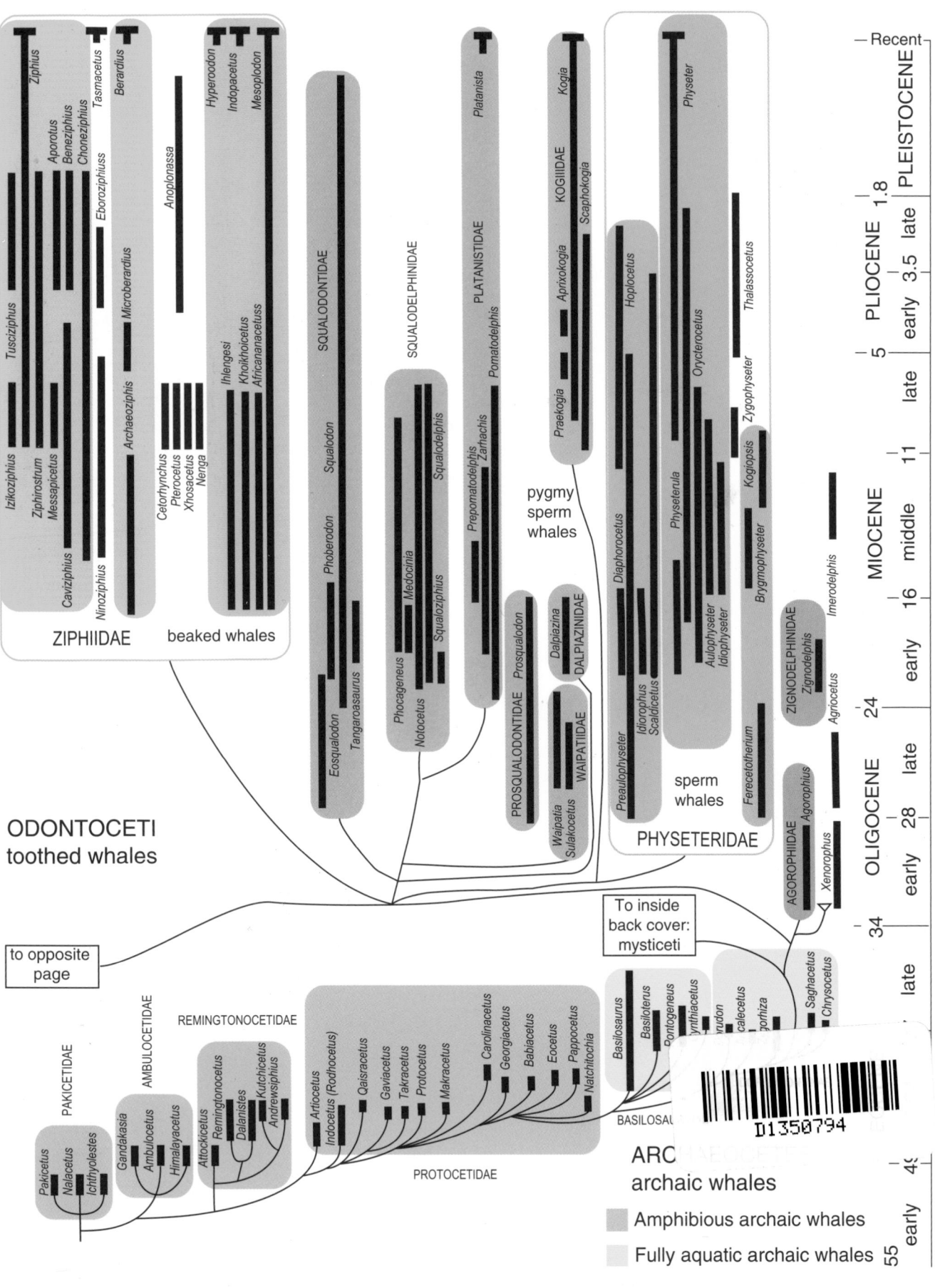

# Encyclopedia of
# MARINE
# MAMMALS

## Cover photographs

(Front cover, top) Antillean manatee in the mangrove channels of Drowned Cayes in the Belize barrier reef lagoon system, Belize, Central America (photo credit: Dr. Leszek Karczmarski, Texas A&M University and University of Pretoria).

(Front cover, middle) Spinner dolphin photographed off Pemba Island, Zanzibar Archipelago, tropical East Africa (photo credit: Dr. Leszek Karczmarski, Texas A&M University and University of Pretoria).

(Front cover, bottom) A false killer whale attacking a mahi-mahi (*Coryphaena hippurus*). Prey sharing in the wild and in captivity is frequently observed for this species. Photograph © Daniel J. McSweeney.

# Encyclopedia of
# MARINE
# MAMMALS

Second Edition

*Editors*

## William F. Perrin
*Southwest Fisheries Science Center, NOAA*
*La Jolla, California*

## Bernd Würsig
*Texas A&M University*
*Galveston, Texas*

## J. G. M. Thewissen
*Northeastern Ohio Universities College of Medicine*
*Rootstown, Ohio*

AMSTERDAM • BOSTON • HEIDELBERG • LONDON • NEW YORK • OXFORD
PARIS • SAN DIEGO • SAN FRANCISCO • SINGAPORE • SYDNEY • TOKYO
Academic Press is an imprint of Elsevier

Academic Press is an imprint of Elsevier
30 Corporate Drive, Suite 400, Burlington, MA 01803, USA
525 B Street, Suite 1900, San Diego, California 92101-4495, USA
360 Park Avenue South, New York, NY 10010-1710, USA
32 Jamestown Road, London NW1 7BY, UK

First edition 2002, Second edition 2009

**Library of Congress Cataloging in Publication Data**
A catalog record for this book is available from the Library of Congress

**British Library Cataloguing in Publication Data**
A catalogue record for this book is available from the British Library

ISBN    978-0-12-373553-9

For information on all Academic Press publications
visit our website at www.elsevierdirect.com

Typeset by Charon Tec Ltd., A Macmillan Company. (www.macmillansolutions.com)

Printed and bound in China

13 14 15   10 9 8 7 6 5 4 3

Working together to grow
libraries in developing countries

www.elsevier.com | www.bookaid.org | www.sabre.org

ELSEVIER    BOOK AID
International    Sabre Foundation

# CONTENTS

# D

# E

# F

# CONTENTS BY SUBJECT AREA

# CONTRIBUTORS

**Alejandro Acevedo-Gutiérrez**
Western Washington University
Bellingham, Washington, USA
*Group Behavior*
*Habitat Use*

**Peter J. Adam**
Northwest Missouri State University
Maryville, Missouri, USA
*Hindlimb Anatomy*

**Alex Aguilar**
University of Barcelona
Barcelona, Spain
*Fin Whale (Balaenoptera physalus)*
*Pollution and Marine Mammals*

**Masao Amano**
Department of Animal Sciences
Teikyo University of Science & Technology
Yamanashi, Japan
*Finless Porpoise (Neophocaena phocaenoides)*

**Paul K. Anderson**
University of Calgary
Alberta, Canada
*Steller's Sea Cow (Hydrodamalis gigas)*

**George A. Antonelis**
National Oceanic and Atmospheric Administration,
National Marine Fisheries Service
Honolulu, Hawaii, USA
*Rookeries*

**Frederick I. Archer II**
Southwest Fisheries Science Center, NOAA Fisheries
La Jolla, California, USA
*Striped Dolphin (Stenella coeruleoalba)*

**Peter W. Arnold**[‡]
Museum of Tropical Queensland
Townsville, Queensland, Australia
*Australian Snubfin Dolphin (Orcaella heinsohni)*

**John P. Y. Arnould**
Deakin University
Melbourne, Australia
*Southern Fur Seals (Arctocephalus spp.)*

**Shannon Atkinson**
University of Alaska Fairbanks Institute of Marine Science
Seward, Alaska, USA
*Endocrine Systems*
*Male Reproductive Systems*

**Whitlow W. L. Au**
Hawaii Institute of Marine Biology,
University of Hawaii
Kaneohe, Hawaii, USA
*Echolocation*

**D. A. Austin**
Parks Canada
Halifax, Nova Scotia, Canada
*Pinniped Ecology*

**Robin W. Baird**
Cascadia Research Collective
Olympia, Washington, USA
*False Killer Whale (Pseudorca crassidens)*
*Risso's Dolphin (Grampus griseus)*

**C. Scott Baker**
Marine Mammal Institute and Department of Fisheries and
    Wildlife
Oregon State University
Newport, Oregon, USA
School of Biological Sciences
University of Auckland
Auckland, New Zealand
*Forensic Genetics*
*Whaling, Modern*

**Lisa T. Ballance**
Southwest Fisheries Science Center,
NOAA Fisheries
La Jolla, California, USA
*Cetacean Ecology*

**Juan A. Balbuena**
University of Valencia
Valencia, Spain
*Parasites*

**John L. Bannister**
The Western Australian Museum
Perth, Western Australia
*Baleen Whales (Mysticetes)*

**Jay Barlow**
Southwest Fisheries Science Center, NOAA Fisheries
La Jolla, California, USA
*Management*
*Population Status and Trends*

**Nélio B. Barros**[‡]
Biology Department
Portland State University

Portland, Oregon, USA
*Diet*

**C. A. Beck**
North Pacific Research Board
Anchorage, Alaska, USA
*Pinniped Ecology*

**Marc Bekoff**
University of Colorado
Boulder, Colorado, USA
*Ethics and Marine Mammals*

**M. Ben-David**
Department of Zoology and Physiology
University of Wyoming
Laramie, Wyoming, USA
*Otters, Marine*

**John L. Bengtson**
National Marine Mammal Laboratory,
Alaska Fisheries Science Center, NOAA Fisheries
Seattle, Washington, USA
*Crabeater Seal (Lobodon carcinophaga)*

**Annalisa Berta**
San Diego State University
San Diego, California, USA
*Mysticetes, Evolution*
*Pinniped Evolution*
*Pinnipedia, Overview*
*Systematics, Overview*

**Martine Bérubé**
Stockholm University
Stockholm, Sweden
*Hybridism*

**Arne Bjørge**
Institute of Marine Research
Oslo, Norway
*Harbor Porpoise (Phocoena phocoena)*

**Nancy A. Black**
Monterey Bay Cetacean Project
Monterey, California, USA
*Pacific White-Sided Dolphin (Lagenorhynchus obliquidens)*

**J. L. Bodkin**
US Geological Survey, Alaska Biological Science Center
Anchorage, Alaska, USA
*Otters, Marine*

**Daryl J. Boness**
University of Maine
Orono, Maine, USA
*Estrus and Estrous Behavior*
*Sea Lions, Overview*

**Asuncion Borrell**
University of Barcelona
Barcelona, Spain
*Pollution and Marine Mammals*

**Peter Boveng**
National Marine Mammal Laboratory,
Alaska Fisheries Science Center, NOAA Fisheries

Seattle, Washington, USA
*Ribbon Seal (Histriophoca fasciata)*

**W. D. Bowen**
Department of Fisheries and Oceans
Bedford Institute of Oceanography
Dartmouth, Nova Scotia, Canada
*Pinniped Ecology*

**Ian L. Boyd**
University of St. Andrews
St. Andrews, Fife, UK
*Antarctic Marine Mammals*
*Pinniped Life History*

**Gillian T. Braulik**
Pakistan Wetlands Programme
Islamabad, Pakistan
Sea Mammal Research Unit,
University of St. Andrews
St. Andrews, Fife, UK
*Susu and Bhulan (Platanista gangetica gangetica and*
*P. g. minor)*

**Jeffrey M. Breiwick**
National Marine Mammal Laboratory,
Alaska Fisheries Science Center, NOAA Fisheries
Seattle, Washington, USA
*Stock Assessment*

**Robert L. Brownell, Jr.**
Southwest Fisheries Science Center, NOAA Fisheries
La Jolla, California, USA
*Minke Whales (Balaenoptera acutorostrata and*
*B. bonaerensis)*
*Whaling, Illegal and Pirate*

**Stephen T. Buckland**
University of St. Andrews
St. Andrews, Fife, UK
*Abundance Estimation*

**John J. Burns**
Fairbanks, Alaska, USA
*Arctic Marine Mammals*
*Harbor Seal and Spotted Seal (Phoca vitulina and*
*P. largha)*

**Douglas S. Butterworth**
University of Cape Town
South Africa
*Competition with Fisheries*

**Claudio Campagna**
Centro Nacional Patagónico
Puerto Madryn, Argentina
*Aggressive Behavior, Intraspecific*

**Humberto Luis Cappozzo**
Museo Argentino de Ciencias Naturales
Buenos Aires, Argentina
*South American Sea Lion (Otaria flavescens)*

**Michael Castellini**
Institute of Marine Science, University of Alaska

Fairbanks, Alaska, USA
*Thermoregulation*

**Susan J. Chivers**
Southwest Fisheries Science Center, NOAA Fisheries
La Jolla, California, USA
*Cetacean Life History*

**Frank Cipriano**
San Francisco State University
San Francisco, California, USA
*Atlantic White-Sided Dolphin (Lagenorhynchus acutus)*

**Phillip J. Clapham**
Alaska Fisheries Science Center, NOAA Fisheries
Seattle, Washington, USA
*Humpback Whale (Megaptera novaeangliae)*
*Whaling, Modern*

**Malcolm R. Clarke**
Lajes de Pico Azores, Portugal
*Diet*

**Rochelle Constantine**
School of Biological Sciences
University of Auckland
Auckland, New Zealand
*Folklore and Legends*

**Lisa Noelle Cooper**
Department of Anatomy
Northeastern Ohio Universities College of Medicine
Rootstown, Ohio, USA
*Forelimb Anatomy*

**Peter Corkeron**
Falmouth, Massachusetts, USA
*Captivity*

**Daniel P. Costa**
Department of Ecology and Evolutionary Biology
University of California
Santa Cruz, California, USA
*Energetics*
*Osmoregulation*
*Pinniped Physiology*

**Daniel F. Cowan**
The University of Texas Medical Branch
Galveston, Texas, USA
*Pathology*

**Enrique A. Crespo**
Centro Nacional Patagónico and Universidad Nacional de la
   Patagonia Blvd
Puerto Madryn, Chubut, Argentina
*Franciscana (Pontoporia blainvillei)*
*South American Aquatic Mammals*

**Daniel E. Crocker**
Department of Biology
Sonoma State University
Rohnert Park, California, USA
*Pinniped Physiology*

**Donald A. Croll**
Ecology and Evolutionary Biology Department,
Center for Ocean Health
University of California
Santa Cruz, California, USA
*Filter Feeding*

**Merel L. Dalebout**
School of Biological, Earth and Environmental Sciences
University of New South Wales
Sydney, Australia
*Forensic Genetics*

**Jim Darling**
Whale Trust
Paia, Hawaii, USA
*Song*

**Stephen M. Dawson**
University of Otago
Dunedin, Otago, New Zealand
*Cephalorhynchus Dolphins (C. heavisidii, C. eutropia,*
   *C. hectori, and C. commersonii)*

**Douglas P. DeMaster**
National Marine Fisheries Service
Seattle, Washington, USA
*Endangered Species and Populations*

**Thomas A. Deméré**
Department of Paleontology
San Diego Natural History Museum
San Diego, California, USA
*Locomotion, Terrestrial*
*Mysticetes, Evolution*

**Lawrence M. Dill**
Department of Biological Sciences
Simon Fraser University
Burnaby, British Columbia, Canada
*Feeding Strategies and Tactics*

**Andrew E. Dizon**
Southwest Fisheries Science Center, NOAA Fisheries
La Jolla, California, USA
*Genetics for Management*

**M. Louella L. Dolar**
Tropical Marine Research for Conservation
San Diego, California, USA
*Fraser's Dolphin (Lagenodelphis hosei)*

**Daryl P. Domning**
Department of Anatomy
Howard University
Washington, DC, USA
*Desmostylia*
*Sirenian Evolution*
*Steller's Sea Cow (Hydrodamalis gigas)*

**Meghan A. Donahue**
Southwest Fisheries Science Center, NOAA Fisheries
La Jolla, California, USA
*Pygmy Killer Whale (Feresa attenuata)*

**G. P. Donovan**
The International Whaling Commission,
Cambridge, UK
*The International Whaling Commission*

**Kathleen M. Dudzinski**
Dolphin Communication Project
Stonington, Connecticut, USA
*Communication*

**Deborah A. Duffield**
Portland State University
Portland, Oregon, USA
*Extinctions, Specific*

**Richard Ellis**
American Museum of Natural History
New York, New York, USA
*Azorean Whaling*
*Whaling, Early, and Aboriginal*
*Whaling, Traditional*

**Sergio Escorza-Treviño**
California State University
Los Angeles, California, USA
*North Pacific Marine Mammals*

**James A. Estes**
Department of Evolutionary Biology and Ecology
University of California
Santa Cruz, California, USA
*Ecological Effects of Marine Mammals*
*Otters, Marine*

**Michael A. Etnier**
Applied Osteology
Bellingham, Washington, USA
*Sustainability*

**Peter G. H. Evans**
Department of Ocean Sciences
University of North Wales
Bangor, Gwynedd, UK
*Habitat Pressures*
*North Atlantic Marine Mammals*

**Michael A. Fedak**
NERC Sea Mammal Research Unit, School of Biological
    Sciences, University of St. Andrews
St. Andrews, Fife, UK
*Reproductive Behavior*

**Mercedes Fernández**
University of Valencia
Valencia, Spain
*Parasites*

**Dagmar Fertl**
Ziphius EcoServices
Plano, Texas, USA
*Albinism*
*Barnacles*
*Fisheries, Interference With*
*Remoras*

**Todd Feucht**
Ocean Embassy
Orlando, Florida, USA
*Training*

**Paul C. Fiedler**
Southwest Fisheries Science Center, NOAA Fisheries
La Jolla, California, USA
*Ocean Environment*

**Frank E. Fish**
West Chester University
West Chester, Pennsylvania, USA
*Streamlining*

**Paulo A. C. Flores**
Centro Nacional de Pesquisa, Conservação e Manejo de
    Mamíferos Aquáticos and Instituto de Pesquisa and Conservação
    de Golfinhos
Florianópolis, South Carolina, Brazil
*Tucuxi and Guiana Dolphin (Sotalia fluviatilis and
    S. Guianensis)*

**Jaume Forcada**
British Antarctic Survey, Natural Environment Research Council
Cambridge, UK
*Antarctic Fur Seal (Arctocephalus gazella)*
*Distribution*
*Monk Seals (Monachus monachus, M. tropicalis, and
    M. schauinslandi)*

**John K. B. Ford**
Fisheries and Oceans Canada, Pacific Biological Station
Nanaimo, British Columbia, Canada
*Dialects*
*Killer Whale (Orcinus orca)*

**R. Ewan Fordyce**
Department of Geology
University of Otago
Dunedin, Otago, New Zealand
*Cetacean Evolution*
*Cetacean Fossil Record*
*Fossil Sites*
*Neoceti*

**Paul H. Forestell**
Long Island University
Brookville, New York, USA
Pacific Whale Foundation
Maui, Hawaii
*Popular Culture and Literature*

**Karin A. Forney**
Southwest Fisheries Science Center, NOAA Fisheries
Santa Cruz, California, USA
*Surveys*

**Charles W. Fowler**
National Marine Mammal Laboratory,
Alaska Fisheries Science Center, NOAA Fisheries
Seattle, Washington, USA
*Sustainability*

**Stuart M. Frank**
Kendall Whaling Museum

Sharon, Massachusetts, USA
*Scrimshaw*

**Adam S. Frankel**
Hawai'i Marine Mammal Consortium
Kamulea, Hawaii, USA
Marine Acoustics, Inc.
Arlington, Virginia, USA
*Sound Production*

**Nicholas J. Gales**
Australian Antarctic Division
Tasmania, Australia
*New Zealand Sea Lion (Phocarctos hookeri)*

**Roger L. Gentry**
National Marine Fisheries Service
Silver Springs, Maryland, USA
*Eared Seals (Otariidae)*
*Northern Fur Seal (Callorhinus ursinus)*

**Joseph R. Geraci**
Aquatic Health Sciences
Stonington, Connecticut, USA
*Health*
*Stranding*

**Tim Gerrodette**
Southwest Fisheries Science Center, NOAA Fisheries
La Jolla, California, USA
*Tuna-Dolphin Issue*

**William G. Gilmartin**
Hawaii Wildlife Fund
Volcano, Hawaii, USA
*Monk Seals (Monachus monachus, M. tropicalis, and*
*M. schauinslandi)*

**R. Natalie P. Goodall**
Museo Acatushún de Aves y Mamíferos Marinos Australes
(AMMA) and Centro Austral de Investigaciones Científicas
(CADIC)
Ushuaia, Tierra del Fuego, Argentina
*Hourglass Dolphin (Lagenorhynchus cruciger)*
*Peale's Dolphin (Lagenorhynchus australis)*
*Spectacled Porpoise (Phocoena dioptrica)*

**Shannon Gowans**
St. Petersburg, Florida, USA
*Bottlenose Whales (Hyperoodon ampullatus and H. planifrons)*

**Justin D. Gregg**
School of Psychology, Áras an Phiarsaigh
Dublin, Ireland
*Communication*

**Ailsa Hall**
Sea Mammal Research Unit
University of St. Andrews
St. Andrews, Fife, UK
*Gray Seal (Halichoerus grypus)*
*Mass Die-Offs*

**Mike O. Hammill**
Maurice Lamontagne Institute

Mont Joli, Quebec, Canada
*Earless Seals (Phocidae)*
*Ringed Seal (Pusa hispida)*

**Philip S. Hammond**
Sea Mammal Research Unit
University of St. Andrews
St. Andrews, Fife, UK
*Mark–Recapture Methods*

**John Harwood**
University of St. Andrews
St. Andrews, Fife, UK
*Mass Mortalities*

**Carolyn B. Heath**
Fullerton College
Fullerton, California, USA
*California, Galapagos, and Japanese Sea Lions (Zalophus*
*californianus, Z. wollebaeki, Z. japonicus*

**M. P. Heide-Jørgensen**
Greenland Institute of Natural Resources
Nuuk, Greenland
*Narwhal (Monodon monoceros)*

**Michael R. Heithaus**
Department of Biological Sciences
Florida International University
North Miami, Florida, USA
*Feeding Strategies and Tactics*

**Louis M. Herman**
University of Hawaii
Honolulu, Hawaii, USA
*Language Learning and Cognitive Skills*

**Roger Hewitt**
Southwest Fisheries Science Center, NOAA Fisheries
La Jolla, California, USA
*Krill and Other Plankton*

**John E. Heyning**[†]
Natural History Museum of Los Angeles County
Los Angeles, California, USA
*Cuvier's Beaked Whale (Ziphius cavirostris)*
*Museums and Collections*

**Mark A. Hindell**
University of Tasmania
Hobart, Australia
*Breeding Sites*
*Elephant Seals (Mirounga angustirostris and*
*M. leonina)*

**A. Rus Hoelzel**
University of Durham
Durham, UK
*Molecular Ecology*

**Aleta A. Hohn**
Southeast Fisheries Science Center, NOAA Fisheries

[†]Deceased

Beaufort, North Carolina, USA
*Age Estimation*

**Sascha K. Hooker**
Sea Mammal Research Unit, University of St. Andrews
St. Andrews, Fife, UK
*Toothed Whales, Overview*

**Joseph Horwood**
Centre for Environment,
Fisheries and Aquaculture Science
Lowestoft, Suffolk, UK
*Sei Whale (Balaenoptera borealis)*

**Erich Hoyt**
Whale and Dolphin Conservation Society
North Berwick, Scotland, UK
*Marine Protected Areas*
*Whale Watching*

**Sara J. Iverson**
Department of Biology
Dalhousie University
Halifax, Nova Scotia, Canada
*Blubber*

**Vincent M. Janik**
School of Biology, University of St. Andrews
St. Andrews, Fife, UK
*Signature Whistles*

**Armando M. Jaramillo-Legoretta**
Instituto Nacional de Ecología
Ensenada, Baja California, México
*Vaquita (Phocoena sinus)*

**F. Javier Aznar**
University of Valencia
Valencia, Spain
*Parasites*

**Thomas A. Jefferson**
Southwest Fisheries Science Center
La Jolla, California, USA
*Clymene Dolphin (Stenella clymene)*
*Dall's Porpoise (Phocoenoides dalli)*
*Rough-Toothed Dolphin (Steno bredanensis)*

**Anne M. Jensen**
UIC Science, LLC
Barrow, Alaska, USA
*Inuit and Marine Mammals*

**Mary Lou Jones**
Cetacean Research Associates
Darnestown, Maryland, USA
*Gray Whale (Eschrichtius robustus)*

**Ronald A. Kastelein**
Harderwijk Marine Mammal Park
Harderwijk, The Netherlands
*Walrus (Odobenus rosmarus)*

**Toshio Kasuya**
Tokyo, Japan
*Giant Beaked Whales (Berardius bairdii and B. arnuxii)*
*Japanese Whaling*

**Hidehiro Kato**
National Research Institute of Far Seas Fisheries
Shimizu, Japan
*Bryde's Whales (Balaenoptera edeni/brydei)*

**Catherine M. Kemper**
South Australian Museum
Adelaide, South Australia
*Pygmy Right Whale (Caperea marginata)*

**Robert D. Kenney**
University of Rhode Island
Narragansett, Rhode Island, USA
*Right Whales (Eubalaena glacialis, E. japonica, and E. australis)*

**Carl Christian Kinze**
CCKonsult
Frederiksberg, Denmark
*White-Beaked Dolphin (Lagenorhynchus albirostris)*

**Gerald L. Kooyman**
Scripps Institution of Oceanography
La Jolla, California, USA
*Diving Physiology*

**Kit M. Kovacs**
Norwegian Polar Institute
Tromsø, Norway
*Bearded Seal (Erignathus barbatus)*
*Hooded Seal (Cystophora cristata)*

**Jeffrey T. Laitman**
Mount Sinai School of Medicine
New York, New York, USA
*Cetacean Prenatal Development*

**André M. Landry**
Department of Marine Biology
Texas A&M University
Galveston, Texas, USA
*Remoras*

**David M. Lavigne**
International Fund for Animal Welfare
Guelph, Ontario, Canada
*Harp Seal (Pagophilus groenlandicus)*

**Rick LeDuc**
Southwest Fisheries Science Center, NOAA Fisheries
La Jolla, California, USA
*Biogeography*
*Delphinids, Overview*

**Jon Lien**
Memorial University of Newfoundland
St. John's, Newfoundland, Canada
*Entrapment and Entanglement*

**John K. Ling**
Clare, South Australia
*Australian Sea Lion (Neophoca cinerea)*

**Jessica D. Lipsky**
Southwest Fisheries Science Center, NOAA Fisheries
La Jolla, California, USA
*Krill and Other Plankton*
*Right Whale Dolphins (Lissodelphis borealis and L. peronii)*

**Thomas R. Loughlin**
National Marine Mammal Laboratory, NOAA Fisheries (Retired)
Seattle, Washington, USA
*Steller Sea Lion (Eumetopias jubatus)*

**Valerie J. Lounsbury**
National Aquarium in Baltimore
Baltimore, Maryland, USA
*Health*

**Lloyd Lowry**
University of Alaska, School of Fisheries and Ocean Sciences
Kailua-Kona, Hawaii, USA
*Ribbon Seal (Histriophoca fasciata)*

**Mary C. Maas**
Northeastern Ohio Universities College of Medicine
Rootstown, Ohio, USA
The University of Texas
Austin, Texas, USA
*Bones and Teeth, Histology Of*

**Stephen A. MacLean**
The Nature Conservancy
Anchorage, Alaska, USA
*Inuit and Marine Mammals*

**Janet Mann**
Georgetown University
Washington, DC, USA
*Parental Behavior*
*Tool Use in Wild Bottlenose Dolphins*

**Lori Marino**
Emory University
Atlanta, Georgia, USA
*Brain Size Evolution*

**Helene Marsh**
James Cook University
Townsville, Queensland, Australia
*Dugong (Dugong dugon)*

**Christopher D. Marshall**
Department of Marine Biology, Texas A&M
University at Galveston
Galveston, Texas, USA
*Feeding Morphology*

**Anthony R. Martin**
NERC Sea Mammal Research Unit, University of St. Andrews
St. Andrews, Fife, UK
*River Dolphins*

**Alla M. Mass**
Russian Academy of Sciences
Moscow, Russia
*Vision*

**Donald F. McAlpine**
New Brunswick Museum
St. John's, Newfoundland, Canada
*Pygmy and Dwarf Sperm Whales (Kogia breviceps and K. sima)*

**Guram A. Mchedlidze**
Georgian Academy of Sciences
Tbilisi, Georgia
*Sperm Whales, Evolution*

**William A. McLellan**
Biology and Marine Biology, University of North Carolina
Wilmington, North Carolina, USA
*Skull Anatomy*

**James G. Mead**
National Museum of Natural History
Washington, DC, USA
*Beaked Whales, Overview (Ziphiidae)*
*Cuvier's Beaked Whale (Ziphius cavirostris)*
*Gastrointestinal Tract*
*Museums and Collections*
*Shepherd's Beaked Whale (Tasmacetus shepherdi)*

**Richard Merrick**
National Marine Fisheries Service, NOAA Fisheries
Woods Hole, Massachusetts, USA
*Endangered Species and Populations*

**Sarah L. Mesnick**
Southwest Fisheries Science Center, NOAA Fisheries
La Jolla, California, USA
*Mating Systems*
*Sexual Dimorphism*

**Edward H. Miller**
Biology Department
Memorial University
St. John's, Newfoundland, Canada
*Baculum*
*Territorial Behavior*

**Nobuyuki Miyazaki**
Tokyo University of Marine Science and Technology,
Tokyo, Japan
*Caspian and Baikal Seals (Pusa caspica and Pusa sibirica)*

**Sue E. Moore**
Alaska Fisheries Sciences Center, NOAA Fisheries
Seattle, Washington, USA
*Climate Change*

**Phillip A. Morin**
Southwest Fisheries Science Center, NOAA Fisheries
La Jolla, California, USA
*Genetics for Management*

**Christian de Muizon**
National d'Histoire Naturelle
Paris, France
*Odobenocetops*
*River Dolphins, Evolutionary History and Affinities*

**William A. Newman**
Marine Biology Research Division,
Scripps Institution of Oceanography
La Jolla, California, USA
*Barnacles*

**Kelly M. Newton**
Ecology and Evolutionary Biology Department,
Center for Ocean Health
University of California
Santa Cruz, California, USA
*Filter Feeding*

**Simon Northridge**
University of St. Andrews
St. Andrews, Fife, UK
*Bycatch*
*Fishing Industry, Effects Of*

**Sirpa Nummela**
Department of Biological and Environmental Sciences
University of Helsinki
Helsinki, Finland
*Hearing*

**Justine K. O'Brien**
SeaWorld and Busch Gardens Reproductive Research Center
San Diego, California, USA
*Captive Breeding*

**Gregory M. O'Corry-Crowe**
Harbor Branch Oceanographic Institution
Fort Pierce, Florida, USA
*Beluga Whale (Delphinapterus leucas)*

**Daniel K. Odell**
Hubbs-SeaWorld Research Institute
Orlando, Florida, USA
*Captive Breeding*
*Marine Parks and Zoos*
*Sirenian Life History*

**Helmut H. A. Oelschläger**
Johann Wolfgang Goethe University
Frankfurt am Main, Germany
*Brain*

**Jutta S. Oelschläger**
Johann Wolfgang Goethe University
Frankfurt am Main, Germany
*Brain*

**Paula A. Olson**
Southwest Fisheries Science Center, NOAA Fisheries
La Jolla, California, USA
*Pilot Whales (Globicephala melas and G. macrorhynchus)*

**Rudy M. Ortiz**
University of California
Merced, California, USA
*Endocrine Systems*

**D. Ann Pabst**
Biology and Marine Biology, University of North Carolina
Wilmington, North Carolina, USA
*Skull Anatomy*

**Debra L. Palka**
Northeast Fisheries Science Center, NOAA Fisheries
Woods Hole, Massachusetts, USA
*North Atlantic Marine Mammals*

**Per J. Palsbøll**
Stockholm University
Stockholm, Sweden
*Genetics, Overview*

**Guido J. Parra**
Cetacean Ecology and Acoustics Laboratory, University of
    Queensland
St. Lucia, Brisbane, Australia
*Humpback Dolphins (Sousa spp.)*

**William F. Perrin**
Southwest Fisheries Science Center, NOAA Fisheries
La Jolla, California, USA
*Atlantic Spotted Dolphin (Stenella frontalis)*
*Blue Whale (Balaenoptera musculus)*
*Bryde's Whales (Balaenoptera edeni/brydei)*
*California, Galapagos, and Japanese Sea Lions (Zalophus*
    *californianus, Z. wollebaeki, Z. japonicus)*
*Coloration*
*Common Dolphins (Delphinus delphis and D. capensis)*
*Elephant Seals (Mirounga angustirostris and M. leonina)*
*Geographic Variation*
*History of Marine Mammal Research*
*Minke Whales (Balaenoptera acutorostrata and B. bonaerensis)*
*Pantropical Spotted Dolphin (Stenella attenuata)*
*Species*
*Spinner Dolphin (Stenella longirostris)*
*South American Sea Lion (Otaria flavescens)*
*Stranding*

**Wayne L. Perryman**
Southwest Fisheries Science Center, NOAA Fisheries
La Jolla, California, USA
*Melon-Headed Whale (Peponocephala electra)*
*Pygmy Killer Whale (Feresa attenuata)*

**Carl J. Pfeiffer**
Virginia Polytechnic Institute and State University
Blacksburg, Virginia, USA
*Whale Lice*

**Robert Pitman**
Southwest Fisheries Science Center, NOAA Fisheries
La Jolla, California, USA
*Indo-Pacific Beaked Whale (Indopacetus pacificus)*
*Mesoplodont Whales (Mesoplodon spp.)*

**Éva E. Plagányi**
University of Cape Town
South Africa
*Competition with Fisheries*

**Paddy P. Pomeroy**
NERC Sea Mammal Research Unit, School of Biological
    Sciences, University of St. Andrews
St. Andrews, Fife, UK
*Reproductive Behavior*

**Paul J. Ponganis**
Scripps Institution of Oceanography
La Jolla, California, USA
*Circulatory System*

**James A. Powell**
Wildlife Trust
St. Petersburg, Florida, USA
*Manatees (Trichechus manatus, T. senegalensis, and T. inunguis)*

**J. Antonio Raga**
University of Valencia
Valencia, Spain
*Parasites*

**Katherine Ralls**
National Zoological Park, Smithsonian Institution
Washington, DC, USA
*Mating Systems*
*Sexual Dimorphism*

**Andrew J. Read**
Nicholas School of the Environment
Duke University
Beaufort, North Carolina, USA
*Telemetry*
*Porpoises, Overview*

**Randall R. Reeves**
Okapi Wildlife Associates
Hudson, Quebec, Canada
*Conservation Efforts*
*Hunting of Marine Mammals*
*Population Status and Trends*
*River Dolphins*

**Joy S. Reidenberg**
Mount Sinai School of Medicine
New York, New York, USA
*Cetacean Prenatal Development*

**Peter J. H. Reijnders**
IMARES, Institute for Marine Resources &
   Ecosystem Studies
Den Burg, The Netherlands
*Pollution and Marine Mammals*

**Julio C. Reyes**
Áreas Costeras y Recursos Marinos
Pisco, Peru
*Burmeister's Porpoise (Phocoena spinipinnis)*

**John E. Reynolds III**
Mote Marine Laboratory
Sarasota, Florida, USA
Marine Mammal Commission
Bethesda, Maryland, USA
*Anatomical Dissection: Thorax and Abdomen*
*Endangered Species and Populations*
*Skeleton, Postcranial*
*Manatees (Trichechus manatus, T. senegalensis,*
   *and T. inunguis)*

**Dale W. Rice**
National Marine Mammal Laboratory, NOAA Fisheries
Seattle, Washington, USA
*Ambergris*
*Baleen*
*Classification*
*Spermaceti*

**W. John Richardson**
LGL Ltd.
King City, Ontario, Canada
*Noise, Effects Of*

**Todd R. Robeck**
Corporate Zoological Operations, Busch Entertainment
   Corporation, SeaWorld
San Diego, California, USA
*Captive Breeding*

**Kelly M. Robertson**
Southwest Fisheries Science Center, NOAA Fisheries
La Jolla, California, USA
*Australian Snubfin Dolphin (Orcaella heinsohni)*

**Tracey L. Rogers**
Australian Marine Mammal Research Centre
Sydney, Australia
*Leopard Seal (Hydrurga leptonyx)*
*Ross Seal (Ommatophoca rossii)*

**Lorenzo Rojas-Bracho**
Instituto Nacional de Ecología
Ensenada, Baja California, México
*Vaquita (Phocoena sinus)*

**Sentiel A. Rommel**
Biology and Marine Biology, University of North Carolina
Wilmington, North Carolina, USA
*Anatomical Dissection: Thorax and Abdomen*
*Skeleton, Postcranial*
*Skull Anatomy*

**Patricia E. Rosel**
Southeast Fisheries Science Center, NOAA Fisheries
Lafayette, LA, USA
*Albinism*

**Graham J. B. Ross**
Australian Biological Resources Study,
Environment Australia
Canberra, Australia
*Humpback Dolphins (Sousa chinensis and S. teuszii)*

**Peter Rudolph**
National Museum of Natural History
Leiden, The Netherlands, UK
*Indo-West Pacific Marine Mammals*

**David J. Rugh**
National Marine Mammal Laboratory, NOAA Fisheries
Seattle, Washington, USA
*Bowhead Whale (Balaena mysticetus)*

**Brooke Sargeant**
Department of Biology
Georgetown University
Washington, DC, USA
*Tool Use in Wild Bottlenose Dolphins*

**Laela S. Sayigh**
Biology Department, Woods Hole Oceanographic Institution
Woods Hole, Massachusetts, USA
*Signature Whistles*

**Bobbi Jo. Schneider**
Department of Biology
University of Akron, Ohio, USA
*Marine Mammal Evolution*

**Michael D. Scott**
InterAmerican Tropical Tuna Commission
La Jolla, California, USA
*Common Bottlenose Dolphin (Tursiops truncatus)*

**Richard Sears**
Mingan Island Cetacean Study, Inc
Longue Pointe de Mingan, Quebec, Canada
*Blue Whale (Balaenoptera musculus)*

**Glenn W. Sheehan**
Barrow Arctic Science Consortium
Barrow, Alaska, USA
*Inuit and Marine Mammals*

**Kim E. W. Shelden**
National Marine Mammal Laboratory
Seattle, Washington, USA
*Bowhead Whale (Balaena mysticetus)*

**Gregory K. Silber**
National Marine Fisheries Service, NOAA Fisheries
Silver Spring, Maryland, USA
*Endangered Species and Populations*

**Vera M. F. da Silva**
Instituto Nacional de Pesquisas da Amazônia
Manaus, Amazonas, Brazil
*Amazon River Dolphin (Inia geoffrensis)*
*Tucuxi and Guiana Dolphin (Sotalia fluviatilis and S. guianensis)*

**Chris Smeenk**
National Museum of Natural History
Leiden, The Netherlands, UK
*Indo-West Pacific Marine Mammals*

**Brian D. Smith**
Wildlife Conservation Society
Bronx, New York, USA
*Irrawaddy Dolphin (Orcaella brevirostris)*
*Susu and Bhulan (Platanista gangetica gangetica and P. g. minor)*

**Fred Spoor**
Department of Anatomy & Developmental Biology
University College London
London, UK
*Balance*

**David St. Aubin**[†]
University of California
Merced, California, USA
*Endocrine Systems*

**Iain J. Staniland**
British Antarctic Survey, Natural Environment Research Council
Cambridge, UK
*Antarctic Fur Seal (Arctocephalus gazelle)*

**S. Jonathan Stern**
San Francisco State University
San Francisco, California, USA
*Migration and Movement Patterns*

**Barbara E. Stewart**
Sila Consultants
Howden, Manitoba, Canada
*Female Reproductive Systems*

**Brent S. Stewart**
Hubbs-SeaWorld Research Institute
San Diego, California, USA
*Diving Behavior*
*Hair and Fur*

**Robert E. A. Stewart**
Department of Fisheries and Oceans
Winnipeg, Manitoba, Canada
*Female Reproductive Systems*

**Ian Stirling**
Canadian Wildlife Service
Edmonton, Alberta, Canada
*Polar Bear (Ursus maritimus)*

**Alexander Ya. Supin**
Russian Academy of Sciences
Moscow, Russia
*Vision*

**Steven L. Swartz**
Cetacean Research Associates
Darnestown, Maryland, USA
*Gray Whale (Eschrichtius robustus)*

**Cynthia R. Taylor**
Wildlife Trust
St. Petersburg, Florida, USA
*Manatees (Trichechus manatus, T. senegalensis, and T. inunguis)*

**Jack Terhune**
University of New Brunswick
St. John, New Brunswick, Canada
*Weddell Seal (Leptonychotes weddellii)*

**Bernie R. Tershy**
Ecology and Evolutionary Biology Department, Long
    Marine Laboratory, Center for Ocean Health
University of California
Santa Cruz, California, USA
*Filter Feeding*

**J. G. M. Thewissen**
Northeastern Ohio Universities College of Medicine
Rootstown, Ohio, USA
*Archaeocetes, Archaic*
*History of Marine Mammal Research*
*Marine Mammal Evolution*
*Musculature*
*Sensory Biology, Overview*

**Jeanette A. Thomas**
Department of Biological Sciences
Western Illinois University-Quad Cities
Moline, Illinois, USA
*Communication*
*Ross Seal (Ommatophoca rossii)*
*Weddell Seal (Leptonychotes weddellii)*

**David Thompson**
Sea Mammal Research Unit
University of St. Andrews
St. Andrews, Fife, UK
*Gray Seal (Halichoerus grypus)*

**Krystal A. Tolley**
South African National Biodiversity Institute
Claremont, South Africa
*Harbor Porpoise (Phocoena phocoena)*

**Fritz Trillmich**
Department of Animal Behavior
University of Bielefeld
Bielefeld, Germany
*Sociobiology*

**Andrew W. Trites**
Marine Mammal Research Unit,
University of British Columbia
Vancouver, British Columbia, Canada
*Predator–Prey Relationships*

**Ted Turner**
Ocean Embassy
Orlando, Florida, USA
*Training*

**Tyler Turner**
Ocean OdysSeas
Aurora, Ohio, USA
*Training*

**Peter L. Tyack**
Woods Hole Oceanographic Institution
Woods Hole, Massachusetts, USA
*Behavior, Overview*
*Mimicry*

**Mark D. Uhen**
United States National Museum of Natural History
Washington, DC, USA
*Basilosaurids*
*Dental Morphology, Evolution Of*

**Koen Van Waerebeek**
Museo de Delfines, Peruvian Centre for Cetacean Research
Pucusana, Peru, USA
*Dusky Dolphin (Lagenorhynchus obscurus)*

**Paul R. Wade**
National Marine Mammal Laboratory, NOAA Fisheries
Seattle, Washington, USA
*Population Dynamics*

**John Y. Wang**
Trent University
Peterborough, Ontario, Canada
FormosaCetus Research and Conservation Group
Thornhill, Ontario, Canada
National Museum of Marine Biology and Aquarium
Checheng, Taiwan
*Indo-Pacific Bottlenose Dolphin (Tursiops aduncus)*
*Stock Identity*

**Gordon T. Waring**
Northeast Fisheries Science Center, NOAA Fisheries
Woods Hole, Massachusetts, USA
*North Atlantic Marine Mammals*

**Douglas Wartzok**
Florida International University
Miami, Florida, USA
*Breathing*

**Mason T. Weinrich**
Whale Center of New England
Gloucester, Massachusetts, USA
*Callosities*

**David W. Weller**
Southwest Fisheries Science Center, NOAA Fisheries
La Jolla, California, USA
*Predation on Marine Mammals*

**Randall S. Wells**
Chicago Zoological Society
Sarasota, Florida, USA
*Common Bottlenose Dolphin*
  *(Tursiops truncatus)*
*Identification Methods*

**Hal Whitehead**
Biology Department
Dalhousie University
Halifax, Nova Scotia, Canada
*Aerial Behavior*
*Culture in Whales and Dolphins*
*Sperm Whale (Physeter macrocephalus)*

**Terrie M. Williams**
Department of Ecology and Evolutionary Biology,
Center for Ocean Health
University of California
Santa Cruz, California, USA
*Swimming*

**Ben Wilson**
Scottish Association for Marine Science
Oban, Argyll, UK
*Reproductive Behavior*

**Loran Wlodarski**
Education Department, SeaWorld Adventure Park
Orlando, Florida, USA
*Marine Parks and Zoos*

**Bernd Würsig**
Texas A&M University
Galveston, Texas, USA
*Aerial Behavior*
*Bow-Riding*
*Dusky Dolphin (Lagenorhynchus obscurus)*
*Ecology, Overview*
*History of Marine Mammal Research*
*Intelligence and Cognition*
*Noise, Effects Of*
*Playful Behavior*

**Alexey V. Yablokov**
Center for Russian Environmental Policy
Moscow, Russia
*Whaling, Illegal and Pirate*

**Tadasu K. Yamada**
National Museum of Nature and Science
Tokyo, Japan
*Omura's Whale (Balaenoptera omurai)*

**Shih-Chu Yang**
FormosaCetus Research and Conservation Group
Hualien, Taiwan
*Indo-Pacific Bottlenose Dolphin (Tursiops aduncus)*

**Pamela K. Yochem**
Hubbs-SeaWorld Research Institute
San Diego, California, USA
*Hair and Fur*

**Joshua H. Yonas**
Department of Biology
San Diego State University
San Diego, California, USA
*Locomotion, Terrestrial*

**Anne E. York**
National Marine Mammal Laboratory
Seattle, Washington, USA
*Abundance Estimation*
*Stock Assessment*

**Kaiya Zhou**
Nanjing Normal University
Nanjing, China
*Baiji (Lipotes vexillifer)*

# PREFACE TO THE SECOND EDITION

Seven years have gone by since the first edition of the Encyclopedia of Marine Mammals was published. The field of marine mammalogy has continued to move at great speed, and significant changes have occurred, even in the roster of our subject animals with the discovery of new species (e.g., Omura's whale), and the disastrous extinction of another (the baiji). One conspicuous difference between the new and the first edition is the use of color throughout the volume, hopefully making it easier and more enjoyable to use. A number of illustrations of marine mammals, kindly provided by Brett Jarrett, are reproduced as they appeared in Jefferson *et al*. "Marine Mammals of the World" (2008) Academic Press.

In addition, authors were given the opportunity to update their chapters, and nearly all took advantage of this to include the latest research results. Several new chapters have been added, covering areas of marine mammal science that have changed significantly, such as those related to climate change and the interface of ecology and conservation. The editors also decided to add conjoined chapters on subjects that were somewhat fragmented in the first edition (e.g., Aerial Behavior) and new chapters to make the volume more comprehensive (e.g., Sense of Balance). We hope that the new edition will get the same positive reception that the first edition enjoyed.

—The Editors

# PREFACE TO THE FIRST EDITION

Marine mammals are awe inspiring, whether one is confronted with the underwater dash of a sea lion, a breaching humpback, or simply the sheer size of a beached sperm whale. It is no surprise that we are fascinated and intrigued by these creatures. Such fascination and curiosity brought us, the editors, to the study of marine mammals at the beginning of our careers, and they keep us excited now. To share the excitement and feed the curiosity of others, scientists or laypersons, we here attempt to summarize the field of marine mammalogy; in a very broad sense, including aspects of history and culture. This was the first reason to compile this encyclopedia.

The science of marine mammals goes back at least to Aristotle, who observed in 400 BC that dolphins gave birth to live young which were nursed with the mother's milk. Observations on the biology of marine mammals expanded throughout the Middle Ages, usually mixing freely with imagination and superstition. Konrad Gesner's *Historia Animalium* (1551), for instance, is a pictorial guide to the animals known in his time. Next to rhinos and seals, it also depicts the unicorn, the fabled mix of a horse and a narwhal. Interest greatly increased with the advent of hunting marine mammals on a large scale. Herman Melville's *Moby Dick* (1851) chronicles nineteenth century Western whaling and displays a curious mix of accurate natural history observations on whales with stubborn misconceptions (such as "whales are fish"). The great whaler/naturalist Charles Scammon accurately described the behavior and aspects of natural history of many species, albeit of necessity from his view behind gun and harpoon.

From these roots, marine mammal science has grown exponentially, especially since the Second World War. Unlike in earlier days, most contemporary research on marine mammals is carried out by observing living animals. Modern marine mammal studies combine aspects of mammalogy, ethology, ecology, animal conservation, molecular biology, oceanography, evolutionary biology, geology, and—in effect—all major branches of the physical and biological sciences, as well as some of the social sciences. This enormous breadth unfortunately necessitates that most marine mammalogists specialize, concentrating on one or a few aspects of marine mammal science and limiting the number of species that they study. Therein lies the second reason for compiling this encyclopedia: we aim to present a summary of the entire field for the scientist who needs information from an unfamiliar subfield.

As editors, we constrained what authors wrote as little as possible, applauding diversity and keeping to minimal guidelines. We consider modern marine mammals to include the mammalian order Cetacea (including whales, dolphins, and porpoises), the order Sirenia (dugongs and manatees), and many members of the order Carnivora: the polar bear, the sea otter and marine otter, and the pinnipeds (true seals, sea lions, fur seals, and walruses). We asked the authors to follow Rice (1998) for the species-level taxonomy and nomenclature of the modern marine mammals (with certain exceptions, as noted in the Marine Mammal Species list), as his work is an excellent, generally accepted listing of diversity.

There is some overlap among the articles. This is not an accident. As in every scientific field, different workers in marine mammalogy have different perspectives on many technical issues and disagree strongly on some of them. We urge the reader to use the cross-indexing to peruse different accounts relating to the same question; on some matters the jury is still very much out, and the range of views is interesting and important.

Ours is an encyclopedia, an alphabetically arranged compilation of articles that are independent and multiauthored, the only such work on marine mammals. However, some other recent books form excellent complements to our work. For example, *Handbook of Marine Mammals* (S. H. Ridgway and R. J. Harrison, Academic Press, 1985–1999) is a series of compendia presenting descriptions of the marine mammal species. *Biology of Marine Mammals* (J. E. Reynolds III and S. A. Rommel, Smithsonian Institution Press, 1999) presents an overview of marine mammals based on a number of long review chapters. *Marine Mammals: Evolutionary Biology* (A. Berta and J. L. Sumich, Academic Press, 1999) presents a current review of the evolutionary aspects of marine mammal science in a textbook format. There are many other authored and edited books, monographs, and research papers, often on more specific topics or particular species. These are listed here in the bibliographies that follow each entry, and the interested reader is encouraged to make use of university libraries, major research libraries (such as in the Smithsonian Institute in Washington, DC, for example), and World Wide Web search engines to find out how to obtain specific reference works. In our modern computer-accessible information era, it is hardly ever appropriate to use the excuse "I cannot find the reference," and we hope that this encyclopedia serves as a text to help point the way.

We hesitated before agreeing to edit this encyclopedia. Marine mammalogy is an exceptionally broad field, ranging across many taxa and across disciplines from molecular genetics and microstructure to whaling history and ethics. We three are all cetologists: we study the evolution and biology of whales, dolphins, and porpoises, and we personally know relatively little about seals, sea cows, or whaling. But we rub shoulders with those who do know much about these things, in our laboratories and universities, in advisory bodies, and at conferences, so we were considered to be in a good position to elicit and edit articles from our colleagues. The project has been fatiguing and sometimes exasperating but elevating nonetheless. We have learned a lot along the way. We owe a great deal to many people. First we thank our editors at Academic Press: Chuck Crumly (the Encyclopedia was his concept and owes its existence to his drive), Gail Rice, and Chris Morris, who all put up bravely with our editing and publishing amateurism and endless missteps and interventions. A very large number of colleagues acted as anonymous peer reviewers for the articles. But the most credit must go to the authors, who gave so freely of their time and expertise. The Encyclopedia is appropriately an international project: articles were authored by scientists in Argentina, Australia, Brazil, Canada, China, Denmark, France, Georgia, Germany, Japan, Mexico, The Netherlands, New Zealand, Norway, Peru, Russia, South Africa, Spain, the United Kingdom, and the United States. The difficulties of such wide participation were eased by the Internet.

We and the authors have engaged in our tasks as a labor of love of our field. We hope that you find not only information in these pages, but also a sense of the excitement of the known and the mystery of the yet-to-be-explored. If this work so affects you, it will have been successful. We also hope that it will help stimulate our growing cadres of young colleagues, naturalists, conservationists, and citizens of earth to contribute to the efforts to save and protect these marvelous creatures of the seas.

W. F. Perrin
B. Würsig
J. G. M. Thewissen

# GUIDE TO THE ENCYCLOPEDIA

The *Encyclopedia of Marine Mammals* is a complete source of information on the subject of marine mammals, contained within a single volume. Each article in the Encyclopedia provides an overview of the selected topic to inform a broad spectrum of readers, from researchers to students to the interested general public.

In order that you, the reader, will derive the maximum benefit from the *Encyclopedia of Marine Mammals*, we have provided this Guide. It explains how the book is organized and how the information within its pages can be located.

## SUBJECT AREAS

The *Encyclopedia of Marine Mammals* presents articles on the entire range of marine mammal study. Articles in the Encyclopedia fall within seven general subject areas, as follows:

- Anatomy and Physiology
- Behavior and Life History
- Ecology and Population Biology
- Evolution and Systematics
- Human Effects and Interactions
- Organisms and Faunas
- Research Methodology

## ORGANIZATION

The *Encyclopedia of Marine Mammals* is organized to provide the maximum ease of use for its readers. All of the articles are arranged in a single alphabetical sequence by title. An alphabetical Table of Contents for the articles can be found beginning on p. v of this introductory section.

As a reader of the Encyclopedia, you can use the alphabetical Table of Contents by itself to locate a topic. Or you can first identify the topic in the Contents by Subject Area and then go to the alphabetical Table to find the page location.

So that they can be more easily identified, article titles begin with the key word or phrase indicating the topic, with any descriptive terms following this. For example, "Noise, Effects Of" is the title assigned to this article, rather than "Effects of Noise" because the specific term *Noise* is the key word.

## ARTICLE FORMAT

Each article in the Encyclopedia begins with an introductory paragraph that defines the topic being discussed and summarizes the content of the article. For example, the article "Baculum" begins as follows:

> The baculum (os penis) is a bone in the penis that occurs in small insectivorous placentals (orders Afrosoricida, Erinaceomorpha, and Soricomorpha), Chiroptera, Primates, Rodentia, and Carnivora. The corresponding element in females is the little-studied clitoris bone (os clitoridis), which has been documented for polar bears and several pinniped species, but presumably is present in all pinnipeds, and in marine and sea otters (it is present in the northern river otter, *Lontra Canadensis*).

Major headings highlight important subtopics that are discussed in the article. For example, the article "Intelligence and Cognition" includes the topics "Brain Size and Characteristics," "Learning," and "Behavioral Complexity in Nature."

## CROSS-REFERENCES

The *Encyclopedia of Marine Mammals* has an extensive system of cross-referencing. References to other articles appear in two forms: as designations within the running text of an article; and as indications of related topics at the end of an article.

An example of the first type, a cross-reference within the running text of an article, is this excerpt from the entry "Baleen Whales:"

> External parasites, particularly WHALE LICE (cyamid crustaceans) and BARNACLES (both sessile and stalked) are common on the slower-swimming more coastal baleen whales such as gray, humpback, and right whales.

This indicates that the items "whale lice" and "barnacles," which are set off in the text by small capital letters, appear as separate articles within the Encyclopedia.

An example of the second type, a cross-reference at the end of the article, can be found in the entry "Forensic Genetics." This article concludes with the statement:

*See Also the Following Articles:*

Classification ■ Molecular Ecology ■ Stock Identity

This reference indicates that these three related articles all provide some additional information about forensic genetics.

## BIBLIOGRAPHY

The Bibliography section appears as the last element of an article, under the heading "References." This section lists recent secondary sources that will aid the reader in locating more detailed or technical information on the topic at hand. Review articles and research papers that are important to a more detailed understanding of the topic are also listed here.

The Bibliography entries in this Encyclopedia are for the benefit of the reader, to provide references for further reading or additional research on the given topic. Thus, they typically consist of a limited number of entries. They are not intended to represent a complete listing of all the materials consulted by the author or authors in preparing the article. The Bibliography is in effect an extension of the article itself, and it represents the author's choice of the best sources available for additional information.

## RESOURCES

The final pages of the *Encyclopedia of Marine Mammals* contain three important resources for the reader.

- *Species List*: This section provides a complete list of living and extinct marine mammal species.

- *Biographies*: This section provides biographical information for more than 50 noted scientists who made important contributions to the field of marine mammal study.
- *Comprehensive Glossary*: This section provides definitions for more than 1000 specialized terms that are used in the articles.

## INDEX

Within the subject index entry for a given topic, references to general coverage of the topic appear first, such as a complete article on the subject. References to more specific aspects of the topic then appear below this in an indented list.

## ENCYCLOPEDIA WEB SITE

The *Encyclopedia of Marine Mammals* maintains its own editorial Web Page on the Internet at:

**http://www.apnet.com/narwhal/**

This site gives information about the Encyclopedia project and features links to many related sites that provide information about the articles of the Encyclopedia. The site will continue to evolve as more information becomes available.

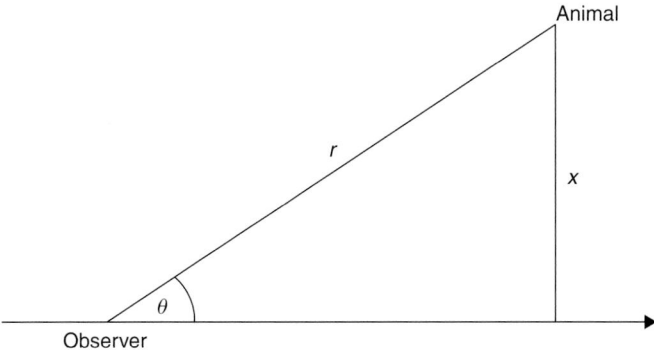

**Figure 1** *The observer records an animal at detection distance* r *and detection angle* θ, *from which the perpendicular distance is calculated as* x = r *sin* θ.

# Abundance Estimation

## Stephen T. Buckland and Anne E. York

Abundance estimation covers the range of techniques by which the size of a population of marine mammals can be estimated. Such population size estimates are often referred to as "absolute" abundance estimates. When it is difficult to estimate absolute abundance with an acceptably low bias, relative abundance indices are often used instead. These are indices that are believed to be proportional to population size, apart from stochastic variation, allowing trends in the population in space and/or time to be assessed. The main techniques for abundance estimation (relative or absolute) are distance sampling, mark–recapture, migration counts, and colony counts.

## I. Distance Sampling

Distance sampling (Buckland *et al.*, 2001, 2004) is the most widely used technique for estimating the abundance of cetaceans. The method is particularly suited to populations of animals that are readily detectable (at least at close quarters) and sparsely distributed over a large area.

The two primary methods of distance sampling are line transect sampling and point transect sampling. The latter method has seldom been applied to marine mammal populations, and we therefore concentrate mostly on line transect sampling. Another distance sampling method is cue counting, which was developed specifically for populations of large whales and the theory for which is closely similar to that for point transects. Distance sampling data may be analyzed using software Distance (Thomas *et al.*, 2006).

### A. Line Transect Sampling

*1. Survey Design* In line transect sampling, the survey design comprises a set of straight lines, randomly or more commonly, systematically spread through the study area for which an abundance estimate is required. For marine mammal surveys, the lines are covered by a team of observers on a ship or boat, or by one or more observers in an aircraft. Because efficiency is improved if lines are placed perpendicular to density contours, a common design for inshore surveys is to have a series of parallel lines as far as possible perpendicular to the coastline. The study area is often divided into geographic blocks or strata, allowing different orientations of the grid of lines in different strata and allowing effort to be greater in high-density strata. For shipboard surveys especially, systematic zig-zag designs are often used because there is then no loss of expensive ship time in traversing off-effort (i.e., not searching for marine mammals) from one line to the next (Buckland *et al.*, 2004). The ship can then be continuously searching for marine mammals during day-light hours.

*2. Assumptions* The following three assumptions should hold:

(1) Animals on or very close to the line are certain to be detected (see later).
(2) Animals are detected before they respond to the presence of the observer, and nonresponsive movement is slow relative to the speed of the observer.
(3) Distances are measured accurately (for ungrouped distance data), or objects are correctly allocated to distance interval (for grouped data).

Bias from nonresponsive movement is generally negligible, provided that the average speed of the animals is less than one-half of the speed of the observer. A fourth assumption is made in many derivations of estimators and variances: whether an object detected is independent of whether any other object is detected. Point estimates are robust to the assumption of independence, and robust variance estimates are obtained by taking the line to be the sampling unit, either by bootstrapping lines or by calculating a weighted sample variance of encounter rates by line.

We do not need to assume that animals are randomly distributed throughout the survey area, provided that lines are placed randomly with respect to the animals. This ensures that objects in the surveyed strip are uniformly distributed on average with distance from the line.

*3. Estimation* Perpendicular distances $x$ are measured from the line to each detected animal. (We will consider the case that animals occur in groups later.) In practice, for shipboard surveys, detection distances $r$ and detection angles $\theta$ are usually recorded, from which perpendicular distances are calculated as $x = r \sin \theta$ (Fig. 1). Suppose there are $k$ lines of lengths $l_1,\ldots,l_k$, with $L = \sum l_j$, and $n$ animals are detected, at perpendicular distances $x_1,\ldots,x_n$. Suppose that animals farther than some distance $w$ from the line are not recorded. Then the "covered area" is $a = 2wL$, within which $n$ animals are detected. However, not all animals within the surveyed area are detected. Let the effective half-width of the strip be $\mu < w$ (so that the proportion of animals within the covered area that are detected is $\mu/w$). Then animal density (number of animals per unit area) is estimated by:

$$\hat{D} = \frac{n}{2\hat{\mu}L} \tag{1}$$

Abundance is estimated as $\hat{N} = A\hat{D}$, where $A$ is the size of the study area. We therefore need an estimate $\hat{\mu}$ of $\mu$. The software Distance provides comprehensive options for these analyses.

Animals often occur in groups, which we term "clusters." If one animal in a cluster is detected, it is assumed that the whole cluster is detected, and the position of the cluster is recorded. Equation (1) then gives an estimate of the density of clusters. To obtain the estimated density of individuals, we must multiply by an estimate of mean cluster size in the population $\hat{E}(s)$:

$$\hat{D} = \frac{n\hat{E}(s)}{2\hat{\mu}L} \tag{2}$$

Because the probability of detection is often a function of cluster size, the sample of cluster sizes exhibits size bias. In the absence of size bias, we can take $\hat{E}(s) = \bar{s}$, the mean size of detected clusters. Several methods exist for estimating $\hat{E}(s)$ in the presence of size bias (Buckland *et al.*, 2001).

These methods assume that once a cluster of animals is detected, it is possible to record the size of that cluster accurately. For shipboard and aerial surveys, this often dictates that at least part of the survey is conducted in "closing mode." After detection, search effort ceases, and the vessel closes with the detected cluster, to allow more accurate estimation of cluster size. This strategy also eases the difficulties of species identification. If "passing mode" is adopted, then underestimation of the size of more distant clusters might be anticipated. Regression methods for correcting size bias also correct for this bias, provided that the sizes of clusters on or close to the transect line are estimated without bias. Where cluster size estimation is problematic, observer training is usually necessary to ensure that bias is not large.

*4. Multiple-Covariate Distance Sampling*   Whether an animal is detected is a function of many factors apart from distance of the animal from the line. Sea state, glare, observer, animal behavior, observation platform, cluster size are a few of the possible factors. When detection on the line is certain, we do not need to model the dependence of detection probability on all these factors, because estimation is pooling robust. However, it is often of interest to estimate how detectability is affected, and the methods to address this issue are available in chapter 3 of Buckland *et al.* (2004) and in Distance (Thomas *et al.*, 2006).

*5. Modeling Density Surfaces*   There is an increasing interest in modeling density surfaces, so that animal density can be related to habitat or environmental variables, or so that abundance for a section of the survey region can be estimated with greater reliability and precision. Figure 2 was obtained using the methods of chapter 4 of Buckland *et al.* (2004) and shows density of minke whales (*Balaenoptera acutorostrata*) in the North Sea, estimated from the 1994 SCANS survey data (Hammond *et al.*, 2002).

*6. Uncertain Detection on the Transect Line*   The standard line transect method assumes that animals on the line are certain to be detected. Double-platform methods in which observers search simultaneously from two platforms are therefore becoming commonplace. This allows extension of the standard methods to the case that animals on the line are not certain to be detected and also, given appropriate field methods, allows adjustment for responsive movement of animals prior to detection by the primary observers. The methods described in chapter 6 of Buckland *et al.* (2004) are available in Distance (Thomas *et al.*, 2006). To reduce the bias arising from unmodeled heterogeneity in the detection probabilities, these

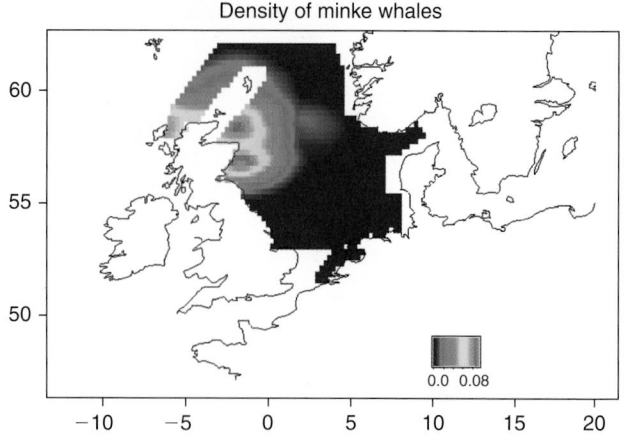

**Figure 2**   *Estimated density of common minke whales in the North Sea.*

methods do not assume full independence between observers, but instead use the weaker assumption of point independence: detections (conditional on recorded covariates) are assumed independent on the line only, as dependence can be expected to be weaker here than away from the line. Double-platform surveys are used widely in cetacean surveys, and have also been used for estimating the abundance of polar bears (*Ursus maritimus*) and seals.

*7. Automated design algorithms and GIS*   Increasingly, survey designs are being developed within Geographic Information Systems (GIS). Automated design algorithms allow the user to avoid the need for designing surveys by hand and allow identification and implementation of more efficient designs. Chapters 7 and 8 of Buckland *et al.* (2004) cover these issues, and Distance (Thomas *et al.*, 2006) has GIS functionality together with a selection of automated design algorithms.

## B. Strip Transect Sampling

Strip transect sampling is a special case of line transect sampling in which it is assumed that all animals out to the truncation distance $w$ are detected. This simplifies analysis, and distances of detected animals from the line need not be measured, except to ensure that they are within distance $w$ of the line. However, the method is seldom efficient for marine mammals; if the strip is narrow enough to ensure that all animals out to $w$ are detected, then many animals are detected beyond $w$, and these observations must be ignored. Abundance of sirenian populations has traditionally been obtained by strip transect methods.

## C. Cue Counting

In cue counting, observers on a ship or in an aircraft cover a sector ahead of their observation platform and record all cues detected within the sector and the distances of the cues from the platform. In principle, the method can be used for any marine mammal, but in practice, it has been used primarily for large whales, for which the cue is the blow. The same design considerations apply as for line transect sampling, although the analysis is essentially the same as for point transect sampling.

If the cues are well defined, such as blows of large whales, then cue counting has the advantage over line transect sampling that the recording unit is the individual cue. Observers need not identify whether different cues are from different animals or how many animals are in a cluster. It also does not matter if some whales stay down so long that they will be undetectable even if they are on the transect line, provided that all cues within the recording sector and very close to the observation platform are detected. Another advantage is that the method requires observer-to-animal distances, which are easier to estimate than perpendicular distances of animals from the line. The main disadvantage is that the method yields estimates of cue density per unit time, which can only be converted into whale density by estimating the cue rate (cues per animal per unit time) from additional costly surveys. The estimated cue rate is prone to bias, both because animals may behave differently when a survey ship is close by and because it is easier to monitor animals that cue frequently, thus biasing the cue rate upward. Additionally, if animals cue more frequently when a ship is bearing down on them, an excess of short distances will be observed in the distance data, biasing the estimation of cue density.

The number of cues per unit area per unit time is estimated by:

$$\hat{D}_c = \frac{2n}{\phi \hat{\rho}^2 T} \tag{3}$$

where $n$ is the number of cues detected in time $T$, $\phi$ is the angle of the sector in radians, and $\hat{\rho}$ is the estimated effective radius of detection. Estimated density is then,

$$\hat{D} = \frac{\hat{D}_c}{\hat{\lambda}}$$

where $\hat{\lambda}$ is the estimated number of cues per animal per unit time (the cue rate). As before, abundance is estimated as $\hat{N} = A\hat{D}$, where $A$ is the size of the study area.

Because cues may be from the same whale, or the same pod of whales, they cannot be assumed independent. This is not a major problem given the robust variance estimation methods provided by software Distance, although model selection tools such as AIC and goodness-of-fit tests are unreliable.

If cues immediately ahead of the vessel might be missed, double-platform methods similar to those for line transect sampling may be used. This has the advantage over those analyses in that it is easier to identify whether a single cue is seen from both platforms, e.g., by recording exact times of cues, than to identify whether a single animal or animal cluster is seen by both platforms, as the two platforms may see different cues from the same animal.

## II. Mark–Recapture

MARK–RECAPTURE tends to be more labor-intensive and more sensitive to failures of assumptions than distance sampling. However, it is applicable to some species that are not amenable to distance sampling methods and, if the marks are long lasting, can yield direct estimates of survival and recruitment rates, which distance sampling cannot do. Mark–recapture methods can be useful for populations that aggregate at some location each year, whereas distance sampling methods are more effective on dispersed populations. The two approaches should therefore be seen as different tools for different purposes. Among marine mammals, mark–recapture has been used most often to estimate the abundance of pinnipeds, usually for the estimation of young of the year. Polar bears have also been the subject of mark–recapture studies. Perhaps the most comprehensive software currently available for analyzing mark–recapture data is MARK (White and Burnham, 1999).

### A. Estimation from a Tagged Subset of Animals

1. *The Petersen Estimator* In its simplest form, mark–recapture consists of marking a sample of $M$ animals from a population of unknown size $N$, returning the animals to the population and then removing, capturing, or observing a sample of $n$ animals. Suppose that, of these $n$ animals, $m$ were marked. We assume that the proportion of marked animals in the second sample is a valid estimate of the proportion of marked animals in the population, giving the following "Petersen" estimator of population size:

$$\hat{N} = nM/m$$

2. *Chapman's Modified Estimator* Inference for the Petersen model is complicated by the fact that the variance of $\hat{N}$ is infinite unless $n + M > N$, in which case $m$ cannot be zero. Chapman's estimator $\hat{N}_c = (n + 1)(M + 1)/(m + 1) - 1$ has finite variance. It also has lower bias (and no bias for $n + M > N$), provided that the assumptions of the estimator are satisfied.

Many variations on this theme have been developed, including extensions to multiple samples, extensions to open populations, "single release" methods, and "single recapture" methods (particularly suited for when the mark is recovered from a dead animal). See "Mark and Recapture Methods", this volume.

3. *Assumptions* The assumptions required for $\hat{N}$ to be a reasonable estimate of population size are that the population of interest is closed over the survey period and that animals are marked and resighted or recaptured at random. In using the ordinary Petersen estimate, it is also assumed that marks are not lost during the survey period and that marking does not affect the probability of resighting or recapturing the animal. Methods have been developed to circumvent these restrictive assumptions and the literature for this topic is rich (see reviews by Pollock, 1991 and Buckland *et al.*, 2000). For most wildlife populations, probabilities of recapture or resighting tend to vary among animals for a variety of reasons. This heterogeneity can be problematic to model and can lead to large bias in abundance estimates, so that the design of a mark–recapture survey should be carefully addressed to minimize heterogeneity.

4. *Estimation of Pinniped Numbers by Mark–Recapture* Mark–recapture techniques have been successfully used to estimate the abundance of young of the year for several species of fur seals. Chapman and Johnson (1968) described the first successful application of this technique for the population of northern fur seals (*Callorhinus ursinus*) on the Pribilof Islands. They marked seals by shearing some hair from their heads and later went back to the colony and counted numbers of marked and unmarked animals within groups of animals. They calculated the Petersen estimate of abundance, which they verified on small colonies where direct counts of young of the year could be made. Resighting was replicated on each colony and several procedures for estimating the variance of the total population size were investigated. These included (1) an empirical estimate calculated as the variance of the mean of replicated estimates for each colony, and the variance of the total calculated by summing the individual colony variances, and (2) a variance for each replicated colony estimate assuming the hypergeometric distribution, with the variance of the mean count for each colony estimated from the variances of the individual counts. They also discussed the use of interpenetrating subsamples to estimate the variance. This procedure is similar in flavor and intent to the bootstrap procedure.

5. *Mark-Recovery Methods* Before the development of line transect methods for estimating the size of populations of large

baleen whales, mark-recovery studies were carried out in which "Discovery" marks were fired into whales, a proportion of which would later be recovered by whalers. Disadvantages of this approach included a requirement for very large sample sizes to ensure an adequate number of recaptures; a long delay before sufficient data accumulated to allow abundance to be estimated; and strong sensitivity of abundance estimates to failures of assumptions. The methods were largely unsuccessful. For a review of the mark–recapture models that are potentially relevant to such data, and of the numerous sources of potential bias in the abundance estimates, see Buckland and Duff (1989). Before the development of mark–recapture or mark–resight techniques for northern fur seal pups, there were many attempts to estimate the population size by tagging pups at birth and recovering the tags in a commercial harvest. This application failed for similar reasons that the use of Discovery tags failed to properly estimate the size of cetacean populations.

### B. Use of Natural Markings

Studies that use natural markings to identify individual animals in a population have become widespread in recent years. These usually rely on photo identification of individuals. A significant milestone in the use of such methods was Hammond *et al.* (1990), which comprises an edited collection of papers from a workshop on the topic. While the technique is undoubtedly of great value, it is important to be aware of its limitations.

Natural markings data can be very effective for estimating survival rates of marine mammals. Abundance estimation is more problematic, as this involves extrapolation from the identified subpopulation. If a high proportion of the population (>80%) can be identified, then abundance estimates are likely to have small bias, especially as there is a tendency to underestimate population size. It is possible to achieve such high rates, e.g., for small coastal populations of bottlenose dolphins (*Tursiops* spp.) and pinniped colonies, provided individuals have distinct markings. The method is then useful because it allows enumeration of almost the whole population without fear of double counting individuals or of seriously underestimating population size. When smaller proportions of animals are identified, estimates of population size can be badly compromised for a variety of reasons. Severe heterogeneity in the "capture" probabilities is common, e.g., because some natural markings are identified much more readily than others or some individuals are more approachable than others. It is notoriously difficult to model such mark–recapture data reliably. Another problem is that the population being estimated is not always well defined, with some animals from elsewhere temporarily entering the population and others temporarily absent. A severe problem for large populations, in which only a small proportion can be identified, is that false positives in the matching procedure, even if they occur only rarely, can lead to a substantial underestimation of population size.

Genetic fingerprinting, if feasible, can reduce this problem substantially. Natural markings studies are invaluable for estimating survival and birth rates, for identifying migration routes, and for detailed studies, including abundance estimation, of a small population. However, they are rarely a cost-effective or reliable method for estimating the size of large populations of marine mammals.

### III. Migration Counts

Many populations of large whales conveniently file past coastal watch points on migration, allowing observers to count a large proportion of the population. This count can then be corrected for animals passing outside watch periods to estimate population size. In practice, further corrections are needed, e.g., to adjust for pods that pass undetected during watch periods, for biased estimation of pod size, for different rates of passage between day and night, and for a component of the population that fails to pass the watch point. Despite the need for various correction terms, migration count data yield very precise estimates of abundance with low bias, provided that the more significant correction factors are estimated reliably. This is unsurprising given that typically 30–40% of the population might be seen by the observers, a much higher fraction than is normal in a distance sampling survey.

The methods usually used for modeling migration counts were developed for the analysis of surveys of the California gray whale (*Eschrichtius robustus*). To estimate the numbers of undetected pods passing during watch periods, two count stations operate independently, and these double-count data are modeled using logistic regression. A polynomial model is used to estimate the rate of passage as it varies through the season, from which numbers of whales passing outside watch periods are estimated. A Bayesian approach is used for analyzing similar data on bowhead whales (*Balaena mysticetus*).

### IV. Colony Counts

Many populations of pinnipeds gather for breeding and pupping at certain times of the year. Researchers often make counts of these populations from cliffs above the colonies, from planes flying overhead, or sometimes from ships passing the colony. Often photographs are taken of the colonies. These are brought back to the laboratory for analysis and form a permanent record of the population. In most pinniped populations, it cannot be assumed that all the animals are on shore at any given time, although in fur seal populations there is a time window in which almost all of the young of the year and breeding males are present, and in certain phocid populations all the young of the year and breeding females are present. Thus, colony counts alone cannot be used to determine absolute abundance of the population size, except for certain classes of animals, and this depends on the reproductive patterns of the population of interest, which must be taken into account when the survey is designed. Serial colony counts can be used to determine the rate of increase of the population if the same proportion of animals is present each year at the colony when the counts are made. This assumption is most likely valid for young of the year. For other segments of the population, this assumption fails if the timing of reproduction changes or if conditions at sea change so that animals need to spend a different amount of time at sea feeding and consequently a different amount of time at the colony.

The size of the harbor seal (*Phoca vitulina*) population in the state of Washington is estimated by combining colony counts made during aerial surveys and mark–recapture to account for animals not present during the aerial surveys. Transponder tags with unique frequencies are attached to animals before the surveys. During the flyovers, animals on shore are counted and radio searches are made to determine the proportion of animals that is ashore. The total population is estimated as $\hat{N}_{tot} = \overline{N}/\hat{p}$ where $\overline{N}$ is the average count of animals on shore and $\hat{p}$ is the estimated fraction of marked animals on shore. The total harbor seal populations in the Gulf of Alaska (Boveng *et al.*, 2003) and in England (Thompson *et al.*, 2005) are estimated from corrected colony counts from aerial surveys. In both cases, the correction factors are estimated from a regression analysis of observed counts on factors that affect the propensity to haul-out on land, such as time of day, tide level, or weather.

A corrected count method is also used to estimate the numbers of southern elephant seal (*Mirounga leonina*) and Antarctic fur seal (*Arctocephalus gazella*) pups on South Georgia. In those surveys, counts of adult females are made from shore or ship along the whole coastline during the pupping season. The counts made at any particular site are then used to estimate the total production for that site based on the adult female haul-out curves and pregnancy/pupping rate estimates from sites that are monitored regularly (twice daily in the case of fur seals) through the breeding season. Similarly, the abundance of northern fur seal pups on the Pribilof Islands is sometimes estimated from mark–recapture estimates on sample colonies coupled with counts of breeding males on all colonies (York and Kozloff, 1987). The ratio of pups to breeding males, estimated on the sampled rookeries, is multiplied by the total count of breeding males on all colonies. Total population size of the stock of northern fur seals has been estimated by multiplying estimated pup numbers by a correction factor derived from a life table assuming a stable age distribution. This method is a very rough tool for estimating the total population and usually no estimates of its bias or variability are computed. In the case of UK gray seals (*Halichoerus grypus*), population dynamics models are fitted to the time series of pup counts, to allow estimation of population size, while accounting for the major sources of uncertainty: observation error, demographic and environmental stochasticity, and model uncertainty (Thomas *et al.*, 2005).

The sizes of colonies of pinnipeds can also be determined using estimates of the area of all colonies coupled with estimates of the density of animals on those colonies. Although this method is often used to estimate the sizes of bird colonies, it has only been used occasionally to estimate pinniped population sizes. The estimates of the areas of the colonies were made from maps of the colonies. Counts or corrected counts, or mark–recapture estimates of the population of interest, are determined on a subsample of colonies. It is assumed that the density of animals in the sampled colonies is representative of the density on all colonies and the total population is estimated by multiplying the total area by the estimated density. Researchers attempted to use this method to estimate the size of the Pribilof northern fur seal population in the late 1940s. At that time, it was thought that the variability of the estimates was too large and efforts were begun to design mark–recapture studies.

Counts, or more often corrected counts, are also sometimes attempted on other marine mammals. For example, because sea otters (*Enhydra lutris*) are difficult to survey by other means, they tend to be counted from a boat. Such counts typically underestimate population size, sometimes substantially (Udevitz *et al.*, 1995). Such counts may be useful as indices if the methods are consistent from survey to survey and if the behavior of animals does not vary in a way that would affect the survey count.

## See Also the Following Articles

Mark–Recapture Surveys

## References

Boveng, P. L., Bengtson, J. L., Withrow, D. E., Cesarone, J. C., Simpkins, M. A., Frost, K. J., and Burns, J. J. (2003). The abundance of harbor seals in the Gulf of Alaska. *Mar. Mamm. Sci.* **19**, 111–127.

Buckland, S. T., and Duff, E. I. (1989). Analysis of the Southern Hemisphere minke whale mark-recovery data. *In* "The Comprehensive Assessment of Whale Stocks: The Early Years" (G. P. Donovan, ed.), pp. 121–143. International Whaling Commission, Cambridge.

Buckland, S. T., Goudie, I. B. J., and Borchers, D. L. (2000). Wildlife population assessment: Past developments and future directions. *Biometrics* **56**, 1–12.

Buckland, S. T., Anderson, D. R., Burnham, K. P., Laake, J. L., Borchers, D. L., and Thomas, L. (2001). "Introduction to Distance Sampling: Estimating Abundance of Biological Populations." Oxford University Press, Oxford.

Buckland, S. T., Anderson, D. R., Burnham, K. P., Laake, J. L., Borchers, D. L., and Thomas, L. (eds) (2004). "Advanced Distance Sampling: Estimating Abundance of Biological Populations." Oxford University Press, Oxford.

Chapman, D. G., and Johnson, A. M. (1968). Estimation of fur seal populations by randomized sampling. *Trans. Am. Fish. Soc.* **97**, 264–270.

Hammond, P. S., Mizroch, S. A., and Donovan, G. P. (eds) (1990). "Individual Recognition of Cetaceans: Use of Photo-Identification and Other Techniques to Estimate Population Parameters." International Whaling Commission, Cambridge.

Hammond, P. S., *et al.* (10 authors) (2002). Abundance of harbour porpoise and other cetaceans in the North Sea and adjacent waters. *J. App. Ecol.* **39**, 361–376.

Pollock, K. H. (1991). "Modelling capture, recapture and removal statistics for estimation of demographic parameters for fish and wildlife populations: Past, present and future." *J. Am. Stat. Assoc.* **86**, 225–238.

Thomas, L., Buckland, S. T., Newman, K. B., and Harwood, J. (2005). A unified framework for modeling wildlife population dynamics. *Aust. N. Z. J. Stat.* **47**, 19–34.

Thompson, D., Lonergan, M., and Duck, C. (2005). Population dynamics of harbour seals *Phoca vitulina* in England: Monitoring growth and catastrophic declines. *J. App. Ecol.* **42**, 638–648.

Thomas, L., *et al.* ( 12 authors) (2006). "Distance 5.0". Research Unit for Wildlife Population Assessment. University of St. Andrews, UK, http://www.ruwpa.st-and.ac.uk/distance/.

Udevitz, M. S., Bodkin, J. L., and Costa, D. P. (1995). Detection of sea otters in boat-based surveys of Prince William Sound, Alaska. *Mar. Mamm. Sci.* **11**, 59–71.

White, G. C., and Burnham, K. P. (1999). Program MARK: Survival estimation from populations of marked animals. *Bird Study* **46**(Suppl), 120–138, Available from http://www.cnr.colostate.edu/~gwhite/mark/mark.htm.

York, A. E., and Kozloff, P. (1987). On the estimation of numbers of northern fur seal, *Callorhinus ursinus*, pups born on St. Paul Island, 1980–86. *Fish. Bull. US* **85**, 367–375.

# Aerial Behavior

## BERND WÜRSIG AND HAL WHITEHEAD

Dolphins and whales (and some pinnipeds, at times) are aerially acrobatic in seemingly exuberant displays of sheer joy. While play may at times be a cause of leaping and other surface activities, there are multiple aerial behavior types and reasons, not all totally understood. We here discuss mainly leaping and breaching, but also mention lunging, spy-hopping, slapping flukes and flippers onto the water surface (lobtailing and flipper slapping), porpoising, and lifting the flukes clear of the water, or fluking.

When large whales leap, the activity is generally termed breaching; when the smaller toothed whales leap, it is generally termed leaping. While the terms can be used interchangeably, we here attempt to stick to those general uses. The breach of a large whale is almost certainly the most powerful action performed by any animal; that of a leaping dolphin rising many body lengths above the surface is one of the most breathtaking. But most breaching and leaping are not immediately functional activities for an aquatic animal, and a question of "Why?" is still only partially answered.

## I. Whale Breaching

Breaching is defined as a jump in which at least 40% of the animal leaves the water. This is distinct from lunging, when less than 40% of the body leaves the water; and lunges are often part of other activity, such as a whale lunging toward another as a part of aggressive display (humpback whale, *Megaptera novaeangliae*, males). The breach (and dolphin leap) may be thought of as an intentional act—leaping out of the water for that express purpose. Dolphins porpoising, to be discussed below, do so as apart of their forward locomotion, and in this special case, the reason for leaping while high-speed swimming is reasonably well understood.

When breaching, sperm whales (*Physeter macrocephalus*) tend to approach the surface vertically from depth (Waters and Whitehead, 1990). Other animals swimming in water less than a few body lengths deep, e.g., humpback and right whales (*Eubalaena* spp.), make a horizontal approach to the breach, gaining speed until, at the last moment, they raise their heads and flukes, pivoting on their flippers, so converting horizontal momentum into vertical motion, and thus rising through the surface (Whitehead, 1985a). To make a full breach, a humpback whale must break the surface at about 15 knots (about 8 m/s), close to its maximum speed. It is likely that some of the most spectacular breaches of other species also represent the full power of the animal.

After breaking the surface, whales have many styles of breaches. In the classic breach of a large whale (such as a humpback or right) the animal emerges from the water at about 20° to 30° from the vertical, twisting so as to land on its back or side, having shown about 90% of its body above the surface at peak emergence (Fig. 1). However, about 20% of the breaches of sperm and humpback whales are "belly flops," with the animal landing ventrally. Breaching whales produce large splashes upon reentry into the water, which can be visible at many kilometers. Frame-by-frame analysis of high-speed photography shows that there are actually two splashes: one is created as the animal falls onto the water surface and initiates a crater of water underneath it, and the other is the secondary splash (and slap sound) produced as the crater collapses on itself, in an act of cavitation.

Breaches are often performed in bouts. Extreme is 130 breaches in 75 min, probably all breaches are performed by the same humpback whale on Silver Bank, West Indies. As such sequences progress, animals tend to show less and less of their bodies, visibly appearing fatigued.

Quantitative breaching rates are only available for a few species and are usually not comparable. However, it is clear that there are substantial differences between species in the rates of this and other forms of aerial activity (Whitehead, 1985b). Frequent breachers include the humpback, right, and sperm whales, as well as virtually all offshore dolphins. In contrast, balaenopterids (blue, *Balaenoptera musculus*; fin, *B. physalus*; sei, *B. borealis*; Bryde's, *B. edeni*; minke, *B. acutorostrata*, *B. bonaerensis*), and most beaked whales (except the northern bottlenose, *Hyperoodon ampullatus*) breach much more rarely.

Breaching rates, then, are not related to size and, at least in the large whales, are inversely related to speed—stouter, slower animals tend to breach more. Instead, interspecifically the best correlate of breaching rate is sociality. Animals found in larger groups, and for whom social structure seems more important, breach more frequently. In sperm whales, the gregarious females breach more often than the much larger, and more solitary, males. Calves of many species breach more frequently than adults.

The circumstances in which breaches occur can provide important clues as to their purpose or function. In some species, different segments of the population breach more frequently than others. For instance, calves of many species breach more frequently than adults. In sperm whales, the gregarious females breach more often than the much larger, and more solitary, males.

Most baleen whales have pronounced seasonal cycles, feeding in winter at high latitudes and breeding in winter nearer the equator. Humpbacks in the western North Atlantic breach about seven times more frequently on their winter breeding grounds in the West Indies than when feeding off Newfoundland.

In a number of species (including humpback, right, and sperm whales) breaching is observed more frequently when groups are merging or splitting. Male humpback whales may breach when they stop singing on the Hawaiian breeding grounds. Breaching is often observed together with lobtailing and/or flipper slapping, with different animals performing different activities at the same time, or one animal switching between the different activities. Studies of humpback and right whales have also shown that breaches by one animal seem to trigger breaches by neighbors. On Silver Bank, breaching humpbacks form clusters about 10 km across. One of the most interesting, and in some ways unexpected, findings that has emerged from several studies of different species is that breaching rates of large whales increase with wind speed. There is currently no generally accepted explanation for this widespread pattern, although it has been postulated that the sound of the breach may serve for better communication in all directions in a surface-noisy ocean.

## II. Dolphin Leaping

The act of dolphins leaping clear of the water is always an energetic and at times a highly acrobatic feat. Just as in large whales breaching, to clear the water the dolphin needs to attain a rapid forward speed and momentum, near the limit of its swimming capability. It generally bends its body abruptly to exit the water and then twists the body mid-air to reenter the water in some structured fashion. Even a noisy "belly flop" after a leap has been designed as such, as multiple similar leaps of the same animals demonstrate. Reentering the water can be head first (unlike in breaching large whales, where it is never

**Figure 1** *A killer whale (*Orcinus orca*) performs a classic breach typical of large whales. The Killer whale is the largest of the delphinids but is not as great a breacher as some of the baleen whales, Photo by Bernd Würsig.*

head-first), creating minimal splash and noise. It can consist of a side, back, or belly splash, resulting in a welter of white water and foam and a considerable percussive (splash) noise in-air and underwater. Finally, there is the "showy" acrobatic leap that consists of spins, somersaults, and various in-air twists. Frame-by-frame analysis of high-speed photographs shows that dolphins control these acrobatics to within split-second timing, affecting muscle movements that allow them to perform the same leap and reentry onto the water again and again. In human terms, a well-trained gymnast or pool diver comes to mind. Dolphin leaps tend to last for 1–2 s, depending on the acrobatics being performed and the size of the leaping individual.

There are three main variations of this leap that tend to create little water disturbance or noise upon reentry. One consists of a "stationary" leap, where the animal comes steeply from depth, usually greater than three body lengths. It leaps in-air, breathes, and tucks its body into a bend to reorient the head downward, then rapidly descends into depth at or very near the original exit point. This leap appears to be executed for the animal to leave whatever it is doing at depth for a minimal time, breathe, and use the in-air weight of its body to regain its position. The need for such an efficient mechanism to breathe becomes clear when we consider that dolphins feeding or mating at depth, e.g., essentially need to interrupt these activities to obtain life-sustaining air. If they can do so rapidly, all the better. The stationary leap is performed singly by dolphins herding food fish below the surface (Würsig, 1979), but often in twos or threes during socializing (Norris *et al.*, 1994) (Fig. 2).

A second head-first reentry leap consists of rapid swimming just below the surface, a very abrupt bend of the body to exit the water, and then a long arcuate in-air leap that may propel the dolphin forward by up to three times of its own body length. While the reentry is head first, there is nevertheless some splashing of water due to the rapidity of the action, kicked up by the body as it exits and again as it enters. This is the "running leap" of dolphins moving at speed, a form of high porpoising. Dolphins propel themselves underwater with several powerful but rapid tail beats and then "sail" through the air, a medium 800 times less dense than water. There is considerable drag generated by crossing the air–water interface, but for an animal that needs to come to the surface to breathe anyway (such as penguins and dolphins), travel efficiency increases above a particular

speed by leaping rather than swimming (Fig. 3). For a 2.5-m-long dolphin, the crossover speed from swimming to leaping is about 4.6 m/s, or 16.6 km/h (Au and Weihs, 1980). Above about 4 m body length (and concomitant body weight), high porpoising is no longer as easy, although killer whales (*Orcinus orca*) moving very rapidly may leap in this manner for short periods of time.

The third head-first reentry leap is designed to gain height. Dolphins, often in twos or threes, leap as high as three times their own body lengths above the surface of the water, usually but not always reentering the water head first. A 2.5-m male pantropical spotted dolphin (*Stenella attenuata*)—spotted dolphins are the champion high leapers—thus leaps about 7 m into the air, or the equivalent of over two apartment stories high. While these leaps may be performed largely for "fun," they may also serve the function of seeing to greater distance by gaining height. Dusky dolphins (*Lagenorhynchus obscurus*) leap in this fashion just before high porpoising toward feeding aggregations with flocking birds some kilometers away (Würsig and Würsig, 1980) (Fig. 4).

Dolphins that noisy leap exit the water in similar fashion as in head-first reentry leaps, but twist the body to reenter with back, side, or belly first. Many noisy leaps end with the dolphin merely falling back onto the water surface. Others are particularly designed to have the animal reenter in a predetermined fashion, and high-speed photography shows subtle tail, flipper, head, or other body readjustments even split seconds before reentering and appearing to be structured to force the body onto the water with a maximal intensity of splash. These

Figure 3   *A bottlenose dolphin* (Tursiop truncatus) *porpoising. Photo by Randall Wells.*

**Figure 2**   *Two dusky dolphins* (Lagenorhynchus obscurus) *perform stationary, or headfirst re-entry, leaps while socializing. Photo by Heidi Pearson.*

**Figure 4**   *Dusky dolphins perform high porpoising leaps while traveling. Photo by Bernd Würsig.*

**Figure 5** *A southern right whale* (Eubalaena australis) *performs a percussive tail slap.*

**Figure 6** *A spinner dolphin* (Stenella longirostris) *spinning.*

observations have led to speculation that noisy leaps are structured for omni-directional communication among dolphins and whales. Indeed, noisy leaps tend to occur more often in higher wind states (when near-surface ambient noise greatly increases), and this observation fits with the hypothesis of communication. Noisy leaps also occur around the periphery of near-surface schools, and in that case, the percussive slaps, as well as the underwater bubble clouds formed by dolphins reentering the water, may serve to frighten fish and cause them to school or aggregate more tightly. Dolphins may at times also stun or debilitate fish prey with the slaps of noisy leaps (as well as with tail slaps), but there is no detailed information on this possibility (Fig. 5).

Some dolphins are especially showy for at least some of their leaps, with spins, somersaults, combinations of flips, head twists, extra tail kicks in-air, and so on. These leaps are almost always associated with an obviously high level of social activity in a school or pod, as evidenced by social rubbing, sexual activity, and a cacophony of whistle and other sounds. Acrobatic leaps usually occur in bouts, with one dolphin (or whale) leaping at least several times. The more social the group, the more leaping dolphins and the longer the individual bouts. These leaps appear graceful to our human eyes and appear particularly structured to be enjoyed in the making and the viewing, like art. However, this may not be the case; we simply do not know.

Spinner dolphins (*Stenella longirostris*) spin by rotating their body rapidly around the long axis up to seven times (usually two to four times) before falling back into the water (Fig. 6). They do so in both

vertical spins and horizontal fashion (Norris and Dohl, 1980; Norris *et al.*, 1994). While twisting of the body in-air may effect some small change in the spin, the angular momentum appears to be almost entirely generated by subsurface corkscrewing in the more dense medium (880 more dense than air!) of water (Fish *et al.*, 2006).

Members of the genus *Lagenorhynchus*, such as Pacific white-sided (*L. obliquidens*) and dusky dolphins, are probably the most aerially acrobatic of all dolphins and whales, with somersaults, twists, and various inventive bends and contortions (Brownell and Cipriano, 1999). Individuals also have the longest bouts of any of the dolphins (some whale breach bouts are as long, and longer; discussed earlier), with up to 36 somersaults having been counted in one dolphin in one about 5-min duration. Interestingly, *Lagenorhynchus* spp. individuals will "never" change leap type during a bout. If a dusky dolphin begins a backward somersault with a half twist to the left and a tail kick just before reentering the water, it will continue this same leap, with no noticeable variation, during that leap bout. Toward the end of the bout, it will tire, muscle action will slow, and the leap will be slightly imperfect. It then quits and breathes while resting at the surface for several minutes. Later, in a different bout, the same individual will leap differently, demonstrating that it knows more than one leap type.

Acrobatic leaps tend to be noisy, but are not structured specifically to make noise. They are structured to be acrobatic, and it is difficult to imagine that they occur for anything but the "fun" (or the art) of it. A more scientifically acceptable explanation may be that acrobatic leaps are not merely an outgrowth of a high level of social activity, but are themselves a call for social activity. Acrobatic leap types may thus serve a social facilitation function that helps to coordinate members of a school or pod. Such facilitation may be especially useful to animals that coordinate finding and aggregating of food and that may need to establish and maintain delicate balances of social and sexual hierarchies. One argument against this stands out: dolphins leaping acrobatically are not being watched by others. They perform their show above the surface while, at any one time of a leap, most or all others are below. Acrobatic leaps may create somewhat different splash sounds from other more simple leaps, but this is not known (Fig. 7).

## III. Other Active Aerial Behaviors

Lunging can be a low form of a would-be breach, and may therefore at times indicate a lowered form of alertness or sociality. But, lunging in whales can also be directed at another individual or individuals, and can signal aggression. Dolphins that play around the eyes and mouths of whales at times apparently do so to elicit lunging from the whales, so that they can ride the "bow-wave" of the lunging behemoth (see Playful Behavior, this volume).

Spy hopping is practiced by both whales and dolphins, and consists of the animal usually quite slowly lifting the head out of the water almost or fully vertical, usually just to the level of the flippers. The term "eye out" might be better for this activity, as we do not know whether the animals are indeed "hopping" out in order to "spy" or look around in-air. At times, spy hopping is attended by a slow rotation of the body, and it does, indeed, appear that the whale or dolphin may be surveying its in-air environment.

Lobtailing or tail slapping consists of forcefully slapping the tail onto the surface of the water (Fig. 5), either ventrum or dorsum up. Large whales are usually oriented more or less vertically in the water column, and the tail stock is sharply bent at the surface in order to produce the tail slap. Smaller toothed whales and dolphins, however, tend to have their bodies oriented parallel to the surface with the

**Figure 7** *A dusky dolphin performing an acrobatic leap. Photo by Jody Weir.*

tail stock not as strongly bent. The slapping action can propel them forward slowly, usually in a semicircle. While the beat frequency in large whales can be quite slow, on the order of several to as many as 10s between slaps, in dolphins it can be up to one or slightly more per second; it is then at times termed "motorboating". Lobtailing produces a loud "crack" sound in air but is not all that loud underwater, and certainly not as loud as many of the vocalizations made by whales and dolphins.

There is high variability of which whale species lobtail and which do so less or not at all. While our review does not pretend to be complete, sperm, right, bowhead (*Balaena mysticetus*), humpback, and gray whales (*Eschrichtius robustus*) are "frequent" lobtailers; but members of the genus *Balaenoptera* tend to lobtail little or not at all. For the smaller cetaceans, Delphinids are champions at lobtailing and other aerial behaviors, but Phocoenids (the porpoises) and river dolphins lobtail (and flipper slap) rarely.

Flipper slapping by whales and dolphins (and by sea lions and fur seals), occurs, as in lobtailing, with both ventrum and dorsum (of the flipper) striking the surface. It also produces a percussive sound in air, and a less loud one underwater.

## IV. Potential Reasons for Aerial Behaviors

It is of course difficult to succinctly summarize our thoughts on reasons for many and diverse gradations of aerial behaviors. Much of the time, especially active behaviors are associated with high levels of alertness, or high levels of sociality. Especially percussive signals may have direct COMMUNICATION function.

A number of authors have suggested that breaching, leaping (when in "noisy" fashion), lobtailing, and flipper slapping may help cetaceans feed by scaring, stunning, herding, or trapping fish or other prey. Although many, probably the majority, of breaches of most species occur in nonfeeding circumstances, the closely related activity of lobtailing is known to sometimes assist feeding, so a direct benefit to food capture cannot be ruled out as a function for some breaches, perhaps especially the "clean-entry leaps" of dolphins.

**Figure 8** *A sperm whale* (Physeter macrocephalus) *in the act of diving and fluking off the coast of New Zealand. Initiation of the dive (top), as the back curves just before lifting the flukes. This fluking sequence shows a particularly high "fluke out" before a fully vertical descent. Photo by B. Würsing*

**A**

Similarly, we cannot completely rule out some rather prosaic potential benefits of breaching and leaping such as stretching, looking above the water, or inhaling water-free air in rough weather. However, these too are unlikely to be important functions of most aerial activities.

One proposed benefit of breaching/leaping that is more consistent with at least some of the evidence is ectoparasite removal. Among the baleen whales, the more heavily infested species tend to be the most frequent breachers, and in Hawaiian spinner dolphins, 44% of spinning dolphins had remoras attached. However, spinners without remoras also breached, and much of the circumstantial evidence points to a very different function—communication.

As noted earlier, the more social species of cetaceans tend to be more aerially active. As sociality is based on communication between members of the same species, the strong inference from these results, then, is that especially percussive aerial activity is a form of communication.

However, there is a paradox in that while breaches, e.g., are excellent at conveying information to visually based human observers above the surface; they are far less prominent for the potential or actual social companions of the breacher. Other whales and dolphins cannot generally see the breacher's body arcing above the surface, and while the reentry makes a noticeable underwater sound, it seems to be less loud than the natural vocalizations of the animals. The paradox may be resolved by considering the theory of "honest signaling": signals are especially useful in animal communication if they convey some important attribute of the signaler that cannot be faked. The distinctive underwater sound or bubble pattern produced by a full breach, while not especially prominent in its own right, is an honest signal of the physical abilities of the breacher (which often seems to leave the surface at close to its maximum speed) and its desire to communicate this by using a significant amount of energy (about 1% of a humpback whale's estimated daily resting metabolic expenditure per breach).

Thus the breach may be a useful signal to nearby potential or actual social companions. What might it be signaling? Suggestions for large whales include aggression, "extreme annoyance" (perhaps with a nearby vessel), an "act of defiance," courtship, or a display of strength by males. Some scientists have suggested that a breach may be used to add emphasis to some other signal, perhaps a vocalization or visual display. By showing the extent of its physical prowess, and expending a significant amount of energy, the whale accentuates the importance of a companion signal. For dolphins, leaps have mainly been considered signals concerning schooling. For instance, it has been suggested that leapers may be used to define the deployment of a school, to recruit dolphins to a cooperative feeding event, or as social facilitators that reaffirm social bonds.

Finally, there is a play. This is probably the most commonly attributed function of breaching, leaping, and much other aerial activity by the general public, but it is also seriously considered by scientists who recognize play as a valid, but hard to define, type of BEHAVIOR. Biological definitions of play usually focus on its lack of immediate biological function, resulting in play becoming a "garbage can" into which activities without an obvious function, such as breaching, get placed in a haphazard fashion. It is likely that many aerial activities described by human observers as "playful" actually function as important signals. However, it is also likely that some, especially those by young animals, provide no immediate benefit, instead, like other forms of play, help to equip the aerially active animal with abilities that may be beneficial in later life.

A discussion of leaping and other aerial activities would be incomplete without mention of other than cetacean marine mammals. Indeed, otariid pinnipeds also high porpoise (probably for the same hydrodynamic efficiency consideration as for dolphins), and especially fur seals and sea otters (*Enhydra lutris*) leap at the surface by rapidly rolling around their own axes. This activity serves to aerate the extremely fine, long, and dense pelage of these marine mammals that use air for insulation. Leaping in pinnipeds and sea otters at times may also consist of play activity, but there is no further information on this point. Especially otariids occasionally flipper slap, at times in apparent play, at times in alarm.

## V. Fluking

Fluking is the act of a whale, dolphin, or porpoise [very rarely, a manatee (*Trichechus* spp.) or dugong (*Dugong dugon*)] raising its tail, or flukes, above the surface of the water during the beginning of a dive (Fig. 8). Usually, whales fluke when diving steeply in water deeper than at least two of their own lengths, although fluking can also occur during shallow submergences. There is great variability in fluking behavior by species: humpback and sperm whales almost always fluke or "fluke out" during the dive; minke and fin whales do so rarely (Leatherwood and Reeves, 1983). Right, bowhead, and gray whales are known to vary the amount and type of fluking depending on whether they are feeding near the surface (no fluking), at moderate depth (occasional fluking), or at depths of 60 m or more ("always" fluking). These species also generally fluke on migration, during the final dive after a series of "near-surface" dives between respirations. Even smaller toothed whales, such as bottlenose dolphins (*Tursiops truncatus*) and pilot whales (*Globicephala* spp.), at times fluke during deep or at least steeply angled dives (Carwardine *et al.*, 1998).

Large whales fluke as a part of bending their bodies as they angle downward, and the fluke out becomes a natural extension of the animal "rolling" forward and down. In smaller delphinids, however, fluking probably has a distinct advantage during the initiation of the dive: the tail and tailstock (or caudal peduncle) held above the surface provide in-air weight to the body and help propel it downward. The effect of this action is quite similar to human skin divers kicking their feet and legs out of water while bending at the waist as they initiate a dive.

Bowhead whales about to dive steeply often stretch their entire bodies so that head and flukes are at the surface, and the belly or mid-part of the body hangs much further below. The spine, in other words, is curved downward. Such a "pre-dive flex" takes about 2 s and occurs just before the last blow before a dive. The pre-dive flex is usually predictive of a fluke-out dive, and pre-drive flex and the flukes out indicate steep, generally deep, diving for feeding or migrating (Würsig and Clark, 1993).

A final form of "flukes out" consists of raising the flukes above water and keeping them there for some time. Dolphins wave their flukes about for at times over 1 min, "headstanding" in apparent play; right, bowhead, and gray whales do so for at times over 10 min. It has been suggested that southern right whales are purposefully sailing by holding their large tails in air during a stiff breeze, but this recreational use of their tails has not been substantiated (Payne, 1995).

Flukes of whales and dolphins have thin trailing edges (similar to the trailing edges of dorsal fins) and are therefore easily tattered or marked by conspecific interactions, getting entangled in lines, or touching other objects. These marks make individual recognition of the flukes possible in those species that habitually fluke out, and fluke-based photographic identification is being practiced for sperm, humpback, and gray whales. In the latter two species, an additional bonus is variable spot or mottling patterns on the flukes (Hammond *et al.*, 1990).

## See Also the Following Articles

Communication ■ Playful Behavior ■ Streamlining ■ Swimming

## References

Au, D., and Weihs, D. (1980). At high speeds dolphins save energy by leaping. *Nature* **284**, 548–550.

Brownell, R. L., Jr., and Cipriano, F. (1999). Dusky dolphins, *Lagenorhynchus obscurus*. *In* "Handbook of Marine Mammals" (S. H. Ridgway, and R. Harrison, eds), 6, pp. 85–104. Academic Press, San Diego.

Carwardine, M., Hoyt, E., Fordyce, R. E., and Gill, P. (1998). "Whales, Dolphins, and Porpoises." Weldon Owen Press, Sydney, Australia.

Fish, F. E., Nicastro, A. J., and Weihs, D. (2006). Dynamics of the aerial maneuvers of spinner dolphins. *J. Exp. Biol.* **209**, 590–598.

Hammond, P. S., Mizroch, S. A., and Donovan, G. P. (eds.) (1990). Individual recognition of cetaceans: Use of photo-identification and other techniques for estimating population parameters. *Rep. Int. Whal. Comn*. Special Issue **12**, 43–52.

Leatherwood, S. L., and Reeves, R. R. (1983). "The Sierra Club Handbook of Whales and Dolphins." Sierra Club Books, San Francisco.

Norris, K. S., and Dohl, T. P. (1980). Behavior of the Hawaiian spinner dolphin, *Stenella longirostris*. *Fish. Bull.* **77**, 821–849.

Norris, K. S., Würsig, B., Wells, R. S., and Würsig, M. (1994). "The Hawaiian Spinner Dolphin." University of California Press, Berkeley.

Payne, R. S. (1995). "Among Whales." Scribner and Sons, NY.

Waters, S., and Whitehead, H. (1990). Aerial behaviour in sperm whales, *Physeter macrocephalus*. *Can. J. Zool.* **68**, 2076–2082.

Wells, R. S., Boness, D. J., Rathbun, G. B., and Behavior, (1999). *In* "Biology of Marine Mammals" (J. E. Reynolds, III, and S. A. Rommel, eds), pp. 324–422. Smithsonian Institution Press, Washington, DC.

Whitehead, H. (1985a). Humpback whale breaching. *Invest. Cetacea* **17**, 117–155.

Whitehead, H. (1985b). Why whales leap. *Sci. Am.* **25**, 84.

Würsig, B. (1979). Dolphins. *Sci. Am.* **240**, 136–148.

Würsig, B., and Würsig, M. (1980). Behavior and ecology of the dusky dolphin, *Lagenorhynchus obscurus*, in the South Atlantic. *Fish. Bull.* **77**, 871–890.

Würsig, B., and Clark, C. W. (1993). Behavior. *In* "The Bowhead Whale" (J. Burns, J. Montague, and C. Cowles, eds), pp. 157–199. Allen Press, Lawrence, KS.

# Age Estimation

## Aleta A. Hohn

### I. Introduction

Age estimation is a tool for obtaining a numerical value of age for animals for which the actual age is not known. Currently, age is estimated primarily from counts of growth layers deposited in several persistent tissues, primarily teeth, less often bone, and in some cases from other layered structures or from chemical signals. Growth layers in the persistent structures are similar in concept to growth rings in trees. Until the use of growth layers became a feasible means of age estimation, relative measures of age, such as tooth wear, pelage or skin color, or fusion of cranial sutures allowed individuals to be placed in age groups; these techniques largely have been replaced with methods that allow for estimation of absolute age by counting growth layers. Marine mammalogists pioneered age estimation from counting growth layers in teeth, initially in pinnipeds (Laws, 1952); this discovery was followed by widespread use for terrestrial mammals as well. Much of the development of this field has focused on how to ensure that age estimates are accurate and precise. That focus has been directed toward verifying the amount of time represented by a growth layer (i.e., calibration or validation), developing increasingly better ways to prepare samples for optimizing counts, and standardizing methods to ensure that growth layer counts are consistent among studies.

Age is fundamental to interpreting and understanding many aspects of the biology of marine mammals. The traditional and most obvious use of age is for estimating parameters used in population dynamics models. Age-specific estimates of fecundity or mortality can be used in these models to project population growth, for example. Estimates of age at sexual maturation are used in absolute terms in population models, while changes in this parameter have been interpreted to reflect changes in population abundance or resource availability and, therefore, indicate a density-compensatory response. Population age structure would also be a useful parameter, although it is rarely known. It is possible, however, to determine the age structure of individuals removed from a population intentionally, such as through directed fisheries, or incidentally, such as in bycatch. This information then can be used to refine estimates of the impact of fisheries on those populations.

The need for accurate and precise estimates of age does not end with traditional population modeling. Of late, there has been increasing concern about the effects of contaminants on the health of marine mammals. Because many of these contaminants bioaccumulate, interpretation of the measured levels of organic or inorganic compounds must be taken as a function of the age, and reproductive condition, of the individual. Furthermore, because indices of health such as blood parameters change naturally with age, understanding the effects of contaminants or other factors on the health of individuals also requires knowing their age. With the recent epizootic events involving morbillivirus (Tautenberger *et al.*, 1996), the ages of individuals infected as well as those with titers indicating previous infections become important in understanding the epidemiology of these outbreaks.

### II. Growth Layer Terminology

In the context of age estimation, the term growth layer is ambiguous. That is because annual increments, as a rule, comprise more than the minimum two growth layers, e.g., a broad layer and a fine layer, needed to differentiate one annual increment from the next. Other layers are usually present; these layers are often referred to as accessory or incremental layers. In the context of age estimation and identifying annual increments, the presence of subannual incremental layers can cause errors and confusion, and have resulted in semantic controversies with regard to the term "growth layer."

To help remedy confusion in terminology, a more descriptive phrase, growth layer group (GLG), was coined at a workshop held in 1977 on estimating age in toothed whales and sirenians (Perrin and Myrick, 1980), predominantly in reference to dentine. Its use has expanded, however, to other marine mammal species and to cement as well as dentine. GLG is a group of layers which occur with cyclical and predictable repetition. Strictly speaking, GLG is a generic term and does not automatically imply deposition that occurred over a 1-year period. It needs to be defined for each species and each

A

use. For practical purposes, however, GLG generally is defined by authors to represent 1 year's deposition, i.e., "annual" is implied. Belows, the term "annual layer" is equivalent to "annual GLG."

## III. Calibration of Annual Layers

Verification that annual layers exist within the complement of visible layers derives from validation or calibration studies. Notably, the first confirmation of annual layers in pinniped teeth occurred soon after teeth were examined for the possibility of age estimation; Scheffer (1950) found external layers (ridges) in the canines of fur seals, *Callorhinus ursinus*, that corresponded to the known age of seals branded as pups and recovered up to 8 years later. Further studies to validate annual layering patterns and to show that patterns are consistent among individuals and species have involved three approaches: (1) examining teeth or bone from animals of known age or known history; (2) examining teeth or bone marked with tetracycline; and (3) comparing growth layers in teeth that have been removed at known intervals (multiple extractions).

For cetaceans, animals of known age or with a known history most often were captive for all or much of their lives. In the latter situation, support for annual layers then hinges on counts of the number of presumed annual layers corresponding to the known age or to the known approximate age of the animal given the length of time it spent in captivity and other data, such as its body length when removed from the wild. Initial encouragement that growth layers in dolphin teeth were annual was from three captive bottlenose dolphins, *Tursiops truncatus* (Sergeant, 1959). Teeth obtained from free-ranging *Tursiops* of known age and known history were significant for confirming and identifying annual layering patterns and determining that annual layers in free-ranging bottlenose dolphins were similar to those in their captive conspecifics (Hohn *et al.*, 1989). Within the pinnipeds, sirenians, and sea otters, numerous studies of free-ranging tagged or individually identified animals have compared the number of growth layers in tooth sections to known ages (e.g., Bowen *et al.*, 1983, and Arnbom *et al.*, 1992). In many of these studies, as in cetacean studies, the actual age of individuals is greater than the "known age" because animals were captured or tagged some time after their birth. Thus, the number of growth layers counted is compared to that minimum age plus an additional number of years estimated as a function of the size of the animal at the time it was first tagged or identified. The most recent and rigorous studies counted growth layers without knowledge of the known ages of specimens in the sample, which eliminates a bias in counting. What is notable about all of these studies is that the authors concluded that they were able to identify annual growth layers (annual GLGs) that correspond to known ages or known approximate ages of the individuals in their samples at least up to some minimum age.

True calibration of growth layer deposition over extended periods of time relative to the lifespan of an animal has not been attempted. To do so would require direct marking of layers, such as through administration of tetracycline, preferably at the same time each year and ideally on the animal's birthday. Tetracycline binds permanently to actively growing mineralized tissue and fluoresces when a bone or tooth section is viewed under ultraviolet light, hence its ability to serve as a marker. Two tetracycline treatments or one treatment followed by extraction have been used to unambiguously identify growth layer deposition over the period of time between the marks, providing a limited calibration; annual layers were determined from this method for several dolphin species, most extensively for spinner dolphins, *Stenella longirostris*, (Myrick *et al.*, 1984) and bottlenose

dolphins (Myrick and Cornell, 1990). Alternatively, multiple extractions of teeth from an individual allow for calibration, but with much restricted sampling opportunities. This method has been used with free-ranging bottlenose dolphins, where two teeth were extracted and growth layer deposition between extractions compared (Hohn *et al.*, 1989). Limited opportunities exist for extensive direct calibration, although captive animals could be used for such studies as could free-ranging populations where individuals are resighted each year and could be caught, administered tetracycline and released.

New techniques continue to be developed as technology becomes increasingly sophisticated. Recently, mass spectrometry was used to identify isotopic signatures of radiocarbon from atomic bomb testing that had been incorporated into teeth of beluga whales (*Delphinapterus leucas*) from Canadian waters (Stewart *et al.*, 2006). Results helped resolve a controversy regarding the number of growth layers deposited annually in beluga teeth.

## IV. Tissues Commonly Used to Obtain Absolute Age Estimates

Given the importance of obtaining age estimates, various tissues and methods have been investigated for elucidating growth layers (Perrin and Myrick, 1980; McCann, 1993; Klevezal', 1996). The most commonly used tissue has been teeth, as they have been for terrestrial mammals (Klevezal' and Kleinenberg, 1967). Fortunately, odontocetes (Figs. 1–3), pinnipeds, sea otters (Fig. 5), and polar bears have teeth that are suitable for use in estimating age. In contrast, teeth cannot be used for baleen whales and manatees, therefore other tissues or methods have been investigated. As alternatives, incremental layers have been found in bone, baleen, and ear plugs. Teeth have several advantages over these other tissues. The normal process of remodeling (resorption and reconstruction) in bone results in resorption of all but the most recent growth layers. For young animals, the number of bone layers may accurately reflect age; otherwise, the number of layers will be less than the age of the animal. The most useful bones are those that show negative allometry, i.e., growing more slowly than the skeleton as a whole (Klevezal, 1996). Growth layers also have been identified in baleen. Unfortunately, baleen abrades fairly quickly during normal use, and relatively few growth layers accumulate. Ear plugs are restricted to just a few species of whales and are challenging to collect.

In the normal course of events, teeth do not remodel, and growth layers continue to be deposited throughout the life of the individual. Teeth are easy to collect, store, and section and have become the preferred means of age estimation for most species with teeth. Within a tooth, two tissues have been used for aging: dentine and cement. New dentine is deposited on the internal surface, i.e., from the pulp cavity side, so that layers deposited when the animal was youngest are found on the outer edges of the tooth or at the crown (Figs. 1–3). Cement or cementum wraps around the outer dentine and functions in anchoring the tooth to its alveolus. In contrast to dentine, new cemental layers are deposited on the external surface (Figs. 3 and 4). In most species of cetaceans, the cemental layer is very thin and the resulting growth layers are so fine that they can be difficult to differentiate. As a result, dentine is primarily used for estimating age. Notable exceptions include the franciscana, *Pontoporia blainvillei*, and the beaked whales, family Ziphiidae, where dentine is useful only for the first few years and then cement, which is extensive, must be used. In addition, for sperm whales, family Physeteridae, and the beluga whale both cement and dentine are well developed and can be used. Because most cetaceans have homodont dentition (the teeth are all approximately the

A

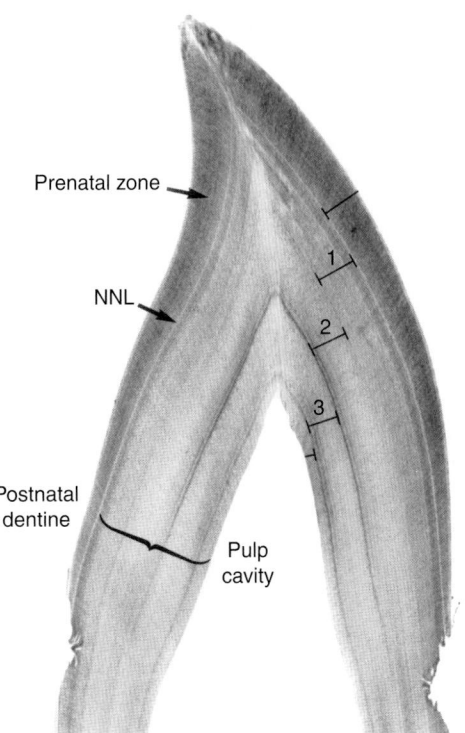

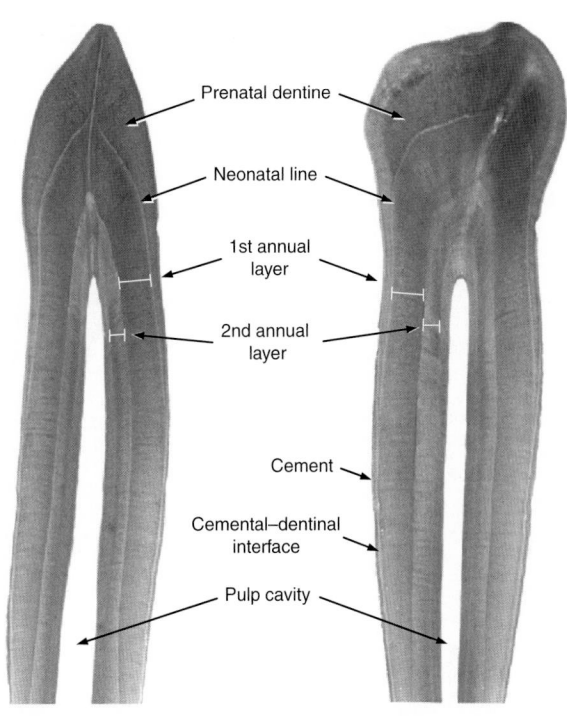

**Figure 1** *Decalcified and stained mid-longitudinal section in the buccal–lingual plane from a free-ranging bottlenose dolphin known to be 3 years of age. This view shows only the upper half of the section. The neonatal line (NNL) represents the time when the animal was born and, therefore, is age "0" for the purpose of estimating age. Dentine external to the neonatal line was deposited before birth and is known as prenatal dentine while the neonatal line and dentine internal to it is postnatal dentine. A thin layer of enamel covered the prenatal dentine but was removed when the tooth was decalcified. The first three complete presumed annual growth layers or GLGs are marked in the sequence they were deposited. Teeth from young dolphins have very little cement and none can be seen in this photograph. Photograph from Hohn* et al. *(1989).*

**Figure 2** *Decalcified and stained mid-longitudinal sections of teeth from a harbor porpoise. Porpoise teeth are spatulate. When sectioned along the buccal–lingual plane, they appear similar to dolphin tooth sections; when sectioned sagitally, the spatulate shape is apparent. The results are comparable in both orientations for this group. A narrow layer of cement occurs externally to the dentine in the part of the tooth that was below the gum line.*

same shape), each tooth contains the same layering pattern, except for the underdeveloped teeth found most anteriorly and posteriorly in the tooth rows.

For pinnipeds, sea otters, and polar bears, cement is most frequently used for age estimation (Garlich-Miller *et al.*, 1993; Bodkin *et al.*, 1997; Figs. 4 and 5), similar to what is done for most terrestrial mammals. For many species, dentine can give accurate age estimates for young animals, but the pulp cavity either becomes occluded or the dentine deposited is too irregular to resolve additional growth layers. Notable exceptions occur in some of the phocids, such as the ringed seal (*Pusa hispida*), Caspian seal (*Pusa caspica*), and the harbor seal (*Phoca vitulina*), where more than 15–20 dentinal layers can be found (Stewart *et al.*, 1996). For these species, which have heterodont dentition, canines are best for counting dentinal layers while the postcanines are better for counting cement.

Although dentine and cement do not remodel like bone, teeth do wear down. When this occurs, generally it is not a problem for age estimation for species whose teeth show limited growth, i.e., do not continue to grow from the root but reach a maximum length when the animal is still relatively young. That is because an important

marker for accurate age estimates is the neonatal line, which is deposited at birth and represents time zero for the purposes of counting growth layers (Figs. 1 and 2). As long as the neonatal line is visible, it is possible to obtain a complete count of growth layers. Initially, the neonatal line extends below the gum line. In species for which tooth growth is limited, even when the tooth wears down above the gum line the neonatal line remains visible in the remaining tooth that was below the gum. In species with continuously growing teeth, such as the walrus (*Odobenus rosmarus*) (including mandibular teeth), bearded seal (*Erignathus barbatus*), beluga whale (*Delphinapterus leucas*), members of the sperm whale family (Fig. 3), and the dugong (*Dugong dugon*), wear continues as the tooth grows up from the root and eventually the neonatal line is worn away. When this occurs, the count of growth layers of dentine or cement is only a minimum. In some species, such as the beluga whale, tooth wear is not uniform and the best estimates of age are made from the least worn tooth.

Manatees (*Trichechus* spp.) present an unusual case for age estimation. In the related dugong, tusks (incisors) and other teeth provide a means for aging using techniques similar to those used for teeth from other species. Manatees lack tusks. Furthermore, they have an indeterminate number of molars that are constantly lost and replaced throughout the life. Therefore, except in young animals, the number of growth layers in a tooth will reflect the age of the tooth but not the age of the individual manatee. As an alternative, it has been demonstrated in manatees that growth layers in tympano-periotic (auditory) bones are annual (Fig. 5) and that resorption occurs at a much slower rate than in other bones, meeting the requirement of a bone with negative allometry. More than 20 annual

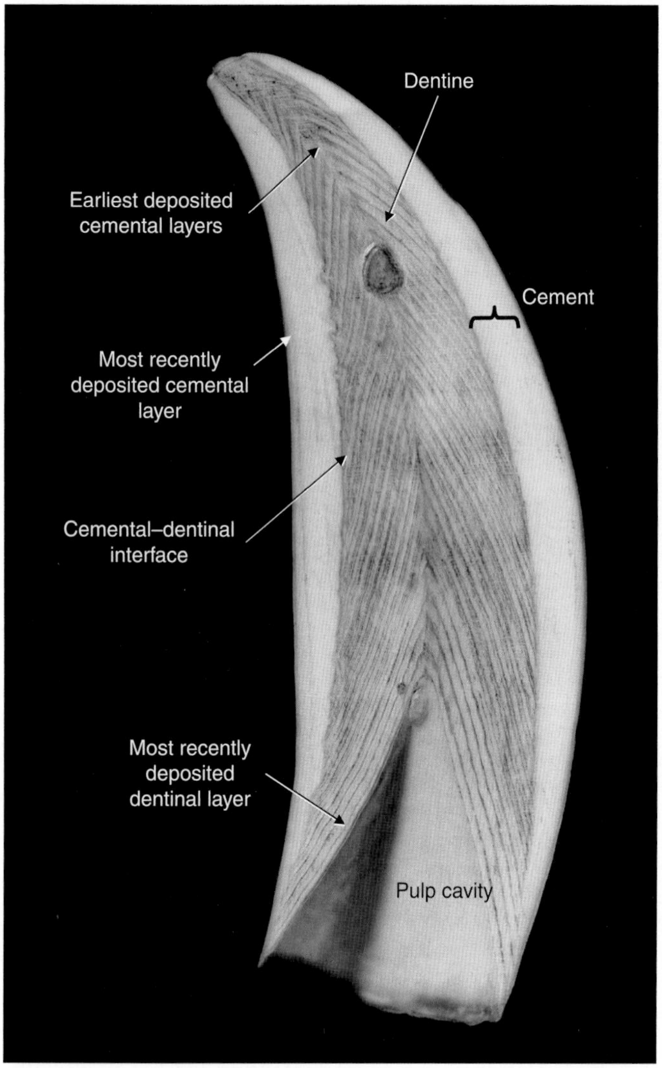

Figure 3 *Mandibular tooth from a sperm whale, cut mid-longitudinally in the buccal–lingual plane, etched in acid, then rubbed with pencil to highlight growth layers in the dentine. Sperm whales have thick cement which from which age can be estimated. In contrast to cement in the dolphin or porpoise tooth, here cement covers all of the dentine. In sperm whales and other species with continuously growing teeth, the tooth adds layers at the bottom (root end) and pushes upward. The cement was deposited when the dentine was still below the gum line. Erupted teeth continuously wear, and in older animals the earliest deposited layers are no longer present. The circular structures are pulp stones that form at the edge of the pulp cavity as globular masses of secondary dentine. Pulp stones are common in some species. Photograph courtesy of the International Whaling Commission.*

layers were found in many specimens and 59 found in a single animal (Marmontel *et al.*, 1996).

Baleen whales also present a special case for age estimation, because they lack teeth. The rorqual whales (family Balaenopteridae) have ear plugs that are deposited in an annual layering pattern throughout life (Laws, 1961; Fig. 6); these are considered accurate for obtaining age estimates. These structures are more difficult to collect and more fragile than teeth or bone. An advantage of ear plugs is that they are not resorbed and do not wear. Other methods of aging have been investigated for balaenopterids as well as the other species of baleen whales. As in manatees, layers occur in the tympanic bullae (auditory bones) in bowhead (*Balaena mysticetus*), gray (*Eschrichtius robustus*), and common minke (*Balaenoptera acutorostrata*) whales (Christensen, 1995), often with no concomitant layers in other cranial or skeletal bones (Klevezal, 1996). Use of tympanic bullae is challenging because extensive effort is required to determine where exactly within the bullae the least amount of resorption and greatest resolution of growth layers will occur. When this region is located, the maximal number of layers will be found. Otherwise, ages will be underestimated. Chemical signals, specifically amino acid racemization, have been used for dolphins and small and large species of whales (Bada *et al.*, 1980), including fin (*Balaenoptera physalus*) and bowhead whales (George *et al.*, 1999). Age is estimated as a function of the proportion of D- and L-isomers of aspartic acid in the lens of the eye.

## V. Collection and Preparation of Tissue for Age Estimation

When the primary tissue to be examined is dentine, especially for old animals, it is critical that a full mid-longitudinal section be obtained. Otherwise, the very fine layers deposited in old animals will be missed. In toothed whales and dolphins (the odontocetes), the possibility of obtaining this mid-longitudinal section is greatly increased if a tooth that is straight in the buccal–lingual plane (cheek to tongue) is used. Generally the largest and straightest teeth occur near the center of the tooth row, and generally teeth are sectioned in the buccal–lingual plane. In some species, sections in the anterior–posterior plane are comparable (Fig. 2). It has become a convention for studies on small odontocetes to use teeth from the center of the left mandible when possible (Perrin and Myrick, 1980). When using specimens from museum collections, often the teeth will have fallen out of the alveoli and so the straightest, largest (in that priority) teeth will be optimal. For studies using cemental layers, postcanines or molars are the generally preferred tissue. In terrestrial mammals, some differences in counts of cemental growth layers among tooth sections from the same individuals have demonstrated that the thickness of the cement influences the deposition pattern, either because the cement is so thin that layers are not readily distinguishable or because the cement is so thick that other incremental layers are apparent and may appear as annual layers (Klevezal, 1996). Differences in cemental thickness can occur both within a molar and between molars (Fig. 4). Ideally, a full investigation of the best site for sectioning can be made to select the optimal tooth and location within that tooth. When that selection has been made, mid-longitudinally sections are more likely than cross-sections to show all of the cemental layers, although cross-sections are commonly used (Klevezal, 1996). As noted above, there is also variability in compact bone thickness in tympanic bullae, resulting in variability in number of growth layers visible; an investigation of the optimal site for sectioning is required. The bone is then cross-sectioned at that site. Ear plugs are sectioned centrally along the long axis of the plug.

Because growth layers are integral to bone and tooth structure, growth layer counts are not sensitive to most of the common ways of storing bones and teeth: cleaned of soft tissue and stored dry such as in museum bone collections, or in alcohol, formalin, or glycerin. It has been suggested that long-term storage in formalin will affect growth layer counts if formalin degrades to formic acid (Perrin and Myrick, 1980). Some teeth will crack at the tip when stored dry, making

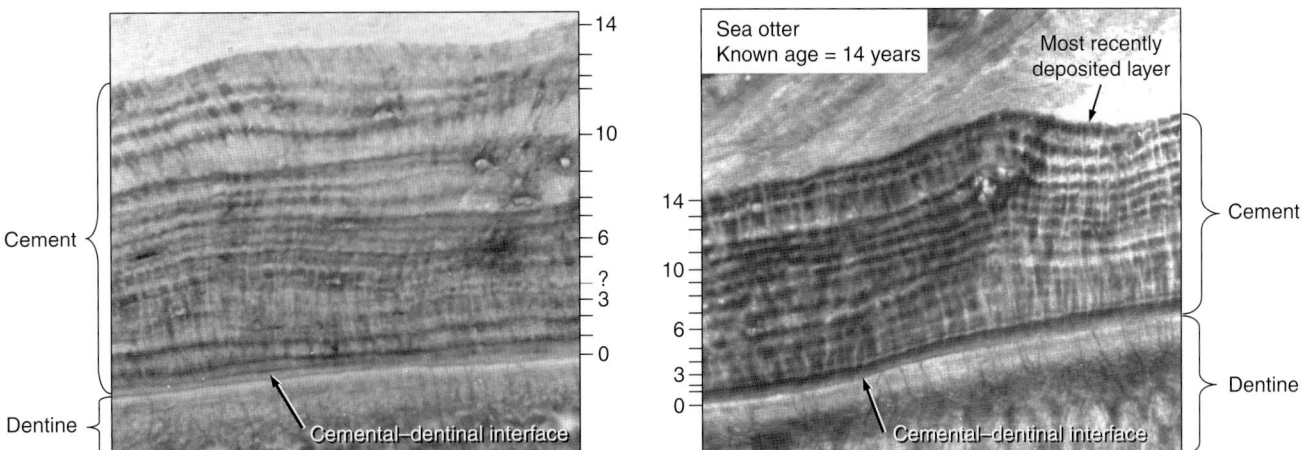

**Figure 4** *Growth layer deposition in cement of a known-age sea otter (14 years). Images are from the same tooth section at different locations. In one location (right image), 14 well-defined, presumably annual layers are visible. These layers are exceptionally clear. In another location (left image), growth layers split and merge: on the right side there appear to be fewer layers, on the left side there appear to be more layers. Presumed annuli are marked on the two images, with the marks on the left image being more subjective and a particularly uncertain layer marked with a dashed line. Counts begin at the interface where the dentine meets the cement; that represents time zero for counting growth layers. Positive identification of annual layers is made by carefully following layers along the tooth to watch for splitting and merging. Photographs of decalcified and stained thin sections courtesy of James Bodkin, USGS, and Gary Matson, Matson's Laboratory, 250X.*

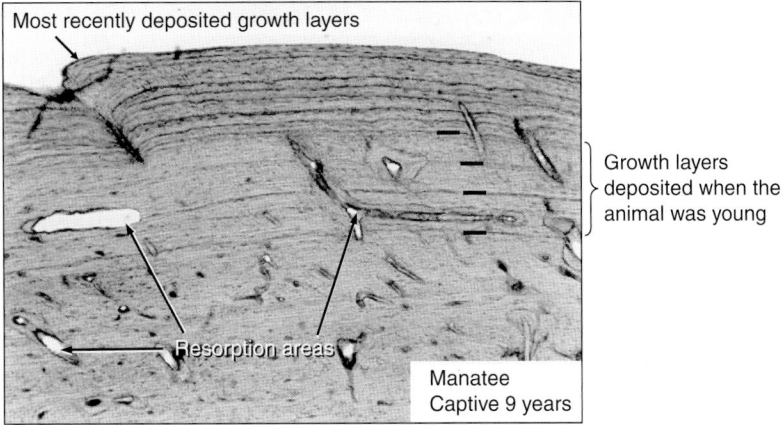

**Figure 5** *Growth layer deposition in the tympano-periotic bone of a manatee that was maintained in captivity for 9 years. Eleven to twelve growth layers can be seen. These layers are primarily on the outer surface of the bone. Even at this age, the bone tissue is being resorbed and beginning the remodeling process. Photograph of decalcified and stained thin section courtesy of Miriam Marmontel and the USGS Sirenia Project.*

sectioning a bit more difficult but not affecting the growth layers. Earplugs are stored in 5–10% buffered formalin (Lockyer, 1984). For amino acid racemization, eye lenses must be collected fresh and immediately frozen (George *et al.*, 1999).

Many creative methods have been tried to obtain the best resolution of growth layers (Perrin and Myrick, 1980). Two of these methods have persisted and become the most widely used: untreated sections (i.e., not decalcified and stained) and decalcified and stained thin sections. The former method generally involves using a low-speed saw with diamond blades to cut a section ranging from 50 to 200 μm thick, depending on the species and tissue, and counting layers directly from that section. The section may or may not

be permanently mounted on a microscope slide. This method was initially the most prevalent one for age estimation from teeth, a little less so for bone, and continues to be widely used because it is fast, easy, and requires little specialized equipment. The increasingly used alternative, decalcified and stained thin sections, requires additional preparation. For this method, whole teeth or thick sections from teeth or bone are decalcified in acid, sectioned on a microtome at from 6 to 25 μm, depending on the species, tissue, and microtome used, then stained in hematoxylin and sometimes counterstained with eosin, two routinely used histological stains. Sections are mounted on a microscope slide. It has become increasingly evident that using the easier method produces inaccurate results

A

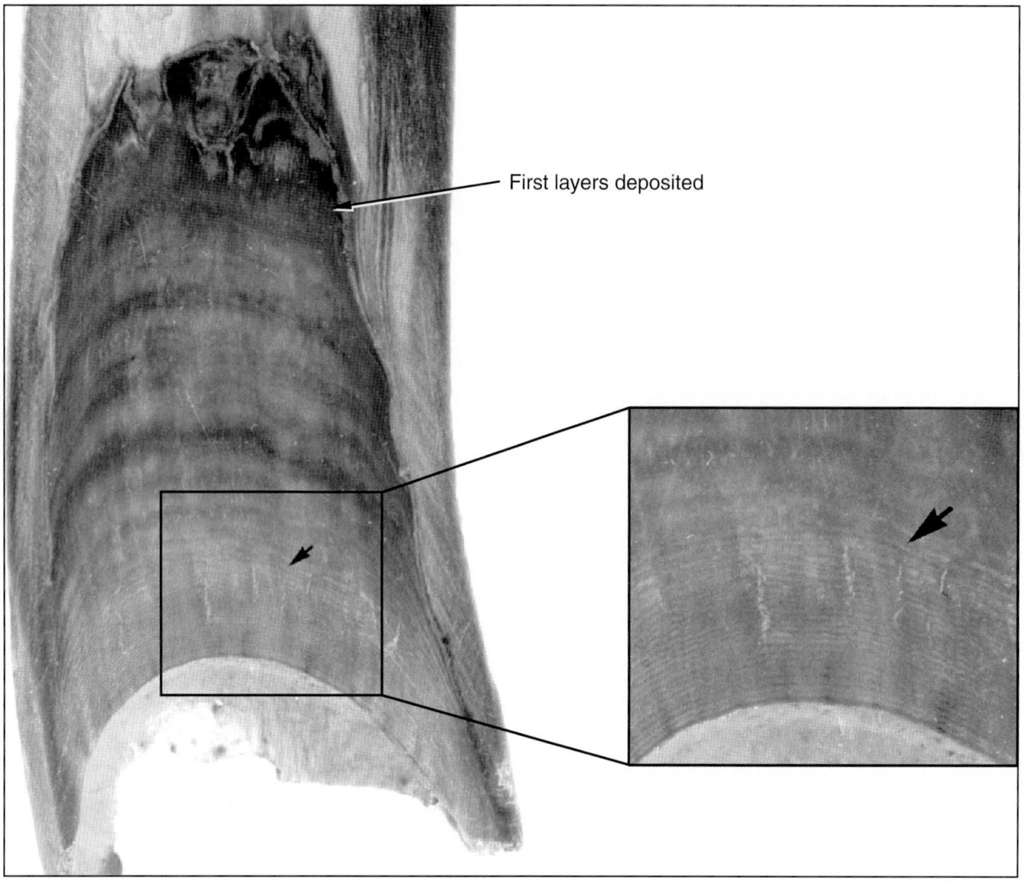

First layers deposited

**Figure 6** *Ear plug from a fin whale, cut mid-longitudinally to expose the growth layers. The arrow denotes a significant and abrupt change in growth layer characteristics that coincides with the transition of the animal from sexually immature to sexually mature. It is called the transition phase. Photograph from Lockyer (1984) and provided courtesy of C. Lockyer and the International Whaling Commission.*

for both bone and tooth sections (Stewart *et al.*, 1996; Hohn and Fernandez, 1999). Stained thin sections allow for much better resolution of growth layers in dentine, cement, and bone to the extent that many layers not apparent in untreated sections are visible and countable in stained sections. This difference is especially apparent in older animals where growth layers become increasingly thinner; staining is required to separate adjacent fine layers. As a result, many estimates of age using untreated sections are underestimates.

## VI. Consistency and Repeatability of Age Estimates

Because annual layers are not the only growth layers present, interpretation of layers is often subjective. Misinterpretation of annual layers or differences in interpretation between investigators or studies leads to errors. For example, is one population but not another actually responding differently to exploitation, or is an apparent difference simply caused by differences in age estimation? Is a population failing to recover because a growth model is incorrect or because the parameters used in that model were incorrect due to misinterpretation of annual layers?

Accuracy and precision are, in large part, influenced by the species being examined. For some species, growth layers are well defined and easily identified, while for other species growth layers are inherently indistinct. Annual layers in polar bear (*Ursus maritimus*)

cement are notably difficult to interpret, at least during the first few years of life. Different areas in the cement have more or less distinct annual layers and accessory layers. Furthermore, different populations within a species may have different growth layer characteristics. For example, harbor porpoise (*Phocoena phocoena*) from the Bay of Fundy have very distinctive growth layers while those from California do not. Within studies it is common to conduct multiple readings of sections by one or more researchers to test the differences between readers or between readings for an individual reader. Measures of precision can be incorporated in models or can be used to evaluate the reliability of ages estimated for a sample.

Consistency and repeatability of age estimates can be increased if the tooth or bone sections are well prepared. Preparing these sections is a multistep process, and at each step the potential for error exists. If the end product is not well done, then the age estimate may be inaccurate or imprecise. For counts using dentine, a large source of error is using a section that is not mid-longitudinal. For all sections, the incorrect stain or degree of staining (light or dark) and under- or over-decalcification also affect the final product in ways that prevent optimal resolution of all growth layers.

Even when sections are perfect, the subjective nature of counting growth layers still results in varying age estimates. Descriptive models of the appearance, size, and complexity of annual layers have been developed to increase consistency, particularly between investigators. These models include photographs with the growth layers interpreted to be annual clearly marked (Hohn, 1990). Such photographs are equally valuable in individual studies to allow other investigators to determine whether the age estimates were obtained using comparable annual layering patterns. To date, such descriptive models have been prepared for bottlenose dolphins (Hohn *et al.*, 1989) and the franciscana (Pinedo and Hohn, 2000). Development of models for other species would increase the accuracy and precision of age estimates. Such models are particularly important and valuable when known-age specimens are available.

## VII. Growth-Layer Tissues as Recording Structures

The annual layering pattern likely is endogenously controlled, while individual subannual growth layers represent events that have a systemic effect on the animal and therefore influence the deposition of the matrix or mineral in teeth, bone, teeth, earplugs, or other tissue. In essence, the annual increments themselves, as well as any layers formed within the annual increments, are recording structures that reflect the physiology of an individual at the time of deposition (Klevezal', 1996). For example, a decrease in thickness of growth layers has been shown to coincide with a general decrease in body growth with age, as well as with maturation, in cetaceans and seals. Following maturation, changes have been shown to occur in the characteristics of growth layers, or in various types of irregularities or anomalies in dentine, reflecting changes in hormone levels associated with maturation, pregnancy, or lactation in females. The appearance of distinctive lines within growth layers (marker lines) has been suggested to indicate parturition in cetaceans and feeding bouts in pinnipeds. Marker lines, as well as anomalies and other tooth characteristics, have demonstrated value in distinguishing animals from various geographic areas, suggesting a means of stock identification (Lockyer, 1999). A decrease in mineral density in growth layers in teeth from Peruvian dusky dolphins (*Lagenorhynchus obscurus*) was found to correspond to an El Niño year (Manzanilla, 1989). Interpreting these structures can serve as another tool for reconstructing life history events for individuals, including changes in growth rates, reproductive events, including number of births, effects of changes in the environment, and distinguishing stocks.

## References

Arnbom, T. A., Lunn, N. J., Boyd, I. L., and Barton, T. (1992). Aging live Antarctic fur seals and southern elephant seals. *Mar. Mamm. Sci.* **8**, 37–43.

Bada, J. L., Brown, S., and Masters, P. M. (1980). Age determination of marine mammals based on aspartic acid racemization in the teeth and lens nucleus. *In* "Age Determination of Toothed Whales and Sirenians" (W. F. Perrin, and A. C. Myrick, Jr., eds), pp. 113–118. International Whaling Commission, Cambridge.

Bodkin, J. L., Ames, J. A., Jameson, R. J., Johnson, A. M., and Matson, G. M. (1997). Estimating age of sea otters with cementum layers in the first premolar. *J. Wildl. Manage.* **61**, 967–973.

Bowen, W. D., Sergeant, D. E., and Øritsland, T. (1983). Validation of age estimation in the harp seal, *Phoca groenlandica*, using dentinal annuli. *Can. J. Fish. Aquat. Sci.* **40**, 1430–1441.

Christensen, I. (1995). Interpretation of growth layers in the periosteal zone of *tympanic bulla* from minke whales *Balaenoptera acutorostrata*. *In* "Whales, Seals, Fish and Man" (A. S. Blix, L. Walløe, and Ø. Ulltang, eds), pp. 413–423. Elsevier Science.

Garlich-Miller, J. L., Stewart, R. E. A., Stewart, B., and Hiltz, E. A. (1993). Comparison of mandibular with cemental growth-layer counts for ageing Atlantic walrus (*Odobenus rosmarus rosmarus*). *Can. J. Zool.* **71**, 163–176.

George, J. C., Bada, J., Zeh, J., Scott, L., Brown, S. E., O'Hara, T., and Suydam, R. (1999). Age and growth estimates of bowhead whales (*Balaena mysticetus*) via aspartic acid racemization. *Can. J. Zool.* **77**, 571–580.

Hohn, A. A. (1990). Reading between the lines: Analysis of age estimation in dolphins. *In* "The Bottlenose Dolphin" (S. Leatherwood, and R. R. Reeves, eds), pp. 575–585. Academic Press.

Hohn, A. A., and Fernandez, F. (1999). Biases in dolphin age structure due to age estimation technique. *Mar. Mamm. Sci.* **15**, 1124–1132.

Hohn, A. A., Scott, M. D., Wells, R. S., Sweeney, J. C., and Irvine, A. B. (1989). Growth layers in teeth from known-age, free-ranging bottlenose dolphins. *Mar. Mamm. Sci.* **5**, 315–342.

Klevezal', G. A. (1996). "Recording Structures of Mammals." A.A. Balkema, Rotterdam.

Klevezal', G. A., and Kleinenberg, S. E. (1967). "Age Determination of Mammals by Layered Structures of Teeth and Bone." Translated from Russian by Israel Progr. Sci. Transl. Jerusalem.

Laws, R. M. (1952). A new method of age determination for mammals. *Nature* **169**, 972–973.

Laws, R. M. (1961). Reproduction, growth and age of southern fin whales, *Balaenoptera physalus*, Linn. *Discov. Rep.* **31**, 327–486.

Lockyer, C. H. (1984). Age determination by means of the earplug in baleen whales. *Rep. Int. Whal. Commn.* **34**, 692–696.

Lockyer, C. H. (1999). Application of a new method to investigate population structure in the harbor porpoise, *Phocoena phocoena*, with special reference to the North and Baltic Seas. *J. Cetacean Res. Manag.* **1**, 297–304.

Manzanilla, S. R. (1989). The 1982–1983 El Niño event recorded in dentinal growth layers in teeth of Peruvian dusky dolphins (*Lagenorhynchus obscurus*). *Can. J. Zool.* **67**, 2120–2125.

Marmontel, M., O'Shea, T. J., Kochman, H. I., and Humphrey, S. R. (1996). Age determination in manatees using growth-layer-group counts in bone. *Mar. Mamm. Sci.* **12**, 54–88.

McCann, T. S. (1993). Age determination. *In* "Antarctic Seals: Research Methods and Techniques" (R. M. Laws, ed.), pp. 199–226. Cambridge University, Cambridge.

Myrick, A. C., Jr., and Cornell, L. H. (1990). Calibrating dental layers in captive bottlenose dolphins from serial tetracycline labels and tooth extractions. *In* "The Bottlenose Dolphin" (S. Leatherwood, and R. R. Reeves, eds), pp. 587–608. Academic Press, San Diego.

Myrick, A. C., Jr., Shallenberger, E. W., Kang, I., and MacKay, D. B. (1984). Calibration of dental layers in seven captive Hawaiian spinner dolphins, *Stenella longirostris*, based on tetracycline labeling. *Fish. Bull. (US)* **82**, 207–225.

Perrin, W. F., and Myrick, A. C., Jr. (eds) (1980). "Age Determination of Toothed Whales and Sirenians." International Whaling Commission, Cambridge.

Pinedo, M. C., and Hohn, A. A. (2000). Growth layer patterns in teeth from the franciscana, *Pontoporia blainvillei*: developing a model for precision in age estimation. *Mar. Mamm. Sci.* **16**, 1–27.

Scheffer, V. B. (1950). Growth layers on the teeth of Pinnipedia as an indication of age. *Science* **112**, 309–311.

Sergeant, D. E. (1959). Age determination in odontocete whales from dentinal growth layers. *Norsk Hvalf. Tid.* **48**, 273–288.

Stewart, R. E. A., Stewart, B., Stirling, I., and Street, E. (1996). Counts of growth layer groups in cementum and dentine in ringed seals (*Phoca hispida*). *Mar. Mamm. Sci.* **12**, 383–401.

Stewart, R. E. A., Campana, S. E., Jones, C. M., and Stewart, B. E. (2006). Bomb radiocarbon dating calibrates beluga (*Delphinapterus leucas*) age estimates. *Can. J. Zool.* **84**, 1840–1852.

Tautenberger, J., Tsai, M., Krafft, A., Lichy, J., Reid, A., Schulman, F., and Lipscomb, T. (1996). Two morbilliviruses implicated in bottlenose dolphin epizootics. *Emerg. Infect. Dis.* **2**, 213–216.

A

# Aggressive Behavior, Intraspecific

## CLAUDIO CAMPAGNA

The heterogeneous phenomena considered as intraspecific, aggressive, or agonistic behaviors represent a conglomerate of social responses, including male disputes over territorial boundaries, female fights to protect an offspring, female harassment and forced copulations, and infant abuse and killing. Agonistic encounters

(a) mediate competition for limited resources economically defendable and valuable to the fitness of an individual (Bartholomew, 1970). Finite resources that can be monopolized would lead to social conflict between individuals of different sexes and generations and of the same sex and similar age class and status. Most often, agonistic confrontation (at least the most conspicuous interactions) involves sexually mature males.

(b) are more common in some social contexts, such as breeding on land in a polygynous mating system, in which competition for resources is typically solved via aggressive disputes. Size and strength (but also agility) correlate positively with winning a contest through exerting dominance over individuals subdued by the costs of rebellion.

(c) have a broad range of costs for actors and recipients, from simple rejection after a ritualized threat display to injury or even death after an overt physical encounter.

The form and frequency of agonistic behavior partially reflect the sophistication of social systems. Aquatic mammals vary widely in the complexity of their societies, thus in the manifestation of agonistic behaviors. The most openly competitive societies characterize the otariids, the walrus, *Odobenus rosmarus*, and phocids that live in crowded conditions (e.g., elephant seals, *Mirounga* spp., and gray seals, *Halichoerus grypus*), a fertile ground for aggressive social interactions (Riedman, 1990). Conversely, polar bears, *Ursus maritimus*, all the mysticetes and river dolphins and some other phocids (e.g., Ross, *Ommatophoca rossii*, and leopard, *Hydrurga leptonyx*, seals) are mostly solitary, except for the mother–calf bond and short periods during reproduction in which breeding males engage in scramble competition over receptive females (Berta and Sumich, 1999). The most complex social systems in the aquatic mammals would characterize some of the odontocete cetaceans such as killer whales, *Orcinus orca*; pilot whales, *Globicephala melas*; bottlenose dolphins, *Tursiops* spp.; or sperm whales, *Physeter macrocephalus* (Connor *et al.*, 1998). These species live in stable social units and show coordinated, cooperative behaviors. The long-term, shared history among individuals of the group would have ritualized many of the overt aggressive responses typical of the polygynous pinnipeds.

## I. Male–Male Competition for Mates

Competition over limited resources to maximize reproductive success would be the most common origin of agonistic encounters. In all the aquatic mammals, males compete for access to reproductive females, either by direct monopolization or, indirectly, through achieving the best place for reproduction or the highest status in a dominance rank. The social and ecological contexts associated with the alternative forms of defense systems would set the stage for the evolution of sexually selected traits, such as dimorphism in size and in special morphological structures (e.g., tusks, manes, elongated snouts).

The behavioral manifestation of conflict directed toward the intimidation of rivals is often referred to as agonistic displays or agonistic social signaling. Behavioral displays include vocal signals, facial expressions, and stereotyped postures and movements, such as static open-mouth threats, open-mouth sparring, foreflipper raise or waving, oblique staring, etc. Overt fighting is commonly a last-resort solution to conflict.

### A. Pinnipeds

The form of male agonistic encounters and its outcome have been described in detail for several pinnipeds. Within the highly polygynous otariids and phocids, there are examples of resource-defense (territorial) and female-defense polygynous systems (Riedman, 1990, and references cited therein). Both types of polygyny may occur in the same species, such as in the South American sea lion, *Otaria flavescens*, as a function of different ecological conditions (Campagna and Le Boeuf, 1988).

The establishment and defense of a territory involves vocal displays, stereotyped postures and movements, and fights. During territorial displays, contenders may rush toward each other with the mouth open or vocalizing, weave heads from side to side, puff out the chest or perform the "oblique stare" posture at one another, but physical contact is carefully avoided. Much of the fighting between otariid males takes place early in the breeding season, when territorial boundaries are being established. When physical contact occurs, it typically lasts a few seconds but may be violent, particularly in the largest sea lions. Fights involve chest-to-chest pushing, vigorous biting of the neck and face, lunging, and slashing at the opponent's flippers, chest, and hindquarters.

In female-defense polygyny, females cannot be sequestered or attracted to a particular place. Males then compete to achieve a position among the females in the breeding colony, and move with the shifting population of females (see Boness and James, 1979, for gray seals and Campagna and Le Boeuf, 1988, for South American sea lions). Association with a particular group of females is not fixed and may change even during the same day due to female redistribution related to the physical environment (high temperatures, variable space due to tidal movements) or to social behaviors (e.g., group raids of ousted males into the colony; Campagna *et al.*, 1988a).

In phocids such as the elephant seals, males aggressively establish a dominance hierarchy, rather than a resource or female defense system. Only the highest ranking individuals have undisturbed reproductive access to females (Le Boeuf and Laws, 1994). During the establishment of hierarchies, males attempt to intimidate each other with vocal displays. If none of the contestants retreat, then a chest-to-chest fight takes place. Fights in elephant seals are violent confrontations and may last half an hour. Each bull throws his weight against the other and slashes at his opponent's face, neck, and back with long canines. Most fights end with multiple lacerations and bloody wounds or even a broken canine tooth; death of one contender does occur on rare occasions.

Vocal threats are a common component of agonistic encounters. Pinnipeds vocalize both in air and underwater (Riedman, 1990). Harp, *Pagophilus groenlandicus*; ringed, *Pusa hispida*; Weddell, *Leptonychotes weddellii*; and bearded, *Erignathus barbatus*; seals and the walrus have a rich underwater vocal repertoire. Males maintain underwater territories and vocalizations seem to be a part of territorial displays. Vocal displays are often complex enough to be called songs, meaning that they occur in predictably similar patterns again and again, as in bird (and human) song. Otariid males, particularly among the fur seals, have a rich variety of airborne threat vocalizations associated with boundary display postures. The California sea lion, *Zalophus californianus californianus*, vocalizes both in air and underwater (several phocids also produce airborne and underwater sounds). The strong airborne calls or barks of California sea lions occur during the breeding and nonbreeding seasons and may serve to advertise dominance. In elephant seals, airborne threat displays consist of loud and directional pulsed sounds that always precede fights.

### B. Cetaceans

There is comparatively little description of agonistic encounters in the rest of the aquatic mammals. Agonistic behaviors to establish dominance relationships were described among dolphins in captivity. Observations of free-ranging cetaceans described a range of behaviors interpreted as agonistic, such as lobtailing, tail, and flipper slaps to the body of other individuals, open-mouth postures, jaw claps, forceful exhalations, chases, body charges and leaps and body slaps, and vocal threat displays (see Wells *et al.*, 1999). Escalated agonistic displays involve striking with flukes, biting, and jousting with tusks.

The scar pattern of some odontocetes has been interpreted as the consequence of tooth marks and violent interactions. Several odontocetes have a conspicuous scar pattern. In Risso's dolphins, *Grampus griseus*, narwhals, *Monodon monoceros*, and in several of the beaked whales, most of the body is covered with scars. At least for the narwhal, scars were associated with intraspecific agonistic encounters (discussed later). Scrape marks are also common in baleen whales. It has been suggested for the southern right whales, *Eubalaena australis*, that males may use the thorny callosities during scramble competition over females.

Males of the humpback whale, *Megaptera novaeangliae*, escort receptive females and vigorously rebuff other males by threatening displays such as thrashing of the flukes. Males also at times violently interact with each other in apparently direct "fights," in so-called social active groups at the surface. The underwater songs of humpback whales on Hawaiian breeding grounds are performed by males, and may serve as communication signals in the context of male competition.

A particularly striking example of male–male competition involves Australian bottlenose dolphins (Connor *et al.*, 1998). Males of this population form stable alliances of 2–3 individuals that cooperate to obtain and control reproductive females. Two alliances occasionally combine efforts to sequester or defend females from another alliance. Alliances in dolphins and group raids in sea lions (discussed later) represent special cases in which competition involves the participation of several individuals simultaneously. Posturing and acoustic threats are common but serious injuries are rare during these male agonistic interactions. One episode of a bottlenose dolphin being rendered unconscious by repeated physical blows to his head was observed during an interaction of a male alliance with a "solo" male, suggesting that some aggressive interactions may be severe (Parsons *et al.*, 2003).

### C. Other Aquatic Mammals

Sea otters, *Enhydra lutris*, polar bears, and sirenians are solitary or live in low-density societies with few direct and prolonged interactions among individuals (Berta and Sumich, 1999). Male sea otters are polygynous, establish breeding territories, and mate in the water.

Females live in low-density areas chosen in relation to the distribution and abundance of food.

During the mating season, polar bear males rove to locate receptive females that are dispersed and solitary. Males access one female at a time. Competition involving physical interactions has been rarely observed but can be inferred from broken teeth and scarring on the head and neck.

Manatees, *Trichechus* spp., form mating groups in which several males compete for access to a receptive female by pushing and shoving each other. Physical competition for females also occurs in dugongs, *Dugong dugon*, with some males presenting scars probably made by the tusks of other males.

## II. Tusks as Special Structures for Aggression?

Two species of marine mammals have extraordinarily developed tusks: the walrus and the narwhal. The two upper canines in both male and female walrus are extraordinarily elongated (Riedman, 1990). The massive tusks of a male can weigh up to 10 pounds and be almost 1 m long. Both sexes use tusks in squabbles, to threaten one another and, perhaps, to establish dominance. Males may force their way to selected places in crowded colonies by pushing and jabbing other walruses with their tusks.

The tusks of narwhals are even more exceptional morphological traits. As a general rule, the left canine in males extends anteriorly into a spiraled tusk to a length that may exceed 2.5 m. Some males have two tusks and a few females also develop one or even two shorter and less robust tusks. It has been suggested that narwhal tusks may be used to disturb or pierce prey, to open-breathing holes in the ice, as defense weapons against predators or as organs of sexual display. Although tusks may be used in more than one context, there is evidence that they largely serve in aggressive encounters (Silverman and Dunbar, 1980; Gerson and Hickie, 1985). Evidence includes direct observations of males crossing tusks and striking them against one another, scar patterns (with adult males having more and larger scars on the head after attaining sexual maturity), a significantly higher incidence of broken tusks in mature males compared to immature individuals or to females, and embedded splinters and tusk tips found in the heads of males. Tusks are also used to spear individuals of other species or, apparently, at rare times even female narwhals.

## III. Sexual Selection and Special Morphological Traits

Pronounced sexual dimorphism in the direction of males being heavier and larger than females is common in all the otariids, the walrus, and some phocids (e.g., elephant and gray seals). This kind of dimorphism often indicates direct physical confrontation among reproductive males involving pushing or strength contests. Dimorphism is not apparent; is slight, or even reversed in most other phocids. Lack or even reverse dimorphism is often accompanied by defense of aquatic territories, aquatic mating, and serial monogamy. Females in these species are often dispersed and breeding occurs over a protracted period. The social and ecological conditions of the latter do not favor frequent direct physical confrontations, but competition does occur and may select in this case for more agile rather than larger individuals.

Among other aquatic mammals, males are much larger than females in some odontocetes, such as killer and sperm whales, whereas dimorphism is reversed in all the mysticetes, possibly due to the female needing as large a body as possible to gestate and nurse her very large calf. Mysticetes may have promiscuous mating systems in which competition for insemination may take place at the level of males displacing each other from the vicinity of a female and of sperm cells displacing or diluting sperm of other males. Gray, *Eschrichtius robustus*, right and bowhead, *Balaena mysticetus*, whales have larger testes than expected based on their body weight, suggesting selection for sperm competition.

Besides dimorphism in body size, males of some species evolved special secondary sexual features that may function in the context of competition for mates. Examples are the enlarged snouts of male elephant seals and gray seals, and the inflatable nasal cavity of hooded seals, *Cystophora cristata*. Hooded seal males can blow a red, balloon-like sac from one nostril that is similar in shape to the long proboscis of elephant seals. These organs may have visual or acoustic effects, and may allow other males and females to judge the quality of a contender or a sexual partner. The developed neck and mane of sea lions with long and thick guard hairs may also have visual effects and serve as a shield that protects internal organs from bites.

## IV. Avoiding Fights

Competition for resources by direct aggression may be a costly experience in species capable of inflicting serious injuries that could lower future fitness of the contestants. Thus, contenders with low chances of success should avoid physical confrontations. Theory predicts that assessment of fighting ability of competitors and of resource value before an escalation of violence, may allow differential adaptive responses on the basis of the perceived asymmetries (Maynard Smith and Parker, 1974). Once a territory or a social hierarchy is established, disputes are asymmetric contests in which territory owners or high-ranking males almost always win. Threat displays may then serve as indicators of quality and motivational state of a contender. Other mechanisms, such as a concept of "dear enemy recognition" in territorial species, have been proposed to prevent escalated contests between neighbors. Individual variation in vocal displays may help territorial males to recognize one another.

In female-defense systems, the proportion of sexually receptive females accessible to a male is variable in space and time. Thus, the level of the asymmetry varies sometimes during the same day within the breeding season. This social context would favor behaviors that are unusual in strict territorial or hierarchical systems, such as group raids in South American sea lions.

## V. Group Raids and Other Forms of Male Harassment of Reproductive Females

Violent behavior toward females that leads to injury and death is common in sea lions and some phocids (e.g., South American sea lions: Campagna *et al.*, 1988a; northern, *Mirounga angustirostris*, and southern, *M. leonina*, elephant seals: Le Boeuf and Mesnick, 1990, Galimberti *et al.*, 2000; New Zealand sea lions, *Phocarctos hookeri*: Chilvers *et al.*, 2005; gray seals, Boness *et al.*, 1995; monk seals, *Monachus* spp., Hiruki *et al.*, 1993).

In the South American sea lion, losers in male–male competition raid the breeding colonies in groups of dozens of individuals (Campagna *et al.*, 1988a). Raiders abduct females from the harems of established males and attempt to mate with them. A male may seize a female in his jaws and hurl her in the air to a spot where he may hold his ground against other males while aggressively keeping her in place. In the process, females are often wounded and can be killed. Group raids may represent a primitive stage of a male alliance or coalition.

Harassed females are injured and sometimes killed by males during mating attempts. Le Boeuf and Mesnick (1990) suggested some social conditions that increase mortality risks to a female during mating: (a) marked male sexual dimorphism, (b) males outnumbering females, (c) use of force or potentially dangerous weapons in mating, and (d) monopolization of mating by a few individuals through direct or indirect control of resources (space, females, food, etc.), with forcible exclusion of the majority of the competitors. All the above traits are common in the majority of polygynous mating systems.

The chief cause of female deaths during the breeding season of elephant seals, the most sexually dimorphic of all the pinnipeds, occurs by traumatic injuries inflicted by males during mating attempts as the females depart the harems for the sea at the end of lactation. Male South American sea lions and elephant seals are 3–5 times heavier than females, have large canines and often bite the neck of the female when copulating. Breeding colonies in the early and late season have a high number of males that intercept departing females and attempt to mate with them. Since the early work by Le Boeuf and Mesnick (1990), increased evidence supports the suggestion that male aggression toward females may be a selective force in shaping female behavior, female choice, maternal performance, and reproductive synchrony (Boness *et al.*, 1995; Galimberti *et al.*, 2000; Chilvers *et al.*, 2005).

## VI. Female Agonistic Behavior

In the polygynous pinnipeds, females are aggressive toward one another and rarely tolerate neighbors close by, which contributes to the regulated density of a site (e.g., Fernández-Juricic and Cassini, 2007). A common context of female agonistic encounters is that of the protection of a pup in a crowded breeding colony. Alien pups are often bitten by females. Aggressive mothers react rapidly and intensively to a threat of their pup originated in a neighbor, and enhance chances of pup survival by decreasing risks of mother–pup separation and pup injury (Christenson and Le Boeuf, 1978).

Females sometimes threaten transient males when the latter approach, or they protest vocally when males mount them. As a result, a harassing male is more likely to be challenged by another male. These challenges generally interrupt a male's approach or mount, and hence a potential copulation. By resisting male copulatory attempts, females can increase the likelihood of mating with a dominant individual, which may be viewed as an indirect form of mate choice.

## VII. Abuse and Killing of the Young

Infanticide is the killing by conspecifics of the young, still dependent on their mothers. Infant or young refers here to a lactating or recently weaned pup, calf, or cub. Infant abuse implies injury of a young either via active violent behaviors or passive neglect. Infant killing and abuse by conspecifics represent a widespread phenomenon in nature. Parental and nonparental infanticide have been reported in almost 100 species of mammals, most of which are terrestrial (Hausfater and Hrdy, 1984; Parmigiani and vom Saal, 1994). Infant killing is the direct outcome of a violent interaction or can result from the indirect neglect of a young or an accident.

This review will focus on violent, nonparental forms of infanticide in aquatic mammals. Parental killing in this group is apparently restricted to the indirect effects of maternal neglect (see Le Boeuf and Campagna, 1994), and will not be treated. Infant abuse is a much more common behavioral occurrence than infanticide. It may imply active violence or passive neglect, and it does not necessarily involve the intended death of the victim. Death in the context of abuse is usually perceived as accidental, a by-product that often follows a process of infection and starvation (Le Boeuf and Campagna, 1994).

Except for otariids and phocids, data on killing and abusing the young are sparse in aquatic mammals. Infanticide is an event that may pass unobserved or unreported. Spotty research coverage, with some species being well known and others virtually unstudied, suggests that the relevance and diversification of abuse and killing of the young may be more widespread than reported here. Explanations of the well-documented cases of abuse and infanticide in aquatic mammals rarely support the adaptive hypotheses that would account for similar episodes in terrestrial species.

## A. Abuse and Killing of the Young by Males

Violent behavior toward young has been described in four of five sea lion species (California and Galapagos, *Zalophus californianus wollebaeki*, sea lions being the exception). Subadult and juvenile males of the South American sea lion abduct (seize), abuse, and kill pups during the breeding season (Campagna *et al.*, 1988b). The behavior was observed in coastal Patagonia (Campagna *et al.*, 1988b), Uruguay (Vaz-Ferreira, 1965), Chile (H. Paves Hernandez and C. Espinoza, pers. comm.), Peru (Harcourt, 1993; Soto *et al.*, 2004), and the Falkland-Malvinas Islands (C. Duck and D. Thompson, pers. comm.). At Punta Norte, Península Valdés, Argentina, 163 successful abductions were recorded in four breeding seasons. More than 20% of the 400 pups born each season were abducted by males. In a typical abduction, a juvenile or subadult male approached the breeding area alone or as part of a group raid (Campagna *et al.*, 1988b), dashed toward a pup and grabbed it. The pups were then abducted away from the breeding group, some were carried out to sea (11% of the abductions), whereas others were released and held close to the abductor. Pups that attempted to escape were shaken violently from side to side, tossed in the air, held crushed against pebble substrate, or submerged. Males defended their abducted pups from other males. Some abductors mounted pups, fully covering them with their massive bodies. About 6% of the pups abducted and 1.3% of the pups born during a season died as a consequence of physical abuse. Dead pups show tooth puncture wounds and extensive hematomas.

Australian, *Neophoca cinerea*, and Hooker's (i.e., New Zealand), *Phocarctos hookeri*, sea lions abduct and abuse pups in a similar fashion described for South American sea lions, with the important difference that adult Hooker's sea lions cannibalize the killed pups. Adult male Australian sea lions grab pups that may be alone or with the mother, and bite, shake, and toss them several times (Higgins and Tedman, 1990). Eight attacks recorded in two breeding seasons resulted in four dead pups (5% of the pups observed) and accounted for 19% of pup mortality in the rookery (Higgins and Tedman, 1990). Adult and subadult male Hooker's sea lions grab pups by the neck, violently thrash them from side to side, and sometimes drown them at sea (Wilkinson *et al.*, 2000). Adult abductors were also observed eating pups. Opportunistic observations on Hooker's sea lions report males abducting pups on two occasions and eating them on nine occasions (Wilkinson *et al.*, 2000). After thrashing the victim repeatedly from side to side, males bit the flesh off the carcass and consumed it. This is the only otariid species for which cannibalism has been described. Immature males apparently do not kill pups, although they may try to keep them under control and occasionally mount them.

Steller sea lions, *Eumetopias jubatus*, may trample, crash into, or push pups over a cliff as an indirect consequence of territorial disputes. In some instances, however, pups are killed as a direct violent action by males.

Episodes of violent behavior toward pups are rare among fur seals, except for the northern fur seals, *Callorhinus ursinus*, (Kiyota and Okamura, 2005). Juvenile, subadult, and adult males of the species may enter the breeding area and sniff, bite, grab, or mount pups. Males also abduct pups of both sexes to other places on land or to sea. Pups are killed by drowning, skull damage, or separation from the mother, the latter leading to emaciation. The behavior closely resembles that of the infanticidal sea lion species.

Among the other pinnipeds, infant abuse and killing have been described in at least four phocids and the walrus. Male northern and southern elephant seals of different age classes kill suckling pups and weanlings (Le Boeuf and Campagna, 1994). Pups are accidentally trampled by bulls in the context of male–male competition and may then die of internal injuries. Weaned pups are abused by pubertal males that attempt to mate with them and, in the process, injure and kill individuals of both sexes (Rose *et al.*, 1991). At the time of first departure to the sea, 30–50% of northern elephant seal weaned pups show signs of having been mounted by a male that range from neck bites, scrapes, cuts, and puncture wounds to deep gashes exposing blubber and profuse bleeding. An adult southern elephant seal male from the Patagonian colony of Península Valdés killed and apparently ate pups. He grabbed weaned pups from the beach, dragged them to sea, kept them underwater until struggling ceased, and then tore off chunks and consumed them. The cannibal returned to the same place at least during two consecutive breeding seasons and killed dozens of weanlings. Male gray seals occasionally shake, toss, bite, mount, and kill pups. There is also evidence of cannibalism in this species (Bédard *et al.*, 1993; Kovacs, 1996). An adult male was involved in killing and eating of pups during three breeding seasons. In a similar modality to the southern elephant seal cannibal, the gray seal male grabbed his victims by the hind flippers, dragged them into the water and drowned them. He later tore off chunks of the pup's body with a biting–shaking action and consumed the blubber, skin, and muscle. Hawaiian monk seal, *Monachus schauinslandi*, males mount pups, and then suffocate or drown them (Hiruki *et al.*, 1993; M. Craig, pers. comm.). Some seals persist in this behavior and may kill many pups. Finally, adult male, female, and immature walruses can jab a pup with their tusks and cause lethal injuries.

In summary, adult, subadult, and juvenile males of several pinniped species injure, abuse or kill suckling and recently weaned pups in the following contexts: (a) accidentally, often as an indirect outcome of trampling and crushing during dominance, female-defense, and territorial disputes; (b) as a direct or indirect consequence of a misdirected sexual assault, such as during abductions and abuse by pubertal males; (c) as a direct or indirect consequence of misdirected aggressive behavior with no clear sexual component, such as attack of pups by territorial males not associated to mounting, herding, or harassing; and (d) as an apparent source of food (cannibalism). The age class involved in the abuse and killing varies with the species. In the Australian and Hooker's sea lions, adults are most aggressive toward pups, but subadult and juvenile males also sequester pups and engage in biting, mounting, and holding them underwater. In the South American sea lions, subadult and juvenile males do most of the abductions; adults are rarely involved in pup abuse. Among phocids, young males seem to be more often involved in abuse than adults; adults may cause pup death or injury as an epiphenomenon of male–male competition.

Reports of violence toward young in the rest of the aquatic mammals are rare. Male polar bears occasionally kill and eat cubs, a behavior that is apparently generalized throughout the Arctic (Taylor *et al.*, 1985). There is indirect evidence suggesting infanticide in the

**A**

common bottlenose dolphin, *Tursiops truncatus* (Patterson *et al.*, 1998). Stranded dolphin calves were found with internal injuries that included contusions around the head and thorax, bone fractures and lacerated organs compatible with violent behavior. The interactions that may have caused the death of the calves were not observed. However, an adult dolphin was observed to interact violently with a dead conspecific calf, and dolphins were also seen to chase and hit harbor porpoises, *Phocoena phocoena*, hard enough to toss them into the air. Stranded harbor porpoises had evidence of trauma similar to that reported for the stranded dolphin calves. Additional indirect evidence of conspecific killing in common bottlenose dolphins is available for the population of the southeast Virginia coast (Dunn *et al.*, 2002). Nine bottlenose dolphins within their first year of life, and thus still dependent upon their mothers, were stranded with multiple skeletal fractures, hematomas, organ lacerations, contusions, and hemorrhages, indicating multidirectional trauma. External signs of trauma were absent, an observation incompatible with predation, boat strike, and fisheries interactions. Lesions were, however, similar to traumatic injuries reported for stranded harbor porpoises and bottlenose dolphins elsewhere and attributed to violent dolphin interactions (Patterson *et al.*, 1998).

### B. Abuse and Killing of the Young by Females

Adult pinniped females repel alien young in the context of aggressive protection of resources intended for their own pup. In the northern elephant seal, females aggressively reject alien pups that approach them (Le Boeuf and Campagna, 1994). They shake, throw, and viciously bite unrelated pups. Attacks may be violent enough to cause extensive wounds or fractures, with subsequent infection and death. Orphan pups that attempt to be nursed from any female are thus particularly vulnerable to attack and injury.

An unusual behavior involving females and resulting in the death of unrelated pups was described for the South American sea lion rookery at Islas Ballestas (Peru; Soto *et al.*, 2004). During the 1997–1998 El Niño (warm water and little food availability) breeding season, virtually all pups starved to death. The following year, only about one-quarter of the females gave birth. These mothers had to defend their newborn pups from the sustained attempts of neighboring females to abduct the latter. It often occurs that otariid females close to parturition attempt to bring alien pups near them. However, the particular breeding context of the post-El Niño year resulted in an unusually high incidence of a behavior that may be related to confusing alien pups with their own pup. Females did not nurse the abducted pups, which were later abandoned. Those that failed to reunite with their mother died from starvation or were killed by young males. In the same El Niño context, a behavior was reported for females who in groups of 4–6 animals abducted and killed pups from other females. The pattern is reminiscent of young male group raids but was considered aberrant and related to low pregnancy rates after the El Niño (Soto *et al.*, 2004).

### C. Male Violent Behavior toward Mature Females and Interspecific Pups

From a behavioral standpoint, abuse and killing of conspecific young by male pinnipeds resemble male violent behaviors directed toward mature females of the same species and toward females and pups of other species. Attributes that allow males to physically overpower competitors would also promote aggressive sexual behaviors

related to achieving access and maintaining control of breeding females. For example, adult and subadult South American sea lion males abduct females from established harems (Campagna *et al.*, 1988a). Abductions involve grabbing, tossing, herding, mounting, and biting. Some females are badly injured and killed in the process. Male harassment of conspecific females may be relatively common in phocids; it was reported at least for both species of elephant seals, the Hawaiian monk seals and the gray seal (Mesnick and Le Boeuf, 1991; Hiruki *et al.*, 1993; Boness *et al.*, 1995).

Strong and large pinniped males with an indiscriminate sexual urge often injure and kill females of other species. Males killing interspecific females during mating attempts were reported in all the sea lions (Miller *et al.*, 1996). South American sea lion males injure and kill South American fur seal, *Arctocephalus australis*, females, and Northern (i.e., Steller) sea lions kill California sea lion females, and even males. Mating attempts with dead females of the same and of a different species occurs in some otariids and phocids, such as the South American sea lion and elephant seals. Abnormal escalation of aggressive sexual behaviors may lead to instances such as one Steller sea lion male killing at least 84 California sea lion females and 12 males over three seasons (Miller *et al.*, 1996), and one southern elephant seal male killing more than 100 Cape fur seal, *Arctocephalus pusillus pusillus*, breeding females over successive breeding seasons (Best *et al.*, 1981).

Sea lion predation of pups of other otariid species typically involves grabbing a pup by the neck, shaking it from side to side, tossing and recovering it, dragging it to sea, submerging and drowning it, biting off flesh, and consuming it. This was described for at least three species. Steller sea lions prey upon northern fur seal neonates (pups under 5 months of age; Gentry and Johnson, 1981). Adult South American sea lions prey on South American fur seal pups (Harcourt, 1993). Hooker's sea lions, a species for which cannibalistic behaviors have been described (Wilkinson *et al.*, 2000), prey on New Zealand fur seals, *Arctocephalus forsteri*, and on Antarctic, A. *gazella*, and subantarctic, A. *tropicalis*, fur seals.

South American sea lions may also injure, abduct, or kill South American fur seal pups without consuming them. Interspecific pup abduction was observed in Peru (Harcourt, 1993), Chile (H. Pavés, pers. comm.), and Uruguay (Vaz Ferreira and Bianco, 1987; Cassini, 1998), where sea lions and fur seals live sympatrically. Males grab a fur seal pup, take it to neighboring beach, and toss and shake the pup, often to death. Young males also defend pups from other sea lions, a behavior that closely resembles the pup abductions within their own species. Fur seal pups may be killed as an indirect consequence of violent treatment, but are not consumed by their abductors. Mothers that attempt to protect their abducted pups typically fail to overpower young males.

In summary, pup killing in some species (e.g., Steller sea lions) is more common in the context of interspecific predation, whereas in others (e.g., South American and Australian sea lions) it occurs more often in a sexual or aggressive social context. In general terms, pup abuse follows a similar pattern to female abuse, with the most aggressive species toward pups being also violent toward conspecific and interspecific females.

### D. The Adaptive Meaning of Infant Abuse and Killing

Several hypotheses have been proposed to account for infanticide in mammals (Hausfater and Hrdy, 1984; Ebensperger, 1998): (1) exploitation or predation, young are killed for nutritional benefits;

(2) resource competition, adult kill unrelated young to increase access to food or breeding space for themselves or their offspring; (3) sexual selection, males kill unrelated offspring to achieve reproductive access to females; (4) parental manipulation, a parent reduces litter size by eliminating all or part of a litter; and (5) social pathology, a maladaptive behavior. Adaptive explanations for the killing of young in aquatic mammals have been suggested for bottlenose dolphins (sexual selection, Connor *et al.*, 1996) and Hooker's sea lions (cannibalism; Wilkinson *et al.*, 2000).

Cannibalism is exceptional among aquatic mammals, and social pathology would be involved in cases such as the cannibal adult male gray seal and the subadult male southern elephant seal described earlier. However, cannibalism in Hooker's sea lions was suggested to fit the food resource hypothesis. Several males kill and eat pups in a fashion similar to that described for the same species preying on fur seals. Pups are easy targets for males and may supply calories in excess of the daily energy requirement of a male, as has been suggested to explain predation of South American sea lions on South American fur seals (Harcourt, 1993). Cannibalism in polar bears appears to occur as carrion feeding and as attacks by males on cubs. There is also evidence of a polar bear male feeding on an adult female, but this is a rare observation of difficult interpretation.

Most instances of infant abuse and killing in pinnipeds are better understood as epiphenomena of indiscriminate sexual and aggressive behaviors (Le Boeuf and Campagna, 1994). Social context, sexual dimorphism, and sexually selected behaviors may set the context for the occurrence of injury and death of the young. Pinniped colonies are often dense, parental investment is limited to females, males are large relative to pups and females, and male movements are frequent in the vicinity of pups. At times during the breeding season, pups may be the most abundant age class in a rookery, increasing the opportunity of social interaction. Reproductive females are aggressive toward conspecifics in general and alien pups in particular. Female aggressive behavior in this context may be explained by the cost of producing milk for individuals that are fasting while nursing. A large proportion of the breeding males do not have sexual access to females, and males have indiscriminate sexual behavior. Pups, particularly those close to be weaned, may be almost as large as young mature females, and are often confused as females. South American sea lion and northern and southern elephant seal males kill pups in the context of misplaced sexual behavior. Abducted South American sea lion pups, e.g., are treated as female substitutes, perhaps as a practice of herding or harem keeping (Campagna *et al.*, 1988b). Pups born in a harem are not likely to be the offspring of the dominant male, since they were sired the previous season. Behavioral mechanisms that can protect pups from direct and indirect violence (e.g., being crushed during male fights) are not then under selective pressure. Infanticide in the Australian sea lions may be the consequence of misdirected aggression. It has been suggested that territorial males may perceive pups as a threat. After killing a pup, males return to their usual position in the territories (Higgins and Tedman, 1990).

It is not yet clear to what extent the abuse and killing of conspecific pups may have a common evolutionary substrate with violent behaviors directed to mature females of the same or other species, and toward the young of other species. Examples among the otariids suggest that a circular gradation may exist from simple predation to infanticide to cannibalism. Steller sea lions kill pups of other species as predators but rarely or never abuse conspecific pups as abductors; South American sea lions prey on pups of other species when adults, and abduct (but do not eat) pups of the same and other species when young; and Hooker's sea lions abduct, abuse, kill and eat conspecific

and interspecific pups. It remains to be determined if this progression is actual. It is possible, however, that the behavioral similarities among these phenomena may indicate a key to understanding the evolutionary origin of abuse.

## See Also the Following Articles

Behavior, Overview ■ Communication ■ Estrus and Estrous Behavior ■ Mating Systems ■ Sexual Dimorphism ■ Sociobiology ■ Territorial Behavior.

## References

Bartholomew, G. A. (1970). A model for the evolution of pinniped polygyny. *Evolution* **24**, 546–559.

Bédard, C., Kovacs, K., and Hammill, M. (1993). Cannibalism by grey seals, *Halichoerus grypus*, on Amet Island, Nova Scotia. *Mar. Mamm. Sci.* **9**, 421–424.

Berta, A., and Sumich, J. L. (eds) (1999). "Marine Mammals: Evolutionary Biology." Academic Press, San Diego.

Best, P. B., Meyer, M. A., and Weeks, R. W. (1981). Interactions between a male elephant seal *Mirounga leonina* and Cape fur seals *Arctocephalus pussilus*. *S. Afr. J. Zool.* **16**, 59–66.

Boness, D. J., and James, H. (1979). Reproductive behavior of the grey seal (*Halichoerus grypus*) on Sable Island, Nova Scotia. *J. Zool. (Lond.)* **188**, 477–500.

Boness, D. J., Bowen, W. D., and Iverson, S. J. (1995). Does male harassment of females contribute to reproductive synchrony in the grey seal by affecting maternal performance? *Behav. Ecol. Sociobiol.* **36**, 1–10.

Campagna, C., and Le Boeuf, B. J. (1988). Thermoregulatory behaviour in the southern sea lion and its effect on the mating system. *Behaviour* **107**, 72–90.

Campagna, C., Le Boeuf, B. J., and Cappozzo, H. L. (1988a). Group raids in southern sea lions. *Behaviour* **105**, 224–249.

Campagna, C., Le Boeuf, B. J., and Cappozzo, H. L. (1988b). Pup abductions, and infanticide in southern sea lions. *Behaviour* **107**, 44–60.

Cassini, M. H. (1998). Inter-specific infanticide in South American otariids. *Behaviour* **135**, 1005–1012.

Chilvers, B. L., Robertson, B. C., Wilkinson, I. S., Duignan, P. J., and Gemmell, N. J. (2005). Male harassment of female New Zealand sea lions, *Phocarctos hookeri*: Mortality, injury, and harassment avoidance. *Can. J. Zool.* **83**, 642–648.

Christenson, T. E., and Le Boeuf, B. J. (1978). Aggression in the female northern elephant seal, *Mirounga angustirostris*. *Behaviour* **64**, 158–172.

Connor, R. C., Richards, A. F., Smolker, R. A., and Mann, J. (1996). Patterns of female attractiveness in Indian Ocean bottlenose dolphins. *Behaviour* **133**, 37–69.

Connor, R. C., Mann, J., Tyack, P. L., and Whitehead, H. (1998). Social evolution in toothed whales. *Trends Ecol. Evol.* **13**, 228–232.

Dunn, D. G., Barco, S. G., Pabst, D. A., and McLellan, W. A. (2002). Evidence for infanticide in bottlenose dolphins of the Western North Atlantic. *J. Wildl. Dis.* **38**, 505–510.

Ebensperger, L. A. (1998). Strategies and counterstrategies to infanticide in mammals. *Biol. Rev* **73**, 321–346.

Fernández-Juricic, E., and Cassini, M. H. (2007). Intra-sexual female agonistic behaviour of the South American sea lion (*Otaria flavescens*) in two colonies with different breeding substrates. *Acta Ethologica* **10**, 1–23.

Galimberti, F., Boitani, L., and Marzetti, I. (2000). Harassment during arrival on land and departure to sea in southern elephant seals. *Ethol. Ecol. Evol.* **12**, 389–404.

Gentry, R. L., and Johnson, J. H. (1981). Predation by sea lions on Northern fur seal neonates. *Mammalia* **45**, 423–430.

Gerson, H. B., and Hickie, J. P. (1985). Head scarring on male narwhals (*Monodon monoceros*): Evidence for aggressive tusk use. *Can. J. Zool.* **63**, 2083–2087.

**A**

Harcourt, R. (1993). Individual variation in predation on fur seals by southern sea lions (*Otaria byronia*) in Peru. *Can. J. Zool.* **71**, 1908–1911.

Hausfater, G., and Hrdy, S. B. (1984). "Infanticide: Comparative and Evolutionary Perspectives." Aldine Press, New York.

Higgins, L. V., and Tedman, R. A. (1990). Effect of attacks by male Australian sea lions, *Neophoca cinerea*, on mortality of pups. *J. Mammal.* **71**, 617–619.

Hiruki, L. M., Gilmartin, W. G., Becker, B. L., and Stirling, I. (1993). Wounding in Hawaiian monk seals (*Monachus schauinslandi*). *Can. J. Zool.* **71**, 458–468.

Kiyota, M., and Okamura, H. (2005). Harassment, abduction, and mortality of pups by nonterritorial male northern fur seals. *J. Mammal.* **86**, 1227–1236.

Kovacs, K. (1996). Grey seal cannibalism. *Mar. Mamm. Sci.* **12**, 161.

Le Boeuf, B. J., and Campagna, C. (1994). Protection and abuse of young in pinnipeds. *In* "The Protection and Abuse of Young in Animals and Man" (S. Parmigiani, and F. vom Saal, eds), pp. 257–276. Harwood Academic Publishers, Chur.

Le Boeuf, B. J., and Laws, R. M. (1994). "Elephant Seals." University of California Press, Berkeley.

Maynard Smith, J., and Parker, G. A. (1974). The logic of asymmetric contests. *Anim. Behav.* **24**, 159–175.

Mesnick, S. L., and Le Boeuf, B. J. (1991). Sexual behavior of northern elephant seals. II. Female response to potentially injurious encounters. *Behaviour* **117**, 262–280.

Miller, E., Ponce de León, A., and DeLong, R. L. (1996). Violent interspecific sexual behavior by male sea lions (*Otariidae*): Evolutionary and phylogenetic implictions. *Mar. Mamm. Sci.* **12**, 468–476.

Parmigiani, S., and vom Saal, F. (1994). "The Protection and Abuse of Young in Animals and Man." Harwood Academic Publishers, Chur.

Parsons, K. M., Durban, J. W., and Claridge, D. E. (2003). Male-male aggression renders bottlenose dolphin (*Tursiops truncatus*) unconscious. *Aquat. Mamm.* **29**, 360–362.

Patterson, I. A. P., Reid, R. J., Wilson, B., Grellier, K., Ross, H. M., and Thompson, P. M. (1998). Evidence for infanticide in bottlenose dolphins: An explanation for violent interactions with harbour porpoises? *Proc. R. Soc. Lond., B, Biol. Sci.* **265**, 1167–1170.

Riedman, M. (1990). "The Pinnipeds: Seals, Sea Lions and Walruses." University of California Press, Berkeley.

Rose, N. A., Deutsch, C. J., and Le Boeuf, B. J. (1991). Sexual behavior of male northern elephant seals: III. The mounting of weaned pups. *Behaviour* **119**, 171–192.

Silverman, H. B., and Dunbar, M. J. (1980). Aggressive tusk use by the narwhal (*Monodon monoceros* L.). *Nature* **284**, 57–58.

Soto, K. H., Trites, A. W., and Arias-Schreiber, M. (2004). The effects of prey availability on pup mortality and the timing of birth of South American sea lions (*Otaria flavescens*) in Peru. *J. Zool.* **264**, 419–428.

Taylor, M. K., Larsen, T., and Schweinsburg, R. E. (1985). Observations of intraspecific aggression and cannibalism in polar bears (*Ursus maritimus*). *Arctic* **38**, 303–309.

Vaz-Ferreira, R. (1965). Comportamiento antisocial en machos subadultos de *Otaria byronia* (lobo marino de un pelo). *Revista de la Facultad de Humanidades y Ciencias, Montevideo* **22**, 203–207.

Vaz Ferreira, R., and Bianco, J. (1987). Acciones interespecíficas entre Arctocephalus australis y Otaria flavescens. *Revista del Museo Argentino de Ciencias Naturales, Zoología* **14**, 103–110.

Wells, R. S., Boness, D. J., and Rathbun, G. B. (1999). Behavior. *In* "Biology of Marine Mammals" (J. E. Reynolds, III, and S. A. Rommel, eds), pp. 324–422. Smithsonian Institution Press, Washington, DC.

Wilkinson, I. S., Childerhouse, S. J., Duignan, P. J., and Gulland, F. M. D. (2000). Infanticide and cannibalism in the New Zealand sea lion. *Phocarctos hookeri. Mar. Mamm. Sci.* **16**, 494–500.

# Albinism

DAGMAR FERTL AND PATRICIA E. ROSEL

Albinism refers to a group of inherited conditions resulting in little or no pigment (hypopigmentation) in the eyes alone or in the eyes, skin, and hair. In humans, all types of albinism exhibit abnormalities in the optic system, including incorrect connections of the optic fibers between the retina and the brain, and incomplete development of the fovea, the area of the retina where the sharpest vision is located (Oetting and King, 1999). It is the presence of these types of eye problems that are best used to define albinism rather than the abnormalities in pigmentation. Inheritance of an altered copy of a gene that does not function correctly is the cause of most types of albinism, and mutations in at least 12 different genes have been identified in different types of albinism in humans (Oetting *et al.*, 2003). Albinistic people most often have white or light skin and hair, and eye color varying from dull gray-blue to brown. In one kind of albinism, only eye color is affected. The "pink" eyes often associated with albinism are due to the reflection from choroid capillaries behind the retina. With pigment lacking in the iris of the eyes, this reflection is visible, similar to red-eye in a flash photograph.

## I. Pigmentation

Mammalian color is almost entirely dependent on presence (or absence) of the pigment melanin in the skin, hair, and eyes. Melanin is produced through a stepwise biochemical pathway in which the amino acid tyrosine is converted to melanin. The enzyme tyrosinase plays a critical role in this pathway, and alterations or mutations in the tyrosinase gene can result in a defective enzyme that is unable to produce melanin, or does so at a reduced rate. Currently, as many as 100 different mutations in this gene have been found associated with albinism (Oetting *et al.*, 2003). At the other end of the spectrum, overproduction of the pigment melanin results in melanism—overly dark animals (Visser *et al.*, 2004). Albinism is differentiated from piebaldism (body pigmentation missing in only some areas) and leucism (dark-eyed anomalously white animals) (Fig. 1). Pigmentation patterns should not be the only criterion used to define albinism, as some mutant phenotypes (pseudoalbinism) may be due to the action of genes at other loci.

**Figure 1**   *Leucistic Antarctic fur seal* (Arctocephalus gazella) *at the isolated subantarctic island, Bouvetøya. Photo by Greg Hofmeyr.*

## II. Albinism and Marine Mammals

Albinism is known to affect mammals, birds, fish, reptiles, and amphibians. In marine mammals, anomalously white individuals have been reported for 21 cetacean species (Fertl *et al.*, 1999; Fertl *et al.*, 2004) and 7 pinniped species (e.g., Rodriguez and Bastida, 1993; Bried and Haubreux, 2000) (Fig. 2). No reports are known of anomalously white sea otters (*Enhydra lutris*) or sirenians. Anomalously white individuals are often presumed to be true albinos. Some of those individuals match the description of true albinism [e.g., there are well-documented reports of albino sperm whales (*Physeter macrocephalus*) and bottlenose dolphins (*Tursiops truncatus*)], but many do not. "Chimo," an anomalously white killer (*Orcinus orca*) captured for display in Canada, was diagnosed postmortem with Chédiak–Higashi Syndrome, (Fig. 3), a type of albinism (Taylor and Farrell, 1973). This inherited disorder is characterized by diluted pigmentation patterns that appear pale gray, eye and white blood cell abnormalities, and a shortened life span. Whales and dolphins also may appear white if extensively scarred, or covered with a fungus, such as Lobo's disease (also known as lobomycosis) (Migaki *et al.*, 1971).

## III. Problems Associated with Albinism

Humans with albinism are often sensitive to light, have limited visual acuity and may display other vision impairments, such as extreme farsightedness, nearsightedness, and astigmatism. There are unpublished reports of apparent vision problems for albino seals, when they are on shore (King, 1983). Costs of this aberrant pigmentation for marine mammals may include reduced heat absorption in colder waters, increased conspicuousness to predators, increased skin and eye sensitivity to sunlight, and impaired visual communication (Hain and Leatherwood, 1982). Despite the costs, some individuals do reach adult age and breeding status.

### *See Also the Following Articles*

Coloration ■ Hair and Fur ■ Vision

### *References*

Bried, J., and Haubreux, D. (2000). An aberrantly pigmented southern elephant seal (*Mirounga leonina*) at Iles Kerguelen, southern Indian Ocean. *Mar. Mamm. Sci.* **16**, 681–684.

Fertl, D., Pusser, L. T., and Long, J. J. (1999). First record of an albino bottlenose dolphin (*Tursiops truncatus*) in the Gulf of Mexico, with a review of anomalously white cetaceans. *Mar. Mamm. Sci.* **15**, 227–234.

Fertl, D., Barros, N. B., Rowlett, R. A., Estes, S., and Richlen, M. (2004). An update on anomalously white cetaceans, including the first account for the pantropical spotted dolphin (*Stenella attenuata graffmani*). *Latin Am. J. Aquat. Mamm.* **3**, 163–166.

A

**Figure 2** *Anomalously white humpback whale* (Megaptera novaeangliae) *sighted off Australia. Photo by Paul Forestell, Pacific Whale Foundation.*

**Figure 3** *An albino killer whale, "Chimo"* (Orcinus orca)*, postmortem diagnosed with Chédiak–Higashi syndrome. Photo by Peter Thomas.*

Hain, J. H. W., and Leatherwood, S. (1982). Two sightings of white pilot whales, *Globicephala melaena*, and summarized records of anomalously white cetaceans. *J. Mammal.* **63**, 338–343.

King, J. E. (1983). "Seals of the World." Cornell University Press, Ithaca.

Migaki, G., Valerio, M. G., Irvine, B., and Garner, F. M. (1971). Lobo's disease in an Atlantic bottle-nosed dolphin. *J. Am. Vet. Med. Assoc.* **159**, 578–582.

Oetting, W. S., and King, R. A. (1999). Molecular basis of albinism: Mutations and polymorphisms of pigmentation genes associated with albinism. *Hum. Mutat.* **13**, 99–115.

Oetting, W. S., Fryer, J. P., Shriram, S., and King, R. A. (2003). Oculocutaneous albinism Type 1: The last 100 years. *Pigment Cell Res.* **16**, 307–311.

Rodriguez, D. H., and Bastida, R. O. (1993). The southern sea lion, *Otaria byronia* or *Otaria flavescens?* *Mar. Mamm. Sci.* **9**, 372–381.

Taylor, R. F., and Farrell, R. K. (1973). Light and electron microscopy of peripheral blood neutrophils in a killer whale affected with Chediak–Higashi syndrome. *Fed. Proc.* **32**, 822.

Visser, I. N., Fertl, D., and Pusser, L. T. (2004). Melanistic southern right-whale dolphins (*Lissodelphis peronii*) off Kaikoura, New Zealand, with records of other anomalously all-black cetaceans. *N.Z. J. Mar. Fresh. Res.* **38**, 833–836.

**Figure 1**   *Male Amazon River Dolphin* (Inia geoffrensis) *in the Rio Negro, Brazil. Characterized by pink body color during adulthood; the largest of the river dolphins. Photo courtesy of Anselmo D'Affonseca.*

# Amazon River Dolphin
## *Inia geoffrensis*

### Vera M.F. da Silva

### I. Characteristics and Taxonomy

The Amazon River dolphin is known as boto or boto-vermelho in Brazil; bufeo colorado in Colombia, Ecuador, and Peru; and toninha and delphin Rosado in Venezuela. It is also known in English as pink dolphin and internationally as boto.

It belongs to the family Iniidae (Mead and Brownell, 2005). The genus *Inia* is currently considered monospecific, with three recognized subspecies: *Inia geoffrensis geoffrensis*, *I. g. boliviensis*, and *I. g. humboldtiana*. However, skull morphology (da Silva, 1994) and genetic analysis (Banguera-Hinestroza *et al.*, 2002) suggest that *I. g. boliviensis* could be a separate species.

The boto (Fig. 1) is the largest of the river dolphins with a maximum recorded body length of 255 cm and body mass of 207 kg for males and 225 cm and approximately 153 kg for females. It is also the most sexually dimorphic of the river dolphins, with males 16% longer and 55% heavier than females (Martin and da Silva, 2006). The body is corpulent and heavy but extremely flexible. Nonfused cervical vertebrae allow movements of the head in all directions. The flukes are broad and triangular; the dorsal fin is long, low, and keel-shaped, extending from the mid-body to the strong, laterally flattened caudal peduncle. The flippers are large, broad, and paddle-like and are capable of separate circular movements. Although most of these characteristics restrict speed during swimming, they allow this dolphin to swim backward and to maneuver between trees and submerged vegetation to search for food in the flooded forest. The rostrum and mandible are prominent, long, and robust. Short bristles along the top of the rostrum persist into adulthood. The melon is small and soft, and the shape can be altered by muscular control. The small, round eyes are functional and the vision is good, both under and above water (Best and da Silva, 1989a,b).

Body color varies with age. Fetuses, neonates, and young animals are dark gray. This pigmentation diminishes in intensity through adolescence to a light gray color. Adult botos become pinker as a consequence of severe scarring on the body surface. Males are pinker than females and more heavily scarred due to intermale aggression (Martin and da Silva, 2006). Adult botos can be dark on the dorsum, but the flanks and underside are pinkish. One albino was captured and maintained in captivity for more than a year in an aquarium in Germany.

### II. Distribution and Abundance

The boto has a wide distribution, occurring almost everywhere it can physically reach without venturing into marine waters. It occurs in six countries of South America—Bolivia, Brazil, Colombia, Ecuador, Peru, and Venezuela—in an area of about 7 million km². It can be found along the entire Amazon River and its principal tributaries, smaller rivers, and lakes, from the mouth near Belém to its headwaters in the Ucayali and Marañon Rivers in Peru. Its principal limits are impassable falls such as those of the upper Xingú and Tapajós Rivers in Brazil, and very shallow waters. A series of rapids and falls along the Madeira River up to the Abuña falls has isolated a population of boto (*I. g. boliviensis*) in the Beni/Mamoré basin in Bolivia, in the southern part of the Amazon Basin. The boto is also found throughout the Orinoco River basin, with the exception of the Caroni and upper Caura Rivers above Para falls in Venezuela. The only connection between the Orinoco and Amazon River Basins is the Cassiquiare Canal, where botos have being sighted. The boto is the most abundant river dolphin. Its current distribution and abundance apparently do not differ from those in the past, although local relative abundance and density are highly seasonal and appear to vary between rivers. During the dry season the dolphins are concentrated in the main channels of the rivers, whereas during the flooded season they disperse into the flooded forest (igapó) and river floodplains (várzea).

Differences in survey methodology used by various workers make comparisons between the results of the different surveys of abundance difficult. Long-distance surveys were carried out on the Solimões–Amazon River, from Manaus to Santo Antonio do Içá-Tabatinga, over a total of 1200 km. The number of sightings averaged 332±55 botos per survey ($n = 9$), and the estimated density was 0.08–0.33 botos/km in the main river and 0.49–0.93 botos/km in the smaller channels. Another boat survey along 120 km of the Amazon River bordering Colombia, Peru, and Brazil carried out by Vidal *et al.* (1997) estimated 345 (CV = 0.12) botos in the study area with a density per km$^2$ of 4.8 in tributaries, 2.7 around islands, and 2.0 along the main banks. In the central Amazon, six multiday visual surveys covering a total trackline distance of 1402 km in strip transect and 810 km in line transect mode were performed throughout the flooding cycle in the Japurá and Amazon Rivers (Martin *et al.*, 2004). The estimated density along river margins across all surveys was 3.7 botos per km$^2$. The density was higher at the river margin, declining with distance toward the center of the river, where the density was lower than for any edge-type habitat along the river margins. In floodplain channels, where a cyclical pattern of boto density followed the water level, Martin and da Silva (2004) estimated peak densities of 18 botos per km$^2$, whereas in rivers and large lakes margins the mean density varied from 1.8 to 5.8 km$^{-2}$. These figures suggest that the boto occurs at higher densities than any other cetacean.

### III. Ecology

The boto is active day and night. The greatest fishing activity occurs between 0600–0900 and 1500–1600 h. It feeds on over 43 species of fish belonging to 19 families. Stomach content analysis has revealed up to 11 fish species in one animal. The mean size of consumed fish is 20 cm (range 5–80 cm), with large fish torn to pieces. In captivity, food sharing has been recorded. Daily consumption is about 2.5% of body weight. The botos diet is unique among cetaceans in that its heterodont dentition allows it to tackle and crush armored prey (da Silva, 1994).

### IV. Behavior and Physiology

The boto is mostly solitary and is not commonly seen in cohesive groups. Group size is generally from one to four individuals. Most groups of two are mother–calf pairs, but mixed groups and groups of males are also common. Large loose aggregations may be seen at confluences and bends of rivers and canals due to large concentrations of fish, or for resting and social purposes. Spatial segregation of the sexes occurs among botos, where the proportion of adult males, at midrising and high water, is higher on the main rivers and lower toward the innermost parts of the flood forest, where most females and calves are found. During low water, all habitats available are equally used by both sexes. Mark–recapture studies carried out in Central Amazon, Brazil, have shown that some individuals are resident in a particular area during the entire year (Martin and da Silva, 2004).

The boto is a slow swimmer with a normal speed of 1.5–3.2 km/h, but bursts of >14–22 km/h have also been recorded. It is capable of strong swimming for some length of time. When surfacing, the melon, tip of the rostrum, and long dorsal keel are out of water simultaneously in a very conspicuous way. The boto does a high-arching roll in which these parts appear sequentially thrust well out of the water. The tail is rarely raised out of the water before a dive. Botos also wave a flipper, show the head or tail above the surface, lob-tail, spy hop, and 'rarely' jump clear of the water.

Studies in captivity indicate that botos are less timid and show less social contact, aggressive behavior, play and aerial behavior than bottlenose dolphins. They are very curious and playful, rarely showing fear of strange objects. However, captivity may not show their true range of behaviors. Wild botos grasp fishermen's paddles, rub against canoes, pull grass under water, throw sticks, and play with logs, clay, turtles, snakes, and fish. They are known to react protectively to injured or captured individuals. Several observers have reported seeing botos in a stationary position, often upside down, with the eyes closed.

### V. Life History

Males attain sexual maturity much later than females at about 200 cm in length. In females, sexual maturity occurs at around 6–7 years of age at body lengths between 175 and 180 cm. Reproductive events are seasonal, and the mating season coincides with low water levels. Gestation time has been estimated at about 11 months, and the calving season is apparently long, with most births occurring at the peak of the river's flood level. The boto's length at birth is about 80 cm. Lactation lasts for more than 1 year and the birth interval is from 2 to 3 years.

### VI. Interactions with Humans

The boto is part of the folklore and culture of Amazonian people, and several legends and myths are commonly known throughout its distribution. Because of these legends, often involving supposed supernatural powers; the boto was protected and respected in the past, although old records mention the use of its oil for illumination. Body parts of incidentally captured animals have been used by local people for medical purposes and as love charms. Recent molecular analysis has revealed that most of the eyes sold in markets in different parts of the Amazonian region, today, as being of *I. geoffrensis* are in reality from *Sotalia guianensis*, which is also known along the coast of Pará and Amapá States as "boto." With increased use of nylon gill nets, machine-made lampara seines, and other fishing techniques, the incidental catching of botos has become more common. Since the mid-1990s a direct catch for use as bait to catch a scavenging catfish, known in Brazil as "piracatinga" (*Callophysus macropterus*), has become the most severe threat to the species. Other threats are the construction of hydroelectric dams on major tributaries affecting the abundance and presence of some species of fish. Dams separate and isolate populations and may reduce the gene pool, thereby increasing chances of local extinction.

*I. geoffrensis* is listed in Appendix II of the Convention on International Trade in Endangered Species of Wild Fauna and Flora (CITES) and is classified by the IUCN as Vulnerable because of serious threats throughout its range.

### See Also the Following Articles

River Dolphins, ■ River Dolphins, Evolutionary History and Affinities

### References

Banguera-Hinestroza, E., Cárdenas, H., Ruiz-Garcýa, M., Marmontel, M., Gaitán, E., Vázquez, R., and Garcýa-Vallejo, F. (2002). Molecular identification of evolutionarily significant units in the Amazon River dolphin Inia sp. (Cetacea: Iniidae). *J. Hered.* **93**, 312–322.

Best, R. C. and da Silva, V. M. F. (1989a). Biology, status and conservation of Inia *geoffrensis* in the Amazon and Orinoco river basin. *In* "Biology and Conservation of the River Dolphins." (W. F. Perrin, R. L. Brownell, Jr.,

K. Zhou, and L. Jiankang, eds.), pp. 23–24, International Union for Conservation of Nature and Natural Resources (IUCN) Species Survival Commission, Gland, Switzerland, Occasional Paper 3.

Best, R. C., and da Silva, V. M. F. (1989b). Amazon River dolphin, Boto. *Inia geoffrensis* (de Blainville, 1817). *In* "Handbook of Marine Mammals" (S. H. Ridgway, and R. J. Harrison, eds), Vol 4, pp. 1–23. Academic Press, London.

Martin, A. R., and da Silva, V. M. F. (2004). River dolphins and flooded forest: Seasonal habitat use and sexual segregation in an extreme cetacean environment. *J. Zool. (Lond.)* **263**, 1–11.

Martin, A. R., and da Silva, V. M. F. (2006). Sexual dimorphism and body scarring in the boto (Amazon River dolphin) *Inia geoffrensis*. *Mar. Mamm. Sci.* **22**, 25–33.

Martin, A. R., da Silva, V. M. F., and Salmon, D. L. (2004). Riverine habitat preferences of botos (*Inia geoffrensis*) and tucuxis (*Sotalia fluviatilis*) in the central Amazon. *Mar. Mamm. Sci.* **20**, 189–200.

Mead, J. G., and Brownell, R. L., Jr. (2005). Order Cetacea. *In* "Mammal Species of the World. A Taxonomic and Geographic Reference." (D. E. Wilson and D. M. Reeder, eds.), Vol. 1, pp. 723–743.

da Silva, V. M. F. (1994). Aspects of the Biology of the Amazonian Dolphins genus *Inia* and *Sotalia fluviatilis*. Ph.D. Thesis, 327pp, University of Cambridge, UK.

Vidal, O., Barlow, J., Hurtado, L. A., Torre, J., Cendon, P., and Ojeda, Z. (1997). Distribution and abundance of the Amazon River dolphin (*Inia geoffrensis*) and the tucuxi (*Sotalia fluviatilis*) in the upper Amazon River. *Mar. Mamm. Sci.* **13**, 427–445.

# Ambergris

## Dale W. Rice

Ambergris is a substance that forms only in the intestines of the sperm whale (*Physeter macrocephalus*). The word comes from the Old French *ambre gris* or "gray amber," as opposed to *ambre jaune*, "yellow amber," which refers to the true, resinous amber. Most ambergris is found in the large intestine or rectum. Probably most lumps of ambergris are eventually voided during defecation, unless they grow too large to pass through the anus. Ambergris is rather rare and may be found in only a few sperm whales. The only fishery in which every whale landed was thoroughly searched for ambergris was that which operated from the island of San Miguel in the Azores from 1934 to 1953; there ambergris was found in only 19 of 1933 whales, or 0.98% (Clarke 2006) (Fig. 1).

Ambergris forms as concretions that usually weigh 0.1–10.0 kg, but rarely much bigger pieces have been recovered; the largest on record, weighing 420 kg, was removed from a 14.9 m bull sperm whale killed in the Southern Ocean on December 21, 1953 (Clarke, 1954). Such huge masses greatly distend the whale's large intestine. Most pieces of ambergris are in the form of an irregular roundish lump, somewhat resembling a potato. Their specific gravity is 0.73–0.95. In consistency they are solid and friable, similar to nearly dry clay. Internally they usually show no laminations, but when broken apart they tend to fracture along concentric cleavage surfaces. In color they are pale yellowish to light gray on the inside, while the outer surface is dark brown with a varnished appearance. The chitinous beaks of cephalopods may almost invariably be found imbedded in the lumps. Fresh ambergris has the highly distinctive pungent odor of sperm whale feces, but aged pieces have an almost pleasant musty or even musky smell.

**Figure 1**   *Lump of ambergris recovered from the rectum of a male sperm whale 16.5 m long (specimen no. DWR 19600189) killed near the Farallon Islands off central California on 11 May 1960. Note the smooth spherical cleavage surface partially exposed on the upper left where several chunks have been broken off. Scale in centimeters.*

Chemically ambergris is a nonvolatile solid consisting mainly of a mixture of waxy, unsaturated, high molecular–weight alcohols. The principal components are epicoprosterol and an ester of ambrein. Epicoprosterol and coprosterol have been found in the feces of other mammals. Ambrein ($C_{23}H_{39}OH$) is the substance which gives ambergris its peculiar properties and odor (Gilmore, 1951). One analysis gave the following chemical composition: ambrein, 25–45%; epicoprosterol, 30–40%; coprosterol, 1–5%; coprostanone, 3–4%; cholesterol, 0.1%; pristane, 2–4%; ketone, 3–4%; free acids, 5–8%; residues insoluble in ether, 10–16% (Berzin, 1971; this analysis was mistranslated in the 1972 English edition of Berzin's book).

The circumstances that induce the production of ambergris are poorly understood. Clarke (2006) hypothesized that the formation of ambergris begins when a mass of indigestible material—mainly squid beaks, which the whale normally vomits—manages to pass through the duodenum. If this mass blocks the intestine, the intestinal wall reacts by absorbing water from the feces-impregnated mass, thus causing it to solidify. As this process continues the mass increases in size by the accretion of additional solid layers.

Contrary to the prevalent notion, ambergris is hardly ever found on beaches; most is recovered directly from whale carcasses. Through the years many people have brought me malodorous globs that they picked up on the seashore in hopes that it was ambergris; none of it ever was. If a suspected specimen of ambergris fits the physical description, the simplest way to confirm its identity is to heat a wire or needle in a flame and thrust it into the sample to a depth of about a centimeter; if the substance is really ambergris it will instantly melt into an opaque fluid the color of dark chocolate. When the needle is withdrawn, the ambergris will leave a tacky residue on it.

Ambergris was known throughout the Moslem world as early as the ninth century. There it was highly valued as an incense, an aphrodisiac, a laxative, a spice, an ingredient in candles and cosmetics, and as a medication for treating a diversity of ailments. Its reputation soon spread around the globe. In those days ambergris was picked up on beaches or found floating on the sea, and its origin remained a complete mystery, thus giving rise to many fanciful and hotly debated theories. In 1574 the Flemish botanist Carolus Clusius was the first author to deduce from the inclusions of squid beaks in ambergris that it was the product of the digestive tract of whales. It was not until after the commencement of the American sperm whale fishery in 1712 that it became generally recognized that ambergris was produced solely by the spermaceti whale (Beale, 1839; Dannenfeldt, 1982). In the ensuing years ambergris was prized mainly as a fixative for fragrances in perfumes. In the twentieth century synthetic chemicals replaced it, so it no longer has much value.

## See Also the Following Articles

Sperm Whale ■ Gastro intestinal tract

## References

Beale, T. (1839). "The Natural History of the Sperm Whale," pp. 1–393. John Van Voorst, London (Reprinted 1973 by The Holland Press, London.).

Berzin, A. A. (1971). "Kashalot," p. 368. Izdatel'stvo "Pishchevaya Promyshlennost", Moscow.

Clarke, R. (1954). A great haul of ambergris. *Norsk Hvalf. Tid.* **43**(8), 286–289.

Clarke, R. (2006). The Origin of ambergris. *Latin Am. J. Aquat. Mamm.* **5**, 7–21.

Dannenfeldt, K. H. (1982). Ambergris: The search for its origin. *Isis* **73**(268), 382–397.

Gilmore, R. M. (1951). The whaling industry: Whales, dolphins, and porpoises. *In* "Marine Products of Commerce" (D. K. Tressler, and J. M. Lemon, eds), pp. 680–715. Reinhold Publishing Corporation, New York.

# Anatomical Dissection: Thorax and Abdomen

## JOHN E. REYNOLDS, III AND SENTIEL A. ROMMEL

The general organization of the postcranial soft tissues does not vary appreciably among mammals. Factors that may influence the relative proportions or positions of organs and organ systems include phylogeny and adaptations to a particular environment or trophic level.

The structure and function of specific organs or organ systems are described in other articles of this encyclopedia. This article provides a "road map" that orients a prosector to the organs and organ systems of marine mammals. For comparative purposes, we focus on the California sea lion (*Zalophus californianus*), Florida manatee (*Trichechus latirostris*), harbor seal (*Phoca vitulina*), and common bottlenose dolphin (*Tursiops truncatus*). Our descriptions are at the gross anatomical level.

To recognize variations on a theme, one must first recognize the theme. Although there is no "typical" mammal, we shall use our own species and the domestic dog as the norms against which to make comparisons. To appreciate human and dog anatomy, we suggest

Hollinshead and Rosse (1985) and Evans (1993), respectively. Anatomy of internal organs of domestic mammals is covered by Schummer *et al.* (1979). For discussions of the anatomy of various types of marine mammals, consult Fraser (1952), Green (1972), Herbert (1987), Howell (1930), King (1983), Murie (1872, 1874), Pabst *et al.* (1999), von Schulte (1916), Slijper (1962), and St. Pierre (1974). Whenever possible, anatomical terms follow the Nomina Anatomica Veterinaria as illustrated by Schaller (1992).

## I. Mammalian Postcranial Landmarks

Marine mammals are generally dissected either ventrally or laterally, but some large, stranded animals must be examined in whatever position they are found. For consistency, we provide figures that describe anatomy in terms of a lateral view, and we discuss organs and organ systems in the order in which they are revealed during necropsy. Although this approach may take some getting used to if one is accustomed simply to the ventral approach, the lateral orientation approximates the living condition more closely.

## A. The Diaphragm

The diaphragm of most marine mammals is generally similar in orientation to that of the diaphragm in both the human and the dog. It lies in a transverse plane and provides a musculotendinous sheet to separate the heart and its major vessels, the lungs and their associated vessels and airways, the thyroid, thymus, and a variety of lymph nodes (all located cranial to the diaphragm) from the major organs of the digestive, excretory, and urogenital systems (all typically caudal to the diaphragm). The diaphragm is generally confluent with the transverse septum (a connective tissue separator between the heart and the liver) and, thus, attaches medially at its ventral extremity to the sternum.

Although the diaphragm separates the heart and lungs from the other organs of the body, the diaphragm is traversed by nerves and other structures such as the aorta (crossing in a dorsal and medial position), the caudal (inferior) vena cava (crossing more ventrally than the aorta, and often slightly right of the midline, although appearing to approximate the center of the liver), and the esophagus (crossing slightly right of the midline, at roughly a midhorizontal level). This approximately transverse orientation exists in most marine mammals, although the orientation of the diaphragm may be more or less diagonal, with the ventral portion being more caudal than the dorsal portion

The West Indian manatee's diaphragm differs from this general pattern of orientation and attachment. The diaphragm and the transverse septum are separate, with the septum occupying approximately the "typical" position of the diaphragm and the diaphragm itself occupying a horizontal plane extending virtually the entire length of the body cavity (Fig. 1B). This apparently unique orientation contributes to buoyancy control (Rommel and Reynolds, 2000). Additionally, there are two separate hemidiaphragms in the manatee (Figs 2B, C). The central tendons attach firmly to the ventral aspects of the thoracic vertebrae, producing two isolated pleural cavities. The position of the manatee diaphragm stands in contrast with the curved, oblique diaphragm (DIA, Fig. 3) of the sea lion, seal, and dolphin.

## B. Regions and Structures Cranial to the Diaphragm

The region cranial to the diaphragm is typically compartmentalized into three sections (1) the pericardium (containing the heart),

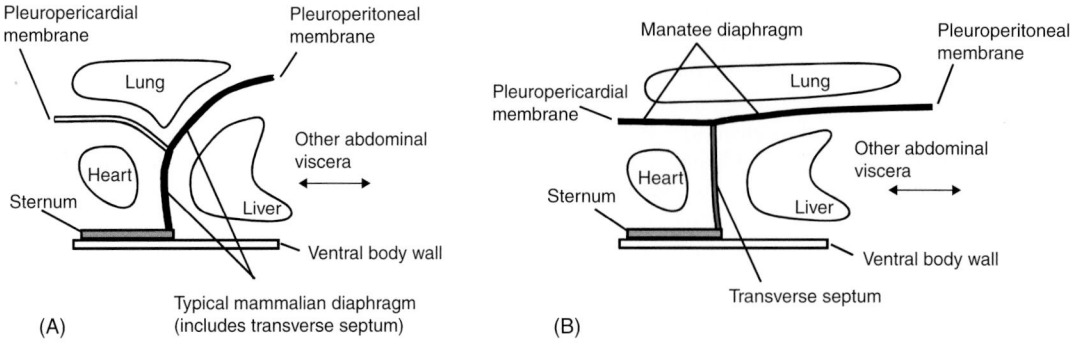

**Figure 1** *Schematic arrangements of mammalian diaphragms (modified after Rommel and Reynolds 2000). (A) The typical mammalian diaphragm extends ventrally from the dorsal midline to attach to the sternum. The typical diaphragm is a separator between the heart and lungs in the front and the liver and other abdominal organs in the back. (B) The manatee diaphragm extends dorsal to the heart and does not touch the sternum. There is a mechanical barrier between the heart and the liver and other abdominal organs but it is a relatively weak barrier called the transverse septum.*

(2) the pleural cavities (containing the lungs), and (3) the mediastinum (Figs 3 and 4).

The pericardium is a fluid-filled sac surrounding the heart (HAR, Fig. 3); in manatees, it often contains more fluid than is found in the pericardia of the typical mammal or in those of other marine mammals. The heart occupies a ventral position in the thorax (immediately dorsal to the sternum), making it easy to see when the overlying muscles, ribs, and sternum are separated. The heart lies immediately cranial to the central portion of the diaphragm (or just the transverse septum in the manatee). Some lungs may embrace the caudal aspect of the heart, separating the heart from the diaphragm. As do the hearts of all other mammals, marine mammal hearts have four chambers, separate routes for pulmonary and systemic circulation, and the usual arrangements of great vessels (vena cavae, aorta, coronary arteries, and pulmonary vessels). Cardiac fat is commonly found in manatees but is typically absent in pinnipeds and cetaceans.

The pleural cavities and lungs of mammals are generally found dorsally and laterally to the heart and are separated along the midline by the heart and mediastinum (discussed later). In the manatee, the lungs are unusual in that they extend virtually the length of the body cavity and remain dorsal to the heart (Rommel and Reynolds, 2000). Lungs of some marine mammals (cetaceans and sirenians) often lack lobes. The size of the lungs of marine mammals varies according to each species' diving proficiency. Marine mammals that make deep and prolonged dives (e.g., elephant seals, *Mirounga* spp.) tend to have smaller lungs than expected (based on allometric relationships) whereas shallow divers (e.g., sea otters, *Enhydra lutris*) tend to have larger than expected lungs.

The mediastinum is typically considered to be the area between the lungs, excluding the heart and pericardium. The mediastinum contains the major vessels leading to and emanating from the heart, nerves (e.g., the phrenic nerve to the diaphragm), and lymph nodes. The thymus, which is larger in younger individuals, is found on the cranial aspect of the pericardium (sometimes extending caudally to embrace almost the entire heart) and may extend into the neck in some species. The thyroid gland is located in the cranial part of the mediastinum along either side of the distal part of the trachea, cranial to its bifurcation into the bronchi (in sea lions, but not in other marine mammals, the bifurcation is cranial to the thoracic inlet).[1] In most marine mammals, the mediastinum is generally not remarkable; in the

manatee, however, the unusual placement of the lungs and the unique diaphragm change how one must define the mediastinum (Rommel and Reynolds, 2000).

One additional structure, located on the cranial aspect of the diaphragm in seals and sea lions, is an atypical mammalian muscular feature associated with the heart. This is the caval sphincter (CAS, Fig. 3), which can regulate the flow of oxygenated[2] blood in the large hepatic sinus to the heart during dives (Elsner, 1969).

### C. Structures Caudal to the Diaphragm

Easy to find landmarks caudal to the diaphragm include a massive liver and the various components of the gastrointestinal (GI) tract. The urogenital organs are generally found only after removal of the GI tract (note that the exception is the uterus of the pregnant female).

1. *The Liver.* Typically, the liver is located immediately caudal to the diaphragm. It is a large, brownish, multilobed organ positioned so that most of its volume/mass is to the right of the midline of the body. Although marine mammal livers are generally similar to the livers of other mammals, in manatees, the organ is displaced somewhat to the left and dorsal relative to its location in most other mammals. The size, color, and "sharpness" of the liver margins can be used to assess the nutritive state and health of individual animals. Bile may be stored in a gallbladder (often greenish in color) located ventrally between the lobes of the liver, although some species (e.g., cetaceans, horses, and rats) lack a gallbladder. Bile enters the duodenum to facilitate the chemical digestion of fats.

2. *The GI Tract.* Most of the volume of the cavity caudal to the diaphragm (the abdominal cavity) is occupied by the various components of the GI tract: the stomach, the small intestine (duodenum, jejunum, and ileum), and the large intestine (cecum,

---

[1]The thoracic inlet is the cranial opening of the thoracic cavity and is bounded by the vertebral and sternal ribs and sternum.

[2]Diving mammals with abundant arteriovenous anastomoses (shunts between arteries and veins that bypass capillary beds) can have high blood pressure and highly oxygenated blood in their veins. One such venous reservoir of oxygenated venous blood is the hepatic sinus of seals (King, 1983).

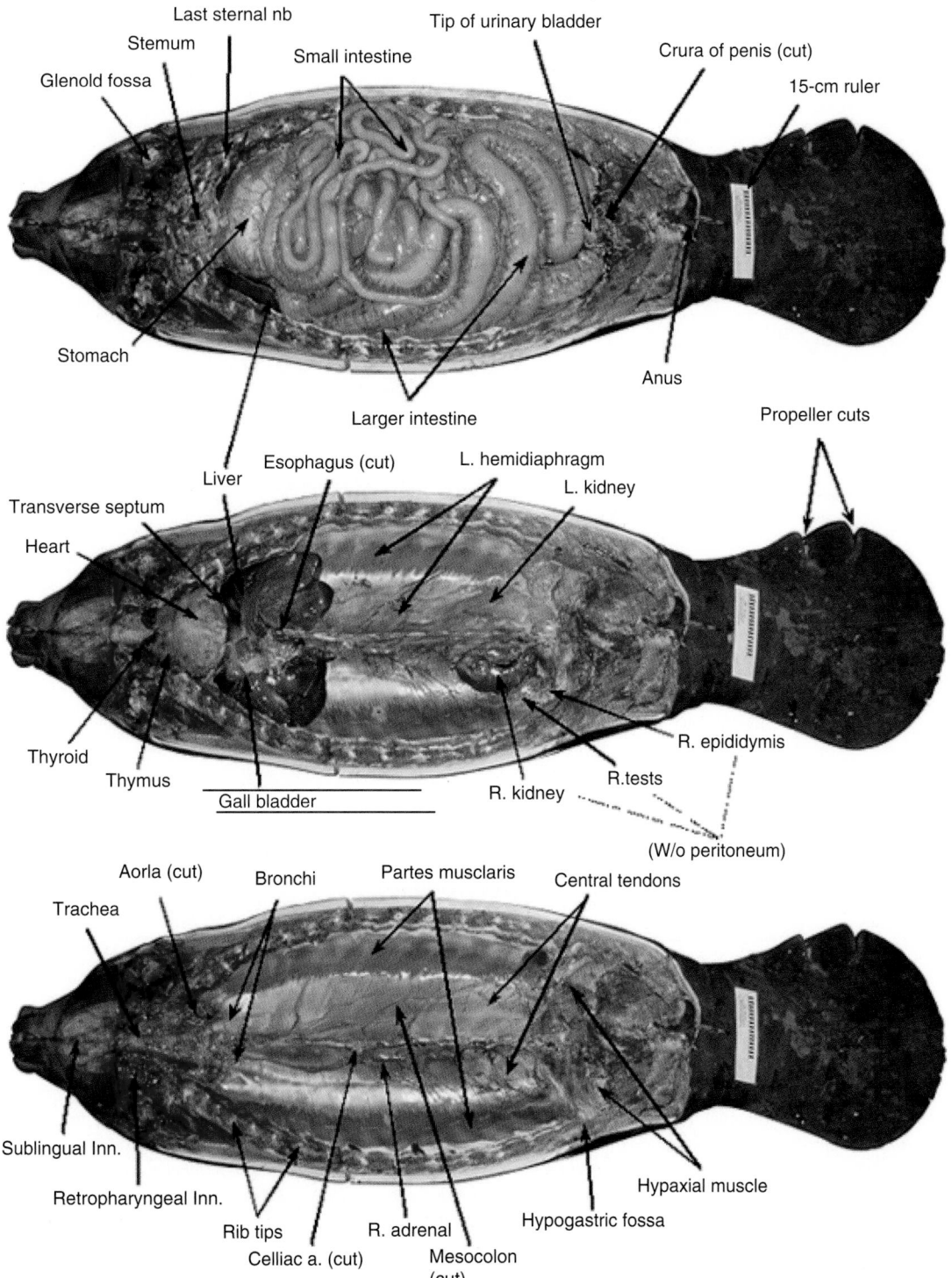

**Figure 2** *Photographs of ventral views of the Florida manatee (modified after Rommel and Reynolds 2000). The ruler is 15-cm long. (A) After removal of the ventral skin, fat, and musculature, the small and large intestines are exposed; the large intestine (with contents) may account for 10% of the total body weight and can measure 20-m long. Portions of the stomach and ventral margins of the liver are visible caudal to the sternum. (B) Removal of the sternum and GI tract reveals the heart, transverse septum, liver, hemidiaphragms, and right kidney (the left kidney was removed to expose that portion of the hemidiaphragm). (C) The two central tendons of the hemidiaphragms attach medially to the ventral aspects of the vertebral column. The diaphragm muscles attach laterally to the ribs. The lungs are flattened, elongate structures dorsal to the hemidiaphragms; when fully inflated, the lungs extend almost the entire length of the region dorsal to the hemidiaphragms. Note the junctions of the central tendon and the pars muscularis of each hemidiaphragm; this approximates the lateral margin of each lung.*

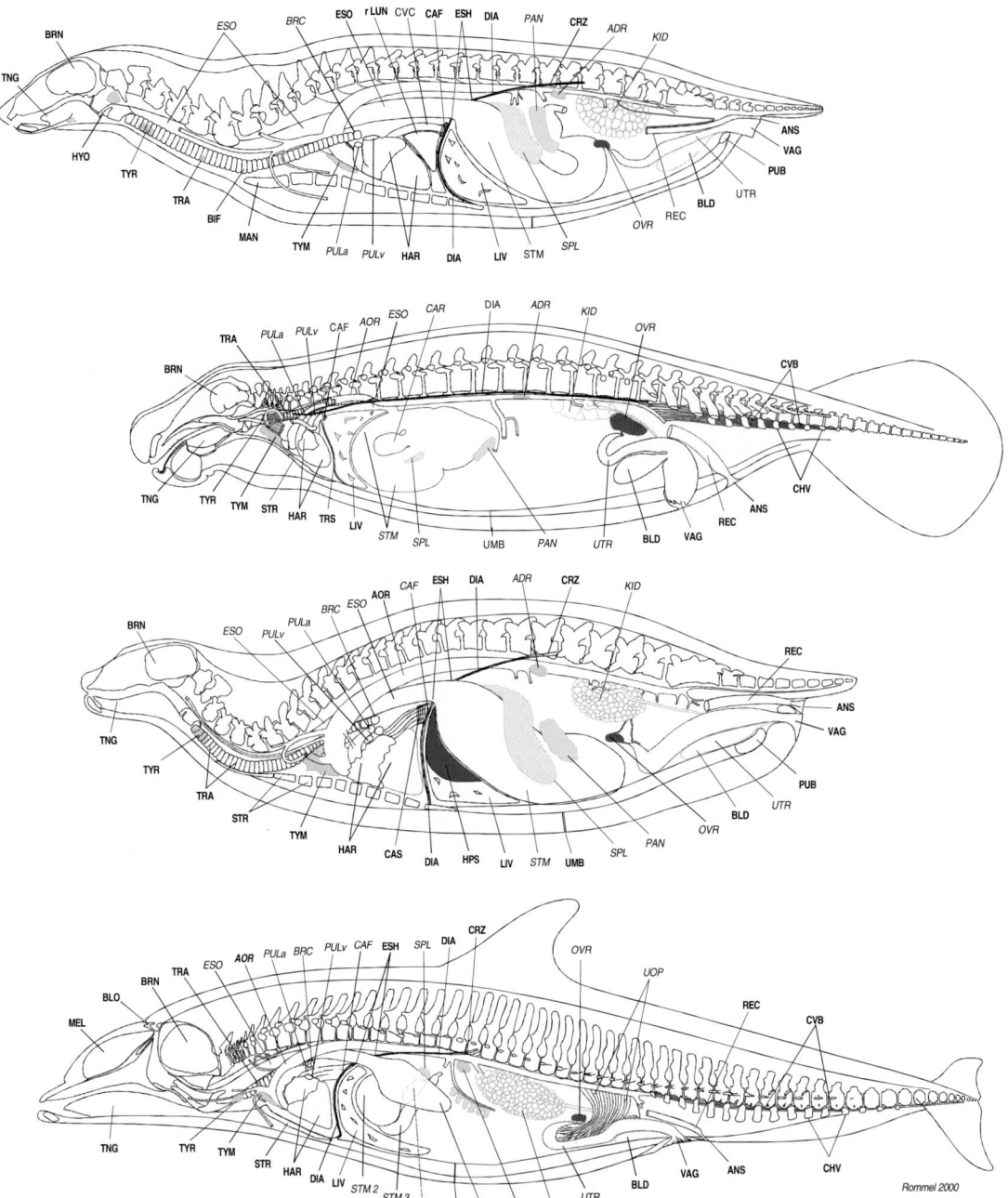

**Figure 3** *Left lateral illustrations of the superficial internal structures and "anatomical landmarks" of the California sea lion* (Zalophus californianus), *Florida manatee* (Trichechus manatus latirostris), *harbor seal* (Phoca vitulina), *and bottlenose dolphin* (Tursiops truncatus) *with the skeleton (minus the distal appendicular elements) superimposed for reference. Our view is a left lateral view, focused on relatively superficial internal structures (labeled in bold) visible from that perspective; the other important bony or soft "landmarks" are not necessarily visible from a left lateral view but they are useful for orientation and are labeled in italics. Skeletal elements are included for reference, but not all are labeled—for these details, consult the figures in the skeleton postcranial and skull chapters. Each drawing is scaled so that there are equivalent distances between the shoulder and the hip; thus, the thoracic and abdominal cavities are roughly equal in length. The shoulder joints are aligned. The left kidney (not visible from this view in the manatee) is illustrated. The relative sizes of the lungs represent partial inflation—full inflation would extend margins to distal tips of ribs (except in the manatee). The following abbreviations are used as labels (structures on the midline are in bold, those off-midline are in italics):* **ANS**, *anus;* **BLD**, *urinary bladder;* **BLO**, *blow hole of dolphin;* **DIA**, *diaphragm, midline extent (except manatee); EYE-eye (note small size in manatee);* **HAR**, *heart;* **ILC**, *iliac crest of the pelvis;* **INT**, *intestines; note the large diameter of the large intestines in the manatee;* **KID**, *left kidney (not visible from this vantage in the manatee);* **LIV**, *liver;* **LUN**, *lung (note that in this illustration, the lung extends under the scapula except in the seal);* **MEL**, *melon, dolphin only;* **OLE**, *olecranon of ulna;* **OVR**, *left ovary;* **PAN**, *pancreas (in this view visible only in seal and sea lion);* **PAT**, *patella;* **PEL**, *pelvic vestige;* **REC**, *rectum;* **SCA**, *scapula;* **SPL**, *spleen;* **STM**, *stomach;* **TRA**, *trachea (not visible in this view of the manatee);* **TYM**, *thymus gland;* **TYR**, *thyroid gland;* **UMB**, *umbilical scar;* **UOP**, *uterovarian plexus in dolphins;* **UTR**, *uterine horn;* **VAG**, *vagina.* © S.A. Rommel.

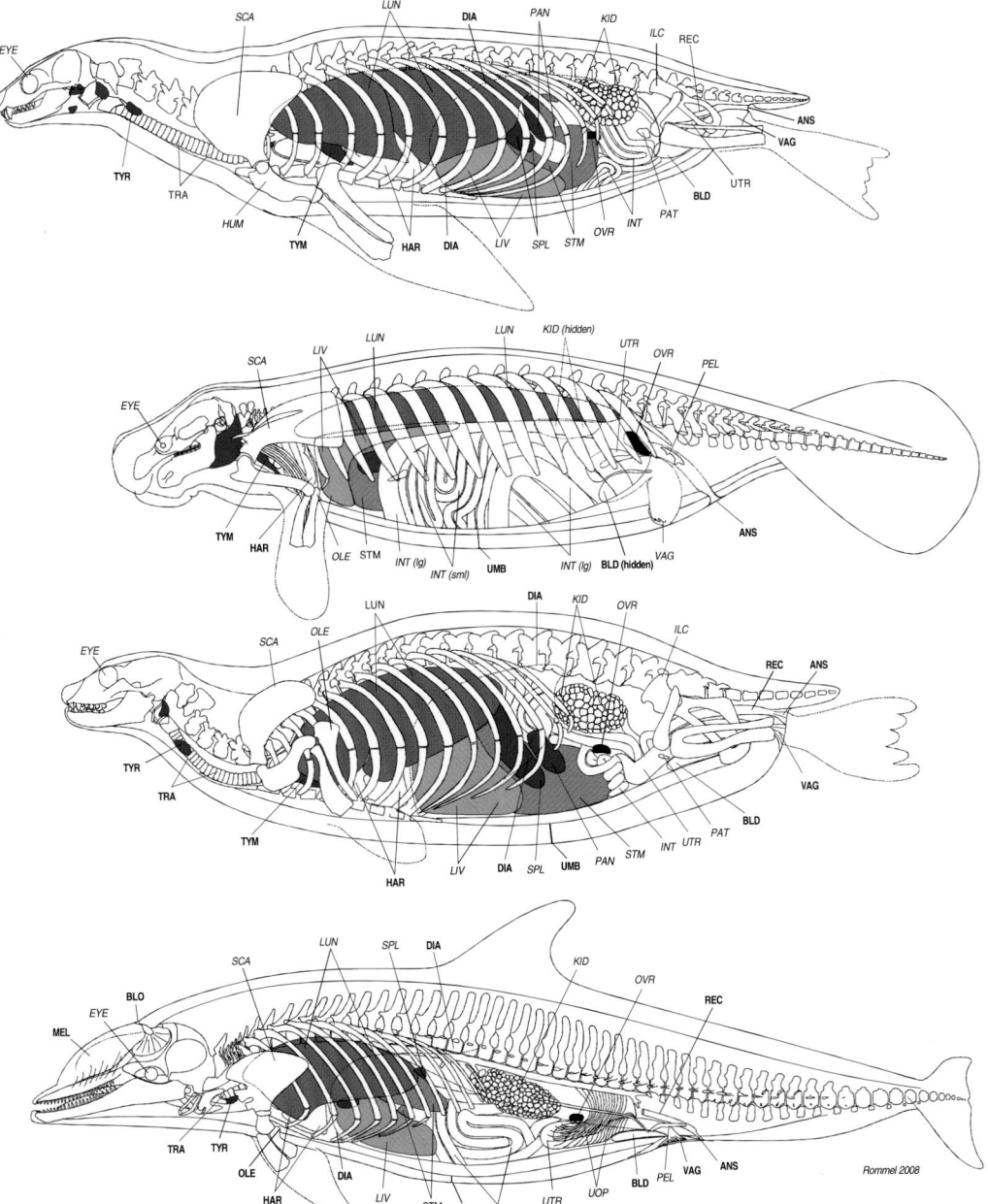

**Figure 4** *A view slightly to the left of the midsagittal plane illustrates the circulation, body cavities, and selected organs of the California sea lion* (Zalophus californianus), *Florida manatee* (Trichechus manatus latirostris), *harbor seal* (Phoca vitulina), *and bottlenose dolphin* (Tursiops truncatus), *with the skeleton for reference. The left lung is removed. Note that the diaphragm separates the heart and lungs from the liver and other abdominal organs. Each drawing is scaled so that there are equivalent distances between the shoulder and the hip; thus, the thoracic and abdominal cavities are roughly equal in length. The shoulder joints are aligned. Note that the manatee's diaphragm is unique and that the distribution of organs and the separation of thoracic structures from abdominal structures require special consideration in these beasts. The following abbreviations are used as labels (structures on the midline are in bold, those off-midline are in italics):* **ADR**, *adrenal gland;* **ANS**, *anus;* **AOR**, *aorta;* **BLD**, *urinary bladder;* **BLO**, *blowhole;* **BRC**, *bronchus;* **BRN**, *brain;* **CAF**, *caval foramen;* **CAR**, *cardiac gland, in manatee only;* **CAS**, *caval sphincter, surrounding the vena cava in the seal and sea lion;* **CHV**, *chevron bones;* **CRZ**, *crus (plural crura) of the diaphragm;* **CVB**, *caudal vascular bundle, in manatee and dolphin;* **DIA**, *diaphragm, cut at midline, extends from crura dorsally to sternum ventrally (except in manatees);* **ESH**, *esophageal hiatus;* **ESO**, *esophagus (to the left of the midline cranially, on the midline caudally);* **HAR**, *heart;* **HPS**, *hepatic sinus within liver, in seals only;* **KID**, *right kidney;* **LIV**, *liver, cut at midline;* **LUN**, *lung, right lung between heart and diaphragm;* **MEL**, *melon, dolphin only;* **PAN**, *pancreas;* **PUB**, *pubic symphysis (seals and sea lions only);* **PULa**, *pulmonary artery, cut at hilus of lung;* **PULv**, *pulmonary vein, cut at hilus of lung;* **REC**, *rectum, straight part of terminal colon;* **SPL**, *spleen;* **STM1**, *fore stomach;* **STM2**, *main stomach* (**STM** *in non-cetaceans);* **STM3**, *pyloric stomach;* **STR**-*sternum, sternabrae;* **TNG**-*tongue;* **TRA**-*trachea;* **TRS**-*transverse septum;* **TYM**-*thymus gland;* **TYR**, *thyroid.* © S.A. Rommel

**A**

colon, and rectum). The proportions and functions of these components reflect the feeding habits and trophic levels of the different marine mammals. Therefore, the gastrointestinal tracts of marine mammals vary considerably.

Food and water travel from the mouth, through a muscular pharynx, and into the esophagus. As noted earlier, the latter pierces the diaphragm to join the stomach, which is typically a single, distensible sac. The distal end of the stomach (the pylorus) is marked by a strong sphincter before it connects with the small intestine (duodenal ampulla in cetaceans). The separation between jejunum and ileum of the small intestine is difficult to distinguish grossly, although the two sections are different microscopically. The junction of the small and large intestines is often (but not in cetaceans) marked by the presence of a cecum (homologous to the human appendix). In manatees, the midgut cecum has two blind pouches called cecal horns. In some marine mammals, the large intestine, as its name implies, has a larger diameter than the small intestine.

The gastrointestinal tracts of pinnipeds and other marine mammal carnivores follow the general patterns outlined earlier, although the intestines can be remarkably long in some species. Cetaceans, however, have some unique specializations (Gaskin, 1978). Cetaceans can have two or three stomachs (usually three) depending on the species being examined. The multiple stomachs of cetaceans function in much the same way as the single stomach found in most other mammals. The first stomach of cetaceans, called the forestomach (essentially an enlargement of the esophagus), is muscular and very distensible, and it acts much like a bird crop, i.e., as a receiving chamber. The second or glandular stomach is the primary site of chemical breakdown among the stomach compartments; it contains the same types of enzymes and hydrochloric acid that characterize a "typical" stomach. Finally, the "U-shaped" third or pyloric stomach ends in a strong sphincteric muscle that regulates the flow of digesta into the duodenum (the duodenal ampulla is sometimes mistakenly called a fourth stomach) of the small intestine. The cetacean duodenum is expanded into a sac-like ampulla. The only other remarkable feature at the gross level is the lack of a cecum, which makes it difficult to tell where the small intestine ends and the large intestine begins. The intestines of some cetaceans may be extremely long (especially in the sperm whale, *Physeter macrocephalus*; Slijper, 1962), but they are not especially long in many other marine mammal species.[3]

Among marine mammals, sirenians have the most remarkably developed gastrointestinal tract. Sirenians are herbivores and hindgut digesters (similar to horses and elephants) so the large intestines (specifically the colon) is extremely enlarged, enabling it to act as a fermentation vat (see Marsh *et al.*, 1977; Reynolds and Rommel, 1996). In horses, the cecum is the region of the large intestine that is enlarged, but in sirenians, the cecum is relatively small and has two "horns." The sirenian stomach is single chambered and has a prominent accessory secretory gland (the cardiac gland) extending from the greater curvature. The duodenum is capacious and has two obvious diverticulae projecting from it. The GI tract and its contents can account for more than 20% of a manatee's weight.

The length and mass of the gastrointestinal tract are impressive and create three-dimensional relationships that can be complex. Simplifying the organization is the fact that tough sheets of connective tissue called mesenteries suspend the organs from the dorsal part of the abdominal cavity and shorter bands of connective tissue (ligaments)[4] hold organs close to one another in predictable arrangements (e.g., the proximal spleen is always found along the

greater curvature of the stomach and is connected to the stomach by the gastrolienal, or gastrosplenic ligament). Also suspended in the mesenteries are numerous lymph nodes and fat.

Accessory organs of digestion include salivary glands (small in most marine mammals but very large in the manatee), pancreas, and liver (where bile is produced and then stored in the gall bladder). The pancreas is sometimes a little difficult to locate because it can be a rather diffuse organ and it decomposes rapidly postmortem; however, a clue to its location is its proximity to the initial part of the duodenum, into which pancreatic enzymes flow. Another organ that is structurally, but not functionally associated with the GI tract is the spleen, which is suspended by a ligament, generally from the greater curvature of the stomach (the first stomach in cetaceans) on the left side of the body. The spleen may be a single organ accompanied by accessory spleens in some species. The spleen is bluish in color and varies considerably in size among species; in manatees and cetaceans it is relatively small but is more massive in some deep-diving pinnipeds (Zapol *et al.*, 1979) in which it acts as a storage region for red blood cells.

3. *Urogenital Anatomy*. The kidneys lie in a retroperitoneal position, typically against the musculature of the back (epaxial muscles) at or near the dorsal midline attachment of the diaphragm (crura). In the manatee, the unusual placement of the diaphragm means that the kidneys lie against the diaphragm, but not against the epaxial muscles. All mammals have metanephric kidneys (i.e., containing cortex, medulla, and calyces). In many marine mammals, the kidneys are specialized as reniculate (multilobed) kidneys, where each lobe (renule) has all the components of a complete metanephric kidney. Why marine mammals have reniculate kidneys is uncertain, but the fact that some large terrestrial mammals also have reniculate kidneys has led to speculation that they are an adaptation associated simply with large body size (Vardy and Bryden, 1981).

The renal arteries of cetaceans enter the cranial poles of the kidneys, whereas in other marine mammals, they enter the hilus (typical of most mammals). Additionally, in manatees, there are accessory arteries on the surface of the kidney. The kidneys are drained by separate ureters, which carry urine to a medially and relatively ventrally positioned urinary bladder. The urinary bladder lies on the floor of the caudal abdominal cavity and, when distended, may extend as far forward as the umbilicus in some species. The pelvic landmarks are less prominent in fully aquatic mammals. In the manatee, the bladder can be obscured by abdominal fat.

Pabst *et al.* (1999) noted that the reproductive organs tend to reflect phylogeny more than adaptations to a particular niche. If one were to examine the ventral side of cetaceans and sirenians before removing the skin and other layers, one would discover that positions of male and female genital openings are different, permitting rather

---

[3]Assessing the length of intestines is fraught with potential bias because it is extremely difficult not to stretch the intestines to unnatural lengths after they are freed from the mesenteries and straightened. Linear measurements of gastrointestinal tract are, therefore, highly subjective.

[4]Ligament has several meanings in anatomy: a musculoskeletal element [e.g., the anterior (cranial) cruciate ligament], a vestige of a fetal artery or vein (e.g., the round ligament of the bladder), the margin of a fold in a mesentery (e.g., broad ligament), and a serosal fold between organs (e.g., gastrolienal ligament).

easy determination of sex in some species without dissection. In all marine mammals, the female urogenital opening is more caudal than the opening for the penis in males. One way to approach dissection of the reproductive tracts is to follow structures into the abdomen from their external openings.

The position and general form of the female reproductive tract in marine mammals are generally similar to those of the female reproductive tracts in terrestrial mammals. The vagina opens cranial to the anus and leads to the uterus, which is bicornuate in marine mammal species. The body of the uterus is found on the midline and is located dorsally to the urinary bladder (the ventral aspect of the uterus rests against the bladder). Although the body of the uterus lies along the midline, it has bilaterally paired, relatively large diameter projections called uterine horns (cornua), which extend laterally. The relatively small-diameter oviducts conduct eggs from the ovaries to the uterine horns where implantation of the fertilized egg and subsequent placental development occur. The dimensions of the uterine horns vary with reproductive history and age. Often the fetus may expand the pregnant horn to the point that it fills a substantial portion of the abdominal cavity. The horns terminate abruptly, narrowing and extending as uterine tubes (fallopian tubes) to paired ovaries. The uterus and the uterine horns are held in place in the abdominal cavity by the broad ligaments. Uterine and ovarian scarring may provide information about the reproductive history of the individual.

The ovaries of mature females may have one or more white or yellow-brown scars, called corpora albicantia and corpora lutea, respectively. Although ovaries are usually solid organs, in sirenians they are relatively diffuse.

Mammary glands are ventral, medial, and relatively caudal in most marine mammals, but they are axillary in sirenians. Many marine mammals have a single pair of nipples; sea lions and polar bears, *Ursus maritimus*, (DeMaster and Stirling, 1981), have two pairs or nipples, and cetaceans have mammary slits (note that some male cetaceans have distinct mammary slits).

The male reproductive tracts of marine mammals have the same fundamental components as the tracts in "typical" mammals, but positional relationships are significantly different. This difference is due to the testicond (ascrotal) position of the testes in most marine mammal species. Sea otters are scrotal (J. Bodkin, personal communication); polar bears are seasonally scrotal (I. Stirling, personal communication); and sea lion testes become scrotal when temperatures are elevated. The testes of some marine mammals are intraabdominal, but in phocids, for example, they lie outside the abdomen, partially covered by the oblique muscles and blubber. The position of marine mammal testes creates certain thermal problems because spermatozoa do not survive well at body (core) temperatures; in some species, these problems are solved by the circulatory adaptations mentioned later.

The penis of marine mammals is retractable and it normally lies within the body wall. The general structure of the penis relates to phylogeny (Pabst *et al.*, 1999).

4. *Adrenal Glands.* The term "suprarenal gland" is often used interchangeably with "adrenal gland." Although the suprarenals often lie immediately atop or very close to the kidneys of terrestrial mammals, adrenals of marine mammals may lie several centimeters cranial to the kidneys, along either side of the median. Adrenal glands can be confused with lymph nodes, but if one slices the organ in half, an adrenal gland is easy to distinguish grossly by its distinct cortex and medulla.

5. *Circulatory Structures.* Blood vessels are often named for the regions they feed or drain. Thus, the fully aquatic marine mammals (cetaceans and sirenians) lack femoral arteries that supply the pelvic appendage. However, most organs in marine mammals are similar to those of terrestrial mammals so their blood supply is also similar. Therefore, readers who want to learn details of typical circulatory anatomy should consult one of the anatomy references cited earlier. The thoracic aorta leaves the heart and lies ventral to the vertebral column, giving off segmental arteries to the vertebrae and epaxial muscles (and in the case of cetaceans and manatees to the thoracic retia). The aorta continues through the aortic hiatus of the diaphragm (between the crura) and into the abdomen as the abdominal aorta and lumbar aorta, which give off several paired (e.g., renal and gonadal) and unpaired (e.g., celiac and mesenteric) arteries. The caudal aorta follows the ventral aspect of the tail vertebrae. In the permanently aquatic marine mammals, there are robust ventral chevron bones that form a canal in which the caudal aorta, its branches, and some veins are protected.

Some of the diving mammals (e.g., seals, cetaceans, and sirenians) have few or no valves in the veins (Rommel *et al.*, 1995); this adaptation simplifies blood collection.[5] Other exceptions to the general pattern of mammalian circulation are associated with thermoregulation and diving. Countercurrent heat exchangers abound, and extensive arteriovenous anastomoses exist to permit two general objectives to be fulfilled (1) regulating loss of heat to the external environment, while keeping core temperatures high and (2) permitting cool blood to reach specific organs (e.g., testes, uteri, and spinal cord) that cannot sustain exposure to high body temperatures (see reviews by Rommel *et al.*, 1998; Pabst *et al.*, 1999).

In mammals, several paths for supplying blood to the brain exist: via the internal carotid, the external carotid, and the vertebral/basilar arteries. Some species use only one, others use two, and manatees use all three pathways. In cetaceans, the path for supplying blood to the brain is unique. The blood destined for the brain first enters the thoracic rete, a plexus of convoluted, small-diameter arteries in the dorsal thorax. Blood leaves the thoracic rete and enters the spinal rete where it surrounds the spinal cord and enters the base of the skull (McFarland *et al.*, 1979). There are two working explanations for this convoluted path of blood to the brain: (1) the elasticity of the retial system allows mechanical damping of the blood pulse pressure wave (McFarland *et al.*, 1979) and (2) the juxtaposition of the thoracic retia to the dorsal aspect of the lungs may provide thermal control of the blood entering the spinal retia. Combined with cooled blood in the epidural veins, the spinal retia may provide some temperature control of the central nervous system.

## II. Overview

Marine mammal postcranial soft tissue anatomy is, in many regards, similar to that of "typical" mammals. However, the relative proportions of and, to some extent, the positions of organs may be somewhat different from the norm.

---

[5]The near absence of valves in the veins of seals and dolphins allows two-way flow to occur, increasing the blood available when venipuncture is used; in contrast, sea lions have numerous valves in the hind flipper veins.

We close with a reminder about orientation: namely the orientation of the prosector relative to the orientation of the specimen and the orientation of the specimen to the orientation of that animal when it was alive. The position of animals during necropsy may be belly-up, obviously not the usual position of the living animals. Thus, gravitational forces make the positional relationships we may observe during necropsy somewhat artificial; we assess "dead anatomy" rather than "living anatomy." We suggest that people examining marine mammal postcranial anatomy bear this fact in mind and try to constantly picture how the structures being observed during necropsy might be arranged in a free-ranging animal. The more the latter perspective can be maintained, the easier it will be to envision dynamic relationships among organs and systems and to relate function (physiology) to structure (anatomy).

## See Also the Following Articles

Female Reproductive Systems ■ Male Reproductive Systems ■ Musculature ■ Forelimb Anatomy ■ Hindlimb Anatomy ■ Skeletal Anatomy ■ Skull Anatomy ■ Gastrointestinal Tract

## References

DeMaster, D. P., and Stirling, I. (1981). *Ursus maritimus. Mamm. Spec.* **145**, 1–7.

Elsner, R. W. (1969). Cardiovascular adjustments to diving. *In* "The Biology of Marine Mammals" (H. T. Andersen, ed.), pp. 117–145. Academic press, New York.

Evans, H. E. (1993). "Miller's Anatomy of the Dog," 3rd Ed. Saunders, Philadelphia.

Fraser, F. C. (1952). "Handbook of R. H. Burne's Cetacean Dissections." Trustees of the British Museum, London.

Gaskin, D. E. (1978). Form and function of the digestive tract and associated organs in cetacea, with consideration of metabolic rates and specific energy budgets. *Oceanogr. Mar. Biol. Annu. Rev.* **16**, 313–345.

Green, R. F. (1972). Observations on the anatomy of some cetaceans and pinipeds. *In* "Mammals of the Sea, Biology and Medicine" (S. H. Ridgway, ed.), pp. 247–297. Thomas, Springfield, IL.

Herbert, D. (1987). "The Topographic Anatomy of the Sea Otter *Enhydra lutris*." Unpublished MS Thesis. Johns Hopkins University, Baltimore, MD.

Hollinshead, W. H., and Rosse, C. (1985). "Textbook of Anatomy." Harper & Row, Philadelphia.

Howell, A. B. (1930). "Aquatic Mammals: Their Adaptations to Life in the Water." Thomas, Springfield, IL.

King, J. E. (1983). "Seals of the World," 2nd Ed. Comstock, Ithaca, NY.

Marsh, H., Heinsohn, G. E., and Spain, A. V. (1977). The stomach and duodenal diverticulae of the dugong (*Dugong dugon*). *In* "Functional Anatomy of Marine Mammals" (R. J. Harrison, ed.), Vol. 3, pp. 271–295. Academic Press, London.

McFarland, W. I., Jacobs, M. S., and Morgane, F. J. (1979). Blood supply to the brain of the dolphin, *Tursiops truncatus*, with comparative observations on special aspects of the cardiovascular supply of other vertebrates. *Neurosci. Biobehav. Rev.* **3**(Suppl. 1), 1–93.

Murie, J. (1872). On the form and structure of the manatee. *Trans. Zool. Soc. Lond.* **8**, 127–202.

Murie, J. (1874). Researches upon the anatomy of the Pinnipedia. 3. Descriptive anatomy of the sealion (*Otatia jubata*). *Trans. Zool. Soc. Lond.* **8**, 501–582.

Pabst, D. A., Rommel, S. A., and McClellan, W. A. (1999). The functional morphology of marine mammals. *In* "Biology of Marine Mammals" (J. E. Reynolds, III, and S. A. Rommel, eds), pp. 15–72. Smithsonian Institution Press, Washington, DC.

Reynolds, J. E., III, and Rommel, S. A. (1996). Structure and function of the Florida manatee, *Trichecus manatus. Anat. Rec.* **245**, 539–558.

Rommel, S. A., and Reynolds, J. E., III (2000). Diaphragm structure and function in the Florida manatee (*Trichecus manatus latirostris*). *Anat. Rec.* **259**, 41–51.

Rommel, S. A., Early, G. A., Matasa, K. A., Pabst, D. A., and McClellan, W. A. (1995). Venous structures associated with thermoregulation of phocid seal reproductive organs. *Anat. Rec.* **243**, 390–402.

Rommel, S. A., Pabst, D. A., and McClellan, W. A. (1998). Reproductive thermoregulation in marine mammals. *Am. Sci.* **86**, 440–448.

Schaller, O. (1992). "Illustrated Veterinary Anatomical Nomenclature." Ferdinand Enke Verlag, Stuttgart.

Schummer, A., Nickel, R., and Sack, W. O. (1979). The viscera of the domestic mammal. *In* "The Anatomy of the Domestic Animals. 2nd Ed." (R. Nickel, A. Schummer, and E. Seiferle, eds), Vol. 2. Verlag Paul Parey, Berlin.

Slijper, E. J. (1962). "Whales." Hutchinson & Co, London.

St. Pierre, H. (1974). The topographical splanchnology and the superficial vascular system of the harp seal *Pagophilus groenlandicus* (Erxleben 1777). *In* "Functional Anatomy of Marine Mammals" (R. J. Harrison, ed.), pp. 161–195. Academic Press, London.

Vardy, P. H., and Bryden, M. M. (1981). The kidney of *Leptonychotes weddellii* (Pinnipedia: Phocidae) with some observations on the kidneys of two other phocid seals. *J. Morphol.* **167**, 13–34.

Von Schulte, H. (1916). Anatomy of a fetus of *Balaenoptera borealis. Mem. Am. Mus. Nat. Hist.* **1**(VI), 389–502, + plates XLIII—XLII.

Zapol, W. M., Liggins, G. C., Schneider, R. C., Qvist, J., Snider, M. T., Creasy, R. K., and Hochachka, P. W. (1979). Regional blood flow during simulated diving in the conscious Weddell seal. *J. Appl. Physiol.* **47**(5), 973–986.

# Antarctic Fur Seal
## *Arctocephalus gazella*

### JAUME FORCADA AND IAIN J. STANILAND

## I. Characteristics and Taxonomy

The genus *Arctocephalus* (G. Saint-Hilaire and F. Cuvier, 1826) derives from the Greek words *arctos* (bear) and *kephale* (head), meaning bear-headed. The species *Arctocephalus gazella* (Peters, 1875) is thought to be named after the German vessel SMS *Gazelle* by Theophil Studer, a zoologist of the Venus Transit Expedition. The species was described from a young female specimen collected at the Kerguelen Islands during the expedition, between 1874 and 1875. The other common name, Kerguelen fur seal, is seldom used. Antarctic fur seals are part of the group Arctocephalinae and evolved to their present form in the past 2–3 million years. There are many similarities between Antarctic fur seals and the other Arctocephalinae, and they can be confused with most other southern fur seals, particularly females.

Both sexes have thick bodies and relatively long necks, which make the head look small (Fig. 1). The snout is smooth and relatively pointed and appears to be flat from the forehead to the nose, especially in males. The skull has a convex forehead and broad and short rostrum, with short nasal bones and a broad flattened palate. There are nine pairs of teeth in the upper jaw and eight pairs in the lower jaw. The tooth rows diverge posteriorly, with a wide diastema between upper postcanines 4 and 5, and 5 and 6. This is a major difference from the other Arctocephalinae, which lack the diastemas. The teeth are small and unicuspid, and the maxillary shelf is long. The upper canines in bulls have obvious external bands that correspond well with single annual growth layers. The lower canines are

**Figure 1** *Antarctic fur seal territory with a dominant bull defending females and their pups at Bird Island, South Georgia.*

**Figure 2** *Different coloration patterns of fur seal pups observed at Bird Island, South Georgia. Approximately 1–2 in 1000 pups are born white, and less than 1 in 5000 have mixed coloration.*

laterally compressed and large; they are used in fights with other bulls for breeding territories.

At birth, mean standard length for pups is 67.4 cm (58–66). Males are born heavier at 5.9 kg (4.9–6.6) compared to females at 5.4 kg (4.8–5.9). This sexual dimorphism is pronounced in adults, with bulls being almost 1.5 times longer and 4 times heavier than females. Bulls' mean standard length is 180 cm (170–200) and mean weight 133 kg (90–197). In females these are 129 cm (117–140) and 34 kg (20–51), respectively. Males have a well-developed scrotum and testicles are external, although they can be concealed, making it especially difficult to distinguish young males from adult females. Younger males lack the heavy manes of bulls but have heavier and larger body foreparts and teeth than females. Long facial vibrissae can extend beyond the pinnae and are usually white. In bulls these can grow up to 48 cm, longer than in any other pinniped.

The body is covered by hair except for the rhinarium, ear tips, and palmar surfaces of the flippers. The hair extends to the base of well-developed nails at the top of the hind flippers, which are often used to groom the hair. The nails on the fore flippers are less well developed. On land the pelage is grizzled dark brown, shading paler below. Color differences are partly due to the length and structure of three different types of hair. From the skin to the outside, the pelt has a dense underlayer of fine fur, which provides thermal insulation, and two distinct types of guard hairs, stouter, and more obvious in the bulls' mane (Bonner, 1968). The pelage of pups is black until the first molt, after which the pelage has the adult coloration. In pups, adult-type guard hairs with white tips may protrude, giving paler coloration, especially in the facial area. Approximately 1–2‰ of the pups are born white, with lack of pigmentation in the guard hairs and much paler underfur and exposed skin, but these are not albinos. Their coloration remains whitish for life. White individuals have only been reported at South Georgia, with incidental records at Bouvetøya Island, King George Island, and Marion Island (De Bruyn et al., 2007). Intermediate colorations, with clumps of black among predominantly white guard hair, are rarer still but are observed in high-density areas (Fig. 2).

## II. Distribution and Abundance

Antarctic fur seals have a very wide distribution and breed primarily in subantarctic and Antarctic locations of the South Atlantic and Indian Ocean sectors of the Southern Ocean. In the South Atlantic, the main breeding populations are south of the Antarctic polar front, on South Georgia Island, South Sandwich Islands, South Orkney Islands, South Shetland Islands (these four archipelagos belong to the Scotia Arc), and Bouvetøya Island. In the Indian Ocean, they are south of the polar front on Heard Island and McDonald Islands and north of the polar front on the Prince Edward Islands, Crozet Islands, Kerguelen Islands, and Macquarie Island. The distribution range widens after the breeding season, when animals can leave the main breeding rookeries. Most records of seals instrumented with telemetry devices suggest unstructured movements or individual dispersal. Bulls can travel very long distances, from the breeding islands to the ice edge and north of the polar front. Seals from South Georgia travel to the Antarctic Peninsula, the Falkland Islands, and southern Argentina, including the Juan Fernández Islands, Tierra del Fuego, and Mar del Plata. A group of Antarctic fur seals were seen at Gough Island, which is mostly populated by subantarctic fur seals, also north of the Polar Frontal Zone (Wilson et al., 2006). Bulls from Kerguelen, Heard and Crozet Islands travel to the ice edge and north of the polar front. During the winter, females disperse at sea with individuals traveling south to the marginal ice zone and north, crossing the polar front and reaching as far as the Mar del Plata region of the Patagonian shelf. Females returning to breed at South Georgia often carry pendunculate barnacles, indicating that they can spend extended periods at sea between their breeding events. Juveniles and bulls are often seen around the breeding beaches throughout the winter. Post-weaning pups remain close to the natal beaches but move to oceanic waters (>500 m) as the winter progresses.

Available population estimates are South Georgia 2,700,000 and pup production (pp) 269,000 (season 1990–1991); Bouvetøya Island 66,000 and pp 15,523 (2001–2002); South Shetland Islands 21,190 and pp 10,100 (2000–2001); Marion Island 3821 and pp 796 (2003–2004); Heard Island 4100 and pp 1278 (2000–2001); Prince Edward Island 2000 and pp 400 (2001–2002). Additional minimum pup production is 6500 at Kerguelen Islands, 295 at Crozet Islands, 350 at South Sandwich Islands, and 1000 at South Orkney Islands. The population of South Georgia probably comprises more than 95% of the world population. However, the total estimate was extrapolated from an estimate of 379,302 (287,363–471,240) breeding females in 1990–1991 (Boyd, 1993) obtained from beach counts and assumed a rate of population increase from 1977 to 1991 of 9.8%. Many females did not return to breed in 1990–1991 because of poor environmental conditions (Boyd, 1993; Forcada *et al.*, 2005) and as a result the counts were low and unrepresentative of the true number of females alive. Since 1990, the number of breeding females at a study site, used to estimate population correction factors and rates of increase, has been significantly declining, whereas circumstantial evidence suggests that seal numbers at other locations of South Georgia have increased. Therefore, reliable recent population estimates and trends are unavailable. Most Antarctic fur seal populations are thought to be increasing at rates well above 5%, although these estimates are not robust and most likely are positively biased.

Modern genetic population structure is known in detail from mitochondrial DNA control region sequences, with 26 haplotypes, including 16 represented in more than one individual, in a study of 145 seals from eight populations (Wynen *et al.*, 2000). The relationship between haplotypes suggests little lineage structure but two genetically distinct regions: a western region including the islands of the Scotia Arc, Bouvetøya, and Marion Islands, and an eastern region including Kerguelen and Macquarie Islands. Seals from Crozet Islands and Heard Island show mixtures of haplotypes from both regions. This suggests that post-sealing populations survived at South Georgia, Bouvetøya, and Kerguelen. South Shetland and Marion would have been recolonized by seals from South Georgia and Bouvetøya, and Macquarie by seals from Kerguelen. The severe reduction of the world population by sealing in the eighteenth and nineteenth centuries could have caused population bottlenecks in most locations. However, present molecular data suggest higher levels of genetic variation at the nuclear and mitochondrial DNA loci than expected from the estimated remnant population levels.

## III. Ecology

The diet of Antarctic fur seals is highly dependent on local prey availability, and comparative differences between sites probably reflect regional differences in prey assemblages rather than differing foraging strategies. For example, the distance from the coast to the shelf break, a proxy for the available shelf habitat, is negatively correlated with the proportion of pelagic fish in the diet of fur seals. In addition, the diet mostly reflects the prey within the seal's narrow depth range. Females are particularly shallow divers and depend on prey migrating into the surface waters, usually at night. Their diet is likely biased toward epipelagic and diurnally migrating mesopelagic species.

The density of Antarctic krill (*Euphausia superba*) is very high in the productive waters of the Atlantic sector of the Southern Ocean, and it dominates in the diet of the seals in this area. In the Indian Ocean sector, euphausids are of minimal importance and are often absent from the diet, which is instead dominated by fish. Other than krill, Antarctic fur seals principally eat myctophids, icefish, and notothenids, although skates and rays are also taken. Squid, a very minor (<1%) part of the diet at South Georgia, can play an important role in some areas, occurring in half the winter scats at Heard Island.

Seasonal differences in diet are reported at most sites where studies have been undertaken, but these are difficult to interpret because prey preference varies with age and sex of fur seals and the composition of the population in an area is very different within and outside of the breeding season. Interannual differences in diet relate to differences in local oceanographic conditions. At South Georgia, increases in myctophid occurrence are closely linked to sea surface temperatures, whereas changes in the consumption of icefish (*Champsocephalus gunnari*) are more closely linked to abundance of krill, its principal prey.

Although normally a very small component of the diet, penguins, especially during fledging, can be an important food source for fur seal bulls, and this may be a significant source of mortality where it occurs. At Marion Island, in a unique situation, the bulls take king penguins on land (Hofmeyr and Bester, 1993).

The dependence on land-based breeding strongly influences the distribution of Antarctic fur seals and their foraging ecology. Females

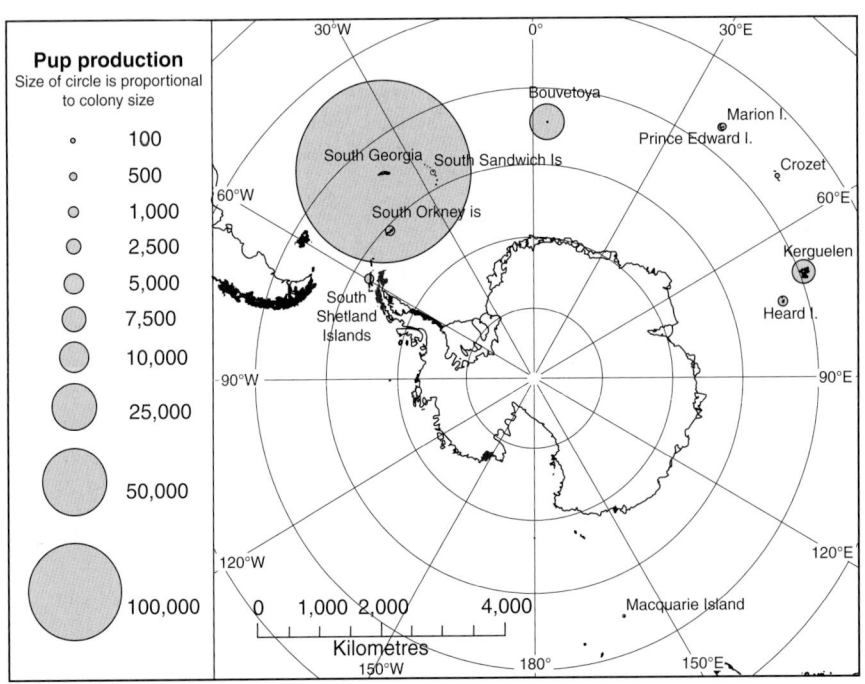

**Figure 3**  *Antarctic fur seal breeding distribution and pup production. Size of circles is proportional to colony size. All named locations have breeding colonies.*

during lactation act as a central place foragers and are thus restricted to foraging in the waters immediately surrounding the breeding beaches, usually around 150km from the pupping location. However, tracking studies have shown that there is large individual variation in the time spent at sea and the distance traveled. Generally female foraging is associated with areas of high gradient, i.e. shelf breaks, in the immediate vicinity. The bathymetry preferences associated with observed foraging patterns are ultimately determined by prey distribution.

Antarctic fur seals preferentially breed on shale and pebble beaches, but they are found on almost all seashore environments, from sandy beaches to exposed rock platforms, and also in areas with vegetation behind landing beaches, such as in the Prince Edward Islands. A few weeks after giving birth, females suckle their pups in the relatively quieter areas behind the breeding beaches. They are generally faithful to a suckling location, which can be quite distant from the water and up to 100m above sea level.

At South Georgia, Antarctic fur seals lay on tussock grass clumps to avoid the often wet and muddy ground. This behavior erodes the top of these clumps and, in areas of high density, kills the tussock, leaving a mound of earth and roots. The destruction of vegetation by recovering populations of fur seals has been a cause of concern, but it may well reflect the return of the environment to its pre-sealing state. This is clear in many areas where sealing artifacts (e.g. try-pots and shelters) are being uncovered by such erosion.

Away from land males are observed hauling out on sea ice and feeding in the marginal ice zone. However they are ill adapted to extreme cold, and if entrapped they can experience blindness through the freezing of their basal tears, often leading to death.

Killer whales are probably the only predator of Antarctic fur seals of all ages, but pups and yearlings are also vulnerable to leopard seal predation. Such predation can be significant, and leopard seals are thought to limit the population growth during winter at Elephant Island, South Shetland Islands. Leopard seals visit Bird Island, South Georgia. Their main prey is pups, although the estimated kill is less than 1% of the island's annual production. Large otariid bulls of other species may also prey on pups where the species co-exist. At Macquarie Island, a single subadult male Hooker's sea lion (*Phocarctos hookeri*) was believed to have killed a total of 54 Antarctic and subantarctic fur seal pups, 43% of the total production.

Antarctic ecosystems are often considered to be isolated and pristine, but Antarctic fur seals regularly cross the polar frontal zone and are therefore a potential vector for diseases. Little is known on the diseases of Antarctic fur seals, but given their similarity to, and mixing with, other fur seal species they are likely to share similar bacteriology and pathology. During the breeding season, the scale of fur seal bull mortality fluctuates between years. The main causes appear to be infections from fighting wounds and pneumonia. Studies have isolated various *Streptococcus* sp., *Staphylococcus* sp., *Bordetella* sp., and *Corynebacterium phocae* from dead animals. Anti-bodies of *Brucella*, which can impair female reproduction, have been found in individuals on the Antarctic Peninsula.

The most obvious external parasitic/commensal organism of Antarctic fur seals is the goose barnacle (*Lepas australis*), typically found on females returning to breeding after extended periods at sea (Setsaas and Bester, 2006). The average infestation of 10 barnacles reported on 4% of returning females has little effect on swimming performance. However, individual seals have been recorded with over half of their fur colonized by the barnacle, which would severely increase the individual's hydrodynamic drag. The barnacles die and drop off within a few days of the seal being ashore. Gastric nematodes are prevalent, and tapeworms have been recorded, but no ticks have been observed.

Antarctic fur seals and elephant seals often breed in the same areas, but the difference in their timing of breeding means that their interactions are limited. Elephant seals occasionally crush fur seal pups and in turn elephant seal weaners are observed with superficial wounds inflicted by fur seals.

Despite penguins being preyed on by fur seals, on land they often breed side by side with the seals. Although fur seals generally ignore their avian neighbors, seal disturbance, usually an individual rushing through a penguin colony, can lead to egg loss and chick mortality.

Fur seals provide resources to many flying birds. At South Georgia, seal carcasses are consumed by giant petrels and skuas and also provide food for pintail ducks and sheathbills. Placentas provide food at the beginning of the breeding season, whereas dead or dying pups provide a steady source of food for scavengers. Increases in the skua and northern giant petrel populations at several subantarctic sites have been attributed to increases in seal carrion, in particular at South Georgia where the species is highly abundant, although this may not be the case in other areas (De Bruyn *et al.*, 2007).

The destruction of tussock stands could potentially impact birds that nest or burrow in this habitat, but this is unlikely to be a major problem, especially compared to the impact of predation from rats and cats where these pests have been introduced. Although Antarctic fur seals share the tussock habitat with larger birds, like wandering albatrosses and giant petrels, there is little evidence of any negative impact on these species.

## IV. Behavior and Physiology

Antarctic fur seal bulls begin to arrive on the breeding beaches in numbers during early November and fight to establish territories. The first females arrive a few weeks later and give birth within a few days of hauling out. The perinatal period lasts 5–8 days, at the end of which the females mate before heading out to sea. Antarctic fur seals are highly polygynous, with bulls on average having "harems" of nine females. However, females exhibit mate choice, and genetic studies have revealed that females from a given male-associated group can travel through a number of neighboring territories in order to mate with a bull who is both heterozygous and unrelated (Hoffman *et al.*, 2007). After mating, males migrate to higher latitudes in January, and their numbers increase on the more southerly islands, on the Antarctic Peninsula and around the ice edge.

Females are income breeders, spending periods ashore nursing the pup (1–2 days) interspersed with foraging at sea (2–11 days). Pups wean in April only after 4 months, the shortest lactation period of any otariids. Only northern fur seals (*Callorhinus ursinus*) have a comparably short lactation.

Extended dives usually relate with feeding or attempts to locate food. Females dive predominately during the night, to shallow depths (0–40m), exploiting diurnally migrating prey within the mixed surface layer. The great sexual dimorphism and the constraints imposed on females by pup rearing lead to obvious sexual segregation. Because of their larger size, males can dive deeper and longer. Maximum dive durations are around 5min for females and 10min for males, reaching maximum depths of 210 and 350m, respectively.

Males foraging around South Georgia prior to breeding are segregated from females both horizontally and vertically. However, their foraging trips are very similar in length to those of breeding females, despite being unconstrained by any offspring demands. Whilst female foraging is concentrated at the shelf break, males mostly forage closer to the breeding colony on the continental shelf. In these shallower waters males dive throughout the day and night and undertake benthic dives (mean depth 200 m), which are not observed in females.

Diving behavior is strongly influenced by prey behavior and differs between foraging locations. Females foraging in shelf waters have a higher proportion of daytime dives and dive deeper than they do in oceanic waters. When males forage in oceanic waters they follow a similar pattern to females foraging there, diving nocturnally within surface waters, but with a deeper dive range (0–60 m).

Bulls perform two main vocalizations. A threatening roar (full-threat call) is generally used in direct aggression against other males or in response to a specific threat. The second type is a "huff-chuff" sound generally used when moving around a territory, interacting with females, or within the area of other seals. This sound appears to act as a display of status, as its use by subordinates can provoke aggression. Females and pups are occasionally heard vocalizing in this way either in aggression toward conspecifics or during play. Although females will huff-chuff and roar, their main vocalizations occur in mother–pup interactions. Mothers and pups establish a bond through both smell and vocalization immediately after birth that is constantly re-enforced throughout the postnatal period. When mothers return from a foraging trip or when they are separated from their pups ashore, the two locate each other by call and response, and once in close proximity identity is confirmed by smell.

Females molt in February–March when clumps of hair can been seen trapped in the nails on their hind flippers. They are still suckling pups at this time and they can continue foraging at sea throughout the molt. Pups begin to molt from their natal coat to adult pelage around early February at about 1–2 months of age. The timing of the bulls' molt is unclear, although identifying paint marks on the fur from behavioral studies can last up to 2 years or more on some individuals, suggesting that any annual molt may only be partial.

## V. Life History

Extreme sexual dimorphism and the species' breeding biology determine completely different life histories for females and males. Mean female age at maturity is 4 years, although first conception occurs as early as 2 and as late as 7 years. By age 6 most females have attained full adult size. Variability in primiparity is largely dictated by body mass, density dependence, and environmental stress. Early age at primiparity may affect survivorship of physically immature females, although this is difficult to observe, because the survival of all young females is highly affected by environmental stress. Reproductive rates increase rapidly from age 2, peak at 0.80 yr at age 8, and remain high, on average 0.75 (0.68–0.77), until the onset of a senescent decline, around age 11. Weaning success increases with age and experience, although it relies heavily on food availability and the ability of mothers to provide. Trauma and lack of food are usually the most common causes of early pup mortality in densely populated areas.

Female longevity is around 20 years; the oldest known female was 24. The mean adult female survival is 0.87 (0.68–0.93); variation is mainly caused by fluctuation of the biological environment. In years with severe climate anomalies, it can be reduced by as much as 15–20%. Female fitness, measured as the asymptotic population growth rate, is most sensitive to changes in the survival of breeders and their propensity to breed. Therefore, with adverse environments, females' body condition may decrease, and they defer or alter breeding rather than put their life at risk. Breeding can be altered by not implanting or reabsorbing the blastocyst, abortion, or by pup abandonment.

The female breeding cycle is highly constrained by their income breeding system and the extreme seasonality of high latitudes. Females can only breed during the short austral summer when suitable pupping habitat is available and the local environment is sufficiently productive to supply constant nourishment for maintenance and pup rearing. This requires a high breeding synchrony that concentrates the birth of 90% of the pups in a 10-day window. This synchrony is highly consistent across the different breeding locations in the Southern Ocean, with only small differences in median birthdate (Hofmeyr et al., 2007) and is mostly affected by local environmental influences.

Despite these constraints, most females breed annually, investing greatly in pupping. They conceive and carry a new fetus although still lactating and are constantly foraging. Because their gestation lasts between 8 and 9 months, annual breeding is achieved by having (1) a diapause period of 3–4 months between conception and implantation of the blastocyst and (2) the shortest lactation among otariids, which is required to cope with the highly seasonal productivity typical of high latitudes. The advantage of diapause is that the pupping interval can be very close to 365 days and is thus adjusted to seasonal breeding; mothers only start investing on the new gestation when the previous pup is weaned or close to weaned. Lactation is arguably the highest breeding cost. Given its heavy reliance on the biological environment, the rapid and successful recovery of the once almost extinct South Georgia population can be explained by the high abundance of its main prey, Antarctic krill. However, the current increased frequency in climate anomalies is reducing the krill supply in the Scotia Sea more regularly, rendering the environment less predictable for breeders. Their vital rates are affected by this fluctuation that reduces the carrying capacity of the environment. This instability is unlikely to support the high rates of breeding success and the population expansion previously seen.

Males reach sexual maturity at ages 3–4, but they are not reproductively active until they reach their adult body size, usually between ages 7 and 8, when they start competing for territories. Territory tenure is highly variable, and with high population density only a few bulls gain access to receptive females. These are the bulls more likely to mate successfully and father most (up to 60%) of the pups born on dense breeding beaches. Higher heterozygosity has been associated with higher bull success, in terms of longer territory tenure and competitive ability. Territory tenure implies greater competition with other bulls and a higher chance of mortality. On average, most bulls live for 8 years, which suggests a high mean mortality rate, possibly 30% higher than that of females. However, reliable estimates are not available for any population.

Antarctic fur seals and subantarctic fur seals (*Arctocephalus tropicalis*) breed sympatrically in at least six locations. At Marion Island hybrids make up at least 0.02% of the island population. At Possession Island, Crozet Islands; Heard Island; and Macquarie Island, Antarctic and subantarctic fur seals occur in breeding groups where New Zealand fur seals (*Arctocephalus forsteri*) also occur. Recent analysis indicates the presence of hybrids involving the three species (17–30% of the pups) and some degree of hybrid reproductive success. Hybrid bulls can hold territories, but their reproductive success is low, with a great proportion of pure-species females in their territories conceiving extra-territorially with conspecifics. This suggests the presence of reproductive isolating mechanisms that limit the frequency of hybrids (Lancaster et al., 2007). The overlap in breeding areas of the different fur seal species is thought to result from recent colonization and the increase and spreading of populations after recovery from intense exploitation.

## VI. Interactions with Humans

Antarctic fur seal populations suffered intense commercial exploitation during the eighteenth and nineteenth centuries, mainly for their pelts. This led to a dramatic reduction of populations worldwide. Large-scale sealing began in the Southern Hemisphere in 1775. Antarctic fur seals were known to be abundant on South Georgia during the eighteenth century; Captain Cook discovered the island in 1774 and found that the beaches "swarmed" with seals. The first record of Antarctic fur seals taken from South Georgia is from 1786. Sealing reached a peak in 1800–1801, when 17 British and American vessels took 112,000 skins. A single ship had a recorded catch of 57,000 seals for that particular season. By 1821 near-extermination was recorded by James Weddell, who calculated a total take of 1.2 million seals. A few more seals were taken in 1838–1839, in the 1870s, and in 1908. Fur-sealing on South Georgia continued until just one seal was found and killed in 1915 (Bonner and Laws, 1964; Headland, 1984). Sealing efforts on the South Shetlands began as soon as they were discovered in 1819, and fur seals were almost completely exterminated in just three seasons. James Weddell calculated that 320,000 were taken during 1821 and 1822.

On South Georgia, the population recovered very rapidly and has now reached very high numbers. A small breeding colony was discovered in 1930s on Bird Island, west of South Georgia, where the recolonization is thought to have started. From 1958 to 1972, the annual rate of increase on that small island was estimated as 16%, with pup production increasing from 10,000 per year in the early 1960s to 90,000 in 1975. The current annual production is thought to be no higher than 60,000 pups per year. The numbers and recovery rates were different at other locations. At Bouvetøya, historical accounts suggest that there were significant numbers of seals present toward the end of sealing (Hofmeyr et al., 2005). At other locations, numbers after sealing are thought to be very small. At Heard Island the first recorded breeding after sealing was in 1963; since then the population is thought to have increased by 12–20% per year (Page et al., 2003). At the South Shetlands, recovery also was fast, with a possible rate of pup production increase of 20% (Hucke-Gaete et al., 2004).

The Antarctic fur seal conservation status is listed as lower risk and Least Concern by the IUCN. However, a number of threats exist. Direct interactions with fisheries have been reported, particularly in the South Georgia area. The problem has not had a significant impact because of good regulation and the use of mitigation measures. A common problem is entanglement in man-made debris, mostly from the fishing industry (Fig. 4). This has been documented since the 1980s, but the effort of removing and reporting entanglements is variable, and there are no good assessments for most areas. The most thorough published study is from 1989 to 1990 on Bird Island, which suggests that 0.1% of the population is affected. Most entanglements were by young males, in polypropylene straps (packaging bands), nylon string, fishing nets, and other materials (Croxall and Boyd, 1990). Up to 30% of the entanglements caused physical injury and less than 20% could be removed easily. The rate of entanglements appears to be increasing, possibly linked to illegal fishing operations, especially in longlining. Entanglements have also been reported on Bouvetøya Island and Marion Island (Hofmeyr et al., 2002), although the rate of occurrence is very low, indicating that it is not a real threat to the current population.

## See Also the Following Articles

Hybridism ■ Southern Fur Seals

**Figure 4** *Young Antarctic fur seal male at Bird Island entangled with a nylon rope loop from a fishing net.*

## References

Bonner, W. N. (1968). The fur seal of South Georgia. *Br. Ant. Surv. Sci. Rep.* **56**, 1–81.

Bonner, W. N., and Laws, R. W. (1964). Seals and sealing. *In* "Antarctic Research" (R. Priestley, R. J. Adie, and G. D. Q. Robin, eds), pp. 163–190. Butterworths, London.

Boyd, I. L. (1993). Pup production and distribution of breeding Antarctic fur seals (*Arctocephalus gazella*) at South Georgia. *Antarct. Sci.* **5**, 17–24.

Croxall, J. P., and Boyd, I. L. (1990). Entanglement in man-made debris of Antarctic fur seals at Bird Island, South Georgia. *Mar. Mamm. Sci.* **6**, 221–233.

De Bruyn, P. J. N., Pistorius, P. A., Tosh, C. A., and Bester, M. N. (2007). Leucistic Antarctic fur seal at Marion Island. *Polar Biol.* **30**, 1355–1358.

Forcada, J., Trathan, P. N., Reid, K., and Murphy, E. J. (2005). The effects of global climate variability in pup production of Antarctic fur seals. *Ecology* **86**, 2408–2417.

Headland, R. (1984). "The Island of South Georgia." Cambridge University Press, Cambridge, UK.

Hoffman, J. I., Forcada, J., Trathan, P. N., and Amos, W. (2007). Female fur seals show active choice for males that are heterozygous and unrelated. *Nature* **445**, 912–914.

Hofmeyr, G. J. G., and Bester, M. N. (1993). Predation on king penguins by Antarctic fur seals. *S. Afr. J. Antarct. Res.* **23**, 71–74.

Hofmeyr, G. J. G., De Maine, M., Bester, M. N., Kirkman, S. P., Pistorius, P. A., and Makhado, A. B. (2002). Entanglement of pinnipeds at Marion Island, Southern Ocean: 1991–2001. *Aust. Mammal.* **24**, 141–146.

Hofmeyr, G. J. G., Krafft, B. A., Kirkman, S. P., Bester, M. N., Lydersen, C., and Kovacs, K. M. (2005). Population changes of Antarctic fur seals at Nyroysa, Bouvetøya. *Polar Biol.* **28**, 725–731.

Hofmeyr, G. J. G., *et al.* (2007). Median pupping date, pup mortality and sex ratio of fur seals at Marion Island. *S. Afr. J. Wildl.Res.* **37**, 1–8.

Hucke-Gaete, R., Osman, L. P., Moreno, C. A., and Torres, D. (2004). Examining natural population growth from near extinction: The case of the Antarctic fur seal at the South Shetlands, Antarctica. *Polar Biol.* **27**, 304–311.

Lancaster, M. L., Bradshaw, C. J. A., Goldsworthy, S., and Sunnucks, P. (2007). Lower reproductive success in hybrid fur seal males indicates fitness costs to hybridization. *Mol. Ecol.* **16**, 3187–3197.

Page, B., Welling, A., Chambellant, M., Goldsworthy, S. D., Dorr, T., and van Veen, R. (2003). Population status and breeding season chronology of Heard Island fur seals. *Polar Biol.* **26**, 219–224.

Setsaas, T. H., and Bester, M. N. (2006). Goose barnacle (*Lepas austra-lis*) infestation of the subantarctic fur seal (*Arctocephalus tropicalis*). *Afr. Zool.* **41**, 305–307.

Wilson, J. W., Burle, M-H., and Bester, M. N. (2006). Vagrant Antarctic pinnipeds at Gough Island. *Polar Biol.* **29**, 905–908.

Wynen, L. P., *et al.* (10 authors) (2000). Postsealing genetic variation and population structure of two species of fur seal (*Arctocephalus gazella* and *A. tropicalis*). *Mol. Ecol.* **9**, 299–314 .

# Antarctic Marine Mammals

### Ian L. Boyd

The Southern Ocean is the ocean subregion surrounding the continent of Antarctica. Its southern boundary is defined by the narrow coastal continental shelf of Antarctica itself. To the north the boundary is defined by an oceanic frontal feature known as the Antarctic convergence or southern polar frontal zone. The zone marks the boundary between cold southern polar waters and temperate northern waters. The ocean temperature can change by as much as 10°C across the front, which may be only a few miles across. The polar front is an important physical feature that determines marine mammal distributions. It defines the normal southern extent of the distributions of most tropical and temperate marine mammals (Fig. 1).

A second feature that is important to marine mammals in the Antarctic is the annual sea ice. The seasonal change in sea ice cover can lead up to 50% of the Southern Ocean being covered in ice during late winter, but by late summer this can have contracted to 10% of the winter maximum. These large seasonal fluctuations in the sea ice have profound implications for the ecology of the Southern Ocean, including that of marine mammals. Many marine mammals, including most cetaceans, migrate north across the polar front in winter.

## I. Antarctic Species

This section deals with true Antarctic species defined as those species whose populations rely on the Southern Ocean as a habitat, i.e., critical to a part of their life history, either through the provision of habitat for breeding or through the provision of the major source of food. Species that inhabit the subantarctic, which is generally seen as including the islands that circle Antarctica in the region of the polar front or the polar frontal zone itself, are not included.

The Southern Ocean accounts for about 10% of the world's oceans but it probably supports >50% of the world's marine mammal

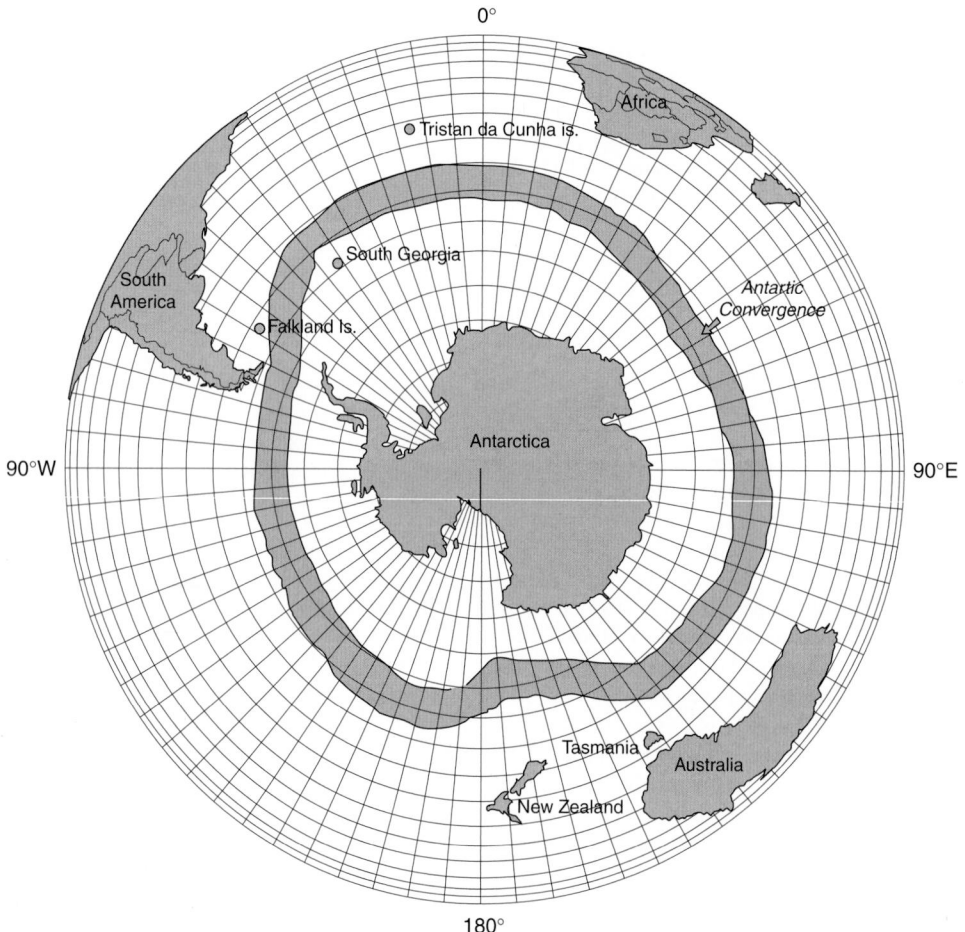

**Figure 1** *Antarctica and the Antarctic Convergence, where waters of warmer northern seas meet colder Antarctic waters. This confluence of waters of different temperatures occurs in some of the roughest seas known.*

biomass, including six species of pinnipeds, eight species of baleen whales, and at least seven species of odontocete whales. Therefore, in terms of the diversity of species, the Antarctic is host to only one-fifth of the world's pinniped and a little less than one-fifth of the world's cetacean species. This low diversity may be attributed partly to the lack of land masses to cause isolation and speciation and also because, although large in its total area, the Southern Ocean does not have the diversity of habitats and prey species seen in other ocean basins.

Among the pinnipeds (Laws, 1984), there is one species from the family Otariidae (eared seals, which include fur seals and sea lions) and there are five species from the family Phocidae (earless or "true" seals), but all of these come from a single subfamily, the Monachinae. This list is as notable as much by its absences as it is for those that are present. For example, there is no representative of the phocid subfamily Phocinae, which contains a diverse collection of species of Northern Hemisphere seals. There are also no representatives of the subfamily Otariinae, which includes all of the sea lions, and there is only one representative of the diverse Southern Hemisphere subfamily Arctocephalinae, which includes the southern fur seals.

Where pinnipeds are concerned, evolutionarily it would appear that there have been only two or three species immigrating into the Antarctic. The main immigration was of an ancestral phocid, possibly related to the nearly extinct tropical phocids of today known as monk seals which gave rise to the four most closely related Antarctic phocids: the crabeater seal (*Lobodon carcinophaga*), Weddell sea (*Leptonychotes weddellii*), Ross seal (*Ommatophoca rossii*), and leopard seal (*Hydrurga leptonyx*). At some later date it is likely that elephant seals (*Mirounga leonina*) arrived. Although these seals extend their distribution into south temperate latitudes, as much as 90% of the world population relies on the Southern Ocean as a critical habitat. These were likely to have been followed, or perhaps preceded, by Antarctic fur seals. The taxonomic status of Southern Hemisphere fur seals, a group within which eight species are currently recognized, is uncertain and it seems probable that many of these are not true species but are instead subspecies. Therefore, the Antarctic fur seal (*Arctocephalus gazella*) may simply be an Antarctic race or subspecies of the southern fur seal.

Among cetaceans (Brown and Lockyer, 1984), there are only three Antarctic species within the highly diverse family Delphinidae, which includes all of the dolphins and porpoises. These three are the hourglass dolphin (*Lagenorhynchus cruciger*), long-finned pilot whale (*Globicephala melas*), and killer whale (*Orcinus orca*). The beaked whales are represented by only three species, because these species are very difficult to identify in the field; it is possible that among the very large number of these individuals that are found in the Southern Ocean, several other species could be present.

## II. Distribution and Abundance

Antarctic marine mammals can be divided ecologically among those associated with fast ice, pack ice, or found in the open ocean. Weddell seals are mostly associated with fast ice, Ross seals with open water or pack ice. Leopard seals are animals mainly of the pack ice zone, but they may also be found feeding at penguin and seal colonies north of the pack ice zone. Crabeater seals travel extensively within the pack ice zone and individuals may have a potential range that extends to the total area of the Antarctic pack. The same may be true for Ross seals, although relatively little is known about the biology of these animals. They have been recorded to migrate north of the pack ice zone. Weddell seals appear to be relatively sedentary, forming more or less isolated populations around the coast of Antarctica.

Elephant seals are known to feed within the pack ice zone on occasion, but they are mainly animals of the open oceans north of the pack ice zone. Antarctic fur seals are sometimes found along the boundary between the pack ice and the open ocean but, again, they are mainly animals of the open ocean. Perhaps up to half of both the Antarctic fur seal and southern elephant seal populations migrate north of the polar front during the winter.

Toothed whales have a stratified distribution within the Southern Ocean relative to the polar front and the edge of the pack ice (Fig. 2). Some species, such as long-finned pilot whales and hourglass dolphins, are more closely associated with the polar front, whereas others, such as killer whales, are more often present close to the pack ice. Bottlenose whales (*Hyperoodon planifrons*) and killer whales are the only cetaceans regularly associated with a distribution within the pack ice zone, but they are also present within the open ocean. Bottlenose whales appear to be able to survive comfortably among almost continuous sea ice cover. Killer whales have been recorded in pack ice in winter, but the only baleen whale found regularly in the pack ice zone is the minke whale (*Balaenoptera bonaerensis*). Other larger species, including the sperm whale (*Physeter macrocephalus*), are restricted to the open ocean, but during the summer they may feed along the boundary between the pack ice and the open ocean. In general, these species are absent from the Southern Ocean during the winter. In the case of the sperm whale, only males are found within the Southern Ocean as females remain north of the polar front throughout the year.

Marine mammal distributions are also affected by bathymetric and oceanographic conditions. Southern right whales (*Eubalaena australis*), which are possibly from the same population that winters at Peninsula Valdes, Argentina, and along the coast of South Africa, spend the summer foraging over the continental shelf of South Georgia within the Southern Ocean. Baleen whale and Antarctic fur seal abundance around South Georgia is also influenced by the local oceanography so that there are regions of predictably high abundance of these marine mammals at specific points along the edge of the continental shelf. Southern elephant seals also appear to migrate from breeding and molting grounds on subantarctic islands to shallow regions along the coast of Antarctica. Most of these types of preferences for different locations are assumed to reflect the distribution of food so that marine mammals migrate to the areas of greatest food abundance.

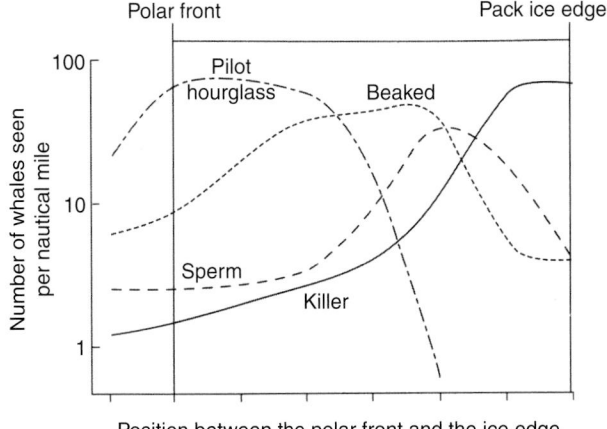

**Figure 2** *Distribution of odontocete whales in the Southern Ocean relative to the southern polar front and the edge of pack ice. Reprinted from Kasamatsu and Joyce (1995) with permission from Cambridge University Press.*

**A**

The crabeater seal is probably the most abundant seal in the world, with population somewhere between 7 and 14 million. There are considerably fewer Weddell seals and leopard seals. Ross seals are rarely seen and the total number is very uncertain, but it is probably the least abundant Antarctic pinniped. The Antarctic fur seal population is more than 3 million and is increasing at about 10% each year. In contrast, the southern elephant seal population within the Antarctic appears to have been relatively stable since the early 1960s, even though the number of elephant seals breeding at sites outside the Antarctic has declined steadily over the same period. The elephant seal population at South Georgia is estimated at 470,000, which probably represents 58% of the world population of the species.

In general, whale populations are in a highly depleted state. Blue whales (*Balaenoptera musculus*) are numbered in the hundreds for the whole of the Antarctic, and the sighting of a blue whale is a rare event. The number of fin whales (*Balaenoptera physalus*) appears to be increasing, as are humpback whales (*Megaptera novaeangliae*) and southern right whales.

Within the Antarctic, there are no significant threats to pinniped species. However, some cetacean populations have been depleted to such a high degree that several are endangered. In particular, blue whales are so rare in the Antarctic that they are possibly close to extinction from the area. Similarly, severely depleted southern right whale and humpback whale populations have very specific migratory routes between summering grounds in the Antarctic and winter grounds in temperate and tropical regions, which make them more vulnerable to threats such as disturbance, habitat loss, and reduced genetic diversity.

## III. Ecology

The presence of a large biomass of marine mammals in the Antarctic is probably a result of the unusual food chain structure of the Southern Ocean. The marine mammals of the Antarctic with large numbers, such as crabeater seals and Antarctic fur seals, rely on krill as their main food source (see Section IV). This is in contrast to marine mammal communities elsewhere that rely mainly on a fish-based diet. Energy enters the food chain through photosynthesis and carbon sequestration by phytoplankton. The relative efficiency with which this energy is passed up the food chain to predators with a krill- or fish-based diet is illustrated in Fig. 3. The efficiency of energy transfer at each step in the food chain can be as low as only a few percent. The fewer steps there are between phytoplankton and marine mammals, the more the energy will be transferred more efficiently to marine mammals. In the Antarctic, there is on average one less step than there is in other oceanic ecosystems, which has led to the very large biomass of marine mammals found in the Southern Ocean.

## IV. Diet

Among seals, there is a progression of dietary specialization from those that mainly eat krill to those that mainly eat fish (Fig. 4). The leopard seal has seabirds and other seals as a major component of its diet, and it is probable that some individuals specialize in feeding on other seals or penguins instead of krill, fish, or squid. Among whales, dietary specializations are divided along taxonomic lines between odontocetes that mainly eat squid and mysticetes that forage primarily on zooplankton.

The crabeater seal is one of the most ecologically specialized of all seals because it feeds almost entirely on Antarctic krill that it

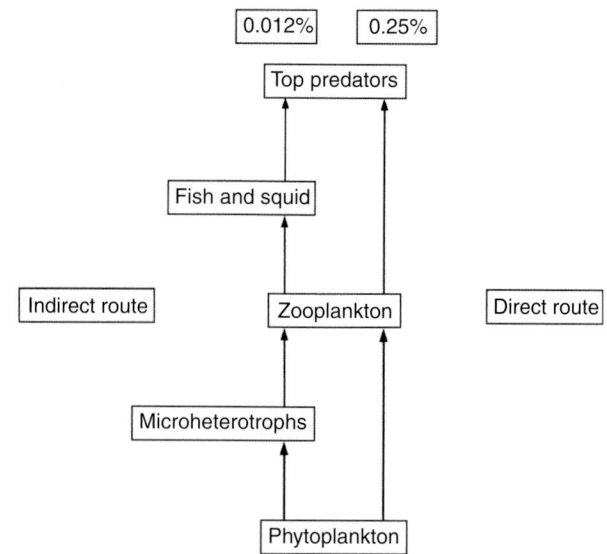

**Figure 3**  *Simplified diagram of energy flow to marine mammals as top predators in marine food chains. This diagram shows the more direct route of energy transfer in the southern ocean, vs the more indirect route elsewhere. The percentage of the energy taken in by phytoplankton that subsequently reaches the top predators is shown at the top of the diagram.*

gathers from the underside of ice floes where the krill themselves feed on the single-celled algae that grow within the brine channels in the ice. Antarctic fur seals also feed on krill to the north of the Antarctic pack ice edge, and many of the other Antarctic seals rely, to varying degrees, on krill as a source of food. Antarctic krill probably sustains more than half of the world's biomass of seals and also sustains a substantial proportion of the biomass of the world's seabirds and whales.

Although the dentition of crabeater seals is modified to help strain krill from the water, the feeding apparatus of the baleen whales is the most highly modified for a diet of plankton. Krill is the major component in the diet of most of the Antarctic baleen whales, although copepods may also be strained from the water, especially by right whales. The Antarctic krill, *Euphausia superba*, often occurs in dense swarms in the open ocean, and the baleen whales have probably evolved to exploit these dense patches of food. Baleen whales eat 30–50 million tons of krill in the Antarctic each year and seals probably eat a similar or slightly lower total amount as whales. Consumption of squid by beaked whales and sperm whales is estimated to be about 14 million tons each year. Killer whales prey on fish and squid but also hunt seals and penguins. Pods of killer whales have been observed tipping over ice floes to push crabeater seals into the water in an effort to catch them.

## V. Exploitation

Throughout the nineteenth and early twentieth centuries, the Antarctic was viewed as an almost limitless source of marine mammals to be hunted for skins, oil, and other products that found expanding markets in Europe and North America. However industrialization of whale and seal hunting brought both greater efficiency and the inevitability that the resources would be exhausted, much to the detriment of the ecology of the Antarctic and its populations of marine mammals.

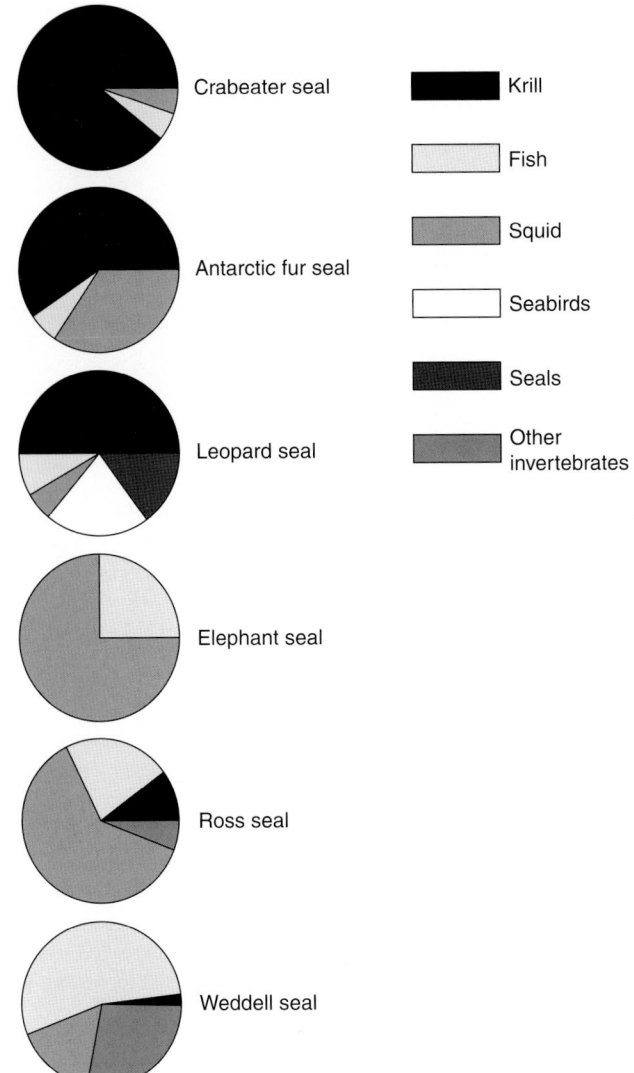

**Figure 4** *Pie charts showing the composition of diets of Antarctic seals.*

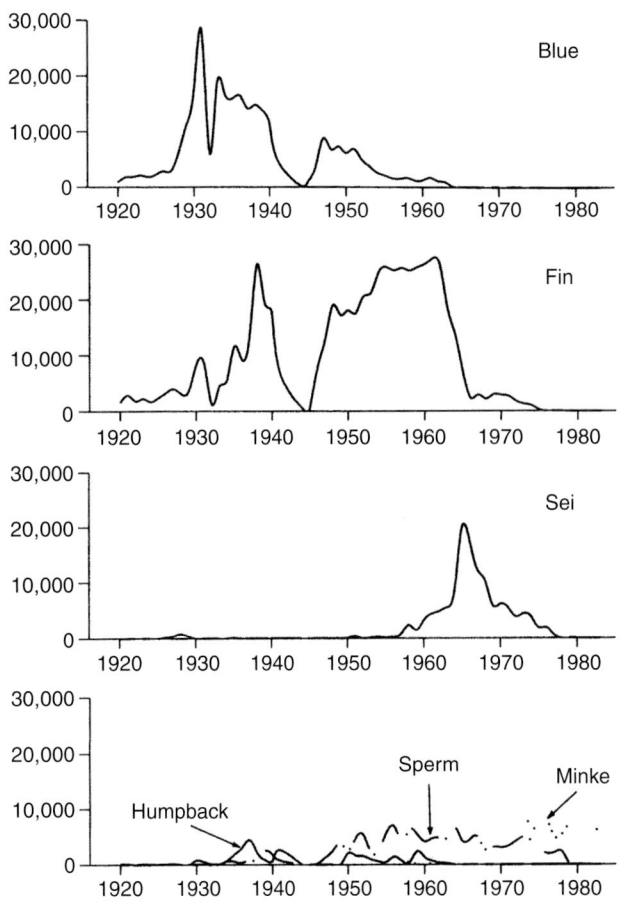

**Figure 5** *Numbers of whales caught in the industrial whale harvest in the Southern Ocean.*

There were three phases of exploitation: exploratory sealing (late eighteenth and early nineteenth centuries), preindustrial sealing and whaling (nineteenth century), and industrial whaling (twentieth century). There are very few records of the exploratory sealing and the preindustrial era. During the exploratory sealing era, exploitation was mainly targeted at fur seals to supply skins for the Chinese market, where there were turned into felt to supply the European market. By about 1830, fur seals in the Antarctic and elsewhere in the Southern Hemisphere had been all but extinguished. In 1825, James Weddell, himself the captain of a sealing vessel, noted that "the number of skins brought from off Georgia cannot be estimated at fewer than 1,200,000." He was referring to South Georgia, where more than 95% of the current world population of Antarctic fur seals resides. This species was considered to be extinct until the early 1920s when whalers saw several individuals at South Georgia. Since then, the numbers have increased rapidly and the population is conservatively estimated to now be on the order of 3 million. The preindustrial era was mainly targeted at whaling and the larger seals, particularly elephant seals for their oil. This activity was mainly undertaken from

sailing vessels. The introduction of steam power to the Antarctic was largely responsible for the transition to industrial whaling.

Industrial whaling began in the early years of the twentieth century. This industry operated for more than 60 years and in that time it removed about 71 million tons of whale biomass involving 1.4 million individual whales from the Antarctic; about 10% of these were taken at South Georgia. Antarctic fur seals feed on krill (Fig. 4), and may have benefited by the reduction in numbers of krill feeding whales and therefore had less competition for their food.

The industry was selective about which species of whales it targeted. The largest and most profitable were selected first, followed by progressively smaller species (Fig. 5). Eventually, the industry became unprofitable because only minke whales were left to exploit and these were too small to be profitable.

## VI. Conservation Measures

Concerns about the effects of industrial whaling on the populations of whales began early in the industrial era. By the early 1920s, the "Discovery Investigations" had been established to determine whale populations mainly around South Georgia. These were funded by a levy on the industry, but they were free from control of the industry. They are one of the first examples of the fledgling field of ecology being used to solve a wildlife management problem. Even though the "Discovery Investigations" made ground-breaking scientific progress and were influential in the introduction of some conservation measures, they came too late to influence the power of the industry and the fate of the populations of whales in the Southern Ocean.

The story of overexploitation of a marine resource in the Southern Ocean repeated itself in the 1960s and 1970s when industrial fisheries targeted the fin fish populations and reduced them to uneconomic levels. This stimulated a renewed effort to ensure that there was proactive conservation of marine living resources in the Southern Ocean. The result was the Convention for the Conservation of Antarctic Marine Living Resources (CCAMLR) and the Convention for the Conservation of Antarctic Seals (CCAS), which came into effect in 1982 and 1978, respectively. One of the unique features of the CCAMLR convention is that it accepts that exploitation has effects on components of the ecosystem far beyond those that are being targeted for exploitation (Kasamatu and Joyce, 1995). This means that any proposals for the exploitation of living resources in the Antarctic must consider the effects that such exploitation is likely to have on marine mammals, whether or not they are the target species. Therefore, even though marine mammals enjoy legal protection in the Antarctic from unregulated exploitation under the environmental protocol within the Antarctic Treaty, they are also protected from other activities within the Southern Ocean ecosystem. Only time will tell if this is sufficient to ensure their long-term survival.

## See Also the Following Articles

Arctic Marine Mammals ▪ Conservation Ecology ▪ Conservation Efforts ▪ Distribution ▪ Krill

## References

Brown, S. G., and Lockyer, C. H. (1984). Whales. *In* "Antarctic Ecology" (R. M. Laws, ed.), Vol. 2, pp. 717–782. Academic Press, London.

Kasamatsu, F., and Joyce, G. G. (1995). Current status of odontocetes in the Antarctic. *Antarct. Sci.* **7**, 365–379.

Laws, R. M. (1984). Seals. *In* "Antarctic Ecology" (R. M. Laws, ed.), Vol. 2, pp. 621–716. Academic Press, London.

Laws, R. M. (ed.) (1993). "Antarctic Seals: Research Methods and Techniques." Cambridge University Press, Cambridge.

# Archaeocetes, Archaic

## J.G.M. THEWISSEN

Archaeocetes is the common name for a group of primitive whales that lived in the Eocene Period (~55–34 million years). Archaeocetes are important because they represent the earliest radiation of cetaceans and because they include the ancestors of the two modern suborders of cetaceans, the Mysticeti (baleen whales) and Odontoceti (toothed whales). Archaeocetes are also the main source of information about the great morphological changes that were associated with the acquisition of aquatic features in cetaceans (Zimmer, 1998). The first archaeocete whales (*Pakicetus* in Fig. 1) looked, externally, nothing like modern whales, instead their bodies resemble wolves with long snouts and powerful tails. Later archaeocetes look more like crocodiles, otters, or sea lions, and it is not until about 39 million years ago that basilosaurid cetaceans

**Figure 1** *Five cetaceans that lived in the Eocene. Clockwise from top: the basilosaurid* Dorudon *(~39 million years old); the ambulocetid* Ambulocetus *(~49 million years old); the pakicetid* Pakicetus *(~50 million years old); the remingtonocetid* Kutchicetus *(~45 million years old); and the protocetid* Rodhocetus *(~45 million years old). Note the increase of aquatic adaptations from* Pakicetus, *the earliest whale, to* Dorudon, *one of the last archaeocetes. Illustration from Thewissen and Williams, 2002.*

can easily be recognized as whales. The archaeocete families document that Eocene cetacean evolution is characterized by increasing aquatic adaptations, documenting amphibious stages that preceded obligate aquatic life in the late Eocene whales (basilosaurids).

Five families of cetaceans are commonly included in archaeocetes: Pakicetidae, Ambulocetidae, Remingtonocetidae, Protocetidae, and Basilosauridae (Williams, 1998). Basilosaurids (also called zeuglodonts) are discussed separately in this volume, and the remaining four families are treated here.

Pakicetidae are only known from the early-to-middle Eocene and lived approximately 50 million years ago in India and Pakistan. Many parts of the skeleton of pakicetids are known, including a number of skulls (Thewissen *et al.*, 2001). Basically, pakicetids varied from fox- to wolf-size. The nasal opening of pakicetids was near the front of the head, and the eyes faced dorsally, similar to crocodiles. Pakicetids had small brains flanked by enormous chewing muscles. The skull and dentition of pakicetids do not resemble those of modern whales and dolphins, but the ear of pakicetids clearly shows that they were cetaceans: there is thick internal lip of bone on the middle ear (the involucrum), and the ear ossicles are pachyostotic and oriented differently from those of other mammals. The limb bones of pakicetids were also very dense, probably as a means to counteract buoyancy, and allow them to wade in water (Gray *et al.*, 2006). Pakicetid fossils are only found in freshwater deposits and most are known from deposits that represent shallow ephemeral streams in an arid climate, it is unlikely that pakicetids were good swimmers. Three genera are included in Pakicetidae: *Pakicetus*, *Ichthyolestes*, and *Nalacetus*.

Ambulocetids are known from middle Eocene rocks in northern India and Pakistan. There are fewer than 10 described ambulocetid fossils, but one of these consists of a nearly complete skeleton of a single individual of *Ambulocetus natans* (Thewissen *et al.*, 1994, 1996). *Ambulocetus* resembled a crocodile in some respects, with short limbs, and a powerful body and tail. It had a large head, with a long snout and eyes that were dorsal on the skull, but faced laterally. The teeth are robust and strongly worn. Skull and vertebrae indicate that the muscles of the head and neck were strong, indicating that *Ambulocetus* was a powerful animal. The shape of the lower jaw of *Ambulocetus*, unlike that of the pakicetids, shows that there was an unusual soft tissue connection between the back of the jaw and the middle ear. In modern odontocetes, this connection consists of a large fat pad that functions as part of the sound-receiving system. This connection is small in *Ambulocetus* and was probably not as important functionally as it is in modern cetaceans. It does show that hearing adaptation arose early in cetacean phylogeny (Nummela *et al.*, 2004). The hindlimbs were relatively short, but the feet were long, and there were four toes. The long paddle-shaped feet indicate that it swam like a modern otter, by swinging its hindlimbs through the water and creating additional propulsive force with its tail (Thewissen and Fish, 1997). The forelimbs were short, with five fingers that each terminated in a short hoof. The hands were much shorter than the feet. The skeleton of *Ambulocetus* indicates that it was probably slow on land. *Ambulocetus* was probably an ambush hunter, attacking prey in or near shallow water. This method of hunting is used by modern crocodiles.

*Ambulocetus* is only known from nearshore marine environments, including estuaries or bays. Geochemical analyses of ambulocetid bones indicate that it drank a mixture of fresh and seawater and that different individuals may have inhabited different microenvironments (Roe *et al.*, 1998). Genera included in Ambulocetidae are *Ambulocetus*, *Gandakasia*, and *Himalayacetus*.

Remingtonocetids are only known from India and Pakistan, from sediments approximately 46 to 43 million years old (Kumar and Sahni, 1980; Bajpai and Thewissen, 1998). Dozens of remingtonocetid fossils have been described, but most of these document only the morphology of skull and lower jaw. Dental and postcranial remains are scarce. The smallest remingtonocetids may have been as small as *Pakicetus*, and the largest may have been close in size to *Ambulocetus*. All early cetaceans had long snouts, but those of remingtonocetids are proportionally even longer than those of other archaeocetes. Skull shape varied between different remingtonocetid genera and possibly reflected different dietary specializations. In *Andrewsiphius* the snout is very narrow and high, and the chewing muscles are weak, suggesting that it may have eaten small, slippery fish. In *Remingtonocetus*, the snout is rounded and robust, and the chewing muscles are large, as would be expected in an animal that attacks larger, struggling prey. No remingtonocetid displays the robust masticatory morphology of *Ambulocetus*. The nasal opening of remingtonocetids is near the front of the skull, similar to pakicetids. The eyes are small, unlike ambulocetids and protocetids. The ear of remingtonocetids is larger than that of pakicetids and ambulocetids, and the connection between the lower jaw and the ear is larger than in ambulocetids. The ears are also set far part, possibly to increase directional hearing. These features are consistent with an increased emphasis on underwater hearing in remingtonocetids. Behind the skull, the remingtonocetid skeleton indicates that the neck was long and mobile and that the hindlimbs were large. Remingtonocetids were certainly able to support their body weight with their limbs, similar to ambulocetids.

The most primitive and oldest remingtonocetid is *Attockicetus*. It is found in the same deposits as *Ambulocetus*. The other remingtonocetids are known from marine, nearshore deposits and may have lived in bays and saltwater swamps. Stable isotope geochemistry indicates that remingtonocetids ingested seawater (Clementz *et al.*, 2006). Remingtonocetid genera include *Remingtonocetus*, *Andrewsiphius*, *Attockicetus*, *Dalanistes*, and *Kutchicetus*.

Protocetids are found in middle Eocene rocks in Indo-Pakistan, Africa, Europe, and North America. Protocetids have never been found at localities with pakicetids or ambulocetids, they are a later radiation, overlapping, partly with remingtonocetids. Protocetids are the oldest whales to disperse across the oceans, although they probably only inhabited the warm seas near the tropics. Many protocetid genera are known, and several of these include several partial skeletons (Gingerich *et al.*, 1994, 2001; Hulbert, 1998). Protocetids are diverse, their average size was similar to that of *Ambulocetus*.

Protocetids had long snouts, large eyes, and their nasal opening was farther caudally than in earlier archaeocetes. This suggests that protocetids could breathe while holding much of their head horizontally, similar to modern cetaceans and foreshadows the origin of the cetacean blowhole. The teeth of protocetids vary between genera, with some showing specializations for crushing hard prey, and others for shearing meat. It is likely that protocetids were active hunters of marine animals, possibly similar to modern pinnipeds. Protocetid locomotor morphology was varied. In general, the tail is well developed and was probably involved in creating propulsive forces. The hind limbs are reduced, and in some species the innominate (pelvis) is not connected by bone to the vertebral column, suggesting that the hind limb did not support the body weight. There are no fossils that document all of protocetid hind limb morphology, but some preserved elements suggest that the hind limbs were short. Indo-Pakistani protocetids inhabited the same environments as the remingtonocetids, and protocetids from other continents are known from shallow marine environments. Known genera of protocetids are *Protocetus*, *Babiacetus*, *Eocetus*, *Georgiacetus*, *Indocetus*, *Natchitochia*, *Pappocetus*, *Rodhocetus*, *Takracetus*, *Artiocetus*, *Rodhocetus*, *Qaisracetus*, *Gaviacetus*, and *Carolinacetus*.

## See Also the Following Articles

Basilosaurids ■ Cetacea, Overview ■ Cetacean Evolution ■ Paleontology

## References

Bajpai, S., and Thewissen, J. G. M. (1998). Middle Eocene cetaceans from the Harudi and Subathu formations of India. *In* "The Emergence of Whales, Evolutionary Patterns in the Origin of Cetacea" (J. G. M. Thewissen, ed.), pp. 213–233. Plenum Press, New York.

Clementz, M. T., Goswami, A., Gingerich, P. D., and Koch, P. L. (2006). Isotopic records from early whales and sea cows: Contrasting patterns of ecological transition. *J. Vertebr. Paleontol.* **26**, 355–370.

Gray, N.-M., Kainec, K., Madar, S., Tomko, L., and Wolfe, S. (2006). Sink or swim? Bone density as a mechanism for buoyancy control in early cetaceans. *Anat. Rec.: Adv. Integr. Anat. Evol. Biol.* **290**, 638–653.

Gingerich, P. D., Raza, S. M., Arif, M., Anwar, M., and Zhou, X. (1994). New whale from the Eocene of Pakistan and the origin of cetacean swimming. *Nature* **368**, 844–847.

Gingerich, P. D., Haq, M., Zalmout, I. S., Khan, I. H., and Malkani, M. S. (2001). Origin of whales from early artiodactyls: Hands and feet of Eocene Protocetidae from Pakistan. *Science* **293**, 2239–2242.

Hulbert, R. C. (1998). Postcranial osteology of the North American middle Eocene protocetid *Georgiacetus*. *In* "The Emergence of Whales, Evolutionary Patterns in the Origin of Cetacea" (J. G. M. Thewissen, ed.), pp. 235–268. Plenum Press, New York.

Nummela, S., Thewissen, J. G. M., Bajpai, S., Hussain, S. T., and Kumar, K. (2004). Eocene evolution of whale hearing. *Nature* **430**, 776–778.

Roe, L. J., Thewissen, J. G. M., Quade, J., O'Neil, J. R., Bajpai, S., Sahni, A., and Hussain, S. T. (1998). Isotopic approaches to understanding the terrestrial to marine transition of the earliest cetaceans. *In* "The Emergence of Whales, Evolutionary Patterns in the Origin of Cetacea" (J. G. M. Thewissen, ed.), pp. 399–421. Plenum Press, New York.

Thewissen, J. G. M., and Fish, F. E. (1997). Locomotor evolution in the earliest cetaceans: Functional model, modern analogues, and paleontological evidence. *Paleobiology* **23**, 482–490.

Thewissen, J. G. M., and Williams, E. M. (2002). The early evolution of Cetacea (whales, dolphins, and porpoises). *Annual Review of Ecology and Systematics* **33**, 73–90.

Thewissen, J. G. M., Hussain, S. T., and Arif, M. (1994). Fossil evidence for the origin of aquatic locomotion in archaeocete whales. *Science* **263**, 210–212.

Thewissen, J. G. M., Madar, S. I., and Hussain, S. T. (1996). *Ambulocetus natans*, an Eocene cetacean (Mammalia) from Pakistan. *Courier Forschungs-Institut Senckenberg* **190**, 1–86.

Thewissen, J. G. M., Williams, E. M., Roe, L. J., and Hussain, S. T. (2001). Skeletons of terrestrial cetaceans and the relationship of whales to artiodactyls. *Nature* **413**, 277–281.

Williams, E. M. (1998). Synopsis of the earliest cetaceans: Pakicetidae, Ambulocetidae, Remingtonocetidae, and Protocetidae. *In* "The Emergence of Whales, Evolutionary Patterns in the Origin of Cetacea" (J. G. M. Thewissen, ed.), pp. 1–28. Plenum Press, New York.

Zimmer, C. (1998). "At the Water's Edge: Macroevolution and the Transformation of Life." Free Press, New York.

# Arctic Marine Mammals

## John J. Burns

## I. Northern Ice-Covered Marine Environments

Traditionally the Arctic is viewed as an ill-defined region around the North Pole that was further subdivided into the high arctic and the low arctic. We are here concerned with much broader, although still poorly defined, areas within which ice-associated bears, pinnipeds, and cetaceans occur. Some freshwater seals are included. It is useful to think in terms of regional climate, oceanography, annual ice dynamics, and life history strategies. For most marine environments, the definitions advanced by Dunbar (1953) are particularly useful. The arctic seas are those in which unmixed polar water from the upper layers of the Arctic Ocean occurs in the upper 200–300 m. A large portion of this zone is ice covered throughout the year. The maritime subarctic includes those seas contiguous with the Arctic Ocean in which the upper water layers are of mixed polar and nonpolar origin. There are, however, some noncontiguous subarctic seas (no water of polar origin) adjacent to terrestrial ecosystems that lie in the subarctic zone. Examples include the Okhotsk Sea, the northern part of the Sea of Japan (Tartar Strait), the Bohai Sea, Lake Baikal in Siberia, and Cook Inlet in Alaska (Fig. 1). In the subarctic, there is a complete annual ice cycle, from formation in autumn to disappearance in summer. Finally, there are areas in the temperate zone where unique climate conditions produce a winter ice cover of relatively short duration. Such areas include the Baltic Sea, the northern Yellow Sea, and the western Sea of Japan.

An estimate advanced in 2005 indicated that over the period 1979–2001, in September, the average annual minimum extent of sea ice was 8 million km², restricted mainly to the Arctic Ocean. The average maximum extent in March was 15 million km², including all of the subarctic seas (or parts thereof), and parts of some in the temperate zone. Most species of the so-called arctic marine mammals are associated with the seasonal ice during the breeding period. They cope with the annual expansion and contraction of the ice cover in a variety of different species-specific ways. Clearly, there are many kinds of ice-dominated habitats formed in response to factors such as regional climate, weather, latitude, currents, tides, winds, land masses, proximity of open seas, and others.

## II. Sea Ice Habitats

Sea ice in the Arctic and subarctic occurs in more complex forms than ice in the Antarctic. This is because of the central location of the Arctic Ocean with its perennial drifting ice, its partially landlocked nature, and the complexity of the subarctic seas encircling it. The annual expansion and contraction of the ice cover provides conditions ranging from the thick and relatively stable multiyear ice of the high latitudes to the transient and highly labile southern pack ice margins that border the open sea. Marine mammals must have regular access to air above the ice, as well as to their food in the ocean below it. During the breeding season, the ice on which pinnipeds haul out must be thick enough and persist long enough for completion of the critical stages of birth, nurture of their young, and, in many cases, completion of the annual molt. Additionally, by virtue of location, behavior, reproductive strategies, and/or physical capabilities, they must be able to avoid excessive predation on dependent and often nonaquatic young. All of the marine mammals must also cope with the great reduction or complete absence of ice during the open water seasons.

There are many different features of the varied types of ice cover that provide marine mammals access to air and allow the pinnipeds to haul out. There are also some features, characteristics, or types of ice that exclude most marine mammals. Important ice features or types include stable land-fast ice (excludes most marine mammals); annually recurring persistent polynyas (irregular shaped areas of open water surrounded by ice); recurrent stress and strain cracks, coastal and offshore lead systems (long linear openings); zones of convergence and compaction (as against windward shores or in constrictions such as narrow straits); zones of divergence (where boundary constraints are eased); the generally labile pack ice of the more southerly seas; and the margins or front zones of broken ice, the characteristics of which are strongly influenced by the open sea (Fig. 2). Ice margins are particularly productive in that ice-edge blooms of phytoplankton and the associated consumers extend many tens of kilometers away from them.

## III. The Role of Sea Ice

There are great differences in how marine mammals exploit ice-dominated environments (Fig. 2). Many are a function of evolutionary constraints imposed on the different lineages of mammals. Polar bears (*Ursus maritimus*) are the most recent arrivals in the high-latitude northern seas, having evolved directly from brown bears (*U. arctos*). They utilize relatively stable ice as a sort of *terra infirma* on which to roam, hunt, den, and rest (Fig. 3). Like their contemporary terrestrial cousins, they are generally not faced with the problem of ice being a major barrier through which they must surface to breathe. Cetaceans are at the other extreme. They live their entire lives in the water and have limited (though differing) abilities to make breathing holes through ice and are therefore constrained to exist where natural openings or thin ice are present. Pinnipeds spend most of their time in the water, but they must haul out to bear their young. Most of them also haul out on ice to suckle their young, to molt, and to rest.

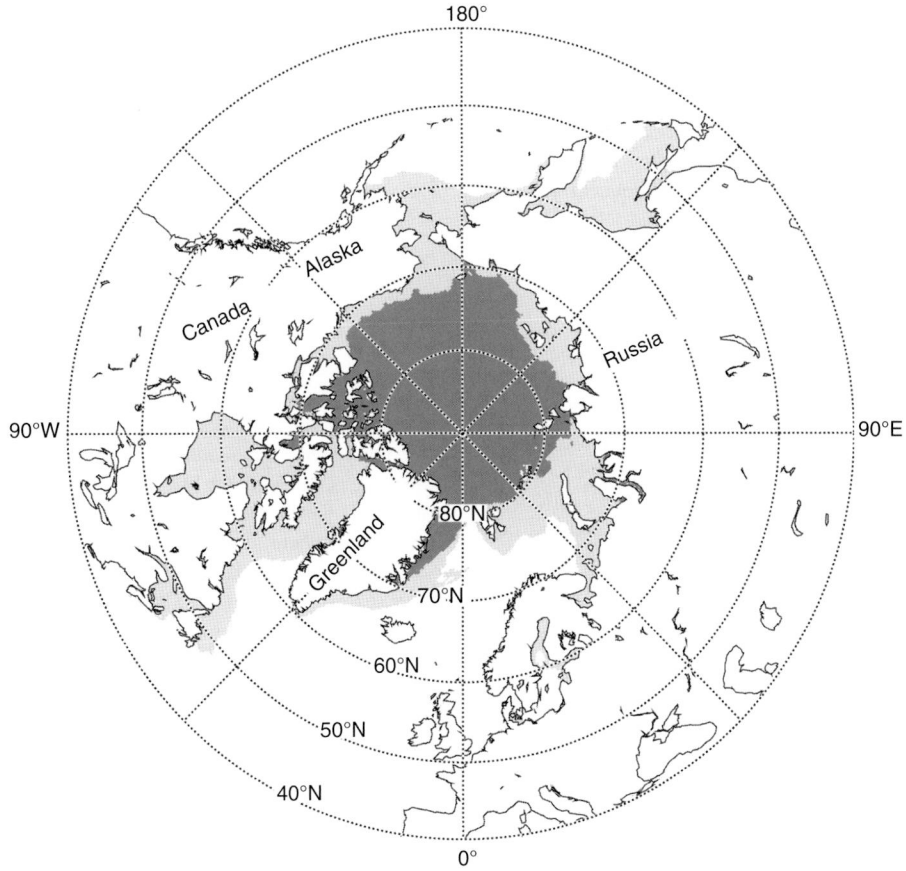

**Figure 1** *Map of the Arctic Ocean and adjacent seas depicting the average maximal (March) and minimal (September) annual extent of sea ice over the period 1979–2001. Courtesy of M. C. Serreze, NSIDC, Boulder, CO).*

**Figure 2** *Typical sea ice of the southeastern Bering Sea front zone in March/April. An aggregation of spotted seals is present.*

**Figure 3** *Tracks of a polar bear (Ursus maritimus) roaming on land-fast ice of the central Beaufort Sea in early spring.*

For cetaceans, obvious benefits are protection from predators, access to ice-associated prey without competition from other animals, and a less turbulent winter environment shielded from perpetual and often storm force winds.

Pinnipeds have flourished in ice-dominated seas both in terms of the number of different species and the number of individuals. All are obliged to haul out either on land or on ice for at least part of the year. As noted by Fay (1974), ice has several special advantages over land, including *isolation* from many predators and other disturbing terrestrial animals; vastly increased *space* away from seashores; a *variety* of different habitats that accommodate more species than does land; easy access to their *food supply*, especially for those that are benthic feeders, or that utilize concentrations of prey associated with ice fronts and polynyas; passive *transportation* to new feeding areas and during migrations; *sanitation* resulting from the ability to avoid or reduce crowding and to haul out on clean ice; and *shelter* among pressure ridges or in snow drifts.

## IV. Ice-Breeding Marine Mammals

Ice-breeding marine mammals in the Northern Hemisphere include eight pinnipeds: gray (*Halichoerus grypus*) (some populations), harp (*Pagophilus groenlandicus*), hooded (*Cystophora cristata*), bearded (*Erignathus barbatus*), ringed (*Phoca hispida*), spotted (*Phoca largha*), ribbon seals (*Histriophora* fasciata), and the walrus (*Odobenus rosmarus*); three cetaceans: narwhal (*Monodon monoceros*), beluga (or belukha) (*Delphinapterus leucas*), and bowhead whale (*Balaena mysticetus*); and one fissiped—the polar bear.

### A. Pinnipeds

A common theme in the ecology of ice-breeding pinnipeds is that of an obligatory, or nearly obligatory, association with ice during the breeding season, which occurs during or shortly after the period of maximal ice extent and relative stability. Seal pups become independent during the spring onset of ice disintegration and retreat. Most species also molt on the ice, after which they disperse to a variety of habitats during the open water season, a few continuing to remain with the diminishing cover. They resume their increasing association with ice during autumn, as it again forms and expands. They haul out on the ice in all seasons during which it is present, although with highly variable frequency depending on species and weather. The maximum number of species and the greatest total number of seals are associated with ice when it is most extensive, and vice versa.

The soft natal fur, or lanugo, of most seals born on ice or in snow lairs, and remaining in one place for long periods of time, is primarily an adaptation for maintaining body heat. Such pups tend to be small, have little insulating blubber, and have a relatively large surface area to body volume ratio at birth. White-coated pups presumably also benefit from the cryptic coloration it provides during the period before they are weaned and begin to enter the water. Prenatal molting occurs in those ice-breeding pinnipeds that are relatively large at birth and can enter the water within hours or days. Detailed discussions of northern ice-breeding seals are presented in the following species accounts, although general comments are noted below.

Gray seals are usually not included in the ice-associated marine mammals category. However, some populations breed on the ice. Gray seals largely inhabit the temperate zone in the North Atlantic region. Their distribution is coastal, often in association with harbor seals (*Phoca vitulina*). There are three populations: those in the Baltic Sea, the eastern North Atlantic, and the western North Atlantic. There is a very wide range in timing of the breeding season. In the eastern Atlantic, pups are born on shore during late autumn to early winter. In both the Baltic and the western Atlantic, however, pups are born during mid- to late winter on ice near shore, or on shore when ice is absent. At birth, gray seals weigh about 15 kg. In all populations almost all pups are born with a silky, whitish coat of lanugo that is retained during the nursing period. They remain on ice or land until after weaning. The late pupping season of the marginally ice-associated breeding populations is thought to be an adaptation to that environment. Grey seals move extensively, although

they are not considered to be migratory. None are associated with sea ice during late spring through autumn.

Spotted seals (or larga seals) occur in continental shelf waters of the Pacific region that are seasonally ice covered. During winter and spring they mainly inhabit the temperate/subarctic boundary areas, occurring in the southern ice front (mainly) of the Bering and Okhotsk seas or in the very loose pack ice of the northern Yellow Sea and Sea of Japan. The birth season is from January through April, depending on latitude. All populations give birth and nurture their pups on the ice, although pups are occasionally born on shore. Newborn pups weigh about 10 kg and have a dense, whitish, wooly lanugo, which is shed toward the end of the month-long nursing period. Seals older than pups usually haul out on the ice to molt, although they also use land when the ice disappears early. As the seasonal ice disintegrates and recedes, all spotted seals disperse, moving to the ice-free coastal zone where they use haulouts on land. The seasonal dispersal can be extensive: in the Okhotsk Sea to its entire perimeter and from the central Bering Sea to most of its perimeter, as well as northward into the northern Chukchi and Beaufort seas. Therefore, some spotted seals reside in the higher latitudes of the subarctic zone during the open water season. They range widely over the continental shelves. There is a close association with sea ice during autumn through spring.

Ribbon seals are animals of the temperate and temperate/subarctic boundary zones in the North Pacific region. Breeding populations are in the Bering and Okhotsk seas and Tatar Strait. During the open water season, they live a completely pelagic existence in the cold temperate waters along and beyond the continental shelves, often far from the locations of their winter habitat. The breeding cycle is similar to that of the spotted seal, and the two occur in relative close proximity to each other during late winter and spring. At the time of pupping and molting, ribbon seals utilize ice of the inner ice front where floes are larger, thicker, more deformed, and more snow covered than in the adjacent ice margin favored by spotted seals. They are noted for hauling out on very clean ice. They pup in late March and April. At birth the pups weigh about 10.5 kg and have a coat of dense, white lanugo. During the nursing period the pups remain on the ice and gradually shed their lanugo. They remain on the ice for some time after they are weaned. In the opinion of this writer, the preference for heavier ice of the inner front, which persists longer than that of the spring ice margin, is because it permits all age classes of these otherwise pelagic seals to haul out until the molt is completed. Ribbon seals do not come ashore unless debilitated. They appear to be the pinniped analog of the Dall's porpoise (*Phocoenoides dalli*) during the pelagic phase of their annual cycle (June through late autumn), dispersing near the shelf breaks and the deeper waters beyond. They have the morphological and physiological attributes of a seal that can dive to great depths and remain submerged for a long time. In the Bering Sea, relatively few move north of their breeding range, except during years of minimal spring ice cover.

Harp seals occur in the North Atlantic region. There are three breeding populations: those of the White Sea, the Greenland Sea and the Gulf of St. Lawrence. They are a gregarious and highly migratory species that lives primarily in the subarctic zone during winter and spring and is broadly distributed in the open sea from the coastal zone to near the ice margin during the open water season. The birth period extends from late January to early April, depending on the region. During the pupping season they form large aggregations in which pups are born in close proximity to each other (often closer than 2.5 m). They prefer large ice fields within the ice front, usually at some distance from the pack ice margins. Here the floes are extensively deformed and ridged, providing shelter to the otherwise exposed pups. At birth the pups weigh about 11.8 kg and have a coat of dense white lanugo. The nursing period lasts from 10 to 12 days and they fast, remaining on the ice floes, for some time after weaning. mating, which occurs after pups are weaned, is followed by the molt. As with the ribbon seal (which is also pelagic after the molt) it seems that the preference of harp seals for the thicker and more stable ice of the inner front zone is because it provides the selective advantage of persisting until the molt is completed. Harp seals make one of the longest annual migrations of any pinniped; some travelling more than 3000 miles from wintering to summering areas. Part of the spring migration is passive as the seals drift on the receding ice.

Hooded seals are a high subarctic, strongly migratory, deep water species that occur in the North Atlantic region and have pups, or whelps, in four different areas: near Jan Mayen, in Davis Strait, off the Labrador coast, and in the Gulf of St. Lawrence. Shifts to heavier ice in the more northerly whelping areas reportedly occur during periods of warmer climate and diminished ice (drift ice pulsations). Pups are mainly born on thick heavily ridged ice floes well within the subarctic pack during late March and early April. At birth, the pups weigh about 22 kg (relatively large) and are comparatively precocious. Their lanugo is shed *in utero* and their birth coat (the blue-back stage) does not resemble the pelage of adults. The nursing period is amazingly brief, averaging 4 days, during which the mothers remain on the ice with their pups. Pups enter the water shortly after weaning, although they spend considerable time on the ice during the postweaning fast. Mating occurs after lactation, and molting after mating. They migrate, both passively on the drifting ice and by swimming, and disperse widely in the open sea (to the Grand Banks), near high latitude shores, and along the edge of the summer pack ice. Extralimital occurrences are common, even to the North Pacific region.

Bearded seals are primarily benthic feeders that have a circumpolar distribution in arctic and subarctic seas. They have evolved in the face of heavy predation pressure by polar bears. Their range broadly overlaps that of all the other ice breeding pinnipeds. They are the least selective of the seals with respect to ice type, provided that it generally overlies water of less than about 200 m deep. Bearded seals are usually solitary and occur from the southern ice margins and fronts (few) to the heavy drifting pack around the rim of the arctic basin, although infrequently in landfast and multiyear ice. Within the heavier pack ice they occur mainly in association with those features that produce open water or thin ice (polynyas, persistent leads, flaw zones, etc.). They are capable of breaking holes in thin ice (<10 cm) and can make or at least maintain breathing holes in thicker ice, with their stout foreclaws. The large pups (about 34 kg) are usually born on the edges of small detached, first year floes very close to the water. The lanugo is shed *in utero*. The pups can swim from birth if necessary, and usually do so, at least in order to move away from the afterbirth. Beyond that they remain on the ice for a day or so. Nursing, which is usually on the ice, lasts 12–18 days, during which time the pups spend a considerable amount of time in the water and begin independent feeding prior to the end of the nursing period. Mating occurs after pups are weaned. The main period of molt is during May and June, and the greatest numbers of all age classes haul out on the ice during that time. However, molting seals are encountered throughout the year. In some areas, such as the Bering and Chukchi seas, the adults and most juveniles migrate to maintain a loose year-round association with ice. They haul out on it throughout the year, although infrequently during winter. In areas where ice disappears during summer (i.e., the Okhotsk Sea) or recedes beyond the continental shelf, they occur in the open sea, in near shore areas, in bays and estuaries, and sometimes haul out on land.

Ringed seals have a circumpolar distribution that includes the arctic and subarctic seas. They have evolved in the face of heavy predation pressure, primarily by arctic foxes, which take pups, and polar bears, which take all age classes. Unique species and subspecies of the subgenus *Pusa* also occur as landlocked populations in Eurasia and include the seals of lakes Baikal, Ladoga, and Saimaa, as well as the Caspian Sea. Ringed seals are the most numerous and widely distributed of the northern ice-associated pinnipeds. During winter to early summer they utilize all ice habitats from the drifting ice margins and fronts (relatively few) to thick stable shore-fast and multiyear ice. Their range extends farther north and includes areas of heavier ice cover than that of any other marine mammal except the polar bear. They occur from shallow coastal waters to the deep of the Arctic Basin. During winter through late spring the adults tend to be solitary and territorial and are most abundant in moderate to heavy pack and shore-fast ice. Ringed seals can make and maintain holes through the ice and crawl out to construct snow lairs above them. In regions where conditions permit, they migrate and maintain a year-round association with ice. In some regions where the pack ice completely or mostly disappears during summer (i.e., the Okhotsk Sea, Baffin Bay, Lake Baikal) they move to nearshore areas and sometimes haul out on land.

Pups are born during late March through April, in snow lairs or cavities in pressure ridges. The pups are small, averaging about 4 kg at birth, and have a thick woolly lanugo, which is usually shed by the end of the nursing period. Lactation lasts 4–6 weeks. Pups mostly remain in the birth lair for the first several days but are soon capable of entering the water and periodically returning to a lair. Mating occurs after the nursing period and is followed by the molt. The peak period of molt in nonpups is during May and June, when the seals haul out above collapsed (melted) lairs, at enlarged breathing holes, or next to natural openings in the ice. Ringed seals are extremely wary when hauled out. During the open water season, depending on the region, they occur in the much reduced pack ice and in open water over a broad area. In some regions they haul out on land.

Walruses are the largest and most gregarious of the ice-breeding northern pinnipeds. They have a discontinuous although nearly circumpolar distribution around the perimeter of the Arctic Ocean and the contiguous subarctic seas. They are benthic feeders mainly restricted to foraging in waters less than 110 m deep. In all areas, their distribution is limited by water depth and in some (i.e., the East Siberian, Laptev, and Kara seas) it is further constrained by severity of ice conditions. In most regions, walruses haul out on ice in preference to land. However, during the open water season, they (mainly males) use land haulouts near the wintering grounds and, in more northerly areas, most come ashore to rest when ice drifts beyond shallow water, as occurs frequently in the Chukchi Sea. During autumn, walruses that migrate southward ahead of the advancing ice also come ashore to rest. All populations are associated with seasonal pack ice during winter to spring/early summer. They mainly use moderately thick floes well into the winter/spring ice cover. The combined requirements for floes low enough to haul out on, but thick enough to support these large animals (usually herds of them) and that are also over shallow productive continental shelves, make walruses particularly dependent on regions within which persistent natural openings are present. They make (batter) holes through ice as thick as 22 cm, using the head, and sometimes maintain them with the aid of their tusks.

Calves are born mainly in early May, which for the Bering Sea population is during the northward spring migration of females, calves, subadults, and some adult males (Fig. 4). Walruses shed their lanugo *in utero*. Calves are born on the ice. They weigh about 60 kg and enter the water from birth, although they haul out frequently.

Cows with young calves often form large nursery herds and migrate passively on the drifting ice, as well as by swimming (Fig. 5). The nursing period lasts more than a year. Walruses haul out in all months of the year.

### B. Cetaceans

The ice-associated cetaceans include two odontocetes (toothed whales), the beluga and the narwhal, and one mysticete (baleen) whale, the bowhead. None have a completely circumpolar distribution. Morphological adaptations to ice seem minimal and include the lack of a dorsal fin in all three and the high "armored" promontory (also termed a "stack") atop which the blowholes of the bowhead are situated. In winter all three species occur in drifting ice where there are persistent natural openings or where the ice cover is thin. Polynyas, shear zones, and leads are important features for them in the regions of heavy pack ice.

The narwhal is a North Atlantic species of the high subarctic and low arctic, which, in winter, consistently occurs in regions of heavy drifting ice over deep water or shelf edges. Adult males have a unique, long unicorn-like tusk which is presumably used in male sexual display. The largest population is that in Davis Strait and Baffin Bay. Seasonal movements of narwhals are directly tied to the advance and retreat of ice. During summer they move to high-latitude, ice-free coastal and nearshore areas, which are often penetrated by deep fjords. Calves are born during the summer, reportedly during July and August, and are nursed for more than a year. This whales' preference for heavy pack ice during winter and spring makes them particularly vulnerable to entrapment during periods of rapid ice formation or when the pack becomes tightly compressed. Most episodes of entrapment are probably brief, though prolonged confinement and rapid ice formation sometimes result in death either by drowning or by polar bears, for which entrapped whales are easy and plentiful prey. Confined whales are also harvested by Inuit hunters whenever they are found.

Beluga whales have a nearly circumpolar distribution that extends from roughly 48°N (the Gulf of St. Lawrence and the northern Sea of Japan) into the summer multiyear pack of the Arctic Ocean. During winter, they are most abundant near the southern ice margins and fronts and as far into the seasonal pack as conditions permit. Again, polynyas, flaw zones, persistent leads, and other features that permit belugas to surface for air are important in the more northerly regions. Belugas often make holes through thin (to about 10 cm) newly formed ice by pushing it up with their head and back. They also surface in openings made by bowheads, with which they often associate during spring migration.

DISTRIBUTION during the open water season is quite variable depending on region. In most cases these whales move into the coastal zone in May to July or early August, where they enter lagoons and estuaries to feed, bear calves in warmer water, and molt. They frequently ascend rivers to feed on seasonally abundant fishes. Telemetry studies have shown that belugas in the Beaufort Sea and the Canadian high arctic spend slightly less than 2 weeks in lagoons, and spend most of their summer feeding in offshore waters (unlike belugas farther south). Some males from the eastern Chukchi and Beaufort Sea stocks are now known to penetrate much farther into the pack ice of the Arctic Ocean during summer than was previously supposed (to beyond 80°N). Other belugas range widely throughout Amundsen Gulf and the Beaufort and northern Chukchi seas during summer and early autumn.

**Figure 4**  *Small herds of walruses* (Odobenus rosmarus) *including females, calves, and subadults on scattered mid-summer ice floes in the eastern Chukchi Sea.*

**Figure 5**  *Part of a huge nursery herd of walruses* (Odobenus rosmarus) *during the northward spring migration through Bering Strait.*

The larger populations include multiple stocks. In the Bering Sea population, the largest of the stocks migrates north through the disintegrating ice cover in spring, and uses both ice-free coastal waters and the summer pack of the Arctic Ocean, ranging from northwestern Canada to northeastern Russia, as the open water season progresses. Most belugas leave the coastal zone by September, although some remain or revisit areas where food is abundant. This habit has resulted in some large and fatal entrapments. Smaller entrapments at sea are not uncommon. All move with the advancing ice in autumn, either migrating southward with it, or moving into it as it forms and expands.

Bowhead whales occur in subarctic waters during winter and spring and, depending on the population, in productive marginal arctic waters during the open water season. These large whales are highly specialized zooplankton feeders and seek areas of high prey abundance. Bowheads may be the slowest growing and latest maturing mammal on earth. Females are thought to become sexually mature between their late teens to mid-twenties (later than humans or elephants). They may live to be well over 100 years old.

The range of bowheads includes the North Atlantic region (three stocks) and North Pacific region (two stocks), with extensive gaps between the two. During winter through early spring they occur from the southern margins of the pack ice to as far into it as persistent natural openings in the ice permit. Large polynya systems are of great importance during winter and spring. In the Pacific sector, the Okhotsk sea stock remains there after the ice has completely disappeared. Most whales of the other stocks migrate northward during spring and southward during autumn. Most whales of the Bering Sea stock maintain a loose association with the summer ice margin, mainly feeding in the open waters south of it. The northward migration begins in late March or early April when they move from the Bering Sea into the eastern Chukchi, and then across the Beaufort Sea through heavy ice in a very long corridor cleaved by a linear system of stress cracks, polynyas, shore leads, and flaw zones. Some migrate into the western Chukchi Sea. Beluga whales commonly migrate with bowheads. Bowheads can stay submerged for long periods and push up through relatively thick ice. These abilities allow them to reside and travel in waters where natural openings in the ice are continually forming and refreezing. Calves are born mainly during April to early June, during the spring migration.

## C. Fissipeds

Two fissipeds roam the high-latitude ice-covered seas: the polar bear and the arctic fox (*Alopex lagopus*). The latter, which rarely enters the water and pups in dens on shore, is not usually considered to be a marine mammal.

Polar bears have a circumpolar distribution in the Arctic and contiguous high subarctic. They are not "marine" in the sense that whales or seals are, but occupy a marine environment in which ice is the substrate on which they live. They prey on other marine mammals, particularly the ringed seal. Depending on the region, they remain with the ice and hunt year round or, where it completely disappears, they come ashore and usually fast or utilize carrion. Exceptions to the latter are some islands (i.e., Wrangel and Herald) where they hunt animals that haul out on shore, particularly walruses, and also feed on the numerous marine mammal carcasses that occur there. On the ice, availability (access) of prey seems to be a more important factor affecting the distribution of bears than is maximum prey abundance. It is difficult for bears to catch marine mammals, except pups, when there are unlimited escape routes and places to surface in a very labile ice cover (*cf.* Fig. 2). For example, few polar bears range south of the northern Bering Sea during winter, even though the majority of other marine mammals (except ringed seals) are south of there. Also, polar bears are not present in the Okhotsk Sea.

Pregnant females make and enter snow dens in early November. These maternity dens can be on the heavy pack ice, on shore-fast ice (relatively few), or on land. The altricial cubs are born in late December or early January, during the arctic winter, and do not emerge with their mothers until late March or early April. Sows that bore their cubs on shore go back to the drifting sea ice after the young emerge from natal dens (Fig. 6). Ringed seal pups, born in lairs beneath the snow starting in late March, are important prey for the sows with cubs.

## V. Possible Effects of Climate Change

It is now well recognized that we are in a phase of accelerated global warming and that the multiyear and seasonal ice cover is being affected. During the period 1979–2006, the average sea ice extent has declined for every month. In September, the usual time of minimal annual extent, the trend of decline is estimated to have been at the rate of −8.6% per decade. The seasonal ice cover is becoming generally less extensive and thinner, and it is forming later and disintegrating earlier than at any time in recorded history. In the Arctic Ocean the multiyear ice cover is also contracting and thinning. Similar changes have occurred in the geological past. The current warming trend, however, seems to be either driven or strongly intensified by anthropogenic inputs (carbon and greenhouse gases) to earth's oceans and atmosphere. In addition to a diminished ice cover, warming conditions also produce rising sea levels, increased ocean circulation, and increased nutrient flow into the northern seas. These changes are likely to have varying effects on the different species of ice-associated marine mammals. At present, we cannot reliably forecast complex changes of the various interacting natural systems that extend from the northern part of the temperate zone to the Arctic.

For some species or populations, e.g., the bowhead whale and spotted seals of northern Beringia, ameliorating conditions might be positive as they would result in more favorable habitat over a broader area then at present. The number of bowhead whales of the Bering Sea population has been increasing steadily and has recently shown a

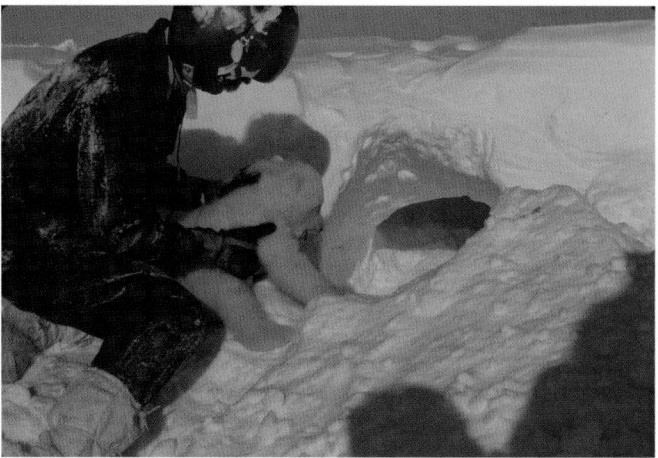

**Figure 6** *NOAA pilot William Harrigan at a temporary snow den on land-fast ice, used by a female polar bear (Ursus maritimus) and her cub, as they traveled from land toward the drifting pack ice in the central Beaufort Sea.*

**A**

remarkable annual increase in calf production. For others, especially those dependent on currently marginal seasonal sea ice habitats, or on heavier and more stable ice habitats of the far north, the changes are likely to have negative impacts. Spotted seals in the Yellow Sea and the Sea of Japan are likely to be negatively affected and, for the true arctic species, there is particular concern about polar bears and their primary prey, ringed seals. At a minimum, global warming will likely result in significant geographic shifts of the seasonal centers of abundance of all ice-associated marine mammals, and populations of some species may decline.

## *See Also the Following Articles*

Antarctic Marine Mammals ■ Biogeography ■ Climate Change ■ Ocean Ecosystems

## *References*

Bo, W. (2006). China's spotted seals face increasing threats. *Pacific Environment*, 4pp. http://www.pacificenvironment.org/article.php?id=1211.

Burns, J. J. (1970). Remarks on the distribution and natural history of pagophilic pinnipeds in the Bering and Chukchi seas. *J. Mammal.* **51**, 445–454.

Burns, J. J. (1981). Ice as marine mammal habitat in the Bering Sea. *In* "The Eastern Bering Sea Shelf: Oceanography and Resources" (D. W. Hood, and J. A. Calder, eds), pp. 781–797. University of Washington Press, Seattle.

Dunbar, M. J. (1953). Arctic and subarctic marine ecology: Immediate problems. *Arctic* **6**, 76–90.

Fay, F. H. (1974). The role of ice in the ecology of marine mammals of the Bering Sea. *In* "Oceanography of the Bering Sea" (D. W. Hood, and E. J. Kelley, eds), pp. 383–399. University of Alaska, Fairbanks.

George, J. C., Bada, J., Zeh, J., Scott, L., Brown, S. E., O'Hara, T., and Suydam, R. (1999). Age and growth estimates of bowhead whales (*Balaena mysticetus*) via aspartic acid racemization. *Can. J. Zool.* **77**, 571–580.

National Research Council (1996). "The Bering Sea Ecosystem." National Academy Press, Washington, DC.

Niebauer, H. J., and Alexander, V. (1989). Current perspectives on the role of ice margins and polynyas in high latitude ecosystems. *In* "Proceedings of the Sixth Conference of the Comité Arctique International" (L. Rey, and V. Alexander, eds), pp. 121–124. E. J. Brill, New York.

Noongwook, G., The Native Village of Savoonga, The Native Village of Gambell, Huntington, H. P., and George, J. C. (2007). Traditional knowledge of the bowhead whale (*Balaena mysticetus*) around St. Lawrence Island, Alaska. *Arctic* **60**, 47–54.

Richard, P. R., Heide-Jørgensen, M. P., and St. Aubin, D. J. (1988). Fall movements of belugas (*Delphinapterus leucas*) with satellite-linked transmitters in Lancaster Sound, Jones Sound, and northern Baffin Bay. *Arctic* **51**, 5–16.

Serreze, M. C., and Barry, R. G. (2005). "The Arctic Climate System." Cambridge University Press, Cambridge.

Serreze, M. C., Holland, M. M., and Stroeve, (2007). Perspectives on the Arctic's shrinking sea-ice cover. *Science* **135**, 1533–1536.

Stirling, I., and Cleator, H. (1981). Polynyas in the Canadian Arctic. Canadian Wildlife Service Occasional Paper 45. Ottawa, Ontario.

Vibe, C. (1967). Arctic mammals in relation to climate fluctuations. *Meddelelser om Grønland*, 1–227.

Wadhams, P. (1990). Evidence for thinning of the arctic ice cover north of Greenland. *Nature* **345**, 795–797.

Weller, G., and Lange, M. (1999). "Impacts of Global Climate Change on the Arctic Regions." Center for Global Change and Arctic System Research, University of Alaska, Fairbanks.

## Atlantic Spotted Dolphin
*Stenella frontalis*

### WILLIAM F. PERRIN

### I. Characters and Taxonomic Relationships

This sturdy spotted dolphin (Figs 1 and 2) is found only in the Atlantic and is commonly seen around the "100-fathom curve" along the southeastern and Gulf US coasts, in the Caribbean, and off West Africa.

The Atlantic spotted dolphin is not always spotted. A large heavy-bodied form found along the coast on both sides of the Atlantic (formerly called *Stenella plagiodon* along the US coast) may be so heavily spotted as to appear white from a distance, but a smaller more gracile form occurring in the Gulf Stream and out into the central North Atlantic can be lightly spotted or entirely unspotted as an adult (Perrin *et al.*, 1987; Viallelle, 1997). A constant diagnostic external feature of *S. frontalis* is a spinal blaze sweeping up into the dorsal cape; this distinguishes it from the very similar pantropical spotted dolphin, *S. attenuata*, also found in the tropical Atlantic. In addition, the peduncle does not exhibit the division into darker upper and lighter lower halves present in *S. attenuata*. The calf of the heavily spotted form is born unspotted, with a three-part color pattern of dark dorsal cape, medium-gray lateral field, and white ventral field. Spots first appear at 2–6 years and increase in size and density up to 16 years (Herzing, 1997). Genetic analyses in correlation with morphology indicate that at least three populations occur in the western Atlantic and the Gulf of Mexico (Adams and Rosel, 2006).

The beak is of medium length (intermediate between those of *Tursiops truncatus* and *S. attenuata*) and sharply demarcated from the melon. The dorsal fin is tall and falcate. Measured adults range from 166 to 229 cm in body length ($n = 106$) and weigh up to 143 kg ($n = 37$) (Perrin *et al.*, 1994a; Nieri *et al.*, 1999). Weight at length is greater than for *S. attenuata* (Perrin *et al.*, 1987).

As in *S. attenuata*, *T. truncatus*, and *T. aduncus*, the skull is characterized by a long rostrum, distal fusion of maxillae and premaxillae

**Figure 1** *Young Atlantic spotted dolphin in the Gulf of Mexico, just developing spots. Spots and blaze below dorsal fin are diagnostic for the species. Photo by R. L. Pitman.*

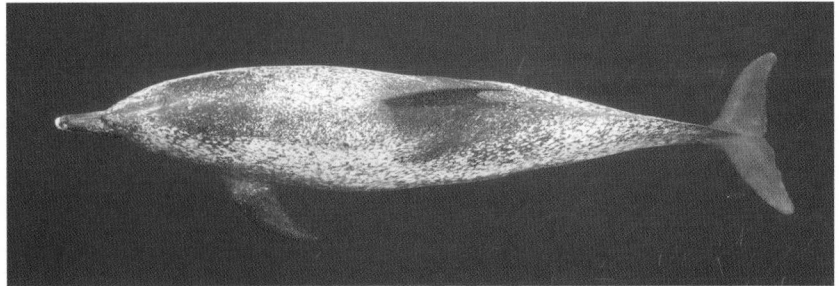

**Figure 2** *Heavily spotted adult Atlantic spotted dolphin in the Gulf of Mexico. Spinal blaze is still visible. Photo by R. L. Pitman.*

in adults, convergent premaxillae, large rounded temporal fossae, and arcuate mandibular rami. Tooth counts are 32–42 in the upper jaw ($n = 115$) and 30–40 in the lower ($n = 107$) vs 35–48 ($n = 315$) and 34–47 ($n = 315$) S. attenuata (Perrin and Hohn, 1994; Perrin *et al.*, 1994a; Nieri *et al.*, 1999). This species and *S. attenuata* overlap in all skull measurements as well as in tooth counts (Perrin *et al.*, 1987). Both species vary greatly geographically. Some specimens of the two species can be identified only with multivariate analysis. However, vertebral counts for the two species do not overlap [67–72 ($n = 52$) in *S. frontalis* vs 74–84 ($n = 75$) in *S. attenuata*].

Taxonomy of the spotted dolphins was long confused, with specimens of this species and the pantropical spotted dolphin (*S. attenuata*) classified or identified under various permutations of the nominal species *S. attenuata*, *S. frontalis*, *S. plagiodon*, *S. froenatus*, *S. pernettyi*, and *S. dubia* (see Hershkovitz, 1966). A revision (Perrin *et al.*, 1987) recognized one pantropical species (*S. attenuata*) and a second species endemic to the tropical Atlantic (*S. frontalis*), both highly variable geographically in size, tooth size, and color pattern. Although the skull of the Atlantic spotted dolphin shows close affinities with that of the pantropical spotted dolphin, the two species did not emerge as sister taxa in a cladistic phylogenetic analysis based on cytochrome *b* mtDNA sequences (LeDuc *et al.*, 1999). *S. frontalis* was imbedded in a strongly supported polytomic clade with *S. coeruleoalba* and *S. clymene* (sister taxa), *Tursiops aduncus*, and *Delphinus* spp. *T. truncatus* was a sister taxon to this clade, with the resulting higher clade imbedded in the five-part polytomic delphinine clade with *S. attenuata*, *S. longirostris*, *Sousa chinensis*, and *Lagenodelphis hosei*. Despite a high degree of cranial similarity, this wide phylogenetic separation suggests that the similarity represents either convergence (homoplasy) or retention of primitive character states (plesiomorphy). The interspecific relationships in color pattern may accord better with the molecular phylogeny, e.g., the pattern of head stripes in *S. frontalis* is closer to those of *T. truncatus* and *T. aduncus* than that of *S. attenuata* (Perrin, 1997). In any case, the existing genus-level taxonomy of the group badly needs revision; *Stenella* is presently polyphyletic and *Tursiops* paraphyletic (LeDuc *et al.*, 1999). A cladistic analysis of morphology (not yet attempted) is in order.

## II. Distribution and Abundance

This species is endemic to the tropical and warm-temperate Atlantic; it is not known to occur in the Pacific or Indian Oceans. The range extends from about 50°N to about 25°S (Jefferson *et al.*, 2007). A discontinuity exists in the range in the western South Atlantic (Moreno *et al.*, 2005). In the western Atlantic, the large heavily spotted form inhabits shallow, gently sloping waters of the continental shelf and the continental-shelf break, usually within or near the 200-m curve but occasionally coming close to shore in pursuit of prey (Perrin

*et al.*, 1994a; Davis *et al.*, 1998; Würsig *et al.*, 2000). In the Southwest Atlantic, they are found within the 1000-m isobath (Moreno *et al.*, 2005). It is usually replaced in nearshore waters by the coastal form of the bottlenose dolphin, *Tursiops truncatus*. There are few estimates of abundance. About 31,000 were estimated to inhabit the northern Gulf of Mexico, but this is thought to be an underestimate (Waring *et al.*, 2006). About 14,000 were estimated from a survey of waters from Maryland to Central Florida (Mullin and Fulling, 2003).

## III. Ecology

Shallow water (6–12 m) over sand flats is utilized as habitat in the Bahamas (Herzing, 1997). A wide variety of prey items has been recorded, including small-to-large epipelagic and mesopelagic fishes and squids and benthic invertebrates; diet may differ between coastal and Gulf Stream forms. Sharks are the only known predators, but it is probably also preyed on by killer whales (*Orcinus orca*) and other small toothed whales.

## IV. Behavior and Physiology

Dives to 40–60 m and lasting up to 6 min have been recorded, but most time is spent at less than 10 m (Davis *et al.*, 1996). Behavior of this dolphin has been studied extensively in the Bahamas (Herzing, 1997; Herzing and Johnson, 1997), where it associates closely with bottlenose dolphins during foraging and traveling. Schools may be segregated by age and sex and fluctuate in size and composition, consisting of up to 100 individuals (Perrin *et al.*, 1994a). In the Azores, Atlantic spotted dolphins join large temporary mixed-species feeding aggregations with tuna, other cetaceans, and seabirds (Clua and Grosvalet, 2001). Echo-location signals in the species recorded in the wild resemble those of other delphinoids in captivity (Au and Herzing, 2003).

## V. Life History

Little is known of the life history of this species. Maximum age in 44 specimens from Brazil was 23 years with maximum length attained by about 20 years (Siciliano *et al.*, 2007). Age at sexual maturation is estimated at 8–15 years in females (Herzing, 1997). First parturition is associated with the mottled phase of spotting development. The average calving interval is about 3 years, with a range of 1–5 years. Nursing has been observed to last up to 5 years. Average first-year natural mortality in a study in the Bahamas was 24%.

## VI. Interactions with Humans

This species does not do well in captivity; most captive animals have died within a year or less, many refusing to eat (Perrin *et al.*, 1994a). It is killed incidentally in fisheries in Brazil, the Caribbean,

**A**

the western North Atlantic, and West Africa (Perrin *et al.*, 1994b; Nieri *et al.*, 1999; Van Waerebeek *et al.*, 2000).

### *See Also the Following Articles*

Bottlenose Dolphins ■ Coloration ■ Geographic Variation ■ Pantropical Spotted Dolphin

### *References*

Adams, L. D., and Rosel, P. E. (2006). Population differentiation of the Atlantic spotted dolphin (*Stenella frontalis*) in the western North Atlantic, including the Gulf of Mexico. *Mar. Biol.* **148**, 671–681.

Au, W. W. L., and Herzing, D. L. (2003). Echolocation signals of wild Atlantic spotted dolphin (*Stenella frontalis*). *J. Acoust. Soc. Am.* **113**, 598–604.

Clua, E., and Grosvalet, F. (2001). Mixed-species feeding aggregations of dolphins, large tunas and seabirds in the Azores. *Aquat. Liv. Res.* **14**, 11–18.

Davis, R. W., Worthy, G. A. J., Würsig, B., and Lynn, S. K. (1996). Diving behavior and at-sea movements of an Atlantic spotted dolphin in the Gulf of Mexico. *Mar. Mamm. Sci.* **12**, 569–581.

Davis, R. W., *et al.* (1998). Physical habitat of cetaceans along the continental slope in the northcentral and western Gulf of Mexico. *Mar. Mamm. Sci.* **14**, 490–507.

Hershkovitz, P. (1966). Catalog of living whales. *US Nat. Mus. Bull.* **246**, 1–259.

Herzing, D. L. (1997). The life history of free-ranging Atlantic spotted dolphins (*Stenella frontalis*): Age classes, color phases, and female reproduction. *Mar. Mamm. Sci.* **13**, 576–595.

Herzing, D. L., and Johnson, C. M. (1997). Interspecific interactions between Atlantic spotted dolphins (*Stenella frontalis*) and bottlenose dolphins (*Tursiops truncatus*) in the Bahamas, 1985–1995. *Aquat. Mamm.* **23**, 85–99.

Jefferson, T. A., Webber, M. A., and Pitman, R. L. (2007). "Marine Mammals of the World: A Comprehensive Guide to Their Identification." Academic Press/Elsevier, San Diego.

LeDuc, R. G., Perrin, W. F., and Dizon, A. E. (1999). Phylogenetic relationships among the delphinid cetaceans based on full cytochrome *b* sequences. *Mar. Mamm. Sci.* **15**, 619–648.

Moreno, I. B., Zerbini, A. N., Danilewicz, D., Santos, M. C. de O., Simões-Lopes, P. C., Brito-Jr, J. L., and Azevedo, A. (2005). Distribution and habitat characteristics of dolphins of the genus *Stenella* (Cetacea: Delphinidae) in the Southwest Atlantic Ocean. *Mar. Ecol. Prog. Ser.* **300**, 229–240.

Mullin, K. D., and Fulling, G. L. (2003). Abundance of cetaceans in the southern US North Atlantic Ocean during summer 1998. *Fish. Bull. US* **101**, 603–613.

Nieri, M., Grau, E., Lamarche, B., and Aguilar, A. (1999). Mass mortality of Atlantic spotted dolphins (*Stenella frontalis*) caused by a fishing interaction in Mauritania. *Mar. Mamm. Sci.* **15**, 847–854.

Perrin, W. F. (1997). Development and homologies of head stripes in the delphinid cetaceans. *Mar. Mamm. Sci.* **13**, 1–43.

Perrin W. F., and Hohn, A. A. (1994). Pantropical spotted dolphin, *Stenella attenuata. In* "Handbook of Marine Mammals." (S. H. Ridgway and R. Harrison, eds.), Vol. 5, pp. 71–98.

Perrin, W. F., Mitchell, E. D., Mead, J. G., Caldwell, D. K., Caldwell, M. C., van Bree, P. J. H., and Dawbin, W. H. (1987). Revision of the spotted dolphins, *Stenella* spp. *Mar. Mamm. Sci.* **3**, 99–170.

Perrin, W. F., Caldwell, D. K., and Caldwell, M. C. (1994a). Atlantic spotted dolphin *Stenella frontalis* (G. Cuvier, 1829). *In* "Handbook of Marine Mammals" (S. H. Ridgway, and R. Harrison, eds), Vol. 5, pp. 173–190. Academic Press, London.

Perrin, W. F., Donovan, G. P., and Barlow, J. (1994b). Gillnets and Cetaceans. *Rep. Int. Whal. Commn, Special Issue* **15**.

Siciliano, S., *et al.* (12 authors) (2007). Age and growth of some delphinids in south-eastern Brazil. *J. Mar. Biol. Assoc. UK* **87**, 293–303 .

Van Waerebeek, K., Ndiave, E., Djiba, A., Diallo, M., Murphy, P., Jallow, A., Camara, A., Ndiave, P., and Tous, P. (2000). A survey of the conservation status of cetaceans in the Senegal, the Gambia and Guinea-Bissau. *CMS WAFCET-1 Report*.

Vialelle, S. (1997). "Dauphins et Baleines des Açores." Espaço Talassa, Lajes do Pico, Azores, Portugal.

Waring, G. T., Josephson, E., Fairfield, C. P., and Maze-Foley, K. (2006). Atlantic and Gulf of Mexico marine mammal stock assessments – 2005. *NOAA Tech. Memo. NMFS-NEFSC* **194**, 346.

Würsig, B., Jefferson, T. A., and Schmidly, D. J. (2000). "The Marine Mammals of the Gulf of Mexico." Texas A&M University, College Station, TX.

# Atlantic White-Sided Dolphin
## *Lagenorhynchus acutus*

### Frank Cipriano

### I. Characteristics and Taxonomy

Atlantic white-sided dolphins are robust and powerful, impressively patterned, and more colorful than most dolphins. A narrow, bright white patch on the side extends back from below the dorsal fin and continues toward the flukes as a yellow-brown blaze above a thin dark stripe (Fig. 1). The back and dorsal fin are black or very dark gray, as are the flippers and flukes, while the belly and lower jaw are white, and the sides of the body a lighter gray. A black eye ring extends in a thin line to the upper jaw, and a very thin stripe extends backward from the eye ring to the external ear. A faint gray stripe may connect the leading edge of the flipper with the rear margin of the lower jaw. The beak is short and grades smoothly into the "melon" (forehead). The upper jaw contains 29–40 and the lower jaw 31–38 small, conical teeth. Molecular analysis has recently been used to examine the evolutionary relationships of *Lagenorhynchus acutus* and the five other currently recognized members of the genus *Lagenorhynchus*. Although formal taxonomic revision awaits a comprehensive review of morphological and molecular characters, the molecular evidence suggests that some of the five are actually more closely related to the right whale dolphins (genus *Lissodelphis*) and some Southern Hemisphere dolphins (genus *Cephalorhynchus*) than they are to *L. acutus* (LeDuc *et al.*, 1999; Harlin-Cognato and Honeycutt, 2006).

### II. Distribution and Abundance

Inhabitants of the cold-temperate North Atlantic, Atlantic white-sided dolphins are usually encountered in waters over the continental shelf and slope, extending into deeper oceanic waters and occasionally into coastal areas (Fig. 2). The southern limit in the western Atlantic is Cape Cod and the submarine canyons south of Georges Bank in the west and Brittany in the east. Groups of Atlantic white-sides are often seen by fishermen and deep-water sailors off the coasts of New England, Nova Scotia and Newfoundland, the western British Isles, Northern Europe and in the Norwegian Sea. There are no records of this species from the inner Baltic Sea, although some sightings and strandings are known from the straits between Denmark, Norway,

**Figure 1**   *The Atlantic white-sided dolphin (C. Brett Jarrett).*

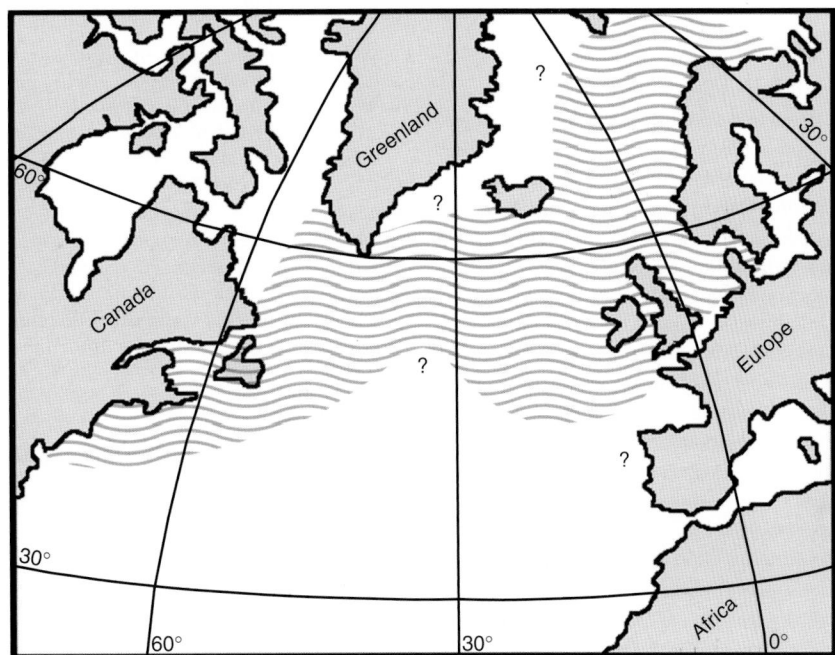

**Figure 2**   *Known distribution limits of the Atlantic white-sided dolphin. The patterned area indicates areas of regular occurrence; question marks indicate uncertainty about occurrence in particular areas.*

and western Sweden. The northern distribution limits are poorly known but extend at least to southern Greenland, southern Iceland, and the south coast of Svalbard Island (Reeves *et al.*, 1999).

Censusing oceanic dolphins is a difficult task, requiring extensive aerial surveys or long observation tracks from survey ships (or both), and then extrapolation of the densities observed to immense ocean areas. Given the wide distribution of Atlantic white-sided dolphins across the northern reaches of the Atlantic, rather wide confidence limits on abundance estimates are to be expected. The best estimate for the western Atlantic is around 50,000 animals (NOAA, 2006), but there is not enough survey coverage for good estimates in the eastern or central Atlantic. For the entire Atlantic there are perhaps 150,000–300,000 (Kaschner, 2004).

## III. Ecology

Analysis of the stomach contents of mass-stranded, incidentally entangled, and drive-caught dolphins is used to assess their diet, since diagnostic "hard parts" (crustacean shells, fish ear bones, and squid beaks) accumulate in stomach chambers. A general indication of the importance of particular prey items can be inferred from the percent contribution of each type, although the number of meals represented by such traces is usually unknown and there may be bias in the retention of different types. Major prey species of Atlantic white-sided dolphins include herring, small mackerel, gadid fishes (codfish and their relatives), smelts and hake, sand lances, and several types of squid (Reeves *et al.*, 1999). Different prey species may predominate at different times of year, representing seasonal movements of prey, or in different areas, indicating prey and habitat variability in the environment. For example, different species of squid are eaten by these dolphins on opposite sides of the Atlantic, while in spring and autumn, sand lance and dolphin distributions in the Gulf of Maine appear to mirror each other. Atlantic white-sided dolphins are probably not deep divers—the maximum dive time recorded from a tagged dolphin was 4 min and most dives were less than a minute in duration.

**A**

## IV. Behavior and Physiology

The number of Atlantic white-sided dolphins observed in a group ranges from a few individuals to several hundred, and mean group size appears to vary with location. In Newfoundland inshore waters 50–60 dolphins in a group are typical; in inshore waters of the British Isles and near Iceland, groups usually contain less than 10 individuals; and off the New England coast group size ranges from a few to around 500, but the usual group size is around 40. Some segregation by sex and age has been suggested from mass stranding records—larger juveniles were absent from some mass-stranded groups that contained many calves, adult males, and pregnant females. Mass strandings involving up to a hundred or more animals at one time are common for this species in the Western North Atlantic (NOAA, 2006).

## V. Life History

Male Atlantic white-sided dolphins are known to reach a maximum body length of about 270 cm and a weight of 230 kg, while adult females reach a maximum size about 20 cm shorter and 50 kg lighter. This is smaller than that well-known oceanarium inhabitant, the common bottlenose dolphin *Tursiops truncatus* (around 380 cm/270 kg maximum) and a bit longer and a lot heavier than the short-beaked common dolphin, *Delphinus delphis* (around 230 cm/75 kg).

Females reach sexual maturity at 200–220 cm, at ages from 6 to 12 years. Males reach sexual maturity at lengths of 215–230 cm, corresponding to ages of 7–11 years. Maximum ages recorded were 22 and 27 years, for males and females, respectively. At birth, Atlantic white-sided dolphins are around 120 cm long, after an approximately 11-month gestation period, and weigh about 25 kg (Perrin and Reilly, 1984). In the western Atlantic, the calving season peaks in mid-summer, while in the eastern Atlantic the calving season may extend several months longer (Weinrich *et al.*, 2001). The lactation period lasts around 18 months, and some stranded individuals were observed to be both pregnant and lactating, suggesting that some individuals may breed annually (Sergeant *et al.*, 1980).

## VI. Interactions with Humans

This species is not currently hunted on a large scale anywhere in its range, although historically many were killed in drive fisheries in Norway and Newfoundland and smaller numbers were taken off Greenland and the Faroe Islands. Incidental mortality has been documented in many areas; these dolphins may be particularly susceptible to entanglement in trawl nets. Recent large catches in pelagic trawl nets have been reported in the Atlantic Frontier off Ireland (Berrow and Rogan, 1997). Within US waters, small numbers of white-sided dolphins have recently been observed caught in sink gill-net, bottom trawl, mid-water trawl, and the Gulf of Maine/Georges Bank herring trawl fisheries (NOAA, 2006). Comprehensive genetic studies and abundance estimates off western Europe are needed to determine the potential impact of mortality in particular areas, since this species seems vulnerable to bycatch from a wide variety of fishing gear types.

## *See Also the Following Articles*

Abundance Estimation ■ Delphinids ■ Overview ■ Genetics for Management ■ North Atlantic Marine Mammals ■ Pacific White-Sided Dolphin ■ White-beaked Dolphin

## *References*

Berrow, S. D., and Rogan, E. (1997). Review of cetaceans stranded on the Irish coast, 1901–95. *Mamm. Rev.* **27**, 51–76.

Harlin-Cognato, A. D., and Honeycutt, R. L. (2006). Multi-locus phylogeny of dolphins in the subfamily Lissodelphininae: Character synergy improves phylogenetic resolution. *BMC Evol. Biol.* **2006**(6), 87.

Kaschner, K. (2004). Modelling and mapping resource overlap between marine mammals and fisheries on a global scale. Ph.D. Thesis, University of British Columbia, Vancouver, BC, Canada.

LeDuc, R. G., Perrin, W. F., and Dizon, A. E. (1999). Phylogenetic relationships among the delphinid cetaceans based on full cytochrome *b* sequences. *Mar. Mamm. Sci.* **15**, 619–648.

NOAA (2006). "U.S. Atlantic and Gulf of Mexico Marine Mammal Stock Assessments – 2006." (G. T. Waring, E. Josephson, C. P. Fairfield, and K. Maze-Foley, eds.), 2nd Ed., *NOAA Tech. Mem.* NMFS-NEFSC-201.

Perrin, W. F., and Reilly, S. B. (1984). Reproductive parameters of dolphins and small whales of the family Delphinidae. *Rep. Int. Whal. Commn.* **Special Issue 6**, 97–133.

Reeves, R. R., Smeenk, C., Brownell, R. L., Jr., and Kinze, C. C. (1999). Atlantic white-sided dolphin *Lagenorhynchus acutus* (Gray, 1828). *In* "Handbook of Marine Mammals" (S. H. Ridgway, and R. Harrison, eds), Vol. 6, pp. 31–56. Academic Press, San Diego.

Sergeant, D. E., St. Aubin, D. J., and Geraci, J. R. (1980). Life history and northwest Atlantic status of the Atlantic white-sided dolphin, *Lagenorhynchus acutus*. *Cetology* **37**, 1–12.

Weinrich, M. T., Belt, C. R., and Morin, D. (2001). Behavior and ecology of the Atlantic white-sided dolphin (*Lagenorhynchus acutus*) in coastal New England waters. *Mar. Mamm. Sci.* **17**, 231–248.

# Australian Sea Lion
## *Neophoca cinerea*

### JOHN K. LING

## I. Characteristics and Taxonomy

The endemic Australian sea lion (Fig. 1) is one of the world's rarest and most unusual seals: rare in terms of very small numbers and unusual in its having a sesquiennial reproductive cycle. It is also a temperate species whose range lies between latitudes 28°S and 38°S around much of the southern part of the island continent (Ling, 1992).

At birth the pups are a dark chocolate-brown to charcoal-gray in color, which changes to the smoky gray (hence the specific name *cinerea*) and cream adult color after the post-natal molt (Walker and Ling, 1981; Ling, 1992). Females retain this coloration throughout life, but males gradually develop a brownish-black coat with increasing age. Males of breeding age have a cream patch on the back of the head and nape of the neck. This species has flattened guard hairs but no underfur—the pelage apparently being adapted to a temperate environment. It also has a relatively thin layer of blubber beneath the skin, about 2 cm thick. Pups measure 62–68 cm in length (nose–tail) and weigh 6.4–7.9 kg at birth, males tending to be heavier than females. Adult females range in length from 132 to 181 cm and weigh between 61 and 105 kg (for a pregnant specimen); males measure up to 200 cm in length and attain weights well in excess of 200 kg.

Marlow and King (1974) summarized the history of the taxonomy of *Neophoca* and concluded that it and the New Zealand sea lion,

Figure 1 *Australian sea lions: adult male (white "cap" on head not visible), two adult females and juvenile (ca. 4 months old) suckling, at North Casuarina Island, off Kangaroo Island, South Australia. Photo courtesy of P.D. Shaughnessy.*

*Phocarctos hookeri*, rightly belonged in different genera, based on skull characteristics. Some common names applied to *Neophoca* are Australian sea lion (preferred), counsellor ("wigged") seal, white-necked or white-capped hair seal or simply hair seal (particularly in early Australian historical times).

## II. Distribution and Abundance

The breeding range of the Australian sea lion extends from Houtman Abrolhos (29°S, 114°E) in Western Australia to the Pages Islands (36°S, 138°E), just east of Kangaroo Island, South Australia, with stragglers reaching central New South Wales on the east coast (Fig. 2).

The most recent study (McKenzie *et al.*, 2005) estimates that there are only 9794 Australian sea lions occupying their wide geographic range in 73 scattered colonies (47 in South Australia and 26 in Western Australia), of which only six produce more than 100 pups in a breeding season. Four-fifths of the population resides in South Australia and a fifth occurs in Western Australia, where more than half the breeding colonies are located, all of which are small. The largest breeding colonies are on Purdie Islands (32°S, 133°E), Dangerous Reef (35°S, 135°E), Seal Bay (36°S, 137°E) on Kangaroo Island, and the two islands of The Pages. Australian sea lions once ranged as far as the eastern end of Bass Strait, but today only stragglers occur there and beyond. The various sea lion colonies are to some extent genetically isolated and members maintain a strong attachment to their respective birth places, particularly adult females.

Australian sea lions were ruthlessly hunted during the sealing era from the late eighteenth to the mid-nineteenth century for their skins and oil, when only a few thousand skins were reported to have been harvested. It is not possible to estimate the number killed for oil, because "seal oil" included fur seal oil and sea lion oil in the cargos. However, there may not have been many sea lions to be taken anyway, compared with the large fur seal populations, which are increasing today after having been almost exterminated by early sealers. Because the first census over most of the sea lion's entire breeding range was carried out only recently (Gales *et al.*, 1990; Dennis and Shaughnessy, 1996), it cannot be determined at this stage whether colonies are growing or declining, except at Seal Bay, where numbers are decreasing. Future surveys of all breeding colonies will need to be undertaken, since counts of live and dead pups provide the most accurate estimates of the size of the total population.

## III. Ecology

After they are weaned, sea lions feed on cephalopods, crustaceans, and fish (Ling, 1992). It is not known how far offshore they forage, but diving appears to begin as soon as females leave the rookery and this takes the sea lions out over the continental shelf. Large prey are generally seized in the mouth and shaken violently at the surface to remove cuttlefish bones, skin or skeletons before swallowing. Depending on size, experimental markers take from 5 to 48 h to pass through the alimentary tract. Australian sea lions are infected by the usual array of external and internal parasites: lice and mites, and acanthocephalans, nematodes, cestodes, and trematodes. Dissections often reveal heavy infestations. Many carry up to several kilograms of pebbles in the stomach, which are thought to aid digestion.

## IV. Behavior and Physiology

The protracted pupping season, during which mating is effected, ensures that there is a high turnover of territories and a breakdown of any harem system (Ling and Walker, 1978; Higgins, 1993; Gales *et al.*, 1994; Gales and Costa, 1997). In contrast to many otariids in which dominant males control small to large numbers of females, *Neophoca* practices what is known a sequential polygyny which still allows males access to several females in a season, but, in general, one at a time. Nevertheless, aggressive encounters do take place between rival breeding males and are a significant cause of mortality among young pups that are unfortunate enough to be attacked or trampled by rampaging bulls. Their lumbering gait resembles something of an ungainly gallop a little above a fast human walking pace and punctuated by frequent rests. Females are most solicitous of their young, and several tourist visitors to Seal Bay and other breeding colonies have received nasty bites when they approached too closely a cow with her pup. When returning from a foraging trip, a female will call from the sea with a soft "moo" and wait for her

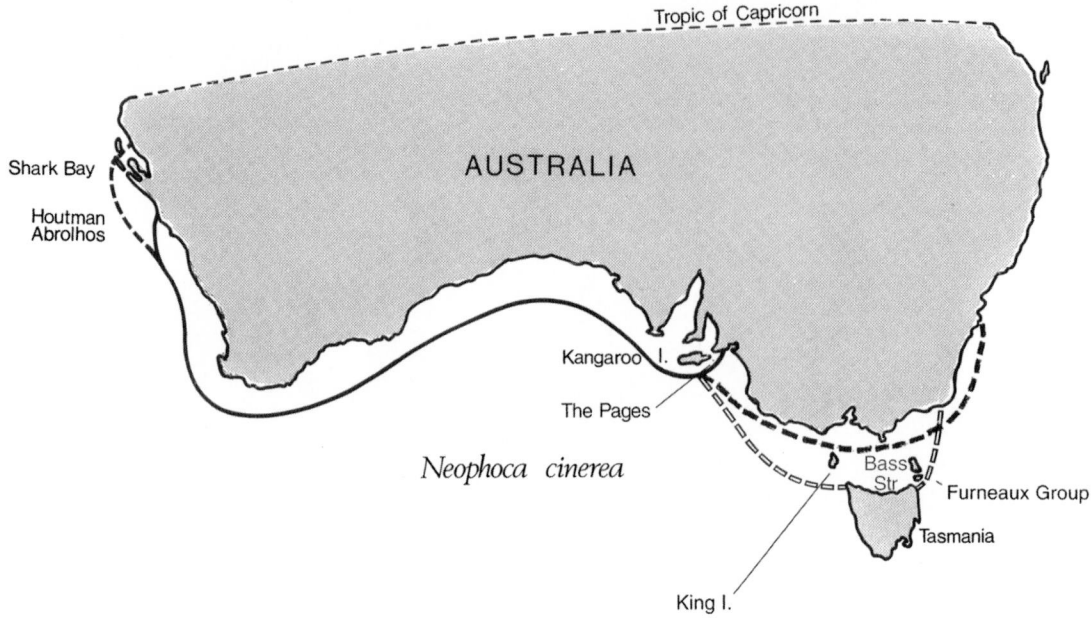

**Figure 2** *Present and past distribution of Australian sea lions. Unbroken solid line depicts current known breeding range, broken solid line depicts seasonal stragglers, and broken open line depicts extent of former breeding range. Courtesy of the Royal Zoological Society of South Australia Inc.*

pup's answering call which resembles a lamb's bleat. When pups are small and site fidelity is very strong, so little searching for each other is necessary. Once the two are reunited, recognition is confirmed by smelling. Sea lions are powerful and skilful swimmers, using their large front flippers to propel them rapidly through the water. They are also excellent surfers and can often be seen riding the waves right into the shallows or "porpoising" along wave crests and troughs farther out to sea.

Large males tend to lie apart from other sea lions, but females and immature animals often lie close together, wriggling, squirming, and scratching constantly. On hot days when the sun temperature may exceed 45°C they will occasionally go into the sea and return a short while later to allow evaporative cooling to take effect. Sea lions may also venture some distance inland to lie under bushes or up steep slopes to find a shelter; they are quite agile on land.

There is a marked increase in concentrations of the hormones progesterone and oestradiol about 3–5 months after the probable mating date, and they reach their highest levels after another 2 months. This suggests that the blastocyst reactivates and implants 3.5–5 months into pregnancy, a similar free blastocyst stage to that of other seals. It also means, however, that post-implantation gestation lasts up to 14 months to fit with the 17.5-month reproductive cycle.

The lipid content of Australian sea lion milk is lower during all stages of lactation than that reported for other middle- and high-latitude eared seals. This may be correlated with the extended lactation period—15–17 months and even longer for some pups—of this species. Lipid and energy content do increase during lactation but the composition varies greatly between and within individuals.

At-sea metabolic rates of Australian sea lions are comparable with those of other eared seals and are up to 6.8 times the predicted basal metabolic rate for terrestrial mammals of similar size. Compared to California sea lions, *Neophoca* works hard during its foraging dives that last from about 3 to 8 min for adult females, at average depths of

67 m and an average maximum depth of 92 m. These are longer dive durations than for other eared (but not true) seals.

## V. Life History

The life history of *Neophoca* is unique in a number of aspects: the approximately 17.5-month aseasonal reproductive cycle, a protracted (i.e. 5 month or longer) pupping season, prolonged (14 month) post-implantation gestation, and lactation lasting almost until the next pup is born or even longer (Ling and Walker, 1978; Gales *et al.*, 1997). In addition, because many of the sea lion colonies are genetically isolated, breeding across their range is asynchronous. Until recently, this unusual reproductive cycle was thought to be an adaptation to the sea lion's environment that was characterized as a nutrient-poor, low-energy, stable *milieu* associated with the eastward flowing Leeuwin Current on the southern coast of Australia. Recent ocean studies have demonstrated significant upwelling in the waters of western Victoria and South Australia as far west as Eyre Peninsula in summer and autumn, which makes this region quite rich in nutrients; this is where about three-quarters of the Australian sea lion population resides.

Pups are born a few days after the females have moved to their breeding sites, to which they are known to return for successive birthings. Viable twins have never been observed, but two aborted fetuses, believed to be twins and estimated to be about 3 months post-implantation, were found on Kangaroo Island in 1985.

Mating takes place 7–10 days post-partum; there is a 3–4-month-free blastocyst (embryonic diapause) stage, followed by a gestation period of up to 14 months duration. The pup is suckled for the next 15–17 months and during this time it learns to forage for food that it will consume in later life. The milk is low in energy (fat) compared with other pinnipeds and its quality may vary according to the foraging success of the mother and stage of lactation. While the pelage is unlikely to be involved in thermoregulation, the flattened hairs overlap each other and provide a smooth but flexible outer surface that reduces turbulence when swimming. Periodic renewal of the hair coat ensures that it functions efficiently in whatever role it has. The timing of pelage renewal or molt is variable. Immature sea lions molt during the breeding season, females begin their molt up to 4 months after parturition (which is about when implantation occurs) and adult males do not start their molt until about 9 months after the breeding season.

The Australian sea lion's unusual life history enables it to survive as a very small population scattered over a wide, in places nutrient-poor longitudinal range. The longer than normal (pinniped) gestation and lactation periods allow the female to nurture a developing fetus and growing pup whilst having to forage. Normal growth rates can also be achieved despite the low energy content of the milk. At the same time, the long maternal association confers many learning and protective advantages on the young sea lion. The protracted, asynchronous pupping season spread out over the wide geographic area again means that food resources can be better shared and there is not a sudden influx of newly independent sea lions, such as occurs with more highly synchronized species that occupy nutrient-rich, higher latitudes.

## VI. Interactions with Humans

At Seal Bay Conservation Park on South Australia's Kangaroo Island humans can approach Australian sea lions quite closely, and could approach even much closer before stricter guidelines were introduced to prevent disturbance to both animals *and* humans! Sea lions at other island colonies are much less tolerant of humans and scatter quickly into the sea when disturbed.

Older males, in particular, occasionally haul out onto Adelaide and other beaches: apparently having left or being driven out of their former breeding colonies. They are sometimes taken into animal care or, more often than not, return to the sea and go elsewhere.

There is some conflict between fishermen or fish-farmers and Australian sea lions which take fish from nets and pens, respectively, rather than having to chase prey themselves. Various scaring devices have been tried in an effort to avoid having to shoot at a species that is threatened with extinction. The main method now uses seal fences consisting of wire strung on stanchions about 1.8 m high.

The Australian sea lion colony on Kangaroo Island is internationally famous because of its proximity to the large city of Adelaide and the public being able to view the animals at close quarters (Robinson and Dennis, 1988). There is also a smaller breeding colony on the mainland at Point Labatt on Eyre Peninsula to the west, which may be viewed from a lookout high above the beach. In February 2005 *Neophoca* was listed as a *Threatened* species, *Vulnerable* category under the Commonwealth of Australia *Environment Protection and Biodiversity Conservation Act 1999* (Shaughnessy, 1999). In view of this and the species' importance to the tourist industry on Kangaroo Island (no other colonies are so easily accessible), the South Australian Government has embarked on an intensive management strategy. The whole Seal Bay area has been designated a *Conservation Park*. Public access is limited to the main beach and only in the company of authorized personnel, but there are also viewing platforms overlooking the beach and other restricted areas. The principal pupping sites in sheltered coves adjacent to the main beach have been declared *Prohibited Areas*. Regular classified censuses are conducted to monitor the status of the Seal Bay colony and enhance its chances of survival and value to tourism.

However, some Australian sea lion colonies appear to be suffering very high pup mortality (30–40%) and decreasing pup production.

Only with a widespread and cooperative research and management effort will the species be perhaps more secure.

Australian sea lions have adapted well to captivity, where there have been at least 45 births around the country since 1981 (Ling *et al.*, 2006). One female produced 8 pups in 11 years. Birth intervals are approximately 17.5 months or multiples thereof, thus mirroring reproduction in the wild. A female, estimated to have been about a year old when caught, lived for 25 years in captivity, and a male, aged approximately 1–2 years when captured, has been captive for more than 22 years. These figures agree closely with maximum ages of tagged sea lions in the natural state. The oldest surviving captive-bred specimen (in January 2007) was aged 21 years 3 months. There have been no attempts to introduce captive-bred Australian sea lions into the wild.

## See Also the Following Articles

Eared Seals (*Otariidae*) ■ Population Status and Trends

## References

Dennis, T. E., and Shaughnessy, P. D. (1996). Status of the Australian sea lion, *Neophoca cinerea*, in the Great Australian Bight. *Wildl. Res.* **23**, 741–754.

Gales, N. J., and Costa, D. P. (1997). The Australian sea lion: A review of an unusual life history. *In* "Marine Mammal Research in the Southern Hemisphere" (M. Hindell, and C. Kemper, eds), Vol. 1, pp. 78–87. Surrey Beatty & Sons, Chipping Norton.

Gales, N. J., Cheal, A. J., Pobar, G. J., and Williamson, P. (1992). Breeding biology and movements of Australian sea lions, *Neophoca cinerea*, off the west coast of Western Australia. *Wildl. Res.* **23**, 405–416.

Gales, N. J., Shaughnessy, P. D., and Dennis, T. E. (1994). Distribution, abundance and breeding cycle of the Australian sea lion *Neophoca cinerea* (Mammalia: Pinnipedia). *J. Zool. (Lond.)* **234**, 353–370.

Gales, N. J., Williamson, P., Higgins, L. V., Blackberry, M. A., and James, I. (1997). Evidence for a prolonged post-implantation period in the Australia sea lion (*Neophoca cinerea*). *J. Reprod. Fertil.* **111**, 159–163.

Higgins, L.V. (1990). Reproductive behavior and maternal investment of Australian sea lions. Ph.D. Thesis, University of California, Santa Cruz.

Higgins, L. V. (1993). The nonannual, nonseasonal breeding cycle of the Australian sea lion, *Neophoca cinerea*. *J. Mammal.* **74**, 270–274.

Ling, J. K. (1992). Neophoca cinerea. *Mammalian Species* **392**, 1–7.

Ling, J. K., and Walker, G. E. (1978). An 18-month breeding cycle in the Australian sea lion? *Search* **9**, 464–465.

Ling, J. K., Atkin, C., Barnes, A., Fischer, A., Guy, M., and Pickering, S. (2006). Breeding and longevity in captive Australian sea lions, *Neophoca cinerea*, at zoos and aquaria in Australia: 1965–2003. *Aust. Mammal.* **28**, 65–76.

McKenzie, J., Goldsworthy, S.D., Shaughnessy, P.D., and McIntosh, R. (2005). Understanding the impediments to the growth of Australian sea lion populations. *SARDI Aquatic Sciences Publication* RD04/0171. Adelaide.

Marlow, B. J., and King, J. (1974). Sea lions and fur seals of Australia and New Zealand—the growth of knowledge. *Aust. Mammal.* **1**, 117–135.

Robinson, A. C., and Dennis, T. E. (1988). The status and management of seal populations in South Australia. *In* "Marine Mammals of Australasia Field Biology and Captive Management" (M. L. Augee, ed.), pp. 87–110. Royal Zoological Society of New South Wales, Sydney.

Shaughnessy, P. D. (1999). "The Action Plan for Australian Seals." Environment Australia, Canberra.

Walker, G. E., and Ling, J. K. (1981). Australian sea lion *Neophoca cinerea* (Peron, 1816). *In* "Handbook of Marine Mammals" (S. H. Ridgway, and R. J. Harrison, eds), Vol.1, pp. 99–118. Academic Press, New York.

**A**

# Australian Snubfin Dolphin

*Orcaella heinsohni*

KELLY M. ROBERTSON AND PETER W. ARNOLD[1]

## I. Characteristics and Taxonomy

The Australian snubfin dolphin is a coastal dolphin species that occurs throughout northern Australia with some evidence for occurrence in Papua New Guinea. This species was previously considered to be a population of Irrawaddy dolphins (*O. brevirostris*). Clear and consistent differences between Asian and Australian *Orcaella* specimens in coloration, cranial and external morphometrics, postcranial morphology, and molecular data are consistent with species-level differences. The Australian snubfin dolphin was formally proposed as a separate species in 2005 (Beasley *et al.*, 2005).

Recent morphological and genetic studies place the genus *Orcaella* in the family Delphinidae, with the closet relative possibly being the killer whale *Orcinus orca*. The Australian snubfin (Fig. 1) dolphin, resembles the Irrawaddy dolphin in appearance and is closely related to it genetically.

**Figure 1** *Australian snubfin dolphin* (Orcaella heinsohni) *from Cleveland Bay, Queensland, Australia. Courtesy Guido J. Parra.*

The name "snubfin dolphin" was first suggested as an alternative common name for Irrawaddy dolphins in 1981. This name highlights a diagnostic external character, is appropriate to all populations, and has been included in lists of common names and general field guides. The proposed common name for this species, the Australian snubfin dolphin, further reflects the fact that the majority of known specimens and morphological work are based on Australian populations.

The species was named for George E. Heinsohn, recognizing his pioneering work on northeastern Australian odontocetes, including the collection and initial analysis of Australian snubfin dolphin specimens.

Total length reaches 230 cm in females and 270 cm in males. Mass of three adults (2.14–2.25 m long) was recorded as 114–133 kg (Arnold and Heinsohn, 1996). The head is rounded in lateral view and lacks a beak. It is usually bounded by a distinct neck crease (Fig. 2) situated about half way between the eye and the anterior insertion of the flipper. The species lacks the dorsal groove that is present on the Irrawaddy dolphin, a distinct indentation from the base of the skull to the anterior edge of the dorsal fin. The body has a subtle three-tone color pattern: a distinct dark brown dorsal cape, light brown lateral field, and white abdominal field. The small variably shaped dorsal fin (from rounded to slightly falcate) is situated in the latter half of the body. A mid-ventral crease runs along the belly from the flippers to the genital slit. The flippers are broad, paddle-like, and highly mobile.

The adult skull retains neotenic features. The number of nasal bones/depressions on each side of the skull vertex varies from 0 to 6, with a mean of 2.9. The nasal bones are nodular in appearance, often with at least two nasals on each side of the vertex. The mesethmoid plate is thin and poorly developed, leaving much of the frontal bone on the anterior face of the vertex exposed. The shallow postnarial pit is filled by a supernumerary bone. The temporal fossa height is greater in the snubfin dolphin than in the Irrawaddy dolphin, with a mean of 61.2 mm vs 45.8 mm (Fig. 3). There are 11–22 teeth in each half of the upper jaw and 14–19 teeth in each lower row.

## II. Distribution and Abundance

The species is confirmed to occur from Broome, Western Australia, north to the Northern Territory and along the Queensland coast as far south as the Brisbane River. Cranial morphological

**Figure 2** *Australian snubfin dolphin displaying neck crease and lack of dorsal groove. From Beasley* et al., *2005.*

---

[1]A posthumous contribution.

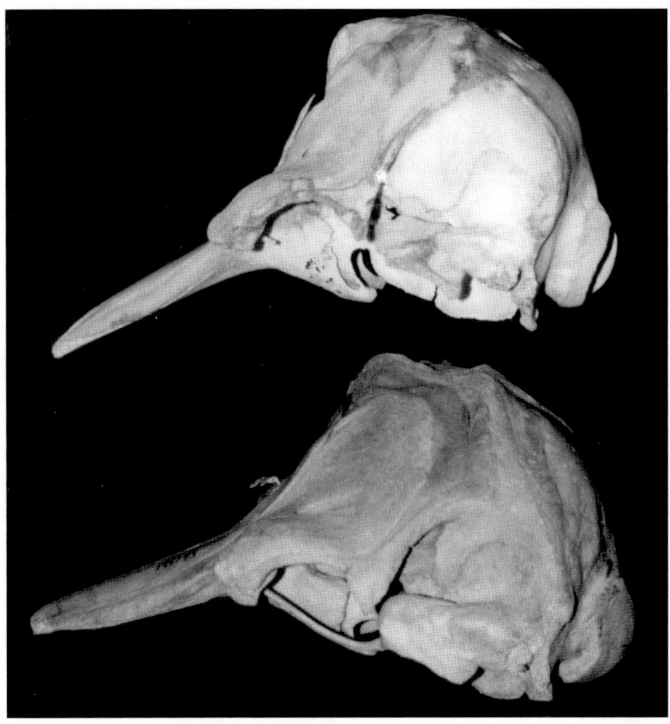

Figure 3 *Skulls showing differences in temporal fossa height between the Australian snubfin dolphin (top) and the Irrawaddy dolphin. From Beasley* et al., *2005.*

features of one Papua New Guinea specimen from Daru were consistent with those of the Australian snubfin dolphin, and it is likely that the species occurs in localized areas around Papua New Guinea. Further study is needed on *Orcaella* in Papua New Guinea and neighboring regions (particularly Indonesia) to confirm the species' respective ranges.

Based on the current known distribution, the Australian snubfin dolphin occurs on the Sahul shelf of Australia/Papua New Guinea, whereas the distribution of the Irrawaddy dolphin corresponds to the Sunda shelf of South and Southeast Asia. These areas are separated by deep oceanic waters and remained separate even during periods of lowered sea levels in the Pleistocene Ice Ages.

Little is known about the current population status of the species. An estimate of about 1000 animals was calculated in 1989 (Freeland and Bayless, 1989) based on aerial survey records. However, the low number of sightings from surveys conducted along the Queensland coast from 1987 to 1995 indicates that the estimate may be high (Parra *et al.*, 2002).

## III. Ecology

Sighting records indicate that the Australian snubfin dolphin occurs mainly in protected, shallow (<15 m deep), coastal waters, especially adjacent to river and creek mouths, with a preference for sea grass beds. It has been observed to co-occur with the Indo-Pacific humpbacked dolphin (*Sousa chinensis*) throughout most of its range in Australian waters. Shark wounds have been observed on individuals in the field (Parra, 2006).

Australian snubfin dolphins have an opportunistic diet, taking a wide range of fishes as well as cephalopods (squid, cuttlefish,

octopus) and shrimps. Fish from the following families have been identified as prey: Clupeidae, Engraulidae, Apogonidae, Chirocentridae, Anguillidae, Pomadasydae, Sillaginidae, Hemirhamphidae, Terapontidae, and Leiognathidae.

## IV. Behavior and Physiology

Australian snubfin dolphins are generally found in small groups of 2–6 animals; however, groups of up to 14 animals have been observed (Parra *et al.*, 2002). Surfacing is generally unobtrusive, with a low roll showing little of the back. Given this, the dolphins' shy behavior, and the small size of the dorsal fin, it is often easy to miss in the field.

Interactions with Indo-Pacific humpback dolphins have been observed to be aggressive and sexual in nature with the humpback dolphins demonstrating dominance. Vocalizations have been recorded by Van Parijs *et al.* (2000) and include broadband clicks (>22 kHz), three or more types of pulsed sounds (>22 kHz), and two types of whistles (1–8 kHz). These vocalizations occur during both socializing and foraging.

## V. Life History

Very little is known about the life history of the species. Most of the information available on the life history of *Orcaella* is based on the Asian species, the Irrawaddy dolphin. It is thought that the life history parameters of the Australian snubfin dolphins and Irrawaddy dolphins are probably very similar. Reproductive seasonality of the snubfin dolphin is unknown. It is thought that gestation may be approximately 14 months. One record of a near-term fetus exists from Townsville, Australia, collected in the month of August. A small number of snubfin dolphins from northeastern Australia were aged using dentinal growth layer groups in teeth. It was estimated that dolphins reached adult size (2.1 m) at 4–6 years, and maximum life span was considered to be about 30 years.

## VI. Interactions with Humans

The current conservation status of the snubfin dolphin in Australia is listed as "Insufficiently Known" by the Action Plan for Australian Cetaceans and "Near Threatened" in the IUCN Red List of Threatened Species.

Previously, the major known threat to the species was accidental capture in nets used to control the number of sharks off popular swimming beaches. Currently, nearshore fishery gillnets are a threat, as they are set in creeks, rivers, and estuaries, the preferred habitat of the snubfin dolphin. Rezoning of the Great Barrier Reef Marine Park in 2004 has provided areas where gillnetting is banned, limited by permit, or allowed for only several months of the year. However, these areas provide very little coastal coverage; many areas outside the Marine Park are not regulated, which is a concern for the conservation of this species.

Concern has also recently been raised about habitat reduction and degradation that may be caused by human population growth. Suggested effects include reduced food supplies (from habitat degradation and overfishing), increased industrial and urban pollution of coastal sites and disturbance from increased vessel traffic. Such potential threats are largely unquantified; it is hard to predict whether they will cause or have already caused fragmentation of dolphin populations leading to potential decline of the species. More research is needed throughout the range in order to help determine proper management decisions to ensure the conservation of the species.

**A**

## See Also the Following Articles

Irrawaddy Dolphin (*Orcaella brevirostris*) Indo-West Pacific Marine Mammals

## References

Arnold, P. W., and Heinsohn, G. E. (1996). Phylogenetic status of the Irrawaddy dolphin *Orcaella brevirostris* (Owen in Gray): A cladistic analysis. *Mem. Queensl. Mus.* **39**, 141–204.

Beasley, I., Robertson, K. M., and Arnold, P. (2005). Description of a new dolphin, the Australian snubfin dolphin *Orcaella heinsohni* sp. n. (Cetacea. Delphinidae). *Mar. Mamm. Sci.* **21**, 365–400.

Freeland, W. J., and Bayliss, P. (1989). The Irrawaddy river dolphin (*Orcaella brevirostris*) in coastal waters of the Northern Territory, Australia: Distribution, abundance and seasonal change. *Mammalia* **53**, 49–57.

Parra, G. J. (2006). Resource partitioning in sympatric delphinids: Space use and habitat preferences of the Australian snubfin and Indo-Pacific humpback dolphins. *J. Anim. Ecol.* **75**, 862–874.

Parra, G. J., Azuma, C., Preen, A. R., Corkeron, P. J., and Marsh, H. (2002). Distribution of Irrawaddy dolphins, *Orcaella brevirostris*, in Australian waters. *Raffles Bull. Zool. Suppl.* **10**, 141–154.

Van Parijs, S. M., Parr, G. J., and Corkeron, P. J. (2000). Sounds produced by Australian Irrawaddy dolphins, *Orcaella brevirostris*. *J. Acoust. Soc. Am.* **108**, 1938–1940.

# Azorean Whaling

## Richard Ellis

They called it the "Western Islands" Grounds because it was an island-group west of Europe, closer to Europe than America. These islands were among the earliest areas visited by the sperm whalers of New England in the eighteenth century. Early in the history of fishery that was to dominate much of New England's commercial history, it was discovered that the waters around the Azores were among the favorite haunts of sperm whales. It soon became apparent that there were other attractions in these islands: food and water could be obtained, and more importantly, men willing to sign aboard could be expeditiously recruited.

From his observatory at Sagres at Cape Saint Vincent in Portugal, the westernmost point in continental Europe, Henry the Navigator (1394–1460) inspired the explorers who would extend the boundaries of the known world. Infante Dom Henrique (as he is known to the Portuguese) was the patron of Gil Eannes, the first man to round the previously untried Cape Bojador on the western hump of Africa, opening the South Atlantic for the voyages of such heroes as Bartolomeo Dias, Vasco da Gama, and Ferdinand Magellan. The pilot Diogo de Sevilha, possibly on a return voyage from Madeira, reached the islands that would eventually be named the Azores, but it was Goncalo Vetho Cabral who is credited with the official discovery of the islands in 1431. Cabral first claimed Santa Maria, then So Miguel, Terceira ("third"), Sao Jorge, Graciosa, Pico, and Faial. The seven islands were named 11-has dos Acores ("Isles of Hawks"), and within a year the Portuguese had settled in. While searching for the nonexistent island of Antilia, Diogo de Tieve discovered the remaining two islands in 1452.

The peaks of submerged volcanoes rising from the Mid-Atlantic Ridge, the nine islands of the Azores archipelago, sizzled out of the North Atlantic some 900 miles east of Portugal. The water around these islands can be over a mile deep; a perfect locale for the deep-feeding sperm whale, or cachalot. After Cabral found the islands, the whales would have only another three centuries—the blink of an eye in cetacean history—to swim unmolested in Azorean waters. The whalers were coming.

As they extended their horizons, the Yankee whalers sailed first to the Bahamas and the West Indies, then out into the Atlantic. The currents of the North Atlantic circulate in a roughly clockwise fashion, but there are enough subcurrents, drifts, and gyres to make sailing less than easy. The same surface ocean movements that allowed Columbus to sail in a south-westerly arc to reach the Caribbean also assisted the whalers as they hitched a ride onto the northerly segment of the Atlantic gyre which pushed them toward Spain and Portugal, and eventually south to Africa. (Although Ponce de Leon is believed to have been the first to describe the Gulf Stream during the early sixteenth century, the whalers recognized its benefits early in the eighteenth century only. These benefits were illustrated and published in 1786 by Benjamin Franklin, the cousin of a Nantucket whaler named Timothy Folger.) Perhaps the whales were not in evidence when they first explored Azorean waters, but the Nantucket whalers first sighted the West African whales in 1773, and 5 years later, they discovered the Western Islands Grounds.

The usual route for whaling in the Atlantic—only the broadest of generalizations, since the whales rarely appeared where or when they were supposed to—would consist of a southward bearing in the spring, to the Carolinas and the West Indies, thence to the Azores, the Cape Verdes, and the coast of Africa in the summer. Eventually, the whalers would re-cross the South Atlantic, and work the Brazil Banks or the Falklands. The ships would return to New England in July and after refitting, sail for the Grand Banks to the north. "Plum-pudding" whaling was the way these short, relatively safe Atlantic voyages were described. It would not be until 1789 that the British whaler *Emelia* would round the Horn and initiate the era of round-the-world whaling voyages.

Because the British whaling fleet was active in Greenland waters, the Atlantic was available to the colonists. During the mid-eighteenth century, French and Spanish privateers and pirates roamed the Atlantic, adding yet another threat to an already hazardous pursuit. The dogged Yankee whalers persevered, however, and continued to visit the Western Islands for sperm whales, because the islands had the reputation of being the home of particularly large whales. (In an account of Azorean whaling, Trevor Housby (1971) describes the capture of a 61-footer, one of the largest bull sperm whales ever measured.) Even though there were large whales to be found there, however, the Azores were only a way-station on the way to such places as the Cape Verde Islands and the whaling grounds of southern Africa.

During the height of square-rigged whaling, the whalers would plunder the waters of the Western Islands for whales and the lands for whalers. Since the days of Vasco da Gama, Portuguese sailors have demonstrated an inordinate desire to go down to the sea in ships. While the Portuguese proved to be brave and competent whalers, however, early New England chauvinism relegated the Azoreans to the same class as anyone who was not a "full-blooded Yankee"—whatever that was supposed to mean in 1820. In later years, Hohman (1928) would write, "as the better type of American forsook the forecastles, their bunks were filled by criminal or lascivious adventurers, by a motley collection of South Sea Islanders known as Kanakas, by cross-bred negroes and Portuguese from the Azores and the Cape Verdes, and by the outcasts and renegades from all the merchant services of both the Old World and the New." When Clifford Ashley, the writer and painter, shipped aboard the *Sunbeam* in 1904, he described the crew in detail, concluding, "The South Sea Islands, East Indies, Cape Verdes, Azores and Canaries, all were liberally represented on our list. Profane, dissolute and ignorant they were, yet, on the whole, as courageous and willing a lot as one could desire."

Most narratives of Atlantic whaling include a visit to the Azores; among the arrivals was the *Bruce*, which J. Ross Browne named *Styx* in his *Narrative of a Whaling Cruise*. Here he describes his first sight of the islands in 1842:

Terceira is a remarkably picturesque island, beautifully laid out in farms, which at this season of the year have a rich golden hue that bespeaks abundant crops. The coast is broken and rugged, and in many places so steep as to preclude the possibility of ascent. Part of the island seems to have been ingulfed by an earthquake, which accounts *for* the rugged appearance of the coast. It is visited at certain seasons of the year by heavy gales and rains, especially in October and November, when there is frequently danger in approaching it. While we lay off and on, awaiting a suitable opportunity of running in, we had hard, shifting winds, and it rained almost incessantly. Mount Brazil, and other elevated portions of the island, were covered most of the time with white, misty clouds.

Browne describes the Azorean whalers as wearing "sennet hats with sugar-loaf crowns, striped bed-ticking pantaloons patched with duck, blue shirts, and knives and belt. They were all barefooted…"

It is impossible to determine when the Azoreans began whaling on their own, but the Portuguese seemed to have maintained a sperm whale fishery, which "they had learned from the New Englanders and carried on upon the coast of Brazil" as early as 1785. The islands sustained international fishery for perhaps a 100 years, but by 1870, the only whalers operating out of the Azores were the Azoreans themselves.

They fitted out their own whaleships, but they were never particularly successful. Their first attempt was the *Cidade da Horta*, a brig that had been abandoned in the islands by the French as not being seaworthy. They probably never sent out more than 10 of their own vessels because their economy was never strong enough to lay out the considerable sums required to build, pay for and man a full-rigged ship. Instead, the Azoreans would sign aboard foreign vessels. Nevertheless the islands later developed a technique that would not be duplicated anywhere else in the world: shore whaling for sperm whales.

In shore whaling, which involves spotting whales from lookouts, the prey has almost invariably been the relatively placid right whale and less frequently the humpback. The reasons for these choices are obvious. Both the right and the humpback are inshore creatures; slow-swimming, passive animals that, more often than not, rolled over and died when they had been lanced. The cachalot, however, is a dangerous threat, given to smashing whaleboats in its death-throes, and less frequently to attacking whaleships and sinking them. It was indeed a courageous whaler who chose to approach the most fearsome of all the great whales in a fragile little cockleshell.

It is possible that the Azoreans learned shore whaling from the Basques, who may have called at the islands as they extended their right-whale fishery to Newfoundland in the sixteenth century. (The Basque term *vigz'a*, which means a lookout, is still in use in the Azores today, and the word *cachalote* is also of Basque origin.) The village of

Horta (also known as Porto Pim) on the island of Faial is believed to have been the site of the first shore station in the Azores, sometime around 1832. The Azorean records are scanty, but it is known that the American consul, a man named Dabney, set up a tryworks at Horta in 1850. From Faial, the industry spread to the other islands, and soon there were stations on São Jorge, Graciosa, Terceira and São Miguel. The Pico islanders began whaling around 1853, following an outbreak of phylloxera that almost totally wiped out the vineyards which had been their main source of revenue. By 1898 there were no less than 29 whaling companies working in the Azores.

Originally, the whaleboats had been imported from New Bedford, but around the turn of the century, a whaleman named Machado built the first boat at Pico. Shortly thereafter, the laborious method of rowing or sailing out to the whaling grounds was abandoned in favor of motorboats, which towed the killing boats out to sea. Although this greatly improved the Azoreans' efficiency by allowing them to go to the whales without the endless hours of backbreaking rowing or time-consuming tacking, the innovation was one of the few attempts at modernization that the Azoreans made. Curiously, at the same time that they adopted motorized launches, the Azoreans abandoned the hand-held harpoon guns which they had been using—somewhat uneasily—since around 1885. They also introduced two-way radios to facilitate communication between the *canoas*. With the exception of the radios and the towing boats, which replaced the whaleship in putting the whalers close to the whales, Azoreans continued to kill and process cachalots in a manner that almost precisely replicated that of the Yankee whalers. Despite the anachronistic nature of the fishery, its economy allowed the technology to be exported. (By 1900, most of the world's whaling was being conducted with exploding harpoons and steam- or diesel-powered catcher boats.) Open-boat whaling was introduced to Madeira in 1941, and although they had only a brief time there, Azorean whalers established a similar fishery in Brazil in 1950.

From the *vigias* on the cliffs, the lookouts stood watching from dawn to dusk, every day of the year. They used powerful binoculars, which they claimed enabled them to spot whales at a distance of 30 miles. When blows were sighted, a rocket was set off to alert all the whalers, who then set out in pursuit. (Another vestige of the Yankee whaling industry was the introduction of English terms into Portuguese. They cried *bloz!* or *baleia!* when a whale was sighted, called the bull whale a *brilo*, the boom a *bûme*, and the junk the *janco*.) The whaleboats, known as *canoas*, were 38-ft long (10 ft longer than the average American whaleboat), and as graceful and seaworthy as the Yankee whaleboats that Clifford Ashley had called "the most perfect water craft that have ever floated." They were smooth-sided, or carvel-built, unlike the Yankee boats, which were clinker-built (Fig. 1). (The Azoreans believed that the acute hearing of the sperm whale enabled it to hear the slap of the water on the strakes.) Where their Yankee predecessors employed a crew of six, the Azorean double-enders shipped a crew of seven. The Azorean harpooner and the steersman did not make the dangerous and awkward change of places after the whale had been struck, so that the harpooner both made fast to the whale and lanced it. Like almost everything else in the Azorean fishery, the harpoons employed followed the New England fashions of the mid-nineteenth century, with a "Temple" toggle head that pivoted to a right angle when plunged into the whale. The boats were equipped with a gaff mainsail and a jib, and if possible, the boat was sailed tight onto the whale for the harpooning. Often the Azorean whalers would paddle up to the whale under sail, using canoe paddles that Robert Clarke described as "betraying their Red Indian origin by their shape and

**Figure 1** *In the swallows, Azorean whalers section a sperm whale before winching the pieces onto shore. The small rowboat is used only for this cutting; the* canoas *(whaleboat) were more than 30-ft long. Photography courtesy of William Dawbin.*

**Figure 2** *The harbor at Horta on the Island of Fayal in the Azores, with American sperm whalers picking up provisions. Credit: New Bedford Whaling Museum.*

the way they were used." The rowing oars were 16–18 ft long, and the steering oar was about 23 ft in length. Clarke wrote, "In the history of seafaring trades there can scarcely be a more remarkable survival than the present use in the Azores of hand weapons to take and kill great whales." The harpoon was not the killing instrument, but was used to make the whale fast to the boat. After the whale towed the boat (which might consume several hours), the whalers threw the lance, a spearlike projectile which was driven deep into the body of the exhausted whale. A towing strap was inserted into the whale's upper jaw so that it could be brought back to shore, sometimes a distance of 25 or 30 miles. The toggle was reeved into the head rather than the tail because a whale normally moves forward through the water (Fig 2).

When John Huston was filming *Moby Dick* in 1955, he sent a crew to film actual whaling in Madeira. The early scenes depicting

whalers chasing and harpooning sperm whales show better than any text the process and the excitement of the chase. The white whale, unavailable for *filming*, was represented by several 90-foot steel, wood, and latex models that were eventually lost at sea off Ireland to the bewilderment of cruising sailors. Whaling in Madeira, some 500 miles southeast of the Azores, is a smaller version of the Azorean fishery and was founded by Azoreans. From 1941 to 1949, almost 1000 whales were taken by 102 Madeiran whaleboats. The last factory was closed in 1981.

Dead whales were usually brought to the stations in the late afternoon and processed the following day. Before the steam-powered whaling station was built at Lages do Pico in 1950, the whales were beached on the rocks at the entrance to the harbor and worked up there. First the head was cut off with a razor-edged blubber spade, then the carcass was stripped of its blubber. Formerly, only the teeth and the blubber were saved, but in later years, the meat was used in the manufacture of fertilizer and livestock feed.

The statistics of the number of whales obtained are not available, but from 1895 to 1897 some 480,000l of whale oil were exported from the Azores. Up to the opening years of the twentieth century, the Azorean fishery had flourished, but by the time of World War I, it had begun to flag. Sperm oil had been used in England and the United States primarily for the manufacture of fine candles, but by 1910 paraffin was substituted and candles became cheaper. Sperm oil had only a limited application in the manufacture of cosmetics and medicinal salves, and because the market was diminishing, the catches decreased as well. In 1910, the Azores accounted for some 73% of all sperm whales caught in the world, but by 1915, the figure had fallen to a depressing 3.8%. World War II saw the return of the factory fleets to the high seas, and their pursuit of sperm whales in the North Atlantic reduced the Azorean catch. In 1949, there were only 125 *canoas* operating out of 19 stations, and the total catch was some 500 whales. As the whaling industry declined and the economy of the islands plummeted, there was a mass evacuation. Whaling was perceived as a dangerous occupation (in 1974 two men were killed when a whale smashed a *canoa*), and it became increasingly difficult to interest young men in this line of work. Many Azoreans crossed the Atlantic to take up residence in New England, and the large Portuguese-speaking enclaves in Massachusetts and Rhode Island are the results of that emigration.

Sperm whale teeth, stored in the Azores, have been finding their way to New England where they are carved into scrimshaw and sold illegally to unsuspecting collectors. There is still a cottage scrimshaw industry in the Azores, but with the passage of the Marine Mammal Protection Act in 1972, it has become illegal to bring whale products into the United States, and the European Economic Community has also imposed strict prohibitions on the import of whale products. With the disappearance of the Azorean markets, the whaling industry has ground to a halt.

Although Portuguese observers attended the meetings of the International Whaling Commission for many years, the country never applied for membership in the commission, perhaps because the government realized that participation would result in sanctions against her whaling. By 1966 sperm whales had been placed in the "protected" category, which meant that they could not be legally killed anywhere. The Azoreans continued to fish in a sporadic fashion, but like so many other whaling operations, theirs was an ecological and economic anachronism, doomed to obsolescence.

In a 1976 *National Geographic* article, Don Moser (1976) wrote that "whaling is dying out in the Azores," and quoted harpooner Almerindo Lemos as saying he can make more money working on a tuna boat. "But I have a craving," says Lemos, "I have an addiction." In 1976, only 200 whales were killed, and since then the number has dropped. In 1982 the boats were still visible, and there were huge piles of dried-out skulls and bones, but it was obvious that the industry, if not over, was on its last legs.

Although commercial whaling officially ceased in 1984, the Azorean Department of Fisheries issued a permit for five male sperm whales to be taken in 1987 in an attempt to stimulate the Azorean economy. Three whales were harpooned and brought to shore, but since the whaling factories had closed down, there were minimal facilities for processing them. A tractor was used to strip off the blubber, and some of the meat was sold for fish bait and fertilizer.

The rest of the carcasses were towed out to sea and discarded, and the teeth were made into scrimshaw trinkets. In their 1988 IWC report, Deimer *et al.*, (1988) wrote:

> The killing of whales led to a debate and protest both inside and outside the Azores. The member of the European Parliament for the Azores, Prof. Vasco Garcia, was prominent amongst those opposing the whaling and proposed that other ways, such as whale-watching, should be found to exploit the region's cetacean resources. The Azores' position as a semi-autonomous part of a new member of the EEC added a further complication to the situation. It appears that whaling is still permitted within the archipelago.

The idea of initiating a whale-watching business in the islands as a means of "exploiting the region's cetacean resources" was a brilliant one, and a decade after Azorean whaling officially ended, a number of companies are offering voyages to the whales. Sperm whales, the very creatures that formed the basis of Azorean whaling, are now the prime attraction there, because, as one website puts it, "the archipelago of Azores is one of the best sites in Europe for the observation of cetaceans. It is one of the few places on earth where it is possible to meet sperm whale pods of females with their offspring." Off São Miguel and Pico islands, watchers might espy blue whales, fin whales, humpbacks, sei whales, and many different species of dolphins including killer whales (which are really large dolphins), false killer whales (ditto), common dolphins, bottlenoses, and Atlantic spotted dolphins. Whaling has ended in the Azores, after a century of intensive exploitation, and the *Museu dos Baleeiros* in Lages do Pico, with its whaleboats, harpoons, and scrimshaw, exists as a reminder of the glorious days of Azorean whaling. The switch from whale-killing to whale-watching shows that it is possible for a society to recognize that whales do not have to be killed to provide jobs and income for locals.

## See Also the Following Articles

Whaling, Early and Aboriginal ■ Whaling, Traditional

## References

Ashley, C. W. (1938). "The Yankee Whaler," Reverside Press, Cambridge, USA.

Browne, J. R. (1841). "Etchings of a Whaling Cruise, with Notes on a Sojourn on the Island of Zanzibar." Harper & Bros. (Reprinted 1968), Harvard Univ. Press, Cambridge, USA.

Clarke, R. (1952). Sperm whaling from open boats in the Azores. Norsk Hvalf-Tid. **42**, 373–385.

Clarke, R. (1954). Open boat whaling in the Azores. Disc. Rep. **26**, 281–354.

Deimer, P., J. Gordon and T. Arnbom. (1988). Sperm whales killed in the Azores during 1987. *International Whaling Commission Report* SC/40/Sp5.

Hohman, E. P. (1928). "The American Whaleman," Reissued 1972. Augustus M. Kelley, New Jersey.

Housby, T. (1971). "The Hand of God: Whaling in the Azores." Abelard-Schuman.

Moser, D. (1976). The Azores: Nine Islands in search of a future. *Natl. Geogr.* **149**(2), 261–288.

Venables, B. (1969). "Baleia! Baleia! Whale Hunters of the Azores. Knopf, New York.

A

# B

## Baculum

### Edward H. Miller

The baculum (os penis) is a bone in the penis that occurs in small insectivorous placentals (orders Afrosoricida, Erinaceomorpha, and Soricomorpha), Chiroptera, Primates, Rodentia, and Carnivora (Burt, 1960). Among marine mammals, it is present in Ursidae (polar bear, *Ursus maritimus*), all Mustelidae

[including the marine otter, *Lontra felina* (undescribed but presumed) and the sea otter, *Enhydra lutris*], and Pinnipedia. The baculum is absent in Cetacea and Sirenia. The corresponding element in females is the little-studied clitoris bone (os clitoridis), which has been documented for polar bears and several pinniped species, but presumably is present in all pinnipeds, and in marine and sea otters (it is present in the northern river otter, *Lontra canadensis*; Mohr, 1963; Fay, 1982).

The baculum is one of several so-called heterotopic bones in mammals, like the kneecap (patella), which form through ossification in connective tissue. In rodents, the bacular shaft is true bone, and includes hemopoietic tissue in the enlarged basal portion. In the caniform Carnivora (which includes bears, otters, and pinnipeds) bacular development has been detailed only in the dog (*Canis familiaris*) but is probably similar in other Caniformia. The dog baculum develops in the proximal portion of the penis, in association with the fibrous septum between the paired corpora cavernosa penis, or in their fibrous non-cavernous portion; centers of ossification on left and right sides fuse early in development. The developing baculum grows dorsally above the urethra, and thickens. The bacular base becomes firmly attached to the corpora cavernosa and to the fibrous tunica albuginea which surrounds them.

The urethral groove in the baculum is deep in the dog but is shallow to absent in bacula of marine mammals (Fig. 1A lower, 1B lower),

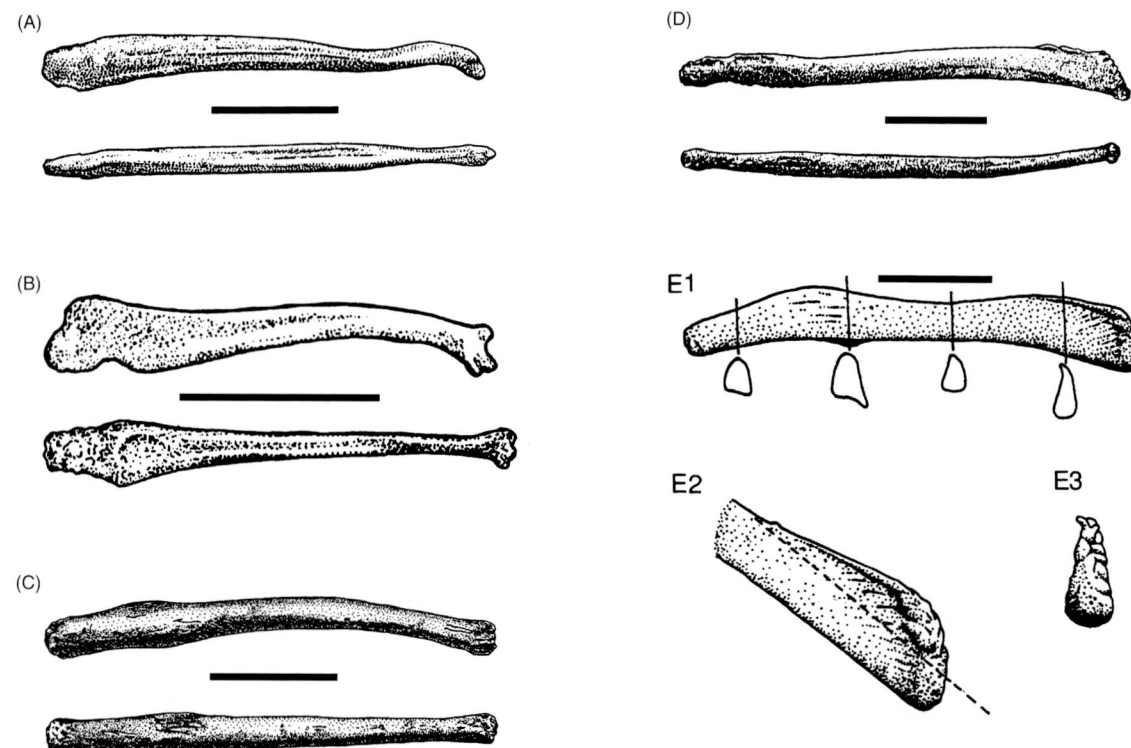

**Figure 1**  *Bacula of marine mammals are large, but most are morphologically simple: (A) polar bear* (Ursus maritimus); *(B) subantarctic fur seal* (Arctocephalus gazella); *(C) Mediterranean monk seal* (Monachus monachus); *(D) crabeater seal* (Lobodon carcinophagus); *(E) Weddell seal* (Leptonychotes weddellii). *All scale bars, 5 cm (no scale bars for E2, E3). Bacula in (A)–(D) are shown in right lateral (upper) and ventral (lower) views. E1: Baculum in right lateral view (note cross-sectional shapes at the indicated points). E2: Oblique view (right side) of the bacular apex (same specimen); dashed line indicates how much growth occurs in the crest (above the line), following sexual maturity. E3: Apical view (dorsal surface above; same specimen). A from R. Didier (1950; Mammalia **14**, 78–94); B from R. Didier (1952; Mammalia **16**, 228–231); C from P. J. H. van Bree (1994; Mammalia **16**, 228–231); D from R. Didier (1953; Mammalia **17**, 21–26); E from G. V. Morejohn (2001; Journal of Mammalogy **81**, 877–881).*

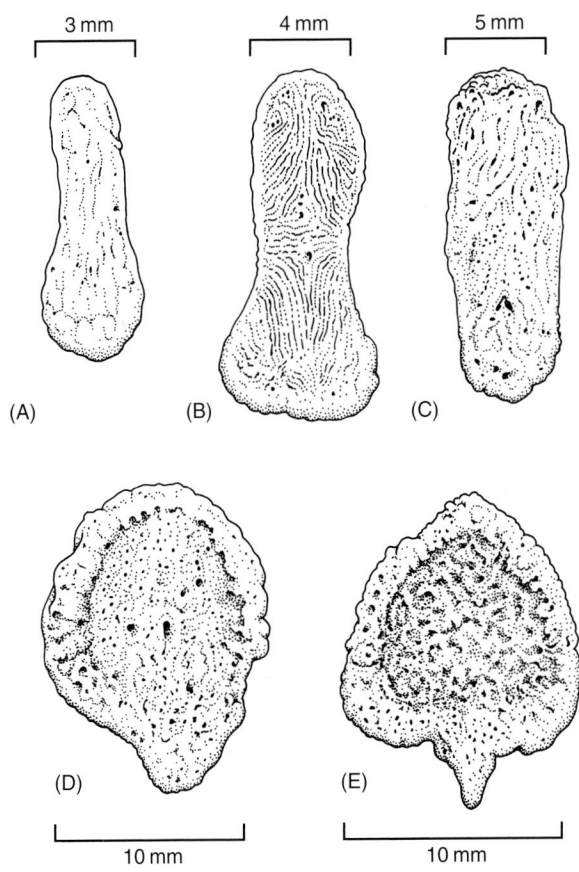

(A)    (B)    (C)

(D)    (E)

**Figure 2** *The bacular apex is morphologically complex and interspecifically diverse in Otariidae. The apex is shown in apical view (dorsal surface up) for (A) unknown species of* Arctocephalus *fur seal; (B) northern fur seal* (Callorhinus ursinus); *(C) California sea lion* (Zalophus californianus); *(D) Australian sea lion* (Neophoca cinerea); *and (E) Hooker's sea lion* (Phocarctos hookeri). *From G. V. Morejohn (1975; Rapports et Proces-verbaux des Reunions, Conseil International pour l'Exploration de la Mer* **169**, *49–56).*

although is likely present terminally in the undescribed baculum of the marine otter, because this is the pattern in the northern river otter (Baryshnikov *et al.*, 2003). Bacula of polar bears and phocid seals are fairly simple, being more or less straight or slightly curved (arched dorsally) structures, and lacking elaborate apices (Fig. 1). In at least some phocids, the bacular apex has a prominent cartilaginous cap (e.g., hooded seal, *Cystophora cristata*). Cross-sectional shapes of phocid bacula vary considerably among species, and a prominent crest develops on the anterior dorsal surface in some Antarctic seals (Fig. 1E). The bacular apex is larger and more elaborate in otariids than phocids, in keeping with the close proximity of the apex to (beneath) the glans penis in otariids where apical size and shape may be functionally important during copulation (Fig. 2). Mustelids possess some of the most diverse and morphologically elaborate elaborate bacula within the Caniformia, although that of the sea otter is relatively simple (Fig. 3; Baryshnikov *et al.*, 2003). Within species, bacula are variable in size, shape, cross-section, and specific structural features, even among individuals of the same age. For example, a dorsal keel may be present or absent in southern elephant seals (*Mirounga leonina*); processes on the shaft near the apex are variably present in California sea lions (*Zalophus californianus*); and bacula may be bilaterally asymmetrical or slightly twisted (Fig. 1D).

Bacula of Carnivora are fairly large (Dixson, 1995; Lariviére and Ferguson, 2002; Ramm, 2007). Bacular length is approximately 6% of body length in otariids, but relatively longer in polar bears (~8%) and phocids (8% in hooded seals; 10% in harp seals, *Pagophilus groenlandicus*); the baculum is also much thicker in phocids than otariids (Mohr, 1963; Scheffer and Kenyon, 1963). In pinnipeds, and indeed among all mammals, the walrus (*Odobenus rosmarus*) has the largest baculum both absolutely (to 62.4 cm in length and 1040 g in mass) and relatively (18% of body length; Fay, 1982). Interspecific differences in bacular size in mammals have been linked to diverse selective pressures: reproductive isolation between species; aquatic vs terrestrial copulation; copulatory duration or pattern; sexual selection and mating system; climate; and risk of fracture (Scheffer and Kenyon, 1963; Eberhard, 1985; Dixson, 1995; Lariviére and Ferguson, 2002; Ramm, 2007). Fractures result from accidents (e.g., falls in walruses), sudden movements during intromission (e.g., in aquatically mating Caspian

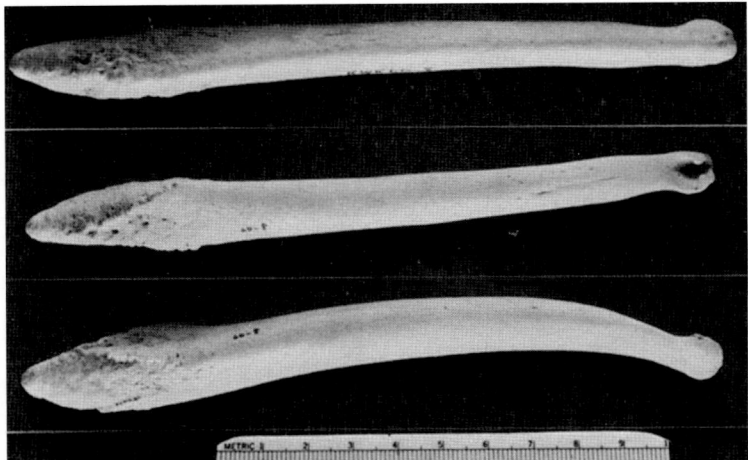

**Figure 3** *The baculum of the sea otter* (Enhydra lutris) *is fairly simple, except for the apex (to the right). Top: dorsal view; center, ventral view; bottom, right lateral view. Scale is in centimeters. From K. W. Kenyon (1969; North American Fauna* **68**, *1–352).*

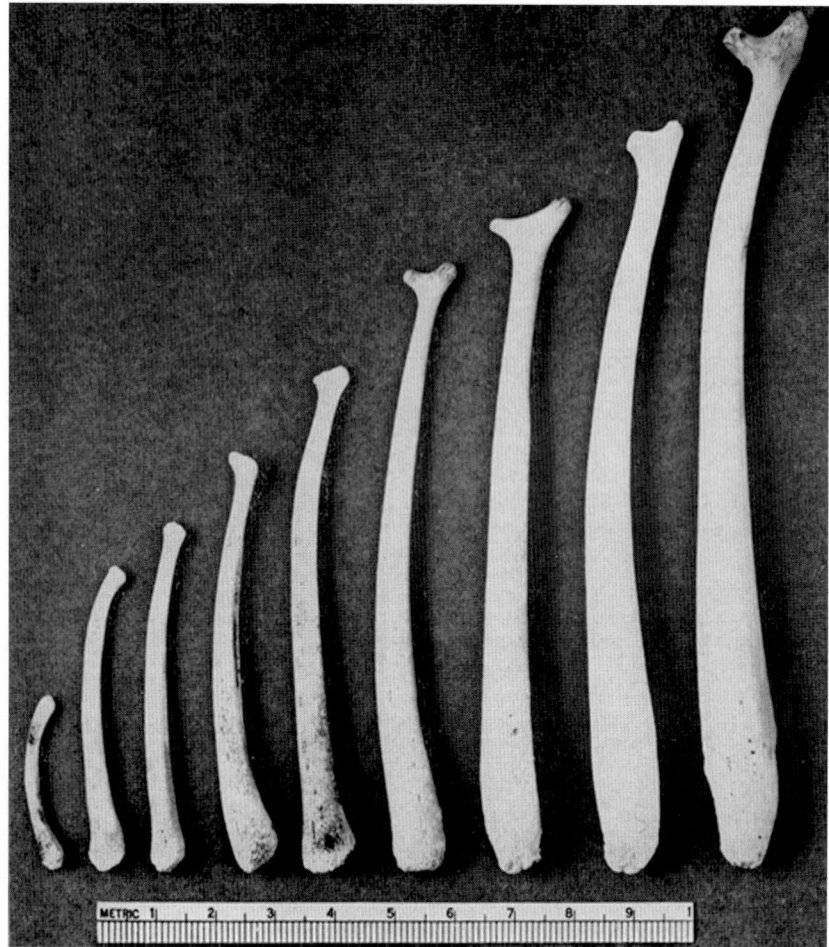

**Figure 4** *Developmental changes in bacular size and shape, illustrated by representative specimens from northern fur seals* (Callorhinus ursinus), *ranging in age from newborn (left) to 8 years of age (right). Specimens are shown in right lateral view, with bacular apex at the top. Scale is in centimeters. From V. B. Scheffer (1950;* Journal of Mammalogy **31***, 384–394).*

seals, *Pusa caspica*), and aggressive social interactions (e.g., fights in adult male sea otters). Healed fractured bacula have been documented for several species. Bacula likely serve several functions: as a mechanical aid in copulation (especially in the absence of full erection), or maintenance of intromission, in aquatic copulations; and to initiate or engage neural or endocrinological responses in females. Bacular size may be limited by adverse effects on females: a female sea otter and a harbor seal (*Phoca vitulina*) pup died from perforation of the vagina during forced copulations with male sea otters. Bacular form and diversity reflect multiple functions, and hence likely have multiple adaptive explanations within and across species.

In Carnivora, bacula grow throughout life in thickness and mass (particularly at the proximal or basal end), but not in length (Fig. 4). Bacular growth is most rapid around puberty. Differential growth occurs in different parts of the baculum (e.g., bacular apex, shaft, and base, in Steller's sea lion, *Eumetopias jubatus*; Miller *et al.*, 2000).

The baculum is anatomically complex and species-specific in many groups, so has been used extensively in mammalian systematics. In addition, bacular growth has been investigated in furbearers and game animals, because it can be informative about age and time of puberty. More recently, the baculum has been studied in the context of mate-choice and sexual-selection theories. In Alaska, the

**Figure 5** *Genitals of African fur seal* (Arctocephalus p. pusillus) *drying under a work table at a seal processing facility in Luderitz, South Africa (1994). Photo: ©International Fund for Animal Welfare.*

Figure 6 *Cooked seal genitals prepared as a meal in the Guolizhuang Penis Restaurant, Beijing, China (September 7, 2007). These were advertised as Canadian seal, so probably were from a harp seal* (Pagophilus groenlandicus), *and killed in the commercial hunt in Quebec or Newfoundland and Labrador. Photo: Feng Li/Gettyimages.*

U.S. Fish & Wildlife Service requires that hunters leave the baculum attached to the hide of sea otters and polar bears, to confirm sex. The seal baculum forms most of the mass of the male genitals that are taken illegally and legally [e.g., in commercial hunts of African fur seals (*Arctocephalus p. pusillus*) in Namibia, and harp seals in Canada], and are usually dried, then sold (mainly in Asia) whole or ground, for use as supposed aphrodisiacs or in traditional medicine (Fig. 5); they are also exported frozen, and served as putatively aphrodisiac-containing food (Fig. 6). Sexual maturation and reproduction may be affected by pollutants, so bacular size and form also may be informative in studies on pollution biology.

## *See Also the Following Articles*

Male Reproductive Systems ■ Mating Systems

## *References*

Baryshnikov, G. F., Bininda-Emonds, O. R. P., and Abramov, A. V. (2003). Morphological variability and evolution of the baculum (os penis) in Mustelidae (Carnivora). *J. Mammal.* **84**, 673–690.

Burt, W. H. (1960). Bacula of North American mammals. *Misc. Pub. Mus. Zool., Univ. Michigan* **113**, 1–76, + 25 plates.

Dixson, A. F. (1995). Baculum length and copulatory behaviour in carnivores and pinnipeds (Grand Order Ferae). *J. Zool. (Lond.)* **235**, 67–76.

Eberhard, W. G. (1985). "Sexual Selection and Animal Genitalia." Harvard University Press, Massachusetts, Cambridge.

Fay, F. H. (1982). Ecology and biology of the Pacific walrus, *Odobenus rosmarus divergens* Illiger. *N. Am. Fauna* **74**, 1–279.

Lariviére, S., and Ferguson, S. H. (2002). On the evolution of the mammalian baculum: Vaginal friction, prolonged intromission or induced ovulation? *Mam. Rev.* **32**, 283–294.

Miller, E. H., Pitcher, K. W., and Loughlin, T. W. (2000). Bacular size, growth, and allometry in the largest extant otariid, the Steller sea lion (*Eumetopias jubatus*). *J. Mammal.* **81**, 134–144.

Mohr, E. (1963). Os penis und Os clitoridis der Pinnipedia. *Z. Säugetierkund.* **28**, 19–37.

Ramm, S. A. (2007). Sexual selection and genital evolution in mammals: A phylogenetic analysis of baculum length. *Am. Nat.* **169**, 360–369.

Scheffer, V. B., and Kenyon, K. W. (1963). Baculum size in pinnipeds. *Z. Säugetierkund.* **28**, 38–41.

# Baiji
## *Lipotes vexillifer*

### Kaiya Zhou

## I. Characteristics and Taxonomy

The baiji or Yangtze river dolphin is endemic to the middle and lower reaches of the Yangtze River in China. It is a relict species and the only living representative of a whole family of mammals. It was described early in the ancient dictionary, *Erh Ya*, published as long ago as 200 BC.

The baiji is a graceful animal with a very long, narrow and slightly upturned beak. It can be easily identified by the rounded melon, longitudinally oval blow hole, very small eyes, low triangular dorsal fin, and broad rounded flippers (Fig. 1). The color is generally bluish gray or gray above and white or ashy white below. Females are larger than males. Maximum recorded length for females is 253 cm and for males is 229 cm (Zhou, 1989). Significant differences between the sexes in external proportions were demonstrated in nine characters, and the skull size is also sexually dimorphic (Gao and Zhou, 1992a, b). The mouth is lined with 31–36 teeth in each tooth row. The crown of the tooth is conspicuously inclined labially and is slightly compressed antero-posteriorly. Its upper half recurves interiorly. The lower half of the lingual side of the crown is a broad, rounded cingulum. The enamel of the entire crown is ornamented with irregular vertical striae and ridges which present a reticular appearance. The lower end of the root widens to form slight anterior and posterior projections (Zhou *et al.*, 1979a). The structure of the stomach is unique in cetaceans. The forestomach is lacking, and the main stomach is divided into three compartments. The connecting channel between the main stomach and the pyloric stomach is absent (Zhou *et al.*, 1979b). The skull is characterized in having an extremely long slender rostrum and mandible (Fig. 2). The rostrum length exceeds two thirds of the condylobasal length. The rostrum bows slightly upward, bends left at the anterior end and is constricted transversely posterior to the end of the tooth row. The total number of vertebrae ranges from 41 to 45. The seven cervical vertebrae are unfused. The costal facets of the second to fifth thoracic vertebrae are located on the posterior edge of the centrum. The facet on the posterior edge of the sixth thoracic disappears or is vestigial, and that of the seventh thoracic sits on the anterior edge of the vertebral body (Fig. 3). The position of the costal facet on the thoracic vertebrae in baiji is unique and is opposite to that in the boto, *Inia geoffrensis*. This feature is one of the morphological bases for favoring rejection of close relationship between the two taxa.

The largest brain weighed was 590 g (Chen, 1979). The largest cranial capacity measured was 590 cm (Zhou *et al.*, 1979a). Comparing brain weight with that in delphinids of similar body size, the former is only about half of the latter. The cerebral hemispheres are short, wide, and highly convoluted. No trace of olfactory bulbs, tracts, or olfactory nerves has been found. The Yangtze River is turbid. The visibility from the surface downward is about 25–35 cm in April and 12 cm in August. A corresponding regression has taken

**B**

**Figure 1**  *Carcass of 2.45-m adult female baiji with a notch in its dorsal fin; found drifting down river near Jiangyin on 15 January 1996. From Zhou et al. (1998).*

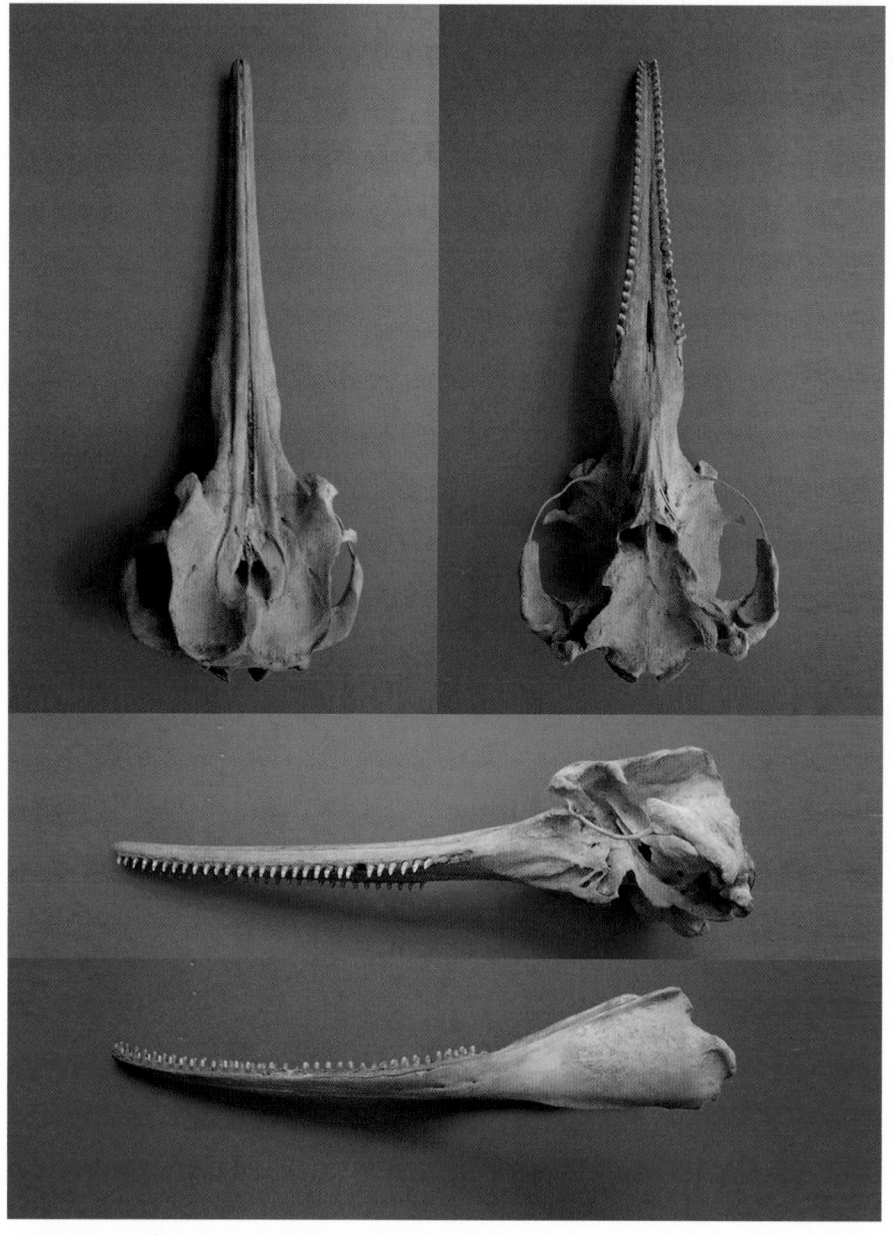

**Figure 2**  *Adult skull of* Lipotes vexillifer, *specimen NJNU 7907, female, CBL 616 mm, collected in 1979 from Guichi City, Anhui Province, China.*

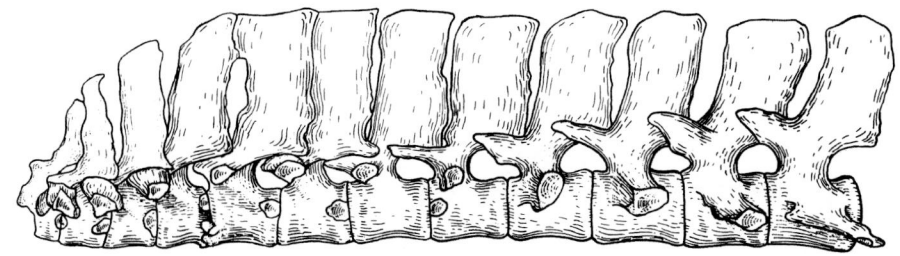

**Figure 3** *Seventh cervical vertebra and thoracic vertebrae of* Lipotes vexillifer. *From Zhou (1989).*

place in the eye of the baiji. The eyes are much smaller and placed much higher than those of marine dolphins. The retinal ganglion cell and optic fiber numbers in baiji are 23,800 and 21,000, respectively (Gao and Zhou, 1987; Gao and Zhou, 1992a, b). Both counts are much lower than those in marine odontocetes. However, the eye of the baiji is functional and objects on the surface or near the surface directly in front of the eye can be distinguished (Zhou, 1989).

Since the baiji was nominated as a member of the Iniidae in 1918 (Miller, 1918), its systematic status and phylogenetic position among the four genera of classical river dolphins and other odontocetes has remained controversial for almost a century (Yan *et al.*, 2005). The species was previously classified as either in the family Platanistidae or in the family Iniidae (Brownell and Herald, 1972). Zhou *et al.* (1978) established the new monotypic family Lipotidae based on osteological studies and anatomy of the stomach. Barnes (1985) recognized a fossil, *Parapontoporia*, as morphologically intermediate between *Pontoporia* and *Lipotes* and placed *Lipotes* and the fossil taxon in a subfamily of Pontoporiidae. Conversely, Muizon (1988) placed the fossil taxon with *Lipotes* in Lipotidae, and Rice (1998) following Muizon to rank Lipotidae as a family. Molecular phylogenetic studies based on mitochondrial and nuclear genes, short interspersed elements (SINEs), and the complete mitochondrial genome since 2000 have strongly supported the referral of *Lipotes* to a separate family Lipotidae (Cassens *et al.*, 2000; Hamilton *et al.*, 2001; Nikaido *et al.*, 2001; Yang *et al.*, 2002; Yan *et al.*, 2005).

The only fossil placed close to *L. vexillifer* is *Prolipotes yujiangensis* (Zhou *et al.*, 1984). The fossil is a fragment of mandible with teeth, including the posterior part of the symphysial portion and initial part of the free portion of the mandibular rami. It was collected from the bank of the Yujiang River in China and considered roughly as Miocene in age. This indicates that the baiji is a relict species.

Parts of eastern and southern China are low-lying deltaic regions formed of sediments deposited by the area's river systems, such as the Yangtze and the Yujiang. Significant sea-level rise would transform these regions into shallow waterways of mixed fluvial and marine origin. This scenario is consistent with the geographical occurrence of the mandibular fragment in southern China (Hamilton *et al.*, 2001).

## II. Distribution and Abundance

Baiji were in recent decades mainly in the mainstream of the middle and lower reaches of the Yangtze River (Zhou *et al.*, 1977). They did occur historically in Dongting and Poyang Lakes; both were appended water bodies of the Yangtze during intense flooding (Zhang *et al.*, 2003). About the turn of the twentieth century, Hoy and Pope collected specimens separately from Dongting Lake and near its mouth, where the lake joins the Yangtze. Dongting Lake was thus incorrectly reported to be the only habitat. The presence of this dolphin in the Yangtze River is noted in documents going back about 2000

years, when it was known only by the ancient name "Ji." In the 1940s, the uppermost records in the Yangtze River were at Huanglingmiao and Liantuo in the Three Gorges area, approximately 50 km upstream of the Gezhouba Dam near Yichang (Zhou *et al.*, 1977). It could be found up to Yichang in the 1960s, which is about 1700 km up from the mouth of the river. However, the range was no farther upstream than Zhicheng in the 1970s and then Jingzhou (formerly called Shashi) in the 1990s (~170 km downstream of the dam site). In the lower part of the river, specimens were obtained in the Yangtze estuary, off the eastern end of Chongming Island, Shanghai, in the 1950s and 1960s (Fig. 4). The range has been no farther downstream than Liuhe since the 1970s. No dolphins were found downstream of Jiangyin, located 256 km upstream of the mouth during surveys in 1997–1999 (Zhang *et al.*, 2003). Some individuals were seen in Fuchun River, immediately south of the Yangtze, during the great flood of 1955, but they disappeared after the construction of the Xinanjiang Hydropower Station in 1957 (Zhou *et al.*, 1977).

The first rough estimate of abundance based on quantitative survey data (1979–1981) was only about 400 animals (Zhou, 1982). On the basis of surveys conducted in 1985–1986, Chen and Hua (1989) made an educated guess that the total population was around 300. Surveys between 1982 and 1986 suggested that there were 100 baiji in a 770-km segment of the lower Yangtze from Hukou to the river mouth (Zhou and Li, 1989). An estimate made by another research group based on surveys in the same segment in 1985–1986 was 78–79 dolphins (Chen and Hua, 1989). Repeated surveys of a 500-km segment of the lower Yangtze (Nanjing-Hukou) in 1989–1991 produced a maximal count of 12 individuals, leading Zhou *et al.* (1998) to infer a total abundance of about 30 baiji in that river segment. The authors reasoned that if the species still inhabited its historical range of about 1700 linear km of river, with a density similar to that found in their study area, the total population in the early 1990s would have been only about 100. Attempted comprehensive surveys of the entire species' range in 1997–1999 resulted in a maximal count (November 1997) of 13 dolphins (including one calf), leading to the generally accepted view that abundance had continued to decline and that the total population was by that time very small. The sighting rate in the 3 years of surveys declined at an annual rate of about 10% (Zhang *et al.*, 2003). Informed guesses in the early 2000s were that there could be only "a few dozen" and "very likely … less than a hundred" (Reeves *et al.*, 2003) baiji left. An intensive 6-week multi-vessel visual and acoustic survey carried out in November–December 2006, covering the entire historical range of the baiji in the main Yangtze channel, failed to find any evidence that the species survives. Although a few undocumented sightings have been reported since 2004, the last authenticated records were that of a stranded pregnant female found in 2001 and a live animal photographed in 2002. The baiji is now thought to be probably extinct (Turvey *et al.*, 2007). In other words, there is no chance for survival of the species.

B

**Figure 4** *Map of China showing distribution of baiji,* Lipotes vexillifer, *and collection locality for the fossil,* Prolipotes yujiangensis. *(▲) Huanglingmiao and Liantuo, the uppermost records in the 1940s; (△) Yichang, the uppermost records in the 1960s; (□) Zicheng, the uppermost records in the 1970s; (●) Jingzhou, the upstream limit of distribution in the 1990s; (○) Distribution in the 1990s; (▼) Yangtze estuary, the lowest records in the 1950s and 1960s; (▽) Liuhe, the downstream limit of distribution in the 1970s; (◎) Tonglu and Fuyang, records in the 1950s in the Fuchun River; (⊕) Guiping, locality where the fossil* Prolipotes yujiangensis, *was found.*

## III. Ecology

The river sections inhabited by the baiji were characterized by having one to several sandbars. Baiji were usually sighted immediately upstream or downstream of a sandbar, where a tributary enters the river. They were generally found in eddy countercurrents below meanders and channel convergences. These areas of former baiji occurrence are also prime fishing areas, set with nets, traps, and hooks (Zhou and Li, 1989).

The baiji appears to have taken any available species of freshwater fish, the only selection criterion being size. The fish could not be so large that it could go down the throat. Sometimes, dead fish were seen floating on the Yangtze with patches of scales torn off. They are believed to have been prey of baiji. At times, a baiji tried a number of times to swallow a larger fish, but in vain and finally let go (Zhou and Zhang, 1991). Occasionally, baiji and Yangtze finless porpoises (*Neophocaena phocaenoides asiaeorientalis*) groups appeared to feed together for short periods (Zhou *et al.*, 1998).

## IV. Behavior and Physiology

Baiji lived in small groups. In the 1980s, the most common group size was 3–4 animals; the largest group observed was about 16 individuals in the middle reaches of the Yangtze River. These larger groups were probably temporary aggregations of several groups. The baiji usually surfaced without causing white water and breathed in a smooth manner (Fig. 5). It had a sequence of several short breathing

**Figure 5** *Baiji in the Yangtze River near Tongling. From Zhou and Zhang (1991).*

intervals (10–30 sec) alternating with a longer one, the longest one up to 200 sec (Zhou *et al.*, 1994). Photographic identifications and sighting records showed that baiji groups made both local and long-range movements. The largest recorded range of a recognizable baiji was more than 200 km from the initial sighting location (Zhou *et al.*, 1998).

Three kinds of behavior, individual behavior, social behavior, and rhythmic behavior, were observed. Clockwise swimming occurred in night, and counterclockwise swimming took place mostly in the daytime. The total duration of either type of swimming in the course of 24 h was nearly equal. Typical resting behavior occurred at night, with peaks in January and May–August (Liu and Wang, 1989). The baiji likely had two main kinds of signals, communication signals and

**Figure 6** *Baiji taken from the Yangtze River near Nanjing in 1982, with 103 hook scars and 5 ulcers in the skin.*

echolocation signals. The former fell into several categories: call signals, bellows of short duration, and squeaks (whistles). Echolocation signals were composed of one pulse train. The bandwidth of these signals was wide and the energy mainly in ultrasonic frequencies (Wang *et al.*, 1989).

## V. Life History

The baiji probably bred and gave birth in the first half of the year. The peak calving season appeared to be February–April. Body length at birth was estimated to be 91.5 cm for both males and females. Calves were closely accompanied by adults and remained alongside and slightly behind the dorsal fin of their presumed mother (Zhou and Li, 1989). They had about the same growth rate until they were about 4 years old based on estimates of dentinal growth layer groups (GLGs), which was the age at sexual maturation for males. After sexual maturation, males grew more slowly than females. The females attained sexual maturation at about 6 years. They continued to grow until about age 8. The smallest mature male and female described were 180 cm and 200 cm long, respectively. The oldest animal that was age-determined was a 242-cm-long female of 24 years of age, and a 21-year-old male was estimated to be about 214 cm in body length (Gao and Zhou, 1992a, b).

## VI. Interactions with Humans

Unlike the case for most historical-era extinctions of large-bodied animals, the baiji was the victim not of active persecution but of incidental mortality resulting from massive-scale human environmental impacts (Turvey *et al.*, 2007). The range contraction and the decline in abundance were caused by a combination of factors: possibly some level of direct exploitation historically; incidental mortality from interactions with fisheries (Fig. 6); vessel traffic; management of navigation channels, and harbor construction; and loss or degradation of habitat by water development, land use practices, and pollution (Chen and Hua, 1989; Zhou *et al.*, 1998).

The Yangtze River is one of the chief river systems of the world, next only to the Nile and the Amazon in length. Nearly one-third of the population of China or almost 10% of the entire world population lives along the Yangtze Valley. It is suffering massive degradation. The banks of the Yangtze have been extensively modified to prevent destructive flooding of agricultural areas, thus reducing the floodplain area. Construction of dams and other barriers along the river and its tributaries has led to changes in fish abundance and distribution. Waste water volume discharged into the Yangtze is about 15.6 billion cubic meters per year. Approximately 80% of the waste waters are discharged directly into the environment without treatment. Up to 1949, approximately 500 domestic commercial vessels operated on the river (Zhou *et al.*, 1998). River traffic increased drastically when China launched the free-market reforms in the 1980s. During the whole-range survey in 2006, a minimum of 19, 830 large shipping vessels (>1 vessel per 100 m of river surveyed) and 1175 fishing vessels were seen between Yichang and Shanghai (Turvey *et al.*, 2007).

The baiji is classified in the World Conservation Union (IUCN) Red List as Critically Endangered—Possibly Extinct. It is designated in the Chinese national First Category of National Key Protected Wildlife Species and has full legal protection throughout its range. Unfortunately, the major threats are continuing, and conservation efforts were unable to save the baiji from extinction. The baiji is the first cetacean species known to have been driven to extinction by human activity.

## See Also the Following Articles

River Dolphins ■ River Dolphins: Evolutionary History and Affinities

## References

Barnes, L. G. (1985). Fossil pontoporiid dolphins (Mammalia: Cetacea) from the Pacific coast of North America. *Nat. Hist. Mus. Los Angeles County Contrib. Sci.* **363**, 1–34.

Brownell, R. L., Jr., and Herald, E. S. (1972). *Lipotes vexillifer. Mamm. Spec.* **10**, 1–4.

Cassens, I., *et al.* (2000). Independent adaptation to riverine habitats allowed survival of ancient cetacean lineages. *Proc. Natl. Acad. Sci. U.S.A.* **97**, 11343–11347.

Chen, P., and Hua, Y. (1989). Distribution, population size and protection of *Lipotes vexillifer. Occ. Papers IUCN SSC* **3**, 81–85.

Chen, Y. (1979). On the cerebral anatomy of the Chinese river dolphin, *Lipotes vexillifer* Miller. *Acta Hydrob. Sin.* **6**, 365–372.

Gao, A., and Zhou, K. (1987). Studies on the ganglion cells of *Neophocaena* and *Lipotes. Acta Zool. Sin.* **33**, 316–322.

Gao, A., and Zhou, K. (1992a). Sexual dimorphism in the baiji, *Lipotes vexillifer. Can. J. Zool.* **70**, 1484–1493.

Gao, G., and Zhou, K. (1992b). Fiber analysis of optic and cochlear nerves of small cetaceans. *In* "Marine Mammal Sensory Systems" (J. A. Thomas, R. Kastelein, and A. Supin, eds), pp. 39–52. Plenum Publishing Corporation, New York.

Hamilton, H., Caballero, S., Collins, A. G., and Brownbell, R. L., Jr. (2001). Evolution of river dolphins. *Proc. R. Soc. Lond. B* **268**, 549–558.

Liu, R., and Wang, D. (1989). The behavior of the baiji, *Lipotes vexillifer*, in captivity. *Occ. Papers IUCN SSC* **3**, 141–145.

Miller, G. S., Jr. (1918). A new river-dolphin from China. *Smiths. Misc. Coll.* **68**, 1–12.

Muizon, C. de. (1988). Les relations phylogénétiques des Delphinida (Cetacea, Mammalia). *Ann. Paléontol. (Vert. Invert.)* **74**, 157–227.

Nikaido, M., *et al.* (2001). Retroposon analysis of major cetacean lineages: The monophyly of the toothed whales and the paraphyly of river dolphins. *Proc. Natl. Acad. Sci. U.S.A.* **98**, 7384–7389.

Reeves, R. R., Smith, B. D., Crespo, E. A., and di Sciara, G. N. (2003). "Dolphins, Whales and Porpoises: 2002–2010 Conservation Action Plan for the World's Cetaceans." IUCN/SSC Cetacean Specialist Group, IUCN, Gland, Switzerland, UK, Cambridge.

Rice, D. W. (1998). Marine mammals of the world, systematics and distribution. *Soc. Mar. Mammal. Spec. Publ.* **4**.

Turvey, S. T., *et al.* (2007). First human-caused extinction of a cetacean species? *Biol. Lett.* **3**, 537–540.

Wang, D., Liu, W., and Wang, Z. (1989). A preliminary study of the acoustic behavior of the baiji, *Lipotes vexillifer. Occ. Papers IUCN SSC* **3**, 137–140.

Yan, J., Zhou, K., and Yang, G. (2005). Molecular phylogenetics of "river dolphins" and the baiji mitochondrial genome. *Mol. Phylogenet. Evol.* **37**, 743–750.

Yang, G., Zhou, K., Ren, W., Ji, G., Liu, S., Bastida, R., and Rivero, L. (2002). Molecular systematics of river dolphins inferred from complete mitochondrial cytochrome *b* gene sequences. *Mar. Mamm. Sci.* **18**, 20–29.

Zhang, X., *et al.* (2003). The Yangtze River dolphin or baiji (*Lipotes vexillifer*): Population status and conservation issues in the Yangtze River, China. *Aquat. Conserv. Mar. Freshwat. Ecosyst.* **13**, 51–64.

Zhou, K. (1982). On the conservation of the baiji, *Lipotes vexillifer. J. Nanjing Nor. Coll. (Nat. Sci.)* **4**, 71–74.

Zhou, K. (1989). Review of studies of structure and function of the baiji, *Lipotes vexillifer. Occ. Papers IUCN SSC* **3**, 99–113.

Zhou, K., Ellis, S., Leatherwood, S., Bruford, M., and Seal, U. (1994). "Baiji (*Lipotes vexillifer*) Population and Habitat Viability Assessment Report." IUCN/SSC Conservation Breeding Specialist Group, Apple Valley, MN.

Zhou, K., and Li, Y. (1989). Status and aspects of the ecology and behavior of the baiji, *Lipotes vexillifer. Occ. Papers IUCN SSC* **3**, 86–91.

Zhou, K., Li, Y., and Qian, W. (1979a). The osteology and the systematic position of the baiji, *Lipotes vexillifer. Acta Zool. Sin.* **25**, 58–74.

Zhou, K., Li, Y., and Qian, W. (1979b). The stomach of the baiji, *Lipotes vexillifer. Acta Zool. Sin.* **25**, 95–100.

Zhou, K., Qian, W., and Li, Y. (1977). Studies on the distribution of baiji, *Lipotes vexillifer. Acta Zool. Sin.* **23**, 72–79.

Zhou, K., Qian, W., and Li, Y. (1978). Recent advances in the study of the baiji, *Lipotes vexillifer. J. Nanjing Nor. Coll. (Nat. Sci.)* **1978**, 8–13.

Zhou, K., Sun, J., Gao, A., and Würsig, B. (1998). Baiji (*Lipotes vexillifer*) in the lower Yangtze River: Movements, numbers, threats and conservation needs. *Aquat. Mamm.* **24**, 123–132.

Zhou, K., and Zhang, X. (1991). "Baiji, the Yangtze River dolphin and other endangered animals of China." Stone Wall Press, Washington, DC.

Zhou, K., Zhou, M., and Zhao, Z. (1984). First discovery of a Tertiary platanistoid fossil from Asia. *Sci. Rep. Whal. Res. Inst., Tokyo* **35**, 173–181.

# Balance

## Fred Spoor

### I. Introduction

The sense organ of balance for the perception of movement and spatial orientation is part of the inner ear, together with the organ of hearing. The mammalian inner ear is housed inside the petrous part of the temporal bone, in a complex-shaped space known as the bony labyrinth. The organ of balance, or vestibular system, consists of two types of motion sensors. The first one, two otolith organs in the membranous utricle and saccule, informs the

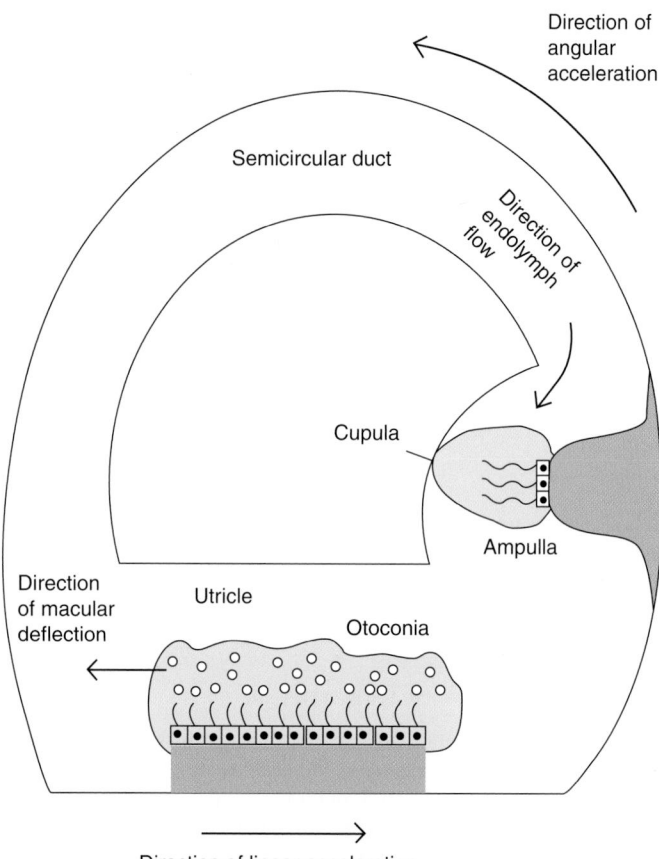

**Figure 1** *Schematic representation of the semicircular canal and otolith systems, only showing the utricle and one membranous duct. The respective inputs, rotational and linear acceleration, and inertia-based responses of the endolymph and macula are indicated.*

brain about changes in gravitational direction and other forms of linear acceleration of the head. In each organ resides a matrix of gel with embedded dense crystals (a macula with otoliths or otoconia), which deforms by inertia in response to such head movements, and this is detected by an associated bed of hair cells (Fig. 1). Little is known of the mammalian otolith system in a comparative context in general, and in relation to life in marine environments in particular. Hence, this part of the organ of balance will not be considered further here (see Spoor and Thewissen, 2008, for details).

The second type of motion sensor concerns the semicircular canal system, which perceives rotational (angular) head movements. It comprises of three membranous ducts, anterior, posterior and lateral, that run inside the semicircular canals of the bony labyrinth. Each duct is connected with the membranous utricle inside the vestibule of the bony labyrinth, and thus forms a fluid circuit filled with endolymph (Fig. 1). The ampulla, a swollen section at one end of each duct, contains a cupula, a gelatinous structure which seals the duct. Changes in head rotation, acceleration or deceleration, will cause the endolymph to lag behind by inertia, slightly deforming the cupula. This is detected by haircells with cilia embedded in the cupula, and provides the neural output, which encodes the head rotation (NB: instant mechanical integration by the system makes that the neural output is directly proportional to rotational velocity, rather than acceleration). The three semicircular ducts of each ear are oriented in approximately orthogonal planes, and any head rotation will thus be sensed by at least one duct. Moreover, the output from both ears is integrated to provide comprehensive representation of self-rotation. This information, combined with otolithic, visual and proprioceptive input, helps coordinate posture and body movements during locomotion, including the reflex stabilization of the head and eyes.

Empirically a relationship has been found among mammals between the length of the semicircular ducts of a species and its specific locomotor repertoire (Spoor *et al.*, 2007). Species that are agile and have fast, jerky locomotion have significantly larger-arced, and thus longer ducts relative to body mass than those that move more cautiously. Presumably this is because longer ducts render the canal system more sensitive, and adjusting length is a way of fine-tuning the sensitivity to match the requirements of particular locomotor behaviors. Relatively long, more sensitive ducts that characterize agile animals can instantly resolve small changes in head rotations and this is thought to be important for precise body coordination during fast and highly maneuverable locomotion. Other functionally important aspects, such as the lumen size (cross-section) of the ducts, the viscosity of the endolymph, or the structure of the cupula, have not been studied comprehensively in a comparative context.

## II. Comparative morphology

Differences between aquatic and terrestrial mammals are largely known from studies using computed tomography (CT) to assess the bony semicircular canals rather than the enclosed membranous ducts. Valid functional information can nevertheless be obtained because the length of the duct along its arc is reliably reflected by that of the canal.

All cetacean species examined thus far have fully developed semicircular canals. However, taking into account the species' overall size they are remarkably small compared with other mammals and in particular with terrestrial species showing rapid and agile locomotion (Fig. 2). This can be demonstrated by measuring the average arc size of the canals, expressed by their radius of curvature (Fig. 1), and plotting it against body mass (as a measure of overall size). Among cetaceans the canal arc size varies with body size in the same way as other mammals (Fig. 3: regression lines are parallel). However, all cetacean species fall way below other mammals, as their canals are on average three times smaller for their body mass. In broad terms these differences mean that the canals of the blue whale (*Balaenoptera musculus*) are just smaller than those of humans, and those of the bottlenose dolphin (*Tursiops truncatus*) are smaller than those of a brown rat (*Rattus norvegicus*).

The semicircular canal arc size in fossil cetaceans that document the transition from land to water reach the small, modern proportions in the middle Eocene (Spoor *et al.*, 2002). Only the earliest cetaceans (pakicetids) have canals with the proportions of land mammals.

Sirenian semicircular canals have been studied in the dugong and the manatee (*Dugong dugon, Trichechus inunguis*). They are small in arc size, at the lower end of the range of non-cetacean mammals (Fig. 3).

Pinnipeds of which the canal arc sizes have been examined include three phocids, four otariids, and *Odobenus rosmarus*. The latter is not significantly different in arc size from terrestrial carnivores (27 species). On the other hand, all three canals are larger in the phocids, whereas the otariids have smaller anterior and posterior canals. See Spoor and Thewissen (2008) for details.

## III. Functional Interpretation

The semicircular canal system of cetaceans is distinctly different from that of all other mammals, by having strongly reduced arc and lumen sizes. The regular pattern of this reduction suggests a functional adjustment of the system, rather than a vestigial condition marked by degeneration and redundancy. The hypothesis explaining this phenomenon, while being fully consistent with the pattern of canal variation seen in other mammals, is based on two key characteristics of cetaceans. The first one is that extant cetaceans, freed from the restrictions of gravitational pull and the need for substrate contact, are particularly agile and acrobatic when compared with terrestrial animals of similar body size (e.g., compare the killer whale *Orcinus* with the African elephant *Loxodonta*, or the dolphin *Tursiops* with the larger bovid species). The second characteristic is that cetaceans have integrated their head and trunk to streamline the body, and in most species the strongly shortened and frequently fused cervical vertebrae allow little neck motility. This has important implications because a motile neck isolates the head from body movements during locomotion. Head rotations are reduced, both passively by inertia, and actively via compensatory neck movements generated by the vestibulo-collic reflex. The canals supply this reflex, and by stabilizing the head thus keep their own input signal within limits. This feedback loop allows the semicircular canals of agile species to be sensitive (i.e., large-arced),

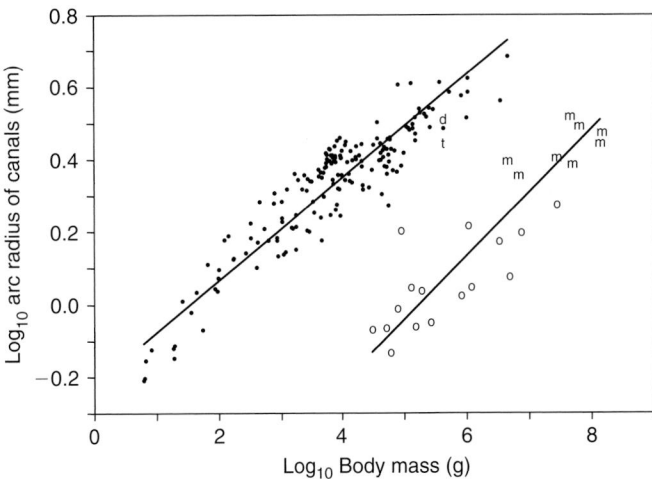

**Figure 3**  *Average arc size of the three semicircular canals (radius of curvature, indicating duct length) plotted against body mass. Reduced major axis regressions are given for the non-cetacean mammal sample (174 species, •) and for the cetaceans (16 odontocetes, o, and 8 mysticetes, m). The dugong (Dugong dugon, d) and manatee (Trichechus inunguis, t) are labeled individually.*

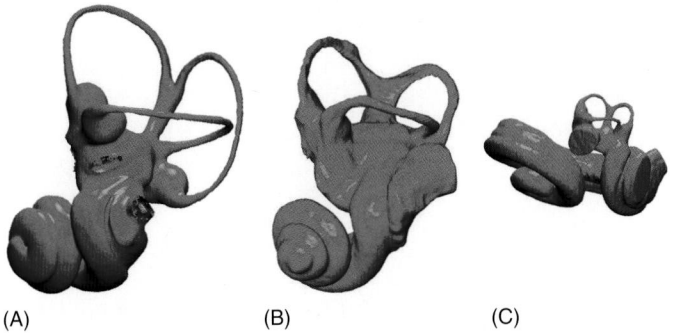

(A)                              (B)                              (C)

**Figure 2**  *Infero-lateral view of the bony labyrinth of (A) a typically agile primate Galago senegalensis, (B) the artiodactyl Hippopotamus amphibius, a sister taxon of extant cetaceans, and (C) the dolphin Tursiops truncatus. Images rendered from CT scans and corrected for body mass.*

without the risk of overstimulation. In cetaceans, on the other hand, little neck motility and ineffective head stabilization combined with acrobatic locomotion implies that the semicircular canal system is likely to experience substantially stronger rotatory input (resulting from movements of the entire body) than in terrestrial mammals of similar body size. The small arc size of cetacean canals may therefore reduce the sensitivity to match the high levels of uncompensated rotations, and avoid overstimulation of the canal system. The loss of canal sensitivity, in response to streamlining of the body, is arguably less critical in an aquatic environment than in, for example, an arboreal setting where less accurate sensory clues easily impair locomotor control. Moreover, less effective stabilization of the eyes is not critical in cetacean navigation which is driven by sonar rather than visual clues. Kinematic analyses of cetacean head motion in comparison with terrestrial mammals of similar size will be essential to test the hypothesis that the cetacean canal system experiences stronger rotatory input.

Sirenians do show reduced neck motility, but not the extreme semicircular canal reduction of cetaceans. However, they are slow and cautious in their swimming, so that fast and effective head stabilization is not a factor of importance. Their canal size is at the lower end of the non-cetacean mammalian variation, as are terrestrial species that are slow and cautious in their locomotion.

Among pinnipeds the semicircular canals of phocids and otariids are different in arc size from terrestrial carnivores, but none show the dramatic size reduction seen in cetaceans. This is expected because they all have motile necks enabling effective head stabilization. Likewise, that phocids have larger canals than terrestrial carnivores is expected, as they are particularly agile in their swimming, and thus follow the normal pattern seen among non-cetacean mammals. However, the smaller anterior and posterior canals of otariids are more difficult to understand. Otariids use a different mode of propulsion than phocids, a bird-like forelimb flight stroke, as opposed to bilateral hind limb undulation, and with a longer neck their center of gravity is located further forward. However, it is not clear how the otariid's smaller canals with reduced mechanical sensitivity relates to either their locomotor pattern or their body plan.

## See Also the Following Article

Sense Organs, Overview

## References

Spoor, F., and Thewissen, J. G. M. (2008). Balance: Comparative and functional anatomy in aquatic mammals. *In* "Senses on the Threshold" (J. G. M. Thewissen and S. Nummela, eds), Chapter 16. University of California Press, Berkeley, California, USA.

Spoor, F., Bajpai, S., Hussain, S. T., Kumar, K., and Thewissen, J. G. M. (2002). Vestibular evidence for the evolution of aquatic behavior in early cetaceans. *Nature* **417**, 163–166.

Spoor, F., Garland, Th., Krovitz, G., Ryan, T. M., Silcox, M. T., and Walker, A. (2007). The primate semicircular canal system and locomotion. *Proc. Nat. Acad. Sci* **104**, 10808–10812.

# Baleen

## DALE W. RICE

The term baleen (also called whalebone) is a mass noun that refers collectively to the series of thin keratinous plates ("baleen plates," Fig. 1) that make up the filtering apparatus in the mouth of a baleen whale. The word derives from the Classical Latin *Balaena* and ultimately from the Greek Φάλλαινα [phallaina], "whale."

Baleen plates are suspended from the whale's palate and are arranged in a row down each side of the mouth, extending from the tip of the rostrum back to the esophageal orifice. The left and right sides are separated by a prominent longitudinal ridge along the midline of the palate. In the rorquals, the two sides are continuous around the tip of the palate, but in the other species the two rows are not confluent. Depending on the species, each "side" of baleen may contain anywhere between 140 and 430 plates. The plates are transversely oriented, and are spaced 1 or 2 cm apart, leaving a narrow gap or slot between adjacent plates. The plates are roughly triangular, with their horizontal basal edges embedded in the palate, their near-vertical labial edges facing outward, and their oblique, fringed lingual edges facing the inside of the mouth. Each plate is slightly curved, with its convex side facing forward, so that its labial edge is directed slightly backward; when the whale is swimming forward,

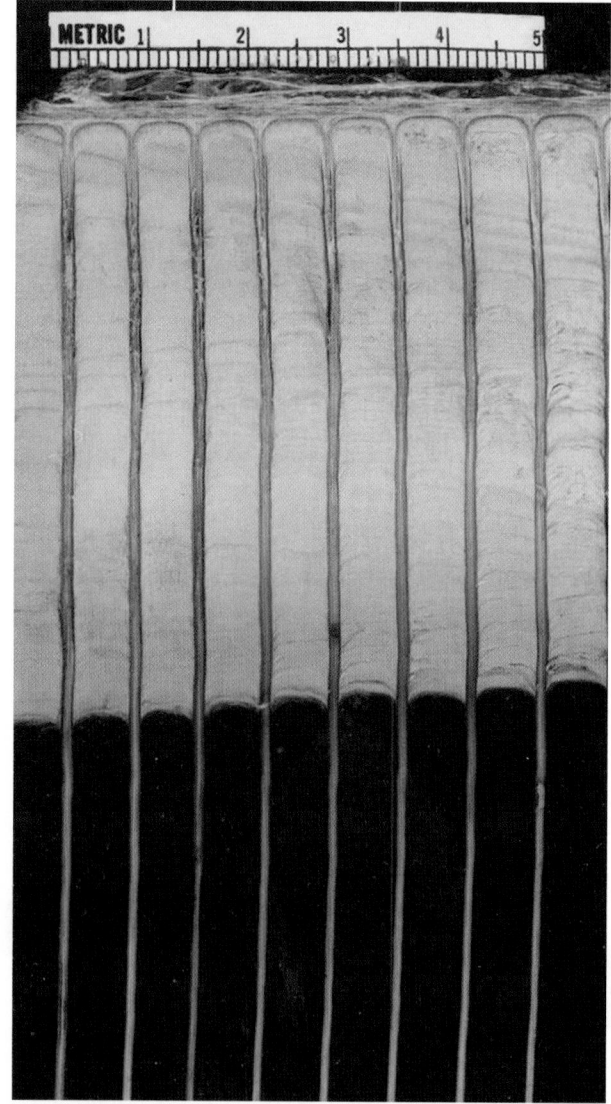

**Figure 1** *Parasagittal section through the palate of a sei whale* (Balaenoptera borealis), *at about midlength of the rostrum, showing the bases of several baleen plates. Anterior is to the right. See Fig. 2 for details.*

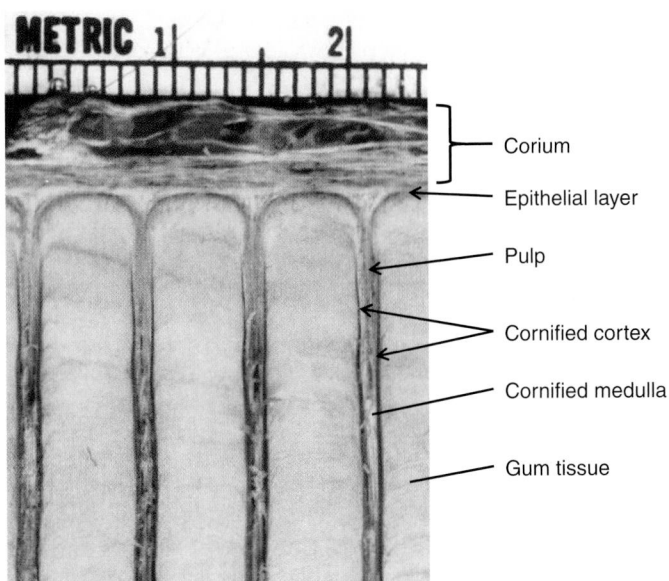

Corium

Epithelial layer

Pulp

Cornified cortex

Cornified medulla

Gum tissue

**Figure 2** *Magnified section of the specimen in Fig. 1, showing the structure of the roots of the baleen plates.*

this arrangement helps to direct the flow of water through the inter-plate gaps from the mouth cavity to the exterior side of the baleen row. The sizes of the plates are smoothly graded, with the longest ones half to two-thirds of the way back from the tip of the rostrum, and only rudimentary ones at the anterior and posterior ends of the row (Williamson, 1973; Pivorunas 1976, 1979).

Each baleen plate is made up of a middle layer, the medulla, which is sandwiched between the thin, smooth outer layers, or cortex (Fig. 2). The medulla consists of a mass of fine, hollow, hairlike keratinous tubules which run parallel to the labial side of the plate, and terminate along the lingual side; the tubules are embedded in and cemented together by a horny matrix.

Evolutionarily, plates presumably originated by modification of the transverse ridges present on the palates of many terrestrial mammals. In whale fetuses the baleen first appears as a series of crosswise ridges along each side of the palate. The palatal tissue of baleen whales is arranged in three layers. The basal layer, several centimeters thick, is the corium. This is overlain by a thin epithelial layer only a few millimeters thick. The outermost epidermal layer, several centimeters thick, is simply called the gum tissue. The corium gives rise to, and is continuous with, the medulla of each baleen plate, whereas the adjacent epithelial layer is deflected downward to produce the cortical layers of each plate. The dense, rubbery gum tissue does not contribute to the formation of the plates, but simply fills the spaces between their bases, where it provides them a firm support. As each plate grows downward, its cortical layers become cornified sooner than the medulla does. This leaves the first few centimeters of the base of the plate with a layer of soft, highly vascular, corial tissue sandwiched between the keratinous outer layers; this soft layer is often called the pulp, by analogy with the pulp in mammalian teeth (Utrecht, 1965). In life, baleen plates are extremely tough and flexible, but once removed they soon become brittle and are easily fractured.

Throughout the life of the whale its baleen plates grow continuously at their base, and wear away along their lingual margin. The cortex and the matrix of the medulla erode away first, freeing the ends of the fibrous tubules for a distance of about 10–20 cm. The freed tubules form a hairy fringe along the entire lingual side of the plate. The fringes of each plate lie back across the lingual edges of the plates

immediately behind them, the whole forming a dense hairy mat that covers the internal apertures to the gaps between the plates. This mat effectively filters out the food organisms while allowing the water to flow out of the whale's mouth through the gaps.

Like human fingernails, the thickness of the baleen plates varies with the nutritional state of the whale. Alternating periods of summer gorging and winter fasting leave a regular series of visible growth zones on the surfaces of the plates. These zones have been used to infer the ages of whales, but because of the constant wear, it is rare for more than five or six zones to remain in a plate (Ruud, 1945). A claim that evidence of individual ovulations could be detected in the growth patterns of baleen plates was never confirmed (Utrecht-Cock, 1965).

The number of baleen plates per side, and their maximum size, shape, color, and other physical attributes are diagnostic for each species of whale. The right whales (family Balaenidae) with their narrow, highly arched rostrum have 250–390 narrow and extremely long plates, about 0.15–0.25 m. wide and up to 2.50 m. long in the black right whales (*Eubalaena* spp.) and 4.00 m. in the bowhead whale (*B. mysticetus*); they are black with a fine whitish fringe. The pygmy right whale (*Caperea marginata*; family Neobalaenidae) has about 230 narrow, short plates up 0.70-m long and 0.12-m wide; they are white with a black labial margin. The gray whale (*Eschrichtius robustus*; family Eschrichtiidae) has 140 thick but narrow and short plates, up to 0.10-m wide and 0.50-m long; they are white or ivory in color, with a coarse white fringe that resembles excelsior. The rorquals (family Balaenopteridae) with their wide, flat rostrum, have 270–430 plates with a basal width 50–95% of their length, which varies from about 0.20 m in the small minke whales to 1.00 m in the huge blue whale. Each species of rorqual has a different color-pattern on its baleen plates: humpback (*Megaptera novaeangliae*)—black with dirty-gray fringe; northern minke (*Balaenoptera acutorostrata*)—white, sometimes with a narrow black stripe along labial margin; Antarctic minke (*B. bonaerensis*) and Omura's (*B. omurai*)—white with a wide black stripe along labial margin; Bryde's (*B. edeni*)—black with light gray fringe; sei (*B. borealis*)—black with fine, silky, white fringe; fin (*B. physalus*)—gray and white longitudinal bands, with fringe the same colors; blue (*B. musculus*)—solid black with black fringe. All of the species of *Balaenoptera* except the blue whale usually have at least a few all-white baleen plates at the tip of the rostrum, mostly on the right side; this asymmetry is most prominent in the fin whale and Omura's whale.

In the nineteenth century, the long baleen plates of the bowhead and right whales were much in demand for uses where a tough but limber material was needed, so they were the most valuable product of the whale fishery. Landings of whalebone at United States ports reached their highest in 1853, with 5,652,300 pounds worth $1,950,000. The last year that any baleen reached the commercial market was 1930. Much of it was made into umbrella ribs, corset busks, and hoops for skirts. The fibrous fringes were used for brooms and brushes (Stevenson, 1907).

## References

Pivorunas, A. (1976). A mathematical consideration on the function of baleen plates and their fringes. *Sci. Rep. Whales Res. Inst.* **28**, 37–55.

Pivorunas, A. (1979). The feeding mechanisms of baleen whales. *Am. Sci.* **67**, 432–440.

Ruud, J. T. (1945). Further studies on the structure of the baleen plates and their application to age determination. *Hvalrådets Skrifter* **29**, 1–69.

Stevenson, C. H. (1907). Whalebone: Its production and utilization. *Bur. Fish. Doc.* **626**, 1–12.

**B**

van Utrecht, W. L. (1965). On the growth of the baleen plate of the fin whale and the blue whale. *Bijdr. Dierk.* **35**, 1–38.

van Utrecht-Cock, C. N. (1965). Age determination and reproduction of female fin whales, *Balaenoptera physalus* (Linnaeus, 1758) with special regard to baleen plates and ovaries. *Bijdr. Dierk.* **35**, 39–100.

Williamson, G. R. (1973). Counting and measuring baleen and ventral grooves of whales. *Sci. Rep. Whales Res. Inst.* **25**, 279–292.

# Baleen Whales (Mysticetes)

### JOHN L. BANNISTER

## I. Characteristics and Taxonomy

The baleen or whalebone whales (Mysticeti) comprise one of the two recent (non-fossil) cetacean suborders. Modern baleen whales differ from the other suborder (toothed whales, Odontoceti), particularly in their lack of functional teeth. Instead they feed, on relatively very small marine organisms, by means of a highly specialized filter-feeding apparatus made up of baleen plates ("whalebone") attached to the gum of the upper jaw. Other differences from toothed whales include the baleen whales' paired blowhole, symmetrical skull, and absence of ribs articulating with the sternum.

Baleen whales are generally huge (Fig. 1). In the blue whale, the largest known animal grows to more than 30-m long and weighing more than 170 tons. Like all other cetaceans, baleen whales are totally aquatic, and like most of the toothed whales, they are all marine. Many undertake very long migrations, and some are fast swimming. A few species come close to the coast at some part of their life cycle and may be seen from shore; however, much of their lives is spent far from land in the deep oceans. Baleen whale females grow slightly larger than the males. Animals of the same species tend to be larger in the Southern than in the Northern Hemisphere.

Within the mysticetes are four families: (1) right whales (Balaenidae, balaenids); (2) pygmy right whales (Neobalaenidae, neobalaenids); (3) gray whales (Eschrichtiidae-eschrichtiids); and (4) "rorquals" (Balaenopteridae, balaenopterids). Within the suborder, 14 species are now generally recognized. Their relationships, including their relationship to terrestrial ungulates, are indicated in Fig. 2.

Right whales (Balaenidae) are distinguished from the other three families by their long and narrow baleen plates and arched upper jaw. Other balaenid features include, externally, a disproportionately large head (approximately one-third of the body length), long thin rostrum, and huge bowed lower lips; they lack multiple ventral grooves. Internally, there is no coronoid process on the lower jaw and cervical vertebrae are fused together. Within the family are two distinct groups—the bowhead (*Balaena mysticetus*) of northern polar waters (formerly known as the "Greenland" right whale), and the three "black" right whales, *Eubalaena* spp. of more temperate seas, so called to distinguish them from the "Greenland" right whale. All balaenids are robust.

Pygmy right whales (*Caperea marginata*) have some features of both right whales and balaenopterids. The head is short (approximately one quarter of the body length), although with an arched upper jaw and bowed lower lips, and there is a dorsal fin. The relatively long and narrow baleen plates are yellowish white, with a dark outer border, quite different from the all-black balaenid baleen plates. Internally, pygmy right whales have numerous broadened and flattened ribs.

Gray whales (*Eschrichtius robustus*) are also somewhat intermediate in appearance between right whales and balaenopterids. They have short narrow heads, a slightly arched rostrum, and between two and five deep creases on the throat instead of the balaenopterid ventral grooves. The body is robust. There is no dorsal fin, but a series of 6–12 small "knuckles" along the tail stock. The yellowish-white baleen plates are relatively small.

Balaenopterids comprise the seven whales of the genus *Balaenoptera* (blue, *B. musculus*; fin, *B. physalus*; sei, *B. borealis*; Bryde's, *B. edeni*; Omura's, *B. omurai*; common minke, *B. acutorostrata*, Antarctic minke, *B. bonaerensis*), and the humpback whale (*Megaptera novaeangliae*). All have relatively short heads, less than a quarter of the body length. In comparison with right whales, the baleen plates are short and wide. Numerous ventral grooves are present, and there is a dorsal fin, sometimes rather small. Internally, the upper jaw is relatively long and unarched, the mandibles are bowed outwards and a coronoid process is present; cervical vertebrae are generally free. All eight balaenopterids are often known as "rorquals" (from the Norse "rørkval, whale with pleats in its throat"). Strictly speaking, the term should probably be applied to the seven *Balaenoptera* species, recognizing the rather different humpback in its separate genus, but many authors now use it for all eight balaenopterids.

Baleen whales are sometimes called "great whales." Despite their generally huge size, some of the species are relatively small, and it seems preferable to restrict the term to the larger mysticetes (blue, fin, sei, Bryde's, Omura's, humpback) together with the largest odontocete (the sperm whale, *Physeter macrocephalus*).

Reviewing the systematics and distribution of the world's marine mammals, Rice (1998) drew attention to the derivation of the Latin word Mysticeti, and clarified the status of a variant, Mystacoceti. He described the former as coming from Aristotle's original Greek *mustoketos*, meaning "the mouse, the whale so-called" or "the mouse-whale" (said to be an ironic reference to the animals' generally vast size). Mystacoceti means "moustache-whales," and although used occasionally in the past (and more obviously appropriate for whales with baleen in their mouths) has been superseded by Mysticeti.

Within the suborder, 14 species are now generally recognized. Although Rice believed that all right whales belong with the bowhead in the genus *Balaena*, recent genetic analyses have recognized three separate right whale species, in the genus *Eubalaena*: in the North Atlantic (*E. glacialis*); in the North Pacific (*E. japonica*); and in the Southern Hemisphere (*E. australis*). Indeed, *Eubalaena* is the only mysticete genus where separate species are recognized in each hemisphere.

The taxonomic status of Bryde's whales is confused. Currently one species is recognized (*B. edeni*) but it has several forms, at least one of which may be a separate species. The "ordinary" form has two distinct sub-forms—offshore and inshore. Another animal, *B. brydei*, was described from specimens taken off South Africa, but subsequently accepted as the same species as *B. edeni*. The situation has not been helped because the location of the type specimen of *edeni* was uncertain until recently and its genetic make-up has yet to be determined. A further similar but smaller species, Omura's whale, *B. omurai*, was described in 2003, and recently accepted (Sasaki *et al.*, 2006 following genetic analysis, but it is not closely related to Bryde's whales, lying outside the clade formed by blue, sei and Bryde's whales (see Fig. 2). Subspecies have been described for several mysticete taxa, but only three are at present commonly in use. They are all blue whales, *B. musculus*: the Antarctic, sometimes known as the "true," blue whale, *B. m. intermedia*; the North Atlantic and North Pacific blue whales (*B. m. musculus*); and the Southern Hemisphere, mainly

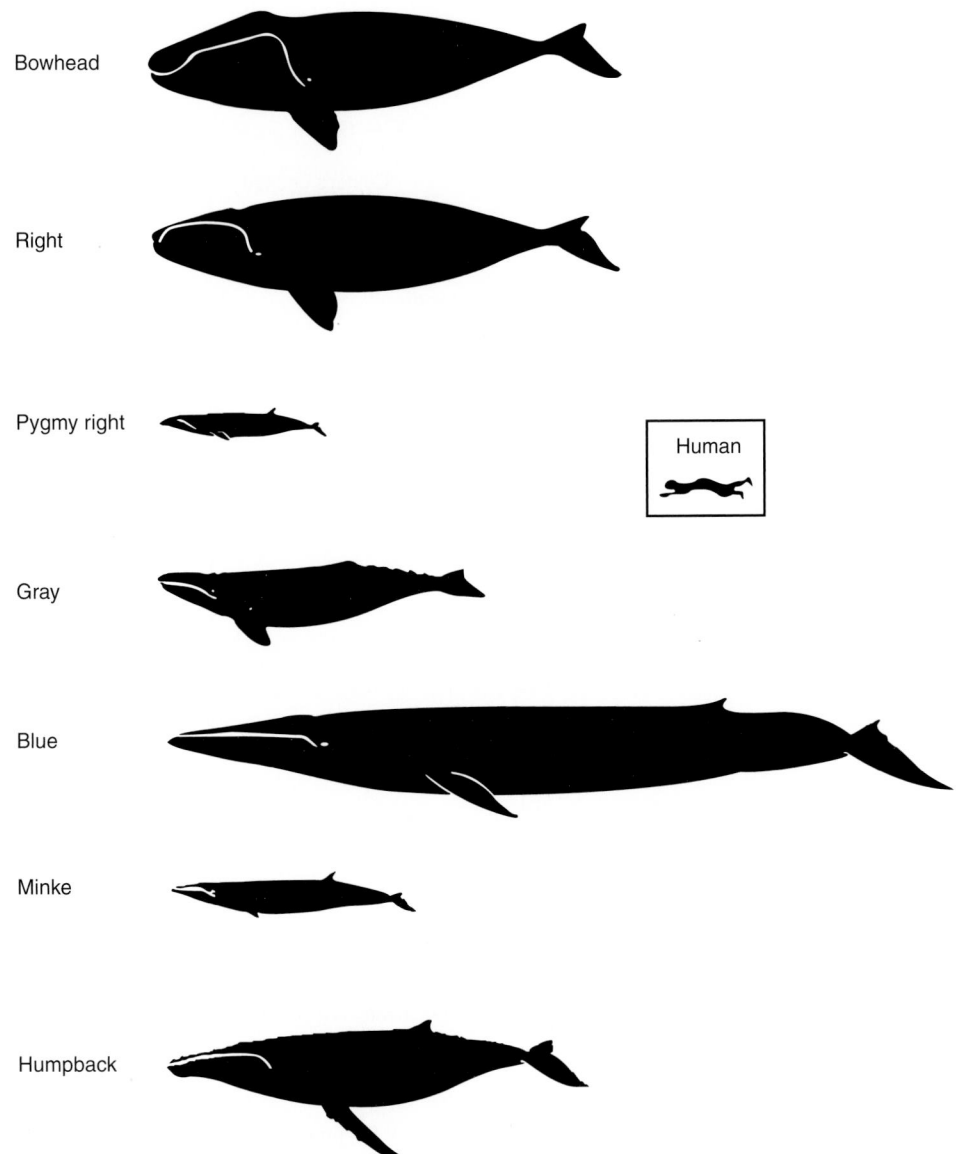

**Figure 1**  *Lateral profiles of representative baleen whales, with a human figure, to scale.*

Indian Ocean, pygmy blue whale, *B. m. brevicauda*. A "dwarf" form of the common minke (*B. acutorostrata*) occurs generally in lower latitudes of the Southern Hemisphere; it has yet to be formally described.

## II. Distribution and Abundace

In addition to the subspecies mentioned in Section I, many stocks or populations have been recognized, some mainly for management purposes, based on more or less valid biological grounds. Some significant examples include:

1. *Bowhead whales*. As well as the currently most abundant population (the Bering-Chukchi-Beaufort Seas stock), four others are recognized—Hudson Bay-Foxe Basin; Davis Strait-Baffin Bay; Svalbard-Barents Sea; Okhotsk Sea.
2. *Right whales*. In the North Atlantic, two populations are currently recognized, western and eastern, with calving grounds off the southeast United States and northwest Africa. The latter may now represent only a relict population. In the North Pacific, there well may once have been two or more stocks, based on feeding ground

information; at least one is now centered in summer on the Sea of Okhotsk and another, though currently in very small numbers, summers in the Gulf of Alaska and Bering Sea.

In the southern right whale, there are several populations, defined by currently occupied calving grounds, but these cover only a proportion of the many areas known from historical whaling records to have once been occupied by right whales. Up-to-date information is available on presumed discrete populations wintering off eastern South America, South Africa, southern Australia and subantarctic New Zealand.

3. *Gray whales*. A western North Atlantic population may have persisted until the seventeenth or eighteenth centuries but is now extinct. The species now survives only in the North Pacific, where, in addition to a flourishing "eastern" stock, wintering on the coast of Baja California and summering in the Bering Sea, a very much smaller western sub-population (the "western" gray whale) summers in the Okhotsk Sea.
4. *Humpback whales*. In the North Atlantic, two major populations have been recognized, one based on animals wintering in

B

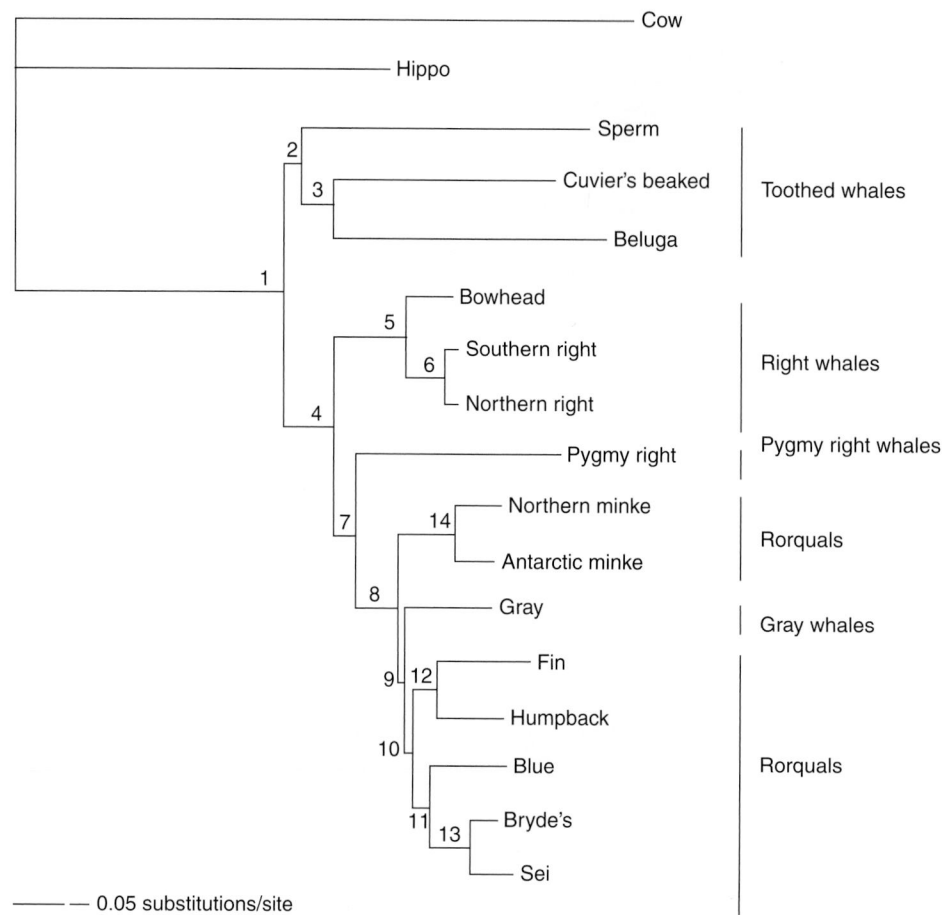

**Figure 2**   *Phylogenetic tree of baleen whales (adapted from Rychel et al., 2004). B. omurai, not included here, lies between the clades for blue, Bryde's and sei whales and for fin and humpback whales (Sasaki et al., 2006). Also, three species of right whale (North Atlantic, North Pacific and southern) are now generally recognized.*

the West Indies, the other, now possibly only a relict population, wintering around the Cape Verde Islands. In the North Pacific, three discrete wintering grounds have been recorded: around the Bonin Islands, Ryuku Islands, and Philippines in the west; around the Hawaiian Islands in the centre; and off Mexico in the east.

In the Southern Hemisphere, seven populations are currently recognized. Six are well-defined, based on calving (wintering) grounds either side of each continent (one off eastern Australia is closely related to animals wintering off New Caledonia and Fiji), and a seventh in the central Pacific. In the northwest Indian Ocean, there is a separate resident population which seems not to migrate.

Baleen whales thus occupy a wide variety of habitats, from open oceans to continental shelves and coastal waters, from the coldest waters of the Arctic and Antarctic, through waters of both hemispheres and into the tropics.

Most specialized is the bowhead, *Balaena*, restricted to the harsh cold and shallow seas of the Arctic and subarctic. The "black" right whales (*Eubalaena*) are more oceanic and prefer generally temperate waters, but come very close to coasts in winter to give birth, particularly in the Southern Hemisphere. Once believed not to penetrate much further south than the Polar Front (Antarctic Convergence, ~50–55°S) there have been recent records in the Antarctic proper, south of 60°S. Whether this is a new phenomenon is unclear: a

report by Sir James Clark Ross of many "common black" (i.e., right) whales in the Ross Sea (eastern Antarctic) at 63°S in December 1840 was discounted when their presence there later that century could not be confirmed. It has been suggested that the currently greatly reduced population of the western North Atlantic right whale, now wintering off the south eastern United States and summering in coastal waters north to the Bay of Fundy (~45°N), may represent the peripheral remnant of a more widely distributed stock, formerly summering north to Labrador and southern Greenland, i.e., to at least 60°N.

The pygmy right whale (*Caperea*) is restricted to Southern Hemisphere temperate waters, mainly between 30°S and 55°S; it can be found coastally in winter in some areas, and year-round in others.

Gray whales (*Eschrichtius*) are the most obviously coastal baleen whales. The long coastal migration of the eastern gray, from Mexico to Alaska, supports a major whale watching industry from November to March. In spring, the animals migrate through the Bering Strait into the more open waters of the Bering Sea, but still favoring more shallow waters.

Among the balaenopterids, fin and sei whales are probably the most oceanic, the former penetrating into colder waters than the latter in summer. Blue whales can be found closer inshore, but often associated with deep coastal canyons, e.g., off central and southern

California and western Australia. The Southern Hemisphere pygmy blue whale (subspecies *B. m. brevicauda*) is restricted to more temperate waters than the Antarctic blue whale (*B. m. intermedia*), not often being found much beyond 55°S. The coastal balaenopterid is the humpback (*Megaptera*), with long migrations between temperate/tropical breeding grounds and cold water feeding grounds. In the Southern Hemisphere, much of its journeys occur along the east and west coasts of the three continents. In the Northern Hemisphere, humpbacks are rather more oceanic, but still coastal at some stage in their migration: in the North Pacific they can be found wintering off the Hawaiian Islands and summering off Alaska; in the western North Atlantic they winter in the Caribbean and summer between New England, the west coast of Greenland, Iceland, and North Norway.

Minke whales are wide ranging, from polar to tropical waters in both hemispheres. In the Southern Hemisphere they can, with blue whales, be found closest to the ice edge in summer. Elsewhere minkes can often occur near shore, in bays and inlets. Their migrations are less well-defined and predictable than the other migratory balaenopterids; in some regions they may be present year-round.

The most localized balaenopterids are Bryde's whale and its close relatives. They are the only balaenopterids restricted entirely to tropical/warm temperate waters, and probably do not undertake long migrations. The two "ordinary" forms of *B. edeni*—inshore and offshore, in several areas—can differ in their movements. Off South Africa, for example, the inshore form is thought to be present throughout the year, whereas the offshore form appears and disappears seasonally, presumably in association with movements of its food, shoaling fish.

## III. Ecology

Although they include the largest living animals, baleen whales feed mainly on very small organisms, and while strictly carnivorous, on zooplankton or small fish. In "filter-feeding"-sieving the sea-baleen whales are quite different from toothed whales, where the prey is captured individually.

Filter-feeding has been described as requiring, in addition to a supply of food in the water, three basic features (1) a flow of water to bring prey near the mouth, (2) a filter to collect the food but allow water to pass through, (3) and a means of removing the filtered food and conveying it to the stomach for digestion. Baleen whales meet those requirements by (a) seeking out areas where their food concentrates, (b) either swimming open-mouthed through food or gulping it in, (c) possessing a highly efficient filter formed by the baleen plates, and (d) forcing the water containing the food out through the baleen plates, and then transferring the trapped food back to the gullet and hence to the stomach. In (d) the tongue is presumed to be involved; in balenopterids the process is aided by the distensible throat and the ability to open the lower jaw to almost 90°.

Although all baleen whales have a filter based on baleen plates, two rather different systems—essentially "skimming" and "gulping"—have evolved to filter a large volume of water containing food. Each relies on the series of triangular baleen plates, borne transversely on each upper jaw. The inner, longer border (hypotenuse) of each plate bears a fringe of fine hairs, forming a kind of filtering "doormat." Quite unrelated to teeth (which appear as early rudiments in the gums of fetal baleen whales), baleen consists of keratin and is close in composition to hair and fingernails. In right whales, filtration is achieved with very long and narrow plates in the very large mouth, in the very large head. The plates, up to 4-m long in bowheads and 2.7 m in other right whales, are accommodated in the mouth by an arched upper jaw, and

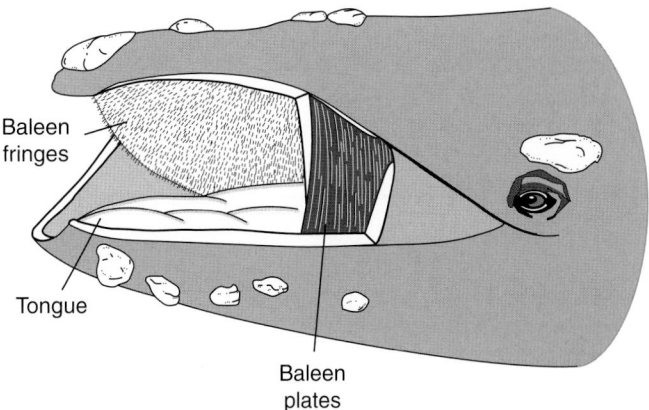

**Figure 3** *Head of a right whale, showing the arrangement of the filter-feeding apparatus. The anterior section of the baleen plates on the left side of this whale were removed to show fringes of right baleen plates and tongue. From Bonner (1980).*

enclosed in massively enlarged and upwardly bowed lower lips (see Fig. 3). There is a gap between the rows at the front of the mouth, and the whole arrangement allows the whale to scoop up a great quantity of water while swimming slowly forwards. In balaenopterids, with their much smaller heads, the baleen plates are shorter and broader and the rows are continuous at the front. Taking in a large volume of water and food is usually achieved by swimming through a food swarm and gulping, while simultaneously greatly enlarging the capacity of the mouth by extending the ventral grooves, depressing the tongue and widely opening the lower jaw, almost to 90° from the body axis. The two systems allow, on the one hand, the relatively slow-swimming balaenids to concentrate their rather sparse food over a period, and on the other, the faster-swimming balaenopterids to take in large amounts of their highly concentrated prey over a shorter time.

Typically, baleen whales feed on zooplankton, mainly euphausiids or copepods, swarming in polar or subpolar regions in summer. That is particularly so in the Southern Hemisphere, where the summer distributions of several balaenopterids depend on the presence of *Euphausia superba* (known to whalers by the Norwegian word "krill") in huge concentrations in the Antarctic. In the Northern Hemisphere, with a more variable availability of food, balaenopterids are more catholic in their feeding. Humpbacks and fin whales, for example, feeding almost exclusively on krill in the south, often commonly take various species of schooling fish in the north.

The variety of organisms taken by the various baleen whale species in different regions is listed in Table I. Although most feeding occurs in colder waters, baleen whales may feed opportunistically elsewhere. All baleen whales but the one, the gray whale, feed in surface waters, generally within 100 m of the surface, and consequently, unlike many toothed whales, do not dive very deep or for long periods. Gray whales feed primarily on bottom-living organisms, almost exclusively amphipods, in shallow waters.

The baleen plate structure, particularly the inner fringing hairs, to some extent mirrors the food organisms taken, or (in the case of *E. superba*) different size classes. Thus there is some correlation between decreasing size of prey and fineness of baleen by species, viz. gray, blue, fin, humpback, minke, sei, and right whales. Where food stocks are very dense, e.g. around subantarctic South Georgia, fin, blue, and sei whales may all overlap in their feeding on *E. superba*.

Baleen whale food consumption per day has been calculated as some 1.5–2.0% of body weight, averaged over the year. Given that

**B**

TABLE I
Baleen Whale Food Items

| Species | Sub-species | Common name | Food items | |
|---------|-------------|-------------|------------|--|
| | | | North hemisphere | South hemisphere |
| B. mysticetus | | Bowhead whale | Mainly calanoid copepods; euphausiids; occasional mysids, amphipods, isopods, small fish | |
| E. glacialis | | Northern right whale | Calanoid copepods; euphausiids | |
| E. australis | | Southern right whale | | Copepods; post-larval *Munida gregaria*; *Euphausia superba* |
| Caperea marginata | | Pygmy right whale | | Calanoid copepods |
| E. robustus | | Gray whale | Gammarid amphipods; occasional polychaetes, small fish | |
| M. novaeangliae | | Humpback whale | Schooling fish; euphausiids | *Euphausia superba* (Antarctic); euphausiids, post-larval *Munida gregaria*, occasional fish (ex-Antarctic) |
| B. acutorostrata | | Common minke (North Atlantic) | Schooling fish; euphausiids | |
| | | Common minke (North Pacific) | Euphausiids; copepods; schooling fish | |
| | B. a. ?spp. | Dwarf minke | | ?Euphausiids, schooling fish |
| B. bonaerensis | | Antarctic minke | | *Euphausia superba*, *E. crystallorophias* |
| B. edeni | | Bryde's whale | Pelagic crustaceans including euphausiids | Schooling fish; euphausiids |
| B. omurai | | Omura's whale | ?Euphausiids | ?Euphausiids |
| B. borealis | | Sei whale | Schooling fish | Copepods incl. *Calanus*; *Euphausia superba* |
| B. physalus | | Fin whale | Schooling fish; squid; euphausiids; copepods | *Euphausia superba* (Antarctic); other euphausiids (ex Antarctic) |
| B. musculus | B. m. musculus | Blue whale | Euphausiids; calanoid copepods; occasional amphipods, squid | |
| | B. m. intermedia | Antarctic blue | | *Euphausia superba* (Antarctic); other euphausiids (ex Antarctic) |
| | B. m. brevicauda | Pygmy blue | | Euphausiids, incl. *E. vallentini*, *E. recurve, Nyctiphanes australis* |

feeding occurs mainly over about 4 months in the summer in the larger species, the food intake during the feeding season has been calculated at some 4% of body weight per day, approximately 4000 kg per day for a large blue whale. To survive the enormous drain of pregnancy and lactation, it has been calculated that a pregnant female baleen whale needs to increase its body weight by up to 65%. The ability to achieve such increase in only a few months' feeding indicates the great efficiency of the baleen whales' feeding system, as well as the considerable nutritive value of the food.

Apart from humans, the most notable baleen whale predator is the killer whale (*Orcinus orca*). Minke whales have been identified as a major diet item of killer whales in the Antarctic, and off British Columbia, Canada. Killer whale attacks have been reported on blue, sei, bowhead, and gray whales, although their frequency and success are unknown. Humpbacks often have killer whale tooth marks

on their bodies and tail flukes. Humpback and right whale calves in warm coastal waters are susceptible to attack by sharks. There are anecdotal reports of calving ground attacks on humpbacks by false killer whales (*Pseudorca*).

A form of harassment occurs on right whales on calving grounds off Peninsula Valdes, Argentina. Kelp gulls have developed the habit of feeding on skin and blubber gouged from adult southern right whales' backs as they lie at the surface. Large white lesions can result. The whales react adversely to such gull-induced disturbance; calf development may be affected and the whales may be driven elsewhere.

External parasites, particularly "whale lice" (cyamid crustaceans) and barnacles (both sessile and stalked) are common on the slower-swimming more coastal baleen whales such as gray, humpback, and right whales. In the latter, aggregations of light-colored cyamids on warty head callosities have facilitated research using callosity-pattern photographs for individual identification. External parasites are much less common on the faster swimming species, although whale lice have been reported on minke whales (in and around the ventral grooves and umbilicus). The highly modified copepod *Pennella* occurs particularly on fin and sei whales in warmer waters; its presence on those species in colder waters, e.g., at South Georgia in the South Atlantic, has been taken to indicate migration from the north. A commensal copepod *Balaenophilus unisetus* often infests baleen plates in warm waters, especially on sei and pygmy blue whales.

A variety of internal parasites has been recorded, although some baleen whales seem less prone to infection than others. They appear, for example, to be less common in blue whales, but prevalent in sei whales. Records include stomach worms (*Anisakis* sp.), cestodes, kidney nematodes, liver flukes, and acanthocephalans ("thorny-headed" worms) of the small intestine.

The cold water diatom *Cocconeis ceticola* often forms a brownish-yellow film on the skin of blue and other baleen whales in the Antarctic. Since the film takes about a month to develop, its extent can be used to judge the length of time an animal has been there. Its presence led to an early common name for the blue whale—"sulfur bottom."

For many years, the origin of small scoop-shaped bites on baleen whale bodies in warmer waters remained a mystery, until they were found to be caused by the small "cookie-cutter" shark, *Isistius brasiliensis*. Some species are highly prone to such attacks. In Southern Hemisphere sei whales the overlapping healing scars can impart a galvanised-iron sheen to the body.

## IV. Behavior and Physiology

*1. Sound production.* Unlike toothed whales, baleen whales are not generally believed to use sound for echolocation, although bowheads, for example, are thought to use sound reflected from the undersides of ice floes to navigate through ice fields. However, sound production for communication, for display, establishment of territory or other behavior, is well developed in the suborder. Blue whales produce the loudest sustained sounds of any living animal. At up to nearly 190 decibels, their long (half minute or more), very low frequency (<20 Hz) moans may carry for hundreds of kilometers or more in special conditions. Fin whales produce similarly low-pulsed sounds (20 Hz). Minke whales also produce a variety of loud sounds. Right whales produce long low moans; bowhead sounds, recorded on migration past hydrophone arrays in nearshore leads, have been used in conjunction with sightings to estimate population size off northern Alaska. Southern right whales, at least, seem to use sound to communicate with their calves.

Humpbacks produce the longest, most complex sound sequences in "songs," described as an array of moans, groans, roars and sighs to high-pitched squeaks and chirps, lasting ten or more minutes before repetition, sometimes over hours. It seems that only the adult males sing, generally only in or close to the breeding season. In any one breeding season, all the males sing the same song, changing slightly over successive seasons. Different populations have different songs; so much so, for example, that those off western Australia have a distinctly different song—less complex, less "chirpy"—than that heard on breeding grounds separated by the Australian continent, off the east coast. But an example of "cultural evolution" involving humpback songs has recently been reported where typical Australian east coast humpback song switched to a new, west coast version over a very short period, only 3 years, between 1995 and 1998, it seems as a result of the influence of a few male singers from the west coast (Noad *et al.*, 2000). "Songs" may also be heard in migrating humpbacks, but less so on the cold water feeding grounds, where, if songs occur at all, they appear generally only as "snatches" or isolated segments.

*2. Swimming and migration.* With their streamlined bodies, rorquals include the fastest swimming baleen whales. Sei whales have been recorded at around 35 knots (>60 km/h) in short bursts; minke and fin whales are also known as fast swimmers, the latter up to 20 knots (37 km/h). Blue whales are among the most powerful swimmers, able to sustain speeds of over 15 knots (28 km/h) for several hours. On migration, humpbacks and gray whales average about 3–4 knots (5–7 km/hr), and bowheads only about 2.7 knots (5 km/hr). Migration speeds for southern right whales are not known, but medium range coastal movements off southern Australia indicate 1.5–2.3 knots (2.7–4.2 km/h) over 24 h for cow/calf pairs.

Baleen whales undertake some of the longest migrations known. Gray whales may cover some 5000 nautical miles (~9000 km) on the round trip between the Baja California breeding grounds and Alaskan feeding grounds. Southern hemisphere humpbacks may cover as much as 50° of latitude either way between breeding and feeding grounds, a round trip of some 6000 nautical miles (~11,000 km); recent records of humpbacks migrating between the Antarctic Peninsula and Costa Rica involved a single trip of ca 8400 km (4500 nautical miles), the longest recorded migration of any mammal. Not all baleen whale migrations are so well marked. The bi-annual movements of Bering Sea bowheads are governed by the seasonal advance and retreat of sea ice, which vary from year to year. Although Southern Hemisphere blue and fin whales all feed extensively in the Antarctic in summer, the locations of their calving grounds are not known. Sei whale migrations are relatively diffuse, and can vary from year to year in response to changing environmental conditions. By comparison, Bryde's whales hardly migrate at all, presumably being able to satisfy both reproductive and nutritional needs in tropical/warm temperate waters. Even among such migratory animals as humpbacks, it may be that not all animals migrate every year: recent studies off eastern Australia have indicated that a proportion of adult females may not return to the calving grounds each year, and individuals have been reported in summer further north. However, Southern Hemisphere migrating humpbacks show segregation in the migrating stream: immatures and females accompanied by yearling calves are in the van of the northward migration, followed by adult males and non-pregnant mature females; pregnant females bring up the rear. A similar pattern occurs on the southward journey, with cow/calf pairs traveling last. Very similar segregation is recorded among migrating gray whales. However, not only is there segregation within some (possibly all) species, the species themselves may arrive on and leave the feeding grounds at different times. At South Georgia, where they were once common, fin whales tended to leave the feeding grounds after blue whales but before sei whales, the latter having arrived there last.

Baleen whale migrations have generally been regarded as taking place in response to the need to feed in colder waters and reproduce in warmer waters. Explanations for such long-range movements have included direct benefits to the calf (better able to survive in calm, warm waters), evolutionary "tradition" (a leftover from times when continents were closer together), and the possible ability of some species to supplement their food supply from plankton encountered on migration or on the calving grounds. Others have rejected these explanations, suggesting there may be a major advantage to migrating pregnant female baleen whales in reducing the risk of killer whale predation on newborn calves in low latitudes. More recently Rasmussen *et al.* (2007) have suggested that calf development in warm water may lead to larger adult size and greater reproductive success.

**B**

*3. Social activity.* Large aggregations of baleen whales are generally uncommon. Even on migration, in those species where well-defined migration paths are followed (e.g., gray whales and humpbacks), individual migrating groups are generally small, numbering only a few individuals. It has been stated that predation is a main factor in the formation of large groups of cetaceans, for example of open ocean dolphins. Given the large size of most adult baleen whales, predation pressure is low, and group size can be correspondingly small.

Blue whales are usually solitary or in small groups of two to three. Fin whales can be single or in pairs; on the feeding grounds they may form larger groupings, up to 100 or more. Similarly, sei whales can be found in large feeding concentrations, but in groups of up to only about six elsewhere. The same is true for minke whales, found in concentrations on the feeding grounds, but singly or in groups of two or three elsewhere. Social behavior has been studied intensively in coastal humpbacks, e.g., on the calving grounds. Male humpbacks compete for access to females by singing and fighting. The songs seem to act as a kind of courtship display. Males congregate near a single adult female, fighting for position. Such aggression can involve lunging at each other with ventral grooves extended, hitting with the tail flukes, raising the head while swimming, fluke and flipper slapping, and releasing streams of bubbles from the blowhole. As a result of such encounters, individuals can be left with raw and bleeding wounds caused by the sharp barnacles. Among southern right whales, surface active groups (known as SAGs) are often observed on the coastal calving grounds in winter, involving a tight group with a number of males pursuing an adult female, but with less aggression. As for humpbacks, it is not yet certain whether such behavior results in successful mating, although intromission in right whale SAGs is often observed.

Feeding balaenopterids have often been reported as circling on their sides through swarms of plankton or fish. It has been suggested that gray whales feed on their right sides, those baleen plates being more worn down, presumably through contact with the seabed. The most remarkable behavior, however, is reported from humpbacks. In the Southern Hemisphere, on swarms of krill, they may feed in the same "gulping" way as other balaenopterids. In the Northern Hemisphere, two methods are commonly reported—"lunging" and "bubble netting." In the former, individuals emerge almost vertically at the surface with their mouths partly open, closing them to force the enclosed water out through the baleen. In the latter, an animal circles below the food swarm; as it swims upwards, it exhales a series of bubbles forming a "net" encircling the prey. It then swims upwards through the prey with its mouth open, as in lunging.

## V. Life History

Young baleen whales, particularly the fetus and the calf, grow at an extraordinary rate. In the largest species, the blue whale, fetal weight increases at a rate of some 100 kg/day towards the end of pregnancy. The calf's weight increases at a rate of about 80 kg/day during suckling. During that 7-month period of dependence on the cow's milk, the blue whale calf will have increased its weight by some 17 tonnes, and increased in length from around 7 to 17 m. Blue whales attain sexual maturity at between 5 and 10 years, at a length of around 22 m, and live for about 80–90 years. Adult female blue whales give birth every 2–3 years, pregnancy lasting some 10–11 months.

Other balaenopterids follow the same general pattern (Fig. 4). Mating takes place in warm waters in winter, birth following some 11 months later. A 7–11 month lactation period may be followed by a year "resting," or almost immediately by another pregnancy. Most adults are able to reproduce from between 5 and 10 years of age, and reach maximum growth after 15 or more years. The smallest balaenopterid, the minke whale, is born after a pregnancy of some 10 months, at a length of just under 3 m. Weaning occurs at just under 6 m, after 3–6 months. The adult female can become pregnant again immediately following birth, but the resulting short calving interval is generally uncommon in baleen whales: 2–3 years is the norm, although humpbacks can achieve a similar birth rate, enabling their stocks to recover rapidly after depletion.

Right whales follow a similar general pattern, but there are some differences. In northern and southern right whales, gestation lasts about 11 months, weaning for about another year. Females are able to reproduce successfully from about 8 years (there are records of successful first pregnancies from 6 years), but the calving interval is usually a relatively regular 3 years. For bowheads, while growth is

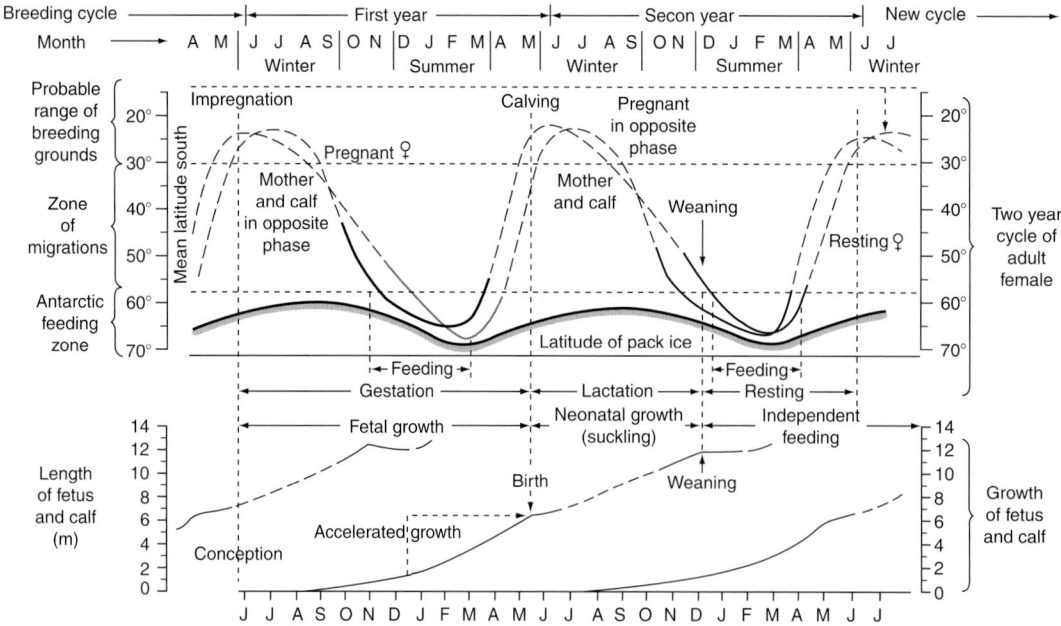

**Figure 4** *"Typical" life cycle of a southern baleen whale. As modified by Bonner (1980), from Mackintosh (1965).*

very rapid during the first year of life (from ~4.5 m), it may be followed by a period of several years with little or no growth. Sexual maturity occurs at 13–14 m: at the reduced growth rate that would not be reached until 17–20 years. Similarly, there is evidence of considerable longevity in this species: harpoon heads and an unexploded bomb-lance found in harvested whales and last known to be used off Alaska in the late nineteenth or early twentieth centuries indicate individual animals can be over 100-years old.

## VI. Interactions with Humans

For centuries, baleen whales have borne the brunt of human greed, for products and profit. Only the sperm whale, largest of the toothed whales, has rivaled them as a whaling target. Black right whales (*Eubalaena*) were taken in the Bay of Biscay at least from the twelfth century, with the fishery extending across the North Atlantic by the sixteenth century. Attention then shifted to the Greenland whale (*Balaena*) near Spitzbergen (Svalbard), and later off southern and western Greenland. Both species' numbers were reduced to only small remnants, and in several areas (e.g., Spitzbergen and Greenland for *Balaena*, the north east Atlantic and the North Pacific for *Eubalaena*) the stocks were virtually exterminated. That destruction was undertaken using the "traditional" whaling method, with open boats and hand harpoons, on the "right" species-"right" because they were relatively easy to approach, floated when dead, and provided huge quantities of product (oil for lighting, lubrication and soap, and baleen ("whalebone") for articles combining flexibility with strength such as corset stays, umbrella spokes, and fishing rods.

Development of the harpoon gun and steam catcher, from 1864, greatly increased the rate of catching, but also allowed attention to turn to the largest baleen whales, the blue and fin whales, whose size and speed, and tendency to sink when dead, had prevented capture by the old methods. From its beginning in the North Atlantic, then, by the end of the century, in the North Pacific, "modern" whaling's next and most intensive phase moved south, first in 1904 at South Georgia, just within the Antarctic zone. Initially on humpbacks, (up to 12,000 were taken in 1 year, 1912, leading to very rapid stock decline) then on blue and fin whales, southern whaling based on such land stations—in the Antarctic in summer and the tropics in winter—was overtaken from the late 1920s by the great development of pelagic whaling, using floating factory ships. Huge annual Southern Hemisphere catches resulted—a maximum of over 40,000 in 1931—averaging around 30,000 animals per year in the later 1930s, and again after the World War II until 1965. Whereas blue whales had been the preferred target in the 1930s, their great reduction in numbers led to a shift in attention progressively over the years to fin whales, to sei whales in the 1960s, and finally to minke. With depletion of stocks and more stringent conservation measures, (killing of humpbacks, blue, and fin whales was banned from the mid-1960s, despite illegal catching until the early 1970s), catches fell to between 10,000 and 15,000 per year in the 1970s. The "old" whaling story had virtually repeated itself—enormous reductions through overfishing of one species or stock leading to exploitation of other species and stocks until, apart from minke whales, only remnants were left. From 1989, a moratorium on all commercial whaling eliminated that pressure, with the exception of limited whaling carried out under exemption for scientific research, and, since 1993, some commercial catching of minke whales in the eastern North Atlantic, and, since 2006, of fin whales off Iceland. Some "aboriginal" whaling has also continued in the Northern Hemisphere, on bowheads off northern Alaska, gray whales in the Bering Sea, on fin and minke whales off Greenland, and on humpbacks in the Caribbean.

Despite the great scale of the kill in "old" and "modern" whaling, no whale species has become extinct through whaling, although a number of individual stocks have been reduced greatly; at least one, the North Atlantic gray whale, has disappeared within the past 200–300 years. In its most recent "Red List" of threatened animals, the World Conservation Union (IUCN, 1996 (Table II) includes no baleen whale species or stocks as either *Extinct*, or *Critically Threatened* (the latter within the *Threatened* category). Within the threatened category, seven taxa—three species, one subspecies, and three stocks—are listed as *Endangered* (*E*); four taxa—one species and three stocks—are *Vulnerable* (*V*). Six taxa—two species, one subspecies, and three stocks—are listed as at *Lower Risk* (*LR*), and two taxa—one species and one subspecies—as *Data Deficient* (*DD*).

Those species under greatest threat (*E*), as determined in 1996, are the Northern Hemisphere right whales, sei, and fin whales, together with the Antarctic blue subspecies, two of the five bowhead stocks (Svalbard-Barents Sea, Okhotsk Sea), and the north west Pacific (western) gray. Next most threatened (*V*) are the humpback, two bowhead stocks (Hudson Bay-Foxe Basin, Davis Strait-Baffin Bay) and the North Atlantic blue. At lower risk (*LR*) are the southern right and Antarctic minke, one bowhead stock (Bering–Chukchi–Beaufort Seas), the North Atlantic (common) minke, north east Pacific (eastern) gray and North Pacific blue; all but one are further qualified as *conservation dependent* (*cd*, not Vulnerable because of specific conservation efforts). The exception is the North Atlantic (common) minke, listed as *near threatened* (*nt*, not Conservation Dependent but almost qualifying as Vulnerable). The two taxa for which insufficient information is currently available (*DD*) are Bryde's whale and the pygmy blue.

The International Whaling Commission's Scientific Committee, responsible for assessments of such stocks' current status, has reported encouraging recent reversals of stock decline for some stocks of some species. One, the eastern gray whale, has recovered under protection from commercial whaling (but with aboriginal catches up to some 150 per year) to at or near its "original" (pre-whaling) state, at about 26,000 animals. Similarly, the Northwest Atlantic humpback and several Southern Hemisphere humpback populations have been showing marked increase. The latest estimate of the North Atlantic stock, some 11,600 animals in 1992–1993, reflects substantial recovery since protection in the 1950s, while several Southern Hemisphere stocks (off southeastern Africa, eastern and western Australia) have been increasing, off Australia at around 10% per year since the early 1980s. Three southern right whale stocks (off eastern South America, South Africa, and southern Australia) have been increasing since the late 1970s at around 7% per year, although all are still well below their "original" stock size. Even the Antarctic blue whale, whose future has been of considerable concern, with estimates for the late 1980s at fewer than 500 animals for the whole Antarctic, has shown recent encouraging signs. Based on a series of Antarctic sightings cruises—mainly for minke whales but including other large whales—the most recent calculations give a population size of some 1700 in 1996, increasing at around 7% per year. At that rate, the population size might now be some four thousand, although still only a small fraction of its original population size, recently estimated at some 240,000.

One species or stock for which there is considerable concern is the western North Atlantic right whale. At very low absolute abundance (only some 300–350 animals), and while recently showing signs of increased reproduction, it is subject to considerable threat from human-induced mortality such as ship strikes and entanglement in fishing gear.

**B**

TABLE II
IUCN Red List Categories for Baleen Whales (1996), abbreviations in text

| Species | Subspecies | Common name | EX | EW | CR | EN | VU | LR | DD | NE |
|---|---|---|---|---|---|---|---|---|---|---|
| | | | | | Threatened | | | | | |
| *B. mysticetus* | | Bowhead whale | | | | °a | °b | ° (cd)c | | |
| *E. glacialis,* *E. japonica* | | North Atlantic and North Pacific right whales | | | | ° | | | | |
| *E. australis* | | Southern right whale | | | | | | ° (cd) | | |
| *Caperea marginata*d | | Pygmy right whale | | | | | | | | |
| *E. robustus* | | Gray whale | | | | °e | | ° (cd)f | | |
| *M. novaeangliae* | | Humpback whale | | | | | ° | | | |
| *B. acutorostrata* | | North Atlantic (common) minke | | | | | | ° (nt) | | |
| | *B. a.* ?spp. | Dwarf minke | | | | | | | | |
| *B. bonaerensis* | | Antarctic minke | | | | | | °(cd) | | |
| *B. edeni* | | Bryde's whale | | | | | | | ° | |
| *B. borealis* | | Sei whale | | | | ° | | | | |
| *B. physalus* | | Fin whale | | | | ° | | | | |
| *B. musculus* | *B. m. musculus* | Blue whale | | | | | °g | °(cd)h | | |
| | *B. m. intermedia* | Antarctic blue | | | | ° | | | | |
| | *B. m. brevicauda* | Pygmy blue | | | | | | | ° | |

[a] Svalbard-Barents Sea bowhead whale; Okhotsk Sea bowhead whale.
[b] Hudson Bay-Foxe Basin bowhead whale; Davis Strait-Baffin Bay bowhead whale.
[c] Bering-Beaufort-Chuckchi Seas bowhead whale.
[d] Pygmy right whale removed from 1996 Red List.
[e] Western gray whale.
[f] Eastern gray whale.
[g] North Atlantic blue whale.
[h] North Pacific blue whale.

It has been calculated (Laws, 1977) that the great reduction of baleen whales by whaling, for the Antarctic to around one-third of original numbers and one-sixth in biomass, must have left a large surplus of food—some 150 million tonnes per year—available for other consumers such as seals, penguins, and fish. (In a different way, earlier whaling in the North Atlantic, particularly on right whales, is believed to have influenced the spread of one sea bird—the fulmar—by providing food in the form of discarded whale carcasses.) In response to an increase in available food, there may well have been increases in growth rates, earlier ages at maturity and higher rates of pregnancy in some baleen whale species. However, the evidence is equivocal, as it is for competition between individual whale species. For some, for example right whales and sei, it has been suggested that an increase in one (right whales) could be inhibited by competition with another (sei whales). In the North Pacific, both sei and right whales can feed on the same prey—copepods—and sei whales can at times be "skimming" feeders, like right whales. However, evidence that they actually compete, on the same prey, in the same area, at the same time, and even on the same prey patch, is lacking. Similarly, there has been much debate and speculation on whether the recovery of the Antarctic blue whale has been inhibited by an apparent increase in minke whales. In that case, there may in fact be very little direct competition for food where the common prey is not limited in abundance (as in the Antarctic) and is available in large patches. The recent increases at substantial annual rates for several stocks of Southern Hemisphere humpbacks and right whales, as well as the recent increase in numbers for the Antarctic blue whale (and possibly even for the fin whale), suggest that such competition is unlikely, at least where, as in the Antarctic, food supplies are abundant.

There is, however, increasing concern over the possible effects of climate change, with reductions in sea ice and rise in sea surface temperature. The former has been considered a factor in reducing the amount of krill available, and the latter has been shown to affect reproductive capacity—elevated sea surface temperature off South Georgia has recently been found (Leaper *et al.*, 2006) to affect conception and subsequent calving rates in the South American population of southern right whales.

## See Also the Following Articles

Toothed Whales, Overview ■ Krill and Other Plankton ■ Conservation Efforts

## References

Bonner, W. N. (1980). "Whales." The Blandford Press, Poole, Dorset, 278 pp.

IUCN (1996). "1996 IUCN Red List of Threatened Animals." IUCN, Gland, Switzerland, 368 pp. plus annexes.

Laws, R. M. (1977). Seals and whales of the Southern Ocean. *Phil. Trans. R. Soc. Lond. B* **279**, 81–96.

Leaper, R., Cooke, J., Trathan, P., Reid, K., Rowntree, V., and Payne, R. (2006). Global climate change drives southern right whale (*Eubalaena australis*) population dynamics. *Biol. Lett.* **2**, 289–292.

Mackintosh, N. A. (1965). "The Stocks of Whales." Fishing News (Books) Ltd, London, 232 pp..

Rice, D. W. (1998). *Marine mammals of the world: Systematics and distribution.* Special Publication Number 4, the Society for Marine Mammalogy. Lawrence, Kansas. 231 pp.

Rychel, A. L., Reeder, T. W., and Berta, A. (2004). Phylogeny of mysticete whales based on mitochondrial and nuclear data. *Mol. Phylogenet. Evol.* **32**, 892–901.

Sasaki, T., Nikaido, M., Wada, S., Yamada, T. K., Cao, Y., Hasegawa, M., and Okada, N. (2006). *Balaenoptera omurai* is a newly discovered baleen whale that represents an ancient evolutionary lineage. *Mol. Phylogenet. Evol.* **41**, 40–52.

# Barnacles

## Dagmar Fertl and William A. Newman

"Barnacle" is the common name for over 1000 marine species of the subclass Cirripedia. Barnacles are unique among crustaceans in being permanently attached as adults to a variety of inanimate and animate objects. They occur in polar, tropical, and temperate waters, being found from high on the shore to the depths of the ocean. The principal superorder is Thoracica, consisting of stalked (order Pedunculata) and sessile (order Sessilia) barnacles (Newman, 1996). Perhaps as many as 20 living barnacle species have some association with marine mammal species, primarily cetaceans (Newman and Ross, 1976). Barnacles attached to marine mammals are often referred to as ectoparasites; however, in actuality, they do not feed on their hosts, but use them as a moving substratum from which they can strain plankton from the passing water. As a result, "epizooitic" is often considered a more appropriate term describing the barnacle's lifestyle. This has been described as an example of symbiosis, usually commensalism, but barnacles create drag and can cause irritations. Therefore, they are perhaps best termed "semiparasitic," since they survive and perpetuate themselves at the host's expense. On the other hand, some marine mammals eat barnacles or their larvae.

## I. Life History

Barnacles were described by Louis Agassiz and T. H. Huxley as nothing more than "a little shrimp-like animal, standing on its head in a limestone house and kicking food into its mouth" (Hoover, 2006). The barnacle's life cycle usually includes six free-swimming planktonic naupliar stages that feed while progressing by molts to the cypris or cyprid stage, which searches for a place to settle. When settling, to anchor itself, the cyprid secretes cement from its antennules, from glands located in their base, and metamorphoses by molting into a juvenile, which begins to secrete adult cement and the calcareous plates that usually constitute its home. In the case of barnacles that attach directly to cetacean skin, a chemical cue from the host tissue likely induces larval settlement (Nogata and Matsumura, 2006).

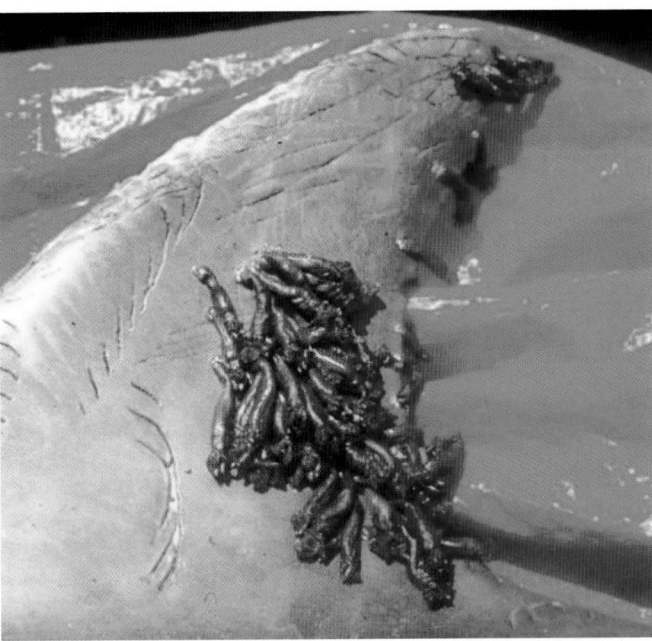

**Figure 1** *The pseudo-stalked sessile barnacle* Xenobalanus *attached to the dorsal fin of a bottlenose dolphin. Courtesy of V. Thayer and K. Rittmaster, North Carolina Maritime Museum.*

Through an opening between the plates, six pairs of feathery thoracic limbs (cirri) can emerge and spread out to sweep through the water like a net to entrap planktonic organisms. Most barnacles are hermaphrodites (i.e., individuals possess the reproductive structures of both sexes). The breeding season of barnacles that cling to whales is probably largely synchronous with that of the whales' breeding season.

## II. Sessile Barnacles

The Sessilia, or sessile barnacles, are stalkless, the usually well-articulated shell wall being attached directly to the substratum. Because of their superficial resemblance to acorns of oak trees, they are called acorn barnacles. Marine mammals host species of *Amphibalanus*, *Balanus*, *Cetopirus*, *Chelonibia*, *Coronula*, *Cryptolepas*, *Platylepas*, *Tubicinella*, and *Xenobalanus*. *Xenobalanus* superficially resembles a stalked barnacle since it has developed an aberrant pseudo-stalk, but it is nonetheless a sessile barnacle (Fig. 1).

## III. Stalked Barnacles

The pedunculate, or stalked, barnacles are more primitive than the sessile barnacles. The terminal sac housing the appendages is called the capitulum. It is supported by a flexible, muscular stalk or peduncle attached to the substratum. The capitulum is usually protected by shell plates. *Conchoderma* spp., the goose barnacles (*Lepas* spp.), and rarely, the leaf barnacles (*Pollicipes* spp.), attach to marine mammals.

## IV. Barnacles and Marine Mammals

Barnacles appear to settle in greatest numbers on large baleen whales, in contrast to toothed whales. Striped dolphins (*Stenella*

**B**

*coeruleoalba*) involved in a mass mortality event in the Mediterranean had an unusual abundance of barnacles likely due to the reduced movement and/or an impaired immune function of the skin of sick individuals (Aznar *et al.*, 2005). Orams and Schuetze (1998) demonstrated that *Xenobalanus* spp. were more prevalent on young than adult bottlenose dolphins (*Tursiops* spp.), presumably because they are less active and/or less resistant.

*Cryptolepas rhachianecti*, considered to be host-specific to the gray whale (*Eschrichtius robustus*), has been found on a killer whale (*Orcinus orca*) stranded in southern California and on belugas (*Delphinapterus leucas*) housed in San Diego Bay. *Xenobalanus globicipitus*, while world-wide in distribution, are almost always found on the trailing edges of the dorsal and pectoral fins and on the flukes of at least 27 cetacean species (Kane *et al.*, 2006; Fig. 1). What may remain of their wall in the skin of the host after death superficially resembles the

**Figure 2** *Humpback whale with the acorn barnacle* Coronula diadema *and a few stalked barnacles (arrows),* Conchoderma auritum *attached to them. Also visible are white-rim scars from acorn barnacles that have dropped or have been knocked off. Courtesy of Y. Ogino, off California, 1999.*

wall of platylepas; this may account for a report (Mead and Potter, 1990) of platylepas on a bottlenose dolphin. *Tubicinella major*, which lives within a columnar shell opening at the surface of its host's skin, and usually found among callosities of southern right whale (*Eubalaena australis*), was once collected from the flank of a stranded northern bottlenose whale (*Hyperoodon ampullatus*). *Coronula* spp., the most generalized of the sessile whale barnacles, are large and generally attach to the skin of baleen whales (Scarff, 1986). *C. reginae* and *C. diadema* (Fig. 2) are commonly epizooites of humpback whales (*Megaptera novaeangliae*), attached to flukes, flippers, ventral grooves, genital slit, and the head (Clarke, 1966). *Cetopirus complanatus* closely resembles *C. reginae*, and both occur on the right whale (Scarff, 1986). Humpback males scrape each other with their barnacle-encrusted flippers (analogous to "brass knuckles") on the breeding grounds; one individual caught during whaling operations was reported to have as much as 450 kg of *Coronula* removed from it. On the other hand, various forms of grooming, including flipper-body grooming (Sakai *et al.* 2006), would likely remove freshly settled larvae and young juveniles; this may account for the lack of barnacles on the bodies of most small toothed whales.

Of the pelagic pedunculate barnacles, *Conchoderma auritum* and *C. virgatum* are commonly recorded from cetaceans, though *Pollicipes polymerus*, a rocky shore barnacle, was recorded on a humpback whale (Clarke, 1966). *Conchoderma* spp. require a hard surface for attachment. *Conchoderma auritum*, identified by its' rabbit ear-like appendages, may be found at a site where teeth are exposed and unprotected (Soto, 2001), such as on erupted teeth of adult male beaked and bottlenose (*Hyperoodon* spp.) whales (Fig. 3), or because of a malformation (including bone injury) in the jaw. *Conchoderma* spp. are less commonly found on baleen plates and were once collected from the penis of a stranded sperm whale (*Physeter macrocephalus*). *C. auritum* is often found attached to *Coronula* spp. (most commonly to *C. diadema*). *C. virgatum*, although sometimes attaching directly to a cetacean, is usually epizootic on other barnacles, and then most often on *C. auritum*. *C. virgatum* has been found on the parasitic copepods *Pennella* spp. and on the cyamid amphipod, *Neocyamus physeteris*, which crawls about on cetaceans and their barnacles (Clarke, 1966; Oliver and Trilles, 2000). *Lepas* spp. usually occur on floating objects, yet *L. pectinata* and *L. hillii* have been found between the teeth of some Mediterranean striped dolphins.

**Figure 3** Conchoderma auritum *(arrow) attached to the teeth of a Blainville's beaked whale* (Mesoplodon densirostris) *off Hawaii. Photo by Alice Mackay, courtesy Cascadia Research.*

There are comparatively few published records of barnacles on pinnipeds, yet *Lepas pacifica*, *L. australis*, and *L. hillii*, as well as *Conchoderma auritum* and *C. virgatum*, are recorded from their dorsal body surface, attached to hair or even directly to the skin of various species, including both species of elephant seals (*Mirounga* spp.) (Best, 1971; Setsaas and Bester, 2006; Fig. 4).

Manatees (*Trichechus* spp.) may acquire acorn barnacles when in brackish or seawater, but when they enter fresh water the barnacles die and drop off, leaving temporary scars. The common barnacle found embedded in the skin of West Indian (*Trichechus manatus*) and West African (*T. senegalensis*) manatees is *Chelonibia manati* (Cintrón De Jesús, 2001), a close relative of it's congeners on turtles. Moreover, turtle barnacles *Platylepas hexastylos* and *P. decorata* have been found on the dugong (*Dugong dugon*) and West Indian manatee. The brackish water species, *Amphibalanus amphitrite*, *A. eburneus*, *A. reticulatus*, and *A. improvisus*, and the marine species, *Balanus trigonus*, attach to the *Chelonibia* spp. on the manatees, rather than to their skin.

It is not surprising that some baleen whales eat barnacle larvae (Mayo and Marx, 1990) since the experimentally estimated filtering efficiency of 95% for plankton larger than 333 μm for the right whale (Mayo *et al.*, 2001) would include the larvae of pelagic and some coastal barnacles.

Sea otters (*Enhydra lutris*) in California and Alaska will eat the large acorn barnacles *Balanus nubilus* and *Semibalanus cariosus*. Faurot *et al.* (1986) reported otters feeding on *Pollicipes polymerus*, suggesting that they may be intentionally ingesting it if not simply being incidental to their take of mussels.

## See Also the Following Articles

Callosities ■ Health ■ Parasites ■ Plankton

## References

Aznar, F. J., Perdiguero, D., Del Olmo, A. P., Repulles, A., Agusti, C., and Raga, J. A. (2005). Changes in epizoic crustacean infestations during cetacean die-offs: The mass mortality of Mediterranean striped dolphins *Stenella coeruleoalba* revisited. *Dis. Aquat. Org.* **67**, 239–247.

Best, P. B. (1971). Stalked barnacles *Conchoderma auritum* on an elephant seal: Occurrence of elephant seals on South African coast. *Zool. Afr.* **6**, 181–185.

Cintrón de Jesús, J. (2001). Barnacles associated with marine vertebrates in Puerto Rico and Florida. Master's thesis, University of Puerto Rico.

Clarke, R. (1966). The stalked barnacle *Conchoderma*, ectoparasitic on whales. *Nor. Hval.-tid.* **55**, 153–168.

Faurot, E. R., Ames, J. A., and Costa, D. P. (1986). Analysis of sea otter, *Enhydra lutris*, scats collected from a California haulout site. *Mar. Mamm. Sci.* **2**, 223–227.

Hoover, L. P. (2006). "Hawai'i's Sea Creatures," 3rd edn. Mutual Publishing, Honolulu, Hawai'i.

Kane, E., Olson, P., and Gerrodette, T. (2006). The commensal barnacle *Xenobalanus globicipitis* Steenstrup, 1851 (Crustacea: Cirripedia) and its relationship to cetaceans of the eastern tropical Pacific. NMFS-SWFSC Administrative Report LJ-06-03.

Mayo, C. A., and Marx, M.K. (1990). Surface foraging behavior of the North Atlantic right whale,, *Eubalaena glacialis*, and associated zooplankton characteristics. *Can. J 2001.*68, 2214–2220.

Mayo, C. A., Letcher, B. H., and Scott, S. (2001). Zooplankton filtering efficiency of the baleen of a North Atlantic right whale, *Eubalaena glacialis*. *J. Cet. Res. Manag.* (Spec. Issue 2), 225–230.

Mead, J. G., and Potter, C.W. (1990). Natural history of bottlenose dolphins along the central Atlantic Coast of the United States. *In* "The Bottlenose Dolphin" (S. Leatnerwood, and R. R. Reeves, eds), pp. 165–195. Academic Press, San Diego.

Newman, W. A. (1996). Cirripedia; suborders Thoracica and Acrothoracica. *In* "Traité de Zoologie Tome VII, Crustacés, Fascicule" (J. Forest, ed.), Vol. 2, pp. 453–540. Masson, Paris (in French).

Newman, W. A., and Ross, A. (1976). Revision of the balanomorph barnacles; including a catalog of the species. Memoirs of the San Diego Society of Natural History, No. 9.

Nogata, Y., and Matsumura, K. (2006). Larval development and settlement of a whale barnacle. *Biol. Lett.* **2**, 92–93.

Oliver, G., and Trilles, J. P. (2000). Crustacean parasites and epizoa of the sperm whale, *Physeter catodon* Linnaeus, 1758 (cetacea, Odontoceti), in the Gulf of Lion (Western Mediterranean). *Parasite* **7**, 311–321.

Orams, M. B., and Schuetze, C. (1998). Seasonal and age/size-related occurrence of a barnacle (*Xenobalanus globicipitis*) on bottlenose dolphins (*Tursiops truncatus*). *Mar. Mamm. Sci.* **14**, 186–189.

Sakai, M., Hishii, T., Takeda, T. S., and Kohsima, S. (2006). Flipper rubbing behaviors in wild bottlenose dolphins (Tursiops aduncus). *Mar. Mamm. Sci.* 22, 966–978.

Scarff, J. E. (1986). Occurrence of the barnacles *Coronula diadema*, *C. reginae* and *Cetopirus complanatus* (Cirripedia) on right whales. *Sci. Rep. Whal. Res. Inst., Tokyo* **37**, 129–153.

Setsaas, T. H., and Bester, M. N. (2006). Goose barnacle (*Lepas australis*) infestation of the Subantarctic fur seal (*Arctocephalus tropicalis*). *Afr. Zool.* **41**, 305–307.

Soto, J. M. R. (2001). First record of a rabbit-eared barnacle, *Conchoderma auritum* (Linnaeus, 1767) (Crustacea, Cirripedia), on the teeth of the La Plata dolphin, *Pontoporia blainvillei* (Gervais & D'Orbigny, 1844) (Cetacea, Platanistoidea). *Mare Magnum* **1**, 172–173.

**Figure 4** *Goose barnacle* (Lepas australis) *attached among the hairs of a Subantarctic fur seal* (Arctocephalus tropicalis). *Courtesy of M. N. Bester.*

# Basilosaurids

## Mark D. Uhen

Basilosaurids are a paraphyletic group of archaeocete cetaceans known from the late middle to early late Eocene of all continents except Antarctica. The family includes 11 species in 8 genera in 2 subfamilies, although some authors elevate the subfamilies

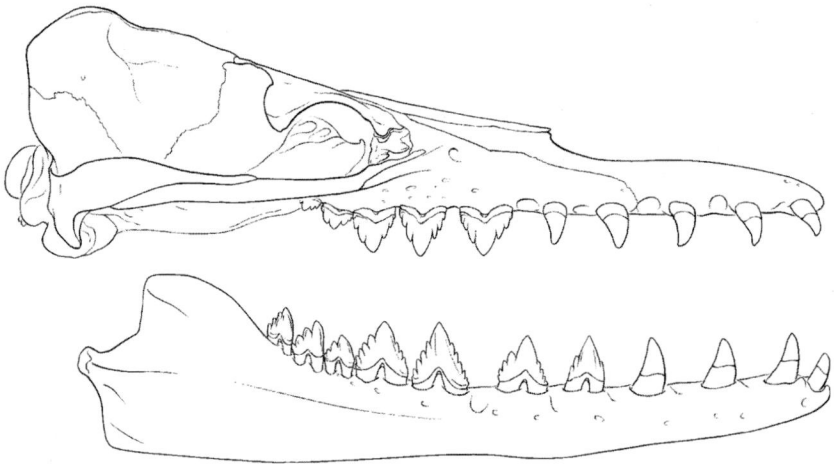

**Figure 1** *Skull and lower jaw of* Dorudon atrox, *lateral view. This drawing is a composite drawn from specimens of* D. atrox *at the University of Michigan Museum of Paleontology by Bonnie Miljour.*

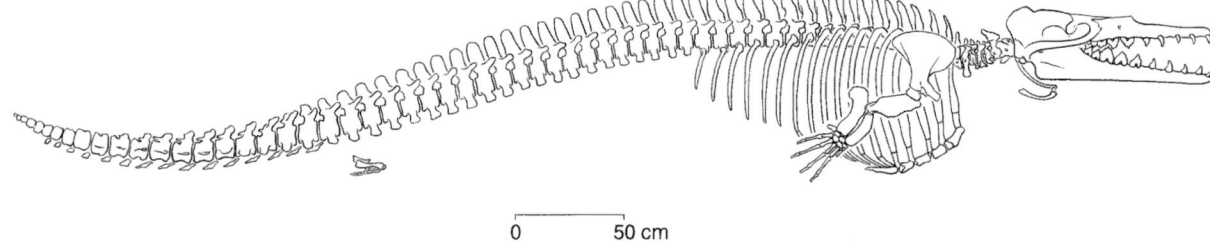

**Figure 2** *Skeleton of* Dorudon atrox, *lateral view. This drawing is a composite drawn from specimens of* D. atrox *at the University of Michigan Museum of Paleontology by Bonnie Miljour.*

to familial rank. They range in size from around 4 m (*Saghacetus osiris*) to around 16 m (*Basilosaurus cetoides*). Basilosaurids are probably the earliest fully aquatic cetaceans (Uhen, 1998) and are thought to have given rise to modern cetaceans (Barnes *et al.*, 1985; Uhen, 1998).

## I. Basilosaurid Characteristics

Like all archaeocetes, basilosaurids lack telescoping of the skull like that seen in modern mysticetes or like that seen in modern odontocetes (see Fig. 1; Miller, 1923). In addition, basilosaurids are diphyodont (have two tooth generations: milk and adult teeth), lack polydonty (11 or fewer teeth per jaw half), and retain a heterodont dentition, in which incisors, canines, premolars, and molars are easy to distinguish based on their morphologies (Kellogg, 1936; Uhen, 1998).

Basilosaurids also share a number of characteristics that distinguish them from other archaeocetes. All basilosaurids lack upper third molars, and the upper molars lack protocones, trigon basins, and lingual third roots. In addition, the cheek teeth of basilosaurids have well-developed accessory denticles on the cheek teeth (Fig. 1). The hind limbs of basilosaurids are greatly reduced (see Fig. 2; Gingerich *et al.*, 1990; Uhen and Gingerich, 2001) and lack a bony connection to the vertebral column. Basilosaurids also lack sacral vertebrae, although vertebrae that are likely to be homologs of sacral vertebrae are identifiable (Kellogg, 1936; Uhen, 1998).

Other characteristics may be found only in basilosaurids (within archaeocetes) but are currently not known from other archaeocetes. For instance, basilosaurid forelimbs had broad, fan-shaped scapulae with the distal humerus, radius, and ulna flattened into a single plane (Fig. 2). In addition, the elbow joint motion was restricted to the same plane, and pronation and supination of the forelimb was not possible based on the articular surfaces of the distal humerus, proximal radius, and proximal ulna. Since forelimbs are poorly known in more derived protocetids, it is unclear whether these features are found only in basilosaurids, or whether they are characteristic of a larger group.

Some of the characteristics of basilosaurids can be seen in some protocetid archaeocetes, like *Georgiacetus*. Although the innominate of *Georgiacetus* is large, it does not appear to have been connected to the vertebral column. None of the vertebrae is fused into a sacrum, yielding a condition similar to that seen in basilosaurids. In addition, the cheek teeth of *Georgiacetus* have small accessory denticles, somewhat different from those in basilosaurids, but certainly larger than any of the serrations seen in other non-basilosaurid archaeocetes.

## II. Taxonomy

Taxonomy for the family Basilosauridae is after Uhen (1998). Basilosaurinae and Dorudontinae are included here in the single family Basilosauridae following Miller (1923) because a single character state (elongate trunk vertebrae) distinguishes basilosaurines from dorudontines (Uhen, 1998). The names "zeuglodonts" and "zeuglodons" are often colloquially used to refer to basilosaurids or archaeocetes in general. These common names are derived from the disused generic name *Zeuglodon* (discussed below), and their usage should be avoided.

## A. Basilosauridae Cope 1868

*1. Basilosaurinae Cope 1868* Basilosaurines are basilosaurids with elongate posterior thoracic, lumbar, and anterior caudal vertebrae. All known basilosaurines are also considerably larger than all known dorudontines, with the exception of *Cynthiacetus* (see below).

A. BASILOSAURUS Harlan 1834 *Basilosaurus* was the first archaeocete whale named. The name was first coined in 1834 by Richard Harlan, who mistook the large vertebrae for those of a large marine reptile, thus the misnomer of "king lizard" for a cetacean. This mistake was pointed out by Richard Owen in 1842, when he attempted to rename the animal *Zeuglodon cetoides*. *Zeuglodon* is considered a junior subjective synonym of *Basilosaurus*, but it has been applied to so many archaeocete whales as to have become a common name for archaeocete or basilosaurid. *Basilosaurus* contains three species from the late middle and early late Eocene: *B. cetoides* is known from the southeastern United States, *B. isis* is known from Egypt and Jordan, and *B. hussaini* is known from Pakistan (Uhen, 1998).

B. BASILOTERUS Gingerich *et al.* 1997 *Basiloterus* can be distinguished from *Basilosaurus* based on its anteroposteriorly long neural arch and more anteriorly projecting metapophyses. *Basiloterus* contains a single species, *B. drazindai*, based on a single lumbar vertebra from the late middle Eocene of Pakistan.

*2. Dorudontinae Miller 1923*

A. DORUDON Gibbes 1845 The genus *Dorudon* was erected in 1845 by Gibbes for a specimen of a small archaeocete that he dubbed *D. serratus*. This specimen is of a juvenile individual with deciduous teeth, making it difficult to compare to other specimens of adult individuals. Nonetheless, the number of species in *Dorudon* grew when Kellogg (1936) removed a number of species from the genus *Zeuglodon* and placed them in *Dorudon*. Subsequently, many of these species were synonymized and/or placed in other genera. Only *D. serratus* from the late Eocene of the southeastern United States, and *D. atrox* (formerly *Prozeuglodon atrox*) from the late Eocene of Egypt remain (Uhen, 1998).

B. ZYGORHIZA True 1908 The genus *Zygorhiza* was erected in 1908 by True for specimens of a small archaeocete from North America that he felt were different from *Dorudon serratus*. Some of these specimens had been part of Koch's *Hydrarchos* and had been called by many different names (Kellogg, 1936). *Z. kochii* can be distinguished from all other dorudontines by the presence of well-developed cuspules on the cingula of the upper premolars. *Zygorhiza* currently includes *Z. kochii* from the late Eocene of the southeastern United States and *Zygorhiza* sp. from New Zealand, as European specimens assigned to *Zygorhiza* are more appropriately identified as Dorudontinae incertae sedis (Uhen, 1998).

C. CYNTHIACETUS Uhen 2005 The genus *Cynthiacetus* was erected by Uhen in 2005 for a large dorudontine that had a skull similar in size and morphology to the skull of *Basilosaurus*, but which lacked the elongate trunk vertebrae of *Basilosaurus*. Previously, animals such as this had often been called *Pontogeneus*, which Uhen designated a nomen dubium. Some of the vertebrae of Koch's *Hydrarchos* were large in size, but not elongate like those of *Basilosaurus*, and were given the name *Zeuglodon brachyspondylus*. Kellogg (1936) suggested that *Zeuglodon brachyspondylus* and Leidy's *Pontogeneus priscus* were the same, and used Leidy's generic name and the specific epithet *brachyspondylus* for the new combination. Neither of these previously named taxa is based on diagnostic type specimens. *Cynthiacetus* is found in both the late Eocene of the southeastern United States and Egypt.

D. SAGHACETUS Gingerich 1992 The generic name *Saghacetus* was coined in 1992 to subsume the former species *Dorudon osiris*, *Dorudon zitteli*, *Dorudon sensitivius*, and *Dorudon elliotsmithii* within a single species, *Saghacetus osiris*. *S. osiris* can be distinguished from other dorudontines based on its small size, and its slightly elongate lumbar and anterior caudal vertebrae. *S. osiris* is known only from the late Eocene of Egypt.

E. ANCALECETUS Gingerich and Uhen 1996 *Ancalecetus* includes one species, *A. simonsi*, which is similar to *Dorudon atrox* but has greatly modified forelimbs that were highly restricted in their range of motion. *A. simonsi* is known from the late Eocene of Egypt (Uhen, 1998).

F. CHRYSOCETUS Uhen and Gingerich 2001 *Chrysocetus* includes one species, *C. healyorum*, which differs from all other dorudontines in the smoothness of the tooth enamel, height of the premolar crowns, and the eruption of its adult teeth in a skeletally juvenile state. *Chrysocetus* is also the only dorudontine for which the innominate is known. *Chrysocetus* is known from the late Eocene of South Carolina (Uhen and Gingerich, 2001).

### B. Questionable Basilosaurids

Excluded from this list is the genus *Gaviacetus*, which was referred to the Basilosauridae by Bajpai and Thewissen (1998). The identification of *Gaviacetus* as a basilosaurid was based on the likely absence of upper third molars in both the type specimen of *Gaviacetus razai* and the type specimen of *Gaviacetus sahnii* (Bajpai and Thewissen, 1998). In addition, Bajpai and Thewissen (1998) referred some elongate vertebrae to *G. sahnii* further supporting their placement of *Gaviacetus* in the Basilosauridae. Since no specimen of *Gaviacetus* clearly shows that the upper third molar is absent, or that any of the cheek teeth have accessory denticles, and since reference of postcrania to unassociated cranial material has proven problematic in the past, I prefer to leave *Gaviacetus* in the Protocetidae as it was originally described until it can be clearly shown to have basilosaurid synapomorphies.

Species that may not be basilosaurids are *Basilosaurus hussaini* and *Basiloterus drazindai*. These species (as well as the genus *Basiloterus*) are based solely on one and two vertebrae respectively. These vertebrae are thought to represent basilosaurines because they are elongate, like the vertebrae of *Basilosaurus*. Although this feature is a distinguishing characteristic of Basilosaurinae within Basilosauridae, it is clear that vertebral elongation is not restricted to basilosaurids. *Eocetus*, a protocetid from Egypt and North America, also has elongate vertebrae, although they are not as elongate as those of *Basilosaurus*. It is possible that *B. hussaini* and *B. drazindai* are also protocetids. Once cranial or dental material associated with vertebrae is found, it will be obvious whether these taxa should be retained in Basilosauridae.

### III. Phylogenetic Relationships

The phylogenetic relationships among basilosaurids, and their relationships to other archaeocetes, mysticetes, and odontocetes are shown in Fig. 3. Many of the character state transformations that occur between basilosaurids and protocetids are associated with the adoption of a fully aquatic existence; such as presence of pterygoid air sinuses, extreme reduction of the hind limb, loss of the sacrum, increase in the number of trunk vertebrae, and the presence of dorsoventrally flattened posterior caudal vertebrae (Uhen, 1998). Other, such as the loss of $M^3$, loss of lingual roots on the upper molars, and the development of accessory denticles on the cheek teeth, have to do with changes in feeding that are not as easy to interpret.

B

**Figure 3** *Cladogram of basilosaurids, selected non-basilosaurid archaeocetes, mysticetes, and odontocetes. Mysticetes and odontocetes not included in Durodontinae.*

Within Basilosauridae, basilosaurines are united by the presence of elongate trunk vertebrae, which dorudontines lack. *Pontogeneus* may be the sister taxon to Basilosaurinae based on its large size. Each genus of dorudontine is distinguishable from the other genera based on the presence of autapomorphies, but it is difficult to confidently link any of the genera based on any clear synapomorphies. The result is a polytomous relationship among the genera or an imbalanced tree with Mysticeti + Odontoceti nested well within Dorudontinae. *Chrysocetus* is preferred as the sister taxon to Mysticeti + Odontoceti based on based on the interpretation of it and early mysticetes and odontocetes as monophyodont. Hopefully, some of the relationships among basilosaurids will become more secure as more of the anatomy of more of the species becomes known.

### See Also the Following Articles

Archaeocetes, Archaic ■ Cetacean Evolution

### References

Bajpai, S., and Thewissen, J. G. M. (1998). Middle Eocene cetaceans from the Harudi and Subathu Formations of India. *In* "The Emergence of Whales" (J. G. M. Thewissen, ed.), pp. 213–234. Plenum Press, New York.

Barnes, L. G. (1985). Evolution, taxonomy and antitropical distributions of the porpoises (Phocoenidae, Mammalia). *Mar. Mamm, Sci.* 1, 149–165.

Gingerich, P. D., Smith, B. H., and Simons, E. L. (1990). Hind limbs of Eocene *Basilosaurus*: Evidence of feet in whales. *Science* **249**, 154–157.

Kellogg, R. (1936). "A Review of the Archacoceti" **482**. Carnegie Institution of Washington Special Publication, 1–366.

Miller, G. S., Jr. (1923). The telescoping of the cetacean skull. *Smiths. Misc. Coll.* **76**(5), 1–71.

Uhen, M. D. (1998). Middle to late Eocene Basilosaurines and Dorudontines. *In* "The Emergence of Whales" (J. G. M. Thewissen, ed.), pp. 29–61. Plenum Press, New York.

Uhen, M. D., and Gingerich, P. D. (2001). New genus of dorudontine archaeocete (Cetacea) from the middle-to-late Eocene of South Carolina. *Mar. Mamm. Sci.* **17**(1), 1–34.

# Beaked Whales, Overview
## *Ziphiidae*

### James G. Mead

Beaked whales belong to the odontocete family Ziphiidae. They are medium-sized cetaceans, adults ranging from 3 to 13 m. They are characterized by a reduced dentition, elongate rostrum, accentuated cranial vertex and enlarged pterygoid sinuses. There are currently 21 recognized species in 5 genera. They are all pelagic, living in the open oceans and feeding on deep-water squid and fish.

### I. Classification and Nomenclature

| | |
|---|---|
| Family **Ziphiidae** | |
| Subfamily **Ziphiinae** | |
| *Berardius arnuxii* | Arnoux's beaked whale |
| *Berardius bairdii* | Baird's beaked whale |
| *Tasmacetus shepherdi* | Shepherd's beaked whale |
| *Ziphius cavirostris* | Cuvier's beaked whale |
| Subfamily **Hyperoodontinae** | |
| *Hyperoodon ampullatus* | Northern bottlenose whale |
| *Hyperoodon planifrons* | Southern bottlenose whale |
| *Indopacetus pacificus* | Longman's beaked whale |
| *Mesoplodon bidens* | Sowerby's beaked whale |
| *Mesoplodon bowdoini* | Andrews' beaked whale |
| *Mesoplodon carlhubbsi* | Hubbs' beaked whale |
| *Mesoplodon densirostris* | Blainville's beaked whale |
| *Mesoplodon europaeus* | Gervais' beaked whale |
| *Mesoplodon ginkgodens* | Ginkgotoothed beaked whale |
| *Mesoplodon grayi* | Gray's beaked whale |
| *Mesoplodon hectori* | Hector's beaked whale |
| *Mesoplodon layardii* | Straptoothed whale |
| *Mesoplodon mirus* | True's beaked whale |
| *Mesoplodon perrini* | Perrin's beaked whale |
| *Mesoplodon peruvianus* | Peruvian beaked whale |
| *Mesoplodon stejnegeri* | Stejneger's beaked whale |
| *Mesoplodon traversii* | Spade-toothed whale |

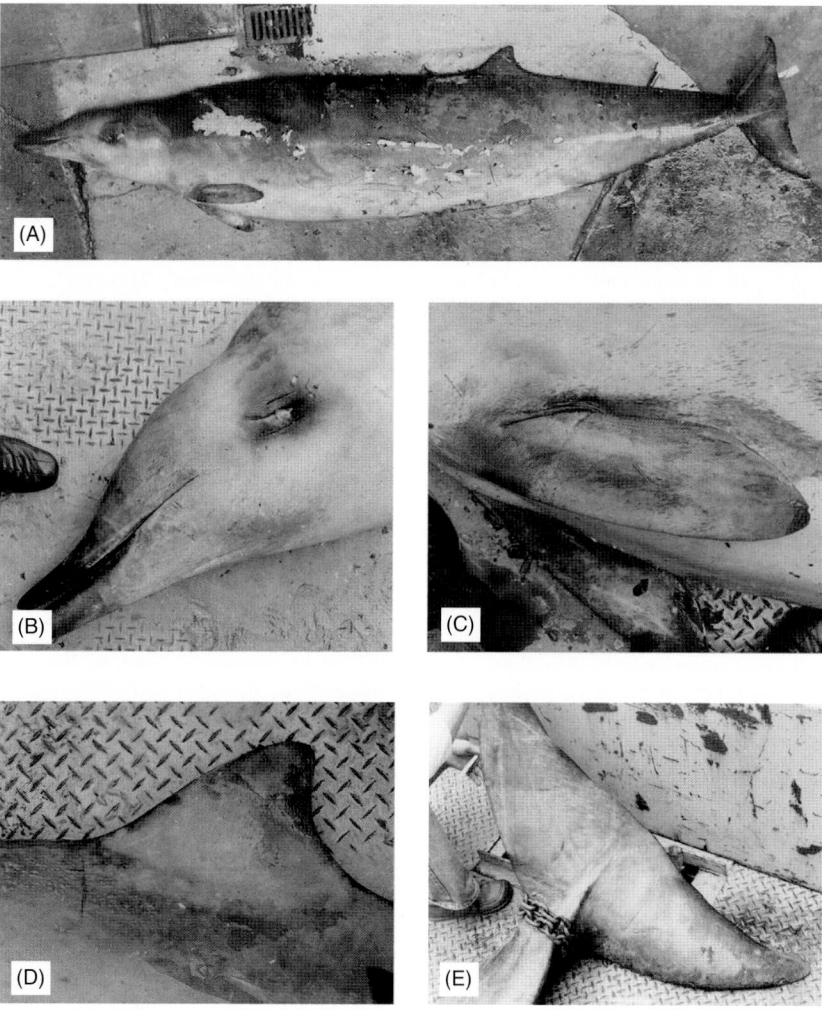

**Figure 1** *Details of the external morphology of an adult male* Mesoplodon mirus *(USNM 504612). (A) Lateral view of the whole animal; (B) lateral view of head; (C) lateral view of flipper; (D) lateral view of dorsal fin; (E) oblique ventral view of flukes.*

The concept of the beaked whales as a separate group of cetaceans became common in the 1860s and 1870s as Gray uses the family Ziphiidae in his *Catalogue of Seals and Whales in the British Museum* (1866) as do Van Beneden and Gervais, in their epic *Ostéographie des Cétacés* (1868–1879). True (1910) studied ziphiid systematics.

The common name of the family, beaked whales refers to their pronounced rostrum or beak. The rostrum of beaked whales is, admittedly, relatively shorter than in most dolphins but relatively longer than most "whales." Most beaked whales are encountered rarely enough that they do not have "common names" but rather "vernacular names," that were coined by scientists.

The only beaked whales that are seen on a regular basis by fishermen (and whalers) are the northern bottlenose whale (*Hyperoodon ampullatus*) and Baird's beaked whale (*Berardius bairdii*). The English name "bottlenose whale" was actually in common use as were the Norwegian name *nebhval* or *naebhval* and the Danish and German name *dögling* or their derivatives in other northern European languages. The name *tsuchi-kujira* or just *tsuchi* is the Japanese common name for Baird's beaked whale (*B. bairdii*).

## II. Anatomy

Living beaked whales are characterized externally by a pronounced rostrum (beak) which blends into a high forehead (or melon) without a break (Fig. 1); a pair of throat grooves; relatively small flippers with short fingers and relative long arm bones; small triangular dorsal fin that is placed far back on the body; and lack of fluke notches. Internally they have a reduction in teeth; fusion of the bones of the rostrum and development of extremely dense rostral elements in males; expansion of the pterygoid air sinus and elimination of its lateral bony wall; and elevation of the bones associated with the nose into a bony protuberance called the vertex (Fig. 2).

Several similarities between beaked whales and sperm whales became evident early. Partly these were due to retention of ancestral characters and partly due to similarities in ecology. Both groups of whales feed at considerable depth and are specialized to feed on squid.

Ziphiids in general have reduced their teeth to the point that teeth in the upper jaw are vestigial or absent and teeth in the lower jaw are reduced to one or two pairs that usually erupt only in adult males. The only exception to this is Shepherd's beaked whale

**B**

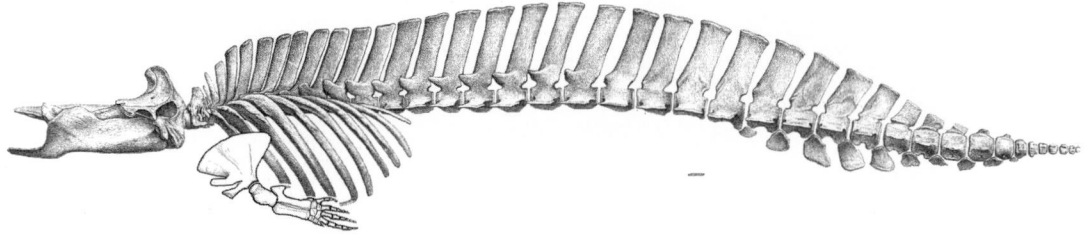

Figure 2    *Skeleton of an adult male* Mesoplodon densirostris *in the Australian Museum, Sydney (after Van Beneden and Gervais, 1868-79:Pl. XXII, Figure 9), Forelimb and pelvic rudiment are from an adult male of the same species in the American Museum of Natural History (after Raven 1942).*

(*Tasmacetus shepherdi*) which has a full dentition in both jaws (see account under *Tasmacetus* for illustration).

The pronounced rostrum results from an anterior extension of the rostral and palatal elements of the skull, the maxilla, premaxilla, and vomer, coupled with a lateral compression to form a beak. Normally these bones are moderately extended in cetaceans to form pincer-like beak, and, in fact, the relative length of the rostra of some of the toothed whales, like the river dolphins, exceeds that of beaked whales.

Beaked whales have a high forehead, which sets off the long rostrum. This forehead is composed of the soft tissue, which forms the facial apparatus and the elevated cranium on which it rests. This soft tissue is responsible for sealing the nasal passages against water and modifying the emitted sound. The blowhole is crescent-shaped with the horns of the crescent pointing anteriorly, except in the genus *Berardius* where they point posteriorly. The forehead merges with the rostrum without a break or groove that is characteristic of other toothed whales, except the rough toothed dolphin (*Steno bredanensis*) which is similar to ziphiids (Fig. 1B).

Beaked whales have a pair of throat grooves which are in the shape of a "v" with its apex pointing forward. The anterior end of the throat grooves lies posterior to the symphysis of the lower jaws and anterior to the jaw joint (i.e., in the throat region). Throat grooves are present in gray whales and sperm whales but absent in all other species. They are not to be confused with the ventral grooves or pleats, which are much longer, stretching from the tip of the jaw back to the umbilicus in rorquals.

Beaked whales have relatively small, unspecialized flippers. They consist of a relatively large forearm (radius and ulna) portion followed by a short phalangeal (finger) portion (Fig. 1C). This also occurs in porpoises (Phocoenidae) and rorquals and appears to be a primitive cetacean character. The dorsal fin of beaked whales is small and triangular, not falcate. The dorsal fin is located on the posterior third of the body, usually over the anus at the junction of the abdomen and tail. The position of the dorsal fin in beaked whales correlates with a relatively long thorax and abdomen and short tail (Fig. 1D).

Beaked whales normally do not have fluke notches and the trailing edge of the flukes is unbroken. Embryologically the fluke notch is formed when the trailing edge of the flukes moves back beyond the end of the caudal vertebrae. The caudal vertebrae anchor the midline of the trailing edge resulting in a notch (Fig. 1E).

The reduction in teeth has proceeded to the point where all function teeth are lost in females and immature males and the dentition is only represented by a single pair of teeth in the lower jaw of males. Females and immature males have a pair of vestigial teeth. The dentition is apparently only used in male, intraspecific, aggression. The two exceptions to this are the genus *Berardius*, which has two pairs of mandibular teeth and *Tasmacetus*, which has a normal odontocete

dentition in both the upper and lower jaws. In *Tasmacetus*, the apical pair of mandibular teeth is enlarged, which suggests that the single pair of teeth in all other ziphiids represents the apical pair. A row of vestigial teeth is sometimes present in the gums of both the upper and lower jaws of some beaked whales, particularly *Mesoplodon grayi* and *Ziphius*.

Fusion of the bones of the rostrum in some males takes place with increasing age. As part of the fusion, the mesorostral canal is filled in by dorsal expansion of the vomer and the individual rostral elements fuse together. This is accompanied by an increase in density of the rostrum. The density of the core of the rostrum has been measured at 2.4 gm/cc in a male of *Mesoplodon carlhubbsi* and 2.6 gm/cc in a male *M. densirostris*.

The pterygoid air sinus is enlarged in the ziphiids, but remains confined to the pterygoid bone and lost its lateral wall. The anterior sinus is not developed in ziphiids. The vertex of the skull has been expanded both laterally and vertically beyond what occurs in all other odontocetes. The vertex is composed of the posterodorsal ends of the maxilla and premaxilla, the nasals and the medial ends of the frontals. The dorsal tip of the vertex has expanded laterally and anteriorly like a mushroom. This region is deeply involved with sound production and modification.

## III. Fossil Record

Ziphiids first appeared in the fossil record in the early Miocene (Muizon, 1991). These early ziphiids had long rostra, full dentitions with the first mandibular tooth often hypertrophied, an elevated synvertex with a premaxillary crest, strong development of the pterygoid sinus with reduction of the lateral wall of the pterygoid and increase in the hamular process, and an auditory region of the skull that has minimal fenestration.

By the middle Miocene fossil ziphiids were abundant. This is a period of maximum diversity of the entire order Cetacea and certainly was for the ziphiids. It is unclear how the Miocene genera relate to the modern genera. There are about 14 genera of fossils currently recognized as ziphiids. Of these 14 genera there are at least 28 species that are based solely on rostral fragments. Critical work has demonstrated that two genera are based upon non-diagnostic fragments and have been regarded as nomina dubia. With further study, particularly of the genera that are based on rostral fragments, there is bound to be a lot of demonstrated synonymy.

Muizon (1991) classified modern and fossil Ziphiidae into three subfamilies (1) the Hyperoodontinae, which contains *Hyperoodon* and *Mesoplodon*; (2) the Ziphiinae, which contains *Ziphius*, *Berardius*, *Tasmacetus* and the fossil genera *Choneziphius*, *Ziphirostrum*, *Cetorhynchus*, and *Ninoziphius*; and (3) the Squaloziphiinae, which currently contains only *Squaloziphius* (Fig. 3).

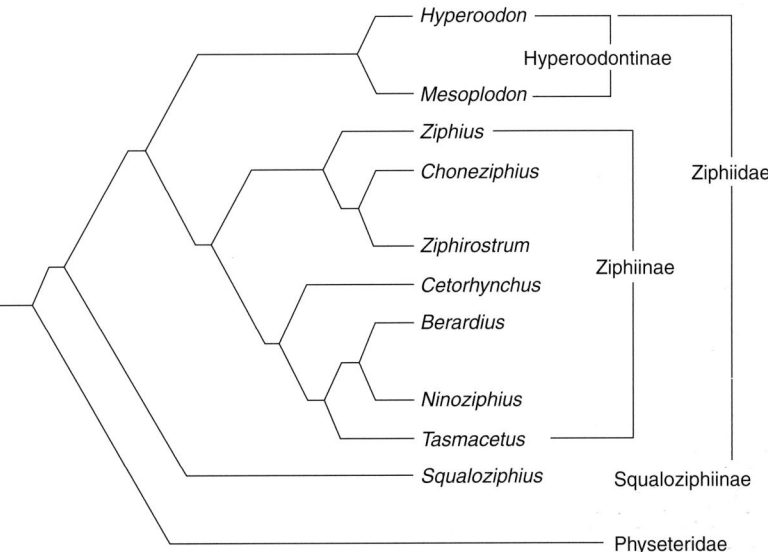

**Figure 3** *Cladogram of the Ziphiidae (after Muizon 1991).* Indopacetus *is included in* Mesoplodon.

## IV. Interactions with Humans

Because of their pelagic habits and general lack of concentrated populations, ziphiids have not had much contact with humans. The only fisheries that had ziphiids as a target species were the bottlenose whale fishery in the North Atlantic and the *Berardius* fishery in the North Pacific.

The bottlenose whale was hunted from the middle of the nineteenth century by Norwegian and British whalers. The catches of the bottlenose whale were part of a multi-species small whale fishery, where catches of one species may serve to subsidize catches of another when the population of the second species has fallen to such a point that fishing of it would not be economical. As a result the population was over-exploited and protected by the International Whaling Commission in the late 1970s.

*Berardius* was hunted primarily by the Japanese who fished it out of shore stations on the northeast coast of Japan since at least the seventeenth century. It was taken incidentally by other nations in the process of whaling for other species. The Japanese market was local to the whaling stations and would sometimes take *Ziphius cavirostris* and the occasional *Mesoplodon*. In the Southern Hemisphere, whalers rarely took the southern forms of *Berardius* (*B. arnuxii*) and *Hyperoodon* (*H. planifrons*). There were no fisheries based on them as the target species.

Ziphiids were moderately large, difficult to find and catch and had habits (deep diving) that did not suit them to captivity. The occasional live stranded animals were sometimes maintained in captivity in hopes of rehabilitating them and learning something of their behavior. The rehabilitation attempts were never successful and the animals always died quickly. One *Mesoplodon* calf that stranded in California in 1989 lived for 25 days in an aquarium.

## See Also the Following Articles

Giant Beaked Whales ■ Mesoplodont Whales ■ Skull Anatomy

## References

Gray, J. E. (1866). "Catalogue of Seals and Whales in the British Museum," 2nd edn. British Museum, London.

Lacépède, B. G. E. (1804). "Histoire Naturelle des Cétacés." Plassan, Paris.

Mead, J. G. (2007). Stomach anatomy and use in defining systematic relationships of the cetacean family Ziphiidae (beaked whales). *Anat. Rec.* **290**(6), 581–595, Figures 1–13.

Moore, J. C. (1968). Relationships among the living genera of beaked whales with classifications, diagnoses and keys. *Fieldiana Zool.* **53**(4), 209–298.

de Muizon, C. (1991). A new Ziphiidae (Cetacea) from the early Miocene of Washington state (USA) and phylogenetic analysis of the major groups of odontocetes. *Bulletin du Musee National d'Histoire Naturelle, Paris, 4e serie, section C* **12**(3–4), 279–326.

Ridgway, S. H., and Harrison, R. (1989). "Handbook of marine mammal's vol. **4**. Academic Press, San Diego.

True, F. W. (1910). An account of the beaked whales of the family Ziphiidae in the collection of the United States National Museum, with remarks on some specimens in other American museums. *U.S. Natl Mus. Bull.* **73**(89), 42 pls.

Van Beneden, P. J., and Gervais, P. (1868–79). "Ostéographie des cétacés vivants et fossiles, comprenant la description et l'iconographie du squelette et du système dentaire de ces animaux, ainsi que des documents relatifs a leur histoire naturelle." A. Bertrand, Paris.

# Bearded Seal
*Erignathus barbatus*

### Kit M. Kovacs

## I. Characteristics and Taxonomy

**B**earded seals are the largest of the northern phocid seals (Fig. 1). Adults are 2–2.5 m long and are gray-brown in color; some individuals have irregular light-colored patches. The weight of bearded seals varies dramatically on an annual cycle, but an average weight for adults is 250–300 kg. Females are somewhat larger than males in this species and can weigh in excess of 425 kg in the spring. The sexes are not easily distinguished. Pups are approximately 1.3-m

**Figure 1** *(A) Bearded seal pup, 2-days old. (B) Adult male bearded seal close-up showing the elaborate vibrissae. (C) An adult bearded seal in typical habitat.*

long at birth and weigh an average of 33 kg. They are born with a partial coat of fuzzy gray-blue fur but have already commenced molting into a smooth dark-gray coat, with a light belly, that is their pelt by the time they are a few weeks old (Kovacs *et al.*, 1996). Their shed fetal hair is formed into disks (similar to hooded seals) that are passed with the placenta. Similar to adults, young animals often have irregular light patches here and there. Pups faces have white cheek patches and white eyebrow spots that give them a "bandit" or "teddy-bear" appearance. Yearlings look very similar to pups, but the facial patterns are somewhat less distinct and they often have dark spots on their bellies.

Bearded seals have several distinctive physical features. Their body shape is very rectangular. Their heads appear to be small compared to their body size, similar to monk seals. They have very square-shaped fore flippers (with the longest toe being the middle one) which bear

very strong claws. Inuit people in the Canadian Arctic refer to this seal as "square-flippers" because of the shape of their front flippers. They also have extremely elaborate, smooth, facial whiskers that tend to curl when dry; this trait gives them their other common name— bearded seal. Females have four mammary glands (another characteristic shared with the monk seal lineage), unlike the other northern phocids, which have only two. Genetically and morphologically, the bearded seal is an intermediate form, sharing characteristics of the two Phocidae subfamilies (Phocinae and Monachinae). Bearded seals are the only species within the genus *Erignathus*.

The dentition of bearded seals is typical for phocid seals: I 3/2 C 1/1 PC 5/5 though anomalies in number are common and it is also not uncommon for the teeth of older animals to be worn to the gum-line.

## II. Distribution and Abundance

Bearded seals have a patchy distribution throughout much of the Arctic and subarctic (Fig. 2; Burns, 1981). Their preferred habitat is drifting pack ice in areas over shallow water shelves. They are often found in coastal areas. Some populations are thought to be resident throughout the year, whereas others follow the retraction of the pack ice northward during the summer and southward once again in the late fall and winter (Kelly, 1988). They can maintain holes in relatively thin ice, but avoid heavy ice areas. During winter they concentrate in areas that contain polynyas or in areas where leads in the ice tend to be a regular feature, or along the outside of pack-ice areas. Juvenile animals wander quite broadly and can be found far south of the normal adult range. A neonate equipped with a satellite tag in Svalbard traveled south to Jan Mayen and then almost to the Greenland coast within the month following weaning, when the tag ceased to transmit. Based on the fact that bearded seals can be locally extirpated quite easily via hunting, it is thought that significant subpopulation structure exists across the Arctic; this is currently the subject of a genetics investigation.

It is not possible to provide accurate abundance estimates for bearded seals because they occur at low density, are spread over a very wide range, are difficult to survey logistically, and receive relatively little research attention. But, this species probably numbers in the hundreds of thousands globally.

## III. Ecology

Bearded seals are a pack ice species. Generally, they are found in ice-filled waters throughout the year. But, levels of primary productivity and benthic biomass as well as sea ice have positive influence on abundance in a given area (Bengtson *et al.*, 2005). Bearded seals are known to come ashore in a few locales to rest, particularly at the time of peak molting in midsummer. Bearded seals are not deep divers; they feed in shallow, often coastal, areas and hence normally are not required to dive to depths more than 100 m. Pups dive to much greater depths during their first year (>450 m), but older, experienced animals remain in shallow water where most of their benthic prey resides (Gjertz *et al.*, 2000). Most dives are less than 10 min in duration, although they can dive for up to 20–25 min.

Bearded seals eat a wide variety of different types of prey, but they are predominantly benthic feeders, eating clams, shrimps, crabs, squid, fishes, and a variety of other small prey that they find near, on, or in the ocean floor. They can search soft-bottom sediments using their whiskers to find hidden prey that they get at using a combination of water jetting and suction (Marshall *et al.*, 2008). Some bearded seals in Svalbard have rust-colored faces and fore flippers. This coloration

**Figure 2** *Map showing the circumpolar, Arctic distribution of bearded seals (pink—the white area over the Arctic Ocean depicts the area that is usually quite consolidated sea ice).*

is the result of iron-compounds from soft-bottom substrates sticking to the hairs while the animals feed and then chemically reacting with oxygen when brought to the surface. The rust material is actually stuck onto the hairs rather than in them.

Polar bears are the main predator of bearded seals, but walruses, killer whales, and Greenland sharks may also take bearded seals, particularly pups and juveniles. They are important as traditional food for humans in parts of their range.

## IV. Behavior and Physiology

Bearded seals are largely solitary, although it is not usual to see them hauled out together in small groups along leads or at holes in the spring or early summer. It is quite unusual to see a bearded seal on land; they prefer to haul out on moving ice. They are rarely more than a body length from the water, and usually face toward the water. However, they are not wary in a general sense—in some areas such as Svalbard, Norway, they are very tame and can be approached by humans to within meters by boat without reaction.

The time of breeding appears to vary somewhat geographically, with peaks occurring sometime between late March and mid-May depending on the locality. Females give birth in a solitary fashion, on small drifting floes in areas of shallow water. Pups are born with a thin layer of subcutaneous blubber, which is thought to be an adaptation to entering the water shortly after birth. Bearded seal pups swim with their mothers when they are only hours old. This precocial entry into the sea is likely a mechanism to avoid polar bear predation. Neonatal swimming skills develop quickly in this species, and pups can dive to depths more than 90 m and remain submerged for periods in excess of 5 min when they are only a few weeks old. They spend approximately half of their time in the water during the nursing period, which lasts a total of 18–24 days and commence foraging on solid food while still accompanied by their mother. Female bearded seals spend little time on the surface with their pups, beyond that which is necessary for nursing. Most of the time, they attend the pups from the water next to the floe on which the pup is resting at a given time. Females do leave their pups unattended for periods to forage during the lactation period (Krafft *et al.*, 2000). Mother–pup pairs tend to remain in an area for some days at a time, but can also move tens of kilometers from one day to the next. Pups grow quickly during the nursing period, gaining about 3.3 kg per day while drinking more than 7.5 l of milk per day. The fat content of the milk is quite stable through lactation, at about 50%. Bearded seals pups have leaner bodies at the time of weaning than less active phocid pups, but they still have significant blubber stores and a body composition that is about one-third fat (Lydersen and Kovacs, 1999).

**B**

Pups are about 110 kg when they are weaned. Weaning does not appear to be as abrupt as it is in most phocid species.

Mating takes place around the time that females leave their off-spring. Male bearded seals perform vocal displays underwater to attract females and they also fight with other males during the breeding season. Their beautiful, but slightly melancholy, underwater songs are composed of a downward spiraling trill that can be heard for many kilometers in calm conditions (Cleator *et al.*, 1989). This behavioral trait of bearded seals is the most studied aspect of their biology. The onset of vocal displays (at least in captivity) is coincident with the onset of sexual maturity (Davies *et al.*, 2006). In the wild, males defend small patches of ocean with elaborate bubble displays, where they sing their songs intensively and repeatedly over a period of some weeks. Bearded seal calls exhibit marked geographic variation in call dialects, whereas repertoires of calls seem to be quite stable regionally (Risch *et al.*, 2007). Relatively little is known regarding the specifics of mating behavior of this species because pairing takes place in the water, but individual territorial males are known to occupy the same areas from 1 year to the next for at least several years (Van Parijs *et al.*, 2003), whereas transient males behave somewhat like "floaters" in the system (Van Parijs *et al.*, 2001).

Bearded seals shed their hair much more diffusely than other phocid seals, losing hair most of the year. But, they do have a concentrated period of molting in June when they prefer not to go into the water. At this time of year there is not a lot of ice available in coastal areas, so bearded seals can be seen in small groups on the available ice. Modestly dense aggregations can occur at this time of year, particularly in poor ice years.

The most notable sensory adaptation of bearded seals beyond their highly developed acoustic system is the extreme development of their facial vibrissae. They have approximately 244 highly sensitive, active-touch receptors within their facial whisker pads, which are among the most sensitive in the animals world with 1300 mylinated axons ("nerve-endings") associated with each whisker (Marshall *et al.*, 2006). The extreme development of the sensitivity of the whiskers of bearded seals is presumably an adaptation to their benthic feeding habit.

### V. Life History

Neonatal growth is fast, similar to all phocid seals, but growth over the rest of the first year of life is minimal by comparison. Female bearded seals reach sexual maturity when they are about 5-years old, whereas males are a bit older, usually 6 or 7 years when they reach maturity. Females give birth annually, similar to other phocid seals. Bearded seals normally live to an age of 20–25 years.

Some populations appear to follow an annual pattern of movement that follows the sea ice retreat in spring and expansion in the fall, whereas in other areas bearded seals seem to be quite stationary. This is likely primarily dependent on the availability of ice to haul out on during the summer season; calving glaciers for example often create ideal areas for bearded seals during summer in some coastal areas.

### VI. Interactions with Humans

Bearded seals are an important subsistence resource for coastal people throughout much of the Arctic. Animals are harvested for use as human food, dog food, and for their thick leather, which is important for various traditional articles of clothing and for making skin boats in Alaska, the Canadian Arctic, and in Greenland. Russia is the only country that has had a commercial-scale harvest of bearded seals. Soviet ships took catches that exceeded 10,000 animals in some years during the 1950s and 1960s. Quotas were introduced to limit the harvests of the declining populations in the Okhotsk and Bering seas, and the catch dropped to a few thousand bearded seals annually through the 1970s and 1980s. This hunt provided food for people and dogs and also fur-farm animal feed. Sinking losses are a serious problem when hunting bearded seals. During much of the year they sink when shot in open water or too close to edges; sinking loss is estimated to be as high as 50%.

Bearded seals fed on a wide variety of food types, many of them low in the food chain, so they tend to have low toxic chemical loads. The most obvious threat to this species, beyond overexploitation at a very local level, is climate change (Kovacs and Lydersen, 2008). Sea ice predictions suggest that the breeding habitat of bearded seals will decline dramatically in the decades to come.

This species has been kept in captivity in only one public aquarium, Polaria, in Tromsø, Norway.

### References

Bengtson, J. L., Hiruki-Raring, L. M., Simpkins, M. A., and Boveng, P. L. (2005). Ringed and bearded seal densities in the eastern Chukchi Sea, 1999–2000. *Polar Biol.* **28**, 833–845.

Burns, J. J. (1981). Bearded seal *Erignathus barbatus* Erxleben, 1977. *In* "Handbook of Marine Mammals" (S. H. Ridgway, and R. J. Harrison, eds), Vol. A, pp. 145–170. Academic Press, London.

Cleator, H. J., Stirling, I., and Smith, T. G. (1989). Underwater vocalizations of the bearded seal (*Erignathus barbatus*). *Can. J. Zool.* **67**, 1900–1910.

Davies, C. E., Kovacs, K. M., and Lydersen, C. (2006). Development of display behavior in young captive bearded seals. *Mar. Mamm. Sci.* **22**, 952–965.

Gjertz, I., Kovacs, K. M., Lydersen, C., and Wiig, Ø. (2000). Movements and diving of bearded seals (*Erignathus barbatus*) mothers and pups during lactation and post-weaning. *Polar Biol.* **23**, 559–566.

Kelly, B. P. (1988). Bearded seal. *In* "Selected Marine Mammals of Alaska: Species Accounts with Research and Management Recommendations" (J. W. Lentfer, ed.), pp. 77–94. U.S. Marine Mammal Commission, Washington, DC.

Kovacs, K. M., and Lydersen, C. (2008). Climate change impacts on seals and whales in the North Atlantic Arctic and adjacent shelf seas. *Sci. Prog.* **91**, 117–150.

Kovacs, K. M., Lydersen, C., and Gjertz, I. (1996). Birth-site characteristics and prenatal molting in bearded seals (*Erignathus barbatus*). *J. Mammal.* **77**, 1085–1091.

Krafft, B. A., Lydersen, C., Kovacs, K. M., Gjertz, I., and Haug, T. (2000). Diving behaviour of lactating bearded seals (*Erignathus barbatus*) in the Svalbard area. *Can. J. Zool.* **78**, 1408–1418.

Lydersen, C., and Kovacs, K. M. (1999). Behaviour and energetics of ice-breeding, North Atlantic phocid seals during the lactation period. *Mar. Ecol. Prog. Ser.* **187**, 265–281.

Marshall, C. D., Amin, H., Kovacs, K. M., and Lydersen, C. (2006). Microstructure and innervation of the mystacial vibrissal follicle-sinus complex in bearded seals, *Erignathus barbatus* (Pinnipedia: Phocidae). *Anat. Rec.* **288A**, 13–25.

Marshall, C. D., Kovacs, K. M., and Lydersen, C. (2008). Feeding kinematics, suction, and hydraulic jetting capabilities in bearded seals (*Erignathus barbatus*). *J. Exp. Biol.* **211**, 699–708

Risch, D., Clarke, C. W., Cockeron, P. J., Elepfandt, A., Kovacs, K. M., Lydersen, C., Stirling, I., and Van Parijs, S. M. (2007). Vocalizations of male bearded seals, *Erignathus barbatus*: Classification and geographical variation. *Anim. Behav.* **73**, 747–762.

Van Parijs, S. M., Kovacs, K. M., and Lydersen, C. (2001). Spatial and temporal distribution of vocalizing male bearded seals—implications for male mating strategies. *Behaviour* **138**, 905–922.
Van Parijs, S. M., Lydersen, C., and Kovacs, K. M. (2003). Vocalizations and movements suggest alternative mating tactics in male bearded seals. *Anim. Behav.* **65**, 273–283.

# Behavior, Overview

## PETER L. TYACK

Marine mammalogists often divide behavioral research into categories defined by mode of study: "acoustics" is studied by recording underwater sounds with a hydrophone, "behavior" is often informally defined as that which can be seen by an observer watching animals, and "diving" is often studied by attaching tags to animals. This method-oriented view of behavior may be convenient for sorting different research traditions, but it obscures the integrated whole of behavior as it has been shaped by evolution. Each method yields its own view of behavior, but no one views alone can provide a complete picture.

Most behavioral ecologists divide behavior along functional lines, i.e., what is the problem the behavior has evolved to solve (Alcock, 1998)? The following is a short list of such problems:

- Foraging behavior: how to find, select, and process prey
- Predator avoidance or defense: the flip side of foraging from the prey's point of view
- Dispersal and migration
- Competition and agonistic behavior
- Sexual behavior: how to find, court, and choose mates
- Parental behavior
- Social behavior and social relationships

This functional taxonomy of behavior is mirrored by Bradbury and Vehrencamp's (1998) functional analysis of animal communication. A receiver can often be viewed as paying attention to a signal to answer a question related to one of these behavioral problems. When the receiver detects one signal out of a larger signal set, the signal can potentially help the receiver to reduce uncertainty about the correct answer. Bradbury and Vehrencamp (1998) suggest that a receiver's questions can be divided into three categories (1) sender identity, (2) sender location, and (3) behavioral context. Depending on the problem, the receiver may be interested in different levels of recognition of the signaler: species, group, sex, age, or individual. Receivers usually need to know something about the location of the signaler: how far away is it? Is it within the receiver's territory? Is it approaching or moving away? The behavioral contexts of animal communication bear a striking resemblance to the functional behavioral problems listed previously: conflict resolution, territory defense, sexual interactions, parent–offspring interactions, social integration, and environmental contexts such as those related to prey and predators. This article discusses marine mammal examples for each of these basic problems in the behavioral ecology of all animal species, with special emphasis on how the marine environment may affect adaptations of marine mammals.

## I. Foraging Behavior: How to Find, Select, and Process Prey

The earliest studies of foraging in marine mammals focused on the stomach contents of dead animals in order to define what kinds of organisms were in the DIET of marine mammals. The best that observers could do in early field studies of living marine mammals was to identify behavior associated with feeding, where feeding was linked to observation of prey at the surface or chases, and so on. However, these observations do not do justice to the complex process by which animals find, select, and handle their prey. Increased efforts in foraging theory to identify the kinds of decisions faced by a foraging individual have focused attention on a more detailed view of the stages of foraging, and new techniques such as tags that can record behavior (Davis *et al.*, 1999; Johnson and Tyack, 2003) have improved our ability to collect the required data. This section discusses the various phases of foraging.

Marine mammals use every sensory modality available to find and select their prey. The optimal senses for solving a particular foraging problem depend on the setting. For example, vision is an excellent distance sense in air but has a limited range underwater. Even though polar bears (*Ursus maritimus*) are classed as marine mammals, they often hunt their prey in air and may use vision in air to search for their pinniped prey. Many seals and dolphins chase fish prey close enough to the surface to be able to use down-welling light to see their prey during daytime. Davis *et al.* (1999) have used video recorders attached to seals to capture images of prey as seals hunt. In many coastal areas, seals can see fish at ranges of 10 m or so. Deep-diving seals such as elephant seals (*Mirounga* spp.) have eyes specially adapted to the wavelengths and low light levels of the deep sea. Many deep-sea organisms have light-producing organs, and researchers have speculated that marine mammals may use vision to find bioluminescent organisms in the dark.

As terrestrial mammals, we humans are accustomed to thinking of vision as the best distance sense, but sound carries much better underwater than light. Some marine mammals have developed sophisticated adaptations to use sound for finding prey. Perhaps the best-known example is the sonar of dolphins. Dolphins and most toothed whales have an auditory system that is specialized for high frequencies, and they can produce a directional beam of intense high-frequency pulses of sound. Most toothed whales echolocate by producing a click and then listening for echoes from surrounding targets. When they are in a search mode, they may produce a slow series of clicks, listening for echoes. Madsen *et al.* (2005) used a tag to record echolocation clicks of foraging Blainville's beaked whales (*Mesoplodon densirostris*) and echoes from prey. These whales typically detect several echoes from each click at varying time delays and ranges. When searching for prey, the beaked whale inter-click intervals are typically 0.4 sec, a two-way travel time corresponding to a range of about 300 m. Sometimes echoes from the seafloor are detected at ranges up to this maximum range. The prey whose echoes are recorded on the tag, however, are typically less than 15 m away from the clicking whale. The foraging whale will pass many of these targets by until it selects one. As the whale approaches to within a few meters of the prey, it changes from slow clicks to a rapid series of pulses with inter-click intervals of about 10 msec. Beaked whales, sperm whales (*Physeter macrocephalus*), and narwhals (*Monodon monoceros*) all show a pattern similar to that of foraging bats, where they search for prey with regular slow clicks and then accelerate clicks into a buzz as they capture prey. Sperm and beaked whales have an increase in angular acceleration at the end of the buzz as they maneuver to

**B**

capture the prey. Together, these results suggest that deep-diving odontocetes use slow series of clicks to monitor several targets at different ranges at the same time. When the whale has selected a prey item, it accelerates the clicks to get more rapid updates on the location of the prey as it maneuvers to capture it.

Although deep-diving toothed whales use echolocation to find prey in the dark depths, there is evidence that dolphins feeding in shallow waters use a combination of senses and cues to detect and select prey. When wild bottlenose dolphins (*Tursiops truncatus*) from inshore waters near Sarasota, Florida, are feeding, they produce echolocation clicks at very low rates (Nowacek, 1999). Dolphins in Sarasota had an overall average click rate of 0.39 click trains/min while foraging and a rate of 0.10 click trains/min while not foraging. Dolphins had the highest click rates and appeared to rely more upon echolocation when they were feeding on fish hiding in seagrass. In contrast, when dolphins were feeding in clear water over sand, they seldom clicked and appeared to rely primarily on vision. Gannon *et al.* (2005) showed that bottlenose dolphins turned toward playback of sounds from fish that are favored dolphin prey, indicating that dolphins may listen for the vocalizations of fish to detect select prey. The dolphins then appeared to use echolocation to pursue and capture the prey.

Dolphins and toothed whales are hunters who chase down individual prey items. Many species feed on highly mobile prey such as schooling fish. When a dolphin charges into a fish school, the fish usually disperse, and this can make it less efficient to find and chase down the remaining fish. Dolphins (Fig. 1) and killer whales (*Orcinus orca*) have been reported to coordinate their feeding so that some individuals keep the fish in a tight school as other individuals feed (Similä and Ugarte, 1993; Benoit-Bird and Au, 2003; Vaughn *et al.*, 2007).

Baleen whales have evolved to capture entire patches of prey in one mouthful. Balaenid whales, such as right (*Eubalaena* spp.) and bowhead whales (*Balaena mysticetus*), are specialized to feed on calanoid crustaceans. When balaenid whales feed, they swim through the prey patch with an open mouth. Because their baleen is very long

and their head has a large cross-sectional area, they catch their prey by engulfing them in the water that flows into the mouth and out through the baleen. The basic problem faced by a feeding balaenid whale is to find a dense enough patch of prey to pay for the time and additional expense of swimming in the open-mouth foraging mode. Their prey move slowly enough that the feeding of balaenid whales is more like grazing then hunting. At times, balaenids coordinate their feeding by swimming in a staggered or "v shape" of up to 13 whales side by side. It is believed that such coordinated feeding keeps prey from escaping to the side, and may therefore more effectively "herd" prey towards each whale mouth (Würsig *et al.*, 1985).

Balaenopterid whales also feed on crustaceans along with fish, but their euphausiid prey are faster and more evasive than calanoid crustaceans. Balaenopterids also may capture schools of baitfish such as capelin, anchovy, sand lance, or even herring. These prey are more mobile than crustaceans, and balaenopterids have evolved a feeding mode that allows them to trap mobile prey. Balaenopterids have accordion-like pleats in the lower jaw, which can expand rapidly. When a balaenopterid feeds, it lunges while opening its mouth, forcing hundreds of gallons of prey and water into the mouth as the pleats expand. The whale then quickly closes its mouth, trapping the prey. The pleats then slowly contract, forcing the water through the baleen and leaving the prey behind. As with toothed whales, when balaenopterids such as humpback whales (*Megaptera novaeangliae*) feed on the most mobile schooling prey, such as herring, they may feed in coordinated groups. Perhaps the most striking reports concern a group of half a dozen or more female humpback whales who associated together each summer for several years in Southeast Alaska. Each individual played a specific role in prey capture, and their movements appeared to be coordinated with a regular series of vocalizations (D'Vincent *et al.*, 1985).

Marine mammals have evolved several different ways to feed on benthic prey that hide submerged in the sediment on the seafloor. Some bottlenose dolphins have been observed to echolocate on small sand dabs buried in the sand. The mustache of the walrus (*Odobenus rosmarus*) is exquisitely sensitive to touch, and walruses use the vibrissae in their mustache to detect prey in the sediment. Trained walruses have demonstrated remarkable abilities to use this mustache to determine the shape of objects, and presumably wild walruses use this ability to identify their favored prey within the sediment. Gray whales (*Eschrichtius robustus*) feed on benthic organisms by rolling more than 45° (usually to the right side), sucking mud and prey into the right side of their mouth and then straining out the prey with their baleen. Gray whales make distinctive pits, measuring about 1-m wide by up to 3-m long and about 1.5-m deep, in the seafloor when they feed in this way and these pits are big enough to be detected by sonar on surface ships.

Most marine mammals just swallow their prey whole, but some species face problems in handling their prey. Dolphins feeding on prey such as catfish with sharp spines may need to learn how to snap off the head and spines before they eat the rest of the fish. False killer whales (*Pseudorca crassidens*) measuring perhaps 3 m in length can capture a large mahi-mahi nearly half their size. It has been reported that one false killer whale may hold the fish while others rip off flesh. The most impressive prey handling among marine mammals involves the sea otter (*Enhydra lutris*), which feeds on shellfish such as abalone (*Haliotis* spp.) and sea urchins. Because these prey are too large or too strong for the otter to break the shell by biting it, most otters use a stone as a tool to open the shells. The sea otter will dive to get a shell, often carrying its stone tool in the axilla. When the otter surfaces, it lies with its stomach up with the stone on its stomach.

**Figure 1** *Dusky dolphins* (Lagenorhynchus obscurus) *exhibit coordinated feeding in Admiralty Bay, New Zealand. "Clean" headfirst re-entry leaps, shown here, are common during coordinated feeding; these leaps facilitate diving while allowing a dolphin to quickly catch a breath. The splash in the background is from an Australasian gannet* (Sula serrator)*, which has just taken a plunge dive; gannets and other seabirds often feed in conjunction with dusky dolphins. Photo by Heidi Pearson.*

It then uses the stone either as a hammer or as an anvil, smashing the shell opens on the stone and then eating the flesh. It must be awkward to carry this stone while diving to feed, and some otters have simplified this by using a bottle for the same task. When the otter dives, it can leave the bottle floating on the surface and can then use it when it surfaces with shellfish.

## II. Avoiding Predators and Defense from Predators

Many marine mammals are top predators and historically may not have faced heavy predation pressure. The primary predators for pelagic marine mammals over evolutionary time are the killer whale and sharks. However, in the last few centuries, humans have been extremely effective predators of marine mammals, driving some species such as the Steller's sea cow (*Hydrodamalis gigas*) to extinction. Seals on ice face the risk of predation from polar bears, and seals hauled out on beaches, especially pups, are at risk from other pinnipeds and terrestrial predators such as foxes. Seals on land are less mobile than at sea and appear to be at a higher risk of predation, for they will usually respond to the approach of a terrestrial predator by entering the water.

The great whales such as baleen and sperm whales are so large that their main predator is the killer whale. Killer whales live in stable groups. They attack large whales in groups and appear to coordinate their attacks in much the same way as a pride of lions (*Panthera leo*) or African wild dogs (*Lycaon pictus*) will attack a herd of ungulates, isolate an individual, and then hold it down to kill it. Baleen and sperm whales use their flukes as a weapon during such an attack, lashing them sideways through the water. Many cetaceans fall silent when a killer whale is detected nearby. After they have been detected, most small odontocetes appear to rely on speed to escape killer whales, whereas some pinnipeds may hide from them, either on the land or on the seafloor.

Early whalers had a predator's view of their marine mammal prey, and their observations make up an unusual body of data on predator defense in some species. While baleen whales often travel in groups, there is little sign of social defense from predators. The young calves of baleen whales may be more vulnerable to predation than adults, but a female with her young calf will tend to be sighted alone rather than with other whales in a group, suggesting that females with young may disperse to reduce the chance a predator will detect them. Sperm whales, however, appear to have a well-developed social defense from dangerous predators such as killer whales and human whalers. Sperm whale calves are born into groups of about 10 adult females with young. Because newborn calves cannot dive deep enough to follow their mothers for their 40- to 50-min foraging dives, they remain nearer the surface when the adults feed. The adults desynchronize their dives, however, so that there is less time that young calves are unattended by an adult. If a predator attacks the group, calves or wounded animals in the group will be surrounded by the rest of the adults. Most adults will face in toward the animal needing protection and will lash their tails facing outwards. This must have been a formidable defense against killer whales but was less successful with human whalers. Whalers knew how predictable this behavior was and would often intentionally injure one animal and leave it. They then could slowly kill each adult attending the injured animal, knowing that adults would be unlikely to abandon an injured group member.

The first step in lowering the risk of predation is to avoid detection. Some seals alter their foraging behavior apparently to avoid visual detection by predators. Female Galapagos fur seals

(*Arctocephalus galapagoensis*) are less likely to make their normal foraging trips when the moon is full. It has been suggested that this avoids the risk that a predator will see them in the moonlight. Because most small cetaceans have little chance to defend themselves from killer whales, they must emphasize strategies to avoid detection by these predators. Dusky dolphins (*Lagenorhynchus obscurus*) mill in the surf zone as killer whales pass by offshore. They will even hide in tidal lagoons and, at times, become stranded in these lagoons until the next tide. Baleen whales also have strategies to avoid detection by killer whales. For example, gray whales that were exposed to experimental playback of the sounds of killer whales fled into shallow water. The surf zone and kelp beds may be particularly good places to hide from an echolocating predator because they absorb and reflect sound, making echolocation more difficult. There is some evidence that killer whales may even have evolved countermeasures to these predator-avoidance strategies. There are two sympatric populations of killer whales in the inshore waters of the Pacific Northwest. One population, called residents, feeds primarily on fish; the other population, called transients, feeds primarily on marine mammals. When residents feed on salmon, a fish with poor hearing, the killer whales make regular series of loud clicks. When transients feed on acoustically sensitive marine mammals such as porpoises, dolphins, or seals have a much stealthier pattern of echolocation (Barrett-Lennard *et al.*, 1996). They produce fewer clicks, and those clicks that are produced are fainter and are produced with an erratic timing. These features appear to be designed to make marine mammal prey less likely to detect and avoid the killer whale.

Cetaceans are also subject to parasitism from animals that bite tissue without causing serious injury. In the tropics, many dolphins are subject to attack from the cookie cutter shark (*Isistius brasiliensis*), which takes bites of skin and blubber about 3 to 5 cm in diameter. Right whales (*E. australis*) in coastal bays in Argentina are subject to attack from seagulls, which peck chunks of skin and blubber from the back of a whale floating at the surface. Although this can evoke a strong behavioral reaction from the whale, right whales do not seem to have an effective defense from this attack, which may be made worse by the growth of seagull populations in areas with human settlements.

Evidence in several cetacean species shows that when an animal has been injured, other members of the group may support them for hours or days. Because marine mammals must breathe air, if they cannot surface on their own, they are at great risk of drowning. This caregiving behavior may cost the caregivers but at potential benefit to the incapacitated member of the group.

## III. Migration and Orientation

Most marine mammals are excellent swimmers, and many species make annual migrations of thousands of kilometers. Most baleen whales have an annual migratory cycle that affects many aspects of their life. These whales are adapted to take advantage of a burst of productivity in polar waters during the summer. Baleen whales store enough energy reserves during their intensive summer feeding season to last for most of the year, and this annual feast/fast cycle helps to select for large size. A humpback whale that is born in the winter in a tropical breeding ground near 20° of latitude will typically migrate in the spring to summer feeding grounds in polar waters near 40–60° of latitude. Humpbacks have traditional feeding grounds and an individual will often visit specific banks or inshore feeding areas of scales of tens of kilometers. Dolphins on both coasts of the United States also show annual migrations of 1000 km

**B**

or more. Off the east coast of the United States, harbor porpoises (*Phocoena phocoena*) and bottlenose dolphins tend to move north in summer and south in winter. It is not known whether the colder temperatures in the north during winter are more important for this seasonal migration than are seasonal changes in prey distribution. Some pinnipeds also have annual migrations of thousands of kilometers. For example, northern elephant seals (*M. angustirostris*) that breed and calve near San Francisco may migrate as far as the Aleutian Islands to feed. Both males and females swim north after the breeding season, following the California Current. Some male elephant seals feed along oceanfronts on the boundary of the Alaska Stream. Very little is known about how marine mammals orient and navigate during migration, and even less is known about how they find oceanographic features such as fronts, which can be important cues for good places to forage.

Other species have more limited annual home ranges. The home ranges of sea otters may be limited to 1–17 km of coastline. Sea otters show strong fidelity to their home range. When 139 sea otters were flown 200 km to an offshore island as part of a reintroduction program, most of the otters left and at least 31 managed to return to the area where they had been captured. Bottlenose dolphins in the inshore waters of Sarasota, Florida, tend to be sighted within a home range of 125 km². "Resident" killer whales in the inshore waters of Puget Sound have seasonal ranges limited to an area several tens of kilometers by about 100 km. Even non-migratory species can be highly mobile. For example, resident killer whales will often swim 100 km or more in a day. Bottlenose dolphins and sea otters may suddenly leave their home ranges and swim 100 km away from the normal range.

## IV. Competition and Agonistic Behavior

When animals are competing for the same resource, they may fight for access. Among animals that exploit a specific substrate, this competition may be for territory. This kind of territorial defense has been well described for many pinnipeds during the breeding season. Female pinnipeds haul out onto beaches or ice to give birth, and many species mate on land as well. This concentration of females creates a valuable resource for males. Males in many species will defend an area of beach from other males and may attempt to monopolize opportunities to mate with females there. For animals that live in the open ocean, resources are not likely to be as tied to a particular location, but rather will move. Animals in this setting are more likely to defend a particular resource at one time than to defend a patch of real estate. For example, a male humpback whale will not defend a specific location during the breeding time, but a male escorting a female will fight other males to limit their access to the female.

This pattern of males competing for access to females, either by defending a group of females or a territory (Fig. 2), is common among mammals and leads to behavioral and morphological adaptations. Males in these species are often larger than females. Some of the most extreme sexual dimorphism among mammals occurs in marine mammals where a successful male may mate with many females in one breeding season. For example, male elephant seals may be 10 times heavier than females (McCann *et al.*, 1989), and mature male sperm whales may be up to 3 times heavier than females (Connor *et al.*, 1998). Some behaviors appear to function to increase the apparent size of a male and may function as visual displays. For example, male humpback whales competing for access to females may lunge with their jaws open, expanding the pleated area under the lower jaw with water. Several observers have suggested this may function to increase the apparent size of a competitor. Males may have larger

**Figure 2**   *A male sea otter* (Enhydra lutris) *patrols his territory in Prince William Sound, Alaska. The function of patrolling is to search for estrous (receptive) females and intruding males. Females (which are 35% smaller than males) may be attracted to resources contained within the territory such as prey, protection from wind and waves, and resting areas. Photo by Heidi Pearson.*

weapons such as teeth or tusks than females. This is particularly striking in beaked whales. In most beaked whale species, the teeth may not erupt at all in females, whereas one or two pairs of teeth erupt in the lower jaw of males at about the time of sexual maturity. Males have scarring patterns, suggesting that these "battle teeth" are used in fights. Males may also have protection such as areas of toughened skin. Male elephant seals, for example, often strike one another on the chest, and this area has thickened and hardened skin.

Fighting often involves a gradually escalating series of threats and responses. Overstrom (1983) presented data suggesting this kind of escalated display for bottlenose dolphins in captivity. The earliest stages of a threat may involve one dolphin directing pulsed sounds toward another. The threat may escalate if the dolphin produces an open-mouth threat display while emitting distinctive bursts of pulses. The longer in duration or louder in sound intensity the pulses are, the stronger the threat may be. As another step in escalation, the animal may accentuate this display with abrupt vertical head movements. One of the most intense threat displays in dolphins is called the jaw clap. A dolphin starts the jaw clap display with an open mouth. The jaw clap consists of an abrupt closure of the gaping jaw, accompanied by an intense pulsed sound. Many of the agonistic visual displays used by bottlenose dolphins are related to movements used to inflict injury. For example, the open-mouth display looks like the first step in preparing to bite.

Some animals live in situations where they interact repeatedly with the same individuals repeatedly. In this setting, animals may develop a predictable hierarchy of who wins and loses in agonistic interactions. Male elephant seals establish a dominance hierarchy on the breeding beaches. The pace of competition is highest before the females appear on the beach. When males are competing using territory or dominance for access to females, they often sort out their competitive relations before the peak of the mating season. Dominance relations have also been studied in captive bottlenose dolphins. The most obvious competitive behaviors are violent fights in which each opponent responds to aggression with an aggressive response. This is not as useful for determining winners or losers as observation of more subtle submissive behaviors. A fight in which each opponent produces aggressive behaviors with no submission does not have an obvious winner or loser, but an animal can be identified as a loser of an interaction if it responds to a neutral or aggressive behavior with a submissive one. Systematic observations of winners and losers in dyadic agonistic interactions reveal that adult males are dominant over adult females. The rate of agonistic interactions is higher in males than in females. The low rate of female agonism means that dominance is rarely contested among females, and female dominance can be stable over years. Two male dolphins in a pool reversed dominance status several times over the years of study. Male dominance relations were characterized by periods of relatively low agonism interspersed with periods of high rates of agonism when one male challenged the other. Little is known about dominance relations among wild cetaceans, but because individuals in many species interact repeatedly with the same conspecifics and can recognize different individuals, dominance relations are likely to be important.

## V. Courtship and Sexual Behavior

Charles Darwin made a distinction between features selected to improve chances of mating and features selected for survival. He called selection for mating sexual selection to discriminate it from natural selection. Darwin defined two kinds of sexual selection: intersexual and intrasexual. Intersexual selection can increase the likelihood that an animal will be chosen by a potential mate; intrasexual selection can increase the likelihood that an animal will outcompete a conspecific of the sex for fertilization of a member of the opposite sex. Reviews that are more recent have included a third mode of sexual selection where a male may attempt to limit the choice of a female by coercing her to mate with him and not to mate with other males.

Differences between male and female mammals alter the costs and benefits of different elements of reproduction. Female mammals all gestate the young internally and are specialized to provide nutrition to the young after birth. In many species, and most marine mammal species, the female provides most of the parental care. Reproduction in most female mammals is limited by the amount of energy and nutrition they can acquire for pregnancy and lactation. Male mammals usually provide much less parental care to their young. This means that reproduction in most male mammals is limited by the number of females with which they can mate. This situation often leads to a polygynous mating system in which there is high variability in the mating success of different males, with some males mating with many different females and other males mating with none. Males in polygynous species often fight other males for access to females. This often leads them to have weapons and to be larger than females; the intensity of polygyny is sometimes estimated by assessing the difference in size of males vs females. As discussed earlier, some of the most extreme cases of sexual dimorphism in mammals occur in marine mammals.

Most traditional discussions of mating systems emphasize male strategies. For example, polygyny occurs where one male mates with more than one female; the number of males with which a female mates is not included in the definition. While the variance of reproductive success is higher in males than in females for most marine mammal species, female reproductive strategies can influence male strategies and impacts other areas of social behavior. Areas in which female reproductive strategies vary include the following:

How seasonal and synchronized is estrus?
Do females have one (monoestrous) or more (polyestrous) estrous cycles per year?
Do females ovulate spontaneously or do they require the presence of a male to ovulate?
How many males are available during estrus?
Can the female select a mate?

If a female mates with more than one male, can she influence which male fertilizes the egg?

There are different patterns for the reproductive strategies of males and females in different polygynous mating systems. This article describes five different categories of male strategy that are used in the literature. The *resource defense strategy* is adopted by males who defend a resource used by females around the time of mating. In this case females do not select a mate but rather select an area for breeding and mate with the male defending this area. The *female defense strategy* is used by males who stay with a female and prevent other males from mating with her while she is receptive. The *sequential defense strategy* differs from the female defense strategy in that a male will defend a female through mating, but then leave in search of other mating opportunities. The distinction between these two male strategies depends in part on whether the female is mono- or polyestrous and on the degree of synchronization of different females. The strategy called by the name "scramble competition" occurs when a male searches for a receptive female, mates with her, and then moves on to search for another female without preventing access for other males. The last three models lie on a continuum of male strategies between pure guarding and pure roving. Whitehead (1990) used modeling to suggest that males should rove between groups of females if the duration of estrus is greater than the time it takes males to swim from group to group. At any one time, a male's decision to leave or stay with a group probably includes other factors, such as his assessment of what other males are doing. A *lekking strategy* occurs when males aggregate in an area with no resources needed by the female and produce displays to attract the female. In leks, males provide no parental care and females select a male for mating.

Some of these male strategies preempt the ability of a female to select a male for mating. In the resource defense model, the female does not select a particular male, but rather will select a particular place with the resource she needs. She will then be most likely to mate with the male who happens to be defending this location. When a female can and does choose a male for mating, she may select a mate based on several different criteria. A female may select a male for inherent qualities based on indicators such as size, age, or an advertisement display. She may assess competition between males and select one based on this performance. In some species, males may compete for access to a particular location, and a female can select a good competitor by mating with a male in such a preferred spot. In some species, a female may mate with several males and allow competition between their sperm to determine which male fertilizes the egg. The males in this system would be likely to devote more resources to sperm production, sperm swimming speed, and so on than species that compete by fighting. Evidence shows that sperm competition may play a role in some cetaceans. Odontocete cetaceans have larger ratios of testis to body weight than most mammals. This contrast is also seen among mysticetes. Balaenid whales form mating groups with multiple males, but there is little sign of fighting between the males. Male right whales have testes weighing more than 900 kg; their testes weigh more than six times what would be predicted for a typical mammal of their body size (Brownell and Ralls, 1986). In contrast, humpback whale males, which fight for access to females, have testes weighing less than 2 kg.

Marine mammals are highly mobile, and in the open ocean it seems unlikely that males could defend a resource in a way that would preempt the ability of a female to use the resource yet mate with another male. Resource defense is much easier to envisage on land. While pinnipeds spend much of their life at sea, they haul out on a

**B**

solid substrate (land or ice) to give birth. Some of these species, such as elephant seals, also mate on land. Females have specific requirements for a place to give birth, and they often return to traditional areas. The selection by females of specific sites for mating and giving birth creates an opportunity for males to defend these sites in order to increase their chances of mating with the females who are selecting the site. In most otariid seals, males appear to employ resource defense strategies for mating. In many of these species, males will arrive before the females and will fight to establish territories that they defend from other males. In some phocid species that mate at sea, males may establish and defend territories just off the beach where females give birth. Genetic analyses of paternity, however, show that the fathers of some pups are not among the territorial males. This suggests that some males have alternate mating strategies.

There are marine mammal species in which males may adopt a strategy of attempting to preempt female choice by guarding a receptive female and preventing her from mating with other males. Northern elephant seal males arrive at breeding beaches before females and compete for dominance status and for position on the breeding beach. A dominant male can guard a group of females and prevent access for other males. If an alpha male can maintain his status, he can prevent access to a group of females for the entire breeding season. This pattern of guarding a group of females is less likely for cetaceans, which are highly mobile. Most male cetaceans would take a shorter time to swim between groups than the duration of female estrus, thus favoring a roving strategy. There is some evidence for sequential female defense in bottlenose dolphins. In field studies of Indian Ocean bottlenose dolphins (*T. aduncus*) in Shark Bay Western Australia and common bottlenose dolphins in Sarasota, Florida, groups of two or three adult male bottlenose dolphins may form consortships with an adult female (Connor *et al.*, 2000). A coalition of males may start such a consortship by chasing and herding a female away from the group in which they initially find her. Some of these consortships appear to be attempts by the males to limit choice of mate by the female, who may try to escape from the males. Males in these alliances may form consortships with several different females during a breeding season.

Many pinnipeds and some baleen whales produce reproductive advertisement displays that may play a role in mediating male–male competitive interactions and may also be used for female choice of a mate. Male humpback whales sing long complex songs during the winter breeding season. Singing males are usually alone and they usually stop singing when joined by another whale (Tyack, 1981). Aggressive behavior is often seen when a male joins a singer; when a female joins, apparent sexual behavior has been observed. Male humpbacks do not seem to be able to defend any resource needed by females on the breeding grounds, so this mating system has been described as a kind of floating lek. Vocal reproductive advertisement displays have also been reported for bowhead whales and many species of seal, including polar ice-breeding seals and harbor seals (*Phoca vitulina*). Most of the phocid seals known to produce songs mate at sea. These seals breed in conditions that foster the development of leks. Females gather to breed on isolated sites, but they mate after they have weaned their pups, so there are few resources males could defend. Females are so mobile that it would be difficult for males to prevent them from gaining access to other males. The females are already concentrated in hot spots around the places where they give birth. This creates an ideal setting for males to cluster near the females, producing advertisement displays to attract females for mating. Some of the songs of whales, of ice-loving seals and the bell-like sounds of the walrus stand as testimony to the power of sexual selection to fashion complex and fascinating advertisement signals.

## VI. Parental Behavior

All mammals have some parental care when the female lactates and suckles the young. The mothering role of the female is critical to mammalian life, and female parental care impacts many aspects of social behavior. There is enormous variability in parental care among marine mammals. Some phocid seals give birth to their young on unstable ice floes, where they cannot count on a stable refuge for the young. The hooded seal (*Cystophora cristata*) has responded to this situation by an intense 4-day period of lactation when the young pup doubles in weight. While female phocid seals generally stay with their young pup and fast while suckling, otariid females will leave their young in order to feed at sea and then they return to suckle the pup. This pattern leads to a large difference in duration of lactation, from 4 days to 2 months in phocids and from 4 months to 2 years in otariids. Most pinnipeds have yearly breeding seasons, so the longest periods of lactation are limited to about 12 months. However, tropical Galapagos sea lions (*Zalophus wollebaeki*) at times nurse their young for more than 12 months and produce young at intervals greater than a year. Phocoenid porpoises and some baleen whale species also have a strong annual breeding cycle. Some baleen whales, such as the blue whale (*Balaenoptera musculus*), wean their young after about 7 months so that the young can start taking solid food during the summer feeding season. All porpoises and baleen whales wean the young within a year. Toothed whales other than the porpoises stand at the other extreme of having very prolonged periods of parental care when the young are dependent. Bottlenose dolphins only 3 m or so in length often suckle the young for 3–5 years, which is remarkably long considering that the 30-m blue whale can wean the young in 7 months.

The longest periods of parental care known among marine mammals involve sperm whales and short-finned pilot whales (*Globicephala macrorhynchus*). In both species, mothers appear to suckle some calves for up to 13–15 years. The young may start to take some solid food by the first few years of life, but this suckling indicates a remarkably long period of dependency for the young. Adult female pilot whales typically start having young by 8–10 years, but by the time they are near 30–40, many cease to reproduce (Marsh and Kasuya, 1984). The ovaries become nonfunctional in these nonreproductive females, showing changes similar to those of human females after menopause. Female pilot whales may live into their 50s, suggesting that they may have a life expectancy 15–20 years after becoming nonreproductive. Most students of life history believe that the life history evolves to maximize lifetime reproductive success. If this has influenced the life history of pilot whales, it suggests that females switch their reproductive effort from having new offspring to parental care of their existing young. The 15- to 20-year duration of this period suggests that 15–20 years of parental care are required for the young to succeed or that these older females are caring for other kin, perhaps in a grandparental role.

## VII. Social Behavior and Social Relationships

Not only do marine mammals show a broad range in the duration of the maternal bond, but also there is great diversity in the duration of social bonds in general, and especially in the importance of individual-specific social relationships. Resident killer whales have the most stable social groups known among mammals: no dispersal of either sex has been described. The only way group composition changes among the resident killer whales of the Pacific Northwest is for an animal to die or for a new animal to be born. The best-known vocalizations from killer whales are group-distinctive repertoires of stereotyped pulsed calls. In contrast, bottlenose dolphins have very fluid social groups. In their fission–fusion society, group composition changes on a minute-by-minute and hour-by-hour basis. However, some individuals may have strong social bonds and be sighted together for years at a time. As was just discussed in the section on parental care, bottlenose dolphin calves suckle for 3–5 years. Adult male bottlenose dolphins may also form coalitions with one to two other unrelated males. Members of a coalition tend to be sighted together 70–100% of the time, and alliances may last for over a decade. It is thought that males form alliances to improve their chances of mating with females, but lone males are also successful breeders. Males within a coalition often have highly coordinated displays, both when feeding and when escorting a female. Each bottlenose dolphin produces an individually distinctive whistle vocalization called a signature whistle, which is probably used for individual recognition (Watwood et al., 2004).

In sperm whales, males have different life history patterns than females. Calves are born into matrilineal groups of females and young. Each matrilineal unit numbers about 10 animals, but often two units associate for days at a time. Males may leave their natal groups when 5–10 years of age, and they then will join bachelor groups. As males grow, they move to higher latitudes and associate in smaller groups of males. As the males approach social and sexual maturity at 20–25 years of age, they are increasingly likely to associate temporarily with female groups during the breeding season, when they may mate with females. The social relationships of males thus change over their lifetime, and adult males appear to have only temporary associations. Young females may stay with their natal groups or may leave, but once they reach sexual maturity at 8–10 years of age, they will tend to associate with the same adult females for decades at a time. Because the matrilineal groups often join with other groups but segregate into the original groups, the females must recognize group members over periods of decades. Sperm whales make rhythmic patterns of sounds called codas. Early reports reported individually distinctive codas, but there are also shared codas that vary with the geographic region. Most of the variation in codas involves differences between groups, and it has been suggested that codas may reaffirm social bonds when a group joins after dispersing to forage. Female sperm whales must have stable social relationships with specific other individuals. These family groups appear to be the basic social unit of sperm whales, with a primary function of vigilance against predators and social defense of calves (Whitehead, 2003).

Baleen whales may feed in groups as do sperm whales, but female baleen whales appear to differ from sperm whales in the importance of group care of young. On the feeding grounds, baleen whales of all sexes are often seen in groups of varying sizes. For humpback whales, the size of the feeding group correlates with the horizontal extent of the prey patch. However, during the breeding season, when a female humpback has a calf, she is extremely unlikely to associate with another adult female. When one or more adults escort a female during the breeding season, the escorts are usually males. In baleen whales there is much less evidence for long-term social bonds than among most toothed whales. Odontocetes with little evidence for stable bonds are species such as the harbor porpoise and delphinids of the genus Cephalorhynchus, which also appear to have fluid groupings with few social bonds more stable than the mother-calf bond, which lasts less than 1 year in the porpoise. However, future research may find social bonds that have not yet been described.

There appears to be a correlation between the social relations of marine mammals and their communication patterns (Tyack, 1986). Baleen whales and pinnipeds with large apparently anonymous breeding aggregations use reproductive advertisement displays to mediate male–male and male–female interactions on the breeding grounds. Killer whales with highly stable groups produce group-specific repertoires of stereotyped calls. Seals and dolphins with strong individual-specific bonds use a variety of different vocalizations for individual recognition, but no such recognition signals are known for porpoises or Cephalorhynchus. Sperm whales appear to use deceptively simple clicks to produce a diverse set of signals consistent with their diverse social groupings.

## VIII. Conclusions

Marine mammals face the same basic problems that have been identified by behavioral ecologists for all animals. However, marine mammals live in an environment that differs in many important ways from the terrestrial environment. Studies since the 1980s have provided ever-growing opportunities for fascinating comparisons between marine mammals and their terrestrial relatives and between the diverse taxa that live in the sea.

## See Also the Following Articles

Communication ■ Feeding Strategies and Tactics ■ Group Behavior ■ Migration and Movement Patterns ■ Predator–Prey Relationships ■ Sexual Dimorphism ■ Territorial Behavior

## References

Alcock, J. (1998). "Animal Behavior: An Evolutionary Approach," 6th edn. Sinauer Associates, Sunderland.

Barrett-Lennard, L. G., Ford, J. K. B., and Heise, K. A. (1996). The mixed blessing of echolocation: Differences in sonar use by fish-eating and mammal-eating killer whales. Anim. Behav. 51, 553–565.

Benoit-Bird, K. J., and Au, W. W. L. (2003). Prey dynamics affect foraging by a pelagic predator (Stenella longirostris) over a range of spatial and temporal scales. Behav. Ecol. Sociobiol. 53, 364–373.

Bradbury, J. W., and Vehrencamp, S. L. (1998). "Principles of Animal Communication." Sinauer Associates, Sunderland.

Brownell, R. L., Jr., and Ralls, K. (1986). Potential for sperm competition in baleen whales. Rep. Int. Whal. Commn Speci. Iss. 8, 97–112.

Connor, R. C., Mann, J., Tyack, P. L., and Whitehead, H. (1998). Social evolution in toothed whales. Trend. Ecol. Evol. 13, 228–232.

Connor, R. C., Wells, R., Mann, J., and Read, A. (2000). The bottlenose dolphin: Social relationships in a fission–fusion society. In "Cetacean Societies: Field Studies of Whales and Dolphins" (J. Mann, R. Connor, P. Tyack, and H. Whitehead, eds), pp. 91–126. University of Chicago Press, Chicago.

Davis, R. W., et al. (8 authors) (1999). Hunting behavior of a marine mammal beneath the Antarctic fast ice. Science 283, 993–996.

D'Vincent, C. G., Nilson, R. M., and Hanna, R. E. (1985). Vocalization and coordinated feeding behavior of the humpback whale in southeastern Alaska. Sci. Rep. Whales Res. Inst., Tokyo 36, 41–48.

Gannon, D. P., Barros, N. B., Nowacek, D. P., Read, A. J., Waples, D. M., and Wells, R. S. (2005). Prey detection by bottlenose dolphins, Tursiops truncatus: An experimental test of the passive listening hypothesis. Anim. Behav. 69, 709–720.

Johnson, M., and Tyack, P. L. (2003). A digital acoustic recording tag for measuring the response of wild marine mammals to sound. IEEE J. Oceanic Eng. 28, 3–12.

Madsen, P. T., Johnson, M., Aguilar de Soto, N., Zimmer, W. M. X., and Tyack, P. L. (2005). Biosonar performance of foraging beaked whales (Mesoplodon densirostris). J. Exp. Biol. 208, 181–194.

Mann, J., Connor, R., Tyack, P. L., and Whitehead, H. (2000). "Cetacean Societies: Field Studies of Whales and Dolphins." University of Chicago Press, Chicago.

Marsh, H., and Kasuya, T. (1984). Changes in the ovaries of the short-finned pilot whale, *Globicephala macrorhynchus*, with age and reproductive activity. *In* "Reports of the International Whaling Commission Special Issue 6: Reproduction of Whales, Dolphins and Porpoises" (W. F. Perrin, R. L. J. Brownell, and D. P. DeMaster, eds), pp. 311–335. International Whaling Commission, Cambridge.

McCann, T. S., Fedak, M. A., and Harwood, J. (1989). Parental investment in southern elephant seals, *Mirounga leonina. Behav. Ecol. Sociobiol.* **25**, 81–87.

Nowacek, D. P. (1999). Sound use, sequential behavior and ecology of foraging bottlenose dolphins, *Tursiops truncatus.* Ph.D. thesis, Woods Hole Oceanographic Institution/Massachusetts Institute of Technology, Massachusetts.

Overstrom, N. A. (1983). Association between burst-pulse sounds and aggressive behavior in captive Atlantic bottlenosed dolphins (*Tursiops truncatus*). *Zool. Biol.* **2**, 93–103.

Pryor, K., and Norris, K. S. (1991). "Dolphin Societies: Discoveries and Puzzles." University of California Press, Berkeley.

Reynolds, J. E., III, and Rommel, S. A. (1999). "Biology of Marine Marine Mammals." Smithsonian Press, Washington, DC.

Similä, T., and Ugarte, F. (1993). Surface and underwater observations of cooperatively feeding killer whales in northern Norway. *Can. J. Zool.* **71**, 1494–1499.

Tyack, P. L. (1981). Interactions between singing Hawaiian humpback whales and conspecifics nearby. *Behav. Ecol. Sociobiol.* **8**, 105–116.

Tyack, P. L. (1986). Population biology, social behavior and communication in whales and dolphins. *Trend. Ecol. Evol.* **1**, 144–150.

Vaughn, R. L., Shelton, D. E., Timm, L. L., Watson, L. A., and Würsig, B. (2007). Dusky dolphin (*Lagenorhynchus obscurus*) feeding tactics and multi-species associations. *N. Z. J. Mar. Freshw. Res.* **41**, 391–400.

Watwood, S. L., Owen, E. C. G., Tyack, P. L., and Wells, R. S. (2004). Signature whistle use by temporarily restrained and free-swimming bottlenose dolphins, *Tursiops truncatus. Anim. Behav.* **69**, 1373–1386.

Whitehead, H. (1990). Rules for roving males. *J. Theor. Biol.* **145**, 355–368.

Whitehead, H. (2003). "Sperm Whales: Social Evolution in the Ocean." University of Chicago Press, Chicago.

Würsig, B., Dorsey, E. M., Fraker, M. A., Payne, R. S., and Richardson, W. J. (1985). Behavior of bowhead whales, *Balaena mysticetus*, summering in the Beaufort Sea: A description. *Fish. Bull.* **83**, 357–377.

# Beluga Whale
## *Delphinapterus leucas*

### GREGORY M. O'CORRY-CROWE

## I. Characteristics and Taxonomy

The beluga whale is a member of the Monodontidae, the taxonomic family it shares with the narwhal, *Monodon monoceros*. Its name, a derivation of the Russian "beloye," meaning "white," appropriately enough captures its most distinctive feature, the pure white color of adults (Fig. 1). The Irrawaddy dolphin, *Orcaella brevirostris*, was considered by some to also be a member of this family. Although superficially similar to the beluga, recent

**Figure 1** Beluga whale, *Delphinapterus leucas*. Flip Nicklin/ Minden Pictures.

genetic evidence strongly supports its position as a member of the family Delphinidae (Lint *et al.*, 1990; LeDuc *et al.*, 1999). The earliest fossil record of the monodontids is of an extinct beluga *Denebola brachycephala* from late Miocene deposits in Baja California, Mexico, indicating that this family once occupied temperate ecozones (Barnes, 1984). Fossils of *D. leucas* found in Pleistocene clays in northeastern North America reflect successive range expansions and contractions of this species associated with glacial maxima and minima.

The beluga whale is a medium-sized toothed whale, 3.5–5.5 m in length and weighing up to 1500 kg. Males are up to 25% longer than females and have a more robust build. As their genus name ("..apterus"—without a fin) implies, they lack a dorsal fin and are unusual among cetaceans in having unfused cervical vertebrae allowing lateral flexibility of the head and neck. They possess a maximum of 40 homodont teeth, which become worn with age. Recent studies have found that beluga whales live much longer than previously thought. Levels of the radioisotope $^{14}$C rose sharply in the marine environment in the late 1950s because of nuclear bomb testing, and researchers were able to detect this increase in growth layers in beluga teeth. Using this increase as a reference point they determined that beluga whales most likely lay down only one growth layer group a year, rather than two (Stewart *et al.*, 2006). As well as doubling the maximum-recorded age from around 40 to 80 years, this discovery has increased the age of first reproduction and necessitated a revision of other life history parameters. Neonates are about 1.6 m in length and are born a gray-cream color that quickly turns to a dark brown or blue-gray. They become progressively lighter as they grow,

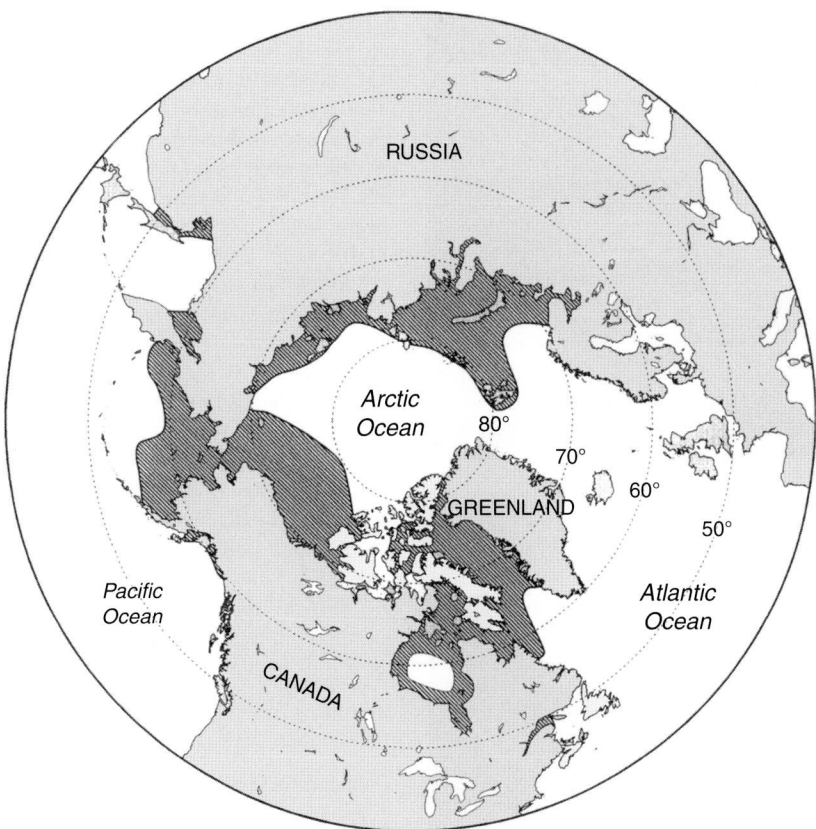

**Figure 2**  *The worldwide distribution of the beluga whale. The northernmost extent of its known range is off Alaska and northwest Canada and off Ellesmere Island, West Greenland, and Svalbard (>80°N). The southern limit of distribution is in the St. Lawrence River in eastern Canada (47–49°N).*

changing to gray, light gray, and finally becoming the distinctive pure white by about age 14 in females and 18 in males (Fig. 3).

Belugas are supremely adapted to life in cold waters. They possess a thick insulating layer of blubber up to 15-cm thick beneath their skin, and their head, tail, and flippers are relatively small. The absence of a dorsal fin is believed by some to be an adaptation to life in the ice or perhaps as a means to reduce heat loss. In its place, belugas possess a prominent dorsal ridge that is used to break through thin sea ice.

## II. Distribution and Abundance

The beluga or white whale inhabits the cold waters of the Arctic and subarctic (Fig. 2). Variation in body size across the species range has been taken as evidence of separate populations. The nonuniform pattern of distribution and predictable return of belugas to specific coastal areas further suggests population structure and has led to the treatment of these summering groups as separate management stocks. Re-sightings of marked or tagged individuals as well as differences in contaminant signatures and limited evidence of geographic variation in vocal repertoire add support to the independent identification of a number of these stocks. Although all are valid to varying degrees, many of these methods used for stock identification have limitations due to incomplete knowledge on year-round distribution, movement patterns, breeding strategies, and social organization. They provide little or no information on rates of individual or genetic

exchange, and although phenotypic differences are highly suggestive, they may not provide evidence of evolutionary uniqueness.

A number of molecular genetic studies confirmed that whales tend to return to their natal areas year after year and that dispersal among many separate summering concentrations is limited, even in cases where there are few geographic barriers (Brown Gladden *et al.*, 1997; O'Corry-Crowe *et al.*, 1997). These molecular findings reveal that knowledge of migration routes and destinations appears to be passed from mother to offspring, generation after generation. Such cultural inheritance of information leads to the evolution of discrete sub-populations, among which there is little dispersal. It is possible that many of these sub-populations may overwinter in a common area and that a certain amount of interbreeding may occur at this time. (deMarch *et al.*, 2002) Regardless of such potential gene flow, in situations where management is concerned with the degree of demographic connectivity among areas, demonstrating that few animals disperse among sub-populations is sufficient evidence to designate them as separate management stocks.

## III. Ecology

The evolutionary history and ecology of belugas are inextricably linked to the extreme seasonal contrasts of the north and the dynamic nature of the sea ice. As well as adaptation to the cold, life in this region has necessitated the evolution of discrete calving and possibly mating seasons, annual migrations, and a unique feature distinguishing it from most other cetaceans, an annual molt.

**B**

**Figure 3** *Beluga whales concentrating near the coast during the brief summer. Note the dark to light gray color of younger animals compared to the white of adults. Flip Nicklin/Minden Pictures.*

**Figure 4** *Aggregations of beluga whales interacting and rubbing on the substrate of a shallow estuary during the summer molt. Flip Nicklin/Minden Pictures.*

As the sea ice recedes in spring, belugas enter their summering grounds, often forming dense concentrations at discrete coastal locations, including river estuaries, shallow inlets, and bays (Fig. 3). Several explanations have been proposed as to why belugas return to these traditional summering areas. In some regions, sheltered coastal waters are warmer, which may aid in the care of neonates. The occupation of estuarine waters also coincides with the period of seasonal molt. Belugas have been observed to actively rub their body surface on nearshore substrates (Smith *et al.*, 1992; Fig. 4), and the relatively warm, low-salinity coastal waters may provide conditions that facilitate molting of dead skin and epidermal regrowth (St. Aubin *et al.*, 1990; Smith and Martin, 1994). Belugas feed on a wide variety of both invertebrate and vertebrate benthic and pelagic prey. In some parts of their range it is clear that belugas are feeding in nearshore waters on seasonally abundant anadromous and coastal fish such as salmon, *Oncorhynchus* spp.; herring, *Clupea harengus*; capelin, *Mallotus villosus*; smelt, *Osmerus mordax*; and saffron cod, *Eleginus gracilis* (Kleinenberg *et al.*, 1964; Seaman *et al.*, 1982). The relative importance of the above factors in determining coastal distribution patterns may vary among regions depending on environmental and biological characteristics (Frost and Lowry, 1990). It is clear, however, that belugas exhibit some degree of dependence on specific coastal areas.

In many areas of the Arctic, belugas soon leave these coastal areas to range widely off shore. Satellite tracking has recorded belugas moving up to 1100 km from shore and penetrating 700 km into the dense polar cap where ice coverage exceeds 90% (Suydam *et al.*, 2001). How these animals find breathing holes in this environment is still a mystery. Analysis of dive profiles suggests that beluga whales may combine the use of sound at depth to find cracks in the ice ceiling overhead. Diving data also indicate that belugas are probably feeding on deepwater benthic prey as well as ice-associated species, including polar cod, *Boreogadus saida*, closer to the surface (Martin *et al.*, 1998; Richard *et al.*, 2001).

Little is known about the distribution, ecology, or behavior of beluga whales in winter. In most regions belugas are believed to migrate in the direction of the advancing polar ice front. However, in some areas belugas may remain behind this front and overwinter in polynyas and ice leads. In the eastern Canadian Arctic some belugas overwinter in the North Water, a large area of open water in northern Baffin Bay (Finley and Renaud, 1980), while in the White, Barents, Kara and Laptev Seas belugas may occur year-round, remaining in polynyas in the deeper water during winter (Kleinenberg *et al.*, 1964; Boltunov and Belikov, 2002).

Killer whales (*Orcinus orca*), polar bears (*Ursus maritimus*) and humans prey on beluga whales. Belugas sometimes become entrapped in the ice where large numbers may perish or be hunted intensively by human.

## IV. Behavior and Physiology

In contrast to the frozen smile of the oceanic dolphins, the ability of belugas to alter the shape of their mouth and melon enables them to make an impressive array of facial expressions. The lateral flexibility of the head and neck further enhances visual signaling and enable beluga whales to maneuver in very shallow waters (1–3 m deep) in pursuit of prey, to evade predators, and generally exploit a habitat rarely used by other cetaceans.

Belugas typically swim in a slow rolling pattern and are rarely given to aerial displays. In nearshore concentration areas, however, such as Cunningham Inlet on Somerset Island in the Canadian High Arctic, belugas may engage in more demonstrative behaviors including spy hopping, tail waving, and tail slapping (Fig. 4).

Studies using satellite-linked transmitters attached to free-swimming whales have confirmed that beluga are capable of covering thousands of kilometers in just a few months, in open water and heavy pack ice alike, while swimming at a steady rate of 2.5–6 km/hr (Lydersen *et al.*, 2001; Suydam *et al.*, 2001). Sensors on these transmitters have also recorded belugas regularly diving to depths of 300–600 m to the sea floor and utilizing different oceanographic regimes (Lydersen *et al.*, 2002). In the deep waters beyond the continental shelf belugas may dive in excess of 1000 m, where the pressure is 100 times that at the surface, and remain submerged for up to 25 min (Martin *et al.*, 1998; Richard *et al.*, 2001)!

Belugas are sometimes seen singly but more commonly occur in groups of 2–10 that may aggregate at times to form herds of several hundred to more than a thousand animals. Single animals are always large adults, while in mixed herds adult males may form separate pods of 6–20 individuals. Adult females form tight associations with newborns and sometimes a larger juvenile, presumably an older calf. These "triads" may join similar groupings to form large nursery groups. At certain times of the year, age, and sex segregation may be more dramatic than at others with males migrating ahead of, or feeding apart from females, young, and immatures. In general, group structure appears to be fluid, with individuals readily forming and breaking brief associations with other whales. Apart from cow–calf pairs there appear to be few stable associations. However, considering the diverse vocal repertoire of beluga whales including individual signature calls, their wide array of facial expressions, and the variety of interactive behaviors observed, as well as the numerous accounts of cooperative behavior, this species appears capable of forming complex societies where group members may not always be in close physical proximity to each other.

In areas of open water beluga whales may divide their days into regular feeding and resting bouts. Belugas appear to predominantly hunt individually, even when within a group, but have also been observed to hunt cooperatively. A typical hunting sequence begins with slow directed movement combined with passive acoustic localization (search mode) followed by short bursts of speed and rapid changes of direction using echolocation for orientation and capture of prey (hunt mode) (Bel'kovitch and Sh'ekotov, 1990).

The beluga possesses one of the most diverse vocal repertoires of any marine mammal and has long been called the "sea canary" by mariners awed by its myriad sounds reverberating through the hulls of ships. Communicative and emotive calls are broadly divided into whistles and pulsed calls and are typically made at frequencies from 0.1 to 12 kHz. As many as 50 call types have been recognized; *groans*, *whistles*, *buzzes*, *trills*, and *roars* to name but a few. Although some geographic variation is apparent, efforts to determine whether there are substantial regional differences or dialects have been hampered by differences among bioacousticians in the categorization of vocalizations. Belugas are capable of producing individually distinctive calls to maintain contact between close kin and can conduct individual exchanges of acoustic signals, or dialogues, over some distance (Bel'kovitch and Sh'ekotov, 1990).

The echolocation system of the beluga whale is well adapted to the icy waters of the Arctic. Its ability to project and receive signals off the surface and to detect targets in high levels of ambient noise and backscatter enables it to navigate through heavy pack ice, locate areas of ice-free water, and possibly even find air pockets under the ice (Turl, 1990).

### V. Life History

Females become sexually mature at age 9–12, males some time later. Gestation is 14–14.5 months with a single calf born in late spring-early summer prior to, or coincident upon entry into warm coastal waters. Mothers produce milk of high caloric content and nurse their young for up to 2 years, the entire reproductive interval averaging 3 years. Little is known about the mating behavior or mating season of beluga whales. Mating is believed to primarily occur in late winter-early spring, a period when most belugas are still on their wintering grounds or on spring migration. Mating behavior, however, has been observed at other times of the year and the question of whether they have delayed implantation is unresolved.

### VI. Interactions with Humans

Because of their predictable migration routes and return to coastal areas, beluga whales have long been an important and reliable resource for many coastal peoples throughout the Arctic and subarctic. However, because past commercial harvesting drove a number of populations to the point of economic extinction, current levels of subsistence take from these populations may not be sustainable. Increasing human activity in the beluga's environment brings with it the threat of habitat destruction, disturbance, and pollution. In areas where there are large commercial fishing operations, belugas, particularly neonates, may be incidentally caught in gill nets. In a number of regions of the Arctic beluga whales exhibit strong avoidance reactions to ship traffic,

whereas in some coastal locations they appear to have developed a certain tolerance to boat traffic. The potential impacts of an emerging whale watching industry in more populated areas are as yet unquantified. In some areas belugas may also be victims of industrial pollution. A high incidence of various pathologies have been found in beluga whales in the St. Lawrence River in Canada and has been linked to high levels of heavy metals and organohalogens found in these whales. Some of these toxins may act by suppressing normal immune response and there is concern that contaminants are adversely affecting population growth (Béland, 1996). Finally, there is concern over the possible downstream effects of hydroelectric dams on estuarine habitats and the environmental and health risks associated with oil and gas development and mining in the Arctic. Beluga whales were one of the first cetaceans to be held in captivity when in 1861 a whale caught in the St. Lawrence River went on display at Barnum's Museum in New York. Today, beluga whales are one of the more common and popular marine mammals in oceanaria across North America, Europe, and Japan. The majority of these animals were wild-caught, but successful breeding programs at a number of facilities are increasing the number of belugas born in captivity. Although the majority of beluga whales in captivity educate and entertain the public, a number of whales have been put to work by the navies of the United States and former Soviet Union.

The large sizes of some Arctic populations and flexible habitat requirements of beluga whales indicate that this species may not be as sensitive to the environmental consequences of current and future climate change as other arctic marine mammals (Laidre *et al.*, 2008). Nevertheless, a number of small, isolated populations at the southern margins of the species range may be vulnerable to continued climate warming, where habitat loss in concert with the genetic and demographic effects of small population sizes may compromise individual fitness and population viability (O'Corry-Crowe, 2008). Furthermore, it is difficult to predict the consequences for beluga whales of increased human activities across the Arctic associated with climate amelioration.

### *See Also the Following Articles*

Arctic Marine Mammals ■ Climate Change

### *References*

Barnes, L. G. (1984). Fossil odontocetes (Mammalia: Cetacea) from the Almejas Formation, Isla Cedros, Mexico. *PaleoBios, Museum of Paleontology, University of California*, **42**, 46 pp.

Béland, P. (1996). The beluga whales of the St. Lawrence River. *Sci. Am.* **May**, 74–81.

Bel'kovitch, V. M., and Sh'ekotov, M. N. (1990). "The Belukha Whale: Natural Behaviour and Bioacoustics." USSR Academy of Science, Moscow, Translated by Woods Hole Oceanographic Institution, 1993.

Boltunov, A. N., and Belikov, S. E. (2002). Belugas (*Delphinapterus leucas*) of the Barents, Kara and Laptev Seas. *In* "Belugas in the North Atlantic and Russian Arctic" (M. P. Heide-Jørgensen, and Ø. Wiig, eds), Vol. 4, pp. 149–168. NAMMCO Sci. Publ.

Brown Gladden, J. G., Ferguson, M. M., and Clayton, J. W. (1997). Matriarchal genetic population structure of North American beluga whales *Delphinapterus leucas* (Cetacea: Monodontidae). *Mol. Ecol.* **6**, 1033–1046.

Finley, K. J., and Renaud, W. E. (1980). Marine mammals inhabiting the Baffin Bay North Water in winter. *Arctic* **33**, 724–738.

Frost, K. J., and Lowry L. F. (1990). Distribution, abundance, and movements of beluga whales, *Delphinapterus leucas*, in coastal waters

**B**

of western Alaska. *In* "Advances in Research on the Beluga Whale, *Delphinapterus leucas*" (T. G. Smith, D. J. St. Aubin, and J. R. Geraci, eds), pp. 39–57. *Can. Bull. Fish. Aquat. Sci.* **224.**

Kleinenberg, S. E., Yablokov, A. V., Bel'kovich, B. M., and Tarasevich, M. N. (1964). "Beluga (*Delphinapterus leucas*): Investigation of the Species." Academy of Sciences of the USSR, Moscow, Translated by Israel Program for Scientific Translations, 1969.

Laidre, K. L., Ferguson, S., Heide-Jørgensen, M. P., Lowry, L., Stirling, I., and Wiig, Ø. (2008). Quantifying the sensitivity of Arctic marine mammals to climate-induced habitat change. *Ecol. Appl.* **18**, S97–S125.

LeDuc, R. G., Perrin, W. F., and Dizon, A. E. (1999). Phylogenetic relationships among the delphinid cetaceans based on full cytochrome *b* sequences. *Mar. Mamm. Sci.* **15**, 619–648.

Lint, D. W., Clayton, J. W., Lillie, W. R., and Postma, L. (1990). Evolution and systematics of the beluga whale, *Delphinapterus leucas*, and other odontocetes: A molecular approach. *In* "Advances in Research on the Beluga Whale, *Delphinapterus leucas*" (T. G. Smith, D. J. St.Aubin, and J. R. Geraci, eds), pp. 7–22. *Can. Bull. Fish. Aquat. Sci.* **224.**

Lydersen, C., Martin, A. R., Kovacs, K. M., and Gjertz, I. (2001). Summer and autumn movements of white whales *Delphinapterus leucas* in Svalbard, Norway. *Mar. Ecol. Prog. Ser.* **219**, 265–274.

Lydersen, C., Nøst, O. A., Lovell, P., McConnell, B. J., Gammelsrød, T., Hunter, C., Fedak, M. A., and Kovacs, K. M. (2002). Salinity and temperature structure of a freezing Arctic fjord-monitored by white whales (*Delphinapterus leucas*). *Geophys. Res. Lett.* **29**, 2119.

deMarch, B. G. E., Maiers, L. D., and Freisen, M. K. (2002). An overview of genetic relationships of Canadian and adjacent populations of belugas (*Delphinapterus leucas*) with emphasis on Baffin Bay and Canadian eastern Arctic populations. *In* "Belugas in the North Atlantic and Russian Arctic" (M. P. Heide-Jørgensen, and Ø. Wiig, eds), 4, pp. 17–38. NAMMCO Sci. Publ.

Martin, A. (1996). "Beluga Whales." Voyager Press, Stillwater.

Martin, A. R., and Smith, T. G. (1992). Deep diving in wild, free-ranging beluga whales, *Delphinapterus leucas*. *Can. J. Fish. Aquat. Sci.* **49**, 462–466.

Martin, A. R., Smith, T. G., and Cox, O. P. (1998). Dive form and function in belugas *Delphinapterus leucas* of the eastern Canadian High Arctic. *Polar Biol.* **20**, 218–228.

O'Corry-Crowe, G. M. (2008). Climate change and the molecular ecology of Arctic marine mammals. *Ecol. Appl.* **18**, S56–S76.

O'Corry-Crowe, G. M., Suydam, R. S., Rosenberg, A., Frost, K. J., and Dizon, A. E. (1997). Phylogeography, population structure and dispersal patterns of the beluga whale *Delphinapterus leucas* in the western Nearctic revealed by mitochondrial DNA. *Mol. Ecol.* **6**, 955–970.

Richard, P. R., Heide-Jørgensen, M. P., Orr, J. R., Dietz, R., and Smith, T. G. (2001). Summer and autumn movements and habitat use by belugas in the Canadian High Arctic and adjacent areas. *Arctic* **54**, 207–222.

Seaman, G. A., Lowry, L. F., and Frost, K. J. (1982). Foods of belukha whales (*Delphinapterus leucas*) in western Alaska. *Cetology* **44**, 1–19.

Smith, T. G., and Martin, A. R. (1994). Distribution and movements of beluga, *Delphinapterus leucas*, in the Canadian High Arctic. *Can. J. Fish. Aquat. Sci.* **51**, 1653–1666.

Smith, T. G., St. Aubin, D. J., and Hammill, M. O. (1992). Rubbing behaviour of belugas, *Delphinapterus leucas*, in a High Arctic estuary. *Can. J. Zool.* **70**, 2405–2409.

St. Aubin, D. J., Smith, T. G., and Geraci, J. R. (1990). Seasonal epidermal molt in beluga whales, *Delphinapterus leucas*. *Can. J. Zool.* **68**, 359–367.

Stewart, R. E. A., Campana, S. E., Jones, C. M., and Stewart, B. E. (2006). Bomb radiocarbon dating calibrates beluga (*Delphinapterus leucas*) age estimates. *Can. J. Zool.* **84**, 1840–1852.

Suydam, R. S., Lowry, L. F., Frost, K. J., O'Corry-Crowe, G. M., and Pikok, D., Jr. (2001). Satellite tracking of eastern Chukchi Sea beluga whales in the Arctic Ocean. *Arctic* **54**, 237–243.

Turl, C. W. (1990). Echolocation abilities of the beluga, *Delphinapterus leucas*: A review and comparison with the bottlenose dolphin, *Tursiops truncatus*. In "Advances in Research on the Beluga Whale, *Delphinapterus leucas*" (T. G. Smith, D. J. St. Aubin, and J. R. Geraci, eds), pp. 119–128. *Can. Bull. Fish. Aquat. Sci.* **224.**

# Biogeography

## RICK LEDUC

### I. Introduction

Biogeography is the study of the patterns of geographic distribution of organisms and the factors that determine those patterns. This discipline plays a critical role in our understanding of marine mammal evolution and adaptation (Berta *et al.*, 2006). Although marine mammals are very mobile, and there is an apparent lack of physical barriers in the world ocean, only *Orcinus orca*, *Physeter macrocephalus*, and perhaps some of the balaenopterids could arguably be considered to have cosmopolitan distributions. Other species have restricted distributions (e.g., coastal South America, Indo-West Pacific), reflecting their ecological requirements and their geographic centers of origin. Because related species tend to have similar ecological requirements and dispersal abilities, the distribution of higher taxa can also show distinct tendencies and restrictions, which reflect the cumulative distributions of their included species. For example, while delphinids, river dolphins, and sirenians have their highest diversity in tropical latitudes, most pinniped, ziphiid, and phocoenid species occur in temperate and polar regions. From a geographic perspective, specific regions can thus be characterized as centers of diversity for these higher taxa, and past global changes in the environment will have influenced their evolutionary history. For example, cooling of the world climates during the Tertiary may have contributed to the radiation of the cold-water adapted pinnipeds and mysticetes.

### II. Types of Distributions

There has been considerable effort in recent years to better document what is known about marine mammal distributions (Rice, 1998; Read *et al.*, 2007). At the species level, distribution patterns can be described at different spatial scales. Broadly speaking, individual species are usually limited to certain latitudinal zones such as tropical, temperate, or polar regions. These descriptions can be further refined into subtropical, cold temperate, and so on, and correlated with patterns of ocean basin or hemisphere endemism. For example, *Stenella clymene* occurs only in the tropical Atlantic, *Eumetopias jubatus* in the cold temperate North Pacific, and *Dugong dugon* in the tropical Indo-West Pacific. On even smaller scales, species may be associated with specific physical features, such as the near-shore coastal areas (e.g., *Sousa* spp.) or the continental slope (e.g., *Berardius bairdii*), or with oceanographic features, such as specific water masses or even bodies of freshwater (e.g., *Lipotes vexillifer* and *Pusa sibirica*).

A few species, notably some of the baleen whales, are highly migratory, summering at high latitudes and spending the winter breeding season at lower latitudes. Some of the migrating rorqual species occupy (at least seasonally) a wide range of latitudes in both hemispheres, although the movements of the Northern and Southern Hemisphere populations are seasonally offset such that they do not normally co-occur in the tropics. At the other end of the spectrum, there are some species (e.g., *Phocoena sinus* and *Monachus schauinslandi*) that have highly restricted ranges. If a formerly wide-ranging species is now limited to a small area, its distribution is considered relict.

There are distributions that are described as pan-tropical (or pan-tropical/temperate), exhibited by many delphinids, ziphiids (e.g., *M. densirostris*), kogiids (e.g., *Kogia breviceps*), and balaenopterids (e.g., *Balaenoptera edeni*). There are a few species and species pairs that occur at higher latitudes in both hemispheres but are absent from tropical waters, the so-called antitropical species and species pairs. These are seen in the families Delphinidae (e.g., *Lissodelphis* spp.), Ziphiidae (e.g., *Hyperoodon* spp.), Phocoenidae (e.g., *Phocoena sinus/P. spinipinnis*), Phocidae (e.g., *Mirounga* spp.), and Otariidae (e.g., *Arctocephalus townsendi/A. philippii*).

## III. Ecology and History Determine Distribution

Beyond these descriptive aspects of biogeography are the factors that determine a given species' distribution. Generally, distributions are determined by the ecology and the history of the species. In some cases, distribution is limited because a species may not be adapted for living in certain environments. For example, tropical delphinids may not range into higher latitudes due to limitations on their abilities to thermoregulate in colder water or find food in different habitats. Tied into this is competition, either from closely related species or from ecologically similar species, which may exclude a species from a particular region in which it could otherwise survive. In the case of South American manatees, it is reasonable to surmise that competition places at least one boundary on the species' ranges. Throughout most of its range, *Trichechus manatus* occurs in both coastal and riverine habitats. However, it does not range into the Amazon River, where the exclusively freshwater *T. inunguis* occurs, although it occupies the coastal areas on either side of the river mouth. Here, the two species are parapatric, and competitive exclusion is likely at work.

The role that history plays in biogeographical patterns should not be overlooked, but it is not always evident from contemporary distributions. The dispersal abilities of organisms may partly explain why species occur in some areas and not in others. For example, the lack of otariids in the North Atlantic is probably not due to the lack of suitable habitat but rather lies in the inability of any North Pacific or South Atlantic species to get there. Of course, one could also tie this into their ecological requirements, in that dispersal to the North Atlantic would be more likely if North Pacific species ranged far enough north for animals to disperse via the Arctic Ocean across northern North America or Eurasia. For some species that have widely separated allopatric populations (e.g., *Cephalorhynchus commersonii*), dispersal from one region to the other is a likely explanation for their distribution. In other cases, vicariance events can explain allopatric distributions. For example, the two subspecies of *Platanista gangetica* occur in different river systems, the Indus and Ganges–Brahmaputra River systems. Although the two forms are not presently in contact, these rivers were all part of a single river until the late Pliocene and probably had sporadic connections through stream capture even until historical times. Therefore, the geographic separation of the populations is from a rather recent vicariance event.

Large-scale changes in the environment can have dramatic influences on species' distributions. For example, in times of global cooling, cold boundary currents in the ocean basins may have extended farther toward the equator. This, in turn, could have enabled temperate species to disperse across the equator to similar habitats in a different hemisphere, giving rise to the antitropical species. Among the antitropical species and species pairs, some tendencies in their distributions are apparent. For example, the northern counterpart occurs in both the North Atlantic and North Pacific in only one species. Although the long-finned pilot whale, *Globicephala melas*, has only been recorded live from the North Atlantic and the Southern Hemisphere, more than 1000-year-old skulls of this species have been unearthed in Japan. For the rest of the seven or so recognized antitropical species and species pairs, all except *Hyperoodon* (which occurs in the North Atlantic) have their northern members limited to the North Pacific. Perhaps the oceanographic and climatic conditions that allow trans-equatorial dispersal for temperate species occur more frequently or become more developed in the Pacific basin than in the Atlantic. The right whales (*Balaena* spp.) present a slightly different scenario, but one that is consistent with this pattern. Now recognized as three distinct species, molecular analyses indicate that the species in the North Pacific (*B. japonica*) and southern ocean (*B. australis*) are more closely related to each other than either is to the North Atlantic species (*B. glacialis*), suggesting a more recent trans-equatorial dispersal in the Pacific basin. The above comparisons do not include the latitudinal migrant species, such as many of the species of balaenopterids. For these, their seasonal occurrence at low latitudes greatly facilitates trans-equatorial dispersal and would not likely require any significant change in oceanographic or climatic conditions.

Latitudinal migrants do, however, present questions regarding the selective advantage to conducting such extensive movements—sometimes covering thousands of miles (e.g., gray whales *Eschrichtius robustus* and humpback whales *Megaptera novaeangliae*). Their occurrence at high latitudes can be explained by the greater abundance of food, but the selective advantage to their seasonal movements to less productive wintering areas is not as apparent. The fact that they occur in high latitudes in the winter season with some regularity means that escape from winter cold may not be a major factor for adults. Calving in warmer climes does make sense, and mating during the same season could lead to wholesale movements of a population. An alternative explanation is that they leave high latitudes in the winter to escape from killer whales, which occur in much higher densities in these areas.

Beyond consideration of the underlying mechanisms of a single species' distribution, it is possible to make inferences about the origins of entire ecological communities. One approach is known as vicariance biogeography (Nelson and Rosen, 1981; Wiley, 1988). Vicariance biogeographers look for congruence between the phylogenetic relationships among species and their geographical distributions. Species distributions can be superimposed on phylogenetic trees to create what are called area cladograms (Fig. 1). If the area cladograms of several unrelated but geographically similar higher taxa are congruent, it is good evidence that a specific sequence of vicariance events operated on all of those taxa as speciation mechanisms. Furthermore, it may allow the researcher to make inferences about the centers of origin for the higher taxa being considered.

If possible, one should try to incorporate the fossil and geologic record when inferring historical mechanisms in biogeography, especially among distantly related taxa. A case in point can be seen in the river dolphins. Among the river dolphins, *Inia* and *Pontoporia* appear to be closest (albeit very distant) relatives among the extant species, the former occupying several South American rivers that flow into the Atlantic, and the latter occurring along the Atlantic coast of South America. However, the closest (albeit even more distant) living

**B**

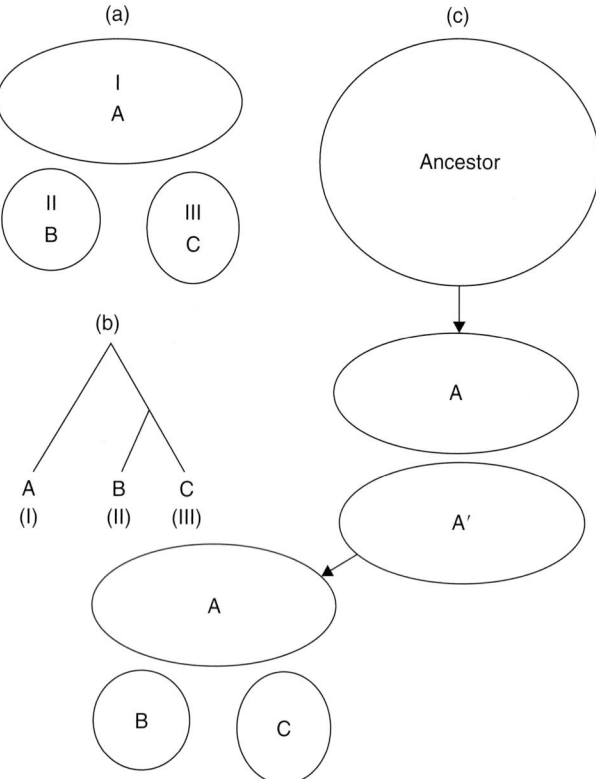

**Figure 1** *In vicariance biogeography, the speciation patterns are determined by vicariance events. The analysis attempts to reconstruct the sequence of vicariance events using the pattern of evolutionary relationships within a group of related species with allopatric distributions. (a) Species "A," "B," and "C" occupy ranges I, II, and III, respectively. (b) If a phylogenetic analysis determines that "B" and "C" are sister species to the exclusion of "A," this pattern of relationships is applied to their respective geographic ranges in an area cladogram. (c) Under this scenario, the range of the ancestral species is first divided by a vicariance event into a northern and a southern half. Populations in these two areas speciate into species "A" and "A'." Species "A'" is the inferred immediate common ancestor to "B" and "C." A later vicariance event divides the range of "A'" into eastern and western halves, giving rise to species "B" and "C." If unrelated species groups occupying these areas show congruent area cladograms, the support for this sequence of vicariance events is strengthened.*

especially for offshore species that are poorly represented in accessible deposits.

In a more recent context, human activities have played and are playing a role in altering species distributions, most often in the form of range reduction. For example, hunting may have played some role in the extirpation of gray whales (*Eschrichtius robustus*) from the North Atlantic and was certainly the primary cause of the complete extinction of the baiji (*Lipotes vexillifer*). More indirect, but just as dramatic, will be the shifts or reductions in species' distributions due to climate change, especially in high latitudes.

## IV. Taxonomic Patterns

As mentioned in the introduction, species within higher taxa share characteristics of their distributions to some degree. It is therefore possible to characterize the distributions of the different groups of marine mammals. The sirenians are primarily a tropical group, with mostly allopatric species occurring in warm coastal waters and some rivers of the Indo-West Pacific and both sides of the Atlantic. The trichechids are represented by two species in the new world (*Trichechus manatus* and *T. inunguis*), and a single congener (*T. senegalensis*) in western Africa, indicating the occurrence of a past trans-Atlantic dispersal event within that lineage. The family Dugongidae, formerly more diverse and widespread, now has only one extant species (*Dugong dugon*) that occurs in the Indian and west Pacific Oceans. One recently extinct species of dugongid (*Hydrodamalis gigas*) had a restricted range in the Commander Islands of the North Pacific, an anomalously cool habitat for a sirenian.

The majority of phocid species inhabit cold temperate and polar regions. Although no species occurs in both the northern and Southern Hemispheres, there are numerous species that are circumpolar either in the Arctic (e.g., *Erignathus barbatus*) or in the Antarctic (e.g., *Lobodon carcinophagus*). In fact, all of the southern phocid species have very broad distributions, their range expansions probably assisted by the oceanic currents that traverse all longitudes in the southern ocean. In the Northern Hemisphere, however, the habitats and ocean currents are much more fragmented by the continental landmasses. In addition to the circumpolar species, there are northern species that have more restricted ranges, either endemic to a single ocean basin (e.g., *Halichoerus grypus*) or limited to landlocked bodies of water (e.g., *Pusa caspica*). In contrast to the rest of the family, the three recent (two extant) species of monk seals (*Monachus* spp.) inhabit(ed) warmer waters of the Mediterranean and eastern Atlantic, Caribbean, and Hawaii. The spread of monk seals to Hawaii must have occurred prior to the rising of the Isthmus of Panama, which has separated the Caribbean and Pacific basins for the past 3 million years.

As a group, the otariids are similar to the phocids in their distribution, although they are less well represented at very high latitudes (near the pack ice) and do not occur in the North Atlantic at all. In addition, individual species tend to have more restricted ranges that are widely allopatric from their congeners. For example, the fur seal genus *Arctocephalus* is very widespread in the Southern Hemisphere, represented by six species (with an additional species endemic to the Galapagos Islands and another to the eastern North Pacific). However, there are only a handful of localities where more than one species occurs together; for the most part the species are allopatric. It appears then that the dispersal abilities of fur seals have allowed them to colonize many areas in the Southern Hemisphere but have not prevented the resulting disjunct populations from speciating. The odobenids are represented by a single circumpolar Arctic species, the walrus (*Odobenus rosmarus*).

relative of this pair is probably *Lipotes*, which is only found in the Yangtze River in China. Considering their freshwater and nearshore habits, it is not obvious from their contemporary distributions how they came to occupy areas a world apart. However, the fossil record has yielded intermediate species from various localities across the North Pacific. Furthermore, the geologic record shows that in the late Pliocene, the major river system in northern South America flowed westward, into what is now the Gulf of Guayaquil in Ecuador. It is thought that the ancestors of *Inia* entered this system from the Pacific. With the uplift of the Andes, much of the river reversed direction and flowed eastward, eventually becoming rivers such as the Amazon and Orinoco. With the North Pacific intermediate forms dying off, and the *Inia* lineage splitting to give rise to the ancestors of *Pontoporia* along the coast, one can see how the present-day species distributions came to be. It should be kept in mind, however, that determining the distributions of fossil taxa is notoriously difficult,

Cetacean species exhibit a wide range of distribution patterns. The family Balaenidae includes one antitropical species group (*Balaena* spp.) and one circumpolar Arctic species (*B. mysticetus*). The gray whale and the various species of balaenopterids are mostly latitudinal migrants in both hemispheres, although the Bryde's-like whales (*Balaenoptera brydei*, *B. omurai*, and *B. edeni*) are restricted to tropical and warm temperate waters, and some primarily migratory species include isolated populations that may be non-migratory (e.g., *Megaptera novaeangliae* in the northern Indian Ocean). In addition to the widespread common minke whale (*Balaenoptera acutorostrata*), the Southern Hemisphere also contains an endemic species of minke whale (*B. bonaerensis*). Similarly, the Southern Hemisphere is also home to two distinct forms (considered subspecies at present) of blue whale (*B. musculus*). In both of these cases, it is not known if the two southern forms represent divergent lineages that arose within the southern ocean or if they were the result of independent dispersal events across the equator.

Sperm whales (*Physeter macrocephalus*) are virtually cosmopolitan, and the kogiids (*Kogia sima* and *K. breviceps*) are worldwide in tropical and warm temperate waters. Beaked whales show a variety of distribution patterns, including pan-tropical species (e.g., *M. densirostris*), antitropical species pairs (*Berardius* spp.), and ocean basin endemics (e.g., *M. bidens*). Some (e.g., *M. peruvianus*) are only known from a few strandings within limited geographic areas. For most species of sperm whales and beaked whales, so little is known about their habits and ecological needs that it is difficult to hypothesize about the mechanisms that have led to their present distributions.

Three of the four species of river dolphins (*Inia geoffrensis*, *Lipotes vexillifer*, and *Platanista gangetica*) occur only in freshwater in specific tropical river systems, with the fourth species (*Pontoporia blainvillei*) having a restricted marine coastal range. The two species of monodontids (*Monodon monoceros* and *Delphinapterus leucas*) are circumpolar in the north, and are among the few resident polar cetaceans, although fossil species of this family occurred as far south as San Diego, California.

Apart from a single Indo-West Pacific coastal species that also ranges into freshwater (*Neophocaena phocaenoides*), the phocoenids are strictly marine and cold temperate to warm temperate in distribution, some with very restricted ranges (e.g., *Phocoena sinus*). Only one phocoenid, *P. phocoena*, has invaded the North Atlantic, becoming very widespread in both oceans of the Northern Hemisphere and even establishing isolated populations in the Black Sea and off West Africa.

The most speciose family of marine mammals, the delphinids, shows a wide variety of distributions, from pan-tropical species (e.g., *Stenella attenuata*) to ocean basin endemics (e.g., *Lagenorhynchus albirostris*) to species with wide-ranging but disjunct populations (e.g., *Delphinus capensis*). Many delphinids are pelagic, although some inhabit coastal waters (e.g., *Cephalorhynchus* spp.) and some even invade freshwater (e.g., *Sotalia fluviatilis*). Only one, *Orcinus orca*, seems to regularly range to the pack ice in the far north and south. For the many pan-tropical/warm temperate species, the continental landmasses effectively separate the populations inhabiting the Indian and Pacific Oceans from those inhabiting the Atlantic Ocean, raising the question of how they came to inhabit all the ocean basins. It has been hypothesized that during warm climatic periods, warm water extended far enough south to allow interchange and range expansion around the Cape of Good Hope. This would enable some species to become pantropical in their distribution, and the subsequent retreat of the warm water and isolation of populations could provide a speciation mechanism for the establishment of the tropical species endemic to the Atlantic Ocean (*S. frontalis* and *S. clymene*).

## V. Conclusion

Why do species live where they do? Answering such a simple question requires the examination of clues from the past as well as the present. Biogeography involves such diverse disciplines as geology, paleontology, ecology, physiology, behavior, and systematics. For marine mammals, studying biogeographical patterns presents real challenges. There is a paucity of information about past distributions and habitats, gaps in our knowledge of contemporary and recent distributions, uncertainties about evolutionary relationships, and a tremendous amount to learn about the basic ecology and physiology of many marine mammals.

## *See Also the Following Articles*

Climate Change ■ Systematics

## *References*

Berta, A., Sumich, J., and Kovacs, K. (2006). "Marine Mammals—Evolutionary Biology," 2nd edn. Academic Press, San Diego.

Brown, J. H., and Gibson, A. C. (1983). "Biogeography." Mosby, St. Louis, Missouri.

Cox, C. B., and Moore, P. D. (1985). "Biogeography," 4th edn. Blackwell Scientific Publications, Cambridge, Massachusetts.

Davies, J. L. (1963). The antitropical factor in cetacean speciation. *Evolution* **17**, 107–116.

Lincoln, R., Boxshall, G., and Clark, P. (1998). "A Dictionary of Ecology, Evolution and Systematics," 2nd edn. Cambridge University Press, Cambridge, UK.

Myers, A. A., and Giller, P. S. (eds) (1988). "Analytical Biogeography." Chapman and Hall, London.

Nelson, G., and Rosen, D. E. (eds) (1981). "Vicariance Biogeography: A Critique." Columbia University Press, New York.

Read, A.J., Halpin, P.N., Crowder, L.B., Best, B.D., Fujioka, E. (eds). (2007). OBIS-SEAMAP: Mapping marine mammals, birds and turtles. World Wide Web electronic publication. http://www.seamap.env.duke.edu, Accessed on October 15, 2007.

Rice, D. W. (1998). "Marine Mammals of the World: Systematics and Distribution." Soc. Mar. *Mamm. Spec. Pub.* **4**., Society for Marine Mammalogy, Lawrence, KS.

Wiley, E. O. (1988). Vicariance biogeography. *Ann. Rev. Ecol. System.* **19**, 513–542.

# Blubber

### Sara J. Iverson

Blubber, a dense vascularized layer of fat beneath the skin, is one of the most well-known and universal characteristics of marine mammals. Although it is not strictly present in polar bears (*Ursus maritimus*) or sea otters (*Enhydra lutris*), all cetaceans, sirenians and pinnipeds have blubber and it may comprise up to 50% of the body mass of some species at certain life stages. Blubber has long been recognized as the primary and most important site of fat, and thus also energy, storage in marine mammals. However, blubber also has a number of other important functions. The blubber layer serves as an insulator in mammals living in often cold marine environments and is thus central to their entire process of thermoregulation.

**B**

Blubber also affects buoyancy and functions as a body streamliner and elastic spring for efficient hydrodynamic locomotion. Although blubber is a dynamic tissue, which can reflect both nutritional state and life history stage of individuals, the tissue itself has likely evolved to best suit the lifestyles, stresses and constraints of specific groups and even individual species of marine mammals. Hence, the study of blubber can provide unique insights not only into phylogenetic relationships and environmental adaptations, but also into aspects of individual feeding habits, foraging ecology, species distribution, and demography that are otherwise difficult to study.

## I. The Structure of Blubber
### A. Tissue Characteristics

Blubber is a specialized subcutaneous layer of fat found only in marine mammals and is different from other types of adipose tissue in that it is anatomically and biochemically adapted to serve as an efficient and adjustable thermal insulator. The blubber layer is almost continuous across the body of marine mammals, lying over but not tightly fixed to the underlying musculature but absent on appendages. Although nearly continuous, the thickness, structure, and biochemical composition of the blubber can vary greatly over the body of an individual in some species and these differences are likely associated with localized differences in function. Consistent with its role as an insulator, there also usually exists a thermal, as well as biochemical, gradient through the depth of the blubber layer. The outer layer (nearest the skin) is usually cooler than the inner layer (nearest the muscle or body core) and activities of individual enzymes in each of these locations may be adapted to function at the different respective temperatures. The polar bear also deposits huge quantities of fat subcutaneously, which likely provides some degree of insulation. However, this superficial adipose tissue does not appear to be a specific anatomical adaptation for that purpose as it does not differ in structure from the superficial fat depots of other large terrestrial carnivores (Pond *et al.*, 1992).

Blubber, like other adipose tissue, is composed of numerous fat cells called adipocytes. Adipocytes develop prior to filling with fat and are composed, like other cells, of mostly protein and water. Once developed, adipocytes can alternately fill and empty with lipid and thus can change greatly in size. Mature adipocytes are generally large and spherical and packed densely into adipose tissue. The cells are surrounded and held in place by a mesh of structural collagen fibers. Although most other types of adipose tissue contain small to moderate amounts of collagen, blubber is distinct in being greatly enriched in collagen and elastic fibers. This gives blubber a firm, tough, and fibrous character from which it derives much of its mechanical and functional properties. The histological structure of the blubber in pinnipeds is relatively uniform throughout its depth. However, in some cetaceans, there is a distinct stratification of the tissue into an inner, middle, and outer layer based on the size, shape, and metabolic characteristics of adipocytes, as well as on the lipid and collagen content of the tissue. Blubber also contains numerous blood vessels and specialized shunts called arterio-venous anastomoses (AVAs), which allow larger and swifter blood flow than would be possible through capillaries alone and are important to the thermoregulation process. The blubber of manatees has been considered unusual in that a layer of muscle is imbedded in the middle of the ventral blubber layer, however, a similar arrangement has also been found at certain body sites in some otariids. A possible functional significance for this arrangement is not known.

### B. Variation in Thickness and Proximate Composition

The thickness of the blubber layer varies among species. In general, because body volume increases more rapidly than body surface area, larger species tend to have greater maximum blubber thickness. Thus, the depth of the blubber layer can be commonly 7–10 cm in adult pinnipeds, 20–30 cm in fin whales (*Balaenoptera physalus*) and up to 50 cm in the bowhead whale (*Balaena mysticetus*). In contrast, in one of the smallest odontocetes, the harbor porpoise (*Phocoena phocoena*), blubber depth generally reaches only 2.5–3.0 cm.

Beyond general species characteristics, the amount, depth, and chemical composition of the blubber also vary with age, nutrition, and reproductive status. The adipose tissue of many newborn mammals is empty of lipid, filling quickly after birth during the lactation period. Proliferation of fat depots in immature mammals is due to an increase in both adipocyte numbers and size. However, in adults, changes in the size of fat depots are primarily due to filling or emptying of adipocytes. The same appears largely true in the case of blubber. Although neonates of large baleen whales are born with a blubber layer that is several centimeters thick, most pinniped neonates are born with very little blubber, at less than 3 mm in depth in some otariids and accounting for less than 5–6% of body mass in most phocids. Most newborn pinnipeds rely instead primarily on fur (otariids) or lanugo (downy hair grown by fetuses, which remains for short periods after birth in many phocids) and delayed entry into the water. For instance, in newborn harp seals (*Phoca groenlandica*), blubber represents less than 6% of body mass and contains only 20% lipid (Worthy and Lavigne, 1983). This rapidly changes during the 12-day lactation period, such that the blubber of a newly weaned harp seal can comprise up to 50% of body mass and contains greater than 90% lipid, representing abundant and replete fat cells. In contrast, during reduced food intake or fasting, lipid is mobilized rapidly from adipocytes and hence undernourished marine mammals are characterized by greatly reduced blubber thickness and lipid content. Likewise, during annual events associated with fasting, such as lactation or molting in some species, blubber is also reduced in depth and lipid content with fat mobilization. For instance, the sternal blubber of female harbor seals (*Phoca vitulina*) changes during the 24-day lactation period from 3.8 to 1.4 cm in depth and from 92.3% lipid, 2.2% protein, and 5.5% water to 76.9% lipid, 5.9% protein, and 17.2% water (Bowen *et al.*, 1992); i.e., the increases in protein and water content reflect the larger proportion of "emptier" fat cells. In a similar manner, lipid content of adipose tissue in polar bears has been shown to reflect reproductive status and likely changes in prey availability (Thiemann *et al.*, 2006).

### C. The Lipids in Blubber

Depot lipid in animals is stored predominantly as triacylglycerols, which consist of three fatty acids esterified, i.e., linked by an ester bond to a glycerol molecule (three-carbon alcohol). The synthesis, storage, and catabolism of fatty acids are the components of lipid energy metabolism. Fatty acids in the marine food web are exceptionally complex and are characterized by high levels of long-chain polyunsaturated fatty acids (PUFA). During digestion of triacylglycerols by monogastric mammals, fatty acids are released from the glycerol backbone but not degraded and they are carried in the bloodstream and taken up by tissues the same way. These fatty acids are then either used for energy or stored as triacylglycerols in adipose tissue. Thus, fatty acids travel up the food chain intact, and because the kinds of fatty acids that can be biosynthesized or modified in mammals are quite limited, most fatty acids found in marine mammal blubber arise from the dietary intake of fish and other prey lipids. Hence, marine mammal blubber lipid is usually characterized by high levels of long-chain PUFA as well as unique fatty acids produced at lower trophic levels of the marine ecosystem.

Marine mammals, like other mammals, can also synthesize some of their own fatty acids from sources such as dietary amino acids consumed in excess of body needs (glucose would be another source but is scarce in diets of marine mammals). These synthesized fatty acids are usually restricted to those with 16 or 18 carbon atoms and usually, at most, one double bond (i.e., 16:0, 16:1$n$–7, 18:0, and 18:1$n$–9). Although these fatty acids are also common in all prey items of marine mammals, some are undoubtedly deposited in marine mammal blubber from biosynthesis.

There are several exceptions to the general characteristics of marine mammal blubber lipids described earlier. In addition to the usual fatty acids that are synthesized by all mammals, one very unusual fatty acid, isovaleric acid, is also found in the blubber of some species of toothed whales (odontocetes) which can arise only from localized biosynthesis. Isovaleric acid is unusual in that it is both very short (five carbons) and branched. When present, it is found in highest concentrations in the outermost layer (nearest the skin) of blubber (Koopman *et al.*, 2003). Additionally, besides the most common form of storage lipid, triacylglycerols, some marine mammals (primarily some odontocetes) store some or all of their fatty acids in blubber as wax esters (Koopman, 2007). A wax ester is a single fatty acid esterified to a long-chain (22–34 carbon) alcohol. In general, wax esters are firm, stable, and resistant to degradation. This is why sperm whale (*Physeter macrocephalus*) oil was popular as an illuminant in the last century.

## II. Role of Blubber in Temperature Regulation: Heat Conservation and Dissipation

As a whole-body envelope of insulation, blubber is central to thermoregulation in marine mammals. Marine mammals, like all mammals, are homeothermic endotherms and hence need to maintain a stable body core temperature of about 37°C in cooler (usually <25°C) and often much colder (−1 to 5°C) fluid environments. Additionally, heat is always lost far more rapidly to water than to air because the thermal conductivity of water is 25 times that of air. There are several ways marine mammals have dealt with this problem. One is to increase body size, which decreases the surface-to-volume ratio and thus provides less surface area per unit volume over which to lose heat. Even the smallest marine mammals are considered large mammals, being one to two orders of magnitude larger than small terrestrial mammals such as rodents and insectivores. Additionally, and perhaps more importantly, large body size generally allows for thicker insulation (be it fur or blubber), which further decreases heat conductance.

Although fur is a far more effective insulator than blubber in air and is used as the sole means of insulation by sea otters, fur acts by trapping pockets of air (a poor thermal conductor) among hairs, which then forms the effective insulative layer. Thus, potentially when fur is wetted, but more significantly when diving under pressure (as most marine mammals do), fur is compressed, expelling the air layer and thus losing its insulative properties. In contrast, blubber does not compress with depth and it is a good insulator because, like air, it has a lower thermal conductivity than water.

Adipose tissue is also less metabolically active than other tissues and thus requires less perfusion by blood, which would otherwise tend to cause heat loss at the body surface. Nevertheless, because blubber is vascularized, circulatory adjustments allow both heat conservation and dissipation as necessary. An important means of regulating heat transfer in marine mammals is by restricting and diverting blood flow through the blood vessels and AVAs in blubber. Restricting blood flow to the blubber's surface (i.e., skin) conserves body heat and allows blubber to act as an effective insulator against cold. Conversely, increasing blood flow into the blubber allows sometimes massive redistribution and dumping of body heat in cases of either very warm water or air or during intense activity (Heath and Ridgeway, 1999).

The effectiveness of blubber as an insulative layer depends on its thickness, lipid content, and lipid composition. As an insulation layer increases in thickness, the lower critical temperature of an animal decreases and thus the animal can accommodate a broader ambient temperature range without having to increase its metabolism for heat production (i.e., to remain thermoneutral). As mentioned earlier, many marine mammals, especially those of larger body size, possess a thick blubber layer, allowing them to remain thermoneutral at most of the temperatures of the world's oceans and, for some pinnipeds, even at air temperatures of −10 to −20°C on polar ice. However, smaller species are limited in the depth of blubber they can carry and also have relatively more surface area over which to lose heat. Hence, most of the smallest cetaceans do not occur at high latitudes. Less insulation increases the lower critical temperature of an animal and requires increased metabolism for heat production. The harbor porpoise is the smallest cetacean species to inhabit temperate waters of the Northern Hemisphere. Although its blubber depth is only several centimeters thick, it is generally twice the thickness and contains more lipid than does a similarly sized dolphin inhabiting tropical waters. These properties appear to confer up to four times greater insulative capacity (Koopman, 2007).

Depletion of lipid from blubber stores will decrease the insulative capacity of the tissue and may seriously compromise an individual, especially if nutritionally stressed. Thus, small species such as the harbor porpoise must feed nearly continually to maintain metabolism and to preserve their blubber's thickness and insulative capacities. In contrast, large whales can fast and mobilize blubber reserves for weeks or months, yet can remain thermoneutral due to a low surface-to-volume ratio as well as the maintenance of a still relatively thick blubber layer. Especially in cetaceans, the thickness, structure, and insulative properties of the blubber may vary across different regions of the body and thus the function of the blubber as an insulator may also vary regionally.

Variation in the lipid composition of blubber may also confer differing insulative capacities. As stated previously, the blubber of marine mammals is composed of large amounts of unsaturated fatty acids. Unsaturated fatty acids have lower melting points than do saturated fatty acids. Thus even when the temperature of the outermost layer of the blubber and skin are near that of cold ambient temperatures, blubber tissue can remain fluid and an effective insulator if the melting point of its fatty acids is low. Saturated and monounsaturated fatty acids abundant in marine mammal blubber (e.g., 16:0, 16:1, 18:0, and 18:1) have melting points of 13–70°C. However, nutritionally important polyunsaturated fatty acids are usually plentiful in marine mammal diets and thus in blubber, conferring an overall melting point in blubber lipid of less than −15°C. Additionally, in some small cold-water cetaceans such as the harbor porpoise, Dall's porpoise (*Phocoenoides dalli*), the bottlenose dolphin (*Tursiops*

**B**

*truncatus*), and the beluga (*Delphinapterus leucas*), high concentrations of the very unusual branched short-chain isovaleric acid are biosynthesized and deposited in blubber (see earlier discussion). Isovaleric acid has an extremely low melting point of $-37.6°C$, which clearly provides fluidity to especially the outer blubber layer of these animals. Although the exact physiological function of isovaleric acid is not understood, its presence may contribute to the superior insulative properties observed previously in harbor porpoise blubber. In contrast to most other marine mammals, while manatees (*Trichechus* sp.) can also store large amounts of blubber, they generally do not tolerate temperatures below $20°C$. As plant eaters, manatees must synthesize the majority of their blubber fatty acids, which would thus be restricted in their degree of unsaturation. However, little is known about the effectiveness of manatee blubber as an insulator in cold temperatures or the role that lipid composition might play in this ability.

### III. Role of Blubber in Energy Storage and Water Balance

Blubber, as a rich energy store, is important in the lives of marine mammals because of the critical role that stored lipid plays in their ecology, reproduction, and survival. Perhaps surprisingly, even though marine mammals obviously live in the environment within which they also forage, reproduction and especially lactation are often spatially and temporally separated from their feeding grounds. For instance, the greatest areas of feeding activity for the large baleen whales are in polar regions during the high primary productivity of summers. However, they migrate in winter to warm tropical waters of low food availability to give birth and nurse their young. In phocid and otariid pinnipeds, parturition and lactation occur on land or ice and thus these activities are also separated from the feeding environment of the lactating female. Female polar bears spend the first 3–4 months of lactation in winter dens, without eating or drinking.

In all female mammals, lactation represents the greatest energetic cost of reproduction, requiring large amounts of nutrient transfer and elevated maternal maintenance costs. Hence, lactation is usually associated with increased maternal food consumption. However, because large energy reserves can be stored in blubber in the form of lipid, baleen whales and large phocid seals are the only mammals (besides holarctic bears) that can complete much or all of lactation without feeding. Again, because a smaller body size constrains the size of blubber stores, the smaller phocids and otariids are able to fast for only portions of lactation. All species of marine mammals produce high fat milks (usually 30–60% fat) to maximize the efficiency of fat transfer from maternal blubber into milk and the subsequent efficiency of neonatal fattening and growth. In species that fast throughout lactation, females switch almost completely to a fat-based metabolism. For instance, during a 16-day lactation period, a gray seal (*Halichoerus grypus*) female draws 97% of the energy required for her own metabolic needs and 90% of the milk energy supplied to her pup solely from her blubber stores. Furthermore, the extent to which she can both maintain lactation and produce a fat pup depends on the size of the blubber layer she starts out with (Mellish *et al.*, 1999). Fasting female polar bears use their extensive subcutaneous adipose tissue in a similar manner during the first months of lactation in winter dens.

Blubber deposition is equally critical to the suckling neonate, both for thermoregulation and to act as an energy reservoir. For example, most newly weaned phocid pups rely on blubber deposited during the suckling period to survive their own subsequent fast of several weeks or months after their mothers have departed the breeding grounds. The energy supplied from blubber is critical to survival of the young while they learn how to forage on their own.

Adult males of many marine mammal species also fast or greatly reduce food intake during the breeding season and during their annual molting period. During these times they rely on stored lipid in blubber as their energy source. Sirenians also use blubber during fasting. For instance in the Amazon, manatees (*Trichechus inunguis*) face dry seasons of up to 6 months at a time, where low waters restrict them to the deep water areas of larger lakes where the aquatic plants they feed upon are unavailable. Hence, food intake during these periods is nil.

Finally, besides being an important fat and energy source for marine mammals, blubber is a critical source of water that is essential to maintaining water balance during fasting. Each kilogram of lipid that is mobilized from blubber and oxidized for energy use by an animal generates a net production of 1.07 kg of metabolic water. In fact, oxidation of blubber yields enough water such that individuals usually do not require an additional external source. This is true even of lactating females that are exporting large quantities of water in milk daily. For instance, a gray seal female exports about 23 kg of water in milk over a 16-day lactation period while fasting and has no external access to water during this time (Iverson *et al.*, 1993). Thus, in most species, blubber functions to maintain both water balance and energy metabolism during periods of fasting.

### IV. Role of Blubber in Locomotion

Several forces act on animals swimming in fluids, and blubber plays a significant role in the way marine mammals deal with these forces. The predominant restrictive force is drag, but the vertical forces of gravity and buoyancy also exist. Drag is the force that resists the movement of a body through a medium and is much greater in seawater than in air due to seawater's higher density and greater viscosity. The single most effective way to reduce both drag and the power required for forward motion through a fluid is to have a smooth streamlined shape. Although all marine mammals tend to be somewhat streamlined in body shape as defined by their musculoskeletal system, blubber provides their form with a smooth sculpted contour. Blubber thickness is often distributed across an animal in a nonuniform manner that ensures this. For instance, the blubber over the hind end of a seal may be thicker than would be necessary for insulative purposes. The blubber layer here instead serves to taper the animal more gradually than would be dictated by the musculoskeleton. In fact, another very important means by which to reduce drag on a body is to be spindle-shaped, i.e., to have a gradually tapering tail end. This acts to reduce the wake left by the animal moving through the water and hence further reduces the forces of drag. Again, blubber creates this effect in cetaceans by a thickening and sculpting of the tailstock (Pabst *et al.*, 1999). This locomotor function may actually constrain the way in which animals utilize their blubber as energy reserves. In large baleen whales as well as the smallest harbor porpoise, blubber may be greatest in depth and fat content, even during nutritional stress, in the posterior dorsal and tail areas of the body (Lockyer, 1987; Koopman, 1998), as blubber in these areas serves important locomotory functions by both streamlining and possibly acting as a biomechanical spring, capable of temporarily storing and releasing elastic strain energy (Pabst, 1996).

Finally, blubber also plays a role in the buoyancy of marine mammals. Buoyancy is the force that acts on a body submerged in water where, if the mass of the body is greater or less than the volume of water it displaces, it will experience either a net downward or net upward force, respectively. In most marine mammals (except the sea otter), buoyancy will be determined primarily by the ratio of its adipose tissue to lean body tissue and body mass. Fat-filled adipose tissue is less dense than seawater, whereas lean tissue is more dense. Thus, the degree to which marine mammals store blubber will affect their buoyancy and thus the energy expended in moving or maintaining position in water. Although some newly weaned phocid pups may be positively buoyant at greater than 43% adipose tissue, most adult marine mammals will not be positively buoyant and thus are not likely to require any counteracting of this force when at the bottom of dives or when feeding at the benthos. However, changes in blubber stores will clearly affect the degree to which they are negatively buoyant. Studies have demonstrated that seals descend faster during diving when they are more negatively buoyant than when they are less negatively buoyant, providing evidence that seals adjust their diving behavior in relation to seasonal changes in buoyancy (Webb *et al.*, 1998; Beck *et al.*, 2000).

## V. Insights from the Study of Blubber

Marine mammals are widely distributed in tropical, temperate, and cold oceans of the world and show a diversity of distributional patterns and apparent physiological adaptations. However, our understanding of these patterns, as well as of the foraging ecology of most marine mammals, is hindered by the difficulties in observing free-ranging animals that spend most or all of their lives at sea. Blubber is clearly of central importance to the structure and function of marine mammals. Due to the fact that blubber has evolved to serve complex functions, and yet the composition and amount of blubber carried by an individual can change rapidly, its study can provide unique insights into the lives of marine mammals as well as the ecosystems in which they live.

The ultrastructure, thickness, and proximate composition of blubber can provide insights into the feeding status of individuals as well as the functional significance of the blubber itself. As stated previously, the proximate composition, especially lipid content, of blubber changes radically in response to feeding and fasting behavior and thus, along with other nutritional indices, may be used to indicate nutritionally stressed vs robust individuals. Because many marine mammals go through predictable annual periods of fasting and fattening, the proximate composition of blubber can also be used to indicate the life cycle stage of an individual. In some cetaceans, the characteristics of blubber differ greatly across sites of the body and thus study of these properties can provide insight into the functions of blubber. For instance, the structure and composition of blubber at specific sites suggest that in some areas on the body (e.g., the thoracic-abdominal area), blubber may play a more important role in insulation and energy storage, whereas at other sites (e.g., the thick ridge posterior to the dorsal fin or at the caudal peduncle) blubber may serve more important roles in maintaining hydrodynamic shape and other locomotory functions (Koopman, 1998; Pabst *et al.*, 1999). Thus, the study of how blubber at these various sites is utilized during times of fat mobilization may provide further insight into adaptations of blubber structure. For example, the finding that blubber in the area of the caudal peduncle is rarely used and always thicker than needed for insulation, even during severe nutritional stress, lends support to the hypothesis that it may be more important in that region for structural support and locomotory functions than as an insulator or energy provider.

Blubber can also provide insight into adaptation and phylogenetic relationships. For instance, the characteristic of storing blubber lipid primarily as wax esters appears to be confined to a group of the odonotocetes (i.e., beaked whales and the sperm whale, *Physeter catodon*). The species in which blubber consists primarily of wax esters, although all closely related, are also all pelagic deep divers. Hence the study of their blubber may provide insight into phylogenetic patterns as well as roles that wax esters may play in deep diving animals (Koopman, 2007). The presence of isovaleric acid is likewise confined to a fairly restricted group of animals, which also may be under special thermal constraints (see earlier discussion). Thus the study of isovaleric acid in blubber may provide clues to its function and potential value in insulation. Additionally, in several species the presence of isovaleric in the outer layer of blubber increases in direct proportion with age, suggesting the possibility of using its level in blubber to estimate ages of unknown individuals in the same population (Koopman *et al.*, 2003).

Finally, the fatty acids in blubber can provide powerful insights into the foraging ecology and diets of both individuals and populations of marine mammals. As stated previously, fatty acids in the marine ecosystem are complex and diverse, fatty acids often travel up the food chain intact, and there are narrow limitations on their biosynthesis in marine mammals. Hence the fatty acids of marine mammal blubber arise in large part from dietary intake and therefore can be used to study aspects of diet and foraging ecology (Iverson, 1993). Given the dynamic nature of lipid mobilization and deposition in marine mammal blubber, fatty acids can provide insight into diets over both time and space. Studies on wild and captive animals demonstrate that there is direct deposition of dietary fatty acids in both marine mammals and their prey and that the influence of dietary fatty acid intake on blubber composition is both substantial and predictable, whether or not rapidly fattening (Iverson *et al.*, 1995; Kirsch *et al.*, 2000). Considered alone, fatty acids stored in a predator can provide powerful qualitative insight into spatial and temporal differences in foraging and diets of individuals and populations. However, recent advances have developed methods that use fatty acids in predators, along with their prey, to quantitatively estimate species composition of predator diets (Iverson *et al.*, 2004). Quantitative fatty acid signature analysis (QFASA) accounts for effects of predator metabolism on fatty acid deposition, and then determines the weighted mixture of prey species fatty signatures that most closely resembles that of the predator's fatty acid stores to thereby infer its diet. QFASA has been validated and used to estimate diets of free-ranging individuals in a number of pinniped species and the polar bear (Iverson *et al.*, 2004, 2006). A blubber biopsy (100–500 mg), or adipose tissue sample, from a free-ranging animal provides relatively non-invasive information about diet that is not dependent on prey with hard parts, nor limited to only the last meal, as are analyses of fecal or stomach contents. This is accomplished most easily in pinnipeds where, using a medical biopsy punch, one can safely obtain a complete sample through the full depth of 5–10 cm. However, in cetaceans, blubber is generally much thicker and layering of fatty acids in the blubber is more pronounced, with dietary fatty acids being most reflected in the inner and middle layers nearest the deep body core. Thus, less work has been done on live animals in these species. Nevertheless, QFASA is now being used to address broad ecosystem-scale processes and is providing new insight into foraging patterns and ecology of free-ranging marine mammals that would otherwise not be possible.

**B**

## VI. Other Specialized Fats

In addition to blubber, several other unusual and specialized fat bodies exist that are unique to a single group of cetaceans, the odontocetes or toothed whales. These fat bodies occur in the forehead tissue (melon) and in and around the mandibles of the lower jaw (mandibular fats) and play important roles in hearing and echolocation. They are composed of a unique array of lipid classes and fatty acids that are likely synthesized with these head tissues (Koopman *et al.*, 2003, 2006). These unusual fats are believed to facilitate sound reception by acting in the melon to focus high frequency sound produced in the nasal passages, while in the mandibular fats, they are organized to form a channel to transmit received sounds to the ear. In all odontocetes examined, short- and branched-chain fatty acids appear to be concentrated in the center of the inner mandibular fat body and immediately adjacent to the earbones. Because sound travels more slowly through these types of fatty acids, this should cause sound entering an odontocete head to bend inwards and be directed to the ears (Koopman *et al.*, 2006). The unique arrangement of lipids within these fat bodies and their direct effect on sound transmission is an important are of current research.

In conclusion, blubber and other fats play a number of major roles in the lives of marine mammals. These fats can also be a powerful tool in trying to understand adaptive solutions of species living in marine environments as wells as insights into their ecology and behavior.

## *See Also the Following Articles*

Skeletal Anatomy ■ Swimming ■ Pinniped physiology

## *References*

Beck, C. A., Bowen, W. D., and Iverson, S. J. (2000). Seasonal changes in buoyancy and diving behaviour of adult grey seals. *J. Exp. Biol.* **203**, 2323–2330.

Bowen, W. D., Oftedal, O. T., and Boness, D. J. (1992). Mass and energy transfer during lactation in a small phocid, the harbor seal (*Phoca vitulina*). *Physiol. Zool.* **65**, 844–866.

Heath, M. E., and Ridgeway, S. H. (1999). How dolphins use their blubber to avoid heat stress during encounters with warm water. *Am. J. Physiol.* **276**, R1188–R1194.

Iverson, S. J. (1993). Milk secretion in marine mammals in relation to foraging: Can milk fatty acids predict diet? *Symp. Zool. Soc. Lond.* **66**, 263–291.

Iverson, S. J., Bowen, W. D., Boness, DJ., and Oftedal, O. T. (1993). The effect of maternal size and milk output on pup growth in grey seals (*Halichoerus grypus*). *Physiol. Zool.* **66**, 61–88.

Iverson, S. J., Oftedal, O. T., Bowen, W. D., Boness, DJ., and Sampugna, J. (1995). Prenatal and postnatal transfer of fatty acids from mother to pup in the hooded seal. *J. Comp. Physiol.* **165**, 1–12.

Iverson, S. J., Field, C., Bowen, W. D., and Blanchard, W. (2004). Quantitative fatty acid signature analysis: A new method of estimating predator diets. *Ecol. Monogr.* **74**, 211–235.

Iverson, S. J., Stirling, I., and Lang, S. L. C. (2006). Spatial and temporal variation in the diets of polar bears across the Canadian arctic: Indicators of changes in prey populations and environment. *Symp. Zool. Soc. Lond.: Conservation Biology Series* **12**, 98–117.

Kirsch, P. E., Iverson, S. J., and Bowen, W. D. (2000). Effect of diet on body composition and blubber fatty acids in captive harp seals (*Phoca groenlandica*). *Physiol. Biochem. Zool.* **73**, 45–59.

Koopman, H. N. (1998). Topographical distribution of the blubber of harbor porpoises (*Phocoena phocoena*). *J. Mammal.* **79**, 260–270.

Koopman, H. N. (2007). Phylogenetic, ecological, and ontogenetic factors influencing the biochemical structure of the blubber of odontocetes. *Mar. Biol.* **151**, 277–291.

Koopman, H. N., Iverson, S. J., and Read, A. J. (2003). High concentrations of isovaleric acid in the fats of odontocetes: Variation and patterns of accumulation in blubber vs. stability in the melon. *J. Comp. Physiol.* **173**, 247–261.

Koopman, H. N., Budge, S. M., Ketten, D. R., and Iverson, S. J. (2006). The topographical distribution of lipids inside the mandibular fat bodies of odontocetes: Remarkable complexity and consistency. *IEEE J. Oceanic Eng.* **31**, 95–106.

Lockyer, C. (1987). Evaluation of the role of fat reserves in relation to the ecology of North Atlantic fin and sei whales. *In* "Approaches to Marine Mammal Energetics" (A. C. Huntley, D. P. Costa, G. A. J. Worthy, and M. A. Castellini, eds), pp. 184–203. Society for Marine Mammalogy Special Publication No. 1., Allen Press, Lawrence, KS.

Mellish, J. E., Iverson, S. J., and Bowen, W. D. (1999). Individual variation in maternal energy allocation and milk production in grey seals and consequences for pup growth and weaning characteristics. *Physiol. Biochem. Zool.* **67**, 677–690.

Pabst, D. A. (1996). Springs in swimming animals. *Am. Zool.* **36**, 723–735.

Pabst, D. A., Rommel, S. A., and McLellan, W. A. (1999). The functional morphology of marine mammals. *In* "Biology of Marine Mammals" (J. E. Reynolds, and S. A. Rommel, eds), pp. 15–72. Smithsonian Institution Press, Washington, DC.

Pond, C. M., Mattacks, C. A., Colby, R. H., and Ramsay, M. A. (1992). The anatomy, chemical composition, and metabolism of adipose tissue in wild polar bears (*Ursus maritimus*). *Can. J. Zool.* **70**, 326–341.

Thiemann, G. W., Iverson, S. J., and Stirling, I. (2006). Seasonal, sexual, and anatomical variability in the adipose tissue composition of polar bears (*Ursus maritimus*). *J. Zool. (Lond.)* **269**, 65–76.

Webb, P. M., Crocker, D. E., Blackwell, S. B., Costa, D. P., and Le Boeuf, B. J. (1998). Effects of buoyancy on the diving behavior of northern elephant seals. *J. Exp. Biol.* **201**, 2349–2358.

Worthy, G. A. J., and Lavigne, D. M. (1983). Changes in energy stores during postnatal development of the harp seal, *Phoca groenlandica*. *J. Mammal.* **64**, 89–96.

# Blue Whale
## *Balaenoptera musculus*

### Richard Sears and William F. Perrin

The blue whale is a baleen whale belonging to the family Balaenopteridae, which includes the group of cetaceans known as rorquals (Fig. 1). Common names are blue whale, sulfur-bottom, Sibbald's rorqual, great blue whale, and great northern rorqual. The largest animal known to have existed on Earth, it is found worldwide, ranging into all oceans (Yochem and Leatherwood, 1985).

## I. Characteristics and Taxonomy

On average, Southern Hemisphere blue whales are larger than those in the Northern Hemisphere. The largest recorded were caught off the South Shetlands and South Georgia and were 31.7–32.6 m (104–107 ft) long. The largest recorded for the Northern Hemisphere was a 28.1-m (92-foot) female reported in whaling statistics from catches in Davis Strait. In the North Pacific females of 26.8 m (88 ft) and 27.1 m (89 ft) have been recorded. A 190-ton female was reported taken off South Georgia in 1947; however, body weights of adults generally range from 50 to 150 tons. For maximum size descriptions, female measurements are used because female baleen whales are larger than males.

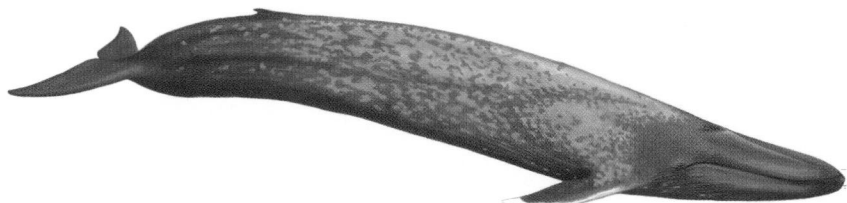

Figure 1 *Blue whale showing the characteristic mottled pigmentation of the species. Drawing by Daniel Grenier.*

Blue whales project a tall (up to 10–12 m) spout, denser and broader than that of the fin whale, *B. physalus*, which in calm conditions can help distinguish between the two species. When surfacing, the blue whale raises its massive shoulder and blowhole region out of the water more than other rorquals. The prominent fleshy ridge just forward of the blowhole, known as the "splash guard," is strikingly large in this species.

When seen from above, blue whales have a tapered elongated shape, with a huge broad, relatively flat, U-shaped head, adorned by a prominent ridge from the splash guard to the tip of the upper jaw or rostrum and massive mandibles. The baleen is black, half as broad as its maximum 1-m length, and 270–395 plates can be found on each side of the upper jaw. There are 60–88 throat grooves or ventral pleats running longitudinally parallel from the tip of the lower jaw to the navel, which enable the throat or ventral pouch to distend when feeding.

The dorsal fin is proportionally smaller than in other balaenopterids and varied in shape, ranging from a small nubbin to triangular and falcate and is positioned far back on the body. The flippers are long and bluntly pointed, slate gray, with a thin white border dorsally and white ventrally; they reach up to 15% of the body length.

In the field, particularly on bright days, blue whales generally appear much paler in coloration than all species of large whale except for the gray whale, *Eschrichtius robustus*, with which it should not be confused due to a great difference in size. Above water, the characteristic mottled pigmentation is a blend of light and dark shades of gray displayed in patches of varying sizes and densities. The underwater color is slate blue on overcast days to silvery or turquoise blue on bright sunny days depending on the clarity of the water. The mottling is found along the body dorsoventrally, occasionally on the flippers, but not on the head and tail flukes. Two prominent pigmentation configurations are found in blue whales, one where a darker, dominant background is mottled with sparser pale patches of pigmentation, while in the other there is a predominantly pale background mottled with sparser dark patches. Blue whale pigmentation can vary, however, from very sparse mottling, where the individual appears uniformly pale or dark, to densely mottled individuals, where the pigmentation is a highly contrasted variegation of spots unique to each whale, which is used in studies involving individual identification. Distinct chevrons curving down and angled back from the apex on both sides of the back behind the blowholes and either pale or dark in tone can be found on some individuals. The tail flukes, sometimes striated ventrally, are predominantly gray above and below; however, some individuals do have white patches of pigmentation on the ventral surface that are used for individual identification (Calambokidis *et al.*, 1990; Sears *et al.*, 1990). The trailing margin of the tail is either straight or curves very slightly from each tip to the median notch. A yellow-green to brown cast, caused by the presence of a diatom (*Cocconeis ceticola*) film, can be seen covering all or part of the body of blue whales found in cold

waters. The yellowish, diatom-induced tint is the reason the "sulfur-bottom" moniker was once used for blue whales.

Three subspecies have been designated: what has been considered the largest, *B. musculus intermedia*, found in Antarctic waters; *B. musculus musculus* in the Northern Hemisphere; and *B. musculus brevicauda*, from the subantarctic zone of the southern Indian Ocean and south western Pacific Ocean, also colloquially known as the "pygmy" blue whale. Although the latter designation is now generally accepted, its validity remains in question.

Our knowledge of the phylogeny of the baleen whales is still in flux. In recent molecular studies, the blue whale has been variously suggested to be the sister taxon of a clade including the Bryde's (*B. edeni*) and sei (*B. borealis*) whales, in combination with them a sister clade of the fin and humpback (*Megaptera novaeangliae*) whales, with gray whales (*Eschrichtius robustus*) the next up the tree (Rychel *et al.*, 2004); in the same arrangement, but with the minke whales (*B. acutorostrata* and *B. bonaerensis*) coming in before the gray whale (Nikaido *et al.*, 2005; Sasaki *et al.*, 2005); again in the same arrangement but with the gray whale not included in the analysis, the balaenids being a sister clade to all the balaenopterids (Nishida *et al.*, 2007); and sister taxon to a clade containing all the other baleen whales except the balaenids (Hatch *et al.*, 2006). In a morphological cladistic analyses, it grouped with the common minke whale in a clade sister to the humpback (Steeman, 2007). Further work is obviously needed.

## II. Distribution and Abundance

Despite having being reduced greatly due to whaling, the blue whale remains a cosmopolitan species separated into populations from the North Atlantic, North Pacific, and Southern Hemisphere (Fig. 2). In the North Atlantic, eastern and western subdivisions are recognized. Photo-identification work from eastern Canadian waters indicates that blue whales from the St. Lawrence, Newfoundland, Nova Scotia, New England, and Greenland all belong to the same population, whereas blue whales photo-identified off Iceland and the Azores appear to be part of a separate population. The best known population in the North Atlantic is that found in the St. Lawrence from April to January, where 435 individuals have been catalogued photographically (Sears *et al.*, 1990). Apart from the Icelandic and Azores sightings, few blue whales have been reported from eastern North Atlantic waters recently. North Atlantic blue whale abundance probably ranges from 600 to 1500 at this time, although more extensive photo-identification and shipboard surveys are needed for more reliable estimates.

In the North Atlantic, blue whales reach as far north as Davis Strait and Baffin Bay in the west, whereas to the east they travel as far north as Jan Mayen Island and Spitzbergen during summer months. Whales sighted recently in winter and spring off the Azores and Canary Islands could be migrating north along the mid-Atlantic ridge to Iceland, where they are seen from May to September.

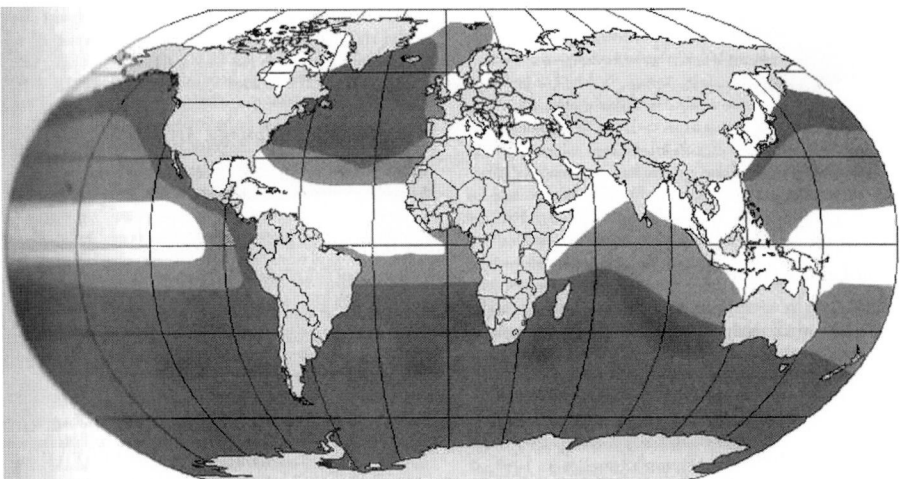

**Figure 2**  *Global distribution of blue whales. Darker gray indicates higher densities.*

Others probably migrate along the European coast, far offshore and out around Ireland north to either Iceland or Norway. It is not clear where the whales winter in the North Atlantic. Some have been observed in the St. Lawrence as late as February; however, acoustic studies have revealed that they are spread out across the North Atlantic basin, south as far as Bermuda and Florida, with concentrations south of Iceland, off Newfoundland and Nova Scotia. The southernmost observations on the eastern side of the North Atlantic are in the waters between the coast of Africa and the Cape Verde Islands.

In the North Pacific, where as many as five subpopulations were thought to exist, acoustic analysis of blue whale vocalizations now indicates there are no more than two. The best known is that from the eastern North Pacific where blue whales can be found as far north as Alaska but are regularly observed from California in summer, south to Mexican and the Costa Rica Dome waters in winter. Abundance estimates of approximately 3000 animals (CV = 0.14) by line-transect methods and 2000 by capture–recapture (photo-identification) have been determined for this population, which has been studied extensively over a good portion of its range (Calambokidis and Barlow, 2004). From late fall to spring, blue whales can be found in the Gulf of California, Mexico, and south to offshore waters of Central America. By April and May they migrate north along the West coast of North America, where a large proportion is found in California waters. From there some reach Canadian waters, and other groups may disperse north to the Gulf of Alaska or west toward the Aleutian Islands.

Few blue whales have been reported recently from the western North Pacific, including the Aleutian Islands, Kamchatka, Kurils, and Japan. They are thought to migrate to Kamchatka or the Kuril Islands and probably farther northeast.

Blue whales are also found in the northern Indian Ocean; however, it is not clear whether these form a distinct population.

In the Southern Ocean, where the blue whale was historically most abundant, it is very rare today, with the abundance estimate at 1700 (95% confidence interval 860–2900) (Branch *et al.*, 2007). A population of 424 (CV = 0.42) has been estimated to frequent the Madagascar Plateau in the austral summer (Best *et al.*, 2003). Although the general population structure in the Southern Ocean is not well understood, evidence shows discrete feeding stocks. A feeding and nursing ground was recently discovered in southern Chile (Hucke-Gaete *et al.*, 2004). Consistent with these feeding areas, the International Whaling Commission has assigned six stock areas for blue whales in the Southern Hemisphere.

## III. Ecology

Food availability probably dictates blue whale distribution for most of the year. Although they can be found in coastal waters of the St. Lawrence, Gulf of California, Mexico, and California, they are found predominantly offshore. They appear to feed almost exclusively on euphausiids (krill) worldwide in areas of cold current upwellings (e.g., in the Southern Hemisphere—Branch *et al.*, 2007). When they locate suitably high concentrations of euphausiids, they feed by lunging with mouth wide open and gulping large mouthfuls of prey and water. The mouth is then almost completely closed and the water is expelled by muscular action of the distended ventral pouch and tongue through the still exposed baleen plates. Once the water is expelled, the prey is swallowed. When they feed just a few meters below the surface, they often surface slowly, belly first, exposing the throat grooves of the ventral pouch, roll to breathe, and evacuate the water before diving to take their next mouthful. If the prey is close to the surface, blue whales lunge vigorously on their sides or lunge up vertically by projecting their cavernous lower jaws 4–6 m up through the surface. Although surface feeding has often been observed during the day, it is more usual for blue whales to dive to at least 100 m into layers of euphausiid concentrations during daylight hours and rise to feed near the surface in the evening, following the ascent of their prey in the water column. In the North Atlantic, blue whales feed on the krill species *Meganyctiphanes norvegica*, *Thysanoessa raschii*, *T. inermis*, and *T. longicaudata*; in the North Pacific, *Euphausia pacifica*, *T. inermis*, *T. longipes*, *T. spinifera*, and *Nyctiphanes symplex*. In Antarctic waters they prey on *E. superba*, *E. crystallorophias*, and *E. vallentini*.

Documentation of natural mortality is rare. The principal predator is the killer whale, *Orcinus orca*, but there is little evidence of attacks on blue whales in the North Atlantic or Southern Hemisphere. However, in the Gulf of California, Mexico, 25% of the blue whales photo-identified carry rake-like killer whale teeth scars on their tails, indicating that attacks occur with some regularity but are probably rarely successful. In the St. Lawrence, ice entrapment, where animals have been crushed, stranded, or suffocated by current and wind-driven ice floes in the late winter-early spring, has been reported.

## IV. Behavior and Physiology

Blue whales are observed most commonly alone or in pairs; however, concentrations of 50 or more can be found spread out in areas of high productivity.

Although not noted for raising their flukes when diving, approximately 18% of blue whales observed in the western North Atlantic and Northeast Pacific do so. This is an individual characteristic, and if the individual is relaxed it will generally raise its flukes high up in the air on each sounding dive. When disturbed, blue whales that raise their flukes when diving will often not raise their tails as high out of the water or not at all and dive more quickly from the surface.

When foraging or feeding at depth, blue whales will generally dive for 8–15 min; 20-min dives are not uncommon. The longest dive recorded was of 36 min; however, dives of more than 30 min are rare. They generally swim at 3–6 km/h when feeding. When traveling, they can attain speeds of 5–30 km/h and when chased by boats, predators, or interacting with other blue whales, they can reach upward of 35 km/h.

Blue whales vocalize regularly throughout the year with peaks from midsummer into winter months. The majority of vocalizations are low frequency or infrasonic sounds of 17–20 Hz, lower than humans can detect. Their sounds, at 188 decibels (re: 1 μPa at 1 m) are one of the loudest and lowest made by any animal. The calls can be heard easily for hundreds of kilometers, thousands of kilometers under optimal oceanographic conditions, and may cover whole ocean basins. The low frequencies are ideal for communication between individuals of a widely dispersed and nomadic species through water without much loss of information. Geographic variation, seasonality, and diel variation in the sounds have been studied intensively in recent years (Stafford, 2003; Širović et al., 2004; Stafford et al., 2004, 2005; Wiggins et al., 2005); the sounds may prove to be useful in delineating populations (McDonald et al., 2006).

Little is known of mating behavior in the species. However, female–male pairings have been noted with regularity in the St. Lawrence from summer into fall, some lasting for as long as 5 weeks. When a female–male pair is approached by a third blue whale, or even a fin whale, vigorous surface displays ensue, where all three animals can be seen racing high out of the water, porpoising forward, causing an explosive bow wave splash, and even at times breaching. Such interactions usually last for 7–25 min.

## V. Life History

Blue whales reach sexual maturity at 5–15 years of age; however, 8–10 years appear to be more usual for both sexes. Length at sexual maturation in females from the Northern Hemisphere is 21–23 m and is 23–24 m in the Southern Hemisphere. Males reach sexual maturity at 20–21 m in the Northern Hemisphere and at 22 m in the Southern Hemisphere. Mating takes place starting in late fall and continues throughout the winter. Females give birth every 2–3 years in winter after a 10- to 12-month gestation period. The calves, which weight 2–3 tons and measure 6–7 m at birth, are weaned when approximately 16 m long at 6–8 months. No specific breeding ground has been discovered for blue whales in any ocean, although mothers and calves are sighted regularly in the Gulf of California, Mexico, in late winter and spring. A portion of the Northeast Pacific Ocean blue whale population could be using this region as a breeding ground.

Longevity is thought to be at least 80–90 years and probably longer. What is certain, however, after extensive photo-identification fieldwork on known individuals in the St. Lawrence and northeast Pacific, is that they live for at least 40 years.

## VI. Interactions with Humans

Because of its great size and the commercial value of the products it yielded, the blue whale was hunted relentlessly beginning in the late 1800s. The greatest number of blue whales was taken from the early 1900s until the late 1930s, with the peak being in the 1930–1931 season when nearly 30,000 were killed. The height of blue whale whaling coincided with the advent of explosive harpoons, steam power vessels, and the construction of factory ships, which could process whale carcasses at sea. The blue whale was severely depleted by whaling, particularly in the Southern Hemisphere, where during the first half of the twentieth century 325,000–360,000 were killed in Antarctic waters alone. A further 11,000 were taken in the North Atlantic, primarily in Icelandic waters, and 9500 in the North Pacific. This unbridled hunt for blue whales, which lasted until its worldwide protection in 1966, brought the blue whale to the brink of extinction and it is still an endangered species today. However, there is evidence for population increase in the Antarctic (Branch, 2004).

Although reports of blue whales approaching vessels are rare, at least 25% of the blue whales photo-identified in the St. Lawrence carry scars that can be attributed to collisions with ships, including whale-watching vessels. This type of scarring has been reported for a few Northeast Pacific blue whales as well. Ship strikes in heavy shipping areas, such as the St. Lawrence and California coast, may have an impact at populations, but data are not available at this point.

Though 12% of blue whales found in eastern Canadian waters carry marks related to contacts with fishing gear, few lethal entanglements have been reported. The size and power of this whale probably enables it to tear through fishing gear relatively unscathed.

Persistent contaminants accumulated over time, such as PCBs commonly found in blue whales from eastern Canadian waters, may have an impact on reproduction and limit the recovery of certain populations.

It has been shown that blue whales react strongly to approaching vessels. The degree of reaction depends on the whale's behavior, as well as the distance, speed, and direction of the vessel at the time of approach. The increasing anthropogenic noise may have an impact on blue whales and their habitat and could also limit recovery of this species.

## See Also the Following Articles

Baleen Whales ■ Cetacean Life History ■ Fluking ■ Noise ■ Effects of Pollution and Marine Mammals

## References

Best, P. B., et al. (2003). The abundance of blue whales on the Madagascar Plateau, December 1996. J. Cetacean Res. Manage. 5, 253–260.

Branch, T. A. (2004). Summary of evidence for increase in Antarctic (true) blue whales. J. Cetacean Res. Manage. (Suppl) 6, 256–258.

Branch, T. A., et al. (2007). Past and present distribution, densities and movements of blue whales Balaenoptera musculus in the Southern Hemisphere and northern Indian Ocean. Mam. Rev. 7, 116–175.

Calambokidis, J., and Barlow, J. (2004). Abundance of blue and humpback whales in the eastern North Pacific estimated by capture-recapture and line-transect methods. Mar. Mamm. Sci. 20, 63–85.

Calambokidis, J., et al. (1990). Sightings and movements of blue whales off central California 1986–88 from photo-identification of individuals. Rep. Int. Whal. Commn 12, 343–348.

Clark, C. W., and Charif, R. A. (1998). Acoustic monitoring of large whales to the West of Britain and Ireland using bottom-mounted hydrophone arrays. October 1996–September 1997. *JNCC Rep.* **281**.

Hatch, L. T., Dopman, E. B., and Harrison, R. G. (2006). Phylogenetic relationships among the baleen whales based on maternally and paternally inherited characters. *Mol. Phylogenet. Evol.* **41**, 12–27.

Hucke-Gaete, R., Osman, L. P., Moreno, C. A., Findlay, K. P., and Llunjblad, D. K. (2004). Discovery of a blue whale feeding and nursing ground in southern Chile. *Proc. R. Soc. Lond. B* **271**(Suppl.), S170–S173.

McDonald, M. A., Mesnick, S. L., and Hildebrand, J. A. (2006). Biogeographic characterization of blue whale song worldwide: Using song to identify populations. *J. Cetacean Res. Manage.* **8**, 55–65.

Nikaido, M., *et al.* (8 authors). The baleen whale phylogeny and a past extensive radiation event revealed by SINE insertion analysis. *Mol. Biol. Evol.* **23**, 866–873.

Nishida, S., Goto, M., Pastene, L. A., Kanda, N., and Koike, H. (2007). Phylogenetic relationships among cetaceans revealed by Y-chromosome sequences. *Zool. Sci.* **24**, 723–732.

Rychel, A. L., Reeder, T. W., and Berta, A. (2004). Phylogeny of mysticete whales based on mitochondrial and nuclear data. *Mol. Phylogenet. Evol.* **32**, 892–901.

Sasaki, T. M., *et al.* (2005). Mitochondrial phylogenetics and evolution of mysticete whales. *Syst. Biol.* **54**, 77–90.

Sears, R., Williamson, J. M., Wenzel, F., Bérubé, M., Gendron, D., and Jones, P. W. (1990). The photographic identification of the blue whale (*Balaenoptera musculus*) in the Gulf of St. *Lawrence, Canada. Rep. Int. Whal. Commn (Spec. Iss.)* **12**, 335–342.

Širović, A., Hildebrand, J. A., Wiggins, S. M., McDonald, M. A., Moore, S. E., and Thiele, D. (2004). Seasonality of blue and fin whale calls and the influence of sea ice in the Western Arctic Peninsula. *Deep-Sea Res. II* **51**, 2327–2344.

Stafford, K. M. (2003). Two types of blue whale calls recorded in the Gulf of Alaska. *Mar. Mamm. Sci.* **19**, 682–693.

Stafford, K. M., Bohnenstiehl, D. R., Tolstoy, M., Chapp, E., Mellinger, D. K., and Moore, S. E. (2004). Antarctic-type blue whale calls recorded at low latitudes in the Indian and eastern Pacific Oceans. *Deep-Sea Res. I* **51**, 1337–1346.

Stafford, K. M., Moore, S. E., and Fox, C. G. (2005). Diel variation in blue whale calls recorded in the eastern tropical Pacific. *Anim. Behav.* **69**, 951–958.

Steeman, M. E. (2007). Cladistic analysis and a revised classification of fossil and recent mysticetes. *Zool. J. Linn. Soc.* **150**, 875–894.

Wiggins, S. M., Oleson, E. M., McDonald, M. A., and Hildebrand, J. A. (2005). Blue whale (*Balaenoptera musculus*) diel call patterns offshore of southern California. *Aquat. Mamm.* **31**, 161–168.

Yochem, P. K., and Leatherwood, S. (1985). Blue whale (*Balaenoptera musculus* Linnaeus, 1758). *In* "Handbook of Marine Mammals" (S. H. Ridgway, and R. Harrison, eds), Vol. 3, pp. 193–240. Academic Press, London.

# Bones and Teeth, Histology of

## Mary C. Maas

The bones and teeth of marine mammals, like those of other vertebrates, consist of both organic and mineral components. Because the mineral component (mostly calcium phosphate) predominates, the constituents of bones (bone and calcified cartilage) and teeth (cementum, dentine, and enamel) are referred as "hard tissues." Each of these hard tissues is distinguished both by its composition and by its microscopic structure. Many of the histological features of marine mammal teeth and bones are typical for mammals, and vertebrates, in general, but others are unique or unusual. Some of these may have evolved in conjunction with their shifts to marine habitats.

## I. Bone

### A. Bone Structure and Composition

Bone consists of highly calcified, intercellular bone matrix, and three types of cells—osteocytes, osteoblasts, and osteoclasts. The outer surface of bone is covered by periosteum, which is bound to bone by bundles of collagen fibers known as Sharpey's fibers, and the inner bone surface is lined with endosteum (Fig. 1). Periosteum is thicker than endosteum, but both consist of fibrous connective tissue lined with osteoprogenitor cells, from which osteoblasts are derived. Osteoblasts are the cells that synthesize bone matrix proteins and are active in bone matrix mineralization. Bone matrix (also known as osteoid) consists of about 33% organic matter (mostly Type I collagen) and 67% inorganic matter (calcium phosphate, mostly hydroxyapatite crystals). The osteoblasts occur as simple, epithelial-like layer at the developing bone surface. As the bone matrix mineralizes, some osteoblasts become trapped in small spaces within the matrix (lacunae). These trapped osteoblasts become osteocytes, the cells responsible for maintenance of the bony matrix. Each lacuna holds only a single osteocyte but is connected with adjacent lacunae by microscopic canaliculi, which house cytoplasmic processes of the osteocytes. Osteoclasts are large, multi-nucleated cells that occur in shallow erosional depressions (Howship's lacunae) on the resorbing bone surface and secrete enzymes that promote local digestion of collagen and dissolution of mineral crystals.

Bone is commonly classified according to its gross appearance as cancellous bone (bone with numerous, macroscopic interconnecting cavities, or trabeculae, also known as spongy or trabecular bone) or compact bone (dense lamellar bone without trabeculae), but both types have the same basic histological structure. In a typical mammalian long bone the diaphysis (shaft) is composed predominantly of compact bone, with cancellous bone confined to the inner surface around a central, medullary cavity (Fig. 1A), while the epiphyses (articular ends) consist mostly of cancellous bone overlain by a thin, smooth layer of compact bone. In short bones a core of cancellous bone is completely surrounded by compact bone, and in the flat bones of the skull, inner and outer plates of compact bone are separated by the diploë, a layer of cancellous bone.

Bone also can be classified histologically, as woven (primary) bone and lamellar (secondary) bone. Woven bone, or primary bone has an irregular structure and is usually replaced in adults by the more highly mineralized lamellar bone. Lamellar bone is organized into thin layers (lamellae), usually 3–7μm thick, which contain parallel collagen fiber bundles. Lacunae containing osteocytes are located between lamellae. There are three types of lamellae: concentric, interstitial, and circumferential (Fig. 1B). Concentric lamellae are arranged in circular layers around a long axis, the haversian canal, which is a vascular channel containing blood vessels, nerves, and connective tissue. Adjacent vertical channels are connected by more horizontally oriented vascular channels (Volkmann's canals). The entire complex consisting of several layers of concentric lamellae around a vascular channel is known as an osteon or haversian system. Interstitial lamellae, which appear as irregularly shaped areas between adjacent osteons, consist of lamellae that are remnants of osteons destroyed during bone remodeling. Circumferential lamellae are arranged parallel to each other and comprise the outer circumferential lamellae laid down next to the periosteum and the inner circumferential lamellae laid down next to the endosteum.

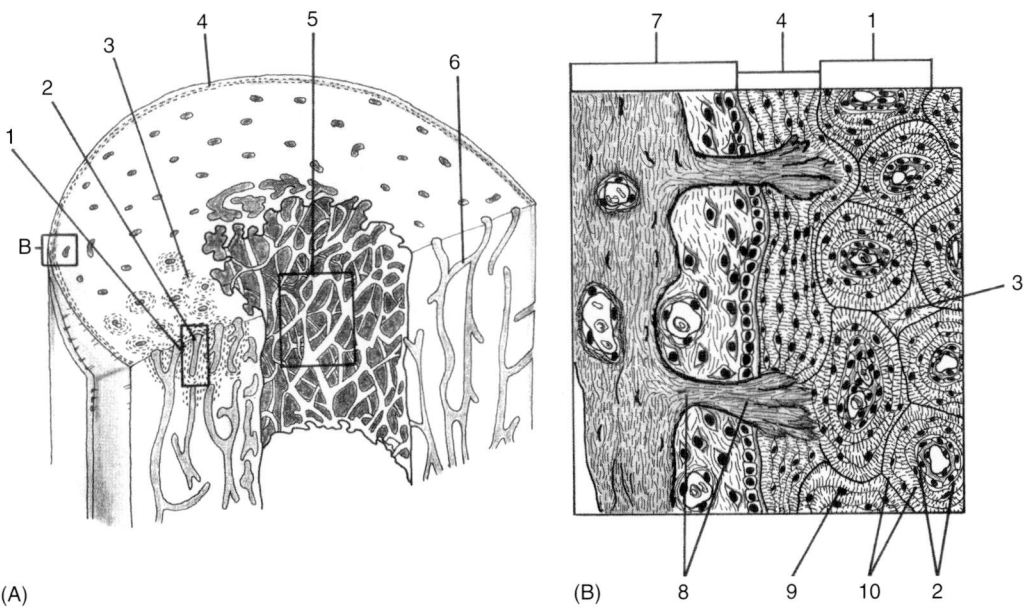

**Figure 1** *(A) Schematic model of the wall of a mammalian long-bone diaphysis, consisting of an outer layer of compact bone and an inner layer of cancellous bone, surrounding a central medullary cavity. Periosteum covers the outer bone surface, and endosteum covers the inner bone surface. (B) Enlarged diagram of periosteum and compact bone in (A). 1, osteon; 2, haversian canal; 3, interstitial lamellae; 4, outer circumferential lamellae; 5, cancellous bone; 6, Volkman's canal; 7, periosteum; 8, Sharpey's fibers; 9, lacuna; 10, concentric lamellae. Adapted from Ten Cate (1989).*

### B. Bone Formation, Growth, and Remodeling

Osteogenesis (bone formation) of membrane bone (intramembranous bone and dermal bone) occurs directly, by mineralization of matrix formed by osteoblasts within condensations of mesenchyme. Osteogenesis of endochondral bone (cartilage bone) occurs indirectly, by deposition and mineralization of bone matrix on a preexisting cartilage matrix (endochondral or cartilage bone). Bone continues to grow through remodeling, as old bone is resorbed through activities stimulated by osteoclasts, and new bone is laid down through the activities of osteoblasts. Influences on bone remodeling include strain and stress imposed by movement and muscle action, hormones, and growth factors. For example, parathyroid hormone is linked to osteoclasts proliferation and activity, and calcitonin, a hormone synthesized within the thyroid, has an inhibitory effect on osteoclast activity (Marks and Popoff, 1988). Modifications of hormonal controls on bone growth and remodeling, in specific parts of the skeleton, are probably responsible for specializations in bone density patterns in some marine mammals (see below). In addition to this typical osteoblastic bone formation, osteogenesis can occur through direct transformation of cartilage or fibrous tissue into bone (metaplastic bone). Metaplastic transformation of chondrocytes into osteoblasts may account for the formation of osseous globules found in the endosteal endochondral bone of some archaeocetes (Buffrénil *et al.*, 1990). Osseous globules are pseudolamellar bone deposited in empty lacunae that once housed chondrocytes, but their mode of origin is controversial.

Because bone growth occurs throughout life, periodic growth marks in skeletal tissue, in particular, periodical deposition of periosteal bone layers, are potentially useful in mammalian age determination. Techniques of skeletal tissue age determination involve the counting of growth layer groups. Growth layer groups are sets of incremental growth lines defined by at least one change in mineral density, such as between more stained and less stained layers or dark and light layers. However, the dynamic nature of bone growth and remodeling limits the accuracy of bone growth layer group counts.

### C. Marine Mammal Bone

Marine mammals show two very different trends in bone architecture and histology, reduced bone density and increased bone density, both of which are associated with their aquatic habits (Wall, 1983). Deep-diving marine mammals, especially Recent cetaceans, have bones that are less dense than homologous elements in terrestrial mammals. They efficiently overcome buoyancy at depth by the active mechanism of lung collapse, while at surface their lighter bones enhance buoyancy, allowing them to float with relatively little expenditure of energy. A pattern of reduced bone density has been thoroughly documented in small to medium-sized odontocetes, some of the large-bodied cetaceans, and some phocids (notably the elephant seal, Mirounga) and is characterized by replacement of cortical bone (the compact bone surrounding medullary cavities) with cancellous bone, which also fills the medullary cavities. This condition is apparently caused by imbalance between bone resorption and redeposition beginning early in ontogeny, and is probably under hormonal control. Increase in cancellous bone in these mammals does not appear to be pathological—the microscopic architecture of cancellous bone in cetacean limbs is significantly more organized than that of typical osteoporotic bone.

In contrast, shallow-diving marine mammals, such as sirenians, overcome buoyancy while diving in large part by the static mechanism of *increased* bone density. Their bones are much denser than typical mammals' bones. This is achieved in different ways; that is, either by osteosclerosis, by pachyostosis, or by a combination of both conditions (pachyosteosclerosis) (Domning and Buffrénil, 1991). Sirenians show pronounced pachyosteosclerosis, especially in the thoracic and occipital regions. Similarly, walruses and some seals have unusually dense limb bones.

Ongoing research in bone histology of extinct marine mammals indicates that both increased bone density and reduced bone density have evolved independently several times in different groups of marine mammals. The earliest sirenians show pronounced pachyosteosclerosis. Likewise, in contrast to Recent cetaceans, some bones of extinct Eocene archaeocetes are osteosclerotic. The earliest cetacean, *Pakicetus* (Pakicetidae), shows a pattern of osteosclerosis similar to that seen in extant semiaquatic mammals, achieved through osteoclast inhibition (Gray *et al.*, 2007). Later, more fully aquatic archaeocetes, including *Ambulocetus* (Ambulocetidae), *Kutchicetus* (Remingtonocetidae), and *Gaviacetus* (Protocetidae) show histological features associated with pachyostosis as well as osteosclerosis. These features are even more pronounced in *Georgiacetus* (Protocetidae) where osteoclast activity is further reduced (Gray *et al.*, 2007) and in the protocetid *Eocetus* where hyperostosis of the periosteal cortex and infilling of the medullary cavity with cancellous bone occurs in ribs and vertebrae (Uhen, 1999). In *Basilosaurus* (Basilosauridae) osteosclerosis of ribs is very pronounced, with total replacement of medullary trabecular bone by compact bone (Buffrénil *et al.*, 1990). In contrast, long bones of some durodontine archeocetes show *reduced* thickness of periosteal compact bone, as in modern cetaceans (Madar, 1998). This, along with greater degree of bone remodeling in ribs of *Zyghoriza* (Durodontinae) than in other archaeocetes suggests that a regional pattern of histological change preceded the systemic osteoporosis of later cetaceans, in conjunction with the functional requirements of the shift from terrestrial to semiaquatic to fully aquatic locomotion (Gray *et al.*, 2007).

The ear region of cetaceans also is characterized by histological specializations of the bone. Both periotic and tympanic bones are noteworthy for their density, compactness, and high mineral content, in comparison with other mammals. This pachyosteosclerosis is due to the hypermineralized state of embryonic woven bone, which is maintained throughout life, rather than later remodeling; its occurrence early in development (in the common dolphin, it begins during the fetal stage and is complete by the first year) probably reflects its role in sound transmission (Buffrénil *et al.*, 2004).

Bone of one marine mammal, the toothed whale *Mesoplodon densirostrus*, exhibits unique histological features. The rostral bone of this odontocete, which is among the densest bone known among tetrapods, is characterized by hypermineralized secondary osteons. These osteons have unusually well-aligned parallel and platy hydroxyapatite crystals and a tubular network of unusually thin collagen fibrils, and thus differ markedly from the structure of haversian systems of typical mammalian lamellar bone (Zylberberg *et al.*, 1998).

Periodic deposition of periosteal bone layers has been used in studies of age determination in mammals, though limited in use by the fact that mammalian bone undergoes remodeling throughout life (Klevezal, 1996). Bone growth layer groups have been studied in a variety of marine mammals including sirenians, pinnipeds, and odontocetes.

## II. Cementum

### A. Cementum Structure and Composition

Teeth of marine mammals, like all mammals, consist of a crown, which extends above the gums, and one or more roots, which extend below the gum line and hold the teeth in bony sockets (alveoli). The roots are covered by cementum (also known as cement), which sometimes extends to cover part of the crown, overlapping the cervical enamel. Cementum, along with the periodontal ligament, comprises

the periodontium, the attachment apparatus of teeth. Cementum is similar in composition to bone. Its mineral component (65% by wet weight) consists of crystals of an impure form of hydroxyapatite similar in shape and size to those of bone. Its organic component (up to 20% of the total tissue) includes cementocytes (cementum cells), ground substance containing proteoglycans, intrinsic collagen fibers, and extrinsic collagen fibers (Sharpey's fibers). The intrinsic fibers and ground substance are the primary constituents of cementum. Intrinsic fibers, like collagen fibers of lamellar bone, are small, on the order of 1–2 μm in diameter. The extrinsic Sharpey's fibers are much larger, typically 3–12 μm in diameter. The intrinsic fibers, ground substance, and cementocytes are derived from cementoblasts, but the extrinsic fibers are derived from fibroblasts of the periodontal ligament.

Cementum is classified according to the relative proportions of the different components, although the different types are gradational. Thus, cementum can be classified as cellular or acellular depending on the relative proportions of cementocytes and ground substance, or it can be classified according to its fiber composition (Fig. 2). Extrinsic fiber cement occurs close to the alveolar bone and is dominated by well-mineralized Sharpey's fibers contained within highly mineralized acellular ground substance. Mixed fiber cement contains intrinsic collagen fibers as well as Sharpey's fibers and ground substance, and may or may not contain cementocytes. Intrinsic fiber cement, which contains only intrinsic collagen fibers, ground substance, and cementocytes, occurs close to roots. In cellular mixed fiber cement and intrinsic fiber cement, cementocytes are contained in lacunae of variable shape that form within the mineralizing ground substance.

Incremental growth layers known as cementing lines or resting lines are sometimes a prominent histological feature of both cellular and acellular cementum. Cementing lines, like the incremental growth layers found in bone, dentine, and enamel, are distinct layers that parallel the developing surface. Due to periodic variation in mineralization during development, they contrast with adjacent layers. Cementum growth layer groups, like those of bone and dentine, can be empirically defined by at least one change in mineral density, such as between translucent and opaque layers, dark and light layers, ridge and groove, or more stained and less stained layers. Empirical studies have shown that cementum growth layer groups record the periodicity of tissue formation, and thus are useful in age determination.

### B. Marine Mammal Cementum

Cementum in marine mammals is for the most part structurally similar to that of other mammals. Cementum growth layers groups are used in conjunction with dentine and bone growth layer groups to estimate age in marine mammals, though their relative clarity varies among species. In some species, cementum formation continues beyond that of dentine, which is an advantage in age determination. In ziphiid whales, where the cementum typically extends over most of the crown and may comprise the bulk of the tooth, cementum growth layer groups are distinguishable without magnification. Ziphiids also have been reported to have an unusual, possibly vascular cementum (Boyde, 1980).

## III. Dentine

### A. Dentine Structure and Composition

Dentine comprises the bulk of the volume of teeth of most mammals. In the crown, dentine is covered by enamel and in the root it is covered by cementum. Circumpulpal dentine surrounds the pulp

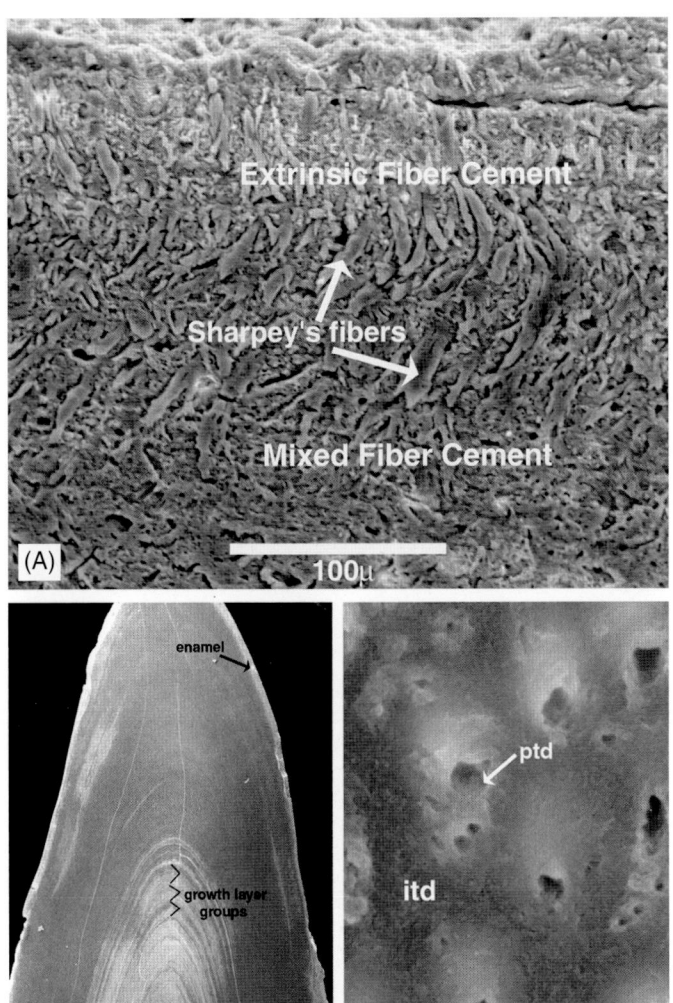

**Figure 2** *Scanning electron micrographs of an isolated tooth of an unidentified delphinoid cetacean (Yorktown Formation, Pliocene, from the Lee Creek Mines, North Carolina, USA). The specimen has been sectioned longitudinally, polished, and etched with dilute HCl. (A) High magnification view of cementum, which grades from extrinsic fiber cement on the outer periphery (top) to mixed fiber cement closer to the cementum-dentine junction (bottom). Classification of cementum depends on the proportion of Sharpey's fibers contained within the matrix. (B) Thin layers of enamel and cementum lie peripheral to dentine of the crown and root. Dentine growth layer groups appear as pairs of dark/light bands. (C) High magnification view of dentine. The walls of cross-sectioned dentine tubules contain hyper-mineralized peritubular dentine (ptd). Less mineralized intertubular dentine (itd) occurs between tubules.*

cavity, which contains connective tissue, nerves, and blood vessels. Circumpulpal dentine is distinguishable histologically from a thin outer layer known as mantle dentine (in the tooth crown) or hyaline dentine (in the root). Dentine tubules, which radiate out from the pulp to the outer dentine surface, are distinctive features of dentine. They are narrow (1–4 μm diameter) tubular structures that form during dentine development around odontoblast (dentine-forming cells) cell processes. In adult teeth tubules contain mostly fluid and amorphous cell debris.

The organic component of dentine consists mainly of very small (on the order of 50 nm in diameter) collagen fibrils. The collagen fibrils in circumpulpal dentine are laid down parallel to the developing dentine surface and perpendicular to the dentine tubules but the mantle dentine contains some large (>1 μm in diameter) collagen fiber bundles known as von Korff fibers. Von Korff fibers are oriented parallel to tubules.

Dentine has 75% mineral (hydroxyapatite). Most of the small (2–3 nm in thickness and probably 20–100 nm in length) hydroxyapatite crystals are aligned parallel to each other and to the small collagen fibrils, but others are radially oriented and form spherical or semispherical structures known as calcospherites. Calcospherites are difficult to distinguish histologically because they typically fuse together. Areas where mineralization is incomplete and calcospherites have not fused are called interglobular dentine. Most mineralization of dentine takes place along the developing dentine front, but dentine deposited in tubule walls (peritubular dentine) undergoes further mineralization (Fig. 2C). In some cases, tubules become occluded by mineralization, forming sclerotic dentine. Denticles (smooth-surfaced, spherical mineralized bodies with a laminar structure) sometimes form by mineralization of collagen fibers within the pulp cavity. These denticles may become attached to the inner surface of the dentine, or embedded in it during continued dentine formation.

### B. Marine Mammal Dentine

The dentine of most marine mammals is structurally similar to that of other mammals, but there are some exceptions. In some, notably the narwhal and sperm whale, the large von Korff fibers are not restricted to the mantle dentine but extend throughout the thickness of dentine, where they are located in the walls of dentine tubules. Denticles have been reported in some odontocetes and sclerotic dentine is found in some marine mammals, especially in seals.

Marine mammal dentine is characterized by prominent incremental growth layers (Fig. 2B) that lie at angles to dentine tubules, and vary in their intensity, both within and among individuals. The finest scale layers are the incremental von Ebner lines, which probably reflect diurnal variation in matrix fiber arrangement. Von Ebner lines appear as alternating dark and light lines in ground sections under polarized light. Other, larger-scale incremental growth layers reflect changes in density due to differences in mineralization. These include the neonatal line, a very prominent growth layer that marks physiological disturbance associated with birth, and other less distinct and consistent growth layers whose physiological bases are uncertain. In some seals, the growth layer groups are accentuated by layers of interglobular dentine. Whatever their origins, there is a regular repetition to growth layer groups that seems to reflect annual or semiannual growth cycles, and counting of dentinal growth layer groups is a primary basis of age determination in pinnipeds, sirenians, and odontocetes.

### IV. Enamel

#### A. Enamel Structure and Composition

Enamel covers the tooth crown in most mammals. It is the most highly mineralized tissue in the body, consisting almost entirely (95% by weight) of highly structured arrangements of hydroxyapatite crystallites. The remaining fraction consists of water and two classes of proteins unique to enamel—enamelins, which predominate in mature (fully mineralized) enamel, and amelogenins, which predominate in developing enamel. The histological structure of enamel reflects the organization of crystallites into units of increasing scale, two of which are enamel prisms and enamel types. This structural organization is determined during enamel development. Unlike bone, cement, and dentine, enamel does not remodel after its initial deposition.

Enamel matrix is secreted by ameloblasts. The activity of these enamel-secreting cells commences at the enamel-dentine junction (EDJ) and continues as ameloblasts retreat outwards, away from the EDJ. Mineral crystals precipitate and grow within the enamel matrix left by the retreating ameloblasts. The orientation and arrangement of the crystallites, and thus the structure of the mature enamel, depends on the shape of the secretory end of the ameloblast. The simplest enamel structure is formed by ameloblasts with flat secretory surfaces. In most mammal teeth, however, the bulk of the enamel is laid down by ameloblasts whose secretory ends form protrusions, called Tomes processes, surrounded by flattened areas called ameloblast shoulders. Because enamel crystallites grow perpendicular to the differently oriented secretory surfaces of the Tomes process and the ameloblast shoulder, there is a regular pattern of discontinuities in crystallite orientations. These discontinuities define the boundaries of enamel prisms and interprismatic enamel.

Enamel prisms are cylindrical bundles of largely parallel hydroxyapatite crystals extending outward from the EDJ towards the outer tooth surface. The prism boundaries are defined by differences in orientations prismatic crystallites and those of the adjacent enamel that fills the spaces between prisms. This enamel is called interprismatic enamel. It is compositionally identical to enamel prisms, but differs in crystallite orientation. The submicroscopic gap produced by the change in crystallite orientations at the prism-interprismatic boundary is known as the prism sheath (Fig. 3). Prism sheaths contain slightly greater concentrations of water and protein than the surrounding enamel, and thus are less dense. This allows prism patterns (the cross-sectional shapes and packing arrangement of prisms and interprismatic enamel) to be distinguished in ground sections or in acid-etched scanning electron microscope preparations. Prisms may have closed, circular cross-sections or open, arc-shaped cross-sections. Prism patterns have been used to distinguish among some mammalian groups, but there is considerable variation within individuals and considerable parallelism among different groups.

Enamel types describe the organization of enamel at a scale greater than individual crystallites or prisms. Common enamel types include parallel crystallite enamel, radial enamel, and decussating enamel. Parallel crystallite enamel, a type of nonprismatic enamel, is a volume of enamel in which hydroxyapatite crystallites are parallel to each other with no discontinuities in orientation and lacking larger-scale structural features, other than incremental lines. Radial enamel refers to a volume of prismatic enamel where prism long axes are parallel to one another and directed radially outward from the EDJ. Decussating enamel is a volume of enamel characterized by layers of parallel prisms, one or more prisms in thickness, whose long axes alternate in orientation with prisms in adjacent layers. Decussating enamel, also known as Hunter-Schreger bands (HSB), includes undulating HSB, where layers of similarly oriented prisms have a gently undulating course from the EDJ to the surface, and zigzag HSB, where the layers undulate with a pronounced vertical amplitude. Differences in enamel types have a phylogenetic component, but also have different mechanical properties that can be important functionally—parallel crystallite enamel may be harder than prismatic enamel, but prismatic enamel, especially decussating enamel, is more resistant to cracks induced by chewing stress. Zigzag enamel is thought to be especially resistant to cracking. Most mammal teeth are composed of more than one enamel type.

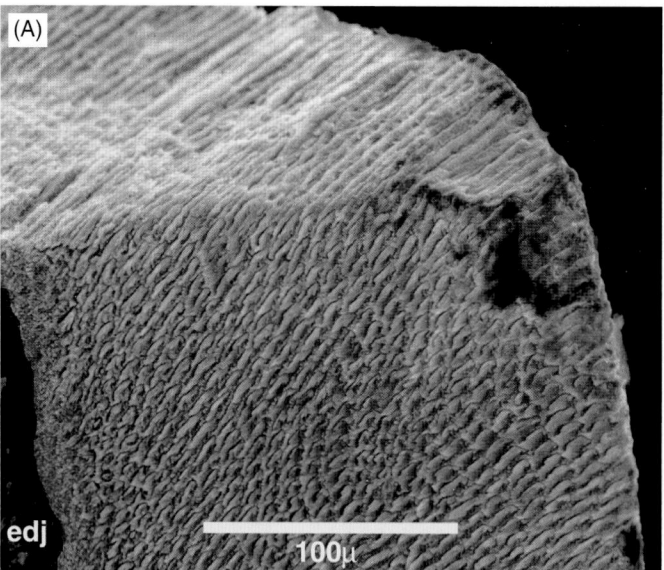

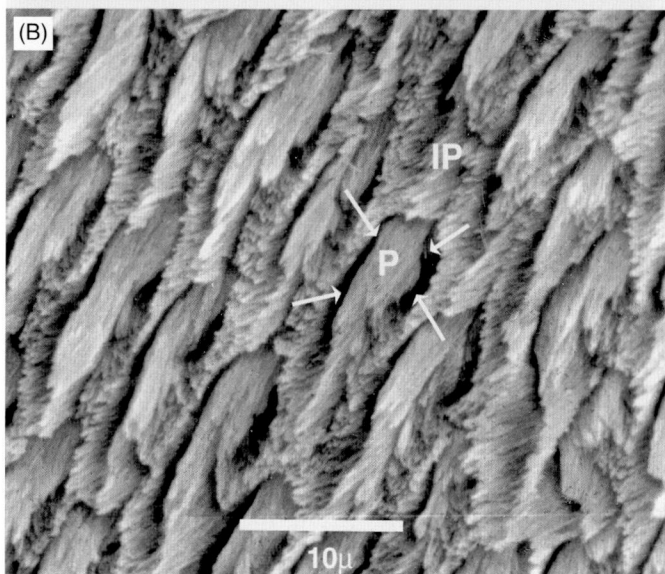

**Figure 3** *(A) Scanning electron micrograph of fractured enamel near tip of tooth (unidentified Pliocene odontocete, Lee Creek Mine, North Carolina, USA). The naturally fractured surface (at top) shows that the prisms take a straight course from the enamel-dentine junction (EDJ) to the outer surface, as is typical in radial enamel. (B) High magnification view showing enamel prisms (P) sectioned oblique to their long axes. Prism crystallites are parallel to each other, but not to crystallites in adjacent interprismatic enamel (IP). Arrows indicate the position of the prism sheath, which has been artificially enlarged in this acid-etched specimen.*

Cross-striations, a record of the daily incremental deposition of enamel, are sometimes evident in both prismatic and nonprismatic enamel. In the scanning electron microscope cross-striations appear as alternating constrictions and varicosities along the length of the prism, suggesting that they reflect variations in rate of enamel secretory activity. More prominent incremental lines, known as brown striae of Retzius, also transect prisms or crystallites. They are oriented parallel to the developing enamel surface, and probably reflect regular interruptions in growth, although their causes and periodicity are not clear.

## B. Marine Mammal Enamel

Although the crowns of most marine mammal teeth are covered with enamel, there is considerable variation in its structural complexity among and within orders. Likewise, prism patterns vary among and within orders, though there is no compelling evidence that prism patterns are diagnostic of particular marine mammal groups.

Most extant cetaceans have thin, structurally simple enamel. In some the enamel consists of a thin layer of radial prismatic enamel with or without an outer layer of nonprismatic parallel crystallite enamel, and in many species the tooth enamel consists entirely of nonprismatic parallel crystallite enamel. In contrast, the most primitive cetacean, the fossil *Pakicetus*, had relatively thick enamel with a more complex structure consisting of parallel crystallite enamel, radial enamel and a thick inner layer of undulating Hunter-Schreger bands. Later archaeocetes show the same arrangement of enamel types, but almost all more derived odontocetes have much less complex enamel. This has led some workers to conclude that the enamel of most extant cetaceans is evolutionary degenerate. Only two extant odontocetes, the Indus dolphin *Platanista* and Amazon dolphin *Inia* have well-developed, undulating HSB. It is unclear whether these were independently acquired in response to functional demands of their diet or a primitive retention from archaeocete ancestors.

Extant sirenians (*Dugong* and *Trichechus*) are reported to have radial enamel with variably circular and arc-shaped prism cross-sections. Similar enamel has been reported for some fossil sirenians, and it is likely that this is primitive for the group. Pinniped enamel has not been described in detail, but enamel of some species appears to be more complex than that of sirenians. *Phoca* has undulating HSB, and walrus enamel shows a transition from undulating HSB to zigzag HSB near cusp tips.

Enamel incremental lines generally are not used in age determination of marine mammals. The thin enamel of many species makes resolution of these lines difficult, and, more importantly enamel only records the period of tooth development during which enamel is laid down, which, in most cases, is before birth.

### *See Also the Following Article*

Age Estimation

### *References*

Boyde, A. (1980). Histological studies of dental tissues of odontocetes. *Rep. Int. Whal. Commn Spec. Iss.*(3), 65–87, Special Issue.

Buffrénil, V.d., Ricqlés, A.d., Ray, C. E., and Domning, D. P. (1990). Bone histology of the ribs of the archaeocetes (Mammalia: Cetacea). *J. Vertebr. Paleontol.* **10**, 455–466.

Buffrénil, V.d., Dabin, W., and Zylerberg, L. (2004). Histology and growth of the cetacean petro-tympanic bone complex. *J. Zool. (Lond.)* **262**, 371–381.

Domning, D. P., and Buffrénil, V.d. (1991). Hydrostasis in the Sirenia: Quantitative data and functional interpretations. *Mar. Mamm. Sci.* **7**, 331–368.

Gray, N.-M., Kainec, K., Madar, S., Tomko, L., and Wolfe, S. C. (2007). Sink or swim? Bone density as a mechanism for buoyancy control in early cetaceans. *Anat. Rec.* **290**, 638–653.

Klevezal, G. (1996). "Recording Structures of Mammals, Determination of Age and Reconstruction of Life History." Translated by M.V. Mina and A.V. Oreshkin. A.A. Balkema, Rotterdam.

Koenigswald, W. V., and Sander, P. M. (1997). "Tooth Enamel Microstructure." A.A. Balkema, Rotterdam.

Maas, M. C., and Thewissen, J. G. M. (1995). Enamel microstructure of *Pakicetus* (Mammalia: Archaeoceti). *J. Paleontol.* **69**, 1154–1162.

Madar, S. I. (1998). Structural adaptations of early archaeocete long bones. *In* "The Emergence of Whales" (J. G. M. Thewissen, ed.), pp. 353–378. Plenum Press, New York.

Marks, S. C., and Popoff, S. N. (1988). Bone cell biology: The regulation of development, structure, and function in the skeleton. *Am. J. Anat.* **183**, 1–44.

Perrin, W. F., and Myrick, A. C., Jr. (eds) (1980). "Age determination of toothed whales and sirenians. Growth of odontocetes and sirenians: Problems in age determination. Report of the International Whaling Commission, Special Issue 3." The International Whaling Commission, Cambridge.

Sahni, A., and Koenigswald, W. V. (1997). The enamel structure of some fossil and recent whales from the Indian subcontinent. *In* "Tooth Enamel Microstructure" (W. V. Koenigswald, and P. M. Sander, eds), pp. 177–191. Balkema, Rotterdam.

Ten Cate, A. R. (1989). "Oral Histology. Development, Structure, and Function," 3rd edn. C.V. Mosby, St. Louis.

Uhen, M. D. (1999). New species of protocetid archaeocete whale, *Eocetus wardii* (Mammalia: Cetacea) from the middle Eocene of North Carolina. *J. Paleontol.* **73**, 512–528.

Wall, W. P. (1983). The correlation between high limb-bone density and aquatic habits in recent mammals. *J. Paleontol.* **57**, 197–207.

Zylberberg, L., Traub, W., Buffrénil, V.d., Allizard, F., Arad, T., and Weiner, S. (1998). Rostrum of a toothed whale: Ultrastructural study of a very dense bone. *Bone* **23**, 241–247.

**B**

# Bottlenose Whales
## *Hyperoodon ampullatus* and *H. planifrons*

Shannon Gowans

## I. Characteristics and Taxonomy

Bottlenose or bottle-nosed whales are large, robust beaked whales (6–9 m) distinguished by their large bulbous forehead and short dolphin-like beak (Fig. 1). They are chocolate brown to yellow in color, being lighter on the flanks and belly. This coloration is believed to be caused by a thin diatom layer. Newborns are gray with dark eye patches and a light-colored forehead. The maxillary crests of males become larger and heavier with age, leading to a change in the shape of the forehead, with mature males having a flat, squared-off forehead whereas females/immature males have a smooth-rounded forehead. The dense bone in the male's forehead may be used for male–male competition, as males head-butt one another (Gowans and Rendell, 1999). Males possess a single pair of conical teeth at the tip of the lower jaw (in females, they remain unerupted); however, these teeth are rarely visible in live animals.

Northern (*H. ampullatus*) and southern (*H. planifrons*) bottlenose whales are the only recognized species within the genus *Hyperoodon* in the family Ziphiidae (Mead, 1989). Recent molecular work on southern bottlenose whales indicates that there may be more than one species (Dalebout *et al.*, 1998). Sightings of a large beaked whale in the tropical Pacific has been identified in the past as a bottlenose whale (either *H. planifrons* or a third, undescribed *Hyperoodon* sp.), however recent evidence suggests these whales are Longman's beaked whale (*Indopacetus pacificus*) (Dalebout *et al.*, 2003).

B

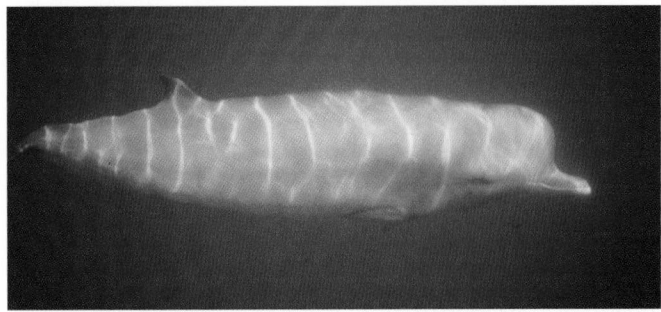

Figure 1   *A northern bottlenose whale underwater.*

## II. Distribution and Abundance

Northern bottlenose whales are found in subpolar and cold temperate waters of the North Atlantic, from the ice edge to the Azores, almost always in waters deeper than 500 m (Wimmer and Whitehead, 2004). They concentrate in submarine canyons, continental shelf edges, and other areas of high relief. A resident year-round population is found in the Gully, a large submarine canyon 200 km off the coast of Nova Scotia, Canada. Gully residents also use nearby canyons but do not appear to make long-distance migrations to other centers of distribution (such as off Labrador, Iceland, Norway, the Faroes, or the Azores). Whales found in different areas of the Atlantic have different length distributions, indicating that there may be geographic isolation between the different whaling grounds. Analysis of both nuclear and mitochondrial DNA indicates reproductive isolation between bottlenose whales in the Gully and Labrador, however there is likely mixing between bottlenose whales from Labrador and Iceland (Dalebout *et al.*, 2006).

Southern bottlenose whales are found throughout the Southern Hemisphere, from the ice edge to 30°S. There are no known areas of concentration, although relatively little effort has been made to identify or census these animals.

## III. Ecology

Bottlenose whales are deep divers feeding predominantly on squid of the genus *Gonatus*, although other species of squid are eaten (Hooker *et al.*, 2001). Adult *Gonatus* are primarily benthic, although juveniles may inhabit the water column. Fish (including herring and redfish) and benthic invertebrates such as starfish and sea cucumbers are occasionally consumed. Time-depth recorders on two northern bottlenose whales in the Gully indicated that these whales were routinely diving to or near the sea floor over 1400 m below the surface (Hooker and Baird, 1999).

Evidence from whaling suggests that northern bottlenose whales migrate north in spring and south in the fall; however, evidence for this is weak and this migration may actually represent a migration of whaling vessels. Stomach contents of stranded animals indicate that both northern and southern bottlenose whales can travel over long distances (~1000 km), although it is not known if these movements are routine. Most individuals in the Gully appear to be year-round residents, although they do routinely spend time outside the canyon, often in nearby canyons. Gully animals do not appear to make long-range migrations.

## IV. Behavior and Physiology

Both northern and southern bottlenose whales are typically found in small groups (one to four individuals), although groups of up to 20 have been observed. Nothing is known about the social organization of southern bottlenose whales, and only the Gully population of northern bottlenose whales has been studied (Gowans *et al.*, 2001). In the Gully, individuals live in fission–fusion groups and most associations are brief (on the order of minutes to a few days). Females form a loose network of associates with most members of the community. However, mature males form long-term companionships with other mature males and these associations last for years. The function of these associations is unknown, but they may be linked to mating and may be similar to male coalitions in bottlenose dolphins.

Northern bottlenose whales are often described as curious, as they will often approach boats. Whalers exploited this behavior to find groups of bottlenose whales, and as healthy whales would often remain near wounded individuals, the entire group was often captured.

## V. Life History

Analysis of whaled carcases off Labrador indicates sexual maturation at 7–11 years for males and 11 years for females (Benjaminsen and Christensen, 1979). A single calf is born after a gestation of at least 12 months. The oldest individual caught during whaling was 37.

## VI. Interactions with Humans

The commercial hunt for northern bottlenose whales began in the 1850s and extended until the 1970s. Over 80,000 whales were captured during this period, and many more were harpooned but not recovered (Benjaminsen and Christensen, 1979). Pre-exploitation numbers are estimated at 40–50,000 whales, although this number is at best a rough guess. There is no current estimate for the size of the North Atlantic population, but it is unlikely that it has fully recovered from whaling (COSEWIC, 2002). Approximately 160 individuals reside in the Gully currently, and this population is likely still recovering from the whaling catch of approximately 60 animals taken from the area in the 1960s. The Scotian Shelf population (including the Gully) is considered to be Endangered under Canadian law.

The study in the Gully represents the first long-term study of live beaked whales. Crews from several documentary films and magazines have visited the Gully, as it is one of the few places where beaked whales can be observed routinely.

## *See Also the Following Article*

Beaked Whales, Overview

## *References*

Benjaminsen, T., and Christensen, I. (1979). The natural history of the bottlenose whale, *Hyperoodon ampullatus* (Forster). *In* "Behavior of Marine Animals" (H. E. Winn, and B. L. Olla, eds), Vol. 3, pp. 143–164. Plenum Press, New York.

COSEWIC. (2002). "COSEWIC Assessment and Update Status Report on the Northern Bottlenose Whale *Hyperoodon ampullatus* (Scotian Shelf population) in Canada" Committee on the Status of Endangered Wildlife in Canada. Ottawa, Vi + 22 pp.

Dalebout, M., Ruzzante, D. E., Whitehead, H., and Øien, N. I. (2006). Nuclear and mitochondrial markers reveal distinctiveness of a small population of bottlenose whales (*Hyperoodon ampullatus*) in the western North Atlantic. *Mol. Ecol.* **15**, 3115–3129.

Dalebout, M., van Heldon, A., van Waerebeek, K., and Baker, C. S. (1998). Molecular genetic identification of southern hemisphere beaked whales (Cetacea: Ziphiidae). *Mol. Ecol.* **6**, 687–692.

Dalebout, M. L., *et al.* (2003). Appearance, distribution and genetic distinctiveness of Longman's beaked whale *Indopacetus pacificus*. *Mar. Mamm. Sci.* **19**, 421–461.

Gowans, S., and Rendell, L. (1999). Head-butting in northern bottlenose whales (*Hyperoodon ampullatus*): A possible function for big heads. *Mar. Mamm. Sci.* **15**, 1342–1350.

Gowans, S., Whitehead, H., and Hooker, S. K. (2001). Social organization in northern bottlenose whales (*Hyperoodon ampullatus*): Not driven by deep water foraging? *Anim. Behav.* **62**, 369–377.

Hooker, S. K., and Baird, R. W. (1999). Deep-diving behaviour of the northern bottlenose whale *Hyperoodon ampullatus* (Cetacea: Ziphiidae). *Proc. R. Soc. Lond. B.* **266**, 671–676.

Hooker, S. K., Iverson, S. J., Ostrom, P., and Smith, S. C. (2001). Diet of northern bottlenose whales inferred from fatty-acid and stable-isotope analyses of biopsy samples. *Can. J. Zool.* **79**, 1442–1454.

Mead, J. G. (1989). Bottlenose whales *Hyperoodon ampullatus* (Forster, 1770) and *Hyperoodon planifrons* (Flowers, 1882). *In* "Handbook of Marine Mammals" (S. H. Ridgway, and R. Harrison, eds), Vol. 4, pp. 321–348. Academic Press, London.

Wimmer, T., and Whitehead, H. (2004). Movements and distribution of northern bottlenose whales, *Hyperoodon ampullatus*, on the Scotian Shelf and adjacent waters. *Can. J. Zool.* **82**, 1782–1794.

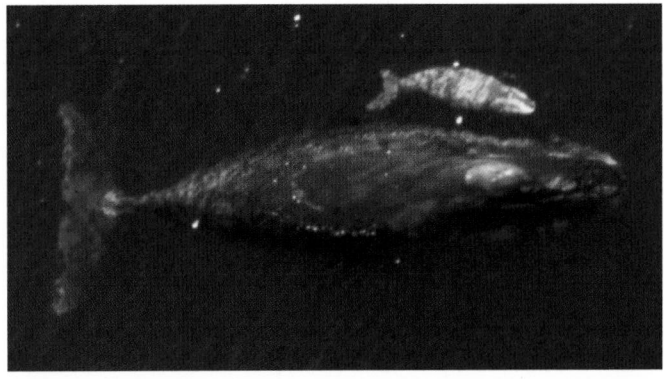

**Figure 1** *Bowhead whales are large, dark cetaceans with various amounts of white on their chins (seen on the far right on the adult in this photo), tail stocks (the paired white spots on the far left), and ventral surfaces (out of sight in aerial photographs). This adult and calf were photographed near Point Barrow, Alaska, during the spring migration from the Bering Sea to the Canadian Beaufort Sea. Photo by David Rugh.*

# Bowhead Whale
## *Balaena mysticetus*

DAVID J. RUGH AND KIM E.W. SHELDEN

## I. Characteristics and Taxonomy

Bowhead whales, sometimes called Arctic right whales, Greenland right whales, great polar whales, or *ahvik*, are members of the family Balaenidae (Mysticeti, Cetacea) that live most of the year associated with sea ice in northern latitudes (Braham *et al.*, 1980; Burns *et al.*, 1993). Bowheads have never occurred in the Southern Hemisphere.

Bowheads are readily identifiable by their large size, rotund shape, lack of a dorsal fin, dark color, white chins, triangular head (in profile), and neck (an indentation between the head and back). They are predominantly black, but most have characteristic white patterns on their chins, undersides, around their tail stocks, and on their flukes (Fig. 1). These patterns distinguish them from the similar-appearing right whales (*Eubalaena* spp.) and are unique to each individual. The white portions of the pattern around the tail and on the flukes increase in extent with age. Most bowheads accumulate distinctive, permanent marks on their backs, perhaps resulting from contact with sea ice or the sea floor when feeding. The bowed appearance of the mouth gives them their name.

These huge marine mammals are among the largest animals on earth, weighing up to 75–100 tons (Reeves and Weatherwood, 1985). Males grow to 14–17 m in length and females 16–18 m, perhaps as long as 20 m. Their flukes are 2–6 m across. The head constitute over a third of the bulk of the body, and the baleen may reach lengths of 4 m (no other whale has baleen longer than 2.8 m) with 230–360 plates on each side of the mouth, making the capacious mouth quite possibly the largest of any animal ever. To insulate them from the icy water, bowheads are wrapped in blubber 5.5–28 cm thick covered by an epidermis up to 2.5 cm thick. This combination of blubber and skin is the thickest of any whale species.

## II. Distribution and Abundance

Currently there are four or five recognized stocks of bowheads defined by geographically distinct segments of the species' total population: the Western Arctic (or Bering–Chukchi–Beaufort stock) found around Alaska, Okhotsk Sea in eastern Russia, Davis Strait and Hudson Bay in northeastern Canada (sometimes considered separate stocks), and Spitsbergen in the North Atlantic (Rugh *et al.*, 2003; Heide-Jørgensen *et al.*, 2006). According to fossil records, bowhead whales may have emerged in the Northern Hemisphere during the Pliocene (roughly 8 million years ago). Genetic mixing among the current stocks was possible during the relatively warm interglacial periods (e.g., AD 1000–1200) when reduction in sea ice meant whales could move between the Atlantic and Pacific Basins. Movement between the Beaufort Sea and Hudson Bay was stopped by icier seas during the "Little Ice Age" (AD 1400–1850). Until recently, temperatures have been cool enough to keep ice across most of the east–west passages of the Arctic, isolating the Western Arctic and Davis Strait–Hudson Bay stocks.

The largest remnant stock, the Western Arctic stock, consists of over 10,000 whales (George *et al.*, 2004) that migrate from the Bering Sea in the winter through the Chukchi Sea to the Beaufort Sea in the summer. This stock is growing at an annual rate of 3%. Prior to commercial whaling, there may have been 10,000–23,000 whales in this stock.

The Davis Strait-Hudson Bay stock is now thought to number at least 7000 bowheads (Cosens *et al.*, 2006), though there may have been over 11,000 prior to commercial whaling. In the Okhotsk Sea, there are now only about 300–400 bowheads where there originally were more than 3000. The Spitsbergen stock, originally at 24,000 bowheads, supported a huge European fishery, but today there are only tens of whales left (Shelden and Rugh, 1995).

## III. Ecology

The only predators of bowhead whales, other than humans, are killer whales (*Orcinus orca*). Scars from killer whale teeth were found on approximately 4–8% of the whales taken in the subsistence hunt by Alaskan Eskimos (George *et al.*, 1994). In part, the bowheads' close association with sea ice may be a way of seeking refuge from killer whales.

As many as 60 species have been found in bowhead stomachs, but the preferred prey are copepods (11 species) and euphausiids (2 species), plus mysids and gammarid amphipods (Lowry, 1993).

## IV. Behavior and Physiology

Bowheads feed throughout the water column, sometimes on the surface (called "skimming") and sometimes at or near the seafloor (as evidenced by mud smeared across their heads and backs, Würsig and Clark, 1993). A bowhead's huge mouth can engulf large volumes of water, including prey, and, as the tongue rises, the water is pushed out, trapping prey on the inside fringed surfaces of the baleen, which acts as a filter all the way around the mouth. The massive tongue (up to 5 m long and 3 m wide) then sweeps the food off the baleen into a very narrow digestive tract. Sometimes a dozen bowheads will feed together in an echelon formation, similar to a line of migrating geese. Perhaps this coordinated effort helps the whales entrap their prey.

Bowheads are well adapted to the risky occupation of being air-breathing mammals in seas often covered with thick ice: they can withstand breaking through ice up to 60 cm thick, and their diving abilities are exceptional—possibly exceeding an hour. These abilities are critical to finding breathing holes when swimming under sea ice. The very low-frequency and very loud calls bowheads produce may help them find mates or assist in following each other when navigating through sea ice. The only other whales commonly found in ice as far north as bowheads are belugas (*Delphinapterus leucas*) and narwhals (*Monodon monoceros*), toothed whales with some of the same characteristics seen in bowheads: smooth backs and relatively thick blubber.

## V. Life History

Bowheads probably mate in late winter or early spring, but sexual activity may occur in any season (Nerini *et al.*, 1984). Acoustics probably play a vital role in reproduction because bowheads are vocally active during the mating season and can hear each other 5–10 km away. Breaching (leaping completely out of the water) and fluke slapping (when the tail smashes down on the water surface) may also play a role in attracting a mate or asserting dominance, but the role of these behaviors is not well understood. Bowhead whales have been observed mating in pairs as well as in larger mating groups. There is a great deal of physical contact and turning in these groups, perhaps as several males attempt to copulate with one female.

More than a year after mating (13–14 months), calves are born, usually during the spring migration between April and June. Calves are about 4 m long at birth. Females have calves 3–4 years apart. The following spring, the young whales, now 6–8 m long, are weaned. After this, juvenile growth is slow compared to other baleen whales. At roughly 15 years of age, when 12–14 m long, females become sexually mature, and males become sexually active when 12–13 m long (Nerini *et al.*, 1984). Bowhead whales may live longer than any other mammals; ancient harpoon points recently discovered in hunted whales and eye lens analysis indicate some bowheads live for more than a century (George *et al.*, 1999).

## VI. Interactions with Humans

Commercial whaling had a profound effect. The bowhead's large size, long baleen, and thick blubber have made them such a valuable commodity that whalers went to great lengths to kill them (Braham *et al.*, 1980; Bockstoce, 1986). Commercial whalers from 17th to 19th centuries were so efficient that they depleted stock after stock of these whales. In fact, even a century after commercial whaling ceased, all bowhead stocks are still considered endangered.

In modern times, Native Alaskans kill about 40 whales per year through quotas set by the International Whaling Commission (IWC). The Chukotka Natives of Siberia have been allotted five bowheads per year from the Alaska quota. Independent of the IWC quota, the Canadian government has allowed a limited hunt of bowheads from the Western Arctic stock (in the eastern Beaufort Sea) and from Davis Strait and Hudson Bay.

## See Also the Following Articles

Baleen ■ Beluga ■ Whale Breaching ■ Filter Feeding ■ Narwhal

## References

Bockstoce, J. R. (1986). "Whales, Ice and Men: The History of Whaling in the Western Arctic." Univ. of Washington Press, Seattle, WA.

Bockstoce, J. R., Botkin, D. B., Philp, A., Collins, B. W., and George, J. C. (2005). The geographic distribution of bowhead whales, *Balaena mysticetus*, in the Bering, Chukchi, and Beaufort Seas: Evidence from whaleship records, 1849–1914. *Mar. Fish. Rev.* **67**, 1–43.

Braham, H. W., Marquette, W. M., Bray, T. W., and Leatherwood, J. S. (1980). The bowhead whale: Whaling and biological research. *Mar. Fish. Rev.* **42**(9–10), 1–96.

Burns, J. J., Montague, J. J., and Cowles, C. J. (eds) (1993). "The Bowhead Whale." Spec. Publ. No. 2., The Society for Marine Mammalogy, Lawrence, KS.

Cosens, S. E., Cleator, H., and Richard, P. (2006). Numbers of bowhead whales (*Balaena mysticetus*) in the Eastern Canadian Arctic, based on aerial surveys in August 2002, 2003 and 2004. Report SC/58/BRG7 submitted to the Sci. Comm., Int. Whaling Commission. 19 pp.

George, J., Philo, L., Hazard, K., Withrow, D., Carroll, G., and Suydam, R. (1994). Frequency of killer whale (*Orcinus orca*) attacks and ship collisions based on scarring on bowhead whales (*Balaena mysticetus*) of the Bering–Chukchi–Beaufort Seas stock. *Arctic* **47**, 247–255.

George, J. C., Bada, J., Zeh, J., Scott, L., Brown, S. E., O'Hara, T., and Suydam, R. (1999). Age and growth estimates of bowhead whales (*Balaena mysticetus*) via aspartic racemization. *Can. J. Zool.* **77**, 571–580.

George, J. C., Zeh, J., Suydam, R., and Clark, C. (2004). Abundance and population trend (1978–2001) of western arctic bowhead whales surveyed near Barrow, Alaska. *Mar. Mamm. Sci.* **20**, 755–773.

Heide-Jørgensen, M. P., Laidre, K. L., Jensen, M. V., Dueck, L., and Postma, L. D. (2006). Dissolving stock discreteness with satellite tracking: Bowhead whales in Baffin Bay. *Mar. Mamm. Sci.* **22**, 34–45.

Lowry, L. F. (1993). Foods and feeding ecology. *In* Burns, J. J., Montague, J. J. and Cowles, C. J. (eds). "The Bowhead Whale." Spec. Pub. 2. Soc. Mar. Mammal., Lawrence, KS.

McCartney, A. P. (ed.). (1995). "Hunting the Largest Animals: Native Whaling in the Western Arctic and Subarctic." The Canadian Circumpolar Institute, Studies in Whaling No. 3, Occasional Publication No. 36.

Moore, S. E., George, J. C., Coyle, K. O., and Weingartner, T. J. (1995). Bowhead whales along the Chukotka coast in autumn. *Arctic* **48**, 155–160.

Nerini, M., Braham, H., Marquette, W., and Rugh, D. (1984). Life history of the bowhead whale, *Balaena mysticetus* (Mammalia: Cetacea). *J. Zool. (Lond.)* **204**, 443–468.

Nicklin, F. (1995). Bowhead whales: Leviathans of icy seas. *Natl. Geograph. Mag.* **188**(2), 114–129.

Reeves, R. R., and Leatherwood, S. (1985). Bowhead whale, *Balaena mysticetus*, linnaeus, 1758. *In* "Handbook of Marine Mammals" (S. H. Ridgway and R. Harrison, eds.) 3, pp. 305–344. Academic Press, New York.

Rugh, D., DeMaster, D., Rooney, A., Breiwick, J., Shelden, K., and Moore, S. (2003). A review of bowhead whale (*Balaena mysticetus*) stock identity. *J. Cetacean Res. Manage.* **5**, 267–279.

Shelden, K. E. W., and Rugh, D. J. (1995). The bowhead whale, *Balaena mysticetus*: Its historic and current status. *Mar. Fish. Rev.* **57**(3–4), 1–20.

Vladimirov, V. L. (1994). Recent distribution and abundance levels of whales in Russian far–eastern seas. *Russ. J. Mar. Biol.* **20**, 1–9.

Würsig, B. (1988). The behavior of baleen whales. *Sci. Am.* **258**, 102–107.

Würsig, B., and Clark, C. (1993). Behavior. *In* Burns, J. J., Mantage, J. J., and Cowles, C. J. (eds). "The Bowhead Whale." Spec. Pub. 2, *Soc. Mar. Mammal.*, Lawrence, KS.

# Bow-Riding

**BERND WÜRSIG**

One of the most fascinating behaviors of dolphins is when they ride the bow pressure waves of boats. Dolphins probably have been bow-riding ever since swift vessels plied the seas, propelled by oar, sail, or very recently in the history of seafaring, motor. The Greeks wrote of bow-riding in the eastern Mediterranean and Aegean Seas by what were most likely bottlenose (*Tursiops truncatus*), common (*Delphinus delphis*), and striped dolphins (*Stenella coeruleoalba*).

Bow-riding consists of dolphins, porpoises, and other smaller toothed whales (and occasionally sea lions and fur seals) positioning themselves in such a manner as to be lifted up and pushed forward by the circulating water generated to form a bow pressure wave of an advancing vessel (Lang, 1966; Hertel, 1969). Dolphins are exquisitely good at bow-riding, able to fine-tune their body posture and position so as to be propelled along entirely by the pressure wave, often with no tail (or fluke) beats needed. Bow-riders at the periphery of the pressure wave do need to beat their flukes, and so do bow-riders of a slowly moving vessel or one with a very sharp cutting instead of pushing bow.

There is often quite a bit of jostling for position at the bow, as dominant animals of a group edge others to a less favorable position, or as one is displaced from the bow by another one approaching (Fig. 1). It is great fun for a person to lean over the bow of a vessel and watch these inter-animal antics, as well as the fine-tuning of positioning, effected by slight body turns and almost imperceptible movements of the flippers. Bow-riding dolphins also tend to emit what sounds to the human listener like a cacophony of underwater whistles and "screams," sounds implicated in high levels of social activity (Brownlee and Norris, 1994). Bow-riding is probably the dolphin behavior most noted, and most enjoyed, by seafaring people the world over.

Of course, riding the bow also makes these animals susceptible to being lanced or harpooned in areas where they are taken by humans. Where this occurs near shore and in apparent smaller populations, dolphins become shy of the bow (Norris, 1974), but on the high seas or in deeper water, probably in larger populations, dolphins often still ride the bow after tens to hundreds of years of (generally small scale) human hunting.

While many species of dolphins, porpoises, and small toothed whale ride the bow, some do not; and in some species, certain populations do not. Bottlenose dolphins are well-known bow-riders the world over, but even they do not ride in some areas (even where they are not hunted) or on some types of vessels. For example, off the shores of Texas in the Gulf of Mexico, they generally do not approach any vessel

**Figure 1** *Common dolphins* (Delphinus delphis) *on the bow of a sailing vessel off Panama. Photo by Bernd Würsig.*

**Figure 2** *Two bottlenose dolphins* (Tursiops truncatus) *leap for a breath between rides on the bow of a shrimp vessel near shore in the Gulf of Mexico. Photo by Bernd Würsig.*

smaller than 15 m long to bow-ride, apparently finding the smaller bows not worth their while. Instead, they "hitch a ride" on the oil tankers and freighters, sometimes larger shrimping vessels while enroute to and from the shrimping grounds, at times bow-riding for 20 or more kilometers at a stretch. Dolphins ride underwater, and must leave their position to breathe, leaping forward and at an angle to the surface before falling back toward the advancing bow in a welter of foam (Fig. 2). Dolphins also ride the stern waves (or wakes) of boats, which present a different hydrodynamic challenge than bow-riding; and in some areas, dolphins that do not approach the bow will nevertheless ride in the influence of a large (or fast small) vessel's wake.

Most oceanic dolphins ride bow waves, with notable exceptions in areas of intensive hunting, such as by tuna vessels of the eastern Tropical Pacific, where vessels chase dolphins in order to net the tuna often affiliated with a dolphin school (Perrin, 1968). However, riding the bow is also "mood dependent"; dusky dolphins (*Lagenorhynchus obscurus*), for example, will not approach vessels when they have not fed for two or more days. These same dolphins will race toward a boat from several kilometers during and after social/sexual activities that take place immediately after bouts of feeding on schooling anchovy (Würsig and Würsig, 1980).

**B**

Why do dolphins bow-ride? It has been proposed that it is a mechanism to efficiently travel from one place to another. However, this is unlikely, for one often sees bow-riding dolphins after some time heading back to whence they picked up the vessel. Instead, it is more likely that riding the bow is done for enjoyment, for the sport of it; in other word, play. This is of great interest to behaviorists, for there are not too many non-domesticated adult mammals that habitually engage in activities just for the fun of them, although the list is growing with detailed observations in nature.

Bow-riding was certainly not "invented" by dolphins as a sport when human-made vessels first came on the scene. Instead, it appears to have been adapted from other wave-riding forms. Dolphins ride on the lee slopes of large oceanic waves and on the curling waves (or surf) that are formed as oceanic waves touch near-shore bottom (these two "rides" are hydrodynamically quite different; Hertel, 1969). Dolphins "body surf" much as do humans, but dolphins are generally much better surfers than humans. Dolphins also ride the bow waves of surging whales such as baleen whales and sperm whales (*Physeter macrocephalus*). Dolphins even "entice" whales to surge ahead by rapidly crossing back and forth a whale's eyes and snout. The whale surges forward in response (and apparent annoyance), often blowing forcefully during the surge. An abrupt bow wave is formed, and the previously heckling dolphins are all lined up in that wave, apparently enjoying its momentary pressure effect. This activity can go on with one whale for 20 min or more, until the whale tires, the bow wave becomes less distinct, and the dolphins abandon it to try with another whale or to go about other activities. They have had their fun, and we are left to wonder what is going on in that large brain during these bouts of quite obvious play.

## See Also the Following Articles

Group Behavior ■ Playful Behavior ■ Aerial Behavior

## References

Brownlee, S. M., and Norris, K. S. (1994). The acoustic domain. *In* "The Hawaiian Spinner Dolphin" (K. S. Norris, B. Würsig, R. S. Wells, and M. Würsig, eds), pp. 161–185. University of California Press, Berkeley.

Hertel, H. (1969). Hydrodynamics of swimming and wave-riding dolphins. *In* "The Biology of Marine Mammals" (H. T. Anderson, ed.), pp. 31–63. Academic Press, New York.

Lang, T. G. (1966). Hydrodynamic analysis of cetacean performance. *In* "Whales, Dolphins, and Porpoises" (K. S. Norris, ed.), pp. 410–434. University of California Press, Berkeley.

Norris, K. S. (1974). "The Porpoise Watcher." Norton Press, New York.

Perrin, W. F. (1968). The porpoise and the tuna. *Sea Front* **14**, 166–174.

Würsig, B., and Würsig, M. (1980). Behavior and ecology of the dusky dolphin *Lagenorhynchus obscurus*, in the south Atlantic. *Fish. Bull.* **77**, 871–890.

# Brain

HELMUT H.A. OELSCHLÄGER AND
JUTTA S. OELSCHLÄGER

Adaptation to aquatic environments is a multiconvergent phenomenon seen in a number of mammalian groups and species (Oelschläger and Oelschläger, 2002). In toothed whales (odontocetes), both the body shape and the morphology of the sensory organs and brain intimate the selective pressures, which may have led to exclusively aquatic life. There are some obstacles, however, in understanding brain evolution in these animals. First, we are only marginally familiar with the brain morphology of very few species, and here we are familiar with mainly the bottlenose dolphin (*Tursiops truncatus*; discussed later). Second, the brain itself does not fossilize; only the outer shape can be studied in natural endocasts, and these are biased covering blood vessels, meninges, and by geological artifacts. Thus, the tracing of brain evolution in fossils is difficult and should be supplemented by phylogenetic reconstruction on the basis of extant relatives. Third, although the comparative consideration of analogous developmental trends (primates) may be useful for the understanding of brain evolution in highly encephalized aquatic mammals, the paucity of data often leads to an overestimation of these analogies.

Among the most fascinating characteristics of toothed whales are the exceptionally large size of their brains, both in absolute and in relative terms, and the extremely dense folding of the neocortex. Whereas dolphins usually have a brain mass of about 200–2000 g, the maximal size is attained in killer whales (*Orcinus orca*) and (giant) sperm whales (*Physeter macrocephalus*) approximating 10,000 g. Basically, odontocete brains show the typical mammalian bauplan and are as complicated morphologically as those of other mammalian groups. To some extent, they parallel the simian and the human brains. In this respect, however, it has to be kept in mind that cetaceans have been subject to profound modifications in brain morphology and physiology during 50 million years of separate evolution in the aquatic environment. Moreover, it is still very difficult to correlate the results of behavioral and physiological research on dolphins with existing neuroanatomical data. Because invasive experimentation is not possible in cetaceans, the functional significance of such data can only be elucidated via comparison with other aquatic or terrestrial mammals.

Most studies during the last decades have focused on the morphology and the potential physiology of the adult toothed whale brain and its functional systems (Jelgersma, 1934; Jansen and Jansen, 1969; Glezer *et al.*, 1988; Ridgway, 1990). Concerning the development of the odontocete brain, the very few recent papers were dedicated to the striped dolphin (*Stenella coeruleoalba*), harbor porpoise (*Phocoena phocoena*; Buhl and Oelschläger, 1986), spotted dolphin (*Stenella attenuata*), narwhal (*Monodon monoceros*), and sperm whale (Oelschläger and Kemp, 1998). Reviews of information on the mammalian brain, including that of marine mammals can be found in Nieuwenhuys *et al.* (1998).

## I. Morphology of the Cetacean Brain

### A. General Appearance

Whereas its development in the embryonal and early fetal period is similar to that of other mammals, the brain of adult whales and dolphins is rather spherical in comparison with that of generalized land mammals (Oelschläger and Oelschläger, 2002) and somehow reminiscent of a boxing glove (Fig. 1). In correlation with the so-called "telescoping" of the skull along the beak-fluke axis, both the cranial vault and the brain are short but wide and even more so in toothed whales (odontocetes) than in baleen whales (mysticetes). In the bottlenose dolphin (Fig. 1), the hemispheres are rounded and high, and the anterior profile is rather steep. In ventral aspect (Fig. 2), the contour of the odontocete forebrain is more trapezoidal, whereas in mysticetes it is more trilobate, with the area of the insula (Fig. 3) being visible externally as an indentation between the orbital and the temporal lobes only (Figs 1 and 2). In comparison with hoofed animals, the telencephalic hemisphere seems to be rotated rostralward and ventralward

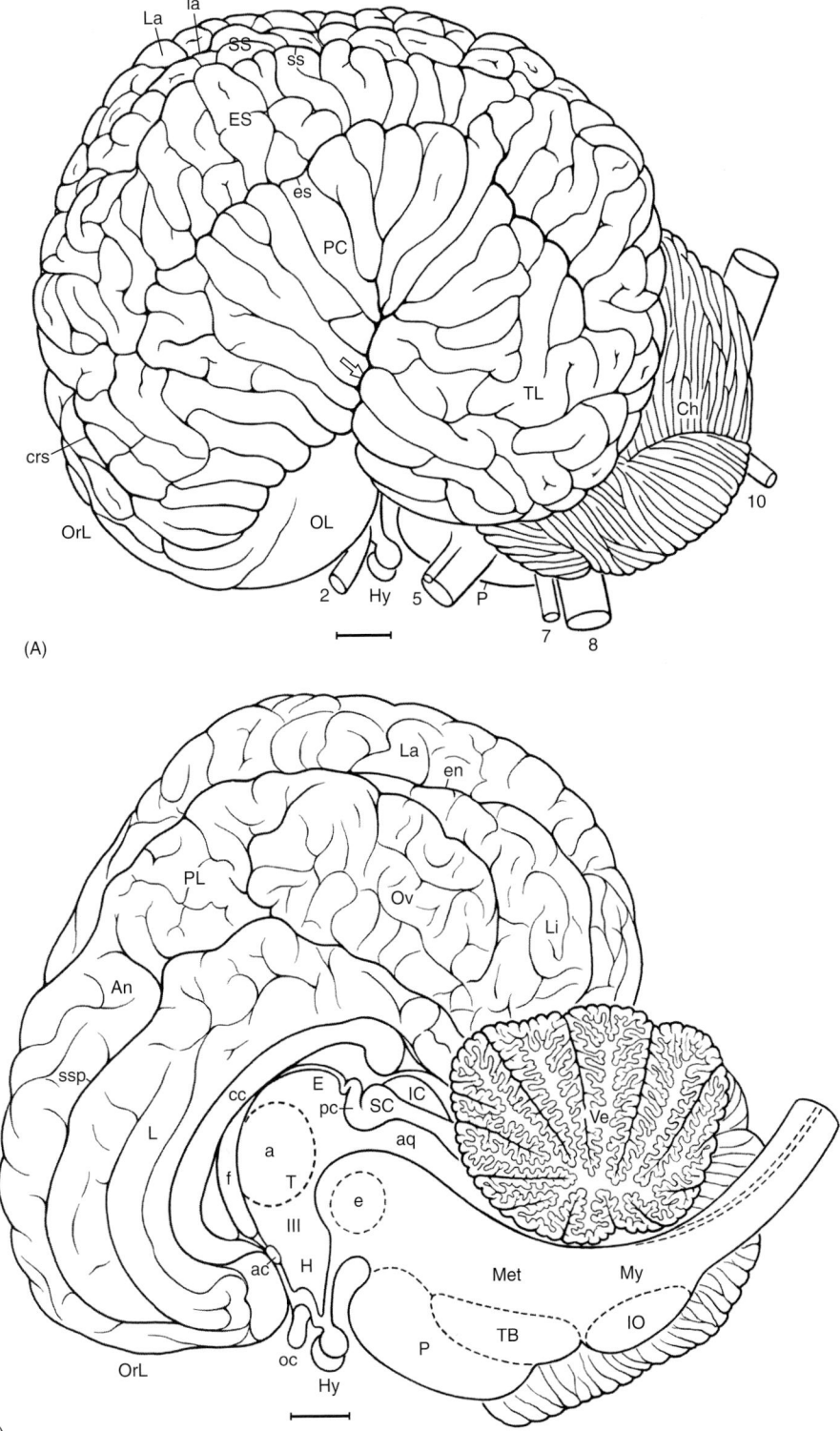

B

(A)

(B)

**Figure 1** *Bottlenose dolphin* (Tursiops truncatus) *brain. (A) lateral aspect, (B) another specimen, mediosagittal aspect. (A) From Langworthy (1932), modified after Morgane and Jacobs (1972) and Pilleri and Gihr (1970); (B) after Morgane and coworkers. Arrow, pointing into sylvian cleft; a, interthalamic adhesion; ac, anterior commissure; An, anterior lobule; aq, cerebral aqueduct; cc, corpus callosum; Ch, cerebellar hemisphere; crs, cruciate sulcus; e, elliptic nucleus; E, epithalamus; en, entolateral sulcus; es, ectosylvian sulcus; ES, ectosylvian gyrus; f, fornix; H, hypothalamus; Hy, hypophysis; IC, inferior colliculus; IO, inferior olive; L, limbic lobe; la, lateral sulcus; La, lateral gyrus; Li, lingual lobule; Met, metencephalon; My, myelencephalon; oc, optic chiasm; OL, olfactory lobe; OrL, orbital lobe; Ov, oval lobule; P, pons; pc, posterior commissure; PC, perisylvian cortex; PL, paralimbic lobe; SC, superior colliculus; ss, suprasylvian sulcus; SS, suprasylvian gyrus; ssp, suprasplenial (limbic) sulcus; T, thalamus; TB, trapezoid body; TL, temporal lobe; Ve, vermis; 2, optic nerve; 5, trigeminal nerve; 7, facial nerve; 8, vestibulocochlear nerve; 10, vagus nerve; III, third ventricle. Scale: 1 cm.*

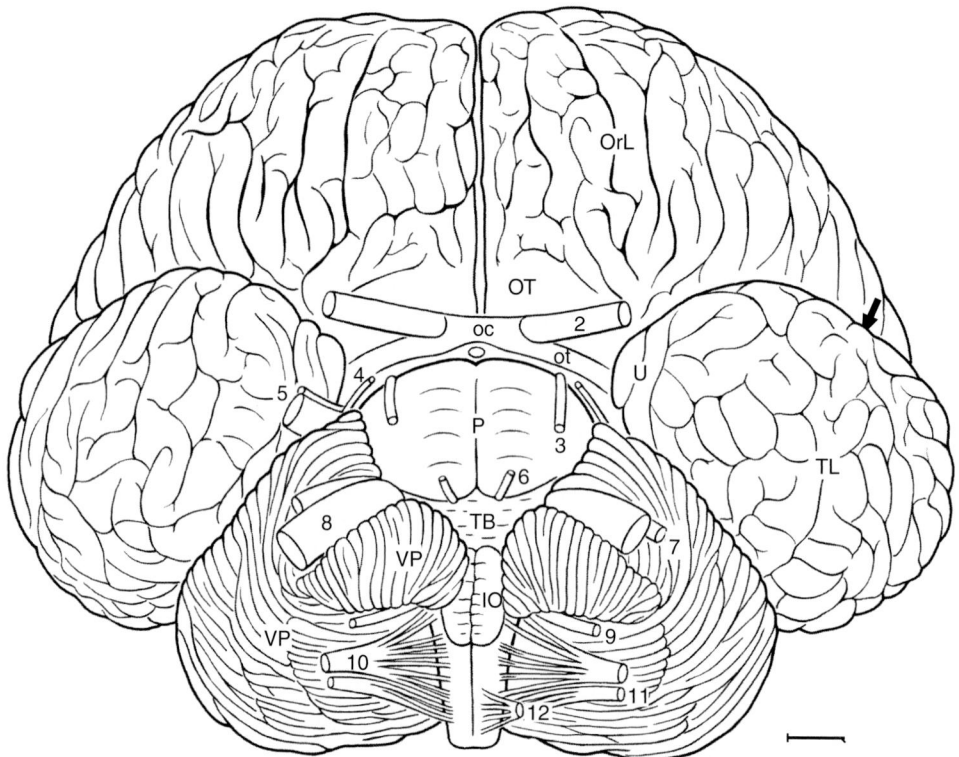

**Figure 2**  *Bottlenose dolphin* (Tursiops truncatus) *brain in basal aspect. From Langworthy (1932), modified after Brauer and Schober (1970), Pilleri and Gihr (1970), and Morgane and Jacobs (1972). Arrow pointing into sylvian cleft. ot, optic tract; OT, olfactory tubercle; U, uncus; VP, ventral paraflocculus; 2–12, cranial nerves; 3, oculomotor nerve; 4, trochlear nerve; 6, abducens nerve; 9, glossopharyngeus nerve; 10, vagus nerve; 11, accessory nerve; 12, hypoglossus nerve. For other abbreviations see previous figure. Scale: 1 cm.*

leading to a subvertical position of the corpus callosum (Fig. 1). In some odontocetes (bottlenose dolphin, sperm whale), the posterior myelencephalon and the anterior spinal cord curve around the cerebellum. Via an S-shaped transition, the spinal cord then continues straight along the body axis, thus accounting for the shortening of the cetacean neck region.

The ventricular system reflects the foreshortening of the brain in the tight coiling of the lateral ventricles, the shortness of the fronto-orbital region (anterior horn), the lack of an occipital pole of the hemisphere (no posterior horn), and the large size of the midbrain (cerebral aqueduct).

### B. Telencephalon

*1. Cortex*  In comparison with generalized tetrapod mammals (Oelschläger and Oelschläger, 2002), the surface of the telencephalic hemispheres is extremely convoluted, particularly in toothed whales (Figs 1–3). Gyrification in baleen whales is presumably less extreme because of the greater width of their cortical layers. It is the neocortex that accounts for the large size of the telencephalon and thus the large size of the brain (percentage of the neocortex: 63% in the generalized Franciscana, *Pontoporia blainvillei*; 87% in the sperm whale).

As in higher primates, most of the cortex of the cetacean olfactory and limbic systems (allocortex) is either restricted to the rostral base of the hemisphere (paleocortex; olfactory system) or located at the inferior horn of the lateral ventricle in the temporal lobe

(archicortex: hippocampus). The archicortex in cetaceans, and particularly in toothed whales (Fig. 3), is much smaller than in terrestrial mammals. This correlates well with the small size of other limbic components, e.g., the fornix as the fiber tract of the hippocampus and the mammillary body as a relay structure, whereas the cortical fields above the corpus callosum ("limbic lobe," Fig. 1) and the entorhinal cortex on the temporal lobe are well developed. As in primates, the cortex of the cetacean limbic lobe presumably does not have an immediate relationship to olfaction. With respect to other large mammals, olfactory components seem to be much reduced or even lacking in adult toothed whales. In contrast to the situation in baleen whales where the nose is small but obviously functional, odontocetes do not exhibit an olfactory part of the nose, olfactory bulb or tract, and the central parts of the olfactory system are moderately developed. In these animals, the mechanical impact of pneumatic sonar signal generation may have led to the elimination of the nasal chemoreceptor systems during evolution (Oelschläger, 2008).

A. SURFACE CONFIGURATIONS  The fissural or gyral pattern of the cetacean cortex, which has been discussed in many papers in the past, bears general resemblance to that of carnivores and ungulates (Figs 1–3). On the convex lateral surface and the vertex of the hemisphere, the main fissures (ectosylvian, suprasylvian, lateral sulcus) run around the Sylvian cleft more or less concentrically. Thus, for example, the ectosylvian gyrus is bordered by the ectosylvian and suprasylvian sulci, the suprasylvian gyrus by the suprasylvian and lateral sulci, and the lateral gyrus by the lateral and entolateral

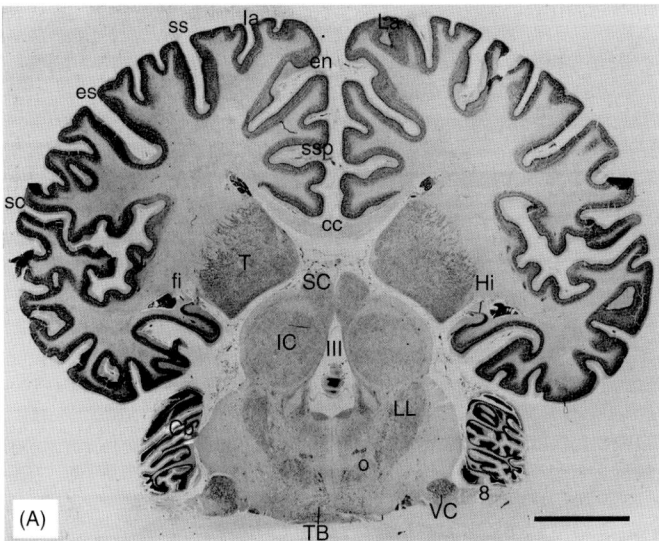

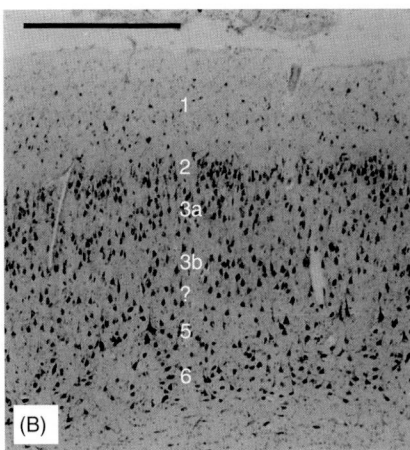

**Figure 3** *Franciscana* (Pontoporia blainvillei), *transverse section of 20-μm thickness through adult brain. Cresyl violet stain. (A) Overview, (B) cortex sample from the lateral gyrus. Cb, cerebellum; fi, fimbria; Hi, hippocampus; J, insula; LL, lateral lemniscus nuclei; O, superior olive; VC, ventral cochlear nucleus. Numbers 1–6 in insert, layers I–VI of the neocortex. For other abbreviations see previous figures. Scale in (A) 1 cm and in (B) 500 μm.*

(paralimbic) sulci. As in other high-encephalized mammals, the insular area (Fig. 3) is covered by so-called "opercula" of the neighboring neocortex, which meet at the lateral hemispheral fossa (Sylvian cleft) and are combined in the term perisylvian cortex (Fig. 1). The medial cortex of the hemisphere is subdivided by the suprasplenial sulcus or limbic cleft and the entolateral or paralimbic cleft. Far rostrally and ventrally, the cruciate sulcus (Fig. 1) separates an anterior motor cortical field from a posterior somatosensory field and is therefore a candidate for homologization with the ansate sulcus in hoofed animals as well as the central sulcus in primates. It is unclear whether the "calcarine sulcus," which originates from the paralimbic cleft and encircles the oval lobule (Fig. 1B), is the homolog of the primate calcarine fissure that houses the primary visual field. Electrophysiological mapping experiments in dolphins have detected visually responding cortical fields in a more anterior and lateral position on the vertex of the hemisphere.

In one of the most plesiomorphic whales (Susu or Indus river dolphin; *Platanista gangetica*), gyrification is still relatively simple. In the dorsal aspect, the main fissures are straight, smooth, and remind of the situation in the mesonychid *Synoplotherium*, a fossil terrestrial relative of the cetaceans. Brain length in the latter still exceeded brain width, the formation of a temporal lobe had only just begun, and the olfactory system was well developed. Archaic fossil cetacean (archaeocete) brains are difficult to interpret morphologically because they apparently had large retia mirabilia on the surface of the brain as is seen in living baleen whales, where they largely conceal the posterior (cerebellar) part. Their telencephalic hemispheres were obviously small and showed no signs of gyrification.

B. LOCALIZATION OF CORTICAL AREAS  Electrophysiological cortical mapping experiments in the bottlenose dolphin located the motor neocortical field in the frontal (orbital) lobe rostral to the paralimbic lobe (Fig. 7). The motor cortex is characterized by the presence of giant pyramidal neurons and gives rise to the pyramidal tract. Laterally and caudally, the motor field is separated from the somatosensory field by the so-called "cruciate" sulcus. The somatosensory field is situated rostral to the visual and auditory fields. The position of the visual fields is somewhat more complicated. Although different in many aspects from other mammalian visual cortices, those of the dolphin are apparently well developed and highly differentiated. All authors place visual cortex in the lateral gyrus, whereas some distinguish a primary visual field near the medial border of the suprasylvian gyrus from a secondary field in the lateral gyrus. Other authors find an additional visual area in the medially adjacent part of the paralimbic lobe (Fig. 7). On the basis of histological analysis, a visual field has also been reported for the borders of the supposed "calcarine" sulcus that separates the oval and lingual sublobules of the paralimbic lobe (Fig. 1). The large primary auditory field lies on the vertex of the hemisphere in the suprasylvian gyrus and lateral to the visual field(s), the secondary auditory field lies more laterally in the medial part of the ectosylvian gyrus (Fig. 7).

Viewed as a whole, the topography of the motor and sensory projection fields of dolphins differ from that in other mammals. However, these primary cortical fields in cetaceans have retained the sequence found in plesiomorphic terrestrial mammals. Therefore, it seems as if the hemisphere would have been expanded to such a degree in a caudal and a ventral direction (huge temporal lobe) that the auditory cortex now extends as a belt along the vertex of the hemisphere and reaches further caudally in the dolphin than the visual field, which is located more in the center of the hemisphere. The lateral surface of the whale hemisphere may be interpreted as a large "association cortex" connecting the auditory fields with other sensory and motor areas.

*2. Commissures*  The size of the individual commissural systems is specific for cetaceans and shows correlations with cortical and nuclear structures throughout the forebrain (Figs 1B, 3, and 4).

The anterior commissure, which links neocortical and paleocortical areas of both temporal lobes, is obviously weak due to the considerable reduction of the olfactory system. The corpus callosum as the main link between the neocortical fields of both hemispheres is rather thin in dolphins relative to total brain mass and in comparison with the situation in other mammals. The cross-sectional area of the cetacean corpus callosum (defined by its area in the midsagittal plane) related to brain mass generally decreases in larger-brained toothed whales, thereby suggesting that increases in brain weight are not necessarily accompanied by increases in callosal linkage between the telencephalic hemispheres. Thus, the corpora callosa of a killer whale

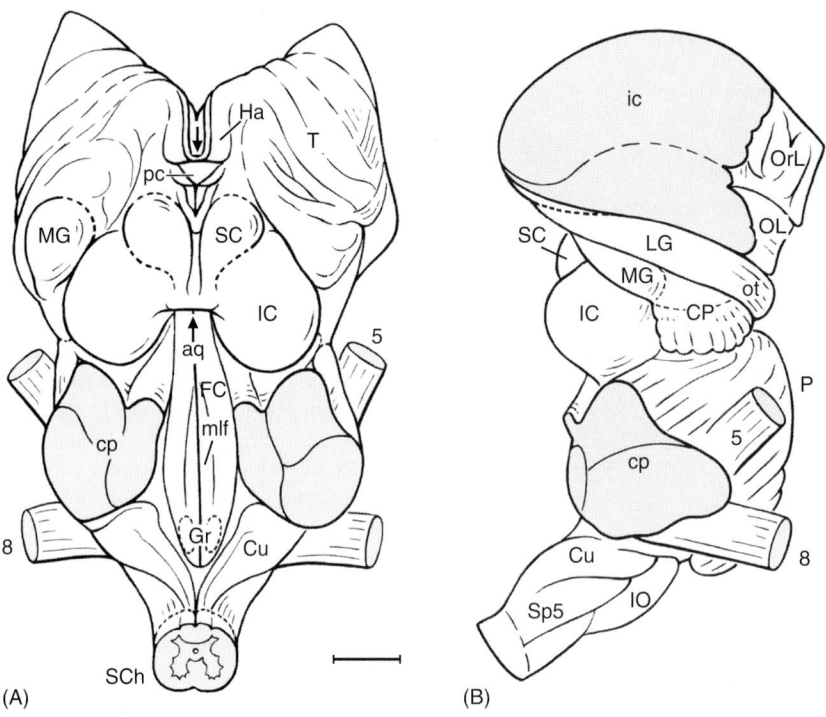

**Figure 4** *Brainstem of the bottlenose dolphin (Tursiops truncatus). (A) Dorsal, (B) lateral aspect. From Langworthy (1931), modified after Pilleri and Gihr (1970), and Morgane and Jacobs (1972). cp, cerebellar peduncles; CP, cerebral peduncle; Cu, cuneate nucleus; FC, facial colliculus; Gr, gracile nuclei; Ha, habenula; ic, internal capsule; LG, lateral geniculate body; MG, medial geniculate body; mlf, medial longitudinal fascicle; SCh, spinal cord; Sp5, spinal nucleus of trigeminal nerve. Arrows in a) pointing into cerebral aqueduct. For other abbreviations see previous figures. Scale: 1 cm.*

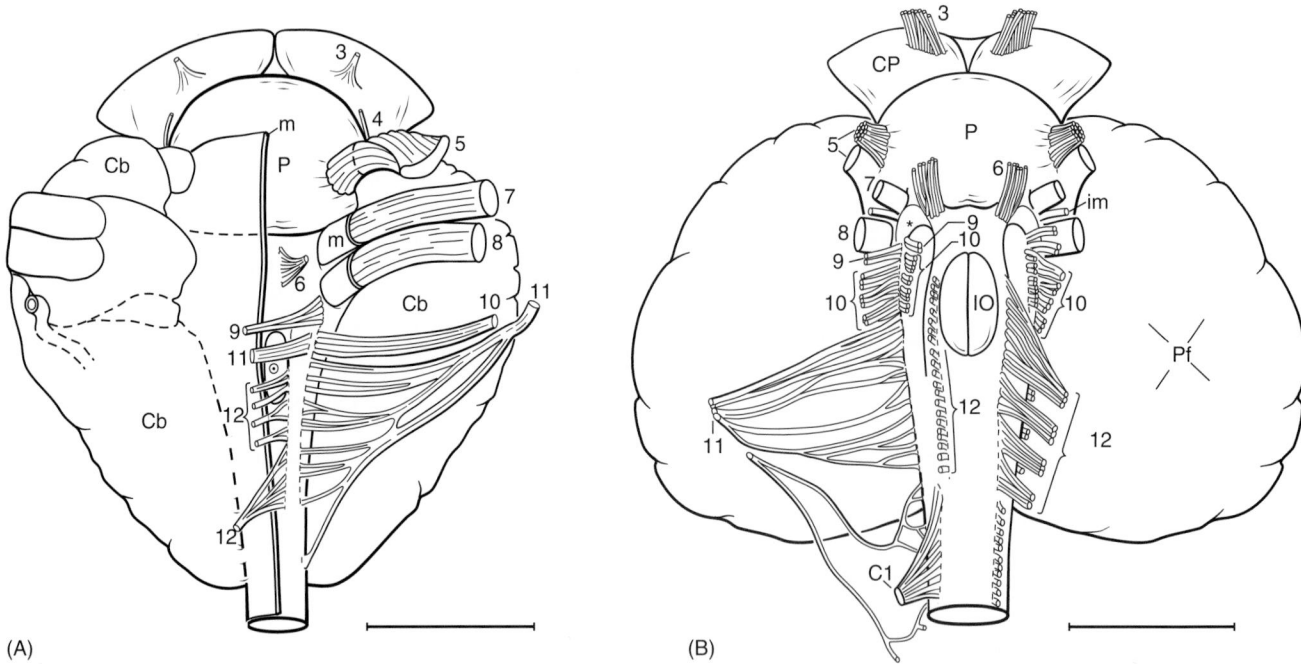

**Figure 5** *Cetacean brainstems in basal aspect. (A) sperm whale (Physeter catodon), after Langworthy (1937), modified. Dotted circle: inferior olive; m, meninx. (B) Fin whale (Balaenoptera physalus), from Jansen (1953), modified after Pilleri and Gihr (1970). Asterisk, facial tubercle; C1, motor root of first cervical spinal nerve; im, intermedius nerve; m, meninx; Pf, paraflocculus. For other abbreviations see previous figures. Scale in (A) and (B): 5 cm.*

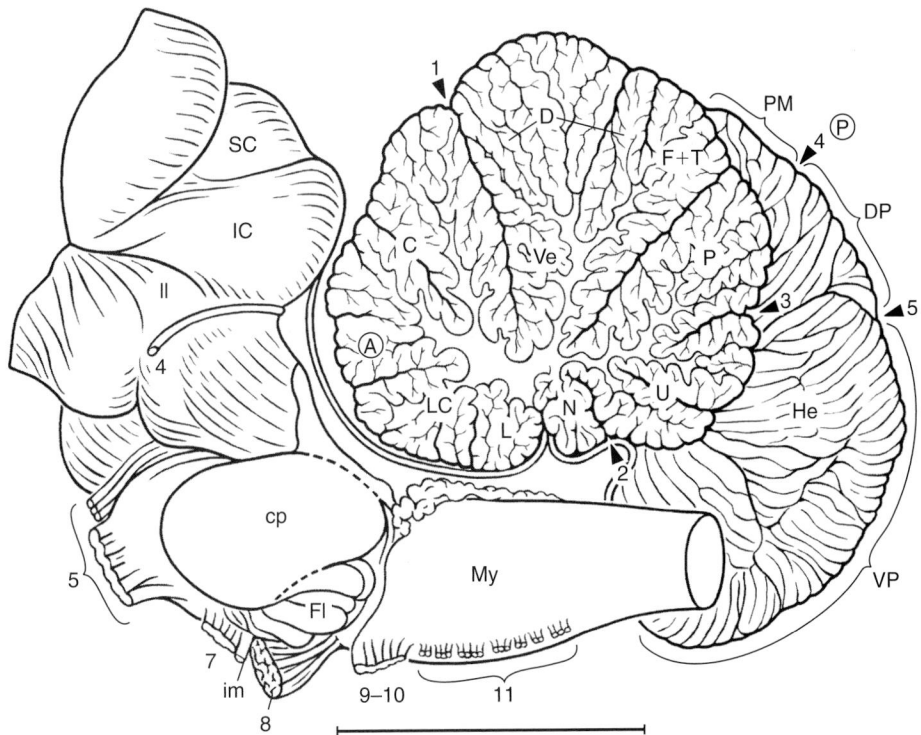

**Figure 6** *Brainstem of the fin whale* (Balaenoptera physalus), *mediosagittal aspect. After Jansen (1953), modified. 1, primary fissure; 2, posterolateral fissure; 3, secondary fissure; 4, parafloccular fissure; 5, intraparafloccular fissure; A, anterior lobe of cerebellum; C, culmen; D, declive; DP, dorsal paraflocculus; Fl, flocculus; F + T, folium and tuber; He, cerebellar hemisphere; L, lingula; LC, lobus centralis; ll, lateral lemniscus; N, nodulus; (P), posterior lobe of cerebellum; P, pyramis; PM, paramedian lobule; U, uvula; VP, ventral paraflocculus. For other abbreviations see previous figures. Scale: 1 cm.*

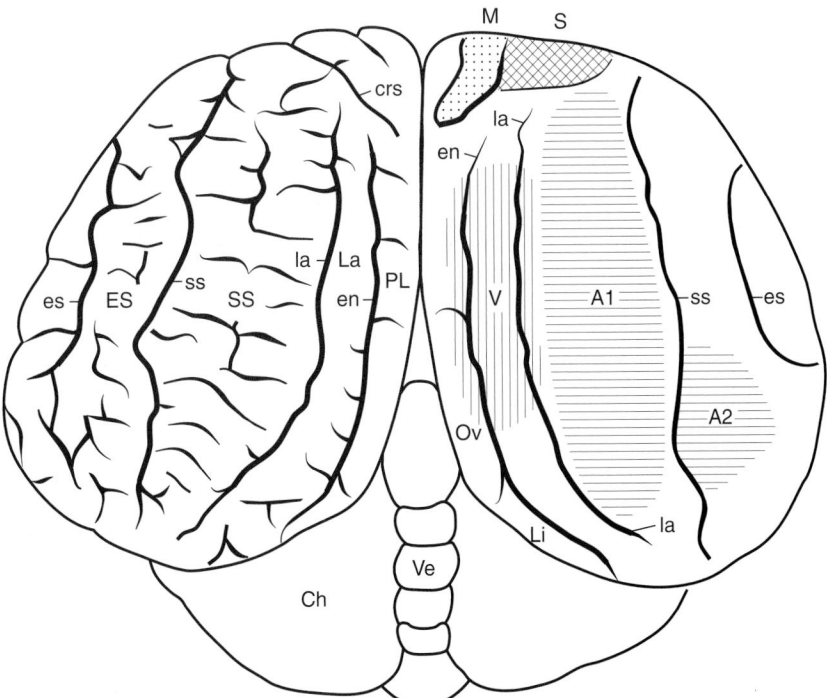

**Figure 7** *Neocortical motor and sensory fields in the bottlenose dolphin* (Tursiops truncatus). *After Morgane et al. (1986). A1, A2, auditory fields; M, motor field; PL, paralimbic lobe; S, somatosensory field; V, visual field. For other abbreviations see previous figures.*

**B**

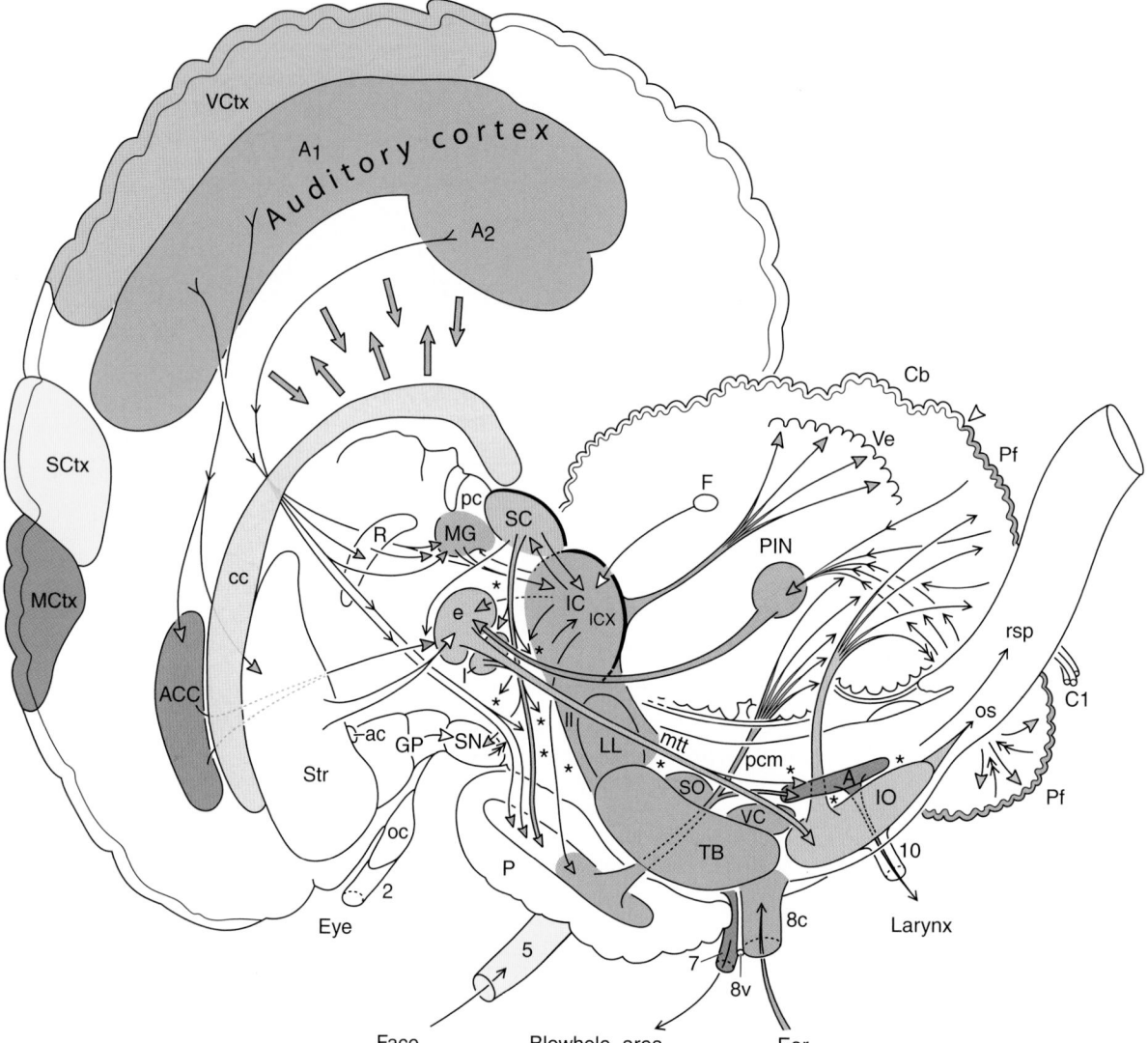

**Figure 8** *Sagittal aspect of dolphin brain with selected sensory and motor structures showing the central position of the auditory system within connective loops between the elliptic nucleus (e), inferior olive (IO), cerebellum (Cb) and including the pons (pontine nuclei, P) presumably involved in phonation and acousticomotor navigation. From Oelschläger (2008), modified. A, nucleus ambiguus; ACC, anterior cingulate cortex; C1, first cervical spinal nerve; F, nucleus fastigii; GP, globus pallidus; I, interstitial nucleus of Cajal; ICX, external cortex (nucleus) of IC; MCtx, motor cortex; mtt, medial tegmental tract; os, olivospinal tract; P, pontine nuclei; pcm, pedunculus cerebellaris medius; Pf, Paraflocculus; PIN, posterior interposed nucleus; R, reticular thalamic nucleus; rsp, reticulospinal tract; SN, substantia nigra; SO, superior olive; SCtx, somatosensory cortex; Str, striatum; TB, trapezoid body and nucleus; VCtx, visual cortex; °°°, periaqueductal gray and reticular formation.*

and a human brain show the same cross-sectional area, with the killer whale brain being some five times heavier than that of the human. Furthermore, this regression of the corpus callosum in larger species obviously has not been compensated by an enlargement of other commissural tracts. In conclusion, the interhemispheric connectivity seems to correlate inversely with brain weight insofar as larger brains possess a lower neocortical neuronal density with the possible result that relatively fewer fibers constitute the corpus callosum. An additional explanation for the regression of the corpus callosum in larger brains could be a smaller percentage of cortical neurons establishing interhemispheric connections and thus a certain independence on the part of both hemispheres. In electroencephalographic experiments, sleeping bottlenose dolphins have been reported to show signs of wakefulness

(low voltage, fast activity waveforms) in one hemisphere and sleep (high voltage, slow wave) in the opposite hemisphere. The posterior commissure (Figs 1, 4, and 8) is very well developed. Provided that its connections to other brain structures are similar to those in terrestrial mammals, the considerable size of the posterior commissure in cetaceans may suggest massive projections from somatosensory relay nuclei of the brain stem and the cerebellum to the contralateral pretectum and thalamus.

*3. Basal Ganglia* All components of the basal ganglia as known from other mammals are present in cetaceans (corpus striatum, globus pallidus, claustrum, amygdaloid complex) and for the most part show the usual topographic relationships to each other and to neighboring structures (Jansen and Jansen, 1969, Morgane and Jacobs, 1972). Moreover, their histological organization corresponds well with that in other mammals. The caudate nucleus as the main part of the large corpus striatum, which bulges distinctly in the area of the "olfactory tubercle" (olfactory lobe) together with the putamen, is largely separated from the latter by a well-developed internal capsule.

Reports regarding the size of the basal ganglia in cetaceans are contradictory in the literature. Quantitative analysis (Schwerdtfeger *et al.*, 1984) has shown that in the generalized Franciscana the corpus striatum, one of the most important centers for locomotion, is large and attains a size index between that of prosimian and simian monkeys. Moreover, as in primates, a size correlation between the striatum and the neocortex seems to be valid for dolphins as well.

The structure of the amygdaloid complex very closely resembles that of other mammals. The size of the amygdala as a whole seems to have been only slightly affected by the reduction of the paleocortex in odontocetes, giving the impression that this nucleus (as in primates) is largely independent of the olfactory system. Its relative size in the Franciscana seems to be larger than in primates presumably on account of its interconnections with the hypertrophied auditory system and the temporal lobe. Particularly the lateral amygdaloid nucleus may bear some relation to auditory function since this nucleus is extremely well developed both in whales and in bats. However, the corticomedial group of the amygdaloid nuclei, which largely depends on the olfactory system functionally, nevertheless occupies the same proportion of the entire amygdaloid complex in the harbor porpoise as in the macrosmatic sheep.

## C. Diencephalon

The relative size of the diencephalon in plesiomorphic dolphins (Franciscana; Susu) is approximately the same as that in monkeys (Figs 1B and 3). There are no reliable data for advanced, delphinid dolphins. The predominant structure is the thalamus. The shape of the diencephalon in mediosagittal aspect is often rather wedge-like in adult cetaceans, with the hypothalamus bending slightly caudalward and tapering in the direction of the hypophysis, particularly in larger toothed whales. In late embryos and early fetuses the floor of the hypothalamus is rather long, whereas it is rapidly foreshortened in later stages during the telescoping process and especially in larger toothed whales. Thus, the transverse interpeduncular fossa between the optic chiasm and pons appears slit-like in adult toothed whales (Figs 1B, 2, and 8).

*1. Epithalamus* The habenular complex is large and the habenular commissure well developed. The pineal organ is reduced or even lacking in cetaceans. A pineal rudiment is present between the habenular and the posterior commissures in embryos and early fetuses of dolphins and the sperm whale, but not in the early fetal narwhal (Oelschläger and Kemp, 1998). In adult whales and dolphins, many observers found the pineal organ to be lacking: it seems plausible that in this case the two fiber tracts unite into a commissural complex (common dolphin; Oelschläger *et al.*, 2008). Rudiments of the pineal organ have also been found in the humpback whale (*Megaptera novaeangliae*) and fin whale (*Balaenoptera physalus*).

*2. Thalamus* Basically, the organization of the large thalamus in cetaceans corresponds well with that in a variety of terrestrial mammals, among them ungulates and primates. There are four groups of nuclei in the dorsal thalamus that constitute about 92% of the thalamus in the bottlenose dolphin: the anterior, medial, ventral, and lateral nuclei. (1) The anterior group of nuclei, which is related to the cortex of the large cetacean limbic lobe, is well developed but constitutes only

a small part of the total dorsal thalamus. The anteroventral nucleus, which projects to the anterior limbic cortex, dominates this group. In contrast, the mammillary body and the interconnecting mammillothalamic tract are comparatively small and thin. (2) In the medial group of the thalamus, the mediodorsal nucleus is remarkably large and merits special interest because of various connections with olfactory and limbic structures as well as a presumed phylogenetic size correlation with the frontal (orbital) lobe of the mammalian telencephalic hemisphere. (3) The ventral group consists mainly of somatosensory nuclei and constitutes a large part of the dorsal thalamus. In mammals, generally, its ventral posterior nucleus (VPN) receives afferents via the medial lemniscus and the spinothalamic and trigeminothalamic tracts and dispatches a main projection to the somatosensory cortex. In the bottlenose dolphin, the ventral posterior nucleus is relatively small and projects to the neocortex anterior to the suprasylvian auditory area (Fig. 7). Compared to its lateral subnucleus, where the body region is represented, the medial subnucleus of the VPN with the head representation is relatively large. In dolphins, the limited somatosensory representation of the body is also reflected in the spinal cord (Oelschläger and Oelschläger, 2002). (4) As in higher primates, the lateral group of thalamic nuclei in cetaceans is dominated by the massive pulvinar, the largest single complex in the thalamus of the bottlenose dolphin. The pulvinar more or less merges in both the strongly protruding medial geniculate nucleus (MG; auditory) and the large lateral geniculate nucleus (LG; visual). The main projection of the inferior pulvinar targets the suprasylvian gyrus, and that of the medial pulvinar the ectosylvian gyrus, whereas the lateral pulvinar projects to the border of the lateral and suprasylvian gyri. The MG is impressively large in cetaceans (Fig. 4) and reflects the outstanding development of the auditory system in these animals. Ventral portions of the MG project to the primary auditory area of the suprasylvian gyrus (Fig. 7), dorsal portions to the "secondary" auditory area in the ectosylvian gyrus as well as to the temporal operculum (perisylvian cortex; Fig. 1). In the bottlenose dolphin, the LG is surprisingly well developed, though less so than the MG. The LG projects to the visually excitable part of the lateral gyrus (Figs 4 and 7), but does not show the laminar organization usually associated with biretinal projection. This may be related to the fact that in cetaceans the fibers in the optic nerve show a complete or almost complete decussation.

*3. Hypothalamus* The basal part of the diencephalon exhibits an organization similar to that encountered in other mammals. The anterior, tuberal, and posterior hypothalamic nuclei are evident but not particularly prominent. The paraventricular and supraoptic nuclei are obvious because of their large hyperchromatic cells, the latter nucleus being especially well formed. As in other mammals, the supraoptic commissure is well developed and well organized. The small size of the mammillary bodies, which in the postnatal animal do not protrude at the brain surface, correlates with the weak development of the hippocampus, postcommissural fornix, and mammillothalamic tract.

## D. Brain Stem and Cerebellum

The percentage of the dolphin midbrain in the total brain volume is relatively low and ranges between that of prosimian and simian primates. Nevertheless, the size index, which is related to the body size and the regression line of basal insectivores (shrews, hedgehogs, tenrecs), shows a remarkable increase of this structure even in plesiomorphic "river dolphins" (*Pontoporia blainvillei*, Schwerdtfeger *et al.*, 1984). This may be attributable to the growth of auditory system components (Figs 2–5 and 7–8). The cerebellum and pons are well developed and the myelencephalon (medulla oblongata) is very large in comparison with that of other mammals. This may be due to the considerable growth of cranial nerve nuclei and their connectivity, particularly those of the trigeminal, auditory, and motor systems.

A. SELECTED NUCLEI   The cetacean brain stem comprises the nuclei known from other mammals; this could be confirmed in our work on the fetal narwhal.

The oculomotor nucleus is the largest eye muscle nucleus, a fact that correlates well with the diameter of the oculomotor nerve. In comparison, the trochlear and abducent nuclei and nerves are rather small and thin in most cetaceans. The sensory trigeminal nuclei (motor, principal, and spinal nucleus) are very well developed, reflecting the large relative size of the cetacean head and the diameter of the trigeminal nerve, which is maximal or submaximal among the cranial nerves. Within cetaceans, the motor nucleus was also the trigeminal is reported to be larger in mysticetes but subdivided more clearly in odontocetes. The sensory principal nucleus was also reported to be larger in baleen than in toothed whales, and its dorsal part giving rise to the well-developed trigeminothalamic tract (Wallenberg). The facial nucleus is very large in cetaceans and often bulges at the ventral surface of the medulla (tuberculum faciale). The nucleus was also divided into a number of cell groups, which can be differentiated from each other cytologically. Each of these cell groups is believed to be responsible for specific muscles or muscular systems, e.g., the dorsal group for muscles of the upper respiratory tract around the blowhole (epicranial complex; Cranford et al., 1996), which are involved in the generation and emission of sonar signals in toothed whales. In comparison with other mammals, the ambiguus nucleus is large in cetaceans, which is similar to some bats (mouse-eared bat; *Myotis myotis*). This nucleus, which is larger in mysticetes than in odontocetes, innervates the muscles of the pharynx, larynx, and the striated muscles of the esophagus via the glossopharyngeus–vagus–accessorius nerve complex. When as in other mammals, it should be involved in cetacean respiration, food processing, and sound production in the larynx. The nucleus of the accessory nerve, which extends into the spinal cord, is moderately developed. This may be related to the extreme foreshortening of the cervical region, restrictions in head and shoulder girdle movability, and the transformation of the forelimb into a steering device (flipper). The hypoglossal nucleus, a derivative of the motor column in the first (occipital) embryonal spinal segments, is well developed, although the flexibility of the tongue is reported to be restricted in most cetaceans. In large baleen whales, the tongue may attain the body mass of a full-size elephant.

Nuclei related or belonging to the extrapyramidal motor system are located in the rostral mesencephalon and in the formatio reticularis throughout the rhombencephalon. The elliptic nucleus (Figs 1 and 8: e, E), which is situated within the central gray rostral or dorsal to the oculomotor nuclear complex, is very conspicuous; in the past it was thought to be unique for the Cetacea until a similar nucleus was found in the elephant. For some time, it was unclear whether the nucleus of Darkschewitsch is integrated into the elliptic nucleus or is even equivalent to this nucleus. Today, the latter opinion is the generally accepted one. In cetaceans, the elliptic nucleus projects via the medial tegmental tract (Fig. 8: mtt) to the rostral medial accessory inferior olive (IO) and correlates with a hypertrophy in cerebellar structures (Cb; discussed later). The red nucleus in cetaceans is little known. In contrast to ungulates, which possess a large rubrospinal tract and lack a spinal pyramidal tract, cetaceans have both weak rubrospinal and corticospinal tracts. The pontine nuclei (Fig. 8: P)

are exceptionally well developed in cetaceans; they receive strong cortical projections and give rise bilaterally to the large brachia pontis (middle cerebellar peduncles, pcm). Thus, the size and the caudal extension of the pons are directly related to the size of the neocortex and the neocerebellum.

The cetacean inferior olive (Figs 1–2, 4–5, and 8: IO) is characterized by an extraordinary development of its medial accessory subnucleus, particularly its rostral portion; in comparison, the principal olive and the dorsal accessory olive appear small. In the two cetacean suborders, there are only minor differences in the relative development of the subnuclei, and both inferior olives join each other in the midline. The rostral part of the medial accessory olive receives massive input from the elliptic nucleus via the medial tegmental tract, and its pronounced development in cetaceans seems to be related to the immense size of the cerebellar posterior interposed nucleus (PIN) and the paraflocculus (Oelschläger et al., 2008). In terrestrial mammals, the medial accessory inferior olive is part of a fiber system involved in directional hearing.

B. CEREBELLUM   The cetacean cerebellum is very large (Figs 1–2, 5–6, and 8), its size obviously being linked phylogenetically with that of the neocortex. In older studies, the relative mass of the cerebellum in baleen whales with respect to total brain mass (average: 20%) was reported to represent a maximal development within the mammalia as a whole. Recently, however, comparative analysis has shown that in relation to body mass the cerebellum of baleen whales is not as voluminous as in larger delphinids such as the killer whale. Concomitantly, it became obvious that the large proportion of the cerebellum in the total brain volume of baleen whales is attributable to the relatively small size of the forebrain. Indeed, in doublelogarithmic regressions, baleen whales rank a little higher than sperm whales, beaked whales, and "river" dolphins but distinctly below the delphinid cetaceans. With respect to the regression line in the "basal insectivores," the cerebellum of the plesiomorphic La Plata dolphin ranks higher than the averages of prosimian and simian monkeys but lower than that of the human. In a group of delphinid species, indices of the total brain mass and cerebellum mass relative to body mass exceeded other groups (sperm whales, river dolphins, baleen whales) by up to three times.

Within cetaceans, the cerebellum of the baleen whales is much better understood owing to ontogenetic histological studies. Only minor external differences, however, seem to exist between the cerebella of toothed and baleen whales. Thus, the mysticete cerebellum is more rounded and slightly hourglass-shaped in the dorsal aspect, whereas the odontocete cerebellum is somewhat more flattened dorsoventrally as a consequence of the stronger telescoping of the brain and resultant overlapping of the cerebellum by the cerebral hemispheres.

The cerebellum consists of two large hemispheres and a comparatively narrow vermis (Figs 1–2, 6, and 8). Two transverse fissures separate three cerebellar lobes: the primary fissure separates the smaller anterior (rostral third) from the large posterior lobe (caudal two-thirds), and the posterolateral fissure separates the posterior lobe from the small flocculonodular lobe. These size relations between the lobes are characteristic for cetaceans. In midsagittal section (Fig. 6), the conventional subdivision of the vermis into nine lobules of the mammalian cerebellum is obvious. In cetacean cerebellar hemispheres, the small size of the anterior lobe may be explained by electrophysiological findings in other mammals indicating that the hemispheral parts of this lobe comprise the cortical representation of the fore and hind limbs that are highly modified or even have vanished in these animals. The caudally adjacent ansiform lobule, which also receives input from the limbs, is similarly small. However, the representation of the head in the simple lobule of the posterior lobe is rather large, and the considerable size of the paramedian lobule (body representation) has been related to the enormous significance of the tail in cetaceans. The paraflocculus, situated between the parafloccular and the posterolateral fissures (Fig. 6), is exceptionally large, particularly the ventral parafloccular lobule. The latter comprises about half of the surface of the cerebellar hemisphere. In mammals, the paraflocculus usually receives climbing fibers from the rostral part of the medial accessory inferior olive (Fig. 8). In cetaceans, both structures are exceptionally large, which strongly indicates a functional relationship between the paraflocculus on the one side and trunk and tail on the other side. The flocculonodular lobe as the principal terminus of primary and secondary vestibulocerebellar fibers ("vestibulocerebellum") is very small in cetaceans, particularly the floccular component. In mammals, generally, the latter is responsible for the regulation of vestibulo-ocular and optokinetic movements as well as compensatory activities of the neck muscles in so-called smooth pursuit movements of the eyes, particularly in carnivorous animals. Dolphins, however, which have limited neck mobility and can use their visual system only during daylight, may have to rely on their auditory system instead to follow their prey effectively.

In the ontogeny of the cetacean cerebellar cortex, which is three-layered, the fundamental mammalian pattern of transverse and longitudinal zones is discernible. These longitudinal zones are obviously related topographically to the development of the cerebellar nuclei (anterior, medial and posterior interposed nuclei, lateral cerebellar nucleus). In cetaceans, the lateral intermediate cortical zone (C2 zone) is enormously developed, occupying about three-fourths of the cerebellar surface (paraflocculus) and correlating with the huge posterior interposed nucleus.

C. MAIN FIBER SYSTEMS: MEDIAL LEMNISCUS   In cetaceans, the afferent spinal system (proprioceptive sensitivity) is moderately developed in accordance with the reduction of the hind limbs and pelvic girdle. In these animals, the dorsal funiculi (gracile and cuneate fascicles) are strikingly small; they are thought to convey input predominantly from the flippers and the tail (sense of position). Nevertheless, cutaneous sensitivity in the trunk was reported to be high. The medial lemniscus is weak in the caudal medulla oblongata, but becomes considerably stronger at more rostral medullary levels, presumably owing to the addition of afferent systems of the head (auditory, trigeminal systems).

*1. Trigeminothalamic Tract (Wallenberg)*   The dorsal part of the principal sensory trigeminal nucleus gives rise to the trigeminothalamic tract. The latter terminates in the medial part of the ventral posterior thalamic nucleus as the main somatosensory thalamic nucleus. As in ungulates and the elephant, where it is extremely well developed, the tract is thought to be responsible for intra- and perioral sensitivity in cetaceans innervating, e.g., tactile bodies on the lips of fin whales and sei whales (*Balaenoptera borealis*) and the epicranial complex (Cranford *et al.*, 1996) in toothed whales.

*2. Medial Tegmental Tract*   The elliptic nucleus (Figs 1B and 8: e), which almost "replaces" the red nucleus in Cetacea, is extraordinarily developed in whales and elephants, and gives rise to the strong medial tegmental tract (mtt) that proceeds to the rostral part of the huge medial accessory inferior olive (IO). The latter nucleus also receives afferents from the spinal cord (spino-olivary tract) and

projects to the lateral intermediate (C2) zone in the huge paraflocculus. The paraflocculus, which was shown in the rat to be the main target of auditory pontocerebellar projections and was estimated in the blue whale (*Balaenoptera musculus*) to receive three-fifths of the pontocerebellar fibers, has a massive projection to the posterior interposed nucleus (PIN) of the cerebellum (Cb). From the latter nucleus, ascending fibers run to the elliptic nucleus and other nuclei at the diencephalic/mesencephalic border which, in turn, project to the inferior olive via the medial tegmental tract (Fig. 8). Similar to the medial accessory olive, the paraflocculus also has been associated with mass movements of the posterior trunk and tail, the only region where the axial skeleton possesses a reasonable range of motion. The medial tegmental tract thus seems to be part of a recurrent circuit (elliptic nucleus, inferior olive, paraflocculus, posterior interposed nucleus, elliptic nucleus), which combines auditory input with locomotor activity and illustrates the dominant position of hearing among sensory systems in whales (acousticomotor navigation; Oelschläger, 2008; Oelschläger *et al.*, 2008).

*3. Pyramidal Tract*   The tract originates in the neocortical motor area rostral to the "cruciate" sulcus (Figs 1A and 7) and runs through the internal capsule. At mesencephalic levels, the localization of the pyramidal tract is difficult, and it is very small at high medullary levels. Typical macroscopical "pyramids" as seen in terrestrial mammals are not present in cetaceans. Here, the pyramidal tracts are weak and situated lateral to the inferior olivary complex, both inferior olives joining each other midsagittally (Figs 2, 4, and 5). Obviously, the extremely well-developed rostral medial accessory olive, which occupies the ventromedial area in the medulla, has pushed the pyramids lateralward. Caudal to the inferior olives, the pyramidal tract disappears in baleen whales. In toothed whales, the pyramidal tracts merge, their crossing (decussation) being described by most authors as indistinct and the pyramidal tract as small and hardly visible (Oelschläger, 2008). Thus, it is likely that pyramidal (corticospinal) fibers do not descend more than a few (cervical) segments in the spinal cord. This pattern resembles much that is seen in hoofed animals and the elephant. In terrestrial mammals, there is an inverse relationship between the development of the pyramidal tract and that of the rubrospinal tract. Perissodactyls and artiodactyls have small pyramidal tracts, whereas their rubrospinal tracts are large. The opposite is seen in primates whose small rubrospinal tract is coexistent with a large spinal pyramid. Such an inverse relationship is not encountered in Cetacea: here, both the rubrospinal and corticospinal tracts are small, which is a speciality in cetaceans.

## II. The Cetacean Spinal Cord

Within the nervous system, the spinal cord (Figs 4A and 8) is responsible for the innervation of the trunk and tail as well as the pectoral and pelvic girdles together with their appendages. Thus, the spinal cord mediates between the locomotory apparatus of the body and the brain by transmitting sensory vs motor information.

All cetaceans are characterized by (1) the subtotal reduction of the pelvic girdle and hind limb and the transformation of the pectoral girdle and fore limb into a steering device (hydrofoil) and (2) the extraordinary development of the axial musculature, thus contributing to the spindle shape of the body required for efficient locomotion by the trunk-and-tail complex. Many of the adaptations in the locomotory apparatus are reflected in the morphology and histology of the spinal cord (for further information *cf.* Oelschläger and Oelschläger, 2002).

B

**B**

## III. Functional Systems

### A. Chemoreceptor Systems

Adult baleen whales still have a small but functional nose probably equipped with an olfactory mucosa and an olfactory nerve, and they possess an olfactory bulb, slender olfactory peduncle, and an olfactory tubercle, a situation resembling that of the human. As a consequence of the adaptation to sonar orientation and communication (high-energy sonar clicks and drastic pressure changes in the upper respiratory system), toothed whales have reorganized their nasal system to such a degree that a short olfactory peduncle is found in the adult animal only very rarely (sperm whale; bottlenosed whale, *Hyperoodon ampullatus*). In general, toothed whale embryos display the anlage of an olfactory bulb, but the latter is reduced in early fetal stages (Oelschläger and Kemp, 1998). Interestingly, however, a remnant of the bulb persists as the large ganglion of the terminal nerve (Buhl and Oelschläger, 1986) that is responsible for the establishment of the hypothalamo-hypophyseal-gonadal axis in prenatal mammals and thus for sexual behavior and reproduction in the adults. In adult dolphins, the terminal ganglia contain the highest number of neurons found within the mammalia (Ridgway *et al.*, 1987). These facts argue both for the non-olfactory nature of the terminal nerve and its possible implications in the control of blood flow in the nose and basal forebrain, in maintenance of the epithelial lining of the upper respiratory tract, and in the sensory control of sonar signal emission as well. A vomeronasal organ (Jacobson's organ) and nerve are absent in cetaceans.

In spite of the strong general reduction of the olfactory system in toothed whales, the olfactory tubercle seems to be well developed even in comparison with that in baleen whales, which possess a small but presumably functional olfactory system. In many cases, the tubercle may not be clearly separated from the neighboring diagonal band and prepiriform cortex, a configuration often called "olfactory lobe" (Fig. 1). In reality, it is the large corpus striatum (basal ganglia) that protrudes here at the basal surface of the brain as an "olfactory" tubercle and is covered by an incomplete layer of thin paleocortex, a situation analogous to that in humans. The remaining paleocortex (diagonal band, piriform cortex, cortical part of amygdala) is moderately developed. The amygdala as a whole, however, is rather large in cetaceans for other reasons (see Section III.E).

Dolphins are clearly sensitive to chemical stimuli (both natural and artifical compounds), and some cetaceans were reported to have functional taste buds. With the olfactory part of the nose disappearing during prenatal development, dolphins still may resort to their taste buds and the trigeminal innervation of the oral cavity for chemoreception.

### B. Visual System

In most cetaceans, the visual system is reported to be fairly well developed (Figs 1–2, 4, 7, and 8). In the adult harbor porpoise, the optic nerve contains 81,700 axons, in the bottlenose dolphin 147,000–390,000 axons compared to 193,000–250,000 axons in the domestic cat and 1,200,000 in the human. The Amazon River dolphin (*Inia geoffrensis*) shows a rather low axon count (15,500), and the optic nerve of the Susu, whose eye lacks a lens and may be capable of serving as a light receptor only, contains a few hundred axons. Large whales have moderate numbers of axons. In baleen whales, the optic nerve may contain approximately the same number of axons (252,000–347,000) as in the bottlenose dolphin, whereas in the sperm whale, with roughly the same body weight and relatively smaller eyes, only 172,000 axons

were counted (Morgane and Jacobs, 1972; *cf.* Oelschläger and Kemp, 1998).

Bottlenose dolphins have a thick retina comprised of rods and cones. The fovea centralis is band-shaped and the neurons in the ganglionic layer very large (up to 150 μm in diameter) with thick dendrites and myelinated axons of up to 9 μm. Cetaceans have laterally placed eyes and, if at all, one small binocular visual field rostrally and ventrally as well as another one dorsally and slightly caudally. The optic fibers show a complete or almost complete decussation, and the lateral geniculate body is not laminated. The superior colliculus is large in most cetaceans, as is the case in many land mammals, including ungulates, carnivores and primates. Dolphins show a definite stratification of the superior colliculi, where layers typical of terrestrial mammals are recognizable. In some baleen whale species, the superior and inferior colliculi are approximately of the same size (length and width), but in others (Southern right whale, *Eubalaena australis*; blue whale, *Balaenoptera musculus*) the inferior colliculi have double the surface of the superior colliculi. In toothed whales, this relation may range from 2:1 to 7:1 (in the Susu).

Electrophysiological mapping studies in dolphins have located the visual cortex not in the dorsocaudal or occipital part of the hemisphere but in a more central position near the midline (Figs 7 and 8). Although well developed, the physiological importance of the cetacean visual system should be by far inferior to that of the auditory system because of functional restrictions in murky water and in darkness.

### C. Auditory System

The auditory system in whales and dolphins basically corresponds to that in terrestrial mammals; however, the various components have been adapted morphologically and physiologically to the specific conditions of hearing under water (Figs 1–5, 7, and 8).

In general, auditory structures are smaller in baleen whales than in toothed whales. According to the impressive diameter of the vestibulocochlear nerve and the number of axons in the cochlear nerve, the cetacean ventral cochlear (VC) nucleus and other auditory centers are very large. The secondary auditory fiber tracts (trapezoid body, acoustic striae) are well developed. In the La Plata dolphin as well as in the harbor porpoise and the common dolphin (*Delphinus delphis*), the absolute volume of the cochlear nuclei is 6–10 times larger than in the cat and the human, in the beluga (*Delphinapterus leucas*) it is 32 times larger, and in the fin whale 20 times larger. Numbers of cochlear neurons range between 583,000 and 1,650,000; i.e., 6–17 times that of the human. The ratio of primary cochlear nerve fibers to secondary cochlear neurons is between 1:5 and 1:8 (in human 1:4). The volume of other auditory nuclei is either very large in comparison or very low, even if we take into account the different body mass of the animals. In the La Plata dolphin, multiples 17 times the volume of the cat's nuclei are found in the nucleus of the trapezoid body and 39 times the intermediate nucleus of the lateral lemniscus. A comparison between volumes of auditory nuclei in the common dolphin and the human is even more impressive. There are multiples for the dolphin VC (16x), lateral superior olive (150x), single nuclei of the lateral lemniscus (up to 200x), and for superior auditory centers such as the inferior colliculus (12x) and laminated medial geniculate nucleus (7x). The cetacean VC is composed of five subunits consisting of specific neuron populations that are also found in terrestrial mammals. This has been proved by cytological investigations in the Franciscana involving the morphology of the axon terminations. The nucleus of the trapezoid body is well developed: a great mass of fibers passes from the VC into the trapezoid body, to the ipsilateral and contralateral nuclei of the lateral lemniscus, and to the inferior colliculus. Interestingly, the dorsal cochlear nucleus (DC) could not be found in many toothed and baleen whales, perhaps because of its reduction to the point of insignificance. Obviously, this nucleus is engaged in the assessment and/or elimination of "auditory artifacts" caused by positional changes of the head and pinnae toward a sound source. In terrestrial mammals (including bats) with "normal" external ears and good movability of the head and pinnae, the DC is well developed and even laminated.

The existence of a medial subnucleus (MSO) apart from the lateral subnucleus (LSO) in the superior olive is discussed in the literature on toothed whales. In some species, the two subnuclei of the superior olive cannot be distinguished, and whether the medial nucleus is very small or even lacking in other species, as has been reported for bats, is not clear at present. In mammals, roughly speaking, the MSO is believed to be engaged in the processing of lower frequencies, the LSO in the processing of higher frequencies including ultrasound. Therefore, it has to be expected that the LSO is larger in toothed whales and the MSO in baleen whales, reflecting the actual neurophysiological adaptation (audiogram) of these animals.

The components of the midbrain tectum differ in size in baleen whales and toothed whales. In most mysticetes the superior colliculi (vision) are larger or at least as large as the inferior colliculi (audition), whereas odontocetes have very large inferior colliculi (Figs 1B, 3, 4, 6, and 8).

The striking dominance of the sense of hearing in adult toothed whales has been emphasized by a number of investigators. Indeed, the auditory system has stamped the morphology, size, and connectivity of the whole odontocete brain (Fig. 8) owing to the necessity of processing vast amounts of acoustic information and on account of the high propagation velocity of sound in water (Ridgway, 1990) and the need of background noise suppression. In the adult sperm whale and other toothed whales, the auditory system seems to be the major source of information. Even when the animals dive to considerable depths and the visual system progressively loses its importance for orientation and the detection of prey, the auditory system might still be functional (navigation by auditory input). Toothed whales scan their surroundings with a sonar system (clicks) and are probably able to integrate the acoustic input together with visual information into two- or even three-dimensional ephemeral images (echolocation imagery) within their extended neocortical auditory projection fields. In addition, communication between individuals by means of acoustic signals clearly is very important for whales and dolphins, especially during hunting activity and/or when vision is reduced. Pelagic dolphins, in particular, tend to live in larger groups. In the open sea, their natural environment is largely represented by the distributional pattern of their kin, which changes more or less continually.

## D. Vestibular System

The four major vestibular nuclei usually found in mammals are also present in cetaceans. The lateral vestibular nucleus of Deiters (magnocellular part) is most conspicuous here and its dimensions are fairly impressive. Most of the other nuclei are rather small. Via the vestibular nerve, the lateral vestibular nucleus receives projections mainly from the maculae in the labyrinth; apart from this, its connections are very similar to those of cerebellar nuclei. In mammals, generally, the other vestibular nuclei receive input from the semicircular canals, but the latter are minute in toothed whales. In these animals Deiters' nucleus, which receives massive input from

the cerebellum, mainly projects to lumbosacral spinal segments via the lateral vestibulospinal tract and seems to be involved in the coordination and modulation of acousticomotor navigation.

## E. Limbic System

In odontocetes, the various cortical and subcortical components of the limbic system show different degrees of development (Figs 1B, 2, 3, and 8). In adult toothed whales, the hippocampus and the mammillary body are unusually small and interconnected by a relatively thin fornix, whereas the anterior thalamic nuclei and the habenulae are better developed than in other mammals. The amygdaloid complex is large in toothed whales (Schwerdtfeger et al., 1984). In contrast to the nucleus of the olfactory tract, which is totally dependent on olfactory input, the amygdala (taken as a whole) is well developed in microsmatic species (baleen whales, human) and even in anosmatic species (toothed whales). This indicates that, apart from olfactory stimuli, input to the amygdala arises from other sources, among them the auditory system. Furthermore, the remarkable development of the cortical limbic lobe (periarchicortex) in the dolphin again points to the largely non-olfactory character of this system.

## F. Cranial Nerves

The eye muscle nerves are thin in smaller cetaceans, particularly in river dolphins with their reduced eyes where the nerves may even be completely lacking. In marine dolphins, the oculomotor nerve comprises the highest number of axons (about one-third that in the human), followed by either the trochlear or the abducent nerves. In large whales (especially baleen whales), the numbers of axons in the eye muscle nerves are distinctly higher and correspond to the situation in the human.

The trigeminal nerve is the thickest cranial nerve in baleen whales and sometimes in the sperm whale as well, whereas in all of the other toothed whales the vestibulocochlear nerve has the maximal diameter (Figs 1A, 2–6, and 8). The diameters of the axons tend to be thin in the trigeminal nerve, whereas cochlear axons rank among the thickest mammalian axons known. In the smaller toothed whales so far investigated, the trigeminal nerve contains 82,000–156,000 axons (human: 140,000). The fact that the trigeminal nerves in the sperm whale and the baleen whales have about the same number of axons (490,000 vs 370,000–500,000) may be explained by the extreme size of the forehead region innervated in these animals.

The vestibulocochlear complex is the largest cranial nerve in most toothed whales, and it is the second largest in baleen whales (Figs 1A, 2, 4, 5A, and 8). In the sperm whale, it contains 215,000 axons, whereas in baleen whales the axon numbers vary from 154,000 to 179,000. Smaller toothed whales have 84,000 (harbor porpoise) to 171,000 axons (beluga), the latter thus ranking near the much larger baleen whales (human: 50,000). The ratio of auditory and vestibular axons within the eighth nerve is disputed in the literature. Some earlier authors found that in the bottlenose dolphin (total: 116,500 fibers) the cochlear fibers comprise about 60% of all the vestibulocochlear fibers (a percentage known from the human). Modern studies report a similar total axon number (113,000), but a much higher fraction of cochlear axons (97%) as opposed to a very low number of vestibular axons. In comparison with the human (19,000 vestibular axons), dolphins possess only one-fifth in vestibular axons, a fact which correlates well with their small semicircular canals, the low number of vestibular ganglion cells, and with the observation that the diameter of the vestibular nerve is barely one-tenth that of the cochlear nerve.

The variation in thickness of the facial nerve and in the number of its axons throughout the cetaceans again sheds some light on biological correlations. Thus, the facial nerve in the sperm whale is nearly as thick as the vestibulocochlear nerve (Fig. 5), containing about 3 times as many axons as in the large baleen whales, 3–8 times more axons than in smaller toothed whales, and 25 times more than the human. Large baleen whales rival sperm whales in body size, and in these giants the head may attain one-third of the total length and mass of the body. In contrast, the absolute and the relative size of the head in other toothed whales is much smaller. Accordingly, it may be speculated that the prominent thickness of the facial nerve and large number of its motor axons in the sperm whale are attributable to the extreme size of the forehead which has to be regarded an oversized "sound machine." Here, the forehead is characterized by the unique amount of acoustic fat tissues and massive blowhole musculature (innervated by the facial nerve) that helps to stabilize the giant fat bodies and to adjust their shape (acoustic lenses) during the emission of sonar signals.

## IV. Neocortex

### A. Layering and Cell Morphology

For a biological interpretation of the cetacean brain, we have to review some existing data about the neocortex, e.g., brain/body mass relationship, absolute and relative cortex mass, and the structure of the neocortex, i.e., the layering, neuron density, synapse number and density, as well as the density of gliocytes. The latter cells nourish, isolate, protect, stimulate/modulate and regenerate the neurons and therefore are as essential for neocortical activity as the neurons themselves.

No other brain surpasses the cetacean brain in richness and complexity of neocortex gyrification. Moreover, toothed whales exhibit encephalization indices which are second to that in the human. Their neocortex shows an extremely tight folding and has a maximal surface area within mammalia. At the same time, dolphins have cortical widths known from ungulates, and even large whales do not seem to attain the average cortical thickness of higher primates (human).

In principle, the cetacean neocortex is six-layered and similar to that in other mammals (Fig. 3). Regional differentiation, however, is indistinct, and cortex lamination is not well expressed, particularly due to the widespread absence of a distinct layer IV in adult toothed whales and the moderate granularization of the cortex, in general. In whales, layer one (molecular layer) can comprise about one-third of the total cortical width, whereas layer II is thin but rich in small pyramidal neurons, the perikarya of which stain intensively with cresyl violet (Nissl stain). This second layer corresponds to the outer granular layer in the human. Its pyramidal cells that border on the molecular layer are mostly "extraverted" in whales, i.e., they show a clear predominance of apical (subpial) dendrites over basal dendrites. Layer III is thick and characterized by a variety of pyramidal cells with comparatively smaller neurons populating the outer half of the layer, whereas larger cells occupy the inner half. In view of the special role played by layer IV (corresponding to the inner granular layer in the human) as a major input layer for thalamocortical specific afferents in advanced land mammals, the precise cytoarchitectonic definition of this layer in cetaceans and its neuronal composition is crucial. It was postulated that, in eutherian mammals, layer IV appears and develops in proportion to the progressive displacement of specific thalamic afferents from the first (molecular) layer to midlevels of the cortex. In evolutionary advanced primates and ungulates, layer IV of the sensory neocortex is relatively wide and

consists mainly of small granular, i.e., non-pyramidal cells mostly of the spiny stellate type. The fact that a lower degree of granularization also exists in insectivores and insectivorous bats and that layer IV is often not discernible here, led to the assumption that the structure of the cetacean neocortex is "primitive" and allegedly shows a considerable number of other plesiomorphic features. These are the lack of definite boundaries, gradual morphological transitions between functionally separate neocortical areas, a poor lamination pattern, a thick layer I, extraverted pyramidal neurons in layer II, a dense band of large pyramidal neurons referred to as layer III$_c$/V, overall weak granularization with a predominance among non-pyramidal neurons on the part of large isodendritic stellate cells, a well-developed layer VI, and a lack of giant pyramidal cells. In some cortical areas of the fin whale brain, however, a narrow zone where nerve cells are sparse or lacking seems to mark the site of layer IV, and also the corresponding external stria of Baillarger was found in a dolphin. Very young postnatal bottlenose dolphins also show a remnant of this layer. The innermost part of the cetacean cortex, presumably corresponding to layers V and VI in terrestrial mammals, usually displays no clear stratification and fades out into the white matter.

According to Deacon (1990), the common ancestor of all cetaceans very likely possessed layer IV granule cells as do the hyraxes or conies (mammals with precursors of genuine hooves), plesiomorphic members of the paenungulate (subungulate) order which also includes the sea cows (sirenians) and elephants. In the highly encephalized elephants, layer IV was found to be lacking in all cortical areas, and their neocortex architecture is strikingly similar to that of the fin whale. The loss of the granule cells in layer IV in cetaceans was therefore regarded a rare derived trait that may be correlated with the shift of the orphaned terminations of the specific thalamic afferents (originally terminating in layer IV) back to layers I and II, thus allowing more neurons in layer II to persist.

However, based on Golgi studies, there is little reason for thinking that the neocortex in the dolphin exhibits less variety than in other mammals, and a fair proportion of true stellate and other non-pyramidal cells exist in the dolphin cortex scattered throughout the layers III and V, which may combine functions of afferent, efferent, and associative layers. Moreover, the distribution of neuronal size classes is rather similar to that occurring in other mammals.

Comparative immunocytochemistry (GABA, calcium-binding proteins) in the primary visual and auditory cortex of dolphins has revealed that the overall quantitative characteristics of the neurons are similar to those in other mammalian orders, being closer to insectivores and rodents in some features and to bats and primates in some other characteristics. Interestingly, in dolphins the laminar distribution of neurons containing neurofilament protein SMI 32 differs from that in primates (monkey), where this protein is mostly found in neurons of layer III which furnish specific corticocortical connections between visual areas and high order association systems. In dolphins, SMI 32 is found exclusively in very large pyramidal neurons situated in layers III$_c$/V. The latter neurons are believed to play a complex role, combining functions of the afferent, efferent, and associative layers of the dolphin neocortex.

### B. Neuron Density

The density of nerve cells is subject to considerable variation throughout the cetacean neocortex not only from area to area but also within the individual layers, although the relatively small and densely packed neurons in layer II account for at least 50% of all the neurons in a given cortical block. Other variations in relevant papers result from different methods of quantitative analysis that have been used (discussed later).

The neuronal density in unspecified cortical areas of two fin whales is around 6800 cells/mm$^3$, similar to an elephant brain (6900/mm$^3$) of about the same size. In comparison with approximately 18,000/mm$^3$ in the human and more than 100,000/mm$^3$ in rat and mouse. This may suggest that there is a direct relationship between cortical neuronal density and brain mass. Other cell counts, however, reported an average of 57,000 neurons/mm$^3$ for the adult human. Neuron counts in a series of toothed whales again revealed some decrease in density with increasing body size and absolute brain mass (harbor porpoise 13,200; bottlenose dolphin 13,000–44,200; beluga 12,300; humpback whale 8300/mm$^3$). Total counts of neurons beneath 1 mm$^2$ of cortical surface (cortical unit) do not take into consideration the different width of the neocortical plate in various mammals and result in 28,500 cells for the adult bottlenose dolphin and 46,400 in an animal of 18 days. Other authors gave a figure of 147,000 was given for all neocortical areas studied in a series of mammals ranging from mouse to human except for the primate visual cortex which harbors about twice as many neurons. A recent investigation within a series of five delphinid species shows that neuron number per cortical unit in three sensory areas (150 μm width, 25 μm thickness, extending from the pial surface to the gray/white transition) is inversely related to brain mass (Poth *et al.*, 2005). Thus, although the larger brains have larger neocortex volumes, the cortical gray matter decreases in favor of the white matter and the decrease in neuron number per cortical unit does not seem to be compensated by an increase in cortical thickness.

The mean size of neuronal perikarya has no definitive relation to brain size. Whereas primates tend to have a similar mean neuronal size and size distribution, other mammals, including cetaceans, show a tendency to increase neuronal size with increasing brain weight. Dolphins have moderate neuron sizes more or less similar to those found in ungulates (sheep, cow, and horse) and in the elephant. Volume measurements reported a maximal neuronal volume of over 20,000 μm$^3$ in the harbor porpoise, which is about double the maximal volume found in ungulates and small- to medium-sized for the human.

### C. Synapses

The synaptic parameters of the visual neocortex in dolphins show many of the qualitative and quantitative features of the generalized mammalian bauplan: For example, the quantitative relationship between the number of synapses contacting different components of cortical neurons such as perikaryon, dendritic shafts, and dendritic spines does not differ significantly from these parameters in most other mammals. The same holds true for the distribution of other synaptic parameters throughout the cortical layers such as the area and form factor of individual synaptic boutons, the length of active zones (densities) of synaptic membranes, and several parameters relating to synaptic vesicles.

In their neocortex, dolphins show a mixture of potential conservative and advanced features at the cortical architectonic level as well as at neuronal and synaptic levels. When looking at the cortex as a whole, the aforementioned so-called conservative features occur mostly in the superficial layers I and II, whereas the deeper layers III–VI are characterized as more or less equivalent to those in advanced terrestrial mammals with the exception of layer IV. To date, adequate comparative data on ungulate groups or other aquatic mammals are not available. Interestingly, cetaceans have a much

lower neuronal density but, at the same time, a much higher synaptic density per volume unit and per neuron than terrestrial mammals. The majority of all synapses (70%) is found in cortical layers I and II. The latter seem to receive the brunt of cortical input as opposed to layer IV in many terrestrial mammals. Thus, in cetaceans, layer II seems to be the main relay element conveying information from the subcortical and intracortical afferents via layer I to the other cortical layers. All of these data, however, are difficult to interpret as to their significance both for function and for the evolution of the cetacean neocortex.

With respect to the total number of synapses in their cortices (0.87 vs $1.3 \times 10^{14}$), the dolphin and the human resemble each other much more closely than other mammalian species. This appears to reflect primarily the generally large volume of the neocortices in both dolphin and human brains as well as the maximal number of synapses per neuron compensating for minimal neuronal density in the dolphin.

### D. Glia

There are only a few data available on gliocytes in the cetacean neocortex. In the bottlenose dolphin, glial density was found to vary in different cortical areas from 28,000 to 93,200 cells/mm$^3$, values rather similar to those in the human (average: 40,000–100,000 cells/mm$^3$). The number of gliocytes per number of neurons (glia/neuron ratio) is species-specific, i.e., varies among the mammalian groups and during ontogenesis owing to changes in neuron density. This basically implies that larger brains have higher glia/neuron ratios. Thus, the ratio rises from small rodents (mouse, rabbit: 0.35) via ungulate species (pig, cow, and horse: 1.1), the human (1.68–1.78) and the bottlenose dolphin (2–3.1), to large whales (fin whale: 4.54–5.85). Also, the glia/neuron ratio within each species increases from birth to maturity, thereby signaling the importance of glia for growing neurons and, thus, for neocortical function. And this implies the necessity of comparing only mature specimens. Accordingly, in species with similar neocortex volumes, the glia/neuron ratio may be interpreted both ways, i.e., in favor of neurons and of glia, respectively, as both are of equal importance.

In summary, the morphology of the dolphin neocortex seems to be equivalent to that of advanced terrestrial mammals, but its specific features are not yet understood. With respect to both ontogenesis and evolution, the allometric process of thinning out the number of neuroblasts and/or neurons via the generation of increasing amounts of glia and neuropil for their connectivity seems to proceed faster in cetaceans than in other mammals. At the same time, the cortical plate seems to spread out more widely than in other mammals, leading to an extremely extended and convoluted neocortex with a minimal neuron density but maximal synaptic density per neuron.

### V. Morphological Trends in Toothed Whales

In the Franciscana (*Pontoporia blainvillei*), one of the least specialized living toothed whales (average brain mass of the adult: 220 g; body mass: 35 kg), the encephalization index (EI: 5.7) ranges above the level of prosimians (lemurs, *Tarsius*) and in the lower echelon of simians (monkeys, apes, and human), whereas the EI level of marine dolphins (13.5) is above that of simian primates but clearly below that of the human. From what we know today, already the Franciscana brain shows all the features typical of toothed whales and of whales, in general, although in a comparatively moderate developmental condition. In progressive pelagic dolphins, these features have reached a maximum as to brain size and structural differentiation: (1) a large telencephalon and, at the same time, the total loss of the rostral olfactory and the whole accessory olfactory system, and the reduction of the remaining paleocortex. The archicortex and some other components of the limbic system are very small, but the limbic lobe cortex seems to be well developed. The neocortex is very much extended and convoluted. As in primates, the neocortex is by far the largest brain structure in river dolphins and other cetaceans. Its size index is higher in river dolphins than the average of prosimians but clearly lower than that of simians. Even higher values for the neocortex can be expected for marine dolphins. Giant cetaceans are difficult to interpret because their brains, although approaching 10 kg in total mass, are dwarfed by their huge bodies. Besides the neocortex, the striatum is one of the most progressive telencephalic structures and indicates a high functional capacity of the motor systems. The amygdala, another component of the basal ganglia and belonging to the limbic system, is considerable in size, although its corticomedial part largely consists of paleocortex. (2) The dinencephalon (thalamus) is large owing to the exceptional volume of different nuclei (medial, dorsal, ventral, medial geniculate, and pulvinar). (3) The large volume of the midbrain is significant due to the extreme size of some components of the auditory system. Thus, for example, the inferior colliculus is much larger than the superior colliculus and the nuclei of the lateral lemniscus are extremely well developed. (4) Another major character is the large size of the cerebellum and particularly of the paraflocculus and posterior interposed nucleus as well as associated structures (elliptic nucleus, pons, inferior olive, and accessory fiber tracts). Interestingly, the central vestibular complex of whales is extremely reduced as are the semicircular canals (Oelschläger, 2008). Deiters' nucleus (lateral vestibular nucleus), however, is very large in toothed whales, presumably in correlation with its functional role as an interface between the cerebellum and the motoneurons of the spinal cord in acousticomotor navigation. (5) Finally, the medulla oblongata is, comparatively speaking, very large due to the outstanding development of the relevant auditory and trigeminal nuclei.

## VI. Cetacean Strategies in Aquatic Adaptation

As in other mammals, the cetacean brain can be regarded as the true center of the body responsible for maintenance of physiological conditions (homeostasis) and survival. By means of the cranial and spinal nerves, the brain and the spinal cord collect all of the sensory information available, evaluating, synthesizing, and transforming it into optimal behavioral responses in manifold aspects: orientation, feeding, defense, communication, and reproduction. Thus, the brain not only represents all parts of the body but also mirrors the ecophysiological situation of the animal within its niche. On account of these tight correlations, evolutionary changes in the environment and in the biology of the species must show in the morphology of the nerves and brain. Rapid transgressions of mammalian groups into totally different habitats have obviously been related to strong selection pressure, i.e., such adaptational processes (in geological terms) unfold rather quickly and may lead to profound changes in brain morphology and function, reflecting changes in the body's periphery.

Whale ancestors had to overcome severe problems with respect to their new environment: As carnivorous plesiomorphic ungulates, they faced high water resistance because of an unfortunate body shape needing hydrodynamic styling. Ancient whales may have been luring amphibious animals comparable with crocodiles. Three-dimensional active hunting and communication underwater in darkness, turbid waters or at greater depths, however, not only

required a spindle-shaped body with reduced or modified limbs but also changes in the phonation and hearing processes. In this respect, the use of high-frequency sounds proved advantageous but necessitated the complete remodeling of the upper respiratory tract as a sound generator and transmitter, the modulation of the ear region as well as the incorporation of the mandible as a secondary sound receiver. Concomitantly, this new life-style on the basis of sonar orientation and communication facilitated the improvement of the central auditory system (as to capacity, versatility, and precision) which can be regarded as the dominant sensory system of toothed whales. The benefits and the success of acousticomotor navigation and the need of adequate brain centers for the analysis of multifaceted and complicated incoming sounds may have spurred the hypertrophy of nuclei along the ascending auditory pathway and expanded neocortical areas (maps) for acoustic detection and memory. A large number of modules in the dolphin neocortex may facilitate the quick scanning and subsequent synthesis of two- or three-dimensional images of their environment and allow intense and diversified social interactions with the members of their group.

## See Also the Following Articles

Brain size ■ Skull Anatomy ■ Vision ■ Hearing

## References

Buhl, E. H., and Oelschläger, H. A. (1986). Ontogenetic development of the N. terminalis in toothed whales. Evidence for its nonolfactory nature. *Anat. Embryol.* **173**, 285–295.

Cranford, T. W., Amundin, M., and Norris, K. S. (1996). Functional morphology and homology in the odontocete nasal complex. *J. Morphol.* **228**, 223–285.

Deacon, T. W. (1990). Rethinking mammalian brain evolution. *Am. Zool.* **30**, 629–705.

Fung, C., Schleicher, A., Kowalski, T., and Oelschläger, H. H. A. (2005). Mapping auditory cortex in the La Plata dolphin (*Pontoporia blainvillei*). *Brain Res. Bull.* **66**, 353–356.

Glezer, I. I., Jacobs, M. S., and Morgane, P. J. (1988). Implications of the "initial brain" concept for the brain evolution in cetacea. *Behav. Brain Sci.* **11**, 75–116.

Jansen, J., and Jansen, J. K. S. (1969). The nervous system of cetacea. In "The Biology of Marine Mammals" (H. T. Andersen, ed.), pp. 175–252. Academic Press, London, New York, San Francisco.

Jelgersma, G. (1934). "Das Gehirn der Wassersäugetiere. Eine anatomische Untersuchung." J.A. Barth, Leipzig.

Manger, P. (2006). An examination of cetacean brain structure with a novel hypothesis correlating thermogenesis to the evolution of a big brain. *Biol. Rev.* **81**, 293–338.

Marino, L. (1998). A comparison of encephalization between odontocete cetaceans and anthropoid primates. *Brain Behav. Evol.* **51**, 230–238.

Morgane, P. J., and Jacobs, M. S. (1972). Comparative anatomy of the cetacean nervous system. In "Functional Anatomy of Marine Mammals" (R. J. Harrison, ed.), pp. 117–244. Academic Press, London, New York, San Francisco.

Nieuwenhuys, R., ten Donkelaar, H. J., and Nicholson, C. (eds) (1998). "The Central Nervous System of Vertebrates," Vol. 3. Springer, Berlin-Heidelberg-New York.

Oelschläger, H. H. A., and Kemp, B. (1998). Ontogenesis of the sperm whale brain. *J. Comp. Neurol.* **399**, 210–228.

Oelschläger, H. H. A., and Oelschläger, J. S. (2002). Brain. In "Encyclopedia of Marine Mammals" (W. F. Perrin, B. Würsig, and J. G. M. Thewissen, eds), pp. 133–158. Academic Press, San Diego.

Oelschläger, H. H. A. (2008). The dolphin brain-a challenge for synthetic neurobiology. *Brain Res. Bull.* doi: 10.1016/j.brainresbull.2007.10.051

Oelschläger, H. H. A., Haas-Rioth, M., Fung, C., Ridgway, S. H., and Knauth, M. (2008). Morphology and evolutionary biology of the dolphin (*Delphinus* sp.) brain—MR imaging and conventional histology. *Brain Behav. Evol.* **71**, 68–86.

Poth, C., Fung, C., Güntürkün, O., Ridgway, S. H., and Oelschläger, H. H. A. (2005). Neuron numbers in sensory cortices of five delphinids compared to a physeterid, the pygmy sperm whale. *Brain Res. Bull.* **66**, 357–360.

Rauschmann, M. A., Huggenberger, S., Kossatz, L. S., and Oelschläger, H. H. A. (2006). Head morphology in perinatal dolphins: A window into phylogeny and ontogeny. *J. Morphol.* **267**, 1295–1315.

Ridgway, S. H. (1990). The central nervous system of the bottlenose dolphin. *In* "The Bottlenose Dolphin" (S. Leatherwood, and R. R. Reeves, eds), pp. 69–97. Academic Press, New York.

Schwerdtfeger, W. K., Oelschläger, H. A., and Stephan, H. (1984). Quantitative neuroanatomy of the brain of the La Plata dolphin, *Pontoporia blainvillei. Anat. Embryol.* **170**, 11–19.

# Brain Size Evolution

### Lori Marino

## I. Introduction

The study of brain size among three major living groups of marine mammals, cetaceans, pinnipeds, and sirenians, is a study in contrasts. Phylogenetic and ecological factors have shaped the course of brain evolution in each group in distinct ways. The resulting diversity of brains provides an illustration of the different successful paths that were taken in the evolution of marine mammals.

## II. The Meaning of Encephalization

Brain size evolution is embodied in the concept of encephalization, which was originally put forth as an Encephalization Quotient (EQ) by anthropologist Harry Jerison. Jerison widely applied this measure to comparisons across different species. EQ is a measure of observed brain size relative to expected brain size derived from a regression of brain weight on body weight for a sample of species. EQ values of one, less than one, and greater than one indicate a relative brain size that is average, below average, and above average, respectively. For example, a species with an EQ of 2.0 possesses a brain twice as large as expected for an animal of its body size. The EQ values reported here are based on a large sample of living mammalian species from Jerison (1973, 1978).

Besides the whole brain, changes in the size of various brain components have also occurred throughout marine mammal evolution and contribute to changes in the overall brain size. Some of these changes are measurable but undoubtedly many changes in the relative size of structures occurred in all marine mammals that are not apparent from the fossil record.

## III. Accessing the Fossil Record

Studies of brain size evolution in fossil marine mammals have depended upon measuring the volume of the endocranial cavity in fossil specimens. Because the specific gravity of brain tissue is nearly the value of water volumetric data have been typically converted to units of weight. In addition to the general problem of finding intact fossil crania, early studies were hampered by difficulties accessing the sediment-filled endocranium. Researchers have taken advantage of the fortuitous occurrence of intact natural endocasts but these are rare. Also, the earlier studies often resulted in overestimates of brain mass from endocranial volume because they did not take into account that total endocranial volume is partly comprised of non-neural, e.g., vascular, components. Cetaceans, for instance, possess a massive endocranial system of blood vessels, called the rete mirabile, which surrounds the brain and can sometimes account for nearly 20% of total endocranial volume. Most recent studies take these vascular structures into account when estimating brain size from endocranial volume. Recently, Computed Tomography (CT) has proven to be an important tool in the study of fossil endocranial features because it is nondestructive and enables more accurate, precise, and reliable measurement of endocranial features than traditional methods.

## IV. Brain Sizes in Fossils and Modern Species

### A. Brain Size Evolution in Archaeocetes

The fossil record of early cetacean brain evolution includes the transition from the immediate land ancestor of cetaceans to the extinct aquatic forms known as archaeocetes. The range of EQs estimated for early, middle, and late archaeocetes is from 0.25 to 0.49 (Table I and Fig. 1) and archaeocetes appear to have experienced no increase in encephalization above their land precursors.

### B. Brain Size Evolution in Odontocetes

The suborder Odontoceti appeared during the early Oligocene and radiated rather dramatically in that epoch and into the Miocene. Data suggest that by the early-mid Miocene odontocetes possessed their present encephalization levels (Table 1 and Fig. 1) and that there was a significant increase in odontocete encephalization levels with their emergence at the Eocene-Oligocene boundary. Those data that do exist suggest that by the early-mid Miocene at least several odontocete species possessed encephalization levels substantially above that of archaeocetes and within the mid-range of living species (Table I and Fig. 1). These data imply that some important changes in brain size occurred during the Oligocene after the turnover from archaeocetes to early odontocetes. Unfortunately, there are no brain size data on odontocetes during the Oligocene. Data on brain size in odontocetes during the Pliocene and Pleistocene are likewise lacking.

The EQs of living odontocetes are generally on a par with nonhuman primates. But some species have achieved a level of encephalization second only to modern humans (EQ ~ 7.0) and equal to or above that of the recent hominid ancestor *Homo habilis* (EQ ~ 4.4). Therefore, a number of odontocete species are significantly more encephalized than other mammals, including nonhuman primates. There is, however, a range of encephalization levels within the odontocetes. The sperm whale (*Physeter macrocephalus*), with an EQ of 0.58, is an example of an odontocete species subject to disproportionate body enlargement for which the measure of EQ is not particularly meaningful. The Delphinidae, however, are the family that contains several species with exceptionally high EQs above 4.0. These include the bottlenose dolphin (*Tursiops truncatus*), the Tucuxi dolphin (*Sotalia fluviatilis*) the Pacific white-sided dolphin (*Lagenorhynchus obliquidens*), and the common dolphin (*Delphinus delphis*).

Odontocete brain evolution was also characterized by increased foreshortening and widening of the brain which coincided with telescoping of the skull. There was a trend towards increased relative size of auditory processing regions such as the acoustic cranial

**B**

### TABLE I
### Estimates of Brain and Body Weight, and EQ for Some Fossil and Living Marine Mammal Species

| Species | Estimated brain weight (g) | Estimated body weight (g) | EQ[a] |
|---|---|---|---|
| Order Cetacea | | | |
| Suborder Odontoceti[b] | | | |
| Family Ziphiidae | | | |
| *Mesoplodon mirus* | 2355 | 929,500 | 1.97 |
| *Mesoplodon europaeus* | 2149 | 732,500 | 2.11 |
| *Mesoplodon densirostris* | 1463 | 767,000 | 1.39 |
| *Ziphius cavirostris* | 2004 | 2,273,000 | 0.92 |
| Family Kogiidae | | | |
| *Kogia breviceps* | 1012 | 305,000 | 1.78 |
| *Kogia simus* | 622 | 168,500 | 1.63 |
| Family Physeteridae | | | |
| *Physeter macrocephalus* | 8028 | 35,833,330 | 0.58 |
| Family Monodontidae | | | |
| *Delphinapterus leucas* | 2083 | 636,000 | 2.24 |
| *Monodon monoceros* | 2997 | 1,578,330 | 1.76 |
| Family Lipotidae | | | |
| *Lipotes vexillifer* | 510 | 82,000 | 2.17 |
| Family Iniidae | | | |
| *Inia geoffrensis* | 632 | 90,830 | 2.51 |
| Family Platanistidae | | | |
| *Platanista gangetica* | 295 | 59,630 | 1.55 |
| Family Pontoporiiadae | | | |
| *Pontoporia blainvillei* | 221 | 34,890 | 1.67 |
| Family Phocoenidae | | | |
| *Phocoena phocoena* | 540 | 51,193 | 3.15 |
| *Phocoenoides dalli* | 866 | 86,830 | 3.54 |
| Family Delphinidae | | | |
| *Tursiops truncatus* | 1824 | 209,530 | 4.14 |
| *Lagenorhynchus obliquidens* | 1148 | 91,050 | 4.55 |
| *Delphinus delphis* | 815 | 60,170 | 4.26 |
| *Grampus griseus* | 2387 | 328,000 | 4.01 |
| *Globicephala melaena* | 2893 | 943,200 | 2.39 |
| *Stenella longirostris* | 660 | 66,200 | 3.24 |
| *Orcinus orca* | 5059 | 1,955,450 | 2.57 |
| *Sotalia fluviatilis* | 688 | 42,240 | 4.56 |
| Suborder Mysticeti[b] | | | |
| Family Eschrichtiidae | | | |
| *Eschrichtius glaucus* | 4305 | 14,329,000 | 0.58 |
| Family Balaenopteridae | | | |
| *Balaenoptera physalus* | 7085 | 38,421,500 | 0.49 |
| *Balaenoptera musculus* | 3636 | 50,904,000 | 0.21 |
| *Megaptera novaeangliae* | 6411 | 39,295,000 | 0.44 |
| Extinct species[c] | | | |
| Family Protocetidae | | | |
| *Rodhocetus kasrani* | 290 | 590,000 | 0.25 |
| Family Remingtonocetidae | | | |
| *Dalanistes ahmedi* | 400 | 750,000 | 0.29 |
| Family Basilosauridae | | | |
| *Basilosaurus isis* | 2520 | 6,480,000 | 0.37 |
| *Saghacetus osiris* | 388 | 350,000 | 0.49 |
| *Dorudon atrox* | 976 | 2,700,000 | 0.40 |
| *Zygorhiza kochii* | 745 | 3,351,000 | 0.26 |
| Family Squalodontidae | | | |
| *Prosqualodon davidi* | 750 | 880,000 | 0.65 |
| Family Physeteridae | | | |
| *Aulophyseter morricei* | 2500 | 1,100,000 | 1.90 |
| Family Argyrocetus | | | |
| *Argyrocetus* sp. | 650 | 72,000 | 3.01 |

(continues)

TABLE I (Continued)

| Species | Estimated brain weight (g) | Estimated body weight (g) | EQ[a] |
|---|---|---|---|
| Family Eurhinodelphidae | | | |
| *Schizodelphis sulcatus* | 368 | 260,000 | 0.72 |
| Order Pinnipedia | | | |
| Family Phocidae | | | |
| *Phoca vitulina* | 250 | 30,000 | 2.08 |
| *Phoca hispida* | 253 | 39,570 | 1.75 |
| *Leptonychotes weddelli* | 520 | 400,000 | 0.76 |
| Order Sirenia | | | |
| Family Trichechidae | | | |
| *Trichechus manatus* | 364 | 756,000 | 0.35 |
| Family Dugongidae | | | |
| *Dugong dugon* | 266 | 281,000 | 0.50 |
| *Hydrodamalis gigas* | 1158 | 7,102,500 | 0.25 |

[a] EQ is based on a reference group of modern mammals from Jerison (1973).
[b] For living species, estimated brain and body weights are averaged across several specimens in most cases.
[c] Body weight estimates for fossil specimens are often general estimates of adult species-specific values and are not necessarily from the same specimen(s) for which brain weight estimates are obtained.

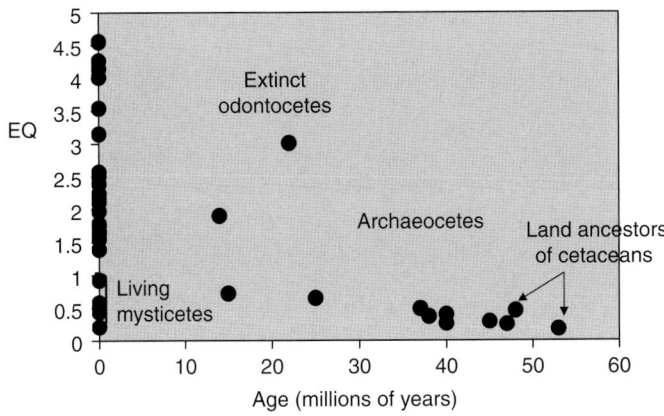

Living odontocetes

**Figure 1** *Pattern of change in encephalization over geological time in archaeocetes and extinct odontocetes, compared with living odontocetes and mysticetes. Encephalization is plotted as EQ where the reference group is a large sample of living mammals.*

nerve and inferior colliculus. In living odontocetes this is evident in the larger relative size of the inferior colliculus to the analogous midbrain visual processing area, the superior colliculus. In addition, structures associated with the processing of olfactory information regressed. Furthermore, the cerebral cortex of odontocetes (and cetaceans in general) has achieved an extremely high level of gyrification. Although surface morphology is not always discernible from fossil endocasts, it is generally thought that this was not a feature of archaeocete brains.

### C. Brain Size Evolution in Mysticetes

The suborder Mysticeti appeared and diversified in the Oligocene and consisted of primitive toothed taxa in addition to the earliest baleen-bearing whales. Extant groups appeared in the mid-late Miocene. There are two problems associated with examining brain size evolution in mysticetes. First, the data on fossil and living mysticete brain size are scarce. This is partly due to the difficulties associated with extracting and measuring such large brains. Second, mysticete brains tend to be smaller than expected relative to body size despite their large absolute size. This is partly due to the enormous body masses achieved by mysticetes.

As in the sperm whale, mysticete bodies are greatly enlarged in ways that do not necessarily require a concomitant increase in neural tissue. EQs of living mysticetes are therefore unrepresentative of actual brain enlargement, with all values falling substantially below 1.0 (Table I and Fig. 1). For this reason, although encephalization has probably occurred throughout mysticete evolution EQ is not an appropriate measure of it in this group, particularly in comparison with terrestrial mammals. In fact, to the extent that disproportionate increases in body size have played a role in body enlargement in any fully aquatic species, EQ will be underestimated relative to terrestrial mammals.

Many of the changes in morphology and size of brain components that occurred in odontocetes also characterize mysticete brain evolution, but to a lesser extent. For instance, unlike in odontocetes, olfactory tracts have remained in some mysticete species and the hypertrophy of the auditory processing regions is not as extreme as in odontocetes.

### D. Brain Size Evolution in Sirenia

Sirenian brain evolution has been markedly conservative with regards to relative brain size. Fossil endocasts of early Eocene sirenians (among the earliest) were small in relation to the skull and already very similar to modern forms. Sirenian encephalization levels are among the lowest of modern mammals. According to the same formula used to derive cetacean encephalization quotients in cetaceans, the Florida manatee (*Trichechus manatus*) possesses an EQ of about 0.35 and the dugong (*Dugong dugon*) about 0.5. The EQ of Steller's seacow (*Hydrodamalis gigas*) was approximately 0.25 (Table I). Body size enlargement explains some of the reason for these low EQs but, given that cetaceans of the same body size possess higher EQs, not all.

Unlike odontocetes, sirenians do possess olfactory bulbs. Perhaps the most striking contrast, however, is the fact that cetacean and sirenian brains anchor the two ends of the spectrum of cortical

**B**

gyrification. Whereas the cetacean cerebral cortex is thin and highly convoluted, the sirenian cortex is unusually thick and almost lissencephalic (smooth). Interestingly, despite these differences, the relative volume of the cerebral cortex in both sirenians and cetaceans is on a par with nonhuman primates.

### E. Brain Size Evolution in Pinnipedia

Pinnipeds diverged from terrestrial carnivores during the early Miocene. This is a relatively more recent date than cetaceans and sirenians diverged from their land ancestors. The pinniped brain, therefore, still bears a number of resemblances to that of terrestrial carnivores. Living pinnipeds possess EQs which hover around the average for terrestrial mammals. For instance, the ringed seal (*Phoca hispida*) possesses an EQ of 1.75, the harbor seal (*Phoca vitulina*) 2.08, and the Weddell seal (*Leptonychotes weddellii*) 0.76 (Table I). These values are fairly representative of pinniped EQ in general. Pinniped olfactory structures are reduced, but not to the same degree as in cetaceans. The cerebral cortex is highly convoluted (and more so than most terrestrial carnivores) but lies somewhere in between the extreme degrees of gyrification and thickness found in cetaceans and sirenians. The pinniped brain is somewhat more spherical in shape than in terrestrial carnivores but did not undergo the dramatic change in overall morphology exhibited in cetaceans.

### V. Discussion

Much more information is needed before we can obtain a complete picture of patterns of brain size evolution in marine mammals. However, what does seem clear is that the different marine mammal groups evolved along distinct paths that led to a great variety of levels of encephalization in modern species. For instance, among odontocetes there was a substantial increase in encephalization in the Oligocene lineages which has led to the existence of a number of dolphin and porpoise species with relative brain sizes challenging only the hominid mammalian line. The relationship between mass and organization must be explored further, as well as the phylogenetic and ecological factors that led to the differential development of brain size among the various marine mammal groups.

*See Also the Following Article*

Brain

*References*

Breathnach, A. S. (1955). Observations on endocranial casts of recent and fossil cetaceans. *J. Anat.* **89**, 533–546.

Gingerich, P. D. (1998). Paleobiological perspectives on Mesonychia, Archaeoceti, and the origin of whales. *In* "The Emergence of Whales" (J. G. M. Thewissen, ed.), pp. 423–449. Plenum Press, New York.

Jerison, H. J. (1973). "Evolution of the Brain and Intelligence." Academic Press, New York.

Jerison, H. J. (1978). Brain and intelligence in whales. *Whales and Whaling: Report of the Independent Inquiry Conducted by Sir Sidney Frost*, **2**, 162–197.

Marino, L. (1998). A comparison of encephalization between Odontocete cetaceans and Anthropoid primates. *Brain Behav. Evol.* **51**(4), 230–238.

Marino, L., McShea, D., and Uhen, M. D. (2004). The origin and evolution of large brains in toothed whales. *Anat. Rec.* **281A**, 1247–1255.

Ridgway, S. H., and Wood, F. H. (1988). Dolphin brain evolution. *Behav. Brain Sci.* **11**, 99–100.

# Breathing

### DOUGLAS WARTZOK

Oxygen is the final electron receptor in the metabolism of marine mammals as it is in all other mammals. Marine mammals obtain oxygen from the air and they breathe at the surface. In contrast, most feeding, mating, and other activities essential to survival occur beneath the surface. Thus most marine mammals minimize the time and they are at the surface and have evolved to load oxygen quickly and use it efficiently. For marine mammals, the breath-holding portion of the breathing cycle is significantly extended compared to the oxygen intake portion.

Because a distinguishing feature of marine mammals is their breath-holding ability, this characteristic has received much more attention than their breathing. However, a number of aspects of marine mammal breathing have necessarily been modified from terrestrial mammals in order to accommodate their submerged lifestyles. As with many modifications from terrestrial mammals, those related to breathing show the greatest difference in those species that dive the deepest and the longest.

In general, cetaceans, otariids, and manatees take only one breath per surfacing. Manatees return to a normal, shallow dive after a single breath. The deeper-diving cetaceans take a series of breaths, each in a subsequent surfacing event, before another dive. Phocid seals remain at the surface for a series of breaths after a dive.

### I. Lung Oxygen Stores

Every inspiration that fills the lungs with air brings in four times as much nitrogen as oxygen. Because nitrogen is neither bound to a carrier in the blood nor metabolized in the tissues, the partial pressure of nitrogen in the blood will equilibrate with that in the lungs. If gas exchange is allowed to take place during a dive, the resulting higher partial pressure of nitrogen in the blood and tissues will result in the formation of nitrogen gas bubbles when the external pressure is reduced as the animal comes to the surface. Thus deep-diving marine mammals limit the exchange of gas from lungs to blood during dives.

Most phocids have been observed to exhale before diving. Weddell seal (*Leptonychotes weddellii*) pups, which are observed to dive after an inspiration, are an occasional exception. At this developmental stage they are shallow divers. California sea lions (*Zalophus californianus*) may dive after a partial inspiration, but then vent air during descent. Dolphins and porpoises making shallow dives routinely dive on inspiration. These breathing behaviors correlate well both with the proportion of total oxygen stores in the lung at the start of a dive and with lung size in proportion to the body size of various marine mammals. Phocids dive with 7% of total oxygen stores in the lung, fur seals with 13%, and delphinids with 22% (Fig. 1; Kooyman, 1973, 1989). The proportionate size of the lung in phocids and manatees is about the same as in terrestrial mammals such as the horse and the human (Fig. 2; Slijper, 1976), whereas the lung size is greater than expected in delphinids and smaller than expected in whales. An outlier in these considerations is the sea otter (*Enhydra lutris*) whose lung is close to three times the expected size for its body mass and accounts for 75% of oxygen stores. The sea otter is not a deep-diving marine mammal, and the relatively large lung in the sea otter may be primarily used for buoyancy when the animal is resting at the surface.

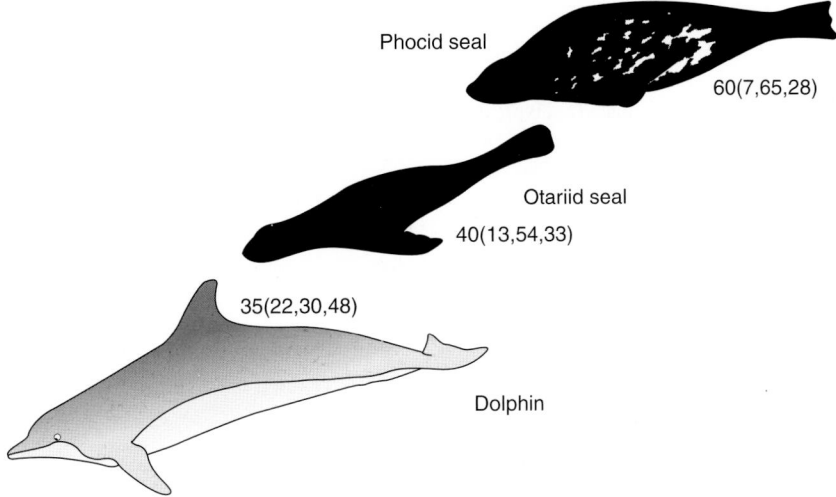

**Figure 1** *Generalized total oxygen storage of major taxa of marine mammals expressed in $O_2 kg^{-1}$. Numbers in parentheses are the percentage of total oxygen store found in the lungs, blood, and muscle, respectively. Modified, with permission, from G. L. Kooyman, 1989 © Diverse Divers: Physiology and Behavior, Springer Verlag.*

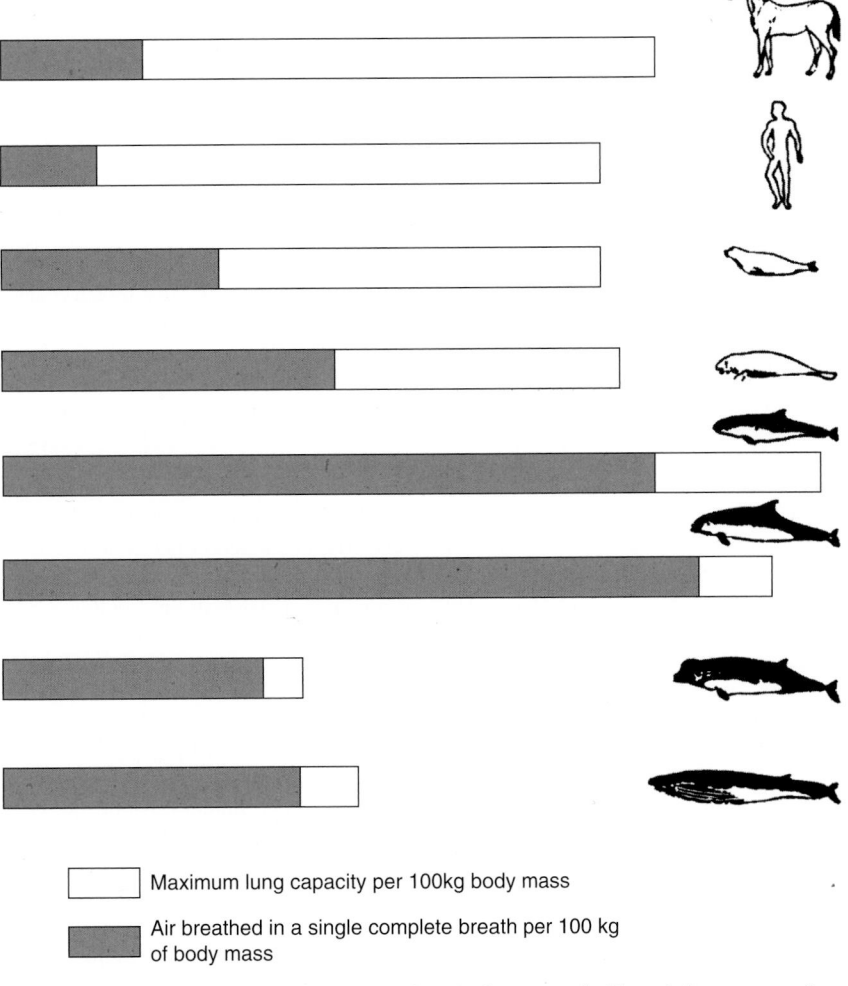

**Figure 2** *The maximum amount of air that the lungs can hold, and the amount of air breathed in and out with each breath, calculated per 100 kg of body mass for a horse, a human, a seal, a manatee, a harbor porpoise, a bottlenose dolphin, a bottlenose whale, and a fin whale. Modified, with permission, from E. J. Slijper, 1976 ©, Whales and Dolphins, University of Michigan.*

**B**

## II. Tidal Volume

The tidal volume (the amount of air breathed in or out during normal respiration) is a larger proportion of the total lung capacity (TLC) in marine mammals than it is of terrestrial mammals. In a typical terrestrial mammal the volume of air inhaled and exhaled in one breath is in the range of 10–15% of TLC. In marine mammals, tidal volume is typically greater than 75% of TLC. The maximum tidal volume or vital capacity (VC) in terrestrial mammals is not more than 75% of TLC, whereas in marine mammals the VC can exceed 90% of TLC. Several factors contribute to the large tidal volume in marine mammals. Marine mammal lungs contain more elastic tissue than those of terrestrial mammals (Kooyman and Sinnett, 1979). The ribs contain more cartilage and are thus more compliant than those of terrestrial mammals. The lung is also more compliant. Marine mammal lungs can collapse and reinflate repeatedly, whereas in terrestrial mammals, lung collapse is a serious situation that requires intervention to reinflate. Although both terrestrial mammals and marine mammals inspire actively and expire passively, the features noted earlier allow a much greater elastic recoil of the lungs, chest cavity, and diaphragm, and thus a greater tidal volume in proportion to TLC.

The terminal portions of the airways in all marine mammals are supported and reinforced by cartilage or muscle. One purpose of this reinforcement is to provide a less collapsible region into which alveolar gases can be forced during a dive to prevent gas exchange with blood at high pressures. This prevents increased nitrogen tensions in the blood and tissues as noted previously. A second purpose of the reinforcement is to keep the terminal airways open even at high-flow rates of gases in and out of the lung during a breath and to allow high expiratory flow rates even as the lung volume decreases. Fig. 3 shows the flow volume profile comparison during exhalation between a harbor porpoise (*Phocoena phocoena*) and a human. There are two striking differences. First, the flow rates are much higher in terms of VC/sec. Second, the flow rates remain very high; even down to a small fraction of the VC. These two factors together allow very rapid exhalation of the full VC. Inspiration takes somewhat longer.

The bottlenose dolphin (*Tursiops truncatus*) completes an exhalation and inhalation cycle in approximately a third of a second. With a tidal volume of 10 l, flow rates through the air passages can be as high as 70 l/sec. In gray whale (*Eschrichtius robustus*) calves the duration of expiration and inhalation is closer to half a second, but the tidal volume can be as great of 62 l, and the maximum flow rate as great as 202 l/sec. Gas flows through the external nares at speeds up to 44 m/sec during inspiration and 200 m/sec during expiration. Cetaceans usually initiate expiration prior to the blowholes breaking the surface. The explosive nature of the expiration creates the small droplets that make the blow visible and clears the upper respiratory passages and the area around the blowholes of any residual water. The time that the blowholes are above the surface is mostly used for inspiration.

The large tidal volume allows for more oxygen loading and greater carbon dioxide unloading during a single breath at the surface. Even in a resting state, the carbon dioxide content of expired air in seals is twice as great as it is in humans. After extended breath holds, alveolar oxygen levels can be as low as 1.5%. The oxygen and carbon dioxide content of expired air after surfacing can provide indirect evidence of physiological adjustments to diving. In bottlenose dolphins, the oxygen content in the first breath after a dive to 200 m is greater than it is in the first breath after an equivalent amount of swimming at 20 m (Ridgway, 1972). The interpretation is that the collapse of lungs in the deeper dive prevented the exchange of oxygen with the exchange of oxygen with the blood during the dive. For the same reason, the content

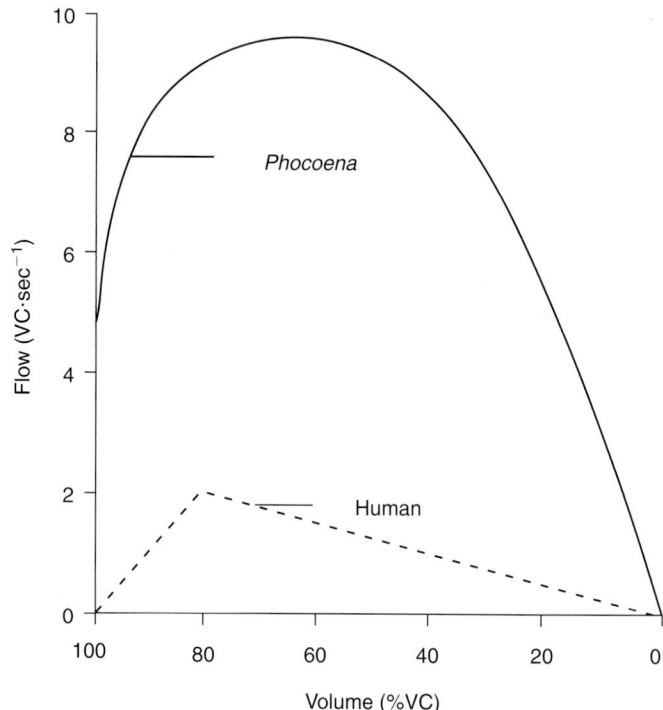

**Figure 3** *Comparison of the flow-volume curves of a human (dashed line) and a harbor porpoise (solid line). Note that in the human, after the volume falls below about 80% of vital capacity, the flow rate declines steadily, but this is not the case in the porpoise. Modified, with permission, from G. L. Kooyman and E. E. Sinnett, 1979 ©, Mechanical properties of the harbor porpoise lung, Phocoena. Res. Physiol.* **36***, 287–300, Elsevier Science.*

of carbon dioxide in the first breath is always less after a dive to depth than after a dive near the surface. In gray seals (*Halichoerus grypus*), the end tidal partial pressure of oxygen in the first exhalation after surfacing is similar to that in the last breath before submergence, again indicating that the lungs were collapsed at depth and there was no gas exchange.

## III. Hyperventilation

Marine mammals often hyperventilate both before and after a dive. Hyperventilation before a dive leads to increased oxygen tensions and reduced carbon dioxide tensions. Hyperventilation arises through an increase in the frequency of breathing and the tidal volume. Although both can initially increase during hyperventilation, lung dynamics requires an eventual reciprocal relationship between tidal volume and frequency of breathing. In Weddell seals, hyperventilation of five to six times resting is accomplished by increasing the tidal volume 1.5–2 times and the respiratory frequency 3 times. If the respiratory frequency rises above 25 breaths/min, the tidal volume, as a proportion of the TLC, is close to that of terrestrial mammals. Harp seals (*Pagophilus groenlandicus*) with respiratory rates of 27 breaths/min and harbor seals (*Phoca vitulina*) with rates of 35 breaths/min have tidal volumes between 20% and 30% of TLC. Because marine mammals normally have such large tidal volumes, they have much less ability to increase total ventilation in response to exercise than terrestrial animals. A human can increase respiration by 4 times and can increase tidal volume by greater than 4 times for an overall increase in ventilation of more than 16 times resting compared to the 5–6 times resting maximum for a Weddell seal.

Marine mammals often exhibit an increase in heart rate on approach to the surface. It has been suggested that this anticipatory tachycardia coincides with the restored perfusion of peripheral tissues so the oxygen levels in the blood drop even lower as the carbon dioxide levels rise. These changes in blood gases increase the gradient between partial pressures in the blood and the lungs and lead to more rapid oxygen uptake and carbon dioxide exhausting during the first breaths at the surface. When gray seals show no anticipatory tachycardia, they do not achieve the maximum rate of oxygen uptake during the first few breaths.

The increased heart rate and breathing on surfacing lead to a rapid restoration of oxygen tensions. In fact, in Weddell seals the blood oxygen partial pressures in the postdive recovery period routinely end up exceeding the resting values. It appears that the purging of carbon dioxide is more critical for determining readiness to dive again than is the replenishing of oxygen. In Weddell seals the partial pressure of carbon dioxide falls to resting levels within a couple minutes after aerobic dives, but may take up to 10 min to reach resting levels after long dives, which rely on anaerobic metabolism. The fact that hyperventilation can continue for an hour after the partial pressures of oxygen and carbon dioxide return to resting levels indicates that prolonged hyperventilation is driven more by lactate-induced changes in the acid–base balance in the blood.

## IV. Ventilation Control

Seals show a more fully developed mammalian reflexive response of increased ventilation to decreases in inspired oxygen concentration or increases in inspired carbon dioxide concentration at an earlier age than terrestrial mammals. However, adult seals show little ventilatory response to decreased oxygen concentration. Instead the adult seals respond by increasing the proportion of time they are at the surface relative to the total time between dives. Apparently the adult seals are able to substitute behavioral control of diving patterns for reflexive control of ventilation.

Breathing hyperoxic gas increases the dive time of Weddell seals, has no effect on the dive time of manatees, and shortens the dive time of hooded seals. A suggested explanation for the latter surprising finding is that the seal breathing a hyperoxic gas mixture dove before it had exchanged all the carbon dioxide and the increased carbon dioxide resulted in the shortened dive time.

Both the nostrils of pinnipeds and the blowhole of cetaceans are normally closed when the controlling muscles are relaxed. The closure in pinnipeds is maintained by muscle tone and pressure of the moustacial pad. Contraction of the nasal and moustacial pad muscles results in a movement of the pad and an opening of the nostrils. In cetaceans, muscles must contract to open the blowhole and to move the nasal so that it is not blocking the airway.

Pinnipeds on land often show a breathing pattern similar to that during diving with breathing periods (eupnea) being shorter than breath hold periods (apnea). The ratio between apnea and eupnea while on land is greatest in those species that normally dive for the longest periods. The periods of apnea also tend to be longer when the animals are asleep than when they are awake. Even the longest bouts of sleep apnea appear to be aerobic: plasma lactate and glucose remain stable even though oxygen tensions drop to very low levels, carbon dioxide tensions increase, and respiratory acidosis occurs. In elephant seal (*Mirounga* spp.) pups, awake apnea does not exceed about 5 min whereas sleep apnea can be as long as 14 min. The pups show a parallel increase in mean sleep apnea and mean dive duration during their first year of life.

Pups of several species have been observed to have a higher breathing rate than adults. Weddell seal pups take breaths/min compared to the 8 breaths/min rate of adults. Pups of Australian (*Neophoca cinerea*) and New Zealand (*Phocarctos hookeri*) sea lions take 13 breaths/min whereas adults of these species typically breathe 3–5 times per minute.

On land and in the water different stages of sleep in pinnipeds are associated with different breathing patterns in different species. During rapid eye movement sleep, breathing is regular and at rates up to 16/min in gray seals, irregular in harp seal pups, and absent in elephant seal pups. Elephant seal pups sleeping in shallow water rise to the surface to breathe without showing brain wave patterns associated with wakefulness. In contrast, Caspian seals (*Pusa caspica*) sleeping below the surface awake prior to surfacing and breathing.

Some species of delphinids show unihemispheric brain waves associated with sleep. Thus there is one cerebral hemisphere that is always awake to control surfacing and breathing. Northern fur seals (*Callorhinus ursinus*) sleeping in the water also sometimes show unihemisperic sleep patterns with one cerebral hemisphere awake to control surfacing and breathing. In contrast to delphinids, no pinniped has shown exclusively unihemispheric sleep brain waves.

## V. Oxygen Loading and Dive Time

Several authors have modeled the diving behavior of marine mammals based on oxygen loading curves at the surface compared with energy expenditure while below the surface. There have been various models based on what the animal may be attempting to maximize, be it time in a deep prey patch, gross energy intake during a dive, net energy intake, energetic efficiency, and so on. All the models conclude that there should be some relationship between the duration of a dive and the time breathing at the surface, either predive or postdive. Although some species do show such relationships over certain dive time intervals, not all species show the expected patterns. The time Weddell seals spend at the surface is independent of preceding dive duration up to the aerobic dive limit. Beyond that time, the surface time increases exponentially with the preceding dive time. However, gray seals (*H. grypus*) show a direct proportionality between dive time and surface time up to dive times of 7 min. Surface time is independent of dive time for dive times greater than 7 min. Surface time of sperm whales (*Physeter macrocephalus*) shows a slight trend to decrease with increasing dive time, but is basically independent of dive time. Elephant seals can maintain, over periods greater than 24 hr, a pattern of long, deep dives followed by surface intervals of 3 min or less. Some of these differences are attributable to species variation among groups with different diving strategies and oxygen loading needs. Additional explanations of the breakdown of models relating surface oxygen uptake to underwater duration and activity include lowered metabolic rates underwater, passive gliding descents, and animals not maximizing any of the foraging related parameters. For example, ringed seals (*Pusa hispida*) appear to be constrained by a risk aversive strategy. Diving under shore fast ice, ringed seals gain access to air only at a few breathing holes. If a seal finds a breathing hole occupied by another seal or detects a polar bear above the hole, it will need oxygen reserves to locate an alternate breathing site.

## VI. Water Conservation during Breathing

Because marine mammals obtain most of their water requirements from their prey and through metabolic production of water, conservation of water is an important adaptation in marine mammals. Renal adaptations for water conservation are discussed elsewhere, but water can also be lost through ventilation. Both pinnipeds and cetaceans exhale air that is not saturated with water vapor. In bottlenose dolphins, the respiratory water loss is reduced by 70% over what it would be compared to a terrestrial mammal breathing dry air. Countercurrent heat exchange and induced turbulence in the nares and nasal sac system allow for the extraction of a majority of the water vapor in the air coming from the lungs. In seals, the bones in the anterior part of the nasal cavity (turbinates) create a very dense mesh through which the expired air must pass. Moist air flowing over this large surface area gives up much of this water before being exhaled.

## VII. Breathing Patterns in Response to Disturbance

Changes in breathing patterns have been used extensively as indicators of disturbance of marine mammals in response to human activity. In many cases, a statistically significant change has been observed in the interbreath interval, total number of breaths during a surfacing, or proportion of time spent at the surface. Although these changes may be statistically significant, it is questionable whether they are biologically significant for an individual animal or indicative of any long-term consequences for the population.

## See Also the Following Articles

Brain ■ Circulatory System ■ Diving Physiology

## References

Butler, P. J., and Jones, D. R. (1997). Physiology of diving of birds and mammals. *Physiol. Rev.* **77**, 837–899.

Kooyman, G. L. (1973). Respiratory adaptations in marine mammals. *Am. Zool.* **13**, 457–468.

Kooyman, G. L. (1989). "Diverse Divers: Physiology and Behavior." Springer-Verlag, Berlin.

Kooyman, G. L., and Sinnett, E. E. (1979). Mechanical properties of the harbor porpoise lung, *Phocoena phocoena. Resp. Physiol.* **36**, 287–300.

Ridgway, S. H. (1972). Homeostasis in the aquatic environment. *In* "Mammals of the Sea: Biology and Medicine" (S. H. Ridgway, ed.), pp. 590–747. Charles C. Thomas, Springfield, IL.

Slijper, E. J. (1976). "Whales and Dolphins." University of Michigan, Ann Arbor.

# Breeding Sites

## Mark A. Hindell

Giving birth to young and the subsequent nursing of those young present unique problems to marine mammals. The strategies that marine mammals employ to deal with these problems can be divided into two broad groups (1) those animals that need to leave the water to breed [seals and polar bears (*Ursus maritimus*)] and (2) those animals that remain at sea to breed (cetaceans, sirenians, and otters). Whichever strategy is used, a crucial component of the reproductive process is the site used for breeding (here, the term breeding is restricted to parturition and suckling and does not include mating). The breeding sites used by marine mammals are quite diverse, both in terms of geography and physical characteristics, ranging from polar to equatorial regions and from sandy beaches to deep ocean basins. Which sites are used for breeding by a particular species is determined by a complicated mixture of factors including evolutionary history, requirements of the young, requirements of the adults, biological characteristics (such as the proximity of prey), and physical characteristics (such as water temperature or beach substrate). Riedman (1990) and Reynolds and Rommel (1999) provide excellent reviews.

## I. Species that Leave the Sea to Breed

All species of seal and polar bears leave the sea to give birth and most also need to be ashore to suckle their young. Some species (predominantly the phocids) remain ashore for most, if not all, of the lactation period, whereas in others (the otariids) the females regularly return to sea to forage during lactation. Only harbor seals (*Phoca vitulina*) and walruses (*Odobenus rosmarus*) have been reported to suckle young at sea. The sites used by these animals for breeding can be divided into two groups: those that breed on land and those that breed on ice.

### A. Marine Mammals that Breed on Land

Pregnant polar bears leave the arctic pack ice with the onset of spring to breed on land. The deteriorating summer pack ice is too unstable for the bears, which have a relatively long period of cub dependence. Females excavate caves in the side of riverbanks or hillsides, thereby by getting close to the permafrost and providing an environment with a stable temperature. As summer progresses and snowfall increases these caves get snowed in and the females need to maintain a cavity with sufficient wall thickness to provide insulation, but also thin enough to allow the passage of air.

Twenty species of seals breed on land. This includes all species of Otariids (fur seals and sea lions) and six of the Phocids (true seals) (Table I). None of these species uses any kind of shelter and all give birth and suckle on the beach exposed to wind, rain (or snow), and waves. Land breeding seals often occur in very large aggregations during the breeding season, which may offer some protection from the weather, but this also puts them at risk of being damaged in fights between adult seals. Land-based breeding sites occur at all but the most extreme polar latitudes.

Some species utilize both land and ice breeding sites. Gray seals (*Halichoerus grypus*) breed on beaches in northern Europe and America, but on pack ice in Northern Canada. Although primarily a land breeding species, harbor seals also breed in pack ice in northern Canada.

Most land breeding seals use islands as their principal breeding sites, with only a few species utilizing mainland beaches. This is likely to be an attempt to avoid large mainland predators, including humans. Generally, species have quite specific requirements of their island breeding sites, and so suitable island breeding sites are often limited. Consequently, those sites that are suitable tend to hold very large numbers of seals, which provides an ideal condition for the evolution of polygyny.

Different species have different substrate requirements. Most species use gradually sloping sandy beaches, such as those used by northern elephant seals (*Mirounga angustirostris*). Fur seals, which are generally more agile than phocids, can also breed on rocky substrates, but again this is quite species-specific. For example, at Macquarie Island where both Antarctic (*Arctocephalus gazella*) and subantarctic (*A. tropicalis*) fur seals breed, *A. gazella* use the open beaches, whereas *A. tropicalis* breeds on nearby rocky headlands.

## TABLE I
### A List of Land Breeding Marine Mammals, also Indicating the Primary Geographic Type of Breeding Site

| Species | Geographic type[a] |
| --- | --- |
| *Otariids* | |
| Antarctic fur seal, *Arctocephalus gazella* | Island |
| Galápagos fur seal, *A. galapagoensis* | Island |
| Guadalupe fur seal, *A. townsendi* | Island |
| Juan Fernández fur seal, *A. philippii* | Island |
| New Zealand fur seal, *A. forsteri* | Island/mainland |
| Northern fur seal, *Callorhinus ursinus* | Island |
| South African/Australian fur seal, *A. pusillus* | Island/mainland |
| South American fur seal, *A. australis* | Island/mainland |
| Subantarctic fur seal, *A. tropicalis* | Island |
| Australian sea lion, *Neophoca cinerea* | Island/mainland |
| California sea lion, *Zalophus californianus* | Island/mainland |
| Galápagos sea lion, *Z. wollebaeki* | Island |
| New Zealand sea lion, *Phocarctos hookeri* | Island/mainland |
| Southern sea lion, *Otaria flavescens* | Island/mainland |
| Steller sea lion, *Eumetopias jubatus* | Island |
| *Phocids* | |
| Monk seal—Hawaiian, *Monachus schauinslandi* | Island |
| Monk seal—*Mediterranean, M. monachus* | Island/mainland |
| Elephant seal—Northern, *Mirounga angustirostris* | Island/mainland |
| Elephant Seal—Southern, *M. leonina* | Island/mainland |
| Harbor seal, *Phoca vitulina* | Island/mainland/ice |
| Gray seal, *Halichoerus grypus* | Island/mainland/ice |

[a]From Riedman (1990).

## TABLE II
### A List of Ice Breeding Marine Mammals, also Indicating the Primary Ice Type

| Species | Ice type[a] |
| --- | --- |
| *Obodenids* | |
| Walrus, *Odobenus rosmarus* | Pack ice |
| *Phocids* | |
| Weddell seal, *Leptonychotes weddellii* | Fast ice |
| Ross seal, *Ommatophora rossii* | Pack ice |
| Crabeater seal, *Lobodon carcinophaga* | Pack ice |
| Leopard seal, *Hydrurga leptonyx* | Pack ice |
| Hooded seal, *Cystophora cristata* | Pack ice |
| Harp seal, *Pagophilus groenlandicus* | Pack ice |
| Ribbon seal, *Histriophora fasciata* | Pack ice |
| Baikal seal, *Pusa sibirica* | Fast ice |
| Ringed seal, *Pusa hispida* | Fast ice |
| Harbor seal, *Phoca vitulina* | Pack ice/land |
| Gray seal, *Halichoerus grypus* | Pack ice/land |
| Largha seal, *Phoca largha* | Pack ice |
| Caspian seal, *Pusa caspica* | Pack ice |
| Bearded seal, *Erignathus barbatus* | Pack ice |

[a]From Riedman (1990).

The reasons for these different preferences are unknown but may arise from resource partitioning.

Proximity to a food source may be another factor that determines the location of breeding sites for some land breeding seals. For example, female otariids need to return to sea to replenish their energy reserves regularly during the lactation period. As their pups are fasting during these trips to sea, faster growth and heavier weaning masses can be achieved if the foraging trips are kept as short as possible. Breeding sites are often located close to the continental shelf break or other oceanographic features that tend to have enhanced primary productivity. In cases where this is not possible, such as the subantarctic fur seal breeding site on Amsterdam Island, female foraging trips are much longer than in other species (or populations) and the pups have correspondingly lower growth rates.

Land breeding phocid seals generally have no requirement for feeding during lactation, and therefore the breeding sites can be located considerable distances from the foraging sites. Southern elephant seals (*Mirounga leonina*), for example, tend to feed in high latitude waters, but breed thousands of kilometers away on subantarctic islands. In this case, the primary requirement of a breeding site is suitable beach structure and perhaps a moderate climate to help the pup in its early life.

### B. Marine Mammals that Breed on Ice

Thirteen species of pinnipeds breed on ice, either floating pack ice or fast ice attached to the land (Table II). Ice breeding seals tend to be monogamous, and this is likely to be a consequence of the breeding habitat. Ice, unlike suitable beaches for land breeding seals, is not a limited resource, and females can haulout anywhere to breed, so large aggregations of females tend not to occur. This limits the opportunities that males have to monopolize access to several females, and the best strategy for them is to find a female and remain with her until estrus. Weddell seals (*Leptonychotes weddellii*) are an exception to this rule, as they breed on fast ice, with limited access to open water. Female Weddell seals therefore tend to aggregate around tide-cracks and other sources of permanent open water.

Most ice breeding seals give birth and suckle their young on the ice. The exceptions to this rule are ringed seals (*Pusa hispida*) and Baikal seals (*P. sibirica*) which can use snow lairs under ice ridges. These lairs afford some protection from the elements and predators, although polar bears and foxes are adept at locating and breaking into these lairs.

Although pack ice is the breeding substrate, many species have preferred geographic regions to which they move for the breeding season. Harp seals (*Phoca groenlandicus*), for example, migrate to the southerly edge of the pack ice, and occupy only a small part of their overall range during the breeding season. The biggest areas are around Newfoundland and off western Greenland. Hooded seals (*Cystophora cristata*) show a similar pattern of migration, moving from widespread northerly feeding areas to more proscribed breeding areas which appear to be associated with the Continental shelf at the southern extent of the summer pack ice. This may provide the adults with access to the shelf for feeding immediately after weaning their pups.

### II. Species that Stay at Sea to Breed

Three groups of marine mammals are sufficiently adapted to a marine existence to give birth or suckle their young in water. These are the cetaceans, the sirenians, and the otters. By giving birth at sea, these animals are no longer constrained to use what is essentially a limiting resource—land. Nonetheless, many species still have quite specific requirements of their breeding sites, and migrate thousands of kilometers to reach their breeding grounds. Such profound

**B**

separation of feeding and breeding sites incurs large energetic costs, so the specific characteristics of these areas must be of considerable importance.

Dugongs (*Dugong dugon*) and manatees (*Trichechus* spp.) have no specialized breeding site requirements. Young are born and remain with their mother while she forages on sea grass beds. Likewise, sea otters (*Enhydra lutris*) give birth at sea and the cubs remain with their mothers. Neither group appears to migrate to specific regions for breeding and remain within their foraging areas.

Cetaceans are the only group of ocean breeding marine mammals in which some species have clear separation of feeding areas and breeding areas. Within the cetaceans, breeding sites can be loosely categorized as (i) coastal, (ii) open ocean, or (iii) non-specific. Species with non-specific breeding sites are those that show no evidence of requiring different environments for breeding. This group contains most of the odontocetes and several of the mysticetes. Many of the smaller odontocete species do make seasonal migrations, but these are not clearly linked to breeding, and seem more related to changes in prey distribution.

Several species of beaked whales appear to have year round high latitude distributions, and do not migrate for breeding, giving birth and suckling their young in polar waters. This behavior is also seen in two mysticete species, the minke (*Balaenoptera acutorostrata*) and the bowhead (*Balaena mysticetus*) whales. Although bowhead whales do seem to move southwards before breeding, they never leave the pack ice zone.

### A. Cetaceans with Coastal Breeding Sites

Humpback whales (*Megaptera novaeangliae*) have several recognized breeding grounds and all are associated with coastal regions. The Southern Hemisphere populations use either the coast of the major southern continents (South Africa, Australia, or South America) or smaller oceanic islands such as Tonga and Fiji, with the preferred sites generally north of 30°. The Northern Hemisphere humpbacks move to the Caribbean, Hawaii, or Cape Verde Islands.

Eastern gray whales (*Eschrichtius robustus*) move south from arctic feeding grounds to breed in Baja California and, as with humpbacks, the breeding sites are close inshore. In fact, they are so inshore that they are close enough to be seen from land and form the basis of a tourist industry.

Of the inshore breeding cetaceans, the right whales (*Eubalaena* spp.) are least migratory. Neither the southern (*E. australis*) or the northern (*E. glacialis* and *E. japonica*) right whale species make long migrations to their breeding sites, but both species favor coastal sites and sheltered bays for giving birth. Individual southern right whales show high levels of breeding site fidelity and use the same bays on several consecutive breeding events.

Only one odontocete species, the beluga (*Delphinapterus leucas*), seems to have an inshore migration so that it can breed in shallow inshore waters.

Why these species seek out inshore waters is not really known. Aside from the thermal advantages common to all migrating whales (Section II.B), the specific advantages associated with inshore breeding are likely to be related to environmental conditions and predator avoidance.

### B. Cetaceans with Oceanic Breeding Sites

Less is known about the characteristics of the breeding sites for these species due to their lack of coastal aggregations, and so much of what is known comes from early tagging studies and whaling records.

Fin whale (*Balaenoptera physalus*) breeding sites are widely spread over oceanic waters in temperate and subtropical waters. There is evidence that some fin whales move toward land and even form loose aggregations, but they do not cross the continental shelf, remaining in deep water. The little information available for blue whales suggests that they use similar breeding sites to fin whales. This may also be the case for sei (*B. borealis*) whales, although they may not penetrate tropical waters. Bryde's (*B. edeni*) whales do not seem to migrate to breed; presumably this is a reflection of this species' largely tropical habitat all year round.

There is on-going debate regarding the reasons for the use of temperate or equatorial breeding sites by mysticete whales. Thermal and energetic advantages to new-born calves are possibilities, but if so, why do some species such as the minke and the bowhead whales, as well as many smaller odontocetes, remain in high latitude waters to breed? An alternative view is that the abundant and predictable food supply of food at high latitudes declines over the dark winter months, but only the species with large body size (and lower mass specific metabolic rates) are able to make the long migrations which allow them to take advantage of warmer waters.

## See Also the Following Articles

Behavior, Overview ■ Cetaceon Life History ■ Estrus and Estrous Behavior ■ Mating Systems ■ Reproductive Behavior

## References

Riedman, M. (1990). "The Pinnipeds: Seals, Sea Lions and Walruses." University of California Press, Berkeley.

Reynolds, J. E., and Rommel, S. A. (1999). "The Biology of Marine Mammals." Smithsonian Institution Press, Washington, DC.

# Bryde's Whales
## *Balaenoptera edeni/brydei*

### HIDEHIRO KATO AND WILLIAM F. PERRIN

### I. Characteristics and Taxonomy

Bryde's whales are the least known of the large baleen whales. We are not even sure how many species are represented in this complex of temperate and tropical whales. They were long confused with sei whales (*Balaenoptera borealis*) because of morphological similarities; this confusion lasted into the 1970s. Bryde's whales were first described by Anderson (1878) based on examination of a stranded balaenopterid on Thaybyoo Creek beach, Gulf of Marataban, Burma. He gave it the scientific name *Balaenoptera edeni* in honor of the British high commissioner in Burma, Sir Ashley Eden. Olsen (1913) found an unrecognized species among "sei whales" caught in Durban, South Africa, and gave it the scientific name *B. brydei* in honor of his sponsor Johan Bryde, a pioneer in South African whaling from the traditional whaling port of Sandefjord. Junge (1950) concluded that the two names were synonymous based on examination of a skeleton collected in Pulu Sugi, Singapore. Further studies by Omura (1959) and Best (1960) supported Junge's view, and their conclusion had been generally accepted until recently, with *B. edeni* having priority as the scientific name. (The common name remained "Bryde's whale" probably due to its wide popularity). However, today it is not clear how many species of Bryde's-whale-like baleen whales are there or what their scientific names should be (Rice, 1998). Wada and Numachi (1991) found that a small form occurring off the Solomon Islands and Java did not accord with other Bryde's whales in allozymes. These results suggesting the existence of at least two species were supported by mtDNA analyses reported by Yoshida and Kato (1999); the Solomon Islands and Java whales were more closely related genetically to sei whales than to "ordinary" Bryde's whales. The potential nomenclature at that point became some permutation of the names *B. brydei*, *B. edeni*, and perhaps a needed third new name. Rice (1998) had proposed provisional recognition of the existence of two species, with the nomenclature unsettled. Subsequently, a specimen of the same apparently undescribed species reported by Wada and Numachi (1991) washed ashore in southern Japan, and based on as well as and the earlier molecular data, Wada *et al.* (2003) described the new species *Balaenoptera omurai* (see chapter Omura's Whale). Sasaki *et al.* (2006) confirmed the genetic distinctness of the new species and its occurrence in the Philippines based on analysis of a bone sample from a whale taken in an artisanal whale fishery (LeDuc and Dizon, 2002). They further suggested that large and small forms of the "ordinary" Bryde's whale should be considered full species, *B. brydei* and *B. edeni*, respectively. However, the degree of differentiation between the two forms is of a level that could be considered consonant with subspecific separation. Moreover, small- and large-form Bryde's whales from various regions around the world have not yet been compared on a global basis. We here straddle the fence and follow Rice's (1998) lead of suggesting that there may be either one or two species of Bryde's whales (aside from Omura's whale). One needs to be aware that the names *B. edeni* and *B. brydei* are used variously by the present workers in the field; e.g., one recent reference listed below (Kanda *et al.*, 2007) refers to "ordinary" large Bryde's whales which is known globally as *B. brydei* (following Sasaki *et al.*, 2006), whereas two others (Heimlich *et al.*, 2005; Murase *et al.*, 2007) refer to the same whales in the western and eastern Pacific as *B. edeni*, the name presently used by the International Whaling Commission (IWC).

A nomenclatural problem remains with Omura's whale. The genetic identity of the holotype specimen of *B. edeni* (in a museum in Calcutta) is unknown. It was at the upper end of the size range for Omura's whale and the small form of the Bryde's whale (the "*edeni*" type) but still physically immature, and it comes from an area of potential overlap of the two forms of Bryde's whale and the new species. If genetic analysis determines that it belongs to the new species (Omura's whale), then that species will bear the name *B. edeni*. If it proves to be a Bryde's whale, then the Bryde's whale (if there is only one species) will bear the name *B. edeni*, *B. brydei* will fall into synonymy, and the name *B. omurai* will stand for Omura's whale. Hopefully a genetic analysis will be carried out on the holotype specimen of *B. edeni* to resolve the situation.

Bryde's whales are medium-sized balaenopterids. Females are larger than males throughout life, by about 2 ft (0.5–0.6 m) at full maturity. It is believed they reach 15.5 m, but most are much smaller. As demonstrated by Best (1977) for South African Bryde's whales, animals from coastal stocks or stocks inhabiting rather areas are generally smaller than those from migratory pelagic stocks (the small-form and large-forms described earlier). Southern Hemisphere animals are also larger than Northern Hemisphere animals. In the South Africa and western North Pacific stocks, body length increases

rapidly until 4–5 years, reaches about 90% of asymptotic length for both sexes at about 10 years, and ceases to increase at about 20–25 years. The length–weight relationship is given by the equation (Ohsumi, 1980):

$$W = 0.0126 \, L^{2.76}$$

where $W$ is body weight in metric tons and $L$ is body length in meters.

If mean lengths at physical maturity in the western North Pacific stock are substituted into the equation, weight estimates are 15.0 (at 13.0 m) and 16.6 (at 13.5 m) tons for males and females, respectively.

Bryde's whales closely resemble sei whales but have a number of distinctive characteristics (Fig. 1). Body color is principally dark smoky gray above and white below, but the dark area extends down to include the throat grooves and the flippers. The boundary between dark and light areas is diffuse. The rostrum is V shaped as in other balaenopterids (Fig. 1). The head occupies about 24–26% of the body.

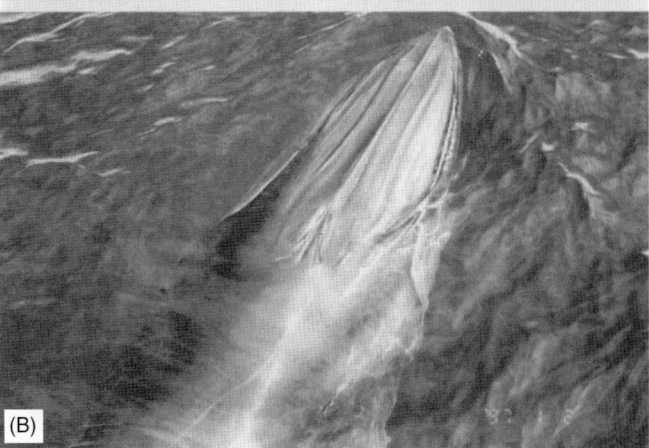

**Figure 1** *(A) Bryde's whale in western North Pacific in summer 1993 (photograph by Tomio Miyashita); (B) head region has a lateral ridge on the rostrum of an animal off South Africa. Photograph by Keiko Sekiguchi, 1997.*

The dorsal fin is extremely falcate with a tapering tip and is located at 25.2–26.6% of body length anterior to the flukes. The flukes are broad (23.5–24.4% of body length), with rather straight posterior margins.

Pelagic Bryde's whales, such as those in the western North Pacific stock or the offshore form off South Africa, bear large numbers of oval pit-like scars from bites by the tropical cookie cutter shark (*Isistius* sp.), which serve as the evidence of migration to tropical waters. This scarring is usually most extensive on the lateral side of the peduncle, leading to an appearance like that of galvanized iron. Such scarring is rare on animals of the coastal form off South Africa; this may also be true for coastal animals in neritic waters off Kochi, southwestern Japan, indicating that the whales do not migrate to tropical waters. Best (1977) further noted that the coastal form has thin scratches on the undersurface of the tail and ventral keel and suggested that these scratches are due to accidental touching of the sea bottom in shallow waters.

The throat grooves extend to or beyond the navel, whereas those of the sei whale do not reach the navel. The number of grooves between the flippers is 54–56.

The most distinctive external character allowing the discrimination of Bryde's whales from other baleen whales is the presence of three prominent ridges on the rostrum (Fig. 1). The ridges run from just behind the tip of the snout to anterior of the blowholes. They are composed of one central ridge and two lateral sub-ridges.

Bryde's whales have 285–350 dark slate-gray baleen plates on each side of the mouth. They are much broader and have coarser bristles than those of sei whales. The longest plate may reach 40 cm in length above the gum. Best (1960, 1977) reported a clear difference in the proportion of length to breadth of the plates in South Africa, with those of the inshore form being more slender than those of the offshore form. Kawamura (1978) found animals in the South Pacific to have finer bristles than those in the western North Pacific stock, probably reflecting a difference in their feeding habits.

The skull occupies about 24.1–25.8% of the total body length. It is relatively broad, short, and flat for a balaenopterid skull (Fig. 2). The rostrum is also relatively short and pointed. Its sides are nearly parallel posteriorly but slightly convexly curved anteriorly. The curved and robust mandible is also conspicuous among balaenopterids. The vertebrae formula is C 7 + D 13 + L 13 + Ca 21–22 = 54–55, for a total slightly lower total count than in sei whales (56–57). Cervical vertebrae are unfused, and thoracic (dorsal) vertebrae have short neural processes. The ribs are relatively thin and broad and are usually numbered

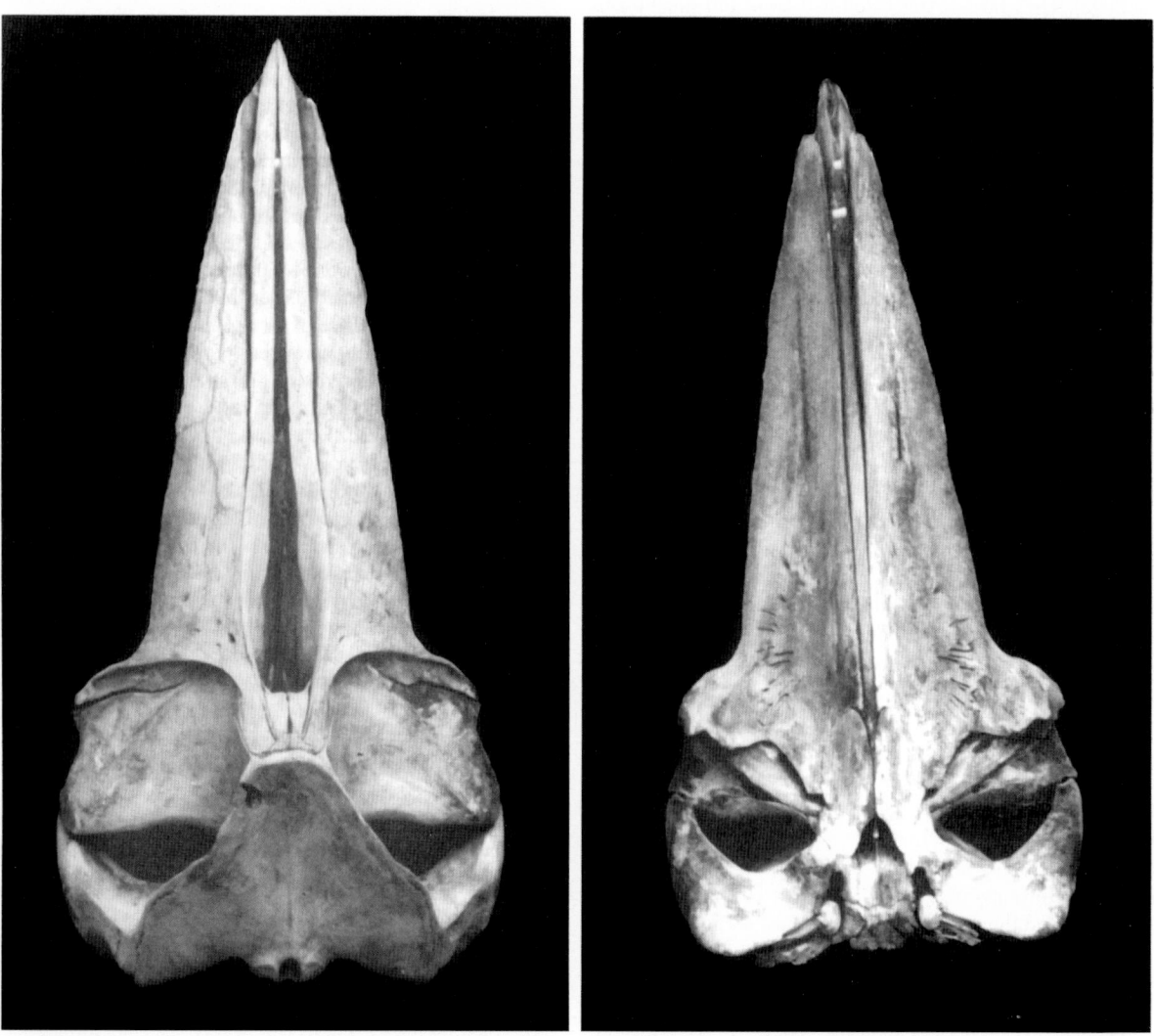

**Figure 2**   *Dorsal and ventral views of a skull of Bryde's whale caught in the western North Pacific and landed at Bonin Island in July 1983. Photograph by Hidehiro Kato.*

13–14. The first rib has a double head, a characteristic shared with the sei whale. The phalangeal formula is I (6), II (5), IV (5), V (3).

## II. Distribution and Abundance

Although there is a general pattern of migration toward the equator in winter and to higher latitudes in summer, Bryde's whales are seen throughout tropical and warm temperate waters of 16.3°C or warmer year round. Their occurrence has been reported from all tropical and temperate waters in the North and South Pacific, Indian Ocean, and South and North Atlantic between 40°N and 40°S (Fig. 3).

In the western North Pacific, Bryde's whales occur in temperate and tropical waters off the Pacific coasts of Japan, Taiwan, and Philippines to 150°W, with the northern limit corresponding approximately to the southern margin of the subarctic boundary at about 40°N and the southern limit at about 2°S. This is the western North Pacific stock, with abundance estimated at about 24,000 (CV = 0.20; IWC, 1997). Bryde's whales inhabit the east China Sea; this stock extends to the coastal waters of the Pacific side of southwest Japan but is restricted to the west of the Kuroshio warm current (Kato *et al.*, 1996). Bryde's whales also occur in eastern tropical Pacific waters, mainly west of 150°W between 20°N and 10°S; abundance is estimated at 13,000 (CV = 0.20). Bryde's whales also inhabit the Gulf of California throughout the year; these are assumed to constitute an independent stock. They are also widely seen in the Southern Hemisphere, occurring continuously from the east coast of Australia to 120°W in a zone between the equator and about 30°S, including northern New Zealand. In the eastern South Pacific, the species is distributed in coastal waters off South America from the equator to 37°S. Occurrence seems to be related somewhat to seasonal upwelling events. Little is known about stock structure in these areas, but it is expected that the coastal stock(s) is genetically different from those of the western South Pacific.

Bryde's whales are common throughout the Indian Ocean, from waters north of 40°S such as the Bay of Bengal and the Arabian Sea. Geographical concentrations occur south of Java to the west coast of Australia (east of 90°E), in the central Indian Ocean (65–90°E), off Madagascar (35–65°E), and off South Africa (east of 35°E). Stock structure in these areas has not been fully examined other than to confirm the existence of inshore and offshore forms off South Africa (discussed later). However, it would not be realistic to assume that one homogeneous stock is distributed over the Indian Ocean.

Two allopatric forms of Bryde's whale known as the inshore and offshore forms are found off the west coast of South Africa (Best, 1977). The inshore form is restricted to within 20 miles from the coast and is seen there throughout the year. The offshore form occurs in waters over 50 miles from the coast and migrates north to the equator in winter. Accurate estimates of population abundance are not available for either stock. Little is known about distribution and stock structure in the rest of the Atlantic, especially in the Northern Hemisphere. However, Bryde's whales have been sighted or stranded from Morocco south to the Cape of Good Hope in the east and from Virginia south to Brazil in the west, including the Gulf of Mexico, the Caribbean, and off Venezuela. Most recently, Best (2001) suggested that there may be three populations off southern Africa, a pelagic population of large whales off the west coast and separate populations of small whales on the east and west coasts.

Genetic analysis of large-form (offshore) Bryde's whales (nominally *B. brydei*) from populations in different ocean basins and hemispheres suggests that there is a little gene flow between the populations and that they should be treated as separate entities for purposes of management (Kanda *et al.*, 2007).

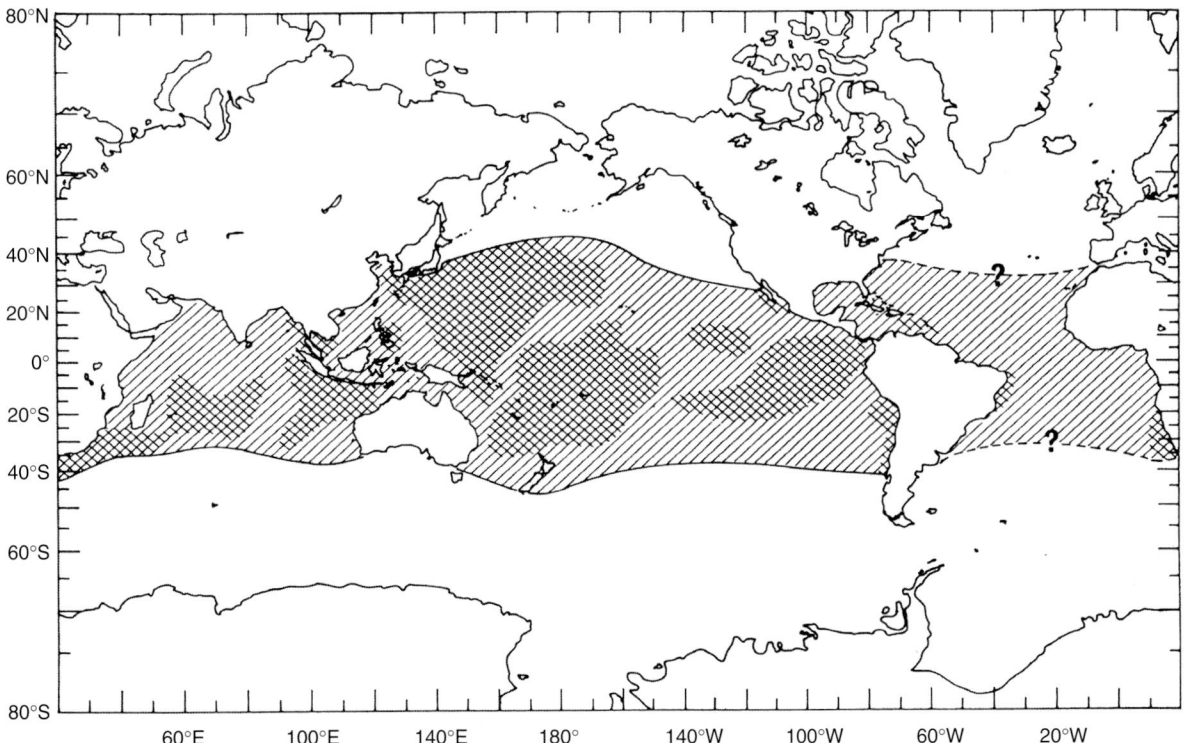

**Figure 3**   *Worldwide distribution of Bryde's whales based on published or available information. Dense hatch represents areas in which higher densities are expected.*

**B**

## III. Ecology

Bryde's whales feed mainly on pelagic schooling fishes such as pilchard, anchovy, sardine, mackerel, herring, and others. They were observed feeding on sardines off Brazil (Siciliano *et al.*, 2004). However, they also feed on small crustaceans such as euphausids and copepods as well as on cephalopods and pelagic red crabs (*Pleuroncodes*) (Best, 1960, 1977; Nemoto and Kawamura, 1977; Ohsumi, 1977a; Kawamura, 1980). They are considered opportunistic feeders, unlike sei whales, which concentrated on copepods. Off South Africa, they tend to be dependent on euphausiids in pelagic waters and feed on schooling fishes in coastal waters; thus feeding habits may be characteristic of stocks (Best, 1977). However, in the pelagic North Pacific, prey selection was shown for Japanese anchovy in 1 year and krill in the next (Murase *et al.*, 2007). Balaenopterids consume about 4% of body weight daily at the peak of the feeding season; this corresponds to approximately 600–660 kg per day for a Bryde's whale.

Bryde's whales are sometimes seen within high-density patches of bonito (*Sarda*) in pelagic waters in the North Pacific. This may be a result of two predators chasing the same prey. Similarly, off the coasts of Kochi and Kasasa, southwestern Japan, it is very common to see Bryde's whales feeding in patches of sardines or juvenile tuna, especially in summer. They have also been observed to utilize bubble net foraging with slow circle swimming under the surface.

## IV. Behavior and Physiology

Bryde's whales do not gather into large groups. They are usually seen singly or in groups of 2–3 in the North Pacific, with a maximum group size of 12. They usually surface steeply like other balaenopterids. The blow is 3–4 m high. The dorsal fin is seen after the blow and then sometimes the dorsal keel is seen. They seldom fluke up before diving. It is generally believed that they usually move at 2–7 km/h but can swim as fast as 20–25 km/h and dive up to 300 m. Bryde's whales breach more often than other balaenopterids. They produce powerful low-frequency tonal and swept calls similar to the calls of other balaenopterid whales (Cummings, 1985). Spatial-temporal variation in frequency and call types may be related to sex, seasonal activity, or population differences (Heimlich *et al.*, 2005; Oleson *et al.*, 2003).

## V. Life History

Although Bryde's whales have a life history similar to that of other balaenopterids, there are species-specific aspects due to the fact that they remain in tropical and temperate waters throughout the year. As for many other migratory large cetaceans, little is known about the breeding grounds, even for inshore or coastal stocks, although it is generally believed that they must be somewhere in lower latitudes for the migratory stocks. In waters off the inshore waters of the Pacific coast of Japan and the East China Sea, females accompanied by small calves sometimes appear in early spring, but there is no direct evidence that they give birth there. In pelagic stocks, peaks of both conception and calving are in winter; although these are much more diffuse than in other migratory balaenopterids. Best (1977) reported that breeding is not seasonally restricted for inshore animals off South Africa; this may be true for other local stocks.

Gestation lasts for about 11 months. Length at birth is about 4 m. The sex ratio at birth is not different from parity (Best, 1977). Lactation lasts about 6 months and calves wean at about 7 m in body

length. Males attain sexual maturity at 11–11.4 m and females at 11.6–11.8 m in the western North Pacific stock. Taking into account bias due to operations and regulation of whaling, Best (1977) found length at sexual maturity for the inshore form to be less than for the offshore form off South Africa, by 1 ft in females and 3 ft in males. The mean age at sexual maturity is slightly less than 7 years. Based on the annual ovulation rate (0.42–0.46) for pelagic Bryde's whale stocks (Best, 1977; Ohsumi, 1977b), the calving interval is about 2 years. Inshore waters off South Africa are very frequent ovulators (1.88 per year), but this does indicate a higher pregnancy rate but rather probably results from the extended breeding season. In summary, Bryde's whales have a 2-year reproductive cycle composed of 11–12 months gestation, 6 months of lactation, and 6 months resting.

## VI. Human Interactions

Bryde's whales were not harvested commercially or substantially until recent times; their value became relatively important in the late 1970s with the shift of whaling to the smaller species. However, commercial harvest of this species has been prohibited by a moratorium imposed by the IWC in 1987.

Because Bryde's whales had been mainly exploited after substantial improvement of IWC stock management procedures adopted in 1975 (the new management procedure or MNP), stocks have been kept relatively stable. A reliable estimate of the population trend for North Pacific Bryde's whales has been available for the western North Pacific stock from a comprehensive assessment conducted by the IWC in 1995 and 1996. According to the assessment (IWC, 1997), the population has been increasing since 1987, and the current population level (mature females in 1996 relative to 1911) ranges from 56.7% to 81.4%.

## *See Also the Following Articles*

Baleen Whales ■ Population Status and Trends ■ Sei Whale ■ Species

## *References*

Anderson, J. (1878). "Anatomical and Zoological Researchers: Comprising an Account of the Zoological Results of the Two Expeditions to Western Yunnan in 1868 and 1875." B. Quaritch, London.

Best, P. B. (1960). Further information on Bryde's whale (*Balaenoptera edeni* Anderson) from Saldanha Bay, South Africa. *Nor. Hval.-tid.* **49**, 201–215.

Best, P. B. (1977). Two allopatric forms of Bryde's whale off South Africa. *Rep. Int. Whal. Commn* (Spec. Iss. 1), 10–38.

Best, P. G. (2001). Distribution and population separation of Bryde's whale *Balaenoptera edeni* off southern Africa. *Mar. Ecol. Prog. Ser.* **220**, 277–289.

Cummings, W. C. (1985). Bryde's whale *Balaenoptera edeni* Anderson, 1878. *In* "Handbook of Marine Mammals" (S. H. Ridgway and R. Harrison, eds), 3, pp. 137–154. Academic Press, London.

Heimlich, S. L., Mellinger, D. K., Nieukirk, S. L., and Fox, C. G. (2005). Types, distribution, and seasonal occurrence of sounds attributed to Bryde's whales (*Balaenoptera edeni*) recorded in the eastern tropical Pacific, 1999–2001. *J. Acoust. Soc. Am.* **118**, 1830–1837.

IWC (International Whaling Commission) (1997). Report of the sub-committee on North Pacific Bryde's whales, Annex G, Report of the Scientific Committee. *Rep. Int. Whal. Commn* **47**, 163–168.

Junge, G. C. A. (1950). On a specimen of the rare fin whale, *Balaenoptera edeni* Anderson, stranded on Pulu Sugi near Singapore. *Zool. Verhand.* **9**, 1–26.

Kanda, N., Goto, M., Kato, H., McPhee, M. V., and Pastene, L. A. (2007). Population genetic structure of Bryde's whales (*Balaenoptera brydei*). *Cons. Genet.* **8**, 853–864.

Kato, H., Shinohara, E., Kishiro, T., and Noji, S. (1996). Distribution of Bryde's whales off Kochi, Southwest Japan, from the 1994/95 sighting survey. *Rep. Int. Whal. Commn* **46**, 429–436.

Kawamura, A. (1978). On the baleen filter area in the South Pacific Bryde's whales. *Sci. Rep. Whales Res. Inst., Tokyo* **30**, 291–300.

Kawamura, A. (1980). Food habits of the Bryde's whales taken in the South Pacific and Indian Oceans. *Sci. Rep. Whales Res. Inst., Tokyo* **32**, 1–23.

LeDuc, R. G., and Dizon, A. E. (2002). Reconstructing the rorqual phylogeny: With comments on the use of molecular and morphological data for systematic study. *In* "Molecular and Cell Biology of Marine Mammals" (C. J. Pfeiffer, ed.), pp. 100–110. Krieger Pub. Co., Malabar, FL.

Murase, H., *et al.* (9 authors) (2007). Prey selection of common minke (*Balaenoptera acutorostrata*) and Bryde's whales (*Balaenoptera edeni*) whales in the western North Pacific in 2000 and 2001. *Fish. Oceanogr.* **16**, 186–201.

Nemoto, T., and Kawamura, A. (1977). Characteristics of food habits and distribution of baleen whales with special reference to the abundance of North Pacific sei and Bryde's whales. *Rep. Int. Whal. Commn* (Spec. Iss. 1), 80–87.

Ohsumi, S. (1977a). Bryde's whales in the pelagic whaling ground of the North Pacific. *Rep. Int. Whal. Commn* (Spec. Iss. 1), 140–150.

Ohsumi, S. (1977b). Further assessment of population of Bryde's whales in the North Pacific. *Rep. Int. Whal. Commn* **27**, 156–160.

Ohsumi, S. (1980). Population study of the Bryde's whale in the Southern Hemisphere under scientific permit in the three seasons, 1976/77–1978/79. *Rep. Int. Whal. Commn* **30**, 319–331.

Olsen, O. (1913). On the external characteristics and biology of Bryde's whale (*Balaenoptera brydei*) a new rorqual from the coast of South Africa. *Proc. Zool. Soc. Lond.* **1913**, 1073–1090.

Olesun, E. M., Barlow, J., Gordon, J., Rankin, S., and Hildebrand, J. A. (2005). Low frequency calls of Bryde's whales. *Mar. Mamm. Sci.* **19**, 407–419.

Omura, H. (1959). Bryde's whale from the coast of Japan. *Sci. Rep. Whales Res. Inst., Tokyo* **14**, 1–33.

Omura, H., Kasuya, T., Kato, H., and Wada, S. (1981). Osteological study of the Bryde's whale from the central South Pacific and eastern Indian Ocean. *Sci. Rep. Whales Res. Inst., Tokyo* **33**, 1–26.

Rice, D. W. (1998). "Marine Mammals of the World, Systematic and Distribution." Society for Marine Mammalogy Special Publication **4**, Society for Marine Mammalogy, Lawrence, KS.

Sasaki, T., Nikaido, M., Wada, S., Yamada, T. K., Cao, Y., Hasegawa, M., and Okada, N. (2006). *Balaenoptera omurai* is a newly discovered baleen whale that represents an ancient evolutionary lineage. *Mol. Phylogenet. Evol.* **41**, 40–52.

Siciliano, S., *et al.* (2004). Strandings and feeding records of Bryde's whales (*Balaenoptera edeni*) in south-eastern Brazil. *J. Mar. Biolog. Assoc. U.K.* **84**, 857–859.

Wada, S., and Numachi, K. (1991). Allozyme analyses of genetic differentiation among the populations and species of the *Balaenoptera*. *Rep. Int. Whal. Commn* (Spec. Iss. 13), 125–154.

Wada, S., Oishi, M., and Yamada, T. K. (2003). A newly discovered species of living baleen whale. *Nature* **426**, 278–281.

Wade, P. R., and Gerrodette, T. (1993). Estimates of cetacean abundance and distribution in the eastern tropical Pacific. *Rep. Int. Whal. Commn* **43**, 477–493.

Yoshida, H., and Kato, H. (1999). Phylogenetic relationships of Bryde's whales in the western North Pacific and adjacent waters inferred from mitochondrial DNA sequences. *Mar. Mamm. Sci.* **15**, 1269–1286.

# Burmeister's Porpoise
## *Phocoena spinipinnis*

### JULIO C. REYES

**B**

## I. Characteristics and Taxonomy

*Phocoena spinipinnis* was described from a specimen captured by fishermen at the mouth of La Plata River, Argentina (Burmeister, 1865; Brownell and Praderi, 1984; Brownell and Clapham, 1999). It belongs to the family Phocoenidae, whose members share the presence of premaxillary eminences in the skull, a reduced posterior extension of the premaxillae not reaching the nasal and, at least in the living species, small spatulated teeth. The Spanish name for this porpoise is "marsopa espinosa" (spiny porpoise) which refers to the series of dermal tubercles present in the dorsal fin. Some vernacular names include "chancho marino," "tonino" (in Peru), and "antonino" (in southern Chile).

The body is stocky with a small, blunt head lacking a beak and proportionally large flippers (Goodall *et al.*, 1995; Figs 1 and 2). The dorsal fin is a diagnostic feature in this species. It is placed far back,

**Figure 1** *Lateral (A) and ventral (B) views of a female Burmeister's porpoise from Peru.*

**Figure 2** *Detail of the head of a Burmeister's porpoise.*

triangular in shape, and canted backward, its leading edge bears a series of 2–4 rows of tubercles ending near the tip of the fin (Fig. 3). The tubercles become larger and sharper in older animals. The flukes are medium sized, with rounded tips and an almost straight trailing edge.

Maximum length reported is 200 cm for specimens from Uruguay. On the Pacific coast of South America, the largest male and female were 182 and 183 cm long, respectively. Maximum known body mass for specimens from the Pacific is 72 kg for a male of 170 cm long ($n = 70$) and 79 kg for a female of 173 cm long ($n = 60$). In the Atlantic, a 191-cm female weighted 105 kg, whereas male of 178 cm length weighed 78 kg.

The coloration of Burmeister's porpoise observed in fresh carcasses and live animals is dark gray, sometimes lead gray, on the back and sides with a light gray to white around the abdominal field. A few animals may present a brownish hue. There is a well-defined

**Figure 3** *Dorsal fin of a Burmeister's porpoise, showing the dermal tubercles characteristic of the species.*

eye patch surrounded by a light gray to white halo. An anterior extension of this eye patch may reach the dark gray lip patch. A wide, dark gray blowhole-to-apex stripe joins the lip patch. The flipper stripe is dark gray, being wide and reaching the lip patch on the right side, and thinner and joining the chin patch on the left side of the head; the flipper stripe may be flanked by thin, light gray stripes. In the abdominal field a pair of stripes extends toward the genital area, splitting into accessory stripes that end in the mammary slits in females and run parallel to the perineal groove in males.

The skull of *P. spinipinnis* (Fig. 4) resembles in several aspects to those of the harbor porpoise, *Phocoena phocoena*, and the vaquita, *Phocoena sinus*, although it has a less antero-posterior compression of the braincase and the dorsal aspect of the supraoccipital is in line with the plane of the rostrum, whereas in *P. sinus* and *P. phocoena*, this portion of the supraoccipital forms an angle with the long axis of the rostrum. Further characteristics differentiating *P. spinipinnis* from *P. sinus* include a longer rostrum, a larger vertex, fewer alveolar teeth, and hamular processes of pterygoids longer than wide and with mesial borders widely separated.

Little information is available on the axial skeleton. Vertebral counts for two specimens were C7, D14, L15, Ca31–35 = 67–68. The first three cervical vertebrae are fused. The first eight pairs of ribs have both capitular and tubercular articulation to the vertebrae. The phalangeal formula is I2, II8, III7, IV4, and V2.

Fossil phocoenids have been found in Peru (de Muizon, 1986; Yáñez *et al.*, 1994), including *Piscolithax longirostris* and *Lomacetus ginsburgi* from the Late Miocene and Early Pliocene in the Pisco Formation. A single incomplete fossil calvarium of *P. spinipinnis* is known from Chile, although locality and date are not known for the specimen. Analysis of the mitochondrial DNA control region and cytochrome *b* gene indicates that *P. spinipinnis* is closely related to

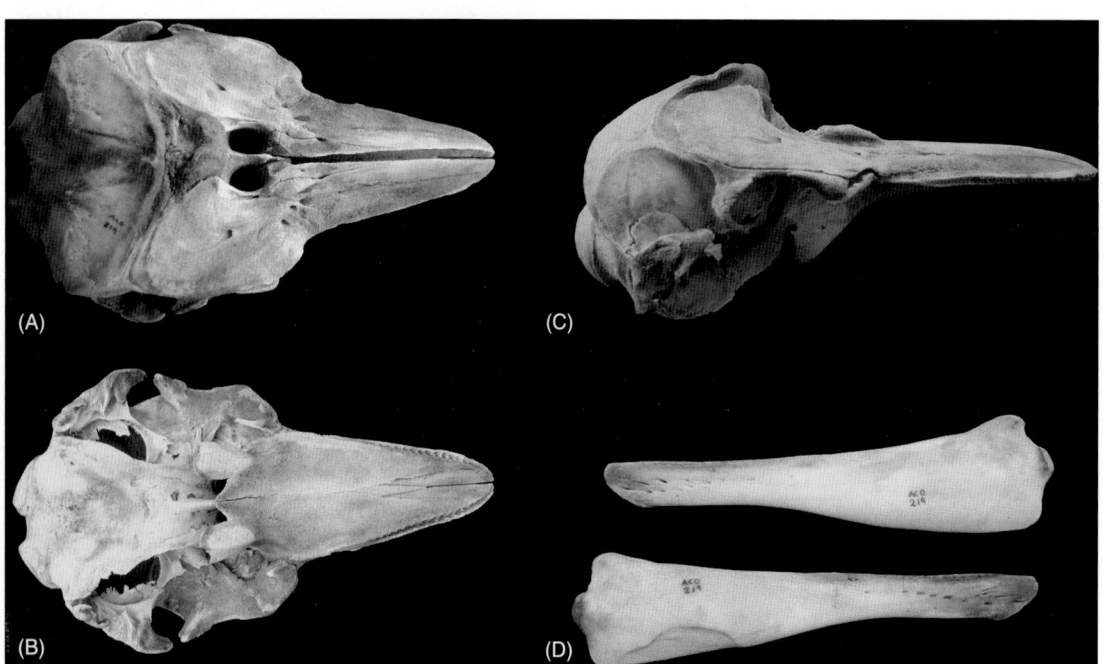

**Figure 4** *Skull of an adult Burmeister's porpoise (ACOREMA-219, male): (A) dorsal, (B) ventral, and (C) lateral views, and (D) mandible.*

*P. sinus* and both are related to the spectacled porpoise (*Phocoena dioptrica*) (Rosel *et al.*, 1995).

## II. Distribution and Abundance

On the west coast of South America, the northernmost record of Burmeister's porpoise is from Paita, northern Peru (5°01'S), at the latitude where the Humboldt Current veers to the west as it is integrated into the South Equatorial Current. From here, the distribution extends south along the Peruvian and Chilean coasts to Valdivia, Chile (39°50'S) (Grimwood, 1969; Clarke, 1962; Goodall *et al.*, 1995; Guerra *et al.*, 1987). There are no confirmed records from the fjords in southern Chile, but the species has been observed in the Magellan Strait, the Beagle Channel, and near Cape Horn. On the east side of the continent, this porpoise is reported from Argentina, Uruguay, and up to Santa Catarina State in Brazil (~28°S) (Pinedo, 1989; Simôes-Lopes and Ximenez, 1989). Based on the data derived from specimen records and analysis of oceanographic variables, it has been proposed that the range of this species is continuous from Paita, Peru, to the La Plata River Basin, Argentina, with records in Uruguay and Brazil related to the intrusion of cold waters associated with the Subtropical Convergence (Sarmiento and Tantaleán, 1991; Molina-Schiller *et al.*, 2005).

Differences in body size between Pacific and Atlantic Burmeister's porpoises led to the proposal of two different stocks of the species (Rosa *et al.*, 2005). Analysis of both mitochondrial and nuclear DNA reinforce the existence of a "Peruvian stock" and a "Chilean-Argentinean stock." Further research is needed based on both morphological and molecular analysis for a better determination of the geographic boundaries of these stocks. There are no abundance estimates for this species.

## III. Ecology

Throughout its range, Burmeister's porpoise is found mainly in coastal waters, 100–1000 m from shore and 5–25 m deep. Off Argentina, however, some animals have been captured in waters 30–60 m deep and 50 km from shore. Water temperatures during sightings ranged from 4°C to 19.5°C. Sightings and captures are reported in all seasons, although in some areas such as southern Chile there seem to be seasonal movements following prey. Seasonal occurrence has also been reported in San Jose Gulf in Argentina. Movements of porpoises following seasonal intrusions of cold waters associated with the Subtropical Convergence may account for the records of this species in Uruguay and Brazil.

Burmeister's porpoise feeds mainly on fishes, the species varying with region. In both Peru and Chile, Peruvian anchovy (*Engraulis ringens*) and hake (*Merluccius gayi*) are by far the main preys consumed, together with other small fish and squids. Off Argentina, sardines, tailed hake, shrimps, and squids are preferred. A few specimens examined in Uruguay and Brazil had mostly fish and squids. There is no information on foraging strategies.

The nematode *Stenurus australis* (Sarmiento and Tantaleán, 1991) is found tightly packed in the cranial sinuses and the inner ear of Burmeister's porpoises over 125 cm in length, suggesting that infestation may be timed with the beginning of ingestion of solid food. The campulid trematode *Nasitrema globicephalae* has been also found in the same location, although in low numbers and only in adult porpoises. In Argentina, *Stenurus minor* has been reported in the cranial sinuses and around the tympanic bulla of animals incidentally taken in nets (Corcuera *et al.*, 1995). The lungs may

be infected with the nematodes *Pseudalius inflexus* and *Halocercus* sp. Gastrointestinal parasites include the trematodes *Synthesium tursionis*, *Pholeter gastrophylus*, and *Braunina cordiformis*; the nematodes *Anisakis typica*, *Anisakis simplex*, and *Pseudoterranova* sp.; and the acantocephalan *Polymorphus* (*Polymorphus*) *cetaceum* (Torres *et al.*, 1992). The latter is present in porpoises from southern Chile and Argentina but absent in Peruvian specimens despite the large sample examined. Parasitic crustaceans of the genus *Isocyamus* have been reported in porpoises from Peru, although prevalence is low (6%). These whale lices are located in fresh wounds, the angle of the mouth, the genital slit, the axillae, and the angle at the base of the dorsal fin. The ectocomensal barnacle *Xenobalanus globicipitis* is found on the tips of flippers, the dorsal fin, and on the flukes of Burmeister's porpoises in Peru, with a maximum of 40 barnacles reported for a single animal. Neither cyamids nor ectocomensal barnacles have been reported from other locations outside Peru.

## IV. Behavior and Physiology

Swimming behavior is described as unobtrusive (Würsig *et al.*, 1977). This porpoise surfaces causing little disturbance in the surface and with little of the body surface exposed and dives with gentle rolls, which combined with the low dorsal fin may account for the difficulty of spotting this species at sea. On occasion it can swim very fast when approached by a boat. There are no reports of "porpoising" or bowriding, although a few animals have been observed riding waves and leaping out of the water. Underwater speed is estimated at 4 km/h, while time under the water surface ranges from 1 to 3 min.

Group size has been usually reported as between two and eight. Based on confirmed sightings, both modal and median group size in this species is 2. From 27 confirmed sightings, mean group size has been estimated at 7.38, with 78% of the sighted groups containing 1–4 animals. Large aggregations of 50–70 animals are sometimes encountered (Van Waerebeek *et al.*, 2002). One of the largest aggregations, comprising nearly 150 porpoises, was sighted off north-central Peru in water 27–31 m deep. The animals were scattered over a large area in small groups of 1–5 individuals (including several mother–calf pairs) forming a loose association. The cause and frequency of these events are unknown, but they may account for sporadic large catches of this species in coastal fisheries.

## V. Life History

In Peru, average length at sexual maturity in females is around 154.8 cm (Goodall *et al.*, 1995a, b; Reyes and Van Waerebeek, 1995;). Pregnancy rate has been estimated at 60%, although considering that some fetuses are too small to be noted, this is thought to be an underestimate. Records of pregnant females simultaneously lactating suggest that annual reproduction may take place. The average length at sexual maturation in males has been estimated as 159.9 cm. There is no evidence of seasonality in the male reproductive cycle.

The size at birth is around 86 cm. The sex ratio in 31 fetuses did not deviate from 1:1. Mating may take place during the austral summer (December through March), with some successful mating occurring outside this season. Gestation may last 11–12 months.

Age data are scarce for this species. Maximum estimated age is more than or equal to 12 growth layer groups (GLGs) for a

196-cm female and more than or equal to 5 GLGs for a 179-cm male. Animals from Argentina may reach physical maturity at a length of 200 cm.

## VI. Interactions with Humans

The largest capture of Burmeister's porpoise occurs off Peru (Read *et al.*, 1988; Van Waerebeek and Reyes, 1990; Reyes and Oporto, 1994; International Whaling Commission, 1994). Published data indicate that 200 to more than 400 porpoises are taken annually in Peruvian gillnet fisheries (targeted for a variety of demersal fishes including sharks and rays) and used for human consumption. Although most takes of this species in Peru are incidental, there is evidence that direct gillnetting in large porpoise aggregations does occur, and it may account for occasional high landings of the species in central Peru. Small numbers of porpoises are incidentally captured in fisheries off northern Chile. A gillnet fishery for rat fish and sciaenids operating in southern Chile takes a few hundred animals each year, which are used as bait in other fisheries (Lescrauwaet and Gibbons, 1994). The number of small cetaceans including Burmeister's porpoises reportedly taken directly for crab bait in southern Chile has shown some decline since 1990, due to several factors modifying the fishery's operations. A few animals may be taken in gillnets in Tierra del Fuego, Argentina; evidence of the use of porpoises as human food in the Beagle Channel area dates back some 6500 years. Off the northern coast of Argentina, an estimated 21–25 Burmeister's porpoises are captured every year in gillnet fisheries (Corcuera *et al.*, 1994). A shark fishery operating off Uruguay took only 8 porpoises in a 16-year period (Praderi, 1990), supporting the idea that the presence of the species off Uruguay and Brazil is related to temporary changes in oceanographic conditions.

Organochlorine compounds have been studied in Burmeister's porpoises from northern Argentina. The levels of pollutants (including DDT metabolites and PCBs) in the blubber are considered very low compared with other marine mammals. Pollutant levels have not been studied in other areas within the species' range.

## See Also the Following Article

Porpoises, Overview

## References

Aguayo, A. (1975). Progress report on small cetacean research in Chile. *J. Fish. Res. Board Can.* **32**, 1123–1143.

Allen, G. M. (1925). Burmeister's porpoise (*Phocoena spinipinnis*). *Bull. Mus. Comp. Zool.* **67**, 251–261.

Brownell, R. L., Jr., and Clapham, P. J. (1999). Burmeister's porpoise *Phocoena spinipinnis* Burmeister, 1865. *In* "Handbook of Marine Mammals" (S. Ridgway, and R. Harrison, eds), Vol VI, pp. 393–410. Academic Press, San Diego.

Brownell, R. L., and Praderi, R. (1982). Status of Burmeister's porpoise, *Phocoena spinipinnis*, in southern South American waters. *In* "Mammals of the Seas," FAO Fisheries Series **5**, 91–96.

Brownell, R. L., and Praderi, R. (1984). *Phocoena spinipinnis*. *Mam. Species* **217**, 1–4.

Burmeister, H. (1865). Description of a new species of porpoise in the Museum of Buenos Aires. *Proc. Zool. Soc. Lond.* **1865**, 228–231.

Clarke, R. (1962). Whale observation and whale marking off the coast of Chile in 1958 and from Ecuador towards and beyond the Galápagos Islands in 1959. *Norsk Hvalfangst. Tid.* **51**, 265–287.

Corcuera, J., Monzón, F., Crespo, E. A., Aguilar, A., and Raga, J. A. (1994). Interactions between marine mammals and coastal fisheries of Necochea and Claromeco (Buenos Aires province, Argentina). *Rep. Int. Whal. Commn Spec. Iss.* **15**, 283–290.

Corcuera, J., Monzón, F., Aguilar, A., Borrel, A., and Raga, A. (1995). Life history data, organochlorine pollutants and parasites from eight Burmeister's porpoises, *Phocoena spinipinnis*, caught in northern Argentine waters. *Rep. Int. Whal. Commn Spec. Iss.* **16**, 365–372.

Goodall, R. N. P., Norris, K. S., Harris, G., Oporto, J. A., and Castello, H. P. (1995a). Notes on the biology of Burmeister's porpoise, *Phocoena spinipinnis*, off southern South America. *Rep. Int. Whal. Commn Spec. Iss.* **16**, 318–347.

Goodall, R. N. P., Würsig, B., Würsig, M., Harris, G., and Norris, K. S. (1995b). Sightings of Burmeister's porpoise, *Phocoena spinipinnis*, off southern South America. *Rep. Int. Whal. Commn Spec. Iss.* **16**, 297–316.

Grimwood, I. R. (1969). Notes on the distribution and status of some Peruvian mammals 1968. *NY Zool. Soc. Spec. Publ.* **21**, 1–86.

Guerra, C., Van Waerebeek, K., Portflitt, G., and Luna, G. (1987). Presencia de cetáceos frente a la Segunda Región de Chile. *Estud. Ocean.* **6**, 87–96.

International Whaling Commission (1994). Report of the workshop on mortality of cetaceans in passive fishing nets and traps. *Rep. Int. Whal. Commn Spec. Iss.* **15**, 1–71.

Lescrauwaet, A.-C., and Gibbons, J. (1994). Mortality of small cetaceans and the crab bait fishery in the Magallanes area of Chile since 1980. *Rep. Int. Whal. Commn Spec. Iss.* **15**, 485–494.

Molina-Schiller, D. M., Rosales, S. A., and De Freitas, T. R. O. (2005). Oceanographic conditions off coastal South America in relation to the distribution of Burmeister's porpoise, *Phocoena spinipinnis*. *Lat. Am. J. Aquat. Mam.* **4**, 141–156.

de Muizon, C. (1986). Un nouveau Phocoenidae (Odontoceti, Mammalia) du Miocène supérieur de la Formation Pisco (Pérou). *C.R. Acad. Sc. París* **16**, 1509–1512, t. 303, Serie II.

Norris, K. S., and McFarland, W. N. (1958). A new harbor porpoise of the genus *Phocoena* from the Gulf of California. *J. Mammal.* **39**, 22–39.

Pinedo, M. C. (1989). Primeiro registro de *Phocoena spinipinnis* (Cetacea, Phocoenidae) para o litoral do Rio Grande do Sul, Brasil, com medidas osteologicas e análise do conteúdo estomacal. *Atlántica* **11**, 85–99.

Praderi, R. (1990). Mortality of dolphins in shark gillnets fisheries off Uruguay. IWC Scientific Committee meeting document SC/090/G1. 22 pp.

Read, A. J., Van Waerebeek, K., Reyes, J. C., McKinnon, J. S., and Lehman, L. C. (1988). The exploitation of small cetaceans in coastal Peru. *Biol. Cons.* **46**, 53–70.

Reyes, J. C., and Oporto, J. A. (1994). Gillnet fisheries and cetaceans in the southeast Pacific. *Rep. Int. Whal. Commn Spec. Iss.* **15**, 467–474.

Reyes, J. C., and Van Waerebeek, K. (1995). Aspects of the biology of Burmeister's porpoise from Peru. *Rep. Int. Whal. Commn Spec. Iss.* **16**, 349–364.

Rosa, S., Milinkovitch, M., Van Waerebeek, K., Berck, J., Oporto, J., Alfaro-Sigueto, J., Van Bressem, M. F., Goodall, N., and Cassens, I. (2005). Population structure of nuclear and mitochondrial DNA variation among South American Burmeister's porpoises (*Phocoena spinipinnis*). *Cons. Genet.* **6**, 431–443.

Rosel, P. E., Haygood, M. G., and Perrin, W. F. (1995). Phylogenetic relationships among the true porpoises (Cetacea: Phocoenidae). *Mol. Phylogenet. Evol.* **4**, 463–474.

Sarmiento, L., and Tantaleán, M. (1991). *Stenurus australis* n.sp. (Nematoda: Pseudaliidae) de *Phocoena spinipinnis* (Burmeister, 1865) (Cetacea: Phocoenidae) de Perú. *Pub. Mus. Hist. Nat., Univ. Nac. May. San Marcos, Ser. Zool.* **36**, 1–4.

Simôes-Lopes, P. C., and Ximenez, A. (1989). *Phocoena spinipinnis* Burmeister 1865, na costa sul do Brasil (Cetacea-Phocoenidae). *Biotemas* **2**, 83–89.

Torres, P., Oporto, J. A., Brieva, L., and Escare, L. (1992). Gastrointestinal helminths of the cetaceans *Phocoena spinipinnis* (Burmeister, 1865) and *Cephalorhynchus eutropia* (Gray, 1846) from the southern coast of Chile. *J. Wildl. Dis.* **28**, 313–315.

Van Waerebeek, K., and Reyes, J. C. (1990). Catch of small cetaceans at Pucusana port, central Peru, during 1987. *Biol. Cons.* **51**, 15–22.

Van Waerebeek, K., Santillán, L., and Reyes, J. C. (2002). An unusually large aggregation of Burmeister's porpoise *Phocoena spinipinnis* off Peru, with a review of sightings from the eastern South Pacific. *Not. Men. Mus. Nac. Hist. Nat., Chile* **350**, 12–17.

Würsig, M., Würsig, B., and Mermoz, J. F. (1977). Desplazamientos, comportamiento general y un varamiento de la marsopa espinosa, *Phocoena spinipinnis*, en el Golfo de San José (Chubut, Argentina). *Physis* **36**, 71–79.

Yáñez, J., Canto, J., and Reyes, J. C. (1994). Cráneo fósil de *Phocoena spinipinnis* (Cetacea: Phocoenidae) hallado en Chile. *Not. Men. Mus. Nac. Hist. Nat., Chile* **324**, 25–29.

# Bycatch

## Simon Northridge

Marine mammals sometimes get caught up and killed in fishing operations. In many cases these deaths are entirely unintended by the fishermen concerned and are incidental to the main fishing operation. They are therefore often referred to as incidental catches. More often they are referred to as "bycatches," although this term is also used to described the capture of some species that, while not the main target of a fishery, still have some value and may therefore be landed. Incidental catches are generally unwanted and discarded. The term bycatch is now commonly used to describe any sort of unintended capture.

Bycatches of marine mammals have probably occurred for as long as people have been putting nets and lines into the water. Most species of marine mammal that occur in places that are heavily fished have been recorded caught in at least one type of fishing gear. Most types of fishing gear have been reported to ensnare marine mammals at one time or another. Some captures seem to defy reason. Large whales, for example, may become caught in a single lobster pot line, and porpoises can get caught in simple fish traps that they are able to find their way into, but not out of. One estimate of global bycatch levels suggests that over 300,000 marine mammals per year are killed in fishing operations globally (Read *et al.*, 2006).

In the past, and indeed in many parts of the world today, bycatch of marine mammals might be treated as a useful bonus and landed for consumption. During the latter half of the twentieth century, however, fishing technology has changed faster and more completely than ever before, which has led to a reappraisal of the issues surrounding bycatch and incidental catch. Nets have become larger and stronger, numerous new fishing techniques have been devised, and fishing intensity throughout the world has increased dramatically, nearly trebling marine fishery landings over a period of just 40 years. Such developments have had unintended negative impacts on non-target species, including marine mammals, so that bycatches have now become a critical issue for some marine mammal populations.

Marine mammals generally reproduce slowly, and their populations are not able to withstand much additional non-natural mortality. The removal of just 1% of the population per year may be more than a marine mammal population can sustain in the longer term. Bycatch is recognized as one of the most important sources of anthropogenic mortality among many species of marine mammals (Reeves *et al.*, 2003; Lewison *et al.*, 2004). For this reason, many nations now legislate to protect marine mammal populations from deliberate or accidental exploitation, and there are several international agreements with the same aim.

Legislation to protect marine mammals from excessive mortality has resulted from a variety of case studies that have uncovered unsustainable levels of incidental capture. Several of these cases have become widely publicized and have generated considerable public attention and debate.

## I. Examples

### A. Eastern Tropical Pacific Tuna Purse Seine Fishery

The first interaction to be recognized as a serious concern for the conservation of marine mammals was the large-scale capture of pelagic delphinids, mainly *Stenella* and *Delphinus* species in the United States tuna purse seine fishery of the eastern tropical Pacific Ocean (ETP). Tuna boat skippers learned in the 1950s that they could catch large tuna by herding dolphin schools with speedboats and then surrounding them with long, deep, purse seine nets. Fishermen were exploiting the curious fact that in the ETP and some other places, large yellowfin tuna *Thunnus albacares* will school under and follow dolphin schools. Once the dolphins and associated tuna are surrounded, the nets can be "pursed," whereby the bottom end of the net is closed off, thereby catching the tuna. At this point the dolphins can still surface to breathe within the encircled net and could escape by jumping over the floats. Pelagic delphinids, however, seem to find it difficult to escape from such an enclosed situation, and many became trapped and died under folds of the surrounding purse seine.

This fishing technique was begun in the 1950s, but was not recognized as a potential conservation problem until the early 1970s, when a monitoring program was established. During much of the 1960s and up to 1972, annual mortalities are thought to have ranged between 200,000 and 500,000. Thereafter a variety of efforts were made to reduce the kill, but tens of thousands of dolphins were still being killed annually throughout most of the 1980s. Pantropical spotted dolphins, *Stenella attenuata*, were the most frequently killed species, and numbers of this species in the ETP were more than halved over the 1960s and 1970s. Populations of other species were also severely impacted.

Largely as a result of public pressure, and the introduction of "dolphin safe" tuna retailing, this practice has now been greatly reduced. New techniques have been devised by the skippers to ensure that a very high proportion of the dolphins used in this way to catch tuna are encouraged to escape from the nets before the fish are removed. Under a training and monitoring scheme run by the Inter-American Tropical Tuna Commission, dolphin mortality had been reduced to less than 1500 animals per year by 2004 (Anon, 2004). Efforts continue to reduce these figures further still. However, despite the great reduction in the kills, the populations have not shown strong signs of recovery; effects of continued large-scale chase and capture may contribute to this failure to rebound (Gerrodette and Forcada, 2005; Wade *et al.*, 2007).

**B**

Throughout the world, since the discovery of the effect of the ETP tuna fishery on dolphin populations, it has become clear that there are numerous other fisheries in which marine mammals are being killed in large numbers. In some cases, populations or species have been threatened with extinction. Two of the most severe cases are those concerning the baiji *Lipotes vexillifer*, which has recently been declared extinct, and the vaquita *Phocoena sinus*.

### B. The Baiji

The baiji (*L. vexillifer*), otherwise known as the Chinese river dolphin, used to inhabit the middle and lower parts of the Yangtze River system in China. The total population size was thought to have numbered a few hundred in the 1980s, and numerous publications warned of its imminent demise throughout the 1990s and into the present millennium. The major source of mortality for this species was snagging in "rolling hook" fishing lines. These are lines equipped with many closely set, sharp, unbaited hooks designed to snag fish foraging on the river bed in the same areas as the Baiji. In one study, 45% of all known Baiji deaths were attributed to snagging in rolling hooks. A recent intensive survey found no remaining baiji and concluded that the species was extinct, probably due to unsustainable bycatch in local fisheries (Turvey *et al.*, 2007).

### C. The Vaquita

The vaquita (*P. sinus*) is a species of porpoise restricted to the upper part of the Gulf of California in Mexico. Population studies suggest that less than 600 animals remained by 1997, that numbers are declining, and that the species is in critical danger of extinction. Again, the major source of mortality is incidental catches in fishing operations, in this case gill nets for fish and shrimps. Gill nets are simple long panels of netting that are set to stand vertically in the water with floats along their top and a weighted rope on their bottom. Depending on the amount of weight added, they either sit on the seabed floating upward or they float at the surface hanging down. They are left to ensnare fish that happen to swim into them, but also catch marine mammals by entangling them. Annual vaquita mortality in gill net fisheries is estimated at around 40–80 per year, which is clearly an unsustainable level of mortality given the size of the population. Progress towards reducing entanglement has been slow in spite of efforts to phase out gill nets in the vaquita's core range, and the development of schemes involving compensation for fishermen (Rojas-Bracho *et al.*, 2006).

## II. Causes for Concern

Although these examples are perhaps the most extreme cases, there are numerous others around the world where significant numbers of marine mammals are killed incidentally in fishing operations. It is usually the smaller species and those that occur in continental shelf waters where most fishing occurs that are impacted most heavily.

Incidental catches do not always impact on entire species. In many instances, marine mammal species may be widespread and in little danger of overall extinction. Nevertheless, incidental catches may be frequent enough to reduce or eliminate a local population (Fig. 1). This is the case for the harbor porpoise, *Phocoena phocoena*. Although they are in no imminent danger of extinction as a species, in several areas including the Gulf of Maine off the US northeast coast, incidental catch rates are or have been high enough to push local populations into decline. In other parts of its range, including

**Figure 1** *A harbor porpoise entangled in a cod gillnet in the North Sea, one of many hundreds dying this way every year in European gillnet fisheries. Photo by Nigel Godden/Sea Mammal Research Unit.*

the Baltic Sea, harbor porpoises have all but disappeared. Although the cause of this disappearance is not known for sure, the use of gillnets in the Baltic has been intense in recent decades, and it is widely believed that bycatch has played a significant role.

Throughout the world, small inshore species such as the harbor porpoise are known to be victims of bycatch in fishing operations, but the level of such bycatches and the likely impacts remain unknown. Monitoring bycatch rates and estimating population sizes are both very expensive. A significant issue in this regard is that there does not need to be a very large number of bycatch kills for the total effect to be significant. When a marine mammal population numbers in the hundreds or even the tens of thousands, a few individuals to a few hundred individuals taken per year may be enough to generate a population decline. Furthermore, even when the marine mammal population is much larger, if the fishery is also large, significant bycatches can occur while still remaining unknown. Generally, bycatches are rare events. Typically in European and North American coastal gillnet fisheries, a capture event only occurs in one or two out of every hundred fishing operations. Such low levels may remain unnoticed, although the aggregate effect over a large number of vessels and operations may be significant. Such low levels of capture also make monitoring more difficult.

Although most attention worldwide has focused on the potential conservation issues that incidental catches of marine mammals raise, animal welfare considerations are also a concern. Whereas some bycatch of marine mammals in fishing operations is an inevitable consequence of fishing, in some nations any large-scale fatalities of marine mammals are publicly unacceptable regardless of whether they are sustainable at a population level.

## III. Attempts to Resolve the Problem

Most of the numerically significant bycatches of marine mammals tend to be in static fishing gear, mainly gill nets. Despite the attention focused on this subject in recent years, it is still not known how or why marine mammals actually become caught in such nets. It is not known, for example, whether mammals are attracted to nets by the curiosity or by the presence of trapped fish, whether they do not notice the netting, or whether they simply do not understand the potential consequence of swimming into it. Despite our ignorance, some progress has been made toward resolving the problem.

One potential solution to the problem of marine mammal capture in gill nets has been developed in North America. Pingers, or acoustic beacons, exploit the sensitive hearing of marine mammals by emitting an intermittent, short, high-pitched noise that most fish cannot hear but that appears to repel or warn off marine mammals. Attached at regular intervals along the length of a gill net, these pocket-sized devices have been shown to reduce the numbers of bycaught marine mammals, mainly harbor porpoises, but also dolphins and sea lions, by up to 90%. Pingers were first developed in Canada, and their use is now mandatory in several US and EU fisheries.

Pingers certainly appear to be useful, but there are still some concerns about their use. If, as seems to be the case, pingers displace animals from an area adjacent to the fishing gear, and if they are used to the very large numbers that would seem to be necessary in some areas, then it is also possible that marine mammals may become excluded from parts of their foraging habitats. Pingers rely on having their batteries replaced or on being recharged, and they rely on people performing this task to ensure that they continue to work. This can be an expensive and time consuming operation that many people might eventually prefer to forget about. Finally, the pingers themselves are expensive, so that the cost of equipping a net with pingers may exceed the cost of the net. In many less-developed countries, it is unlikely that they will ever become widely used for this reason alone.

Issues with mobile fishing gear are somewhat different. There are or have been several initiatives worldwide that aim to keep marine mammals out of towed fishing gear. In the ETP tuna fishery referred to earlier, special techniques, and nets have been developed to help dolphins to escape from the purse seine net once the net has fully encircled the school of tuna. During the "backdown procedure" the skipper reverses the vessel and is able to sink a part of the net float line under the water, enabling the dolphins to escape. This part of the net is also made up with a smaller meshed panel, the Medina panel, reducing the chances of dolphins becoming entangled as they escape. Similarly, in New Zealand, special marine mammal escape devices have been designed and used in squid trawls. A large grid is placed near the rear of the net, set at a 45° angle to the vertical plane. Fish can pass through the grid, but larger animals such as sea lions are forced upward and out the net through an escape hatch. Similar devices have been designed and tested for dolphins in the United Kingdom and France.

In general, the incidental capture of marine mammals is caused by a combination of fishing technique or gear design and the behavior of the marine mammal. Resolving problematic interactions therefore involves some combination of change to fishing gear use or design and the manipulation of marine mammal behavior. Very little is known about the behavior of marine mammals in relation to fishing gear, especially in the context of incidental capture. In part this is because of the difficulties of studying marine mammals underwater and the rarity of such events in most cases, which makes observing their occurrence very difficult. Finding solutions to the problem is therefore a slow and arduous process.

Most fishing practices and gear designs have been adopted by fishing communities because they are effective in catching fish, and making changes may therefore reduce the profitability of a fishery. Effective mitigation measures therefore need to be devised in collaboration with the fishing community to minimize the adverse impacts on fish catches and ensure that bycatch rates once lowered remain low, but they may also require a legislative approach to ensure compliance. In this respect, managing the incidental capture of marine mammals may be seen as part of a much more wide-ranging and ongoing problem of managing a global industry that, in the last 50 years, has outgrown its resource base.

## See Also the Following Articles

Entrapment and Entanglement Fisheries ■ Interference with Tuna–Dolphin Issue ■ Baiji ■ Vaquita

## References

Anon, (2004). "Annual Report of the Inter-American Tropical Tuna Commission." IATTC, La Jolla, CA.

Gerrodette, T., and Forcada, J. (2005). Non-recovery of two spotted and spinner dolphin populations in the eastern tropical Pacific. *Mar. Ecol. Prog. Ser.* **291**, 1–21.

Gosliner, M. L. (1999). The tuna-dolphin controversy. *In* "Conservation and Management of Marine Mammals" (J. Twiss, and R. Reeves, eds), pp. 120–155. Smithsonian Institution Press, Washington, DC.

Kraus, S. D., Read, A. J., Solow, S., Baldwin, K., Spradlin, T., Anderson, E., and Williamson, J. (1997). Acoustic alarms reduce porpoise mortality. *Nature* **388**, 525.

Lewison, R. L., Crowder, L. B., Read, A. J., and Freeman, S. A. (2004). Understanding impacts of fisheries bycatch on marine megafauna. *Trend. Ecol. Evol.* **19**, 598–604.

Northridge, S. P., and Hofman, R. J. (1999). Marine mammal interactions with fisheries. *In* "Conservation and Management of Marine Mammals" (J. Twiss, and R. Reeves, eds), pp. 99–119. Smithsonian Institution Press, Washington, DC.

Perrin, W. F. (1999). Selected examples of small cetaceans at risk. *In* "Conservation and Management of Marine Mammals" (J. Twiss, and R. Reeves, eds), pp. 296–310. Smithsonian Institution Press, Washington, DC.

Perrin, W. F., Donovan, G. P., and Barlow, J. (1995). Gillnets and Cetaceans. *Rep. Int. Whal. Commn* **15**(Spec. Iss.).

Read, A. J., Drinker, P., and Northridge, S. (2006). Bycatch of Marine Mammals in U.S. and Global Fisheries. *Conserv. Biol.* **20**, 163–169.

Reeves, R. R., Smith, B. D., Crespo, E. A., and Notarbartolo di Sciara, G. (2003). "Dolphins, Whales and Porpoises. 2002–2010 Conservation Action Plan for the World's Cetaceans." IUCN, Gland, Switzerland and Cambridge, UK.

Rojas-Bracho, L., Reeves, R. R., and Jaramillo-Legorreta, A. (2006). Conservation of the vaquita *Phocoena sinus*. *Mam. Rev.* **36**, 179–216.

Turvey, S. L., *et al.* (2007). First human caused extinction of a cetacean species? *Biol. Lett.* **3**, 537–540.

Wade, P. R., Watters, G. M., Gerrodette, T., and Reilly, S. B. (2007). Depletion of spotted and spinner dolphins in the eastern tropical Pacific: Modeling hypotheses for their lack of recovery. *Mar. Ecol. Prog. Ser.* **343**, 1–14.

Wickens, P. A. (1995). A review of operational interactions between pinnipeds and fisheries. *FAO Fish. Tech. Pap.* **346**.

B

# California, Galapagos, and Japanese Sea Lions
*Zalophus californianus,*
*Z. wollebaeki,* and *Z. japonicus*

CAROLYN B. HEATH AND WILLIAM F. PERRIN

## I. Characteristics and Taxonomy

The California, Galapagos, and Japanese sea lions are closely related species that together comprise the genus *Zalophus*. They occupy (or occupied, in the case of the extinct Japanese sea lion) widely separated regions of the Pacific, including the temperate western and eastern North Pacific and the tropical Galapagos Archipelago (Fig. 1).

The three sea lions are now regarded as separate species: *Zalophus californianus* (Lesson, 1828), *Z. wollebaeki* (Sivertsen, 1953), and *Z. japonicus* (Peters, 1866), respectively. Previously they were typically considered to be geographically isolated subspecies (*Z. californianus californianus, Z. c. wollebaeki,* and *Z. c. japonicas*), but recent discoveries of substantial morphological and behavioral differences among them led to their reclassification. Most recently (Sakahira and Nimi, 2007) analysis of ancient DNA extracted from skeletal remains of the Japanese sea lion has determined that it is most closely related to the California sea lion, with a 7.02% divergence in nucleotides, a level suggesting that the two forms diverged 2.2 million years ago in the late Pliocene.

California sea lions are highly sexually dimorphic: the weight and length for adult males is about 350 kg and 2.4 m compared to about 100 kg and 1.8 m for adult females. Males in the Gulf of California appear to be smaller than their Pacific counterparts. Newborn California sea lion pups weigh 6–9 kg for females and males, respectively. They are dark brown to black until they molt to a tawny brown color at 4–6 months (Fig. 2). Females remain this color throughout adulthood, whereas male coats typically darken as they age. Adult males usually are a dark brown, but can range from light brown to black. All individuals appear darker when wet. It is difficult to distinguish females from young males until the latter begin developing the chest, dark color, and sagittal crest of adult males (Fig. 3). The sagittal crest, which is unique to *Zalophus*, is usually topped by white fur and is quite conspicuous. Galapagos sea lions are smaller than California sea lions and appear to be much less sexually dimorphic. Females weigh

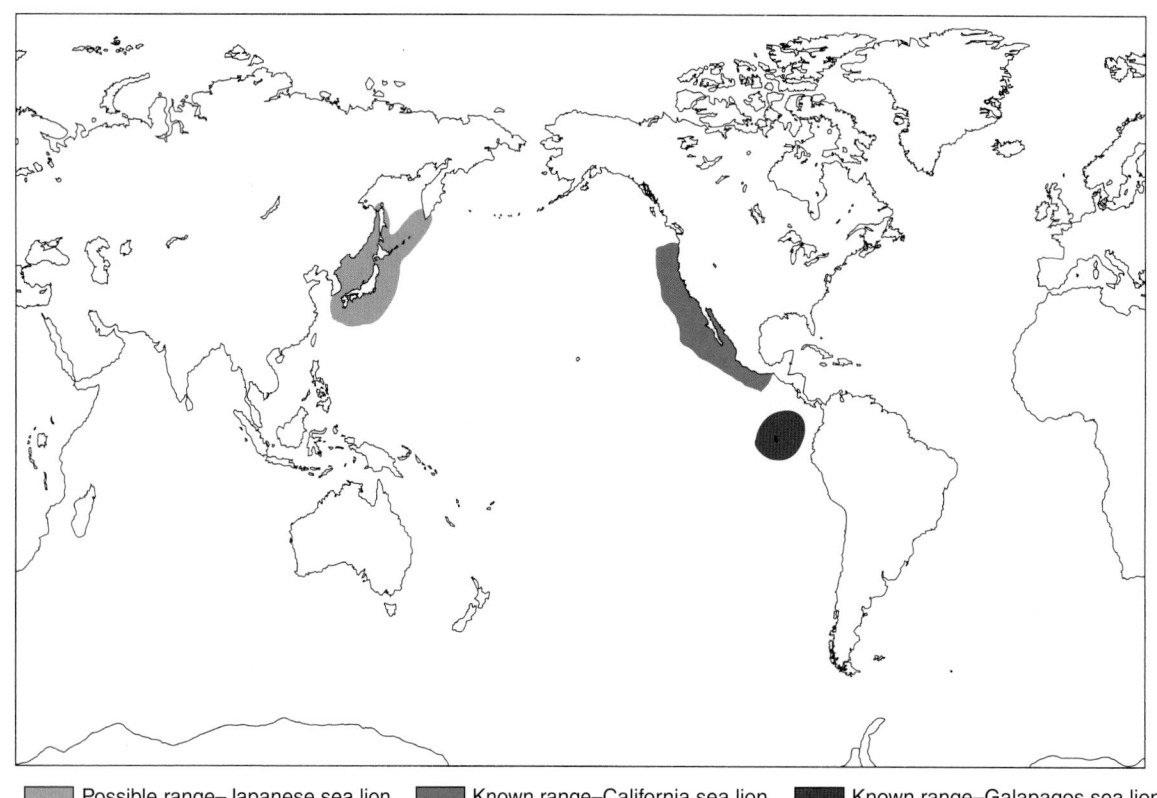

Possible range–Japanese sea lion    Known range–California sea lion    Known range–Galapagos sea lion

**Figure 1**    *Ranges of California, Galapagos, and Japanese sea lions. The Japanese sea lion is extinct.*

**Figure 2** *An adult female California sea lion suckling her pup after returning from a feeding trip at sea (Photo by C. B. Heath).*

**Figure 3** *An adult male California sea lion on his territory, with pups playing in protected shallows. The broad neck and chest are indicative of an adult male; the fat deposits they contain will sustain him while fasting for the duration of his tenure. Galapagos and Japanese sea lions are similar in appearance to California sea lions. Photo by C. B. Heath.*

about 80 kg. No weights are available for males. Young California and Galapagos sea lions are especially playful, spending much time chasing and mock fighting with each other, playing with objects such as kelp and feathers, and bodysurfing. All age classes are fairly vocal; territorial males are exceedingly so. Little is known about the physical characteristics of the Japanese sea lion.

## II. Distribution and Abundance

The main breeding areas of California sea lions include the Channel Islands in southern California and Mexican islands off the Pacific coast of Baja and in the Gulf of California. Rarely, births may occur as far north as the Farallon Islands off central California. Outside the breeding season, animals (mostly males) are common as far north as Vancouver Island. Vagrants occur as far north as Prince William Sound, Alaska, and as far south as Chiapas, Mexico. Females and immatures may disperse from the breeding islands to forage but apparently do not migrate as extensively as males.

Differences in mitochondrial DNA indicate that the Pacific and Gulf of California populations of California sea lions have long been genetically isolated. MHC genetic differences are consistent with either ecologically distinct patterns of selection pressures and/or geographical isolation (Bowen *et al.*, 2006). However, no differences have been found in the cranial morphology of the two populations, and subadult males appear to migrate between colonies in the southern Gulf of California and those along the Pacific Coast of Mexico. The degree of interchange between Mexican and US populations is not known.

Galapagos sea lions breed on all the islands of the archipelago. In 1986, a small rookery was established outside of the Galapagos on Isla de La Plata off the coast of Ecuador. Vagrants have been reported along the mainland coast of Ecuador and on Islas del Coco (Costa Rica) and Gorgona (Columbia).

Although historical and archeological records are incomplete, Japanese sea lions appear to have lived in coastal areas from Kyushu to southern Kamchatka. Their range was likely centered along the west and east coasts of Honshu, off Shikoku and Kyushu, in the Seto Inland Sea, and on islands in the Sea of Japan and the Izu region. Known rookeries include Takeshima and Ullung-do in the southern Sea of Japan, the northwest and also central-eastern coasts of Honshu, and four islands in the Izu region. Vagrants have been noted to southwestern Sakhalin, the Kuril Islands, southern Kamchatka, and the east coast of South Korea.

The US population of California sea lions is currently estimated to be at least 141,842 (Carretta *et al.*, 2007); an additional unknown number were at sea or hauled out at unsampled locations during the census. Annual pup production is currently 40,000–50,000. Maximum population growth rate is estimated at 6.52%; however, massive pup mortality during El Niño years greatly reduces growth. The most recent counts indicates continued growth of the population. A 2000 census of the Mexican Pacific coast resulted in an estimate of about 75,000–87,000 California sea lions (Lowry and Maravilla-Chavez, 2005). Annual growth estimated by two methods was 0.4% and 3.2%. A 1992 census of the Gulf of California population recorded about 23,000 animals; correction factors yielded a population estimate of about 31,000 (Aurioles and Zavala, 1994). Thus, the total California sea lion population is about 248,000.

The Galapagos sea lion has been largely spared the commercial exploitation that depleted the California sea lion population. However, the number of Galapagos sea lions apparently does fluctuate frequently due to El Niño events and epidemics of seal pox. The population was estimated at 20,000–50,000 in 1963. Following the 1997–1998 El Niño, which was accompanied by up to 90% pup mortality and 45% overall mortality, the population was estimated at only about 14,000 individuals (Salazar, 1999).

Between 30,000 and 50,000 Japanese sea lions may have existed in the mid-1800s. Heavy, unregulated hunting depleted the population such that by the 1950s 50 to 60 animals on Takeshima were the only ones reported. The IUCN lists the species as extinct; some scientists believe this status needs to be confirmed by surveys in remote regions.

**C**

## III. Ecology

California and Galapagos sea lions breed on sandy beaches and rocky areas on remote islands. Because females must forage at sea during lactation, breeding areas are restricted to regions of high marine productivity. California sea lion rookeries along the Pacific coast are in a very productive upwelling zone, and productivity in the Gulf of California is also very high due to tide- and, wind-generated upwelling. The waters of the continental margin adjacent to areas once used by breeding Japanese sea lions are quite productive. Although low productivity generally excludes otariids from breeding in the tropics, the islands in the Galapagos Archipelago are bathed in nutrient-rich upwelling currents. This creates an isolated pocket of high productivity, which supports the Galapagos sea lions. The importance of high productivity can be seen in the devastation that occurs during El Niño events, when a plume of warm, nutrient-poor water emanating from the equatorial Pacific decreases the availability of the sea lions' prey. These frequent, but unpredictable events are most severe in the eastern tropical Pacific where the Galapagos sea lion lives, with lesser impacts at the higher latitudes occupied by California and Japanese sea lions. Sea lions living in the Gulf of California, Mexico, may be largely protected from the effects of El Niños; strong tidal mixing there should be able to resupply nutrients to surface waters during an El Niño event.

The breeding habitat occupied by *Zalophus* ranges from temperate to tropical regions. As a result, breeding animals are often subjected to high temperatures while on land. The effects of these high temperatures and of El Niño events are described in detail below and in section IV.

California sea lions eat a wide variety of prey, which is determined to some degree by its relative availability. The most common prey items in southern California are market squid (*Loligo opalescens*), northern anchovy (*Engraulis mordax*), Pacific whiting (*Merluccius productus*), rockfish (*Sebastes* spp.), jack mackerel (*Trachurus symmetricus*), Pacific mackerel (*Scomber japonicus*), and blacksmith (*Chromis punctipinnis*) (Lowry *et al.*, 1986; Lowry and Carretta, 1999). Anchovy, whiting and rockfish are also important in the Mexican Pacific and Gulf populations, as are midshipmen (*Porichthys* spp.). Myctophids, sardines, cutlassfish, alopus, tusk eels, anchoveta, grunts, squids, and bass are frequently prey in various areas of the Gulf (Sanchez, 1992; García-Rodríguez and Aurioles-Gamboa, 2003; Mellink and Romera-Saavedra, 2005). Diet in the Gulf varies greatly among years, seasons, locations; and probably individuals, including with variation in the availability of the Pacific sardine (*Sardinops sagax*). El Niño events cause shifts in the DIET, and species otherwise rarely consumed, such as the pelagic red crab, may become more common in the diet. Feeding can occur at any hour of the day. Dives typically last for about 2 min, but can be as long as 10 min. Dive depth averages 26–98 m but can be well over 200 m (Feldkamp *et al.*, 1991). Sea lions in California may follow dolphins to take advantage of their better food-finding abilities (Bearzi, 2006).

The staple of the Galapagos sea lion diet is sardines. During El Niño events, however, partial shifts to green eyes (1982–1983) and myctophids (1997–1998) have occurred (Trillmich and Dellinger, 1991; Salazar, 1999). Galapagos sea lions forage within a few kilometers of the coast, feeding during the daytime on a near-daily basis. Dive depth averages 37 m but can reach 186 m. There is no information on what the Japanese sea lions ate.

The environmental changes that occur during El Niño events also elicit behavioral responses by *Zalophus*. However, unlike the fairly constant heat stress experienced on rookeries, the environmental

stresses associated with El Niño are unpredictable and only occur every few years. In addition, because the degree of stress varies among events and locations, the sea lions' responses must be somewhat flexible. El Niños cause a reduction in prey availability for *Zalophus* throughout much of its range (Keiper *et al.*, 2005). The potential consequences of this reduction are mitigated somewhat by adaptations that have evolved over the sea lions' long history of coexistence with El Niños. However, the severe impacts of some El Niño events demonstrate the limits of these adaptations. The 1982–1983 El Niño was a particularly strong one, and much is known about its effects on the California sea lions that breed in southern California. Some nonbreeding sea lions in this region responded to local prey depletion by migrating north to more productive areas. Many immatures and some adult females left their normal winter foraging areas and migrated to central California (Huber, 1991). Emigration was thus apparently an option for some individuals to reduce the effects of El Niño. Territorial males in southern California showed no measurable effects from this event, most likely due to their preseason foraging farther north. Adult females, however, appeared to be more tied to the general vicinity of the BREEDING SITES, where prey reduction was more pronounced. The increase in spontaneous abortions during the 1982–1983 winter indicates that some of these females were unable to find adequate prey (Francis and Heath, 1991). Females that did manage to produce full-term pups then faced the greater challenge of nourishing them. Feeding during lactation makes females quite vulnerable to localized decreases in food availability. In southern California they attempted to compensate for decreased prey by increasing their foraging effort while at sea, partially shifting their prey, and by slightly prolonging their feeding trips (Lowry *et al.*, 1986; Ono *et al.*, 1987). These efforts, however, were inadequate to compensate for the strength of the 1982–1983 El Niño. Females apparently made less milk: pups suckled less, grew more slowly, and weighed less at age 2 months (Ono *et al.*, 1987). Pup mortality increased, and pup production decreased by 30–71% at various islands. Fewer of the male pups were weaned by age 1 year, and more of them stayed on their birth island and suckled into their second year. Fewer females mated during the El Niño summer, presumably a sign that they were undernourished. As a result, pup production was still low in the following year. Because pup production took several years to return to pre-El Niño levels, it is possible that there may have been some mortality of breeding females and juveniles associated with this event. In Mexico, pup production on at least one Pacific island decreased by 50% during the 1982–1983 El Niño, while effects appeared to be very weak in the Gulf (Aurioles and Le Boeuf, 1991). An even stronger El Niño occurred in 1997–1998. While not as widely monitored for its effects on California sea lions, they appear to have been even greater.

The reduction of prey during El Niño events is particularly strong in the eastern tropical Pacific. Because Galapagos sea lions are isolated from alternative feeding areas by vast expanses of unproductive tropical waters, emigration to better feeding areas is not an option for them. Mortality has thus been very high for this species during El Niño events. Between 80% and 95% of the pups born in 1982 did not survive their first year of life. Pup production at various rookeries in 1983 was between 3% and 65% of normal years. Adult female mortality was estimated at 20%, and territorial male mortality was particularly severe (Trillmich and Dellinger, 1991). During the 1997–1998 El Niño, pup mortality was close to 90%, and mortality for the overall population was about 45% (Salazar, 1999).

Oceanic conditions also change in Japan during El Niño events, but what effects this may have had on the Japanese sea lions is not known. The oceanographic counterpart to El Niños are Las Niñas, periods of generally cooler ocean temperatures and greater productivity. Little is known of their effects on pinnipeds or any role they might play in the recovery from El Niño events.

The toxin domoic acid produced by blooms of the diatom *Pseudonitzschia australis* has emerged as a cause of massive episodic mortality of California sea lions (Lefebvre *et al.*, 2000; Scholin *et al.*, 2000; Gulland *et al.*, 2002). The majority of rescued affected animals die in captivity or strand again when "cured" and released (Gulland *et al.*, 2002). Domoic acid has also been shown to cause reproductive failure (Brodie *et al.*, 2006).

The growing arsenal of toxic chemicals and waste that makes its way into marine mammals' habitats and prey has generated much concern. This is particularly relevant for the Channel Islands population of California sea lions; their proximity to the major metropolitan areas of southern California exposes them to a great deal of urban and industrial runoff, waste, and debris (Connolly and Glaser, 2002; Kannan *et al.*, 2004; Stapleton *et al.*, 2006). Levels of contaminants are lower in Gulf of California sea lions (Del Toro *et al.*, 2006). Because they are high-level predators, sea lions are vulnerable to compounds such as organochlorines (e.g., DDT and PCBs) that become increasingly concentrated as they move up the food chain. Laboratory studies of such compounds have revealed that they can suppress pinniped immune systems, rendering them more vulnerable to disease. However, establishing such clear cause-and-effect relationships in wild populations exposed to organochlorines is more difficult due to confounding factors. California sea lions were found to have elevated levels of organochlorines associated with increased stillbirths and premature pupping, but the level of contribution of disease to this problem could not be determined. One link that has been shown is between levels of PCBs and mortality from carcinoma (Ylitalo *et al.*, 2005). Although the specific links among chemicals, immune system responses, and disease or mortality are incompletely understood, enough indications of problems exist to warrant caution and further research.

## IV. Behavior and Physiology

In southern California, where California sea lions have been studied extensively, animals can be found on the breeding islands year-round. The number ashore increases rapidly in May with the onset of the breeding season. At this time adult males begin fighting for territories along the shorelines of the rookeries. Most are unsuccessful and retreat to sea or to "bachelor" beaches nearby. Those that establish territories maintain their boundaries with ritualized displays and frequent barking. Territorial males fast throughout their tenure, surviving on fat accumulated during the off-season. Tenure lasts from 1 to 45 days and may end when residents are displaced by another male or when their fat reserves are depleted. Some males maintain territories for multiple breeding seasons. Throughout May and June females give birth to a single pup a few days after coming ashore. Vocal and olfactory imprinting follows birth and is used by mothers and pups to reunite after separations (Fig 4). Mothers spend the first week postpartum with their pup and then begin alternating feeding trips at sea with suckling bouts on land. The feeding trip length is largely determined by the distance to the foraging grounds and the availability of prey. It averages 2–3 days, but varies with location and year. Stays on land average 1–2 days. This pattern continues until the pup is weaned. Females mate about 27 days after giving birth, an unusually long interval for otariids. Not all females breed every year. At many rookeries, females form "milling" groups as a prelude to mating (Fig 5). In these groups of 2–20 females, the females often mount each other and the territorial male. Eventually 1–2 of the females mate and the milling activity ends (Heath, 1989).

**Figure 4** *An adult female California sea lion vocalizes to her newborn pup. Mothers and pups imprint on each others' calls and smell at birth, which helps them recognize each other and reunite after separations. Note remnants of the amniotic sac on the pup.*

**Figure 5** *A "milling" group of California sea lions consiting of several females and one male (large individual on right). These groups form in some areas as a prelude to mating. Females often mount each other or the territorial male while in these groups. Note the male's sagittal crest, which is unique to Zalophus.*

While ashore, both males and females make regular movements to the water to cool off (Peterson and Bartholomew, 1967).

After the breeding season ends in August, most adult and subadult males leave the southern California rookeries and migrate north, where they feed throughout the fall and winter. Females and juveniles appear to disperse to feed in the general vicinity of the breeding islands.

C

California sea lions living in the Gulf of California, Mexico, experience a similar annual cycle with some notable differences. The pupping season lasts 1–2 weeks longer for at least some rookeries in the northern half of the Gulf (Morales, 1990), and the interval between birth and mating appears more variable than in the US population (Heath, 1989). About 40% of the adult males in this region remain around the breeding islands year-round, and the age of weaning appears to be both older and more variable than in southern California (Heath *et al.*, 1996). The breeding season is very protracted in the Galapagos sea lion, and territorial males are observed most of the year. Births occur from June to March, but the peak pupping period varies among rookeries and years. The interval between birth and mating has been estimated at about 3 weeks. Females with new pups interfere aggressively in sibling conflicts, defending the younger pup (Trillmich and Wolf, 2008). Territorial males sometimes go to sea to feed, thus often losing their territory to another male, but sometimes reclaiming it upon their return. Migration within the archipelago is minimal (Wolf and Trillmich, 2007).

California sea lions are highly intelligent and adaptable animals, capable of learning a simple sign language in captivity (Schusterman and Krieger, 1986). Both California and Galapagos sea lions can recognize individuals in the wild through scent, sound, and probably sight. They are also gregarious animals, with much opportunity for social interaction during the breeding season, at least. These traits would seem to dispose them to sustained relationship with other individuals; however, the only obvious social bond is between mothers and their offspring. Perhaps longer term studies of permanently marked individuals will reveal other types of relationships.

Life is a compromise, and pinnipeds have evolved many adaptations in response to the sometimes conflicting pressures of breeding on land and feeding at sea. Certain environmental conditions can increase the costs or benefits of some of these adaptations and bring about compensatory changes in behavior. Two such conditions—high air temperatures on ROOKERIES and decreased prey availability during El Niño events—play particularly important roles in shaping the breeding and foraging behavior of California and Galapagos sea lions. For example, male California sea lions engaged in foraging trips extending over more than twice the distance and lasting 3 times as long during anomalous oceanographic conditions in 2005 as compared to 2004 (Weise *et al.*, 2006).

California sea lion females are very particular about male behavior. Boisterous, overly attentive, or aggressive males are typically abandoned and left to sit alone on their territories. Any interference with female movements is simply not tolerated. Should a male attempt to block a female's path, she needs only to extend her neck out and up, and sway it side to side as she walks. This long-neck display signals males that she requires free passage; the rare male that does not respond to this will likely be subjected to a display of jerky hopping and flipper slapping, which will dissuade him from interfering further. Males seem to have little option but to acquiesce if they are to be successful at breeding.

Why might this situation exist, especially in such apparent contrast to some otariid species where females may be herded, threatened, and even injured by territorial males? The explanation appears to partly lie in the animals' thermoregulatory needs. The particularly warm climate in which *Zalophus* breed increases the cost of moving between marine and terrestrial habitats. The blubber, fur, and large size that insulate against cold ocean waters can lead to overheating while ashore, thus making necessary regular access to wet substrate or water for cooling. The breeding fat of adult males intensifies their thermal stress, thereby limiting the distribution and number of territories. Successful males have territories containing access to water; others must abandon

their territories (and any hopes of mating) during the heat of the day. Breeding females must also have access to cooler shoreline areas on a near-daily basis, and they regularly travel through several males' territories while moving from resting to cooling areas. Furthering this pattern of female movements is the unusually long interval (about 21–27 days) between birth and MATING. During that interval females must also make regular feeding trips to sea to replenish their milk supplies. A male that prevented these premating thermoregulatory and feeding excursions by herding the females would be left with, at best, only severely stressed females, which are not likely to mate. Thus the combination of warm breeding areas and delayed mating together foster a system in which males may control each other, but not the females. Rather, when it is time to mate, many females leave the male's territory in which they have given birth and mate with another male that they may have encountered in their movements about the rookery. Females show a surprising degree of unity in their selection of mates. As a result, many males holding territories during the breeding season never or rarely mate, while a few males mate with many females (Heath, 1989). This dramatically increases the degree of polygyny in California sea lions: males make the first cut by excluding many of their gender from the rookeries, and females make another cut via their selection of mates.

The influence of temperature on behavior can also be seen in the high percentage of copulations that occur in contact with the water. This percentage increases in hotter regions, as does the amount of time females spend cooling off at the shoreline. In the Gulf of California, where air temperatures are very high, nearly all territorial defense and breeding activities are restricted to the water (Heath, 1989; Garcia, 1992).

Galapagos sea lions show a similar response to high air temperatures with their great reliance on shoreline areas for cooling. Females with pups segregate in areas offering the most options for cooling (tidepools, rock surfaces close to the water, shade), whereas males congregate in areas where only shade is available (Wolf *et al.*, 2005).

Environmental conditions in at least part of the former range of the Japanese sea lion are similar to those of its congeners. However, barring the discovery of historical records, we will never know if their behavioral responses were the same as those of the California and Galapagos sea lions.

Standard metabolic rate at the surface in captive California sea lions has been estimated at 1.9 to 3 times that for terrestrial mammals of similar size and decreased by about half during submersion (Hurley and Costa, 2001). Full aerobic diving capacity is not reached until a weight of about 125 kg at 4–6 years of age (Weise and Costa, 2007). However, benthic-feeding pinnipeds in nature may often exceed their aerobic dive limit (Costa *et al.*, 2001).

## V. Life History

Most California sea lion pups are weaned by 10–12 months of age. The number that continue to suckle as yearlings or even 2-year-olds varies among years (Francis and Heath, 1991). Pup mortality is roughly 15–20% for the first 6 months of life. In the Gulf of California, normal pup mortality to age 6 months is about 5%. In the southern Gulf of California, at least, it then increases to about 40% for the next 6 months as pups venture into the water (Aurioles and Sinsel, 1988). Sexual maturity occurs at about 4–5 years of age, although males are not large enough to hold breeding territories for several more years. Longevity is estimated at 15–24 years. Sources of mortality include starvation, infection, sharks, killer whales (*Orcinus orca*), toxic phytoplankton blooms, entanglement, shooting, and disease.

As with all other parameters, we have virtually no information regarding the life history of Japanese sea lions.

## VI. Interactions with Humans

The annual return of pinnipeds to predictable breeding areas makes them particularly vulnerable to exploitation. Subsistence hunting of California sea lions probably occurred for several thousand years without much of an effect on the population. However, commercial harvesting during the 1800s and early 1900s in southern California and Mexico reduced the population to only about 1500 animals by the 1920s (Zavala-González and Mellink, 2000). The harvest, which at various times was for hides, blubber, meat, predator control, or the whiskers and BACULA sold as aphrodisiacs, probably focused on adult males. A floundering market coupled with protective legislation allowed the population to start increasing by the 1940s, although some killing and live collecting continued until the 1970s. The 1972 US Marine Mammal Protection Act and similar legislation in Canada and Mexico greatly facilitated the population's recovery. Like all marine mammals, California and Galapagos sea lions spend a good portion of their lives in remote areas or underwater, hidden from our view. However, the California sea lion is one of the most familiar marine mammals. This is due in part to their being the most commonly used "seal" performer in animal park shows and also to their habituation to human presence in some areas, especially where there is a comfortable dock or buoy to be acquired by this tolerance. These activities are indicative of their intelligence and flexible nature, which itself can sometimes lead to less positive interactions with humans.

Is the sea lion an enviable hunter or lowly thief? The answer to this question is largely a matter of perspective. Certainly, California sea lions are highly skilled at catching their prey of fish and squid, and their growing population consumes many tons of them yearly [e.g., depredation on salmon (Weise and Harvey, 2004) and competition with artisanal fisheries in Mexico (Aurioles-Gamboa et al., 2003)]. As with many predators, they are also flexible and opportunistic in their search for food, as their diet and thus foraging patterns vary with age, location, and environmental changes caused by things such as El Niño and commercial fishing. While this flexibility is partly responsible for the recovery of this species, it at times also brings them into direct COMPETITION WITH HUMAN FISHERIES. Healthy populations of fish, sea lions, and humans have coexisted throughout much of our history; however, the demand for marine resources generated by a rapidly increasing human population, coupled with its increasingly efficient exploitation of those resources, has heightened concerns about competition between humans and other marine predators. This, combined with the highly visible actions of individual sea lions that have learned to exploit the easy take from fishing lines and nets, has led some to view California sea lions as marine pests rather than an integral part of a healthy ecosystem.

Close interactions with fisheries causes entanglement in gear leading often to death for the sea lion. This occurs most often in set and drift gillnet fisheries (Carretta et al., 2007). However, it is estimated that a safe level of incidental take in US waters would be about 8500, an order of magnitude greater than the present estimated or likely take. Mitigation measures that show promise to reduce incidental mortality and depredation are the use of acoustic pingers (Barlow and Cameron, 2003) and use of nets only during the day and when sea lions are not around (Maravilla-Chavez et al., 2006).

Another form of competition occurs when sea lions make themselves at home on docks and other man-made resting areas. Although this can be quite inconvenient, in areas such as Pier 39 in San Francisco, the situation has been converted into a popular tourist attraction.

Entanglement with marine debris is a problem found in all populations of California and Galapagos sea lions. Materials such as packing bands and discarded fishing line or nets can become caught on the animals' necks or flippers, leading to injury, infection, reduced feeding efficiency, or death. Sea lions are also killed incidentally in some fisheries.

The southern California population of California sea lions is currently thriving and is thus apparently quite able to recover from its interactions with humans. The extinction of the Japanese sea lion, however, reminds us that there can be limits to this recovery. This is especially true for smaller populations, such as those of the Galapagos sea lions or the Mexican population of California sea lions. If harmful human activities were to increase substantially or happened to coincide with natural stresses such as epidemics or El Niño event, recovery might not be so rapid or complete.

## See Also the Following Articles

Extinctions, Specific ■ Habitat Pressures ■ Pinniped Ecology ■ Territorial Behavior

## References

Aurioles, G. D., and Le Boeuf, B. J. (1991). Effects of the El Niño 1982–83 on California sea lions in Mexico. *In* "Pinnipeds and El Niño: Responses to Environmental Stress" (F. Trillmich, and K. A. Ono, eds), 88, pp. 112–118. Ecological Studies, Springer-Verlag, Berlin.

Aurioles, G. D., and Sinsel, F. (1988). Mortality of California sea lion pups at Los Islotes, Baja California Sur, Mexico. *J. Mammal.* **69**, 180–183.

Arioles, G. D., and Zavala, G. A. (1994). Ecological factors that determine distribution and abundance of the California sea lion *Zalophus californianus* in the Gulf of California. *Cienc. Mar.* **20**(4), 535–553.

Aurioles-Gamboa, D., García-Rodríguez, F., Ramírez-Rodríguez, M., and Hernández-Camacho, C. (2003). Interaction between the California sea lion and the artisanal fishery in La Paz Bay, Gulf of California, Mexico. *Cienc. Mar.* **29**, 357–370.

Barlow, J., and Cameron, G. A. (2003). Field experiments show that acoustic pingers reduce marine mammal bycatch in the California drift gill net fishery. *Mar. Mamm. Sci.* **19**, 265–283.

Bearzi, M. (2006). California sea lions use dolphins to locate food. *J. Mammal.* **87**, 606–617.

Bowen, L., *et al.* (10 authors) (2006). MHC gene configuration variation in geographically disparate populations of California sea lions (*Zalophus californianus*). *Mol. Ecol.* **15**, 529–533.

Brodie, E. C., *et al.* (8 authors) (2006). Domoic acid causes reproductive failure in California sea lions (*Zalophus californianus*). *Mar. Mamm. Sci.* **22**, 700–707.

Carretta, J. V., *et al.* (11 authors) (2007). US Pacific marine mammal stock assessments: 2007. *NOAA Tech. Memo. NMFS* **15**, NOAA-TM-NMFS-SWFSC-414.

Connolly, J. P., and Glaser, D. (2002). *p,p'*-DDE bioaccumulation in female sea lions of the California Channel Islands. *Cont. Shelf Res.* **22**, 1059–1078.

Costa, D. P., Gales, N. J., and Goebel, M. E. (2001). Aerobic dive limit: How often does it occur in nature? *Comp. Biochem. Physiol. A Mol. Integr. Physiol.* **129**, 771–783.

Del Toro, L., Heckel, G., Camacho-Ibar, V. F., and Schramm, Y. (2006). California sea lions (*Zalophus californianus*) have lower chlorinated hydrocarbon contents in northern Baja California, México, than in California, USA. *Environ. Pollut.* **142**, 83–92.

Feldkamp, S. D., DeLong, R. L., and Antonelis, G. A. (1991). Effects of El Niño 1983 on the foraging patterns of California sea lions (*Zalophus californianus*) near San Miguel Island, California.

**C**

*In* "Pinnipeds and El Niño: Responses to Environmental Stress" (F. Trillmich, and K. A. Ono, eds), 88, pp. 146–155. Ecological Studies, Springer-Verlag, Berlin.

Francis, J. M., and Heath, C. B. (1991). Population abundance, pup mortality, and copulation frequency in the California sea lion in relation to the 1983 El Niño on San Nicolas Island. *In* "Pinnipeds and El Niño: Responses to Environmental Stress" (F. Trillmich, and K. A. Ono, eds), 88, pp. 119–128. Ecological Studies, Springer-Verlag, Berlin.

García, R. M. C. (1992). Conducta territorial del lobo marino *Zalophus californianus* en la lobera Los Cantiles, Isla Angel de la Guarda, Golfo de California, México. Tesis de Licenciatura, Facultad de Ciencias, UNAM, Mexico, D.F.

García-Rodríguez, F. J., and Aurioles-Gamboa, D. (2003). Spatial and temporal variation in the diet of the California sea lion (*Zalophus californianus*) in the Gulf of California, México. *Fish. Bull. US* **102**, 47–62.

Gulland, F. M. D., Haulena, M., Fauquier, D., Langlois, G., Lander, M. E., Zabka, T., and Duerr, R. (2002). Domoic acid toxicity in California sea lions (*Zalophus californianus*): Clinical signs, treatment and survival. *Vet. Rec.* **150**, 475–480.

Heath, C. B. (1989). The Behavioral Ecology of the California Sea Lion. Ph.D. dissertation, University of California, Santa Cruz, CA.

Heath, C. B., Adams, M., and Garcia, M. (1996). Geographic variation in the duration of maternal care in the California sea lion. *In* "Abstracts, Symposium on Otariids." Smithsonian Institution, Washington, DC.

Huber, H. R. (1991). Changes in the distribution of California sea lions north of the breeding rookeries during the 1982–83 El Niño. *In* "Pinnipeds and El Niño: Responses to Environmental Stress" (F. Trillmich, and K. A. Ono, eds), 88, pp. 129–137. Ecological Studies, Springer-Verlag, Berlin.

Hurley, J. A., and Costa, D. P. (2001). Standard metabolic rate at the surface and during trained submersions in adult California sea lions (*Zalophus californianus*). *J. Exp. Biol.* **204**, 3273–3281.

Kannan, K., Kajiwara, N., Le Boeuf, B. J., and Tanabe, S. (2004). Organochlorine pesticides and polychlorinated biphenyls in California sea lions. *Environ. Pollut.* **131**, 425–434.

Keiper, C. A., Ainley, D. G., Allen, S. G., and Harvey, J. T. (2005). Marine mammal occurrence and ocean climate off central California, 1986 to 1994 and 1997 to 1999. *Mar. Ecol. Prog. Ser.* **289**, 285–306.

Lefebvre, K., *et al.* (8 authors) (2000). Domoic acid-producing diatoms: Probable cause of neuroexcitotoxicity in California sea lions. *Mar. Environ. Res.* **50**, 485–488.

Lowry, M. S., and Carretta, J. V. (1999). Market squid (*Loligo opalescens*) in the diet of California sea lions (*Zalophus californianus*) in southern California (1981–1995). *Calif. Coop. Ocean. Fish. Invest. Rep.* **40**, 196–207.

Lowry, M. S., and Maravilla-Chavez, O. (2005). Recent abundance of California sea lions in western Baja California, Mexico and the United States. Proceedings of 6th California Islands Symposium, Ventura, CA, December 1–3, 2003, 485–497.

Lowry, M. S., Oliver, C. W., and Weber, J. B. (1986). The food habits of California sea lions at San Clemente Island, California; April 1983 through September 1985. Southwest Fisheries Science Center Admin. Rept. LJ-86-33, La Jolla, CA.

Maravilla-Chavez, M. O., Hernández-Vazquez, S., Zavala-González, A., and Ortega-Rubi, A. (2006). Reduction of the impact produced by sea lions on the fisheries in Mexico. *J. Environ. Biol.* **27**, 629–631.

Mellink, E., and Romera-Saavedra, A. L. (2005). Diet of California sea lions, *Zalophus californiensis*, at San Jorge Island, northern Gulf of California, Mexico, 1998–1999. *Cienc. Mar.* **31**, 369–377.

Morales, V. J. B. (1990). Parámetros reproductivos del lobo marino en la Isla Angel de la Guarda, Golfo de California, Mexico. Tesis de Maestria, Fac. de Ciencias, UNAM, Mexico, D.F.

Ono, K. A., Boness, D. J., and Oftedal, O. T. (1987). The effect of a natural environmental disturbance on maternal investment and pup behavior in the California sea lion. *Behav. Ecol. Sociobio.* **21**, 109–118.

Peterson, R. S., and Bartholomew, G. A. (1967). "Natural History and Behavior of the California Sea Lion." *Amer. Soc. Mammal.* Special Publ. No. 1. Allen Press, Lawrence, KS.

Sakahira, F., and Nimi, M. (2007). Ancient DNA analysis of the Japanese sea lion (*Zalophus californianus japonicus* Peters, 1866): Preliminary results using mitochondrial control-region sequences. *Zool. Sci.* **24**, 81–85.

Salazar, S. K. (1999). Dieta, tamaño poblacional y interacción con desechos costeros del lobo marino *Zalophus californianus wollebaeki* en las Islas Galápagos. Disertación de Licenciatura, Pontifica Universidad Católica del Ecuador.

Sanchez, A. M. (1992). Contribución al conocimiento de los hábitos alimentarios del lobo marine *Zalophus californianus* en las Islas Angel de la Guarda y Granito, Golfo de California, Mexico. Tesis Profesional, UNAM, Mexico, D.F.

Scholin, C. A., *et al.* (2000). Mortality of sea lions along the central California coast linked to a toxic diatom bloom. *Nature* **403**, 80–84.

Schusterman, R. J., and Krieger, K. (1986). Artificial language comprehension and size transposition by a California sea lion (*Zalophus californianus*). *J. Comp. Physiol.* **100**, 348–355.

Stapleton, H. M., *et al.* (9 authors) (2006). Determination of HBCD, PBDEs and MeO-BDEs in California sea lions (*Zalophus californianus*) stranded between 1993 and 2003. *Mar. Pollut. Bull.* **52**, 522–531.

Trillmich, F., and Dellinger, T. (1991). The effects of El Niño on Galapagos pinnipeds. *In* "Pinnipeds and El Niño: Responses to Environmental Stress" (F. Trillmich, and K. A. Ono, eds), 88, pp. 66–74. Ecological Studies, Springer-Verlag, Berlin.

Trillmich, F., and Wolf, J. B. W. (2008). Parent–offspring and sibling conflict in Galápagos fur seals and sea lions. *Behav. Ecol. Sociobio.* **62**, 363–375.

Weise, M. J., and Costa, D. P. (2007). Total body oxygen stores and physiological diving capacity of California sea lions as a function of age and sex. *J. Exp. Biol.* **210**, 278–289.

Weise, M. J., and Harvey, J. T. (2004). Impact of the California sea lion (*Zalophus californianus*) on salmon fisheries in Monterrey Bay, California. *Fish. Bull. US* **103**, 685–696.

Weise, M. J., Costa, D. P., Kudela, R. M. (2006). Movement and diving behavior of male California sea lion (*Zalophus californianus*) during anomalous oceanographic conditions of 2005 compared to those of 2004. *Geophy. Res. Lett.* **33**, L22S10 (6 pp.).

Wolf, J. B. W., and Trillmich, F. (2007). Beyond habitat requirements: Individual fine-scale site fidelity in a colony of the Galápagos sea lion (*Zalophus wollebaeki*) creates conditions for social structuring. *Oecologia* **152**, 553–567.

Wolf, J. B. W., Kauermann, G., and Trillmich, F. (2005). Males in the shade: Habitat use and sexual segregation in the Galápagos sea lion (*Zalophus californianus wollebaeki*). *Behav. Ecol. Sociobiol.* **59**, 293–302.

Ylitalo, G., *et al.* (10 authors) (2005). The role of organochlorines in cancer-associated mortality in California sea lions (*Zalophus californianus*). *Mar. Pollut. Bull.* **50**, 30–39.

Zavala-González, A., and Mellink, E. (2000). Historical exploitation of the California sea lion, *Zalophus californianus*, in México. *Mar. Fish. Rev.* **62**, 35–40.

# Callosities

## MASON T. WEINRICH

Perhaps no external feature on any baleen whale is as distinctive as the hardened, raised patches of skin, called callosities (pronounced cal-OS-it-ies), found on the head of all extant right whale species (North Atlantic right whale, *Eubalaena glacialis*, North Pacific right whale, *E. japonica*, and Southern right whales,

**Figure 1** *The top of the head of a northern right whale* (Eubalaena glacialis) *showing the most prominent callosity, usually called the bonnet, on its summit. The tip of the snout of the whale is toward the right, and the spray on the left indicates the blowhole. A second callosity is visible near the waterline, on the lower jaw.*

*E. australis;* Fig. 1). These features are unique characteristic of the genus *Eubalaena,* and are immediately notable and visible upon sighting the whale. Old whalers called the most visible callosity, on the tip of the rostrum, the "bonnet"; that name has stuck to the present day.

The term callosity gained acceptance in the first part of the twentieth century. The word extends from the term "callus," which refers to a variety of thickened tissues in many species. Speculation of their origin included the possibility that the callosity was an "excrescence" (a commonly used term in the late nineteenth century) from barnacles found on the head, abrasions from rubbing the head, or that they were irritations from whale lice. A number of whalers and scientists have noted the coincident occurrence between hair clusters on the whale and callosities, and Payne *et al.* (1983) note that callosities also occur in the same locations as facial hair in humans, e.g., above the eyes, behind the nostrils (blowholes) and along the lip (lower and along the ridge above the upper jaw), and on the "chin." Although there are some locations where callosities are found and hairs are not present, large callosities have at least a scattering of hairs over their surface, and smaller ones often have at least a single hair near the center of the callosity.

Callosities are a naturally occurring physical feature of the whale. In late term fetuses and immediately after birth, the areas where callosity tissue will erupt can be seen as lighter gray patches as opposed to darker gray/black surrounding skin, although the skin is still smooth. Shortly after birth, the callosity tissue erupts and acquires a pitted, jagged texture which stays with each animal throughout its life.

Although the callosities maintain their gray color throughout life, they often appear white, yellow, or orange because of the coloration of the whale lice living there. Whale lice (*Cyamid* spp.) are a small amphipod that in right whales have developed a unique relationship. Cyamids are found primarily in the crevices and crannies formed by callosity tissue. Because of reduced laminar flow, they can more easily adhere to the whale in these callosity patches. It appears that the cyamids feed on the dead, sloughing skin of the whale but do not hurt the host on which they live. Herein, they reproduce and form dense colonies, which create the whitish coloration seen highlighting the callosities (and allowing for individual photo-identification of right whales). Three different cyamid species can be found on right whales—*Cyamus ovalis, C. gracilis,* and *C. erraticus*—each with a unique coloration. Young southern right whale calves predominantly have an orange coloration to their head from an infestation of *C. erraticus,* which is typically replaced by *C. ovalis* when the calf reaches 2 months of age and their initial rapid growth slows (Rowntree, 1996). One scientist has even proposed that eruption of callosities in young whales may be facilitated by cyamids which, through eating a portion of the skin which comprises the callosity, create an area of lowered laminar flow in which they could more easily adhere to the whale, although this remains to be confirmed.

The function of callosities remains unknown. Male southern and northern right whales have, on average, a greater portion of the surface area of their head covered by callosity tissue than do females. These may be used by males in mating competition, and observers have reported seeing males in mating groups "deliberately" running the dorsal side of their heads along the backs of other males, with the recipient of the scrape reacting by "twisting and writhing" (Payne and Dorsey, 1983). Given the sensitivity of cetacean skin, it would be likely that contact from the callosity of another animal would be painful. While use in competition may account for the greater amount of callosity tissue in males, it does not explain why callosities are also present in females. Rowntree (1996) hypothesized that callosities may also function to raise sensory hairs from the laminar flow close to the skin, to detect dense swarms of plankton (also suggested for the tubercles and sensory hairs found on the rostrum of humpback whales, *Megaptera novaeangliae*). She further suggested that the ability of the whale to sense the behavior of cyamids in response to the presence of plankton (such as standing in a feeding position) around the sensory hairs may facilitate the whale's detection of its prey.

In the past 30 years callosities have received increased attention from cetologists photo-identifying individual right whales, as the pattern of the callosities varies between individuals. In the southern right whale, individual identification is facilitated by a configuration where the bonnet callosity covers only the front portion of the rostrum, and there are several additional "rostral islands" between the bonnet and the blowholes. This is referred to as a "broken" callosity. Researchers can then use the shape of the bonnet in addition to the number, location, and shape of the rostral islands to identify individuals. In the North Atlantic right whale, however, the callosity can cover the entire

**C**

area between the tip of the rostrum and the blowholes, referred to as a "continuous" callosity (found only rarely in southern right whales). Identification can be confounded since the whale lice on and around the callosities are mobile, sometimes masking the true edge of the callosity. However, by using additional distinctive features, including the three-dimensional configuration of the callosities, additional scars or marks, and crenulations on the lower lip, North Atlantic researchers have still been able to reliably identify each individual. Photographic catalogs of identified right whales, primarily of their callosity patterns, are maintained for the North Atlantic Ocean (by the New England Aquarium, Boston, MA), the North Pacific Ocean (the US National Marine Fisheries Service), and in the numerous places in the Southern Hemisphere, including off of South America near Peninsula Valdez, Argentina and off the coast of Brazil, off the coast of South Africa, and off of New Zealand's sub-Antarctic Auckland Islands (Best *et al.*, 2001). Additional collections of photographs of individual right whales based on callosities and other natural markings exist in various institutions around the world.

## See Also the Following Article

Right Whales

## References

Best, P. B., Bannister, J. L., Brownell, R. L., and Donovan, G. P. (2001). Right whales: Worldwide Status. *Journal of Cetacean Research and Management*(Special Issue 2), 309.

Hamilton, P. K., and Martin, S. M. (1999). "A Catalog of Identified Right Whales from the Western North Atlantic: 1935 to 1997." New England Aquarium, Boston, MA.

Kraus, S. D., Moore, K. E., Price, C. A., Crone, M. J., Watkins, W. A., Winn, H. E., and Prescott, J. E. (1986). The use of photographs to identify individual northern right whales (*Eubalaena glacialis*). Reports of the International Whaling Commission, Special Issue 10, 145–151.

Matthews, L. H. (1938). Notes on the southern right whale, *Eubalaena australis*. Discovery Report, **17**, 169–182.

Omura, H., Ohsumi, S., Nemoto, T., Nasu, K., and Kasuya, T. (1969). Black right whales in the North Pacific. Sci. Rep. Whales Research Institute, Tokyo **21**, 1–78.

Payne, R., and Dorsey, E. M. (1983). Sexual dimorphism and aggressive use of callosities in right whales (*Eubalaena australis*). *In* "Communication and Behavior of Whales" (R. Payne, ed.), pp. 295–329. AAAS Selected Symposium 76, Westview Press, Boulder, CO.

Payne, R. S., and Rowntree, V. J. (1984). Southern right whales: A photographic catalogue of individual whales seen in waters surrounding Peninsula Valdes, Argentina. Unpublished Report available from Center for Long Term Research, Inc., 191 Weston Road, Lincoln, MA 01773.

Payne, R., Brazier, O., Dorsey, E.M., Perkins, J.S., Rowntree, V.J., and Titus, A. (1983). External features in southern right whales (*Eubalaena australis*) and their use in identifying individuals. *In* "Communication and Behavior of Whales" (R. Payne, ed.), pp. 371–445. AAAS Selected Symposium 76, Westview Press, Boulder, CO.

Rowntree, V. (1983). Cyamids: The louse that moored. Whalewatcher (J. American Cetacean Society) **17**(4), 14–17.

Rowntree, V. (1996). Feeding, distribution, and reproductive behavior of cyamids (Crustacea: Amphipoda) living on humpback and right whales. *Can. J. Zool.* **74**, 103–109.

Tomilin, A. G. (1957). Mammals of the USSR, Vol. 8: Cetacea. Translated by the Israel Program for Scientific Translations. Available from National Technical Information Service, 5285 Port Royal Road, Springfield, VA 22161.

# Captive Breeding

TODD R. ROBECK, JUSTINE K. O'BRIEN AND DANIEL K. ODELL

## I. Marine Mammals in Captivity

Animals have been held in captivity in one form or another for hundreds, if not thousands, of years. Private collections turned into "public" collections. A private animal collection at Schloss Schönbrunn, Vienna, Austria, was opened to the public in 1765 and is considered to be one of the first modern zoos. The first marine mammals to be held in captivity may have been polar bears (*Ursus maritimus*) and various pinnipeds. Reeves and Mead (1999) provide an excellent overview of marine mammals in captivity. Harbor porpoises (*Phocoena phocoena*) may have been held as early as the 1400s, polar bears since about 1060, and walruses (*Odobenus rosmarus*) since 1608. As with terrestrial animals, marine mammals were held in private collections.

However, most pinnipeds and some sirenians were not held in captivity until the late 1800s and early 1900s. Being considerably more difficult to capture, transport, and maintain, cetaceans, with few exceptions, have only been held in captivity since the mid-1900s. According to Reeves and Mead (1999), 4 species of sirenians, 33 pinnipeds, 51 cetaceans, the polar bear, and the sea otter (*Enhydra lutris*) have been held in captivity. Of these, 2 species of sirenians, 22 pinnipeds, 15 cetaceans, the polar bear, and the sea otter have reproduced in captivity. Among these, however, only a few species, such as the polar bear, California sea lion (*Zalophus californianus*), harbor seal (*Phoca vitulina*), bottlenose dolphin (*Tursiops truncatus*), and killer whale (*Orcinus orca*), have enough numbers and have been reproductively managed with enough production to be considered part of a successful captive breeding program.

Successful captive breeding of any marine mammal requires a combination of an appropriate habitat, adequate nutrition and a social structure that is conducive toward successful reproduction. It becomes obvious when analyzing the history of successful births and survivorship of these species in captivity that early animal managers had little thought or, in some cases, knowledge of the requirements necessary for the development of successful breeding programs. In contrast, past records of breeding and survivorship with recent trends beginning in the mid-1980s where captive breeding successes have equaled or, in some cases, surpassed the best scientific estimates of wild population breeding and survivorship, and one can see just how far the captive marine mammal community has evolved. Detailed censuses of captive marine mammals in North America (Asper *et al.*, 1990; Andrews *et al.*, 1997) have shown the increasing numbers of captive-bred marine mammals, particularly California sea lions, harbor seals, and bottlenose dolphins. In 1995, 70% of the California sea lions, 56% of the harbor seals, and 43% of the bottlenose dolphins on display in North American facilities were captive-born. This compares with 3%, 4%, and 6%, respectively, in 1975.

## II. Why Breed Marine Mammals in Captivity?
### A. Legal Necessity

In the "early days," when animals were just being displayed as "curiosities," it was easier to collect replacements when animals died. In most countries today this is simply not possible. Despite only a few species of marine mammals being threatened or endangered, they and their habitats are protected by a myriad of national and international laws and regulations. Although a number of countries are considering regulation of the minimum conditions (e.g., pool size and volume, water quality, food quality and handling, medical care) under which marine mammals may be held in zoos, marine parks, or research facilities, apparently only the USA has such regulations in place. Even though the natural habitat of most marine mammals cannot be duplicated in captivity, the trend is toward larger, more complex, habitat-oriented displays and exhibits. Together, the various laws and regulations have reduced the collection of wild marine mammals and have eliminated smaller facilities that did not have the financial resources to adequately provide for their animals as required by law.

The just-mentioned laws and regulations favor captive breeding programs. A successful captive breeding program eliminates costly field expeditions and animal transports.

### B. Maintaining/Enlarging a Captive Population

A successful captive breeding program can provide animals for other institutions with adequate holding facilities but without the financial resources to maintain a breeding colony or (if even possible) to collect wild animals. Captive-born animals have a known medical history and, to some extent, are imprinted on their keepers. Captive breeding programs have, out of necessity, reduced the impact on wild populations of marine mammals. Professional organizations such as the Association of Zoos and Aquariums (AZA) have established studbooks and other programs to assist with the management of captive animal breeding colonies. For example, studbooks track individual animals from birth to death and their reproductive histories. Computer programs are used to pair animals to optimize genetic diversity. In the USA, formal studbooks exist for the beluga whale (*Delphinapterus leucas*), common bottlenose dolphin, Florida manatee (*Trichechus manatus latirostris*), gray seal (*Halichoerus grypus*), harbor seal, northern fur seal (*Callorhinus ursinus*), and polar bear and several others are under development. Similar studbooks are in place on other continents. The AZA also hosts a Marine Mammal Taxon Advisory Group whose function is to promote managed captive breeding of marine mammals.

### C. A Breeding Program as a Scientific Resource

A successful captive breeding program is a unique scientific resource in that it allows one to document, in great detail, various aspects of reproductive behavior, reproductive physiology, and the subsequent birth, growth, and development of the offspring. This is particularly valuable for cetaceans, which are typically difficult to study in great detail in their natural habitat due to various environmental factors and the fact that they spend most of their lives underwater. It is, however, most important to recognize that studies on captive marine mammals, even in the best breeding colonies, do not and cannot replace field studies. Both types of studies (i.e., laboratory and field) are necessary to fully describe the biology of a species.

Routine components of a proper animal husbandry program include regular physical exams, collection of blood, urine, and fecal samples, body measurements, and body weights. These samples and data are virtually impossible to collect from wild marine mammals. Consider, e.g., what it would take to get a daily urine sample from a wild bottlenose dolphin or killer whale! Captive animals are easily conditioned to provide urine samples and to station for blood sampling, body measurements, and so on. Figures 1, 2, and 3 illustrate the kinds of observations that can be made and the kinds of data that can be gathered.

**Figure 1** *Birth of a killer whale* (Orcinus orca) *at SeaWorld Florida. Photo credit: SeaWorld Florida.*

### D. A Breeding Program as a Conservation Resource

A successful captive breeding program may provide the physical and human resources necessary to save some species of marine mammals in imminent danger of extinction (Ralls and Meadows, 2000). These resources may allow us to maintain a viable gene pool until the habitat can be restored or other reasons for endangerment are eliminated. While such an undertaking is certainly honorable and the right thing to do, the magnitude of the job should not be underestimated and there are, at the present time, obvious limits based in good measure on the sheer size of the animals. For example, the population of right whales (*Eubalaena glacialis*) in the western North Atlantic Ocean is about 350 animals and may be decreasing due to human activities. These animals may reach lengths of 18 m and weights on the order of 20 tons. No facility in existence today (or likely to be in the foreseeable future) could hold a breeding group of right whales. However, threatened or endangered marine mammals such as the Hawaiian and Mediterranean monk seals (*Monachus schauinslandi* and *M. monachus*, respectively) and the river dolphins (families Platanistidae, Iniidae, and Pontoporiidae) could be maintained in viable captive breeding colonies. The importance of taking immediate action on behalf of these species is demonstrated by the recent extinction of a small cetacean, the baiji (*Lipotes vexillifer*). If a captive breeding colony had been established, this species would still be present today and this colony may have provided potential animals for reintroduction to the wild.

### III. Assisted Reproductive Technologies (ART)

The domestic animal industry [cattle (*Bos* spp.), pigs (*Sus* spp.), horses (*Equus caballus*)] long ago realized that it is much more efficient to move genetic material (e.g., semen) to different facilities than it is to move the animal. Methodologies for semen collection, preservation, and transportation, along with methods for artificial insemination (AI), induction of superovulation, and embryo collection and transfer, were developed and are in widespread use worldwide today. Some of these techniques have been applied successfully to endangered animals. And recently, AI has been successfully developed in three species of marine mammals; killer whales, bottlenose dolphins, and Pacific white-sided dolphins (*Lagenorhynchus obliquidens*) (Robeck *et al.*, 2003, 2004, 2005).

What then are the needs for the development and application of ART to marine mammals? The most obvious ART that should be

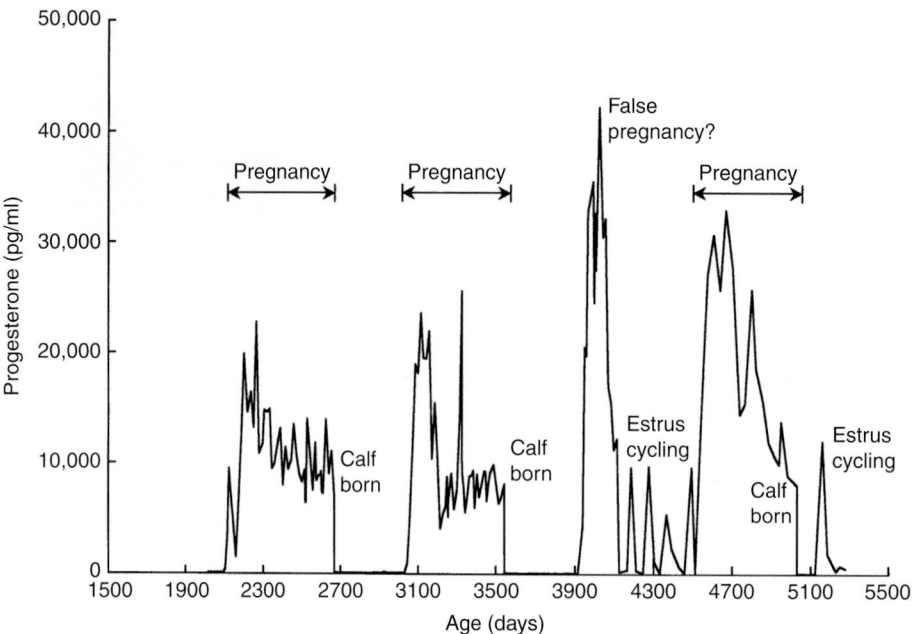

**Figure 2** *Serum progesterone of killer whale* (Orcinus orca) *from birth through sexual maturity and calving and showing estrous cycling and pregnancies. Progesterone levels were "zero" until after 1900 days of age.*

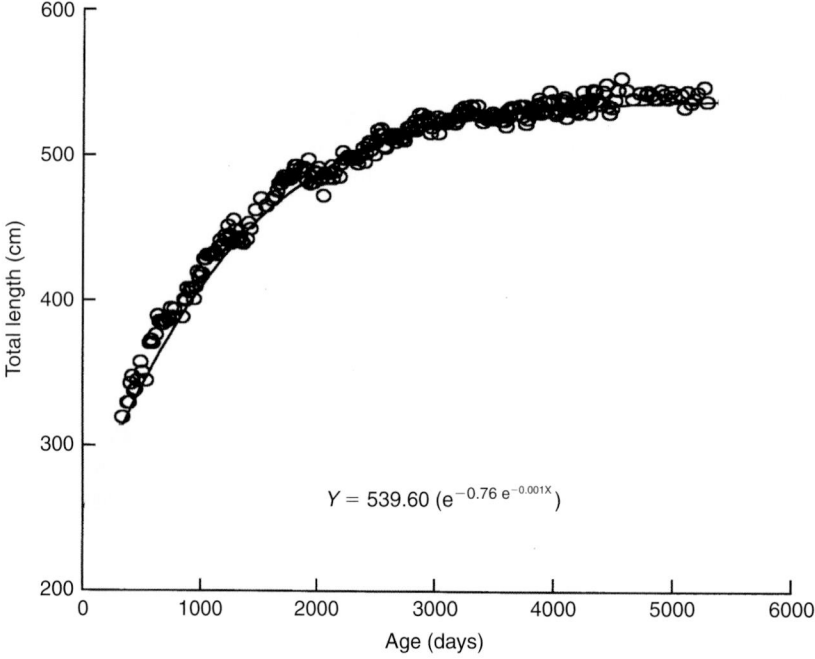

**Figure 3** *Killer whale* (Orcinus orca) *growth (length) curve as an example of data that can be gathered from captive-bred animals.*

developed and the one that is most likely to have an immediate impact on the genetic management of captive marine mammals is AI. Once developed in all captive species, AI can provide an immediate mechanism for marine mammal managers to increase the genetic fitness of their respective populations without having to rely on the transportation of animals between facilities. Wild animal population studies have shown that dolphins develop strong social ties to other animals and that these bonds can be maintained throughout the life of the animals (Connor *et al.*, 2000). To what extent these bonds are important for

the health of these animals can only be speculated, but it seems prudent for managers who have groups of compatible animals to carefully consider the impact of removing or introducing individuals on the population social dynamics. Further, bringing animals from other locations may expose the new population of animals to bacterial or viral organisms, which they have no natural resistance against.

The development of AI, the most common assisted reproductive technology, for commercial use began in the 1950s. The successful application of AI and other ART to domestic species and humans was in part because of their accessibility for research into their reproductive mechanism or reproductive physiology. Critical importance for the successful application of ART in any species is the determination of how reproductive hormones, particularly luteinizing hormone (LH), relate to ovulation (Sorensen, 1979).

Once ARTs were developed in domestic species, it was naively believed that they could easily be transferred to exotic species. However, relatively little success was realized with exotic animals until the late 1970s when an endocrinological breakthrough occurred. This breakthrough was the ability to analyze reproductive hormones in urine. This technology was successfully used to characterize the endocrinological events in a wide range of exotic species, including for the first time in marine mammals; the killer whale (Walker *et al.*, 1988). Other smaller cetacean species proved more difficult to train for urine collection, and detailed endocrinological evaluations would have to wait until a technique was developed for training urine collection in bottlenose dolphins (Lenzi, 2000). This simple technique has since revolutionized urine collection and the subsequent endocrine evaluation of cetaceans (whereas Pinnipeds are yet to undergo such extensive evaluation). As a result, nearly 15 years after the first published report of urinary hormones in killer whales, endocrine characteristics have now been reported in four additional cetacean species.

Although the capability to characterize the endocrine cycle is critical, for it to be useful to predict the timing of the actual insemination, it must first be related to ovulation. For deduction of the temporal relationship between urinary hormones and ovulation in cetaceans, two important techniques had to be developed. This first was an assay system that could rapidly detect the LH surge—a brief and dramatic hormonal surge that in most mammalian species precedes ovulation at a regular interval. The second was to develop a method to allow for consistent observation of ovarian activity using transabdominal ultrasonography. Brook (2001) was the first to develop a technique to accomplish this in the bottlenose dolphin. This simple technique which relies on understanding the anatomical location of the ovaries in relationship to the abdominal musculature has since been successfully utilized to consistently locate ovaries in many other captive marine mammals, including the killer whale, beluga, the Pacific white-sided dolphin (Fig. 4), and the false killer whale (*Pseudorca crassidens*). Finally, by combining this ultrasound technique with rapid LH assay systems, landmark research was completed that allowed characterization of the timing between the LH surge and ovulation (Robeck *et al.*, 2004, 2005). This information then opened the door for the development and application of AI in cetaceans.

Once ovulation timing could be predicted, the optimum timing of insemination had to be systematically determined for each species. However, before any AI attempts could be made the collection of semen had to be developed. Currently, collection of semen on a routine basis has been achieved only in a few marine mammal species, and as one would expect, this list correlates with the species where AI has now been accomplished. The first species where semen collection was successfully trained was the bottlenose dolphin (Keller, 1986). Bottlenose dolphins seem to have consistently elevated libidos and thus were relatively easily trained to provide semen via manual stimulation. Since the success with bottlenose dolphins, four killer whales and three Pacific white-sided dolphins and one beluga have been trained to provide semen on a regular basis. Obviously, to reach the full potential of AI, all genetically valuable males should be trained to provide semen.

Once the semen has been collected, it must be stored temporarily for immediate use or permanently by cryopreservation. Semen cryopreservation has been successfully accomplished by different methods (straws, pellets, or cryovials) in all of the species for which

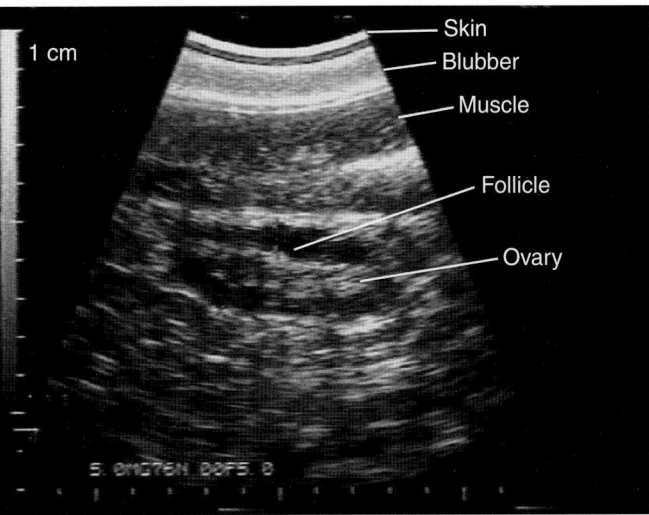

**Figure 4** *Sonographic image of an ovary of a Pacific white-sided dolphin* (Lagenorhynchus obliquidens) *showing a follicle.*

semen has been obtained (Schroeder and Keller, 1990; Robeck and O'Brien, 2004; Robeck *et al.*, 2004; O'Brien and Robeck, 2007). Although cryopreserved semen provides a long-term supply of genetic material from a particular animal and is the only method currently available to store semen for long (greater than a few days) periods of time, when used with AI it has approximately half the fertilizing life span of fresh semen. As a general rule, frozen–thawed semen must be deposited in the uterus and at close proximity to ovulation. Fresh, cooled semen, having a longer fertilizing life span than frozen–thawed semen, can be placed with less accuracy, generally in the true cervix or distal uterine body, and with a greater time interval prior to ovulation. Thus, when attempting to develop AI, first attempts are generally made using fresh cooled semen.

All of the successful AIs in marine mammals have been performed using an endoscope to deposit either fresh or frozen–thawed semen into the uterus (uterine body or the horn). Before this could be accomplished, a thorough understanding of female reproductive tract anatomy was essential. Despite the close phylogenetic relationship between captive delphinids, significant variation exists in their reproductive tract morphology. For example, bottlenose dolphins have a pseudocervical vaginal fold or flap that must be traversed before encountering the true cervix. Knowledge of the anatomy of the bottlenose dolphin would do little to help placement of semen into the uterine body of a closely related member of the delphinid family, the Pacific white-sided dolphin. This species has a cervix which is composed of a series of three annular folds that present anatomical barriers to the placement of semen. The killer whale, another member of the delphinid family, has a completely different arrangement consisting of two longitudinally ridged cervices aligned in series.

Once the semen has been deposited and ovulation has occurred, how do you determine if you have had success? With the application of ultrasound for pregnancy detection (Williamson *et al.*, 1990), more and more marine mammal practitioners have become aware of the fact that all species of captive delphinids can exhibit variable periods of false pregnancies. These periods of elevated progesterone are currently endocrinologically indistinguishable from pregnancy. Thus, pregnancy can only be confirmed with the use of ultrasound.

In the bottlenose dolphin, the conception rate after AI using frozen–thawed semen is 65–70%. AI in this species has been taken a step further by the development of sperm-sorting technology, a

**Figure 5** *The world's first female bottlenose dolphin* (Tursiops truncatus) *born at SeaWorld San Diego (California) after using AI with X-selected spermatozoa. Photo credit: Mike Aguilera, SeaWorld San Diego.*

sex predetermination method (Johnson *et al.*, 1989). The ability to skew the sex ratio through insemination of females with X chromosome-bearing (female) spermatozoa helps to alleviate space and social problems that often occur with multiple adult breeding males. AI of sex-sorted, cryopreserved sperm could also contribute to improved genetic management of captive populations in the same way as conventionally frozen sperm. Further, small populations can be replenished at a faster rate with higher numbers of females than males. Development of this technology for application to bottlenose dolphin population management has led to the world's first cetacean pregnancy using AI with sex selected then cryopreserved sperm (O'Brien and Robeck, 2006). The technology has been integrated into SeaWorld's multi-site reproductive and genetic management program for this species, with five sex-selected (female) calves being born to date (O'Brien and Robeck, unpublished data; Fig. 5). Sex sorting of previously frozen spermatozoa, derived from ejaculates or post-mortem epididymal samples can also be accomplished providing species-specific modifications are developed in controlled studies (O'Brien *et al.*, 2004, 2005). The diverse applications of sperm sorting in conjunction with AI have great implications for the genetic and social management of captive cetaceans, particularly those which are endangered with extinction.

## IV. Challenges for the Future

If self-sustaining populations are to be developed, then all reproductively capable animals must contribute to the gene pool. Producing viable offspring is the primary goal of any captive breeding program, but it must be managed to avoid overpopulation of the facility and to prevent or minimize inbreeding. This may require separate facilities for adult males, preparturient females, and females with new offspring. Breeding may be regulated simply by separating the sexes or by neutering (physically or chemically) both males and females.

Currently, many populations of marine mammals are kept in small groups or same-sex groups. Therefore, a large portion of reproductively mature animals within these groups are not reproducing. These animals are functionally excluded from contributing to the collective captive gene pool. Therefore, valuable financial resources and pool space are being used for a minority of the available genetic lines. This inefficient use of animal resources must be corrected before long-term population stability can be achieved. Judicious use of contraception and continued development of AI in all captive

marine mammal species would help managers optimize utilization of the population's genetic pool. However, if the maximum utilization of genetic resources does not result in a predictable stable population, then genetic infusions from wild stocks will be necessary. ART may provide another answer to this future dilemma if semen (and possibly oocytes) can be collected from wild animals that are incidentally or purposely killed in fisheries or subsistence hunting activities. Alternatively, temporary capture of wild males, followed by semen collection using electro-ejaculation, then release back into the wild after appropriate monitoring, could also represent a means of obtaining valuable genetic material for potential infusion into the captive population using ART. As an alternative to the ART solution or until ARTs such as AI and IVF were perfected, mangers could "borrow" adult males from wild populations for 1–2 years for breeding purposes and then return them. This, of course, involves considerable expense and there is no guarantee that any given animal would breed successfully.

## See Also the Following Articles

Captivity ■ Genetics for Management ■ Marine Parks and Zoos

## References

Andrews, B., Duffield, D. A., and McBain, J. F. (1997). Marine mammal management: Aiming at the year 2000. *IBI Rep.* **7**, 125–130.

Asper, E. D., Duffield, D. A., Dimeo-Ediger, N., and Shell, D. (1990). Marine mammals in zoos, aquaria and marine zoological parks in North America: 1990 census report. *Int. Zoo Yearbook* **29**, 179–187.

Brook, F. M. (2001). Sonographic imaging of the reproductive tract of the female bottlenose dolphin, *Tursiops truncatus aduncas*. *Reproduction* **121**, 419–428.

Connor, R. C., Wells, R., Mann, J., and Read, A. (2000). The bottlenose dolphin: Social relationships in a fission–fusion society. *In* "Cetacean Societies: Field Studies of Whales and Dolphins" (J. Mann, R. Connor, P. Tyack, and H. Whitehead, eds), pp. 91–126. University of Chicago Press, Chicago.

Johnson, L. A., Flook, J. P., and Hawk, H. W. (1989). Sex Pre-selection in Rabbits: Live Births from X- and Y-Sperm Separated by DNA and Cell Sorting. *Biol. Reprod.* **41**, 199–203.

Keller, K. V. (1986). Training of the Atlantic bottlenose dolphins (*Tursiops truncatus*) for artificial insemination. Proceedings of International Association of Aquatic Animal Medicine **14**, 22–24.

Lenzi, R. (2000). Operant conditioning and ultrasound together at work to successfully condition voluntary urine collection. Proceedings of International Marine Animal Trainers Association, **17**.

O'Brien, J. K., and Robeck, T. R. (2006). Development of sperm sexing and associated assisted reproductive technology for sex pre-selection of captive bottlenose dolphins (*Tursiops truncatus*). *J. Reprod. Fertil. Develop.* **18**, 319–329.

O'Brien, J. K., and Robeck, T. R. (2007). Semen collection, characterization and preservation in a beluga (*Delphinapterus leucas*). 1st International Workshop on Beluga Whale Research, Husbandry and Management in Wild and Captive Environments. Valencia, Spain, pp. 8–9.

O'Brien, J. K., Hollinshead, F. K., Evans, K. M., Evans, G., and Maxwell, W. M. C. (2004). Flow cytometric sorting of frozen–thawed spermatozoa in sheep and non-human primates. *J. Reprod. Fertil. Develop.* **15**, 367–375.

O'Brien, J. K., *et al.* (8 authors) (2005). Flow cytometric sorting of fresh and frozen–thawed spermatozoa in the western lowland gorilla (*Gorilla gorilla gorilla*). *Am. J. Primatol.* **66**, 297–315.

Ralls, K., and Meadows, R. (2000). Captive breeding and reintroduction. *In* "Encyclopedia of Biodiversity." Academic Press, New York.

C

Reeves, R. R., and Mead, J. G. (1999). Marine Mammals in Captivity. *In* "Conservation and Management of Marine Mammals" (J. R. Twiss, Jr, and R. R. Reeves), p. 412–436. Smithsonian Institution Press, Washington, D.C.

Robeck, T. R., and O'Brien, J. K. (2004). Effect of cryopreservation methods and pre-cryopreservation storage on bottlenose dolphin (*Tursiops truncatus*) spermatozoa. *Biol. Reprod.* **70**, 1340–1348.

Robeck, T. R., Greenwell, M., Boehm, J. R., Yoshioka, M, Tobayama, T., Steinman, K, and Monfort, S. T. (2003). Artificial insemination using frozen–thawed semen in the Pacific white-sided dolphin (*Lagenorhynchus obliquidens*). Proc. IAAAM May 2003.

Robeck, T. R., Steinman, K. J., Gearhart, S., Reidarson, T. R., McBain, J. F., and Monfort, S. L. (2004). Reproductive physiology and development of artificial insemination technology in killer whales (*Orcinus orca*). *Biol. Reprod.* **71**, 650–660.

Robeck, T. R., *et al.* (10 authors) (2005). Estrous cycle characterization and artificial insemination using frozen–thawed spermatozoa in the bottlenose dolphin (*Tursiops truncatus*). *Reproduction* **129**, 659–674.

Schroeder, J. P., and Keller, K. V. (1990). Artificial insemination of bottlenose dolphins. *In* "The Bottlenose Dolphin" (S. Leatherwood, and R. R. Reeves, eds), pp. 447–460. Academic Press, San Diego.

Sorensen, A. M. (1979). "Animal Reproduction: Principles and Practices." McGraw-Hill, New York.

Walker, L. A., *et al.* (8 authors) (1988). Urinary concentrations of ovarian steroid hormone metabolites and bioactive follicle-stimulating hormone in killer whales (*Orcinus orca*) during ovarian cycles and pregnancy. *Biol. Reprod.* **39**, 1013–1020.

Williamson, P., Gales, N. J., and Lister, S. (1990). Use of real-time B-mode ultrasound for pregnancy diagnosis and measurement of fetal growth in captive bottlenose dolphins (*Tursiops truncatus*). *J. Reprod. Fertil.* **88**, 543–548.

# Captivity

Peter Corkeron

## I. The Debate

The debate over the ethics of marine mammals in captivity is essentially about cetaceans (whales, dolphins, and porpoises) because other marine mammals such as seals and sea lions do not inspire the same passion as whales and dolphins do. The ongoing debate over whether cetaceans should be kept in captivity is relatively recent in contrast to the history of human/marine mammal interactions. Human interest in marine mammals was initially based on the commercial value of killing seals and whales for oil, meat, and hides. The larger animals represented a greater profit, so small marine mammals such as dolphins were mostly considered to be pests to fishermen. Occasional reports of marine mammals being kept in captivity as curiosities are scattered throughout history: polar bears (*Ursus maritimus*) were kept by Scandinavian rulers prior to the Middle Ages; a killer whale (*Orcinus orca*) that had been live stranded was kept and used for sport by Roman guards during the first century ad; and seals were kept in menageries by the eighteenth century. In the mid-1800s, P. T. Barnum displayed belugas (*Delphinapterus leucas*) and common bottlenose dolphins (*Tursiops truncatus*) in his New York museum for a short time, and in the late 1800s, the Brighton Aquarium in England displayed harbor porpoises (*Phocoena phocoena*) for several months. A new era of modern marine mammal exhibits began in the late 1930s

when the Marine Studios at Marineland opened in Florida, USA (Reeves and Mead, 1999).

At first, marine mammal facilities were quite popular. During the 1950s, 1960s, and 1970s, the number of aquaria and zoological parks displaying marine mammals increased rapidly to meet public demand, especially in Europe, North America, and Australasia. Simultaneously, technology and methods for the capture, transport, and maintenance of marine mammals improved with increasing knowledge and experience. Scientists took advantage of the ability of the animals at these facilities to conduct groundbreaking research on dolphin acoustics, human/dolphin communication, dolphin brain function, hearing and echolocation, and behavior. Soon the public became more familiar with dolphins through shows at aquaria and from "Flipper," a popular TV show, in which a dolphin was portrayed as a free-ranging family friend. With this heightened awareness, the public began to understand that dolphins were not large fish, but intelligent and friendly marine mammals. However, the image transformation inspired—at least in part by animals at marine mammal facilities—would soon become a public relations nightmare for the industry that helped to make these animals popular.

## II. The Impact

In developed nations, the boom in aquaria and oceanaria experienced through the 1970s came to a near halt during the mid-1980s due to the growing debate over keeping cetaceans in captivity. Pressure from non-government organizations (NGOs) and changes in public opinion forced the closure of some existing facilities and prevented some new facilities from being opened. While the 1990s saw a decline in the number of facilities for keeping cetaceans in Australia and some parts of Europe, but the number of captive facilities increased in other parts of the world, notably in Asia, Latin America, and the Caribbean, where there was little or no domestic agitation against oceanaria.

Compared with most terrestrial mammals, marine mammals are expensive and difficult to maintain in captivity. They require a good deal of logistical support, such as high-quality food sources, specialized veterinary care, large enclosures, and expensive water-quality maintenance systems. Cetaceans and sirenians (manatees, *Trichechus* spp., and dugongs, *Dugong dugon*]), being wholly aquatic, present greater logistic difficulties than any other marine mammals. A few species predominate at these facilities because they have shown greater success in captivity, and because they are relatively easy to capture. Of these animals, those most often used in public performances include common or Indo-Pacific (*Tursiops aduncus*) bottlenose dolphins, belugas or killer whales, and California sea lions (*Zalophus californianus*), whereas phocids such as harbor seals (*Phoca vitulina*), sea otters (*Enhydra lutris*), and sirenians are more typically maintained in non-performing exhibits.

Most modern facilities in developed nations maintain high standards of operation and animal care. However, many facilities, particularly those in less developed nations, fall well short of such standards and internationally, new captive facilities are opening mostly in less developed nations. There is concern that capture and holding facilities that are effectively unregulated exist, particularly in southeast Asia and parts of Latin America and the Caribbean, supplying animals to support the growth of the aquarium industry in these regions.

Even if only for short periods of time, almost every species of marine mammal other than most of the great whales have been maintained in captivity. As an anomaly, two young gray whales (*Eschrichtius robustus*) were kept in captivity for about 1 year each, and a wealth of

C

behavioral and physiological data was collected. However, most others are dolphins and porpoises, pinnipeds, and several members of the carnivores. Animals are kept captive for different reasons: display in zoos and aquaria, military work, scientific (including military) research, and rehabilitation for injured or sick animals prior to release, although these categories are not mutually exclusive. Recent years have seen the growth of "swim-with" programs, where visitors pay to enter the water with captive cetaceans, and "Dolphin Assisted Therapy," where people with illnesses spend time with captive dolphins. Size and temperament are generally the limiting factors in keeping some marine mammal species for long periods of time. Most are kept in zoos and aquaria, some live in open ocean enclosures. However, many commercial facilities with cetaceans are not traditional zoological gardens but marine parks where there tends to be more emphasis on performances by animals. (Temporary restraint for a few minutes to hours for research purposes is not considered as captivity here.)

## III. Regulations for Collection, Care, and Maintenance

### A. International Regulations

In general, regulations dealing with marine mammals in captivity cover collection, care, and maintenance of animals, and movements of animals between countries. The extent to which existing laws are administered and enforced varies internationally. The International Whaling Commission (IWC) is the central international instrument for the protection of whales; however its effectiveness is debated. National attitudes to whaling, and to hunting small cetaceans, do not necessarily transpose to captivity. Norway, for example, a staunch whaling nation, has a national ban on keeping cetaceans in captivity, although seals are kept for display and scientific research.

The major instrument regulating the international trade in captive cetaceans is the Convention on International Trade in Endangered Species of Wild Fauna and Flora (CITES). Most cetaceans are listed in CITES Appendix II, which provides a means of regulating and monitoring trade for species not threatened with extinction but which are vulnerable to overuse. This listing allows for international trade with properly issued permits. Some cetaceans that are kept in captivity (or due to their size could be kept in captivity) are listed in Appendix I of CITES, indicating that they are either currently threatened with extinction or may be affected by trade. These include the recently deemed to be extinct baiji or Chinese river dolphin (*Lipotes vexillifer*), Ganges river dolphin (*Platanista gangetica gangetica*), Indus River dolphin (*P. g. minor*), tucuxi (*Sotalia fluviatilis*), Indo-Pacific humpback dolphin (*Sousa chinensis*), Atlantic humpback dolphin (*Sousa teuszii*), finless porpoise (*Neophocaena phocaenoides*), and the vaquita (*Phocoena sinus*). With regard to marine animals, the CITES convention includes the requirement that suitable housing and care is available, and for Appendix I listed animals, that they are not to be used for "primarily commercial purposes."

Most nations keeping marine mammals in captivity are signatories to CITES, although this does not guarantee problems cannot occur, as the export of bottlenose dolphins from the Solomon Islands to Mexico in 2003, in dubious circumstances, demonstrated. International agreements can regulate capture, but most trade in wild-caught animals now comes from nations with few or no regulations on capture, or with regulations that are ignored. International pressure on some nations (e.g., Iceland in 1989) has resulted in the closure of their capture industry.

In the USA, Congress passed unprecedented regulatory legislation in 1972 to bring under its protection all marine mammals within the borders of its jurisdiction: The Marine Mammal Protection Act (MMPA)'s intent was to protect marine mammals from human actions (predominantly fishing) that led to extinction. However, the MMPA specifically authorized the collection of free-ranging animals for scientific research and public display and education. Depending on the species involved, the collection of marine mammals is governed by a permit process administered either by National Oceanic and Atmospheric Administration (NOAA) Fisheries or the Fish and Wildlife Service. The standards for the maintenance of marine mammals in research or public display facilities are established and monitored by the Animal and Plant Health Inspection service under the US Animal Welfare Act, and all marine mammal-related activities are monitored by the presidentially appointed Marine Mammal Commission. Although collection is still permitted in the USA, there have not been many bottlenose dolphins collected for US facilities since 1989 due to a self-imposed moratorium observed by members of the Alliance of Marine Mammal Parks and Aquariums on capturing bottlenose dolphins from the Gulf of Mexico. In addition, there is a changing trend regarding holding and breeding bottlenose dolphins in North America. In 1976, 94% of captive bottlenose dolphins held were originally free-ranging, 6% were captive born. By 1996, 44% of bottlenose dolphins were captive born.

Several countries have legislated regulations or guidelines to govern collection and maintenance of marine mammals since 1972. New Zealand passed an MMPA in 1978; and in 1980, Australia passed the Whale Protection Act. In Australia, the state of Victoria put a ban on issuing permits for keeping cetaceans for display or for collecting them for export. Here, legislation does not absolutely preclude issuing a permit to capture free-ranging animals, but general government policy, the legal requirement for public comment on an application for capture, and the need for signed Ministerial approval for capture permits, all mean that it is highly unlikely that permits would be issued. Canada developed guidelines that forbid the capture of killer whales and gives priority to Canadian institutions when considering permits to capture belugas. However, in the past 10 years for which data are available (1997–2006), 28 belugas have been imported into Canada from the Russian Federation (data downloaded from the UNEP-WCMC CITES trade database, http://www.unep-wcmc.org/citestrade/).

Legislation can interact with government policy and public opinion to affect the capture industry. For example, guidelines established in some countries such as the UK do not specifically prohibit the collection and display of marine mammals. However, following a scientific review of appropriate standards for dolphinaria (Klinowska and Brown, 1986), all displays of cetaceans in the UK were closed, as existing facilities could not meet the new standards. The last dolphinarium in the UK closed in the early 1990s. Movements of cetaceans into and within the European Union (EU) are regulated under EU wildlife trade regulation, established to fulfill EU member nations responsibilities under CITES. Trade in animals listed under Annex A of this Regulation (including all cetaceans) is permitted for "research or education aimed at the preservation or conservation of the species," breeding for conservation, and biomedical research (the latter is not relevant for cetaceans), but not simply for commercial use.

In Japan, multi-species drive fisheries combine capturing cetaceans alive for aquaria and dead for food. US legislation requires that captures be conducted humanely, effectively denying animals from the Japanese fishery to institutions in the USA. The World Association of Zoos and Aquariums (WAZA) and the American Zoo and Aquarium Association have issued statements describing drive fisheries as inhumane and calling for their cessation. The WAZA resolved in 2004 (Anonymous, 2004) that taking animals from drive fisheries into captivity was against their code of ethics, to which all members of the Association (including those in Japan) must adhere.

Regional conservation agreements have also been developed. Several nations that are signatory to the Agreement on the Conservation of Small Cetaceans of the Baltic and North Seas (ASCOBANS) are mandated to take part in conservation and research measures. Such measures include preventing the release of potentially harmful substances into the environment, developing fishing gear to reduce bycatch, and reducing the impact of other potentially harmful human activities. They are obliged to prohibit intentional killing of small cetaceans and to release immediately any healthy small cetaceans caught incidentally (Anonymous, 1992).

## B. Care and Maintenance

The first published accounts of the behavior of captive cetaceans were provided by Charles Townsend in the early 1900s when he was the director of the New York Aquarium. His observations of a group of bottlenose dolphins described their social behavior and some sensory capabilities. Townsend understood the importance of developing health care and water treatment regimes to promote long-term survival. Requirements for the care and maintenance of marine mammals can vary dramatically between countries and can vary between jurisdictions within a country. These requirements include regulations regarding pool dimensions and construction materials, food quality and feeding schedules, water quality, air quality, veterinary care, and educational message. There are countries (e.g., the USA) where the agency responsible for overseeing capture and international transport is different from that responsible for care and maintenance.

A core aspect of the argument against maintaining captive cetaceans is that it is impossible to provide an adequate environment for cetaceans in captivity. The basic reasons put forward for this are: that pools can never be of adequate in size; that regardless of size, pool construction is inappropriate; and that it is impossible to keep animals in suitable social groups. Even the larger commercial facilities, e.g., one that includes a complex of four linked (sand-bottomed, rock-lined) pools of 30,000,000l, with a maximum depth of 7.5 m, holding 12 bottlenose dolphins, will be considered inadequate to some, because they see them as an inadequate substitute for the home ranges of free-ranging dolphins. Facilities differ greatly in their resources, and so the quality of their environment.

## IV. Issues

### A. Experiencing Captive Marine Mammals

Zoos and aquaria in North America alone are visited by over a hundred million people each year. There is an assumption by the supporters of such facilities that people, having experienced living marine mammals in close proximity, will be more likely to develop (or enhance) their marine conservation ethic. It is clear that public support for marine mammal conservation increased substantially in the latter half of the twentieth century. It is also clear that marine environmental degradation, particularly through overfishing, also increased substantially at the same time. Putting aside this disconnect, the extent to which commercial oceanaria have contributed to peoples' enhanced affection for marine mammals has yet to be measured and reported in a compelling manner and, not surprisingly, is questioned by some of those calling for closures. Also debated is the extent to which this was historical, and whether peoples' attitudes

today are also influenced. The public view of killer whales seemingly changed radically at the same time as they appeared in captivity, supporting the "captive animals as ambassadors" argument. However, other factors were also in play, including TV programs that presented such animals to a wider public and in a new light.

Unfortunately, as noted, there is little current research (by either side) on the extent to which visiting aquaria affects peoples' conservation ethic. Around four-fifths of respondents to a survey carried out in 1998 (Kellert 1999) believed that facilities should not be permitted to display marine mammals unless there was scientific or educational benefit. Over 90% of respondents felt that facilities should not be allowed to keep marine mammals in captivity unless the animals were well maintained, both physically and mentally.

Some opponents of keeping cetaceans captive suggest that it would be better if people were to view cetaceans through commercial whale watching, and that viewing captive cetaceans enhances an inappropriate world view (e.g., demonstrating human dominance over nature). Nearly 80% of respondents to the survey cited earlier had been to a zoo or an aquarium in the previous 5 years, but less than 20% had been whale watching over the same period (Kellert, 1999). Whether most people who visit aquaria would go whale watching if the aquaria did not exist is unknown, but if they did, it could result in an extra several million people a year going whale watching in US waters. The environmental impact of this has not been estimated, but could be substantial. Another argument that conservation benefit could be achieved through multimedia presentations featuring whales and dolphins in the wild is being developed but remains untested.

Most of this discourse is set in a Western context. The greatest growth in facilities keeping captive marine mammals is occurring in the developing world, and so this discourse may be inappropriate. Public awareness of marine mammals in most other nations is less developed than it is in the West. It may be that zoos and aquaria can make a significant contribution in these countries. However, improving conservation outcomes from experiencing captive cetaceans presupposes that the quality of educational material provided in structured programs at zoos and aquaria is acceptable. One institution in Bangkok maintaining captive cetaceans also advertises an "Orangutan boxing show" (see http://safariworld.com/), suggesting that this presupposition can be misplaced.

### B. Scientific Value of Captive Marine Mammals

The value of studies conducted with captive marine mammals has also come under scrutiny. However, before field studies of living cetaceans burgeoned (after the late 1970s), captive animals were the major means by which scientists collected data on biology and behavior. Some phenomena such as echolocation may remain unproved were it not for studies on captive animals. Even today, the echolocation capabilities of most dolphin species remain unknown, and controlled experiments with captive animals are the main way by which this information is obtained.

Because there are many other questions that are intractable using free-ranging animals, captive marine mammals remain the primary source of data for several fields, including comparative psychology, cognition, and immunology, and captive studies contribute substantially to aspects of acoustics and physiology. However, recent technological improvements have allowed playback experiments to ask acoustic questions of free-ranging animals, although problems of experimental design still plague this field (Deeke, 2006). The remarkable sensory and cognitive abilities of dolphins are, paradoxically but

**C**

understandably, two of the arguments used against their maintenance in captivity.

Most of the opposition to the scientific value of captive marine mammals appears to be based on two considerations: the extent to which inference can be drawn from captive studies to the biology of free-ranging animals, and whether the results of such research provide ethical justification for keeping animals captive. Interplay continues between work on captive and free-ranging animals. For example, attempts to understand the factors leading to the decline of Steller's sea lions (*Eumetopias jubatus*) have been informed by captive studies of diet and physiology (Rea *et al.*, 2007; Tollit *et al.*, 2007), studies of free-ranging animals (Trites *et al.*, 2006; Ban and Trites, 2007), and synthesis by modeling (Guénette *et al.*, 2006).

Regulations controlling the manner in which marine mammals are maintained for scientific research, and ethical limitations on research, vary between countries. These limitations are not necessarily related to other legal controls over keeping marine mammals. For example, in one experiment, thermocouples were surgically implanted (under anesthesia) into the brains of a juvenile harp (*Phoca groenlandica*) and a juvenile hooded (*Cystophora cristata*) seal. The conscious animals were then force-dived to assess the manner in which seals' brains cooled while diving (Odden *et al.*, 1999). Such an experimental protocol would be unlikely to obtain ethics approval in many countries, yet was permitted in Norway, where maintaining cetaceans in captivity is banned completely.

### C. Captive Breeding for Conservation

Breeding animals in zoological gardens has a role in the conservation of some endangered species. It has been suggested that such *ex-situ* conservation provides a justification for maintaining aquaria. However, developing such programs should not be an excuse to ignore our responsibility to implement conservation strategies to protect *in-situ* populations and their habitat. Furthermore, captive breeding claims for cetaceans need to be carefully evaluated. There was a project to maintain baiji in captivity in "semi-natural reserves" as part of the attempt to conserve this species. This project, supported in principle by the World Conservation Union's Cetacean Specialist Group, was unsuccessful. Despite international recognition that *ex-situ* conservation was the strategy with the best chance to conserve baiji, it failed to achieve this important goal. Debate within the scientific community as to how to manage this program continued up to the year prior to the declaration of the baiji's likely extinction (Reeves and Gales, 2006; Wang *et al.*, 2006; Yang *et al.*, 2006).

The baiji example helps illuminate the difficulties confronting those wishing to develop *ex-situ* breeding programs for marine mammals. Captive breeding programs require significant numbers of captive animals to reduce the probability of inbreeding. This appears logistically improbable at present, as animals, money and space seem unlikely to become available. The species most commonly maintained in captivity are not at conservation risk, so expertise is lacking for species that may require conservation measures in the foreseeable future. Artificial insemination is a useful tool to reduce inbreeding among animals held in countries where free-ranging animals are no longer captured, but its value in *ex-situ* conservation seems limited at present. The animals most likely to benefit from *ex-situ* conservation programs are those listed on CITES Appendix I, but the vaquita, the cetacean most likely to go extinct next, is also not likely

to be saved through captive breeding. The Chinese experience keeping finless porpoises in "semi-natural reserves," developed during the baiji project, offer possibilities, but these reserves are yet to demonstrate that they can be self-sustaining. Permits for the capture or importation of individuals from species at risk are unlikely to be issued in countries where the probability of successful captive breeding is highest. Although it seems unlikely that zoos and aquaria will contribute to species conservation through captive breeding, rehabilitating sick or injured animals can have conservation value.

### D. Rehabilitation

At times, free-ranging marine mammals that are ill, injured, or have suffered some misadventure require rehabilitation. In most cases, these are from species that are not at risk of extinction, and so the issue is one of animal welfare rather than conservation. An exception to this is the efforts made to rehabilitate Florida manatees (*Trichechus manatus latirostris*) where rehabilitation and subsequent release of each individual manatee has clear conservation value. By contrast, an attempt to rehabilitate a dugong hand-raised after stranding as a young calf, into the waters of southeast Queensland proved unsuccessful, and the animal, emaciated and injured, was recaptured 6 months after release. The dugong is now held in a purpose-designed facility with educational displays, in an attempt to use it as an "ambassador" for its species.

The contrasting examples of manatees and dugongs suggest that with each rescue comes the possibility of acquiring new knowledge that can benefit all marine mammals, endangered or not. Opposition to the role of commercial captive facilities in rehabilitation tends to be strongest in countries with effective networks to deal with marine mammal strandings and disentanglements, separate from commercial zoos and aquaria. In other places, zoos and aquaria are the organizations with the resources, funding and expertise to handle injured or sick animals effectively, and their role in rehabilitation can be significant.

### E. Release

Perhaps in response to the decrease in collecting free-ranging animals, attention in some countries through the 1990s and early part of this century focused more on releasing those already held captive. Attempts have been made to release captive dolphins back to the waters from which they came. Following the closure of "Atlantis," in Perth, Western Australia, nine bottlenose dolphins were released in 1992. Animals were radiotracked after a gradual release back to the waters from which some had been caught 11 years previously (three were captive born). After a few weeks, three animals in very poor condition were returned to a sea pen, but the fate of most was unknown, due to the failure of the radio tags (Gales and Waples, 1993). The animals that were recovered all died of poisoning in late 1999.

The release program for Keiko, the killer whale that starred in the movie "Free Willy," encapsulates some of the issues regarding captivity and release. Caught as a calf by the Icelandic capture industry, imported to Canada then to Mexico, where he was held in an inadequate pool with no other members of his own species, the public was alerted to his plight by his movie role. NGOs organized his importation to another captive facility in the USA, where his condition improved before he was moved to Iceland, where he was held in a sea pen and trained to catch prey prior to his release, in the summer of 2002. In just under 2 months, he swam to Norway, resumed contact with people, and inadvertently became a tourist attraction.

In concert with Norwegian authorities, those responsible for his release moved him to a more secluded area, intending to re-release him. He died of pneumonia in late 2003, ending the 9-year project, and demonstrating that life does not always imitate the movies.

However, not all release projects have failed. In 1990, two young male bottlenose dolphins, held captive for 2 years, were released into the waters from which they were taken. Both animals were successfully re-integrated into their social group and showed no aberrant interactions with humans (Wells *et al.*, 1998). However, this was a well-designed and relatively brief period of captivity, and the animals were released precisely with the group from which they had been captured.

Just as the conservation value of zoos and aquaria should be questioned, so too should the value of release projects. Do the perceived benefits of the ambassadorial value of the release outweigh the costs (e.g., funding for conservation projects foregone) and risks associated with such projects? Likewise, release programs raise their own ethical issues (Waples and Stagoll, 1997). The legacy of Keiko's project—his ambassadorial value appears to be in demonstrating that high-profile, well-designed, expensive release attempts can fail—on future releases remains to be seen.

### F. Funding

Just as some zoos and aquaria contribute significant funds and resources to rescue and rehabilitation programs, some agencies keeping captive marine mammals also support research programs. For example, most of the research into cetaceans' acoustic faculties has been funded through the US Office of Naval Research (ONR). There have been suggestions that the significant contribution made by the ONR to funding cetacean research affects the capacity of scientists to comment openly on issues related to the US Navy, although this is disputed (Gannon *et al.*, 2004; Weilgart *et al.*, 2005). Some captive facilities and groups opposed to captivity either employ full-time research staff or provide funding for research projects. The relative funding provided by the two groups varies dramatically between countries, and there are places where one or both of the two groups contribute significantly to research effort.

### G. Military Use of Marine Mammals

Humans have a long history of using animals for military purposes. The sensory abilities of some animals are appreciably better than those of humans and our tools, so animals are used to detect enemies or weaponry. Tracker dogs (*Canis lupus* spp.) are one example. The information in the public domain regarding military use of marine mammals suggests that this is the major use of marine mammals by the military. The ability of dolphins, small whales, and sea lions to detect and retrieve objects or people, and their ability to make repeated deep dives without suffering from the bends, make them valuable for military operations. For example, in this century dolphins have been used for mine detection in wars (http://www.spawar.navy.mil/sandiego/technology/mammals/mine_hunting.html).

The major argument against this is that it is inherently unethical to train animals for involvement in warfare. Most military uses involve marine mammals working in the open sea, rather than being maintained in enclosures. This can involve animals spending time in waters to which animals may not be physiologically adapted, which is another area of concern. As marine mammals are relatively expensive to train and keep, very few nations use them for military purposes. After the collapse of the Soviet Union, lack of funding to maintain marine mammals used by the Soviet Navy has led to ex-military animals being sold internationally.

### V. Conclusion

The debate continues over whether marine mammals, especially whales and dolphins, should be kept in captivity. In general, less concern is expressed over keeping other marine mammals. Basically, the debate between pro- and anti-captivity groups is simple: does any benefit achieved by holding animals in captivity outweigh any costs involved—both to the individual animal and to the population from which the animal came—assuming that all parties view "benefit" as contributing to the conservation of free-ranging marine mammals. Both sides believe their position to be valid and rational.

The growth of oceanaria in countries with relatively weak rule of law and increasing captures of free-ranging animals to fill the needs of these facilities has created new challenges for those opposing captivity. There are situations where the concern that taking free-ranging animals will threaten populations seems justified. Indo-Pacific humpback dolphins and Irrawaddy (*Orcaella brevirostris*) dolphins in some parts of southeast Asia have been identified as at risk from the burgeoning, and effectively unregulated, aquarium industry. To prevent this, no captures should be permitted unless stock structure is well understood and takes are demonstrably sustainable, but realities on the water suggest that such regulation is unlikely to be introduced or enforced. However, when compared with other anthropogenic threats to the viability of marine mammal populations, takes for captivity appear to generally represent a localized problem.

There will probably always be differences in peoples' ethical stance regarding whether it is appropriate to keep animals, especially cetaceans, in captivity. But perhaps through comparison with other anthropogenic influences, some common ground in the debate can be found. A dichotomy is generally made between captivity and "the wild." "The wild" implies pristine, or at least healthy, environments, unsullied by human intervention. But the influence of human activities, particularly resulting from fisheries and global warming, on marine environments is so pervasive and profound that "the wild" is becoming a misnomer. The dichotomy is between animals held in captivity, in conditions that vary dramatically between institutions, and animals living in oceans that are also affected by peoples' activities, also to differing degrees. On many issues—marine pollution, overfishing, whaling, marine mammal rescue—some (but by no means all) oceanaria and zoos hold positions that are indistinguishable from organizations that oppose captivity. The WAZA position against drive fisheries is an example.

Attempts to rehabilitate marine mammals that have spent most of their lives in captivity have not met with resounding success, yet breeding programs for some species, particularly bottlenose dolphins, have. So marine mammals in captivity appear a fact of life for the foreseeable future. At the same time, substandard facilities are still opening around the world, and capturing free-ranging animals to stock their tanks. In the USA at least, far more people go to zoos and aquaria than go whale watching, and whale watching can have deleterious impacts (Lusseau, 2004; Bejder *et al.*, 2006). Perhaps common ground could be found between the more responsible members of both sides of the debate, in working to improve conditions at substandard facilities, ensuring that live captures are (at the very least) managed humanely and are sustainable, and improving the educational content of all displays.

*See Also the Following Articles*

Captive Breeding ■ Ethics and Marine Mammals ■ Marine Parks and Zoos

*References*

Anonymous (1992). The Agreement on the Conservation of Small Cetaceans of the Baltic and North Seas. Available at: http://www.service-board.de/ascobans_neu/files/agreement1992.pdf.

Anonymous (2004). 2004 AZA Board Meeting: Minutes. Available at: http://www.aza.org/Publications/2004/12/2004BoardMtg.pdf.

Ban, S., and Trites, A. W. (2007). Quantification of terrestrial haul-out and rookery characteristics of Steller sea lions. *Mar. Mamm. Sci.* **23**, 496–507.

Bejder, L., *et al.* (10 authors) (2006). Decline in relative abundance of bottlenose dolphins exposed to long-term disturbance. *Conserv. Biol.* **20**, 1791–1798.

Deeke, V. B. (2006). Studying marine mammal cognition in the wild: A review of four decades of playback experiments. *Aquat. Mamm.* **32**, 461–482.

Gales, N., and Waples, K. (1993). The Rehabilitation and Release of Bottlenose Dolphins from Atlantis Marine Park, Western Australia. *Aquat. Mamm.* **19**, 49–59.

Gannon, D. P., Johnston, D. W., Read, A. J., and Nowacek, D. (2004). Resonance and dissonance: Science, ethics, and the sonar debate. *Mar. Mamm. Sci.* **20**, 898–899.

Guénette, S., Heymans, S. J. J., Christensen, V., and Trites, A. W. (2006). Ecosystem models show combined effects of predation, competition and ocean productivity on Steller sea lions (*Eumetopias jubatus*) in Alaska. *Can. J. Fish. Aquat. Sci.* **63**, 2495–2517.

Kellert, S. R. (1999). "American Perceptions of Marine Mammals and their Management." The Humane Society of the United States, Washington, D.C.

Klinowska, M., and Brown, S. (1986). "A Review of Dolphinaria." Department of the Environment, London.

Lusseau, D. (2004). The hidden cost of tourism: Detecting long-term effects of tourism using behavioral information. *Ecol. Soc.* **9**(1), 2 [online] URL: http://www.ecologyandsociety.org/vol9/iss1/art2/.

Odden, Å., Folkow, L. P., Caputa, M., Hotvedt, R., and Blix, A. S. (1999). Brain cooling in diving seals. *Acta Physiol. Scand.* **166**, 77–78.

Rea, L. D., Rosen, D. A. S., and Trites, A. W. (2007). Utilization of stored energy reserves varies by age and season in Steller sea lions. *Can. J. Zool.* **85**, 190–200.

Reeves, R. R., and Gales, N. J. (2006). Realities of Baiji conservation. *Conserv. Biol.* **20**, 626–628.

Reeves, R. R., and Mead, J. M. (1999). Marine mammals in captivity. *In* "Conservation and Management of Marine Mammals" (J. R. Twiss, and R. R. Reeves, eds), pp. 412–436. Smithsonian Institute Press, Washington, D.C.

Tollit, D. J., Heaslip, S. G., Barrick, R. L., and Trites, A. W. (2007). Impact of diet-selection and the digestion of prey hard remains on determining the diet of the Steller sea lion (*Eumetopias jubatus*). *Can. J. Zool.* **85**, 1–15.

Trites, A. W., Calkins, D. G., and Winship, A. J. (2006). Diet of Steller sea lions, *Eumetopias jubatus*, in Southeast Alaska, 1993–1999. *Fish. Bull.* **105**, 234–248.

Wang, D., Zhang, X., Wang, K., Wei, Z., Würsig, B., Braulik, G. T., and Ellis, S. (2006). Conservation of the Baiji: No simple solution. *Conserv. Biol.* **20**, 622–625.

Waples, K. A., and Stagoll, C. S. (1997). Ethical issues in the release of animals from captivity. *BioScience* **47**, 115–121.

Yang, G., Bruford, M. W., Wei, F., and Zhou, K. (2006). Conservation options for the Baiji: Time for realism? *Conserv. Biol.* **20**, 620–622.

Weilgart, L., Whitehead, H., Rendell, L., and Calambokidis, J. (2005). Signal-to-noise: Funding structure versus ethics as a solution to conflict-of-interest. *Mar. Mamm. Sci.* **20**, 898–899.

Wells, R. S., Bassos-Hull, K., and Norris, K. S. (1998). Experimental return to the wild of two bottlenose dolphins. *Mar. Mamm. Sci.* **14**, 51–71.

# Caspian and Baikal Seals
## *Pusa caspica* and *Pusa sibirica*

NOBUYUKI MIYAZAKI

## I. Characteristics and Taxonomy

Baikal seals (*Pusa sibirica*) and Caspian seals (*P. caspica*) have common features such as small size, delicate skull, and affinity for ice. Based on mtDNA haplotypes, Caspian seals were derived from the common ancient type of *Pusa* 60 million years ago and were subsequently isolated in the Caspian Sea. Baikal seals were derived from a ringed-seal-like ancestor in the Arctic Ocean and isolated in Lake Baikal 40 million years ago (Sasaki *et al.*, 2003).

Baikal seals, which do not have distinct spots, are uniform dark silver gray dorsally and light yellowish gray ventrally (Fig. 1). Caspian seals are irregularly spotted with brown or black against a light grayish yellow background (Fig. 2).

The Baikal seal possesses enlarged eyes that enable it to find prey in water as deep and clear as that in Lake Baikal (Endo *et al.*, 1998a, b, 1999). The skull of the Caspian seal possesses the same thin frontal bone and dorsoventrally developed zygomatic arch found in the Baikal seal that are required to accommodate the enlarged eyeball in the orbit (Endo *et al.*, 2002). A large zygomatic width, greater length of jugal, and smaller orbital width are correlated with the large orbit in these species. There is a slightly closer morphological affinity between Baikal and ringed seals than between Baikal and Caspian seals. This relationship coincides well with the genetic relationship among these indicated by analysis of mtDNA (Sasaki *et al.*, 2003).

## II. Distribution and Abundance

The population of Baikal seals from 1971 to 1978 was estimated to be between 68,000 and 70,000 animals (Pastukhov, 1978a). In 1987–1988, an outbreak of morbillivirus infection resulted in a large mass mortality of Baikal seals (Grachev *et al.*, 1989). A mass death also occurred in 1998.

The Caspian seal population declined from about 1 million animals early in the twentieth century to 360,000–400,000 by the end of the 1980s (Krylov, 1990). In the spring of 1997, a mass death of several thousand seals occurred.

## III. Ecology

Baikal seals feed mainly on four fish species: the greater golomyanka (*Comephorus baicalensis*), the lesser golomyanka (*C. dybowskii*), the Baikal yellowfin sculpin (*Cottocomephorus grewingki*), and the longfin sculpin (*C. comephoroides*), all of which are not of commercial value. In captivity, an adult Baikal seal consumed up to 5.6 kg of fish per day (Pastukhov, 1969).

Caspian seals in the northern Caspian Sea feed on *Clupeonella engrauliformis*, *C. grimmi*, *C. delicate caspia*, Gobiidae, *Rutilus rutilus caspicus*, *Atherina mochon pontica*, *Lucioperca lucioperca*, other fish species, and crustaceans (Khuraskin and Pochtoyeva, 1997). It is estimated that an adult Caspian seal takes 2–3 kg fish per day, or approximately 1 metric ton of fish per year.

Hazardous chemicals (heavy metals, organochlorine compounds, organotin compounds, radionuclides, etc.) are present in high concentrations in Baikal and Caspian seals, which are long-lived

Figure 1 *Baikal seal. Photo courtesy of Dr. Yuuki Watanabe.*

Figure 2 *Caspian seal on Pearl Island, northwestern North Caspian Sea.*

(>40 year) top predators in the enclosed ecosystems (Dietz *et al.*, 1990; Frank *et al.*, 1992; Nakata *et al.*, 1995; Watanabe *et al.*, 1998, 1999; Kajiwara *et al.*, 2002 Yoshitome *et al.*, 2003; Ikemoto *et al.*, 2004).

## IV. Behavior and Physiology

Movements and dive patterns of Baikal seals appear to be associated primarily with seasonal diet movements of their primary prey, golomyanka and sculpins, and are correlated secondarily with patterns of ice formation and thaw. Based on data obtained by Argos satellite-linked transmitters, most dives of Baikal seals are to depths of 10–50 m, although a few dives exceed 300 m (Stewart *et al.*, 1996). Dives may last between 2 and 6 min, but a few dives exceed 40 min. According to Watanabe *et al.* (2004), who used an advanced Japanese data logger, Baikal seals dived almost continuously, to an average of 68.9 m, with dives deeper than 150 m being concentrated around dusk and dawn. Maximum depth was 245 m. Mean and maximum duration was 6.0 min and 13.5 min, respectively. In the daytime, dives were characterized by higher swimming speeds (mean: 1.2 m/sec) and upward-directed acceleration events. At night, dives were shallower around midnight and characterized by lower speeds (0.9 m/sec) and non-directional deceleration events. Baikal seals actively chased pelagic fishes such as golomyanka during the day and swam upward at around 2 m/sec. at a body angle of 40°, suggesting that the

seals use vision to search and chase for silhouetted prey against the brighter water overhead. Experimental research on stroke-and-glide swimming pattern using lead weights indicated that body density of the seal is 1027–1046 kg/m$^3$, corresponding to 32–41% lipid content, for the weighted condition, and 1014–1022 kg/m$^3$, 43–47% lipid content, for the unweighted condition (Watanabe *et al.*, 2006).

For two adult male Caspian seals using Argos satellite-linked transmitters, most dives were less than 50 m in depth whereas a few exceeded 200 m. Dives were mostly less than 50 sec long but some exceeded 200 sec.

Mass die-off of thousands of Baikal seals in Lake Baikal occurred in 1987–1988, and the virus isolated from the dead Baikal seals was identified as canine distemper virus, which genetically and antigenetically was close to the canine distemper viruses isolated from a dog and a ferret (Grachev *et al.*, 1989; Osterhaus *et al.*, 1989; Mamaev *et al.*, 1995, 1996). According to Ohashi *et al.* (2001), a virus neutralizing test and ELISA (enzyme-linked immunosorbent assay) clearly suggested that a distemper virus epidemic occurred in Caspian seals in early 1997 and that the canine distemper virus infection has continued to occur in Lake Baikal in recent years. The genome of canine distemper virus found in one of the dead Caspian seals and the sequence of the *P* gene in that animal were distinct from those of the Baikal seal virus, laboratory strains, and field isolates from terrestrial animals in other area (Forsyth *et al.*, 1998). According to Kennedy *et al.* (2000), the mass die-off of Caspian seals was caused by a canine distemper virus.

## V. Life History

The maximum known age in both sexes in Baikal seals is 56 years for females and 52 for males (Pastukhov, 1993). According to Amano *et al.* (2000), the oldest age of Baikal seals in samples (*n* = 73) collected in 1992 was 24.5 years for females and 35.5 years for males. In Caspian seals collected from Pearl Island in the western North Caspian Sea (*n* = 118), the oldest age was 43.5 years for females and 33.5 years for males.

Growth in length of Baikal seals appears to cease around the age of 15 years (Amano *et al.*, 2000). The seals may continue to grow for 8–9 years after the age of sexual maturation (6 years for females and 7 years for males). Asymptotic body length is 140 cm in males and 130 cm in females. In Caspian seals, growth appears to cease around the age of 10 years, which is the age of sexual maturation in both sexes. Asymptotic body length is 118 cm in males and 111 cm in females.

Most Baikal seals breed by 6 years for females and 7 years for males (Thomas, 1982). Newborn pups are 65 cm in body length and 4.1 kg in body weight on average. A rather high rate of twinning (4% of annual births) is exhibited compared to other seals (Pastukhov, 1968). Mating may occur underwater in March at about the time mothers wean their pups. Mothers nurse the pups in a birthing lair. The lactation period is estimated at 2–3 months. The mating system is assumed to be polygamous with little or no pair bonding. In winter, when Lake Baikal is covered with ice averaging 80–90 cm in thickness with a maximum of 1.5 m, seals are sighted throughout the lake and adjacent to breathing holes in the ice. In Baikal seals of 7 years or more, 84% of females give birth to a pup yearly (Pastukhov, 1993).

Caspian seal pups are born on the ice from the middle of January to the end of February and are about 60 cm in length. Mating takes place between the end of February and the middle of March. Sexual maturity is attained at 4–6 years in females and 6 years in males (Ognev, 1935; Fedoseev, 1975). The pregnancy rate of Caspian seals over 9 years was 31.3% ($n = 30$) in 1997 and 1998.

## VI. Interactions with Humans

The number of Baikal seals taken annually has varied. Before 1917 about 2000 to 9000 were taken; in 1930 about 6000; and currently between 5000 and 6000 (Pastukhov, 1978b). According to Khuraskin and Pochtoyeva (1997), 115,000–174,000 Caspian seals have been hunted annually since the early nineteenth century. A total of 86,000 animals were killed in 1966. From 1970, seal hunting on the northern ice was limited to a catch of 20,000–25,000 pups.

## See Also the Following Article

Earless Seals (*Phocidae*)

## References

Amano, M., Miyazaki, N., and Petrov, E. A. (2000). Age determination and growth of Baikal Seals (*Phoca sibirica*). *Adv. Ecol. Res.* **31**, 449–462.

Anderson, S. S., Livens, F. R., and Singleton, D. L. (1990). Radionuclides in grey seals. *Mar. Pollut. Bull.* **21**, 343–345.

Dietz, R. C., Nielsen, C. O., Hansen, M. M., and Hansen, C. T. (1990). Organic mercury in Greenland birds and mammals. *Sci. Total Environ.* **95**, 41–51.

Endo, H., Sasaki, H., Hayashi, Y., Petrov, E. A., Amano, M., and Miyazaki, N. (1998a). Macroscopic observations of the muscles of the face, and eye in the Baikal seal (*Phoca sibirica*). *Mar. Mamm. Sci.* **14**, 778–788.

Endo, H., Sasaki, H., Hayashi, Y., Petrov, E. A., Amano, M., and Miyazaki, N. (1998b). Functional relationship between muscles of mastication, and the skull with enlarged orbit in the Baikal seal (*Phoca sibirica*). *J. Vet. Med. Sci.* **60**, 699–704.

Endo, H., Sasaki, H., Hayashi, Y., Petrov, E. A., Amano, M., and Miyazaki, N. (1999). CT examination of the head of Baikal seal (*Phoca sibirica*). *J. Anat.* **194**, 119–126.

Endo, H., Sasaki, S., Arai, T., and Miyazaki, N. (2002). The muscles of mastication in the Caspian seal (*Phoca caspica*). *Anat. Histol. Embryol.* **31**, 262–265.

Fedoseev, G. A. (1975). Principal population indicators of dynamics of numbers of seals of the family Phocidae. *Ekologiya* **5**, 62–70 [Transl. Consultants Bureau, New York, 439–446].

Forsyth, M. A., Kennedy, S., Wilson, S., Eybatov, T., and Barett, T. (1998). Canine distemper virus in a Caspian seal. *Vet. Rec.* **143**, 662–664.

Frank, A., Galgan, V., Roos, A., Olsson, M., Petersson, L. R., and Bignert, A. (1992). Metal concentrations in seals from Swedish waters. *Ambio* **21**, 529–538.

Froehlich, K., Rozanski, K., Povinec, P., Oregioni, B., and Gastaud, J. (1999). Isotope studies in Caspian Sea. *Sci. Total Environ.* **237/238**, 419–427.

Grachev, M. A., *et al.* (1989). Distemper virus in Baikal seals. *Nature* **338**, 209.

Ikemoto, T., *et al.* (2004). Comparison of trace element accumulation in Baikal seals (*Pusa sibirica*), Caspian seals (*Pusa caspica*) and northern fur seals (*Callorhinus ursinus*). *Environmental Pollution* **127**, 83–97.

Kajiwara, N., *et al.* (2002). Organochlorine and organotin compounds in Caspian seals (*Phoca caspica*) collected during an usual mortality event in the Caspian Sea in 2000. *Environ. Pollut.* **117**, 391–402.

Kennedy, S., *et al.* (2000). Mass die-off of Caspian seals caused by canine distemper virus. *Emerg. Infect. Dis.* **6**, 1–5.

Khuraskin, L. S., and Pochtoyeva, N. A. (1997). Status of the Caspian seal population. *In* "Caspian Environment Program" (H. Dumont, S. Wilson, and B. Wazniewicz, eds), pp. 86–94. Proceedings of the First Bio-network Workshop, Bordeaux, November 1997, World Bank, Washington, DC.

Krylov, V. I. (1990). Ecology of the Caspian seal. *Finn. Gamete Res.* **47**, 32–36.

Mamaev, L. V., *et al.* (1995). Characterisation of morbilliviruses isolated from Lake Baikal seals (*Phoca sibirica*). *Vet. Microbiol.* **44**, 251–259.

Mamaev, L. V., *et al.* (1996). Canine distemper virus in lake Baikal seals (*Phoca sibirica*). *Vet. Rec.* **138**, 437–439.

Nakata, H., Tanabe, S., Tatsukawa, R., Amano, M., Miyazaki, N., and Petrov, E. A. (1995). Persistent organochlorine residues and their accumulation kinetics in Baikal seal (*Phoca sibirica*) from Lake Baikal, Russia. *Environ. Sci. Technol.* **29**(11), 189–197.

Ognev, S. I. (1935). "Mammals of USSR and adjacent countries. Vol. III: Carnivora (Fissipedia and Pinnipedia)." Acad. Sci. USSR, Moscow. [In Russian; English translated by A. Birron and Z. S. Coles for Israel Program for Scientific Translation, 1962].

Ohashi, K., *et al.* (2001). Seroepidemiological survey of distemper virus infection in the Caspian Sea and in Lake Baikal. *Vet. Microbiol.* **82**, 203–210.

Osterhaus, A. D. M. E., Groen, J., UytdeHaag, F. G. C. M., Visser, I. K. G., van de Bildt, M. W. G., Bergman, A., and Klingeborn, B. (1989). Distemper virus in Baikal seals. *Nature* **338**, 209–210.

Pastukhov, V. D. (1968). On twins in *Pusa sibirica* Gmel. *Zool. Zhurnal* **47**, 479–482. [English summary.]

Pastukhov, V. D. (1969). Some results of observations on the Baikal seals under experimental conditions. *In* "IV oye Vsesoyuznaya Konferentsiya po Izucheniyu Morskikh Mlekopitayushchikh." Tezisy Dokladov. Can. Fish. Res. Bd. Translation Series, No. 3544.

Pastukhov, V. D. (1978a). Scientific-production experiment on the Baikal seal. *In* "Morskiye Mlekopitayushchiye," pp. 257–258. Moscow, USSR.

Pastukhov, V. D. (1978b). Baikal seal. *In* "Problemy Baikala" (G. I. Galaziy, and K. K. Votintsev, eds), pp. 251–259. Nauka, Sibirskoye, Otdeleniye, Novosibirsk.

Pastukhov, V. D. (1993). "Baikal Seals." Nauka, Moscow, USSR.

Sasaki, H., Numachi, K., and Grachev, M. A. (2003). The origin and genetic relationship of the Baikal seal, *Phoca sibirica*, by restriction analysis of mitochondrial DNA. *Zool. Sci.* **20**, 1417–1422.

Stewart, B. S., Petrov, E. A., Baranov, E. A., Timonin, A., and Ivanov, M. (1996). Seasonal movements and dive patterns of juvenile Baikal seals, *Phoca sibirica*. *Mar. Mamm. Sci.* **12**(4), 528–542.

Thomas, J. (1982). Mammalian species. *Am. Soc. Mamm.* **188**, 1–6.

Yoshitome, S., Kunito, T., Ikemoto, T., Tanabe, S., Zenke, H., Yamauchi, M., and Miyazaki, N. (2003). Global distribution of radionucleides ($^{137}CS$ and $^{40}K$) in marine mammals. *Environ. Sci. Technol.* **37**, 4597–4602.

Watanabe, I., Tanabe, S., Amano, M., Miyazaki, N., Petrov, E. A., and Tatsukawa, R. (1998). Age-dependent accumulation of heavy metals in Baikal seal (*Phoca sibirica*) from the Lake Baikal. *Arch. Environ. Contam. Toxicol.* **35**, 518–526.

Watanabe, M., Tanabe, S., Tatsukawa, R., Amano, M., Miyazaki, N., Petrov, E. A., and Khuraskin, S. L. (1999). Contamination levels and specific accumulation of persistent organochlorines in Caspian seal (*Phoca caspica*) from the Caspian Sea, Russia. *Arch. Environ. Contam. Toxicol.* **37**, 396–407.

Watanabe, Y., Baranov, E. A., Sato, K., Naito, Y., and Miyazaki, N. (2004). Foraging tactics of Baikal seals differ between day and night. *Mar. Ecol. Prog. Ser.* **279**, 283–289.

Watanabe, Y., Baranov, E. A., Sato, K., Naito, Y., and Miyazaki, N. (2006). Body density affects stroke patterns in Baikal seals. *J. Exper. Biol.* **209**, 3269–3280.

# *Cephalorhynchus* Dolphins
## *C. heavisidii, C. eutropia, C. hectori,* and *C. commersonii*

### Stephen M. Dawson

## I. Characteristics and Taxonomy

The four dolphins of the genus *Cephalorhynchus* are small, coastal species. They are blunt-headed (hence the frequent mistake of calling them porpoises), chunky dolphins with rounded, almost paddle-shaped flippers. The most characteristic feature of the genus is the dorsal fin, which is proportionally large, either with a shallowly sloping leading edge and a rounded, convex trailing edge [like a Mickey Mouse ear: Hector's (*C. hectori*), Commerson's (*C. commersonii*), and Chilean dolphin (*C. eutropia*)] or upright and roughly triangular (Heaviside's dolphin, *C. heavisidii*). In color pattern, Chilean dolphins and Hector's dolphins are most similar.

The *Cephalorhynchus* dolphins are among the smallest dolphins. Indeed, in length, Hector's dolphins from New Zealand's South Island (maximum 145 cm; *ca.* 50 kg) and South American Commerson's dolphins (maximum 146 cm; *ca.* 45 kg) are the smallest of all dolphins. Both Hector's and Commerson's dolphins have isolated populations in which individuals grow larger. Commerson's dolphins at the Kerguelen Islands grow to 174 cm (86 kg) and Maui's dolphins (North Island Hector's dolphins) reach at least 152 cm (65 kg). In both Hector's and Commerson's dolphins females are 5–10% larger than males. Far fewer Heaviside's and Chilean dolphins have been measured, so it is not clear whether females are larger in these species also. Heaviside's dolphins reach about 174 cm (*ca.* 75 kg), and Chilean dolphins reach at least 167 cm (*ca.* 63 kg).

Studies of mtDNA suggest that the *Cephalorhynchus* dolphins originated from a common Lissodelphinine ancestor in South Africa. Following the west-wind drift, the genus colonized New Zealand, then South America. Commerson's dolphin and Chilean dolphin are proposed to have speciated along the two coasts of South America during the glaciation of Tierra del Fuego (Pichler *et al.*, 2001).

Two of the *Cephalorhynchus* species have genetically isolated populations. One isolate, a subspecies of Commerson's dolphin (*C. commersonii kerguelenensis*; Robineau *et al.*, 2007) is found at the Kerguelen Islands (see later) and probably arose via a founder event, perhaps ten thousand years ago. The other, a subspecies of Hector's dolphin, (*C. hectori maui*; Baker *et al.*, 2002) probably arose via this species' extreme site fidelity and a population bottleneck. South Island Hector's dolphins (*C. hectori hectori*) are fragmented into three genetically distinctive populations (east coast, west coast, south coast), which could only be maintained if there were very low levels of interchange among them (Pichler *et al.*, 2001). Interestingly, a recent study of mtDNA sequences in Heaviside's dolphin from almost 1000 nmi of South African/Nambian coast showed no evidence of population subdivision (van Vuuren *et al.*, 2002).

The *Cephalorhynchus* dolphins are all shallow-water species, but their radiation shows that they have made exceptional movements establishing new populations in similar habitats—a process perhaps triggered by low water temperatures and extensive glaciation during ice ages.

## II. Distribution and Abundance

*Cephalorhynchus* dolphins are found only in Southern Hemisphere waters (Fig. 1). Heaviside's dolphin occurs off the west coast of South Africa and Namibia. Hector's dolphin is found solely off New Zealand. The Chilean dolphin is found in the coastal waterways of Chile and along the exposed west coast. The remaining species, Commerson's dolphin, has the strangest distribution. Its principal stronghold is in the inshore waters of Argentina and in the Strait of Magellan, but it also occurs at the Falkland Islands, and has an isolated population 8500 km away at the Kerguelen Islands, in the Indian Ocean. The Kerguelen Commerson's dolphins are larger, retain the darker juvenile coloration into adulthood, and show skeletal and genetic differences. All *Cephalorhynchus* species favor waters less than 100 m deep and are often seen in the surf zone, especially in summer.

Of the four species, only Hector's dolphin has had its abundance assessed quantitatively throughout its range. The South Island subspecies is most common along the east and west coasts of the South Island between 41°30'S and 44°30'S, with regions of high abundance at Banks Peninsula and between Greymouth and Westport. Total abundance is estimated at 7270 (CV = 16.2%; Dawson *et al.*, 2004; Slooten *et al.*, 2004). The North Island subspecies is found on the west coast between 36°25'S and 39°S but is regularly seen only between the entrance of Manukau harbor and Port Waikato. Its abundance is estimated at 111 (CV = 44%; Slooten *et al.*, 2006b).

South American Commerson's dolphins are found off the east coast between Rio Negro (40°S) and Cape Horn (55°15'S) down into the Drake Passage (61°50') and also at the Falkland Islands. They are common between Peninsula Valdez and Tierra del Fuego (42°S to 54°S) and very common in the eastern Straits of Magellan, where an aerial survey in summer 1984 indicated a population of around 3200. A more recent survey (June 1996) suggested a much smaller population size (1206; CV = 27%; Lescrauwaet *et al.*, 2000) but it is not clear whether this difference is due to a decline in abundance, differences in survey methods, or seasonal redistribution of the population. Recent aerial surveys suggest a total population of about 21,000. At Kerguelen, Commerson's dolphins are restricted to the immediate vicinity of the islands, where they are most frequently seen on the eastern side in the Golfe du Morbihan.

Heaviside's dolphins are found on the west coast of Southern Africa from 17°09'S on the Namibian coast to around Cape Town (34°13'S). Sightings are frequent between Walvis Bay and Cape Town. Abundance in the southern part of its range, along a 390-km stretch of the west coast north of Cape Town is estimated at 6345

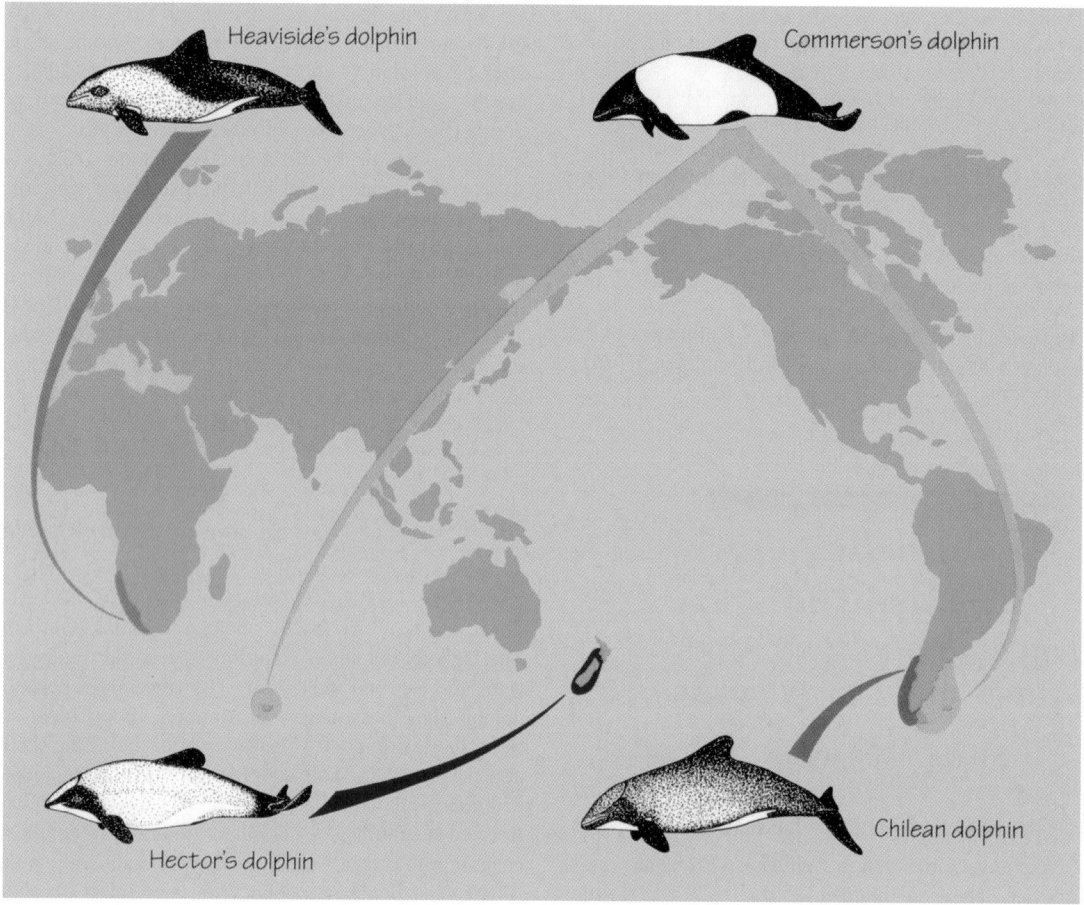

**Figure 1**    *Distribution of the four* Cephalorhynchus *dolphins.*

(CV = 26%; Elwen, 2007). High mtDNA diversity also implies a relatively large population size.

Chilean dolphins have been seen over a very wide latitudinal range, from Valparaíso (33°S) to near Cape Horn (55°15'S), on both open and sheltered coasts. They are seen regularly in only a few places, however, including Valdivia, Golfo de Arauco and at Isla Chiloé. These local populations are likely to be very small. For example, at southern Isla Chiloé, among the places where this species is most commonly seen, a recent detailed study estimated a local population of 59 (CV = 4%; Heinrich, 2006). Sightings in the Chilean channels are very rare. A boat survey covering 1600 nmi in March/April 2006 made only five sightings of Chilean dolphins between Ushuaia and the southern tip of Isla Chiloé. The total population appears to be very small (low thousands at most). Suggestions that the species is becoming very rare are impossible to refute without further dedicated survey work.

As part of work to assess the likely effectiveness of marine protected areas for conserving Hector's dolphin (and Maui's dolphin), the dolphins' offshore distribution has been studied via intensive aerial surveys off the South Island west coast, at Banks Peninsula on the east coast, and the North Island west coast. At Banks Peninsula, where large rivers to the north and south have resulted in broad shallowly shelving underwater plains, Hector's dolphins range much farther offshore than elsewhere (maximum 16.3 nmi in summer; 18.2 nmi in winter; maximum depth = 90 m). In similar surveys off the west coasts of the North and South Islands we have not seen the species beyond 5.3 nmi offshore, nor in water deeper than 60 m

(Rayment *et al.*, unpublished data). In all areas they are very seldom seen in water deeper than 50 m in summer.

## III. Ecology

### A. Diet

All four species feed on a wide variety of coastal prey, focusing on benthic and small pelagic schooling fish and squid. The South American species supplement their fish/squid diet with crustaceans (mysids, euphausiids, and *Munida* spp.) and, strangely, algae (Goodall, 1994a, b). Juvenile hake are the dominant prey item in the diet of Heaviside's dolphins, which, unusually in this genus, also includes a significant component of octopus. The diet of Hector's dolphins is more varied on the South Island east coast, where eight species of fish and squid make up 80% of the diet by mass, than it is on the corresponding west coast (where four species make up 80%; Dawson and Slooten, 1996).

Individual dolphins in this genus appear to be at least seasonally resident in a local area. Long-running, year-round studies on Hector's dolphins have shown that individuals usually range over about 30 km of coastline (Bräger *et al.*, 2002). Chilean dolphins at Isla Chiloé appear to have similar, or even smaller home ranges (Heinrich, 2006). In all species at least some individuals remain in local areas year-round, though inshore abundance is generally reported to be lower in winter. Individual Hector's dolphins have been resighted in the same general area year-round, for over 20 years. In general Hector's dolphins tend to spread out, form smaller groups and range further from shore in winter. This seasonal difference in distribution appears to be greatest at Banks Peninsula, where, on intensive aerial surveys out to 15 nmi offshore, 81% of dolphins seen were within 4 nmi from shore in summer, but only 44% in winter (Slooten *et al.*, 2006a). Similar surveys on the South Island west coast show that fewer dolphins are found within the first mile offshore in winter, but that the limit of offshore extent is very similar in summer (5.3 nmi) and winter (5.2 nmi; Rayment *et al.*, unpublished data).

There is no evidence for large-scale migration in any of the *Cephalorhynchus* species. The largest documented movements are of about 250 km, in Commerson's dolphins (Coscarella, 2005). Despite wide-ranging surveys over more than 20 years, no two sightings of the same individual Hector's dolphin are more than 106 km apart. Small home ranges have been confirmed by satellite tagging in Heaviside's dolphin and Hector's dolphins. The data from Heaviside's dolphins suggest inshore–offshore diurnal movement, most probably driven by movements of their principal prey, juvenile hake (Elwen *et al.*, 2006).

### IV. Behavior and Physiology

Small group sizes are characteristic of this genus. Though occasionally sightings of more than 50 have been made, most sightings are of between 2 and 10 individuals. In high-density areas, there may be several such groups nearby. In Hector's dolphin, these often coalesce to form a large, temporary group of 25 or so. Large groups often show boisterous behavior such as chases, leaps, and lobtailing. Sexual displays and copulation are also more common after groups have joined (Slooten, 1994). Such large groups are usually short-lived; often splitting after 10–30 min. Frequently groups lose, gain or swap members in this process. Associations among adult Hector's dolphins are weak. It is likely that the other species are similar.

Hector's, Commerson's, and Heaviside's dolphins are strongly attracted to boats, and readily bow-ride. Chilean dolphins are commonly reported to be shy and elusive, and this is confirmed by a recent hilltop study of reactions to boats in Yaldad Bay, on Isla Chiloé (Ribiero *et al.*, 2005). Some groups, however, are boat positive, repeatedly bow-riding or wake-surfing (see Fig. 2). Chilean dolphins have been heavily hunted in parts of their range and their usual wariness of boats has probably been acquired.

All *Cephalorhynchus* species are thoroughly at home in turbulent water close to shore and are frequently seen surfing.

The sounds of Hector's dolphin and Commerson's dolphin have been studied in detail. Almost all the sounds are short (*ca.* 140 μsec, roughly 1/7000 sec), high-frequency (*ca.* 125 kHz) narrow-band, ultrasonic clicks (Fig. 3) which are used in "trains" of a few dozen to several thousand (Kamminga and Wiersma, 1982; Dawson and Thorpe, 1990). Clicks may contain one, two or three main pulses and appear to have a role in communication and well as sonar. Hector's dolphins use complex clicks disproportionately in social contexts and a particular

**Figure 2**  *The four* Cephalorhynchus *dolphins (clockwise from upper left): Hector's dolphin (photo by author), Commerson's dolphin (photo by Marine Mammal Lab, CENPAT, Puerto Madryn), Heaviside's dolphin (photo by Simon Elwen, University of Pretoria), and Chilean dolphin (photo by author).*

type of click is used preferentially in feeding (Dawson, 1991). Click rates can be exceptionally high (maximum 1149 clicks/sec). Such high click rates generate an audible tone of the same frequency as the click rate. These signals, which sound like a "cry" or "squeal," are strongly linked to aerial behaviors in Hector's dolphins, and apparently indicate excitement. They have been recorded from all four *Cephalorhynchus* species. It seems that the *Cephalorhynchus* dolphins do not whistle, as other delphinids do.

Preliminary analysis of Maui's dolphin sounds show no obvious differences to those made by South Island Hector's dolphins. Given this, and the close similarity of the sounds of Hector's and Commerson's dolphins, it is likely that the sounds of Heaviside's and Chilean dolphins will be similar also.

## V. Life History

Reproduction has been studied only in Commerson's dolphins and Hector's dolphins, but the limited information available for Heaviside's and Chilean dolphins suggest they are similar. Females bear their first calf between 6 and 9 years, in spring to late summer. Males reach sexual maturity at 5–9 years. Mating and calving occur in spring to late summer, and the gestation period is around 10–11 months. Maximum ages so far found via ageing teeth from dissected dolphins are 20 (Hector's dolphin; Slooten, 1991) and 18 (Commerson's dolphin; Lockyer *et al.*, 1988). At Banks Peninsula, six photographically identified Hector's dolphins were first seen in 1985 and known to be alive in 2006, hence are at least 22 years old.

Studies of photographically identified Hector's dolphins suggest that mature females calve every 2–4 years. Late maturity, relatively short life, and long intervals between calves inevitably result in a low population growth rate. Modeling studies have shown that Hector's

dolphin populations, in the absence of human impact, would be expected to grow by about 2% per year (Slooten and Lad, 1991).

## VI. Interactions with Humans

All four species have, at some stage, been hunted for food or bait. The South American species have fared worst: they have been extensively hunted for bait for crab pots. In Chile, estimates of numbers taken range from 1250 to 4120 dolphins/year (comprising mostly Peale's, Chilean, and Commerson's dolphins; Lescrauwaet and Gibbons, 1994) in the late 1970s and probably reached a maximum between 1980 and 1986. As many as 1300–1500 Chilean dolphins may have been harpooned each year near the Western Strait of Magellan in the late 1970s and early 1980s (Goodall, 1990b). Chilean dolphins are now virtually absent from this area. Direct take for bait seems to have ceased; legislation is now more restrictive and the crab fishery has declined. However, there is no monitoring of cetacean catches (deliberate or incidental), and enforcement of existing legislation in such a remote and convoluted area is practically impossible.

All four *Cephalorhynchus* species suffer some degree of incidental catch in fishing gear. In South America, bycaught animals have been used as bait and for human consumption, but it is suspected that this is now uncommon. Numbers taken currently are unknown for any of the four species. While a wide range of fishing methods are known to have resulted in *Cephalorhynchus* mortality (including gillnetting, purse seining, trawling, lobster potting), gillnets pose a larger problem than any other fishing method.

In the mid-1980s an average of 57 Hector's dolphins were caught each year in gillnets in the Canterbury region of the east coast of New Zealand's South Island. Given the low reproductive rate of this species, it was clear that this level of catch was unsustainable.

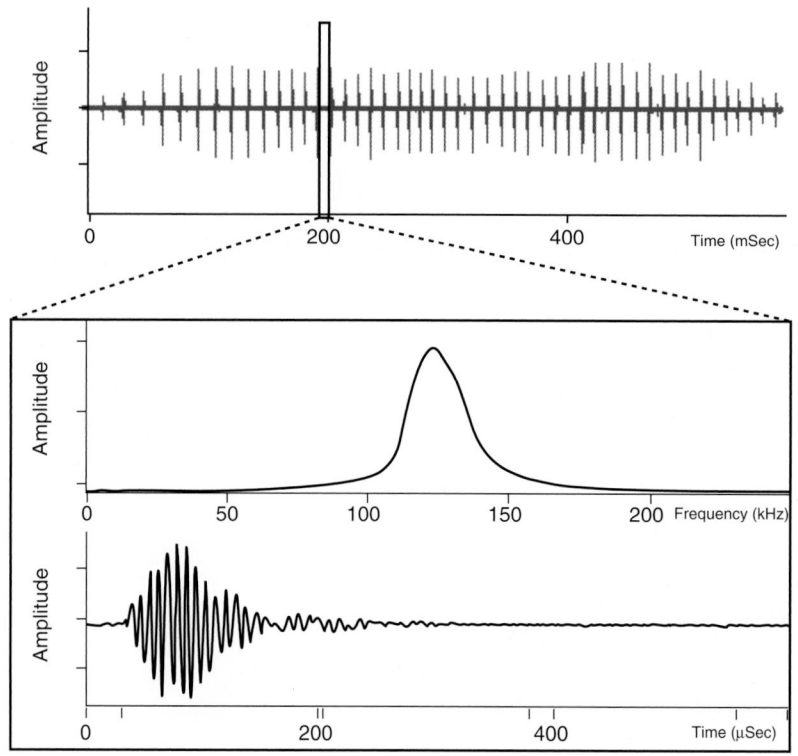

**Figure 3** *A "train" of high-frequency Hector's dolphin clicks, with a "zoomed" section showing the waveform and spectrum shown for an example click.*

An 1170-km² sanctuary, in which commercial gillnetting is effectively illegal and amateur gillnetting restricted to particular times and places was established in 1989. This has reduced, but not solved the bycatch problem in Canterbury (Dawson and Slooten, 2005). Analysis of observed catches in the Canterbury region in 1997/1998 suggest that 18 Hector's dolphins were caught in commercial gillnets to the north, south, and offshore of the current sanctuary. Observer coverage since has confirmed a continuing bycatch problem but has been insufficient to estimate how many are taken.

On the shores of Argentina it is very common to see intertidal gillnets set from the shore. This is an artisanal fishery targeting inshore fish species and king crabs. The nets are normally fixed to the bottom at low tide and take advantage of large tidal ranges to be completely submerged at high tide. Bycatch of Commerson's dolphin on the La Angelena coast and in the Ria Gallegos estuary in the Santa Cruz province has been estimated at 179 for the 1999/2000 fishing season (Iñíguez *et al.*, 2003). Given that this style of gillnetting is widely used in Argentina, and that members of the *Cephalorhynchus* genus have low reproductive rates and very low potential for population growth, bycatch of Commerson's dolphins, at least in this area, is almost certainly unsustainable. Bycatch of Commerson's dolphins is also associated with the rapidly expanding mid-water trawl fisheries along the Atlantic Patagonian coast.

Salmon and mussel farming is becoming widespread in the sheltered inlets of Chile, especially north of 44°S. This development is most intensive in Isla Chiloé, where it overlaps with resident groups of Chilean dolphins. In Yaldad Bay, these dolphins have been shown to avoid mussel farms (Ribiero *et al.*, 2007). In addition to direct competition for space, increased aquaculture-related boat traffic has caused changes in dolphin behavior (Ribiero *et al.*, 2005), and is likely to exacerbate overall impact. There is significant potential for burgeoning aquaculture to displace dolphins from some of these areas.

There appear to be fewer threats to Heaviside's dolphin than to the others (Best and Abernathy, 1994). Its abundance, along with high levels of genetic diversity and apparent lack of genetic population structure are all good signs. It seems that this species, and Commerson's dolphin, despite the high level of impacts on the latter, are the most numerous members of this genus.

The Chilean dolphin needs urgent consideration. Impacts over the last three decades have been severe and the population is substantially reduced. Recent detailed studies at Isla Chiloé confirm high site fidelity (Heinrich, 2006), which appears to be a characteristic of this genus. In these small local populations, even an apparently low level of mortality, such as that caused by increasing use of intertidal gillnets to catch escaped salmon at Chiloé, could be enough to ensure continuing decline. High site fidelity also increases susceptibility to habitat modification and decreases potential for declining local populations to be replenished from elsewhere. Hence there is real risk of population fragmentation, as has happened in Hector's dolphin. Wide-ranging population surveys are urgently needed before we can say with confidence whether the species numbers hundreds (as the most dire warnings imply) or thousands.

South Island Hector's dolphins have received some direct conservation action, via the Banks Peninsula Marine Mammal Sanctuary. Recent surveys and modeling suggest that the Banks Peninsula population is still declining. In part this is because they range far further offshore at Banks Peninsula than elsewhere, and hence the Sanctuary's 4 nmi offshore boundary provides less protection than was assumed (Slooten *et al.*, 2006a).

Maui's dolphins (North Island Hector's dolphins) are now protected from gillnetting along a 210 nmi stretch of open coast (most of the dolphins' current range), out to 4 nmi offshore. Both commercial and amateur gillnetting is allowed within the harbors in this zone, and trawling is permitted on the open coast beyond one nautical mile from shore. This protection was brought about by evidence of (a) apparent contraction in alongshore range over the last 20 years, (b) decline in the number of mtDNA lineages present in current samples, (c) analyses of the level of bycatch in fishing operations, (d) continuing discovery of gillnet-marked Maui's dolphin carcases, and (e) bycatch in gillnets and trawl nets reported during interviews with fishers (Dawson *et al.*, 2001). Maui's dolphins currently number about 111 individuals (CV = 44%; Slooten *et al.*, 2006b) and are probably the most seriously endangered marine mammal.

## See Also the Following Articles

Bycatch ■ Fishing Industry, Effects of

## References

Baker, A. N., Smith, A. N. H., and Pichler, F. B. (2002). Geographical variation in Hector's dolphin: Recognition of new subspecies of *Cephalorhynchus hectori*. *J. R. Soc. NZ* **32**, 713–727.

Best, P. B., and Abernathy, R. B. (1994). Heaviside's dolphin *Cephalorhynchus heavisidii* (Gray, 1828). *In* "Handbook of Marine Mammals," Vol. V, Delphinidae and Phocoenidae (S. H. Ridgway, and R. Harrison, eds), pp. 241—267. Academic Press. New York.

Bräger, S., Dawson, S. M., Slooten, E., Smith, S., Stone, G. S., and Yoshinaga, A. (2002). Site fidelity and along-shore range in Hector's dolphin, an endangered marine dolphin from New Zealand. *Biol. Cons.* **108**, 281–287.

Coscarella, M. A. (2005). Ecología, comportamiento y evaluación del impacto de embarcaciones sobre mandas de tonina overa *Cephalorhynchus commersonii* en Bahía Engano, Chubut. Ph.D. Thesis, Unversidad de Buenos Aires, Argentina.

Dawson, S. M. (1991). Clicks and communication: The behavioural and social contexts of Hector's dolphin vocalisations. *Ethology* **88**, 265–276.

Dawson, S. M., and Slooten, E. (1996). "The Downunder Dolphin: The story of Hector's Dolphin." Canterbury University Press, Christchurch.

Dawson, S. M., and Slooten, E. (2005). Management of gillnet bycatch of cetaceans in New Zealand. *J. Cetacean Res. Manag.* **7**, 59–64.

Dawson, S. M., and Thorpe, C. W. (1990). A quantitative analysis of the acoustic repertoire of Hector's dolphin. *Ethology* **86**, 131–145.

Dawson, S. M., Slooten, E., Pichler, F., Russell, K., and Baker, C. S. (2001). North Island population of Hector's dolphin threatened with extinction. *Mar. Mamm. Sci.* **17**, 366–371.

Dawson, S. M., Slooten, E., Du Fresne, S. D., Wade, P. R., and Clement, D. (2004). Small-boat surveys for coastal dolphins: Line-transect surveys of Hector's dolphins. *Fish. Bull. US* **201**, 441–451.

Elwen, S. H. (2007). The distribution, movements and abundance of Heaviside's dolphins in the nearshore waters of the Western Cape, South Africa. Ph.D. Thesis, University of Pretoria.

Elwen, S., Meÿer, M. A., Best, P. B., Kotze, P. G. H., Thornton, M., and Swanson, S. (2006). Range and movements of female Heaviside's dolphins (*Cephalorhynchus heavisidii*), as determined by satellite-linked telemetry. *J. Mammal.* **87**, 866–877.

Goodall, R. N. P. (1994a). Commerson's Dolphin *Cephalorhynchus commersonii* (Lacépede 1804). *In* "Handbook of Marine Mammals," Vol. V, Delphinidae and Phocoenidae (S. H. Ridgway and R. Harrison, eds), pp. 241–267. Academic Press. New York.

Goodall, R. N. P. (1994b). Chilean Dolphin *Cephalorhynchus eutropia* (Gray, 1846). *In* "Handbook of Marine Mammals," Vol. V, Delphinidae and Phocoenidae (S. H. Ridgway and R. Harrison, eds), pp. 269–287. Academic Press, New York.

Heinrich, S. (2006). Ecology of Chilean dolphins and Peale's dolphins at Isla Chiloé, Southern Chile. Ph.D. Thesis, University of St Andrews, UK.

**C**

Iñiguez, M. A., Hevia, M., Gasparrou, C., Tomsin, A. L., and Secchi, E. R. (2003). Preliminary estimate of incidental mortality of Commerson's dolphins (*Cephalorhynchus commersonii*) in an artisanal setnet fishery in La Angelina beach and Ría Gallegos, Santa Cruz, Argentina. *Lat. Am. J. Aquat. Mamm.* **2**, 87–94.

Kamminga, C., and Wiersma, H. (1982). Investigations on cetacean sonar V. The true nature of the sonar sound of *Cephalorhynchus commersonii*. *Aquat. Mamm.* **9**, 95–104.

Lescrauwaet, A., and Gibbons, J. (1994). Mortality of small cetaceans and the crab bait fishery in the Magallanes are of Chile since 1980. *Rep. Int. Whal. Comm. (Spec issue)* **15**, 485–494.

Lescrauwaet, A., Gibbons, J., Guzman, L., and Schiavini, A. (2000). Abundance estimation of Commerson's dolphin in the eastern area of the Strait of Magallan-Chile. *Rev. Chil. Hist. Nat.* **73**, 473–478.

Lockyer, C., Goodall, R. N. P., and Galeazzi, A. R. (1988). Age and body length characteristics of *Cephalorhynchus commersonii* from incidentally caught specimens off Tierra del Fuego. *Rep. Int. Whal. Comm. (Spec. Issue)* **9**, 71–83.

Pichler, F., Robineau, D., Goodall, R. N. P., Meÿer, M. A., Olivarría, C., and Baker, C. S. (2001). Origin and radiation of Southern Hemisphere coastal dolphins (genus *Cephalorhynchus*). *Mol. Ecol.* **10**, 2215–2223.

Ribeiro, S., Viddi, F. A., and Freitas, T. R. O. (2005). Behavioural responses of Chilean dolphins (*Cephalorhynchus eutropia*) to boats in Yaldad Bay, Southern Chile. *Aquat. Mamm.* **31**, 234–242.

Ribeiro, S., Viddi, F. A., Cordeiro, J. L., and Freitas, T. R. O. (2007). Fine-scale habitat selection of Chilean dolphins (*Cephalorhynchus eutropia*): Interactions with aquaculture activities in southern Chiloé Island, Chile. *J. Mar. Biol. Assoc. UK* **87**, 119–128.

Robineau, D., Goodall, R. N. P., Pichler, F., and Baker, C. S. (2007). Description of a new subspecies of Commerson's dolphin, *Cephalorhynchus commersonii* (Lacépède, 1804), inhabiting the coastal waters of the Kerguelen Islands. *Mammalia* **2007**, 172–180.

Slooten, E. (1991). Age, growth and reproduction in Hector's dolphins. *Can. J. Zool.* **69**, 1689–1700.

Slooten, E. (1994). Behavior of Hector's dolphin: Classifying behavior by sequence analysis. *J. Mammal.* **75**, 956–964.

Slooten, E., and Lad, F. (1991). Population biology and conservation of Hector's dolphin. *Can. J. Zool.* **69**, 1701–1707.

Slooten, E., Dawson, S. M., and Rayment, W. J. (2004). Aerial surveys for coastal dolphins: Abundance of Hector's dolphins off the South Island West Coast, New Zealand. *Mar. Mammal. Sci.* **20**, 477–490.

Slooten, E., Dawson, S. M., and Rayment, W. J. (2006a). Offshore distribution of Hector's dolphins at Banks Peninsula: Is the Banks Peninsula Marine Mammal sanctuary large enough? *NZ J. Mar. Freshw. Res.* **40**, 333–343.

Slooten, E., Dawson, S. M., Rayment, W. J., and Childerhouse, S. J. (2006b). A new abundance estimate for Maui's dolphin: What does it mean for managing this critically endangered species? *Biol. Conserv.* **128**, 576–581.

van Vuuren, B. J., Best, P. B., Roux, J.-P., and Robinson, T. J. (2002). Phylogeographic population structure in the Heaviside's dolphin (*Cephalorhynchus heavisidii*): Conservation implications. *Anim. Conserv.* **5**, 303–307.

# Cetacean Ecology

## Lisa T. Ballance

Ecology is the study of the natural environment and of the relationships of organisms to each other and to their surroundings. From its natural history beginnings in the late 1800s, the field of ecology has blossomed into a diverse discipline, encompassing empirical and theoretical research in fields as diverse as mathematics, conservation, physiology, geography, and behavior. Cetacean ecology describes the relationships between cetaceans and their physical and biological environment, including their interactions with prey, predators, competitors, and commensals.

Studying whales or dolphins in their natural environment is a formidable challenge. This is not only due to the logistical constraints of attempting to study highly mobile, oceanic animals that spend nearly all of their lives underwater, but also because of the political and legal constraints of working on protected species, which include most cetaceans. Early insights into cetacean ecology came largely from anecdotes and observations handed down by early whalers (Herman Melville's *Moby Dick* is a classic example). Although much remains unknown, technological developments over the past decade have greatly facilitated our understanding of cetacean ecology. Insights gained from individuals fitted with devices that transmit data on location and depth of dive through a satellite and to a researcher's desk, from linking movements and distribution with remotely sensed data on surface water properties of the ocean, or from sophisticated shipboard equipment designed to quantify density of prey in the water column are just three examples of technological advancements that have greatly contributed to clarifying cetacean ecology.

Cetaceans include approximately 84 species; new species continue to be described. They are a diverse group (Perrin, 1991). They range in size from less than 1 m long for a newborn vaquita (*Phocoena sinus*), to 33 m in an adult blue whale (*Balaenoptera musculus*); they occupy water ranging in temperature from −2°C to over 30°C; they exhibit a diverse array of life history strategies. Consider the sperm whale (*Physeter macrocephalus*) which can remain beneath the water for over an hour and dive to depths of several thousand meters, the Ganges river dolphin (*Platanista gangetica*) which inhabits fresh water so turbid it is functionally blind, beaked whales of the genus *Mesoplodon* which are so pelagic and so elusive that some have never been seen alive in the wild, the gray whale (*Eschrichtius robustus*) which annually migrates some 15,000 to 20,000 km between breeding and feeding areas, and the bowhead whale (*Balaena mysticetus*) which uses its rostrum to break ice in the Arctic.

One of the challenges of ecology is to search for pattern within diversity. Despite their diversity of form, behavior, and habitat, all cetaceans have some key features in common that underscore the fact that they are secondary marine forms, derived from terrestrial ancestors. That they are all air-breathing, live-bearing homeotherms provides a unifying theme. This chapter provides an overview of cetacean ecology with the ultimate goal being to identify some unifying principles in the ways that cetaceans interact with each other and with their environment.

## I. Habitat

### A. Habitat in the Ocean

Marine habitat is largely about oceanography. Terrestrial habitat is typically defined by the interaction between physical structure (as defined by physiography, and characteristics of primary producers) and meteorological factors, whereas marine habitat is almost entirely defined by hydrographic, physical, and chemical properties of water (e.g., water masses, surface currents, fronts, eddies, island wakes). Marine habitat types can be geographically fixed, when referring to the benthos or coastal ecosystems, e.g., but in most cases, marine habitat is not static in space or time. Instead, habitat types move with the water masses and surface currents that define them. Physical structure can define marine habitat, in the case of kelp forests or coral reefs, e.g., but in most cases, this type of physical structure is completely absent from marine habitats. Light attenuates more quickly and sound travels faster and farther in water as compared to air; thus marine organisms, including cetaceans, rely less on vision and more on the auditory sense than terrestrial mammals. The base of the trophic web in marine systems is formed by planktonic organisms. There is no analogous counterpart in terrestrial systems and marine animals have therefore evolved unique and specialized morphological and behavioral adaptations for taking advantage of this prey base.

On a global scale, cetaceans have invaded a large proportion of the ocean's habitats. They inhabit coastal waters up to and including the surf zone (gray whale, some populations of bottlenose dolphins *Tursiops truncatus*, harbor porpoise *Phocoena phocoena*, Commerson's dolphin *Cephalorhynchus commersonii*), neritic waters over continental shelves (long-beaked common dolphin *Delphinus capensis*, *Lagenorhynchus* spp., *Cephalorhynchus* spp., *Phocoena* spp.), and the most oceanic of systems (sperm whale, Fraser's dolphin *Lagenodelphis hosei*, beaked whales). They are found in tropical waters (pantropical spotted dolphin *Stenella attenuata*), temperate seas (Risso's dolphin *Grampus griseus*), and polar oceans, up to and within pack ice (beluga *Delphinapterus leucas*, bowhead whale). They utilize much of the water column, some being confined to relatively shallow depths (most dolphins and baleen whales), and others diving to thousands of meters (sperm whale, many beaked whales). And they have invaded the world's major river systems (Ganges, Indus, Amazon/Orinoco).

Cetaceans in different habitats might be expected to show differential development of adaptations which reflect selective pressures of the environments in which they function. For example, species in polar seas must conserve heat and so, bowhead whales have relatively large bodies, thick blubber layers, and short appendages. Deep-diving species (sperm and beaked whales) must conserve oxygen and might be expected to have large blood volumes, high hematocrit, and a well-developed diving response. Species that forage in low light conditions (night feeders, deep divers, species living in turbid rivers) should have well-developed echolocation abilities relative to those that function in habitats with greater light levels and better visibility.

The geographic range that a single species occupies runs from cosmopolitan to extremely local. For example, the killer whale (*Orcinus orca*) can be found throughout the world's oceans and, with the exception of humans, is the most wide-ranging mammal on earth. At the other extreme is the vaquita, a tiny porpoise that occupies a few hundred square km in the northern Gulf of California. And some species migrate between widely separate breeding and feeding areas; this pattern is characteristic of most (if not all) of the baleen whales.

### B. Species–Habitat Relationships

The relationship between a species and its habitat is a defining feature of its ecology. Species–habitat relationships form the basis for defining a species' ecological niche, in turn, a driving factor in determining competitive relationships, and a species' role in communities. Species–habitat relationships identify core requirements, critical knowledge for effective management and conservation. And a solid understanding of species–habitat relationships allows for prediction of distribution and abundance (Ferguson *et al.*, 2006), prediction that can facilitate mitigation of anthropogenic impacts such as ocean noise and climate change.

Because quantification of species–habitat relationships by definition involves integration of two very different types of data: species and habitat, and because marine habitat is so often defined by

oceanographic features, the study of species–habitat relationships is not straight forward. Simple correlations between a species and one or two directly measured oceanographic variables form the basis for early understanding of these relationships. More recently, two types of analytical tools have been used (Redfern *et al.*, 2006). Descriptive methods include overlays of species data on maps of oceanographic measures, correlation analysis, goodness of fit metrics, analysis of variance, and ordination. Modeling techniques are more sophisticated analytically, requiring parameter estimation, model selection, uncertainty estimation, and model evaluation. Many of these methods are still being developed.

In many geographic regions and for many species, we are beginning to identify those features that correlate with centers of distribution, thereby possibly identifying what may be called critical habitat. For example, some species associate with ice edges (beluga), some with continental shelf edges or seamounts (beaked whales), and some with shorelines (gray whale, bottlenose dolphin, harbor porpoise). For oceanic species habitat preferences are often defined by less obvious features: physical and chemical characteristics of the water itself, which define water masses and current boundaries. For example, some species associate with cold-water currents (Heaviside's *Cephalorhynchus heavisidii*, Commerson's, Peale's *Lagenorhynchus australis* dolphins). Blue whales in the eastern Pacific are found in relatively cool, upwelling-modified waters with high primary and secondary productivity. And in the eastern tropical Pacific, pantropical spotted dolphins and spinner (*Stenella longirostris*) dolphins segregate from common dolphins (*Delphinus* spp.) according to thermocline depth and strength, sigma-t (a measure of seawater density computed from surface temperature and salinity), and surface water chlorophyll content. These differences are statistically significant and these species-specific distribution patterns track oceanographic variation on a seasonal and interannual basis (reviewed in Ballance *et al.*, 2006).

Prey are likely the drivers of these species–habitat relationships, not the physical variables typically used in these types of analyses. In fact, there are a number of studies which have linked general distribution and movement patterns of cetacean species (humpback *Megaptera novaeangliae*, fin *Balaenoptera physalus*, long-finned pilot whales *Globicephala melaena*, Atlantic white-sided *Lagenorhynchus acutus*, bottlenose, common dolphins) with those of their prey.

## II. Food, Feeding, and Foraging

### A. Cetacean Prey

Most of what is known about the food of cetaceans comes from data collected from dead animals, through directed fisheries, incidental mortality, or strandings. Prey of cetaceans fall into four general categories.

The first prey type consists of small individuals which school at relatively shallow depths (surface to several hundred meters). These are primarily planktonic crustaceans (euphausiids, copepods, amphipods), and small fish [e.g., herring (*Clupea* spp.), sardine (*Sardinops* spp.), anchovy (Engraulidae), sandlance (Ammodytidae)]. They tend to occur in temperate or polar seas or in those tropical latitudes that are associated with high productivity. They generally occur at low trophic levels, have small body sizes, and occur in dense aggregations. Accordingly, the cetaceans feeding upon them capture multiple individuals simultaneously, have large body sizes, and have evolved filtering mechanisms (baleen) to strain prey items from the water. All mysticetes feed on this prey type.

The second prey type is comprised of larger organisms that also school at relatively shallow depths (surface to several hundred meters),

**C**

or migrate up to shallow depths during the night. This includes many pelagic fishes [e.g., hake (*Merluccius* spp.), pollock (*Pollachius* spp., *Theragra* spp.), myctophids (Myctophidae)] and schooling squids (*Loligo* spp., *Dosidicus* spp.) that occur throughout the world's oceans. Because these prey are larger, they generally occupy higher trophic levels and are captured individually. Their cetacean predators typically have smaller body sizes. They include all of the large-schooling dolphins (e.g., dusky *Lagenorhynchus obscurus*, common, striped *Stenella coeruleoalba*, spotted dolphins) and some small-schooling or solitary species (e.g., bottlenose, Commerson's, river dolphins *Platanista* spp.). These cetaceans tend to have a high tooth count, pointed teeth, and pointed snouts, all adaptations for pursuing fast, individual prey.

The third prey type is comprised of large, solitary squid (e.g., *Gonatus* spp.). These are most often found in deep waters throughout the world's oceans. Because of their size and solitary habits, they are captured individually. Cetacean predators of these prey include the sperm whale, dwarf, and pygmy sperm whales (*Kogia* spp.), all of the beaked whales (Ziphiidae), and pilot whales (*Globicephala* spp.). They are deep divers and tend to have reduced dentition, rounded heads, and well-developed melons, the latter perhaps indicative of the importance of echolocation for prey detection in the dark depths.

The final prey type includes species at high trophic levels that are themselves top predators. These include predatory fishes [e.g., tunas (Scombridae), sharks, salmonids], marine birds, pinnipeds, and cetaceans, including the largest of whales [rorquals (*Balaenoptera* spp.) and sperm whales]. Few cetaceans are able to take these prey items. They include the killer whale and, possibly, false killer whale (*Pseudorca crassidens*), and pilot whales. Two distinct forms of killer whales occur in waters off the west coast of North America: those that take fish and those that take mammals and birds. There is some indication that multiple ecotypes are found in Antarctic waters, and perhaps throughout the world's oceans (Pitman *et al.*, 2007).

## B. Capturing Prey

Cetaceans have two main types of feeding apparatus: baleen and teeth. Baleen is used for straining prey items from the water or, in the case of the benthic-feeding gray whale, from the sediment. Teeth are used for catching individual prey items. Species with a high tooth count use them to grasp individual prey; those with a low tooth count tend to be suction feeders.

Most of what is known about prey capture strategies is relevant to cetaceans which feed on small prey that school at relatively shallow depths (the mysticetes). This is because it is relatively easy to observe these animals feeding in the wild. Mysticetes have baleen plates suspended from the roof of their mouths that they use to strain prey items from the water. The number of baleen plates, their length, and the density of baleen fibers per plate vary between species and are correlated with prey size. The Balaenidae (right whales *Eubalaena* spp., bowhead whales) and sei whales (*Balaenoptera borealis*) have the greatest number of plates with the finest filtering strands and feed mainly on tiny copepods. Blue whales and most other rorquals have an intermediate number of plates with coarser filtering strands and feed on larger prey items such as euphausiids and small fishes. Gray whales have the fewest number of plates with the coarsest strands and are largely bottom feeders, sifting benthic infauna from muddy substrate.

In addition to specializing on different prey sizes, baleen whales have specialized feeding methods that also correlate with the morphology of their baleen. "Skimmers," the right whales, swim slowly with their mouths open through dense clouds of slow-swimming copepods. "Gulpers," including most rorquals, lunge into dense schools of euphausiids or fishes with their mouths open, closing them rapidly to trap their prey. All rorquals have throat grooves that run along the ventral surface of the mouth and throat, which allow the buccal cavity to expand during a lunge, taking in huge quantities of water, and with this, prey. A variation on this type of feeding is used by humpback whales when they form "bubble nets": streams of bubbles emitted from the blowhole as the whale swims in a circular pattern toward the surface. The bubbles form an ascending curtain, which concentrates prey inside. Most of these cetaceans are solitary feeders but they regularly aggregate in areas of high prey density and, when prey are extremely dense, will feed co-operatively at times, through bubble-net feeding or in staggered echelon formations.

Cetaceans that feed on larger fish and squid that school at relatively shallow depths capture individual prey items and swallow them whole. High speed is important, as is vision. Typically these predators forage co-operatively, herding prey into tight aggregations and capturing them in turn. Acoustic signaling is presumably important for the co-ordination of schooling activities. Some cetaceans in this group feed as individuals, particularly those found in coastal areas. They show a wide range of prey capture behaviors, including slapping fish with their flukes and deliberately stranding themselves on the beach in pursuit of fishes.

Cetaceans taking large, solitary squid feed at depth, in partial to full darkness. For this reason, not much is known about how they capture prey. They probably do not feed co-operatively because their prey do not school, and because most of these cetaceans occur in small schools and are slow swimming. Most have reduced dentition, and evidence indicates that they are suction feeders, using the gular muscles and tongue in a piston-like action to suck prey into their mouths. How they are able to get close enough to their prey to suck them in remains a mystery. One intriguing idea is that they are able to partially stun prey with echolocation bursts (Norris and Møhl, 1983).

Cetaceans that prey upon top predators show a wide range of prey capture methods and hunt as individuals as well as co-operatively in groups, depending upon prey size and characteristics. For example, killer whales may take pinnipeds by beaching themselves intentionally to grab adults and pups from rookeries but hunt co-operatively to take dolphins and large whales. Co-operative behaviors include prey encirclement and capture, division of labor during an attack, and sharing of prey (Pitman *et al.*, 2001).

## C. Locating Prey

Most cetaceans are visual predators, at least in part. For odontocetes, echolocation is equally important in locating and targeting prey, more so than vision in some species. Although only confirmed for a handful of captive species, all odontocetes are assumed to be able to echolocate and to use this sense extensively when foraging. At present, there is no evidence that mysticetes have the ability to echolocate, although they do produce low frequency sounds that travel long distances (hundreds of km). The long wavelengths of these pulses cannot resolve features finer than the wavelengths themselves (tens of m), and so, it is doubtful that they could be used to locate and target prey patches. The effective range of vision and echolocation is a function of water clarity and the specific echolocation abilities of a species, but both are probably limited to distances on the order of hundreds of meters to a few km.

On a larger spatial scale, patchiness and variability in space and time are characteristic of most marine ecosystems and little is known about how cetaceans locate prey in such environments. Presumably, many species simply travel large distances in a continuous search. This is particularly likely to be the case in regions of low productivity, where prey patches are few and far between, such as the oceanic tropics. Here, schooling may increase the chances of encountering a patch (the more eyes and ears, the better), and dolphin schools have been observed moving through the water in wide line-abreast formations, apparently searching for prey.

There are circumstances under which prey occur predictably in space and time, and it is likely that cetaceans search for and exploit these opportunities. For example, oceanographic features (e.g., boundaries between currents, eddies, and water masses) increase prey abundance or availability by enhancing primary production, by passively carrying planktonic organisms, and by maintaining property gradients (e.g., fronts) to which prey actively respond. Topographic features also (e.g., islands, seamounts) are sites of prey aggregation. Therefore, a good foraging strategy is simply to locate these physical features and many species of cetaceans (right, blue, fin, humpback, sperm, killer whales, spinner, Risso's, common, Atlantic spotted *Stenella frontalis* dolphins) have been found to associate with them.

Many species of cetaceans locate and associate with predictable point sources of prey. For example, killer whales aggregate around pinniped rookeries when young seals and sea lions are weaning. Rough-toothed dolphins (*Steno bredanensis*) associate with flotsam in the oceanic tropics, which serves to aggregate communities of animals at a wide range of trophic levels (Pitman and Stinchcomb, 2002). A wide variety of cetaceans associate with fishing operations to take their discards or their target species.

And there are times when prey are more accessible than others. The pelagic community of fishes and invertebrates, which live at depth during the day but migrate to the surface at night, provides an opportunity for cetaceans to predictably locate prey near the surface, and some dolphins (spotted, spinner, dusky, common) are known to feed on organisms in this community at night.

## III. Cetacean Predators

By far the most important non-human predator of cetaceans is the killer whale. Their pack-hunting behavior allows them to take everything from the fastest dolphins and porpoises to the largest whales, including blue and sperm whales. Other predators known to occasionally prey on smaller or weakened individuals include large sharks, and possibly false killer and pilot whales. Polar bears (*Ursus maritimus*) take cetaceans along the ice edge.

The ecological significance of this predation pressure in the lives of whales and dolphins is difficult to assess, but it may be significant (Corkeron and Connor, 1999; Pitman *et al.*, 2001). Individual large whales often show signs of killer whale tooth rake marks on their flippers, fins, and flukes, and up to one-third of the bottlenose dolphins off eastern Australia bear shark bite scars, suggesting that they regularly encounter predators. It has been hypothesized that large whales may undergo their annual migrations to reach calving grounds in areas of lower killer whale densities (i.e., the tropics). Aggregative behavior is a common defensive strategy among prey species and it is possible that schooling evolved in dolphins primarily as a defense mechanism against predators. These kinds of behavioral adaptations have cascading effects influencing not only distribution and abundance, but also social structure, timing and mode of reproduction, foraging strategies, and speciation patterns. Although its significance has been downplayed, the degree to which predation (top-down forcing) has structured cetacean ecology may have been under-estimated.

Sharks are the other main predator of cetaceans. Large-bodied sharks can maim or kill individual cetaceans but are likely significant predators only for smaller bodied dolphins or perhaps tiny calves of larger bodied cetaceans. The tiny cookie-cutter shark (*Isistius brasiliensis*) can also be considered a cetacean predator. These tropical sharks regularly take scoops of skin and muscle from many species of cetaceans and while these wounds are not fatal, they leave scars for the remainder of an individual's life.

## IV. Schooling

Like many animals, cetaceans form aggregations for two main reasons: feeding and protection. Feeding can bring animals together in passive aggregations in areas of high resource abundance. Alternatively, animals may actively seek others to take advantage of benefits provided by other school members. Schools also serve to protect members from predation, by providing cover for individual members, by confusing predators with synchronized movements of many individuals, by reducing the probability of predation on any one individual, by increasing the chance of detection of a predator, and by providing for co-ordinated defense. Occurring in large groups also increases the potential for social interactions, including reproduction; this may only be a secondary benefit of schooling.

The majority of cetaceans occur in schools, although there are some species that regularly occur solitarily or in very small groups of pairs or trios (many mysticetes, large male sperm whales, most beaked whales, dwarf and pygmy sperm whales, and river dolphins). Most schooling species have characteristic school sizes (although they can vary somewhat area to area). For example, rough-toothed dolphins typically occur in groups of 10–20, pilot whales occur in schools of dozens, and some oceanic dolphins (*Stenella* spp., *Delphinus* spp.) regularly occur in groups of hundreds or thousands (Wade and Gerrodette, 1993).

School size correlates with feeding habits: species that form large schools are almost all shallow-diving species that feed mainly on schooling prey, whereas those which occur in school sizes of 25 or fewer tend to be (a) deep-diving species and feed mainly on larger squids or (b) coastal species feeding on dispersed prey. School size also correlates with predation pressure; large cetaceans, presumably subject to lower predation pressure than small species, occur only in small groups, whereas small cetaceans, subject to higher predation pressure, occur in schools whose size correlates with openness of habitat: the more open, the larger the school size. School size should correlate with resource availability and will affect reproductive strategies, although the nature of these relationships remains largely unexplored.

Although most schools are monospecific, several species regularly occur in mixed-species schools. Some of these associations appear to be opportunistic: bottlenose dolphins, e.g., have been recorded to occur with over 20 different species of whales and dolphins. Other associations appear to be more prescribed: spotted and spinner dolphins regularly occur together in mixed schools. Risso's, Pacific white-sided (*Lagenorhynchus obliquidens*), and northern right whale (*Lissodelphis borealis*) dolphins are commonly found in association. The nature of these interactions (e.g., why these species-specific associations occur, how these species avoid competition) is unknown.

## V. Communities and Coexistence

Studies of communities typically focus on identifying member species and their interactions and then address mechanisms for their coexistence. These kinds of studies comprise a large part of the ecological knowledge for many terrestrial species. In contrast, very little is known about this aspect of cetacean ecology.

There are regularly occurring species assemblages. For example, pantropical spotted and spinner dolphins are frequently found in mixed-species schools in association with yellowfin tuna (*Thunnus albacares*) and are accompanied by large and speciose flocks of seabirds; this association is particularly prevalent in the eastern tropical Pacific, as opposed to other tropical oceans. There are variations in typical co-occurrence patterns. In the Gulf of Mexico, e.g., five species of *Stenella* coexist in a relatively small area, more *Stenella* species than any other tropical ocean. The nature of the interactions between species in these assemblages, why they associate, and the reasons for variations in community membership patterns are almost completely unknown.

Coexisting species, particularly those that are closely related or have similar ecological roles, potentially compete for resources. An often cited example is the southern ocean, where the relative abundances of cetaceans, pinnipeds, and seabirds, all krill consumers, have been reported to have changed between pre- and post-whaling years. One plausible explanation is competitive release: the decrease in biomass of cetacean predators released a huge prey base of krill to pinnipeds and seabirds, both of which were able to increase in abundance (Laws, 1977).

Ecological theory states that stable communities of coexisting species must differ in resource utilization in some way: through prey species or size specialization, differential habitat use, or diet pattern. Such niche partitioning is fairly clear for cetaceans on a broad scale. For example, there are species that feed on fish and those that feed on squid. There are species feeding in shallow water and those that feed at depth. Some cetaceans feed at night and others during the day.

On a smaller scale, one of the best known examples of niche partitioning is for baleen whales. In this group, there is a fair degree of prey specialization that presumably allows for niche partitioning in areas of sympatry. Blue whales feed almost entirely on euphausiids; fin whales and humpbacks feed mainly on fishes but take euphausiids when they are abundant; and right whales and sei whales feed mainly on copepods. Odontocetes provide additional possible examples. Bottlenose, short-beaked common (*Delphinus delphis*), pantropical spotted dolphins, and harbor porpoise exhibit diet specialization among age, sex, and reproductive class, although this diet specialization could be due to differing energy requirements. Aside from these examples, very little is known about how, or if, cetaceans partition resources.

Ultimately, to understand community structure, the mechanisms by which species partition resources, not merely the presence of differences in resource use, are of principal interest. The question then becomes, given that there are differences, what mechanisms can explain them? Community ecologists have identified interference and exploitative competition, mutualism, morphological or physiological factors, and habitat structure as potential mechanisms for maintaining resource utilization differences. This is an area that remains almost completely unexplored for cetaceans and the communities in which they are found.

## VI. The Role of Cetaceans in Marine Ecosystems

What role do cetaceans play in marine ecosystems and what is their significance? Most cetaceans are apex predators. As such, they take tons of prey from the ecosystem. In so doing, it seems likely that cetaceans affect the life history strategies and population biology of their prey, as well as organisms at other trophic levels that interact in various ways with these prey. Little is known about the details of

**Figure 1** *Australasian gannets* (Sula serrator) *feed on small-schooling fish* (e.g., *pilchard*, Sardinops neopilchardus) *which have been herded to the surface by dusky dolphins* (Lagenorhynchus obscurus) *in Admiralty Bay, New Zealand.Photo by Christopher Pearson.*

these dynamics, although this may be the most significant way in which cetaceans impact marine ecosystems.

More specific effects have been documented. For example, benthic feeders such as gray whales alter habitat by regularly turning over substrate (between 9% and 27% of the benthos in the northern Bering Sea) and therefore, significantly affect the species composition of benthic communities. Feeding cetaceans provide feeding opportunities for seabirds by driving prey to the surface, sometimes injuring or disorienting it (Fig. 1). In one study up to 87% of all feeding individuals from four seabird species in the Bering Sea associated with gray whale mud plumes (Obst and Hunt, 1990). Large whales dying at sea may sink to the bottom and provide rare but superabundant food and habitat for deep-water species. There is evidence that mollusc communities may have specialized on these resources for the past 35 million years, and some speculate that whale carcasses may have been instrumental in the dispersal of hydrothermal-vent faunas. Feces of some cetaceans, particularly large whales in areas of low productivity, may play a significant role in nutrient cycling. Cetaceans are host to a variety of commensal or parasitic species; in some cases (Cyamid whale lice), these species are completely dependent on cetaceans through all life stages.

## VII. Macroecology

Macroecology is a branch of ecology that attempts to characterize the relationships between organisms and their environment by increasing the spatial and temporal scale of investigation. Macroecologists ask questions about the relationships between abundance, distribution, and diversity of individuals, populations, or species and incorporate principles of biogeography, paleobiology, systematics, and the earth sciences in their search for pattern. Macroecology is a relatively recent field but has resulted in significant and novel insights. For example, the concept of species richness hotspots, the underlying mechanisms that produce them, and the ecological and conservation implications are all products of the macroecological perspective. These advances pertain largely to terrestrial systems but there are notable exceptions. Worldwide patterns of tuna and billfish diversity indicate that there are clear hotspots, areas of high species richness, that appear to hold in general for other taxa and trophic levels, and are functions of oceanography (Worm *et al.*, 2005). The significance of these patterns and their relevance to cetaceans is unknown, but a promising field of investigation.

## VIII. Concluding Remarks

The field of cetacean ecology is a dynamic discipline. Technological advances and heightened interest have resulted in a great deal of additional knowledge, a trend that will no doubt continue. However, any attempts to make ecological sense of cetaceans as marine organisms and to interpret their distribution patterns, foraging ecology, community structure, and role in ecosystems must take into account the fact that many cetaceans today exist as remnant populations that have been drastically reduced through anthropogenic effects: commercial exploitation, incidental mortality, and habitat destruction. Additionally, the study of cetacean ecology is no longer relevant without incorporating anthropogenic influences into the biological environment; these are pervasive and in many cases, permanent. The recent declaration of extinction of the baiji (*Lipotes vexillifer*) (Turvey *et al.*, 2007) and an almost certain similar fate for vaquita if drastic measures are not immediately taken (Jaramillo-Legorreta *et al.*, 2007) are tragic indications of the state of many marine ecosystems. This means that cetacean ecologists must also add human sociology, economics, and policy to the list of disciplines that will likely affect the search for ecological patterns.

## See Also the Following Articles

Biogeography ■ Distribution ■ Ecology, Overview ■ Geographic variation ■ Habitat use ■ Ocean Ecosystems ■ Ocean Environment ■ Pinniped Ecology

## References

Ballance, L. T., Pitman, R. L., and Fiedler, P. C. (2006). Oceanographic influences on seabirds and cetaceans of the eastern tropical Pacific: A review. *Prog. Oceanogr.* **69**, 360–390.

Corkeron, P. J., and Connor, R. C. (1999). Why do baleen whales migrate? *Mar. Mamm. Sci.* **15**, 1228–1245.

Ferguson, M. C., Barlow, J., Fiedler, P., Reilly, S. B., and Gerrodette, T. (2006). Spatial models of delphinid (family Delphinidae) encounter rate and group size in the eastern tropical Pacific Ocean. *Ecol. Modell.* **193**, 645–662.

Jaramillo-Legorreta, A., Rojas-Bracho, L., Brownell, R. L., Jr., Read, A. J., Reeves, R. R., Ralls, K., and Taylor, B. L. (2007). Saving the vaquita: immediate action, not more data. *Conserv. Biol.* doi:10.1111/j.1523-1739.2007.00825.x. Published online.

Laws, R. M. (1977). Seals and whales of the Southern Ocean. *Philos. Trans. R. Soc. Lond. B Biol. Sci.* **27**, 81–96.

Norris, K. S., and Møhl, B. (1983). Can odontocetes debilitate prey with sound? *Am. Nat.* **122**, 85–104.

Obst, B. S., Jr., and Hunt, G. L. (1990). Marine birds feed at gray whale mud plumes in the Bering Sea. *Auk* **107**, 678–688.

Perrin, W. F. (1991). Why are there so many kinds of whales and dolphins? *BioScience* **41**, 460–461.

Pitman, R. L., and Stinchcomb, C. (2002). Rough-toothed dolphins (*Steno bredanensis*) as predators of mahimahi (*Coryphaena hippurus*). *Pac. Sci.* **56**, 447–450.

Pitman, R. L., Ballance, L. T., Mesnick, S. L., and Chivers, S. J. (2001). Killer whale predation on sperm whales: Observations and implications. *Mar. Mamm. Sci.* **17**, 494–507.

Pitman, R. L., Perryman, W. L., LeRoi, D., and Eilers, E. (2007). A dwarf form of killer whale in Antarctica. *J. Mammal.* **88**, 43–48.

Redfern, J. V., *et al.* (19 authors) (2006). Techniques for cetacean–habitat modeling. *Mar. Ecol. Prog. Ser.* **310**, 271–295.

Turvey, S. T. *et al.* (16 authors) (2007). First human-caused extinction of a cetacean species? *Biol. Lett.* doi:10.1098/rsbl.2007.0292. Published online.

Wade, P. R., and Gerrodette, T. (1993). Estimates of cetacean abundance and distribution in the eastern tropical Pacific. *Reports of the International Whaling Commission* **43**, 477–493.

Worm, B., Sandow, M., Oschlies, A., Lotze, H. K., and Myers, R. A. (2005). Global patterns of predator diversity in the open oceans. *Science* **309**, 1365–1369.

# Cetacean Evolution

### R. Ewan Fordyce

## I. Patterns of Evolution

Fossils show that cetaceans arose from terrestrial ancestors more than 50 million years ago. They have evolved to become the dominant group of marine mammals in terms of taxonomic and ecological diversity and geographic range. Structural, genetical, and ecological patterns in living species are used widely to infer cetacean evolution. Fossils provide the only direct evidence of extinct species and past structures; they indicate extinct clades and ecologies, revealing patterns not seen today. Patterns from modern and fossil species have, since 2000, hugely expanded understanding of evolution, although high-level/deep-time phylogenies yet show limited agreement. Discussion of evolution depends on an agreed phylogenetic approach. Here, concepts of crown and stem group are fundamental, especially when comparing fossil vs molecular evidence for deep-time events. For any clade (monophyletic group), the crown group comprises all the living species plus all descendants of their most recent common ancestor. This monophyletic crown group may contain only one species; it may also contain extinct species. The stem group comprises those species, all extinct, that are closer to the crown group than to any other clade.

There is little conclusive evidence on evolutionary rates, modes, and mechanisms of ancestor-to-descendant transitions. Evolutionary processes at the species level have likely included natural selection, sexual selection, coevolution, founder effects, vicariance, and hybridization. At larger scales, changes in ocean structure/ocean ecosystems correlate with some changes in structural/ecological evolution in Cetacea, suggesting that the distribution and abundance of global food resources have played major roles in evolution.

## II. Ancient Ecology

Ecology is the sum of interactions that species have with their environment, biological, physical, and chemical. Paleoecology explores such relationships of the past. On geological timescales, evolution and paleoecology are inextricably linked, for evolution occurs through natural selection and adaptation to the environment. An appreciation of paleoecology thus helps understand the history of Cetacea. Several ecological factors have been considered repeatedly in accounts of cetacean evolution (Fig 1), including feeding and predator–prey interactions, migration, thermal adaptations, and habitat shifts.

For extinct species, lifestyle can be inferred in several ways. The traditional approach of "taxonomic uniformitarianism" assumes that a fossil had similar ecological strategies to its close living relatives: the present is applied to the past. Thus, e.g., fossil sperm whales

C

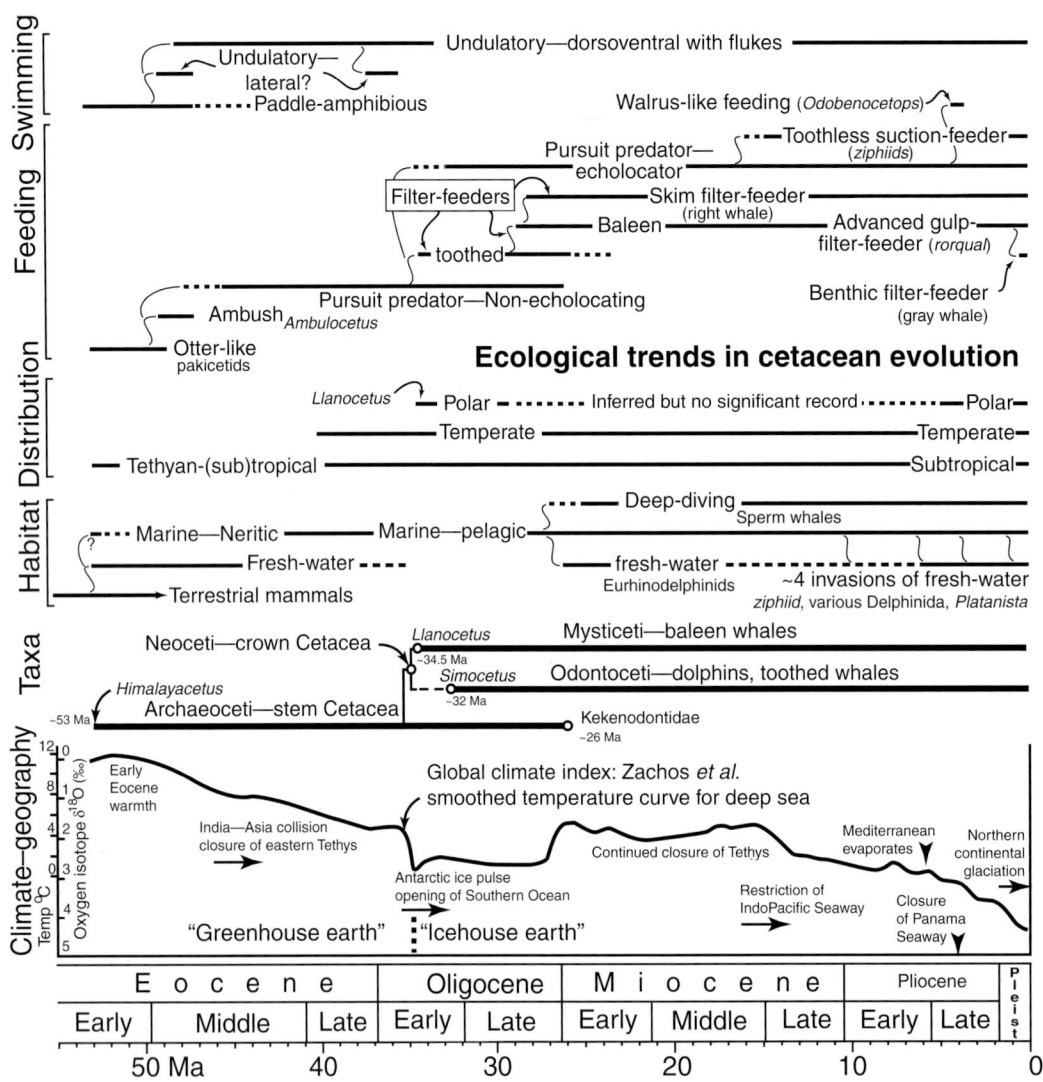

**Figure 1** *Summary of inferred ecological strategies important in cetacean evolutionary history plotted against geological time, inferred global climate, geographic/oceanic change, and taxon history. Selected taxa are given as examples. Slightly revised from Figure 1, in Cetacean Evolution, R. E. Fordyce, "Encyclopedia of Marine Mammals" (W. F. Perrin et al., eds), Copyright Elsevier 2002.*

presumably were deep divers. However, fossils show that cetaceans have evolved through long-term change in structure, with natural selection providing the adaptive fine-tuning. Structural change, then, implies changes in ecology. Clearly, taxonomic uniformitarianism should be supported by other methods. The method of extant phylogenetic bracketing (Witmer, 1995) proposes that some attribute present in crown (living) species can be inferred to have occurred in an extinct species or extinct clade within the crown group. For example, extinct species within the crown Odontoceti can be inferred to have echolocated because phylogenetically adjacent living species echolocate.

For some fossils, ancient ecologies may be inferred by analogy. The bizarre extinct Peruvian *Odobenocetops* species, or walrus whales, have no structural equivalents among living Cetacea, but their jaw form was remarkably similar to that of the unrelated living walrus, *Odobenus rosmarus*. A comparable style of suction feeding on molluscs is inferred. This approach to paleoecology can

be expanded by studying functional morphology, which includes reconstructing soft tissues onto fossils using known modern correlations of muscle, nerve, and vessel to bone. Finally, geological evidence (e.g., from sediments and isotopes) can help identify the ancient environments in which cetaceans lived, sometimes indicating dramatic or novel ancient ecologies.

### III. Major Radiations

Changes in structure and diversity of fossils reveal three major radiations—times of functional, ecological, and taxonomical diversification—in cetacean history. First, cetaceans diversified in near-tropical shallow waters of the Tethys seaway between India and Asia about 45–53 million years before the present (Ma), spreading into temperate waters by 40 Ma. This early phase of evolution was influenced by local marine productivity (Gingerich, 2005), although the initial shift to fresh waters may have been for protection rather than feeding (Thewissen *et al.*, 2007). Up to 20 genera of Archaeoceti (stem or basal Cetacea) were present in the Lutetian (geological) stage, and over 40 species have been reported for the interval 34–53 Ma (Uhen and Pyenson, 2007; see also pbdb.org). This initial radiation of archaeocetes involved shifts from riverine and near-shore marine settings to fully oceanic habits, accompanied by changes in locomotory and hearing mechanisms (Gingerich, 2005). In terms of structural variation and, by inference, ecology, these early cetaceans were comparable with living mysticetes and did not show the wide disparity seen among living odontocetes. Archaeocete diversity dropped late in Eocene times, foreshadowing the rise of Neoceti.

The second major radiation, involving the Neoceti or crown group Cetacea, occurred early in the Oligocene (Fordyce, 2003; see also FOSSIL RECORD). Echolocating odontocetes and filter-feeding mysticetes appeared about 35 Ma and diversified rapidly in about 5 M years after originating from Eocene archaeocetes which neither echolocated nor filter-fed. Concurrent events included the final breakup of Gondwana, opening of the Southern Ocean, cooling and increased tropics-to-polar temperature gradients, and changes in ocean ecosystems and productivity. Probably, cetaceans radiated in direct response to new ecological opportunities in rapidly changing oceans (Lindberg and Pyenson, 2007; Berger, 2007). In the Late Oligocene, 23.5–29 Ma, there were 16 families or higher clades and probably approximately 50 species, with some species ranging into polar and fresh waters. A few relict Oligocene archaeocetes are known.

Third, the middle and the late Miocene 10–12 Ma, saw the start of a marked radiation of "modern" mysticetes and odontocetes and a decline of ancient groups. Crown balaenopterids and Delphinida radiated to become major components of fossil assemblages, with crown Delphinoidea well established in the Pliocene (see FOSSIL RECORD). Two extinct delphinoid families (Odobenocetopsidae, Albireonidae) appeared briefly. Some archaic cetacean groups, including the Eurhinodelphinidae, Squalodontidae, and other members of the Platanistoidea, declined to extinction. Many crown genera appeared by the start of the Pleistocene, 2 Ma, but evolutionary patterns during the following ice ages are not clear because of a patchy fossil record. It has been suggested that many north–south species pairs evolved in the later Pleistocene but molecular studies point to earlier north–south splits.

## IV. Evolutionary Processes

Darwin's original evolutionary mechanism, natural selection, is the process that leads to the adaptation or "fine-tuning" of organisms to the environment through beneficial structure or behavior. The result is differential reproductive success. Intra- or interspecific competition for limited resources, especially food, is a fundamental part of natural selection. Some cetologists have indicated that competition for food is important among species of baleen whales and between baleen whales and other plankton-eating vertebrates, with implications for the origins of the species involved. However, the structure of the feeding complexes of baleen whales, mode behavior of feeding, and geographic distribution indicate that, in spite of sympatry, niche overlap and thus competition are limited between species (Pastene *et al.*, 2007). Perhaps previous competition among ancestral species led to taxonomic and ecological divergence seen today. Among fossils, platanistoid dolphins declined in diversity as delphinoids diversified in the middle to late Miocene; this pattern of extinction in one group and radiation in another could indicate that delphinoids out competed platanistoids (Fordyce and Muizon, 2001). However, differences in jaw and skull structure imply little ecological overlap between the two groups. Perhaps changing

oceanic circulation and climate at about this time caused the extinction of platanistoids and allowed the radiation of delphinoids.

Darwin identified sexual selection as a mechanism in evolution. Here, one sex chooses a specific member of the other sex as a mating partner. Sexual selection plausibly accounts for sexual dimorphism, particularly in structures involved in display, and has been linked with polygamous mating systems. Examples of possible sexually selected structures in male Cetacea include the large dorsal fin in *Orcinus orca*, conspicuous mandibular teeth in many ziphiids, prominent foreheads in species of *Hyperoodon* and *Globicephala*, and the prominent tusks in *Monodon monoceros* and the extinct *Odobenocetops leptodon*. Perhaps the size dimorphism in some species also results from sexual selection; examples include male *Physeter macrocephalus*—with its grossly enlarged forehead—and female *Balaenoptera* species.

Hybridization is potentially important in the origin of new species, and cetacean hybrids are known, e.g., in *Balaenoptera*, between *Phocoenoides* and *Phocoena*, and between several genera of Delphinidae. However, no convincing cases for speciation by hybridization have been identified in Cetacea.

Coevolution, especially involving mimicry, predator prey, and host–parasite interactions, is an important phenomenon. For cetaceans, mimicry is seen in pygmy and dwarf sperm whales, *Kogia* spp.; these have a remarkably shark-like form complete with underslung jaw and pigmented false gill slit. Presumably, to look like a predator will lessen the chance of being preyed upon. Predation may have other roles in evolution. The presence of predators, such as the killer whale, *Orcinus orca*, has been used to explain species-specific distribution patterns, including the distinctive and supposedly ancient migration patterns of mysticetes from poles to tropics (Corkeron and Connor, 1999).

Convergent evolution occurs when species show similarity not inherited from a common ancestor. Among Cetacea, the "river dolphins," long discussed as if they form a real group, encompass species in four different families. *Platanista*, however, is convergent in its riverine habits, small body size, and long rostrum with the three delphinoid relatives *Inia*, *Pontoporia*, and *Lipotes*. The southern delphinid *Cephalorhynchus hectori* is similar in body form and some aspects of ecology and behavior to the unrelated porpoise *Phocoena phocoena* (Phocoenidae). Unrelated groups have convergently reduced and lost teeth, as seen in living *Mesoplodon* (Ziphiidae) and the Pliocene ziphiid-mimic delphinid *Australodelphis*.

There is a wide interest in the developmental mechanisms that allow rapid evolution of the structure. Consider heterochrony, which involves a change in timing or rate of development of structures relative to the equivalent processes in an ancestor. Some features might evolve when juvenile structures persist into adult stages (pedomorphosis), whereas others arise when structures develop "beyond" that of the ancestral adult stage (peramorphosis). Apparent pedomorphic features in Neoceti include the shortened intertemporal region and longer vomer and mesorostral groove, the rounded cranium and persistent interparietal bone on the skull of many Delphinidae and Phocoenidae, and the down-turned rostrum and relatively symmetrical skull in Phocoenidae. Possible peramorphic features include extra body parts, e.g., increase in number of vertebrae, as in Dalls porpoise, *Phocoenoides dalli*, or phalanges generated through a delayed halt in development.

## V. Evolution and Geography

Darwin realized that speciation is often related to geography. The range of a species may either be split by a change in physical habitat (namely, split by a vicariant event) or be expanded by dispersal beyond normal limits. Populations that become geographically isolated through such events can diverge and, via allopatric speciation, may become new species. Most discussion of such ideas focuses on terrestrial habitats, but there are clear marine parallels. During 50 M years of cetacean history, geographic changes have included the closure or opening of some straits and ocean basins, dramatic swings in continental shelf habitat area through sea level fluctuations, and major shifts in current systems, upwellings, and latitudinal water masses. Oceanic temperature regimes changed in parallel. This physical evolution of the oceans probably influenced cetacean evolution at many levels.

The distributions of modern and fossil Cetacea indicate an important role for geography in evolution (Fig. 2). Some living species have obvious northern and southern populations or closely related north–south species pairs, but do not occur in the tropics. Such bipolar or antitropical distributions probably arose allopatrically when populations became isolated either side of the tropics through changing sea temperatures or current regimes, sometimes leading to speciation. Antitropical distributions are marked (Rice, 1998) in populations of rorquals *Balaenoptera*, species of right whales *Eubalaena*, some beaked whales (Ziphiidae species of *Berardius*, *Hyperoodon*, and *Mesoplodon*), and dolphins (Delphinidae, e.g., *Lissodelphis* spp.).

Among delphinids, molecular studies reveal that six antitropical species of *Lagenorhynchus* split at varying times, not just during the geologically recent (~2 Ma) ice ages, and that these species represent two or more different lineages (LeDuc *et al.*, 1999). At least one genus appears to have speciated around an ocean: four species of the delphinid genus *Cephalorhynchus* occur around the circum-Antarctic Southern Ocean. There is some fossil evidence for allopatric species pairs: closely related species of the small dolphin *Kentriodon* that occur in Miocene strata in California and Maryland perhaps evolved from an ancestor that ranged through the central American seaway before the uplift of Panama.

For porpoises, the endangered vaquita *Phocoena sinus*, from the Gulf of Mexico, perhaps originated when a few of Burmeisters porpoise, *Phocoena spinipinnis*, crossed the equator only tens of thousands of years ago. The vaquita, with its limited distribution and low genetic variability, illustrates founder effects; a new population, e.g., in an isolated region beyond normal range limits, may be established by only a few original founders which are not genetically representative of the original population.

Sea level fluctuations have changed the extent of continental shelf habitat dramatically, especially during cycles of glaciation and cooling over the last 2 M years. At peak glaciation, sea level was 100+ m lower than at present, leading to fragmentation or loss of shelf habitat for long intervals, along with colder conditions. Such habitat change could lead to extinctions. However, if species durations typically exceed 100,000 years, as is plausible, then most living cetaceans have survived several of these fluctuations. The record of Late Pliocene and Early Pleistocene Cetacea includes a few species from shallow-water habitats (e.g., the dolphin *Parapontoporia*) that might have later disappeared because of Pleistocene climate changes and/or habitat loss.

Some cetacean groups show significant change in range over time. The beluga and narwhal (Monodontidae) are now restricted to cold north polar waters, but during warmer Pliocene times (4 Ma), monodontids occurred far south in subtropical waters. Further evidence for changing geographic ranges comes from archaeocetes, which, for about 10 M years after their origin, were subtropical to tropical, later spreading to temperate regions. The peak modern diversity for Cetacea is in temperate waters.

Allopatric speciation results from lineage splitting or cladogenesis. New species could also evolve by anagenesis, involving transformation from an ancestral to descendant species without lineage splitting. Anagenetic change would produce a fossil record with two or more species in succession and might be expected in lineages containing geographically wide-ranged species. Species with limited ranges seem more likely to experience allopatric speciation. Among living Cetacea, the sperm whale *Physeter macrocephalus* represents a genus with a single abundant and widely distributed species with no immediate relatives (sister species), and anagenesis might be considered in explaining its history.

Freshwater settings were important in the early transition of cetaceans from land to sea in the Tethys. Early archaeocetes were amphibious, with well-developed hindlimbs, and they include species from freshwater strata. Freshwater habits for some archaeocetes are indicated by oxygen isotopes from teeth. Fossil Neoceti also show several reinvasions of freshwaters. For example, one unnamed Late Oligocene species of Eurhinodelphinidae occurs in lake sediments of central Australia. The dolphin shows no obvious structural adaptations to a lacustrine habitat, but the presence of several specimens of slightly differing geological age indicates long-term occupation of the habitat. A single large ziphiid fossil is known from Miocene freshwater sediments of Kenya. All the living "river dolphins" arose from marine ancestors, some well known from fossils, which invaded freshwaters in Asia and South America in up to four separate events. Mysticetes seem never to have occupied river systems.

## VI. Life History Traits

In terms of life history traits, cetaceans appear to be K-strategists (Estes, 1979). That is, compared with many other mammals, cetaceans are large animals that have slow reproductive rates, produce a single offspring, show significant parental of young, have long reproductive lives, and have relatively low mortality rates. This reproductive strategy has been linked, in an evolutionary sense, to the nutritional requirements of both the young and the parents and thus to food availability. The fossil record of cetaceans, which shows major evolutionary change linked to oceanic change, supports the idea that food resources have been fundamental in cetacean history. The exact roles of physical vs biological effects, namely, bottom-up or top-down drivers, in cetacean evolution are contentious.

## VII. Taxonomic Longevity

How long do species and genera persist? The fossil record shows that few, if any, species have ranges longer than one geological stage (a stage is a time unit commonly 4–5 M years) (Uhen and Pyenson, 2007). Amongst extinct Cetacea, the close-spaced succession of Eocene archaeocetes points to species durations of 1–2 M years. Living species can be traced back into the Pleistocene but not reliably longer, further implying durations of approximately 2 M years. Molecular studies can estimate dates for separation of lineages of crown species, e.g., 1.9 to 3 and 3.8 to 9.6 M years for species of *Lagenorhynchus* and greater than 5 M years for *Balaenoptera musculus* and *B. physalus*. Dates for lineage separation, however, are not necessarily the same as dates for species durations. Beyond the species, some crown genera have reliable fossil records back to the Early Pliocene or Late Miocene, e.g., approximately 6 Ma for *Balaenoptera*, and approximately 5 Ma for *Delphinus*. The supposed origin for crown right whales, *Balaena* and *Eubalaena*, pre-dating *Morenocetus* at 20 Ma, is surprisingly long and needs verifying. Similarly, some molecular predictions for

C

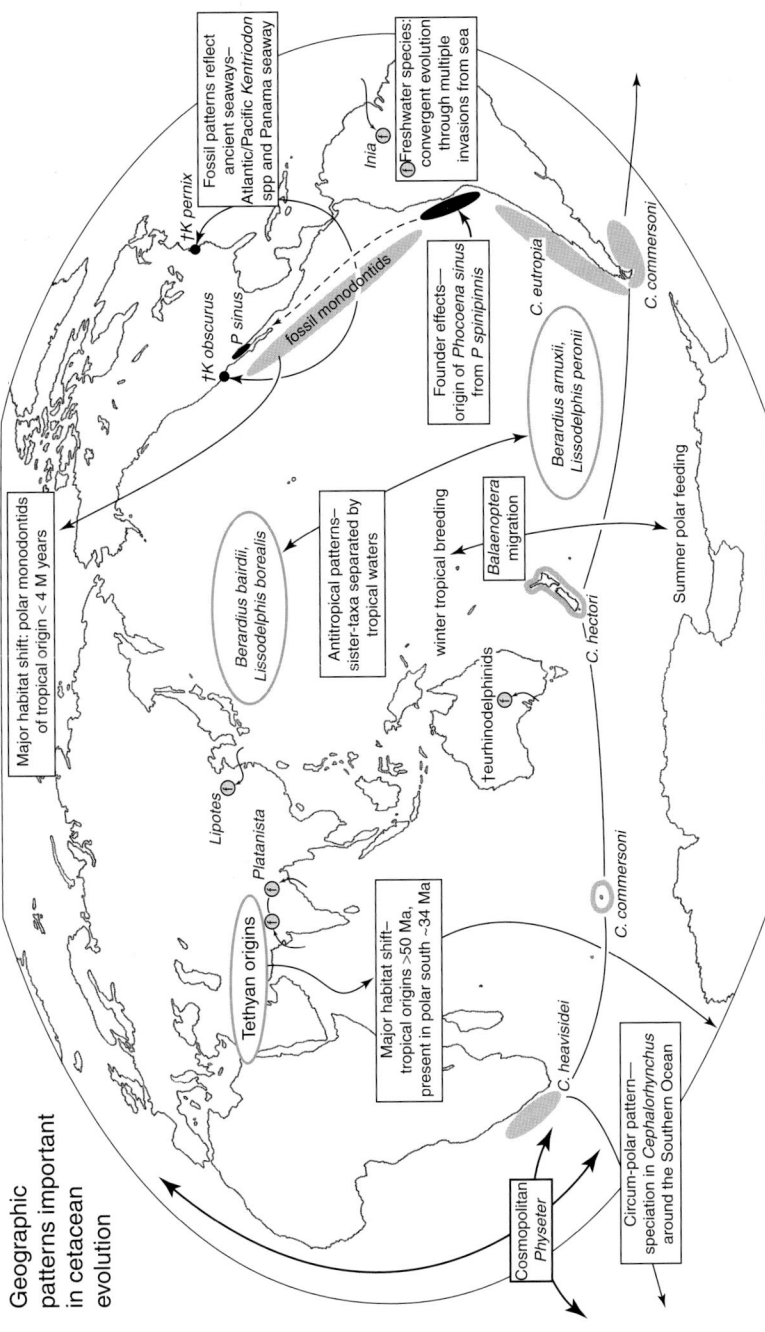

Geographic patterns important in cetacean evolution

Major habitat shift: polar monodontids of tropical origin < 4 M years

†K pernix

Fossil patterns reflect ancient seaways— Atlantic/Pacific *Kentriodon* spp and Panama seaway

*Inia* ⊕ ⊕Freshwater species: convergent evolution through multiple invasions from sea

†K obscurus
*P sinus*

fossil monodontids

†K obscurus
*P sinus*

Founder effects— origin of *Phocoena sinus* from *P spinipinnis*

*C. eutropia*

*C. commersoni*

*Berardius arnuxii,* *Lissodelphis peronii*

*Berardius bairdii,* *Lissodelphis borealis*

Antitropical patterns— sister-taxa separated by tropical waters

winter tropical breeding

*Balaenoptera* migration

Summer polar feeding

*C. hectori*

†eurhinodelphinids

*Lipotes* ⊕
*Platanista* ⊕

Tethyan origins ⊕

Major habitat shift— tropical origins >50 Ma, present in polar south ~34 Ma

*C. commersoni*

*C. heavisidei*

Cosmopolitan *Physeter*

Circum-polar pattern— speciation in *Cephalorhynchus* around the Southern Ocean

**Figure 2** *Distribution patterns illustrating the possible role of geography in cetacean evolution. Patterns indicated include antitropical and circum-Antarctic distributions, allopatric species pairs between oceans, founder effects, convergent origins for various freshwater dolphins, habitat expansion in general for Cetacea tropical origins, followed later by spread as far as the poles, and geologically recent changes in habitat. Selected taxa are given as examples. Fossils are marked with a dagger. Slightly revised from Figure 3, in Cetacean Evolution, R. E. Fordyce, "Encyclopedia of Marine Mammals" (W. F. Perrin et al., eds), Copyright Elsevier 2002.*

origins of crown genera are markedly longer than indicated by the fossil record.

Little is known about rates of evolution in Cetacea. Family-level patterns in the fossil record, and some molecular studies, imply phases of rapid radiation, and presumably rapid evolution, for the Oligocene. Unlike the situation for some terrestrial mammals, the cetacean fossil record is too sparse to reveal quantifiable change in structure over time. For living species, no speciation event has been dated reliably enough to clearly reveal evolutionary rate, but there is some evidence that the vaquita, *Phocoena sinus*, evolved fast, over tens, rather than hundreds, of thousands of years.

## VIII. Diversity and Disparity

Diversity is the number of species within a taxon (such as a genus or a family) and is effectively an index of taxonomic richness, whereas disparity indicates the variation in structure or basic design within a taxon. Diversity is easy to assess, particularly now that advances in the philosophy and practice of systematics have produced a generally accepted species level classification of living cetaceans, but the study of disparity is still developing. A comparison of the two living clades Odontoceti and Mysticeti reveals quite different patterns of diversity (Fig. 3). Mysticetes

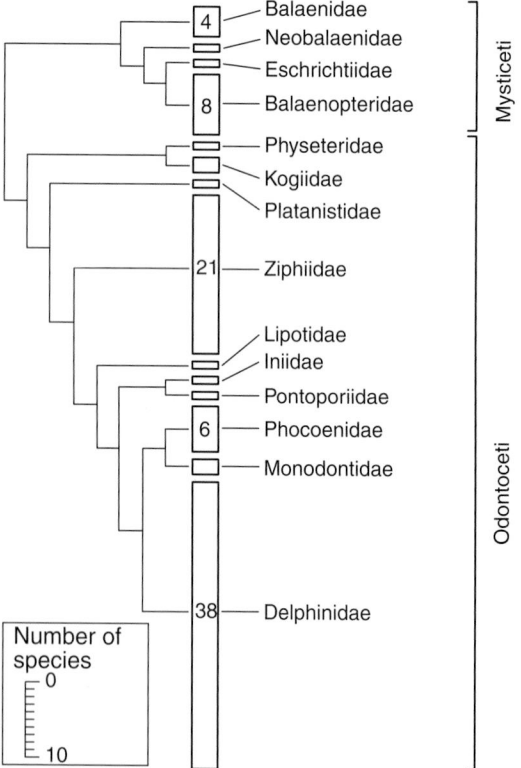

**Figure 3** *Phylogenetic patterns of species diversity at the family level among living Cetacea, using currently understood higher-level relationships. Traditionally, families are clusters of species and genera with a similar body plan; thus, families usefully reveal disparity at high levels within the Cetacea. Mysticetes are less speciose and less disparate than odontocetes; delphinids and then ziphiids are the most speciose and most disparate of the odontocetes.*

include 14 species in four families. The Balaenopteridae is most speciose, with 8 species in two genera (*Balaenoptera*, 7; *Megaptera*, 1). Broadly speaking, species of *Balaenoptera* vary in size, distribution, and behavior, but species boundaries may be blurred. Species are distinguished mainly on aspects of the feeding apparatus baleen size and spacing, size and shape of the upper jaw, and skeletal differences are rather minor. Thus, disparity appears low. The gray whale, *Eschrichtius robustus*, is structurally quite different (disparate) from other mysticetes and is generally placed in its own monotypic family, although some molecular studies place it within the Balaenopteridae. Fossils show the gray whale lineage to extend at least back to the Pliocene, approximately 4 Ma.

Odontocetes include at least 74 species in 10 families. With 38 species, Delphinidae is the most diverse family of cetaceans, and disparity seems much higher than, e.g., within Balaenopteridae. Among delphinids, there is a great variation in body size and proportion, skull form, proportions of the feeding apparatus and teeth, and distribution of air sinuses in the skull base. Among beaked whales (Ziphiidae; 21 species), the genus *Mesoplodon* has 14 rather similar species in which only adult males are separated easily. Disparity here appears low and awaits explanation in terms of evolutionary ecology within the genus. Among other odontocetes, 4 species of small long-beaked "river dolphins" each represent a single family: Iniidae, Pontoporiidae, Lipotidae, and Platanistidae. Although these "river dolphins" are superficially similar externally, they differ markedly in skull form—hence the common separation into 4 families. The most intriguing is *Platanista gangetica* (family Platanistidae), which appears to be the last of the ancient and once diverse superfamily Platanistoidea. For crown *Platanista*, Platanistidae, and Platanistoidea, diversity is low (one living species of *Platanista*), with no fossil record, but many stem platanistoid lineages include diverse species ranging back to 28–29 Ma.

As a group, odontocetes are structurally diverse. Huge variation is seen in the skull, involving the shape, size, and sometimes basic construction of the feeding apparatus and teeth, facial region, including bony origin for nasofacial muscles implicated in echolocation, nasal passages, acoustic system, and air sinuses in the skull base. Disparity might be viewed as caused by reduced constraints on body form, allowing specialization in many different directions. This explanation for disparity, however, does not explain the ultimate origin of structural diversity. Future study of odontocete structural diversity could involve constructional morphology, which offers an alternative to the strongly selectionist/adaptationist accounts of structure and evolution in modern literature.

## XI. Extinction

Extinction, the disappearance of lineages, is the fundamental complement to evolution. Clearly, a species goes extinct when the number of individuals and geographic range drops to nil. The fossil record reveals that extinction is inevitable and, for lineages, usually terminal; few species go extinct by evolving into descendants. Among different styles of extinction, there is no evidence that cetaceans have been involved in mass extinction comparable to that affecting dinosaurs. Taxonomic extinction (the disappearance of a clade) has occurred, as shown by the fossil record of such well-defined clades as the odontocetes Eurhinodelphinidae and Squalodontidae, and the *Cetotherium* group of mysticetes. Environmental change might explain extinction through loss of habitat or food or through climate change. Most extinction has probably involved the piecemeal extinction of single species, but the cetacean fossil record is too patchy to expect pattern or cause to be clear. Species susceptible to extinction are those in low-diversity clades, e.g., one or two species in a genus, with no close relatives, occurring in geographically limited physical settings that are unstable over geological time. For Cetacea, this means particularly the "river dolphins." Conversely, widely distributed oceanic species would seem resistant to extinction.

Patterns of extinction beg a question that ecologists might consider unthinkable: Are there vacant modern niches that formerly were occupied by Cetacea? For example, stem Platanistidae lived in shallow marine settings until about 10 Ma, but *Platanista* now occurs only in freshwaters, and species of Squalodontidae and Eurhinodelphinidae were widely distributed before their demise in the later Miocene. Judging from the functional complexes seen in the latter fossils, there are no modern equivalents to these groups: some morphotypes and lifestyles have disappeared.

## See Also the Following Articles

Archaeocetes, Archaic ■ Mysticete Evolution ■ Basilosaurids

## References

Berger, W. H. (2007). Cenozoic cooling, Antarctic nutrient pump, and the evolution of whales. *Deep Sea Res. II* **54**, 2399–2421.

Corkeron, P. J., and Connor, R. C. (1999). Why do baleen whales migrate? *Mar. Mamm. Sci.* **15**, 1228–1245.

Estes, J. A. (1979). Exploitation of marine mammals: r-selection of K-strategists? *J. Fish. Res. Board Can.* **36**, 1009–1017.

Fordyce, R. E. (2003). Cetacea evolution and Eocene–Oligocene oceans revisited. *In* "From Greenhouse to Icehouse. The Marine Eocene–Oligocene Transition" (D. R. Prothero, L. C. Ivany, and E. Nesbitt, eds), pp. 154–170. Columbia University Press, New York.

Fordyce, R. E., and Muizon, C. de. (2001). Evolutionary history of whales: A review. *In* "Secondary Adaptation of Tetrapods to Life in Water. Proceedings of the International Meeting, Poitiers, 1996" (J.-M. Mazin, and V. de Buffrénil, editors), pp. 169–234. Verlag Dr Friedriech Pfeil, München.

Gingerich, P. D. (2005). Cetacea. *In* "Placental Mammals: Origin, Timing, and Relationships of the Major Extant Clades" (K. D. Rose, and J. D. Archibald, eds), pp. 234–252. Johns Hopkins University Press, Baltimore.

Leduc, R. G., Perrin, W. F., and Dizon, A. E. (1999). Phylogenetic relationships among the delphinid cetaceans based on full cytochrome b sequences. *Mar. Mamm. Sci.* **15**, 619–648.

Lindberg, D. R., and Pyenson, N. D. (2007). Things that go bump in the night: Evolutionary between cephalopods and cetaceans in the Tertiary. *Lethaia* **40**, 335–343.

Pastene, L. A., Goto, M., Kanda, N., Zerbini, A. N., Kerem, D. A. N., Watanabe, K. B. Y., Hasegawa, M., Nielsen, R., Larsen, F., and Palsboll, P. J. (2007). Radiation and speciation of pelagic organisms during periods of global warming: The case of the common minke whale, *Balaenoptera acutorostrata*. *Molecular Ecol.* **16**, 1481–1495.

Rice, D. W. (1998). "Marine Mammals of the World. Systematics and Distribution." Society for Marine Mammalogy, Lawrence, KS.

Thewissen, J. G. M., Cooper, L. N., Clementz, M. T., Bajpai, S., and Tiwari, B. N. (2007). Whales originated from aquatic artiodactyls in the Eocene epoch of India. *Nature* **450**, 1190–1194.

Uhen, M. D., and Pyenson, N. D. (2007). Diversity estimates, biases, and historiographic effects: Resolving cetacean diversity the Tertiary. *Palaeontol. Electronica* **10**, 11A [1–22].

Witmer, L. M. (1995). The extant phylogenetic bracket and the importance of reconstructing soft tissues in fossils. *In* "Functional Morphology in Vertebrate Paleontology" (J. J. Thomason, ed.), pp. 19–33. Cambridge University Press, Cambridge.

# Cetacean Fossil Record

## R. Ewan Fordyce

## I. Introduction

The fossil record of Cetacea—whales, dolphins extends back more than 50 million years (Fig. 1). Hundreds of species are known, based on fossils from near-shore to deep-ocean marine strata and, occasionally, freshwater sediments. Remains vary from less common near-complete skeletons through skulls and teeth to abundant single and usually undiagnostic bones (Fordyce and Muizon, 2001). The taxonomy at family level is adequate to review the diversity and spatio-temporal distribution of fossil Cetacea, although cladistic relationships of families and other taxa to one another, and clade names, are volatile, and cetologists should use cladistic results with caution. Standard zoological techniques are used in taxonomy, classification, and analysis of function. Routine geological techniques are used to date fossils, to interpret sedimentary environments, and to extract geochemical signals such as isotopes from fossils. The fossil record shows patterns of evolution and extinction that link strongly with biological and physical environmental change in the oceans. Fossils provide the only direct means of dating clade origins and thus calibrating molecular clocks.

## II. Occurrence, Environment, and Age

Fossil cetaceans occur in sedimentary rocks. Originally, remains accumulated in mud, silt, sand, or gravel which, as flesh decayed, was buried and turned to rock through compaction and/or deposition of cementing minerals. Sedimentary rocks are recognized as discrete formations (genetically unified bodies of strata), and are named formally, e.g., the Calvert Formation, Maryland. Marine mammals come from strata including sandstone, mudstone, limestone, greensand, and phosphorite, most of which are marine rocks now exposed on land. Rare fossils have been recovered from the sea floor. Because broadly similar rock types may form at different times and places, sedimentary rocks must be dated to establish their time relationships.

Two correlated timescales, relative and absolute, are used for the fossil record. The relative timescale has named intervals (epochs; Fig. 1) in an agreed international sequence: Eocene, Oligocene, Miocene, Pliocene, and Pleistocene. These epochs are usually subdivided into early, middle, and late. Stages (e.g., Aquitanian of Fig. 1) may provide finer subdivision. Typically, distinct age-diagnostic fossils are used to recognize time intervals. The most reliable dates are based on oceanic microfossils with short-time ranges, such as foraminifera, which allow correlation between ocean basins. Because of compounded errors of long-distance correlation, ages are rarely accurate to within 1 million years, and many fossils can be placed only roughly within a stage. Beyond the relative timescale, absolute dates in millions of years are needed to understand rates of processes in phylogeny (involving, e.g., molecular clocks) and in geology (involving, e.g., rates of sediment accumulation or of climate change). Absolute dates are usually obtained from radiometric analysis of grains of volcanic rock interbedded with fossiliferous strata.

## III. The Fossil Record

Fossil cetaceans are classified on the evidence of skeletons (Fig. 2). No other significant body parts preserve, and fossils have not yet produced biomolecules useful in molecular taxonomy. Skulls are by far the most versatile and thus important elements in classification,

C

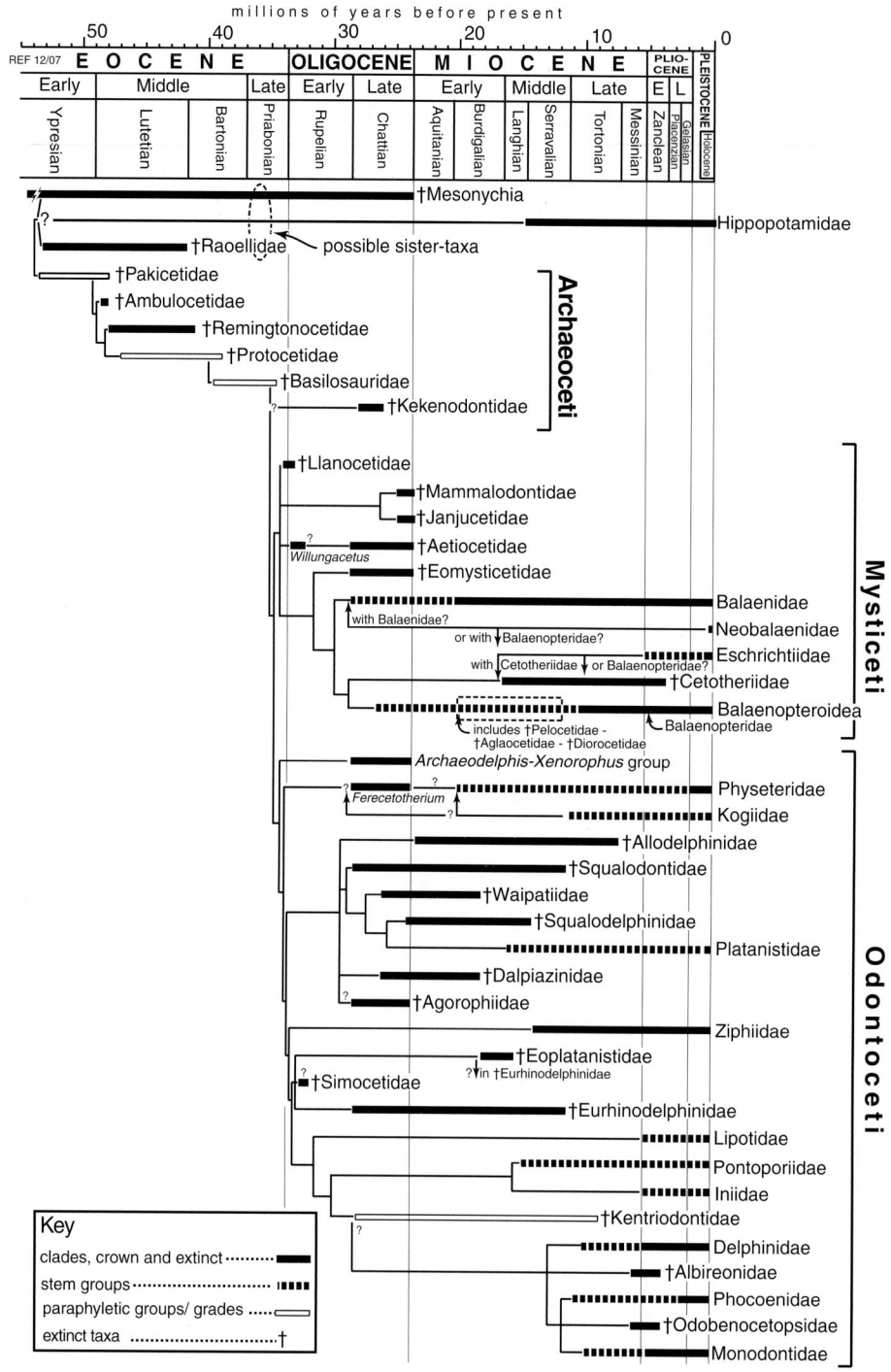

**Figure 1** *Stratigraphic record and inferred relationships of family-level clades of Cetacea. Timescale shows absolute time, Epochs (e.g., Eocene) and their subdivisions (e.g., early, middle), and Stages (e.g., Priabonian). Bars show age ranges for family-level cetacean taxa: crown clades and wholly extinct clades (infilled bars), stem clades (heavy dash), and grades (open bars). Accuracy of ranges varies between different groups and different time intervals. Inferred relationships follow literature cited in text. Some, but not all, major regions of uncertainty are indicated.*

but teeth and, rarely, other bones (vertebrae, limb elements) have been used at times (Fordyce and Muizon, 2001).

    Methods of handling, recording and interpreting specimens are ever-changing; consider, e.g., the development of CT and laser scanning, of computer-aided phylogenetics and shape analysis, and of geochemistry. Computer databases have become fundamental basic tools which allow rapid access to information worldwide (e.g., relevant here: http://pbdb.org). An important conceptual advance since

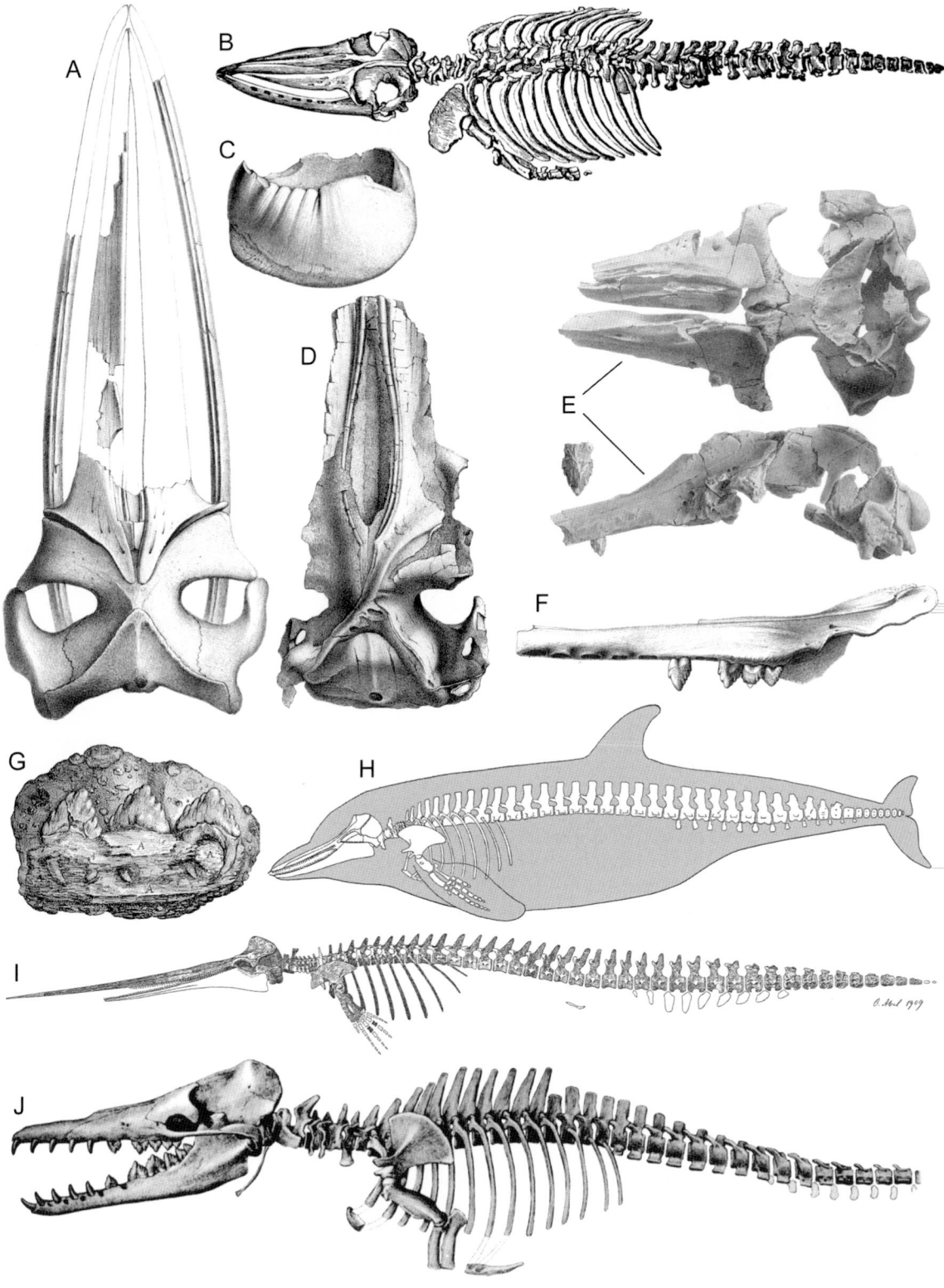

**Figure 2** *Historically important fossil cetaceans figured in early paleontological and systematic literature. (A) Cetotherium rathkii (Mysticeti), skull; from Van Beneden and Gervais 1868–1880. (B) Protororqualus cuvieri (Mysticeti), skeleton; from Zittel 1925. (C) Balaena primigenia (Mysticeti), tympanic bulla; from Van Beneden and Gervais 1868–1880. (D) Metopocetus vandelli (Mysticeti), skull; from Van Beneden and Gervais 1868–1880. (E) Agorophius pygmaeus (Odontoceti), skull and tooth; from True 1907. (F) Squalodon gratelupi (Odontoceti), part of upper jaw; from Van Beneden and Gervais 1868–1880. (G) Squalodon melitensis (Odontoceti), teeth; from Scilla 1670. (H) Kentriodon pernix (Odontoceti), skeleton; from Kellogg 1927. (I) Schizodelphis longirostris (Odontoceti), skeleton; from Abel 1931. (J) Saghacetus osiris (Archaeoceti), skeleton; from Stromer 1908.*

2000 is the gradual introduction of stem- and crown-group terminology into cetacean taxonomy and phylogeny, although this approach is not yet used widely.

Cladistic analyses abound, particularly for the Neoceti, using variable combinations of modern anatomy, molecules, and fossils. Molecular phylogenies give a fairly consistent higher level pattern which is reflected in some but not all morphological cladograms. Contradictory cladograms are inevitable, because many clades are sampled poorly in terms of quality of specimens, number of species, and spectrum of crown-to-stem species. Few analyses of Neoceti agree on, e.g., the number of suprageneric clades and their content. There are implications for crown- and stem-group nomenclature, sister-group relationships, and the known vs predicted times of origins for genera and families. The use of new suprageneric clades for each new rearrangement of often long-known and sometimes fragmentary taxa, without adding major new data from revealing new specimens, is a dubious advance. Where possible, the summary below uses "traditional" family names to convey current understanding.

### A. Archaeoceti

Knowledge of basal whales has expanded dramatically since 1980, giving new insights into phylogeny, ecology, and distribution (Gingerich, 2005). For many years, basal archaeocetes were known only from *Protocetus atavus* (Mokattam Formation, Middle Eocene, ~46.0 Ma; Egypt-Tethys) (Kellogg, 1936). Since the 1980s, new finds, especially in the eastern Tethys, have greatly increased the diversity of species, genera, and families (Thewissen, 1998; Gingerich, 2005). Basal archaeocetes are placed in the Pakicetidae, typified by the small *Pakicetus inachus*, from non-marine redbeds of the Kuldana Formation (49–~49.5 Ma), Pakistan. Skull structure indicates limited underwater hearing capabilities, and the teeth are simpler than those of many later forms. *Pakicetus* has been cited as evidence that the earliest cetaceans radiated slowly in productive shallow waters of the Tethys seaway between Asia and India. The fragmentary jaw of *Himalayacetus subathuensis* from India has been used to push the pakicetid (and cetacean) record back to approximately 53.5 Ma. Isotopes from *Himalayacetus* indicate marine foraging habits, but other pakicetids were probably fresh water (Clementz *et al.*, 2006). Other pakicetids include *Ichthyolestes* and *Nalacetus*, also from the eastern Tethys. It was long thought that pakicetids arose from the terrestrial artiodactyl group Mesonychia, but molecular recognition of the hippopotamus as the sister taxon to Cetacea called mesonychid links into question. Later studies have supported links with hippopotamus, with small deer-like raoellids, and still with mesonychids (Geisler and Uhen, 2003; O'Leary and Gatesy, 2007; Thewissen *et al.*, 2007).

The family Protocetidae, now expanded beyond *Protocetus*, is an Early to Middle Eocene grade for species in which the skull has an enlarged supraorbital shield, the mandible has a large mandibular foramen, and hindlimbs are reduced. Protocetids lack the complex teeth and pterygoid sinuses of younger cetaceans. Many protocetids have been named since 1990, with 14 genera in three subfamilies forming a comb-like sequence of genera rather than a single clade. Some protocetids occur in the western Tethys to western central Atlantic, including the large *Eocetus* (Egypt, North Carolina), *Pappocetus* (Nigeria; North Carolina) *Natchitochia* (Louisiana), apparently *Protocetus* (Texas), and *Carolinacetus* (South Carolina). Protocetids have a high diversity in the eastern Tethys, judging from the range of teeth and skulls from Pakistan and India. *Babiacetus* (~43.5 Ma) is known from teeth and jaws, while partial skulls represent the slightly older (45.5–46-Ma) *Gaviacetus*, *Takracetus*, and

*Indocetus*. *Rodhocetus kasrani* (Domanda Formation, 46–46.5 Ma; Pakistan) has a skeleton strikingly intermediate between that of land mammals and later whales (Gingerich, 2005). Cetacean features include the short neck vertebrae and more-posterior vertebrae adapted for dorsoventral oscillation, but *Rodhocetus* retains a femur and sacrum. *Rodhocetus* is from deep- rather than shallow-water deposits, implying early colonization of offshore habitats. Another protocetid, *Georgiacetus vogtlensis* (McBean Formation, 40–41 Ma; Georgia) shows derived features in the cheek-teeth, an incipient pterygoid sinus in the skull base, and a reduced link between sacrum and pelvis, all of which presage features seen in the more-crownward Basilosauridae.

Two other rather specialized archaeocete families are reported only from the Early and Middle Eocene of the eastern Tethys (Thewissen, 1998). Firstly, *Ambulocetus natans* (Kuldana Formation, 48–49 Ma; Ambulocetidae) includes a substantially complete skeleton with a long-snouted skull and well-developed fore- and hind limbs. *Ambulocetus* perhaps swam using pelvic paddling and dorsoventral undulations of the tail, comparable in style to some modern otters. A crocodile-like mode of predation in water is possible, but locomotion on land was probably clumsy. Secondly, the family Remingtonocetidae includes specialized long-snouted Middle Eocene species of *Remingtonocetus* and four other genera, known from at least partial skulls (43.5–~45 Ma). Despite previous suggestions, remingtonocetids are unrelated to odontocetes.

Basilosaurids, from the later Middle and Late Eocene, ranged well beyond the Tethys. Opinions differ as to whether there is one family (with two subfamilies, Basilosaurinae and Dorudontinae), or two families, based around *Basilosaurus* and *Dorudon*. These archaeocetes are typified by the 15 m long *Basilosaurus cetoides*, first described, and named as a fossil reptile, from Louisiana (Jackson Formation, Late Eocene, 36–~39 Ma; western North Atlantic; Kellogg, 1936). The large size of *Basilosaurus*, and its elongate vertebrae, are specialized features used to recognize a clade Basilosaurinae. Also included is *Basilosaurus isis* (~39 Ma; Egypt, central Tethys), which has small but functional hind limbs of ungulate-like character (Gingerich, 2005). Large later Eocene archaeocetes, presumably basilosaurines, have been reported from scattered localities worldwide (e.g., Northeastern Atlantic, proto-Southern Ocean, Southwest Pacific and eastern Tropical Pacific), indicating a wide range for Cetacea in Eocene time.

The second subfamily of basilosaurids, the Dorudontinae, is a grade which includes smaller, more generalized and somewhat dolphin-like species of *Dorudon*, *Saghacetus*, and others. The genera are rather similar to one another, and are diagnosed on size, tooth form, and limb form. The typical species *D. serratus* is fragmentary, but other dorudontines include some magnificent fossils (e.g., *Dorudon atrox*, Birket Qarun Formation, 39–40 Ma; Egypt; Uhen, 2004). The small *Saghacetus osiris* (Fig. 2j) ranges into the latest Eocene, Priabonian, or 33.7–37.0 Ma (Gingerich, 2005), and thus overlaps in age with the oldest of the Neoceti. The formally named dorudontines are from the Tethys and western Central Atlantic, but fragmentary late middle Eocene specimens from New Zealand indicate an early geographic spread.

Basilosaurids differ from more basal archaeocetes in having cheek-teeth with complex denticles, and expanded basicranial air sinuses. These features, which imply more sophisticated feeding and hearing capacities, link basilosaurids closely with early odontocetes and mysticetes. Some dorudontines have been identified as sister taxa to the Odontoceti + Mysticeti (Uhen, 2004; Gingerich, 2005), while basilosaurines seem too specialized, in terms of large size and elongate vertebral bodies, to be directly ancestral to living cetaceans.

Fragmentary Oligocene fossils with archaeocete-like features, e.g., *Kekenodon onamata* (New Zealand, Kokoamu Greensand, Late Oligocene, 27–~28 Ma; Kekenodontidae), have been known since the 1800s. For a while, such animals were thought to be basal toothed mysticetes. By 2004, it was apparent that the *Kekenodon*-like animals are indeed archaeocetes in Kellogg's classical sense; they fall cladistically between dorudontines and the Neoceti. *Kekenodon* and unnamed New Zealand specimens occur in the same late Oligocene strata as early crown Mysticeti and Odontoceti.

## B. Mysticeti

Since the 1960s, the fossil record of mysticetes has expanded to reveal diverse toothed and toothless Oligocene species which effectively "bridge the gap" between archaeocetes and baleen-bearing crown Mysticeti. The latter is the clade containing extant rorquals (Balaenopteridae), right whales (Balaenidae) and the more enigmatic gray whale (Eschrichtiidae) and pygmy right whale (Neobalaenidae). The crown clade is conceptually different from group Chaeomysticeti, or toothless baleen whales (Mitchell, 1989). This summary reviews families: at crown level, the Balaenopteridae, Balaenidae, Eschrichtiidae, and Neobalaenidae, and for the fossil record, Aetiocetidae, Llanocetidae, Mammalodontidae, Janjucetidae, Cetotheriidae, Diorocetidae, Aglaocetidae, Pelocetidae, Eomysticetidae, and Cetotheriopsidae. Mysticetes, fossil and living, are generally considered filter feeders, although feeding mode is contentious for some toothed groups.

Pivotal in discussion of mysticete origins is *Aetiocetus cotylalveus* (Aetiocetidae; Yaquina Formation, Late Oligocene) from Oregon. Initially, this small cetacean was identified as an archaeocete because it has teeth, but other features, including the flattened triangular rostrum, indicate that it is an archaic mysticete. Aetiocetids are moderately diverse in their reported North Pacific range, and include species of *Chonecetus* and *Morawanocetus* from Japan (Barnes *et al.*, 1995) and perhaps the fragmentary early Oligocene (~32 Ma) *Willungacetus* of Australia. From Victoria, on the margins of the proto-Southern Ocean, are two other notable toothed species each in its own family: the short-snouted *Mammalodon colliveri* and *Janjucetus hunderi* (about Oligocene–Miocene boundary, 23–~24 Ma; Mammalodontidae and Janjucetidae; Fitzgerald, 2006). *Mammalodon* and *Janjucetus* show some specialized features that are highly disparate from other mysticetes, such as tiny skull size, rostrum remarkably shorter than the cranium, big orbits, and prominent robust occluding teeth. These animals are relicts that represent basal Mysticeti. It is debatable how their structure and inferred habits, including raptorial macrophagy, truly represent the early stages of mysticete history. Perhaps *Mammalodon* and *Janjucetus* were highly specialized, pedomorphic, geographically restricted forms that lived in extreme marginal settings in isolated seas not used by contemporaneous crown Mysticeti and Odontoceti, having abandoned the filter-feeding habits otherwise typical of mysticetes.

The oldest reported archaic toothed mysticete is *Llanocetus denticrenatus* (Llanocetidae), named from a fragmentary toothed jaw and a brain cast from the La Meseta Formation, latest Eocene (~35 Ma) of Seymour Island, Antarctica (Mitchell, 1989). A nearly complete large skull is now known; the skull was originally about 2 m long, and has a broad rostrum with widely spaced denticulate teeth that occluded with the lower teeth. The exact role of the teeth in feeding is uncertain, but it is possible that spaces between the teeth were filled with proto-baleen. The posterior of the *Llanocetus* skull is superficially quite like that of basilosaurid archaeocetes. A smaller

*Llanocetus*-like toothed mysticete is known from the basal Oligocene of New Zealand.

Remains of toothless baleen-bearing mysticetes, the Chaeomysticeti of Mitchell (1989), are common in Miocene and younger strata worldwide. Whitmore and Sanders (1977) reported few Oligocene baleen whales, but since 1977, the late Oligocene record back to 28–29 Ma has expanded significantly. Many fossils, particularly those older than late Miocene (~12 Ma) are more archaic than living baleen whales, and have long been placed in the family Cetotheriidae (Fordyce and Muizon, 2001). Cladistic analyses allow the traditional use of a paraphyletic/ polyphyletic grade Cetotheriidae to be abandoned (Bouetel and Muizon, 2006; Steeman, 2007). Cetotheres are typified by the small Middle Miocene (12–13 Ma) *Cetotherium rathkii* from Ukraine (Paratethys), which differs from living mysticetes in that the upper jaw (rostrum) thrusts back into bones of the braincase with a sharp narrow triangular apex, almost obscuring the nasal bones (Fig. 2a). The related early Pliocene *Herpetocetus sendaicus* (Yushima Formation, Japan) has a similar rostral structure and, further, shows the strange superficially reptilian jaw articulation that led Van Beneden to propose *Herpetocetus* in 1872. *Piscobalaena*, *Nannocetus*, and *Metopocetus* (Fig. 2d) are also in the extinct clade Cetotheriidae, which ranges from middle Miocene to early Pliocene (4–16 Ma) (Bouetel and Muizon, 2006; Steeman, 2007). It is not clear where other named archaic mysticetes—former "cetotheres"—belong, or whether the Cetotheriidae lies in the stem Mysticeti, or with gray whale (*Eschrichtius*, Eschrichtiidae) within the crown Mysticeti.

Two other families are reported only from the Oligocene, which was generally a time of great diversification of Mysticeti. The Eomysticetidae contains the long narrow-skulled species of *Eomysticetus*. Eomysticetids, including associated skulls and skeletons, are important members of late Oligocene cetacean assemblages from New Zealand; "*Mauicetus*" *lophocephalus* is a widely recognized example. Whether the Cetotheriopsidae, containing *Cetotheriopsis* and *Micromysticetus*, is distinct is debatable; the fossils are too fragmentary to elucidate overall structure and likely lifestyle. Eomysticetids and cetotheriopsids reportedly form a clade basal to crown Mysticeti (Geisler and Sanders, 2003).

The family Balaenopteridae comprises the fast swimming, gulp feeding, rorquals (*Balaenoptera*) and humpback whale (*Megaptera*). Many fossils have been assigned to the family since the 1800s (Fig. 2b), although there has long been debate on how to distinguish archaic balaenopterids from cetotheres. Cladistic analysis allows crown and stem Balaenopteridae to be differentiated, with the crown group including fossil species in *Parabalaenoptera* and nominally *Megaptera*; living *Megaptera novaeangliae* is the most basal crown species (Deméré *et al.*, 2005). The oldest crown balaenopterid identified is the late Miocene (~10–11 Ma) "*Megaptera*" *miocaena*, which is notably younger than some of the crown divergence times predicted from molecular phylogenies (Sasaki *et al.*, 2005). There are many records of later Miocene, Pliocene and Pleistocene *Balaenoptera* fossils, but none has resolved the time of origin of any one living species. The middle Miocene *Eobalaenoptera* is of uncertain position, not clearly the oldest known balaenopterid. A unifying feature of the crown balaenopterids is the distinctive structure above the orbits on the skull, where the frontal bone above the eye is depressed to house the large muscles that close the lower jaw.

Cladistic studies disagree on which fossil genera are closer to crown Balaenopteridae than to other family clades of mysticetes. Perhaps some genera, including former "cetotheres," belong in the Balaenopteroidea, as extinct clades: the Diorocetidae (*Diorocetus* and others), Aglaocetidae (*Aglaocetus* and others) and Pelocetidae

C

(*Pelocetus* and others) (Steeman, 2007). These are mostly animals with skulls 1 to 2 m long, representing species comparable in size to minke whale. Fossils come from Miocene and Pliocene localities mainly around Europe and the Americas. It is not clear that such fossils deserve separate family status, given the highly variable nature of type or reference material, and reliance on earbones to define clades. Also debatable is the position of the Late Oligocene (~24 Ma) *Mauicetus*, a minke whale-sized gulp-feeder: the most basal genus in the Balaenopteroidea (Steeman, 2007) or, alternatively, a stem balaenopterid? Other *Mauicetus*-like fossils from New Zealand are older, at approximately 26 Ma, emphasizing the antiquity of rorquals.

The living gray whale *Eschrichtius robustus* has a fossil record extending about 0.5 M years into the Pleistocene. Historic gray whale bones indicate that these animals occurred in the North Atlantic, where they were probably exterminated by hunting. Of older putative gray whales, a fragmentary, but convincing, Late Pliocene eschrichtiid is known from Teshio, Japan (Ichishima *et al.*, 2006). *Archaeschrichtius* (Italy, late Miocene) is based on a mandible. On the origin of eschrichtiids, morphological cladistic analyses variously place *Eschrichtius* close to balaenids, to balaenopterids and, unexpectedly, to cetotheres.

Right whales are known from many fragmentary fossils (Fig. 2c), but few that include informative material such as skulls (Bisconti, 2005). Modern right whales are large, slow moving, skim feeding mysticetes which are skeletally so conservative as to have long been placed in one genus, *Balaena*, although *Eubalaena* is now used for other than bowhead whale. The oldest named fossil right whale is the early Miocene *Morenocetus parvus* (~20 Ma) from Patagonia, but an older unnamed stem balaenid has been reported from New Zealand (Kokoamu Greensand, ~28 Ma). Perhaps *Morenocetus* is cladistically closer to *Balaena* than to *Eubalaena*, with the crown Balaenidae originating before approximately 20 Ma (Bisconti, 2005); an early divergence, approximately 17 Ma (Sasaki *et al.*, 2005), is predicted for the molecular split between *Balaena* and *Eubalaena*. Such divergence times imply extremely slow structural evolution in living right whales. Notable younger fossil balaenids include the small early Pliocene (4–~5 Ma) *Balaenella brachyrhynus* (Kattendijk Sand, Belgium), and the contemporaneous large *Balaena ricei* (early Pliocene, Yorktown Formation, Atlantic Coastal Plain).

There is no firm published fossil record of Neobalaenidae to indicate the origins of the enigmatic pygmy right whale, *Caperea marginata*. *Caperea* has long been presumed to belong close to the Balaenidae, a position supported by most anatomical cladistic studies (Steeman (2007). However, molecular analyses generally place *Caperea* closer to *Eschrichtius* and balaenopterids. The skull of *Caperea* is quite disparate from balaenids, suggesting that the molecular relationships should be considered seriously. Useful fossils, however, are lacking.

## C. Odontoceti

Odontocetes are much more diverse in terms of taxa (families, genera) and structure than are mysticetes. Living odontocetes all appear to be echolocating raptors, and a similar lifestyle is inferred for most extinct species. Late Eocene odontocetes have been reported from Washington State, but are not formally named. The oldest named odontocete is the small *Simocetus rayi* (Simocetidae; Alsea Formation, Oregon; Fordyce, 2002), of early Oligocene age (perhaps 32 Ma). The skull in *Simocetus* is generally archaic, but the animal has a highly specialized downward arched upper jaw (rostrum) with a toothless

tip; perhaps this was a sediment-grubbing slurry-feeder. Facial structures suggest that *Simocetus* could echolocate. Basicranial features hint at links with enigmatic species of *Squaloziphius* and *Argyrocetus* (both provisionally Eurhinodelphiidae, within crown Odontoceti). *Simocetus* clearly is not close to the late Oligocene *Agorophius pygmaeus* (Cooper Marl, Late Oligocene, >24 Ma, South Carolina; Whitmore and Sanders, 1977). *Agorophius* typifies the Agorophiidae, a family widely used until 1980 as a grade for archaic odontocetes. Where *Agorophius* belongs within the Odontoceti is uncertain for lack of information on the species, as the holotype skull is lost, and the one useful published figure is a lithograph from the 1860s (Fig. 2e). There is no compelling reason to regard the Agorophiidae as a basal clade of odontocetes.

The most basal named odontocete is *Archaeodelphis patrius*, based on a fragmentary skull of uncertain origin and possible Oligocene age. Uniquely, this enigmatic species barely shows evidence of the naso-facial muscles which, in other odontocetes, are implicated in echolocation. Despite its reputedly ancestral position, because its orbit (lacrimal) and skull base are somewhat specialized, *Archaeodelphis* is not directly on a lineage leading to living species. Almost as archaic is *Xenorophus*, a bizarre Late Oligocene genus containing one or two species with specialized facial structures superimposed on an archaic skull (Whitmore and Sanders, 1977).

Sperm whales (*Physeter macrocephalus*, and *Kogia* spp.) appear to be the most basal living odontocetes. The oldest fossils of *P. macrocephalus*, marking the crown Physeteridae, are geologically young fragmentary specimens of uncertain Pleistocene to possibly latest Pliocene (~2 Ma). Skulls of stem physeterids are rare but informative; they are pivotal in cladistic studies (Bianucci and Landini, 2006), and help to recognize the subfamilies Physeterinae, Hoplocetinae, and Aulophyseterinae. Early Miocene sperm whales, such as *Diaphorocetus* and *Idiorophus* from the South Atlantic, show the characteristic basined facial bones of later sperm whales, while features of the brain case link them firmly with other odontocetes. Many named fossil sperm whales are based on isolated teeth which, however, reveal little about the actual animal or about relationships. The fragmentary *Ferecetotherium kelloggi* is reportedly late Oligocene (23 + Ma; Maikop Formation, Paratethys-Caucasus), providing the earliest record for stem physeterids. There is no reliable fossil evidence for the age of the crown *Kogia* (living dwarf and pygmy sperm whales, variously termed Kogiidae or Kogiinae), but molecular studies on *K. sima* imply crown origins well before 4 Ma (Chivers *et al.*, 2005). The oldest stem kogiids are later Miocene species from the eastern Pacific, including *Praekogia* from Isla Cedros (Almejas Formation, ~6 Ma), and the large narrow-skulled *Scaphokogia* from the Pisco Formation of Peru (Muizon, 1988a). The *Kogia* clade probably arose amongst stem physeterids/physeteroids.

Fossil beaked whales, Ziphiidae, are reported widely as isolated and worn rostra ("beaks" or upper jaws) and rarely as more complete skulls from unconformities, and from modern oceanic phosphate sediments (Bianucci *et al.*, 2007). Such fossils may be reworked from older sediments, and difficult to date. The oldest well-dated ziphiid is the middle Miocene (13–~15 Ma) *Archaeoziphius microglenoideus*, which is close to the crown genera *Berardius* and *Tasmacetus* (Lambert and Louwye, 2006). Like most modern species, crown-ziphiid fossils have high crania and robust rostra with reduced teeth. Anatomical cladistic studies recognize the subfamilies Hyperoodontinae, Ziphiinae and Berardiinae, all with both living and extinct species. There are no firm reports of stem Ziphiidae. The widely discussed *Squaloziphius emlongi* (Clallam Formation, Oligocene–Miocene, ~23 Ma, Washington; Muizon, 1991) is not clearly a beaked whale, and other ancient supposed ziphiids (e.g., *Notocetus*) are platanistoids. Perhaps the most unusual occurrence of a fossil cetacean is that of a large ziphiid in fresh water Miocene strata of Kenya.

The content and relationships of the Platanistoidea are contentious. Is the crown clade monotypic (*Platanista* only), or polytypic (all living "river dolphins"), and do platanistoids lie between physeterids and ziphiids or nearer to delphinoids (Muizon, 1991; Nikaido *et al.*, 2001; Geisler and Sanders, 2003)? This summary views Platanistoidea as comprising Platanistidae and diverse extinct relatives: Squalodelphinidae, Waipatiidae, Squalodontidae, Allodelphinidae, and the less certainly recognized Prosqualodontidae, Patriocetidae and Dalpiazinidae. Geisler and Sanders (2003), however, excluded most of the latter from Platanistoidea and also from crown Odontoceti.

Crown Platanistidae contains only the bizarrely disparate riverine *Platanista gangetica* from India and Pakistan, with no fossil record. Notable stem platanistids are known, all Miocene (6–~16 Ma) long-beaked shallow marine forms from the western North Atlantic and Paratethys (Barnes, 2006): *Prepomatodelphis*, *Zarhachis*, *Pomatodelphis*. These dolphins have facial crests which place them close to *Platanista*, but because of the dorsoventrally flattened jaws, Barnes classified them in the Pomatodelphininae. The record suggests that *Platanista* invaded fresh water late in its history.

Muizon (1991) recognized the extinct Squalodelphinidae as sister taxon to the Platanistidae. Squalodelphinids are typified by the long-beaked *Squalodelphis* (Early Miocene, Mediterranean); they have heterodont teeth and asymmetrical crania, but no facial crests. Southern squalodelphinids include several species of *Notocetus* from the latest Oligocene–early Miocene (18–~24 Ma; New Zealand, Patagonia) bordering the Southern Ocean. A small heterodont marine dolphin with a slightly asymmetrical skull, *Waipatia maerewhenua* (late Oligocene, ~25 Ma; New Zealand) typifies a late Oligocene–early Miocene family Waipatiidae, basal to the Platanistidae + Squalodelphinidae (Fordyce, 1994). As with other early odontocetes, *Waipatia* shows skull features which indicate an ability to echolocate. Other previously enigmatic odontocetes, such as *Sulakocetus* and *Microcetus* from the Tethys-North Atlantic, also may be waipatiids.

Probably the best recognized of the extinct platanistoids are the shark-toothed dolphins, Squalodontidae. These are geographically widespread medium to large odontocetes with long rostra and conspicuous, large, triangular heterodont teeth (Rothausen, 1968) (Fig. 2f, g); there are no comparable living long-jawed robust-toothed raptors. Squalodontids were long considered central in the phylogeny of living odontocetes, but recently they have been recognized as platanistoids (Muizon, 1991). Squalodontids are typified by *Squalodon gratelupi* (Early Miocene, ~20 Ma, eastern North Atlantic) (Fig. 2f). Other species with more-complete skulls, from both northern and Southern Hemisphere locations, are important in diagnosing the group. The squalodontids include at least *Squalodon* and *Phoberodon*, but exclude many taxa, originally based on teeth, now known to belong elsewhere including Mysticeti and Archaeoceti. Notable undescribed late Oligocene species are known from South Carolina and New Zealand, while the youngest squalodontids are late middle to early late Miocene (10–12 Ma, Rothausen, 1968). It is not clear whether the robust broad-beaked *Prosqualodon* (latest Oligocene-early Miocene (18–~24 Ma, marginal Southern Ocean) deserves its own family, Prosqualodontidae, or belongs in Squalodontidae. Likewise, the late Oligocene *Patriocetus*, with a wide, almost shelf-edged, rostrum, has been placed in a monotypic Patriocetidae but equally could be a disparate squalodontid. *Patriocetus* lived in isolated waters of the Austrian Paratethys.

The Allodelphinidae encompasses the extinct marine *Allodelphis* (Jewett Sand, Early Miocene, ~20 Ma; California) and a few other Northeast Pacific species as young as late Miocene. Allodelphinids are basal platanistoids with a unique premaxillary structure (Barnes, 2006). The most enigmatic of the reputed platanistoids is the extinct Dalpiazinidae of Muizon (1988b), established for the early Miocene fragmentary *Dalpiazina* ["*Acrodelphis*"] *ombonii*. Formerly, this and other Miocene species with dorsoventrally flattened rostra, reminiscent of *Pomatodelphis* (discussed earlier), were placed in the widely cited Acrodelphidae which, however, is no longer used because of questionable typology.

Species in the extinct Eurhinodelphinidae are small- to medium-sized dolphins with dramatically long rostra on which the long premaxillae are toothless (Fig. 2i). Species of *Schizodelphis*, *Xiphiacetus* and others were significant members of early, middle and perhaps late Miocene cetacean faunas in the Atlantic, and the family ranged into the Pacific (Lambert, 2005). There appear to be no modern ecological equivalents. The early record is patchy; Oligocene forms comprise the poorly known *Iniopsis* (a fragmentary skull, Paratethys-Caucasus) and an unnamed Late Oligocene species from lake strata of central Australia which marks an early invasion of fresh waters. Relationships are uncertain; eurhinodelphinids could lie with delphinoids, with platanistoids, with ziphiids, or between physeteroids and other crown Odontoceti (Muizon, 1991; Geisler and Sanders, 2003; Lambert, 2005). Another extinct group, the monotypic early Miocene Eoplatanistidae, is close to eurhinodelphinids.

The remaining odontocetes include the ocean dolphins (Delphinidae) and close relatives Phocoenidae and Monodontidae. The exact relationships of other dolphin groups are more contentious; Muizon (e.g., 1988c) placed all in the Delphinida, as followed here. Dolphins—(Delphinida) generally dominate odontocete fossil assemblages from the late Miocene on. Most early or archaic fossil dolphins have been placed in the Kentriodontidae, variously used as a grade or clade. Kentriodontids are geographically widespread small to medium-sized long-beaked animals with more or less symmetrical skulls and near-homodont teeth. *Kentriodon* includes *Sotalia*-sized dolphins (Fig. 2h that were widespread in the early Miocene, as indicated by fossils from the northwest Atlantic, and around the Pacific. Many other genera (e.g., *Delphinodon*, *Tagicetus*, *Pithanodelphis*, *Atocetus*) have been reported from northern and a few southern localities. Rare *Kentriodon*-like fossils are known from the late Oligocene (>23 Ma). The oldest alleged Delphinidae are late Miocene (5–11 Ma), but it seems that no well-dated, well-preserved, named delphinid is older than Pliocene. Nevertheless, stem delphinids presumably evolved by 10–11 Ma, given the known record of their likely sister taxon Phocoenidae. Named delphinids are major components of, especially, Pliocene assemblages (2–5 Ma). For example, fossils from northern Italy (Bianucci, 1996) include skulls, teeth, earbones referred to *Stenella*, *Tursiops*, *Orcinus*, *Hemisyntrachelus*, and *Astadelphis*. Fossil delphinids include some unusual morphotypes, such as the Antarctic Pliocene *Australodelphis*, which is toothless and convergent in form with beaked whales.

Like delphinids, porpoises (Phocoenidae) and white whales (Monodontidae) range back to the late Miocene. Phocoenids have a better record than monodontids, with the oldest stem taxon, *Salumiphocaena stocktoni*, from the eastern North Pacific (Monterey Formation, 10–11, California; Barnes, 1985). *Salumiphocaena* shows characteristic phocoenid features, such as premaxillary eminences and a frontal boss. Rosel *et al.* (1995) predicted an origin for phocoenids at 12–16 Ma, and suggested a radiation for most crown species

**C**

2–3 Ma. There is no older record of well-dated firmly identified crown porpoises, and the diverse extinct genera, such as *Piscolithax*, *Lomacetus*, and *Haborophocoena*, are unrevealing about predicted crown origins. However, the Mio-Pliocene *Piscolithax* is a stem phocoenid, originating before the crown radiation (Fajardo-Mellor *et al.*, 2006). Judging from fossils, most of phocoenid history occurred in the north to eastern tropical Pacific.

A meager record of monodontids gives little clue about crown or stem origins. Fossils of the living species of *Delphinapterus* and *Monodon*, from mid-latitude North Atlantic shores, have not been reported firmly to be older than Pleistocene, 2 Ma, and most are likely late Pleistocene, less than 0.5 Ma. In the Atlantic, an early Pliocene *Delphinapterus* occurs in North Carolina (Yorktown Formation, ~4.5 Ma; Whitmore, 1994). The archaic (presumed stem) monodontid *Denebola brachycephala* is from Isla Cedros, equatorial eastern Pacific (Almejas Formation, late Miocene, >5 Ma), and other unnamed late Miocene and Pliocene monodontids have been reported from the Pacific coast of North America and Peru (Barnes, 1984).

Two extinct families of presumed delphinoids are reported from the late Miocene-early Pliocene of the eastern Pacific. Barnes (1984) based the Albireonidae on the porpoise-like *Albireo whistleri*, from Cedros Island. Two species of tusked "walrus whales" (*Odobenocetops*: Odobenocetopsidae) are from the Pisco Formation, Peru (Muizon and Domning, 2002). The latter are uncannily like the living walrus, *Odobenus*, in skull form. Some cetologists have doubted that *Odobenocetops* is a cetacean because of its highly disparate form. Neither genus, *Albireo* or *Odobenocetops*, has been subject of detailed published cladistic analysis; they have been linked with Phocoenidae and Monodontidae respectively.

Three families of "river-dolphin" are sister taxa to the Delphinoidea in both anatomical and molecular phylogenetic analyses: the Pontoporiidae + Iniidae, and more basally, the Lipotidae. Each family is represented by a single living species with no noteworthy fossil record, but each is known from stem representatives. Stem pontoporiids occur in freshwater strata in South America and Mio-Pliocene marine rocks bordering the eastern North Pacific and the North Atlantic. The small short-beaked *Brachydelphis mazeasi* (Pisco Formation, 12–~15 Ma) from Peru is so disparate that it has been put in its own subfamily (Pontoporiidae: Brachydelphininae; Muizon, 1988a) related to the long-beaked Peruvian marine fossil *Pliopontos littoralis* (4–~5 Ma), the large marine fossil *Pontistes rectifrons*, and the living paralic-neritic franciscana (*Pontoporia blainvillei*). Early Pliocene (~4.5 Ma) *Pontoporia*-like fossils are known from North Carolina (Yorktown Formation), and there are significant late Miocene pontoporiids from Europe (Pyenson and Hoch, 2007).

In the early 1900s, many fossils were placed in the family Iniidae, with the living *Inia geoffrensis*, but most of these belong elsewhere. Stem iniids (the Ischyrorhynchinae) include the South American freshwater *Ischyrorhynchus* and *Saurodelphis* (late Miocene, "Mesopotamiense" horizon, >5 Ma; Cozzuol, 1996), and the fragmentary marine *Goniodelphis hudsoni* (Early Pliocene, 4–5 Ma) from Florida. Stem iniids must have arisen with their sister taxon Pontoporiidae; iniids thus have a significant ghost lineage at least back to approximately 12–15 Ma.

The baiji, *Lipotes vexillifer*, apparently now extinct, was a riverine descendant of a marine lineage known from long-beaked species of *Parapontoporia* from Isla Cedros (Almejas Formation, ~5 Ma) and southern California (San Diego Formation, ~2–3 Ma) (Barnes, 1984; Pyenson and Hoch, 2007). According to Muizon (1988c), neither genus is close to *Pontoporia*. The problematic *Prolipotes*, based on a

fragment of possibly Miocene lower jaw from freshwater deposits in China, is too incomplete to confirm a relationship with *Lipotes*.

## See Also the Following Articles

Mysticete Evolution ■ Sperm whales, Evolution ■ Basilosaurids

## References

Barnes, L. G. (1984). Fossil odontocetes (Mammalia: Cetacea) from the Almejas Formation, Isla Cedros, Mexico. *Paleobios* **42**, 1–46.

Barnes, L. G. (1985). Evolution, taxonomy and antitropical distributions of the porpoises (Phocoenidae, Mammalia). *Mar. Mamm. Sci.* **1**, 149–165.

Barnes, L. G. (2006). A phylogenetic analysis of the Superfamily Platanistoidea (Mammalia, Cetacea, Odontoceti). *Beiträge zur Paläontologie* **30**, 25–42.

Barnes, L. G., Kimura, M., Furusawa, H., and Sawamura, H. (1995). Classification and distribution of Oligocene Aetiocetidae (Mammalia; Cetacea; Mysticeti) from western North America and Japan. *Isl. Arc.* **3**, 392–431.

Bianucci, G. (1996). The Odontoceti (Mammalia, Cetacea) from Italian Pliocene. Systematics and phylogenesis of Delphinidae. *Palaeontographia Italia* **83**, 73–167.

Bianucci, G., and Landini, W. (2006). Killer sperm whale: A new basal physeteroid (Mammalia, Cetacea) from the Late Miocene of Italy. *Zool. J. Linn. Soc.* **148**, 103–131.

Bianucci, G., Lambert, O., and Post, K. (2007). A high diversity in fossil beaked whales (Mammalia, Odontoceti, Ziphiidae) recovered by trawling from the sea floor off South Africa. *Geodiversitas* **29**, 561–618.

Bisconti, M. (2005). Skull morphology and phylogenetic relationships of a new diminutive balaenid from the lower Pliocene of Belgium. *Palaeontology* **48**, 793–816.

Bouetel, V., and Muizon, C.de. (2006). The anatomy and relationships of *Piscobalaena nana* (Cetacea, Mysticeti), a Cetotheriidae s.s. from the early Pliocene of Peru. *Geodiversitas* **28**, 319–395.

Chivers, S. J., Leduc, R. G., Robertson, K. M., Barros, N. B., and Dizon, A. E. (2005). Genetic variation of *Kogia* spp. with preliminary evidence for two species of *Kogia sima*. *Mar. Mamm. Sci.* **21**, 619–634.

Clementz, M. T., Goswami, A., Gingerich, P. D., and Koch, P. L. (2006). Isotopic records from early whales and sea cows: Contrasting patterns of ecological transition. *J. Vertebr. Paleontol.* **26**, 355–370.

Cozzuol, M. A. (1996). The record of the aquatic mammals in southern South America. *Münchner Geowissenschaftliche Abhandlungen. A, Geologie und Paläontologie* **30**, 321–342.

Deméré, T. A., Berta, A., and McGowen, M. R. (2005). The taxonomic and evolutionary history of fossil and modern balaenopteroid mysticetes. *J. Mammal. Evol.* **12**, 99–143.

Fajardo-Mellor, L., Berta, A., Brownell, R. L., Boy, C. C., and Goodall, R. N. P. (2006). The phylogenetic relationships and biogeography of true porpoises (Mammalia: Phocoenidae) based on morphological data. *Mar. Mamm. Sci.* **22**, 910–932.

Fitzgerald, E. M. G. (2006). A bizarre new toothed mysticete (Cetacea) from Australia and the early evolution of baleen whales. *Proceedings of the Royal Society B – Biological Sciences* **273**, 2955–2963.

Fordyce, R. E. (1994). *Waipatia maerewhenua*, new genus and new species (Waipatiidae), an archaic Late Oligocene dolphin (Cetacea: Odontoceti: Platanistoidea) from New Zealand. *Proceedings of the San Diego Museum of Natural History* **29**, 147–176.

Fordyce, R. E. (2002). *Simocetus rayi* (Odontoceti: Simocetidae) (new species, new genus, new family), a bizarre new archaic Oligocene dolphin from the eastern North Pacific. *Smithsonian Contrib. Paleobiol.* **93**, 185–222.

Fordyce, R. E., and Muizon, C. de. (2001). Evolutionary history of whales: A review. *In* "Secondary Adaptation of Tetrapods to Life in

Water. Proceedings of the International Meeting, Poitiers, 1996" (J.-M. Mazin, and V. de Buffrénil, eds), pp. 169–234. Verlag Dr Friedriech Pfeil, München.

Geisler, J. H., and Sanders, A. E. (2003). Morphological evidence for the phylogeny of Cetacea. *J. Mamm. Evol.* **10**, 23–129.

Geisler, J. H., and Uhen, M. D. (2003). Morphological support for a close relationship between hippos and whales. *J. Vertebr. Paleontol.* **23**, 991–996.

Gingerich, P. D. (2005). Cetacea. *In* "Placental Mammals: Origin, Timing, and Relationships of the Major Extant Clades" (K. D. Rose, and J. D. Archibald, eds), pp. 234–252. Johns Hopkins University Press, Baltimore.

Ichishima, H., Sato, E., Sagayama, T., and Kimura, M. (2006). The oldest record of Eschrichtiidae (Cetacea: Mysticeti) from the Late Pliocene, Hokkaido, Japan. *J. Paleontol.* **80**, 367–379.

Kellogg, A. R. (1936). A review of the Archaeoceti. *Carnegie Inst. Washington Publ.* **482**, 1–366.

Lambert, O. (2005). Review of the Miocene long-snouted dolphin *Priscodelphinus cristatus* du Bus, 1872 (Cetacea, Odontoceti) and phylogeny among eurhinodelphinids. *Bulletin De L'Institut Royal Des Sciences Naturelles De Belgique, Sciences De La Terre* **75**, 211–235.

Lambert, O., and Louwye, S. (2006). *Archaeoziphius microglenoideus*, a new primitive beaked whale (Mammalia, Cetacea, Odontoceti) from the Middle Miocene of Belgium. *J. Vertebr. Paleontol.* **26**, 182–191.

Mitchell, E. D. (1989). A new cetacean from the Late Eocene La Meseta Formation, Seymour Island, Antarctic Peninsula. *Can. J. Fish. Aquat. Sci.* **46**, 2219–2235.

Muizon, C.de. (1988a). Les vertébrés fossiles de la Formation Pisco (Pérou). Troisième partie: les odontocètes (Cetacea, Mammalia) du Miocène. *Institut Français D'Études Andines Mémoire* **78**, 1–244.

Muizon, C.de. (1988b). Le polyphylétisme des Acrodelphidae, Odontocètes longirostres du Miocène européen. *Bull. Mus. Nat. Hist. 4 Série*, **10C**, 31–88.

Muizon, C.de. (1988c). Les relations phylogénétiques des Delphinida (Cetacea, Mammalia). *Annales De Paléontologie* **74**, 159–227.

Muizon, C.de. (1991). A new Ziphiidae (Cetacea) from the Early Miocene of Washington State (USA) and phylogenetic analysis of the major groups of odontocetes. *Bull. Mus. Nat. Hist. 4 Série* **12C**, 279–326.

Muizon, C.de., and Domning, D. P. (2002). The anatomy of *Odobenocetops* (Delphinoidea, Mammalia), the walrus-like dolphin from the Pliocene of Peru and its palaeobiological implications. *Zool. J. Linn. Soc.* **134**, 423–452.

Nikaido, M., Matsuno, F., Hamilton, H., Brownell, R. L., Cao, Y., Ding, W., Zuoyan, Z., Shedlock, A. M., Fordyce, R. E., Hasegawa, M., and Okada, N. (2001). Retroposon analysis of major cetacean lineages: The monophyly of toothed whales and the paraphyly of river dolphins. *Proc. Natl. Acad. Sci. USA*, **98**, 7384–7389.

O'Leary, M. A., and Gatesy, J. (2007). Impact of increased character sampling on the phylogeny of Cetartiodactyla (Mammalia): Combined analysis including fossils. *Cladistics* **1**, 1–46.

Pyenson, N. D., and Hoch, E. (2007). Tortonian pontoporiid odontocetes from the Eastern North Sea. *J. Vertebr. Paleontol.* **27**, 757–762.

Rosel, P. E., Haygood, M. G., and Perrin, W. F. (1995). Phylogenetic relationships among the true porpoises (Cetacea: Phocoenidae). *Mol. Phylogenet. Evol.* **4**, 463–474.

Rothausen, K. (1968). Die systematische Stellung der europaischen Squalodontidae (Odontoceti: Mamm.). *Paläontologische Zeitschrift* **42**, 83–104.

Sasaki, T., Nikaido, M., Hamilton, H., Goto, M., Kato, H., Kanda, N., Pastene, L. A., Cao, Y., Fordyce, R. E., Hasegawa, M., and Okada, N. (2005). Mitochondrial phylogenetics and evolution of mysticete whales. *Syst. Biol.* **54**, 77–90.

Steeman, M. E. (2007). Cladistic analysis and a revised classification of fossil and recent mysticetes. *Zool. J. Linn. Soc.* **150**, 875–894.

Thewissen, J. G. M. (ed.) (1998). "The emergence of whales: Evolutionary patterns in the origin of Cetacea." Plenum, New York.

Thewissen, J. G. M., Cooper, L. N., Clementz, M. T., Bajpai, S., and Tiwari, B. N. (2007). Whales originated from aquatic artiodactyls in the Eocene epoch of India. *Nature* **450**, 1190–1194.

Uhen, M. D. (2004). Form, function and anatomy of *Dorudon atrox* (Mammalia: Cetacea): An archaeocete from the middle to late Eocene of Egypt. *Univ. Mich. Pap. Paleontol.* **34**, 1–222.

Uhen, M. D., and Pyenson, N. D. (2007). Diversity estimates, biases, and historiographic effects: Resolving cetacean diversity the Tertiary. *Palaeontol. Electronica* **10**, 11A [1–22].

Whitmore, F. C. (1994). Neogene climatic change and the emergence of the modern whale fauna of the North Atlantic Ocean. *Proceedings of the San Diego Society of Natural History* **29**, 223–227.

Whitmore, F. C., and Sanders, A. E. (1977). Review of the Oligocene Cetacea. *Syst. Zool.* **25**, 304–320.

**C**

# Cetacean Life History

## Susan J. Chivers

A species' life-history strategy is defined by parameters that describe how individuals allocate resources to growth, reproduction, and survival. The allocation of resources presumably results from natural selection maximizing the reproductive fitness of individuals within a species. Biologists studying life-history strategies collect data to answer questions about how long individuals of a species live, the ages at which they become sexually mature and first reproduce, and where and when they travel to find sufficient food to survive (Fig. 1). In search of answers, these biologists may be found in a laboratory estimating an individual animal's age, at a computer modeling growth rates, or at sea observing animals in their natural habitat.

Among cetacean species, the life-history strategies are diverse and differ markedly between the two cetacean suborders: the baleen whales (suborder Mysticeti) and the toothed whales (suborder Odontoceti). This diversity demonstrates the range of successful strategies that have evolved and enable cetaceans to live in a completely aquatic environment as well as the influence of their phylogeny on adapting them to a particular niche. Reviewing the strategies of species within each suborder reveals that the baleen whales share more similar life-history characteristics. All species are large and long lived, and all of the baleen whales, except the bowhead whale (*Balaena mysticetus*) and Bryde's whale (*Balaenoptera edeni*), make long-range annual migrations between breeding grounds in tropical waters and feeding grounds in temperate or polar waters. However, the life-history patterns observed among odontocetes are more varied. These species range in size from the small, relatively short-lived (<24 years) harbor porpoise (*Phocoena phocoena*) to the large, relatively long-lived (>70 years) sperm whale (*Physeter macrocephalus*) and occupy diverse habitats, ranging from pelagic and coastal ocean waters to estuarine and fresh waters.

The life-history strategies for relatively few cetacean species are known in detail. Most of the biological data available for baleen whales were collected during whaling operations, whereas odontocetes have been studied from animals incidentally taken during fishery operations, taken in directed fisheries, found stranded on the

- Mating strategy
- Age at first birth
- Birthing interval
- Parental care
- Senescence

Reproduction    Life History    Survival

- Longevity
- Predation
- Movement patterns
- Feeding ecology

Growth

- Rates/patterns of growth
- Maximum size or length
- Size-at-birth
- Morphology

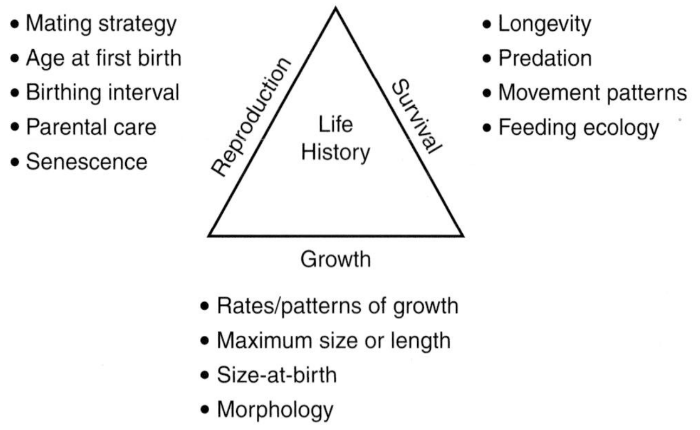

**Figure 1** *Studies of cetacean life-history integrate data describing an individual animal's allocation of resources to growth, reproduction, and survival. Compiling data from many individuals allows the parameters listed next to each category to be estimated, which in turn describes the growth, reproduction, and survival strategies of a species. Life-history data may be collected by observing individual animals in directed photo-identification, tagging, marking or telemetry studies, or by necropsying animals to collect teeth for aging, body length measurements for quantifying growth rates, gonads for determining reproductive condition, and skin for estimating individual relatedness or determining "local" adaptations using molecular genetic techniques.*

beach, or observed in the wild or in captivity. Among the most well-known cetacean life histories are the humpback whale (*Megaptera novaeangliae*), fin whale (*B. physalus*), common bottlenose dolphin (*Tursiops truncatus*), and killer whale (*Orcinus orca*). Three of these species—the humpback whale, common bottlenose dolphin, and killer whale—have each been the subject of long-term studies, which have provided unique data about the natural variability of a species' life-history strategy based on the observed demographics of individual animals. Our knowledge about the life-history strategies of cetacean species is still incomplete, particularly for rarely encountered species, but our knowledge is expanding rapidly as more specimens are collected and new techniques are developed.

## I. Characteristics of Cetacean Life Histories

Although diverse life-history strategies are exhibited by cetacean species, there are a few common characteristics that are likely necessary adaptations to live successfully in a completely aquatic environment. All species give birth to single, large, and precocial young. The presence of multiple fetuses or multiple births has been documented only rarely, and there are no known cases of successfully reared multiple offspring. Gestation times are approximately a year. Among the baleen whales, the estimates for gestation range from 10 to 12 months, and among the odontocetes, estimates range from 10 months for the harbor porpoise (Gaskin *et al.*, 1984) to 14 or 15 months for the sperm whale (Best *et al.*, 1984) and 17 months for the killer whale (Olesiuk *et al.*, 1990; Baird, 2000). Most of the small delphinids (e.g., *Stenella* spp.) have gestation periods of 11 to 12 months (Perrin and Reilly, 1984). The length of the gestation period in part balances the cost of producing a large neonate. Additionally, all cetaceans are relatively long lived. Among baleen whales, estimates of longevity range from 6 decades for the minke whale (*B. acutorostrata*) up to 10 decades for the fin whale and more than

10 decades for the bowhead whale, while among odontocetes, estimates of longevity range from approximately 2 decades for the harbor porpoise up to 7 decades for the sperm whale. Bowhead whales are the cetacean species known to live the longest, and a weapon fragment recovered from a whale in 2007 indicated that these whales may live between 115 and 130 years. Additional generalizations about life-history strategies are presented in Section III.

## II. Methods of Studying Life History

Longitudinal and cross-sectional studies of cetacean species have provided data necessary for understanding their life-history strategies. Longitudinal studies are rare but valuable, because they provide unique data on the variability of individual demographics. Three species, the humpback whale (Clapham, 1996), the common bottlenose dolphin (Wells and Scott, 1990), and the killer whale (Olesiuk *et al.*, 1990), have been the subject of ongoing studies that originated during the 1970s. These studies are possible because individuals are relatively accessible and easily distinguishable in the field by natural markings. These studies have quantified individual variability in reproduction and survival through time and have provided unique insights into the species' life-history strategy by incorporating observations of the species' social behavior and ecology. However, most of our knowledge about cetacean life-history strategies is the result of cross-sectional studies. In these studies, data are collected from individual animals sampled primarily from directed or incidental takes. The primary advantage of these studies is that a complete suite of morphological and biological data can be collected, which allows explicit determination of reproductive and physical maturity as well as an estimate of age. Estimates of age for most cetaceans are made from the layering patterns evident in the ear plugs or ear bones of baleen whales (Lockyer, 1984; Christensen, 1995) and in the teeth of odontocetes (Perrin and Myrick, 1980). One exception is bowhead whales, which have been aged using an aspartic acid racemization technique and the recovery of weapon fragments from carcasses (George *et al.*, 1999). Accurate determination of reproductive maturity in both sexes requires examination of the gonads. In females, the presence of one corpus or more in the ovaries indicates sexual maturity, and in males, the presence of spermatazoa and large seminiferous tubules in histologically prepared testes tissue indicates sexual maturity (Perrin and Reilly, 1984). Physical maturity is determined in both sexes by examining the vertebral column for evidence of fusion. That is, when the vertebral epiphyses are fused with the centrum, an animal is considered physically mature.

Life-history studies are designed to collect data on body size, age, and reproductive and physical maturity from many individuals to estimate parameters that characterize a species' allocation of resources to growth, reproduction, and survival. Estimated parameters may include age-specific growth and pregnancy rates, the average age at attainment of sexual maturity, calving interval and longevity. Age is the primary independent variable for all studies, because age explicitly demonstrates the tradeoff in resource allocation to growth and reproduction during an individual animal's life. The expected pattern of resource allocation from birth through attainment of sexual maturity is primarily for growth and then for reproduction once sexual maturity is attained. Also the probability of an individual surviving to the next age class increases with increasing age after weaning until sexual maturity is attained and then remains high throughout the individual's reproductive years. Data on age-specific growth and reproductive rates, combined with estimates of age-specific survival rates, are essential to comparing and contrasting the life-history strategies of different species.

## III. Cetacean Life-History Patterns

The neonates of all cetacean species are relatively large when compared to other mammal species. In fact, neonate size ranges from approximately 29% of the female's asymptotic total body length in most of the baleen whales to between 40% and 48% of the female's length in the odontocetes. The large size of neonates, combined with their ability to swim and grow rapidly immediately after birth, increases their probability of survival. The lactation period for the baleen whales lasts only about 6 months, and the young grow rapidly during that period because the fat content of the milk is high. However, the calves of odontocetes grow more slowly, and the lactation period lasts approximately a year or more. The difference in calf growth rates between the two suborders of cetaceans is probably due to the transfer of energy to the young through the milk. Oftedal (1997) has estimated that the energy output through milk ranges from 0.40 to 1.06 MJ/kg$^{0.75}$ for mysticetes and from 0.09 to 0.17 MJ/kg$^{0.75}$ for odontocetes. For species with a lactation period of more than a year, the additional investment likely further increases the calf's probability of survival by facilitating the learning of social behaviors (e.g., common bottlenose dolphin, short-finned pilot whale, *Globicephala macrorhynchus*).

Patterns of growth differ between the sexes of many cetacean species, resulting in some degree of sexual dimorphism. Both males and females have high growth rates while suckling, but growth slows after weaning and again after reaching sexual maturity. However, the sex that grows largest tends to grow for a longer period of time and may have higher growth rates after weaning. Among baleen whales, females attain lengths that are generally 5% larger than the males. Similarly, among odontocetes, the females are slightly larger than the males in the porpoises and river dolphins. However, for other odontocetes, males are larger than females. Sexual dimorphism is most marked in sperm whales, in which males are 60% larger than females. Among the smaller delphinids, such as the common bottlenose dolphin, pantropical spotted dolphin (*Stenella attenuata*), and common dolphins (*Delphinus* spp.), males are approximately 2–10% larger than females.

The breeding cycle for all cetacean species has three parts: a gestation period, a lactation period, and a resting, or anestrous, period. This cycle is 2 years or more for most cetacean species. The exceptions are the minke whale and harbor porpoise, which can breed annually. The breeding cycle of the blue (*B. musculus*), Bryde's, humpback, sei (*B. borealis*), and gray (*Eschrichtius robustus*) whales includes an 11-month gestation period, a 6- to 7-month lactation period and a 6- to 7-month resting, or anestrous, period for a minimum of a 2-year cycle, while the breeding cycle for the bowhead and right whales (*Balaena glacialis*) is 3 to 4 years starting with a 10- to 12-month gestation period. Furthermore, the breeding season of baleen whales is synchronized with their migration cycle. These species travel long distances to breed in tropical waters. The exceptions are the Bryde's whale and the pygmy Bryde's whale, which spend all year in tropical waters and do not breed synchronously (Lockyer, 1984). Several hypotheses have been proposed for the adaptive significance of the large-scale migrations of baleen whales. Although the phenomenon remains unexplained, hypotheses of increased survival rates for neonates in tropical waters by reducing thermoregulatory demands or the risk of predation by killer whales have been proposed (Corkeron and Connor, 1999).

Similar to other life-history characteristics, the breeding cycle for odontocetes is more variable than that of mysticetes. Porpoises have the shortest breeding cycle, which is approximately 1 year and includes a 10-month gestation. In fact, annual breeding among the porpoises has been well documented for the harbor porpoise. The

**Figure 2**  *Mother and calf pantropical spotted dolphin* (Stenella attenuata) *in the eastern tropical Pacific. Photographed by R. L. Pitman.*

smaller delphinid species seem to have 2- to 3-year calving intervals, which include an 11- to 12-month gestation and a 1- to 2-year lactation period. However, the larger odontocetes such as the killer whale, short- and long-finned (*G. melas*) pilot whales, and sperm whale have calving intervals of more than 3 years, which includes a 12- to 17-month gestation period and a 2- to 3-year, or longer, lactation period. Breeding synchrony also varies among odontocetes. Species inhabiting temperate waters like the harbor porpoise have been found to have more synchronous breeding seasons than species inhabiting tropical waters. For example, studies of the pantropical spotted dolphin (Fig. 2) and the striped dolphin (*S. coeruleoalba*), which inhabit tropical waters in the Pacific Ocean, found that young are born throughout the year, although most births occur during the spring and fall (Perrin and Reilly, 1984).

Age at attainment of sexual maturity is delayed in all cetacean species as would be expected for large, long-lived mammals. However, the range of ages is quite broad and reflects the unique set of adaptations that characterize the life-history strategy of each species. The range in age of sexual maturity among the baleen whales is from approximately 4 years for humpback whales to approximately 10 years for the fin, sei, and Bryde's whales (Lockyer, 1984) and to 25 years for bowhead whales (George *et al.*, 1999). Among odontocetes, the range in age at attainment of sexual maturity is about the same as that observed for the baleen whales and seems to be correlated to a degree with longevity and body size. The youngest age at attainment of sexual maturity is 3 years for harbor porpoise, which is the smallest odontocete and is estimated to live approximately two decades (<24 years). However, many of the larger odontocetes reach sexual maturity at ages of 10 years or more and live for four or more decades (Perrin and Reilly, 1984).

Reproductive success varies throughout the life of female cetaceans. Initially, reproductive success is relatively low, peaks several years after the age at attainment of sexual maturity and then declines as the female ages. This phenomenon is also characteristic of large terrestrial mammals and is probably due in part to a tradeoff in costs

C

between reproduction and growth that must occur because physical maturity is attained several years after sexual maturity and to learning to care for young. Evidence for low reproductive success among newly matured females has been documented in the common bottlenose dolphin and the fin whale. Lower reproductive rates for older females have also been documented in the common bottlenose dolphin as longer interbirth intervals for older females that include a 3- to 8-year lactation period. Postreproductive females with senescent ovaries have been identified in only a few odontocetes, including the short-finned pilot whale (Marsh and Kasuya, 1986) and the pantropical spotted dolphin (Myrick *et al.*, 1985), but senescence has not yet been identified in any of the baleen whales. The adaptive significance of senescence is not yet understood but likely contributes to increased reproductive success. For example, several species that exhibit senescence also have fairly complex social structures (e.g., sperm whale, short-finned pilot whale), and the role of postreproductive females in their societies may be associated with increased survival rates of the young by these females participating in the care of young that are not their own.

## IV. Characteristics of Male Life Histories

The life-history characteristics of males are less well known than those of females, primarily because this knowledge is less critical to understanding a species' reproductive potential and population dynamics. In this sense, females are the limiting sex. However, knowledge about the life-history strategies of males provides a more complete picture of a species population dynamics and provides information about the species' breeding strategy and social structure.

One of the major differences between the life-history strategies of male and female cetaceans is the age at attainment of sexual maturity. In species with the greatest degree of sexual dimorphism, the difference in age at attainment of sexual maturity for males and females is greatest. This difference reflects the additional time required to grow to about 85% of their asymptotic length, which is the approximate size at which all mammals become sexually mature. For example, sperm whale males reach sexual maturity at a much later age than females. The estimated age at attainment of sexual maturity for the female sperm whale is from 7 to 13 years and for males, is approximately 20 years (Best *et al.*, 1984; Rice, 1989). The difference is similar in the killer whale and the short- and long-finned pilot whale (Lockyer, 1993; Baird, 2000). However, the smaller delphinid species that show less sexual dimorphism reach sexual maturity at more similar ages. In fact, the difference in age between the sexes is about 3 years for the common bottlenose dolphin, pantropical spotted dolphin and spinner dolphin (*S. longirostris*) with males reaching sexual maturity at the older age (Perrin and Reilly, 1984).

Sexual dimorphism has been used as a predictor of cetacean mating systems. For example among odontocetes, the degree of sexual dimorphism exhibited by the sperm whale, short-finned pilot whale, and killer whale has been hypothesized to indicate male–male competition in a polygynous mating system. The presence of scars inflicted by other males provides evidence of male–male competition in sperm whales. However, fairly recent data that the short-finned pilot whale and killer whale have limited male dispersal from natal pods and likely breed promiscuously suggests that sexual dimorphism may have evolved for reasons other than mating. One interpretation of these data is that the presence of large males in their natal pod enhances their reproductive fitness by improving, e.g., the foraging efficiency of the pod (Wells *et al.*, 1999).

## V. Life-History Parameters and Demography

Knowledge of a species' life-history strategy provides the foundation for understanding the species' demography because their life-history characteristics reflect the species' adaptations to a particular niche, which is bounded by constraints of the environment as well as their morphology and physiology. While life-history studies primarily focus on individual variability in traits that express these adaptations for a species, each study can usually only focus on a particular group of animals within the species. The comparison of studies made on different groups of animals within a species' range, however, reveals variability in the average expression of life-history traits among demographically isolated populations. For example, pantropical spotted dolphins north and south of the equator have different breeding seasons, and the estimates of asymptotic length for animals in the western Pacific are 4 to 7 cm longer than those from the eastern Pacific (Perrin and Reilly, 1984). Similar examples exist for other cetacean species. There are also examples in the literature of cetacean populations responding to changes in the availability of resources through time. This is called density dependence. For example, changes in the age at attainment of sexual maturity for fin, sei, and minke whales through time have been reported and are presumed to be a response to increased per capita resource availability following reductions in population abundance that resulted from commercial whaling (Lockyer, 1984). Similarly for the striped dolphin and the spinner dolphin, changes in the age of sexual maturity and pregnancy rates have been reported and explained as responses consistent with increased resource availability that resulted from decreased population abundance (Perrin and Reilly, 1984). Ultimately, these responses are reflected as changes in the population's growth rate. In addition to understanding a species' life-history strategy and its inherent variability, recognition of these types of population-level responses is important to consider when developing conservation and management plans.

Estimates of age-specific reproductive rates and survival rates are critical to quantifying a species' demography. However, for nearly all cetacean species, age-specific survival rates are unknown and are likely to remain so. Because demographic studies must include age-specific survival rates, unique solutions have been sought to allow the estimation of survival rates based on imperfect knowledge (Barlow and Boveng, 1991). Longitudinal studies like those of the common bottlenose dolphin and the humpback whale provide the only source of data to estimate survival rates, and these data are generally used as a guide for estimating survival rates for other species with similar, but less well known, life histories.

## VI. Life-History Studies and the Future

Several new technologies are being actively applied to studies of cetacean species and are contributing to our knowledge about their life-history strategies. Specifically, the expansion of molecular genetic techniques, the development of satellite and VHF (Very High Frequency) tracking technology, and the collection of high-resolution vertical aerial photographs have allowed more detailed data collection on individual animals. For example, the application of molecular genetic markers as tags for individuals has been successfully demonstrated

**Figure 3** *A pantropical spotted dolphin* (Stenella attenuata) *wearing a radio tag and time-depth recorder to study diving behavior. Photographed by M. D. Scott.*

with the humpback whale data set (Palsbøll, 1999). Application of this technique to cetacean species whose individuals cannot be readily recognized by natural marks may facilitate life-history studies for those species. Additionally, the results of molecular genetic studies on several cetaceans, including the beluga whale (*Delphinapterus leucas*) and Dall's porpoise (*Phocoenoides dalli*), have confirmed hypotheses of male-biased dispersal (O'Corry-Crowe *et al.*, 1997; Escorza-Treviño and Dizon, 2000). Although not a particularly surprising result, because male-biased dispersal is common among large terrestrial mammals, molecular genetics provided the tool to examine large data sets to address this question for cetacean species. Two other technologies are contributing to our understanding of cetacean breeding patterns. These are the development of molecular techniques to identify pregnant individuals by measuring hormone concentrations in the blubber, which will allow pregnancy rates to be estimated for populations of wild cetaceans, and the collection of high-resolution vertical aerial photographs to identify breeding seasons of cetacean populations using measurements of animal size (Perryman and Lynn, 1993). Similarly, the development of satellite and VHF tracking technology is continuing and providing new insights about how cetacean species live their lives. There have been notable successes documenting movement patterns of beluga whales in the Arctic (Martin *et al.*, 1998) and blue whales in the North Pacific (Mate *et al.*, 1999) using satellite tags, and documenting the diving behavior of a number of cetacean species using VHF tags (Hooker and Baird, 2001) (Fig. 3). As these and other new technologies continue to be developed and applied, new insights about the life-history strategies of all cetacean species will be provided that will complement and expand knowledge obtained from more traditional life-history study methods.

## See also the Following Articles

Age estimation ■ Female reproductive system ■ Pinniped life history ■ Population dynamics ■ Sexual dimorphism

## References

Baird, R. W. (2000). The killer whale: Foraging specializations and group hunting. *In* "Cetacean Societies: Field Studies of Dolphins

and Whales" (J. Mann, R. C. Connor, P. L. Tyack, and H. Whitehead, eds), pp. 127–153. University of Chicago Press, Chicago.

Barlow, J., and Boveng, P. (1991). Modeling age-specific mortality for marine mammal populations. *Mar. Mamm. Sci.* **7**, 50–65.

Best, P. B., Canham, P. A. S., and MacLeod, N. (1984). Patterns of reproduction in sperm whales, *Physeter macrocephalus*. *Rep. Int. Whal. Comm. Spec. Issue* **6**, 51–79.

Christensen, I. (1995). Interpretation of growth layers in the periosteal zone of *tympanic bulla* from minke whales *Balaenoptera acutorostrata*. *In* "Whales, Seals, Fish and Man" (A. S. Blix, L. Walloe, and O. Ulltang, eds), pp. 413–423. Elsevier Science, Amsterdam.

Clapham, P. J. (1996). The social and reproductive biology of humpback whales: An ecological perspective. *Mamm. Rev.* **26**, 27–49.

Corkeron, P. J., and Connor, R. C. (1999). Why do baleen whales migrate? *Mar. Mamm. Sci.* **15**, 1228–1245.

Escorza-Treviño, S., and Dizon, A. E. (2000). Phylogeography, intraspecific structure, and sex-biased dispersal of Dall's porpoise, *Phocoenoides dalli*, revealed by mitochondrial and microsatellite DNA analyses. *Mol. Ecol.* **9**, 1049–1060.

Gaskin, D. E., Smith, G. J. D., Watson, A. P., Yasui, W. Y., and Yurick, D. B. (1984). Reproduction in the porpoises (Phocoenidae): Implications for management. *Rep. Int. Whal. Comm. Spec. Issue* **6**, 135–148.

George, J. C., Bada, J., Zeh, J., Scott, L., Brown, S. E., O'Hara, T., and Suydam, R. (1999). Age and growth estimates of bowhead whales (*Balaena mysticetus*) via aspartic acid racemization. *Can. J. Zool.* **77**, 571–580.

Hooker, S. K., and Baird, R. W. (2001). Diving and ranging behaviour of odontocetes: A methodological review and critique. *Mamm. Rev.* **31**, 81–105.

Lockyer, C. (1984). Review of baleen whale (*Mysticeti*) reproduction and implications for management. *Rep. Int. Whal. Comm. Spec. Issue* **6**, 27–50.

Lockyer, C. (1993). A report on patterns of deposition of dentine and cement in teeth of pilot whales, genus, *Globicephala*. *Rep. Int. Whal. Comm. Spec. Issue* **14**, 137–161.

Marsh, H., and Kasuya, T. (1986). Evidence for reproductive senescence in female cetaceans. *Rep. Int. Whal. Comm. Spec. Issue* **8**, 57–74.

Martin, A. R., Smith, T. G., and Cox, O. P. (1998). Dive form and function in belugas *Delphinapterus leucas* of the eastern Canadian High Arctic. *Pol. Biol.* **20**, 218–228.

Mate, B. R., Lagerquist, B. A., and Calambokidis, J. (1999). Movements of North Pacific blue whales during the feeding season off southern California and their southern fall migration. *Mar. Mamm. Sci.* **15**, 1246–1257.

Myrick, A. C., Jr., Hohn, A. A., Barlow, J., and Sloan, P. A. (1985). Reproductive biology of female spotted dolphins, *Stenella attenuata*, from the eastern tropical Pacific. *Fish. Bull.* **84**, 247–259.

O'Corry-Crowe, G. M., Suydam, R. S., Rosenberg, A., Frost, K. J., and Dizon, A. E. (1997). Phylogeography, population structure and dispersal patterns of the beluga whale *Delphinapterus leucas* in the western Nearctic revealed by mitochondrial DNA. *Mol. Ecol.* **6**, 955–970.

Oftedal, O. T. (1997). Lactation in whales and dolphins: Evidence of divergence between baleen- and toothed-species. *J. Mamm. Gland Biol. Neoplasia* **2**, 205–230.

Olesiuk, P. F., Bigg, M. A., and Ellis, G. M. (1990). Life history and population dynamics of resident killer whales (*Orcinus orca*) in the coastal waters of British Columbia and Washington State. *Rep. Int. Whal. Comm. Spec. Issue* **12**, 209–243.

Palsbøll, P. J. (1999). Genetic tagging: Contemporary molecular ecology. *Biol. J. Linn. Soc.* **68**, 3–22.

Perrin, W. F., and Myrick, Jr., A. C., (eds) (1980). Age determination of toothed whales and sirenians. *Rep. Int. Whal. Comm. Spec. Issue* **3**.

Perrin, W. F., and Reilly, S. B. (1984). Reproductive parameters of dolphins and small whales of the Family Delphinidae. *Rep. Int. Whal. Comm. Spec. Issue* **6**, 97–133.

C

Perryman, W. L., and Lynn, M. S. (1993). Identification of geographic forms of common dolphin (*Delphinus delphis*) from aerial photogrammetry. *Mar. Mamm. Sci.* **9**, 119–137.

Rice, D. W. (1989). Sperm whale, *Physeter macrocephalus* Linnaeus, 1758. *In* "Handbook of Marine Mammals" (S. H. Ridgway, and R. Harrison, eds), Vol. 4, pp. 177–233. Academic Press, London.

Wells, R. S., and Scott, M. D. (1990). Estimating bottlenose dolphin population parameters from individual identification and capture–release techniques. *Rep. Int. Whal. Comm. Spec. Issue* **12**, 407–415.

Wells, R. S., Boness, D. J., and Rathbun, G. B. (1999). Behavior. *In* "Biology of Marine Mammals" (J. E. Reynolds, III, and S. A. Rommel, eds), pp. 324–422. Smithsonian Institution Press, Washington.

**C**

# Cetacean Prenatal Development

JOY S. REIDENBERG AND
JEFFREY T. LAITMAN

Very little is known about the specifics of intrauterine growth and development in cetaceans. Indeed, the precise time intervals of such development, the basic genetic determiners, and any distinctive growth trajectories are basically unknown. What is known about cetacean prenatal development is that, as they are mammals, it is to be expected that the same basic stages of early cell division, pattern formation, organogenesis, and growth and differentiation will also be similar. For example, the "embryonic" period is usually defined as the time frame within which an animal's body plan and its organs and organ systems (i.e., integument, skeletal, muscular, nervous, circulatory, respiratory, digestive, urinary, and reproductive) are established. Once all organs form, the "fetal" period of growth and distinctive development commences. Cetacean prenatal development will similarly follow this course. It is also to be expected that the absolute time of these periods will differ both from terrestrial species, between odontocetes and mysticetes, and among the different species therein.

Many studies that have noted aspects of cetacean prenatal development (most in passing rather than by detailed, systemic analysis) have used terms such as "embryo" or "fetus" in a seemingly imprecise manner. Adding to this complexity is the fact that the precise gestation periods for many cetacean species are not known. In light of the earlier discussion, our use of the terms embryo and fetus (or embryonic and fetal periods) should be taken as representing approximate guides to stages of development rather than as a precise descriptor of an absolute time frame.

It is important to remember in discussing cetacean prenatal development that most current knowledge derives from observations on embryonic or fetal specimens discovered in pregnant cetaceans either found stranded or taken aboard whaling ships. In many cases, only a length or weight is recorded (if at all) with an occasional description of external appearance. It is usually impossible to distinguish the age of the specimen, as the date of conception and the length of gestation cannot be known with any certainty. As most breeding and calving seasons are known, however, some approximations are available and have been provided. In this chapter, we focus on changes occurring in the late embryonic through fetal periods.

We refer to the review articles by Štěrba *et al.* (1994) and Thewissen and Heyning (2007) for a discussion of the embryonic period.

## I. Integument and External Characteristics

The overall coloring of the embryo appears light pink, due to the transparency of the skin (integument) allowing the underlying tissues perfused with blood to be visible (Fig. 1). The skin consists of the epidermis (which has four layers), dermis, and hypodermis, which increase in thickness throughout the embryonic period (Meyer *et al.*, 1995). Skin coloration begins during the early fetal period. In mysticetes, dark coloration occurs initially along the rostrum bordering the opening of the oral cavity. As the fetus grows, dark patches appear along the dorsum of the thorax and the abdomen, and on the pectoral flippers, tail flukes, and dorsal fin (Fig. 2). The separate and irregularly shaped patches fuse and grow into a more uniform pattern. For many species, this is usually a countershaded pattern of dark dorsum and light ventrum that resembles the adult's coloration (Fig. 3).

Hairs can be found along the surfaces of the upper jaw. In odontocetes, hairs appear on the lateral aspect near the tip of the rostrum, whereas in mysticetes, they are found both laterally and dorsally on the broad rostrum. In some cetacean species, these hairs can also be found on the margins of the lower jaw. These hairs appear to have some tactile properties and may derive from the vibrissae of terrestrial mammals. Although most odontocetes will lose these hairs shortly after birth (except perhaps platanistids), they are retained into adulthood in some species of mysticetes.

External ears (pinnae) do not develop, thus maintaining a streamlined surface contour in the ear region of cetaceans. Only a remnant of the external auditory canal is visible as a small hole present in the skin behind the eye.

Mammary glands (mammae) are epidermal organs derived from modified sweat glands. In terrestrial mammals, and presumably cetaceans, the mammae develop along a mammary ridge (the "milk line"), which extends bilaterally from the axilla (where the forelimb joins the thorax) to the inguinal region (where the hindlimb joins the pelvis). The position of the mammae that eventually develop varies in different species: thoracic (e.g., primates, sirenians), thoraco-abdominal (e.g., felids), thoraco-inguinal (e.g., canids, suids), and inguinal (e.g., ungulates). Cetacea, like their ungulate relatives, only develop inguinal mammae. In females, the teats (nipples) of the mammae are internalized, being withdrawn into the mammary slits (which are positioned with one on either side of the genital slit). This internal location helps streamline the body contour and thus reduce drag during locomotion.

It is difficult to sex the cetacean embryo or early fetus, as they only display a small genital tubercle (Fig. 1). As the genital tubercle develops, however, it is directed cranially in males and caudally in females (Amasaki *et al.*, 1989b). Although the penis/clitoris may be totally exposed in an earlier fetus, the external genitalia are usually not completely visible in the full-term fetus as they are withdrawn into the genitoanal slit. In a postmortem specimen, the penis usually protrudes through the slit due to relaxation of the retractor penis muscle (Fig. 2). The genitoanal slit opens into a common vestibule occupied caudally by the anus and rostrally by the urogenital openings. In males, the urethra is contained in the penis; in females, the urethra is independent of the clitoris, and there is a separate opening for the vagina. In males, the genitoanal slit is elongated, reaching almost to the navel. In comparison, the genitoanal slit of females is very short, appearing only between the two mammary slits. Both

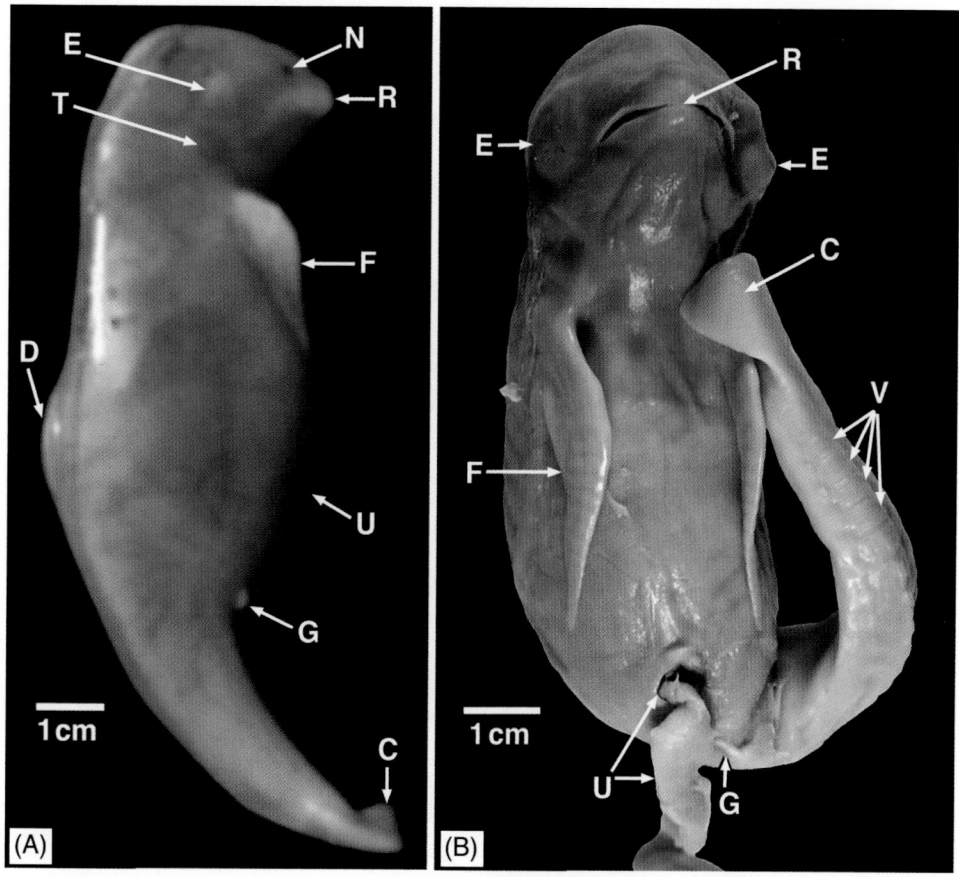

**Figure 1** *Two early fetal long-finned pilot whales* (Globicephala melaena) *obtained postmortem from pregnant, beach-stranded whales on the shores of Cape Cod, Massachusetts, USA. (A) Lateral view of a fresh specimen of a very small, unpigmented fetus that appears pink. This is due to the transparency of the thin skin allowing the color of the blood-perfused tissues to show through. Based upon its external appearance, this appears to be a very early fetus, probably very close to the transition between the embryonic and the fetal periods of development. Note the prominent rostrum and the rudimentary development of a dorsal fin, tail flukes, and genital tubercle. The dark spot above the mouth and anterior to the eye appears to be the left nostril. (B) Ventral view of an early fetus preserved in alcohol. The unpigmented fetal skin (which was pink) is now discolored to a tan hue. The tail is folded laterally to the left side. Due to desiccation, the skin is shrunken against the skeleton revealing the shapes of the skull, ribs, individual segments of the flipper phalanges, and the caudal vertebrae. Note the large eyes, prominent rostrum, attached umbilical cord, and genital tubercle just caudal to the umbilicus. C, caudal fluke; D, dorsal fin; E, eye; F, flipper; G, genital tubercle; N, nostril; R, rostral tip; T, tympanic (ear/pinna) area; U, umbilicus/umbilical cord; V, vertebrae.*

males and females have a streamlined external shape, as the penis/clitoris is withdrawn into the genitoanal slit and there are no scrotal (testes are intra-abdominal) or labial skin protrusions, thus further reducing drag during swimming.

## II. Musculo-Skeletal System

The forelimb extremities of whales are called pectoral flippers. Although cetaceans are derived from a quadrupedal ancestor, adult whales do not possess hindlimbs. During the embryonic period, both forelimb and hindlimb buds are present as paddle-shaped projections, with the forelimb developing before the hindlimb

(Amasaki *et al.*, 1989c). The rudimentary hindlimb buds form skeletal element anlagen, vascular plexes, and nerves (Sedmera *et al.*, 1997a), but are completely absorbed by the fetal period (Thewissen *et al.*, 2006). By birth, the only remaining vestige of the hindlimb is a skeletal remnant of the femur embedded into the lateral body wall, and a rudimentary pelvis that is not attached to the vertebrae. The forelimbs, however, continue developing during the embryonic and the fetal periods. Early on, they assume the elongated shape of a typical mammalian arm and forearm, with grooves separating the digits apparent toward the distal edge. The skin overlying the flippers matures faster than the skin over the trunk (Meyer *et al.*, 1995). The stalk-like arm and the forearm foreshortens into one functional

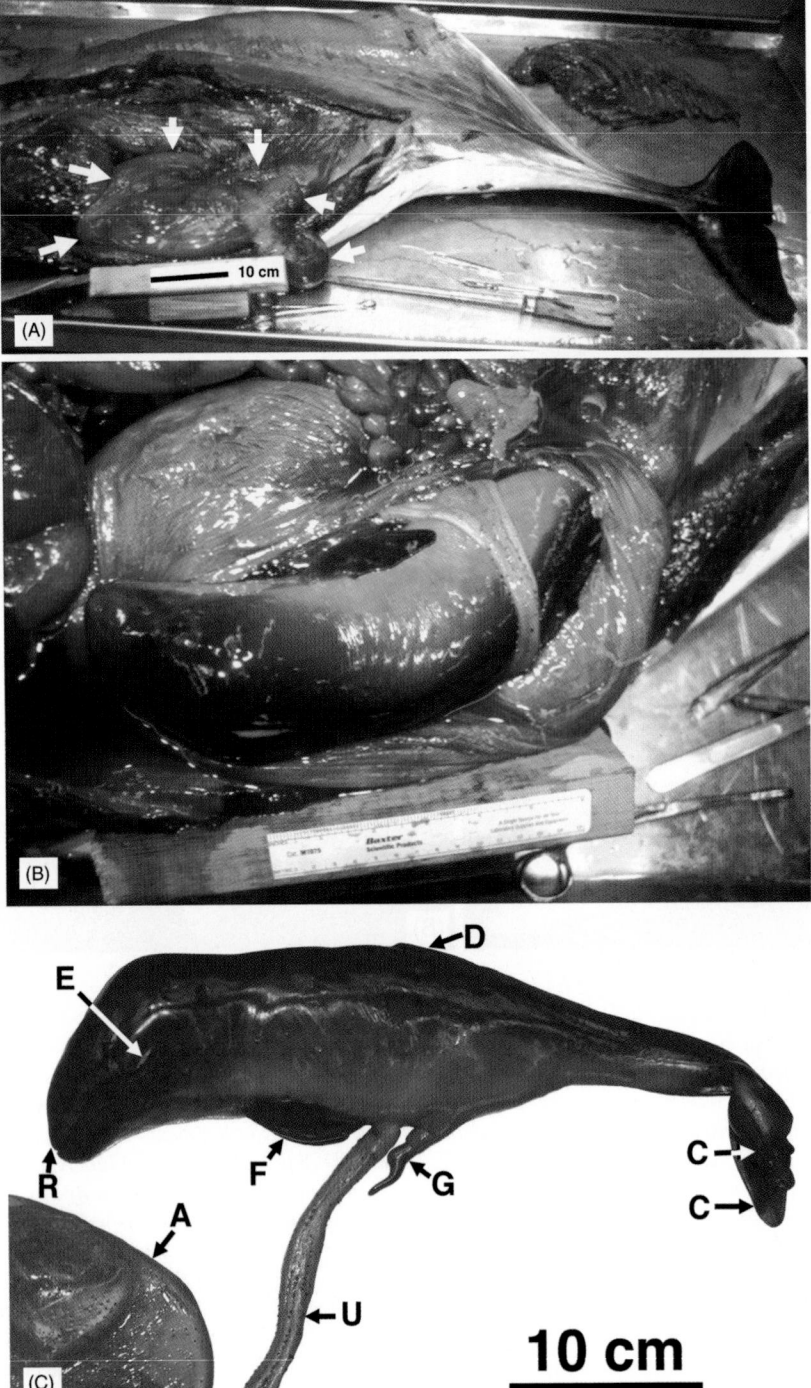

**Figure 2** *Pregnant harbor porpoise* (Phocoena phocoena) *and her fetus, obtained postmortem from a beach-stranding in Cape Cod, Massachusetts, USA. (A) Note the pregnant uterus in situ as revealed during a necropsy (outlined by white arrowheads). The fetus occupies only one horn of the bicornuate uterus. The other uterine horn and ovary are displaced out of the maternal abdominal cavity to the right of a wooden block with a scale-bar. (B) This is a close-up showing the uterus opened to reveal the fetus. Note the fetal head is directed toward the maternal abdomen, whereas the fetal tail is directed toward the maternal vagina. The umbilical cord can be seen wrapped around the abdominal region of the fetus. (C) The fetus after removal from the uterus. The umbilical cord is still attached and the amniotic sac has been removed and is visible in the lower left area of this image. Note the presence of amniotic "pearls" (i.e., black dots present on the umbilical cord and the amniotic sac). The penis is fully extruded due to postmortem muscle relaxation. The dorsal fin is still folded flat against the back. Black skin pigmentation is present. The fetus is not yet large enough to warrant curling, and thus no fetal folds are found. A, amniotic sac; C, caudal fluke; D, dorsal fin; E, eye; F, flipper; G, penis protruding from genital slit; R, rostral tip; U, umbilical cord.*

unit. The skeletal elements (humerus, radius, ulna, and carpal bones) lose their mobility at the elbow and the wrist joints, maintaining flexibility only at the shoulder joint.

During the fetal period, the manus of the pectoral flipper fuses into a paddle-shape (the distal portion never separates into individual digits, and the interdigital grooves disappear). Odontocete flippers contain five digits—a pattern reminiscent of a terrestrial ancestry. The number of digits within the flipper varies in mysticetes; members of the Balaenidae and Eschrichtiidae families retain all five digits, whereas the rorquals (members of the family Balaenopteridae) have reduced that number to four. The tip of the flipper elongates in the caudal direction as differentiation of the phalangeal cartilages progresses proximo-distally. The central digits exhibit hyperphalangia (or polyphalangia), i.e., the number of phalangeal elements expands beyond the maximum of three found in most terrestrial mammals (Calzada and Aguilar, 1996). The degree of hyperphalangia varies greatly between species. For example, the second and the third digits of *Globicephala melas* have 14–15 and 11 phalanges respectively, whereas there are seven elements in each of these two digits in *Stenella attenuata* and only five for each of these digits in *Physeter macrocephalus*. Expansion in the number of phalangeal elements, rather than in the lengths of the elements, probably helps support the elongated form of the flipper while retaining some small degree of flexibility that is reminiscent of fin function in fish. Hyperphalangy and elongated pectoral flipper form may also relate functionally to increasing/decreasing aspect ratio (i.e., relationship between length and width), hydrodynamic form (streamlining effects), or locomotor function (limited to steering, braking, and lift in most species, but can include increased maneuverability or propulsion, e.g., humpback whales) (Cooper *et al.*, 2007a, b).

The tail flukes do not appear until the fetal period, after the hind-limbs have regressed. The midline of the tail enlarges dorsally and ventrally in the vertical plane to form the slender and hydrodynamic tail-stock. The number of caudal vertebrae may increase above that typically seen in terrestrial mammals (perhaps up to 24 in mysticetes and perhaps up to 48 in odontocetes, compared with up to 21 in ungulates). Note that the actual number of caudal vertebrae is difficult to determine with accuracy, as there are no clear anatomical landmarks to separate the caudal region from the lumbar region. The caudal tip develops two horizontal plates of tissue that do not contain any skeletal elements. These plates form the tail flukes. As the fetus nears full term, the tail flukes curl ventrally at their caudal tips so that they are directed rostro-medially (Fig. 3). This curling of the flukes makes the tail tip more compact and easier to present through the vagina during birth (see later).

At about the same time that the tail flukes appear, a bulge develops along the midline of the back in the region where the dorsal fin will form (Fig. 1). The bulge-shape is then modified to a species-specific shape (e.g., falcate, triangular, rounded, ridge). When sexual dimorphism in fin height is seen (e.g., *Orcinus*), it does not occur prenatally. The vertebrae of the back unfold from the embryonic curvature (ventrally concave), to a horizontally aligned column in the early fetal period. In the late fetal period, however, the growing fetus folds again, only this time the body curves laterally. This flexibility may be possible, in part, due to the lack of a sacrum and lengthening of the vertebral column. There are additional lumbar vertebrae in most cetaceans (perhaps up to 15 in mysticetes and perhaps up to 29 in odontocetes), compared with the usual six of ungulates or five of humans. Again, this number is difficult to determine with accuracy, as there is no sacrum or pelvis, and rib articulations can vary. As the side of the fetal head approximates the tail, the dorsal fin folds

flat against the concave side of the body (Fig. 3). Dorsal fin folding facilitates vaginal delivery (see later).

The ribs of odontocetes are hinged along the lateral aspect, giving each rib two osseous elements joined by a synovial joint. Postnatally, this will facilitate thoracic cavity collapse during diving (as pressure increases with depth, the volume of air in the lungs will decrease).

## III. Head and Neck

The large embryonic head lies in the typical mammalian pose with the face directed ventrally at 90° to the long axis of the body. The maxillary and the mandibular regions form a ventrally projecting, conical rostrum that curves slightly caudally. This projection resembles a parrot's beak, being rather thick at the base. In the early fetal period, the rostrum elongates, particularly in the long beaked species (e.g., *Stenella longirostris*, *Platanista gangetica*). In the mid-fetal period, the head and neck junction straightens into the adult position, aligned horizontally with the body. The neck region is shortened and stiffened and in many species (e.g., *Globicephala macrorhynchus*) all the seven cervical vertebrae become cranio-caudally compressed and fuse together (Ogden *et al.*, 1981). This enables a smoother transition in form between the head and the thorax, and a midline head position relative to the body's longitudinal axis. The shortened neck enhances streamlining and fusion of the cervical vertebrae improves head stability during locomotion. Vertebral fusion limits lateral or rotational head motion, leaving only dorso-caudal head movements (which help begin the propulsive body wave) at the large joint between the first cervical vertebra and the skull's enlarged occipital condyles.

The hyoid apparatus is derived from the second and the third branchial arches. The single basihyal and paired thyrohyals form the large "U"-shaped plate to which the muscles of the tongue, larynx, and sternum attach, and the paired epihyals, ceratohyals, stylohyals, and tympanohyals form the osseous chains bilaterally connecting the basihyal with the skull (Reidenberg and Laitman, 1994).

The mandible (jaw) forms around a cartilaginous precursor (Meckel's cartilage) derived from the first branchial arch. The cetacean mandible is largely comprised of a horizontal body, with very little (if any) vertical projection forming the ascending ramus. The condylar process is short, and the condylar head may appear to rest directly superior to the caudal portion of the mandibular body. In many odontocetes, the condylar head migrates with fetal development to the caudal aspect of the mandible, whereas in some mysticetes, the condylar head occupies the dorso-caudal edge of the mandible. Although the ascending ramus develops most of its vertical height postnatally in many terrestrial mammals, the ascending ramus of cetaceans remains practically nonexistent through the adult stage.

The caudal portion of the first branchial arch contributes to the formation of the upper portions of the first two ear ossicles (malleus and incus). The caudal portion of the second branchial arch contributes to the lower portion of these same two ear ossicles as well as the body of the third ossicle (stapes, except for the footplate, which derives from the otic placode).

In terrestrial mammals, and presumably cetaceans, the skull is derived from two types of bones: chondrocranial (that which preforms in cartilage and then ossifies), and desmocranial (that which does not form a cartilaginous stage, but rather directly ossifies in mesenchyme). The portion preformed in cartilage (the skull base) tends to be less plastic in its shape than that which ossifies from

C

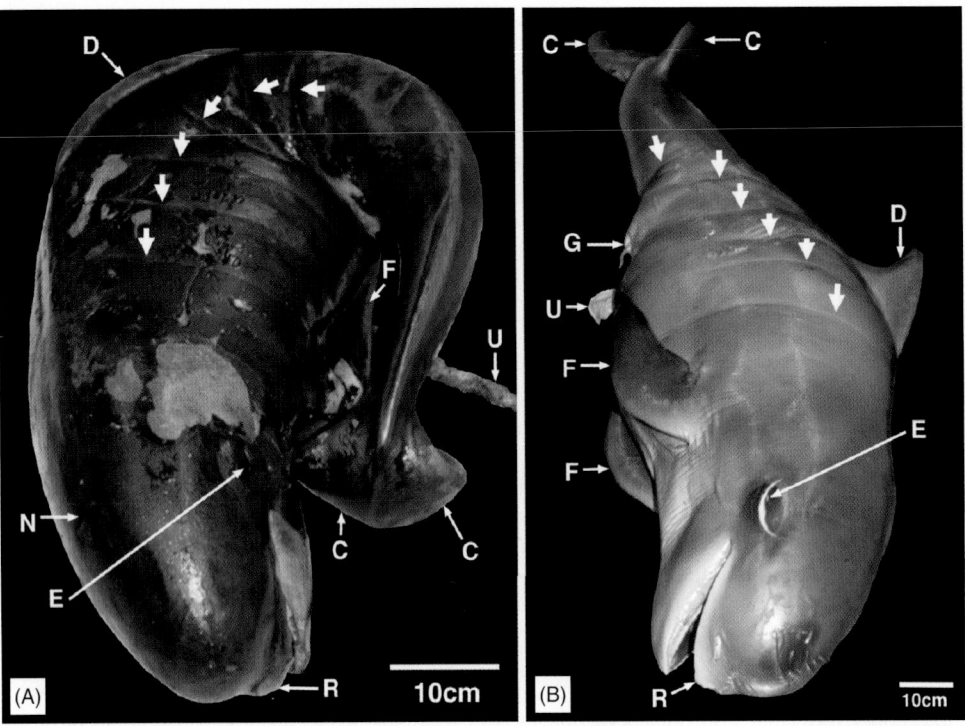

**Figure 3** *Full-term fetal long-finned pilot whales* (Globicephala melaena) *obtained postmortem from pregnant, beach-stranded whales on the shores of Cape Cod, Massachusetts, USA. Both whales were stored frozen and then thawed. Note the darker gray to black pigmentation of the skin, and the well-developed melon that now overrides the rostrum. The lighter patches are areas where the skin has sloughed off postmortem. (A) Near full-term fetus curled in the fetal position as it was found in utero, with the tail folded against the left side. The umbilical cord is evident under the distal tail-stock. The left lateral surface of the thorax and the abdomen shows a number of dorso-ventral fetal folds (white arrowheads; grooves between skin folds located mainly on the lateral abdominal wall of only one side). Note the prominent crimping along the pronounced concavity at the beginning of the tail-stock. The dorsal fin lies flat against the body and the tail flukes are curled ventrally. (B) The near full-term fetus has been uncurled from its fetal position, and the flukes and the dorsal fin have been extended. The fetal tail was originally curled to the right, as evidenced by the prominent fetal folds along the right lateral abdominal surface (white arrowheads). The tip of the penis can be seen protruding through the genital slit, just caudal to the stalk of the umbilical cord seen near the right flipper tip. The crescent-shaped tear in the fragile skin over the right eye and the wrinkles on the melon (forehead region) are freeze-thaw artifacts. C, caudal fluke; D, dorsal fin; E, eye; F, flipper; G, genital slit with protruding penis; N, single nasal opening (blowhole); R, rostral tip; U, umbilical cord.*

membranes (the cranial vault). Cetaceans appear to be no exception to this rule. In fact, they are an excellent example of the plasticity of the cranial vault, as this region is grossly modified compared with terrestrial mammals.

In the fetal period, the elements of the cranial vault begin to shift their relative positions so that the maxilla approximates or meets the occipital dorsally. This process of bony overlapping (called "telescoping") creates a layered appearance to the skull, where portions of bone are buried on the inner surface. In odontocetes, the parietals are depressed laterally and the premaxillary and the maxillary bones overlap the frontal bone dorsally, whereas in mysticetes, the premaxilla slides over the frontal and the base of the maxilla moves under the frontal bone. The cranial vault thus changes shape from dolichocephalic (longer than wide) to brachycephalic (wider than long). The ear ossicles begin to rotate into their adult position during the

early fetal period. No paranasal air sinuses (i.e., maxillary, ethmoid, sphenoid, or frontal) form within the skull either prenatally or postnatally (a diving adaptation that prevents injury from expanding/contracting the volume of an enclosed air space during depth/pressure changes). The bony nares migrate caudally to the near adult position on the dorsum of the head.

The structures of the nasal region are forming in the early fetal period, but asymmetry is not detected yet. Nasal conchae (bony plates that project from the nasal septum and walls in terrestrial mammals) never form. The nasal plugs (the tissues that close off the airway) are present, and may derive from the tissue that forms the upper lip in terrestrial mammals. The melon, which may also derive from this same tissue, has not yet formed the characteristic bulge in the forehead region. The nasal air sacs, diverticulae of the nasal tract, begin to bud off the soft tissues of the nasal passageways.

The nasal apertures, which appear initially on the dorsum of the rostrum (Fig. 1), begin to migrate caudally toward the adult position at the top of the head. They can be found near the junction of the rostrum and the swelling containing the forebrain. The nasal skull grossly transforms so that the nasal floor projects ahead of the nasal passageways into the rostrum, whereas the lateral parts (which form the walls of the nasal passages) shift from horizontal to vertical (Klima, 1999). There are two separate bony nasal passageways in all cetaceans. The soft tissues above the skull that surround the nasal passageways are maintained as two separate tubes in mysticetes. In odontocetes, however, the two soft tissue passageways fuse near their exit at the skin into one common blowhole opening. There are further differences within odontocetes in the development of the nasal skull. *Phocoena* has the most posteriorly positioned nares, whereas *Physeter* has the most anteriorly positioned nares. There are additional specializations in *Physeter* related to the unique forehead containing the spermaceti organ, including asymmetrically sized and positioned narial openings in the skull and soft tissue pathways through the head.

The larynx (voice-box) forms from cartilage elements of the fourth through sixth branchial arches. Its position in cetaceans is similar to that found in terrestrial mammals. The front part (epiglottis) overlaps its ventral surface with the dorsal surface of the soft palate, creating a bridge to channel air from the nasal region into the trachea and the lungs. In odontocetes, the larynx undergoes elongation of its rostral portion, forming a "goose beak" shape that inserts into the nasal region. The epiglottis elongates during the mid-fetal period. The posterior cartilages (corniculates) are still shorter than the epiglottis, and will not reach their full proportions (i.e., exceed the epiglottis in height) until the fetus reaches full term. The laryngeal "goose beak" of odontocetes inserts into a muscular sphincter derived from the palatopharyngeal arch of the soft palate. Postnatally, this interlock will keep the rostral opening of the larynx connected with the posterior nasal cavity. This connection imparts circumferential protection from the digestive tract, allowing air to flow between the nasal region and the lungs for sound production while prey is swallowed whole underwater. Internally, the odontocete larynx develops a midline fold (which is bifurcated in *Kogia breviceps*) that appears homologous to the vocal folds of terrestrial mammals (Reidenberg and Laitman, 1988). The mysticete larynx more closely resembles that of terrestrial mammals, except that there is a large and muscular sac attached ventrally in the midline (Reidenberg and Laitman, 2007). In *Caperea marginata*, this sac lies on the right side.

Late in the fetal period, both odontocetes and mysticetes form tooth buds. Odontocete teeth are single cusped and usually conical in shape with species-specific variations, e.g., narwhal (long spiral tusk), platanistids (needle-shaped), beaked whales (flattened and broad), and porpoises (flattened, spade-shaped). Postnatally, teeth are used primarily for grasping and aggression. As there are no incisors, canines, or molars (odontocetes are "homodonts"—all teeth have the same shape), the task of breaking up food is passed onto the stomach (see later). The tooth buds of fetal mysticetes are sometimes multicusped, resembling the teeth of related terrestrial mammals (Slijper, 1979). The mysticete tooth buds are more numerous in the upper jaw than in the lower jaw, but all are usually resorbed before birth. The formation of rudimentary baleen plates, which occurs concurrently with tooth bud degradation, may be induced by the process of tooth bud resorption (Ishikawa and Amasaki, 1995). As the mysticete fetus grows, longitudinal baleen ridges form in the gums of the upper jaw. These longitudinal ridges develop transverse divisions and rows of papillae comprised of epidermal folds that become cornified. The cornified papillae, which are tubular in shape, elongate and coalesce with their neighbors to form baleen plates (Slijper, 1979).

Throat grooves are a series of parallel, longitudinal folds found on the external, ventral surface of the head in rorqual mysticetes that enable expansion of the oral cavity. In other mysticetes and some odontocetes, a single pair of throat grooves can be found at the base of the jaw, and may indicate expansion of this region during tongue and hyoid depression (see later). In rorqual mysticetes, throat grooves begin forming in the fetal period, appearing initially between the umbilicus and the pectoral flipper. A second set of ventral throat grooves appears next near the tip of the mandible. Toward the end of gestation, the two sets of throat grooves join to form one complete set running from the mandibular tip to the umbilicus. These throat grooves enable expansion of the floor of the mouth to engulf prey during feeding.

In both mysticetes and odontocetes, lingual papillae develop along the lateral border of the tongue during the fetal period. Since newborn cetaceans lack lips, these papillae probably play an important role postnatally during nursing in grasping the teat, creating a seal for suction, and forming a channel for milk to flow into the oral cavity. These papillae attain maximal size in the early postnatal period of odontocetes, but can sometimes be found persisting in adults.

The tongues of mysticetes and odontocetes differ greatly in their construction, and this difference is evident in the fetal stage. Odontocete tongues are more closely related to the tongue of terrestrial mammals, being very muscular. Their tongues have large insertions on the broad bones of the hyoid apparatus (Reidenberg and Laitman, 1994). This arrangement helps depress the tongue into the throat like a piston, thereby creating enough negative pressure to draw in prey—a mechanism referred to as "suction feeding." The mysticete tongue (particularly in rorqual whales) is unusual in its structure because it can be flattened against the floor of the mouth and expanded laterally along with the throat pleats during prey engulfing. In addition, there is a fibrocartilage structure in the ventral throat region of rorqual whales that may be related to attachment of the mylohyoid muscle. This structure may aid jaw mechanics and support the tongue and floor of the mouth during expansion/contraction of the throat grooves.

## IV. Internal Organs

The tracheal rings usually develop as "O"-shaped rings, unlike the "C"-shaped rings of most terrestrial mammals. A bronchus leading to the right lung develops above the carina (tracheal bifurcation). As this bronchus emerges directly from the trachea above the primary (main stem) bronchi, it is termed a "tracheal bronchus." A right tracheal bronchus is a feature also found in the closely related artiodactyls.

The lungs mature from the embryonic glandular stage, to the fetal canalicular stage (see Drabek and Kooyman, 1983 for more information on stages of lung development). Next, muscular sphincters develop around the terminal bronchioles. Since this feature is not found in terrestrial mammals, it may be an adaptation for diving. The next phase of development is the alveolar stage. By the mid-fetal period, cartilaginous rings develop in the terminal bronchioles. This is another feature not found in terrestrial mammals that may also be an adaptation for diving, since the cartilage rings may keep the terminal airways patent under high pressures and during lung collapse at depth.

During the embryonic period, the heart is visible and has probably undergone a similar differentiation as in other mammals. The

**C**

heart begins as a straight tube, but during the late embryonic period, it folds and forms septa that eventually divide it into four chambers found in all mammalian hearts. The cetacean heart, however, shows differences in shape from terrestrial mammals. In both odontocetes and mysticetes, it is laterally (transversely) broad and cranio-caudally compressed, with the apex being formed by both ventricles. The cetacean heart has specializations which may be adaptive for diving, such as anastomoses between the dorsal and the ventral interventricular arteries, and hypertrophy of the right ventricle (Tarpley *et al.*, 1997). Diving adaptations also occur in the great vessels, such as an expandable aortic arch.

The internal carotid artery, which is a major supplier of blood to the brain in terrestrial mammals, tapers dramatically in the neck and terminates under the skull base at the carotid canal before reaching the brain. This reduction of the internal carotid artery probably occurs in all cetaceans that exhibit cervical retia mirabila (see later) as this is the only structure it appears to supply. Interestingly, the internal carotid artery is also reduced or absent in many artiodactyl species. The ductus arteriosus (a fetal vascular connection between the aorta and the pulmonary artery) was thought to remain patent postnatally, but a study in adult pilot whales showed that it does close (Johansen *et al.*, 1988). This is no different from terrestrial mammals, and is probably the condition in other cetaceans as well.

The fetus develops complex networks of anastomosing, coiled blood vessels called retia mirabila. These vessel masses are found in regions surrounding the dorsal thoracic cage, the region near the foramen magnum, and the spinal cord. Although the functions of retia mirabila are not known, it is thought that they are adaptations to diving and resurfacing. Their vessel structure may compensate for the rapid pressure changes of descent and ascent with a slow and a sustained response that moderates blood flow. By dampening oscillations in blood pressure, sensitive tissues, such as heart muscle or the brain and spinal cord, continue to receive steady perfusion, thus avoiding oxygen debt and lactic acid build up. As these vessels appear to store blood near vital tissue (e.g., brain, spinal cord, heart), they may thus function as a reservoir, distributing blood to these oxygen sensitive tissues when normal circulation is affected (e.g., as pressure increases during diving, or metabolism is slowed). A less widely held hypothesis for the retia mirabila's function is in trapping the nitrogen bubbles (emboli), which may come out of solution in the bloodstream during ascent from a prolonged dive. These bubbles are potentially fatal, as they can block smaller blood vessels and therefore interrupt blood flow in the capillary beds of organs (a condition known in human divers as decompression sickness, or caisson disease).

The fetus also develops a complex network of vessels that supply and drain the testes and the uterus. These vessels are arranged in a plexus to enable thermoregulatory countercurrent exchange. This conserves heat where needed and allows extra heat to be drawn away from these organs. Thus, the male can keep the testes cool and the pregnant female can keep the fetus in the uterus from overheating, despite their internal location under the insulating blubber (Rommel *et al.*, 1993).

In mammals, gut development begins with a single, relatively straight gastrointestinal tube that is suspended in the midline of the coelom. As the embryo develops, the gut tube differentiates into the foregut and the hindgut, and as each section further develops its specific shapes, individual regions of the gut tube begin to rotate into different positions within the coelomic cavity. Toward the end of the embryonic period, the thoraco-abdominal wall is distended. This is probably because the stomach is developing its multiple chambers and intestinal rotation is occurring. Cetaceans develop

a multichambered stomach (see Tarpley *et al.*, 1987, on *Balaena mysticetus*; Mead, 2007, on beaked whales) much like that found in closely related ruminant artiodactyls, the closest group of living land mammals to the cetaceans. The divisions of the cetacean stomach include, from proximal to distal: forestomach, main stomach, and pyloric stomach. As it does in ruminants, the cetacean forestomach arises from the stomach bud rather than the esophagus (Amasaki *et al.*, 1989a), but is not divided into the three small chambers (rumen, reticulum, psalterium) found in, e.g., the cow. The size of the forestomach may be dependent upon the consistency of the prey. In odontocetes, a large and a muscular compartment may signify a function in breaking down whole fish or crustaceans, whereas a smaller compartment possibly relates to a diet of soft prey such as cephalopods. In mysticetes, the forestomach is smaller than the main stomach, perhaps relating to the relatively small size of their prey. The cetacean main stomach and pyloric stomach (which can have up to 12 chambers, e.g., beaked whales) is equivalent to the cow's single rennet stomach (Slijper, 1979).

The process of intestinal rotation probably resembles that of other mammals, involving temporary herniation (protrusion) into the umbilicus, rotation and folding, and then return of the contents back to the abdomen where they lie more compactly. Thus, by the early fetal period the abdomen is no longer distended from the process of intestinal rotation. The cecum and large intestines then further differentiate, developing circular folds that divide the intestines into multiple connected chambers resembling the haustra (sacculations) of terrestrial mammals (Amasaki *et al.*, 1989a).

In embryos of terrestrial mammals, and presumably cetaceans, the earliest kidney is the mesonephros, comprised of ducts and tubules. The embryonic metanephric duct, which buds off the mesonephric duct, becomes the ureter. As the mesonephros regresses, a second kidney structure, the metanephros, develops around the metanephric duct and is retained as the final kidney.

The fetal kidney develops as a cluster of many small and relatively independent kidney units called renicules, which will be retained in the adult. An adult kidney divided into renicules or lobes is not unusual in mammals (e.g., ox, otter), and may indicate persistence of the fetal condition. The apparent functional advantage of a kidney divided into renicules in large mammals appears to be related to a maximum size for the length of the tubules, which might otherwise be too long for proper function in a large single kidney.

The urogenital sinus (derived from the embryonic cloaca) becomes the urethra. The urinary bladder develops from the proximal portion of the allantois.

The remnants of the mesonephric duct become the efferent ductules, epididymis, and deferent duct for sperm transport in males. The gonads (ovaries and testes) develop from gonadal ridges, which are paired thickenings of the coelomic epithelium. In females, paramesonephric (Mullerian) ducts develop simultaneously with the mesonephric ducts. The paramesonephric ducts become the bicornuate uterus and oviducts in females, but degenerates in males except for the prostatic sinus.

In males, the testes are intra-abdominal, i.e., they do not descend as in most terrestrial mammals, and thus there is no scrotal sac. The internalization of these structures helps streamline the body shape, thus reducing drag during swimming. Interestingly, cetaceans develop a gubernaculum (which functions in testicular descent in terrestrial mammals), but do not develop the peritoneal out-pocketing that occurs with testicular descent (the vaginal process) during the fetal period (van der Schoot, 1995). A complex vascular plexus supplies the testes (see later), functioning as a countercurrent heat exchanger to keep the testes cool despite their internal location under the insulating blubber.

The genital tubercle gives rise to the penis or clitoris (Amasaki *et al.*, 1989b). The external genitalia are not usually visible externally in the full-term fetus as they are withdrawn into the genital slit (see earlier discussion on genitoanal slit and contents).

## V. Nervous System

Brain development in the embryonic period resembles that of other terrestrial mammals. The brain is comprised of three main sections: prosencephalon (forebrain), mesencephalon (middlebrain), and rhombencephalon (hindbrain). The corticospinal tract does not develop to the same degree as terrestrial mammals, probably owing to the loss of the hindlimbs and the reduced role of the forelimbs in propulsion. The cochlea is enlarging, whereas the vestibular system is rudimentary in size—a disparity that will remain in the adult. Olfactory bulbs and nerves are present in both odontocete and mysticete embryonic brains.

The rate of brain growth and degree of encephalization differs for different species (Pirlot and Kamiya, 1975). In the early fetal period, typical cetacean features begin to develop. For example, the olfactory bulbs and nerves disappear in odontocetes. In mysticetes, however, the olfactory bulbs and nerves are retained. Since adult mysticetes retain olfactory mucosa, it is presumed that they use a sense of smell to help locate plankton, particularly swarms of krill. There is some dispute, however, as to the existence of a vomeronasal organ (a chemoreceptive organ that functions mainly in detecting sexual pheromones in terrestrial mammals). Although it had been thought that adult whales had a vomeronasal organ, fetal studies of mysticetes and odontocetes show both the vomeronasal organ and the nerve to be absent (Oelschläger *et al.*, 1987). The function of a cetacean vomeronasal organ is purely speculative, but may include detecting the presence and mating status of other whales and perhaps even the odor of food in the mouth. The terminal nerve (a sensory, but not chemosensory, derivative of the olfactory placode sometimes called "cranial nerve 0") persists and may function in autonomic innervation of intracranial arteries and mucous epithelium of the nasal air sacs.

In the mid-fetal period, the head and the neck regions align horizontally. During this process, the cervical section of the spinal cord, which was previously flexed ventrally, must now arch under and around the cerebellum to join the thoracic spinal cord (the dorsal aspect is thus concave). As the cervical vertebrae are compressed, much of the cervical spinal cord is contained within the skull.

## VI. Gestational Length

Since whales are related to the terrestrial ungulates, it is not surprising that their gestation is of a similar length (Table I). Horses, for example, have a gestation of 11 months (compared with 9 months for a human's pregnancy and 22 months for an elephant's gestation). Gestation in mysticetes lasts for 10–14 months. The length of gestation in odontocetes, however, is more variable, ranging 9 to 17 months. The length of the gestation is not correlated with body size (e.g., although the sperm whale is the largest odontocete, its gestation period is nearly the same as the smaller pilot whale and less than the Baird's beaked whale).

The length of gestation in cetaceans may be related to food supply and migration. Mysticetes mate in warm waters, migrate to cold waters to feed, and then migrate back to warm waters to calve. This behavioral cycle, which takes 1 year, thus appears to be related to gestational length. Interestingly, since most feeding occurs in the colder waters, a pregnant mysticete whale may well be fasting while simultaneously spending energy in migratory locomotion and nourishing a growing embryo/fetus. In this regard, it is significant that the first half of the pregnancy (largely embryo development) takes place during the migration to the feeding areas, whereas the second half of the pregnancy (largely fetal growth) takes place during the migration back to the calving areas. A whale migrating to the feeding areas is not carrying a large load of stored energy, compared with a whale returning to the calving areas from the feeding areas. In some species, pregnant whales may increase food intake by 50–60% above normal during the last 6 months of gestation. Thus, the energy demands on a whale in the early pregnancy stages may be smaller than that of a whale in the later stages of pregnancy, when the fetus is growing at a rapid rate. In odontocetes, these energy constraints appear to have less of a temporal impact on gestational periods. This may be due, in part, to a more constant energy supply (year-round access to a food supply) for those species that migrate, or the lack of migration in other species.

Since the calf must be able to swim, see, hear, and vocalize immediately after birth, the nervous and the muscular systems of the calf must be well developed. This also translates to a fairly long gestation, with as much development as possible occurring prenatally (compared with the human baby, which completes much of its neuromuscular development postnatally). The long gestation also enables calves to grow to a large size before birth, reaching approximately 1/4 to 1/3 of the mother's size. Once the fetus has attained a near adult form, the most dramatic changes appear to be mainly in the overall size of the fetus. As the fetal period progresses, the growth rate rapidly increases. For example, the blue whale (*Balaenoptera musculus*) gains approximately 100 kg/day in the last 2 months alone. Large newborns are also common among the whales' closest land relatives, the ungulates, which also have well-developed neural and muscular systems at birth. An additional advantage of a large calf is a smaller surface-to-volume ratio (which helps the calf conserve heat). Thus, since whales have relatively large calves it is not surprising that multiple births are a rarity.

## VII. Maternal Uterus, Placenta, and Umbilical Cord

Cetaceans have a bicornuate uterus (two horns joined in a Y-shape). The fetus usually develops in one horn (either horn for mysticetes, but most frequently the left horn for odontocetes), whereas the other horn is generally occupied by the allantois (one of the embryonic membranes) and placenta. The cetacean uterus has a complex vascular plexus that functions in countercurrent heat exchange (Rommel *et al.*, 1993). This keeps the fetus from overheating, despite its insulated location under the maternal blubber and adjacent to the locomotor musculature of the maternal abdomen.

The placenta is epithelio-chorial (or cotyledon), which means that the maternal and the fetal tissues do not fuse into one tissue, as in humans. Rather, their vascular systems remain separated by two epithelial layers with separate capillary beds. This arrangement ensures that the two layers separate relatively easily at birth, thus minimizing the inevitable loss of blood. Not surprisingly, this type of placenta is also found in the ungulates, the group of terrestrial mammals most closely related to cetaceans.

The umbilical cord is short and thick, with "amnion pearl" knobs on the outer surface that appear to regulate development

TABLE I
Cetacean Gestations and Newborn Calf vs Adult Measurements

|  | Gestation (in months) | Newborn calf length, weight | Mature adult length, weight |
|---|---|---|---|
| **Mysticetes** | | | |
| Right whale (*Balaena glacialis*) | 12 | 4.5–6 m, 1,000 kg | 10.09–17.7 m, 9,055–80,000 kg |
| Bowhead whale (*B. mysticetus*) | 13–14 | 3.6–4.5 m, 1,000 kg | 11.5–18 m |
| Pygmy right whale (*Caperea marginata*) | 10 | 1.5–2 m | 5.47–6.45 m, 3,100–3,500 kg |
| Gray whale (*Eschrichtius robustus*) | 13–14 | 4.5–5 m, 500–800 kg | 13–15.2 m, 14,000–35,000 kg |
| Humpback whale (*Megaptera novaeangliae*) | 11–11.5 | 4–5 m, 900–1,500 kg | 11.5–19 m, 25,000–48,000 kg |
| Minke whale (*Balaenoptera acutorostrata*) | 10 | 2.4–3 m, 300–400 kg | 6.9–10.7 m, 4,000–13,500 kg |
| Bryde's whale (*B. brydei, B. edeni*) | 11–12 | 3.95–4.3 m, 900 kg | 11.6–15.6 m, 16,000–25,000 kg |
| Sei whale (*B. borealis*) | 11.5–12 | 4.5 m, 780 kg | 13–18.3 m, 20,000–25,000 kg |
| Fin whale (*B. physalus*) | 11 | 6.4–6.5 m, 1,750–1,800 kg | 17.5–27 m, 30,000–90,000 kg |
| Blue whale (*B. musculus*) | 11–12 | 7–8 m, 2,000–3,000 kg | 19–31 m, 100,000–200,000 kg |
| **Odontocetes** | | | |
| Sperm whale (*Physeter macrocephalus*) | 14–16 | 3.5–4.5 m, 1,000–1,016 kg | 8.3–20.5 m, 16,000–57,000 kg |
| Pygmy sperm whale (*Kogia breviceps*) | 11 | 1.2 m, 30 kg | 2.42–3.7 m, 400 kg or less |
| Dwarf sperm whale (*K. sima*) | 9–12 | 1–1.04 m, 14 kg | 1.97–2.86 m, 210–303 kg |
| Cuvier's beaked whale (*Ziphius cavirostris*) | 12 | 2.5–3 m | 5.1–7.5 m, 3,000 kg |
| Baird's beaked whale (*Berardius bairdii*) | 17 | 4.5–4.8 m | 10–12.8 m, 11,000 kg |
| Northern bottlenosed whale (*Hyperoodon ampullatus*) | 12 | 3–3.5 m | 6–9.8 m |
| Southern bottlenosed whale (*H. planifrons*) | 12? | 2.5–2.9 m, 150–200 kg | 5.7–7.8 m |
| Hector's beaked whale (*Mesoplodon hectori*) | ? | 1.8–2.1 m | 4–4.5 m, 800 kg or less |
| True's beaked whale (*M. mirus*) | ? | 2.2–2.3 m, 136 kg | 4.8–5.3 m, 1,394 kg |
| Gervais' beaked whale (*M. europaeus*) | ? | 1.6–2.1 m, 49 kg or more | 3.7–5.2 m, 1,178 kg or more |
| Sowerby's beaked whale (*M. bidens*) | 12 | 2.4–2.7 m, 185 kg | 5.05–5.5 m |
| Gray's beaked whale (*M. grayi*) | ? | 2.1–2.42 m | 4.74–5.64 m, 1,075–1,100 kg |
| Andrews'/deepcrest beaked whale (*M. bowdoini*) | 10? | 2.2 m | 4.5 m or less |
| Hubb's beaked whale (*M. carlhubbsi*) | ? | 2.5 m | 5.3 m, 1,432 kg |
| Strap-toothed whale (*M. layardii*) | 9–12? | 0.76–2.8 m | 5–6.2 m |
| Blainville's beaked whale (*M. densirostris*) | ? | 1.9 m, 60 kg | 4.5–5.8 m, 1,033 kg |
| Ganges river dolphin (*Platanista gangetica*) | 10.5 | 0.67–0.9 m | 1.7–2.5 m, 69–85 kg |
| Amazon river (boto) dolphin (*Inia geoffrensis*) | 9–12 | 0.75–0.8 m, 7–8 kg | 2–2.6 m, 100–160 kg |
| Chinese river dolphin (*Liptotes vexillifer*) | ? | 0.57–0.95 m, 10 kg or less | 2.1–2.5 m, 125–160 kg |
| Franciscana (*Pontoporia blainvillei*) | 10–11 | 0.7–0.8 m, 7.3–8.5 kg | 1.4–1.74 m, 25 kg-53 kg |
| Beluga whale (*Delphinapterus leucas*) | 14–14.5 | 1.5–1.6 m, 79–80 kg | 3–5.5 m, 400–1,500 kg |
| Narwhal (*Monodon monoceros*) | 14–15.3 | 1.5–1.6 m, 80 kg | 3.4–4.7 m, 1,000 kg |
| Commerson's dolphin (*Cephalorhynchus commersonii*) | 11–12 | 0.65–0.75 m | 1.25–1.75 m, 35–86 kg |
| Hector's dolphin (*C. hectori*) | ? | 0.5–0.7 m | 1.2–1.8 m, 50–60 kg |
| Rough-toothed dolphin (*Steno bredanensis*) | ? | 1 m | 2.2–2.3 m, 122 kg |
| Humpback dolphin (*Sousa chinensis*) | ? | 0.9–1.08 m | 2.1–3 m, 85–284 kg |
| Tucuxi/gray river dolphin (*Sotalia fluviatilis*) | 10–10.2 | 0.7–0.92 m | 1.3–1.9 m, 35–40 kg |
| Bottlenosed dolphin (*Tursiops truncatus*) | 12 | 0.9–1.3 m, 32 kg | 1.9–4 m, 90–650 kg |
| Indian Ocean/Red Sea bottlenose dolphin (*T. aduncus*) | 12 | 1.03 m, 13.8 kg | 1.8–2.43 m, 176–200 kg |
| Pan-tropical spotted dolphin (*Stenella attenuata*) | 11.2–11.5 | 0.8–0.9 m | 1.8–2.57 m, 100–119 kg |
| Atlantic spotted dolphin (*S. frontalis*) | 12 | 0.76–1.20 m | 1.9–2.3 m, 110–143 kg |
| Spinner dolphin (*S. longirostris*) | 9–11 | 0.7–0.8 m | 1.3–2.16 m, 26.5–75 kg |
| Striped dolphin (*S. coeruleoalba*) | 12–13 | 0.8–1 m | 2.16–2.4 m, 131–156 kg |
| Common dolphin (*Delphinus delphis*) | 10–11.5 | 0.76–0.86 m | 0.93–2.6 m, 70–163 kg |

(*continues*)

TABLE I (Continued)

|  | Gestation (in months) | Newborn calf length, weight | Mature adult length, weight |
|---|---|---|---|
| Fraser's dolphin (*Lagenodelphis hosei*) | 10–12? | 0.95–1 m | 2.25–2.65 m, 200 kg or more |
| White-beaked dolphin (*Lagenorhynchus albirostris*) | ? | 0.95–1.6 m, 40 kg or more | 2–3 m |
| Atlantic white-sided dolphin (*L. acutus*) | 10–12 | 1.08–1.22 m | 2–2.8 m |
| Pacific white-sided dolphin (*L. obliquidens*) | 10–12 | 0.8–1.24 m | 1.7–2.5 m, 75–181 kg |
| Dusky dolphin (*L. obscurus*) | 11 | 0.55–0.70 m, 3.7 kg | 1.6–2.1 m, 40–80 kg or more |
| Northern right whale dolphin (*Lissodelphis borealis*) | ? | 0.8–1 m | 2–3.1 m, 115 kg |
| Southern right whale dolphin (*L. peronii*) | ? | 0.86 m, 5 kg? | 2.18–2.5 m |
| Risso's dolphin (*Grampus griseus*) | 12–14 | 1.1–1.5 m, 59 kg | 2.6–4.3 m, 500 kg |
| Melon-headed whale (*Peponocephala electra*) | ? | 1 m | 2.2–2.75 m, 160–275 kg |
| Pygmy killer whale (*Feresa attenuata*) | ? | 0.8 m | 2–2.6 m, 150–225 kg |
| False killer whale (*Pseudorca crassidens*) | 15.1–15.6 | 1.2–1.93 m, 80 kg | 3.5–6 m, 700–2,200 kg |
| Killer whale (*Orcinus orca*) | 12–17 | 2.06–2.5 m, 180 kg | 4.6–9.75 m, 2,600–10,500 kg |
| Long-finned pilot whale (*Globicephala melas*) | 12–16 | 1.38 m, 55–85 kg | 3.8–6.3 m, 280–1,750 kg |
| Short-finned pilot whale (*G. macrorhynchus*) | 14.9–15 | 1.4–1.85 m, 55 kg | 3.01–7.2 m, 600–3,950 kg |
| Irrawaddy dolphin (*Orcaella brevirostris*) | 14 | 0.9–1 m, 12.3–12.5 kg | 2.1–2.75 m, 85–150 kg |
| Finless porpoise (*Neophocaena phocaenoides*) | 11–12 | 0.6–0.9 m | 1.2–1.9 m, 35 kg |
| Harbor porpoise (*Phocoena phocoena*) | 11 | 0.7–0.9 m, 5–9 kg | 1.4–2 m, 40–90 kg |
| Burmeister's porpoise (*P. spinipinnis*) | 11–12 | ? | ? |
| Spectacled porpoise (*P. dioptrica*) | ? | 0.46–0.8 m? | 1.8–2.4 m |
| Dall's porpoise (*Phocoenoides dalli*) | 11–12 | 0.95–1 m, 25 kg | 1.7–2.2 m, 123–200 kg |

? = missing or estimated datum

(cornification) of fetal skin (Fig. 2). It contains two arteries and two veins, as well as an allantoic duct. When the calf is born, the umbilical cord breaks off at the fetal end, allowing the calf to swim unhindered to the surface. Since the mother does not appear to bite off the umbilical cord, it is presumed to break with little force at the moment of birth. The umbilical ring contains invaginations that probably weaken the connection between the fetal epithelium and the umbilical cord. The umbilical arteries and veins are also constricted and weak where they enter the fetal abdomen. The umbilical cord attaches midway along the length of the fetus (unlike the more caudal attachment found in fetuses of other mammals, in which the neck contributes more to fetal length than the tail). Thus, the umbilical cord will be stretched taut to the same degree regardless of whether the head or the tail is delivered first. The stretch from the delivered fetus pulling taut the umbilical cord, which is attached via the placenta to the mother's uterus, may cause its rupture at the umbilicus.

## VIII. Fetal Position and Birth

Birth takes place underwater. In most observed captive births in odontocetes, the fetal tail emerged first through the vaginal opening. This tail-first presentation may appear unusual, particularly when compared with the usual head-first presentations of most terrestrial mammals. Interestingly, captive manatees have also been observed to deliver their young tail-first underwater. Although births in the wild have been less frequently documented, they appear to be more commonly tail-first presentations in odontocetes and may be equally tail-first or head-first in mysticetes. As the pelvis in whales is rudimentary, it appears to have little, if any, effect on passage of the fetus during birth. In fact, the large size of the cetacean brain at birth may be possible, in part, because of the ease with which the large head of the fetus can be delivered through this rudimentary pelvis. Since there is no significant bony constriction at the pelvic outlet, there does not appear to be a physical need for a head-first delivery as in most large terrestrial mammals.

The higher frequency of tail-first presentations also may be explained by the shape and the intrauterine position of the fetus. The cetacean fetus has a fusiform shape, with the rostrum and the tailstock both being relatively small in diameter. The tail flukes, dorsal fin, and pectoral flippers are very pliable, and are flattened against the body (fin and flippers) or curled back toward the midline to form a small knob (flukes). This folding and curling not only helps keep the fetal body within the smallest dimensions, but it also enables the fetus to maintain a relatively smooth exterior contour with no protrusions to inhibit delivery through the vagina. In addition, the whole fetus is laterally flexed into a U-shape, with the tail recurved toward the head so that the flukes are positioned adjacent to the rostrum. While this fetal folding reduces the intrauterine volume needed for carrying the fetus, it also leaves the fetus with both its rostrum and its tail flukes directed toward the maternal tail. However, since the uterine horn is also folded, only one end of the fetus can thus be directed toward the cervix. In odontocetes, it is most commonly the tail. Since the tail flukes are smaller than the cetacean head, they can therefore slip out the vagina more easily and thus are more likely to emerge first.

The head being directed away from the cervix before parturition may be a function of either fetal shape or fetal weight. Since both the center of gravity and the largest diameter is closer to the fetal head, its "rest" position may more likely be with the heavier fetal head nearer the center of gravity of the mother (Fig. 2). This places the fetal head in the more distensible part of the mother's abdomen, and away from the more mobile tail-stock (which, due to locomotor constraints, may have less capability for expansion).

Once the fetus is settled into this birth position, continued growth appears to cause it to recurve caudally to fit within the mother's abdomen. Unlike terrestrial mammals, the head is not flexed ventrally in the late-term cetacean fetus because there is practically no neck and the cervical vertebrae are largely fused. Rather, the fetus folds in half laterally to conserve space. The curved midsection of the fetus takes up more room in the maternal abdomen than the fetal head and tail. Thus, the fetal abdomen is placed cranially in the mother, where there is more room for expansion, whereas the fetal head and tail are directed caudally near the less expandable maternal tail-stock. Although the fetal head is directed caudally, it is positioned at the tip of the uterine horn (which is thus also folded to face caudally) and not adjacent to the cervix. In this folded position, it is unlikely that the fetus can reposition itself to completely switch from a tail-first to a head-first presentation. As the fetus is delivered, its body must unfold. Thus, midway through parturition, the fetal head will again face toward the maternal head as the fetal body straightens. The newborn calf bears light colored bands and shallow vertical grooves, called "fetal folds," along the skin of the lateral abdomen (Fig. 3). These markings indicate the concave side of the fetus as it was folded *in utero*.

## See Also the Following Article

Baleen Whales

## References

Amasaki, H., Daigo, M., Taguchi, J., and Nishiyama, S. (1989a). Morphogenesis of the digestive tract in the fetuses of the southern minke whale, *Balaenoptera acutorostrata. Anat. Anz.* **169**, 161–168.

Amasaki, H., Ishikawa, H., and Daigo, M. (1989b). Development of the external genitalia in fetuses of the southern minke whale, *Balaenoptera acutorostrata. Acta. Anat.* **135**, 142–148.

Amasaki, H., Ishikawa, H., and Daigo, M. (1989c). Developmental changes of the fore- and hind-limbs in the fetuses of the southern minke whale, *Balaenoptera acutorostrata. Anat. Anz.* **169**, 145–148.

Calzada, N., and Aguilar, A. (1996). Flipper development in the Mediterranean striped dolphin (*Stenella coeruleoalba*). *Anat. Rec.* **245**, 708–714.

Cooper, L. N., Berta, A., Dawson, S. D., and Reidenberg, J. S. (2007a). Evolution of hyperphalangy and digit reduction in the cetacean manus. *Anat. Rec.* **290**, 654–672.

Cooper, L. N., Dawson, S. D., Reidenberg, J. S., and Berta, A. (2007b). Neuromuscular anatomy and evolution of the cetacean forelimb. *Anat. Rec.* **290**, 1121–1137.

Drabek, C. M., and Kooyman, G. L. (1983). Terminal airway embryology of the delphinid porpoises, *Stenella attenuata* and *S. longirostris. J. Morphol.* **175**, 65–72.

Ishikawa, H., and Amasaki, H. (1995). Development and physiological degradation of tooth buds and development of rudiment of baleen plate in southern minke whale, *Balaenoptera acutorostrata. J. Vet. Med. Sci.* **57**, 665–670.

Johansen, K., Elling, F., and Paulev, P. E. (1988). Ductus arteriosus in pilot whales. *Jpn. J. Physiol.* **38**, 387–392.

Klima, M. (1999). Development of the cetacean nasal skull. *Adv. Anat. Embryol. Cell Biol.* **149**, 1–143.

Meyer, W., Neurand, K., and Klima, M. (1995). Prenatal development of the integument in Delphinidae (Cetacea: Odontoceti). *J. Morphol.* **223**, 269–287.

Mead, J. G. (2007). Stomach anatomy and use in defining systemic relationships of the cetacean family ziphiidae (beaked whales). *Anat. Rec.* **290**, 581–595.

Oelschlager, H. A., Buhl, E. H., and Dann, J. F. (1987). Development of the nervus terminalis in mammals including toothed whales and humans. *Ann. NY. Acad. Sci.* **519**, 447–464.

Ogden, J. A., Lee, K. E., Conlogue, G. J., and Barnett, J. S. (1981). Prenatal and postnatal development of the cervical portion of the spine in the short-finned pilot whale *Globicephala macrorhyncha. Anat. Rec.* **200**, 83–94.

Pirlot, P., and Kamiya, T. (1975). Comparison of ontogenetic brain growth in marine and coastal dolphins. *Growth* **39**, 507–524.

Reidenberg, J. S., and Laitman, J. T. (1988). Existence of vocal folds in the larynx of odontoceti (toothed whales). *Anat. Rec.* **221**, 884–891.

Reidenberg, J. S., and Laitman, J. T. (1994). Anatomy of the hyoid apparatus in Odontoceti (toothed whales): Specializations of their skeleton and musculature compared with those of terrestrial mammals. *Anat. Rec.* **240**, 598–624.

Reidenberg, J. S., and Laitman, J. T. (2007). Discovery of a low frequency sound source in Mysticeti (baleen whales): Anatomical establishment of a vocal fold homolog. *Anat. Rec.* **290**, 745–760.

Rommel, S. A., Pabst, D. A., and McLellan, W. A. (1993). Functional morphology of the vascular plexuses associated with the cetacean uterus. *Anat. Rec.* **237**, 538–546.

van der Schoot, P. (1995). Studies on the fetal development of the gubernaculum in cetacea. *Anat. Rec.* **243**, 449–460.

Sedmera, D., Misek, I., and Klima, M. (1997a). On the development of Cetacean extremities: I. Hind limb rudimentation in the spotted dolphin (*Stenella attenuata*). *Eur. J. Morphol.* **35**, 25–30.

Slijper, E. J. (1979). "Whales," 2nd English Ed. Cornell University Press, Ithaca, NY.

Štěrba, O., Klima, M., and Schlidger, B. (1994). Embryology of dolphins: Staging and ageing of embryos and fetuses of some cetaceans. *Adv. Anat. Embryol. Cell Biol.* **157**, 1–133.

Tarpley, R. J., Hillmann, D. J., Henk, W. G., and George, J. C. (1997). Observations on the external morphology and vasculature of a fetal heart of the bowhead whale, *Balaena mysticetus. Anat. Rec.* **247**, 556–581.

Thewissen, J. G. M., Cohn, M. J., Stevens, L. S., Bajpai, S., Heyning, J., and Horton, W. E., Jr. (2006). Developmental basis for hind-limb loss in dolpins and origin of the cetacean bodyplan. *Proc. Natl. Acad. Sci. USA* **103**, 8414–8418.

Thewissen, J. G. M., and Heyning, J. (2007). *In* "Reproductive Biology and Phylogeny of the Cetacea: Whales, Dolphins, and Porpoises" (D. L. Miller, ed.), pp. 307–329. Science Publishers, Enfield NH.

# Circulatory System

## Paul J. Ponganis

### I. Introduction

Although the circulatory systems of marine mammals follow the general mammalian plan, they are most notable for features associated with the diving response, thermoregulation, and large body mass. This chapter will emphasize anatomical and functional aspects of the circulatory system in these animals.

The cardiovascular reflexes and adjustments which occur during diving are reviewed in Chapter Diving Physiology. Specific features of the circulatory system vary with orders, families, and species. These adaptations include large blood volumes, large capacitance structures (spleens and venous sinuses), venous sphincter muscles, vascular adaptations for thermoregulation, aortic windkessels, and vascular retia. The heart, arterial and venous systems, and blood volume will be discussed first. Then specific structural adaptations of the circulation in various groups will be considered.

## II. General Anatomy

### A. Heart

The basic structure and size of hearts in pinnipeds and cetaceans are typical of mammals. The four-chambered heart, with right ventricular outflow to the lungs, and left ventricular output to the systemic circulation, weighs 0.5% to 1% of body mass in most pinnipeds and small cetaceans (Slijper, 1962; King, 1983). In the great whales, relative heart mass is smaller, about 0.3% to 0.5% of body mass. Chamber size, stroke volume, and resting cardiac output and heart rate (where measured) are also in the general mammalian range, and in agreement with mammalian allometric equations. Both the foramen ovale and ductus arteriosus are closed in adult seals and cetaceans as in other mammals. Therefore, utilization of an intermittent fetal circulatory pathway does not appear to be a mechanism to bypass a potential increase in pulmonary vascular resistance during diving.

### B. Arterial/Venous Systems

General aspects of the arterial and venous systems in marine mammals are remarkable for several features (Elsner and Gooden, 1983; Butler and Jones, 1997). First, dense sympathetic nerve innervation of proximal as well as distal arteries in seals may represent a mechanism by which the intense sympathetic vasoconstriction of the dive response can be maintained independent of local tissue metabolite induced vasodilatation in the periphery. Angiography during forced submersions of seals supports this model. Second, venous capacitance is highly developed, especially in phocid seals and whales. This includes a large hepatic sinus and posterior vena cava,

the latter of which in seals has been estimated to be capable of storing a fifth of the seal's blood volume. Presumably, this large venous capacitance is related to the large blood volume of seals.

In some species, the spleen appears to be a significant storage organ for red blood cells (Butler and Jones, 1997; Kooyman and Ponganis, 1998). Increased splenic volumes in several pinniped species, and extensive sympathetic nerve innervation and smooth muscle development in the splenic capsule, are consistent with this storage role. Fluctuations in hematocrit between resting and diving states, or anesthetized and stressed states also support such a role for the spleen in seals. It has been estimated that 30% of the blood volume can be stored in the spleen in Weddell seals.

Another feature of the venous system, again well developed in both seals and whales, is the extradural venous system (Harrison and Tomlinson, 1956). These veins, located within the vertebral canal and above the spinal cord, receive blood flow from the brain, back, and pelvic regions (Fig. 1). They are linked with both the posterior and anterior vena cava via paravertebral communicating veins. In seals and cetaceans, the extradural veins are the primary venous drainage of the brain; the internal jugular vein is poorly developed or absent. The function of such a prominent vertebral venous system in these animals is unclear. In humans, it has been noted that extradural vein flow may participate in brain temperature regulation, and that, in the upright posture, the vertebral veins, kept open by attachment to the bony walls of the vertebral canal, are the primary cerebral venous drainage since blood flow decreases in the jugular veins due to venous collapse in the upright posture (Gauer and Thron, 1965). The direction and magnitude of flow within the extradural vein vary with the respiratory cycle (Ronald *et al.*, 1977). During a breath hold (apnea), extradural vein flow is low and has been reported to vary in direction (i.e., rostrally or caudally; Fig. 2). However, during breathing (eupnea), extradural vein flow is increased and rostral in direction (Fig. 3). It has been estimated that the extradural vein may contribute as much as 20% of eupneic venous return to the heart via its intrathoracic connections to intercostal veins, the enlarged right azygous vein, and the anterior vena cava (Ponganis *et al.*, 2006). These intrathoracic connections allow transmission of negative inspiratory pressures within the chest to the extradural vein. This enhances rostral blood flow during the inspiration since the ligamentous attachments of this vein to the vertebral canal prevent

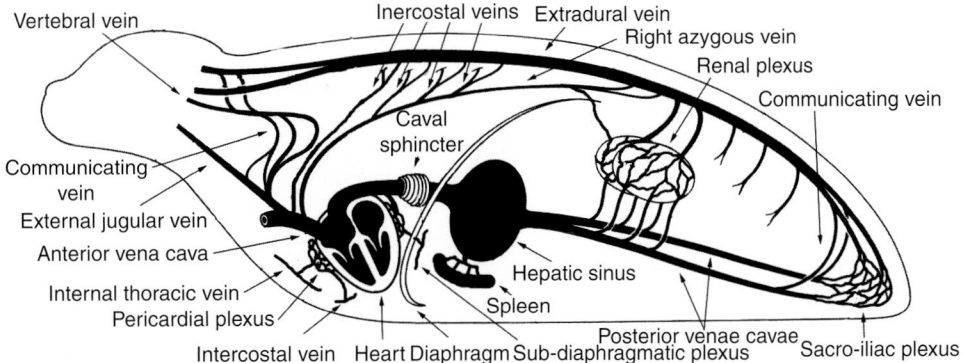

**Figure 1** *Venous anatomy of the phocid seal (modified from King, 1983). Notable features include the prominent extradural vein, large hepatic sinus, vena caval sphincter muscle, and numerous intrathoracic venous connections (pericardial venous plexus, intercostals veins, right azygous vein). Structures are not drawn to scale.*

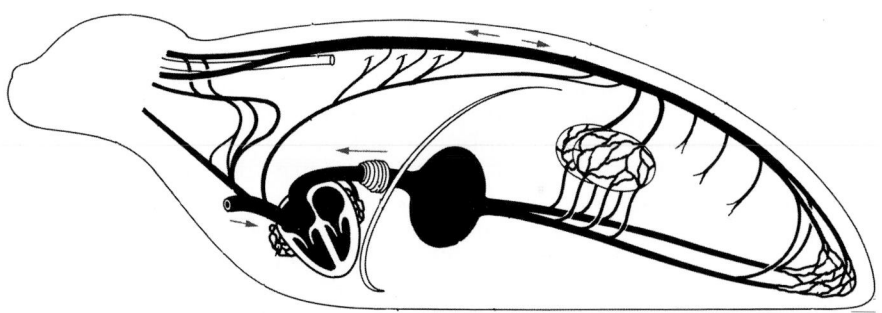

**Figure 2**   *Routes of venous return of blood to the heart during a breath hold (modified from Ponganis et al., 2006; after King, 1983). Blood flow in the extradural vein is low in magnitude and variable in direction. Primary control of venous return is via the vena caval sphincter. Blood also returns to the heart via the anterior vena cava.*

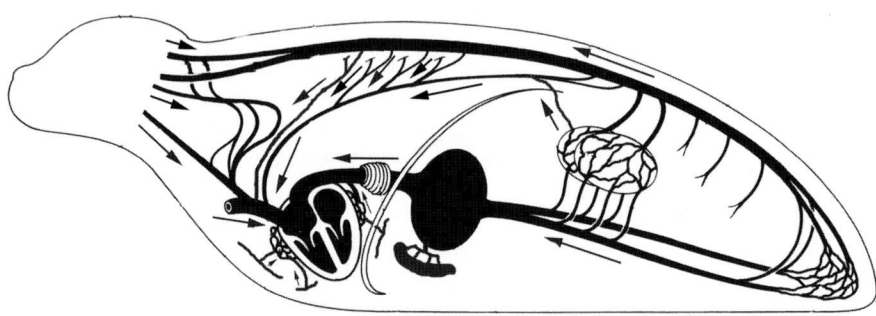

**Figure 3**   *Routes of venous return of blood to the heart during breathing (modified from Ponganis et al., 2006; after King, 1983). Primary return is again via the posterior vena cava, and the anterior vena cava. It is postulated that extradural vein flow may also contribute as much as 20% of venous return during inspiration via connections to the intercostal veins, right azygous vein, and pericardial plexus. (Arrows do not indicate magnitude of flow at any site).*

collapse of the vein. Therefore, at least one important function of the prominent extradural vein in seals may be its contribution to eupneic venous return to the heart.

### C. Blood Volume

Blood volumes are elevated and contribute to increase blood oxygen stores in marine mammals (Butler and Jones, 1997; Kooyman and Ponganis, 1998). On a mass specific basis, most measurements indicate that blood volumes are 2–3 times the 70 ml/kg human value, and that blood volume is greater in more active, and in longer diving species. The largest blood volumes (200–260 ml/kg) have been found in some of the best divers, including elephant seals, Weddell seals, and sperm whales.

### III. Structural Adaptations

#### A. Vascular Thermoregulatory Adaptations

The parallel pattern of counter-flowing arteries and veins, characteristic of counter current exchange units, is present in the flukes and flippers of cetaceans (Scholander and Schevill, 1955). Such arrangements, characteristic of blood vessel patterns in the limbs of many animals, are considered to conserve body heat by transferring heat from warm, out-going arterial blood to cool venous blood returning

from the limb. A superficial venous system, which does not return in conjunction with out-going arteries, also occurs in the skin. These veins, which have well-developed muscular walls, are considered to represent a route by which heat can be dissipated to the environment during periods of thermal stress.

Another structural adaptation observed in pinnipeds is the presence of numerous arterio-venous (a-v) anastomoses in the skin (Bryden and Molyneux, 1978). These structures represent a mechanism by which blood bypasses tissue capillary networks, and instead shunts directly from the arterial to venous system. The a-v anastomoses are distributed uniformly over the body surface of phocid seals, but, in otariids, are found in greater densities in the flippers. It is presumed that flow through these vessels allows heat exchange at the skin surface.

More recently, counter current anatomy has been observed in the reproductive organs of dolphins and pinnipeds (Rommell et al., 1995). It has been proposed that return of blood from the skin via vascular anastamoses allows relatively cool venous blood to prevent overheating of these organs. Temperature patterns along the length of the colon in the dolphin have been consistent with this hypothesis.

### B. Aortic Bulbs/Windkessels

In pinnipeds, again particularly in phocid seals, the aortic root (ascending aorta) is dilated, forming the so-called aortic bulb. The bulb can accommodate the stroke volume ejected by the heart, and it is more distensible than the distal aorta. It has been proposed that the aortic bulb acts as a windkessel: gradual contraction of the bulb due to elastic fibers within its wall contributes to maintenance of blood flow especially to the brain and heart during diastole (relaxation phase of the cardiac cycle).

The ascending aorta, aortic arch, and proximal carotid arteries of whales are also very compliant, and have also been hypothesized to act as a windkessel and preserve blood flow during diastole (Shadwick and Gosline, 1994). This is especially important in whales because long diastoles accompany slow heart rates. Low heart rates can occur in whales due to both their large body masses and the cardiovascular responses which occur during diving.

Maintenance of blood flow and pressure during a long diastole is of course critical to the brain, but also to the heart. This is because coronary perfusion occurs during diastole when the heart is relaxed. Myocardial flow is dependent on the diastolic blood pressure as the driving pressure. Thus, species which are either large or have more profound diving responses are likely to have some form of an aortic windkessel.

A compliant ascending aorta may also contribute to a reduction in the impedance that the left ventricle must pump against during the peripheral vasoconstriction of the diving response. This reduction in afterload will decrease the work and oxygen consumption of the heart, which is of course beneficial to a diver with a limited oxygen supply.

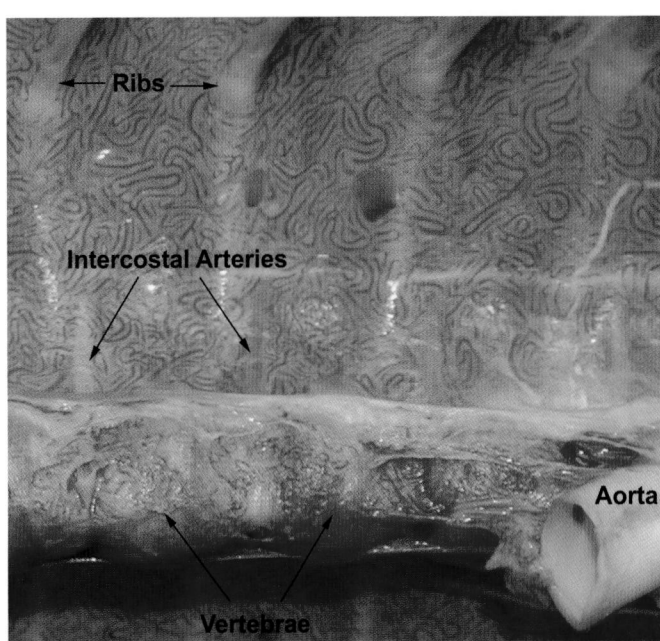

**Figure 4** *Ventrol view of thoracic rete mirabile (red swirls between ribs) in the right chest of a common dolphin,* Delphinus delphis. *Resection of the aorta reveals intercostal arteries at base of spinal column. These vessels extend into the rete. Dorsal is on bottom.*

### C. Vascular Retia

The *retia mirabilia* (wonderful nets) of cetaceans have long been noted by anatomists (Slijper, 1962; McFarland *et al.*, 1979). These plexuses of anastamosing arteries and veins occur along the vertebrae and base of skull, and are especially prominent in the thorax. The vascular retia are well developed in dolphins, in fact, more so than in large whales; they are also found in sirenians.

The thoracic rete is supplied by vessels from the aorta, which anastamose to form a complex, spongiform structure beneath the dorsal thoracic wall (Fig. 4). This vascular tissue extends around the vertebrae into the vertebral canal, and forms the primary arterial blood supply to the brain in cetaceans. The carotid arteries are vestigial or absent. The spinal meningeal artery in dolphins extends from the rete to the brain. Although mean blood pressure in the spinal meningeal artery of dolphins is equal to the aortic pressure, it is notable that the pressure is non-pulsatile; there is no systolic peak or diastolic trough. Thus, the cetacean brain appears to receive nonpulsatile blood flow. The significance of such a flow pattern as well as the function of the retia are unknown.

Slijper also reports the presence of large venous retia in the abdomens of whales. Hypotheses about the role of the retia have included windkessel functions, intrathoracic vascular engorement to prevent "lung squeeze" during diving, thermoregulation, and modification of composition of the blood.

### D. Inferior Vena Caval Sphincter

In most pinnipeds, the posterior vena cava is associated with a striated muscle sphincter at the level of the diaphragm (Fig. 5). Again, this is most well developed in phocid seals (Harrison and Tomlinson, 1956; King, 1983). The sphincter is innervated by the right phrenic

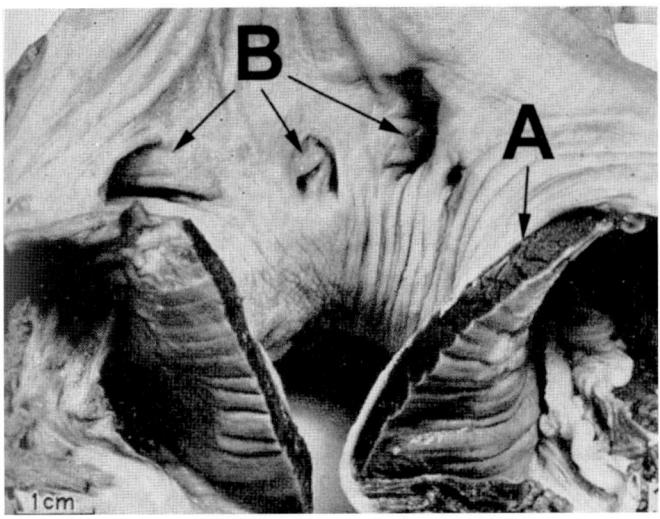

**Figure 5** *The posterior vena caval sphincter muscle of an elephant seal (modified from Harrison and Tomlinson, 1956). A: Transected muscle of the vena caval sphincter and B: openings of the pericardial venous plexus into the thoracic portion of the posterior vena cava.*

nerve, and is located cranial to the large hepatic sinus and inferior vena cava. Relaxation/contraction of the sphincter has been observed angiographically during forced submersions, and it is assumed that this is a mechanism to regulate venous return to the heart during diving bradycardias (Harrison and Tomlinson, 1956; Ronald *et al.*, 1977; Elsner and Gooden, 1983). Vena caval sphincters are also described in whales (Slijper, 1962); they presumably regulate blood return from the large venous capacitance vessels in the abdomen (posterior vena cava and venous rete).

C

Another venous structure, especially developed in phocid seals, is the pericardial venous plexus (Harrison and Tomlinson, 1956). This extensive venous network is connected to intercostal veins, internal thoracic veins, and a sub-diaphragmatic venous plexus; empties into the posterior vena cava just cranial to the vena caval sphincter (Figs 1 and 5). It is especially developed in the better diving seals such as elephant seals. Its significance is unknown, although it has been reported to be associated with brown fat and has been hypothesized to function in thermoregulation (Blix *et al.*, 1975). It may also be part of the pathway for eupneic venous return.

## *See Also the Following Articles*

Diving Physiology ■ Brain

## *References*

Blix, A. S., Grav, H. J., and Ronald, K. (1975). Brown adipose tissue and the significance of the venous plexuses in pinnipeds. *Acta Physiol. Scand.* **94**, 133–135.

Bryden, M. M., and Molyneux, G. S. (1978). Arteriovenous anastomoses in the skin of seals II. The California sea lion (*Zalophus californianus*) and the northern fur seal (*Callorhinus ursinus*) (Pinnipedia: Otariidae). *Anat. Rec.* **191**, 253–260.

Butler, P. J., and Jones, D. R. (1997). The physiology of diving of birds and mammals. *Physiol. Rev.* **77**, 837–899.

Elsner, R. W., and Gooden, B. (1983). "Diving and Asphyxia: A Comparative Study of Animals and Man." Cambridge University Press, Cambridge.

Gauer, O. H., and Thron, H. L. (1965). Postural changes in circulation. *In* "Handbook of Physiology: Circulation" (W. F. Hamilton, and P. Dow, eds), Vol. III, pp. 2409–2439. American Physiological Society, Washington, DC.

Harrison, R. J., and Tomlinson, J. D. W. (1956). Observations on the venous system in certain Pinnipedia and Cetacea. *Proceedings of the Zoological Society of London* **126**, 205–233.

King, J. E. (1983). "Seals of the World." Cornell University Press, Ithaca, NY.

Kooyman, G. L., and Ponganis, P. J. (1998). The physiological basis of diving to depth: Birds and mammals. *Annu. Rev. Physiol.* **60**, 19–32.

McFarland, W. L., Jacobs, M. S., and Morgane, P. J. (1979). Blood supply to the brain of the dolphin, *Tursiops truncatus*, with comparative observations on special aspects of the cerebrovascular supply of their vertebrates. *Neurosci. Biobehav. Res.* **3**, 193.

Ponganis, P. J., Stockard, T., Levenson, D. H., Berg, L., and Baranov, E. A. (2006). Intravascular pressure profiles in elephant seals: A hypothesis on the extradural vein and venous return to the heart. *Comp. Biochem. Physiol. A* **145**, 123–130.

Rommell, S. A., Early, G. A., Matassa, K. A., Pabst, D. A., and McLellan, W. A. (1995). Venous structures associated with thermoregulation of phocid seal reproductive organs. *Anat. Rec.* **243**, 390–402.

Ronald, K., McCarter, R., and Selley, L. J. (1977). Venous circulation of the Harp seal (*Pagophilus groenlandicus*). *In* "Functional Anatomy of Marine Mammals" (R. J. Harrison, ed.), pp. 235–270. Academic Press, New York.

Scholander, P. F., and Schevill, W. E. (1955). Counter-current vascular heat exchange in the fins of whales. *J. Appl. Physiol.* **8**, 279–282.

Shadwick, R. E., and Gosline, J. M. (1994). Arterial mechanics in the fin whale suggest a unique hemodynamic design. *Am. J. Physiol.* **267**, R805–R818.

Slijper, E. J. (1962). "Whales." Hutchinson and Co, London.

# Classification (Overall)

## DALE W. RICE

Four clades of placental mammals (class Mammalia: cohort Placentalia) independently evolved adaptations for life in the oceans. These are the still-living pinnipeds (sea lions, walruses, and seals), cetaceans (whales, dolphins, and porpoises), and sirenians (manatees and dugongs), and the extinct desmostylians. The pinnipeds are amphibious animals capable of terrestrial locomotion, and must haul out on shore to give birth. The cetaceans and sirenians (except for a few primitive Eocene species) are totally aquatic, having lost their hind limbs and evolved huge muscular tails with terminal flukes for swimming. The extinct desmostylians were quadrupedal amphibious creatures. The pinnipeds and cetaceans are carnivorous, the sirenians and desmostylians are herbivorous. Although primarily oceanic, several members of each of the three living groups have secondarily invaded freshwater habitats. The systematics of all of these sea mammals, living and fossil, is a flourishing field of research; many details are currently contested, and several paraphyletic groupings await resolution, so changes in the prevailing classification (Table I) may be anticipated.

## I. Pinnipeds

The pinnipeds were long classified as order Pinnipedia, separate from but closely related to the terrestrial carnivores of the order Carnivora. In recent years cladistic analyses of both morphological and molecular data have clearly shown them to be members of the suborder Caniformia of the order Carnivora.

Two strongly differentiated groups of living pinnipeds were long recognized: the eared seals, or sea lions and fur seals (family Otariidae), and the earless, or true, seals (family Phocidae); the walruses (family Odobenidae) were usually associated with the former group. Some taxonomists maintained that pinnipeds are a diphyletic assemblage, and that the eared seals shared a common ancestry with the bears (family Ursidae), while the true seals were most closely related to the weasel group (family Mustelidae), or more specifically the otters (subfamily Lutrinae). Those authorities allocated the pinnipeds to two superfamilies: Phocoidea for the family Phocidae, and Otarioidea, which included the extinct families Enaliarctidae and Desmatophocidae, and the extant Otariidae and Odobenidae. The Enaliarctidae included several late Oligocene and early Miocene genera that were postulated to have given rise to the other three families of otariids in the Miocene (Repenning and Tedford, 1977).

With the advent of cladistic methods, a different picture emerged. All of the molecular analyses and most of the morphological analyses have supported the hypothesis that all pinnipeds shared a common ancestry. However, the position of the pinnipeds within the suborder Caniformia is still disputed. Although some investigators favor a sister-group relationship to the Ursidae, other studies (Bininda-Emonds and Russell, 1996; Dragoo and Honeycutt, 1997) do support a closer affinity to the Mustelidae. One analysis placed the genus *Kolponomos* as the sister taxon of the pinnipeds. *Kolponomos*, an amphibious bear-like creature that lived along the coasts of Washington and Oregon during the Miocene, is currently listed in the family Amphicynodontidae, a paraphyletic group from which the Ursidae descended (Tedford *et al.*, 1994). With the present preponderance of evidence, the most appropriate classification is that of McKenna and Bell (1997), who rank the

## TABLE I
### Classification and geologic ranges of the living and fossil families of marine mammals.[1]

Order CARNIVORA (in part)
  Suborder CANIFORMIA (in part).
    Superfamily Phocoidea.
      †Family Enaliarctidae°. L. Olig.-M. Mioc.
      Family Otariidae (fur-seals and sea-lions). M. Mioc.-Rec.
      Family Odobenidae (walruses). L. Mioc.-Rec
      †Family Desmatophocidae°. E.-M. Mioc.
      Family Phocidae (true seals).L. Mioc.-Rec.

Order CETARTIODACTYLA (in part)
  Infraorder CETACEA
    †Stem groups [="Suborder Archaeoceti"]
      †Superfamily Protocetoidea
        Family Pakicetidae. E. Eoc.
        †Family Protocetidae°. M. Eoc.
        †Family Ambulocetidae. E.-M. Eoc.
      †Superfamily Remingtonocetoidea
        †Family Remingtonocetidae. M. Eoc.
      †Superfamily Basilosauroidea
        †Family Basilosauridae° (zeuglodonts). M.-L. Eoc.

Parvorder MYSTICETI
  Stem-groups
    †Family Llanocetidae. L. Eoc. or E. Olig.
    †Family Janjucetidae. L. Olig.
    †Family Aetiocetidae. L. Olig.
    †Family Mammalodontidae. L. Olig.
    †Family Kekenodontidae. L. Olig.
  †Superfamily Eomysticetoidea
    †Family Eomysticetidae. L. Olig.
    †Family Cetotheriopsidae. L. Olig.-E. Plioc.
  Superfamily Balaenoidea
    Family Neobalaenidae. Rec.
    Family Balaenidae. E. Mioc.-Rec.
  Superfamily Cetotherioidea.
    Family Eschrichtiidae. L. Mioc.-Rec.
    †Family Cetotheriidae. M. Mioc.-E. Plioc.
  Superfamily Balaenopteroidea
    †Family Pelocetidae. M. Mioc.
    †Family Aglaocetidae. E. Mioc.-M. Mioc.
    †Family Diorocetidae. M. Mioc.-L. Mioc.
    Family Balaenopteridae (rorquals). L. Mioc.-Rec.

Parvorder ODONTOCETI.
  Stem-groups.
    †Family Xenorophidae. L. Olig.

      †Family Agorophiidae. L. Olig.
      †Family Simocetidae. L. Olig.
    Superfamily Physeteroidea.
      Family Physeteridae (sperm whales). L. Olig.-Rec.
      Family Kogiidae (pygmy sperm whales). L. Mioc.-Rec.
    Superfamily Ziphioidea.
      Family Ziphiidae (beaked whales). M. Mioc.-Rec.
    Superfamily Platanistoidea.
      †Family Prosqualodontidae. E. Mioc.
      †Family Squalodontidae (shark-toothed dolphins). L.
      Olig.-M. Mio.
      †Family Patriocetidae. L. Olig.
      Family Platanistidae (Indian river-dolphins).
      M. Mioc.-Rec.
      †Family Squalodelphinidae. E. Mioc.
      †Family Dalpiazinidae. E. Mioc.
      †Family Waipatiidae. L. Olig.
    †Superfamily Eurhinodelphinoidea
      †Family Eurhinodelphinidae (long-snouted dolphins).—
      E.-L. Mioc.
      †Family Eoplatanistidae. E. Mioc.

    Superfamily Inioidea.
      Family Pontoporiidae (La Plata dolphins).—M. Mioc.-Rec
      Family Iniidae (Amazon river-dolphins). L. Mioc.-Rec.
    Superfamily Lipotoidea
      Family Lipotidae (Chinese river-dolphins). L. Mioc.-Rec
    Superfamily Delphinoidea.
      †Family Kentriodontidae°. L. Olig.-L. Mioc.
      †Family Albireonidae.—L. Mio.-E. Plio.
      Family Delphinidae (dolphins). M. Mioc.-Rec.
      Family Phocoenidae (porpoises). L. Mioc.-Rec.
      Family Monodontidae (belugas and narwhals).—
      Mioc.-Rec.
      †Family Odobenocetopsidae. E. Plio.

Order SIRENIA.
  †Family Prorastomidae. E.-M. Eoc.
  †Family Protosirenidae. M. Eoc.
  Family Dugongidae°(dugongs). M. Eoc.-Rec.
  Family Trichechidae (manatees). M. Mioc.-Rec.

†Order DESMOSTYLIA.
  †Family Paleoparadoxiidae. L. Olig.-M. Mioc.
  †Family Desmostylidae. L. Olig.-M. Mioc.

Extinct taxa are marked with a dagger (†), and taxa that appear to be paraphyletic are marked with an asterisk (°). Abbreviations: E = early; M = middle; L = late; Eoc. = Eocene; Olig. = Oligocene; Mioc. = Miocene; Plioc. = Pliocene; Pleist. = Pleistocene; Rec. = Recent
[1]Sources: McKenna and Bell 1997 (pinnipeds); Gingerich 2005 (archeocetes); Fitzgerald 2006 and Steeman 2007 (mysticetes); Fordyce and Muizon (1991) with updates (odontocetes); Domning 1994 and Gheerbrant et al., 2005 (Sirenians); Inuzuka 2000 (desmostylians).

pinnipeds as the superfamily Phocoidea, and the bear-like terrestrial carnivores as the superfamily Ursoidea, both under the parvorder Ursida, and place the mustelids in parvorder Mustelida; those authors also included all the genera of amphicynodonts as unallocated stem groups of the Phocoidea.

The cladistic studies have also reopened the question of interfamilial relationships of pinnipeds. A total-evidence analysis indicated the following "phyletic sequence": Desmatophocidae—Phocidae—Odobenidae—Otariidae (Dragoo and Honeycutt, 1997). The traditional pairing of the Odobenidae with the Otariidae was likewise

supported by molecular and morphological analyses of the living taxa. However, a comprehensive morphological analysis (Berta and Wyss, 1994) affirmed the paraphyletic nature of the Enaliarctidae, and arranged the other families of pinnipeds in the following phyletic sequence: Otariidae—Odobenidae—Desmatophocidae (paraphyletic).

Mention must be made of three other fossil genera of "otter-like seals" or "seal-like otters." These are *Potamotherium* with two species from freshwater Oligocene and Miocene deposits in Europe and North America, *Semantor macrurus* from freshwater Miocene or Pliocene deposits of Kazakhstan, and *Necromites nestoris* from

C

a marine Pliocene stratum of Azerbaijan. The latter two are known only by the hinder halves of their skeletons. All three have been proposed as primitive pinnipeds or phocids, but their phylogenetic placement remains totally problematic.

## II. Cetaceans

Whales differ so much from other placental mammals that their evolutionary relationships long remained conjectural. Gregory (1910) ranked Cetacea as a superorder—one of seven into which he divided all placental mammals. Simpson (1945) ranked Cetacea as an order, but made it the sole member of "cohort" Mutica, one of his four cohorts of placental mammals. Both of these classifications implied that the Cetacea had no close affinity with any of the other orders of mammals.

The marked anatomical dissimilarities between the baleen whales and the toothed whales led a few earlier cetologists to question the monophyly of the Cetacea, but none of them ever proposed an explicit hypothesis of diphyly. However, all recent studies, both morphological and molecular, overwhelmingly confirm the monophyletic origin of cetaceans.

During the past several decades a remarkable series of Eocene fossil cetaceans has been unearthed, mostly near the shores of the ancient Tethys Sea in Pakistan, India, and Egypt (Thewissen, 1998). These finds document the rapid evolutionary transition of the cetaceans from amphibious quadrupeds to fully aquatic forms during the interval from 34 to 54 million years ago (Ypresian through Priabonian). The phylogenetic relationships among these primitive Eocene cetaceans have yet to be fully resolved. For the interim they are allocated to five families, some of which are paraphyletic (Uhen, 2004). The entire assemblage has traditionally been included in the paraphyletic suborder Archaeoceti.

Early cladistic analyses of morphological data from fossil and living taxa showed the cetaceans (order Cetacea) and the extinct mesonychids (order Acreodi or Mesonychia) as monophyletic sister groups, which together constituted the sister group to the monophyletic Artiodactyla (O'Leary and Geisler, 1999). The mesonychids were cursorial, wolf-like creatures whose feet had five toes that bore hoof-like claws; they lived throughout the Holarctic from the early Paleocene to the early Oligocene. This apparent close relationship between the mesonychids and cetaceans became the generally accepted hypothesis (Luo and Gingerich, 1999), and led McKenna and Bell (1997) to classify both in order Cete, and to reduce Acreodi and Cetacea to subordinal rank. They further divided the Cetacea into two infraorders, Archaeoceti and Autoceta [sic], with the latter including the Mysticeti (baleen whales) and Odontoceti (toothed whales) as parvorders.

However, a contradictory classification resulted when cladistic analyses were based solely on the molecular data. In those studies the Cetacea consistently appeared within the Artiodactyla (even-toed hoofed mammals), and now a sister–taxon relationship between the cetaceans and hippos (family Hippopotamidae) is strongly supported (Nikaido et al., 1999; Shimamura et al., 1999). Especially strong support for this hypothesis comes from the presence of SINEs (short interspersed elements) in the genomes. SINEs are unique nucleotide sequences of 70 to 500 base pairs, more than 1000 copies of which are inserted throughout the genome. Because of these features, convergence of SINE sequences between any two lineages, or complete loss of the complement of SINEs, is virtually impossible. Thus they are near-perfect characters for phylogenetic analyses. Many authors then classified the living orders Artiodactyla and Cetacea under a supraordinal taxon Cetartiodactyla. Subsequent molecular studies are converging toward a consensus classification of the living Cetartiodactyla:

CETARTIODACTYLA Montgelard, Catzeflis, and Douzery (1999)
   TYLOPODA Illiger (1811) ................................................Camels
   ARTIOFABULA Waddell, Okada, and Hasegawa (1999)
   SUINA Gray (1868).................................................................. Pigs
   CETRUMINANTIA Waddell, Okada, and Hasegawa (1999)
     CETANCODONTA Arnason et al. (2000)[1]
      ANCODONTA Matthew (1929).......................Hippopotamuses
     CETACEA Scopoli (1777)
      MYSTICETI Cope (1869)................................. Baleen whales
      ODONTOCETI Flower (1867).......................Toothed whales
    RUMINANTIA Scopoli (1777)
     TRAGULINA Flower (1883)...............................Chevrotains
     PECORA Linnaeus (1758).........................................Pecorans

At first paleontologists were skeptical of these molecular results. Despite the rich fossil record of the artiodactyls, there appeared to be no evidence that would support the derivation of cetaceans from hippos or any other subclade within the Artiodactyla. There was a gap in the fossil record between the early Eocene, when cetaceans arose, and the late Miocene, when the hippos first appeared. However the latter objection would be weakened if Boisserie et al. (2005) are correct in claiming that hippos are simply late-surviving anthracotheres. The anthracotheres were mostly amphibious pig-like creatures that lived from the middle Eocene to the Miocene.

Are cetaceans the sister group to the Mesonychia, or did they arise from within the Artiodactyla? This controversy was finally resolved with the recovery of the heel bones (astragali) of three species of ancestral cetaceans which had functional hind legs: Pakicetus attocki and Icthyolestes pinfoldi of the Pakicetidae and Rodhocetus balochistanensis of the Protocetidae (Gingerich et al., 2001; Thewissen et al., 2001). These three cetaceans were found to have had "double-pulley" heel bones, a character diagnostic of the Artiodactyla. On this type of astragalus the articular facets on both the proximal (tibial) end and the distal (navicular) end are trochleated, or shaped like the wheel of a pulley. Subsequent studies revealed other morphological similarities between whales and hippos (Geisler and Uhen, 2003).

Price et al. (2005) calculated a fully resolved "supertree" for all living species of Cetartiodactyla, based on all available published morphological characters. Their results were congruent with the classification based on molecular data, noted above. The most recent and comprehensive phylogenetic analysis of both living and fossil taxa, based on morphological characters, again was largely congruent with the molecular results as far as the living taxa are concerned (Geisler et al. 2007). In this tree the Cetacea, Hippopotamidae, and the extinct Raoellidae appear as an unresolved trichotomy; the anthracotheres appear as the sister group to this clade. Thewissen et al. (2007) made a detailed comparison between the earliest cetaceans and the Raoellidae, both of which lived in the same area during the Eocene, and concluded that the two taxa are sister groups.

---

[1]Arnason et al. (2000) proposed the name CETANCODONTA as a replacement for WHIPPOMORPHA Waddell, Okada, and Hasegawa (1999). This change was made to avoid confusion with the name HIPPOMORPHA Wood (1937), which is currently in use as a suborder of the order Perissodactyla.

The raoellids were raccoon-sized amphibious creatures—Thewissen *et al.* (2007) likened their habitus to that of the living water chevrotains (*Hyemoschus aquaticus*) of central Africa.

The more advanced post-Eocene cetaceans are postulated to have descended from an archeocete, most plausibly a member of the family Basilosauridae, subfamily Dorudontinae. The monophyly of each of the two modern suborders, Mysticeti and Odontoceti, is strongly corroborated by a suite of complex morphological synapomorphies. The oldest fossil cetacean allocated to the suborder Mysticeti is *Llanocetus denticrenatus* (family Llanocetidae) from the end of the Eocene of the Antarctic Peninsula. This and several Oligocene genera of the families Aetiocetidae, Janjucetidae, Kekenodontidae, and Mammalodontidae all posessed teeth rather than baleen. In the late Oligocene appeared the first toothless, baleen-bearing cetaceans; they were long assigned to the family Cetotheriidae, a paraphyletic assemblage from which the four living families descended. In a recent cladistic analysis, Steeman (2007) resolved the old Cetotheriidae into five monophyletic families: Cetotheriidae (*sensu stricto*), Cetotheriopsidae, Pelocetidae, Aglaocetidae, and Diorocetidae.

The earliest members of the suborder Odontoceti appeared in the late Oligocene. During that epoch there lived a number of peculiar genera whose phylogenetic relationships remain unresolved. One distinctive superfamily, the Eurhinodelphinoidea, or long-snouted dolphins, diversified and then died out during the Miocene. All of the living odontocetes other than the peculiar river-dolphins clearly fall into three superfamilies, Physeteroidea (sperm whales), Ziphioidea (beaked whales), and Delphinoidea (dolphins, porpoises, etc.), all of which first appeared in the late Oligocene (Muizon, 1988, 1991). Beaked whales were long thought to be closely related to sperm whales, but a majority of recent cladistic analyses suggest that they are closer to the delphinoids. Studies of the river-dolphins have resulted in an emerging consensus that the Platanistidae are only distantly related to the others, and are closer to the family Squalodontidae, or shark-toothed porpoises, that lived during the Miocene. The Iniidae, Lipotidae, and Pontoporiidae appear to constitute one or more branches from the ancestral lineage of the Delphinoidea.

## III. Sirenians

Because of their superficially whale-like physique, many nineteenth century naturalists classified the sirenians as the "herbivorous cetacea." Modern studies have revealed that the Sirenia, along with the extinct Desmostylia, are marine members of a supraordinal group called the Tethytheria, which also embraces the elephants (order Proboscidea) and several other extinct groups (Ray *et al.*, 1986). The earliest sirenians had four limbs and were capable of terrestrial locomotion (Domning, 2001). The early to middle Eocene *Prorastomus* (family Prorastomidae) probably swam with only its hind limbs, but the middle Eocene *Protosiren* (family Protosirenidae) probably used its well-developed tail as well. The latter genus thus foreshadowed the still-living Dugongidae and Trichechidae, fully aquatic forms which have lost their hind limbs and swim by means of caudal flukes (Domning, 1994).

## IV. Desmostylians

The affinities of the Oligocene and Miocene desmostylians long remained problematic because they were known only from skulls recovered from the North Pacific rim, but many early authors classified them as a suborder of the Sirenia. Discovery of complete skeletons finally showed them to be quadrupedal hippopotamus-like animals, sufficiently different from sirenians to be ranked as a separate order. The most recent cladistic analysis places the Desmostylia

closer to the Proboscidea than to the Sirenia (Ray *et al.*, 1994). McKenna and Bell (1997) demoted the Tethytheria to a suborder of their new order Uranotheria, under which they ranked Sirenia as one infraorder, and also included the Desmostylia and Proboscidea as parvorders under the infraorder Behemota, but almost all recent authors still refer to Desmostylia as an order. Two families are recognized, Paleoparadoxiidae and Desmostylidae (Inuzuka, 2000).

## V. Other Marine Species

Several species of mammals that belong to otherwise terrestrial groups have become facultative or obligate members of the marine ecosystem (Rice, 1998). The polar bear (*Ursus maritimus*; family Ursidae) and the arctic fox (*Vulpes lagopus*; family Canidae) range widely over the north polar pack-ice. Among the 10 or so species of otters (family Mustelidae: subfamily Lutrinae), the sea otter (*Enhydra lutris*) of the North Pacific and the marine otter (*Lutra felina*) of western South America are strictly marine, and local populations of at least six other species feed in coastal marine waters. Two species of bats (order Chiroptera) also catch fish in coastal waters, the greater bulldog bat (*Noctilio leporinus*; family Noctilionidae) of the neotropics and the fishing bat (*Myotis vivesi*; family Vespertilionidae) of the Gulf of California. Finally, the most unexpected marine mammals were five species of large ground sloths of the genus *Thalassocnus* (order Phyllophaga: family Nothrotheriidae) which lived along the coast of Peru from the late Miocene to the late Pliocene (Pujos and Salas, 2004). They evidently grazed on algae or seagrasses.

## *See Also the Following Articles*

Cetacean Evolution ■ Systematics, Overview ■ Pinniped Evolution

## *References*

Berta, A., and Wyss, A. R. (1994). Pinniped phylogeny. *Proc. San Diego Soc. Nat. Hist.* **29**, 33–56.

Bininda-Emonds, O. R. P., and Russell, A. P. (1996). A morphological perspective on the phylogenetic relationships of the extant phocid seals (Mammalia: Carnivora: Phocidae). *Bonner Zool. Monogr.* **41**, 1–256.

Boisserie, J. R., Lihoreau, F., and Brunet, M. (2005). The position of Hippopotamidae within Cetartiodactyla. *Proc. Natl. Acad. Sci.* **182**(5), 1537–1541.

Domning, D. P. (1994). A phylogenetic analysis of the Sirenia. *Proc. San Diego Soc. Nat. Hist.* **29**, 177–189.

Domning, D. P. (2001). The earliest known fully quadrupedal sirenian Pezosiren portelli Prorastomidae (new gen. and sp.) from early middle Eocene of Jamaica. *Nature* **413**, 625–627.

Dragoo, J. W., and Honeycutt, R. L. (1997). Systematics of mustelid-like carnivores. *J. Mammal.* **78**(2), 426–443.

Fitzgerald, E. M. G. (2006). A bizarre new toothed mysticete (Cetacea) from Australia and the early evolution of baleen whales. *Proc. R. Soc. B* **273**, 2955–2963.

Fordyce, E., and Muizon, C. de (2001). Evolutionary history of cetaceans: A review. In "Secondary Adaptation of Tetrapods to Life in Water" (J.-M. Mazin, and V. de Buffrénil, eds), pp. 169–233. Verlag Dr. Friedrich Pfeil, Munchen.

Geisler, J. A., and Uhen, M. D. (2003). Morphological support for a close relationship between hippos and whales. *J. Vertebr. Paleontol.* **23**(4), 991–996.

Geisler, J. H., Theodor, J. M., Uhen, M. D., and Foss, S. E. (2007). Phylogenetic relationships of cetaceans to terrestrial artiodactyls. *In* "The Evolution of Artiodactyls" (D. R. Prothero, and S. E. Foss, eds), pp. 19–31. Johns Hopkins University Press, Baltimore.

**C**

Gheerbrant, E., Domning, D. P., and Tassy, P. (2005). Paenungulata (Sirenia, Proboscidea, Hyracoidea, and relatives). In "The Rise of Placental Mammals" (K. D. Rose, and J. D. Archibald, eds), pp. 84–105. Johns Hopkins University Press, Baltimore.

Gingerich, P. D. (2005). Cetacea. In "The Rise of Placental Mammals" (K. D. Rose, and J. D. Archibald, eds), pp. 234–252. Johns Hopkins University Press, Baltimore.

Gingerich, P. D., Haq, M. U., Zalmout, I. S., Khan, I. H., and Malkani, S. (2001). Origin of whales from early artiodactyls: Hands and feet of Eocene Protocetidae from Pakistan. *Science* **293**, 2239–2242.

Gregory, W. K. (1910). The orders of mammals. *Bull. Am. Nat. Hist.* **27**, 1–524.

Inuzuka, N. (2000). Primitive Late Oligocene Desmostylians from Japan and phylogeny of the Desmostylia. *Bull. Ashoro Mus. Paleontol.* **1**, 91–123.

Luo, Z., and Gingerich, P. D. (1999). Terrestrial Mesonychia to aquatic Cetacea: Transformation of the basicranium and evolution of hearing in whales. *Mus. Paleontol. Univ. Mich. Pap. Paleontol.* **31**, 1–98.

McKenna, M. C., and Bell, S. K. (1997). "Classification of mammals above the species level." pp. 631. Columbia Univ. Press, New York.

Muizon, C. de. (1988). Les relations phylogénétiques des Delphinida (Cetacea, Mammalia). *Ann. Paléontol. (Vertebr. Invertebr.)* **74**(4), 157–227.

Muizon, C.de. (1991). A new Ziphiidae (Cetacea) from the Early Miocene of Washington State (USA) and phylogenetic analysis of the major groups of odontocetes. *Bull. Mus. natl. Hist. nat., Paris (4e sér.)* **12**, 279–326, sect. C (3–4).

Nikaido, M., Rooney, A. P., and Okada, N. (1999). Phylogenetic relationships among cetartiodactyls based on insertions of short and long interspersed elements: Hippopotamuses are the closest extant relatives of whales. *Proc. Natl. Acad. Sci.* **96**, 10261–10266.

O'Leary, M. A., and Geisler, J. H. (1999). The position of Cetacea within Mammalia: Phylogenetic analysis of morphological data from extinct and extant taxa. *Syst. Biol.* **48**(3), 455–490.

Price, S. A., Bininda-Emonds, O. R. P., and Gittleman, J. L. (2005). A complete phylogeny of the whales, dolphins and even-toed hoofed mammals (Cetartiodactyla). *Biol. Rev.* **80**, 445–473.

Pujos, F., and Salas, R. (2004). A systematic reassessment and paleogeographic review of fossil Xenarthra from Peru. *Bull. Inst. Fr. Etudes andines* **33**(2), 331–377.

Ray, C. E., Domning, D. P., and McKenna, M. C. (1994). A new specimen of *Behemotops proteus* (order Desmostylia) from the marine Oligocene of Washington. *Proc. San Diego Soc. Nat. Hist.* **29**, 205–222.

Repenning, C. A., and Tedford, R. H. (1977). Otarioid seals of the Neogene. *Geological Survey Professional Paper* **992**, 1–93.

Rice, D. W. (1998). *Marine mammals of the world: Systematics and distribution* **4**, 1–231, Soc. Mar. Mammal. Spec. Publ.

Shimamura, M., Abe, H., Nikaido, M., Ohshima, K., and Okada, N. (1999). Genealogy of families of SINEs in cetaceans and artiodactyls: The presence of a huge superfamily of tRNA^Glu-derived families of SINEs. *Mol. Biol. Evol.* **16**(8), 1046–1060.

Simpson, G. G. (1945). The principles of classification and a classification of mammals. *Bulletin of the American Museum of Natural History* **85**(i–xvi), 1–350.

Steeman, M. E. (2007). Cladistic analysis and a revised classification of fossil and recent mysticetes. *Zoological Journal of the Linnean Society* **150**(4), 878–894.

Tedford, R. H., Barnes, L. G., and Ray, C. E. (1994). The early Miocene ursoid carnivoran Kolponomos: Systematics and mode of life. *Proc. San Diego Soc. Nat. Hist.* **29**, 11–32.

Thewissen, J. G. M. (ed.) (1998). "The emergence of whales: Evolutionary patterns in the origin of Cetacea," p. 477. Plenum Press, New York.

Thewissen, J. G. M., Williams, E. M., Roe, L. G., and Hussain, S. T. (2001). Skeletons of terrestrial whales and the relationship of whales to artiodactyls. *Nature* **413**, 277–281.

Thewissen, J. G. M., Cooper, L. N., Clementz, M. T., Bajpai, S., and Tiwari, B. N. (2007). Whales originated from aquatic artiodactyls in the Eocene epoch of India. *Nature* **450**, 1190–1194.

Uhen, M. D. (2004). Form, function, and anatomy of *Dorudon atrox* (Mammalia, Cetacea): An archaeocete from the middle to late Eocene of Egypt. *Mus. Paleontol. Univ. Mich. Papers on Paleontology* **34**, 1–122.

# Climate Change

## SUE E. MOORE

Global climate change is shifting the state of the world ocean toward a future of increased acidity, reduced productivity and sea ice cover, higher sea levels and loss of marine biodiversity and ecosystem function (IPCC, 2007). Reduction in ocean productivity over the past decade has been driven chiefly by the warming and stratification of low latitude waters, which blocks nutrients necessary for phytoplankton growth (Behrenfeld *et al.*, 2006). Simultaneously, warming has resulted in dramatic reductions in sea ice at sub-polar and polar latitudes, with concomitant shifts anticipated in marine ecosystem structure (Bluhm and Gradinger, 2007). The impacts of these changes on marine mammals will be mainly indirect, often mediated through alteration of physical habitat and predator–prey dynamics. The concept of ecological scale, described as the interface between population biology and ecosystem science (Levin, 1992), is used here to interpret how climate change may affect marine mammals (Moore, 2005). Because many species of marine mammals migrate between feeding and breeding areas, the concept of phenology, or the relation between climate and periodic biological phenomena, is also invoked (Root *et al.*, 2003; Durant *et al.*, 2007). Finally, potential synergies between climate change and the rate and extent of marine mammal exposure to disease or natural toxins are considered (Harvell *et al.*, 2002; Van Dolah, 2005).

## I. Ecological Scale

The effects of climate change on a given species will vary with the ecological scale on which that species exists. Ecological scale is determined by intrinsic life history characteristics and, for marine mammals, can extend from years to centuries in time and from tens to thousands of kilometers in space (Fig. 1). While individuals of some species roam across ocean basins for decades to centuries, others live shorter but more productive lives (in terms of number and frequency of offspring) within small freshwater, estuarine or coastal home ranges. This breadth of scale can confound attempts to predict and describe the effects of climate change on marine mammals as a group. A basic tenet of ecology is that community structure is influenced by (a) disturbance events that physically alter habitats and (b) competition among species for resources in those altered habitats. On the temporal scale, it is possible that marine mammals can adapt to disturbance introduced by climate change, given that most extant species have evolved over roughly the past 10 million years. However, it may be that ice-obligate species may not adjust as rapidly as more recent invading migrants that are evolved to more open water conditions. While description of the actual evolutionary steps that led to existing marine mammal fauna is outside the bounds of

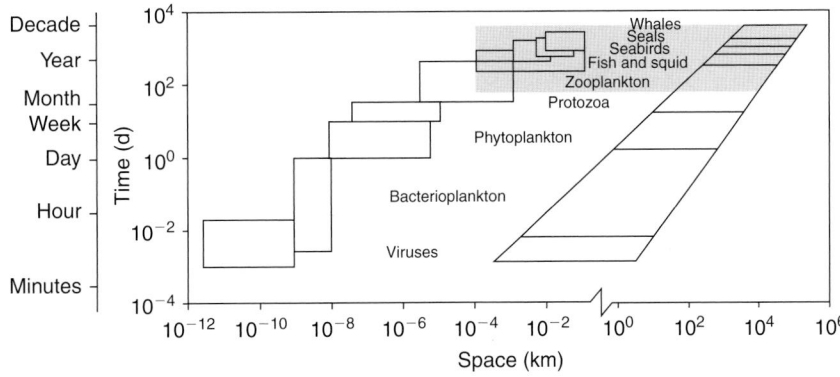

**Figure 1**  *Marine mammal ecological scale (shaded area) is determined by a species' intrinsic life history characteristics and, for marine mammals, can extend from years to centuries in time and from tens to thousands of kilometers in space. Reproduced from Moore (2005), with permission.*

scientific measurement, impacts on marine mammals of anticipated warming over the next 50–100 years can be addressed with an eye toward the effects of disturbance anticipated across latitudinal zones.

### A.  Marine Mammals in Polar Regions

In polar regions, the loss of sea ice and glacial disintegration are the clearest signals of disturbance to the marine environment attributable to climate change. In the Arctic, polar basin multiyear sea ice has disappeared at a rate of 9% per decade over the past 30 years. If that rate continues or increases due the absorption of heat by larger expanses of dark sea water (the albedo effect), the Arctic Ocean will be ice-free in September by sometime between 2040 and 2060. Of note, the rate of loss of seasonal sea ice measured to date varies among Arctic sub-regions and most models predict retention of sea ice during winter. The effects of the loss of seasonal sea ice, including declines in total ice cover and thickness as well as earlier breakup, will depend on the degree to which a species uses sea ice for basic life functions. Classifying the various species as ice-obligate, ice-associated, or seasonally migrant provides a conceptual model (Fig. 2). Clearly, the loss of sea ice will be most detrimental to those species, such as the polar bear (*Ursus maritimus*), the walrus (*Odobenus rosmarus*), and ice seals, that rely on ice as a platform for hunting and giving birth. For some ice seals, warmer temperatures and increased precipitation will further degrade sea ice habitat, including subnivean lairs that are required by ringed seals for rearing pups. Diminished sea ice could have cascading effects on the prey of ice-associated whales and seals. In some cases, loss of sea ice could be beneficial if it enhances productivity along the ocean basin slopes and shelves. In other cases, the decline of sea ice may result in the loss of preferred prey items, and the predators that cannot switch to alternative food will decline in numbers and perhaps range. The loss of sea ice loss may be most advantageous to seasonally migrant species, as it will give them opportunities to forage in Arctic waters earlier and stay later in the feeding season. Because humans live and exploit resources in the Arctic, changes in access as a result of climate change will complicate and compound the effects of sea ice loss on marine mammals. Importantly, all of the linkages shown in Fig. 2 are dynamic and interconnected. Sea ice loss will precipitate complex and cascading interactions among physical and biological components of the Arctic ecosystem. As top predators in this system, marine mammals are well positioned to function as well-observable indicators of climate change (Laidre *et al.*, 2008; Moore and Huntington, 2008).

Changes in marine mammal distribution and movements may also signal ecosystem shifts accompanying the loss of Antarctic sea ice. The dynamic relationship between sea ice and krill will mediate the impact of climate warming on whales, seals, and seabirds in the Southern Ocean. The density of krill, the primary prey of migratory baleen whales and many Antarctic seals, is positively correlated with extensive sea ice. Off the Antarctic Peninsula, years of extreme ice extent have been associated with high krill biomass and "good" years for penguins and seals. Conversely, reduced sea ice is associated with high salp biomass, few krill, and poor survivorship for young seals and penguins. Although the production–prey–predator dynamics among ice, krill, and whales remain poorly understood, feeding opportunities for large whales, at least in some regions of the Antarctic, are likely to decrease with the loss of sea ice.

### B.  Marine Mammals in Temperate and Tropical Regions

In temperate and tropical regions, ocean acidification and coral bleaching are the clearest signals of disturbance to the marine environment from climate change. Although these signs of environmental perturbation are alarming in their own right, the ultimate effects of these changes on marine mammals are not immediately evident. Although acidification will have cascading effects on trophic structure in the oceans, responses at the level of marine mammal prey cannot yet be predicted. Many climate models predict increases in sea level concomitant with thermal expansion of the warming oceans and melting of glaciers and ice caps. Rising sea levels can mean a loss of habitat for seals and sea lions that rely on low-lying coastal areas for rest, molting, pup birth and rearing, and courtship/mating. In the case of endangered and endemic fauna, such as Hawaiian monk seals (*Monachus schauinslandi*) in the Northwestern Hawaiian Islands, an evaluation of potential effects of sea level rise by 2100 found that maximum projected habitat loss ranged from 65% to 75% under median and maximum scenarios of sea level rise, respectively. For small isolated populations, such as those of monk seals (*Monachus* spp.), this loss of habitat increases extinction risk. Some climate change models predict increasing storm events, which may deepen the mixed layer and thereby increase nutrient availability in the upper ocean and ultimately enhance production of marine mammal prey. Storm effects are predicted to be most evident in temperate and sub-polar waters, and could result in more prey for marine mammals in these domains. At the same time, storms can severely

C

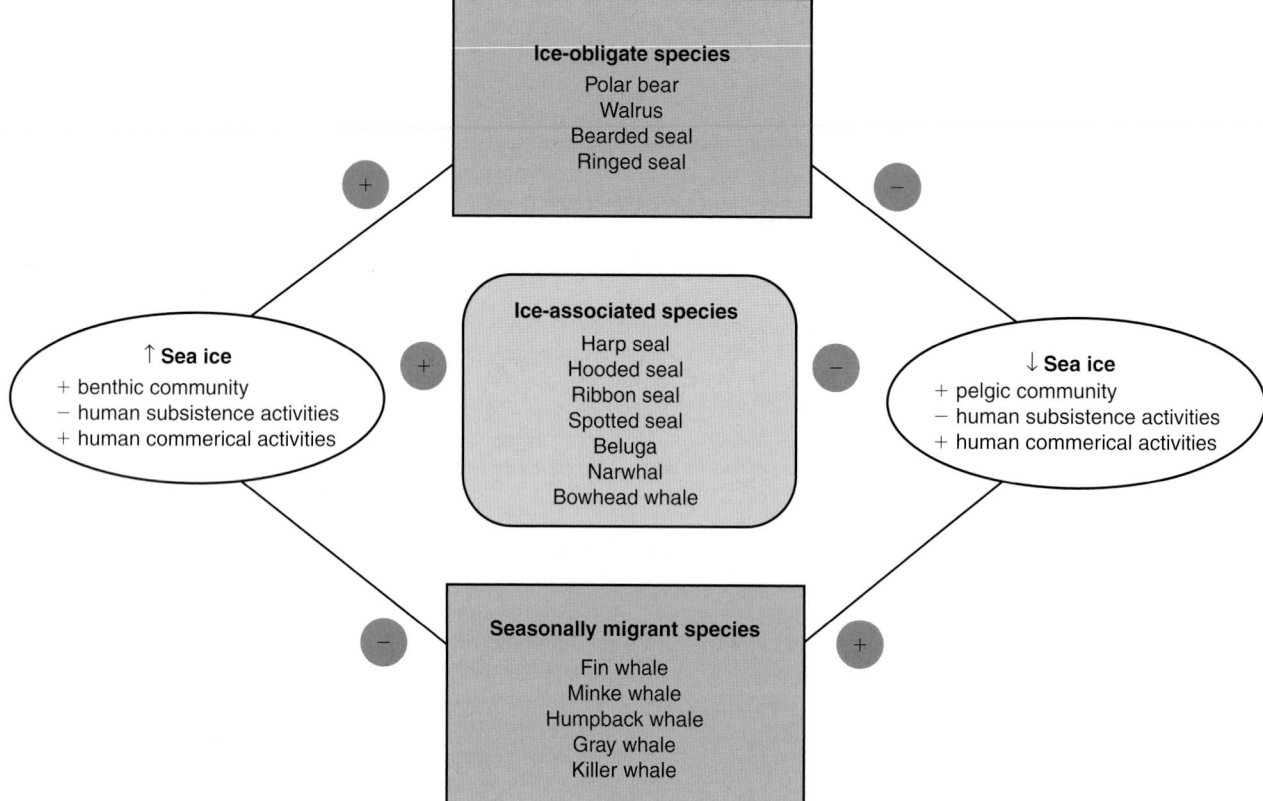

**Figure 2** *A conceptual model of sea ice impacts on ice obligate, ice associated and seasonally migrant marine mammal species. Dashed lines indicate uncertainty regarding potential impact of sea ice gain or loss for ice-associated species. Marine mammals will be affected by anticipated changes in productivity in both benthic and pelagic prey communities, and by changes in human subsistence and commercial activities in the Arctic. From Moore and Huntington (2008), with permission.*

reduce the survival of young pinnipeds on land or ice, and have catastrophic effects on sirenian feeding habitat.

## II. Phenology

Although marine mammal species have responded to climatic changes throughout their evolutionary history, the rates of those changes have been far slower than that measured during the past 100 years. Responses to recent global warming across a range of species from grasses and trees to mollusks and mammals have included poleward shifts in distribution and changes in the timing (phenology) of life history events such as migration, flowering, or reproduction. The broad-scale annual migrations by marine mammals between feeding and breeding areas have evolved to maximize foraging, reproductive success, and offspring survival. Although the environmental cues that initiate migration are not well understood, there is evidence to suggest that climate change is altering the timing of migrations in some marine mammal species. Such alteration in timing, or shifts in seasonal changes, to physical habitat, can lead to a mismatch between predator requirements and prey availability on the feeding grounds.

### A. Migration Timing

Long-term data on migratory timing are available for only a few marine mammal species. One of the best records is that for the migration of Eastern North Pacific gray whales (*Eschrichtius robustus*) between feeding areas offshore Alaska and Siberia and breeding areas in the lagoons and coastal waters of Baja California, Mexico. The southbound migration for this population has been documented from

a census site in central California over the past 40 years, providing a rare opportunity to examine migratory timing. In doing so, researchers were able to define a week delay in migration that coincided with, and appeared to be a step-response to, the strong El Niño (periodic ocean warming) event that occurred in the North Pacific during 1998/1999. Prior to 1998/1999, the overall median date for gray whales passing southbound was 8 January, but since then the overall median date has shifted to 15 January. This shift in migration timing was accompanied by reports of more calves seen offshore California, well north of the Mexican calving areas. These observations suggest that gray whales have modified their migration timing, and possibly their breeding range, in response to a climate event in the ocean. In another less-well-documented case, belugas (white whales, *Delphinapterus leucas*) that migrate along the Alaskan coast in summer now arrive near the village of Point Lay roughly 2 weeks earlier than they did in the 1980s. Subsistence hunters have noted this change and altered their activities accordingly. Conversely, in the case of bowhead whales (*Balaena mysticetus*) that migrate between the Bering and Beaufort seas, the timing of the migration has not changed but whales appear to linger longer along the route, with more animals seen feeding near Barrow, Alaska, through early summer. This latter case also could be attributable to a growing whale population simply expanding its foraging range, and not necessarily to climate change.

### B. Feeding

There is no doubt that the temporal aspects of feeding are important for marine mammals, especially in polar regions where the productive season is short. In many cases, animals must consume a yearlong supply of food over the course of a few months because reproductive periods necessitate fasting, especially for species that haul-out on land, and those large whales that feed in high latitude waters and migrate to lower latitudes to mate/calve. The effects of climate warming on polar bears is a clear case where access to prey such as ringed seals (*Pusa hispida*) has been disrupted by earlier breakup and later formation of sea ice in some areas of the eastern Canadian Arctic. Long-term records show declines in population size and body condition for bears in Western Hudson Bay and Baffin Bay due to the extended fasting imposed on the bears by longer ice-free periods. There is no other species of marine mammal for which such a long-term record exists to investigate the potential effects of climate change on feeding opportunities. However, anecdotal observations suggest that rapid diminution of seasonal sea ice in the Arctic could also be affecting feeding opportunities and recruitment in Pacific walruses. In summer, female walruses and their calves ride the retreating sea ice north from the Bering Sea to the northern Chukchi Sea. In the recent past, at maximum recession the sea ice edge as approximately at the edge of the Chukchi Sea continental shelf and adult female walruses could make easy forays there to feed, then return to suckle calves while hauled out on sea ice. However, in recent years, sea ice has retreated rapidly and far into the deep water of the Canadian Basin. Adult walruses must undertake long swims (*ca*. 100+ km) between sea ice haul-outs and shallow-water feeding areas, at considerable energetic cost. In addition, the report of at least nine walrus pups separated from adult females in deep water habitat suggests that young may become separated from their mothers before they are weaned. Without the food or protection of their mothers, these pups would almost certainly die. Whether this situation is having a measurable effect on the Pacific walrus population is unknown, as population size and trend for these pagophilic (ice-associated) animals have not been accurately determined.

### III. Disease and Toxins

Infectious diseases can cause rapid declines in wildlife populations. Rates of pathogen development, disease transmission, and host susceptibility are all influenced by climate, with a greater incidence of disease anticipated with warming. Marine mammal health and reproductive success are also adversely affected by toxins associated with harmful algal blooms (HABs). Marine mammal deaths associated with HABs appear to have increased over the past three decades, as have the frequency and geographic distribution of the events. Although coincident with climate warming, these apparent increases in HAB's may also reflect improvements in the ability to detect HABs and in the capability to identify algal toxins in marine mammal tissues. While marine mammals may face greater risk of mortality due to disease outbreak or exposure to toxins in a warming ocean, the magnitude of these threats and their relationships to climate are difficult to judge. Fortunately, diagnostic tools to monitor and measure the effects of disease and HABs on marine mammal populations are in rapid development, and combined with access to satellite imagery of the oceans and geo-spatial modeling, there is some hope for rapid advances in this field.

### See Also the Following Articles

Biogeography ■ Ecology, Overview ■ Ocean Environment

### References

Behrenfeld, M. J., *et al.* (2006). Climate-driven trends in contemporary ocean productivity. *Nature* **444**, 752–755.

Bluhm, B., and Gradinger, R. (2008). Regional variability in food availability for Arctic marine mammals. *Ecol. Appl.*

Durant, J. M., Hjermann, D.Ø., Ottersen, G., and Stenseth, N. Chr. (2007). Climate and the match or mismatch between predator requirements and resource availability. *Clim. Res.* **33**, 271–283.

Harvell, C. D., Mitchell, C. E., Ward, J. R., Altizer, S., Dobson, A. P., Ostfeld, R. S., and Samuel, M. D. (2002). Climate warming and disease risks for terrestrial and marine biota. *Science* **296**, 2158–2162.

IPCC (2007). Intergovernmental Panel on Climate Change (IPCC), Fourth Assessment Report (http://www.ipcc.ch/).

Laidre, K. L., Stirling, I., Lowry, L. F., Wiig, Ø, Heide-Jørgensen, M. P., and Ferguson, S. H. (2008). Quantifying the sensitivity of arctic marine mammals to climate-induced change. *Ecol. Appl.* **18**, S97–S125.

Levin, S. A. (1992). The problem of pattern and scale in ecology. *Ecology* **73**, 1943–1967.

Moore, S. E. (2005). Long-term environmental change and marine mammals. *In* "Marine Mammal Research: Conservation beyond Crisis" (J. E. Reynolds, III, W. F. Perrin, R. R. Reeves, S. Montgomery, and T. J. Ragen, eds). John Hopkins University Press, Baltimore.

Moore, S. E., and Huntington, H. P. (2008). Article marine mammals and climate change: impacts and resilience. *Ecol. Appl.* **18**, S157–S165.

Root, T. L., Price, J. T., Hall, K. R., Schneider, S. H., Rosenzweig, C., and Pounds, J. A. (2003). Fingerprints of global warming on wild animals and plants. *Nature* **421**, 57–60.

Van Dolah, F. M. (2005). Effects of harmful algal blooms. *In* "Marine Mammal Research: Conservation beyond Crisis" (J. W. Reynolds, III, W. F. Perrin, R. R. Reeves, S. Montgomery, and T. J. Ragen, eds). John Hopkins University Press, Baltimore.

# Clymene Dolphin
## *Stenella clymene*

### THOMAS A. JEFFERSON

### I. Characteristics and Taxonomy

The Clymene dolphin is a small, but rather stocky dolphin with a moderately long beak, separated from the melon by a distinct crease (Fig. 1). The dorsal fin is tall and nearly triangular to slightly falcate, and the flippers and flukes are typical of dolphins of the genera *Stenella* and *Delphinus*. The body shape is probably most similar to that of the striped dolphin, but coloration is very different. The color pattern is distinctly tripartite, with a white belly, light gray flanks, and dark gray cape (Fig. 1). The cape dips below the dorsal fin, somewhat lower than in the spinner dolphin. There is an eye stripe that runs forward to the upper beak and connects with a dark gray stripe running down the length of the top of the beak. The most distinctive feature is a black "mustache" marking of variable extent and intensity on the top of the beak. The lips are dark. Often there is a dark, indistinct band between the white belly and gray sides. With the exception of the mustache, most of the species' external characters are very similar to those of the spinner dolphin. This is one of the reasons why the Clymene dolphin was not fully recognized as a distinct species until 1981 (Perrin *et al.*, 1981).

Not many individuals of this species have been examined in detail and measured, so data are limited. These small dolphins probably do not reach much over 2.0 m in length, with males somewhat larger and heavier than females (Jefferson, 1996). Adult-sized females have been between 171 and 190 cm, and males between 176 and 197 cm (Perrin and Mead, 1994; Jefferson, 1996). The maximum

C

**Figure 1** *A Clymene dolphin leaps at the bow wave of a research vessel in the northern Gulf of Mexico, showing the species' diagnostic characteristics. The dark vertical streak on the tail stock is not typical coloration. Photo by R. L. Pitman.*

**Figure 2** *Three Clymene dolphins ride the waves produced by a research vessel in the offshore Gulf of Mexico. These dolphins are avid bowriders and are very active, often leaping and breaching alongside vessels. Photo by T. Pusser.*

known weight is about 80 kg, but considering the few specimens that have been weighed, they probably reach somewhat greater weights than this.

The skull of this species is very similar to that of *Stenella longirostris* and *S. coeruleoalba* (especially the latter). It can be distinguished by its small size (CBL <415 mm), combined with a short, broad rostrum. Tooth counts range from 39–52 (upper) and 39–48 (lower) (Jefferson and Curry, 2003). Total vertebral counts for the small number of specimens examined so far have been 70–76.

The species was named after the Greek sea nymph, Clymene (daughter of Oceanus and Tethys), and therefore Clymene should always be capitalized in the common name (Jefferson and Curry, 2003). Other English common names include short-snouted spinner dolphin and helmet dolphin. Taxonomically, *S. clymene* is considered to be most closely related to *S. longirostris* and *S. coeruleoalba* (Perrin *et al.*, 1981; Perrin and Mead, 1994). However, genetic studies indicate that its cytochrome *b* sequence is actually closer to that of *S. coeruleoalba* (LeDuc *et al.*, 1999). Specimens thought to be hybrids between *S. clymene* and *S. longirostris* have been observed in the southwestern Atlantic (Silva *et al.*, 2005).

## II. Distribution and Abundance

The Clymene dolphin is found only in the Atlantic Ocean, in tropical to warm-temperate waters (Fertl *et al.*, 2003). The exact range is not well documented, especially in South Atlantic, mid-Atlantic, and West African waters. Presumably it occurs continuously across the Atlantic Ocean. Recently, the known range off the coast of West Africa has been extended south to Angola (Weir, 2006). Most sightings have been in deep, offshore waters, although Clymene dolphins are sometimes observed very close to shore where deep water approaches the coast (such as around some islands of the Caribbean). It is present year-round in at least the northern Gulf of Mexico and probably throughout much of its tropical range.

No estimates of overall abundance exist, although there are estimated to be over 17,000 Clymene dolphins in the northern Gulf of Mexico (Mullin and Fulling, 2004). Considering this, it seems likely that the global abundance of the species is over 100,000 dolphins, and the species is not considered to be in danger of extinction. However, despite this, little is actually known about the status of any stock.

## III. Ecology

There is very little known about the feeding ecology of this species, as very few stomachs have been examined. It apparently feeds mostly on mesopelagic fishes and squids, including some species that are vertical migrators (Jefferson and Curry, 2003).

External parasites include barnacles on appendages and whale lice in lesions and body grooves. Internal parasites have not been well studied but include various worms and flukes in the blubber and muscle, respiratory system, digestive system and brain and in the mammary glands of females. They can cause disease and have been implicated in the deaths of some animals.

Clymene dolphins associate with dolphins of other species on occasion, in particular spinner dolphins. Associations with tuna are known to occur off the West African coast (Cadenat and Doutre, 1958). Many Clymene dolphins bear bite marks and scars from cookie-cutter sharks on their bodies, and large sharks and killer whale are probable predators (although actual predation events have not been documented).

## IV. Behavior and Physiology

Schools of this species are often moderately large, although most appear to consist of less than a few hundred individuals. In the Gulf of Mexico, where most information on school size comes from, the average group size is 42 dolphins (Mullin *et al.*, 1994). Schools may be segregated by age and sex class, as evidenced by several mass stranded herds that were composed largely of individuals of one or the other sex (Jefferson *et al.*, 1995).

Clymene dolphins are active bow riders, sometimes approaching ships from a distance for a free ride (Fig. 2). They are also often aerially active and they do spin on their long axes like spinner dolphins (something that only a few species of dolphins do), although apparently not as frequently or as elaborately as the spinner dolphin. Cooperative foraging techniques have been observed in the Gulf of Mexico (Fertl *et al.*, 1997).

Although there has been little work done on acoustic behavior (Mullin *et al.*, 1994), these animals often appear to be quite vocal, with whistles in the frequency range of 6–19 kHz (Wang, 1993). Virtually nothing is known about the species' physiology.

## V. Life History

There have been no published studies on the life history of this species based on large samples of specimens. Most of what we know is based on scant information from strandings, mostly from the Gulf of Mexico. Both males and females appear to reach sexual maturity by a length of 180 cm (Jefferson and Curry, 2003). Nothing is known of other life history parameters, but they are thought to be broadly similar to those of other members of the genus *Stenella*.

## VI. Interactions with Humans

Clymene dolphins have not been held captive, except for occasional animals that were kept temporarily after stranding alive. No major conservation problems are known for this species, but it is likely that some undocumented problems exist. Some dolphins are known to be killed in directed fisheries in the Caribbean, and others incidentally in nets throughout most parts of the range. This may be one of the species involved in the tuna purse seine fishery in the Gulf of Guinea area of West Africa (Maigret, 1981). It is possible that large, but undocumented, incidental catches may occur there, as they have in the eastern tropical Pacific. There has been almost no work on environmental contaminants in this species (Jefferson and Curry, 2003).

## *See Also the Following Articles*

Spinner Dolphin ▪ Striped Dolphin ▪ Bow-riding

## *References*

Cadenat, J., and Doutre, M. (1958). Notes sur les Delphinidés ouest-africaines. I.-Un *Prodelphinus*? indéterminé des côtes du Sénégal. *Bull. de l'Inst. Français d'Afr. Noire* **20A**, 1483–1484.

Fertl, D., Schiro, A. J., and Peake, D. (1997). Coordinated feeding by Clymene dolphins (*Stenella clymene*) in the Gulf of Mexico. *Aquat. Mamm.* **23**, 111–112.

Fertl, D., Jefferson, T. A., Moreno, I. B., Zerbini, A. N., and Mullin, K. D. (2003). Distribution of the Clymene dolphin *Stenella clymene*. *Mamm. Rev.* **33**, 253–271.

Jefferson, T. A. (1996). Morphology of the Clymene dolphin (*Stenella clymene*) in the northern Gulf of Mexico. *Aquat. Mamm.* **22**, 35–43.

Jefferson, T. A., and Curry, B. E. (2003). *Stenella clymene. Mamm. Spec.* **726**, 1–5.

Jefferson, T. A., Odell, D. K., and Prunier, K. T. (1995). Notes on the biology of the Clymene dolphin (*Stenella clymene*) in the northern Gulf of Mexico. *Mar. Mamm. Sci.* **11**, 564–573.

LeDuc, R. G., Perrin, W. F., and Dizon, A. E. (1999). Phylogenetic relationships among the delphinid cetaceans based on full cytochrome b sequences. *Mar. Mamm. Sci.* **15**, 619–648.

Maigret, J. (1981). Rapports entre les Cétacés et la pêche thonière dans l'Atlantique tropical oriental. *Notes Africaines* **171**, 77–84.

Mullin, K. D., and Fulling, G. L. (2004). Abundance of cetaceans in the oceanic northern Gulf of Mexico, 1996–2001. *Mar. Mamm. Sci.* **20**, 787–807.

Mullin, K. D., Higgins, L. V., Jefferson, T. A., and Hansen, L. J. (1994). Sightings of the Clymene dolphin (*Stenella clymene*) in the Gulf of Mexico. *Mar. Mamm. Sci.* **10**, 464–470.

Perrin, W. F., and Mead, J. G. (1994). Clymene dolphin *Stenella clymene* (Gray, 1846). *In* "Handbook of Marine Mammals" (S. H. Ridgway, and R. Harrison, eds), Vol. 5, pp. 161–171. Academic Press, San Diego.

Perrin, W. F., Mitchell, E. D., Mead, J. G., Caldwell, D. K., and van Bree, P. J. H. (1981). *Stenella clymene*, a rediscovered tropical dolphin of the Atlantic. *J. Mammal.* **62**, 583–598.

Silva, J. M., Jr., Silva, F. J. L., and Sazima, I. (2005). Two presumed interspecific hybrids in the genus *Stenella* (Delphinidae) in the tropical West Atlantic. *Aquat. Mamm.* **31**, 468–472.

Wang, D. (1993). Whistle structures: a preliminary comparison between four delphinid species in the Gulf of Mexico. Ph.D. Thesis, Institute of Hydrobiology, The Chinese Academy of Sciences.

Weir, C. R. (2006). First confirmed records of Clymene dolphin, *Stenella clymene* (Gray, 1850), from Angola and Congo, south-east Atlantic Ocean. *Afr. Zool.* **41**, 297–300.

**C**

# Coloration

## WILLIAM F. PERRIN

Marine mammals are not as colorful as birds or fishes or reptiles, but many have striking and distinctive coloration patterns that are useful in their taxonomy, presumably have function and adaptive value, and can vary individually and with age, sex, geographic region, and even time of the year.

## I. Terminology

A number of schemes have been proposed for naming the elements of color patterns in cetaceans; the usage here follows Perrin (1973, Perrin, 1997) and Perrin *et al.* (1991). In delphinids and phocoenids (Fig. 1), the *bridle* is composed of the *blowhole stripe* running from the blowhole to the apex of melon and the *eye stripe* from the eye to the apex of melon. Both stripes may have complex internal structure. An *eye spot* may be visible, and there may also be a small *ear stripe* or *spot*. The *eye-to-anus stripe* runs from the eye to the anal/genital region and may have *accessory stripes*. The *flipper stripe* runs forward from the base of the flipper variously to the eye (e.g., in spinner dolphin, *Stenella longirostris*), corner of the mouth (e.g., pantropical spotted dolphin, *S. attenuata*), or forward along the rostrum to join the *lip mark* ventrolaterally (common dolphins, *Delphinus* spp.).

The overall color pattern in at least some delphinids can be analyzed in terms of interacting independent components (Fig. 2). A basic *cape* is covered with a *dorsal overlay* of varying extent and intensity and may not be visible except in fetal or anomalously pigmented specimens. A crisscross of the boundaries of these two elements in *Delphinus* spp. yields a complex four-part pattern of a dark-gray *dorsal field* (cape and overlay combined), buff or yellowish *thoracic patch* (cape alone), light-gray *flank patch* (overlay alone), and white *ventral field* (outside both cape and overlay). In some anomalous individuals, the overlay may be absent, yielding a simplified pattern of cape only [e.g., in *Delphinus delphis* (Perrin *et al.*, 1995) and *Stenella longirostris* (Perrin, 1973)]. Spotting appears to be yet another independent component that develops with maturation in some species.

In pinnipeds, coloration can be a property of different *pelages* or pelage elements, through a range from white to silver, gray or bluish gray, brown, and black. The *lanugo* is a fetal pelage that develops and can be lost before birth, although in many species it is shed a few days or weeks after birth. Juveniles may undergo additional *molts* and changes of color. The coarse *guard hairs* can differ in color from

C

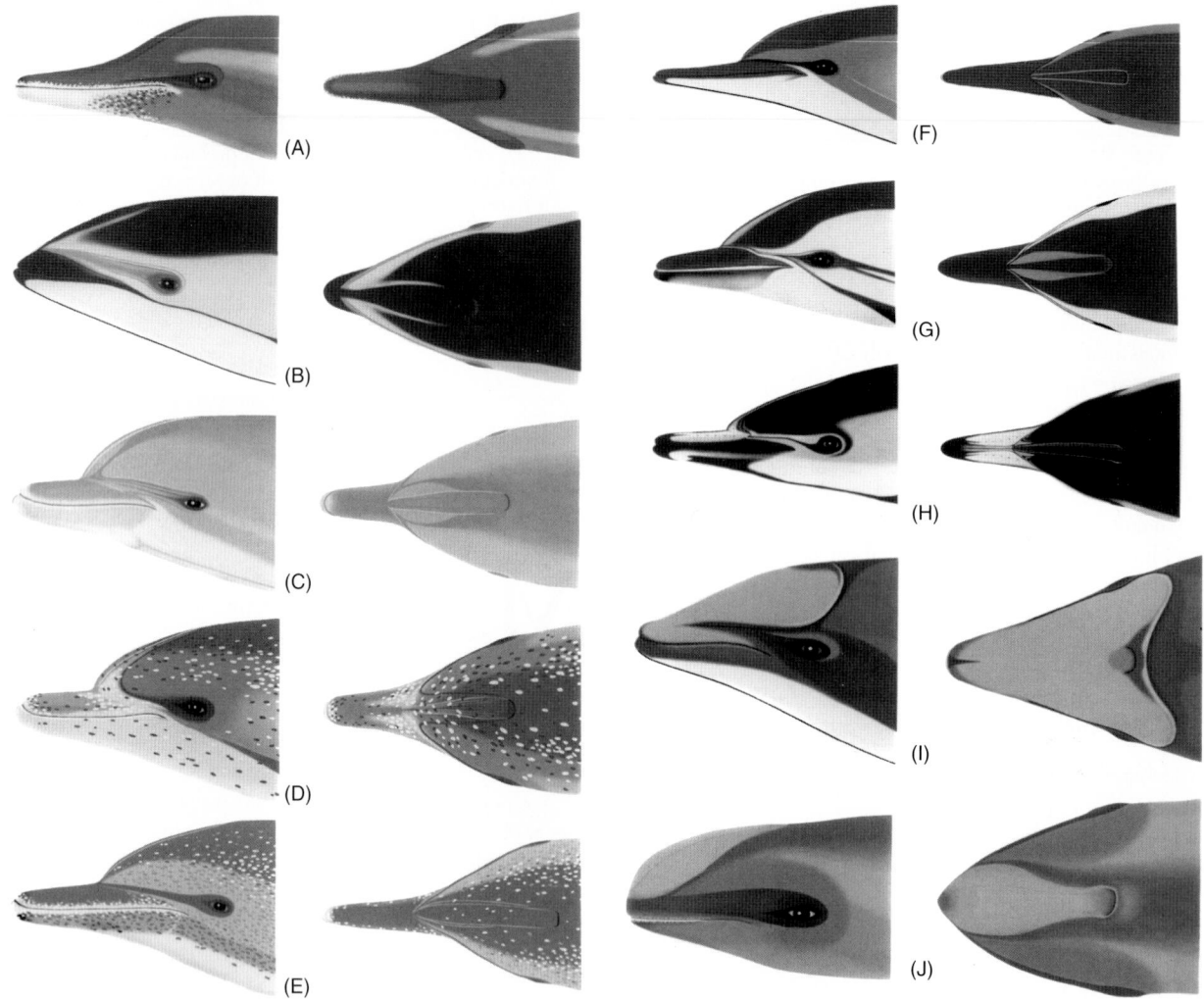

Figure 1  *Typical appearance of bridle in 10 delphinid species:* (A) Steno bredanensis, (B) Lagenorhynchus obliquidens,
(C) Tursiops truncatus, (D) Stenella frontalis, (E) S. attenuata, (F) S. longirostris, (G) S. coeruleoalba, (H) Delphinus delphis,
(I) Cephalorhynchus eutropia, *and* (J) Peponocephala electra. *From Perrin (1997).*

the hairs of the *undercoat*. Many seals are simply uniformly colored
or counter-shaded, but some have bold patterns, such as the harp
seal (*Pagophilus groenlandicus*) and ribbon seal (*Histriophoca fasciata*), and others are spotted.

## II. Development

Coloration typically changes between birth and adulthood. In
some cases appearance changes radically, whereas in others the
change is more in contrast and distinctness of pattern elements. Only
a few examples are discussed here.

The spotted dolphins, *Stenella attenuata* and *S. frontalis*, are
unspotted at birth. Small dark spots appear in large juveniles in the
throat region and spread over the ventral surface, enlarging as maturity approaches. Light spots appear on the back and spread in a similar fashion, although not in an even distribution over the back. In
*S. attenuata* the dark ventral spots fuse and lighten to yield a light
gray, faintly dappled ventral surface. In *S. frontalis*, both ventral and
dorsal spots persist into maturity.

The beluga, *Delphinapterus leucas*, is dark gray at birth but lightens as it grows; adults are white. A similar trend is seen in Asian populations of the Indo-Pacific humpbacked dolphin, *Sousa chinensis*.
The reverse of this trend is seen in many other cetaceans, such as
pilot whales and beaked whales; calves are lighter at birth and darken
with age, although neonates of the bottlenose dolphin and the finless
porpoise, *Neophocaena phocaenoides*, in some regions are darker
than juveniles and adults.

The development of coloration tends to take opposite courses
in different groups of pinnipeds (Bonner, 1990). Otariids are born
dark and become lighter as juveniles, most darkening again as adults.
However, most Northern Hemisphere phocids (the phocinines),
including *Phoca largha*, *Pusa* spp., *Pagophilus groenlandicus*,
*Histriophoca fasciata*, and *Halichoerus grypus*, are born with a white
or yellowish lanugo, which is shed at 2–5 weeks. This molt exposes
either the adult pattern of spots or other marks or a juvenile counter-
shaded coloration that later changes to the adult patterned state.
The harbor seal, *Phoca vitulina*, and the hooded seal, *Cystophora*

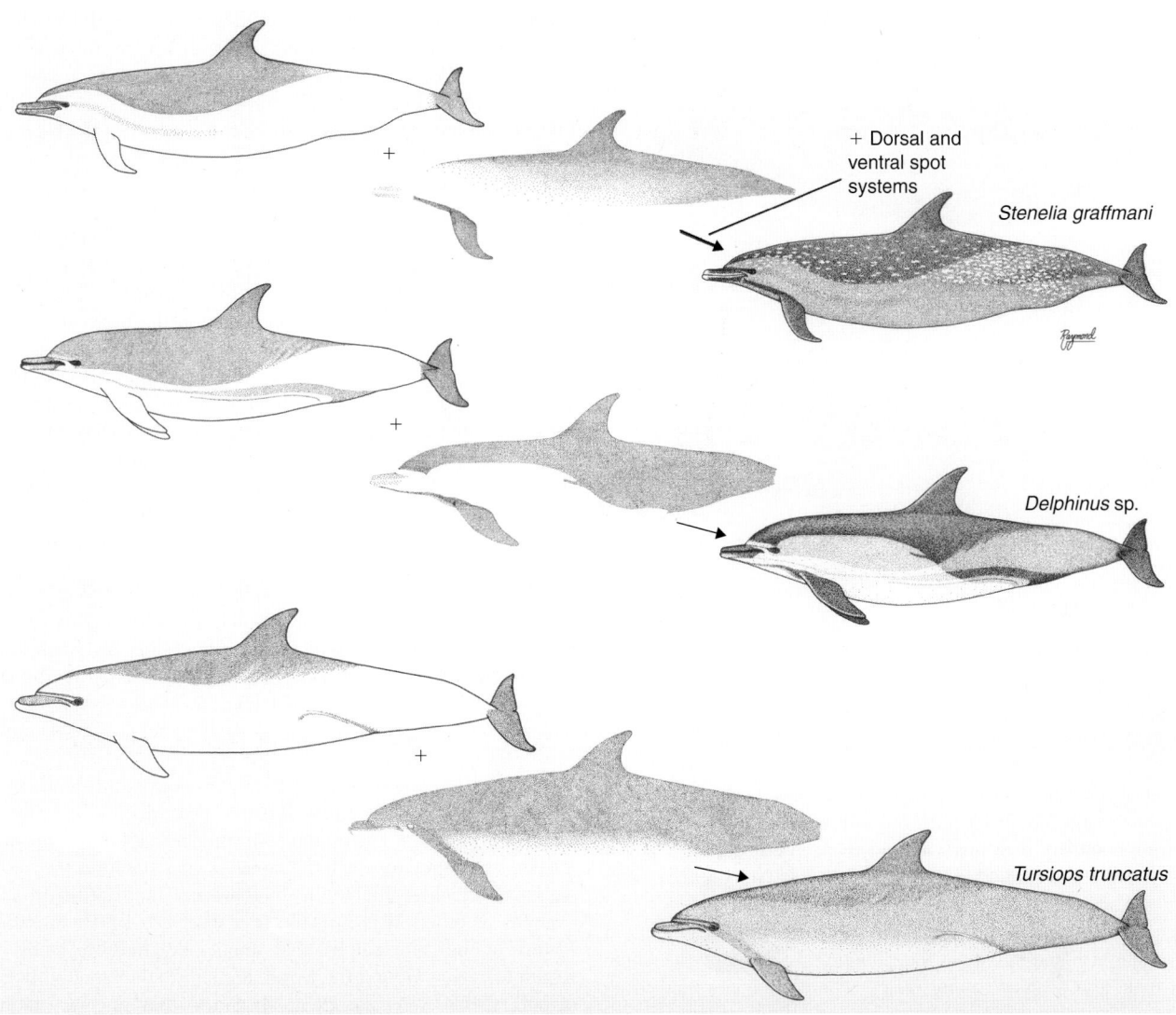

+ Dorsal and
ventral spot
systems

*Stenelia graffmani*

*Delphinus* sp.

*Tursiops truncatus*

**Figure 2** *Component analyses of color patterns of* Stenella attenuata, Delphinus delphis, *and* Tursiops truncatus *(from left to right): basic cape plus dorsal overlay yields a complex color pattern. From Perrin (1973).*

*cristata*, are unusual in that the lanugo is molted before birth (always in the hooded seal and usually in the harbor seal). The bearded seal, *Erignathus barbatus*, is born with a grayish-brown lanugo (with white muzzle and white blotches), and in the monk seals (*Monachus* spp.) and elephant seals (*Mirounga* spp.) the lanugo is black. In the southern phocids, the Weddell seal (*Leptonychotes weddellii*), Ross seal (*Ommatophoca rossii*), crabeater seal (*Lobodon carcinophaga*), and leopard seal (*Hydrurga leptonyx*), pups are born with pale-gray to brownish-gray coats; in the leopard seal the birth coat resembles the adult state in color and pattern. The walrus (*Odobenus rosmarus*) molts the lanugo *in utero* and has sparse whitish, yellowish, or silver-gray coat at birth. The lanugo, whether light or dark, is usually thick and wooly, and it has been suggested that it functions in heat conservation until a BLUBBER layer accumulates. Another suggested function, at least for the white lanugo, is camouflage against predators on the ice, although the southern ice-breeding monachines do not have white coats at birth. It has been posited that intrauterine loss of the lanugo in the harbor seal is a secondary adaptation to breeding on land since its descent from an ice-breeding ancestor, implying a camouflage function. Why some birth coats are light and others dark is still a matter for speculation.

## III. Sexual Dimorphism

In delphinid cetaceans, sexually dimorphic color-pattern elements are typically associated with the genital region. For example, a black tear-drop-shaped patch surrounding the genital slit in Commerson's dolphin, *Cephalorhynchus commersonii*, has its apex directed posteriorly in adult males and anteriorly (sometimes with a posterior invagination around the genital slit) in females (Fig. 3). A lateral stripe extending from the eye to the genital region in Fraser's dolphin, *Lagenodelphis hosei* is broader and darker in adult males than in females (Jefferson *et al.*, 1997). Adult male beaked whales (Ziphiidae) of many species tend to develop white areas on the head; in some species the entire head becomes white, whereas in others the white area may be confined to the front or top of the head or to the rostrum (Ridgway and Harrison, 1981–1999).

C

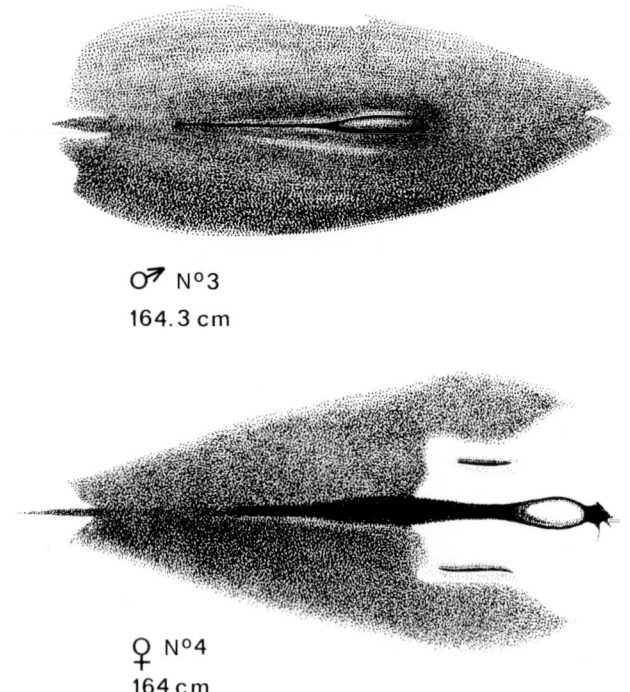

♂ N°3
164.3 cm

♀ N°4
164 cm

**Figure 3**  *Sexual dimorphism in the form of a genital patch in* Cephalorhynchus commersonii. *From Robineau (1984).*

Ventral coloration is dimorphic in the Mediterranean monk seal, *Monachus monachus*, from birth onward; as in some dolphins, the posterior boundary of a mark around the genital opening is arcuate in females and straight in males (Badosa and Grau, 1998). Adult males are also much darker than females. In other phocids, the color pattern is usually more distinct and with darker elements in males than in females, e.g., in the ribbon seal and harp seal. Some adult male gray seals are almost black, whereas females tend to be lighter colored. However, some species, e.g., the bearded seal and Weddell seal, are not noticeably dimorphic in coloration (Ridgway and Harrison, 1981). For walruses, the pattern seen in many phocids is reversed; old adult males tend to be lighter colored than females. The pattern is variable in the otariids; in some, e.g., the California sea lion, *Zalophus californianus*, and northern fur seal, *Callorhinus ursinus*, the adult male is darker than the female, whereas in others, e.g., the Steller sea lion, *Eumetopias jubatus*, there is no apparent dimorphism.

## IV. Geographic and Individual Variation

Color patterns vary among adults in marine mammal species, both individually and geographically. Biologists use individual variations in coloration and other natural marks as "tags" in studies of abundance, movements, and life history. "Mark and recapture" studies using natural marks can estimate population size; this has been applied to a number of cetacean species, including killer whales (*Orcinus orca*), minke whales (*Balaenoptera acutorostrata*), fin whales (*B. musculus*), blue whales (*B. physalus*), humpback whales (*Megaptera novaeangliae*), right whales (*Eubalaena* spp.), bowhead whales (*Balaena mysticetus*), bottlenose dolphins, and others (Hammond *et al.*, 1990). Long-range movements of migratory whales have been documented using natural marks, clarifying migratory cycles and stock structure and affiliations. Longitudinal studies of individuals and groups over

generational time have been vital in arriving at estimates of such important life history parameters as age at first reproduction, calving interval, and survivorship, and these have been made possible by the use of natural marks to keep track of individuals (for the Atlantic spotted dolphin, *Stenella frontalis*, in the Caribbean, the killer whale in the Pacific Northwest, and a variety of other small cetaceans). Examples of coloration features that have been or could be used as natural marks include shape of the dorsal saddle and postocular spot in the killer whale (Visser and Mäkäinen, 2000), details of the eye and blowhole stripes in dolphins and porpoises, color and patterning of the underside of the flukes in the humpback whale, and patterns of spots, blotches and white areas on the body in right, gray (*Eschrichtius robustus*), bowhead, blue, and minke whales. Some naturally imposed marks, such as scars from infraspecific fighting and from bites by predators or cookie cutter sharks (*Isistius* spp.), have also proven useful, although factors such as fading and acquisition of new marks during a study must be taken into consideration (Blackmer *et al.*, 2000). For some drably colored marine mammals, e.g., manatees (*Trichechus* spp.), scars (including those from collisions with boats and fishing gear) are the only marks available for use.

The use of natural marks in individually identifying pinnipeds from their patterns of spots and blotches has been complicated by the difficulty of photographing seals from a standard angle or in a standard posture. This problem has been approached by the development of computer-aided matching of images using a three-dimensional model to correct for orientation and posture (Hammond *et al.*, 1990).

As for other morphological characters, coloration tends to vary geographically most in those features that vary individually most within a population. For example, in spinner dolphins in the eastern Pacific, the degree to which the dorsal overlay obscures the underlying cape is highly variable; in some animals the cape is prominent whereas in others it is invisible (Perrin *et al.*, 1991). It is in this feature of the color pattern that spinner dolphins vary most from region to region around the world. Similarly, adult spotted dolphins of both species (*Stenella attenuata* and *S. frontalis*) vary individually in the degree of spotting, and average spotting varies geographically as well; the offshore Gulf Stream form of *S. frontalis* is all but unspotted (Perrin *et al.*, 1987). The *truei* and *dalli* color morphs of Dall's porpoise (Fig. 4), originally described as different species, typify regional populations but include the range of coloration in a single population (Ridgway and Harrison, 1981–1999), and an additional color-pattern morph has been recently discovered (Amano *et al.*, 2000). Baleen whales vary geographically in coloration of the baleen and in details of color pattern of the body and appendages, which also vary greatly within populations. Spotting is highly variable among individual harbor seals, and differences can be found between populations even on different islands in an archipelago (Hammond *et al.*, 1990). Humpback whales of the Southern Hemisphere tend to have more white pigmentation in their bodies than their Northern Hemisphere counterparts.

## V. Genetics

Little is known of the genetic basis of coloration in marine mammals. Albinism is possibly an autosomal dominant trait as in humans. Analysis of populational data for the North Atlantic right whale, *Eubalaena glacialis*, suggests that the presence of white ventral skin patches is an autosomal recessive trait (Schaeff and Hamilton, 1999); it is not evident from data that the trait is subject to selection pressure.

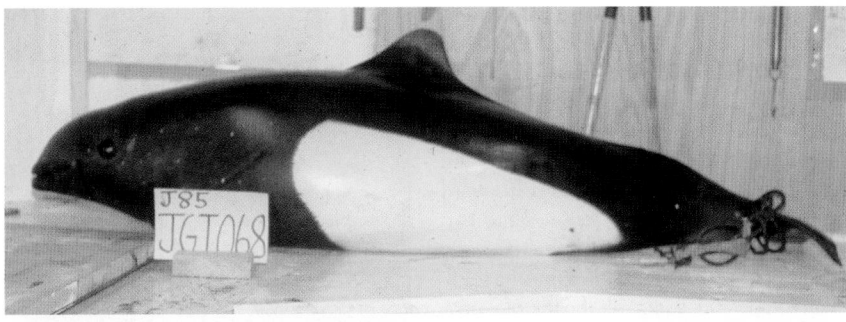

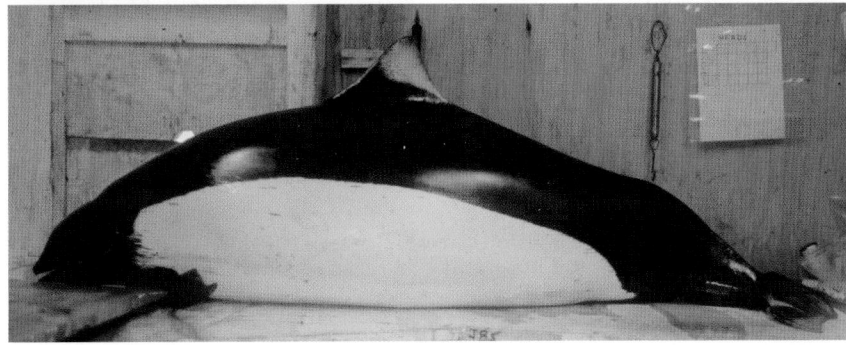

**Figure 4** *Two color morphs of Phocoenoides dalli: dalli type (A) and truei type (B).* *Photos courtesy of T. A. Jefferson.*

## VI. Microanatomy

Pigmentation in cetaceans is limited to the occurrence of melanin in the epidermis. The distribution of melanin in the skin of *Delphinus delphis* (Gwinn and Perrin, 1975) is described here as an example. The pigment is usually concentrated around the bases of the dermal papillae and extends in bands from their apices. Portions of the skin that appear white show very small amounts of diffuse pigment (particles unresolvable at 1250 × magnification) and small granules (<5 mm in diameter). The buff color characteristic of the thoracic patch is associated with an equal prominence of diffuse and small granules or higher density of diffuse pigment. The gray flank patch has some small granules but mostly large granules (5 mm). The black regions of the back and flukes have the highest density of large granules. It may be hypothesized that the type of melanin that produces the buff color is of a composition that does not allow further polymerization but favors its combination with a protein instead of aggregation into granules. In gray and black areas of the color pattern, the aggregation of particles proceeds, and melanocytes containing them migrate toward the surface of the epidermis until diffuse pigment is largely replaced by granular pigment.

## VII. Function and Evolution

An early analysis of coloration in cetaceans (Yablokov, 1963) proposed several functions: acquisition of prey, protection from predators, and communication with conspecifics. Cetacean patterns were divided into three types: (1) uniform or finely spotted, adapted to planktonic feeding or feeding in murky water or great depths where vision is not important; (2) strongly spotted, striped, or patterned, for intraspecific recognition; and (3) counter shaded, as camouflage against predators in animals foraging near the surface. However, spots would seem also to be useful in camouflage, as the surface of the sea appears dappled when seen from below. Notably, spotted dolphins are strongly counter shaded at birth and only begin to develop spots at about the age when they begin to forage on their own; in this case the likely function of the spots is camouflage against prey. This pattern of development is the reverse of that in many species of deer, which are spotted at birth and lose their spots as they become self-sufficient; there the function is camouflage against predators. The stripes and marks on the typical delphinid head (eye stripe, blowhole stripe, flipper stripe, eye spot, and lip mark) may also serve as camouflage against prey, obscuring the eye and mouth as in many terrestrial mammals and other tetrapods, including reptiles, amphibians, birds, and fishes. Many cetaceans have prominent white color-pattern elements or patches, often bordered with dark pigmentation; these may function in species recognition or serve to signal the positions of school mates in low-light conditions. Similar functions have been suggested for the bold color patterns seen in some phocid pinnipeds, together with possible uses in signaling sex and age. The patterns may also serve in disruptive coloration for camouflage against predator or prey.

The fin whale and dwarf minke whale (*Balaenoptera acutorostrata* unnamed subspecies) are unusual among marine mammals in being asymmetrically patterned. In the fine whale, the left anterior third of the body and the baleen on the left side are dark, but the lower jaw and the anterior baleen on the right side are white (Ridgway and Harrison, 1981–1999). The white area has been proposed to function to maintain counter shading when the whale rolls on its side during feeding or to startle prey during prey herding. A similar but lesser asymmetry obtains in the dwarf minke whale (Arnold *et al.*, 2005).

A simple counter-shaded pattern has been proposed to be the most primitive and generalized for the delphinid cetaceans (Mitchell, 1970) because it is a pattern shared by many taxa of marine organisms inhabiting near-surface waters and used for concealment through counter lighting. The crisscross pattern of *Delphinus* spp. is perhaps the most derived (as the most complex), with a possible function of obscuring the presence of a small calf swimming side by side with the mother.

**C**

The color pattern can evolve through paedomorphosis, the retention of fetal or juvenile characteristics into adulthood. The dark "hoods" on the heads of some dolphins and small-toothed whales may have evolved this way. The system of stripes on the head (eye stripe and blowhole stripe) in delphinid and phocoenid cetaceans develops from a single mark across the back of the head behind the blowhole in small fetuses (Perrin, 1997). A progressive forward invagination of the mark on each side of the head creates the two stripes, which are initially broad and then narrow to varying degrees. The stripes vary among species in width and definition, but in some species, e.g., *Cephalorhynchus* spp. and some of the globicephalinine species such as *Peponocephala electra*, a fetal condition is retained and the "blowhole stripe" effectively covers the entire top of the head (Fig. 1). It is interesting to note that the *Cephalorhynchus* species are also paedomorphic in their osteology, convergent on the phocoenids in body and skull size and shape.

Paedomorphosis may account for geographic variation in the expression of color-pattern elements in some species. For example, the cape is visible in the fetus of the killer whale but is usually not expressed in the postnatal animal. However, the cape is distinctly visible in all whales in an Antarctic population (Pitman and Ensor, 2003). Similarly, the adult of the Kerguelen Islands form of Commerson's dolphin, *Cephalorhynchus commersonii*, retains a grayish portion of the color pattern seen only in calves in the South American population (Robineau, 1984).

## VIII. Coloration as a Taxonomic Character

Color-pattern features are useful in taxonomy, even lending themselves to cladistic analysis (Perrin, 1997 for Delphinoidea; Arnold *et al.*, 2005 for Mysticeti). In some cases, what were thought to be color variants of a single species have proved to be distinct species [e.g., the common dolphins *Delphinus delphis* and *D. capensis* (Heyning and Perrin, 1994); and the minke whales, *Balaenoptera acutorostrata* and *B. bonaerensis* (Best, 1985)]. However, in other cases, species defined on the basis of color-pattern differences have been subsequently lumped [e.g., the striped dolphin, *Stenella coeruleoalba*, and its nominal synonyms (Fraser and Noble, 1970) and the *dalli* and *truei* forms of Dall's porpoise, *Phocoenoides dalli* (Ridgway and Harrison, 1999)], demonstrating that the same caution must be used in employing color-pattern characters as in the use of any other morphological characters. The contributions of ontogenetic, individual, and geographic variation must be delineated before species-level differences and higher level taxonomic relationships emerge.

## IX. Ephemeral and Anomalous Variation

Real and apparent changes in coloration can be ephemeral or environmentally induced. For example, some dolphins (and old walruses) are at some times pink and other times white. The Amazon river dolphin, *Inia geoffrensis*, is also known as "*bufeo colorado*" because of its pink color when seen in its natural habitat; in captivity in a temperate-latitude aquarium it is white. Some adult individuals of the Indo-Pacific humpback dolphin, *Sousa chinensis*, in the waters of Hong Kong have been observed to be bright pink (Fig. 5) whereas some other tropical dolphins sometimes exhibit pink bellies. This fact is even incorporated in the scientific name of one dolphin, the dwarf spinner dolphin of Southeast Asia, *Stenella longirostris roseiventris* (Perrin *et al.*, 1999, 2007). The pink color is due to dilation of subcutaneous blood vessels, presumably for purposes of thermoregulation; dolphins everywhere must dispose of excess heat generated by metabolism,

**Figure 5** *Adult Chinese humpback dolphin, illustrating pink coloration in response to excess heat load. Photo courtesy of T. A. Jefferson.*

which is a greater problem in warm tropical water. Even the Arctic walrus, *Odobenus rosmarus*, must "dump" heat, and palely pigmented old males may appear "rosy" during relatively warm weather. In cetaceans, the pink coloration is visible only in animals or parts of animals that are normally white (lacking melanin), although light-gray dolphins, such as immature individuals of *Sousa chinensis*, may sometimes appear purplish because of the subcutaneous suffusion with blood. When a "pink" walrus enters the water, the skin becomes ischemic (deprived of blood) and the animal appears white.

Other pinnipeds change color when wet; e.g., the brown California sea lion becomes almost black. Seasonally molting pinnipeds change color, becoming drabber as the molt approaches. Reddish pelage in harbor seals can be the result of deposition of iron oxide precipitates on the hair shaft. Even conditions such as water color, cloud cover, and angle of sun can cause cetaceans to apparently change color in the water. The striped dolphin, also known as the blue-and-white dolphin, can appear brown and white in turbid water or under overcast skies. A sojourn in high latitudes can lead to an accumulation of a coat of yellowish diatoms in the blue whale, *Balaenoptera musculus*, leading to one of its other common names, "sulfurbottom." In one minke whale taken in the Antarctic, the areas normally white were pinkish and the baleen, blubber, and connective tissue were orange; the cause of the "carotenoid" coloration was not apparent (Kato, 1979) but may have been metabolic. Cetaceans in captivity in shallow tanks can become darker when exposed to the sun. The intensity of coloration has been reported to vary seasonally in common dolphins in the Black Sea. In some cetacean species, apparent stripes or spots can be scars. Adult male beaked whales inflict long parallel rakes on each other with their teeth (the behavior also occurs in species in which the teeth erupt in females, although scarring is still usually more prevalent in males, e.g., in Amazon river dolphins, Martin and da Silva, 2006). Most cetaceans that frequent tropical waters bear oval or star-shaped scars from the bites of cookie cutter sharks; these can be so numerous as to give the animal an overall spotted appearance.

Anomalous conditions such as albinism (see chapter Albinism), melanism (Visser *et al.*, 2004) and pie-bald coloration have been recorded for many cetacean and pinniped species. In one anomalous color-pattern variety that has been seen in the short-beaked common dolphin in widely separated parts of the world, the dorsal overlay is missing, eliminating the typical crisscross pattern (Perrin *et al.*, 1995).

Coloration in cetaceans changes with death. Subtle elements of color pattern disappear quickly in a dead stranded animal as the skin dries, and a dolphin with a complex color pattern can turn solid black when long dead or frozen (careful thawing in water can sometimes bring back some of the pattern). The outer layers of the skin can be abraded away during stranding, which has resulted in more than one report of a stranded "white whale." The accounts of coloration in some original species descriptions of cetaceans were based on long-dead specimens and thus are defective (e.g., Heaviside's dolphin, *Cephalorhynchus heavisidii*, is not uniformly black as originally described, but boldly patterned). Much remains to be learned about the color pattern of the living animal for many species.

## See Also the Following Articles

Albinism ■ Geographic Variation ■ Mark and Recapture ■ Sexual Dimorphism ■ Species

## References

Amano, M., Marui, M., Guenther, T., Ohizumi, H., and Miyazaki, N. (2000). Re-evaluation of geographic variation in the white flank patch of dalli-type Dall's porpoises. *Mar. Mamm. Sci.* **16**, 631–636.

Arnold, P. W., Birtles, R. A., Dunstan, A., Lukoschek, V., and Matthews, M. (2005). Colour patterns of the dwarf minke whale *Balaenoptera acutorostrata* sensu lato: Description, cladistic analysis and taxonomic implications. *Mem. Queensland Mus.* **51**, 277–307.

Badosa, E., and Grau, E. (1998). Individual variation and sexual dimorphism of coloration in Mediterranean monk seal pups (*Monachus monachus*). *Mar. Mamm. Sci.* **14**, 390–393.

Best, P. B. (1985). External characters of southern minke whales and the existence of a diminutive form. *Sci. Rep. Whales Res. Inst. Tokyo* **36**, 1–33.

Blackmer, A. L., Anderson, S. K., and Weinrich, M. T. (2000). Temporal variability in features used to photo-identify humpback whales (*Megaptera novaeangliae*). *Mar. Mamm. Sci.* **16**, 338–354.

Bonner, W. N. (1990). "The Natural History of Seals." Facts On File, New York.

Fraser, F. C., and Noble, B. A. (1970). Variation of pigmentation pattern in Meyen's dolphin, *Stenella coeruleoalba* (Meyen). *Invest. Cetacea* **2**, 147–163, pl. 1–7.

Gwinn, S., and Perrin, W. F. (1975). Distribution of melanin in the color pattern of *Delphinus delphis* (Cetacea: Delphinidae). *Fish. Bull. US* **73**, 439–444.

Hammond, P. S., Mizroch, S. A., and Donovan, G. P. (eds) (1990). Individual recognition of cetaceans: Use of photo-identification and other techniques to estimate population parameters. *Rep. Int. Whal. Commn. spec. issue* **12**.

Heyning, J. E., and Perrin, W. F. (1994). Evidence for two species of common dolphins (genus *Delphinus*) from the eastern North Pacific. *Nat. Hist. Mus. Los Angeles County Contr. Sci.* **442**, 1–35.

Jefferson, T. A. (1988). Phocoenoides dalli. *Mamm. Species* **319**, 1–7.

Jefferson, T. A., Pitman, R. L., Leatherwood, S., and Dolar, M. L. L. (1997). Developmental and sexual variation in the external appearance of Fraser's dolphins (*Lagenodelphis hosei*). *Aquat. Mamm.* **23**, 145–153.

Kato, H. (1979). Carotenoid colored minke whale from the Antarctic. *Rep. Whales Res. Inst. Tokyo* **31**, 97–99, pl. 1.

Martin, A. R., and da Silva, V. M. F. (2006). Sexual dimorphism and body scarring in the boto (Amazon river dolphin) *Inia geoffrensis*. *Mar. Mamm. Sci.* **22**, 25–33.

Mitchell, E. (1970). Pigmentation pattern evolution in delphinid cetaceans: An essay in adaptive coloration. *Can. J. Zool.* **48**, 717–740.

Perrin, W. F. (1973). Color pattern of spinner porpoises (*Stenella cf. S. longirostris*) of the eastern tropical Pacific and Hawaii, with comments on delphinid pigmentation. *Fish. Bull. US* **70**, 983–1003.

Perrin, W. F. (1997). Development and homologies of head stripes in the delphinoid cetaceans. *Mar. Mamm. Sci.* **13**, 1–43.

Perrin, W. F., Mitchell, E. D., Mead, J. G., Caldwell, D. K., Caldwell, M. C., van Bree, P. J. H., and Dawbin, W. H. (1987). Revision of the spotted dolphins, *Stenella* spp. *Mar. Mamm. Sci.* **3**, 99–170.

Perrin, W. F., Akin, P. A., and Kashiwada, J. V. (1991). Geographic variation in external morphology of the spinner dolphin *Stenella longirostris* in the eastern Pacific and implications for conservation. *Fish. Bull. US* **89**, 411–428.

Perrin, W. F., *et al.* (1995). An anomalously pigmented form of the short-beaked common dolphin (*Delphinus delphis*) from the southwestern Pacific, eastern Pacific, and eastern Atlantic. *Mar. Mamm. Sci.* **11**, 241–247 (14 authors).

Perrin, W. F., Dolar, M. L. L., and Robineau, D. (1999). Spinner dolphins (*Stenella longirostris*) of the western Pacific and Southeast Asia: Pelagic and shallow-water forms. *Mar. Mamm. Sci.* **15**, 1029–1053.

Perrin, W. F., Aquino, M. T., Dolar, M. L. L., and Alava, M. N. R. (2007). External appearance of the dwarf spinner dolphin *Stenella longirostris roseiventris*. *Mar. Mamm. Sci.* **23**, 464–467.

Pitman, R. L., and Ensor, P. (2003). Three forms of killer whales (*Orcinus orca*) in Antarctic waters. *J. Cetacean Res. Manage.* **5**, 131–139.

Ridgway, S. H., and Harrison, R. J. (eds) (1981–1999). "Handbook of Marine Mammals," Vols 1–6. Academic Press, San Diego.

Robineau, D. (1984). Morphologie externe et pigmentation du dauphin de Commerson, *Cephalorhynchus commersonii* (Lacépède, 1804), en particulier celui des Îles Kerguelen. *Can. J. Zool.* **62**, 2465–2475.

Schaeff, C. M., and Hamilton, P. K. (1999). Genetic basis and evolutionary significance of ventral skin color markings in North Atlantic right whales (*Eubalaena glacialis*). *Mar. Mamm. Sci.* **15**, 701–711.

Visser, I. N., and Mäkäinen, P. (2000). Variation in eye-patch shape of killer whales (*Orcinus orca*) in New Zealand waters. *Mar. Mamm. Sci.* **16**, 459–469.

Visser, I. N., Fertl, D., and Pusser, L. T. (2004). Melanistic southern right-whale dolphins (*Lissodelphis peronii*) off Kaikura, New Zealand, with records of other anomalously all-black cetaceans. *NZ J. Mar. Freshw. Res.* **38**, 833–836.

Yablokov, A. V. (1963). Types of color of the Cetacea. *Byul. Morsk. Obshch. Ispytat. Pri. (Ot. Biol.)* **68**(6), 27–41 [In Russian].

# Common Bottlenose Dolphin
## *Tursiops truncatus*

RANDALL S. WELLS AND MICHAEL D. SCOTT

### I. Characters and Taxonomy

Bottlenose dolphins (*Tursiops truncatus*) are arguably the best known of all cetaceans. They figured prominently in the legends of the ancient Greeks and Romans and were described in the writings of Aristotle, Oppian, and Pliny the Elder. Several books for scientific and public audiences have focused on this species (Caldwell and Caldwell, 1972; Shane, 1988; Leatherwood and Reeves, 1990; Thompson and Wilson, 1994; Reynolds *et al.*, 2000), and a number of comprehensive review articles have been produced as well (Tomilin, 1957; Leatherwood and Reeves, 1982; Shane *et al.*, 1986; Wells and Scott, 1999). The name *Tursiops* can be translated as "dolphin-like," deriving from the Latin *Tursio* ("dolphin") and the Greek suffix *-ops* ("appearance"); *truncatus* derives from the Latin

**Figure 1** *Lateral view of an adult male bottlenose dolphin* (Tursiops truncatus, *photo by R. S. Wells*).

**Figure 2** *Ventral view of a bottlenose dolphin* (Tursiops truncatus, *photo by R. S. Wells*).

*trunco-* ("truncated"), apparently referring to the flattened teeth used by Montagu (1821) as an identifying characteristic. The common English name is "common bottlenose dolphin," distinguishing this species from the Indo-Pacific bottlenose dolphin, *T. aduncus*. It is still often referred to as "porpoise" in the southeastern USA.

Though no conclusive fossil evidence of the origin of *Tursiops* exists, fossil records extend back several million years (Barnes, 1990). The geographical distribution of the fossils falls within the range of the modern animals. Anatomical features suggest that *Tursiops* evolved from some ancestral group of extinct fossil Delphininae, perhaps related to the subfamily Steninae, which might have evolved from the Kentriodontidae.

Common bottlenose dolphins are cosmopolitan in distribution, and demonstrate a great deal of geographical variation in morphology. *T. truncatus* is found in most of the world's warm temperate to tropical seas, in coastal as well as offshore waters. They are recognizable by their generalized appearance—a medium-size, robust body, a moderately falcate dorsal fin, and dark coloration, with a sharp demarcation between the melon and the short rostrum (Figs 1, 2). Adult lengths range from about 2.5 m to about 3.8 m, varying by geographic location (Mead and Potter, 1990; Read *et al.*, 1993). Body size appears to vary inversely with water temperature in many parts of the world, but not the eastern Pacific. Bottlenose dolphins are colored light gray to black dorsally and laterally, with a light belly (Fig. 2). A light blaze or brush

marking is sometimes observed on their sides. A distinct cape may be visible or may be obscured when the color pattern is very dark.

Variation in size, coloration, and cranial characteristics associated with feeding have led to descriptions of at least 20 nominal species of *Tursiops* (Hershkovitz, 1966; Rice, 1998). Recognition of the polymorphic nature of *Tursiops* and the existence of clinal variation had led to general agreement for many years that *Tursiops* was a single-species genus (Tomilin, 1957; Mitchell, 1975; Honacki *et al.*, 1982). However, recent genetic, morphologic, and physiologic studies suggest that revision of the genus may be necessary to acknowledge significant differences between forms from different oceans, as well as differences between forms in inshore vs offshore habitats within ocean basins (Hersh and Duffield, 1990; LeDuc *et al.*, 1999; Mead and Potter, 1995; Rice, 1998). Inshore bottlenose dolphins in the Atlantic and some other regions tend to be smaller, lighter in color, have proportionately larger flippers, and differ in hematologic and mitochondrial DNA features from offshore forms (Hersh and Duffield, 1990; LeDuc *et al.*, 1999); however, eastern Pacific offshore bottlenose dolphins are smaller and darker than inshore forms. The taxonomic status of *Tursiops* is made even more confusing by observations of hybridization with several other odontocete species (Sylvestre and Tanaka, 1985).

## II. Distribution and Abundance

Common bottlenose dolphins are found in temperate and tropical marine waters around the world, with an estimated 600,000 animals world-wide (Fig. 3). In the North Pacific, they are commonly found as far north as the southern Okhotsk Sea, the Kuril Islands, and central California. In the North Atlantic, they are seen inshore during summer months off New England, offshore as far north as Nova Scotia, and they have been recorded off Norway and the Lofoten Islands. Bottlenose dolphins occur as far south as Tierra del Fuego, South Africa, Australia, and New Zealand. Limits to the species' range appear to be temperature related, either directly or indirectly, through distribution of prey. Off the coasts of North America they tend to inhabit waters with surface temperatures ranging from about 10°C to 32°C. At the northern limit of the species' range in the western North Atlantic, they are seasonally migratory, with a more southerly distribution in the winter.

## III. Ecology

*Tursiops* inhabits most warm temperate and tropical shorelines, adapting to a variety of marine and estuarine habitats, even ranging into rivers. Common bottlenose dolphins are primarily coastal, but are also found in pelagic waters, near oceanic islands, and over the continental shelf, especially along the shelf break. In the Indian Ocean, *T. truncatus* tends to inhabit offshore waters, whereas *T. aduncus* is the more-common coastal species.

The diets of common bottlenose dolphins have been described from many regions (Barros and Odell, 1990). A large variety of fish and/or squid forms most of the diets, although bottlenose dolphins seem to show a consistent preference for sciaenids, scombrids, and mugilids. Most fish prey are bottom-dwellers, but some surface-dwellers or pelagic fish are also represented in the diets. Noise-producing fish make up a large part of the *Tursiops* diet, presumably because sound helps the dolphins to locate prey (Barros and Wells, 1998; Gannon *et al.*, 2005). Differences in diets have been found where both inshore and offshore *Tursiops* ecotypes have been identified. Across a population, common bottlenose dolphins may appear to be generalists with regards to prey, but individuals within the

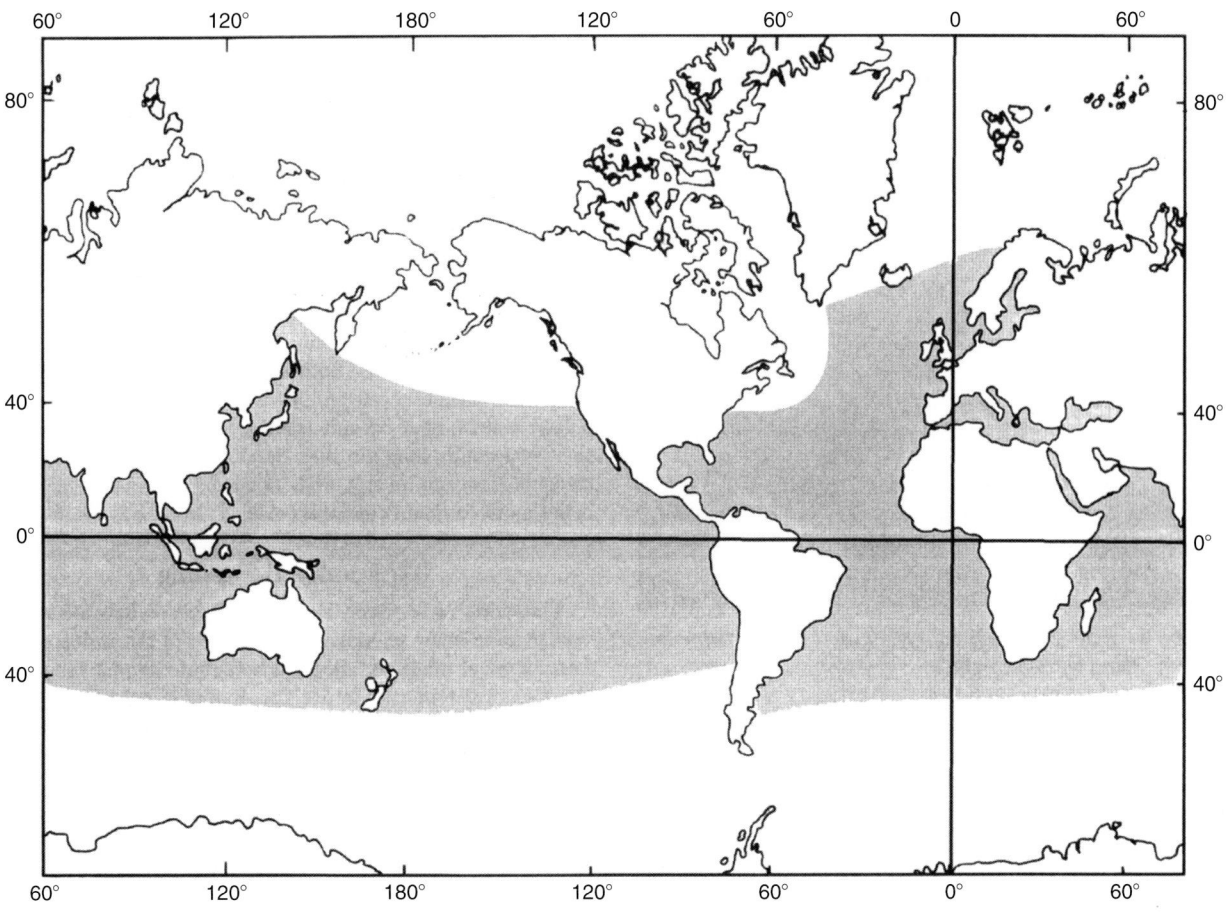

**Figure 3**   *Species range of the common bottlenose dolphin (*Tursiops truncatus*).*

population may show some degree of specialization. In some cases bottlenose dolphin groups feed in different areas depending on sex and size, with lactating females and their calves frequenting and feeding in the near-shore zone, adolescents feeding slightly farther offshore, and resting females and adult males feeding farther still.

Sharks are probably the most important predators of bottlenose dolphins, although killer whales (*Orcinus orca*) may also occasionally prey on them as well. Mutual tolerance during encounters between sharks and dolphins is probably typical, but as many as half of all bottlenose dolphins bear shark-bite scars as evidence of occasional encounters, depending on the region. In at least some areas, *Tursiops* appears to be a relatively minor and occasional part of the diets of sharks. Most wounds and scars from sharks tend to be found on the posterior and ventral regions of the dolphins, suggesting that the dolphins were ambushed from behind and below; some attacks may have been something other than a predation attempt (e.g., sharks defending a territory). The primary shark predators of common bottlenose dolphins are the bull shark (*Carcharhinus leucas*), tiger shark (*Galeocerdo cuvier*), great white shark (*Carcharodon carcharias*), and dusky shark (*Carcharhinus obscurus*) (Wood *et al.*, 1970). Observations of captive dolphins suggest that they may recognize certain species of sharks as potential threats (McBride and Hebb, 1948; Irvine *et al.*, 1973).

Anecdotal accounts describe common bottlenose dolphins attacking sharks by butting them with their rostra or by striking them with their flukes (summarized by Wood *et al.*, 1970). Defense may explain the apparently high survival rate indicated by the shark-bite scars on

living dolphins. The relatively infrequent occurrence of shark-bite scars on young dolphins indicates either that the calves are well protected by their mothers, or that attacks on young dolphins are generally fatal (Wells *et al.*, 1987; Cockcroft *et al.*, 1989a).

Stingrays are an increasing source of mortality for common bottlenose dolphins in some areas (Walsh *et al.*, 1988). The dolphins were wounded externally, or internally from ingestion of small rays, and deaths resulted from physical trauma as the barb migrated and penetrated vital organs, causing infection or toxicosis.

## IV. Behavior and Physiology

Coastal common bottlenose dolphins exhibit a full spectrum of movements, including seasonal migrations, year-around home ranges, periodic residency, and a combination of occasional long-range movements and repeated local residency (Shane *et al.*, 1986; Wells and Scott, 1999). Much less is known about the ranging patterns of pelagic bottlenose dolphins. In some places, coastal dolphins living at the high-latitude or cold-water extremes of the species' range may migrate seasonally, as is the case along the Atlantic coast of the USA. Long-term residency has been reported from many parts of the world and may take the form of a relatively permanent home range or repeated occurrence in a given area over many years. For example, the year-round residents of several dolphin communities along Florida's west coast have maintained relatively stable, slightly overlapping home ranges during more than 37 years of observations and through at least five generations; seasonal changes in habitat

**C**

use may occur within the ranges (Scott *et al.*, 1990; Wells and Scott, 1999). Nearby ranges sometimes can be distinguished by genetic differences (Duffield and Wells, 1991, 2002; Sellas *et al.*, 2005; Parsons *et al.*, 2006). Home range bounds are often demarcated by physiographic features such as passes or abrupt changes in water depth. Some dolphins may use seasonal home ranges joined by a traveling range.

Longer-distance movements have been reported for some coastal common bottlenose dolphins, including range shifts of several hundred kilometers in apparent response to environmental changes such as an El Niño warm-water event (Wells *et al.*, 1990) and a 600-km roundtrip for several identifiable dolphins in Argentina (Würsig and Würsig, 1979). Average daily movements of 33–89 km, monitored through travel distances of as much as 4200 km, have been reported for bottlenose dolphins in offshore waters (Tanaka, 1987; Wells *et al.*, 1999a).

Common bottlenose dolphins are typically found in groups of 2–15 individuals, although groups of more than 1000 have been reported (Leatherwood and Reeves, 1982; Shane *et al.*, 1986; Scott and Chivers, 1990). In general, bottlenose dolphins in bays and estuaries tend to form smaller groups than those in offshore waters, but the trend does not continue linearly with increasing distance from shore (Wells *et al.*, 1999). Group composition tends to be dynamic, with sex, age, reproductive condition, familial relationships, and affiliation histories appearing to be the most important determining factors (Wells, 2003). Subgroupings may be stable or repeated over periods of years. Basic social units include nursery groups, mixed sex groups of juveniles, and adult males as individuals or strongly bonded pairs (Wells *et al.*, 1987; Owen *et al.*, 2002; Rogers *et al.*, 2004). Females bear calves sired by multiple males over the course of a lifetime; long-term monogamous bonds have not been observed (Duffield and Wells, 2002; Wells, 2003).

Dominance hierarchies have been observed in captivity, with large adult male common bottlenose dolphins dominating all other pool-mates, females forming a less-rigid hierarchy, with the largest females dominant over smaller animals. Aggressive behaviors including contact and posturing are used to establish and maintain hierarchies. Serious agonistic interactions have been noted in the wild between male conspecifics (Parsons *et al.*, 2003) as well as with other species of dolphins (Herzing *et al.*, 2003).

Bottlenose dolphins in the wild appear to be active both during the day and at night, interspersing bouts of feeding, traveling, socializing, and idling or resting (Shane *et al.*, 1986; Wells *et al.*, 1999b). The duration and frequency of activities are influenced by such environmental factors as season, habitat, time of day, and tidal state, and by physiological factors such as reproductive seasonality. Bottlenose dolphins feed in a large variety of ways and habitats, primarily as individuals, but cooperative herding of schools of prey fish also occurs (Leatherwood, 1975). Individual prey capture involves behaviors as diverse as high-speed chases with a pin-wheeling capture at the surface, "fishwhacking" in which a fleeing fish is struck with the dolphin's flukes and often knocked clear of the water, pushing fish onto shore and then partially beaching to capture them, creating bubble bursts to drive prey to the surface, and herding and perhaps disorienting fish with percussive leaps and tail lobs referred to as "kerplunking." Calves apparently learn foraging specializations from their mothers, and patterns may spread through a population from observation, as an indication of cultural transmission of knowledge (Mann and Sargeant, 2003; Wells, 2003).

Common bottlenose dolphins in shallow habitats tend to make relatively brief dives, surfacing on average twice every minute. In deep water habitats, such as the Bermuda Pedestal, dives to more than 500 m and for longer than 5 min have been documented, correlating with reported nightly vertical migrations of mesopelagic prey (Klatsky *et al.*, 2007).

Common bottlenose dolphins produce three categories of sounds: whistles, echolocation clicks, and burst-pulse sounds. Dolphins produce a large variety of whistles, including largely stereotypic "signature whistles" that are individually specific and appear to be used to communicate identity, location, and possibly emotional state (Caldwell *et al.*, 1990; Sayigh *et al.*, 1990; Janik *et al.*, 2006). Once the signature whistle develops in neonates, it remains stable for many years. The signature whistles of many male calves are similar to the whistles of their mothers while those of female calves are not. Recent studies have demonstrated that common bottlenose dolphins spread over distances of hundreds of meters to kilometers may remain in acoustic contact with one another through whistles (Quintana-Rizzo *et al.*, 2006). Dolphin echolocation involves the production of "clicks," with peak frequencies of about 40–130 kHz (Au, 1993). Echolocation is hypothesized to be used in navigation, foraging, and predator detection, among other possible functions (Nowacek, 2005; Quintana-Rizzo *et al.*, 2006). Burst-pulses ("squawks") tend to be produced during social interactions.

## V. Life History

Analyses of dentinal and cemental growth layer groups in teeth (Hohn *et al.*, 1989) have shown that female common bottlenose dolphins can live to more than 57 years, and males up to 48 years (Wells and Scott, 1999). Calves achieve most of their growth during the period of suckling, i.e., the first 1.5 to 2 years of life. Females typically reach sexual and physical maturity before males, leading to sexual dimorphism in some regions. Age at sexual maturity varies by region, but in general females usually reach sexual maturity at 5 to 13 years. Sexual maturity for males tends to occur at 9 to 14 years; paternity testing of common bottlenose dolphins in Sarasota Bay, Florida has shown that males 13 to 40 years of age have sired offspring (Duffield and Wells, 2002; Wells, 2003).

Although births have been reported from all seasons, calving tends to be diffusely seasonal, with peaks during spring–summer months (Urian *et al.*, 1996; Thayer *et al.*, 2003). Hormonal monitoring of captive common bottlenose dolphins indicates that females are spontaneous sporadic ovulators, ovulating repeatedly during a given season, while males may be active throughout the year with prolonged elevation of testosterone concentrations over the months that different females may be ovulating. The reproductive lifespan is prolonged; females up to 48 years of age have successfully given birth and raised young (Wells and Scott, 1999). Calves are born after a gestation period of about 1 year and range in length from about 84 to 140 cm (Perrin and Reilly, 1984), depending on the geographic region.

Lactation is the primary source of nutrition for the first year of life and may continue for several more years. Solid food has been found along with milk in the stomachs of calves as young as 4 months old. Maternal investment for free-ranging calves typically extends for about 3–6 years, with separation often coinciding with the birth of the next calf (Wells and Scott, 1999). Simultaneously pregnant and lactating females have been noted on occasion.

Common bottlenose dolphin mortality results from both natural and anthropogenic sources (Wells and Scott, 1999). Natural causes include old age, failure to thrive as a calf, intra-specific agonistic interactions, predation, stingray barb wounds, disease, and biotoxins from harmful algal blooms such as red tides. Threats of human origin include entanglement in nets, entanglement in or ingestion of recreational fishing gear, pollution, boat collisions, noise, and drive or harpoon fisheries. In some cases, the distinction between natural and anthropogenic causes of death become blurred, e.g., when it is suspected that exposure to toxic environmental contaminants may increase a dolphin's susceptibility to natural pathogens or biotoxins.

## VI. Interactions with Humans

Common bottlenose dolphins take advantage of human activities to facilitate prey capture in a variety of ways. In Mauritania and Brazil, dolphins regularly drive schools of mullet toward fishermen wading with nets in shallow water, and in many parts of the world dolphins collect discarded fish from behind shrimp trawls and small purse seines, or steal fish from various types of fishing gears.

Common bottlenose dolphins were publicly displayed first at the Brighton Aquarium in 1883, then at the New York Aquarium in 1914, and have been a regular attraction at Marineland of Florida since 1938. *Tursiops* continues to be the most common dolphin species maintained in collections and displayed throughout the world. According to a May 2000 National Marine Fisheries Service inventory, 35 US facilities held 392 common bottlenose dolphins; in addition, several hundred bottlenose dolphins were held in at least 16 other countries. Within the US, approximately 70% of the dolphins are held primarily for public display, whereas the remainder are used primarily for research or military purposes. Improved facilities and increased knowledge about the requirements for care of dolphins have led to increasing success in the long-term maintenance of the animals, to the point where birth and survivorship rates at the better facilities approach and, possibly in a few cases, surpass those of wild populations (DeMaster and Drevenak, 1988; Wells and Scott, 1990).

The largest of the historical fisheries for common bottlenose dolphins involved several countries surrounding the Black Sea, where dolphins were caught for oil, meat, and leather. Because of declines in dolphin populations, these countries have since outlawed the fishery. Directed takes still occur in other parts of the world, such as Peru, Sri Lanka, Faroe Islands, and Japan for human consumption, to reduce the perceived competition with commercial fisheries, or for bait (Wells and Scott, 1999). Live-capture fisheries for dolphins for public display have existed for more than 100 years. More than 1500 *Tursiops* were removed from the waters of the US, Mexico, and the Bahamas by 1980 for display, research, or military applications in many parts of the world (Leatherwood and Reeves, 1982). Though no bottlenose dolphins have been collected in US waters since 1989, some live-capture fisheries continue in Cuba, the Solomon Islands, Russia, and Japan.

Incidental catches of small numbers of *T. truncatus* have been reported for several fisheries, including purse-seine fisheries for tunas, sardines, and anchovetas (Wells and Scott, 1999). In some cases dolphins have been killed by fishermen to prevent damage to their fishing gear or stealing of the catch or bait (Leatherwood and Reeves, 1982). In the US, entanglement in, or ingestion of recreational fishing gear is resulting in increasing numbers of common bottlenose dolphin mortalities.

Though the impacts of habitat alteration and pollution on dolphins have not been studied systematically, anecdotal accounts suggest that human-caused degradation may have led to declines in some dolphin populations (Wells and Scott, 1999). Extremely high concentrations of chlorinated hydrocarbon residues have been found in the tissues of *Tursiops* in many parts of the world, with males accumulating higher concentrations than females with age (O'Shea, 1999; Wells et al., 2005). Cockcroft et al. (1989b) suggested that

first-born calves of South African bottlenose dolphins, identified by the authors as *T. truncatus*, received 80% of their mother's body burden of contaminant residues (polychlorinated biphenyls and dieldrin), perhaps leading to increased neonatal mortality but also reducing levels of contaminants in the mothers. Studies along the west coast of Florida monitoring contaminant concentrations in identifiable resident dolphins supported these findings, with high concentrations in first-born calves and nulliparous females and lowered levels on females while lactating (Wells et al., 2005). Accumulation of contaminants in tissues of males have reached levels that theoretically could impair testosterone production and thus reduce reproductive ability. Preliminary findings suggest that even relatively low levels of PCBs and DDT metabolites can result in a decline in bottlenose dolphin immune system function, health, and reproductive success (Lahvis et al., 1995; Schwacke et al., 2002). Other anthropogenic chemical contaminants, such as perfluoroalkyl compounds, are of emerging concern (Houde et al., 2006).

Responses to other human use of dolphin habitat through dolphin feeding and swimming with dolphins are receiving increased attention (Samuels and Bejder, 2004; Cunningham-Smith et al., 2006). It is clear that common bottlenose dolphins suffer mortality and serious injury from collisions with boats (Wells and Scott, 1997), relative abundance may decline with increased dolphin-watching tours (Bedjer et al., 2006), and behaviors such as dive patterns, heading, synchronicity, aerial behaviors, and communications may change in the presence of vessels (Janik and Thompson, 1996; Nowacek et al., 2001; Hastie et al., 2003; Buckstaff, 2004; Goodwin and Cotton, 2004; Mattson et al., 2005; Lusseau, 2006).

Although there are many threats operating on local populations, *T. truncatus* is widespread and abundant, and as a species it does not appear to merit concern for major global population decline. Therefore, the common bottlenose dolphin has been classified by the IUCN as "Least Concern."

## See Also the Following Articles

Delphinids, Overview ■ Indo-Pacific ■ Bottlenose Dolphin

## References

Au, W. L. (1993). "Sonar of Dolphins." Springer Verlag, New York.
Barnes, L. G. (1990). The fossil record and evolutionary relationships of the genus *Tursiops*. In "The Bottlenose Dolphin" (S. Leatherwood, and R. R. Reeves, eds), pp. 3–26. Academic Press, San Diego.
Barros, N. B., and Odell, D. K. (1990). Food habits of bottlenose dolphins in the southeastern United States. In "The Bottlenose Dolphin" (S. Leatherwood, and R. R. Reeves, eds), pp. 309–328. Academic Press, San Diego.
Barros, N. B., and Wells, R. S. (1998). Prey and feeding patterns of resident bottlenose dolphins (*Tursiops truncatus*) in Sarasota Bay, Florida. *J. Mammal.* **79**, 1045–1059.
Bedjer, L., et al. (10 authors) (2006). Decline in relative abundance of bottlenose dolphins exposed to long-term disturbance. *Conserv. Biol.* **20**, 1791–1798.
Buckstaff, K. C. (2004). Effects of watercraft noise on the acoustic behavior of bottlenose dolphins, *Tursiops truncatus*, in Sarasota Bay, Florida. *Mar. Mamm. Sci.* **20**, 709–725.
Caldwell, D. K., and Caldwell, M. C. (1972). "The World of the Bottlenosed Dolphin." J. B. Lippincott Co, Philadelphia.
Caldwell, M. C., Caldwell, D. K., and Tyack, P. L. (1990). Review of the signature-whistle hypothesis for the Atlantic bottlenose dolphin. In "The Bottlenose Dolphin" (S. Leatherwood, and R. R. Reeves, eds), pp. 199–234. Academic Press, San Diego.

Cockcroft, V. G., Cliff, G., and Ross, G. J. B. (1989a). Shark predation on Indian Ocean bottlenose dolphins *Tursiops truncatus* off Natal, South Africa. *South Afr. J. Zool.* **24**, 305–309.

Cockcroft, V. G., De Kock, A. C., Lord, D. A., and Ross, G. J. B. (1989b). Organochlorines in bottlenose dolphins *Tursiops truncatus* from the east coast of South Africa. *South Afr. J. Mar. Sci.* **8**, 207–217.

Cunningham-Smith, P., Colbert, D. E., Wells, R. S., and Speakman, T. (2006). Evaluation of human interactions with a provisioned wild bottlenose dolphin (*Tursiops truncatus*) near Sarasota Bay, Florida, and efforts to curtail the interactions. *Aquat. Mamm.* **32**, 346–356.

DeMaster, D. P., and Drevenak, J. K. (1988). Survivorship patterns in three species of captive cetaceans. *Mar. Mamm. Sci.* **4**, 297–311.

Duffield, D. A., and Wells, R. S. (1991). The combined application of chromosome, protein and molecular data for investigation of social unit structure and dynamics in *Tursiops truncatus. In* "Genetic Ecology of Whales and Dolphins" (A. R. Hoelzel, ed.) pp. 155–169. *Rep. Int. Whal Commn Spec. Iss.* No. 13, Cambridge.

Duffield, D. A., and Wells, R. S. (2002). The molecular profile of a resident community of bottlenose dolphins, *Tursiops truncatus. In* "Molecular and Cell Biology of Marine Mammals" (C. J. Pfeiffer, ed.), pp. 3–11. Krieger Publishing Company, Melbourne.

Gannon, D. P., Barros, N. B., Nowacek, D. P., Read, A. J., Waples, D. M., and Wells, R. S. (2005). Prey detection by bottlenose dolphins (*Tursiops truncatus*): An experimental test of the passive-listening hypothesis. *Anim. Behav.* **69**, 709–720.

Goodwin, L., and Cotton, P. A. (2004). Effects of boat traffic on the behaviour of bottlenose dolphins (*Tursiops truncatus*). *Aquat. Mamm.* **30**, 279–283.

Hastie, G. D., Wilson, B., Tufft, L. H., and Thompson, P. M. (2003). Bottlenose dolphins increase breathing synchrony in response to boat traffic. *Mar. Mamm. Sci.* **19**, 74–84.

Hersh, S. L., and Duffield, D. A. (1990). Distinction between Northwest Atlantic offshore and coastal bottlenose dolphins based on hemoglobin profile and morphometry. *In* "The Bottlenose Dolphin" (S. Leatherwood, and R. R. Reeves, eds), pp. 129–139. Academic Press, San Diego.

Hershkovitz, P. (1966). "Catalog of Living Whales." *US Natl. Mus. Bull.* 246. Smithsonian Institution, Washington, DC.

Herzing, D. L., Moewe, K., and Brunnick, B. J. (2003). Interspecies interactions between Atlantic spotted dolphins, *Stenella frontalis* and bottlenose dolphins, *Tursiops truncatus*, on Great Bahama Bank, Bahamas. *Aquat. Mamm.* **29**, 335–341.

Hohn, A. A., Scott, M. D., Wells, R. S., Sweeney, J. C., and Irvine, A. B. (1989). Growth layers in teeth from known-age, free-ranging bottlenose dolphins. *Mar. Mamm. Sci.* **5**, 315–342.

Honacki, J. H., Kinman, K. E., and Koeppl, J. W. (1982). "Mammal Species of the World." Allen Press, Inc. and the Association of Systematics Collections, Lawrence.

Houde, M., *et al.* (8 authors) (2006). Biomagnification of perfluoroalkyl compounds in the bottlenose dolphin (*Tursiops truncatus*) food web. *Environ. Sci. Technol.* **40**, 4138–4144.

Irvine, B., Wells, R. S., and Gilbert, P. W. (1973). Conditioning an Atlantic bottle-nosed dolphin, *Tursiops truncatus*, to repel various species of sharks. *J. Mammal.* **54**, 503–505.

Janik, V. M., and Thompson, P. M. (1996). Changes in surfacing patterns of bottlenose dolphins in response to boat traffic. *Mar. Mamm. Sci.* **12**, 597–602.

Janik, V., Sayigh, L. S., and Wells, R. S. (2006). Signature whistle shape conveys identity information to bottlenose dolphins. *Proc. Natl. Acad. Sci. USA* **103**, 8293–8297.

Klatsky, L. J., Wells, R. S., and Sweeney, J. C. (2007). Offshore bottlenose dolphins (*Tursiops truncatus*): Movement and dive behavior near the Bermuda Pedestal. *J. Mammal.* **88**, 59–66.

Lahvis, G. P., Wells, R. S., Kuehl, D. W., Stewart, J. L., Rhinehart, H. L., and Via, C. S. (1995). Decreased lymphocyte responses in free-ranging bottlenose dolphins (*Tursiops truncatus*) are associated with increased concentrations of PCBs and DDT in peripheral blood. *Environ. Health Perspect.* **103**, 67–72.

Leatherwood, S. (1975). Some observations of feeding behavior of bottlenosed dolphins (*Tursiops truncatus*) in the northern Gulf of Mexico and (*Tursiops cf.* gilli) off southern California, Baja California, and Nayarit, Mexico. *Mar. Fish. Rev.* **37**, 10–16.

Leatherwood, S., and Reeves, R. R. (1982). Bottlenose dolphin (*Tursiops truncatus*) and other toothed cetaceans. *In* "Wild Mammals of North America: Biology, Management, Economics" (J. A. Chapman, and G. A. Feldhamer, eds), pp. 369–414. Johns Hopkins University Press, Baltimore.

Leatherwood, S., and Reeves, R. R. (1990). "The Bottlenose Dolphin." Academic Press, San Diego.

LeDuc, R. G., Perrin, W. F., and Dizon, A. E. (1999). Phylogenetic relationships among the delphinid cetaceans based on full cytochrome *b* sequences. *Mar. Mamm. Sci.* **15**, 619–648.

Lusseau, D. (2006). The short-term behavioral reactions of bottlenose dolphins to interactions with boats in Doubtful Sound, New Zealand. *Mar. Mamm. Sci.* **22**, 802–818.

Mann, J., and Sargeant, B. (2003). Like mother, like calf: The ontogeny of foraging traditions in wild Indian Ocean bottlenose dolphins (*Tursiops* sp.). *In* "The Biology of Traditions: Models and Evidence" (D. Fragaszy, and S. Perry, eds), pp. 236–266. Cambridge University Press, Cambridge.

Mattson, M. C., Thomas, J. A., and St. Aubin, D. (2005). Effects of boat activity on the behavior of bottlenose dolphins (*Tursiops truncatus*) in waters surrounding Hilton Head Island, South Carolina. *Aquat. Mamm.* **31**, 133–140.

McBride, A. F., and Hebb, D. O. (1948). Behavior of the captive bottle-nose dolphin, *Tursiops truncatus. J. Comp. Physiol. Psychol.* **41**, 111–123.

Mead, J. G., and Potter, C. W. (1990). Natural history of bottlenose dolphins along the central Atlantic coast of the United States. *In* "The Bottlenose Dolphin" (S. Leatherwood, and R. R. Reeves, eds), pp. 165–195. Academic Press, San Diego.

Mead, J. G., and Potter, C. W. (1995). Recognizing two populations of the bottlenose dolphin (*Tursiops truncatus*) off the Atlantic coast of North America: Morphological and ecological considerations. *Int. Biol. Res. Inst. Rep.* **5**, 31–43.

Mitchell, E. (1975). Report of the meeting on smaller cetaceans, Montreal, April 1–11, 1974. *J. Fish. Res. Board Can.* **32**(7), 889–983.

Nowacek, D. P. (2005). Acoustic ecology of foraging bottlenose dolphins (*Tursiops truncatus*), habitat-specific sound use of three sound types. *Mar. Mamm. Sci.* **21**, 587–602.

Nowacek, S. M., Wells, R. S., and Solow, A. R. (2001). Short-term effects of boat traffic on bottlenose dolphins, *Tursiops truncatus*, in Sarasota Bay, Florida. *Mar. Mamm. Sci.* **17**, 673–688.

O'Shea, T. J. (1999). Environmental contaminants and marine mammals. *In* "Biology of Marine Mammals" (J. E. Reynolds, III, and S. A. Rommel, eds), pp. 485–563. Smithsonian Institution Press, Washington, D.C.

Owen, E. C. G., Hofmann, S., and Wells, R. S. (2002). Ranging and social association patterns of paired and unpaired adult male bottlenose dolphins, *Tursiops truncatus*, in Sarasota, Florida, provide no evidence for alternative male strategies. *Can. J. Zool.* **80**, 2072–2089.

Parsons, K. M., Durban, J. W., and Claridge, D. E. (2003). Male–male aggression renders bottlenose dolphin (*Tursiops truncatus*) unconscious. *Aq. Mamm.* **29**, 360–362.

Parsons, K. M., Durban, J. W., Claridge, D. E., Herzing, D. L., Balcomb, K. C., and Noble, L. R. (2006). Population genetic structure of coastal bottlenose dolphins (*Tursiops truncatus*) in the northern Bahamas. *Mar. Mamm. Sci.* **22**, 276–298.

Perrin, W. F., and Reilly, S. B. (1984). Reproductive parameters of dolphins and small whales of the family Delphinidae. *In* "Reproduction in Whales, Dolphins and Porpoises" (W. F. Perrin, R. L. Brownell and D. P. DeMaster, eds), pp. 97–133. *Rep. Int. Whal. Commn* Spec. Iss. No. 6, Cambridge.

Quintana-Rizzo, E., Mann, D. A., and Wells, R. S. (2006). Estimated communication range of social sounds used by bottlenose dolphins (*Tursiops truncatus*). *J. Acoust. Soc. Am.* **120**, 1671–1683.

Read, A. J., Wells, R. S., Hohn, A. H., and Scott, M. D. (1993). Patterns of growth in wild bottlenose dolphins, *Tursiops truncatus. J. Zool. (Lond.)* **231**, 107–123.

Reynolds, J. E., III, Wells, R. S., and Eide, S. D. (2000). "Biology and Conservation of the Bottlenose Dolphin." University of Florida Press, Gainesville.

Rice, D. W. (1998). "Marine Mammals of the World: Systematics and Distribution." Special Publication No. 4, Society for Marine Mammalogy, Allen Press, Lawrence.

Rogers, C. A., Brunnick, B. J., Herzing, D. L., and Baldwin, J. D. (2004). The social structure of bottlenose dolphins, *Tursiops truncatus*, in the Bahamas. *Mar. Mamm. Sci.* **20**, 688–708.

Samuels, A., and Bejder, L. (2004). Chronic interaction between humans and free-ranging bottlenose dolphins near Panama City Beach, Florida, USA. *J. Cetacean Res. Manage.* **6**, 69–77.

Sayigh, L. S., Tyack, P. L., Wells, R. S., and Scott, M. D. (1990). Signature whistles of free-ranging bottlenose dolphins *Tursiops truncatus*: Stability and mother–offspring comparisons. *Behav. Ecol. Sociobiol.* **26**, 247–260.

Schwacke, L. H., Voit, E. O., Hansen, L. J., Wells, R. S., Mitchum, G. B., Hohn, A. A., and Fair, P. A. (2002). Probabilistic risk assessment of reproductive effects of polychlorinated biphenyls on bottlenose dolphins (*Tursiops truncatus*) from the southeast United States coast. *Environ. Toxicol. Chem.* **21**, 2752–2764.

Scott, M. D., and Chivers, S. J. (1990). Distribution and herd structure of bottlenose dolphins in the eastern tropical Pacific Ocean. *In* "The Bottlenose Dolphin" (S. Leatherwood, and R. R. Reeves, eds), pp. 387–402. Academic Press, San Diego.

Scott, M. D., Wells, R. S., and Irvine, A. B. (1990). A long-term study of bottlenose dolphins on the west coast of Florida. *In* "The Bottlenose Dolphin" (S. Leatherwood, and R. R. Reeves, eds), pp. 235–244. Academic Press, San Diego.

Sellas, A. B., Wells, R. S., and Rosel, P. E. (2005). Mitochondrial and nuclear DNA analyses reveal fine scale geographic structure in bottlenose dolphins (*Tursiops truncatus*) in the Gulf of Mexico. *Conserv. Genet.* **6**, 715–728.

Shane S. H. (ed.) (1988). The Bottlenose Dolphin in the Wild. Felton, CA: Hatcher Trade Press.

Shane, S. H., Wells, R. S., and Würsig, B. (1986). Ecology, behavior and social organization of the bottlenose dolphin: A review. *Mar. Mamm. Sci.* **2**, 34–63.

Sylvestre, J. P., and Tanaka, S. (1985). On the inter-generic hybrids in cetaceans. *Aquat. Mamm.* **11**, 101–108.

Tanaka, S. (1987). Satellite radio tracking of bottlenose dolphins *Tursiops truncatus*. *Nippon Suisan Gakkaishi* **53**, 1327–1338.

Thayer, V. G., *et al.* (11 authors) (2003). Reproductive seasonality of western Atlantic bottlenose dolphins off North Carolina, USA. *Mar. Mamm. Sci.* **19**, 617–629.

Thompson, P., and Wilson, B. (1994). "Bottlenose Dolphins." Voyageur Press, Stillwater.

Tomilin, A. G. (1957). Zveri SSSR i prilezhashchikh stran. IX: Kitoobraznye [Mammals of the USSR and Adjacent Countries. Vol. IX: Cetacea] (V. G. Heptner, ed.). Nauk USSR, Moscow. English translation, 1967, Israel Program for Scientific Translations, Jerusalem.

Urian, K. W., Duffield, D. A., Read, A. J., Wells, R. S., and Shell, E. D. (1996). Seasonality of reproduction in bottlenose dolphins, *Tursiops truncatus. J. Mammal.* **77**, 394–403.

Walsh, M. T., Beusse, D., Bossart, G. D., Young, W. G., Odell, D. K., and Patton, G. W. (1988). Ray encounters as a mortality factor in Atlantic bottlenose dolphins (*Tursiops truncatus*). *Mar. Mamm. Sci.* **4**, 154–162.

Wells, R. S. (2003). Dolphin social complexity: Lessons from long-term study and life history. *In* "Animal Social Complexity: Intelligence, Culture, and Individualized Societies" (F. B. M. de Waal, and P. L. Tyack, eds), pp. 32–56. Harvard University Press, Cambridge.

Wells, R. S., and Scott, M. D. (1990). Estimating bottlenose dolphin population parameters from individual identification and capture–release techniques. *In* "Individual Recognition of Cetaceans: Use of Photo-identification and Other Techniques to Estimate Population Parameters" (P. S. Hammond, S. A. Mizroch, and G. P. Donovan, eds), pp. 407–415. *Rep. Int. Whal. Commn* Spec. Iss. No. 12.

Wells, R. S., and Scott, M. D. (1997). Seasonal incidence of boat strikes on bottlenose dolphins near Sarasota, Florida. *Mar. Mamm. Sci.* **13**, 475–480.

Wells, R. S., and Scott, M. D. (1999). Bottlenose dolphin *Tursiops truncatus* (Montagu, 1821). *In* "Handbook of Marine Mammals" (S. H. Ridgway, and R. Harrison, eds), pp. 137–182. Academic Press, San Diego, 6, the Second Book of Dolphins and Porpoises.

Wells, R. S., Scott, M. D., and Irvine, A. B. (1987). The social structure of free-ranging bottlenose dolphins. *In* "Current Mammalogy" (H. H. Genoways, ed.), 1, pp. 247–305. Plenum Press, New York.

Wells, R. S., Hansen, L. J., Baldridge, A., Dohl, T. P., Kelly, D. L., and Defran, R. H. (1990). Northward extension of the range of bottlenose dolphins along the California coast. *In* "The Bottlenose Dolphin" (S. Leatherwood, and R. R. Reeves, eds), pp. 421–431. Academic Press, San Diego.

Wells, R. S., Rhinehart, H. L., Cunningham, P., Whaley, J., Baran, M., Koberna, C., and Costa, D. P. (1999a). Long-distance offshore movements of bottlenose dolphins. *Mar. Mamm. Sci.* **15**, 1098–1114.

Wells, R. S., Boness, D. J., and Rathbun, G. B. (1999b). Behavior. *In* "Biology of Marine Mammals" (J. E. Reynolds, III, and S. A. Rommel, eds), pp. 324–422. Smithsonian Institution Press, Washington, DC.

Wells, R. S., *et al.* (10 authors) (2005). Integrating life history and reproductive success data to examine potential relationships with organochlorine compounds for bottlenose dolphins (*Tursiops truncatus*) in Sarasota Bay, Florida. *Sci. Total Environ.* **349**, 106–119.

Wood, F. G., Caldwell, D. K., and Caldwell, M. C. (1970). Behavioral interactions between porpoises and sharks. *Invest. Cetacea* **2**, 264–277.

Würsig, B., and Würsig, M. (1979). Behavior and ecology of the bottlenose dolphin, *Tursiops truncatus*, in the South Atlantic. *Fish. Bull.* **77**, 399–412.

# Common Dolphins
*Delphinus delphis* and *D. capensis*

WILLIAM F. PERRIN

## I. Characteristics and Taxonomy

The short-beaked common dolphin (*Delphinus delphis*) is the most abundant dolphin in offshore warm-temperate waters in the Atlantic and Pacific, often coming from a distance to join a boat and ride the bow wave. The closely related long-beaked common dolphin (*D. capensis*) is seen less often in most regions and is difficult to distinguish from its congener at sea. The very-long-beaked endemic Indian Ocean subspecies of the long-beaked common dolphin (*D. capensis tropicalis*) is poorly known; its taxonomic status has only recently been clarified (Jefferson and Van Waerebeek, 2002). There is also a distinct short-beaked form in the Black Sea, currently thought to be a subspecies, *D. delphis ponticus* (Amaha, 1994).

C

Figure 1 *Short-beaked common dolphins*, Delphinus delphis. *Photograph by Robert Pitman.*

Figure 2 *Long-beaked common dolphin*, D. c. capensis, *off California. Photograph by Robert Pitman.*

Figure 3 *Extremely-long-beaked common dolphin in Indonesia; may be* D. c. tropicalis. *Photo courtesy of Danielle Kreb.*

All the common dolphins are slender and have a long beak sharply demarcated from the melon (Figs. 1 & 2). The dorsal fin is high and moderately falcate, although in some areas it may be erectly triangular. The two species are distinguished from other dolphins by a unique crisscross color pattern formed by interaction of the dorsal overlay and cape, resulting in distortions of the usual delphinine lateral and ventral fields. The lower margin of the dorsal overlay passes high anteriorly and dips to cross the ventral margin of the low-riding cape, yielding a four-part pattern of dark gray to black uppermost portion or spinal field (cape under dorsal overlay), buff to pale yellow anterior portion or thoracic patch (undiluted cape), light to medium gray posterior portion or flank patch (undiluted dorsal overlay/lateral field), and white abdominal field. A dark flipper-to-anus stripe may parallel the lower margin of the cape and extend to the genital region. A dark flipper stripe runs forward to join the lip patch on the underside of the beak. In the short-beaked species, the color pattern is crisper, the thoracic patch is more yellowish, the sub-cape stripe tends to be anteriorly narrow and faint, and the flipper stripe tends to be narrow and pass low below the corner of the gape. In the long-beaked species including the Indian Ocean subspecies (Fig. 3), the pattern is more muted, the spinal field may be grayish, the thoracic patch tends to be pale buff, the flipper-to-anus stripe tends to be broad anteriorly and may be pronounced and contiguous with the flipper stripe, and the flipper stripe tends to wander toward the corner of the gape before passing ventrally to join the lip patch mark. In the short-beaked species, the dorsal fin and flippers may be all white

or have white centers. These relative features may not be evident in juveniles.

Full species status for the two forms has not been widely recognized until recently (Evans, 1994; Heyning and Perrin, 1994; Rosel *et al.*, 1994; Rice, 1998), and data on size and weight for the three species have been commingled in the literature for some parts of the world. Data exist for well-documented series of adults of two species from California waters. In the short-beaked species, males were 172–201 cm long (*n* = 28) and females were 164–193 cm (*n* = 37) vs 202–235 cm (*n* = 15), and 193–224 cm (*n* = 10), respectively, for the long-beaked species. The short-beaked species ranged to about 200 kg and the long-beaked species to about 235 kg. However, this pattern of differential size may not hold globally; a geographic form of the short-beaked species in the eastern tropical Pacific ranges in length to 235 cm, as large as the long-beaked species in California waters. Also, it has been impossible or difficult to identify definitively common dolphins to species in some areas based on morphology, e.g., South Africa (Samaai *et al.*, 2005) and the eastern North Atlantic (Murphy *et al.*, 2006), and the monophyly of *D. capensis* has been questioned on genetic grounds; various disjunct regional populations may have originated from *D. delphis* separately (Natoli *et al.*, 2006), possibly making the present taxonomy incorrect. Work remains to be done on the alpha taxonomy of the genus; there may be more than two species.

The SKULLS of *Delphinus* spp. are different from those of all the other delphinines (*Stenella* spp., *Tursiops* spp., *Sousa* spp., and *Lagenodelphis hosei*) in the combination of long narrow rostrum and deep palatal grooves. They are similar to the skulls of *Stenella longirostris*, *S. clymene*, *S. coeruleoalba*, and *L. hosei* in having a strongly dorsoventrally flattened rostrum with distally splayed teeth, about 40–60 teeth in each row, relatively small temporal fossae, and sigmoid mandibular rami. The two species in sympatry differ in proportional length of the rostrum in adults: the ratio of rostral length to zygomatic width in *D. delphis* is 1.21 to 1.47 and in *D. c. capensis* is 1.52 to 1.77 (Heyning and Perrin, 1994). Upper/lower tooth counts in California were 42–54/41–53 (*n* = 49/47) and 47–60/47–57 (*n* = 53/51), respectively. Vertebral counts were 74–80 (*n* = 80) in *D. delphis* and 77–80 (*n* = 25) in *D. capensis*. The osteology of *D. c. tropicalis* has not been well described, but the rostral/zygomatic ratio of the holotype specimen is 2.06 and the tooth counts 65–65/57–58, both well beyond the upper range of *D. capensis* for California. Six *Delphinus* skulls from the Arabian Peninsula ranged continuously from 1.72 to 1.94 in

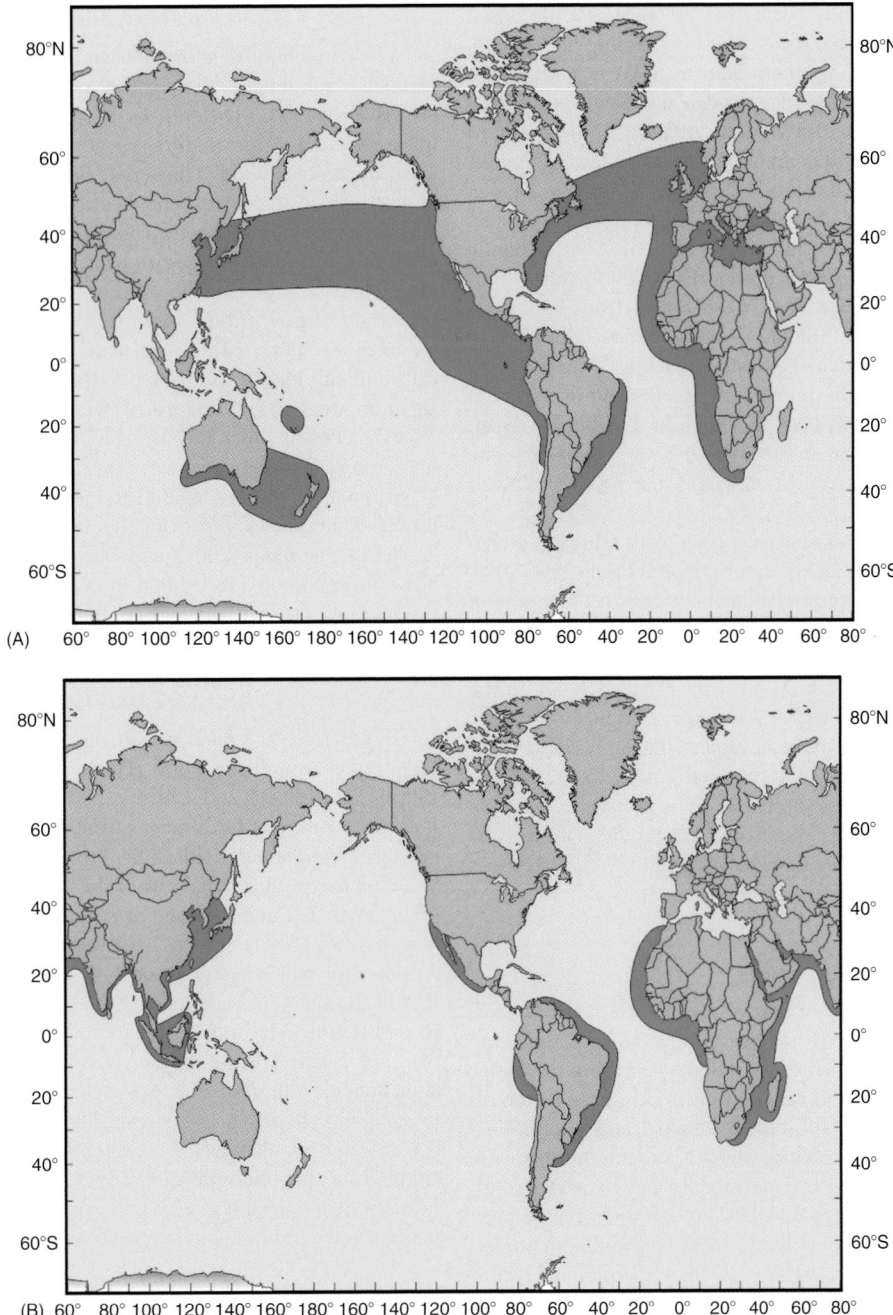

**Figure 4**  *Distributions of (A) short-beaked and (B) long-beaked common dolphins.*

rostral/zygomatic ratio. Three (with the lowest ratios) had been identified as "*D. capensis*" and the others as "*D. tropicalis*"; the range overlaps that of *D. capensis* from California (Smeenk *et al.*, 1996).

The common dolphins are members of the delphinid subfamily Delphininae *sensu stricto* (LeDuc *et al.*, 1999). In a cladistic phylogenetic analysis based on cytochrome *b* mtDNA, they share a strongly supported polytomic clade with *Stenella coeruleoalba* and *S. clymene* (sister taxa), *S. frontalis*, and *Tursiops aduncus*. *D. tropicalis* was basal in the *Delphinus* clade, with California *D. delphis,* Black Sea *D. delphis*, and *D. capensis* in a subsidiary three-part polytomy, making *D. delphis* seemingly paraphyletic. However, apparent paraphyly among terminal taxa can result from incomplete lineage sorting in newly evolved species (LeDuc *et al.*, 1999); reciprocal monophyly has not yet had a chance to evolve in such cases. As noted earlier, *Delphinus* shares several skull features with *S. longirostris, S. coeruleoalba, S. clymene,* and *L. hosei,* but that grouping is not supported by the molecular results to date.

## II. Distribution and Abundance

Common dolphins occur in warm-temperate and tropical waters worldwide from about 40–60°N to about 50°S (Jefferson *et al.*, 2007), but because the three species were considered to be forms of one until recently, many DISTRIBUTION records and much documentation of range have not been identified to species (Rice, 1998). Based on records with diagnostic characters (Fig. 4; Heyning and Perrin, 1994), *D. delphis* occurs from southern Norway to West Africa in the eastern Atlantic (including the Mediterranean and Black Seas), from Newfoundland to Florida in the western Atlantic, from southern Canada to Chile along the coast and pelagically in the eastern Pacific; in the central North Pacific (but not in Hawaii); from central Japan to Taiwan and around New Caledonia, New Zealand, and Tasmania in the western Pacific; and is possibly absent from the South Atlantic and Indian Oceans. *D. c. capensis* occurs disjunctly in warm-temperate and tropical coastal waters in West Africa, from Venezuela to Argentina in the western Atlantic, from southern California to central Mexico, and in Peru in the eastern Pacific, around Korea, southern Japan, and Taiwan in the western Pacific, and in waters of Madagascar, South Africa, and possibly Oman in the Indian Ocean. The identification of the Oman specimens is in some doubt (Smeenk *et al.*, 1996). *D. tropicalis* is known only from the northern Indian Ocean and Southeast Asia.

*Delphinus delphis* is a very abundant species, with estimates available for several regions (IUCN, 2008): eastern tropical Pacific, 2,963,000; US west coast, 400,000; European continental waters, 63,400; portion of European offshore waters, 273,000; western North Atlantic, 121,000; western Mediterranean, 19,400. There has been a major decline in the Mediterranean during the last 30–40 years. Population size in the Black Sea is unknown. Only a few local abundance estimates exist for *D. capensis* (IUCN, 2008): off California, around 10,000; Pacific coast of Mexico, 55,000; Gulf of California, 69,000; South Africa, 15,000–20,000. There are no estimates for the subspecies *D. c. tropicalis*. Multiple stocks of *D. delphis* exist in the eastern tropical Pacific, identified by differences in size, shape, coloration and life-history characteristics (Perryman and Lynn, 1993; Danil and Chivers, 2006).

## III. Ecology

*D. delphis* and *D. capensis* are narrowly sympatric in some near-shore waters; schools of the two species may be seen in the same general area on the same day. However, *D. capensis* seems to prefer shallower and warmer water and generally occurs closer to the coast. In the eastern North Pacific, substantial seasonal and inter-annual changes in abundance of *D. delphis* suggest migrations that vary with oceanographic conditions; the movements may be north–south and/or inshore–offshore (Forney and Barlow, 1998). In the tropical eastern Pacific, *D. delphis* occupies primarily upwelling-modified habitats with less tropical characteristics than surrounding water masses (Ballance *et al.*, 2006); separate northern, central, and southern stocks associated with different upwelling areas are recognized in the management of incidental mortality in tuna fisheries (Perrin *et al.*, 1985). *D. delphis* prefers coastal and upper slope waters n the Mediterranean (Bearzi *et al.*, 2003) and migrate between offshore and shallow coastal waters in the Black Sea (Birkun, 2006).

As would be expected for a species occupying a wide range of habitats, common dolphins feed on a variety of prey, including small mesopelagic fishes and squids found in the deep scattering layer and epipelagic schooling species such as small scombroids, clupeoids, and market squids (Evans, 1994; Ohizumi *et al.*, 1998; De Pierrepont *et al.*, 2005; Pusineri *et al.*, 2007). The diet varies taxonomically with region; in the western Atlantic, oceanic common dolphins consume more squid than neritic ones do (Lahaye *et al.*, 2005). Foraging dives to 200m have been recorded. Diet varies with season as well as with region (Evans, 1994). FEEDING habits of the two species have not been compared.

## IV. Behavior and Physiology

Schools of hundreds or thousands are thought to be composed of smaller subunits of about 20–30 perhaps closely related individuals (Evans, 1994). Differences between schools in cranial measurements and color pattern suggest that schools have temporal integrity at some level. There may be segregation in schools by age and sex. Association with schools of pilot whales (*Globicephala* spp.) and other dolphins (*Lagenorhynchus* spp., *Stenella coeruleoalba*, *Grampus griseus*) have been observed (Evans, 1994; Frantzis and Herzing, 2002), as has "bow riding" on mysticete whales, possibly the origin of bow riding on vessels. Common typical aerial behavior includes "pitch poling," in which the dolphin leaps high vertically and falls lengthwise back into the water to create a large splash. In some areas in the eastern tropical Pacific, schools of *D. delphis* "carry" yellowfin tuna and are chased and captured by tuna fishermen using purse seines.

Seasonal migrations related to environmental change have been reported for *D. delphis* from the Black Sea (Birkun, 2006), New Zealand (Neumann, 2001) and the eastern tropical Pacific (Danil and Chivers, 2006). Individual long-beaked common dolphins may exhibit long-term local residency (Bernal *et al.*, 2003).

## V. Life History

Gestation in *D. delphis* as estimated at 10–11.7 months for several populations around the world (Perrin and Reilly, 1984; Murphy and Rogan, 2006; Danil and Chivers, 2007; Westgate and Read, 2007). Estimated length at birth ranges from 80 to 93cm. The calving interval varies from about 1 (Black Sea) to about 3 years (eastern Pacific). Weaning occurs at 5–6 months in the Black Sea, but possibly later in other areas. Estimates of age at sexual maturation vary with region from 3 years (Black Sea) to 7–12 years (eastern Pacific and western Atlantic) for males and 2–4 and 6–8 years, respectively, for females. Part of the range of variation may be a density-dependent effect due to exploitation. Maximum estimated age was 30 years for both sexes (western North Atlantic). Other possible factors explaining the variation include differences in age-determination methods and biases in sampling. Reproductive seasonality also varies with region, with cooler-water populations exhibiting more synchrony of breeding. The degree of sexual dimorphism and very large testes (translation of one Japanese name for the species is "testis dolphin") suggest sperm competition and a promiscuous breeding system (Murphy *et al.*, 2005).

## VI. Interactions with Humans

Short-beaked common dolphins have died in large numbers in tuna purse seines in the eastern tropical Pacific (e.g., 24,307 in 1986; IATTC, 2006) and, with long-beaked and Indian Ocean common dolphins, in gillnet, trawl and purse-seine fisheries around the world (Perrin *et al.*, 1994; IUCN, 2008); the impact of these kills on the populations is for the most part unknown. Common dolphins are also killed in nets enclosing tuna feedlots in Australia (Kemper and Gibbs, 2001). Use of pingers on gillnets has reduced gillnet mortality to an insignificant level in the eastern North Pacific (Carretta *et al.*, 2006). Direct take for human consumption and shark bait is a threat to *D. capensis* in Peru (IUCN, 2008). Prey depletion is thought to be a threat to the subspecies of *D. delphis* in the Black Sea (IUCN, 2008).

Common dolphins have been kept successfully in captivity for up to 8 years in eastern European oceanaria and reproduced successfully in captive facilities in the USA, New Zealand, and Russia

(Evans, 1994). *D. capensis* has hybridized with *T. truncatus* in captivity, producing viable offspring of unknown fertility (T. Goff, unpublished data). *D. delphis* is among the species exploited in whale watching in Europe and elsewhere (Hoyt, 2003).

## See Also the Following Articles

Coloration ■ Delphinids ■ Hybridism ■ Incidental Catches ■ Tuna-Dolphin Issue

## References

Amaha, A. (1994). Geographic variation of the common dolphin, *Delphinus delphis* (Odontoceti Delphinidae). Ph.D. Thesis, Tokyo University of Fisheries.

Ballance, L. T., Pitman, R. L., and Fiedler, P. C. (2006). Oceanographic influences on seabirds and cetaceans of the eastern tropical Pacific: A review. *Prog. Oceanog.* **60**, 360–390.

Bearzi, G., Reeves, R. R., Notarbartolo di Sciara, G., Politi, E., Cañadas, A., Frantzis, A., and Mussi, B. (2003). Ecology, status and conservation of short-beaked common dolphins *Delphinus delphis* in the Mediterranean Sea. *Mamm. Rev.* **33**, 224–252.

Bernal, R., Olavarria, C., and Moraga, R. (2003). Occurrence and long-term residency of two long-beaked common dolphins, *Delphinus capensis* (Gray 1828), in adjacent small bays on the Chilean central coast. *Aquat. Mamm.* **29**, 396–399.

Birkun, A. (2006). Short-beaked common dolphin (*Delphinus delphis ponticus*): Black Sea subspecies. *In* "The Status and Distribution of Cetaceans in the Black Sea and Mediterranean Sea" (R. R. Reeves, and G. Notarbartolo di Sciara, eds), pp. 16–22. IUCN Centre for Mediterranean Cooperation, Malaga, Spain.

Carretta, J. V., Forney, K. A., Muto, M. M., Barlow, J., Baker, J., Hansen, B., and Lowry, M. S. (2006). U.S. Pacific marine mammal stock assessments: 2005. *NOAA Tech. Memo. NOAA-TM-NMFS-SWFSC* **388**, 317.

Danil, K., and Chivers, S. J. (2006). Habitat-based spatial and temporal variability in life-history characteristics of female common dolphins *Delphinus delphis* in the eastern tropical Pacific. *Mar. Ecol. Prog. Ser.* **318**, 277–286.

Danil, K., and Chivers, S. J. (2007). Growth and reproduction of female short-beaked common dolphins, *Delphinus delphis*, in the eastern tropical Pacific. *Can. J. Zool.* **85**, 108–121.

De Pierrepont, J. F., Dubois, B., Desermonts, S., Santos, M. B., and Robin, J. P. (2005). Stomach contents of English Channel cetaceans stranded on the coasts of Normandy. *J. Mar. Biol. Assoc. UK* **85**, 1539–1546.

Evans, W. E. (1994). Common dolphin, white-bellied porpoise *Delphinus delphis* Linnaeus, 1758. *In* "Handbook of Marine Mammals" (S. H. Ridgway and R. Harrison, eds), 5, pp. 191–224.

Forney, K. A., and Barlow, J. (1998). Seasonal patterns in the abundance and distribution of California cetaceans, 1991–1992. *Mar. Mamm. Sci.* **14**, 460–489.

Frantzis, A., and Herzing, D. L. (2002). Multi-species associations of striped dolphins (*Stenella coeruleoalba*), short-beaked common dolphins (*Delphinus delphis*), and Risso's dolphins (*Grampus griseus*) in the Gulf of Corinth (Greece, Mediterranean Sea). *Aquat. Mamm.* **28**, 188–197.

Heyning, J. E., and Perrin, W. F. (1994). Evidence for two species of common dolphins (genus *Delphinus*) from the eastern North Pacific. *Nat. His. Mus. Los Angeles County Contrib. Sci.* **442**, 35.

Hoyt, E. (2003). "The Best Whale Watching in Europe." Whale and Dolphin Conservation Society, Unterhaching, Germany, 60pp.

Inter-American Tropical Tuna Commission (IATTC). (2006). *Ann. Rep.* 2004.

IUCN. (2008). The IUCN Red List of Threatened Species (http://www.iucnredlist.org).

Jefferson, T. A., Webber, M. A., and Pitman, R. L. (2007). "Marine Mammals of the World: A Comprehensive Guide to Their Identification." Academic Press/Elsevier, San Diego.

Jefferson, T. A., and Van Waerebeek, K. (2002). The taxonomic status of the nominal dolphin species *Delphinus tropicalis* van Bree, 1971. *Mar. Mamm. Sci.* **18**, 787–818.

Kemper, C. M., and Gibbs, S. E. (2001). Dolphin interactions with tuna feedlots at Port Lincoln, South Australia and recommendations for minimizing entanglements. *J. Cetacean Res. Manag.* **3**, 283–292.

Lahaye, V., Bustamante, P., Spitz, J., Dabin, W., Das, K., Pierce, G. J., and Caurant, F. (2005). Long-term dietary segregation of common dolphins *Delphinus delphis* in the Bay of Biscay, determined using cadmium as an ecological tracer. *Mar. Ecol. Prog. Ser.* **305**, 275–285.

LeDuc, R. G., Perrin, W. F., and Dizon, A. E. (1999). Phylogenetic relationships among the delphinid cetaceans based on full cytochrome *b* sequences. *Mar. Mamm. Sci.* **15**, 619–648.

Murphy, S., and Rogan, E. (2006). External morphology of the short-beaked common dolphin, *Delphinus delphis*: Growth, allometric relationships and sexual dimorphism. *Acta. Zool.* **87**, 315–329.

Murphy, S., Collet, A., and Rogan, E. (2005). Mating strategy in the male common dolphin (*Delphinus delphis*): What gonadal analysis tells us. *J. Mammal.* **86**, 1247–1258.

Murphy, S., Herman, J. S., Pierce, G. J., Rogan, E., and Kitchener, A. C. (2006). Taxonomic status and geographical cranial variation of common dolphins (*Delphinus*) in the eastern North Atlantic. *Mar. Mamm. Sci.* **22**, 573–599.

Natoli, A., Cañadas, A., Peddemors, V. M., Aguilar, A., Vaquero, C., Fernández-Piqueras, P., and Hoelzel, A. R. (2006). Phylogeography and alpha taxonomy of the common dolphin (*Delphinus* sp.). *J. Evol. Biol.* **19**, 943–954.

Neumann, D. R. (2001). Seasonal movements of short-beaked common dolphins (*Delphinus delphis*) in the north-western Bay of Plenty, New Zealand: Influence of sea surface temperature and El Nino/La Nina. *NZ J. Mar. Freshw. Res.* **35**, 371–374.

Ohizumi, H., Yoshioka, M., Mori, K., and Miyazaki, N. (1998). Stomach contents of common dolphins (*Delphinus delphis*) in the pelagic western North Pacific. *Mar. Mamm. Sci.* **14**, 835–844.

Perrin, W. F., and Reilly, S. B. (1984). Reproductive parameters of dolphins and small whales of the family Delphinidae. *Rep. Int. Whal. Commn. Spec. Iss.* **6**, 97–133.

Perrin, W. F., Scott, M. E., Walker, G. J., and Cass, V. L. (1985). Review of geographical stocks of tropical dolphins (*Stenella* spp. and *Delphinus delphis*) in the eastern Pacific. *NOAA Tech. Rep. NMFS* **28**.

Perrin, W. F., Donovan, G. P., and Barlow, J. (eds). (1994). Gillnets and Cetaceans. *Rep. Int. Whal. Commn.* Spec. Iss. 15, 629.

Perryman, W. L., and Lynn, M. S. (1993). Identification of geographic forms of common dolphin (*Delphinus delphis*) from aerial photography. *Mar. Mamm. Sci.* **9**, 119–137.

Pusineri, C., Magnin, V., Meynier, L., Spitz, J., Hassani, S., and Ridoux, V. (2007). Food and feeding ecology of the common dolphin (*Delphinus delphis*) in the oceanic Northeast Atlantic and comparison with its diet in neritic areas. *Mar. Mamm. Sci.* **23**, 30–47.

Rice, D. W. (1998). Marine mammals of the world. *Soc. Mar. Mamm. Spec. Pub.* **4**, 231.

Rosel, P. E., Dizon, A. E., and Heyning, J. E. (1994). Genetic analysis of sympatric morphotypes of common dolphins (genus *Delphinus*). *Mar. Biol.* **119**, 159–167.

Samaai, T., Best, P. B., and Gibbons, M. J. (2005). The taxonomic status of common dolphins *Delphinus* spp. in South African waters. *South Afr. J. Mar. Sci.* **27**, 458–459.

Smeenk, C., Addink, M. J., van den Berg, A. B., Bosman, C. A. W., and Cadée, G. C. (1996). Sightings of *Delphinus cf. tropicalis* van Bree, 1971 in the Red Sea. *Bonn. Zool. Beitr.* **46**, 389–398.

Westgate, A. J., and Read, A. J. (2007). Reproduction in short-beaked common dolphins (*Delphinus delphis*) from the western North Atlantic. *Mar. Biol.* **150**, 1011–1024.

C

# Communication in Marine Mammals

Kathleen M. Dudzinski, Jeanette A. Thomas and Justin D. Gregg

Communication is a process by which a sender produces a signal, which alters the probability of a subsequent behavior in a receiver(s). Often, but not always, communication facilitates social behavior. Given the highly social behavior found in many marine mammals, the study of communicative behavior is essential to understanding the role that signaling plays in regulating social interactions for these species. To understand communication in a given species, it is important to view the mode of the signal (i.e., visual, acoustic, tactile, gustatory, or olfactory), medium in which the signal is transmitted (air and/or water), mechanisms of signal production (anatomical and/or physiological), function(s) of the signal (e.g., aggression/submission, mate attraction, parental care, TERRITORIAL defense), and whether signals are multi-modal.

This chapter is a brief overview of communication in marine mammals. Even with more than 40 years of focused studies on the social lives of marine mammals, relatively little is understood about the majority of species within marine mammal groups—cetaceans, pinnipeds, sirenians, sea otters (*Enhydra lutris*), and the polar bears (*Ursus maritimus*). Behavioral characteristics and social relationships are adapted to each species' unique ecology. Marine mammals are either amphibious or totally aquatic. Each life mode imposes different constraints on signaling and communication. A paucity of studies on communication exists for many marine mammals, especially polar bears, sea otters, dugongs (*Dugong dugon*), and manatees (*Trichechus* spp.). Even less is known about the marine otter (*Lontra felina*), so we do not discuss this species. The majority of research on communication has been conducted on pinnipeds (such as Weddell seals, *Leptonychotes weddellii*; and California sea lions, *Zalophus californianus*) and cetaceans (particularly bottlenose dolphins, *Tursiops* spp.; killer whales, *Orcinus orca*; and humpback whales, *Megaptera novaeangliae*). Thus, our discussion highlights and compares species predominantly represented in the literature.

## I. Definition

A clear definition of communication is needed to facilitate consistency among studies and to avoid ambiguities in methods. Bradbury and Vehrencamp (1998) provided this definition: "communication involves the provision of information (via a signal) by a sender to a receiver, and subsequent use of this information by the receiver in deciding how or whether to respond." The signal is the vehicle by which the sender and receiver exchange information. Both the sender and the receiver rely on signals to meet individual challenges within a group setting, such as reproduction, predator defense, territory defense, foraging, maintenance of social bonds, and parental care. Signals are mechanisms or "tools" specialized over time to be informative, salient to interactions among individuals, and adapted for optimum transmission in their environment(s). In mammals, sensory channels can include chemical (i.e., taste and olfaction), mechanical (i.e., tactile and acoustic), photic (visual), and electromagnetic modes (Herman, 1980; Reynolds and Rommel, 1999). While most terrestrial mammals evolved signals in each

of these sensory channels (Hauser, 1997), marine mammals have not, primarily because of limitations of the aquatic environment. Amphibious marine mammals tend to use sensory modes similar to terrestrial mammals, but the strictly marine mammals have limited abilities for olfaction because water is not a good medium for long-term, site-specific use of scent. Likewise, marine mammals use vision only for short distances because water movement, plankton blooms, murky water, or darkness at depth limits the range and applicability of vision in water. In totally aquatic mammals, communication is achieved primarily through acoustic and tactile modes.

## II. Chemical Communication

Chemoreception is common among terrestrial mammals, but little is understood about how marine mammals sense chemical signals in the water (Reynolds and Rommel, 1999). The olfactory sense and anatomy are not suited for communication in water and this sensation declines with greater adaptation to an aquatic lifestyle (Reynolds and Rommel, 1999). Jansen and Jansen (in Anderson, 1969) found that adult odontocetes lack olfactory nerves, bulbs, and tracts; the same are reduced greatly in adult mysticetes. Furthermore, cetaceans, pinnipeds, sea otters, and sirenians all close their nasal openings while in water, thus preventing smell. Chemoreception in water may be more taste than smell. The possibility that scents are pheromone based in nature and function has not been examined in marine mammals, although anecdotal accounts exist. It has been suggested that belugas (*Delphinapterus leucas*) release a "pheromone" when alarmed. Belugas react to blood in the water by either quickly escaping or becoming unusually excited. Trails of both feces and urine deposited by schools of dolphins could contain sexual pheromones. At times, spinner dolphins (*Stenella longirostris*) appear to swim deliberately through dispersing excrements deposited by schoolmates (Norris *et al.*, 1994).

*1. Cetacea* Whether under water or at the surface, cetaceans keep their blowholes closed, except during brief respirations at the surface. Most studied cetaceans have the ability to taste, although with somewhat different receptive qualities than terrestrial mammals (Reynolds and Rommel, 1999). Taste buds have been documented behaviorally and physiologically for both cetaceans and sirenians (Herman, 1980; Schusterman *et al.*, 1986; Reynolds and Rommel, 1999). Experiments on the taste sensitivities of common bottlenose dolphins (*Tursiops truncatus*) have shown that they can discriminate sour, sweet, bitter, and salty solutions. They were least sensitive to different salt concentrations, which seems adaptive given they live in a marine environment.

*2. Pinnipedia* Whether on land or under water, pinnipeds keep their nares closed, except to respire. The olfactory anatomy of pinnipeds is variably reduced, more for phocids and for odobenids than for otariids (Reynolds and Rommel, 1999). Pinnipeds employ scents to exchange or gain information about colony members; e.g., male northern fur seals (*Callorhinus ursinus*) sniff the hindquarters of females to assess their estrous state (Reynolds and Rommel, 1999). The largest glands of pinnipeds are around the vibrissae and could play a role in mother–pup recognition. Mothers and pups maintain a great deal of nose-to-nose contact and use odor cues for recognition in air. Little work has been conducted on pinniped taste sensations. Gustatory abilities also have been demonstrated in Steller sea lions (*Eumetopias jubatus*) and California sea lions. Both studies found similar taste abilities with some sensitivity to acidic, basic, and salty solutions, but not to sweet.

*3. Sirenia* Similar to cetaceans, sirenians keep their nasal openings closed under water or at the surface, except to quickly breathe at the surface. Sirenians have a rudimentary olfactory system (Reynolds and Rommel, 1999) and likely rely, to a limited degree, on chemicals for signal exchange among conspecifics. However, because aquatic plants are known to have different tastes and smells, manatees and dugongs could use this sense in foraging. No information is available on taste abilities in sirenians.

*4. Sea otter* Unlike other mustelids, known for their musky smell, the sea otter has no scent glands. This is likely a result of the aquatic environment in which scent marking would have limited usefulness. When under water, sea otters close their nares. Kenyon (1975) reported that sea otters in water commonly surface and sniff the air, and the male sea otter smells the genital area of an estrous female during pre-copulatory behavior. The common "nosing" behavior observed between sea otters is thought to involve scent recognition or chemoreception (Riedman and Estes, 1990). It is assumed that sea otters and polar bears have taste abilities similar to their terrestrial counterparts, although the exact extent of this sense is unknown for these species.

*5. Polar bear* Polar bears have a keen sense of smell, especially useful in foraging. While little is known about how olfaction is used for communication among polar bears, patterns are likely similar to those observed in other ursids and used by males to find potential mates (Ovsyanikov, 1996; Stirling, 1999). No studies are available on taste abilities of polar bears.

## III. Visual Communication

Behavioral displays are well documented for many marine mammals with visual detection and acuity levels being good both above and under water for all species studied (Reynolds and Rommel, 1999). Under water, VISION is limited by light levels, the concentration and type of organic matter suspended in the water column, and depth (for a thorough discussion of light in the ocean and visual adaptations by marine mammal species for visual detection and acuity both above and below the water; Reynolds and Rommel, 1999; Mass and Supin, 2007). Visual displays can be simple, such as sexually dimorphic features, body postures or coloration patterns, or they can be elaborate sequences of behaviors that indicate a context, species, age, sex, or reproductive condition. Movements and postures often are highlighted in species with conspicuous color patterns. In clear water, visual signals provide cetaceans and other marine mammals a close-range alternative to acoustic signaling; however, displays could inadvertently alert predators or prey. Some marine mammals have adaptations for vision (e.g., large eyes, tapetum) that allow them to see and potentially communicate via visual signals in low light conditions. The anatomical adaptations for vision in water vary greatly among marine mammals (Mass and Supin, 2007).

*1. Cetacea* Visual displays for odontocetes include behaviors, coloration, and morphological traits. Several species possess distinct visual characters that might or might not be considered secondary sexual characteristics [e.g., male spinner dolphins have a forward sloping dorsal fin and bulging ventral keel, male Dall's porpoise (*Phocoenoides dalli*) have a pronounced ventral hump, male narwhals (*Monodon monoceros*) have long spiral tusks and in several species of beaked whales males have lower teeth that protrude outside the mouth]. These sexually dimorphic characteristics may be used to regulate social signaling, and possibly mating. Recurrent body color patterns (spots, saddle patches, capes, and longitudinal striping) are evident in several delphinids, especially species living in clear water where surface reflections may be important to social signaling. Disruptive coloration has likely evolved for social signaling, or to deceive prey and predators. Pelagic species tend to live in large, inter-specific schools and also seem to possess the most complex color patterns among small cetaceans. This complexity may be important for species and individual recognition, as well as social signaling (Thomas and Kastelein, 1990). Overt actions and gestures such as open-jaw threat displays, aerial leaps, tail lobs, flared pectoral fins, and S-shaped postures form the majority of behavioral visual displays expressed by cetaceans. Changes in posture can be used to communicate to conspecifics, predators, and prey. Posture and behavioral signaling can also be used to synchronize actions among individuals or groups as a signal for group coordination or for social interaction (e.g., synchrony between male coalitions of bottlenose dolphins when herding a female). Whalers early on recognized that the shape and height of the "blow" associated with respiration at the water surface is quite distinctive in some species of odontocetes and mysticetes. Similarly, the blow could be used as a social signal among cetaceans, especially to indicate location and species.

*2. Pinnipedia* Several pinnipeds incorporate body coloration or postures into visual displays; e.g., TERRITORIAL BEHAVIOR of a Weddell seal patrolling the water underneath a crack in the fast ice consists of loud trill and teeth chatter sounds accompanied by an S-shaped posture that thrusts the chest forward and the hind flippers downward. When approached on the pack ice, Ross seals (*Omatophoca rossii*) assume a head-up posture that displays the stripes on their chest, an open mouth of teeth, and this is accompanied by noisy sounds. Interestingly, this species has been called the "singing seal," but the postures and sounds indicate aggression toward an intruder.

*3. Sirenia* Sirenians have poor color vision and poor visual acuity for near-field objects. Often considered solitary, manatees may congregate at well-defined, traditional locations called "rendezvous sites" in Florida where tactile contact seems to be the primary form of close-range communication (Reynolds and Rommel, 1999). Little information is available on signal exchange for visual communication among manatees or dugongs.

*4. Sea otter* Riedman and Estes (1990) described a "head jerk" movement commonly seen in sea otters. This rapid side-to-side head movement is a visual display that may be involved in communicating social status, reproductive status, or other information. In general, little is known about the visual displays of sea otters.

*5. Polar bear* Polar bears exhibit visual displays on land, like other ursids, including bearing of teeth, upright sparring, chasing, and wrestling between males.

## IV. Tactile Communication

Visual displays are useful for close-range communication among marine mammals and, because of close proximity, visual displays may readily become tactile signals. Extensive touching and rubbing occurs in both captive and free-ranging animals during PLAY, sexual, maternal, and social contexts using the nose or rostrum, flippers, pectoral fins, dorsal fin, flukes, abdomen, and the entire body. Tactile contacts often are observed during AGGRESSIVE BEHAVIOR, but are characterized by more overt actions, such as biting, raking, ramming, wrestling, and butting. Tactile signals can be modified to increase the information content—who, where, and how animals touch, as well as the intensity of a touch, factor into the signal content. Often, tactile signals combine with other signals and grade into each other; e.g., a chase, then wrestle, teeth chatter, then nip, can escalate into a full biting and sparring match. The advantage of a graded signal is that the sender or receiver can choose to withdraw at any point along the progression.

## A. Responsiveness to Touch

*1. Cetacea*   An inclination for tactile responsiveness has been noted in studies of wild and captive individuals of all cetaceans. Among mysticetes, the "friendly" gray whales (*Eschrichtius robustus*) of San Ignacio Lagoon, Mexico are noted for approaching and rubbing under small boats and for tolerance of petting by tourists. In the wild, both Atlantic spotted (*Stenella frontalis*) and bottlenose dolphins rub body parts into the sand or along rocky edges, and are in frequent contact with each other (Fig 1). Gentle contact behavior between conspecifics (petting, stroking, nuzzling) has been recorded in many cetacean species [e.g., humpback and North Atlantic right whales (*Eubalaena glacialis*)] and is common among mothers and their calves. However, there is no evidence of allogrooming or mother grooming of a calf in cetaceans. Contact swimming bouts, where one dolphin lays its pectoral fin on the flank of a conspecific, have been recorded in bottlenose dolphins in Shark Bay Australia, with one bout reportedly lasting over 30 min (Connor *et al.*, 2006). All odontocetes in captivity seek and are receptive to gentle body contact. Mild tactile stimulation (e.g., rubbing of gums, flippers, or dorsal fin) serves as an effective re-enforcer in training of most odontocetes. Trainers suggest that tactile stimulation is reinforcing, and perhaps rubbing among dolphins might also be rewarding. Rubbing and touching serves a secondary function to help remove dead skin that continually sloughs in cetaceans. In addition, the flow of water across the body may help cetaceans judge swim speed or water depth. The bow-riding behavior exhibited by cetaceans in the family Delphinidae likely provides a tactile sensory experience.

*2. Pinnipedia*   Pinnipeds vary in their degree of gregarious behavior and thus tolerance for tactile stimulation by conspecifics. Leopard seals (*Hydrurga leptonyx*) are solitary predators and rarely seen in close proximity. In contrast, Weddell seals congregate in breeding colonies, but each mother–pup pair maintains an individual space. The more polygynous pinnipeds, such as walrus (*Odobenus rosmarus*) and California sea lions, often crowd onto beaches, piling on top of each other, with little regard for "personal space." This tolerance of body contact may provide a thermoregulatory advantage, as well.

Regardless of adult spacing, in pinnipeds a mother and pup maintain close tactile communication. Young pinniped pups often crawl over their mothers, and sleep touching their mother. There is, however, no maternal grooming of the young in pinnipeds.

*3. Sirenia*   In Crystal River, Florida, some manatees seek physical contact with divers, whereas others avoid divers. Florida manatees (*Trichechus manatus latirostris*) sometimes "body surf" on currents generated below dams when floodgates are partially opened. This surfing can last for up to an hour, with manatees repeatedly riding the currents in parallel formation. Often, nuzzling and vocalizations accompany manatee body surfing (Reynolds and Odell, 1991). When not eating (nearly 8h/day), manatees curiously investigate objects, socialize by mouthing and rubbing against each other, and play together (Reynolds and Odell, 1991). The fact that manatees have a green tinge from the algae that grows on their skins suggests that they do not slough their skin often, but perhaps rubbing helps keep the amount of algae in check. Manatees are often seen mouthing or bathing in a water stream from a hose running off boats. Mothers do not groom their calves.

*4. Sea otter*   Sea otters possess thick layers of fur for warmth and protection, and thus grooming is part of their social structure, as with many social terrestrial mammals. Unlike other marine mammals, sea otters do not have subcutaneous fat, they must rely on keeping their underfur dry and therefore groom themselves repeatedly to spread waterproofing squalene oil over the surface of their fur. Because grooming is essential to keeping warm, sea otters spend a large part of their day grooming. Like terrestrial mammals, sea

**Figure 1**   *Two Atlantic spotted dolphins* (Stenella frontalis) *rubbing their bodies in the sand.*

otter mothers groom their pups by licking their fur. They are the only marine mammal to have the ability to hold and manipulate their young for grooming. Mother sea otters float on their back and hold the pup on their abdomen and spend a great deal of time grooming their pup and remove feces and urine by licking the pup's urogenital region. Sea otter grooming behavior is probably at least partially hygienic in function; however, in other mammal species (e.g., primates, canids), grooming behavior signals affection, appeasement, or reconciliation. A variety of tactile behavior relating to grooming (rubbing, shaking, stroking) and other social interactions (shoving, pawing, wresting) have been documented for sea otters (Riedman and Estes, 1990). When sleeping or resting, sea otters float on their back at the water surface; however, the water movement can "wash them out to sea" (Fig. 2). In Alaska, sea otters often synchronize their sleeping time and form a raft of bodies that bob together. Sea otters in groups often "hold paws" or otherwise keep close body contact to maintain the raft.

*5. Polar bear* More data are required to better understand how polar bears use touch in communication. As in other ursids, mother polar bears likely have close tactile contact with their young for nursing and grooming in the den. Adult males are seen in intense fights grasping each other with "bear holds," nose-to-nose open-mouth threats, growls, and biting. Small cubs are often seen sleeping together, perhaps for comfort and warmth.

## V. Acoustic Communication

Marine mammals use both vocal and non-vocal acoustic communication. Richardson *et al.* (1995) summarized the hearing and acoustic abilities of marine mammals. Because of the ease with which sound travels in water and the large area over which sound can be transmitted, as opposed to in air, underwater acoustic signals evolved to be the principal mode of information transmission for fully aquatic mammals and a predominant mode of communication for amphibious marine mammals. Recording and analyzing sounds from marine mammals is relatively easy, but determining the

context and function of sounds is not. Deecke (2006) published a recent summary of studies using playback techniques to examine the function of sounds in marine mammals. The impact that anthropogenic underwater NOISE (Southall *et al.*, 2007) may have on the communicative signals of marine mammals has been highlighted in recent studies of many odontocetes (Foote *et al.*, 2004), endangered mysticete species (Croll *et al.*, 2002), sirenians (Miksis-Olds *et al.*, 2007), and pinnipeds (Southall *et al.*, 2000).

### A. Non-vocal Communication

*1. Cetacea* Non-vocal communication can include noise from flukes or flippers striking the water surface, as well as the percussive sounds of jaw claps, teeth gnashing, or bubble emissions. Breaches, leaps, tail slapping, and chin slapping produce sounds under water that likely carry a communicative message. Most cetaceans are known to leap vigorously into the air, called breaching. A breach produces airborne and underwater sounds upon reentry that carry for several kilometers. Breaching could be a spacing mechanism or help cetaceans remain in acoustic contact. Breaching often indicates general excitement or arousal deriving from any of several causes, including sexual stimulation, location of food, or a response to injury or irritation. Calves and their mothers also breach on occasion and sometimes in unison. Clearly, there can be many immediate causes of breaching (e.g., parasite or dead skin removal) and further study is needed to clarify and understand the multiple contexts in which breaching occurs. Dusky dolphins (*Lagenorhynchus obscurus*) are well known for the three leap types they produce in association with three stages of cooperative feeding: head-first re-entry leaps, noisy leaps, and social, acrobatic leaps. The latter two create sounds that function to signal peers or, as for the noisy leaps, could act as a sound "barrier" to disorient prey and keep them tightly schooled. Upon water re-entry after a spin, a breach, a back slap, or a head or tail slap, spinner dolphins generate omni-directional noise that propagates over short to intermediate distances. Spinner dolphins' aerial behavior seems designed to

**Figure 2** *Typical behavior of a sea otter* (Enhydra lutris)*-floating on its back with a shell.*

**Figure 3** *A group of Atlantic spotted dolphins* (Stenella frontalis) *socializing and exchanging vocal and behavioral signals modified by various postures.*

produce noise, as many leaps are common at night, when visual contact is limited, or in daytime occur in fully alert, but dispersed schools (Fig. 3). Visual signals during leaps likely convey position information to schoolmates and could facilitate aerial inspection for feeding sites or for detecting environmental features.

Most observers agree that tail slaps often convey a threat or distress. Tail slapping, which produces extensive, low-frequency underwater and aerial sound, often occurs dozens of times in succession. It is likely that tail slaps among mysticetes and odontocetes often, but not always, have an agonistic component. Pectoral fin slapping is observed predominantly in humpback whales, although other baleen whales and some smaller delphinids also exhibit this behavior. The exact communicative nature of pectoral fin slapping is not clear,

although it may signal frustration or irritation, maintain individual spacing, or serve to invite PLAY or socializing.

The blow or sounds associated with respiration at the water surface are quite distinctive in both odontocetes and mysticetes. It is likely these sounds may be used incidentally by conspecifics to locate each other.

*2. Pinnipedia*    Pinnipeds do not use non-vocal communication as much as cetaceans, i.e., pinnipeds exhibit fewer hind-flipper or fore-flipper slaps or breaches. However, harbor seals (*Phoca vitulina*) and Baikal (*Phoca siberica*) seals slap their fore flippers against their body when disturbed. This sound seems to warn intruders and is quite loud, especially from a wet seal. The most common example of non-vocal communication in pinnipeds is teeth chattering, which provides both an acoustic and a visual aggressive sign. Many pinnipeds also produce a loud snort or hiss as they exhale, especially after a long dive. This exhalation could have communication significance because it is more forceful in some situations, such as when an intruder approaches while they are hauled out.

*3. Sirenia*   Manatees are known to slap the water surface with their tail fluke as a form of communication.

*4. Sea otter*   Riedman and Estes (1990) described sea otter behaviors involving contacting the water's surface ("splashing" and "porpoising"), although it is not known if this behavior is intended as non-vocal communication. Sea otters produce a distinctive and loud "tap, tap, tap" sound as they use a rock on their chest as an anvil to crack shellfish, but it is not known whether this may serve as a deliberate form of communication at times.

*5. Polar bear*   Little is known about non-vocal auditory communication in polar bears, but as a top level predator it may be adaptive to minimize non-vocal sounds.

## B. Vocal Communication

*1. Mysticete cetaceans*   Generally, sounds of baleen whales are very different from those of odontocetes, with a wide range of types and quantity of phonation across mysticete species. Social functions proposed for mysticete sounds include long-range contact, assembly calls, sexual advertisement, greeting, spacing, threat, and individual identification; however, only rarely has a specific sound been associated with a given behavioral event. It is probable that sounds produced by mysticetes serve to synchronize biological or behavioral activities in listeners that promote subsequent feeding or breeding. Known and examined baleen whale sounds seem to fall into three basic categories: low-frequency moans, short thumps or knocks, and chirps and whistles. Additionally, the "songs" of humpback whales have been described in some detail.

A. MYSTICETE LOW-FREQUENCY MOANS   Low-frequency moans are from 1 to 30 sec in duration, with dominant frequencies between 20 and 200 Hz. These sounds can be either pure tones, as in the second-long, 20-Hz sounds of fin whales (*Balaenoptera physalus*), or more complex tones with a strong harmonic structure. Theoretically, these low-frequency, long-wavelength sounds are ideal for long-range communication. A 20-Hz moan from a fin whale has a wavelength of almost 75 m, which means that it passes unimpeded over most obstacles, only bouncing off something large, like a seamount, the water surface, or the ocean bottom. These sounds could travel hundreds of kilometers to reach conspecifics for signaling. Payne and Webb (1971) predicted that theoretically the low-frequency, high-amplitude signals of mysticetes could travel from pole to pole if it were not for interfering water surface and oceanic bottom topography.

Low-frequency sounds (20 Hz) of blue whales (*Balaenoptera musculus*) are recorded across ocean basins at distances of several hundred kilometers. Blue whales are the largest creatures to inhabit the earth; they traverse large expanses in a relatively short time. It is no wonder then that the social structure of these animals reflects a scale that we are only beginning to comprehend. However, tracking the distinctive vocal behavior of this species may provide important clues. "Old blue," a single blue whale, was tracked for nearly 80 days using its distinctive, repeated, 20-Hz signal received by bottom mounted SOSUS (Navy hydrophone) arrays off the coast of the eastern US.

B. MYSTICETE SHORT THUMPS OR KNOCKS   Short thumps or knocks are less than 200 Hz, less than 1 sec long, and are currently known to be produced by right (*Eubalaena* spp.), bowhead (*Balaena mysticetus*), gray, fin, and minke whales (*Balaenoptera acutorostrata*). Clark (1983) recorded and studied southern right whale (*Eubalaena australis*) sounds in relation to behavior and found that their sounds were not random, but were related to social context and activity. Resting whales were least soniferous, whereas mildly social groups produced the most varied suite of sounds, including high, hybrid, and pulsive calls, body and flipper slaps, and forceful blows. Clark's work showed that "up calls" functioned as a request for contact between whales: lone swimming whales often produced up calls that were returned by other whales in the vicinity prior to joining.

C. MYSTICETE CHIRPS AND WHISTLES   Mysticete chirps and whistles tend to be >1 kHz, but change frequency rapidly and are less than 0.10 sec in duration. These pure tones involve harmonics and seem to be produced by most baleen whales.

D. MYSTICETE SONG   Humpback whale songs are probably the most recognized and well known of mysticete vocalizations. Humpback whale males produce what is considered true song because they use elements repeated in phrases and phrases repeated in themes. Songs are very long (up to 30 min), vary at different breeding grounds throughout the world, and change from year to year at a breeding ground. Only males sing while solitary and all sing the same song during each season in the same breeding ground. While males rarely sing outside of the breeding season, the song remains relatively constant from the end of one breeding season to the start of the next. The song could advertise each male's fitness as a mate and control male spacing when advertising to females. For whatever specific purpose, humpback songs represent an evolved signal used by males to communicate information about their internal (e.g., reproductive condition or fitness) and external (e.g., location, proximity) state to conspecifics, likely both females and other males.

*2. Odontocete cetaceans*   Odontocete sounds can be divided broadly into two signal types: pulsed and narrow-band tonal sounds. Some pulsed sounds (clicks) are implicated in echolocation and can be of broad spectral composition as in the bottlenose dolphin, or of narrow-band composition as in the narwhal (Thomas *et al.*, 2004). Other burst-pulsed sounds, described in the literature as barks, squawks, squeaks, blats, buzzes, and moans, have social functions. Narrow-band tonal sounds are continuous signals called whistles. Limitations in audio equipment led to the suggestion that whistles, or frequency-modulated (FM) pure tones, were limited to the human mid-to-upper sonic range of frequency (5–15 kHz), and were of 0.5–2.0 sec in duration. Improvements in technology yielding a more complete bandwidth for recording dolphin sounds indicate that dolphins produce FM pure tones across a broad-frequency range, from 5 kHz to at least 85 kHz. Other FM tonal sounds include screams and chirps.

Research on sound communication in bottlenose dolphins and other delphinids has centered on whistle sounds for pragmatic reasons. The sonic range of whistles is recorded and analyzed easily. Also, whistles are produced by the most common captive species, the bottlenose dolphin, and appear to have no function other than communication. Because the number of non-whistling species, such as the harbor porpoise (*Phocoena phocoena*) and Commerson's dolphin (*Cephalorhynchus commersonii*), is relatively large, it is premature to regard whistles as the principal means for sound communication among odontocetes.

Stereotypical calls produced by members of a social group that vary among populations have been termed DIALECTS, and have been described in at least two species of odontocetes. In British Columbia, matrilineal groups of killer whales have repertoires of call types that are unique to each pod. Cultural transmission has been implicated in the development of orca dialects (Deecke *et al.*, 2000). In sperm whales (*Physter macrocephalus*), codas are stereotyped sequences of 3–40 broad-band clicks usually lasting less than 3 sec in total (Watkins and Schevill, 1977). Rendell and Whitehead (2003) categorized these codas into six acoustic "clans" for populations in the South Pacific and the Caribbean. These vocal clans have ranges that span thousands of kilometers, are sympatric, and contain many thousands of whales. Like killer whale dialects, the codas produced by these clans may result from cultural transmission.

A. ODONTOCETE-PULSED SOUNDS All recorded toothed cetaceans produce pulsed underwater sounds. These sounds can be used for echolocation or communication (Herman, 1980; Au, 1993). They can be divided into two subclasses: pulse trains and burst-pulse sounds. Pulse trains, also called click trains, are sequences of acoustic pulses repeated over time. Individual pulses are about 50 μsec, with varying peak frequencies of 5–150 kHz. The repetition rate of pulses within a click train can vary from 1–2 to several hundreds per second. Click trains are thought to function mainly for echolocation (Thomas *et al.*, 2004).

Burst-pulse sounds can be defined as high repetition rate pulse trains where the interpulse interval is less than 5 μsec, which are similar in shape to echolocation pulses. Because of the high repetition rate in burst pulses, these sounds are not perceived as discrete sequences of sounds by the human ear, but are heard as a continuous sound. Their peak frequencies vary among species from 20 kHz in killer whales to above 100 kHz in Commerson's dolphins. Burst-pulse sounds are proposed as functioning primarily for communication, and have been linked to the social interactions of some species (Herzing, 2004; Blomqvist *et al.*, 2005). The directional characteristics of many pulsed sounds, the relative ease with which they can be localized, their variability, and possibly the intensity with which they can be produced enhance their potential value as communication signals. Indeed, in situations described as alarm, fright, or distress, broad-band high-intensity squeaks have been heard from bottlenose dolphins and harbor porpoises. River dolphins in the family Platanistidae, some members of the family Delphinidae (i.e., killer whales, and Commerson's dolphins), as well as Physeteridae and Phocoenidae, which do not whistle, most likely communicate via pulsed sounds.

B. ODONTOCETE NARROW-BAND TONAL SOUNDS Narrow-band tonal sounds, i.e., whistles, are produced over a range of 5–20 kHz, are FM, and can last from milliseconds to a few seconds. These sounds sometimes have a rich harmonic content that extends into the ultrasonic range of frequencies up to 70–80 kHz for some dolphin species. Whistles vary greatly in contour from simple up-or-down sweeps to FM warbles to U-loops and inverted U-loops. Whistles often grade from one type to another. Whistles are thought to function only for communication, but are not produced by all odontocetes. In at least two odontocete species, false killer whales (*Pseudorca crassidens*) and belugas, whistle frequency shifts upward to avoid low-frequency ambient NOISE.

Of relatively low frequency, whistles travel longer distances in WATER than pulsed sounds. Although less directional than pulsed sounds, whistles probably are localized easily by cetaceans. Bottlenose dolphins and probably other whistling species can produce whistles and clicks simultaneously. Given these attributes, whistles provide a potential vehicle for maintaining acoustic communication and coordination during food search by echolocation. Also, whistles possess little overlap with the major portion of the echolocation frequency spectrum, minimizing potential masking effects. If whistles have species, regional, or individual specificity, this would at least allow for the identification of schoolmates or familiar associates or aid in the assembly of dispersed animals and in the coordination, spacing, and movements of individuals in rapidly swimming, communally foraging herds.

Many different situations can elicit whistling. Whistling could appear as a simple phono-reaction in response to hearing another animal's whistle. Mimicry of whistles or of artificial sounds has been documented in bottlenose dolphins by Peter Tyack (1986) and Louis Herman and colleagues (Richards *et al.*, 1984), and in belugas (Ramirez, 1999), revealing the plasticity of the sound production system (Kuczaj and Yeater, 2006). A correlation between whistling and feeding was noted among wild and captive delphinids; e.g., pods of false killer whales produce more whistles while feeding than during traveling. Dolphins accidentally captured in tuna seine nets whistle intensely. Captive bottlenose dolphins newly introduced into a tank or temporarily separated from a familiar pool mate whistle nearly continuously.

There have been several attempts to inventory the whistle repertoire of wild and captive delphinids. This is often difficult because the whistle contours are so variable. The size of a repertoire, including both whistles and pulsed sounds, is probably limited to fewer than 40 discrete types. However, it is possible that whistles are graded, rather than discrete signals. In a graded system, several basic types of signals transition to one another through a series of intermediate forms.

In 1965, the Caldwells presented the idea of SIGNATURE WHISTLES from observations indicating that each dolphin in a captive bottlenose group tended to produce whistles that were individually distinctive, stereotyped in certain acoustic features, and therefore called "signature" whistles. In the late 1980s and early 1990s, Tyack and colleagues proposed the hypothesis that dolphins use "signature whistles" to refer to each other and themselves (Caldwell *et al.*, 1990 in Leatherwood and Reeves, 1990; Tyack, 1986).

3. *Pinnipedia* Pinnipeds typically produce FM and pulsed sounds. Except for male walruses, pinnipeds do not whistle. The number of vocalizations produced by pinnipeds is correlated with their MATING SYSTEM and whether mating occurs under water or on land (Stirling and Thomas, 2003). Phocids tend to be more vocal under water, especially the true seals that mate under water. In general, otariids are much more vocal on land, often obtaining high densities that result in highly soniferous colonies. Polar pinnipeds are much more vocal under water than temperate or tropical pinnipeds. Early polar explorers reported hearing "eerie, ghost-like sounds from underneath the water." Because of polar bear predation, Arctic pinnipeds are essentially silent while hauled out. In contrast, Antarctic pinnipeds are vocal when they haul out. Comparing the vocal repertoire size and MATING SYSTEM of three species of Antarctic phocids, Stirling and Thomas (2003) found distinctive differences. The Weddell seal congregates in colonies up to 100 mothers with pups, whereas males establish underwater territories beneath the fast ice that are vigorously patrolled and defended with an elaborate repertoire of 34 sounds. Mating in Weddell seals is polygamous; males mate with as many females as will enter their territory. Presumably, the 6-week period that males defend their underwater territory assists females, hauled out on the ice above, in mate selection. The polygynous, but solitary, leopard seal has an intermediate number of underwater vocalizations (9–12, depending on the region), apparently used to establish short-term underwater territories in the pack ice and attract females to mate. This large pinniped predator produces a surprisingly musical repertoire of sounds, but primarily during mating. This seal also

exhibits geographic variations in sounds around the Antarctic. The crabeater seal (*Lobodon carcinophaga*) is seasonally monogamous: a male hauls out on an ice floe with a female, guards her and her pup against attacks from predators, and then conveniently is available for mating when the pup weans. It is unlikely that this male is the pup's father, and the pair bond is well established for the season. Consequently, this pinniped has a single monotonous call.

In all pinnipeds, mothers and pups exchange vocalizations that are important in pup recognition and reuniting the mother and pup after she returns from a foraging bout. Recognition of one's pup is especially important in some otariid mothers that go to sea to forage for up to 7 days before returning to nurse their pup. In many pinnipeds, the vocal repertoire of the mother and pup is unique and distinct from their sounds during other social activities or their underwater sounds. This repertoire occurs mostly while hauled out, but is also used by mothers coaxing their pups into the water or to haulout.

The majority of documented pinniped sounds are within the range of human hearing. Only one study on a captive leopard seal has examined ultrasonic frequencies, with underwater sounds up to 164 kHz (Awbrey *et al.*, 2004). Some species are nearly silent, whereas others have large repertoires that vary by season, sex, age, and whether the animal is in the air or water. Pinniped calls have been described as grunts, rasps, rattles, growls, creaky doors, warbles, trills, chirps, chugs, clicks, and whistles. Clicks are produced, but experimental attempts to demonstrate echolocation have not been successful. These studies, however, were on California sea lions and harbor seals and some researchers suggest that ECHOLOCATION, if present in pinnipeds, would more likely occur in the polar pinnipeds, which live in ice-covered waters and total darkness during the polar winters (Thomas *et al.*, 2004).

Phocid calls are primarily between 100 Hz and 15 kHz, with peak spectra less than 5 kHz. Typical source levels of underwater sounds are 130 dB re 1 μPa, but are as high as 193 dB re 1 μPa in a territorial Weddell seal. Northern elephant seals (*Mirounga angustirostris*) are reported to produce infrasonic vibrations while vocalizing in air (Fig. 4).

One of the most elaborate repertoire of sounds is from the Weddell seal, which has a separate repertoire of sounds for communicating while hauled out, from its sounds for underwater contexts (Terhune *et al.*, 2001). At the Hutton Cliffs colony in the Antarctic, Weddell seals had 34 types of underwater sound, including trills, chugs, chirps, guttural glugs, and knocks. Eleven types of trills are used exclusively by males for territorial advertisement and defense and could be used in a graded context to convey the degree of warning, i.e., shortest, quietest trills are just a reminder, but long, loud trills are an emphatic warning to an intruder. This species also uses prefixes and suffixes with main call types, seeming to warn or emphasize a message. The trills are as long as 75 sec. The repertoire on the opposite sides of the Antarctic (Palmer Peninsula and near Davis Station) shows geographic variations, including some unique usage of "mirror-image" calls, i.e., an upsweep followed by the mirror-image, downsweep. Male pups as young as 2 months try to perfect the long, loud trills, using comical, voice-cracking sweeps reminiscent of adolescent humans.

Otariid airborne sounds range from 1 to 4 kHz, with harmonics up to 6 kHz. Barks in water are slightly louder than in air, and both center around 1.5 kHz. Individual California sea lion sounds have unique variations suggesting signal components for identity. Odobenid sounds are low in frequency, 500 Hz, with a peak of 2 kHz. Under water, walruses have a unique bell-like sound, but also produce clicks and whistles. Recent studies indicate that territorial male walruses have their own distinctive sound patterns.

*4. Sirenia* Sounds of sirenians are low in amplitude and probably only propagate short distances. From field observations of mother–calf manatee pairs, it appears that vocalizations play a key role in keeping the mother and calf together. Some researchers even describe this vocal exchange as dueting, where the mother and calf exchange chirps (Reynolds and Odell, 1991). Another example of communication in manatees included a mother–calf pair on opposite sides of a flood control gate. For nearly 3 h, the mother placed her head in the narrow opening and vocalized to the calf until the gate opened enough for the calf to swim through. Although most evidence is anecdotal, sirenians (at least manatees) seem to use sounds to communicate with conspecifics.

Dugongs are highly social, occurring in groups up to several hundred animals. Sound probably plays the most important role in communication. Vocalizations of dugongs are low frequency, ranging from 1 to 8 kHz and seem to be especially important in maintaining the mother–calf bond. Studies in the clear waters of Shark Bay Australia suggest that dugong males establish territories to attract estrous females. Reynolds and Odell (1991) suggested that low-frequency vocalizations play a role in mate attraction in dugongs.

Manatee and dugong sounds are described as chirps, whistles, squeals, barks, trills, squeaks, and frog-like calls. West Indian manatee (*Trichechus manatus* spp.) sounds range from 0.6 to 5.0 kHz, whereas Amazonian manatees (*Trichechus inunguis*) produce sounds reaching 10 kHz, with distress calls having harmonic structure up to 35 kHz (Reynolds and Rommel, 1999). Manatee vocalizations appear to be stereotypical, with little variability between individuals and subspecies. Vocalizations consist primarily of short tonal harmonic complexes with small-frequency modulations at the beginning and end (Nowacek *et al.*, 2003). Nonetheless, there is evidence that variation in Amazonian manatee vocalizations could allow for individual recognition (Sousa-Lima *et al.*, 2002). Dugongs produce calls between 0.5 and 18.0 kHz with maximum energy between 1 and 8 kHz (Reynolds and Rommel, 1999).

*5. Sea otter* In sea otters, social interactions, pup care, and mating occur at the water surface; still, little is known about sea otter vocal behavior. No underwater vocal sounds have been reported for sea otters. The inter-tidal zone is a noisy, churning environment that would not be a good environment for exchange of sounds and would make recording difficult. In addition, sea otters forage singly and probably do not need to communicate while foraging. Kenyon (1975)

**Figure 4** *Bull northern elephant seals* (Mirounga angustirostris) *competing for territory.*

**C**

provided a detailed summary of sea otter sounds heard in air. He described their sounds as (1) baby cry—a sharp, high-pitched "waah-waah" sound used by pups in distress situations or when wanting to attract mother's attention; (2) scream—given by adults in distress or when a female has lost her young (it is an "ear-splitting" version of the pup cry detectable 0.5 miles away in the wild); (3) whistle or whine ("whee-whee")—a high-pitched sound resembling a human whistle (denotes frustration or mild distress, given by captive sea otters when feeding is delayed, detectable 200 m away in the wild); (4) coo ("ku-ku-ku")—produced by females before and after mating or while eating a "particularly pleasing food," detectable only 34 m away; (5) snarl or growl—originating deep in the throat produced by a newly captured sea otter, audible only a few meters away; (6) hiss—short, explosive cat-like hiss used in startle situations; (7) grunt—soft groaning sound produced while eating; (8) bark—a staccato bark trailing off into a whistle, indicative of frustration produced by a young male; and (9) cough, sneeze, and yawn as in other mammals. A subsequent study by McShane *et al.* (1995) confirmed many of these vocal categories, and indicated that sea otters use a variety of graded signals that likely enhance the ability to share detailed and highly variable information between known individuals. Furthermore, the structural characteristics of the calls could easily allow for individual recognition.

*6. Polar bear*    Polar bears have a variety of sounds used in different contexts. Growls serve as a warning to others bears and defense of a food source. Hissing, snorting, loud roars, and groans or grunts are aggressive sounds. Chuffing was documented as a response to stress, whereas mother polar bears will produce soft chuff sounds or low growls when scolding cubs (Ovsyanikov, 1996).

## VI. Conclusions

Animals live in an ever-changing world. Reactions and responses to environmental and social variables must be flexible and adaptive for survival and reproduction. Examination of signaling behavior and subsequent receiver responses provides a window into nonhuman minds, as well as to the social complexity of other species. It can be assumed that evolutionary processes are at work on signals to keep them informative and useful to individuals. Ecological factors, coupled with social relationships and interactions, provide the principal force in the evolution of communication systems (Hauser, 1997). Foraging, mating, and parental strategies are examples of components that influence signaling behavior. In marine mammals, coastal or oceanic species living in relatively clear water may be more likely to use visual signals (e.g., postures, coloration patterns) than species inhabiting riverine or turbid environments. Similarly, amphibious species require a suite of signals useful both in air and under water. Differential communication is also evidenced in the foraging methods of several delphinid species. Communal foragers have more complex signals compared with more solitary hunters. Frequent interactions with conspecifics necessitate a higher rate of information exchange than for solitary species. Observing and examining the social and ecological differences among individuals and groups will help elucidate the mechanisms underlying the use and evolution of different signals to exchange information among individuals, i.e., to communicate.

### *See Also the Following Articles*

### *References*

Anderson, H. T. L (1969). "The Biology of Marine Mammals". Pergamon Press, New York.

Au, W. W. L. (1993). "The Sonar of Dolphins." Springer-Verlag, New York.

Awbrey, F. T., Thomas, J. A., and Evans, W. E. (2004). Ultrasonic sound production in a captive leopard seal (Hydrurga leptonyx). *In* "Echolocation in Bats and Dolphins," pp. 535–541. University of Chicago Press, Chicago.

Blomqvist, C., Mello, I., and Amundin, M. (2005). An acoustic play-fight signal in bottlenose dolphins (*Tursiops truncatus*) in human care. *Aquat. Mamm.* **31**, 187–194.

Bradbury, J. W., and Vehrencamp, S. L. (1998). "Principles of Animal Communication." Sinauer, Sunderland.

Clark, C. W. (1983). Acoustic communication and behavior of the southern right whale (*Eubalaena australis*). *In* "Communication and Behavior in Whales" (R. Payne, ed.), pp. 163–198. Westview Press, Boulder.

Connor, R. C., Mann, J., and Watson-Capps, J. (2006). A sex-specific affiliative contact behavior in Indian Ocean bottlenose dolphins, *Tursiops* sp. *Ethology* **112**, 631–638.

Croll, D. A., Clark, C. W., Acevedo, A., Tershy, B., Flores, S., Gedamke, J., and Urban, J. (2002). Only male fin whales sing loud songs. *Nature* **417**, 809.

Deecke, V. B. (2006). Studying marine mammal cognition in the wild: A review of four decades of playback experiments. *Aquat. Mamm.* **32**, 461–482.

Deecke, V. B., Ford, J. K., and Spong, P. (2000). Dialect change in resident killer whales: Implications for vocal learning and cultural transmission. *Anim. Behav.* **60**, 629–638.

Foote, A. D., Osborne, R. W., and Hoelzel, A. R. (2004). Environment: Whale-call response to masking boat noise. *Nature* **428**, 910.

Hauser, M. D. (1997). "The Evolution of Communication." MIT Press, Cambridge.

Herman, L. M. (1980). "Cetacean Behavior: Mechanisms and Functions." Wiley, Inc, New York.

Herzing, D. L. (2004). Social and nonsocial uses of echolocation in free-ranging *Stenella frontalis* and *Tursiops truncatus*. *In* "Echolocation in Bats and Dolphins" (J. A. Thomas, C. Moss, and M. Vater, eds), pp. 404–409. University of Chicago Press, Chicago.

McShane, L. J., Estes, J. A., Riedman, M. L., and Staedler, M. M. (1995). Repertoire, structure, and individual variation of vocalizations in the sea otter. *J. Mammal.* **76**, 414–427.

Leatherwood, S., and Reeves, R. R. (1990). "The Bottlenose Dolphin." Academic Press, New York.

Kenyon, K. (1975). "The Sea Otter in the Eastern Pacific Ocean." Dover Publications, New York.

Kuczaj, S. A., and Yeater, D. B. (2006). Dolphin imitation: Who, what, when and why? *Aquat. Mamm.* **32**, 413–422.

Mass, A., and Ya Supin, A. (2007). Adaptive features of aquatic mammals' eye. *Anat. Rec.* **290**, 701–715.

Miksis-Olds, J., Donaghay, P. L., Miller, J. H., Tyack, P. L., and Nystuen, J. A. (2007). Noise level correlates with manatee use of foraging habitats. *J. Acoust. Soc. Amer.* **121**, 3011–3020.

Norris, K. S., Würsig, B., Wells, R. S., and Würsig, M. (1994). "The Hawaiian Spinner Dolphin." University of California Press, Berkeley.

Nowacek, D. P., Casper, B. M., Wells, R. S., Nowacek, S. M., and Mann, D. A. (2003). Intraspecific and geographic variation of West Indian manatee (*Trichechus manatus* spp.) vocalizations. *J. Acoust. Soc. Am.* **114**, 66–69.

Ovsyanikov, N. (1996). "Polar Bear: Living with the White Bear." Voyageur Press, Stillwater.

Payne, R., and Webb, D. (1971). Orientation by means of long range acoustic signaling in baleen whales. *Ann. NY Acad. Sci.* **188**, 110–142.

Ramirez, K. (1999). "Animal Training: Successful Animal Management through Positive Reinforcement." Shedd Aquarium, Chicago.

Rendell, L. E., and Whitehead, H. (2003). Vocal clans in sperm whales (Physeter macrocephalus). *Proc. R. Soc. Lond. B Biol. Sci.* **270**, 225–231.

Reynolds, J. E., III, and Odell, D. K. (1991). "Manatees and Dugongs." Facts on File, New York.

Reynolds, J. E., and Rommel, R. (1999). "Biology of Marine Mammals." Smithsonian Institution Press, Washington, DC.

Richardson, W. J., Greene, C. R., Malme, C. I., and Thomson, D. H. (eds) (1995). "Marine Mammals and Noise." Academic Press, New York.

Richards, D. G., Wolz, J. P., and Herman, L. M. (1984). Vocal mimicry of computer-generated sounds and vocal labeling of objects by a bottlenosed dolphin, *Tursiops truncatus*. *J. Comp. Psychol.* **98**, 10–28.

Riedman, M. L., and Estes, J. A. (1990). The sea otter (*Enhydra lutris*): Behavior, ecology, and natural history. *US Fish Wildlife Serv., Biol. Rep.* **90**(14).

Schusterman, R. J., Thomas, J. A., and Wood, F. G. (eds) (1986). "Dolphin Cognition and Behavior: A Comparative Approach." Lawrence Erlbaum Associates, Hillsdale.

Sousa-Lima, R. S., Paglia, A. P., and da Fonseca, G. A. B. (2002). Signature information and individual recognition in the isolation calls of Amazonian manatees, *Trichechus inunguis* (Mammalia: Sirenia). *Anim. Behav.* **63**, 301–310.

Southall, B. L., Schusterman, R. J., and Kastak, D. (2000). Masking in three pinnipeds: Underwater, low-frequency critical ratios. *J. Acoust. Soc. Am.* **108**, 1322–1326.

Southall, B., *et al.* (13 authors) (2007). Marine mammal noise exposure criteria: Single sources and single individual. *Aquat. Mamm.* **33**, 411–521.

Stirling, I. (1999). "Polar Bears." University of Michigan Press, Ann Arbor.

Stirling, I., and Thomas, J. A. (2003). Relationships between underwater vocalizations and mating systems in phocid seals. *Aquat. Mamm.* **29**, 227–246.

Terhune, J. M., Healey, S. R., and Burton, H. R. (2001). Easily measured call attributes can detect vocal differences between Weddell seals from two areas. *Bioacoustics* **11**, 211–222.

Thomas, J. A., and Kastelein, R. A. (1990). "Sensory Abilities of Cetaceans: Laboratory and Field Evidence. NATO Life Science Series", vol. 196. Plenum Press, New York.

Thomas, J. A., Moss, C. F., and Vater, M. (2004). "Echolocation in Bats and Dolphins." University of Chicago Press, Chicago.

Tyack, P. L. (1986). Whistle repertoires of two bottlenose dolphins, *Tursiops truncatus*: Mimicry of signature whistles? *Behav. Ecol. Sociobiol.* **18**, 251–257.

Vauclair, J. (1996). "Animal Cognition: An Introduction to Modern Comparative Psychology." Harvard University Press, Cambridge.

Watkins, W. A., and Schevill, W. E. (1977). Sperm whale codas. *J. Acoust. Soc. Am.* **62**, 1485–1490.

# Competition with Fisheries

## Éva E. Plagányi and Douglas S. Butterworth

### I. Introduction

From an ecological perspective, competition is a situation where the simultaneous presence of the two competitors is mutually disadvantageous. This review focuses on biological interactions (also known as trophic interactions), and specifically the competition for food and fishery resources between marine mammals

and fisheries, in contrast to operational interactions in which marine mammals damage or become entangled in fishing gear with negative consequences for both the fishery and the animals (Northridge, 1991, for a review).

Interactions due to bycatches in fisheries constitute one of the major threats to marine mammals (see Fisheries, interference with). These two forms of conflict are sometimes difficult to separate because e.g., animals may damage fishing gear in the process of removing fish therefrom. A third important marine mammal–fishery interaction concerns anisakid nematodes whose larvae use commercial fish and squid for transmission to marine mammals (see Parasites), but this is not of direct relevance to the current topic.

Competitive interactions between marine mammal populations and fisheries can either be "direct" or "indirect." In the former case, the two groups share a common prey species whereas in the latter case, e.g., a marine mammal may prey on a species that is also an important component of the diet of a commercial fish species.

Perceived conflicts between marine mammals and humans in pursuit of common sources of food have come increasingly to the fore in recent years. Escalating pressures on shared resources are expected in the future because of both the increasing marine mammal populations and an increasing human population. Reductions in directed takes in response to recognition that several populations of marine mammals were heavily over-exploited in the nineteenth and earlier part of the twentieth century, as well as a widespread change in people's perceptions of whether marine mammals should still be regarded as renewable resources available for harvest, have meant that several marine mammal populations are currently on the increase, sometimes by as much as 5–10% per annum (Bowen *et al.*, 2006). From the human population perspective, the Food and Agriculture Organization of the United Nations (FAO, 2006) has estimated that over 2.6 billion people worldwide currently rely on fish and shellfish for more than 20% of their animal protein. Marine capture fisheries seem unlikely to much exceed the present global level, so that ability to meet the demands from an increasing human population will be heavily dependent on a continuation of the recent rapid increase in aquaculture production.

Commercial fisheries and marine mammals frequently target the same fish species, so that faced with possible shortages in marine food production in the future, it is likely that the possible impacts of growing marine mammal populations on the sustainable harvest of commercial fisheries will be vigorously questioned. Concerns about the consequences for fisheries of an increasing marine mammal population have already been expressed in southern Africa, e.g., where in 1990, Cape fur seals (*Arctocephalus pusillus pusillus*) were estimated to consume some 2 million tons of food a year. Considering that this amount was about the same as the annual human catch of fish in the region, and that the fur seal population was anticipated to increase further, the reasons for concerns and potential for conflict are obvious. A second example concerns the Pacific Ocean, where marine mammals are estimated to consume about 150 million tons of food per annum, which is some 3 times the current annual fish harvest by humans.

This chapter first presents a brief summary of some specific examples which address the question of whether marine mammal populations have negatively impacted the potential yields from fisheries through competition. Examples of perceived competitive interactions are included because the evidence is generally inconclusive. Secondly, some examples pertinent to the reverse—whether fisheries negatively impact marine mammals—are summarized.

## II. Detrimental Effects of Marine Mammals on Fisheries

### A. Pinnipeds (Seals, Sea Lions and Walruses)

In the early 1990s, catastrophic collapses occurred in the cod (*Gadus morhua*) fisheries on the East coast of Canada. Although several hypotheses have been posited to explain this, the most likely cause was overfishing. Harp seal (*Pagophilus groenlandicus*) populations off Newfoundland and Labrador increased at an estimated rate of 5% per annum over the 1980s and 1990s and are known to consume a substantial tonnage of juvenile cod. The socio-economic implications of the collapse of the cod fishery were huge, with some 40,000 fishers rendered out of work, and there is an obvious temptation to argue a causal relationship between the failure of the cod population to recover as rapidly as expected after its protection and the increase in harp seal abundance. Although the results of at least one ecosystem modeling study support the hypothesis that the recovery of these cod populations is being retarded to some extent by the increased abundance of harp seals, ecosystem models generally have poor predictive reliability, largely because of data limitations.

Demonstrating that either a fishery or a marine mammal will be adversely affected as a result of an increase in the removals by one party of a limited resource is not simple. Inferences based on assumptions of a linear relationship between predator and prey abundance are often incorrect because of the complex nonlinear interactions in an ecosystem. For example, off the west coast of South Africa, seals consume almost as much hake as is taken by the commercial fishery (Fig. 1). However, the commercially valuable hake consists of two species, a shallow-water (*Merluccius capensis*) and a deep-water species (*M. paradoxus*), with the larger of the shallow-water species eating the smaller individuals of the deep-water species. The results of multispecies models suggest that the net effect of a seal cull would be less hake overall because fewer seals would mean more shallow-water hake, and hence more predation on small deep-water hake (Punt and Butterworth, 1995). This study highlights the complexity of predation,

food-fish and fishery interactions and hence the difficulties of demonstrating conclusively that marine mammals are in direct competition with humans for food fish, as may superficially appear to be the case.

### B. Whales

Numerous multispecies modeling studies have been employed to investigate the direct and indirect effects of common minke whales (*Balaenoptera acutorostrata*) on cod, herring (*Clupea harengus*), and capelin (*Mallotus villosus*) fisheries in the Greater Barents Sea. Minke whales are abundant in this region and prey on all three species, prompting the question of whether or not fishermen could expect greater catches if the populations of these marine mammals were reduced. The indications of these studies are that there is competition between the whales and the fishers in this region, and that the fisheries are likely to respond linearly to changes in whale abundance. The studies estimate that each minke whale reduces the potential annual catches of both cod and herring by some 5 metric tonnes. Similarly, studies off Iceland suggest that the piscivorous minke, humpback (*Megaptera novaeangliae*), and fin (*B. physalus*) whales may be having a considerable impact on the region's cod stock. The cod fishery is of key importance to the Icelandic economy, and the rebuilding of the cod population and catches are recognized as an important economic consideration. It is therefore not surprising that arguments have been put forward that there is a need to reduce whale populations to permit commercial fisheries to increase.

Marine mammals are thought to exert relatively minor influences on systems such as the North Sea and Baltic Sea, whereas they have been identified as potentially serious competitors off, e.g., the northeastern USA, a region that includes important fishery areas such as the Gulf of Maine and Georges Bank. The latter region exemplifies the conflicts that can arise between fishery management plans tasked with rebuilding prey populations and prescriptions, by the USA Marine Mammal Protection Act in this case, to facilitate an increase in the abundance of marine mammal predators.

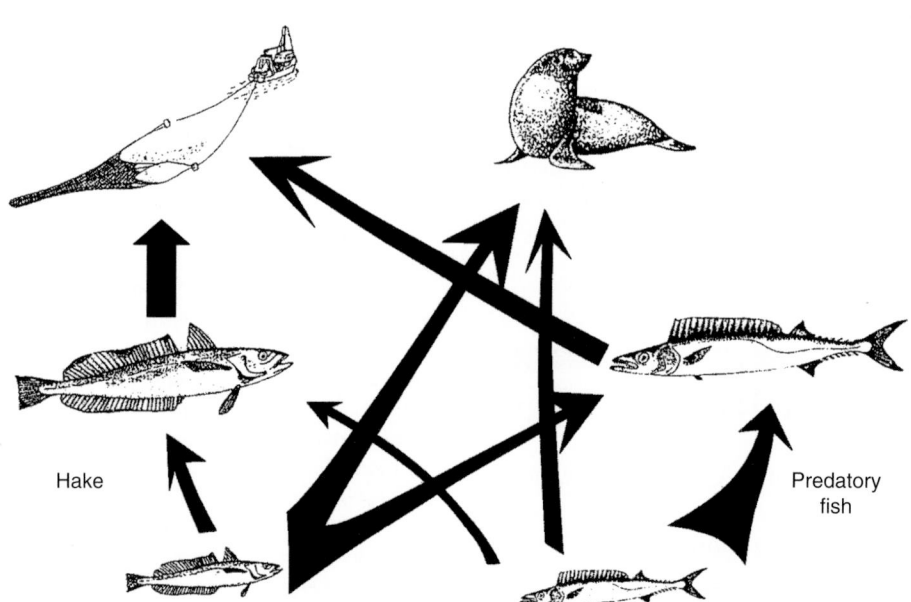

**Figure 1** *Schematic showing the complexities of predation, food-fish and fishery interactions as summarized in a minimal realistic model of Cape fur seals and Cape hake interactions off South Africa. Reproduced with permission from Punt (1994).*

## C. Small Cetaceans

In many areas of the world, coastal fishermen consider dolphins as serious competitors, although retaliation by the fishermen usually has only minor effects on the populations or on their perceived damage. Common dolphins (*Delphinus delphis*) in the Mediterranean have often been perceived as a threat to the purse-seine and trawl fisheries operating in these waters, and as a result have been deliberately caught in direct retaliation. Declines in this population have been attributed to both direct and incidental catches by the fishery.

The largest hunt designed to reduce the perceived level of competition with fisheries took place in the Black Sea from 1870. In the mid-1900s, tens of thousands of dolphins and porpoises were killed every year as a result of fishing industry claims of competition. Other examples of cetacean kills due to perceived competition include the bombing of belugas (*Delphinapterus leucas*) from the air by the Quebec government in the 1920s and 1930s, the commissioning by the Icelandic government in 1956 of a US naval vessel to kill killer whales (*Orcinus orca*), and the use of explosives and firearms by Alaskan fishers in the mid-1980s to eliminate local killer whales.

## D. Sea Otters

Sea otters prey on a variety of marine invertebrates such as urchins and abalone. Off southern California, southern sea otters (*Enhydra lutris nereis*) have been labeled by some as responsible for the decline of the abalone (*Haliotis* spp.) fishery, but there is little direct evidence to support this notion. The commercial abalone fishery in California was closed in late 1997, and factors such as commercial fishing, poaching, disease, and changing environmental conditions are all thought to have contributed to the decline of these commercially valuable shellfishes. Although the southern sea otter population is listed as federally threatened, southward movements of otters increased the overlap between otter fishing grounds and abalone fishing areas, and several conflicts exist with commercial and recreational abalone fisheries. Southern sea otters in some areas likely exert a greater impact on red abalone (*H. rufescens*) populations than human harvests.

## III. Detrimental Effects of Fisheries on Marine Mammal Populations

### A. Pinnipeds

The western Alaska population of Steller sea lions (*Eumetopias jubatus*) declined by 75% between 1976 and 1990, with subsequent continuing declines of the western stock resulting in its listing in 1997 as an endangered species under the US Endangered Species Act. Several groups have argued that this decline is due in part to the large fishery harvest of walleye pollock (*Theragra chalcogramme*), simultaneously a key source of food for sea lions and the most important US commercial fishery. There is considerable evidence to suggest that nutritional stress played a role in reducing both the recruitment and the reproductive rates of Steller sea lions (DeMaster and Atkinson, 2002). Measures to reduce the perceived competition between sea lions and fisheries for ground fish stocks include the establishment of "buffer" (no-trawl) areas to include important locations where the sea lions breed, feed, and rest, as well as specifying a pollock harvest which is more evenly distributed over the remaining areas and spread throughout the year. However, the results of modeling studies indicate that the observed sea lion population decline cannot be explained solely through trophic interactions, and rather is more likely linked to inadequate recruitment and shifts in environmental conditions which lead to changes in the favored complex of species

(DeMaster and Atkinson, 2002; Cornick *et al.*, 2006; Fay and Punt, 2006). Moreover, studies such as Trites *et al.* (1999) highlight the difficulties of predicting the direction and magnitude of a change in an ecosystem arising from a reduction in predation or fishing pressure. They posit that, paradoxically, Steller sea lion and northern fur seal (*Callorhinus ursinus*) populations might realize greater benefits if adult pollock and large flatfish were more heavily fished. This competitive release effect may result because, e.g., pollock are cannibalistic and hence decreased adult pollock abundance as a result of heavier fishing may result in increased numbers of juvenile pollock available to marine mammals.

## B. Whales

Competition effects are difficult to quantify, but it has been proposed by Whitehead and Carscadden (1985) that the collapse of the eastern Canadian capelin fishery in the 1970s had a negative effect on fin whales. They suggest that a shortage of capelin might have allowed humpback whales to out-compete fin whales because the latter rely principally on capelin as a prey source.

If competitive predation between a marine mammal and fishery occurs, it implies that the marine mammal population is limited by food availability and hence it should be possible to demonstrate a response of some vital population parameters to a change in food availability. Past probable population increases of several krill-eating marine mammals, such as minke whales, crabeater seals, and fur seals, have been attributed by some investigators to a likely large increase in the availability of krill (*Euphausia superba*) in Antarctic waters (Mori and Butterworth, 2006). Following the substantial reduction through overexploitation of large whale populations during the early twentieth century, some 50–150 million tons of "surplus" krill is argued to have become available annually to other predators. This "krill surplus" hypothesis (Laws, 1977) is yet to be universally accepted, and questions remain concerning potential corroborative evidence, e.g., whether trends in the mean age at maturity of minke whales and crabeater seals (*Lobodon carcinophaga*) are fully consistent with the changes in food availability that the hypothesis suggests.

Trites *et al.* (1999) suggested that marine mammal populations can be quickly reduced through reductions in prey abundance but show a generally slow recovery when abundant food becomes available.

## C. Small Cetaceans

Dolphin populations that have restricted or localized coastal distributions may be particularly susceptible to competition with fishers for limited food resources. Prey depletion is considered of primary or secondary importance in causing habitat degradation and loss of at least four small cetacean species in the Mediterranean and Black Seas (Notarbartolo di Sciara *et al.*, 2002). Prey depletion is the most likely proximate cause of declines in short-beaked common dolphins in the Mediterranean, with observed malnutrition in other marine mammals, such as bottlenose dolphins, also being attributed to overfishing of their prey stocks and intensive trawling (Bearzi *et al.*, 2003). Reduced prey availability due to fishing may also play a more indirect role in compromising animal health, which is suggested to have led to the large die-off of Mediterranean striped dolphins in 1990–1992 (Aguilar, 2000). As stocks of some preferred fished species, such as sardines and anchovies in the Mediterranean, are depleted, there is a concern that fishing will refocus on other small pelagic fish that are simultaneously the prey of common dolphins and important in meeting the growing demands of the aquaculture industry (Bearzi *et al.*, 2003).

**C**

**C**

### D. Sea Otters

Recent declines—in excess of 50% since the mid-1980s—in northern sea otter (*Enhydra lutris kenyoni*) populations in Alaska have recently led to its listing as "Threatened." Although the reasons for this decline are still subject to considerable debate, one hypothesis suggests the declines are indirectly linked to competition with fisheries (Doroff *et al.*, 2003). As discussed above, fishing is argued to be one of the factors contributing to the decline of pinniped populations (harbor seals *Phoca vitulina* and Steller sea lions) in some of Alaska's Aleutian Islands. Killer whales preferentially feed on pinnipeds in this region, but as a result of the decline in pinnipeds, they have switched to sea otters as prey. Estes *et al.* (1998) argueed that reduced populations of fish prey responsible for providing high caloric and nutritive value to pinnipeds, may impact not only directly on pinniped populations but also indirectly on killer whale and sea otter populations.

### E. Sirenians (Dugongs and Manatees)

Although direct kills and incidental capture in fishery gear are problems, these mammals feed mostly on vascular aquatic plants so that there is no direct competition with humans for a shared food resource.

### IV. Assessing the Competitive Effects

Commercial fishermen in many parts of the world perceive marine mammals as serious competitors for a scarce resource, whereas others argue that marine mammals are being used simply as scapegoats for failed fisheries management policies. Scientific evidence is therefore increasingly being sought to settle these disputes, but it is becoming increasingly appreciated that the scientific methodologies required to address them are complex, time consuming, data hungry, and beset with difficulties.

Initial attempts to quantify the impact of consumption by marine mammals on fish catches used a simple approach. They took account of the fact that, particularly for pinnipeds, the sizes of fish eaten tend to be smaller than are taken by commercial fisheries (Beddington *et al.*, 1985). Thus 1 ton of a commercially desired fish species eaten by seals say, does not translate exactly into 1 ton less in the allowable catch for fishers. This is because although a fish eaten by a seal would have grown larger by the time it became vulnerable to fishing, it might also have died before reaching the size as a result of other sources of natural mortality.

It is now acknowledged that such computations, which essentially treated marine mammals as the equivalent of another fishing fleet, are likely to be inadequate because of oversimplification. There are three complicating factors which need to be addressed in performing more realistic computations, while still accepting that both data and computing power limitations necessarily restrict the degree of complexity that is viable to incorporate in multispecies models. The first concerns how many of the large number of interacting species in any ecosystem need to be considered. Secondly, do age-structure effects need to be taken into account? One instance where this can become important is when one species that predates on the small juveniles of a second, finds itself the prey of the larger adults of that same species. For example, whiting *Merlangius merlangus* feeds extensively on the youngest (0+ and 1+) age classes of the commercially valuable cod, in turn an important predator on the smaller individuals of whiting. Finally, the customary modeling assumption that species interactions occur homogeneously over space may well be sufficiently

flawed to invalidate results. Moreover, the distribution of seal breeding and resting sites does not necessarily reflect their feeding distributions. Modern animal tagging technology has demonstrated, e.g., that gray seals (*Halichoerus grypus*) and southern elephant seals (*Mirounga leonina*) may travel hundreds of kilometers to a preferred feeding site.

However, with the development of several recent models (Koen-Alonso and Yodzis, 2005; Boyd *et al.*, 2006) as well as generalized multispecies modeling tools such as Ecosim and Ecospace (Walters *et al.*, 1997), groundwork is being laid to provide a more reliable basis for scientific evaluation of these competitive effects (Plagányi, 2007, for a review).

### V. Considering the Influence of Fish and Krill Harvesting on the Ecosystem

The adoption of the Convention for the Conservation of Antarctic Marine Living Resources (CCAMLR) was a watershed in international fishing agreements in that it was the first to acknowledge the importance of maintaining the ecological relationships between harvested, dependent, and related populations of marine resources. Krill is the primary food source of a number of marine mammal species, and concern has been expressed that the rapidly expanding krill fishery might negatively impact or retard the recovery of previously over-exploited populations such as the large baleen whales of the Southern Hemisphere. Thomson *et al.* (2000) calculated, within a modeling procedure, the level of krill fishing intensity that would reduce krill availability, and hence the population of a predator to a particular level. Moreover, research is currently in progress regarding the subdivision of the precautionary catch limit for krill among 15 small-scale management units (SSMUs) in the Scotia Sea, to reduce the potential impact of fishing on land-breeding predators.

In general, initiatives such as these pursued under CCAMLR recognize the need to balance the needs of predators with the socio-economic pressures underlying fishery harvests and represent a realistic step forward in resolving some of the management quandaries resulting from competition for limited marine resources.

### VI. Food Web Competition

Trites *et al.* (1997) and Kaschner *et al.* (2001) assessed the competition between fisheries and marine mammals for prey and primary production in the Pacific and North Atlantic Oceans respectively, concluding that marine mammals in these areas collectively consume about 3 times as much food as humans harvest. Kaschner (2004) presented similar arguments based on a global analysis of catch and food consumption by marine mammals and fisheries. In the Northern Hemisphere, the greatest overlaps occur with pinnipeds and dolphins and porpoises, whereas in the Southern Hemisphere overlaps between baleen whales and large toothed whales are the most substantial. The dietary overlap between the prey items of marine mammals and fisheries is however less than the foregoing might seem to suggest, because specialized feeding habits mean, e.g., that some of the targeted prey are either unfit for human consumption or are not currently viable for commercial harvest.

Trites and others argue that whilst direct competition between fisheries and marine mammals for prey appears limited, indirect competition for primary production may be a cause for concern. Such so-called food web competition may occur if there is overlap between the trophic flows supporting the two groups (see Fig. 2). Evidence in support of food web competition between marine mammals and fisheries

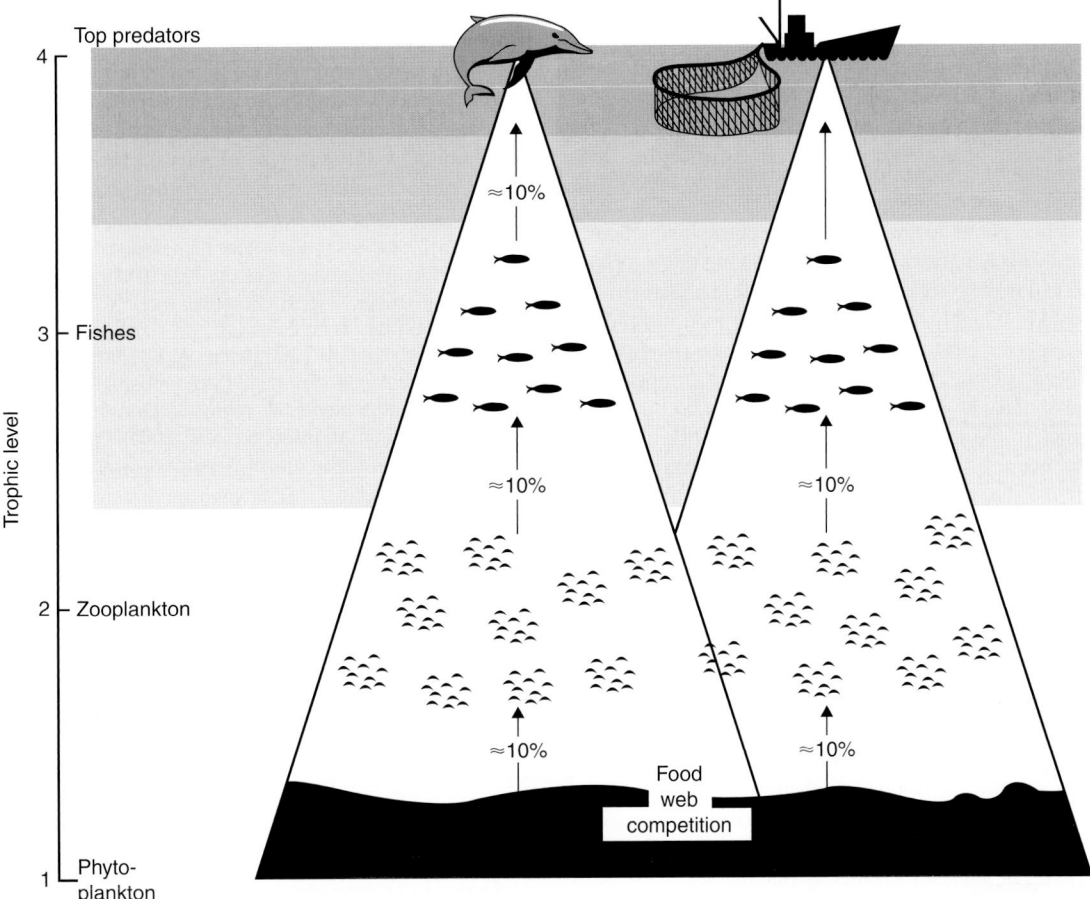

**Figure 2** *Schematic example of indirect competition for food by marine mammals and fisheries. The representation shows how top predators, such as marine mammals, may be affected by fisheries because of limits on the primary productivity available to support the two groups. Thus even though the mammals' prey and species taken by fisheries may not overlap, so-called food web competition occurs at the base of the food pyramids. Reproduced with permission from Trites et al. (1997).*

is provided by a negative correlation between estimates of primary production required to support fisheries catches and to support the number of marine mammals estimated in the different FAO Statistical Areas in the Pacific Ocean.

## VII. Additional Indirect Interactions

There are a number of additional instances, as summarized in Plagányi and Butterworth (2005), where fishing has impacted (or is likely to directly impact) marine mammals by damaging critical habitats upon which they depend, by altering the structure of ecosystems or by otherwise altering marine mammal population dynamics and/or population parameters. For example, trawling may have important effects on the fish populations upon which both fisheries and marine mammals depend. Fisheries-generated habitat destruction may impact most heavily on species such as the gray whale *Eschrichtius robustus*, which feeds primarily on benthic organisms such as amphipods, and the walrus which feeds on molluscs and other bottom-dwelling invertebrates. Other destructive fishing practices such as the use of explosives or cyanide in coral areas may seriously damage the habitat, with consequent repercussions for fish and dependent marine mammal populations.

Noise pollution from fishing vessels has been implicated in affecting marine mammals negatively both by interfering with the sensitive echolocation systems of toothed cetaceans and thereby indirectly reducing their foraging efficiency and by routing marine mammals away from preferred feeding areas (Roussel, 2002).

Fishing changes the overall size distribution of fish, and such changes over time may either increase the competitive overlap between fishers and marine mammals or may prove positive for marine mammals with a preference for smaller fish. However, a shift to preying on smaller fish may have negative effects on the bioenergetics of some species.

Several cetacean and pinniped species are known to feed in association with trawlers, as the resulting disturbance may bring prey species higher in the water column where they are easier targets for marine mammals, or alternatively trawlers might beneficially concentrate food (Fertl and Leatherwood, 1997). Although discards associated with trawling provide ready forage for several species (including dolphins and seals), the advantages of opening new feeding niches for marine mammals are likely offset by other negative impacts of trawlers and alterations in marine mammal foraging strategies.

Fishing may play yet another indirect role in increasing the mortality rates of a marine mammal species by forcing animals to either increase their foraging time or to forage further afield in areas where they are themselves at higher predation risk. However, some shark fisheries may have a positive indirect effect on marine mammals because of the associated reduced predation on (in particular young) animals.

## VIII. Summary

Despite a persistent notion worldwide that there is a mass-for-mass equivalence in the prey of marine mammals and the yields available to fishers, the evidence points to much more complicated situations in which this is hardly likely to be the case. Furthermore, the complexity of ecosystems could well be such that the response to a marine mammal cull, e.g., could be highly diffused through the food web, involving many other species (Yodzis, 2000). In some cases, competition effects are reduced because, e.g., one of the putative competitors in fact reduces the abundance of a predatory fish species, in turn affecting the abundance of the target prey species. It is worth noting that although marine mammals are the most obvious scapegoat of fishers because of their visibility, there is typically greater competitive overlap in the feeding "niches" of fish predators and fishermen.

Because of the difficulties of providing definitive scientific advice on such questions, scientists often equivocate. It is currently virtually impossible to wholly substantiate claims that predation by marine mammals is adversely impacting a fishery or *vice versa*. In the absence of definitive answers, fisheries managers are increasingly applying the "Precautionary Principle," which requires that "where there are threats of serious or irreversible damage, lack of full scientific certainty shall not be used as a reason for postponing cost-effective measures to prevent environmental degradation." But this has been argued both ways in this context: either that marine mammal culls should not take place in the absence of clear evidence that they will benefit fisheries or alternatively that marine mammals should be culled in the absence of clear evidence that their consumption of fish will not possibly damage fisheries.

As more and better information on marine mammal diets becomes increasingly available, one of the key uncertainties in resolving questions as to the degree of competitive overlap between marine mammals and fisheries relates to limited understanding at present of the feeding strategies of marine mammals. There is a need to quantify not only spatial and temporal variability in diet but also the conditions under which predators switch to alternative prey species as the abundances of the various species change. It is important also to bear in mind that some marine mammals which are highly specialized (or conversely, highly specialized fishers) are most vulnerable to competition effects because they cannot readily change their diet in response to overfishing of a vital food source.

## See Also the Following Articles

Fishing Industry ■ Effects of Fisheries ■ Interference with Hunting of Marine Mammals Incidental Catches

## References

Aguilar, A. (2000). Population biology, conservation threats and status of Mediterranean striped dolphins (*Stenella coeruleoalba*). *J. Cetacean Res. Manage.* **2**, 17–26.

Bearzi, G., Reeves, R. R., Notarbartolo-di-Sciara, G., Politi, E., Cañadas, A., Frantzis, A., and Mussi, B. (2003). Ecology, status and conservation of short-beaked common dolphins (*Delphinus delphis*) in the Mediterranean Sea. *Mamm. Rev.* **33**, 224–252.

Beddington, J. R., Beverton, R. J. H., and Lavigne, D. M. (1985). "Marine Mammals and Fisheries." Allen & Unwin, London, 354 pp.

Bowen, W. D., Beck, C. A., Iverson, S. J., Austin, D., and McMillan, J. I. (2006). Linking predator foraging behaviour and diet with variability in continental shelf ecosystems: Grey seals of eastern Canada. *In* "Top Predators in Marine Ecosystems – Their Role in Monitoring and Management" (I. Boyd, S. Wanless, and C. J. Camphuysen, eds). Cambridge University Press, Cambridge.

Boyd, I., Wanless, S., and Camphuysen, C. J. (eds) (2006). "Top Predators in Marine Ecosystems – Their Role in Monitoring and Management." Cambridge University Press, Cambridge.

Butterworth, D. S. (1999). Do increasing marine mammal populations threaten national fisheries? *In* "Issues Related to Indigenous Whaling, Tonga" (M. R. Freeman, ed.). World Council of Whalers, Brentwood Bay, B.C., Canada.

Cornick, L. A., Neill, W., and Grant, W. E. (2006). Assessing competition between Steller sea lions and the commercial groundfishery in Alaska: A bioenergetics modelling approach. *Ecol. Mod.* **199**, 107–114.

DeMaster, D., and Atkinson, S. (eds) (2002). "Steller Sea Lion Decline: Is It Food II." University of Alaska Sea Grant, Fairbanks, AK-SG-02-02, 80 pp.

DeMaster, D. P., Fowler, C. W., Perry, S. L., and Richlen, M. F. (2001). Predation and competition: The impact of fisheries on marine mammal populations over the next one-hundred years. *J. Mamm.* **82**, 641–651.

Doroff, A. M., Estes, J. A., Tinker, M. T., Burn, D. M., and Evans, T. J. (2003). Sea otter population declines in the Aleutian archipelago. *J. Mamm.* **84**, 55–64.

Estes, J. A., Tinker, M. T., Williams, T. M., and Doak, D. F. (1998). Killer whale predation on sea otters linking oceanic and nearshore ecosystems. *Science* **282**, 473–476.

Fay, G. and Punt, A. E. (2006). Modeling spatial dynamics of Steller Sea Lions (*Eumetopias jubatus*) using maximum likelihood and Bayesian methods: evaluating causes for population decline. *In* "Sea lions of the world" (A. W. Trites, S. K. Atkinson, D. P. DeMaster, L. W. Fritz, T. S. Gelatt, L. D. Rea, and K. M. Wynne, eds), pp. 425–433. Alaska Sea Grant Program.

Fertl, D., and Leatherwood, S. (1997). Cetacean interactions with trawls: A preliminary review. *J. Northw. Atl. Fish. Sci.* **22**, 219–248.

Food and Agriculture Organization of the United Nations (FAO) (2006). "The State of World Fisheries and Aquaculture." FAO, Fisheries Department, Rome.

Gales, N., Hindell, M., and Kirkwood, R. (eds) (2003). "Marine Mammals – Fisheries Tourism and Management Issues." CSIRO Publishing.

Kaschner, K. (2004). "Modelling and Mapping of Resource Overlap between Marine Mammals and Fisheries on a Global Scale. Ph.D. Thesis, MMRU, Fisheries Centre, Department of Zoology, University of British Columbia, Vancouver, Canada.

Kaschner, K., R. Watson, V. Christensen, A. W. Trites and Pauly, D. (2001). Modeling and mapping trophic overlap between marine mammals and commercial fisheries in the North Atlantic. *In*: "Fisheries impacts on North Atlantic ecosystems: catch, effort and national/regional datasets" (D. Zeller, and R. D. Pauly, eds). *Fish. Cent. Res. Rep.* **9**, 35–45.

Koen-Alonso, M., and Yodzis, P. (2005). Multispecies modelling of some components of the marine community of northern and central Patagonia, Argentina. *Can. J. Fish. Aquat. Sci.* **62**, 1490–1512.

Laws, R. M. (1977). The significance of vertebrates in the Antarctic marine ecosystem. *In* "Adaptations within Antarctic ecosystems," Third Symposium on Antarctic Biology (G.A. Llano, ed.), Scientific Committee for Antarctic Research.

Mori, M., and Butterworth, D. S. (2006). A first step towards modelling the krill–predator dynamics of the Antarctic ecosystem. *CCAMLR Sci.* **13**, 217–277.

Northridge, S. P. (1991). An updated world review of interactions between marine mammals and fisheries. *FAO Fish. Tech. Pap.* **251**(Suppl. 1).

Notarbartolo di Sciara, G., Aguilar, A., Bearzi, G., Birkun, A. Jr., and Frantzis, A. (2002). Overview of known or presumed impacts on the different species of cetaceans in the Mediterranean and Black Seas.

*In* "Cetaceans of the Mediterranean and Black Seas: state of knowledge and conservation strategies" (G. Notarbartolo di Sciara, ed.). A report to the ACCOBAMS Secretariat, Monaco, February 2002. Section 17.

Plagányi, É. E. (2007). Models for an ecosystem approach to fisheries. *FAO Fish. Tech. Pap.* **477**, Rome, FAO.

Plagányi, É. E., and Butterworth, D. S. (2005). Indirect Fishery Interactions. *In* "Marine Mammal Research: Conservation Beyond Crisis" (J. E. Reynolds, W. F. Perrin, R. R. Reeves, S. Montgomery, and T. J. Ragen, eds), pp. 19–46. John Hopkins University Press, Baltimore, Maryland, 240 pp.

Punt, A. E. (1994). Data analysis and modelling of the seal–hake biological interaction off the South African West Coast. Report submitted to the Sea Fisheries Research Institute, Cape Town, South Africa.

Punt, A. E., and Butterworth, D. S. (1995). The effects of future consumption by the Cape fur seal on catches and catch rates of the Cape hakes. 4: Modeling the biological interaction between Cape fur seals Arctocephalus pusillus pusillus and the Cape hakes Merluccius capensis and M. paradoxus. *South Afr. J. Mar. Sci.* **16**, 255–285.

Roussel, E. (2002). Disturbance to Mediterranean cetaceans caused by noise. *In* "Cetaceans of the Mediterranean and Black Seas: state of knowledge and conservation strategies" (G. Notarbartolo di Sciara, ed.). A report to the ACCOBAMS (Agreement on the Conservation of Cetaceans in the Black and Mediterranean Sea) Secretariat, Monaco.

Schweder, T., Hagen, G. S., and Hatlebakk, E. (2000). Direct and indirect effects of minke whale abundance on cod and herring fisheries: A scenario experiment for the Greater Barents Sea. *In* "Minke Whales, Harp and Hooded Seals: Major Predators in the North Atlantic Ecosystem" (G. A. Víkingsson, and F. O. Kappel, eds), 2, pp. 120–132. NAMMCO Sci. Pub, Tromsø, Norway.

Tamura, T. (2003). Regional assessments of prey consumption and competition by marine cetaceans in the world. *In* "Responsible Fisheries in the Marine Ecosystem" (M. Sinclair, and G. Valdimarsson, eds). FAO and CABI Publishing, UK.

Thomson, R. B., Butterworth, D. S., Boyd, I. L., and Croxall, J. P. (2000). Modeling the consequences of Antarctic krill harvesting on Antarctic fur seals. *Ecol. Appl.* **10**, 1806–1819.

Trites, A. W., Christensen, V., and Pauly, D. (1997). Competition between fisheries and marine mammals for prey and primary production in the Pacific Ocean. *J. Northw. Atl. Fish. Sci.* **22**, 173–187.

Trites, A. W., Livingston, P. A., Mackinson, S., Vasconcellos, M. C., Springer, A. M., and Pauly, D. (1999). Ecosystem change and the decline of marine mammals in the Eastern Bering Sea. testing the ecosystem shift and commercial whaling hypotheses. *Fish. Cent. Res. Rep.* **7**, 1–106.

Walters, C. J., Christensen, V., and Pauly, D. (1997). Structuring dynamic models of exploited ecosystems from trophic mass-balance assessments. *Rev. Fish. Biol. Fish.* **7**, 139–172.

Whitehead, H., and Carscadden, J. E. (1985). Predicting inshore whale abundance – whales and capelin off the Newfoundland coast. *Can J. Fish. Aquat. Sci.* **42**, 976–981.

Yodzis, P. (2000). Diffuse effects in food webs. *Ecology* **81**, 261–266.

# Conservation Efforts

## RANDALL R. REEVES

Efforts to conserve marine mammals began early in the twentieth century. The impetus for these efforts came from the recognition that populations of several highly valued species—fur seals and the sea otter (*Enhydra lutris*)—had been nearly extirpated by hunting. In most instances, self-regulation through market feedback had been the only thing that prevented extinctions.

In other words, as the animal populations were reduced by overkill, it became increasingly difficult to hunt them profitably, so the hunting effort declined. This mechanism was clearly inadequate to protect the stocks of whales because modern whaling was a multispecies enterprise. As right whales (Balaenidae, *Eubalaena* spp.) and blue whales (*Balaenoptera musculus*) became scarce, the fleets simply redirected their attention to humpback, fin, and sei whales (*Megaptera novaeangliae, Balaenoptera physalus*, and *B. borealis*, respectively), but any right or blue whale encountered would still be killed. By the late 1920s and 1930s, the whaling industry had begun to place limits on oil production and had given some protection to the depleted right whales and gray whales (*Eschrichtius robustus*). Eventually, international agreements emerged to manage the industry on terms more favorable to conservation. It was not until the 1970s, however, that the multispecies problem in commercial whaling had been addressed properly. In fact, few serious efforts to conserve marine mammals for reasons other than as a response to stock depletion or exhaustion were made until the late 1960s and early 1970s.

A discussion of marine mammal conservation can be organized in a number of ways—according to different types of threat (e.g., directed hunting, bycatch in fisheries, chemical pollution), on a species or population basis, by geographical region, or chronologically (Twiss and Reeves, 1999; Whitehead *et al.*, 2000; Evans and Raga, 2001; Reeves and Reijnders, 2002; Reynolds *et al.*, 2005). The first part of this chapter is organized according to levels of governance. Conservation efforts have been and should be made at many different levels, from global international agreements all the way "down" to actions by local communities and individual citizens. Therefore, some efforts to conserve marine mammals at the international, regional, national, and local levels are reviewed, and this is followed by a discussion of some of the principal threats and how they are being addressed. Next is a brief overview of the geography of marine mammal conservation, which considers regional differences in the seriousness of threats and in how they are being addressed. Finally, an attempt is made to identify the most threatened marine mammal species and populations.

## I. What Is "Conservation"?

"Conservation" is defined here as the preservation of wild populations so that they continue to replicate themselves in a natural context for an indefinite, but long, time into the future (i.e., at least hundreds of generations). This means that not only the animals themselves, but also the environments (habitats and "ecosystems") that sustain them and the biotic communities to which they belong, need to be preserved. Neither the maintenance of a few individuals in zoo-like conditions, nor the preservation of frozen DNA, constitutes a conservation endpoint. Either of those approaches, however, can be part of a broader effort to achieve conservation goals.

The unit of conservation has traditionally been the species, classically defined as a group of interbreeding natural populations that is isolated reproductively from other such groups. In practice, conservation biologists generally agree that it is insufficient to be concerned only with preserving species. They argue that it is also important to preserve the natural variety within species, including genetic and behavioral variants. One way of achieving this more ambitious objective is by ensuring the survival of local or geographical populations ("stocks"). There is a substantial and growing body of literature on the "stock" concept as it applies to marine mammals (Taylor, 2005).

The term "conservation" has a long history and is often cast in three different perspectives: biocentric, economic, and ecologic.

C

*Biocentric* conservation emphasizes the intrinsic value of all life forms and is rooted in religious or philosophical beliefs that place humans on the same plane as other organisms. Although the concept of "animal rights" shares similar roots, it differs from biocentric conservation in that it focuses on the importance of individuals rather than on populations or genomes. Concerns about animal welfare and humane treatment also focus on individuals and are not always central to conservation, as defined here. *Economic* conservation regards wild animal populations as resources to be used for human benefit. A central tenet is sustainability: killing or other forms of extractive, or consumptive, use are allowed and perhaps even encouraged, but only on the condition that such use does not compromise the ability of a wild population to regenerate itself. Finally, *ecologic* conservation places a premium on the maintenance of natural systems and processes. Individuals, populations, and species derive importance from their functional relationships with the communities of which they are a part.

The term "conservation" is sometimes used as though it were synonymous with "protection," particularly in the anti-whaling/pro-whaling debate. As used here, conservation does not rule out killing or other forms of use as long as the central goal—population persistence—is assured. Other terms that have tended to replace "conservation" in many forums over the last quarter-decade are "sustainable development" and "sustainable use" (Lavigne, 2006), which assume (questionably) that (a) there is no inherent conflict or incompatibility between human wishes and wildlife needs, (b) wild organisms *must* be used if they are to be preserved, (c) scientific understanding is sufficient to define sustainability, and (d) mechanisms exist to ensure against "over-use."

## II. International Conservation Efforts

Organized conservation efforts at the international level are carried out mainly by intergovernmental organizations established by treaties or conventions (Table I). A few non-governmental organizations (NGOs) also operate on a global basis. Some, such as the International Union for the Conservation of Nature (IUCN, actually a combination of inter- and non-governmental), International Fund for Animal Welfare (IFAW), and World Wide Fund for Nature (WWF), address a wide range of environmental issues.

The scale of any particular effort depends on the geographical distribution of the organisms (or phenomena) being conserved or the threat being addressed. Relatively few international conservation instruments focus solely on marine mammals. The best known is the International Whaling Commission (IWC), established under the International Convention for the Regulation of Whaling (ICRW) signed in Washington, DC, in 1946 (Gambell, 1999). A global conservation body was clearly necessary to manage the exploitation of the great whales, customarily defined as the baleen whales (Fig. 1) plus the sperm whale (*Physeter macrocephalus*). Most of these animals migrate over long distances and have been hunted on a truly worldwide scale for centuries (Reeves and Smith, 2006).

The IWC's authority as the body responsible for managing whaling worldwide has been challenged in recent years, and there is ongoing controversy about its scope and reach. Some member states (e.g., Japan, Denmark, and Russia) have traditionally insisted that "small cetaceans," meaning all toothed species except the sperm whale and the "bottlenose" whales (defined in the IWC schedule as the northern and southern bottlenose whales, *Hyperoodon ampullatus* and *H. planifrons*, respectively, Arnoux's beaked whale, *Berardius arnuxii*, and Baird's beaked whale, *B. bairdii*), are not covered by the ICRW

and that their exploitation and conservation are national, or at most regional, concerns. This interpretation of the Commission's competence ignores the fact that many populations of small cetaceans move seasonally across national borders or onto the high seas. It also fails to acknowledge the close biological relationships among the cetaceans, which mean that they face common threats (e.g., bycatch in fisheries, bioaccumulation of pollutants) and are similarly vulnerable to over-exploitation (Fig. 2). In the absence of IWC oversight, various bilateral and multilateral instruments have been developed to manage takes of small cetaceans (see Section III), and national programs of full protection or managed exploitation are typical (see Section IV).

Any international agreement is effective only if the parties ensure compliance and enforcement. Typically, sovereign states are unwilling to accede to a convention unless they are allowed to opt out of provisions with which they disagree. Under the ICRW, e.g., member countries have 90 days to consider their options before any amendment to the regulations comes into effect. Once an objection has been lodged, the measure is no longer binding on the objecting country. On this basis, Norway has continued commercial whaling for common minke whales (*Balaenoptera acutorostrata*) despite the IWC's global moratorium established in 1986. Japan and Iceland have used another "loophole" to continue (or resume) whaling. The ICRW allows contracting governments to grant special permits to take whales for scientific research. Although the IWC's scientific committee reviews and comments on permit proposals, its advice is non-binding. Japan has kept its commercial whaling industry viable by issuing permits to kill hundreds of common and southern minke whales (*B. bonaerensis*), plus a growing array of other species (Bryde's [*B. edeni/brydei*], sei, fin, and sperm whales) each year, allegedly for research purposes (but see, Gales *et al.*, 2005).

The United Nations Convention on the Law of the Sea (UNCLOS) was ratified in 1982. Rather than strengthening efforts to conserve marine mammals, however, this framework convention has tended to provide states with a rationale for opting out of agreements such as the ICRW. Under the convention, the idea that countries have exclusive sovereign rights to manage resources within 200 nautical miles of their coastlines became firmly entrenched. This has been interpreted to mean that the hunting of coastal stocks of marine mammals should not be subject to international oversight and regulation. Also, although Article 65 calls for member states to "work through the appropriate international organizations" for the conservation, management, and study of cetaceans, it leaves governments with considerable latitude to interpret what that means. Canada, e.g., withdrew from the IWC in 1982, arguing that a bilateral commission with Greenland sufficed as an "appropriate international organization" to manage the hunting of belugas (*Delphinapterus leucas*) and narwhals (*Monodon monoceros*) (see Section III) and that the obligation of "working through" an appropriate international body to manage the hunting of bowhead whales (*Balaena mysticetus*) could be discharged by occasionally sending experts and observers to IWC meetings.

The Antarctic, an important seasonal feeding ground for migratory whale populations and home to several endemic seal species, is a global commons. As such, it requires its own international regime of protection and conservation (Kimball, 1999). The Antarctic Treaty system consists of four separate instruments: the initial framework treaty signed in 1959 (entered into force in 1961), the seals convention of 1972, the marine ecosystem-oriented Convention on the Conservation of Antarctic Marine Living Resources (CCAMLR) of 1980, and a 1988 convention on mineral resources. As a whole, this system is nearly comprehensive, particularly taking into account the overlapping responsibilities of the IWC and other instruments

### TABLE I
#### Current International Conservation Conventions and Institutions

| Name of Entity | Year of Initiation | Location of Secretariat or HQ | Primary Mandate or Responsibility in Relation to Marine Mammals | Comments on Effectiveness |
|---|---|---|---|---|
| International Convention for the Regulation of Whaling (ICRW); International Whaling Commission (IWC) | Signed 1946, entered into force 1948; IWC established 1951 | Cambridge, UK | Conservation of whale stocks (officially concerned only with baleen whales, sperm and bottlenose whales) | Very strong scientific component; controversial but highly effective in 1970s–1980s; suffered loss of credibility and authority in 1990s. |
| Convention on International Trade in Endangered Species of Wild Fauna and Flora (CITES) | Signed 1973, entered into force 1975 | Geneva, Switzerland | Regulation and monitoring of international trade in products from species and populations classified as threatened | Highly politicized and rancorous, but continued through 1990s to be largely effective. |
| International Union for the Conservation of Nature (IUCN) | Established 1948 | Gland, Switzerland (with country offices) | Maintains Red List of Threatened Species, sponsors specialist groups (e.g., Cetacean, Seal, Sirenia, Polar Bear, Otter), provides advice to CITES and IWC | Specialist groups provide scientific expertise, promote and coordinate conservation research. |
| World Wide Fund for Nature (WWF) | Established as World Wildlife Fund in 1961 | Gland, Switzerland (with many national affiliates) | Lobbies for conservation, supports conservation research, and participates in international conservation fora | Influences policies of IWC and CITES, many national affiliates conduct local or regional marine mammal research and conservation programs (e.g., Philippines, USA, Canada, Hong Kong, Malaysia). |
| TRAFFIC Network (trade monitoring program of IUCN and WWF) | 1976 | Cambridge, UK (with regional or national offices) | Monitoring international trade in wildlife, works in close cooperation with CITES Secretariat | Important role in documentation of trade, with emphasis on threatened species. |
| Convention on the Conservation of Migratory Species of Wild Animals (CMS; or Bonn Convention) | Signed 1979, entered into force 1983 | Bonn, Germany | Conservation of "entire populations or any geographically separate part of the population of any species or lower taxon…, a significant proportion of whose members cyclically and predictably cross one or more national boundaries." | HAS RECOGNIZED CETACEANS, BUT NOT PINNIPEDS OR SIRENIANS, AS HIGHLY MIGRATORY SPECIES; see Table 2 for relevant regional agreements. |
| Convention for the Conservation of Antarctic Seals | 1972, entered into force 1978 | None, but scientific advice is provided by Scientific Committee on Antarctic Research's Group of Specialists on Seals, based in Cambridge, UK | Conservation of Antarctic seals, regulation of sealing, facilitation of scientific research on seals | First international conservation agreement to be established *prior to* the initiation of exploitation. |
| Convention on the Conservation of Antarctic Marine Living Resources (CCAMLR) | 1980, entered into force 1982 | Hobart, Tasmania, Australia | Facilitation of recovery of depleted whale stocks; prevention of further irreversible human-caused changes in Antarctic ecosystem | Krill monitoring program, ecosystem focus, strong scientific base. |
| United Nations General Assembly Drift-net Resolution 46/215 | 1991, took effect end of 1992 | None | Elimination of large-scale (longer than 2.5km), high-seas drift net fishing (and thus elimination of the large associated bycatch of marine mammals) | More than 1000 vessels were withdrawn from this type of fishing, but drift netting continues inside national 200nmi Exclusive Economic Zones (and probably to some extent illegally in international waters). |

C

C

Figure 1 *(A) A fin whale* (Balaenoptera physalus) *is butchered at a whaling station in Iceland on July 11, 1988. (B) A young Icelander poses with baleen. Iceland used a scientific rationale to justify continued whaling operations for a few years after the International Whaling Commission's moratorium took effect in 1986. Later, in 1992, Iceland withdrew from the commission, only to rejoin in 2002. Along with Norway and Japan, Iceland has been a strong advocate of resumed commercial whaling and the reopening of international trade in whale products. Photographs by Steve Leatherwood.*

such as the 1972 Convention on the Prevention of Marine Pollution by Dumping of Wastes and Other Matter (the London Dumping Convention) and the 1973–1978 Convention for the Prevention of Pollution from Ships (the MARPOL Convention). It should provide an adequate legal basis for protecting Antarctic marine mammal populations. What it cannot do is reverse the devastation of the southern stocks of baleen whales caused by the whaling industry. Nor can it protect the seal and whale populations from the ongoing (and worsening) effects of climate change (see Section VIII).

Figure 2 *(A) Sri Lankans begin butchering a Risso's dolphin* (Grampus griseus; *in August 1985). A diverse array of dolphins and whales are killed in Sri Lanka, partly as a bycatch of net fisheries and partly by direct harpooning. Estimates of the total annual kill of small cetaceans in Sri Lanka during the 1980s were in the tens of thousands. (B) Harbor porpoise* (Phocoena phocoena) *skin and meat are sold, along with other local wildlife, for domestic consumption in West Greenland. The annual reported catch of harbor porpoises in Greenland averaged about 2300 between 2000 and 2005. Photographs by Steve Leatherwood.*

The Convention on International Trade in Endangered Species of Wild Fauna and Flora (CITES) has been in force since 1975. Although CITES has little relevance when the products of exploitation are only for domestic consumption or when animals are captured alive and placed in institutions within the country of origin, it becomes highly relevant when the animals or their products or derivatives cross international borders. The economic stakes can be high, and this is certainly true of the trade in whale meat and blubber, which are in demand by Japan. Under CITES, species and geographical populations can be listed in one of three appendices. Appendix I species or populations are threatened with extinction, and trade in their products for primarily commercial purposes is prohibited. Those in Appendix II are not considered to be in immediate danger of extinction but may become so unless trade is strictly regulated. The third appendix includes species or populations that are subject to national regulation and for which multilateral cooperation is necessary to avoid over-exploitation. The goals of monitoring and regulation are achieved through a system of permits and certificates for export or import issued by national governmental authorities. As of 2007, when the 14th Meeting of the Conference of the Parties took place in The Hague, all of the commercially valuable baleen whales were listed in CITES Appendix I, as were some of the odontocetes, including the sperm whale. Also, a series of resolutions, consolidated in 1979 and calling for CITES member states to honor IWC restrictions on whaling by prohibiting the trade of whale products, remained in force even though it had been challenged forcefully by Japan and Norway. In addition to its role in the whaling arena, CITES has been used to limit the trade in live cetaceans for public display and research (Fisher and Reeves, 2005).

The United Nations ban on pelagic drift nets was a major conservation achievement. In the early 1990s, at least 40 million nontarget fish, sharks, seabirds, marine mammals, and sea turtles were being killed annually by the Japanese drift net fishery for squid. Altogether, the high-seas squid drift net fisheries in the North Pacific were killing approximately 15,000–30,000 northern right whale dolphins (*Lissodelphis borealis*), 11,000 Pacific white-sided dolphins (*Lagenorhynchus obliquidens*), and 6000 Dall's porpoises (*Phocoenoides dalli*) each year. Drift nets set for salmon, tuna, and billfish were taking thousands more dolphins, porpoises, whales, and pinnipeds each year. The threats to populations of pelagic cetaceans, pinnipeds, seabirds, and many other organisms from large-scale, high-seas drift nets were judged sufficiently severe and widespread to necessitate action by the United Nations. A resolution passed by the General Assembly in 1991 called on member nations to enact a moratorium on such drift netting by the end of 1992. This global ban was a valuable step and undoubtedly helped avert catastrophic declines in some marine animal populations. However, the UN decree could not affect the use of these nets inside the 200 nautical mile limit of coastal states. Consequently, pinnipeds and cetaceans continue to be killed in large numbers by drift nets deployed in coastal waters. Also, the global proliferation of pelagic long lining (at times as a replacement for drift netting) has brought new problems, including both depredation (when mammals damage or remove caught fish) and bycatch (when mammals are hooked or entangled in the gear) (Read, 2005).

## III. Regional and Bilateral Conservation Efforts

In cases involving species or populations with well-defined distributions that cross several national boundaries, multilateral regional bodies have sometimes been established to monitor and manage exploitation (Table II). Included among these are some "international"

instruments that are in fact regional because their scope is defined by the limited geographical ranges of the animals involved. For example, membership in the Inter-American Tropical Tuna Commission (IATTC), which is involved in managing the incidental mortality of dolphins in purse seines, has been geographically diverse since it was created in 1949 under a treaty between the USA and Costa Rica. Only states with an interest in fishing for tuna in the eastern tropical Pacific Ocean have joined the commission, and over the years this has included France, Japan, Vanuatu, the USA, and Mexico in addition to a number of Central and South American countries. Thus, while the commission is international in the sense of having a geographically varied membership, its purview is distinctly regional. Similarly, the International Agreement on the Conservation of Polar Bears and their Habitat involves only the northern circumpolar countries where polar bears occur, and it is therefore treated here as a regional agreement.

The 1911 Treaty for the Preservation and Protection of Fur Seals (often referred to as the North Pacific Fur Seal Convention) involved four countries: Great Britain (signing on behalf of Canada), the USA, Russia, and Japan. This was essentially an agreement among the states involved in the exploitation of northern fur seals (*Callorhinus ursinus*), which are endemic to the North Pacific Ocean. Pelagic sealing was banned, and as part of the agreement, Japan and Canada were allocated a portion of the profits from the controlled killing (mainly of "surplus" male seals) on the Pribilof (USA) and Commander (Russia) islands. This treaty lapsed in 1941, when Japan withdrew, and was replaced in 1957 by the Interim Convention on the Conservation of North Pacific Fur Seals. The northern fur seal is frequently cited as a conservation success story. Elimination of pelagic sealing, in combination with regulations limiting the kill at breeding rookeries, allowed the seal population to make a strong recovery from its depleted state in the early 1900s. The population reached about 2 million in the 1950s but had dipped below a million by the early 1980s. Numbers in US waters stood at about three-quarters of a million in 2006. The 1957 interim convention, having lapsed in 1984, has not been replaced.

The polar bear (*Ursus maritimus*) treaty mentioned earlier is often cited as an example of an effective international agreement. Discussions among the range states—Canada, the USA, Denmark (on behalf of Greenland), Norway, and the Soviet Union (now Russia)—began in the mid-1960s, when the future of the polar bear was of great concern because of overhunting and habitat deterioration (Lyster, 1985). The three main objectives of the agreement, which was signed in 1973 and took effect in 1976, were to ensure that appropriate restrictions were placed on hunting, that polar bear habitat was preserved, and that needed research was conducted in a coordinated fashion. The Polar Bear Specialist Group of the IUCN Species Survival Commission has served as a *de facto* scientific committee, meeting every few years to share information, discuss research needs, and assess the state of polar bear conservation. At its 14th working meeting in 2005, the group concluded that the total population of wild polar bears was between about 20,000 and 25,000, more than half of them in Canada. It stressed the serious implications of climate change and pollution, the need for better regulation of hunting and ship traffic, and the importance of identifying and protecting critical habitat.

The multinational hunt for harp and hooded seals (*Pagophilus groenlandicus* and *Cystophora cristata*, respectively) in the northern North Atlantic proceeded without meaningful regulation until the late 1950s, when Norway and the Soviet Union established a bilateral commission to set quotas for commercial catches of harp and hooded seals as well as walruses (*Odobenus rosmarus*) in the northeastern Atlantic. The reach of this agreement has been interpreted to include large areas of the Greenland and Barents seas, Denmark Strait, and waters

C

## TABLE II
### Regional or Bilateral Conservation Agreements Currently in Effect

| Name of Entity | Year of Initiation | Location of Secretariat or HQ | Mandate or Objectives | Comments on Effectiveness |
|---|---|---|---|---|
| Inter-American Tropical Tuna Commission (IATTC) | 1949 | La Jolla, California | Initially to document and manage tropical tuna fisheries; since then, expanded to include documentation, mitigation, and regulation of dolphin mortality incidental to fishing operations in eastern tropical Pacific | Operates programs to: place observers aboard tuna vessels, reduce dolphin mortality through diagnosis and solution of gear problems, and training for captains and crews; provides mechanism for linking tuna industry with government agencies and environmental NGOs. |
| North Atlantic Marine Mammal Commission (NAMMCO) | 1992 | Tromsø, Norway | Sustainable use and management of marine mammals in the North Atlantic Ocean | Emphasis on ecological interactions (e.g., rationales for culling marine mammals to protect fish stocks), hunting rights of coastal communities, forum for scientific information exchange. |
| Canada/Greenland Joint Commission on Narwhal and Beluga | 1989 | None (Ottawa, Canada; Nuuk, Greenland) | Cooperative research and management related to "shared" stocks of narwhals and white whales | Forum for bilateral studies and sharing of information, with management measures left to national authorities and local "co-management" bodies. |
| Agreement on the Conservation of Small Cetaceans of the Baltic and North Seas (ASCOBANS) | Signed 1991 (concluded under CMS, the Bonn Convention — see Table 1), entered into force 1994 | Bonn, Germany | Cooperation to achieve and maintain a "favourable conservation status" for small cetaceans in the region | Most effort has been directed at estimating abundance and incidental takes of harbor porpoises and dolphins and at seeking ways to reduce bycatch. |
| Agreement on the Conservation of Cetaceans of the Black Sea, Mediterranean Sea and Contiguous Atlantic Area (ACCOBAMS) | 1996 (concluded under CMS, the Bonn Convention — see Table 1), entered into force 2001 | Monaco | Cooperation to achieve and maintain a "favourable conservation status" for cetaceans in the region, including the complete prohibition of deliberate taking and establishment of a network of "specially protected areas to conserve cetaceans" | Differs from ASCOBANS in that scope includes all cetaceans, not only "small" species; focuses on bycatch, disturbance and injury by recreational and industrial vessel traffic, prey depletion; planning basin-wide population surveys. |
| International Agreement on the Conservation of Polar Bears and their Habitat | Signed 1973, entered into force 1976 | None; follows rotating chairmanship of IUCN Polar Bear Specialist Group | To prevent polar bear populations from becoming endangered because of hunting or other human activities | Provides a framework for communication and cooperation among circumpolar countries, emphasis on research and monitoring; signatory states are supposed to "enact and enforce such legislation and other measures as may be necessary to give effect to the Agreement." |
| U.S.-Russia Agreement on Cooperation in the Field of Environmental Protection | 1972 | None (Washington and Moscow) | Marine Mammal Project, under Area V of the Agreement, provides for information exchange, coordination of research activities, and joint or cooperative research | Annual scientific meetings, formerly focused on Bering and Chukchi seas regions, now also considers, e.g., Caspian seals, bycatch in Japanese salmon drift nets operating within Russian EEZ. |

near the island of Jan Mayen. A similar bilateral agreement pertaining to the northwestern Atlantic was signed by Canada and Norway in 1971. A series of regional bodies have become involved in monitoring the North Atlantic seal hunt and assessing the harp and hooded seal populations to provide management advice. Starting in the 1960s, a

Sealing Panel of the International Commission for Northwest Atlantic Fisheries (later the Northwest Atlantic Fisheries Organization, or NAFO) recommended overall quotas and other conservation measures (e.g., opening and closing dates for sealing from ships) related to hunting of the western populations. In recent years, scientific advice on harp and hooded seal stocks has come from a working group convened jointly by the International Council for the Exploration of the Sea (ICES) and NAFO. This group's advice is presented to the North Atlantic Marine Mammal Commission (NAMMCO), which in turn offers management advice to its members. Canada decides unilaterally on sealing quotas in the western North Atlantic, whereas Norway and Russia continue to allocate quotas in the West Ice (Jan Mayen) and East Ice (White Sea) on a bilateral basis.

NAMMCO is a regional body established in 1992 by several countries that had become frustrated by the IWC's unwillingness to allow the resumption of commercial whaling. Its membership consists only of Iceland, Norway, Greenland, and the Faroe Islands, the latter two belonging to the Kingdom of Denmark but with "home rule" governments. NAMMCO has devoted much of its attention to species for which there is little or no direct conflict with the IWC, notably harp and hooded seals, ringed seals (*Pusa hispida*), gray seals (*Halichoerus grypus*), walruses, long-finned pilot whales (*Globicephala melas*), and northern bottlenose whales.

An international treaty, the Convention on the Conservation of Migratory Species of Wild Animals (or Bonn Convention), provides a mechanism for developing regional conservation agreements. Three that deal explicitly with cetaceans have been concluded thus far. The Agreement on the Conservation of Small Cetaceans in the Baltic and North Seas (ASCOBANS) entered into force in 1994, with a membership that initially included Belgium, Denmark, Germany, the Netherlands, Poland, Sweden, and the UK (Finland, France, and Lithuania have joined since then). A major achievement has been the instigation, planning, and completion of two large-scale abundance surveys of cetaceans throughout the Baltic and North Seas, one of them (in 2005) extending westward to include a large area of the contiguous northeastern Atlantic. A second CMS-generated instrument, the Agreement on the Conservation of Cetaceans of the Black Sea, Mediterranean Sea and Contiguous Atlantic Area (ACCOBAMS), was concluded in 1996 and entered into force in 2001. Both of those agreements have tried to address a wide range of issues, including incidental mortality in fisheries, chemical pollution, and underwater noise. The development, and especially the implementation, of conservation plans for populations of special concern (e.g., harbor porpoises, *Phocoena phocoena*, in the Baltic; short-beaked common dolphins, *Delphinus delphis*, in the Mediterranean; Fig. 3) has been a major challenge. The third agreement is a memorandum of understanding for the Conservation of Cetaceans and their Habitats in the Pacific Islands Region. It was opened for signatures in 2006 and the first meeting was held in early 2007. As is true of most multilateral instruments, these ultimately depend on the willingness of sovereign state parties to enact and enforce any proposed measures.

Canada and Greenland have a bilateral agreement to manage the hunting of transboundary stocks of white whales and narwhals. In 1989, the two governments signed a memorandum of understanding that recognized the importance of hunting to the Inuit and called for "the rational management, conservation and optimum utilization of living resources of the sea" as reflected in the UNCLOS (see section II). The Joint Commission on Conservation and Management of Narwhal and Beluga meets annually, as does its Scientific Working Group. In addition to management advice directed at government agencies and, in Canada, the Nunavut Wildlife Management Board (a "co-management" body established under an aboriginal land-claims agreement), the commission's Scientific Working Group plans and undertakes collaborative research on narwhals and belugas.

**Figure 3** *Short-beaked common dolphins* (Delphinus delphis) *in Kalamos, western Greece, where they were plentiful as recently as the mid-1990s but are now rarely seen. The Mediterranean population of common dolphins is officially listed by IUCN as endangered. A conservation plan commissioned by UNEP's Agreement on the Conservation of Cetaceans of the Black Sea, Mediterranean Sea and Contiguous Atlantic Area has been available since 2004, but no significant implementation of measures outlined in the plan has occurred. Photograph by Giovanni Bearzi/Tethys Research Institute.*

## IV. National Conservation Efforts

In the USA, the Marine Mammal Protection Act was passed by Congress in 1972, and it has been the cornerstone of a massive domestic commitment to conservation (Baur *et al.*, 1999). Although the MMPA is not the only such law in the world (many countries confer full legal protection to marine mammals), it is undoubtedly the most sweeping of its kind. At the time of the act's passage, a preeminent concern was the annual slaughter of more than 100,000 young harp and hooded seals on the spring pack ice off Newfoundland and in the Gulf of St. Lawrence. Public outrage at film footage of seal pups being clubbed to death was probably the most influential single factor in forcing Congress to pass legislation. In addition, however, there was growing concern about the deplorable condition of the world's stocks of large whales, especially the blue whale. Moreover, controversy swirled around the killing of pelagic dolphins by the American tuna fleet in the eastern tropical Pacific (Gosliner, 1999). The estimated kill from 1960 through 1971 had been more than 370,000 dolphins annually (totaling nearly 4.5 million by that time), and environmentalists were understandably outraged.

The resulting legislation was both comprehensive and innovative. An immediate embargo was placed on the importation of marine mammal products, with only a few specified exceptions. Deliberate taking was banned, although Alaskan Eskimos and other aboriginal people were allowed to continue hunting marine mammals for food, skins, and other products as long as the main purpose was to meet basic community needs. A strong emphasis in the act was placed on research, and science was accorded a prominent role in influencing how decisions were to be made. Existing federal departments were given the responsibility of implementing the new law, with the Department of Commerce to manage cetaceans and most pinnipeds and the Department of the Interior to manage polar bears, sea otters, sirenians, and walruses. In addition, however, an entirely new and independent federal agency, the Marine Mammal Commission, was established to oversee implementation.

C

C

The goals of management, as set forth in the Marine Mammal Protection Act, were to achieve and maintain "optimum sustainable populations" of marine mammals and to reduce incidental mortality from fishing operations (including tuna seining) to "insignificant levels approaching zero." An optimum sustainable population has been defined operationally as having a lower bound at the maximum net productivity level and an upper bound at the unexploited population (carrying capacity) level. By defining population status in terms of productivity, the act emphasized the health and stability of ecosystems rather than economic yield. The ambitious and lofty goals of the act have been pursued over the past 35 years with what appears to be an undiminished national commitment to the conservation of marine mammals.

The European Union has used a more selective approach to achieve certain objectives related to marine mammal conservation. In 1983, the European Community (as it was then called) established a controversial ban on the importation of products from seal pups. The explicit goal was to stop the clubbing of young white-coated harp seals and blue-backed hooded seals, a concern related primarily to animal welfare. In combination with closure of the US import market from 1972, the European ban effectively destroyed the profitability of sealskin production in North America and Greenland, with serious unintended economic and social consequences in Eskimo communities where the hunting of ringed seals was a major source of income. Canada was forced to stop the commercial hunt for unweaned harp and hooded seals, essentially bringing the large-scale, ship-based sealing industry to a halt. For more than a decade, the populations of harp and hooded seals were allowed to increase. Since the mid-1990s, however, Canada's commercial sealing industry has been reinvigorated with government subsidies and aggressive product marketing, particularly focused on the export of seal penises to China and other Asian countries, so the kill of harp and hooded seals has returned to levels not seen since the early 1970s.

The European Community also effectively banned the importation of whale products in 1982–1983 by declaring that all cetaceans would be treated as though they were in Appendix I of CITES (no commercial trade allowed). Greenland, with its special relationship to Denmark, was exempted from the ban, meaning that narwhal tusks could be imported to EU countries under the normal provisions of CITES Appendix II. Thus, although the EU measure caused a steep decline in the value of narwhal ivory from Canada (Fig. 4), it had comparatively little effect on the market for tusks from Greenland, which was traditionally centered in Europe. Having lost access to the American market and much of the European market for narwhal ivory, entrepreneurs in Canada found new buyers in Asia, a pattern similar to that seen with seal products.

National conservation efforts are often influenced by international law or governance, and such influence can be for either good or ill. The US government, in implementing the Marine Mammal Protection Act, has had to take account of what are sometimes conflicting commitments under international agreements. For example, the USA has always belonged to the IATTC, whose primary goal is to maximize tuna catches. A sometimes uneasy alliance has been forged over the years between the IATTC and the US National Marine Fisheries Service, the agency responsible for pursuing the "zero mortality rate" goal for dolphins mandated by the Marine Mammal Protection Act. Although substantial progress had been made toward that goal by the late 1980s, animal protection groups continued to mount legal challenges, insisting that the procedure of setting purse seines around dolphins to catch tuna should cease altogether. Their efforts led to the "dolphin-safe" labeling of canned tuna and

**Figure 4** *The long, spiraled tusk of the male narwhal* (Monodon monoceros) *has commercial value and is traded in the global marketplace. Here, Inuk hunters on northern Baffin Island, Canada, August 1975, have dragged a carcass onto a beach and are preparing to remove the valuable tooth, which will later be sold to a non-Inuk who will likely export it to Europe. The 1972 ban on imports of marine mammal products into the USA, and measures taken by the European Union since the early 1980s, have limited the international commerce in narwhal ivory, but the trade remains lucrative thanks to outlets in the Middle and Far East. Photographs by Randall Reeves.*

to embargoes on US imports of tuna from countries continuing to "fish on dolphins." The IATTC took the position that by redirecting all fishing effort away from "dolphin sets" and toward "school sets" and "log sets" (neither of which involve dolphin encirclement), the bycatch of other species (e.g., billfish and turtles) and the proportion of undersized tuna in the catch would both increase. In general, the Fisheries Service has tended to assign a higher priority to dolphin protection than has the IATTC, and the relationship between the two agencies has been strained because of this and other differences. To make matters worse, Mexico mounted a challenge to the tuna embargoes under the General Agreements on Tariffs and Trade (GATT), insisting that they were unwarranted and unacceptable impediments to free trade. The dispute-resolution panel ruled that the embargoes were indeed inconsistent with GATT provisions, and the USA consequently had to seek a balance between its commitment to marine mammal protection and its support for the principle of free trade (Gosliner, 1999).

In a more positive vein, the IWC has managed to influence the conservation of small cetaceans in Japan, despite Japan's refusal to recognize the commission's authority to impose measures related to dolphins, porpoises, and smaller species of toothed whales. A variety of small cetaceans have been hunted in Japanese coastal waters for many decades (Fig. 5). The IWC Scientific Committee's standing Subcommittee on Small Cetaceans meets annually to review the status of species, particular threats, and technical approaches to eliminating or managing threats. In its reviews of stocks, the subcommittee has repeatedly found evidence of over-exploitation by Japanese coastal cetacean fisheries. As a result, the government of Japan has been forced, through international pressure from governments and NGOs, to implement research programs and management measures.

The most glaring (but not only) example is the striped dolphin (*Stenella coeruleoalba*). A drive hunt for striped dolphins, in which entire schools are herded toward shore and killed *en masse*, has

**Figure 5** *Japanese fishermen have hunted small cetaceans for many decades, often driving hundreds of animals toward shore where they are killed en masse. Rough-toothed dolphins* (Steno bredanensis), *shown here, are rarely taken in drive hunts, and even more rarely is a photographer on hand to record the carnage. Photograph by Rusty White, courtesy of Hubbs Marine Research Institute.*

**Figure 6** *A harbor porpoise* (Phocoena phocoena) *that had been trapped in a herring weir in Passamaquoddy Bay, New Brunswick, Canada (summer 1980), being lifted from a seine net before tagging and release. Efforts to extricate marine mammals from fishing gear and return them to the wild typically require that fishermen and scientists work together cooperatively. Photograph by Randi Olsen.*

taken place annually on Japan's Izu Peninsula for more than a century. Catches of as many as 22,000 animals occurred in some years during the 1940s and 1950s. By the 1980s, when the hunters introduced a voluntary catch limit of 5000 dolphins, the annual average catch had declined to less than 3000/year, presumably because the dolphin population was seriously depleted. Finally, in 1989, quotas were imposed. Although the hunt should have stopped entirely to allow the population to recover, the striped dolphin example at least helped demonstrate that stronger measures by the national government in Japan would be needed to prevent further over-exploitation.

## V. Local and Individual Conservation Efforts

Top-down approaches to resource management have often failed. The cost of policing human actions is likely to be unacceptably burdensome when local people assign little or no legitimacy to the management regime. It is generally agreed that the greater the local or community involvement, the more likely it is that conservation efforts will succeed in the long run (Mangel *et al.*, 1996). Marine mammal hunting communities in the Arctic and in Australia [where dugongs (*Dugong dugon*) are the principal prey species] have forged cooperative management ("co-management") agreements with government agencies. Ideally, such agreements recognize the interests and rights of local people, and the broader national and international concerns are represented by the central government.

A prominent example is the Alaska Eskimo Whaling Commission (AEWC), which was established in 1977 by whalers in northern Alaska in reaction to the IWC's controversial decision to ban bowhead whaling. After several years of difficult negotiations, marked by threats, lawsuits, and even a grand jury investigation into violations of the agreed bowhead quota in 1980, a cooperative agreement was reached between the AEWC and the National Oceanic and Atmospheric Administration (NOAA), the federal body directly responsible by law for implementing IWC decisions within the USA. Under this agreement, the AEWC assumes responsibility for managing the hunt, monitoring compliance with the quota and other regulations, and reporting each year's results. Quotas on the number of strikes and landings of bowheads are still negotiated through the IWC.

Singling out the contributions of individuals to the cause of conservation is an arbitrary undertaking. The conscientious daily efforts of bureaucrats, scientists, writers, educators, fishermen, engineers, veterinarians, lobbyists, lawyers, and political activists all merit recognition. Several individuals are mentioned here, but with the caution that their work, while it may be exemplary, is not necessarily exceptional.

As discussed further later (under Section VII), the rescue and rehabilitation of injured or otherwise incapacitated individual marine mammals may have little or no conservation value. Nevertheless, in some circumstances, especially when an endangered species is involved, intervention can be important. Jon Lien, a professor at Memorial University in Newfoundland, began working with fishermen in the late 1970s to devise ways of extricating whales from fishing gear. The problem of entrapment and entanglement was a concern of conservationists because, at the time, humpback whales in the North Atlantic were considered endangered (their status has improved since then). It was of concern to fishermen because of the economic losses associated with damaged gear and lost fishing time, as well as the personal danger involved when dealing with these large animals at sea. Lien gained the confidence of fishermen and developed a successful program for assisting in the safe release of entrapped or entangled whales. Subsequently, Charles Mayo, David Mattila, and their associates at the Center for Coastal Studies on Cape Cod began rescuing whales from fishing gear on the US coast, with an emphasis on endangered right whales (*Eubalaena glacialis*). Disentanglement teams are now integral to right whale recovery efforts in the eastern USA and southeastern Canada, thanks to the pioneering efforts of Lien, Mayo, and Mattila. A similarly successful program centered on rescuing harbor porpoises, minke whales, and occasionally right whales trapped in herring weirs has been in operation in Canada's Bay of Fundy for more than two decades (Fig. 6).

One of the greatest obstacles to conservation can be the difficulty of defining and demonstrating the significance of a threat.

C

Marine debris pollution provides a clear example (Laist *et al.*, 1999). Although it is widely accepted today that marine debris, such as derelict fishing gear and plastic packaging material, is a menace to wildlife, the problem's seriousness was not recognized until the early 1980s. Charles Fowler, a scientist with the National Marine Fisheries Service in Seattle, was engaged in research to determine the cause of the continuing decline of North Pacific fur seals (see Section III). Despite bureaucratic resistance and the skepticism of scientific colleagues, Fowler pressed ahead with the task of marshaling data to test the hypothesis that entanglement in debris was a major cause of juvenile mortality in fur seals. His painstaking compilation of evidence, together with mathematical models, finally convinced others that at least this one marine mammal species was being affected at the population level. Fowler's work provided the impetus for a chain of events, from beach clean-up campaigns to the signing of international treaties, intended to reduce the ocean's burden of debris and therefore lessen the risks to seals and cetaceans, to say nothing of seabirds, turtles, and other marine wildlife.

One final example of an individual's ability to change the course of conservation policy again relates to dolphin mortality in the eastern tropical Pacific tuna fishery. By the late 1980s, many conservationists had forgotten about this issue, assured that dolphin mortality had been reduced substantially as a result of changes in fishing techniques and the imposition of annual quotas on the number of dolphins from each species that could be killed before fishing would have to cease. In 1987–1988, however, Sam LaBudde, who described himself as an "itinerant biologist," spent 5 months aboard a Panama-registered tuna boat. Although he had signed on as an ordinary seaman and cook, he carried a video camera and clandestinely recorded grisly footage of dolphins being killed. When the scenes were aired on national television, it galvanized public support within the USA for strong measures to be taken against the non-American tuna fleet. While LaBudde's actions can be viewed as either heroic or deceitful, depending on one's point of view, there is no doubting his courage or his influence on the course of conservation.

## VI. Protected Areas

The designation of specially protected areas (e.g., reserves, sanctuaries, parks) is a tool increasingly used to achieve conservation goals. Of the many such areas around the world, relatively few exist for the explicit purpose of benefiting marine mammals (Hoyt, 2005). Mexico declared Scammon's Lagoon (Laguna Ojo de Liebre) a refuge for gray whales in 1971, and San Ignacio Lagoon was given similar status in 1979. Also, Mexico established a Biosphere Reserve in the upper Gulf of California in 1993 mainly to protect the highly endangered vaquita (*Phocoena sinus*) and the totoaba, an endangered sea bass (*Totoaba macdonaldi*). New Zealand created the Banks Peninsula Marine Mammal Sanctuary in 1988 to protect Hector's dolphins (*Cephalorhynchus hectori*), like the vaquita an endemic coastal species, from entanglement in gill nets. In 1999, the parliament of the North German state of Schleswig-Holstein established a sanctuary for small cetaceans off the islands of Sylt and Amrum in the North Sea, intended to protect harbor porpoises from the dangers associated with gillnet fishing, jet skiing, and high-speed motor boating. In the USA, the Hawaiian Islands Humpback Whale Sanctuary was declared in 1993, and several other marine sanctuaries were established in large part because of public interest in the marine mammals that use them for feeding, breeding, or both (e.g., the California Channel Islands, Gulf of the Farallones, Monterey Bay, and Stellwagen Bank sanctuaries).

In addition to those areas explicitly created to benefit marine mammals, there are many small sites in the Antarctic and on the sub-Antarctic islands that are designated as specially protected areas or sites of special scientific interest under the Agreed Measures for the Conservation of Antarctic Fauna and Flora (1964), or as nature reserves under national legislative instruments, many of which protect vital haul-out habitat for pinnipeds. Norway's Svalbard (Spitsbergen) archipelago in the northeastern North Atlantic was the site of some of the worst excesses of early whaling and walrus killing, yet since 1973 about half of the land area has been declared to be inside nature reserves and national parks, and the Svalbard population of walruses is expanding rapidly.

All too often, protected areas are created in response to a public outcry, but without an accompanying ongoing commitment to enforce meaningful restrictions on human activities within them. The marine sanctuary program in the USA, e.g., has failed to meet the public's high expectations, largely because no serious attempt has been made to regulate fishing within the sanctuaries. To the program's credit, though, the dumping of wastes and the exploration for oil and gas have been strictly regulated, and this may be seen as having conservation value for marine mammals and other organisms. So-called "paper parks" and "paper reserves" can be counterproductive for conservation because they provide false assurance that space and resources have been set aside for wildlife.

## VII. Strategies to Enhance the Survival and Reproduction of Individuals

At times, human intervention can improve the chances for individual marine mammals to survive and reproduce. Organized programs for rescuing marine mammals that strand (come ashore) alive or that are injured and debilitated do manage to release some animals after rehabilitation. However, the success rate is low, and the conservation value of such programs has often been called into question. Many strandings represent "natural" mortality. Thus, while intervention may be justified as a humane gesture intended to improve the welfare of the stranded animals, it can also be argued that natural processes should be allowed to proceed without human interference. Only in a few special cases can rescue, rehabilitation, and release efforts be considered to have made a clear, positive difference for a marine mammal population.

Most of the rivers in southern Asia inhabited by river dolphins are partitioned by irrigation dams (called barrages). When dolphins on the upstream side of such dams get too close to the intake structures of adjacent canals, they run the risk of becoming marooned in the canals, unable to return to the safety of the main river channel. Wildlife officers and conservationists sometimes attempt to locate and rescue these ill-fated dolphins. Between January 2000 and July 2007, at least 46 Indus dolphins (*Platanista gangetica minor*) were trapped in canals near Sukkur Barrage in Pakistan (Gill Braulik and WWF-Pakistan/Uzma Khan, personal communication, 21 July 2007). The majority of them were successfully captured and returned to the river. Sindh Wildlife Department and WWF-Pakistan are developing a systematic procedure for notification and response, and are refining a protocol for rescuing river dolphins that enter irrigation canals. Also, on several occasions Ganges dolphins (*Platanista gangetica gangetica*) have become trapped in isolated pools, shallow streams, or rice paddies, and have been successfully captured and released into safer areas (Fig. 7).

In Florida, several facilities that display captive marine mammals have been collaborating for many years with the US. Fish

Figure 7 *Interest in marine mammals, and concern about their conservation, is not limited to wealthy, countries. In Bangladesh, e.g., young conservationists are eager to contribute, and they collaborate with local and international scientists to study and conserve freshwater and coastal cetaceans (A). Here, a Ganges river dolphin (Platanista gangetica) is returned to the Sundarbans Delta after having been found stranded, its long beak fouled with monofilament gill netting (B). Photographs by Mowgliz, courtesy of Brian Smith.*

Rescue and rehabilitation programs can contribute to conservation in less direct ways, too. For example, as John Reynolds (1999) has pointed out, "Educating people about manatee conservation as they watch recuperating animals in a zoo setting can make a strong impression that may do more to encourage actual conservation than reading an article or watching a documentary about manatees." The whale and river dolphin rescue efforts mentioned earlier also serve to heighten awareness, educate people about conservation issues, and inspire actions to prevent further entanglement and entrapment. Reynolds also points out that manatees in captivity have allowed scientists to study their species' reproduction, osmoregulatory capabilities, and sensory abilities. Knowing more about manatee biology and physiology is important for conservation.

Finally, rescue and rehabilitation programs offer opportunities to instrument and monitor animals after release. This can lead to new discoveries about the animals and allow researchers to test new study methods. There are many examples, but one in particular stands out. In 1997 an adult male bottlenose dolphin (*Tursiops truncatus*), nicknamed "Gulliver," stranded in Florida. He was treated for a variety of ailments and, after about 4 months in captivity, released far offshore bearing a satellite-linked transmitter. Gulliver's travels were impressive. After a week moving northward along the continental shelf, he headed southeast, swimming against the North Equatorial Current. He traversed waters more than 5000 m deep and reached an area northeast of the Virgin Islands before his transmitter stopped working, having covered 4200 km in 47 days. This study showed that bottlenose dolphins can be extremely mobile and that previous assumptions about the distributional limits of pelagic stocks needed to be reconsidered.

Another example of human intervention to enhance survival comes from the northwestern Hawaiian Islands, where biologists from the US National Marine Fisheries Service have captured and translocated endangered monk seals (*Monachus schauinslandi*). In one program on Laysan Island, they caught some adult males that had been seen participating in "mobbing," or collective attacks on adult females and juvenile seals. The males were moved by ship to Johnston Island, some 600 miles south of Laysan, and released in the hope that they would survive but not return to carry on their destructive behavior toward other monk seals. In another program, called "Headstart," female pups at Kure Atoll have been collected after weaning and kept in a fenced beach enclosure for several months. The watered portion of the enclosure is kept well stocked with fish taken from nearby reefs, and the young seals have a chance to learn to forage in safety from large sharks, adult male monk seals, and hazardous fishing gear—all potential causes of mortality. The idea is that by the time they are released, they will have survived a critical stage in the life cycle and be ready for independence.

Captive breeding, with the intention of using captive-born young to reestablish a species in its former range or to supplement and reinvigorate a depleted wild population, is sometimes employed as a conservation strategy when necessary and feasible. However, only one serious attempt has been made to restock a wild population of marine mammals with animals that were conceived, born, and reared in captivity. A number of captive-born harbor seals (*Phoca vitulina*) were released into the Dutch Wadden Sea, where their species had been depleted (although harbor seals are not threatened globally). The released seals were monitored with telemetry devices, and early results suggested that they had survived and adapted reasonably well.

Although captive breeding programs have been discussed in relation to Yangtze River dolphins, or baiji (*Lipotes vexillifer*), and Mediterranean monk seals (*Monachus monachus*), both gravely endangered, none of these programs have come to fruition.

and Wildlife Service to rehabilitate injured or orphaned manatees (*Trichechus manatus*). The animals are cared for and either maintained permanently in educational exhibits or, if judged healthy enough, released back into the wild. A single facility (Sea World in Orlando) was reported to have responded to 160 requests for assistance with distressed manatees from 1976 to 1995. More than half of the animals brought into captivity died, but nearly 60 individuals were eventually returned to the wild. Virtually all of the injuries to manatees in Florida waterways are caused by human activity (mainly boating), so the rehabilitation program is almost entirely compensatory in the sense of helping to offset human-caused mortality.

A much-publicized "seminatural reserve" was established for river dolphins in a Yangtze River oxbow, but this facility was stocked primarily with finless porpoises (*Neophocaena phocaenoides*) rather than dolphins. The single female baiji introduced to the reserve became entangled in fishing gear and died. No attempt was made to place this female with the lone male baiji in captivity (which itself later died), so there was no prospect of captive breeding. Efforts to capture additional baiji for captivity or for stocking the seminatural reserve failed, and a recent survey led investigators to conclude that the baiji is likely extinct and that further capture and translocation efforts would be pointless (Turvey *et al.*, 2007).

Translocation efforts played a role in the sea otter's reoccupation of parts of its original range. More than 700 otters were taken from high-density areas in Alaska during the 1960s and early 1970s and released at unoccupied sites in British Columbia, Washington, and Oregon. Populations are now well established in British Columbia and Washington. A controversial attempt was made during the 1980s to establish a new population of sea otters in the California Channel Islands in view of the risk that an oil spill could destroy the mainland population. More than 135 otters were captured and translocated to San Nicolas Island, but their numbers did not increase as expected and by the mid-2000s only about 30–35 remained.

## VIII. Reduction of Environmental Pollution (Chemical and Acoustic)

The role of POLLUTION in impairing the productivity and survival of marine mammals was first realized in the 1970s, when a correlation was found between the rate of reproductive failure (premature births, still births, and abortions) in California sea lions (*Zalophus californianus*) and elevated tissue levels of DDT (Vos *et al.*, 2003). Also during the 1970s, studies of seals in the Baltic and North seas provided suggestive evidence that organochlorine pollutants pose serious risks to the health and reproductive potential of marine mammals. The production and use of DDT, PCBs, and some other dangerous persistent organochlorine chemicals began to be restricted in North America and western Europe in the 1970s, and there has been a general trend toward further restrictions since then. Unfortunately, however, the problem is far from solved. For example, India continued to produce 4000 metric tons of DDT at least as recently as the mid-1990s, and at least some of the former Soviet states have continued to manufacture and use PCBs. Moreover, the persistent nature of these chemicals means that they continue to be present in the environment, either temporarily sequestered in sediments or recycling in food webs, and therefore marine mammals continue to be vulnerable to their effects. While it must be acknowledged that the principal motivation for banning the release of harmful substances into the environment has had less to do with protecting marine mammals than with protecting human health (and birds, in the case of DDT), there is no doubt that reports of high levels of contaminants in marine mammals have contributed to public concern (O'Hara and O'Shea, 2005).

Acoustic pollution is thought to be especially damaging to cetaceans, as they depend heavily on sound for information about their environment, for foraging, and for communication. Military sonar has been implicated in numerous mass strandings of cetaceans, particularly beaked whales (Hildebrand, 2005). Noise associated with the offshore exploration, development, and transport of oil and gas has been a particular source of concern, and many millions of dollars have been invested in studies of effects (Richardson *et al.*, 1995). In some instances, notably those involving seismic and drilling noise in the

**Figure 8** *The franciscana* (Pontoporia blainvillei) *is one of several marine mammal species with a restricted distribution (coastal waters of eastern South America between approx. 18°30'S and 41°10'S) that experience substantial incidental mortality in fisheries. Although progress has been made toward assessing the impact of such mortality on franciscana populations in some areas, nowhere has significant progress been made at reducing the bycatch rate. This image shows franciscanas killed incidentally in the coastal gill net fishery for demersal fish (sciaenids) based in the port of Rio Grande, southern Brazil, 1994. Photograph by Eduardo Secchi.*

Arctic, steps have been taken to minimize the exposure of whales and seals to high-energy sounds. In some countries, government agencies and companies have conducted monitoring programs to determine when marine mammals are present in an area so that operations can be suspended or moved to protect them. In a similar vein, the sites and timing of military exercises have, in a few instances, been planned with the safety of marine mammals and other marine wildlife as a primary consideration.

It has become increasingly clear that human-induced changes in global climate will have (and probably already have had) significant effects on marine mammal populations. The effects will be most obvious for ice-associated species: the phocid seals in the Arctic and Antarctic and the walrus and polar bear in the Arctic. These animals use sea ice as a platform for resting, giving birth, or, in the case of polar bears, hunting. As the extent and thickness of pack ice decrease from global warming, these species will lose critical habitat. Once again, as in the case of toxic chemical pollution, the primary motivation for taking steps to reduce emissions of greenhouse gases and ozone-depleting substances has been concern about human welfare rather than a desire to conserve marine mammals.

## IX. Reduction of Conflicts with Fisheries

Fishery policies are the key to many of the most pressing marine mammal conservation problems. While there are examples of effective action to reduce marine mammal mortality in fishing gear, such as the UN ban on high-seas drift netting, the seasonal or permanent closure of certain areas to gill netting, and the development and implementation of deterrence programs using pingers and similar devices, the sad truth is that many critical situations simply continue to deteriorate (Fig. 8). For example, although some legal limits have been placed on gill netting and commercial fishing in a portion of the northern Gulf of California, there has been little effective enforcement, and vaquitas remain in jeopardy (Rojas-Bracho *et al.*, 2006; Jaramillo-Legorreta *et al.*, 2007). In China's Yangtze River, it is illegal to fish with electricity and explosives, yet there is almost no enforcement, and river dolphins (if any survive, which seems unlikely) and finless porpoises continue to be killed and injured as unintended victims (Turvey *et al.*, 2007).

## X. Reduction of Disturbance and Direct Harm from Vessel Traffic

Manatees living in Florida's motorboat- and barge-infested waterways are frequently struck and injured, if not killed outright, by watercraft (also see section VII). On average, about 50 Florida manatees are killed by boat collisions each year, and many more are injured and harassed by vessel traffic. Although this problem had long been recognized, it was not until the late 1970s and early 1980s that serious efforts were made to reduce the risk of collisions and disturbance. More than 20 areas have been designated as protection zones for manatees, where vessel speed is regulated and signs warn visitors of the need to exercise caution. In some key manatee congregation areas, all waterborne human activity, including diving, boating, and swimming, is prohibited.

Another marine mammal species that is clearly threatened by ship strikes is the North Atlantic right whale. Where thousands of right whales were present in the past, all that remains is a small population of about 350 to 400 centered along the east coast of Canada and the USA. Several right whales are killed by ship collisions each year. This mortality, combined with that caused by entanglement in fishing gear, is considered sufficient to have stalled population recovery (Kraus *et al.*, 2005). Efforts have been made in both Canada and the USA to map the seasonal distribution and movements of right whales and to caution vessel captains to watch for and avoid them. Both countries have also modified the official shipping lanes into key ports in the hope that this will reduce the risk of ship strikes.

It remains to be seen whether manatees and right whales will be able to withstand the effects of human activities in the coastal and inshore waters we share with them. Thus far, our own species' recreational and commercial use of the marine environment has been regarded as sacrosanct, and the few gestures made to accommodate the needs of these other species have had to overcome strenuous resistance from boaters, the shipping industry, military authorities, and others.

## XI. Giving Economic Value to Living Wild Marine Mammals

In the 1950s, a few nature enthusiasts in southern California began venturing into near-shore waters to watch gray whales. At the time, scientists were just beginning to document the remarkable recovery of this whale population—a result of the protection from whaling afforded by the IWC and, in recent years, Mexico's protection of the breeding lagoons in Baja California. Interest in watching whales grew steadily, and by the mid-1970s, conservationists were suggesting that the "non-consumptive" use of whales as objects of tourism might eventually rival whaling in economic value. The 1980s and 1990s saw the rapid proliferation of tour enterprises for observing whales and dolphins. Even in the whaling countries of Norway, Iceland, and Japan, whale watching has become a popular and remunerative form of recreation. In eastern Canada, helicopter tours to the pack-ice pupping grounds of harp seals have been encouraged by

animal-welfare groups as a way of demonstrating that seals also can generate income (tourism revenues) without having to be killed. In the Antarctic and Arctic, opportunities to observe marine mammals are an important aspect of nature-oriented tourism.

## XII. Zoogeography of Marine Mammal Conservation

Threats to marine mammal species and populations are relatively well understood and are being addressed to some degree in North America, Europe, South Africa, Australia, and New Zealand. However, even in those parts of the world, serious problems remain. In fact, North Atlantic and North Pacific (*Eubalaena japonica*) right whales, Hawaiian monk seals, northern (or Steller) sea lions (*Eumetopias jubatus*), and sea otters in US waters, Mediterranean monk seals and some local populations of bottlenose dolphins, harbor porpoises, and short-beaked common dolphins in Europe, dugongs in parts of Australia, and Hector's dolphins in New Zealand are still in trouble. Elsewhere in the world, marine mammal populations are slipping away even before there has been a chance to document their distribution and abundance, or to elucidate their ecological roles.

Table III lists 20 of the world's most threatened marine mammal taxa. The list is by no means authoritative, or exhaustive. Some species, such as the franciscana dolphin (*Pontoporia blainvillei*) and Caspian seal (*Pusa caspica*), might merit inclusion except for the fact that their total numbers are still believed to be in the tens of thousands. For other species, such as the West African manatee (*Trichechus senegalensis*) and Amazonian manatee (*T. inunguis*), we have very little understanding of how many there are or the extent to which their distribution has been reduced by over-exploitation, incidental mortality in fishing gear, and habitat deterioration. For these and many other situations, there has been little or no active conservation. Another concern is that by limiting the list to recognized species and subspecies, geographical populations are left out. In particular, numerous geographically isolated populations that are known to be in serious trouble are missing simply because they have not been accorded a subspecies designation. Among the more obvious examples are several freshwater populations of Irrawaddy dolphins (*Orcaella brevirostris*), the western Pacific population of gray whales, and the Cook Inlet (Alaska) population of belugas, all of which number in the tens or low hundreds and are listed by IUCN as Critically Endangered.

Endemism is a feature that is often associated with vulnerability. Many of the species and subspecies in Table III are on the list because they occur in only one place. For example, the baiji was confined for the last several decades of its existence (it is likely now extinct) to the main stem of the Yangtze River, and the Indus river dolphin to the main stem of the Indus. The vaquita is limited to the upper portion of the Gulf of California, and the Saimaa, Ungava, and Ladoga ringed seals occur only within single networks of freshwater rivers and lakes. The effects of endemism are, of course, scale dependent—the smaller the range, the more vulnerable the population tends to be. A species or population that ranges throughout, or on both sides of, an ocean basin is usually less vulnerable than one limited to a single stretch of coastline or a single river or lake. However, an extensive range and great mobility also mean that management for conservation (e.g., protection from hunting, entanglement in fishing gear, exposure to ship strikes, and other threat factors) must be pursued on a large spatial scale and, often, across multiple jurisdictions.

The conservation challenges that lie ahead are truly endless. As the global economy becomes more integrated and as the human appetite (and capacity) for consuming our planet's resources expands,

C

## TABLE III
### Twenty of the world's most threatened marine mammal taxa, including recently extirpated ones
(subspecies taxonomy follows Rice, 1998, except as noted)

| Taxon | Range States | Approx. Abundance | Main Threats |
|---|---|---|---|
| 1. Caribbean monk seal, *Monachus tropicalis* | Mexico, USA, Bahamas, Jamaica, Cuba, Haiti, Dominican Republic, Guadeloupe, and other Caribbean states | Probably extinct | Deliberate killing, loss of habitat due to development |
| 2. Baiji or Yangtze river dolphin, *Lipotes vexillifer* | China | Probably extinct | Fishery bycatch, loss and degradation of habitat due to development, contamination and depletion of prey resources, vessel strikes and disturbance |
| 3. Vaquita, *Phocoena sinus* | Mexico | 100s | Fishery bycatch |
| 4. Mediterranean monk seal, *Monachus monachus* | Turkey, Greece, Italy, Mauritania, Morocco, Western Sahara, Libya, Madeira (Portugal) | 500 | Fishery bycatch, shooting by fishermen, loss of pupping and pup-rearing habitat |
| 5. North Atlantic right whale, *Eubalaena glacialis* | Canada, USA, Iceland, Norway, UK, Spain, Portugal, France | 350–400 | Ship strikes, fishery bycatch, possibly effects of small population size (depletion from past over-exploitation) |
| 6. North Pacific right whale, *Eubalaena japonica* | Russia, Japan, Korea, China, Canada, USA, Mexico, Canada | Mid to high 100s | Ship strikes, fishery bycatch, possibly effects of small population size (depletion from past over-exploitation) |
| 7. Hawaiian monk seal, *Monachus schauinslandi* | USA (Hawaiian archipelago) | 1,000 | Fishery bycatch, disturbance on pupping beaches, debris entanglement, possibly prey depletion by commercial fisheries |
| 8. Hector's dolphin, *Cephalorhynchus hectori* | New Zealand | 7,000 | Fishery bycatch, vessel strikes |
| 9. Japanese sea lion, *Zalophus californianus japonicus* | Japan, Korea, Russia | Probably extinct | Deliberate killing, fishery bycatch |
| 10. Maui's (North Island Hector's) dolphin, *Cephalorhynchus hectori maui* (new subspecies recognized in 2002) | New Zealand | 100 | Fishery bycatch |
| 11. Saimaa ringed seal, *Pusa hispida saimensis* | Finland | 280 | Fishery bycatch, changes in habitat due to water management policies, chemical contamination |
| 12. Ungava harbor seal, *Phoca vitulina mellonae* | Canada | Low to mid 100s | Loss and fragmentation of habitat due to water management policies, hunting |
| 13. Red Sea dugong, *Dugong dugon hemprichii* | Egypt, Saudi Arabia, Yemen, Eritrea, Sudan | Probably low 1,000s at most, possibly only 100s | Fishery bycatch, hunting |
| 14. Bhulan or Indus river dolphin, *Platanista gangetica minor* | Pakistan | 1,200 | Loss and fragmentation of habitat due to water management policies, accidental movement into canals and other unsafe areas, fishery bycatch, chemical contamination |
| 15. Yangtze River finless porpoise, *Neophocaena phocaenoides asiaeorientalis* | China | 1,500–2,000 | Fishery bycatch; loss, degradation, and fragmentation of habitat due to water management policies and sand mining; possibly contamination and depletion of prey resources |
| 16. Southern sea otter, *Enhydra lutris nereis* | USA | 2,800–3,000 | Fishery bycatch, human-mediated disease |
| 17. Susu or Ganges river dolphin, *Platanista gangetica gangetica* | India, Bangladesh, Nepal | At least low 1,000s | Fishery bycatch, deliberate hunting, loss and fragmentation of habitat due to water management policies, accidental movement into canals and other unsafe areas, chemical contamination |
| 18. Florida manatee, *Trichechus manatus latirostris* | USA, Bahamas (occasionally) | 3,500 | Vessel strikes, fishery bycatch, exposure to toxic organisms (probably related to human activities), habitat modifications due to water management and energy policies |
| 19. Antillean manatee, *Trichechus manatus manatus* | Caribbean and Atlantic mainland coastal states from Mexico to Brazil, Cuba, Puerto Rico, Trinidad, Dominican Republic, and other Caribbean island states | Unknown but probably 1,000s | Fishery bycatch, deliberate hunting and trapping |
| 20. Ladoga ringed seal, *Pusa hispida ladogensis* | Russia | About 5,000 | Fishery bycatch, disturbance at haul-out sites |

marine mammals will inevitably experience new threats, even while long-standing ones persist. We are in danger of losing numerous populations, some species, and a few genera (e.g., *Monachus*). One entire family of cetaceans, the Lipotidae, appears to have been lost very recently (Turvey *et al.*, 2007). Another river dolphin family, the Platanistidae, is far from secure, particularly given the ever-mounting pressure on the freshwater systems inhabited by the two extant subspecies in southern Asia. However impressive the array of conservation efforts may seem on paper, it is far from adequate (Bearzi, 2007). Only with a genuine, broad-scale change in how we value the remnants of the world's natural variety and abundance, and thus in how we use and care for the Earth's precious resources, can we hope to head off a cascade of marine mammal EXTINCTIONS in the coming decades.

## See Also the Following Articles

Captive Breeding ▪ Competition with Fisheries ▪ Conservation Ecology ▪ Distribution ▪ Fishing Industry ▪ Effects of Illegal and Pirate Whaling ▪ Pollution and Marine Mammals

## References

Baur, D. C., Bean, M. J., and Gosliner, M. L. (1999). The laws governing marine mammal conservation in the United States. *In* "Conservation and Management of Marine Mammals" (J. R. Twiss, Jr., and R. R. Reeves, eds), pp. 48–86. Smithsonian Institution Press, Washington, D.C.

Bearzi, G. (2007). Marine conservation on paper. *Conserv. Biol.* **21**, 1–3.

Evans, P. G. H., and Raga, J. A. (eds) (2001). "Marine Mammals: Biology and Conservation." Kluwer Academic, New York.

Fisher, S. J., and Reeves, R. R. (2005). The global trade in live cetaceans: Implications for conservation. *J. Int. Wildl. Law Policy* **8**, 315–340.

Gales, N. J., Kasuya, T., Clapham, P. J., and Brownell, R. L., Jr. (2005). Japan's whaling plan under scrutiny. *Nature* **435**, 883–884.

Gambell, R. (1999). The International Whaling Commission and the contemporary whaling debate. *In* "Conservation and Management of Marine Mammals" (J. R. Twiss, Jr., and R. R. Reeves, eds), pp. 179–198. Smithsonian Institution Press, Washington, D.C.

Gosliner, M. L. (1999). The tuna-dolphin controversy. *In* "Conservation and Management of Marine Mammals" (J. R. Twiss, Jr., and R. R. Reeves, eds), pp. 120–155. Smithsonian Institution Press, Washington, D.C.

Hildebrand, J. (2005). Impacts of anthropogenic sound. *In* "Marine Mammal Research: Conservation Beyond Crisis" (J. E. Reynolds, III, W. F. Perrin, R. R. Reeves, S. Montgomery, and T. J. Ragen, eds), pp. 101–123. Johns Hopkins University Press, Baltimore.

Hoyt, E. (2005). "Marine Protected Areas for Whales, Dolphins and Porpoises: A World Handbook for Cetacean Habitat Conservation." Earthscan, London.

Jaramillo-Legorreta, A., Rojas-Bracho, L., Brownell, R. L., Jr., Read, A. J., Reeves, R. R., Ralls, K., and Taylor, B. L. (2007). Saving the vaquita: Immediate action, not more data. *Conserv. Biol.* **21**, 1653–1655.

Kimball, L. A. (1999). The Antarctic Treaty system. *In* "Conservation and Management of Marine Mammals" (J. R. Twiss, Jr., and R. R. Reeves, eds), pp. 199–223. Smithsonian Institution Press, Washington, D.C.

Kraus, S. D., *et al.* (16 authors) (2005). North Atlantic right whales in crisis. *Science* **309**, 561–562.

Laist, D. W., Coe, J. J., and O'Hara, K. J. (1999). Marine debris pollution. *In* "Conservation and Management of Marine Mammals" (J. R. Twiss, Jr., and R. R. Reeves, eds), pp. 342–366. Smithsonian Institution Press, Washington, D.C.

Lavigne, D. M. (ed.) (2006). "Gaining Ground: In Pursuit of Ecological Sustainability." International Fund for Animal Welfare, Guelph.

Lavigne, D. M., Scheffer, V. B., and Kellert, S. R. (1999). The evolution of North American attitudes toward marine mammals. *In* "Conservation and Management of Marine Mammals" (J. R. Twiss, Jr., and R. R. Reeves, eds), pp. 10–47. Smithsonian Institution Press, Washington, D.C.

Lyster, S. (1985). "International Wildlife Law: An Analysis of International Treaties Concerned with the Conservation of Wildlife." Grotius Publications Limited, Cambridge.

Mangel, M., *et al.* (42 authors) (1996). Principles for the conservation of wild living resources. *Ecol. App.* **6**, 338–362.

Northridge, S. N., and Hofman, R. J. (1999). Marine mammal interactions with fisheries. *In* "Conservation and Management of Marine Mammals" (J. R. Twiss, Jr., and R. R. Reeves, eds), pp. 99–119. Smithsonian Institution Press, Washington, D.C..

O'Hara, T. M., and O'Shea, T. J. (2005). Assessing impacts of environmental contaminants. *In* "Marine Mammal Research: Conservation beyond Crisis" (J. E. Reynolds, III, W. F. Perrin, R. R. Reeves, S. Montgomery, and T. J. Ragen, eds), pp. 63–83. Johns Hopkins University Press, Baltimore.

Perrin, W. F. (1999). Selected examples of small cetaceans at risk. *In* "Conservation and Management of Marine Mammals" (J. R. Twiss, Jr., and R. R. Reeves, eds), pp. 296–310. Smithsonian Institution Press, Washington, D.C..

Read, A. J. (2005). Bycatch and depredation. *In* "Marine Mammal Research: Conservation Beyond Crisis" (J. E. Reynolds, III, W. F. Perrin, R. R. Reeves, S. Montgomery, and T. J. Ragen, eds), pp. 5–17. Johns Hopkins Univeristy Press, Baltimore.

Reeves, R. R., and Reijnders, P. J. H. (2002). Conservation and management. *In* "Marine Mammal Biology: An Evolutionary Approach" (A. R. Hoelzel, ed.), pp. 388–415. Blackwell Science, Oxford.

Reeves, R. R., and Smith, T. D. (2006). A taxonomy of world whaling: Operations and eras. *In* "Whales, Whaling, and Ocean Ecosystems" (J. A. Estes, D. P. DeMaster, R. L. Brownell, Jr., D. F. Doak, and T. M. Williams, eds), pp. 83–102. University of California Press, Berkeley.

Reeves, R. R., Smith, B. D., Crespo, E. A., and Notarbartolo di Sciara, G. (2003). "Dolphins, Whales and Porpoises: 2002–2010 Conservation Action Plan for the World's Cetaceans." IUCN, Gland.

Reynolds, J. E., III (1999). Efforts to conserve the manatees. *In* "Conservation and Management of Marine Mammals" (J. R. Twiss, Jr., and R. R. Reeves, eds), pp. 267–295. Smithsonian Institution Press, Washington, D.C.

Reynolds, J. E., III, Perrin, W. F., Reeves, R. R., Montgomery, S., and Ragen, T. J. (eds) (2005). "Marine Mammal Research: Conservation beyond Crisis." Johns Hopkins University Press, Baltimore.

Rice, D. W. (1998). Marine Mammals of the World: Systematics and Distribution" Society for Marine Mammalogy, Special Publication No. 4.

Richardson, W. J., Greene, C. R., Jr., Malme, C. I., and Thomson, D. H. (1995). "Marine Mammals and Noise." Academic Press, San Diego.

Rojas-Bracho, L., Reeves, R. R., and Jaramillo-Legorreta, A. (2006). Conservation of the vaquita, *Phocoena sinus*. *Mamm. Rev.* **36**, 179–216.

Taylor, B. L. (2005). Identifying units to conserve. *In* "Marine Mammal Research: Conservation Beyond Crisis" (J. E. Reynolds, W. F. Perrin, R. R. Reeves, S. Montgomery, and T. J. Ragen, eds), pp. 149–162. Johns Hopkins University Press, Baltimore.

Turvey, S. T., *et al.* (16 authors) (2007). First human-caused extinction of a cetacean species? *Biol. Lett.* **3**, 537–540.

Twiss, J.R. Jr. and Reeves, R. R. (eds), (1999). Conservation and Management of Marine Mammals. Smithsonian Institution Press, Washington, D.C.

Vos, J. G., Bossart, G. D., Fournier, M., and O'Shea, T. J. (2003). "Toxicology of Marine Mammals." Taylor & Francis, London.

Whitehead, H., Reeves, R. R., and Tyack, P. L. (2000). Science and the conservation, protection, and management of wild cetaceans. *In* "Cetacean Societies: Field Studies of Dolphins and Whales" (J. Mann, R. C. Connor, P. L. Tyack, and H. Whitehead, eds), pp. 308–332. University of Chicago Press, Chicago.

# Crabeater Seal
## *Lobodon carcinophaga*

### John L. Bengtson

## I. Characteristics and Taxonomy

The crabeater seal may be the most abundant pinniped in the world, existing in the millions around Antarctica. The scientific name, *Lobodon carcinophaga* (Hombron and Jacquinot, 1842), is derived from Greek, and means "lobed tooth" (*Lobodon*) "crab eater" (*carcinophaga*). Crabeater seals have finely divided, lobed teeth, presumably an adaptation to their specialized diet of krill. The multiple cusps of upper and lower postcanine teeth interlock to form a sieve that can be used to filter crustaceans from seawater. A bony protrusion on the lower jaw behind the most posterior postcanine tooth fills the gap in this sieve so that prey cannot escape at the rear of the mouth.

Adult crabeater seals are generally about 205 to 240 cm long, with some older male and female individuals reaching lengths of up to 264 and 277 cm, respectively (Laws *et al.*, 2003). During the summer molting period, adults typically exhibit average weights of about 200 kg (males) and 215 kg (females). Pups weigh about 35 kg at birth but can grow to more than 100 kg by the time they are weaned.

The pelts of crabeater seals usually have medium brown to silver hair over most of their body, although darker coloration and spotting is not uncommon on the front and rear flippers and flanks (Fig. 1). The hair fades in color throughout the year, so recently molted seals may appear darker than those about to begin their molt, whose pelts can appear silvery-white. The body form is relatively slender compared to other phocids, and crabeater seals' faces have a somewhat pointed snout. Crabeater seals have a high incidence of obvious scarring on their bodies, mostly caused by leopard seal (*Hydrurga leptonyx*) attacks (Fig. 2). Adults typically also have small scars from bites around their front and rear flippers (both sexes) and around their lower jaw and throat (mostly males) from intra-specific interactions during the breeding season.

Crabeater seals are highly mobile on ice, and when disturbed often raise their heads and arch their backs. They can move surprisingly quickly over ice and snow, and on a cold day (i.e., when not subject to overheating) they may be capable of outrunning a fit human.

## II. Distribution and Abundance

Crabeater seals have a circumpolar Antarctic distribution, spending the entire year in the pack ice zone as it advances and retreats seasonally. Occasionally crabeater seals are found along the southern fringes of South America, Australia, New Zealand, and Africa, but such sightings or strandings are rare. Genetic analyses suggest that the circumpolar crabeater seal population is panmictic; there are no known subspecies of crabeater seals.

Crabeater seals migrate over large distances in association with the annual advance and retreat of the pack ice. Although they can be found anywhere within the pack ice zone, it is typical to find higher densities of crabeater seals over and at the edge of the continental shelf, as well as in the marginal ice zone (Burns *et al.*, 2004; Southwell *et al.*, 2005). Crabeater seals sometimes congregate in large groups (i.e., hundreds of individuals), which may be associated with migration or foraging.

**Figure 1** *Crabeater seal head and shoulders, illustrating spotting around front flippers. Photo by J. L. Bengtson.*

**Figure 2** *Nearly all crabeater seals possess long, raking scars on their torsos resulting from attacks in their first year of life by leopard seals. Photo by J. L. Bengtson.*

There is presently no reliable estimate of the total abundance of crabeater seals. Past estimates have ranged from 2 to 75 million individuals, although a population estimate in the range of 5–10 million is likely to be more reasonable. The observed densities of crabeater seals censused in the 1980s were lower than densities observed in the late 1960s and early 1970s (4.3 vs 11.4 seals/nm$^2$ in the Weddell Sea and 1.9 vs 4.9 seals/nm$^2$ in the Pacific Ocean Sector, respectively) (Erickson and Hanson, 1990). However, it is unclear whether these differences in densities reflected a change in population abundance or a shift in distribution within the sea ice zone. An international research initiative, the Antarctic Pack Ice Seals (APIS) Program, is evaluating survey data to refine estimates of the abundance and distribution of crabeater seals (Southwell *et al.*, 2008).

## III. Ecology

In their first year, crabeater seals experience a surprisingly high mortality rate that may be as high as 80%, which is perhaps double that which might normally be expected (Boveng and Bengtson, 1997). For the approximately 20% of crabeater seals that survive past their first birthday, as many as 78% exhibit large, raking scars on their bodies resulting from attacks by leopard seals (Fig. 2), suggesting that leopard seals may have a significant negative impact on crabeater seal populations. Most attacks by leopard seals on crabeater seals occur in the crabeater seals' first year; fresh wounds, indicating a recent attack, are rarely seen on crabeater seals that are older than 1 year (Siniff and Bengtson, 1977).

Studies of crabeater seal diet have shown that these seals depend almost exclusively on Antarctic krill (*Euphausia superba*). Most investigators have reported that krill comprise over 95% of the crabeater diet, with the remainder being made up of small quantities of fish and squid (Øritsland, 1977). As specialist krill predators, crabeater seals do not appear to switch their prey seasonally.

## IV. Behavior and Physiology

In a peculiar behavioral twist, crabeater seals likely hold the record for any pinniped wandering inland from the coast. Carcasses have been found up to 113 km from open water and as high as 1100 m above sea level. Seals that wander inland, become lost, and die; they may eventually become mummified in the cold, dry, Antarctic air, and can remain in this "freeze-dried" state for many decades or centuries (Stirling and Kooyman, 1971).

Similar to other Antarctic pack ice seals, crabeater seals exhibit a daily haulout pattern in summer that generally involves hauling out on ice floes during the middle of the day (Bengtson and Cameron, 2004). However, usually less than 80% of crabeater seals haul out simultaneously on the ice, even during the height of the molting period in January and February. Haulout patterns also vary markedly among seasons, with as few as 40% of seals hauling out at the peak of daily haulout during winter months.

During daily foraging periods in summer, which normally occur during the night, crabeater seals dive nearly continuously for periods of up to 16 h. In one study, a single crabeater seal continued diving for 44 h without interruption. Although crabeater seals have been recorded diving to depths of over 600 m, most dives are less then 100 m deep and less than 5 min in duration (Bengtson and Stewart, 1992; Nordøy *et al.*, 1995; Burns *et al.*, 2004; Wall *et al.*, 2007). Foraging dives made during crepuscular periods are often deeper than those made during the darkest hours, suggesting that the seals may prefer dark conditions when catching their principal prey, Antarctic krill.

## V. Life History

During the breeding season, crabeater seals form "family groups," consisting of a female, her pup, and an attendant male who guards the female from other males until she completes lactation (Siniff *et al.*, 1979). The peak of pupping is in mid- to late-October, with pups still observed with adults as late as mid-December (Southwell *et al.*, 2003b; Southwell, 2004). Pups are born with a light brown lanugo that is molted about 2 weeks later. Following weaning at about 2–3 weeks of age, the attendant male and the female form a "mated pair" and remain together for an estimated 1 to 2 weeks. Estrus, ovulation, and copulation occur approximately 4 days after the pup weans (Laws *et al.*, 2003b). Females without pups also form mated pairs as they come into estrus. Crabeater seals can live up to 40 years, but adults dying at about 20–25 years is more typical.

A large group of crabeater seals experienced an incident of mass mortality in 1955 in the vicinity of an Antarctic base where sledge dogs were active (Laws and Taylor, 1957). Up to 97% of mixed-age aggregations died during that event. It was speculated that a viral infection may have been associated with the die-off, and circumstantial evidence suggests that it may have been caused by a distemper-like virus. Blood samples taken from crabeater seals in the late 1980s confirmed that populations of crabeater seals along the Antarctic Peninsula had antibodies similar to those related to canine distemper and phocine distemper viruses that were responsible for major epi-zootic die-offs of harbor seals (*Phoca vitulina*) in the Northern Hemisphere in the late 1980s (Bengtson *et al.*, 1991).

## VI. Interactions with Humans

Crabeater seals were harvested commercially twice during the past century: in 1964/1965 by Norway, and in 1986/1987 by the former Soviet Union. In both cases, the sealing ventures were judged to be economically unsuccessful. However, the concern generated by the earlier harvest was sufficient to mobilize an international effort to prevent potential over-exploitation of the seals. This concern resulted in the Convention for the Conservation of Antarctic Seals, which came into effect in 1978, and provides international oversight for the conservation and management of crabeater seals throughout their range.

## *See Also the Following Articles*

Antarctic Marine Mammals ■ Earless Seals (Phocidae)

## *References*

Adam, P. J. (2005). Lobodon carcinophaga. *Mamm. Species* **772**, 1–14.

Bengtson, J. L., and Cameron, M. F. (2004). Seasonal haulout patterns of crabeater seals (*Lobodon carcinophaga*). *Polar Biol.* **27**, 344–349.

Bengtson, J. L., and Laws, R. M. (1985). Trends in crabeater seal age at maturity: An insight into Antarctic marine interactions. *In* "Antarctic Nutrient Cycles and Food Webs" (W. R. Siegfried, P. R. Condy, and R. M. Laws, eds), pp. 670–675. Springer-Verlag, Berlin.

Bengtson, J. L., and Siniff, D. B. (1981). Reproductive aspects of female crabeater seals (*Lobodon carcinophagus*) along the Antarctic Peninsula. *Can. J. Zool.* **59**, 92–102.

Bengtson, J. L., and Stewart, B. S. (1992). Diving and haulout behavior of crabeater seals in the Weddell Sea, Antarctica, during March 1986. *Polar Biol.* **12**, 635–644.

Bengtson, J. L., Boveng, P., Franzén, U., Have, P., Heide-Jørgensen, M. P., and Härkönen, T. J. (1991). Antibodies to canine distemper virus in Antarctic seals. *Mar. Mamm. Sci.* **7**, 85–87.

Bengtson, J. L., Hill, R. D., and Hill, S. E. (1993). Using satellite telemetry to study the ecology and behavior of Antarctic seals. *Korean J. Polar Res.* **4**, 109–115.

Boveng, P. L., and Bengtson, J. L. (1997). Crabeater seal cohort variation: Demographic signal or statistical noise? *In* "Antarctic Communities: Species, Structure, and Survival" (B. Battaglia, J. Valencia, and D. W. H. Walton, eds), pp. 241–247. Cambridge Univ. Press, Cambridge.

Burns, J. M., *et al.* (2004). Winter habitat use and foraging behavior of crabeater seals along the Western Antarctic Peninsula. *Deep-Sea Res. (II Top. Stud. Oceanogr.)* **51**, 2279–2303 (9 authors).

Erickson, A. W., and Hanson, M. B. (1990). Continental estimates and population trends of Antarctic ice seals. *In* "Antarctic Ecosystems: Ecological Change and Conservation" (K. R. Kerry, and G. Hempel, eds), pp. 253–264. Springer-Verlag, Berlin.

Gales, N. J., Fraser, W. R., Costa, D. P., and Southwell, C. (2004). Do crabeater seals forage cooperatively? *Deep-Sea Res. (II Top. Stud. Oceanogr.)* **51**, 2305–2310.

Kooyman, G. L. (1981). Crabeater seal, *Lobodon carcinophagus* (Hombron and Jacquinot, 1842). *In* "Handbook of Marine Mammals" (S. H. Ridgeway, and R. J. Harrison, eds), pp. 221–235. Academic Press, London.

Laws, R. M. (1984). Seals. *In* "Antarctic Ecology" (R. M. Laws, ed.), Vol. 2, pp. 621–715. Academic Press, London.

Laws, R. M., and Taylor, R. J. F. (1957). A mass dying of crabeater seals, *Lobodon carcinophagus* (Gray). *Proc. Zool. Soc. Lond.* **129**, 315–324.

Laws, R. M., Baird, A., and Bryden, M. M. (2003a). Breeding season and embryonic diapause in crabeater seals (*Lobodon carcinophagus*). *Reproduction* **126**, 365–370.

C

Laws, R. M., Baird, A., and Bryden, M. M. (2003b). Size and growth of the crabeater seal *Lobodon carcinophagus* (Mammalia: Carnivora). *J. Zool.* **259**, 103–108.

Nordøy, E. S., Folkow, L., and Blix, A. S. (1995). Distribution and diving behaviour of crabeater seals (*Lobodon carcinophagus*) off Queen Maud Land. *Polar Biol.* **15**, 261–268.

Øritsland, T. (1977). Food consumption of seals in the Antarctic pack ice. *In* "Adaptations Within Antarctic Ecosystems" (G. A. Llano, ed.), pp. 749–768. Smithsonian Institution, Washington.

Siniff, D. B., and Bengtson, J. L. (1977). Observations and hypotheses concerning the interactions among crabeater seals, leopard seals, and killer whales. *J. Mamm.* **58**, 414–416.

Siniff, D. B., Stirling, I., Bengtson., J. L., and Reichle, R. A. (1979). Social and reproductive behavior of crabeater seals (*Lobodon carcinophagus*) during the austral spring. *Can. J. Zool.* **57**, 2243–2255.

Southwell, C. (2004). Satellite-linked dive recorders provide insights into the reproductive strategies of crabeater seals (*Lobodon carcinophagus*). *J. Zool.* **264**, 399–402.

Southwell, C., Kerry, K. R., and Ensor, P. H. (2005). Predicting the distribution of crabeater seals *Lobodon carcinophaga* off east Antarctica during the breeding season. *Mar. Ecol. Prog. Ser.* **299**, 297–309.

Southwell, C., Kerry, K. R., Ensor, P., Woehler, E. J., and Rogers, T. (2003). The timing of pupping by pack-ice seals in East Antarctica. *Polar Biol.* **26**, 468–652.

Southwell, C., Paxton, C. G. M., Borchers, D., Boveng, P., and de la Mare, W. (2008). Taking account of dependent species in management of the Southern Ocean krill fishery: Estimating crabeater seal abundance off east Antarctica. *J. Appl. Ecol.* **45**.

Stirling, I., and Kooyman, G. L. (1971). The crabeater seal (*Lobodon carcinophagus*) in McMurdo Sound, Antarctica, and the origin of mummified seals. *J. Mamm.* **52**, 175–180.

Wall, S. M., Bradshaw, C. J. A., Southwell, C. J., Gales, N. J., and Hindell, M. A. (2007). Crabeater seal diving behaviour in eastern Antarctica. *Mar. Ecol. Prog. Ser.* **337**, 265–277.

# Culture in Whales and Dolphins

## Hal Whitehead

Evidence is growing that culture is an important determinant of the behavior of whales and dolphins (Rendell and Whitehead, 2001). Among the many definitions of culture, one that is commonly used by evolutionary biologists and is useful when studying the phenomenon in whales and dolphins, is behavioral variation between sets of animals maintained and transmitted by social learning (Laland and Hoppitt, 2003). As clearly shown in the case of humans, when culture becomes an important determinant of behavior, then evolution and ecology are channeled into unusual paths (Richerson and Boyd, 2004).

There are two principal approaches to the study of nonhuman culture (Laland and Hoppitt, 2003). Because some scientists will only ascribe culture to a behavioral pattern if it can be proved to be transmitted between animals by imitation or teaching, they investigate transmission mechanisms experimentally. Others, who use a broader definition of culture encompassing any form of social learning, not just imitation or teaching (Whiten and Ham, 1992), look for patterns of behavioral variation in wild populations that cannot be explained by either genetic factors or environmental differences plus individual learning. This has been called the "ethnographic" approach to the study of culture.

The bottlenose dolphin, *Tursiops* spp., has been shown experimentally to possess sophisticated social learning abilities, including vocal and motor imitation (Herman, 2002), but these have not been closely tied to observed patterns of behavior in the wild. Although social learning of other cetacean species has not been studied experimentally, there is observational evidence for imitation and teaching in some other whales and dolphins, especially killer whales, *Orcinus orca* (Guinet and Bouvier, 1995).

Using the second, ethnographic, approach to the study of culture, there is good evidence for cultural transmission in several cetacean species. Most notable are the complex and stable vocal (call dialects) and behavioral (foraging patterns and techniques) cultures of sympatric pods, clans, communities, and types of killer whales (Boran and Heimlich, 1999; Yurk *et al.*, 2002; see KILLER WHALES). The parallel slow evolution of calls in neighboring pods not only demonstrates cultural evolution, but also indicates that killer whales actively delineate their social structures using cultural markers (Deecke *et al.*, 2000).

The sperm whale, *Physeter macrocephalus*, another large odontocete with a matrilineally based social system, also has important matrilineally transmitted cultures. In the South Pacific, social units of female sperm whales are members of clans that are distinguished by the types of coda vocalizations that they use (Rendell and Whitehead, 2003). Although clans are sympatric, units form groups only with other units from their clan. The clans, which span several thousands of kilometers and might contain ten thousand members, have distinctive movement patterns, foraging success, and seem to differ in their reproductive rates (Whitehead and Rendell, 2004; Marcoux *et al.*, in press).

Sympatric cultural variants seem to be quite common in odontocetes. In Shark Bay, Australia, there are at least 13 foraging strategies that are not used equally by individual bottlenose dolphins, at least some of which are likely socially learned (Mann and Sargeant, 2003; Krützen *et al.*, 2005). In several parts of the world, inshore and riverine odontocetes have developed fishing cooperatives with local human populations, in which both humans and dolphins benefit (Simões-Lopes *et al.*, 1998).

Perhaps most remarkable of all cetacean cultures is the song of male humpback whales, *Megaptera novaeangliae*. All males in any ocean basin sing nearly the same song, but it evolves over periods of months and years (Payne, 1999). This evolution is usually gradual, but over a 2-year period, the males off eastern Australia adopted the radically different western Australian song, which they had heard from a few itinerant males, making this the first known instance of a non-human cultural revolution (Noad *et al.*, 2000; Fig. 1).

The case that culture is the sole cause of these differences in cetacean behavior varies in strength, being perhaps strongest for humpback whale songs and weakest for bottlenose dolphin foraging strategies (Rendell and Whitehead, 2001; Laland and Janik, 2006). While it is very likely that culture has a role in all cases, ecological or genetic influences may also be present, making the exclusionary approach of studying culture in the wild ineffective. New techniques are being developed that, instead of only ascribing culture when other influences can be ruled out, apportion behavioral variation to genetic, ecological and cultural causes (Laland and Janik, 2006), and these may be particularly important in the future study of cetacean culture.

Several factors may be implicated in the apparent importance of cultural transmission of behavior among cetaceans. Long lives, prolonged parental care, and substantial cognitive abilities are often associated with the evolution of cultural faculties, and these are generally characteristic of cetaceans, as well as other cultural animals, females which is known only from humans among non-cetaceans (Rendell and Whitehead, 2001).

Culture potentially affects ecology and population biology in a variety of ways. It can increase feeding success, the diversity of diet, and either promote or inhibit population structure. Thus, cetacean culture may need to be included into considerations of the conservation and management of whales and dolphins (Whitehead *et al.*, 2004). As an example, the bottlenose dolphin population in Hervey Bay, Australia, includes two sympatric communities, one that feeds extensively on the discards from trawlers and one that does not (Chilvers and Corkeron, 2001). Management of the fishery will have complex repercussions for this population.

The focused study of culture in whales and dolphins is just beginning. Despite denials from those who demand experimental proof of imitation or teaching before attributing culture, there are strong indications that, in common with humans (*Homo sapiens*) and chimpanzees (*Pan troglodytes*), much of the behavioral repertoire of many cetaceans is learned socially and constitutes culture. Culture may also be an important attribute of other marine mammals, with the foraging techniques of sea otters (*Enhydra lutris*) perhaps forming the clearest example (Estes *et al.*, 2003).

The cultures of whales and dolphins possess features that have no known parallel outside humans. Examples are the sympatric multifaceted cultures of killer whales, and the giant physical scales such as ocean-wide songs of humpback whales and clans of sperm whales. They represent an independent evolution of cultural faculties outside the primate line, and thus help us understand both how culture evolved and what its consequences may be.

## See Also the Following Articles

Behavior, Overview ■ Communication ■ Dialects ■ Ethics and Marine Mammals ■ Intelligence and Cognition ■ Tool Use

## References

Boran, J. R., and Heimlich, S. L. (1999). Social learning in cetaceans: Hunting, hearing and hierarchies. *Symp. Zool. Soc., Lond.* **73**, 282–307.
Chilvers, B. L., and Corkeron, P. J. (2001). Trawling and bottlenose dolphins' social structure. *Proc. Roy. Soc. London B* **268**, 1901–1905.
Deecke, V. B., Ford, J. K. B., and Spong, P. (2000). Dialect change in resident killer whales: Implications for vocal learning and cultural transmission. *Anim. Behav.* **40**, 629–638.
Estes, J. A., Riedman, M. L., Staedler, M. M., Tinker, M. T., and Lyon, B. E. (2003). Individual variation in prey selection by sea otters: Patterns, causes and implications. *J. Anim. Ecol.* **72**, 144–155.
Guinet, C., and Bouvier, J. (1995). Development of intentional stranding hunting techniques in killer whale (*Orcinus orca*) calves at Crozet Archipelago. *Can. J. Zool.* **73**, 27–33.
Herman, L. H. (2002). Vocal, social, and self-imitation by bottlenosed dolphins. *In* "Imitation in animals and artifacts" (K. Dautenhahn, and C. L. Nehaniv, eds), pp. 63–108. MIT Press, Cambridge.
Krützen, M., Mann, J., Heithaus, M. R., Connor, R. C., Bejder, L., and Sherwin, W. B. (2005). Cultural transmission of tool use in bottlenose dolphins. *Proc. Natl. Acad. Sci. USA* **102**, 8939–8943.
Laland, K. N., and Hoppitt, W. (2003). Do animals have culture? *Evol. Anthrop.* **12**, 150–159.
Laland, K. N., and Janik, V. M. (2006). The animal cultures debate. *Trends Ecol. Evol.* **21**, 542–547.
Mann, J., and Sargeant, B. (2003). Like mother, like calf: The ontogeny of foraging traditions in wild Indian ocean bottlenose dolphins (*Tursiops* sp.). *In* "The biology of traditions; models and evidence" (D. M. Fragaszy, and S. Perry, eds), pp. 236–266. Cambridge University Press, Cambridge.

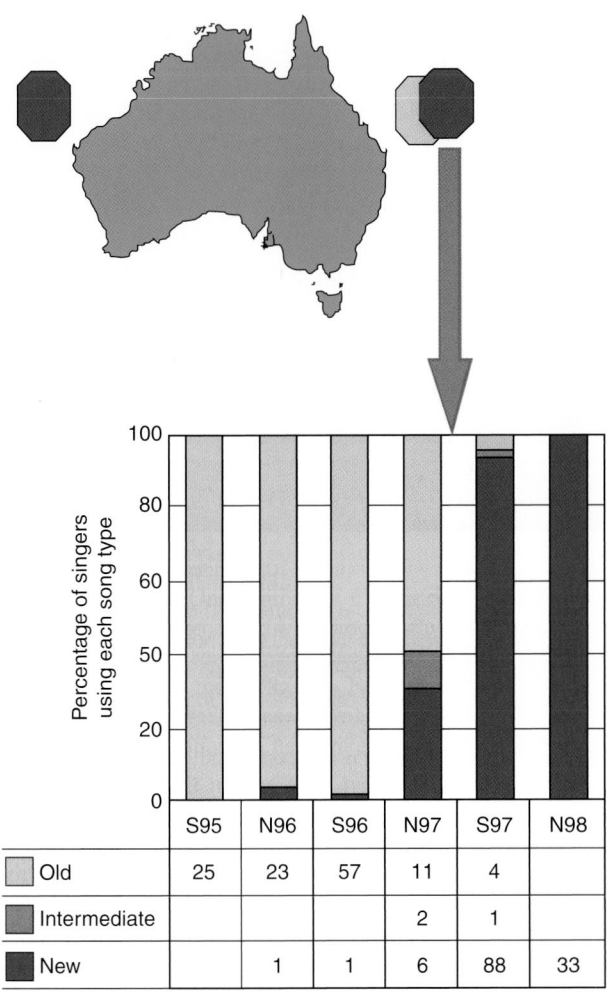

**Figure 1** *The revolution in humpback* (Megaptera novaeangliae) *song type off the east coast of Australia during 1995–1998, as males from the east coast population adopted the west coast song (adapted from Noad* et al., *2000).*

such as primates and some birds. The wide movements of cetaceans and the greater variability of the marine biotic environment relative to that on land, as well as the stable matrilineal social groups of some species, are potentially important factors in the evolution of some of the more unusual aspects of cetacean culture (Rendell and Whitehead, 2001).

Culture can affect the evolution of other aspects of the lives of animals. There have been a number of suggestions for gene-culture coevolution in cetaceans, and culture may be implicated in some of their unusual behavioral and life history traits. For instance, it has been proposed that the separation between "resident," fish-feeding and "transient," mammal-feeding forms of killer whales (which now show morphological and genetic differences) was originally driven by culture (Boran and Heimlich, 1999). Another suggestion is that cultural selection may have caused the remarkably low diversity of mitochondrial genes found in matrilineal whales, as these genes and beneficial cultural traits may have been inherited in parallel by daughters from their mothers (Whitehead, 1998). Culture may also be implicated in mass strandings, as well as in the pronounced menopause shown by killer and pilot whale (*Globicephala* spp.)

C

Marcoux, M., Rendell, L., and Whitehead, H. (2007). Indications of fitness differences among vocal clans of sperm whales. *Behav. Ecol. Sociobiol.* **61**, 1093–1098.

Noad, M. J., Cato, D. H., Bryden, M. M., Jenner, M. N., and Jenner, K. C. S. (2000). Cultural revolution in whale songs. *Nature, Lond.* **408**, 537.

Payne, K. (1999). The progressively changing songs of humpback whales: A window on the creative process in a wild animal. *In* "The origins of music" (N. L. Wallin, B. Merker, and S. Brown, eds), pp. 135–150. MIT Press, Cambridge.

Rendell, L., and Whitehead, H. (2001). Culture in whales and dolphins. *Behav. Brain Sci.* **24**, 309–324.

Rendell, L., and Whitehead, H. (2003). Vocal clans in sperm whales (*Physeter macrocephalus*). *Proc. Roy. Soc. London B* **270**, 225–231.

Richerson, P. J., and Boyd, R. (2004). "Not by genes alone: How culture transformed human evolution." University of Chicago Press, Chicago.

Simões-Lopes, P. C., Fabián, M. E., and Menegheti, J. O. (1998). Dolphin interactions with the mullet artisanal fishing on southern Brazil: A qualitative and quantitative approach. *Revta Bras. Zool.* **15**, 709–726.

Whitehead, H. (1998). Cultural selection and genetic diversity in matrilineal whales. *Science* **282**, 1708–1711.

Whitehead, H., and Rendell, L. (2004). Movements, habitat use and feeding success of cultural clans of South Pacific sperm whales. *J. Anim. Ecol.* **73**, 190–196.

Whitehead, H., Rendell, L., Osborne, R. W., and Würsig, B. (2004). Culture and conservation of non-humans with reference to whales and dolphins: Review and new directions. *Biol. Conserv.* **120**, 431–441.

Whiten, A., and Ham, R. (1992). On the nature and evolution of imitation in the animal kingdom: Reappraisal of a century of research. *Adv. Study Behav.* **21**, 239–283.

Yurk, H., Barrett-Lennard, L., Ford, J. K. B., and Matkin, C. O. (2002). Cultural transmission within maternal lineages: Vocal clans in resident killer whales in southern Alaska. *Anim. Behav.* **63**, 1103–1119.

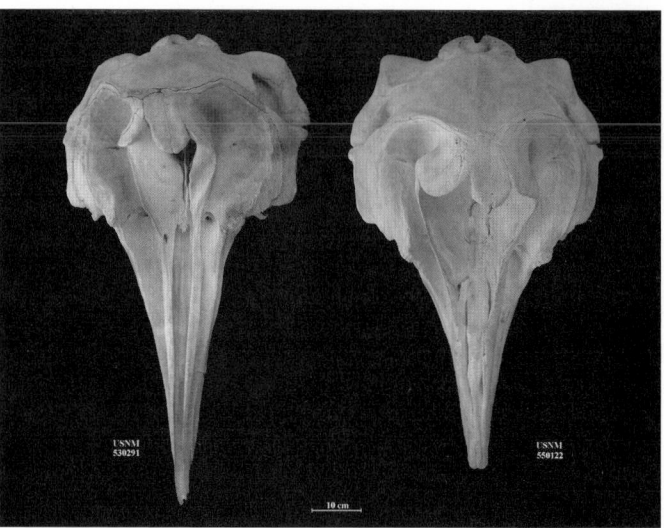

**Figure 1**  *As males of* Ziphius cavirostris *become sexually mature, they begin to resorb bone in front of the nasal passages, creating over time a distinct cavity or prenarial basin seen easily in this dorsal view (right). Adult females and immature males lack this basin (left).*

# Cuvier's Beaked Whale
## *Ziphius cavirostris*

### JOHN E. HEYNING AND JAMES G. MEAD

### I. Characteristics and Taxonomy

The original description of *Ziphius cavirostris* is based on a partial cranium collected near the village of Fos, France, in 1804. In his species description, Cuvier mistakenly identified the specimen as a fossil because he thought the skull was "petrified." The specimen actually represents part of a skull, including a densely ossified rostrum, with a well-developed prenarial basin of an adult male. This basin, or cavity, on the top of the skull just anterior to the bony nares led to the trivial name of *cavirostris*. The densely ossified rostrum is found in adult males of Cuvier's beaked whale and mesoplodont beaked whales (Fig. 1). The function of this rock-hard snout is unknown, but it has been postulated to either reinforce the skull when males fight or serve as a sound conduit. The most common English names for *Z. cavirostris* are Cuvier's beaked whale and the goose-beaked whale, both of which are in wide usage.

The general body shape of *Z. cavirostris* is similar to that of other beaked whales with a rather robust, cigar-shaped body (Heyning,

1989). The falcate dorsal fin is relatively small and set approximately two-thirds of the body length posterior to the rostrum. The flippers are also relatively small, narrow, and can be tucked into a slight depression or pocket along the body wall. This flipper pocket is also found in other ziphiids and is assumed to allow the flippers to be held tight against the body while swimming. As with other ziphiids, the flukes are proportionately large and, as a rule, lack the distinctive medial notch found in all other cetaceans. The head is rather blunt in profile with a small, poorly defined rostrum that grades into a gently sloping melon region. There is no significant difference in total length between sexes for *Z. cavirostris*, with an average adult size of 6.1 m. There are several reports of specimens that exceed 7.0 m in length, but virtually all of these appear to be either estimates of lengths or based on misidentified animals. The largest accurately measured specimen of *Z. cavirostris* is an adult male that measured 6.93 m from the Falkland Islands. There is one pair of throat grooves that converge, but do not meet anteriorly. Beaked whales feed primarily by suction and these grooves allow the throat region to expand as they slurp in their prey.

The pigmentation pattern for male *Z. cavirostris* is a dark slate gray over most of its body with a distinctively white head. This white coloration continues slightly posterior along the dorsum. This appears to be the pattern found in most mature males (Fig. 2). Adult females tend to vary in general color from a dark gray to a reddish-brown, with a slight lightening of the skin on the head. This is not as dramatic a contrast as in males and does not appear to extend posteriorly on the dorsal aspect of the body. Usually there are some distinctive patterns of dark pigment on the head of adult females. The eye is typically dark, and there is a highly variable pair of dark crescents surrounding the eye, one anteriorly and one posteriorly. Newborns are dark black or bluish-black above and lighter below. This pigmentation pattern is very similar to that found on young beaked whales of the genus *Mesoplodon* and may represent the primitive pigmentation pattern for the calves of many ziphiids. Light oval patches and linear marks are quite common on the skin of *Z. cavirostris*, which can give an animal a mottled appearance.

**Figure 2** *This dead, stranded* Z. cavirostris *shows lime whitish head characteristic of adult males. Photo by W. Perrin, National Marine Fisheries Service.*

The oval patches on ziphiids have been attributed to scars left by lampreys or cookie-cutter sharks of the genus *Isistius*. Linear marks have been attributed to scars resulting from the teeth of males raking along the skin during intraspecific fighting. The only erupted teeth are the apical pair in adult males, and linear scars are most prevalent in mature males.

## II. Distribution and Abundance

As with most uncommon cetaceans, the distribution of *Z. cavirostris* is known primarily from strandings. This type of information may be somewhat biased, especially with regard to the abundance for a particular area. Stranding records indicate that *Z. cavirostris* is the most cosmopolitan of the beaked whales and is distributed in all oceans and most seas except in the high polar waters (Heyning, 1989). Strandings of *Z. cavirostris* are the most numerous of all beaked whales, indicating that they are probably not as rare as originally thought. Observations of live animals in the field reveal that the blow of *Z. cavirostris* is low, diffuse, and directed forward, making sightings more difficult, and there is some evidence that they avoid vessels by diving. These two facts may explain why these whales are rarely seen at sea. Single animals are frequently observed with pods ranging in size up to seven animals. There are several records of mass strandings.

## III. Ecology

Although most general accounts of *Z. cavirostris* list squid as the primary prey item, very few actual stomach contents have been analyzed, and care must be invoked in any interpretation. Stomach contents from *Z. cavirostris* caught off Japan varied consistently, with a predominance of squid from animals taken in waters slightly under 1000 m in depth, but fish are the most abundant prey item found in animals in deeper waters. This evidence has been interpreted to suggest that *Z. cavirostris* is somewhat opportunistic in its feeding habits. Most of the prey items listed are open ocean, mesopelagic, or deep-water benthic organisms, concurring with the idea that *Z. cavirostris* is an offshore deep-diving species. Ectoparasites that have been reported include the barnacles *Xenobalanus* sp. from the flukes and dorsal fin and *Conchoderma* sp. on the erupted apical teeth. The following internal parasites have been reported from *Z. cavirostris*;

Nematoda, *Anisakis* sp., *Crassicauda boopis*, and *Crassicauda crassicauda*; and Cestoda, *Phyllobothrium* sp.

## IV. Behavior and Physiology

Recent studies have demonstrated that *Ziphius* can dive to 1888 m and stay submerged for up to 85 min while foraging using echolocation clicks and buzzes.

## V. Life History

The minimum length at sexual maturity is 527 cm for females, which average 580 cm at sexual maturity (Heyning, 1989). Males average 550 cm at sexual maturity. There are no detailed studies of age, though females have been estimated at 30 years of age and males at 36 years. The mean length at birth is 270 cm.

## VI. Interactions with Humans

In the past, there have been few small cetacean fisheries that have taken *Ziphius*. In the Japanese *Berardius bairdii* fishery, *Z. cavirostris* have been taken on an opportunistic basis with catches of varying from 3 to 35 animals taken yearly. Although the *Berardius* fishery continues, there has been no takes of *Z. cavirostris* in recent years. It is probable that killer whales occasionally prey on *Z. cavirostris*.

Association of mass stranding of *Ziphius* and naval sonar exercises leads one to speculate that repeated deep dives lead to supersaturation of the blood with nitrogen and abnormal behavioral responses resulting in stranding (Cox *et al.*, 2006). In addition, ship noise may disrupt their behavior.

## *See Also the Following Articles*

Barnacles ▪ Beaked Whales ▪ Overview ▪ Mesoplodont Whales

## *References*

Cox, T. M., *et al.* (2006). Understanding the impacts of anthropogenic sound on beaked whales. *J. Cetacean Res. Manag.* **7**, 177–187.

Heyning, J. E. (1989). Cuvier's beaked whale, *Ziphius cavirostris*. *In* "Handbook of Marine Mammals" (S. H. Ridgway, and R. Harrison, eds), Vol. 4, pp. 289–308. Academic Press, London.

# Dall's Porpoise
## *Phocoenoides dalli*

### THOMAS A. JEFFERSON

### I. Characteristics and Taxonomy

Typical of the porpoise family, Dall's porpoise has a stocky body, and it has a short, wide-based, triangular dorsal fin (Fig. 1). The dorsal fin is slightly falcate at the tip, but the entire fin may be canted forward in adult males (Jefferson, 1990). The tail stock is deepened, especially in adult males, and males also have a prominent post-anal hump of connective tissue. There is an extremely short, poorly defined beak. The flippers and flukes are small, and the fluke blades may also be canted forward in older individuals.

The color pattern is diagnostic. "Dall's" are largely dark gray to black with a large, ventrally continuous white patch that extends up about halfway on each flank. In addition, there is light gray to white frosting, or trim, on the upper part of the dorsal fin and on the trailing edges of the flukes. Some other light patches may exist, particularly around the base of the tail stock. Newborn animals are muted in color and do not have the fluke and dorsal fin frosting, which develops with age in older animals.

There are two major color morphs, one with a flank patch that extends forward to about the level of the dorsal fin (*dalli* type) and the other with a flank patch extending to about the level of the flippers (*truei* type). These forms were variously considered as separate species and subspecies in the past, but most recent work suggests that they are in fact color variants, with few or no other phenotypic differences. However, genetic analyses have confirmed that they do form separate populations (Escorza Trevino *et al.*, 2004), and Rice (1998) still considered them as separate subspecies. Other color types (all-black, all-white, and forms intermediate between *dalli* and *truei* types) have also been observed.

Dall's porpoises reach maximum known lengths and weights of about 239 cm and 200 kg. Males grow longer and heavier than females, and adult males have secondary sexual characteristics (as discussed earlier). There is a great deal of geographical variation; size, shape, and coloration differences have been documented among different areas of the species' range (Amano and Miyazaki, 1992).

The SKULL of Dall's porpoise is larger than that of most other phocoenids and may reach 340 mm in length. The rostrum is wide at the base and relatively short. There are prominent "maxillary shields" that make an angle of about 130° with the rostrum axis (Houck and Jefferson, 1999). Tooth counts are highly variable, but generally number 21–28 per tooth row. The TEETH are shaped like grains of rice (not strongly spatulate, as in *Phocoena*) and are extremely small, the smallest of any species of cetacean. They often do not rise above the level of the gums and are considered by many to be rudimentary. Dall's porpoises have an unusual skeleton, with extremely long, slender dorsal, and lateral processes on the vertebrae. Total vertebral counts generally number 92–98.

Recent studies of mtDNA and morphology suggest that the previous classification of Dall's porpoise and the spectacled porpoise (*Phocoena dioptrica*) in the same subfamily was erroneous (Rosel *et al.*, 1995; Fajardo-Mellor *et al.*, 2006). These two species do not appear to be closely related, and their similarities may be the result of CONVERGENT EVOLUTION.

Intergeneric hybrids between Dall's porpoises and harbor porpoises (*Phocoena phocoena*) have been examined and described (Baird *et al.*, 1998; Willis *et al.*, 2004). Free-ranging hybrids are regularly observed around Vancouver Island, British Columbia, suggesting that such hybridization events may not be all that rare and supporting the hypothesis of a close relationship between these two species.

The species was named after William H. Dall, who collected the type specimen in Alaska in 1873. Other English common names include True's porpoise, white-flank porpoise, and spray porpoise.

### II. Distribution and Abundance

Dall's porpoise is found only in the North Pacific Ocean and adjacent seas (Bering Sea, Okhotsk Sea, and Sea of Japan), from about 32–35°N (southern California and southern Japan) in the south to about 63°N (central Bering Sea) in the north. When water temperatures are unseasonably cold, they may extend down to around Scammon's Lagoon, Baja California, Mexico (Morejohn, 1979). On rare occasions, they may also go through the Bering Strait into the Chukchi Sea. Up to 10 different stocks are recognized, based on studies of morphology, genetics, and ecological parameters (see review in Houck and Jefferson, 1999). Sex-biased dispersal and migration patterns have been elucidated from molecular genetic analyses (Escorza Trevino *et al.*, 2004).

Dall's porpoise is a cold-water species, avoiding tropical/subtropical waters. This is an oceanic species that is found in deep offshore waters, but also in deeper nearshore and inshore waters along the west coast of North America. There are seasonal inshore–offshore and north–south movements in both the eastern and western North Pacific (Forney and Barlow, 1998; Houck and Jefferson, 1999), but in most areas these are poorly defined.

Current global abundance is not well established (due to lack of survey data for some areas and probable biases in available estimates),

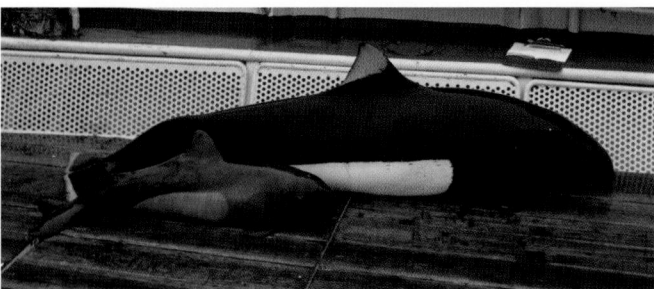

**Figure 1** *Dall's porpoise mother and calf killed in a Japanese salmon driftnet south of the Aleutian Islands in the mid-1980s. Such large-scale kills have now been much reduced, due to a United Nations ban on oceanic driftnet fishing.*

but it was thought to number over 1.2 million individuals in the 1980s (Buckland *et al.*, 1993). There are thought to be around 104,000 along the Pacific coast of Japan, 554,000 in the Okhotsk Sea, 83,000 in Alaska, and 100,000 along the US west coast.

## III. Ecology

Prey of Dall's porpoise include a wide variety of small fishes and cephalopods (several dozen species have been identified—see Houck and Jefferson, 1999). The most common prey items include schooling fishes (such as herrings, anchovies, mackerels, and sauries), mesopelagic fishes (such as myctophids and deep-sea smelts), and squids. KRILL, decapods, and shrimps have been found in some stomachs, but these are not considered to be common prey items. Amano and Miyazaki (1992) found that the skulls of Dall's porpoises grew to larger sizes in areas with higher productivity, suggesting that food availability may affect growth. Dall's porpoises in some areas appear to feed preferentially at night on vertically migrating fish and squid associated with the deep scattering layer (DSL).

Dall's are thought to be deep divers and capable of feeding at great depths; however, very few dives have been measured directly. The first dive depth data were obtained from a single individual in the transboundary area between British Columbia and Washington State. Seventeen dives were recorded, the deepest to 94 m (Hanson and Baird, 1998). However, Dall's porpoises are probably capable of much deeper dives.

Several internal parasites have been identified from various areas of the body of this species, and these parasites can cause disease and pathology which can even lead to death (see review in Houck and Jefferson, 1999). The trematode fluke, *Nasitrema*, appears to be particularly serious and has been implicated in the stranding and subsequent deaths of several specimens. Whale lice have also been found on the external surface of the body.

Large sharks, and especially killer whales, are predators. Several attacks by killer whales on Dall's porpoises have been observed in recent years, and it appears that killer whales may be major predators on this species, at least in Alaskan waters.

## IV. Behavior and Physiology

Small groups are most common, although large aggregations of several hundred to about one thousand have been reported on occasion (Houck and Jefferson, 1999). Groups of over 20–30 are rather uncommon. Very little is known of the group structure of this species, except that group composition appears to be quite fluid. Recently, evidence of mate guarding behavior, whereby a male maintains longer associations with females (presumably to exclude other males from mating with them) has been documented (Willis and Dill, 2007). This is probably related to a polygynous mating system.

Dall's porpoises are very fast swimming and active animals. They are often seen moving very quickly, slicing along the surface, creating a sloppy, V-shaped splash. These are called rooster tail splashes (Fig. 2). However, aerial behavior, such as breaching and leaping out of the water, is virtually nonexistent. Dall's porpoises are willing and capable bow riders and will converge on the bow of a fast-moving boat from all around. They have even been seen to "snout ride" on bow waves pushed forward by the heads of large whales (like blue and humpback whales). When moving more slowly, they roll at the surface in a subdued behavior more typical of other species of porpoises (Fig. 3).

The physiology of this species has not been extensively studied, but some work mostly related to diving physiology has been done.

**Figure 2** *Most often when Dall's porpoises are seen at sea, they are swimming very fast and "roostertailing," as this porpoise is doing while riding the bow wave of a research vessel in Southeast Alaska.*

**Figure 3** *A Dall's porpoise slow rolls in Monterey Bay, California. This is an adult male, based on the forward-canted dorsal fin (the animal is moving left to right).*

Dall's have a thin blubber layer, large skeletal muscle mass, thick tracheal cartilage rings, deeply folded vestibular sacs, fatty pads lining the lungs, relatively small brain, and large adrenal and thyroid glands. They also have relatively high blood oxygen content. In addition, they appear to have a relatively high metabolic rate and interestingly, captive animals were never seen to sleep (see review in Houck and Jefferson, 1999).

## V. Life History

Growth and reproductive parameters have been estimated for several populations in the central and western North Pacific, based on large samples of specimens killed in various fisheries. Length at birth is about 100 cm. Estimates of length and age at sexual maturity range from about 172 to 187 cm and 4–7 years for females, and from 175 to 196 cm and 3.5–8 years for males (Ferrero and Walker, 1999; see Houck and Jefferson, 1999 for a review). Gestation lasts about 10–12 months. The length of lactation is not well known (but is most likely <1 year). The calving season is in the summer for all populations that have been studied to date, although sometimes calves may be born outside of the main calving season (Jefferson, 1989). There appears to be significant geographic variation in growth and life history parameters.

## VI. Interactions with Humans

Small numbers of Dall's porpoises have been kept in captivity in oceanaria and research institutes in the United States and Japan, but most individuals have not survived very long. Due to their poor record of survival in captivity and their apparent intractability, they are not currently a species that is sought after for captive display.

A number of human-caused threats to Dall's porpoise populations have been identified, including environmental POLLUTION and habitat alteration. However, the most serious threats are clearly the various fishery kills of this species (review in Houck and Jefferson, 1999). These include directed kills in Japanese waters and INCIDENTAL CATCHES in various fisheries (most prominently several drift gillnet fisheries) throughout the range. The most heavily impacted populations were those in the central and western North Pacific. Between 1981 and 1990, over 45,000 Dall's porpoises were killed in Japanese driftnet fisheries, and between 1981 and 1994, more than 247,000 were directly taken in a Japanese harpoon fishery (Houck and Jefferson, 1999). Fortunately, these kills have been greatly reduced in recent years.

### *See Also the Following Articles*

Bow-Riding ■ Geographic Variation ■ North Pacific Marine Mammals ■ Porpoises ■ Overview

### *References*

Amano, M., and Miyazaki, N. (1992). Geographic variation and sexual dimorphism in the skull of Dall's porpoise, *Phocoenoides dalli*. *Mar. Mamm. Sci.* **8**, 240–261.

Baird, R. W., Willis, P. M., Guenther, T. J., Wilson, P. J., and White, B. N. (1998). An intergeneric hybrid in the family Phocoenidae. *Can. J. Zool.* **76**, 198–204.

Buckland, S. T., Cattanach, K. L., and Hobbs, R. C. (1993). Abundance estimates of Pacific white-sided dolphin, northern right whale dolphin, Dall's porpoise and northern fur seal in the North Pacific, 1987–1990. *Int. N. Pac. Fish. Comm. Bull.* **53**, 387–407.

Escorza Trevino, S., Pastene, L. A., and Dizon, A. E. (2004). Molecular analyses of the *truei* and *dalli* morphotypes of Dall's porpoise (*Phocoenoides dalli*). *J. Mammal.* **85**, 347–355.

Fajardo-Mellor, L., Berta, A., Brownell, R. L., Jr., Boy, C. C., and Goodall, R. N. P. (2006). The phylogenetic relationships and biogeography of true porpoises (Mammalia: Phocoenidae) based on morphological data. *Mar. Mamm. Sci.* **22**, 910–932.

Ferrero, R. C., and Walker, W. A. (1999). Age, growth, and reproductive patterns of Dall's porpoise (*Phocoenoides dalli*) in the central North Pacific Ocean. *Mar. Mamm. Sci.* **15**, 273–313.

Forney, K. A., and Barlow, J. (1998). Seasonal patterns in the abundance and distribution of California cetaceans, 1991–1992. *Mar. Mamm. Sci.* **14**, 460–489.

Hanson, M. B., and Baird, R. W. (1998). Dall's porpoise reactions to tagging attempts using a remotely-deployed suction-cup tag. *MTS J.* **32**, 18–23.

Houck, W. J., and Jefferson, T. A. (1999). Dall's porpoise *Phocoenoides dalli* (True, 1885). *In* "Handbook of Marine Mammals" (S. H. Ridgway, and R. Harrison, eds), Vol. 6, pp. 443–472. Academic Press, San Diego.

Jefferson, T. A. (1988). *Phocoenoides dalli*. *Mamm. Spec.* **319**, 7 pp.

Jefferson, T. A. (1989). Calving seasonality of Dall's porpoise in the eastern North Pacific. *Mar. Mamm. Sci.* **5**, 196–200.

Jefferson, T. A. (1990). Sexual dimorphism and development of external features in Dall's porpoise *Phocoenoides dalli*. *Fish. Bull.* **88**, 119–132.

Morejohn, G. V. (1979). The natural history of Dall's porpoise in the North Pacific Ocean. *In* "Behavior of Marine Animals, Volume 3: Cetaceans" (H. E. Winn, and B. L. Olla, eds), pp. 45–83. Plenum Press, New York.

Rice, D. W. (1998). Marine mammals of the world: Systematics and distribution. *Soc. Mar. Mamm. Spec. Pub.* **4**, 231 pp.

Rosel, P. E., Haygood, M. G., and Perrin, W. F. (1995). Phylogenetic relationships among the true porpoises (Cetacea: Phocoenidae). *Mol. Phy. Evol.* **4**, 463–474.

Willis, P. M., and Dill, L. M. (2007). Mate guarding in male Dall's porpoises (*Phocoenoides dalli*). *Ethology* **113**, 587–597.

Willis, P. M., Crespi, B. J., Dill, L. M., Baird, R. W., and Hanson, M. B. (2004). Natural hybridization between Dall's porpoises (*Phocoenoides dalli*) and harbour porpoises (*Phocoena phocoena*). *Can. J. Zool.* **82**, 828–834.

# Delphinids, Overview

### RICK LEDUC

## I. Introduction

For most species of delphinids, basic aspects of their evolution, physiology, ecology, behavior, and population structure are virtually unknown. Even abundance estimates for many species are very imprecise. For the biologist, dolphin research presents challenges and opportunities in trying to understand individual species and how they fit into marine ecosystems.

## II. Taxonomic Overview

The family Delphinidae is one of three extant families (with Phocoenidae and Monodontidae) in the cetacean superfamily Delphinoidea (which also includes two extinct families, Kentriodontidae and Albireonidae). Delphinids likely arose in the mid- to late Miocene (11–12 mya) from kentriodontid-like ancestors and quickly radiated into many different morphological and ecological types. This early radiation produced precursors of many modern forms; many of the early delphinid fossils can be assigned to extant genera, particularly *Tursiops*. Today the Delphinidae is the most speciose family of marine mammals, with 33–35 recognized extant species arranged into 17–19 genera. At present, there is much uncertainty about the evolutionary relationships among the species of delphinids. Of the many recent classifications that have been proposed, two are depicted here. One represents a more traditional view of dolphin taxonomy (Table Ia), and the other is a revised classification based on various recent molecular and morphological phylogenetic analyses (Table Ib). Some of these analyses have led to changes in the number of species in certain genera (*Orcaella* and *Sotalia*) (Beasley *et al.*, 2005; Caballero *et al.*, 2007). In fact, some of the best known genera, such as *Orcinus*, may undergo revision in the number of recognized species (Pitman and Ensor, 2003). There is still uncertainty surrounding the deeper relationships, such as those among the different subfamilies. Also, there will no doubt be additional revisions proposed in the future at the genus level, especially involving the apparently paraphyletic genera *Stenella* and *Tursiops*. In part, this changing nature of delphinid taxonomy is due to the new molecular and analytical tools available to researchers, but it also reflects the uncertainties about evolutionary relationships that have long been recognized by morphological systematists but have yet to be addressed.

**D**

**TABLE I**

**Two Classifications of the Family Delphinidae**

| (a) A classification reflecting a traditional view of species interrelationships. | (b) Revised classification based on molecular systematic analysis. Adapted from LeDuc *et al.* (1999) and incorporating the results of Beasley *et al.* (2005) and Caballero *et al.* (2007). |
|---|---|
| Family Delphinidae<br>Subfamily Stenoninae<br>    *Steno bredanensis*<br>    *Sousa chinensis*<br>    *S. teuszii*<br>    *Sotalia fluviatilis*<br>Subfamily Delphininae<br>    *Lagenorhynchus albirostris*<br>    *L. acutus*<br>    *L. obscurus*<br>    *L. obliquidens*<br>    *L. cruciger*<br>    *L. australis*<br>    *Grampus griseus*<br>    *Tursiops truncatus*<br>    *Stenella frontalis*<br>    *S. attenuata*<br>    *S. longirostris*<br>    *S. clymene*<br>    *S. coeruleoalba*<br>    *Delphinus delphis*<br>    *D. capensis*<br>    *Lagenodelphis hosei*<br>Subfamily Lissodelphininae<br>    *Lissodelphis borealis*<br>    *L. peronii*<br>Subfamily Cephalorhynchinae<br>    *Cephalorhynchus commersonii*<br>    *C. eutropia*<br>    *C. heavisidii*<br>    *C. hectori*<br>Subfamily Globicephalinae<br>    *Peponocephala electra*<br>    *Feresa attenuata*<br>    *Pseudorca crassidens*<br>    *Orcinus orca*<br>    *Globicephala melas*<br>    *G. macrorhynchus*<br>Subfamily Orcaellinae<br>    *Orcaella brevirostris* | Family Delphinidae<br>Subfamily Stenoninae<br>    *Steno bredanensis*<br>    *Sotalia fluviatilis*<br>    *S. guianensis*<br>Subfamily Delphininae<br>    *Sousa chinensis*<br>    *Stenella clymene*<br>    *S. coeruleoalba*<br>    *S. frontalis*<br>    *S. attenuata*<br>    *S. longirostris*<br>    *Delphinus delphis*<br>    *D. capensis*<br>    *Tursiops truncatus*<br>    *T. aduncus*<br>    *Lagenodelphis hosei*<br>Subfamily Lissodelphininae<br>    *Lissodelphis borealis*<br>    *L. peronii*<br>    *Cephalorhynchus heavisidii*<br>    *C. hectori*<br>    *C. eutropia*<br>    *C. commersonii*<br>    *Sagmatias obscurus*<br>    *S. obliquidens*<br>    *S. cruciger*<br>    *S. australis*<br>Subfamily Globicephalinae<br>    *Feresa attenuata*<br>    *Peponocephala electra*<br>    *Globicephala melas*<br>    *G. macrorhynchus*<br>    *Pseudorca crassidens*<br>    *Grampus griseus*<br>Subfamily Orcininae<br>    *Orcinus orca*<br>    *Orcaella brevirostris*<br>    *O. heinsohni*<br>*Incertae sedis*<br>    *Lagenorhynchus albirostris*<br>    *Leucopleurus acutus* |

## III. Morphology

Dolphins have the typical morphological characteristics of toothed whales, such as spindle-shaped bodies, single external blowholes, telescoping of the skull such that the maxillary bones overlap the frontals in the supraorbital region, left-skewed cranial asymmetry, polydonty, and homodonty (in most). In some species, there has evolved a secondary reduction in the number of teeth, often seen as an adaptation for feeding on squid. The evolution of the delphinoid lineage saw the development of elaborate systems of pterygoid sinuses and better-isolated ear bones, probably increasing their ability to echolocate and perhaps giving them an advantage over some of the other odontocete groups of the Miocene. The presumed ancestors of the delphinids, the kentriodontids, were small dolphins with short to medium length rostra and, unlike most modern delphinoids, had symmetrical cranial vertices. The development in the Delphinidae of asymmetry in the cranial vertex and in the premaxillary bones suggests a further refinement of their echolocation capabilities and may partly explain their evolutionary success. The most noticeable difference between delphinids and their closest relatives, the phocoenids (true porpoises), is that the latter have spade-shaped teeth while delphinids have conical or peg-like teeth, as do most other odontocetes (toothed whales). They also differ from phocoenids in the shape of the facial region of the skull, including having a more distinct vertex. Within the Delphinidae, the most obvious variation among species relates to the feeding apparatus—the development of the rostrum, jaws, and teeth. There is a broad spectrum of rostrum lengths and widths, tooth counts, and tooth sizes, reflecting the range of ecological niches occupied by the different species. For example, total tooth counts range from less than 10 in *Grampus griseus* (which has no teeth in the upper jaw) to 250 in *Stenella longirostris*.

Delphinids also show wide variation in their external morphology. Only a few species (e.g., *Orcinus orca*, *Globicephala* spp.) are dramatically sexually dimorphic, although many others may have subtler dimorphism in body size and shape, coloration, and dorsal fin shape. In size they are small to medium cetaceans, with adults ranging from less than 1.5 m (*S. longirostris* from the Gulf of Thailand, some species of *Cephalorhynchus*) to over 9 m (*O. orca*). External rostrum (beak) length varies widely, from very long on some (e.g., *Delphinus delphis tropicalis*) to very short on others (e.g., *Lagenorhynchus albirostris*). The external beak is completely absent in a number of delphinids with short bonyrostrum in the skull, particularly in *Orcinus*, *Globicephala*, *Feresa*, *Pseudorca*, and *Peponocephala*. Those species that lack an external rostrum often have heads that are rounded or even bulbous in profile. In most delphinid species, the dorsal fin is pointed and falcate, although it is triangular in some subspecies of *S. longirostris* and in male *O. orca*, round in *Cephalorhynchus hectori*, and even forward canted in males of *S. longirostris orientalis*. The dorsal fin is completely missing in *Lissodelphis* spp., and *Sousa* spp. has a pronounced hump at the base of its dorsal fin. The color patterns of delphinids are similarly varied, from bold black and white patterns (e.g., *O. orca* and some of the *Cephalorhynchus* spp.) to complex patterns of black, white, and gray (e.g., *Delphinus* spp. and *Stenella* spp.) to rather simple patterns of black (e.g., *Globicephala* spp.) or gray (e.g., *Sousa* spp.). The complex color patterns exhibited by some delphinids are composites of various elements, including stripes, capes, overlays, spots, and blazes (Perrin, 1973). Regardless of their overall color and pattern, all dolphins tend to have a countershaded aspect, where the ventral surface is lighter than the dorsum. This countershading reduces an animal's visibility in the marine environment, where the ambient light comes from above. Besides being a mechanism for species recognition, the more complex patterns may also play a role in camouflage against waves at the surface or against the dappling of light penetrating the water.

### IV. Distribution and Habitat

As a group, the family Delphinidae reaches its highest diversity in tropical and warm-temperate latitudes. There are numerous species with pantropical distributions and others that occur in tropical waters but are limited to one or two ocean basins (Rice, 1998). For example, *Stenella frontalis* and *S. clymene* are limited to the lower latitudes of the Atlantic Ocean, while *Tursiops aduncus* and *Orcaella brevirostris* only occur in the Indian and west Pacific Oceans. In colder areas one can find species in the genera *Globicephala*, *Cephalorhynchus*, *Lissodelphis*, and *Lagenorhynchus* (including *Sagmatias* and *Leucopleurus* of Table Ib). Interestingly, genetic evidence suggests that the majority of these cold-temperate species appear to be closely related (subfamily Lissodelphininae in the classification of Table Ib). Only one recognized species (*Globicephala melas*) has an antitropical distribution, although some antitropical species pairs [e.g., *Lissodelphis* spp., *Lagenorhynchus* (or *Sagmatias*) *obscurus/obliquidens*] have been hypothesized as being single species with antitropical populations. Only one species, *O. orca*, ranges into high latitudes near the polar ice. Indeed, occurring also in mid- and low latitudes, this species is probably the most cosmopolitan of all the cetaceans.

Within this broad range of geographic distributions, delphinids occupy an equally diverse array of habitats. Many species occur far

offshore in deep water, where the specifics of their ecological requirements are poorly known. In fact, in tropical seas, only the cetacean fauna of the eastern tropical Pacific has been extensively and systematically studied; here some differences in the species composition have been observed in different water masses (Au and Perryman, 1985). Areas with a stable mixed layer and a shallow thermocline are frequented by *Stenella attenuata*, *S. longirostris*, and *Steno bredanensis*, while areas with more variable conditions and some amount of upwelling contain species such as *Globicephala macrorhynchus*, *Delphinus delphis*, *Stenella coeruleoalba*, and *Peponocephala electra*. In any ocean, some of the offshore species may also range closer to the coast (e.g., *D. delphis*), or even have populations or sister species that are restricted to the coastal waters or the nearshore habitat (e.g., *D. capensis*, coastal populations/species of *Tursiops*). In a few cases, coastal populations may ascend a short distance up rivers, but only two species (*Sotalia fluviatilis* of South America and *O. brevirostris* of the Asian Indo-West Pacific) regularly occur far upstream. These two species also present a contrast in patterns of divergence. The former has a sister species occurring in coastal waters of middle and South America on the Atlantic side (*S. guianensis*), while the latter has shown no such divergence between coastal and riverine species, instead having a sister species in Australia and New Guinea (*O. heinsohni*). These cases illustrate how speciation can occur across ecological boundaries or across discontinuities in geographical distribution.

### V. Social Organization and Behavior

All dolphin species are social to some degree. However, characteristic group sizes for the different species range from small pods of just a few individuals to large schools numbering in the thousands. Due to the difficulty of observing dolphins in the wild, especially those occupying offshore habitats, very little is known about the behavior and social organization of most species. The populations that have been studied over longer time scales (e.g., *O. orca* in the northeast Pacific and *Tursiops truncatus* in the western Atlantic) are those that form relatively stable and small social groups within a short distance of the coast. By using photos to identify individuals via fin markings and color patterns, and by supplementing observations with genetic data, associations of individuals and their genealogical relatedness have been recorded and monitored over generations. In these populations, some long-term associations and patterns have been noted. For example, it appears that bonds between individuals in a pod and/or between individuals and a particular area (philopatry) tend to be stronger for females than males; the society can even be considered matrilineal for *O. orca*. A similar pattern was inferred from molecular data on *G. melas* in the North Atlantic. However, one must be cautious in extrapolating these social patterns to other delphinid species. For example, some species that occur on the high seas are found in schools that number in the thousands, and these associations appear much more fluid in their composition. In fact, social patterns observed for inshore groups of species like *T. truncatus* and *O. orca* may not even reflect the organization of offshore populations of the same species. What little is known about the social organization of offshore dolphins comes from direct observations of school sizes and from life history data collected from mass strandings and fishery kills. In the few pelagic species studied (primarily *S. attenuata* and *S. longirostris*), there is evidence for promiscuity, strong mother–calf bonds, and some segregation by age and sex both within and between schools.

In addition to their intraspecific social organization, most dolphin species are seen at least occasionally in the company of other species. One famous association is between *S. attenuata* and *S. longirostris* in the eastern tropical Pacific, an aggregation that also includes yellow-fin tuna (*Thunnus albacares*) and numerous species of seabirds. Other associations are frequently observed, such as *P. electra* associated with *Lagenodelphis hosei*, or *T. truncatus* with *G. macrorhynchus*. A few species, such as *Lissodelphis borealis*, have been observed with a wide variety of marine mammals, including mysticetes and pinnipeds.

Most of the information on the behavior and cognition of individual dolphins comes from studies of captive animals, primarily of *T. truncatus*. Apart from humans, dolphins have the highest ratio of brain size to body mass of any animal. Their intelligence and behavioral versatility is legendary and is still being explored. Using controlled experiments, dolphins have been shown to be capable of understanding complex commands, including the incorporation of abstract concepts and variations of syntax, of devising novel behaviors of their own volition, and of self-recognition. Although behavioral activities in the wild are more difficult to interpret, there are well-documented observations of cooperative hunting (e.g., *O. orca*), play behavior (e.g., *T. truncatus*), and even tool use (*T. truncatus*). In fact, *T. truncatus* has been observed surfing in many coastal areas, and many dolphin species are avid riders of the bow-wakes of ships and large whales. Some species regularly perform aerial maneuvers such as high leaps and flips and, in the case of *S. longirostris*, spins. At the present time, the function of many of these behaviors can only be guessed at. Dolphins also have an array of vocalizations such as clicks, whistles, and squeals, which are used in part for their well-developed echolocation and in part for communication. However, only those species kept in captivity have been extensively studied. For the rest, due to the logistic difficulties of collecting data on fast-swimming dolphins in the pelagic realm, the vocalizations of those species that have been recorded are difficult to understand in terms of functionality. There has been recent progress made in developing methods to use recorded vocalizations to identify dolphin species, but the complexities and similarities of the calls of certain species (e.g., *Stenella* and *Delphinus* spp.) make this a difficult endeavor.

## VI. Feeding

Ecologically, dolphins have also radiated dramatically. There are species that forage on fish, squids, and/or other invertebrates. A few species (e.g., *O. orca*, *Pseudorca crassidens*) even take mammalian prey, including large whales and pinnipeds. While some have fairly specific diets, a few species have rather broad tastes. In those species that eat a wide variety of prey items, one type of food (e.g., fish vs squid) may predominate. For some species (e.g., *S. coeruleoalba*), the preferred food type varies among populations, while in others (e.g., *S. attenuata*), it may even vary among individuals within a population depending on their sex and life history stage. Perhaps the most dramatic segregation of foraging strategies is seen in the populations of *O. orca* in the eastern North Pacific, where two distinct groups exist in sympatry, one specializing on mammals and the other on fish (mostly one species of salmon). Individual species of delphinids usually forage in a particular part of the water column, specializing in epipelagic prey (e.g., *S. attenuata*), mesopelagic prey (e.g., *Lissodelphis* spp.), or even benthic prey (e.g., some species of *Cephalorhynchus*). The few measurements of swimming speeds (often anecdotal) suggest that some may exceed 20 knots in short bursts, although prolonged cruising speeds are generally on the order of 5–9 knots (9–17 km/h). Direct data on diving depths for wild delphinids are practically nonexistent. However, *S. longirostris*, a relatively small species, is thought to dive to at least 200–300 m, based on an analysis of prey items. Greater depths are no doubt possible for larger species (e.g., *Globicephala* spp.) that feed mainly on larger squid. In fact, a trained pilot whale

(*Globicephala* sp.) is thought to have reached 610 m in an experimental situation. At the other extreme, certain populations of two dolphin species have been known to intentionally beach themselves in pursuit of prey. In one case, groups of *T. truncatus* drive and pursue schooling fish onto the beach and in the other, *O. orca* beach themselves to take pinnipeds hauled out near the water's edge. The behavioral adaptability of dolphins may be best illustrated by those species that incorporate human activities into their foraging, such as the *Sousa chinensis* that feed on trawl discards or the *O. orca* that raid longlines for their catch. Along the Gulf of Mexico coast of the United States, there are even some *T. truncatus* that feed within shrimp trawl nets, many of whom are "caught," released, and netted again.

## VII. Reproduction

As with many aspects of dolphin biology, there is much that is unknown about the reproductive biology of most species, although many of the parameters seem to be correlated with body size. Estimated ages at attainment of sexual maturity range from about 6 (*D. delphis*) to 16 (*O. orca*). Like most mammals, when the age at sexual maturation differs between the sexes, it is usually the females that reach maturity at a younger age. Like other cetaceans, dolphins bear single young. Gestation periods are rarely well documented and are thought to range from about 9 months to 16 months, although most last less than a year (16 months was an estimate for *G. melas*). Most species appear to show at least some seasonality in their breeding, although this varies in degree. Estimates of calving intervals similarly vary, even among populations within species or among studies, ranging from just over a year for many species to approximately 8 years in one study of *O. orca* from the eastern North Atlantic. Reliable estimates of longevity are quite rare, but range from around 20 years for the smaller species up to about 60 years for females of the larger species. In the large species, which tend to be more sexually dimorphic, females may live 15–20 years longer than males. The causes of natural mortality, when they can be ascertained, are usually parasites, pathogens, or predation by killer whales or sharks.

## VIII. Abundance and Conservation

Some delphinid species are no doubt the most numerous of cetaceans, occurring in schools of thousands in large portions of the world ocean. Reliable global estimates for most species do not exist, but in the eastern tropical Pacific alone, some species (*S. attenuata*, *D. delphis*) number in the low millions. At the other end of the spectrum, species with restricted ranges (e.g., *C. hectori*) may have total populations of only a few thousand. In spite of (and in some cases because of) their general abundance, dolphins face numerous anthropogenic threats. They are still hunted in some parts of the world by harpoon, drive fisheries, or nets. Two well-known examples are the drive fisheries for *S. coeruleoalba* in Japan and for *G. melas* in the Faeroe Islands. There are many other smaller dolphin fisheries still in operation, mostly in developing countries, often in spite of the protection of assorted laws and treaties. Usually the meat from intentional takes is for human consumption, although in some areas it is used as bait for crab pots or shark longlines. In addition to the mortality from directed fisheries, many dolphins are also taken incidentally in the course of other fishing operations. The dolphin mortality from the eastern tropical Pacific tuna purse-seine fishery has been well studied and greatly reduced in recent years. However, many dolphins of a variety of species are still caught in coastal and offshore gillnets all over the world (Perrin *et al.*, 1994). It is difficult to accurately assess the severity of these threats to dolphin populations, given that the extent of the mortality can only be roughly guessed at in most cases and the sizes of the affected populations are largely unknown. Nevertheless, mortality from fisheries bycatch may present major threats to some dolphin populations.

Compounding the impact of these direct kills are indirect effects from human activities. In some areas, large-scale fishing operations may adversely affect dolphin populations, either by direct competition for prey or by alteration of a region's ecology. Pollution also undoubtedly takes its toll on dolphin health, particularly in coastal areas, either by direct poisoning effects or by making the animals more susceptible to pathogens and parasites. Subtle effects on fitness like decreases in reproductive capacity or shortened lifespans are almost impossible to detect, but large-scale mortalities are difficult to ignore and are not at all unusual. There have been mass strandings in recent years along the Mediterranean coast, in the southeastern United States, and in the Gulf of California in Mexico. These large-scale die-offs often involve multiple species (although one species usually predominates) and may occur over periods of several months. It is difficult to determine if such mass mortalities are increasing in frequency in recent years because historical events may not have been adequately documented. It is also difficult to assess their impact on dolphin populations. Nevertheless, they are a cause for concern if for no other reason than as indicators of the declining health of marine ecosystems.

## See Also the Following Articles

Coloration ■ Cetacean Life History ■ Mass Mortalities

## References

Au, D. W. K., and Perryman, W. L. (1985). Dolphin habitats in the eastern tropical Pacific. *Fish. Bull.* **83**, 623–643.

Beasley, I., Robertson, K. M., and Arnold, P. (2005). Description of a new dolphin, the Australian snubfin dolphin *Orcaella heinsohni* sp. N. (Cetacea, Delphinidae). *Mar. Mamm. Sci.* **21**, 365–400.

Caballero, S. *et al.* (11 authors). (2007). Taxonomic status of the genus *Sotalia*: Species-level ranking for "tucuxi" (*Sotalia fluviatilis*) and "costero" (*Sotalia guianensis*) dolphins. *Mar. Mamm. Sci.* **23**, 358–386.

Krützen, M., Mann, J., Heithaus, M. R., Connor, R. C., Bejder, L., and Sherwin, W. B. (2005). Cultural transmission of tool use in bottlenose dolphins. *PNAS* **102**, 8939–8943.

Leatherwood, S., and Reeves, R. R. (eds) (1990). "The Bottlenose Dolphin." Academic Press, San Diego, CA.

LeDuc, R.G., Perrin, W.F. and Dizon, A.E. (1999). Phylogenetic relationships among the delphinid cetaceans based on full cytochrome *b* sequences. *Mar Mamm. Sci.* 15, 619–648.

Lincoln, R., Boxshall, G., and Clark, P. (1998). "A Dictionary of Ecology, Evolution and Systematics," 2nd Ed. Cambridge University Press, Cambridge.

Mann, J., Connor, R. C., Tyack, P. L., and Whitehead, H. (eds) (2000). "Cetacean Societies: Field Studies of Dolphins and Whales." University of Chicago Press, Chicago, USA.

Perrin, W. F. (1973). Color patterns of spinner porpoises (*Stenella* cf. *S. longirostris*) of the eastern tropical Pacific and Hawaii, with comments on delphinid pigmentation. *Fish. Bull., US* **70**, 983–1003.

Perrin, W. F. and Reilly, S. B. (1984). Reproductive parameters of dolphins and small whales of the family Delphinidae. *In* "Reproduction in Whales, Dolphins and Porpoises," (W. F. Perrin, R. L. Brownell, and D. P. Demaster, eds.), *Rep. Int. Whal. Commn. Spec. Iss.* **6**, 97–133.

Perrin, W. F., Donovan, G. P., and Barlow, J. (eds) (1994). Gillnets and Cetaceans. *Rep. Int. Whal. Commn. Spec. Iss.* (15).

Pitman, R. L., and Ensor, P. (2003). Three forms of killer whale (*Orcinus orca*) in Antarctic waters. *J. Cetacean Res. Manage.* **5**, 131–139.

Rice, D. W. (1998). "Marine Mammals of the World: Systematics and Distribution." *Soc. Mar. Mamm. Spec. Pub.* **4**. Society for Marine Mammalogy, Lawrence, KS.

Ridgway, S. H., and Harrison, R. (eds) (1994). "Handbook of Marine Mammals. Volume 5: The First Book of Dolphins." Academic Press, London.

Ridgway, S. H., and Harrison, R. (eds) (1999). "Handbook of Marine Mammals. Volume 6: The Second Book of Dolphins and the Porpoises." Academic Press, London.

# Dental Morphology, Evolution of

## MARK D. UHEN

**M**odern marine mammals exhibit some of the most highly derived dentitions in all of Mammalia. Among cetaceans, living odontocetes exhibit a wide variety of conditions from the polydont delphinids to the complete lack of teeth in some ziphiids. Living mysticetes completely lack teeth as adults. Add to this even more divergent conditions in fossil mysticetes, odontocetes, and their early forebears, the archaeocetes, and the result is a broad range of conditions in which is embedded the story of how modern cetacean dentitions arose from the more typically mammalian teeth of early artiodactyls (Fig. 1).

The herbivorous Sirenia and Desmostylia are distinct but related orders. Modern dugongids lack functional teeth as adults and modern trichechids have a highly derived pattern of serial dental eruption. Desmostylian teeth are also highly derived, but in a very different fashion when compared to sirenians (Fig. 1, 2).

Pinniped dentitions (Fig. 2) are general, simplified compared to other those of their relatives, terrestrial carnivores, and they lack the carnassial pair typical of carnivoran dentitions (fourth upper premolar, P4/, and the lower first molar, M/1). Different pinniped dentitions have become specialized for various feeding modes from the large peg-like teeth of molluscivorous odobenids to the delicate, denticulate cheek teeth of the crabeater seal for filter feeding on krill.

## I. Cetacea

### A. Archaeocetes

Archaeocetes are diphyodont and heterodont with their incisors in line with the cheek teeth and separated by diastemata (a gap between teeth). The anterior teeth are conical and single rooted, and the premolar series forms a morphological gradient from the anterior teeth to the posterior premolars, and usually smaller molars. The molars have basins, where teeth occlude during chewing, but these are smaller than in related land mammals. Archaeocetes often exhibit vertical wear facets on the buccal (cheek side) surfaces of the lower molars that resulted from contact with the lingual (tongue side) surfaces of the upper cheek teeth during chewing (O'Leary and Uhen, 1999). Clementz *et al.* (2003) studied diet using isotopes.

*1. Non-Basilosaurid Archaeocetes* Non-basilosaurid archaeocetes have a dental formula of 3.1.4.3/3.1.4.3 (three incisors, one canine, four premolars, and three molars per half jaw), that they share with early artiodactyls and other early mammals and that distinguishes them from the later basilosaurids. Non-basilosaurid archaeocetes have two

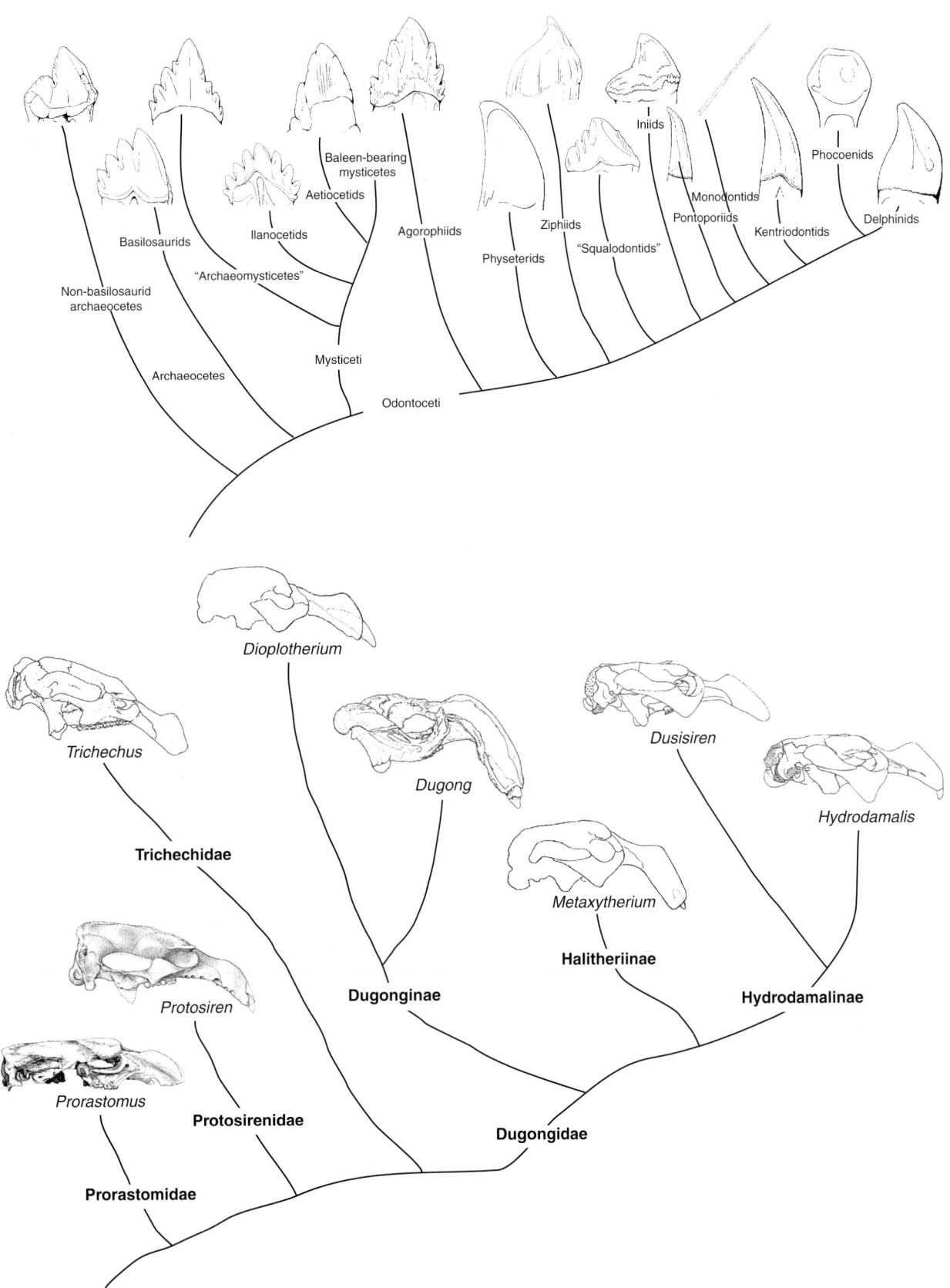

**Figure 1**  *Phylogenetic relationships of cetaceans with examples of dental morphology. For those cetaceans with heterodont teeth, molars (posterior cheek teeth) are shown (top). Phylogenetic relationships of Sirenia (after Domning, this volume; bottom).*

**D**

cusps on the buccal side and one on the lingual side. Between these cusps (elevated areas) are basins (depressed areas) in which a cusp from the opposite side (upper or lower) occludes. Ambulocetids have a protocone lobe on the lingual side of the upper molars that is smaller than that in pakicetids, and larger than that in remingtonocetids and protocetids (Thewissen *et al.*, 1996). Remingtonocetids have labiolingually narrow incisors and premolars; upper molars large and narrow, sometimes lacking a lingual third root, and crenulations on the cutting edges of the teeth, which distinguishes them from other non-basilosaurid archaeocetes (Gingerich *et al.*, 1995; Kumar and Sahni, 1986). Protocetids have robust teeth with upper molars having three roots and the cheek teeth lacking accessory denticles.

*2 Basilosauridae* Basilosaurids differ from other archaeocetes in that upper and lower first premolars are replaced in dorudontine basilosaurids, but it is unclear how broadly this character is distributed within the archaeocetes (Uhen, 2000). Basilosaurids lack M$^3$, giving them a dental formula of 3.1.4.2/3.1.4.3. The cheek teeth of basilosaurids have accessory denticles (small cusps) along their crests. It is unclear whether any (or all or none) of the accessory denticles are homologous with more primitive mammalian cusps. Basilosaurid molars lack basins in which teeth occlude.

### B. Neoceti

The fossil record of archaeocetes ends in the late Eocene. Mysticeti and Odontoceti can be distinguished from archaeocetes. Even though modern mysticetes lack teeth as adults, they develop embryonic tooth buds that are later resorbed (Karlsen, 1962). In addition, early fossil mysticetes have well-developed teeth as adults (Emlong, 1966; Mitchell, 1989; Barnes and Sanders, 1996). Even a late-occurring archaeocete has been described that may be monophyodont (Uhen and Gingerich, 2001). Early mysticetes and odontocetes are similar to archaeocetes in that they are heterodont, with a morphological gradient from conical, single-rooted anterior teeth to triangular (or rounded) multi-rooted cheek teeth with accessory denticles.

*1. Odontoceti* Modern odontocetes can be generally characterized as being polydont, monophyodont, with single-rooted teeth that grow throughout life, but there are numerous counter examples to these generalizations, in both modern and fossil cetaceans discussed below. The earliest odontocetes show only limited polydonty, with *Agorophius* having eight, and *Xenorophus* with ten (Fordyce, 1982 and references therein). Polydonty could have originated by intercalation of deciduous and permanent teeth, and could have been further increased in later odontocetes by the addition of supernumerary teeth, but it would be difficult to explain why no odontocete is know that exceeds three teeth in the premaxilla under this scenario (Fordyce, 1982). The description of early mysticetes with one extra molar in each quadrant (Barnes and Sanders, 1996) may indicate that polydonty arose prior to the split of Mysticeti and Odontoceti by terminal addition of teeth, although many other toothed mysticetes are not polydont (Emlong, 1966; Fordyce, 1982).

All modern and fossil odontocetes are thought to be monophyodont (Fordyce, 1982). It is unclear whether the teeth of modern odontocetes are homologous with the deciduous or adult teeth (or both), of archaeocetes. One archaeocete that appears to possess adult teeth in a skeletally juvenile individual suggests that the teeth of all Mysticeti + Odontoceti may be adult teeth (Uhen and Gingerich, 2001), but Karlsen (1962) suggested that the embryonic

tooth buds of mysticetes were homologous with the deciduous dentition of more primitive mammals.

The early odontocetes Agorophiidae (including *Xenorophus*), Simocetidae, Waipatiidae, Squalodontidae are all considerably heterodont. Their anterior teeth are conical with long, single roots. These teeth grade into teeth with a more triangular shape with multiple accessory denticles on the anterior and posterior edges of the teeth and two roots. The most posterior lower cheek teeth may lack accessory denticles on their anterior edges. These differences in tooth form are reminiscent of the differences along the tooth rows of archaeocetes, but the teeth are smaller compared to the size of the skull.

Modern physeteroids lack upper teeth (although some anomalous specimens have tiny upper teeth that do not erupt), but some fossil representatives (like *Scaldicetus* have similarly sized teeth in the lower and the upper jaws (Hirota and Barnes, 1995). Ziphiidae are thought to have secondarily reduced the dentition from a primitively polydont condition. One modern species, *Tasmacetus shepherdi*, and additional fossil species are polydont. In other modern species, the female usually lack teeth entirely, and males have one tooth on each side of the lower jaw, and none in the upper jaw. Iniidae have rugose heterodont teeth that are expanded lingually in the posterior portion of the jaw. Monodontidae have very reduced dentitions. Male narwhals (*Monodon monoceros*) have a single, long, spiral tusk, usually on the left side with a small unerupted tooth on the right. Females usually have two unerupted teeth. One fossil monodontid, *Odobenocetops*, has a walrus-like skull with one large tusk-like tooth, and another shorter tusk on the opposite side (de Muizon, 1994). Members of the extinct Kentriodontidae have small conical teeth that are generally homodont, a feature shared with one of their likely descendants, the Delphinidae, but not with another, the Phocoenidae, which have numerous small spatulate teeth that are embedded in the gums.

*2. Mysticeti* Despite the lack of teeth in modern mysticetes they are thought to have evolved from toothed ancestors in part because they develop tooth buds as embryos that are later resorbed (Karlsen, 1962) and many tooth-bearing early mysticetes are known from the Oligocene epoch (Fordyce and Barnes, 1994). All post-Oligocene mysticetes are baleen bearing and lack teeth (Fordyce and Barnes, 1994).

One group of toothed mysticetes includes species with archaeocete-like teeth (Fordyce, 1989; Barnes and Sanders, 1996). *Llanocetus denticrenatus* from the late Eocene of Antarctica (Mitchell, 1989) has teeth that are similar in size to those of archaeocetes and the cheek teeth are similarly double rooted and have accessory denticles. Some species have an extra molar in each quadrant over the count of basilosaurids, yielding three upper and four lower molars (Barnes and Sanders, 1996). These teeth may have been used to filter food from seawater, since the teeth generally lack wear resulting from tooth/food contact (Mitchell, 1989). Fitzgerald (2006) suggested that the large-toothed mysticetes (archaeomysticetes, *Mammalodon*, and *Janjucetus*) were not filter feeders, but were macrophagous predators.

The second group of toothed mysticetes includes species that all have tiny teeth that are conical anteriorly and have tiny denticles posteriorly, but are all single rooted, the Aetiocetidae. Some aetiocetids have the plesiomorphic number of teeth for archaeocetes (11), while others are polydont. Aetiocetids may have had some form of proto-baleen, and may have used this baleen and their small teeth for filter feeding.

The earliest baleen-bearing mysticetes that lack teeth are the Eomysticetidae, from the early Oligocene of South Carolina. All

D

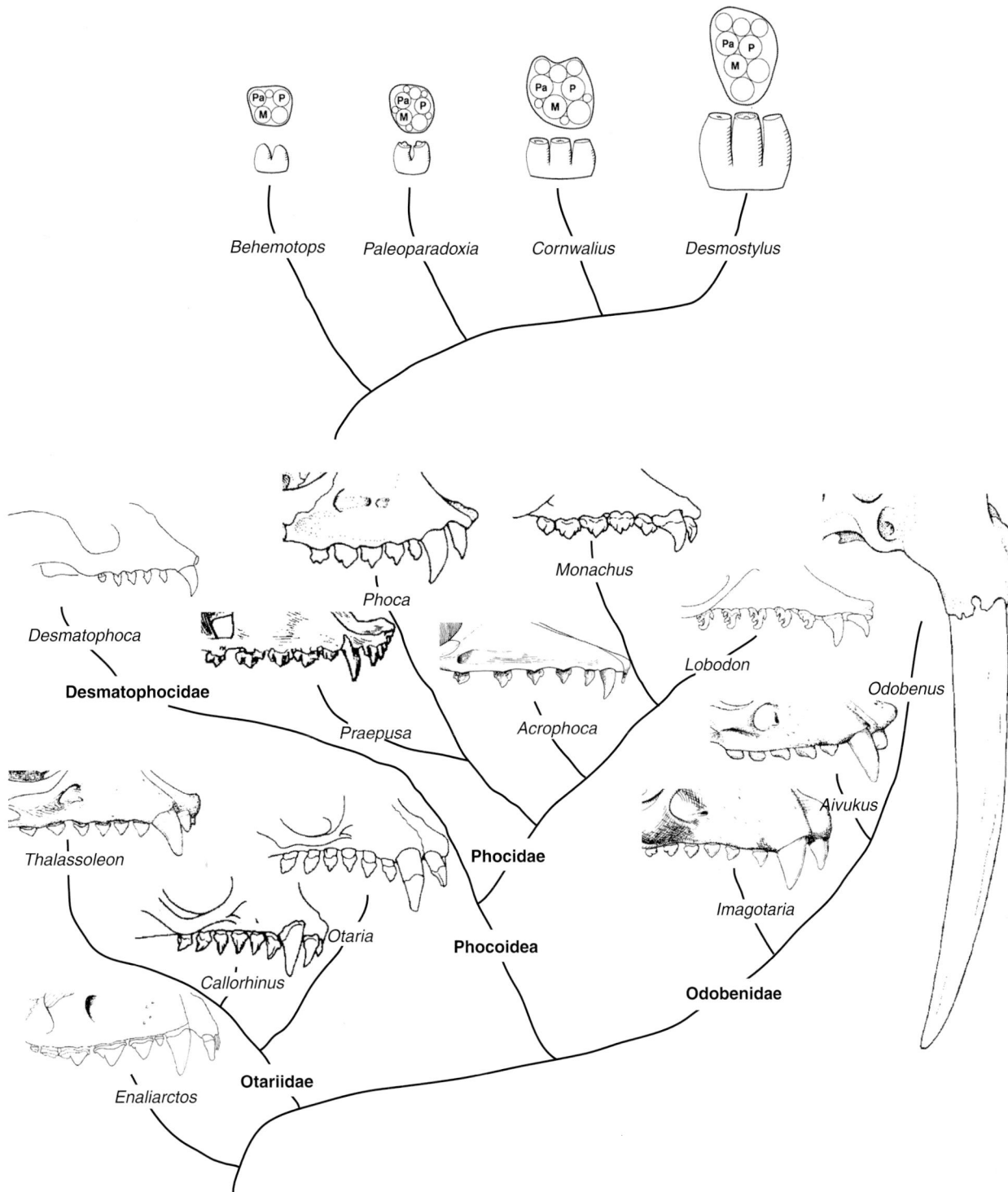

**Figure 2**  *Phylogenetic relationships of Desmostylia with examples of molar dental morphology. (after Domning, this volume; top).*
*Phylogenetic relationships of pinnipeds (after Berta, this volume; bottom).*

baleen-bearing toothless mysticetes belong to the Chaeomysticeti. All known post-Oligocene mysticetes belong to the Chaeomysticeti, and are thus toothless. Modern mysticetes use several modes of feeding, and have different sizes and shapes of baleen plates, but they are all bulk feeders, and use their baleen for prey capture.

## II. Sirenia

All Eocene sirenians have a dental formula of 3.1.5.3/3.1.5.3. Unfortunately, sirenian cheek teeth are relatively similar and not diagnostic of specific clades. These teeth are low crowned with two rows of large, rounded cusps (see SIRENIAN EVOLUTION). Some sirenians, particularly within the Dugongidae, developed tusks which were used for feeding on seagrass rhizomes, even though modern *Dugong* uses its tusks primarily for interactions with other individuals. (Domning and Beatty, 2007).

The Dugongidae diversify and some show a tendency to evolve large, blade-like incisor tusks (Domning, 2001). Modern *Dugong* is unusual in that it has reduced its cheek teeth to open-rooted, ever-growing pegs whose enamel crowns wear off, leaving flat surfaces of dentin. Modern *Dugong* has 5 or 6 functional cheek teeth per quadrant, and the teeth are replaced from the rear, with new teeth erupting in the rear of the tooth row, and the anteriormost tooth being resorbed and lost (Husar, 1978a). Steller's sea cow, *Hydrodamalis gigas* completely lacked teeth, and instead had keratinous pads for mastication.

Trichechidae (Manatees) are derived from basal dugongids that were diphyodont. In contrast, modern *Trichechus* have 5–7 functional cheek teeth per quadrant at one time. Teeth are replaced in the rear, and migrate forward, yielding a significant number of supernumerary teeth, but it is not clear how many teeth eventually erupt (Husar, 1978b). The earliest fossil form to exhibit this type of tooth replacement is *Ribodon*, from South America and North Carolina (Domning, 2001).

### III. Desmostylia

Desmostylia are diphyodont and heterodont with a large diastema separating the anterior teeth from the cheek teeth. The procumbent incisors are caniniform and the canines are tusk-shaped in some desmostylians. The premolars become more complex and molariform from anterior to posterior. The cheek teeth are bunodont, with large cusps and thick enamel. As the teeth wear, the tips of the cusps wear away, leaving a ring of thick enamel surrounding an area of dentine (Domning, 2002), and eventually evolving in teeth that look as an agglomeration of a large number of conical enamel tubes (Fig. 2; Inuzuka *et al.*, 1995).

The dental formula of desmostylians changes over their history. *Behemotops*, has a dental formula of 3.1.4.3 (Inuzuka *et al.*, 1995); *Paleoparadoxia* has a dental formula of 3.1.3.3; *Desmostylus japonicus* has a dental formula of 0.1.2.3/2.1.2.3 (Kohno, 2002); and *Desmostylus hesperus* has a dental formula of 0.0.2-1.3/2.1.1.3 (Kohno, 2002). Thus, desmostylians undergo a pattern of upper incisor loss over their evolutionary history.

### IV. Pinnipedia

The earliest known pinniped is *Enaliarctos* (Berta *et al.*, 1989). *Enaliarctos* has a dental formula of 3.1.4.2/3.1.4.2, which represents a loss of $M^3$ compared to basal Carnivora, as well as basal Arctoidea. *Enaliarctos* also has small upper molars relative to the premolars when compared to other arctoids, and to other pinnipeds as well (Berta, 1991). *Enaliarctos* shares with other pinnipeds a reduction and/or loss of the $M_1$ entoconid, reduction and/or loss of the $M_1$ hypoconid, and reduction and/or loss of the upper molar cingulum, suggesting a reduction in the carnassial function of the teeth (Berta, 1991).

Crown group Pinnipedia (to the exclusion of *Enaliarctos* and *Pteronarctos*) is characterized by the loss of the embrasure pit between $P^4$ and $M^1$, loss of the $P^4$ protocone, and the loss of $M_2$, indicating a further reduction in the carnassial function of the teeth (Berta, 1991).

Otariidae have a dental formula of 3.1.4.0-3/2.1.4.1-2, including a molar count that varies both within and among the genera (Kubota

and Togawa, 1964; Repenning and Tedford, 1977; Nowak, 1999). Otariid third incisors are enlarged and caniniform (Nowak, 1999). The premolars and molars are similar in morphology. Cheek teeth in *Thalassoleon* are double rooted (Repenning and Tedford, 1977), as they are in some modern otariids (Repenning, 1977). All otariids have relatively homodont, generally conical cheek teeth (Repenning, 1977).

Phocoidea (Phocidae + Desmatophocidae) have a dental formula of 2-3.1.4.0-2/1-2.1.4.0-2 (Nowak, 1999; Deméré and Berta, 2002). Most phocids retain more complex tooth crowns than most other pinnipeds, with at least three distinct cusps on each crown (Nowak, 1999). The earliest known named phocid, *Devinophoca*, retains an $M^1$ with three roots, which is similar to the condition in the $M^1$ of *Enaliarctos* (Koretsky and Holec, 2002). Long-snouted phocids such as *Acrophoca* have relatively small cheek teeth with large diastemata between them, which also sport embrasure pits (de Muizon, 1981). Modern *Lobodon* has elongate, recurved cusps on the cheek teeth which it uses to filter feed krill from sea water when the lower and upper dentitions are occluded (Adam, 2005).

Odobenidae are characterized by a progressive reduction and simplification of the dentition. More basal odobenids, such as *Imagotaria*, retain a plesiomorphic dental formula (3.1.4.2/3.1.4.1 or 2) and a plesiomorphic dental morphology, with single roots on $P_{2-4}$, which are however, bilobed. (Repenning and Tedford, 1977). *Aivukus* has a more reduced dental formula (2.1.4.1/2.1.4.1) and more simplified, peg-like teeth as well. All of the lower teeth are circular in cross-section, single-rooted, solid columns of dentine with a heavy layer of cementum (Repenning and Tedford, 1977). *Aivukus* also possessed an enlarged canine which eventually closed its root, so was not ever growing (Repenning and Tedford, 1977). Modern *Odobenus* has a variable number of teeth, with the most common dental formula being 1.1.3.0/0.1.3.0. The teeth are reduced to dentine pegs except for the greatly enlarged canine tusks. The tusks can reach 100-cm long in males and 80-cm long in females (Nowak, 1999).

### *See Also the Following Articles*

Pinniped Evolution ■ Desmostylia ■ Sirenian Evolution

### *References*

Adam, P. J. (2005). Lobodon carcinophaga. *Mamm. Spec.* **772**, 1–14.

Barnes, L. G. (1984). Whales, dolphins and porpoises: Origin and evolution of the Cetacea, Mammals: Notes for a short course, University of Tennessee Department of Geological Sciences Studies in Geology. **8**, 139–154.

Barnes, L. G., and Sanders, A. E. (1996). The transition from Archaeoceti to Mysticeti: Late Oligocene toothed mysticetes from South Carolina, U.S.A. *J. Vertebrate Paleontol.* **16**(3), 21A.

Berta, A. (1991). New *Enaliarctos* (Pinnipedimorpha) from the Oligocene and Miocene of Oregon and the role of "enaliarctids" in pinniped phylogeny. *Smith. Contr. Paleobiol.* **69**, 1–33.

Berta, A., and A. R. Wyss. (1994). Pinniped phylogeny. Proceedings of the San Diego Society of Natural History, 29 (Contributions in Marine Mammal Paleontology Honoring Frank C. Whitmore Jr.): 33–56.

Berta, A., Ray, C. E., and Wyss, A. R. (1989). Skeleton of the oldest known pinniped, *Enaliarctos mealsi*. *Science* **244**, 60–62.

Clementz, M. T., Hoppe, K. A., and Koch, P. L. (2003). A paleoecological paradox: The habit and dietary preferences of the extinct tethythere *Desmostylus*, inferred from stable isotope analysis. *Paleobiology* **29**(4), 506–519.

Deméré, T. A., and Berta, A. (2002). The Miocene pinniped *Desmatophoca oregonesis* Condon, 1906 (Mammalia:Carnivora), from the Astoria formation, Oregon. *Smith. Contr. Paleobiol.* **93**, 113–147.

Domning, D. P. (2001). Evolution of the Sirenia and Desmostylia. *In* "Secondary Adaptation of Tetrapods to Life in Water" (J.-M. Mazin, and V.d. Buffrénil, eds), pp. 151–168. Verlag Dr. Friedrich Pfeil, München, Germany.

Domning, D. P. (2002). Desmostylia. *In* "Encyclopedia of Marine Mammals" (W. F. Perrin, B. Würsig, and J. G. M. Thewissen, eds), pp. 319–322. Ed. 1. Academic Press, San Diego.

Domning, D. P., and Beatty, B. L. (2007). Use of tusks in feeding by dugongid sirenians: Observations and tests of hypotheses. *The Anatomical Record: Advances in Integrative Anatomy and Evolutionary Biology* **290**(6), 523–538.

Emlong, D. R. (1966). A new archaic cetacean from the Oligocene of northwest Oregon. *Bull Oreg. Univ. Mus. of Natural History* **3**, 1–51.

Fitzgerald, E. M. G. (2006). A bizarre new toothed mysticete (Cetacea) from Australia and the early evolution of baleen whales. Proceedings of the Royal Society: 1–9 online pages.

Flynn, J. J., Finarelli, J. A., Zehr, S., Hsu, J., and Nedbal, M. A. (2005). Molecular phylogeny of the Carnivora (Mammalia): Assessing the impact of increased sampling on resolving enigmatic relationships. *Syst. Biol.* **54**(2), 317–337.

Fordyce, R. E. (1982). Dental anomaly in a fossil squalodont dolphin from New Zealand, and the evolution of polydonty in whales. *N.Z. J. Zool.* **9**, 419–426.

Fordyce, R. E. (1989). Problematic early Oligocene toothed whale (Cetacea, ?Mysticeti) from Waikari, north Canterbury, New Zealand. *N.Z. J. Geol. Geophys.* **32**, 395–400.

Fordyce, R. E. (1992). Cetacean evolution and Eocene/Oligocene environments. *In* "Eocene–Oligocene Climatic and Biotic Evolution" (D. R. Prothero, and W. A. Berggren, eds), pp. 368–381. Princeton University Press, Princeton, New Jersey.

Fordyce, R. E., and Barnes, L. G. (1994). The evolutionary history of whales and dolphins. *Annu. Rev. Earth Planet. Sci.* **22**, 419–455.

Gingerich, P. D., Arif, M., and Clyde, W. C. (1995). New archaeocetes (Mammalia, Cetacea) from the middle Eocene Domanda Formation of the Sulaiman Range, Punjab (Pakistan), Contributions from the Museum of Paleontology, The University of Michigan, **29**(11), 291–330.

Hirota, K., and Barnes, L. G. (1995). A new species of middle Miocene sperm whale of the genus Scaldicetus (Cetacea; Physteridae) from Shiga-mura, Japan. *Island Arc* **3**(4), 453–472.

Husar, S. L. (1978a). Dugong dugon. *Mamm. Spec.* **88**, 1–7.

Husar, S. L. (1978b). Trichechus manatus. *Mamm. Spec.* **93**, 1–5.

Inuzuka, N., Domning, D. P., and Ray, C. E. (1995). Summary of taxa and morphological adaptations of the Desmostylia. *Island Arc* **3**, 522–537.

Karlsen, K. (1962). Development of tooth germs and adjacent structures in the whalebone whale (*Balaenoptera physalus* (L.)). *Hvalrådets Skrifter* **45**, 1–56.

Kohno, N. (2002). Dental formula and tooth replacement pattern in Desmostylus as revealed by high-resolution X-ray CT. *J. Vertebrate Paleontol.* **22**(3), 75A.

Koretsky, I. A., and Holec, P. (2002). A primitive seal (Mammalia: Phocidae) from the early middle Miocene of central Paratethys. *Smith. Contr. Paleobiol.* **93**, 163–178.

Kubota, K., and Togawa, S. (1964). Numerical variations in the dentition of some pinnipeds. *Anat. Rec.* **150**(2), 487–502.

Kumar, K., and Sahni, A. (1986). *Remingtonocetus harudiensis*, new combination, a middle Eocene archaeocete (Mammalia, Cetacea) from western Kutch, India. *J. Vertebrate Paleontol.* **6**(4), 326–349.

Mitchell, E. D. (1989). A new cetacean from the late Eocene La Meseta Formation, Seymour Island, Antarctic Peninsula. *Can. J. Fish. Aquat. Sci.* **46**, 2219–2235.

de Muizon, C. (1981). Deux nouveaux Monachinae (Phocidae, Mammalia) du Pliocène de Sud-Sacaco. Éditions Recherche sur les Civilisations, Mémoire, 6(Les vertébrés fossiles de la formation Pisco (Pérou)): 1–150.

de Muizon, C. (1994). Are the squalodonts related to the platanistoids? Proceedings of the San Diego Society of Natural History, 29 (Contributions in Marine Mammal Paleontology Honoring Frank C. Whitmore Jr.): 135–146.

Nowak, R. M. (1999). "Walker's Mammals of the World," 6th Ed. The Johns Hopkins University Press, II, 1919 p.

O'Leary, M. A., and Uhen, M. D. (1999). The time of origin of whales and the role of behavioral changes in the terrestrial–aquatic transition. *Paleobiology* **25**(4), 534–556.

Repenning, C. A. (1977). Adaptive evolution of sea lions and walruses. *Syst. Zool.* **25**(4), 375–390.

Repenning, C. A., and Tedford, R. H. (1977). Otarioid seals of the Neogene. *Geological Survey Professional Paper* **992**, 1–93.

Thewissen, J. G. M., Madar, S. I., and Hussain, S. T. (1996). *Ambulocetus natans*, an Eocene cetacean (Mammalia) from Pakistan. *Courier Forschungsinstitut Senckenberg* **191**, 1–86.

Uhen, M. D. (2000). Replacement of deciduous first premolars and dental eruption in archaeocete whales. *J. Mammal.* **81**(1), 123–133.

Uhen, M. D., and Gingerich, P. D. (2001). New genus of dorudontine archaeocete (Cetacea) from the middle-to-late Eocene of South Carolina. *Mar. Mamm. Sci.* **17**(1), 1–34.

# Desmostylia

## DARYL P. DOMNING

The Desmostylia are the only completely extinct order of marine mammals. They were hippopotamus-like amphibious herbivores (Fig. 1) that were confined to the North Pacific Ocean and are known only as fossils of Oligocene and Miocene age. [For comprehensive references to the published literature on desmostylians, see Domning (1996).]

## I. Desmostylian Relationships, Origins, and Distribution

Desmostylians have undergone more of a taxonomic odyssey than perhaps any other mammals. Although they were placed initially (and afterward, most commonly) among either the Sirenia or the Proboscidea, various authors later assigned them to the Monotremata, Marsupialia, Multituberculata, or an order of their own. The latter view prevailed only in 1953. Today they are classified in the supraordinal group Tethytheria together with the living Sirenia and Proboscidea; the Proboscidea are probably their sister group. The teeth of the most primitive desmostylians closely resemble those of anthracobunids, early tethytheres which are sometimes considered true proboscideans (Domning *et al.*, 1986; Gheerbrant *et al.*, 2005).

Since the most primitive known desmostylians, as well as other early tethytheres, are found in Asia, this continent, bordering the eastern part of the ancient Tethys Sea, is likely to have been the area where this order arose, probably during the Paleocene or Eocene. However, desmostylians do not appear in the fossil record until the early Oligocene, about 30 million years ago. By that time they had already spread to the North American shore of the Pacific; and from then to their extinction some 20 million years later, they inhabited the North Pacific littoral from Japan to Baja California.

D

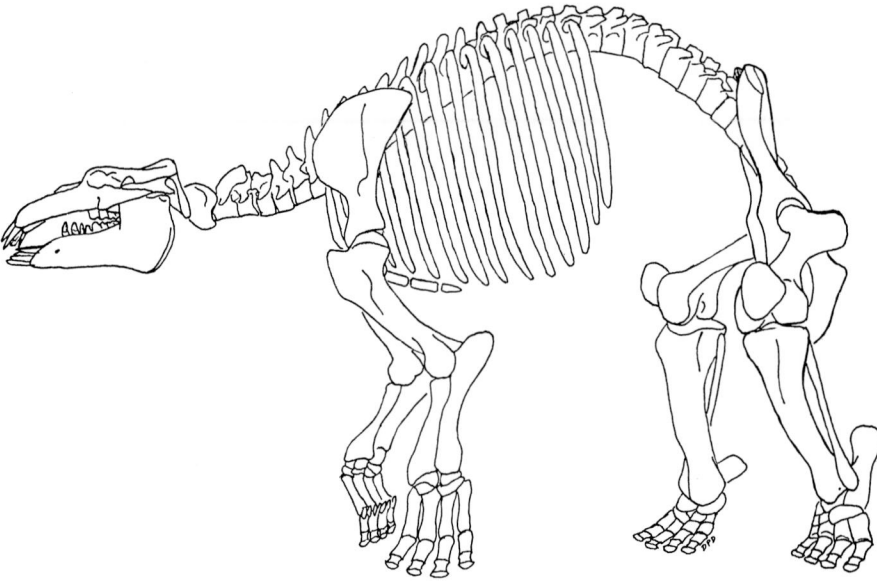

**Figure 1**  *Skeleton of* Paleoparadoxia tabatai, *a Miocene desmostylian, in terrestrial pose. Total length about 2.2 m. Note hyperextension and anterolateral direction of front toes, anterior direction of hind toes, and strong abduction of knees. From Domning (2002), reproduced with permission of the Smithsonian Institution Press.*

## II. Anatomy and Mode of Life

The desmostylian skull (Fig. 2) features a more or less long, narrow, and little-deflected rostrum; dorsally protruding orbits; a stout paroccipital process; an external auditory meatus nearly enclosed ventrally by contact of the posttympanic and postglenoid processes; and a large epitympanic sinus opening into the temporal fossa. The dental formula is primitively 3.1.4.3, with procumbent, transversely aligned incisors and canines; the lower canine is especially enlarged. The fourth lower deciduous premolar is primitively trilobed. The body (Fig. 1) is stout and compact, with a relatively short neck, a deep thorax, a broad sternum with paired plate-like sternebrae, a strongly arched lumbar spine, and a very reduced tail. The limbs are robust, with considerable torsion in the tibia and an ankle joint that is oblique to the tibial shaft (conditions also seen to varying degrees in many noncursorial land mammals). The metacarpals are longer than the metatarsals. The bones show some osteosclerosis (increased density) but no pachyostosis (increased volume).

Controversy over desmostylians' structure and posture has even surpassed the disagreements over their classification (Inuzuka *et al.*, 1995). Once complete skeletons were discovered, showing that they possessed four stout limbs and were capable of some sort of locomotion on land, paleontologists and artists created a startling variety of reconstructions, including ones resembling hippopotami, tapirs, sea lions, crocodiles, and creatures unlike anything else (Inuzuka, 1982). The interpretations presented most recently and defended in the most detail have portrayed them as hauling out in the manner of sea lions (Repenning; see Repenning and Packard, 1990), as walking with the limbs sprawled in a "herpetiform" stance (Inuzuka; e.g., Inuzuka *et al.*, 1995), or as keeping the legs under the body in a more conventional land-mammal fashion, with resemblances to ground sloths (Domning, 2002; Fig. 1).

In contrast, their aquatic behavior has occasioned little argument. Their style of swimming is generally agreed to have been like that of polar bears; i.e., alternate paddling with the forelimbs while the hind limbs were used for steering. Because desmostylian fossils are found exclusively in marine deposits, there has also never been any doubt expressed that they were strictly marine mammals, despite the lack of clearly aquatic specializations in their skeletons.

The peculiar, heavily enameled tooth structure of the more highly derived desmostylians has occasionally led to suggestions that they fed on molluscs or other shelled prey. However, these teeth tend to be high crowned, like those of grazing ungulates, rather than low, broad, and pavement-like as seen in animals that do crush shellfish. Moreover, the teeth of earlier desmostylians closely resemble those of undoubted herbivores. Hence a diet of marine plants for all desmostylians is now generally conceded.

## III. Diversity

Only about seven genera and fewer than a dozen species are currently recognized in this order (Table I). They are currently grouped into two families, Paleoparadoxiidae (*Behemotops*, *Paleoparadoxia*) and Desmostylidae (*Ashoroa*, *Cornwallius*, *Kronokotherium*, *Desmostylus*, and *Vanderhoofius*). All these taxa, so far as is known, had broadly similar postcranial skeletons, and they are distinguished mainly by details of skull and dentition.

The most primitive form named so far is the Oligocene *Behemotops*, an animal nearly the size of a Nile hippopotamus and with low crowned, anthracobunid-like teeth (Ray *et al.*, 1994). The Miocene *Paleoparadoxia* is similar, but has a more retracted nasal opening, fewer premolars, and a long postcanine diastema. *Cornwallius*, another late Oligocene genus, shows a tendency for the molar cusps to become more columnar. This trend continued

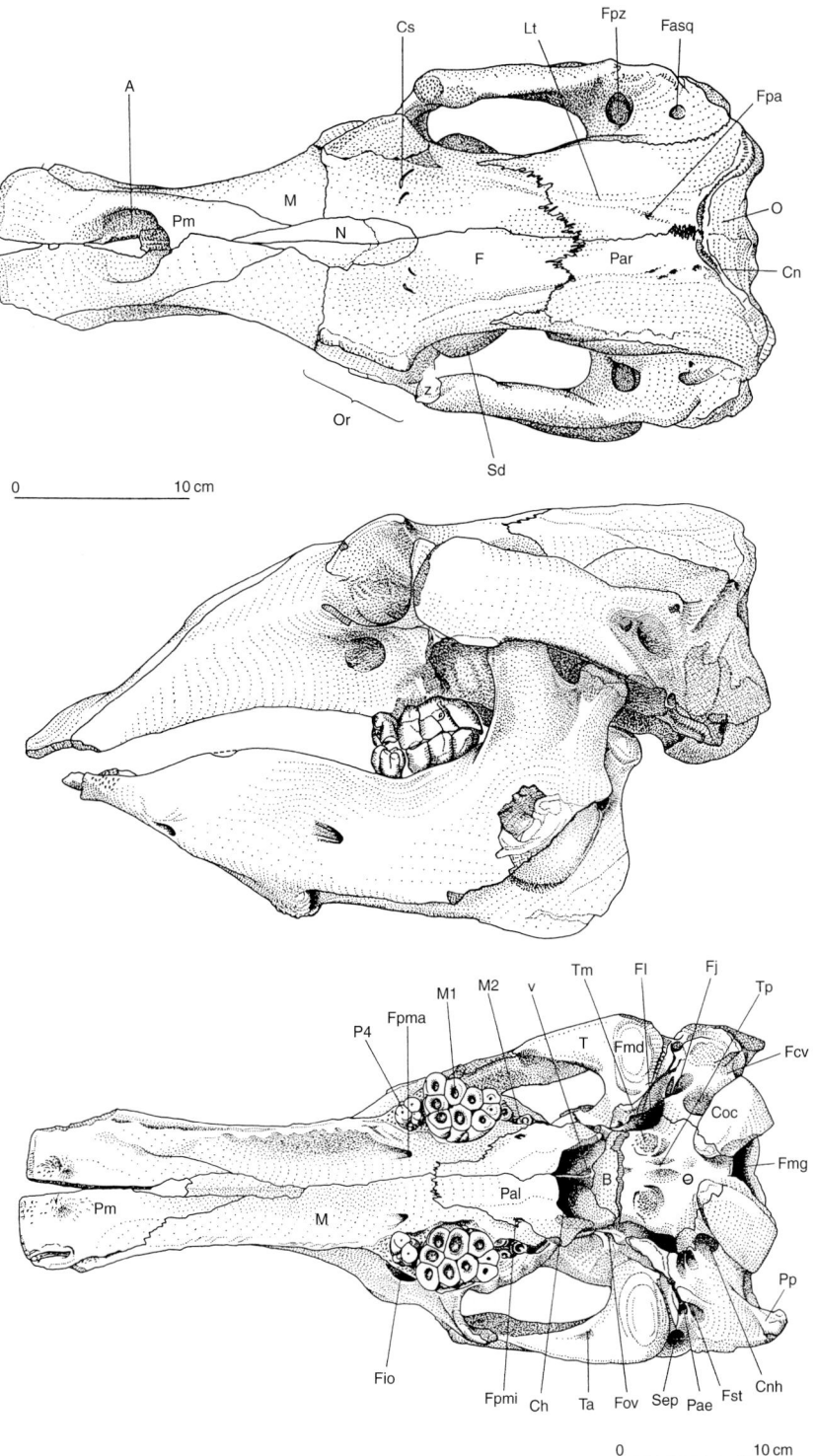

**D**

**Figure 2** *Skull and mandible of* Desmostylus hesperus, *a Miocene desmostylian, in dorsal, lateral, and ventral views (immature specimen). The left mandibular angle is broken and the large right dental capsule is visible. Note the columnar cusps of the molars. A, nasal aperture; B, basisphenoid; Ch, choana; Cn, nuchal crest; Cnh, hypoglossal canal; Coc, occipital condyle; Cs, supraorbital canal; F, frontal; Fasq, anterior squamosal foramen; Fio, infraorbital foramen; Fj, jugular foramen; Fl, foramen lacerum; Fmd, mandibular fossa; Fmg, foramen magnum; Fov, foramen ovale; Fpa, parietal foramen; Fpma, greater palatine foramen; Fpmi, lesser palatine foramen; Fpz, postzygomatic foramen; Fst, stylomastoid foramen; Lt, temporal crest; M, maxilla; M1, M2, upper first and second molars; N, nasal; O, occipital; Or, orbit; P4, upper fourth premolar; Pae, external auditory meatus; Pal, palatine; Par, parietal; Pm, premaxilla; Pp, paroccipital process; Sd, dental capsule; Sep, epitympanic sinus; T, squamosal (temporal); Ta, articular tubercle; Tm, muscular tubercle; Tp, pharyngeal tubercle; V, vomer; Z, zygomatic (jugal). From Inuzuka (1988), reproduced with permission of the Geological Survey of Japan.*

**D**

TABLE I
Genera of Desmostylia and Their Temporal Range

| Genus | Range |
| --- | --- |
| *Behemotops* | Early to late Oligocene |
| *Paleoparadoxia* | Early to late Miocene |
| *Ashoroa* | Late Oligocene |
| *Cornwallius* | Late Oligocene |
| *Kronokotherium* | ?Early to ?middle Miocene |
| *Desmostylus* | Early to late Miocene |
| *Vanderhoofius* | Middle Miocene |

in the Miocene with the genera *Kronokotherium*, *Vanderhoofius*, and *Desmostylus*; the latter name, meaning "bundle of columns," expresses the appearance of a molar in these most highly derived and characteristic members of the group (Fig. 2). The procumbent incisors and canines are variously reduced, sometimes leaving only the large lower canines as digging organs, and the adult cheek dentition comprises only molars.

The diversity of desmostylians was constrained by their limited geographic range and dietary opportunities. Apparently they were adapted to cooler climates than the tropical sirenians, but they seem to have succumbed to competition from these more fully aquatic herbivores as soon as hydrodamaline sirenians evolved the ability to spread into the cool home waters of the desmostylians.

## See Also the Following Article

Sirenian Evolution

## References

Domning, D. (1996). Bibliography and index of the Sirenia and Desmostylia. *Smith. Contr. Paleobiol.* **80**, 1–611.

Domning, D. (2002). The terrestrial posture of desmostylians. *Smith. Contr. Paleobiol.* **93**, 99–111.

Domning, D., Ray, C., and McKenna, M. (1986). Two new Oligocene desmostylians and a discussion of tethytherian systematics. *Smith. Contr. Paleobiol.* **59**, 1–56.

Gheerbrant, E., Domning, D., and Tassy, P. (2005). Paenungulata (Sirenia, Proboscidea, Hyracoidea, and relatives). *In* "The Rise of Placental Mammals: Origin and Relationships of the Major Extant Clades" (K. Rose, and J. Archibald, eds), pp. 84–105. Johns Hopkins Univ. Press, Baltimore.

Inuzuka, N. (1982). Atlas of reconstructed desmostylians. *Abstracts, 36th Annual Meeting, Association for the Geological Collaboration in Japan*, Saitama, 44–61. [In Japanese.]

Inuzuka, N. (1988). The skeleton of *Desmostylus* from Utanobori, Hokkaido. I. Cranium. *Bull. Geol. Surv. Japan* **39**, 139–190. [In Japanese.]

Inuzuka, N., Domning, D., and Ray, C. (1995). Summary of taxa and morphological adaptations of the Desmostylia. *Island Arc* **3**, 522–537.

Ray, C., Domning, D., and McKenna, M. (1994). A new specimen of *Behemotops proteus* (Order Desmostylia) from the marine Oligocene of Washington. *Proc. San Diego Soc. Nat. Hist.* **29**, 205–222.

Repenning, C., and Packard, E. (1990). Locomotion of a desmostylian and evidence of ancient shark predation. *In* "Evolutionary Paleobiology of Behavior and Coevolution" (A. Boucot, ed.), pp. 199–203. Elsevier, Amsterdam.

# Dialects

## John K.B. Ford

Consistent variations in the underwater vocalizations of marine mammals can exist among individuals, groups, or populations. Differences in vocal patterns between geographically isolated populations have been described in several species of pinnipeds and cetaceans. Examples are regional variations in the vocal repertoires of the Weddell seal, *Leptonychotes weddellii* (Thomas and Stirling, 1983), the trill vocalizations of the bearded seal, *Erignathus barbatus* (Risch *et al.*, 2007), and the songs of humpback whales, *Megaptera novaeangliae*, in different ocean areas (Payne and Guinee, 1983). Although these types of variations have occasionally been referred to as *dialects*, they are more appropriately termed *geographic variations* (Conner, 1982). Such differences may result from long-term geographic and, therefore, reproductive isolation leading to genetic distinctions among populations. However, cultural variations may also contribute to acoustic distinctions between marine mammal populations. Pinnipeds and cetaceans are among the few mammalian orders in which vocal mimicry and learning has been documented (Janik and Slater, 1997, 2000). Transmission of vocal patterns across generations may thus depend more on cultural than on genetic mechanisms. Copying errors and other forms of cultural mutation and drift may be responsible for at least a portion of the vocal differences between distant populations. Geographic variation in vocal patterns most likely represents epiphenomena, or byproducts, of social and genetic isolation.

Dialects, which are consistent differences in the vocal repertoires of local, neighboring populations, or groups that can potentially mix and interbreed, likely arise primarily through social learning. Generally rare among mammals, dialects have been described in two cetacean species to date, killer whales (*Orcinus orca*) and sperm whales (*Physeter macrocephalus*). Among pinnipeds, there is evidence for vocal dialects between neighboring sites in Weddell seals (Morrice *et al.*, 1994) and harbor seals, *Phoca vitulina* (Van Parijs *et al.*, 2003).

In coastal waters of the northeastern Pacific, matrilineal kinship groups, or pods, of "resident" killer whales have repertoires of 7–17 call types that vary among pods (Ford, 1991; Yurk *et al.*, 2002). All pods have distinctive features in their call repertoires, and thus each has a unique dialect. Certain pods share a portion of their call repertoires with others, and these are considered to belong to the same acoustic clan; each clan is acoustically distinct. Pods belonging to different clans have overlapping ranges and interact frequently, despite having very different call repertoires. New pods form by gradual fission of older, larger pods, along maternal lines. This process appears to be accompanied by divergence of common dialects, thus dialects reflect the historical matrilineal genealogy of pods within clans. Differences in fine-scale features of call structure exist among closely related matrilines within pods (Deecke *et al.*, 2000; Miller and Bain, 2000). Dialects likely serve as acoustical "badges" that help maintain cohesion and integrity of matrilineal groups and may also serve as an inbreeding avoidance mechanism (Ford, 1991; Barrett-Lennard, 2000). Group-specific dialects have also been found among killer whale groups off Norway (Strager, 1995) and Kamchatka (Filatova *et al.*, 2006).

The other cetacean species with group-specific dialects, the sperm whale, tends also to live in matrilineal groups, although these lack the long-term stability seen in "resident" killer whales. Sperm whale groups were found to have repertoires that consistently varied in the proportional usage of different coda types and classes (Weilgart and Whitehead, 1997). As in killer whales, groups of sperm whales with distinct dialects regularly interact. Geographic variations in coda repertoires was also noted in different oceans and in different areas within oceans (Whitehead *et al.*, 1998), but such variations were weaker than those observed in group-specific dialects within local regions. In the South Pacific, sperm whale groups were grouped into six sympatric acoustic clans based on coda sharing (Rendell and Whitehead, 2003). These clans differed in patterns of movement, habitat use patterns, and feeding success (Whitehead and Rendell, 2004).

## References

Barrett-Lennard, L. G. (2000). Population structure and mating systems of northeastern Pacific killer whales. Ph.D. Dissertation, University of British Columbia, Vancouver, BC.

Conner, D. A. (1982). Dialects versus geographic variation in mammalian vocalizations. *Anim. Behav.* **30**, 297–298.

Deecke, V. B., Ford, J. K. B., and Spong, P. (2000). Dialect change in resident killer whales: Implications for vocal learning and cultural transmission. *Anim. Behav.* **60**, 629–638.

Filatova, O. A., Burdin, A. M., and Hoyt, E. (2006). Vocal dialects and population structure in killer whales of eastern Kamchatka. In 20th Annual Conference of the European Cetacean Society, Gdynia, Poland. April 2–7, 2006, pp. 43–44.

Ford, J. K. B. (1991). Vocal traditions among resident killer whales (*Orcinus orca*) in coastal waters of British Columbia. *Can. J. Zool.* **69**, 1454–1483.

Janik, V. M., and Slater, P. J. B. (1997). Vocal learning in mammals. *Adv. Study Behav.* **26**, 59–99.

Janik, V. M., and Slater, P. J. B. (2000). The different roles of social learning in vocal communication. *Anim. Behav.* **60**, 1–11.

Miller, P. J. O., and Bain, D. E. (2000). Within-pod variation in the sound production of a pod of killer whales, *Orcinus orca. Anim.Behav.* **60**, 617–628.

Payne, R., and Guinee, L. N. (1983). Humpback whale, *Megaptera novaeangliae*, songs as an indicator of 'stocks'. *In* "Communication and Behavior of Whales" (R. Payne, ed.), pp. 333–358. Westview Press, Boulder, Colorado, USA.

Rendell, L. E., and Whitehead, H. (2003). Vocal clans in sperm whales (*Physeter macrocephalus*). *Proc. R. Soc., Biol. Sci.* **270**, 225–231.

Risch, D., *et al.* (2007). Vocalizations of male bearded seals, *Erignathus barbatus*: classification and geographical variation. *Anim. Behav.* **73**, 747–762.

Strager, H. (1995). Pod-specific call repertoires and compound calls of killer whales, *Orcinus orca* Linnaeus, 1758, in waters of northern Norway. *Can. J. Zool.* **73**, 1037–1047.

Thomas, J. A., and Stirling, I. (1983). Geographic variation in the underwater vocalizations of Weddell seals (*Leptonychotes weddelli*) from Palmer Peninsula and McMurdo Sound, Antarctica. *Can. J. Zool.* **61**, 2203–2212.

Van Parijs, S. M., *et al.* (2003). Patterns in the vocalizations of male harbor seals. *J. Acoust. Soc. Am.* **113**, 3403–3410.

Weilgart, L., and Whitehead, H. (1997). Group-specific dialects and geographical variation in coda repertoire in South Pacific sperm whales. *Behav. Ecol. Sociobiol.* **40**, 277–285.

Whitehead, H., and Rendell, L. E. (2004). Movements, habitat use and feeding success of cultural clans of South Pacific sperm whales. *J. Anim. Ecol.* **73**, 190–196.

Whitehead, H., Dillon, M., Dufault, S., Weilgart, L., and Wright, J. (1998). Non-geographically based population structure of South Pacific sperm whales: dialects, fluke-markings and genetics. *J. Anim. Ecol.* **67**, 253–262.

Yurk, H., Barrett-Lennard, L., Ford, J. K. B., and Matkin, C. O. (2002). Cultural transmission within maternal lineages: Vocal clans in resident killer whales in southern Alaska. *Anim. Behav.* **63**, 1103–1119.

# Diet

## Nélio B. Barros and Malcolm R. Clarke

The ancestors of present-day marine mammals moved into water, possibly to escape competition on land for food resources, escape predation, or take advantage of relatively abundant food supplies in the seas. Most likely, it was a combination of factors. Eventually, the development of echolocation capabilities in odontocetes (toothed whales) and physiological adaptations for deep and prolonged dives allowed for the exploration of deep waters in search for food. Some groups of cetaceans, such as beaked whales and several phocid seals were able to evolve the ability to dive to great depths and take advantage of food resources unavailable to other predators. From a presumed terrestrial insectivore diet, marine mammals switched largely to fish, squid, and shrimp as main prey (in addition to other crustaceans, mollusks, and zooplankton organisms) (Bowen and Siniff, 1999).

### I. Methods of Study

Various methods have been used to gain insight into what marine mammals eat. Their aquatic lifestyle usually limits direct observations of feeding, except in shallow waters in geographical areas where water visibility is good. The following methods have been used to draw inferences on marine mammal diet.

### A. Direct Observations of Feeding

This method has been limited to what can be observed above the surface, from a vessel, a vantage point on land, or the air. Although much can be learned from these observations, especially when made systematically in a particular area, subsurface feeding behavior is generally not observed, and the picture obtained on feeding is incomplete, at best. Prey that present aerial behavior (e.g., mullet, flying fish) tend to be overrepresented, whereas bottom-dwelling species (e.g., toadfish, flatfish, octopus) are underrepresented. Modern "crittercam" video cameras mounted on marine mammals (quite successfully on several pinniped species) promise to give us much more direct information on food ingested.

### B. Traditional Methods

The traditional method to study marine mammal food habits has been the analysis of food remains present in vomit or scat from living animals, and the stomachs and intestines of stranded animals. This method relies on the finding and identification of structures representing a typical meal, e.g., fish bones and the jaws of cephalopods, often referred to as "beaks" due to their superficial resemblance to beaks of parrots. Fish ear stones (or otoliths), in particular, are diagnostic structures in the identification of prey because their size and shape vary considerably from species to species (Fig. 1) (Härkönen, 1986; Nolf, 1993).

Fish otoliths are calcareous structures (their primary composition is calcium carbonate) and are more resistant to digestion than bones. In life, they are housed in capsules inside the fish's skull, and their main function in a fish is to provide information on balance and sound reception. Each species of fish has three pairs of otoliths (the sagitta, lapillus, and asteriscus). With a few exceptions (e.g., catfish), the largest pair of otoliths (the sagitta) is also the one most distinctive and recognizable, and the one used in species identification (Smale *et al.*, 1995; Furlani *et al.*, 2007). Similarly, cephalopod beaks possess

D

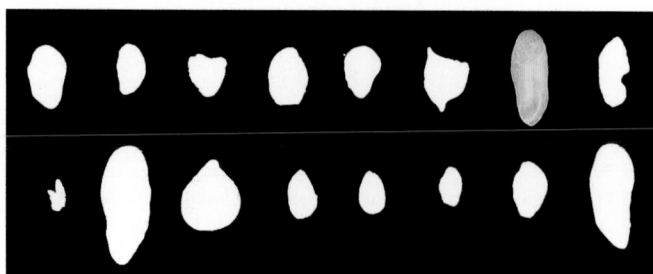

**Figure 1** *Fish ear stones (otoliths) of different sizes and shapes retrieved from stomachs of bottlenose dolphins stranded in Florida.*

**Figure 2** *Cephalopod lower (left) and upper (right) jaws or "beaks" from a giant squid* (Architeuthis).

morphological features that vary and can be used in the identification of species (Fig. 2). Although the upper beaks have some taxonomic value, the lower beaks are generally the ones used in prey identification (Clarke, 1986a). Beaks are composed of chitin, a similar material to our fingernails and mammal horns, and are not dissolved by digestive processes. Otolith and cephalopod beak lengths correlate well with the length and weight of the animal from which they came; relationships with their weights are described by a power curve. These allow reconstruction of the original meal by weight. Estimates can then be made of consumption of particular prey species by single mammals and, sometimes, their populations. The relationships relating otolith/beak length to animal length are best described by straight lines, which indicate the target lengths of prey species.

The advantages of the otolith/beak method are (1) knowledge of prey composition and size classes allows for understanding spatial and temporal distribution of predators; (2) studies of predator–prey dynamics are possible; (3) prey species may be very poorly sampled by humans using other methods, and diets can give considerable information on the species in an area available to predators; (4) changes in diet during growth and over time can be monitored;

(5) analysis requires low cost or little equipment; and (6) samples can be collected from carcasses in an advanced stage of decomposition.

The disadvantages of this method are (1) prey with no hard parts (e.g., invertebrates) will be underrepresented; (2) different digestion rates of prey can make calculations of reconstructed meal sizes complicated (fish otoliths can last for about 1 day in the gastrointestinal tracts of marine mammals, whereas squid beaks may accumulate for several days or longer); (3) there can be potential bias in using feeding data gathered from stranded (possibly sick) animals, as they may not be representative of the population at large; (4) analyses of food matter present in fecal material (e.g., scats) may present biases, as prey structures can be partially or completely digested through passage in the gastrointestinal tract; correction factors may be necessary to account for such limitations (Bowen, 2000); and (5) a comprehensive reference collection of hard structures (fish otoliths and cephalopod beaks) of the most common prey in a particular area is of great advantage in species identification, but is not always available.

### C. Use of Novel Tools (New Technology) to Understand Feeding Ecology

Several new methods of studying diets are likely to become more important with time (Read, 1998).

*1. Stable Isotopes* The principle of this method is that ratios of heavier vs lighter isotopes of particular elements (carbon, nitrogen, oxygen, sulfur) in tissues of predators can be traced to those of their prey as they are assimilated through the diet (Latja and Michener, 1994; Walker and Macko, 1999).

The following are the advantages of this method: (a) it is ideal to detect shifts in diet; (b) different tissues of the predator yield information reflecting the feeding history relative to the last days, months, or the entire life of the animal; these data can be used to gain insights into the distribution, movements, and migratory habits of the animal; (c) isotopic ratios of carbon reflect those of the primary producers in the area; isotopic ratios of nitrogen are indicative of the trophic level occupied by the organism, which are helpful in understanding habitat utilization and trophic relationships; and (d) historical reconstruction of values through time can yield intraannual and interannual variability in feeding. The disadvantages of this method are that a reference database for the isotopic signature of prey is needed and the cost of the equipment used in the analyses is high.

*2. Fatty Acids* The fatty composition of a prey is species specific and, as these compounds are assimilated through the diet and accumulated in the fatty tissues of predators (e.g., blubber of marine mammals), they can be used as tracers of diet (Iverson et al., 1997; Wetzel and Reynolds, 2004; Budge et al., 2006). Fatty acid analysis can be useful in (a) reconstructing diets in time and space; (b) population studies of various marine mammal species using different feeding grounds; (c) studies of energetic transfers between mother and their offsprings; and (d) the application of the technique to free-ranging animals, which can be done by the relative noninvasive collection of tissue (e.g., through biopsy darts). Similarly in stable isotopes, this method requires a reference database for the chemical signature of the various prey species, and the cost of the equipment is high. In addition, there is stratification of fat in the outer and inner blubber layer of marine mammals, and incomplete sampling of the blubber layer may yield misleading results of dietary information. Additional variability may be associated with from what part of the body the sample is taken, making interstudy comparisons difficult.

*3. Molecular Identification of Prey*   This method involves the genetic identification of material from scats, stomach contents, and gut bacteria, which must be separated prior to analysis. The disadvantages of this method are that a reference database for the genetic signature of prey is needed and there might be biases associated with prey DNA survival during digestion (Deagle and Tollit, 2007). However, the increased application of this method to marine mammal dietary studies (Purcell *et al.*, 2004; Casper *et al.*, in press; Dunshea *et al.*, in press) reflects the potential of this tool.

*4. Video-Taping Studies (Using "Crittercams") of Animals Feeding at Depth*   This method has the following advantages: (a) it documents the actual feeding behavior of the predator, and the identity of the prey species can be verified by the images; (b) prey behavior during detection and capture can be documented; and (c) different feeding strategies can also be observed (e.g., cooperative feeding). Among the disadvantages (a) only captures of a few species can be observed; (b) ambient light must be relatively bright and the water must be clear; (c) there is difficulty in the attachment and recovery of the equipment (video camera); and (d) there are high costs associated with the equipment and its operation.

*5. Acoustics*

A. DIGITAL ACOUSTIC RECORDING TAGS (D-TAGS)   Placing of acoustic tags in deep-diving cetaceans, such as beaked and sperm whales, has allowed the recording of ultrasonic clicks and the returning echoes of potential prey (Johnson and Tyack, 2003; Johnson *et al.*, 2004; Madsen *et al.*, 2005; Zimmer *et al.*, 2005; Watwood *et al.*, 2006). The computerized instruments record whale movements, depth, and sounds on dives, which shed light into whales' foraging behavior at great depths.

B. PREY TARGET STRENGTH   This method estimates the abundance of prey in the wild by measuring the target strength of the individual organisms that potentially comprise the prey spectrum (Benoit-Bird and Au, 2001; Benoit-Bird *et al.*, in press). The advantages of acoustic sampling vs other direct methods, such as trawling, are cost-effectiveness, a less biased assessment of overall biomass and relative composition of the prey field, and the potential of examining prey variability in larger spatial and temporal scales.

In summary, although identifying and measuring items in vomit, scats, and stomach contents have many disadvantages, it provides more information at considerably less cost than other methods and cannot be replaced effectively by any other method at present.

## II. Diets in General

Marine mammals, all together, eat a great variety of animals from minute crustaceans, less than 1-mm long, to giant squids, over 15 m in length. These disparate prey live in a wide range of habitats, from the shallow shelf seas and estuaries to over 2000-m deep in the deepest oceans, from near the water surface to the ocean bottom. The animals consumed vary in texture from soft-skinned and gelatinous octopods to hard-scaled, muscular fish and vary in mobility from sedentary clams to jet-propelled squids. The three species of manatees (*Trichechus* spp.) and the dugong (*Dugong dugon*) of the order Sirenia rely on sea grasses and river plants; they are grazers in the true sense.

All this variety in food organisms has led to many specializations in structure. Most obviously, the mouth has developed a great variety of tooth numbers, sizes, and shapes or, for those eating very small organisms, a special filter made of horny plates, frayed on one edge, the baleen. Various species dive to greater depths, thus permitting

an extension of their feeding grounds from the continental shelves, down the continental slopes, to depths exceeding 2000 m. Similarly, thickening of their fatty blubber layer has permitted further extension of their feeding grounds into the cold waters of the Antarctic and the Arctic. The diet of any species reflects its adaptations; its mouth adaptations make it possible for it to catch certain types and sizes of prey. The actual species that compose the diet depend on its own and on the prey's depth and geographic distribution.

Another property of the food is quality. Crustaceans, fish, and cephalopods vary in protein, fat, and mineral constituents and proportions. Marine mammal species can sometimes shift between these three groups during a year, as supply fluctuates or during their own migrations, but this may drastically affect their physical condition and health. Even within one of these major groups, the protein content, for instance, may vary greatly so that change to another species of prey as the main food item might markedly affect the predator. This is likely to be important when the mammal shifts from a diet of shellfish, which are very muscular, to deep-living oceanic fish, which are generally lower in protein content. Similarly, shelf-living cephalopods are mainly soft-bodied, have weak muscles, and are low in protein so that twice as much has to be eaten. The high acidity and presence of ammonium salts in high concentration in some soft-bodied squids also probably require special adaptations of digestive processes.

The food of rarer marine mammals, and species that have not been caught for their oil, is not well known. Knowledge depends on information from occasional strandings, and stomach contents are often difficult to collect because of the size of the carcass. Debris from the stomachs only occasionally includes complete, readily identifiable, prey animals. Usually information on diet has to be obtained from hard pieces, mainly fish otoliths and cephalopod beaks.

### A. Cetaceans

*1. Baleen Whales (Mysticeti)*   These possess baleen plates, and all but the gray whales (*Eschrichtius robustus*) collect swarming animals by skimming through the water or by gulping. They therefore primarily eat shoaling plankton or small nekton together with a few larger animals, such as fish and squids, caught with the shoals (Kawamura, 1980).

A. RIGHT WHALES (BALAENIDAE AND NEOBALAENIDAE)   These five species (three right whales, *Eubalaena* spp., the bowhead whale, *Balaena mysticetus*, and the pygmy right whale, *Caperea marginata*) have very long baleen plates hanging from the roof of the mouth, whose finely frayed inner edges can trap very small plankton. They mainly eat small crustaceans ranging from minute copepods less than 1-mm long, favored by the bowhead whale of the northern seas and the pygmy right whale to small euphausiid crustaceans called "krill" as much as 25-mm long, eaten by the southern right whale (*Eubalaena australis*). The bowhead is also known to eat a small molluscan called a pteropod.

B. GRAY WHALE   This species has very tough baleen plates that become worn, particularly at the right side, by rubbing on the sea floor from which it principally sucks, by piston action of its tongue, bottom amphipod crustaceans and mollusks and bristle worms.

C. RORQUALS (BALAENOPTERIDAE)   These species have shorter baleen plates than right whales and generally favor larger prey than copepods. Blue whales (*Balaenoptera musculus*) eat midwater crustaceans, mainly krill, in the Antarctic and other euphausiid species in the North Pacific and the North Atlantic. The fin whale

D

**D**

(*B. physalus*) eats krill in the Antarctic but in the North Pacific it broadens its diet to include fish such as clupeids, muscular squids, and a copepod. It eats the fish capelin in the North Atlantic. Humpback whales (*Megaptera novaeangliae*) eat mainly krill or "lobster krill" in the Southern Hemisphere, but mainly anchovies and cod in the Northern Hemisphere. Assorted squid are also eaten by humpback whales. The sei (*B. borealis*) whale eats 20 species of densely shoaling midwater crustaceans, including krill and copepods, in addition to anchovy, cod, and assorted oceanic squids. Minke whales (*B. acutorostrata* and *B. bonaerensis*) eat assorted crustaceans in the Arctic and the Antarctic seas, and also fish, including anchovy, in the North Pacific and herring in the North Atlantic. They also take assorted midwater squid in the south tropical seas and appear to rely more on fish than other baleen whales. Bryde's whales (*B. brydei* and *B. edeni*) eat crustaceans, including krill, but also various fish, including mullet and anchovy in the Southern Hemisphere and anchovy in the North Pacific.

*2. Toothed Whales (Odontoceti)* Within this group, comprising 7 families and 52 species, teeth are developed and the main prey items are fish and cephalopods. The cetacean species that live on the continental shelf eat muscular fish, such as herring (*Clupea* spp.), pilchards, whiting, and soles; muscular cephalopods, such as inshore squid, cuttlefish, and octopods, are occasionally taken as well. Odontocetes living in the deep ocean eat mainly lantern fishes and soft-bodied, often gelatinous, squids. Around oceanic islands, both lantern fishes and more muscular species, such as horse mackerel and trumpeter fish, are often taken.

A. DOLPHINS (DELPHINIDAE)  In 45% of the species in this family, cephalopods comprise over 75% and fish less than 25% of the diet (Clarke, 1986b). In 24% of the species, cephalopods comprise 50–75% and fish 25–50%. Depending on the species of dolphin, the diet can be muscular species living on the continental shelf, as for many populations of the short-beaked common dolphin (*Delphinus delphis*) and the common bottlenose dolphin (*Tursiops truncatus*); or as in spinner (*Stenella longirostris*) and spotted dolphins (*S. attenuata* and *S. frontalis*) the diet may include many soft-bodied oceanic species. However, on the whole, dolphins favor muscular squid rather than soft-bodied ones, even in oceanic waters; this also applies to pilot whales (*Globicephala* spp.). A few species include other prey groups in their diet, e.g., Commerson's dolphins (*Cephalorhynchus commersonii*) eat some krill and killer whales (*Orcinus orca*) also prey on seals and other cetaceans.

B. PORPOISES (PHOCOENIDAE)  In half of the species in this family, cephalopods comprise over 75% and fish less than 25% of the diet. In the other half of the species, cephalopods comprise 50–75% and fish 25–50%. Being inshore cetaceans, food consists of common (often economically important to humans) species including muscular inshore squid, cuttlefish, and octopus, as well as fish such as herrings, whitings, and bottom-living soles.

C. BEAKED WHALES (ZIPHIIDAE)  In over half the species of beaked whales, more than half of the food is cephalopod and the rest is fish. These whales are deep divers and at least one species favors soft-bodied squids. The number of teeth is much reduced in this family and, in one species, the strap-toothed beaked whale (*Mesoplodon layardii*), the two teeth in the lower jaw grow over the upper jaw and limit it to a narrow gape. It is remarkable that this does not seem to inhibit capture of oceanic squids.

D. NARWHAL (*MONODON MONOCEROS*) AND BELUGA (*DELPHINAPTERUS LEUCAS*), MONODONTIDAE  The diet of the narwhal includes fish such as Greenland halibut (*Reinhardtius hippoglossoides*) and polar cod (*Boreogadus saida*), muscular squid, and shrimp. The suggestion that the long tooth of male narwhals is regularly used to stir prey from the mud is unlikely to be true. Beluga feed on fish such as capelin (*Mallotus villosus*) and sand lance, as well as larger species such as cod and flounder. Sand- and bottom-living worms show that they probably feed on the bottom as well as in midwater.

E. SPERM WHALES (PHYSETERIDAE AND KOGIIDAE)  All three species feed mainly on squids, although a few fish are taken, including large sharks. Sperm whales (*Physeter macrocephalus*) eat mainly deep-living oceanic squids, and most of these are soft-bodied or gelatinous, luminous, and weak swimmers. Contrary to common belief, the average weight of their prey is not great, varying from 0.5 kg off South Africa to 7 kg in the Antarctic, although some large sperm whales can eat squids over 15 m in length. Pygmy and dwarf sperm whales (*Kogia breviceps* and *K. sima*) eat some of the same species as their larger relative but, because they spend some time on continental shelves, they also include muscular, shelf-living squids and octopods (Clarke, 1996).

F. RIVER DOLPHINS (INIIDAE, PONTOPORIIDAE, LIPOTIDAE, PLATANISTIDAE)  This group includes four species of dolphins, three of which inhabit the freshwater systems of major rivers in South America, China (but recently declared extinct), and the Indian subcontinent. The fourth species has a marine distribution and is found in coastal waters of the Atlantic coast of South America. The riverine species feed on a variety of freshwater fish (including sharks) and prawns, and occasionally also prey on other groups, such as freshwater turtles. As cephalopods are strictly marine in distribution, they are not part of the diet of riverine dolphins. The marine species in this group (the "franciscana," *Pontoporia blainvillei*) eats mainly bottom-dwelling fish, coastal species of cephalopods, and several species of shrimp.

## B. Pinnipeds [Seals, Sea Lions, Walruses (*Odobenus rosmarus*)]

All but 2 of the 36 species of pinnipeds (seals and sea lions) probably include both fish and cephalopods in their diet. The exceptions inhabit freshwater systems where cephalopods do not occur. Most pinniped species inhabit coastal regions or seas close to oceanic islands, which partly influences their choice of diet (Klages, 1996).

*1. Fur Seals and Sea Lions (Otariidae)* Of the 16 species, 10 take benthic cephalopods and 11 eat midwater squids. At least 14 eat muscular cephalopods and 3 of these eat oily squids whereas only 1 eats soft-bodied squids. Three species consume all these on the continental shelf, whereas 8 eat them in oceanic waters.

A. NORTHERN FUR SEAL (*CALLORHINUS URSINUS*)  On the continental shelf this species eats primarily small shoaling fish and muscular squids. In offshore waters, they eat mainly muscular oceanic squids.

B. GUADALUPE FUR SEAL (*ARCTOCEPHALUS TOWNSENDI*)  They eat oceanic cephalopods and lantern fish.

C. JUAN FERNANDEZ FUR SEAL (*A. PHILIPPII*)  They apparently eat muscular, oceanic squids.

D. GALÁPAGOS FUR SEAL (*A. GALAPAGOENSIS*)  This species mainly eats muscular, oceanic squids.

E. CAPE AND AUSTRALIAN FUR SEALS (*A. PUSILLUS*)  They eat both shelf and oceanic fish and cephalopods, as well as both midwater and bottom species. They favor muscular rather than soft-bodied cephalopods.

**F. NEW ZEALAND FUR SEAL** (*A. FORSTERI*)  This species eats not only midwater fish and muscular cephalopods, but also takes penguins.

**G. ANTARCTIC FUR SEAL** (*A. GAZELLA*)  They eat mainly krill in the Antarctic, but further north take oceanic fish, as well as muscular and soft-bodied squids.

**H. SUBANTARCTIC FUR SEAL** (*A. TROPICALIS*)  They eat oceanic squids, both muscular and soft-bodied.

**I. STELLER SEA LION** (*EUMETOPIAS JUBATUS*)  They eat muscular, bottom octopods as well as oily oceanic squids, and polar cod.

**J. CALIFORNIA, GALÁPAGOS, AND JAPANESE SEA LIONS** (*ZALOPHUS CALIFORNIANUS, Z. WOLLEBAEKI, AND Z. JAPONICUS*)  These species eat shelf-living, muscular squids, and octopods, as well as muscular, oceanic squids.

**K. SOUTH AMERICAN SEA LION** (*OTARIA FLAVESCENS*)  They eat mainly bottom and midwater shelf fish and also take some shelf cephalopods.

**L. AUSTRALIAN SEA LION** (*NEOPHOCA CINEREA*)  This species eats shelf octopods and cuttlefish.

**M. NEW ZEALAND SEA LION** (*PHOCARCTOS HOOKERI*)  This species eats shelf fish, cephalopods, and crustaceans.

*2. Earless (true) Seals (Phocidae)*  This group eats a variety of fish, cephalopods, and crustaceans of both inshore and oceanic species, depending on their locality. Of the 19 species, at least 15 eat muscular cephalopods, 5 eat oily species, and 4 eat soft-bodied squids. Some slight deviations from this pattern are given.

**A. HARP SEAL** (*PAGOPHILUS GROENLANDICUS*)  This species eats mainly fish and crustaceans, especially amphipods and euphausiids, although both bottom and midwater oceanic cephalopods are also taken.

**B. BEARDED SEAL** (*ERIGNATHUS BARBATUS*)  They eat mainly bottom shelf invertebrates, such as clams, and also fish, in addition to a few species of octopods.

**C. GRAY SEAL** (*HALICHOERUS GRYPUS*)  Gray seals eat schooling fish, squids, octopods, and occasionally sea birds.

**D. CRABEATER SEAL** (*LOBODON CARCINOPHAGA*)  They eat krill almost exclusively.

**E. ROSS SEAL** (*OMMATOPHOCA ROSSII*)  Ross seals eat oceanic species of fish and squids of both muscular and soft-bodied species.

**F. LEOPARD SEAL** (*HYDRURGA LEPTONYX*)  This species eats krill, fish, soft-bodied squids, and occasionally mammals.

**G. WEDDELL SEAL** (*LEPTONYCHOTES WEDDELLII*)  They eat mainly cephalopods, including muscular and soft-bodied species, and bottom octopods.

**H. ELEPHANT SEALS** (*MIROUNGA ANGUSTIROSTRIS AND M. LEONINA*)  These species eat oceanic species, including muscular, soft-bodied, and oily species and, seasonally shelf squids.

*3. Walrus*  These are benthic feeders in shallow Arctic seas at depths less than 100 m. Their main food is clams but they also eat a small quantity of bottom octopods and have been known to attack other seals.

### C. Sea Otter (*Enhydra lutris*)

These generally eat bottom invertebrates on the continental shelf, usually very close to shore, including clams, sea urchins, and other invertebrates. However, especially in the northern distributions of their range, they also feed on fishes.

### D. Polar Bear (*Ursus maritimus*)

This species eats mainly harp seals and also feeds on other seals, young beluga and narwhal, young walrus or sick animals, and fish such as Arctic char (*Salvelinus alpinus*). Polar bears also feed on terrestrial species of mammals and birds, on carcasses of bowhead and gray whales, and occasionally on humans. Polar bear males are known to kill and eat cubs of their own kind, possibly in part to incite the female to come into estrus again rapidly.

### E. Sirenians

The manatees and dugong feed on tropical grasses and roots and rhizomes in nearshore areas in saline environments and on water hyacinths (*Eichhornia* spp.), water lilies, and other vegetation in rivers and lakes. The extinct Steller's sea cow (*Hydrodamalis gigas*) was a cold-adapted species, last found off the Kamchatcka Peninsula in far east Russia; it fed on cold-water kelp.

## *See Also the Following Articles*

Baleen Feeding Strategies ■ Tactics Filter Feeding Predator–Prey Relationships.

## *References*

Benoit-Bird, K. J., and Au, W. W. L. (2001). Target strength measurements of animals from the Hawaiian mesopelagic boundary community. *J. Acoust. Soc. Am.* **110**, 812–819.

Benoit-Bird, K. J., Gilly, W. F., Au, W. W. L., and Mate, B. (2008). Controlled and *in situ* target strengths of the jumbo squid *Dosidicus gigas* and identification of potential acoustic scattering sources. *J. Acoust. Soc. Am.* **123**, 1318–1328.

Bowen, W. D. (2000). Reconstruction of pinniped diets: Accounting for complete digestion of otoliths and cephalopod beaks. *Can. J. Fish. Aquat. Sci.* **57**, 898–905.

Bowen, W. D., and Siniff, D. B. (1999). Distribution, population biology, and feeding ecology of marine mammals. *In* "Biology of Marine Mammals" (J. E. Reynolds, III, and S. A. Rommel, eds), pp. 345–384. Smithsonian Institution Press, Washington, DC.

Budge, S. M., Iverson, S. J., and Kooper, H. N. (2006). Studying trophic ecology in marine ecosystems using fatty acids: A primer in analysis and interpretation. *Mar. Mamm. Sci.* **22**, 759–801.

Casper, R. M., Jarman, S. N., Gales, N. J., and Hindell, M. A. (2007). Combining DNA and morphological analyses of faecal samples improves insight into trophic interactions: A case study using a generalist predator. *Mar. Biol.* 152, 815–825.

Clarke, M. R. (1986a). "A Handbook for the Identification of Cephalopod Beaks." Clarendon Press, Oxford.

Clarke, M. R. (1986b). Cephalopods in the diet of odontocetes. *In* "Research on Dolphins" (M. M. Bryden, and R. Harrison, eds), pp. 281–321. Clarendon Press, Oxford.

Clarke, M. R. (1996). Cephalopods as prey III. Cetaceans. *Philos. Trans. R. Soc. Lon. B.* **351**, 1053–1065.

Deagle, B. E., and Tollit, D. J. (2007). Quantitative analysis of prey DNA in pinniped faeces: Potential to estimate diet composition? *Conserv. Genet.* **8**, 743–747.

Dunshea, G., Barros, N. B., Wells, R. S., Gales, N., and Jarman, S. (2008). Pseudogenes and DNA-based diet analyses: A cautionary tale from a relatively well sampled predator-prey system. *Bull. Entomol. Res.* 98, 239–248.

Furlani, D., Gales, R., and Permberton, D. (2007). "Otoliths of Common Australian Temperate Fish: A Photographic Guide." CSIRO Publishing, Melbourne.

Härkönen, F. (1986). "Guide to the Bony Fishes of the Northeast Atlantic." Danbiu ApS, Hellerup.

Iverson, S. J., Frost, K. J., and Lowry, L. F. (1997). Fatty acid signatures reveal fine scale structure of foraging distribution of harbor seals and their prey in Prince William Sound, Alaska. *Mar. Ecol. Prog. Ser.* **151**, 255–271.

Kawamura, A. (1980). A review of the food of balaenopterid whales. *Sci. Rep. Whales Res. Ins.* **32**, 155–197.

Klages, N. T. W. (1996). Cephalopods as prey. II. Seals. *Philos. Trans. R. Soc. Lon. B* **351**, 1045–1052.

Johnson, M. P., and Tyack, P. L. (2003). A digital acoustic recording tag for measuring the response of wild marine mammals to sound. *IEEE J. Oceanic Eng.* **28**, 3–12.

Johnson, M. P., Madsen, P. T., Zimmer, W. M. X., Aguilar de Soto, N., and Tyack, P. L. (2004). Beaked whales echolocate on prey. *Proc. R. Soc. Lond., B, Biol. Sci.* **271**, S383–386.

Latja, K., and Michener, R. H. (1994). "Stable Isotopes in Ecology and Environmental Science." Blackwell Scientific Publications, Oxford.

Madsen, P. T., Johnson, M., Aguilar de Soto, N., Zimmer, W. M. X., and Tyack, P. L. (2005). Biosonar performance of foraging beaked whales (*Mesoplodon densirostris*). *J. Exp. Biol.* **208**, 181–194.

Nolf, D. (1993). A survey of perciform otoliths and their interest for phylogenetic analysis, with an iconographic synopsis of the Percoidei. *Bull. Mar. Sci.* **52**, 220–239.

Perrin, W. F. (1975). Variation of spotted and spinner porpoise (genus *Stenella*) in the eastern tropical Pacific and Hawaii. *Bull. Scripps Inst. Oreanogr.* 21, 206 pp.

Purcell, M., Mackey, G., LaHood, E., Huber, H., and Park, L. (2004). Molecular methods for the genetic identification of salmonid prey from Pacific harbor seal (*Phoca vitulina richardsi*) scat. *Fish. Bull.* **102**, 213–220.

Read, A. J. (1998). Possible application of new technologies to marine mammal research. *Mar. Mamm. Comm. Contr. Rep.*, 1–36, T30919695.

Smale, M. J., Watson, G., and Hecht, T. (1995). Otolith Atlas of Southern African Marine Fishes. *Icthyological Monographs of the J.L.B. Smith Institute of Ichthyology* **1**, 1–253.

Walker, J. L., and Macko, S. A. (1999). Dietary studies of marine mammals using stable carbon and nitrogen isotopic ratios of teeth. *Mar. Mamm. Sci.* **15**, 314–334.

Watwood, S. L., Miller, P. J. O., Johnson, M., Madsen, P. T., and Tyack, P. L. (2006). Deep-diving foraging behavior of sperm whales (*Physeter macrocephalus*). *J. Anim. Ecol.* **75**, 814–825.

Wetzel, D. L., and Reynolds, J. E., III (2004). Definitive identification of fatty acids constituents in marine mammal tissues. *Can. J. Fish. Aquat. Sci.* **61**, 554–560.

Zimmer, W. M. X., Johnson, M. P., Madsen, P. T., and Tyack, P. L. (2005). Echolocation clicks of free-ranging Cuvier's beaked whales (*Ziphius cavirostris*). *J. Acoust. Soc. Am.* **117**, 3919–3927.

# Distribution

## Jaume Forcada

## I. Distribution Patterns and Preferences

Marine mammals are found in almost all the different marine environments, and their distribution varies according to the physical, chemical, and biological characteristics of the water masses they use. The effects of oceanographic phenomena, wind-induced movements (e.g., water currents, local divergence, and upwelling areas and water fronts, thermocline depth), and the topography can be used to characterize distribution. In the case of pinnipeds, the haul-out and breeding and molting habitats on land or ice also characterize their distribution. In the polar bear, breeding and cub-rearing terrestrial habitats are also relevant.

In freshwater environments, marine mammals are found in rivers and lakes. Examples of riverine species are the river dolphins (Platanistidae, Iniidae, Lipotidae, and Pontoporiidae) and the manatees. A few Phocidae live in freshwater inland lakes: the Saimaa seal (*Pusa hispida saimensis*) in Finland, the Caspian seal (*Pusa caspica*) in the Caspian Sea, the Baikal seal (*Pusa sibirica*) in lake Baikal, and the Ungava (common) seal (*Phoca vitulina mellonae*) in freshwater lakes of the Hudson Strait.

In marine environments, distribution can be generally described as coastal (in estuarine or near shore waters), neritic (in waters on the continental shelf), or oceanic (in waters beyond the continental slope, in the open seas or oceans). Examples of marine mammals that reside primarily in coastal waters are populations of bottlenose dolphins (*Tursiops* spp.), sea otters (*Enhydra lutris*), and dugongs (*Dugong dugon*). Primarily neritic species include gray whales (*Eschrichtius robustus*), harbor porpoise (*Phocoena phocoena*), and California sea lions (*Zalophus californianus*). Primarily oceanic species include the sperm whale (*Physeter macrocephalus*) and beaked whales (family Ziphiidae). These generalizations should be used with caution, as many species occur in multiple habitats. Some species shift from one habitat to another seasonally, such as the switch from neritic feeding grounds to coastal migratory routes and breeding grounds by gray whales. Some species have populations that reside in a variety of habitats, such as the bottlenose dolphin, which occurs in coastal, neritic, oceanic, and, occasionally, riverine habitats.

Marine mammal distribution can also be classified according to general geographic areas. These are characterized by latitudinal bands and average water temperatures. Thus, marine mammals have tropical and/or subtropical, temperate, Antarctic, or Arctic distributions. Some species can be strictly included in just one of these categories, such as exclusively Arctic species [bowhead whales (*Balaena mysticetus*), polar bear (*Ursus maritimus*), narwhal (*Monodon monoceros*), and beluga (*Delphinapterus leucas*)], but, again, other species often have multiple classifications. A clear example is the baleen whales that migrate from cold high latitudes to tropical low latitudes. Some species, such as the killer whale (*Orcinus orca*) are found in all the marine waters of the world, from the equator to the Arctic and the Antarctic. Finally, similar and closely related species may occupy different latitudinal (hemispheres, ocean basins) or longitudinal (different oceans and seas) ranges. Examples of pairs of similar species that occur in different hemispheres are the northern (*Hyperoodon rostratus*) and southern (*Hyperoodon planifrons*) bottlenose whales and the northern (*Mirounga angoustirostris*) and southern (*Mirounga leonina*) elephant seals. An example of very similar cetacean species with different distribution preferences within the same ocean basin are the long- and short-finned pilot whales (*Globicephala melas* and *G. macrorhynchus*).

Detailed data on distribution are provided in the species account of this encyclopedia and therefore only overall patterns by taxa are given in this section to avoid redundancy. Additional detailed description of marine mammal distribution can be found in the chapters for cetaceans, the sea otter (*E. lutris*), pinnipeds, sirenians, and the polar bear.

### A. Cetaceans

Cetaceans live permanently in aquatic environments. They can be found in all the oceans and most of the seas of the world, and distribution patterns vary between and within families. The Balaenidae, the Balaenopteridae, the gray whale, the sperm whale, and the killer whale are found in polar, temperate, and tropical waters. They are found in the Northern and Southern Hemispheres, except gray and bowhead whales, which are only found in the Northern Hemisphere. As noted earlier, other strictly northern and also Arctic species are the narwhal and beluga. The pygmy right whale (*Caperea marginata*) is only found in the Southern Hemisphere. Most delphinids live in tropical and temperate waters of both hemispheres. More tropical Delphinidae are *S. attenuata*, *S. longirostris*, *S. frontalis*, *Steno bredanensis*, *Sotalia fluviatilis*, *Globicephala macrorhynchus*, *Pseudorca crassidens*, *Peponocephala electra*, and *Feresa attenuata*. Other tropical odontocetes are the pygmy (*Kogia breviceps*) and dwarf (*K. sima*) sperm whales, Irrawaddy dolphin (*Orcaella brevirostris*), and many Ziphiidae. Most Phocoenidae live in temperate or subtropical waters with some species exclusively in the Northern Hemisphere (*P. phocoena*, *P. sinus*, and *Phocoenoides dalli*) and some exclusively in the Southern Hemisphere (*P. spinipinnis* and *P. dioptrica*). All Delphinidae of the genus *Cephalorhynchus* live in temperate waters of the Southern Hemisphere. Of the river dolphins, the Amazon River dolphin, *Inia geoffrensis*, lives in the large lakes and tributaries of the Amazon and Orinoco basins. The franciscana (*Pontoporia blainvillei*) lives in the coastal central Atlantic waters of South America, but is commonly found in the mouth of the rivers and ocean waters surrounding estuaries. Similarly, the tucuxi is distributed in both fresh and marine waters. The two subspecies of the family Platanistidae (*Platanista gangetica gangetica* and *P. gangetica minor*) live in the major rivers of India and Pakistan, the Indus and Ganges, and the baiji (*Lipotes vexillifer*), at present considered extinct, lived in the Yangtze river and formerly lived in some of the lakes along this extremely large inland river system.

## B. Pinnipeds

Pinnipeds are amphibious mammals and spend most of their life in aquatic environments. However, they must return to land or ice for breeding (giving birth and rearing their offspring) and molt after breeding. Other possible reasons for hauling out are resting, thermoregulation, and escape from predators. Some common characteristics of nonaquatic habitats are space availability, isolation from predators, and proximity to food supply. Pinnipeds with tropical and temperate distributions find these conditions in isolated rookeries or beaches of remote places, which often are on islands. In the Mediterranean monk seal (*Monachus monachus*), terrestrial habitats are secluded spaces and caves with preferably underwater entrance. Ice characteristics condition the distribution and activity patterns of pinnipeds; pack ice offers a more constant substrate than fast ice, which varies highly seasonally in extent. Some pinnipeds reproduce in fast ice, such as the hooded seal (*Cystophora cristata*), and the leopard seal (*Hydrurga leptonyx*) and the duration of lactation and rearing of their young strongly depend on ice conditions. In general, seasonal changes in oceanographic conditions and ice cover condition the distribution of pinnipeds in the pack ice.

Among the Phocidae, geographical or latitudinal distributions include the Arctic, subarctic, and temperate areas, subtropical and tropical areas, and subantarctic and Antarctic areas. Antarctic seals are the Weddell (*Leptonychotes weddellii*), crabeater (*Lobodon carcinophaga*), leopard, and Ross (*Ommatophoca rossii*) seals. A subantarctic and Antarctic seal is the southern elephant seal. In the Northern Hemisphere, tropical and subtropical species are the Hawaiian (*Monachus schauinslandi*), Mediterranean, and the extinct Caribbean (*Monachus tropicalis*) monk seals. Subarctic and temperate-water seals are the gray (*Halichoerus grypus*), harbor (*Phoca vitulina*), and northern elephant seals. Arctic and subarctic seals are the harp (*Pagophilus*

*groenlandicus*), hooded, bearded (*Erignatus barbatus*), ringed ribbon (*Histriophoca fasciata*), spotted (*Phoca largha*), Baikal (*Pusa sibirica*), and Caspian seal. Among phocids, harp, hooded, bearded, ribbon, spotted, Ross, and leopard seals breed in the pack ice; the crabeater and ringed seals breed in pack and fast ice; the southern elephant seal breeds on land and fast ice; the Baikal and Caspian seals on fast ice; the harbor and gray seals on land and ice; and the northern elephant and the monk seals breed on land. Phocids with coastal and continental shelf distribution are the harp, harbor, gray, bearded, ringed, ribbon, spotted, Weddell, crabeater, leopard, and Mediterranean monk seals. Continental slope and oceanic seals are the Hawaiian monk, northern and southern elephant, Ross, and hooded seals. The walrus (*Odobenus rosmarus*, family Odobenidae) breeds on the pack ice and occurs in waters of the continental shelf. All the Otariidae breed and rear their offspring on land. Most of them disperse after breeding and therefore have neritic and oceanic distributions depending on season and reproductive status. Many Otariidae have subtropical or tropical distributions, such as the California (*Z. californianus*) and Galapagos (*Z. wollebaeki*) sea lions and the Guadalupe (*Arctocephalus townsendi*) and Galapagos (*A. galapagoensis*) fur seals. The Steller sea lion (*Eumetopias jubatus*) is found from Arctic to temperate waters of the eastern North Pacific. The other sea lions are distributed in tropical and subantarctic waters in the Southern Hemisphere: the Australian (*Neophoca cinerea*), New Zealand (*Phocarctos hookeri*), and South American (*Otaria flavescens*) sea lions. In the Southern Hemisphere, all the fur seals are found in temperate or subantarctic waters: the New Zealand (*Arctocephalus forsteri*), South African (*A. pusillus pusillus*), subantarctic (*A. tropicalis*), Australian (*A. pusillus doriferus*), Juan Fernandez (*A. philippii*), and South American (*A. australis*) fur seals. Only the Antarctic fur seal (*A. gazella*) can be strictly considered subantarctic and Antarctic. In the Northern Hemisphere, the northern fur seal (*Callorhinus ursinus*) is found in the subarctic and temperate North Pacific.

## C. Sirenians

All the Sirenia are found in tropical or subtropical waters. The manatees have restricted ranges in different oceans and river systems. The West Indian manatee (*Trichechus manatus*) is found from southern North America and the Caribbean to northern South America in the western Atlantic, and the Amazon manatee (*T. inunguis*) in the Amazon drainage. In the eastern Atlantic, the African manatee (*T. senegalensis*) is found in western Africa, from Senegal to Angola. Manatees are coastal, although they may be found in continental shelf waters, transiting between islands, in the Caribbean. The dugong (*Dugong dugon*) is the most widely distributed sirenian, in the Indian and the western Pacific oceans, with a preference for shallow coastal bays.

## D. Polar Bear and Sea Otter

The polar bear has a circumpolar distribution, mostly above the Arctic circle. It uses coastal, neritic waters, and breeds and rears its offspring on ice, inside snow lairs. Ice is also important for polar bears as a platform to travel, especially in the ice floes, between foraging areas and areas where they give birth and rear their young and as a substrate to hunt seals. The time of spring sea-ice breakup is important for polar bear feeding. After hybernation, bears start looking for food for them and their offspring, and an important component of the diet are seals. Sea ice is essential habitat for seals. Without it the seals move to other places, restricting the bear's food supply. The sea otter is found in the Pacific coasts of North America and Russia, essentially in temperate and subarctic waters. It lives

D

near shore and comes ashore on Aleutian Islands. Its distribution is conditioned by predators (e.g, killer whales) and food availability, such as the prey they usually use in kelp forests (e.g., sea urchins and abalone).

## II. Factors Affecting Marine Mammal Distribution

Marine mammal distributions are affected by demographic, evolutionary, ecological, habitat related, and anthropogenic factors. Demographic factors include the abundance, age, and sex structure of the populations and the reproductive status and life cycle of individuals. Evolutionary factors include morphological, physiological, and behavioral aspects of the species' adaptations. Ecological factors include biological production and use of prey, distribution of prey and predators, and competitors. Habitat includes factors such as water temperature, salinity, density, thermocline depth, and the type of substrate and the bathymetry. Anthropogenic factors are the human effects that alter the natural distribution of marine mammals, including pollutants, human-induced sounds, habitat encroachment, modification by global warming (sea-ice habitat), and incidental and direct kills. Distribution is the product of factors that act in a parallel or interactive way over different scales of space and time on each species, and sometimes on groups of species. As an example, baleen whale distribution depends on their ability to exploit planktonic organisms (evolutionary), the oceanographic characteristics of the water masses where they feed (habitat), and the trophic level they exploit (ecological).

### A. Demographic Factors

The dynamics of marine mammal populations can determine distribution changes and patterns. The number of individuals that live in particular areas depends on the capacity of those areas to sustain their biological requirements. In general, the most critical requirements are prey availability and energy. The depletion of food resources by marine mammal populations influences the movement or dispersal to other areas. The age and sex structure of marine mammal aggregations also affect the distribution patterns. Habitat requirements for breeding females or females with offspring are not the same as those of adult males. In the case of odontocetes, females with calves may require coastal areas with locally abundant food resources and protection from predators. Adult males, not having to care for their offspring, are less limited in movements and distribution range. In offshore dolphins, large cohesive aggregations may be required by breeding females for protection in the open ocean, and foraging distances will be greater due to patchiness of their prey. In the case of pinnipeds, distributional differences according to age and sex classes and reproductive status are related to the seasonality of their life cycles, their adaptation to aquatic feeding, and their need to periodically return to land to breed. In breeding colonies, individuals will gather seasonally to mate, give birth, and nurse their pups over variable periods of time according to species. After the breeding season, pinnipeds often display age- and sex-related differences in habitat use and foraging areas. Dispersal according to age and sex classes is often associated with these characteristics.

### B. Evolutionary Factors

All factors related to the secondary aquatic adaptation of marine mammals influence their distribution to some extent. Diving capacities in terms of duration and maximum depths allow particular species to exploit different habitats. In sperm whales and elephant seals, deep diving allows access to prey unavailable to the shallower diving dolphins or porpoises. Hence, their distributions are associated with deep canyons and other deep ocean areas. In sperm whales, this ability also requires complex social systems that insure the protection of newborns or youngsters, particularly while mothers spend long times underwater in search of prey. Another notable physiological adaptation is thermoregulation, which allows marine mammals to extend their distribution ranges from the warm equatorial waters to the coldest high latitudes. Efficient insulation and body temperature regulation systems allow the polar bear and the sea otter to spend a substantial part of their life at sea and survive in cold waters. The relative inability to regulate body temperature adequately in colder water of neonates is a hypothetical factor that leads baleen whales to migrate from the cold feeding grounds to the warmer calving grounds. Morphological adaptations, such as the feeding apparatus of baleen whales, also influence their distribution. As active filter feeders, they can capture planktonic (e.g., copepods, krill) or schooling (e.g., sand lance, capelin, herring) prey, which are abundant in the particular areas where whales distribute. Finally, the cohesiveness of large dolphin schools and the sensorial integration of individuals allow them to range in offshore areas, find food actively and efficiently, and obtain protection from predators.

### C. Ecological Factors

Marine mammal distribution is in great measure related to prey distribution. The ability to exploit different trophic levels and resources classifies different marine mammals from top predators, such as the killer whales or leopard seals, to low-trophic level feeders, such as northern right whales or manatees. Marine mammals can be considered as either specialists or generalists, and these two aspects imply differentiated distribution patterns, although it is very likely that it is an abundant prey source like copepods for northern right whales and Antarctic krill for many whales migrating seasonally to the Southern Ocean that renders them specialists. Manatees, being specialist feeders, have restricted distributions where sea grass meadows provide continued food. Despite often being categorized as specialists, odontocetes or phocids tend to use a wide range of prey items. Thus, they can be distributed over wider ranges and change their distribution seasonally according to the availability of their prey. The killer whale, as a species, has a broad diet, yet different populations have more specialized diets, transient killer whales feed mainly on pinnipeds and other marine mammals but must range widely to maintain this diet, whereas resident killer whales feed on large fishes such as salmon. In both instances, the distributions of the whales are synchronized to the life cycles of their prey. In the eastern North Pacific, the transients concentrate seasonally near pinniped rookeries, whereas residents live near the mouth of salmon-spawning rivers. In other cases, marine mammals tend to use the same home range, such as coastal bottlenose dolphins, feeding on different prey species that change their distribution seasonally. In this case, distribution patterns must be studied and interpreted at a finer scale. Interspecific competition is an additional ecological factor determinant of variable distribution. Violent attacks on harbor porpoises by bottlenose dolphins have been reported in their common range in the North Sea. Finally, predation plays an important role in the selection of habitats and distribution areas by marine mammals, especially those of smaller size, such as ringed seals. This species appears to select the fast ice to avoid predation by polar bears.

### D. Habitat-Related Factors

Marine mammals are usually found in waters with high densities of principal prey species. These waters are characterized by the physical conditions that facilitate the accumulation of the prey. Relevant oceanographic variables characterizing marine mammal habitats are water temperature, salinity, density, chlorophyl concentration, and thermocline depth. These characteristics are related to upwelling fronts, often related to differences in species distribution. As an example, spinner and spotted dolphins range in the same areas of the eastern tropical Pacific, often traveling in the same schools. They occur in the same overall ocean area as common and striped dolphins but appear to have preferences for water masses of different oceanographic characteristics (Reilly, 1990). Ocean topography and bathymetry are often related to local oceanographic phenomena that influence marine mammal distribution. Underwater canyons, marine ridges, and irregular topographies concentrate prey for deep divers such as sperm whales or elephant seals. In contrast, mysticetes often have preferences for shallow waters (Elwen and Best, 2004) and sometimes with high topographic variation. In these waters their prey accumulates at frontal interfaces between mixed and stratified waters. Temperature is also an important factor limiting waters of specific characteristics. For instance, warm water bounderies like convergence zones are important for northern right whales (Keller *et al.*, 2006). The ice is also a critical habitat element for marine mammals; the seasonal and highly dynamic changes of ice cover determine their patterns of change in distribution. It provides shelter during reproduction for pinnipeds, access to seasonally abundant food, and also delimits the distribution ranges of some cetaceans, such as the bowhead whale.

## E. Anthropogenic Factors

Human alteration of habitats can change marine mammal distributions significantly. Marine mammals that haul out on land are particularly affected by habitat encroachment by human development. The three species of monk seals have suffered substantial changes in their original distributions, and one of them, the Caribbean monk seal, became extinct because of this. In the case of the Mediterranean monk seal, a major change in habitat preferences occurred as a result of human development but also of deliberate kills for human uses. The seals changed their haulouts from open beaches to difficult-to-access caves, often with underwater entrance. This has created severe habitat fragmentation. Commercial exploitation has also affected marine mammal distributions greatly. Whale stocks were reduced to the point that many original distribution areas are not used anymore. Overfishing of prey items has led to changes in marine mammal distributions. Pollution of coastal areas has degraded many original marine mammal habitats, thus affecting their original distributions. Human-induced changes in local water temperature have changed the seasonal distribution of the Florida manatee (Reynolds and Odell, 1991). This population previously migrated to warmer waters in winter but now uses the thermal vents in waters close to power plants and has changed migration patterns substantially. Expanding sources of sound in the ecosystem (e.g., large ship traffic, naval experiments) and pollution may also affect marine mammal distribution. A recently new indirect human impact is global warming, which is altering the distribution of species very attached to critical habitats. For example, the dependence of polar bear in certain areas of the seasonal sea ice to access their prey has been affected by changes in the air-temperature–sea-ice relationships. These have caused an earlier spring sea-ice breakup and late sea-ice formation in autumn. In addition, the more extensive melting of sea ice in certain regions may extend the range of predatory species like killer whales, giving them access to new prey sources. In contrast, the contraction of sea ice brings about reductions in prey,

like the Antarctic krill in the South Atlantic sector of the Southern Ocean, which have repercussions for distribution and abundance of Antarctic fur seals and crabeater and leopard seals, and possibly many baleen whale species.

## III. Movements and Seasonality

The distribution of marine mammals changes seasonally as their biological and ecological requirements change. Marine mammals respond to changes in the environment, such as in temperature, ice coverage, and prey availability. Daily requirements in terms of energy or protection against predators depend on the reproductive status and the season; these are clearly not the same for females with nursing offspring as for solitary males. Movements are a response to changes in the environment and the biological requirements of a species. In tropical areas, movements are expected to vary according to the patchiness of the environment. Distances covered in short periods of time may vary depending on the conditions, but a very marked seasonality is not commonly found. In high latitudes, changes during the cold winter affect the distribution of marine mammals, their tolerance to physical conditions, and their life history requirements. Thus, seasonality is more marked.

Movements can be classified as migration, dispersal, and daily travel. Migration is the seasonal change between two geographic locations that is related to species reproductive cycle, changes in the physical environment, like water temperature or ice coverage and extent, and prey availability. Dispersal is the movement from the place of birth to other areas in which individuals reach a feeding area, join a breeding population, or find another group of individuals with which to spend the next stage of its life. The classification of movements may be somewhat arbitrary because marine mammals do not always follow strict periodic patterns. They instead respond to the limitations of the environment in providing constant food or other requirements. Short-scale movements are difficult to detect and must be put in the context of the species life cycle before being classified. A typical example of migration is the one of baleen whales; humpback, right, and gray whales undertake long-distance travels, often thousands of kilometers, between the tropical calving grounds in winter and the high-latitude feeding grounds in summer. In contrast, most otarids have dispersal movements from their birth colonies toward different feeding areas or other breeding colonies when they reach sexual maturation. In any case, movement patterns vary among individuals, according to their age, sex, and reproductive condition. A prereproductive young whale may delay its departure from the high-latitude feeding grounds to extend the feeding season, whereas a pregnant female must leave for the low-latitude calving grounds to give birth to its offspring.

### A. Cetaceans

Cetaceans spend their entire life in aquatic habitats and are in constant movement. Understanding their seasonal distribution is more difficult than in pinnipeds for technical and logistical reasons, the manipulation and tagging of animals is less efficient and more expensive. Thus, classifying movements as dispersal or migration is even more confusing, except in some well-studied populations of baleen whales. Their life patterns and cycles make the concept of dispersal a little ambiguous, however, because seeming residency, site fidelity, or habitat discreteness may be just apparent, short-term attributes of their distribution. Only migration in large whales is known from long-term studies. Studies on migration range from the examination of catch statistics of whaling operations to the use of modern telemetry technology.

**D**

Contrary to classic accounts of whale migration, the most recent studies show how movements vary across whale populations and species. Mysticetes appear to have periodical migrations with relatively consistent patterns over the years. Seasonal movements in odontocetes are far less consistent over time, including those of the sperm whale, which has been classified as a migratory species with marked seasonal patterns. However, only males appear to be involved in the long latitudinal migrations between the tropics and the poles, while females, calves, and young stay year round in tropical areas. In general, as in other marine mammals, factors inducing migration are the biological cycle, greatly determined by reproductive needs, and factors in the environment (e.g., prey availability, changes in water temperature). These factors may trigger the start of seasonal movements, although not all individuals will respond in the same way.

Annual migrations are best known for species with more coastal ranges, such as the gray, right, or humpback whales. However, virtually all mysticete species are known to migrate. No data are available for the pygmy right whale. Most mysticetes have latitudinal migrations, from tropical breeding grounds to high(er) latitude feeding grounds. In breeding grounds, mating and calving take place. Migratory species are right, blue (*Balaenoptera musculus*), fin (*B. physalus*), sei (*B. borealis*), humpback, and gray whales. Bowhead whales also migrate, but their longitudinal movements are equal to or greater than their latitudinal changes, and they never leave Arctic waters. Bryde's (*Balaenoptera edeni*), common minke (*B. acutorostrata*) and Antarctic minke (*B. bonaerensis*) whales, however, have less clear movement patterns. Bryde's whales, often confused at sea with sei whales, spend most of the year in warm tropical waters and calving does not have the same marked seasonality seen in other balaenopterids. This indicates a possibly different reproductive cycle, in which whales feed and mate year round. In this case, whale movements are more similar to those of many odontocete species, in constant search for food, with variable utilization of prey, and different prey types through the year. Among the best known migrations are those of the gray and humpback whales. Gray whales migrate annually from feeding grounds in the Arctic to their calving areas in the lagoons of Baja California in Mexico. Interannual changes in the timing and numbers reaching the different migratory destinations have been observed. The migration of humpback whales is also very well studied, and male sperm whales are the best example of long-range migration in odontocetes.

Movements in odontocetes have different scales depending on geographical areas, family, and species. It is generally accepted that most movements are in response to prey availability, and the largest movements, often called migration, occur in oceanic odontocetes. In the eastern tropical Pacific, movements are reported to be wide. Several species of *Stenella* had daily movements of 53 km/day and hundreds of kilometers over months, and these reflected seasonal changes in distribution. Dolphins moved inshore, toward the American continent, in fall and winter and offshore in spring and summer. Other methods, such as line transect surveys in California, have shown how several dolphin species have different patterns of abundance and distribution depending on the season. Pacific white-sided dolphins (*Lagenorhynchus obliquidens*), Risso's dolphins (*Grampus griseus*), common dolphins (*Delphinus* spp.), and northern right whale dolphins (*Lissodelphis borealis*) were less abundant in summer than in winter, and significant north/south shifts in distribution were reported for Dall's porpoises and common and Pacific white-sided dolphins. Significant inshore–offshore differences were found for the northern right whale dolphin. Some dolphin species show variable distribution patterns, such as bottlenose dolphins or killer whales. Difference in patterns has been attributed to different varieties or ecotypes of the same species. In the case of bottlenose dolphins, a well-studied coastal population (Scott *et al.*, 1990) showed a year-round residency with slight seasonal changes within the population home range. It has been argued that dolphins in Florida follow the mullet migrations into the Gulf of Mexico during the fall. Short-term movements have also been observed in a resident bottlenose population in east Scotland. In contrast, Atlantic offshore bottlenose dolphins, described as a possible different form, have wider movements and a broader distribution (Wells *et al.*, 1999).

## B. Pinnipeds

Migration is not uncommon in pinnipeds, and the advent of new telemetry has helped describe the migratory movements of several species. Dispersal is very common in pinnipeds and depends on the abundance of prey, its energy content, and the seasonality of prey distribution. In addition, their reproductive cycle mandates that individuals return to land or ice to give birth, nurse, and rear their offspring and molt. Pinnipeds also haul out for resting, thermoregulation, and to escape predators, among other reasons. If the environment provides constant food resources, such as in some tropical areas, there will not be a clear need to disperse. In contrast, pinnipeds living in high latitudes will be more dependent on ice cover, availability of seasonally changing prey, reproduction, and population size. These will create density-dependent effects, such as dispersal and distributional changes. Thus, dispersal can vary with latitude, based on the stability of prey resources.

Phocids appear to migrate more than otariids, as they generally live in higher latitudes, where the environment (e.g., the ice cover) is more variable. For otariids that live in tropical areas with a more constant environment, habitat regulates the growth of the standing colonies and conditions of dispersal. Otariids also have longer lactations and rearing periods than true seals. Their breeding behavior and requirements in terms of habitat are also different, allowing them to stay longer in the breeding colonies. Periodic events that lead to drastic changes in food availability or other environmental limitations, such as the El Niño southern oscillation (Trillmich and Ono, 1981), also favor dispersal. In Antarctic fur seals of the South Atlantic, the lagged effects of El Niño-Southern Oscillation (ENSO) reduce prey availability and affects the life history of breeding mothers. This promotes breeding abstention, low return of breeders, and causes low survivorship among mothers after severe El Niños. Among otariids, only the northern fur seal has a well studied and distinctive migration.

Both elephant seal species and the hooded and the harp seals are good examples of migratory seals. Northern and southern elephant seals spend between 8 and 10 months at sea each year, with long-distance migrations between breeding and molting sites. Both species have two long migration trips between postbreeding and postmolting areas. The northern elephant seal migration, of between 18,000 and 21,000 km, is the longest reported for any mammal.

Harp and hooded seals have interannually variable distribution patterns, dependent on the time of the year, the geographic location, and the density of individuals in the breeding colonies. Harp seals live in colonies in the subarctic pack ice, where breeding takes place. The largest population of this species is in Newfoundland. Individuals from this population start their southward migration in late September along the coast of the Baffin Islands and go eastward through the Hudson Straits, reaching Labrador between October and December. There are variations in migration timing and patterns between age classes.

Gray seals from numerous colonies in the British Isles and gray seal pups disperse widely during the first year. Adult seals show high variability in their movements along the coasts of Scotland, especially in postbreeding periods and after the molt, from March to May. Although long-distance travel by adults occurs, short travel at close range from particular haul outs is more common. Juveniles tend to spend longer periods at sea (Hammond *et al.*, 1993).

### C. Sirenians

The best-studied movements by sirenians are of manatees in waters of Florida. Water temperature is a major determinant of seasonal movements of Florida manatees, and dispersal is higher in warmer months. In winter, manatees tend to aggregate in areas of warmer waters, such as natural freshwater springs or the outfalls of power plants (Reynolds and Odell, 1991).

### D. Polar Bear

Seasonal movements in polar bears have been reported in all their distribution range. Long-range movements also occur and are mostly related to the ice cover and extent. Predation on seal pups also influences movements, and bears disperse more during pinniped pupping seasons. In summer, when ice melts in many areas, bears move to land, where they remain for a few months, before leaving in November–December. Pregnant females stay longer on land than males.

## IV. Study of Marine Mammal Distribution

The study of distribution depends on each species' habitat and its abundance, so that scale is a significant factor. Distribution changes have to be interpreted in space and time, and different methods are to be used according to species range and density. The distribution of a species occupying an extensive ocean area is best studied by air or shipboard surveys following systematically placed transects. In surveys, visual and/or acoustic data on species are collected according to predetermined protocols. Oceanographic variables and data on position of individuals can be incorporated in spatial modeling (Hedley *et al.*, 1999). Results of this modeling on repeated surveys can be compared to study seasonal patterns and changes over time. In species that live in fragmented habitats and that are not abundant, knowledge of the location of animal aggregations is essential. In these cases, the best possible information is obtained from telemetry studies using high-frequency radio tags, satellite-linked radio tags, and geolocation time-depth recorders, GPS tags, and geolocation tags. The use of telemetry devices is also essential in understanding seasonal movements and patterns. The life of the batteries and permanence of the tags in the animals are critical to the duration of the studies. The study of habitat, an integral part of distribution, changes with the species life time because of the above-mentioned factors. It is difficult to monitor a cohort of animals, ideally tagged since birth, because of the long average life span of marine mammals. In practice, the general distribution patterns of a marine mammal population are the sum of the individual specific movements over space and time. Monitoring just a few animals over a restricted time duration (e.g., that of a telemetry device battery) produces partial information on the overall patterns and may show a high variability between individuals. Therefore, inferences at the population level must be made cautiously to avoid biased perceptions of the species distribution.

## *See Also the Following Articles*

Cetarean Ecology ■ Pinniped Ecology

## *References*

Bowen, W. D., and Siniff, D. B. (1999). Distribution, population biology, and feeding ecology of marine mammals. *In* "Biology of Marine Mammals" (J. E. Reynolds, and S. A. Rommel, eds), pp. 423–484. Smithsonian Institution Press, Washington.

Elwen, H. S., and Best, P. B. (2004). Environmental factors influencing the distribution of southern right whales (*Eubalaena australis*) on the south coast of South Africa II: within bay distribution. *Mar. Mamm. Sci.* **20**, 583–601.

Forney, K. A., and Barlow, J. (1998). Seasonal patterns in the abundance and distribution of California cetaceans, 1991–1992. *Mar. Mamm. Sci.* **14**, 460–489.

Gaskin, D. E. (1982). "The Ecology of Whales and Dolphins." Heineman Educational Books, London.

Hammond, P. S., McConnell, B. J., and Fedak, M. A. (1993). Grey seals off the east coast of Britain: Distribution and movements at sea. *Symp. Zool. Soc. (Lond.)* **66**, 211–224.

Hedley, S. L., Buckland, S. T., and Borchers, D. L. (1999). Spatial modelling from line transect data. *J. Cetacean Res. Manage.* **1**, 255–264.

Keller, C. A., Ward-Geiger, L. I., Brooks, W. B., Slay, C. K., Taylor, C. R., and Zoodsma, B. J. (2006). North Atlantic right whale distribution in relation to sea-surface temperature in the southeastern United States calving grounds. *Mar. Mamm. Sci.* **22**, 4216–4445.

Reilly, S. B. (1990). Seasonal changes in habitat and distribution differences among dolphins in the eastern tropical Pacific. *Mar. Ecol. Prog. Ser.* **66**, 1–11.

Reynolds, J. E., III., and Odell, D. K. (1991). "Manatees and Dugongs." Facts on File, New York.

Riedman, M. L. (1990). "The Pinnipeds: Seals, Sea Lions, and Walruses." University of California Press, Berkeley, CA.

Riedman, M. L., and Estes, J. A. (1988). A review of the history, distribution and foraging ecology of sea otters. *In* "The Community Ecology of Sea Otters" (G. R. Van Blaricom, and J. A. Estes, eds), pp. 4–21. Springer–Verlag, Heidelberg.

Scott, M. D., Wells, R. S., and Irvine, A. B. (1990). A long-term study of bottlenose dolphins in the west coast of Florida. *In* "The Bottlenose Dolphin" (S. Leatherwood, and R. R. Reeves, eds), pp. 235–265. Academic Press, San Diego.

Trillmieh, F., and Ono, K. A. (1981). "Pinnipeds and El Niño: Responses to Environmental Stress," Ecological Studies 88. Springer–Verlag, Berlin.

Wells, R. S., Rhinehart, H. L., Cunningham, P., Whaley, J., Baran, M., Koberna, C., and Costa, D. P. (1999). Long distance offshore movements of bottlenose dolphins. *Mar. Mamm. Sci.* **15**, 1098–1114.

Wiig, O., Born, E., and Garner, E. W. (1995). "Polar Bears: Proceedings of the Eleventh Working Meeting of the IUCN/SSC Polar Bear Specialist Group." IUCN, Cambridge, UK.

# Diving Behavior

## BRENT S. STEWART

Except for the polar bear (*Ursus maritimus*), all marine mammals feed exclusively in aquatic environments, and mostly in the world's oceans (Reeves and Stewart, 2003). The depths at which they hunt for and capture prey and the time spent submerged vary among pinnipeds, cetaceans, sea otters, and sirenians as a function of physical and physiological adaptations among these taxa, environmental conditions (e.g., coastal or pelagic, tropical or polar, season), and body size, age, and health of individuals. All are ultimately tied to the sea surface to periodically breathe, yet natural selection has operated to minimize the time needed there and to maximize the amount of time that can be spent submerged hunting and capturing prey. What has become known in recent years is that these animals spend substantial parts of their lives moving within

**D**

the water column to relatively great depths and some over vast geographic areas in search of food. Among the amphibious pinnipeds, these aquatic foraging bouts can extend, with minor interruptions, for several weeks to several months, punctuated by periods of several days to weeks on land or ice when no feeding occurs when these animals rest, molt, or breed. For the less amphibious sea otters (*Enhydra lutris*), diving and foraging periods may be separated by periods spent sleeping or resting at the sea surface rather than on land. Among the wholly aquatic cetaceans and sirenians, foraging bouts may last several hours or perhaps days, interrupted by periodic resting periods at the sea surface.

Individuals of some species, particularly sperm whales (*Physeter macrocephalus*) and many mysticete cetaceans, evidently fast during migrations or in particular breeding areas. Although the diving performance and the patterning of individual dives or sequences of dives vary among species, what has become apparent for all marine mammals is that little time is spent at the surface between successive dives to exchange gases (i.e., unload carbon dioxide from tissue and blood and restore tissue oxygen stores). This allows for sustained, repetitive diving and hunting, and is made possible by physiological adaptations for conserving heat and oxygen and by anatomical adaptations that promote effective movement in the aquatic environment (e.g., reducing drag through streamlining and efficient propulsion mechanisms; Reeves and Stewart, 2003).

## I. Methods of Studying Diving Behavior

The simplest method for studying the diving behaviors of marine mammals is direct observation of the timing and location of appearances of individuals at the sea surface, the number of breaths taken there, and the duration of the animal's disappearance under water before reappearing. With some assumptions and strong inference, much can be deduced about what animals are doing while hidden beneath the ocean surface. Indeed, most early knowledge of diving, feeding, traveling, and migratory behaviors was based on such interpretations.

Other techniques for documenting diving behavior have used radio transmitter and telemetry instruments, operating at various radio frequencies. Sonic transmitters, operating at relatively low frequencies or wavelengths, allow the tracking of animals when they are submerged by placing a microphone (hydrophone) beneath the sea surface to listen for and orient to these signals. Higher frequencies are used for in-air detection and tracking but generally yield less detailed observations, mostly when an animal reached the surface and how long it spent there. Durations of dives are inferred from periods of radio silence, as transmissions that occur when the animal is submerged will rapidly be attenuated in salt water. When vocalizing underwater, some marine mammals may also be tracked with hydrophones to detect and localize those sounds. All of these techniques require constant tracking and observers must be within a few hundred meters (surface observers) or kilometers (observers in aircraft), as the signals attenuate quickly.

During the past several decades, and in particular since the early 1990s, an enormous amount of information has been added to those simple observations due to technological developments and their application to free-ranging marine mammals. For example, in the late 1960s and the early 1970s, an encapsulated mechanical device was used in the Antarctic to study the diving patterns of Weddell seals (*Leptonychotes weddellii*). That instrument provided a continuous trace on photographic film of the depth of the seal vs time. The

spooled film was pulled at a known rate past a small radioactive particle, which rested on a pressure-sensitive arm. Thus a two-dimensional record was made on the film of depth vs time. From these records came the first long-term (about 7 days continuous, based on film capacity) data on the vertical movements of free-ranging marine mammals (Kooyman, 2006). Those instruments, called time-depth recorders (TDRs), were later deployed on a number of species of fur seals and some sea lions to study the effects of variation in body size and environment on the foraging patterns of lactating females. However, because the instruments were rather large and because they were attached with harnesses, they likely had some influence on the recorded durations of dives because of the effects of drag on swimming that they imposed, particularly for fur seals. Other simple instruments used capillary tubes with pressure sensors attached to record the maximum depth of a single dive or the maximum depth achieved during a period of diving.

Mechanical instruments were replaced in the late 1980s with much smaller electronic instruments, armored to keep seawater out under extreme hydrostatic pressures. These instruments could collect and store substantially more data on depth and duration of dives and also had less impact on behavior. Indeed, today most of these instruments weigh less than 50 g and can be glued (Figs 1 and 2) to the hair or fur of pinnipeds for long-term (up to a year) monitoring, attached to the dorsal fin of small cetaceans (Fig. 3), attached to the skin surface of large whales with subdermal anchors or deeply embedded into their blubber, or attached with suction cups to the skin of cetaceans for shorter term (to several days) study. Because these instruments may now also collect data other than just water depth as a function of time (e.g., swim speed, ambient light level, compass bearing, seawater temperature, salinity), they are called time-data recorders. These instruments are generally controlled by small microprocessors that can be programmed to record measures of various parameters at particular intervals that are then stored in electronic memory for several months or more. Thus, detailed records (e.g., at 1-sec intervals) of a marine mammal's position in the water column, in addition to other environmental and behavioral data, can be collected continuously for months or more.

Even more recently, technological developments and improvements have involved remote sensing of diving patterns and geographic movements of marine mammals using radio transmitters that communicate with earth-orbiting satellites, most notably the

**Figure 1** *A satellite-linked data recorder (SLDR) glued to the dorsal pelage of an adult male ribbon seal* (Histriophoca fasciata). *Photo by B. S. Stewart.*

Figure 2 *A satellite-linked data recorder (SLDR) glued to the dorsal pelage of a northern fur seal* (Callorhinus ursinus). *Photo by B. S. Stewart.*

Figure 3 *A satellite-linked data recorder attached to the dorsal fin of a short-beaked common dolphin* (Delphinus delphis). *Photo by B. S. Stewart.*

Figure 4 *Gluing a satellite-linked data recorder to a physically restrained ribbon seal* (Histriophoca fasciata). *Photo by B. S. Stewart.*

two polar-orbiting satellites of the ARGOS Data Collection and Location System (DCLS). These transmitters are known as platform transmitter terminals (PTTs) and packaged instruments that include microprocessor-based data recorders in addition to the PTT are known as satellite relay data recorders (SRDRs). They allow animals to be located several times each day, and also allow small amounts of behavioral and environmental data to be transmitted through the DCLS. Further continuing improvement and miniaturization of film and digital video equipment are allowing the underwater diving, social, and hunting behaviors of marine mammals to be visually documented (Parrish and Littnan, 2008).

Most of what is now known and summarized below on the diving behaviors of marine mammals is based on two-dimensional (i.e., depth vs time) data from electronic TDRs, which are occasionally supplemented by geographic locations of the animals at the sea surface. Some data have been collected recently on the movements of animals in a three-dimensional ocean space beneath the sea surface for several hours to a couple of days (Harcourt *et al.*, 2000; Simpkins *et al.*, 2001). But the seductiveness of representations of a single spatial vector (depth) vs time as a trace in a two-dimensional, linear

spatial format has led some researchers to infer the geographical form of large numbers of dives in three-dimensional space. Moreover, some researchers have extended inferences even further to assign physiological and behavioral function to those guessed three-dimensional spatial forms. Though those inferences and conclusions of function have yet to be substantively validated, they are nonetheless interesting hypotheses for further rigorous inspection (*cf.* Brillinger and Stewart, 1997; Fedak *et al.*, 2001).

A substantial amount of information has been collected on diving patterns of a number of pinniped species compared to relatively little progress in the study of cetacean diving patterns. The primary reason for the difference in quantity and quality of data between these taxa is principally due to the greater difficulty of keeping instruments attached to cetaceans compared to the long-term attachment of instruments, up to 1 year, to pinnipeds by gluing them to their hair (Fig. 4). Regardless, the dive patterns of virtually all species were limited to particular times of the year and even to particular classes of individuals (e.g., lactating female pinnipeds). Nearly year-round monitoring of northern and southern elephant seals (*Mirounga angustirostris* and *M. leonina*, respectively) has been the exception. Consequently, any discussion of diving patterns is conditioned on these important constraints. Moreover, it has not been confirmed for all cases whether hunting or feeding occurs whenever animals are submerged and diving. The incorporation of additional environmental and physiological sensors to TDRs and PTTs will likely help refine studies of diving patterns to more rigorously evaluate spatial form and function of subsurface movements and to enhance the summaries of dive patterns presented here.

## II. Pinnipeds

### A. Otariids

California sea lion females (*Zalophus californianus*) dive mostly to depths of around 75 m for about 4 min during summer and then deeper and longer the rest of the year (maximum depth of 536 m and longest dive of 12 min). When at sea for several days at a time and up to 1–2 weeks at some seasons, California sea lions dive virtually continually and rest at the surface for only about 3% of the time.

Juvenile Steller sea lions (northern sea lion, *Eumetopias jubatus*) dive to average depths of 21 m (maximum 200 m). Most dives last less than 2 min. They are generally shallower at night and deeper in spring and summer than in winter. Adult females dive deeper than juveniles and dives are deeper in winter than in autumn (Pitcher *et al.*, 2005).

Southern sea lions (*Otaria flavescens*) dive mostly at night, apparently to the sea bed, where they hunt at depths down to 250 m. While at sea near the Falkland Islands, these sea lions dive virtually continually. Near Patagonia along the Argentina coast, over half of the time that lactating females are at sea they are diving. Their dives are mostly to depths of 19–62 m (maximum of 97–175 m) and for 2–3 min (maximum of 4.4–7.7 min). Diving is continuous during these bouts, and time spent at the surface between successive dives is brief, around 1 min.

Lactating New Zealand sea lions (Hookers sea lions, *Phocarctos hookeri*) also dive almost continually when at sea, averaging about 7.5 dives per hour, varying little with time of day. Dive depths average about 123 m (maximum of 474 m) and last between 4 and 6 min (maximum of 11.3 min). Most dives are evidently to the sea bed to forage on demersal and epibenthic fish, invertebrates, and cephalopods.

A few lactating Australian sea lion (*Neophoca cinerea*) females were reported to repeatedly forage on the sea bottom (~150 m deep) on the continental shelf of South Australia within 30 km of the coast.

Northern fur seals (*Callorhinus ursinus*) can be at sea continuously for several months or more from autumn through spring, but their diving behavior has not been studied then. Most data come from lactating female fur seals that are foraging near ROOKERIES in the Bering Sea in summer. Then they forage in bouts that mostly occur at night. Seals mostly make shallow dives to depths of 11–13 m, lasting around 1–1.5 min (Baker and Donohue, 1999). These dives tend to be at night when seals are in pelagic habitats.

Depths and durations of dives of Galapagos fur seals (*Arctocephalus galapagoensis*) increase as they get older. Six-month-old seals dive to depths of around 6 m for up to 50 sec and dives occur at all hours. One-year-old seals reach depths of 47 m and durations average 2.5 min. Most of those dives occur at night. When 18 months old, seals are at sea mostly at night, diving continually for periods lasting around 3 min and reaching depths of 61 m (Horning and Trillmich, 1997).

Lactating Juan Fernandez fur seal (*Arctocephalus philippii*) females dive mostly at night to depths of 50–90 m, although most dives are shallower than 10 m. They last, on average, 1.7–2.0 min (longest 3.46 min).

Lactating female New Zealand fur seals (*Arctocephalus forsteri*) dive as deep as 274 m, and their longest dives have been measured at around 11 min. Median dive depths are around 5–10 m. They occur in bouts with the longest bouts at night. The deepest dives occur around dawn and dusk. Dives are shallowest (30 m) and shortest (1.4 min) in summer and get progressively deeper and longer through autumn (54 m, 2.4 min) and winter (74 m, 2.9 min; Mattlin *et al.*, 1998).

Most dives of female Australian fur seals (*Arctocephalus pusillus doriferus*) are to the sea bed on the continental shelf at depths of 65–85 m. The median depth of one foraging male fur seal was 14 m and the median duration of dives was 2.5 min. The deepest dive was to 102 m and the longest was 6.8 min, and the seal spent about one-third of its time at sea diving and foraging, with little variation in activity with time of day.

Lactating female subantarctic fur seals (*Arctocephalus tropicalis*) at Amsterdam Island dive predominantly at night. These foraging dives get progressively deeper and longer from summer (10–20 m and about 1-min long) through winter (20–50 m and about 1.5-min long). The deepest dive recorded was 208 m and the longest was 6.5 min (Georges *et al.*, 2000).

Lactating female Antarctic fur seals (*Arctocephalus gazella*) dive mostly at night when they are at sea for periods of 3–8 days at a time in summer. Those dives are shallower at night (to ~30 m) than dives made during the day (40–75 m), closely matching the vertical distribution of krill. Maximum depths and durations of dives have been measured at 82–181 m and 2.8–10 min, respectively, for individual females. Seals apparently adjust their diving behavior to maximize the proportion of time that they spend at depth. Young pups dive mostly to depths of about 14 m, depending on their body size, for mean durations of 20 sec. Their diving abilities continue to develop during their first couple of months of life, and by the time they are weaned at around 4 months of age, they are able to dive to the same depths and for about the same amount of time as adult females.

## B. Odobenids

The diving patterns of the walrus (*Odobenus rosmarus*) are not well studied. It is known, however, that its dives may last 20 min or more, although most may be less than 10 min and not exceed 100 m. The longest dive recorded lasted about 25 min and the deepest was to 133 m. Most dives are likely shallower than about 80 m, as its benthic prey of mollusks are generally found in relatively shallow coastal or continental shelf habitats. Near northeast Greenland, walruses may be submerged about 81% of the time when they are at sea and are presumably diving and foraging most of that time.

## C. Phocids

Phocid seals generally are at sea continually for weeks to months and appear to dive, and perhaps forage, virtually constantly (Reeves *et al.*, 1992; Reeves and Stewart, 2003). Elephant seals are perhaps the best studied of marine mammal divers. The dives of weaned southern elephant seal pups are to about 100 m for about 6 min and they dive virtually continuously when at sea for several months. Heavier pups dive deeper (to ~130 m) and longer (7 min) than smaller pups (88 m and 5 min). Dives of juvenile southern elephant seals last around 15.5 min (maximum of 39 min) to depths averaging 416 m (deepest 1270 m) and they spend about 90% of their several months at sea diving. Intervals between dives are brief, rarely lasting more than 2 min. Adult southern elephant seal dives on average 400–600 m and 19–33 min (deepest 1444 m and longest 113 min) and also occur continuously while they are at sea for up to 7–8 months (Campagna *et al.*, 1999).

Northern elephant seals also dive continually when at sea for several months or more, with only brief periods at the surface (1–3 min) between dives. Dives of adults are to modal depths of 350–400 m and 700–800 m (maximum of 1567 m), and average 20–30 min (maximum of 77 min). Depths and durations of dives differ between adult males and females depending on season and geographic location (Stewart and DeLong, 1995). Generally, these seals feed on pelagic fish and squid, although some seals may also dive to and feed near the sea floor near the coastlines of continents and islands.

Dives of Hawaiian monk seals (*Monachus schauinslandi*) are between 3- and 6-min long and mostly shallow, between 10- and 40-m deep, where the seals forage near the sea bed on epibenthic fish, cephalopods, and other invertebrates. Adults may occasionally dive to greater depths of up to 550 m when foraging outside of the shallow atoll lagoons of the northwestern Hawaiian Islands (Stewart *et al.*, 2006).

Weddell seals forage for much, if not most, of the year beneath the unbroken fast ice and the more open pelagic pack ice zones of the Antarctic. Diving and foraging occur in bouts of about 40–50

D

consecutive dives over a several hour period, usually to depths of 50–500 m. Dives of young pups are relatively shallow and brief but get progressively deeper and longer as pups age. They plateau when the pups are weaned when about 6–8 weeks old. The dives of 1 year olds are somewhat shallower, to around 118 m, compared to adult females (163 m). The deepest dive recorded is about 750 m and the longest over 73 min. Dives are shallower (350–450 m) in spring (October–December) than in summer (January; 50–200 m) evidently reflecting a shift in preferred hunting depths.

Among the Antarctic pack ice seals, Ross seals (*Ommatophoca rossii*) are also relatively deep divers. One female that was monitored near the Antarctic Peninsula in summer dove exclusively at night, mostly to depths of 110 m (maximum of 212 m) and for about 6.4 min (longest 9.8 min). Diving was continual while the seal was in the water with about 1 min between dives. The deepest dives (175–200 m) occurred near twilight and the shallowest (~75–100 m) at midnight (Bengtson and Stewart, 1997).

In summer, crabeater seals (*Lobodon carcinophaga*) dive primarily at night and haul out on pack ice during the day, although some diving bouts may last up to 44 h without interruption. Most dives are 4–5 min long to depths of 20–30 m, with maximum depths and durations of 430 m and 11 min, respectively. Dives near twilight are deepest and those near midnight shallowest (Bengtson and Stewart, 1992). Diving patterns in other areas of the Antarctic are similar (*cf.* Nordoy *et al.*, 1995; Wall *et al.*, 2007).

Baikal seals (*Pusa sibirica*) apparently dive continually from September through May, when the freshwater Lake Baikal is frozen over, and haul out only infrequently then (Stewart *et al.*, 1996). Most dives are 10–50 m deep in the middle of Lake Baikal where the water depth is around 1000–1600 m. Occasionally, seals descend to more than 300 m. Dives last between 2 and 6 min but some have been measured at more than 40 min.

Dives of another closely related freshwater seal, the Saimaa seal (*Pusa hispida saimaansis*) of Lake Saimaa in eastern Finland, last about 6 min in spring and increase to about 10–11 min by autumn. In summer and autumn, long series of sequential dives lasting more than 10 min each may occur over 3 h or more. The longest dive recorded is about 23 min, when the seal may actually have been resting on the bottom rather than feeding.

Modal dive depths for breeding age, male ringed seals (*P. hispida hispida*) are 10–45 m and for subadult males and postpartum females 100–145 m. Durations of dives for adult males are around 4 min and around 7.5 min for adult females.

Harbor seal (*Phoca vitulina*) diving behaviors have been studied in several areas throughout their range in the North Pacific and North Atlantic Oceans. Dives in the Wadden Sea (northeast Atlantic) average from 1–3 min (maximum of 31 min) with little variation between night and daytime behavior. When in the water, about 85% of their time is spent diving. In the western Atlantic, foraging dives of adult males are mostly deeper than 20 m but are shallower during the mating period, when they are defending aquatic territories or searching for females to mate with instead of foraging (Bowen *et al.*, 1999). Dives of lactating females are 12–40 m and occur in bouts lasting several hours, mostly during the day. In southern California, dives are as deep as 446 m. Most, however, are to modal depths of 10, 70, or 100 m with an occasional mode at around 280. Harbor seals near Monterey dive to and forage at depths between 5 and 100 m for up to 35 min (Eguchi and Harvey, 2005).

Bearded seal (*Erignathus barbatus*) adult females near the coast of Spitzbergen, Norway, dive mostly at night to depths of around 20 m (deepest at 288 m) and for 2–4 min (longest 19 min). Nursing

pups may dive to around 10 m (maximum of 84 m) for about 1 min (maximum of 5.5 min). Pups spend about 40% of their time in the water diving. Depths and durations of dives increase as the pups age (Kraft *et al.*, 2000).

Most dives of lactating female gray seals (*Halichoerus grypus*) are to the sea floor and last about 1.5–3 min (maximum of 9 min). Most foraging dives of juvenile gray seals in the Baltic Sea are to depths of 20–40 m (Sjoberg and Ball, 2000).

Lactating female harp seals (*Pagophilus groenlandicus*) dive about 40–50% of the time that they are at sea. Dives average about 3 min (maximum of 13 min) to depths of up to 90 m.

Hooded seals (*Cystophora cristata*) repeatedly dive to depths of 1000 m or more and for 52 min or longer. Most feeding dives appear to be to depths of 100–600 m.

## III. Cetaceans
### A. Odontocetes

Limited data for odontocete cetaceans so far indicate that short-beaked common dolphins (*Delphinus delphis*) may forage at depths of up to 260 m for 8 min or more, although most dives are around 90-m deep, last about 3 min, and are mostly at night. Pantropical spotted dolphins (*Stenella attenuata*) dive to at least 170 m; most of their dives are to 50–100 m for 2–4 min and most feeding appears to occur at night. Atlantic common bottlenose dolphins (*Tursiops truncatus*) near Grand Bahama Island off southeastern Florida often dive to the ocean bottom (7–13 m depth) and burrow into the sediment ("crater-feeding") to catch fish dwelling or hiding there. Long-finned pilot whales (*Globicephalas melas*) dive to over 500–600 m for up to 16 min. Northern bottlenose whales (*Hyperoodon ampullatus*) regularly dive to the sea bed at depths of 800–1500 m for more than 30 min per dive and occasionally for 2 h (Hooker and Baird, 1999).

Harbor porpoise (*Phocoena phocoena*) near Japan have dived almost continuously when observed for short periods. Maximum dive depths are around 70–100 m, although about 70% of dives may be less than 20 m. These porpoises descend to and ascend from depth at greater rates when diving deeply than when the dives are shallow. In waters near Denmark, porpoises dive as deeply as 84 m and for up to 7 min from spring through late autumn.

Female beluga (white) whales (*Delphinapterus leucas*) dive more often between 2300 h and 0500 h than during the day, although males may dive at the same rate at all hours. Dive rates and time spent at the surface decline whereas dives deepen and lengthen from early through late autumn. Most dives are deep (400–700 m), with the deepest recorded at 872 m, and last about 13 min on average (maximum of 23 min). Dive duration increases with body size (Martin and Smith, 1999).

Narwhals (*Monodon monoceros*) regularly dive to more than 500 m and occasionally deeper than 1000 m, but most dives are to depths of 8–52 m and last less than 5 min, although as long as 20 min on occasion. The rate of diving varies between adult males and females. When diving shallowly, narwhals descend and ascend relatively slowly (<0.05 m/sec) compared with deeper, longer dives (1–2 m/sec) where substantially more time is spent at maximum depth.

Killer whales (*Orcinus orca*) along the northern coast of North Island, New Zealand, dive to the ocean bottom (~12-m depth or less) after stingrays and perhaps probe into the sediment to catch them.

Sperm whales are deep and long-duration divers. Near Kaikoura, New Zealand, the average duration of dives is about 41 min with about 9 min spent at the surface between dives. Both durations and surface intervals are longer in summer than in winter. Males spend little time at the surface compared to females (Jaquet *et al.*, 2000). Average dive durations have been measured at 36 min near Sri Lanka and about 55 min near the Azores. Sperm whales in the Caribbean were reported to make dives averaging 22–32 min during the day (longest 79 min) and 32–39 min at night (longest 63 min). Off Japan, sperm whales dive to around 550–650 m along the Kumano coast with no differences between day and night patterns, whereas dive patterns are strongly diurnal to around 470 m at night and to around 850 m at night near the Ogasawara Islands, evidently related to differences in behaviors of their prey (Aoki *et al.*, 2007).

## B. Mysticetes

As yet there is no evidence for a taxonomic relationship between body size and maximum dive depths for mysticete cetaceans, although preliminary correlations have been reported between maximum dive durations and body size.

While in shallow coastal lagoons during the spring breeding season, gray whales (*Eschrichtius robustus*) dive for about 1–5 min (maximum of 28 min) to average depths of 4–10 m (maximum recorded of 20.7 m). It is not clear what the function of these dives may be other than perhaps subsurface resting, as breeding whales are presumed to fast. In the Bering Sea in summer, when whales are feeding on the bottom, dives average 3–4 min at depths ranging from less than 10 to 79 m (Würsig *et al.*, 1986).

Fin whales (*Balaenoptera physalus*) in the Ligurian Sea dive repeatedly to depths around 180 m (maximum 474 m) for around 10 min (longest 20 min) while they prey on deep-dwelling krill. Elsewhere, fin whale dives have been reported to last about 5 min near Iceland and about 3 min in the North Atlantic and near Long Island, New York, in summer. Elsewhere, fin whales dive to around 98 m for 6 min when foraging and to around 59 m for 4 min when not foraging (Croll *et al.*, 2001).

When chased by commercial whalers, dives of blue whales (*Balaenoptera musculus*) lasted up to 50 min. Blue whales off central and southern California otherwise spend about 94% of their time submerged. Dives lasting longer than 1 min are 4.2–7.2 min, on average (longest 18 min), and to around 105 m (deepest 150–200 m). When foraging, blue whales dive to around 140 m for 8 min and to 68 m for 5 min when not foraging (Croll *et al.*, 2001). Dives of pygmy blue whales (*B. musculus brevicauda*) have been measured to average 9.9 min (longest 26.9 min).

Humpback whales (*Megaptera novaeangliae*) in Frederick Sound, Alaska, make rather brief (most less than 3 min) and shallow (60 m or less) dives, although some may exceed 120 m on occasion.

When on summer feeding grounds in the Beaufort Sea, dives of bowhead whales (*Balaena mysticetus*) last 3.4–12.1 min and some are to the relatively shallow sea bed. Dives of calves are very short compared to adults and they also spend more time at the surface between dives. Most dives of juveniles last about 1 min (longest 52 min) to depths of around 20 min. Longer dives, up to 80 min, have been observed for bowhead whales that were harpooned and being chased by whalers. Dives lasting longer than 1 min ("sounding dives") average between 7 and 14 min. Dives made while whales are migrating through heavy pack ice are deeper and longer than those made while in open water. Lactating females dive less often and for shorter periods than other adult whales.

Dives of North Atlantic right whales (*Eubalaena glacialis*) near Cape Cod, Massachussetts, last around 2.1 min.

## IV. Other Marine Mammals

Manatees (*Trichechus manatus*, *T. inunguis*, and *T. senegalensis*) feed on floating and submerged vegetation in shallow nearshore habitats, so it is unlikely that their dives often exceed 25–30 m. Direct observations of free-ranging animals have shown that most dives are less than 5 min, although a few have been timed at more than 20 min. These longer dives may have periods of rest at the bottom rather than feeding activity. Dugongs (*Dugong dugon*) also feed on submerged vegetation, most often in coastal and offshore seagrass beds either on the sea bottom to depths of 20 m or in surface canopies. The longest foraging dives observed are around 6 min, but most have been reported to last only between 2 and 4 min.

Sea otters dive and forage mostly on the seafloor in shallow nearshore waters. Dives may be in bouts lasting several hours during the day and night, interrupted by periods at the surface to groom, process food, or rest. Juvenile males often dive in deeper water, for longer periods, and further from shore than juvenile and adult females (Ralls *et al.*, 1995). In southeast Alaska, sea otters spend about 11–12 h each day diving, about 9 h of that actively foraging. Adult males in recently occupied habitats spend less time foraging than do adult females and also less time foraging than do sea otters in less recently colonized habitats (Bodkin *et al.*, 2007).

Polar bears are powerful swimmers and probably make some dives while moving among ice floes, the fast-ice edge, or coastlines, but nothing is known of the details of such diving performance. They prey mostly on ringed seals and whale carcasses on the surface of the ice or along shorelines and also on white whales and narwhals that they may attack and kill at the sea surface and then drag out of the water to consume.

## *See Also the Following Articles*

Feeding Strategies and Tactics ■ Swimming ■ Telemetry

## *References*

Aoki, K., Amano, M., Yoshioka, M., Mori, K., Tokuda, D., and Miyazaki, N. (2007). Diel diving behavior of sperm whales off Japan. *Mar. Ecol. Prog. Ser.* **349**, 277–287.

Baker, J. D., and Donohue, M. J. (1999). Ontogeny of swimming and diving in northern fur seal (*Callorhinus ursinus*) pups. *Can. J. Zool.* **78**, 100–109.

Bengtson, J. L., and Stewart, B. S. (1992). Diving and haulout behavior of crabeater seals in the Weddell Sea, Antarctic during March 1986. *Polar Biol.* **12**, 635–644.

Bengtson, J. L., and Stewart, B. S. (1997). Diving patterns of a Ross seal (*Ommatophoca rossii*) near the eastern coast of the Antarctic Peninsula. *Polar Biol.* **18**, 214–218.

Bodkin, J. L., Monson, D. H., and Esslinger, G. G. (2007). Activity budgets derived from time-depth recorders in a diving mammal. *J. Wildl. Manage.* **71**, 2034–2044.

Bowen, W. D., Boness, D. J., and Iverson, S. J. (1999). Diving behaviour of lactating harbour seals and their pups during maternal foraging trips. *Can. J. Zool.* **77**, 978–988.

Brillinger, D. R., and Stewart, B. S. (1997). Elephant seal movements: Dive types and their sequences. *In* "Modelling Longitudinal and Spatially Correlated Data: Methods, Applications and Future Directions," pp. 275–288. Springer, Berlin.

Campagna, C., Fedak, M. A., and McConnell, B. J. (1999). Post-breeding distribution and diving behavior of adult male southern elephant seals from Patagonia. *J. Mammal.* **80**, 1341–1352.

Croll, D. A., Acevedo-Gutierrez, Tershy, B., and Urban-Ramirez, J. (2001). The diving behavior of blue and fin whales: Is dive duration shorter than expected based on oxygen stores? *Comp. Biochem. Physiol.* **129A**, 797–809.

Eguchi, T., and Harvey, J. T. (2005). Diving behavior of the Pacific harbor seal (*Phoca vitulina richardii*) in Monterey Bay. *Mar. Mamm. Sci.* **21**, 283–295.

Fedak, M. A., Lovell, P., and Grant, S. M. (2001). Two approaches to compressing and interpreting time-depth information as collected by time-depth recorders and satellite-linked data recorders. *Mar. Mamm. Sci.* **17**, 92–110.

Georges, J. Y., Tremblay, Y., and Guinet, C. (2000). Seasonal diving behavior in lactating subantarctic fur seals on Amsterdam Island. *Polar Biol.* **23**, 59–69.

Harcourt, R., Hindell, M., Bell, D., and Waas, T. (2000). Three-dimensional dive profiles of free-ranging Weddell seals. *Polar Biol.* **23**, 479–487.

Hooker, S. K., and Baird, R. W. (1999). Deep-diving behaviour of the northern bottlenose whale, *Hyperoodon ampullatus* (Cetacea: Ziphiidae). *Proc. R. Soc. Lond., B* **266**, 671–676.

Horning, M., and Trillmich, F. (1997). Ontogeny of diving behaviour in the Galapagos fur seal. *Behaviour* **134**, 1211–1257.

Jaquet, N., Dawson, S., and Slooten, E. (2000). Seasonal distribution and diving behaviour of male sperm whales off Kaikoura: Foraging implications. *Can. J. Zool.* **78**, 407–419.

Kooyman, G. L. (2006). Mysteries of adaptation to hypoxia and pressure in marine mammals. *Mar. Mamm. Sci.* **22**, 507–526.

Kraft, B. A., Lydersen, C., Kovacs, K. M., Gjertz, I., and Haug, T. (2000). Diving behaviour of lactating bearded seals (*Erignathus barbatus*) in the Svalbard area. *Can. J. Zool.* **78**, 1408–1418.

Lagerquist, B. A., Stafford, K. M., and Mate, B. R. (2000). Dive characteristics of satellite-monitored blue whales (*Balaenoptera musculus*) off the central California coast. *Mar. Mamm. Sci.* **16**, 375–391.

Martin, A. R., and Smith, T. G. (1999). Strategy and capability of wild belugas, *Delphinapterus leucas*, during deep, benthic diving. *Can. J. Zool.* **77**, 1783–1793.

Mattlin, R. H., Gales, N. J., and Costa, D. P. (1998). Seasonal dive behaviour of lactating New Zealand fur seals (*Arctocephalus forsteri*). *Can. J. Zool.* **76**, 350–360.

Nordoy, E. S., Folkow, L., and Blix, A. S. (1995). Distribution and diving behaviour of crabeater seals (*Lobodon carcinophagus*) off Queen Maud Land. *Polar Biol.* **15**, 261–268.

Parrish, F. A., and Littnan, C. L. (2008). Changing perspectives in Hawaiian monk seal research using animal-borne imaging. *Mar. Technol. Soc. J.* **41**(4), 30–34.

Pitcher, K. W., Rehbert, M. J., Raum-Suryan, K. L., Gelatt, T. S., Swain, U. G., and Sigler, M. F. (2005). Ontogeny of dive performance in pup and juvenile Steller sea lions in Alaska. *Can. J. Zool.* **83**, 1214–1231.

Ralls, K., Hatfield, B. B., and Siniff, D. B. (1995). Foraging patterns of California sea otters as indicated by telemetry. *Can. J. Zool.* **73**, 523–531.

Reeves, R. R., and Stewart, B. S. (2003). Marine mammals of the world: An introduction. *In* "Walker's Marine Mammals of the World," pp. 1–64. The Johns Hopkins University Press, Baltimore, Maryland.

Reeves, R. R., Stewart, B. S., and Leatherwood, J. S. (1992). "The Sierra Club Handbook of Seals and Sirenians." Sierra Club Books, San Francisco.

Simpkins, M. A., Kelly, B. P., and Wartzok, D. (2001). Three-dimensional movements with individual dives by ringed seals (*Phoca hispida*). *Can. J. Zool.* **79**, 1455–1464.

Sjoberg, M., and Ball, J. P. (2000). Grey seal, *Halichoerus grypus*, habitat selection and haulout sites in the Baltic Sea: Bathymetry or central-place foraging. *Can. J. Zool.* **78**, 1661–1667.

Stewart, B. S., and DeLong, R. L. (1995). Double migrations of the northern elephant seal, *Mirounga angustirostris*. *J. Mammal.* **76**, 196–205.

Stewart, B. S., Petrov, E. A., Baranov, E. A., Timonin, A., and Ivanov, M. (1996). Seasonal movements and dive patterns of juvenile Baikal seals, *Phoca sibirica*. *Mar. Mamm. Sci.* **12**, 528–542.

Stewart, B. S., Antonelis, G. A., Baker, J. D., and Yochem, P. K. (2006). Foraging biogeography of Hawaiian monk seals in the northwestern Hawaiian Islands. *Atoll Res. Bull.* **543**, 131–145.

Wall, S. M., Bradshaw, C. J. A., Southwell, C. J., Gales, N. J., and Hindell, M. A. (2007). Crabeater seal diving in eastern Antarctica. *Mar. Ecol. Prog. Ser.* **337**, 265–277.

Würsig, B., Wells, R. S., and Croll, D. A. (1986). Behavior of gray whales summering near St. Lawrence Island, Bering Sea. *Can. J. Zool.* **64**, 611–621.

**D**

# Diving Physiology

## GERALD L. KOOYMAN

## I. Introduction

Ever since humankind has lived by and gone down to the sea, we have been awestruck by the creatures that make it their home. First we feared them, later we ate them, and now we try to emulate them with humble attempts to set "world" diving records. At present the record for a descent-assisted dive is at 214 m during a breath hold which lasted a little less than 4.5 min. Many marine mammals exceed that depth within the first few months of life. Premier divers such as elephant seals (*Mirounga* spp.) and sperm whales (*Physeter* macrocephalus) will occasionally dive to depths beyond a kilometer (Table I). The spectacular abilities of marine birds and mammals to dive deep and for long periods of time are a source of interest and curiosity for marine scientists and amateurs alike.

When marine mammals descend below the sea's surface they leave behind the thin skin of the earth's atmosphere with one of its essential ingredients to all vertebrate life—oxygen. They begin a journey that is incredible in diverse ways. The magnitude of incredulity varies according to the species, but for all, even the most humble of marine mammals such as the sea otter (*Enhydra lutris*), much if not most of the experience is beyond our imagination. Unlike flying, in which our technology now enables us to fly faster, higher, and further than any bird, bat, or pterosaur ever has or did, marine mammals, particular those that dive to great depths, explore and exploit a realm that overwhelms much of our technology and which enables us to gain only fleeting glimpses of what their environment is like. Recently we have enlisted the animals themselves to help us discover more about this cold, dark world without oxygen, where awesome hydrostatic pressures always prevail. However, "crittercams" will only give us fleeting glimpses, under very special conditions, with those few species that lend themselves to the attachment of these cameras. Life in the deep blue remains a mystery. So too do the means that enable diving mammals to exploit this habitat.

This chapter discusses some of what is known about adaptations to breath holding and overcoming the crushing effects of pressure. These adaptations are unique among vertebrates. Even after our primordial fish ancestors overcame great obstacles to adapt to the terrestrial environment, and eventually to spread throughout all land habitats of the world, the sea continued to be a rich habitat that would bring great success to those species that exploited it. Some air-breathing vertebrates are doing just that. In fact, this has occurred several times in the history of vertebrates as they reinvaded the sea. The marine reptiles of the Mesozoic were diverse, abundant, and no doubt very capable divers. They had at least one major advantage over marine mammals, a small brain. The brains of mammals require a substantial share of the oxygen being supplied to the body, and it is an obligate need with very little reserve for those times when supply is interrupted. Within 3 min after blood flow and oxygen transport to the human brain is interrupted, there is irreversible damage. This sensitivity of large, complex brains to a grave need for oxygen makes it seem a contradiction that animals who routinely breath hold many times every day are all so smart. Proportionately in terms of brain size relative to body size, several cetacean species have some of the largest brains of mammals. Despite this "handicap" marine mammals have been an extremely successful group that are found in all the world's oceans, in extremely large numbers, and have the biomass of some species matching that of any of the formerly abundant terrestrial mammals of the world.

What is the secret of their success? Some routinely dive to depths of several hundred meters, and a few species may occasionally descend from 1 to 2 km (Table I). Although these depths may seem just a superficial range compared to the ocean limit of 11 km, with an average depth of 3.5 km, the range used by most marine mammals is in the zone of greatest oceanic life. Nevertheless, this region of cold, dark waters requires special adaptations enabling the animal to endure low temperatures and find prey in the "dark." Marine mammal diving skill provides a dramatic contrast to human capacities. On average we can dive to a few meters for about 30 sec. The super athletes, who make a career of setting records such as the record breath-hold dive of 214 m, require mechanical aids of weights, pulleys, and drop lines. To extend our depth beyond these few meters humans have gone to costly extremes in mechanical devices. Most deep submersibles are usually limited to several hundred meters depth, but ALVIN, the workhorse of the scientific submersibles, can go as deep as 4500 m.

Adaptations of marine mammals to the marine environment are diverse in order for them to become successful marine predators. They involve many systems in and out of the body, ranging from external body shape to overcome the high density and viscosity of water to the sensory systems necessary to find their way and to detect prey and predator. Space will allow for only a few of the numerous adaptations necessary for a successful marine mammal. The following paragraphs discuss adaptations to hypoxia and pressure. These paragraphs address pelagic, offshore deep divers in which the adaptations are the most extreme.

## II. Adaptations to Hypoxia
### A. Oxygen Stores and their Distribution

An increased total body $O_2$ store is considered an essential factor in the breath-hold capacity of diving mammals. The oxygen consumed by body metabolism during a breath hold is stored in three compartments, the respiratory system, the blood, and the body musculature. The theoretical maximum amount of oxygen available in each compartment is a function of several criteria. The respiratory oxygen store is dependent on lung volume and the concentration of oxygen in the lung at the start of a breath hold. The blood and muscle oxygen stores are dependent on blood volume and muscle mass, and the concentration of the oxygen-binding proteins of hemoglobin in blood, and myoglobin in muscle. From the measurements of myoglobin concentration in the muscles of many species of divers it is clear that one of the most consistent hallmarks of oxygen storage in all marine mammals that dive to depth is an elevated myoglobin concentration (Kooyman and Ponganis, 1997; Kooyman *et al.*, 1999). This trait is more characteristic of deep divers than any changes in blood volume, hemoglobin concentration, or respiratory volumes. However, increased blood volume and hemoglobin concentration often contribute to elevated oxygen storage.

As the distribution of oxygen stores vary among species, so do the ranges of the total oxygen store (Table I). In humans the total store is 20 ml $O_2$/kg body mass, which is about a fifth of the nearly 100 ml $O_2$/kg body mass in elephant seals (*Mirounga* spp.). Using the seal as our basic model it is noted that most of its oxygen is in blood and muscle. The large amount relative to terrestrial animals, using the human average as a standard, is a result of a blood volume 3 times, a hemoglobin concentration 1.5 times, and a myoglobin concentration approximately 10 times the human value. In seals the lung is a minor source of oxygen, as it is in most other marine mammals. It is less than 5% of the total in part because seals exhale to 50% of their total lung capacity just before diving. Furthermore, at depth the lung is collapsed and does not exchange gas.

### TABLE I
Distribution and Quantity of Oxygen Stores, Maximum and Routine Diving Depths, and Durations for Some Marine Mammals

| Species | Body mass (kg) | Total store (ml/kg) | Lung | Blood (%) | Muscle | Routine depth (m) | Maximum depth (m) | Routine duration (min) | Maximum duration (min) |
|---|---|---|---|---|---|---|---|---|---|
| Human | 70 | 20 | 24 | 57 | 15 | 5 | 214 | 0.25 | 6 |
| Weddell seal | 400 | 87 | 5 | 66 | 29 | 200 | 741 | 15 | 93 |
| Elephant seal | 400 | 97 | 4 | 71 | 25 | 500 | 1,653 | 25 | 120 |
| California sea lion | 100 | 40 | 21 | 45 | 34 | 40 | 275 | 2.5 | 10 |
| Bottlenose dolphin | 200 | 36 | 34 | 27 | 39 | | 535 | | |
| Cuvier's beaked whale | 3,000 | | | | | 1,070 | 1,888 | 58 | 85 |
| Sperm whale | 10,000 | 77 | 10 | 58 | 34 | 500 | 2,035 | 40 | 73 |

Note: There is an extensive list of diving capabilities of many species of diving animals at: http://polaris.nipr.ac.jp/~penguin/penguiness/index.html

## B. Cardiovascular Responses

The cardiovascular response to breath holding falls into at least two categories of whether the dive is extended or of routine duration for that species. Measurements of cardiovascular and metabolic responses under these circumstances are very limited for any species and most measurements are from seals. Diving mammals are arrhythmic breathers with pauses between each series of breaths. The resting maintenance heart rate is probably most closely reflected in the rate during the respiratory pause or apnea. Using the heart rate during apnea as a basis of comparison for heart rates during a routine dive, the heart rates during the dive are lower than the rate of a resting apneusis, and this occurs despite the fact that the mammal is swimming. When an extended dive is performed, the heart rate is even lower than that during routine diving. Because no measurements of blood flow distribution have been directly measured during dives of marine mammals, it is by extrapolation from indirect measures of other organ functions that allude to what may be occurring. During routine dives it is likely that gastric, renal, and hepatic functions are reduced to a small amount, but no more than what can be compensated for by higher than normal performance during the short, breathing intervals at the surface. Muscle may utilize a small part of the circulating blood oxygen, but it probably relies on its internal store of oxygen bound to myoglobin for much of aerobic metabolic needs.

Extended dives, those that are 3–5 times the routine dives, are uncommon. They are most likely to occur because of some urgent need such as a Weddell seal (*Leptonychotes weddellii*) searching for a new hole under sea ice, or an elephant seal hiding at depth to escape notice from a passing pod of killer whales (*Orcinus orca*) near the surface. The cardiovascular response in these extreme cases may be a limitation of blood flow to obligate aerobic tissues, the most conspicuous of which is the brain. Having no internal store of oxygen, and a need to be at full functional capacity, a constant supply of oxygen and other metabolites provided by the blood, as well as transport of waste products of metabolism from the brain, means that constant blood flow is essential. There is a lesser need for transport of oxygen to the heart because of a reduced work load (the slower heart rate) and small store of internal oxygen. Blood flow to muscle is reduced to a trickle as it draws from the large oxygen store within the muscle and the internal store of glycogen for the production of the high energy compounds of adenosine triphosphate (ATP). The high concentration of myoglobin in *all* mammals that dive to depths greater than about 100 m indicates that myoglobin is a key adaptation for diving. Blood flow would be a liability since the affinity of myoglobin for oxygen is much higher than is the affinity of hemoglobin for oxygen. Consequently, any flow to muscle that had utilized much of its oxygen store would strip oxygen from the circulating blood and deprive more vital organs such as the brain from oxygen. A reduced blood flow to muscle also decreases cardiac output needs and, hence, the work of the heart and its oxygen consumption. Thus, the degree of muscle blood flow reduction during long and short dives is key to understanding the management of oxygen stores. Unfortunately, little is known about this crucial topic. Unlike other organs, muscle is widely distributed in the body, and the vascularity is diffuse. Consequently, it is an intractable problem which has not lent itself to study.

Muscle also has a great capacity for anaerobic metabolism and tolerance for high concentrations of the metabolic end product of lactic acid which is stored in the form of lactate. Nevertheless muscle must continue to function for locomotion either continuously as a Weddell seal swims below the sea ice, or intermittently as in an elephant seal as it drifts in the depths, but in the end must call upon muscle to provide the locomotion to return to the surface. In contrast, the splanchnic organs may shut down or greatly reduce function until the diving mammal returns to the surface.

## C. Metabolic Responses

The cessation of metabolic function in the splanchnic organs will reduce metabolic rate substantially since these organs functioning at normal rates account for nearly 50% of the total resting metabolism of the animal. In addition the heart is beating more slowly and performing less work, which may also be the case for striated muscle. In the cold environment at depth some tissues may also be cooling which would result in an additional savings in energy consumption. The final result is to lower the overall metabolic rate to below the resting level during these short and metastable conditions.

## D. Anaerobic Metabolism

Dominating the many factors that affect how long an animal may breath hold is the amount of oxygen available and its rate of utilization. Through oxygen supported metabolic pathways 18 times more high energy ATP is produced from glucose than through anaerobic processes. Furthermore, carbon dioxide and water, the end products of oxygen supported catabolism are less polluting to the cells and circulation than those of anaerobic catabolism. Finally, nerve cells, especially within the brain, are completely dependent on aerobic metabolism. Therefore, the duration of time an animal may breath hold is most strongly affected by the availability of oxygen, with subsidiary support from anaerobic glycolysis and creatine phosphate catabolism. Although an animal may extend its dive considerably by relying on anaerobic glycolysis, the subsequent recovery is in turn extensive because of the time required to process lactic acid and restore the acid base balance of the cells and circulatory system. For routine dives that occur in sequence over many hours aerobic metabolism is the only practical option. Oxygen supported metabolic pathways are also the only means of producing ATP that is derived from catabolism of fat and protein.

## E. Aerobic Diving Limit

The only diving mammal in which there has been a detailed correlation of the diving duration and the postdive blood lactate concentration is the Weddell seal (Fig. 1) (Kooyman *et al.*, 1980; Ponganis *et al.*, 1993). The source of the lactate has been shown to be from muscle, in which it accumulates rapidly as muscle oxygen is depleted. After the seal surfaces, there is increased blood flow to muscle and much of the lactate is flushed into the circulation and gradually disappears over several minutes. If the seal should dive again before all of the blood lactate is processed, it will continue to decline over the course of the dive unless that dive exceeds what has been termed the aerobic diving limit (ADL). The ADL is defined as the diving duration beyond which there is a net increase in lactate production (Kooyman, 1985). This rise in lactate concentration first occurs primarily in muscle, and eventually diffuses from the organ into the circulation where it can be measured easily. It has been proposed that this threshold be called the diving lactate threshold (DLT) to avoid confusion about the numerous ways that the ADL has been derived since the first measurements were made on Weddell seals (Butler and Jones, 1997). From those measurements it was also shown that with some reasonable accuracy the ADL could be obtained from the quotient of the $O_2$ store divided by metabolic rate, the calculated ADL (cADL). Because this limit predicts basic information

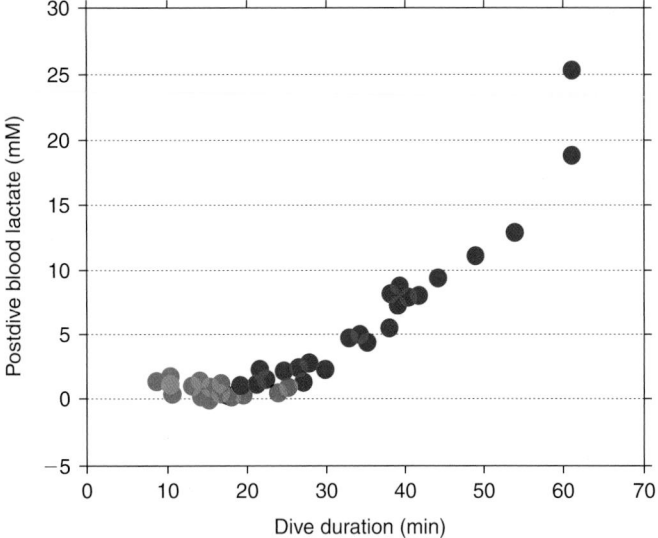

**Figure 1** *Peak concentration of lactate in arterial blood after dives of different duration in adult Weddell seals. The inflection represents the transitions from completely aerobic dives and is considered the aerobic diving limit (ADL) or diving lactate threshold (DLT). Red circles reflect blood values of dives in which there is no net production of lactate, and blue circles are those in which there was a net production.*

about the foraging behavior of diving animals as well as clarifying physiological responses and models to breath holding, it has been calculated for many diving species. Some of the most intriguing calculations have been made for the elephant seal, a continuous diver who appears to allow no surface time for recovery from dives exceeding the cADL. There are several possibilities that could resolve this puzzle. A recent computer simulation of oxygen store depletion provides a physiological model of aerobic diving that may clarify this problem, and provide direction for further studies.

This comprehensive, numerical model uses as its data source the Weddell seal because there is extensive information on this species. The calculations in the model are based on or derived from available data on cardiac output, $O_2$ depletion rates of blood and muscle, blood flow patterns in various organs, and the diving metabolic rates that may occur in the Weddell seal. The model demonstrates how the matching or mismatching of oxygen transport and regional oxygen consumption can affect the ADL. This theoretical treatment of the ADL goes a long way in understanding how oxygen must be managed during a dive, and in particular, explains how 31% of the body $O_2$ store remains unconsumed under the most optimal cardiac output conditions. The model also shows that only 49% of the muscle oxygen provision comes from the internal store during the longest possible aerobic dive. In regard to recovery from an extended dive, the oxygen replenishment rate is much more rapid than the reconversion of lactate to glycogen (Davis and Kanatous, 1999).

Assuming that the model can be applied to other aquatic species, it may help to explain the enigma of the serial dives in the elephant seal which exceed previous cADL's. However, the model does not take into account the influence of creatine phosphate to support the few dives that may appear to exceed the ADL, and for which some have invoked some unusual hypometabolic responses. Because creatine phosphate concentration is 15–20 mmol/kg in mammals, this is enough to have a significant effect on the magnitude of the ADL

and the production of energy without oxygen, but before measurable amounts of lactate are produced (Butler and Jones, 1997).

## III. Adaptations to Pressure

Once a marine mammal descends below the surface it not only must deal with the lack of oxygen but also with the effects of pressure. This is one of the most imposing physical variables to which vertebrates must adjust (Kooyman, 1989). We become especially sensitive to pressure during the most modest dive to depth because our airspaces such as the middle ear and facial sinuses make us acutely aware of any difference between the ambient pressure and our internal pressure. More subtle is the effect of pressure on the lung. For humans the lung is an important oxygen store, but in deep-diving mammals the lung is not an important oxygen store. Over a long period of evolution the main function of the vertebrate lung became the exchange of gases between blood and air. During the descent to depth, this function is diminished in deep-diving marine mammals. As the transfer of gases between the lung and blood slackens or ceases, the rise in nitrogen partial pressure within the lung is not matched in the blood. The lack of gas exchange also results in the avoidance of nitrogen narcosis and oxygen toxicity. Even with this adaptation there is still the pure physical effect of hydrostatic pressure on the nervous system. In terrestrial mammals pressure causes over stimulation or uncoordinated nerve conduction and dysfunction called high pressure nervous syndrome (HPNS). How do marine mammals manage to avoid these problems that are manifested in their terrestrial relatives?

The pressure within all airspaces must closely match that of the ambient pressure or suffer damage to the membranes and blood vessels lining the space and a breakdown in normal function. There are at least three major airspaces within most mammals that are liabilities for diving. First are the facial sinuses. Any experienced diver or airline passenger is aware that flying or diving during or soon after having a head cold is a bad idea. The blockage that may ensue during rapid pressure changes can cause extreme pain and serious damage to tissues and blood vessels lining the walls of these cavities. Marine mammals do not have this problem because they have no facial sinuses. Thus, one problem is dealt with by the absence of the airspace.

Similar to all other mammals there is a cavity that forms the middle ear. This is a rigid structure that has little or no compressibility. A pressure differential is prevented, at least for seals and sea lions, because of a complex vascular sinus lining the wall of the middle ear cavity. As the pressure within the middle ear cavity begins to fall below that of the vascular tree, the blood sinuses volume increases. This is a result of the close match between ambient pressure and blood pressure transferred from one fluid (sea water) to another (blood). Another problem is resolved by the reduction of an airspace by hydraulic compression through the vascular system.

Third, the largest airspace of all, and potentially the most problematic is the lung. Volume pressure curves of the chest wall and lung of the ribbon seal, *Histriophoca fasciata*, show that both the chest and the lung are nearly limitless in the degree of compression collapse that they can tolerate. This must be so for other diving mammals as well, and in less detailed studies it has been shown that there is exceptional compressibility in other seals, sea lions, and dolphins. Dolphins and other toothed whales show the most extreme modifications within the lung among marine mammals, or any other mammal. Most notable is the reinforcement of peripheral airways, the loss of respiratory bronchioles, and the presence of a series of bronchial sphincters. Sea lions also have robust cartilaginous airway reinforcement extending to the

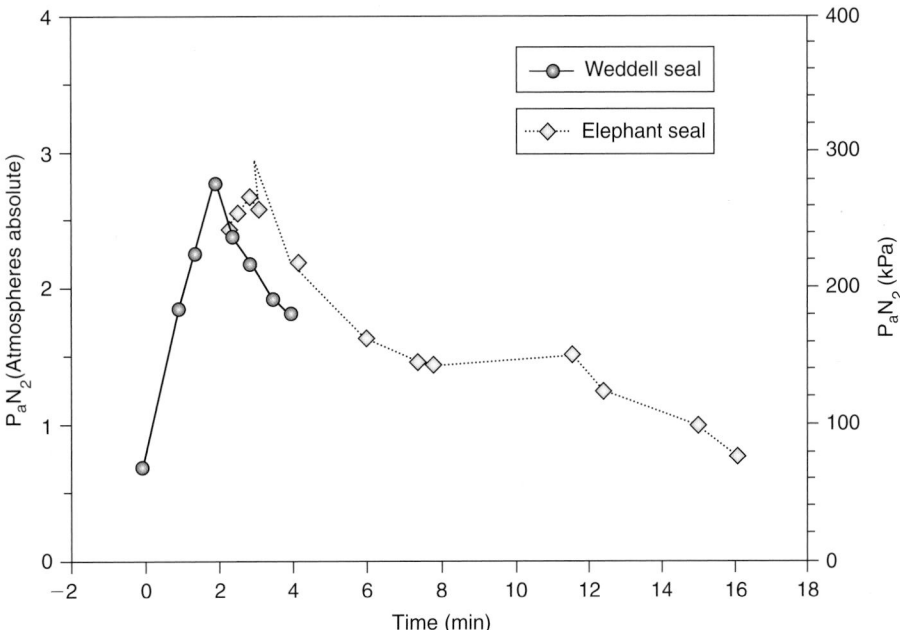

**Figure 2** *Arterial N₂ tensions in elephant seals and a Weddell seal. The elephant seal submersion (yellow diamonds) were simulated in a water filled hydraulic compression chamber to pressures equivalent to a sea water depth of 136 m. Compression began at zero; at 11 min pressure was released and at 14 min the submersion ended. The Weddell seal (red circles) made a free dive to 89 m under Antarctic ice. Maximum depth was reached at 5 min, and the dive ended after 8 min.*

alveolar sac, but there are no bronchial sphincters. In seals there is no cartilage in the terminal airway, but the walls are thickened by connective tissue and smooth muscle, which reduces their compliance to less than that of the alveoli. Hence, during compression the more compliant alveoli collapse first and the gases within these alveoli are squeezed into the upper airway spaces.

These airways enable a graded collapse of the lung to occur during a dive to depth. The result is that most of the lung air is forced into the upper airways where gas exchange with the blood ceases. It has been shown that blood $PN_2$ in seals only rises slightly, no matter how deep the dive. During simulated and actual dives to depth, the $P_aN_2$ of the elephant seal, *Mirounga angustirostris*, peaked at 300 kPa and equilibrated to 200 kPa where it was approximately the same as venous $PN_2$ (Kooyman *et al.*, 1972). This was independent of the ambient pressure from 30 to 136 m (1460 kPa). Similar values were obtained for Weddell seals diving voluntarily to depths as great as 230 m (2400 kPa) (Fig. 2) (Falke *et al.*, 1985). These small increases in $P_aN_2$ indicate that lung collapse in both species occurred between 20–50 m. The early occurrence of lung collapse in seals makes the lung almost useless as an $O_2$ store, but it limits $N_2$ absorption during the dive. These $N_2$ values are below the minimum $PN_2$ of 330 kPa found to be necessary for bubble formation in cats, and it is assumed that a similar threshold for bubble formation prevails in marine mammals. An additional benefit of early lung collapse is that it eliminates the likelihood of nitrogen narcosis. This condition is often experienced by SCUBA divers descending to depths greater than 30 m. At these depths tissue nitrogen level is at least 399 kPa; greater than the $P_aN_2$ measured in seals. A final thought is the intriguing condition of elephant seals at sea when they spend 90% of their time underwater and at depths greater than 100 m. At these times the lung does not do what it was originally evolved to do, i.e., to exchange gas with the blood and with the atmosphere. Instead it is collapsed to

a solid organ, and the alveoli become unavailable for gas exchange. Nevertheless, there are recent incidences of Cuvier's (*Zipius cavirostris*) and Blainville (*Mesoplodon densirostris*) beaked whale strandings that are shrouded in mystery. Fatal strandings of these species were associated with US Navy sonar experiments. Postmortem dissections of some of the beaked whales showed the presence of gas bubbles that appeared to have occurred *in vivo* (Jepson *et al.*, 2003) . We know that these species make extreme dives while hunting for prey (Tyack *et al.*, 2006). Is it possible that gas bubbles could have formed during abnormal diving behavior related to the sonar tests, which resulted in decompression sickness?

Finally, in humans and other nonaquatic animals descending to depths of more than 100 m and at rates of 100 m/min the mechanical compression on nervous tissue can cause HPNS. The symptoms are modest to severe tremors throughout the body that can become so severe as to be incapacitating. The range of depths and rate of descent are modest compared to that of some deep-diving marine mammals. Surely a well-adapted mammal does not experience HPNS, which leads to the compelling question of what the difference is between the neural makeup of a marine mammal that protects it from experiencing HPNS and that of a terrestrial mammal that is susceptible to HPNS. We must conclude that the structure of the nervous system is modified in a unique way for a life at high pressure.

## IV. Epilogue: Mysteries of the Deep

Our understanding of the physiology of diving in marine mammals is still elementary. These animals have adapted to some of the most extreme conditions on the planet. What we can learn from them about hypoxia and pressure is of much intellectual interest as well as of clinical significance. In addition to the brief summary presented, there is a host of other adaptations not mentioned and

D

in some cases not studied. Some of these are, blood and muscle interactions under extreme hypoxia, acoustical and visual sensing under the extreme conditions of depth where pressure is intense, the cold is penetrating, and light from the surface is at times nil. Many organisms have adapted to a life under these various conditions, but marine mammals commute to the depths, and where they excel is in their adaptability to rapid changes in these extremes as they move from the conditions at the interface of air and water to those of several hundred meters to even a few kilometers beneath the surface.

## References

Butler, P. J., and Jones, D. R. (1997). Physiology of diving of birds and mammals. *Physiological. Rev.* **77**, 837–899.

Davis, R. W., and Kanatous, S. B. (1999). Convective oxygen transport and tissue oxygen consumption in Weddell seals during aerobic dives. *J. Exp. Biol.* **202**, 1091–1113.

Falke, K. J., Hill, R. D., Qvist, J., Schneider, R. C., Guppy, M., Liggins, G. C., Hochachka, P. W., and Elliot, Z. (1985). Seal lungs collapse during free diving: Evidence from arterial nitrogen tensions. *Science* **229**, 556–558.

Jepson, P. *et al.* (17 authors) (2003). Gas-bubble lesions in stranded cetaceans. *Nature* **425**, 575–576.

Kooyman, G. L. (1985). Physiology without restraint in diving mammals. *Mar. Mamm. Sci.* **1**, 166–178.

Kooyman, G. L. (1989). Diverse Divers: Physiology and Behavior. (K. Johansen, and D. S. Farner, eds.), *Zoophysiol. Ser.*, Vol. 23, 216pp. Springer-Verlag, New York.

Kooyman, G. L., and Ponganis, P. J. (1997). The challenges of diving to depth. *Am. Sci.* **85**, 530–539.

Kooyman, G. L., Schroeder, J. P., Denison, D. M., Hammond, D. D., Wright, J. J., and Bergman, W. P. (1972). Blood $N_2$ tensions of seals during simulated deep dives. *Am. J. Physiol.* **223**, 1016–1020.

Kooyman, G. L., Wahrenbrock, E., Castellini, M., Davis, R. W., and Sinnett, E. (1980). Aerobic and anaerobic metabolism during voluntary diving in Weddell seals: Evidence of preferred pathways from blood chemistry and behavior. *J. of Comp. Physiol.* **138**, 335–346.

Kooyman, G. L., Ponganis, P. J., and Howard, R. S. (1999). Diving animals. *In* "The Lung at Depth" (C. E. G. Lundgren, and J. N. Miller, eds), pp. 587–620. Dekker, Inc., New York.

Ponganis, P., Kooyman, G., and Castellini, M. (1993). Determinants of the aerobic dive limit of Weddell seals: Analysis of diving metabolic rates, postdive end tidal $PO_2$ and blood and muscle oxygen stores. *Physiol. Zool.* **66**, 732–749.

Tyack, P. L., Johnson, M., Soto, N. A., Sturlese, A., and Madsen, P. T. (2006). Extreme diving of beaked whales. *J. Exp. Biol.* **209**, 4238–4253.

# Dugong
## *Dugong dugon*

### HELENE MARSH

The dugong (*Dugong dugon*) looks rather like a cross between a rotund dolphin and a walrus. Its body, flippers, and fluke resemble those of a dolphin without a dorsal fin. Dugongs can be difficult to distinguish from dolphins in the wild, especially as they often occur in muddy water. They surface very discreetly, often with only their nostrils showing above the water. Dugongs tend to move more slowly than dolphins and lack a dorsal fin. The Dugong's head looks somewhat like that of a walrus without the long tusks. Growing to a length of up to about 3 m, the dugong is the only extant plant-eating mammal that spends all its life in the sea. The other sea cows (or sirenians), the three species of manatee, all use fresh water to varying degrees (Reynolds and Odell, 1991).

## I. Characteristics and Taxonomy

Adults are gray in color but often appear brown from the air or from a boat. Older "scarback" individuals may have a large area of unpigmented skin on the back above the pectoral fins. The dugong's head is distinctive with the mouth opening ventrally beneath a broad, flat muzzle. The tusks of mature males and some old females erupt on either side of the head. The eyes are small and not prominent. Externally the ears consist of only small openings, one on either side of the head. The flippers are short and, unlike those of the West Indian and West African manatees, lack nails. There are two mammary glands, each opening via a single teat situated in the "armpit" or axilla. The mammaries are somewhat reminiscent of the breasts of human females, a similarity which probably explains the legendary links between mermaids and sirenians. Hindlimbs are absent. Unlike manatees, which have a paddle-shaped tail, the tail of the dugong is triangular like that of a whale (Fig. 1).

There is one species of Dugong, *Dugong dugon*, in the family Dugongidae. The only other recent (but extinct) dugongid is *Hydrodamalis*, Steller's sea cow. Dugongidae and Trichechidae (manatees) are the two modern families of Sirenia.

## II. Distribution and Abundance

The dugong has a large range. Its extent of occurrence is some 140,000 km of coastline across more than 40 countries and includes tropical and subtropical coastal and island waters from East Africa to Vanuatu, between about 26° and 27° north and south of the equator (Marsh, 2006). Timed depth recorders show that dugongs spend most of their time feeding in shallow water less than 10 m deep (Chilvers *et al.*, 2004), suggesting a potential area of occupancy of more than 125,000 km². The dugong's historic distribution was broadly coincident with the tropical Indo-Pacific distribution of its seagrass food plants. It is believed that throughout most of its range outside Australia and the Arabian region, the dugong is currently represented by relict populations separated by large areas where it is close to extinction or extinct. The degree to which dugong numbers have dwindled, and their range fragmented, is not known. It is encouraging that dugongs still seem to be present at the high latitude limits to their range, Okinawa Japan, Mozambique, Shark Bay, and Moreton Bay Australia, New Caledonia, and Vanuatu.

Over most of its range, the dugong is known only from incidental sightings, accidental drownings, and the anecdotal reports of fishermen. However, within Australia, extensive aerial surveys have resulted in a more comprehensive knowledge of dugong distribution. A significant proportion of the world's dugongs is found in northern Australian waters from Moreton Bay in the east to Shark Bay in the west. Dedicated aerial surveys of dugong populations in Australian waters indicate that dugongs are the most abundant marine mammal in the inshore waters of northern Australia. Some areas of suitable habitat have not been surveyed. Nonetheless, the available population estimates sum to about 85,000 dugongs (Marsh *et al.*, 2002, 2003). This accuracy of these estimates is unknown as there is still uncertainty about the correction factor used for the number of animals that are not available to observers due to water turbidity.

**Figure 1** *Gentle vegetarians, dugongs often frequent muddy near-shore and estuarine waters that make it hard to positively identify them. Slow moving and lacking any hint of a dorsal fin, dugongs can be confused with dolphins, especially at a distance.*

## III. Ecology

Dugongs are mostly seagrass specialists, uprooting whole plants when they are accessible, leaving long serpentine furrows depleted of seagrass in seagrass meadows. When the whole plant cannot be uprooted, dugongs feed only on the leaves. Dugongs forage for the rhizomes of some seagrasses, presumably because seagrass rhizomes tend to have higher starch concentrations than seagrass leaves, which tend to be higher in nitrogen. Dugongs switch between species of seagrasses as their availability changes. The preferred seagrasses are relatively low in fiber, high in available nutrients such as nitrogen and starch, and easily masticated. Experiments simulating dugong grazing indicate that feeding dugongs alter both the species composition and the nutrient qualities of seagrass communities. Like large terrestrial herbivores, dugongs can maintain grazing lawns. If dugongs become locally extirpated, seagrass meadows are likely to deteriorate as dugong habitat. Dugongs also eat invertebrates, both incidentally and deliberately, especially in winter at the high latitude limits to their range.

Dugongs frequent coastal waters. Major concentrations of dugongs tend to occur in wide shallow protected bays, wide shallow mangrove channels, and in the lee of large inshore islands where there are sizable seagrass beds. Dugongs are also regularly observed in deeper water further offshore in areas where the continental shelf is wide, shallow, and protected. This distribution reflects that of deepwater seagrasses. Dugong feeding scars have been observed at depths of up to 33 m in seagrass beds off northeastern Queensland, however, evidence from timed depth recorders deployed on 15 dugongs caught in northern Australia, indicate that dugongs spend about half of their daily activities within 1.5 m of the sea surface and more than 70% of their time less than 3 m from the sea surface.

Dugongs may be short of food for several reasons, including habitat loss, seagrass dieback, decline in the nutrient quality of available seagrass, or a reduction in the time available for feeding due to disturbance from boat traffic. Experience from various parts of northern Australia suggests that episodic losses of hundreds of square kilometers of seagrass are associated with extreme weather events such as some cyclones and floods. For example, approximately 1000 km² of seagrass was lost in Hervey Bay in Queensland in 1992–1993, because of high turbidities resulting from flooding of local rivers and runoff turbulence from a cyclone. Such events can cause extensive damage to seagrass communities through severe wave action, shifting sand, adverse changes in salinity, and light reduction. Dugongs respond to large-scale seagrass dieback by using at least two strategies (1) move away from the affected area, and/or (2) postpone breeding. Episodic seagrass diebacks occur relatively frequently in some regions and this frequency may be increased by climate change. Thus it will be imperative to consider seagrass diebacks in marine planning and dugong management.

## IV. Behavior and Physiology

Knowledge of the social behavior of dugongs is rudimentary. The habits and habitats of dugongs make them difficult to observe and the lack of distinct size classes or obvious sexual dimorphism limits the data obtained from direct observation. The only definite long-lasting social unit is the cow and her calf. Most dugongs are sighted in groups of one or two animals. Large aggregations of up to several hundred animals are regularly seen at some locations but the composition of such groups appears fluid. Despite this gregariousness, little is known of the structure and function of dugong herds. Recent observations of focal animals in dugong herds in relatively clear water using a blimp-mounted video camera indicated that dugongs change their nearest neighbor every few minutes. One of the functions of herds seems to be maintaining the seagrass meadow at the stage favored by dugongs.

As with many other mammals, the mating behavior of dugongs seems to vary with location. Mating herds have been observed at several locations along the Queensland and Northern Territory coasts. In these herds, splashing and fighting precede mating. The presumed female swims, turns, twists, and thrashes as she attempts to escape from her persistent entourage. Members of this entourage engage in bouts of violent fighting before attempting to mount the female in a form of pack rape Fig. 2). In contrast, in South Cove in Shark Bay, Western Australia, presumed male dugongs defend mutually exclusive territories in which unique behaviors are displayed in order to attract females. It is not known whether this behavior occurs elsewhere in the dugong's range.

Seventy dugongs were fitted with satellite PTTs and/or GPS transmitters in subtropical and tropical waters of Queensland and the Northern Territory, Australia (Sheppard et al., 2006). Most movements have been localized to the vicinity of seagrass beds and seem to be commuting movements dictated by the tide. Twenty-eight of the 70 dugongs were also fitted with timed depth recorders. The dugongs were tracked for periods ranging from 15 to 551 days and exhibited a large range of individualistic movement behaviors; 26 individuals were relatively sedentary (moving <15 km) while 44 made large-scale movements (>15 km) of up to 560 km from their capture sites. Male and female animals, including cows with calves, undertook large-scale movements. The body lengths of these dugongs ranged from 1.9 to 3 m. At least some of the movements were return movements to the capture location, suggesting that such movements are ranging rather than dispersal movements. Solitary individuals or small groups of dugongs are occasionally recorded in the waters of isolated oceanic islands, such as the Seychelles and Cocos (Keeling) Islands after years of apparent absence, suggesting that dugongs are capable of crossing ocean trenches.

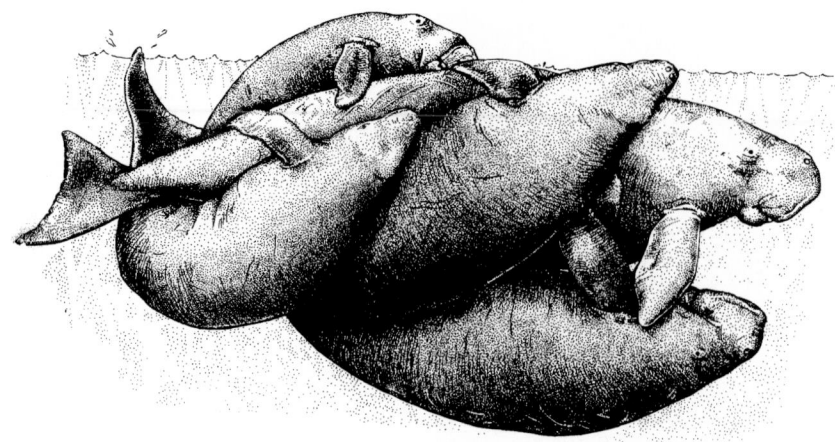

**Figure 2** *Part of a mating herd of Dugongs. Several males are attempting to embrace the female and mate with her. Drawing by Lucy Smith, reproduced from Bryden* et al. *(1993), with permission.*

## V. Life History

Dugongs are long lived with a low reproductive rate, long generation time, and a high investment in each offspring. Like the teeth of other marine mammals, dugong tusks accumulate "growth layer groups" that are used to estimate age, rather like the growth rings of a tree. The oldest dugong whose tusks have been examined for age determination was estimated to be 73 years old when she died. Females do not bear their first calf until they are at least 6 and up to 17 years old. Gestation is approximately 13 months. The usual litter size is one. The calf suckles for 18 months or so and the period between successive births is very variable; estimates range from 3 to 7 years. Dugongs start eating seagrasses soon after birth and grow rapidly during the suckling period. Population simulations indicate that a dugong population is unlikely to be able to increase more than 5% per year, making the dugong highly susceptible to overexploitation by humans.

## VI. Interactions with Humans

Dugongs are vulnerable to human impacts because of their life history and their dependence on seagrasses that are restricted to coastal habitats, which are often under pressure from human activities. The sustainable level of anthropogenic mortality is likely to be in the order of about 2% of the female population per year. This rate will be lower in areas where their reproductive rate has been reduced by food shortage.

Accidental entangling in gill and mesh nets set by commercial fishers is considered to be a major, but largely unquantified, cause of dugong mortality in most countries in the dugong's range. Shark nets set for bather protection have been another source of dugong mortality in Queensland, Australia. Between 1962 and 1999, shark nets set on swimming beaches in Queensland netted some 800 dugongs. The rate of decline in the catch per beach in the shark nets over this 40-year period averaged about 8% per year. If the shark net catches are a reliable index of the overall decline in the dugong population from all causes, the population of dugongs along the urban coast of Queensland in the mid-1990s was only a small fraction of that in the 1960s.

Triggered by the evidence of dugong decline, their statutory responsibilities and more generic conservation concerns, the relevant management agencies are attempting to address all known human impacts on dugongs along the urban coast of Queensland between about 15° 30'S and 28°S by a comprehensive series of management interventions. These initiatives include generic initiatives to protect the marine environment such as the establishment and rezoning of an extensive series of large-scale multiple use marine protected areas, especially the 348,000 km² Great Barrier Reef World Heritage Area, and interventions to improve water quality. In addition, there have been a series of more dugong-relevant initiatives including (1) banning the commercial dugong oil industry in the 1960s; (2) partnerships between management agency staff and indigenous traditional owners to develop a mutually acceptable, legal framework for sustainable dugong hunting at culturally relevant scales; (3) the replacement of shark nets by baited hooks at most locations; (4) the establishment of Dugong Protection Areas in the Great Barrier Reef Region and Hervey Bay in 1997; (5) voluntary vessel lanes and/ or speed restrictions to protect dugongs from vessel strikes in several major dugong habitats; (6) a review of the use of the herbicide diuron, which has been detected in both dugong tissues and the sediments associated with seagrass beds; (7) a marine wildlife carcass salvage program; and (8) a dugong research strategy which includes a dugong monitoring program based on aerial surveys. The results of 20 years of aerial surveys suggest that the management intervention has been successful and that dugong numbers are now stable at the spatial scale of the entire urban coast of Queensland although numbers in individual bays fluctuate in response to changes in the quality of the seagrass habitat.

Dugong meat tastes like beef or pork. Dugong hunting for food and oil was once widespread throughout the dugong's range and still occurs in at least 31 countries. Today the dugong is legally protected in most countries. Members of Aboriginal and Torres Strait Islander communities in northern Australia and villagers in Western Province in Papua New Guinea are still permitted to hunt and modeling evidence suggests that the level of hunting is not sustainable in Torres Strait between Australia and Papua New Guinea and along the adjacent remote coastal waters of Cape York. The indigenous peoples of northern Australia and Papua New Guinea consider dugong hunting to be an important expression of their cultural identity and dugong meat is still an important food source in some remote regions of northern Australia and Papua New Guinea. Dugong oil extracted by boiling the parts of the dugong not used for food, such as the head, is used as a panacea for aches, pains, and many illnesses. A dugong yields about 35% of its body weight in usable meat and fat and on average approximately 18 l of oil.

It is increasingly recognized that arrangements to insure that indigenous hunting of dugongs is sustainable will require the active participation of relevant indigenous communities. The Australian Government has recently formalized this approach in a 2005 policy entitled "Sustainable harvest of marine turtles and dugongs in Australia—A national partnership approach."

The World Conservation Union (IUCN) classifies the dugong as vulnerable to extinction at a global scale on the basis of anecdotal evidence of declines in its abundance throughout most of its range. In recent years—there has been an encouraging and widespread increase in interest in dugong biology and management. Nonetheless, the difficulties of ameliorating adverse influences on dugongs in the highly populated developing countries that comprise most of its range emphasize the importance of the remote regions of tropical Australia to dugong conservation.

It is vital that dugong management is coordinated at a biologically appropriate scale, which dugong movements suggest is hundreds of kilometers. The October 2007 signing of a memorandum of understanding concerning the conservation and management of dugongs and their habitats throughout their range is a very encouraging first step to resolving the mismatch between the biological and the geopolitical scales at which dugong management has occurred to date.

## See Also the Following Articles

Indo-West Pacific Marine Mammals ▪ Manatees ▪ Sirenan Life History

## References

Aragones, L., Lawler, I. R., Foley, W. J., and Marsh, H. (2006). The effects of dugong grazing and turtle cropping on nutritional characteristics of tropical seagrasses. *Oecologia* **149**, 635–647.

Bryden, M., Marsh, H., and Shaughnessy, P. (1993). "Dugongs, Whales, Dolphins and Seals: A Guide to Sea Mammals of Australia." Allen and Unwin, Sydney.

Chilvers, B. L., Delean, S., Gales, N. J., Holley, D. K., Lawler, I. R., Marsh, H., and Preen, A. R. (2004). Diving behaviour of dugongs, *Dugong dugon*. *J. Exp. Mar. Biol. Ecol.* **304**, 203–224.

Heinsohn, R., Lac, R. C., Lindenmayer, D. B., Marsh, H., Kwan, D., and Lawler, I. R. (2004). Unsustainable harvest of dugongs in Torres Strait and Cape York (Australia) waters: Two case studies using population viability analysis. *Anim. Conserv.* **7**, 417–425.

Marsh, H. (2006). *Dugong dugon*. In: "IUCN Red List of Threatened Species "<www.iucnredlist.org>. Downloaded

Marsh, H., Penrose, H., Eros, C., and Hugues, J. (2002). *Dugong: Status reports and action plans for countries and territories*. United Nations Environment Programme. Early Warning and Assessment Report Series 1, 162pp. URL: http://www.tesag.jcu.edu.au/dugong

Marsh, H., Penrose, H., and Eros, C. (2003). A future for the dugong. *In* "Marine Mammals: Fisheries, Tourism and Management Issues" (N. Gales, M. Hindell, and R. Kirkwood, eds), pp. 383–399. CSIRO Publishing, Victoria.

Marsh, H., Lawler, I. R., Kwan, D., Delean, S., Pollock, K., and Alldredge, M. (2004). Aerial surveys and the potential biological removal technique indicate that the Torres Strait dugong fishery is unsustainable. *Anim. Conserv.* **7**, 435–443.

Reynolds, J. E., III, and Odell, D. K. (1991). "Manatees and Dugongs", Facts on File. New York.

Reynolds, J. R., and Rommell, S. A. (1999). "Biology of Marine Mammals." Smithsonian Institution Press, VA.

Sheppard, J., Preen, A. R., Marsh, H., Lawler, I. R., Whiting, S., and Jones, R. E. (2006). Movement heterogeneity of dugongs, *Dugong dugon* (Müller) over large spatial scales. *J. Exp. Mar. Biol. Ecol.* **334**, 64–83.

# Dusky Dolphin
## *Lagenorhynchus obscurus*

KOEN VAN WAEREBEEK AND BERND WÜRSIG

## I. Characteristics and Taxonomy

The dusky dolphin was described by John W. Gray in 1828 from a stuffed skin and a single skull shipped to the British Museum from the Cape of Good Hope. No subspecies are recognized. Because of their morphological similarities, the stocky dusky dolphin from the Southern Hemisphere and the Pacific white-sided dolphin, *Lagenorhynchus obliquidens*, from the northern Pacific Ocean are considered phylogenetically closely related species despite the absence of a fossil record. Some researchers suggested that *L. obliquidens* could almost equally well be regarded as a subspecies of the dusky dolphin. However, a close scrutiny of morphological and life history parameters does not support this premise. Cytochrome *b* sequence analysis is consistent with the "sister species" hypothesis; the divergence date is estimated at 1.9–3.0 million years ago (Cipriano, 1997; Cassens *et al.*, 2003).

The smallish dusky dolphin can be recognized by its short beak and the bluish black to dark gray of the dorsal field contrasting with the white belly, as well as the light gray of the thoracic patch and two-pronged flank patch. The dark lips and eye patch also stand out. The falcate dorsal fin is two-toned with a dark leading edge (Fig. 1). Unlike in *L. obliquidens*, the linear dorsal flank blaze does not extend farther anteriorly than about mid-body. Heavily pigmented specimens are found off Peru and Argentina. The holotype of *Delphinus fitzroyi* (Waterhouse, 1838) caught off Argentina from Darwin's ship *Beagle* was such a melanistic form. Various cases of anomalous, piebald pigmentation are probably equivalent to so-called partial albinism. Tooth counts range for both upper and lower half jaw 26–39, with some differences between populations.

Both males and females off Peru reach sexual maturity at about 175cm; average adult size is approximately 185cm, while the largest two known individuals measured 211 (male) and 205cm (female), based on 693 specimens. Dusky dolphins rarely exceed 100kg in weight. No significant dimorphism is present in morphology, including coloration patterns, except that in adult males the dorsal fin is more curved, has a broader base, and a greater surface area than in females, presumably a secondary sexual characteristic. Dusky dolphins from southwestern Africa and New Zealand are some 8–10cm shorter than Peruvian specimens of both sexes, which supports a conclusion of discrete populations based on cranial variability and molecular genetics studies. Mature skulls from South West Africa and New Zealand are on average 8.5% (3.1cm) shorter than skulls from Peru and Chile (Van Waerebeek, 1992a; 1993a, b; Cassens *et al.*, 2003, 2005).

Possible hybrids have been described between a dusky dolphin and a southern right whale dolphin, *Lissodelphis peronii* (Yazdi, 2002) and between a dusky dolphin and a long-beaked common dolphin, *Delphinus capensis* (Reyes, 1996). Analysis of nine microsatellite loci and two mitochondrial gene fragments suggest that dusky dolphins from Argentina and South West Africa recently separated (*ca.* 2000 generations ago) from an ancestral Atlantic population and since then diverged without considerable gene flow (Cassens *et al.*, 2005). There are low levels of genetic differentiation among most populations. Only the Peruvian dusky dolphin stock is highly differentiated, especially at mitochondrial loci, suggesting that major

D

**Figure 1** *A dusky dolphin in a somersault leap off New Zealand. This acrobatic leap type indicates a high behavioral social level by the individual, and very likely its group. Photo by B. Würsig*

fluctuations in its population size have led to an increased rate of genetic drift. No genetic subdivision was detected within the Peruvian and African stocks (Cassens *et al.*, 2005). Nor was any subdivision among regions found in New Zealand, where at least one, if not two, historical population expansions were proposed (Harlin *et al.*, 2003).

## II. Distribution and Abundance

Dusky dolphins are distributed around South America, from northern Peru south to Cape Horn and from southern Patagonia north to about 36°S, including the Falkland Islands (Malvinas); off southwestern Africa from False Bay, South Africa to Lobito Bay, Angola; and in New Zealand waters, including off Chatham and Campbell Islands. Populations of unknown size inhabit waters surrounding oceanic islands of the Tristan da Cunha Archipelago in the mid-Atlantic, the Prince Edward Islands, and Crozet and Amsterdam Island in the southern Indian Ocean (Van Waerebeek *et al.*, 1995). The species has been recently confirmed from southern Australia but seems to be rare there (Gill *et al.*, 2000). What appeared to be a possible distribution gap between 36–46°S off Chilean Patagonia (Van Waerebeek, 1992b) can be attributed to minimal observer effort, as sightings are now confirmed from that region. There is increasing evidence that dusky dolphins may occasionally venture into waters of the Southern Ocean South of the Antarctic Polar Front (Van Waerebeek *et al.*, in press).

In a 15,000 nmiles² fishing area off Argentina, abundance was estimated as 7252 dolphins (Dans *et al.*, 1997a). No abundance estimates are available at the scale of an entire population. However, dusky dolphins are the most common small cetaceans on the Patagonian Shelf (Schiavini *et al.*, 1999) and one of three most abundant species in Peru's coastal waters.

Sightings and morphological and molecular genetic evidence suggest a disjunct distribution across oceans and the existence of discrete stocks confined to continental shelves and oceanic island groups, with indication of some level of male dispersal along coasts (Van Waerebeek *et al.*, 1995; Cassens *et al.*, 2003, 2005).

## III. Ecology

Dusky dolphins inhabit predominantly neritic waters above continental shelves and slopes but can also be found over deep water if close to continents or islands. These coastal habits explain the discontinuous distribution across the temperate southern ocean and the reproductive isolation of populations. They seem to prefer waters with sea-surface temperatures of 10–18°C (Brownell and Cipriano, 1999), but lower temperatures have been reported.

Dusky dolphins over the extensive continental shelf off Argentina forage cooperatively on small schooling fishes during the day. On the east coast of New Zealand, where deep oceanic waters reach close to shore, animals typically feed at night on prey associated with the deep scattering layer (Würsig *et al.*, 1997). Some of these same dolphins, identified by natural marks on their bodies, have been found to feed during the day in the Marlborough Sounds on the northern edge of the South Island of New Zealand, reinforcing a generally held belief that these animals adapt behaviorally relative to habitat and prey availability patterns (Markowitz *et al.*, 2003).

Prey items include a wide variety of fish and squid species. Dominant are small schooling fishes such as anchovies, lantern fishes, pilchards, and sculpins, but also hakes, horse mackerel, hoki, red cod, and a large number of squid species, i.e., of the genera *Notodarus*, *Todaroides*, and *Loligo* amongst others. The 80% anchovy diet component (by weight) found by McKinnon (1994) in Peru, suggesting an almost exclusive dependence on this prey, is now understood to have been biased by localized and short-term sampling. New evidence indicates that Peruvian dusky dolphins are opportunistic feeders that will readily feed on other prey species where and when anchovy is not abundant. In New Zealand, dusky dolphins feed on a variety of fish and squid species; lantern fish, hoki, and squid are predominant (Cipriano, 1992; Würsig *et al.*, 1997). In a small sample from Argentina, anchovy represented 46% by weight of the diet (Alonso *et al.*, 1998).

Killer whales (*Orcinus orca*) are occasional predators, and dusky dolphins within about 5 km of shore, when killer whales approach, will enter very shallow water to avoid them (Würsig and Würsig, 1980). In Golfo Nuevo, Argentina, mothers with calves and smaller groups, as well as groups exhibiting resting behavior, were found in the shallowest waters, supporting the hypothesis that movement to shallower water is related to increased safety for individuals (Garaffo, 2007). This is also the case for dusky dolphins in open ocean New Zealand waters (Weir, 2007).

Prevalence of helminth parasites in dusky dolphin is reported for Peruvian and Argentine stocks only. *Anisakis simplex*, *Nasitrema* sp., *Phyllobothrium delphini*, *Braunina cordiformis*, *Hadwenius* sp., and *Pholeter gastrophylus* are common, while *Crassicauda* sp., *Halocercus* sp., and *Corynosoma australe* are rare, with some significant differences between the Pacific and Atlantic populations (Van Waerebeek *et al.*, 1993; Dans *et al.*, 1999). The epidemiology of diseases has been studied in some detail in Peruvian dusky dolphins. Endemic presence was established for poxvirus, papillomavirus, a herpes-like virus, and *Brucella* sp. Genital diseases in a large sample (N = 502) included ovarian cysts, ovarian, and uterine tumours, vaginal calculi, abscesses of the broad ligament and testicular lesions of unknown aetiology (Van Bressem *et al.*, 1994, 2000, 2006; Van Bressem and Van Waerebeek, 1996).

## IV. Behavior and Physiology

Although dusky dolphins can move over great distances—a range of 780 km is confirmed—no well-defined seasonal migration patterns are apparent. However, the Argentina and New Zealand populations exhibit inshore–offshore movements both on a diurnal and on a seasonal scale (Würsig *et al.*, 1997).

Surface feeding activity typically happens in large groups, accompanied with extensive aerial display. It is believed that such aerial activity helps synchronize cooperative foraging and after-feeding social (including sexual) activities (Würsig and Würsig, 1980; Würsig *et al.*, 1989). Individual dolphins alter their foraging strategies and prey types on a seasonal/regional basis (Markowitz *et al.*, 2003; Würsig *et al.*, 2007).

In New Zealand, during the morning, dolphin groups are spread out, leaping often, and are closest to shore. Around midday, the dolphins tend to rest in tight groups. They then spread out again and move offshore in late afternoon/early evening. Groups of hundreds of individuals exhibit coordinated travel and noisy leaps. At night, the dolphins forage (and presumably feed) in small subgroups at the shallowest dive depths but individually dive down to as deep as 130 m (Würsig *et al.*, 2007). In coastal Peru, dusky dolphins are commonly seen together with long-beaked common dolphins in large feeding aggregations of many hundreds or thousands of individuals, which in turn attract thousands of guano-producing seabirds (cormorants, pelicans, boobies) that dive among the dolphins. These spectacular aggregations can be observed from shore-based vantage points, especially in winter. Alternatively, large numbers of dolphins may be spread out over a huge area, in what appears to be a single meta-group travelling in the same direction. However, circumstances of gillnet entanglement suggest that Peruvian dusky dolphins may also engage in nocturnal foraging. Dusky dolphins sometimes associate with short-beaked common dolphins, *Delphinus delphis*, off New Zealand and with southern right whale dolphins, *L. peronii*, in Namibia. School sizes vary from only 3–5 individuals to the more common dozen or so, and occasionally up to about 2000 (the latter off Kaikoura, New Zealand, in April–May; Markowitz, 2004). Most of the species' echolocation signals have bimodal frequency spectra with a low-frequency peak between 40 and 50 kHz and a high-frequency peak between 80 and 110 kHz. The wave form and spectrum of the echolocation signals are similar to those of other dolphins measured in the field (Au and Würsig, 2004).

## V. Life History

In Peru and New Zealand, most births occur in late winter (August, September, and October), but a few neonates appear during other seasons. Growth rate of fetuses is 0.261 cm/day. Neonatal length of Peruvian dusky dolphins averages 91 cm at 9.6 kg. Gestation lasts for 12.9 months, followed by 12 months of lactation and a resting period of 3.7 months. The size of adult testes increases, reaching a maximum in September and October, in synchrony with the peak period of conception (Van Waerebeek and Read, 1994). The largest single-testis mass was 5120 g, proportionately among the highest of any mammal. Very large testes size, relative lack of sexual dimorphism, and apparent lack of aggressive behavior between males suggest a promiscuous mating system and sperm competition (Van Waerebeek, 1992a).

In Peru, sexual maturity for females is estimated at 4.3–5 years and for males at 3.8–4.7 years. The lower values, which are the most recent, possibly indicate a density-dependent response to heavy exploitation (Chávez-Lisambarth, 1998) or an adaptation to an unstable environment with recurring El Niño-Southern Oscillation (ENSO) events. Anomalous dentine deposition during El Niño years

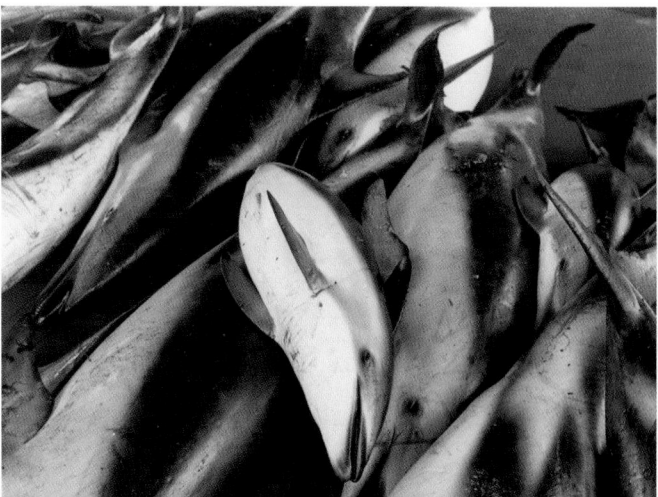

**Figure 2**  *In the early 1990s thousands of dusky dolphins were landed in Peruvian ports each year. Photo by K. Van Waerebeek*

off Peru may reflect low foraging success. Female Argentinean dusky dolphins mature at 6.3 years of age and, in contrast with Peru and New Zealand, summer is the prime birth season (Dans *et al.*, 1997b).

## VI. Interactions with Humans

In the absence of global abundance estimates, the status of the dusky dolphin remains indeterminate. Unknown numbers are caught in gill nets in New Zealand waters, although current catches have dropped from those of the 1970s and 1980s. Some 200 dusky dolphins were incidentally captured off Kaikoura in 1984 (Brownell and Cipriano, 1999). Off Argentina, from 1982–1994, a variable number of Patagonian dusky dolphins, typically a few hundred each year, died in midwater trawls (Dans *et al.*, 1997a). Arguably of greatest concern are fishery-related mortality levels in Peruvian coastal waters, where both directed and accidental entanglements in artisanal drift nets, as well as harpooning, have killed thousands each year since 1985. In the 1991–1993 period, an estimated 7000 animals per annum were captured (Fig. 2), an exploitation thought to be unsustainable (Van Waerebeek and Reyes, 1994). In 1994, in a single port an estimated 1272 (SE 227) dusky dolphins were landed in a 7-month period (Van Waerebeek *et al.*, 1997). Although the second (1994) ban on the landing and sale of dolphins is believed to have reduced direct captures and landings in Peru, no measures were adopted to address bycatch, and unknown numbers of dolphins continue to be killed in fisheries. Indications are that the trend of a significant decline in the proportion of dusky dolphins in the overall cetacean take from 77.5% (in 1985–1990) down to 52.8% (1991–1993) has continued. Nowadays, captured dolphins are routinely used as shark bait in both longline and drift gillnet fisheries, and in at least in some ports dolphin meat is landed clandestinely and sold in public markets. While legislation has helped to curb visible trade, the lack of inspections, especially on boats at sea, has failed to stop the exploitation.

Only a few individuals have been exhibited, in aquaria in New Zealand, South Africa, and Australia, largely because dusky dolphins adapt rather poorly and have failed to reproduce successfully in captivity. There is increasing concern about impacts of dolphin-based tourism in New Zealand and Argentina, and several studies of the problem have been conducted. A moratorium on dusky dolphin tourism in New Zealand, established in 1999, is currently under review (Würsig *et al.*, 2007).

D

## See Also the Following Articles

Delphinrds, Overview ∎ Pacific White-Sided Dolphin

## References

Alonso, M. K., Crespo, E. A., García, N. A., Pedraza, S. N., and Coscarella, M. A. (1998). Diet of dusky dolphins, *Lagenorhynchus obscurus*, in waters off Patagonia, Argentina. *Fish. Bull., US* **96**, 366–374.

Au, W. W. L., and Würsig, B. (2004). Echolocation signals of dusky dolphins (*Lagenorhynchus obscurus*) in Kaikoura, New Zealand. *J. Acoust. Soc. Am.* **115**, 2307–2313.

Brownell, R. L., and Cipriano, F. (1999). Dusky dolphin— *Lagenorhynchus obscurus* (Gray, 1828). *In* "Handbook of Marine Mammals." (S. H. Ridgway, and S. R. Harrison, eds.), Vol. 6, "The Second Book of Dolphins and Porpoises," 85–104.

Cassens, I., Van Waerebeek, K., Best, P. B., Crespo, E. A., Reyes, J., and Milinkovitch, M. C. (2003). The phylogeography of dusky dolphins (*Lagenorhynchus obscurus*): A critical examination of network methods and rooting procedures. *Mol. Ecol.* **12**, 1781–1792.

Cassens, I., Van Waerebeek, K., Best, P. B., Tzika, A., van Helden, A., Crespo, E. A., and Milinkovitch, M. C. (2005). Evidence for male dispersal along the coasts but no migration in pelagic waters in dusky dolphins (*Lagenorhynchus obscurus*). *Mol. Ecol.* **14**, 107–121.

Chávez-Lisambarth, L. E. (1998). Age determination, growth and gonad maturation as reproductive parameters of dusky dolphin *Lagenorhynchus obscurus* (Gray, 1828) from Peruvian waters. Thesis Dr. rer. nat., University of Hamburg.

Cipriano, F. W. (1992). Behavior and occurrence patterns, feeding ecology, and life history of dusky dolphins (*Lagenorhynchus obscurus*) off Kaikoura, New Zealand. Unpublished Ph.D. Dissertation, University of Arizona, Tucson, USA. 216 p.

Cipriano, F. (1997). Antitropical distributions and speciation in dolphins of the genus *Lagenorhynchus*: A preliminary analysis. *In* "Molecular Genetics of Marine Mammals," pp. 317–338. Society for Marine Mammalogy Special Publication **3**.

Crespo, E. A., *et al.* (1997). Distribution and school size of dusky dolphins, *Lagenorhynchus obscurus* (Gray, 1828), in the southwestern South Atlantic Ocean. *Rep. Int. Whal. Commn.* **47**, 693–697.

Dans, S. L., Crespo, E. A., Garcia, N. A., Reyes, L. M., Pedraza, S. N., and Alonso, M. K. (1997a). Incidental mortality of Patagonian dusky dolphins in mid-water trawling: Retrospective effects from the early 1980s. *Rep. Int. Whal. Commn.* **47**, 699–703.

Dans, S. L., Crespo, E. A., Pedraza, S. N., and Alonso, M. K. (1997b). Notes on the reproductive biology of the female dusky dolphins *Lagenorhynchus obscurus* (Gray, 1828) off the Patagonian coast. *Mar. Mamm. Sci.* **13**, 303–307.

Dans, S. L., Reyes, L. M., Pedraza, S. N., Raga, J. A., and Crespo, E. A. (1999). Gastrointestinal helminths of the dusky dolphin, *Lagenorhynchus obscurus*, (Gray, 1828) off Patagonian coasts, in the Southwestern Atlantic Ocean. *Mar. Mamm. Sci.* **15**, 649–660.

Garaffo, G. V., Dans, S. L., Pedraza, S. N., Crespo, E. A., and Degrati, M. (2007). Habitat use by dusky dolphin in Patagonia: How predictable is their location? *Mar. Biol.* **152**, 165–177.

Gill, P. C., Ross, G. B., Dawbin, W. H., and Wapstra, H. (2000). Confirmed sightings of dusky dolphins (*Lagenorhynchus obscurus*) in southern Australian waters. *Mar. Mamm. Sci.* **16**, 452–459.

Harlin, A. D., Markowitz, T., Baker, C. S., Würsig, B., and Honeycutt, R. L. (2003). Genetic structure, diversity, and historical demography of New Zealand's dusky dolphin. *J. Mammal.* **84**, 702–717.

Markowitz, T. (2004). Social organisation of the New Zealand dusky dolphin. Ph.D. Dissertation, Texas A&M University, Texas, USA. 278pp.

Markowitz, T., Harlin, A. D., Würsig, B., and McFadden, C. J. (2003). Dusky dolphin foraging habitat: Overlap with aquaculture in New Zealand. *Aquat. Conserv. Mar. Freshw. Ecosyst.* **14**, 133–149.

McKinnon, J. (1994). Feeding habits of the dusky dolphin, *Lagenorhynchus obscurus*, in the coastal waters of central Peru. *Fish. Bull., US* **92**, 569–578.

Reyes, J. C. (1996). A possible case of hybridism in wild dolphins. *Mar. Mam. Sci.* **12**, 301–307.

Schiavini, A., Pedraza, S. N., Crespo, E. A., González, R., and Dans, S. L. (1999). Abundance of dusky dolphins (*Lagenorhynchus obscurus*) off north and central Patagonia, Argentina, in spring and a comparison with incidental catch in fisheries. *Mar. Mam. Sci.* **15**, 828–840.

Van Bressem, M. F., and Van Waerebeek, K. (1996). Epidemiology of a poxvirus in small cetaceans from the eastern South Pacific. *Mar. Mam. Sci.* **12**, 371–382.

Van Bressem, M. F., Van Waerebeek, K., García-Godos, A., Dekegel, D., and Pastoret, P. P. (1994). Herpes-like virus in dusky dolphins, *Lagenorhynchus obscurus*, from coastal Peru. *Mar. Mam. Sci.* **10**, 354–359.

Van Bressem, M-F., Van Waerebeek, K., Siebert, U., Wünschmann, A., Chávez-Lisambarth, L., and Reyes, J. C. (2000). Genital diseases in the Peruvian dusky dolphin (*Lagenorhynchus obscurus*). *J. Comp. Pathol.* **122**, 266–277.

Van Bressem, M. F., Van Waerebeek, K., and Bennett, M. (2006). Orthopoxvirus neutralising antibodies in small cetaceans from the Southeast Pacific. *LAJAM* **5**, 49–54.

Van Waerebeek K. (1992a). Population identity and general biology of the dusky dolphin *Lagenorhynchus obscurus* (Gray, 1828) in the Southeast Pacific. Doctoral Dissertation, Institute for Taxonomic Zoology, University of Amsterdam. 160pp.

Van Waerebeek, K. (1992b). Records of dusky dolphins *Lagenorhynchus obscurus* (Gray, 1828) in the eastern South Pacific. *Beaufortia* **43**, 45–61.

Van Waerebeek, K. (1993a). External features of the dusky dolphin *Lagenorhynchus obscurus* (Gray 1828) from Peruvian waters. *Estud. Oceanol.* **12**, 37–53.

Van Waerebeek, K. (1993b). Geographic variation and sexual dimorphism in the skull of the dusky dolphin *Lagenorhynchus obscurus* (Gray 1828). *Fish. Bull., US* **91**, 754–774.

Van Waerebeek, K., and Read, A. J. (1994). Reproduction of dusky dolphins *Lagenorhynchus obscurus* from coastal Peru. *J. Mammal.* **75**, 1054–1062.

Van Waerebeek, K., and Reyes, J. C. (1994). Post-ban small cetacean takes off Peru: A review. *Reports of the International Whaling Commission (Spec. Iss.)* **15**, 503–520.

Van Waerebeek, K., van Bree, P. J. H., and Best, P. B. (1995). On the identity of *Prodelphinus Petersii* Lütken, 1889 and records of dusky dolphins *Lagenorhynchus obscurus* (Gray, 1828) from the Indian and southern mid-Atlantic Oceans. *S. Afr. J. Mar. Sci.* **16**, 25–35.

Van Waerebeek, K., Reyes, J.C. and Alfaro, J.O. (1993). Welminth parasites and phoronts of dusky dolphins *Lagenorhynchus obscurus* (Gray, 1828) from Peru. *Aquat. Mamm.* 19, 159–169.

Van Waerebeek *et al* (8 authors) (In press.) Odontocetes of the Southern Ocean Sanctuary *J. Cetacean Res, Manage*.

Van Waerebeek, K., *et al.* (9 authors) (1997). Mortality of dolphins and porpoises in coastal fisheries off Peru and southern Ecuador in 1994. *Biol. Conserv.* **81**, 43–49.

Waterhouse, G. R. (1838). On a new species of the genus *Delphinus*. *Proc. Zoo. Soc. Lond.*

Weir, J. (2007). Dusky dolphin nursery groups off Kaikoura, New Zealand. Unpublished MS Thesis. Texas A&M University, Texas, USA. 75pp.

Würsig, B., and Würsig, M. (1980). Behavior and ecology of the dusky dolphin, *Lagenorhynchus obscurus*, in the South Atlantic. *Fish. Bull.* **77**, 871–890.

Würsig, B., Würsig, M., and Cipriano, F. (1989). Dolphins in different worlds. *Oceanus* **32**(1), 71–75.

Würsig, B., Cipriano, F., Slooten, E., Constantine, R., Barr, K., and Yin, S. (1997). Dusky dolphins (*Lagenorhynchus obscurus*) off New Zealand: Status of present knowledge. *Rep. Int. Whal. Commn. Sci.* **47**, 715–722.

Würsig, B., Duprey, N., and Weir, J. (2007). Dusky dolphins (*Lagenorhynchus obscurus*) in New Zealand waters, present knowledge and research goals. *DOC. Res. Dev. Ser.* **270**, 1–28.

Yazdi, P. (2002). A possible hybrid between a dusky dolphin (*Lagenorhynchus obscurus*) and the southern right whale dolphin (*Lissodelphis peronii*). *Aquat. Mamm.* **28**, 211–217.

# Eared Seals
## *Otariidae*

### ROGER L. GENTRY

The Otariidae, or "eared seals," evolved around the rim of the North Pacific Ocean 11–12 million years ago and subsequently spread into the Southern Hemisphere. Around three million years ago, sea lions diverged from the fur seals that preceded them. The two are distinguishable by their pelage. Sea lions have a single coat of hair whereas fur seals have stiff outer guard hairs emerging from a dense, thick layer of fine underfur, which is waterproof, thereby providing thermoregulation.

Otariids comprise 16 species (Table I). All breeding sites they use are between 60°S and 55°N latitude. Eight species breed in temperate latitudes, two in subpolar, and two in equatorial regions (King, 1989). The sea lions use only temperate or equatorial regions, but fur seals use all three. No species of otariids mate on ice, but at least two encounter snow during breeding and pup rearing. Four species mate on the shores of deserts where the climate is ameliorated by fog or cool ocean winds. Only two species of fur seals and three species of sea lions still inhabit the North Pacific waters where otariids evolved. Otariids never occupied the North Atlantic Ocean, perhaps because the Central American seaway was closed when the family dispersed into the Southern Hemisphere.

Otariids use both island and mainland sites, but at present, island sites predominate worldwide. Islands offer a combination of freedom from terrestrial predators (including humans), cooling winds for THERMOREGULATION, and closer access to offshore prey concentrations than most mainland sites provide. The preferred substrate for mating is rock or sand, rarely soil or mud. All species mate on open ground except some small populations of Hooker's sea lions (see Table I for Scientific names), which use forested sites.

## I. Unique Traits

Otariids are truly amphibious in that they feed at sea but, mate and rear their young on land. Therefore, their anatomy strikes a balance between functioning in air and underwater.

Unlike other pinnipeds or cetaceans, otariids have an external ear flap (hence their scientific and common names), an air filled auditory canal, and a middle ear structure very similar to a terrestrial ear. Otariids hear best in the frequency range of 2–12 kHz. In contrast, dolphins and porpoises that rely on a sophisticated echolocation system to feed hear best from 8 to 90 kHz or higher. This difference suggests that otariids either do not echolocate or do so in a rudimentary way.

Laboratory tests have demonstrated that sea lions have excellent visual acuity at low-light levels. Large eyes, an all-rod retina, and a well-developed reflective layer (tapeta lucidum) account for this ability (see Vision, this volume). These features suggest that eared seals feed visually in the dark (at night, in deep water, or both). However, they maintain good visual acuity in air in bright light by being able to close down their pupil to a pinhole. Well developed, highly enervated vibrissae may provide tactile cues that combine with visual cues for feeding in the dark.

Locomotion also points out the balance between their terrestrial and aquatic lives. Unlike phocids, otariids can rotate their rear flippers forward and walk or run using all four limbs (see Locomotion, Terrestrial, this volume). They can outrun a human over slippery rocks. Using their chin and their long, strong front flippers, they can climb steeply sloping surfaces, which means they can take advantage of broken, rocky terrain for BREEDING SITES. They swim using the

TABLE I
Otariids, Eared Seals

| Common name | Scientific name | Location | Approximate numbers | Trend |
|---|---|---|---|---|
| Northern fur seal | *Callorhinus ursinus* | Subpolar | 1,400,000 | Stable |
| Antarctic fur seal | *Arctocephalus gazella* | Subpolar | 3,000,000 | Increasing |
| Sub-antarctic fur seal | *A. tropicalis* | High temperate | >310,000 | Increasing |
| Juan Fernandez fur seal | *A. philippii* | Temperate | 18,000 | Increasing |
| Guadalupe fur seal | *A. townsendi* | Temperate | >7000 | Increasing |
| Cape fur seal | *A. pusillus pusillus* | Temperate | 1,700,000 | Increasing |
| Australian fur seal | *A. pusillus doriferus* | Temperate | >60,000 | Stable |
| South American fur seal | *A. australis* | Temperate | >285,000 | Increasing |
| New Zealand fur seal | *A. forsteri* | Temperate | 135,000 | Increasing |
| Galapagos fur seal | *A. galapagoensis* | Temperate | 40,000 | Fluctuating |
| Galapagos sea lion | *Zalophus wollebaeki* | Temperate | 40,000 | Fluctuating |
| California sea lion | *Z. californianus* | Temperate | >188,000 | Increasing |
| Japanese sea lion | *Z. japonicus* | Temperate | Extinct | |
| Steller sea lion | *Eumetopias jubatus* | High temperate | >76,000 | Decreasing |
| Hooker's sea lion | *Phocarctos hookeri* | Temperate | 13,000 | Stable |
| Australian sea lion | *Neophoca cinerea* | Temperate | 12,000 | Stable |
| Southern sea lion | *Otaria flavescens* | Temperate | 275,000 | Decreasing |

front flippers powered by well-developed pectoral muscles, while the rear flippers trail behind and are used only in turning and stopping. Otariids are quick, graceful swimmers that can leap free of the water and "porpoise" to breathe while swimming fast. They can execute a forward somersault while surfing.

Laboratory tests have also shown that otariids have well-developed cognitive abilities. They can perform complex learning tasks and have excellent memories, abilities that are perhaps related to their role as high-level consumers which depend on patchy prey to survive.

Otariids must reduce heat loss to cold ocean waters, but simultaneously have the ability to lose body heat while on land in the sun. A BLUBBER layer (abetted by the underfur in fur seals) reduces heat loss to water. To promote heat loss on land, otariids use shade, immersion in tide pools or spray, flipper waving or panting. At least one species urinates on its rear flippers for evaporative cooling. Only the flippers are naked, so evaporative cooling from sweat is ineffective. A few animals drink seawater, but usually only at the beginning of fasting at the start of the breeding season. As in VISION and HEARING, the primary adaptations for thermoregulation and water balance seem to be for life at sea, with secondary adaptations for life on land.

## II. Diet

Otariids tend to be generalist feeders, taking a wide variety of prey. For example, the northern fur seal takes at least 63 species of prey over its full range. Fur seals and sea lions tend to feed on dissimilar prey. Fur seals often feed in deep water beyond the continental shelf break on small squid or fish, especially myctophids (lantern fish). The Antarctic fur seal takes krill as well as fish, and the New Zealand fur seal supplements fish with rock lobster. Sea lions tend to feed on or near the continental shelf and tend to take larger or more mature stages of prey than are used by fur seals (e.g., adult halibut). Hooker's and Australian sea lions specialize in squid and octopus, which are hunted on the bottom and under rocks. Several species of sea lions are known to eat the young of other seals, one species exhibits cannibalism, and several are known to eat penguins or other birds. For an as yet unexplained reason, otariids (especially sea lions) intentionally swallow fist-sized rocks.

## III. Maternal Strategy

Otariids differ from phocids in that the females of all species continue to feed while they are lactating (see Pinnipedia Physiology, this volume). Otariid mothers capture prey and within a few days time transfer the energy it contains to their young on shore in the form of milk fat. They make a series of brief, regular nursing visits to shore from birth to weaning, an interval that varies with latitude. In contrast, most phocid mothers gather energy from prey for several months, lay it down as extensive blubber reserves, and then deliver it to their young in a single, prolonged nursing bout that may last a few weeks at most.

Feeding during lactation has widespread implications for otariid natural history. Eared seal mothers are restricted in how long they may be absent for foraging by the limited fasting abilities of their newborn pups. To meet this restriction, mothers must find abundant prey of a given type close enough to the colony that they can commute between the two and still experience a net energetic gain for themselves and their young. The number of sites that meet these conditions and that have the proper terrain, cooling winds, absence of predators, and other factors are relatively few. Otariids tend to gather in large numbers (up to hundreds of thousands) on the few sites in the world that meet these specific restrictions and to form dense aggregations there. Therefore, largely because of maternal commuting, otariids tend to breed in a few large dense colonies that are most often on

islands. Most phocids do not commute for foraging and therefore tend to breed dispersed as pairs or in small numbers on a large number of dispersed sites. Among phocids, only elephant seals (*Mirounga* spp.) form colonies that superficially resemble those of otariids (few sites, large dense groups).

The details of otariid maternal strategies vary considerably according to the local foraging environment. As a generalization, the duration of the maternal feeding trip, duration of nursing visits, and fat content of the milk all increase with increasing distance to feeding locations. However, the amount of milk fat and its delivery schedule are both constrained by the need to wean the young at a particular age. At high latitudes this age is 4 months, which allows mother and young time to migrate to lower latitudes before winter begins. At temperate latitudes the age is 9–12 months, which allows females to bear young annually without having to support two simultaneously. The two species at the equator are not limited by season, but by the rate at which mothers can bring their pups to the size needed for independent foraging. In good years, Galapagos fur seal mothers may achieve this in 12 months. However, when periodic El Nino events disrupt the food supply, they may not achieve it until age 3 years.

## IV. Reproductive Adaptations

Otariids and phocids both possess a postpartum estrus (Riedman, 1990). The uterus is Y shaped with two "horns," one of which holds the full-term fetus while the other prepares to receive the new blastocyst soon after birth. In all but one species of otariid, females enter estrus and mate during the 4- to 11-day perinatal nursing period. (The exception is the California sea lion, which mates about 23 days postpartum, well beyond its perinatal nursing period.) Because the perinatal period may be the females' longest single shore visit of the year, estrus at that time gives adult males the longest uninterrupted chance to mate with them. The process is quite efficient; pregnancy rates may exceed 93% for some age groups of females.

Dense gatherings have produced the most striking feature of the otariids, namely sexual dimporhpism in body size and appearance (Fig. 1). In each species the adult male is 2–4.5 times the size of the female. Looking across species, there is more dissimilarity in the appearance of males as a group than there is among females as a group. These differences in body size and appearance probably result from the dissimilar selection pressures that the two sexes are under. Unlike females, in which fitness is measured by the quality of the single young they bear, male fitness is measured by the number of offspring produced. This means that males attempt to obtain the largest possible number of mates but females do not. Increased size not only gives males an advantage in fights related to obtaining mates, it also allows them to fast longer and thus remain among estrous females longer than males of smaller size. Male otariids weigh from 200 to 1000 kg and can fast for 12 weeks, although a fast of 4–6 weeks is more common.

## V. Mating System

In all species, males defend space on the land sites that females use for giving birth and nursing their young (Riedman, 1990). In two species (California sea lions and Juan Fernandez fur seals) males also partition some of the aquatic areas where females gather. Defending space gives males exclusive reproductive access to all females on that site. Failure to obtain space among females means having low reproductive success because of the characteristics of estrus (see later). In very few cases do males defend the actual females on a site rather than the boundaries around the site they use.

**Figure 1**    *Sexual dimorphism in otariids, note the large male and smaller female* (C. Brett Jarrett)

Adult female

Pup

Adult male

E

Defended space in otariids should be referred to as territories not as leks. The reason is that, except possibly in the California sea lion, otariid females do not actively choose their mates as in a lek system. Instead, females choose their parturition site and then mate with any nearby male when estrus occurs. Females may use the exact same parturition site for 10 years or more, whereas the males that defend these sites change every breeding season or two.

All otariids are polygynous; average adult sex ratios of up to 10 females per male are common. Although the sex ratio at birth is near unity, males have higher mortality rates during maturation than females, and some males that reach adulthood are excluded from mating by the territorial system, which leaves more mating females than males. The average adult sex ratio is difficult to determine because a variable number of females will be at sea on any given day.

Sexual receptivity in female otariids is usually terminated by the physical act of coitus. For that reason, receptivity may last only minutes per year, and most females (85–90%) copulate only once. No second estrus is known to occur later in the year, even for females that fail to mate immediately postpartum. The advantage to females of a single copulation is that they can dispense with mating and quickly resume feeding to support their young. The advantage to males is that they can inseminate more females with fewer copulations in a shorter period of time. Most virgin females mate on the same grounds as more mature females, but at the end of the breeding season. For these reasons, males that do not acquire territory on traditional pieces of ground at predictable times of year find many fewer mates than males that do.

Female otariids have an open society that features a loose, size-related dominance system. Larger, older, more experienced females generally have more access to favored rest sites, water, or shade than younger, smaller females. However, no evidence has been found for hierarchies of individuals. Also, no evidence has been found for any social bond between adult females, including those between mothers and their previous female offspring. The society is open in the sense that females can enter and leave it frequently without loss of social status. It thus fits with, and may have resulted from, the need of females to forage specific to the needs of their nursing offspring.

Females maintain close spacing while on shore. All sea lions plus the South African fur seal are thigmotactic (seek full body contact with others); spacing in the other species varies seasonally and with the radiant load (less contact on hot days). Despite their close spacing on shore, females are simultaneously aggressive toward each other.

Fights that draw blood are uncommon, but females threaten each other frequently in various ways.

The offspring of otariids are precocial. Some can swim on the day of birth but most defer swimming for a month or more. Sexual dimorphism is evident at birth. In one species (northern fur seal) it exists as early in embryonic development as the sexes can be distinguished. During the lactation period, otariid pups tend to form dense aggregations and to avoid contact with adult males, which may bite and kill any pup that approaches them. Pups engage in play bouts that feature many of the components of adult aggressive behavior, sexual behavior, and prey handling in mixed order. These patterns appear to be innate in that pups born in captivity display behavior that they cannot have witnessed. Sneak suckling, especially in starvelings, occurs probably in all otariids. In some species (Steller's sea lion), some females that lose their pups may actually foster a foreign pup.

Weaning is difficult to observe and document. It has been studied well in only two species (northern and Antarctic fur seals). In both species, most pups leave shore before their mothers, thereby weaning themselves at 4 months, just prior to migrating. In non-migratory species, varying degrees of mother–young conflict exist at weaning, depending on environmental conditions, age at weaning, and presence of younger offspring.

Outside the breeding season, when no adult males are present, weaned juveniles may reside among the females and young. Juvenile females may remain there during the breeding season, but juvenile males are usually excluded from breeding areas and may gather on nearby all-male landing areas. In some areas of the world, several species of seals may mix on these landing areas. Juvenile males spend much time play fighting and occasionally make brief running forays into breeding areas or wait offshore to intercept females departing on foraging trips. Copulations sometimes occur in these circumstances.

## VI. Foraging Behavior

Otariids usually only dive deeply when foraging. This summary is based on the dive records for females, which have been studied more thoroughly than males. Most otariids feed in the water column, sometimes over very deep water. Dive depths measured in the field reflect the vertical distribution of the prey more than the physiological limits of the divers. These depths are usually less than 450 m, and dives to more than 200 m are uncommon. Dive durations are usually less than 12 min, and dives of more than 5–7 min are uncommon. Most dives of otariids (more than 85% for individuals) are aerobic, i.e., the duration of most dives does not exceed the estimated oxygen that animals take down stored in hemoglobin or myoglobin. Diving occurs at all hours, but nighttime diving predominates. The dive depth may change with time of day, suggesting that the otariid is following the daily migration of prey toward the surface at dusk and toward deep water at dawn. There is as yet no evidence of cooperative feeding by otariids (such as group attacks by killer whales), but coordinated group diving has been observed.

## VII. Population Trends

Starting in the sixteenth century, otariids were harvested for furs, hides, blubber, various organs, or, in the case of Steller sea lions, vibrissae (for cleaning opium pipes). By the end of the nineteenth century, sealers had obliterated many stocks and reduced many species to near extinction. Stocks recovered throughout the twentieth century in varying degrees. After a long lag time, the Antarctic fur seal recovered at a rate of 10% per year, the highest known for any otariid population. Recovery for most species was nearer 5–7%. The Japanese sea lion is probably extinct. Some populations (Hooker's sea lion, Australian sea lion, see Table I) are small but apparently stable, whereas others are small but growing (Guadalupe and Juan Fernandez fur seals). Two species (Galapagos fur seal and sea lion) experience intermittent declines and recoveries related to El Nino events. The northern fur seal has been declining at various rates from 1956 to the present. The only otariid species presently of concern to managers is the Steller sea lion, which is rapidly declining for as yet unknown reasons. Worldwide, fur seal populations tend to be increasing faster than sea lion populations and outnumber them by an order of magnitude (nearly 7 million fur seals worldwide compared to just over 600,000 sea lions, Table I). The diet and place of foraging differ between the two groups and may explain these differences. It could be the habit of feeding beyond the continental shelf, like a fur seal, that has accounted for the present rapid growth of the California sea lion population.

## See Also the Following Articles

Pinniped Ecology ■ Pinniped Life History ■ Pinniped Physiology ■ Rookeries ■ Sea Lions, Overview

## References

King, J. E. (1983). "Seals of the World." Cornell University Press, 240 pp.
Riedman, M. (1990). "The Pinnipeds: Seals, Sea Lions, and Walruses." University of California Press, Berkeley, CA.

# Earless Seals
## *Phocidae*

### MIKE O. HAMMILL

## I. Systematics

The family Phocidae, consists of the earless or "true" seals. They are distinguished from sea lions and fur seals (family Otariidae), by the absence of external visible ear pinnae, internal testes, generally larger size and the inability to draw their hind limbs forward under their body when on land (King, 1983).

This latter character, the absence of tusks, and a notched tongue also distinguishes them from the family Odobenidae (walruses).

There has been considerable debate as to whether pinnipeds were diphyletic or monophyletic in origin. The diphyletic view suggests that odobenids and otariids are related to the bears (Ursidae), and that phocids were more closely linked to the otters, weasels, and skunks (Mustelidae). However, a re-evaluation of morphological evidence and the application of molecular techniques supports the monophyletic hypothesis, with pinnipeds descending from arctoid carnivores, a group which includes the bears (Arnason *et al.*, 1995).

The Phocidae can be divided into two sub-families: the Monachinae, with $2n = 34$ chromosomes, consisting of the southern phocids, the southern and northern elephant seals and the monk seals; and the Phocinae, or northern seals, with $2n = 34$ chromosomes (bearded seal, *Erignatus barbatus* and hooded seal, *Cystophora cristata*) and $2n = 32$ chromosomes (in the remaining seven species; Table I). The separation between these two groups has been confirmed in molecular studies, but the relationships among members within the sub-families are uncertain. More recent work indicates that the gray seal (*Halichoerus grypus*) may not be sufficiently separated from the ringed seal to warrant its own genus, and the closer relationship between ribbon and harp seals (*Histriophoca fasciata* and *Pagophilus groenlandicus*, respectively) than of either to the *Phoca* group (Carr and Perry, 1997). This implies that the harp seal should be moved from the genus *Phoca* and retain its old name *Pagophilus groenlandicus*. Support has also been found for grouping the ribbon, harp and hooded together, but the use of karyotypic data as a diagnostic landmark in the phylogenetic analysis, may be sufficient to maintain the separation (Table I).

## II. Distribution

Phocids are found throughout all of the world's major oceans except for the Indian Ocean. Twelve species breed on ice, six species breed on land, including the West Indies monk seal, which is probably extinct. The gray seal breeds on both land and on ice (Fig. 1). Among the ice-breeding seals, eight species breed primarily on the pack ice. Four species breed on land-fast ice. Phocids in the Northern Hemisphere have also colonized freshwater areas; these include the harbor seal (*Phoca vitulina mellonae*) in freshwater lakes of northern Quebec; the ringed seal (*Pusa hispida ladogensis* and *Pusa hispida saimensis*) in Lakes Lagoda and Saimaa in Russia and Finland, respectively; and the Baikal seal (*Pusa sibirica*) in Lake Baikal in Siberia. Ringed seals are also frequent in Nettelling Lake on Baffin Island in northern Canada.

Pinnipeds are adapted to marine foraging but must haul out on land or ice for parturition and successful rearing of offspring. Marine adaptations include a thick blubber layer for insulation, modifications in limbs and body shape to improve hydrodynamics and agility, and anatomical and physiological changes to improve diving performance.

## III. Ecology

Although young seals do not have to deal with the energetic costs associated with reproduction they must learn to identify and capture sufficient prey for survival and to support growth. Young harbor seals are capable of following their mothers in the water soon after birth, but are unable to duplicate the diving performance of the older female (Beck *et al.*, 2003). As harbor seal pups develop, they develop the ability to control and reduce their heart rate to conserve oxygen while diving. Blood oxygen stores increase quickly, but by the time animals are weaned muscle oxygen stores are still much lower

## TABLE I
### Members of the Family Phocidae, General Distribution and Breeding Habitat

| Common name | Latin name | Subspecies | Distribution | Breeding habitat |
|---|---|---|---|---|
| **Northern Hemisphere** | | | | |
| Gray seal | *Halichoerus grypus* | | North Atlantic | Land, ice breeder |
| Harp seal | *Pagophilus groenlandicus* | | North Atlantic | Pack-ice breeder |
| Harbor seal | *Phoca vitulina* | *P.v. vitulina* | Atlantic, Pacific Oceans. Arctic regions | Land breeder |
| | | *P.v. concolor* | | |
| | | *P.v. stejnegeri* | | |
| | | *P.v. richardsi* | | |
| | | *P.v. mellonae* | | |
| Spotted seal | *Phoca largha* | | North Pacific, Chukchi Sea | Pack-ice breeder |
| Caspian seal | *Pusa caspica* | | Caspian Sea | Fast-ice breeder |
| Ringed seal | *Pusa hispida* | *P.h. hispida* | Arctic regions, Baltic Sea | Fast-ice breeder |
| | | *P.h. botnica* | | |
| | | *P.h. ochotensis* | | |
| | | *P.h. krascheninikovi* | | |
| | | *P.h. saimensis* | | |
| | | *P.h. ladogensis* | | |
| Hooded seal | *Cystophora cristata* | | North Atlantic | Pack-ice breeder |
| Bearded seal | *Erignathus barbatus* | *E.b. barbatus* | Arctic | Pack-ice breeder |
| | | *E.b. nauticus* | | |
| Baikal seal | *Pusa sibirica* | | Lake Baikal, Siberia | Fast-ice breeder |
| Northern elephant seal | *Mirounga angustirostris* | | North Pacific | Land breeder |
| Ribbon seal | *Histriophoca fasciata* | | North Pacific (Chukchi, Bering and Okhotsk Seas) | Pack-ice breeder |
| Hawaiian monk seal | *Monachus schauinslandi* | | Pacific Ocean (Hawaiian Islands) | Land breeder |
| Mediterranean monk seal | *Monachus monachus* | | Mediterranean Sea, Black Sea, Atlantic (Northwest African coast) | Land breeder |
| West Indian monk seal | *Monachus tropicalis* | | Caribbean Sea–Gulf of Mexico area | Land breeder |
| **Southern Hemisphere** | | | | |
| Southern elephant seal | *Mirounga leonina* | | Subantarctic, Antarctic, southern South America | Land breeder |
| Weddell seal | *Leptonychotes weddellii* | | Antarctic | Fast-ice breeder |
| Ross seal | *Ommatophoca rossi* | | Antarctic | Pack-ice breeder |
| Leopard seal | *Hydrurga leptonyx* | | Antarctic | Pack-ice breeder |
| Crabeater seal | *Lobodon carcinophagus* | | Antarctic | Pack-ice breeder |

**Figure 1** *Gray seal pup in white coat. This fur is called laguno.*

than those of yearlings. Thus newly weaned animals are limited in their physiological capacity to forage compared to older animals. Consequently, it would be expected that young animals would prefer shallower areas compared to older animals. However, there is evidence to suggest that juvenile Hawaiian monk seals and gray seals forage in deeper water, while adults, particularly females forage in shallower areas. This exclusion of juveniles to peripheral areas might be one mechanism affecting population growth, since juveniles with their smaller size and diving capabilities would have more difficulties finding sufficient prey to meet their energy requirements, than would the larger more experienced adults.

Pup mortality during the lactation period is normally quite low. Among land breeding species, trampling and wounds caused by interactions between adults may encourage infection, which may result in death of the pup. This problem is more aggravated in crowded colonies where the number of interactions would increase. In Arctic ice breeding species, predation by bears, foxes, and birds such as ravens (*Corvus* sp.) and gulls (*Larus* sp.) in the case of ringed seal pups are an important source of pup mortality, although the

E

effects of predation are reduced by the use of lairs. Ringed seal pups are also quite active and are capable swimmers at a very young age. However, swimming incurs a high metabolic cost owing to the minimum blubber thickness. Repeated disturbance may affect growth and survival. Surface predators are not present in the southern polar regions, but leopard seals and killer whales are important marine predators. In the pack ice, pups may drown or be crushed as a result of storms causing the breakup and rafting of the ice.

## IV. Behavior and Physiology

Compared to otariids, phocids generally spend more time at sea, swim more slowly, and dive deeper and for longer periods. Southern elephant seals may dive to over 1200 m and remain below the surface for up to 120 min (Le Boeuf and Laws, 1994). Other deep-diving phocids include northern elephant seals (*Mirounga* sp., 1500 m), Weddell seals (*Leptonychotes weddellii*, 700 m), and recent hooded seal dives to 1400 m have been documented. Phocids have adopted what is sometimes referred to as a "slow-lane" strategy to reduce energy use during diving (Burns *et al.*, 2007). This is achieved through a combination of (1) apnea with exhalation upon initiation of diving to minimize buoyancy- and pressure-related problems; (2) an enhanced oxygen carrying capacity which is accomplished by a greater blood volume, an increase red cell mass (hematocrit) within the blood cell volume, a greater hemoglobin concentration in the red blood cells, and possibly a higher content of oxygen-carrying myoglobin in the muscles; (3) a generally larger body size to maximize oxygen carrying abilities while minimizing mass-specific energy demands while diving.

The characteristic phocid lactation strategy consists of building-up energy reserves throughout the year, fasting during lactation and lactating for a short period. The utilization of fat reserves to satisfy their own energy requirements and the costs of providing milk for her pup has favored selection for large body size, because energy stores scale to $Mass^{1.0-1.19}$, whereas metabolic requirements scale to $Mass^{0.75}$. Thus increased body size increases energy storing capabilities at a greater rate than increasing mass-specific energy requirements. This has lead to larger body size in phocid females than among otariid females (phocid females: mean = 229 kg, median = 141 kg; otariid females mean = 80 kg, median = 55 kg). The need to build up energy reserves over the year probably adds about 12% to the daily energy requirements of a female phocid, but it also allows the spatial and temporal separation of feeding and reproduction.

To minimize the metabolic costs associated with lactation, phocids have shortened the lactation period to 4–50 days instead of the months seen in otariids and odobenids. This is achieved by remaining beside the pup, providing more opportunities for the pup to suckle, and producing a very fat-rich milk, which increases the energy transfer per volume of milk consumed. In the elephant seals the fat content is very low at the beginning of lactation (≈10%), but increases to about 50% fat by mid-lactation. In harp, gray, and Weddell seals, the milk fat content increases from around 40% to between 50% and 65% fat by mid lactation, while in hooded seals there is relatively little change in fat content of the milk (55–68%) over the short 4-day lactation period.

## V. Life History

Phocids whelp on the ice, isolated islands, or inaccessible beaches. This makes it more difficult for terrestrial predators to approach seals undetetected. Some ice-breeding species, such as the Baikal seal and the ringed seal, are also protected by using small caves or lairs under the snow to such predators as humans, polar bears (*Ursus maritimus*), Arctic foxes (*Alopex lagopus*), and birds (*Corvus sp.*; *Larus* sp.). These lairs also provide shelter from cold ambient temperatures. Current global warming trends will likely result in the reduction of suitable ice habitat for many phocids, particularly in the more temperate regions of their distribution. This will impact not only seal populations, but also predators that rely on seals as food.

Sexual maturity is delayed in phocids. Some females are sexually mature at the age of 3+ years, but the mean age of sexual maturity is normally around 4–6 years, although it may vary with the population size and availability of resources. In Northwest Atlantic harp seals the mean age of sexual maturity among females may vary from 5.8 to 4.1 years. These changes have mirrored changes in population size due to exploitation. As the population has increased, the mean age of sexual maturity has also increased. It is currently around 5.3 years. Normally about 80% of the adult females are pregnant, but some interannual variability in adult reproductive rates can occur and in recent years the age-specific pregnancy rates among adults have been less than 75%. Extremely low (≈60%) adult reproductive rates have been documented in some years among ringed seals in the Beaufort Sea and Hudson Bay areas of northern Canada. These changes may be related to changes in ice conditions and availability of food resources.

Males become sexually mature around the same time or slightly later than females, but recently mature males appear to be incapable of defending access to females until 2 or more years after they are sexually mature. Sexually monomorphic species have the longest life expectancies. For example, ringed seals aged 45 years old have been reported. The life expectancy of sexually dimorphic species is much shorter particularly among males. In elephant seals, which show perhaps the most extreme level of sexual dimorphism, males are sexually mature at about 5 years of age, but they seldom achieve any rank within the colony until the age of 8 years. The greatest reproductive success occurs between the ages of 9 and 12 years and males die by the age of 14 years.

The female phocid reproductive cycle is characterized by parturition, a short lactation period (4–50 days), copulation near the end of lactation, embryonic diapause (≈3 months), and active fetal growth (≈9 months). In most temperate species implantation occurs during late summer–early autumn when light levels are decreasing, and pupping occurs during the spring, when light levels are increasing. However, an irregular pattern is seen among gray seals. Gray seals in the United Kingdom give birth during the fall and implantation occurs during the spring. Implantation of the embryo occurs after the molt, when female energy reserves are at a minimum. Embryonic diapause provides females a means of terminating reproduction if conditions are poor before her investment becomes too costly. It also leads to synchronization of reproductive activity.

Stability of the whelping habitat and vulnerability to predation appear to be two important factors that have influenced the duration of lactation within the phocid group. Pups on the unstable, drifting pack ice nurse for only 4 days in the hooded seal, to as much as 30 days among the crabeater and Ross seals (*Lobodon carcinophaga* and *Ommatophoca rossii*; mean = 18 days). In contrast, the longest lactation periods are found among the fast ice breeding Weddell, Baikal, and ringed seals (mean = 57 days). The lengthy lactation period among the very small ringed and Baikal seals may be related to their small size and their relative inability to store and deliver energy quickly to their offspring. The Weddell seal, which weighs around 450 kg, is almost 4 times heavier than the diminutive ringed seal. The lengthy nursing period of Weddell seal pups may keep the young away from the fast ice edges where their exposure to aquatic predators such as leopard seals and killer whales would be greater. The duration of lactation among land-breeding phocids is intermediate to that of the fast ice and pack ice animals with an average of 32 days.

The "fasting strategy" is certainly characteristic of lactation in the largest phocids such as the elephant seals and in the hooded seal. Facultative feeding occurs among female Weddell, bearded, harp, and gray seals, while feeding appears to be obligatory among the smaller phocids (≤100 kg), such as the harbor and ringed seals. Feeding may be necessary because females are unable to store sufficient energy to satisfy their own energy requirements, plus energy required for lactation. The need to continue foraging during lactation means that females must leave their pups in a safe area while they forage or the pup must accompany the female. In both instances, the spatial separation between whelping sites and foraging areas is limited. In the ringed seal, females scrape out lairs beneath the snow to haul out in. Females leave their pups to feed under the ice, but remain close enough to help the pup move to an alternate lair if a predator approaches. Among harbor seals, foraging activity is restricted during the early stages of lactation. In large groups, as lactation advances, some females leave their pups unattended while they forage. In areas where only small groups are seen, the pups may follow the females over extensive distances of as much as 30 km away from the original haulout site.

Fewer studies have examined male energy expenditures during breeding owing in part to the difficulties associated with handling large, dangerous, and aggressive animals, or their inaccessibility in the case of males that spend much of their time in water. Daily energy expenditures of males during the breeding season are much lower than those of females, because they do not incur the costs of milk production. However, once females wean their pups they resume feeding. This contrasts with males, who continue fasting to maximize their access to successively receptive females as the breeding season advances. Elephant, gray, and hooded seals appear to terminate breeding activity when fat levels have declined to levels similar to those observed in females. This suggests that the overall reproductive effort is similar between the two sexes.

At birth, phocid neonates are larger than otariid neonates. When female body size is taken into account, phocid and otariid mass at birth represents about 10% of their mother's mass. The pups are quite lean at birth with a fat content of 5–8%. Notable exceptions are hooded, harbor, and bearded seals which are larger relative to other species with a mass at birth equal to about 12–13% of the maternal mass. Harbor, hooded, and bearded seal pups also differ from otariid and other phocid pups by the presence of a thin blubber layer at birth. Fat content represents 11–14% of the total body mass in harbor and hooded seal pups at birth. Phocid pups gain weight rapidly, achieving a weaning mass 2–5 times their birth mass in a period of 4–50 days, but mass at weaning is relatively constant across species, being equivalent to 25–30% of the mother's mass. Among hooded, harp, and gray seal pups, 65–75% of the milk energy is deposited primarily as fat. Elephant, hooded, harp, ringed, and gray seal pups are 40–50% fat at weaning, while species with very active pups such as Weddell, bearded, and harbor seals may contain only 34–37% fat at weaning. The rate of mass gain is inversely related to the duration of the lactation period when expressed relative to the female's metabolic mass. The lowest rates of relative mass gain are seen among Hawaiian monk seals, bearded, harbor, and ringed seals, species where the pups are very active and begin entering the water early during lactation. Ice-breeding neonate phocids such as harp, ringed, ribbon, Caspian, Baikal, and Largha seals are born with a white, relatively long fur called lanugo (Fig. 1). The white color may provide some protective camouflage on the white snow from predators. However, the lanugo may be more effective in its role as an insulator, particularly to the very young pup who has not yet developed a thick layer

**Figure 2** *Gray seal pups that are molting, shedding their laguno. Such partially molted or ragged-jacket and weaned animals are called beaters. Adult female on right.*

of blubber for insulation. The structure and color of this fur permits short-wave energy received from the sun to be transmitted through the fur, where it is absorbed by the dark skin. It also traps heat energy radiated from the animal's skin and thus acts like a greenhouse, heating up the air trapped within the fur, but limiting heat loss to the outside air. Among species where the young enter the water very soon after birth (e.g., harbor seals, bearded seals, and hooded seals), the young are born with a thin layer of blubber, and the lanugo is shed or molted within the uterus. This is because blubber acts as a much better insulator in water than fur. Among harbor seals, 5–30% of the pups are born with a white lanugo, depending on the region, but this is quickly replaced by a grayish pelage similar to that of adults. In bearded seal the shedding of the lanugo begins *in utero*, but is completed after birth. In the hooded seal, the fetal fur is expelled in small clumps on the ice with the placenta, which was referred to as silver dollars by commercial sealers. At birth the pups are covered by blue (dorsal) and silver (ventral) fur that is much thicker and longer than the adult fur and at this stage the pups are known as bluebacks. Hooded seal pups differ from harbor and bearded seals in that they do not enter the water until they are weaned. However, in this species lactation lasts for only 4 days. The remaining species that whelp on the ice in the Northern Hemisphere tend to give birth to pups with a white lanugo, which may afford some protection against surface predators. Gray seals have their pups on both the land and on the ice, but the pups are born with a white lanugo, which is eventually molted to leave a dark spotted fur in the case of males or a silvery spotted fur in the case of females (Fig. 2). Neonates of the southern ice-breeding seals, such as Weddell, crabeater, leopard, and Ross seals, are born with a gray, brown, or grayish-brown pelage. Southern ice breeding phocids are not exposed to surface predators other than man and hence the white lanugo may not be required. Elephant and monk seal pups are born on land with a black pelage.

The study of mating systems among phocids has relied heavily on behavioral observations of animals in the breeding areas. Male reproductive success has been evaluated by a male's ability to monopolize females or by the number of copulations observed. However, DNA techniques have suggested that the evaluation of reproductive success may be more complex, with the existence of alternative mating strategies within a population. In captive harbor seals, behavioral observations during courtship were not reliable indices of paternity, while in one population of gray seals, large males sired significantly fewer pups than would otherwise have been indicated from their observed mating opportunities. Females tended to produce several

E

**Figure 3**   *Two male hooded seals fighting on pack ice in the Gulf of St. Lawrence, Canada. Note the inflatable sac on the male on the left.*

pups fathered by the same male, who in many cases was not the large attendant male.

Phocids breeding on land prefer islands or isolated beaches, where threats from predation are reduced. The combination of habitat limitations and synchronization of reproductive activity encourages the aggregation of females. Males are not involved in caring for the pup. Therefore, the best strategy for males is to copulate with as many females as possible, while the strategy of females is to successfully rear her pup. Males will attempt to prevent other males from having access to females. In phocids, this may involve defending a geographical area, but if the female moves, the male will follow to defend a new space around the female. The number of females that can be defended is limited by habitat features and the skills of defending males. Large, open beaches are more difficult for a male to control than more topographically irregular sites, where geographical barriers will aid established males to limit the approach of intruding males. Reproductive success will also be affected by the fighting and signaling ability of males and how long they can remain beside females without leaving to feed. In male hooded seals a series of displays involving the inflatable nasal sac and nasal septum are often associated with the approach of other males (Figs 3 and 4). Not all approaches result in combat, suggesting some signaling occurs. The gradual evolution toward large size observed in females, which permitted the separation of reproduction and feeding, would also have operated on males as well. Larger males could spend more time ashore fasting. Larger males would also be favored over small males in combat, although experience and individual skill development would also be contributing factors. The greatest degree of polygyny seen among phocids is exhibited by the elephant seals, who may defend harems containing upward of 100 females. Southern elephant seal males may weigh up to 3700 kg, while northern elephant seal males may reach 2300 kg. In both species the males are typically 5–6 times larger than the females. Males arrive on the whelping area just before the females and the most successful males remain on the beaches fasting until the last females leave about 3 months later.

In other species the development of polygyny is more variable. Gray seals copulate on land and occasionally in water. Males at 350 kg are about 50% larger than females. In some areas in the British Isles, they control access to as many as 7.5 females, while

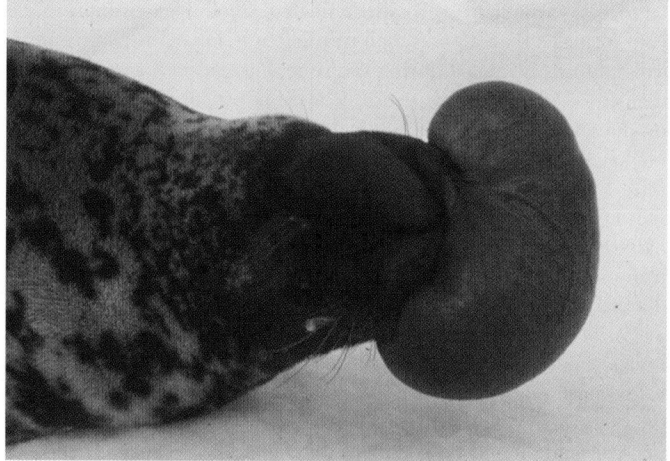

**Figure 4**   *Male hooded seal with inflated sac. The walls of the sac are actually part of the nasal septum.*

in the open beaches and ice breeding areas in Canada, males are only able to control access to 1.3–2.6 females. The development of polygyny would be limited by the three-dimensional nature of the marine environment. Among hooded seals the marked sexual dimorphism observed in many terrestrial mating species is also observed. Male hooded seals weigh up to 440 kg, while females may reach about 290 kg. The hooded seal male begins to defend one female about midway through the short 4-day lactation period. Once the pup is weaned, he accompanies the female into the water, mates with her, and then returns to the ice and will attempt to establish himself beside another female. Since pupping occurs over a 2- to 3-week period, males have opportunities for multiple matings during that period. Mating success will be affected by his ability to defend access to females against other males and by the amount of time he can spend on the ice before his energy reserves are depleted. Less is known about the structure of mating systems of phocids that copulate primarily in water. The development of polygyny would be limited by the three-dimensional nature of the marine environment. Among the phocids that copulate in the water, little difference in body size is seen between males and females, or in some cases such as bearded seals and Weddell seals the females appear to be slightly larger. In these cases, smaller size may favor underwater agility. The mating system of harp seals has been referred to as promiscuous, but little information is available. Males do haul out, but no displays or fighting are observed on the ice. Extensive vocal activity within harp seal whelping patches has been recorded, and groups of males are observed often patrolling leads, vocalizing and diving. It is possible that male harp seals are displaying to females, suggesting more of a lek-type system, but there is insufficient data to comment further. Among ringed seals, the presence of predators in the fast ice such as bears, foxes, and humans select against aggregation, while the need to continue feeding during lactation, in an area generally considered to have low productivity would also be a contributing factor. In this species, some underwater vocal activity has been recorded, but unlike the underwater vocalizations of the widely dispersed bearded seal that can be heard over distances of 25 km, the underwater vocalizations of the ringed seal are relatively weak, limiting their use as a signal to potential mates. However, the males emit a strong odor during the breeding season owing to the enlargement and increased secretion activity of sebaceous and apocrine glands in the muzzle region. The strong odor may serve as a signal to inform both females and other males that a breathing hole is used or belongs to a particular male. In harbor seals, the females move with their pups between haulout sites and foraging areas. Males are unable to defend females or sites against other males. As the time that females will become receptive approaches, males reduce the size of the range that they occupy, but remain in the water, making repeated short dives that are associated with underwater vocal displays. Some males establish themselves near haulout sites, some near foraging areas used by females, while others appear to establish themselves on transit routes between haulout sites and foraging areas. It has been suggested that a "lekking" type of mating system occurs in this species. In the Antarctic Weddell seal, males appear to defend underwater territories around breathing holes and cracks.

Many early studies relied on stomach content material from hunted animals, fecal collections, and entrapments in fishing gear to provide information on diving and foraging activity. During the last decade, major technological advances have provided researchers with satellite transmitters, time-depth recorders, stomach temperature probes, and video recorders to study diving and foraging activity.

Phocids feed on a wide variety of prey, including invertebrates such as amphipods, mysids, squid, and krill, and vertebrate prey such as fish. Birds have been recorded in the diet of some species such as the leopard seal (*Hydrurga leptonyx*) and harp seal. Leopard seals also prey on other seal species and cannibalism has been reported in gray seals. Diet composition may change seasonally, geographically, and with age. Newly weaned pups of ringed and harp seals begin foraging on zooplankton in their initial attempts to forage independently and then become more piscivorous as their skills develop. Phocids feed primarily on smaller prey that can be consumed whole, but large prey may be taken. Under certain conditions where prey are very abundant or accessible, e.g., when fish are caught in nets, seals may consume only pieces from fish.

Little is known about factors affecting prey choice. Research has indicated that in harp and harbor seals feeding preferences occur for particular types and sizes of prey that may be independent of local abundance. Harp seals digest capelin more efficiently than they do most other prey, and throughout their range capelin forms an important component of their diet, while other species such as commercially important Atlantic cod (*Gadus morhua*) form only a very minor

component in the overall diet. While foraging phocids must balance their intake of oxygen and the distribution of oxygen for locomotion, body maintenance, and processing of food (specific dynamic action or heat increment of feeding). Their approach to balancing these sometimes conflicting needs will influence their foraging strategy. A seal may process (digest) food while actively swimming and foraging. If it uses only aerobic metabolism, then the consumption of oxygen required to process the food will reduce the amount of time spent diving and collecting food. If the seal attempts to maintain the duration of diving, then the switch to anaerobic conditions will force the animal to rest at the surface until lactic acid levels are reduced. A second strategy is to forage and then spend time resting at the surface, hauled out on the ice or on land until food processing is completed. However, resting at the surface increases vulnerability to surface predators such as sharks or killer whales, while returning to shore or ice involves time lost due to transit and may limit the distances that foraging can occur away from haulout sites. A third strategy involves foraging and then if successful, reducing locomotory costs by drifting during the surfacing phase of the dive, allowing food processing to occur. Many phocids may utilize these first two strategies. Weddell, ringed, harp, and hood seals are often seen hauled out on the ice and occasionally on land. Harbor and gray seals may forage in offshore areas, but rarely spend more than a few days away from haulout sites. The third strategy appears to be utilized by elephant seals, which spend almost 8–9 months of the year at sea.

Male and female phocids have different reproductive strategies, which carry over into different foraging strategies as well. Among phocids, mating occurs at the end of lactation. Females begin foraging almost immediately and some recovery in mass has been observed prior to the moult which occurs 1–4 months later. Since females must support a developing fetus, it might be important that a minimum threshold be attained to ensure implantation and continued fetal development. After implantation females initially gain mass quickly. Then, the rate of mass gain appears to be slow, possibly to avoid increasing energy costs of transport associated with gaining mass too early in the season. About 4 months prior to birth, the rate of mass gain increases once again. This coincides with the period when fetal support costs are highest and females would store energy to carry them through lactation. Many males continue to lose mass throughout the breeding season until just after the molt. Since males only need to rebuild energy reserves in time for the subsequent breeding season, the rate of post-molt mass gain is slow. Gray seals, elephant seals, and hooded seals are three species where important differences in foraging activity have been observed. Among gray seals, females undertake periods of intensive foraging postbreeding and postmolting, whereas males replenish reserves predominantly during the few months preceding the breeding season; females perform longer dives, spend more time in more restricted feeding areas, and are more selective and feed on higher quality pelagic prey than males. Northern elephant seals, differ slightly in that males and females have similar postbreeding foraging patterns, with both sexes accumulating body reserves following the breeding season. However, in northern elephant seals, females are dispersed over a broad geographical area across the northeastern Pacific from the coast to as far west as 150°W, but tend to remain between 44°N and 52°N. Foraging occurs both offshore and during transit between inshore and offshore areas. Diurnal changes in diving depths are observed, indicating that they are foraging on vertically migrating prey in the pelagic and mesopelagic environment. Their principal prey are mesopelagic squid and fish. Males utilize a different foraging strategy by foraging little while en route to particular foraging areas along

E

E

continental margins of Washington State to as far north as the Aleutian Islands. Once on site, repeated, uniform flat-bottomed dives predominate diving, with little diurnal variability, suggesting intensive foraging activity, possibly on benthic prey such as energy-rich elasmobranchs and cyclostomes (sharks, skates, ratfish, and hagfish). While at sea both species dive continuously, spending almost 90% of their total time at sea submerged, leading to the suggestion that they should be called surfacers instead of divers. At the opposite end of the size spectrum, a very different strategy is seen among harbor seals. Although capable of diving to depths of 500 m, harbor seals rarely dive deeper than 65 m and in some studies an average of 65% of diving activity occurs at depths of less than 4 m. Visual and telemetry data, indicate that harbor seals in some areas spend most of their time very close to the coast in shallow water areas. Foraging distances rarely exceed 50 km away from haulout sites. Little difference is seen between males and females in dive depths and foraging distances away from haulout sites outside of the breeding season.

## See Also the Following Articles

Pinniped Physiology ■ Pinniped Life History

## References

Arnason, U., Bodin, K., Gullberg, A., Ledje, C., and Mouchaty, S. (1995). A molecular view of pinniped relationships with particular emphasis on the true seals. *J. Mol. Evol.* **40**, 78–85.

Beck, C. A., Bowen, W. D., McMillan, J. L., and Iverson, S. J. (2003). Sex differences in the diving behaviour of a size-dimorphic capital breeder: The grey seal. *Anim. Behav.* **66**, 777–789.

Boyd, I. L. (1993). "Marine Mammals: Advances in Behavioural and Population Biology," Symposia of the Zoological Society of London, No. 66. Clarendon Press, Oxford, p. 404.

Burns, J. M., Lestyk, K. C., Folkow, L. P., Hammill, M. O., and Blix, A. S. (2007). Size and distribution of oxygen stores in harp and hooded seals from birth to maturity. *J. Comp. Physiol.-B* **177**, 687–700.

Carr, S. M., and Perry, E. A. (1997). Intra- and interfamilial systematic relationships of phocid seals as indicated by mitochondrial DNA sequences. *In* "Moleculare Genetics of Marine Mammals" (A. E. Dizon, S. J. Chivers, and W. F. Perrin, eds), pp. 277–290. The marine Mammal Society. Allen Press Inc, Lawrence, KS, Special Publication Number 3.

Hochachka, P. W., and Mottishaw, P. D. (1998). Evolution and adaptation of the diving response: Phocids and otariids. *In* "Cold Ocean Physiology" (H. O. Portner, and R. C. Playle, eds), pp. 391–431. Cambridge University Press.

King, J. E. (1983). "Seals of the World," 2nd Ed. Cornell University Press, Ithaca, NY.

Le Boeuf, B. J., and Laws, R. M. (eds) (1994). "Elephant Seals: Population Ecology, Behavior and Physiology." University of California Press, Los Angeles, CA, 414 pp.

Lydersen, C., and Kovacs, K. M. (1999). Behaviour and energetics of ice-breeding, North Atlantic phocid seals during the lactation period. *Mar. Ecol. Prog. Ser.* **187**, 265–281.

Parrish, F. A., Marshall, G. J., Littnan, C. S., Heithaus, M., Nja, S., Ecker, B., Raun, R. B., and Antonelis, G. (2005). Foraging of juvenile monk seals at French Frigate Shoals, Hawaii. *Mar. Mamm. Sci.* **21**, 93–107.

Renouf, D. (1991). "The Behaviour of Pinnipeds." Chapman and Hall Ltd., New York, 410 pp.

Smith, T. G., Hammill, M. O., and Taugbol, G. (1991). A review of the developmental, behavioural and physiological adaptations of the ringed seal, *Phoca hispida*, to life in the Arctic winter. *Arctic* **44**, 124–131.

# Echolocation

## WHITLOW W.L. AU

Echolocation is the process in which an animal obtains an assessment of its environment by emitting sounds and listening to echoes as the sound waves reflect off different objects in the environment. In a very general sense any animal that can emit sounds may be able to hear echoes from large obstacles (i.e., humans yelling in a canyon); however, such a process is not considered echolocation. The term "echolocation" is reserved for a specialized acoustic adaptation by animals who utilize this capability on a regular basis to forage for prey, navigate, and avoid predators. Echolocating dolphins are often searching for entities that are considerably smaller than themselves and must make fine discrimination of these objects. Over the eons of time, this specialized adaptation has been continually refined under evolutionary pressures.

Echolocation in bats was suspected as early as 1912, but it was not until 1938 when G. Pierce and D. Griffin that provided evidence of bats emitting ultrasonic pulses using an ultrasonic detector that the concept began to gain acceptance. Echolocation in dolphins was suspected around 1947 as was evidenced in the personal notes of A. McBride (1956), the first curator of Marine Studio (later Marineland) in Florida. However, it was not until 1960 that Norris and colleagues (1961) performed the first unequivocal demonstration of echolocation in dolphins by placing rubber suction cups over the eyes of an Atlantic bottlenose dolphin (*Tursiops truncatus*) and observing that the animal was able to swim and avoid various obstacles. Ultrasonic pulses were also detected as the blindfolded dolphin swam and avoided obstacles, such as a vertical maze of suspended pipes.

Since the Norris demonstration, considerable progress has been made in our understanding of the echolocation capabilities of dolphins. Most of the research has been done with the Atlantic bottlenose dolphin, the most common dolphin in captivity. Research in dolphin echolocation can be divided into the following areas: (1) sound production mechanism and propagation in the dolphin's head, (2) sound reception and auditory capabilities, (3) sound transmission and the characteristics of echolocation signals, (4) target detection capabilities, (5) target discrimination capabilities, (6) auditory nervous system function and capabilities, and (7) signal processing modeling. This article addresses each of the first six areas, providing the most recent findings in most cases along with some fundamental capabilities.

## I. Sound Production Mechanism and Propagation in the Dolphin's Head

The head of a dolphin shown in Fig. 1 is a very complex structure with unique air sacs and special sound-conducting fats. Once of the most perplexing issues that has eluded researchers since the discovery of echolocation has been the location and mechanism of sound production in dolphins. In the mid-1980s, Cranford (1988) began using modern X-ray computer tomography (CT) and magnetic resonance imaging techniques to study the internal structure within a dolphin's head. These non-invasive techniques allowed Cranford to study the relative position, shape, and density of various structures in the dolphin's head and helped him to conclude that a structure called the monkey lip-dorsal bursae (MLDB) in the dolphin nasal complex was the location of the sound generator.

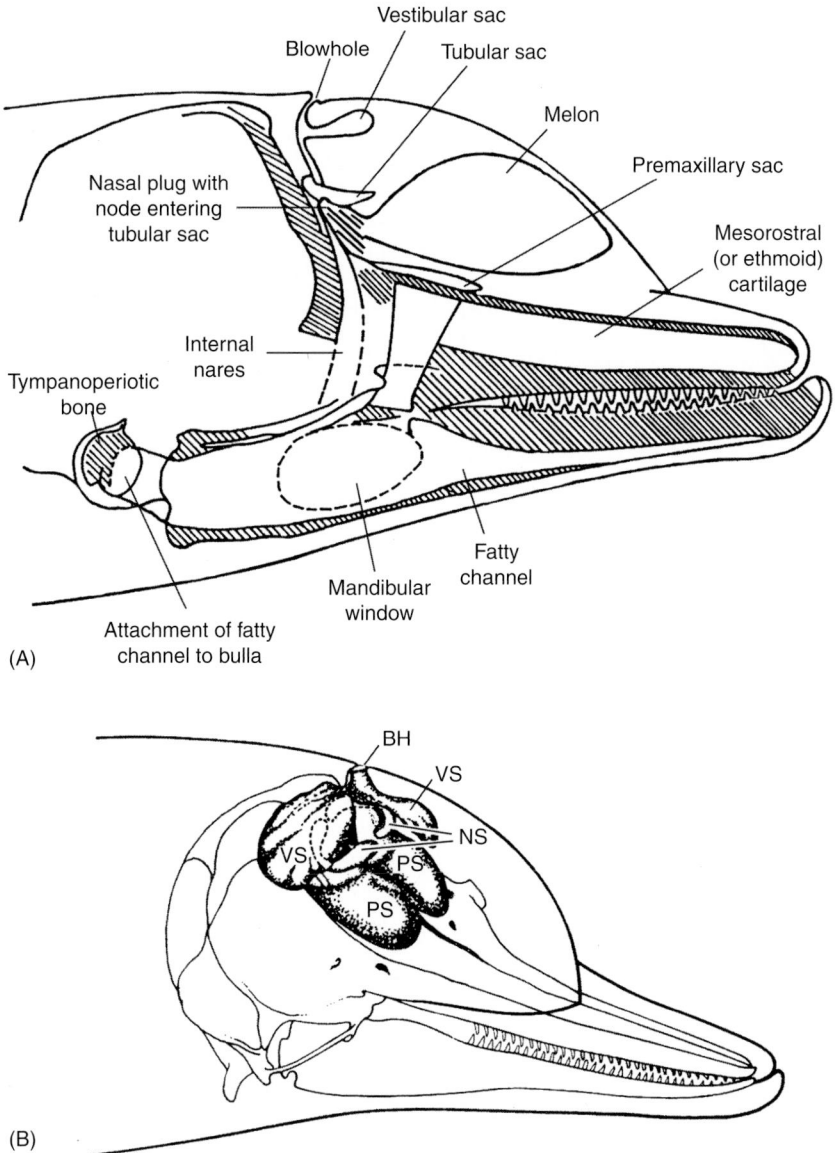

**Figure 1** *(A) Schematic of a dolphin's head (adapted from Norris, 1968) and (B) three-dimensional diagram of the air sacs in a dolphin's head. PS, premaxillary sac; VS, vestibular sac; NS, nasofrontal (tubular) sac; AS, accessory sac. Adapted from Purves and Pilleri (1983).*

Eventually, Cranford *et al.* (1997) used high-speed video of the phonic lips (previously referred to as the monkey lips) with simultaneous hydrophone observations of echolocation signals. There are two sets of phonic lips, associated with the two nares in the dolphin's nasal complex. Cranford and colleagues have obtained additional high-speed video observation of movements in both sets of phonic lips during the production of echolocation signals and whistles.

The numerical simulation of sound propagation in the head of a dolphin by Aroyan (2001) has provided considerable understanding of the role of the air sacs, skull, and melon in the propagation of sounds in a dolphin's head. One of the interesting problems Aroyan considered was that of a plane wave propagating toward the head of a dolphin (as depicted in Fig. 2A); to determine where sounds would focus in the dolphin's head in a similar manner to the process used by geologists to determine the epicenter of an

earthquake. He numerically solved the three-dimensional wave equation (also shown in Fig. 2) using a finite-difference technique and a supercomputer. The density and sound velocity structure of the dolphin's head were estimated from the CT scan results of Cranford. The grid points represent a pictorial illustration of how the head of a dolphin may be mathematically subdivided so that the solution of the wave equation is numerically determined at each grid point. The results for the geometry depicted in Fig. 2A are shown in Fig. 2B, with focal regions at the two auditory bullas and at the MLDB region of the nasal system, supporting Cranford's earlier suspicion of the MLDB being the site of the sound generator for echolocation sounds.

Aroyan then placed a hypothetical sound source at the MLDB region and numerically solved the three-dimensional wave equation as the sound propagated through the dolphin's head into the water. He

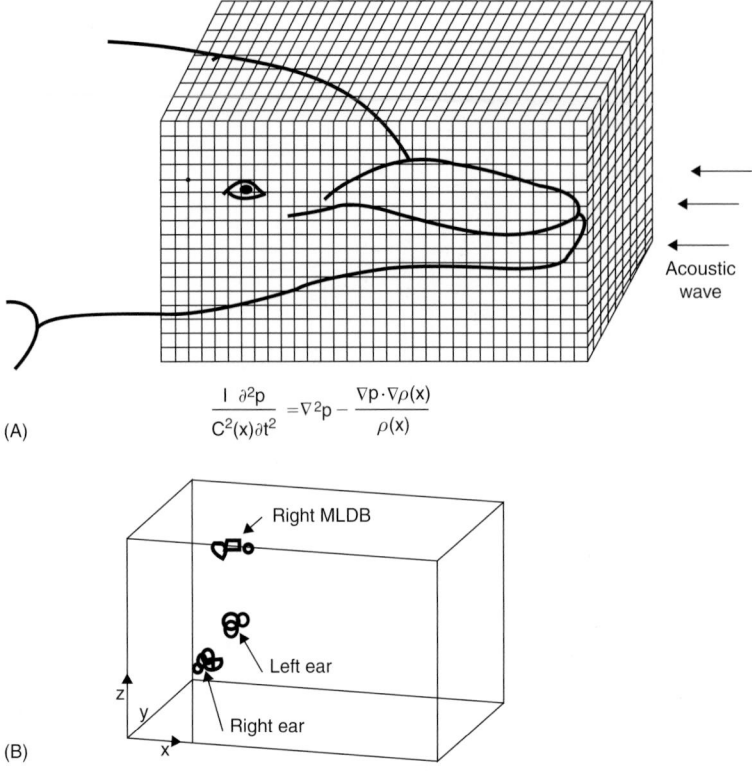

$$\frac{1}{C^2(x)} \frac{\partial^2 p}{\partial t^2} = \nabla^2 p - \frac{\nabla p \cdot \nabla \rho(x)}{\rho(x)}$$

(A)

(B)

**Figure 2** *(A) Configuration of numerical simulation of sound propagation in the head of a dolphin to determine acoustic focal regions and (B) results of numerical simulation for the geometry depicted. After Aroyan (2001).*

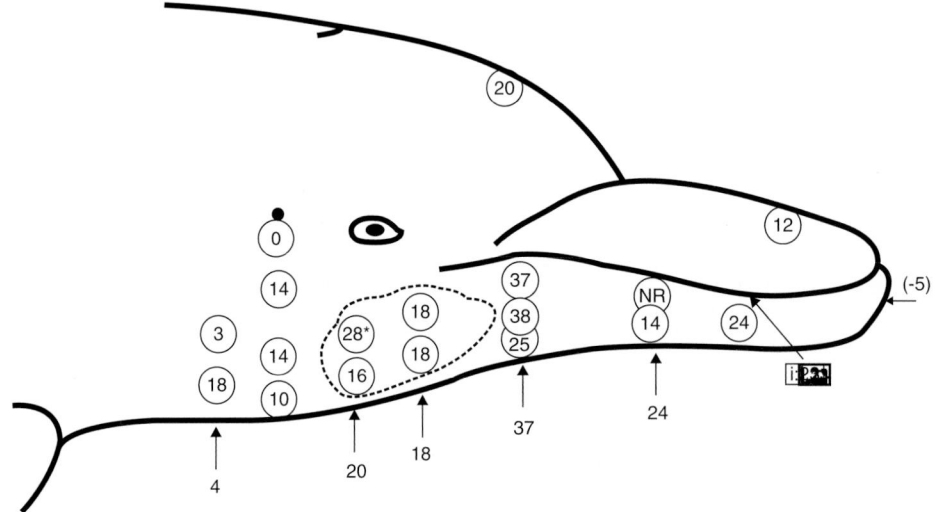

**Figure 3** *Relative reception sensitivity of hearing at different locations around a dolphin's head. The higher the number, the more sensitive the region. Adapted from Møhl et al. (1999).*

found that the skull, the various air sacs, and the non-homogeneous melon all played important roles in forming the beam in which sounds are transmitted into the free field. The specific characteristics of this beam are discussed in Section II. He also showed that if a hypothetical sound source was placed at the larynx, the resulting beam in the free field was not compatible to the actual beam measured for dolphins, therefore, essentially eliminating the larynx as a possible site for a sound generator.

## II. Sound Reception and Auditory Capabilities

### A. Auditory Capabilities

Dolphins do not have pinnae and their external auditory meatus is but a pinhole with vestigial fibrous tissues connecting the surface to the tympanic structure. Kenneth Norris was the first to postulate that sound enters the dolphin's auditory system through the thin posterior portion of the mandible (see Fig. 3) and is transmitted via a fat-filled canal to the tympano-periotic bone, which contains the middle and inner ears. Electrophysiological measurements, such as the one by Bullock *et al.* (1968), were conducted that provided evidence to support Norris' theory. However, the acoustic conditions for both experiments were less than ideal: the subjects were confined to small holding tanks and their heads were held near the surface to keep the electrodes from shorting out by the water. The acoustic propagation for such a situation can be extremely variable, with the sound pressure level changing drastically because of multipath propagation, on the order of 10–20 dB.

Møhl *et al.* (1999) took a slightly different electrophysiological approach by measuring the brain stem-evoked potential of a bottlenose dolphin that was trained to beach itself on a rubberized mat.

A special suction cup hydrophone having a water interface between the piezoelectric element and the skin of the dolphin was positioned at different locations on the dolphin's head. By performing the measurement in air, the point at which sound from the piezoelectric element enters into the dolphin could be firmly established. Acoustic energy will only propagate toward the dolphin's skin and energy propagating in any other direction will be reflected back at the boundary of the suction cup. Møhl and colleagues positioned the suction cup hydrophone at different locations around the dolphin's head, and at each location, the amount of attenuation needed to obtain the evoked potential threshold was determined. Their results are shown in Fig. 3, where the circles indicate the different positions of the suction cup and the number within each circle represents the amount of attenuation needed to achieve threshold. Therefore, a larger number is indicative of a more sensitive region of sound reception. The dashed line indicates the area of the pan bone or mandibular window shown in Fig. 1A. These results indicated that the area just forward of the pan bone area of the dolphin's lower jaw is the most sensitive area of sound reception, which seems to be inconsistent with Norris' pan bone theory. However, the numerical simulations of acoustic propagation by Aroyan suggest that sounds that enter the dolphin's lower jaw just forward of the pan bone actually propagate below the skin surface to the pan bone and enter into the lower jaw through the pan bone.

The hearing sensitivity of a dolphin was first measured accurately in 1967 by Dr. Scott Johnson in a pioneering experiment. Johnson found that the upper limit of hearing of an Atlantic bottlenose dolphin was 150 kHz. Since Johnson's research, audiograms have been determined for the harbor porpoise (*Phocoena phocoena*), Amazon River dolphin (*Inia geoffrensis*), beluga whale (*Delphinapterus leucas*), false killer whale (*Pseudorca crassidens*), Chinese river dolphin (*Lipotes vexillifer*), Risso's dolphin (*Grampus griseus*), Tucuxi (*Sotalia fluviatilus*), and killer whale (*Orcinus orca*). The audiograms for these odontocetes are shown in Fig. 4.

One of the most interesting features of these audiograms is the high upper frequency limit of hearing extending beyond 100 kHz. This is rather remarkable when the wide range of sizes of the animals depicted in Fig. 4 is considered. The largest animal represented in Fig. 4 is the killer whale, which weighed about 3,600 kg and was about 5 m in length compared to the smallest animal, the harbor porpoise, which typically weighs about 33 kg and is about 1.3 m in length. The typical rule of thumb in mammalian hearing is that larger animals tend to have limited high-frequency hearing capabilities. A killer whale is considerably larger than the smallest dolphin, yet its upper limit of hearing is approximately 120 kHz, and may therefore represent an exception to the norm. Another interesting feature of Fig. 4 is the fact that the maximum sensitivity (lowest point in the graph) for the different odontocetes is very similar within 10 dB.

Although dolphins do not have pinnae, the auditory system of the dolphins is directional. Sounds are received best when the source is directly in front of an animal. The receiving beam pattern of a *T. truncatus* in both horizontal and vertical planes is shown in Fig. 5. Several features of the beam patterns are worth pointing out. First, the beam becomes wider as the frequency decreases. Planar transducers also behave in a similar fashion with their beam becoming narrower as the frequency increases. Second, the major axis of the beam in the vertical plane is pointed upward with respect to the tooth line by about 5–10°. Third, the major axis in the horizontal plane is pointing directly in front of the animal parallel to the longitudinal axis of the dolphin. The receiving beam pattern can also be discussed in terms of the spatial variation in the hearing sensitivity of the dolphin. Therefore, the dolphin has the best high-frequency hearing sensitivity when sounds approach from the front and poorer sensitivity as the sound sources move to other locations about the animal's head.

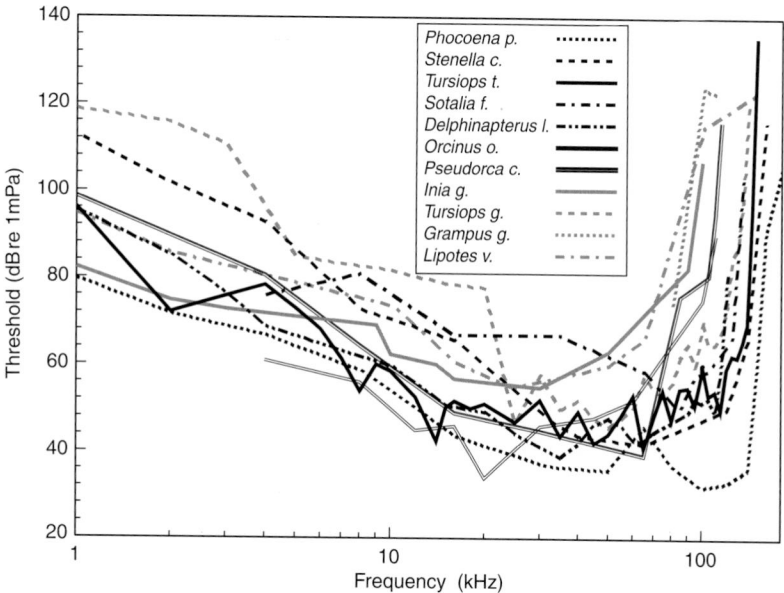

**Figure 4**  *Audiograms for 11 species of odontocetes.*

## III. Sound Transmission and the Characteristics of Echolocation Signals

There is a distinct difference in the echolocation signals used by odontocetes that produce whistle signals and those that do not whistle. Whistling dolphins project short, almost exponentially decaying signals with durations of 40–70 μsec and bandwidths of tens of kilohertz. Non-whistling dolphins and porpoises project signals with longer durations of 120–200 μsec and with narrow bandwidths that are typically less than 10 kHz. An example of a typical echolocation signal produced by an Atlantic bottlenose dolphin is shown in Fig. 6, along with a typical echolocation signal produced by a harbor porpoise (a non-whistling animal). Whether riverine dolphins produce whistles is still an open question; however, these dolphins emit signals that are of the broadband, short duration variety. Most

odontocete species produce whistles, and only a few, such as the harbor porpoise, Commerson's dolphin (*Cephalorhynchus commersonii*), Hector's dolphin (*C. hectori*), Dall's porpoise (*Phocoenoides dalli*), and pygmy sperm whale (*Kogia breviceps*), are known to not whistle.

The amplitudes of the echolocation signals also are very different between whistling and non-whistling odontocetes. Whistling dolphins, such as *T. truncatus*, *P. crassidens*, and *D. leucas*, can project echolocation signals with peak-to-peak amplitudes as high as 225 dB re 1 mPa. The center frequency of the signals used by whistling dolphins is affected by the level of the outgoing signal. The center frequency of clicks varies almost linearly with the peak-to-peak amplitude. Non-whistling dolphins and porpoises, such as *P. phocoena* and *Phocoenoides dalli*, emit signals that normally do not exceed 170 dB re 1 mPa. Peak-to-peak source level measurements for *P. phocoena* by,

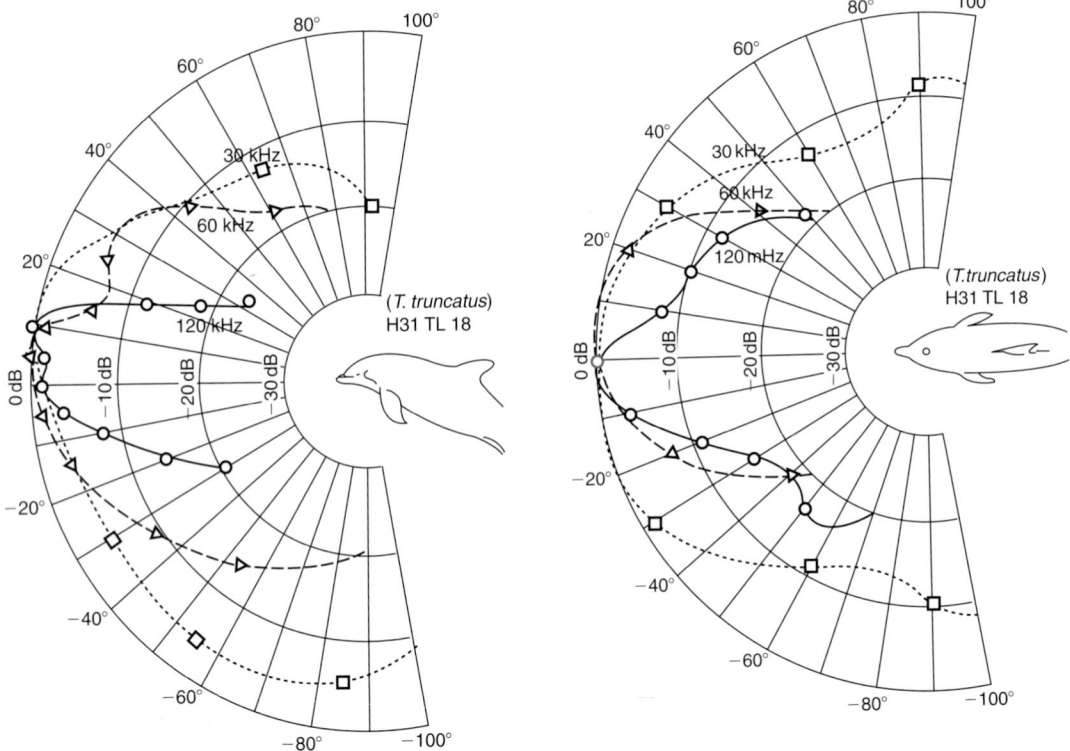

**Figure 5**   *The receiving beam patterns in the horizontal and vertical planes for different frequencies. Adapted from Au (1993).*

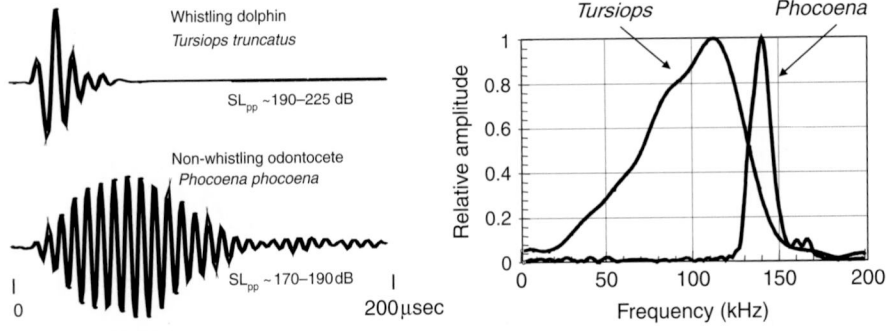

**Figure 6**   *Typical echolocation signal emitted by* T. truncatus *(a whistling dolphin) and* P. phocoena *(a non-whistling porpoise). The source level (SL) is the peak-to-peak sound pressure level referenced to 1 mPa at 1 m.*

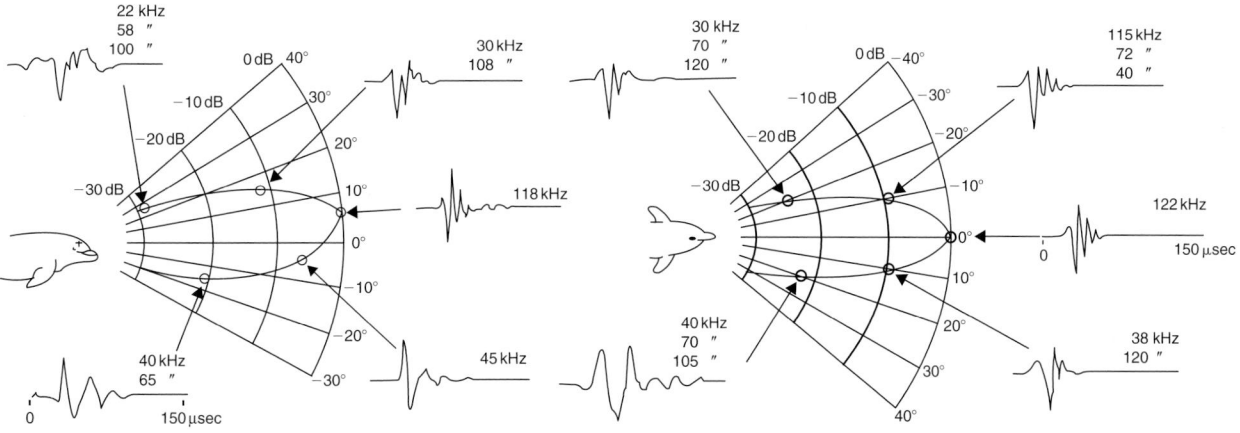

**Figure 7** *The transmission beam pattern in horizontal and vertical planes. The signals shown with each beam pattern are all the same signal captured simultaneously by five hydrophones located about the dolphin's head. After Au (1993).*

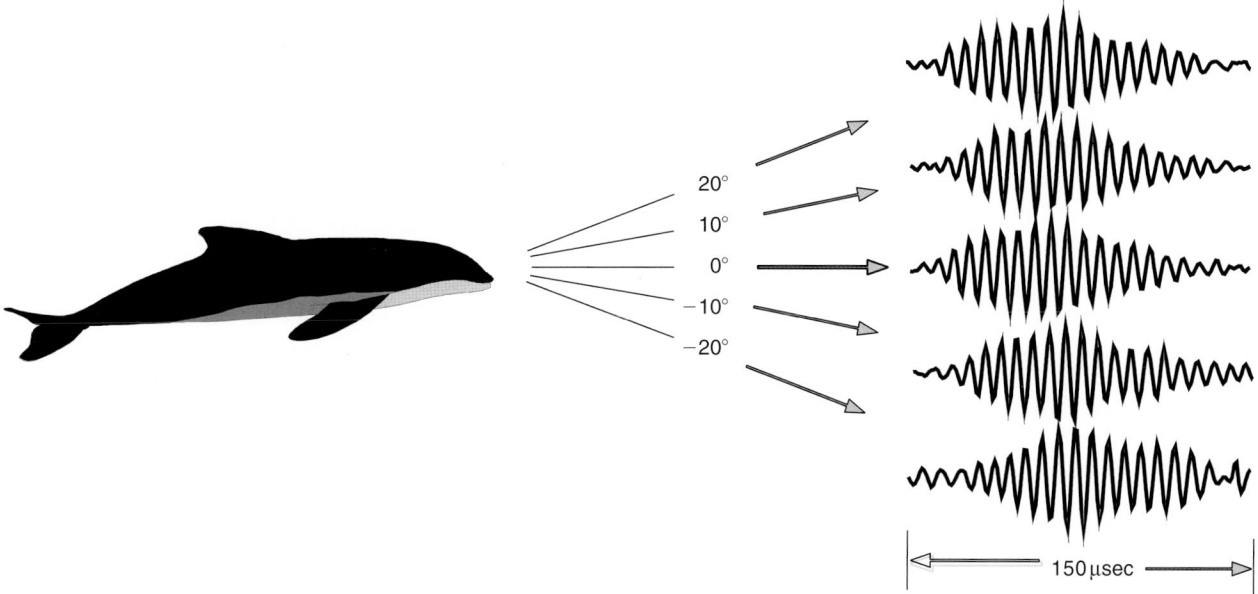

**Figure 8** *The waveform of an echolocation signal detected by hydrophones spaced about the head of a porpoise* (P. phocoena). *After Au et al. (1999).*

while the animal was performing a target detection task, indicated an average peak-to-peak source level of only 160 dB, which is considerably smaller than the 210–225 dB used by *T. truncatus*, *P. crassidens*, and *D. leucas*. The center frequency of the *P. phocoena* signal, which is typically between 120 and 145 kHz, does not depend on the level of the projected sonar signals.

Echolocation signals are projected from a dolphin's head in a beam. An example of the transmitting beam pattern for a *T. truncatus* in both horizontal and vertical planes is shown in Fig. 7. The signal shown at different angles about the animal's head is the same signal captured by an array of hydrophones. Note that only the signal traveling along the major axis of the beam is undistorted. This phenomenon occurs in horizontal and vertical planes. The numbers above each signal are the maxima in the frequency spectra of the signals, in order of descending amplitude. The further away from the major axis of the beam, the more the signal is changed. This property of the beam makes it very difficult to measure echolocation signals in the wild. Occasionally,

dolphins in the wild may actually swim directly toward a hydrophone so that relatively true measurements can be made.

Beam pattern measurements have also been conducted for *D. leucas*, *P. crassidens*, and *P. phocoena*. The signals from all of these animals, with the exception of *P. phocoena*, exhibit changes in frequency content when the measuring hydrophone is located away from the major axis. However, in the case of *P. phocoena*, the signals detected by hydrophones located away from the major axis are not distorted, as can be seen in Fig. 8. Distortion does not occur because the signal has a relatively narrow bandwidth.

One of the most fundamental properties of a sonar system is its maximum detection range. A simple way to determine the maximum target detection range of sonar is to gradually move a specific target away from the sonar until the target can no longer be detected. Au (1993) used a 7.62-diameter water-filled stainless-steel sphere as the target to determine the maximum detection range of *T. truncatus*. The target was moved progressively away from an echolocating

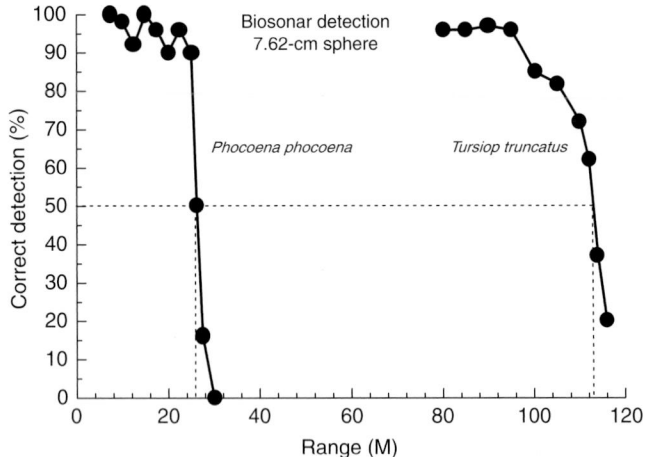

**Figure 9** *Target detection performance as a function of range for* T. truncatus *and* P. phocoena. *After Au (1993) and Kastelein* et al. *(1999).*

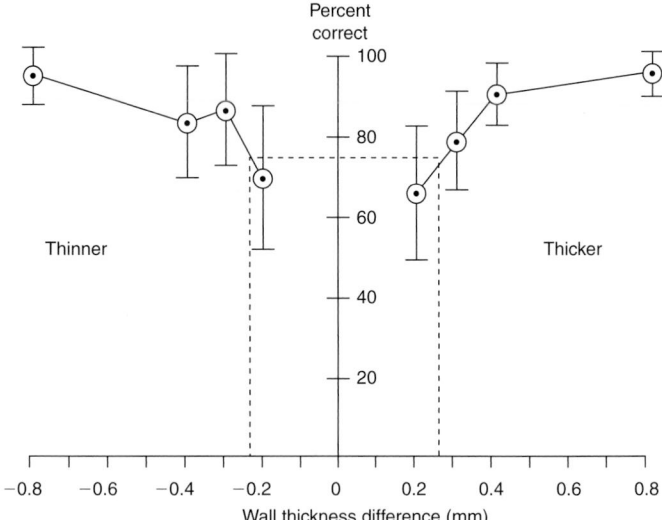

**Figure 10** *Performance of an echolocating dolphin in the wall thickness discrimination experiment. Adapted from Au (1993).*

dolphin until the animal could no longer detect its presence. Stringent psychophysical techniques were used and many sessions were conducted in order to stabilize performance and to determine the probability of detection as a function of range. Kastelein *et al.* (1999) used the same type of target (7.62-cm diameter water-filled stainless-steel sphere) as Au and colleagues to determine the sonar detection capability of *P. phocoena*. The results of both experiments are shown in Fig. 9.

The 50% correct detection threshold for the bottlenose dolphin occurred at a range of 113 m. The 50-percent correct detection threshold for the harbor porpoise was approximately 26 m. An experiment by Au and Snyder took place in Kaneohe Bay, Hawaii, a bay that has a high level of noise and the bottlenose dolphin's hearing was masked by the background noise level. The harbor porpoise target detection experiment was performed at a location in the Netherlands, where the ambient background noise was moderate, between 100 and 150 kHz. Therefore, the harbor porpoise was not masked by background noise and yet its detection threshold was considerably shorter than *Tursiops*. The difference in the two-way propagation losses for 113 and 26 m is 36 dB. The bottlenose dolphin typically produces clicks that are 50–60 dB greater than that of the harbor porpoise. Therefore, the large difference in the levels of projected signals can account for most, but not all, of the very large difference in the detection threshold ranges of both animals. If the target detection experiment with *T. truncatus* were conducted in a body of water with a low ambient background noise, the dolphin's target detection range would be much longer and, in that case, the difference in the two-way transmission would probably match the difference between the source levels used by the two different species.

## IV. Target Discrimination Capabilities

There have been many target discrimination experiments involving echolocating dolphins. Unfortunately, in many of these experiments the reflection characteristics of the targets were not measured or were measured with tone-burst signals instead of with a simulated dolphin-like signal. The experiment involving wall thickness discrimination by an echolocating dolphin is one that provided appropriate echo characteristics of the targets (Au, 1993). In this experiment, the

dolphin was presented with two hollow aluminum targets separated by 20° at a range of 8 m. The standard target had a wall thickness of 0.63 cm and the comparison targets had wall thicknesses that were different than the standard by ±0.8, ±0.4, ±0.3, and ±0.2 mm. All the targets had a length of 12.7 cm. On any given trial, the standard and comparison were introduced into the water separated by ±20° about the center axis. The dolphin was required to swim into a hoop and echolocate the targets when a screen was lowered out of the way and then touch a paddle that was on the same side of the center line as the standard target. Two sets of targets were available so that the position of the standard could be switched on any given trial.

The results of the wall thickness experiment are shown in Fig. 10. The dolphin performed very well with correct responses in the mid-90 percentile. The animal's correct response performance became progressively worse as the difference in wall thickness decreased. The 75% correct performance threshold was 0.27 mm for the case in which the comparison targets were thinner than the standard target and 0.23 mm when the comparison targets were thicker than the standard.

Echoes from the standard target and the 20.3 mm comparison target are shown in Fig. 11. There are several cues that the animal might have used in order to perform this discrimination. One cue is the difference in the time delay between the first and the second echo component for both the standard and the comparison target. If this cue was used, it suggests that the dolphin could discriminate differences of about 0.5–0.6 μsec. Another cue could be the difference in the time-separation pitch (TSP). When humans are presented with two correlated broadband acoustic signal separated by time $T$, a TSP equal to $1/T$ can be perceived. TSP stimuli also have a frequency spectrum that is rippled. The third possible cue is the difference in the frequency spectra of the echoes. The frequency spectrum of the echo from −0.3 mm comparison target is shifted between 2 and 3 kHz from the spectrum of the standard target. If the dolphin was using this cue, it suggests that the animal was able to perceive a frequency difference of 2–3 kHz in the broadband echoes.

The fact that echolocating dolphins can produce signals with peak frequencies between 100 and 135 kHz implies a relatively fast nervous system response because the time between periods of

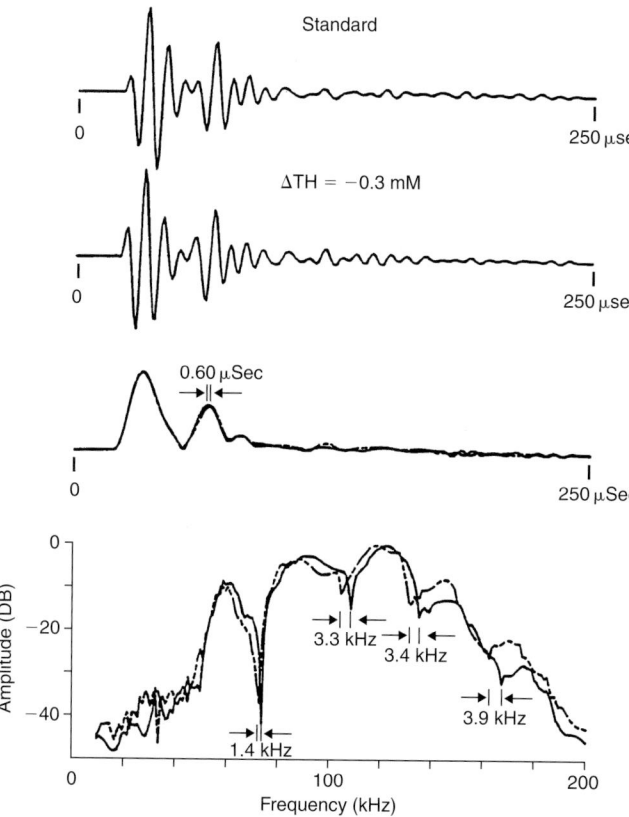

**Figure 11** *Echoes from the standard and the −0.3 mm targets. The top two traces are the echo waveforms, the middle trace is the envelope of the echo waveforms overlayed upon each other, and the bottom curve is the spectra of the echoes. Adapted from Au (1993).*

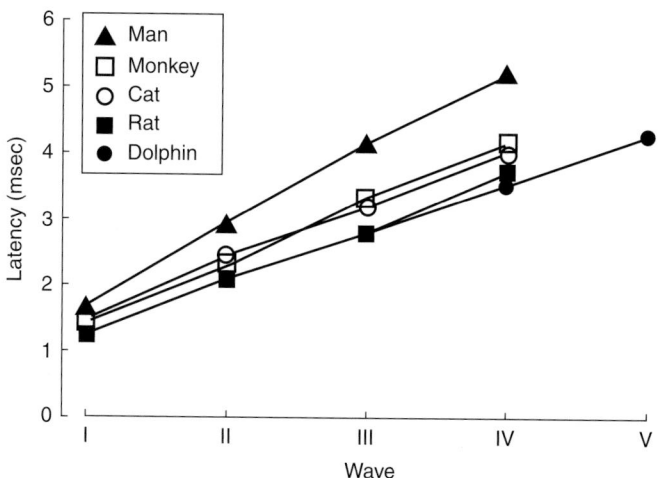

**Figure 12** *Brain stem-evoked potential latency for different animals. Adapted from Ridgway (1983).*

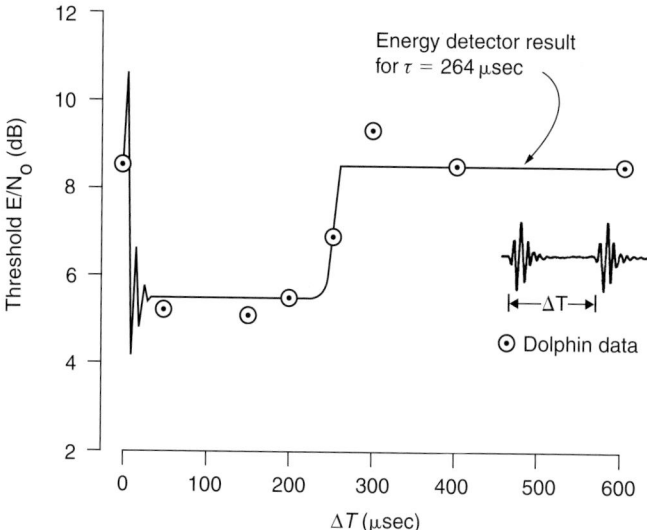

**Figure 13** *Results of the phantom echo experiment. After Au (1993).*

sound transmission and the auditory response can be short, on the order of 7–10 μsec. The speed of the auditory response can be determined by performing an electrophysiological experiment in which the time difference between the projection of the stimulus and the onset of the brain stem response is measured. This time difference is typically referred to as the latency of response. A comparison of the latency for a variety of mammals is shown in Fig. 12. In order to

fully appreciate Fig. 12, it is important to understand that the brain stem-evoked potential consists of several waves (indicated by Roman numerals) that arrive at the measuring electrodes at slightly different times. From Fig. 12, it is obvious that dolphins have an extremely fast auditory nervous response, faster than that of a rat. The acoustic stimulus must travel into the inner ear where the cochlea nerves discharge and send the electrical pulses to higher auditory centers and eventually to the brain stem. What is very astonishing is the fact that a rat's head is considerably smaller than that of dolphins, yet dolphins have shorter responses than the rat.

The response time of the auditory system of a dolphin can be estimated by performing an integration time experiment. Target detection experiments use phantom echoes to determine this (Au, 1993). The phantom echo generator would digitize the outgoing signal, which was detected by a hydrophone 1.5 m in front of the dolphin stationing in a hoop. The digitized signal was stored in the memory of a personal computer, and at the appropriate time delay representing the two-way transit time for a target at 20 m, the "echo," was sent back to the dolphin via a small transducer located 2 m in front of the animal. In the initial phase, a single echo was sent to the dolphin and the dolphin's hearing threshold for the single echo was determined by varying the amplitude of the echo in a staircase fashion. In the second phase, two echoes were separated by a variable spacing that was sent back to the dolphin. The dolphin's threshold was obtained by varying the amplitude of the whole echo in a staircase fashion for various separation times between the two echoes.

The results of the phantom echo experiment are shown in Fig. 13. For an echo consisting of a single click, the threshold is shown on the ordinate of the curve. Then two echoes were sent back to the animal with a separation time of 50 μsec. The threshold for the two-click echo at 50 μsec was approximately 3 dB lower than for the single click threshold. This was expected because the two-click echo had twice the amount or 3 dB more energy than the single click echo, and the dolphin auditory system, like most mammals, behaves as an energy detector. As the time separation between the two clicks increased to 200 msec, the dolphin's threshold remained very constant. However, when the time separation increased to 250 μsec and beyond, the dolphin's threshold began to move toward the threshold for a single click echo. The solid line in Fig. 13 represents the output

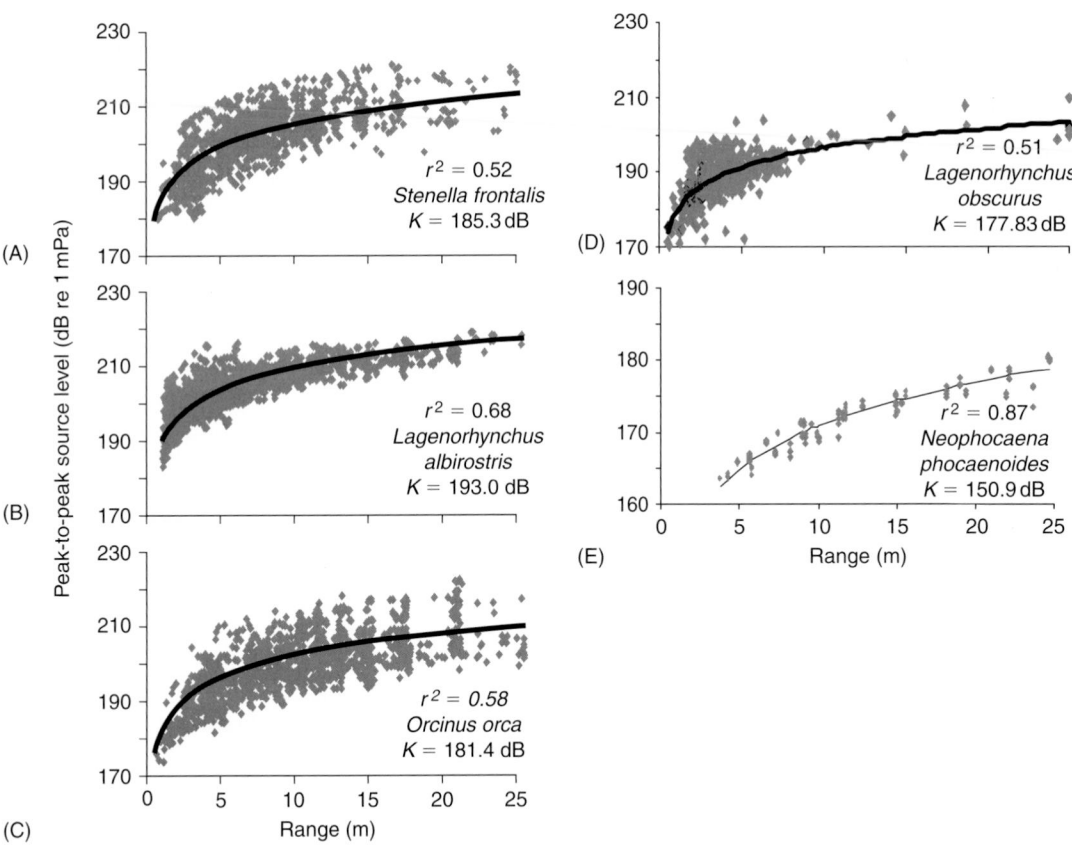

**Figure 14**  *Source level of echolocation signals from five species of odontocetes as a function of range. The data in (A)–(C) are from Au and Benoit-Bird (2003), in (D) from Au and Wursig (2004), and (E) from Li* et al. *(2006).*

of an energy detector having an integration of 264 μsec. The energy detector response with a 264 μsec integration time best fits the animal results. This integration time is extremely small compared to the integration time of any other mammal.

## V. Echolocation Signals of Free-Swimming Dolphins

It has only been recently that detailed studies of dolphin echolocation in the wild have been undertaken using a multiple hydrophone array so that the echolocating animal could be localized. One remarkable finding is that the echolocation signals measured with five different species of odontocetes in widely separate locations, suggest that these animals use a form of time-varying gain or automatic gain control. Time-varying gain is a technique used in technological sonar by which the gain of the receiver increased with time as a function of 40 log R, where R is the range in meters, to compensate for two-way spherical spreading losses that the signal and echo experience while propagating the target and back. Dolphins instead, vary the peak-to-peak source level (sound pressure level reference to 1 m in front of any projector) so that the further away a target is the greater the output level of the echolocation signal or conversely, the closer the target the lower the output level. Such variation in the output level of five species of free swimming dolphins are shown in Fig. 14. The solid line is the curve $K + 20 \log R$, that best fit the data with the $r^2$ value shown for each species. Not only are the data shown in Fig. 14 from five different species of odontocetes, but the measurements were made at widely separated locations. The data for

*Stenella frontalis* were obtained in the waters of the Bahama Islands, *Lagenorhynchus alborostris* in the waters of Iceland, *Orcinus orca* in the waters of Vancouver Island, British Columbia, *Lagenorhynchus obcurus* in the waters of Kailoura, New Zealand, and *Neophocaena phocaenoides* in the Yangtze River of China.

The dynamic control of the echolocation source level is probably not the result of a cognitive process, but rather a natural consequence of how echolocation clicks are produced. Dolphins typically emit echolocation clicks at a rate that allow the echoes to return to the animal before the next click is emitted. Consequently, the repetition rate increases as an animal closes on a target. The clicks are produced within the nasal system of the dolphins by manipulating the air flow through the phonic lips, previously referred to as the MLDB complex. Dolphins initially pressurize their nasal system and then emit a click train with the clicks occurring at relatively low repetition rate and the animal continually adjusting the rate as targets are located. If the dolphin chooses to keep the amount of acoustic energy emitted relatively constant (or within certain limits for each pressurization cycle), then the amplitude of the signal can be high when the repetition rate is low but must continually decreases as the repetition rate increases. The data are consistent with the notion that there is a dynamic coupling between reception of echoes, repetition rate, and source level of transmitted signals.

## VI.  Conclusions

Dolphins have very keen echolocation capabilities, much keener than any man-made sonar, especially in a shallow water environment.

The use of relatively short broadband echolocation signals by whistling dolphins is probably the most important factor in the dolphin's good discrimination capabilities. The broad frequency range of hearing extending over 10 octaves and the good peak sensitivity of 30–40 dB re 1 mPa are certainly contributing factors in the dolphin's echolocation capabilities. Another feature of the dolphin's auditory system that contributes to its good echolocation capabilities is the extremely rapid response of its auditory nervous system. The auditory nervous system of the dolphin probably responds faster than that of any other animal if the relative dimensions of the auditory system are taken into account. Finally, dolphins are extremely mobile and can investigate objects at different aspects and angles to maximize the amount of echo information from objects and thus enhance their echolocation capabilities.

## See Also the Following Articles

Brain ■ Hearing ■ Song ■ Sound Production

## References

Aroyan, J. L. (2001). Three-dimensional modeling of hearing in *Delphinus delphis. J. Acoust. Soc. Am.* **110**, 3305–3318.

Au, W. W. L. (1993). "The Sonar of Dolphins." Springer-Verlag, New York.

Au, W. W. L., and Benoit-Bird, K. J. (2003). Automatic gain control in the echolocation system of dolphin. *Nature* **423**, 861–863.

Au, W. W. L., and Würsig, B. (2004). Echolocation signals of dusky dolphins (*Lagenorhynchus obscurus*) in Kaikoura, New Zealand. *J. Acous. Soc. Am.* **115**, 2307–2313.

Au, W. W. L., Kastelein, R. A., Rippe, T., and Schooneman, N. M. (1999). Transmission beam pattern and echolocation signals of a harbor porpoise (*Phocoena phocoena*). *J. Acoust. Soc. Am.* **106**, 3699–3705.

Bullock, T. H., Grinnell, A. D., Ikezono, E., Kameda, K., Katsuki, Y., Nomoto, M., Sato, O., Suga, N., and Yanagisawa, K. (1968). Electrophysiological studies of the central auditory mechanisms in cetaceans. *J. Compar. Physiol. A.* **59**, 117–156.

Cranford, T. W. (1988). The anatomy of acoustic structures in the spinner dolphin forehead as shown by X-ray computed tomography and computer graphics. *In* "Animal Sonar: Processes and Performance" (P. E. Nactigall, and P. W. B. Moore, eds), pp. 67–77. Plenum Publishing, New York.

Cranford, T. W., Van Bonn, W. G., Chaplin, M. S., Carr, J. A., Kamolnick, T. A., Carder, D. A., and Ridgway, S. H. (1997). Visualizing dolphin sonar signal generation using high-speed video endoscopy. *J. Acoust. Soc. Am.* **102**, 3123.

Kastelein, R. A., Au, W. W. L., Rippe, T., and Schooneman, N. M. (1999). Target detection by an echolocating harbor porpoise (*Phocoena phocoena*). *J. Acoust. Soc. Am.* **105**, 2493–2498.

Li, S., Wang, D., Wang, K., and Akamatsu, T. (2006). Sonar gain control in echolocating finless porpoises (*Neophocaena phocaenoides*) in open waters. *J. Acoust. Soc. Am.* **120**, 1803–1806.

MacBride, A. F. (1956). Evidence for echolocation in cetaceans. *Deep Sea Res.* **3**, 153–154.

Møhl, B., Au, W. W. L., Pawloski, J. L., and Nachtigall, P. E. (1999). Dolphin hearing: relative hearing as a function of point of application of a contact sound source in the jaw and head region. *J. Acoust. Soc. Am.* **105**, 3421–3424.

Norris, K. S. (1968). The echolocation of marine mammals. *In* "The Biology of Marine Mammals" (H. T. Andersen, ed.), pp. 391–423. Academic Press, New York.

Norris, K. S., Prescott, J. H., Asa-Dorian, P. V., and Perkins, P. (1961). An experimental demonstration of echo-location behavior in the porpoise, *Tursiops truncatus. Biol. Bull.* **120**, 163–176.

Purves, P. E., and Pilleri, G. (1983). "Echolocation in Whales and Dolphins." Academic Press, London.

Ridgway, S. H. (1983). Dolphin hearing and sound production in health and illness. *In* "Hearing and Other Senses: Presentations in Honor of E. G. Wever" (R. R. Fay, and G. Gourevitch, eds), pp. 247–296. Amphora Press, Gronton.

# Ecological Effects of Marine Mammals

## James A. Estes

There are two ways in which marine mammals and their ecosystems can interact. One encompasses the effects of the ecosystem on marine mammals; the other, the effects of marine mammals on their ecosystems (Fig. 1). Ocean scientists in general and marine mammalogists in particular often consider their worlds from the former perspective. However, the latter perspective should also be of interest for two main reasons. First, there is a large and growing body of evidence from diverse ecosystems for the ecological importance of large vertebrate consumers, including several marine mammal species (Pace *et al.*, 1999; Shurin *et al.*, 2002). And second, significant ecological effects of marine mammals are implied from their great abundance, high trophic status, high metabolic rates, and the resulting fact that some of these consumers co-opt significant proportions of their ecosystem's primary production (Estes *et al.*, 2006). Many marine mammal species have been depleted through overexploitation for protein, oil, and other products. A few others have increased dramatically in recent years due to protection from human harassment or perhaps other factors. If marine mammals are important drivers of ecosystem structure and function, the ecological effects resulting from these changes in their abundance could be substantial. It follows that the structure and function of future world oceans may depend critically on the way in which the distribution and abundance of marine mammals are managed.

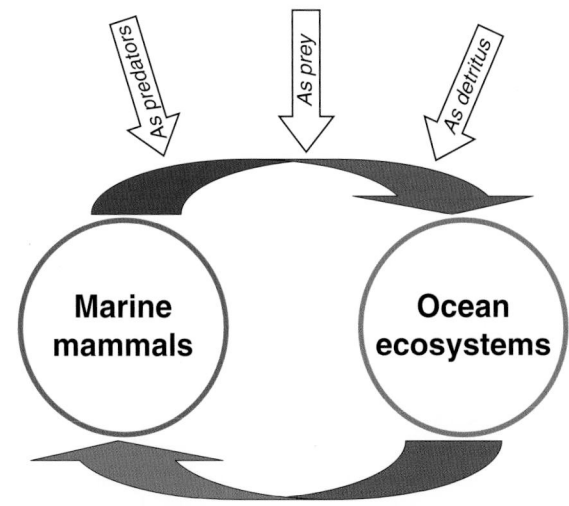

**Figure 1** *Cartoon drawing illustrating the various interactions between marine mammals and their ecosystems.*

## I. Conceptualizing and Understanding Interaction Web Processes

### A. The Nature of Species Interactions

Before considering the ecological roles of marine mammals in marine ecosystems, it is important to reflect on the diverse ways in which species interact with one another and how these processes can be understood. The influence of any one species on another can be categorized broadly as positive, negative, or neutral. The most ubiquitous and important species interactions probably are those that occur between consumers and their prey. Consumers are fueled by the things they eat and thus the influences of prey on their consumers are usually positive. Prey are killed by the things that eat them and thus the influences of consumers on their prey are usually negative. Competitive interactions between any two species are defined by reciprocal negativity. These interactions can be weak or strong, symmetrical or asymmetric, and manifested through exploitation of a shared resource or behavioral interference. Mutualisms are defined by reciprocal positive relationships. Like competitive interactions, these can be weak or strong, symmetrical or asymmetric. Unlike competitive interactions, mutualisms take many forms.

### B. Bottom-Up vs Top-Down Forcing

Bottom-up forcing occurs when the distribution and abundance of species are influenced by production and the efficiency of energy transfer upward through the food web. Top-down forcing occurs when the distribution and abundance of species are influenced by the impacts of consumers on their prey (Hunter and Price, 1992). These processes are not mutually exclusive nor do they act independently from one another. Bottom-up effects are ubiquitous in nature whereas top-down effects may not be. The important questions for this chapter are whether or not top-down forcing processes influence the structure and organization of marine mammal-dominated ecosystems and if so, the relative degree to which variation in bottom-up vs top-down forcing is responsible for changes in the distribution and abundance of marine mammals and their prey.

### C. Direct Vs Indirect Effects

Species may interact with one another directly or indirectly. The direct effects of one species on another are those that occur in the absence of intervening species. Indirect effects, in contrast, include one or more intervening species. The indirect effect of predators on lower trophic status species through top-down forcing is known as a trophic cascade (Paine, 1980). It is important to recognize that for any community of species there are vastly more potential indirect than direct interactions (Estes, 2005).

### D. Alternate Stable States

Contrary to a widely held view that underlies much of natural resource management and conservation, the same assemblage of species in similar physical settings does not necessarily organize itself in a single manner. In fact, there is growing evidence for alternate stable-state systems, among which the nature of species interactions and the abundances and distributions of species can differ substantially (Scheffer et al., 2001). There are numerous specific causes for this seemingly peculiar feature of ecosystem behavior (Doak et al., 2008). Cultural differences in foraging behavior can develop among individuals within a species as the result of serendipitous events that may be entirely lost to specific explanation. In general, alternate stable states arise within similar species assemblages because the transition vectors (i.e., the forces that drive population change following a perturbation) are commonly strongly non-linear and the pathways of historical context are highly unpredictable. Given this well-known feature of ecosystem dynamics on the one hand and the fact that so many marine mammal populations and their ecosystems have been extensively perturbed by various human-induced influences on the other, there is strong likelihood that the recovery of these systems will lead to alternate states.

## II. Approaches to Understanding

While the distribution and abundance of species are relatively easy to observe and measure, species interactions are invisible and therefore difficult to grasp. Two approaches (philosophies) have been employed in efforts to observe and measure the strength of species interactions.

### A. Perturbation Approaches

Perturbation-based approaches derive from the assumption that species interactions maintain ecosystems in equilibrium or quasi-equilibrium states, and thus that the interaction web pathways and their strengths can be observed and measured from responses to single species perturbations. Perturbation experiments have been done in three general ways. One of these is through purposeful manipulations, ideally conducted in a controlled and replicated manner. Properly executed over adequate scales of space and time, experiments of this nature provide the least equivocal evidence for species interactions. Most marine mammals cannot be purposefully manipulated for logistic, ethical, and legal reasons. Thus natural experiments based on fortuitous perturbations is another approach that has been used to understand the ecological roles of these and other large, mobile species. Yet another approach is to use historical records to infer the nature of species interactions through the retrospective analysis of patterns of covariation in the abundance of species over time. The scales of such historical analyses have varied from decades or centuries in the case of records obtained by modern humans, to millions of years for those based on the geological record.

### B. Constructionist Approaches

Finding informative perturbations and using them to answer specific questions can be challenging. An alternative approach to understanding species interactions is to construct model interaction webs from information on species distributions, abundances, diets, metabolic rates, life histories, behaviors, etc., and then to observe how the model system responds to perturbations of interest. The Ecopath/Ecosim/Ecospace family of mass balance models is one well-known example of this approach (Pauly et al., 2000), although the coupling of energetics and demography to estimate the strength of both bottom-up and top-down trophic effects and a variety of other methods have also been used. By and large, these constructionist approaches are more useful in evaluating the plausibility of particular hypotheses than in predicting population or ecosystem change.

## III. Case Studies

### A. Otters

Sea (Enhydra lutris) otters and the nearshore habitats in which they live provide the clearest and most compelling evidence for the ecological effects of a marine mammal. One reason for the utility of this particular system is that the Pacific maritime fur trade perturbed sea otter populations in such a way that their ecological effects could be observed on appropriate scales of space and time. Another helpful attribute is that many of the key species in the sea otter's coastal ecosystem are easy to observe, measure, and experimentally manipulate. Studies built around the decline and recovery of sea otters have shown that they limit the size, abundance, and distribution of their benthic invertebrate prey in both soft-sediment and rocky-reef habitats. These and other studies have further demonstrated that the control of herbivorous sea urchins by sea otter predation helps to maintain kelp forests on shallow reefs across much of the North Pacific Ocean (Estes and Duggins, 1995). This trophic cascade from otters to sea urchins to kelp indirectly affects other species and ecosystem processes through increased production, the creation of three-dimensional habitat (the kelp forest), and reduced water flow. The interactions resulting from sea otter predation may have exerted strong selective influences on various other species (Steinberg et al., 1995; Estes et al., 2005).

North American otters also have important ecosystem-level effects, which are founded on a characteristic high latitude sea-to-land production gradient. By foraging at sea and defecating at traditional land-based sprainting sites, North American otters vector nutrients from the sea to the land, thereby increasing secondary production, altering plant species composition, and generating habitat heterogeneity across coastal landscapes. In areas where sea otters and North American otters coexist, sea otters may enhance the effects of North American otters by increasing production and fish abundance in the coastal marine ecosystem.

The direct and indirect effects of sea otters and North American otters are discussed in the chapter on OTTERS, MARINE.

## B. Sirenians

Like sea otters, sirenians live in coastal marine systems where many of the species with which they interact can be observed and manipulated. Because sirenians are exclusively herbivores, their direct impacts are on aquatic plant assemblages.

Dugongs (*Dugong dugon*) in the tropical southwest Pacific Ocean provide the clearest evidence for the ecological effects of sirenians. Foraging dugongs uproot seagrasses, reducing their overall biomass, and creating heterogeneous habitats. These effects are clearly evident in seagrass meadows and have even been used to determine the presence or absence of dugongs in particular areas. By generating organic detritus, disturbing the substrate, and suspending sediments, dugong foraging has numerous effects on seagrass species composition and succession, as well as on associated species of invertebrates and fishes.

The intensity of dugong foraging on seagrass meadows varies seasonally in Western Australia because of changes in dugong habitat utilization in response to the risk of tiger shark (*Galeocerdo cuvier*) predation. When large tiger sharks are present during the warm season, dugongs tend to forsake the shallower seagrass-dominated habitats in favor of deeper channels where presumably they are better able to avoid shark attacks. Dugongs also change the way they feed in seagrass meadows (more cropping and less substrate excavation) when sharks are abundant. Large sharks therefore reduce the overall impact of dugong foraging in seagrass-dominated systems.

Although the kelp-eating Steller's sea cow (*Hydrodamalis gigas*) has been extinct for 250 years, the possible role of this species as an herbivore in kelp forest ecosystems is the subject of long-standing interest and speculation. Grazing sea cows may have created light gaps in the surface canopy-forming kelps, thereby increasing benthic illumination and releasing the competitively subordinate under story kelps from light limitation. Steller's sea cows are believed by most experts to have succumbed to extinction from direct human exploitation. However, others have argued that sea cows may only have been able to persist through an indirect asymmetric mutualism with sea otters, and thus that the sea cow's demise may have been caused or at least facilitated by kelp forest collapses resulting from human overexploitation of sea otters.

Because manatees (*Trichechus* spp.) tend to inhabit highly turbid tropical rivers and streams, relatively little is known of their effects in these systems.

## C. Pinnipeds

The ecological effects of pinnipeds are poorly known, due in part to the perspective from which they have been studied and in part to the absence of opportunity. With several notable exceptions, the interactions between pinnipeds and their prey have commonly been viewed from a bottom-up perspective. Furthermore, the areas of the ocean where pinnipeds feed are intrinsically more difficult to study than those described above for marine otters and sirenians.

Despite these limiting perspectives and the limited opportunities for empirically based studies, there are reasons to suspect that pinnipeds are strong interactors in some ecosystems. Pinnipeds have relatively high field metabolic rates and live at high population densities compared with terrestrial carnivores of comparable body size. Pinniped foraging ranges are often limited by distance from shore, thus concentrating their potential ecological effects in relatively narrow bands of coastal habitat.

The most compelling case study of a consumer-induced effect by pinnipeds involves harbor seals (*Phoca vitulina*) trapped in the lakes of eastern Canada. As freshwater lakes were formed following the retreating Pleistocene ice sheet, harbor seals survived in some of these but not others, thus providing the opportunity to contrast lake systems with and without seals. Such contrasts suggest that seal predation affected the size, species composition, and life history patterns of salmonids (Power and Gregoire, 1978). Smaller scale studies have been done on the influences of benthic feeding walruses in the western Arctic. Walruses not only reduce prey biomass, they create pits in the substrate that accumulate detritus, thus facilitating a detritivore-based food web. The only other substantive efforts to document the ecological effects of pinnipeds come from constructionist approaches, most of which have focused on competition between pinnipeds and fisheries.

The distribution, abundance, and behavior of several pinniped species may be influenced by the realization or risk of predation. Various recent population declines have been attributed to predation by sharks and killer whales (*Orcinus orca*). Striking behavioral contrasts in the reaction of pagophilic (ice-associated) pinnipeds to humans between the Northern (fright reactions) and Southern Hemispheres (generally tame) are purportedly due in large measure to the differential risk of predation from polar bears (*Ursus maritimus*) and humans in the Northern Hemisphere.

## D. Cetaceans

The potential influence of cetaceans on marine ecosystems is intriguing because of the antiquity of cetacean evolution, the diversity of foraging modes employed by various mysticetes and odontocetes, and because cetaceans comprise far more consumer biomass than other marine mammal groups (Estes *et al.*, 2006). This latter feature suggests not only large effects of cetaceans on their associated ecosystems, but a release from any such effects following the overexploitation of whales by industrial whaling. As is generally true for pinnipeds, the ecological effects of small cetaceans are mostly unexplored, for likely the same reasons.

The great whales are thought to be ecologically important because of their influences as predators, as prey, and as detritus (Fig. 1).

*1. As Predators*   Gray whale (*Eschrichtius robustus*) foraging on the Bering Sea shelf probably provides the clearest example of consumer-induced effects by a large cetacean. Gray whales consume various epibenthic and infaunal invertebrates, in the process re-suspending large quantities of sediment and nutrients. Gray whale feeding pits contain 50% less invertebrate biomass, and a fauna dominated more by free-living scavenger amphipods compared with the tube-building amphipods and polychaetes that characterize unexploited sites. The rate of sediment re-suspension from gray whale feeding is substantial, equaling or exceeding the rate of sediment input from the major rivers that enter the Bering Sea. This latter process must have a wide range of influences on marine fishes, birds, and mammals in the highly productive Bering Sea Ecosystem.

Other consumer-induced effects by great whales are known or suspected. The depletion of once abundant krill-feeding mysticetes by industrial whaling in the Southern Ocean purportedly led to more abundant krill, in turn resulting in improved body condition, earlier reproduction, and population increases by other krill-feeding species, including the smaller-bodied mysticetes and odontocetes, penguins, and various pinnipeds. Similar kinds of effects likely occurred in squid-based food webs following the reduction of sperm whales. Time series analyses further suggest that predatory fishes and piscivorous or semi-piscivorous whales once competed for what may have been a jointly limiting forage-fish prey base in both the North Atlantic and North Pacific Oceans.

*2. As Prey*   Because of their large size, great abundance, and high energy density, the great whales represent a valuable nutritional resource for both human and animal consumers. At least one large predator, the killer whale, is known to attack and consume great whales. These behemoths therefore were quite possibly an important prey resource for killer whales, especially prior to industrial whaling. Modern industrial whaling may have facilitated this interaction by dispatching the living whales, advertising their locations through the sounds produced by exploding harpoons and preventing the carcasses from sinking by injecting them with gas, thereby greatly extending the periods of time carcasses were available on the ocean's surface to scavenging by killer whales. The reduction of great whales through industrial whaling and the sudden elimination of harvested carcasses as a food resource for killer whales at the end of the industrial whaling era, may have caused transient killer whales to expand their diets to include smaller marine mammal species, ultimately resulting in population declines of some of these. Although there are numerous records of attacks by killer whales on various great whales, and living whales commonly have rake marks on their flukes from failed killer whale attacks, the importance of consumer–prey interactions between killer whales and large whales is much contended.

*3. As Detritus*   Dead whales that are not immediately consumed either wash ashore or sink to the sea floor. Although whale falls only constitute an estimated 0.3% of the particulate organic flux to the seafloor, these materials are highly concentrated so that the area in the immediate vicinity of a fallen carcass receives the equivalent of several thousand years of organic carbon input in a single pulse. A diverse assemblage of species (approximately 370 in the North Pacific Ocean alone) utilizes these carcasses, some of which appear to be obligate associates of whale falls.

Whale falls typically undergo a succession of stages upon reaching the sea floor. These include an initial *mobile-scavenger stage*, in which organisms like sleeper sharks (*Somniosus* spp.), hagfish, crabs, and amphipods remove flesh from the carcass; an *enrichment-opportunist stage* in which invertebrates and heterotrophic bacteria colonize the organic carbon-rich skeleton and surrounding sediments; a *sulfophilic stage* in which chemoautotrophic organisms exploit the sulfide emitting anaerobic decomposition of skeletal lipids; and a *reef stage* in which organisms exploit the physical structure of inorganic skeletal remains. Because the first three of these stages may require decades to run their course, the effects of whaling on deep sea assemblages are perhaps only now becoming manifest.

Marine mammals in general and large whales in particular provide important nutritional resources to various terrestrial vertebrates, including bears, foxes, eagles, and condors. The demise of the California condor (*Gymnogyps californianus*) may have been facilitated by the whaling-induced reduction in stranded carcasses.

## IV. Density-Mediated Vs Trait-Mediated Effects

The aforementioned examples all are of what have been referred to as *density-mediated effects*, or those for which the interaction strength is determined by mass–energy relationships between consumers and prey. Consumers can also influence their ecosystems through *trait-mediated effects* or behavioral responses to the risk of being eaten. Trait-mediated effects have been referred to under the rubric of "the ecology of fear." Although the study of trait-mediated effects is still in its infancy, various and sundry examples illustrate their potential importance to the dynamics of consumer–prey interactions involving marine mammals (Box 1). Density- and trait-mediated effects usually are complementary rather than antagonistic, with trait-mediated effects likely being the more important of the two in some instances.

## V. Future Directions

Although marine mammals clearly can have important and far-reaching effects on marine ecosystems, at this juncture the support for this contention comes mostly from theory, analogy with other species and ecosystems, and a smattering of case studies (Bowen, 1997). What might scientists do to better understand the ecological roles of marine mammals in the sea? One useful approach would be to document associated changes in the ecosystem as marine mammal populations grow or decline, keeping in mind and controlling for the potentially confounding influences of other environmental factors. Another potentially useful approach would be to use theory and interdisciplinary synthesis to better define the limits of ecosystem behavior. Modeling approaches involving demography, energetics, and behavior could be used to answer such important questions as whether killer whale/marine mammal assemblages are sustainable without killer whale predation on great whales, and if marine mammal population changes are more sensitive to bottom-up or top-down forcing. These latter approaches cannot provide definitive answers, but they can establish the plausibility of hypothesized processes, which is an important step in the search. Finally, marine mammalogists should continue to expand their conceptual visions and conduct their research in the company of interdisciplinary collaborators.

---

**Box 1**

Trait-mediated effects of consumers on their prey may complement density-mediated effects, establishing qualitatively new pathways of important species interactions and even exceeding or overriding the more traditionally understood density-mediated effects in some systems (Wirsing *et al.*, in press). The following select examples illustrate the range of known or suspected trait-mediated effects in marine mammal-dominated systems.

- Great whale migrations from high-latitude foraging areas to low-latitude breeding and calving sites were once thought to function primarily as a means of energy conservation through reduced heat loss in the warmer tropical or subtropical oceans. Subsequent analyses, and the fact that not all large whales migrate toward the equator to reproduce, cast doubt on this explanation. An alternative (and still contended) hypothesis is that large whales migrate to low-productivity tropical waters to reduce the risk of predation by killer whales on the highly vulnerable newborn calves.

- Dugongs spend more time feeding in shallow seagrass meadows during the cool seasons, when large sharks are rare, than during the warm seasons when large sharks are relatively abundant. This behavioral response to the risk of attack by sharks reduces the intensity of disturbance and herbivory by dugongs just as though they were actually being eaten by the predator.

- As sea otters re-colonized long unoccupied habitats in British Columbia, they foraged on the abundant red sea urchins (*Strongylocentrotus franciscanus*) in kelp-deforested ecosystems that had developed in their absence. The damaged urchin tests and other uneaten remains were dropped to the seafloor where they elicited a flight response by healthy conspecifics. Kelps re-colonized areas from which the urchins had fled just as though they had been directly removed by sea otter predation.

---

### See Also the Following Articles

Biogeography ■ Distribution ■ Ecology, Overview ■ Habitat Use

### References

Bowen, W. D. (1997). The role of marine mammals in aquatic ecosystems. *Mar. Ecol. Prog. Ser.* **158**, 267–274.

Doak, D. F. *et al.* (14 authors) (2008). Understanding and predicting ecological dynamics: Are major surprises inevitable? *Ecology*: **89**, 952–961.

Estes, J. A. (2005). Carnivory and trophic connectivity in kelp forests. *In* "Large Carnivores and the Conservation of Biodiversity" (J. C. Ray, K. H. Redford, R. S. Steneck, and J. Berger, eds), pp. 61–81. Island Press, Washington, DC.

Estes, J. A., and Duggins, D. O. (1995). Sea otters and kelp forests in Alaska: Generality and variation in a community ecological paradigm. *Ecol. Monogr.* **65**, 75–100.

Estes, J. A., Lindberg, D. R., and Wray, C. (2005). Evolution of large body size in abalones (*Haliotis*): Patterns and implications. *Paleobiology* **31**, 591–606.

Estes, J. A., DeMaster, D. P., Brownell, R. L., Jr., Doak, D. F., and Williams, T. M. (2006). Retrospection and review. *In* "Whales, Whaling and Ocean Ecosystems" (J. A. Estes, D. P. DeMaster, D. F. Doak, T. M. Williams, and R. L. Brownell, Jr., eds), pp. 388–393. University of California Press, Berkeley, CA.

Hunter, M. D., and Price, P. W. (1992). Playing chutes and ladders: Heterogeneity and the relative roles of bottom-up and top-down forces in natural communities. *Ecology* **73**, 724–732.

Pace, M. L., Cole, J. J., Carpenter, S. R., and Kitchell, J. G. (1999). Trophic cascades revealed in diverse ecosystems. *Trends Ecol. Evol.* **14**, 483–488.

Paine, R. T. (1980). Food webs: Linkage, interaction strength, and community infrastructure. *J. Anim. Ecol.* **49**, 667–685.

Pauly, D., Christensen, V., and Walters, C. (2000). Ecopath, Ecosim, and Ecospace as tools for evaluating ecosystem impact of fisheries. *ICES J. Mar. Sci.* **57**, 697–706.

Polis, G.E., Power, M.E. and Huxel, G.R. (eds), (2004). Food Webs at the Landscape Level. University of Chicago Press, Chicago, IL.

Power, G., and Gregoire, J. (1978). Predation by fresh water seals on the fish community of Lower Seal Lake, Quebec. *J. Fish. Res. Board Can.* **35**, 844–850.

Scheffer, M., Carpenter, S., Foley, J. A., Folke, C., and Walker, B. (2001). Catastrophic shifts in ecosystems. *Nature* **413**, 591–596.

Shurin, J. B., *et al.* (8 authors) (2002). A cross-ecosystem comparison of the strength of trophic cascades. *Ecol. Lett.* **5**, 785–791.

Steinberg, P. D., Estes, J. A., and Winter, F. C. (1995). Evolutionary consequences of food chain length in kelp forest communities. *Proc. Natl. Acad. Sci. USA* **92**, 8145–8148.

Wirsing, A. A., Heithaus, M. R., Frid, A., and Dill, A. M. (2008). Seascapes of fear: Evaluating sublethal predator effects experienced and generated by marine mammals. *Mar. Mamm. Sci.* **24**, 1–15.

Wootton, J. T., and Emmerson, M. (2005). Measurement of interaction strength in nature. *Ann. Rev. Ecol. Evolution Syst.* **36**, 419–444.

E

---

# Ecology, Overview

## BERND WÜRSIG

Marine mammals have entered just about all ocean habitats, and several mighty rivers and inshore seas as well. Only the deep abyss is foreign to them, but—remarkably—elephant seals (*Mirounga* spp.), sperm whales (*Physeter macrocephalus*), and several other toothed whales can "easily" dive to depths that exceed 1000 m, where it is cold and dark and where the pressure is 100 times and more what we experience on land. Perhaps just as remarkable is the fact that some of these divers, the pinnipeds, are also able to live on land, where they mate, give birth, and molt.

Morphologic, physiologic, and behavioral adaptations to the environments of marine mammals are largely driven by their food and the habitats of their prey. Although there are various ways that ecological adaptations can be divided, this article does so by several broad-based general habitat types: open ocean, semipelagic, coastal, and riverine feeding and breeding habitats and—for pinnipeds and the polar bear (*Ursus maritimus*)—their obligatory stint on land to breed.

**E**

## I. Open Ocean

There are two major types of open ocean marine mammals: "surface dwellers" and "deep divers."

### A. Surface Dwellers

In most of the open ocean or pelagic zone, smaller toothed whales and dolphins spend their entire lives within about 200 m of the surface. The near-surface environment is low in primary and secondary productivity except in latitudes higher than about 40° north and south of the equator. Therefore, these pelagic cetaceans travel great distances in search of food, often in large herds of hundreds to thousands. The large herds may be for better detection of prey, possible cooperative prey herding, and enhanced detection of predators such as deep water sharks and the larger cousins of dolphins, killer whales (*Orcinus orca*). All of these capabilities may be enhanced by several species traveling together, in so-called multispecies aggregations. An example in the eastern tropical Pacific (ETP), where a dolphin herd may travel over 1000 km in 1 week, is the co-occurrence of spinner (*Stenella longirostris*), pantropical spotted (*S. attenuata*), and common (*Delphinus* spp.) dolphins. These dolphins are slim-bodied (or "sleek"), built for speed and long-distance endurance. They do not have the thick blood (packed with red blood cells) so characteristic of deep divers. Instead, they feed on sporadically encountered near-surface fishes and squid, or at night on animals that rise to within several hundred meters of the surface in association with the deep scattering layer (DSL). Their occurrence in large schools has another function: the school is the social, breeding, and calf caregiving unit, and these nomadic wanderers tend to be within their "complete" society at all times (Norris and Dohl, 1980). Exceptions are when young males, e.g., may form separate bachelor herds or bands or when adult males move among breeding herds (as in sperm whales).

While several species of baleen whales migrate through deep water, they tend to feed on rich areas of invertebrates and fishes that are found more often close to shore. However, others habitually feed in open ocean waters. As is the case for the surface-dwelling odontocetes, baleen whales most often feed within about 200 m of the surface, as none of them are exceptionally deep divers. The balaenopterid whales termed rorquals, blue (*Balaenoptera musculus*), fin (*B. physalus*), sei (*B. borealis*), and Bryde's and Eden's whales (*B. brydei* and/or *B. edeni*) are good examples of oft-pelagic, near-surface feeders. Blue and fin whales tend to feed on euphausiid crustaceans, or krill; whereas sei and Bryde's whales feed more commonly on shoals of fishes. All of them lunge through their food rapidly. The balaenid right (*Eubalaena* spp.) and bowhead whales (*Balaena mysticetus*) often surface-skim feed in the open ocean of productive high latitudes, whereas gray whales (*Eschrichtius robustus*) feed on ampeliscid (tube-dwelling) amphipods. The latter do so in waters less than 200 m deep, both near shore, and far away from land in the Bering, Chukchi, and Beaufort Seas. Although rorquals are built for speed so that they can lunge into food rapidly, balaenid and gray whales tend to a more rotund body shape, with great blubber storage capability for long fasts (Berta and Sumich, 1999).

Many pinnipeds also feed near the surface and, at times, up to several hundred kilometers from shore. The smaller true seals (such as ringed seals, e.g., *Pusa hispida*) and all of the eared seals are not deep divers and therefore stay near the surface in those generally higher latitude waters where they find themselves in the open sea. Near-surface feeding pinnipeds are not likely to be out on truly oceanic seas further than several hundred kilometers from land.

Northern fur seals (*Callorhinus ursinus*), however, are often found in deep pelagic waters of the north Pacific (Costa, 1993).

### B. Deep Divers

Many of the larger toothed whales and a few true seals dive "deeply," or below about 500 m. Sperm whales are likely to be the champion divers. They routinely feed at depths around 500 m on fishes and squid, but can also dive to 2000 m and more in search of the larger truly pelagic squid. Although we know little of the dive capabilities of other deep-diving odontocetes—pilot whales (*Globicephala* spp.), beaked whales (family Ziphiidae), dwarf and pygmy sperm whales (family Kogiidae), and the false killer whale (*Pseudorca crassidens*) are good examples—it is likely that all of them are capable of greater than 500-m dives as they feed largely on mid-sized deep water fishes (often of the family Myctophidae) and squids. Curiously, the largest dolphin-like (or delphinid) cetacean, the killer whale, appears to feed without diving deeply. It is possible, but remains unproved, that some smaller toothed whales can evade killer whales by diving down.

The champion pinniped divers are northern (*Mirounga angustirostris*) and southern (*M. leonina*) elephant seals as well as the Weddell (*Leptonychotes weddellii*) and probably several other large true (or phocid) seals. They can (but do not often) dive down to 1000 m and beyond. They feed on fishes and squid at these depths, but it has been surmised that at least some deep dives are "resting dives" as the animals conserve energy while their metabolism is largely shut down at depth. Such possible rest (or "sleep") may even help them evade detection by predators such as most active sharks, who are not deep divers, and killer whales. Because elephant seals spend only about 15% of their time at the surface, it is not really correct to call them "divers." Their life is underwater and they are indeed "surfacers" who come up only for life-sustaining air (LeBoeuf and Laws, 1994).

All deep divers have adapted physiologically and morphologically for the task. Blood and muscles have changed to hold as much oxygen as possible, and peripheral vasoconstriction and shutting off non-vital body functions during a dive take place.

## II. Semipelagic

Quite a few marine mammals habitually occur in the zone between shallow and deep water, often at the edge of the continental shelf or some other underwater feature. There is high productivity there, caused by upwelling or current systems as sea meets land, and it makes sense that this is a major point of aggregation. Sperm whales off Kaikoura, on the South Island of New Zealand, feed in such a zone near the deep Hikurangi Trench about 10 km out. However, the sperm whales are often within 1–5 km from shore, in productive waters 200–600 m in the deep, shore side of the trench. Blue whales of Monterey Bay, California, do so as well, as they take advantage of large stands of krill to enter the area in late summer. Dall's porpoises (*Phocoenoides dalli*) are also found in some abundance in Monterey Bay, not as frequently in very shallow nor very deep waters, but on the edge of the productive Monterey Canyon. Dozens of species and hundreds of geographic examples could be cited as those that occur in such productive "neither nearshore nor open ocean" zones (Evans, 1987; Reynolds and Rommel, 1999).

Several dolphins are "semipelagic" in another sense. They seek out deep productive waters in areas close to shore so that they can feed in the open sea yet retreat to the shallows, often into bays and inlets or onto expansive shoals during rest. Spinner dolphins of the tropical islands of the Pacific have such a habit. During the day, they rest and socialize within island bays and lagoons, even entering atolls through narrow passes in some areas. It is believed that nearshore rest is to avoid large oceanic swells and trade winds, as well as predation by large oceanic sharks. At night, these dolphins head out to sea, often only 1–5 km from land off abruptly rising volcanic islands or coral atolls. The dolphins meet the DSL as it comes to within several hundred meters of the surface at night and thus have a food resource available that these only-average divers could not obtain during the day, when the DSL is 600 m or more below. Atlantic spotted dolphins (*Stenella frontalis*) appear to do the same, but have daytime rest over an expansive shallow area: the Grand Banks of the Bahamas, only 6–10 m deep.

Pinniped females that go on foraging dives in between nursing their young on land, such as Galapagos fur seals (*Arctocephalus galapagoensis*), also use the productive shelf and drop-off waters to feed while—in their case—needing to return to land to take care of their young.

### III. Coastal

Many marine mammals can be termed "coastal," and because all of the various taxonomic orders and suborders have coastal representatives, one to several examples of each group are given here.

The most coastal baleen whale is undoubtedly the gray whale, for it feeds in shallow waters of the Bering Sea, usually but not always near coasts; travels on its immense migration from the Bering Sea to Baja California, Mexico—and back—along the coast; and mates and calves near and in coastal lagoons of the subtropics. It is likely that this rather slow cetacean hugs the coastline for safety (mainly, one surmises, for its young) against shark and killer whale predation. It probably also uses the coast to navigate. It would not be surprising, although present information is not clear on this point, if gray whales use the depth contours, rocky outcroppings of headlands, and other near-coastal features as signs of location as surely as we find our way to and from the supermarket. The coastline also allows them to find clouds of mysids, small aggregating invertebrates, among kelp beds, and to occasionally feed on stands of in-benthic invertebrates while on migration. A second "coastal" animal is the humpback whale (*Megaptera novaeangliae*), for it feeds in bays and inlets, breeds near islands, and only uses deep oceanic waters to get to and from these ends of migration. In the northeast Pacific, humpbacks feed in the fjord-like bays of southern Alaska and breed around the Hawaiian and Mexican Revillagigedo islands.

Odontocete cetaceans have many coastal representatives, with the best studied of them being the bottlenose dolphin (*Tursiops truncatus*). While separate populations of this highly adaptable species can exist in deep oceanic waters as well, it is the coastal form that has taken our fancy and makes for one of the better captive animals, presumably because it feels at home in small groups and with confines of cliffs, rocks, bayous, and channels. Bottlenose dolphins variably nose and poke their way among rocks to feed; feed on schooling fishes in the nearshore, at times trapping schools against a beach or cliff; feed on the bottom; and encircle prey as a cooperating group in the open coastal sea. Dolphins of the *Cephalorhynchus* genus of Southern Oceans tend toward coastal living, as do the humpback dolphins of the genus *Sousa*, harbor porpoises (*Phocoena phocoena*), and beluga whales (*Delphinapterus leucas*) of the Arctic. Interestingly, these animals tend toward fission–fusion societies, traveling in subgroups of variable size from day to day. It is likely that they aggregate in small groups for greatest efficiency in hunting, and that the social or breeding unit consists of all of the small groups of an area that get together at some time throughout a year, but never

all at once. Most but not all coastal waters are turbid as well, and it may be that echolocation and communicative sounds are particularly well developed in these animals.

Many pinnipeds have coastal representatives, especially for the physically smaller species. California sea lions (*Zalophus californianus*) rest on the shore and feed in the coastal zone, hardly ever venturing further than several kilometers from land. Harbor seals (*Phoca vitulina*) and the two endangered tropical monk seals (*Monachus* spp.) do so as well.

Sea otters (*Enhydra lutris*), marine otters of Chile (*Lontra felina*), and the sirenians are all coastal shallow-water feeders. Otters feed on invertebrates on the bottom or on kelp-associated fishes. While many populations of sea and marine otters do not frequently haul out on land, they use kelp beds as resting stations and perhaps as a means to hide from sharks and killer whales (Estes and Duggins, 1995). The West Indian manatee (*Trichechus manatus*) and the dugong (*Dugong dugon*), the latter largely of the nearshore Indian Ocean, feed on seagrasses and are thereby restricted to the shallows (Reynolds and Odell, 1991).

### IV. Riverine

While the term "marine mammals" is meant for mammals that take all or most of their food from the sea, several species are included that have gone to a largely freshwater environment. Because these have close taxonomic affiliations to several other marine mammals, this inclusion makes sense.

There are several obligate river dolphins: the susu and bhulan [now listed as subspecies within one species (*Platanista gangetica*)] of the Indian subcontinent; the baiji (*Lipotes vexillifer*) of the Yangtze River of China (recently declared likely to be extinct, Turvey *et al.*, 2007); and the boto or Amazon river dolphin (*Inia geoffrensis*) that also occurs in the Orinoco basin of South America. These dolphins live their lives in mighty rivers, feeding on invertebrates and fishes, generally in small groups numbering fewer than about six animals. Their eyes have adapted to the less saline environment and their kidneys do not need to process the salty foods of the ocean. It is likely that they would not survive in salt water. The Amazonian manatee (*Trichechus inunguis*) is also restricted to the extensive freshwater system of the Amazon basin.

In addition to obligate river dolphins and the Amazonian manatee, there are several mammals that are facultative, those who have populations that occur in rivers and those who go in and out of rivers to the adjacent ocean. Of the first type are finless porpoises (*Neophocaena phocaenoides*) that occur throughout nearshore waters of southern Asia and a bit of the Indian Ocean, but have a thoroughly freshwater population in the Yangtze River. Recent work shows that the freshwater form has eyes, skin, and kidneys that are adaptively different from their ocean-going conspecifics. As well, the diminutive tucuxi dolphin (*Sotalia fluviatilis*) occurs nearshore along much of the tropical Atlantic Central and South American coast, but as separate populations in the Amazon River basin. Of the second type of marine mammals, where some go in and out of rivers as members of the same population, are Irrawaddy dolphins (*Orcaella brevirostris*), bottlenose dolphins, belugas, West Indian manatees, and West African manatees (*Trichechus senegalensis*). To date, there are no well-defined morphologic or physiologic differences between those who are frequent in freshwaters more than others, and it is assumed that this wide salinity tolerance is itself an adaptation that allows exploitation of food resources in ecologically diverse realms. Belugas seem to enter rivers more often during a concentrated period of skin sloughing or molt; these are the only whales with a well-defined molting period, although all whales and dolphins are "constantly" shedding their outer skin.

Almost all pinnipeds are generally tied to the sea to feed, but a form of the harbor seal and the Asian Lake Baikal (freshwater) and Caspian Sea (somewhat salty) seals (*Pusa* spp.) occur in land-locked areas. They occur in remnants of areas that were once connected to oceans.

## V. Life on Land

Polar bears do considerable feeding on land or ice, pinnipeds all need to come to land to give birth, and sea otters do so variably by population. Off California, sea otters give birth in the water, but are usually surrounded and buoyed by *Macrocystis* spp. giant kelp fronds. While some pinnipeds, such as the walrus (*Odobenus rosmarus*) and Weddell seals, mate in water, others do so on solid land or ice, and all females need to come to solid substrates to give birth and to suckle their young. Indeed, most newborn pinnipeds (and the polar bear) are not yet a marine mammals and would become overexposed rapidly and die if they were to be dunked into water. The natal pelt of most true or phocid seals is a downy fur, or lanugo, that holds insulating hair but is not waterproof; they have brown fat, a type of lipid that breaks down rapidly to generate heat; and they instinctively huddle near mother and each other to stay warm (Stirling, 1988; Berta and Sumich, 1999, on polar bears).

Pinnipeds have shortened and greatly changed fore and hind flippers, modified beautifully for swift and precise movement in water. However, they have had to compromise their morphology to keep a bit of it—so very necessary for procreation—available for life on land. It is now known that early cetaceans lived a similarly dual existence, with morphologic, physiologic, and likely behavioral compromises to survive in both realms. It is tempting to speculate whether pinnipeds, given another 20 million years of evolution, could make the same total transition to the sea.

## See Also the Following Articles

Cetacean Ecology ▪ Diving Physiology ▪ Ocean Environment ▪ Pinniped Ecology

## References

Berta, A., and Sumich, J. L. (1999). "Marine Mammals: Evolutionary Biology." Academic Press, San Diego.

Costa, D. P. (1993). The relationship between reproductive and foraging energetics and the evolution of the Pinnipedia. *Symp. Zool. Soc. Lond.* **66**, 293–314.

Estes, J. A., and Duggins, D. O. (1995). Sea otters and kelp forests in Alaska: Generality and variation in a commuity ecological paradigm. *Ecol. Monogr.* **65**, 75–100.

Evans, G. P. H. (1987). "The Natural History of Whales and Dolphins." Facts on File Press, New York.

LeBoeuf, B.J., and Laws, R.M. (eds), (1994). "Elephant Seals". University of California Press, Berkeley, CA.

Norris, K. S., and Dohl, T. P. (1980). The structure and function of cetacean schools. *In* "Cetacean Behavior: Mechanisms and Functions" (L. M. Herman, ed.), pp. 305–316. Wiley-Interscience Press, New York.

Reynolds, J. E., III, and Odell, D. E. (1991). "Manatees and Dugongs." Facts on File Press, New York.

Reynolds, J.E., III, and Rommel, S.A. (1999). "Biology of Marine Mammals". Smithsonian Institution Press, Washington, DC.

Stirling, I. (1988). "Polar Bears." University of Michigan Press, Ann Arbor.

Turvey, S. T., *et al.* (16 authors) (2007). First human-caused extinction of a cetacean species? *Biol. Lett.* **3**, 537–540.

# Elephant Seals
## *Mirounga angustirostris* and *M. leonina*

MARK A. HINDELL AND WILLIAM F. PERRIN

## I. Characteristics and Taxonomy

The northern elephant seal (*Mirounga angustirostris*) and the southern elephant seal (*M. leonina*) are the largest pinnipeds (Ling and Bryden, 1981; McGinnis and Schusterman, 1981). The most striking characteristic of both species is the pronounced sexual dimorphism, with males weighing 8–10 times more than females. Male southern elephant seals have been recorded to weigh 3700 kg, whereas females only weigh between 400 and 800 kg. This makes the elephant seal the most sexually dimorphic mammal (Fig. 1). There are other pronounced secondary sexual differences in morphology, all of which are related to the highly polygynous mating strategy of the species. Most notable of these is the large proboscis of the male that plays a key part in dominance displays with other males (Sanvito *et al.*, 2007a).

The evolutionary origins of the species are unclear, with estimates of the divergence from a common ancestor ranging from as little as 10,000 years ago to as far back as the Pleistocene. Morphological differences between the species are, however, quite distinct and they are readily differentiated. The southern species is larger, with northern elephant seals rarely reaching more than 2300 kg. In both species, adult females exhibit a considerable range in body weight and there are no clear differences between them in this feature. Adult males of the northern species have a longer proboscis than the southern species. Northern elephant seals also have a more developed chest shield, a region of the neck, chest, and shoulders of thickened and scared skin associated with fighting. In the northern species, this has a distinctive red coloration. Females lack the proboscis and chest shields, but northern females are distinguished by a noticeably narrower and flatter nose than in the southern species.

## II. Distribution and Abundance

Despite their physical similarities, the two species have very different geographic distributions, with at least 8000 km separating them. Southern elephant seals have a more extensive range, with breeding sites on islands scattered right around the subantarctic (Fig. 2). Very occasionally, pups are even born on the Antarctic mainland. The range extends north to Patagonia and the Falkland Islands (Las Malvinas) (Lewis *et al.*, 2006a). Studies have indicated that when not ashore during the breeding season or for their annual molt, southern elephant seals utilize most of the Southern Ocean ranging from waters north of the Antarctic Polar Front (sometimes called the Antarctic Convergence) to the high Antarctic pack ice. There is some separation of feeding areas between the sexes, with males tending to feed in the more southerly waters associated with the Antarctic continental shelf.

Northern elephant seals have a limited breeding distribution, pupping at approximately 15 colonies between Point Raines in northern California to the Baja California Peninsula in Mexico (Fig. 3). Most of the colonies occur on offshore islands, but a small number occur on the mainland coast. As with southern elephant seals, the

**Figure 1**  *Northern elephant seal* (C. Brett Jarrett).

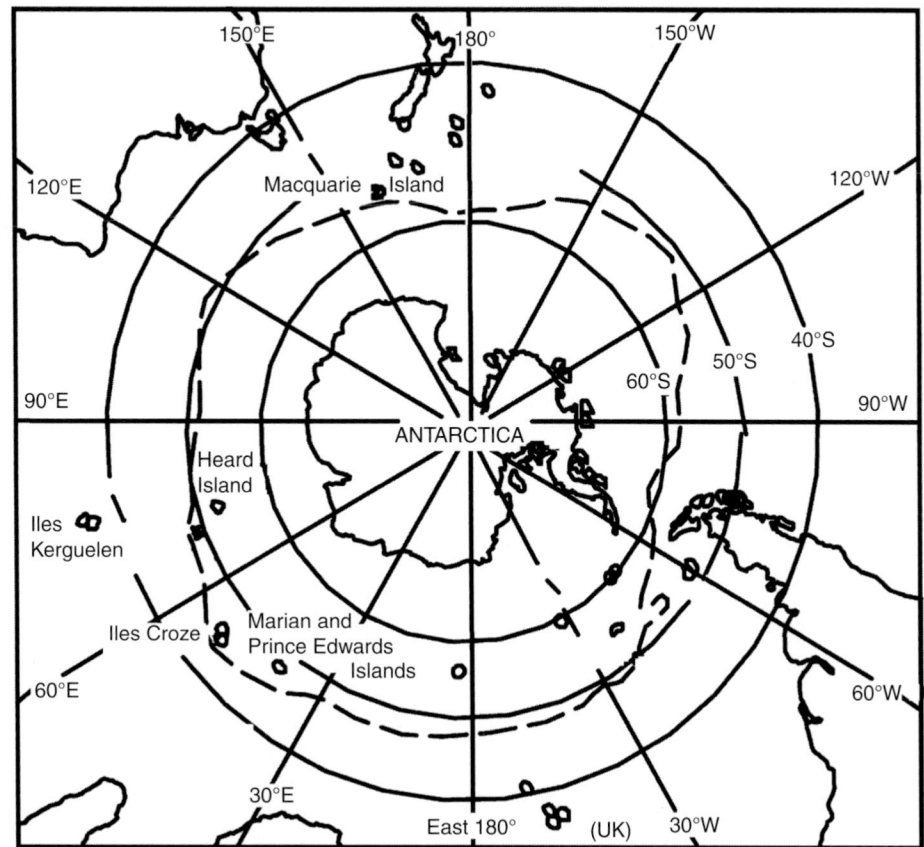

**Figure 2**  *Map of the Southern Ocean indicating the major southern elephant seal breeding islands* (named in map).

northern species disperses widely during the non-breeding phase of its annual cycle. Many individuals travel northward along the North American coast to feed in the Gulf of Alaska and the Aleutian Islands, which is a round trip of more than 10,000 km. This MIGRATION is even more remarkable as many individuals make it twice per year, returning to their southern breeding grounds to molt. The northern species also exhibits sexual differences in foraging areas, with males tending to use the more northerly areas and females heading in a more northwesterly direction and feeding in deep oceanic waters of the North Pacific Ocean.

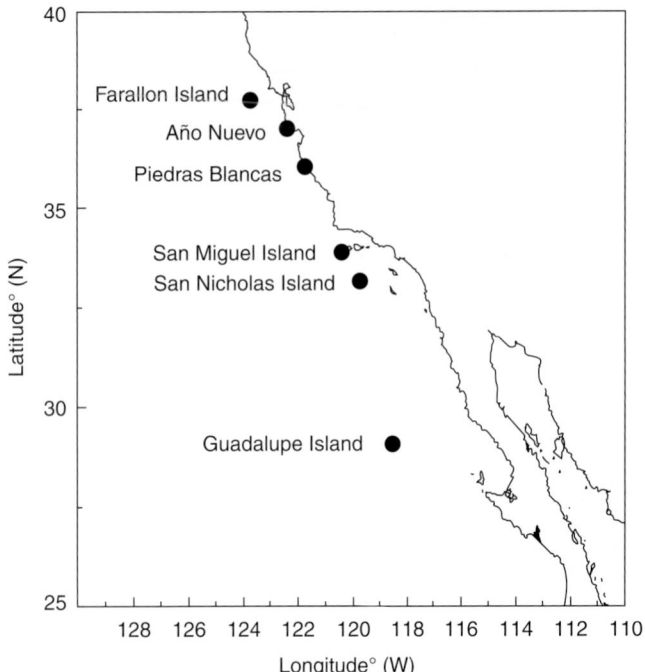

**Figure 3** *Map of the West coast of North America indicating the major northern elephant seal breeding sites.*

Northern elephant seals are presently undergoing a rapid population increase and range extension, whereas the southern species is currently experiencing a decline in two of its three major populations.

Population declines in the order of 50% have been recorded in both the southern Indian and the southern Pacific Ocean populations since the 1950s and 1960s, whereas the populations in the South Atlantic are stable or increasing. For example, the relatively small breeding population at Prince Edward Island declined at least 66.3% between 1977 and 2004 (Bester and Hofmeyr, 2005). The estimated total population size for southern elephant seals in 2000 was 640,000. Modeling suggests that the declines are due to relatively small changes in fecundity and survival (McMahon *et al.*, 2005a), but the underlying causes are presently unclear. Two plausible hypotheses are that they are related to interspecific competition or to environmental change causing change in food availability (McMahon *et al.*, 2005b; Pistorius *et al.*, 2005).

Conversely, northern elephant seal populations are currently increasing at a rate of approximately 6% per annum. This is the latest phase in one of the most remarkable population recoveries of any mammal. The total number of northern elephant seals in 1890 is thought to have been less than 100 individuals after 50 years of intensive and indiscriminate hunting by sealers. The last published estimates of the population put the total population at 127,000 in 1991. The expansion may have slowed in some regions (e.g., no change in abundance was detected at one of the San Benito Islands in Mexico between 1980 and 2001) (García-Aguilar and Morales-Bojorquez, 2005).

As a direct consequence of the extensive hunting in the nineteenth century, northern elephant seals passed through an intense genetic bottleneck, which has seen almost all genetic variation removed from the population. The relatively recent expansion onto several islands has not yet resulted in discernible genetic variation between the breeding groups.

In contrast, southern elephant seals have quite a clear genetic structure, with four distinct stocks: the southern Pacific Ocean, the south Atlantic, the southern Indian Ocean, and a small, but increasing population at Peninsula Valdes in Argentina. The integrity of the subpopulations appears to be maintained by the extremely low interchange rates between populations. Although genetically distinct, animals from the subpopulations are indistinguishable from each other in external features.

## III. Ecology

Elephant seals spend more than 80% of their annual cycle at sea, making long migrations to favorable foraging areas and feeding intensively to build up the blubber stores required to support them during breeding and molting haulouts (van den Hoff, 2001). The seals prey on deep-water squid and fish (Daneri and Carlini, 2002; van den Hoff *et al.*, 2003) and, as a result, have developed the remarkable ability to dive to depths in excess of 1500 m and for as long as 120 min (Le Boeuf and Laws, 1994). While these values are the extremes of those recorded, even the average values are impressive. Adult females routinely make dives of 20 min and reach depths of 400–800 m. Paradoxically, although the males generally dive for longer, about 30 min, they often do not go as deep. This is a reflection of their tendency to feed over continental shelves, whereas females use deeper open water. Age-related shifts occur (Field *et al.*, 2007), which may accomplish niche expansion and resource partitioning (Lewis *et al.*, 2006b; Field *et al.*, 2007). For example, in the Antarctic, juvenile males remain within the pack ice to forage on the shelf (Bailleul *et al.*, 2007); similar behavior is observed for juvenile males on the Patagonian shelf (Campagna *et al.*, 2007).

Foraging activity and success vary with oceanography features such as sea-surface temperature (SST) and events in both hemispheres (Bradshaw *et al.*, 2004; Campagna *et al.*, 2006; Simmons *et al.*, 2007), as do the prey species taken (Piatkowski *et al.*, 2002; Crocker *et al.*, 2006).

## IV. Behavior and Physiology

Elephant seals spend as much as 90% of the time submerged, the majority of it hunting for food, but other behaviors, such as traveling from place to place, and apparently even resting, take place at depths of more than 200 m. A number of morphological and physiological adaptations and behaviors make this, and prolonged fasting ashore, possible (Hindell *et al.*, 2000; Lester and Costa, 2006; Davis *et al.*, 2007).

Monachine seals all have an unusual annual molt, which entails the shedding of epidermal tissue in addition to the hair. The rich supplies of blood required at the body surface for the new skin and hair require the animals to leave the water in order to conserve body heat. The seals therefore spend 3–5 weeks fasting ashore during this time, once again relying on stored blubber to supply their energy requirements (Champagne *et al.*, 2006). Primiparous females tend to molt and haul out earlier than other breeding females (Kirkman *et al.*, 2004); timing of molting also varies over other age/sex classes (Kirkman *et al.*, 2003).

Both species of elephant seal are highly polygynous, with large, dominant males (alpha males) presiding over large aggregations of females known as "harems." Competition between males for the alpha position is intense and leads to spectacular fights. Successful males will have almost exclusive access to harems consisting of up to 100 females and so the reproductive benefits of success are very high. This has led to the evolution of the pronounced secondary sexual characteristics of immense body size and exaggerated proboscis. Vocalizations also contribute to the struggle for females; they have been found to convey significant information on age, size, and resource-holding potential (Sanvito *et al.*, 2007b).

### V. Life History

Age at first breeding in females ranges from 3 to 8 years, with average values varying with species, population status, and environmental changes (Sydeman and Nur, 1994). There are tradeoffs between fecundity and survival. In one study of the northern elephant seal, maternal survival was inversely related to age of primiparity, although this was not found in a similar study of the southern species (Pistorius *et al.*, 2004).

The annual breeding cycle begins when the largest males haul out on deserted beaches (in August for *M. leonina* and in December for *M. angustirostris*). Pregnant females then haul out in large numbers, aggregating into harems, and giving birth to their single pup 2–5 days after arriving. The females stay with their pup throughout the ensuing lactation period, never feeding and relying on their thick BLUBBER layer to sustain them and to supply many liters of milk required by the rapidly growing pup. At birth the pups weigh between 30 and 40 kg in both species, but by the time they wean, southern elephant seal pups weigh approximately 120–130 kg and northern elephant pups weigh approximately 140–150 kg. The difference in weaning weight is due to the slight difference in the duration of lactation in the two species, with southern elephant seals weaning at 23–25 days and northern elephant seals at 26–28 days.

Several days before weaning their pups, the females come into estrus and are mated by the dominant males. Although fertilization takes place at this time, the blastocyst does not implant until several months later. This ability, known as delayed implantation, is common to many pinnipeds. Once the pup is weaned, the females depart to sea, leaving the pups to fend for themselves. The pups spend the next 4–6 weeks teaching themselves to swim and hunt, during which time they rely heavily on the large reserves of blubber that they got from their mothers while suckling. When the pups eventually leave their natal beaches they spend the next 6 months at sea. This is a difficult time for the pups and as many as 30% of them die at this time.

### VI. Human Interactions

Today, both species of elephant seal are relatively free of adverse interactions with humans. Southern elephant seals are only rarely captured in the nets of Southern Ocean fishing fleets, and this has never been reported for the northern species. There are some grounds for concern that some large-scale fisheries may be competing with the seals for preferred prey species, but this is difficult to quantify given the current paucity of information on the diet of both species. Present impacts may be minimal, but increasing fishing activity may result in significant competition in future (Hindell *et al.*, 2003).

However, both species have a long history of direct exploitation by humans as they were hunted extensively during the 1800s for their blubber, which yielded an unusually high-quality oil. In the case of the northern elephant seal, this hunting was so intense that the populations were reduced to a small group breeding on a single island by 1890. The more widespread and southerly distribution of the southern species meant that the exploitation was less intensive. Nonetheless, the seals were reduced dramatically at all of their major breeding sites.

The exploitation of southern elephant seals also continued longer than that of northern elephant seals, with commercial operations continuing until 1919 at Macquarie Island and until 1964 at South Georgia.

Anthropogenic climate change may affect elephant seals; evidence of correlations between population–dynamics parameters and ENSO may provide insights into the likely impacts of the change (McMahon and Burton, 2005).

### *See Also the Following Articles*

Blubber ■ Mating Systems ■ Pinniped Evolution ■ Earless Seals ■ Pinniped life history

### *References*

Bailleul, F., Charrassin, J. B., Ezraty, R., Girard-Ardhuin, F., McMahon, C. R., Field, I. C., and Guinet, C. (2007). Southern elephant seals from Kerguelen Island confronted by Antarctic sea ice. Changes in movements and diving behaviour. *Deep-Sea Res. (2 Top. Stud. Oceanogr.)* **54**, 343–355.

Bester, M. N., and Hofmeyr, G. J. G. (2005). Numbers of elephant seals at Prince Edward Island, Southern Ocean. *S. Afr. J. Wildl. Res.* **35**, 85–88.

Bradshaw, C. J. A., Higgins, J., Michael, K. J., Wotherspoon, S. J., and Hindell, M. A. (2004). At-sea distribution of female southern elephant seals relative to variation in ocean surface properties. *ICES J. Mar. Sci.* **61**, 1014–1027.

Campagna, C., Piola, A. R., Marin, M. R., Lewis, M., and Fernández, T. (2006). Southern elephant seal trajectories, fronts and eddies in the Brazil/Malvinas Confluence. *Deep-Sea Res. (1 Oceanogr. Res. Pap.)* **53**, 1907–1924.

Campagna, C., Piola, A. R., Marin, M. R., Lewis, M., Zajaczkovski, U., and Fernández, T. (2007). Deep divers in shallow seas: Southern elephant seals on the Patagonian shelf. *Deep-Sea Res. (1 Oceanogr. Res. Pap.)* **54**, 1792–1814.

Champagne, C. D., Houser, D. S., and Crocker, D. E. (2006). Glucose metabolism during lactation in a fasting animal, the northern elephant seal. *Am. J. Physiol. Regul. Integr. Comp. Physiol.* **291**, R1129–R1137.

Crocker, D. E., Costa, D. P., Le Boeuf, B. J., Webb, P. M., and Houser, D. S. (2006). Impact of El Niño on the foraging behavior of female northern elephant seals. *Mar. Ecol. Prog. Ser.* **309**, 1–10.

Daneri, G. A., and Carlini, A. R. (2002). Fish prey of southern elephant seals, *Mirounga leonina*, at King George Island. *Polar Biol.* **25**, 739–743.

Field, I. C., Bradshaw, C. J. A., Burton, H. R., Sumner, M. D., and Hindell, M. A. (2005). Resource partitioning through oceanic segregation of foraging juvenile southern elephant seals (*Mirounga leonina*). *Oecologia* **142**, 127–135.

Field, I. C., Bradshaw, C. J. A., Corey, J. A., van den Hoff, J., Burton, H. R., and Hindell, M. A. (2007). Age-related shifts in the diet composition of southern elephant seals expand overall foraging niche. *Mar. Biol.* **150**, 1441–1452.

García-Aguilar, M. D. L., and Morales-Bojorquez, E. (2005). Estimating the haul-out population size of a colony of northern elephant seals (*Mirounga angustirostris*) in Mexico, based on mark-recapture data. *Mar. Ecol. Prog. Ser.* **297**, 297–302.

Hindell, M. A., Bradshaw, C. J. A., Sumner, M. D., Michael, K. J., and Burton, H. R. (2003). Dispersal of female southern elephant seals and their prey consumption during the austral summer: Relevance to management and oceanographic zones. *J. Appl. Ecol.* **40**, 703–715.

Kirkman, S. P., Bester, M. N., Hofmeyr, G. J. G., Jonker, F. C., Pistorius, P. A., Owen, ., and Strydom, N. (2004). Variation in the timing of the breeding haulout of female southern elephant seals at Marion Island. *Austral. J. Zool.* **52**, 379–388.

E

Le Boeuf, B. J., and Laws, R. M. (1994). "Elephant Seals: Population Ecology, Behavior and Physiology." University of California Press, Berkley, CA.

Lester, C. W., and Costa, D. P. (2006). Water conservation in fasting northern elephant seals (*Mirounga angustirostris*). *J. Exp. Biol.* **209**, 4283–4294.

Lewis, R., Campagna, C., Marin, M. R., and Fernández, T. (2006a). Southern elephant seals north of the Antarctic polar front. *Ant. Sci.* **18**, 213–221.

Lewis, R., O'Connell, T. C., Lewis, M., Campagna, C., and Hoelzel, A. R. (2006b). Sex-specific foraging strategies and resource partitioning in the southern elephant seal (*Mirounga leonina*). *Proc. Roy. Soc. B.* **273**, 2901–2907.

Ling, J. K., and Bryden, M. M. (1981). Southern elephant seal— *Mirounga leonina*. *In* "Handbook of Marine Mammals" (S. H. Ridgway, and R. J. Harrison, eds), Vol. 2, pp. 297–328. Academic Press, London, Seals.

McGinnis, S. M., and Schusterman, R. J. (1981). Northern elephant seal—*Mirounga angustirostris*. *In* "Handbook of Marine Mammals" (S. H. Ridgway, and R. J. Harrison, eds), Vol. 2, pp. 229–350. Academic Press, London, Seals.

McMahon, C. R., and Burton, H. R. (2005). Climate change and seal survival: Evidence for environmentally mediated changes in elephant seal, *Mirounga leonina*, pup survival. *Proc. Roy. Soc. B.* **272**, 923–928.

McMahon, C. R., Bester, M. N., Burton, H. R., Hindell, M. A., and Bradshaw, C. J. A. (2005a). Population status, trends and a re-examination of the hypotheses explaining the recent declines of the southern elephant seal *Mirounga leonina*. *Mamm. Rev.* **35**, 82–100.

McMahon, C. R., Hindell, M. A., Burton, H. R., and Bester, M. N. (2005b). Comparison of southern elephant seal populations, and observations of a population on a demographic knife-edge. *Mar. Ecol. Prog. Ser.* **288**, 273–283.

Piatkowski, U., Vergani, D. F., and Stananelli, Z. B. (2002). Changes in the cephalopod diet of southern elephant seal females at King George Island, during El Niño-La Niña events. *J. Mar. Biol. Ass. UK* **82**, 913–916.

Pistorius, P. A., Bester, M. N., Lewis, M. N., Taylor, F. E., and Kirkman, S. P. (2004). Adult female survival, population trend, and the implications of early primiparity in a capital breeder, the southern elephant seal (*Mirounga leonina*). *J. Zool.* **263**, 107–119.

Pistorius, P. A., Bester, M. N., and Taylor, F. E. (2005). Pubescent southern elephant seal males: Population changes at Marion Island and the food limitation hypothesis. *S. Afr. J. Wildl. Res.* **35**, 215–218.

Sanvito, S., Galimberti, F., and Miller, E. H. (2007a). Having a big nose: Structure, ontogeny, and function of the elephant seal proboscis. *Can. J. Zool. Rev. Can. Zool.* **85**, 207–220.

Sanvito, S., Galimberti, F., and Miller, E. H. (2007b). Vocal signaling of male southern elephant seals is honest but imprecise. *Anim. Behav.* **73**, 287–299.

Simmons, S. E., Crocker, D. E., Kudela, R. M., and Costa, D. P. (2007). Linking foraging behaviour of the northern elephant seal with oceanography and bathymetry at mesoscales. *Mar. Ecol. Prog. Ser.* **346**, 265–275.

Slade, R. W., Moritz, C., Hoelzel, A. R., and Burton, H. R. (1998). Molecular population genetics of the southern elephant seal *Mirounga leonine*. *Genetics* **149**, 1945–1957.

Stewart, B. S. (1997). Ontogeny of differential migration and sexual segregation in northern elephant seals. *J. Mammal.* **78**, 1101–1116.

Sydeman, W. J., and Nur, N. (1994). Life history strategies of female northern elephant seals. *In* "Elephant Seals: Population Ecology, Behavior and Physiology" (B. J. Le Boeuf, and R. M. Laws, eds), pp. 137–153. University of California Press, Berkley, CA.

van den Hoff, J. (2001). Dispersal of southern elephant seals (*Mirounga leonina*) marked at Macquarie Island. *Wildl. Res.* **28**, 413–418.

van den Hoff, J., Burton, H., and Davies, R. (2003). Diet of male southern elephant seals (*Mirounga leonina* L.) hauled out at Vincennes Bay, East Antarctica. *Polar Biol.* **26**, 27–31.

# Endangered Species and Populations

RICHARD MERRICK, GREGORY K. SILBER, DOUGLAS P. DEMASTER, AND JOHN E. REYNOLDS, III

## I. Introduction

Utilization of marine mammals was an integral part of life in early human coastal communities, but humans probably had little effect on most marine mammal populations until commercial exploitation began in the late 1700s for seals and a century later for whales. Vessels traveled to progressively more distant locations, hunting and processing efficiency improved, and commercial applications of marine mammal products expanded. The peak of the southern sealing industry occurred in 1821, when *Lloyd's Register* included 164 sealing vessels. However, by 1830, most seal populations had been depleted, and *Lloyd's Register* only showed one full-time sealing vessel. Commercial whaling reached a zenith (at least for large whales) in the 1930–1940s, and declined markedly in the 1970s because of reduced marine mammal population sizes, development of (synthetic) substitutes for some whale products, and international conservationist pressures.

By the beginning of the twentieth century, many marine mammal populations had reached perilously low levels. Indeed, human exploitation resulted in the extinction of the Steller's sea cow (*Hydrodamalis gigas*) (in 1768, only 27 years after its discovery by Russian explorers), the North Atlantic population of the gray whale (*Eschrichtius robustus*), and the Atlantic sea mink (*Mustela macrodon*). The Caribbean monk seal (*Monachus tropicalis*) and the Japanese sea lion (*Zalophus californiaus japonicus*) also became extinct in the twentieth century as a result of human interactions, although not necessarily hunting. Sadly, we must report that, in the short time since our writing for the first issue of this encyclopedia, another species, the "baiji" or Chinese river dolphin (*Lipotes vexillifer*), has been declared extinct (Turvey *et al.*, 2007). Several other species, including the vaquita (*Phocoena sinus*), Hawaiian (*Monachus schauinslandi*) and Mediterranean (*M. monachus*) monk seals, West Indian manatee (*Trichechus manatus*), and North Pacific (*Eubalaena japonica*) and North Atlantic (*E. glacialis*) right whales are now in as precarious a state as the "baiji" faced during the years prior to its extinction. For these species, conservation decisions in the next decade will be critical to long-term persistence.

Eschricht and Reinhardt (1861) and Scammon (1874) foresaw the extinction crisis for marine mammals and warned that certain species or populations of marine mammals had reached dangerously low levels. The increasing recognition of this crisis led to specific measures designed to recover these populations, and also to heightened worldwide concern to prevent the extinction of marine mammals. Some specific efforts included: (1) the 1911 North Pacific Fur Seal Convention (protecting northern fur seals, *Callorhinus ursinus* and sea otters, *Enhydra lutris* (see www.intfish.net/treaties/furseals11.htm); (2) the 1931 League of Nation's "Convention for the Regulation of Whaling" banning the whaling of right whales in all oceans (see www.intfish.net/treaties/whales31.htm); (3) the 1972 "Convention for the Conservation of Antarctic Seals"; (4) the 1991 "Agreement on the Conservation of Small Cetaceans of the Baltic and North Seas (ASCOBANS)" and "Agreement on the Conservation of Cetaceans in the Black Sea, Mediterranean Sea, and contiguous Atlantic area (ACCOBAMS)"; and (5) the International Whaling Commission's global moratorium on commercial whaling in 1986. Other efforts have been more inclusive, such as the US Marine Mammal Protection Act (US MMPA) which protects all marine mammals (endangered or not) within USA territorial waters. Still other efforts, such as those by the International Union for the Conservation of Nature and Natural Resources (IUCN), the Convention on International Trade in Endangered Species of Wild Fauna and Flora (CITES), the US Endangered Species Act (ESA), and the Canadian Species At Risk Act (SARA) have focused on recovering marine mammal species or populations threatened with extinction on a global scale.

Because of these activities, most marine mammal harvests have ceased, and many formerly endangered species or populations are recovering, including the Southern Hemisphere population of fur seals (*Arctocephalus* spp.), the Northern and Southern Hemisphere populations of elephant seals (*Mirounga* spp.), the western Arctic bowhead whale (*Balaena mysticetus*) population, and the Northern Hemisphere populations of humpback whales (*Megaptera novaeangliae*). In 1994, the eastern North Pacific population of gray whales became the first marine mammal species, subspecies, or distinct population segment (DPS) to be formally declared recovered under the US List of Endangered and Threatened Wildlife and Plants. It is important to focus on such recoveries, and apply the lessons learned to promote the recovery of other depleted populations and species.

Most conservation efforts have been directed at reversing the impacts of commercial exploitation. Unfortunately, as harvests have declined or even ceased, other anthropogenic threats have arisen, including those having direct (e.g., fishery by-catches, vessel strikes) or indirect (e.g., loss of prey to commercial fisheries) impacts. These threats have kept some marine mammal populations at low levels and slowed the recovery of others. Although attempts have—and are—being made to reduce bycatch, ship strikes, and other human threats, only time will tell if these can be mitigated. Furthermore, new challenges are now emerging, which will have to be addressed by future generations. These include anthropogenic activities related to habitat destruction, degradation of water quality, disturbance of animals (e.g., anthropogenic noise), bioaccumulation of toxins, and global climate change (including ocean acidification). These impacts are less obvious than direct mortalities and are not easily studied or measured, but may be equally as deadly.

The greatest future threat to all marine species is global climate change. For some marine mammal species, climatic impacts will be profound. For species such as bowhead whales, polar bears (*Ursus maritimus*), and pagophilic seals which depend on ice, changes in distributions are already occurring due to the retreating ice sheets (Derocher *et al.*, 2004). Retreating ice also renders polar areas more accessible to humans (e.g., trans-polar shipping), exposing species in these regions to anthropogenic additional impacts (e.g., noise disturbance) (US Marine Mammal Commission, 2006). For temperate species, the impacts from global climate change are less clear. Rising sea levels and increases in the number or severity of storms, while potentially catastrophic to coastal human communities, would likely have little effect on most marine mammal populations, particularly pelagic ones. However, changes in ocean productivity resulting from sea temperature increases, changes in currents, ocean acidification,

increases in storms, and greater river run-off may have significant biotic impacts. These could change an entire ecosystem's carrying capacity, affecting the system's ability to support marine mammals at pre-exploitation abundance levels.

In this chapter, we focus on marine mammals that require protection in order to survive well into the twenty-first century and beyond. We give reasons why some marine mammals are in critical condition today, and we identify lessons and trends that may help explain why some groups recover and others do not. We provide definitions of criteria under which such groups are classified by legislation, conventions, or other approaches such as the IUCN Red List of Threatened Animals, CITES, the MMPA, and the US ESA. Finally, we highlight the types of data and information that, if collected, would be helpful in conservation and recovery efforts.

## II. Why Do Marine Mammal Populations Become Endangered or Fail to Recover?

What makes a species endangered? One can define an endangered species based solely on demographic characteristics; i.e., the abundance of the species is so low *or* abundance is declining so precipitously *or* the species range has retracted so greatly that the species is in danger of becoming extinct. Note that while an endangered species may be synonymous with a taxonomic species, it can also be defined as a DPS of the species. An alternate approach is to consider a species endangered because of threats which could lead to its extinction. Ultimately, what makes a species endangered is some marked change to the species itself or to its ecosystem (e.g., increased exploitation, loss of habitats, etc.). Many, if not most, marine mammal species considered to be at risk of extinction reached this situation as a result of human activity (i.e., harvesting). Only in recent decades have humans affected marine mammal habitat sufficiently to place species at risk. These more recent habitat impacts now place more marine mammal species at risk of extinction, even those not formerly at risk due to harvesting.

Various aspects of marine mammal demography also contribute to their extinction vulnerability. Marine mammals generally exhibit low reproductive potential, maturing at a late age and giving birth to few offspring. This strategy means that once a population is reduced, it will be slow to recover. Moreover, even pristine populations of some marine mammal species or subspecies were fairly small [e.g., killer whales (*Orcinus orca*), monk seals (*Monachus* spp.)]. Such low abundance would certainly make the species vulnerable to extinction and incapable of withstanding significantly increased predation (e.g., the Steller's sea cow). This innate vulnerability, coupled with various environmental threats, has led to the precarious status of some species and populations.

### A. Life History

Life-history attributes are biological characteristics of a species that maximize the fitness, and include such traits as age at sexual maturity, age-specific survival, sex- and age-specific growth rates, reproductive interval (or its inverse, the reproductive rate), and longevity. Life-history attributes dictate the potential for population growth. Biologists identify two extreme types of life history strategies among species: the "*r*-strategists" in which the ability to reproduce quickly is critical (and which typically have a high fecundity, small body size, short generation time, and wide dispersal of offspring, each of whom has a low probability of survival to adulthood), and the "K-strategists" in which the ability to compete successfully for limited resources is critical (and which typically exhibit a large body size, long-life expectancy, and produce fewer offspring that require extensive parental care until they mature). For further details on r and K strategies, see Pianka (1970) and Reynolds *et al.* (2000).

The life histories of plant and animal species lie along a continuum between these two extremes. As a group, marine mammals are *extremely* close to the K end of the spectrum. K-strategists generally adapt poorly to changing conditions, so human impacts on their environments can severely compromise population recovery. When the abundance of marine mammal populations is low, these species are vulnerable to extinction from events (e.g., epizootics and demographic and environmental stochasticity) that would not otherwise be threatening at higher population levels. Consider the North Atlantic right whale, in which longevity may approach or even exceed a century and in which reproduction may not occur until age 10 and thereafter only once every 3–5 years. After right whale populations were markedly reduced due to commercial exploitation, it was biologically impossible for the species to quickly rebound. Moreover, at low population levels, competition for copepod prey with r-selected finfish species can make recovery even less likely or further delayed. Right whales are simply not adapted to rapid recovery (from either overharvesting or from anthropogenic impacts to their habitats), and when population abundances are low, extinction vulnerability is extremely high. Likewise for other marine mammal species, life-history strategies provide points of vulnerability.

## B. Environmental Threats

Although commercial harvests peaked some time ago for most marine mammals, harvesting of seals, small cetaceans, and even the great whales (i.e., fin, *Balaenoptera physalus* and minke whales, *Balaenoptera acutorostrata*) continues today. Even with regulations in place, there have been occasional blatant violations. The Russian government has made available modern whalers' log books, which show that quotas and protective regulations for some populations of endangered species, including the North Pacific right whale, were ignored as late as the 1970s (Danilov-Danil'yan and Yablokov, 1995), and it would not be surprising if other nations also ignored some whaling regulations. Thus, some species or populations thought to have been protected for decades were only protected on paper.

Both natural and anthropogenic habitat alterations can affect the abundance of marine mammals. Frequently, changes to coastal and marine environments occur in subtle and diverse ways, making it difficult to tease apart the various possible impacts to marine mammals. A telling example involved the die-off in 1987–1988 of Atlantic coastal bottlenose dolphins (*Tursiops truncatus*) along the southeastern coast of the United States. At least 740 animals died, prompting the National Marine Fisheries Service to list the population as depleted under the MMPA. Cause of death was first suggested to be ingestion of brevetoxin; later, scientists indicated that high-contaminant loads were involved; and later still, other scientists noted the presence of morbillivirus in preserved tissues of dolphins from the die-off. As noted by Reynolds *et al.* (2000), the precise interplay between the natural toxins, anthropogenic toxicants, viral infections, immune dysfunction, opportunistic infections, and death are still unclear.

The dolphin die-off illustrates that habitat change (whether natural or human induced) may compromise population health and make large numbers of animals susceptible to natural pathogens. In a relatively large population, such as coastal bottlenose dolphins, the problem is serious. For a species like the Mediterranean monk sea (*Monachus monachus*), which suffered a disastrous die-off in 1997

due to as-yet-undetermined causes (perhaps either saxitoxin poisoning or morbillivirus, or some combination of these or other causes), the problem becomes critical when an already small population is further reduced over a matter of weeks.

Changes in prey availability also seriously impact the viability of marine mammal populations. In two cases involving endangered marine mammals—the western population of the Steller sea lion (*Eumetopias jubatus*) and the French Frigate Shoals Hawaiian monk seal—the commercial harvesting of the preferred prey of these species has been suggested as an important cause of their population declines. Natural fluctuations in prey availability (e.g., El Niño Southern Oscillation events) can precipitate marine mammal population declines, but these events have only a transient impact on species with robust population sizes.

It is relatively easy to count how many animals are killed by hunters or through incidental takes. However, it is exceedingly difficult (due to both the variety and magnitude of effects and to potential synergisms) to quantify the impacts of chemical and noise pollution, harvesting of marine mammal prey, and other effects on marine mammal habitats which may compromise, or at least retard, the recovery of populations.

## III. Which Marine Mammals Are Endangered?

Although there are a number of lists of protected species both regionally and globally, we provide three widely accepted lists of protected marine mammals. Table I lists endangered and threatened (from the US ESA) and depleted (from the MMPA) species and populations. Table II lists marine mammals included in Appendices I and II of CITES. Table III lists those marine mammals classified by the IUCN as Endangered (E), Critically Endangered (CE), and Vulnerable (V). Similar listings exist under the Canadian SARA.

Of the species listed in Tables I–III, the status of some is more critical than others. The western North Atlantic population of humpback whales, which numbers more than 10,000 individuals, is far more abundant than the populations of vaquitas (perhaps 150 individuals left; Jaramillo-Legorreta *et al.*, 2007), North Atlantic right whales (somewhat more than 300), and Mediterranean monk seals (around 300 left in the wake of an epizootic in 1997, which killed over half of the members of the largest colony in northwestern Africa). Species or populations which are critically endangered based on a very small population size (hundreds of animals), and therefore require immediate, effective conservation efforts include:

- *Cetaceans*: Indus river dolphin (*Platanista gangetica minor*), vaquita, North Pacific and North Atlantic right whales, several populations of blue whales (*Balaenoptera musculus*), western North Pacific gray whale, Cook Inlet and St. Lawrence River beluga whales (*Delphinapterus leucas*), and North Atlantic Arctic bowhead whales.
- *Pinnipeds*: Mediterranean monk seal, Saimaa ringed seal (*Pusa hispida saimensis*), and several populations of Atlantic walrus (*Odobenus rosmarus rosmarus*).
- *Sirenians*: Several populations of dugongs (*Dugong dugon*), as well as West African (*Trichechus senegalensis*) and West Indian manatees.

Objectively classifying populations according to their precise level of vulnerability is very difficult as predicting extinction probabilities is fraught with uncertainty. However, the above species are relatively simple to classify because their population sizes are so very low. Apart from population size, a variety of qualitative and quantitative

## TABLE I
Marine Mammal Species and Populations Listed Under the United States Endangered Species Act (ESA) and Marine Mammal Protection Act (MMPA) as Endangered (E), Threatened (T), and Depleted (D)

| Order/species | Common name | Range | Status |
|---|---|---|---|
| **Cetacea** | | | |
| *Lipotes vexillifer* | Baiji | Changjiang (Yangtze) River, China | E/D |
| *Platanista minor* (=*P. gangetica minor*) | Indus river dolphin | Indus River and tributaries, Pakistan | E/D |
| *Phocoena sinus* | Vaquita | Northern Gulf of California, Mexico | E/D |
| *Stenella attenuata* | Northeastern offshore spotted dolphin | Eastern tropical Pacific Ocean | D |
| *Stenella longirostris* | Eastern spinner dolphin | Eastern tropical Pacific Ocean | D |
| *Tursiops truncatus* | Mid-Atlantic coastal bottlenose dolphin | Atlantic coastal waters, New York to Florida | D |
| *Eubalaena glacialis* (=*Balaena glacialis glacialis*) | Northern right whale | North Atlantic, North Pacific Oceans; Bering Sea; Sea of Okhotsk | E/D |
| *Eubalaena australis* (=*Balaena glacialis australis*) | Southern right whale | South Atlantic, South Pacific, Indian, and Southern Oceans | E/D |
| *Balaena mysticetus* | Bowhead whale | Arctic Ocean and adjacent seas | E/D |
| *Megaptera novaeangliae* | Humpback whale | Oceanic, all oceans | E/D |
| *Balaenoptera musculus* | Blue whale | Oceanic, all oceans | E/D |
| *Balaenoptera physalus* | Finback or fin whale | Oceanic, all oceans | E/D |
| *Eschrichtius robustus* | Western North Pacific gray whale | Western North Pacific Ocean | E/D |
| *Balaenoptera borealis* | Sei whale | Oceanic, all oceans | E/D |
| *Physeter macrocephalus* | Sperm whale | Oceanic, all oceans | E/D |
| **Carnivora** | | | |
| *Lutra feline* | Marine otter | Western South America; Peru to southern Chile | E/D |
| *Enhydra lutris nereis* | Southern sea otter | Central California coast | T/D |
| *Monachus schauinslandi* | Hawaiian monk seal | Hawaiian Archipelago | E/D |
| *Monachus tropicalis* | Caribbean monk seal | Caribbean Sea and Bahamas (probably extinct) | E/D |
| *Monachus monachus* | Mediterranean monk seal | Mediterranean Sea; northwest African coast | E/D |
| *Arctocephalus townsendi* | Guadalupe fur seal | Baja California, Mexico, to southern California | T/D |
| *Callorhinus ursinus* | Northern fur seal | North Pacific Rim from California to Japan | D |
| *Eumetopias jubatus* | Western North Pacific Steller sea lion | North Pacific Rim from Japan to Prince William Sound, Alaska (east of 144°W longitude) | E/D |
| *E. jubatus* | Eastern North Pacific Steller sea lion | North Pacific Rim from Prince William Sound, Alaska, to California (east of 144°W longitude) | T/D |
| *Phoca hispida saimensis* (=*Pusa hispida saimensis*) | Saimaa seal | Lake Saimaa, Finland | E/D |
| **Sirenia** | | | |
| *Trichechus manatus* | West Indian manatee | Caribbean Sea and North Atlantic from southeastern United States to Brazil; and Greater Antilles Islands | E/D |
| *Trichechus inunguis* | Amazonian manatee | Amazon River basin of South America | E/D |
| *Trichechus senegalensis* | West African manatee | West Africa coasts and rivers; Senegal to Angola | T/D |
| *Dugong dugon* | Dugong | Northern Indian Ocean from Madagascar to Indonesia, Philippines, Australia, southern China, Palau | E/D |

*Note*: Species listed under the ESA as E or T are also listed under the MMPA as D. However, there are a few species listed under the MMPA as depleted that are not listed under the ESA. Equivalent species names used by Rice (1998) appear in parentheses.

approaches are used to assess extinction risk. The IUCN approach (www.iucnredlist.org/info/categories_criteria2001) is the most widely used; under this approach, extinction vulnerability is based on population size (or population trends), or on the extent of historical habitat occupied, or on a population viability analysis. However, other approaches are also used around the world and it is important to understand which approach is being used when reference is made to in danger of extinction.

## IV. Recovery and Non-recovery of Species and Populations: Lessons and Trends

One intriguing question is why some marine mammal populations have recovered from low population sizes (even when effective conservation measures have been absent) while others have remained low and in danger of extinction (Clapham *et al.*, 2008). In most cases, the answer lies in differences in life histories or differences

E

## TABLE II
### Marine Mammals Listed by CITES Under Appendices I and II

| Order/species | Common name | Appendix |
|---|---|---|
| **Cetacea** | | |
| *Balaena mysticetus* | Bowhead | I |
| *Balaenoptera acutorostrata* | Minke whale | I |
| *B. borealis* | Sei whale | I |
| *B. edeni* | Bryde's whale | I |
| *B. musculus* | Blue whale | I |
| *B. physalus* | Fin whale | I |
| *Berardius* spp. | Beaked whale | I |
| *Caperea marginata* | Pygmy right whale | I |
| *Eschrichtius robustus* | Gray whale | I |
| *Eubalaena* spp. (=*Balaena glacialis* spp.) | Right whale | I |
| *Hyperoodon* spp. | Bottlenosed whale | I |
| *Lipotes vexillifer* | Chinese river dolphin white flag dolphin | I |
| *Megaptera novaeangliae* | Humpback whale | I |
| *Monodon monoceros* | Narwhal | II |
| *Neophocaena phocaenoides* | Finless porpoise | I |
| *Phocoena sinus* | Vaquita; Gulf of California harbor porpoise | I |
| *Physeter macrocephalus* | Sperm whale | I |
| *Pontoporia blainvillei* | La Plata River dolphin | II |
| *Sotalia fluviatilis* | Tucuxi | I |
| *Sousa* spp. | Humpbacked dolphins | I |
| **Carnivora** | | |
| *Arctocephalus australis* | Southern fur seal | II |
| *A. galapagoensis* | Galapagos fur seal | II |
| *A. philippii* | Juan Fernandez fur seal | II |
| *A. townsendi* | Guadalupe fur seal | I |
| *Enhydra lutris nereis* | Southern sea otter | I |
| *Mirounga leonina* | Southern elephant seal | II |
| *Monachus* spp. | Monk seals | I |
| *Ursus maritimus* | Polar bear | II |
| **Sirenia** | | |
| *Dugong dugon* (except in Australia) | Dugong | I |
| *Dugong dugon* (Australia) | Dugong | II |
| *Trichechus inunguis* | Amazonian manatee | I |
| *T. manatus* | West Indian manatee | I |
| *T. senegalensi* | West African manatee | II |

*Note*: From (Federal Register, 1999). Equivalent species names used by Rice (1998) appear in parentheses.

## TABLE III
### Marine Mammals Listed by the IUCN (1996) as Critically Endangered (CE), Endangered (E), or Vulnerable (V)

| Order/species | Common name | Category |
|---|---|---|
| **Cetacea** | | |
| *Balaena mysticetus* | Bowhead | E/V |
| *Balaenoptera borealis* | Sei whale | E |
| *B. musculus* | Blue whale | E/V |
| *B. physalus* | Fin whale | E |
| *Cephalorhynchus hectori* | Hector's dolphin | V |
| *Delphinapterus leucas* | Beluga | V |
| *Eschrichtius robustus* | Gray whale | E |
| *Eubalaena glacialis* (=*Balaena glacialis glacialis*) | Northern right whale | E |
| *Inia geoffrensis* | Boto, Amazon river dolphin | V |
| *Lipotes vexillifer* | Baiji, Yangtze river dolphin | CE |
| *Megaptera novaeangliae* | Humpback whale | V |
| *Neophocaena phocaenoides* | Finless porpoise | E |
| *Phocoena phocoena* | Harbor porpoise | V |
| *Phocoena sinus* | Vaquita | CE |
| *Physeter catodon* (=*Physeter macrocephalus*) | Sperm whale | V |
| *Platanista gangetica* (=*P. gangetica gangetica*) | Ganges river dolphin | E |
| *P. minor* (=*P. gangetica minor*) | Indus river dolphin | E |
| **Carnivora** | | |
| *Arctocephalus galapagoensis* | Galapagos Island fur seal | V |
| *A. philippii* | Juan Fernandez fur seal | V |
| *A. townsendi* | Guadalupe fur seal | V |
| *Callorhinus ursinus* | Northern fur seal | V |
| *Eumetopias jubatus* | Steller seal lion | E |
| *Halichoerus grypus* | Gray seal | E |
| *Lutra feline* | Marine otter | E |
| *Monachus monachus* | Mediterranean monk seal | CE |
| *M. schauinslandi* | Hawaiian monk seal | E |
| *Phoca caspica* (=*Pusa caspica*) | Caspian seal | V |
| *P. hispida botnica* (=*Pusa hispida botnica*) | Baltic seal | V |
| *P. h. ladogensis* (=*Pusa hispida ladogensis*) | Ladoga seal | V |
| *P. h. saimensis* (=*Pusa hispida saimensis*) | Saimaa seal | E |
| *Phocarctos hookeri* | Hooker's sea lion | V |
| *Zalophus californianus japonicus* (=*Z. japonicus*) | Japanese sea lion | Extinct? |
| *Zalophus californianus Wollebaeki* (=*Z. wollebaeki*) | Galapagos sea lion | V |
| **Sirenia** | | |
| *Dugong dugon* | Dugong | V |
| *Trichechus inunguis* | Amazonian manatee | V |
| *T. manatus* | West Indian manatee | V |
| *T. senegalensis* | West African manatee | V |

*Note*: Where more than one classification category is given for a particular species in this table, it means that different populations or populations of that species are threatened at different levels of severity. Similarly, a particular classification does not necessarily mean that a species is threatened range wide at that level; rather the classification may reflect the status of only one population or population. Equivalent species names listed by Rice (1998) appear in parentheses.

in proximity to (and extent of) anthropogenic impacts. A number of examples exist of divergent recovery trajectories in (a) closely related species; (b) sympatric and ecologically similar, but distantly related species; and (c) populations of the same species occurring in different regions.

Both the North Atlantic and eastern North Pacific right whale populations are small, probably numbering fewer than 500 individuals combined, and population growth in each is negligible or non-existent. In contrast, right whale populations in the Southern Hemisphere have increased at estimated rates of 7–8% per annum for a number of years and these populations now total over 7500 individuals.

Both the Southern and Northern Hemisphere right whale populations were severely reduced by commercial whaling in the nineteenth and early- to mid-twentieth centuries. However, the responses of these populations to the cessation of harvesting have been very different. As these populations represent closely related species which are likely ecological equivalents (in marginally different ecosystems) and display similar life-history traits, the differences in life-history patterns are not likely to be responsible for the differences in recovery ability. Changes in competitive relationships within their respective ecosystems might also be operative, but these are very difficult to identify. A more likely explanation for the lack of recovery of North Atlantic right whales is continued levels of anthropogenic serious injury and mortality. NOAA Fisheries has recorded 38 ship strike-related mortalities and injuries of North Atlantic right whales during 1978–2002 and six right whales were known to be seriously injured or died between 2000 and 2004. The actual number is almost certainly higher because not all carcasses are reported. Because of low population abundance and low reproductive ability, the North Atlantic right whale population cannot sustain even this level of mortality. The right whale's recovery in the North Atlantic may also be constrained by its diet, which is highly focused on only one or two species of copepods; other large whale species have broader diets and seem more able to utilize additional prey types when necessary. Large increases during the last decade in the abundance of Atlantic herring (*Clupea harengus*) and mackerel (species which also consume copepods) may mean that the Northwest Atlantic ecosystem can no longer support as many right whales as it did historically. Southern Hemisphere right whales do not appear to be exposed to the same level of anthropogenic mortality, and it is unclear whether these populations experience any significant competition for prey.

Sympatric marine mammal populations of distantly related species can also show divergent population growth rates. Humpback whales, which occur sympatrically with right whales in the North Atlantic, exhibit a positive recovery trajectory (possibly as high as 3–6% per year) while right whale abundance remains extremely low and relatively stable (Waring *et al.*, 2007). The north–south distribution of the two species is relatively similar (although the southern breeding ground of humpback whales is further south than the southern limit of the right whale's distribution). However, right whales generally occur closer to shore than humpbacks, perhaps exposing the former to greater human impacts from commercial fishing and shipping. Although right and humpback whales feed at different trophic levels, the recovery of humpbacks suggests that human activities, rather than marine habitat changes, are affecting recovery of the North Atlantic right whale population.

Gray whales provide an example of a recovery in one population and a concurrent lack of recovery in a geographically distant population of the same species. The eastern North Pacific gray whale population has increased during the past 30 years to pre-exploitation levels (i.e., 15,000–20,000 animals) (although a recent publication by Alter *et al.* (2007) has questioned this assessment). As a result of this increase, the population was removed from the US List of Endangered and Threatened Wildlife in 1994. In contrast, the western North Pacific gray whale population is small and remains listed as endangered. Again, the discrepancies in growth rates between the two populations are difficult to interpret. Gray whales migrate very close to shore and are likely subject to the same types of human threats. Both populations are ecologically similar and have nearly identical life history patterns. So far, scientists have been unable to identify what is hampering the recovery of the western North Pacific population. It is likely, however, to be some form of human activity (such as entanglements in fishing gear, ship strikes, directed and illegal hunting, pollution, or other types of habitat degradation) or an unidentified ecological change.

## V. Improving the Recovery of Species and Populations

To recover an endangered species from extinction, it should be recognized that conservation measures needed to halt (and reverse) declines in abundance will not be effective without an understanding of (1) the life history of the species or population; (2) the species or population structure; and (3) the effects of human activities on the species or population.

### A. Information on Life Histories of Endangered Species

Information on age-specific rates of birth and survival is critical to assessing recovery potential. It is important to determine whether recovery is being hindered by inadequate reproduction, inadequate recruitment to the adult population, and/or low adult survival. For some populations, both reproduction and survival will be found to be inadequate to support recovery. However, once reliable data on life-history parameters are available, it should be possible to identify the proximate cause of the reduced survival or reproduction, and then implement strategies to rectify this situation.

For example, juvenile mortality was found to be very high in Hawaiian monk seals within three of their six extant breeding colonies. Researchers determined that part of this mortality was due to adult male monk seals mobbing and killing adult females and pups. In response to this situation, researchers removed many (but not all) of the adult male monk seals in one area where the mobbing was most severe. The result was an immediate and almost total elimination of additional deaths caused by mobbing in that colony.

Another example involves California's southern sea otter population, which in the late 1970s and early 1980s was reported to be in decline due to low survival of juveniles and adults. Using observer data obtained from several commercial fisheries operating along the central California coast, it was determined that the incidental mortality of sea otters captured in the set gill net fishery for Pacific halibut was too high. The State of California subsequently passed legislation which moved this fishery further offshore, where interactions with sea otters were thought unlikely. Following this management action, population recovery ensued.

Sorting out the proximate and ultimate causes of a decline is rarely this simple. Although a tremendous amount can be known about the life history of an affected group (e.g., all evidence about the Steller sea lion decline pointed toward reduced juvenile and adult survival as the ultimate cause), this knowledge does not always help to direct management efforts. In such cases, management agencies are required (although they may fall short) to manage in a way that "errs on the side of the animal." This means that managers are required to be cautious in authorizing human activities which may have a non-negligible probability of adversely affecting listed species or the habitats upon which they depend.

E

In summary, it is possible to take actions that directly promote the recovery of listed populations. However, without knowledge of the life-history parameters impeding recovery (as well as knowledge about the underlying cause of the population bottleneck), it is often impossible to know the appropriate conservation measures. Unfortunately, the requisite research to determine rates of reproduction and survival in wild populations is expensive and difficult. As well, it often takes several years of collecting such data to obtain estimates precise enough to evaluate different hypotheses about population recovery or to test the effectiveness of proposed or enacted management measures. Nonetheless, without such information, most recovery efforts are severely hampered.

## B. Information on the Population Structure of Endangered Species

Although national legislation such as the US ESA and Canada's SARA (see Introduction) would, by their very names, appear to approach conservation at the species level, the actual intent of the US Congress in passing the ESA was that management should be local or regional, and directed at populations or subpopulations. Management at the species level can lead to loss of biological diversity if local populations are extirpated, even though the species as a whole is healthy. The importance of management at the population level is emphasized by the case of Steller sea lions. In 1990, the species was designated as threatened under the US ESA. By the late 1990s, however, scientists had demonstrated that (a) there were two discrete populations of the Steller sea lions; (b) the eastern population was smaller than the western population, but was stable and possibly increasing in numbers; and (c) the western population was in precipitous decline, with animal counts at some sites declining by 90% or more since the late 1970s. Accordingly, the ESA listing was modified in 1997 so that the western population was reclassified as endangered, whereas the eastern population remained listed as threatened.

The Steller sea lion example illustrates some important points. First, with endangered marine mammals (and other organisms as well), "distinct population segments" (or "distinct vertebrate populations," e.g., as defined below according to the US ESA) need to be identified so that management and conservation efforts can focus on the most critical groups and areas. However, even with published guidelines for DPS designations, the actual designation is often more of an art than a science; a more standardized approach to genetic analyses would probably provide for better DPS definitions (Fallon, 2008). Secondly, even though many populations and species may be accorded special protection, taxonomic subunits are often more in need of management intervention than the species as a whole. Unfortunately, in some cases, management still occurs at the species level. Worse, the general public has frequently developed serious misconceptions about management practices and equated the precarious status of a species (e.g., the North Atlantic right whale) with the status of all species in a taxonomic group (e.g., the baleen whales).

## C. Information on the Effects of Human Activities on Endangered Species

Recovery of a species (or DPS) requires that threats be removed or, at the least, mitigated. If the threat is as simple as over-harvesting, then the required management actions are usually obvious. However, the causes for population declines or lack of recovery are,

in many cases, obscure or difficult to address. Firstly, the complexity of marine systems often makes it difficult to know what threats are causing the declines. Secondly, it is difficult to gauge whether a threat has the potential to significantly impact a marine species. Finally, translating the threat into specific effects on life-history traits may be all but impossible. Nonetheless, current efforts to implement ecosystem approaches to management (Merrick *et al.*, 2007) and initiatives for monitoring marine mammal health (http://www.nmfs.noaa.gov/pr/health) provide some hope that such threats may be ameliorated.

## VI. Laws and Legislation to Recover Endangered Species

Endangered marine mammal species are currently protected by a variety of domestic and international laws and treaties, and collaborative efforts by nations, individuals, and organizations to reduce takes and prevent extinctions. Some efforts have been inclusive, protecting ALL marine mammals, endangered or not, within a country's territorial waters (i.e., the US MMPA). Other efforts, by groups such as the IUCN (now called the World Conservation Union) and the CITES have focused on species or populations threatened with extinction on a global scale. In almost all cases, at least four criteria have been considered in prioritizing the focus of these activities: (1) population size and demography; (2) the extent to which human activities adversely affect the animals of concern, either directly or indirectly; (3) the adequacy and protection of habitat deemed necessary for survival; and (4) the extent of markets and trade in products from the populations and species of concern. Recently, quantitative criteria have begun to be developed which should reduce the subjectivity involved in categorizing species as endangered or threatened (IUCN, 2000; DeMaster *et al.*, 2004).

The US ESA has weathered significant challenges since its enactment over 30 years ago. Although amended, the Act and its intent remain largely unchanged. The Act aims to "provide a means whereby the ecosystems upon which endangered species and threatened species depend may be conserved, [and] to provide a program for the conservation of such endangered species and threatened species…" Currently, 34 marine mammal species or populations are identified as "depleted" under either the ESA List of Endangered and Threatened Wildlife and Plants or the MMPA (Table I). Recovery plans have been prepared for 22 of these, and a number of plans are in the process of being drafted or updated. As noted earlier, only one marine mammal population (the eastern North Pacific population of gray whales) has been removed from the list.

This record may be viewed as not particularly good by some. However, relative lack of successes (i.e., de-listings) cannot be attributed to (a) inefficiencies of the ESA, (b) those charged with implementing the Act, or (c) lack of efforts by managers or conservation advocates. Challenges exist in collecting adequate data on populations and threats, on newly arising threats, and on conflicts with other human interests. In addition, as previously mentioned, even unimpeded recovery rates for most marine mammal species are relatively slow compared to many other taxa.

Other nations are taking related steps to recover marine mammal populations at risk of extinction. However, in many parts of the world, protective efforts and measures are minimal or inconsistent. Lack of firm enforcement of existing legislation or the total lack of legislation in many countries may be a key factor hampering worldwide efforts to recover endangered species.

## Acknowledgments

Our grateful thanks to Dr. Fred Serchuk and Chris Uyeda of NOAA for the editing and helpful comments provided on an earlier draft of this manuscript.

## *References*

Alter, S. E., Rynes, E., and Palumbi, S. R. (2007). DNA evidence for historic population size and past ecosystem impacts of gray whales. *Proc. Natl. Acad. Sci. USA* **104**, 15162–15167.

Brewer, R. (1988). "The Science of Ecology." Saunders, Philadelphia.

Clapham, P. J., Aguilar, A., and Hatch, L. T. (2008). Determining spatial and temporal scales for management: Lessons from whaling. *Mar. Mamm. Sci.* **24**, 183–201.

Danilov-Danil'yan, V. F., and Yablokov, A. V. (1995). "Soviet Antarctic Whaling Data (1947–1972)." Center for Russian Environmental Policy, Moscow.

DeMaster, D. *et al.* (10 authors) (2004). "Recommendations to NOAA Fisheries: ESA Listing Criteria by the Quantitative Working Group, 10 June 2004." US Department of Commerce, NOAA Tech. Memo. NMFS F/SPO-67.

Derocher, A. E., Lunn, N. J., and Stirling, I. (2004). Polar bears in a warming climate. *Integr. Comp. Biol.* **44**, 163–176.

Eschricht, D. F., and Reinhardt, J. (1861). Om Nordhvalen (*Balaena mysticetus* L.) navnlig med Hensyn til dens Udbredning i Fortiden og Nutiden og til dens ydre og indre Saerkjender. K. Danske Videnskabernes Selskabs Skrifter. Series 5. *Naturvidenskabelig og Mathematisk Afdeling* **5**, 433–590.

Fallon, S. (2008). Genetic data and the listing of species under the US Endangered Species Act. *Conserv. Biol.* **21**, 1186–1195.

Federal Register (1999). Code of Federal Regulations. 50 CFR, Section 23.23.

IUCN (2000). "IUCN Red List Categories." Available at http://www.iucn.org.

Jaramillo-Legorreta, A., Rojas-Bracho, L., Brownell, R. L., Jr., Read, A. J., Reeves, R. R., Ralls, K., and Taylor, B. L. (2007). Saving the vaquita: Immediate action, not more data. *Conserv. Biol.* **26**, 1653–1655.

Merrick R. *et al.* (8 authors) (2007). "Report of the Protected Species SAIP Tier III Workshop, 7–10 March 2006, Silver Spring, MD." US Department of Commerce, NOAA Tech Report. NMFS F/SPO-78.

Pianka, E. R. (1970). On r and K selection. *Am. Nat.* **104**, 592–597.

Reynolds, J. E., III, Wells, R. S., and Eide, S. D. (2000). "The Bottlenose Dolphin—Biology and Conservation." University Press of Florida, Gainesville, FL.

Rice, D. W. (1998). "Marine Mammals of the World. Systematics and Distribution," Special Publication No. 4. Society for Marine Mammalogy, Lawrence.

Scammon, C. M. (1874). "The Marine Mammals of the Northwestern Coast of North America (1968 edition)." Dover, New York.

Turvey, S. T., *et al.* (16 authors) (2007). First human-caused extinction of a cetacean species? *Biol. Lett.* **3**, 537–540.

US Fish and Wildlife Service (1996). "Endangered Species Act of 1973 as Amended Through the 100th Congress." US Government Printing Office 414-990/50033.

US Marine Mammal Commission (1995). "The Marine Mammal Protection Act of 1972 as amended February 1995." Available at www.mmc.gov/legislation/mmpa.html

US Marine Mammal Commission (2006). "Advisory Committee on Acoustic Impacts on Marine Mammals," Report to the Marine Mammal Commission. Available at www.mmc.gov/sound/committee/pdf/soundFACAreport.pdf

World Conservation Monitoring Centre (2000). "IUCN Red List Categories." Available at www.wcmc.org.uk

# Endocrine Systems

SHANNON ATKINSON, DAVID ST. AUBIN,
AND RUDY M. ORTIZ

## I. Introduction

Endocrine systems function by regulating and integrating physiological processes to meet specific needs of the organism and facilitate adaptation to dynamic and chronic environmental changes or perturbations. Internally, hormone systems are constantly changing in response to environmental cues such as photoperiod, temperature, energetic demands, food and water availability, and reproductive status or season. Hormones are the chemical substances that are typically produced and released into circulation by specialized cells that are localized in small glands or organs. Because of their great potency and ability to broadly influence bodily functions, hormones are regulated by an exquisite set of negative and positive feedback loops that may link several organs. For the most part, endocrine systems in marine mammals follow the basic organization and chemical characteristics of other mammals. Nevertheless, it is intriguing to examine how these systems allow marine mammals to meet the peculiar challenges imposed by their environment. The following sections will review and highlight our current understanding of endocrine systems and how they respond to either natural or artificially manipulated environments to enhance our knowledge of hormone functions in marine mammals.

## II. Neuroendocrine Perception of Environmental Changes

Many species of marine mammals inhabit highly variable ecosystems that possess dramatic seasonal changes in environmental variables such as air and water temperature, salinity, photoperiod, and prey resources. Some marine mammals experience migrations that are associated with similar effects. For example, mysticetes exploit productive cold waters at high latitudes during the long days of summer but retreat to tropical habitats in the fall to bear calves under less challenging conditions. Daylength appears to be an important cue for many life history traits, including the initiation of migration patterns as well as several aspects of reproduction. Pinnipeds, particularly those in polar environments, seasonally partition activities such as breeding and molting to take advantage of favorable conditions that will increase the survival of offspring or allow recovery from fasts necessitated by long periods ashore (Atkinson, 1997). West Indian manatees regularly move in and out of freshwater and marine habitats that vary greatly in salinity without any apparent consequence on their ability to regulate body water and electrolytes (Ortiz *et al.*, 1998). These are but a few examples that illustrate the dynamic environmental conditions that stimulate a host of different endocrine systems to allow the animals to properly adapt and thrive.

While limited work has been done on the neuroendocrinology of marine mammals, it is clear that the sensory systems that predominate are vision, hearing, olfaction, and likely gustation. All of these systems link directly to the brain stimulating the hypothalamus, pineal, and hippocampus (St. Aubin, 2001). Most of the hypothalamic effects are transferred to the pituitary gland or hypophysis, a small, compound structure located at the base of the brain. The glandular portion (adenohypophysis) produces a set of hormones (Fig. 1) that

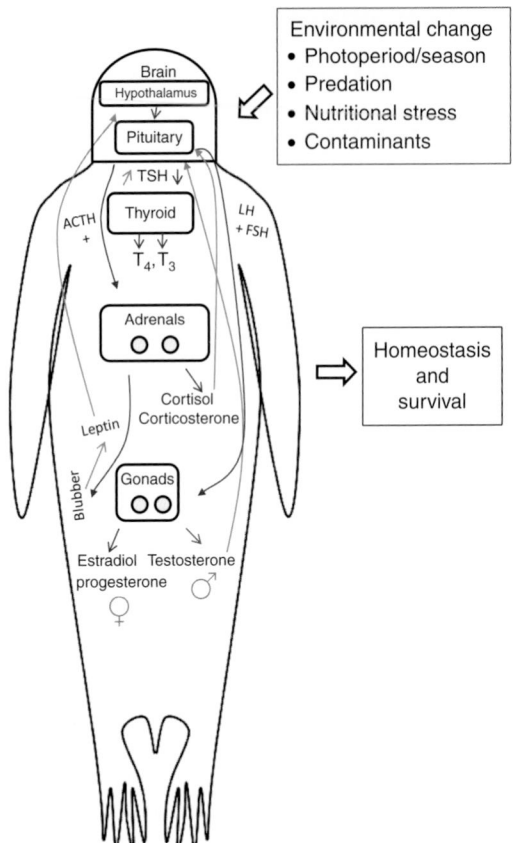

**Figure 1** *An overview of endocrine organs and hormones in a marine mammal.*

either directly elicit tissue responses (e.g., growth hormone (GH), prolactin, and gonadotropins) or modulate the activity of other endocrine glands (e.g., adrenocorticotropic hormone (ACTH) and thyrotropin or thyroid-stimulating hormone (TSH)). Early investigators on whaling vessels were impressed by the size of the pituitary in mysticetes (up to 53.5 g in a blue whale, *Balenoptera musculus*), although on a body weight basis, it is unremarkable. Nevertheless, the abundance of the tissue afforded the opportunity for extensive studies to extract and characterize the principal hypophyseal hormones. Fin whale, *B. physalus*, ACTH was found to be identical to that of humans, whereas blue whale TSH more closely resembled that of non-primates. Studies on beluga whales (*Delphinapterus leucas*) and bottlenose dolphins, (*Tursiops truncatus*), as well as Steller sea lions (*Eumetopias jubatus*), harbor seals (*Phoca vitulina*), and ringed seals (*Phoca hispida*) have shown that synthetic ACTH and bovine TSH are capable of eliciting expected responses from the target glands, demonstrating at least some homology in the structure of these hormones among various groups of mammals (St. Aubin and Dierauf, 2001; Mashburn and Aktinson, 2004, 2007a).

## III. Thyroid Hormones

Early investigators were as impressed by the size of the cetacean thyroid as they were by the pituitary, to the degree that large whales were considered as a possible commercial source for thyroxine (T₄), the principal hormone synthesized by the gland. It was not only the size, but the proportions of the thyroid that called attention to the importance of this gland in cetaceans (St. Aubin, 2001); beluga whales have 3 times more thyroid tissue per unit body weight than a thoroughbred horse, and bottlenose dolphins have nearly twice as much thyroid as do humans (400 vs 250 mg/kg).

Stimulation of the thyroid gland follows a somewhat typical pattern of hormone biosynthesis and secretion in that the initial stimulus comes from the hypothalamus in the form of thyrotropic releasing hormone (TRH). TRH acts on the pituitary to produce and secrete thyroid stimulating hormone (TSH). TSH is carried via the circulatory system to the thyroid gland whereby it stimulates the production of the thyroid hormones (TH). Two major forms of TH exist, thyroxine ($T_4$) and triiodothyronine ($T_3$), with both present in the bioactive free form ($fT_4$ and $fT_3$) as well as the globulin-bound form. Free$T_3$ is the physiologically active TH and $T_4$ is converted to $T_3$ by monodeiodination. Most assays measure either the free form or total form ($TT_4$ or $TT_3$), the latter being the sum of the free plus bound forms (Oki and Atkinson, 2004). In Steller sea lions, $T_4$ constitutes the primary form (97%) in circulation with only 3% in $TT_3$ (Myers *et al.*, 2007). Additionally, the free forms only account for 0.02% and 0.01% for $fT_4$ and $fT_3$, respectively. Thyroid hormones play an important role in metabolism and in regulating homeostasis of the body. Both $T_4$ and $T_3$ effect metabolism by calorigenic and thermogenic actions related to lipid, protein, and carbohydrate metabolism. $T_4$ stimulates oxidative phosporylation and mitochondrial respiration, which increase the animal's metabolic rate and subsequent metabolic heat production (St. Aubin *et al.*, 1996; Oki and Aktinson, 2004; Myers *et al.*, 2007). TH are also necessary for normal growth and development, especially in critical developmental periods. Early researchers compared terrestrial mammals to aquatic mammals and concluded that the metabolic rate of aquatic mammals must be elevated to counteract the heat loss from living in cold water. The size of the thyroid gland in cetaceans was used as evidence for this physiological adaptation. Subsequent studies failed to demonstrate a difference in metabolic rates of terrestrial mammals and aquatic mammals. In pinnipeds, circulating TH levels and the basal metabolic rate are comparable to those in terrestrial mammals, but surprisingly, the blood TH concentrations are lower than in West Indian manatees, (*Trichechus manatus*), a species with a distinctly lower metabolism (Ortiz *et al.*, 2000). Assumptions about the metabolic rate cannot be based solely on circulating TH levels.

Concentrations of TH tend to be highest in pups of all pinniped species examined to date (Engelhardt and Ferguson, 1980; Haulena *et al.*, 1998; Myers *et al.*, 2006). Circulating levels in neonatal pinnipeds are understandably high in view of the need for metabolically derived heat until their lanugo coat gains insulating ability and blubber reserves are established. Thereafter, seasonal fluctuations in the blood levels of $T_4$ and $T_3$ have been correlated with changing metabolic needs. Among cetaceans, the seasonality of TH activity in beluga whales is characterized by marked elevation in circulating levels of $T_4$ and $T_3$ and by histological evidence of intense cellular activity in the thyroid gland during the summer (St. Aubin, 2001). Colloid reserves are depleted by columnar follicular cells, in contrast to the quiescent appearance of the gland in spring and fall when low cuboidal cells surround abundant stores of TH. The implications of this burst of thyroid activity are broad, favoring mobilization of blubber, and promoting the effects of other agents such as GH. It also coincides with a unique event in beluga epidermis, cell production is enhanced, presumably under the influence of TH, and superficial turnover is accelerated by the relatively warm (10–15°C) freshwater environment. Taken together, these events constitute a true molt, lacking only the production of a hair coat to be fully analogous to the process in a pinniped. No comparable transformation has been described in any other cetacean; only the rubbing behavior of killer whales (*Orcinus orca*), on the cobbles of Telegraph Cove, British Columbia, hints of a seasonal pulse in epidermal growth. Studies on circulating levels of TH in Atlantic bottlenose dolphins have not revealed significant annual variation, even though the 15°C range of water temperatures experienced by these animals is equivalent to that encountered by belugas (St. Aubin, 2001).

As in terrestrial fur-bearing mammals, TH are seasonally elevated to promote hair growth in pinnipeds during the annual molt (Ashwell-Erickson *et al.*, 1986; Boily, 1996; Oki and Atkinson, 2004). Controversy in the literature regarding the degree of association between visible molt and thyroid activity in pinnipeds may derive from the difficulty in recognizing when hair growth is actually stimulated by elevated TH. Hair loss, the overt sign of molting, may be enhanced by increased levels of cortisol at a time when $T_4$ and $T_4$ levels are low. Cortisol suppresses the secretion of TSH from the pituitary and inhibits the deiodination of $T_4$ to $T_3$. The fluctuations in circulating levels of metabolically potent substances such as cortisol and TH during the molt draw attention to how intensely pinnipeds are physiologically affected at that stage of their annual cycle.

## IV. Substrate Metabolism

Marine mammals exhibit a number of different behaviors that can produce highly variable responses in substrate-level metabolism that are regulated by various endocrine systems. Phocid seals experience prolonged periods (months) of food and water deprivation as a natural component of their life history to which they are well adapted. Otariids also exhibit natural periods of food deprivation, but the durations are not as long. Because the strictly water-borne animals such as cetaceans and sirenians commonly migrate great distances, it is thought that they will exhibit periods of intermittent caloric restriction. Thus, whether the period of food deprivation or fasting is natural or imposed, almost all marine mammals experience caloric restriction, necessitating the stimulation of robust physiological regulation of substrates to maintain homeostasis. While it is well-recognized that TH play a critical role in regulating metabolism, this section will focus on other hormones as TH were discussed previously. Hundreds of hormones contribute to substrate metabolism in vertebrates either directly or indirectly; however, only a few have been studied in marine mammals. In most mammals, glucose is the primary substrate contributing to the animal's metabolic rate; however, in phocids and likely all marine mammals, carbohydrates contribute very little (<10%) (Keith and Ortiz, 1989; Champagne *et al.*, 2006). Thus, it is not surprising that for those marine mammals examined, insulin does not appear to contribute to the regulation of glucose as levels are relatively low (Kirby and Ortiz, 1994; Ortiz *et al.*, 2003b; Champagne *et al.*, 2005). The bolus infusion of glucose (glucose tolerance test; GTT) led to only a transient stimulation of insulin secretion suggesting that the pancreas of marine mammals is insensitive to a high-carbohydrate load (Kirby and Ortiz, 1994). Because of the lack of carbohydrate in the diet, this finding should not be that surprising, but rather more indicative of the adaptive mechanisms these animals have evolved to deal with this environmental constraint. Furthermore, the relatively slow clearance of glucose from circulation following a bolus infusion is indicative of insulin resistance (Kirby and Ortiz, 1994), which combined with a chronic state of hyperglycemia in many species makes these animals

appear, clinically like diabetic/obese humans (Ortiz *et al.*, 2003b; Champagne *et al.*, 2005). While changes in plasma glucagon or insulin: glucagon ratios have been reported in fasting elephant seals suggesting that it may contribute to glucose metabolism (Ortiz *et al.*, 2003b; Champagne *et al.*, 2005), these levels have yet to be definitively shown to contribute to glucose metabolism. In lactating female elephant seals, the changes in endogenous glucose production were not associated with expected changes in insulin:glucagon ratios or glucagon concentrations (Champagne *et al.*, 2006). The possibility exists that these studies failed to measure the concentrations at the key periods when the dynamic changes are occurring to contribute to glucose metabolism; however, the paucity of available data would suggest that the counterregulatory insulin–glucagon mechanism does not contribute appreciably to carbohydrate metabolism.

Lipids or fatty acids are the primary substrates contributing to the metabolism of marine mammals. During periods commonly associated with elevated levels of fatty acid mobilization such as prolonged fasting (postweaning, molt, breeding, etc.), increased levels of cortisol, GH, and ghrelin have been observed. A synchronous pattern of free fatty acid and GH concentrations was reported in harp seals. In addition, a significant correlation between plasma GH and free fatty acids in fasting elephant seal pups was detected suggesting that GH contributes to fatty acid metabolism (Ortiz *et al.*, 2003b). The gene encoding GH in the finback whale (*Balaenoptera physalus*) has been cloned and characterized with the data on this suggesting that the GH gene in cetaceans appears to be evolving more slowly than in other artiodactyls (Tsubokawa *et al.*, 1980, Wallis *et al.*, 2005).

The glucocorticoid, cortisol, consistently increases during fasting conditions in pinnipeds and likely contributes to the mobilization of fatty acids to support the animal's metabolism (Ortiz *et al.*, 2001, 2003a, b). As its name implies, cortisol can also stimulate gluconeogenesis and thus likely contributes to carbohydrate metabolism. A bolus injection of the catecholamine, epinephrine, induced an increase in plasma glucose within 25 min in Weddell seals (*Leptonychotes weddellii*) suggesting that catecholamines stimulate glycolysis in marine mammals (Hochachka *et al.*, 1995).

Leptin is a relatively newly discovered hormone that has received considerable attention in marine mammals recently. Leptin is a product of the *ob* gene in mammals and is primarily produced in adipocytes. In humans and other mammals, its plasma concentration is correlated with the animal's body mass or lipid stores. Its primary role in terrestrial mammals is to signal the brain to inhibit food intake and to provide an index of the body's energy reserves. For these reasons, it is not surprising that leptin has received considerable attention in recent studies. However, the data from those marine mammals studied suggest that leptin does not stimulate typical mammalian responses with respect to energy balance. In fasting elephant seal pups (*Mirounga angustirostris*), plasma leptin concentrations were not correlated with body fat and fasting did not affect plasma concentrations (Ortiz *et al.*, 2001; Ortiz *et al.*, 2003a, b). No consistent changes in serum leptin concentrations were observed in acutely fasted Steller sea lions and levels were also not correlated with body fat. In contrast, plasma leptin decreased over 5 days of fasting in Antarctic fur seals (*Arctocephalus gazelle*) (Arnould *et al.*, 2002). Serum leptin decreased between the reproductive/breeding and molting seasons in adult southern elephant seals suggesting that leptin may serve as a permissive hormone contributing to the reproductive effort of marine mammals as it does in humans (Guilherme *et al.*, 2004). Under natural conditions, adult southern elephant seals have the highest reported average concentrations (9.5 ng/ml) of any marine mammal. Interestingly, serum leptin is stimulated by adrenocorticotropin (ACTH), with mean levels

**E**

E

approaching 15 ng/ml (Mashburn and Atkinson, 2008). Whether ACTH directly stimulates the *ob* gene or promotes its stimulatory actions via another hormone is unknown; however, this indicates that the *ob* gene is inducible in marine mammals and that it can be stimulated exogenously to increase circulating levels in a relatively short period of time (<30 min). Unfortunately, these studies do not elucidate the physiological role of these hormones, they do provide an index of their potential contribution to regulating substrate metabolism and a basis from which to design future studies to assess their physiological function.

## V. Adrenal Hormones

The adrenal gland in mammals is comprised of two distinct tissue layers, the cortex and medulla. The medulla is the primary site of catecholamine (epinephrine and norepinephrine) synthesis and release. The cortex is divided into three distinct regions or zones characterized by distinct cortical cell types: (1) zona glomerulosa, (2) zona fasciculata, and (3) zona reticularis. Cells of the zona glomerulosa are primarily responsible for synthesizing and releasing the mineralocorticoid, aldosterone, which directs the reabsorption of $Na^+$. The zona glomerulosa is particularly well developed in young seals suggesting that a fully functional system for the regulation of $Na^+$ is intact by the time these animals are born (Bryden, 1994). Cells of the zona fasciculata are primarily responsible for synthesizing and releasing the glucocorticoids, cortisol, corticosterone, and cortisone. The positive correlations among adrenal steroids (aldosterone, cortisol, and corticosterone) exhibited by bottlenose dolphins suggest that an active pituitary–adrenal axis is present in these animals (Ortiz and Worthy, 2000). Cells that comprise the zona reticularis serve as the border between the cortex and medulla, but can produce and release small amounts of androgens. This zone's androgenic capacity in marine mammals has not been evaluated; however, it is well recognized that the glomerulosa and fasciculata produce the aforementioned steroids in marine mammals. Pinnipeds have been reported to produce the glucocorticoids, corticosterone, cortisone, and cortisol (St. Aubin and Geraci, 1986; Liggins *et al.*, 1993; Boily, 1996; Gardiner and Hall, 1997; Ortiz *et al.*, 2001; Engelhard *et al.*, 2002; Guinet *et al.*, 2004; Mashburn and Atkinson, 2004; Oki and Atkinson, 2004; Mashburn and Atkinson, 2008), while corticosterone and cortisol have been detected in dolphins (St. Aubin *et al.*, 1996; Ortiz and Worthy, 2000), and cortisol in manatees (Ortiz *et al.*, 1998). The precise physiological function of glucocorticoids in marine mammals has not been delineated, but it is expected that they contribute in a typical mammalian fashion to glucoregulation. This is supported by a number of studies in which changes in circulating glucose are associated with changes in plasma cortisol. Phocid seals have the highest cortisol production rate (when corrected for surface area) reported for any species studied (Liggins *et al.*, 1993), in addition to relatively high-metabolic clearance rates. The relatively high levels of cortisol found in most phocid seals can be attributed to production rates that far exceed the metabolic clearance rates (Liggins *et al.*, 1993). Under normal conditions, the levels of cortisol found in otariids are more similar to those observed in terrestrial mammals. Thus, it has been suggested that the relative hypercortisolemia is adaptive to facilitate the deep and extensive dives that are more common in phocids.

Because adrenal hormones are primarily stimulated in response to "stressful" cues, their primary functions are to support metabolism during stressful events by liberating glucose and fatty acids. In those marine mammals examined, the bolus infusion of ACTH produces a predictable increase in cortisol and corticosterone (Mashburn and

Atkinson, 2008). While acute handling and restraint does not appear to influence plasma cortisol b and aldosterone concentrations in seals and dolphins (St. Aubin *et al.*, 1996; Ortiz and Worthy, 2000; Ortiz *et al.*, 2001; Engelhard *et al.*, 2002), prolonged restraint and encirclement was associated with elevated plasma aldosterone concentrations in dolphins (St. Aubin *et al.*, 1996). Repeated handling also had no effect on circulating plasma cortisol in southern elephant seals suggesting that these animals can become rapidly desensitized to frequent handling procedures in the field, which is important since this can be a common research practice in many studies involving some marine mammals (Engelhard *et al.*, 2002).

Because routine handling and manipulation of animals could potentially disrupt biochemical and hormonal homeostasis in some of these animals, if they have not become desensitized to these procedures, alternative methods of sampling animals are required. Recently, considerable attention has been dedicated to evaluating fecal hormone levels, primarily adrenal steroids (but also androgens and estrogens) to circumvent the need to restrain or even handle an animal to obtain a blood/tissue sample (Mashburn and Atkinson, 2004; Petrauskas *et al.*, 2006). This is particularly important for field applications or when working with very large animals that cannot be easily restrained or immobilized. Protocols to reproducibly measure urinary steroid (and protein) levels in smaller captive animals have been implemented with great success and usefulness (Hong *et al.*, 1982; Ortiz *et al.*, 1998, 2002a, b, 2003a, b). Protocols using salivary samples have also been used with equal success (Pietraszek and Atkinson, 1994; Petrauskas *et al.*, 2006). These alternative methods of sample collections will expand our abilities to enhance our understanding of the impacts of dynamically changing environmental cues on not only adrenal hormones but hormones from other endocrine systems as well.

## VI. Osmoregulation

In mammals, a large number of hormones are known to contribute to water and electrolyte balance, but hormones most commonly attributed to osmoregulation are angiotensin (Ang I, II, or III), atrial natriuretic peptide (ANP), aldosterone, and vasopressin (AVP). Renin converts angiotensinogen to angiotensin I (Ang I), which is quickly converted to Ang II (the most potent angiotensin) by angiotensin-converting enzyme (ACE). Subsequently Ang II stimulates the release of aldosterone from the zona glomerulosa in the adrenal gland, which in turn induces the reabsorption of $Na^+$ from the distal tubule of the nephron and the colon resulting in a decrease in excreted $Na^+$. Collectively, renin, Ang II, and aldosterone comprise the renin–angiotensin–aldosterone system (RAAS). Functional and responsive RAAS have been identified in pinnipeds (Ortiz *et al.*, 2000b, 2002a, b; Ortiz *et al.*, 2006), dolphins (Malvin *et al.*, 1978), and manatees (Ortiz *et al.*, 1998). The sensitivity of RAAS in West Indian manatees is much greater than in pinnipeds and bottlenose dolphins, which would be expected since manatees are commonly found in $Na^+$ depleted environments and do not drink seawater. Oral intubation of West Indian manatees resulted in a neuroendocrine stress response, producing an increase in RAAS (Ortiz *et al.*, 1998).

Probably the first published account of a bioassay using extracted cetacean tissues was performed by Eichelberger and his colleagues (1940). Extracted renin from the kidney of a bottlenose dolphin was injected into dogs at two doses (concentrations not reported), 4 and 2 ml. In the initial trial, 4 ml of extract induced an immediate increase (48%) in blood pressure (BP), and remained 38% higher than control pressure for 10 min. In a separate trial, an initial 2 ml dose of extract induced a 21% increase in BP; however, it was not sustained. A second 2 ml dose resulted in a 27% increase in BP, which also was not sustained. However, the two doses collectively were able to induce a 9% increase in BP that was sustained. A third dose of 2 ml produced a 20% increase in BP that sustained an additional 4% increase in BP. These experiments demonstrated that (1) renin in dolphins (and probably all cetaceans) causes a vascular pressor effect as in other mammals; (2) the pressor effect is dose dependent; and (3) the response is graded until a threshold is reached, which results in a sustained elevation in BP. These important early studies shed light on the sensitivity and kinetics of the renin-induced increase in BP and suggest that RAAS regulates BP as well as water and electrolyte balance in marine mammals as in terrestrial mammals.

Although most seals live in salt water, they likely don't drink and captive animals residing in freshwater may be susceptible to hyponatremia (reduced plasma $Na^+$), which appears to be reconciled via the pituitary–adrenal axis. Early infusion studies with Baikal (*Phoca sibirica*) and ringed seals demonstrated an increase in fractional clearance of $Na^+$ following infusion of hyperosmotic saline indicating that tubular $Na^+$ reabsorption was reduced (and likely aldosterone) (Hong *et al.*, 1982). However, a negative correlation between excreted aldosterone and excreted $Na^+$ was not observed prompting a question of the role of RAAS in seals. More recently, similar studies with northern elephant seal pups (*Mirounga angustirostris*) demonstrated a similar increase in fractional excretion of $Na^+$, but in the presence of elevated aldosterone suggesting that an alternative, and likely non-hormonal, mechanism mediated the increased fractional excretion of $Na^+$ (Ortiz *et al.*, 2002b). In this case, increased glomerular filtration rate (GFR) overcame the aldosterone-mediated retention of $Na^+$ suggesting that renal hemodynamics play a critical role in regulating water and electrolytes in pinnipeds, especially during conditions of increased $Na^+$ load, which may be observed during feeding.

While aldosterone and Ang II can also serve to reabsorb water, the most potent anti-diuretic agent is vasopressin, which stimulates the synthesis of water channels (aquaporins) in the collecting duct of the kidney. Detectable levels of AVP have been reported for pinnipeds (Hong *et al.*, 1982; Skog and Folkow, 1994; Zenteno-Savin and Castellini, 1998; Ortiz *et al.*, 2000a, b, 2002a, b, 2006), cetaceans (Ortiz and Worthy, 2000), and manatees (Ortiz *et al.*, 1998). Although not conclusive, a number of studies provide compelling evidence to suggest that tubular water reabsorption is mediated by AVP in seals as in terrestrial mammals. The intravenous infusion of pitressin (synthetic AVP) in a water-loaded harbor seal resulted in an immediate decrease in urine flow rate along with concomitant increases in urinary electrolyte concentrations. Pitressin infusions induced an increase in urine osmolality and in osmotic clearance, which was greater than urine volume suggesting that free water reabsorption was increased (or free water clearance was reduced). More recently, a bolus infusion of AVP in fasting elephant seal pups paradoxically induced a diuresis associated with increased osmotic clearance (primarily natriuresis) and not reduced free water clearance (Ortiz *et al.*, 2003c). Additionally, this infusion of AVP acutely increased plasma cortisol and aldosterone in the presence of reduced plasma renin activity suggesting that AVP possesses natriuretic and neuronendocrine capabilities (similar to some terrestrial mammals). Under force-fasted conditions, Baikal and ringed seals exhibited an increase in excreted AVP associated with a concomitant decrease in urine flow rate and increase in urine osmolality (Hong *et al.*, 1982). A positive and significant correlation between urine osmolality and excreted AVP was also demonstrated in Baikal and

ringed seals further suggesting that the observed increase in urine osmolality was attributed to an increase in tubular water reabsorption via AVP stimulation. Furthermore, in gray seals (*Halichoerus grypus*), force-fasted conditions induced an increase in urine osmolality in conjunction with an increase in plasma osmolality and AVP (Skog and Folkow, 1994). However, in naturally fasting, postweaned northern elephant seal pups urine osmolality increased despite decreased plasma AVP. These latter data were refuted recently, using a much larger sample of fasting pups to demonstrate that plasma AVP concentrations are constant and relatively low with no change in plasma osmolality (Ortiz *et al.*, 2000b). Also, under conditions of isotonic and hypertonic saline infusion, increased urinary AVP excretion was negatively correlated with reduced free water clearance and positively correlated with urinary cAMP excretion suggesting that AVP mediates free water reabsorption via a cAMP-mediated mechanism in the collecting duct as in terrestrial animals during stimulatory conditions such as feeding (Oritz *et al.*, 2002b). The only other known bioassay of cetacean tissue was that of plasma-extracted AVP on water retention in water-loaded rats by measuring the electrical conductivity of the urine. In most of the dolphin samples measured by bioassay, AVP was not detected. For those samples that showed detectable amounts of AVP, concentrations were very low, as were the amounts from pituitary extracts. Also, the lack of a correlation between urine flow rate and plasma AVP concentrations in fasting dolphins led researchers to contend that AVP does not significantly regulate urine volume and thus water retention in these animals. However, significantly greater concentrations of AVP have been measured recently in free-ranging dolphins, which may be the result of advances in assay techniques for hormone measurements (Ortiz and Worthy, 2000). Increased plasma osmolality appears to stimulate AVP release in manatees, like in terrestrial mammals; however, the anti-diuretic actions of AVP in manatees have yet to be elucidated (Ortiz *et al.*, 1998).

Atrial distention, induced by increased cardiac pressure (volume), is the primary mechanism for the release of ANP. The actions of ANP oppose those of Ang II and aldosterone by inhibiting the synthesis and release of renin, thereby resulting in an increase in excreted $Na^+$. Levels of ANP have only been reported for pinnipeds (Zenteno-Savin and Castellini, 1998; Ortiz *et al.*, 2002a, b). A bolus infusion of AVP in fasting elephant seal pups increased the urinary excretion of the natriuretic factor, urodilatin, that may partially explain the observed natriuresis in this case (Ortiz *et al.*, 2003c). Beyond this, the natriuretic effects of ANP and its related peptides have not been further examined in marine mammals.

## VII. Diving

For logistical reasons, examining the effects of diving on the various endocrine systems is difficult, and trying to elucidate the function of these systems during diving would be even more so. Nonetheless, a few studies have evaluated the effects of diving or sleep apnea. Diving has profound effects on blood flow to the kidneys and glomerular filtration (Murdaugh *et al.*, 1961; Zapol *et al.*, 1979; Davis *et al.*, 1983), resulting in a decrease in renal activity, which could alter the response of vasoactive hormones (i.e., Ang II, AVP, ANP). Voluntary bouts of sleep apnea in elephant and Weddell seal pups reduced heart rate associated with a decrease in Ang II and AVP (vasoconstrictors), and an increase in ANP (vasoconstrictor inhibitor) (Zenteno-Savin and Castellini, 1998). This increase in ANP was attributed to an increase in cardiac pressure, which is a known stimulus of ANP release. The changes in vasoactive hormones observed during sleep apnea are likely to occur during breath-hold diving as well, based on the available data on renal blood flow during diving and the direction of change in vasoactive hormones during sleep apnea.

Diving is also associated with an increase in epinephrine that likely contributes to diving bradycardia, peripheral vasoconstriction, and inhibition of pancreatic insulin release to maintain circulating plasma glucose concentrations (Hochachka *et al.*, 1995). The inhibition of insulin results in a reduced insulin:glucagon ratio that is also important for the post-dive recovery to facilitate lactate clearance and maintain glucose for aerobic metabolism. The hormone erythropoietin (EPO) stimulates the production and release of red blood cells in response to tissue hypoxia (Richmond *et al.*, 2005). Thus, it is not surprising that age-related increases in hematocrit and hemoglobin are correlated with increased concentrations of EPO suggesting that EPO contributes to the development of blood oxygen stores in marine mammals. While leptin does not appear to contribute to the regulation of substrate-level metabolism or energy balance, it may contribute to pulmonary surfactant production and thus, may play a role in diving physiology (Hammond *et al.*, 2005).

## VIII. Reproduction

As with many of the endocrine systems, reproductive activity is controlled by the hypothalamus secreting gonadotrophic releasing hormone (GnRH) which acts on the anterior pituitary. The pituitary in turn releases the gonadotrophins, luteinizing hormone (LH) and follicle stimulating hormone (FSH). The two gonadotrophins collectively act on the ovaries and testes to stimulate follicle development and spermatogenesis, respectively (Atkinson, 1997). Through follicular development the granulosa cells of the antral follicles produce estrogen and upon luteinization of these cells (and subsequent ovulation), progesterone is produced. Estrogen promotes cellular hydration and proliferation, while progesterone promotes fat deposition and smooth muscle relaxation. In the testes, testosterone is produced by the Leydig cells and acts on the seminiferous tubules to stimulate spermatogenesis (Atkinson and Yoshioka, 2007).

Collectively, the sex steroids—estrogen, testosterone, and progesterone—are measured in low concentrations (typically <0.5 ng/ml for each steroid) in the circulation of prepubertal mammals and fluctuate very little. The onset of sexual maturity is detected by seasonal fluctuations of each of the sex steroids.

Reproduction in most marine mammals is a highly seasonal (Atkinson and Yoshioka, 2007). In particular, the high-latitude pinnipeds have tight synchrony of various reproductive events and the associated behaviors. With the exception of one species of pinniped (the Australian sea lion, *Neophoca cinerea*), all seals are annual breeders with their estrous cycles synchrononized by parturition (Atkinson, 1997). Most pinnipeds are seasonally monoestrous with ovulation occurring once per year. The duration of estrus, or the period of sexual receptivity, is a function of elevated estrogen. Steroid hormone peaks in circulating concentrations of estrogen followed 1–2 days later by elevated progesterone have been characterized for several species of pinniped and reflects preovulatory follicular maturation and subsequent corpus luteum formation (Pietraszek and Atkinson, 1994). Once ovulation has occurred progesterone concentrations in the circulation are elevated regardless of whether or not conception takes place (Atkinson *et al.*, 1999). The conceptus of pinnipeds undergoes an embryonic diapause of variable length, with 1–5 months being reported. While the reactivation of the blastocyst signals at the end of diapause, the exact cue remains somewhat elusive (Boyd, 1991). As the size of the corpus luteum increases just before the embryonic reactivation, it is thought that the ovary is functional in cuing the end of the diapause. However, the scientific literature on the subject also implicates photoperiod as being an environmental cue. Active or placental gestation is sustained by progesterone that is produced by the corpus luteum which atrophies at parturition. Lactation is extremely variable in pinnipeds ranging from 4 days to up to 36 months, with the phocid seals having shorter more defined lactational periods than the otariids (Atkinson, 1997). Prolactin is important in development of mammary glands in anticipation of lactation and in maintaining early lactation. Prolactin concentrations are high before parturition and reach a peak 0–3 days after birth declining to non-detectable concentrations before the end of lactation. As in terrestrial mammals, prolactin is under inhibitory dopaminergic control. Prolactin does not decline precipitously when laction is stopped abruptly and when prolactin is inhibited, the postovulatory rise in progesterone is suppressed, indicating prolactin may play a controlling role in ovulation and luteal development in pinnipeds.

In male pinnipeds circulating concentrations of testosterone exhibit seasonal elevations preceding the breeding season (Atkinson and Gilmartin, 1992). The seasonal elevations appear to be tighter with increasing latitude. The increase in testosterone prior to the breeding season is followed by an increase in cortisol, at least in Weddell seals and Steller sea lions.

LH and FSH have been measured in both odontocetes and mysticetes, with females having higher concentrations than males (Atkinson and Yoshioka, 2007). As with the pinnipeds, prepubertal cetaceans are characterized by low circulating concentrations of the sex steroids. Reproductive cycles in female cetaceans are highly variable, with some falling into relatively narrow time frames, such as in belugas, whereas others show very little seasonality. Odontocetes such as bottlenose dolphins and killer whales, which have been studied extensively in captivity (West *et al.*, 2000), and false killer whales (*Pseudorca crassidens*) demonstrate some of the variability possible within a species. Some adult female whales may be seasonally polyestrous, with up to seven cycles per year, whereas others may be anestrous for a year or more. In addition, elevated progesterone concentrations up to 10 months in a non-pregnant captive female false killer whale were recorded.

Estrous cycle lengths have been best described for killer whales (Robeck *et al.*, 1993). They exhibit a 41-day estrous cycles, divided into a 7-19-day luteal phase, dominated by progesterone, and a follicular phase. Progesterone is increasingly the predominant hormone responsible for sustaining pregnancy, and is commonly used as a diagnostic indicator for pregnancy detection. As many cetaceans exhibit annual reproduction cycles, it is clear that conception can occur during lactation. It has also been reported for many cetaceans that extended periods of anestrus can occur; these are characterized by low or undetectable levels of the sex steroids followed by measureable ovarian activity.

Endocrinology of male cetaceans has not been studied extensively when compared to their female counterparts. As with female cetaceans, male sex steroids (i.e., testosterone) are low in immature animals. Once sexual maturity is reached, testosterone is correlated with testes weight in several of species (Atkinson and Yoshioka, 2007). As with most of the pinnipeds, testosterone concentrations in circulation are highest prior to peak sperm density and the peak period of the female breeding activity. Testosterone levels in male bottlenose dolphins tend to be higher in spring and fall, roughly coinciding with calving peaks (gestation is approximately 12 months), although individual males show varying patterns from year to year and are capable of impregnating females in almost any month. Seasonal constraints on breeding would appear to be less critical in tropical and subtemperate species than in those exploiting more polar habitats. An exception would be river dolphins exposed to drastic seasonal fluctuations in habitat associated with dry and rainy seasons. Synchrony of reproductive activity, and by inference reproductive hormones, is likely an important consideration.

The challenges associated with conducting endocrine studies on cetaceans are largely logistical and several studies have been conducted on both captive and free-ranging cetaceans to develop non-invasive or novel methods of sample collection. Monitoring of estrous cycles and the detection of pregnancy has been attempted in saliva, feces, ocular secretions, milk, and blubber in addition to the standard serum or plasma (Pietraszek and Atkinson, 1994; Theodorou and Atkinson, 1998; West *et al.*, 2000). Most of these media were successful in the detection of progesterone, with the exception of saliva and ocular secretions from false killer whales (Atkinson *et al.*, 1999).

Numerous factors have been implicated in altering the endocrine status of marine mammals. Social suppression of reproduction in both male and female cetaceans has been suggested (Atkinson and Yoshioka, 2007). Anthropogenic chemicals such as organochlorines and polybrominated compounds, have been shown to have endocrine disrupting effects that are detrimental to normal reproduction. It is likely large-scale changes in environmental (e.g., Global warming) may also impact reproduction through several mechanisms, including nutritional stress and immunosuppression. Factors that alter normal reproductive functioning are likely to be the topics of enhanced research in the future.

## IX. Circadian Patterns and Melatonin

The biological clock controlling the circadian pattern of behavior and many physiological processes is a neural network located in the suprachiasmatic nucleus (SCN) of the basal hypothalamus. The SCN is responsible for the circadian rhythm of hormones in mammals. There is a direct link between photoperiod, pituitary, and pineal gland hormone secretion.

Melatonin is the hormone most commonly associated with photoperiodism in mammals (St. Aubin, 2001). It is produced primarily by the pineal gland (epiphysis) located above the third ventricle of the brain. Daylight suppresses the production of melatonin, which regulates the activity of the hypothalamus, pituitary, and, indirectly, the gonads, adrenals and thyroid. Melatonin secretion ceases in Weddell seals under continuous natural light. Unfortunately, the hormone and its activity have received relatively little attention in the marine mammal literature. This deficit is ironic, as the pineal gland of the newborn southern elephant seal can weigh over 9g, which is the size of the brain of a hamster, the subject of so much of the research on melatonin.

The exceedingly high plasma levels of melatonin in newborn seals draw attention to the important role of this hormone in the survival of animals born under harsh environmental conditions. Concentrations (69,000ng/ml) in southern elephant seals at birth, but fall steadily during the ensuing 7–10 days. A similar pattern has been observed in harp (*Pagophilus groenlandicus*), gray, Weddell, and northern elephant seals (Stokkan *et al.*, 1995). It has been postulated that the hormone promotes the generation of the TH and cortisol, which in turn accelerate metabolism and provides the heat necessary to withstand the extreme conditions experienced at the time of birth. Cortisol in

Alaskan harbor seals did not show a diurnal rhythm in winter, only in summer (Oki and Atkinson, 2004).

Marked seasonality is evident in the size of the pineal in southern elephant seals. The gland is largest in the winter dark period and the high circulating levels of melatonin at that time presumably suppress gonadal function. In terrestrial mammals melatonin acts in the mediobasal hypothalamus through dopaminergic receptors that inhibit GnRH. Melatonin levels are uniformly low during the spring breeding season. A circadian rhythm in melatonin secretion is evident during the winter, but equivocal during the summer.

The literature on melatonin in cetaceans is sparse when compared to pinnipeds. The very presence of a discernible pineal gland had been uncertain, but was eventually identified in a number of cetacean species. Extrapineal sources of melatonin, such as from the retina, likely augment the role of the small gland in integrating photoperiod with metabolic functions. One might expect that polar cetaceans such as belugas and narwhals (*Monodon monoceros*), need to entrain their endocrine physiology with seasonal changes marked by day length, similar to high-latitude pinnipeds.

## X. Conclusions

The ecosystems in which marine mammals reside have required the morphology and physiology of these species to evolve to be able to cope with the challenging environments. One of the major systems that allows the body to maintain homeostasis is the endocrine system. The sensory systems of marine mammals, like terrestrial mammals, link directly to the brain and the master glands, the hypothalamus and pituitary.

The metabolism of marine mammals is driven by the oxidation of fatty acids and appears to be regulated by the glucocorticoid, cortisol, and the protein hormone, growth hormone. During diving conditions, the adrenal products, cortisol and epinephrine, may contribute to glucose metabolism. The counterregulatory insulin–glucagon mechanism typical of terrestrial mammals does not appear to contribute appreciably to the regulation of carbohydrate, which should not be surprising since carbohydrate contribution to the animals metabolism is minimal. Leptin also does not appear to contribute to energy balance in those animals studied, but may play a role in reproduction and diving. The TH are essential in regulating metabolism and thermogenesis, but also in early growth and development. All groups of marine mammals possess an active and functional RAAS that contributes to renal $Na^+$ reabsorption and likely blood pressure. Under natural conditions, AVP is relatively low and likely not contributing to renal water reabsorption in pinnipeds and dolphins; however, during isotonic and hypertonic infusion, AVP appears to mediate renal water reabsorption in seals. While not definitive, it likely plays a similar role in manatees. Reproduction in many marine mammals is highly synchronized, requiring tight endocrine controls to ensure success. Also linking to the synchrony of seasons is the pineal gland and its production of melatonin.

Important advances in analytical techniques such as development of fecal, urinary, salivary, and tissue hormone measurement protocols; increased sensitivities of RIAs and EIAs; and improved antibody production have enhanced our ability to detect low levels of hormones and expanded our ability to measure a greater variety of hormones. All of these advancements will increase our understanding of the endocrine systems in marine mammals and their needs to adapt to changing environments.

E

E

## See Also the Following Articles

Pinniped Physiology ■ Diving

## References

Arnould, J. P., Morris, M. J., Rawlins, D. R., and Boyd, I. L. (2002). Variation in plasma leptin levels in response to fasting in Antarctic fur seals (*Arctocephalus gazella*). *J. Comp. Physiol.* **B172**, 27–34.

Ashwell-Erickson, S., Fay, F. H., Elsner, R., and Wartzok, D. (1986). Metabolic and hormonal correlates of molting and regeneration of pelage in Alaskan harbor and spotted seals (*Phoca vitulina* and *Phoca largha*). *Can. J. Zool.* **64**, 1086–1094.

Atkinson, S. (1997). Reproductive biology of seals. *Rev. Reprod.* **2**, 175–194.

Atkinson, S., and Gilmartin, W. G. (1992). Seasonal testosterone pattern in Hawaiian monk seals (*Monachus schauinslandi*). *J. Reprod. Fert.* **96**, 35–39.

Atkinson, S., and Yoshioka, M. (2007). Endocrinology of reproduction. *In* "Reproductive Biology and Phylogeny of Cetacea" (B. G. M. Jamieson, ed.), Vol. 7, pp. 171–192. Science Publishers, Enfield, NH.

Atkinson, S., Combelles, C., Vincent, D., Nachtigall, P., Pawloski, J., and Breese, M. (1999). Monitoring of progesterone in captive female false killer whales (*Pseudorca crassidens*). *Gen. Comp. Endo.* **115**, 323–332.

Boily, P. (1996). Metabolic and hormonal changes during the molt of captive gray seals (*Halichoerus grypus*). *Am. J. Physiol.* **270**(5 Pt 2), R1051–R1058.

Boyd, I. L. (1991). Changes in plasma progesterone and prolactin concentrations during the annual cycle and the role of prolactin in the maintenance of lactation and luteal development in the Antarctic fur seal (*Arctocephalus gazella*). *J. Reprod. Fert.* **91**(2), 637–647.

Bryden, M. M. (1994). Endocrine changes in newborn southern elephant seals. *In* "Elephant Seals: Population Ecology, Behavior and Physiology" (B. J. LeBoeuf, and R. M. Laws, eds), pp. 387–397. University of California Press, Berkeley, CA.

Champagne, C. D., Houser, D. S., and Crocker, D. E. (2005). Glucose production and substrate cycle activity in a fasting adapted animal, the northern elephant seal. *J. Exp. Biol.* **208**, 859–868.

Champagne, C. D., Houser, D. S., and Crocker, D. E. (2006). Glucose metabolism during lactation in a fasting animal, the northern elephant seal. *Am. J. Physiol. Regul. Integr. Comp. Physiol.* **291**, R1129–R1137.

Davis, R. W., Castellini, M. A., Kooyman, G. L., and Maue, R. (1983). Renal glomerular filtration rate and hepatic blood flow during voluntary diving in Weddell seals. *Am. J. Physiol. Regul. Integr. Comp. Phsyiol.* **245**, R743–R748.

Eichelberger, L., Leiter, L., and Geiling, E. M. K. (1940). Water and electrolyte content of dolphin kidney and extraction of pressor substance (renin). *Proc. Soc. Exp. Biol. Med.* **44**, 356–359.

Engelhardt, F. R., and Ferguson, J. M. (1980). Adaptive hormone changes in harp seals (*Phoca groenlandica*), and gray seals (*Halichoerus grypus*), during the postnatal period. *Gen. Comp. Endocrinol.* **40**, 434–445.

Engelhard, G. H., Brasseur, S. M., Hall, A. J., Burton, H. R., and Reijnders, P. J. (2002). Adrenocortical responsiveness in southern elephant seal mothers and pups during lactation and the effect of scientific handling. *J. Comp. Physiol. B* **172**, 315–328.

Gardiner, K. J., and Hall, A. J. (1997). Diel and annual variation in plasma cortisol concentrations among wild and captive harbor seals (*Phoca vitulina*). *Can. J. Zool.* **75**, 1773–1780.

Guilherme, C., Bianchini, A., Martinez, P. E., Robaldo, R. B., and Colares, E. P. (2004). Serum leptin concentration during the terrestrial phase of the Southern elephant seal *Mirounga leonine* (*Carnivora: Phocidae*). *Gen. Comp. Endocrinol.* **139**, 137–142.

Guinet, C., Servera, N., Mangin, S., Georges, J. Y., and Lacroix, A. (2004). Change in plasma cortisol and metabolites during the attendance period ashore in fasting lactating subantarctic fur seals. *Comp. Biochem. Physiol. A Mol. Integr. Physiol.* **137**, 523–531.

Hammond, J. A., Bennett, K. A., Walton, M. J., and Hall, A. J. (2005). Molecular cloning and expression of leptin in gray and harbor seal blubber, bone marrow, and lung and its potential role in marine mammal respiratory physiology. *Am. J. Physiol. Regul. Integr. Comp. Physiol.* **289**, R545–R553.

Haulena, M., St. Aubin, D. J., and Duignan, P. J. (1998). Thyroid hormone dynamics during the nursing period in harbour seals, *Phoca vitulina*. *Can. J. Zool.* **76**, 48–55.

Hochachka, P. W. *et al.* (9 authors) (1995). Hormonal regulatory adjustments during voluntary diving in Weddell seals. *Comp. Biochem. Physiol. B Biochem. Mol. Biol.* **112**, 361–375.

Hong, S. K., Elsner, R., Claybaugh, J. R., and Ronald, K. (1982). Renal functions of the Baikal seal (*Pusa sibirica*) and ringed seal (*Pusa hispida*). *Physiol. Zool.* **55**, 289–299.

Houser, D. S., Crocker, D. E., Webb, P. M., and Costa, D. P. (2001). Renal function in suckling and fasting pups of the northern elephant seal. *Comp. Biochem. Physiol.* **A129**, 405–415.

Keith, E. O., and Ortiz, C. L. (1989). Glucose kinetics in neonatal elephant seals during postweaning aphagia. *Mar. Mamm. Sci.* **5**, 99–115.

Kirby, V. L., and Ortiz, C. L. (1994). Hormones and fuel regulation in fasting elephant seals. *In* "Elephant Seals: Population, Ecology, Behavior, and Physiology" (B. J. Le Boeuf, and R. M. Laws, eds), pp. 374–386. University California Press, Berkeley, CA.

Liggins, G. C., France, J. T., Schneider, R. C., Knox, B. S., and Zapol, W. M. (1993). Concentrations, metabolic clearance rates, production rates and plasma binding of cortisol in Antarctic (*phocid seals*). *Acta. Endocrinol. (Copenh)* **129**, 356–359.

Malvin, R. L., Ridgway, S., and Cornell, L. (1978). Renin and aldosterone levels in dolphins and sea lions. *Proc. Soc. Exp. Biol. Med.* **157**, 665–668.

Mashburn, K. M., and Atkinson, S. (2004). Evaluation of adrenal function in serum and feces of Steller sea lions (*Eumetopias jubatus*): Influences of molt, gender, sample storage, and age on glucocorticoid metabolism. *Gen. Comp. Endo.* **136**, 371–381.

Mashburn, K., and Atkinson, S. (2007). Seasonal and predator influences on adrenal function in adult Steller sea lions: Gender matters. *Gen. Comp. Endoc.* **150**, 246–252.

Mashburn, K. L., and Atkinson, S. (2008). Variability in leptin and adrenal response in juvenile Steller sea lions (*Eumetopias jubatus*) to adrenocorticotropic hormone (ACTH) in different seasons. *Gen. Comp. Endocrinol.* **155**, 352–358.

Murdaugh, H. V., Jr., Schmidt-Nielsen, B., Wood, J. W., and Mitchell, W. L. (1961). Cessation of renal function during diving in the trained seal (*Phoca vitulina*). *J. Cell. Comp. Physiol.* **58**, 261–265.

Myers, M. J., Rea, L. D., and Atkinson, S. (2006). The effects of age, season and geographic region on thyroid hormones in Steller sea lions (*Eumetopias jubatus*). *Comp. Biochem. Physiol. Part A* **145**, 90–98.

Oki, C., and Atkinson, S. (2004). Diurnal patterns of cortisol and thyroid hormones in the harbor seal (*Phoca vitulina*) during summer and winter seasons. *Gen. Comp. Endo.* **136**, 289–297.

Ortiz, R. M., and Worthy, G. A. J. (2000). Effects of capture on plasma adrenal steroids and vasopressin levels in free-ranging bottlenose dolphins (*Tursiops truncatus*). *Comp. Biochem. Physiol.* **125A**, 317–324.

Ortiz, R. M., Worthy, G. A. J., and MacKenzie, D. S. (1998). Osmoregulation in wild and captive West Indian manatees (*Trichechus manatus*). *Physiol. Zool.* **71**, 449–457.

Ortiz, R. M., Worthy, G. A. J., and Byers, F. M. (1999). Estimation of water turnover rates of captive West Indian manatees (*Trichechus manatus*) held in fresh and salt water. *J. Exp. Biol.* **202**, 33–38.

Ortiz, R. M., MacKenzie, D. S., and Worthy, G. A. J. (2000a). Thyroid hormone concentrations in captive and free-ranging West Indian manatees (*Trichechus manatus*). *J. Exp. Biol.* **203**, 3631–3637.

Ortiz, R. M., Wade, C. E., and Ortiz, C. L. (2000b). Prolonged fasting increases the response of the renin–angiotensin–aldosterone system, but not vasopressin levels, in postweaned northern elephant seal pups. *Gen. Comp. Endocrinol.* **119**, 217–223.

Ortiz, R. M., Wade, C. E., and Ortiz, C. L. (2001a). Effects of prolonged fasting on plasma cortisol and TH in post weaned northern elephant seal pups. *Am. J. Physiol.* **280**, R790–R795.

Ortiz, R. M., Noren, D. P., Litz, B., and Ortiz, C. L. (2001b). A new perspective on adiposity in a naturally obese mammal. *Am. J. Physiol. Endocrinol. Metabol.* **281**, E1347–E1351.

Ortiz, R. M., Wade, C. E., Costa, D. P., and Ortiz, C. L. (2002a). Renal effects of fresh water-induced hypo-osmolality in a marine adapted seal. *J. Comp. Physiol. B* **172**, 297–307.

Ortiz, R. M., Wade, C. E., Costa, D. P., and Ortiz, C. L. (2002b). Renal responses to plasma volume expansion and hyperosmolality in fasting seal pups. *Am. J. Physiol. Regul. Inter. Comp. Physiol.* **282**, R805–R817.

Ortiz, R. M., Houser, D. S., Wade, C. E., and Ortiz, C. L. (2003a). Hormonal changes associated with the transition between nursing and natural fasting in northern elephant seals (*Mirounga angustirostris*). *Gen. Comp. Endocrinol.* **130**, 78–83.

Ortiz, R. M., Noren, D. P., Ortiz, C. L., and Talamantes, F. (2003b). Ghrelin and growth hormone increase with fasting in a naturally adapted species, the northern elephant seal. *J. Endocrinol.* **178**, 83–89.

Ortiz, R. M., Wade, C. E., Ortiz, C. L., and Talamantes, F. (2003c). Acutely elevated vasopressin increases circulating concentrations of cortisol and aldosterone in fasting northern elephant seal (*Mirounga angustirostris*) pups. *J. Exp. Biol.* **206**, 2795–2802.

Ortiz, R. M., Crocker, D. E., Houser, D. S., and Webb, P. M. (2006). Angiotensin II and aldosterone increase with fasting in breeding adult male northern elephant seals (*Mirounga angustirostris*). *Physiol. Biochem. Zool.* **79**, 1106–1112.

Petrauskas, L., Tuomi, P., and Atkinson, S. (2006). Noninvasive monitoring of stress hormone levels in a female Steller sea lion (*Eumetopias jubatus*) pup undergoing rehabilitation. *J. Zoo. Wildl. Med.* **37**(1), 75–78.

Pietraszek, J., and Atkinson, S. (1994). Concentrations of estrone sulfate and progesterone in plasma and saliva, vaginal cytology, and bioelectric impedance during the estrous cycle of the Hawaiian monk seal (*Monachus schaunslandi*). *Mar. Mamm. Sci.* **10**, 430–441.

Richmond, J. P., Burns, J. M., Rea, L. D., and Mashburn, K. L. (2005). Postnatal ontogeny of erythropoietin and hematology in free-ranging Steller sea lions (*Eumetopias jubatus*). *Gen. Comp. Endocrinol.* **141**, 240–247.

Robeck, T. R., Schneyer, A. L., McBain, J. F., Dalton, L. M., Walsh, M. T., Czekala, N. M., and Kraemer, D. C. (1993). Analysis of urinary immunoreactive steroid metabolites and gonadotropins for characterization of estrous cycle, breeding period and seasonal estrous activity of captive killer whales (*Orcinus orca*). *Zoo Biol.* **12**, 173–187.

Skog, E. B., and Folkow, L. P. (1994). Nasal heat and water exchange is not an effector mechanism for water balance regulation in grey seals. *Acta Physiol. Scand.* **151**, 233–240.

St. Aubin, D. J. (2001). Endocrinology. *In* "CRC Handbook of Marine Mammal Medicine" (L. A. Dierauf, and F. M. D. Gulland, eds), 2nd Ed, pp. 165–192. CRC Press, Boca Raton, FL.

St. Aubin, D. J., and Geraci, J. R. (1986). Adrenocortical function in pinniped hyponatremia. *Mar. Mamm. Sci.* **2**, 243–250.

St. Aubin, D. J., and Dierauf, L. A. (2001). Stress and marine mammals. *In* "CRC Handbook of Marine Mammal Medicine" (L. A. Dierauf, and F. M. Gulland, eds), 2nd Ed, pp. 253–269. CRC Press, Boca Raton, FL.

St. Aubin, D. J., Ridgway, S. H., Wells, R. S., and Rhinehart, H. (1996). Dolphin thyroid and adrenal hormones: Circulating levels in wild and semidomesticated *Tursiops truncatus*, and influence of sex, age, and season. *Mar. Mamm. Sci.* **12**, 1–13.

Stokkan, K. A., Vaughan, M. K., Reiter, R. J., Folkow, L. P., Martensson, P. E., Sager, G., Lydersen, C., and Blix, A. S. (1995). Pineal and thyroid functions in newborn seals. *Gen. Comp. Endocrinol.* **98**, 321–331.

Theodorou, J., and Atkinson, S. (1998). Monitoring of total androgen concentrations in saliva from captive Hawaiian monk seals (*Monachus schuainslandi*). *Mar. Mamm. Sci.* **14**(2), 304–310.

Tsubokawa, M., Kawauchi, H., and Li, C. H. (1980). Isolation and partial characterization of growth hormone from fin whale pituitary glands. *Int. J. Biochem. (Tokyo)* **88**, 1407–1412.

Wallis, O. C., Maniou, Z., and Wallis, M. (2005). Cloning and characterization of the gene encoding growth hormone in finback whale (*Balaenoptera physalus*). *Gen. Comp. Endocrinol.* **143**, 92–97.

West, K. L., Atkinson, S., Carmichael, M. J., Sweeney, J. C., Krames, B., and Krames, J. (2000). Concentrations of progesterone in milk from bottlenose dolphin during different reproductive status. *Gen. Comp. Endo.* **117**, 218–224.

Yoshioka, M., Mohri, E., Tobayama, T., Aida, K., and Hanyu, I. (1986). Annual changes in serum reproductive hormone levels in the captive bottlenosed dolphins. *Bull. Japan Soc. Sci. Fish.* **52**, 1939–1946.

Zapol, W. M., Liggins, G. C., Schneider, R. C., Qvist, J., Snider, M. T., Creasy, R. K., and Hochachka, P. W. (1979). Regional blood flow during simulated diving in the conscious Weddell seal. *J. Appl. Physiol.* **47**, 968–973.

Zenteno-Savin, T. and Castellini, M. A. (1998). Changes in the plasma levels of vasoactive hormones during apnea in seals. *Comp. Biochem. Physiol.* **119C**, 7–12.

E

# Energetics

### DANIEL P. COSTA

## I. Introduction

Energetics provide a method to quantitatively assess the effort animals spend acquiring resources, as well as the relative way in which they allocate those resources. Energy flow models are analogous to cost–benefit models used in economics. Costs take the form of energy expended to acquire and process prey, and to maintain body functions. The energetic benefits are manifest as food energy used for growth and reproduction. Measurement of energy acquisition and allocation provide a quantitative assessment of how animals organize their daily or seasonal activities, and how they prioritize their behaviors. Thus, energy flow can be described as what goes into the animal as food and what comes out in the form of growth, reproduction, repair, waste, or metabolic work. Survival requires a positive balance between the costs of maintenance and the acquisition of food energy. If a marine mammal cannot compensate for decreases in energy acquisition, it must either reduce its overall rate of energy expenditure or utilize stored energy reserves. Conversely, in order to grow and reproduce, animals must obtain more energy than is needed to survive. Marine mammals undergo profound variations in this feast or famine dynamic equilibrium as they can gain significant amounts of food energy while feeding in highly productive environments, followed by prolonged negative energy balance while fasting during migration or reproduction (Brodie, 1975; Costa, 1993; Lockyer, 1993) (Fig. 1).

The balance of how energy acquisition and expenditure is achieved differs for individual species and environments. For

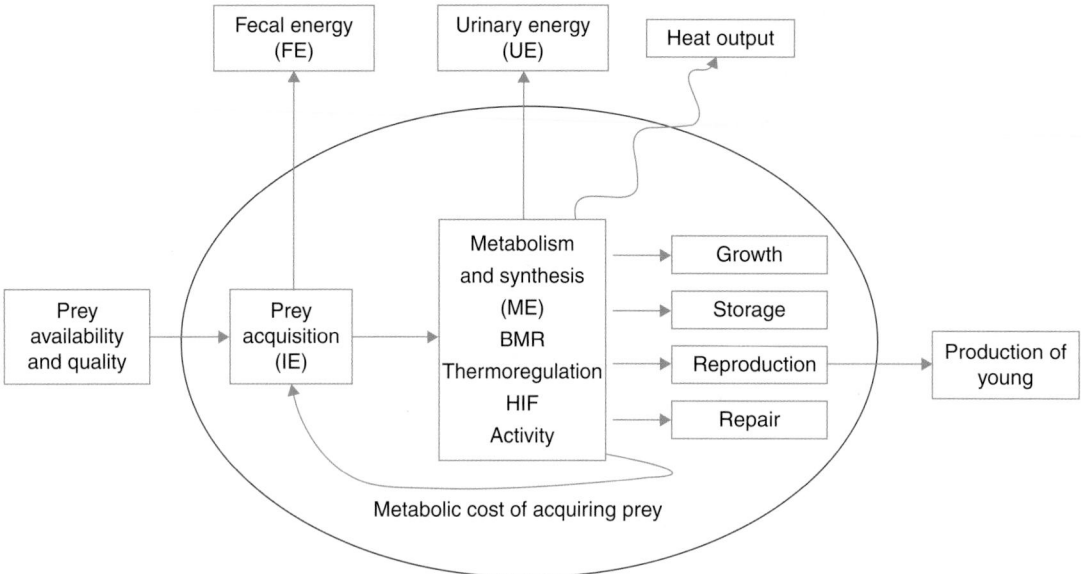

**Figure 1** *A conceptual diagram of energy flow through a typical mammal. The outer oval represents the animal. Anything that passes through that envelope is material (straight line) or energy (waved line) entering or leaving the animal. IE, ingested energy; ME, metabolic energy; FE, fecal energy; and UE, urinary energy.*

**Figure 2** *A minke whale,* Balaenoptera acutorostrata, *feeding in the Antarctic ice. Photograph by Dan Costa.*

some species, sea otters (*Enhydra lutris*), sea lions, and fur seals (Otariidae), very high rates of energy expenditure are met by high rates of energy acquisition (Costa and Williams, 2000). These animals preferentially live in nearshore environments or upwelling regions where food is abundant (Costa, 1993). Sirenians represent the opposite extreme. These marine mammals exhibit comparatively low existence costs and are able to survive on a low-quality diet. They have adapted to a diet of grasses that is in high abundance but of low quality, energetically. They are able to do this because; they live in the climatically benign tropics where maintenance costs are low (Gallivan and Best, 1980; Gallivan *et al.*, 1983; Irvine, 1983).

The seasonal migrations of large cetaceans demonstrate this interrelationship between energetic demand, energy availability, and local productivity (Brodie, 1975). Although maintenance costs may be elevated in polar regions, the ability to take advantage of the seasonally high productivity associated with the sea ice during the polar summer more than compensates (Fig. 2). When confronted with the high energetic costs of reproduction and of winter conditions,

the mysticete whales opt for the more benign tropics. Further, their large body size makes the cost of migration extremely low (see Locomotion, this volume). While prey availability may be low in the tropics, so too are the existence costs, especially for a large animal that is able to utilize energy reserves stored in the blubber.

A conceptual diagram of the relationship between energy acquisition and allocation is shown in Fig. 1. The rate of energy consumed by the animal is referred to as Ingested Energy (IE). The energy that remains after the losses associated with the production of feces and urine is the Metabolic Energy (ME). This is the energy available for maintenance and repair, growth or reproduction. Energy expended for maintenance includes key processes such as basal metabolism, digestion (heat increment of feeding, HIF), thermoregulation, and activity (locomotion, grooming, feeding, etc.). The rate of prey energy acquired is directly related to the availability and quality of prey. As prey becomes less available, the cost of finding it increases and the animal spends a greater proportion of its time and therefore energy searching for it. Eventually, there is a threshold when more energy is spent searching for prey than is obtained and the animal goes into negative energy balance (Winship *et al.*, 2002; Rosen and Trites, 2005). While the best situation is to have access to abundant high energy prey, in some scenarios low-quality prey that is more abundant may be more optimal than searching for high-quality prey that is difficult to find.

## II. Energy Acquisition

Not all of the ingested material consumed is digestible. Food energy remaining after digestion and elimination of Fecal Energy (FE) is known as the Apparently Digested Energy (ADE). The proportion of ADE to IE is called the assimilation efficiency and ranges from 88% to 97.9%, for a diet of fish to 72.2% for invertebrate prey with a high chitin content (Martensson *et al.*, 1994; Lawson *et al.*, 1997; Costa and Williams, 2000; Rosen and Trites, 2000). The assimilation efficiency decreases as the rate of prey intake increases, but is greater when a diet composed of different species of fish with different proximate compositions is consumed (Trumble and Castellini, 2005). Sirenians extract less energy from their food than other marine mammals (84.6%), because plant material, which contains cellulose and requires bacterial fermentation to digest, is harder to digest. However, they are more efficient than other hindgut fermenters, such as horses, *Equus caballus* (45–59%) (Burn, 1986).

Chemical energy lost as urea and other metabolic end products in the urine is defined as Urinary Energy (UE). Metabolizable Energy (ME) is the net energy remaining after fecal and urinary energy loss, and represents the energy available for growth or reproduction and for supporting metabolic processes such as work (locomotion) and respiration (thermoregulation, maintenance metabolism, HIF). The ME for pinnipeds varies between 78.3% for a squid diet to 91.6% for an anchovy diet (Costa and Williams, 2000).

## III. Energy Expenditure

### A. Cost of Maintenance Functions

Maintenance costs are those associated with homeostasis and include basal metabolism, HIF, repair (molt, fighting disease, and/or parasites), thermoregulation, and activity (see Thermoregulation and Locomotion, this volume).

*1. Basal Metabolism* It has generally been assumed that the basal metabolic rates of aquatic mammals are elevated when compared to terrestrial mammals of similar size. The current view of basal metabolic rates of marine mammal is more complex as many studies did not conform to standardized criteria for measurements of basal metabolism (Lavigne *et al.*, 1986). These criteria require that the subjects be adults, resting, thermoneutral, non-reproductive, and post-absorptive. This has been further confused by the expectation that all marine mammals should employ the same metabolic response. Specialization for marine living has occurred independently in three mammalian orders: the sirenians, cetaceans, and carnivores. Further, within the carnivores there are three separate transitions to a marine existence: pinnipeds, sea otters, and polar bears, *Ursus maritimus*. Based on this diversity, we might expect different metabolic adaptations between the groups (Fig. 3). West Indian manatees, *Trichechus manatus*, have BMRs lower than values predicted, while phocid seals have BMRs closer to those of similar sized terrestrial mammals. Conversely, sea otters, otariids, and odontocetes appear to have BMRs greater than terrestrial mammals of equal size.

The BMR of an animal is not constant, but varies seasonally (Rosen and Renouf, 1995; Williams *et al.*, 2007), with the animal's nutritional state (Rosen and Trites, 1999), as well as with the animal's body composition (Rea and Costa, 1992). Some species such as sirenians and walrus, *Odobenus rosmarus*, have dense bone, whereas seals may be composed of as much as 50% fat. When metabolic rates are expressed relative to body mass, a disproportionate amount of fat or particularly dense bone will lower the apparent metabolic rate. This is due to the low metabolic rates of bone and adipose tissue in comparison to lean tissue. Many marine mammals undergo prolonged fasts that are accompanied by profound changes in body composition. Most of the mass change during fasting is due to loss of adipose tissue with a comparatively smaller change in lean tissue. For example, northern elephant seal females, *Mirounga angustirostris*, loose 42% of their initial mass, but of this only 14.9% comes from lean tissue with 57.9% coming from adipose tissue (Costa *et al.*, 1986). This results in an overall change in body composition of 39% fat at parturition to 24% fat at weaning (see Pinnipedia Physiology, this volume). Since lean mass is the primary contributor to whole

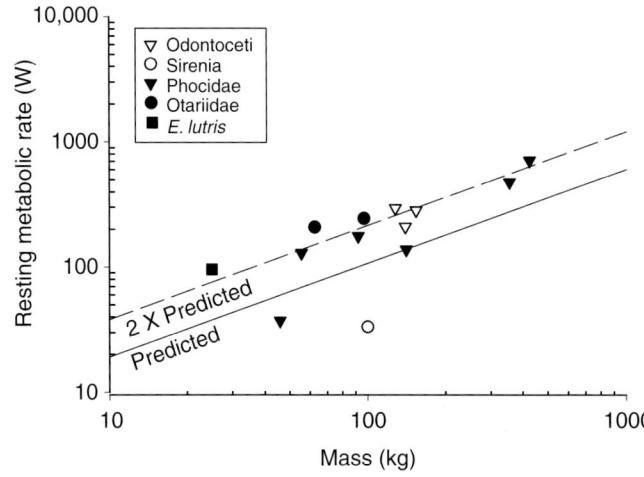

**Figure 3** *Resting metabolic rate (RMR) of marine mammals in relation to body mass. Measurements were made for the animals resting in water. The solid line denotes the predicted metabolic rate for equal sized terrestrial animals; the dashed line represents 2 times the predicted levels. Species included are: Odontoceti*—Stenella attenuata, Tursiops truncatus; *Phocidae*—Phoca vitulina, Phoca hispida, Leptonychotes weddellii, Pagophilus groenlandicus, Haliochoerus grypus; *Otariidae*—Zalophus californianus; *Sirenia*—Trichechus manatus.

animal metabolism, the animals whole body metabolism is likely to change little even though there has been a major change in its body mass (Rea and Costa, 1992).

The ability to digest and process the greater amount of prey associated with the higher metabolic rates of marine mammals may have also required changes in their morphology. Specifically, all carnivorous marine mammals, regardless of their ancestry (carnivore or herbivore) have comparatively longer small intestines than similarly sized terrestrial carnivores. Further, there is a high correlation between small intestine length and BMR in mammals (Williams *et al.*, 2001).

*2. Heat Increment of Feeding* When food is consumed, the animal's metabolic rate increases over fasting levels. The HIF, also known as the Specific Dynamic Action (SDA), may be considered the "tax" that is required to process food energy for conversion to Metabolizable Energy (ME). The magnitude of energy allocated to HIF varies between 5% and 17% of the ME (Costa and Kooyman, 1984; Markussen *et al.*, 1994; Rosen and Trites, 1997). In addition, the duration of HIF following a meal will depend on the amount of food consumed and its composition. In many mammals, the HIF is considered excess or waste heat. However, sea otters incorporate the additional heat produced from HIF to meet their high thermoregulatory costs associated with their small size (they are the smallest marine mammal) (Costa and Kooyman, 1984). While grooming, feeding, and swimming sea otters use the heat produced from activity to supplement their thermoregulatory needs, while at rest sea otters incorporate the heat produced from HIF to augment their thermoregulatory needs (Fig. 4).

*3. Fur vs Blubber* It is important to consider the potential differences in the energy budgets of animals that use fur or blubber for insulation. The overall time–energy budget of an animal that uses fur (fur seal or sea otter) is fundamentally different from an animal that uses blubber (sea lion, seal, or dolphin). Although fur is not a living

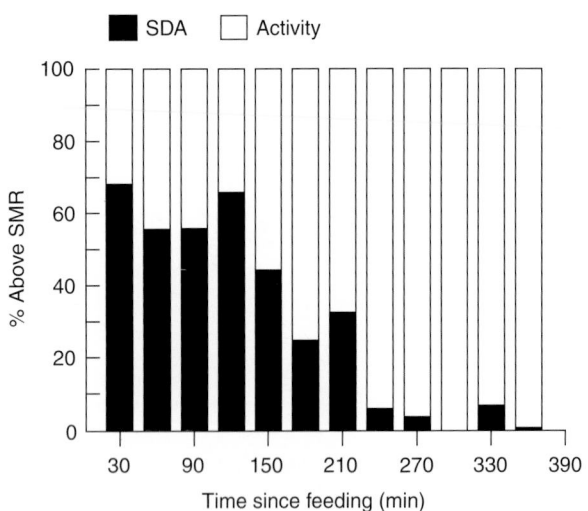

Figure 4    *The thermal budget of sea otters relies on heat production above BMR (or SMR). The contribution of HIF (or SDA) to the overall metabolism is highest immediately after feeding and reaches zero within 390 min after feeding. As HIF decreases the animal becomes more active and thus compensates for the decrease in heat production from HIF (Costa and Kooyman, 1984).*

tissue, it requires the maintenance of an air layer, which is done by frequent grooming. Sea otters spend up to 16% of their day grooming. While sea lions, seals and dolphins, spend no time grooming, they must take in sufficient food to lay down a thick blubber layer, which is a living tissue and must be supplied with blood. Furthermore, blubber serves dual roles as an insulator and as an energy store. During fasting or periods of low food availability a marine mammal must balance its utilization of blubber for energy needs with the potential loss of the blubber layer as an insulator. For pinnipeds a way of increasing blood flow to the skin and keeping thermoregulatory costs low while ashore is to huddle together. This reduces their effective surface area that is exposed to the cold (Fig. 5). Such behavior is commonly observed in cold climates and can change during the day. For example, in the morning when it is cold, animals clump together and as it gets warmer they separate. Finally, huddling behavior is less common in fur seals, as there isn't much of an incentive to huddle if you use fur for insulation.

## B. Cost of Growth and Reproduction

For growth and reproduction to occur, an animal must acquire energy and nutrients in excess of that required for supporting maintenance functions. These additional energetic costs vary with the species of marine mammal, the sex, and reproductive pattern. In pinnipeds, polar bears, sea otters (and probably mysticetes and sirenians) the cost of reproduction in males is limited to the cost of finding and maintaining access to estrous females. Evolution favors a pattern of energy expenditure that maximizes reproductive success in males. The costs associated with reproduction in aquatic and terrestrially breeding males is quite similar when normalized for differences in body mass. Larger body size is preferred in terrestrially breeding male pinnipeds since it confers both an advantage in fighting and allows the male to maintain terrestrial territories longer (Fig. 6) (see Pinnipedia Physiology, this volume). In addition, larger animals can fast longer because they have a lower mass specific metabolic rate than smaller animals (Costa, 1993). In species that compete

Figure 5    *A group of California sea lions,* Zalophus californianus, *huddling on a California beach. A behavior typical in the cool winter months or during the cool mornings. Photograph by Dan Costa.*

Figure 6    *The extreme difference in body size between a male (on top) and female (underneath) northern elephant seal (*Mirounga angustirostris*). The animals are copulating in this picture. Photograph by Dan Costa.*

for females in the water, males are comparatively smaller than the species that breed on land. For the aquatic breeders, underwater agility is more important than large size when competing for mates.

The cost of reproduction for females can be broken down into the energetic requirements of gestation and lactation. The cost of gestation is small relative to the cost of lactation. Even given the strikingly different reproductive patterns in marine mammals, there is little variation in fetal mass at birth among marine mammals, but as a group they appear to invest more energy into fetal birth mass (and thus more into gestation) than terrestrial mammals (Fig. 7). This higher investment in gestation by all marine mammals except polar bears is associated with the production of precocial young (Fig. 8). The young of cetaceans, sirenians, pinnipeds, and sea otters need to be capable of dealing with life in the water or on a crowded rookery within seconds of birth. As a group marine mammals exhibit considerable variation in both the duration and pattern of maternal investment (Fig. 9). Phocid seals and mysticete whales have extremely short lactation durations, which are compensated for by higher rates of energy transfer that enable the young to grow rapidly (Fig. 10).

Although phocid pups are weaned early, they still rely on maternally derived energy, stored as blubber, for weeks or months after weaning. The disadvantage of this rapid growth is that most of the mass and energy is stored as fat with proportionately little protein. The advantage of longer lactation is that young get more protein and other nutrients allowing greater growth of lean tissue. However, longer lactation is energetically more expensive (Costa, 1991a, b, 1993).

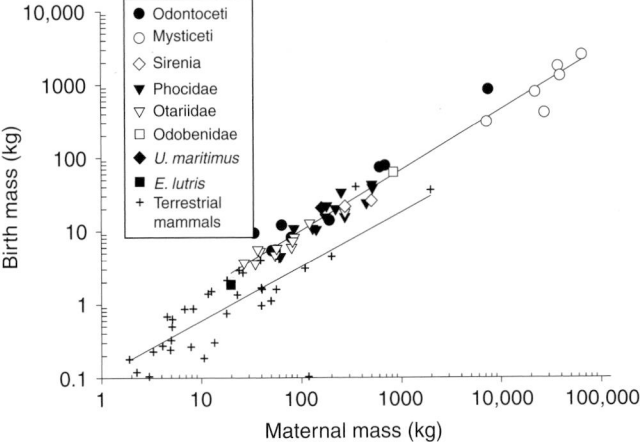

**Figure 7** *Birth mass plotted in relation to maternal mass for marine and terrestrial mammals. Species of marine mammals included are: Odontoceti*—Inia geoffrensis, Pontoporia blainvillei, Stenella attenuata, Globicephala melaena, Physeter macrocephalus, T. truncatus, Phocoena spinnipinnis, P. phocoena, Delphinapterus leucas; *Mysticeti*—Balaenoptera musculus, B. physalus, B. acutorostrata, B. borealis, Megatpera novaengliae, Eschrictius robustus; *Phocidae*—Mirounga angustirostris, M. leonina, Cystophora cristata, Phoca vitulina, P. hispida, Leptonychotes weddelli, Monachus schauinslandi, Pagophilus groenlandicus, Erignathus barbatus, Lobodon carcinophagus, Histriophoca fasciata, Haliochoerus grypus; *Otariidae*—Arctocephalus gazella, A. forsteri, A. galagagoensis, A. tropicalis, A. pusillis, Callorhinus ursinus, Zalophus californianus, Neophoca cinerea, Eumatopias jubatus, Otaria byronia; *Sirenia*—Dugong dugon, Trichechus manatus.

**Figure 8** *A harbor seal* (Phoca vitulina)*, mother and pup on a California beach. The pup was recently born and shows the extreme level of precociality typical for marine mammals. In harbor seals the pup is born with the adult pelage and it can go to sea within an hour of birth. Photograph by Dan Costa.*

*1. Variation in Milk Composition* The rapid growth of marine mammal young is made possible by the ingestion of extremely lipid rich milk. With a few exceptions, terrestrial animals produce milk that is low in fat; cows, *Bos taurus* and humans, *Homo sapiens*, produce milk that contains 3.7% and 3.8% milk fat, respectively. Lipid-rich milk allows the mother to transfer high levels of energy in a very short period. Hooded seals, *Cystophora cristata*, are most impressive with a 4-day lactation interval and a milk fat of 65% lipid (Bowen *et al.*, 1985). In view of this, it is not surprising that marine mammals with the highest growth rates produce milk with the highest lipid content.

Lactation also enables mothers to optimize the delivery of energy to their young. The energy content of the milk is independent of the type or quality of prey consumed, or the distance or time taken to obtain it. Although milk is ultimately derived from the prey consumed, a mother can process, concentrate, or utilize stored reserves to produce milk. For example, some species feed on fish, while others feed on fish or squid. Yet, all of these species provision their offspring with milk of significantly greater energy density than the prey consumed (Costa, 1991b).

*2. Body Size and Maternal Resources: The Role of Maternal Overhead* Fasting during lactation is a unique component of the

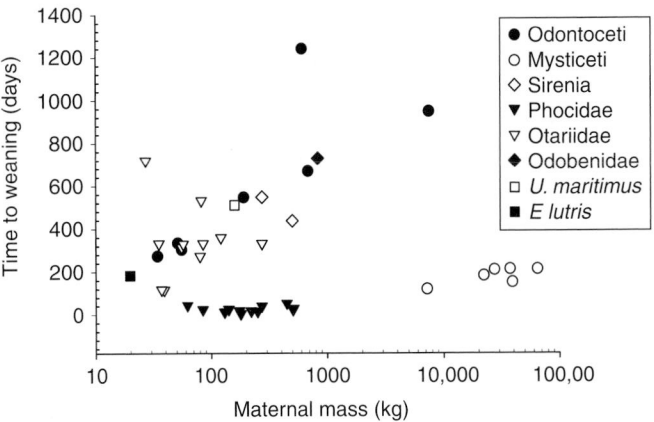

**Figure 9** *Time to weaning plotted as a function of maternal mass for marine mammals. Lactation durations of phocid seals and mysticete whales are shorter than in all other marine mammals. Species are same as in Fig. 3.*

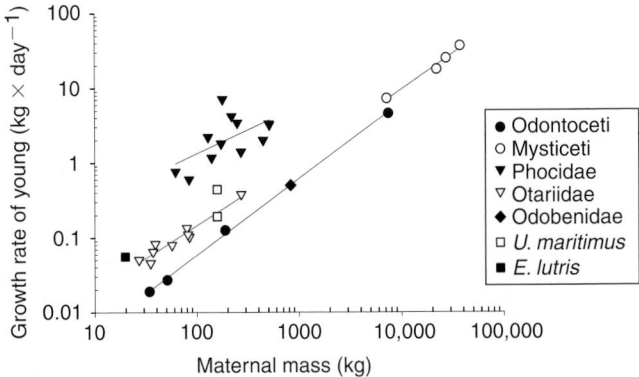

**Figure 10** *Growth rate of suckling marine mammals as a function of maternal mass. Lines represent least squares regressions for each taxonomic group. Species are same as in Fig. 3.*

life history pattern of marine mammals (Costa, 1993; Oftedal, 2000). With the exception of bears, no other mammal is capable of producing milk without feeding. By undertaking this energetic challenge, mysticetes and pinnipeds are able to separate where and when they feed from where and when they breed. In mysticete whales, this allows them to feed in the highly productive polar regions of the world's oceans, but retain the thermal advantage of breeding in the calm tropical regions (Fig. 2) (Brodie, 1975). Migrating to warmer waters for parturition reduces the thermal demands on the newborn calf and additional thermal savings for the mother.

**Figure 11** *A recently born northern elephant seal* (Mirounga angustirostris) *pup suckles from its' mother (below) is compared to a recently weaned pup (30 days old). The mother does not eat or drink during the 26- to 28-day lactation interval, and after weaning the pup fasts for 2–3 months before going to sea. Elephant seals, like many true seals, fast during the entire lactation interval. Photograph by Dan Costa.*

Among pinnipeds, the separation of feeding from lactation is necessary to allow for terrestrial parturition (Bartholomew, 1970). Most phocids store sufficient energy reserves for the entire lactation period, whereas all otariids must feed during lactation (see Pinnipedia Physiology and Pinniped Reproduction, this volume) (Costa, 1991a, b). A phocid mother typically remains on or near the rookery continuously from the birth of the pup until it is weaned; whereas milk is produced from body reserves stored prior to parturition (Fig. 11). Although some phocids feed during lactation, most of the maternal investment is derived from body stores. Their reproductive pattern is less constrained by the time it takes to travel and exploit distant prey, which allows utilization of a more dispersed or patchy food resource (Costa, 1993). By spreading out the acquisition of prey energy required for lactation over many months at sea, northern elephant seal females only need to increase their daily food intake by 12% to cover the entire cost of lactation.

The ability of a marine mammal female to fast while providing milk to her offspring is related to the size of her energy and nutrient reserves and the rate at which she utilizes them. When food resources are far from the breeding grounds, as may occur for some phocids and large mysticete whales, the optimal solution is to maximize the amount of energy and nutrients provided to the young and to minimize the amount of energy expended on the mother. The term "metabolic overhead" refers to the amount of energy a female expends on herself while onshore (seals) or while in the calving grounds (whales). Larger females have a lower metabolic overhead than smaller females. This is because maintenance metabolism scales as mass$^{0.75}$, and fat stores scale as mass$^{1.0}$. As body size increases, energy reserves increase proportionately faster than maintenance metabolism.

*3. Energy Investment and Trip Duration*   Many phocids fast throughout the lactation interval, whereas otariid females feed intermittently between suckling bouts onshore (Fig. 12) (Costa, 1991a, b). Otariid mothers modify the timing and patterning of energy and nutrient investment to optimize energy delivery to their young (Boyd, 1998; Trillmich and Weissing, 2006). Otariid mothers making short feeding trips that provide their pups with less milk energy than mothers that make long trips. In comparison to otariids, phocids may have a reproductive pattern that is better suited for dealing

**Figure 12**   *A Galapagos sea lion* (Zalophus wollebaeki) *female suckling her pup on left, and on a trip to sea on right. Fur seals and sea lions intermittently suckle their pup on shore between trips to sea to forage. Photograph by Dan Costa.*

with dispersed or unpredictable prey, or prey that is located at great distances from the rookery (Costa, 1993). However, fasting during lactation places a limit on the duration of investment and this limits the total amount of energy that a phocid mother can invest in her pup.

Phocids are buffered from short-term fluctuations in prey availability due to their unique reproductive pattern. In phocids, reproductive performance (maternal investment) during a given season reflects prey availability over the preceding year and represents the mother's foraging activities over a much larger spatial and temporal scale than the foraging activities of otariids (Costa, 1993). It follows that the weaning mass of a phocid pup is an indicator of the mother's foraging success over the previous year, whereas the subsequent post-weaning survival of the pup is related to both its weaning mass (energy reserves provided by the mother) and the resources available to the pup after weaning.

## C. Field Metabolic Rates

A number of approaches have been used to study the metabolic rate of animals at sea. One approach, time budget analysis, sums the daily metabolic costs associated with various activities (Williams *et al.*, 2004). Other methods rely on predictive relationships between physiological variables and metabolic rate. For example, metabolic costs can be indirectly assessed by measurements of changes in body mass and composition, variations in heart rate or ventilation rate, or with the dilution of isotopically labeled water (Folkow and Blix, 1992; Boyd *et al.*, 1999; Costa and Gales, 2000).

Field metabolic rates (FMR) provide insight into the energetic strategies used by marine mammals (Costa, 1993; Costa and Gales, 2003). The best data exist for pinnipeds and the common bottlenose dolphin, *Tursiops truncatus* (Fig. 13), and indicate that foraging otariids and bottlenose dolphins expend energy at 6 times the predicted basal metabolic level (Costa and Williams, 2000). In contrast, the metabolic rate of diving elephant seals (*Mirounga* spp.) and Weddell, *Leptonychotes weddellii*, seals are only 1.5–3 times the predicted basal rate (Castellini *et al.*, 1992). The lower diving metabolic rate of phocid seals contributes to their superb diving ability (Costa, 1993) (see Diving Physiology, this volume). The importance of the thermal environment on field metabolic rate can also be seen in Galapagos fur seals, *Arctocephalus galapagoensis*, which due to the warm equatorial climate have a substantially reduced field metabolic rate compared to other otariids (Trillmich and Kooyman, 2001). An interesting consequence of the high metabolic rate of marine mammals is that the presence of a few foraging individual can have a significant impact on community structure (Estes *et al.*, 1998; Springer *et al.*, 2003).

FMR are quite variable both between and within species (Fig. 13). Such variation is thought to be associated with year-to-year changes in both the abundance and availability of prey (Costa, 2007). In response to reduced availability of prey, fur seals and sea lions mothers increased their foraging effort in an effort to keep the duration of their foraging trip the same. However, there reaches a point where they can no longer increase their foraging effort and have to spend more time at sea to obtain the same amount of prey energy. If a mother spends more time to deliver the same amount of energy, the offspring receives less overall energy. As a result, more of the offspring's energy is spent on maintenance and its growth will slow and in the worst case the pup will eventually die.

*1. Energetics of Prey Choice*  The amount of work, and therefore energy expenditure that an animal puts into locating prey varies

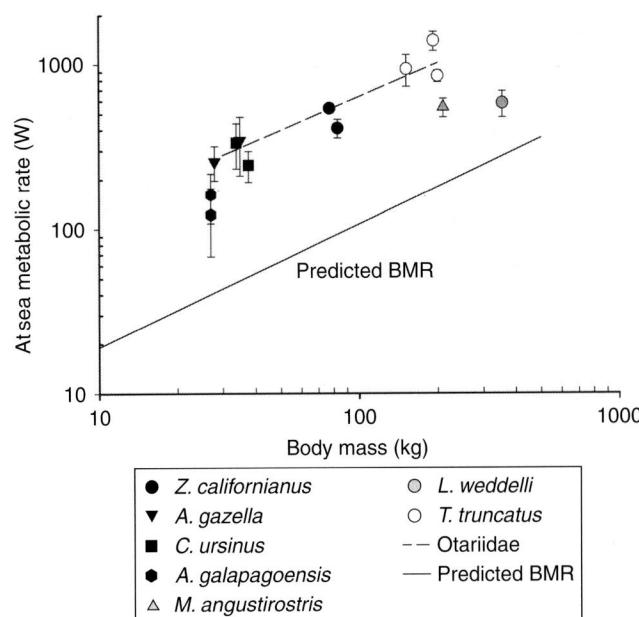

**Figure 13**  *At sea metabolic rate measurements determined from the O-18 doubly labeled water method as a function of body mass. Data on Weddell seals (*Leptonychotes weddellii*) were measured using open circuit respirometry on seals diving from an ice hole. The solid line represents the predicted basal metabolic rate for a terrestrial animal of equal size; the dashed line is the best-fit linear regression for the otariidae with the exception of the data from* Arctocephalus galapagoensis ($r^2 = 0.53$). *Error bars represent ± one standard deviation. Multiple points for each species reflect measurements taken over different years and show the range of interannual variation within a species.*

as a function of the energy content, availability, and location of the prey both geographically as well as its depth in the water column. Both size and proximate composition (fat, carbohydrate, protein, and water content) affect the energy content of prey. Prey availability varies as a function of the absolute abundance of prey (amount of prey per unit of habitat) and its distribution in the environment. A predator is more efficient when foraging on prey that is clumped than on prey that is evenly dispersed. Similarly, prey that is near the surface is easier to obtain than prey located at depth. Marine mammals forage in areas where prey has been concentrated as a result of oceanographic processes like eddies, fronts, and upwelling regions associated with bottom topography.

Sea otters provide an excellent example of the factors that determine the energetics of prey choice (Fig. 14) (Riedman and Estes, 1990). In recently occupied areas, sea otters feed on preferred prey items like clams, abalone, *Haliotis* spp., or sea urchins. In such environments they find large, energy-rich, abundant prey that is easy to handle, consume, and digest. In such situations, lower quality prey items (turban snails, sea stars, mussels, chitons) are generally not eaten. These items may be abundant, but they are energy poor, and difficult to eat and digest. As the abundance and size of their preferred prey declines, sea otters switch to less preferred but more accessible prey like turban snails, kelp crabs, and in some cases, chitons and sea stars. Some sea otters specialize on different types of prey and are more efficient predators than non-specialists (Estes *et al.*, 2003).

E

**Figure 14** *A California sea otter feeds on a crab. Photograph by Dan Costa.*

Polar bears represent another example of optimal prey choice and its relation to the prey energy quality. Feeding predominately on ring seals, *Phoca hispida*, polar bears eat the energy-rich blubber layer and leave behind the lean "core" of the carcass (Stirling and McEwan, 1975). Due to its high lipid content the blubber has a per unit mass energy content almost 10 times greater than that of the lean tissue of the ring seal. Thus, polar bears consume the most energy dense part of the ring seal and then move on to find another kill.

*2. Variations in Foraging Energetics* Different foraging behaviors are associated with different metabolic costs. For example, in sea lions benthic foraging is more expensive than epipelagic or near surface feeding (Costa and Gales, 2000, 2003). The gulping behavior of blue, *Balaenoptera musculus*, fin, *B. physalus*, and other whales of the family Balaenopteridae also appears to be quite costly due to the tremendous drag created as they open their enormous mouths to engulf entire schools of prey (Croll *et al.*, 2001). The blue whale finds a school of krill, and opens its mouth engulfing the entire school of krill. The whale then expels the water through its baleen plates, retaining the krill in its mouth. Consider how much drag, and thus increased effort, it takes to swim with an open mouth the size of a blue whale through the water!

## *See Also the Following Articles*

Diving Physiology ▪ Pinniped Physiology ▪ Thermoregulation

## *References*

Bartholomew, G. A. (1970). A model for the evolution of pinniped polygyny. *Evolution* **24**, 546–559.

Bowen, W. D., Oftedal, O. T., and Boness, D. J. (1985). Birth to weaning in 4 days: Remarkable growth in the hooded seal, *Cystophora cristata. Can. J. Zool.* **63**, 2841–2846.

Boyd, I. L. (1998). Time and energy constraints in pinniped lactation. *Am. Nat.* **152**, 717–728.

Boyd, I. L., Bevan, R. M., Woakes, A. J., and Butler, P. J. (1999). Heart rate and behavior of fur seals: Implications for measurement of field energetics. *Am. J. Physiol. Heart Circ. Physiol.* **45**, H844–H857.

Brodie, P. F. (1975). Cetacean energetics, an overview of intraspecific size variation. *Ecology* **56**, 152–161.

Burn, D. M. (1986). The digestive strategy and efficiency of the West Indian manatee, *Trichechus manatus. Comp. Biochem. Physiol.* **85A**, 139–142.

Castellini, M. A., Kooyman, G. L., and Ponganis, P. J. (1992). Metabolic rates of freely diving Weddell seals—correlations with oxygen stores, swim velocity and diving duration. *J. Exp. Biol.* **165**, 181–194.

Costa, D. P. (1991a). Reproductive and foraging energetics of pinnipeds: Implications for life history paterns. *In* "Behaviour of Pinnipeds" (D. Renouf, ed.), pp. 300–344. Chapman and Hall Ltd., London.

Costa, D. P. (1991b). Reproductive and foraging energetics of high latitude penguins, albatrosses and pinnipeds: Implications for life history patterns. *Am. Zool.* **31**, 111–130.

Costa, D. P. (1993). The relationship between reproductive and foraging energetics and the evolution of the Pinnipedia. *In* "Marine Mammals: Advances in Behavioural and Population Biology" (I. L. Boyd, ed.), 66, pp. 293–314. Oxford University Press, Symposium Zoological Society of London, Oxford.

Costa, D. P. (2007). A model of the variation in parental attendance in response to environmental fluctuation: Foraging energetics of lactating sea lions and fur seals. *Aquat. Conserv.: Mar. Freshw. Ecosys.* **17**, S44–S52.

Costa, D. P., and Kooyman, G. L. (1984). Contribution of specific dynamic action to heat balance and thermoregulation in the sea otter *Enhydra lutris. Physiol. Zool.* **57**, 199–203.

Costa, D. P., and Williams, T. M. (2000a). Marine mammal energetics. *In* "The Biology of Marine Mammals" (J. Reynolds, and J. Twiss, eds), pp. 176–217. Smithsonian Institution Press, Washington, DC.

Costa, D. P., and Gales, N. J. (2000b). Foraging energetics and diving behavior of lactating New Zealand sea lions, *Phocarctos hookeri. J. Exp. Biol.* **203**, 3655–3665.

Costa, D. P., and Gales, N. J. (2003). Energetics of a benthic diver: Seasonal foraging ecology of the Australian sea lion, *Neophoca cinerea. Ecol. Monogr.* **73**, 27–43.

Costa, D. P., Le Boeuf, B. J., Huntley, A. C., and Ortiz, C. L. (1986). The energetics of lactation in the northern elephant seal, *Mirounga angustirostris. J. Zool.* **209**, 21–34.

Croll, D. A., Acevedo-Gutierrez, A., Tershy, B. R., and Urban-Ramirez, J. (2001). The diving behavior of blue and fin whales: Is dive duration shorter than expected based on oxygen stores? *Comp. Biochem. Physiol., Part A Mol. Integr. Physiol.* **129**, 797–809.

Estes, J. A., Tinker, M. T., Williams, T. M., and Doak, D. F. (1998). Killer whale predation on sea otters linking oceanic and nearshore ecosystems. *Science* **282**, 473–476.

Estes, J. A., Riedman, M. L., Staedler, M. M., Tinker, M. T., and Lyon, B. E. (2003). Individual variation in prey selection by sea otters: Patterns, causes and implications. *J. Anim. Ecol.* **72**, 144–155.

Folkow, L. P., and Blix, A. S. (1992). Metabolic Rates of minke whales (*Balaenoptera acutorostrata*) in cold water. *Acta Physiol. Scand.* **146**, 141–150.

Gallivan, G. J., and Best, R. C. (1980). Metabolism and respiration of the Amazonian manatee (*Trichechus inunguis*). *Physiol. Zool.* **53**, 245–253.

Gallivan, G. J., Best, R. C., and Kanwisher, J. W. (1983). Temperature regulation in the Amazonian Manatee *Trichechus inunguis. Physiol. Zool.* **56**, 255–262.

Irvine, A. B. (1983). Manatee metabolism and its influence on distribution in Florida. *Biol. Conserv.* **25**, 315–334.

Lavigne, D. M., Innes, S., Worthy, G. A. J., Kovacks, K. M., Schmitz, O. J., and Hickie, J. P. (1986). Metabolic rates of seals and whales. *Can. J. Zool.* **64**, 279–284.

Lawson, J. W., Miller, E. H., and Noseworthy, E. (1997). Variation in assimilation efficiency and digestive efficiency of captive harp seals (*Phoca groenlandica*) on different diets. *Can. J. Zool.* **75**, 1285–1291.

Lockyer, C. (1993). Seasonal changes in body fat condition of northeast Atlantic pilot whales, and their biological significance. *In* "Reports of the International Whaling Commission Special Issue, 14. Biology of Northern Hemisphere Pilot Whales" (G. P. C. H. L. Donovan, and A. R. Martin, eds), pp. 325–350. Cambridge University Press, Cambridge.

Markussen, N. H., Ryg, M., and Øritsland, N. A. (1994). The effect of feeding on the metabolic rate in harbor seals (*Phoca vitulina*). *J. Comp. Physiol. [B]* **164**, 89–93.

Martensson, P. E., Nordoy, E. S., and Blix, A. S. (1994). Digestibility of krill (*Euphausia superba* and *Thysanoessa* sp.) in minke whales (*Balaenoptera acutorostrata*) and crabeater seals (*Lobodon carcinophagus*). *Br. J. Nutr.* **72**, 713–716.

Oftedal, O. T. (2000). Use of maternal reserves as a lactation strategy in large mammals. *Proc. Nutr. Soc.* **59**, 99–106.

Oftedal, O. T., Boness, D. J., and Tedman, R. A. (1987). The behavior, physiology, and anatomy of lactation in the Pinnipedia. *Curr. Mammal.* **1**, 175–245.

Rea, L. D., and Costa, D. P. (1992). Changes in standard metabolism during long-term fasting in northern elephant seal pups (*Mirounga angustirostris*). *Physiol. Zool.* **65**, 97–111.

Riedman, M. L., and Estes, J. A. (1990). The sea otter *Enhydra lutris* behavior ecology and natural history. *US Fish Wildl. Serv. Biol. Rep.* **90**, 1–126.

Rosen, D., and Renouf, D. (1995). Variation in the metabolic rates of captive harbour seals. *In* "Whales, Seals, Fish and Man" (A. S. Blix, L. Walloe, and O. Ulltang, eds), pp. 393–399. Elsevier Science B.V, Amsterdam.

Rosen, D. A., and Trites, A. W. (1997). Heat increment of feeding in Steller sea lions, *Eumetopias jubatus*. *Comp. Biochem. Physiol. A* **118**, 877–881.

Rosen, D. A. S., and Trites, A. W. (1999). Metabolic effects of low-energy diet on Steller sea lions, *Eumetopias jubatus*. *Physiol. Biochem. Zool.* **72**, 723–731.

Rosen, D. A. S., and Trites, A. W. (2000). Digestive efficiency and dry-matter digestibility in Steller sea lions fed herring, pollock, squid, and salmon. *Can. J. Zool.* **78**, 234–239.

Rosen, D. A., and Trites, A. W. (2005). Examining the potential for nutritional stress in young Steller sea lions: Physiological effects of prey composition. *J. Comp. Physiol. [B]* **175**, 265–273.

Springer, A. M., *et al.* (8 authors) (2003). Sequential megafaunal collapse in the North Pacific Ocean: An ongoing legacy of industrial whaling? *Proc. Natl. Acad. Sci. USA* **100**, 12223–12228.

Stirling, I., and McEwan, E. H. (1975). The caloric value of whole ringed seals (*Phoca hispida*) in relation to polar bear (*Ursus maritimus*) ecology and hunting behavior. *Can. J. Zool.* **53**, 1021–1027.

Trillmich, F., and Kooyman, G. L. (2001). Field metabolic rate of lactating female Galapagos fur seals (*Arctocephalus galapagoenis*): The influence of offspring age and environment. *Comp. Biochem. Physiol., Part A Mol. Integr. Physiol.* **129**, 741–749.

Trillmich, F., and Weissing, F. J. (2006). Lactation patterns of pinnipeds are not explained by optimization of maternal energy delivery rates. *Behav. Ecol. Sociobiol.* **59**, 1–13.

Trumble, S. J., and Castellini, M. A. (2005). Diet mixing in an aquatic carnivore, the harbour seal. *Can. J. Zool.* **83**, 851–859.

Williams, T. M., Haun, J., Davis, R. W., Fuiman, L. A., and Kohin, S. (2001). A killer appetite: Metabolic consequences of carnivory in marine mammals. *Comp. Biochem. Physiol., Part A Mol. Integr. Physiol.* **129**, 785–796.

Williams, T. M., Fuiman, L. A., Horning, M., and Davis, R. W. (2004). The cost of foraging by a marine predator, the Weddell seal *Leptonychotes weddellii*: Pricing by the stroke. *J. Exp. Biol.* **207**, 973–982.

Williams, T. M., Rutishauser, M., Long, B., Fink, T., Gafney, J., Mostman-Liwanag, H., , and Casper, D. (2007). Seasonal variability in otariid energetics: Implications for the effects of predators on localized prey resources. *Physiol. Biochem. Zool.* **80**, 433–443.

Winship, A. J., Trites, A. W., and Rosen, D. A. S. (2002). A bioenergetic model for estimating the food requirements of Steller sea lions *Eumetopias jubatus* in Alaska, USA. *Mar. Ecol. Prog. Ser.* **229**, 291–312.

# Entrapment and Entanglement

## Jon Lien

**F**ishermen use a variety of techniques to capture fish. A common method is the use of gill nets that hang passively in the water, like curtains, and ensnare fish that blunder into them, or with barriers such as cod traps that direct the fish into traps that hold them until they are removed.

Because fishing nets are an unusual barrier, cryptic, and hard to detect, they on occasion catch marine mammals. When non-target species are accidentally caught in nets, they are termed *bycatch*. Bycatches of some species of marine mammals, such as harbor porpoises, *Phocoena phocoena*, are common in several areas. Because of the strength of modern materials now used in constructing nets, even larger species of cetaceans are sometimes captured incidentally in fishing gear. Such entrapments seriously threaten the North Atlantic right whale, *Eubalaena glacialis*, population (Johnson *et al.*, 2007). Any species can be captured in nets. However, humpback whales, *Megaptera novaeangliae*, perhaps because of their abundance in coastal waters where nets are commonly used or because of the many barnacles they carry, seem extremely vulnerable to entanglement in fishing gear. I present here a case history of whale entrapment from Newfoundland and Labrador, Canada; there are other areas of the world with entrapment and entanglement problems of many species of marine mammals (Lien, 1995).

In the late 1970s, humpback whales were seen in greater numbers in inshore waters of Newfoundland and Labrador as the bait fish capelin, *Mallotus villous*, which is their major prey, was seriously depleted offshore on the Grand Banks. The humpbacks moved inshore to feed on spawning capelin that occurs in the same areas where fishermen place their nets. Inevitably, this meant trouble. Fishermen began to report whale collisions which left nets badly damaged. On occasion, humpback whales would actually be caught and held in the nets. Initially, about 50% of the animals that were caught died. Because this was a new anthropogenic source of mortality in this recovering humpback population, it was a serious conservation concern.

Because of the problem, a program was established by the Whale Research Group of Memorial University of Newfoundland to aid both the whales and the fishermen when incidental captures occurred. Fishermen anywhere along Newfoundland and Labrador's 17,000 km of coastline that accidentally caught a whale could call the Entrapment Assistance Program by a toll-free phone number. A trained team would be dispatched to release the whale alive and to save as much of the fishing gear as possible. Because there were real benefits for fishermen in minimizing gear losses and lost fishing time, they cooperated very well with the program, and it benefited whales as well. Humpback mortality as a result of entrapment was reduced to about 10% of the animals that were captured. Those fewer whales died before help could reach the animal.

However, during the 1980s, groundfish populations were being seriously depleted by overfishing. To make a living, fishermen responded by fishing more nets, and inshore effort increased dramatically. With more barriers, the frequency of collisions and entrapments by whales increased. In one fishing season, the Entrapment Assistance Program received over 150 reports of entrapped humpback whales. In addition, other species, such as common minke whales, *Balaenoptera acutorostrata*, were also being caught. The program to release whales was extremely busy and effective. However, it was apparent that something was required to prevent collisions.

It was not practical to expect fishermen to stop fishing or to substantially modify where or how they fished. Instead, because cetaceans are acoustic specialists, scientists experimented with electronic devices that could be placed on nets. The hypothesis was that noisier nets would better alert the whales to their presence so they could avoid them. The alarms emitted higher frequency sounds that cetaceans could hear but were not detected by groundfish. Thus, fish catches would not be similarly reduced. Such devices were used successfully in areas where the likelihood of collisions was high. Acoustic alarms reduced collisions by about 80%. This was good news for both whales and fishermen. Such devices have now been used in many parts of the world and relative to all manner of marine mammals, with variable success (Barlow and Cameron, 2003).

Other news was not good for fishermen, however. By 1992, groundfish populations were so seriously depleted by fishing that a moratorium on fishing was established. All nets were removed from the water. Collisions and entrapments of whales were reduced to near zero. The moratorium on groundfish fishing continued until 1998, when small quotas were once again established. Closures and greatly reduced quotas on other species of groundfish followed by 1994.

The quotas that were established were very small compared to historical levels, so far fewer nets were used than before. The reduction in the number of nets kept accidents with humpbacks low. In addition, quotas were allocated in shares to individual fishermen. Thus, each fisherman did not have to fish competitively but was assured of a fair portion of fish that they could catch when they wanted, usually when they could realize the best prices for fish. In Newfoundland, best prices occur in the fall, and most fishing occurred then, a period when whale abundance inshore is relatively low. This combination of lower total fishing effort and a shift in fishing effort to a different season had kept collisions and entrapments to a very low level.

The Entrapment Assistance Program continues to be available to release whales carefully and to aid fishermen in retrieving fishing gear. In areas where whale abundance is high, some fishermen continue to use alarms on nets. However, at present, the incidental capture of humpback whales in traditional inshore nets of Newfoundland and Labrador is a relatively minor problem for both whales and fishermen. Because inshore fisheries are dynamic and changing activities that importantly affect the environment, they must be monitored continuously.

For example, in Newfoundland waters, recently developed fisheries for crab, turbot, and flounder are causing trouble for whales, often quite far from shore. Humpback whales can be caught easily, and the weight of the long strings of fishing nets or pots are able to hold these large animals. A dockside education program has been developed to ensure that fishermen know how to properly release an animal they catch.

### *References*

Barlow, J., and Cameron, G. A. (2003). Field experiments show that acoustic pingers reduce bycatch in the California drift gillnet fishery. *Mar. Mamm. Sci.* **19**, 265–283.

Johnson, A. J., Kraus, S. D., Kenney, J. F., and Mayo, C. A. (2007). The entangled lives of right whales and fishermen: Can they co-exist? *In* "The Urban Whale" (S. D. Kraus, and R. M. Rolland, eds), pp. 380–408. Harvard University Press, Cambridge.

Lien, J. (1994). Entrapments of large cetaceans in passive inshore fishing gear in Newfoundland and Labrador (1979–1990). *In* "Gillnets and Cetaceans." (W. F. Perrin, G. P. Donovan, and J. Barlow, eds.), pp. 149–157. Rep. Int. Whal. Comm. Spec Issue 15.

# Estrus and Estrous Behavior

## Daryl J. Boness

Estrus is a state of sexual receptivity during which the female will accept the male and is capable of conceiving. This behavioral state is under hormonal regulation involving the ovary and pituitary gland, and precedes or coincides with ovulation (i.e., production of an egg). The existing anatomical and physiological evidence suggests that the physiological process underlying estrus in marine mammals is comparable to that in other mammals.

Our knowledge about estrus, estrous behavior, and the estrous cycle in marine mammals is highly variable. Among cetaceans, most of it is derived from studies of the ovaries and reproductive tracts of animals killed during whaling or collected from beached and stranded specimens. The little behavioral information that is available comes mostly from studies of captive animals although some comes from observations of free-ranging animals. Among pinnipeds, we know far more from behavioral observations and physiological studies, but this is concentrated on those species that mate on land. There is almost a complete void of information on aquatically mating species, which comprise about half the pinnipeds and the majority of the phocids or true seals. Estrous cycles in marine mammals are relatively long compared to terrestrial mammals, usually a year or more, and are seasonal and synchronous within a species or population.

### I. Hormones and Anatomy of Estrus

As noted earlier, the hormonal cycle associated with estrus and reproduction in marine mammals appears to be similar to what happens in other mammals. High plasma concentrations of estrogens at the time of parturition decline rapidly and then begin a sharp rise again (Fig. 1; Boyd, 1991) as the cells surrounding the follicle, known as the theca interna, secrete estrogen. This rise in estrogen, along with a decline in progesterone, which inhibits follicle growth, is responsible for the rapid follicular growth in the ovary and the onset of estrous behavior. At the same time the follicle grows, the epithelium of the ovary thins and eventually ruptures, marking ovulation.

In the marine carnivores [pinnipeds, sea otters (*Enhydra lutris*), and polar bears (*Ursus maritimus*)], the period of anestrus, following ovulation and fertilization of the egg, involves a delay in implantation of the fertilized egg (Daniel, 1981). This phenomenon does not appear to occur in cetaceans, and is thought to have an adaptive function that allows females to disassociate the time of mating from parturition. The effect of the failure to implant is that it suspends development for a period of time and can subsequently work to help produce highly synchronized births, which is an important component of the mating system of pinnipeds especially. The physiology of delayed implantation is not well understood. Efforts to try to

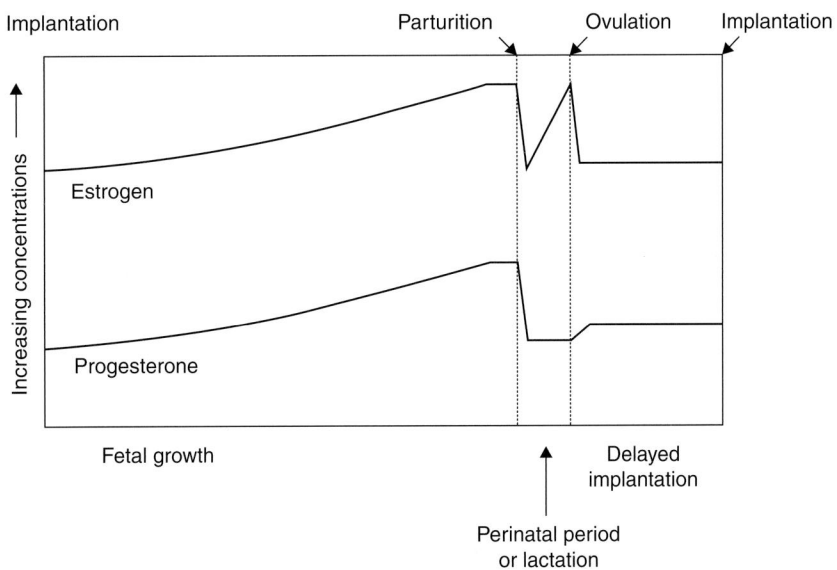

**Figure 1** *A schematic of estrogen and progesterone levels in the circulatory system of female seals in relation to the timing of various reproductive events. Modified from Boyd* et al. *(1999).*

artificially induce implantation using ovarian steroids failed, suggesting these hormones are not what trigger implantation. However, a recent hormonal study on free-ranging northern fur seals (*Callorhinus ursinus*) found that androgens, "male hormones," which are produced by females, might play a role in delayed implantation in female fur seals (Browne *et al.*, 2006).

All marine mammals have a reproductive tract that has two uterine horns (bicornuate) terminating at the ovaries, which are of equal size and have a common blood source. In pinnipeds, reproduction alternates between the right and left ovary and horn in successive reproductive periods. This is not the case in cetaceans, however. Among baleen whales, both ovaries are equally functioning, whereas in many odontocetes the left ovary appears to function early in life and the right one later in life (Boyd *et al.*, 1999).

## II. Timing of Estrus and Estrous Cycles

In pinnipeds, most of our knowledge about the timing of estrus comes from behavioral observations of mounting and mating behavior. Most species are monestrous, with the exception being the walrus (*Odobenus rosmarus*), which has two estrous periods within a year. The double estrus in walruses is undoubtedly linked to the 15-month gestation. With this long gestation, parturition occurs when there are few reproductively active males available, yet there is a postpartum estrus. The second estrus occurs 9–10 months after birth, during the peak of male sexual activity. Otariid seals show relatively little variation in the timing of estrus, with it occurring early in lactation during what has been called the perinatal period and within 4–14 days following parturition for most species. In a few species (California sea lions, *Zalophus californianus*; Steller sea lions, *Eumetopias jubatus*; and Galapagos fur seals, *Arctocephalus galapagoensis*), estrus occurs after maternal foraging trips begin, which may be as much as a month after parturition. Although a postpartum estrus may be an ancestral condition, the nearly yearlong lactation in otariids, along with the dispersal of males away from breeding sites after mating, would have selected for estrus to be early in lactation

and shortly after parturition. Linked to this ecological pattern of spatial separation between breeding on land and feeding at sea is the existence of delayed implantation in seals, which tends to synchronize parturition, and thus estrus (Wells *et al.*, 1999). The benefit to otariid females being receptive shortly after parturition would then be having access to the most competitive and highest quality males because they are present on land.

In phocid seals the timing of estrus is more variable with respect to parturition, ranging from about 5 days to between 1 and 2 months after parturition. In fact, for phocids estrus tends to occur in late lactation or after pups have been weaned, although this may not be the case for Mediterranean monk seals (*Monachus monachus*), which appears to have a 5-month lactation period and little seasonality in breeding. This accompanies a very different lactation pattern from that of otariids. All phocids, with the exception of the Mediterranean monk seal, have a short lactation ranging from 4 to 60 days. Although those phocids that have extremely short lactation periods will come into estrus within 2 weeks of parturition like most otariids, those species that lactate for a month or more have shifted estrus away from being juxtaposed to parturition if this was the ancestral pattern. Reasons for this difference between otariids and phocids are unclear. It may relate to phocid females being in estrus for longer periods of time, providing for greater opportunities to mate with multiple males and thereby increasing the quality of males fertilizing their offspring through sperm competition.

The duration of estrus in otariids appears to be very short based on observed copulatory behavior. In most species, females are observed mating once, although about 30% of females in some species are seen mating twice, and in several more recent studies DNA analyses indicate that fertilization of a female was by a male other than the one observed mating with her (see later for a possible explanation). The observed multiple matings still appear to occur within a 2-day period. We know nothing about the duration of estrus in the aquatically mating phocids, but for all three of the species that mate on land, estrus lasts for several days. Based on the spread of observed copulations, it may last up to 13 days in the northern elephant seal (*Mirounga angustirostris*) and 5–7 days in the southern elephant

(*M. leonina*) and gray seals (*Halichoerus grypus*). However, there is evidence in some of these phocids that females may use copulation as a strategy to avoid injury and harassment as they try to leave the breeding grounds (Mesnick and Le Boeuf, 1991). Thus it is possible that estrus has ended sooner than indicated by observed copulations.

An estrous period several days long is common in the sea otter as well, a species in which females are known to mate with more than one male (Riedman and Estes, 1990). Sea otters have an annual cycle like most pinnipeds. Female sea otters typically become receptive within a few days of weaning their pup and receptivity lasts 3–4 days. It is possible that ovulation is induced by copulation in this species, as other mustelids exhibit induced ovulation. There is evidence of polyestrus in females that are not impregnated successfully during an initial estrus.

Little is known about the details of the reproductive cycle of polar bears. There is a breeding season between March and May, when females come into estrus (Stirling, 1998). Anecdotal evidence suggests that estrus occurs over multiple days with "courtship behavior" that may be necessary to stimulate ovulation. Individual females have extended periods of maternal care and do not become receptive annually. Intervals between estrus may be as low as 2 years but more typically average over 3 years.

The cycles of cetaceans are much more variable than the cycles of other marine mammals. Many of the mysticete whales have a 2-year cycle, including the fin (*Balaenoptera physalus*), blue (*Balaenoptera musculus*), sei (*Balaenoptera borealis*), humpback (*Megaptera novaeangliae*), and gray (*Eschrichtius robustus*) whales. The fin whale, e.g., becomes receptive in the winter in low latitude, warm water, where mating occurs, and then migrates to cooler high latitudes where it feeds through the summer. In the fall it again migrates to low latitudes to give birth, and after a second migration to cooler water for feeding and lactation the female weans the calf and has a period of anestrus for 5–6 months before becoming receptive again, having made another migration to warm, low-latitude waters. The Northern (*Eubalaena glacialis*) and Southern (*E. australis*) right whales, and bowhead whale (*Balaena mysticetus*) may have an even longer interval of 3–4 years between estrous periods, whereas the minke whale (*Balaenoptera acutorostrata*) appears to have an annual cycle (Perrin *et al.*, 1984).

Odontocetes have an annual breeding season with most mating occurring over a diffuse 2- to 5-month peak for many species studied (Mann *et al.*, 2000). However, the estrous cycles of individual animals are not likely to be annual in many species because of multiyear lactation periods, which may inhibit cycling. Lactation periods in excess of a year are common among odontocetes and may last up to 3.5 years in some species. Unfortunately, more behavioral studies in which individual animals are followed are needed. However, even this has its drawbacks in that copulatory behavior in some dolphins may have a non-sexual social component that could make it difficult to obtain a clear understanding of the estrous cycle from behavior. Data from reproductive tracts of dead odontocetes suggest that females of most or even all species are polyestrus.

Captive studies have provided some useful insights into the estrous cycle of non-lactating odontocetes. For example, one study showed that bottlenose dolphins (*Tursiops* spp.) ovulate spontaneously and have a 21- to 42-day estrous cycle. Also, a 3-year study with two short-beaked common dolphin (*Delphinus delphis*) females and with four common bottlenose dolphin (*Tursiops truncatus*) females found somewhat similar results with respect to periods of estrus and anestrus. The common dolphin females exhibit extended anestrus of 1–2 years but also cycled as many as 7 times within a year. There

was no seasonality to these cycles. The bottlenose dolphin females also had a yearlong anestrus followed by polyestrus, although the maximum number of cycles within a year was only three. Along with this reduced number of cycles compared to the common dolphin the cycles were seasonal, occurring only in April and May.

Sirenians (manatees, *Trichechus* spp. and dugongs, *Dugong dugon*) appear to follow a similar pattern as the cetaceans with a multiyear lactation that results in an estrous cycle of 2 or more years (Wells *et al.*, 1999). Among sirenians, after weaning a calf, females may also undergo polyestrus before becoming impregnated. Breeding may not be continuous, as 33% and 56% of a sample of ovaries from manatees and dugongs, respectively, showed that females were neither pregnant nor lactating. Little is known about the length of estrus in sirenians. However, given that males and females tend to be solitary and need to search out each other for reproduction, it is likely that estrus is relatively long and ovulation might be induced. Mating or estrus groups, consisting of one female and multiple males trying to mate with the female, in manatees and dugongs last for variable periods of time, from a few hours to several days.

## III. Estrous Behavior and Signals

The behavior of female gray and elephant seals gives one the impression that they do not come into estrus, except that they ultimately do mate before departing from the breeding colony. Females generally show no soliciting behavior and appear to protest mounting attempts by males right up to the time they leave. However, upon closer observation, subtle differences in the behavior of females do become apparent. For example, females may reduce the duration of their protests when they are in estrus, but they may do this selectively to "potentially high-quality" males. In southern elephant seals, males reportedly will move from female to female, placing their head across the female's back and waiting for her reaction before either moving on or subsequently attempting to try to mount her. In northern elephant seals near the end of lactation, a female will not respond vocally to the harem master when he places his foreflipper over her back and she may even show a lordosis response by spreading her flippers and raising her tail end.

Most likely there are cues that signal the sexual status of these phocid females. For example, in the gray seal, attempts by males to mount females are not indiscriminate. They begin about a week before females are first observed copulating and increase in frequency linearly during that period (Fig. 2; Boness and James, 1979). As gray or elephant seal males do not routinely sniff the head or tail end of females prior to attempting to mount them, the existence of olfactory signals is less clear than for otariids, in which sniffing behavior preceding mounting is typical (see later). Possible explanations for the selective mounting by gray seal males could be that they use either the body shape of females that are becoming depleted from fasting or the increased rotundness of their pups as a signal of their sexual status.

Another interesting example of behavior that likely indicates signaling (probably chemical) of estrus occurs in polar bears. As noted earlier, males and females are solitary and must find each other for reproduction. Males have been observed detecting the tracks of a female and following them up to 100 km to find the female (Stirling, 1998). Manatee and dugong females may increase the range of their movements when they become estrus to increase the likelihood of encountering males.

Our best information about estrous behavior in pinnipeds comes from some elegant experimental work by Roger Gentry in his 19-year

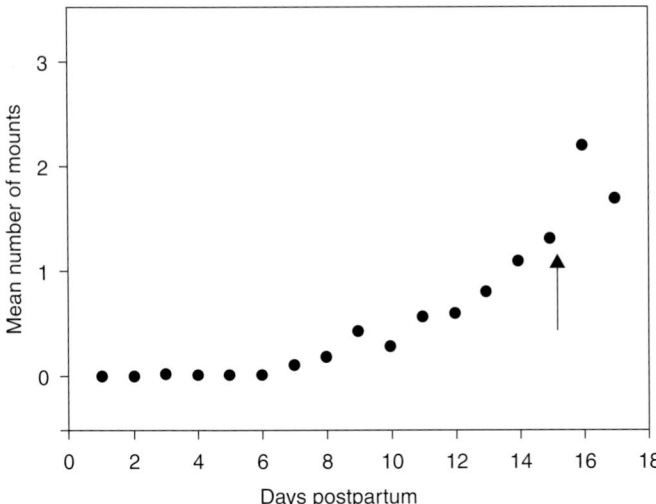

**Figure 2** *Mean number of attempted copulations with females by gray seal males as a function of the number of days postpartum for the females. The arrow represents the average time of onset of estrus, defined by the first observed copulation. Modified from Boness and James (1979).*

study of the northern fur seal in Alaska (Gentry, 1998). These studies involved capturing groups of females, controlling their access to males, manipulating the size and age of males to which the females were given access, and using surrogates in place of males. Field observations found that prior to estrus, males engaged in sniffing the nose and mouth of females and that sniffing the tail region sometimes followed this. Investigation of the latter often led to a lordosis response. With this in mind, the studies were designed to determine the role of the male in triggering estrus, terminating it, determining the change in behavior of females, confirming the importance of olfactory cues, and understanding the mechanism of how estrus is terminated.

Gentry's studies showed that females prevented from copulating did not prolong estrus nor recycle and receptivity lasted an average of 34h. A non-estrous female's behavior changed dramatically within as short as a 5-min period, going from being aggressive to a male to emitting low-level estrous vocalization and showing lordosis. The change was reflexive in that either artificial means or a male elicited it. Copulation by a territorial male usually terminated estrous behavior within 15min, whereas a juvenile or senescent male copulation did not terminate estrus quickly. All copulation led to fertilization, however. Males that actively suppressed the aggressive behavior of females appeared to enhance the speed with which females became receptive. Finally, there appears to be a "vaginal code" based on male size and thrusting pattern for the termination of estrus following copulation.

We are only at the very early stages of beginning to understand about the pre-estrous and estrous behavior of cetacean females. In one study of captive spinner dolphins (*Stenella longirostris*), an attempt was made to correlate behavioral and hormonal patterns, examining behaviors that could have a sexual function. The behaviors included genital-to-genital contact, beak-to-genital contact, other-to-genital contact, ventral presentation, chases, and non-genital contact. The only behavior that showed any relationship with hormonal states indicative of estrus was beak-to-genital contact. This involved females inserting her beak into the genital slit of the male, most often during the follicular phase of her estrous cycle. One question that remains is to what extent such behavior in captivity represents normal behavior in the wild.

Studies of bottlenose dolphins in Shark Bay, Australia have shown that males form consortships with females that are receptive and try to control access to them. It is unclear how males detect the sexual status of these females, but the consortships are not formed with females that are non-receptive. Females may not have a choice in these consortships, which often involve coordinated activity by male coalitions to control the female.

There is growing evidence from behavioral observations of whales in the wild that females may choose which males that mate or that they may incite competition, as has been demonstrated in some pinnipeds. For example, individual female right whales will move into shallow water or raise their tales to avoid mating by certain males. They also mate with more than one male, providing the opportunity for sperm competition. Female humpback whales lead competing males on high-speed chases. Humpback whales are noted for their song, which is only sung by males. There is some evidence to suggest that females may use these male songs as a basis upon which to choose males.

## IV. Reproductive Senescence

Reproductive senescence can be a decline in maternal performance or a complete cessation of reproduction, i.e., no estrus cycling involving ovulation. A number of studies of pinnipeds have shown a decline in maternal performance by older females, but females continued to cycle and produce offspring until they died (Bowen *et al.*, 2006). In contrast, physiological data has shown at least one species of cetacean, the short-finned pilot whale (*Globicephala macrorhynchus*), that ceases cycling and producing offspring more than 10 years before the end of their life expectancy (Mann *et al.*, 2000). In several other cetacean species markedly reduced fertility rates in older females also suggest senescence in estrus cycling may occur. The reasons why senescence would evolve in marine mammals or other mammals, for that matter, are unclear. However, two hypotheses that have been proposed are that females in species that have high levels of maternal investment may benefit by continuing to rear their last and preceding offspring still under their care, or by enhancing their offsprings' success by helping to rear grand-offspring.

While gradual progress is being made in our understanding of estrus and estrous behavior in marine mammals, there is still much to be learned, especially in those species where sexual behavior is entirely aquatic. Some of our greatest opportunities may come from new technologies at hand that are being used on animals in the wild, such as the animal-borne video systems or tags that can record and store vocalizations made by an animal or by surrounding animals. We also can expect to learn more from detailed studies of animals in captivity.

*See Also the Following Articles*

Behavior, Overview ■ Mating Systems

*References*

Boness, D. J., and James, H. (1979). Reproductive behaviour of the grey seal (*Halichoerus grypus*) on Sable Island, Nova Scotia. *J. Zool. Lond.* **188**, 477–500.

E

Bowen, W. D., Iverson, S. J., McMillan, J. I., and Boness, D. J. (2006). Reproductive performance in grey seals: Age related improvement and senescence in a capital breeder. *J. Anim. Ecol.* **75**, 1340–1351.

Boyd, I. L. (1991). Environmental and physiological factors controlling the reproductive cycles of pinnipeds. *Can. J. Zool.* **69**, 1135–1148.

Boyd, I. L., Lockyer, C., and Marsh, H. D. (1999). Reproduction in marine mammals. *In* "Biology of Marine Mammals" (J. E. Reynolds, III, and S. A. Rommel, eds), pp. 218–286. Smithsonian Institution Press, Washington, DC.

Browne, P., Conely, A. J., Spraker, T., Ream, R. R., and Lasley, B. L. (2006). Sex steroid concentrations and localization of steroidogenic enzyme expression in free-ranging female northern fur seals (*Callorhinus ursinus*). *Gen. Comp. Endocrin.* **147**, 175–183.

Daniel, J. C. (1981). Delayed implantation in the northern fur seal (*Callorhinus ursinus*) and other pinnipeds. *J. Reprod. Fert. (Suppl.)* **29**, 35–50.

Gentry, R. L. (1998). "Behavior and Ecology of the Northern Fur Seal." Princeton University Press, Princeton, NJ.

Mann, J., Connor, R. C., Tyack, P. L., and Whitehead, H. (eds) (2000). "Cetacean Societies." University of Chicago Press, Chicago, IL.

Mesnick, S. L., and Le Boeuf, B. J. (1991). Sexual behavior of male northern elephant seals. 2. Female response to potentially injurious encounters. *Behaviour* **117**, 262–280.

Perrin, W. F., Brownell, R. L., and Demaster, D. P. (eds) (1984). "Reproduction in Whales, Dolphins and Porpoises." International Whaling Commission, Cambridge.

Riedman, M. L., and Estes, J. A. (1990). The sea otter (*Enhydra lutris*): Behavior, ecology and natural history. US Fish & Wildlife Service, Biol. Rep. 90.

Stirling, I. (1998). "Polar Bears." University of Michigan Press, Ann Arbor.

Wells, R. S., Boness, D. J., and Rathbun, G. B. (1999). Behavior. *In* "Biology of Marine Mammals" (J. E. Reynolds, III, and S. A. Rommel, eds), pp. 324–422. Smithsonian Institution Press, Washington, DC.

# Ethics and Marine Mammals

### Marc Bekoff

The whale in the sea, like the wolf (*Canis lupus*) on land, constituted not only a symbol of wildness but also a fulcrum for projecting attitudes of conquest and utilitarianism and, eventually, more contemporary perceptions of preservation and protection.

Stephen R. Kellert R(1996)

## I. Humans and Other Animals: Multidimensional Encounters

Humans are a curious lot, and our intrusions, intentional, and inadvertent, have significant impacts on other people, non-human animal beings ("animals") plants, water, the atmosphere, and inanimate landscapes. Often our influence is subtle and long term. Our relationship with other animals is a complex, ambiguous, challenging, and frustrating affair, and we must continually reassess how we should interact with our non-human kin. Often we live with deep contradictions about what we do and what we believe we morally should do.

There are many important and difficult issues that demand serious consideration in discussions of the ethics of how human beings interact with animals. Their complexity is compounded because highly charged subjective opinions and passions run high. This article highlights just how complex and multidimensional these issues are. It is meant to be a starting point for a discussion of different perspectives; none is more important than others. What is important is that we all agree that ethics is an essential element in any discussion of human interactions with other animals.

Since I wrote the original version of this essay for the first edition of this encyclopedia (2002), things have gotten worse for numerous aquatic animals including marine mammals. Human-induced (anthropogenic) assaults on aquatic ecosystems and on individuals due to climate change, recreation, and over-fishing have increased globally (Bekoff, 2007a and references therein; see also For cod's sake, act now, 2006; Mackenzie, 2006; Osinga and 't Hart, 2006; Raloff, 2006). The Food and Agricultural Organization of the United Nations (FAO) noted in February 2006 that their "…most recent global assessment of wild fish stocks found that out of the almost 600 major commercial species groups monitored by the Organization, 52 percent are fully exploited while 25 percent are either overexploited (17%), depleted (7%) or recovering from depletion (1%). Twenty percent are moderately exploited, with just three percent ranked as underexploited" (Food and agricultural organization of the United Nations, 2006). There are also problems with non-target species getting caught due to fishing activities. For example, in 1990, about 42 million marine mammals and sea birds were caught in drift nets as squid and tuna were being harvested (Fox, 1997). About 129,000 Olive Ridley turtles (*Lepidochelys olivacea*) have died over the past 13 years because they suffocate in the nets of fishing boats not using mandatory turtle-excluder devices. Experts fear that the movement of giant ships and artificial illumination would put the turtles in even deeper trouble in the years ahead. Whales also are non-target victims of fishing nets. In 2003 the World Wildlife Fund reported that nearly 1000 whales, dolphins, and porpoises drowned daily after becoming entangled in fishing nets and other equipment (Verrengia, 2003). Annually, more than 300,000 individuals may perish because of fishing activities. And while there is a global moratorium on commercial whaling since 1986, Japan and Iceland continue to hunt as part of what they call "scientific programs." Norway has objected to the moratorium and runs commercial whaling operations.

Here I focus on human–dolphin and human–whale encounters (e.g., hunting, keeping animals in captivity, swimming programs, using them for entertainment), for ethical questions that arise when considering these types of highly visible interactions can be used as illustrations for human encounters with other marine mammals, including the pinnipeds [seals, sea lions, and the walrus (*Odobenus rosmarus*)], manatees (*Trichechus* spp), and polar bears (*Ursus maritimus*). Understandably, there is much and growing public interest in these rendezvous. Individuals of these species are sentient beings (capable of experiencing pleasure and pain) and sentience must influence how we interact with them (Bekoff, 2006a, b, c, 2007b). In scientific research there are always surprises. New scientific data appear that force us to rethink what we know and to revise our stereotypes. For example, spindle cells, which were long thought to exist only in humans and other great apes, have been discovered in humpback whales (*Megaptera novaeangliae*), fin whales (*Balaenoptera physalus*), killer whales (*Orcinus orca*), and sperm whales (*Physeter macrocephalus*) in the same area of their brains as spindle cells in human brains (Coghlan, 2006). This brain region is linked with social organization, empathy, and intuition about the feelings of others, as well as rapid gut reactions. Spindle cells are important in processing emotions. There is also a growing database showing that fish are sentient beings that experience pain and suffering (Sneddon, 2003; Moccia and Duncan, 2004).

Dolphins, whales, and other marine mammals have often been fabricated to be the animals we want them to be. Most descriptions of dolphins and other cetaceans picture them as highly intelligent and capable of experiencing pleasure and pain with remarkable social and cognitive skills. Indeed, dolphins and other marine mammals seem to fulfill some criteria of "personhood" in that they are alive, aware of their surroundings, sentient, and may have a sense of self. Why, then, do some people feel comfortable intruding into their worlds if it will cause pain and suffering? Toni Frohoff (1998, p. 84) has poignantly noted: "Currently we are walking a fine line in our relationship with cetaceans. The same attraction that motivates us to protect them from harm is also what drives us to be close to them, to have them 'within reach.'" It is because dolphins and other marine mammals are thought to be attractive, harmless, endowed with mystical qualities, or to be of economic value as commodities for show or food that we seek them out. However, we may bring much harm to them in our efforts to include them in our lives, even in ways that do not involve killing them.

Many issues that are pertinent when considering marine mammals are also raised when discussing other mammals such as terrestrial CARNIVORES. Wolves and whales have been among the most persecuted of animals during the last three centuries, wolves because they were feared predators and whales because of their economic value. The near EXTINCTION of both wolves and some whales was important in setting environmental policies. However, while many people swim and wade with dolphins, few if any truly dance or howl with wolves. Thus, the close contact that many humans have with some marine mammals leads to other questions that are unique to these encounters.

## A. What Should We Do?

Human impacts on other animals, intentional and inadvertent, are universal. A major question in need of serious debate is should we ever interfere in other animals' lives—when might human interference be permissible? Thus, should we let other animals be and not intentionally interfere in their lives? Or, should we hunt them for food whenever we so wish? Should we hunt only when there are no alternative food sources? Should we interfere in other animals' lives when we have spoiled their habitats or when they are sick, provide food when there is not enough food to go around, provide care to young if a parent does not, stop aggressive encounters, stop predators in their tracks, or translocate individuals from one place to another, including zoos, wildlife and marine theme parks, and aquariums? Should human interests always trump those of other animals? If not, then when should the interests of other animals trump our own?

The question of when humans *should* intrude is a difficult one. However, just because we *can* do something does not mean that we *should* or *have* to do it. Furthermore, just because some intrusions may be *relatively* less injurious than others, this line of reasoning places us on a very slippery slope and can, in the end, lead to narrow or selfish anthropocentric claims. Even in situations when humans have good intentions, those intentions are not always enough.

This article discusses some basic principles that underlie the use and exploitation of marine mammals, especially whales and dolphins, presents a few general questions, and discusses some representative examples. Definitive answers to these and other questions are quite elusive, but open discussion can provide guidelines for proactive decision-making. All too often we are left in the position of trying to rectify messy and difficult situations that we have created; proactivity needs to become the *modus operandi* for future actions. For many questions about how animals should be treated by humans there are no "right" or "wrong" answers. However, there are better and worse answers. Perhaps it will turn out that in some cases what we think is the "right" action is not when the big picture is carefully analyzed.

It is essential to remember that even if wild or captive marine mammals develop close social bonds with humans, these animals are socialized or habituated individuals, but they are not domesticated animals. Often people remark that individuals who interact closely with humans have become domesticated. However, domestication does not happen to an individual during his or her lifetime. Domestication is a long-term evolutionary process during which humans selectively breed animals for desirable traits. "Domesticated" and "socialized" or "habituated" are not synonyms.

## B. Human Responsibilities

It is generally accepted that humans have unique responsibilities to other living organisms (and to inanimate environments). Our unique responsibilities stem from our (at least most of us) being moral agents who are responsible for our actions. It is usually assumed that neither animals nor young human infants and mentally impaired adults are moral agents. Rather, they are moral patients, not usually held responsible for their actions. They do not know "right" from "wrong" or "good" from "bad." Nature, too, cannot be good or bad, although the consequences of natural acts can be better (good) or worse (bad). Nature simply is. We do not have to apologize for nature's ways. Nor should her ways—her supposed cruelty and ruthlessness—be used as excuses for what we do to other animals (including humans).

It is important to stress that most, if not all, other animals depend on our goodwill and mercy. Individuals can choose to be intrusive, abusive, or compassionate. We do not have to do something because someone else wants us to do it. We do not have to do something just because we can do it. Each of us is responsible for our choices.

## II. The Moral Status of Animals: Animal Rights and Animal Welfare

In current discussions about the moral status of animals, there is an obvious "progressive" trend for greater protection for wild and captive animals. (This might be due in part to an increasing number of people moving from farms and rural areas to more urban environments.) This is clearly the case for marine mammals (Kellert, 1999). In a survey of American's perception of marine mammals, most respondents were opposed to commercial WHALING, often for ethical reasons. Concern was also expressed for the commercial exploitation of seals, sea otters (*Enhydra lutris*), walruses, and polar bears. Most Americans also objected to commercial whaling by native peoples or the resumption of killing gray whales (*Eschrichtius robustus*). A majority of Alaskans opposed oil and gas development if it injured or killed marine mammals. There was also an unsuccessful effort prior to the reauthorization of the Marine Mammal Protection Act in 1988 to prohibit any invasive research involving marine mammals unless that research would directly benefit the subject of the research.

In recent years, philosophers and scientists have devoted increasing attention to questions about the moral status of animals. Many people support a position called the *rights* view. To say that an animal has a right to have an interest protected means that the animal has a claim, or entitlement, to have that interest protected even if it would benefit us to do otherwise. Humans have an obligation to honor that claim for other animals (just as they do for humans who cannot protect their own interests). Thus, if a wild dolphin has a

E

right to feed, then humans have an obligation to allow her to do so and not do anything to interfere with her feeding activities. Likewise, if a dolphin has a right to life, she cannot be used in war games, warfare, or other activities in which death is possible.

Animal rights advocates stress that animal's lives are valuable in and of themselves (they have inherent value) and that their lives are not valuable because of what they can do for humans (their utility) or because they look or behave like us. Animals are not property or "things," but rather they are living organisms that are worthy of our compassion, respect, friendship, and support. Animals are not "lesser" or "not as valuable" as humans; they are not property that can be abused or dominated. Human benefits are irrelevant for determining how animals should be treated.

Many people believe that the rights view and the *animal welfare* view are identical. They are not. Animal welfarists focus on individual's usefulness to humans. They practice *utilitarianism*, in which the general rule of thumb is that the right actions are those that maximize utility summed over all those who are affected by the actions. Often welfarists/utilitarians are called "wise users." They believe that while humans should not abuse or exploit animals, as long as we make the animals' lives comfortable, physically and psychologically, we are taking care of them and respecting their welfare. Welfarists are concerned with the quality of animals' lives. However, welfarists do not believe that animals' lives are valuable in and of themselves. Many conservation biologists and environmentalists are utilitarians who are willing to trade off individuals' lives for the perceived good of higher levels of organization such as populations, species, or ecosystems.

The welfarists' rule of thumb, and it is not a moral rule, is that it is permissible to use animals if the relationship between the costs to the animals and the benefits to the humans is such that the costs are less than the benefits. Welfarists believe that if animals experience comfort, appear happy, experience some of life's pleasures, and are free from prolonged or intense pain, fear, hunger, and other unpleasant states, then we are fulfilling our obligations to them. If individuals show normal growth and reproduction and are free from disease, injury, malnutrition, and other types of pain and suffering, they are doing well. Thus, welfarists argue that using animals in experiments, slaughtering them for human consumption, and using them for treating human disorders (e.g., dolphin-assisted therapy, DAT, programs) are permissible as long as these activities are conducted in a humane way. Welfarists do not want animals to suffer from any unnecessary pain, but they sometimes disagree among themselves about what pain is necessary and what humane care really is. Welfarists agree that the pain and death animals experience are sometimes justified because of the benefits that humans derive. The ends—human benefits—justify the means—the use of animals.

## III. Hunting Whales

Whale HUNTING brings to light numerous issues that reflect utilitarian thinking. Whales are frequently viewed as commodities, whether they are hunted centers or whether it is economical. Rarely are the costs to the individuals entered into the equation. In the past, many people thought that whales were an inexhaustible resource and historically there were few restrictions on killing them. When whale watching became popular, whales were more valuable alive than dead. They went from being a consumptive to a non-consumptive resource.

Political and sociocultural motives also play a role in whale hunting. Various indigenous people (e.g., the Makah in Washington State) want to be able to hunt whales (in the Makah case, gray whales) because their ancestors did so, because it was part of their cultural heritage. They claim that the tradition of whale hunting defines "who they are."

The revival of aboriginal whaling is controversial in various parts of the world and involves species other than the gray whale. When the target species is on the brink of extinction, few argue that any type of whaling is permissible. Likewise, when killing whales is essential for food, few argue against the practice. However, when the whales are not endangered, people disagree about continuing to kill whales or reestablishing this practice. Some argue that whale hunting is permissible because it is part of the heritage of a given indigenous group (it is cultural revival), whereas others argue that there are other cultural practices that are no longer followed and little effort to regain them. Why is whaling hunting so important if it is not essential for sustenance?

Methods of killing whales also are controversial. For example, the Makah used a rifle to kill a whale who had been wounded by a harpoon. A majority of Americans oppose the use of weaponry. Hunting whales produces much pain and suffering. Chasing and stalking individuals compromises their physical and psychological well-being and death usually is not instantaneous, frequently taking upward of 10 min. Furthermore, family groups are broken up. All in all, hunting whales and other animals, including such marine mammals as seals, raises numerous difficult ethical issues.

## IV. Keeping Animals in Captivity
### A. Swimming with Dolphins

"Swim with dolphin" and "petting pool" programs are very controversial. Such proffered reasons as "it's fun," "aren't the animals cute," or "it's a spiritual experience" are insufficient to justify these practices. Much attention has been given to the question of whether human encounters with dolphins may have negative effects on the dolphins. Human–dolphin interactions may be noisy and stressful. One study reported that captive dolphins showed enlarged adrenals, especially those individuals exposed to humans on numerous occasions. The long-term effects of swim programs on dolphin behavior and well-being still need to be studied systematically, but evidence shows that the stress associated with these programs may have long-term effects on the dolphins.

Swim programs are risky to humans. Dolphins are large and strong animals. While higher risks seem to be experienced more in non-controlled swims, there are also serious risks in controlled swims that might be fatal. "Controlled" refers to situations when all interactions are directed by trainers who give the dolphins commands at all times. "Non-controlled" means interactions are allowed to occur spontaneously.

It is also important to know if DAT programs truly work. While some researchers claim that DAT is an effective therapeutic intervention for several disorders (e.g., depression, autism, cerebral palsy, mental retardation) others disagree. Criticisms center on the use of improper statistical methods and the lack of controlled studies. It is often very difficult to assess experimentally the positive effects of animals on people. In many programs, no other animals, including such domesticated species as dogs, were used as controls to see if they might be as or more effective than dolphins.

Another question that is also important to consider is whether programs that involve interactions with captive dolphins help to educate people about these and perhaps other animals. More research is needed to determine if contacts with dolphins actually change people's attitudes about them. Intuitions are not enough. To date, there is no solid evidence that interactive captive programs with dolphins are more effective educationally than non-interactive programs. Indeed, some marine biologists fear that these programs may send the message that it is permissible to take animals from the wild and bring them into captivity and keep them in small tanks where they are bored, deprived, and needlessly die. There are also serious concerns about the fate of dolphins once they are too old or aggressive to partake in swim programs. Yet another concern centers on the possibility that these programs may teach people to expect the same kinds of interactions from free-ranging wild animals.

While there have been attempts to regulate swimming programs, little has actually been accomplished. In the United States, federal regulations controlling these programs, mandated in 1998 (after a delay of 4 years before finalizing them) by the US Department of Agriculture's Animal and Plant Health Inspection Service (APHIS) were suspended in April 1999 soon after they were invoked because, according to their press release: "It has come to our attention that the language in the new regulations may be confusing to some. Therefore, we are suspending enforcement of the regulations in order to take a closer look at the language and make it more understandable." (www.aphis.usda.gov/lpa/press/1999/04/dolphin.txt) During this process, there are no regulations for these popular programs. Some people believe the federal regulations were suspended because of pressure from the lucrative industries that exploit dolphins.

## B. Petting and Feeding Programs

Petting and feeding programs allow people to pet and feed captive dolphins. Many of these programs may not be adequately supervised or monitored. There are some major concerns with these programs, including that dolphins may be unable to avoid encounters with humans and may be highly stressed, the water in which dolphins and humans interact is often heavily chlorinated and may be unhealthy for dolphins (and humans), dolphins may be fed foreign objects that can harm them, and there seems to be little, if any, education value to these programs. There are few data that speak to these and other concerns and this information is needed to determine if petting and feeding programs can be properly regulated. One of the main questions is whether dolphins can be accessible to people in these programs and still be protected from harm.

While feeding and harassing wild dolphins is illegal in the United States, and there are severe penalties for engaging in these activities, this is not so for other countries. There are documented instances of wild dolphins being fed firecrackers, golf balls, plastic objects, balloons, and fish baits with hooks (so that hooked dolphins can be caught). Provisioning dolphins with fish has been associated with a change in the social behavior of free-ranging Indo-Pacific bottlenose dolphins (*Tursiops aduncus*) in Monkey Mia, Australia. Dolphins who have been fed also change their foraging behavior and frequent heavily trafficked harbors and marinas. Some get struck by boats. People have also been seriously injured trying to feed wild dolphins. The National Marine Fisheries Service and other organizations are mounting highly visible campaigns to stop the feeding and harassment of wild dolphins. It also has been noted that some problems associated with feeding terrestrial mammals (changes in foraging patterns and hunting skills) are relevant to concerns about the feeding of dolphins.

Clearly, much more information is needed concerning petting and feeding programs for captive and wild animals. Especially needed are data concerning the effects of these programs on dolphin mortality,

health, and psychological and emotional well-being. It is also important to counter the possibility that feeding captive dolphins may send the message that it is permissible to feed wild individuals.

## C. Zoos: Aquariums and Marine Theme Parks

The existence of zoos, including aquariums and marine theme parks, raises many important and difficult ethical questions. Certainly, numerous people are interested in exotic animals, including marine mammals. Kellert found that a majority of Americans objected to the captive display of marine mammals in zoos and aquariums if there were no demonstrated educational and scientific benefits. They were concerned with the care given to captive individuals. To date, no unequivocal data show that there are any significant educational and scientific benefits that help the animals, despite beliefs that such benefits accrue. An average zoo visitor spends only about 30 sec–2 min at a typical exhibit and only reads some signs about the animals. A number of surveys have shown that visiting zoos to be entertained was the predominant reason people went to the zoo. In one study at the Edinburgh Zoo in Scotland, only 4% of zoo visitors went there to be educated, and no one specifically stated they went to support conservation. To date, very few empirical data support the notion that much educational information is learned *and* retained that helps the animals in the future. Indeed, some people worry that keeping animals in captivity for humans to view carries the message that it is all right to do so.

Many questions center on how individuals are captured, transported, and kept in various types of captive situations. Animals often are injured and otherwise stressed during capture and transport. Family groups may be broken up and the social structure of populations decimated. The effects of changing the social structure of wild populations are little known. Well-intentioned people often argue that the lives of captive animals are better, of higher quality, than those of wild relatives, but available data for marine mammals suggest that this claim is not well supported. From an ethical perspective, one must consider whether this claim is even relevant, for keeping animals in captivity radically alters numerous behavior patterns that have evolved over millennia. Predation, starvation, and disease are part of what it is to be wild. Is a longer unnatural life in captivity better than a shorter natural life in the wild?

Breeding surplus animals for profit (e.g., polar bears who become the center of media parades and then are moved to other zoos when their resource value or utility is fully exploited) also demands serious discussion. Similarly, the trading, donating, or loaning of unwanted or surplus animals who cannot be released into the wild—treating them as property—also raises numerous ethical questions.

The benefits of keeping marine mammals in captivity, to the animals themselves, remain unknown. Because the social and physical environments of marine mammals are virtually impossible to replicate in captivity, ethical questions arise when these animals are maintained in unnatural environments. There can be little doubt that the quality of life is compromised. In captivity, evolved patterns of foraging, care giving, and migrating are lost as are natural patterns of social organization (group size and composition). In captivity, for practical reasons, group sizes may be much smaller than those observed in wild relatives. Stereotyped behaviors often result from conditions of captivity, as do self-mutilation and unusually high levels of aggression. Furthermore, individuals often cannot escape from the glaring eye of the public and cannot choose when and where to rest.

E

**E**

There also seems to be higher mortality (spontaneous abortions, still-births) in captive vs wild individuals (especially killer whales). There is also higher mortality for adult killer whales in captivity. Limited data on annual survival rates suggest that there is high mortality during acclimation to captivity and differences in annual survival rates among different species and age classes within species.

There is little evidence that people leave zoos or aquariums with any long-lasting sentiments or knowledge that benefit either the animals they have seen or their wild relatives. Furthermore, few zoos are engaged in conservation efforts for marine mammals. The Association of Zoos and Aquariums (AZA) which oversees zoos in the United States and grants accreditation if they meet certain standards, admits in its *own* executive summary that "Little to no systematic research has been conducted on the impact of visits to zoos and aquariums on visitor conservation knowledge, awareness, affect, or behavior."

Many ethical concerns are also raised because first and foremost, zoos are businesses and their bottom line centers on money. It costs an enormous amount of money to bring marine mammals into captivity and to keep them there. It has been suggested that the money used to capture, transport, and keep animals in captivity would be better used to do research in the wild. Also, much money is spent on public relations and not on the animals themselves. Some feel that the images of nature that are represented to the public are a manufactured corporate point of view that centers more on what the public wants than what is good for the animals. Witness the existence of numerous "Flippers" (the prototypical dolphin) and "Shamus" (the model killer whale) whose lives do not resemble even closely the lives of free-living conspecifics or relatives.

Similar questions are raised when considering research on captive animals. Certainly, information may be gathered about various aspects of their lives (e.g., maternal behavior, self-recognition, social behavior, communication, and cognitive capacities). However, research on captive animals is being increasingly carefully scrutinized by some researchers, philosophers, many universities, and various funding agencies. Some relevant questions include: Is it ever permissible to keep individuals in captivity regardless of their utility, is the knowledge that is gained by studying captive individuals justified by keeping them in cages or tanks, and could more reliable data be collected under more natural conditions? Very little still is known about the life histories of most marine mammals. For many people it is the benefits that the captive individuals and other members of their (or other) species might accrue that is central, not the benefits that humans might gain. However, rarely are results used to benefit the animals other than in learning about medical treatments and husbandry to make their lives in captivity better. Rarely do wild individuals benefit from work done on captive relatives.

### V. Research Ethics

In addition to questions concerned with how humans treat other animals, the study of ethics also considers questions dealing with such areas as (1) the context of research (where it is done, are local people involved when researchers "go into the field" in countries other than their own, are local customs and beliefs about native fauna respected); (2) scientific integrity (researchers' responsibility for integrity in data collection, analysis, and dissemination); (3) the ownership of data (do data "belong" to any single person or to the team that is involved in their collection, analysis, and dissemination); (4) authorship (whose names should appear on a publication and in what order); and (5) individual responsibility for the integrity of a project as a whole and for the integrity of the results. A good deal of

trust is involved in all research, and questions that arise in these general areas require, and are receiving, much attention in the scientific community. Studies of marine mammals often require large teams of people, some of whom have never met, and it is important for all individuals to realize that each is responsible not only for his or her involvement, but for the composite product that is generated.

Another area of concern, some aspects of which are included in this volume, is research methodology (trapping, marking, tracking, and observing animals; experimentally manipulating social groups, food supply, and habitat). Often, human intrusions have major effects on animals' behavior even if they are unintentional. For example, the mere handling of individuals can influence their behavior and their acceptance back into a group, as can fitting individuals with various telemetric devices. Tracking or stalking animals can lead to changes in their activity patterns so that they spend more time avoiding humans than feeding or giving care. Most data come from animals other than marine mammals, and future studies of the effects of various methods are needed. Ethical considerations require that we learn about the effects of research methodologies and attempt to avoid them. In some cases the methods used may preclude collecting data relevant to the questions at hand.

### VI. Ecotourism

Ecotourism (whale watching, swimming with wild dolphins, photographing animals, visiting pinniped rookeries), some aspects of which are discussed in this volume, also raises numerous ethical questions concerning human intrusions into the lives of other animals. When, if ever, this activity is justified requires serious debate. People often try to interact with wild marine mammals but do not attempt to pet wild zebras (*Equus* spp.) or lions (*Panthera leo*). What principles underlie these differences in attitude?

Whether ecotourism is less intrusive on the lives of marine mammals than various research practices awaits further study. Indeed, there are observations of humans causing seal pups to stampede and being trampled and of humans striking and injuring animals with boats. It is essential to educate the public of possible negative effects of ecotourism.

### VII. The Future: Being Proactive

Kellert's study of American perceptions of marine mammals and their management shows clearly that most people support the various goals of the US Marine Mammal Protection Act. Most are willing to "render significant sacrifices to sustain and enhance marine mammal populations and species … These findings clearly indicate that marine mammals possess considerable aesthetic, scientific, and moral support among the great majority of Americans today" (1999, pp. iv–v).

It is in the best traditions of science to ask questions about ethics; it is not anti-science to question what we do when we interact with other animals. Ethics can enrich our views of other animals in their own worlds and in our different worlds and help us to see that their lives are worthy of respect, admiration, and appreciation. Indeed, it is out of respect, admiration, and appreciation that many humans seek out the company of whales, dolphins, polar bears, and other marine mammals.

Many ethical issues are extremely difficult to reconcile and generate highly charged and deep emotional and passionate responses. Achieving a win-win situation for animals and humans will be very difficult. However, it is clear that the increasingly detailed attention being given to various sorts of human–marine mammal interactions is showing that there are innumerable negative effects on the lives of the animals. While many negative influences have been anticipated or are not surprising, the severity of human influences has not been fully appreciated. We must be careful not to love these animals to their (or our) deaths. Humans are indeed dangerous to marine mammals and they are dangerous to us.

The study of ethics can also broaden the range of possible ways in which we interact with other animals without compromising their lives. Ethical discussions can help us see alternatives to past actions that have not served us or other animals well. Thus, the study of ethics can be enriching to other animals and to us. If we believe that ethical considerations are stifling and create unnecessary hurdles over which we must jump in order to get done what we want to accomplish, then we will lose rich opportunities to learn more about other animals and also ourselves. Our greatest discoveries come when our ethical relationships with other animals are respectful and not exploitive.

Allowing human interests always to trump the interests of other animals is not the best strategy if we are to solve the numerous and complex problems at hand. We need to learn as much as we can about the lives of wild animals. Our ethical obligations also require us to learn about the innumerable ways in which we influence animals' lives when we study them in the wild and in captivity and what effects captivity has on them. As we learn more about how we influence other animals, we will be able to adopt proactive, rather than reactive, strategies.

The fragility of the natural order requires that people work harmoniously so as not to destroy nature's wholeness, goodness, and generosity. The separation of "us" (humans) from "them" (other animals) engenders a false dichotomy, the result of which is a distancing that erodes, rather than enriches, the possible numerous relationships that can develop among all animal life.

Public education is critical. However, it does not always work (Cunningham-Smith *et al.*, 2006). To disseminate information about what is called the "human dimension," administrators of zoos, wildlife theme parks, aquariums, and areas where animals roam freely should inform visitors of how they may influence the behavior of animals they want to see. Tourism companies, nature clubs and societies, and schools can do the same. By treading lightly, humans can enjoy the company of other animals without making them pay for our interest in their fascinating lives. Our curiosity about other animals need not harm them.

Many marine mammals are closely linked to the wholeness of many ecosystems, and how they fare is tightly associated with how communities and ecosystems fare. By paying close attention to what we do to them, and why we do, what we do, where, and when we do it, we can help maintain the health of individuals, species, populations, and ecosystems. Concerning animal welfare, we can always do better. Quite often, "good welfare" is not "good enough."

## Acknowledgments

I thank Toni Frohoff, Robert Hofman, Dale Jamieson, Naomi Rose, and Trevor Spradlin for comments on a previous draft of this paper. Trevor Spradlin kindly sent me voluminous material dealing with human–dolphin interactions, much of which I can only summarize here. Support for summary statements can be found, for the most part, in J. R. Twiss, Jr., and R. R. Reeves (eds.) "Conservation and Management of Marine Mammals," Smithsonian Institution Press, Washington, DC, and in other sections of this encyclopedia.

## See Also the Following Articles

Captivity ■ Hunting of Marine Mammals ■ Marine Parks and Zoos ■ Whale Watching

## References

AZA, Executive summary: Visitor learning in zoos and aquariums. http://www.aza.org/ConEd/VisitorLearning/Documents/VisitorLearning ExecutiveSummary.pdf

Beck, A., and Katcher, A. (1996). "Between Pets and People: The Importance of Animal Companionship (revised edition)." Purdue University Press, Lafayette.

Bekoff, M. (1998). "Encyclopedia of Animal Rights and Animal Welfare." Greenwood, Westport.

Bekoff, M. (2001). Human–carnivore interactions: Adopting proactive strategies for complex problems. *In* "Carnivore Conservation" (J. L. Gittleman, S. M. Funk, D. W. Macdonald, and R. K. Wayne, eds). Cambridge University Press, London.

Bekoff, M. (2006a). "Animal Passions and Beastly Virtues: Reflections on Redecorating Nature." Temple University Press, Philadelphia.

Bekoff, M. (2006b). Animal emotions and animal sentience and why they matter: Blending "science sense" with common sense, compassion and heart. *In* "Animals, Ethics, and Trade" (J. Turner, and J. D'silva, eds), pp. 27–40. Earthscan Publishing, London.

Bekoff, M. (2006c). The public lives of animals: A troubled scientist, pissy baboons, angry elephants, and happy hounds. *J. Consc. Stud.* **13**, 115–131.

Bekoff, M. (2007a). Aquatic animals, cognitive ethology, and ethics: Questions about sentience and other troubling issues that lurk in turbid water. *Dis. Aquat. Org.* **75**, 87–98.

Bekoff, M. (2007b). "The Emotional Lives of Animals: A Leading Scientist Explores Animal Joy, Sorrow, and Empathy—and Why They Matter." New World Library, Novato.

Bekoff, M., and Jamieson, D. (1991). Reflective ethology, applied philosophy, and the moral status of animals. *Perspect. Ethol.* **9**, 1–47.

Bekoff, M., and Jamieson, D. (1996). Ethics and the study of carnivores. *In* "Carnivore Behavior, Ecology, and Evolution" (J. L. Gittleman, ed.), pp. 16–45. Cornell University Press, Ithaca.

Coghlan, A. (2006). Whales boast the brain cells that 'make us human'. *New Scientist* **27**, November. http://www.newscientist.com/article/dn10661-whales-boast-the-brain-cells-that-make-us-human.html

Cunningham-Smith, P., Colbert, D. E., Wells, R. S., and Speakman, T. (2006). Evaluation of human interactions with a provisioned wild bottlenose dolphin (*Tursiops truncatus*) near Sarasota Bay, Florida, and efforts to curtail the interactions. *Aquat. Mamm.* **32**, 346–356.

Davis, S. G. (1997). "Spectacular Nature: Corporate Culture and the Sea World Experience." University of California Press, Berkeley, CA.

Food and agricultural organization of the United Nations 2006. http://www.fao.org/newsroom/en/news/2006/1000239/index.html

For cod's sake, act now (2006). *New Scientist* **11**, November, p. 5.

Fox, M. W. (1997). "Eating with Conscience." NewSage Press, Troutdale, OR.

Francione, G. L. (1999). "Introduction to Animal Rights: Your Child or the Dog?." Temple University Press, Philadelphia.

Frohoff, T. G. (1998). In the presence of dolphins. *In* "Intimate Nature: The Bond Between Women and Animals" (L. Hogan, D. Metzger, and B. Peterson, eds), pp. 78–84. Ballantine, New York.

Frohoff, T. G., and Packard, J. M. (1995). Human interactions with free-ranging and captive bottlenose dolphins. *Anthrozoös* **8**, 44–53.

Herzing, D. L., and White, T. I. (1998). Dolphins and the question of personhood. *Etica Anim.* **9**, 64–84.

Iannuzzi, D., and Rowan, A. N. (1991). Ethical issues in animal-assisted therapy programs. *Anthrozoös* **4**, 154–163.

E

Jamieson, D., and Regan, R. (1985). Whales are not cetacean resources. *In* "Advances in Animal Welfare Science, 1984" (M. W. Fox, and L. Mackley, eds), pp. 101–111. MartinusNijhoff, The Hague.

Kellert, S. R. (1996). "The Value of Life: Biological Diversity and Human Society." Island Press, Washington, DC.

Kellert, S. R. (1999). "American Perceptions of Marine Mammals and their Management." Humane Society of the United States, Washington, DC.

Kirkwood, J. K., Bennett, P. M., Jepson, P. D., Kuiken, T., Simpson, V. R., and Baker, J. R. (1997). Entanglement in fishing gear and other causes of death in cetaceans stranded on the coasts of England and Wales. *Vet. Rec.* **141**, 94–98.

Lavigne, D. M., Scheffer, V. B., and Kellert, S. R. (1999). The evolution of North American attitudes toward marine mammals. *In* "Conservation and Management of Marine Mammals" (J. R. Twiss, Jr., and R. R. Reeves, eds), pp. 10–47. Smithsonian Institution Press, Washington, DC.

Mackenzie, D. (2006). Glimmer of hope for "doomed" fish. *New Scientist* **11**, 10, November.

Marino, L., and Lilienfeld, S. O. (1998). Dolphin-assisted therapy: Flawed data, flawed conclusions. *Anthrozoös* **11**, 194–200.

Moccia, R. D., and Duncan, I. J. H. (2004). Investigating fear in domestic rainbow trout, *Oncorhynchus mykiss*, using an avoidance learning task. *Appl. Anim. Behav. Sci.* **87**, 343–354.

Nathanson, D. W. (1998). Reply to Marino and Lilienfeld. *Anthrozoös* **11**, 201–202.

Nollman, J. (1999). "The Charged Border: Where Whales and Humans Meet." Holt, New York.

Osinga, N., and 't Hart, P. (2006). Fish-hook ingestion in seals (*Phoca vitulina* and *Halichoerus grypus*): The scale of the problem and a non-invasive method for removing fish-hooks. *Aquat. Mamm.* **32**, 261–264.

Raloff, J. (2006). New estimates of the shark-fin trade. *Science News*, 4 November. http://www.sciencenews.org/articles/20061104/food.asp

Rose, N., and Farinato, R. (1995/1999). "The Case Against Marine Mammals in Captivity." Humane Society of the United States, Washington, DC.

Samuels, A., and Spradlin, T. R. (1995). Quantitative behavioral study of bottlenose dolphins in swim-with-dolphin programs in the United States. *Mar. Mamm. Sci.* **11**, 520–544.

Sneddon, L. U. (2003). The evidence for pain in fish: The use of morphine as an analgesic. *Appl. Anim. Behav. Sci.* **83**, 153–162.

Verrengia, J. (2003). Nearly 1,000 whales drowning daily in fishing nets. www.eurocbc.org/bycatch

# Extinctions, Specific

### Deborah A. Duffield

Extinction of species has been occurring since the dawn of life on Earth. Sometimes this process is a gradual one, with a few species disappearing over a long period of time. Occasionally, the process appears to be quite rapid and widespread; such mass extinctions often result in the disappearance of many of the flora and fauna on Earth in a relatively short period of time. Such a mass extinction is currently taking place, only this time the cause is not climactic, tectonic, or cosmic, but rather is the result of the activities of one species, *Homo sapiens* (Domning, 1999). It is well established that many species of animals have disappeared over the last 400 years because of humans, including such well-known examples as the passenger pigeon, dodo, quagga, and great auk. Many other lesser-known species of animals, along with many representatives of the plant

kingdom, have also died out because of human activities. It is thought that a total of 1190 species of plants and animals have gone extinct since 1600 (Smith *et al.*, 1993). When this is combined with the countless number of fungi, protists, and bacteria that have disappeared without notice, it is easy to see why this human decimation of the Earth's living organisms may easily rank among one of the most major mass extinctions that are a part of our planet's history.

Until recently, the effect of human activities has been primarily felt by those species whose members are relatively large and conspicuous. The great Pleistocene extinctions of mammoths and horses in North America, along with the demise in more recent times of many species of large, flightless birds, are examples of this. Being relatively large, it would be expected that marine mammals would have suffered the same fate as their terrestrial counterparts. The size and remoteness of the habitat of most marine mammals, however, spared them the slaughter and resultant slide toward extinction until more recently. In the last few hundred years, technology gave humans the ability to seek out and kill these animals throughout the vast expanses of the ocean, and many marine mammal species have been brought to the verge of extinction. Lack of profitability from increased costs inherent in trying to find fewer animals, coupled with increased public awareness of and resistance to the destruction, has brought a halt to the exploitation of many of these marine mammals, and in some cases the numbers of certain species are on the rise. This fortunate development came too late for three species of marine mammal: Steller's sea cow (*Hydrodamalis gigas*), the Caribbean monk seal (*Monachus tropicalis*), and the Japanese sea lion (*Zalophus japonicus*). Most recently the river dolphin of the Yantee River (the baiji, *Lipotes vexillifer*) has become extinct. In addition, one population has become extinct in recent times, the Atlantic gray whale (*Eschrichtius robustus*). Direct human destruction was the primary cause of these extinctions, and in all cases the story is one of human ignorance, shortsightedness, and greed.

## I. Steller's Sea Cow

In the late autumn of 1741, a Russian exploratory vessel, the *Saint Peter*, wrecked on a bleak, uninhabited island near the western end of the Aleutian chain (Ford, 1966). This island was completely isolated, save for smaller nearby Copper Island, and was over 200 miles from the nearest Russian settlement on the Kamchatka peninsula. As food supplies were practically non-existent in the winter, it was a joyous occasion when it was discovered that the nearshore waters around the island were inhabited by huge, slow-moving, previously unknown marine mammals. These were Steller's sea cows, later named after the naturalist accompanying the voyage, Georg Wilhelm Steller. Upon killing, this animal provided large quantities of beef-like meat and almond-tasting oil. Throughout the winter, the crew members of the *Saint Peter* slaughtered sea cows. When in the summer of 1742, the men who survived the winter reached Kamchatka, they spread the word of the wealth in furs to be had in the Bering Sea and Alaska and of the huge sea cow that would provide food necessary for the required long voyages. From then on, fur traders and hunters would stop at Bering Island and Copper Island to slaughter these animals and stock their vessels with meat and oil. In addition, parties hunting fur-bearing animals would winter on these islands and dine on sea cow meat (Stejneger, 1887). By 1754, the sea cow was gone from Copper Island. With its disappearance 14 years later from Bering Island, *Hydrodamalis gigas* was extinct (Haley, 1978).

What little is known about this species comes from the observations and written accounts of Georg Steller and from bones found on Bering and Copper Islands. Members of this species were large, fusiform animals with a relatively small head and a horizontally flattened tail. Individuals grew to about 25 ft in length and probably weighed upward of 4 tons. They had short forelimbs with no digits, a tough hide, and thick layers of fat for insulation. These characteristics permitted the animals to survive in the frigid, storm-tossed waters of the north Pacific, far from the tropical areas inhabited by their nearest living relatives, manatees, and dugongs (Domning, 1978). Like other SIRENIANS, these animals were herbivorous, grazing on the large kelp beds that grew in the shallow waters around the Commander Islands. They had no teeth, but rather two flat, horny chewing plates, one in each jaw, between which the animals mashed kelp. Steller observed that they spent most of their day feeding, head down, with their backs exposed. They lived in shallow, sandy areas, preferring the mouths of running streams with their input of freshwater. They may not have been capable of diving (or if so, only near the surface), a characteristic that would have restricted them to the intertidal area outward to the seaward edge and to the surface of this zone (Reeves *et al.*, 1992).

This shallow restricted range, coupled with their slow-moving and docile habits, made Steller's sea cows easy prey for human hunters, even those with relatively primitive weapons and boats. Grazing animals could be easily approached, harpooned, dragged ashore, and butchered (Stejneger, 1936). Often a wounded animal would escape from its killers and swim out to sea. It has been estimated that only one out of every five sea cows killed was actually utilized for food. Even though this species apparently lived all along the rim of the North Pacific into the late Pleistocene, by the early eighteenth century it was confined as a relict population to the shallow waters around the Commander Islands. This reduction of numbers was thought to be possibly a result of hunting by indigenous peoples (Whitmore, 1977). In 1741, the population was estimated to be no more than 2000 individuals, and relentless hunting pressure, together with the wasteful harvest method, led the complete extermination of Steller's sea cow by 1768 (Stejneger, 1887). Even though there have been reports of sightings in the North Pacific of animals reputed to be sea cows, some as recently as 1977, it is almost certain that *H. gigas* was driven to extinction over 200 years ago.

## II. Caribbean Monk Seal

Nearly 300 years before the Bering expedition and almost half a world away, the demise of another species of marine mammal at the hands of Europeans began. In 1494, on his second voyage to the Caribbean, Columbus discovered the Caribbean monk seal (*Monachus tropicalis*) on the coast of Santo Domingo (Hispaniola). He ordered his men to kill eight of these "sea wolves" for food, and from that time on the killing never ceased (King, 1983). Probably confined to sandy beaches on remote islands and atolls because of centuries of aboriginal hunting on large islands and the mainland, the monk seals were sought out and ruthlessly slaughtered by fishermen, turtle hunters, buccaneers, and organized sealers. In the Bahamas, up to 100 seals per night were known to be killed (Maxwell, 1967). By the mid-nineteenth century, many zoologists thought that the Caribbean monk seal was already extinct. However, in 1886 a small herd was discovered near the Yucatan peninsula, and because so little was known about the species, North American museums sent expeditions out to gather specimens. One of these expeditions killed 49 seals, 5 of which were pregnant females with near-term fetuses (Ward, 1887). For a species in such precarious circumstances, such activities of science were also of no help. The last confirmed sighting was reported by C. Bernard Lewis, Director of the Institute of Jamaica, who

said that until 1952 a small colony existed at Seranilla Bank, halfway between Jamaica and Honduras (Rice, 1973). Since that time, although some pinnipeds have been sighted in the Caribbean (these probably being individuals of other seal or sea lion species outside of their normal range), no confirmable sighting of a monk seal in the Caribbean has been made. In 1996, the International Union for the Conservation of Nature declared *M. tropicalis* extinct.

Even though known to Europeans for over 400 years, little scientific information was obtained concerning the Caribbean monk seal before its disappearance. The species belongs to the family Phocidae, and with the Mediterranean monk seal (*Monachus monachus*) and the Hawaiian monk seal (*Monachus schauinslandi*) comprise the genus *Monachus*. The Caribbean species was apparently very similar to the other two monk seals, with males reaching 6–7 ft in length, and females somewhat smaller. The fur of adults was brown on back with a gray tinge and yellowish on the underside. The fur of pups was dark to black. Pupping probably peaked in December, a time different from that of the other two species (Kenyon, 1981). At the time of Columbus, the range of the seal included the shores and islands of the Caribbean Sea and Gulf of Mexico, eastward to the Bahamas and Florida Keys and westward to the Yucatan Peninsula and the Central American coast. The most primitive of all the seals, the three monk seal species probably evolved in areas remote from land predators and never learned to flee nor defend themselves from animals such as humans. In effect, they were "genetically tame" and could easily be clubbed to death by human exploiters (Kenyon, 1981). Like all monk seals, *M. tropicalis* would have been very sensitive to human disturbance, a fact that usually leads to poor reproduction and pup survival (Bonner, 1990). As humans spread through the Caribbean, these characteristics would ensure that the species declined quickly. It became so scarce by 1887 that it was referred to as an "almost mythical species" (Allen, 1887). Its persecution and demise continued, however, and sometime after 1900 the species went into its inexorable slide to extinction. Although it probably survived until the 1950s, represented mostly by old individuals past reproductive age, the species was doomed due to human slaughter and habitat destruction (Kenyon, 1977). Hoping that the species might somehow have survived in some remote area, extensive surveys were conducted of suitable monk seal habitat in both 1973 and 1984. Sadly, no evidence of the Caribbean monk seal was found, only signs of human habitation and use (LeBoeuf *et al.*, 1986). Although reports of sightings have surfaced into the last decade, the species is almost certainly extinct. As stated by biologist Karl Kenyon (1977): "My conclusion from the 1973 survey is that the Caribbean monk seal has been extinct since the early 1950s. The fact that I saw no monk seals was not as important as the fact of ubiquitous human presence … Even if a few old Caribbean monk seals had survived to the 1970s, all available evidence leads me to believe that there is no hope that the species can recover. Man has now dominated its environment."

## III. Japanese Sea Lion

The Japanese sea lion, formerly considered to be a subspecies of the California sea lion, has now been recognized as a separate species, *Zalophus japonicus* (Rice, 1998). Currently considered extinct by the IUCN, it is possible that a small number of Japanese sea lions may still exist in Korean waters, although the animals reported in Korea may actually be Steller sea lions, *Eumetopias jubatus* (Nishiwaki, 1973). The Japanese sea lion formerly inhabited the southern sea of Japan and coastal waters of Japan from Hokkaido to Kyushu. No individuals of this species have been seen in Japanese waters for over 50 years.

E

Little is known concerning the species, but it is felt to have been morphologically similar to its counterpart, the California sea lion (*Zalophus californianus*). The Japanese sea lion has probably disappeared due to direct hunting pressure and habitat destruction.

## IV. Extinct Population

### A. Atlantic Gray Whale

The gray whale (*Eschrichtius robustus*) is best known as a coastal dwelling cetacean that migrates along the westerncoast of North America from Alaska to Baja California. Almost exterminated by whalers, this population has rebounded as a result of being completely protected, except for aboriginal subsistence hunting (Reilly, 1984). Another population is found in the western North Pacific, where it migrates between the Okhotsk Sea and southern South Korea. This population has been reduced to very low numbers due to overexploitation and is currently highly endangered. A third population in the western North Atlantic appears to have existed until the seventeenth century and was probably rendered extinct by early whaling activity (Mead and Mitchell, 1984). Accounts from this period, although somewhat confusing, describe a whale known as the "scrag" whale, which bears a strong resemblance to the gray whale. Earlier Icelandic accounts describe a "sandloegja," which is also felt to have been a gray whale (Fraser, 1970). Subfossil remains of gray whales have been found in eastern North America, and radioactive dating has shown the latest to have been from approximately 1675 AD (Mead and Mitchell, 1984). The population went extinct shortly thereafter. Subfossil remains of gray whales have also been found in Europe, but are apparently much older. Therefore, an eastern Atlantic population of gray whales probably also occurred in historical times, apparently going extinct sometime before 500 AD, quite possibly at the hands of early European whalers (van Deinse and Junge, 1937).

## V. Prospects for the Future

Even though few species of marine mammals have gone extinct at the hands of humans, many have come very close. Elephant seals, fur seals, monk seals, walruses, and sea otters all have narrowly escaped extinction. Among cetaceans, the great rorquals, the gray whale, and the right whales were all nearly exterminated. Some have rebounded, some appear to be slowly increasing; others apparently are not increasing and may never recover. The history of sealing and whaling makes for depressing reading. Although these industries have mostly disappeared in today's world (with some notable exceptions), some species of marine mammals are still highly endangered. The Mediterranean monk seal is down to less than 1000 individuals and is being forced into tiny pockets of habitat by an explosion of tourism. The vaquita (*Phocoena sinus*), a small porpoise from the northern Gulf of California, exists as a small population under pressure from unintentional destruction from fisheries (Rojas-Bracho and Taylor, 1999). Many other species are also endangered, although they are not in as precarious a state as these. As much as may be desired, it will never again be possible to observe the "sea wolves" of Columbus lounging on the tropical beaches of the Caribbean or Steller's sea cow rising out of the northern mists.

## See Also the Following Articles

California, Galapagos and Japanese Sea Lions ■ Endangered Species and Populations ■ Monk Seals ■ Steller's Sea Cow

## References

Allen, J. A. (1887). The West Indian seal (*Monachus tropicalis*, Gray). *Bull. Amer. Mus. Nat. Hist.* **2**, 1–34.

Bonner, W. N. (1990). "The Natural History of Seals." Facts on File, New York, p. 197.

Domning, D. P. (1978). Sirenian evolution in the North Pacific Ocean. *In* "University of California Publications in Geological Sciences", Vol. 118. University of California Press, Berkeley, CA, p. 176.

Domning, D. P. (1999). Endangered species: The common denominator. *In* "Conservation and Management of Marine Mammals" (J. R. Twiss, Jr., and R. R. Reeves, eds), pp. 332–341. Smithsonian Institution Press, Washington, DC.

Ford, C. (1966). "Where the Sea Breaks Its Back." Little, Brown, Boston, p. 206.

Fraser, F. C. (1970). An early 17th century record of the California gray whale in Icelandic waters. *Invest. Cetacea* **Vol. 2**, 13–20.

Haley, D. (1978). Steller sea cow. *In* "Marine Mammals" (D. Haley, ed.), pp. 236–241. Pacific Search Press, Seattle.

Kenyon, K. W. (1977). Caribbean monk seal extinct. *J. Mammal.* **58**, 97–98.

Kenyon, K. W. (1981). Monk seals. *In* "Handbook of Marine Mammals" (S. H. Ridgway, and R. J. Harrison, eds), Vol. 2, pp. 195–220. Academic Press, London.

King, J. E. (1983). "Seals of the World." Cornell University Press, Ithaca, NY, p. 240.

LeBoeuf, B. J., Kenyon, K. W., and Villa-Ramirez, B. (1986). The Caribbean monk seal is extinct. *Mar. Mamm. Sci.* **2**, 70–72.

Maxwell, G. (1967). "Seals of the World." Constable, London, p. 153.

Mead, J. G., and Mitchell, E. D. (1984). Atlantic gray whales. *In* "The Gray Whale" (M. L. Jones, S. L. Swartz, and S. Leatherwood, eds), pp. 33–53. Academic Press, Orlando, FL.

Nishiwaki, M. (1973). Status of the Japanese sea lion. *In* "Seals," pp. 80–81. New Series Supplementary Paper No. 39. IUCN, Morges, Switzerland.

Reeves, R. R., Stewart, B. S., and Leatherwood, S. (1992). "The Sierra Club Handbook of Seals and Sirenians." Sierra Club Books, San Francisco, CA, p. 359.

Reilly, S. B. (1984). Assessing gray whale abundance: A review. *In* "The Gray Whale" (M. L. Jones, S. L. Swartz, and S. Leatherwood, eds), pp. 203–223. Academic Press, Orlando, FL.

Rice, D. W. (1973). Caribbean monk seal (*Monachus tropicalis*). *In* "Seals," pp. 98–112. New Series Supplementary Paper No. 39. IUCN, Morges, Switzerland.

Rice, D. W. (1998). "Marine Mammals of the World: Systematics and Distribution," Special Publication 4. Society for Marine Mammals, Lawrence, KS, p. 231.

Rojas-Bracho, L., and Taylor, B. L. (1999). Risk factors affecting the vaquita (*Phocoena sinus*). *Mar. Mamm. Sci.* **15**, 974–989.

Smith, F. D. M., May, R. M., Pellew, T. H., Johnson, T., and Walter, K. (1993). How much do we know about the current extinction rate? *Trends Ecol. Evol.* **8**, 375–378.

Stejneger, L. (1887). How the great northern sea cow (*Rytina*) became exterminated. *Am. Nat.* **21**, 1047–1054.

Stejneger, L. (1936). "Georg Wilhelm Steller, the Pioneer of Alaskan Natural History." Harvard University Press, Cambridge, MA, p. 623.

van Deinse, A. B., and Junge, G. C. A. (1937). Recent and older finds of the California gray whale in the Atlantic. *Temminckia* **2**, 161–188.

Ward, H. L. (1887). Notes on the life history of *Monachus tropicalis*, the West Indian seal. *Amer. Nat.* **21**, 257–264.

Whitmore, F. C., Jr., and Gard, L. M., Jr. (1977). "Steller's Sea Cow (*Hydrodamalis gigas*) of the Late Pleistocene Age from Amchitka, Aleutian Islands, Alaska." US Government Printing Office, Washington, DC, p. 20.

# False Killer Whale
## *Pseudorca crassidens*

### Robin W. Baird

## I. Characteristics and Taxonomy

The false killer whale is one of the larger members of the family Delphinidae, with adult males reaching lengths of almost 6 m and females reaching up to 5 m. The common name comes from similarity not in external appearance to the killer whale (*Orcinus orca*) but rather in skull morphology of these two species. In fact, the two species do not appear to be closely related; based on genetic similarity, false killer whales appear to be most closely related to the Risso's dolphin (*Grampus griseus*), melon-headed whale (*Peponocephala electra*), pygmy killer whale (*Feresa attenuata*), and pilot whales (*Globicephala* spp.). There is evidence of geographic variation in skull morphology (Kitchener *et al.*, 1990), but no subspecies are currently recognized.

Largely black or dark gray in color (usually with a lighter blaze on the ventral surface between the flippers), it is easily recognizable with its rounded head, gracile shape (Fig. 1), small falcate

dorsal fin located at the midpoint of the back, and distinctive flippers (with a bulge on the leading edge). Scars from inter- and intraspecific interactions eventually are re-pigmented, unlike in the closely related Risso's dolphin. False killer whales are slightly sexually dimorphic, with the melon of males protruding farther forward than in females. Their teeth are large and conical, with 7–11 in each of the upper jaws and 8–12 in each lower jaw.

## II. Distribution and Abundance

False killer whales are found in all tropical and warm temperate oceans of the world, and occasional records of their presence in cold temperate waters have also been documented. Although they are typically characterized as pelagic in habits, they do approach close to shore and utilize shallow waters around oceanic islands. These oceanic habits have hindered the study of this species in the wild, and most of what is known comes from stranded individuals, captive animals, and limited observations of groups around oceanic islands. In the Pacific there is evidence of limited gene flow, and the population around the main Hawaiian Islands is demographically isolated from the rest of the tropical Pacific (Chivers *et al.*, 2007). No estimates of worldwide population size are available, although false killer whales appear to be naturally uncommon throughout their range. Regional estimates for the Hawaiian Islands Exclusive Economic Zone suggest a small population size, in the low hundreds of individuals. No information on population trends is available.

## III. Ecology

False killer whales are one of the handful of species that regularly mass strand, with the largest stranding recorded of 835 individuals. The diet appears to be diverse, in terms of both species and size of prey (Fig. 2). In general they feed on a variety of oceanic squid and fish but have also been documented feeding on smaller delphinids being released from tuna purse-seines in the eastern tropical Pacific. One case of predation on a humpback whale (*Megaptera novaeangliae*) calf has also been recorded, and they have been documented attacking sperm whales (*Physeter macrocephalus*). Nonaggressive

**Figure 1** *The highly acrobatic false killer whale (*Pseudorca crassidens*) leaping while chasing prey. The false killer whale does not resemble the killer whale (*Orcinus orca*) in external appearance, although the skulls of the two species are quite similar. Photograph © Robin W. Baird.*

**Figure 2** *A false killer whale attacking a mahi-mahi (*Coryphaena hippurus*). Prey sharing in the wild and in captivity is frequently observed for this species. Photograph © Daniel J. McSweeney.*

interspecific associations with bottlenose dolphins (*Tursiops truncatus*) and rough-toothed dolphins (*Steno bredanensis*) have also been reported. No predators of false killer whales have been reported, although large sharks and killer whales likely take some individuals.

## IV. Behavior and Physiology

False killer whales are considered to be extremely social, usually traveling in groups of 20 to 100 individuals. Long-term (15 years) associations among individuals have been documented in Hawaiian waters, and analyses of associations of photo-identified individuals indicate strong bonds among individuals (Baird *et al.*, 2008). Such bonds are also evident from their propensity to strand *en masse*, and by the affiliative behavior of stranded animals. False killer whales are active during the day, and food sharing in the wild has been regularly recorded. Little is known about the diving behavior of this species; one tagged animal dove for up to 12 min and to depths of over 230 m.

## V. Life History

Life history information comes entirely from stranded individuals. Because the deposition rate of growth layer groups in the teeth has not been calibrated, there is some uncertainty in life history parameters. Both sexes are thought to mature between about 8 and 14 years of age, although there is some suggestion that males may mature later. Maximum longevity has been estimated at 57 years for males and 62 years for females (Kasuya, 1986). Calving interval for one population has been reported as almost 7 years, and calving may occur year-round, with a peak in late winter.

## VI. Interactions with Humans

A number of types of interactions between humans and false killer whales have been documented. In Hawaii they are regularly encountered by commercial whale- or dolphin-watching vessels 2nd often bowride. They have been maintained in captivity in a number of aquaria around the world, including in Japan, the United States, the Netherlands, Hong Kong, and Australia. They have been successfully bred in captivity in several locations, and there they have produced viable interspecies hybrids with bottlenose dolphins. False killer whales are one of several species of odontocetes that occasionally steal fish from both commercial and recreational fishermen, with these types of interactions noted in Japan, Hawaii, the Indian Ocean, and the Gulf of Mexico.

Conflicts with fisheries have resulted in direct killing in Japan. Small numbers have been occasionally taken in fisheries, both directly and incidentally as bycatch. In Hawaiian waters the number killed or seriously injured incidentally in the longline fishery is greater than the population is thought to be able to sustain. They are one of a growing list of species that has been recorded ingesting discarded plastic, and high levels of toxins have been documented in tissues collected from stranded animals. It is unknown, however, whether such toxins contribute to immunosuppression in this species.

### See Also the Following Articles

Delphinids ■ Indo-West Pacific Marine Mammals

### References

Acevedo-Gutierrez, A., Brennan, B., Rodriguez, P., and Thomas, M. (1997). Resightings and behavior of false killer whales (*Pseudorca crassidens*) in Costa Rica. *Mar. Mamm. Sci.* **13**, 307–314.

Baird, R.W. *et al.* (10 authors). (2008). False Killer Whales (*Pseudorca crassidens*) around the main Hawaiian Islands: long-term site sidelity, inter-island movements, and association patterns. *Mar. Mamm. Sci.* **24**, 591–612.

Brown, D. H., Caldwell, D. K., and Caldwell, M. C. (1966). Observations on the behavior of wild and captive false killer whales, with notes on associated behavior of other genera of captive delphinids. *Los Angeles County Mus. Contrib. Sci.* **95**, 1–32.

Chivers, S. J., Baird, R. W., McSweeney, D. J., Webster, D. L., Hedrick, N. M., and Salinas, J. C. (2007). Genetic variation and evidence for population structure in eastern North Pacific false killer whales (*Pseudorca crassidens*). *Can. J. Zool.* **85**, 783–794.

Kasuya, T. (1986). False killer whales. *In* "Report of Investigation in Search of Solution for Dolphin-Fishery Conflict in the Iki Island Areas" (T. Tamura, S. Ohsumi, and S. Arai, eds). Japan Fisheries Agency, Tokyo.

Kitchener, D. J., Ross, G. J. B., and Caputi, N. (1990). Variation in skull and external morphology in the false killer whale, *Pseudorca crassidens*, from Australia, Scotland and South Africa. *Mammalia* **54**, 119–134.

Koen Alonso, M., Pedraza, S. N., Schiavini, A. C. M., Goodall, R. N. P., and Crespo, E. A. (1999). Stomach contents of false killer whales (*Pseudorca crassidens*) stranded on the coasts of the Strait of Magellan, Tierra del Fuego. *Mar. Mamm. Sci.* **15**, 712–724.

Odell, D. K., and McClune, K. M. (1999). False killer whale *Pseudorca crassidens* (Owen, 1846). *In* "Handbook of Marine Mammals" (S. Ridgway, ed.), Vol. 6, pp. 213–243. Academic Press, New York.

Palacios, D. M., and Mate, B. R. (1996). Attack by false killer whales (*Pseudorca crassidens*) on sperm whales (*Physeter macrocephalus*) in the Galapagos Islands. *Mar. Mamm. Sci.* **12**, 582–587.

Purves, P. E., and Pilleri, G. (1978). The functional anatomy and general biology of *Pseudorca crassidens* (Owen) with a review of hydrodynamics and acoustics in Cetacea. *Invest. Cetacea* **9**, 67–227.

Stacey, P. J., and Baird, R. W. (1991). Status of the false killer whale, *Pseudorca crassidens*, in Canada. *Can. Field-Nat.* **105**, 189–197.

Stacey, P. J., Leatherwood, S., and Baird, R. W. (1994). *Pseudorca crassidens. Mamm. Sp.* **456**, 1–6.

# Feeding Morphology

CHRISTOPHER D. MARSHALL

## I. Functional Morphology

Functional morphology is a diverse field of biology that integrates anatomy, biomechanics, and behavior. It is the study of structure, its relationship to function, and organismal adaptation. Marine mammals, and their adaptations to the aquatic environment, have interested functional morphologists for long time. Accordingly, our knowledge of the anatomy of marine mammals is extensive for many species, and new data continues to compile quickly. However, direct experimental investigations are largely lacking relative to terrestrial mammals. This has been due to the difficulty of working with large mammals in an aquatic environment, and the lack of technology that can be taken in the field. As a result morphology has been used extensively to predict function and behavior of marine mammals. However, experimental work regarding functional and behavioral performance of marine mammals is beginning to flourish as electronics, and other technologies, become smaller, portable, and less expensive. When integrated, morphological inference and direct empirical measurements of function greatly enrich our knowledge of how marine mammals interact with their environment.

A basic necessity for survival of any organism is to feed and forage. The aquatic environment imparts strong selection pressures on feeding adaptations of marine mammals, as well as other life history attributes. Marine mammal adaptations for feeding are especially divergent relative to terrestrial mammals, and this will be the focus of this chapter. As in any comparative study, the phylogenetic history of the organism of interest is of paramount importance. This is particularly true for functional morphology and comparative biomechanics. The fact that "marine mammals" are a diverse collection of nonrelated mammals that have returned to the sea is not only an important consideration for functional studies but also it makes marine mammals an interesting study group regarding convergent evolution of form and function, and phylogenetic constraint.

## II. Using Morphology to Predict Behavioral Performance

The skulls of modern cetaceans are among the most derived among mammals. In contrast to the dog (commonly used as a "typical" mammal and a morphological baseline), cetaceans have drastically remodeled the morphology of the rostrum, nares, cranium, ear bones (petrosal bones), and mandible (Fig. 1). The maxilla and premaxilla elongate the facial region but the cranium is shortened because of overlapping (telescoping) of cranial bones. This disparity of facial length vs cranial length is variable among cetaceans (e.g., the short blunt face of *Globicephala* and the long narrow rostra of platanistids). Odontocete jaws can be virtually edentulous (e.g., ziphiids) or filled with several hundred simple homodont teeth (e.g., platanistids). The condylar processes of odontocete mandibles (articulation of the mandible with the skull) are simple (Fig. 1) and allows for only a simple dorsal–ventral motion; most mammals can also move their jaws side-to-side (herbivores excel at this). In general, delphinid muscles of mastication (temporalis, masseter, and pterygoids) are relatively small. The pterygoid muscles are the dominant muscle group. Orcas (*Orcinus orca*) are exceptions to this generalization; these delphinids possess dominant temporalis muscles, perhaps related to the requirements of taking large prey. Compared to the dog, skull attachments for the temporalis and masseter muscles of most delphinids are also reduced (i.e., zygomatic arch, and coronoid and angular processes; Fig. 1). Such comparative anatomical insights are far from esoteric. Functional inferences from skull anatomy and biomechanical measurements, such as simple lever mechanics, can be used to predict feeding behavior, and indirectly trophic ecology.

LEVERAGE—Levers are simple devices that transmit forces from one place to another using a pivot. For example, a person might want to move a large rock that is too heavy to pick up. With the aid of a rigid beam and a smaller rock, one could build a simple Type I lever by placing the tip of the beam under the large rock and pivoting the beam on top of the small rock. The large rock can be moved by imposing force on the beam at the end opposite to the rock. This force is known as the in-force ($F_{in}$), the small rock is the fulcrum, and the force generated to move the load is the out-force ($F_{out}$). The distance between the fulcrum and the in-force is the in-lever arm ($L_{in}$) and the distance between the fulcrum and the load is the out-lever arm ($L_{out}$). The directions of $F_{in}$ and $F_{out}$ are the lines of action.

In the simplest case, which provides the greatest leverage, the lines of action of $F_{in}$ and $F_{out}$ are parallel to each other and at right angles to the lever arms. Mechanical advantage (MA) is the ratio of the in-lever arm to out-lever arm and also the ratio of the out-force to the in-force (MA = $L_{in}/L_{out} = F_{out}/F_{in}$). A lever system in which an $F_{in}$ of 5 N results in an $F_{out}$ of 10 N would have a mechanical advantage of 10/5 or 2. Greater mechanical advantage is attained with levers that have a long $L_{in}$ a and short $L_{out}$. Another component to levers is velocity ($v$). Each lever arm possesses a velocity ($v_{in}$ and $v_{out}$), and the length of the lever arms influences the velocity such that $v_{in}L_{out} = v_{out}L_{in}$. This relationship is the reciprocal of the force–lever arm relationship ($F_{out}L_{out} = F_{in}L_{in}$). Therefore, there is a trade-off between mechanical advantage and velocity; powerful levers are slow and move only a short distance, whereas fast levers are not powerful, but move over longer distances. This latter type is characteristic of most biological levers.

*Biological Levers*—The musculoskeletal system of vertebrates is comprised of numerous levers, often in series, that are comprised of muscles, tendons, bones, and joints. Muscles are contractile elements that are attached to bones and cross bony joints by tendons; shortening of muscles produces movement at the joint. Biological levers tend to possess short in-lever arms and long in-lever arms, which impart rapid movement (high $v_{out}$) with large excursions of distance, but with reduced mechanical advantage. Other lever configurations include Type II and Type III levers. Both of these lever configurations are characterized by having the load and the in-force on the same side of the fulcrum. Type II levers place the load between the fulcrum and the in-force. Type III levers place the in-force between the fulcrum and the load. Mandibles are excellent examples of Type III levers. Lever mechanics of mandibles are useful for predicting feeding behavior. For example, consider the lever mechanics between a dog (*Canis familiaris*) and a bottlenose dolphin (*Tursiops truncatus*; Fig. 2). The temporal mandibular joint (TMJ) forms the fulcrum of the lever. The temporalis, masseter, and pterygoid muscles comprise the $F_{in}$ and the functions to close the lower jaw. For simplicity, only the temporalis muscle will be used in this example. The temporal fossa, the recessed space in which the temporalis muscle resides and attaches, can be used to estimate the muscular line of action. The temporalis muscle originates in this temporal fossa, but crosses the TMJ to insert on the mandible. The in-lever arm is the distance from this muscle insertion on the mandible to the TMJ. The distance from this muscle insertion to the tip of the lower jaw is the out-lever arm. The $F_{out}$ is the bite force produce during jaw closure. Notice that for both the dog and the bottlenose dolphin in Fig. 2, the line of action for the $F_{in}$ is not perpendicular to the out-lever arm. Instead both are angled posteriorly, which has the affect of reducing the $F_{in}$ by the *sin* of the angle ($\theta$) from the perpendicular. Note that this angle is much greater for the bottlenose dolphin than for the dog. Also notice that the temporalis fossa, which reflects the size and mass of the temporalis muscle, is smaller in the bottlenose dolphin than in the dog, which again reduces the magnitude of the $F_{in}$ in the bottlenose dolphin. In addition, the proportion of the in-lever arm to the out-lever arm is smaller in the bottlenose dolphin than in the dog, resulting in a lower mechanical advantage. Even qualitatively, and holding body sizes equal, it is obvious that dogs should have a greater bite force than dolphins based on temporalis muscle mechanics alone. However, since force and velocity are inversely proportional, the bottlenose dolphin jaw would be much faster than the dog. Such an arrangement is advantageous to a piscivore, which requires a fast snapping jaw with only enough force

F

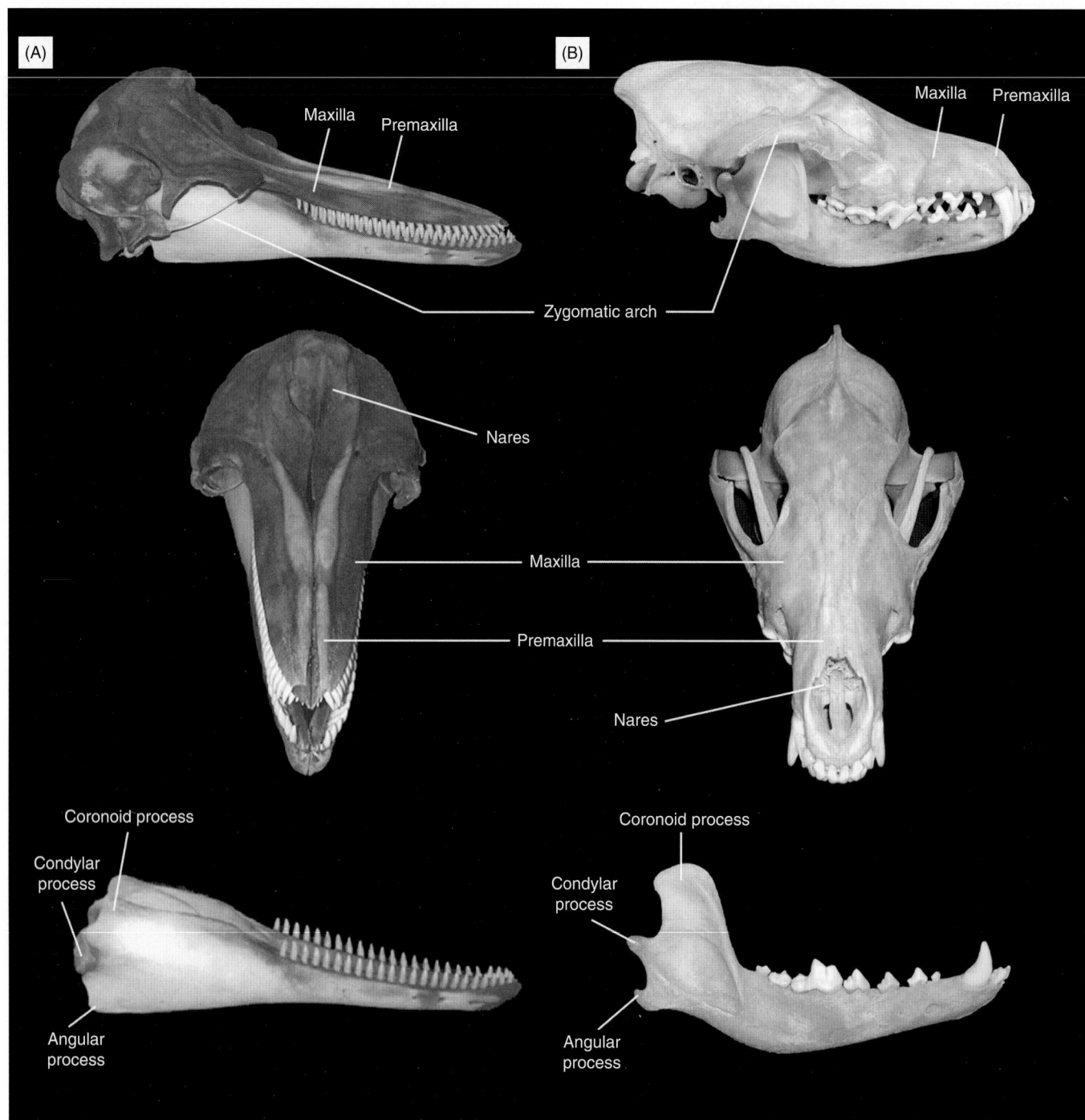

**Figure 1**  *Comparison of the skull morphology of a typical odontocete, (A) a bottlenose dolphin with (B) a dog. Note the elongation of the premaxillary and maxillary bones, thin zygomatic arch, and simple mandible of the bottlenose dolphin relative to the dog.*

to grasp and hold onto a fish before it is swallowed whole, without mastication. If muscles were available, a calculation of physiological cross-section of jaw muscles could provide a quantitative calculation of the $F_{in}$, and therefore bite force. These morphological predictions of behavioral performance of biting can provide meaningful insights in the capability of the feeding apparatus of organisms that are not available for direct studies. The diversity of feeding capabilities among marine mammals can be appreciated by examining the diversity of marine mammal skulls (Fig. 3). These types of morphological inferences can become more powerful if validated by measuring behavioral performance in live animals. Understanding how morphology constrains behavioral performance can provide indirect inference into an organism's ecology, and its ability to exploit the resources of its environment.

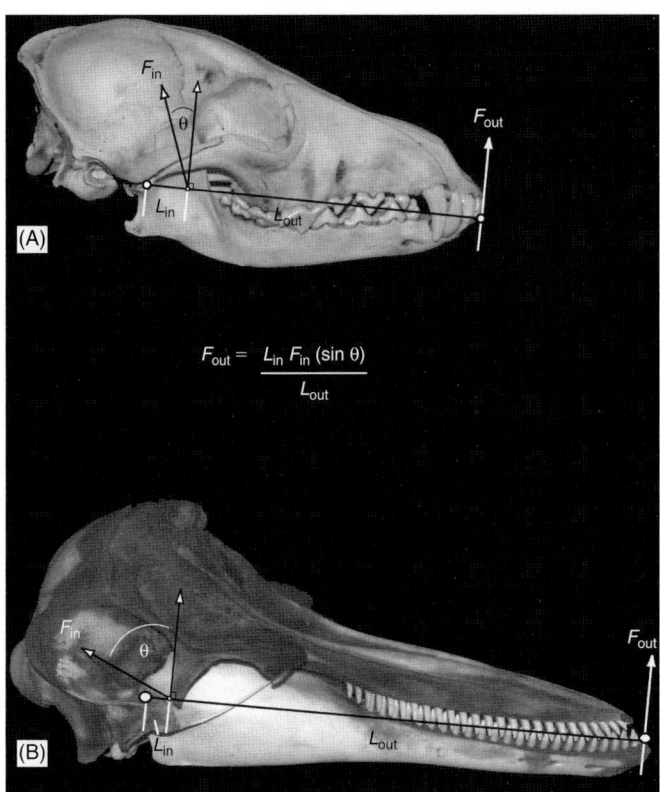

$$F_{out} = \frac{L_{in} F_{in} (\sin \theta)}{L_{out}}$$

**Figure 2** *Comparison of Type III jaw levers in (A) a dog and (B) a bottlenose dolphin . White lines indicate limits of lever arms (L), and squares indicate theoretical perpendicular lines of action. Arrows indicate forces (F).*

## III. Cetacean Functional Feeding Morphology

Cetaceans have developed some of the most specialized and varied feeding mechanisms among mammals. This should not be a surprise since cetaceans exhibit an amazing amount of ecological diversity and inhabit a diverse number of habitats ranging from the tropics to the polar regions. These feeding mechanisms can be generalized to four categories that are specific to the major cetacean groups. The feeding modes of odontocetes range from ram and raptorial feeding to suction feeding, while, mysticetes are generally categorized as skimmers and engulfers.

*Odontocetes*—Although the oral apparatus of odontocetes is not as derived as that of mysticetes, they are still greatly modified compared to terrestrial mammals. One explanation for derived mandibles of odontocetes is that they exhibit dual roles in feeding and sound reception. However, even among odontocetes there is a wide range of morphological and functional diversity. This is reflected in the number of teeth found among odontocetes; some species may possess 200–300 homodont teeth (e.g., river dolphins) within long narrow jaws, whereas other odontocetes are characterized by drastic reduction in tooth number, tooth function, and possess blunt rostra.

The delphinid jaw has often been called a pincer jaw, which refers to a raptorial method of prey capture (Pabst *et al.*, 1999; Werth, 2000a). As demonstrated in the lever example, relatively low mechanical advantage is a modification for quickly grasping and capturing prey, which are then swallowed whole without mastication. River dolphins are exemplary examples of this feeding strategy and are convergent with gharials, specialized crocodilian piscivores. Although "odontocete" often invokes an image of a long snouted dolphin

with a mouthful of teeth, this does not reflect the diversity of the feeding apparatus among this group. In fact, the breadth of morphological diversity is found among the non-delphinid odontocetes. The use of suction to feed is likely common among this group, but has only been demonstrated for a few species. Suction feeding requires a mechanism that rapidly lowers the intraoral pressure relative to the surrounding water. A rapid increase in the volume of the oral cavity can create this pressure difference. Aquatic vertebrates that use suction typically have a suite of anatomical features such as a smooth flat palate, and a robust hyoid apparatus with large lingual and hyoid musculature (e.g., genioglossus, styloglossus, and hyoglossus) which are used to depress a piston-like tongue, or the floor of the mouth. Cetaceans presumed to use suction to feed possess many of these morphologies and may generate negative intraoral pressures in the same way as other aquatic vertebrates. Suction has been observed in live stranded beaked whale (*Mesoplodon carlhubbsi*) calves; investigative palpation suggested that the motion was originating in the region of the throat grooves. Manipulation of ziphiid cadavers demonstrated that the tongue could easily be retracted toward the hyoid apparatus by the extrinsic tongue muscles. Retraction of the tongue by manipulation of these muscles resulted in the distention of the throat grooves. Werth (2000b) provided the first quantitative kinematic feeding investigation of odontocetes using pilot whales (*Globicephala melas*). This study not only validated the use of suction in this species but also demonstrated a four-phase feeding cycle that was divergent from the feeding cycle of terrestrial mammals. Furthermore, experimental work demonstrated varying suction capability in three odontocetes of differing head shape (common dolphin, *Delphinus delphis*; Atlantic white-sided dolphin, *Lagenorhychus acutus*; and harbor porpoise, *Phocoena phocoena*) using cadavers to directly measure pressure change at three locations in the oral cavity during simulations of hyoid depression (Werth, 2006a). Pressure data was incorporated into a biomechanical model to predict suction capability. Although the largest negative pressures were found within the posterior mouth cavity in all three species, the greatest suction capability was found in harbor porpoises. Such experimental work provides support for a suite of traits associated with presumed suction feeders, which includes a short broad rostrum, reduction or loss of the teeth, adaptations that occlude lateral gape, and a primarily teuthophagous diet. A comparative investigation of feeding kinematics in captive pygmy and dwarf sperm whales (*Kogia breviceps*, and *K. sima*—kogiids), and bottlenose dolphins verified the presumption (based on morphology) that kogiids also use suction as their primary feeding mode and that bottlenose dolphins (in this study) are primarily ram feeders, as long suspected (Bloodworth and Marshall, 2005). A ram-suction index (RSI) calculated for each species verified these observations (*Kogia* = 0.67±0.29; *Tursiops* = 0.94±0.11). Overall the feeding cycle duration of kogiids was significantly shorter than bottlenose dolphins (470±139 ms vs 863±337 ms). *Kogia* mean maximum gape angle (39.8±18.9°), and mean maximum opening and closing gape angle velocities (293±261·deg./sec and 223±121·deg./sec, respectively) were significantly greater than the mean bottlenose dolphin maximum gape angle (24.8±6.6°) and mean maximum opening and closing gape angle velocities (84±56·deg./sec and 120 ± 54·deg./sec, respectively). Negative RSI values in kogiids were correlated with increasing maximum gular depression and retraction, wide gape angle, and rapid opening gape angle velocity. The rapid jaw opening velocity in kogiids likely contributed to suction generation, in addition to hyoid depression. To understand the underlying mechanism of the observed feeding kinematics in kogiids and bottlenose dolphins, a morphological and biomechanical investigation

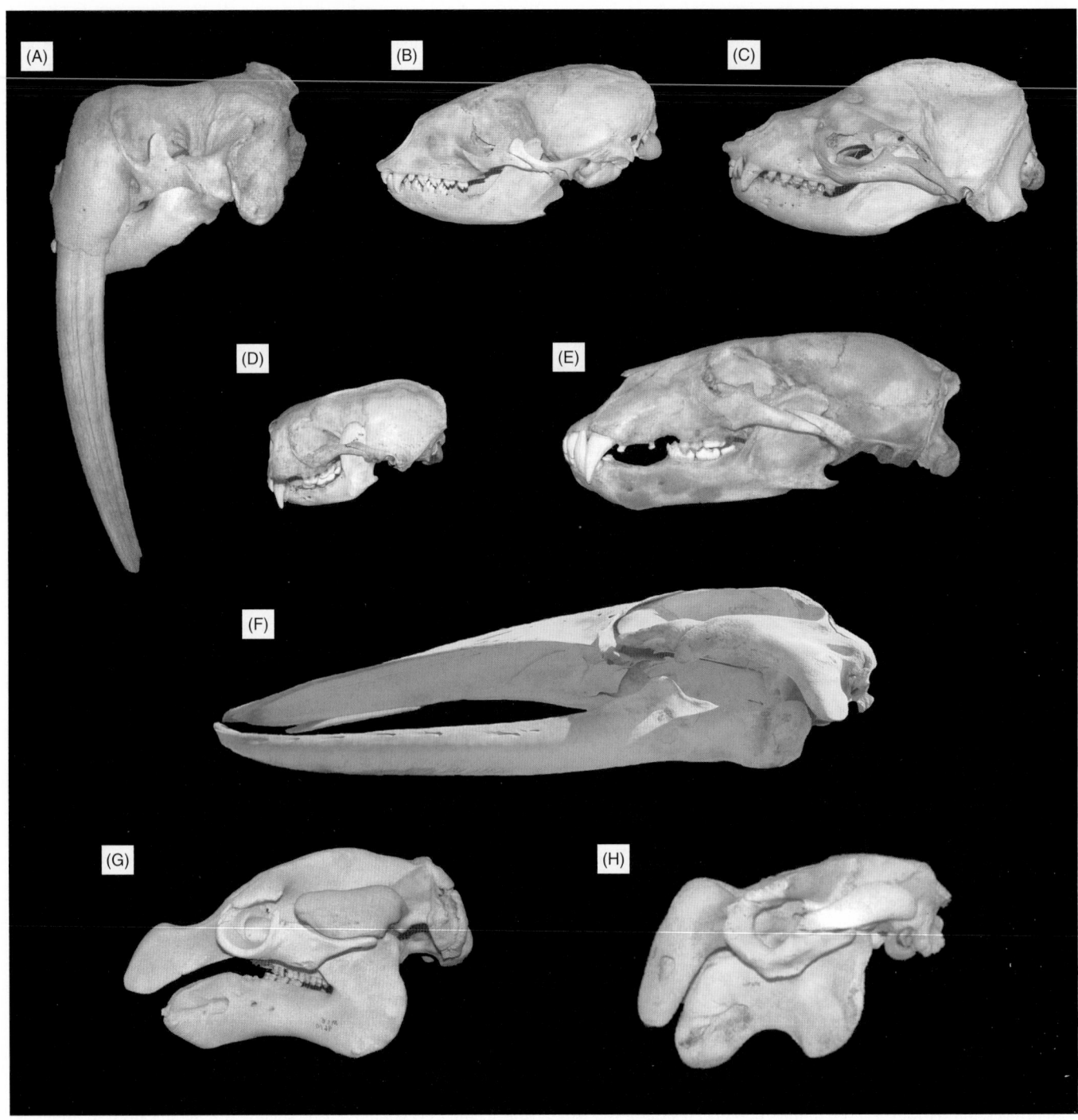

**Figure 3** *Diversity of skull morphology among marine mammals. (A) walrus, (B) harbor seal, (C) California sea lion, (D) sea otter, (E) polar bear, (F) gray whale, (G) West Indian manatee, and (H) dugong.*

of the hyoid and tongue musculature of both kogiids and bottlenose dolphins was conducted (Bloodworth and Marshall, 2007). The hypothesis that suction feeders possess a more massive hyolingual apparatus that can generate increased maximum theoretical muscle tension contraction, which would presumably result in greater intraoral negative pressures, was tested. Surprisingly, no significant differences in hyolingual maximum theoretical muscle tension were demonstrated between kogiids and bottlenose dolphins. Instead, all evidence pointed to modifications of orofacial morphology, including increased breadth of the mandible, tongue shape, and lateral gape occlusion, for increased suction feeding performance. A recent morphometric analysis of odontocete mandibles from all living species supports these conclusions; kogiids possess the bluntest mandibles of all odontocetes. (Werth, 2006a)

*Mysticetes*—All mysticetes possess baleen and feed upon plankton (Slijper, 1962). However the diet of some rorquals, or balaenopterids, is predominantly fish and squid, while gray whales (*Eschrichtius robustus*) feed upon marine amphipods and fish in addition to plankton. The jaws of mysticetes are edentulous; teeth have been replaced by baleen. Baleen are plates of keratinized epidermis (much like hair, nails, and horns) that are used to filter water. They have a straight outer edge, and a rounded inner edge, which is lined by a fringe of hair. The multiple overlapping plates result in the intertwining of the hairs into a woven mat, which separates plankton as water flows into the oral cavity. The differing morphologies of the baleen plates, and the maxilla that suspends the plates, reflect differing feeding mechanics employed by balaenids, balaenopterids, and gray whales.

In general, balaenids are skimmers and balaenopterids are engulfers (Pabst *et al.*, 1999; Werth, 2000a). The maxilla of all mysticetes arch higher above the mandible compared to odontocetes (Figs. 2 and 3). However, the maxilla of balaenids exhibits a markedly distinctive higher arch relative to balaenopterids, which functions to accommodate their taller, narrower, and greater number of baleen plates relative to balaenopterids. This increased number and height of baleen increase the effective surface area for straining plankton from water. Balaenids feed by swimming slowly at the surface with their mouths slightly agape. Water flows into the mouth, through the baleen plates where plankton are filtered, and out the corner of the mouth. Periodically the mouth is closed and plankton are removed from the baleen by the tongue, and ingested. A model of hydrodynamic flow through the oral cavity of bowhead whales (*Balaena mysticetus*) suggests that the unique morphological structures and actions of the oral apparatus, such as the subrostral gap, orolabial sulcus, curvature of baleen, mandibular rotation, and lingual mobility, permit the steady flow of water through the baleen and may improve the efficiency of filtration.

Instead of skimming, balaenopterids lunge and engulf large quantities of water, which is then strained through the shorter, wider, and fewer plates of baleen. This feeding strategy relies on the momentum of the body to operate. Despite their enormous size, rorquals are slender, streamlined, and faster than the relatively bulky and slow swimming balaenids. Although it is likely true for many vertebrates, mysticetes body shape is linked with feeding strategy due to related morphological specializations for hydrodynamic performance. Balaenopterids actively fluke and increase their speed, or lunge, just prior to opening their jaws. High-resolution digital electronic tags attached to free-ranging fin whales (*Balaenoptera physalus*), which included dual-axial accelerometers and hydrophones (to measure fluid flow noise), have verified and characterized this behavior in great detail (Goldbogen *et al.*, 2006). Balaenopterids allow the forward momentum and sudden increase of inertial drag to open the large jaws and cause the throat grooves to expand. Throat grooves in balaenopterids are pleats of blubber which undergo large-scale deformation to provide an expansive cavity (cavum ventrale), which enables these whales to engulf enormous quantities of water. The throat blubber of fin whales can expand to as much as four times its resting length circumferentially, and up to 1.5 times its resting length longitudinally. This extensibility is a direct result of the material properties of the grooved blubber (Orton and Brodie, 1987). The small and weak mandibular articulation with the skull is not capable of preventing the mandible from being detached from the head during the forceful engulfment event. The frontomandibular stay apparatus is a strong ligament extending from the supraorbital process of the frontal bone of the skull to the coronoid process of the mandible. This innovation allows the mouth to open to 90° while protecting the jaw joint from over-extension and damage. The stay apparatus also allows each side of the mandible to rotate ventrodorsally along its long axis (the dorsal surface rotates medially, the ventral surface rotates laterally). This is possible because of the loose articulation of the mandible with the skull and the unfused and flexible mandibular symphysis. Mandibular rotation further increases the amount of water the whale is able to engulf and may assist in the expansion of the throat grooves. After the mouth is closed, water is forced through the baleen, separating prey items, which are ingested. This water movement is powered by the retraction of the elastic throat grooves, contraction of muscles deep to the grooved blubber, and the return of the tongue to its original position. This behavior is supported by more recent kinematic data of wild feeding rorquals.

The oral apparatus of gray whales possesses characteristics of both balaenids and rorquals (Nerini, 1984). Instead of filtering water, gray whales filter sediment for marine amphipods, creating troughs in the sea floor. Feeding is not completely understood but is thought to be accomplished with the animal swimming on its side, and either using its jaws to excavate long troughs on the sea floor, or using suction to introduce prey laden sediment into the oral cavity. The latter behavior appears more likely and is supported by observations of feeding in a captive gray whale calf. The baleen of gray whales is unusually thick and sturdy, and it is presumed that amphipods are strained through it in a manner similar to balaenids straining plankton from water. Baleen on the right side show considerable more wear than the left side suggesting that this is the preferred side of the mouth to feed. Digital acoustic tags, with 3D accelerometers, attached to gray whales feeding off the central British Columbia Coast demonstrated that feeding occurred on the right side 97.2% of the time and 98.5% of the time at an angle greater than 45° (Woodward and Winn, 2006). Gray whales also skim the surface water for plankton in the same manner as balaenids and are also capable of engulfing prey in a manner similar to balaenopterids. This diversity in feeding modes allows for dietary flexibility so that alternate food sources can be used when prey distribution changes.

## IV. Sirenian Functional Feeding Morphology

Sirenians are unusual among marine mammals because of their herbivorous nature, which transcends all aspects of their natural history (Hartman, 1979; Reynolds and Odell, 1991; Reep and Bonde, 2006). The sirenian skull is also derived relative to other mammals. Like cetaceans, the facial region also dominates the skull. Whereas cetaceans have evolved a relatively enlarged maxilla, sirenians exhibit pronounced and expanded premaxillary bones (Fig. 3). This is particularly true of dugongs (*Dugong dugon*). The large anterior bones of both manatees and dugongs increase the surface area for the attachment of large facial muscles that form a muscular hydrostat integral for feeding; the large narial basin allows for greater movement of these facial muscles. Due to their herbivorous diet, sirenians do masticate. The temporalis muscle is well developed, but the masseter muscle is relatively simple (unlike terrestrial herbivores). The coronoid process of the mandible is expanded and modified. The head of the condylar process is small and flat, and the corresponding mandibular fossa is shallow allowing for a great mobility of the TMJ. The mandible is large and heavy; only cheek teeth are present (with the exception of tusks in male dugongs). Unusual skull characters include the massive pterygoid processes, which may provide a second jaw articulation (ptyergoid-mandibular) that could allow the lower jaw to rotate and produce transverse (side-to-side) movements during mastication. Such an articulation would displace the fulcrum of the lower jaw from the condyles to the ptyergoid processes—an unusual situation among mammals.

Deflection of the rostrum is correlated with sirenian feeding ecology. Dugongs are benthic foraging specialists that target the rhizomes of sea grasses. Their rostrum is accordingly strongly downturned (~70°),

F

which is advantageous for benthic feeding. Alternatively, Amazonian and West African manatees (*T. inunguis* and *T. senegalensis*) possess the least deflected rostra (~30° and 26° respectively), and this also reflects their trophic ecology. For example, Amazonian manatees are restricted to the freshwater rivers, lakes, and floodplains of the Amazonian basin where they feed upon food high in the water column such as inundated vegetation of the vàrzea, igapó, floating meadows, and emergent grasses (Gramineae; Best, 1981; Rosas, 1994). The diet of West African manatees is not well known, but the murky waters of their habitat do not support extensive submerged aquatic vegetation, and West African manatees most likely rely upon natant, emergent aquatic, and semi-aquatic vegetation (Best, 1981). West Indian manatees (*Trichechus manatus*) exhibit rostral deflections that are intermediate between those of dugongs and other manatees. They are generalist feeders and feed anywhere in the water column. They also feed on a greater diversity of freshwater, brackish, and marine plants.

All manatees possess an unusual form of tooth replacement (Domning and Hayek, 1984). Only cheek teeth, which number 6–8 at any one time, are present. Teeth erupt in the back of the mouth and move anteriorly as they wear. The bony septa between each tooth are reabsorbed in front of a tooth and re-deposited behind it. This allows the teeth to move through the bone of the mandible. Teeth wear as they progress forward. When the teeth reach the anterior-most portion of the tooth row and are completely worn, the roots are absorbed and the tooth is shed. A new tooth erupts in the posterior tooth row that continues the conveyer-like process. Manatees are apparently able to produce an unlimited number of cheek teeth. In contrast, dugongs possess simplified peg-like teeth that are open rooted. Enlarged horny pads on the upper and lower palate play an important role in mechanical reduction and processing of sea grasses rather than teeth (Marsh *et al.*, 1999; Lanyon and Sanson, 2006a, b).

All sirenians possess muscular snouts that are covered by modified vibrissae, or bristles. Vibrissae are specialized hairs that transmit tactile information from the environment to the central nervous system. Although, these bristles are homologous with mystacial and mental vibrissae of the dog, they are short, thick, and robust. The expanded lips of all sirenians move fields of bristles to manipulate vegetation, and introduce it into the mouth (Hartman, 1979; Marshall *et al.*, 2003). The use of vibrissae by sirenians to manipulate objects in their environment is a departure from the classical sensory function of mammalian vibrissae. For example, many other mammals use vibrissae to detect tactile cues. Pinnipeds employ whisking movements for more directed tactile exploration. The sniffing behavior and related vibrissal movement during exploration by rodents involve sweeping of the mystacial vibrissae forward and backward in conjunction with protraction and retraction of the rhinarium and head. The modification of manatee bristles to actively manipulate food and other objects appears to be unique to sirenians. Facial muscles of all mammals, including those that move vibrissae (whisking or manipulative behaviors), are innervated by Cranial Nerve VII, and the cell bodies of the axons of this nerve are located in the facial motor nucleus within the brainstem. The size and morphology of the facial motor nucleus among mammals can be used to infer functional importance of facial muscles and vibrissal use. The facial muscles in Florida manatees are innervated by a large and prominent facial motor nucleus, with clear subdivisions, particularly within the lateral subdivision that is known to innervate the rostral-most facial muscles (Marshall *et al.*, 2007). Such neuroanatomical data supports the important function role that the muscular-vibrissal complex serves in manatees and likely all sirenians. The use of neuroanatomical data is an important tool for predicting function from morphology.

## V. Pinniped Functional Feeding Morphology

Pinnipeds (sea lions, seals, and walruses) represent a diverse group of marine mammals with varying life histories. Among marine mammals, their skulls are most similar to that of the dog (Figs. 1 and 3), with notable exceptions such as walruses (*Odobenus rosmarus*; Fig. 3). Pinnipeds are characterized by a large rounded cranium, short snout, large orbits, and narrow interorbital distance. In general, the skulls of otariids (sea lions) are less variable than those of phocids (seals), although some species show sexually dimorphic characteristics. The orbits in small phocids (and therefore the eyes) encompass a greater proportion of the skull than that in larger phocids. The narial basin of elephant seals (*Mirounga angustirostris*) and hooded seal skulls (*Cystophora cristata*) are enlarged; as in sirenians it is presumed that this allows greater movement of their mobile proboscises. It is presumed that the use of suction feeding is more prevalent among pinnipeds (King, 1983). Pinnipeds that are presumed, or known, to employ suction feeding (e.g., walruses, bearded, crabeater, ringed, and harp seals) tend to have short, wide rostra with jaws that have scoop-like anterior ends, and a long mandibular symphysis, or a mandible in which the ventral borders are angled toward the oral cavity.

Relative to the dog, tooth number in pinnipeds is reduced, and the cheek teeth are relatively uniform in cusp number and shape—virtually homodont. These changes in the cheek teeth are likely adaptations related to eating fish. At least two species, the leopard seal (*Hydrurga leptonyx*) and crabeater seal (*Lobodon carcinophagus*), have specialized teeth. The leopard seal primarily feeds on large vertebrates, such as penguins and other seals. The distinctive cheek teeth possess three long shearing cusps (King, 1983). Crabeater seals actually feed on krill, not crabs. The cusps of their cheek teeth are complicated and modified to form a sieve. These seals swim into a krill patch with mouth open, sucking in water. When the mouth is closed, water is strained through the sieve, and the krill are consumed.

The skull of walruses differs significantly from other pinnipeds (Fig. 3), which is related to the presence of tusks (the dominant feature of the skull), and their specialized feeding behavior (Fay, 1982). The maxillary bones are enlarged to accommodate and anchor the tusks to the skull. The short, wide rostrum is advantageous for benthic feeding, and increases the surface area for their numerous and highly mobile whiskers used to detect infaunal prey. The posterior head is flat and broad, providing a large surface area for attachment of neck flexor muscles. The enlargement of the maxillary bones to anchor the tusks to the skull and large regions for attachment of neck muscles are important for hauling out behavior. Walruses commonly use their tusks to pull and lift their bodies from the water. Hence the derivation of their Latin name *Odobenus* (tooth walker); the tusks are not used for feeding. Walruses excel at suction feeding (Fay, 1982; Kastelein *et al.*, 1994). Powerful intraoral pressures are generated by the piston-like tongue and design of the oral cavity. Suction is used alternatively with hydraulic jetting (the forceful ejection of water from the mouth and the opposite behavior of suction) to excavate bivalve prey and then remove them from their shells. Recent direct observations of feeding walruses in Greenland demonstrate that some populations use their flippers to assist in excavation of prey.

Prey capture by pinnipeds has not been widely investigated (King, 1983, Reidman, 1999). However, kinematic investigations of bearded seal feeding demonstrated that these benthic foraging specialists employ an excavation behavior similar to walruses. Direct measurements of suction generation in bearded seals demonstrated that their suction capability is comparable to that of walruses. In general, vibrissae appear to be important for prey capture and discrimination. Vibrissal tactile discrimination by harbor seals has been shown to be as sensitive as the hands of monkeys. The vibrissae in harbor seals form a hydrodynamic receptor system that is tuned to the frequency of water movement made by swimming fish (Dehnhardt, 1998; Dehnhardt *et al.*, 2001). It is presumed that peripheral nerves within the vibrissae that terminate on mechanoreceptors are involved; however, the mechanism is still unknown. Compared to terrestrial mammals the number of nerves innervating mystacial vibrissae of pinnipeds is much higher. The number of axons in ringed (*Pusa hispida*) and bearded seals (*Erignathus barbatus*) are around 1500 and around 1650, respectively (Hyvärinen, 1989; Marshall *et al.*, 2006), whereas most mammals possess approximately 200 axons per whisker.

## VI. Sea Otter and Polar Bear Functional Feeding Morphology

The functional morphology of sea otters and polar bears has not received much attention. Sea otters forage on the bottom, in waters as deep as 40 m (Kenyon, 1969; VanBlaricom and Estes, 1988; Reidman and Estes, 1990). Their diet is varied but shellfish and urchins comprise a large portion. Otters use their forepaws to excavate clams, and to pry shellfish and urchins from the rocky substrate, sometimes using tools. Food is usually consumed at the surface, and behavioral observations suggest that otters do not use their teeth underwater, even when feeding on fish. Upon surfacing, fish are killed by a bite to the head. A rock or some other tool is usually carried in a flap of skin in the axilla region (under the arm) and is used to pound open shellfish. The spines of urchins are simply bitten off, and the test (shell) of the urchin is crushed with the cheek teeth. Their cheek teeth are broad, flat, and covered with thick enamel. The shearing cusps of the carnassial teeth have been lost; sea otters are adapted for crushing their food (Kenyon, 1969). Polar bears grasp their prey with their mouths and break the neck or skull of their prey with their large masticatory muscles and robust dentition. Their masticatory apparatus appears to resemble a robust version found in other bears.

## *See Also the Following Article*

Skull Anatomy

## *References*

Best, R. C. (1981). Foods and feeding habits of wild and captive Sirenia. *Mamm. Rev.* **11**, 3–29.

Bloodworth, B., and Marshall, C. D. (2005). Feeding Kinematics of *Kogia* and *Tursiops* (Odontoceti: Cetacea): Characterization of suction and ram feeding. *J. Exp. Biol.* **208**, 3721–3730.

Bloodworth, B., and Marshall, C. D. (2007). A functional comparison of the hyolingual complex in pygmy and dwarf sperm whales (*Kogia breviceps & K. sima*), and bottlenose dolphins (*Tursiops truncatus*). *J. Anat.* **211**, 78–91.

Dehnhardt, G. (1998). Seal whiskers detect water movements. *Nature* **394**, 235–236.

Dehnhardt, G., Mauck, B., Hanke, W., and Bleckmann, H. (2001). Hydrodynamic trail-following in harbour seals (*Phoca vitulina*). *Science* **293**, 102–104.

Domning, D. P., and Hayek, L. A. C. (1984). Horizontal tooth replacement in the Amazonian manatee (*Trichechus inunguis*). *Mammalia* **48**, 105–127.

Fay, F.H. (1982). Ecology and Biology of the Pacific Walrus, *Odobenus rosmarus divergens* Illiger. North American Fauna No. 74, United States Department of the Interior Fish and Wildlife Service, Washington DC, 279pp.

Goldbogen, J. A., Calambokidis, J., Shadwick, R. E., Oleson, E. M., McDonald, M. A., and Hildeband, J. A. (2006). Kinematics of foraging dives and lunge-feeding in fin whales. *J. Exp. Biol.* **209**, 1231–1244.

Hartman, D. S. (1979). "Ecology and Behavior of the Manatee in Florida. Special Publication Number 5." American Society of Mammalogists, Lawrence, Kansas, USA.

Hyvärinen, H. (1989). Diving in darkness: Whiskers as sense organs of the ringed seal. *J. Zool. (Lond.)* **218**, 663–678.

Kastelein, R. A., Muller, M., and Terlouw, A. (1994). Oral suction of a Pacific walrus (*Odobenus rosmarus divergens*) in air and under water. *Z. Säugetierkd.* **59**, 105–115.

Kenyon, K. W. (1969). The sea otter in the eastern Pacific Ocean. *North American Fauna* **68**, 1–352.

King, J. E. (1983). "Seals of the World," 2nd Ed. Cornell University Press, Ithaca.

Lanyon, J. M., and Sanson, G. D. (2006a). Degenerate dentition of the dugong (*Dugong dugon*), or why a grazer does not need teeth: Morphology, occlusion and wear of mouthparts. *J. Zool.* **268**, 133–152.

Lanyon, J. M., and Sanson, G. D. (2006b). Mechanical disruption of seagrass in the digestive tract of the dugong. *J. Zool.* **270**, 277–289.

Marsh, H., Beck, C. A., and Vargo, T. (1999). Comparison of the capabilities of dugongs and West Indian manatees to masticate seagrasses. *Mar. Mamm. Sci.* **15**, 250–255.

Marshall, C. D., Maeda, H., Iwata, M., Furuta, M., Asano, A., Rosas, F., and Reep, R. L. (2003). Orofacial morphology and feeding behaviour of the dugong, Amazonian, West African and Antillean manatees (Mammalia: Sirenia): Functional morphology of the muscular–vibrissal complex. *J. Zool.* **259**, 1–16.

Marshall, C. D., Amin, H., Kovacs, K., and Lydersen, C. (2006). Microstructure and innervation of the vibrissal follicle–sinus complex in the Bearded Seal, *Erignathus barbatus* (Pinnipedia: Phocidae). *Anat. Rec.* **288A**, 13–25.

Marshall, C. D., Vaughn, S. D., Sarko, D., and Reep, R. L. (2007). "Topographical Organization of the Facial Motor Nucleus in Florida Manatees (*Trichechus manatus latirostris*)." *Brain Behav. et al. Evol.* **70**, 64–173.

Nerini, M. (1984). A review of gray whale feeding ecology. In "The Gray Whale Eschrichtius Robustus" (M. L. Jones, S. L. Swartz, and S. Leatherwood, eds), pp. 423–449. Academic Press, Inc, Orlando.

Orton, L. S., and Brodie, P. F. (1987). Engulfing mechanisms of fin whales. *Can. J. Zool.* **65**, 2898–2907.

Pabst, D. A., Rommel, S. A., and McLellan, W. A. (1999). The functional morphology of marine mammals. In "The Biology of Marine Mammals." Smithsonian Press, Washington, DC.

Reep, R. L., and Bonde, R. K. (2006). "The Florida Manatee: Biology and Conservation." University of Florida Press, Gainesville, 190pp.

Reidman, M. L. (1999). "Pinnipeds: Seals, Sea Lions, and Walruses." University of California Press, Berkley, 439pp.

Reidman, M.L., and. J.A. Estes. (1990). The Sea Otter (Enhydra lutris): Behavior, Ecology, and Natural History. US Fish and Wildlife Report 90, 1–126.

Reynolds, J. E., and Odell, D. K. (1991). "Manatees and Dugongs." Facts on File, New York, 192pp.

Rosas, F. C. W. (1994). Biology, conservation, and status of the Amazonian manatee *Trichechus inunguis*. *Mamm. Rev.* **24**, 49–59.

Slijper, E. J. (1962). "Whales." Basic Books, New York.

VanBlaricom, G. R., and Estes, J. A. (1988). "The Community Ecology of Sea Otters. Ecological Studies, volumes 65. Springer-Verlag, Berlin, 247pp.

Werth, A.J. (2000a). Feeding in Marine Mammals. In "Feeding" (K. Swenck, ed.), pp. 487–526, Academic Press, San Diego.

Werth, A. J. (2000b). A kinematic study of suction feeding and associated behavior in the long-finned pilot whale, *Globicephala melas* (Traill). *Mar. Mamm. Sci.* **16**, 299–314.

Werth, A. J. (2006). Odontocete suction feeding: Experimental analysis of water flow and head shape. *J Morphol* **267**, 1415–1428.

Werth, A.I. (2006b). Mandibular and dental variation and the evolution of suction feeding in Odontoreti. *J. Mammal.* 87, 579–588.

F

Woodward, B. L., and Winn, J. P. (2006a). Apparent lateralized behavior in gray whales feeding off the central British Columbia coast. *Mar. Mamm. Sci.* **22**, 64–73.

# Feeding Strategies and Tactics

Michael R. Heithaus and Lawrence M. Dill

## I. Introduction

Marine mammals are found in a wide range of habitats including the open ocean, coastal waters, rivers, lakes, and even on ice floes and land. They feed on a variety of prey species from aquatic plants to microscopic zooplankton to the largest marine mammals, and a diverse array of strategies and tactics is used to locate and capture these prey. Some marine mammals consume huge numbers of prey items at a time (batch feeding) while others attack and consume prey items singly (raptorial feeding). Many marine mammals forage in large groups while others feed alone. In this chapter, we will consider the wide range of marine mammal foraging behaviors and the circumstances and habitats that led to the adoption of particular feeding strategies and tactics.

Before embarking upon a review of marine mammal foraging, it is important to make a distinction between a strategy and a tactic, terms which have specific meanings in the field of Behavioral Ecology. To put simply, a strategy is a genetically based decision rule (or set of rules) that results in the use of particular tactics. Tactics are used to pursue a strategy and include behaviors (Gross, 1996). Tactics may be fixed or flexible, in the latter case they depend on the condition of the individual or characteristics of the prey or environment. For example, a humpback whale's (*Megaptera novaeangliae*) strategy may be to use that tactic which will maximize energy intake at any particular time. The whale may pursue this strategy by switching between the tactics used to capture fish and those used to catch krill, depending upon the relative abundance of these two prey types.

Our understanding of marine mammal foraging is hampered by the difficulty of studying these animals. They live in an environment where observations are difficult (often beneath the surface), our presence can disturb their foraging behavior, and feeding events often occur quickly and are easy to miss. Despite this, and thanks to many emerging technologies, much is known. We will begin our review by considering ways that marine mammals find and capture their prey, continue with a discussion of group foraging, then conclude with a discussion of the causes of variation in feeding strategies and tactics.

## II. Finding Prey

The first step in foraging is locating prey. This may be done over many temporal and spatial scales and can involve migrations of thousands of kilometers or switching between habitats separated by only a few meters to forage in prey-rich locations. Then, once a marine mammal is in a prey-rich area it still must locate prey.

### A. Habitat Use

One way that marine mammals can increase their chances of encountering prey is to spend time foraging in those habitats with high prey abundance. There is evidence that a variety of marine mammals tend to aggregate in areas with high food concentrations. For example, the highest densities of polar bears (*Ursus maritimus*) are found along floe-edges and on moving ice, habitats that contain the highest density of seals; resident killer whales (*Orcinus orca*) are most abundant in Johnston Strait, British Columbia when salmon migrate through the strait. Also, the distribution of humpback whales in the Gulf of Maine appears to reflect the availability of fish prey, and humpback whale distribution in southeast Alaska may partially be determined by krill abundance. When there are a variety of habitats available to marine mammals in a restricted area (such as nearshore environments), a theoretical model predicts that, if the main concern of the animals is to maximize energy intake, they should be distributed proportional to the amount of food available in each habitat (Tregenza, 1995). Testing this hypothesis is difficult since marine mammal prey availability is often hard to quantify. However, the distribution of bottlenose dolphins (*Tursiops* spp.) in Shark Bay, Western Australia, conforms to this hypothesis and matches that of their fish prey in winter months at scales of 100s of meters to kilometers. Similarly, Hawaiian spinner dolphins (*Stenella longirostris*) appear to match the availability of their vertically migrating prey at scales of 20 m to kilometers. Humpback and minke (*Balaenoptera acutostrata*) whales do not appear to conform to this hypothesis. Instead, they appear to show a threshold response to prey availability, only using a habitat once prey density has reached a particular level, but above this threshold there is a tight relationship between zooplankton abundance and whale abundance. Of course, prey availability is not the only factor that might influence habitat use of marine mammals, and this will be considered in detail later.

Many marine mammals forage over great distances, and they may have limited knowledge of the distribution of prey patches, especially in pelagic habitats. In these situations, marine mammals, including pinnipeds and cetaceans, may adopt movement tactics that should maximize the probability of encountering prey. Displacement rates are relatively high, and movements relatively linear, in areas of low prey abundance, but animals exhibit low displacement rates and high turning rates ("area restricted searches") when they encounter rich patches.

### B. Migration

When suitable habitats for a marine mammal are widely spaced, movements between them are considered migrations. There is thus a continuum between habitat use decisions and migrations. Some migrations appear to be driven primarily by variation in food availability. Unlike baleen whales, sperm whales (*Physeter macrocephalus*) cannot fast for long periods of time, and female groups use migrations up to 1100 km as part of a strategy for surviving in a variable habitat with low local food abundance and poor foraging success. In fact, this tactic may be the reason that female sperm whales are found in permanent social groups as they may benefit from the experience of old females during migrations.

Migration frequently involves trade-offs between feeding and another factor, like reproduction. Baleen whales and some pinnipeds feed only for a relatively short period of time in high productivity high latitude waters, then fast during the rest of the year while moving to, and spending time at, low latitude breeding grounds. For example, northern elephant seals (*Mirounga angustirostris*) forage along the entire North Pacific, then migrate to a few beaches on the California coast to breed and molt. Also, some humpback whales in the Pacific Ocean reproduce in warm, low productivity, Hawaiian waters, then move to the more productive waters of the north Pacific to feed during the summer months. However, some humpback whales remain in the southeast Alaska feeding grounds yearround, and individuals that do not consume enough prey during the feeding season may forego migration to continue feeding.

## C. Searching and Diving

The way in which an animal moves through its environment can influence its encounter rate with prey, and many animals exhibit stereotyped search patterns. Marine mammals that forage on concentrated prey may continually patrol through areas where they expect to encounter concentrations. For example, leopard seals (*Hydrurga leptonyx*) will patrol along ice edges where departing and returning penguins congregate and killer whales patrol nearshore areas in search of seals. When groups of marine mammals forage, they often spread out into widely spaced subgroups and/or move forward in a line abreast formation (e.g., dusky dolphins [*Lagenorhynchus obscurus*], pilot whales [*Globicephala* spp.], Risso's dolphins [*Grampus griseus*], bottlenose dolphins, killer whales). Spreading out in such fronts may either reduce foraging competition among individuals or increase the probability that prey is detected so the subgroups can converge to feed.

Once a marine mammal has selected a habitat for foraging, it must execute a strategy that will optimize its net energy intake rate, often with respect to trade-offs and constraints. For a diving animal this means that it must balance the energetic costs of diving with the energetic gains of foraging. The costs of diving vary greatly among marine mammals. Polar bears, sea otters (*Enhydra lutris*), and most pinnipeds are divers—they spend most of their time above water or have long surface intervals between food gathering dives. In contrast, most cetaceans and sirenians can best be thought of as surfacers—they spend the majority of their time submerged, and make trips to the surface only to breathe (see Boyd, 1997).

Theoretical studies of optimal diving suggest that as the depth at which prey are located increases, both dive times and surface times should increase (Kramer, 1988), and the type of dive a marine mammal executes will depend on the depth and the distribution of prey. Some predictions of optimal diving theory are supported by several studies of marine mammals, and both dive times and surface times increase with dive depth in pinnipeds, cetaceans, and sirenians. Because a diving individual should behave in a manner that optimizes its net energy intake, marine mammals may exceed aerobic limits when the energetic pay-off is sufficient. In addition to energetic considerations, predation risk may influence the diving behavior of marine mammals, and therefore result in deviations from optimal diving predictions, based on energetic currencies alone (Frid *et al.*, 2006).

There is a great deal of variation in the depths to which marine mammals dive. Some, like sea otters, nearshore odontocetes, and otariids, tend to be shallow divers. Others, including sperm whales, elephant seals (*Mirounga* spp.), and beaked whales are extremely deep-divers, sometimes foraging over 1000 m from the surface. Some species minimize the depths to which they must dive, and thus the costs, by modifying their diel pattern of foraging. For example, some dolphins and pinnipeds are nocturnal foragers on prey whose daily movements bring them closer to the surface at night (e.g., spinner dolphins, northern fur seals [*Callorhinus ursinus*], Antarctic fur seal [*Arctocephalus gazella*]). The diving tactics of beluga whales (*Delphinapterus leucas*) may be influenced by competition with pinnipeds, which are superior divers. The belugas generally forage over the deepest waters and, because of their body size, are able to gain access to benthic areas that the smaller pinnipeds cannot. Although the time spent at the bottom decreases with increasing depth, belugas compensate by increasing their ascent and descent rates as dive depth increases, a result also found in narwhals (*Monodon monoceros*).

## D. Prey Detection

Marine mammals have many ways to detect their prey including vision, various types of mechanoreception, echolocation, and hearing. Most marine mammals appear to rely on vision to at least some extent. The large, forward pointing eyes of pinnipeds suggest that vision is an important method for detecting prey. Even species that dive to extreme depths, like the elephant seal, are capable of using vision to find prey in dark waters at their foraging depth. Vision may be less important in other taxa. For example, river dolphins (*Platanista gangetica*) of the Indian subcontinent have eyes that are greatly reduced and may be mostly blind. Sea otters can use their forepaws to find food and discriminate prey items without the aid of vision, and many pinnipeds are found in turbid waters, making vision a poor method of prey detection. However, they are able to use their vibrissae (whiskers) to detect prey through active touch or through minute water movements caused by their prey.

Odontocete cetaceans have a method of prey detection not available to other marine mammals—echolocation. In controlled situations, odontocetes can detect relatively small objects at a considerable distance. For example, a bottlenose dolphin can detect a 7.62-cm diameter sphere from over 100 m. However, it is still unclear how efficient echolocation is under natural conditions. It is likely to be less efficient than suggested by laboratory and controlled experiments (as has been shown for bats), and may vary greatly depending on environmental conditions such as noise.

Echolocation is not always an effective way to detect prey. While most fish cannot hear echolocation calls, clupeid fish and other marine mammals can. Therefore, odontocetes foraging on prey that can detect their echolocation may have to use tactics other than echolocation for detecting prey. This difference in the ability of potential prey to detect echolocation is reflected in the foraging behavior of fish-eating ("resident") and mammal-eating ("transient") killer whales off British Columbia. While resident whales commonly use echolocation during foraging, transients do not. Also, when transients echolocate; their pulses are of low intensity and are irregular in timing, frequency, and structure; a pattern that may be difficult for prey to detect. Instead of echolocation, mammal-eating killer whales appear to use passive listening to detect their prey. Other marine mammals probably use passive listening opportunistically, especially bottlenose dolphins that feed on a variety of noisy fish species. Elephant seals and other pinnipeds also have good hearing abilities in water and may use passive listening to find prey.

## III. Capturing and Consuming Prey

A diverse array of tactics is used by marine mammals to capture and consume their prey once they have located it (Fig. 1). The most widespread tactic of raptorial predators is to simply chase down individual prey items that they have encountered. However, there are many other, more unique tactics employed by marine mammals.

## A. Stalking and Ambushing

Marine mammals often hunt prey that are non-sessile, fast-moving, and have good sensory abilities and, thus, could avoid predators if their approach were too obvious. For example, seals can avoid polar bears by diving back through the ice, and penguins can avoid leopard seals by hauling out, as can pinnipeds approached by killer whales near land. When hunting elusive prey, a predator must rely on either stalking or ambushing. A stalking predator attempts to conceal its identity or presence until it approaches its prey close enough

F

Figure 1    *Whales employ a diverse array of foraging tactics. Art by Pieter A. Folkens/Higher Porpoise DG.*

for a sudden, successful attack. In contrast, an ambush predator conceals itself and lies in wait, leaving the approach to the prey.

Polar bears use both stalking and ambush methods when hunting seals hauled out on the ice near breathing holes. In terrestrial stalking, bears creep forward and use ice for cover to closely approach their intended prey. Bears also stalk seals by swimming circuitously through interconnected channels or even under the ice, occasionally surfacing through holes to breathe and monitor their prey. However, an ambushing tactic, where a bear lies, sits, or stands next to a breathing hole waiting for a seal to surface, is the most energy-efficient and most commonly used foraging tactic.

Leopard seals also use both stalking and ambush tactics when foraging. Stalking leopard seals may swim under the ice below a penguin, then break through to capture the bird, or they may swim submerged near a fur seal beach and lunge at pups when they get close enough. Alternatively, leopard seals may ambush their prey by hiding between ice flows near a penguin landing beach. Sea otters will stalk birds by swimming underwater and grabbing them from below, a tactic similar to that used by Steller sea lions (*Eumetopias jubatus*) hunting northern fur seal pups and leopard seals stalking Adelie penguins (*Pygoscelis adeliae*).

Another behavior that could be considered stalking is wave riding and intentional beaching used to capture young pinnipeds and penguins near the water's edge. This tactic is commonly used by killer whales and occasionally by Steller sea lions and leopard seals. This may be a particularly dangerous foraging tactic, especially for young killer whales that may not be able to return to the water if they strand too high on the beach.

Some stalking predators make detours that involve moving away from the prey and potentially losing visual contact temporarily before making another approach. Polar bears will make detours from their prey while stalking aquatically, and dolphin subgroups may detour away from a school of fish to attack it from opposing sides. Weddell seals (*Leptonychotes weddellii*) have also been observed making detours when stalking cod under fast ice. These detours allow the seal to remain out of the fish's view and to attack from very close range below the fish.

## B. Prey Herding and Manipulation

To capture them more efficiently, marine mammals may actively manipulate the behavior of their prey. In other words, marine mammals take advantage of normal prey behaviors to enhance their ability to capture them. These manipulations may help a marine mammal flush prey from hiding, capture an individual prey item, or increase the density of prey aggregations so as to increase the forager's energetic intake rate. Prey herding is a common tactic used by dolphins, porpoises, whales, and pinnipeds and may be considered prey manipulation when they take advantage of natural schooling and flight behavior of their prey. Dolphin and porpoise groups and individuals have been observed herding prey against shorelines or other barriers, reducing the number of escape routes. Dolphins use shorelines for more than herding fish. Bottlenose dolphins inhabiting salt marshes sometimes form small groups that rush at fish trapped against a mudbank. The wave created by the rapid swim causes fish to strand on the mudbank, and the dolphins slide up the bank and pick fish off the mud before sliding back into the water. A similar behavior is performed by both individuals and groups of humpback dolphins (*Sousa* spp.) foraging around sandbanks off Mozambique.

Marine mammals also herd fish in open waters. When schools are at the surface, dolphins may split into groups to attack from different directions, herding the fish into a ball between subgroups. Other times, fish may be herded up from deeper waters, and trapped between circling individuals and the surface. During a herding event, individuals swim around the fish school, and below it, preventing its escape. Fish herding in open waters has been reported in many dolphin species, porpoises, and sea lions. Sea lions are also found feeding on schools of fish that are herded to the surface by dusky dolphins, but it is unclear if the sea lions aid in fish herding. The tactic of herding fish is found in a variety of marine predators. Although there are no reports of prey herding for many species of pelagic dolphins, it is probably a wide-spread tactic employed by marine mammals feeding on schooling fish.

During a prey-herding event, many different tactics may be used to cause the fish to move into a tight ball and to capture fish in these balls. Splashing at the surface causes fish schools to compact. Dusky dolphins perform leaps at the edge of fish schools that they are herding, and spotted dolphins (*Stenella* spp.) have been observed tail-slapping and splashing at the edge of a fish school when it started to break apart or move in a different direction, but the function of these behaviors is still unclear. Killer whales in Norway and humpback whales in the northwest Atlantic also use tail-slaps when they near schools of prey. Tail-flicks by humpbacks may also be used to concentrate schooling euphausiid prey in southeast Alaska, though this may simply be a hydro-mechanical effect.

Another tactic that marine mammals can use to herd prey is flashing light-pigmented areas of their body toward a fish school. Killer whales herding herring swim under the school and flash their white undersides to keep the school from diving, and a similar behavior has been noted in spotted dolphins. Humpback whales in southeast Alaska may also use flashes to help concentrate prey by rotating their elongated pectoral flippers while they herd herring, thereby showing the highly visible white undersides.

Fish show strong avoidance responses to bubbles and are reluctant to cross barriers composed of them. Not surprisingly, marine mammals take advantage of this response. The use of bubbles during foraging has been observed in many odontocetes, mysticetes, and pinnipeds. Spotted dolphins use bubbles to isolate individual fish, pulling them away from the school with the water disturbance created by the passing bubble, so they can be consumed, and Weddell seals blow bubbles into ice crevices where fish are hiding to flush them out. Killer whales also use bubbles to flush prey, and blow large bubbles toward rays buried in the sediment, causing them to move. However, bubbles are primarily used to concentrate and contain schools of fish. For example, killer whales blow large bubbles near the surface to keep fish in a tight ball. Humpback whales are the best-known bubble users and bubble feeding may be conducted by individual whales or in large groups. Whales deploy bubbles in a variety of formations including columns, curtains, nets, and clouds, with the tactic used dependent on the characteristics of the prey aggregations.

Sound and pressure waves may also be used to manipulate prey behavior. For example, bubble-netting humpback whales in southeast Alaska produce loud "feeding calls" as they rise to the surface, presumably herding prey up into bubble nets which are meters above the herring schools. Similarly, Icelandic killer whales may use low-frequency calls to herd herring schools into tighter groups. Bottlenose dolphins off Australia and Florida use tail-slaps known as "kerplunks" while foraging in shallow seagrass habitats. The kerplunk displaces a significant amount of water, creates a plume of bubbles, and causes a low-frequency sound. Kerplunks may cause startle responses in fish and help the dolphin locate and flush their prey, while the bubbles may provide a barrier to contain the fish. Humpback whales in the western Atlantic may flush burrowing fish (sand lance) from the bottom by scraping the substrate with their head, then feed on the fish once they have entered the water column.

### C. Prey Debilitation

Marine mammals sometimes debilitate their prey before they consume it. Killer whales attacking mysticetes often swim onto their backs when the prey tries to surface, and in some cases the victim may drown instead of dying from its wounds. While killer whales are herding herring, individuals thrash their tail through the school, stunning fish with the physical impact of their flukes; they then feed on the stunned and injured fish. The whales probably use this tactic because it is energetically more efficient than whole body attacks. Bottlenose dolphins strike fish with their tails ("fish whacking") when foraging alone or in groups, sometimes knocking the fish through the air. Also, there is evidence that walruses (*Odobenus rosmarus*) may use their tusks to kill or stun intended seal prey. Recent studies suggest that the hypothesis that odontocetes use sound to debilitate prey cannot be supported at this time.

### D. Tool Use

"Tool use is the external employment of an unattached environmental object to alter more efficiently the form, position or condition of another object, another organism, or the user itself when the user holds or carries the tool during or just prior to use and is responsible for the proper and effective orientation of the tool" (Beck, 1980). Tool use by marine mammals is reviewed in detail elsewhere in this volume, but more marine mammals use tools during foraging than is generally appreciated, so this behavior deserves brief mention here.

Sea otters are the best-known marine mammal tool users and will pick up rocks from the bottom and place them on their chest to use as an anvil for crushing mussels, crabs, or urchins, or use them to smash or dislodge abalone (*Haliotis* spp.) off rocks. In some cases, the rocks are retained between foraging dives to be reused. Certain bottlenose dolphins carry sponges on their rostra, apparently as a tool to aid foraging (Fig. 2). Also, there are popular accounts of polar bears throwing blocks of ice at basking seals to injure or trap them, and polar bears in captivity are often observed throwing large objects, raising the possibility of tool use in the wild. Another behavior that might be considered tool use involves killer whales creating waves to wash hauled out seals into the water. Finally, the use of bubbles to concentrate schooling fish or aid in flushing fish from hiding or a school (discussed earlier) fits Beck's definition of tool use.

F

**Figure 2** *Bottlenose dolphins* (Tursiops aduncus) *use sponges as tools to aid in foraging. Photograph by Michael R. Heithaus.*

### E. Benthic Foraging

While most marine mammals pursue their prey in the water column, several species forage on benthic organisms. There are three basic methods that marine mammals use to obtain prey from the bottom: collecting, extracting, and engulfing. Epibenthic prey is simply collected by foraging marine mammals. Sea otters collect echinoderms (mostly sea urchins), crabs, and other benthic organisms with their forepaws.

Infaunal prey items must be extracted from the substrate and require the predator to excavate in some manner. Sea otters use their forepaws to dig for clams in soft-sediment areas and may produce large pits over the course of several dives, occasionally surfacing with a clam. Harbor seals (*Phoca vitulina*) dig for prey in sandy habitats with their foreflippers or snouts while narwhals and belugas use water jets to dislodge mollusks buried in the sea floor. Walruses use a combination of tactics to obtain buried bivalves including digging with their snouts (not tusks) and hydraulic jetting. Walruses make multiple excavations on each dive and have been recorded consuming at least 34 clams on a single dive. Killer whales, in New Zealand, engage in benthic foraging on rays and have been observed pinning them to the bottom and may also be digging for them. Bottlenose dolphins in the Bahamas also dig for infaunal prey ("crater feeding"), and once a burrowing fish has been located, the dolphin will dive into the soft sand and use its flukes to drive deeper, almost up to the flippers, to catch the fish.

If many small infaunal prey items are consumed in a single feeding event, they may be engulfed while still in the sediment. Gray whales (*Eschrichtius robustus*) feeding near the bottom use suction to pull sediment and prey into their mouths, then filter the sediment and water out through their baleen.

### F. Batch Feeding

Batch feeding is a tactic employed to consume a large number of prey items in a single feeding event. While mysticetes are obligate batch feeders, some pinnipeds facultatively use this tactic. There are two basic types of batch feeding: skimming and engulfing. Skimmers, most notably the right whales (*Eubalaena* spp.) and bowhead whale (*Balaena mysticetus*), swim through concentrations of zooplankton, either at the surface or in the water column, with their mouths open, filtering water through their fine baleen plates which traps prey.

Engulfers include the rorqual whales and several pinnipeds. These species engulf large amounts of water and prey, then filter the water back through their baleen plates or teeth. Rorquals have a suite of adaptations, including expandable gular pleats and a lower jaw that can disarticulate from the upper jaw, that allow them to engulf huge volumes of water, and fish or crustacean prey, in each feeding attempt. "Lunge feeding" is one of the most common tactics of rorqual whales feeding near the surface and may take several forms. During a typical lunge, a whale surfaces with its mouth open to capture prey near the surface. Lunge feeding may be done singly or in groups, and in combination with many of the prey concentration tactics.

All Antarctic seals (crabeater [*Lobodon carcinophaga*], Weddell, Ross [*Ommatophoca rossii*], leopard seals) include zooplankton in their diet, as do some Arctic seals (ringed [*Pusa hispida*], ribbon [*Histriophoca fasciata*], harp [*Pagophilus groenlandicus*], largha [*Phoca largha*], and harbor seals). Of these, the crabeater seal is the most specialized batch feeder and zooplankton may comprise up to 94% of its diet. The cheek teeth of crabeater and some other seals are modified for straining krill, which are probably sucked into the mouth when the seal depresses its tongue, then trapped against the cheek teeth as the water is expelled.

### G. Ectoparasitism, Kleptoparasitism, and Scavenging

Predators kill their prey in the course of consuming it (Ricklefs, 1990). While most marine mammal foraging is predatory, there are several ways that animals may forage which do not involve killing their own prey. For example, an animal may gouge mouthfuls of flesh from a "host" without killing it (sometimes referred to as ectoparasitism). Although marine mammals fall victim to such ectoparasites (small sharks), there are no concrete examples of marine mammals using this tactic. However, killer whales may effectively ectoparasitize large whales as some attacks do not kill the victim. Kleptoparasitism (food stealing) has been observed only in otters and polar bears but may occur in other species. For example, pilot whales were observed harassing sperm whales until they regurgitated and the pilot whales consumed the regurgitated food. Scavenging is a common foraging tactic, but it does not appear to be widespread among marine mammals. However, it may be an important tactic for polar bears and some pinnipeds. Also, several odontocete species that feed on trawler discards or longline catches could be considered facultative scavengers.

### H. Herbivory

Sirenians (manatees [*Trichechus* spp.] and dugongs [*Dugong dugon*]) are the only marine mammals that routinely feed on plants, and both manatees and dugongs may be found foraging individually or in large groups. Manatee feeding appears to be more flexible than that of dugongs as the former will consume either floating or rooted vegetation and sometimes leaves from overhanging branches or vegetation along banks. Dugongs feed almost exclusively on seagrasses but may also intentionally consume benthic invertebrates. While manatees tend to crop vegetation, dugongs often dig up rhizomes and leave large feeding trails through seagrass beds, which can have a large impact on seagrass biomass, both above and in the sediment (Fig. 3), and even on invertebrate communities of the seagrass.

### I. Prey Preparation and Consumption

While some marine mammal prey can be consumed immediately after capture, others require extensive handling before they are eaten, and some are only partially consumed. Sea otters remove the heads of birds that they capture and strip the muscle from the breast,

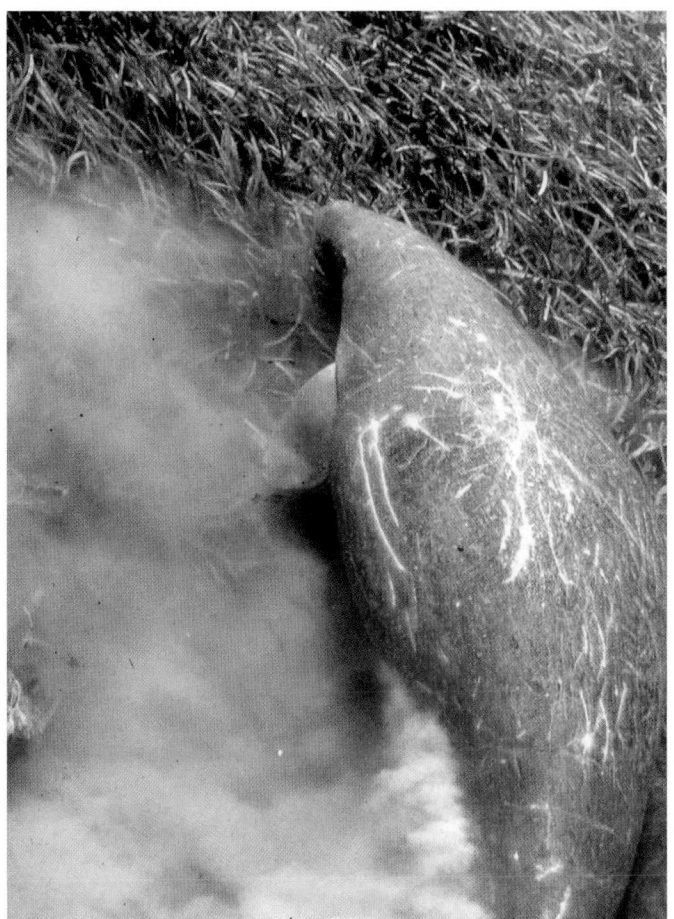

**Figure 3** *Dugongs* (Dugong dugon) *preferentially forage on below-ground portions of many seagrass species, creating a cloud of sediment during foraging activity. Photograph by Michael R. Heithaus.*

neck, and legs. Many dolphins and sea lions remove the heads from large fish before consuming them, and bottlenose dolphins will strip flesh from spiny fish. Head and spine removal may reduce the probability that a predator is injured while consuming prey, but it may also be a mechanism to reduce the intake of bony material that provides no nutritional value. Dolphins do not always consume their prey correctly, and sharp spines have been implicated in deaths of bottlenose dolphins stranded in Texas and Florida.

Odontocetes cannot chew prey and must spend considerable time handling large prey items. For example, bottlenose dolphins will drag large fish along sandy bottoms until pieces that are small enough to swallow are broken off. Killer whales are well known for their extensive handling of prey, especially pinnipeds, after capture. Killer whales often breach upon sea lion and seal prey as well as slap them with their tails. The function of these behaviors is unclear, but they may tenderize the prey, aid in training calves in hunting techniques, or even debilitate dangerous prey. Selective feeding on energy-rich portions of prey is common in both killer whales and polar bears. Killer whales will selectively eat the blubber and tongue of whales that they kill and polar bears prefer the blubber and muscle of seals and narwhals over the internal organs. Finally, harbor seals will regurgitate after feeding on sandlance to remove sand from their stomachs. The regurgitated fish are re-swallowed before they sink to the bottom.

## IV. Group Foraging

Many foraging tactics are executed by groups of marine mammals. Sometimes these groups are merely aggregations of animals attracted to the same resource, and there appears to be little interaction among individuals as they pursue prey individually. Other group foraging behaviors, like herding of prey, appear to be highly coordinated efforts and may involve animals cooperating with each other to increase their net energy intake rate. It is not always easy to determine whether group foraging marine mammals are cooperative or not. For example, the echolocation rate of an individual resident killer whale decreases as group size increases, suggesting that there may be information transfer. While this could represent cooperative information sharing, it is also possible that individual whales are parasitizing the information of others, as shown in bats. In many cases it is difficult to assess whether marine mammals are foraging cooperatively because group-living may be selected for by factors other than food, and group-foraging, whether cooperative or non-cooperative, is therefore simply a necessary epiphenomenon. One important consideration in studies of cooperation is whether groups are kin based as individuals in kin groups are more likely to engage in cooperative behavior to increase their inclusive fitness.

### A. Cooperative Foraging, Food Sharing, and Cultivation

Cooperation can be defined as "an outcome that—despite individual costs—is "good" in some appropriate sense for the members of the group … and whose achievement requires collective action" (Mesterson-Gibbons and Dugatkin, 1992). Most cooperation is achieved through a mechanism of by-product mutualism, in which an individual acts selfishly to benefit itself, and its actions incidentally benefit other individuals, but not all by-product mutualisms are cooperative. A possible example of by-product mutualism involves bowhead whales skim-feeding in groups with whales staggered in an inverse "V" formation. This formation may aid a whale in prey capture by using adjacent whales as a wall to trap prey or to catch prey that have escaped from the whale in front. These groups sometimes appear to be coordinated, with whales changing direction and leadership. All individuals probably act selfishly, but their presence may benefit other whales.

The above definition of cooperation requires three things be shown to support the hypothesis that a group is cooperative. First, individuals acting cooperatively must realize a short-term cost. This cost may include having to share food with other individuals or an opportunity cost by not attacking prey immediately while herding. Next, energy intake rate of individuals benefiting from cooperation must be higher than what they would have gained without cooperation. Finally, collective action must be required for the hunt to be successful. It is worth noting that in cooperative groups not all group members are required to receive equal benefits, and in groups that appear to be cooperative, a number of individuals may be non-cooperative (Packer and Ruttan, 1988).

There are many possible examples of cooperative foraging in the marine mammal literature involving mysticetes, odontocetes, pinnipeds, and sirenians. However, most anecdotal accounts of possible cooperative foraging behavior do not provide enough detail to determine whether these groups were truly cooperative. For example, many dolphin species are known to break into subgroups that spread out across a large front when foraging or to travel in line abreast formation. Generally, when one subgroup finds fish, other subgroups join to feed. This behavior has often been considered cooperative foraging. However, none of the three criteria for cooperation outlined previously have been shown to apply to these cases. Furthermore, although some authors have assumed that joining subgroups were recruited, they may simply be converging once they determine that another group located food. Fish herdings by dolphins, porpoises, whales, and sea lions have all been cited as examples of cooperative foraging. In these cases, there does appear to be a cost involved as individuals do not start foraging immediately but wait until the school has been herded to the surface ("temporary restraint"). Larger groups of dusky dolphins forage on a single fish school for longer periods of time than do small groups. Some authors have suggested that this indicates an increase in individual intake and that herding requires collective action. However, it is important to measure individual intake rates because longer foraging durations of large groups may simply be the result of larger schools being herded (and increased time until school depletion) or of increased foraging interference in large groups. More studies are required to support the hypothesis that such groups are cooperative.

Deliberate prey sharing provides strong evidence for cooperative hunting but must be viewed with caution as some apparent food sharing may represent intense competition for large prey items or kleptoparasitism (Packer and Ruttan, 1988). Prey-sharing has been observed in few marine mammal species but has been documented in both mammal-eating and fish-eating killer whale populations. Also, an apparent case of prey-sharing has been documented in leopard seals when two seals killed penguins, but one individual released its penguin to be consumed by the other. Prey sharing also has been documented in false killer whales (*Pseudorca crassidens*).

There are a few other examples of marine mammal foraging which appear to represent cooperative foraging. Leopard seals have been observed hunting in a coordinated fashion, with one seal driving penguins toward a second seal hiding behind an ice flow. The process was repeated several times, and both seals caught penguins each time, sharing prey in one instance. Collective action is required if killer whales are to capture large or swift prey and, in general, larger groups are seen when transient killer whales attack such prey.

There may even be a division of labor during their hunts. Also, there is a cost as prey are divided among group members. Bubble-netting humpback whales feeding on herring in southeast Alaska represent another potential example of cooperative foraging. In these groups, one whale deploys a bubble net, starting at a depth shallower than the herring schools. The whales then apparently drive the prey up into the bubble net and simultaneously lunge through the herring trapped against the surface (Fig. 4). Although there are apparently costs to this behavior and coordination is probably required, no data exists on intake rates in these groups. Finally, it is possible that large groups of synchronously diving crabeater seals cooperatively herd krill, but future studies are needed to verify this possibility.

One study has suggested that dugongs cultivate seagrass as they forage in large groups moving among seagrass banks. Although dugong grazing changes seagrass communities to stands of more profitable species, for deliberate cultivation to occur, cooperation among dugongs would be required. In general, true cultivation (gardening) is favored to evolve only when the individual that cultivates an area realizes the benefits of that action (Branch *et al.*, 1992). This implies both a fixed and a defendable feeding site (Branch *et al.*, 1992). Cooperative cultivation by dugongs is unlikely since individual dugongs that moved to a previously cultivated area, before the cultivating individuals, would benefit from reduced foraging competition. Also, there do not appear to be any mechanisms to prevent such cheating. A more likely explanation for the observed pattern of dugong foraging is "traplining" where dugong groups rotate among the most profitable seagrass meadows, and the changes in seagrass communities are an incidental by-product of dugong foraging on rhizomes.

### B. Optimal Group Size

The question of why particular group sizes are observed has been raised several times. For some species, group size has been suggested to be that which maximizes the intake rate of individuals in the group (optimal group size). However, this may not generally be the case. When it is difficult for a group to exclude joiners (e.g., when foraging

**Figure 4** *Humpback whales* (Megaptera novaeangliae) *in southeast Alaska cooperate to catch herring. They use a variety of tactics to manipulate the behaviour of herring, including the use of bubbles as tools. Photograph by Michael R. Heithaus.*

on a large fish school), the observed group size will often be greater than that which maximizes intake of each group member, since individuals will continue to join a foraging group until the average energy intake in the group approaches that of a solitary forager (stable group size; Giraldeau, 1988). Also, the benefits of defending resources may be low in large groups since individuals that do not defend the resource will realize higher intake rates than those individuals that try to defend against joining individuals. Finally, group size is likely to be larger than that which is optimal for foraging considerations if there are other benefits of being in a group (e.g., mating opportunities, protection from predators). Therefore, it is likely that most dolphins feeding on large schools of fish are in groups larger than those that would maximize energy intake of each group member. However, some marine mammals may be found in groups that are of optimal size for maximizing energy intake. For example, killer whales feeding on marine mammals may be able to regulate group size as individual prey items are easily defended, and groups (which are kin based) may be able to exclude other individuals before foraging commences. This may explain why the modal group size of three individuals observed in foraging transient killer whales is the group size that maximizes its members' energy intake.

## V. Variation in Feeding Strategies and Tactics

Marine mammals show a high degree of variability and flexibility in their foraging tactics. Individuals may be flexible in their foraging tactics depending on their state or circumstances, and this flexibility may lead to variation in foraging tactics among populations, individuals, and age/sex classes. Variation in feeding tactics may also arise from differences in the ways individuals solve cost-benefit trade-offs. Some of these differences among individuals may be genetically based and thus considered strategic variation.

### A. Trade-offs

Evolution favors strategies that maximize fitness (usually by maximizing lifetime reproductive success). For example, animals may pursue a strategy that maximizes their expected lifetime energy intake, which may involve a trade-off between maximizing short-term energy intake and minimizing predation risk because habitats that are prey rich are often the most dangerous (Lima and Dill, 1990, Fig. 5). Therefore, marine mammals may sometimes accept lower energetic returns to forage in safe habitats (see Wirsing *et al.*, 2008). For example, bottlenose dolphins in Shark Bay, Western Australia, match the distribution of their prey when their primary predator, the tiger shark (*Galeocerdo cuvier*), is absent but shift to forage mostly in low-risk, low-food areas when sharks are abundant. Dugongs in Shark Bay also reduce their use of food-rich but dangerous shallow habitats as tiger shark abundance increases. In addition, dugongs primarily excavate seagrass rhizomes when tiger sharks are scarce but switch almost exclusively to cropping seagrass leaves, which allows greater vigilance, when sharks are abundant. Finally, female polar bears with cubs often select habitats with lower food abundance to avoid potentially infanticidal adult males, and trade-offs between predation risk to calves and food availability at high latitudes may have led to the evolution of seasonal migrations in baleen whales.

Trade-offs between feeding and predation may also result in habitat use patterns that vary with behavior. For example, spinner dolphins rest in shallow nearshore coves with sandy bottoms during the day, possibly to reduce the probability of shark attack, then move offshore to feed on deep scattering layer organisms at night. Similarly, bottlenose dolphins in Shark Bay rest almost exclusively in safer, relatively deep waters, but will sometimes move into higher risk but more productive shallow habitats to feed.

Trade-offs between feeding and reproduction also may influence foraging patterns. For example, most phocid females fast during lactation and must consume sufficient food before the breeding season while female otariids make foraging trips of variable duration throughout lactation (see Wells *et al.*, 1999 for a review).

Prey selection can be viewed as the result of another type of trade-off. Each potential prey item differs in the energy required to capture it and the amount of energy the predator will gain from eating it. This trade-off sometimes results in selective foraging where one prey type is favored over others irrespective of its relative abundance. For example, harp seals always preferentially feed on capelin (*Mallotus villosus*) and select Arctic cod (*Arctogadus glacialis*) only in nearshore waters. Prey preferences have also been shown in resident killer whales. Off Alaska, resident killer whales prefer coho salmon (*Oncorhynchus kisutch*), while those off British Columbia prefer chinook salmon (*O. tshawytscha*) that are energy rich but relatively scarce. Prey-selection may also take the form of capturing a particular size of prey. For example, harbor seals in Scotland feed primarily on the most abundant fish species but prefer fish of 10–16cm in length. Changes in the relative costs and benefits of particular prey items may lead to prey switching, which has been observed in some marine mammals.

### B. Ontogenetic Variation

There are often distinct differences in the foraging behaviors of marine mammals of different ages. Such differences may be the result of changing physiological or foraging abilities, the relative importance of energy intake and survival at different life history stages, or differences in experience if a learning period is required for the successful use of a particular foraging tactic. Diving by young seals and sea lions is constrained by physiological development, and they typically make shorter and shallower dives than do adults (e.g., Steller and Australian [*Neophoca cinerea*] sea lions, Weddell and elephant seals). During their first trip to sea, elephant seal pups make a transition from short, shallow dives to a pattern similar to adult seals, with longer deeper

**Figure 5** *Foraging decisions made by individuals can be influenced by the presence of predators. Some individuals may forage in areas where they are more likely to be attacked by predators if the energy gain in these habitats is sufficient. Photograph by Michael R. Heithaus.*

F

dives that show diel fluctuations. This transition appears to be related to both changes in the physiology of young seals and possibly prey distribution. Young seals of different sizes may adopt different diving tactics. For example, larger yearling Weddell seals engage in relatively shallower dives to forage on benthic prey compared to small yearlings which make deeper dives to forage on energy rich prey. However, the cause of this variation is unclear.

Learning and cultural transmission of foraging tactics play important roles in the acquisition of foraging tactics in cetaceans. For example, there is a long period of practice required for young killer whales to become adept at using the intentional stranding tactic to capture pinnipeds. This period of learning may involve calves preferentially associating with the female pod-members (not necessarily their mother) that engage in this tactic most frequently. Similarly, it appears that sponge-carrying by bottlenose dolphins is socially passed within matrilines, especially to female offspring. Finally, sea otters tend to display the diet preferences of their mothers.

## C. Inter-Individual Variation

Within many marine mammal populations substantial differences exist among individuals in the foraging tactics that they employ. Northern fur seal females perform two distinct types of foraging dives: shallow dives, which seem to be directed towards vertically migrating prey, and deep dives to feed near the bottom. Shallow dives are made only between dusk and dawn while deep dives occur both at night and during daylight hours. Some individual seals specialize in one dive type or the other while other individuals use a mix of tactics. Southern sea lion (*Otaria flavescens*) individuals differ in their propensity to hunt fur seal pups. In Alaska, only juvenile male Steller sea lions prey upon fur seal pups while in Peru, most hunting is done by just a few adult males, and there are large differences in the success rates of different individuals. Similarly, sea otter predation on birds appears to be largely restricted to a few individuals, and a few individual sea lions have learned to wait at fish ladders and at the mouths of freshwater streams to take advantage of spawning steelhead. Leopard seals also vary in their hunting tactics. For example, a single individual was responsible for all ambushing attacks on Adelie penguins observed in Prydz Bay, Antarctica. In gray seals (*Halichoerus grypus*), there are distinct differences between sexes in diet and foraging behavior (Beck *et al.*, 2007). Finally, individual variation in the prey species consumed by sea otters may be a result of differences in diving tactics as juvenile males forage further offshore and make longer dives than other age/sex classes.

Cetaceans also show individual variation in feeding tactics. In Shark Bay, many unique tactics including kerplunking, sponge-carrying, and extreme shallow water foraging are restricted to a small number of individual bottlenose dolphins. Adult female killer whales perform most of the intentional strandings to catch elephant seals, and within a pod individual females differ in their use of this tactic. Most individual minke whales around the San Juan Islands specialize in either lunge feeding or feeding in association with birds. These two tactics are usually observed in different regions with individual whales showing inter- and intra-seasonal site fidelity. Individual humpback whales differ in their use of various types of lunge-feeding and bubble-netting tactics that may relate to dietary specializations on either krill or herring (*Clupea* spp.) and to the distribution of these prey items. Finally, reproductive state may influence the foraging tactics of cetaceans as lactating female bottlenose, common (*Delphinus* spp.), and pantropical spotted dolphins (*S. attenuata*) consume different prey items than do other dolphins.

## D. Intra-Individual Variation

Individual marine mammals can switch among foraging locations and tactics depending on their age, body condition, group size, and prey distribution and abundance. For example, pinnipeds can change their diving behavior in response to increased foraging costs as seals make shallower dives and dive at a steeper angle to maximize their time at a foraging depth. Individuals that encounter different habitats often switch among tactics depending on their location. For example, humpback whales may switch between foraging in large bubble-netting groups and engaging in individual lunges to capture krill. Sperm whale foraging behavior is linked to foraging success, and foraging is more common when prey availability is high or the energetic cost of capturing prey is relatively low. Also, sea otters change the number of prey items they collect on each foraging dive depending on the average prey size available. Offshore of Sable Island, Nova Scotia, harbor seals switch between pursuit and benthic foraging tactics depending on prey type, and Baikal seals (*Pusa sibirica*) shift their foraging tactics between day and night. They use visual cues to feed on pelagic fishes during the day but move to shallow waters, likely to feed on crustaceans using tactile cues, at night.

The flexibility of marine mammals is highlighted by their ability to take advantage of human activities. Many odontocetes, pinnipeds, and sea otters have learned to steal fish from nets. Seal lions will even jump into encircling nets to feed or will follow fishing vessels for days to take advantage of the abundant food resources offered by fishing operations. Bottlenose dolphins are well known for foraging behind trawlers and feeding on discarded fish or fish in nets. Some individual bottlenose dolphins also have learned to take advantage of direct handouts of fish offered by people, and many species of odonotocetes remove either bait or fish from fishing lines. In the Bering Sea and off Southern Brazil, killer whales may damage over 20% of the fish captured by longline fisheries.

Both the diversity of habitats in which marine mammals live and the flexibility of individuals has led to the wide variety of foraging tactics exhibited by the group. However, further studies of these tactics are still of great interest, especially systematic investigations of the function and use of particular tactics and the circumstances in which they are employed. For example, current studies are beginning to use a multivariate approach to teasing apart the roles of cultural transmission, genetics, and environmental factors on inter-individual variation in the use of foraging tactics within a population. Such detailed studies will improve the ability to predict influences of anthropogenic changes to marine habitats and prey availability on marine mammals and aid in efforts to conserve them.

## See Also the Following Articles

Behavior, Overview ■ Feeding Morphology ■ Filter Feeding ■ Toel Use

## References

Beck, B. B. (1980). "Animal Tool Behavior: The Use and Manufacture of Tools by Animals." Garland Press, New York.

Beck, C. A., Iverson, S. J., Bowen, W. D., and Blanchard, R. (2007). Sex differences in grey seal diet reflect seasonal variation in foraging behaviour and reproductive expenditure: Evidence from quantitative fatty acid signature analysis. *J. Anim. Ecol.* **76**, 490–502.

Boyd, I. L. (1997). The behavioural and physiological ecology of diving. *Trend. Ecol. Evol.* **12**, 213–217.

Branch, G. M., Harris, J. M., Parkins, C., Bustamante, R. H., and Eekhout, S. (1992). *In* "Plant-Animal Interactions in the Marine Benthos" (D. M. John, S. J. Hawkins, and J. H. Price, eds) pp. 405–423. Clarendon Press, Oxford.

Frid, A., Heithaus, M. R., and Dill, L. M. (2006). Dangerous dive cycles and the proverbial ostrich. *Oikos* **116**, 893–902.

Giraldeau, L. A. (1988). The stable group and the determinants of foraging group size. *In* "The Ecology of Social Behavior" (C. N. Slobodchikoff, ed.), pp. 33–53. Academic Press, New York.

Gross, M. R. (1996). Alternative reproductive strategies and tactics: Diversity within sexes. *Trend. Ecol. Evol.* **11**, 92–98.

Kramer, D. L. (1988). The behavioral ecology of air breathing by aquatic animals. *Can. J. Zool.* **66**, 89–94.

Lima, S. L., and Dill, L. M. (1990). Behavioral decisions made under the risk of predation: A review and prospectus. *Can. J. Zool.* **68**, 619–640.

Mesterson-Gibbons, M., and Dugatkin, L. A. (1992). Cooperation among unrelated individuals: Evolutionary factors. *Q. Rev. Biol.* **67**, 267–281.

Packer, C., and Ruttan, L. (1988). The evolution of cooperative hunting. *Am. Nat.* **132**, 159–198.

Ricklefs, R. F. (1990). "Ecology." W. H. Freeman & Co., New York.

Tregenza, T. (1995). Building on the ideal free distribution. *Adv. Ecol. Res.* **26**, 253–307.

Wells, R. S., Boness, D. L., and Rathbun, G. B. (1999). Behavior. *In* "Biology of Marine Mammals" (J. E. Reynolds, and S. A. Rommel, eds), pp. 324–422. Smithsonian Institution Press, Washington, DC.

Wirsing, A. J., Heithaus, M. R., Frid, A., and Dill, L. M. (2008). Seascapes of fear: Methods for evaluating sublethal predator effects experienced and generated by marine mammals. *Mar. Mamm. Sci.* **24**, in press.

# Female Reproductive Systems

## R.E.A. Stewart and B.E. Stewart

## I. Introduction

The female reproductive system in marine mammals is composed of the basic mammalian reproductive organs: ovary, oviduct, uterus, cervix, vagina, clitoris, and vaginal vestibule. Under the control of endocrine system, these organs are engaged in the reproductive cycle of ovulation, fertilization, implantation, fetal growth, and parturition. Ancillary to reproduction are the mammary glands and lactation. Some variation in anatomy, morphology, and physiology of the reproductive organs, and in reproductive cycles, exists among orders of marine mammals. Species-specific differences within orders also exist, reflecting both phylogeny and the variety of environments inhabited by marine mammals. Variation also exists in how marine mammals use their basic mammalian anatomy in different marine habitats. This is discussed in other articles such as those on reproductive strategies, life history, lactation, and behavior. Here the gross anatomical and morphological characteristics of female reproductive systems are described and the functional adaptations are noted.

## II. Anatomy and Morphology

The ovary is the organ where eggs or ova mature and are released during ovulation. Usually, there are two functional ovaries suspended from the abdominal or pelvic cavity by a short mesentery, the mesovarium, which attaches to the dorsal side of the broad ligament. Dugong (*Dugong dugon*) ovaries are also attached to the diaphragm by peritoneal folds that form pouches in the dorsal abdominal wall.

The ovaries are surrounded by the ovarian bursa, a fold of mesosalpinx which forms a peritoneal capsule. There is considerable variation in development and in the extent to which the bursa communicates with the celomic cavity. The ovarian bursa of odontocetes develops *in utero* whereas in mysticetes it develops after birth. In polar bears (*Ursus maritimus*) and other carnivores, the periovarian space between the ovary and peritoneal lining of the bursa communicates with the peritoneal cavity by a narrow passage which may become distended at estrus with fluid of unknown origins. In all marine mammals, and mammals in general, the function of the bursa is to ensure that the ova pass into the oviduct where fertilization occurs.

Marine mammal ovaries vary in size and shape. Quiescent dugong ovaries are small, flattened ovoids or spheres. Ovaries in the Amazonian manatee (*Trichechus inunguis*) are broad and flattened against a short mesovarium. Sea otters (*Enhydra lutris*) have lenticulate, compressed oval ovaries. The odontocete ovary is more or less spherical to ovoid in shape, with a smooth surface in the resting condition whereas in mysticetes ovaries are flat and elongated. Phocid ovaries are ovoid and smooth in the resting state. In some species of phocids, (e.g., gray seals, *Halichoerus grypus*), fetal hypertrophy of the ovaries exists through hormonal influence of the pregnant female. This condition may be less pronounced in otariids.

Typically, eggs ripen and ovulate alternately between the ovaries in successive reproductive cycles and the ovaries are of similar size. However, in some odontocetes there is a prevalence of activity in the left ovary (e.g., 70% in pilot whales, *Globicephala*) and the left ovary is larger than the right (Slijper, 1966). The right ovary may become active later in life.

The mammalian ovary is covered by germinal epithelium (Fig. 1) below which lies connective tissue (*tunica albuginea*) of varying thickness. Germinal epithelium is often invaginated into the *tunica albuginea*, forming small folds, pits, or subsurface crypts. These invaginations are particularly well developed in pinnipeds and form surface fissures in sea otter ovaries. Below the *tunica albuginea* is a layer of follicles and *corpora* that are derived from them. The ovary also contains stromal and connective tissue, interstitial tissue, vascular, nervous and lymphatic tissues, and embryological remnants. The interstitial cells of cetaceans are less numerous and less prominent than those in some other mammalian orders, such as rodents. Understanding the maturation process of the follicles, and development and subsequent regression of the luteal bodies for each species allows researchers to assess the reproductive status of females (immature, ovulating, etc.).

Follicular maturation (Fig. 1) proceeds through a series of changes characterized by two phases. In the first phase, there is a rapid increase in the size of the oocyte and a slow increase in the size of the follicle. Second, there is slow growth of the oocyte and a rapid increase in the size of the follicle which can be seen macroscopically. In dugongs, mature follicles may be just visible as translucent bodies or they may protrude from the ovarian surface. In West Indian manatees (*Trichechus manatus*) the mature follicles appear as large masses of bead-like spherules in the ovary. Similarly, mysticete ovaries may appear grape-like with protruding follicles. Maturing follicles of odontocetes and pinnipeds tend to be more widely dispersed in the ovary.

Oocytes develop within the ovary during fetal development but are dormant until puberty is reached. After puberty, and partly in

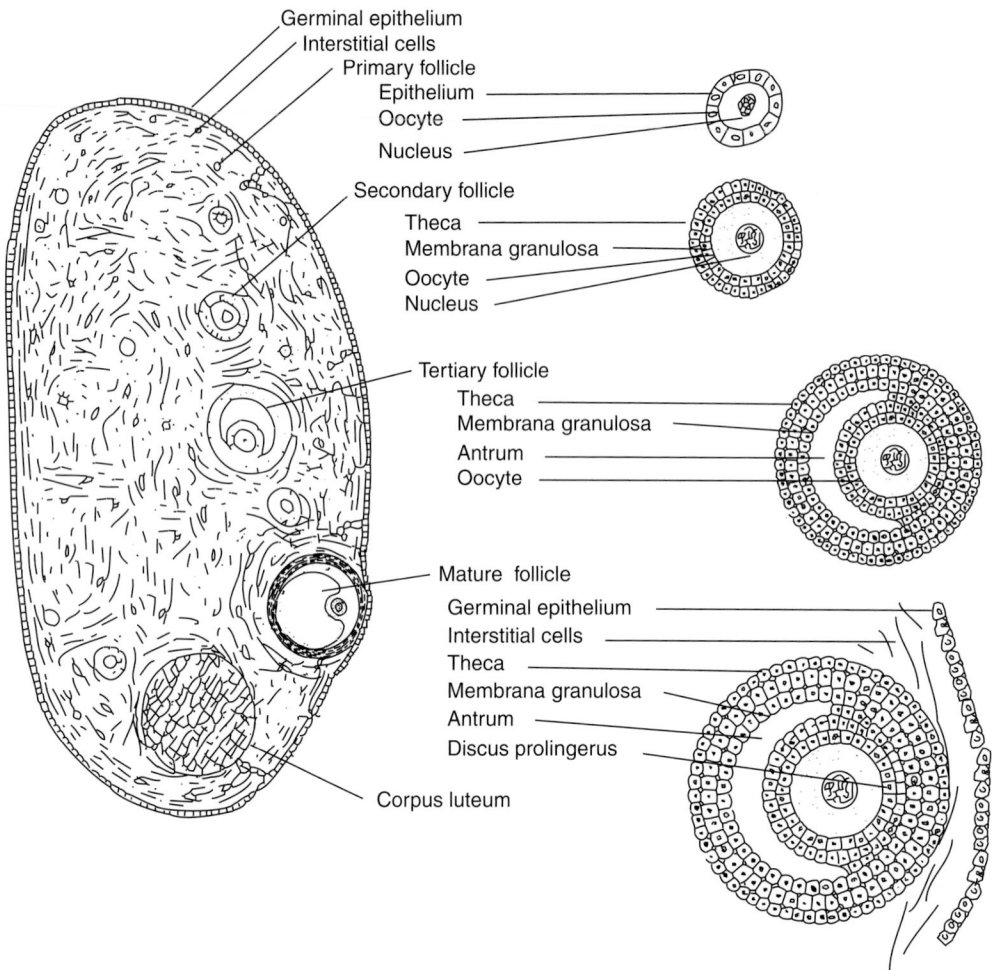

**Figure 1** *The main structures of a marine mammal ovary showing stages of follicular development. Modified from Penny and Waern (1965).*

concert with the annual reproductive cycle, they become primary follicles (Fig. 1) with a single layer of flattened epithelial cells surrounding each oocyte. These follicles lack connective tissue or thecal investment. As primary follicles increase in size they sink deeper into the cortex of the ovary toward the medulla or central area.

Secondary follicles are formed as the single layer of flattened epithelial cells around the oocyte thickens and becomes cuboidal or columnar, forming a distinct *membrana granulosa* (granular membrane or layer). This granulosa layer quickly becomes several layers thick while being encapsulated in an outer sheath or *theca* derived from the *stroma*. The *theca* divides into two layers. The inner layer, the *theca interna*, is glandular and well vascularized while the outer layer, the *theca externa*, is composed of connective tissue. The oocyte is now surrounded by a membrane, the *zona pellucida*, which is jelly-like, contains large amounts of polysaccharides, and lies between the plasma membrane of the oocyte and the granulosa cells.

Tertiary follicles rapidly increase in size because of an increased number of granulosa and thecal cells. Then, one or more cavities form in the granulosa and as the cavities enlarge they coalesce into an antrum which fills with follicular fluid (*liquor folliculi*). The fluid-filled follicle is now surrounded by a wall except at the point of attachment of the oocyte.

Mature or Graafian follicles (Fig. 1) contain an oocyte which is surrounded by an irregular cluster of granulosa cells. These granulosa cells eventually form the *corona radiata* which in turn is attached to cells forming the *discus prolingerus* or *cumulus oophorus*. There is mortality among developing follicles such that only a fraction of those primary follicles that start to develop will ever mature.

Under endocrine control, the mature follicle releases the ovum (ovulation). After ovulation, the *corpus luteum* (yellow body) develops from cellular components of the follicle. Luteinization is the process of transformation of follicular granulosa cells into luteal cells which contain carotenoid luteins (yellow pigments). There is a significant increase in cell size during luteinization. This volumetric growth is in contrast with the accretional growth of thecal cells in tertiary follicles. The *corpus luteum* (CL) is considered to be a gland and several types exist, the nomenclature being based on the morphology or function of the CL. For example, a CL of pregnancy develops when ovulation is followed by fertilization; accessory *corpora lutea* (plural) develop by luteinization of unruptured follicles (common in cetaceans) (Perrin *et al.*, 1984).

If fertilization of the ovum occurs, the CL of pregnancy persists through gestation. There are several phases of development in this *corpus luteum gravidatitis*: a short post-ovulatory phase when the CL is small, poorly vascularized, and minor changes occur in the luteal cells; a phase during delayed implantation when the CL is smaller than its initial size, vascularization is still poor, and the luteal cells show marked cytoplasmic vacuolation; a short phase related to implantation when vacuoles disappear, the CL enlarges with resumed glandular activity, and vascularization increases; a phase post-implantation and during early pregnancy characterized by minor cell adjustments such as fluid accumulation, more obvious intercellular spaces, appearance of small vacuoles, increase of connective tissue, and thickened walls of blood vessels; the phase of the duration of pregnancy when the *corpus luteum* size is maintained; and the post-parturient phase as a *corpus albicans* (CA). In cetaceans, accessory CLs may also form in the ovary of pregnancy.

In beluga, the *corpora lutea gravidatitis* were about 3–4 cm in diameter and weighed approximately 22 g. CL diameters in blue whales (*Balaenoptera musculus*) average approximately 14 cm; in minke whales (*B. acutorostrata*) they averaged about 7 cm and 160 g. In some odontocetes, such as bottlenosed dolphins (*Tursiops truncatus*), the CL *gravidatitis* may protrude far out from the general outline of the ovary and is connected by a stalk (pedunculated).

The process of degeneration of the *corpus luteum* into the *corpus albicans* is similar regardless of the type of *corpus luteum* (CL of ovulation, pregnancy, pseudopregnancy, or lactation) that is regressing. There are four patterns of degeneration: fibrohyalin invasion, lipoid degeneration, slow necrobiosis, and fast necrobiosis. Regardless of the regression pattern, glandular elements are lost, lutein granules vanish, and the size of the body diminishes until the white or gray scar-like *corpus albicans* (CA) is formed.

In most mammals, *corpora albicantia* are assimilated either relatively quickly post-partum or after one or two reproductive cycles, as in the sea otter. In marine mammals that cycle every 2 or 3 years, such as the walrus (*Odobenus rosmarus*), the CA may persist for some time. In cetaceans, *corpora albicantia* are thought to persist throughout a female's lifetime (Perrin *et al.*, 1984), a consequence of the large amount of connective tissue present and its poor vascularization, leading to a slow rate of regression. Some attempts have been made to characterize various types of cetacean *corpora albicantia*, but a definitive way to distinguish those bodies derived from ovulation from those of pregnancy has not been established.

An ovarian structure that appears in at least some pinnipeds is the *hilar rete* (Boyd, 1984). These glandular cells are most abundant during delayed implantation but their function is not clear. They also occur in a number of terrestrial mammals including carnivores, primates, rodents, and hyraxes.

The oviduct, uterus, and vagina are all derivatives of the Müllerian duct system. The oviduct, fallopian tube, or uterine tube is generally highly convoluted and is enclosed in the ovarian bursa. The anterior end forms a funnel or infundibulum near the ovary and the posterior end of the oviduct enters the uterus. In eutherian mammals with a bicornuate uterus (e.g., cetaceans), the isthmus can be straight or convoluted but it has a thick wall and a narrow lumen. Dugong uterine tubes lack a mesosalpinx and exist as a 4-cm long cord-like convoluted tube which lies dorsal to the peritoneum. Generally, the oviduct is lined by simple columnar epithelium that are ciliated, and have occasional goblet cells. There is an inner circular layer and an outer longitudinal layer.

The uterus classification scheme in mammals is based on progressive fusion of the caudal ends of the oviducts. Four major types are recognized (Table I, Fig. 2) of which three types are represented in marine mammals; no marine mammal has a simplex uterus. All types

**TABLE I**
**Uterine Types**

| Type of uterus | Uterine horns | Cervix | Example |
|---|---|---|---|
| Duplex | Two completely separate horns | 2 | Walrus |
| Bipartite | Two horns separated internally by a septum but sharing a small common area near the cervix | 1 | Phocids |
| Bicornuate | Two horns with no internal septum, forming a single body of the uterus | 1 | Cetaceans, sirenians, mustelids, ursids |
| Simplex | No horns, one uterine body without compartments | 1 | Humans |

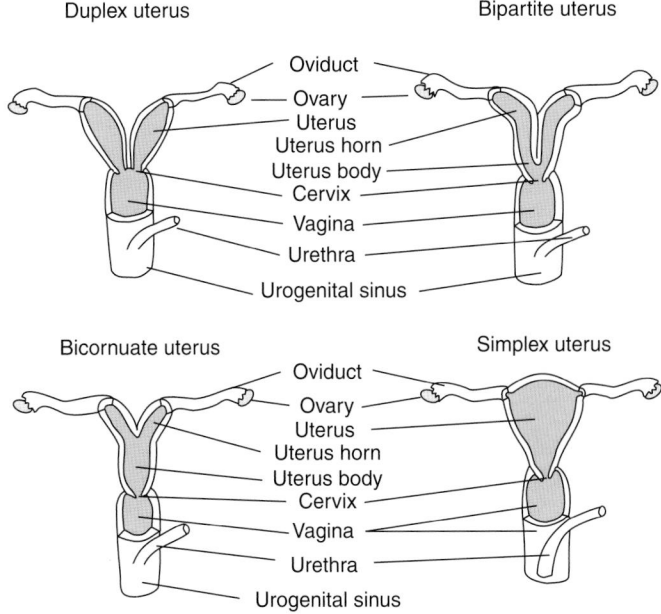

**Figure 2** *Uterus types found in placental mammals. The simplex type is not found in marine mammals. Modified from Romer (1962).*

of uteri are supported by broad ligaments, have two oviducts and deliver into a single vagina.

The uterine wall has three layers: on the outside, the serous membrane; in the middle, the myometrium, which contains the internal circular muscle and the external longitudinal muscle separated by the vascular layer; and the inner lining of the uterus, the endometrium, composed of an epithelial lining of the lumen, a glandular layer, and some connective tissue. All uterine types exhibit changes in the layers of the uterine wall that precede implantation and development of the placenta. During the luteal phase of the follicle (post-ovulatory) the endometrium increases in thickness and the glands become extremely branched and convoluted.

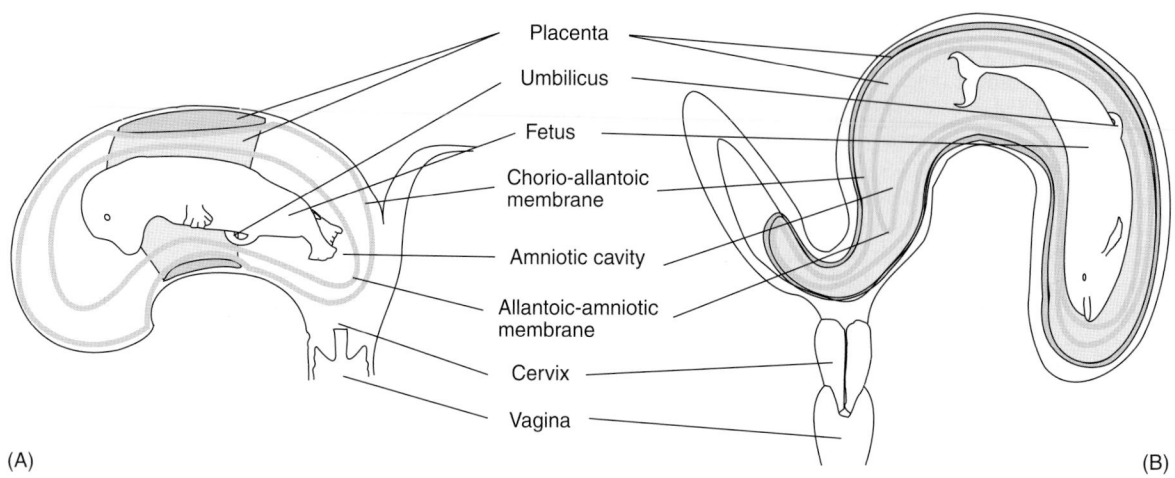

Placenta

Umbilicus

Fetus

Chorio-allantoic
membrane

Amniotic cavity

Allantoic-amniotic
membrane

Cervix

Vagina

(A)

(B)

**Figure 3** *(A) Zonary and (B) diffuse placentas found in marine mammals (B modified from Slijper, 1966).*

Embryonic membranes develop before the embryo implants. The yolk sac forms first as endoderm surrounds the nutrient uterine fluid. The chorion and amnion develop as a double layer, originating from embryonic ectoderm. The innermost layer is the amnion which forms the fluid-filled amniotic sac encompassing the fetus. The outer layer of cells of this sac wall is the chorion. The allantois develops as an extension of the embryonic hind gut and forms the allantoic cavity. The allantois fuses with the chorion forming a small round area, the allanto-chorion. This region becomes the placenta (Fig. 3).

The embryonic membranes persist after the placenta is fully developed. In whales, the amniotic sac and allantois extend into the contralateral uterine horn. The chorion has many folds that mesh with the uterine mucosa. In pinnipeds, the chorion extends beyond the zonary placenta. The large allantois almost completely surrounds the amnion. In sirenians, the allantois is also large and nearly fills the chorionic sac.

The placenta is comprised of maternal and embryonic tissue in close union. It allows nutritional, respiratory, and excretory exchange between the maternal and fetal circulatory systems by diffusion across the placental membranes. Additionally, the placenta functions as a protective barrier to bacteria and other large molecules, produces some food materials, and synthesizes hormones required to maintain the pregnancy. The umbilical cord connects the placenta to the ventral surface of the embryo, and is formed of mesoderm and blood vessels. This connection is broken at birth.

There are three types of placentas seen in mammals, defined by the fetal tissue that adheres to the uterine wall. They are the chorionic placenta, the yolk-sac placenta, and the chorio-allantoic placenta which are found in all marine mammals. The chorio-allantoic placenta is the most advanced placenta type in its ability to provide rapid diffusion between the uterine and fetal circulatory systems. As the blastocyst implants and sinks into the uterine endometrium, chorionic villi grow quickly and push further into the endometrium. This process is accompanied by a breakdown of uterine tissue. The degraded debris is called embryotroph. The blastocyst absorbs the nourishing embryotroph until the villi are fully developed and the embryonic vascular system becomes functional.

Although all placentas are derived from both maternal and fetal tissues, the degree of separation of maternal and fetal circulatory systems is variable and is a function of the type of placenta that has developed. Chorio-allantoic placentas can be further subdivided based on the fetal and maternal cell layers that are in contact. In epitheliochorial placentas (cetaceans), the epithelium of the chorion is in contact with the uterine epithelium. The villi rest in endometrial pockets. In endotheliochorial placentas, (pinnipeds, mustelids, ursids), degradation of the maternal endometrium is more pronounced and the epithelium of the chorion is in contact with the endothelial lining of the uterine capillaries. Hemochorial placentation (sirenians), is characterized by destruction of the endothelium of the uterine blood vessels, allowing blood sinuses to develop in the endometrium. There is direct contact, therefore, between the chorionic villi and maternal blood. Hemoendothelial placentas are not found in marine mammals but occur in some rodents and in lagomorphs.

Placenta shape is characterized by the pattern in which villi are distributed over the chorion. Two shapes occur in marine mammals—diffuse and zonary (Fig. 3). In the diffuse placenta (cetaceans), the villi occur over the entire chorion providing a large surface area for exchange. In a zonary placenta (all other marine mammals), there is a continuous band of villi covering the equator of the chorion. In walrus and sirenians, the zonary placenta leaves detectable scars on the uterus. Both diffuse and zonary placentas are also found in terrestrial mammals.

When the placenta is expelled from the uterus postpartum as afterbirth, a maternal component may or may not be lost in this process. Epitheliochorial placentas have villi which pull out of the uterine pits easily and no endometrium is pulled away. Therefore, no bleeding occurs at birth and this placenta is referred to as nondeciduous. The other types of placentas allow for closer association of maternal and fetal circulatory systems through degradation of the endometrium and extensive intermingling of uterine and chorionic tissue. At birth then, part of the uterine component of the placenta is torn away and bleeding occurs. These are termed deciduous placenta. Bleeding is arrested quickly by collapse of the uterus, contractions of the myometrium constricting blood vessels, and blood clotting. The subsequent development of uterine scars at the bleeding site can result in persistent features that can be used in the interpretation of reproductive history in a female (e.g., walrus, Fay, 1982), although their persistence is variable among species.

The cervix is a well-muscled sphincter that marks the transition between the uterus and the vagina. The West Indian manatee has a rounded cervix. In the dugong, the long, thin-walled vagina has a keratinized shield originating in the vault region. This shield surrounds the cervix and extends along with the ventral wall of the vagina. In cetaceans, the cervix is long with a thick wall and a narrow, sinuous lumen. The portion of the cervix projecting into the vagina (*portio vaginalis uteri*) is species specific in length, e.g., very short in narwhal (*Monodon monoceros*) but long in common porpoise (*Phocoena phocoena*). Vagina length in pinnipeds equals or slightly exceeds the urogenital canal in length.

Many cetaceans have several folds in the upper part of the vagina which are not found in any other mammal. There is a valve-like arrangement to these 4–12 circular folds which project distally and look like a chain of funnels with their mouths directed toward the cervix. This configuration may act to retain sperm but the function of these folds is unclear. Harrison (1969) noted the folds appeared capable, anatomically, of a pumping action and speculated that they may relate to the formation of vaginal plugs. Vaginal plugs are congealed masses of semen that occur in some mammalian orders. They are thought to assist sperm retention and discourage subsequent matings by competing males. There is some suggestion of vaginal plugs in *Tursiops* and *Delphinus* but little other evidence. Vaginal calculi are masses of organic and inorganic material that have been found in *Delphinus*, *Stenella*, and *Lagenorhynchus*. Calculi contain material identical in composition to mammalian bone and some contained recognizable embryonic bones (Perrin *et al.*, 1984). "Vaginal plugs" therefore may represent small or deteriorating vaginal calculi.

The remainder of the female reproductive system consists of the clitoris, urethra, and vaginal vestibule. For all pinnipeds, the presence of the *os clitoridis* has been recorded but it is generally small (<1 cm) and its appearance is irregular. In cetaceans, the clitoris projects from a strand of fibrous connective tissue on the anterior border of the vulva. Dugongs possess a clitoris with a large conical glans which has prominent fissures dividing it into lobes.

The urethra also opens into the vaginal vestibule, draining the bladder which lies ventral to the vagina. The urethra in pinnipeds opens by a large urinary papilla just caudal to the hymen. Pinnipeds have long urogenital canals compared to other mammals and there is a prominent hymenal fold which is smaller in otariids. Dugongs have a urethra which is surrounded by prostate-like tissue which also surrounds the narrow distal portion of the vagina.

In pinnipeds, the vaginal vestibule opens to the exterior just ventral to the anus in a common furrow. In cetaceans, the vulva is a slit-like aperture with *labia majora* and *labia minora* which may be poorly developed. The slit is positioned just anterior to the anus.

Considerable variation exists in mammary gland configuration and nipple location in marine mammals. Phocids have distinct mammary glands enclosed in connective sheaths lying under the blubber layer. In otariids, the mammary glands coalesce to form a sheet-like layer under the blubber over most of the ventral surface of the body. Cetacean mammary glands are elongate, narrow, flat organs that extend in the subcutaneous connective tissue at both sides of the ventromedial line. They extend from a little posterior of the umbilicus to slightly anterior to the anus.

Cetaceans have two nipples in elongated recesses, one on either side of the midline in the urogenital slit. Sirenians have two pectoral nipples. Sea otters have only two functional teats on the lower abdomen, compared to six or more in most other mustelids. Polar bears have four teats, two on either side of the midline of the belly slightly posterior of the axillae and two about 15 cm posterior to the anterior pair. Otariids, odobenids, and two genera of phocids (*Erignathus* and *Monachus*) have four teats while the other phocids have two that correspond to the posterior pair of those in otariids. Pinniped nipples are retracted beneath the level of the body surface when a pup is not nursing and become erect during suckling.

## III. Reproductive Cycle

In marine mammals, ovulation can be either spontaneous or induced. Spontaneous ovulators release an egg even in the absence of breeding (phocids, cetaceans). Induced ovulators release an egg only in response to coital stimulation (polar bear, sea otter).

The ova normally are fertilized in the oviduct within 24 h of breeding and the zygote's first cell divisions occur there. At the 8–16 cell stage movement to the uterus occurs with assistance of muscle contractions in the oviduct. There, further differentiation into the blastocyst and subsequent implantation takes place.

Implantation of the blastocyst may be immediate (within 1–2 weeks) or after some protracted delay of several months. This delay, also known as embryonic diapause, occurs in the carnivores and pinnipeds and is considered to be obligate. Its duration is species-specific. Facultative delayed implantation can occur in other species of mammals if a female is nursing a large litter at the time of insemination (e.g., some marsupials, some insectivores). The length of active gestation is generally related to the body size of the female and a delay in implantation is thought to allow the young to be born at an advantageous time. For example, harp seals (*Phoca groenlandica*) have a delay of about 3 months followed by an active gestation of about 8.5 months. The delay produces a cycle that is nearly 1 year long, allowing births and breeding to occur during large spring aggregations when pack-ice conditions are suitable. There may be some flexibility in the duration of the delay, and in sea otters, this flexibility may lead to the variation seen in estimates of total gestation, ranging from 6 to 8 months.

Obligate delayed implantation is characterized by ovulation, fertilization, and differentiation up to the blastocyst stage which creates a hollow ball of 100–400 cells surrounding a fluid-filled cavity. Further differentiation of the blastocyst then stops and implantation in the endometrium does not occur. The blastocyst is free-floating in the uterus and is covered by a *zona pellucida*, a noncellular protective layer, for the period of dormancy. Resumption of blastocyst differentiation occurs prior to implantation. During delayed implantation, pregnancy is indicated, macroscopically, by the presence of a *corpus luteum*, and increased diameter of a uterine horn which shows marked surface vascularization, smoothening of the endometrial folds, and development of a nidation chamber where implantation will occur.

Usually, implantation of the blastocyst occurs in the uterine horn corresponding to the active ovary (ipsilateral). However, in odontocetes with only one active ovary, there is a tendency for trans-uterine migration of the blastocyst which will implant in the other (contralateral) horn. The fetal membranes project into the horn opposite the implantation site. In mysticetes, and most other mammals, there is a slight prevalence (60%) of implanted fetuses in the right horn, reflecting a similar rate of ovarian activity in the right ovary (60%) (Slijper, 1966).

Once implanted, the blastocyst begins to differentiate tissues and organs, and remains in the uterus for the duration of its fetal phase. Nourishment and protection in utero allows for relatively high survival rates of the fetus.

At birth, powerful and rhythmic contractions of the uterine myometrium aided by the abdominal muscles expel the fetus. Continued contractions force the placenta from the uterus and vagina. In cetaceans, birth underwater must be rapid to prevent drowning of the neonate. The newborn swims unaided or is pushed to the surface by its mother or attendants to breath for the first time. Other birthing platforms include land fast ice (ring seal—*Phoca hispida*, Weddell seal—*Leptonychotes weddellii*), pack ice (harp seal, crabeater seal—*Lobodon carcinophagus*), and terrestrial sites (harbor seal—*Phoca vitulina*, polar bears, otariids). Few births have been observed in sea otters and may take place both on shore and in the water. Sirenian births are also rarely seen but are thought to occur in shallow water, although there is some evidence they may also calve on low sand bars.

Cetacean and pinniped neonates have relatively large body sizes, approximately 8–10% maternal weight, compared to mammals in general (Slijper, 1966, Kovacs and Lavigne, 1992) a benefit to these animals that need to swim at birth or shortly afterward, and to maintain homeothermy in cold water. Most marine mammals usually give birth to a single offspring. Twin live births are exceedingly rare in pinnipeds and the sea otter and have never been documented in cetaceans, although multiple fetuses have been observed. Some twins have been reported among sirenians. It is thought that multiple births are incompatible with the production of newborns that are large relative to maternal body size, as is found in these animals. It may also be difficult for a marine mammal mother to properly tend more than one offspring in the marine environment.

Multiple births are the norm in polar bears, however ursids have extremely small young, relative to adult female body weight, that are born in dens and emerge with mother after considerable time when substantial postnatal growth has occurred. Most ringed seals and Baikal seals (*Phoca siberica*) are also born in dens, excavated in snow drifts from a hole scratched in the sea ice by the mother.

All marine mammals suckle their young with milk exclusively before a transition to solid food items and complete weaning is made. The period of lactation is again species-specific and can be relatively short (4 days in hooded seals—*Cystophora cristata*, 10–12 days in harp seals), or more protracted (up to 2 years in sirenians, some cetaceans, walrus), although the young may start to eat solid food before weaning is completed. Marine mammal milk is typically high in fat (40–50%), high in protein (7–19%), and low in lactose (trace 5%) compared to terrestrial mammals.

Milk may be forcefully ejected from the teats or may be sucked from the teats by the neonate. In cetaceans, forceful ejection of milk is required because neonates cannot suck with their lips and must hold their breath during underwater nursing bouts. Indeed, young *Tursiops* can only remain underwater for less than a minute, so they nurse two or three times an hour over an entire 24-h period. Cetacean nipples become protruded during nursing and the milk is expelled under pressure, likely due to contraction of either the cutaneous muscles or the myoepithelial cells surrounding the alveoli. Walrus may suckle young in water and the teats are surrounded by sphincter-like folds of skin which suggest that milk may be squirted into the calves' mouth. Young sea otter pups nurse while lying on the female's chest while older ones lie in the water perpendicular to her. Sirenian calves nurse at the surface with their nostrils in the air or just below the surface. All other marine mammals nurse their young on ice or on land and neonates actively suck milk.

Weaning can be abrupt by abandonment (most phocids and otariids) or extend over some time (walrus, sirenians). The extended care of the young during lactation, and sometimes beyond in a period of learning, further increases survivorship of the offspring beyond the high rate of fetal survival. It also increases the efficiency of reproduction in that maternal energy expended towards young results in a high rate of offspring that reach reproductive maturity, consistent with other K-selected life history traits.

Age of maturation varies by species and, within a species, can be influenced by environmental factors that affect growth and fattening. Breeding success is often lower in younger breeders, but lifetime reproductive success, the number of descendants produced, of those that breed young and survive, can be high. Diminished reproductive frequency (reproductive senescence) has been described for some marine mammals (walrus, polar bear, some fur seals, and some cetaceans). Although short-finned pilot whales (*Globicephala macrorhynchus*) of advancing age become senescent with no follicular activity, they may still lactate, nursing not only their own previous young but also other young in the pod.

## See Also the Following Articles:

Cetacean Life History ▪ Cetacean Prenatal Development ▪ Endocrine System ▪ Estrus and Estrous Behavior ▪ Polar Bears ▪ Pinniped Life History ▪ Reproductive Behavior ▪ Sirenian Life History

## References

Berta, A., and Sumich, J. L. (1999). "Marine Mammals: Evolutionary Biology." Academic Press, San Diego.

Boyd, I. L. (1984). Occurrence of hilar rete glands in the ovaries of grey seals (*Halichoerus grypus*). *J. Zool. (Lond.)* **204**, 585–588.

Boyd, I. L., Lockyer, C., and Marsh, H. D. (1999). Reproduction in marine mammals. *In* "Biology of Marine Mammals" (J. E. Reynolds, III, and S. A. Rommel, eds), pp. 218–286. Smithsonian Institution Press, Washington.

Fay, F. H. (1982). "Ecology and Biology of the Pacific Walrus *Odobenus rosmarus divergens* Illiger," North American Fauna, Number 74. Department of the Interior, Washington, DC.

Harrison, R.-J. (1969). Reproduction and reproductive organs. *In* "The Biology of Marine Mammals" (H.T. Anderson, ed.), pp. 253–348. Academic Press, London.

Harrison, R. J., and King, J. E. (1965). "Marine Mammals." Hutchinson & Co. (Publishers) Ltd, London.

Harrison, R. J., Brownell, R. L., Jr., and Boice, R. C. (1972). Reproduction and gonadal appearances in some Odontocetes. *In* "Functional Anatomy of Marine Mammals" (R. J. Harrison, ed.), Vol. 1, pp. 361–429. Academic Press, London.

Kovacs, K. M., and Lavigne, D. M. (1992). Maternal investment in otariid seals and walurses. *Can. J. Zool.* **70**, 1953–1964.

Leatherwood, S. and Reeves, R.R eds. (1990). The Bottlenose Dolphin, Part VII Husbandry and Captive Breeding. Academic Press, San Diego.

Penny, D. A., and Waern, R. (1965). "Biology, An Introduction to Aspects of Modern Biological Science." Sir Isaac Pitman (Canada) Limited, Toronto.

Perrin, W. F., R. L. Brownell Jr., and D. P DeMaster, (eds) (1984). Reproduction in whales, dolphins and porpoises. Reports of the International Whaling Commission Special Issue 6. International Whaling Commission, Cambridge.

Romer, A. S. (1962). "The Vertebrate Body." W. B. Saunders Company, Philadelphia.

Slijper, E. J. (1966). Functional morphology of the reproductive system in Cetacea. *In* "Whales, Dolphins and Porpoises" (K. S. Norris, ed.), pp. 277–319. University of California Press, Berkeley and Los Angeles.

van Tienhoven, A. (1968). "Reproductive Physiology of Vertebrates." W. B. Saunders, Philadelphia.

Vaughn, T. A. (1972). "Mammalogy." W. B. Saunders, Philadelphia.

Zuckerman, L., and Weir, B. J. (1977). "The Overy. Vol. 1 General Aspects,". Physiology. Academic Press, New York.

# Filter Feeding

Donald A. Croll, Bernie R. Tershy and Kelly M. Newton

## I. Filter Feeding and the Marine Environment

A fundamental necessity for any organism is acquiring sufficient food for maintenance, growth, and reproduction. This search for food likely drove the return of mammals to the ocean where they were able to exploit highly productive coastal waters. With their return to the sea, marine mammals evolved a number of foraging techniques. Filter feeding, found in the mysticete whales and three species of pinnipeds (crabeater seals, *Lobodon carcinophaga*; leopard seals, *Hydrurga leptonyx*; and Antarctic fur seals, *Arctocephalus gazella*) is the most unique of these adaptations for feeding, and is not found in any terrestrial mammals.

Filter feeding allows these marine mammals to exploit extremely abundant, but small schooling fish and crustaceans by taking many individual prey items in a single feeding event. This adaptation arose in response to the unique patterns of productivity and prey availability in marine ecosystems. Low standing biomass and high turnover of small-sized primary producers that respond rapidly to nutrient availability characterize marine food webs. Due to spatial differences in the physical dynamics of marine ecosystems, productivity tends to be more patchy and ephemeral than in terrestrial systems. Consequently, marine grazers (e.g., schooling crustaceans and fish) often occur in extremely high densities near these patches of high primary production. Most marine mammals are primary carnivores and feed on these dense, patchily distributed aggregations of schooling prey. The spatial and temporal patchiness of this prey means that marine mammals must often travel long distances to locate prey, and the larger body size of marine mammals likely plays an important role (Croll *et al.*, 2005).

Initially, thermoregulatory requirements selected for larger body sizes as mammals returned to the ocean. However, once dependent upon marine prey, large body size also provided a buffer for the patchy and ephemeral distribution of marine prey. Thus, larger individuals could endure longer periods and travel longer distances between periodic feeding events on patchy prey. While adaptive for exploiting patchy prey resources, a consequence of larger body size is a higher average daily prey requirement. For marine mammals that feed on patchy and ephemeral resources, this requires individuals to take in large quantities of prey during the short periods of time it is available (Berta and Sumich, 1999; Bowen and Siniff, 1999).

Filter feeding is a foraging strategy that allows individuals to capture and process large quantities of prey in single mouth full, thus allowing them to acquire energy at high rates when small prey are aggregated. Indeed, for mysticetes, large body size is probably a prerequisite for attaining a sufficiently large surface area for filter feeding. Thus, the interaction of availability of prey resources, high concentrations of prey in schools, and selection for large body size likely led to the evolution of filter feeding. Ultimately, large body size and filter feeding allowed some marine mammals to exploit the extremely high densities of schooling prey that develop at high latitudes during the spring and summer, but fast during the winter when these resources disappear. Large body size provided an energy store for wintering and long distance migration without feeding (Berta and Sumich, 1999).

Due to this dependency on patchy but extremely productive food resources, it is not surprising that filter-feeding whales are believed to have first evolved and radiated in the southern hemisphere during the Oligocene at the initiation of the Antarctic Circumpolar Current (ACC). It is generally agreed that the initiation of the ACC led to cooling of the southern oceans, increased nutrient availability and thus increased productivity. This increased productivity provided a rich resource of zooplankton that could be effectively exploited through filter feeding (Berta and Sumich, 1999). Recently, the discovery of a late Oligocene fossil archaic mysticete that was a macrophagous predator casts doubt on the suggestion that the initial radiation of mysticetes was linked to the evolution of filter feeding (Fitzgerald, 2006).

Present-day filter-feeding marine mammals concentrate their foraging in polar regions and highly productive coastal upwelling regions. The southern ocean is still the most important foraging area for filter-feeding marine mammals. Prior to their exploitation by humans, the highest densities of mysticetes occurred in highly productive southern waters. Crabeater seals, Antarctic fur seals, and leopard seals are found primarily in the southern oceans where seasonally dense aggregations of krill develop (Berta and Sumich, 1999).

## II. Diet, Filter-Feeding Structures, and Prey Capture

All filter-feeding species feed on prey that form dense aggregations (primarily pelagic schooling fish and crustaceans or densely aggregated benthic amphipods). Two feeding adaptations have evolved to allow the exploitation of these dense aggregations: baleen (mysticete whales) and modified dentition (seals).

### A. Seals—Diet, Feeding Morphology, and Behavior

Unlike mysticetes, pinnipeds evolved in the Northern Hemisphere where krill was not likely an important component of their diet, and adaptations for filter feeding are not nearly as extensive in pinnipeds as in mysticetes.

Only three pinniped species regularly filter feed: crabeater seals, leopard seals, and Antarctic fur seals (Riedman, 1990). When filter feeding, all the three species feed almost exclusively on Antarctic krill, *Euphausia superba* in the Southern Ocean where it is large in size, abundant, and forms extremely dense aggregations. Of the three species, crabeater seals are most highly specialized with krill comprising up to 94% of their diet, while krill comprises approximately 33% of the diet of leopard seals and Antarctic fur seals. The most remarkable adaptation for filter feeding in pinnipeds is found in the dentition of crabeater and leopard seals. In both species elaborate cusps have developed on the postcanines in both the upper and lower jaws (Fig. 1) (Berta and Sumich, 1999). Once the mouth closes around a small group of krill, water is filtered out through the cusps, trapping krill in the modified teeth. Little detailed information is available on the behavior used by filter-feeding pinnipeds to capture prey. However, data from Antarctic fur seals and crabeater seals indicate that they track the diel migration of krill: shallow dives are performed during the night and deeper dives during the day (Boyd and Croxall, 1992).

### B. Mysticetes—Diet and Feeding Morphology

Most mysticetes feed primarily on planktonic or micronectonic crustaceans (copepods and krill) and pelagic schooling fish found in shallow waters. Gray whale, *Eschrichtius robustus*, diet consists primarily of benthic gammarid amphipods, although they can forage

F

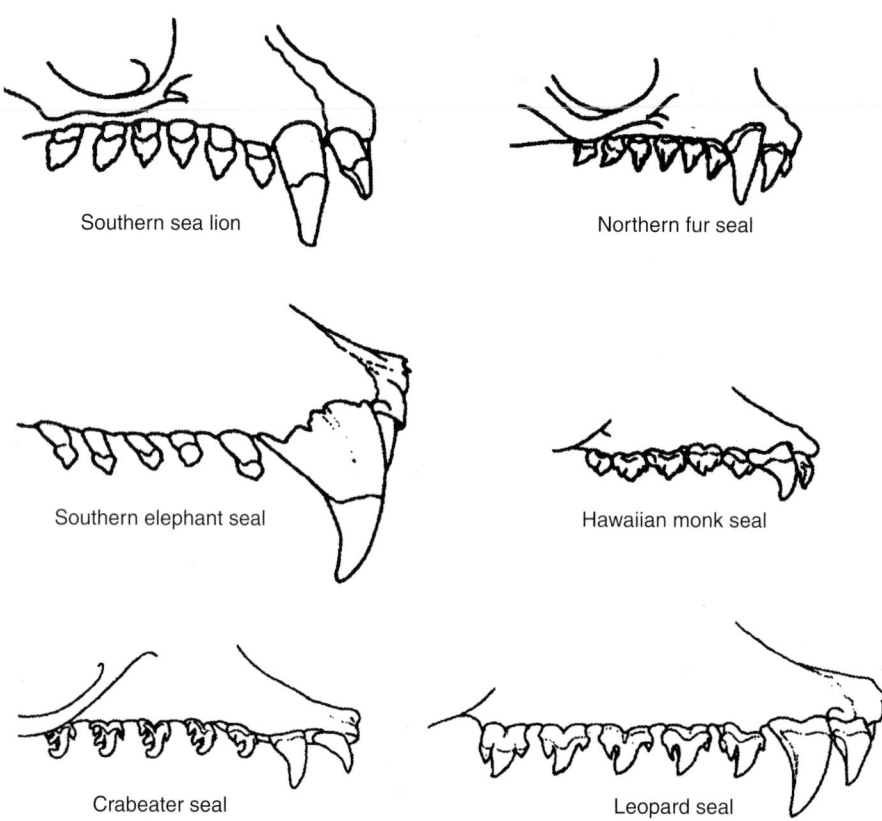

Southern sea lion

Northern fur seal

Southern elephant seal

Hawaiian monk seal

Crabeater seal

Leopard seal

**Figure 1**   *Dentition patterns in pinnipeds. Note modified cusps in postcanine teeth in filter-feeding crabeater and leopard seals. From Berta and Sumich (1999).*

on a wide variety of prey, including schooling mysids in some areas. Right, *Eubalaena* spp., and bowhead, *Balaena mysticetus*, whales primarily feed on copepod crustaceans of the genus *Calanus*. All of the rorquals feed on euphausiids (krill) to some extent, and blue whales, *Balaenoptera musculus*, feed almost exclusively upon euphausiids (see section on krill). The other rorquals have a more varied diet that includes copepods (sei whales, *Balaenoptera borealis*), and schooling fish (minke, *B. acutorostrata*, Bryde's; *B. edeni*, humpback; *Megaptera novaeangliae*; and fin whales, *B. physalus*) (Berta and Sumich, 1999).

All present-day mysticetes lack teeth and instead have rows of baleen plates made of keratin that project ventrally from the outer edges of the palate. Similar to fingernails, the plates grow continuously from the base, but are worn by the movements of the tongue. As the edges of the plates wear, hair-like fibrous strands emerge as fringes. The outer fibers of these fringes are coarser while the inner fibers form a tangled fringe that overlaps with fringes on adjacent baleen plates. Rows of baleen plates form an extended filtering surface along each side of the palate.

The coarseness of the hair-like fibrous fringes, the density of fibers (number of fibers/cm$^2$), number of baleen plates, and length of baleen plates varies between species, and is related to the prey species captured in the filtering mechanism. Because gray whales feed primarily upon sediment-dwelling benthic amphipods, they have the coarsest filtering mechanism, made up of about 100, 1-m long individual plates with very coarse fibers. This coarse filtering structure allows them to separate amphipods from bottom sediments. In contrast, right whales that feed on small copepods have a fine filtering mechanism composed of more than 350 baleen plates that can

exceed 3 m in length. The fibers of right whale baleen are very fine, forming a dense mat capable of capturing copepods that are less than 5 mm (Berta and Sumich, 1999).The strong, flexible, and light characteristics of baleen plates made them commercially important in the nineteenth century where they served some of the roles of today's plastics.

Mysticetes have evolved three types of filter feeding: sediment-straining (gray whales), skimming (right and bowhead whales), and lunging or gulping (rorquals). The morphology of mysticetes reflects these different strategies. Gray whale heads are straight and relatively short, contain short, coarse baleen, and their throat regions possess only a few grooves (3–5) in the gular region that allows limited distension for taking in bottom sediment, water, and amphipods. Right and bowhead whale's heads have a strongly arched rostrum that allows them to have very long- and fine-textured baleen within a relatively blunt mouth. They have no throat grooves for distension and instead feed by swimming slowly (3–9 km/h) with their jaws held open for long periods while skimming prey from the water. The shape of their baleen minimizes the pressure wave in front of the whale that develops while swimming slowly through prey and enhances prey entry into the mouth (Werth, 2004; Lambertsen *et al.*, 2005). Rorqual heads are large and contain enormous mouths that extend posteriorly nearly half of the total body length. Their mouths contain relatively short baleen that ranges from fine (sei whales) to medium texture (blue, fin, humpback, and minke whales). The heads and bodies of rorquals are much more streamlined than the other mysticetes, allowing them to swim rapidly into a prey school to gulp large quantities of water and schooling prey. One of the most remarkable adaptations for feeding is the presence, in rorquals, of

**Figure 2**   *Skim-feeding in right and bowhead whales. From Berta and Sumich (1999).*

70–80 external throat grooves. During gulping, these grooves open like pleats to allow the mouth cavity to expand up to 4 times in circumference, taking in a volume of water equivalent to about 70% of the animals' body weight or greater (Croll *et al.*, 2001; Acevedo-Gutierrez *et al.*, 2002; Goldbogen *et al.*, 2006; Goldbogen *et al.*, 2007). The filter-feeding strategy of Balaenids appears to focus upon enhanced filter area whereas Balaenopterid strategy allows for greater filter pressure.

### C. Mysticetes—Feeding Behavior

Observations of feeding gray whales in the Arctic and Bering Sea have shown that the whales roll to one side and suck benthic invertebrate prey and bottom sediments, with some distension of the mouth cavity through the expansion of the throat grooves. Water and mud are expelled through the side of the mouth (Berta and Sumich, 1999). A similar behavior is used by gray whales that do not migrate as far north where they feed on a variety of benthic invertebrates and schooling mysids. This benthic foraging behavior creates scrapes of 1–5-m deep in the ocean floor, and several studies have shown that the disturbance is an important factor in the ecology of soft-bottom benthic communities of the Arctic and Bering Seas. Observational and direct measurement studies have shown that most gray whales and rorquals exhibit a strong right-side rolling preference while filter feeding (Woodward and Winn, 2006). For gray whales this right-side preference has been identified by shorter baleen and fewer parasitic barnacles on the right side.

Right and bowhead whales forage by skimming with their mouths open through concentrations of crustaceans near the surface and deeper in the water column. As the whale swims, water and prey enter through a gap between the two baleen plates in the front of the mouth and water exits along the sides of the mouth. Prey are swept into the back of the mouth by the dynamically controlled flow of water through the mouth and the side-to-side sweeping action of the large muscular tongue (Fig. 2). When the mouth is opened, the large lower lip abducts to create a gutter-like channel to direct water flow along the outside of the baleen to draw water out via negative hydrodynamic pressure through the baleen (Werth, 2004; Lambertsen *et al.*, 2005). While right and bowhead whales generally feed singly, at times they may feed alongside one another—a V-formation of 14 bowhead whales has been observed.

Rorqual lunge feeding has been described as the largest biomechanical event that has ever existed on earth (Croll *et al.*, 2001). Rorquals capture food by initially swimming rapidly (3–5 m/s in fin whales) at a prey school and then decelerating while opening the mouth to gulp vast quantities of water and schooling prey (Fig. 3). To maximize the opening, the lower jaw opens to almost 90° of the body axis. This is possible because the lower jaw has a well-developed coronoid process. This process is where the large temporalis muscle inserts, and provides an anchor and mechanical advantage for control of the lower jaw while maximizing the gape for prey capture. It is not developed in other whale species, and a tendinous part of the temporalis muscle, the frontomandibular stay, enhances and strengthens the mechanical linkage between the skull and the lower jaw (Lambertsen *et al.*, 1995). The expansion of the mouth during each lunge greatly increases drag and brings the body of the whale almost to a stop. As a result, it appears that filter feeding in rorquals is an energetically costly behavior (Croll *et al.*, 2001; Goldbogen *et al.*, 2006; Goldbogen *et al.*, 2007).

With the mouth open, the onrush of water and prey are accommodated by the distending ventral pleats. The tongue invaginates to form a hollow sac-like structure (cavum ventrale) which lines the inside of the gular region and the ventral pleats distend fully. After engulfing entire schools of prey, the lower jaw is closed. The tongue and the elastic properties of the ventral walls of the throat act in concert to force water out through the baleen (Fig. 3) (Lambertsen *et al.*, 1995; Goldbogen *et al.*, 2006).

Although the process described above is fundamentally the same in all rorquals, some species exhibit modifications and additional adaptations. Sei whales skim-feed in a manner similar to right whales, as well as feeding by lunging. Fin and blue whales often feed in pairs or trios that have a consistent echelon configuration. Humpback whales have a diverse diet and a wider variety of feeding behaviors. They have been observed bottom feeding, and while feeding on schooling fishes have been observed to produce a cloud of bubbles and feed cooperatively to assist in prey capture.

Laboratory experiments have shown schooling fish to react to bubbles by aggregating more densely. Humpback whales appear to take advantage of this as one member of a group of foraging whales that form long-term associations produce a net of bubbles. The bubble cloud serves to aggregate and confuse the prey. Members of the group dive below the bubble cloud and surface together—one whale immediately adjacent to another. The location of the whales in the surfacing group appears to be fairly constant through time. Humpbacks thus likely enhance prey capture success by both using bubbles and foraging cooperatively. A variation of bubble cloud feeding has been observed in humpback whales feeding on sand lance off New England. Here the bubble-cloud feeding is followed by a tail slap—believed to cause the sand lance to aggregate more densely.

### D. Mysticetes—Feeding Ecology

All filter-feeding whales exhibit distinct migration patterns linked to seasonal patterns in prey abundance. Seasonally dense aggregations of prey are probably necessary for successful filter feeding. For example,

F

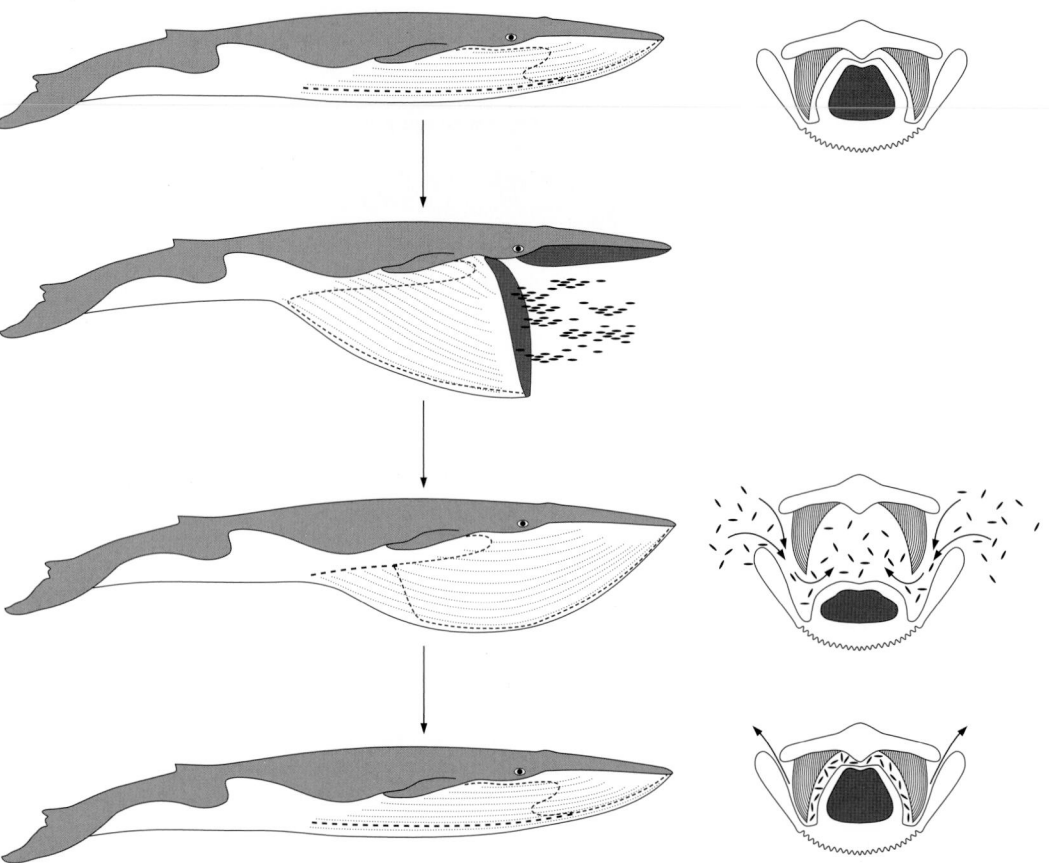

**Figure 3** *Lunge feeding in rorqual whales, demonstrating expansion of the throat pleats in invagination of the tongue. From Berta and Sumich (1998).*

gray whales undergo the longest migration of any mammal—foraging during the summer and fall in the Bering Sea and Arctic Ocean when dense aggregations of benthic amphipods become available with the seasonal increase in productivity. Humpback whales seasonally migrate from breeding areas to higher latitude foraging areas where schooling fish and krill become seasonally abundant (Berta and Sumich, 1999). The timing of coastal migration patterns of the California blue whale appears to be linked to annual patterns in coastal upwelling and krill development patterns (Croll *et al.*, 2005).

Studies of the diving behavior and daily movement patterns of right whales has shown that they track dense aggregations of copepods that in turn track oceanographic features such as fronts. Zooplankton densities in regions where right whales foraged in the southwestern Gulf of Maine were approximately three times the mean densities in the region (whale feeding densities averaged 3.1–5.9 g/m$^3$, compared to 1.1–3.6 g/m$^3$ where whales were not foraging). In a related study using hydroacoustic surveys, zooplankton densities where right whales were foraging were 18–25 g/m$^3$ (compared to 1–5 g/m$^3$ where whales were not foraging). Whale diving behavior is related to the depth of prey aggregations. In a year where copepods did not undergo diel migrations, dive depths averaged 12 m, with no dives exceeding 30 m throughout the day and night. In contrast, in a year where copepods showed strong diel shifts in depth (near the surface at night, deeper during the day), whale dive depths were significantly longer during the day (Mayo and Marx, 1990; Baumgartner and Mate, 2003; Baumgartner *et al.*, 2003).

Rorquals also track seasonal and diel patterns in the abundance and behavior of their prey. In general, the distribution and movement patterns of most rorquals consist of a seasonal migration from high latitudes where foraging takes place to low latitudes where they mate and give birth. However, data from blue whales in the Pacific indicate that feeding also takes place at low latitude, "upwelling-modified" waters, and data from both the Pacific and the Indian Oceans indicate that some blue whales may remain at low latitudes year-round. Fin and blue whales foraging on krill off the coast of North America concentrate their foraging effort on dense aggregations of krill deep (150–300 m) in the water column during the day, and may cease feeding when krill becomes more dispersed near the surface at night (Croll *et al.*, 1998; Croll *et al.*, 2005).

Rorqual foraging appears to only occur in regions of exceptionally high productivity, often associated with fronts, upwelling centers, and steep topography. It has been estimated that fin whales require prey concentrations of at least 17.5 g/m$^3$ to meet daily energy requirements. Krill densities where humpback whales were foraging in southeast Alaska have been estimated at 910 individuals/m$^3$, and minimum required prey densities for humpbacks were about 50 individuals/m$^3$ (Dolphin, 1987a, b). Krill densities in schools where blue whales were foraging in Monterey Bay, California were estimated at 145.3 g/m$^3$ compared to an overall mean density of zooplankton of 1.3 g/m$^3$ in the area (Croll *et al.*, 2005).

### III. Summary

Filter feeding in marine mammals is an adaptation that allows individuals to take in large quantities of prey in one mouth full. This is particularly adaptive in marine ecosystems where prey are relatively small and often densely aggregated, but patchy and ephemeral in space and time. Most filter-feeding species feed on schooling fish and crustaceans. The large body size of marine mammals and particularly mysticetes facilitates filter feeding by providing the ability to have a large filtering area relative to body volume. In addition, large body size likely provides an energetic buffer for animals that must move long distances between dense prey patches and endure long periods of fasting between foraging events.

## See Also the Following Articles

Baleen Whales ■ Feeding Morphology ■ Feeding Strategy and Tectics ■ Krill and Other Plankton

## References

Acevedo-Gutierrez, A., Croll, D.A., and Tershy, B.R. (2002). High feeding costs limit dive time in the largest whales. *J. Exp. Biol.* **205**(12), 1747–1753.

Baumgartner, M. F., and Mate, B. R. (2003). Summertime foraging ecology of North Atlantic right whales. *Marine Ecology-Progress Series* **264**, 123–135.

Baumgartner, M. F., Cole, T.V.N., Clapham, P.J., and Mate, B.R. (2003). North Atlantic right whale habitat in the lower Bay of Fundy and on the SW Scotian Shelf during 1999–2001. *Marine Ecology-Progress Series* **264**, 137–154.

Berta, A., and Sumich, J. L. (1999). "Marine Mammals: Evolutionary Biology." Academic Press, San Diego.

Bowen, W. D., and Siniff, D. B. (1999). Distribution, population biology, and feeding ecology of marine mammals. *In* "Biology of Marine Mammals" (J. E. Reynolds, and S. A. Rommel, eds), pp. 423–484. Smithsonian Institution Press, Washington.

Boyd, I. L., and Croxall, J. P. (1992). Diving Behavior of lactating Antarctic fur seals. *Can. J. Zool.* **70**(5), 919–928.

Croll, D. A., (12 authors) (1998). An integrated approach to the foraging ecology of marine birds and mammals. *Deep-Sea Res. II* **45**, 1353–1371.

Croll, D. A., Acevedo-Gutierez, A., Tershy, B.R., and Urban-Ramirez, J. (2001). The diving behavior of blue and fin whales: Is dive duration shorter than expected based on oxygen stores? *Com. Biochem. Physiol., Part A Mol. Integr. Physiol.* **129A**(4), 797–809.

Croll, D. A., Marinovic, B., Benson, S., Chavez, F.P., Black, N., Ternulb, R. and Tershey, B.R. (2005). From wind to whales: Trophic links in a coastal upwelling system. *Mar. Ecol. Prog. Ser.* **289**, 117–130.

Dolphin, W. F. (1987a). Dive behavior and estimated energy expenditures of foraging humpback whales in southeast Alaska. *Can. J. Zool.* **65**, 354–362.

Dolphin, W. F. (1987b). Prey densities and foraging of humpback whales, *Megaptera novaeangliae. Experientia* **43**(4), 468–471.

Fitzgerald, E. M. G. (2006). A bizarre new toothed mysticete (Cetacea) from Australia and the early evolution of baleen whales. *Proc. Roy. Soc. Lond. B* **273**, 2955–2963.

Goldbogen, J. A., Calambokidis, J., Shadwick, R.E., Oleson, E.M., McDonald, M.A., and Hildebraud, J.A. (2006). Kinematics of foraging dives and lunge-feeding in fin whales. *J. Exp. Biol.* **209**(7), 1231–1244.

Goldbogen, J. A., Ryenson, N.D., and Shadwick, R.E. (2007). Big gulps require high drag for fin whale lunge feeding. *Mar. Ecol. Prog. Ser.* **349**, 289–301.

Lambertsen, R., Rasmussen, K.J., Lancaster, W.C., and Hintz, R.J. (1995). Frontomandibular stay of Balaenopteridae-a mechanism for momentum recapture during feeding. *J. Mammal.* **76**(3), 877–899.

Lambertsen, R. H., Ulrich, N., and Straley, J. (2005). Functional morphology of the mouth of the bowhead whale and its implications for conservation. *J. Mammal.* **82**(2), 342–353.

Mayo, C. A., and Marx, M. K. (1990). Surface foraging behavior of the North Atlantic right whale, *Eubalaena glacialis*, and associated zooplankton characteristics. *Can. J. Zool.* **68**(10), 2214–2220.

Riedman, M. (1990). "The Pinnipeds Seals, Sea Lions, and Walruses." University of California Press, Berkeley.

Werth, A. J. (2004). Models of hydrodynamic flow in the bowhead whale filter feeding apparatus. *J. Exp. Biol.* **207**, 3569–3580.

Woodward, B. L., and Winn, J. P. (2006). Apparent lateralized behavior in gray whales feeding off the Central British Columbia coast. *Mar. Mamm. Sci.* **22**(1), 64–73.

# Fin Whale
## *Balaenoptera physalus*

### ALEX AGUILAR

## I. Characters and Taxonomic Relationships

Fin whales were initially described by Frederik Martens in 1675 and then by Paul Dudley in 1725. From these descriptions, Linnaeus created his *Balaena physalus* in 1758, which was later designated by Lacépède as *Balaenoptera physalus*.

The fin whale is very close to the other balaenopterids, particularly its congeners and shares with them the same chromosome number of $2n = 44$. It appears to be particularly close to the blue whale (*B. musculus*), from which it diverged between 3.5 and 5 million years ago and with which several hybrids have been described. Although the hybridization rate between these two species has not been properly assessed, it may be in the range of one for every 500–1000 fin whales. At least in one case, a female hybrid was pregnant (Bérubé and Aguilar, 1998).

The fin whale (Fig. 1) is SEXUALLY DIMORPHIC, with females being about 5–10% longer than males (Gambell, 1985). In the Southern Hemisphere, the average body length of adults is about 26 m for females and 25 m for males; in the Northern Hemisphere the corresponding lengths are 22.5 and 21 m. The fin whale is a slender balaenopterid, its maximum girth being between 40% and 50% of the total length. The rostrum is narrow, with a single, well-developed longitudinal ridge. BALEEN plates number 350–400 in each row and their maximum length is up to 70 cm. The dorsal fin is falcate and located at 75% of the total length; it is higher than that of blue whales, but lower than in sei whales (*B. borealis*) or Bryde's whales (*B. edeni*). The ventral grooves are numerous and extend from the chin to the umbilicus. The pigmentation of the cephalic region is strikingly asymmetrical, whereas the left side, both dorsally and ventrally, is dark slate, the right dorsal cephalic side is light gray and the right ventral side is white. This asymmetry also affects the baleen plates: those on the whole left side and the rear two-thirds of the right side are gray, whereas those on the front third of the right maxilla are yellowish. Particularly in adults, the skin of the flanks in the rear trunk is often covered by small round scars and stripes attributed to the attachment of lampreys and remoras. The white ventral region of whales inhabiting cold waters may have a yellowish layer produced by an infestation of diatoms.

The body mass of adult individuals typically ranges from 40 to 50 metric tons in the Northern Hemisphere and from 60 to 80 metric tons in the Southern Hemisphere. A general formula for estimating body weight (*W*) from body length (*L*) is $W = 0.0015 \, L^{3.46}$. If the

**Figure 1**  *Fin whales feeding in the North Atlantic (Photo by Brenda Rone, courtesy of Northeast Fisheries Science Center).*

girth at the level of the navel ($G$) is available, a more precise formula is $W = 0.0469\ G^{1.23}L^{1.45}$.

The relative mass of body tissues varies seasonally according to nutritive condition (Lockyer and Waters, 1986). Average mass relative to total body weight is 18.4±3.3% for blubber, 45.3±4.4% for muscle, 15.5±2.4% for bone, and 9.8±2.1% for viscera. The liver is large, usually weighing 230–600 kg. The heart is similar in relative size to that of terrestrial mammals but larger than in odontocetes and weighs 130–290 kg. Kidneys are large and weigh 50–110 kg. The right lung is about 10% heavier than the left, with each one weighing 100–160 kg. The spleen weighs 2–7 kg and sometimes has accessory bodies of smaller size.

The rostrum of the fin whale is sharply pointed, without the lateral curvature typical of blue whales. The zygomatic width is about 50–55% of the condylo-premaxillary length, the width of the rostrum at midlength is approximately 30–35% of its basal width, and the whole skull measures about 20–25% of the total body length. Ribs usually number 16 pairs, with the last pair being smaller and not attached to the vertebral column. The number of vertebrae ranges from 60 to 63, with a typical formula of C, 7; D, 15 (16); L, 14 (13–16); and Ca, 25 (24–27). The sternum is broad and variable in shape, usually in the form of a cross or a trefoil. As in other balaenopterids, the scapulae are fan shaped with a convex upper margin. The flippers lack a third digit. The typical digit formula is I, 3–4; II, 6; IV, 5–6; and V, 3–4.

Because the fin whale makes seasonal migrations (see later), which follow alternate schedules in each hemisphere, the northern and southern populations do not appear to come into contact. This has led to genetic isolation and, as a consequence, two forms, sometimes accepted as subspecies, are recognized: *B. physalus physalus* inhabiting the Northern Hemisphere and *B. physalus quoyi* in the Southern Hemisphere. In addition to the difference in body size mentioned earlier, the flippers of whales from the Northern Hemisphere are shorter and broader than those of their southern counterparts. Small variations in body proportions and coloration between fin whales from different locations in the Northern Hemisphere have also been described. The occurrence off the western coast of South America of a pygmy form, dark in color, and possibly with black baleen, has been proposed though not demonstrated to be genetically distinct (Clarke, 2004).

## II. Distribution and Abundance

The fin whale is a cosmopolitan cetacean. It is found in most large water masses of the world, from the Equator to the polar regions (Fig. 2). However, in the most extreme latitudes it may be absent near the ice limit; thus, it has not been reported off northern Novaya Zemlya or Franz Josef Land in the Northern Hemisphere or in the Weddell or Bellinghausen Seas in the Antarctic. It is also absent from the Black Sea and is very rare in the eastern Mediterranean, the Baltic Sea, the Persian Gulf, the Red Sea, and most equatorial regions. The largest concentrations are usually located in temperate and cold waters. Within these, fin whales are thought to segregate

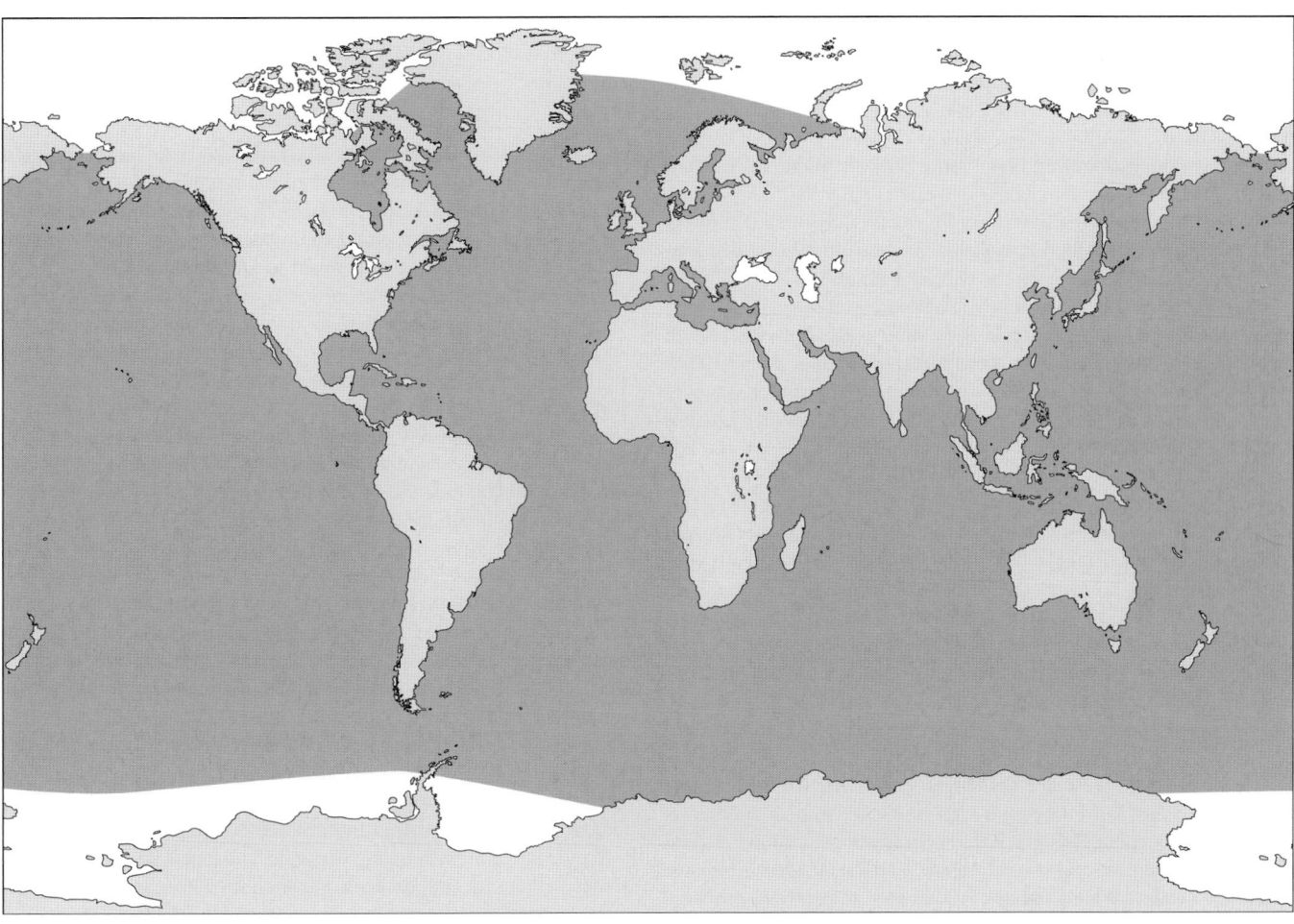

**Figure 2**   *Distribution of fin whales (dark-shaded area).*

into independent breeding units, the identity of which is often a matter of discussion (Bérubé *et al.*, 1998). Overall, the fin whale density tends to be higher outside the continental slope than inside it.

The estimated population numbers in recent years are 15,200 in the Antarctic, south of 307°S; 25,800 in the central North Atlantic (East Greenland-Iceland, Jan Mayen and the Faeroes); 4100 in the northeastern North Atlantic (North and West Norway); 17,400 in the Spain–Portugal–British Isles area; 1700 off West Greenland; 1000 off Newfoundland; 2800 off the east coast of North America south of the Gulf of St Lawrence; 5700 in the Bering Sea, Aleutian Islands, and Gulf of Alaska; and 3300 off the west coast of the United States.

## III. Ecology

The fin whale feeds on a wide variety of organisms, depending on availability (Kawamura, 1980). Possibly, diet varies with season and locality. Preferred prey in the Northern Hemisphere seems to be krill composed of the euphausiid *Meganyctiphanes norvegica*, although other species of planktonic crustaceans (*Thysanoessa inermis, Calanus finmarchicus*), schooling fishes such as capelin (*Mallotus villosus*), herring (*Clupea harengus*), mackerel (*Scomber scombrus*), and blue whiting (*Micromesistius poutassou*), and even small squids are also consumed. In the Southern Hemisphere, the diet is almost exclusively krill, mostly the euphausiid *Euphausia vallentini* but also other planktonic crustaceans such as *Euphausia superba, Parathemisto gaudichaudii*, or *Calanus tonsus* in smaller

proportions. Similar to other balaenopterids, the fin whale feeds in summer, when an adult whale is estimated to consume up to 1 ton of euphausiids per day, and fasts in winter (see later).

Because the distribution range and the DIET of fin whales overlap with those of other balaenopterid whales, interspecific competition is likely to occur. This is especially likely in the case of the blue whale, which is often found forming mixed schools with fin whales. This association, together with the evolutionary proximity of the two species, makes blue fin hybrids relatively common (see earlier discussion).

Because they are large and swim fast, fin whales do not have significant predators, with the exception of the killer whale (*Orcinus orca*). In certain regions where this odontocete is abundant, signs of past attacks of killer whales can often be seen on the flippers, flukes, and flanks of fin whales.

## IV. Behavior and Physiology

In the Southern Hemisphere, fin whales engage in north–south seasonal migration, feeding in the summer and breeding and fasting in the winter. A similar latitudinal movement has been generally proposed for the Northern Hemisphere, although in many areas this pattern is not clear. For example, in the North Atlantic, although several summer grounds have been identified in medium to high latitudes (northern Morocco, Gibraltar Straits, northeastern Spain, Scotland, northern and western Norway, Newfoundland, the Faroe Islands, and Iceland), no definite wintering grounds are known.

It has been suggested that in this ocean the latitudinal movements of the species are short because of the influence of the Gulf Stream, which would make the higher latitudes suitable as wintering grounds. An alternative explanation is that individuals that concentrate near the coast in the feeding season tend to disperse into open waters during the winter, therefore being more difficult to detect. Indeed, there is evidence that some whales remain in higher latitudes during the coldest months of the year. Also, fin whales may remain at lower latitudes year-round if food is available, as shown by the lack of seasonality of historical catches in the Gibraltar Straits (36°N).

The migration route apparently follows areas of low geomagnetic intensity and gradient and, similarly to other balaenopterids, not all components of the population move together (Walker *et al.*, 1992). Pregnant females are usually the first to initiate the seasonal movement and are soon followed by adult males and resting females. Lactating females and juvenile individuals of both sexes are the last to migrate.

Compared to other large cetaceans, fin whales are relatively fast swimmers. Normal cruising speed is 5–8 knots but may increase up to 15 knots for short bursts. The SWIMMING is usually smooth, and breaching is very rare, except when the whale is harassed. Dives are limited to 100–200 m and usually last 3–10 min. The species is not gregarious and its only social bond appears to be that of cows with their nursing calves, a link that vanishes at weaning. School size is small; the fin whale usually swims single or in groups of two to seven, but transitory large aggregations may occur in highly productive areas.

The sounds of fin whales are simple and mostly consist of low-frequency moans and grunts and high-frequency pulses, apparently with a social function. Regional differences have been found between the Gulf of California and several Atlantic and Pacific Ocean regions. Fin whale moans are loud and can be heard for at least tens, probably hundreds of kilometers.

## V. Life History

Growth during the postnatal period is rapid, and 95% of the maximum body size is reached when whales are 9–13 years old (Aguilar and Lockyer, 1987). Males grow faster than females but stop growing sooner. Physical maturity, as determined by the degree of ossification of the vertebral column, is reached at about the age of 25 in both sexes. Longevity has not been determined precisely, but individuals of up to 80–90 years old are known.

Reproductive behavior is poorly known (Lockyer, 1984), but taking into account the reduced mass of their testes, their mating system probably relies on competition between males for monopolizing females and reduced sperm competition. In the Northern Hemisphere, sexual maturity is attained at an approximate length of 17.5 m in males and 18.5 m in females; corresponding figures for the Southern Hemisphere are 19 m in males and 20 m in females. These lengths typically correspond to an age of about 6–7 years in males and 7–8 years in females.

In the Northern Hemisphere, the mating period is December–February, whereas in the Southern Hemisphere it is May–July. Gestation lasts about 11 months, at the end of which a calf about 6–7 m long and weighing 1–1.5 metric tons is born. Only one calf is usually produced per pregnancy, although twinning has been reported. Weaning occurs when the calf is about 6–7 months old and measures 11- to 13-m long.

Weaning is followed by a 6-month resting period, at the end of which mating takes place. Therefore, the reproductive cycle is completed in about 2 years. Gross pregnancy rates (number of pregnant females in relation to that of adult females) are typically estimated at 38–49%.

## VI. Interaction with Humans

Second in size after the blue whale, the fin whale has been much sought after by whalers; it was one of the target species when modern operations started in the early 1870s, and its exploitation continued until the moratorium on commercial whaling came into force in 1985. At the beginning, it was taken in large numbers off northern Norway where catches were recorded as early as 1876; overexploited, this population soon decreased in abundance and the whaling stations had to close in 1904 because of insufficient catches (Tønnessen and Johnsen, 1982). The fishery then extended to other areas of the North Atlantic and from there to the Antarctic, the Pacific, and the Indian Oceans. Captures peaked during the period of 1935–1970, when the fin whale formed the largest part of worldwide catches, with about 30,000 individuals taken annually. In most localities, such a level of exploitation far exceeded the sustainable yield of the populations, and one after another the operations collapsed.

Fortunately, however, management measures introduced by the INTERNATIONAL WHALING COMMISSION in the 1960s and 1970s (IWC, 2007) prevented populations from being reduced to such a low level as its predecessor in the whale exploitation history, the blue whale. In the 1980s, when the moratorium was approved, large-scale commercial fishing of fin whales was only taking place off northwestern Spain and Iceland. Both operations were discontinued, the first in 1985 and the second in 1989. Today, the species is exploited in West Greenland, where about 19 individuals are struck per year by the "aboriginal" whale fishery; in the Antarctic Ocean, where since 2005, 3–10 individuals are taken per year under the Japanese Special Permit Research Program; and off Iceland, where in 2006 commercial whaling was reinitiated with a take of seven individuals. After discontinuation of whaling, populations have either stopped their decline or slowly recovered from losses.

Apart from exploitation, no other severe conservation threats have been identified (Klinowska, 1991). Collisions of fin whales with vessels sometimes occur in locations where boat traffic is very intense, but their frequency does not appear to be significant at the population level. Incidental catches in fishing gear do occur but are uncommon. Given the oceanic distribution of fin whales and their lower position in the food chain, pollution or reduction of food supply by overfishing do not have a significant impact. Overall levels of contaminants are low or extremely low as compared to other marine mammals (O'Shea and Brownell, 1994; Law, 1996). In the North Atlantic and the Mediterranean Sea, where the highest levels of organochlorine compounds have been detected, average concentrations of DDTs in blubber ranged from 0.2 to 12 ppm and those of PCBs ranged from 0.5 to 8 ppm. In fin whales off Chile and South Africa, concentrations were much lower, often below analytical detection limits. A number of heavy metals have also been found in tissues of fin whales, although usually at low concentrations. Levels in the muscle of North Atlantic individuals are in the range of 10–90 ppb of zinc, 0.2–2 ppb of copper, 0.1–2 ppb of cadmium, and 2–3 ppm of mercury.

## *References*

Aguilar, A.y., and Lockyer, C. H. (1987). Growth, physical maturity and mortality of fin whales (Balaenoptera physalus) inhabiting the temperate waters of the northeast Atlantic. *Can. J. Zool.* **65**(2), 253–264.

Bérubé, M., and Aguilar, A. (1998). A new hybrid between a blue whale, *Balaenoptera musculus*, and a fin whale, *B. physalus*: Frequency and implications of hybridization. *Mar. Mamm. Sci.* **14**(1), 82–98.

Bérubé, M., *et al.* (9 authors). (1998). Population genetic structure of North Atlantic, Mediterranean Sea and Sea of Cortez fin whales, *Balaenoptera physalus* (Linnaeus, 1758): Analysis of mitochondrial and nuclear loci. *Mol. Ecol.* **7**, 585–599.

Clarke, R. (2004). Pygmy fin whales. *Mar. Mamm. Sci.* **20**(2), 329–334.

Gambell, R. (1985). Fin whale, *Balaenoptera physalus* (Linnaeus, 1758). *In* "Handbook of Marine Mammals" (S. H. Ridgway, and R. Harrison, eds), vol. 3, pp. 171–192. Academic Press, London, UK.

International Whaling Commission (2007). Report of the Sub-committee on the Revised Management Procedure. *J. Cetacean Res. Manag.* **9**(Suppl.), 9–12.

Kawamura, A. (1980). A review of food of balaenopterid whales. *Sci. Rep. Whales Res. Inst. Tokyo,* **32**, 155–197.

Kawamura, A. (1994). A review of baleen whale feeding in the southern ocean. *Rep. Int. Whal. Commn.* **44**, 261–271.

Klinowska, M. (1991). "Dolphins, Porpoises and Whales of the World." The IUCN Red Data Book. IUCN, Gland, Switzerland, 429pp.

Law, R. J. (1996). Metals in marine mammals. *In* "Environmental Contaminants in Wildlife: Interpreting Tissue Concentrations" (W. N. Beyer, G. H. Heinz, and A. W. Redmon-Norwood, eds), pp. 357–376. CRC Press Inc., Boca Raton, USA.

Lockyer, C. H. (1984). Review of baleen whale (Mysticeti) reproduction and implications for management. *Rep. Int. Whal. Commn.*(Special Issue 6), 27–50.

Lockyer, C., and Waters, T. (1986). Weights and anatomical measurements or northeastern Atlantic fin (*Balaenoptera physalus*, Linnaeus) and sei (*B. borealis*, Lesson) whales. *Mar. Mamm. Sci.* **2**(3), 169–185.

O'Shea, T. J., and Brownell, R. L. (1994). Organochlorine and metal contaminants in baleen whales; a review and evaluation of conservation implications. *Sci. Total Environ.* **154**, 179–200.

Tomilin, A. G. (1967). "Mammals of the USSR and Adjacent Countries. Cetacea." Israel program for Scientific Translations, Jerusalem, 756pp.

Tønnessen, J. N., and Johnsen, A. O. (1982). "The History of Modern Whaling." C. Hurst and Company, London, UK, 798 pp.

Walker, M. M., Kirschvink, J. L., Ahmed, G., and Dizon, A. E. (1992). Evidence that fin whales respond to the geomagnetic field during migration. *J. Exp. Biol.* **171**, 67–78.

# Finless Porpoise
## *Neophocaena phocaenoides*

### Masao Amano

## I. Characteristics and Taxonomy

The finless porpoise is a small phocoenid cetacean lacking a dorsal fin (Fig. 1). Instead of the dorsal fin, a ridge sprinkled with horny tubercles runs down the middle of the back. Although these tubercles seem to be homologous to tubercles on the leading edge of the dorsal fin of other phocoenids, their function is not obvious. It is plausible that they serve as a sensory organ or organ for mutual contact, as many nerve endings occur in the tubercles (Kasuya, 1999). The species has a rounded head without an apparent beak. Its color is uniformly dark to pale gray and somewhat lighter on the ventral side. The teeth are spatulate in shape like those of most other porpoises of the family phocoenidae.

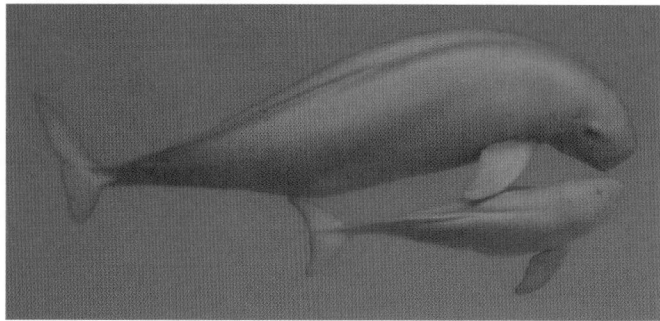

**Figure 1** *Mother and calf pair of Yangtze finless porpoises. Photo courtesy of the Institute of Hydrobiology, the Chinese Academy of Sciences.*

The finless porpoise was originally described by G. Cuvier as *Delphinus phocaenoides* based on a skull supposedly from the Cape of Good Hope. However, there have been no further records of this species from the southwest coast of Africa, and the true type locality is now considered to be the Indian coast. The genus names *Neomeris*, which is not valid since preoccupied by a polychaete, and *Meomeris*, which is a misspelled name and not available, were formerly used for the finless porpoise (Rice, 1998). Other English common names are finless black porpoise and Indian black porpoise.

Currently, three subspecies are generally recognized: *Neophocaena phocaenoides phocaenoides* from the Indian Ocean to the South China Sea, *N. p. asiaeorientalis* in the Yangtze River, and *N. p. sunameri* from the East China Sea and coasts of Japan. *N. p. phocaenoides* has a broader (3–10 cm) and lower dorsal ridge, and *N. p. asiaeorientalis* and *N. p. sunameri* have a narrower (0.2–1.2 cm) and higher dorsal ridge. *N. p. asiaeorientalis* has the narrowest (<0.8 cm) dorsal ridge. Differences among three subspecies exist in external skull and postcranial morphology (Gao and Zhou, 1995a, b, c; Jefferson, 2002). The darkness and ontogenetic change of color pattern vary among subspecies. *N. p. phocaenoides* is born with lighter color and darkens with growth to be almost black in adults. However, neonates of *N. p. asiaeorientalis* and *N. p. sunameri* are black to dark gray and get lighter with age (Kasuya, 1999, Jefferson and Hung, 2004). The border between the range of *N. p. phocaenoides* and *N. p. sunameri* seems to be around the Taiwan Strait, where both the subspecies have been observed, but there is still no evidence of genetic isolation (Jefferson, 2002). The border between the less differentiated *N. p. asiaeorientalis* and *N. p. sunameri* may be around the mouth of the Yangtze River, but this is less certain.

Five local populations are clearly identified in Japanese waters based on skull morphology and mtDNA variability (Yoshida, 2002). mtDNA analysis also revealed population differentiation in the Yangtze River (Zheng *et al.*, 2005). Similarly, finless porpoises are considered to likely live in relatively small isolated local populations in other areas.

The only known fossil of *Neophocaena* is a skull from the bottom of the Seto Inland Sea, Japan. It is considered to be from the late Pleistocene and resembles the skull of recent *N. p. sunameri* (Kimura and Hasegawa, 2005). There are conflicting views of whether the finless porpoise is the most primitive or derived member among the phocoenids. A genetic study suggested the species is the oldest lineage in the family (Jefferson and Hung, 2004).

F

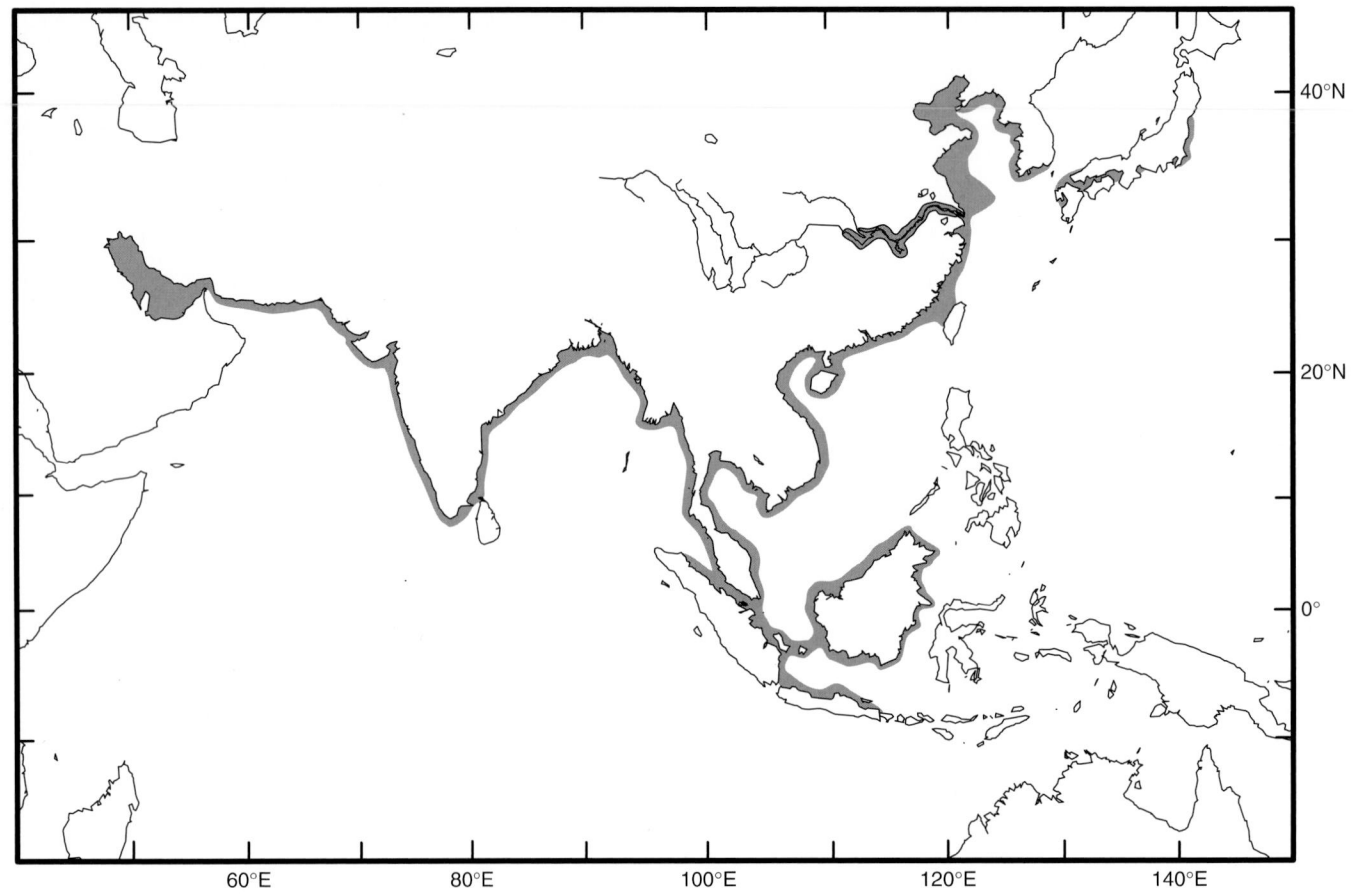

**Figure 2** *Distributional range of the finless porpoise.*

## II. Distribution and Abundance

Finless porpoises inhabit shallow coastal waters and some rivers in the Indo-Pacific region (Fig. 2). Their range is from the Persian Gulf in the west and along the coasts of India, Indochina Peninsula, China, and Korea to northern Japan. The southernmost limit of the range is the northern coast of Java. In the Yangtze River, finless porpoises are found to about 1600 km upstream from the sea (Jefferson and Hung, 2004).

Total abundance of the finless porpoise is unknown. There are some estimates for several local populations (Biodiversity Center, 2002; IWC, 2006): about 19,000 in Japanese waters (total of 5 populations), about 27,000 off the western coast of South Korea (IWC, 2006), and at least 220 in Hong Kong and adjacent waters. The number inhabiting the Yangtze River is considered to be less than 2000 (Xia *et al.*, 2005).

## III. Ecology

Finless porpoises prefer the mangrove zone in tropical waters. They are also found in estuaries and large river systems such as the Indus, Ganges, and Yangtze Rivers. In Japan, habitat of the finless porpoise is strictly defined by depth (<50 m) and sandy or soft bottom (Shirakihara *et al.*, 1992).

Finless porpoises feed on various prey species from surface to bottom dwellers, including schooling fishes, squids, octopuses, shrimps, and prawns, and they are considered to be opportunistic feeders (Jefferson and Hung, 2004). The known predators apart from

humans are killer whales (*Orcinus orca*) and large sharks, including the great white shark (Jefferson and Hung, 2004).

## IV. Behavior and Physiology

Finless porpoises are observed year-round in most localities, but seasonal fluctuation in density suggesting some movements has been reported in some areas.

Finless porpoises rarely form large schools and are usually found in pairs. Kasuya (1999) concluded that a mother and calf or an adult pair is the basic social unit. Finless porpoises usually swim quietly and do not leap or make splashes. They tend to avoid vessels and do not bow ride. However, they can be easily observed in some areas, and commercial porpoise-watching trips are conducted in a few places in Japan.

## V. Life History

Finless porpoises grow to around 170 cm in length and 70 kg in weight, but animals over 200 cm and 100 kg have been reported. Males grow slightly larger than females. Size seems to be different geographically and generally greater in the northern population.

Information on reproduction is limited to that of porpoises from Chinese and Japanese waters (Kasuya, 1999; Jefferson *et al.*, 2002). The oldest specimen was a 33-year-old female from the South China Sea, but most aged specimens were younger than 25 years in all areas. Age at sexual maturation is similar among areas and ranged from 4 to 6 years in both sexes. The gestation period is estimated at around 11 months. The peak of the calving season is different among localities: October–January in the South China Sea, April–May in the Yangtze River, November–December and March in western Kyushu, and April–June in the Seto Inland Sea and on the Pacific coast of Japan. Females are thought to calve every 2 years. The lactation period is estimated to be about 7 months, which is short as in other phocoenids.

## VI. Interaction with Humans

Because of its nearshore habitat, the finless porpoise tends to be threatened by many human activities. Although there is no fishery that takes finless porpoises directly, incidental catches mainly by gillnets occur throughout its range (Jefferson and Hung, 2004; IWC, 2006). Habitat degradation by land reclamation and deforestation of mangrove areas are also serious problems. It has been suggested that intensive sand dredging is responsible for severe decline of the species in the central-eastern Seto Inland Sea, Japan (Kasuya *et al.*, 2002; Shirakihara *et al.*, 2007). Moreover, very high levels of toxic contaminants, including organochlorine and butyltin compounds and heavy metals, have been reported in finless porpoises from eastern Asia (Jefferson and Hung, 2004; Ramu *et al.*, 2005). In the Yangtze, the combined threats of bycatches, habitat degradation including heavy traffic and construction, and pollution, which in concert caused extinction of the baiji, are also suspected to have contributed to the recent significant decline of the finless porpoise in the river (Wang *et al.*, 2005).

The distribution seems to be separated into relatively small local subpopulations by unsuitable habitats of deep water or rocky bottoms. Such local subpopulations are easily depleted, contributing to the decline of genetic diversity of the larger regional metapopulations.

Despite these concerns, our ability to assess the current status and population trends of the species is still very limited, especially for the populations in the Indian Ocean and the southeast Asia.

## *See also the Following Articles*

Baiji ■ Indo-West Pacific Marine Mammals ■ Pollution and Marine Mammals ■ Porpoises, Overview

## *References*

Biodiversity Center (2002). "Survey of Marine Animals (Finless Porpoises)." Biodiversity Center, Ministry of the Environment, Fujiyoshida, Japan, 136pp. [In Japanese].

Gao, A., and Zhou, K. (1995a). Geographical variation of external measurements and three subspecies of *Neophocaena phocaenoides* in Chinese waters. *Act. Ther. Sin.* **15**, 81–92 [In Chinese with English summary].

Gao, A., and Zhou, K. (1995b). Geographical variations of skull among the populations of *Neophocaena* in Chinese waters. *Act. Ther. Sin.* **15**, 161–169 [In Chinese with English summary.].

Gao, A., and Zhou, K. (1995c). Geographic variations of postcranial skeleton among the populations of *Neophocaena* in Chinese waters. *Act. Ther. Sin.* **15**, 246–253 [In Chinese with English summary.].

IWC. (2006). Report of the Scientific Committee, Annex L, Report of the sub-committee on small cetaceans. *J. Cetacean Res. Manage.* **8** (Suppl.), 221–240.

Jefferson, T. A. (2002). Preliminary analysis of geographic variation in cranial morphometrics of the finless porpoise (*Neophocaena phocaenoides*). *Raffles Bull. Zool.* **10**(Suppl.), 3–14.

Jefferson, T. A., and Hung, S. K. (2004). *Neophocaena phocaenoides*. *Mamm. Sp.* **746**, 1–12.

Jefferson, T. A., Robertson, K. M., and Wang, J. Y. (2002). Growth and reproduction of the finless porpoise in southern China. *Raffles Bull. Zool.* **10**(Suppl.), 105–113.

Kasuya, T. (1999). Finless porpoise *Neophocaena phocaenoides* (G. Cuvier, 1829). *In* "Handbook of Marine Mammals" (S. H. Ridgway, and R. Harrison, eds), Vol. 6, pp. 411–442. Academic Press, San Diego.

Kasuya, T., Yamamoto, Y., and Iwatsuki, T. (2002). Abundance decline in the finless porpoise population in the Inland Sea of Japan. *Raffles Bull. Zool.* **10**(Suppl.), 57–65.

Kimura, T., and Hasegawa, Y. (2005). Fossil finless porpoise, *Neophocaena phocaenoides*, from the Seto Inland Sea, Japan. *Bull. Gunma Mus. Nat. Hist.* **9**, 65–72 [In Japanese with English Summary.].

Ramu, K., Kajiwara, N., Tanabe, S., Lam, P. K. S., and Jefferson, T. A. (2005). Polybrominated diphenyl ethers (PBDEs) and organochlorines in small cetaceans from Hong Kong waters: Levels, profiles and distribution. *Mar. Poll. Bull.* **51**, 669–676.

Rice, D. W. (1998). "Marine Mammals of the World." Special publication No. 4, the Society for Marine Mammalogy, Lawrence, KS.

Shirakihara, K., Yoshida, H., Shirakihara, M., and Takemura, A. (1992). A questionnaire survey on the distribution of the finless porpoise, *Neophocaena phocaenoides*, in Japanese waters. *Mar. Mamm. Sci.* **8**, 160–164.

Shirakihara, K., Shirakihara, M., and Yamamoto, Y. (2007). Distribution and abundance of finless porpoise in the Inland Sea of Japan. *Mar. Biol.* **150**, 1025–1032.

Wang, D., Hao, Y., Wang, K., Zhao, Q., Chen, D., Wei, Z., and Zhang, X. (2005). The first Yangtze finless porpoises successfully born in captivity. *Env. Sci. Poll. Res.* **12**, 247–250.

Xia, J., Zheng, J., and Wang, D. (2005). *Ex situ* conservation status of an endangered Yangtze finless porpoise population (*Neophocaena phocaenoides asiaeorientalis*) as measured from microsatellites and mtDNA diversity. *ICES J. Mar. Sci.* **62**, 1711–1716.

Yoshida, H. (2002). Population structure of finless porpoise (*Neophocaena phocaenoides*) in coastal waters of Japan. *Raffles Bull. Zool.* **10**(Suppl.), 35–42.

Zheng, J., Xia, J., He, S., and Wang, D. (2005). Population genetic structure of the Yangtze finless porpoise (*Neophocaena phocaenoides asiaeorientalis*): Implications for management and conservation. *Biochem. Gen.* **43**, 307–320.

# Fisheries, Interference With

### DAGMAR FERTL

**M**arine mammal interactions with fisheries may be characterized as biological (ecological) or operational (IUCN, 1981). This section focuses on operational interactions. "Operational" interactions are those in which marine mammals take fish from fishing operations (depredation), disturb fishing, or damage fishing gear. Interference with fisheries might negatively affect fisheries by resulting in loss of bait, damage to fishing gear, decreased catches, reduced fish weight (in the case of fish farms), or increased time spent during fishing operations. Difficult to assess is hidden damage, i.e., the amount of fish wholly removed from nets without a trace and the catch losses due to the presence of marine mammals.

## I. Depredation

Evidence from some fisheries shows that cetaceans, pinnipeds, sirenians, and even the sea otter (*Enhydra lutris*) are attracted to fishing gear and attempt to remove bait and catches during

F

commercial and recreational fisheries. Read (2005) noted that more information is needed about depredation in artisanal fisheries. Marine mammals also feed at mariculture (i.e., fish farm) enclosures and fish aggregated at natural and artificial constraints in river systems, such as falls or fish ladders. Gear likely to have the most interactions with (and interference from) marine mammals are purse seines, trawls, gill nets, traps, and hooks and lines. In fact, baited longlines have been described as "loaded lunchboxes" and shrimp trawlers as "smorgasbords".

Fishing operations concentrate food of interest, decreasing energy expenditure associated with foraging by marine mammals. Nursing females may especially benefit from this feeding technique. Fishing operations may permit marine mammals to select food of higher caloric value. Some food niches, not otherwise available to some marine mammals, may be opened up, making prey easier to access that might be normally more difficult (e.g., because of depth required to dive).

Feeding in association with fisheries is likely a learned behavior, with increasingly more individual marine mammals seeking out fishing gear for an easy meal. Very often, acoustic aspects of the fishing activity (e.g., cavitation noise from changes in the propeller speed of ship engines) act as a "dinner bell." As noted by Königson et al. (2006), an earlier foraging experience could encourage individuals to return to feeding grounds that were previously successful. It has been suggested that this type of feeding behavior is also passed from generation to generation by observation and participation (Fertl and Leatherwood, 1997).

Marine mammals may cause abrasions and wounds to fish during unsuccessful capture attempts or while "playing" with fish during fishing operations, even when they are satiated. It is not known how the number of injured fish relates to the total number killed. Marine mammals may take portions of fish or the entire fish, rendering them nonmarketable. Most estimates of loss vary between <1% and 8% of the total catch. There are exceptions, such as in some longline and salmon gill net fisheries. Various species of cetaceans (e.g., killer whale [*Orcinus orca*], false killer whale [*Pseudorca crassidens*], sperm whale [*Physeter macrocephalus*], pilot whale [*Globicephala* spp.]) and pinnipeds (e.g., South American sea lion [*Otaria flavescens*] and California sea lion [*Zalophus californianus*]) depredate longline-caught fish (Gilman et al., 2006). Flesh may be torn from hooked fishes or fish may be removed completely (often leaving only the head or lips) as lines are hauled back to the fishing vessel. Estimates of lost catch per attack by killer whales on longlines off Alaska range from 25% to 100%; of the catch (Yano and Dahleim, 1995), while world-wide catch rate reductions of only 1–3% are reported for sperm whales with longline fisheries (Sigler et al., 2008). Depredation rates may be related to a variety of factors, including low availability of fish in the area at the time or presence of particular individual marine mammals causing problems. Longline fishermen off Alaska report that sperm whale depredation is more common when fishery offal (discarded heads and internal organs) is unavailable (Sigler et al., 2008).

## II. Disturbance

When the mere presence (i.e., attendance) of a marine mammal(s) can cause catches to be reduced and time wasted, fishing operations are disturbed. Cetaceans and pinnipeds may disturb fishing activity by causing fish shoals to disperse or sound (dive rapidly); and thus escape being trapped by the net. Marine mammals might also cause a large decrease in prey abundance in a fishing area; e.g., during acoustic surveys of Spanish sardine (*Sardinella aurita*) in

Venezuela, Fréon and Misund (1999) observed that fish disappeared when dolphins arrived in the area. In the case of fish farms, predators may attack and harass fish through the pen walls, thus stressing, scarring, and wounding the fish and resulting in lower product quality through reduced value or reduced fish weight.

Disturbance of fishing activities even occurs in dolphin–human fishing cooperatives, such as one in Brazil, with "bad" dolphins (*ruim* in Portuguese) occasionally interfering with the fishing by dispersing fish and damaging nets and netted fish (Pryor et al., 1990).

Once a marine mammal has located fishing gear, fishermen might have to move to another site or the marine mammal would continue to take substantial amounts of the catch. This disturbance results in cost to the fishermen because they have to move and relocate to fish.

Fishing operations may be impeded in other ways. Seals may be caught in the fish pump (used to remove fish from the net) and die. Fishing operations may be disrupted if live seals are brought abroad in nets or otherwise make their way onto fishing boats, especially if they manage to get below deck. Crew injuries may result from direct interactions with seals, including people being bitten and nylon burns or cut fingers from tug-of-wars between a fisherman and a seal.

## III. Gear Damage

When a marine mammal attempts to remove fishery catches entangled in a net's mesh or hooked on a line, holes may be torn or hooks removed. Marine mammals may become incidentally entangled during these encounters; netting and rope may be lost in trying to free live or dead animals. Damaged gear may not fish as efficiently, and a loss of catch may result. The visible gear damage such as holes and tears in the nets is only a small part of the total economic losses. Indirect expenses include costs for new material, time for repairing fishing gear and reduction in gear durability, increased time and fuel consumption due to emptying the gear more often should also be considered. Further, repairs may be costly, and the time spent repairing fishing gear may be significant in a seasonal fishery where most of the catch is taken in a period of a few weeks. Additionally, animals become injured, with fish hooks in their mouths or dorsal fin disfigurements that occur during the struggle by the animal to free itself (Baird and Gorgone, 2005).

There are numerous examples of marine mammals (particularly pinnipeds) damaging fishing gear; there are more reports of damage to static gear (e.g., fish traps, gill nets, longlines) than moving gear (e.g., trawls, purse seines). Gear damage may also result from accidental collision with fishing gear. For example, humpback whales (*Megaptera novaeangliae*) in Newfoundland in the 1970s caused several hundred thousand dollars of damage annually as a result of colliding with cod traps while feeding on capelin (*Mallosus villosus*).

Cetaceans and pinnipeds are often blamed for damage to gear or catches that should actually sometimes be attributed to other predators. For example, bottlenose dolphins (*Tursiops truncatus*) often are blamed for damage to trawl and gill nets when sharks are often the real culprits. Killer whales and sharks are both known to feed on longline catches off Brazil (Dalla Rossa and Secchi, 2007). Although South African fur seals (*Arctocephalus australis*) are blamed for taking lobsters from traps, clear evidence shows that octopus are to be blamed (Wickens. 1996). Birds may also be seen taking bait off hooks as they are cast into the water during line fishing.

### A. Pinniped Interference with Fisheries

Seals, sea lions, and fur seals take caught fish from nets, hooks, trawls, seines (Fig. 1), or traps, and attack fish that are being raised

**Figure 1** *A gray seal* (Halichoerus grypus) *raiding a gillnet with caught herring* (Clupea spp.) *in the northern Baltic Sea. About 200 herring were set on the net, and 20 min later the net was empty. Photo by Sara Königson.*

**Figure 2** *Sperm whale* (Physeter macrocephalus) *near a longline vessel in the Gulf of Alaska. Photo by Heather Vukelic, SEASWAP, permit number 473-170-01.*

in aquaculture pens. Pinnipeds may also be attracted by discarded bycatch; e.g., Nitta and Henderson (1993) observed a Hawaiian monk seal (*Monachus schauinslandi*) feeding on discarded fish. The impact of pinnipeds on fisheries is of particular concern through depredation and gear damage during gill netting on the west coasts of North America, Japan, Britain, Scandinavia, and Chile; through depredation, net damage, and disturbance at fish farms on the west coast of North America, Chile, and the United Kingdom (Nash *et al.*, 2000), and depredation from trawls, depredation and gear loss from hand lines (Meÿer *et al.*, 1992), and disturbance of purse seining in South Africa (Wickens, 1995). Estimates of the consumption by seals from fishing operations in South Africa show it to be a minimal percentage of fishery catches and a small proportion of the total predation by seals (Meÿer *et al.*, 1992). Preliminary calculations of overall economic losses resulting from these seals' interference show this to be small (0.3%) in comparison with the wholesale value of the catches.

Some pinnipeds converge on areas where anadromous fish stocks aggregate or where the movements of fish are constrained naturally or artificially (bottleneck or "choke points" where salmonids aggregate in response to human-made structures or natural river physiography, such as fish ladders or below falls, respectively). Seals may be attracted to a farm by escaped fish and oil slicks from feed, or increased wild fish outside the pens. The most thoroughly studied pinniped/salmonid conflict is California sea lion predation on winter steelhead (*Oncorhynchus mykiss*) at Washington State's Ballard Locks. The severe decline in salmon is considered primarily a direct result of human activities (Fraker and Mate, 1999); however, much concern has been voiced that the expanding populations of seals and sea lions may be causing a further decline (or impeding the recovery) of various salmon runs in the Pacific Northwest.

## B. Cetacean Interference with Fisheries

The following is not meant to serve as a comprehensive list, but illustrates some of the better known cetacean–fishery interactions. Bottlenose dolphins and "blackfish" (e.g., killer whales, false killer whales, pilot whales) are notorious fish stealers, and there are widespread reports of catch and gear damage by these species. Cetaceans may feed directly on a fishery's target species, such as killer whales and sperm whales feeding on sablefish (black cod, *Anoplopoma fimbria*) in the North Pacific longline fishery (Yano and Dahlheim, 1995). Interactions between sperm whales and killer whales with longline fisheries (Fig. 2) also have been well documented in the Southern Ocean (in particular, South Georgia, the Kerguelen Islands, and Southern Chile). Such interactions include entanglement in gear, following vessels for days, and observed feeding off gear. Killer whales also feed in association with bottom set fisheries off eastern Russia, interfere with the tuna fishery in the Strait of Gibraltar, and even take salmon off lines in recreational fisheries. Zollett and Read (2006) documented depredation by bottlenose dolphins in the Florida king mackerel (*Scomberomorus cavalla*) troll fishery. Long-finned pilot whales (*Globicephala melas*) in Newfoundland frequent traps to remove the target species squid. Bottlenose dolphins in Belize have been observed retrieving fish from local, homemade fish traps. Even baleen whales are known to interact with fisheries, following boats and taking fish from nets; an example is fin whales (*Balaenoptera physalus*) off Archipelago Campano feeding in association with trawls and encircling nets (Mussi *et al.*, 1999). Cetaceans also prey on fish confined in mariculture (i.e., fish farm) enclosures (Kemper *et al.*, 2003; Díaz López, 2006).

Cetaceans may also feed on fish that are ancillary to the catch, as in the case of bottlenose dolphins feeding on bycatch from trawl fisheries for shrimp and prawn (Fertl and Leatherwood, 1997). Whales and dolphins may interfere with traps or pots (Fig. 3), such as bottlenose dolphins in the Indian River Lagoon in Florida interacting with the crab pot fishery, apparently to steal bait fish (Noke and Odell, 2002).

## C. Sirenian Interference with Fisheries

Fishermen in Jamaica and Sierra Leone have complained about damage caused to gill nets by "net robbing" West Indian and West

**Figure 3** *Bottlenose dolphin* (Tursiops truncatus) *investigating fish trap utilized by the Haitian fishermen, in 70ft of water at the Northwest Point of Navassa National Wildlife Refuge. The trap is an Antillean Z-trap constructed of meshed/woven bamboo (3–4cm mesh size) with wooden cross supports and corners and opposing funnel entrances. Photo by Amy V. Uhrin, NOAA Center for Coastal Fisheries and Habitat Research, Beaufort, North Carolina.*

African manatees (*Trichechus manatus* and *T. senegalensis*) (Powell, 1978, Reeves *et al.*, 1988). Manatees have been described as stripping the flesh off fish entangled in gill nets and leaving the bones. Fishermen in Puerto Rico have noticed manatees circling gill nets, picking out fish.

### D. Sea Otter Interference with Fisheries

When a sea otter investigates a lobster pot, Dungeness crab trap, or live-fish trap, the individual itself can become trapped and die (Newby, 1975).

### E. Toward Solutions

In response to presumed or real interference with fishing operations, fishermen use various means to deter marine mammals in an attempt to safeguard their catches and gear. Lethal methods have been attempted, including shooting at or killing the marine mammal with a variety of objects and methods, sometimes involving poison. Sometimes these practices are illegal. Seals have been persecuted much more intensely than cetaceans. Lethal methods have not been found to be a consistently effective means of keeping pinnipeds from interacting with fishing operations. The idea is that if problems are caused by a few rogue seals, then removal of these animals should eliminate the problem. However, this method removes individuals that are then often replaced by others.

A diversity of nonlethal methods has been attempted (Werner *et al.*, 2006). At the most basic level, fishermen throw stones or bait to distract the predator. Other methods used include firing gunshots (nonlethal and lethal), using explosives (such as firecrackers and seal bombs), acoustic deterrent devices (ADDs), gear modifications, gear switching, physical barriers (nets), vessel chase (hazing), tactile

harassment (e.g., rubber bullets), visual signals, and taste aversion (baiting fish using a chemical to induce vomiting).

The most commonly used deterrents are ADDs that have been widely used to attempt to reduce depredation on fish. These include pingers, acoustic harassment devices (AHDs), passive acoustic devices, predator sounds, and banging pipes. Marine mammals are difficult to deter by acoustic methods, and the acoustic signal of the AHDs over time can be a "dinner bell" effect, alerting animals to the presence of a fish pen or trap. New high-intensity AHDs appear to be more effective but have a greater potential for causing hearing damage, as well as affecting nontarget species. In some cases, a problem may be eased by changing the location of the fishing effort.

The most successful mitigation measures appear to be changes to fishing gears or fishing methods where a particular change may reduce or exclude problems, thereby resulting in a permanent solution. Implementation of anti-predator cages around fish farms, physical barriers at the entrance of fish ladders, and the change to synthetic twine in gill nets are some examples. Other measures include exclusion devices in nets to mitigate bycatch (Suuronen *et al.*, 2006). The Southeast Alaska Sperm Whale Avoidance Project (SEASWAP) is testing changes to the longline fishery, such as circle hauls that minimize engine cycling (which as noted earlier, appears to attract whales), and changing the time of year the fishermen deploy their gear.

Capture and relocation of "problem" pinnipeds has proven ineffective, with the animals returning to the problem area. California sea lions have been captured at the Ballard Locks and placed in temporary captivity and released after the steelhead run. This proved ineffective in the long term, as did permanent captivity, which eliminates the "problem" sea lions without having to kill them but is limited by the availability of facilities that can hold the sea lions and the costs involved in capturing and holding the animals.

Past efforts have been unsuccessful in finding effective, long-term, nonlethal approaches to eliminating or reducing marine mammal–fishery conflicts. Some nonlethal deterrence measures appear to be effective initially or effective on "new" animals but become ineffective over time or when used on "new" animals in the presence of "repeat" animals that do not react to deterrence. Further research on the development of new technologies and techniques is needed.

### See Also the Following Articles

Competition with Fisheries ■ Feeding Strategies and Tactics ■ Incidental Catches ■ Management ■ Noises, Effects of ■ Parasites.

### References

Baird, R. W., and Gorgone, A. M. (2005). False killer whale dorsal fin disfigurements as a possible indicator of long-line fishery interactions in Hawaiian waters. *Pac. Sci.* **59**, 593–601.

Dalla Rossa, L., and Secchi, E. R. (2007). Killer whale (*Orcinus orca*) interactions with the tuna and swordfish longline fishery off southern and south-eastern Brazil: A comparison with shark interactions. *J. Mar. Biol. Assoc. UK* **87**, 135–140.

Díaz López, B. (2006). Bottlenose dolphin (*Tursiops truncatus*) predation on a marine fin fish farm: Some underwater observations. *Aquat. Mamm.* **32**, 305–310.

Fertl, D., and Leatherwood, S. (1997). Cetacean interactions with trawls: A preliminary review. *J. Northw. Atl. Fish. Sci.* **22**, 219–248.

Fraker, M. A., and Mate, B. R. (1999). Seals, sea lions, and salmon in the Pacific Northwest. *In* "Conservation and Management of Marine Mammals" (J. R. Twiss, Jr., and R. R. Reeves, eds), pp. 156–178. Smithsonian Institution Press, Washington, D.C.

Fréon, P., and Misund, O. A. (1999). "Dynamics of Pelagic Fish Distribution and Behaviour: Effects on Fisheries and Stock Assessment." Fishing News Books, Oxford.

Gilman, E., Brothers, N., McPherson, G., and Dalzell, P. (2006). Review of cetacean interactions with longline gear. *J. Cet. Res. Manag.* **8**, 215–223.

IUCN (1981). Report of IUCN workshop on marine mammal/fishery interactions, La Jolla, California, 30 March–2 April 1981. International Union for the Conservation of Nature and Natural Resources, Gland.

Kemper, C. M., *et al.* (8 authors) (2003). Aquaculture and marine mammals: Co-existence or conflict? *In* "Marine Mammals: Fisheries, Tourism and Management Issues" (N. Gales, M. Hindell, and R. Kirkwood, eds), pp. 208–225. CSIRO Publishing, Melbourne.

Königson, S. J., Lundström, K. E., Hemmingsson, M. M. B., Lunneryd, S. G., and Westerberg, H. (2006). Feeding preferences of harbour seals (*Phoca vitulina*) specialised in raiding fishing gear. *Aquat. Mamm.* **32**, 152–156.

Meÿer, M. A., Kotze, P. G. H., and Brill, G. W. (1992). Consumption of catch and interference with linefishing by South African (Cape) fur seals *Arctocephalus pusillus pusillus. South Afr. J. Mar. Sci.* **12**, 835–842.

Mussi, B., Miragliuolo, A., Monzini, E., Diaz Lopez, B., and Battaglia, M. (1999). Fin whale (*Balaenoptera physalus*) feeding ground in the coastal waters of Ischia (Archipelago Campano). *Eur. Res. Cet.* **13**, 330–335.

Nash, C. E., Iwamoto, R. N., and Mahnken, C. V. W. (2000). Aquaculture risk management and marine mammal interactions in the Pacific Northwest. *Aquaculture* **183**, 307–323.

Newby, T. C. (1975). A sea otter (*Enhydra lutris*) food dive record. *Murrelet* **56**, 7.

Nitta, E. T., and Henderson, J. R. (1993). A review of interactions between Hawaii's fisheries and protected species. *Mar. Fish. Rev.* **55**, 83–92.

Noke, W. D., and Odell, D. K. (2002). Interactions between the Indian River Lagoon blue crab fishery and the bottlenose dolphin, *Tursiops truncatus. Mar. Mamm. Sci.* **18**, 819–832.

Powell, J. A., Jr. (1978). Evidence of carnivory in manatees (*Trichechus manatus*). *J. Mammal.* **59**, 442.

Pryor, K., Lindbergh, J., Lindbergh, S., and Milano, R. (1990). A dolphin–human fishing cooperative in Brazil. *Mar. Mamm. Sci.* **6**, 77–82.

Read, A. J. (2005). Bycatch and depredation. *In* "Marine Mammal Research: Conservation Beyond Crisis" (J. E. Reynolds, III, W. F. Perrin, R. R. Reeves, S. Montgomery, and T. J. Ragen, eds), pp. 4–17. Johns Hopkins University Press, Baltimore.

Reeves, R. R., Tuboku-Metzger, D., and Kapindi, R. A. (1988). Distribution and exploitation of manatees in Sierra Leone. *Oryx* **22**, 75–84.

Sigler, M. F., Lunsford, C. R., Straley, J. M., and Liddle, J. B. (2008). Sperm whale depredation of sablefish longline gear in the northeast Pacific Ocean. *Mar. Mamm. Sci.* 24, 16–27.

Suuronen, P., Siira, A., Kauppinen, T., Riikonen, R., Lehtonen, E., and Harjunpää, H. (2006). Reduction of seal-induced catch and gear damage by modification of trap-net design: Design principles for a seal-safe trap-net. *Fisheries* **79**, 129–138.

Werner, T., Kraus, S., Read, A., and Zollett, E. (2006). Fishing techniques to reduce the bycatch of threatened marine animals. *Mar. Tech. Soc. J.* **40**, 50–68.

Wickens, P. A. (1995). "A Review of Operational Interactions between Pinnipeds and Fisheries." FAO Fisheries Technical Paper, No. 346.

Wickens, P. A. (1996). Fur seals and lobster fishing in South Africa. *Aquat. Conserv.: Mar. Freshw. Ecosys.* **6**, 179–186.

Yano, K., and Dahlheim, M. E. (1995). Killer whale, *Orcinus orca*, depredation on longline catches of bottomfish in the southeastern Bering Sea and adjacent waters. *Fish. Bull.* **93**, 355–372.

Zollett, E. A., and Read, A. J. (2006). Depredation of catch by bottlenose dolphins (*Tursiops truncatus*) in the Florida king mackerel (*Scomberomorus cavalla*) troll fishery. *Fish. Bull.* **104**, 343–349.

# Fishing Industry, Effects Of

## Simon Northridge

The fishing industry probably represents the single area of human activity that has the most profound effects on marine mammals. These effects can be categorized broadly as "operational effects" and "biological effects." Operational effects include the accidental capture of marine mammals in fishing gear, a problem that has brought about the extinction of one cetacean species in recent years, and threatens several other populations too. Although accidental capture usually results in the death of the animal concerned, there are also instances where marine mammals are injured or affected in some way during fishing operations so that their survival probability or reproductive potential is compromised. Not all operational interactions have a negative effect on marine mammals. In some cases the effect of the fishing operations may be positive for the marine mammal where, e.g., they feed on discarded fish or take fish that have been caught before these can be retrieved onto the fishing vessel. In a few cases there are even mutually beneficial collaborative efforts between fishermen and marine mammals, with marine mammals assisting in fish capture and being rewarded with a portion of the catch.

Biological effects encompass all the consequences of the large-scale removal of animal biomass from the marine ecosystem through fishing activities, including, although not limited to, possible competition for resources between fisheries and marine mammals. Competitive interactions can be direct or indirect. Direct competition occurs where the mammal and the fishery are both taking the same kind of fish. Indirect competition includes situations where the fishery and the marine mammal population are taking two different types of fish, but where the removal of one of these fish influences the availability of the other through some competitive or predatory link. Indirect interactions need not be competitive, and sometimes the effect of the fishing industry may be to increase the abundance of marine mammal prey items through indirect ecological interactions. Sometimes fisheries may physically alter a habitat and so change the composition and abundance of the fish community to the detriment or advantage of marine mammals and other predators.

Operational effects cover interactions between fisheries and marine mammals that relate to the mechanical process of fishing. Several fisheries have well-documented problems with unwanted entrapment of marine mammals. In some cases the numbers of animals involved are large enough to seriously endanger the marine mammal populations concerned; in one case (baiji, *Lipotes vexillifer*) a species has been driven to extinction, largely due to fishery interactions, while it is feared the same fate awaits the vaquita (*Phocoena sinus*). Examples considered cover gill net fisheries, pelagic trawls, and purse seine fisheries.

Gill nets are a widely used fishing gear with a long history of use in many parts of the world. Their use has become more widespread since the 1950s or 1960s with the introduction of nylon as a netting material during the 1950s. They represent a fuel-efficient means of fishing and, when set on the seabed, provide a fishing method that can be used to exploit areas of rough ground that cannot be fished easily by towed gear. When used in surface waters, they are usually left to drift with the wind and tide and are effective in targeting

**F**

**Figure 1** *Young gray seal caught and drowned in a skate tangle net set in the English North Sea. Photo by Dave Sanderson, SMRU.*

dispersed fish schools. They are usually left to fish unattended, and fishing times may range from a few hours to several weeks, but 24 h would be a typical soak time. It has been suggested that in contrast to the traditional nets that were made of cotton and other natural fibers, the use of stronger nylon twines has contributed to an increased rate of marine mammal entanglement. This, coupled with a dramatic increase in their usage since the 1950s, has led to some serious conservation and animal welfare problems with respect to marine mammals.

Small cetaceans, such as porpoises (Phocoenidae), and some species of seals seem especially prone to becoming entangled in gill nets. In some instances, this does not present a conservation problem. In Britain, e.g., gray seals *Halichoerus grypus* are frequently caught and drowned accidentally in gill net and tangle net fisheries (Fig. 1). In a seal tagging program run by the Sea Mammal Research Unit in the North Sea, over 20,000 gray seals have been tagged soon after birth since the 1950s. Returns of tags by fishermen indicate that at least 10% of known subsequent deaths are because of net entanglement, and at least 1.5% of all pups tagged were recovered dead in fishing nets. Tag loss over the months and years after tagging and failure to return tags from entangled seals are two reasons why this latter figure must be an underestimate of total mortality rates due to entanglement. Despite such mortality in fishing gear, gray seal numbers have more than doubled in British waters in the past 30 years.

In other cases, accidental catches in fishing gear *can* lead to conservation problems. An isolated population of Hector's dolphin (*Cephalorhynchus hectori*) that lives off the west coast of the North Island of New Zealand now numbers around 100 individuals, and bycatches in gillnets continue to occur despite a ban on gillnetting in part of the range of these animals. There are concerns that this population may be driven to extinction before long (Slooten *et al.*, 2006; Slooten, 2007)

A porpoise species that is endemic to Mexico, the vaquita, is threatened with extinction through accidental catches in gill nets. In this case, the species has a restricted range, in the upper Gulf

of California in Mexico, where there are large numbers of small boats using gill nets and tangle nets to catch a wide variety of fishes. Although the population is thought to be less than 200, one estimate suggests that at least 40 animals drown in gill nets every year. This is clearly an unsustainable rate of mortality and this species' future therefore seems bleak (Rojas-Bracho, Reeves & Jaramillo-Legorreta, 2006).

The North Atlantic right whale (*Eubalaena glacialis*), once one of the more commonly seen whales in the North Atlantic, is now reduced to a population of around 300 animals in that area. The population is declining and it is estimated that at current rates it will be extinct within 200 years (Caswell, Fujiwara & Brault, 1999). Most of this population migrate along the eastern seaboard of the United States every year, where they too are vulnerable to entanglement in gill nets and also in lobster pot lines. Lobster pots are usually set in "strings" of several pots joined by ropes, with each end of the string marked by a surface-floating buoy attached by a rope to the pot string on the seabed. Although only three right whale deaths have been attributed directly to this cause since 1970, evidence of entanglement-related scars on live animals has been identified in around 60% of the population. It has been suggested that some entangled whales die and are dragged to the seabed without being recovered. Others may suffer injuries that lead to subsequent death by other causes, while at least two right whales, encumbered with fishing gear, have been fatally wounded in collisions with ships. These high levels of entanglement may therefore be a major factor of the decline in this population of right whales, even though the vast majority of the entanglements are not immediately fatal.

Marine mammal entanglements in gill nets may occur for a variety of possible reasons. Some people maintain that the animals do not detect the netting and swim into it before realizing it is there. It may be that the animals do detect the netting but that they do not recognize it as something dangerous and attempt to swim through it as though it were some natural obstacle, such as seaweed. Another possibility is that the animals are fully aware of the netting, and the danger it poses, but they simply make mistakes and become entangled while feeding close to the net due to inattention.

There has been much attention given to means of reducing the numbers of marine mammals that become caught in gill nets and tangle nets because of the conservation problems that such entanglements represent. So far, the only effective means of reducing bycatch that has been found is the use of pingers. Pingers are small battery-powered devices that emit a brief high-pitched noise every few seconds. They are attached to the float line or lead line of the gill net and are of similar size and shape as a net float so as to avoid tangling the net when it is being set or hauled. They are effective in reducing the entanglement rates of several marine mammal species in gill nets, although exactly why they work is not clear. They have been developed in collaboration with the fishing industry and are currently being used in several major gill net fisheries around the world.

Pelagic trawling is another fishing method that has increased in recent decades. Trawling dates back for more than a century and initially performed with low-opening nets dragged along the seabed. As various aspects of technology have improved, so trawling techniques have been refined, and trawls have been used to catch fish above the seabed and even near the surface of the water. The development in the 1950s of acoustic fish finders and net sounders that enable the skipper to control the position of the net with respect to that of a fish school has been a key technological development.

Initially, during the 1950s, pelagic trawls were used to catch fish like herring that form dense schools. Typical nets might have had an opening of around $2000 \, m^2$, perhaps 50-m wide and 40-m high. Since then, both nets and trawlers have grown in size, as other pelagic fish species have been targeted, some of which may form much more dispersed schools. Net openings at least 10 times as great are now common, with horizontal and vertical openings of up to 200 m. There are numerous records of marine mammals becoming caught in pelagic trawls, sometimes in large enough numbers to present a conservation problem, although the nature and scale of the problem remain obscure in most fishing areas.

Hooker's sea lion (*Phocarctos hookeri*) is endemic to the sub-Antarctic islands of New Zealand, around which a pelagic trawl fishery for squid developed in the 1980s. Observations of fishing activity between 1988 and 1995 suggested annual mortality of between 20 and 140 Hooker's sea lions, at a rate of about 1 animal every 340 trawl tows. The total population of this species was only around 13,000 animals in the mid-1990s, and the accidental catches were therefore considered significant. There is currently a quota system in operation in this fishery to limit accidental catches (Wilkinson, Burgess & Cawthorn, 2003).

Dolphins have also been reported caught in pelagic trawl fisheries in several areas, including the United States, Europe, and New Zealand. The capture of dolphins in pelagic trawls appears to be very variable, depending on the area and probably the type of trawl being used. In one US pair-trawl fishery, common bottlenose (*Tursiops truncatus*), short-beaked common (*Delphinus delphis*), and Risso's (*Grampus griseus*) dolphins were observed accidentally caught in an average of one in every five trawl tows, although this fishery has subsequently been closed. In one New Zealand midwater trawl fishery, rates of common and bottlenose dolphin catches averaged about one animal in every nine trawl tows observed. In European Atlantic waters, one study recorded common and Atlantic white-sided dolphins (*Lagenorhynchus acutus*) caught in sea bass (*Dicentrarchus labrax*) and albacore tuna (*Thunnus alalunga*) pelagic trawl fisheries at rates of one in every trawl tow to 1 in every 15 trawl tows. In other pelagic trawl fisheries for anchovy (*Engraulis encrasicolus*), pilchard (*Sardina pilchardus*), and mackerel (*Scomber scombrus*) in the same area, no dolphin deaths were recorded.

The reasons why some pelagic trawl fisheries have relatively high levels of accidental dolphin catches and others have low or zero levels are not yet clear. It is possible that such factors as the dimensions of the net, the towing speed and duration, and the foraging activity of dolphins around the nets could all be important. There are several accounts of dolphins taking advantage of trawling activity by feeding on fish escaping through the meshes of the trawl. Such behavior may increase chances of dolphin entanglement, but these interactions remain relatively poorly investigated.

As with gill nets and pelagic trawls, technical innovations during the 1950s enabled the development and expansion of purse seine fisheries around the world. Purse seines are used to catch pelagic fish and work by first encircling a school of fish with a long net, hanging from the surface down to depths sometimes of several hundred meters. The bottom edge of the net can be pursed so as to prevent any escape under the netting once it has been set in a circle. The major technical innovations that allowed this fishery to develop globally were the introduction of nylon as a netting material, enabling much larger and stronger nets to be constructed, and the development of the power block, with which large amounts of netting could be hoisted up out of the water.

American fishermen working in the eastern tropical Pacific Ocean during the late 1950s worked out a way of using dolphin schools in conjunction with purse seine nets to catch yellowfin tuna (*Thunnus albacares*). They discovered that tuna would aggregate under dolphin schools, even if the dolphins were chased. This meant that by using speedboats to corral dolphins they could exploit this behavioral characteristic of the fish to round up tuna schools that would otherwise be invisible below the surface. By corralling a dolphin school and setting a purse seine net around it, a school of tuna would normally also be encircled. The dolphins were not intentionally killed in this activity, but many died as nets were being hauled in and fish were being brought on board. The scale of the fishery meant that in some years hundreds of thousands of dolphins drowned as a result of this fishing activity.

The eastern spinner dolphin subspecies (*Stenella longirostris orientalis*) was depleted to 44% of its original pre-1959 level, whereas the northeastern offshore stock of the pantropical spotted dolphin stock (*Stenella attenuata*) was reduced to about 20% of its pre-exploitation level. From the 1970s onward, in the face of public concerns over the issue, the fishery developed and implemented means of reducing this toll. By encouraging the dolphins out of the net before trying to remove the fish, annual mortality was reduced to a few thousand per year. Despite this, populations of these two species do not appear to have recovered, and it has been suggested that there may be additional, sublethal, impacts of the fishing technique that might be hampering population recovery (Wade *et al.*, 2007).

Normally, interactions between marine mammals and hook and line fisheries have few negative impacts on the marine mammals. In the case of the baiji or Chinese river dolphin (*Lipotes vexillifer*), however, the situation was different. The baiji was endemic to the Yangtze River system in China and is now thought to be extinct after years in which the population was in decline. One of the most important causes of mortality is animals becoming ensnared in "rolling hooks." This type of fishing involves using many sharp unbaited hooks on a line set on the bottom of the river to snag bottom-dwelling fish. Dolphins foraging near such hooks sometimes become snagged too, occasionally causing death. Among the recovered river dolphin carcasses, over 50% of the dolphins had died as a result of such entanglements or other fishing-related injuries (Turvey *et al.*, 2007).

Not all operational interactions have a negative impact on marine mammals. In some cases marine mammals are able to exploit fishing activities to their own advantage. In several parts of the world cetaceans, often killer whales (*Orcinus orca*) and false killer whales (*Pseudorca crassidens*), have been reported to remove fish from hooks during longline fishing operations. This can make fishing in certain areas with longlines unprofitable, and many methods have been tried to eliminate such behavior (Donoghue *et al.*, 2003). There has been remarkably little success reported in trying to prevent such predation, and sometimes the boats involved have had to switch gear or move to other areas.

Other types of fishery where marine mammals actively benefit from fishing activities include trawling, purse seining, and lobster potting. Several species, including bottlenose and white-beaked dolphins (*Lagenorhynchus albirostris*), gray seals, and South African fur seals (*Arctocephalus pusillus*), have been reported to remove fish from fishing gear (Fertl and Leatherwood, 1997). Dolphins typically take undersized fish as they come through the cod end of a trawl. Fur seals swim into purse seine nets by climbing over the float line once a school of fish has been encircled and make a meal of the trapped fish. Gray seals have been observed removing bait from baited lobster pots. There are numerous other examples of marine mammals taking advantage of fishing activities in similar ways, often taking marketable fish, which may provoke considerable resentment on the part of the fishing crews.

In a few places in the world, including Burma (Myanmar), Mauritania, and Brazil, fishermen and dolphins have learned to collaborate in the capture of fish, usually by dolphins driving fish toward fishermen waiting with nets. Some of the catch is then given back to the dolphins as a reward (Busnel, 1973; Thien, 1977; Pryor *et al.*, 1990).

The more widespread but less well-understood interactions between fisheries and marine mammals are ecosystem-level effects, where fisheries may cause fundamental changes to the species composition of the marine environment (see Competition with Fisheries).

Every year the fishing industry removes about 80–90 million tons of fish and other marine organisms from the world's oceans. Another 7–8 million tons of unwanted animal biomass may be caught but discarded prior to landing every year. It has been suggested that fisheries may account for 8% of the global primary productivity of the oceans, and in some of the more heavily exploited areas as much as 35% of local primary productivity may be required to sustain fishery catch levels. It is clear that such levels of fishing activity are likely to have profound effects on marine ecosystems, especially on the top predators, such as marine mammals, as fish populations are reduced and restructured on a very large scale. Therefore in theory, fisheries may compete with marine mammals by depleting their food.

There is of course another side to this concern. Whereas fisheries may cause a depletion of the food resource for marine mammals, marine mammals are also accused of consuming large amounts of fish, thereby reducing the amount of food available for people to eat (Tamura, 2003). As a result of this latter concern, in some parts of the world there are frequent calls for marine mammals to be culled as unwanted competitors.

In both cases, however, it has proved extremely difficult to demonstrate any clear competitive interaction between marine mammals and fisheries. This is mainly because of the complexities of the marine ecosystem, which make it very hard to predict how changes in one fish stock will affect either their predators or their prey. Some brief examples will illustrate this point.

The North Sea and adjacent areas are among the most heavily fished sea areas of the world, with annual landings of all species of over 2 million metric tons. One of the most numerous marine mammals in this region is the gray seal (*Halichoerus grypus*), which feeds on a range of fish species, including Atlantic cod, haddock, whiting, saithe, sand eels, sole, plaice, Atlantic herring, and sprats (*Gadus morhua*, *Melanogrammus aeglefinus*, *Merlangius merlangus*, *Pollachius virens*, *Ammodytes* spp., *Solea solea*, *Pleuronectes platessa*, *Clupea harengus*, and *Sprattus sprattus*). Fisheries also target all of these species, and most of the fish stocks concerned are designated as fully exploited or overexploited. Despite these facts, the gray seal population has more than doubled in size since the early 1980s to somewhere between 130,000 and 200,000 individuals during which time fishing pressure in the region has been intense. There are at least two good reasons for this apparent paradox. First, the most important food of the gray seal is the sand eel (primarily *Ammodytes marinus*) or "sand lance" in the United States. The sand eel fishery was taking over 1 million metric tons of sand eels per year in the late 1990s and was the single largest fishery in terms of the amount landed in the North Sea. Despite this fishing pressure, sand eels were extremely abundant and appeared to have increased in abundance since the 1950s. The proposed reason for this is that sand eel numbers have increased in response to massive declines in the abundance of Atlantic herring and mackerel as a result of intense herring and mackerel fishing in the 1960s and 1970s. Paradoxically,

therefore, heavy fishing pressure may have led to an increase in a less commercially valuable species that is a major food of gray seals.

Alongside this, although gray seals and fisheries consume other more marketable species in common, gray seals typically consume smaller-sized individuals of around 15–20 cm in length, whereas commercial fisheries generally concentrate on fish of 30 cm and larger. As commercial fishing pressure intensifies, larger fish have become scarcer, but smaller fish of the same species may not be affected or at least not until there are two few large fish left to generate sufficient eggs to replenish the stock. Indeed, in many of the commercially useful gadid cod family fishes, small fish are consumed in large numbers by bigger fish of the same species. Thus, reduction of the numbers of larger fish may actually boost the numbers of smaller fish (Daan *et al.*, 2005).

It would seem that although gray seals are feeding in an area that is very heavily fished because they exploit a niche that is not directly in competition with fisheries, they are able to thrive. Indeed their population expansion may have been assisted by fishery-induced changes to the species and size structure of the system they inhabit.

Although pinniped numbers may have increased with increasing fishing pressure in one area, this is by no means the norm. In the Gulf of Alaska and the Bering Sea, several pinniped populations have undergone dramatic declines over periods when fishing activity has been increasing. The Pribilof population of Northern fur seals (*Callorhinus ursinus*) declined from 1.25 million animals in 1974 to around 877,000 animals in 1983. At around the same time, harbor seal (*Phoca vitulina richardii*) numbers in the Gulf of Alaska and southeastern Bering Sea declined, with one major haul-out site at Tugidak Island recording an 85% reduction in numbers between 1976 and 1988. Populations of both of these species now seem to be stable or increasing, but numbers of Steller sea lions *Eumetopias jubatus* in western Alaska started to decline in the 1970s and numbers declined until at least 2000. Declines of up to 80% have been recorded in some areas.

The reasons for the declines of these three species over much the same time period are not known, but there is some agreement that food availability, especially for younger animals, seems to be a key issue. For all of these three species, Alaska pollock (*Theragra chalcogramma*) is an important prey item. Alaska pollock is also the target of one of the largest single-species fisheries in the world. The fishery for Alaska pollock increased greatly during the 1970s with over a million tons being landed annually from the eastern Bering Sea alone throughout most of the 1980s and 1990s.

An obvious explanation is that the fishery has deprived the pinnipeds of their food. A closer inspection, however, reveals that the situation is more complex. Overall, the pollock biomass has stayed remarkably buoyant throughout this time period. Numbers of pollock, especially numbers of the larger or older age classes that are the target of the fishery, have not declined until relatively recently. Smaller pollock are consumed by the pinnipeds and are not targeted by the fishery but are cannibalized by the larger fish. Numbers of smaller pollock have declined. Furthermore, for Steller sea lions at least, pollock does not appear to be a favored food item. In other parts of the Steller sea lion's range, such as southeastern Alaska, where the population is not in decline, pollock makes a smaller contribution to the diet, after oily fish such as herring.

It may be that the three pinniped species have all suffered from some change in the relative abundance of their preferred diet items. It has not yet been possible to determine whether such ecosystem-level changes have been the result of long-term oscillations in oceanographic conditions or whether the pollock fishery has in some way altered the abundance of the pinniped's preferred prey items through the cascading effects of restructuring the pollock population. Controversy still surrounds the issue (Fritz & Hinckley, 2005; Trites *et al.*, 2007), but it serves to illustrate that indirect or competitive interactions between fisheries and marine mammals are always difficult to interpret or understand.

The effects of fisheries on marine mammals do not necessarily have to be mediated through changes in their food supply. Changes to predation on marine mammals could also arise through the effects of fishing. On the Atlantic coast of Canada, also a heavily fished region, gray seal numbers have also been increasing steadily for more than two decades for unknown reasons. One suggested reason has been that fisheries may have greatly reduced the number of large sharks in coastal waters of Atlantic Canada, and as gray seals are known to be subject to predation by certain sharks, such a reduction might be one of the factors contributing to the increase in gray seal numbers (Brodie & Beck, 1983).

Conversely, recent increases in predation on sea otters (*Enhydra lutris*) in the Aleutian Islands have been attributed to behavior changes in killer whales in the region, some of which now seem to be preying more heavily on sea otters than in previous decades. The possibility has been raised that this change in behavior has been caused by a decline in other more favored food items, including sea lions (Springer *et al.*, 2003). Such declines could, as has been suggested earlier, be at least partly due to the effects of fishing on sea lion food items.

All of these hypotheses demonstrate the complex ways in which fishery-induced changes to the marine ecosystem may affect marine mammals, although none has yet proved testable. In almost all cases where some form of competition is perceived, any closer scrutiny of the situation reveals that the complex predatory interrelations of the marine ecosystem make it extremely difficult to predict the results of any proposed management action. The extent to which the fishing industry competes with marine mammals is therefore still very much an open question.

## See Also the Following Articles

Bycatch ■ Competition with Fisheries ■ Entrapment and Entanglement ■ Forensic Genetics ■ Tuna-Dolphin Issue ■ Vaquita.

## References

Brodie, P., and Beck, B. (1983). Predation by sharks on the grey seal (*Halichoerus grypus*) in eastern Canada. *Can. J. Fish. Aquat. Sci.* **40**, 267–271.

Busnel, R. G. (1973). Symbiotic relationship between man and dolphins. *Trans. N Y Acad. Sci. II* **35**, 112–131.

Caswell, H., Fujiwara, M., and Brault, S. (1999). Declining survival probability threatens the North Atlantic right whale. *Proc. Natl. Acad. Sci USA* **96**, 3308–3313.

Daan, N., Gislason, H., Pope, J. G., and Rice, J. C. (2005). Changes in the North Sea fish community: Evidence of indirect effects of fishing? *ICES J. Mar. Sci.* **62**, 177–188.

Donoghue, M., Reeves, R.R., and Stone, G.S. (2003). "Report of the Workshop on Interactions Between Cetaceans and Longline Fisheries, Apia Samoa November 2002." New England Aquarium, Apia, Samoa.

Fertl, D., and Leatherwood, S. (1997). Cetacean interactions with trawls: A preliminary review. *J. Northw. Atl. Fish. Sci* **22**, 219–248.

Fritz, L. W., and Hinckley, S. (2005). A critical review of the regime shift - "Junk Food"- nutritional stress hypothesis for the decline of the western stock of Steller sea lion. *Mar. Mamm. Sci.* **21**, 476–518.

Gosliner, M. (1999). The tuna–dolphin controversy. *In* "Conservation and Management of Marine Mammals" (J. Twiss, and R. Reeves, eds), pp. 120–155. Smithsonian Institution Press, Washington, DC.

Mangel, M., and Hofman, R. J. (1999). Ecosystems: Patterns, processes and paradigms. *In* "Conservation and Management of Marine Mammals" (J. Twiss, and R. Reeves, eds), pp. 87–98. Smithsonian Institution Press, Washington, DC.

Marine Mammal Commission. (2000) "Annual Report to Congress, 1999." Marine Mammal Commission, Washington, DC.

Merrick, R. L., Cumbley, M. K., and Bryd, G. V. (1997). Diet diversity of Steller sea lions *Eumetopias jubatus* and their population decline in Alaska: A potential relationship. *Can. J. Fish. Aquat. Sci.* **54**, 1342–1348.

Northridge, S. P. (1991). An updated world review of interactions between marine mammals and fisheries. *FAO Fish. Tech. Pap.* **251**, Suppl. 1. Food and Agriculture Organisation of the United Nations, Rome.

Northridge, S. P., and Hofman, R. J. (1999). Marine mammal interactions with fisheries. *In* "Conservation and Management of Marine Mammals" (J. Twiss, and R. Reeves, eds), pp. 99–119. Smithsonian Institution Press, Washington, DC.

Perrin, W. F., Donovan, G. P., and Barlow, J. (1994). Cetaceans and gillnets. *Rep. Int. Whal. Commn., Spec. Iss.* **15**.

Pryor, K., Lindbergh, J., Lindbergh, S., and Milano, R. (1990). A dolphin–human fishing cooperative in Brazil. *Mar. Mamm. Sci.* **6**, 77–82.

Rojas-Bracho, L., Reeves, R. R., and Jaramillo-Legorreta, A. (2006). Conservation of the vaquita *Phocoena sinus*. *Mam. Rev.* **36**, 179–216.

Slooten, E. (2007). Conservation management in the face of uncertainty: Effectiveness of four options for managing Hector's dolphin bycatch. *Endang. Spec. Res.* **3**, 169–179.

Slooten, E., Dawson, S., Rayment, W., and Childerhouse, S. (2006). A new abundance estimate for Maui's dolphin: What does it mean for managing this critically endangered species? *Biol. Cons.* **128**, 576–581.

Springer, A. M., *et al.* (8 authors). (2003). Sequential megafaunal collapse in the North Pacific Ocean: An ongoing legacy of industrial whaling? *Proc. Proc. Nat. Acad. Sci. USA* **100**, 12223–12228.

Tamura, T. (2003). Regional assessments of prey consumption and competition by marine cetaceans in the world. *In* "Responsible Fisheries in the Marine Ecosystem" (M. Sinclair, and G. Valdimarsson, eds), p. 448. CABI Publishing, Beijing, China.

Thien, U. T. (1977). The Burmese freshwater dolphin. *Mammalia* **41**, 233–234.

Trites, A. W., *et al.* (2007). Bottom-up forcing and the decline of Steller sea lions (*Eumetopias jubatus*) in Alaska: Assessing the ocean climate hypothesis. *Fish. Oceanogr.* **16**, 46–67.

Turvey, S. L., et al. (16 authors) (2007). First human caused extinction of a cetacean species? *Bio. Lett.* **3**, 537–540.

Wade, P. R., Watters, G. M., Gerrodette, T., and Reilly, S. B. (2007). Depletion of spotted and spinner dolphins in the eastern tropical Pacific: Modeling hypotheses for their lack of recovery. *Mar. Ecol. Prog. Ser.* **343**, 1–14.

Wilkinson, I., Burgess, J., and Cawthorn, M. W. (2003). New Zealand sea lions and squid: Managing fisheries impacts on a threatened marine mammal. *In* "Marine Mammals: Fisheries, Tourism and Management Issues" (N. J. Gales, M. Hindell, and R. Kirkwood, eds), pp. 192–207. CSIRO, Collingwood, Victoria.

# Folklore and Legends

## ROCHELLE CONSTANTINE

Folklore and legends are usually traditional stories popularly regarded as the telling of historical events. When in the form of myths, they often involve some form of the supernatural.

They have been with us for thousands of years and, because of this, folklore and legends form the basis of many religious beliefs, value systems, and the way we perceive our place in the world and our interaction with other animals. Man has long revered whales and dolphins in legends. For thousands of years they have been aligned with the gods, mythologized, and celebrated in art.

Some of the earliest legends about dolphins were told in Greek mythology, where it was believed that the sun god Apollo assumed the form of a dolphin when he founded his oracle at Delphi on the edge of Mount Parnassus. The ancient Greeks also believed Orion was carried into the sky riding on the back of a dolphin and was gifted three stars by the gods. This constellation is now known as Orion's Belt.

Many cultures, both ancient and recent, revered dolphins and believed them to be messengers from the gods. The pre-Hellenic Cretans appeared to have honored dolphins, and the ancient Greek, Oppian (*ca.* ad 180), wrote of godly intervention in the dolphins' move to the sea. It was believed that by the devising of Dionysus, the Greek god of wine, dolphins exchanged their life on land for life at sea and took on the form of fishes. Even though they changed form, it was believed that they retained the righteous spirit of man and, because of this, they preserved their human thoughts and deeds. Oppian also wrote of dolphins stranding themselves to die, so mortals could bury them and thereby remember the dolphins' gentle friendship. This was seen as an example of how magnificent dolphins were (McIntyre, 1974). The close alignment with man meant that ancient Greeks held dolphins in extremely high regard and that killing a dolphin was tantamount to killing a person. Both crimes were punishable by death.

Pliny the Elder (AD 23–79), a Roman philosopher, told the story of a peasant boy in the Mediterranean Sea who developed a relationship with a solitary dolphin he named Simo. The legend tells that the boy fed the dolphin and, in return, the dolphin gave him rides across the bay on its back. The boy became ill and died and, according to local knowledge, the dolphin returned to their meeting place for many days until it was believed that it died of a broken heart. Many of the Roman legends involved close human/dolphin contact and may seem fanciful, but in more recent times these bonds between solitary dolphins and humans have been well documented.

Drawings carved into rocks in northern Norway of killer whales (*Orcinus orca*) and other local animals are the earliest known artwork portraying dolphins. These drawings have been estimated at 9000 years old. The most detailed and colorful ancient art work was done by the ancient Greek and Minoan (Crete 3000–1500 BC) people. Dolphins were portrayed on frescoes, mosaic floors, coins, vases, and in sculpture. One of the earliest known pieces is a dolphin fresco painted *ca.* 1600 BC on the wall of the queen's bathroom in the Minoan palace of Knossos. The Dionysus cup dated 540 BC shows the Greek wine god with dolphins and grapes. Coins portraying dolphins have been found in Syracuse, Greece, *ca.* 480 BC, and the Romans also had dolphin coins in Second century BC.

It appears in many legends that the great whales were not necessarily held with such high regard as the dolphins. Whales were typically described as monsters of the sea, their great size to be feared by all. Oppian (*ca.* AD 180) told of the hunt of a whale—its monstrous size and unapproachable limbs a terrible sight to behold. In biblical times, the story of Jonah and the whale was well known, and it is popular even today. The story tells of Jonah who fled from the lord by boat to Tarshish. When the ship was underway, the lord caused a great storm. In fear of their lives, Jonah asked the mariners to cast him into the sea so the lord would again make the sea calm and spare the mariners' lives. Once Jonah was in the sea, however, the lord prepared a "great fish" to swallow him. He was in the belly of the whale for 3 days and 3 nights where he prayed and vowed salvation to the lord. Upon his vow the lord spoke to the whale and it vomited Jonah onto dry land and spared his life. Although today we know that it is unlikely that this event truly occurred, the story displayed the power of the lord and what he was capable of doing to those who defied him (Unsworth, 1996).

In his novel "Moby Dick" (1851), Herman Melville described a white sperm whale (*Physeter macrocephalus*) of uncommon magnitude, capable of great ferocity, cunning, and malice. Melville's novel summarized the fears of Yankee whalers that the tables would be turned and the whale would become the attacker (Melville, 1851).

Not all folklore portrays whales as fearsome beasts. Maori folklore of the Ngati Porou people tells of their ancestors being carried safely across the Pacific to New Zealand on the back of a whale (Ihimaera, 1987). The Ngai Tahu people consider the sperm whales off the coast of the South Island as taonga (treasures). If a whale strands, prayers are said to return its spirit to Tangaroa, the Maori god of the sea. After this, the lower jawbone is removed for ceremonial carving and placement on the marae, the tribes' traditional meeting grounds.

The north Alaska Inuit people have for over 1000 years relied on whale products for their survival. As with many traditional hunting societies, ceremonies accompany the hunt to ensure good luck, and many hunters take charms or amulets to ensure their luck and safety. Some believe the skull of the dead whale must be returned to the sea to assure the immortality and reincarnation of the whale, thereby protecting future hunting success (Lowenstein, 1994).

The Haida people of northwestern North America tell of an evil ocean people who used killer whales as canoes. The Haida turned a chief into a killer whale, and they believe that this whale now protects them from attacks by the ocean people.

The Tlingit (pronounced "Kling-kit") people of southeastern Alaska immortalize killer whales in their beliefs and folklore. Images of killer whales appear in many of their masks, carvings, totems, and blankets. At gatherings, the Tlingit tell stories, including one about the origin of killer whales. They believe a man from the Seal people carved many killer whales from wood but only the one carved from yellow cedar would swim. The legend says he carved many more from cedar and they swam up the inlet where he taught them how to hunt and what to hunt for. He also taught them not to hurt people. The Tlingit in return do not hunt killer whales and they believe that because of this the killer whales look after them (Fig. 1).

On Mornington Island in the Gulf of Carpentaria, northern Australia, a tribe of Aborigines have been in direct communication with Indian Ocean bottlenose dolphins (*Tursiops aduncus*) for thousands of years. They have a medicine man who calls the dolphins and "speaks" to them telepathically. By these communications he assures that the tribes' fortunes and happiness are maintained.

Many people who live on the banks of the Amazon River believe that river dolphins or botos (*Inia geoffrensis*) have the ability to transform themselves into handsome young men to woo women during fiestas and times of ceremony. So strong is this belief that some children are believed to have been fathered by these dolphins (Sangama de Beaver and Beaver, 1989).

A Japanese legend tells of a gigantic whale that challenges a sea slug to a race after boasting that he is the greatest animal in the sea. The sea slug accepts and arranges for his friends to wait at different beaches along the chosen course. On the day of the race, the whale surges ahead, but when he arrives at the first beach he is astonished to find the sea slug already there. So he challenges it to another race, only to have the sea slug win again. This happens many times until the whale admits defeat. This legend is analogous to the European legend of the tortoise and the hare but shows the Japanese peoples' close relation to the sea and its inhabitants and their use in teaching moral lessons.

**Figure 1** *Killer whale* (Orcinus orca) *images on the front of a Tlingit dance house. Photograph courtesy of Alaska State Library, Vincent Soboleff Photograph Collection.*

Perhaps such legends and folklore serve the purpose of helping people understand their past or to help society learn valuable lessons. In many societies today we revere whales and dolphins, and this will continue to develop our folklore into the future.

## See Also the Following Articles

Ethics and Marine Mammals ■ Popular Culture and Literature

## References

Ihimaera, W. (1987). "The Whale Rider." Reed Books, Auckland.
Lowenstein, T. (1994). "Ancient Land; Sacred Whale: The Inuit Hunt and its Rituals." Farrar Straus and Giroux, New York.
McIntyre, J. (1974). "The Mind of the Dolphin." Charles Scribner's Sons, New York.
Melville, H. (1851). "Moby Dick," 2003 Ed. Penguin Books, New York.
Sangama de Beaver, M., and Beaver, P. (1989). "Tales of the Peruvian Amazon." AE Publications, Largo.
Unsworth, B. (1996). "Classic Sea Stories." Random House, London.

# Forelimb Anatomy

## Lisa Noelle Cooper

Marine mammals are descended from terrestrial mammals whose forelimbs were weight-bearing appendages specialized for terrestrial locomotion. In the transition to an aquatic lifestyle, most marine mammals evolved a flipper by encasing the forelimb in soft tissue (Fig. 1). Most living marine mammals have a flipper, and flipper shape and the morphology of the underlying bony structures greatly affect the function of marine mammal forelimbs.

## I. Cetaceans

From the gracile and crescent-shaped flippers of a pilot whale, to the thick and door-like flippers of right whales, cetacean flippers come in lots of shapes and sizes (Figs. 2 and 3) (Howell, 1930; Benke, 1993). Most delphinids have small and thin flippers, except the broad and thick flippers of the killer whale (*Orcinus orca*). Killer whales display sexual dimorphism in that the male flippers are larger compared to female flippers. Beaked whales and the pygmy sperm whale can tuck the flippers into an indentation in the body wall during deep dives. Bowhead and right whales have large, broad flippers, while pygmy right whales and rorqual whales have elongated and very thin flippers. Intermediate between the forelimb morphologies seen in right whales and rorqual whales, the gray whale has a broad and elongated flipper. The most unusual flipper shape is seen in humpback whales as they have longest flippers of any cetacean and the leading edge of the flipper is scallop shaped by the presence of large tubercles.

Cetacean flippers function to stabilize the body and aid in turns (Woodward *et al.*, 2006). Large bowhead and right whale flippers are useful when the whale is turning at slow speeds. Gray whales make long migrations and breed in shallow lagoons, and as such their flippers have a broad surface area useful for turning at slow speeds, but the elongate flipper is also useful for generating lift while migrating. Most rorqual whales use their tiny flippers to stabilize and aid in turns. The humpback whale flipper is the exception among rorquals. The flipper is slapped on the water surface during mate displays and social touching. While swimming, the flipper moves in alternating dorsal–ventral strokes, and deforms into concave and convex arc shapes. Leading edge tubercles increase flipper area to maintain laminar flow, hinder generation of tip vortices, and allow a greater generation of lift.

All modern cetaceans have a remarkably shortened humerus, and the radius and ulna are flattened (Fig. 1A) (Howell, 1930). In most small-sized odontocetes, the carpal bones of the wrist have distinct bony articular facets, but almost all mysticetes and large-bodied odontocetes have burr-shaped carpal bones that lack articular facets and these carpals are immersed in a block of cartilage (Fig. 1A).

Cetacean metacarpals and phalanges, the bones of the digits, have unique characteristics compared to other mammals. These bones are hourglass shaped and lack all articular facets such that the dorsal and palmar, and proximal and distal surfaces are unidentifiable. Differentiating characteristics between metacarpals and phalanges are lacking, and identification of these elements can only be certain in articulated limbs. Cetaceans also have epiphyses on both the proximal and distal surfaces of the metacarpals and phalanges. Cetaceans are the only mammals to have a greater number of phalanges per digit (hyperphalangy) than the standard mammalian condition of two phalanges in the thumb, followed by three phalanges in the other digits (Table I). Odontocetes have the greatest number of phalanges in digits II and III, while mysticetes display the greatest number of phalanges in digits III and IV. Most cetaceans have five digits, but three families of mysticetes (Neobalaenidae, Eschrichtiidae, and Balaenopteridae) lack metacarpal I and all the phalanges of digit I (Fig. 1A).

Compared to other marine mammals, the cetacean flipper has a distinct lack of muscular and soft tissue structures (Howell, 1930). The triceps muscle complex is reduced with only the heads originating from the scapula being functional. The elbow is locked and the triceps

**F**

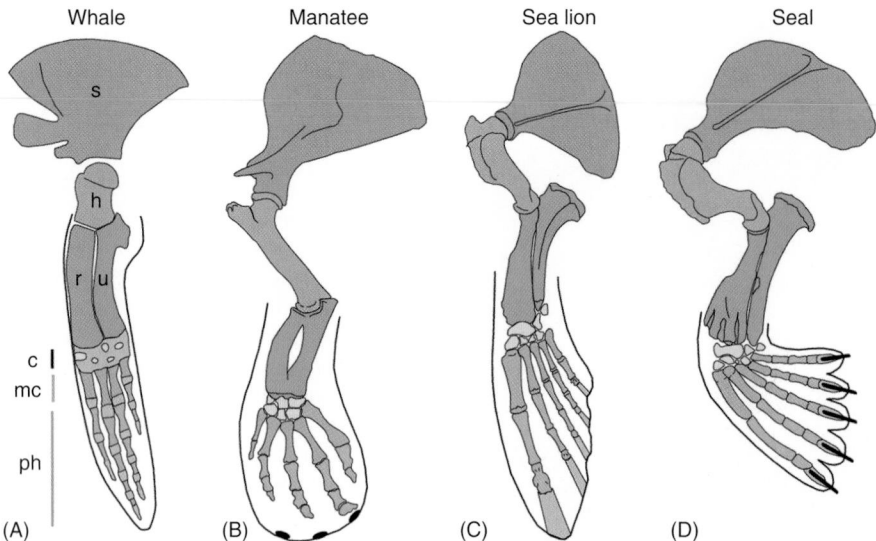

**Figure 1** *Left forelimbs of marine mammals: (A) blue whale, (B) manatee, (C) sea lion, (D) seal (Howell, 1930). Cartilage indicated by gray, claws shown in black. s, scapula; h, humerus; r, radius; u, ulna; c, carpals; mc, metacarpals; ph, phalanges.*

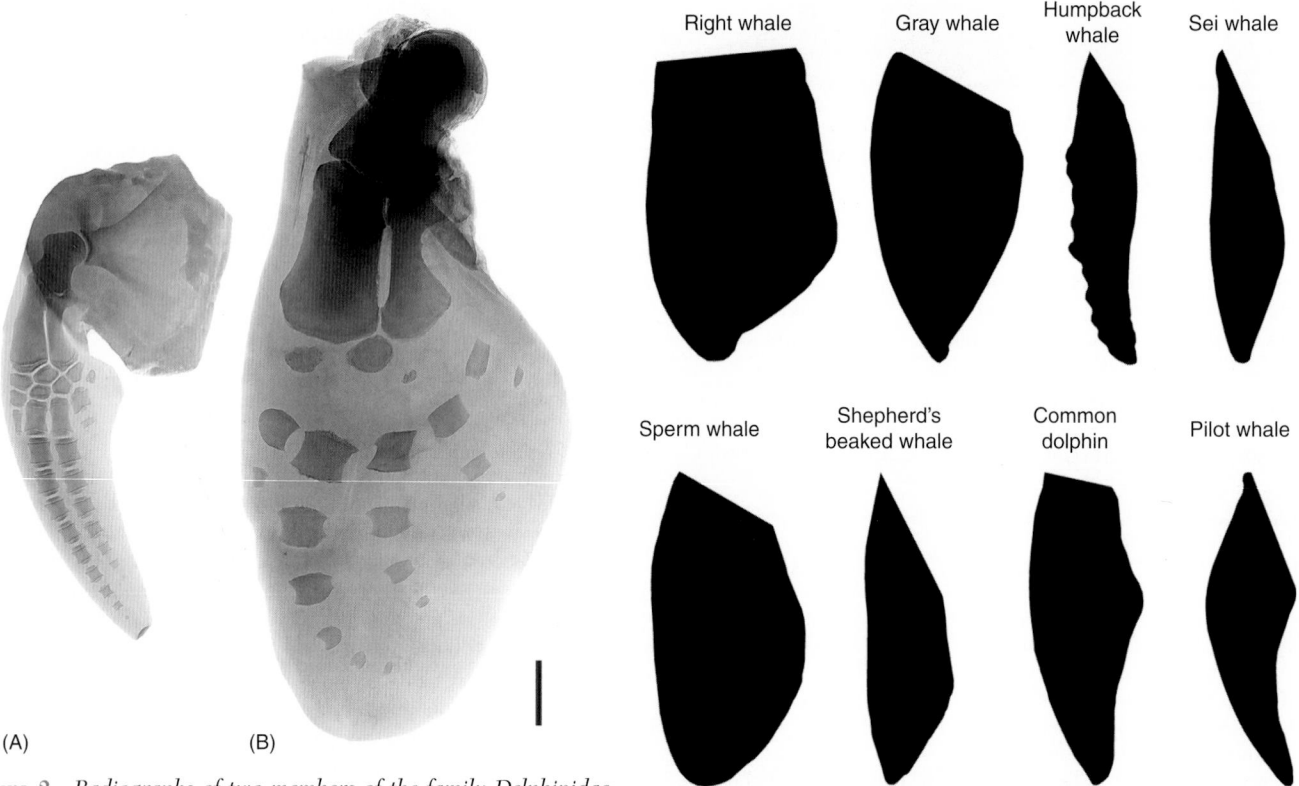

**Figure 2** *Radiographs of two members of the family Delphinidae showing variation in flipper size and shape: (A) Risso's dolphin (Grampus griseus), (B) killer whale (Orcinus orca). Scale bar 10 cm (Jacobsen, 2007).*

**Figure 3** *Flipper shapes of some cetaceans. Top row representative mysticetes, bottom row representative odontocetes. The leading edge of the flipper is to the left.*

humeral heads are reduced. Flexor and extensor muscles are prominent in most mysticetes (except *Megaptera*), and sperm whales and beaked whales (physeterids, kogiids, and ziphiids), but are lacking in other families of odontocetes (monodontids, phocoenids, and delphinids). Intrinsic manus muscles (interossei, lumbricals, abductors, adductors) are absent in most cetaceans, with exceptions found in physeterids and kogiids.

Fossil evidence indicates cetaceans increased surface area for muscles acting on the shoulder joint, immobilized the elbow and wrist, and elongated the manus (Uhen, 2004). The earliest Eocene archaeocetes used their forelimbs for terrestrial locomotion and their limbs appeared similar to those of Eocene artiodactyls. In the late Eocene, basilosaurid archaeocetes were fully aquatic and developed a wider

TABLE I

Phalangeal Formulas for Some Cetaceans. The Greatest Number of Phalanges Are Seen in Digits II and III in Odontocetes, and Digits III and IV in Mysticetes

| Taxon | Species | Digit I | Digit II | Digit III | Digit IV | Digit V |
|---|---|---|---|---|---|---|
| Odontocetes | | | | | | |
| sperm whale | *Physeter* | 0–1 | 3–6 | 4–5 | 2–4 | 0–3 |
| pygmy sperm whale | *Kogia breviceps* | 1–2 | 7–8 | 6–7 | 5–6 | 3–5 |
| Gervais' beaked whale | *Mesoplodon europaeus* | 0 | 5–6 | 5–6 | 4 | 2–3 |
| Susu | *Platanista* | 1–2 | 3–5 | 3–4 | 3–5 | 3–5 |
| Beluga | *Delphinapterus* | 1 | 3–6 | 3–4 | 2–4 | 1–3 |
| Harbor porpoise | *Phocoena phocoena* | 0–2 | 5–9 | 5–8 | 2–4 | 0–2 |
| Long-finned pilot whale | *Globicephala melas* | 2–3 | 12–13 | 8 | 2 | 0–1 |
| Atlantic white-sided dolphin | *Lagenorhynchus acutus* | 1–2 | 7–10 | 5–6 | 2–4 | 0–2 |
| Killer whale | *Orcinus* | 0–2 | 4–6 | 3–4 | 3 | 0–2 |
| Bottlenose dolphin | *Tursiops truncatus* | 0–1 | 5–8 | 5–6 | 2–4 | 0–2 |
| Mysticetes | | | | | | |
| Bowhead whale | *Balaena* | 0–2 | 3 | 4–5 | 3 | 2 |
| Southern right whale | *Eubalaena australis* | 0–2 | 3–4 | 4–5 | 4 | 3 |
| Northern right whale | *Eubalaena glacialis* | 1–2 | 4 | 4–5 | 2–3 | 2–3 |
| Pygmy right whale | *Caperea* | Absent | 2–4 | 3–5 | 3–4 | 1–3 |
| Gray whale | *Eschrichtius* | Absent | 2–3 | 4–5 | 3–4 | 2–3 |
| Minke whale | *Balaenoptera acutorostrata* | Absent | 3 | 6–7 | 5–6 | 3 |
| Sei whale | *Balaenoptera borealis* | Absent | 3–4 | 5–7 | 4–7 | 2–4 |
| Blue whale | *Balaenoptera musculus* | Absent | 3–4 | 5–8 | 5–7 | 3–4 |
| Humpback whale | *Megaptera* | Absent | 2 | 7–8 | 6–7 | 2–3 |

scapula, allowing for greater areas of origin for the infraspinatus and supraspinatus muscles. Basilosaurids also developed a strong deltopectoral crest on the humerus for insertion of the deltoid muscle. This crest was lost in Oligocene cetaceans as the insertion of the deltoid muscle shifted to the distal and dorsal surface of the humerus. The elbow joint lost mobility as the distal end of the humerus evolved a v-shaped articular surface that locked the radius and ulna in place. Fossil evidence indicates elbow mobility was lost about 29 million years ago. It is currently unknown when wrist mobility was lost. Fossil evidence indicates the process of digital elongation, indicated by hyperphalangy, began at least 7–8 million years ago, although it may have started much earlier during the Oligocene.

## II. Sirenians

Manatees do not use their flippers as control surfaces while the animal is swimming; instead forelimbs mostly function to orient the animal and make small corrective movements during feeding, rest, or socializing. The forelimbs are the main sources of propulsion while the animal is in contact with the sea floor, in which manatees may "walk" on the sea floor by placing flippers one in front of another, or propel themselves by paddling (Hartman, 1979). Forelimb movements are supported by abundant musculature and large, rounded tendons throughout the proximal and distal limb (Murie, 1872). Manatee digits are immersed in thick connective tissue and lack the ability to abduct and adduct, but retain intrinsic muscles of the manus (abductors and interossei).

Modern sirenians (dugongs and manatees) have a slightly modified mammalian forelimb (Howell, 1930). The elbow is mobile in sirenians, and this joint motion is stabilized by a proximally fused radius and ulna. The wrist is highly mobile and lacks a pisiform carpal bone. Dugongs have three carpal elements in the wrist, while manatees have six. The manatee manus is slightly modified as digit I lacks one phalanx, digit IV is the longest and most robust, and phalanges on the ends of the digits are irregularly shaped and flattened (Fig. 1AB). Manatees also have a number of broad and flat nails on the surface of the flipper (Fig. 1b), and some captive manatees increase the number of nails.

## III Marine Carnivores

### A. Pinnipeds

Pinnipeds, unlike other marine mammals, have pairs of flippers on both the forelimbs and hindlimbs. This discussion will only address the foreflippers.

Pinnipeds are unique in that their flippers are utilized mostly in aquatic locomotion and have limited utility on land. Odobenid (walrus) forelimbs act as paddles or rudders for steering (Gordon, 1981) and are used to remove sediment when searching for prey. While on land, walrus forelimbs support the trunk by placing the digits flat and bending the wrist at a right angle. This bent forelimb morphology makes terrestrial locomotion akward. Otariid pinnipeds (English, 1976, 1977) have elongated and thin flippers that flap like

F

a bird wing to produce thrust underwater, and are used to support the trunk on land. Phocid forelimbs function solely in steering while underwater but are usually held flush with the body wall, and are not a significant source of propulsion. On land, most phocids do not use their forelimbs as a weight-bearing appendage.

Walrus (*Odobenus*) flippers are short compared to other pinnipeds, but are very broad and have tiny nails on the dorsal surface. Otariids have elongate and thin flippers with a slight crescent of skin at the ends of each digit. Phocid flippers are divided between some of the digits, and long thin nails extend beyond the dorsal surface of all five digits.

The digits of pinnipeds also have unique characteristics (Howell, 1930). All pinnipeds have elongated the digits by developing bars of cartilage at the ends of each digit. These cartilaginous extensions are longest in otariids, slightly shorter in the walrus and shortest in some phocids. Metacarpal I is longer and thicker than metacarpal II in all pinnipeds except phocines.

Pinnipeds also display large and complex forelimb muscles. The walrus has large and powerful muscles, with relatively the same sized muscle bellies as otariids.

Otariids isolate more than half of the forelimb musculature in the proximal portion of the forelimb. The triceps muscle complex is relatively large, and allows for elbow retraction. Muscles acting on the otariids wrist create palmar flexion, which is the main source of propulsion. Otariids also have muscles acting on the digits: interossei, digital abductors and adductors, and in some specimens a single lumbrical. Phocids have an enlarged triceps muscle complex.

The earliest fossil pinniped, *Enaliarctos mealsi*, already had forelimbs modified as flippers. No fossils indicate the transition between terrestrial carnivores and aquatic pinnipeds (Berta *et al.*, 1989).

## B.  Sea Otters

Sea otters (*Enhydra*) do not use their forelimbs while swimming. The forelimbs are specialized in movements requiring great dexterity: prey manipulation, grooming, and caring for young (Howard, 1973).

Sea otter forelimbs are small and retractable claws extend from each of the digits (Fig. 4). The digits cannot act individually as they are connected by soft tissue webbing. Thick pads line the palmar surfaces of digits. Forelimb musculature is well developed.

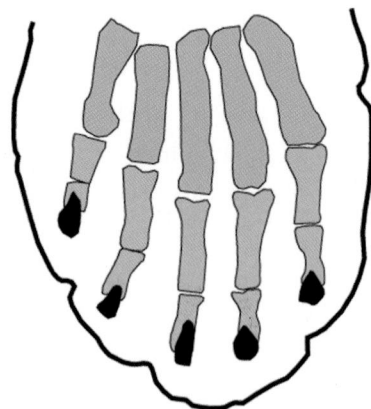

Figure 4    *Sea otter foreflipper (Howard, 1973).*

The giant extinct sea otter *Enhydritherium* was propelled by its forelimbs, but all modern sea otters are pelvic paddlers with enlarged hindlimbs.

## C.  Polar Bears

Polar bears are powerful swimmers but also walk on ice or land. The forelimbs are incredibly strong and are the main sources of propulsion while swimming, killing prey, fighting, and hauling out of the water. Alternating strokes of forelimb flexion generate propulsion while swimming and the hindlimbs trail and remain motionless. While fighting another, polar bears will stand on their hind limbs, wrap forelimbs around another and bite. To haul out of the water, the polar bear pulls itself mostly out of the water with its stong forelimbs, and uses the hindlimbs after most of the body mass is out of the water. While walking on ice or land, polar bears place the whole hand flat on the substrate.

Polar bear forelimbs are similar to other bears, except that the scapula has a narrow postscapular fossa. This fossa gives origin to the subscapularis muscle.

## *See Also the Following Article*

Skeletal anatomy

## *References*

Benke, H. (1993). Investigations on the osteology and the functional morphology of the flipper of whales and dolphins (Cetacea). *Invest. Cetacea* **24**, 9–252.

Berta, A., Ray, C. E., and Wyss, A. R. (1989). Skeleton of the oldest known pinniped, *Enaliarctos mealsi*. *Science* **244**, 60–62.

Cooper, L. N., Berta, A., Dawson, S. D., and Reidenberg, J. (2007). Evolution of digit reduction and hyperphalangy in the cetacean manus. *Anat. Rec.* **290**, 654–672.

Davis, D. D. (1949). The shoulder architecture of bears and other carnivores. *Field. Zool.* **31**, 285–305.

English, A. W. M. (1976). Functional anatomy of the hands of fur seals and sea lions. *Am. J. Anat.* **147**, 1–17.

English, A. W. M. (1977). Structural correlates of forelimb function in fur seals and sea lions. *J. Morphol.* **151**, 325–352.

Fish, F. E., and Battle, J. M. (1995). Hydrodynamic design of the humpback flipper. *J. Morphol.* **225**, 51–60.

Gordon, K. R. (1981). Locomotor behavior of the walrus (*odobenus*) *J. Zool. Lond*, **195**, 349–357.

Hartman, D. S. (1979). "Ecology and behavior of the manatee (*Trichechus manatus*)." American Society of Mammalogists, Special Publication No. 5.

Howard, L. D. (1973). Muscular anatomy of the forelimb of the sea otter (*Enhydra lutris*). *Proc. Cal. Acad. Sci.* **XXXIX**, 411–500.

Howell, A. B. (1930). "Aquatic mammals: Their adaptations to life in the water." Charles C. Thomas Press, Springfield.

Jacobsen, J. K. (2007). "Radiographs from the Humboldt State University Vertebrate Museum." Humboldt, California.

Murie, J. (1872). On the structure of the manatee (*Manatus americanus*). *Trans. Zool. Soc. London* **8**, 127–202.

Shulte, H.von W., and Smith, M.de F. (1918). The external characters, skeletal muscles, and peripheral nerves of *Kogia breviceps* (Blainville). *Bull. Amer. Mus. Nat. Hist.* **37**, 7–72.

Woodward, B. L., Winn, J. P., and Fish, F. E. (2006). Morphological specializations of baleen whales associated with hydrodynamic performance and ecological niche. *J. Morphol.* **267**, 1284–1294.

Uhen, M. D. (2004). Form, function, and anatomy of *Dorudon atrox* (Mammalia, Cetacea): An archaeocete from the middle to late Eocene of Egypt. *Univ. Mich., Pap. Paleontol.* **34**, 1–222.

# Forensic Genetics

C. Scott Baker and Merel L. Dalebout

## I. Introduction

Molecular genetics provide a powerful tool for the conservation and management of cetaceans and other marine mammals—the identification of products derived from hunting, strandings, and fisheries bycatch (Baker and Palumbi, 1994; Cipriano and Palumbi, 1999; Dalebout *et al.*, 2002a). Such products include soft tissue such as meat, organs, blubber, skin, and blood, as well as teeth, bone, baleen, and hair. The examples given here are for cetaceans, but the approaches can be applied to carnivores and sirenians. Although the species origins of these products may be impossible to determine on the basis of appearance, they contain DNA that can be amplified, sequenced, and compared to sequences from known specimens. With advances in molecular methods over the last decade, DNA can now be recovered from almost any biological source, even products that have been preserved, cooked, or canned (Asensio Gil, 2007). With a comprehensive reference library of homologous sequences, such as the control region or cytochrome *b* gene of the mitochondrial (mt) DNA, a product of unknown origin can be attributed in most cases to one of the approximately 90 accepted or proposed species of cetaceans (Baker *et al.*, 2003). If a comprehensive archive of tissue is maintained as part of a regulated hunt or documented fisheries bycatch, it is also possible to trace the origins of a product to a specific individual by matching of nuclear DNA genotypes (Cipriano and Palumbi, 1999; Dalebout *et al.*, 2002b; Palsbøll *et al.*, 2006). Although many of the applications of these methods are not intended for criminal prosecution, they share the common methodology of wildlife forensic genetics (US Fish and Wildlife Service, 2001) and the broader objectives of improving controls over trade and exploitation of protected species.

The forensic use of molecular genetic methods is of particular interest to the International Whaling Commission (IWC), as it attempts to develop a Revised Management Scheme (RMS) for the regulation of any future commercial whaling, and to the Conventional on International Trade in Endangered Species of Wild Fauna and Flora (CITES), as it attempts to implement a verifiable system for controlling trade in cetacean products. An important application of forensic genetics has been to identify the species and, in some cases, geographic origins of whale, dolphin, and porpoise products sold in two countries with active commercial markets: Japan and the Republic of (South) Korea (Baker and Palumbi, 1994; Baker *et al.*, 1996; Grohman *et al.*, 1999; Simmonds *et al.*, 2002). Of particular, concern has been the sale of protected species or populations (stocks) derived from illegal, unreported, and unregulated (IUU) exploitation (Baker *et al.*, 2000b, c; Baker *et al.*, 2002). Other applications include identifying stranded individuals and fisheries bycatch, particularly for poorly described species such as beaked whales (Henshaw *et al.*, 1997; Dalebout *et al.*, 1998; Dalebout *et al.*, 2004), and monitoring of trade in pinniped penises sold as aphrodisiacs (Malik *et al.*, 1997). Most recently, molecular identification of species and capture–recapture analysis of DNA genotyping from individual products have been used to estimate the true level of bycatch for some species sold in commercial markets (Baker *et al.*, 2006; Baker *et al.*, 2007).

## II. Molecular Taxonomy and Identification of Cetacean Species

The methods for molecular identification of species in trade developed initially from basic research on species-level phylogenetic relationships and the genetic structure of populations (Baker and Palumbi, 1994 ; Baker *et al.*, 1996, 1994; DeSalle and Birstein, 1996; Malik *et al.*, 1997; Roman and Bowen, 2000; Shivji *et al.*, 2002). More recently, there has been an explosion of interest in the systematic application of these techniques to basic organismal taxonomy (Hebert *et al.*, 2003; Tautz *et al.*, 2003; Blaxter, 2004), including cetaceans (Baker *et al.*, 2003; Ross *et al.*, 2003; Dalebout *et al.*, 2004). Now referred to as "molecular taxonomy" or "DNA taxonomy," the objective of identifying known species from a designated homologous gene sequence differs from the usual goal of molecular phylogenetics, which is more concerned with hierarchical relationships above the species level. For species identification of cetaceans and other animal species, the molecular marker of choice has usually been mitochondrial DNA (mtDNA). In general, mtDNA offers two important advantages over nuclear genetic markers. First, because of its maternal inheritance and absence of recombination, the phylogenetic relationship of mtDNA sequences reflects the history of maternal lineages within a population or species. (If hybridization is encountered, nuclear markers are required to identify the paternal species; discussed later). Second, all else being equal, the effective population size of mtDNA genomes is one-fourth that of autosomal nuclear genes and their rate of random genetic drift is proportionately greater. This results in more rapid differentiation of mtDNA lineages among populations, compared to nuclear genes, and consequently greater sensitivity in the detection of recent historical demographic or speciation events. The ability to detect population differentiation is also be enhanced by the rapid pace of mtDNA evolution, which is generally estimated to be five to ten times faster than nuclear coding DNA in most species of mammals.

Although one approach to molecular taxonomy has advocated a universal "DNA barcode of life" for all animal species based on the mtDNA cytochrome *c* oxidase I gene (COI) (Hebert *et al.*, 2003), it is not clear that this locus is the most sensitive or reliable for identification of cetaceans (Amaral *et al.*, 2007). Instead, species-level identification of marine mammals has relied primarily on the phylogenetic reconstruction of sequences from the mtDNA control region (sometimes referred to as the D-loop) or the cytochrome *b* gene. The control region of the mtDNA does not code for a protein or RNA and, in the absence of these constraints, accumulates mutational substitutions more rapidly than other regions. For this reason, it has become the marker of choice for most studies of the population structure of cetaceans and pinnipeds. The cytochrome *b* gene, a protein region of the mtDNA, has also been used widely for species-level phylogenetics (Arnason and Gullberg, 1996; LeDuc *et al.*, 1999) and in some cases, for population structure of marine mammals (Lento *et al.*, 1997). Because of the large number of reference sequences available on public databases such as *GenBank* and *EMBL*, both loci have been used for species-level identification of marine mammals.

The basic steps involved in the phylogenetic identification of an unknown specimen or market product are illustrated in Fig. 1. First, mtDNA is extracted from the product in question, such as a flensed piece of skin and blubber from a commercial market. Second, a fragment of the mtDNA control region is amplified from the product via PCR (Saiki *et al.*, 1988)—a cyclic, *in vitro* enzymatic reaction that results in the exponential replication of a small targeted fragment

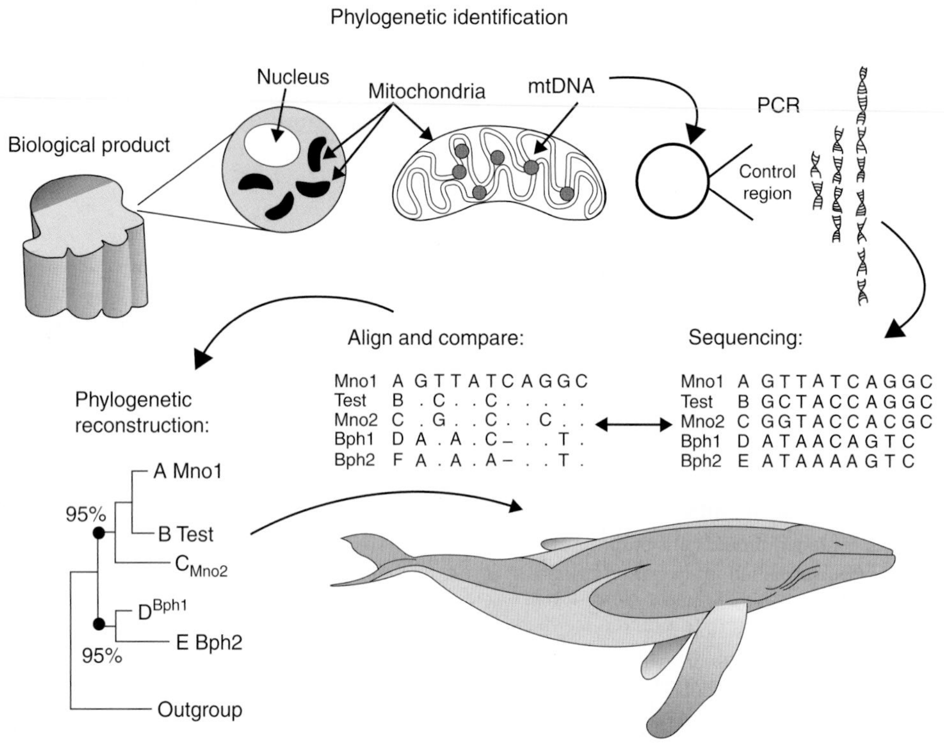

Phylogenetic identification

**Figure 1** *The basic steps involved in species identification of an unknown cetacean product using of nucleotide sequences amplified, by PCR, from the mtDNA control region. The "test sequence" derived from an unknown species origin is aligned and compared to a comprehensive reference database of sequences from specimens of known provenance, such as that available on the web-based program, www.DNA-surveillance.*

of mtDNA (usually <1000 base pairs). Third, the exact nucleotide sequence of the amplified fragment is determined using a dideoxy-terminator sequencing reaction, followed by electrophoresis through an acrylamide gel. In most laboratories, this step is now automated with a computer-assisted laser scanner. Fourth, the sequence of the product, now referred to as the "test sequence," is aligned to and compared with the sequences from reference samples. For protein coding regions, such as cytochrome *b* or COI, alignment is unambiguous among cetaceans (or other mammals) because of the absence of insertions or deletions (indels) in the codon sequences. For the noncoding CR, however, the presence of numerous single- and multiple-base indels often requires an automated alignment procedure and manual confirmation to optimize identification.

Finally, the "test sequence" is grouped with the most closely related reference sequences in the reference database using phylogenetic reconstruction methods, such as minimum evolution (neighbor joining), maximum parsimony, or maximum likelihood. The reconstruction is usually represented as a "tree," with closely related sequences forming neighboring branches. This approach allows a hierarchical comparison to establish first the suborder and family derivation using a small number of reference sequences from each of a large number of species. Once the family or subfamily is established, the phylogenetic reconstruction of the test sequence can be repeated using a larger number of reference sequences to better represent the diversity within each of the smaller number of species at this taxonomic level. A close relationship or match with a "reference" sequence provides evidence for identification of the species origin of the product. One or more "out-groups" (i.e., distantly related species) are used to protect

against a misclassification error. The strength of support for an identification or phylogenetic grouping is evaluated by "bootstrap" resampling of the sequence data. The relative support for a grouping or branch in the tree is shown as the percentage agreement from a large number (>1000) of bootstrap simulations.

As a conservative approach to forensic identification, a species identification should be considered "confirmed" only if the test sequence is "nested" within the range of reference sequences for a given species. This is necessary because the molecular systematics of some marine mammals, particularly the cetaceans, are not fully described (Reeves *et al.*, 2004; discussed later). If a test sequence is intermediate between two groups of reference sequences, rather than nested within one or the other, it could be a related species or subspecies not included in the reference database. Where a reference database is considered to be comprehensive in regards to known species, the finding of a particularly divergent sequence could be evidence of an unknown or unrecognized species or subspecies (Baker *et al.*, 1996; Dalebout *et al.*, 2004; Dalebout *et al.*, 2007).

When a large set of reference sequences is available from the range of a single species, it is possible to use intraspecific variation to evaluate the geographic origin of a sample (Baker *et al.*, 2000a). In many cases, the management of marine mammals is based on geographic populations or stocks (Dizon *et al.*, 1992). Catch quotas and limits of incidental mortality from fisheries bycatch are usually set according to such stock definitions, as well as according to species. Hunting may be allowed in an abundant stock but prohibited in another stock of the same species that is depleted from past exploitation. However, the ability to identify or estimate the geographic origin of a specimen or product is determined by the genetic distinctiveness of the recognized stocks, as well as by the comprehensiveness of the reference samples.

## III. Web-Based Species Identification with www.DNA-surveillance

To assist in the genetic identification of whales, dolphins, and porpoises, an interactive application for phylogenetic identification has been developed and is accessible through the website, http://www.DNA-surveillance (Ross *et al.*, 2003). *DNA Surveillance* (2008) implements phylogenetic methods for identification of species within a particular taxonomic group, such as the currently available data-sets for whales, dolphins, and porpoises. The application automates the procedure of species identification by aligning a user-submitted gene sequence of unknown origin against a set of validated reference sequences. The evolutionary distances between the unknown or "test" sequence and each of the reference sequences is computed and a phylogenetic tree displays the affinity of the unknown sequence to the reference sequences.

*DNA Surveillance* differs in several important ways from the blast search options available on the website of the international genetic database, *GenBank* (Ross and Murugan, 2006). The problems associated with using *GenBank* for species identification are particularly relevant to cetaceans, where the primary taxonomic identification of the voucher specimen can be ambiguous or incorrect (Henshaw *et al.*, 1997; Dalebout *et al.*, 1998). Sequences entered in *GenBank* are not curated and often are not associated with identifiable reference or voucher specimen material. The taxonomic representation of a BLAST search is difficult to judge because of the large number of redundant gene sequences for some species, the absence of sequences from other closely related species and the nature of the pair-wise alignment and search algorithm. BLAST and related search engines seek only locally maximal matches in pair-wise comparisons. The extreme (E) value associated with each sequence hit in a BLAST search is not a rigorous measure of evolutionary distance or genetic similarity and is dependent on the size of the database being searched (Karlin and Altschul, 1990). Inconsistent application of keywords also reduces the power of searching *GenBank* by fields, impeding effective data mining. By contrast, *DNA Surveillance* is designed specifically for species identification of selected taxonomic groups. The reference sequences in *DNA Surveillance* are prealigned at each hierarchical level of the database, using a mixture of algorithmic and manual methods, to create an optimized alignment. The sequences in *DNA Surveillance* were chosen to reflect known phylogenetic diversity at the species and population level (where available). The genetic distances and trees in *DNA Surveillance* are calculated using standard phylogenetic algorithms, as implemented in the Phylogenetic Algorithms Library (Drummond and Strimmer, 2001).

The reference data sets mounted on *DNA-surveillance* comprise sequences from both the mtDNA control region and cytochrome *b* gene. Reference sequences were selected to reflect the generic, specific, or geographic diversity observed at a taxonomic level and to maximize the discriminatory power of the analysis. In an effort to validate the dataset, most sequences were derived from specimens that had been identified by experts and for which diagnostic skeletal material or photographic records were collected (Dizon *et al.*, 2000). Data sets are arranged hierarchically, allowing initial family-level identification of cetaceans and subsequently more detailed analysis within the suborders Mysticeti (baleen whales) and Odontoceti (toothed whales). The datasets currently mounted on *DNA-surveillance* at this writing are taxonomically comprehensive, with a total of 399 control region sequences and 264 cytochrome *b* sequences representing 88 species (DNA-surveillance, 2008). Sequences from documented specimens represent all of the 83 species recognized by Rice (Rice, 1998), with two exceptions: the Atlantic hump-backed dolphin, *Sousa teuszii*, and the Indian hump-backed dolphin *S. plumbea* (the latter of which has not been accepted by IWC). The datasets also includes one species found in an alternate species listings, *Platanista minor* (IUCN Red Book), and seven species accepted, revised, or proposed in publications since Rice (1998): *Balaenoptera omurai* (Wada *et al.*, 2003), *Eubalaena australis* and *E. japonicus* (Rosenbaum *et al.*, 2000), *Mesoplodon perrini* (Dalebout *et al.*, 2002b), *M. traversii* (van Helden *et al.*, 2002), *Orcaella heinsohni* (Beasley *et al.*, 2005), and *Sotalia guianensis* (Caballero *et al.*, 2007). Both the control region and cytochrome *b* datasets include sequences from multiple specimens for most species.

In a typical analysis, the user copies a DNA sequence (in FASTA or text format) into a data input window and chooses the appropriate reference data set. The test sequence is aligned by a simple profile alignment against the pre-aligned data set of reference sequences. The user can also choose that a computationally more intensive full alignment of the test and reference sequences be performed as part of the analyses. A neighbor-joining (NJ) tree is built from the table of evolutionary distances (Saitou and Nei, 1987) and rooted using an out-group appropriate for each data set. The phylogenetic tree, in both graphic and text format, and a table of distances are displayed and can be downloaded to the user. An optional bootstrap analysis (Felsenstein,1985) can be performed to assess the robustness of the resulting phylogenetic tree.

## IV. Taxonomic Uncertainties and Species Identification

Problems in molecular identification of cetacean species can occur if taxonomic sampling is incomplete (missing species) or within-species sampling is not sufficiently representative of diversity. In cases of deep intraspecific diversity or shallow interspecific divergence, an unknown test sequence could group with the next most closely related species as a result of such sampling error. For this reason, it is important that levels of genetic diversity within, and divergence between species in a group of interest are assessed as part of the development of a molecular taxonomy (Dalebout *et al.*, 2007). As a conservative approach, Baker *et al.* (1996) suggested that identification of a test sequence should be considered conclusive only if it nests within the diversity of reference sequences for a given species. In practice, this is less crucial if phylogenetic support for the species-level grouping is strong and the taxonomic sampling is known to be complete.

More problematic are cases where the phylogenetic reconstruction of mtDNA sequences is not concordant with accepted species taxonomy described from morphological characteristics. Instead of the expected pattern of species-specific monophyly, where all mtDNA lineages (sequences) from a given species group with each other, some lineages group with another species (paraphyly) or fail to form species-specific groups (polyphyly). The reasons for species-level paraphyly or polyphyly of mtDNA are varied and include recent hybridization as well as incomplete lineage sorting due to recent speciation (Funk and Omland, 2003). If species or species complexes are truly paraphyletic, it is unlikely that a phylogenetic approach to "species" identification will be successful. Instead, molecular identification can only to be made with confidence to a higher taxonomic

F

rank within which mtDNA lineages are monophyletic (e.g., genus or subfamily). Although species-level monophyly of mtDNA has been demonstrated for many species of cetaceans, including the beaked whales (Dalebout *et al.*, 2004; Dalebout *et al.*, 2007) and baleen whales (Baker *et al.*, 1993; Rosenbaum *et al.*, 2000; Wada *et al.*, 2003), apparent paraphyly is reported for some species of the family Delphininae, particularly in the genera *Stenella*, *Tursiops*, and *Delphinus* (Dizon *et al.*, 2000; Reeves *et al.*, 2004).

## V. Individual Identification and a Diagnostic "DNA Register"

An alternative to the species identification of an unknown specimen or product is individual identification by DNA genotyping or "profiling" using variable nuclear markers. As in human forensic genetics, a combination of variable nuclear markers (such as microsatellites, Single Nucleotide Polymorphisms (SNPs), or nuclear introns) can be used to establish individual identity with high probability (or to exclude identity with certainty, barring experimental error). The DNA profile of the market product can be compared to that from archived tissue collected in a regulated hunt or documented bycatch for verification of trade records. One of the first efforts to track the individual identity of a whale in trade involved a product from the Japanese market, identified initially as a blue whale, *Balaenoptera musculus*. The mtDNA sequence from this product matched closely with the published sequence of a blue/fin (*B. physalus*) hybrid killed during a scientific whaling program by Iceland. Because mtDNA is maternally inherited, it cannot, by itself, identify a product as a hybrid. Subsequent comparison of variable nuclear DNA introns from tissue archived during the Icelandic whaling program supported the assumption that this product was derived from this hybrid individual (Cipriano and Palumbi, 1999).

In response to concerns about the continued sale of protected species and the poor control of whale-meat markets, the Government of Norway initiated a program to DNA-profile all whales taken in its commercial hunt (IWC, 1998). The DNA profiles of each individual whale are stored on an electronic database, forming a "DNA register" of all products intended for the market. If the register is comprehensive or "diagnostic," a match with a market product would confirm the legality of the product (Dizon *et al.*, 2000). A product that did not have a match in the register would be illegal. Further genetic investigation would then be required to determine the species and geographic origin of illegal products. The Governments of Japan has also committed to the development of DNA registers as part of its ongoing programs of scientific whaling and intends to include the bycatch of baleen whales destined for sale in commercial markets (IWC, 2005b). The effectiveness of the Norwegian DNA register was tested recently with products from North Atlantic common minke whales (*Balaenoptera acutorostrata*) purchased at Norwegian markets. The results demonstrate the matching of the test profiles to the register, confirming the potential power of the DNA registers, but highlighted a number of methodological problems that need to be addressed to ensure successful implementation for control of trade (Palsbøll *et al.*, 2006).

Individual identification of market products can also be a powerful tool for describing market dynamics even in the absence of a DNA register or official tissue archive. Dalebout *et al.* (2002b) used mtDNA sequences to identify minke whale products sold in Japanese and Korean markets and subsequent DNA profiling to identify replicate products derived from the same individual. Many of these products in both countries were derived from whales taken as unregulated bycatch (discussed later). Individual identification provided information

on distribution of products and a minimum "census" of the true number of takes in this bycatch. More recently, Baker *et al.* (2007) expanded on this work in the Korean market, using a modified capture–recapture model based on DNA profiles, to estimate the true number of whales in trade (reflecting the true number killed) over a 5-year period from 1999 to 2003 (discussed later).

## VI. Monitoring of Commercial Markets in Whale, Dolphin and Porpoise Products

In recognition of historic patterns of over-exploitation, the International Whaling Commission (IWC) voted in 1982 to impose a global moratorium on commercial whaling. Although the moratorium took effect in 1986, whaling never actually stopped. IWC member nations continue to hunt some species of whales for scientific research or for aboriginal and subsistence use. Whales killed for scientific research can be sold legally to domestic consumers and traded to other member nations of the IWC (subject to CITES permits), thereby sustaining a commercial market for meat, skin, blubber, and other whale products. Small cetaceans are also hunted or taken as fisheries bycatch and sold for consumption in many parts of the world (Clapham and Van Waerebeek, 2007). Although the IWC regulates only hunting of large whales, international trade in all cetaceans is subject to CITES. When some species are protected by an international prohibition against hunting or trade but similar species are not, it is crucial to identify the origin of products that are actually sold in retail markets.

In an effort to monitor the sale and trade of cetaceans products, molecular methods have been used to identify the species and geographic derivation of products sold in two countries with active commercial markets: Japan and the Republic of (South) Korea. Whale meat is widely available in retail markets of both countries despite the international moratorium on commercial whaling (Chan *et al.*, 1995; Mills *et al.*, 1997; Kang and Phipps, 2000). Japan sustains a legal market for whale products through its growing scientific whaling programs in the Southern Hemisphere and the North Pacific Ocean (Gales *et al.*, 2005). South Korea has no program for scientific hunting but reports a substantial fisheries bycatch of cetaceans each year, including minke and other baleen whales (Mills *et al.*, 1997). Products from this unregulated incidental mortality are sold in local markets but their international trade is prohibited by CITES.

Surveys of whale-meat markets conducted, since 1993 in Japan and 1994 in Korea, have employed both species identification and individual identification to detect the sale of protected species and assess the true take of species in unregulated bycatch or by illegal hunting. As summarized in the year 2000 (Baker *et al.*, 2000b), surveys of Japanese markets have revealed numerous cases of protected species of large whales including sperm whales (*Physeter macrocephalus*), fin whales, blue/fin whale hybrids, two species of Bryde's whales (*B. edeni* and *B. brydei*, following the taxonomy of Wada *et al.*, 2003), sei whales (*B. borealis*), humpback whales (*Megaptera novaeangliae*), and gray whales (*Eschrichtius robustus*). With the expansion of the Japanese scientific programs since 2000, however, some formerly protected species are now included in this hunt and regularly available on commercial markets.

Market surveys have also provided information on the diversity of small cetacean products available for sale. In Japan (Endo *et al.*, 2005), a total of 160 "small cetacean" products sold for human consumption in markets from 2000 to 2003 were identified as originating from seven species of the family Delphinidae, one species of beaked whale (*Berardius bairdii*) and one species of porpoise

(*Phocoenoides dalli*). In Korea (Baker *et al.*, 2006), a total of 357 whale-meat products, purchased from late 2003 to early 2005, were identified as originating from 15 species of cetaceans: three baleen whales (North Pacific minke, common form Bryde's and humpback), three beaked whales (Stejneger's beaked whale, *Mesoplodon stejnegeri*; Cuvier's beaked whale, *Ziphius cavirostris*; and Blainville's beaked whale, *Mesoplodon densirostris*), seven species of the family Delphinidae (bottlenose dolphin, *Tursiops truncatus*; Risso's dolphin, *Grampus griseus*; short-beaked common dolphin, *Delphinus delphis*; Pacific white-sided dolphin, *Lagenorhynchus obliquidens*; false killer whale, *Pseudorca crassidens*; killer whale, *Orcinus orca*; short-finned pilot whale, *Globicephala macrorhynchus*), and two porpoises (harbour porpoise, *Phocoena phocoena*; finless porpoise, *Neophocaena phocaenoides*).

Detailed comparisons of mtDNA sequences and individual identification by DNA profiling have provided information on high levels of unregulated exploitation of minke whales in coastal water of Japan and Korea. The North Pacific minke whale forms at least two stocks with marked differences in frequencies of mtDNA haplotypes(Goto and Pastene, 1997): the "J" stock found in the Sea of Japan/East Sea, and the "O" stock found in the North Pacific to the east of Japan. Although the "O" stock is subject to legal scientific hunting by Japan and is reported to be relatively abundant, the "J" stock was depleted by commercial hunting before 1986 and is considered a "Protection Stock" by the IWC. Using molecular methods and mixed-stock analysis, market surveys from 1993 to 1999 showed a large proportion of products from Japan were derived from the protected "J" stock despite relatively low numbers in official reports of fisheries bycatch (Baker *et al.*, 2000a).

Surveys of Korean markets have raised similar concerns about exploitation of the "J" stock minke whales. The sale of minke whales reportedly taken as incidental bycatch supports a thriving trade in whale products concentrated in three coastal cities along the southeastern coast of the Korean peninsula: Busan, Ulsan, and Pohang (Kang and Phipps, 2000). As trade in whale products is unregulated, the dynamics of market distribution are not well described (IWC, 2006a). Available information suggests that fishermen negotiate the sale of bycatch informally through a network of perhaps 10 wholesalers operating in these three cities. Whale products are sold in numerous small shops and restaurants in or around Busan, Ulsan, and Pohang, including speciality whale-meat restaurants and large fisheries markets (IWC, 2006). Given the high commercial value of whale and dolphin products (reportedly up to US$100,000 wholesale for an adult minke whale), there is considerable incentive to enhance the potential for bycatch through modified fishing practices, similar to that of traditional "net whaling". Although the Government of South Korea reports relatively large numbers of minke whale as bycatch in its annual progress report to the IWC, market surveys indicate that these records are incomplete, perhaps due to substantial levels of illegal hunting (IWC, 2005a). A capture–recapture analysis of individual market products purchased during market surveys from 1999 to 2003 (discussed earlier, Baker *et al.*, 2007) estimated that the true number of whales entering trade across the 5-year survey period was 827 individuals (CV = 0.24), significantly greater than the officially reported bycatch of 458 whales for this period. Considering results from surveys of both Japanese and Korean markets, the estimated true levels of illegal, unreported, or unregulated exploitation has serious implications for the survival of this genetically distinct coastal population.

## VII. Conclusions

Molecular methods have great power to detect trade in protected species and to monitor or estimate unregulated or undocumented trade in whales, dolphins, and porpoises. Efforts to improve monitoring and detection of IUU exploitation of cetaceans and control trade in cetacean products would be enhanced greatly by the establishment of diagnostic "DNA registers" (Dizon *et al.*, 2000). The Governments of Japan and Norway have both committed to the development of DNA registers as part of their ongoing programs of scientific or commercial whaling, and in the case of Japan, the effort is intended to include the bycatch of baleen whales destined for sale in commercial markets (IWC, 2005b). Korea has made efforts to improve the collection of biological samples from the bycatch of baleen whales (IWC, 2006), although it has not committed to develop a DNA register. No country has yet committed to develop a register for small cetaceans, even though products from directed hunting and bycatch of these species are often destined for commercial markets (Clapham and Van Waerebeek 2007).

Given the commitment to DNA registers by Japan and Norway, it is puzzling that the governments of both countries oppose implementation of market surveys as a component of any system of observation and monitoring of future whaling. Formal statements by both countries claim that the IWC has no competency in market monitoring (IWC, 2001a, b), although methods for market surveys to estimate bycatch and other human induced mortality have been under discussion at the IWC for several years (IWC, 2003). Assuming a continuation of this political opposition, it is likely the future market surveys will have to follow an "empirical" approach, similar to that advocated for surveys of wild-meat markets (Fa *et al.*, 2004) and including methods of estimation typically used in the molecular ecology of living populations Baker (2008). For countries that regulate hunting or keep official records of bycatch, an empirical approach is likely to require more effort and to yield less precise estimates than combining market surveys with a diagnostic DNA register (IWC, 2006). For countries such as China, Taiwan, Indonesia, and the Philippines, where trade in whale and dolphin products is known or suspected (Barnes,1991; Dolar *et al.*, 1994; Mills *et al.*, 1997) but which keep few records of hunting or bycatch, an empirical approach will be the only option available for monitoring exploitation in the foreseeable future.

## See Also the Following Articles

Hunting of Marine Mammals ■ Genetics, Overview ■ International Whaling Commission

## References

Amaral, A. R., Sequeira, M, and Coelho, M. M. (2007). A first approach to the usefulness of cytochrome *c* oxidase I barcodes in the identification of closely related delphinid cetacean species. *Mar. Freshw. Res.* **58**, 505–510.

Árnason, U., and Gullberg, A. (1996). Cytochrome *b* nucleotide sequences and the identification of five primary lineages of extant cetaceans. *Mol. Biol. Evol.* **13**, 407–417.

Asensio Gil, L. (2007). PCR-based methods for fish and fishery products authentication. *Trends Food Sci.Tech.* **18**, 558–566.

Baker, C.S. (2008). Invited review: A truer measure of the market: the molecular ecology of fisheries and wildlife trade. *Mol. Ecol.* in press.

Baker, C. S., and Palumbi, S. R. (1994). Which whales are hunted? A molecular genetic approach to monitoring whaling. *Science* **265**, 1538–1539.

Baker, C. S., *et al.* (13 authors) (1993). Abundant mitochondrial DNA variation and world-wide population structure in humpback whales. *Proc. Nat. Acad. Sci. USA*, **90**, 8239–8243.

Baker, C. S., Cipriano, F., and Palumbi, S. R. (1996). Molecular genetic identification of whale and dolphin products from commercial markets in Korea and Japan. *Mol. Ecol.* **5**, 671–685.

Baker, C. S., Lento, G. L., Cipriano, F., and Palumbi, S. R. (2000a). Predicted decline of protected whales based on molecular genetic monitoring of Japanese and Korean markets. *Proc. Roy. Soc. Lond. B* **267**, 1191–1199.

Baker, C. S., Lento, G. M., Cipriano, F., Dalebout, M. L., and Palumbi, S. R. (2000b). Scientific whaling: Source of illegal products for market? *Science* **290**, 1695–1696.

Baker, C. S., Lento, G. M., Cipriano, F., and Palumbi, S. R. (2000c). Predicted decline of protected whales based on molecular genetic monitoring of Japanese and Korean markets. *Proc. Roy. Soc. Lond. B* **267**, 1191–1199.

Baker, C. S., Dalebout, M. L., Lento, G. M., and Funahashi, N. (2002). Gray whale products sold in commercial markets along the Pacific Coast of Japan. *Mar. Mamm. Sci.* **18**, 295–300.

Baker, C.S., Dalebout, M. L., Lavery, S. and Ross, H. A. (2003). www. DNA-surveillance: Applied molecular taxonomy for species conservation and discovery. *Trend. Ecol. Evol.* **18**, 271–272.

Baker, C. S., Lukoschek, V., Lavery, S., Dalebout, M. L., Yong-un, M., Endo, T., and Funahashi, N. (2006). Incomplete reporting of whale, dolphin and porpoise "bycatch" revealed by molecular monitoring of Korean markets. *An. Cons.* **9**, 474–482.

Baker, C. S., *et al.* (8 authors) (2007). Estimating the number of whales entering trade using DNA profiling and capture-recapture analysis of market products. *Mol. Ecol.* **16**, 2617–2626.

Barnes, R. H. (1991). Indigenous whaling and porpoise hunting in Indonesia. *In* "Cetaceans and Cetacean Research in the Indian Ocean Sanctuary" (S. Leatherwood, and G. P. Donovan, eds), 3, pp. 99–106. UN Env. Prog., Nairobi, Mar. Mamm. Tech. Rep.

Beasley, I., Robertson, K. M., and Arnold, P. (2005). Description of a new dolphin, the Australian snubfin dolphin *Orcaella heinsohni* sp. n. (Cetacea, Delphinidae). *Mar. Mamm. Sci.* **21**, 365–400.

Blaxter, M. L. (2004). The promise of a DNA taxonomy. *Phil. Tran. Roy. Soc. B* **359**, 669–679.

Caballero, S., *et al.* (2007). Taxonomic status of the genus *Sotalia*: Species level ranking for "tucuxi" (*Sotalia fluviatilis*) and "costero" (*Sotalia guianensis*) dolphins. *Mar. Mamm. Sci.* **23**, 358–386.

Chan, S., Ishihara, A., Lu, D. J., Phipps, M., and Mills, J. A. (1995). Observations on the whale meat trade in East Asia. *Traffic Bull.* **15**, 107–115.

Cipriano, F., and Palumbi, S. R. (1999). Genetic tracking of a protected whale. *Nature* **397**, 307–308.

Clapham, P. J., and Van Waerebeek, K. (2007). Bushmeat and bycatch: The sum of the parts. *Mol. Ecol.* **16**, 2607–2609.

Dalebout, M. L., Van Helden, A., Van Waerebeek, K., and Baker, C. S. (1998). Molecular genetic identification of southern hemisphere beaked whales (Cetacea: Ziphiidae). *Mol. Ecol.* **7**, 687–695.

Dalebout, M. L., Lento, G. M., Cipriano, F., Funahashi, N., and Baker, C. S. (2002a). How many protected minke whales are sold in Japan and Korea? A census by DNA profiling. *An. Cons.* **5**, 143–152.

Dalebout, M. L., Mead, J. G., Baker, C. S., Baker, A. N., and Van Helden, A. L. (2002b). A new species of beaked whale *Mesoplodon perrini* sp. n. (Cetacea: Ziphiidae) discovered through phylogenetic analyses of mitochondrial DNA sequences. *Mar. Mamm. Sci.* **18**, 577–608.

Dalebout, M. L., Baker, C. S., Cockroft, V. G., Mead, J. G., and Yamada, T. K. (2004). A comprehensive molecular taxonomy of beaked whales (Cetacea: Ziphiidae) using a validated mitochondrial and nuclear DNA database. *J. Hered.* **95**, 459–473.

Dalebout, M. L., *et al.* (9 authors) (2007). A divergent mtDNA lineage among Mesoplodon beaked whales: Molecular evidence for e new species in the tropical Pacific? *Mar. Mamm. Sci.* **23**, 954–966.

DeSalle, R., and Birstein, V. J. (1996). PRC identification of black caviar. *Nature* **381**, 197–198.

Dizon, A. E., Lockyer, C., Perrin, W. F., DeMaster, D. P., and Sisson, J. (1992). Rethinking the stock concept: A phylogeographic approach. *Cons. Biol.* **6**, 24–36.

Dizon, A., Baker, S., Cipriano, F., Lento, G., Palsbøll, P., and Reeves, R. (2000). Molecular genetic identification of whales, dolphins, and porpoises: Proceedings of a workshop on the forensic use of molecular techniques to identify wildlife products in the marketplace. *US Dep. Com. NOAA Tech. Mem. NOAA-TM-NMFS-SWFSC-***286**.

DNA Surveillance. (2008). http://www.DNA-surveillance.auckland. ac.nz:9000/.

Dolar, M. L. L., Leatherwood, S. J., Wood, C. J., Alava, M. N. R., Hill, C. L., and Aragones, L. V. (1994). Directed fisheries for cetaceans in the Philippines. *Rep. Int. Whal. Commn.* **44**, 439–449.

Drummond, A., and Strimmer, K. (2001). PAL: An object-oriented programming library for molecular evolution and phylogenetics. *Bioinformatics* **17**, 662–663.

Endo, T., Haraguchi, K., Hotta, Y., Hisamichi, Y., Lavery, S., Dalebout, M. L., and Baker, C. S. (2005). Total mercury, methyl mercury and selenium levels in the red meat of small cetaceans sold for human consumption in Japan. *Env. Sci. Tech.* **39**, 5703–5708.

Fa, J. E., Johnson, P. J., Dupain, J., Lapuente, J., Koster, P., and Macdonald, D. W. (2004). Sampling effort and dynamics of bushmeat markets. *An. Cons.* **7**, 409–416.

Felsenstein, J. (1985). Confidence limits on phylogenies: An approach using the bootstrap. *Evolution* **39**, 783–791.

Funk, D. J., and Omland, K. E. (2003). Species-level paraphyly and polyphyly: Frequency, causes, and consequences, with insights from animal mitochondrial DNA. *Ann. Rev. Ecol. Evol. Syst.* **34**, 397–423.

Gales, N. J., Kasuya, T., Clapham, P. J., and Brownell, L. R., Jr. (2005). Japan's whaling plan under scrutiny. Useful science or unregulated commercial whaling? *Nature* **435**, 883–884.

Goto, M., and Pastene, L. A. (1997). Population structure in the western North Pacific minke whale based on an RFLP analysis of the mtDNA control region. *Rep. Int. Whal. Commn.* **47**, 531–538.

Grohman, L., *et al.* (1999). Whale meat from protected species is still sold on Japanese markets. *Naturwissenschaften* **86**, 350–351.

Hebert, P. D. N., Cywinska, A., Ball, S. L., and DeWaard, J. R. (2003). Biological identifications through DNA barcodes. *Proc. Roy. Soc. Lond. B* **270**, 313–321.

van Helden, A. L., Baker, A. N., Dalebout, M. L., Reyes, J. C., Van Waerebeek, K. V., and Baker, C. S. (2002). Resurrection of *Mesoplodon traversii* (Gray, 1874), senior synonym of *M. bahamondi* Reyes, Van Waerebeek, Cárdenas and Yañez, 1995 (Cetacea: Ziphiidae). *Mar. Mamm. Sci.* **18**, 609–621.

Henshaw, M. D., LeDuc, R. G., Chivers, S. J., and Dizon, A. E. (1997). Identifying beaked whales (Family Ziphiidae) using mtDNA sequences. *Mar. Mamm. Sci.* **13**, 487–495.

IWC (1998). Report of the Scientific Committee, Annex Q. Report of the working group on proposed specifications for a Norwegian DNA database register for minke whales. *Rep. Int. Whal. Commn.* **49**, 287–289.

IWC (2001a). Annex I: Report of the working group on stock definition, Appendix 2 Statement from the Government of Japan concerning DNA identification and tracking of whale products. *J. Cetacean Res. Manag.* **3**(Supp.), 237.

IWC (2001b). Annex I: Report of the working group on stock definition, Appendix 3 Statement from the Government of Norway concerning DNA identification and tracking of whale products. *J. Cetacean Res. Manag.* **3**(Supp.), 237.

IWC (2003). Report of the Scientific Committee, Annex M: Report of the Sub-Committee on estimation of bycatch and other human-induced mortality. *J. Cetacean Res. Manag.* **5**(Supp), 392–401.

IWC (2005a). Annex I, report of th e Infractions Sub-Committee: Appendix 3, Summary of infractions reports received by the Commission in 2003. *Ann. Rep. Int. Whal. Commn.* **2004**, 110.

IWC. (2005b). Report of the Specialist Group on the DNA Register/Market Sampling Scheme Approach (SGDNA). IWC meeting document IWC/M05/RMSWG 5.

IWC (2006). Report of the Scientific Committee, Annex J: Working group on estimation of bycatch and other human-induced mortality (BC). *J. Cetacean Res. Manage.* **8**(Supp.), 177–184.

Kang, S., and Phipps, M. (2000). "A survey of whale meat markets along South Korea's coast." TRAFFIC East Asia.

Karlin, S., and Altschul, S. F. (1990). Methods for assessing the statistical significance of molecular sequence features by using general scoring schemes. *Proc. Nat. Acad. Sci. USA* **87**, 2264–2268.

LeDuc, R. G., Perrin, W. F., and Dizon, A. E. (1999). Phylogenetic relationships among the delphinid cetaceans based on full cytochrome *b* sequences. *Mar. Mamm. Sci.* **15**, 619–648.

Lento, G. M., Haddon, M., Chambers, G. K., and Baker, C. S. (1997). Genetic variation, population structure and species identity of Southern Hemisphere fur seals, *Arctocephalus* spp. *J. Hered.* **88**, 202–208.

Malik, S., Wilson, P. J., Smith, R. J., Lavigne, D. M., and White, B. N. (1997). Pinniped penises in trade: A molecular genetic investigation. *Cons. Biol.* **11**, 1365–1374.

Mills, J., Ishirhara, A., Sakaguchi, I., Kang, S., Parry-Jones, R., and Phipps, M. (1997). "Whale Meat Trade in East Asia: A Review of the Markets in 1997." TRAFFIC International, Cambridge, UK.

Palsbøll, P. J., Berube, M., Skaug, H. J., and Raymakers, C. (2006). DNA registers of legally obtained wildlife and derived products as means to identify illegal takes. *Cons. Biol.* **20**, 1284–1293.

Reeves, R.R., Perrin, W. F., Taylor, B. L., Baker, C. S. and Mesnick, M. L. (2004). Report of the Workshop on Shortcomings of Cetacean Taxonomy in Relation to Needs of Conservation and Management, April 30-May 2, 2004. *US Dep. Comm. Tech. Mem. NOAA-TM-NMFS-SWFSC*-**363**.

Rice, D. W. (1998). "Marine Mammals of the World: Systematics and Distribution." Society for Marine Mammalogy, Lawrence, KS.

Roman, J., and Bowen, B. W. (2000). The mock turtle syndrome: Genetic identification of turtle meat purchased in the south-eastern United States of America. *An. Cons.* **3**, 61–65.

Rosenbaum, H. C., *et al.* (2000). World-wide genetic differentiation of *Eubalaena*: Questioning the number of right whale species. *Mol. Ecol.* **9**, 1793–1802.

Ross, H. A., *et al.* (2003). DNA surveillance: Web-based molecular identification of whales, dolphins and porpoises. *J. Hered.* **94**, 111–114.

Ross, H. A., and Murugan, S. (2006). Using phylogenetic analyses and reference datasets to validate the species identities of cetacean sequences in GenBank. *Mol. Phy. Evol.* **40**, 866–871.

Saiki, R. K., *et al.* (1988). Primer-directed enzymatic amplification of DNA with a thermostable DNA polymerase. *Science* **239**, 487–491.

Saitou, N., and Nei, M. (1987). The neighbor-joining method: A new method for reconstructing phylogenetic trees. *Mol. Biol. Evol.* **4**, 406–425.

Shivji, M., Clarke, S., Pank, M., Natanson, L., Kohler, N., and Stanhope, M. (2002). Genetic identification of pelagic shark body parts for conservation and trade monitoring. *Cons. Biol.* **16**, 1036–1047.

Simmonds, M. P., Haraguchi, K., Endo, T., Cipriano, F., Palumbi, S. R., and Troisi, G. M. (2002). Human health significance of organochlorine and mercury contaminants in Japanese whale meat. *J. Tox. Env. Health A* **65**, 1211–1235.

Tautz, D., Arctander, P., Minelli, A., Thomas, R. H., and Vogler, A. P. (2003). A plea for DNA taxonomy. *Trends Ecol. Evol.* **18**, 70–74.

US Fish and Wildlife Service (2001). A quality assurance manual. National Fish and Wildlife Forensic Laboratory, Oregon.

Wada, S., Oishi, M., and Yamada, T. K. (2003). A newly discovered species of living baleen whale. *Nature* **426**, 278–281.

# Fossil Sites, Noted

### R. Ewan Fordyce

## I. Introduction

Fossil marine mammals—Cetacea, Sirenia, Desmostylia, Pinnipedia and other aquatic carnivores—are known from hundreds of sites worldwide (Fig. 1). Localities span from modern tropics to poles, in both north and south and on all major continents, but with most in northern temperate regions. Usually, sites preserve marine sedimentary rocks, which have been exposed on land through sea level fall and/or uplift, followed by erosion. There are a few records (dredgings) from the deep ocean, and there are some important fresh water sites for secondarily nonmarine species. Fossils give only a general guide to former distributions in ancient oceans. Sites vary from rich localized concentrations at sites a few tens of meters across, to scattered occurrences across many kilometer which become significant at the regional level, and they range in age from Eocene to Pleistocene (Fig. 2). The case studies below, given in sequence from oldest to youngest, span all the major time intervals and oceans.

## II. The Role of Geological Processes

Marine mammal history has been affected by geological changes in oceans and climates over millions (M) of years (Fordyce and Muizon, 2001). These changes ultimately reflect global tectonic processes: continental drift and the rearrangement of land and sea. Continents are now relatively more emergent than for much of the past 50 million years, with less continental shelf and less extensive shallow continental sea than in the past. Most continents preserve coast-parallel strips of ancient marine rock now exposed on land. These may be extensive and a notable source of fossils (e.g., Atlantic Coastal Plain, eastern USA), or limited (e.g., most of Africa). Sometimes extensive shallow epicontinental seas onlapped the continents, as in northern Europe and the Paratethys. Major drops in sea level occurred about 30 million years ago (Ma) and, associated with widespread glaciation and global cooling, since 2 Ma (major fluctuations; Ma, million years).

When the first cetaceans and sirenians appeared, beyond 50 Ma, the extensive shallow Tethys Sea stretched from the Pacific to about the modern Mediterranean. By the end of the Eocene, India had moved northwards to collide with Asia, closing much of the Tethys. More-western remnants of the Tethys, through what is now southern Eurasia, were eliminated in the Miocene, when Africa collided with Eurasia. Later, the Mediterranean dried out completely about 6 Ma, with dramatic consequences for the biota.

In the south, Australia moved north away from Antarctica opening part of the Southern Ocean by the end of Eocene time (34 Ma). Later, Antarctica and South America separated in the Oligocene about 30–23 Ma, to open the Drake Passage, allowing west-to-east flow of a newly developed Antarctic Circumpolar Current. This current isolated Antarctica thermally, and probably allowed the Antarctic icecap to expand, global climates to cool, and global oceans to become more heterogeneous. Australia continued to drift north, so that in about Middle Miocene (~15 Ma) it closed the Indopacific seaway between Australia and Asia and restricted equatorial circulation between the Indian and Pacific Oceans. In the middle Pliocene (~3–4 Ma), the Panama Seaway closed, cutting Caribbean–Pacific

F

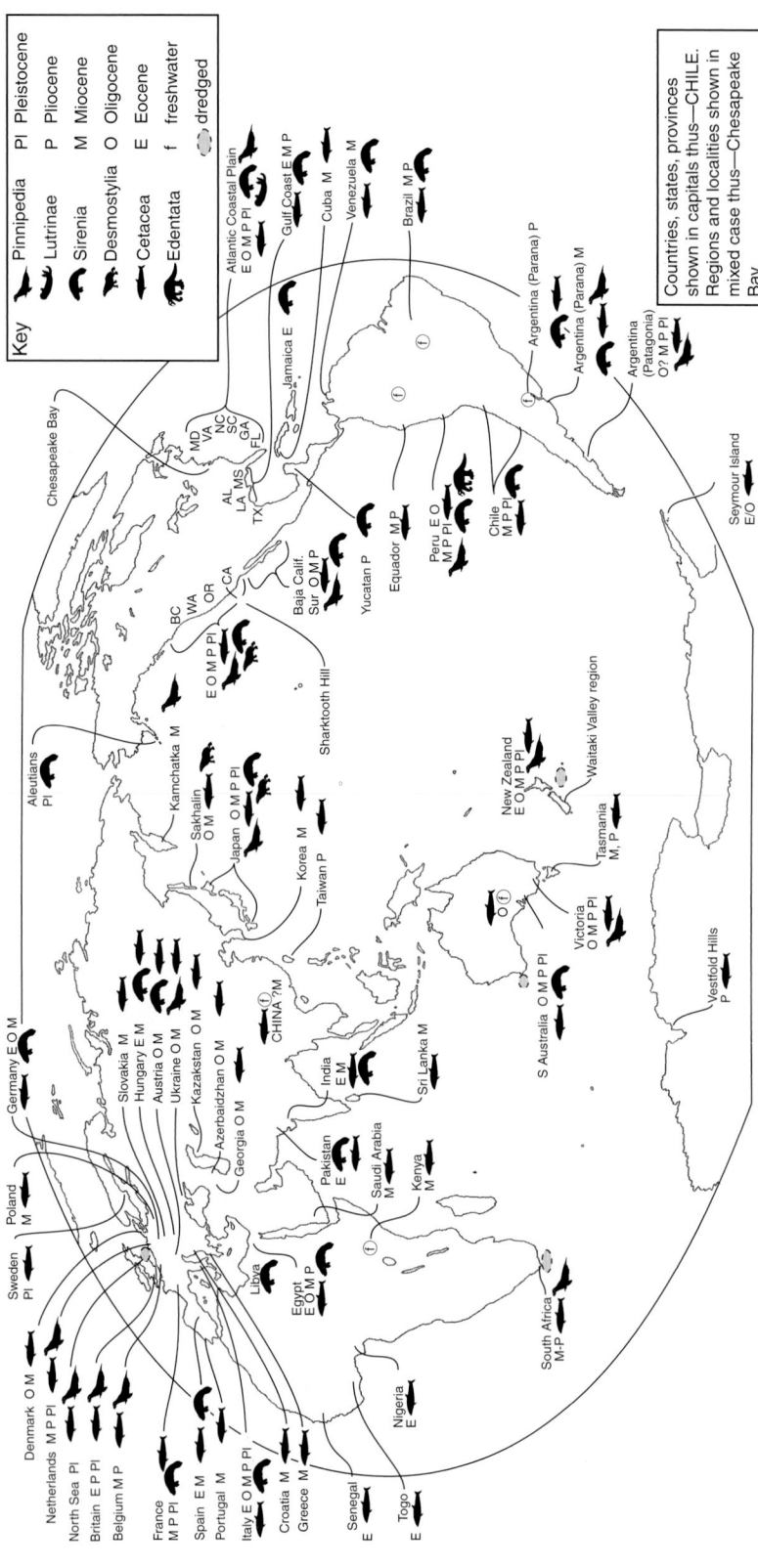

**Figure 1** *Selected localities for fossil marine mammals. Slightly revised from figure 1, in Fossil Sites, R. E. Fordyce, "Encyclopedia of Marine Mammals,"* W.F. Perrin et al. (eds). © Elsevier 2002.

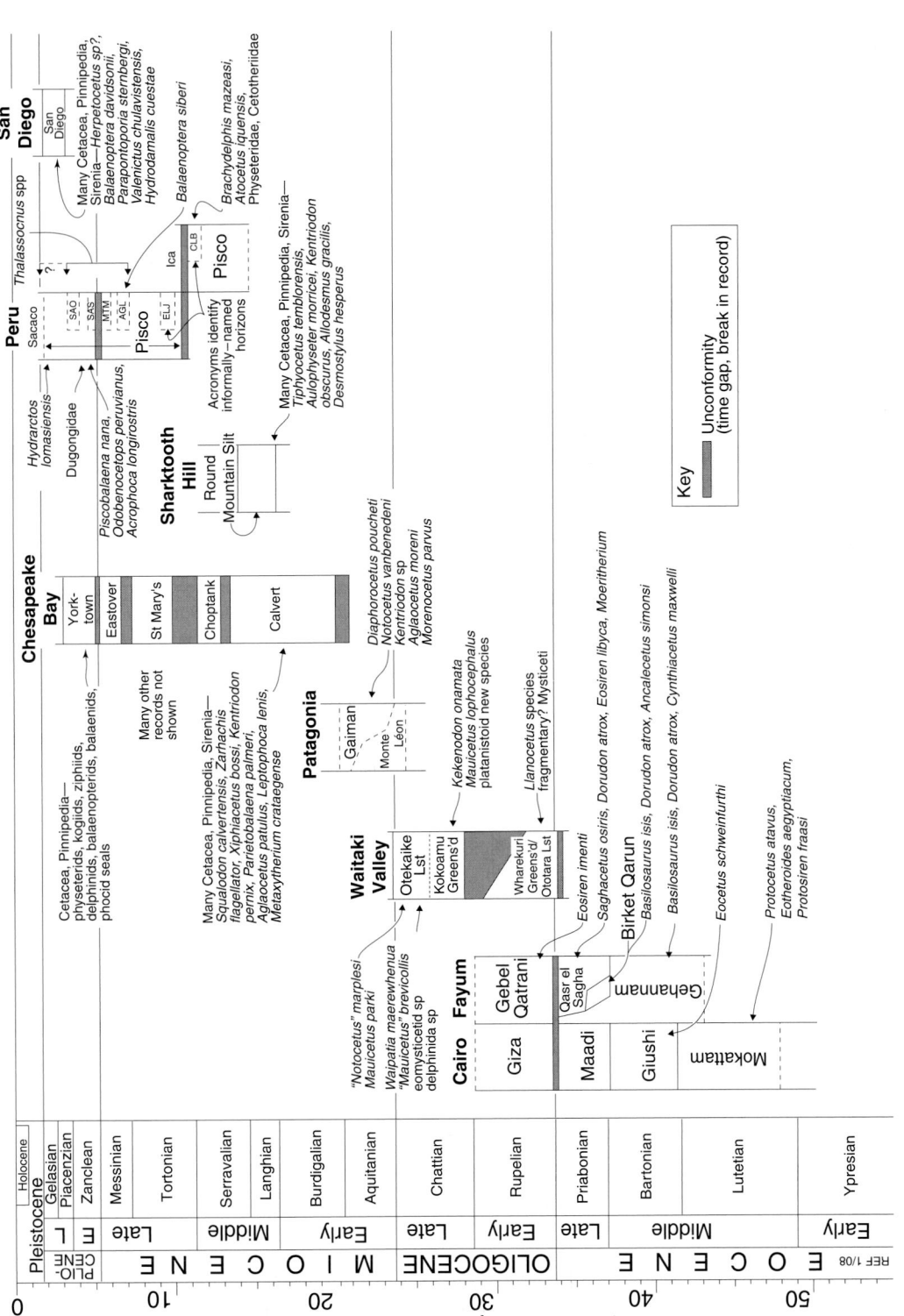

**Figure 2** *Geological age ranges for key localities for fossil marine mammals discussed under case studies. Revised from figure 2, in Fossil Sites, R. E. Fordyce, "Encyclopedia of Marine Mammals," W.F. Perrin et al. (eds), © Elsevier 2002.*

F

links. The closure of the Panama Seaway correlates closely with the start of Northern Hemisphere continental glaciation.

## III. A Global Summary of Localities

Important localities occur in marine sequences around the modern Mediterranean, which is a remnant of the formerly extensive Tethys sea and its now-vanished northeast arm, the Paratethys. Cetaceans, pinnipeds, and sirenians are notable. Italy has many sites of Pliocene to Oligocene age, while the most significant localities along the southern Mediterranean are in the Egyptian Eocene (discussed later). Paratethyan localities to the northeast include some in Austria, Hungary, Slovakia, Croatia, Romania, Ukraine, and several in the Caucasus mountains and borders of the Caspian Sea including Georgia, Azerbaijan, and Kazakstan.

Eastern North Atlantic cetaceans and pinnipeds have come from Miocene–Pleistocene and, rarely, Eocene-Oligocene sequences bordering the North Sea, in Denmark, northern Germany, Poland, Sweden, Netherlands, Belgium, Britain, and North Sea dredgings. Eocene to Pliocene fossils from the western North Atlantic include many from the Atlantic Coastal Plain (Delaware to Florida) and Gulf Coast. The Caribbean-tropical Atlantic has few reported sites, but southwest Atlantic fossils from Argentina include latest Oligocene, and Miocene–Pleistocene, with some from Miocene–Pliocene nonmarine strata. A few sites in the eastern tropical Atlantic have produced Eocene cetaceans (Nigeria, Senegal, Togo), the tip of South Africa has rich Pliocene bone-beds including cetaceans and pinnipeds, and abundant cetacean bones dredged from offshore.

Eocene basal Cetacea from Pakistan and India lived in the now-obliterated Tethys Sea. Otherwise, there are sparse reports of fossil Neoceti from around the Indian Ocean. From Kenya comes a Miocene apparent freshwater beaked whale (Ziphiidae), and ziphiid rostra have been dredged from off Western Australia.

A few regions around the Pacific, which was the largest ocean during cetacean history, have received concentrated attention. Japanese Mio-Pliocene cetaceans and pinnipeds are well documented, and studies of Oligocene cetaceans are underway. There are only sporadic records of fossil marine mammals from further north in the Pacific (Sakhalin and the Aleutian chain). From the Eastern North Pacific (Mexico to British Columbia) come notable Miocene–Pleistocene assemblages; of hundreds of known Oligocene fossils, only a few species are described. Significant assemblages of Mio-Pliocene cetaceans and pinnipeds come from Peru and, largely unstudied, northern Chile. New Zealand assemblages, from the Southern Ocean margin, span from the Eocene to Quaternary, including important Oligocene material; most fossils are cetacean but for a few geologically young pinnipeds. A scattered Oligocene to Neogene record of cetaceans and pinnipeds and fragmentary sirenians from Australia also hints at the composition of Southern Ocean faunas. One Eocene and one Pliocene site are reported from Antarctica.

## IV. Eocene: Mediterranean/Tethys (Northern Egypt)

Eocene strata in northern Egypt, near Cairo and southwards at Fayum, have produced fundamentally important archaeocete cetaceans and sirenians. For nearly a century, the cetaceans were the most archaic known (Kellogg, 1936). Assemblages have modest diversity, involving a few species of sometimes excellent preservation; fossils are locally abundant. The Cairo sequence, at Gebel Mokattam, is slightly older than that of Fayum. Patterns of fossiliferous strata reflect changing sea levels and shallow marine deposition at a passive continental margin (Gingerich, 1992). Marked unconformities,

or breaks in the record, were caused by lowered sea level. Paleoenvironments range from riverine (with sirenians) through estuarine, active shoreface, barrier bar, and shallow shelf.

At Gebel Mokattam, Cairo, the Mokattam Formation comprises limestones deposited in a shallowing marine shelf setting from approximately 48–41 Ma. Key fossils are the protocetid archaeocete *Protocetus atavus* and early sirenians such as *Eotheroides aegyptiacum* (Dugongidae; named by famous anatomist Richard Owen), *Eosiren abeli*, and *Protosiren fraasi* (Protosirenidae)—all about 46 Ma. Another protocetid, *Eocetus schweinfurthi*, is younger (~41–37 Ma), from the overlying shallow marine Giushi Formation.

To the south, Fayum marine strata span the Middle and Late Eocene (~43–35 Ma). Gehannam Formation limestone has yielded fossils of the sirenians *Eotherioides* and *Protosiren*, and associated skeletons of archaeocetes: the 4.5- to 5-m long *Dorudon* (formerly *Prozeuglodon*) *atrox* and larger ~16-m *Basilosaurus isis*, the latter known to have hind limbs (Uhen, 2004; Gingerich, 2005). Archaeocete skeletons are abundant at horizons that probably represent low-stands of sea level. *Moeritherium*, a pig-sized estuarine stem-proboscidean, is present. The top of the Gehannam Formation varies in age according to locality. It is succeeded in places by barrier beach sands of the Birket Qarun Formation, a unit with many archaeocete skeletons in the base, including the dorudontine *Ancalecetus simonsi*. The youngest marine rocks at Fayum are lagoonal strata of the Qasr el Sagha Formation (~35–37 Ma), a unit that has produced the sirenian *Eosiren libyca*, two dorudontine archaeocetes—the small *Saghacetus osiris* and larger *Dorudon atrox*—and *Moeritherium*. Above this marine sequence, the sirenian *Eosiren imenti* occurs in Oligocene riverine strata of the Gebel Qatrani Formation.

The Egyptian localities are important sources of early archaic cetaceans and sirenians, including significant type specimens (basic standards of reference). The geologically youngest Egyptian archaeocetes, from the latest Eocene, were contemporaneous with the oldest crown-group Cetacea, Neoceti.

## V. Oligocene: Southwest Pacific (Waitaki Valley, New Zealand)

Thin marine Oligocene strata (34–23 Ma) in the Waitaki Valley area of New Zealand, at 45°S, are a source of early Neoceti, or "modern" Cetacea (Fordyce and Muizon, 2001; Fordyce, 2006). Important localities include those near Oamaru, Duntroon, Wharekuri, and Hakataramea Valley, spanning some tens of kilometer. Field studies started in the 1880s; Benham and Marples were notable researchers, both on archaic mysticetes referred to *Mauicetus*. Many specimens represent unnamed new species.

The Wharekuri Greensand and Ototara Limestone have produced some of the few Early Oligocene (~33 to ~30 Ma) cetaceans reported worldwide. The strata are truncated above by a major unconformity probably caused by sea level fall at approximately 30 Ma, and are succeeded by the Kokoamu Greensand, a thin (usually <5 m), burrowed, shelly, sediment-starved unit. Greensand grades up over several meter into the massive Otekaike Limestone (up to 50-m thick), which spans the later Late Oligocene (~25 Ma) and earliest Miocene (~23 Ma). Sediments accumulated in a quiet setting below storm wave base, in mid-to-outer shelf depths. Associated vertebrates include penguins, sharks, and bony fish, but no other marine mammals. International correlation is based on planktic foraminifera, coccoliths, and strontium dating. Cetaceans occur both as isolated bones and as skeletons. The source units are often cemented, producing resistant outcrops so that excavations need pneumatic tools and saws, are limited in scope, and mostly recover partial rather than whole skeletons.

Basal Oligocene cetaceans from the Wharekuri Greensand include two presumed mysticetes: a large species with severe vertebral pathology, and a neonate. Elsewhere, fragments from the Ototara Limestone include a small *Llanocetus*-like toothed archaic mysticete. The overlying Kokoamu Greensand has produced baleen-bearing whales including Marples' widely cited "*Mauicetus*" *lophocephalus* and other unnamed eomysticetids, presumed stem-balaenopterids, and a stem-balaenid. There are no definite records of toothed basal mysticetes. Sparse odontocetes from the lower to middle Kokoamu Greensand are mostly Platanistoidea, including species of Squalodontidae, and provisionally identified Waipatiidae and Dalpiazinidae, but no true dolphins (Delphinida) are reported. In the early 2000s, enigmatic toothed cetaceans, long known from fragments, were finally identified as late persisting archaeocetes. A pivotal specimen is a skull, which is placed cladistically between Neoceti and the archaeocete group Basilosauridae; the fragmentary and enigmatic *Kekenodon onamata* is probably related.

Cetaceans are better preserved in the upper Kokoamu Greensand and overlying Otekaike Limestone. Mysticetes include the stem balaenopterid *Mauicetus parki* and relatives, and eomysticetids. Odontocetes are mainly rare but well-preserved platanistoids, including Squalodontidae, the small *Waipatia* (Waipatiidae), "*Microcetus*" and *Notocetus* (both Squalodelphinidae), and undescribed Dalpiazinidae. Tantalizing fragments of an archaic species of Delphinida and a sperm whale are known. There are notable absences, including the mysticete groups Aetiocetidae, Mammalodontidae and Janjucetidae. In contrast to South Carolina, there are no archaic odontocetes comparable to *Agorophius* and *Xenorophus*.

Small-to-medium-sized mysticetes dominate assemblages, with small odontocetes also conspicuous. The shallow broad seas could have been a breeding ground for mysticetes from the recently developed Southern Ocean ecosystem to the south. Why platanistoids are common but delphinoids and sperm whales are rare is not clear. Perhaps the shallow seaway suited neritic rather than more-pelagic species. The assemblages give the impression that the Oligocene was a time of structural/ecological experiment. Species were spread fairly evenly amongst diverse family-level taxa, whereas, for extant cetaceans, a few families account for most species diversity. Some of the better preserved fossils will help resolve cladistic relationships of extant Cetacea, thus providing an independent standard against which to compare molecular classifications.

## VI. Miocene: Southwest Atlantic (Patagonia)

Strata of the "Patagonian" marine stage in southern Argentina have produced Early Miocene Cetacea, which include basal (stem) members of modern families. These fossils are from the Gaiman and Monte Leon Formations of Santa Cruz, Chubut, and Rio Negro Provinces, and represent the Leonian local stage, Early Miocene and possibly latest Oligocene. Other marine mammals—pinnipeds, sirenians—have not been reported. The strata represent shallow-water settings, in which age-diagnostic microfossils are uncommon, and exact international correlations are uncertain. The Chubut River region has provided the main described material. More than a century of basic taxonomy includes work by Moreno, Lydekker, True, and Muizon. Cabrera (1926) and Cozzuol (1996) summarized the fossils.

Amongst the Early Miocene Cetacea, the small *Morenocetus* is an early right whale. The rostrum is not known, but other skull features are consistent with balaenid affinities. Some studies, surprisingly, place this whale in the crown Balaenidae, rather than in the stem. Of less certain relationships are the stem-balaenopterids *Aglaocetus* and "*Plesiocetus*," also known from variably complete skulls. Skull form is quite different between these mysticetes, indicating quite different habits and affinities.

Odontocetes are notably more diverse, and include some of the best-known early sperm whales: the small *Diaphorocetus*, and the much larger *Idiorophus*. Both are archaic in appearance, retaining upper teeth with obvious enamel and a narrow anterior on the rostrum. Archaic platanistoid odontocetes include the large and long-beaked shark-toothed dolphin *Phoberodon* (Squalodontidae) and the short-beaked *Prosqualodon*. *Prosqualodon* is also present in New Zealand and Australia; it has been placed variously in the Squalodontidae or its own group Prosqualodontidae, in turn dubiously allied with the true dolphins (Delphinida). A third group of platanistoids, the Squalodelphinidae, is represented by *Notocetus*, in turn important in refining the higher classification of the Platanistoidea. *Notocetus*-like odontocetes also occur in New Zealand pointing, as for *Prosqualodon*, to a Southern Ocean distribution. *Notocetus* and other squalodelphinids range into the North Atlantic and Mediterranean. The eurhinodelphinid *Argyrocetus* shows extreme lengthening of the rostrum, particularly the toothless premaxillae; another Early Miocene species in this genus has been reported from California. True dolphins (Delphinida) are rare, represented by the kentriodontid *Kentriodon*.

The Patagonian cetaceans indicate that "modern" ecological structuring (e.g., deep-diving sperm whales, skim-feeding balaenids) was established by the Early Miocene. Some species probably had circum-Southern Ocean distributions as also seen amongst living cetaceans.

## VII. Miocene: NorthWest Atlantic (Chesapeake Bay)

Shallow-dipping strata of the Chesapeake Group exposed around the western shores of Chesapeake Bay, in Maryland and Virginia, have been an important source of Miocene and Pliocene marine mammals for over 150 years (Gottfried *et al.*, 1994). Fossil cetaceans, sirenians, and phocid seals are present, along with land mammals, other vertebrates, and rich assemblages of marine invertebrates. Early studies on these fossils were made in the 1800s by the pioneering vertebrate paleontologists Harlan, Leidy, and Cope. Later, True (early 1900s) and Kellogg (1920s–1960s) produced such detailed descriptions of taxa such as *Delphinodon* and *Parietobalaena* that the Chesapeake fossils have become international standards of comparison. Significant collections are held in the Smithsonian Institution.

Marine mammals occur in five formations of the Chesapeake Group: Calvert (Early to early Middle Miocene), Choptank (Middle Miocene), St. Mary's (later Middle Miocene), Eastover (Late Miocene) and Yorktown (Pliocene). These are mainly fine grained, shallow water strata deposited in climate regimes ranging from subtropical (early Miocene, ~19 Ma) to near-modern (Pliocene, ~4.5 Ma) (Weems and Edwards, 2007). Strata are soft, so that fossils can be collected easily. The fossils are often preserved well, with fine sutural detail present and only limited crushing. Material ranges from single worn bones to nearly-complete skeletons.

Fossils from the Calvert Formation are important in revealing faunal composition around the Early to Middle Miocene boundary, approximately 16 Ma. Amongst the diverse cetaceans, odontocetes include species in three families of Platanistoidea: the shark-toothed dolphins, Squalodontidae (*Squalodon*), Squalodelphinidae (e.g., *Phocageneus*), and stem-Platanistidae (*Zarhachis*, possibly *Pomatodelphis*, and the long-enigmatic *Araeodelphis*). The extinct long-snouted Eurhinodelphinidae include two Calvert species of *Schizodelphis*, and two of *Xiphiacetus* including the former "*Eurhinodelphis*" *bossi*. True dolphins, Delphinoidea, are placed in the archaic grade family Kentriodontidae (e.g., *Kentriodon*, *Hadrodelphis*). There is a small sperm whale, Physeteridae (*Aulophyseter*; =*Orycterocetus* of older accounts), and extinct dolphins of uncertain affinities (*Tretosphys*, *Pelodelphis*).

Mysticetes represent stem balaenopterids or balaenopteroids (e.g., *Aglaocetus*, *Diorocetus*, *Parietobalaena*; Steeman, 2007); there are no firm reports for the Cetotheriidae or Balaenidae, groups that appear higher in the Chesapeake Group. Other marine mammals are markedly rarer in the Calvert Formation. Phocid seals, known mostly from isolated elements but also from rare partial skeletons, form the basis for the phocine *Leptophoca*, and a species of the monachine *Monotherium*. Sea cows include the extinct dugong *Metaxytherium crataegense* which also occurs in the Pacific (Montera Formation, Peru), indicating movement presumably through the Central American Seaway.

Marine mammal assemblages from the Chesapeake Group reveal marked faunal change over time. Archaic cetaceans from the Calvert Formation, such as the eurhinodelphinids, squalodontids, and squalodelphinids, are rare or absent in overlying (younger) units, while extant families such as the Delphinidae and Balaenopteridae become significant by the Pliocene. Concurrently, species-level diversity dropped. Ecological shifts are indicated by the absence of squalodontids and long-snouted eurhinodelphinid and stem-platanistid/platanistoid dolphins.

South of Chesapeake Bay, strata of the lower Yorktown Formation (Early Pliocene, ~4.5 Ma) at Lee Creek Mine, North Carolina, have produced thousands of isolated marine mammal bones (Whitmore, 1994), expanding the fauna beyond the Yorktown at Chesapeake Bay. Fossils include ziphiids, physeterids, kogiids, delphinids, balaenopterids, balaenids, and phocid seals. Unusual elements include monodontids (now found only at high latitudes) and pontoporiids (now restricted to the South Atlantic). Similar assemblages occur in the San Diego and Pisco Formations.

Strata of the Chesapeake Group, especially the Calvert Formation, are one of the world's richest sources of fossil marine mammals, particularly Cetacea. The abundance of young animals could reflect favorable conditions for calving, or, equally, a high mortality for young individuals.

## VIII. Miocene: Northeast Pacific (Sharktooth Hill, California)

The Sharktooth Hill Bonebed of Kern County, in the foothills of the Sierra Nevada, California, is a key horizon for marine mammal fossils in the Pacific Basin. The assemblage includes more than 100 marine vertebrates (Kellogg, 1931, Barnes, 1977). Cetaceans, pinnipeds, rare desmostylians, turtles, marine birds, chondrichthyans, teleosts, and terrestrial vertebrates occur abundantly in, and less commonly above, a thin (10–30 cm) dense and geographically widespread layer in the Round Mountain Silt (Temblor Formation, Middle Miocene). This bone-bed formed probably over hundreds to thousands of years about 16 Ma, a little after a peak time of global warmth; it represents a mix of organisms from pelagic, neritic, and terrestrial environments. Such a deposit containing spectacularly concentrated fossils is termed a lagerstätte (fossil deposit of exceptional abundance or quality). Most of the fossils are isolated elements dominated by mammalian postcrania and by shark teeth, but sometimes there are natural associations of bones from one individual. Preservation varies; bones may be finely preserved, or eroded, and sometimes carry marks caused by scavengers or predators. The deposit has been studied extensively, mainly by research groups from Los Angeles, San Diego, and Berkeley.

Cetaceans dominate the marine mammals (Barnes, 1977). Mysticetes include presumed stem balaenopteroids *Aglaocetus*, *Parietobalaena*, *Peripolocetus* and *Tiphyocetus*, known from partial skulls, and previously regarded as Cetotheriidae. A possible right whale is reported, but Cetotheriidae in the strict sense appear absent. Amongst odontocetes, archaic sperm whales include the moderate-sized

*Aulophyseter*. The taxonomy of the smaller odontocetes is less certain, for Kellogg, who named most of the species, based on several new species and genera on isolated periotics which, in a few cases, have been linked to later-discovered skulls. Of note are a long-beaked allodelphinid dolphin ("*Squalodon*" *errabundus*; Platanistoidea), and species in the archaic delphinoid group Kentriodontidae (e.g., *Kentriodon*). Several named odontocetes are still enigmatic (e.g., *Oedolithax*, *Lamprolithax*). In terms of ecological composition, the cetaceans are comparable to faunas in Californian waters today.

Otariid (or otarioid) pinnipeds are significant and include the types for some species or genera. They are the small archaic imagotariine *Neotherium*, three species of the large *Allodesmus* (including the rare type-species *A. kernensis* and more common *A. gracilis*), the large rare *Pelagiarctos*, and two unnamed "desmatophocines." No phocids have been described. Of other marine mammals, sirenians have not been reported, but fragmentary desmostylians occur.

There are conspicuous faunal similarities with the slightly older assemblage from the Calvert Formation of Maryland and Virginia, western North Atlantic; e.g., *Parietobalaena* and *Kentriodon* are reported for both places. Differences are also marked; e.g., phocoid seals, sirenians, and squalodontid and eurhinodelphinid odontocetes are absent from the Sharktooth Hill Bonebed, while otarioid seals and desmostylians are absent from the Calvert.

## IX. Mio-Pliocene: Southeast Pacific (Peru)

Since the early 1980s, Muizon and others have documented diverse later Miocene and Pliocene marine mammals, including some quite surprising ecotypes, from the sandy Pisco Formation of southern Peru. Rich localities are around Sacaco, near Lomas, where well-preserved skeletons of cetaceans and pinnipeds occur, and Ica, to the north. Pisco strata include diatomite, sandstone, siltstone, and occasional conglomerate, sometimes volcanic-rich, with marine invertebrates and bone-rich horizons. Settings are intertidal, barrier bar, lagoonal, and shallow embayment; these environments were probably protected from the open Pacific, perhaps by rocky promontories. The sequence of vertebrate faunas involves: Cerro la Bruja (base), El Jahuay, Aguada de Lomas, Montemar, Sud-Sacaco, and Sacaco (top). The exact age for each horizon is uncertain because the Pisco Formation mostly lacks fossils useful in international correlation.

Odontocete cetaceans are the most diverse marine mammals (Muizon, 1988). The oldest productive Pisco horizon, Cerro la Bruja, is probably Middle Miocene, approximately 15–12 Ma; taxa include kentriodontids (e.g., *Atocetus*) and an early record for stem Pontoporiidae (the peculiarly short-beaked *Brachydelphis*). Other younger Miocene horizons at El Jahuay (~9 Ma) and Aguada de Lomas (8–7 Ma) yield early porpoises (stem Phocoenidae, e.g., *Australithax*), archaic sperm whales (narrow-skulled *Scaphokogia*), archaic dolphins (Kentriodontidae, e.g., *Atocetus*), beaked whales (stem Ziphiidae), and sperm whales (stem Physeteridae) (Muizon, 1988). Younger assemblages, from the latest Miocene Montemar (~6 Ma), and Pliocene Sud-Sacaco (4–5 Ma) and Sacaco (3–4 Ma) localities for the Pisco Formation, are similar in content. Of note are stem pontoporiids (*Pliopontos*), porpoises (*Piscolithax*), ziphiids (*Ninoziphius*), sperm whales, and kentriodontids, but also more-modern dolphins (Delphinidae) and an unnamed beluga (Delphinapterinae). The most peculiar Pliocene odontocetes are two species of the extinct tusked walrus-mimic *Odobenocetops*. These animals, reported only from the Pisco Formation, have secondarily lost many of the distinctive facial features of odontocetes (Muizon and Domning, 2002).

Mysticeti from the Pisco Formation include cetotheres and up to six balaenopterids: a small cetothere, *Piscobalaena*, the larger *Piscocetus* (a possible Cetotheriid) and an extinct *Balaenoptera* (Balaenopteridae). *Piscobalaena* includes excellent specimens important in understanding the Cetotheriidae (Bouetel and De Muizon, 2006). Some of the mysticetes preserve baleen plates *in situ* on the skull (Brand *et al.*, 2004).

Up to nine species of phocid seals are present. Two lobodontines, the *Monachus*-like *Piscophoca pacifica* and long-skulled *Acrophoca longirostris*, are known from articulated skeletons. From the relatively barren upper Pisco Formation comes the unique specimen of an extinct fur seal, *Hydrarctos* (Otariidae: Arctocephalinae) which is probably Late Pliocene. A sirenian, probably close to *Dugong*, is from Early Pliocene lagoonal deposits. Despite the unexpected occurrence, it seems certain that the extinct species of Pisco sloth *Thalassocnus* really were marine mammals. Marine sloths range from 7–8 Ma to 4–3 Ma and possibly 3–1.5 Ma at Sacaco (Muizon *et al.*, 2003), they are abundant, there are no other putative land mammals in the vertebrate assemblage, and the adjacent coast was a desert.

Pliocene assemblages from the Pisco Formation include some cetaceans and pinnipeds similar to those from the Yorktown Formation of the Chesapeake Group, North Atlantic, indicating contact through the Central American Seaway before the uplift of Panama. Several genera also occur at Isla Cedros, Baja California Sur. When considered with roughly contemporaneous assemblages from elsewhere in the eastern Pacific (Isla Cedros and San Diego), and contrasted with modern communities, it seems that there must have been considerable faunal turnover late in the Pliocene or in the Pleistocene.

## X. Plio-Pleistocene: Northeast Pacific (San Diego, California)

Well-preserved geologically young marine mammals—from the later Pliocene and earlier Pleistocene—are rare, but important in revealing marine mammal ecology before the dramatic climate shifts and sea level change of the Pleistocene glaciations. One notable sequence is the ~84 m of San Diego Formation at and near San Diego, Southern California (Deméré, 1994). This sandy to gravelly unit was deposited late in the Pliocene (3–2 Ma) and possibly Pleistocene (>1.5 Ma), in settings mostly from shoreface to mid- and outer shelf. The lower finer strata with abundant fossils are marine, while the coarser and sparsely fossiliferous upper strata are partly nonmarine. The formation contains many marine mammals that are now extinct, including some whose descendants live in quite different settings today.

Pinnipeds, Cetacea, and Sirenia are present (Barnes, 1977; Deméré, 1994), including many complete enough to identify to species level. The pinnipeds include Otariidae, e.g., an extinct species of *Callorhinus*, and Odobenidae, e.g., the extinct long-tusked walrus *Valenictus* and *Dusignathus*. There are no phocids.

Three families of mysticetes include two species of the cetothere *Herpetocetus*, providing one of the last records of Cetotheriidae *sensu stricto*. The extinct minke whale *Balaenoptera davidsonii* is one of five rorquals (Balaenopteridae), and there are two right whales (Balaenidae). Notable amongst odontocetes are the long-beaked *Parapontoporia* (a marine stem Lipotidae), two porpoises (Phocoenidae), a beluga-like animal (Delphinapterinae), and two dolphins (Delphinidae, including *Stenella* or *Delphinus*). The huge sirenian *Hydrodamalis cuestae* appears to be a direct ancestor to the recently exterminated *Hydrodamalis gigas*, Steller's sea cow, of the cold North Pacific.

These fossils imply major shift in geographic range and/or ecology in geologically recent times: walruses (now only one species, of *Odobenus*) and belugas currently live in cold northern waters, as *Hydrodamalis* also did until a few 100 years ago. Descendants of *Parapontoporia* have left the eastern Pacific, and cetotheres are extinct.

## See Also the Following Articles

Cetacean Evolution ■ Cetacean Fossil Record

## References

Barnes, L. G. (1977). Outline of eastern North Pacific fossil cetacean assemblages. *Syst. Zool.* **25**, 321–343.

Bouetel, V., and De Muizon, C. (2006). The anatomy and relationships of *Piscobalaena nana* (Cetacea, Mysticeti), a Cetotheriidae s.s. from the early Pliocene of Peru. *Geodiversitas* **28**, 319–395.

Brand, L. R., Esperante, R., Chadwick, A. V., Poma Porras, O., and Alomía, M. (2004). Fossil whale preservation implies high diatom accumulation rate in the Miocene–Pliocene Pisco Formation of Peru. *Geology* **32**, 165–168.

Cabrera, A. (1926). Cetáceos fosiles del Museo de la Plata. *Rev. Mus. La Plata* **29**, 363–411.

Cozzuol, M. A. (1996). The record of the aquatic mammals in southern South America. *Münchner Geowissenschaftliche Abhandlungen. A, Geologie und Paläontologie* **30**, 321–342.

Deméré, T. A. (1994). Two new species of fossil walruses (Pinnipedia, Odobenidae) from the upper Pliocene San Diego formation, California. *Proc. San Diego Soc. Nat. Hist.* **29**, 77–98.

Fordyce, R. E. (2006). A southern perspective on cetacean evolution and zoogeography. *In* "Evolution and Biogeography of Australasian Vertebrates" (J. R. Merrick, M. Archer, G. Hickey, and M. S. Y. Lee, eds), pp. 755–778. AusSciPub, Sydney.

Fordyce, R. E. and Muizon, C. de. (2001). Evolutionary history of whales: A review. *In* "Secondary Adaptation of Tetrapods to Life in Water. Proceedings of the International Meeting, Poitiers, 1996" (J.-M. Mazin and V. de Buffrenil, eds), pp. 169–234. Verlag Dr Friedriech Pfeil, München.

Gingerich, P. D. (1992). Marine mammals (Cetacea and Sirenia) from the Eocene of Gebel Mokattam and Fayum, Egypt: Stratigraphy, age and paleoenvironments. *Univ. Mich., Pap. Paleontol.* **30**, 1–84.

Gingerich, P. D. (2005). Cetacea. *In* "Placental Mammals: Origin, Timing, and Relationships of the Major Extant Clades" (K. D. Rose, and J. D. Archibald, eds), pp. 234–252. Johns Hopkins University Press, Baltimore.

Gottfried, M. D., Bohaska, D. J., and Whitmore, F. C. (1994). Miocene cetaceans of the Chesapeake Group. *Proc. San Diego Soc. Nat. Hist.* **29**, 229–238.

Kellogg, A. R. (1931). Pelagic mammals from the Temblor formation of the Kern River region, California, series 4. *Proc. Californian Acad. Sci.* **19**, 217–297.

Kellogg, A. R. (1936). A review of the Archaeoceti. *Carnegie Inst. Washington Publ.* **482**, 1–366.

Muizon, C.de. (1988). Les vertébrés fossiles de la Formation Pisco (Pérou). Troisième partie: Les odontocètes (Cetacea, Mammalia) du Miocène. *Inst. Fran. d'Étud. Audines Mem.* **78**, 1–244.

Muizon, C.de., and Domning, D. P. (2002). The anatomy of *Odobenocetops* (Delphinoidea, Mammalia), the walrus-like dolphin from the Pliocene of Peru and its palaeobiological implications. *Zool. J. Linn. Soc.* **134**, 423–452.

Muizon, C.de., Mcdonald, H. G., Salas, R., and Urbina, M. (2003). A new early species of the aquatic sloth *Thalassocnus* (Mammalia, Xenarthra) from the late Miocene of Peru. *J. Vertebr. Paleontol.* **23**, 886–894.

Steeman, M. E. (2007). Cladistic analysis and a revised classification of fossil and recent mysticetes. *Zoological Journal of the Linnean Society* **150**, 875–894.

Uhen, M. D. (2004). Form, function and anatomy of *Dorudon atrox* (Mammalia: Cetacea): An archaeocete from the middle to late Eocene of Egypt. *Univ. Mich., Pap. Paleontol.* **34**, 1–222.

**F**

Weems, R. E., and Edwards, L. E. (2007). The age and provenance of "*Eschrichtius*" *cephalus* Cope (Mammalia : Cetacea). *J. Vertebr. Paleontol.* **27**, 752–756.

Whitmore, F. C. (1994). Neogene climatic change and the emergence of the modern whale fauna of the North Atlantic Ocean. *Proc. San Diego Soc. Nat. Hist.* **29**, 223–227.

# Franciscana Dolphin
## *Pontoporia blainvillei*

### Enrique A. Crespo

## I. Characteristics and Taxonomy

Franciscana (*Pontoporia blainvillei*) is also known as the La Plata River dolphin. In Uruguay and Argentina it is called *franciscana*, whereas in Brazil it is called *toninha* or *cachimbo*. Although both this species and the Yangtze river dolphin, *Lipotes vexillifer*, were until recently regarded as of the family **Pontoporiidae**, the franciscana is now the sole member of this family. The franciscana is the only one of the five river dolphins living in the marine environment. It is one of the smallest dolphins and has an extremely long and narrow beak and a bulky head. The franciscana is brownish to dark gray above, turning lighter to the flanks and belly (Fig. 1). The number of teeth in the upper and lower jaws ranges from 53 to 58 and from 51 to 56, respectively.

### A. Fossil Record

Three records have been related to the franciscana and assigned to the Family **Pontoporiidae**: *Brachidelphis mazeasi*, a middle miocene fossil from the Pisco Formation (Perú), *Pontistes rectifrons*, a late miocene fossil found in the Paraná Formation (Argentina), and *Pliopontos littoralis*, a pliocene fossil closely related to the living species described from the Pisco Formation (Perú).

### B. Geographic Variation

Skull morphology, genetic markers and parasites have been used to identify stocks. The existence of two potential populations was tested by means of the differences in skull morphology. A northern (smaller) form was proposed between Rio de Janeiro and Santa Catarina and a southern (larger) form for Rio Grande do Sul, Uruguay, and Argentina. The existence of differences between populations was confirmed some years later, using mtDNA from samples collected at Rio de Janeiro and Rio Grande do Sul (Secchi *et al.*, 1998). It was found that six exclusive haplotypes were present in the northern population and five in the southern one, indicating some degree of segregation between the stocks. Recent work on mtDNA and radio tracking carried out at Bahía Samborombón and Bahía Anegada (Argentina) reveal significant genetic division at the regional level, fine-scale structure within the study area, limited movement patterns, a small home range, and a high degree of isolation (Bordino *et al.*, 2007). Gastrointestinal parasites were also used as bioindicators to study the existence of stocks. The parasites seem to indicate segregation into two functional or ecological stocks between southern Brazil–Uruguay and Argentina. Three species of parasites were recommended as biological tags (*Hadwenius pontoporiae*, *Polymorphus cetaceum*, and *Anisakis typica*). On the base of the present information, at least three stocks or populations could exist.

## II. Distribution and Abundance

The species is endemic in southwestern Atlantic waters. Based on the distribution of sightings and catches, the franciscana lives in a narrow strip of coastal waters beyond the surf to the 30-m isobath

**Figure 1**   *A live-stranded franciscana in a tank. Photo by R. Bastida.*

(Fig. 2). The complete range known for the franciscana extends from Itaúnas (18°25′S, 39°42′W) in Espirito Santo, Brazil, to the northern coast of Golfo San Matías (41°10′S) in northern Patagonia, Argentina. Recent surveys carried out in Argentina showed that franciscana is also found up to the 50-m isobath. However, density declines with distance from the coast. In the strip between the 30- and the 50-m isobaths, density is half that between the coast and the 30-m isobath. With regard to abundance estimates, only two surveys were carried out for franciscanas. One survey was at Rio Grande do Sul State coast, southern Brazil, a region where there are current data on annual incidental mortality. At Rio Grande, the density was estimated to be 0.657 dolphins per km$^2$, with a population estimation of 42,000 individuals in 64,000 km$^2$ between the coast and the 30-m isobath. In Argentina, the second area where the franciscanas were surveyed, density was lower than in southern Brazil (0.304–0.377 dolphins per km$^2$) and abundance was estimated to be 15,000 individuals between the coast and the 50-m isobath in 50,000 km$^2$.

## III. Ecology

Little is known about the northern stock or population between Espirito Santo and Santa Catarina, a region that is under the influence of the Brazil tropical current. Between southern Brazil and Golfo San Matías, the franciscana lives in a transition zone in which the surface circulation of the southwestern Atlantic is dominated by the opposing flows of subtropical and subantarctic water masses. The coastal marine ecosystem is characterized by continental runoffs with a high discharge of high-nutrient river flows (e.g., Lagoa dos Patos, Río de la Plata). Juvenile sciaenids, the most important prey of the franciscana, are typically associated with those continental runoffs and the influence of subtropical shelf waters. The franciscana feeds mostly near the bottom on fishes of several families, such as sciaenids, engraulids, batrachoidids, gadids, carangids, and atherinids. However, sciaenids account for most of the fish species. The diet also includes squids, octopus, and shrimps. The franciscana feeds on the most abundant species in the region and seems to change its diet according to seasonal prey fluctuations. A comparison of results between two studies carried out 15 years apart in Rio Grande do Sul showed shifts in prey composition in which important prey of the former period were depleted in artisanal fisheries. Among predators, remains of franciscanas were found in stomach contents of killer whales (*Orcinus orca*) and several species of sharks.

## IV. Behavior and Physiology

Very little is known about behavior of free-ranging franciscanas, in part because they are difficult to observe in the wild and in part as a consequence of low sighting effort. The franciscana is thought to be

**Figure 2** *Distribution range of the franciscana dolphin* (Pontoporia blainvillei) *in the Southwestern Atlantic Ocean. The shaded area represents approximately the 30-m isobath.*

solitary or not gregarious. Herd size may range from 2 to 15 individuals. In aerial surveys carried out in southern Brazil with the objective of estimating abundance, 37 sightings gave a mean herd size of 1.19 (SD: 0.47, range: 1–3). In aerial surveys conducted in Argentina, 101 franciscanas were observed in 71 sightings with an average of 1.43 (SD = 0.85, range: 1–5) individuals per group. A study of wild behavior at Bahía Anegada in southern Buenos Aires Province showed a seasonal pattern with cooperative feeding, with traveling activities increasing during winter and high tide. The mean swimming speed was estimated in 1.3 m/sec (+/− 0.09) with a maximum of 1.8 m/sec, and mean dive duration was estimated in 21.7 sec (+/− 19.2) (range from 3 to 82 sec). The average at the surface was estimated to be 1.2 sec.

## V. Life History

Females are larger than males. Adult females range between 137 and 177 cm in total length, whereas males range between 121 and 158 cm. The weight of the mature females range between 34 and 53 kg and that of males range between 29 and 43 kg. Neonates in Uruguay range in size between 75 and 80 cm, whereas in southern Brazil they range between 59 and 77 cm (some of the smaller neonates could be near term fetuses). Neonates weigh around 7.3–8.5 kg. Age at sexual maturity is estimated to be 2.7 years, and the gestation period is between 10.5 and 11.1 months. Females give birth around November and lactation lasts for 9 months. However, calves take solid food around the third month, sizing between 77 and 83 cm. Mating seems to occur in January and February. The calving interval is around 2 years; nevertheless, few females are lactating and pregnant at the same time. Reproductive capacities and life span are low for the species, which is a problem for the population to sustain the mortality rates caused by fisheries. Longevity has been estimated to be close to 15 years for males and 21 for females, fairly low when compared to most of the small cetaceans. Few individuals attain ages over 10 years. Three types of acoustic signals have been recorded, including low, high, and ultra high frequency clicks.

## VI. Interactions with Humans

Incidental catches in gillnets, mostly of juvenile individuals, became a serious problem for the species throughout its distribution range, probably since the end of World War II. At that time, many artisanal fisheries for sharks developed in the region for Vitamin A production, which was exported to Europe. During the 1970s, gillnet mortality in Uruguay was estimated at above 400 individuals/year and fell to around 100 individuals/year in the last few years for economic reasons. Nevertheless, minimum mortality rates were always estimated at several thousands of individuals throughout the distribution range. At present, higher mortality rates are shown by the fisheries at Rio Grande do Sul and Buenos Aires Province, where no less than 700–1000 and 500–800 are, respectively, incidentally taken. The estimated mortality for the whole distribution range could be around 1200–1800 individuals per year. Due to the variability found in mortality rates and abundance estimates, it is not known if those mortality rates are sustainable. In gross numbers, the upper limits of abundance estimations cannot account for the lowest estimates of mortality. Therefore, more precise estimates are needed along with conservation measures to preserve the species. Other threats to the franciscana include habitat degradation. A large proportion of the distribution range is subject to pollution from several sources, especially the agricultural use of land and heavy industries between São Paulo in Brazil and Bahía Blanca in Argentina. The coastal zone is also intensely used for boat traffic, tourism, and artisanal and industrial fishing operations.

## References

Andrade, A., Pinedo, M. C., and Pereira, J., Jr. (1997). The gastrointestinal helminths of the franciscana, *Pontoporia blainvillei*, in southern Brazil. *Rep. Int. Whal. Commn.* **47**, 669–674.

Barnes, L. G. (1985). Fossil pontoporiid dolphins (Mammalia: Cetacea) from the Pacific Coast of North America. *Cont. Sci. Nat. Hist. Mus. Los Angeles County* **363**, 1–34.

Barreto, A. S., and Rosas, F. C. W. (2006). Comparative growth analysis of two populations of *Pontoporia blainvillei* on the Brazilian coast. *Mar. Mamm. Sci.* **22**, 644–653.

Bassoi, M. (1997). Avaliação da dieta alimentar de toninha, *Pontoporia blainvillei* (Gervais and D'Orbigny, 1844), capturadas acidentalmente na pesca costeira de emalhe no sul do Rio Grande do Sul. Dissertação de Bacharelado. Fundação Universidade do Rio Grande. Rio Grande-RS. 68pp.

Bordino, P., Thompson, G., and Iñiguez, M. (1999). Ecology and behaviour of the franciscana dolphin *Pontoporia blainvillei* in Bahía Anegada, Argentina. *J. Cetacean Res. Manag.* **1**, 213–222.

Bordino, P., Wells, R. S., and Stamper, M. A. (2007) Site Fidelity of Franciscana Dolphins *Pontoporia blainvillei* off Argentina. Abstract accepted at 17th Biennial Conference on the Biology of Marine Mammals, South Africa.

Brownell, R. L., Jr. (1975). Progress report on the biology of the franciscana dolphin *Pontoporia blainvillei* in Uruguayan waters. *J. Fish. Res. Board Can.* **32**, 1073–1078.

Brownell, R. L., Jr. (1984). Review of reproduction in platanistid dolphins. *Rep. Int. Whal. Commn.* (Special Issue 6), 149–158.

Busnel, R. G., Dziedzic, A., and Alcuri, G. (1974). Études préliminaires de signaux acoustiques du *Pontoporia blainvillei* Gervais et D'Orbigny (Cetacea, Platanistidae). *Mammalia* **38**, 449–459.

Corcuera, J., Monzón, F., Crespo, E. A., Aguilar, A., and Raga, J. A. (1994). Interactions between marine mammals and coastal fisheries of Necochea and Claromecó (Buenos Aires Province, Argentina). *Rep. Int. Whal. Commn.* (Special Issue 15), 283–290.

Cozzuol, M. A. (1996). Contributions of southern South America to vertebrate paleontology. *Münchner Gewissensch. Abh.* **30**, 321–342.

Crespo, E. A., Harris, G., and Gonzalez, R. (1998). Group size and distributional range of the franciscana *Pontoporia blainvillei*. *Mar. Mamm. Sci.* **14**, 845–849.

Crespo. E. A., Pedraza, S. N., Grandi, M. F., Dans, S. L., and Garaffo, G. Abundance and distribution of endangered Franciscana dolphins (*Pontoporia blainvillei*) in Argentine waters and conservation implications. Manuscript submitted to *Mar. Ecol. Prog. Ser.*

Danilewicz, D., Claver, J. A., Pérez Carrera, A. L., Secchi, E. R., and Fontoura, N. F. (2004). Reproductive biology of male franciscanas (*Pontoporia blainvillei*) (Mammalia: Cetacea) from Rio Grande do Sul, southern Brazil. *Fish. Bull.* **102**, 581–592.

Kasuya, T., and Brownell, R. L., Jr. (1979). Age determination, reproduction and growth of franciscana dolphin *Pontoporia blainvillei*. *Scient. Rep. Whales Res. Inst.* **31**, 45–67.

Méndez, M., Rosenbaum, H. C., and Bordino, P. (2007). Conservation genetics of the franciscana dolphin in Northern Argentina: Population structure, by-catch impacts, and management implications. *Conserv. Genet.*, DOI 10.1007/s10592007-9354-7.

Muizon, C. De. (1988). Les Vertebrés fossiles de la Formation Pisco (Pérou) Triosieme partie: Les Odontocétes (Ceacea, Mammalia) du Miocene. Recherche sur les Grandes Civilisations, Institut Française d'Etudes Andines. *Mémoire* **78**, 1–244.

Pérez Macri G., and Crespo, A. (1989). Survey of the franciscana, *Pontoporia blainvillei*, along the Argentine coast, with a preliminary evaluation of mortality in coastal fisheries. *In* "Biology and Conservation of the River Dolphins" (W. F. Perrin, R. L. Brownell, Jr., K. Zhou and J. Liu, eds), pp. 57–63. Occasional Papers of the IUCN Species Survival Commission (SSC) 3.

Pinedo, M. C. (1982). Analises dos contudos estomacais de *Pontoporia blainvillei* (Gervais and D'Orbigny, 1844) e *Tursiops gephyreus* (Lahille, 1908) (Cetacea, Platanistidae e Delphinidae) na zona estuarial e costeira de Rio Grande, RS, Brasil. M.Sc. Thesis, Universidade do Rio Grande do Sul, Brasil. 95pp.

Pinedo, M. C. (1991). Development and variation of the franciscana, *Pontoporia blainvillei*. Ph.D. Thesis, University of Californa, Santa Cruz. 406pp.

Pinedo, M. C., Praderi R., and Brownell, Jr. R. L. (1989). Review of the biology and status of the franciscana *Pontoporia blainvillei*. In "Biology and Conservation of the River Dolphins" (W. F. Perrin, R. L. Brownell, Jr., K. Zhou and J. Liu, eds). pp. 46–51. Occasional Papers of the IUCN Species Survival Commission (SSC) 3.

Praderi, R., Pinedo, M. C., and Crespo, E. A. (1989). Conservation and management of *Pontoporia blainvillei* in Uruguay, Brazil and Argentina. *In* "Biology and Conservation of the River Dolphins" (W. F. Perrin, R. L. Brownell Jr., K. Zhou and J. Liu, eds), pp. 52–56. Occasional Papers of the IUCN Species Survival Commission (SSC) 3.

Secchi, E. R., Zerbini, A. N., Bassoi, M., Dalla Rosa, L., Moller, L. M., and Roccha-Campos, C. C. (1997). Mortality of franciscanas, *Pontoporia blainvillei*, in coastal gillneting in southern Brazil: 1994–1995. *Rep. Int. Whal. Comm.* **47**, 653–658.

Secchi, E. R., Wang, J. Y., Murray, B., Roccha-Campos, C. C., and White, B. N. (1998). Populational differences between franciscanas, *Pontoporia blainvillei*, from two geographical locations as indicated by sequences of mtDNA control region. *Can. J. Zool.* **76**, 1622–1627.

Secchi, E. R., Kinas, P. G., and Muelbert, M. (2004). Incidental catches of franciscana in coastal gillnet fisheries in the Franciscana Management Area III: Period 1999–2000. *Latin Am. J. Aquat. Mamm.* **3**, 61–68.

**Figure 1** *Fraser's dolphins in Verde Island Passage, Philippines. (A) Group of females and calves and (B) an adult male, showing the "bandit mask." Photographs by M.L.L. Dolar.*

# Fraser's Dolphin
## *Lagenodelphis hosei*

### M. Louella L. Dolar

## I. Characteristics and Taxonomy

Fraser's dolphin was described in 1956 based on a skeleton collected by E. Hose from a beach in Sarawak, Borneo in 1895. F.C. Fraser gave it the genus name *Lagenodelphis*, due to what appeared to him a similarity of the skull to those of *Lagenorhynchus* spp. and *Delphinus delphis*. The external appearance of this species was not known until 1971 when specimens were found in widely separated areas: near Cocos Island in the eastern tropical Pacific, South Africa, and southeastern Australia (Perrin *et al.*, 1973).

Fraser's dolphin is easily identified by its stocky body, short but distinct beak, and small, triangular, or slightly falcate dorsal fin; the flippers and flukes are also small (Fig. 1; Jefferson *et al.*, 1993; Jefferson and Leatherwood, 1994; Perrin *et al.*, 1994). The color pattern is striking and varies with age and sex (Jefferson *et al.*, 1997). For example, a distinct black head stripe or "bridle" is absent in calves, variable in females, and extensive in adult males, where it merges with the eye-to-anus stripe to form a "bandit mask." Color pattern in the genital region may also be sexually dimorphic. The back is brownish gray, the lower side of the body is cream colored, and the belly is white or pink. Other features that appear to vary with age and sex are dorsal fin shape and the post-anal hump. With some variability, the dorsal fin is slightly falcate in calves and females and more erect or forward canted in adult males. Similarly, the post-anal hump is either absent or slight in females and young of both sexes and well developed in adult males. From a distance,

the eye-to-anus stripe makes Fraser's dolphin look similar to the striped dolphin, *Stenella coeruleoalba*. However, the distinctive body shape of Fraser's dolphin rules out confusion with other species. The largest male recorded was 2.7 m long and the largest female 2.6 m with males over 10 years old significantly larger than females. Large males could weigh up to 210 kg. Based on a limited number examined, it is tentatively proposed that Fraser's dolphins in the Atlantic are larger than those in the Pacific.

Fraser's dolphin belongs to the subfamily Delphininae. Based on cytochrome *b* mtDNA sequences, it is more closely related to *Stenella*, *Tursiops*, *Delphinus*, and *Sousa* than it is to *Lagenorhynchus* (LeDuc *et al.*, 1999). Morphologically, the skull (Fig. 2) structure shows close similarity with that of the common dolphin, *D. delphis*, in terms of the presence of deep palatal grooves, and with those of *S. longirostris*, *S. coeruleoalba*, and the Clymene dolphin, *Stenella clymene*, in several other characteristics.

## II. Distribution and Abundance

Fraser's dolphin is a tropical species, distributed between 30°N and 30°S. Strandings outside this limit, such as in southeastern Australia, Brittany, United Kingdom and Uruguay, are considered unusual and are probably influenced by temporary oceanographic events. Density and abundance are known only for a few areas: eastern tropical Pacific, 289,300 with CV 0.34 (Wade and Gerrodette, 1993); eastern Sulu Sea, 13,518 with CV 0.26 and density 0.58/km² (Dolar *et al.*, 2006); Hawaii,

F

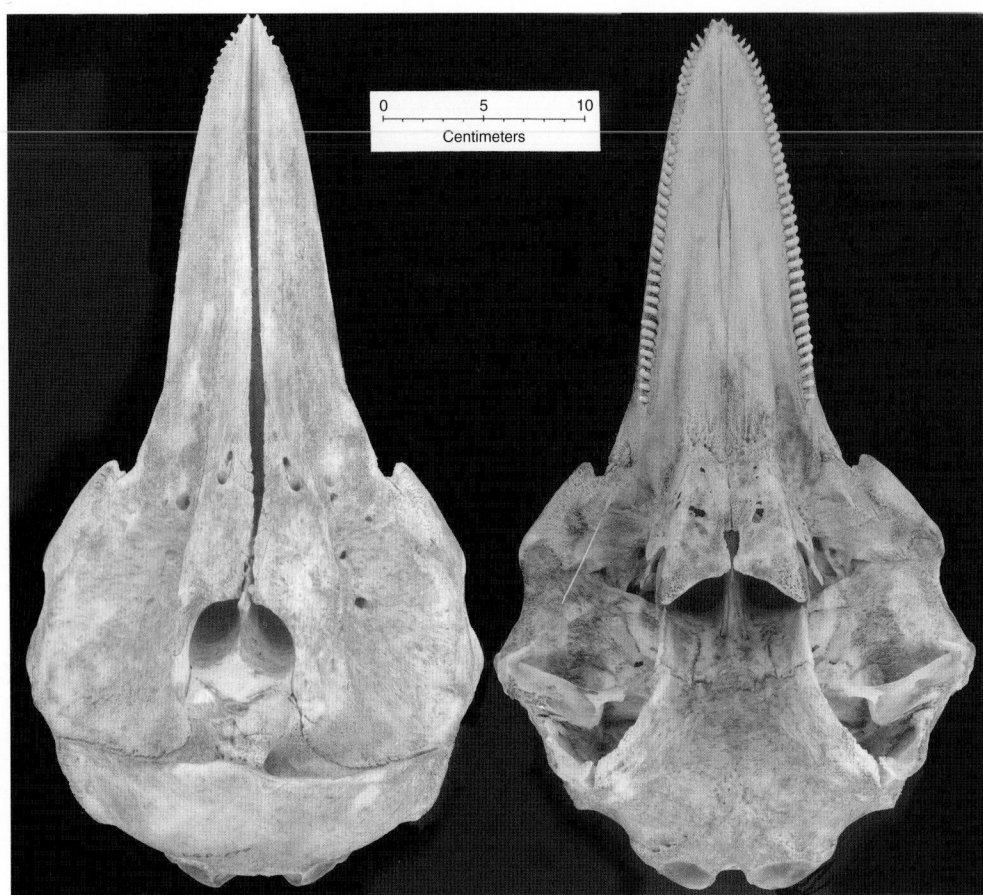

**Figure 2**   *Skull of Fraser's dolphin. Photograph by W. F. Perrin.*

10,226 with CV 1.16 and density 0.0042/km² (Barlow, 2006). Populations in Japan and the Philippines differ morphologically (Perrin *et al.*, 2003).

## III.  Ecology

Fraser's dolphins are typically an oceanic species, except in places where deep water approaches the coast such as the Philippines, Indonesia, and Lesser Antilles, where Fraser's dolphins can be observed within 100 m from shore (Balance and Pitman, 1998; Dolar *et al.*, 2006). In the eastern tropical Pacific, they were observed to occur at least 15 km from the coast where water depth was between 1500 and 2500 m. In the Sulu Sea, Philippines, high sighting rates and large school sizes were in waters 700 to 3,500 m, peaking at about 1000 m, although some animals were observed in shallower waters adjacent to the continental shelf. In the Gulf of Mexico, sightings have been in waters around 1000 m deep, and the animals appear to be more common in the Gulf than anywhere else in the North Atlantic (Würsig *et al.*, 2000). Affinity to deep waters can be explained by the type of prey eaten by Fraser's dolphins, which is composed of mesopelagic fish, crustaceans, and cephalopods (Robison and Craddock, 1983; Dos Santos and Haimovici, 2001; Dolar *et al.*, 2003). It is also suggested that compared to other pelagic dolphins, Fraser's dolphins feed selectively on larger prey that inhabit deeper waters. In the eastern tropical Pacific and the Sulu Sea, the most common fishes in the diet are the Myctophidae and the Chauliodontidae, and the most common crustaceans are the Oplophoridae. Mesopelagic cephalopods also comprised a significant amount of the DIET in the Sulu Sea animals, i.e., about 30% by volume.

In southern Brazil, cephalopods, fishes, and crustaceans were found in the stomachs of four stranded Fraser's dolphins. Based on prey composition, it was hypothesized that Fraser's dolphins in the eastern tropical Pacific feed at two depth horizons: the shallowest level of no less than 250 m and the deepest no less than 500 m. In the Sulu Sea, Fraser's dolphins appear to feed from near surface to deeper than 600 m. Myoglobin (Mb) concentrations in the skeletal muscles of Fraser's dolphin support the distribution and feeding habits of this species (Dolar *et al.*, 1999). However, in South Africa and in the Caribbean, Fraser's dolphins were observed FEEDING near the surface (Sekiguchi *et al.*, 1992; Watkins *et al.*, 1994).

Although no PREDATION has been reported, Fraser's dolphins may be preyed upon by killer whales (*Orcinus orca*), false killer whale (*Pseudorca crassidens*), and large sharks. Cookie cutter sharks (*Isistius brasiliensis*) are thought to inflict circular wounds.

An external PARASITE, *Xenobalanus* sp., and internal parasites (*Phyllobothrium delphini*, *Monorhygma grimaldi*, *Anisakis simplex*, *Tetrabothrius* sp., *Bolbosoma* sp., *Strobicephalus triangularis*, *Campula* sp., and *Stenurus ovatus*) have been observed in Fraser's dolphins.

Dolphin morbillivirus was found present in samples of Fraser's dolphins from the southwest Atlantic with an indication that the virus might be endemic to these animals (Van Bressem *et al.*, 2001).

## IV.  Behavior and Physiology

Fraser's dolphins often swim in tight fast-moving schools of 100 to 1000 individuals with the members of the school "porpoising in low-angle, splashy leaps" and have been reported to swim away from vessels in the eastern tropical Pacific. In the Philippines, they were observed to ride the bow at boat speed less than three knots but were often displaced by melon-headed whales (*Peponocephala electra*).

In the eastern tropical Pacific and in the Gulf of Mexico, Fraser's dolphins are often found together with melon-headed whales (Perryman *et al.*, 1994; Würsig *et al.*, 2000). Although also sometimes seen with melon-headed whales (when in relatively shallow waters), Fraser's dolphins in the Sulu Sea are often seen with short-finned pilot whales, *Globicephala macrorhynchus* (Dolar *et al.*, 2006). Other species Fraser's dolphins have not been seen with are the false killer whale, Risso's dolphin (*Grampus griseus*), spinner dolphin, pantropical spotted dolphin (*S. attenuata*), bottlenose dolphin (*Tursiops truncatus*), and sperm whale (*Physeter macrocephalus*). In the western tropical Indian Ocean, Fraser's dolphins have not been seen with any other species.

Fraser's dolphins are deep divers. Based on the vertical distribution of the prey they eat, they could dive down to 600 m or deeper. Mb concentration in epaxial muscles averages at 7.1 g Mb/100$^g$ muscle and is comparable to those of the very best divers such as the Weddell seal (*Leptonychotes weddellii*), northern bottlenose whale (*Hyperoodon ampullatus*), and sperm whale.

## V. Life History

The oldest specimen recorded from a sample of 16 dolphins from southeastern Brazil (Siciliano *et al.*, 2007) was 19 years old. An asymptotic length of 231.2 cm predicted for this group using the Gompertz model occurred at about 7–8 years old. In Japan, the oldest specimen from a sample of 108 dolphins was estimated to be 17.5 years old with the males reaching sexual maturity at about 7–10 years at 220–230 cm and the females at 5–8 years at 210–220 cm (Amano *et al.*, 1996). The school showed a mixed-age group and a ratio of 1:1 between males and females. Mating may be promiscuous. The annual ovulation rate is about 0.49, and the gestation period is about 12.5 months. The calving interval is approximately 2 years. In Japanese waters, calving appears to peak in spring and fall. Limited samples from South African waters suggest that calving occurs in summer. Length at birth is estimated to be about 100–110 cm.

## VI. Interaction with Humans

Fraser's dolphins are caught in drive nets in Japan and by harpoon in the Lesser Antilles, Indonesia, and (before they became protected) in the Philippines (Dolar, 1994; Dolar *et al.*, 1994). They are also caught incidentally in purse seines in the eastern tropical Pacific and the Philippines, in trap nets in Japan, in gill nets in South Africa and Sri Lanka, in anti-shark nets in South Africa, and in drift nets in the Philippines.

## References

Amano, M., Miyazaki, N., and Yanagisawa, F. (1996). Life history of Fraser's dolphin, *Lagenodelphis hosei*, based on a school captured off Pacific coast of Japan. *Mar. Mamm. Sci.* **12**, 199–214.

Balance, L. T., and Pitman, R. B. (1998). Cetaceans of the tropical western Indian Ocean: Distribution, relative abundance, and comparisons with cetacean communities of two other tropical ecosystems. *Mar. Mamm. Sci.* **14**, 428–459.

Barlow, J. (2006). Cetacean abundance in Hawaiian waters estimated from summer/fall survey in 2002. *Mar. Mamm. Sci.* **22**, 446–464.

Bones, M., Neill, B., and Reid, B. (1998). Fraser's dolphin (*Lagenodelphis hosei*) stranded in South Uist: First record in UK waters. *J. Zool.* **246**, 460–461.

Dolar, M. L. L. (1994). Incidental takes of small cetaceans in fisheries in Palawan, central Visayas and northern Mindanao in the Philippines. *Rep. Int. Whal. Commn* **15**(Spec. Issue), 355–363.

Dolar, M. L. L., Leatherwood, S., Wood, C., Alava, M. N. R., Hill, C., and Aragones, L. V. (1994). Directed fisheries for cetaceans in the Philippines. *Rep. Int. Whal. Commn* **44**, 439–449.

Dolar, M. L. L., Suarez, P., Ponganis, P., and Kooyman, G. L. (1999). Myoglobin in pelagic small cetaceans. *J. Exp. Biol.* **202**, 227–236.

Dolar, M. L. L., Walker, W. A., Kooyman, G. L., and Perrin, W. F. (2003). Comparative feeding ecology of spinner dolphins (*Stenella longirostris*) and Fraser's dolphins (*Lagenodelphis hosei*) in the Sulu Sea. *Mar. Mamm. Sci.* **19**, 1–19.

Dolar, M. L. L., Perrin, W. F. P., Taylor, B. L., and Kooyman, G. L. (2006). Abundance and distributional ecology of cetaceans in the central Philippines. *J. Cetacean Res. Manage.* **8**, 93–111.

Dos Santos, R. A., and Haimovici, M. (2001). Cephalopods in the diet of marine mammals stranded or incidentally caught along southeastern and southern Brazil (21–34°S). *Fish. Res.* **52**, 99–112.

Jefferson, T. A., and Leatherwood, S. (1994). Lagenodelphis hosei. *Mam. Spec.* **470**, 1–5.

Jefferson, T. A., Leatherwood, S., and Weber, M. A. (1993). "FAO Species Identification Guide: Marine Mammals of the World." FAO, Rome.

Jefferson, T. A., Pitman, R. L., Leatherwood, S., and Dolar, M. L. L. (1997). Developmental and sexual variation in the external appearance of Fraser's dolphins (*Lagenodelphis hosei*). *Aquat. Mamm.* **23**, 145–153.

LeDuc, R. G., Perrin, W. F., and Dizon, A. E. (1999). Phylogenetic relationships among the delphinid cetaceans based on full cytochrome *b* sequences. *Mar. Mamm. Sci.* **15**, 619–648.

Perrin, W. F., Best, P. B., Dawbin, W. H., Balcomb, K. G., Gambell, R., and Ross, G. J. B. (1973). Rediscovery of Fraser's dolphin *Lagenodelphis hosei*. *Nature* **241**, 345–350.

Perrin, W. F., Leatherwood, S., and Collet, A. (1994). Fraser's dolphin, *Lagenodelphis hosei* Fraser, 1956. *In* "Handbook of Marine Mammals" (S. H. Ridgway, and R. Harrison, eds), Vol. 5, pp. 225–240. Academic Press, San Diego.

Perrin, W. F., Dolar, M. L. L., Amano, M., and Hayano, A. (2003). Cranial sexual dimorphism and geographic variation in Fraser's dolphin, *Lagenodelphis hosei*. *Mar. Mamm. Sci.* **19**, 484–501.

Perryman, W. L., Au, D. W. K., Leatherwood, S., and Jefferson, T. A. (1994). Melon-headed whale, *Peponocephala electra* Gray, 1846. *In* "Handbook of Marine Mammals" (S. H. Ridgway, and R. Harrison, eds), Vol.5, pp. 363–386. Academic Press, San Diego.

Robison, B. H., and Craddock, J. E. (1983). Mesopelagic fishes eaten by Fraser's dolphin, *Lagenodelphis hosei*. *Fish. Bull. US* **81**, 283–289.

Sekiguchi, K., Klages, N.T., and Best, P.B. (1992). Comparative analysis of the diets of smaller odontocete cetaceans along the coast of Southern Africa. *S. Afr. J. Mar. Sci.* 12, 843–861.

Siciliano, S., *et al.* (2007). Age and growth of some delphinids in southeastern Brazil. *J. Mar. Biol. Assoc. UK* **87**, 293–303.

Wade, P. R., and Gerrodette, T. (1993). Estimates of cetacean abundance and distribution in the eastern tropical Pacific. *Rep. Int. Whal. Commn* **43**, 477–493.

Watkins, W. A., Daher, M. A., Fristrup, K., and Notobartolo di Sciara, G. (1994). Fishing and acoustic behavior of Fraser's dolphin (*Lagenodelphis hosei*) near Dominica, southeast Caribbean. *Carib. J. Sci.* **20**, 76–82.

Würsig, B., Jefferson, T. A., and Schmidly, D. J. (2000). "The Marine Mammals of the Gulf of Mexico." Texas A & M Univ. Press, College Station, TX.

Van Bressem, M. F., *et al.* (2001). An insight into the epidemiology of dolphin morbillivirus worldwide. *Vet. Microbiol.* **81**, 287–304.

# Gastrointestinal Tract

### James G. Mead

The gastrointestinal tract consists of all structures derived from the primitive gut tube and distal to the esophagus. As such, the gastrointestinal tract includes the stomach, small intestine, large intestine, and those accessory structures that have formed from that part of the gut (liver, gall bladder, pancreas,

hepatopancreatic duct, anal tonsils). The posterior boundary is the lower part of the anal canal where the mucous membrane of the gut ends and the epidermis begins.

The anatomy of the gastrointestinal tract has long fascinated workers. Grew (1681) is the earliest worker who dealt with that topic exclusively. Tyson (1680), in his marvelous treatment of the anatomy of the harbor porpoise (*Phocoena phocoena*), went extensively into the gastrointestinal tract. Owen dissected the dugong (*Dugong dugon*) (1838) and then summarized the information on the digestive system of mammals in his magnum opus on comparative anatomy (1868). William Turner (1912) did extensive studies of the stomach of cetaceans which are summarized in his catalog of the specimens of marine mammals in the Anatomical Museum of the University of Edinburgh. Langer (1988) and Reynolds and Rommel (1996) discussed the gastrointestinal tract of the Sirenia in some detail.

The parts of the gastrointestinal tract are described starting with the stomach and progressing distally, using the terminology of Chivers and Langer (1994). The major features of the gastrointestinal tract are summarized in Table I. Some dimensions are presented in this chapter but taking consistent measurements of the gastrointestinal tract, both length and volume, is difficult due to the elasticity of the organs. At death, the muscles lose their tonus and the length and volume can double or triple (Slijper 1962).

TABLE I

**Comparative Morphology of the Gastrointestinal System of Marine Mammals**

| Taxon | Stomach | | | | | Small intestine | | | |
|---|---|---|---|---|---|---|---|---|---|
| | Stomach type | Forestomach | Main stomach | Connecting chambers | Pyloric stomach | Cardiac gland | Duoenum | Duodenal ampulla | Duodenal diverticula |
| **Pinnipedia** | | | | | | | | | |
| Phocid | Unilocular | Absent | Present | Absent | Absent | Absent | Present | Absent | Absent |
| Otariid | Unilocular | Absent | Present | Absent | Absent | Absent | Present | Absent | Absent |
| Odobenid | Unilocular? | Absent? | Present? | Absent? | Absent? | Absent? | Present | Absent | Absent? |
| **Sirenia** | | | | | | | | | |
| Dugonid | Unilocular | Absent | Present | Absent | Present | Present | Present | Present | Present |
| Trichechid | Unilocular | Absent | Present | Absent | Present | Present | Present | Present | Present |
| **Cetacea** | | | | | | | | | |
| **Mysticete** | | | | | | | | | |
| Balaenopterid | Plurilocular | Present | Present | Present | Present | Absent | Present | Present | Absent |
| Eschrichtiid | Plurilocular | Present | Present | Present | Present | Absent | ? | ? | Absent |
| Balaenid | Plurilocular | Present | Present | Present | Present | Absent | ? | ? | Absent |
| Neobalaenid | Plurilocular | Present | Present | ? | Present | Absent | ? | ? | Absent |
| **Odontocete** | | | | | | | | | |
| **Delphinoid** | | | | | | | | | |
| Delphinid | Plurilocular | Present | Present | Present | Present | Absent | Present | Present | Absent |
| Phocoenid | Plurilocular | Present | Present | Present | Present | Absent | Present | Present | Absent |
| Monodontid | Plurilocular | Present | Present | Present | Present | Absent | Present | Present | Absent |
| Platanistoid | Plurilocular | Variable | Hyper. | Variable | Variable | Absent | Present | Present | Absent |
| Physeteroid | Plurilocular | Present | Present | Present? | Present | Absent | Present | Present | Absent |
| Ziphioid | Plurilocular | Absent | Variable | Hyper. | Present | Absent | Present | Present | Absent |

hyper. = hypertrophied
undiff. = undifferentiated
? = unknown

## I. Major Organs

### A. Stomach

The stomach is a series of compartments starting with the cardiac, fundic and ending with the pyloric. The boundary of the stomach with the esophagus is determined by the epithelial type; stratified squamous for the esophagus, columnar for the stomach. The distal boundary is marked by the pyloric sphincter.

*1. Pinnipedia* The stomach in pinnipeds is relatively uncomplicated when compared to the rest of marine mammals. The stomach in the California sea lion (*Zalophus californianus*) (Green, 1972, p. 286) consists of a simple cardiac chamber into which the esophagus enters, followed by a narrowing into the pyloric chamber. There is a prominent pyloric sphincter. The pyloric end of the stomach is strongly recurved of the cardiac portion. The stomach in the walrus, southern sea lion (*Otaria flavescens*) and Weddell seal (*Leptonychotes weddellii*) do not differ from the California sea lion. The stomach of the walrus (*Odobenus rosmarus*), although it is not described in any detail, does not appear to differ markedly from that of the other pinnipeds. Pinnipeds follow the carnivore plan of a relatively simple single-chambered (monolocular) nonspecialized stomach.

*2. Sirenia* The stomach in the dugong is moderately complex. Externally it is a simple oval organ with the esophageal opening in the center. Internally there is a ridge (gastric ridge) that divides the stomach into two compartments, the cardiac and the pyloric portions. There is a development of a powerful sphincter up to 4-cm thick at the esophageal/gastric junction (Owen, 1868). The stomach wall is highly muscular. There is a cardiac gland that is roughly spherical and about 15cm in diameter in adults. The cardiac gland opens into the first compartment, where the esophagus also opens. The mucosa in the cardiac gland is packed with the gastric glands that are distinguishable from the glands in the main stomach compartment. The glands consist of chief and parietal cells in a ratio of about 10:1. The mucosa in the cardiac glands is arranged in a complex plicate structure. The pyloric aperture is in the second compartment. The cardiac region of the stomach extends for several centimeters from the esophageal junction. The stomach is lined by gastric glandular epithelium with a particular abundance of goblet cells and mucus-secreting gastric glands.

The stomach of the dugong appears to be modified to secrete mucus to aid in lubricating the ingested material and prevent abrasion to the mucosa. It is interesting that the salt content of the dugong diet is high; the sodium is about 30 times and the chloride about 15 times that of terrestrial pasture plants.

The stomach in the recently extinct *Hydrodamalis* (Steller's sea cow) was apparently very large. According to Steller it was 6-ft long and 5-ft wide when distended with masticated sea-weed. The stomach in the manatees (*Trichechus* spp.) is very similar to that in the dugong. The stomach is divided by a muscular ridge into cardiac and pyloric regions. A single cardiac gland opens into the cardiac region of the stomach.

**G**

| Jejunum | Ileum | Large intestine | | Accessory organs | | | | |
| | | Caecum | Colon | Liver | Gall bladder | Pancreas | Hepato-pancreatic duct | Anal tonsils |
|---|---|---|---|---|---|---|---|---|
| Undiff. | Undiff. | Present | Present | Multilobed | Present | Present | **Absent** | Absent? |
| Undiff. | Undiff. | Present | Present | Multilobed | Present | Present | Present | Absent? |
| Undiff. | Undiff. | Present | Present | Multilobed | Present? | Present | Present | Absent? |
| | | | | | | | | |
| Undiff. | Undiff. | **Hyper.** | Present | Multilobed | Present | Present | **Absent** | Absent? |
| Undiff. | Undiff. | **Hyper.** | Present | Multilobed | Present? | Present | **Absent** | Absent? |
| | | | | | | | | |
| Undiff. | Undiff. | Present | Present | Bilobed | **Absent** | Present | Present | Absent? |
| ? | ? | Present? | Present? | Bilobed | **Absent** | Present | Present | **Present** |
| ? | ? | **Absent** | Undiff. | Bilobed | **Absent** | Present | Present | Absent? |
| ? | ? | Present? | Present? | Bilobed | **Absent** | Present | Present? | Absent? |
| | | | | | | | | |
| Undiff. | Undiff. | **Absent** | Undiff. | Bilobed | **Absent** | Present | Present | Absent |
| Undiff. | Undiff. | **Absent** | Undiff. | Bilobed | **Absent** | Present | Present | Absent? |
| Undiff. | Undiff. | **Absent** | Undiff. | Bilobed | **Absent** | Present | Present | Absent? |
| Undiff. | Undiff. | **Variable** | Undiff. | Bilobed | **Absent** | Present | Present | **Variable** |
| Undiff. | Undiff. | **Absent** | Undiff. | Bilobed | **Absent** | Present | Present | **Present?** |
| Undiff. | Undiff. | **Absent** | Undiff. | Bilobed | **Absent** | Present | Present | Absent? |

*3. Cetacea* The cetacean stomach is a diverticulated composite stomach, consisting of regions of stratified squamous epithelium, fundic mucosa, and pyloric mucosa. The stomach, as typified by a delphinids, consists of four chambers. These have been referred to by various anatomical terms: forestomach (first, esophageal compartment, paunch) main stomach (second, cardiac, fundus glandular, proximal), connecting chamber (third, fourth, "narrow tunneled passage," "conduit ètroit," intermediate, connecting channel, connecting division), and pyloric stomach (third, fourth, fifth, pyloric glandular, distal).

A. Forestomach There is no full consensus about the homology of the forestomach in Cetacea. It is lined with stratified squamous epithelium, such as the esophagus, and there was reason to believe that it was just an esophageal sacculation. Embryological work in the minke whale (*Balaenoptera acutorostrata*) demonstrated that the forestomach was formed from the stomach bud, but that the esophagus was not. This indicates that the cetacean esophagus is homologous to the forestomach of ruminants.

The forestomach of delphinoids (also called paunch) is lined with stratified squamous non-keratinized epithelium. The epithelial lining is white in freshly dead animals and is thrown into a series of longitudinal folds when empty. Similar to the other chambers in the stomach it is variable in size. It is pyriform and on the order of 30-cm long in an adult *Tursiops truncatus* (280cm total length). The forestomach is highly muscular but has no glandular functions. The forestomach/main stomach aperture is a wide opening (3–5cm in adult *Tursiops*) in the wall of the forestomach near the esophageal end. The forestomach functions as a holding cavity analogous to the crop of birds or the forestomach of ungulates. Because the communication with the main stomach is so wide, there is a reflux of digestive fluids from the main stomach and some digestion takes place in the forestomach. The same general relationships hold in *Phocoena*, *Delphinapterus*, and *Monodon*.

The forestomach of platanistoids is unusual in *Inia geoffrensis* and *Platanista gangetica* in that the esophagus runs directly into the main stomach and the forestomach branches off the esophagus. In the two other genera of platanistids the forestomach is lacking entirely.

The forestomach is present in *Physeter catodon*, where it was approximately 140 by 140cm and lined with yellowish-white epithelium in a 15.6m male. The forestomach is absent in all ziphiids, and present in all species of mysticetes.

B. Main Stomach The main stomach has a highly vascular, glandular epithelium which is grossly trabeculate. The epithelium of the main stomach is dark pink to purple. The main stomach is the compartment which secretes most of the digestive enzymes and acids and in which place digestion commences. It has also been known as the fundic stomach. It is present in all cetaceans.

In delphinoids, the main stomach is approximately spherical and on the order of 10–15cm in adult *Tursiops*. The same general relationships hold in *Phocoena*, *Delphinapterus*, and *Monodon*. In *Platanista* there is a constricting septum of the main stomach which forms a small distal chamber, through which the digesta must pass. *Lipotes vexillifer* presents an unusual situation in having three serially arranged main stomach compartments. The second and third compartments are very much smaller than the first and are topographically homologous with the connecting chambers. However they are lined by epithelium that has fundic glands, typical of the main stomach.

There is nothing remarkable about the main stomach of physeteroids, but in some ziphiids there is a subdivision in their main

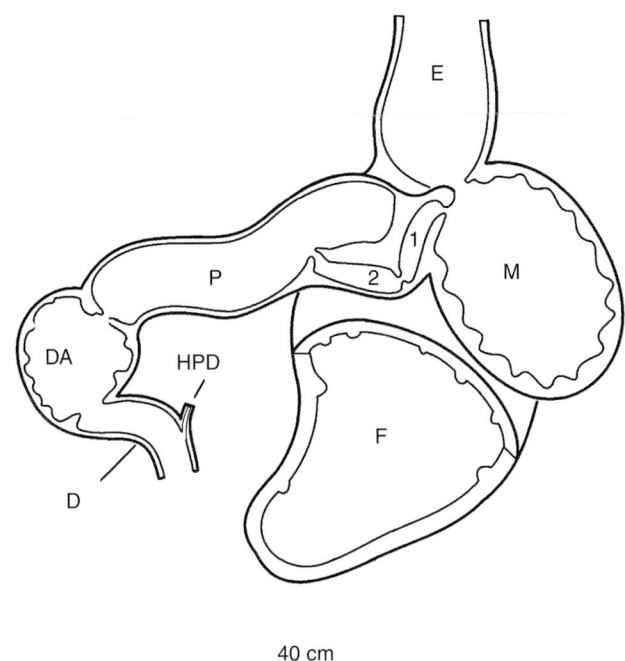

**Figure 1** *Stomach of a spinner dolphin,* Stenella longirostris, *in ventral view. D, duodenum; DA, duodenal ampulla; E, esophagus; F, forestomach; HPD, hepatopancreatic duct; M, main stomach; P, pyloric stomach; 1, 2, compartments of connecting chambers. After Harrison* et al. *(1970).*

stomach. There is an incipient constriction in the main stomach of *Berardius bairdii* and *Mesoplodon bidens* that divides the stomach into two compartments. The connecting chambers exit off the second compartment. Another type of stomach modification has occurred in *Mesoplodon europaeus* and *M. mirus*, where a large septum has developed forming a blind diverticulum in the main stomach. An additional septum has developed in the diverticulum in *Mesoplodon europaeus* subdividing it Fig. 2. There is nothing remarkable about the main stomach in mysticetes.

C. Connecting Chambers The connecting chambers, also called the connecting channel, the intermediate stomach, or the third stomach, are present in all Cetacea. They are lined with pyloric epithelium and are easily overlooked in dissections. They are small in most cetaceans but have been greatly developed in ziphiids. Because of their proliferation in ziphiids, where they seem to function as something more than channels between the main and pyloric stomachs, their name was changed from connecting channels to connecting chambers.

The connecting chambers in a typical delphinoid consist of two narrow compartments lying between the main stomach and the pyloric stomach. The diameter of the connecting chambers is 0.8 cm in adult *Tursiops* and the combined length is 7–9cm. The epithelial lining is very similar to the pyloric stomachs. In some species the compartments are simple, serially arranged; in others they may have diverticulae. The same general relationships hold in *Phocoena*, *Delphinapterus*, and *Monodon*.

Connecting chambers occur in all the species of platanistids, with the exception of *Lipotes*. In that species, the compartments lying between the main stomach and the pyloric stomach (second and third compartments of the main stomach) are lined with epithelium containing fundic and mucous glands in the first compartment and fundic glands in the second compartment. This would make them subdivisions of the main stomach.

The connecting chambers in ziphiids are globular compartments, ranging in number from 3 to 11. They are separated by septa and communicate by openings in the septa. The openings are sometimes central in the septa, sometimes peripheral. The connecting chambers are lined with pyloric epithelium. The connecting chambers in specimens of adult *Mesoplodon* (~5 m long) are about 10 cm in diameter.

Many workers have described the connecting chambers in a number of species of *Balaenoptera* (blue, *B. musculus*; fin, *B. physalus*; sei, *B. borealis*; minke, *B. acutorostrata* and *B. bonaerensis*). The connecting chambers in common minke whales were 10–30 cm in length. The inflated connecting chambers in an 8.5-m female *Balaena mysticetus* were 5 cm in diameter and 17 cm combined length. The presence of connecting chambers was not mentioned in dissections of right whales. The connecting chambers are relatively large in a newborn *Eschrichtius*.

D. PYLORIC STOMACH  The pyloric stomach in delphinoids is a simple tubular cavity lined by typical mucous producing pyloric glands. The epithelium is in many ways similar to the epithelium of the small intestine. The pyloric stomach is about 20-cm long and 4 cm in flat diameter in an adult *Tursiops*. The same general relationships hold in *Phocoena*, *Delphinapterus*, and *Monodon*.

The pyloric stomach in *P. gangetica* is a single chamber about 12-cm long and contains abundant large tubular pyloric glands. The pyloric stomach is comparable in *Inia* and *Pontoporia* but differs markedly in *Lipotes*. In that species it is differentiated into a proximal bulbous compartment and a smaller distal compartment. The epithelial lining in *Lipotes* is similar to all other Cetacea.

The available data on the pyloric stomach of physeteroids is scanty. The pyloric compartment is present and there is no reason to assume that it is different from the rest of the cetaceans. The pyloric stomach in a newborn *Ziphius* was a simple spherical compartment that measured about 10 cm in diameter. It was lined with smooth pyloric epithelium and communicated with the duodenum through a strong pyloric sphincter. This is also the case in *Hyperoodon*, *Tasmacetus*, and some species of *Mesoplodon* (*M. densirostris*, *M. hectori*, and *M. stejnegeri*). In *B. bairdii* the main pyloric compartment has expanded in volume to where it is nearly the size of the main stomach, and has developed a small distal accessory chamber. The pyloric compartments are in series, accessory chamber lies between it and the duodenum. In all other species of *Mesoplodon* examined to date (*M. bidens*, *M. europaeus*, and *M. mirus*), a blind diverticulum has developed. The diverticulum comes off the proximal side of the pyloric stomach and lies along the distal connecting chambers. The accessory pyloric stomach communicates with the pyloric stomach through a wide opening.

In all of the balaenopterid species examined (*B. acutorostrata*, *B. borealis*, *B. musculus*, and *B. physalus*), the pyloric stomach is smaller than the main stomach. The pyloric stomach contained 8.5–12.1% of the total inflated stomach volume (18–39). It is lined with smooth pyloric epithelium. In balaenids and newborn *Eschrichtius*, the pyloric stomach appears to be similar to that of balaenopterids.

## B. Small Intestine

The small intestine starts at the pyloric sphincter. Digestion continues in the small intestine and absorption of the nutrients takes place here via absorbtive villi in the mucosa. The small intestine consists of duodenum, ileum, and jejunum. The hepatic and pancreatic ducts open into the duodenum. The duodenum is short and has longitudinal folds. In the jejunum, the folds are circular (plicae circulares). Circular folds gradually disappear and are replaced by longitudinal folds toward the end of the ileum. The diameter of the intestine increases where it ends at the ileocolic orifice, opening into the colon. The ileocolic orifice is usually provided with a sphincter permitting partial closure.

*1. Pinnipedia*  The demarcation between pylorus and duodenum is sharply marked by position of the duodenal (Brunner's) glands in *Leptonychotes weddelli*. The duodenum is 1 or 2 ft in length. Small plicae circulares and short irregular villi were present in the duodenum. Jejunum and ileum are hard to differentiate. *Phoca vitulina* has a small intestine of "great length," 40 ft in a seal 3 ft long (snout → end of flippers). An adult male *Mirounga leonina* (4.80 m length) had a small intestine length of 202 m. *Otaria byronia* lacks plicae circulares, villi being arranged on delicate transverse linear folds. *Eumetopias* has small intestine length of 264 ft (~80 m). Owen (1853) described the intestine in passing in his description of a young walrus. The small intestine was 75 feet (~23 m) long, the cecum was 1.5 in. (3.8 cm), and the large intestine was 1 foot (~30 cm) in length.

*2. Sirenia*  The duodenum of the dugong and manatee has two duodenal diverticula that are crescentic in shape and about 10–15 cm long, measured in the curve. They communicate via a common connecting channel with the duodenum. The lining of the diverticulae is similar to the pyloric region of the stomach and contains mucous glands. The duodenum is about 30 cm in length, similar to other medium-sized mammals. Both the duodenum and the diverticulae contain prominent plicae circulares. There is a weak sphincter at the distal end of the duodenal ampulla. The length of the small intestine is from 5.4 to 15.5 m, 4–7 times the body length of the animal. Brunner's glands are present in the duodenum and the diverticulae. Paneth cells are absent in contrast to most domestic terrestrial herbivores. The diverticulae appear to enlarge the surface of the proximal duodenum which would allow a larger volume of digesta to pass from the stomach at one time.

*3. Cetacea*  In delphinoids, there is no cecum and no marked differentiation between large and small intestine. Intestine length ranged between 8.85 and 16.80 m in specimens of *Tursiops*, *Delphinus*, and *Stenella* (total lengths from 160 to 230 cm). There is a duodenum about 30-cm long but they lack differentiation between jejunum and ileum. Examination of the small intestines by light microscopy revealed a lack of well-developed villi in delphinids.

The length of the small intestine in a 204 cm *I. geoffrensis* was 4.15 m and the duodenum is approximately 20-cm long, with the jejunum not differentiated from the ileum. A prominent longitudinal fold begins at the opening of the hepatopancreatic duct in the duodenum and continues throughout the small intestine. The small intestine varies in diameter from 0.7 to 0.8 cm. The small intestine grades into a "smooth-walled portion" which is 1 cm in diameter. The boundary between small and large intestine is indistinct, with the "smooth-walled portion" was 80 cm in length and graded into the colon distally.

There are no plicae circulares or typical villi in the intestine of *Pontoporia*. A distinct continuous longitudinal fold occurs in the small intestine. There are abundant plicae circulares in the proximal part of the intestine in *P. gangetica*, which change to longitudinal

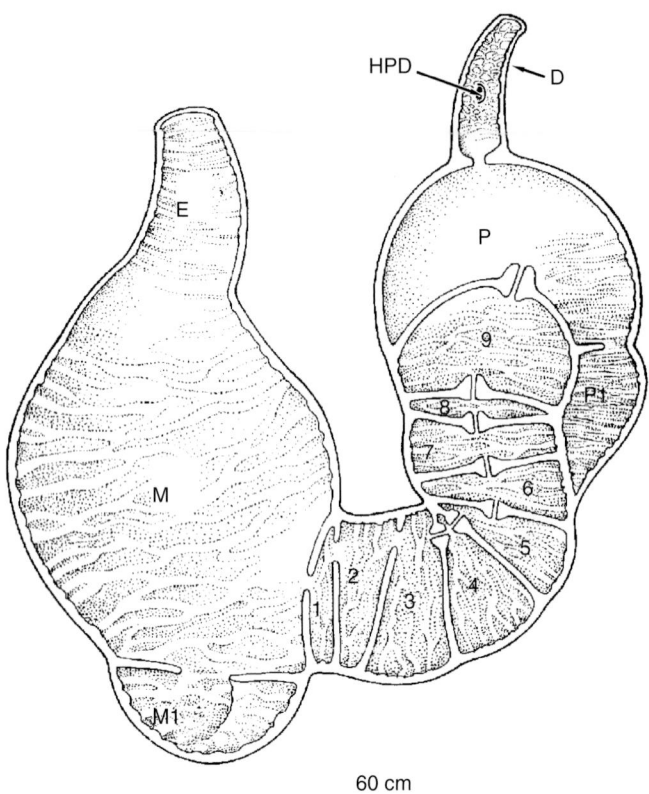

**Figure 2** *Stomach of a Gervais' beaked whale,* Mesoplodon europaeus, *in dorsal view. D, duodenum; E, esophagus; HPD, hepatopancreatic duct; M, main stomach; M1, accessory main stomach; P, pyloric stomach; P1, accessory pyloric stomach; 1–9, compartments of connecting chambers. Drawing by Trudy Nicholson.*

folds in the last meter or two. There is a prominent cecum and an ileocolic sphincter. The small intestine in most platanistids is extremely long. The ratio of small to large intestine length is between 50% and 60% in *Pontoporia*, 50% in *Inia* but only around 9% in *Platanista*.

The total intestinal length in *Physeter* adults can range up to 250 m. The plicae circulares are unusual in that they appeared to be spiral. There is no cecum and the transition between the small and the large intestine is gradual. In beaked whales, the intestine is also long, and may reach 6-times body length in *Hyperoodon*. There is a unique vascular rete (mirabile) intestinale associated with the large and small intestine in at least *Ziphius and Berardius* and there is no cecum.

The mean ratio of the length of the small intestine to body length in minke whales (*B. acutorostrata*) was rather small (3.92) and averaged 36 m in length. The minke whale possessed a duodenal ampulla, but there was no indication of differentiation of the jejunum and the ileum.

## C. Large Intestine

The large intestine consists of the colon (ascending, transverse and descending), cecum, vermiform appendix, rectum, anal canal and anus. The cecum is a diverticulum of the proximal end of the colon, near the ileocolic juncture. The vermiform appendix is the narrowed apex

of the cecum. The colon functions to absorb water and consolidate the fecal material. Most mammalian colons have their longitudinal muscle fibers arranged in groups into bands called taenia coli. The colon has a series of lateral bulges called haustra, and fatty projections (appendices epiploicae). The rectum is the straight portion of the large intestine that transverses the pelvis. The anal canal is the specialized terminal portion of the large intestine. The anal canal has many lymph nodes and glands and the anal sphincter controls excretion of fecal wastes.

1. *Pinnipedia* In pinnipeds the cecum is short and blunt or round and an appendix is not present. The large intestine is relatively short and not much larger in diameter than the small intestine. No taenia coli, plicae semilunares, haustra, and appendices epiploicae are present.

The colon is about 6-ft long (183 cm) in an adult *Leptonychotes*. The colon grades into the rectum, which begins at the pelvic inlet and ends at the anal canal. Throughout the length of the rectum the lining is thrown into large irregular transverse rectal plicae. Toward the distal portion of the rectum, the plicae become organized into five longitudinal anal columns that continue into the anal canal. The anal canal is much smaller in diameter than the rectum. Small coiled tubular rectal glands were present. The anal canal ends where the mucosa changes into a pigmented cornified stratified squamous epithelium (epidermis). There are circumanal glands, which are coiled tubular structures, representing modified sweat glands, and these are confined exclusively to this region. In *Leptonychotes* there is no evidence of other anal glands, sacs, or scent glands.

2. *Sirenia* The cecum in the dugong is conical and was about 6-in. long and 4-in. wide at the base in the half-grown specimen that Owen dissected (Owen 1838, 1868) A sphincter is present in the ileocecal juncture. There is no constriction between the cecum and the colon. The epithelia lining of the cecum is smooth and its walls are muscular. Owen (1868, p.457) hypothesizes that the cecum acts as a pump, "giving a first powerful impulse to the long column of vegetable 'magma' usually distending the colon."

The colon in the dugong is thinner-walled than the small intestine and is between 4 and 11 times the total body length (5.8–30.0 m). There are no taeniae coli (Hill, 1945). The lining of the colon is smooth, with the exception of irregular folds that are present at the wider terminal portion. The lining of the rectum is provided with longitudinal folds which become finer and more numerous in the anal canal. The lining of the anus is grayer and harder than the lining in the rectum. The anal canal is about 5-cm long. At the distal end of the canal the longitudinal folds become higher and terminate in globular swelling which occlude the lumen and which has termed "anal valves."

3. *Cetacea* In delphinoids, there is no cecum and no marked differentiation between large and small intestines in delphinoids. The colon in *I. geoffrensis* is 40-cm long, followed by a 5-cm rectum and a 3-cm anal canal. The proximal and distal portions of the colon are 1 cm and 1.5 cm respectively. There is a pronounced cecum that is 5–9 cm (2–3.5 in.) long in *P. gangetica*. The large intestine is short, 60 cm in adults. The lengths of the large intestine (cecum, colon, rectum, and anus) in 4 specimens that ranged between 76 and 127 cm total length, ranged from 25.5 to over 58 cm (the 127 cm specimen was lacking the cecum). There was no trace of taeniae coli. There is no cecum in *Pontoporia*. The longitudinal fold in the small intestine of *Pontoporia* divides to become two distinct longitudinal folds. Taeniae and haustra coli were not found.

In large adult *Physeter* the large intestine can be up to 26-m long. The mucosa of the large intestine in *Physeter* is not folded, and there is no cecum. The diameter of the descending colon is markedly increased in *Kogia* spp. There is no cecum and the transition between large and small intestines is gradual in ziphiids.

Mysticetes have a very short cecum except in right whales where it is absent altogether. There is a marked difference between the diameter of the large and small intestines in right whales (*Eubalaena* spp.). In the blue whale (*Balaenoptera musculus*), the taeniae consists of three longitudinal muscular bands. The mean ratio of the large intestine to body length in minke whales (*B. acutorostrata*) is 40%. The mean ratio of cecum length to body length is 4%; the cecum varies between 30 and 50 cm.

## II. Accessory Organs

The liver is derived from a diverticulum of the embryonic duodenum. As it grows, the liver expands to become the largest internal organ. The liver functions in storage and filtration of blood, in the secretion of bile, which aids in the digestion of fats, and is concerned with the majority of the metabolic systems of the body.

The liver is multilobed in pinnipeds, and has 7 or 8 lobes in *Otaria*. In Sirenia, the diaphragm is oriented in the dorsal plane instead of the transverse plane. The liver in the dugong and manatee are flattened against the dorsally oriented diaphragm. The liver is comprised of four lobes, the normal central, left and right, and the fourth, Spigelian lobe that lies on the dorsal border of the liver and is closely associated with the vena cava.

The liver in cetaceans is divided into two lobes by a shallow indentation. Occasionally, there is a third intermediate lobe. The cetacean liver is greater than would be expected for a mammal of its size.

The gallbladder is located on the posterior side of the liver where the hepatic duct issues, it forms in embryology from the same diverticulum as the liver. The gall bladder stores and concentrates the bile that is secreted by the liver. The gall bladder is universally present in pinnipeds and tends to be pyriform and located in a fossa of one of the subdivisions of the right lobe of the liver. The gall bladder is small in the dugong and is strongly sigmoid in shape. It lies on the ventral surface of the central lobe where the falciform and round ligaments attach. The gall bladder is absent in all members of the order Cetacea. The duct systems for bile produced in the liver are increased in diameter, suggesting that bile is stored here.

The pancreas also develops out of outgrows of the embryonic duodenum. It consists of two developmental bodies, the dorsal and ventral pancreas, which may empty into either the hepatic duct or directly into the duodenum. The pancreas secretes enzymes that are discharged into the duodenum and insulin that is discharged directly into the blood. The pancreas in marine mammals appears to have no remarkable differences from other mammals.

Tonsils are bodies of organized lymphatic tissues around crypts which they use to communicate to the lumen of whatever system they are in. In some cetaceans, but not in pinnipeds and sirenians, clusters of lymphatic tissue that can be called tonsils occur in the anal canal.

Anal tonsils have not been reported in most delphinoids, balaenids, eschrichtiids, and neobalaenids, and *Pontoporia*. Anal tonsils were found in the anal canal of *P. gangetica* and *Stenella coeruleoalba*, and lymphoid tissue also occurred in the anal canal of *I. geoffrensis*. Anal tonsils also occur in sperm whales. In the gray whale, anal tonsils consist of masses of lymphatic tissue that communicated with the anal canal via crypts. They lie near the boundary of the anal canal with the rectum, 30–40 cm from the anal orifice.

## See Also the Following Articles

Anatomical Dissection, Thorax and Abdomen ■ Diet ■ Energetics

## References

Chivers, D. J., and Langer, P. (eds) (1994). "The Digestive System in Mammals: Food, Form and Function." Cambridge University Press, Cambridge.

Green, R. F. (1972). Observations on the anatomy of some cetaceans and pinnipeds. *In* "Mammals of the Sea—Biology and Medicine." (S. H. Ridgway, ed.), Chapter 4, pp. 247–297. Charles C Thomas, Springfield.

Grew, N. (1681). Musaeum regalis societatis. or a catalogue & description of the natural and artificial rarities belonging to the Royal Society and preserved at Gresham College. . . . . . . whereunto is subjoyned the comparative anatomy of stomachs and guts. London, 4to, il. 7, 386 pp., l. 1 + il. 2, pp. 142, 31 pls.

Harrison, R. J., Johnson, F. R., and Young, B. A. (1970). The oesophagus and stomach of dolphins (*Tursiops, Delphinus, Stenella*). *J. Zool.* **160**, 377–390.

Langer, P. (1988). "The Mammalian Herbivore Stomach: Comparative Anatomy, Function and Evolution." Gustav Fischer, Stuttgart and New York.

Langer, P. (1996). Comparative anatomy of the stomach of the Cetacea. Ontogenetic changes involving gastric proportions—mesenteries—arteries. *Zeitschrift für Säugetierkunde* **61**, 140–154.

Mead, J. G. (2007). Stomach anatomy and use in defining systematic relationships of the cetacean family Ziphiidae (beaked whales). *Anat. Rec.* **290**(6), 581–595, Figures 1–13.

Olsen, M. A., Nordboy, E. S., Blix, A. S., and Mathieson, S. D. (1994). Functional anatomy of the gastrointestinal system of northeastern Atlantic minke whales (*Balaenoptera acutorostrata*). *J. Zool.* **234**, 55–74.

Owen, R. (1838). [On the anatomy of the Dugong.] Proceedings of the Zoological Society of London,pp. 28–45.

Owen, R. (1868). On the anatomy of vertebrates. Volume III (1868), Mammals. Longmans, Green, and Co., London, x + 915 pp., [613 figs] [1868].

Reynolds, J. E., and Rommel, S. A. (1996). Structure and function of the gastrointestinal tract of the Florida manatee, *Trichechus manatus latirostris*. *Anat. Rec.* **245**, 539–558.

Slijper, E. J. (1962). "Whales." Hutchinson and Co, London, 475 pp.

Turner, W. (1912). "The Marine Mammals in the Anatomical Museum of the University of Edinburgh." Macmillan and Co, London, xv + 207 pp.

Tyson, E. (1680). [1980] - PHOCAENA or the Anatomy of a Porpess dissected at Gresham College. B. Tooke, London.

**G**

# Genetics for Management

## Phillip A. Morin and Andrew E. Dizon

Certain kinds of genetic information are particularly well suited to assist in designing strategies to protect human-impacted marine mammals. What sort of genetic information is required depends on the particular conservation goals wildlife managers seek to achieve when protecting specific species, or populations within species. For example, is the goal to prevent extinction of the species as a whole or to prevent extirpation of local, but not necessarily genetically unique, populations? For most developed nations, these goals are codified in laws presumably reflecting, at least in democratic societies, the will of the public. To achieve these goals, managers often choose between controversial and conflicting strategies, such as various limits on the species and numbers of marine mammals that can be incidentally killed during certain fishing operations. Relaxed limits favor the fishermen but may put a population of marine mammals at risk; stringent limits are less risky but may put an unsupportable burden on fishermen by restricting their fishing options. Obviously, the kind and the quality of biological data, genetic or otherwise, informing this choice are critical. Decisions have to be based on the current scientific information available, or they will be challenged in the courts. Although most scientific information on impacted populations is of value, certain kinds of information are much more important for the management process. If only limited data are available (molecular or other), biased or misleading conclusions can result in inappropriate decisions being made, eventually imperiling the population needing protection in the first place. Biological data on marine mammals, especially cetaceans, are difficult and consequently expensive to obtain. By consuming limited conservation funds, even good but irrelevant studies can impede the conservation effort. To insure that genetic studies proposed are relevant for management needs requires an understanding of the policy (the conservation goals) before doing the science (the information gathering) (Taylor and Dizon, 1999).

One advantage that genetic analyses have over "whole animal" studies is that data are easier to collect and few constraints are put on the quality of a sample or its origin. DNA is a relatively tough molecule, and adequate samples can be obtained from tiny amounts of a variety of tissues such as skin, blood or blood stains, hair follicles, placenta, excrement, baleen, modern or ancient bone, or, in some circumstances, formalin-preserved tissues. For instance, adequate amounts of mtDNA from *ca.* 1000-year-old bowhead whale (*Balaena mysticetus*) bones have been obtained. More recent historical samples of bone and baleen from St Lawrence Island in the Bering Sea have been used for both mtDNA and SNP analysis (Morin and McCarthy, 2007). For live animals, projectile biopsying (crossbow, firearm, or lance) has been used successfully for all but the smallest and shyest cetaceans (see chapter on Genetics, Overview).

## I. The "Conservation Unit"

Today, defining the population segment on which to focus conservation efforts is the primary use of genetic information. The US Marine Mammal Protection Act of 1972 (MMPA), the US Endangered Species Act of 1973 (ESA), the relevant legislation of some other nations, and the Revised Management Procedure of the International Whaling Commission (IWC) all direct that management efforts must be focused on populations below the species level. Although most other countries have not necessarily established laws codifying the conservation unit, biologists are generally in agreement that species comprise a collection of semi-isolated populations (i.e., species-wide panmixia is the exception) and that those semi-isolated populations should be the focus of management. However, the devil is in the details, and there is much controversy on the precise definition of these units. Besides having obvious biological consequences

for getting the groupings correct, there can be economic ones as well. For instance, quotas on harvest or incidental take are calculated as some allowable fraction of the overall abundance within the chosen conservation unit. A small conservation unit is the most *biologically* risk-averse because quotas are then necessarily small, and there is a greater likelihood that removals will be equally distributed over the whole unit. However, a large conservation unit is the most *economically* risk-averse because the quotas are larger, and there is the potential that excessive removals in one part of the range (the sink) will be compensated for by immigration from outside of the exploited region (the source).

Policy tries to provide managers with guidance to balance conservation and economic issues by defining the management unit (MU). For instance, the US ESA seeks to prevent the extinction of distinct population segments that are evolutionarily unique. The policy addresses last-ditch efforts to rescue populations whose abundances are so low, or whose abundances will become so low in the near future, that if something is not done immediately, they will likely go extinct. These so-called evolutionarily significant units (ESUs) are defined in the statutes as (1) being "substantially" reproductively isolated from other population segments of the same species and (2) representing an important component in the evolutionary legacy of the species. The first criterion speaks to the rate of exchange between the population segment and other segments. The second speaks to the time the population segment has been isolated. In contrast, the US MMPA seeks to maintain viable populations across their historical ranges at 50% of their historical population size. This act addresses maintenance of abundance. The MMPA conservation units could be characterized as demographically independent populations (DIPs) to contrast them with ESUs. Some use the term "management unit" to refer to a DIP, but because both DIPs and ESUs are MUs in the strict sense, it is important to distinguish them. Genetic data are useful for defining both. However, the policy goals are different and, consequently, the details of genetic studies directed toward either must take slightly different approaches.

## A. The Evolutionarily Significant Unit

Because the ESA is concerned with conservation units that are characterized as being "evolutionarily" different, the genetic methodology employed must be sensitive to evolutionary distances between taxa. Indeed, the traditional academic use of genetic data is employed to reconstruct common ancestry and to group taxa based on common ancestry. No restriction is based on the taxon level examined (subspecies, species, genus, family, etc.), except that the taxa are assumed to be reproductively isolated and that sufficient time has passed so that measurable genetic differences have accrued between every individual in one taxa and every individual in another. For higher level taxonomic relationships, the grouping derives *a priori* from a particular classification based on morphological distinctiveness. For groupings below the species level, the grouping often derives *a priori* from geographical clustering; some have termed this phylogeography to contrast it to traditional phylogenetics.

Regardless, the key to ESU status is still reproductive isolation and time. Using DNA sequence data to test these *a priori* groupings to see if they are genetically accurate, an investigator demonstrates that all the individuals of each *a priori* stratum fall into exclusive genetic clusters (Waples, 1991; Hillis *et al.*, 1996; Ross *et al.*, 2003). If so, ESU status can be presumed for the groupings. The evidence addresses the policy that protection should be offered to a population segment that is first of all "substantially" reproductively isolated. If they were not isolated, it would be impossible to demonstrate the presence of exclusive genetic clustering. The genetic evidence is usually presented in the form of a branching diagram representing the evolutionary pathways leading to mutually exclusive genetic clusters (Fig. 1A). In cetaceans, several species have been defined almost exclusively on the basis of genetic evidence for reproductive isolation, in the form of substantial genetic differentiation from other animals previously thought to be of the same species (Kingston and Rosel, 2004; Dalebout *et al.*, 2007).

If animals are commonly moving between groups and interbreeding, the groups would not be reproductively isolated from one another and would share genetic material. As a result, the genetic analysis would not find unique groupings of individuals corresponding to each population, and no ESUs could be defined.

### B. The Demographically Independent Population

Consider, however, if the individuals in the sample fail to fall into exclusive genetic clusters that are congruent with the *a priori* classification. For example, what is happening if some of the individuals sampled in the Northern Hemisphere cluster genetically with those in the South (Fig. 1B)? This situation can be the result of (1) insufficient time having elapsed from when the populations were split to purge ancestral shared alleles or haplotypes from the populations, (2) a degree of gene flow that exists or has existed recently (e.g., a few adventuresome individuals immigrated to the south or vice versa to breed), or (3) a combination of the two. It also means that the populations under consideration do not meet ESU criteria. Nevertheless, the populations may be genetically distinguishable if there are significant frequency differences in alleles or haplotypes between the groups. These populations would be characterized as DIPs and the definition would pertain to an intermediate situation between complete, long-term isolation of the ESUs and free gene flow between geographically distinct populations (panmixia).

It is in the range of dispersal rates between the virtual isolation of the ESU and complete panmixia where the interpretation of genetic information requires an understanding of policy. The logical thread goes as follows: e.g., the US MMPA establishes, albeit somewhat obliquely, that populations be maintained at 50% of their historical capacity as functioning elements of their ecosystems. This is interpreted to mean that adequate population levels shall be maintained across their historical ranges. It would forbid management action that resulted in extirpation in one portion of the range, even if such extirpation would not reduce the overall species abundance to below 50% of historical levels.

What happens if anthropogenic mortality occurs at different levels in different parts of the range, e.g., there is heavy incidental take in the southern part of the range because it overlaps with a gill net fishery, but none at all in the central and the northern part of the range? For example, consider a temperate, coastal species that inhabits waters from northern California through Canada, the Aleutian Peninsula, to Japan. Due to the large distances involved, distinct habitat differences, and the coastal behavior of this species, complete panmixia is not very likely and some population structure, i.e., dispersal between certain population segments, is reduced. Say samples are available from each of five putative population groupings (defined *a priori*) in the US Pacific northwest waters. An extensive genetic analysis using both mtDNA and microsatellites is performed, and initial analyses using phylogenetic methods demonstrate no

**Phylogeographic concordant clustering**

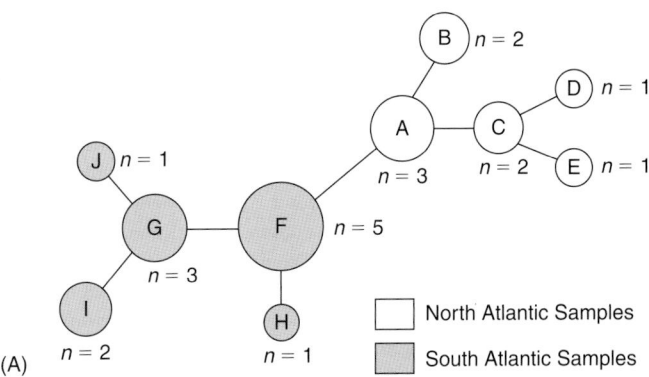

(A)

**Phylogeographic modal clustering**

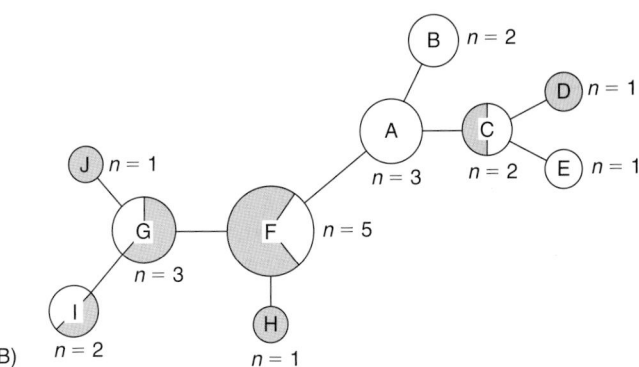

(B)

**Figure 1** *Hypothetical genetic evidence representing two different evolutionary histories presented in the form of branching diagrams representing the evolutionary pathways leading to mtDNA haplotypes observed in a sample of marine mammals. The size of circles is proportional to the number of individuals in the sample exhibiting the particular haplotype, and each haplotype differs from a connected neighbor by a 1-bp difference. (A) North Atlantic and South Atlantic stocks have been isolated for a sufficient amount of time so that there are no haplotypes common to both. Geographic strata are concordant with genetic ones. (B) The isolation of the two stocks is (1) recent so that common haplotypes (C, F, G, and I) have not yet been purged via genetic drift from the North Atlantic, the South Atlantic, or both or (2) the isolation is incomplete, and there is a degree of continual interchange between the stocks. Even though the geography and the genetics are not strictly concordant, the distribution of haplotypes within each of the two stocks in this example is modally different.*

striking genetic clustering concordant with the geographic groupings. However, proximal populations were observed to share haplotypes and microsatellite alleles, and statistical analysis showed that significant frequency differences for the mtDNA haplotypes and for many of the microsatellite loci distinguish the populations. The inference here is that dispersal is sufficiently limited among the five populations so that some genetic differentiation has occurred among them. The populations are isolated but cannot be considered ESUs because the "evolutionary legacy" criterion is not met. They should be considered DIPs because dispersal between them is sufficiently reduced to warrant managing them separately [e.g., establishing individual quotas for incidental kills ("take") for each population]. Moritz described

**G**

such populations as MUs, representing "populations connected by such low levels of gene flow that they are functionally independent" (Moritz, 1994).

This recommendation can actually be made with confidence because of the shape of the curve that relates genetic differentiation and dispersal (Fig. 2). The strength of the result is reflected in the left-hand portion of the graph, genetic differentiation is detectable only when exchange rates between the putative populations are virtually nonexistent from a demographic or management point of view. This is in the range of a few dispersers per generation. However, the weakness of genetic analyses comes from how rapidly genetic differentiation declines as dispersal increases only slightly. Genetic differentiation disappears at dispersal rates that still might be considered insignificant from a demographic point of view, say a few percent per year. In other words, it is very difficult to demonstrate statistically significant genetic differentiation if dispersal between strata is more than a few dispersers per year.

So by demonstrating genetic differentiation, the geneticist has confidently demonstrated demographically insignificant exchange rates. The management consequences are that any anthropogenic mortality within the strata must be compensated for by production from within rather than dispersal from adjacent, perhaps less impacted, units. Under this circumstance, which is actually common in coastal populations, mistakenly assuming that adjacent populations will serve as a source for the losses within the impacted population can result in destruction of the impacted population and failure to maintain it as a functioning element of its ecosystem. Disregarding

the geneticist's recommendation may mean that the manager will have failed to meet a policy goal stipulated in the US MMPA.

However, it is not a "symmetrical" situation. What happens when genetic evidence fails to establish significant demographic isolation between units? Because there was no evidence of population subdivision and hence restricted dispersal, a manager may be tempted to use this negative evidence to infer that the putative populations could be coalesced into one larger MU. Coalescence of two or more small populations into one larger MU would allow the manager to establish a larger incidental take quota and avoid the inevitable economic and political consequences of restricting fishing effort to reduce the incidental fishing mortality. The manager argues that high levels of take in one localized portion of the range (the sink) will be compensated for by production in and dispersal from less exploited portions of the range (the source).

This would turn out to be an appropriate decision if the failure to find evidence of population subdivision was due to demographically high levels of exchange between the exploited and the unexploited regions. However, the decision may have serious biological consequences if the failure to find genetic differences was simply because the experimental design of the genetic study lacked statistical power to discriminate subdivision (e.g., too few samples tested, too little portion of the genome tested, or an insufficiently variable portion of the genome tested), or if genetic isolation of populations is recent. In reality, although undetected, in this case the populations were demographically isolated, and it would be unlikely that adjacent populations could replenish losses due to incidental take in the exploited region. Because exchange between populations may be high enough to prevent detection genetically but not high enough for demographic replenishment, failure to discriminate the subdivision genetically should not at present be used as a scientific rationale for coalescing smaller populations into larger MUs in the absence of sufficient evidence for statistical power to detect such subdivision. Such evidence can be obtained from simulations of the populations and genetic data, as in the program POWSIM (Ryman and Palm, 2006).

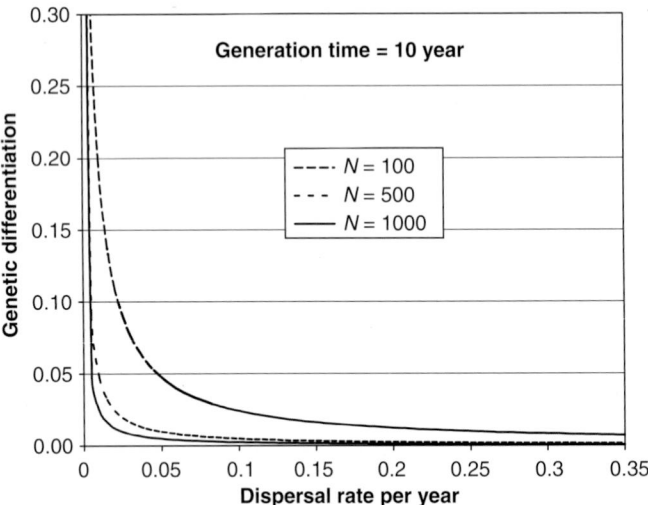

**Figure 2** *The idealized relationship between the degree of genetic differentiation (fixation index), dispersal rate expressed as the average dispersal rate per year, and population size expressed as the number of breeding animals, or breeding females in the case of mtDNA analyses (effective population size). The fixation index ranges between 1 (no common alleles or haplotypes) and 0 (no differences in allelic or haplotypic distribution). Demographically insignificant rates of exchange (e.g., 1% per year) in anything but the smallest effective population sizes probably result in an inability to subdivide populations with any degree of statistical confidence. Perhaps more importantly, because the curve is so flat at this point and higher, genetic data have little resolution to accurately estimate dispersal rate in this range.*

## II. Molecular Markers

Currently, management-oriented genetic studies use primarily (1) genotypes from microsatellite loci within the $3 \times 10^9$ or so base pairs (bp) of the mammalian nuclear genome or (2) DNA sequence data from a portion of the $1.6 \times 10^4$ bp of the mitochondrial genome; the subsequence is also known as a haplotype (Fig. 3). Mitochondrial (mt)DNA is a multicopy, circular, cytoplasmic DNA that in marine mammals is inherited intact from the mother. In contrast, microsatellites are part of the nuclear genome and are inherited biparentally. They are short stretches of repeated DNA that are distributed abundantly in the nuclear genome and show exceptional variability in most species. Newer markers that are rapidly gaining ground in molecular ecology studies of marine mammals include single nucleotide polymorphisms (SNPs) (Morin *et al.*, 2004), amplified fragment length polymorphisms (AFLPs) (Kingston and Rosel, 2004), and sequencing of nuclear genes (Palumbi and Cipriano, 1998). These are all methods for assessing sequence variation primarily in nuclear DNA, and primarily at the level of individual nucleotide changes (though insertions/deletions may also be assessed). Because single nucleotide changes are the most common type of variation in the genome, methods for assessing large numbers of polymorphic sites, such as SNP and AFLP genotyping, provide good statistical

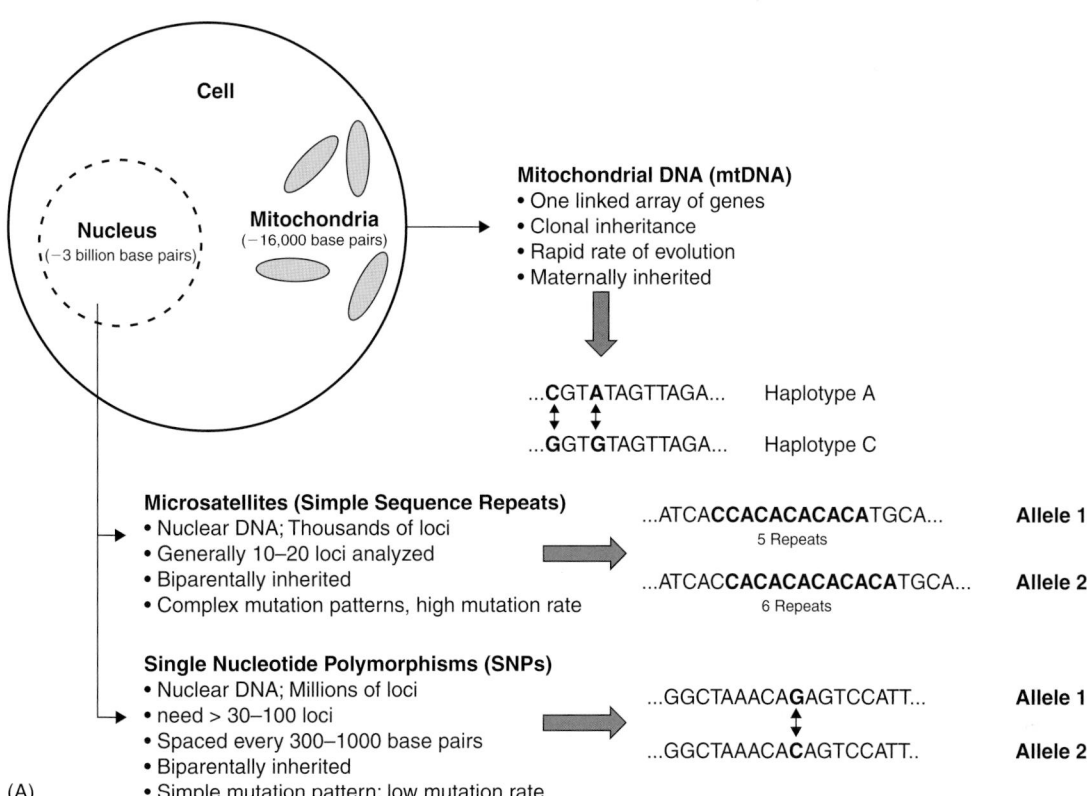

**Figure 3** *Currently, data for most management genetic studies primarily consist of microsatellite DNA, mitochondrial DNA, or both, but SNPs are becoming more common. (A) Microsatellites are short tandem repeats (two, three, or four base repeats) of nucleotides, e.g., CACACACACA ..., ATGATGATG ..., or GATAGATAGATA .... Microsatellite data consist of n pairs of alleles for each individual at m number of microsatellite loci within the $3 \times 10^9$ or so base pairs (bp) of the mammalian nuclear genome. There is estimated to be a microsatellite region every 3000 or so base pairs. Microsatellites are part of the nuclear genome and are inherited biparentally. Mitochondrial data consist of n DNA subsequences (haplotypes) at some locus within the $1.6 \times 10^4$ bp of the mitochondrial DNA genome. mtDNA is a multicopy cytoplasmic DNA that, in vertebrates, is inherited intact from the mother. Each mitochondrion may have 5–10 DNA molecules, and there may be from 100 to 1000 mitochondria per cell. For mitochondrial DNA, a sequenced portion of 12 bp of the 16,000-bp molecule is shown. (B) Sample 1 is heterozygous at microsatellite locus A having a pair of alleles that have five and six CA repeats, and nucleotides C and T at SNP locus A. Sample 1 also possesses an "A"-type mitochondrial haplotype that, e.g., differs by 2 bp from the "C"-type. For actual studies, the number of microsatellite loci examined might typically range from 8 to 20, and the size of the mitochondrial sequence examined might range from 350 to 1200 bp.*

power for genetic assessment of populations, while also assaying more of the genome. With SNPs, the possibility for looking at patterns of variation in individual genes opens the possibility of directly or indirectly assessing genetic variation under selection, or variation associated with known phenotypes. One often-used example of this is genetic determination of sex.

## III.  Focusing on the Individual

In the previous sections, the focus was on a population of animals united by some characteristic, e.g., geographic locale. In this section, the focus is on the individual and what information genetic studies can provide to management.

### A.  Illegal Traffic and Trade

Two sorts of questions are usually asked

1. Did sample X come from the same individual as sample Y? Nuclear marker analysis is used to establish an individual's genetic fingerprint; this is also known as genotyping.
2. What is the provenance of sample X, i.e., what species or geographic population characterizes the sample? For this, sequence analyses are generally employed at higher levels of differentiation, and genotypes are used for assignment to a group or population.

Question 1 is much like placing crime suspects at the crime scene via something the suspect has left behind (e.g., clothing fibers, fingerprints, hair, DNA), and genotyping is a highly reliable means of answering it. The genetic profile of a piece of meat in a market of unknown provenance could be compared with the genetic profiles in a database of "legally" harvested whales or, alternatively, the sample could be compared with the genetic profiles in a database of biopsied, protected ones.

Question 2 is more general and deals with establishing that the sample came from an animal that belonged to a certain group or taxon. Genetic analyses can help determine whether a given market sample came from a proscribed or a permitted taxon. For example, a particular market sample is humpback whale (*Megaptera novaeangliae*). The unknown sample is compared genetically with samples whose taxon identity is known. Because the genetic differences between taxa above the species level are so large, assignment analyses are almost infallible (e.g., did the sample come from a whale or a cow?). In most situations, assignment is accurate at the species level (e.g., did the sample come from a minke whale *Balaenoptera acutorostrata/bonaerensis*, or a blue whale *B. musculus*?). However, there are exceptions, such as discriminating species among the genera *Delphinus*, *Stenella*, *and Tursiops* using only mtDNA control-region sequence. Accurate assignment of an individual sample to its geographic origin is more difficult [e.g., did the sample come from a gray whale (*Eschrichtius robustus*) harvested off the eastern Pacific Ocean or from the Okhotsk Sea?]. Although there are exceptions to this rule, in general the lower the taxonomic division the greater the difficulty in distinguishing provenance of an individual sample. At these lower levels, relatively large numbers of microsatellites or SNPs may provide sufficient power for statistical assignment tests. An example of this was the use by Kingston and Rosel (2004) of hundreds of variable AFLP loci to identify clear genetic differentiation of coastal and pelagic *Tursiops truncatus* populations in the western North Atlantic and differentiation between two sympatric species of *Delphinus* with relatively low mtDNA sequence divergence.

### B.  Other Uses of Individual-Oriented Genetic Information

Genetic mark-recapture and monitoring methods based on genotyping can be substituted for traditional tagging methods, i.e., Discovery tags, for estimating population size, dispersal rate, and migration pathways (Palsbøll, 1999; Schwartz *et al.*, 2007). The management value of such data is obvious. However, if populations are large, the number of "recaptures" is likely to be small, and the cost of genetic analysis of many samples can be high. Nevertheless, such methods have been used to estimate population size and level of reproductive isolation (Garrigue *et al.*, 2004) and to complement ongoing photographic identification projects. Besides re-identification of individuals, genotyping can be used to reliably identify parent–offspring relationships, although large numbers of microsatellite loci must be examined to do this accurately. It is probably worth the effort because by doing so, dispersal can be measured over two generations rather than over the lifetime of single individuals. For conservation decisions, inter- rather than intragenerational movement (i.e., geneflow) is probably a more important parameter than movements of a single individual. Another important demographic parameter that emerges from a study of parent–offspring relationships is the fraction of mature animals enjoying reproductive success. In other words, what is the particular breeding structure of the population, and how does that influence effective population size, inbreeding, and geneflow between populations (e.g., if a small proportion of males actually reproduce, dispersal does not accurately reflect gene flow)?

Finally, determining sex provides a means to examine geographical segregation by sex and whether males or females are the dispersers. It is a common situation with many marine mammal species that females tend to be strongly philopatric, returning year after year to specific feeding or breeding sites. Female philopatry can be demonstrated by examining genetic population subdivision separately in males and in females. If only females are strongly philopatric, mtDNA subdivision should be apparent among the females but not the males. When males are the dispersers but not females, nuclear marker subdivision should be nonexistent because the males of breeding age serve as a "conduit" to homogenize the alleles between populations. If there are data on age, it is sometimes possible to demonstrate that the likelihood of dispersal increases with age of the males. There are policy implications in demonstrating female philopatry. Although this sort of population structuring would not qualify the population as an ESU, it does qualify it as a DIP worthy of management. If the animals from a particular feeding or breeding area are extirpated (males and females), recolonization will not likely take place. The strongly philopatric females from other breeding or feeding grounds would not recolonize the depopulated region, and the dispersing males would not likely return to an area with no females. Thus, if policy deliberately excluded populations based on female philopatry, there could likely arise a situation where take could reduce or fragment ranges.

### C.  The Hidden Power of Molecular Genetics

In addition to providing answers to population subdivision, dispersal, individual identities, and breeding behavior, molecular genetic analyses present a previously unexploited opportunity for gaining understanding of marine mammals via remote, nonlethal sampling. Some of these data can have direct relevance for management. Consider that a skin sample contains the entirety of the individual's genetic blueprint. The ability to read this blueprint is progressing at an astounding rate, and although most of the progress is within the human genome, around 70% of the cetacean genes are homologous, and tools developed for medical research can be utilized for marine mammals. For example, DNA sequence information extracted from the genes of skin cells can provide data about expressed characteristics of other tissues or organs. Sequencing visual pigment genes from skin is a good example. Levenson *et al.* (2006) have shown that, with collateral data about visual performance of particular photoreceptors via behavioral or physiological testing, it is possible to extrapolate from the DNA sequence to the spectral sensitivity. Understanding the visual abilities of cetacean could aid in the design of fishing nets with increased color contrast, making them more visible to marine mammals, thereby reducing entanglement rates while sustaining the catch rate of the target species.

## IV. Conclusion

Although examination of genetic material offers unparalleled insights into many biological aspects of an animal's life, certain sorts of genetic information provide data that are directly relevant to the management process. The most important is the definition of the conservation unit. By common sense and by law in many countries, this unit is created out of the understanding that the vast majority of species (marine mammal or otherwise) are not panmictic. Species are subdivided geographically into isolated and semi-isolated groupings. Genetic analyses can measure this directly and provide the main avenue whereby the geneticist can provide information to facilitate management decision-making. Other genetic information on impacted populations is certainly of high value. This chapter has provided some examples. Regardless of the sort of genetic information collected, to insure that genetic studies and information will be useful for management requires a clear understanding of the conservation policy that the studies are designed to help implement.

## *See Also the Following Articles*

Conservation Efforts ■ Forensic Genetics ■ Genetics Overview ■ Molecular Ecology ■ Stock Identity

## *References*

Amos, W. (1997). Marine mammal tissue sample collection and preservation for genetic analysis. *In* "Molecular Genetics of Marine Mammals" (A. Dizon, S. Chivers, and W. Perrin, eds), pp. 107–116. The Society for Marine Mammalogy, Lawrence, KS.

Avise, J. C. (1998). The history and purview of phylogeography: A personal reflection. *Mol. Ecol.* **7**, 371–380.

Dalebout, M. L., *et al.* (2007). A divergent mtDNA lineage among *Mesoplodon* beaked whales: Molecular evidence for a new species in the tropical Pacific? *Mar. Mamm. Sci.* **23**, 954–956.

Garrigue, C., Dodemont, R., Steel, D., and Baker, C. S. (2004). Organismal and gametic capture-recapture using microsatellite genotyping confirm low abundance and reproductive autonomy of humpback whales on the wintering grounds of New Caledonia. *Mar. Ecol. Prog. Ser.* **274**, 251–262.

Hillis, D. M., Moritz, C., and Mable, B. K. (1996). "Molecular Systematics." Sinauer Associates, Sunderland, Massachusetts.

Kingston, S. E., and Rosel, P. E. (2004). Genetic differentiation among recently diverged delphinid taxa determined using AFLP markers. *J. Hered.* **95**, 1–10.

Levenson, D. H., Ponganis, P. J., Crognale, M. A., Deegan, J. F., Dizon, A., and Jacobs, G. H. (2006). Visual pigments of marine carnivores: Pinnipeds, polar bear, and sea otter. *J. Comp. Physiol. A. Neuroethol. Sens. Neural. Behav. Physiol.* **192**, 833–843.

Morin, P. A., and McCarthy, M. (2007). Highly accurate SNP genotyping from historical and low-quality samples. *Mol. Ecol. Notes* **7**, 937–946.

Morin, P. A., Luikart, G., Wayne, R. K., and Grp, S. W. (2004). SNPs in ecology, evolution and conservation. *TREE* **19**, 208–216.

Moritz, C. (1994). Defining evolutionarily significant units for conservation. *TREE* **9**, 373–375.

Palsbøll, P. (1999). Genetic tagging: Contemporary molecular ecology. *Biol. J. Linn. Soc.* **68**, 3–22.

Palumbi, S. R., and Cipriano, F. (1998). Species identification using genetic tools: The value of nuclear and mitochondrial gene sequences in whale conservation. *J. Hered.* **89**, 459–464.

Ross, H. A., *et al.* (9 authors) (2003). DNA surveillance: Web-based molecular identification of whales, dolphins, and porpoises. *J. Hered.* **94**, 111–114.

Ryman, N., and Palm, S. (2006). POWSIM: A computer program for assessing statistical power when testing for genetic differentiation. *Mol. Ecol. Notes* **6**, 600–602.

Schwartz, M. K., Luikart, G., and Waples, R. S. (2007). Genetic monitoring as a promising tool for conservation and management. *TREE* **22**, 25–33.

Taylor, B. L., and Dizon, A. E. (1999). First policy then science: Why a management unit based solely on genetic criteria cannot work. *Mol. Ecol.* **8**, S11–S16.

Waples, R. S. (1991). Pacific salmon, *Onycorhynchus* spp., and the definition of species under the endangered species act. *Mar. Fish. Rev.* **53**, 11–22.

**G**

# Genetics, Overview

### PER J. PALSBØLL

## I. Introduction

Genetics is the study of the transmission of and variation in hereditary traits. In the case of genetic analyses of natural animal populations at the level of organisms or above (e.g., populations or phyla), most studies draw their inferences from the relative degree of difference in consanguinity (i.e., kinship or relatedness) among individuals, populations, and species. The confidence with which such inferences can be relied upon depends on the accuracy of the genetic estimates derived from the collected genetic data, which in turn is linked to the amount of genetic data as well as the underlying assumptions made during the analysis of the data.

In principle, the relative degree of relatedness among organisms is estimated from, and positively correlated with, the proportion of shared inherited characters. It is possible to use any hereditary trait in an organism toward this end; however, the farther removed from the *locus* that encodes the trait under study (i.e., the DNA itself), the higher the chance that external factors may have altered the phenotypic expression of the hereditary trait. Consequently, while relatedness may be estimated from morphological characters, and a single morphological character might represent the expression of many loci, the phenotypic expression might have been altered by extrinsic factors, such as environmental or physiological variables, to an unknown extent, thereby masking the genetic underpinnings of the morphological trait. In contrast, the composition of most cellular components is usually not susceptible to such extrinsic variation, and thus the interpretation of the observed variation of such cellular components may be directly linked to the state of the encoding locus, the genotype. This observation explains why biochemical/molecular methods were so readily adopted in place of morphological methods to estimate genetic and phylogenetic relationships when efficient methods to detect biochemical changes emerged in the mid-1960s. Until the 1980s, the biochemical/molecular methods applied to natural populations were mainly indirect in the sense that they did not detect differences in the DNA sequence of the encoding locus itself. For instance, the most widely employed biochemical method, allozyme electrophoresis, detects differences in the overall electric charge of enzymes caused by amino acid replacements. An important limitation of allozyme electrophoresis is that only a small part of the genome consists of genes encoding enzymes, and only a small subset of the possible amino acid substitutions result in a change of the overall electrical charge of the enzyme. In addition, homoiothermic organisms (birds and mammals) have reduced level of isozyme variation compared to poikilothermic animals and plants.

Despite these limitations, a large number of studies have been conducted based upon allozyme electrophoresis, providing novel and valuable insights. Interested readers should consult the works of Wada and Danielsdóttir, both of whom have undertaken extensive allozyme-based studies of various cetacean species.

The most basic level of genetic organization is the DNA sequence of the genome itself, which became accessible in a practical manner due to a series of technical advances during the 1980s culminating with the development of the polymerase chain reaction (PCR) by Mullis and coworkers in 1987. The PCR technique permits simple and robust *in vitro* amplification of any specific nucleotide sequence if the nucleotide sequence of the flanking regions is known. Once amplified, the exact nucleotide sequence of the locus is readily determined. PCR-based analysis of DNA sequences has become the predominant method used in genetic studies of marine mammals. For this reason I will rely upon examples based upon analysis of DNA sequences rather than allozymes or morphological characters in this chapter.

## II. Obtaining Tissue Samples

A prerequisite for DNA-based methods is, naturally, DNA. The most common source of genomic DNA is from soft tissue samples. Soft tissue samples are readily available from dead animals, e.g., stranded or killed specimens. However, it is often scientifically or ethically desirable to obtain samples from free-ranging, live animals. The advantage of PCR-based techniques is that only a minute amount of target DNA is required, and hence adequate amounts of DNA are readily obtained from skin biopsies, sloughed skin, hair, and even feces, which can be collected from free-ranging marine mammals with relative ease.

The high sensitivity of PCR-based methods also enables the use of historical samples, such as hair from old furs, baleen or even dried blood obtained from old logbooks. However, the quality of DNA extracted from such historical samples is usually inferior and obtained in much lower concentrations than DNA extracted from fresh tissue samples. The same is usually true for DNA extracted from fecal or similar degraded samples. The low concentration and often highly degraded DNA obtained from historical samples necessitates additional precautionary measures to prevent cross-contamination among samples as well as repeated analyses to insure that a correct genotype is obtained due to various artifacts potentially occurring during PCR, such as allelic dropout and spurious alleles.

Tissue samples can be collected from free-ranging animals by invasive and noninvasive techniques, each with respective advantages and disadvantages. Invasive techniques, such as the collection of skin biopsies, enable a directed sampling scheme. This implies that, conditions permitting, skin biopsies can be collected from those individuals relevant to the specific objective of the study and a biopsy can usually be linked to a specific individual. Multiple biopsy systems have been developed to collect skin biopsies from marine mammals, all principally consisting of a delivery unit, such as a crossbow or gun, and a projectile unit, usually an arrow (called a bolt in the case of a crossbow). The projectile unit carries the biopsy tip and a stop to limit the depth of penetration, which may act as a float as well. The biopsy tip is typically a simple hollow tube of stainless steel with one or more barbs retaining the sample when the projectile unit is retracted after hitting the target animal. Systems of various kinds and ranges have been developed; the currently most powerful system was developed by Finn Larsen with which a skin biopsy was collected from a blue whale (*Balaenoptera musculus*) at a distance of approximately 70 m (~210 ft, Fig. 1). Skin biopsies from pinnipeds or smaller odontocetes are usually collected when the animals haul out on land or bow-ride using a hand pole on which a biopsy tip is mounted. Invasive sampling techniques are under some circumstances viewed as intrusive and thus undesirable. In order to investigate such concerns, data have been collected during biopsy sampling to assess possible adverse effects. To date the only discernable effects appear to be short term and may be equally attributable to the multiple close approaches of the boat toward the target animal while attempting to collect a sample. Although the resolution of such studies is typically low, given the pervasive use of skin biopsy sampling today in e.g., baleen whales, any substantial side effects would likely have been detected by now.

The alternative, noninvasive sampling methods are typically of a more opportunistic and random nature, which may prohibit the

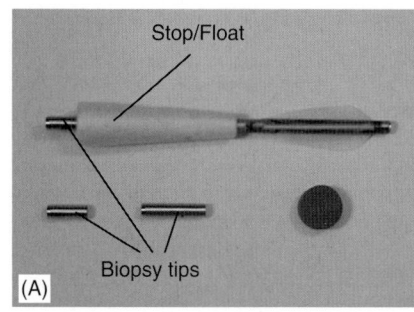
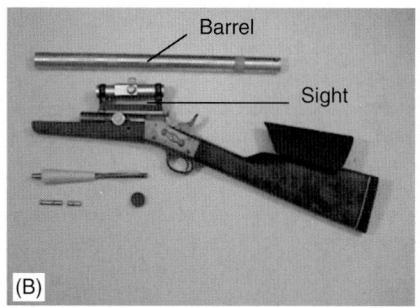

**Figure 1**   *The "Larsen" long-range skin biopsy system. (A) The projectile unit with biopsy tips and concave stop, which acts as a float as well. (B) The delivery system (a Remington rolling block system rifle), complete with barrel and sighting aid. Pictures courtesy of Finn Larsen.*

pursuit of some research objectives. For cetaceans, the most common kind of noninvasive samples are from sloughed skin. The outer epidermis in cetaceans differs from that of most other mammals by the lack of dead keratinized cells and consists mostly of live cells complete with nuclei and mitochondria, the two cellular compartments with genomes in mammals. There is considerable variation among cetacean species in terms of the amount and how often the skin is sloughed. Sperm whales (*Physeter macrocephalus*) have been observed to slough massive amounts of skin, whereas other species, such as fin (*Balaenoptera physalus*) and minke (*B. acutorostrata* and *B. bonaerensis*) whales rarely slough any skin. The main disadvantage when collecting samples such as sloughed skin in a noninvasive manner is the opportunistic nature of the samples and the difficulty in linking a specific sample to a particular individual during multi-individual sightings, which may influence the pursuit of some research objectives. In addition, the quality and quantity of DNA extracted from such samples are more variable than those of samples from skin biopsies.

Some projects have used skin swabs, where skin is scraped off the target individual without penetrating the epidermis. Such methods also require multiple close encounters, just as is the case when collecting skin biopsies, and may thus be equally invasive in terms of the degree of disturbance to the target individual.

Genomic DNA has also been successfully extracted from samples collected from fecal plumes in the water column, from dugongs, dolphins, or baleen whales, which contain epithelial cells from the intestinal tract. Among pinnipeds, the most common noninvasive samples are fecal samples (scats), typically collected from haul-out sites on land. In bears, hair has proven an excellent source of noninvasive samples, where the DNA is extracted from the root cells. In order to insure that hair samples contain the root (which is where most of the DNA is contained), the hair must be pulled out of the skin. In the case of bears (and other terrestrial animals with fur), a simple, highly effective sampling scheme has been utilized based upon "hair-traps" with scent lures to attract bears and barbed wire that passively collect hair samples.

Samples are usually preserved by freezing with or without some conservation buffer. Commonly used conservation solutions are 70–96% ethanol or distilled water saturated with sodium chloride and 20% DMSO, both of which enable storage at ambient temperatures, although it is usually recommended to freeze samples at –18°C or below.

### III. Commonly Analyzed Genetic Markers

As mentioned earlier, genetic analyses of different taxa, e.g., individuals, populations, or species, are in essence about estimating the relative degrees of consanguinity among the included taxa. Put simply, the higher the proportion of shared traits/characters between two entities, the higher the degree of relatedness, or, in the case of nucleotide sequences, the more mutations (i.e., differences in the nucleotide sequence) at the same locus separating two different entities the less related they are inferred to be.

In principle, there are two kinds of mutations in nucleotide sequences, substitutions or insertions/deletions of one or more nucleotides. The latter kind of mutations is commonly observed at microsatellite and minisatellite loci and short interspersed elements (SINEs), all of which have been employed in genetic analyses of marine mammals. The most commonly analyzed loci of this kind, at this time, are microsatellite loci. Most DNA sequence changes at microsatellite loci consist of additions or deletions of one or more DNA sequence repeats. These kinds of mutations are likely due to single-strand slippage when the parental template and the new native DNA strand misalign in the repeated DNA sequence during DNA replication. This mode of mutation is termed a step-wise mutation model (Fig. 2). Mutation rates at microsatellite loci are often high and have been estimated at $10^{-4}$–$10^{-5}$, which is several-fold higher compared to single-nucleotide substitutions. The high mutation rate often results in multiple alleles at each locus and consequently high levels of heterozygosity. Microsatellite loci are therefore well suited as genetic markers for the estimation of close relationships, such as parent–offspring relations. In contrast, microsatellite loci are less well suited to estimate more distant relationships due to high levels of allozygosity due to the high rate and step-wise mode of mutation. Alleles at a microsatellite locus will differ solely by the number of repeats, and two copies of the same allele (i.e., the same number of repeats) may be allozygous or autozygous (Fig. 2). This aspect has to be taken into account during the data analysis, and several estimators of genetic distance have been developed specifically for microsatellite loci. However, accounting for the step-wise mutation model in the estimation also introduces additional variance in the estimation of the degree of genetic divergence, which in turn reduces the precision of the estimate. Although the probability of allozygosis is low among closely related individuals, such as members of the same population, it increases with the degree of genetic divergence and becomes an issue as individuals, populations, and species become more genetically divergent. Many assessments have also demonstrated that a simple step-wise mutation model is inadequate because of additional mutational constraints acting upon microsatellite loci, such as limits on the number of repeats, rare multi-repeat mutations, and microvariants (mutations involving partial repeats), all of which affect the usability of microsatellite loci for the estimation of distant evolutionary relationships.

Similar to microsatellite loci, minisatellite loci are composed of tandem repeated DNA sequences but with a larger repeat size (>30 nucleotides long). Alleles are discriminated based upon the number of repeats as for microsatellite loci, but the individual fragments

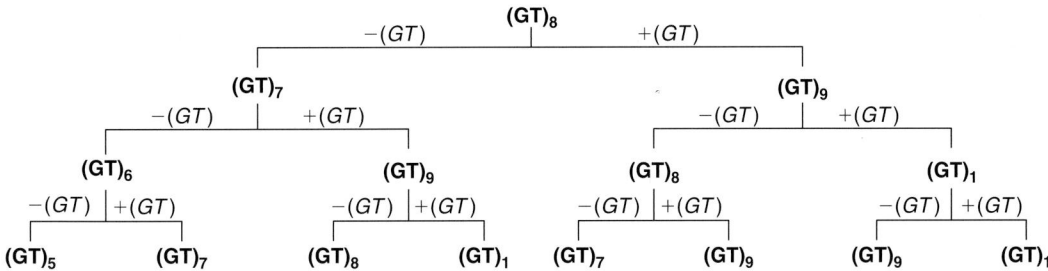

**Figure 2**  *The step-wise mutation mode at microsatellite loci. +/− (GT) denotes a mutation by single-strand slippage, i.e., addition or deletion of a single GT repeat unit.*

are much longer. The process of altering the number of repeats at minisatellite loci is known as unequal crossover, when recombination occurs during meiosis. Analyses based upon minisatellite loci became popular in the late 1980s but have since largely been replaced by microsatellite analysis, which are technically simpler and more readily standardized within and among laboratories.

SINEs are irreversible insertions into the genomic DNA of reverse-transcribed RNA probably of viral origin. The random and irreversible nature of these insertions implies low or no reversal of "mutations" making these kinds of these genetic markers ideal for estimating more distant genetic relationships. Not surprisingly they have been used successfully in resolving the phylogenetic relationships within Cetacea and the placement of cetaceans within Artiodactyla.

The rate of nucleotide substitutions is typically several-fold lower compared to insertion and deletions at microsatellite loci. Substitutions are detected either by whole-sequence analysis or as single nucleotide polymorphisms (SNPs), and in some cases as amplified fragment length polymorphisms (AFLPs). The two latter methods are more efficient compared to whole-sequence analysis, but each method has important drawbacks relative to whole-sequence analyses.

In a SNP analysis, the detection method typically targets only a single polymorphic nucleotide position and detects only the two different bases (e.g., guanine or cytosine) that have been characterized previously at that specific position. Such an analytical system results in a biallelic locus, with very low mutation rates. This in turn yields low statistical power and low accuracy compared to whole-sequence and microsatellite analysis. The chief advantages of SNP analyses are the ability to fully and easily automate data generation, and that data are readily comparable among laboratories. SNP analyses also target very small DNA fragments and are therefore ideally suited for old samples or samples of poor quality. To date the application of dedicated SNP analysis in marine mammals has been limited, but the use will likely increase (Morin *et al.*, 2004).

AFLP analyses are somewhat similar to the use of minisatellites in terms of the nature of the final data and the associated difficulties in standardization of data, thereby hampering comparisons between laboratories (and even among analyses within the same laboratory). The main advantages of AFLPs are the large amount of data (loci) generated in a single analysis, and that generic PCR primers are used to generate the data. In contrast, SNP, SINE, whole-sequence, and microsatellite analysis requires the development of PCR primers that target a specific locus, which is both cost- and time-consuming to develop. AFLP data are treated as dominant alleles, prohibiting the use of more traditional population genetic assessments, which may be critical for some studies.

The lower rate of single-nucleotide substitutions implies lower autozygosity, and single-nucleotide substitutions are therefore in many ways better suited to estimate more distant evolutionary relationships than microsatellite loci. However, the rate of single-nucleotide substitutions differs among and within loci due to varying (often unknown) selective pressures. An example is codons in exons. In most cases a single amino acid is encoded by at least four different codons. The different codon sequences encoding for the same amino acid typically differ at the third position, at times at the first and only rarely at the second codon position. Hence, nucleotide substitutions at the third position are usually synonymous and not subject to selective constraints. In contrast, the majority of nucleotide substitutions at the first and the second codon positions are nonsynonymous. The selective constraints are thus higher at the first and the second codon positions, and the substitution rate is usually lower than that at the third codon position.

Because of the different selective pressures relative to codon position, phylogenetic analyses usually stratify nucleotide sequence data according to codon position. There are, however, multiple exceptions to this rule of thumb, such as MHC genes, which evolve primarily by amino acid substitutions.

In mammals, the vast majority of the genome does not encode enzymes and was thus presumed to be under little or no selection pressure. However, as more and more genomic data becomes available, the large variations in mutation rates observed among such putatively noncoding DNA sequences indicate the existence of selective constraints acting on these DNA sequences as well. Possible explanations are aspects such as chromosome pairing during meiosis, replication and transcription rates, chromosomal stability, and numerous other phenomena.

A prerequisite for the estimation of the relative degree of genetic divergence among taxa is a model that adequately captures the underlying mutational mechanisms. One important assumption in most evolutionary models is the absence of homoplasy. The commonly employed infinite-site mutation model assumes that new mutations always occur at a new site in the nucleotide sequence that has not previously been subject to a mutation. The infinite-allele model differs slightly in that multiple mutations at the same position can occur, but no allozygous alleles have identical nucleotide sequences. The consequence of either model is that identical nucleotide sequences are all assumed to be autozygous. Although these idealized models probably are applicable to closely related taxa, multiple mutations do occur, especially at fast evolving nucleotide sequences such as the mitochondrial control region, which need to be accounted for as taxa become genetically divergent.

The earlier mentioned variance in mutation rates among loci is in fact an advantage, as it enables the researcher to pick loci with mutation rates that are appropriate for the level of genetic divergence of the taxa under study. Usually the goal is to uncover sufficient amounts of variation to facilitate accurate estimations, while keeping the amount of homoplasy as low as possible.

Mammalian cells contain two different genomes. The cell nucleus harbors two full complements of chromosomes, one paternally and the other maternally inherited. Secondly, the mitochondria, in the cell cytoplasm, possess a small genome, consisting of a circular DNA molecule of approximately 16,500 nucleotides in length in mammals. During the formation of the zygote, the sperm cells appear not to contribute any mitochondria to the resulting zygote in mammals, although rare cases of paternal leakage of mitochondrial DNA have been reported. Thus in principle and for all practical purposes, the offspring inherits only the maternal mitochondrial genome.

## IV. Analyses of Individuals

In the case of marine mammals, genetic methods have been applied to identify individuals and parent–offspring relations as well as full-siblings for a number of different purposes.

Identifying marine mammals by traditional tagging methods is often not feasible. In many instances, marine mammals are simply too large, have too-wide ranges, and live in a too-dense medium to make traditional tagging practical. Tag attachments are usually relatively short-lived (although this is changing), in part because of the significant drag caused by the water, unless the tag is attached to solid structures, such as the tusk of a male narwhal. Although individual identification from natural markings has been applied successfully to a number of marine mammal species, this approach is limited to species with sufficient levels of natural variation among individuals.

In contrast, individuals from most species are readily identified by "genetic fingerprinting," even species with much reduced levels of genetic variation such as northern elephant seals (*Mirounga angustirostris*). Coworkers and I set out to verify if "genetic tagging" was feasible for a wide-ranging cetacean species. Our study included 3068 skin biopsy samples collected over a period of 8 years (from 1988 to 1995) from humpback whales (*Megaptera novaeangliae*) across the North Atlantic (Palsbøll *et al.*, 1997). Each humpback whale was identified by its composite genotype collected from six hypervariable microsatellite loci. The main issue in individual identification from a genetic profile is the probability of identity. The probability of identity is readily estimated from the population allele frequencies for all degrees of relationship raging from unrelated individuals to first-order relatives, such as parent–offspring pairs, and decreases rapidly with the number of loci. The difficulty lies in determining the proportion of each degree of relationship in a sample, which in turn determines the expected number of individuals that have identical genetic profiles by chance; these different animals may be mistakenly be assumed to represent the same individual. Although the probability of identity is positively correlated with the degree of relatedness, the proportion of pairs of a specific degree of relation decreases with as the degree of relatedness increases. In the case of the humpback whale study mentioned earlier, the probability of identity and expected numbers of different individuals with identical composite genotypes were estimated assuming that all samples came from unrelated individuals only, first for each maternally related feeding aggregation and subsequently for the entire population. The expected number of pairs of different individuals with identical genetic profiles by chance in the total sample of 3068 was estimated to be less than one. Consequently, skin biopsy samples with identical genetic profiles were inferred as originating from the same individual. In total, 698 such samples with duplicate genetic profiles were detected. In all cases samples inferred as originating from the same individual had the same sex and mitochondrial control region sequence.

In a few cases, samples had been collected from the same individual humpback whale as far apart as 7500 km. The overall pattern of re-sightings within and among sampling areas was in agreement with two decades of sighting records of individual humpback whales identified by their natural markings. The genetic "tags" were also used to estimate the abundance of humpback whales on the breeding grounds in the West Indies using mark-recapture techniques. Since the sex of each individual whale had been determined by genetic analysis as well, separate estimates of male and female abundance were calculated. Unexpectedly, the study yielded a significantly higher estimate of males at 4894 (95% confidence interval, 3374–7123) relative to that of females at 2804 (95% confidence interval, 1776–4463). The reason for this apparent underrepresentation of females on the breeding range (the sex ratio among the calves and all whales on the feeding grounds has previously been estimated at 1:1) could not be resolved on the basis of the data collected during the study. However, the authors suggested either spatial or temporal segregation among females as the source of the difference between the two abundance estimates.

Pomilla and Rosenbaum (2005) identified the same single female humpback whale by genetic tagging first in the Indian Ocean off Madagascar in 2000 and later again in the South Atlantic off Gabon (i.e., on the other side of the African continent) in 2002, demonstrating the tremendous migration potential migrating whales posses.

An aspect of marine mammal biology where genetic methods are especially useful is determination of parentage, e.g., to study breeding strategies and to assign individual reproductive fitness. Paternal reproductive success can be assessed in several ways, either by

determination of specific parentage or by the level of paternal variation among the offspring. The former approach is relatively straightforward, as individuals that are related as parent and offspring will have at least one allele in common at each locus. However, as is the case for individual identification (as discussed earlier), two individuals that are not related as parent and offspring may also share the minimum of one allele at each locus by chance. The probability that two individuals not related in parent–offspring manner share one or two alleles at each locus by chance decreases with the number and variability of loci genotyped. Hence, confident assignment of parentage requires that a relatively large number of variable loci are genotyped. In addition to a sufficient number of genetic markers, an adequate set of samples is required in order to insure that parent and offspring pairs are among the collected samples. To date only a few studies have attempted assignment of paternity in marine mammals, e.g., in gray seals, (*Halichoerus grypus*) or harbor seals (*Phoca vitulina*) either by analysis of microsatellite loci or "multi-locus" DNA fingerprinting, as in the case of the northern elephant seal where genetic diversity is exceptionally low.

Hoelzel *et al.* (1999) compared reproductive success of northern and southern (*M. leonina*) male elephant seals estimated as the proportion of pups fathered by the α-male in his own harem. Previous behavioral observations indicated a higher level of competition for matings among male northern elephant seal compared to male southern elephant seals, leading to the hypothesis that northern elephant seal α-males on average are less successful than their southern conspecifics. The genetic analysis corroborated this hypothesis, finding that southern elephant seal α-males sired a significantly higher proportion of pups in their own harem than did northern elephant seal α-males.

Multi-locus DNA fingerprinting differs from microsatellite analysis mainly by the fact that the alleles from multiple loci are detected simultaneously and is in essence similar to AFLP methods. The simultaneous detection of multiple loci prevents assignment of individual alleles to loci, which is why the degree of relatedness usually is estimated from the proportion of bands shared between individuals. However, the relationship between the degree of band sharing and relatedness is not straightforward, which is why the degree of band sharing is usually calibrated with a sample of individuals of known relationship, i.e., parent–offspring pairs.

Amos *et al.* (1993) employed multi-locus fingerprinting as well as microsatellite loci to study the pod structure of long-finned pilot whales (*Globicephala melas*). The whales are found in groups known as pods. Pilot whale pods appear to consist of mature animals as well as immature animals, presumably calves of the mature females. However, the genetic analyses revealed that also the adult males within a pod were closely related to the mature females in the same pod, indicating that males stay within their natal pod even after they become mature. The genetic analyses revealed that mature males had not sired the calves in their own pod. Curiously, calves of the same cohort in a pod shared paternal alleles, indicating that a single or few closely related males sired calves of the same age. The authors proposed that mature males leave their natal pod briefly and mate with receptive females when pods meet during the breeding season. This hypothesis would explain why no males appeared to have sired calves within their own pod. Mature males of different ages within a pod would then also be maternally related and in fact live "with Mom" for their entire life!

Cerchio *et al.* (2005) used microsatellite loci to assess the reproductive skew among male humpback whales. Humpback whales congregate on tropical breeding grounds where groups of males of up to 25–30 individuals compete intensely for access to mate a single female that presumably is in estrus. The studies detected an unexpected low degree of skew in reproductive success among males.

Individual-based analyses such as the above examples have the potential to address new issues with genetic methods that previously were infeasible. Traditional population genetic analyses (see later) yield evolutionary estimates of genetic divergence/diversity and may thus be of limited relevance to contemporary management and conservation issues. However, identifying individuals and parent–offspring relations provides a "real-time" insight into population structure and dispersal at a time scale relevant to management and conservation purposes.

The collection of multi-locus genotypes at highly variable loci, such as microsatellite loci, from each individual enables so-called assignment tests. If one assumes panmixia, then the probability of observing an individual's multi-locus genotype may be estimated from the population allele frequencies. If the probability is substantially higher for one population compared to the remainder candidate populations, the individual is assigned to that population. By this approach it is also possible to estimate "real-time" gene flow and recent events. Gaggiotti *et al.* (2002) used such an approach to determine the origin of individuals in newly established gray seal rookeries off the British Isles, demonstrating density-dependent dispersal in this species from the older established rookeries.

## V. Analyses of Populations

A large number of genetic studies of marine mammals have been undertaken for the purpose of identifying population structure and mechanisms of intra-specific evolution. In practical terms, the aim is to determine if individuals belonging to the same partition are more closely related to each other than with individuals from other partitions, which is expected if partitions represent different entities (e.g., pods, population, or species). In numerical terms this objective translates into estimation of the degree of genetic heterogeneity among subpopulations, traditionally estimated as the relative increase in homozygosity due to population subdivision, e.g., Wright's $F$ statistics. The increase in homozygosity due to population structure is a product of random genetic drift. Random genetic drift denotes the random changes in allele frequencies resulting from the sampling of alleles for each new generation from the parental generation. If one assumes panmixia within each subpopulation with respect to the locus under study (which is likely to be the case in most instances) and the absence of any selection, the offspring generation can then be viewed as a random sample of the parental alleles. As with any random sampling process, such sampling is subject to stochastic variation, i.e., alleles are not resampled in exactly the same proportions as those found in the parental generation, and the allele frequencies will thus oscillate between generations unless the population is very large. The long-term consequence of random genetic drift in a finite-sized population is that all but one allele will be lost from the population in the absence of introduction of new alleles by gene flow and mutation. In other words, due to random genetic drift, alleles are lost from a population (thereby increasing the homozygosity) at a rate depending on the population size as well as the rate of introduction of new alleles either by mutation or gene flow from other subpopulations. Since the process is random, it follows that different alleles will increase/decrease in frequency due to random genetic drift in different populations. Overall the effect of random genetic drift is that we find more homozygotes among the sampled individuals than expected from the overall allele frequencies estimated from all populations combined. Gene flow homogenizes allele frequencies among populations by transferring alleles from one population to other populations. If there are no major fluctuations in effective population size, gene flow, or mutation rates, an equilibrium state is reached where the rate of divergence in allele frequencies due to random genetic drift and mutation is equivalent to the rate of homogenization due to gene flow. Even very low levels of gene flow (e.g., 10 individuals per generation) among populations will homogenize allele frequencies among populations to an extent that no effect of random genetic drift and mutation can be detected. Neither the mutation rate nor the effective population size is usually known in natural populations. For instance, two populations may have a similar level of genetic variation (e.g., estimated as the heterozygosity) but differ in terms of population sizes and mutation rates. For instance, the degree of heterozygosity estimated among samples collected from a small population at loci with high mutation rates may be similar to that estimated from a large population at loci with low mutation rates. As the level of genetic variation depends upon the combination of effective population size and mutation rate (and these are typically unknown), it is common to simply combine both in the composite parameter $\theta$ (called theta), where $\theta = 4N_e\mu$ (for a diploid locus). $N_e$ denotes the effective population size and $\mu$ the mutation rate. The advantage of this approach is that $\theta$ can be estimated from population genetic data, i.e., from the number of alleles, heterozygosity, polymorphic nucleotide positions, and the variance in allele size (for microsatellite loci). Comparisons of estimates of $\theta$ are used to draw inferences regarding differences in mutation rates among loci within single populations or differences in effective population size among populations as well as estimates of genetic divergence.

Several recent studies (Roman and Palumbi, 2003; Alter *et al.* 2007) have utilized population-specific estimates of $\theta$ from both haploid mitochondrial and nuclear autosomal DNA sequences to infer the effective population size in North Pacific gray whales (*Eschrichtius robustus*), as well as North Atlantic minke, fin, and humpback whales. The genetic estimates of effective population sizes in these populations, inferred from the amount of current degree of genetic diversity, were then equated to census population sizes, which in all cases were much higher than the abundance estimated in these populations today by use of other methods. For instance, the estimate of abundance arrived at in this manner for North Atlantic humpback whales was 240,000 individuals, much more than the best estimate of 10,600 individuals derived from mark-recapture data. The authors argued that their estimates reflect historical, pre-whaling abundance in these populations, as there is a time lag between the reduction in abundance (e.g., caused by whaling) and the corresponding decrease in genetic diversity in a population. These genetically derived estimates of abundance are subject to a number of assumptions about gene flow, sampling of all relevant populations, demographic changes, unknown mutation rates, and finally whether evolutionary estimators may be readily applied to contemporary populations (i.e., what is the effect of whaling, which for these species began some 250 years ago or later). The conservation and management implications of these findings are substantial; not only would conservation targets need be raised considerably, but the results also question our fundamental understanding of the "natural" state of our oceans. Given the many underlying assumptions and the large effects of violations of these assumptions, it remains open whether, for instance, these genetic estimates of abundance reflect local (e.g., North Atlantic) humpback whales, or global abundance.

As mentioned earlier, many population genetic studies of marine mammals have employed analysis of microsatellite loci. In addition, the nucleotide sequence of the maternally inherited mitochondrial control region is usually determined as well. The mitochondrial control region constitutes the only major noncoding region of the mitochondrial genome, with mutation rates well above those for the remainder of the mitochondrial genome. Usually the sequence of the first 300–500 nucleotides in the mitochondrial control region is determined, which constitutes the most variable part of the mitochondrial control region. Because the mitochondrial genome is maternally inherited, any results from this locus estimate only the degree of maternal relation among samples. Most microsatellite loci, however, are of autosomal origin and thus inherited in a Mendelian manner.

The different modes of transmission of the mitochondrial and nuclear genome imply that each may reflect a different evolutionary relationship for the same set of samples. Palumbi and Baker (1994) investigated this aspect in 1994 in a study of humpback whales. In addition to mitochondrial control region sequences, the study also included data collected from the first intron in the nuclear protein-encoding locus actin. A phylogenetic analysis of the actin intron I allele nucleotide sequences revealed the existence of two main evolutionary lineages with no apparent geographic distribution. The two lineages could be distinguished by a SNP, which was detected by digestion with the restriction endonuclease MnlI to yield a biallelic locus. This detection method was subsequently employed in the analysis of samples collected off Hawaii and western Mexico, both winter breeding grounds for eastern North Pacific humpback whales. Although the distribution of mitochondrial control region alleles was highly heterogeneous between the same two population samples (the Hawaiian sample being almost entirely monomorphic), no significant level of heterogeneity was detected in the distribution of the two actin intron I alleles. These "contrasting" results, i.e., little or no gene flow at the mitochondrial locus but indications of high levels of gene flow at the nuclear actin intron I locus, were interpreted as the result of male-mediated gene flow, different rates of random genetic drift at each of the two genomes or a combination of both. A subsequent study (Baker et al., 1998) revealed significant levels of heterogeneity also at nuclear loci (mainly microsatellite loci) among samples collected from humpback whales off California and Alaska, which winter off Mexico and Hawaii, respectively. The simplest explanation for the seemingly discrepant outcome of the two studies is likely an increase in statistical power due to larger sample sizes and the inclusion of additional nuclear loci in the analysis (actin intron I as well as four microsatellite loci). However, the results do not eliminate the possibility of some contribution from male biased gene flow to the level of heterogeneity; more work is required to reach an affirmative conclusion.

The issue mentioned earlier, i.e., different degrees of male and female gene flow, is highly relevant when studying marine mammals. This has been clearly demonstrated in several population genetic analyses of species such as the North Atlantic humpback whales as well as northern right whales (Eubalaena glacialis) and belugas (Delphinapterus leucas). North Atlantic humpback whales summer at several high-latitude feeding grounds off the eastern sea border of North America, West Greenland, Iceland, Jan Mayen, and Bear Island in the Barents Sea. Whales from these distinct feeding grounds all appear to congregate on common winter grounds in the West Indies. The winter constitutes the breeding and the mating season. Calves are born during the winter and follow the mother during the spring migration to a high-latitude feeding ground and later on during the autumn migration back to the West Indies. At the end of their first year the calves separate from their mother. The calf will, however, continue to migrate back to the same high-latitude feeding ground in subsequent summers to which it went with its mother during the first summer. The population genetic consequence of this maternally directed migration pattern is that North Atlantic humpback whale summer feeding grounds can be viewed as a single panmictic population with respect to nuclear loci, but structured in terms of mitochondrial loci. The latter is due to the maternal transmission of the mitochondrial genome in combination with the maternally directed site-fidelity to the high-latitude summer feeding grounds. Nuclear alleles are exchanged when humpback whales from different summer feeding grounds mate in the West Indies. However, the calves only inherit their maternal mitochondrial genome, and thus there is in principle no exchange of mitochondrial DNA among summer feeding grounds, if calves keep returning to their maternal high-latitude summer feeding ground. Several population genetic studies have analyzed North Atlantic humpback whales and in conclusion found what was expected from the earlier work (Palsbøll et al., 1995). However, low levels of heterogeneity have also been detected at nuclear loci when comparing western and eastern North Atlantic high-latitude summer feeding grounds, indicating that some eastern North Atlantic humpback whales may winter and breed elsewhere than in the West Indies.

On a much more detailed scale, (Hoelzel et al., 1998; Hoelzel et al., 2007) determined the genotype at multiple microsatellite loci and the nucleotide sequence in the variable part of the mitochondrial control region in samples collected from pods of killer whale (Orcinus orca) observed in Puget Sound in the northeastern Pacific. Two kinds of killer whale pods are found in Puget Sound, resident and transient pods. The latter pods spend only part of the year in Puget Sound. Although the resident pods seem to feed almost exclusively on fish, the diet of transient pods is mainly composed of marine mammals. The two kinds of pods also differ in average number of individuals and vocalizations. The genetic analysis revealed significant levels of heterogeneity between resident and transient killer whales not only at the mitochondrial locus but also at the nuclear loci as well. This result was interpreted as evidence of a highly restricted degree of gene flow between two different kinds of foraging specialists, and in fact it might be that this feeding specialization drives the genetic divergence between the two sympatric groups of killer whales.

All the above-mentioned examples assume the absence of selection on the marker, but one could well envision natural selection affecting the degree and distribution of genetic variation among and within subpopulations.

One such possibility is the sperm whale (P. macrocephalus) for which very low levels of variation have been detected in the mitochondrial control region on a worldwide scale. This observation prompted Whitehead (1998) to propose cultural transmission of adaptive traits in matrilineal whale species as the cause of the low levels of variation at maternally inherited mitochondrial loci. The basic principle proposed by Whitehead is that long-term association between females and their offspring facilitates an efficient cultural transmission of behavioral traits, e.g., feeding behaviors. If a maternal lineage adapts more efficient behaviors, which in turn increases that lineage's reproductive success, such maternal lineage will eventually increase in proportion within the population. The model is similar to genetic inheritance of adaptive traits, i.e., natural selection, the only difference being that transmission across generations is culturally mediated as opposed to genetically. Since the mitochondrial genome is maternally transmitted it will thus "hitchhike" along with maternal cultural transmission of advantageous behavioral traits. The study reported low levels of genetic variation at mitochondrial loci in species which were classified as matrilineal by the author, i.e., species with pods presumably consisting of females and their offspring, such as pilot whales and sperm whales. In contrast, the nucleotide

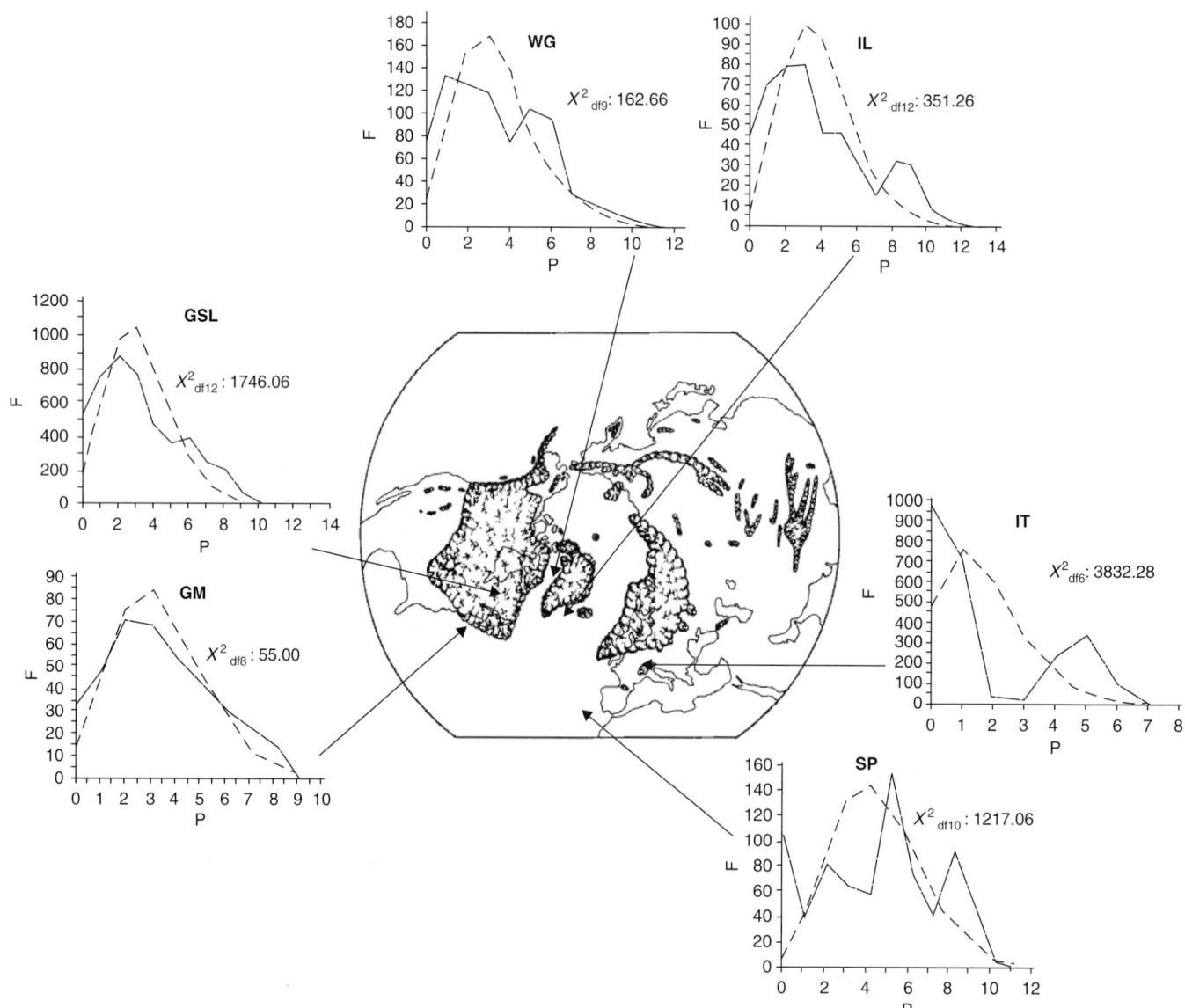

**Figure 3** *Indication of postglacial expansions on western North Atlantic fin whale, Balaenoptera physalus, populations from genetic data. Observed (solid line) and expected (dashed line) frequency distributions of pair-wise differences among mitochondrial control region nucleotide sequences in North Atlantic fin whale populations under a model of exponential expansion (see text for details). A close match between the observed and the expected distribution suggests that the samples were obtained from an exponentially expanding population. The marked areas on the map of the Northern Hemisphere indicate the presence of solid ice sheets during the last Pleistocene glaciation. Center part of figure from Pielou (1979), copyright © 1979 John Wiley and Sons, Inc. Reprinted by permission of John Wiley and Sons, Inc.*

diversity was on average 10-fold higher in species classified in the study as nonmatrilineal. The author demonstrated, by use of computer simulations, that maternal cultural transmission of advantageous behavioral traits could indeed reduce the nucleotide sequence variation at mitochondrial loci significantly if the cultural transmission was efficient and the selective advantage relatively high (~0.1). Although there have been no objections to the hypothesis of cultural transmission of adaptive behavioral traits occurring in cetaceans, others have pointed toward other evolutionary models, such as continued selection and fluctuating population sizes, as equally compatible with the observed data.

The environment inhabited by marine mammals is relatively devoid of physical barriers in comparison to the terrestrial environment. In addition, many marine mammal species have wide ranges, and thus there is a high potential for dispersal. In spite of this, most genetic studies of marine mammals have detected population structure in the distribution of genetic variation within as well as between ocean basins. The lack of physical barriers to dispersal indicates that intrinsic factors may play a role in generating population structure, such as foraging specialization and maternally directed site-fidelity. Even for species where no obvious behaviors limiting dispersal have been observed, population genetic structure was detected, such as in the case of polar bears (*Ursus maritimus*) and fin whales (*Balaenoptera physalus*). In these two instances, it appears that the availability of prey is, at least in part, responsible for generating population genetic structure. In the case of polar bears,

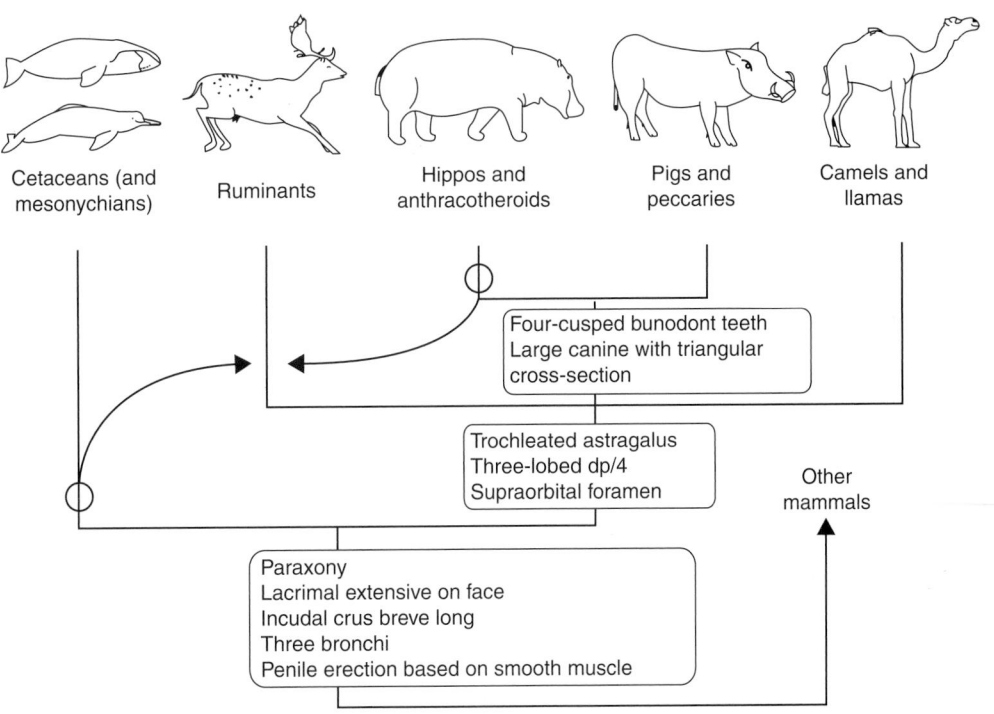

Cetaceans (and mesonychians)  Ruminants  Hippos and anthracotheroids  Pigs and peccaries  Camels and llamas

Four-cusped bunodont teeth
Large canine with triangular cross-section

Trochleated astragalus
Three-lobed dp/4
Supraorbital foramen

Other mammals

Paraxony
Lacrimal extensive on face
Incudal crus breve long
Three bronchi
Penile erection based on smooth muscle

**Figure 4** *The changes to the traditional artiodactyl phylogeny suggested by the findings of Shimamura* et al. *(1997) (see text for details). Reprinted by permission from Nature (Milinkovitch and Thewissen, 1997). Copyright © (1997) Macmillan Magazines Ltd.*

Paetkau *et al.* (1999) analyzed 16 different microsatellite loci in a total of 473 polar bears collected from all areas of the Arctic. The study detected a pattern of genetic divergence among subpopulations that was consistent with the distribution of active annual sea-ice, which in turn relates to the abundance of ringed seals, which is their main prey. A study of North Atlantic and Mediterranean Sea fin whales (Bérubé *et al.*, 1998) was based upon analyses of mitochondrial control region sequences as well as six microsatellite loci in 309 specimens. The population structure revealed from the genetic analyses was consistent with an isolation-by-distance model, which could be explained by a distribution described as a "patchy-continuum" previously suggested by Sergeant and based upon the overall distribution of prey. Interestingly the fin whale study also revealed the possible effect of major geological events, in this case glaciations, upon the present day levels and distribution of genetic variation. The frequency distribution of mitochondrial nucleotide sequences suggested that fin whale population in the western North Atlantic had undergone rapid expansion in abundance most probably from a small postglacial founder population (Fig. 3).

## VI. Analyses of Interspecific Relationships

A well-founded phylogenetic description of marine mammals is fundamental to our understanding of the unique evolution and adaptations observed in this group of mammals. Phylogenetic studies have been conducted at several levels, e.g., among cetaceans as well as at higher levels, such as the relation of cetaceans to ungulates.

The latter question has attracted much attention, as molecular data is emerging complementing earlier morphological estimates of the phylogenetic affinities of marine mammals. The results emerging from the molecular data are, at the moment, inconsistent with the morphological findings as well as among the different molecular data sets themselves with regard to the branching order in several parts of the evolutionary tree. There are multiple possible explanations for such incongruence, such as incomplete taxonomic sampling, inadequate model of change (molecular and morphological), insufficient choice, and number of out-groups as well as incomplete fossil records. As mentioned earlier, the level of homoplasy increases with genetic divergence, which complicates the interpretation of nucleotide sequence data. Instead of DNA nucleotide sequences, the more common sort of data employed in phylogenetic analyses, Shimamura and coworkers (1997) used SINEs. The random and irreversible nature of SINEs make these loci ideal loci for phylogenetic analyses, with a very simple mutation model that is devoid of many of the problems, such as homoplasy, codon position, transition/transversion ratio, and the like, which introduce variability in analyses of single-nucleotide substitutions. The SINE-based study found support for the notion that Artiodactyla is a paraphyletic group, in that cetaceans did not constitute a sister group but originated within Artiodactyla (Fig. 4). Earlier studies, based upon a sequence analysis of nuclear loci encoding milk proteins by Gatesy and coworkers also arrived at the same conclusion, i.e., artiodactyls are a paraphyletic group, also from the position of the cetacean branch. The paraphyly of Artiodactyla was subsequent supported in a comprehensive phylogenetic estimation conducted by involving data from several nuclear and mitochondrial loci. Given the highly specialized cetacean morphology, comparisons of morphological characters with terrestrial mammals is not a trivial matter, making molecular-based phylogenies attractive.

A perhaps more controversial study is an early phylogenetic analysis that estimated the phylogenetic position of the sperm whales within Cetacea from mitochondrial nucleotide sequences. Conventional taxonomy based upon morphological characters places this distinct and old lineage of cetaceans among the odontocetes, as sperm whales share many morphological characters with other

**G**

odontocetes, the presence of teeth and echolocation being the most obvious traits. In contrast, the new study found that sperm whales were significantly more closely related to the baleen whales than to the remainder of the odontocetes (Milinkovitch *et al.*, 1993). The result of this study has since been the subject of numerous additional analyses and reanalyses and in many ways become a case study of phylogenetic estimation from genetic characters and revealed many different (problematic) aspects of single-locus phylogenetic estimation. Such studies have shown that estimation of taxonomic relationships from nucleotide sequences is sensitive to such factors as choice of out-groups, taxonomic sampling, sequence alignment, and long branches. Subsequent analyses based upon nuclear and mitochondrial loci showed a strong support among nuclear genes for the traditional odontocete affinity of the sperm whales and less strong support for the alternate view among the mitochondrial nucleotide sequences. Later studies based upon Y-chromosome specific sequences have also confirmed the "traditional" cetacean phylogeny (Nishida *et al.*, 2007).

The same studies have also illuminated rapid radiation events leading to the extant baleen whales and within the toothed whales, in particular the dolphins, which explains why phylogenetic estimates based upon mitochondrial genes (all belonging to the same locus) have proven inadequate to resolve the phylogenetic relationships or yielded erroneous results.

In several instances interspecific analyses of DNA sequences have revealed "cryptic" species in cetaceans. Initial work on mitochondrial DNA sequences, which was later confirmed by analyses of 13 nuclear loci, suggested three species of right whales; a North Atlantic (*E. glacialis*), a southern right whale (*E. australis*), and a North Pacific (*E. japonica*) (Rosenbaum *et al.*, 2000).

The above examples illustrate that our understanding is still far from satisfactory and that additional work is necessary before we have a more thorough and definitive understanding of the evolution of this highly specialized group of mammals and the underlying molecular mechanisms which we employ in our inferences.

## *See Also the Following Articles*

Forensic Genetics ■ Genetics for Management

## *References*

Alter, S. E., Rynes, E., and Palumbi, S. R. (2007). DNA evidence for historic population size and past ecosystem impacts of gray whales. *Proc. Natl. Acad. Sci. USA* **104**, 15162–15167.

Amos, B., Schlötterer, C., and Tautz, D. (1993). Social structure of pilot whales revealed by analytical DNA profiling. *Science* **260**, 670–672.

Baker, C. S., *et al.* (1998). (10 authors) Population structure of nuclear and mitochondrial DNA variation among humpback whales in the North Pacific. *Mol. Ecol.* **7**, 695–707.

Bérubé, M., *et al.* (9 authors) (1998). Population genetic structure of North Atlantic, Mediterranean Sea and Sea of Cortez fin whales, *Balaenoptera physalus* (Linnaeus 1758): Analysis of mitochondrial and nuclear loci. *Mol. Ecol.* **7**, 585–599.

Cerchio, S., Jacobsen, J. K., Cholewiak, D. M., Falcone, E. A., and Merriwether, D. A. (2005). Paternity in humpback whales, *Megaptera novaeangliae*: Assessing polygyny and skew in male reproductive success. *Anim. Behav.* **70**, 267–277.

Gaggiotti, O. E., Jones, F., Lee, W. M., Amos, W., Harwood, J., and Nichols, R. A. (2002). Patterns of colonization in a metapopulation of grey seals. *Nature* **416**, 424–427.

Hoelzel, A. R., Dahlheim, M., and Stern, S. J. (1998). Low genetic variation among killer whales (*Orcinus orca*) in the Eastern North Pacific and genetic differentiation between foraging specialists. *J. Hered.* **89**, 121–128.

Hoelzel, A. R., Le Boeuf, B. J., Reiter, J., and Campagna, C. (1999). Alpha-male paternity in elephant seals. *Behav. Ecol. Sociobiol.* **46**, 298–306.

Hoelzel, A. R., Hey, J., Dahlheim, M. E., Nicholson, C., Burkanov, V., and Black, N. (2007). Evolution of population structure in a highly social top predator, the killer whale. *Mol. Biol. Evol.* **24**, 1407–1415.

Milinkovitch, M. C. and Thewissen, J. G. M. (1997). Even-toed fingerprints on whale ancestry. *Nature* **388**, 622–623.

Milinkovitch, M. C., Orti, G., and Meyer, A. (1993). Revised phylogeny of whales suggested by mitochondrial ribosomal DNA sequences. *Nature* **361**, 346–348.

Morin, P. A., Luikart, G., Wayne, R. K., and SNP Working Group (2004). SNPs in ecology, evolution and conservation. *Trends Ecol. Evol.* **19**, 208–216.

Nielsen, R., Mattila, D. K., Clapham, P. J., and Palsbøll, P. J. (2001). Statistical approaches to paternity analysis in natural populations and applications to the North Atlantic humpback whale. *Genetics* **157**, 1673–1682.

Nishida, S., Goto, M., Pastene, L. A., Kanda, N., and Koike, H. (2007). Phylogenetic relationships among cetaceans revealed by Y-chromosome sequences. *Zool. Sci.* **24**, 723–732.

Paetkau, D., *et al.* (1999). Genetic structure of the world's polar bear populations. *Mol. Ecol.* **8**, 1571–1584.

Palsbøll, P. J., *et al.* (1995). Distribution of mtDNA haplotypes in North Atlantic humpback whales: The influence of behaviour on population structure. *Mar. Ecol. Prog. Ser.* **116**, 1–10.

Palsbøll, P. J., *et al.* (1997). Genetic tagging of humpback whales. *Nature* **388**, 676–679.

Palumbi, S. R., and Baker, C. S. (1994). Contrasting population structure from nuclear intron sequences and mtDNA of humpback whales. *Mol. Biol. Evol.* **11**, 426–435.

Pomilla, C., and Rosenbaum, H. C. (2005). Against the current: An interoceanic whale migration event. *Biol. Lett.* **1**, 476–479.

Roman, J., and Palumbi, S. R. (2003). Whales before whaling in the North Atlantic. *Science* **301**, 508–510.

Rosenbaum, H. C., *et al.* (20 authors) (2000). World-wide genetic differentiation of *Eubalaena*: Questioning the number of right whale species. *Mol. Ecol.* **9**, 1793–1802.

Shimamura, M., *et al.* (9 authors) (1997). Molecular evidence from retroposons that whales form a clade within even-toed ungulates. *Nature* **388**, 666–670.

Whitehead, H. (1998). Cultural selection and genetic diversity in matrilineal whales. *Science* **282**, 1708–1711.

# Geographic Variation

## William F. Perrin

### I. The Nature of Geographic Variation

Mammals vary from place to place, in size, shape, coloration, osteology, and genetic features, including chromosomes, enzymes, and DNA sequences. They also vary in sounds produced, other behavior, life history, breeding system, parasites, contaminant loads, biochemical features such as fatty acids, and other characters. This chapter focuses on geographic variation in morphology. When morphological variation and range are discontinuous, i.e., the populations or metapopulations are allopatric and can be diagnosed from one or, more commonly, a few characters, they are usually recognized as species, with the inference that they have diverged irrevocably in their evolutionary paths. When this is not true and groups differ from each other on average (modally) rather than absolutely, the variation is considered to be *geographic variation* within a species, and the form is recognized as a *subspecies*, *race*, or *geographic form*.

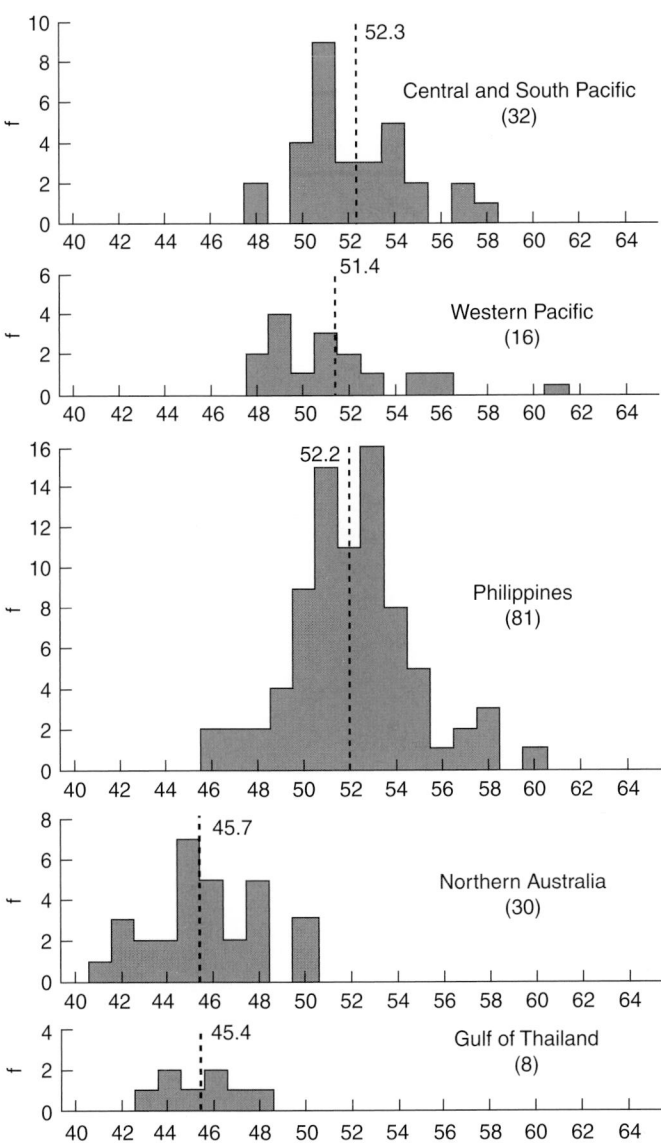

**Figure 1** *Geographic variation in the number of teeth in spinner dolphins from five regions. Average and frequency distribution are shown; sample sizes are shown in parentheses. By permission from Perrin et al. (1999).*

Mammal species tend to vary geographically most in those features that vary most within a population. If, as for most mammals, body size varies broadly within a population, then geographic variants will usually differ in average body size. In another example, odontocete cetaceans are unusual among mammals in that they vary greatly in the number of teeth and vertebrae within a species, and, as expected, these features differ sharply between geographic forms (Fig. 1).

Geographic variation may be *discordant*, i.e., the geographic pattern may differ among characters. For example, spinner dolphins (*Stenella longirostris*) in the eastern tropical Pacific vary differentially in color pattern, dorsal fin shape, fluke shape, and skull features (Perrin *et al.*, 1991), creating a complex mosaic of subspecies (see later) and varying zones of intergradation, depending on what feature is being looked at.

It is likely that most geographic variation in morphology (and the underlying genetic basis) in marine mammals is due to differential selection (ecological divergence) rather than genetic drift. By saying that two populations belong to the same species, we are implying that there is, or recently has been, gene flow between them. Populations can diverge sharply morphologically in the presence of even substantial gene flow if the ecologically engendered differential selection is strong enough (Orr and Smith, 1998). However, modeling studies have indicated that social behavioral characteristics, such as female phylopatry and polygynous breeding systems, both common in marine mammals, can lead to the sequestering of variation due to drift within populations; this may accelerate evolutionary divergence (Storz, 1999).

Neutral genetic differences can accumulate across populations due to drift, and markers for this geographic variation are used extensively in defining marine mammal populations for purposes of assessment and management.

## II. Subspecies

*Subspecies* are formally named or otherwise recognized geographic variants within a species. Subspecies are currently recognized for 29 of the 127 or so marine mammal species (Table I).

The situation for subspecies is not as tidy as might be implied from Table I. Many of these subspecies were poorly described and may prove to be invalid; others may turn out to be full species. Some probably include multiple distinctive populations that deserve subspecific status but are as yet poorly understood. As indicated, some geographic forms have been recognized but not yet named. Some workers would disagree with certain of these subspecific designations and perhaps recognize others. Many additional subspecies were described in the past but have since been discounted (Rice, 1998). As for all of taxonomy, progress in classification at this level (*beta taxonomy*) is uneven and iterative.

In a recent workshop on cetacean taxonomy (Reeves *et al.*, 2004), participants agreed that the subspecies concept may be construed as broad enough to contain two types of entities, populations that are not quite far enough along the continuum toward full specieshood to be judged as species, and populations that may well be species but for which not quite enough evidence is available to justify their designation as such. They also agreed that while the subspecies concept has been largely based on morphology, it should be extended to include genetic evidence, especially as many marine species are more susceptible to genetic sampling than to collection of full osteological specimens. The workshop arrived at the following criteria advised to be applied in defining subspecies:

> The subspecies concept should be understood to embrace groups of organisms that appear to have been on independent evolutionary trajectories (with minor continuing gene flow), as demonstrated by morphological evidence or at least one line of appropriate genetic evidence. Geographical or behavioral differences can complement morphological and genetic evidence for establishing subspecies. As such, subspecies could be geographical forms or incipient species.

## III. Cetaceans

For the odontocetes, in every case where adequate samples of specimens from different regions have been available for examination, geographic variation has been found (Amano and Miyazaki, 1992; Perrin and Brownell, 1994; Archer and Perrin, 1999; Perrin *et al.*, 2003; Jefferson and Van Waerebeek, 2004; Lazaro *et al.*, 2004), so it can be expected to be universal.

Body size tends to be larger in open waters than in closed seas. For example, in the contiguous eastern North Atlantic, Mediterranean, and Black Sea, the short-beaked common dolphin, *Delphinus delphis*, is largest in the North Atlantic, smallest in the Black Sea, and intermediate

**TABLE I**
**Currently Recognized Subspecies[a,b]**

| Species | Subspecies |
| --- | --- |
| Cetaceans | |
| *Balaenoptera acutorostrata* | *B. a. acutorostrata* (North Atlantic) |
| | *B. a. scammoni* (North Pacific) |
| | *B. a.* subsp. (Southern Hemisphere) |
| *B. borealis* | *B. b. borealis* (North Atlantic and North Pacific) |
| | *B. b. schlegellii* (Southern Hemisphere) |
| *B. physalus* | *B. p. physalus* (North Atlantic and North Pacific) |
| | *B. p. quoyi* (Southern Hemisphere) |
| *B. musculus* | *B. m. musculus* (North Atlantic and North Pacific) |
| | *B. m. indica* (northern Indian Ocean) |
| | *B. m. brevicauda* (Southern Hemisphere)[c] |
| | *B. m. intermedia* (Southern Hemisphere)[c] |
| *Platanista gangetica* | *P. g. gangetica* (Ganges and Brahmaputra rivers) |
| | *P. g. minor* (Indus river system) |
| *Inia geoffrensis* | *I. g. geoffrensis* (Amazon below Bolivia) |
| | *I. g. boliviensis* (Rio Madeira, Bolivia) |
| | *I. g. humboldtiana* (Orinoco river system) |
| *Delphinus delphis* | *D. d. ponticus* (Black Sea) |
| *Delphinus capensis* | *D. c. capensis* (warm-temperate Atlantic and Pacific Oceans) |
| | *D. c. tropicalis* (Indian Ocean) |
| *Stenella attenuata* | *S. a. attenuata* (pelagic tropical waters)[d] |
| | *S. a. graffmani* (eastern Pacific coastal) |
| *S. longirostris* | *S. l. longirostris* (pelagic tropical waters) |
| | *S. l. orientalis* (eastern Pacific offshore) |
| | *S. l. centroamericana* (eastern Pacific coastal) |
| | *S. l. roseiventris* (inner Southeast Asia) |
| *Lagenorhynchus obscurus* | *L. o. obscurus* (southern Africa) |
| | *L. o. fitzroyi* (southern South America) |
| | *L. o.* subsp. (New Zealand) |
| *Cephalorhynchus commersonii* | *C. c. commersonii* (South America and Falkland Islands) |
| | *C. c. kerguelenensis* (Kerguelen Islands) |
| *Cephalorhynchus hectori* | *C. h. hectori* (South Island, New Zealand) |
| | *C. h. maui* (North Island, New Zealand) |
| *Globicephala melas* | *G. m. melas* (North Atlantic) |
| | *G. m. edwardii* (Southern Hemisphere) |
| *Orcinus orca* | *O. o. orca* (cosmopolitan) |
| | *O. o.* subspp. (transients and residents, sympatric in Northeast Pacific) |
| *Neophocaena phocaenoides* | *N. p. phocaenoides* (Indian Ocean to southern China Sea) |
| | *N. p. sunameri* (western North Pacific) |
| | *N. p. asiaeorientalis* (Yangtze River) |
| *Phocoena phocoena* | *P. p. phocoena* (North Atlantic) |
| | *P. p. vomerina* (eastern North Pacific) |
| | *P. p.* subsp. (western North Pacific) |
| | *P. p. relicta* (Black Sea) |
| *Phocoenoides dalli* | *P. d. dalli* (North Pacific) |
| | *P. d. truei* (Kuril Peninsula, northern Japan) |
| Carnivores | |
| *Arctocephalus pusillus* | *A. p. pusillus* (southern Africa) |
| | *A. p. doriferus* (Australia) |
| *A. australis* | *A. a. australis* (Falkland Islands/Malvinas) |
| | *A. a. gracilis* (southern South America) |
| *Odobenus rosmarus* | *O. r. rosmarus* (Atlantic Arctic) |

(continues)

TABLE I (continued)

| Species | Subspecies |
|---|---|
| | *O. r. laptevi* (Kara Sea to eastern Siberia) |
| | *O. r. divergens* (Pacific Arctic) |
| *Erignathus barbatus* | *E. b. barbatus* (Atlantic Arctic) |
| | *E. b. nauticus* (Laptev Sea to Pacific Arctic) |
| *Phoca vitulina* | *P. v. vitulina* (eastern North Atlantic) |
| | *P. v. concolor* (western North Atlantic) |
| | *P. v. mellonae* (freshwater, eastern North America) |
| | *P. v. stejnegeri* (western North Pacific) |
| | *P. v. richardii* (eastern North Pacific) |
| *Pusa hispida* | *P. h. hispida* (Arctic Ocean and Bering Sea) |
| | *P. h. botnica* (Baltic Sea) |
| | *P. h. lagodensis* (Lake Ladoga, Russia) |
| | *P. h. saimensis* (freshwater lakes in Finland) |
| | *P. h. ochotensis* (Sea of Okhotsk) |
| *Halichoerus grypus* | *H. g. grypus* (western and eastern Atlantic) |
| | *H. g. macrorhynchus* (Baltic Sea) |
| *Pagophilus groenlandicus* | *P. g. groenlandicus* (North Atlantic) |
| | *P. g. oceanicus* (White and Barents Seas) |
| *Ursus maritimus* | *U. m. maritimus* (Atlantic Arctic) |
| | *U. m. marinus* (Pacific Arctic) |
| *Enhydra lutris* | *E. l. lutris* (western North Pacific) |
| | *E. l. kenyoni* (Aleutians to Washington State) |
| | *E. l. nereis* (California to Mexico) |
| *Sirenians* | |
| *Trichechus manatus* | *T. m. manatus* (South American mainland) |
| | *T. m. latirostris* (southeastern United States and Caribbean) |
| *Dugong dugon* | *D. d. dugon* (Indian and western Pacific Oceans) |
| | *D. d. hemprichii* (Red Sea) |

[a]From Amaha (1994), Rice (1998), Perrin *et al.* (1999), Baker *et al.* (2002), Jefferson and Van Waerebeek (2002), Krahn *et al.* (2004), Viaud-Martínez *et al.* (2007), Robineau *et al.*, 2007.
[b]Approximate ranges in parentheses.
[c]Relative winter (breeding) ranges of *B. m. brevicauda* and *B. m. intermedia* unknown.
[d]Combines "subsp. A" and "subsp. B."

in average size in the Mediterranean (Perrin, 1984). The common bottlenose dolphin, *Tursiops truncatus*, shows the same pattern, being largest in the open Atlantic and smallest in the Black Sea, as does the beluga, *Delphinapterus leucas*, in the Canadian Arctic (Stewart and Stewart, 1989; Doidge, 1990). Body size also varies inshore/offshore. In the eastern tropical Pacific, the coastal subspecies of the pantropical spotted dolphin, *Stenella attenuata graffmani*, is on average larger than the offshore form, *S. a. attenuata*. It also has larger teeth; it may prey on larger tougher, benthic fish species, whereas the offshore form feeds primarily on small epipelagic fishes and squids. The pattern is repeated in the Atlantic spotted dolphin; the coastal form is larger than the offshore form in the Gulfstream (Perrin *et al.*, 1987). However, in the bottlenose dolphin in the western North Atlantic, the pattern is reversed; the offshore form is larger than the coastal form (Hoelzel *et al.*, 1998), in correlation with different stomach contents and parasite loads. Variation in size can also be latitudinal; short-beaked common dolphins in the eastern Pacific are longest in the Central Stock off Central America and shorter to the north and south (Perryman and Lynn, 1993). Although it has been suggested that some of this variation in body size could be ecophenotypic (due, e.g., to differential nutrition across areas of varying productivity), it is thought to most likely be determined genetically.

The dorsal fin is another feature that varies markedly with region in some odontocetes. A dramatic example of this can be seen in the tropical Pacific; whereas the fin in spinner dolphins in the Hawaii and the South Pacific is slightly falcate and subtriangular, typical of the species around the world, in large adult males in the far eastern Pacific (*Stenella longirostris orientalis* and *S. l. centroamericana*) the fin is canted forward, with a convex posterior margin (Fig. 2). Animals in a broad zone of HYBRIDIZATION or intergradation are intermediate. A similar variation is present in short-beaked common dolphins; large adult males from the equatorial offshore eastern Pacific have more erect, triangular dorsal fins than in other regions. In both species the more erect (or forward-canted) dorsal fin is correlated with the development of a post-anal ventral hump (of unknown function).

Color pattern also varies within a species. In the *truei* form of Dall's porpoise (*Phocoenoides dalli truei*) in the western Pacific the ventrolateral white field is greatly enlarged from that in *P. d. dalli*. The just-described geographic variation in dorsal fin shape in the spinner dolphin is correlated with variation in color pattern; the dorsal overlay in the eastern spinner is extensive and dark, obscuring the cape and giving the animal a monochromatic rather than a tricolor appearance. In killer whales in the Antarctic, the cape is visible; in

G

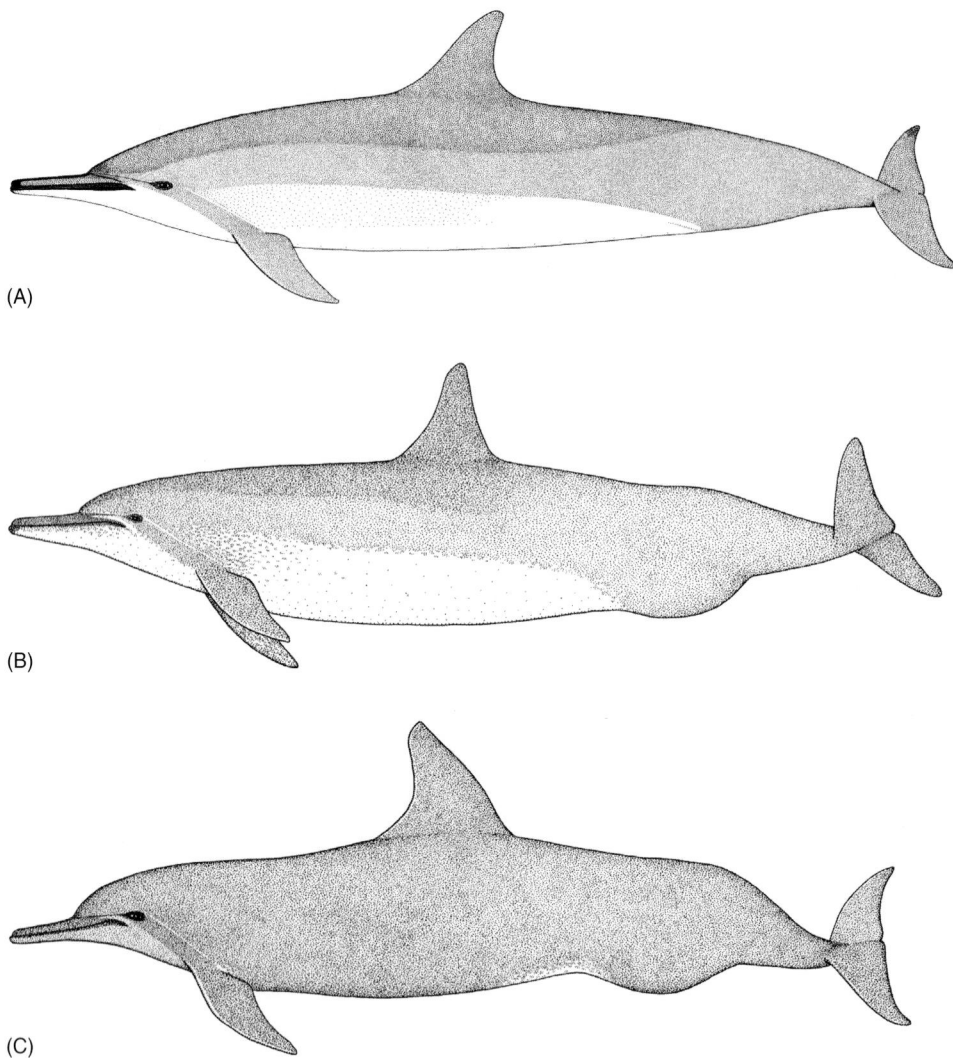

**Figure 2** *Geographic variation in shape and color pattern in spinner dolphins from the eastern and central Pacific: (A) Hawaii, (C) far eastern Pacific, and (B) intermediate form from region far offshore in eastern Pacific. By permission from Perrin (1990).*

other areas of the world it is not (Evans *et al.*, 1982). The degree of spotting in the Atlantic spotted dolphin (*Stenella frontalis*) varies from intense along the US east coast to slight or none in animals in the offshore Gulf Stream to the northeast (Perrin *et al.*, 1987).

The most extensive studies of geographic variation in odontocetes have dealt with cranial features, characters that can be measured in collections of museum specimens. Within a species, variation has been found to be greatest in elements involved in feeding: size and number of teeth, length and breadth of the rostrum, and size of the temporal fossa. This implies that much geographic variation must be associated with factors in trophic ecology such as available forage, foraging techniques, and competition. Cranial variation in the offshore spotted dolphin has been shown to be correlated with environmental parameters such as water depth, solar insulation, sea-surface temperature, surface salinity, and thermocline depth (Perrin *et al.*, 1994), and the distribution of two forms of the spinner dolphin in the eastern Pacific is associated with different water masses (Fiedler and Reilly, 1993). Different geographic forms or subspecies may also exhibit different patterns in life history parameters, such as age and size at attainment of sexual maturity, fecundity, and survival, but these differences can be due to differential population status as well as genetic factors (Chivers and DeMaster, 1994).

Mysticetes have not been as well studied because of their large size and a paucity of museum series of specimens (Reeves *et al.*, 2004). Subspecies and populations have been recognized mainly on the basis of DISTRIBUTION and, more recently, genetic differences. No adequate morphological comparisons of the recognized populations of bowhead whales, *Balaena mysticetus*, have been carried out; the same is true for the two extant populations of the gray whale, *Eschrichtius robustus*, and the several populations of humpback whales, *Megaptera novaeangliae* (apart from fluke coloration). It is only in the rorquals that progress has been made in documenting geographic variation in morphology; this has been due to the availability of large series of specimens in factory-ship whaling operations. A dwarf form of the minke whale, *Balaenoptera acutorostrata*, exists in the Southern Hemisphere. Minke whales from the Sea of Japan ("J Stock") and Pacific coast of Japan ("O Stock") may differ modally in body proportions and baleen and flipper coloration (Kato *et al.*, 1992). Small coastal and large offshore forms of Bryde's whales, *B. eden/brydei*, have been described from South Africa and Japan. The pygmy blue whale, *B. musculus brevicauda*, is shorter (by 2 m in the North Pacific; Gilpatrick *et al.*, 1997) and heavier than other blue whales. (Thus the heaviest animal on earth is called "pygmy.")

## IV. Carnivores

Similar to cetaceans, pinnipeds can vary geographically in body size (e.g., *Arctocephalus tropicalis* among Amsterdam, Gough, and Marion Islands; *Odobenus rosmarus* between the Atlantic and the Pacific Arctic; *Mirounga leonina* between Macquarie Island and South Georgia; *Pusa hispida* between the Baltic and the Sea of Okhotsk; Rice, 1998), coloration (*Phoca vitulina* between different islands off California; Yochem *et al.*, 1990), and cranial features (*Phoca vitulina* among several subspecies; *P. hispida* between pack ice and shore-fast ice and between the freshwater populations; *Halichoerus grypus* between the two sides of the Atlantic and the Baltic; *Histriophoca fasciata* between the western and the eastern parts of the Bering Sea; and *Monachus monachus* between the Mediterranean and the Atlantic; Rice, 1998).

The three subspecies of sea otter are distinguished on the basis of body size and cranial characters (Wilson *et al.*, 1991) and the two subspecies of polar bears on the basis of skull size (Rice, 1998).

Although for some small cetaceans efforts to find genetic markers concordant with geographic morphological variation have failed (see Dizon *et al.*, 1991 for *Stenella longirostris*), the reverse is true for pinnipeds; genetic differences or reproductive isolation have been found between populations that cannot be distinguished morphologically in *Arctocephalus forsteri*, *Zalophus californianus*, *Eumetopias jubatus*, and *P. hispida* (Hoelzel, 1997; Loughlin, 1997; Rice, 1998). Factors causing this may be polygyny and strong phylopatry (promoting accumulation of neutral variation due to drift) combined with relatively uniform ecological selection promoting morphological homogeneity over the range of the species.

## V. Sirenians

A molecular study of *Trichechus manatus* (García-Rodríguez *et al.*, 1998) found variation not in accordance with the presently recognized subspecies based on cranial characters; *T. m. latirostris* is closely linked to the Caribbean population of *T. m. manatus*, whereas the phylogenetic distances among the Caribbean, Gulf of Mexico, and South American populations of *T. m. manatus* are comparable to that between *T. manatus* and the Amazonian manatee, *T. inunguis*. These results were supported by further genetic investigations (Vianna *et al.*, 2006). As for many other marine mammal taxa, the taxonomy is ripe for revision based on both morphological and molecular characters.

Geographic variation in cranial morphology has been found within *Dugong d. dugon*, between Australia and Tanzania and between the Gulf of Carpentaria and Queensland in Australia (Rice, 1998).

## *See Also the Following Articles*

Biogeography ■ Coloration ■ Genetics for Management ■ Morphology ■ Functional Species

## *References*

Amaha, A. (1994). Geographic variation of the common dolphin, *Delphinus delphis* (Odontoceti Delphinidae). Ph.D. Thesis, Tokyo University of Fisheries.

Amano, M., and Miyazaki, N. (1992). Geographic variation and sexual dimorphism in the skull of Dall's porpoise, *Phocoenoides dalli*. *Mar. Mamm. Sci.* **8**, 240–261.

Archer, F. I., II, and Perrin, W. F. (1999). *Stenella coeruleoalba*. *Mamm. Spec.* **603**, 9.

Baker, A. N., Smith, A. N. H., and Pichler, F. B. (2002). Geographical variation in Hector's dolphin: Recognition of new subspecies of *Cephalorhynchus hectori*. *J. Roy. Soc. New Zeal.* **32**, 713–727.

Chivers, S. J., and DeMaster, D. P. (1994). Evaluation of biological indices for three eastern Pacific dolphin species. *J. Wildl. Manag.* **58**, 470–478.

Dizon, A. E., Southern, Š. O., and Perrin, W. F. (1991). Molecular analysis of mtDNA types in exploited populations of spinner dolphins (*Stenella longirostris*). *Rep. Int. Whal. Commn. Spec. Iss.* **13**, 183–202.

Doidge, D. W. (1990). Age-length and length-weight comparisons in the beluga, *Delphinapterus leucas*. *In* "Advances in Research on the Beluga Whale." (T. G. Smith, D. J. St. Aubin, and J. R. Geraci, eds). *Can. Bull. Fish. Aquat. Sci.* **224**, 59–68.

Evans, W. E., Yablokov, A. V., and Bowles, A. E. (1982). Geographic variation in the color pattern of killer whales. *Rep. Int. Whal. Commn.* **32**, 687–694.

Fiedler, P. C., and Reilly, S. B. (1993). Interannual variability of dolphin habitats in the eastern tropical Pacific. II. Effects on abundances estimated from tuna vessel sightings 1975–1990. *Fish. Bull. US* **92**, 451–463.

García-Rodríguez, A. I., *et al.* (1998). Phylogeography of the West Indian manatee (*Trichechus manatus*): How many populations and how many taxa? *Mol. Ecol.* **7**, 1137–1149.

Gilpatrick, J. W., Perryman, W. L., Brownell, R. L., Jr., Lynn, M. S., and DeAngelis, M. L. (1997). Geographic variation in North Pacific and Southern Hemisphere blue whales (*Balaenoptera musculus*). IWC Sci. Comm., SC/49/O9, 32. Available from International Whaling Commission, The Red House, 135 Station Road, Impington, Cambridge CB4 9NP, UK.

Hoelzel, A. R. (1997). Molecular ecology of pinnipeds. *In* "Molecular Genetics of Marine Mammals." *Soc. Mar. Mamm. Spec. Pub.* **3**, 147–157.

Hoelzel, A. R., Potter, C. W., and Best, P. B. (1998). Genetic differentiation between parapatric nearshore and offshore populations of the bottlenose dolphin. *Proc. R. Soc. Lond., B* **265**, 1177–1183.

Houck, W. J., and Jefferson, T. A. (1999). Dall's porpoise *Phocoenoides dalli* (True, 1885). *In* "Handbook of Marine Mammals" (S. H. Ridgway, and R. Harrison, eds), Vol. 3, pp. 443–472. Academic Press, London.

Jefferson, T. A., and Van Waerebeek, K. (2002). The taxonomic status of the nominal dolphin species *Delphinus tropicalis* van Bree, 1971. *Mar. Mamm. Sci.* **18**, 787–818.

Jefferson, T. A., and Van Waerebeek, K. (2004). Geographic variation in skull morphology of humpback dolphins (*Sousa* spp.). *Mar. Mamm. Sci.* **30**, 3–17.

Kato, H., Kishiro, T., Fujise, Y., and Wada, S. (1992). Morphology of minke whales in the Okhotsk Sea, Sea of Japan and off the east coast of Japan, with respect to stock identification. *Rep. Int. Whal. Commn* **42**, 437–453.

Krahn, M. M., *et al.* (2004) (11 authors). 2004 status review of Southern Resident Killer Whales (*Orcinus orca*) under the Endangered Species Act. *NOAA Tech. Mem. NMFS-NWFSC-62.*

Lazaro, M., Lessa, E. P., and Hamilton, H. (2004). Geographic genetic structure in the franciscana dolphin (*Pontoporia blainvillei*). *Mar. Mamm. Sci.* **20**, 201–214.

Loughlin, T. R. (1997). Using the phylogeographic method to identify Steller sea lion stocks. *In* "Molecular Genetics of Marine Mammals." *Soc. Mar. Mamm. Spec. Pub.* **3**, 159–171.

Orr, M. R., and Smith, T. B. (1998). Ecology and speciation. *TREE* **13**, 502–505.

Perrin, W. F. (1984). Patterns of geographical variation in small cetaceans. *Acta Zool. Fennica* **172**, 137–140.

Perrin, W. F. (1990). Subspecies of *Stenella longirostris* (Mammalia: Cetacea: Delphinidae). *Proc. Biol. Soc. Wash.* **103**, 453–463.

Perrin, W. F. (1998). *Stenella longirostris*. *Mamm. Spec.* **599**, 7.

Perrin, W. F., Akin, P. A., and Kashiwada, J. V. (1991). Geographic variation in external morphology of the spinner dolphin *Stenella longirostris* in the eastern Pacific and implications for conservation. *Fish. Bull. US* **89**, 411–428.

Perrin, W. F., and Brownell, (1994). A brief review of stock identity in small marine cetaceans in relation to assessment of driftnet mortality in the North Pacific. *Rep. Int. Whal. Commn. Spec. Iss.* **15**, 393–401.

Perrin, W. F., Dolar, M. L. L., and Robineau, D. (1999). Spinner dolphins (*Stenella longirostris*) of the western Pacific and Southeast Asia: Pelagic and shallow-water forms. *Mar. Mamm. Sci.* **15**, 1029–1053.

Perrin, W. F., Mitchell, E. D., Mead, J. G., Caldwell, D. K., Caldwell, M. C., van Bree, P. J. H., and Dawbin, W. H. (1987). Revision of the spotted dolphins, *Stenella* spp. *Mar. Mamm. Sci.* **3**, 99–170.

**G**

Perrin, W. F., Schnell, G. D., Hough, D. J., Gilpatrick, J. W., Jr., and Kashiwada, J. V. (1994). Reexamination of geographic variation in cranial morphology of the pantropical spotted dolphin, *Stenella attenuata*, in the eastern Pacific. *Fish. Bull. US* **92**, 324–346.

Perrin, W. F., Dolar, M. L. L., Amano, M., and Hayano, A. (2003). Cranial sexual dimorphism and geographic variation in Fraser's dolphin, *Lagenodelphis hosei*. *Mar. Mamm. Sci.* **19**, 484–501.

Perryman, W. L., and Lynn, M. S. (1993). Identification of geographic forms of common dolphin (*Delphinus delphis*) from aerial photogrammetry. *Mar. Mamm. Sci.* **9**, 119–137.

Reeves, R. R., Perrin, W. F., Taylor, B. L., Baker, C. S., and Mesnick, S. L. (2004). Report of the Workshop on Shortcomings of Cetacean Taxonomy in Relation to Needs of Conservation and Management, April 30–May 2, 2004 La Jolla, California. *NOAA Technical Memorandum NOAA-TM-NMFS-SWFSC-363.* 94pp.

Rice, D. W. (1998). Marine mammals of the world. *Soc. Mar. Mamm. Spec. Pub.* **4**, 231.

Robineau, D., Goodall, R. N. P., Pichter, F., and Baker, C. S. (2007). Description of a new subspecies of Commerson's dolphin, *Cephalorhynchus commersonii* (Lacépède, 1804), inhabiting the coastal waters of the Kerguelen Islands. *Mammalia* **2007**, 172–180.

Stewart, B. E., and Stewart, R. E. A. (1989). *Delphinapterus leucas. Mamm. Spec.* **336**, 8.

Storz, J. F. (1999). Genetic consequences of mammalian social structure. *J. Mammal.* **80**, 553–569.

Vianna, J. A., *et al.* (2006). Phylogeography, phylogeny and hybridization in trichechid sirenians; implications for manatee conservation. *Mol. Ecol.* **15**, 433–447.

Viaud-Martínez, M., *et al.* (10 authors) (2007). Morphological and genetic differentiation of the Black Sea harbour porpoise *Phocoena phocoena. Mar. Ecol. Prog. Ser.* **338**, 281–294.

Wilson, D. E., Bogan, M. A., Brownell, R. L., Jr., Burdin, A. M., and Maminov, M. K. (1991). Geographic variation in sea otters, *Enhydra lutris. J. Mammal.* **72**, 22–36.

Yochem, P. K., Stewart, B. S., Mina, M., Zorin, A., Sadovov, V., and Yablokov, A. V. (1990). Non-metrical analyses of pelage patterns in demographic studies of harbor seals. *Rep. Int. Whal. Commn Spec. Iss.* **12**, 87–90.

# Giant Beaked Whales
## *Berardius bairdii and B. arnuxii*

### Toshio Kasuya

### I. Characteristics and Taxonomy

These two species are the largest members of the family Ziphiidae (Fig. 1). Arnoux's beaked whale, *Berardius arnuxii* Duvernoy, 1851, was described based on a skull from New Zealand.

A specimen of similar characters found in the Bering Sea was the basis for another species, Baird's beaked whale, *B. bairdii* Stejneger, 1883. Due to uncertainty in morphological difference, the validity of the two species has long been questioned (Balcomb, 1989), but they are now considered to represent distinct species based on mitochondrial DNA (Dalebout *et al.*, 2004).

Currently recognized morphological differences between the two species are slight and limited to smaller adult size in Arnoux's beaked whale (8.5–9.75 m vs 9.1–11.1 m) and possible differences in flipper size and in the shape of nasal bones and vomer. Condylobasal lengths of skulls of adult Arnoux's beaked whales range 1174–1420 mm, and those of Baird's beaked whale are 1343–1524 mm. Other measurements in percentage of condylobasal length are (both species combined), length of rostrum 60.7–69.5%, width of rostrum at base 64.4–82.3%, and breadth across zygomatic processes of squamosals 47.1–56.5%. Nasal bones are large but do not overhang the superior nares. Among Ziphiidae, their skulls show the least bilateral asymmetry and are distinguished by the greatest nasal area on the vertex of skull, followed by frontals and premaxillae in decreasing order (Dalebout *et al.*, 2003). A pair of large teeth erupt on the anterior end of the lower jaw at around sexual maturity and abrade rapidly. The tooth is flat, triangular in shape (about $8 \times 8 \times 3$ cm), and has elements of rudimental enamel, thin dentine, massive secondary dentine filling the pulp cavity, and thick cementum that covers the root.

The vertebral formula of 3 Arnoux's beaked whales was C7, T10-11, L12-13, Cd17-19, total 47–49, and that of 49 Baird's beaked whales off Wadaura, Japan, C7, T9-11, L12-14, Cd17-22, total 47–52 (mean 48.9); most (41) had either 48 or 49 vertebrae (Kasuya unpublished). There are five phalanges in the manus.

The stomach lacks an esophageal compartment, and the glandular stomach has up to nine segments. The cecum is absent. The nasal tract has three pairs of sacs.

The entire body is dark brown. The ventral side is paler and has irregular white patches. Tooth marks of conspecifics are numerous on the back, particularly on adult males.

These are the least sexually dimorphic species in the Ziphiidae. The body is slender with a small head, a low falcate dorsal fin, and small flippers that fit into depressions on the body. A pair of throat grooves and some accessory ones contribute to expand the oral cavity at suction feeding. The equation $W = (6.339 \times 10^{-6})L^{3.081}$ expresses the relationship between body weight ($W$, in kg); and body length ($L$, in cm) off Japan. The blowhole is crescent shaped with the concavity directing anteriorly. The melon is small and its front surface is almost vertical, with a slender projecting rostrum.

### II. Distribution and Abundance

Arnoux's beaked whales inhabit vast areas of the Southern Hemisphere outside of the tropics, from the Ross Sea at 78°S to Sao

**Figure 1**    *Baird's beaked whale (C. Brett Jarrett).*

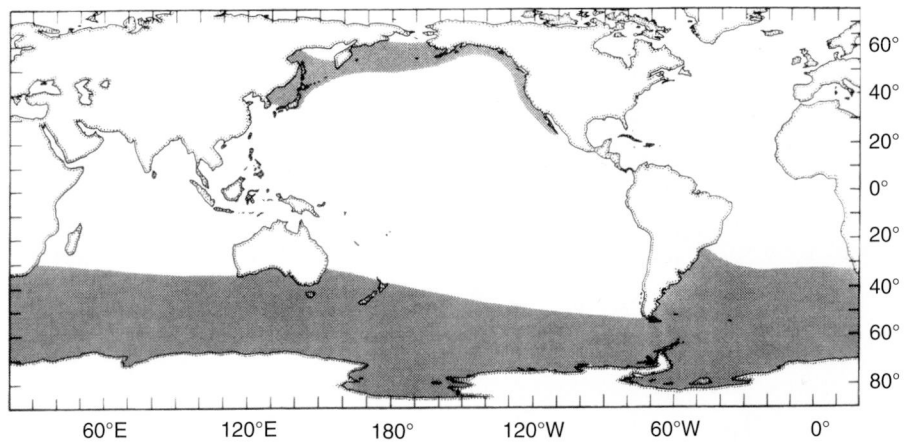

**Figure 2** *Giant beaked whales in the genus Berardius are distributed disjunctly.* B. arnuxii *occurs in waters around the Antarctic, reaching northward to the shores of the Southern Hemisphere continents.* B. bairdii *ranges across the northern Pacific from Japan, throughout the Aleutians, and southward along the coast to the southern tip of Baja California.*

G

Paulo (24°S), northern New Zealand (37°S), South Africa (31°S), and southeastern Australia (29°S) (Fig. 2). Baird's beaked whales inhabit the temperate North Pacific and adjacent seas, mainly deep waters over the continental slope. The northern limits are at Cape Navarin (62°N) in the Bering Sea and in the central Okhotsk Sea (57°N), where they occur even in shallow waters of 200–1000 m. On the American side they usually occur north of northern Baja California (30°N), but there are records from La Paz (24°) in the southern Gulf of California. The southern limits on the Asian side are at 36°N on the Japanese coast in the Sea of Japan and at 34°N on the Pacific coast. They occur year-round in the Okhotsk Sea and the Sea of Japan, including the drift ice area of the former. A vagrant was taken off the Chinese coast (30°N) (Wang, 1999).

Off the Pacific coast of Japan, the whales appear in May in waters over the continental slope at depths of 1000–3000 m. Their numbers increase toward summer when hunting commences and then decrease toward October. During this period they are almost absent in waters farther offshore. Factors determining the range are unknown and their wintering ground has not been identified (Kasuya and Miyashita, 1997).

Abundance has been estimated only for Japanese waters, 5029 for the Pacific coast, 1260 for the eastern Sea of Japan, and 660 for the southern Okhotsk Sea, with 95% confidence intervals of about 50% on both sides of the mean (International Whaling Commission, 1992, 1994).

## III. Ecology

For Baird's beaked whales off the Pacific coast of Japan (35°N), 81.8% of the food in number of individuals eaten was benthopelagic fish (Moridae and Macrouridae comprised 81.3%) and 18.0% various cephalopods (Gonatidae and Cranchiidae comprised 11.7%), but among individuals in the southern Okhotsk Sea (44–45°N) fish made up only 12.9% (Moridae and Macrouridae comprised 9.2%) and most of their food was cephalopods (87.1%) (Gonatidae and Cranchiidae comprised 86.7%) (Walker *et al.*, 2002). These data agree with earlier studies of Baird's beaked whales off California, that ate mainly Moridae, Macrouridae, and cephalopods with occasional ingestion of pelagic fish, and of an Arnoux's beaked whale that had squid beaks in the stomach, and suggest that they are opportunistic feeders.

Cyamids attach on teeth and skin, stalked barnacles on teeth, and diatoms on skin. Wounds attributable to the cookie-cutter shark,

*Isistius brasiliensis*, are common in Japan. Scars from killer whale teeth are common on flippers and tail flukes, suggesting predation. Internally, they are heavily parasitized in the stomach, liver, blubber, and kidney, with extensive kidney pathology due to the nematode *Crassicauda giliakiana*.

## IV. Behavior and Physiology

Information comes from observation of Baird's beaked whales off Japan (Kasuya, 1986; Kasuya and Miyashita, 1997; Kasuya *et al.*, 1997). When traveling, they form tight schools of 1–30 individuals (mean 5.9) (Fig. 3). Schools of 2–9 individuals constitute 64% of the encounters and singletons 14%. Diving lasts up to 67 min (mean 18.2), 39% last less than 11 min, 27% 11–20 min, and 18% 21–30 min. Time at the surface is 1–14 min (mean 3.9) and tends to be greater after a longer dive. During surface schooling, individuals blow continuously while swimming slowly and are easily identifiable from vessels.

## V. Life History

Age is determined using growth layers in the tooth cementum. The gestation time is unknown, although 17 months is suggested from interspecies relationship among toothed whales. Neonates are about 4.6 m long. Females first ovulate at age 10–15 years when they are 9.8–10.7 m and live to about 54 years.

The testis is histologically mature when it weighs 1.5 kg, which corresponds to age 6–11 years or a body length of 9.1–9.8 m, but continues growing until age 30, when it reaches 3–9 kg. Males live to about 84 years. Lack of behavioral data inhibits judgment of the age at which males begin to participate in reproduction.

Physical maturity, determined from the fusion of vertebral epiphyses to the centrum, is attained before 15 years and within 5 years after attainment of sexual maturity. Mean body lengths of whales 15 years or older are 10.45 m (SD = 0.31, n = 22) in females and 10.10 m in males (SD = 0.35, n = 66).

The sex ratio is about 1:1 before sexual maturity (44% males at age 3–9 years), after which males increase with age to reach 100% at 55 years and over. Seventeen percent of 170 individuals taken off Japan were males at ages 55–84 years, which were older than the oldest females (54 years). This explains the high male proportion (67.6%)

**Figure 3** *A school of Baird's beaked whales in the southern Okhotsk Sea (44° 15′N, 145° 30′) on 26 September 1994. Copyright Asahi Shinbun.*

the operation moved to outer seas. After World War II the fishery expanded to the entire northern Pacific, reported a maximum annual catch of over 300, and subsequently declined. A question exists about the reliability of the statistics (Balcomb and Goebel, 1977). In 2007 the industry operated with a quota of 10 for the Sea of Japan, 4 for the southern Okhotsk Sea, and 52 for the Pacific coasts.

There are no identified threats to the species except for takes by Japanese whaling from three putative stocks off Japan (Okhotsk Sea, Sea of Japan, and Pacific). Effect of the present catch should be evaluated together with consideration of effects of the past whaling operations on the stocks, using newer abundance estimates.

## See Also the Following Articles

Beaked Whales, Overview ■ Japanese Whaling

## References

Balcomb, K. C. (1989). Baird's beaked whale *Berardius bairdii* Stejneger, 1883: Arnoux's beaked whale *Berardius arnuxii* Duvernoy, 1851. *In* "Handbook of Marine Mammals" (S. H. Ridgway, and R. Harrison, eds), 4, pp. 261–288. Academic Press, San Diego.

Balcomb, K. C., and Goebel, C. A. (1977). Some information on a *Berardius bairdii* fishery in Japan. *Rep. Int. Whal. Commn* **27**, 485–486.

Dalebout, M. L., *et al.* (2003). Appearance, distribution, and genetic distinctiveness of Longman's beaked whales, *Indopacetus pacificus*. *Mar. Mamm. Sci.* **19**, 421–461.

Dalebout, M. L., Baker, C. S., Mead, J. G., Cockroft, V. G., and Yamada, T. K. (2004). A comprehensive and validated molecular taxonomy of beaked whales, family Ziphiidae. *J. Hered.* **95**, 459–473.

International Whaling Commission (1992). Report of the sub-committee on small cetaceans. *Rep. Int. Whal. Commn* **42**, 108–119.

International Whaling Commission (1994). Report of the sub-committee on small cetaceans. *Rep. Int. Whal. Commn* **44**, 178–228.

Kasuya, T. (1986). Distribution and behavior of Baird's beaked whales off the Pacific coast of Japan. *Sci. Rep. Whales Res. Inst., Tokyo* **37**, 61–83.

Kasuya, T., and Miyashita, T. (1997). Distribution of Baird's beaked whales off Japan. *Rep. Int. Whal. Commn* **47**, 963–968.

Kasuya, T., Brownell, R. L., Jr., and Balcomb, K. C., III (1997). Life history of Baird's beaked whales off the Pacific coast of Japan. *Rep. Int. Whal. Commn* **47**, 969–979.

Walker, W. A., Mead, J. G., and Brownell, R. L. (2002). Diet of Baird's beaked whales, *Berardius bairdii*, in the southern Sea of Okhotsk and off the Pacific coast of Honshu, Japan. *Mar. Mamm. Sci.* **18**, 902–919.

Wang, P. (1999). "Chinese Cetaceans." Ocean Enterprises, Hong Kong [In Chinese].

in the sample. Such a sex ratio imbalance is common among whaling samples from Japan (Sea of Japan, Okhotsk Sea, and Pacific), Russia (Kuril and Aleutian Islands), and Canada (off Vancouver Island) and is believed to reflect a lower natural mortality rate of males.

The proportion of females among sexually mature individuals was 23%. This is increased to 37% with the assumption that males attain reproductive capacity at 30 years when testicular growth ceases. Ovulation occurs throughout life at an approximate rate of once every 2 years. The apparent high fecundity and shorter longevity of females combined with greater male longevity have invited speculations on their social structure including a possible male contribution in rearing calves.

## IV. Interactions with Humans

Exploitation has not been reported for the Arnoux's beaked whale. Past hunting of Baird's beaked whales by the USSR, Canada, and the United States was at low levels. Only Japan currently hunts them. This started in the early seventeenth century at the entrance of Tokyo Bay. The annual catch was less than 25 before 1840 and then declined. In 1891, whaling cannons were introduced and

---

# Gray Seal
*Halichoerus grypus*

## AILSA HALL AND DAVID THOMPSON

### I. Characteristics and Taxonomy

The gray seal is the only member of the genus *Halichoerus* a member of the family of the true seals or Phocidae. Its species name, *grypus*, means hook nosed, referring to the Roman nose profile of the adult male. *Halichoerus* means sea pig in Greek. This species exhibits SEXUAL DIMORPHISM with the mature males

Figure 1   *Female gray seal, Halichoerus grypus.*

Figure 2   *Male gray seal, Halichoerus grypus.*

weighing between 170 and 310 kg and adult females between 100 and 190 kg. Individuals from the population in the western Atlantic are significantly larger than those from the eastern Atlantic; males can weigh over 400 kg and females over 250 kg. Genetic studies suggest that the western and the eastern Atlantic populations are distinct and diverged approximately 1 million years ago (Boskovic *et al.*, 1996).

Morphological differences between the sexes can be seen in Figs. 1 and 2. The neck and chest of the male are wrinkled and often scarred, whereas females are much sleeker. Both have the convex nose and wide muzzle, which are very pronounced in the male. Pelage patterns are highly variable, but many of the females are slate gray in color with a distinctive cream/off-white underside, particularly around the neck, spotted with black blotches. Males are usually more uniformly dark when mature, but subadults can have similar cream-colored patches on the neck and the side of the face. Females mature at between 3 and 5 years old and males around 6 years, although they are probably not socially mature until 8 years old.

## II.  Distribution and Abundance

Figure 3 shows the geographic range of the gray seal. Breeding rookeries are usually on remote beaches and uninhabited islands but they also breed on fast ice in the Baltic and the Gulf of St Lawrence. Table I shows the relative size of the various gray seal populations worldwide. Because the relationship between the number of pups born each year (pup production) and the total population size is not well known, the pup production estimates for each population are given. The total population will probably lie somewhere between 3 and 4.5 times the pup production depending on the local survival and fecundity patterns.

## III.  Ecology

Gray seals are highly successful predators of the North Atlantic. They feed on a variety of fish species and cephalopods (Hammond *et al.*, 1994a, b). However, a large proportion of their diet is sand eels or sand lance (*Ammodytidae*), which can make up over 70% of the diet at some locations and in some seasons. Other prey include whiting, cod, haddock, saithe, and flatfish (plaice and flounder). They are largely demersal or benthic feeders, and foraging trips lasting between 1 and 5 days away from a haul-out site are frequently focused on discrete areas that are within 40 km of a haul-out site (McConnell *et al.*, 1999).

## IV.  Behavior and Physiology

The females give birth, on land or on ice, to a single white-coated pup between September and March. The earliest breeding colonies are those in the south of the United Kingdom and Ireland. Further north around the British Isles the breeding season is later, between October and November. In Canada, peak pupping is not until January and in the Baltic it occurs in late February–early March.

At birth the pup weighs between 11 and 20 kg and, over the lactation period, lasting an average of 18 days, can quadruple in weight to over 40 kg. The mothers' milk is very fat rich (around 50–60% lipid) and is mobilized from her blubber stores. The pup's white coat, known as the lanugo, is shed at weaning. The pup then undergoes a postweaning fast on land for a period lasting, on average, approximately 25 days (Bennett *et al.*, 2007), during which it loses approximately 0.5 kg per day. The reason for this fasting period is not fully understood, but physiological changes during this time suggest that it is related to the development of diving ability.

Toward the end of lactation the female comes into estrus and mates. On some colonies there may be as many as 10 females to 1 male, whereas on rookeries, where access is not restricted by narrow gullies, the sex ratio may be 2 females to 1 male. Males compete for access to females but do not defend discrete territories, and matings may occur in the water as females return to the sea, as well as on land. Females fast during the breeding season and may lose up to 40% of their initial body weight as they do not feed during this time. The gestation period is 8 months, and to achieve a 12-month breeding cycle the fertilized egg is not planted until 4 months after conception. This occurs around the time of the annual molt when animals spend longer time hauled out on land. Gray seals generally return to their natal site to breed and show a high degree of site fidelity, often returning to within meters of their previous pupping sites (Pomeroy *et al.*, 1994).

On average, gray seal dives are generally short, lasting between 4 and 10 min with a maximum recorded duration of about 30 min. Gray seal foraging is mainly confined to the shallow continental shelf waters. Typically, animals dive down to the sea bed, in relatively shallow waters 60–100 m in depth but are capable of routinely diving to 200 m in some areas. Dives to more than 300 m have been recorded.

## V.  Interactions with Humans

Hunted throughout much of its range until the mid twentieth century it is now protected under national and international legislation. Small scale controlled hunting continues in parts of its range and

G

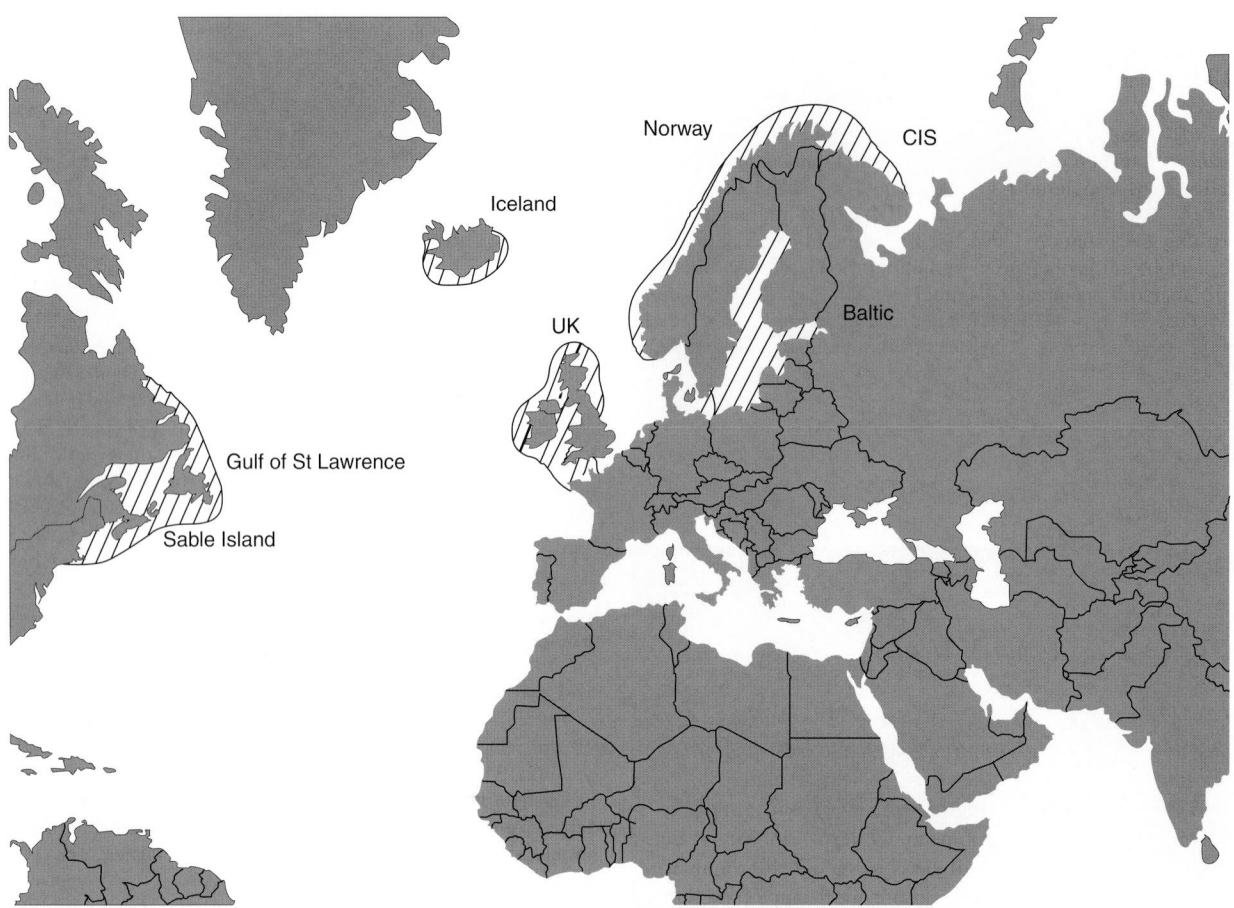

Figure 3   *Geographic distribution of the gray seal.*

TABLE I

**Relative Sizes of Gray Seal Populations. Pup Production Estimates are Used to Avoid Uncertainty in Overall Population Estimates**

| Region | Pup production | Years when latest information was obtained | Possible population trend |
| --- | --- | --- | --- |
| Scotland | 40,600 | 2006 | Stable or slowly increasing |
| NE England | 2,700 | 2006 | Increasing |
| SW Britain | 1,800 | 2006 | Increasing |
| UK | 45,100 | | Increasing |
| Ireland | 300 | 1998 | Unknown[a] |
| Wadden Sea | 200 | 2004 | Increasing[b] |
| Norway | 1200 | 2003 | Unknown[b] |
| Russia | 800 | 1994 | Unknown[b] |
| Iceland | 1200 | 2002 | Declining[b] |
| Baltic | 4,000[c] | 2003 | Increasing[b] |
| Europe excluding UK | 7,700 | | Increasing |
| Canada – Sable Island | 41,500 | 2004 | Increasing[c] |
| Canada – Gulf St Lawrence | 6,100 | 2000 | Declining[b] |
| WORLD TOTAL | 100,400 | | Increasing |

Baltic pup production estimate based on mark-recapture estimate of total population size and an assumed multiplier of 4.7

[a]Kiely, O., and Myers, A. A. (1998). Grey seal pup production at The Inishkea Island Group, Co. Mayo and Blasket Islands, Co. Kerry. Proceedings of the Royal Irish Academy, **98b**, 113–122.

[b]Data summarized in Grey Seals of the North Atlantic and the Baltic (2007). (T. Haug, M. Hammill, and D. Olafsdottir (eds.). NAMMCO Scientific publications Vol. 6.

[c]Bowen, W. D., McMillan, J. I., and Blanchard, W. (2007). Reduced population growth of gray seals at Sable Island: Evidence from pup production and age of primiparity. Mar. Mammal Sci., **23**(1), 48–64.

some animals are killed as part of control measures to reduce interactions with fisheries. Gray seal populations are intensively monitored and regular counts throughout its range suggest that its population is increasing in most areas.

## References

Bennett, K. A., Speakman, J. R., Moss, S. E., Pomeroy, P., and Fedak, M. A. (2007). Effects of mass and body composition on fasting fuel utilisation in grey seal pups (*Halichoerus grypus*): An experimental study using supplementary feeding. *J. Exp. Biol.* **210**, 3043–3053.

Boskovic, R., Kovacs, K. M., Hammill, M. O., and Whilte, B. N. (1996). Geographic distribution of mitochondrial DNA haplotypes in grey seals (*Halichoerus grypus*). *Can. J. Zool.* **74**, 1787–1796.

Hammond, P. S., Hall, A. J., and Prime, J. H. (1994a). The diet of grey seals around Orkney and other island and mainland sites in northeastern Scotland. *J. Appl. Ecol.* **31**, 340–350.

Hammond, P. S., Hall, A. J., and Prime, J. H. (1994b). The diet of grey seals in the Inner and Outer Hebrides. *J. Appl. Ecol.* **31**, 737–746.

McConnell, B. J., Fedak, M. A., Lovell, P., and Hammond, P. S. (1999). Movements and foraging areas of grey seals in the North Sea. *J. Appl. Ecol.* **36**, 573–590.

Pomeroy, P. P., Anderson, S. S., Twiss, S. D., and McConnell, B. J. (1994). Dispersion and site fidelity of breeding female grey seals (*Halichoerus grypus*) on North Rona, Scotland. *J. Zool. (Lond.)* **233**, 429–447.

# Gray Whale
## *Eschrichtius robustus*

### MARY LOU JONES AND STEVEN L. SWARTZ

### I. Characteristics and Taxonomy

The gray whale (*Eschrichtius robustus*, Lilljeborg, 1861) is the only living species in the family Eschrichtiidae (Ellerman and Morrison-Scott, 1951) (=Rhachianectidae; Weber, 1904). The genus name honors Danish zoologist Daniel Eschricht; *robustus* means strong or oaken in Latin. The evolutionary history of the modern genus is not well known. The fossil record of *E. robustus* includes recent and Pleistocene subfossils from both sides of the North Atlantic and a Pleistocene fossil (200–300 Ka) from California (Deméré *et al.*, 2005). A specimen from Japan that Ichishima *et al.* (2006) assigned to *Eschrichtius* sp. extends the earliest fossil record of the modern genus to the late Pliocene (1.8–3.5 Ma).

Recently the content of the family Eschrichtiidae has expanded to include two extinct genera based on specimens from Italy. "*Balaenoptera*" *gastaldii* † (Strobel, 1875), formerly assigned to Balaenoptera by Portis (1885), instead appears to be a new genus of basal eschrichtiid from the late Pliocene (1.8–3.5 Ma) (Deméré *et al.*, 2005). The new genus (and species) *Archaeschrichtius ruggieroi* † (Bisconti and Varola, 2006) was formally described based on a specimen (7.5–11 Ma) from the late Miocene and represents the oldest eschrichtiid identified to date. The two new taxa are characterized as intermittent suction feeders like the living gray whale. Bisconti and Varola (2006) propose an age of divergence for Eschrichtiidae of about 10 Ma and the Mediterranean Basin as the center of origin of the family.

Questions remain regarding the phylogenetic position of Eschrichtiidae among baleen whales. Recent phylogenetic reconstructions of extant mysticetes based on morphology indicate Eschrichtiidae + Balaenopteroidae (6 rorquals and the humpback, *Megaptera novaeangliae*) are included within a monophyletic Balaenopteroidea (Geisler and Sanders, 2003). However, analyses based on molecular characters (Rychel *et al.*, 2004) and "total evidence" support two competing hypotheses concerning relationships within the Balaenopteroidea (1) the eschrichtiids and balaenopterids as sister taxa, and (2) eschrichtiids nested within a paraphyletic Balaenopteridae (Deméré *et al.*, 2005).

The gray whale (Fig. 1) is a slow-moving sturdy mysticete, slimmer than right whales and stockier than most rorquals. It attains a maximum length of 15.3 m (50 ft). The skin is a mottled light to dark gray with whitish blotches and heavily infested with barnacles and cyamids, or "whale lice," especially on the head (Fig. 2). Instead of a dorsal fin, the back has a hump followed by a series of fleshy knobs, or "knuckles," along the tailstock. The baleen is cream to pale yellow, the shortest (5–40 cm), coarsest, and with fewest plates (130–180 per side) of any whale. There are typically 2–7 short, deep, longitudinal creases on the throat rather than the numerous long ventral pleats of balaenopterids. The narrow triangular head (seen from above) is moderately arched downward (seen from the side) and relatively small (~20% of skeletal length) (Fig. 3). The overall skull structure is less telescoped than in other extant mysticetes. Unique to grays is a bulging "tailstock cyst" (10–25 cm wide) of unknown function on the ventral surface of the caudal peduncle. The flukes are 3–3.6 m across and frequently lifted before a deep dive. The blow is 3–4 m high, heart-shaped, bushy, or columnar. The behavioral ecology of the gray whale is unique among mysticetes, it is the most coastal; makes the longest migration; calves in warm bays, lagoons, and coastal areas; and is an intermittent suction feeder that regularly forages on benthos (organisms living within, at, or near to the sea floor), but also feeds opportunistically on plankton and nekton by gulping and skimming.

### II. Distribution and Abundance

Once found throughout the Northern Hemisphere, the gray whale became extinct in the Atlantic and now is a relict species confined to the productive neritic and estuarine waters of the North Pacific Ocean and adjacent waters of the Arctic Ocean (Fig. 4). Mixing between Atlantic and Pacific populations was possible during warm interglacial periods (such as 1–12 Ka). Whales could have moved between the Beaufort Sea and the Hudson Bay until the Little Ice Age when temperatures cooled enough for ice to form across most of the Northwest passage of the Arctic (1400–1850) isolating the populations. There are two extant Pacific populations of gray whales. The eastern population (also called the *American*, *California*, or *Chukchi* stock) occurs in the eastern North Pacific and Amerasian Arctic Oceans (between North America and Asia). The remnant western population (also called the *Asian*, *Korean*, or *Okhotsk stock*) occurs in the western North Pacific (off Asia). Genetic studies show the groups are differentiated at the population level.

### A. North Atlantic Population(s) (Extinct)

The modern gray whale once existed in the North Atlantic in the coastal waters of Europe, Iceland, and North America. Subfossils from Europe have been found off the Baltic coast of Sweden,

G

**Figure 1** *The head of the gray whale is typically covered with patches of barnacles and whale lice (A). The blow is heart-shaped if viewed from directly in front or behind (B). Instead of a dorsal fin, grays have a low hump followed by a series of bumps (C). The flukes often are lifted above the surface before a deep dive (D).*

**Figure 2** *Dense clusters of barnacles surrounded by whale lice (cyamids) develop shortly after birth. Barnacles leave white scars on the skin, which slowly re-pigment over time.*

Belgium, the Netherlands, and England (the most recent dated 1650), and along the coast of the United States from New Jersey to Florida (the most recent from colonial times about 1675). Based on written accounts, the last few gray whales in the North Atlantic were killed in the late seventeenth or the early eighteenth century by early Basque, Icelandic, and Yankee whalers. Whether coastal whaling was solely responsible for or only hastened the extinction

of an already generally moribund North Atlantic population(s) is unknown (Fig. 4).

### B. Western North Pacific Population (Critically Endangered)

The western gray whale is now a remnant population close to extinction that occurs off Russia, Japan, Korea, and China. Although it once utilized coastal feeding sites throughout the northern Sea of Okhotsk, today the core of the population feeds primarily from June to November off the northeastern coast of Sakhalin Island, Russia. During 2004–2006, six western gray whales (~5% of the population) that were sighted in feeding areas off Sakhalin Island also fed along the eastern coast of Kamchatka Peninsula, Russia. The Sakhalin Island foraging habitat, which is vital especially for pregnant and lactating females, lies within a region that is undergoing massive oil and gas development (Fig. 4; Reeves, 2005). The known parts of the population's current north–south migratory route include eastern Kamchatka, the eastern shores of Sakhalin Island and mainland Russia, possibly the Korean Peninsula, and the east and west coast of Japan. The mating and winter calving areas have yet to be determined. Speculation that they winter off the south coast of Korea is not supported with any observations. Evidence suggests whales traverse the East China Sea into the South China Sea to tropical waters off southeastern China (18–20°N) at least as far south as Hainan Island in winter. Before the twentieth century, offshoots of the population migrated off eastern and western Japan (Omura, 1984), but today sightings are rare.

The western north Pacific gray whale is one of the most critically endangered populations of whales (CR in IUCN Red List), and its

**Figure 3**  *Gray whales commonly spy hop.*

G

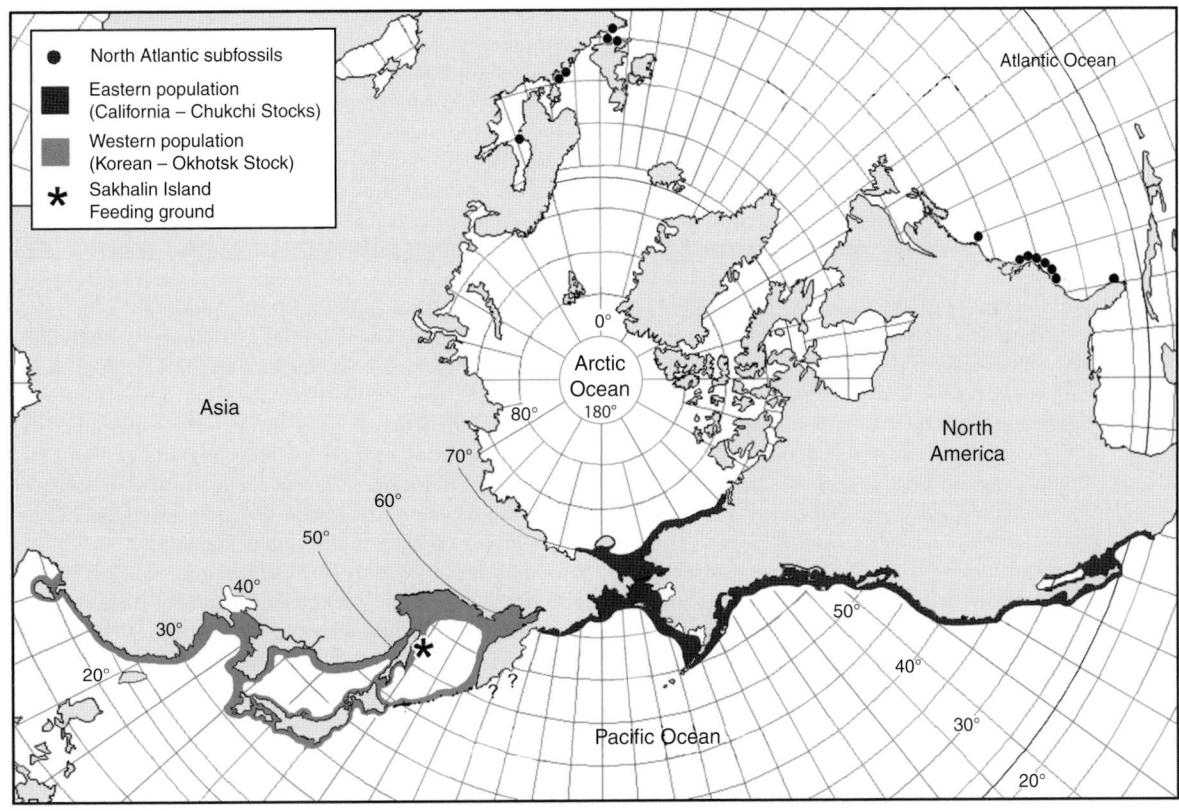

**Figure 4**  *Distribution of the extant populations of gray whale.*

survival is in jeopardy (http://www.iucnredlist.org). This stock was reduced to such low numbers by heavy international exploitation off Russia, Korea, and Japan from the 1840s to the mid-1960s that it was nearly extinct. Estimates of prior abundance are highly speculative, from 1500 to 10,000 whales. In 2007, the population size was estimated to be only 121 whales (90% CI = 112–130) (Cooke *et al.*, 2007). Constraints affecting population recovery include the low estimated number of reproductive females (28; 90% CI = 24–33), low juvenile survival, a male-biased sex ratio, the alarming advent of

thin and emaciated whales (~5–51% from 1999–2007), and genetic isolation from the eastern gray whale population. Of great concern is the number of females that have recently died along the coast of Japan. During 2005–2007, four female gray whales were entrapped and died in net fisheries (three off northeastern Honshu and one in Tokyo Bay; Cooke *et al.*, 2007). Surviving whales continue to be threatened by occasional exploitation by Japanese fishermen, entanglement in fishing gear, collisions with ships, and oil and gas exploration/development. Even the loss of one additional female per year (about the death rates in recent years) would be sufficient to drive the population to extinction. Killer whales also at times prey on western gray whales, especially calves.

## C. Eastern North Pacific Population

From late May to early October, the core of the eastern population is on its feeding range in the shallow coastal and shelf waters between Alaska and Russia and along their northern coastlines. Until the late 1990s, the northern Bering Sea, especially the Chirikov Basin, as well as the southern Chukchi Sea were the primary feeding grounds and supported high biomass of benthic amphipod prey and large numbers of feeding gray whales. However, in recent years a major ecosystem shift has been underway in the northern Bering Sea. A warming trend from arctic to subarctic conditions is resulting in higher air temperatures, lower winter ice cover, and increases in seawater temperature. Primary productivity has declined by about 70% from 1988–2004, reducing food supply to benthic communities in the Chirikov Basin. The ice-dominated, shallow ecosystem favoring benthic amphipod communities is being replaced by one dominated by pelagic fish. The ecosystem reorganization no longer favors benthic predators. Gray whales have responded by shifting their foraging distribution northward from the Chirikov Basin into the southern Chukchi Sea. Today, the primary feeding grounds are located only in the Chukchi Sea. Ongoing environmental changes should be expected to affect a much larger portion of the Pacific-influenced sector of the Arctic Ocean.

Secondary feeding sites are located in the western Bering Sea (off the Koryak coast), the southeastern Bering Sea (primarily along the eastern Alaska Peninsula and mainland coast), and in the Beaufort Sea (east to 130°W off northern Canada). Since 1980, there has been a westward opening of the range into the East Siberian Sea (179°E in 2007) as the pack ice and Arctic ice cap recede due to global warming. The northern boundary of the range is the Arctic ice-edge (72–75°N as of 2007). The foraging limits, however, are thought to be determined by ocean currents (originating with the Green Belt in the Bering Sea) that deliver high nutrient concentrations and high primary production biomass at broad frontal zones to benthic communities with resultant high secondary production which gray whales exploit (Highsmith et al., 2006). The southern limit of the range along the Asiatic coast of the Bering Sea is approximately 60°N. Since 1979, a few whales have foraged annually off southeastern Kamchatka Peninsula (51–55°N). During 2004–2006, six were identified as western gray whales indicating the feeding ranges of both populations overlap in this area. The first ever winter-long acoustic study in 2003–2004 detected gray whale calls in the Beaufort Sea, suggesting whales are extending residence times in the Arctic as the region warms, although it cannot be certain they have not wintered in the Arctic previously.

This population makes the longest annual migration of any mysticete, 15,000–20,000 km (roundtrip). Spanning up to 55° of latitude, the migration connects Arctic feeding grounds with southern mating regions, calving, and assembling grounds in temperate and in subtropical coastal waters in winter, which are presumably safer from killer whale predation. In the fall, whales start the southward migration with females in late pregnancy going first, followed by adults and immature females, then immature males. They exit the Bering Sea via Unimak Pass, Alaska, with approximately 90% filing through from mid-November to late December, and mainly follow the coast to Mexico. The trip averages 2 months. Mating occurs mainly during the middle of the migration, but courtship/mating also happens in the winter assembly area.

Winter grounds extend from central California south along Baja California, the Gulf of California, and mainland coast of Mexico south to Bahía de Banderas, Jalisco (20°45′N, 105°34′W). Pregnant females carrying near-term fetuses begin to arrive by late December to early January. By mid-February, the bulk of the population has arrived. The calving areas are coastal California and the west coast of the Baja California Peninsula—San Diego Bay (possibly occupied historically, but no longer used), Laguna Ojo de Liebre (Scammon's Lagoon), Laguna Guerrero Negro (no longer used), Laguna San Ignacio, and Bahía Magdalena and adjacent waters (from Estero Las Animas to Bahía Almejas), and eastern shore of the Gulf of California— the open coast of Yavaros (Sonora) and Bahía Reforma (Sinaloa) (Rice and Wolman, 1971). Most calves are born near to or within the lagoons and Bahía Magdalena region. The majority of whales, except most mothers with newborns, occur outside of the lagoons and estuaries in the waters of Bahía Sebastián Vizcaíno, Bahía de Ballenas, and elsewhere along the coastline. El Niño (warm) and La Niña (cold) events cause variability in the Mexican Pacific, which influences gray whale distribution. During El Niño, the migration of whales to Bahía Magdalena diminishes, while it increases during La Niña.

The north migration to the feeding grounds occurs as two phases. The first phase (late January through March) consists of newly pregnant females who go first to maximize feeding time in the Arctic, followed by adult females and males, then juveniles. The second phase (April through May) consists primarily of mothers and calves. They remain in the breeding area 1–1.5 months longer, allowing calves to strengthen and rapidly increase in size before the north migration. Cows and calves travel very close to shore (90% are within 200 m) mostly alone or in pairs.

The north migration culminates in the dispersal of the population throughout the Arctic feeding grounds. An exception is pockets of whales that feed opportunistically south of the Bering Sea during summer, or perhaps much of the year. Roughly, 1000 whales (~0.5% of the population of ~18,000 in 2007) forage along Kodiak Island, southeast Alaska, British Columbia, and areas southward to Baja California. Many may migrate through these areas, but some stop to feed for the summer, perhaps longer (Darling, et al., 1998; Moore et al., 2007). In Mexico, a few occur year-round in Bahía San Quintin and Cabo San Lorenzo, on the Pacific coast of Baja California, and Bahía de Los Angeles in the Gulf of California.

Commercial whaling of eastern gray whales began in 1846. Estimates of historic pre-whaling abundance differ widely. Those derived from whaling statistics range from 15,000–20,000 whales (Henderson, 1972) to 30,000–40,000 (Scammon, 1874), while a recent genetic analysis of DNA variables suggests a much larger pre-whaling population averaging approximately 96,000 individuals (76,000–118,000), but this estimate likely measures the eastern and western stocks together (Alter et al., 2007). The population was hunted to very low numbers by 1939 (models suggest possibly a few thousand). The stock finally received limited protection from commercial whaling in 1937 and full protection in 1946 under international treaty, except for a small aboriginal hunt. It is not known how many whales remained. The population made a dramatic comeback to 21,000 whales, and was removed from the US government's endangered species list in 1994. Growth continued through 1997/1998, when abundance reached a high of 29,758 whales (Rugh et al., 2005). At that time, there were indications that the population might be approaching carrying capacity and might begin to level off, but it has since suffered a major mortality event.

In 2000/2001 the abundance estimate dropped alarmingly by 35% to 19,448, and in 2001/2002 it was 39% less at 18,178 whales, indicating there had been a large die-off of about 11,500 whales over a 4-year period (Rugh *et al.*, 2005). In addition, the number of calves produced annually during 1999–2001 was about 70% lower than during the previous 5-year period. The population die-off was coincident with a spike in the number of stranded dead whales in 1999 (274 whales) and 2000 (368 whales), about 10 times greater than the annual average during the previous decade. The emaciated condition of many indicated they had starved. During this time, many living gray whales were also thin or emaciated.

Food shortage is the likely cause of these events, although contributing factors such as disease cannot be ruled out (Moore *et al.*, 2007). A large downturn in benthic amphipod biomass occurred during the 1990s in the Chirikov Basin, a primary feeding area, due to ecosystem changes in the northern Bering Sea, or overgrazing of the prey base that could no longer sustain the growing population, or both. Since malnourishment was also occurring in the western population, an alternative hypothesis posited that more global or ocean-wide changes are influencing the availability of, or access to, primary prey for the species.

Preliminary analysis of the latest 2006/2007 census suggests that both population size and calf abundance are similar to the low levels observed in 2001/2002, and some of the whales are "skinny." In San Ignacio Lagoon, a major breeding area, approximately 11–13% of the whales were "skinny" rather than robust during 2007, indicating those whales were nutritionally stressed. The future of the population will hinge on the whales' ability to adjust quickly to the ongoing ecological transitions associated with global climate warming.

## III. Ecology

### A. Diet and Feeding

The feeding ecology of gray whales is unique and complex. The diet consists of a wide variety of benthic organisms (infaunal, epibenthic, and hyperbenthic) but also includes planktonic and nektonic organisms (midwater and sea surface) and perhaps some plants. Gray whales use three foraging methods, they typically rely on intermittent suction as their primary mode and also opportunistically employ gulping and skimming to capture midwater and sea surface species. They are able to switch techniques to exploit the most optimum prey species, or assemblage of species at any one location within their summer–fall feeding range and elsewhere in the migratory and the wintering areas. Foraging activity occurs at water depths of 4–120 m, but mostly at 50 m or less. Water is obtained from their food (most fish and invertebrate prey consist of 60–80% water) and metabolically derived water.

Most gray whales concentrate their feeding during 5 months, from about May through October (eastern population) and June through November (western population) when they are in high-latitude high-productivity waters where food resources are patchy but dense. They forage primarily on or near the ocean floor and appear to feed continuously, 24 h a day. Gray whales preferentially forage on aggregations of crustaceans and invertebrates. At least 60 benthic amphipod species, 80–90 other benthic invertebrate species, and small fish occur in the northern diet. Principal prey in soft-bottom habitats include infaunal amphipods (tube-dwelling ampeliscid amphipods and burrowing pontoporeid amphipods dominate), polychaete worms, and bivalves. In some areas eastern gray whales also feed on locally abundant swarming species such as cumaceans, mysids, shrimp, krill, mobile amphipods, and shoals of sardines and anchovy. Energy might also be obtained from plant material (e.g., algae, kelp, sea grass) which is deliberately ingested in some cases, probably more than has usually been assumed, rendering the gray whale a partial herbivore, but almost nothing is known about the role of plants in their feeding ecology (Nerini, 1984). During the feeding season an adult eastern gray whale might consume approximately 220,800 kg of food (using 1200 kg/day and 184 feeding days).

During most or all of the 6–7 months when gray whales are migrating and on the winter grounds, they primarily fast and rely on stored lipid in body fat and blubber as the prime energy source. When whales return to northern feeding grounds, they will have lost 16–30% of body weight and must single-mindedly forage to replenish fat reserves. Lactation represents the greatest cost of reproduction, and pregnant females put on 25–30% more weight than other whales (exclusive of fetus). Lipids are also a critical source of water essential to maintaining water balance during fasting or greatly reduced food intake. Exceptions are small lactating females, which probably resume feeding on the north migration, and juvenile animals that tend to feed opportunistically throughout the year.

Some eastern gray whales (~1000 or 0.5% of the population) do not complete the annual migration to the feeding grounds and remain south of the Aleutian Islands during summer to forage opportunistically and sporadically in localized areas on infauna, swarming benthic invertebrates, and planktonic preys (e.g., amphipods, isopods, mollusks, cumaceans, shoaling mysids, shrimp, crabs, herring, eggs, and larvae).

To suction feed, the gray whale uses gular muscles and the tongue in a piston-like action to suck prey into the mouth. When foraging on infauna (prey living within sediments), whales roll on their side with the head just above the bottom and swim slowly while suctioning prey and sediment into the side of the mouth in pulses and filtering the prey with their baleen. This creates a series of large excavations, or "feeding pits," in a single dive (pits are ~3 m long, 1 m wide, and 1/2 m deep), and whales often trail clouds of sediment, or "mud plumes," in their wake (Fig. 5). This foraging on infauna significantly affects the habitat through sediment disruption and resuspension, and removal of the benthic prey assemblages. Gray whales also commonly use intermittent suction to feed on swarming organisms in near-bottom water and shoaling prey in the water column. In these instances, suction feeding does not result in seafloor excavations or mud plumes. When viewed from above the sea surface, indications of suction feeding in the water column include defecating whales, fecal slicks, whales trailed by seabirds, and animals "working" (diving repeatedly) in an area.

As a feeding generalist and flexible forager, the gray whale is responsive to feeding opportunities along their entire range and varies its foraging method accordingly. This provides insights into gray whale survival over the millennia. During Pleistocene glacial advances, the most recent of which ended 10–12 Ka, sea level was approximately 75 m lower than now. Consequently, areas that are currently major feeding grounds were above sea level, and marine access to the Arctic was blocked by the Bering land bridge. The ability to use alternative prey, feeding modes, and locations may have been critical to the species during periods of glacial maxima when continental shelf areas were above sea level (Highsmith *et al.*, 2006).

### B. Predators and Parasites

The killer whale (*Orcinus orca*) is the only predator of gray whales. Tooth rakes often occur on living whales, thus many attacks are not fatal. A reduced risk of calf mortality by killer whales (more

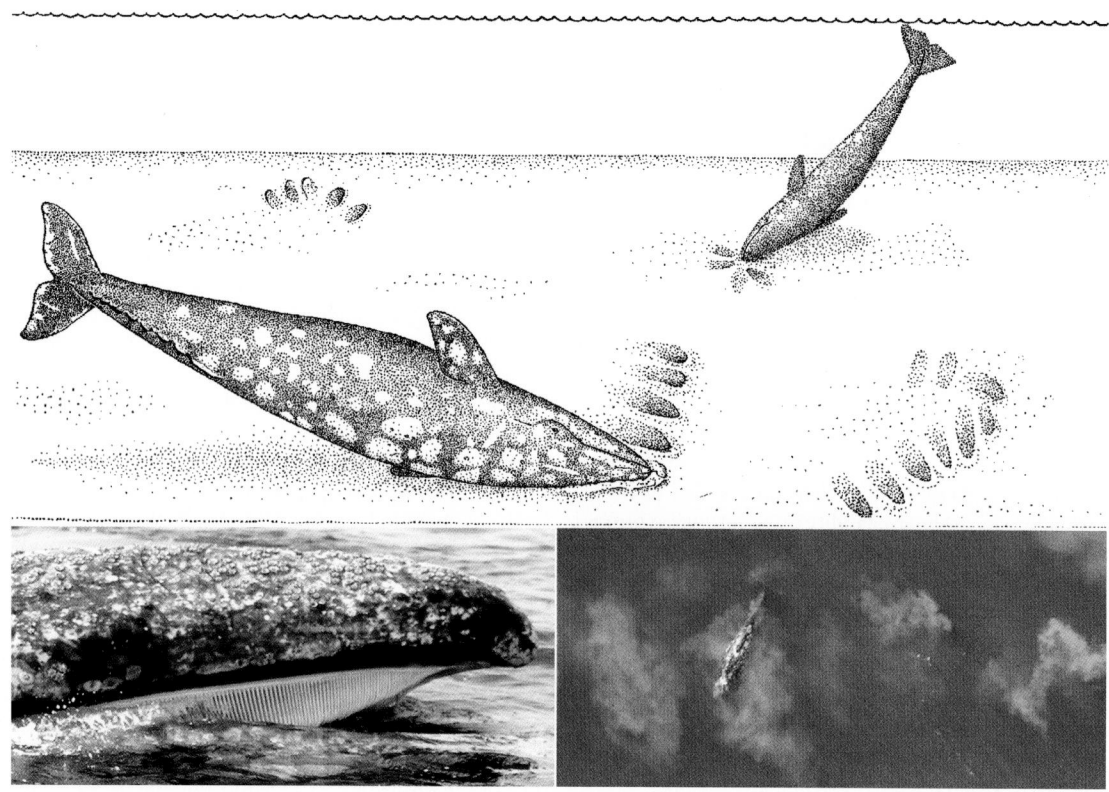

Figure 5    *Feeding gray whale.*

abundant at high latitudes in colder coastal seas) is proposed as the primary benefit to females migrating to give birth in the subtropics. Even so, calves remain prime targets during the north migration. Gray whales attempt to escape from killer whales by swimming into shallow water, often inside the surf zone.

Gray whales have heavy infestations of external parasites and commensals, more than any other cetacean. A host-specific barnacle, *Cryptolepas rhachianecti*, forms large colonies deeply embedded in the skin. Three species of cyamids (whale lice), *Cyamis scammoni*, *Cyamus kessleri*, and *Cyamus ceti* feed on skin around barnacles, blowholes, skin folds, and swarm into wounds (Fig. 2). These are not known to be harmful and may be more accurately regarded as mutualists or commensals rather than parasites, although whales in poorer nutritional condition tend to have heavier infestations. Apparently, gray whales are less prone to internal parasites than other cetaceans. These include trematodes (liver), nematodes (stomach), and cestodes and acanthocephalans (small intestine). Some require fishes as intermediate hosts (Rice and Wolman, 1971).

### C. "Skinny" and "Stinky" Whales

From 1999 to 2007, gray whales in poor body condition, unusually thin or emaciated and referred to as "skinny" whales, have been present in both populations. The proximate cause is nutritional stress, but underlying cause(s) remain unknown. Explanations include natural or human-caused changes in prey availability or habitat quality, physiological changes due to stress, disease, or a combination of these.

Since the 1960s, indigenous Russian whalers have reported eastern gray whales exhibiting a strong, foul-smelling medicinal odor, called "stinky" whales. The meat is malodorous and inedible (even village dogs will not eat it), and such whales have severe halitosis. The problem only exists in the region of Chukotka where similar phenomena occur in the meat of seals, walrus, fish, and murre eggs. The odiferous condition may result from altered metabolism due to a limited food supply, new or different foods consumed, disease, biotoxins, or abnormal metabolic pathways resulting from genetic anomalies.

## IV. Behavior and Physiology

### A. Social Organization

Gray whales form no long-lasting associations and apparently have little social cohesion. The bond between mother and calf stops after weaning. They migrate solo or in transitory pairs and small groups. On the feeding grounds, whales are usually single or in small groups and widely dispersed, not surprising given the large benthic area needed to supply the energy requirement of each whale. Large aggregations can occur on feeding grounds and breeding grounds but are in constant flux. There is no territoriality or aggression toward conspecifics. Care-giving behavior to aid young in the calving areas and joint defense against killer whale attacks occur but are rare. Whalers report standing-by behavior in which whales support or refuse to leave an injured companion, making them easy targets.

### B. Sensory Perception

Underwater sound rather than vision is the gray whale's primary sensory modality. They create a variety of phonations, which are mostly low frequency broadband signals that range from about 100 Hz to 4 kHz but may go up to 12 kHz. Use of mostly low frequency sounds might be an adaptive strategy for grays to circumvent the high levels of natural background noise prevalent in the coastal environment (e.g., waves, bubbles, currents, ice movement) by producing sounds that are generally at frequencies below it. Unfortunately, manmade noise is a rapidly increasing pollutant in the ocean. Much of it occurs in the lower frequency range and has a high level of output (e.g., high-intensity air guns for seismic exploration, military and civilian sonar systems, ship-shock trials, offshore drilling and construction, industrial activities, supertankers, icebreakers). Anthropogenic noise can interfere with whale communication, reduce ability to hear natural sounds that aid in foraging and navigation, and may damage hearing. Grays try to circumvent this noise by increasing call types, rates, and loudness to enhance signal transmission and reception.

Gray whales see moderately well both in air and in water, but the visual system likely is of inferior importance to that of the auditory system because of functional restrictions in turbid water and darkness. The position of the eyes suggests that they have stereoscopic vision forward and downward permitting efficient estimation of distance. The eyes are adapted for heightened sensitivity to dim light and for improving contrast and resolution underwater. Grays possess a tiny presumably functional olfactory system, but are microsmatic at best. The sense of touch is very well developed. Taste buds occur at the back of the tongue.

### C. Swimming, Breathing, and Diving

On migration, eastern gray whales mostly progress in one direction and travel at the same speed day and night. The southward trip to the winter grounds averages 55 days, at approximately 7–9 km/h (144–185 km/day). Northbound grays average approximately 4.5 km/h (88–127 km/day). Mothers and calves travel approximately 96 km/day; speed is about the same as other whales, but they pause to rest and nurse. Calves position themselves in a "drafting" position alongside their mothers, and the hydrodynamic effect is that the calf can gain thrust to swim while saving energy at traveling speed. If pursued, grays reach 13 km/h but can only maintain that speed for a few hours. In extreme duress, speed can surge to 16 km/h for short bursts. Breaching, spyhopping, and lobtailing are common during migration and on the breeding grounds (Fig. 6). Maximum dive depth is approximately 170 m. Maximum duration of breath holding is approximately 26 min., associated with resting and hiding. Blow pattern varies depending on age (calf vs adult) and behavior.

## V. Life History

Gray whales appear to have a promiscuous mating system, where both sexes may copulate with several partners during the breeding season. Multiple inseminations likely occur, suggesting that sperm competition is a feature in reproduction. Sexual maturity is attained from 6 to 12 years (average is 8 for both sexes), at a mean length of 11.7 m in females (called cows) and 11.1 m in males (called bulls). Mating and calving are strongly seasonal and synchronized with the migratory cycle. Bulls have a peak of spermatogenesis in late autumn or in early winter correlated with the time cows come into estrus. Thus mating occurs mainly during the middle of the south migration, but courtship/mating activity also continues during January and February as whales travel into and socialize throughout the winter assembly range where most females calve (eastern population).

Bulls can mate annually. Females usually have one estrous cycle per 2 years (rarely they may ovulate twice), thus at most only half the reproductive females are available annually for mating. Ovulation usually occurs in late November and December within a 3-week period during the south migration (while the eastern population is still north of central California). Females usually conceive following their first ovulation but may undergo another estrous cycle about 40 days later if they fail to conceive (Rice and Wolman, 1971).

Copulation is belly to belly. Pairs or trios of whales sometimes court and mate quite gently. However, in larger groups of up to 20 consorting adults, there is a high level of activity marked by rolling, splashing, and energetic cavorting (Fig. 7). Estimates of the gestation period vary from 11 to 13 months. Cows bear one calf at intervals of 2 years, but longer intervals of three or more years occur. Birth season for the eastern population lasts from late December to early March (median birth date is January 27) when near-term cows are in or near the Mexican calving grounds, although some are born off California.

A mother's bond with her calf is very close; they are unusually affectionate, protective, and will fight fiercely, even to the death, to defend young from danger (Fig. 7). Calves consume about 189 l of rich milk per day (53% fat, greater than any other cetacean, 6% protein). Weaning occurs at 7–8 months around August when calves are 7.6–9.5 m long. Cows then have a 3–4 month resting period until November–December when estrus begins anew.

**Figure 7** *Gray whales mate with multiple partners, often in large, energetic courting groups (A). Newborn calves have more uniformly dark skin and are supported on their mothers' backs for their first few breaths (B).*

**Figure 6** *Gray whales breach frequently while migrating and on the winter range. One animal was observed to breach 40 consecutive times.*

Most neonates are 4.6–4.9 m long and weigh about 680–920 kg. Adults weigh 16,000–45,000 kg and stop growing at about 40 years, when the average female is 14.1 m long and the average male is 13.0 m. The largest female recorded was 15.3 m and the largest male 14.6 m long. After birth, females are slightly bigger than males at all ages, but there is no significant difference in their appearance (the distance from the genital slit to the anus is longer in males). Longevity is unknown (age is calculated from growth layers in the waxy earplugs that fill the auditory canal), but estimates vary from at least 40 to over 80 years.

## IV. Interactions with Humans

### A. Whaling

Although the International Whaling Commission (IWC) banned commercial whaling of gray whales in 1946, it allows aboriginal whaling of the eastern population for cultural and subsistence purposes. For the years 2008–2012, the quota for aboriginal kills off the Chukotka Peninsula (Russian Federation) is set at 620 whales, with a maximum of 140 in any year. Whale oil, meat, bones, hide, and baleen are used.

There has been illegal hunting of gray whales in violation of the IWC moratorium on whaling, by its member nations as well as pirate whaling by fleets acting beyond national jurisdiction. In 2000 it was revealed that this prohibited species was killed "at every sighting" by the former Soviet Union from 1961–1979. Occasionally a western gray is taken incidentally by Japanese fishermen (e.g., in 2005, three females were killed in fishing nets). The IWC prohibits killing them deliberately, but sale as "bycatch" occurs in markets in Korea, Japan, and elsewhere. DNA profiling of whale meat suggests the true magnitude of intentional "net whaling" (deliberate entanglement in fishing nets and gear) of western gray whales sold as bycatch in Asia is larger than reported. If this mortality continues in such a small population, the population is projected to decline toward extirpation.

### B. Whale-Watching Industry and Friendly Whales

The eastern population of gray whales supports a major whale-watching industry. Whalers dubbed the gray whale the "devil fish" for its ferocity when harpooned, yet it is a gentle species if unmolested. Known today for approaching boats curiously and letting whale watchers pet it, the gray is popularly called the "friendly whale" (Fig. 8). From a conservation perspective, tourism poses both risks and benefits to gray whales. Risks arise from the potential for vital behavior patterns and essential habitat to be degraded by too much attention. Benefits come from a better-educated public more likely to highly value gray whales and to provide support for their protection and conservation of their habitats.

### C. Oil and Gas Development and Exploration

Discovery of extensive oil and gas resources on the Sakhalin Shelf has placed the Critically Endangered western gray whales in peril. The feeding ground off Sakhalin Island lies within the region which is now the site of ongoing large-scale oil and gas development by several consortia of Russian and multinational companies. Although some measures are being taken to mitigate deleterious effects, the oil and gas production, in addition to associated extensive shipping and aircraft traffic, may damage the habitat, stress or disturb the whales, or displace the population, which is dependent on the Sakhalin Shelf for its primary feeding ground (Fig. 9).

### D. International and National Protection

Gray whales received protection from commercial whaling under the 1937 *International Agreement for the Regulation of Whaling*, to which most whaling nations concurred, and more comprehensive protection under the 1946 *International Convention for the Regulation of Whaling*, to which the Soviet Union and Japan also adhered. In the United States, two statutes provide legal protection, the *Marine Mammal Protection Act*, passed in 1972; and the *Endangered Species Act*, which became law in 1973. The eastern population recovered and in 1994 the US Department of Interior removed it from the *List of Endangered and Threatened Wildlife and Plants* (under the US Endangered Species Act). The World Conservation Union (IUCN) reclassified it from *Endangered* to *Lower Risk: conservation dependent@* in the *1996 IUCN Red List of Threatened Animals*. However, these actions had no bearing on the status of the western gray whale population, which remained *Endangered* throughout its range. It was reclassified in the 2000 IUCN Red List from *Endangered* to *Critically Endangered* (under 1996 categories and criteria) (http://www.iucnredlist.org). The gray whale is listed in Appendix 1 of CITES.

**Figure 8** *A "friendly" gray whale calf and mother allow whale watchers to pet them (note the tip of the cow's rostrum in the foreground).*

**Figure 9** *The critically endangered western gray whale's core feeding ground off Sakhalin Island, Russian Federation, in the Okhotsk Sea is located within a multi-national oil and gas development area.*

Mexico recognized the importance of the breeding lagoons to the recovery of the gray whale and it is the only nation to provide important habitat protection for the eastern population. In 1972, it established Ojo de Liebre Lagoon (the principle calving and nursery area) as the world's first whale refuge. In 1979, San Ignacio Lagoon became a *Whale Refuge and Maritime Attraction Zone*. In 1980, reserve status extended to Laguna Manuela and Laguna Guerrero Negro. All lie within the *El Vizcaíno Biosphere Reserve*, created in 1988. In 1993, the United Nations Educational, Scientific, and Cultural Organization (UNESCO) made Ojo de Liebre and San Ignacio Lagoons World Heritage Sites. Lastly, in 2002, all Mexican territorial seas and EEZ were declared a refuge to protect large whales.

## See Also the Following Articles

Baleen Whales (Mysticetes) ■ Whaling, Early and Aboriginal ■ Whaling, Traditional

## References

Andrews, R. C. (1914). Monographs of the Pacific Cetacea. I. The California gray whale (*Rhachianectes glaucus* Cope). *Mem. Am. Mus. Nat. Hist.* **1**, 227–287.

Alter, S. E., Rynes, E., and Palumbi, S. R. (2007). DNA evidence for historic population size and past ecosystem impacts of gray whales. *Proc. Natl. Acad. Sci.* **104**, 15162–15167.

Bisconti, M., and Varola, A. (2006). The oldest eschrichtiid mysticete and a new morphological diagnosis of Eschrichtiidae (gray whales). *Riv. Ital. Paleo. Strat.* **112**, 1–11.

Cooke, J. G., Weller, D. W., Bradford, A. L., Burdin, A. M., and Brownell, R. L., Jr. (2007). Population assessment of western gray whales in 2007. IWC Scientific Committee meeting document SC/59/BRG41.

Darling, J. D., Keogh, K. E., and Steeves, T. E. (1998). Gray whale (*Eschrichtius robustus*) habitat utilization and prey species off Vancouver Island, BC. *Mar. Mamm. Sci.* **14**, 692–720.

Dedina, S. (2000). "Saving the Gray Whale: People, Politics, and Conservation in Baja California." University of Arizona Press, Tucson.

Deméré, T. A., Berta, A., and McGowen, M. R. (2005). The taxonomic and evolutionary history of fossil and modern balaenopteroid mysticetes. *J. Mammal. Evol.* **12**, 99–143.

Geisler, J. H., and Sanders, A. E. (2003). Morphological evidence for the phylogeny of Cetacea. *J. Mammal. Evol.* **10**, 23–129.

Henderson, D. A. (1972). "Men and Whales at Scammon's Lagoon." Dawson's Book Shop, Los Angeles.

Highsmith, R. C., Coyle, K. O., Bluhm, B. A., and Kona, B. (2006). Gray whales in the Bering and Chukchi Seas. *In* "Whales, Whaling, and Ocean Ecosystems" (J. A. Estes, D. P. DeMaster, D. F. Doak, T. M. Williams, and R. L. Brownell, eds). University of California Press.

Ichishima, H., Sato, E., Sagayama, T., and Kimura, M. (2006). The oldest record of Eschrichtiidae (Cetacea: Mysticeti) from the late Pliocene, Hokkaido, Japan. *J. Paleontol.* **80**, 367–379.

Jones, M. L., Swartz, S. L., and Leatherwood, S. (eds) (1984). "The Gray Whale *Eschrichtius robustus*." Academic Press, Orlando, Florida, USA.

Moore, S. E., Wynne, K. M., Kinney, J. C., and Grebmeier, J. M. (2007). Gray whale occurrence and forage southeast of Kodiak Island, Alaska. *Mar. Mamm. Sci.* **23**(2), 419–428.

Nerini, M. (1984). A review of gray whale feeding ecology. *In* "The Gray Whale *Eschrichtius robustus*" (M. L. Jones, S. L. Swartz, and S. Leatherwood, eds). Academic Press, Orlando, Florida, USA.

Omura, H. (1984). History of gray whales in Japan. *In* "The Gray Whale *Eschrichtius robustus*" (M. L. Jones, S. L. Swartz, and S. Leatherwood, eds). Academic Press, Orlando, Florida, USA.

Reeves, R. R. *et al.* (2005). Report of the Independent Scientific Review Panel on the Impacts of Sakhalin II Phase 2 on Western North Pacific Gray Whales and Related Biodiversity. IUCN, Gland, Switzerland. [Available from http://www.iucn.org].

Rice, D. W., and Wolman, A. A. (1971). Life History and Ecology of the Gray Whale (*Eschrichtius robustus*). *Am. Soc. Mamm. Spec. Pub.* **3**.

Rugh, D. J., Roderick, C. H., Lerczak, J. A., and Breiwick, J. M. (2005). Estimates of abundance of the eastern North Pacific stock of gray whales (*Eschrichtius robustus*) 1997–2002. *J. Cetacean Res. Manag.* **7**, 1–12.

Rychel, A., Reeder, T., and Berta, A. (2004). Phylogeny of mysticete whales based on mitochondrial and nuclear data. *Mol. Phylogenet. Evol.* **32**, 892–901.

Scammon, C. M. (1874). "The Marine Mammals of the Northwestern Coast of North America Together with an Account of the American Whale-Fishery." John H. Carmany and Co., San Francisco.

Tomilin, A. G. (1957). "Mammals of the U.S.S.R. and adjacent countries. Vol. IX. Cetacea." Akad. Nauk, SSSR, Moscow (transl. by Israel Program for Sci. Transl., Jerusalem, 1967).

Urbán-R., J., Rojas-Bracho, L., Pérez-Cortés, H., Gómez-Gollardo, A., Swartz, S., Ludwig, S., and Brownell, R. L., Jr. (2003). A review of gray whales (*Eschrichtius robustus*) on their wintering grounds in Mexican waters. *J. Cetacean Res. Manag.* **5**, 281–295.

G

# Group Behavior

## ALEJANDRO ACEVEDO-GUTIÉRREZ

Many animals spend part or all of their lives in groups. Their size and composition have diverse effects on morphology and behavior including relative brain size and extent of sexual dimorphism. A group may be viewed as any set of individuals, belonging to the same species, which remain together for a period of time interacting with one another to a distinctly greater degree than with other conspecifics. Thus, the study of group-living is the study of social behavior, and marine mammal societies can be remarkably diverse (Fig. 1).

Groups can be classified based both on the amount of time individuals interact with each other and on the benefits that individuals receive. Schools last for periods of minutes to hours while groups last for months to decades. Aggregations (or non-mutualistic groups) do not provide a larger benefit to individuals than if they were alone, while groups (or mutualistic groups) do provide such benefit to their members. Aggregations are formed because a nonsocial factor, e.g., food, attracts individuals to the same place; groups are formed because they provide a benefit to their members.

Recent studies have highlighted the challenges of defining marine mammal groups in nature, particularly those of cetaceans. For instance, most scientists determine whether individual dolphins belong to the same group based on the distance separating individuals (usually ≤10 m) or the radius comprised by the group (usually ≤100 m), and/or by whether individuals are engaging in the same behavior or not. However, a study of bottlenose dolphins (*Tursiops truncatus*) in Sarasota Bay, Florida, indicates that the communication range of social sounds between females and dependent calves could be at a minimum 487 m and reach up to 2 km or more (Quintana-Rizzo *et al.*, 2006). Hence, the traditional distinctions of a group—practical, replicable, and undoubtedly useful in advancing our understanding of group behavior—are likely not meaningful to a cetacean. Because cetaceans rely on acoustic communication to maintain group cohesion, the study

Figure 1 *(A) Blue whales (Balaenoptera musculus) are usually found alone or in small numbers. (B) South American sea lions (Otaria flavescens) aggregate in large numbers during the breeding season. Photos by A. Acevedo-Gutierez.*

highlights the importance of understanding communication range to define cetacean groups.

To describe the social structure of a population, it is essential to measure how much time individuals spend together (association patterns) and the rate at which individual associations changes over time (lagged association rates). However, the amount of time that animals spend together depends both on genuine social affiliations and on how much individual home ranges overlap. For instance, two individuals may be observed together because they have a similar home range, forming then an aggregation, rather than because they are genuinely affiliated, which would then be a group. Employing network analyses, association analyses, and estimates of lagged association rates at different spatial scales, a study of coastal bottlenose dolphins in eastern Scotland shows that the population is composed of two social units with restricted interactions as a result of social affiliation (Lusseau *et al.*, 2006). The study highlights the importance of network analyses and the use of different temporal (lagged association rates) and spatial scales to understand the organization of social marine mammals.

## I. Theory of Group-Living

There appear to be three conditions under which group-living will evolve, the benefits to the individual outweigh the costs, the costs outweigh the benefits but strong ecological constraints prevent dispersal from the natal territory (for instance, lack of high-quality breeding openings explains within-population dispersal decisions and family groups in birds), and the area where the group lives can accommodate additional individuals at no cost.

### A. Benefits and Costs of Group-Living

Group-living is usually explained in terms of benefits to the individual group members via direct or indirect fitness. Increases in direct fitness include mechanisms such as direct benefits of group-living, direct and indirect reciprocity, and mutualism. Increases in indirect fitness are achieved via kin selection. It has been argued that when competition occurs at the level of groups rather than individuals, group-living is best explained in terms of benefits to the groups themselves, group selection. However, many scientists consider that whenever interactions occur at a local spatial scale, and dispersal is limited, then interactions occur among genetic relatives, and thus kin selection rather than group selection is operating (Nowak, 2006).

Kin selection is perhaps the most frequently employed argument to explain benefits of group-living. For instance, kin selection explains the generalities of cooperative breeding in mammals and birds (Brown, 1987; Jennions and Macdonald, 1994), and the evolution of cooperation among male chimpanzees (*Pan troglodytes*) (Morin *et al.*, 1994). Further, mammalian female kin (including several odontocete species) spend more time in close proximity and are more likely to help each other. Females may allosuckle or gain higher reproductive success by forming coalitions with kin. However, explanations based on kin selection are in some cases inadequate and some behaviors are best explained in terms of direct fitness via diverse mechanisms. Direct benefits from early detection of danger explain the sentinel behavior of meerkats (*Suricata suricatta*) (Clutton-Brock *et al.*, 1999) and delayed direct benefits to the subordinate male explain the occurrence of dual-male courtship displays in long-tailed manakins (*Chiroxiphia linearis*) (McDonald and Potts, 1994). By-product mutualism explains territorial coalitions in Australian fiddler crabs (*Uca mjoebergi*), which assist other crabs in defending their neighboring territories; in this manner, the neighbor keeps its territory and the ally pays to retain an established neighbor rather than renegotiate boundaries with a new neighbor (Backwell and Jennions, 2004).

Benefits to increase foraging efficiency and reduce predation, and the number of individuals that can be supported by the available local resources have been typically viewed as important factors shaping group-living. These benefits apparently apply to all social organisms; for instance, the ability to disperse and exploit new food patches appears to favor the aggregation of solitary slime mold into a multicellular organism known as a slug (Kuzdzal-Fick *et al.*, 2007). However, increased foraging efficiency and reduction of predation can be accomplished through a myriad of different mechanisms (Table I). In addition, increased foraging and reduce predation are sometimes inadequate to explain group-living, in African lions (*Panthera leo*), female-grouping patterns are best explained as facilitating cooperative defense of cubs against infanticidal males and defense of territory against other females, not as increasing foraging efficiency (Packer *et al.*, 1990). Group-living can also impose several costs to individuals, including increased competition over access to resources and mating opportunities, exposure to infection, and conspicuousness to predators (Table II). In general, it is believed that for mammals the main benefit of sociality is protection against predators whereas the main cost is increased competition for resources.

**TABLE I**
**Benefits of Group Living**

*Reduction of predation°*
  Enhanced ability to detect predators: sensory integration.
  Enhanced ability to deter predators, even larger than group members.°
  Enhanced ability to escape, including predator confusion and coordinated evasion behavior.°
  Reduced individual probability of being selected as prey:
    By associating with conspecifics: dilution effect.°
    By hiding behind conspecifics: selfish herd.°

*Allocation of time to other activities*
  Reduced individual vigilance time
    Because of group vigilance (many eyes).°
    Because of decreased individual predation risk.°
  Increased foraging time for mothers by having babysitters.°

*Enhanced detection and capture of prey*
  Foraging in risky, but profitable, areas.
  Finding prey or reducing variation in food intake through cooperative searching: sensory integration.°
  Following more knowledgeable animals in the group to a food source: information transfer.°
  Following other species with more specialized senses to a food source.°
  Joining resources uncovered by others, also known as conspecific attraction, kleptoparasitism, area copying, scrounging, or tolerated theft.°
  Acquisition of innovative feeding behaviors from another group member:
    Social learning through social facilitation (contagion of motivational states).
    Directing attention to particular locations or objects: local enhancement.
    Imitation of knowledgeable tutors.°
    Information sharing and cultural transmission.°
  Increased diversity and size of prey that is captured:
    Due to more individuals foraging.°
    Due to prey flushed by movements of group members.°
    Due to individuals with different skills or abilities foraging together: skill pool effect.
  Increased food intake as a result of communal foraging.°
  Lower risks of injury while hunting.°

*Acquisition or defense of resources*
  Large groups defend, occupy, or displace small groups from better territories.
  Large groups acquire or defend localized food sources, including carcasses, from conspecifics or other species.°

*Improved reproduction*
  Caring and protection of offspring.°
  Learning to be a parent.°
  Finding mates in isolated or vast areas.°
  Enhanced reproductive synchrony.°
  Enhanced survival when there is prevention of dispersal to neighboring territories.
  Males benefit from cooperative displays, subdominant males receive the payoff later in time.
  Males in large groups gain access to females.°

*Reduction of parasitism*
  When number of hosts in a group increases more rapidly than the number of mobile parasites, reduced individual probability of being parasitized by associating with conspecifics: dilution effect.

*Other*
  Huddling to survive cold temperatures.°

°Suggested or documented costs in marine mammals.

The magnitude of the costs of sociality may be important in shaping group-living. It has been suggested that in some species differences in group size may be related to the differences in their costs of locomotion. Under this ecological-constraints model, large group sizes must travel farther each day because they deplete food patches more rapidly or require searching larger areas. Supporting the model, day range, and group size are positively correlated in various primate and carnivore species (Wrangham *et al.*, 1993). Animals that travel further spend more energy and reproduce less efficiently, hence a negative relationship between group size and reproductive performance is predicted within species. However, marine mammals reduce the costs of locomotion by developing energy-conserving swimming behaviors such as routine transit speeds, wave-riding, porpoising, and gliding. Hence, one would expect that group size in marine mammals is unrelated with distance traveled. Although I am unaware if such correlative study has been conducted, it has been hypothesized that reduced cost of locomotion coupled with a lack of restriction to a particular territory has allowed some populations of killer whales (*Orcinus orca*), and possibly long-finned pilot whales (*Globicephala melas*), to develop societies in which females and males remain with their natal group for life (Fig. 2). In this manner, males traveling with their mothers can have large home ranges and thus find potential mates.

---

TABLE II
Costs of Group Living

---

*Increased predation*
    Large groups more attractive to predators.
    Larger groups more likely to be detected by a predator: encounter effect.

*Reduced foraging efficiency*
    Increased amount of food needed for group.°
    Increased energy spent, distance traveled, or area covered to find food for group.°
    Increased conspicuousness: prey able to detect predators sooner than if predators are alone.
    Reduction in food intake due to sharing of prey, scramble competition, scrounging, and individual discrepancies in foraging success.
    Reduction in food intake due to interference by the behavior of other individuals.°
    Reduce ability to learn innovative foraging skills due to scroungers in the group.

*Increased conflicts for resources due to presence of more conspecifics or other species*
    Individuals from other groups or species following social parasitism.°
    Individuals from other groups or species attracted to feeding parties: local enhancement.°

*Reduced reproduction, increased competition for mates, or other limited resources*
    Individual discrepancies in number or quality of mates obtained.°
    Extrapair copulations and loss of fertilizations to other members of group.°
    Increased intra-specific competition for limited resources.°
    Increased infant mortality.°
    Increased risk of exploitation of parental care by conspecifics.
    Theft of nest material.

*Increased risk of infection*
    Increased contagious parasitism.
    Increased disease transmission.

---

°Suggested or documented costs in marine mammals.

**Figure 2** *In certain populations male and female killer whales (Orcinus orca) remain with their natal groups throughout their lifetime. Photo by Christopher Pearson*

Recent studies have documented novel strategies followed by individuals living in groups, including the complexity of intragroup and intergroup interactions. Female African lions cooperate to defend their territory from intruders; however, some individuals consistently lead the approach whereas other individuals lag behind without being punished by the leaders (Heinsohn and Packer, 1995). One potential explanation for this tolerance is that females need to defend their territories against other groups and their success depends in part on the number of defending females even if some individuals never lead the charge. Pinyon jays (*Gymnorhinus cyanocephalus*) draw sophisticated inferences about their own dominance status relative to that of strangers that they have observed interacting with known individuals (Paz-y-Miño *et al.*, 2004). That is, they make judgments about relationships on the basis of indirect evidence rather than by learning through direct interactions with other individuals. The study is the first experimental demonstration of transitive inference in animals and implies that such cognitive capabilities are widespread among social species.

Complex social behaviors have also been reported in marine mammals. In a breeding colony of gray seals (*Halichoerus grypus*) at the island of North Rona, Scotland, a few large males monopolize matings on the breeding beaches, however females over the years give birth to full siblings not sired by the dominant male (Amos *et al.*, 1995). The fathers of the pups are nondominant males that mate with the same females in different seasons. Thus behavioral polygyny and genetic fidelity seem to operate simultaneously in this colony. It has been suggested that this strategy of partner fidelity is maintained in the population because it may diminish aggressive interactions between dominant males and thus reduce the pup mortality originated by these clashes.

Perhaps one of the most complex behaviors described in marine mammals is the formation of alliances among male Indo-Pacific bottlenose dolphins (*Tursiops aduncus*) in Shark Bay, Australia, in which complexity is only matched by humans (Connor *et al.*, 1999; Connor, 2007). Males form strong and stable bonds for over 10 years with one or two other males, males in these first-order alliances cooperate to form aggressively maintained consortships with individual females.

Each first-order alliance forms moderately strong bonds with one or two other alliances, these second-order alliances do not endure for more than a few years and males cooperate to take or defend females from other alliances. Recently, it has been described that some second-order alliances associate regularly and cordially with other groups, suggesting a third level of alliance formation. A different strategy is for males to form a large but loose superalliance that competes with the smaller and more stable first-order or second-order alliances. Members of the superalliance split into smaller alliances of pairs and trios that are constantly changing but are always comprised of males from the superalliance. These pairs and trios join conflicts involving members of the superalliance and are always victorious. It is hypothesized that the large size of the superalliance allows individuals to compete with the smaller alliances and that the fluidity of individual associations within the superalliance allows males to maintain affiliative bonds. Because alliance formation is a strategy to obtain access to females, it has been hypothesized that alliances increase male reproductive success. Supporting this hypothesis, a recent study suggests that the vast majority of paternities are achieved by males involved in some form of alliance. However, within first-order alliances a few males dominate fertilizations, indicating skew in reproductive success among males in stable alliances (Krützen *et al.*, 2004). Given that males in stable first-order alliances formed small second-order alliances are more related to their allies than expected by chance, the reproductive skew among males may be explained in terms of kin selection (Krützen *et al.*, 2003). Interestingly, superalliances, where no reproductive skew has been described, are not formed by closely related males. Finally, the observation that some juvenile males without an alliance may achieve paternities suggests that alternative male tactics other than alliance formation exist in the population. What favors alliance formation among males? A recent model predicts that males will form alliances except when the number of males competing for a receptive female is very few, when there are substantial costs to being in an alliance, or when alliances do not out-compete single males.

## B. Female Social Behavior

One previously neglected area of research is the study of female social behavior. Females and males frequently have different interests, as a result female relationships are important in understanding social evolution independently of the behavior of males. For instance, dominant female chimpanzees have a higher reproductive success than subordinate ones, apparently because they are able to establish and maintain access to good foraging areas, competing in extreme cases as intensely as males (Pusey *et al.*, 1997). Females may also influence behaviors that affect the interests of males, female bird song appears to have evolved in part to compete for males, however this behavior has the potential consequence of preventing polygyny by deterring rival females (Langmore, 1998). In mammals, most females remain within their natal area or group throughout their lives. This female philopatry facilitates the formation of social groups through kin selection. Hence, understanding social behavior in many mammal species requires studying female social behavior. For instance, theoretical and empirical evidence indicates that females live in groups that often exceed the optimal group size (Silk, 2007). This discrepancy is associated with mechanisms regulating group size, such as trying to oust other group members or exclude immigrants, actions that are costly to the individuals performing them.

The study of females is also essential to understand group-living in marine mammals. Captive female bottlenose dolphins

**Figure 3** *Alloparenting behavior apparently allows sperm whale (Physeter macrocephalus) mothers to make deep foraging dives. Photo by A. Acevedo-Gutierrez.*

maintain dominance hierarchies and also compete aggressively against each other. However, unlike chimpanzees, it is unknown if female dominance hierarchies in free-ranging dolphins translate into differences in reproductive success. The preference and fidelity of female gray seals at North Rona toward nondominant males undermines the polygynous strategy of dominant males and results in a different mating system from that inferred by behavioral observations (Amos *et al.*, 1995). The large number of females in colonies of certain pinnipeds, such as northern elephant seals (*Mirounga angustirostris*), has permitted the existence of alloparenting and the appearance of a distinct suckling strategy by calves, milk-stealing. Male and female sperm whales (*Physeter macrocephalus*) have different grouping strategies, females appear to spend their entire lives within their natal group, forming strong matrilineal societies. Adult males are less social, leaving their nursing group when they reach puberty and after they have reached their late 20s roam among nursery groups looking for mates. The function of the female groups is to provide care for calves that are too young to follow their mothers during their deep foraging dives (Fig. 3). It has been suggested that this alloparenting reduces the period in which the calf is unaccompanied and thus provides protection from predators and also perhaps provides communal nursing (Whitehead, 1996; Mann *et al.*, 2000a). Thus key features of the sperm whale society are explained solely by the behavior of females.

## II. Social Behavior of Marine Mammals

There are several differences between terrestrial and marine environments that have allowed the evolution of distinctive strategies in marine mammals. Drag, heat loss, and density of the water generate differences in scaling and costs of locomotion, allowing many marine mammals to have large body sizes and large home ranges. Sound is the form of energy that best propagates in water, not surprisingly marine mammals employ it for social communication and many species navigate via echolocation. Marine mammals must find food that is for the most part dispersed and patchy, thus they appear to have no territories outside of the breeding season. Due to the global effects of the atmosphere and the ocean in the marine environment, marine mammals are affected by both global and local processes as

**Figure 4** *In certain species, pinniped females cluster during the breeding season and males are able to monopolize access to them. Photo by A. Acevedo-Gutierrez.*

G

exemplified by the impact of El Niño Southern Oscillation events on different populations.

### A. General Strategies

Sirenians, sea otters (*Enhydra lutris*), and polar bears (*Ursus maritimus*) are solitary animals that have few social interactions beyond mating and mother/offspring pairs. The time that these pairs remain together is 1–1.5 years in sirenians, 5–7 months in sea otters, and 2.5 years in polar bears. When a female becomes receptive sirenians form aggregations that have as many as 17 males physically competing for access to the female or defending display territories. During the breeding season male sea otters establish territories that include the areas occupied by several females, whereas male polar bears mate with only one partner because females have a dispersed distribution.

Most pinnipeds aggregate in colonies during the breeding season, a major factor influencing the size of these colonies is the distribution of habitat available for parturition. Pinnipeds give birth out of the water and thus the areas favored for parturition are oceanic islands, ice, or isolated mainland regions not easily accessible to terrestrial predators. When available space is limited, females become densely aggregated in large colonies that favor mating systems in which males defend either aggregations of females or areas occupied by females, or aggregate and display before aggregations of females (Fig. 4). However, when parturition space is dispersed, females are isolated, males usually have access to only one female, and no colonies are formed. At small spatial scales (within colonies), lack of suitable habitat might also explain high density of females in many cases; however, in various populations there is plenty of unoccupied space and females are still clustered. Hence, other factors need to be invoked to explain this clustering. For instance, females reduce the individual probability of being harassed by less competitive males by clustering (Trillmich and Trillmich, 1984). Female Galápagos sea lions (*Zalophus wollebaeki*) avoid overheating by clustering along the wet shoreline, which in turns determines the distribution of dominant and subdominant males (Wolf *et al.*, 2005). The strongest association found in pinnipeds is formed by a mother and her offspring, and lasts from less than 1 week to almost 3 years, depending on the species. Pinnipeds haul out together outside of the breeding season.

Although this non-reproductive social behavior is poorly known, there is evidence that it increases vigilance for predators in harbor seals (*Phoca vitulina*). It is believed that hauling out together also allows pinnipeds to rest, avoid predators, molt or warm themselves. For instance, walruses (*Odobenus rosmarus*) in large numbers may decrease the rate of body heat loss, particularly in calves, when on land or on ice.

The complexity of cetacean societies appears to be related to amount of time invested in lactating and in rearing their calf after weaning. Baleen whales are found in schools of varying size, from single individuals to more than 20 whales. Pairs of mothers and their offspring form stable associations that last less than 1 year. It is currently unclear if long-term associations exist among adult whales. Most females give birth every 2–3 years, and have the potential to produce more than 20 calves throughout their lifetime. Schools of baleen whales have been observed in both feeding and breeding grounds. For instance, feeding humpback whales (*Megaptera novaeangliae*) forage alone, in aggregations, or as a group, depending on prey type, while aggregations of breeding males display acoustically or compete directly for access to females. Odontocetes are the most social marine mammals and have different types of societies as suggested by the large variation in school size between species (Table III). Short-term associations between adults characterize porpoises. Associations between mothers and their offspring last for 8–12 months. Females breed every 1 or 2 years and may give birth to 15 calves or more during their life span.

It is believed that medium-sized dolphin species live in fission–fusion societies with fluid group membership. Yet, a recent study indicates that spinner dolphins (*Stenella longirostris*) around Midway Atoll live in stable bisexually bonded societies of long-term associates with strong geographic fidelity, no obvious fission–fusion, and limited contacts with other populations (Karczmarski *et al.*, 2005). It is hypothesized that the geographic isolation and small size of the remote atoll favor long-term group fidelity and social stability over the fluidity of the fission–fusion society is replaced. Bottlenose dolphins live in fission–fusion societies that are believed to reduce feeding competition by allowing individuals to disperse. Associations between adults are varied, they last a short amount of time in some individuals and several years in others. In certain populations males form relatively stable groups and rove among female groups. Females give birth at least every 3 years and may produce close to 10 calves throughout their lifetime. Calves remain with their mothers 2–11 years (Fig. 5). Adult females form strong bonds with their calves as well as stable, moderate-level associations with other females within social clusters named bands. Bands tend to be composed of female relatives, but they can also include unrelated females. It has been hypothesized that reproductive condition (e.g., females with same-aged calves) determines associations within bands while kinship determines band membership. Why do female bottlenose dolphins form groups? Two leading hypothesis are protection from predators and defense against sexual coercion by males.

Little is known about the social structure of beaked whales (Ziphiidae) given their pelagic and deep-diving behavior. However, the northern bottlenose whale (*Hyperoodon ampullatus*) society in the northwestern Atlantic appears to comprise roving strong male–male bonds and weaker female–female bonds (Gowans *et al.*, 2001). Female and immature whales form a loose network of associations, showing neither preferential associations with particular individuals nor long-term bonds. Although males form many short-term associations, associations between some males last for several years. This social organization is reminiscent of that observed in some bottlenose

### TABLE III
#### School Sizes of Odontocetes

| Species[a] | Average school size | Maximum school size |
|---|---|---|
| *Phocoenoides dalli* (7) | 2.3–7.4 | 5–500 |
| *Neophoca phocaenoides* (1) | 2.0 | 13 |
| *Phocoena phocoena* (6) | 1.2–5.7 | 15–100 |
| *P. sinus* (1) | 1.9 | 7 |
| *P. spinipinnis* (1) | 4.5 | 10 |
| *Cephalorhynchus commersonii* (1) | 6.9 | 110 |
| *Lissodelphis borealis* (2) | 9.9–110.2 | 60–2000 |
| *Delphinus* sp. (4) | 46.8–385.9 | 650–4000 |
| *Grampus griseus* (9) | 6.3–63 | 20–500 |
| *Lagenodelphis hosei* (1) | 394.9 | 1500 |
| *Lagenorhynchus acutus* (1) | 53.2 | ? |
| *L. obliquidens* (3) | 10.8–88 | 50–6000 |
| *L. obscurus* (3) | 9.5–86 | 24–1000 |
| *Sotalia fluviatilis* (1) | 2.5 | 10 |
| *Sousa plumbea* (1) | 6.6 | 25 |
| *Stenella attenuata* (5) | 26.0–360.0 | 148–2400 |
| *S. clymene* (1) | 41.6 | 100 |
| *S. coeruleoalba* (3) | 60.9–302 | 500–2136 |
| *S. frontalis* (2) | 6.0–10.0 | 50–65 |
| *S. longirostris* (4) | 37.6–134.1 | 95–1700 |
| *Steno bredanensis* (2) | 14.7–40.0 | 53 |
| *Tursiops aduncus* (2) | 10.2–140.3 | 80–1000 |
| *T. truncatus* (29) | 3.1–92.0 | 18–5000 |
| *Feresa attenuata* (1) | 27.9 | 70 |
| *Globicephala macrorhynchus* (2) | 12.2–41.1 | 33–230 |
| *G. melas* (3) | 9.3–84.5 | 220 |
| *Orcinus orca* (10) | 2.6–12.0 | 5–100 |
| *Pseudorca crassidens* (1) | 18 | 89 |
| *Peponocephala electra* (2) | 135.3–199.1 | 400 |
| *Delphinapterus leucas* (3) | 3.8–32.9 | 100–500 |
| *Monodon monoceros* (1) | 3 | 50 |
| *Inia geoffrensis* (2) | 1.6–2.0 | 8–10 |
| *Lipotes vexillifer* (1) | 3.4 | 10 |
| *Platinista gangetica* (1) | 1.4 | 3 |
| *Kogia sima* (1) | 1.7 | ? |
| *Physeter macrocephalus* (6) | 3.7–22.1 | 17 |
| *Berardius bairdii* (1) | 7.2 | 25 |
| *Ziphius cavirostris* (1) | 2.3 | 7 |

[a]Values in parentheses indicate number of studies.

dolphin populations foraging in shallow, enclosed bays rather than that of sperm whale populations, which forage in deep water canyons as northern bottlenose whales do. Baird's beaked whales (*Berardius bairdii*) apparently employ a novel social strategy. Males live longer than females and thus there is an excess of mature males over

**Figure 5** *Bottlenose dolphin (Tursiops truncatus) calves remain with their mother for up to 8 years.*

females (Kasuya, 1995). It has been hypothesized that these traits indicate a society in which males provide significant parental care by rearing weaned calves, protecting them from predators and teaching them foraging skills.

In the case of the sperm whale and large-sized delphinids (pilot whales and some populations of killer whales), females appear to spend their entire lives within their natal group, forming strong matrilineal societies. Females usually breed every 3–6 years and may give birth to about 5 calves throughout their lifetime, more in the case of long-finned pilot whales. Females may live over 20 years past their post-reproductive years. It has been suggested that this strategy allows old females to transmit and store cultural information, and provide alloparental behavior. In the case of short-finned pilot whales (*Globicephala macrorhynchus*), it is possible that non-reproductive females even provide alloparental nursing (Kasuya, 1985). Male sperm whales and perhaps male short-finned pilot whales leave their nursing group when they reach puberty. However the former, after they have reached their late 20s, roam among nursery groups looking for mates; the latter appear to join a different nursery group and remain in it, engaging in few clashes with other males, apparently because they are able to engage in non-reproductive mating with old females, as it apparently occurs in bonobos (*Pan paniscus*). Male killer whales in some populations, and perhaps male long-finned pilot whales, remain in their natal group for life but mate with females from other groups when they meet, hence avoiding inbreeding. It is important to explain the absence of male dispersal because in the majority of social mammals males disperse from their natal group and do not interact with relatives (in a few species it is the females who disperse). The accepted explanation is that this sexually dimorphic dispersal and lack of interaction with relatives avoids inbreeding in mammals. The lack of male dispersal in killer whales has been explained in terms of the benefits that male apparently provide to the offspring of related females, such as assistance in hunting and teaching (Mann *et al.*, 2000a).

Among vertebrates, female killer whales, short-finned pilot whales, humans (*Homo sapiens*), and probably sperm whales, spend a substantial part of their adult life reproductively sterile and helping their close relatives. As such, the females undergo menopause and the species can be viewed as eusocial (McAuliffe and Whitehead, 2005). Cetacean menopause is believed to be adaptive, where the benefits of assisting kin outweigh the costs of reproductive cessation. Similar to human grandmothers, cetacean grandmothers appear to help by storing and providing information to the other members of their matrilines. This informative role of grandmothers might be the primary motor of eusociality and also supports the growing evidence of culture among cetaceans (Whitehead, 1998, 2007; Whitehead *et al.*, 2004). An important impact of culture can be found in social learning among matrilineal odontocetes, whereby learned behaviors passed on to family members are being conserved within matrilines and affecting the course of genetic evolution. For instance, it has been suggested that the division of sympatric resident and transient killer whales off the west coast of Washington State and Canada was originally cultural; however, they show enough differences in feeding behavior, vocalizations, social systems, morphology, and genetics that they may be incipient species. In another example, indirect measures of the reproductive success of groups of sperm whales vary according to differences in culture between the groups. Although modification of the course of genetic evolution through culture has only been demonstrated in humans, further studies in more species and longer datasets on well-studied species will shed more light into the impact of culture on cetacean evolution. Why do some cetacean species engage in social learning? Apparently the prevalence of social learning and culture in cetaceans is related to patterns of environmental variation (Whitehead, 2007). Under this scenario, social learning is advantageous in environments where variation in biotic and abiotic factors is large over long time scales, such as marine ecosystems.

### B. Foraging

Increased foraging efficiency is considered to be one of the principal roles of group-living in cetaceans. However thus far transient killer whales provide the only clear example supporting the argument that marine mammals live in groups because of foraging benefits (Baird and Dill, 1996). Transient killer whales live in the Pacific Northwest and prey on harbor seals and other small marine mammals. Individuals maximize their caloric intake if they feed in groups of three, which is the size of the group in which they live. The small size of these groups is apparently maintained by the departure of all female offspring and all but one male offspring from their natal group.

Two benefits of group-living through foraging efficiency are the ability to search for prey as a group and to forage communally (Fig. 6). Searching for prey as a group allows individuals to combine their sensory efforts, which should be an advantage when prey has a dispersed and a patchy distribution. Communal foraging allows individuals to combine efforts to pursue and capture prey. This behavior has been reported in dolphins, baleen whales, including blue whales (*Balaenoptera musculus*) and bowhead whales (*Balaena mysticetus*), and pinnipeds, such as fur seals and sea lions. However, in some instances it is unclear whether individuals combine efforts to pursue and capture prey, or merely aggregate in an area where food is concentrated. A particular type of communal foraging behavior, termed prey herding, has been observed when feeding on shoaling fish. Individuals encircle shoals of fish and thus create a tight, motionless ball of prey from which they can grab individual fish with their mouths, in some cases individuals release bubbles to further tighten the ball of prey. This herding of prey has been well described in humpback whales, dusky dolphins (*Lagenorhynchus obscurus*), and killer whales. However, it has also been reported in other species, such as bottlenose dolphins, common dolphins (*Delphinus* spp.), clymene dolphins (*Stenella clymene*), and Atlantic spotted dolphins (*S. frontalis*). It is difficult to document this behavior, and no study has yet quantified the success of cetaceans in herding prey.

**Figure 6**   *Communal foraging allows dolphins to combine pursuing efforts.*

### C. Predation

Reduction of predation is considered to be another principal function of group-living in cetaceans, certain shark species and some large delphinids attack cetaceans, and calves suffer higher mortality than adults do. However, pinnipeds apparently also form groups in response to predation. Walruses sometimes form groups lasting throughout the year in the water and on haul-out sites. It has been suggested that this may be a female strategy for pup defense against predation by polar bears.

Thus far no conclusive evidence shows that group-living in cetaceans is driven because of benefits in reduction of predation, although it has been suggested that this could be the case in sperm whales. Nonetheless group-living may provide several benefits to reduce predation. Groups are able to mob and chase away predators, as has been observed in hump-backed dolphins (*Sousa* spp.) when attacked by a shark. It is believed that other dolphins also employ this antipredatory strategy. Sperm whales, and perhaps humpback whales, employ the marguerite formation, in which adults surround young individuals by having their heads toward the center (horizontal formation) or toward the surface (vertical formation). Adults have their flukes toward the periphery and employ them to slap at the predators, which in the majority of observations have been killer whales.

Group-living appears to be mostly related to food and predation in terrestrial and marine mammals. Thus it has been argued that the variation of group sizes among dolphin species is related to food availability, related to prey habitat, or to the need to defend from predators. For instance, the reproductive success of female Indo-Pacific bottlenose dolphins is highest in shallow waters, either because calves and their mothers are able to detect and avoid predators or because prey density is highest (Mann *et al.*, 2000b).

I compiled data on the average school size from 24 species of the family Delphinidae (Table III). Because definitions of school vary among researchers, I attempted to make values comparable by selecting only studies with at least 30 observations throughout a season and that defined schools as the number of individuals engaged in similar activities regardless of distance between them. I averaged the values from species belonging to the same genus and related them to crude measures of predation pressure and prey habitat, measures that were obtained from the literature. Results indicate that regardless of the body weight of the genus, average school sizes are larger when predation pressure is high rather than low, and when prey

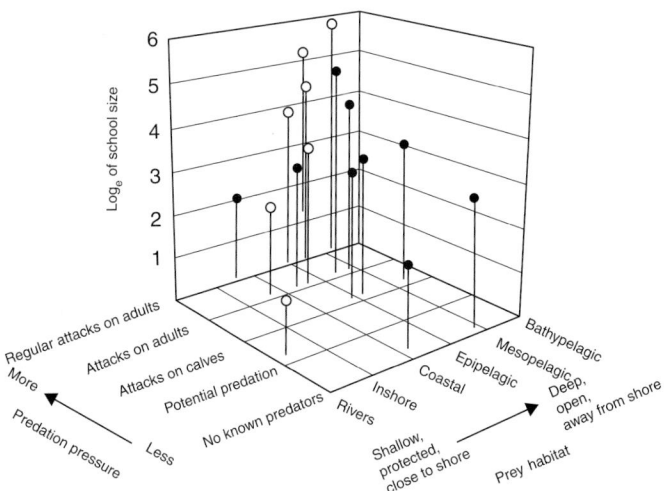

**Figure 7** *Relationship between predation pressure, prey habitat, and average school size of 16 genera of the family Delphinidae. Open circles indicate small genera (females weigh less than 150kg), solid circles indicate large species (females weigh more than 150kg).*

**Figure 8** *Leaps and slaps at the water may attract other dolphins to feeding events.*

is found in open rather than enclosed waters (Fig. 7). School sizes are largest when both predation pressure is high and prey lives at depth in oceanic waters. Thus it appears that the average school size of dolphin genera is related both to the predation they experience and the habitat where their prey lives. However, comparative research of group size in Delphinoidea (Delphinidae, Phocoenidae, Monodontidae)—including phylogeny, physical environment, diet, predation pressure, and life history—indicated that phylogeny explained most of the observed variation in group size (Gygax, 2002a, b). Although group size also increased with openness of the habitat and showed a U-shaped relationship with temperature, the simplest interpretation of the study is that group size resulted from a random process and has been marginally shaped by direct selection.

### D. Resource Defense

Interspecific contests over food are thought to also influence the group size and the group composition of predators. In the case of marine mammals, one study has documented the influence of competitive interactions with sharks on dolphin food intake and apparently on group size (Acevedo-Gutiérrez, 2002). As Isla del Coco, an oceanic island off Costa Rica, underwater observations indicate that bottlenose dolphins eat less food if there are many silky sharks (*Carcharhinus falciformis*) converging on the same fish school. The observations also indicate that the interactions between sharks and dolphins represent contests over food between these two similarly sized species, and not predation on the dolphins by the sharks. Dolphin groups of moderate size (around 10 individuals) are able to chase sharks away from the shoal and monopolize it. However, dolphins are not always found in such numbers when feeding because individual dolphins eat less as the number of dolphins increases. (There are fewer cookies available per person the more people are at the party.) As such, dolphins appear to increase group size when sharks are present but not when they are absent by leaping and/or by producing sounds to attract other dolphins. Dolphins leap out of the water and slap the water

with their flukes and body more often while feeding than while engaged in any other type of behavior (Fig. 8). They also increase whistle production in the presence of sharks but not when sharks are absent. Given the patchy and ephemeral nature of food resources in the open ocean, it is expected that further observations from species living in pelagic zones will indicate that interspecific contest over food are relatively common and are more important than currently viewed in influencing group-living in marine mammals.

### III. Conclusion

Group-living involves benefits and costs, and the resultant society represents a balance between the different interests of all group members. The aquatic environment has allowed marine mammals to pursue complex and sometimes unique social strategies. At the same time, the basic needs of finding food, insuring reproduction and evading predators are also found in terrestrial environments. This convergence provides interesting parallels between the social strategies of marine mammals and those of terrestrial mammals, chimpanzees and bottlenose dolphins, elephants and sperm whales. Not surprisingly, much insight on group-living in marine mammals is gained by examining the societies of other taxa, most notably birds and terrestrial mammals. Of particular usefulness are comparative studies examining traits that are both consistent across taxa and common in some groups but not in others. These comparisons will allow us to examine specific hypotheses and test predictions regarding sociality as well as assist us in identifying data gaps and research needs.

It is clear that many questions about the group behavior of marine mammals remain unanswered and that much work remains to be done. For instance, the relationship between group-living and fitness remains to be described. However, studies on previously neglected topics such as female social behavior have increased in the last years and have improved our understanding of marine mammal societies.

As is the case of studies on terrestrial mammals, long-term studies of free-ranging populations have provided the most critical information to understand the evolution of group-living of marine mammals. Examples include bottlenose dolphins in Florida, Australia and, recently, Scotland; killer whales in the Pacific Northwest; sperm whales in the Galapagos Islands; humpback whales in their breeding and foraging grounds; northern elephant seals in California;

G

Antarctic fur seals (*Arctocephalus gazella*) and, recently, southern elephant seals (*Mirounga leonina*), and gray seals in the United Kingdom and Canada. A cursory review of the literature indicates that many more such studies are being started in other populations and, most importantly, on other taxa. It seems certain that our understanding of marine mammal group-living in the next 10 years will be not only more thorough but more inclusive.

## See Also the Following Articles

Behavior, Overview ■ Communication ■ Sociobiology

## References

Acevedo-Gutiérrez, A. (2002). Interactions between marine predators: Dolphin food intake is related to number of sharks. *Mar. Ecol. Prog. Ser.* **240**, 267–271.

Alexander, R. D. (1974). The evolution of social behavior. *Annu. Rev. Ecol. Syst.* **5**, 325–383.

Amos, B., Twiss, S., Pomeroy, P., and Anderson, S. (1995). Evidence for mate fidelity in the gray seal. *Science* **268**, 1897–1899.

Backwell, P. R. Y., and Jennions, M. D. (2004). Coalition among male fiddler crabs. *Nature* **430**, 417.

Baird, R. W., and Dill, L. M. (1996). Ecological and social determinants of group size in transient killer whales. *Behav. Ecol.* **7**, 408–416.

Berta, A., Sumich, J. L., and Kovacs, K. M. (2006). "Marine Mammals. Evolutionary Biology," 2nd Ed. Academic Press, San Diego.

Brown, J. L. (1987). "Helping and Communal Breeding in Birds: Ecology and Evolution." Princeton University Press, Princeton.

Clutton-Brock, T. H., O'Riain, M. J., Brotherton, P. N. M., Gaynor, D., Kansky, R., Griffin, A. S., and Manser, M. (1999). Selfish sentinels in cooperative mammals. *Science* **284**, 1640–1644.

Connor, R. C. (2007). Dolphin social intelligence: Complex alliance relationships in bottlenose dolphins and a consideration of selective environments for extreme brain size evolution in mammals. *Philos. Trans. R. Soc. Lond., B, Biol. Sci.* **362**, 587–602.

Connor, R. C., Heithaus, M. R., and Barré, L. M. (1999). Superalliance of bottlenose dolphins. *Nature* **397**, 571–572.

Gowans, S., Whitehead, H., and Hooker, S. K. (2001). Social organization in northern bottlenose whales, *Hyperoodon ampullatus*: Not driven by deep-water foraging? *Anim. Behav.* **62**, 369–377.

Gygax, L. (2002a). Evolution of group size in the dolphins and porpoises: Interspecific consistency of intraspecific patterns. *Behav. Ecol.* **13**, 583–590.

Gygax, L. (2002b). Evolution of group size in the superfamily Delphinoidea (Delphinidae, Phocoenidae and Monodontidae): A quantitative comparative analysis. *Mamm. Rev.* **32**, 295–314.

Heinsohn, R., and Packer, C. (1995). Complex cooperative strategies in group-territorial African lions. *Science* **269**, 1260–1262.

Hoelzel, A. R. (ed.) (2002). "Marine Mammal Biology. An Evolutionary Approach." Blackwell Publishing, Oxford.

Jennions, M. D., and Macdonald, D. W. (1994). Cooperative breeding in mammals. *Trends Ecol. Evol.* **9**, 89–93.

Karczmarski, L., Würsig, B., Gailey, G., Larson, K. W., and Vanderlip, C. (2005). Spinner dolphins in a remote Hawaiian atoll: Social grouping and population structure. *Behav. Ecol.* **16**, 675–685.

Kasuya, T. (1995). Overview of cetacean life histories: An essay in their evolution. *In* "Developments in Marine Biology" (A. S. Blix, L. Walloe, and O. Ultang, eds), Vol. 4, pp. 481–497. Elsevier Science Publishers B. V., Amsterdam.

Krebs, J. R., and Davies, N. B. (eds) (1978–1997). "Behavioural Ecology. An Evolutionary Approach," 1st.–4th. Eds. Blackwell Scientific Publishing, Oxford.

Krützen, M., Sherwin, W. B., Connor, R. C., Barré, L. M., Van de Casteele, T., Mann, J., and Brooks, R. (2003). Contrasting relatedness patterns in bottlenose dolphins (*Tursiops* sp.) with different alliance strategies. *Proc. R. Soc. Lond., B, Biol. Sci.* **270**, 497–502.

Krützen, M., Barré, L. M., Connor, R. C., Mann, J., and Sherwin, W. B. (2004). O father: Where art thou?—Paternity assessment in an open fission–fusion society of wild bottlenose dolphins (*Tursiops* sp.) in Shark Bay, Western Australia. *Mol. Ecol.* **13**, 1975–1990.

Kuzdzal-Fick, J. J., Foster, K. R., Queller, D. C., and Strassmann, J. A. (2007). Exploiting new terrain: An advantage to sociality in the slime mold *Dictyostelium discoideum*. *Behav. Ecol.* **18**, 433–437.

Langmore, N. E. (1998). Functions of duet and solo songs of female birds. *Trends Ecol. Evol.* **13**, 136–140.

Lusseau, D., *et al.* (8 authors) (2006). Quantifying the influence of sociality on population structure in bottlenose dolphins. *J. Anim. Ecol.* **75**, 14–24.

Mann, J., Connor, R. C., Tyack, P. L., and Whitehead, H. (eds) (2000a). "Cetacean Societies. Field Studies of Dolphins and Whales." University of Chicago Press, Chicago.

Mann, J., Connor, R. C., Barré, L. M., and Heithaus, M. R. (2000b). Female reproductive success in bottlenose dolphins (*Tursiops* sp.): Life history, habitat, provisioning, and group-size effects. *Behav. Ecol.* **11**, 210–219.

McAuliffe, K., and Whitehead, H. (2005). Eusociality, menopause and information in matrilineal whales. *Trends Ecol. Evol.* **20**, 650.

McDonald, D. B., and Potts, W. K. (1994). Cooperative display and relatedness among males in a lek-mating bird. *Science* **266**, 1030–1032.

Morin, P. A., Moore, J. J., Chakraborty, R., Jin, L., Goodall, J., and Woodruff, D. S. (1994). Kin selection, social structure, gene flow, and the evolution of chimpanzees. *Science* **265**, 1193–1201.

Nowak, M. A. (2006). Five rules for the evolution of cooperation. *Science* **314**, 1560–1563.

Packer, C., Scheel, D., and Pusey, A. E. (1990). Why lions form groups: Food is not enough. *Am. Nat.* **136**, 1–19.

Paz-y-Miño, C., Bond, A. B., Kamil, A. C., and Balda, R. P. (2004). Pinyon jays use transitive inference to predict social dominance. *Nature* **430**, 778–781.

Pusey, A., Williams, J., and Goodall, J. (1997). The influence of dominance rank on the reproductive success of female chimpanzees. *Science* **277**, 828–831.

Quintana-Rizzo, E., Mann, D. A., and Wells, R. S. (2006). Estimated communication range of social sounds used by bottlenose dolphins (*Tursiops truncatus*). *J. Acoust. Soc. Am.* **120**, 1671–1683.

Silk, J. B. (2007). The adaptive value of sociality in mammalian groups. *Philos. Trans. R. Soc. Lond., B, Biol. Sci.* **362**, 539–559.

Trillmich, F., and Trillmich, K. G. K. (1984). Mating systems of pinnipeds and marine iguanas: Convergent evolution of polygyny. *Biol. J. Linn. Soc. Lond.* **21**, 209–216.

Whitehead, H. (1996). Babysitting, dive synchrony, and indications of alloparental care in sperm whales. *Behav. Ecol. Sociobiol.* **38**, 237–244.

Whitehead, H. (1998). Cultural selection and genetic diversity in matrilineal whales. *Science* **282**, 1708–1711.

Whitehead, H. (2007). Learning, climate and the evolution of cultural capability. *J. Theor. Biol.* **245**, 341–350.

Whitehead, H., Rendell, L., Osborne, R. W., and Würsig, B. (2004). Culture and conservation of non-humans with reference to whales and dolphins: Review and new directions. *Biol. Conserv.* **120**, 431–441.

Wolf, J. B. W., Kauermann, G., and Trillmich, F. (2005). Males in the shade: Habitat use and sexual segregation in the Galápagos sea lion (*Zalophus californianus wollebaeki*). *Behav. Ecol. Sociobiol.* **59**, 293–302.

Wrangham, R. W., Gittleman, J. L., and Chapman, C. A. (1993). Constraints on group size in primates and carnivores: Population density and day-range as assays of exploitation competition. *Behav. Ecol. Sociobiol.* **32**, 199–209.

# Habitat Pressures

## Peter G. H. Evans

Like other animals, marine mammals may have preferred locations in which they spend the majority of time or where they engage in particular important life history activities such as giving birth, calf rearing, or feeding. The array of physical and oceanographic features that typify those locations forms the habitat of a species or local population. Often these are difficult to define. An ice-breeding seal clearly depends on pack ice upon which to give birth and that constitutes its' breeding habitat, and a gray whale (*Eschrichtius robustus*) may seek out a sheltered tropical lagoon to calve, but for a large open-ocean baleen whale like a fin whale (*Balaenoptera physalus*) or blue whale (*B. musculus*), identifying its habitat requirements for breeding can be a difficult task. The same applies to feeding habitats: manatees (*Trichechus* spp.) and dugongs (*Dugong dugon*), e.g., require specific habitats such as shallow seagrass beds for feeding, but oceanic dolphins may range the high seas in pursuit of shoaling (schooling) fishes, making it difficult to identify whether they have specific habitat requirements. Human activities impinge upon the lives of marine mammals if they damage or destroy those habitats which may be important to them. Our knowledge of habitat pressures facing marine mammals is therefore limited to particular species, and especially to locations nearshore where animals have been studied more intensively and their ecological requirements are better defined.

Habitats formed by eddies, thermoclines, and fronts, particularly if they are driven by currents or wind, may shift from one locality to another during the life span of a marine mammal, leading to shifts in their geographic distributions. Habitats determined by geomorphological features such as depth, topography, available haulout, or den sites [in the case of pinnipeds and polar bears (*Ursus maritimus*), respectively], are relatively stable over time in relation to location. Strong site fidelity may lead a population to have difficulty in adjusting to changes in local food availability.

Habitat pressures upon marine mammals from anthropogenic influences may be grouped into five categories: (1) physical damage to their environment: a river or seabed and its constituent communities; (2) contamination from chemical pollutants; (3) direct removal of important prey through fisheries; (4) disturbance from human activities either by the introduction of sound into the environment or through ship strikes; and (5) physical and oceanographic effects from global climate change.

## I. Physical Damage

Human population pressures frequently lead to direct changes to coastal and riverine environments. Estuaries are turned into industrial harbors, wetlands are drained for agricultural purposes or for tourism, and coastal waters are modified often irreversibly by dredging of the seabed and input of a wide variety of pollutants. Some of the most obvious detrimental changes to a habitat come from alteration of rivers inhabited by particular dolphin species (Reeves and Smith, 1999). Water is often taken out of rivers for other uses, such as for drinking, flood control, or irrigation agriculture. In Pakistan, e.g., most of the annual flow of the Indus River is diverted into canals, and this, along with dam construction, has resulted in the Indus river dolphin (*Platanista gangetica minor*) losing probably at least half of its historical range (Reeves *et al.*, 1991). Dams modify water flow and affect the sedimentation of rivers; they also block traditional movement patterns of marine mammals that can lead to population fragmentation. The construction of large dams (such as the Ghezouba Dam and the Three Gorges Dam) along the Yangtze River system had serious consequences upon the already endangered baiji (*Lipotes vexillifer*) (Reeves and Leatherwood, 1994), which is now believed to be extinct. It may also restrict movements of more widespread species such as the Amazonian manatee (*Trichechus inunguis*) in Brazil (Rosas, 1994).

On land, one of the greatest habitat pressures leading to mass extinctions of fauna and flora is that of deforestation, particularly in the tropics. In the 1980s, Latin American countries are estimated to have eliminated 7.4 million hectares of tropical forests annually, with Brazil sustaining the greatest annual loss with 3.2 million hectares per year. This deforestation directly affects the fresh water habitats of the boto or Amazon river dolphin (*Inia geoffrensis*), as well as the Amazonian manatee (Rosas, 1994; Reeves and Smith, 1999; Reynolds, 1999).

After centuries of direct exploitation, pinnipeds have largely sought sites remote from human activities to give birth to their pups. They, therefore, are less likely to experience direct physical damage to those breeding habitats.

## II. Chemical Pollution

Nearshore environments in particular are exposed to a potential wide range of pollutants as a result of industrial and agricultural activities. Those pollutants may concentrate in the food web, and either degrade the habitat by removing important prey populations or cause health deficiencies in the local populations of marine mammal species. Although high levels of potentially damaging pollutants have frequently been detected in marine mammals, particularly seals and coastal small cetaceans inhabiting nearshore environments, direct causal links with health status have rarely been demonstrated. Baltic ringed (*Pusa hispida*) and gray (*Halichoerus grypus*) seals during the 1970s had lesions of the reproductive system attributed to high PCB and DDT levels in their tissues. By the late 1980s and through the 1990s, as levels in those pollutants declined, the proportion with lesions had declined substantially, along with an increase in their pregnancy rate (O'Shea, 1999; Reijnders *et al.*, 1999). In an experimental study with harbor seals (*Phoca vitulina*), females fed with fish from the heavily polluted Dutch Wadden Sea had poorer reproductive success than those fed less contaminated fish from the North Atlantic. The effects were attributed to PCBs or their metabolites, and seals with the highest PCB intake were found to have reduced blood levels of thyroid hormones and vitamin A, both of which are known to be important in reproduction, including spermatogenesis.

Belugas (*Delphinapterus leucas*) in the highly polluted St. Lawrence Estuary in North America had a high prevalence of tumors which had been attributed to carcinogenic compounds such as polycyclic aromatic hydrocarbons (PAHs) and other toxic compounds such as PCBs (Martineau *et al.*, 1999; Michaud and Béland, 2001). These were thought to account for low reproductive success in this population. However, although both sets of compounds occurred at high levels in this population, a direct link has not been clearly demonstrated, and the population in fact appeared to have increased since hunting ceased in 1979 (Kingsley, 2001).

Stranded harbor porpoises (*Phocoena phocoena*) from around the British Isles had PCB concentrations sufficiently high to cause adverse physiological effects, and mortality identified as from infectious diseases was considered to be associated with chronic exposure to these chemicals (Jepson *et al.*, 2005). Mass mortalities of striped dolphins (*Stenella coeruleoalba*) in the Mediterranean, bottlenose dolphins (*Tursiops truncatus*) in the eastern United States, harbor seals in the North and Baltic Seas, and Baikal seals (*Pusa sibirica*) in Lake Baikal also have showed significantly high concentrations of PCBs, which were thought to have reduced resistance to disease, thus making these populations more susceptible to virus infection.

Despite examples like these of apparent links between contamination and health status, the biological significance and nature of effects generally remains uncertain, and it has been impossible to demonstrate conclusively that demographic changes to a population can be attributed to pollution. The only exceptions are where pollution can be shown to lead directly to mortality. After the Exxon Valdez tanker went aground in Prince William Sound, Alaska, in 1989, releasing large volumes of crude oil, several thousand sea otters (*Enhydra lutris*) and about 300 harbor seals died as a result of the oiled pelts losing their vital insulation properties (Loughlin, 1994).

## III. Competition with Fisheries

Habitats compromise animal and plant communities in an often complex web of interaction. When one or more members of the community are removed in large numbers, this can have repercussions throughout the food web, altering predator–prey relationships and competition for resources. Following the intense exploitation of large baleen whales in the Southern Ocean during the first half of the twentieth century, it was estimated that their overall biomass was reduced from 43 million tons to about 6.6 million tons, and that this made available a "surplus" of about 153 million tons of krill (Laws, 1985). These massive changes to the food web of the Southern Ocean had important effects on the remaining members with individual whales growing faster, reaching sexual maturity at an earlier age, and exhibiting increased pregnancy rates. Similar changes in life history parameters were seen in other marine species like the Antarctic crabeater seal (*Lobodon carcinophaga*) and several seabird species.

During the twentieth century, fisheries around the world intensified to such an extent that major changes in fish stocks were observed for many species. Rarely, however, has it been possible to show that prey depletion had reduced the numbers of a particular marine mammal species. Many marine mammals have catholic (broad or species rich) diets, and appear to respond by switching prey. The relative ease of capture and nutritive contents of different prey species may vary, but it has scarcely ever been possible to demonstrate that these have affected reproductive or survival rates, and hence led to a decline in that population. More often than not, the species appears to respond by shifting its distribution.

On both sides of the North Atlantic, fishing activities have markedly reduced the stocks of Atlantic mackerel and herring (*Clupea* spp.), resulting in other fish (upon which they prey) such as sand lance, sprat, and gadoid species becoming locally very abundant. Not only did some cetacean species like harbor porpoises and humpback whales (*Megaptera novaeangliae*) switch their diets to include those prey in greater amounts, but some also showed geographic shifts in distribution. Gray seals, feeding largely on sand lance, increased in number in the North Sea at around 7% per year, while right whales (spp.), feeding largely on plankton (the prey of sand lance) in the North-west Atlantic showed local declines. When some local sand lance and sprat populations crashed a few years later, further changes were witnessed. In the Gulf of Maine, e.g., fish-eating humpback and fin whales were replaced by plankton-eating right and sei (*Balaenoptera borealis*) whales, harbor porpoises moved nearer shore, and Atlantic white-sided dolphins (*Lagenorhynchus acutus*) became abundant and white-beaked dolphins rare (*L. albirostris*) (Kenney *et al.*, 1996).

In the Bering Sea and Gulf of Alaska, substantial declines in the numbers of Steller sea lions (*Eumetopias jubatus*), harbor seals, and northern fur seals (*Callorhinus ursinus*), as well as several species of fish-eating birds, have occurred since the 1970s. Although other factors may also be involved, most of these declines have been attributed to a decline in food availability resulting from the development of the Walleye pollock (*Pollachius pollachius*) fishery, a key prey species for many of these marine mammals following the demise of local herring stocks (Reeves and Reijnders, 2002). Similarly, the collapse of productivity of the Barents Sea ecosystem, brought on partly from excessive fishing mortality, has had far-reaching effects on a range of species from seabirds through to marine mammals (Bjørge, 2002).

## IV. Disturbance

Sounds are introduced into marine and fresh water environments from a wide variety of sources: motor-powered vessel traffic of various sizes; active sonar for object detection including fish-finding and submarines; seismic exploration and subsequent drilling and production for oil and gas; explosions from military exercises and ocean science studies; and marine dredging and construction (Richardson *et al.*, 1995; Würsig and Evans, 2002; Nowacek *et al.*, 2007). Most of the sounds produced are concentrated between 10 and 500 Hz frequency. However, speedcraft of various types generate noise mainly between 2 and 20 kHz by cavitation of the propeller, and sidescan and military sonar generate sounds between 2 and 500 kHz (particularly in the lower-frequency range) (Evans, 1996).

Among cetaceans, baleen whales have rather different hearing sensitivities to those of toothed whales and dolphins. The former are most sensitive at low frequencies below 5 kHz and the latter above 1 kHz. Thus, baleen whales are likely to be most vulnerable to large vessels, oil and gas activities, marine dredging and construction, whereas toothed whales and dolphins may be more susceptible to recreational speedboats and most forms of active sonar.

Changes in behavior (e.g., movement away from the sound, increased dive times, clustering behavior) are often recorded in the vicinity of loud sounds. Few experimental studies have been conducted to test the nature and duration of negative responses. One such study in relation to low-frequency regular ATOC (Acoustic Thermometry of Ocean Climate project) sound pulses was conducted west of California (Calambokidis *et al.*, 1998). Aerial surveys showed no significant differences in numbers of marine mammals of any species between control and experimental surveys, but humpback and sperm whales (*Physeter macrocephalus*) were on average further from the sound source during the experimental periods. Although many other studies have reported negative reactions, there is very little information concerning the long-term impact of sound disturbance. In Hawaii, humpback whale mothers with their calves are thought to have shifted their distribution offshore in response to the high volume of recreational traffic. Whale and seal watching itself can impose pressures upon marine mammals, disturbing seals from haulout or breeding sites, and whales (and dolphins) from favored feeding areas. These have even been found to have long-term consequences upon reproductive success, as in one bottlenose dolphin population from Australia (Bejder *et al.*, 2006).

Besides those indirect effects where sound disturbance may interfere with or frighten marine mammals, there is some evidence that loud sounds can cause physical damage. Temporary or permanent shifts in hearing thresholds may occur which could affect auditory acuity, and post-mortem examination of humpback whales found dead in the vicinity of drilling operations has revealed ear damage. Most notably, mid-frequency sonar (mainly between 2 and 10 kHz) used in military activities has been linked to mass strandings of beaked whales, and there have been a number of recent such events (e.g., in the Bahamas and the Canaries) (Evans and Miller, 2004; Cox *et al.*, 2006). In those cases, however, it has not been entirely clear whether the strandings have resulted from direct acoustic trauma or some behavioral change leading to gas bubble formation.

A new concern has arisen within Europe with the rapid expansion of offshore wind farms, mainly in shallow areas where harbor porpoises and harbor seals occur. Pile-driving activities during the construction phase in particular situations appear to have negative impacts on harbor porpoises, which can be relatively long-lasting (Carstensen *et al.*, 2006).

Powered vessels pose an obvious threat to marine and fresh water mammals through direct damage. Collisions have been reported in a wide variety of species, and in some, such as the Florida manatee (*Trichechus manatus*) and the North Atlantic (*Eubalaena glacialis*) and North Pacific (*E. japonica*) right whale, they are regarded as the major threat to their survival. With the advent of high-speed ferries in many parts of the world, ship strikes are being reported with increasing frequency, especially affecting some of the larger baleen whales like fin whales and the slower swimming toothed whale species like sperm whales and pilot whales (*Globicephala* spp.).

## V. Climate Change

As a result of emissions by humans of substances which deplete the ozone layer, our increasing use of hydrocarbons for energy and fuel, and large-scale deforestation and desertification, the world is experiencing climate change such that it is predicted that, in the next hundred years, temperatures will rise by 1.0–3.5°C and overall sea level will rise by anywhere from 15 to 95 cm. Obvious consequences will be the melting of polar ice, drowning of coastal plains, and changes to shallow seas. Other less direct implications include an increase in the frequency and velocity of storms, and more extreme seasonal fluctuations in local climate (including, e.g., El Niño Southern Oscillation events). Shifts in areas of primary productivity may lead to distributional changes for many marine mammal species, but some such as the polar bear, land-breeding pinnipeds, and coastal cetaceans and sirenians may find it difficult to adjust to the loss of important feeding or breeding habitat (Würsig *et al.*, 2002). Already there is concern that less stable ice in some parts of the Arctic has reduced the availability of ringed seals to polar bears,

thus reducing the breeding success of the bears, which in those areas depend upon this species for food (Tynan and DeMaster, 1997).

During recent El Niño events, there has been reproductive failure in many seabird populations and some colonies of fur seals. During the 1982 El Niño, e.g., all Galapagos fur seal (*Arctocephalus galapagoensis*) females lost their pups due to starvation (Trillmich and Dellinger, 1991). However, many pelagic toothed whales and dolphins, being less tied to a particular locality, simply shifted their distributions: short-finned pilot whales (*Globicephala macrorhynchus*), e.g., left southern Californian waters following the departure of a species of squid, their main prey. Such changes can affect other members of the ecosystem. When the squid returned some years later, the temporarily vacant niche became occupied by another cetacean species, the Risso's dolphin (*Grampus griseus*) (Shane, 1995).

Despite the many pressures upon their habitats, marine mammals appear to be remarkably resilient, often living in highly modified coastal and riverine environments. Of course, because demographic changes may be slow and difficult to detect, we rarely know whether these are nonetheless having negative effects. In the case of small local populations of endangered species like the northern right whales, vaquita (*Phocoena sinus*), various river dolphins, monk seals (*Monachus* spp.), and manatee populations, the dangers of habitat pressures are all too obvious. Howvever, even for other species, a precautionary approach would be prudent, and there is scope for the establishment of protective areas where human activities can be zoned.

## See Also the Following Articles

Endangered Species and Populations ■ Entrapment and Entanglement ■ Fishing Industry, Effects of.

## References

Bejder, L., *et al.* (10 authors) (2006). Relative abundance of bottlenose dolphins (*Tursiops* sp.) exposed to long-term anthropogenic disturbance. *Conserv. Biol.* **20**, 1791–1798.

Bjørge, A. (2002). How persistent are marine mammal habitats in an ocean of variability? Habitat use, home range and site fidelity in marine mammals. *In* "Marine Mammals: Biology and Conservation" (P. G. H. Evans, and J. A. Raga, eds), pp. 63–91. Plenum Press/Kluwer Academic, New York.

Calambokidis, J., Chandler, T. E., Costa, D. P., Clark, C. W., and Whitehead, H. (1998). Effects of the ATOC sound source on the distribution of marine mammals observed from aerial surveys off Central California. *The World Marine Mammal Science Conference*, January 1998 (Abstract).

Carstensen, J., Henriksen, O. D., and Teilmann, J. (2006). Impacts of offshore wind farm construction on harbour porpoises: Acoustic monitoring of echolocation activity using porpoise detectors (T-PODs). *Mar. Ecol. Prog. Ser.* **321**, 295–308.

Cox, T. M., *et al.* (36 authors) (2006). Understanding the impacts of anthropogenic sound on beaked whales. *J. Cetacean Res. Manag.* **7**, 177–187.

Evans, P. G. H. (1996). Human disturbance of cetaceans. *In* "The Exploitation of Mammals—Principles and Problems Underlying Their Sustainable Use" (N. Dunstone, and V. Taylor, eds), pp. 376–394. Cambridge University Press, Cambridge, MA.

Evans, P. G. H., and Miller, L. A. (2004). Active sonar and cetaceans. *Proceedings of Workshop held at the ECS 17th Annual Conference*, Las Palmas, Gran Canaria, March 8, 2003. European Cetacean Society, Kiel, Germany.

Jepson, P. D., Bennett, P. M., Deaville, R., Allchin, C. R., Baker, J. R., and Law, R. J. (2005). Relationships between PCBs and health status

H

in UK-stranded harbour porpoises (*Phocoena phocoena*). *Environ. Toxicol. Chem.* **24**, 238–248.

Kenney, R. D., Payne, P. M., Heinemann, D. W., and Winn, H. E. (1996). Shifts in northeast shelf cetacean distributions relative to trends in Gulf of Maine/Georges Bank finfish abundance. *In* "The Northeast Shelf Ecosystem: Assessment, Sustainability, and Management" (K. Sherman, N. A. Jaworski, and T. J. Smayda, eds), pp. 169–196. Blackwell, Oxford.

Kingsley, M. C. S. (2001). Beluga surveys in the St Lawrence: A reply to Michaud and Béland. *Mar. Mamm. Sci.* **17**, 213–218.

Laws, R. M. (1985). The ecology of the Southern Ocean. *Am. Sci.* **73**, 26–40.

Loughlin, T. R. (1994). "Marine Mammals and the Exxon Valdez." Academic Press, San Diego.

Martineau, D., Lair, S., DeGuise, S., Lipscomb, T. P., and Béland, P. (1999). Cancer in beluga whales from the St. Lawrence estuary, Quebec, Canada: A potential biomarker of environmental contamination. *J. Cet. Res. Manag.* (Special Issue 1), 249–265.

Michaud, R., and Béland, P. (2001). Looking for trends in the endangered St. Lawrence beluga population. A critique of Kingsley, M.C.S. 1998. *Mar. Mamm. Sci.* **17**, 206–212.

Nowacek, D. P., Thorne, L. P., Johnston, D. W., and Tyack, P. L. (2007). Responses of cetaceans to anthropogenic noise. *Mamm. Rev.* **37**, 81–115.

O'Shea, T. J. (1999). Environmental contaminants and marine mammals. *In* "Biology of Marine Mammals" (J. E. Reynolds, III, and S. A. Rommel, eds), pp. 485–563. Smithsonian Institution Press, Washington, DC.

Reeves, R. R., and Leatherwood, S. (1994). Dams and river dolphins: Can they co-exist? *Ambio* **23**, 172–175.

Reeves, R. R., and Smith, B. D. (1999). Interrupted migrations and dispersal of river dolphins: Some ecological effects of riverine development. *In* "Proceedings of the CMS Symposium on Animal Migration," Gland, Switzerland, April 13, 1997 (UNEP/CMS, Ed.), pp. 9–18. Convention on Migratory Species, Technical Series Publication No. 2. United Nations Environment Programme, Bonn/The Hague.

Reeves, R. R., and Reijnders, P. J. H. (2002). Conservation and management. *In* "Marine Mammal Biology: An Evolutionary Approach" (A. R. Hoelzel, ed.), pp. 388–415. Blackwell Publishing, Oxford.

Reeves, R. R., Chaudhry, A. A., and Khalid, U. (1991). Competing for water on the Indus plain: Is there a future for Pakistan's river dolphins? *Environ. Conserv.* **18**, 341–350.

Reijnders, P. J. H., Aguilar, A., and Donovan, G. P. (1999). Chemical Pollutants and Cetaceans. *J. Cet. Res. Manag.* (Special Issue 1).

Reynolds, J. E., III (1999). Efforts to conserve manatees. *In* "Conservation and Management of Marine Mammals" (J. R. Twiss, Jr, and R. R. Reeves, eds), pp. 267–295. Smithsonian Institution Press, Washington, DC.

Richardson, W. J., Greene, C. R., Jr., Malme, C. I., and Thomson, D. H. (1995). "Marine Mammals and Noise." Academic Press, San Diego.

Rosas, F. C. W. (1994). Biology, conservation and status of the Amazonian manatee *Trichechus inunguis*. *Mamm. Rev.* **24**, 49–59.

Shane, S. H. (1995). Relationship between pilot whales and Risso's dolphins at Santa Catalina Island, California, USA. *Mar. Ecol. Prog. Ser.* **123**, 5–11.

Trillmich, F., and Dellinger, T. (1991). The effects of El Niño on Galapagos pinnipeds. *In* "Pinnipeds and El Niño: Responses to Environmental Stress" (F. Trillmich, and K. A. Ono, eds), pp. 66–74. Springer-Verlag, Berlin.

Tynan, C. T., and DeMaster, D. P. (1997). Observations and predictions of Arctic climate change: Potential effects on marine mammals. *Arctic* **50**, 308–322.

Würsig, B., and Evans, P. G. H. (2002). Cetaceans and humans: Influences of noise. *In* "Marine Mammals: Biology and Conservation" (P. G. H. Evans, and J. A. Raga, eds), pp. 555–576. Plenum Press/Kluwer Academic, New York.

Würsig, B., Reeves, R. R., and Ortega-Ortiz, J. G. (2002). Global climate change and marine mammals. *In* "Marine Mammals—Biology and Conservation" (P. G. H. Evans, and J. A. Raga, eds), pp. 589–608. Kluwer Academic/Plenum Publishers, New York.

# Habitat Use

### ALEJANDRO ACEVEDO-GUTIÉRREZ

## I. Introduction

### A. Temporal and Spatial Scales in Ecology

Ecology is the study of interactions between organisms and their environment, and the distribution and abundance of organisms resulting from these interactions. The environment of any organism includes abiotic factors—non-living chemical and physical factors such as temperature and light—and biotic factors—living organisms with which any individual interacts. For instance, other organisms may compete with an individual for food and resources, prey upon it, or change its physical and chemical environment. At the core of both ecology and conservation biology are questions that examine the relative importance of various environmental components in determining the distribution and abundance of organisms.

Habitat use studies attempt to describe, explain, and predict the distribution and abundance of organisms. In these studies, identifying the factors that influence distribution and abundance at different spatial and temporal scales is fundamental. This concept can be illustrated by examining the distribution and abundance of blue whales (*Balaenoptera musculus*) from the California/Mexico populations and fin whales (*B. physalus*) from the Gulf of California population. Blue whales and to a lesser extent fin whales depend on krill (Euphausiacea) as a prey item. Krill form large, dense swarms during the day and have their largest concentrations below 100 m in depth. At night, krill come near the surface but are scattered over large areas. During winter in Bahía de Loreto, Gulf of California, México, blue and fin whales engage in deep foraging dives during the day while at night they perform shallow dives, very few of which appear to be foraging dives. During the day, krill swarms are found around underwater edges, where depth diminishes rapidly, and blue and fin whales concentrate their movements and feeding in those areas. Both blue and fin whales move out of Loreto around early spring. Bahía de Loreto is thus a short-term feeding site for both whales, which behavior and movements closely match those of krill.

By combining the information from Loreto to that from other studies, the following general picture of the California/Mexico population of blue whales emerges: In late spring, blue whales move north to feed during summer and early fall along the California coast in the Farallones Islands, Cordell Banks, and Monterey Bay, on large swarms of krill. The whales move back south in fall, feeding around the Channel Islands and perhaps off Bahia Magdalena, Mexico. During winter, whales are back in the Gulf of California, including Bahía de Loreto. However, there is a large degree of variation, and many whales may winter in the Costa Rica Dome, an oceanographic feature in the Pacific Ocean. Although the picture of the Gulf of California population of fin whales is less complete, we know that they feed during the spring in the southern region of the Gulf of California, including Loreto. During the summer, a time of year in which krill are less abundant in the gulf, they move further north into the gulf to prey on schooling fish.

This example illustrates the importance of defining temporal and spatial scales in an ecological study, and documents how the distribution of marine mammals is influenced by the environment at different spatial and temporal scales. At scales of days and tens of kilometers, blue whales are found during the day along canyon edges, feeding on krill swarms. At scales of months and hundreds to thousands of kilometers, blue whales move to different coastal areas to exploit krill swarms. In the case of fin whales, at scales of days and tens of kilometers, they are also found during the day along canyon edges feeding on krill swarms. However, at scales of months and hundreds to thousands of kilometers, they move within the same oceanographic area (the Gulf of California) and switch prey items, from krill to schooling fish. Even closely related species of marine mammals can make different decisions regarding their distribution: blue whales move out of the Gulf of California and look for the same prey item; fin whales remain in the Gulf of California and switch prey items. Given that marine mammals are generally long lived and that their cost of locomotion in the water is relatively low, understanding their distribution and abundance at multiple temporal and spatial scales is even more crucial than for shorter-lived or less mobile organisms.

## B. Research on Marine Mammal Habitat Use

Marine mammals are highly mobile, tend to cover large areas, move in three spatial dimensions, and spend the vast majority of their lives under water. Hence, controlled experiments are next to impossible to conduct, and describing, explaining, and predicting distribution and abundance present unusual challenges to researchers. In general, studies are unable to show a causal explanation between a factor or factors and the observed distribution and abundance of a marine mammal population; rather, scientists rely on quantitative correlations that are indicative of potential causal factors. For instance, several studies document that during the summer belugas (*Delphinapterus leucas*) in Alaska are distributed near coastal mud flats and river mouths; however, it is unclear whether the observed distribution is caused by prey availability, breeding, calving, molting, or shelter from predators (Goetz *et al.*, 2007).

Our understanding of marine mammal habitat use has been improved by employing remote-sensing techniques and sophisticated statistical analyses. Remote-sensing techniques allow scientists to correlate marine mammal distribution with dynamic environmental variables that take into account spatial or temporal scales. For example, in the Gulf of St. Lawrence, the distribution of blue whales, fin whales, minke whales (*Balaenoptera acutorostrata*), and humpback whales (*Megaptera novaeangliae*) is highly correlated with thermal fronts, which were described from sea-surface-temperature satellite images (Doniol-Valcroze *et al.*, 2007). Remote-sensing techniques also allow scientists to describe the three-dimensional space distribution of marine mammals and correlate it with environmental factors. For instance, the amount of time that Weddell seals (*Leptonychotes weddellii*) in Antarctica spend at the bottom phase of a dive correlates with an index of prey abundance (Mitani *et al.*, 2003; Watanabe *et al.*, 2003; Mitani *et al.*, 2004). These results were obtained by attaching recorders to individual seals, which recorded dive behavior, acceleration, geomagnetic intensity, and digital still pictures. This methodology allows scientists to describe three-dimensional spatial use of seals, which spent their time under water on a small region with a steep bottom contour, apparently searching for bentho-pelagic prey throughout the water column (Mitani *et al.*, 2004).

Sophisticated statistical analyses allow scientists to simultaneously correlate the distribution and abundance of marine mammals with many different environmental factors. In the Bahamas, the occurrence of Blainville's beaked whales (*Mesoplodon densirostris*) is correlated in decreasing order of importance with seabed aspect (facing direction), seabed gradient (slope), and water depth, an analysis conducted with generalized additive modeling (GAM). Blainville's

whales occur in areas with a northeast aspect, intermediate gradients, and depths between 200 and 1000 m where bottom topography forces the Deep Western Boundary Current toward the surface (MacLeod and Zuur, 2005). The authors hypothesize that prey are concentrated in these same areas. In the Faroe-Shetland Channel north of the United Kingdom, a GAM analysis of dolphin sounds indicates that dolphin distribution is best predicted by a combination of water noise level, time of day, month, and water depth (Hastie *et al.*, 2005).

Despite our inability to determine causality and hence fully explain the relationship between marine mammal distribution and abundance, and several biotic and abiotic factors, several tools such as remote-sensing techniques and GAM analysis allow scientists to describe, predict, and partially explain the determinants of such relationships.

## C. Habitat Use and Evolution

Understanding the distribution and abundance of marine mammals is important not only to ecologists, conservation biologists, environmentalists, managers, and tour operators, but also to evolutionary biologists. This is because the interactions of organisms with their environment that occur over a long period of time are important causes of evolutionary change. Lake Apoyo, a volcanic crater lake in Nicaragua, was seeded only once by the ancestral benthic species *Amphilophus citrinellus* from which the new limnetic species *A. zaliosus* evolved within less than 10,000 years by exploiting a different habitat (Barluenga *et al.*, 2006). These two species are both reproductively isolated and eco-morphologically distinct; thus providing a convincing example of habitat use as an agent of evolutionary change via sympatric speciation. Sympatric speciation is a contentious concept in evolutionary biology, for which few convincing examples exist worldwide. Hence, documenting evolutionary change in marine mammals due to habitat use has been extremely difficult. The apparent incipient speciation of sympatric resident and transient killer whales (*Orcinus orca*) off the west coast of Washington State and Canada may be such an example (Baird *et al.*, 1992). However, it has been suggested that the division was originally cultural (Whitehead *et al.*, 2004).

## II. Intrinsic Factors in Habitat Use

Most explanations on habitat use by marine mammals refer to environmental factors, such as prey availability, predation, or temperature, which are extrinsic to the organisms. However, traits intrinsic to the organisms themselves may affect their ability to exploit certain habitats and hence determine their distribution and abundance.

## A. Body Size

Body size affects many important traits in organisms, including morphology, metabolic rate, and reproductive costs. Species with large body size and large amounts of fat stored in blubber are able to travel far or to exploit very patchy resources. An example of the relationship between habitat use and large body size is the northern elephant seal (*Mirounga angustirorstris*). This species makes a double migration each year: one after molting and one after breeding, with individual annual movements of 18,000–21,000 km (Stewart and DeLong, 1993, 1995). Adults stay at sea for 8–9 months of the year to forage, using the California Current as a corridor to foraging areas further north that are related to water masses and the distribution of squid. While at sea, both sexes dive almost continuously, remaining submerged for about

90% of the total time. Large body size has also predisposed marine mammals to long dives, because body size augments oxygen stores and diminishes specific metabolic rate (use of oxygen per unit of mass) (Hoelzel, 2002). Because deep dives require longer dives than shallow dives, large marine mammals such as the sperm whale (*Physeter macrocephalus*) are able to exploit deep-water habitats. However, given the many different physiological adaptations for diving found in marine mammals, body size alone cannot predict the vertical distribution of marine mammals.

### B. Age

Given that marine mammals are long-lived predators, the habitat that they are able to exploit may change over time to reflect increased physiological capabilities as body size increases and by increased learning. Dive and depth duration in Australian sea lions (*Neophoca cinerea*) increase with age; however, such development is slow (Fowler *et al.*, 2006). Pups at 6 months of age show minimal diving activity, they are weaned at about 17 months of age, and as 23-month-old juveniles they tend to dive to 40–50 m depths (62% the depth of adults). Pup and juveniles do not reach adult depth or durations and hence occupy shallower habitats than those exploited by adults. Adult New Zealand fur seals (*Arctocephalus forsteri*) utilize continental shelf waters and deep waters over the shelf break, where presumably high densities of fishes and cephalopods are found, while juveniles use pelagic waters up to 1000 km from the habitats used by adults (Page *et al.*, 2006). It is hypothesized that due to their small body size, juveniles cannot efficiently utilize prey in the same habitats as adults because they do not have the capacity to spend enough time under water at the greater depths. Hence, adult male and female New Zealand fur seals are large enough to engage in benthic feeding in shelf breaks, whereas the smaller-sized juveniles are constrained to epipelagic feeding at night.

### C. Sex

Many marine mammal species segregate by sex. The harbor seal (*Phoca vitulina*) provides a good example of such segregation (Boness *et al.*, 1994; Coltman *et al.*, 1997). Females and males have similar body sizes (females weigh about 85 kg; males about 110 kg) and mate at sea. Females nurse their pups for about 24 days, fasting for about 1 week and then having to take regular foraging trips while lactating. Most males forage early in the season and in doing so most individuals maintain or increase body mass during this period. During the latter part of the breeding season, males rarely forage and spend time reproducing when females are receptive. Hence, the habitat occupied by both sexes is different: during the first 10 days after birth, females are on land while males are at sea diving to depths exceeding 60 m; between 10 and 20 days after birth, females make trips at sea and dive to 50–60 m while males are also at sea in areas where females move but diving to only 20 m. It appears that the different habitat use in which females and males engage represents a balance between foraging and reproduction to maximize reproductive success.

The relationship between habitat use and sex is also related to body size in sexually dimorphic marine mammal species. Such body size differences may require the sexes to use different habitats. For instance, gray seals (*Halichoerus grypus*) are sexually dimorphic in size (Breed *et al.*, 2006). At Sable Island, Nova Scotia, males and females utilized different habitats, differences that were most pronounced just before and immediately after breeding. Females mainly used mid-shelf regions whereas males primarily used areas along the continental shelf break. It is hypothesized that these differences maximize fitness

by reducing intersexual foraging competition. Southern elephant seals (*Mirounga leonina*) from Kerguelen Island travel to the Antarctic shelf (Bailleul *et al.*, 2007). As the ice expands during winter, females appear to shift from benthic to pelagic foraging, while males continue to forage almost exclusively benthically over the continental shelf. It is hypothesized that this difference in habitat use is related to the different energetic requirements between the two sexes, or to the need for females to return to Kerguelen in the spring to give birth, whereas males can remain in the ice.

### D. Individual Variability

Differences in habitat use may also be related to individual variability. For instance, there is significant variation between individual female Antarctic fur seals (*Arctocephalus gazella*) in trip durations and the maximum distance reached from the breeding beach (Staniland *et al.*, 2004). Apparently, there is a strong individual component to where a seal forages, especially in terms of the distance traveled. The authors suggest that once the foraging area is selected by an individual seal, the dive behavior within that area is determined by the area itself, perhaps related to the spatial and temporal distribution of the prey within it, and not by the individual seal.

### E. Life History

Life history refers to the patterns of resource allocation to maintenance (survival), growth, and reproduction. Life history traits appear also to influence habitat use in marine mammals. As described earlier, blue whales migrate from the Gulf of California to the California coast searching for krill aggregations, while fin whales remain in the Gulf of California and switch prey items. In this case, the blue whale pattern is to move to another body of water and feed on the same prey; the fin whale pattern is to remain in the same body of water and feed on different prey. Along the Scandinavian coast, harbor porpoises (*Phocoena phocoena*) experience different ecological regimes during the year and shift from pelagic prey species in deep waters to more coastal and/or demersal prey in relatively shallow waters (Fontaine *et al.*, 2007). In this case, the harbor porpoise pattern is similar to that of fin whales: they both adapt their foraging to local oceanographic conditions rather than perform an extensive migration.

Larger body size implies a longer dive time. However, whales of the family Balaenopteridae (rorquals) dive less than expected based on body size because their foraging strategy of lunging is costly (Acevedo-Gutiérrez *et al.*, 2002). Apparently, the effort needed to accelerate a large mass increases the costs of feeding and reduces time under water. Despite engaging in behaviors to reduce such costs—such as gliding gaits during dive descent, accelerating at the beginning of a lunge and gliding throughout the rest of the lunge—rorquals do not exploit the deep waters that smaller species use. In this case, the rorqual pattern is to exploit relatively shallow habitats due to the constraints imposed by their foraging behavior.

### III. Extrinsic Factors in Habitat Use

Most habitat use studies attempt to explain the distribution and abundance of marine mammals in relationship to external biotic and abiotic factors. Two important extrinsic factors influencing the distribution of a species are food availability and predation risk. In general, marine mammals should exploit areas of high prey density and avoid areas of high predator density. However, it is also important to understand the temporal and spatial scales, given that the predictability of prey distribution tends to decrease with the spatial scale.

## A. Prey Availability

Prey availability is the most frequently factor invoked to explain the distribution and abundance of marine mammals, regardless of the spatial and temporal scale of the study. However, understanding the mechanism influencing prey availability itself has proved as challenging as determining the causality of marine mammal distributions. Croll *et al.* (2005) took advantage of a relatively straightforward system: blue whales feed exclusively upon dense but patchy schools of pelagic krill; hence, understanding krill distribution will assist in understanding blue whale distribution. By employing remote-sensing techniques and concurrent measurements, they examined the temporal and spatial linkages between intensity of upwelling, primary production, distribution of krill, and distribution and abundance of blue whales in Monterey Bay, California. The study indicated that seasonally high primary production supported by coastal upwelling combined with topographic breaks off California maintained high densities to allow exploitation by blue whales. Blue whales appeared in the area in late summer and early fall and fed exclusively upon adult krill *Thysanoessa spinifera* and *Euphausia pacifica* aggregations, diving to depths between 150 and 200 m on the edge of the Monterey Bay Submarine Canyon. High krill densities were supported by high primary production between April and August and a submarine canyon that provided deep water down-current from an upwelling region. Peak krill densities occurred in late summer and early fall, lagging the seasonal increase in primary production by 3–4 months, due to the growth to adulthood of krill spawned around the spring-increase in primary production, and to decreased upwelling in late summer. It is predicted that the annual migratory movements of the California blue whale population reflect seasonal patterns in productivity in other foraging areas in the Northeast Pacific. The annual increase in the abundance of blue whales was linked to wind-driven upwelling, but these linkages occurred through a sequence of bottom-up biological processes that lagged in time. Consequently, models that attempt to predict the distribution and abundance of marine mammals need to include bottom-up processes and temporal scales.

Another example of the importance of understanding the spatial and temporal distribution of prey to describe, explain, and predict marine mammal distribution is found in dugongs (*Dugong dugon*). Like other herbivores, dugongs must select quality food plants to optimize their nutrient intake. Across multiple spatial scales, they appear to prefer some seagrass pastures and avoid others. At medium spatial scale remote sensing, it was confirmed that a 24 km² seagrass meadow in Hervey Bay, Australia, is an important dugong habitat due to the presence of five species of seagrasses, which covered 91% of the total habitat area (Sheppard *et al.*, 2007). However, at a small spatial scale, dugong use within the meadow is still not well understood because the influence of seagrass food quality on dugong grazing patterns and nutritional ecology is poorly understood. Consequently, understanding the dynamics of seagrass communities is essential for predicting patterns of habitat use by dugongs.

## B. Predation Risk

Predation risk is an important factor explaining the distribution and abundance of marine mammals regardless of the spatial and temporal scale of the study. For instance, tiger shark (*Galeocerdo cuvier*) predation risk correlates well with the habitat use of Indo-Pacific bottlenose dolphins (*Tursiops aduncus*) in Shark Bay, Western Australia (Heithaus and Dill, 2002, 2006). The biomass of dolphin prey is greater in shallow habitats than in deeper ones; however, when tiger sharks are present in the area, their density is highest in shallow habitats. It is believed that shallow habitats are also inherently risky because shark detection apparently decreases as dolphin echolocation efficiency and visual detection of sharks camouflaged over seagrass diminish in shallow habitats. Hence, shallow habitats are the best places to forage for dolphins, but are also the most risky. As a result, in seasons of high shark abundance, dolphins foraged much less in the productive but risky shallow habitats than expected if food was the only relevant factor. These results suggest that dolphin habitat use reflects a trade-off between predation risk and prey availability. Besides showcasing the importance of predation risk in explaining habitat use, this study also indicates that the distribution and abundance of marine mammals is simultaneously affected by more than one factor. Further, because the distribution and abundance of tiger sharks are influenced by species other than dolphins, the distribution of the primary prey of the sharks may indirectly influence dolphin habitat use. Hence, as also exemplified by the studies described in the section on prey availability, it is important to consider the community context in studies of habitat use.

## C. Intraspecific Competition

In many species, differences in habitat use between sexes are apparently a consequence of social interactions. A recent study of the Galápagos sea lion (*Zalophus wollebaeki*) indicates that sexual segregation on land was high both during the reproductive and non-reproductive periods (Wolf *et al.*, 2005). A generalized linear model of habitat use showed that adult males frequented habitat types that adult females used much less, with males being most abundant in suboptimal inland habitats. It is hypothesized that this habitat segregation resulted as a by-product of social processes, primarily intrasexual competition and female avoidance of male harassment.

## D. Human Influence

Human activities, also termed anthropogenic influences, are an important extrinsic factor affecting the distribution and abundance of marine mammals. Boat traffic is an activity with many documented cases of impact on marine mammal habitat use. In the short term, this activity may cause marine mammals to temporarily abandon or avoid a particular site. For instance, the number of harbor seals hauled out in a particular site may diminish dramatically in relation to boat traffic (Suryan and Harvey, 1999; Johnson and Acevedo-Gutiérrez, 2007). In New Zealand, the frequency of bottlenose dolphin (*Tursiops truncatus*) visits to Milford Sound has diminished as a result of boat traffic; additionally, when dolphins visit the fjord they remain at the entrance, away from tour boats (Lusseau, 2005). In the long term, boat traffic may create a permanent abandonment of areas visited by marine mammals and hence creating a permanent shift in distribution. For example, boat traffic may cause harbor seals to abandon haulout sites where alternative haulout locations are limited (Suryan and Harvey, 1999). In Shark Bay, Australia, the abundance of Indo-Pacific bottlenose dolphins has declined in areas operated by two or more dolphin-watching boats compared to areas with no boats or with only one boat (Bejder *et al.*, 2006).

Human activities may also cause marine mammals to visit rather than leave a particular area. For instance, two sympatric communities of Indo-Pacific bottlenose dolphins are found in Moreton Bay, Australia (Chilvers *et al.*, 2007). The non-trawler community does not associate with trawler vessels, whereas the trawler community associates with trawlers to feed on flushed prey. While the distribution of the non-trawler community is explained by season and tide,

H

the distribution of the trawler community is explained by the distribution of trawler boats.

## IV. Conclusion

The habitat use of marine mammals is affected by abiotic and biotic factors, including intrinsic and extrinsic factors. The scientists' goal is to describe, explain, and understand the relative importance of each factor in the distribution and abundance of marine mammals. To reach this goal, the temporal and spatial scales of the study system need to be clearly defined. Given the challenges inherent in studying marine mammals, the use of sophisticated remote-sensing technologies and statistical models has been very successful in gathering and integrating data on habitat use. Long-term studies and studies in new regions are fundamental to answering questions on habitat use. However, the most promising line of work is to conduct integrative studies that consider community and ecosystem structure at different spatial and temporal scales, such as the study on the California/Mexico population of blue whales described throughout this chapter.

## *See Also the Following Articles*

Cetacean Ecology ■ Distribution ■ Pinniped Ecology.

## *References*

Acevedo-Gutiérrez, A., Croll, D., and Tershy, B. (2002). High feeding costs limit dive time in large whales. *J. Exp. Biol.* **205**, 1747–1753.

Bailleul, F., Charrassin, J.-B., Ezraty, R., Girard-Ardhuin, F., McMahon, C. R., Field, I. C., and Guinet, C. (2007). Southern elephant seals from Kerguelen Islands confronted by Antarctic sea ice. Changes in movements and in diving behaviour. *Deep Sea Res. II* **54**, 343–355.

Baird, R. W., Abrams, P. A., and Dill, L. M. (1992). Possible indirect interactions between transient and resident killer whales: Implications for the evolution of foraging specialization in the genus *Orcinus. Oecologia* **89**, 125–132.

Barluenga, M., Stölting, K. N., Salzburger, W., Muschick, M., and Meyer, A. (2006). Sympatric speciation in Nicaraguan crater lake cichlid fish. *Nature* **439**, 719–723.

Bejder, L., *et al.* (10 authors) (2006). Decline in relative abundance of bottlenose dolphins exposed to long-term disturbance. *Conserv. Biol.* **20**, 1791–1798.

Berta, A., Sumich, J. L., and Kovacs, K. M. (2006). "Marine Mammals. Evolutionary Biology," 2nd Ed. Academic Press, San Diego.

Boness, D. J., Bowen, W. D., and Oftedal, O. T. (1994). Evidence of a maternal foraging cycle resembling that of otariid seals in a small phocid, the harbor seal. *Behav. Ecol. Sociobiol.* **34**, 95–104.

Breed, G. A., Bowen, W. D., McMillan, J. I., and Leonard, M. L. (2006). Sexual segregation of seasonal foraging habitats in a non-migratory marine mammal. *Proc. R. Soc. Lond., B, Biol. Sci.* **273**, 2319–2326.

Chilvers, B. L., Corkeron, P. J., and Puotinen, M. L. (2007). Influence of trawling on the behaviour and spatial distribution of Indo-Pacific bottlenose dolphins (*Tursiops aduncus*) in Moreton Bay, Australia. *Can. J. Zool.* **81**, 1947–1955.

Coltman, D. W., Bowen, W. D., Boness, D. J., and Iverson, S. J. (1997). Balancing foraging and reproduction in male harbour seals: An aquatically mating pinniped. *Anim. Behav.* **54**, 663–678.

Croll, D. A., Marinovic, B., Benson, S., Chavez, F. P., Black, N., Ternullo, R., and Tershy, B. R. (2005). From wind to whales: Trophic links in a coastal upwelling system. *Mar. Ecol. Prog. Ser.* **289**, 117–130.

Doniol-Valcroze, T., Berteaux, D., Larouche, P., and Sears, R. (2007). Influence of thermal fronts on habitat selection by four rorqual whale species in the Gulf of St. Lawrence. *Mar. Ecol. Prog. Ser.* **335**, 207–216.

Fontaine, M. C., Tolley, K. A., Siebert, U., Gobert, S., Lepoint, G., Bouquegneau, J. M., and Das, K. (2007). Long-term feeding ecology and habitat use in harbour porpoises *Phocoena phocoena* from Scandinavian waters inferred from trace elements and stable isotopes. *BMC Ecol.* **7**, doi:10.1186/1472-6785-7-1.

Fowler, S. L., Costa, D. P., Arnould, J. P. Y., Gales, N. J., and Kuhn, C. E. (2006). Ontogeny of diving behaviour in the Australian sea lion: Trials of adolescence in a late bloomer. *J. Anim. Ecol.* **75**, 358–367.

Goetz, K. T., Rugh, D. J., Read, A. J., and Hobbs, R. C. (2007). Habitat use in a marine ecosystem: Beluga whales *Delphinapterus leucas* in Cook Inlet, Alaska. *Mar. Ecol. Prog. Ser.* **330**, 247–256.

Hastie, G. D., Swift, R. J., Slesser, G., Thompson, P. M., and Turrell, W. R. (2005). Environmental models for predicting oceanic dolphin habitat in the Northeast Atlantic. *ICES J. Mar. Sci.* **62**, 760–770.

Heithaus, M. R., and Dill, L. M. (2002). Food availability and tiger shark predation risk influence bottlenose dolphin habitat use. *Ecology* **83**, 480–491.

Heithaus, M. R., and Dill, L. M. (2006). Does tiger shark predation risk influence foraging habitat use by bottlenose dolphins at multiple spatial scales? *Oikos* **114**, 257–264.

Hoelzel, A. R. (ed.) (2002). "Marine Mammal Biology. An Evolutionary Approach." Blackwell Publishing, Oxford.

Johnson, A., and Acevedo-Gutiérrez, A. (2007). Regulation compliance and harbor seal (*Phoca vitulina*) disturbance. *Can. J. Zool.* **85**, 290–294.

Lusseau, D. (2005). Residency pattern of bottlenose dolphins *Tursiops* spp. in Milford Sound, New Zealand, is related to boat traffic. *Mar. Ecol. Prog. Ser.* **295**, 265–272.

MacLeod, C. D., and Zuur, A. F. (2005). Habitat utilization by Blainville's beaked whales off Great Abaco, northern Bahamas, in relation to seabed topography. *Mar. Biol.* **147**, 1–11.

Mann, J., Connor, R. C., Tyack, P. L., and Whitehead, H. (eds) (2000). "Cetacean Societies. Field Studies of Dolphins and Whales." University of Chicago Press, Chicago, IL.

Mitani, Y., Sato, K., Ito, S., Cameron, M. F., Siniff, D. B., and Naito, Y. (2003). A method for reconstructing three-dimensional dive profiles of marine mammals using geomagnetic intensity data: results from two lactating Weddell seals. *Polar Biol.* **26**, 311–317.

Mitani, Y., Watanabe, Y., Sato, K., Cameron, M. F., and Naito, Y. (2004). 3D diving behavior of Weddell seals with respect to prey accessibility and abundance. *Mar. Ecol. Prog. Ser.* **281**, 275–281.

Page, B., McKenzie, J., Sumner, M. D., Coyne, M., and Goldsworthy, S. D. (2006). Spatial separation of foraging habitats among New Zealand fur seals. *Mar. Ecol. Prog. Ser.* **323**, 263–279.

Sheppard, J. K., Lawler, I. R., and Marsh, H. (2007). Seagrass as pasture for seacows: Landscape-level dugong habitat evaluation. *Estuar. Coast. Shelf Sci.* **71**, 117–132.

Staniland, I. J., Reid, K., and Boyd, I. L. (2004). Comparing individual and spatial influences on foraging behaviour in Antarctic fur seals *Arctocephalus gazella. Mar. Ecol. Prog. Ser.* **275**, 263–274.

Stewart, B. S., and DeLong, R. L. (1993). Double migrations of the northern elephant seal, *Mirounga angustirostris. Symp. Zool. Soc. Lond.* **66**, 179–194.

Stewart, B. S., and DeLong, R. L. (1995). Seasonal dispersion and habitat use of foraging northern elephant seals. *J. Mammal.* **76**, 196–205.

Suryan, R. M., and Harvey, J. T. (1999). Variability in reactions of Pacific harbor seals, *Phoca vitulina richardsi*, to disturbance. *Fish. Bull.* **97**, 332–339.

Watanabe, Y., Mitani, Y., Sato, K., Cameron, M. F., and Naito, Y. (2003). Dive depths of Weddell seals in relation to vertical prey distribution as estimated by image data. *Mar. Ecol. Prog. Ser.* **252**, 283–288.

Whitehead, H., Rendell, L., Osborne, R. W., and Würsig, B. (2004). Culture and conservation of non-humans with reference to whales and dolphins: Review and new directions. *Biol. Conserv.* **120**, 431–441.

Wolf, J. B. W., Kauermann, G., and Trillmich, F. (2005). Males in the shade: Habitat use and sexual segregation in the Galápagos sea lion (*Zalophus californianus wollebaeki*). *Behav. Ecol. Sociobiol.* **59**, 293–302.

**Figure 1** *Catastrophic-type molt in the northern elephant seal* (Mirounga angustirostris), *where the upper epidermis and hairs are shed in large patches within a few weeks. Photograph by B. S. Stewart.*

**H**

# Hair and Fur

PAMELA K. YOCHEM AND BRENT S. STEWART

## I. Structure and Function

The presence of hair is one of the characteristics that distinguishes mammals from other vertebrates. Hair consists of keratinized epidermal cells, formed in hair follicles located in the dermal layer of the skin. Adaptations to an aquatic or amphibious lifestyle are apparent in marine mammal skin and hair (Ling, 1974; Williams *et al.*, 1992; Pabst *et al.*, 1999; Reeves *et al.*, 2002). Pinniped and sea otter (*Enhydra lutris*) hairs are flattened in cross-section rather than round as in other carnivores. This is evidently an adaptation for enhancing streamlining of the body and reducing drag during swimming. Pinnipeds and sea otters have diffuse smooth muscle in their dermis, but they lack true arrector pili muscles. Pinnipeds, sea otters, and polar bears (*Ursus maritimus*) possess sebaceous glands and sweat glands, but these are absent in cetaceans and sirenians. Cetacean skin is hairless except for a few vibrissae or bristles occurring mostly on the rostrum or around the mouth. These are usually lost before or soon after birth. Sirenians have widely scattered hairs. The integument of pinnipeds, sea otters, and polar bears generally has two layers of hair. The outer protective layer consists of long, coarse guard hairs and the inner layer is composed of softer intermediate hairs or underfur. Polar bear, sea otter, and otarid guard hairs are medullated (having a sheath), whereas phocid and walrus hairs (*Odobenus rosmarus*) are not. The hairs typically grow in groups or clumps, with a single guard hair emerging cranial to one or more underfur hairs. Each hair grows from a separate follicle, but the underfur follicles feed into the guard hair canal so that all hairs in a particular clump emerge from a single opening in the skin. Some pinnipeds have a relatively sparse hair coat [walrus, elephant seals (*Mirounga* spp.), and monk seals (*Monachus* spp.) with a single guard hair per canal], whereas others have a lush, thick coat (fur seals, with dozens of underhair or fur follicles feeding into each guard hair canal). Sea otters have the densest fur of any mammal, with approximately 130,000 hairs/cm², about twice as dense as that of northern fur seals (*Callorhinus ursinus*). Albinism and other skin and hair color anomalies have been reported in pinnipeds and cetaceans (Fertl *et al.*, 1999; Bried and Haubreux, 2000).

The appendages of some pinnipeds and the pads of sea otters are hairless, allowing these species to readily lose excess body heat by conduction to the environment. Although most marine mammals rely on blubber for insulation, a layer of air trapped within the hair or fur serves as the primary insulator in fur seals and sea otters and keeps the skin dry when the animals are submerged. Sea otter pelage is coated with squalene, a hydrophobic lipid that aids in waterproofing the fur. Skin secretions in pinnipeds also assist in waterproofing, and provide defense against microbial infections (Meyer *et al.*, 2003).

## II. Molt

Many phocid seals possess a white lanugo coat *in utero*; this may be lost before birth, or may persist for several weeks (as in some arctic and antarctic species). This pelage provides insulation for neonates of ice-breeding seals until they develop a blubber layer and also may serve as camouflage or protective coloration. Other examples of distinct neonatal pelage include the wooly black coat of elephant seals, which is replaced by a silvery hair coat after the pup is weaned, and the fluffy buff-colored pelage of sea otter pups, which persists for several months. The signals for initiation and control of the annual pelage cycle are not known for most species but are thought to include endocrine (thyroid, adrenal, and gonadal hormones), thermal, and nutritional influences (Ling, 1970; Ashwell-Erickson *et al.*, 1986). Molt is generally seasonal, beginning shortly after breeding. Sea otters may molt year-round, although more hairs are generally replaced in summer than in winter. The duration of molt in pinnipeds ranges from a very rapid and "catastrophic" shedding of large patches of superficial epidermis and associated hairs (elephant seals, monk seals) (Fig. 1) to the more gradual pattern seen in otariids, with hairs replaced over several months. A disruption of the molt process, resulting in breakdown of the protective skin barrier, appears to underly Northern Elephant Seal Skin Disease (Beckmen *et al.*, 1997; Yochem, 2008), an ulcerative dermatopathy affecting primarily yearling northern elephant seals (*Mirounga angustirostris*).

## *See Also the Following Articles*

Blubber ■ Energetics ■ Pinniped Physiology ■ Streamlining ■ Thermoregulation

## *References*

Ashwell-Erickson, S., Fay, F. H., and Elsner, R. (1986). Metabolic and hormonal correlates of molting and regeneration of pelage in Alaskan harbor and spotted seals (*Phoca vitulina* and *Phoca largha*). *Can. J. Zool.* **64**, 1086–1094.

Beckmen, K. B., Lowenstine, L. J., Newman, J., Hill, J., Hanni, K., and Gerber, J. (1997). Clinical and pathological characterization of northern elephant seal skin disease. *J. Wildl. Dis.* **33**, 438–449.

Bried, J., and Haubreux, D. (2000). An aberrantly pigmented southern elephant seal (*Mirounga leonina*) at Iles Kerguelen, southern Indian Ocean. *Mar. Mamm. Sci.* **16**, 681–684.

Fertl, D., Pusser, L. Y., and Long, J. J. (1999). First record of an albino bottlenose dolphin (*Tursiops truncatus*) in the Gulf of Mexico, with a review of anomalously white cetaceans. *Mar. Mamm. Sci.* **15**, 227–234.

Ling, J. K. (1970). Pelage and molting in wild mammals with special reference to aquatic forms. *Quart. Rev. Biol.* **45**, 16–54.

Ling, J. K. (1974). The integument of marine mammals. *In* "Functional Anatomy of Marine Mammals" (R. J. Harrison, ed.), Vol. 2, pp. 1–44. Academic Press, London.

Meyer, W., Seegers, U., Herrmann, J., and Schnapper, A. (2003). Further aspects of the general antimicrobial properties of pinniped skin secretions. *Dis. Aquat. Org.* **53**, 177–179.

Pabst, D. Ann, Rommel, S. A., and McLellan, W. A. (1999). The functional morphology of marine mammals. *In* "Biology of Marine Mammals" (J. E. Reynolds, II, and S. A. Rommel, eds), pp. 15–72. Smithsonian Institution Press, Washington DC.

Reeves, R. R., Stewart, B. S., Clapham, P. J., and Powell, J. A. (2002). "National Audubon Society Guide to Marine Mammals of the World." Alfred A. Knopf, New York, 527 pp.

Williams, T. D., Allen, D. D., Groff, J. M., and Glass, R. L. (1992). An analysis of California sea otter (*Enhydra lutris*) pelage and integument. *Mar. Mamm. Sci.* **8**, 1–8.

Yochem, P. K. (2008). The molting process in Northern Elephant seals: Cellular and biochemical correlates. Ph.D. Dissertation, University of California, Davis, CA.

**Figure 1**   *The harbor porpoise.*

shape is a characteristic that distinguishes porpoises from the dolphin family, which have conical teeth. Another characteristic feature of harbor porpoises is the presence of tubercles or small hard bumps on the leading edge of the dorsal fin. The function of these is not yet known.

Although the fossil record containing porpoises is poor, recent genetic investigations have made it possible to reconstruct the most probable relationships among the porpoises. Early morphological studies suggested that harbor porpoises were related to Burmeister's porpoise (*Phocoena spinipinnis*) and the vaquita (*Phocoena sinus*), and therefore these three species have been placed in the same genus. However, genetic information suggests close relation to the Dall's porpoise (*Phocoenoides dalli*), a species endemic to the Pacific Ocean (Rosel *et al.*, 1995).

# Harbor Porpoise
## *Phocoena phocoena*

### ARNE BJØRGE AND KRYSTAL A. TOLLEY

### I. Characteristics and Taxonomy

The harbor porpoise is a small odontocete inhabiting coastal temperate and boreal waters of the Northern Hemisphere (Bjørge and Donovan, 1995). It derives its common English name from the Latin for pig (*porcus*) and is sometimes referred to as the "puffing pig" in parts of Atlantic Canada. The Norwegian common name "nise" is derived from an old Norse word for sneeze and refers to the sound the porpoises make when they surface to breathe.

Harbor porpoises have a short, stocky body (Fig. 1) resulting in a rotund shape, an adaptation that helps them limit heat loss in the cold northern climes (McLellan *et al.*, 2002). The dorsal side of the harbor porpoise and the tail flukes are dark gray, almost black. The chin and underbelly are contrasting light gray, almost white. The head and sides are shaded gray, and darker gray stripes originate near the back of the mouth and run back toward the flippers, which are dark gray. Individual differences in the shading patterns occur. The triangular-shaped dorsal fin makes this species easily recognizable at sea, as does its characteristic forward rolling behavior when it surfaces.

Harbor porpoises have small spade-shaped teeth, about 22–28 pairs in the upper jaw and 21–25 pairs in the lower jaw. The spade

### II. Distribution and Abundance

Harbor porpoises are distributed throughout the coastal waters of the North Pacific, the North Atlantic, and the Black Sea (Fig. 2). The porpoises in each of these ocean basins are reproductively isolated, resulting in division of the species into subspecies: *Phocoena phocoena phocoena* in the Atlantic Ocean and *Phocoena phocoena vomerina* in the Pacific Ocean. Most likely there is an additional, yet unnamed subspecies in the western North Pacific. The harbor porpoises of the Black Sea are classified as a separate subspecies *Phocoena phocoena relicta*. These subspecies differ from each other morphologically and genetically. Atlantic harbor porpoises have larger skulls but shorter jaws than Pacific harbor porpoises. Some morphological differences and variation in pigmentation are observed within the Atlantic Ocean porpoises.

Within the ocean basins the subspecies are divided into several genetically distinct population units. Thirteen population units have been suggested for the North Atlantic. Recent genetic studies indicate that the population structure might be more complex and that the current population units should be revised (Andersen, 2003). Several population units are described from the North Pacific.

The global population size of harbor porpoises is at least 700,000. The North Sea is a particularly important porpoise habitat housing about 335,000 porpoises (Hammond *et al.*, 2002). Other important harbor porpoise habitats are the Gulf of Maine—Bay of Fundy area, the US west coast, and Alaska (Bjørge and Donovan, 1995). The abundance is declining in some areas, mainly due to human-induced factors. The entire subspecies in the Black Sea numbers only about 10,000 porpoises and is possibly declining. Baltic Sea porpoises number a few hundred and are critically endangered.

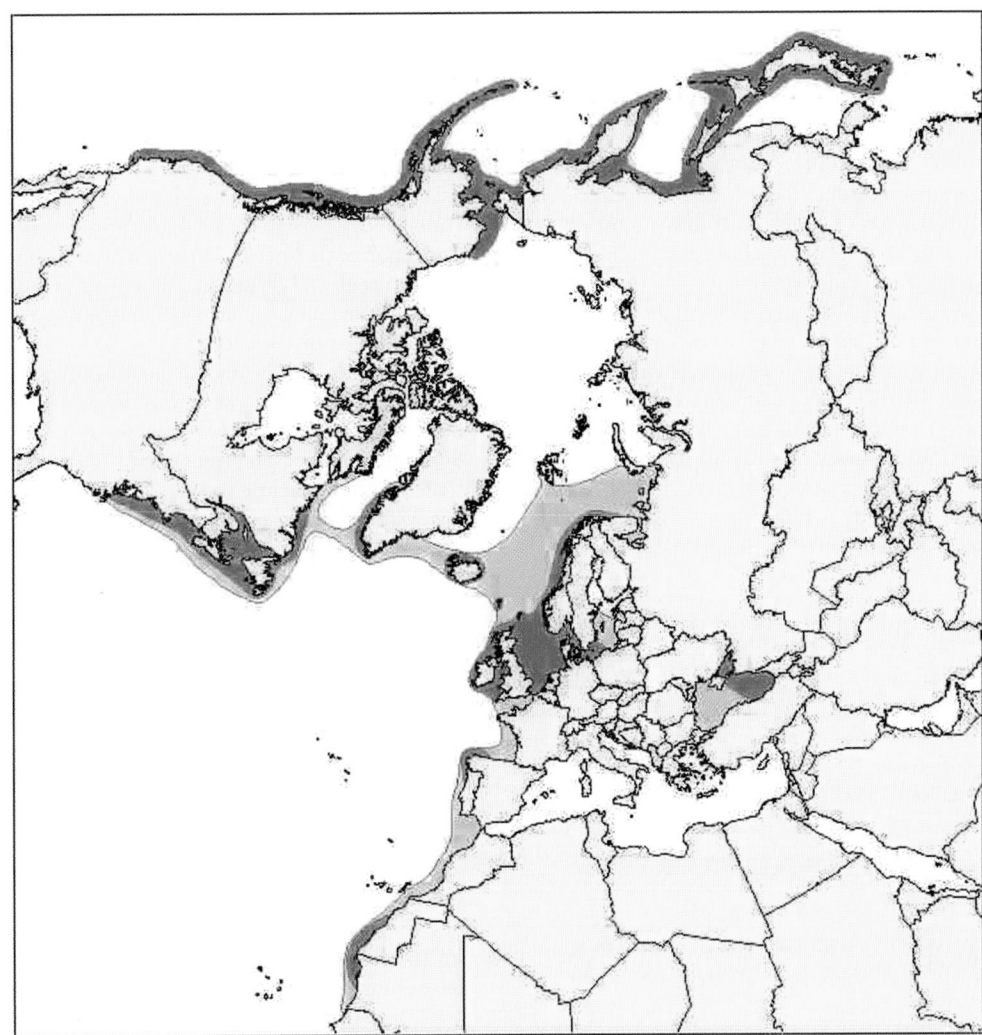

**Figure 2** *The regular worldwide distribution of the harbor porpoise is shown in dark green. The distribution where harbor porpoises occasionally occur is shown in light green.*

### III. Ecology

Harbor porpoises are primarily fish feeders, but in some areas they prey on squid and crustaceans. Small pelagic schooling fishes with high lipid content such as herring, sprat, and anchovy and a range of bottom-dwelling fishes are common prey species. Harbor porpoises usually forage near the sea bottom in waters less than 200 m depth. They also are known to forage close to the surface, e.g., on sprat. When harbor porpoises occur in deeper waters their diet may include mid-water species such as pearlsides. These mid-water fishes become available to the harbor porpoises when they migrate toward the surface at night. Although they generally feed independently, groups of porpoises have been observed collaborating to keep schools of fish closely together and herding them to the surface. The harbor porpoises possibly use the sea bottom and the sea surface as back walls when they are chasing fish.

### IV. Behavior and Physiology

Some studies have demonstrated that harbor porpoises may reside within an area for an extended period of time. However, onshore/offshore migrations and movements parallel to the coast are known to occur. Two large-scale surveys in the North Sea and adjacent waters in 1994 and 2005 showed a stable population size but a shift in distribution from north to south within the North Sea over this period. Such shifts may mirror changes in distribution and availability of important prey species. In some coastal areas, harbor porpoises migrate offshore to avoid ice during winter. In the long and narrow fjords of Norway, where porpoises live year-round, input from rivers form a fresh water top layer, which may freeze within a few hours. Under such circumstances, porpoises can be fatally trapped. The water input from rivers during winter has increased in recent decades due to climate change, and increased risk of ice entrapment of porpoises may be an unexpected effect of higher winter temperatures.

Dive telemetry data have shown that porpoises can dive to at least 220 m. The majority of extended dives are about a minute long, but dives of over 5 min have been recorded (Westgate *et al.*, 1995). The most cost-effective swim speed of harbor porpoises is estimated to about 1.4 m/sec (Otani *et al.*, 1998, 2001). The aerobic dive limit (ADL) depends on swim speed and for porpoises swimming at about 1.5 m/sec or less is about 4 min. Free ranging porpoises usually swim at a speed of slightly less than 1 m/sec and most of their dives are less than 4 min. They, therefore, spend most of their time working aerobically.

Harbor porpoises have extremely thick blubber compared to body size, an adaptation that aids in thermoregulation of the relatively small body in cold waters. Calves have thicker blubber and are more rotund than adults, providing them with an excellent capacity to conserve heat. It is the outer blubber layer that is most stable and important for conserving heat. The inner layer is more metabolically active and serves as an efficient energy store.

Harbor porpoises are normally found in small groups of 1–3 animals often consisting of a mother–calf pair. Larger groups of 6–8 animals are not uncommon and on rare occasions they may form much larger aggregations. Their swimming and surfacing movements are quick, but they rarely leap out of the water. When surfacing, their dorsal side is exposed for a few seconds in the characteristic forward roll movement. When the porpoises occasionally rest at the surface for extended periods, the body is tilted slightly backward with the blowhole as the most elevated part of the body.

## V. Life History

Harbor porpoises at birth are usually about 70–75 cm long and weigh 5 kg (Lockyer, 2003). The calving season varies from region to region, but in most areas calving takes place from May to August. Mating takes place approximately a month and a half after the calving season. The gestation period is approximately 10.5 months. The calves are weaned before they are 1-year old but may begin to catch small solid food items (e.g., euphausiids) when they are just a few months old. Calves grow rapidly, with males reaching about 120 cm and females 125 cm in length at the end of their first year. On average, adult females reach 160 cm in length and weigh around 60 kg. Males are smaller than females, growing only to about 145 cm and 50 kg. The largest recorded size for this species was from a female which was over 200 cm and 70 kg. They become sexually mature between 3 and 4 years of age but are not physically mature until they are about 5 (males) and 7 (females) years old. Harbor porpoises have an average life span of about 8–10 years, although some have been documented to live longer than 20 years. The oldest documented harbor porpoise was 23 years old.

In the Atlantic, most sexually mature females produce a new calf every year, but in the Pacific it appears that the calving interval may be 2 years. Harbor porpoises likely have a promiscuous mating system, whereby each individual mates with several other individuals. Further, they are thought to be "sperm competitors" because males produce large quantities of sperm, presumably in order to mate with several females. The testes undergo large seasonal change in size: they increase up to 800 g just prior to the mating season but regress in winter to a total weight of about 200 g.

## VI. Interactions with Humans

The harbor porpoise as a species is not threatened and IUCN has listed it as LC (least concern). However, the populations in the Baltic Sea and in the Black Sea are listed as CR (critically endangered) and EN (endangered), respectively. Harbor porpoises inhabit coastal waters and are therefore exposed to many negative environmental effects of modern society; eutrophication, chemical pollution, noise, ship traffic, and overfishing of prey are just a few of the

human-induced disturbances of this species (Jepson *et al.*, 1999; Beineke *et al.*, 2007; Lahaye *et al.*, 2007).

In the past, harbor porpoises were harvested for their meat and blubber in many areas, e.g., in Puget Sound, the Bay of Fundy, Gulf of St. Lawrence, Labrador, Newfoundland, Greenland, Iceland, Black Sea, the Danish Belt Seas, and the Baltic Sea. Most of these fisheries are now closed, but hunting still occurs and is increasing in Greenland, where the annual take now exceeds 2000. However, legal protection of the species in most areas does not protect the porpoises against accidental deaths in fishing nets. Throughout its range there are high incidental bycatches of harbor porpoises in fishing gear, and in many areas the bycatch is above sustainable levels. Entanglement in fishing nets is currently the most significant human-induced threat to the porpoises. Modifications in fishing practices are urgently needed to ensure the long-term survival of some porpoise population units. Harbor porpoises spend time close to the surface when they breathe and close to the sea bottom when they forage in shallow waters. Porpoises are therefore exposed both to driftnets hanging from the surface (e.g., driftnets set for salmon) and to bottom-set gill nets. The porpoises emit click trains and use echoes of their own sound to find fish and navigate. Therefore, they are able to detect nets before they are in physical contact with the nets. However, the range within which they can detect the netting of gill nets is less than 10 m. Mean click-train interval of about 12 sec is common in harbor porpoises. However, about 4% of all click-train intervals are longer than 50 sec. Therefore, a swim speed of slightly less than 1 m/sec makes porpoises very susceptible to entanglement in fishing gear during periods of prolonged click-train intervals. The use of modified, more detectable nets, or the use of "pingers" (devices that emit warning sounds) on gill nets may assist in mitigating bycatches. In some areas, knowledge of porpoise movements and habits has aided in setting fishing regulations designed to help protect the species.

The harbor porpoise is a high-trophic-level predator exposed to bio-accumulated pollutants. Some of the bio-accumulated pollutants (e.g., PCBs) suppress the immune system when they enter mammals in elevated concentrations. It can be assumed that chronic PCB exposure predisposes harbor porpoises to infectious disease mortality. Baltic Sea porpoises have about 250% higher levels of PCBs than North Sea porpoises.

North Sea harbor porpoises have a high burden of mercury, and it appears that mercury burden is associated with prevalence of parasitic infection and certain pathological diseases such as pneumonia. About 15% of stranded porpoises died from pneumonia.

The recent development of offshore windmill parks in shallow waters is a possible new threat that could displace harbor porpoises from their preferred habitat. These windmills emit sounds with yet unknown long-term effect on porpoises. The number of offshore windmill parks is currently growing in the North Sea region in an attempt to generate renewable, environment-friendly energy.

## See Also the Following Articles

Fisheries, Effects of Porpoises, Overview

## References

Akamatsu, T., *et al.* (2007). Comparison of echolocation behaviour between coastal and riverine porpoises. *Deep Sea Res. II* **54**, 290–297.

Andersen, L. W. (2003). Harbour porpoises (*Phocoena phocoena*) in the North Atlantic: Distribution and genetic population structure. *NAMMCO Sci. Pub.* **5**, 11–29.

Beineke, A., Siebert, U., Müller, G., and Baumgartner, W. (2007). Increased blood interleukin-10 mRNA levels in diseased free-ranging harbor porpoises (*Phocoena phocoena*). *Vet. Immunol. Immunopath.* **115**, 100–106.

Bjørge, A., and Donovan, G. P. (eds.) (1995). Biology of the phocoenids. *Rep. Int. Whal. Commn Spec. Iss.* **16**.

Fontaine, M. C., *et al.* (2007a). Rise of oceanographic barriers in continuous populations of a cetacean: The genetic structure of harbour porpoises in Old World waters. *BMC Biol.* **5**(30), doi:10.1186/1741-7007-5-30.

Fontaine, M. C., Tolley, K. A., Siebert, U., Gobert, S., Lepoint, G., Bouquegneau, J.-M., and Das, K. (2007b). Long-term feeding ecology and habitat use in harbour porpoises *Phocoena phocoena* from Scandinavian waters inferred from trace elements and stable isotopes. *BMC Ecol.* **7**(1), doi:10.1186/1472-6785-7-1.

Hammond, P. S., *et al.* (2002). Abundance of harbour porpoise and other cetaceans in the North Sea and adjacent waters. *J. Appl. Ecol.* **39**, 361–376.

Haug, T., Desportes, G., Vikingsson, G. A., and Witting, L. (eds) (2003). "Harbour Porpoises in the North Atlantic," Vol. 3. NAMMCO Scientific Publications.

Jepson, P. D., *et al.* (8 authors) (1999). Investigating potential associations between chronic exposure to polychlorinated biphenyls and infectious disease mortality in harbour porpoises from England and Wales. *Sci. Tot. Env.* **243–244**, 339–348.

Lahaye, V., *et al.* (13 authors) (2007). Biological and ecological factors related to trace element levels in harbour porpoises (*Phocoena phocoena*) from European waters. *Mar. Env. Res.* **64**, 247–266.

Lockyer, C. (2003). Harbour porpoises (*Phocoena phocoena*) in the North Atlantic: Biological parameters. *NAMMCO Sci. Pub.* **5**, 71–89.

McLellan, W. A., *et al.* (8 authors) (2002). Ontogenetic allometry and body composition of harbour porpoises (*Phocoena phocoena*, L.) from the western North Atlantic. *J. Zool.* **257**, 457–471.

Otani, S., Naito, Y., Kawamura, A., Kawasaki, M., Nishiwaki, S., and Kato, A. (1998). Diving behaviour and performance of harbor porpoises, *Phocoena phocoena*, in Funka Bay, Hokkaido, Japan. *Mar. Mamm. Sci.* **14**, 209–220.

Otani, S., Naito, Y., Kayo, A., and Kawamura, A. (2001). Oxygen consumption and swim speed of the harbor porpoise *Phocoena phocoena*. *Fish. Sci.* **67**, 894–898.

Read, A. J. (1999). Harbour porpoise *Phocoena phocoena* (Linneaus, 1758). *In* "Handbook of Marine Mammals" (S. H. Ridgway, and R. Harrison, eds), Vol. 6, pp. 323–355. Academic Press, London.

Read, A. J., and Hohn, A. A. (1995). Life in the fast lane: The life history of harbor porpoises from the Gulf of Maine. *Mar. Mamm. Sci.* **11**, 423–440.

Rosel, P. E., Haygood, M. G., and Perrin, W. F. (1995). Phylogenetic relationships among the true porpoises (Cetacea: Phocoenidae). *Mol. Phyl. Evol.* **4**, 463–474.

Tolley, K. A., and Heldal, H. E. (2002). Inferring ecological separation from regional differences of radioactive caesium in harbour porpoises (*Phocoena phocoena*). *Mar. Ecol. Prog. Ser.* **228**, 301–309.

Tolley, K. A., and Rosel, P. E. (2006). Population structure and historical demography of eastern North Atlantic harbour porpoises inferred through mtDNA sequences. *Mar. Ecol. Prog. Ser.* **327**, 297–308.

Westgate, A. J., and Tolley, K. A. (1999). Geographical variation in organochlorine contaminants in harbour porpoises (*Phocoena phocoena*) from the western North Atlantic. *Mar. Ecol. Prog. Ser.* **177**, 255–268.

Westgate, A. J., Read, A. J., Berggren, P., Koopman, H. N., and Gaskin, D. E. (1995). Diving behaviour of harbour porpoises, *Phocoena phocoena*. *Can. J. Fish. Aquat. Sci.* **52**, 1064–1073.

## Harbor Seal and Spotted Seal
### *Phoca vitulina* and *P. largha*

### JOHN J. BURNS

The harbor seal (*Phoca vitulina*) is also widely known as the common seal. It occurs over a great latitudinal range and in many different coastal and insular habitats around the rims of both the North Atlantic and North Pacific regions (King, 1983). Spotted seals (*P. largha*), in contrast, occur only in seasonally ice-covered seas of the Western Hemisphere (Burns, 1986). The name "larga seal" is sometimes used for the spotted seal and is derived from *largha*, which is part of the scientific name. These two sibling species are the most closely related members of the subfamily *Phocinae* and are fascinating examples of adaptations to vastly different environments. Most harbor seals occur in habitats that are sea ice free throughout the year, or at least where their coastal haulout and rookery sites are clear of sea ice during the breeding season. Spotted seals, conversely, utilize sea ice during the breeding season. In this context, it is important to distinguish between sea ice and fresh water icebergs calved from tidewater glaciers. Both species are of medium size. In some areas of the North Pacific their distributions overlap.

## I. Characteristics and Taxonomy
### A. Appearance

Based on external appearances, harbor and spotted seals older than weaned pups are not readily distinguishable from each other. Body size of spotted seals falls within the range of that for all but the largest harbor seals.

The pelage pattern and coloration of harbor seals is variable (Fig. 1). Background color ranges from yellowish or yellowish-gray (light phase) to blackish (dark phase). Light phase seals are usually paler on the flanks and belly than on the back, are covered with small black spots, and often show small pale rings, usually on the slightly darker dorsum. Dark phase harbor seals also have dark spots that are largely masked by the background coloration. Usually the dark seals show obvious light rings, especially on the dorsum. Seals of intermediate coloration are common. Throughout their broad range there are regions within which a particular pelage type predominates. Ungava seals (*P. v. mellonae*) are of the dark phase, as are most western Pacific harbor seals. Spotted seals are more uniform in color and pattern (Fig. 2). They tend to resemble light-phase harbor seals, which has contributed to the confusion about these two species.

### B. Size

The average length of harbor seals varies among populations. The smallest and largest seals occur in the North Pacific region and therefore they bracket the size of animals from other regions. Those from the northern Gulf of Alaska are the smallest. The average standard length and weight of adult males from that area is about 160 cm and 87 kg, while that of adult females is about 148 cm and 65 kg. Newborn pups average 82 cm and 10 kg (Pitcher and Calkins, 1979). The largest seals are from the Aleutian Islands and northern Japan. Length and weight of adult males ranges from 174 to 186 cm and 87 to 170 kg and

**Figure 1**   *Adult harbor seals* (Phoca vitulina) *on Año Nuevo Island, CA.*

**Figure 2**   *An adult female spotted seal* (Phoca largha) *(right) with her lanugo-clad pup (center, partially concealed), and an attending male (left), in the Bering Sea.*

that of adult females from 160 to 169 cm and 60 to 142 kg. Newborn pups were up to 98 cm and 19 kg (Naito and Nishiwaki, 1972; Burns and Gol'tsev, 1984).

Spotted seals are about the same size as most harbor seals and there are slight differences among populations. Adult males from the Bering Sea range from 161 to 176 cm and 85 to 110 kg. Adult females are 151 to 169 cm and 65 to 115 kg (Burns, 1986). Near-term fetuses and newborn pups from the Okhotsk Sea are 78 to 92 cm long and 7 to 12 kg (Trukhin, 2005). Healthy pups usually double and sometimes triple their birth weight during the 3- to 4-week nursing period.

## C. Diagnostic Characters

There are genetic (O'Corry-Crowe and Westlake, 1997), morphological, ecological, and behavioral differences between harbor and spotted seals. The breeding habitat of harbor seals is coastal and insular.

They give birth mainly on shore rookeries, although in some parts of Alaska they utilize icebergs calved from tidewater glaciers in protected fjords. Spotted seals use seasonal sea ice, mostly far from shore. During the breeding season, harbor seals occur in herds with no obvious social organization. Spotted seals occur as widely scattered adult pairs, usually with a pup (triads). In the areas where they occur together, harbor seals breed about 2 months later than spotted seals (reproductive separation). The pelage of newborn harbor seals is like that of adults because the lanugo is shed before birth (*in utero*). Occasionally, especially in the northern parts of their range, or in the case of premature pups, the lanugo is retained for up to a few days after birth. The pups usually enter the water shortly after birth, often within an hour. Spotted seal pups retain their whitish wooly lanugo, which is important for thermoregulation, for about 4 weeks. After the lanugo is shed the pelage resembles that of adult animals (Figs. 3 and 4). They remain on the ice during the nursing period and are abruptly weaned (abandoned).

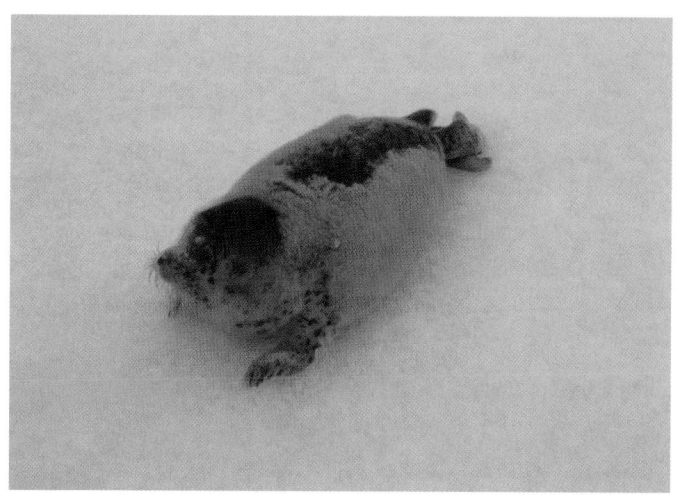

**Figure 3** *Partially molted, weaned, spotted seal pup* (Phoca largha).

**Figure 4** *Completely molted spotted seal pup* (Phoca largha). *Note the adult-like pelage.*

There are no individual cranial measurements that consistently separate harbor from spotted seals. As with body size, skull measurements are within the range of those in harbor seals. The ratios of some measurements are useful for differentiating between the two species (Chapskii, 1969; Shaughnessy and Fay, 1977, Burns *et al.*, 1984). Those ratios include jugal length/condylobasal length; nasal length from maxillo-frontal suture/condylobasl length; and interorbital width/mastoid width. Several non-metrical characters, used in combination, do permit differentiation. In harbor seals the skull is more massive, the bullae are more flattened and angular, the premolar teeth of adults are mostly obliquely set (straight in spotted seals), the posterior margin of the jugal bone is mostly angular (as opposed to rounded), the glenoid fossa is more flattened and angular, the bony process of the external auditory meatus is mostly straight and blade like (as opposed to mostly blunt and rounded), the shape of the posterior edge of the bony palate is mostly acute (as opposed to mostly rounded), and the hyoid arch is incomplete, having abbreviated stylohyals that are not attached to the bullae (as opposed to complete and attached in spotted seals). Unfortunately, none of these diagnostic characteristics are useful for differentiating live seals under field conditions.

Very experienced observers can distinguish between these two seals, even those with similar pelage, based on behavior when hauled out together on land, on general facial features of adults, and on behavior when frightened into the water.

## D. Nomenclature

There are five presently recognized subspecies of harbor seals: *P. v. vitulina* (Linnaeus, 1758); *P. v. concolor* (DeKay, 1842); *P. v. mellonae* (Doutt, 1942); *P. v. richardii*[1] (Grey, 1864); and *P. v. stejnegeri* (Allen, 1902). The spotted seal (Pallas, 1811) is considered to be a monotypic species. The different subspecies of harbor seal were originally recognized on the basis of geographical separation and skeletal morphology. Recent studies of their genetics sustain those conclusions. Boundaries between the eastern and western subspecies within both the North Atlantic and North Pacific Oceans are not known with certainty.

## II. Distribution and Movements

The distribution of harbor and spotted seals is shown in Fig. 5. Harbor seals occur over a latitudinal range from about 30°N to 80°N in the eastern Atlantic region and about 28°N to 62°N in the eastern Pacific region. They have the broadest distribution and occur in most different habitats of any other pinniped. Although the centers of abundance (greatest numbers of breeding animals) are in the northern temperate zone, breeding colonies of these seals occur north or south of that zone, depending on the presence of required environmental conditions created by regional oceanographic and climatic conditions. The high-latitude distribution in the Atlantic region is due to relatively warm oceanographic features, including the so-called North Water in Baffin Bay (eastern Canada–west Greenland) and the strong influence of warm water carried across the Atlantic to northern Europe by the Gulf Stream and associated gyres.

In the eastern Atlantic, *P. v. vitulina* normally occurs from the French coast bordering the English Channel, throughout the North Sea and northward to Finmark on the Barents Sea; including into the southern Baltic Sea and waters of Ireland and Great Britain. Stragglers occur to Portugal in the south and to the eastern Barents Sea in the northeast. The northernmost breeding population (here assumed to be *P.v. vitulina*) is in western Svalbard (Spitsbergen) at 78°30′N.

The boundary between *P. v. vitulina* of the eastern Atlantic and *P. v. concolor* of the western Atlantic is not known. However, harbor seals extend across the North Atlantic as a series of widely separated populations that occur at Svalbard, the Faeroe Islands (uncertain), Iceland, southern East Greenland, and West Greenland northward to about Upernavik (72°N). In Greenland the seals are considered to be *P.v. concolor*, as are those in most of eastern North America.

In the western Atlantic region the normal range of *P. v. concolor* extends from about 40°N (New Jersey) to about 73°N (northern Baffin Island, Canada); including into Hudson Bay and southern Foxe Basin. Stragglers have occurred as far south as Florida. The resident fresh water seal of the Ungava Peninsula in eastern Canada, *P. v. mellonae*, was first described and recognized as a separate subspecies by Doutt (1942), mainly on the basis of skull features and apparent isolation. It occurs in several drainage systems that empty into eastern Hudson Bay, where *P. v. concolor* is found. The subspecific designation of the

---

[1]Editorial protocol for this book requires the nomenclature of Rice (1998), which is *P.v. richardii*. The correct nomenclature, in my opinion, is *P.v. richardsi* in accordance with the explanations in Shaughnessy and Fay (1977). The person in whose honor this subspecies of seal was named was Capt. Richards, not Capt. Richard.

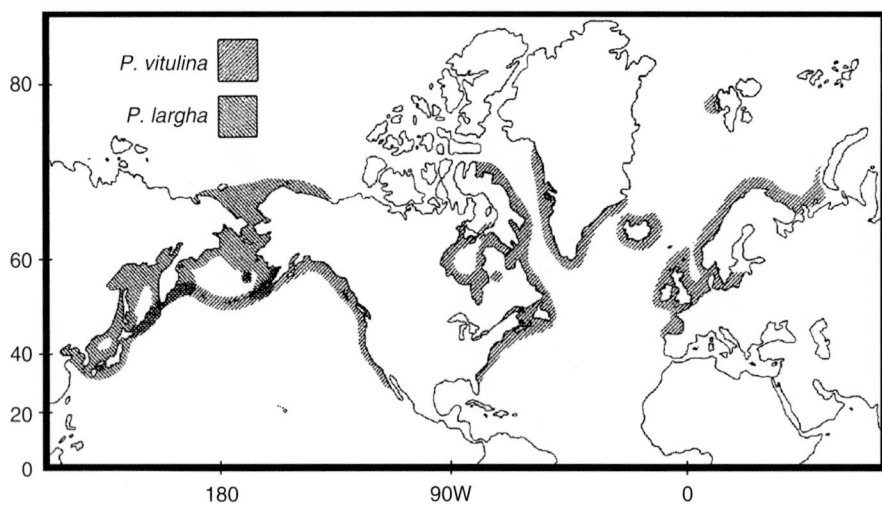

**Figure 5** *The distribution of harbor seals* (Phoca vitulina) *(five subspecies) and spotted seals* (Phoca largha).

Ungava seals was in doubt for several reasons, including their close proximity to salt water harbor seals; the fact that in general harbor seals occur frequently in rivers and lakes; and because the fresh water drainages in which it occurs flow into Hudson Bay. However, passage to and from salt water is blocked by numerous obstacles resulting from isostatic uplifting (rebound) of the peninsula since the last Pleistocene glaciation. The distinct status of this rare fresh water seal has been upheld on the basis of genetic differences (Smith, 1999).

In the North Pacific region the distribution of harbor seals extends from Cedros Island near the west-central coast of Baha California, Mexico (about 28°N), northward to the Gulf of Alaska and southeastern Bering Sea including the Pribilof Islands, across the entire Aleutian Ridge (the Aleutian and Commander islands) to the Kamchatka Peninsula of eastern Russia, southward in the Kuril Islands and beyond to Hokkaido Island in northern Japan. *P. v. richardii* is the subspecies of the eastern North Pacific region and *P. v. stejnegeri* occurs in the western Pacific. The boundary between these two subspecies is currently thought to be in the western Alaska Peninsula–eastern Aleutian Islands, although uncertainty about that question still exists. The northernmost pupping colonies in the Pacific region are in Prince William Sound, Alaska, at about 61°13′N. That is some 1920 km farther south than the northernmost breeding group in the Atlantic region.

Great distances separate the Atlantic and Pacific forms. There are no breeding colonies between Baffin Island in northeastern Canada and the Pribilof Islands of southeastern Bering Sea, nor between northern Norway and the Pribilof Islands.

Seasonal and annual movements of harbor seals are quite varied depending on the environments in which they occur. They are usually considered to be relatively sedentary, with a high degree of fidelity to one or a few haulout sites. This view, although perhaps applicable to some populations, is a gross oversimplification. It is now recognized that they move, in some cases quite extensively. Generalizations are inappropriate in view of this seals' wide distribution and differences in stock sizes, population dynamics, and the varied environments they occupy. In most instances, some individuals are likely more sedentary and show stronger site fidelity than others. Kinds of movement include migrations, juvenile dispersal, seasonal shifts, shifts related to breeding activity, responses to seasonal habitat exclusion, responses to acute or chronic disturbance, and immigration/emigration, occasionally on a relatively large scale.

The spotted seal was, until recently, considered to be a subspecies of harbor seal. It is now recognized as a distinct species (Shaughnessy and Fay, 1977), that includes several widely separated breeding populations. The centers of abundance during the breeding season are mainly in the temperate/subarctic boundary regions. The seal is well adapted to exploit the "front" and broken ice zones of seasonal sea ice that overlies continental shelves during winter and spring. Spotted seals resort to haulouts on land during ice-free seasons of the year. There are great seasonal expansions and contractions of range, commensurate with the annual cycle of sea ice advance and retreat. Their distribution in all areas is most restricted during the period of maximum ice cover. They occur in the Bering, Chukchi (in summer), Beaufort (in summer), and Okhotsk seas, Tartar Strait, the Sea of Japan, and the northern Yellow Sea/Bo Hai (Bohai Sea), and adjacent embayments that border China, and in summer the eastern Korean Peninsula. The most southern breeding populations (about 38°N) are in the Sea of Japan and the Yellow Sea. Their occurrence at these southern latitudes is because of a cold winter climate, dominated by the so-called Siberian High Pressure system that results in a limited sea ice cover during mid-winter.

In all areas, as the seasonal ice cover recedes and disintegrates, spotted seals expand their range and haulout on land. Some animals of the population that winters and pups on ice in the Bering Sea migrate northward into the Chukchi and Beaufort seas during the ice-free months. Their summer-early autumn distribution extends as far north as 71°30′N near Point Barrow, Alaska, and to about 70°N on the northern shores of Chukotka, Russia. Thus, the total range of the Bering Sea population extends over 15° of latitude or about 1665 km.

Spotted and harbor seals are sympatric (have overlapping ranges) in the southeastern and southwestern Bering Sea, on the Kamchatka Peninsula, in the Kuril Islands, and northern Japan. Similarities in general appearance and occurrence on land (sometimes in close proximity) have long contributed to the confusion about these two different species.

## A. Abundance

Population sizes of harbor and spotted seals have fluctuated due to both natural and anthropogenic causes, including hunting, incidental taking, competition with commercial fishers for food, habitat alteration, disturbance, protective measures, diseases, climate regime shifts, and other factors. Some populations are small and isolated; persisting in what may be marginal habitat. These may be the ones most vulnerable to changes in environmental factors and to direct exploitation. In general, direct exploitation has now been reduced greatly. Most populations are currently protected from hunting except under terms of special licenses or in areas where they are taken by indigenous peoples for subsistence purposes. Population estimates for some regions are fragmentary and in several cases outdated. For others, they are compilations of surveys in different subregions, on changing populations, sometimes several years apart. Nonetheless, they provide useful indications of regional abundance, and further illustrate this seals' coastal and insular distribution, primarily in the northern temperate zone.

In the middle 1980s there were perhaps 98,000 eastern Atlantic harbor seals (*P. v. vitulina*) (Reijnders *et al.*, 1993). By then populations had recovered after prolonged and sometimes intensive hunting and control programs. The largest numbers were and still are around the rim of the North Sea and Iceland. Areas of greatest abundance were in Great Britain (up to 47,000), Iceland (28,000), the Wadden Sea (10,000), and Kattegat/Skagerrak (6000). The smallest known populations are in the Baltic Sea (perhaps 200) and around Svalbard (500–600 in 1990). In 1988 a large proportion of some populations died from a viral epidemic: up to 48% in parts of southeastern Great Britain and an estimated 60% in the Wadden Sea and Kattegat/Skagerrak (Dietz *et al.*, 1989). These affected populations recovered rapidly and by 1992 there were an estimated 7250 seals in the Wadden Sea and perhaps 5200 in Kattegat/Skagerrack (Reijnders, *et al.*, 1993).

The current number of all western Atlantic harbor seals (*P. v. concolor*) is not known with certainty, but may be around 115,000. Data for Canadian waters are fragmentary and based largely on information from Sable Island, reported in Boulva and McLaren (1979). That population has declined. Baird (2001) summarized available information and suggested that perhaps 12,700 harbor seals, not including those of the Ungava Peninsula, may be in Canadian waters. In parts of eastern Canada, as well as in Greenland these seals are harvested for meat and for their beautiful hides, which are made into traditional clothing and other articles of Native handicraft (Teilmann and Dietz, 1994). About 100,000 seals occur in US waters (Waring *et al.*, 2007). Almost all are in Maine. That population has been increasing and, in 2001, was estimated to be about 99,300 (Gilbert *et al.*, 2005). There is no population estimate for Greenland. They were never abundant, and have apparently undergone a long-term decline. In the past they were more widely distributed, mainly in West Greenland. Fewer than 40 a year were harvested in Greenland in the late 1980s (Teilmann and Dietz, 1994).

The number of fresh water seals of the Ungava Peninsula is low and the estimate of 120 to perhaps 600 animals is uncertain (Reijnders *et al.*, 1993). The actual number is probably closer to the lower value. This includes the entire subspecies. Ungava seals are considered to be possibly endangered, vulnerable, and rare. That designation is based on a lack of information, a very limited range, low numbers, and potential threats from proposed development in the region. This subspecies may well be a relict, persisting in habitat that has been altered drastically and unfavorably by the natural geological processes of post-glacial land rebound (uplift) during Holocene and Recent times (Smith, 1999).

In the eastern Pacific region, harbor seals are abundant although as elsewhere numbers have fluctuated greatly. There is no estimate for the number in Mexican waters. In waters from California to Alaska the total may approach 350,000 of which 74,000 occur from California to Washington State (Carretta *et al.*, 2007), 75,000–88,000 in British Columbia in 1988 (Olesiuk *et al.*, 1990; Baird, 2001), and about 180,000 in Alaska (Angliss and Outlaw, 2007). The reported breakdown of those estimates for US waters is: 34,283 in California in 2005; 24,732 in Oregon and coastal Washington in 2003; 14,612 in inland waters of Washington in 2003 (Jeffries *et al.*, 2003); 112,391 in southeastern Alaska in 2006; 45,000 in the Gulf of Alaska including the Aleutian Islands in 2006; and 20,109 in southeastern Bering Sea in 2006. From California to southeastern Alaska they have increased over many years. In recent years the rate of increase has slowed or stopped, perhaps as these populations approach or reach an equilibrium level. In the Gulf of Alaska region the trend has been the opposite of that farther south, with an 85% decline between 1976 and 1988. These opposite trends in the southern and northern parts of their range are apparently mainly responses to a major shift in Pacific climate that began about 1976 (Ebbesmeyer *et al.*, 1991).

The range of the western Pacific harbor seal extends across the Aleutian Ridge to Asia. These seals are predominately of the dark color phase, they tend to occur in very small groups (as opposed to large aggregations), and they mainly occupy rocky islands and shorelines. Regional estimates of numbers are: Aleutian Islands, about 3400 in 1994 (Withrow and Loughlin, 1995); Commander Islands, 1500; Kamchatka Peninsula, 200; Kuril Islands, 1900; and northern Japan, 300. Estimates for the latter four areas are from the early 1990s (Reijnders *et al.*, 1993). They are classified as rare in Japanese and Russian waters, and are now protected from hunting.

There are no reliable estimates of the present size of spotted seal populations, except perhaps in the Bohai Sea. They are, however, common within all parts of their known normal range in the Okhotsk, Bering, and Chukchi seas. Various estimates, all of which have high levels of uncertainty, are noted by Trukhin (2005), including that of 290,000 in the 1990s, inclusive of all populations combined. Indirect and anecdotal information from Native subsistence hunters suggests that over the past two decades spotted seals of the Bering Sea may have slowly declined. If the perceived decline is real, it may be a response to changed food-web dynamics, and a greater frequency of less favorable sea ice conditions during late winter/early spring. This would be in accord with the numerical trend for harbor seals in the Gulf of Alaska and parts of the southeastern Bering Sea. Spotted seals are an important resource to American-Native subsistence hunters in coastal areas of the Bering and Chukchi seas, where an estimated 5265 a year may be taken (Angliss and Outlaw, 2007). There is no confidence interval associated with that estimate and it is questionably high. There is essentially no subsistence hunting in the Okhotsk Sea. Ship-based commercial hunting in both the Okhotsk and Bering seas was reduced during the late 1980s and finally ceased in 1995. The size of populations in the Yellow Sea, and the Sea of Japan, noted in Trukhin (2003), were 4000–4500 and perhaps up to 8000+, respectively. According to Dong and Shen (1991), in the Bohai Sea, including Liaodong Bay, there were an estimated 4000–4500 in 1990; the estimate subsequently included in Trukhin (2005). This compares with estimates of >7000 in the 1930s, >8100 in 1940, and 2269 in 1979, after a period of intensive harvesting. There is a report of as few as 1000 at the present time (Bo, 2006), although no substantiating information was included. In China, these seals were accorded protection from hunting in the 1980s.

## III. Ecology
### A. Foods and Feeding

Feeding forays of harbor seals can be close to haulout sites, or many miles distant, either along the coast (including rivers) or seaward. They are capable of feeding at considerable depths (to 500+ m) and are generalists that prey mainly on abundant and easily available foods, with diets varying by season and region. There are long-term changes in foods that are associated with environmental changes, and therefore dynamic changes in the abundance of different prey species. Primary food items are small- to medium-size fishes (or age classes), such as various members of the codfish family, hake, mackerel, herring (*Clupea* spp.), sardines, smelts, shad (*Alosa* spp.), capelin (*Mallotus villosus*), sand lance, sculpins, a variety of flatfishes, salmonids, and many others. Their propensity for cod, salmons, and other commercially important species has resulted in long-standing conflicts with fishermen in many areas. Cephalopods (squid and octopus) are usually reported as being next important after fishes, followed by crustaceans including mainly shrimps and crabs. Several studies have reported that shrimp may be particularly important to recently weaned pups.

Although there is great diversity in foods, a few items usually comprise the majority of seasonal diets in an area. As examples, in Atlantic Canada, 23 different food items were identified but 4 accounted for 84% of the estimated biomass of prey consumed (Boulva and McLaren, 1979). In the Gulf of Alaska, fishes comprised 73.8% of the diet and 27 different species were eaten. The four most important foods were walleye pollock (*Theragra chalcogramma*), cephalopods, capelin, and flatfishes (Pitcher, 1980). In the western Aleutian Islands the main food items, at least in 1958 and 1962, were Atka mackerel (*Pleurogrammus monopterygius*) and octopus (Kenyon, 1965). The main foods of the Ungava seals are thought to be resident brook (*Salvelinus fontinalis*) and lake (*S. namaycush*) trout Smith, 1999). Seals in Lake Iliamna, Alaska, feed on the variety of salmonids [charr (*Salvelinus* spp.), trout, and salmon] that occur there in large numbers (traditional local knowledge).

The food habits of spotted seals are noted in several publications including Tikhomirov (1966); Lowry and Frost (1981); and Trukhin (2005). These seals are also generalist feeders, although in continental shelf waters, they primarily utilize similar types of abundant fishes, crustaceans, and cephalopods. Because they have a pelagic distribution in winter–spring and a different coastal and pelagic distribution during ice-free months, there are major seasonal and regional differences in food habits. Additionally, there are age-related differences. Most reports about food habits are based on seals examined during spring (mainly April and May) when they are associated with sea ice. A few samples are from animals collected in the coastal zone during autumn, and there are anecdotal observations of summer feeding, especially in areas where subsistence and commercial fishing activities occur. There are few data from the late autumn and winter months, although in the Bering and Okhotsk seas these seals occur where pollock, herring, eelpout, flounders, shrimp, and crabs are abundant.

Independent feeding by spotted seal pups begins after they are abruptly weaned. During the time of fasting and early independent feeding they live on their accumulated fat reserves and loose between 18% and 25% (sometimes up to 30%) of their weight (Burns, 1986). The first food consumed is frequently small amphipods or euphausiids. Abundant schooling fishes are the main foods of older seals and, in the Okhotsk Sea, occurred in 89% of seals 1–4 years old and 70% of seals >5 years old. Cephalopods were next in importance, followed by decapods. Amphipods were still consumed by the 1- to 4-year olds but were not found in older animals. The frequency of occurrence of cephalopods was higher in older age animals. Spotted seals were reported to feed more in the morning and evening than at other times of the day. During spring the main food items in the Bering Sea, depending on the region, were pollock,

arctic cod (*Boreogadus saida*), sand lance, and capelin. In the Okhotsk Sea, pollock were most important. In Peter the Great Bay (Sea of Japan) the dominant fishes were saffron cod (*Eleginus gracilis*), flounders, and rockfish, and in Tartar Strait (between mainland Russia and Sakhalin Island) they were saffron cod, flounders, and salmon. In all areas, crustaceans and cephalopods were also important. There has been little sampling in coastal habitats during summer when anadromous and coastal spawning fishes such as charr, salmon, capelin, smelt, herring, flounders, saffron cod, and other species are abundant. According to traditional local knowledge, those foods are utilized intensively by the seals.

### B. Haulouts

As already noted, harbor and spotted seals are superficially quite similar in appearance. Harbor seals haul out mainly on land, although in some areas of mainland Alaska they use icebergs calved from tidewater glaciers. Also, in the northern parts of their range, where labile sea ice occurs to or very near shore, they haul out on it until the land sites are accessible, usually long before the pupping season. They use haulouts throughout the year, although most frequently and in greatest numbers during the pupping and molting seasons. Regardless of season, haulout activity is strongly affected by the stage of the tide, air temperature, wind speed, precipitation, and time of day. They lie close to the water when hauled out, and usually flee when disturbed. Habituation is not uncommon where regular and continuous, non-threatening, human activities occur, and they are not unduly harassed. The substrate at natural haulouts on land is diverse and includes mud flats, sand and gravel bars and beaches, rocks, glacial icebergs, and occasionally sea ice. Depending on the region, haulouts can be on lakes, rivers, estuaries, bays, ocean shorelines, islands, islets, ledges, and any other setting where the seals can rest, undisturbed, with immediate access to deep water. They may, on occasion, haul out on man-made structures such as docks, floats, and log rafts.

Spotted seals use sea ice starting with its formation in autumn (Burns, 1970). They often concentrate in large numbers on the early ice that forms near river mouths and estuaries (fresh water freezes at a higher temperature than seawater) and feed on autumn spawning fishes. As the ice thickens, becomes attached to land, and extends farther from shore, spotted seals move seaward into the drifting ice. Their association is mainly with the highly labile marginal areas and they move (southward in the Chukchi/Bering sea region) to maintain an association with that habitat. During the cold weather of winter they rarely haul out. Peak haulout on the ice is during the pupping and molting season.

As the sea ice cover retreats and disintegrates in late spring–early summer, spotted seals again move shoreward and, in the Bering Sea, northward. Again, large aggregations can often be seen close to shore on the last remnants of former shore-fast ice and on ice flushed from rivers and estuaries. At this time of year they feed extensively on the dense schools of spawning herring and smelt. They haul out on shore when the ice is gone. Between haulout bouts on land, some seals travel long distances in the open sea, even between Alaska and Siberia, and use multiple haulouts (Lowry *et al.*, 1998). Shore haulouts are mostly on isolated mud, sand, or gravel beaches, or on rocks close to shore. They are often on river bars, tidal flats, and barrier islands. Spotted seals are especially vigilant on land, where they are subjected to attack from a variety of predators. Their association with sea ice starts again as soon as it begins to form in the autumn.

## IV. Behavior and Physiology

### A. Mating System

In both harbor and spotted seals, mating occurs in the water at about the time that pups are weaned, although females mating for the first time or that have not given birth in a specific year may breed outside of the peak period of the post-parturient animals. There is intermale competition for females.

Harbor seal males use vocal behavior and display dives, within aquatic territories, for mate attraction and for male–male competition (Hayes *et al.*, 2004), that often involves vigorous fighting (Burns and Gol'tsev, 1984). The territories are near haulouts or along female traffic corridors, to maximize their exposure to estrous females, and they are polygynous. Thus, the locations are often fairly fixed over time. Males exhibit interannual fidelity to their acoustic display stations, which are within their larger territories (Hayes *et al.*, 2004).

The mating system of spotted seals is quite different than that of harbor seals (Burns, 1970, 1986). They begin to form pairs prior to or early in the pupping season, before females obtain estrus. They are considered to be annually monogamous and territorial. Triads consisting of a female, her pup, and an attending male can be seen on the ice, or with the attending male close by (Fig. 6). These triads are widely spaced, although there are regions of high abundance. In an environment of drifting and shifting ice floes the spacing suggests that aquatic vocal behavior by males is important, both to initially attract a mate and to maintain a shifting territory around her. Interestingly, adult male spotted seals do not have the assorted types of aggression wounds that are common on harbor seals during the mating season. Females attend their pups on the ice, especially during the early nursing period, and the males stay with the females. Pairs that include an adult female that did not pup are also formed. In the Bering and Okhotsk Seas, such pairs are seldom seen on the ice in early April (prior to the molt in adults), probably because there is no pup for the female to attend.

### B. Molt

In harbor seals, the molt generally occurs during mid-summer to early autumn, within 2 or 3 months of the pupping season (Bigg, 1981). During the molt, seals haul out more frequently than at any other time of the year except for the pupping season. There are differences in timing among age and sex cohorts. Usually yearlings begin and end the molt earliest, followed by subadults, then adult females, and last, adult males. There is overlap among these general age groups. Throughout their extensive range the molt occurs after cessation of the breeding season. Accordingly, it occurs later in the year in the late breeding populations such as those in Europe, British Columbia, and Puget Sound.

Spotted seals of the Okhotsk and Bering seas molt mainly in late spring (Burns, 1970; Trukhin, 2005). Pups, as mentioned, have the color and pelage pattern of adults after their lanugo is shed. Older seals begin the molt immediately after the breeding season and show an overlapping age-related sequence similar to that of harbor seals. The period of intensive molt is during May and June, during which time the sea ice is retreating rapidly and deteriorating. In areas where the ice disappears early, or in minimal ice years, the molt is completed on shore haulouts and at sea.

## V. Life History

### A. Mortality Factors

Seal control programs, commercial hunting (now discontinued), large-scale die-offs due to epizootic diseases, and natural long-term population changes are known to occur in harbor seals. As examples, in late 1979 and 1980 an estimated 500 seals died along the New England coast, from an influenza virus of avian origin. Another less severe disease-caused die-off occurred in the same area in 1982 (Geraci *et al.*, 1982). The largest known incident of mass deaths occurred in northern Europe, during 1988–1989, when an estimated

**Figure 6** *A triad of spotted seals* (Phoca largha) *during breeding season, in Bering Sea: lanugo clad pup (right), the mother (center), and attending adult male (left, in the water).*

18,000 harbor seals died due to a viral infection that rapidly spread among some colonies in the North Sea region. In all areas the populations had previously reached high levels after cessation of control programs (Dietz *et al.*, 1989). In the eastern North Pacific, south of the northern Gulf of Alaska, there has been a sustained long-term increase in numbers. Farther north they declined about 85% between 1976 and 1988 (Baird, 2001; Angliss and Outlaw, 2007). These changes in the eastern Pacific are probably mainly related to natural large-scale ocean regime shifts now known to have occurred. Nothing is known about natural fluctuations in spotted seals, although it is probable that they have also been affected by climate change and therefore changes in carrying capacity of their more remote environment.

The reported predators of harbor seals include killer whales (*Orcinus orca*), sharks, Steller sea lions (*Eumetopias jubatus*), eagles, ravens (*Corvus* spp.), and gulls. Spotted seals are preyed on by those same animals and also by walruses (*Odobenus rosmarus*), polar bears (*Ursus maritimus*), and arctic foxes (*Alopex lagopus*), as well as by shore-based predators such as brown bears (*Ursus arctos*), wolves (*Canis lupus*), wolverines (*Gulo gulo*), and red foxes (*Vulpes vulpes*). Indigenous peoples still harvest small numbers of both species of seals for subsistence uses. They are also occasionally taken incidentally in the course of other activities, particularly in commercial gill net fisheries for salmon.

### B. Reproduction

In both harbor and spotted seals, mating occurs at about the time that pups are weaned (Bigg, 1969; Boulva and McLaren, 1979; Burns, 1986). As with all other pinnipeds, fertilization is followed by a prolonged period of delayed implantation (embryonic diapause) that lasts up to about 2.5 months, after which the embryo implants and resumes development. The total gestation period, from fertilization to birth, is about 10.5 months. In most populations, pregnancy rates exceed 85%; in other words, most sexually mature females bear a pup every year.

In general, female harbor seals reach sexual maturity at ages 3–4 years, and physical maturity by age 6 or 7 years. Males obtain sexual maturity at 4–5 years and physical maturity by 7–9 years. The maximum life span is around 35 years, although few animals live that long in the wild. They are reproductively active throughout their lives. All harbor seal populations have a similar reproductive cycle. However, over their very broad range the specific timing of events varies. Depending on the region in question, births occur in late winter to summer. Within a specific population the peak of pupping can shift slightly over time, apparently in response to significant environmental change. Additionally, there is some interannual variability. In general, the pupping season extends over a period of up to about 10 weeks, within which there is about a 2-week peak. Females bear a single pup although twinning has been recorded. In most regions, pups are born on land, usually between the high- and low-tide water lines. In some parts of Alaska pups are born on floating icebergs calved from tidewater glaciers in protected fjords.

Newborn pups can and do enter the water, often being forced to do so by tidal inundation of birth sites or because of disturbance by birds scavenging afterbirth. Mother–pup bonding is a critical phase of behavior within the first hour of birth, as mutual recognition is required to locate and/or remain with each other on rookeries and in the water. Young pups often cling to their mothers' backs in the water. Mothers feed during the approximately 4-week nursing period (some reports indicate as long as 6 weeks). Pups start to catch their

own food during the late stages of the nursing period, and some maintain an association with their mothers after weaning.

To put timing of the generalized reproductive cycle into a regional context, the peak period of pupping can be used as the benchmark event. For harbor seals of the European coast, most pupping occurs during late June and early July. In most of eastern Canada and Greenland, births are mainly during mid-May to mid-June, slightly later at higher latitudes (Boulva and McLaren, 1979; Gjertz and Børset, 1992). However, the Ungava seals reportedly pup during late April or early May (Smith, 1999). There are considerable differences among populations of the Pacific region (Temte *et al.*, 1991). Births occur during early February in Mexican waters; in March–April in southern California; in May along the outer Washington coast; between late June and September in Puget Sound and southern British Columbia; during May to late June in northern British Columbia, most of Alaska and Japan; and early June to late July (peak around July 1) in the Aleutian, Pribilof, and Kuril islands.

Spotted seals have the same basic reproductive cycle as harbor seals, although timing of events is directly related to the most favorable sea ice conditions at the time of birth through weaning (Burns, 1970). Those events have evolved to coincide with the average period of greatest extent and stability of seasonal ice (births), and the subsequent onset of its seasonal disintegration (weaning). The timing of these conditions varies by region.

Pups are born exposed on the ice and, during the first 2 or 3 weeks, are more like land mammals. They spend most of the time on ice floes, without benefit of snow lairs, until weaned (*cf.* Fig. 2). Their only protection from wind is that provided by their mothers or the shelter of ice ridges. The exposed and relatively immobile pups are not subjected to significant predation by polar bears or arctic foxes because the labile marginal ice in which they occur during spring is well south of the normal range of those predators. Polar bears and arctic foxes do not occur in the Okhotsk Sea or farther south.

Unseasonably early destruction and disintegration of the ice front zone, caused occasionally by severe southerly storms, probably results in a high mortality of nursing pups. During early life the dense coat of lanugo provides the required insulation for maintaining body heat, although that important function is assumed by the rapidly increasing blubber layer acquired during the 4-week nursing period. At weaning, most pups are heavier than at any other time during their first year of life. They are so fat and buoyant that they are poor divers. This large energy store provides sustenance during the early stages of adjustment to independent life.

Weaning, which is abrupt, coincides with the normal seasonal onset of ameliorating spring weather and disintegration of the seasonal ice cover. The use of sea ice as a platform on which to bear and nurture pups is central to the ecology of spotted seals. These events (birth, dependence during the nursing period, weaning, and early independence) are more restricted in time than is the case with harbor seals. Pups are born earlier in the more southerly parts of this species' range. In the Yellow Sea the peak period is during late January (Dong and Shen, 1991); in the Sea of Japan it occurs during February and March, and in both the Okhotsk and Bering seas the peak is during the first half of April. Mothers feed during the nursing period, although the pups remain on the ice, sometimes wriggling over brash ice to move between closely adjacent ice floes.

Pupping on land has been reported to occur on shores of the Bohai Sea and occasionally elsewhere. It occurs during years when, or in regions where, suitable sea ice is insufficient or absent (Trukhin, 2005; Bo, 2006). It may become more frequent as a result of global warming. Spotted seal pups born and nurtured on land might sustain a higher incidence of mortality, due to their relatively long period of dependence, and slow development of aquatic capabilities.

## VI. Interactions with Humans

### A. Conservation Concerns

There are similar conservation concerns relevant to both harbor and spotted seals. The general problem of pollution from military, agricultural, and/or industrial activities (including coastal and off-shore oil and gas development) is of particular concern because of its direct and indirect effects on seals and the foods they eat. Oil spills are and will continue to be a chronic problem. Major disease outbreaks may have been intensified because of suppressed immune responses caused by a variety of pollutants. The role of increased density dependent disease-caused mortality is not understood, but it may become more evident in populations that achieve or exceed the carrying capacity of available habitat. Hunting may still be an important factor in limiting or reducing some of the small breeding populations of harbor seals in Greenland and northeastern Canada. Fishing activities can affect both species adversely by causing incidental mortalities, and by competing for fish the seals depend upon for food. Fishery interactions are probably limiting any increase of the small populations of harbor seals in northern Japan and parts of Greenland. Development projects can alter or eliminate important habitat or displace seals by increased disturbance near haulouts. This would be most likely for spotted seals, as there are no known instances of habituation. Disturbance at shore haulouts, together with a low level of illegal hunting, are considered to be problems for small population of spotted seals in the Bohai Sea. They use summer haulouts in increasingly industrialized coastal areas of China and the eastern Korean Peninsula. The small relict population of Ungava seals may be particularly vulnerable to proposed hydroelectric projects within their limited range.

### B. Climate Change

Climate change, specifically warming, will have major impacts on harbor and spotted seals. The contentious aspect of that important issue is the extent to which natural change is being exacerbated by anthropogenic effects. Climate has changed many times in the past and has been an important force affecting zoogeography, population fluctuations, extirpation, and extinction. Global warming is definitely causing later formation and earlier breakup, and reducing the extent and thickness of seasonal sea ice. It might well result in an increase of suitable habitat for harbor seals in the north, and an overall decrease of spotted seal habitat, especially in the southern parts of its range.

## See Also the Following Articles

Earless Seals ■ Mass Die-offs ■ Migration and Movement Patterns ■ Pinniped Ecology ■ Skull Anatomy

## References

Angliss, R. P., and Outlaw, R. W. (2007). "Alaska Marine Mammal Stock Assessments, 2006." US Department of Commerce, NOAA Technical Memorandum, NMFS-AFSC-168.

Baird, R. W. (2001). Status of harbour seals, *Phoca vitulina*, in Canada. *Can. Field-Natur.* **115**, 663–675.

Bigg, M. A. (1969). The harbour seal in British Columbia. *In* "Bulletin of the Fisheries Research Board of Canada." Vol. 172, Ottawa, Canada.

Bigg, M. A. (1981). Harbour seal *Phoca vitulina* Linnaeus, 1758 and *Phoca largha* Pallas, 1811. *In* "Handbook of Marine Mammals" (S. H. Ridgway, and R. J. Harrison, eds), Vol. 2, pp. 1–27. Academic Press, London.

Bo, W. (2006). China's spotted seals face increasing threats. *Pacific Environment*, http://www.pacificenvironment.org/article.php?id = 1211.

Boulva, J., and McLaren, J. A. (1979). Biology of the harbor seal, *Phoca vitulina*, in eastern Canada. *In* "Bulletin of the Fisheries Research Board of Canada." Vol. 200, Ottawa, Canada.

Burns, J. J. (1970). Remarks on the distribution and natural history of pagophilic pinnipeds in the Bering and Chukchi seas. *J. Mammal.* **51**, 445–454.

Burns, J. J. (1986). Ice seals. *In* "Marine Mammals of Eastern North Pacific and Arctic Waters" (D. Haley, ed.), 2nd Ed., pp. 216–229. Pacific Search Press, Seattle.

Burns, J. J., and Gol'tsev, V. N. (1984). Comparative biology of harbor seals, *Phoca vitulina* Linnaeus, 1758, of the Commander, Aleutian and Pribilof islands. *In* "Soviet-American Cooperative Research on Marine Mammals." (F. H. Fay, and G. A. Fedoseev, eds.), Vol. 1, pp. 17–24. NOAA Technical Report NMFS 12.

Burns, J. J., Fay, F. H., and Fedoseev, G. A. (1984). Craniological analysis of harbor and spotted seals of the North Pacific region. *In* "Soviet-American Cooperative Research on Marine Mammals." (F. H. Fay, and G. A. Fedoseev, eds.), Vol. 1, pp. 5–16. NOAA Technical Report NMFS 12.

Carretta, J. V., Forney, K. A., Muto, M. M., Barlow, J., Baker, J., Hanson, B., and Lowry, S. (2007). "U.S. Pacific Marine Mammal Stock Assessments: 2006." US Department of Commerce, NOAA Technical Memorandum, NMFS-SWFSC-398.

Chapskii, K. K. (1969). Taxonomy of seals of the genus *Phoca, sensu stricto* in light of contemporary craniological data. *In* "Marine Mammals" (V. A. Arsen'ev, B. A. Zenkovich, and K. K. Chapskii, eds), pp. 294–304. Science Publisher, Moscow [In Russian. English translation Can. Dept. Secr. State, Transl. Bur., 1972].

Dietz, R., Heide-Jørgensen, M. P., and Härkönen, T. (1989). Mass deaths of harbor seals (*Phoca vitulina*) in Europe. *Ambio* **18**, 258–264.

Dong, J., and Shen, F. (1991). Estimates of historical population size of harbour seal in Liaodong Gulf. *Mar. Sci.* **3**, 40–45.

Doutt, J. K. (1942). A review of the genus *Phoca*. *Ann. Carnegie Mus.* **29**, 61–125.

Ebbesmeyer, C. C., Cayan, D. R., McLain, D. R., Nichols, F. H., Peterson, D. H., and Redmond, K. T. (1991). 1976 step in the Pacific climate: Forty environmental changes between 1968–1975 and 1977–1985. *In* "Proceedings Seventh Annual Pacific Climate Workshop," pp. 115–126. California Department of Water Research, Asilomar, CA.

Fay, F. H. (1974). The role of ice in the ecology of marine mammals of the Bering Sea. *In* "Oceanography of the Bering Sea" (D. W. Hood, and E. J. Kelley, eds), pp. 383–389. University of Alaska Fairbanks, Institute of Marine Science, Occas. Publ. 2.

Geraci, J. R., *et al.* (12 authors) (1982). Mass mortality of harbour seals: Pneumonia associated with influenza A virus. *Science* **215**, 1129–1131.

Gilbert, J. R., Waring, G. T., Wynne, K. M., and Guldager, N. (2005). Changes in abundance of harbor seals in Maine, 1981–2001. *Mar. Mamm. Sci.* **21**, 519–535.

Gjertz, I., and Børset, A. (1992). Pupping in the most northerly harbor seal (*Phoca vitulina*). *Mar. Mamm. Sci.* **8**, 103–109.

Hayes, S. A., Costa, D. P., Harvey, J. T., and LeBouef, B. J. (2004). Aquatic mating strategies of the male Pacific harbor seal (*Phoca vitulina richardii*): Are males defending the hotspot? *Mar. Mamm. Sci.* **20**, 639–656.

Heide-Jørgensen, M. P., and Härkönen, T. J. (1988). Rebuilding seal stocks in the Kattegat-Skagerrak. *Mar. Mamm. Sci.* **4**, 231–246.

H

Hoover, A. A. (1988). Harbor seal, *Phoca vitulina*. *In* "Selected Marine Mammals of Alaska: Species Accounts with Research and Management Recommendations" (J. W. Lentfer, ed.), pp. 125–187. Marine Mammal Commission, Washington, DC.

Jeffries, S., Huber, H., Calambokidis, J., and Laake, J. (2003). Trends and status of harbor seals in Washington State: 1978–1999. *J. Wildl. Manage* **67**, 208–219.

Kenyon, K. W. (1965). Food of harbor seals at Amchitka Island, Alaska. *J. Mammal.* **46**, 103–104.

King, J. E. (1983). "Seals of the World," 2nd Ed. Comstock Publishing Associates, Ithaca.

Lowry, L. F., and Frost, K. J. (1981). Feeding and trophic relationships of phocid seals and walruses in the eastern Bering Sea. *In* "The Eastern Bering Sea Shelf: Oceanography and Resources" (D. W. Hood, and J. A. Calder, eds), Vol. 2. University of Washington Press, Seattle.

Lowry, L. F., Frost, K. J., Davis, R., DeMaster, D. P., and Suydam, R. S. (1998). Movements and behavior of satellite-tagged spotted seals (*Phoca largha*) in the Bering and Chukchi seas. *Pol. Biol.* **19**, 221–230.

Lowry, L. F., *et al.* (8 authors) (2000). Habitat use and habitat selection by spotted seals (*Phoca largha*) in the Bering Sea. *Can. J. Zool.* **78**, 1959–1971.

Naito, Y. (1974). The hyoid bones of two kinds of harbour seals in the adjacent waters of Hokkaido. *Sci. Rep. Whales Res. Inst.* **26**, 313–320.

Naito, Y., and Nishiwaki, M. (1972). The growth of two species of harbour seal in adjacent waters of Hokkaido. *Sci. Rep. Whales Res. Inst.* **24**, 127–144.

O'Corry-Crowe, G. M., and Westlake, R. L. (1997). Molecular investigations of spotted seals (*Phoca largha*) and harbor seals (*P. vitulina*), and their relationship in areas of sympatry. *In* "Molecular Genetics of Marine Mammals" (A. E. Dizon, S. J. Chivers, and W. F. Perrin, eds), pp. 291–304. Special Publications Number 3, The Society for Marine Mammalogy.

Olesiuk, P. F., Bigg, M. A., and Ellis, G. M. (1990). Recent trends in abundance of harbour seals, *Phoca vitulina*, in British Columbia. *Can. J. Fish. Aquat. Sci.* **47**, 992–1003.

Pitcher, K. W. (1980). Food of the harbor seal, *Phoca vitulina richardsi* in the Gulf of Alaska. *Fish. Bull.* **78**, 544–549.

Pitcher, K. K., and Calkins, D. (1979). "Biology of the harbor seal, (*Phoca vitulina richardsi*), in the Gulf of Alaska." "US Deptartment of Commerce, NOAA, OCSEAP Final Report. Vol. 19, pp. 231–310.

Quakenbush, L. T. (1988). Spotted seal, *Phoca largha*. *In* "Selected Marine Mammals of Alaska: Species Accounts with Research and Management Recommendations" (J. W. Lentfer, ed.), pp. 107–124. Marine Mammal Commission, Washington, DC.

Reijnders, P. J. H., *et al.* (9 authors) (1993). "Seals, Fur Seals, Sea Lions and Walruses: Status of Pinnipeds and Conservation Action Plan." IUCN, Gland.

Rice, D. W. (1998). "Marine mammals of the World. Systematics and Distribution." Special Publication No. 4. The Society for Marine Mammalogy, Lawrence. Kansas, USA.

Rugh, D. J., Shelden, K. E. W., and Withrow, D. E. (1997). Spotted seals, *Phoca largha*, in Alaska. *Mar. Fish. Rev.* **59**, 1–17.

Shaughnessy, P. D., and Fay, F. H. (1977). A review of the taxonomy and nomenclature of North Pacific harbour seals. *J. Zool. Lond.* **182**, 385–419.

Smith, R. J. (1999). The Lacs des Loups Marins harbour seal, *Phoca vitulina mellonae* Doutt 1942: Ecology of an isolated population. Ph.D. Dissertation, The University of Guelph, Guelph, Ontario, Canada.

Teilmann, J., and Dietz, R. (1994). Status of the harbour seal, *Phoca vitulina*, in Greenland. *Can. Field Nat.* **108**, 139–155.

Temte, J. L., Bigg, M. A., and Wiig, Ø. (1991). Clines revisited: The timing of pupping in the harbour seal (*Phoca vitulina*). *J. Zool. Lond.* **224**, 617–632.

Tikhomirov, E. A. (1966). Certain data on the distribution and biology of the harbor seal in the Sea of Okhotsk during the summer–autumn period and hunting it. *In* "Soviet Research on Marine Mammals in the Far East," Vol. 58, pp. 105–115. Izvestia TINRO [In Russian]. Vladivostok, Russia.

Trukhin, A. M. (2005). "Spotted Seal." Russian Academy of Sciences, Dalnauka, Vladivostok [In Russian].

Waring, G. T., Josephson, E., Fairfield, C. P., and Maze-Foley, (2007). "US Atlantic and Gulf of Mexico marine mammal assessments-2006," 2nd Ed. US Department of Commerce, NOAA Technical Memorandum, NMFS-NE-201.

Withrow, D. E., and Loughlin, T. R. (1995). Abundance and distribution of harbor seals (*Phoca vitulina richardsi*) along the Aleutian Islands during 1994. *In* "Marine Mammal Assessment Program, Status of Stocks and Impacts of Incidental Take 1994," pp. 173–205. National Marine Mammal Laboratory, Seattle.

# Harp Seal
## *Pagophilus groenlandicus*

### DAVID M. LAVIGNE

## I. Characteristics and Taxonomy

The harp seal is one of the most abundant and best known of all pinniped species. Referred to by most scientists by the Latin name, *Pagophilus groenlandicus* (Erxleben, 1777), which means the ice-lover from Greenland, it is still sometimes called *Phoca groenlandica*, the Greenland seal. Its common names, the harp or saddleback seal, come from the black wish-bone or harp-shaped marking found on the backs of adults (Fig. 1). The faces of adults are also black, whereas the remainder of the body appears silvery-gray when dry. Young pups, which have a characteristic white pelt at birth, are known as "whitecoats" (Fig. 1).

Adult harp seals are about 1.7 m (5.6 ft) in length, with females being marginally smaller than males. Adults weigh about 130 kg (288 lbs) early in the pupping season, but their mass varies considerably throughout the year and from one year to the next. The adult dental formula is: I 3/2, C 1/1, PC 5/5 = 34.

## II. Distribution and Abundance

The harp seal inhabits the North Atlantic and Arctic Oceans from northern Russia in the east to Newfoundland and the Gulf of St. Lawrence (Canada) in the west (Fig. 2). Its annual range is essentially tied to the southern and northern limits of pack ice and is largely coincident with the subarctic region of the North Atlantic.

Although some scientists recognize two subspecies, it is more common to refer to three distinct populations or stocks, based on geographic distribution and small morphological, genetic and behavioral differences. One population, found largely in the Barents Sea, reproduces on the "East Ice" in the White Sea off the coast of Russia. This population is designated by some Russian scientists as *P. g. oceanis* (Rice, 1998). A second population lives off the east coast of Greenland and breeds on the "West Ice" in the Greenland Sea near the island of Jan Mayen. The third lives in the Northwest Atlantic off the east coast of Canada and breeds in two locations: on the "Front" off the coast of Newfoundland and Labrador, and in the Gulf of St. Lawrence. The latter two populations are assigned to *P. g. groenlandicus* by some scientists (Rice, 1998).

Figure 1    *Adult female harp seal with "whitecoat" pup. Photograph by N. Lightfoot.*

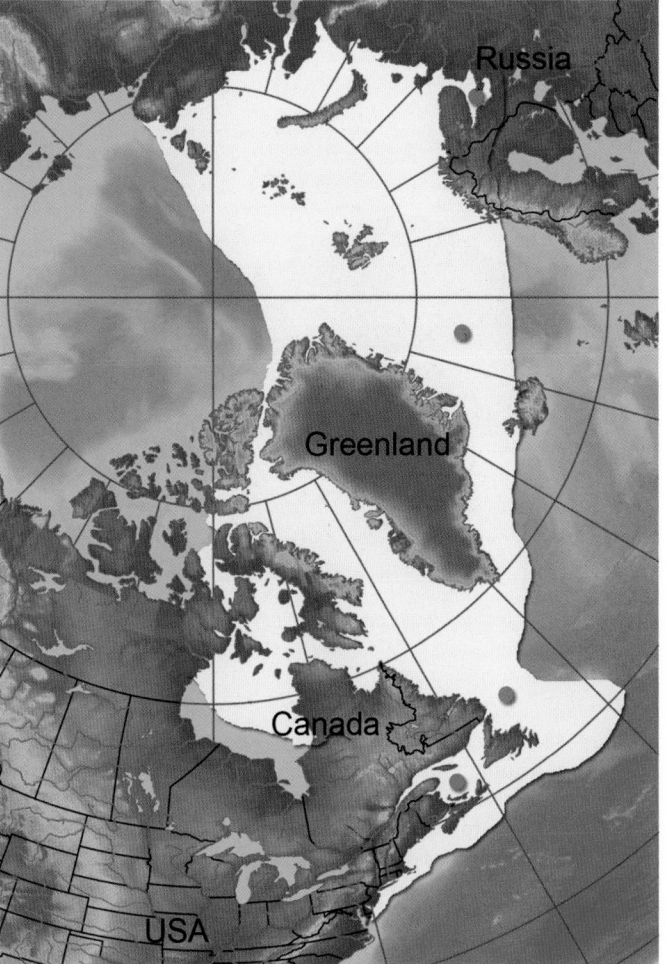

Figure 2    *Harp seal distribution in the North Atlantic Ocean. The four red circles indicate major pupping areas (see text for details).*

The Northwest Atlantic harp seal population is the largest of the three. Population size was estimated to be 5.5 million in 2007 (95% CI = 3.8–7.1 million), not significantly different from the 2005 population estimate of 5.8 million (95% CI = 4.1–7.6 million)

(Department of Fisheries and Oceans, 2007). The most recent estimate of population size for the West Ice population is 634,960 (95% CI = 425,140–844,860) animals aged 1 year and older (1+) and 106,710 (± 35,041) pups in 2005 (ICES, 2006).

Population size for the White Sea population was estimated at 2,064,600 (95% CI = 1,496,520–2,633,480) 1+ animals, and 360,880 (± 62,279) pups, also in 2005 (ICES, 2006).

## III. Ecology

Harp seals exhibit catholic feeding habits, which vary with age, season, location, and year. While at least 67 species of fin fish and 70 species of invertebrates have been recorded in their stomachs (Wallace and Lawson, 1997), harp seals tend to concentrate on smaller fishes such as capelin (*Mallotus villosus*), arctic cod (*Boreogadus saida*), and polar cod (*Arctogadus glacialis*), and a variety of invertebrates, including euphausids (*Thysanoessa* sp.). Harp seals rarely eat Atlantic cod, *Gadus morhua*, the biomass of which has been reduced by more than 99% in the Northwest Atlantic since the early 1960s, largely due to overfishing (Hutchings, 2006).

Before the collapse of cod, predation by harp seals in the northern Gulf of St. Lawrence accounted for 1% of large cod mortality, whereas fishing accounted for 46%. Morissette *et al.* (2006) concluded that seals play an important role in maintaining the structure of the ecosystem and, overall, have a positive impact on marine ecosystems.

Harp seals are prey for polar bears (*Ursus maritimus*), killer whales (*Orcinus orca*), and sharks (e.g., *Somniosus microcephalus*). Their major predator, however, is *Homo sapiens*.

## IV. Behavior and Physiology

Harp seals are highly migratory animals that spend most of the year at sea, traveling and feeding, sometimes in groups (Fig. 3), sometimes alone. They are particularly gregarious during the breeding season, when most adult females haul out on ice to give birth and nurse their pups. Adult males congregate nearby, waiting to participate in mating once the pups are weaned. Harp seals exhibit little sexual dimorphism and appear to have a promiscuous breeding system.

Harp seals maintain a thick blubber layer that not only provides insulation against the heat-draining properties of cold water but supplies a rich source of energy that can be used during fasts and when food is scarce. Blubber also rounds out the body contours to streamline the seal's body and reduce drag when swimming. In addition to providing propulsion in water, the flippers serve to regulate heat loss by means of countercurrent heat exchangers. Harp seals also have brown fat that can be used to warm cool blood returning from the periphery, just as neonatal harp seals use brown fat for rapid heat production. Among the behavioral means of reducing heat loss, harp seals on ice can keep their fore flippers held against their bodies and their hind flippers pressed together to reduce the exposed surface area and thereby conserve heat.

Such adaptations seem adequate for maintaining homeothermy, even in cold climates. The evidence is that harp seals, like other marine mammals, do not need (or have) elevated metabolic rates or huge appetites to meet their energy demands, either on land or in water (Lavigne *et al.*, 1986; Innes *et al.*, 1987).

Harp seals are modest divers. Average maximum dive depth is 370 m (1214 ft) and mean dive duration is about 16 min (Schreer and Kovacs, 1997).

Vision is the harp seal's primary sense. The harp seal eye is relatively large, covered by a cornea that is constantly lubricated by tears

**H**

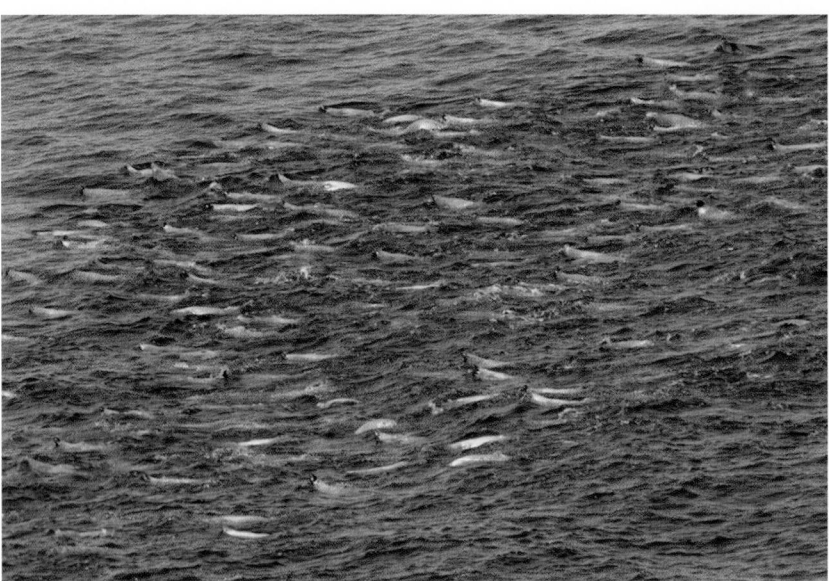

Figure 3 *A group of harp seals moving northward in the Gulf of St. Lawrence. Photograph by IFAW/S. Cook.*

produced by lacrimal glands to protect it from salt water. Unlike the eyes of terrestrial mammals, the seal's eye lacks tear ducts to drain away the tears, which explains why harp seals often appear to be crying. The harp seal eye has a large spherical lens to provide additional focusing power under water and a highly mobile pupil which contracts in bright light to produce a reasonably focused image, even when the animal is on land or ice. Visual sensitivity in dim light is enhanced by the harp seal's rod-dominated retina, backed by a reflective tapetum lucidum, reminiscent of that found in cats. The rod visual pigment exhibits maximum sensitivity in the blue-green region of the spectrum, consistent with Lythgoe and Dartnall's "sensitivity hypothesis" (Lavigne and Ronald, 1975). Harp seals also have cone photoreceptors (Nagy and Ronald, 1970) for vision in bright light, which possibly allow some form of color vision. Like other ice-dependent seals, the harp seal cornea tolerates the high levels of ultraviolet radiation found in bright, snowy environments, thereby avoiding "snow-blindness" (ultraviolet damage to the cornea), an affliction that plagues humans in such environments.

Harp seal hearing is also adapted to function both in air and under water. In air, the seal's hearing range is similar to that of humans although not quite as sensitive. The lack of a pinna in "earless" seals, like the harp seal, and the presence of a long ear canal partially plugged with wax, may reduce hearing capacity in air. Under water, the hearing range is extended to more than 60,000 Hz and, over those frequencies heard in both air and water, the harp seal actually hears better under water. Even more remarkably, seals possess good directional hearing both in air and under water.

Harp seals are known for their varied vocalizations during the breeding season. While their vocalizations on ice are relatively few (e.g., "bawling" vocalizations of hungry pups calling for their moms, "mumbling" sounds observed in playing pups, and threatening growls and "warbling" associated with agonistic behaviors in older animals; Kovacs, 1987), at least 19 call types have been identified in wild harp seals under water during courtship and mating (Perry and Terhune, 1999). The extent to which wild harp seals vocalize outside the breeding season remains unknown. Despite their "high-frequency" hearing under water, there is no evidence that harp seals are capable of echolocation.

Figure 4 *A female harp seal identifies her pup. Photograph by IFAW/S. Cook.*

Harp seal females often use their sense of smell to identify their own offspring during the nursing period (Fig. 4). Their sense of smell may also allow harp seals to detect approaching predators when on ice. In water, however, seals keep their nostrils tightly closed and their sense of smell is of no use.

Harp seals also have well developed, beaded whiskers, called vibrissae, arranged in horizontal rows on both sides of the snout. There is some evidence that the vibrissae provide tactile information, as they do in other carnivores, including cats. Their vibrissae are also sensitive to low frequency, waterborne vibrations and may function to detect the movement of fish and other aquatic organisms.

## V. Life History

Female harp seals congregate near the southern limits of their range (Fig. 2) from late February to mid-March (depending on location) to give birth to their white-coated pups. Each female gives birth

**Figure 5** *Molted harp seal pup or "beater," aged about 4 weeks. Photograph by IFAW/S. Cook.*

to a single pup, although twin fetuses have been reported. Newborn pups weigh about 11 kg (24 lbs) at birth and lack the thick insulating layer of blubber found in older seals. Pups are nursed on a fat-rich milk for about 12 days, during which time they deposit a thick (*ca.* 5 cm or 2 in.) blubber layer and grow at about 2.2 kg (5 lbs) per day. During this time they pass quickly through a number of recognizable developmental stages. At weaning the pups weigh, on average, 36 kg (80 lbs). At this stage they are known as "graycoats" because their spotted, gray juvenile pelage has grown in and can be seen beneath the white neonatal coat. Shortly thereafter, the white coat becomes loose and, within a few days, it begins to fall out. Once the white coat is completely molted, exposing the black-spotted, silvery-gray pelt of the young harp seal, the animal is called a "beater" (Fig. 5).

At the age of about 13–14 months, young harp seals undergo their second molt; the "beater" pelt is replaced by a similar spotted pelt and the animals are renamed "bedlamers." Bedlamers retain the spotted pelt through successive annual molts until the spots begin to disappear and the dark, harp-shaped pattern of the adult coat begins to emerge. Older harp seals with a combination of the spotted bedlamer pelt and the distinct adult "harp" are called "spotted harps."

The transition from the bedlamer pelt to the adult pelage begins with the onset of sexual maturity. In 2001, the mean age of sexual maturity for females was about 5.3 years (Sjare *et al.*, 2004). Most male harp seals, which possibly mature later than females, develop the black "harp" marking abruptly, whereas in females the transition is more gradual and may take many years. Some female harp seals never lose all their spots or develop a complete "harp" (Lavigne and Kovacs, 1988). Any seal with a complete harp and black face may be aged anywhere from about 5 to 30+ years—the life expectancy of the species.

Weaning in harp seals is abrupt. The adult females simply leave their pups on the ice and turn their attention to mating. Mating usually occurs in the water (Lavigne and Kovacs, 1988) but has also been photographed on the ice. The fertilized egg divides several times, forming a spherical embryo that floats freely in the womb for more than 3 months before implanting in the wall of the uterus. This type of suspended development—known as delayed implantation—ensures that all females give birth to their pups at the same time each year when the short-lived pack ice is available as a whelping platform.

Meanwhile, the weaned pups remain on the ice and undergo a post-weaning fast. This fast may last upward of 6 weeks, during which the weaned seals may lose up to half of their body mass as they draw on their thick blubber layer for sustenance. Eventually,

they enter the water, or the ice melts beneath them, and they begin swimming and feeding on their own.

After mating, older harp seals congregate once again on pack ice farther north to undergo the annual molt. Following the molt, in which the pelt and surface layers of skin are replaced, the seals continue their migration to summer feeding areas in subarctic and arctic waters to the north. All three populations exhibit similar patterns of annual migration, although the timing of specific events, such as pupping, mating, and molting, varies slightly among populations.

## VI. Interactions with Humans

All three harp seal populations have been hunted by humans for centuries, and all three have undergone documented declines in numbers as a result of over-exploitation. And all three continue to be hunted today.

The commercial hunt for Northwest Atlantic harp seals, which occurs mainly after the spring whelping season, remains the largest hunt for any marine mammal population in the world. Over 1 million seals were killed between 2004 and 2007, inclusive, of which more than 95% were pups under the age of 3 months. The Total Allowable Catch (TAC) for Canada's 2007 commercial seal hunt was 270,000, which was higher than the estimated replacement yield (the number of seals that can be removed from the population without causing it to decline from one year to the next) of 165,000. This population is also hunted in the eastern Canadian Arctic (some 800 animals per year) and off the west coast of Greenland (currently about 70,000–90,000 animals per year) during the summer months.

For harp seals breeding on the West Ice, the TAC for 2004 was set at 15,000 1+ animals, or an equivalent number of pups (where one 1+ animal = two pups), almost double the estimated sustainable catch level for this population of 8200 1+ animals. Catches in 2004 and 2005 were 9895 (including 8288 pups) and 5808 (including 4680 pups), respectively (ICES, 2006).

For the White Sea population, the 2004 TAC was set at 45,000 1+ animals (or an equivalent number of pups, where one 1+ animal = 2.5 pups). The reported catch for this population in 2005 was 22,474 (including 15,420 pups) (ICES, 2006).

Over-exploitation, particularly in the Northwest Atlantic, and the largely unregulated trade in seal products (especially seal skins, seal oil, and penises) in the absence of a precautionary management approach (e.g., Johnston *et al.*, 2000; but see Hammill and Stenson, 2007), remain potential threats to the species. Other frequently cited threats involve proposals to cull harp seal populations ostensibly to benefit commercial fisheries, including recently proposed "seal exclusion zones;" incidental catches in fishing gear; environmental contaminants; and reduced food availability due to overfishing.

Today, the most serious threat is global warming and its potential effects on ice-breeding harp seals (Johnston *et al.*, 2005). In 10 of the 12 years between 1996 and 2007, there has been below average ice cover off Canada's east coast. The lack of suitable ice, combined with violent storms, and early ice break-up, disrupts the harp seals' normal pupping season. This can result in increased abortions if female seals do not find ice upon which to give birth, or increased mortality of newborns if the ice breaks up before the end of nursing. In 2002, Canadian government scientists estimated that 75% of pups born in the southern Gulf of St. Lawrence died even before the annual seal hunt began; in 2007, almost all the pups born in that region may have suffered a similar fate. Such effects, in any given year, result in reduced cohort (year-class) size and have longer-term implications for population size and trends.

If warm years with reduced ice coverage become the norm, as appears to be happening, there will be additional effects. These include effects on timing of reproduction and the loss of critical breeding habitat. They also include effects on the distribution and abundance of fish and invertebrates, leading to further changes in availability of prey for harp seals, and ripple effects on seal condition, growth, reproductive success, and survival.

## See Also the Following Articles

Earless seals (Phocidae) ■ Hunting of Marine Mammals

## References

Department of Fisheries and Oceans (2007). *A review of ice conditions and potential impact on harp seal neonatal mortality in March 2007.* DFO Canadian Science Advisory Secretariat Science Response 2007/2008.

Hammill, M. O., and Stenson, G. B. (2007). Application of the precautionary approach and conservation reference points to management of Atlantic seals. *ICES J. Mar. Sci.* **64**, 702–706.

Hutchings, J. A. (2006). Ecological and fisheries sustainability: Common goals uncommonly achieved. *In* "Gaining Ground: In Pursuit of Ecological Sustainability" (D. M. Lavigne, ed.), pp. 101–112. International Fund for Animal Welfare and University of Limerick, Guelph, Canada and Limerick, Ireland.

ICES (2006). Report of the ICES/NAFO Working Group on Harp and Hooded Seals (WGHARP). August 30–September 3, 2005. ICES Advisory Committee on Fishery Management. *ICES C.M. 2006/ACFM* 06, D,G. 50 pp.

Innes, S., Lavigne, D. M., Earle, W. M., and Kovacs, K. M. (1987). Feeding rates of seals and whales. *J. Anim. Ecol.* **56**, 115–130.

Johnston, D. W., Meisenheimer, P., and Lavigne, D. M. (2000). An evaluation of management objectives for Canada's commercial harp seal hunt, 1996–1998. *Cons. Biol.* **14**, 729–737.

Johnston, D. W., Friedlaender, A. S., Torres, L. G., and Lavigne, D. M. (2005). Variation in sea ice cover on the east coast of Canada, 1969 to 2002: Climate variability and implications for harp and hooded seals. *Clim. Res.* **29**, 209–222.

Kovacs, K. M. (1987). Maternal behaviour and early behavioural ontogeny of harp seals, *Phoca groenlandica. Anim. Behav.* **35**, 844–855.

Lavigne, D. M., and Ronald, K. (1975). Pinniped visual pigments. *Comp. Biochem. Physiol.* **52B**, 325–329.

Lavigne, D. M., and Kovacs, K. M. (1988). "Harps and Hoods: Ice-breeding Seals of the Northwest Atlantic." University of Waterloo Press, Waterloo, Ontario, Canada.

Lavigne, D. M., Innes, S., Worthy, G. A. J., Kovacs, K. M., Schmitz, O. J., and Hickie, J. P. (1986). Metabolic rates of seals and whales. *Can. J. Zool.* **64**, 279–284.

Morissette, L., Hammill, M. O., and Savenkoff, C. (2006). The trophic role of marine mammals in the Northern Gulf of St. Lawrence. *Mar. Mamm. Sci.* **22**, 74–103.

Nagy, A. R., and Ronald, K. (1970). The harp seal, *Pagophilus groenlandicus* (Erxleben 1777). VI. Structure of the retina. *Can. J. Zool.* **48**, 367–370.

Perry, E. A., and Terhune, J. M. (1999). Variation of harp seal (*Pagophilus groenlandicus*) underwater vocalizations among three breeding locations. *J. Zool. Lond.* **249**, 181–186.

Rice, D. W. (1998). "Marine Mammals of the World. Systematics and Distribution." Society for Marine Mammalogy, Special Publication Number 4, Lawrence, Kansas, USA.

Schreer, J. F., and Kovacs, K. M. (1997). Allometry of diving capacity in air-breathing vertebrates. *Can. J. Zool.* **75**, 339–358.

Sjare, B., Stenson, G. B., and Healy, B. (2004). Changes in the reproductive parameters of female harp seals (*Pagophilus groenlandicus*) in the Northwest Atlantic. Canadian Science Advisory Secretariat Research Document 2004/107.

Wallace, S.D., and Lawson, J.W. (1997). A review of stomach contents of harp seals (*Phoca groenlandica*) from the Northwest Atlantic: An update. *IMMA Tech. Rep.* 97-01. International Marine Mammal Association, Inc., Guelph, Ontario, Canada.

# Health

JOSEPH R. GERACI AND VALERIE J. LOUNSBURY

The health of an animal is linked to age, behavior, and environment. Like terrestrial species, marine mammals are subject to infection, injury, and metabolic disturbances. Our understanding of marine mammal health is impeded not only by the difficulties inherent in studying these species in the wild, but also by their unique biology. Today, the challenge is compounded by human impacts on the health of marine mammals and their environment.

## I. Adaptations to Life at Sea

Cetaceans, sirenians, pinnipeds, and sea otters (*Enhydra lutris*), all taxonomically distant, have evolved similar biological mechanisms to cope with a marine existence (Elsner, 1999). These include biological and behavioral strategies for controlling body temperature, diving, maintaining salt and water balance, and promoting reproductive success—adaptations vital to health and survival.

### A. Temperature Balance

Other than tropical waters, the sea is always colder than a mammal's body temperature. Water conducts thermal energy about 20 times faster than air, so even a few degrees difference can be enough to drain vital heat reserves. To counter this, marine mammals have evolved numerous mechanisms, of which blubber arguably tops the list. More or less a fatty envelope, blubber in cetaceans and certain pinnipeds offers mechanical protection, warmth, buoyancy, nutrients when food is scarce, and fresh water in reserve. Otariid pinnipeds, having thinner blubber and less body fat than phocids or the walrus (*Odobenus rosmarus*), are thus less tolerant of cold and depend to a certain extent on their pelage for insulation. This is especially true for otariid pups, which may not acquire an adult coat or adequate fat until they are about 3 months old and, in the meantime, are prone to hypothermia when they become wet. Species with less blubber rely on other strategies. The sea otter depends entirely on a high metabolic rate (and high caloric intake) to generate heat and on its dense, well-groomed fur to prevent heat loss. The living sirenians, with low metabolic rates and little ability to control surface heat loss, are narrowly restricted to tropical and subtropical waters.

Environmental temperature has more than a subtle bearing on health. To survive in a cold climate, a marine mammal must be robust, appropriately insulated, and have all surface heat control mechanisms operating. If not, the only recourse is to increase metabolic rate and either eat more or borrow fat from vital fat reserves. Ironically, as a last measure to conserve heat, pinnipeds and sea otters may haul out on land where the prospect of feeding is hopeless.

Can a marine mammal be too warm? Not if it is healthy and in the right place. In a warming environment, a whale may eat less and metabolize blubber, which effectively reduces insulation, and shed excess heat by increasing blood flow to the skin, particularly of the extremities. On land a wet seal or sea otter may get some comfort from evaporative cooling, and to avoid overheating when dry, it relies mostly on circulatory and behavioral adaptations (e.g., seeking shade, sleeping, moving to the surf zone). These strategies work to a point. A sea otter out of water can become distressed at air temperatures as low as 10°C and die within hours at 21°C. A cetacean stranded on a sunny beach can literally cook inside its own blubber.

## B. Breathing and Diving

Marine mammals forage at all depths. Sea otters and sirenians, which feed in relatively shallow waters, have little need to dive deeply or for more than just a few minutes. Some species of phocids and odontocete cetaceans, however, make sustained dives to 600–1000 m or more, where they must cope with extraordinary pressures, periods of a half hour or more without oxygen, and the potential accumulation of deadly nitrogen. How do they adapt?

During a prolonged dive, circulation to the skin and viscera may almost cease, allowing oxygen to be channeled to organs that need it most, such as the heart and brain. A long, deep dive—whether to feed, explore, or escape—requires a shift to anaerobic metabolism, which is physiologically exhausting and therefore needs to be highly managed. So for the most part, marine mammals are adapted to feed within the depth and time limit of their aerobic diving capacity. That, in turn, is governed by the size, age, and health of the individual. Because of their relatively greater capacity to store oxygen, large animals tend to be better divers. It is not surprising that juveniles may find it difficult to reach prey that is easily accessible to adults.

## C. Salt and Water Balance

The osmotic concentration of the sea is nearly 4 times greater than that of mammalian body fluids. Chemical equilibrium thus favors both loss of body fluids into the sea and encroachment of salts into the animal. Marine mammals have strategies that stop this from happening: (1) external surfaces are impermeable to seawater; (2) body water is highly conserved—sweat glands are either reduced or absent and the kidneys efficiently concentrate urine; (3) they drink little seawater and acquire most of their fresh water from food (water makes up about 70% of a fish, 80% of a squid, and over 90% of aquatic plants, and each gram of dietary fat or blubber they metabolize yields close to its weight in fresh water). In pinnipeds and cetaceans, the physiological response to stress is also designed to conserve water. During stress, the adrenal gland produces aldosterone, a hormone that induces the kidney to draw salt and water from what would otherwise form urine, back into the body. Maintaining this delicate salt and water balance depends, among other factors, on adequate blubber, well-functioning kidneys, proper hormonal balance, an intact epidermis, and—above all—good health.

## D. Strategies for Rearing Young

The social, physical, and biological conditions that together create a healthy environment are especially critical during the period of an animal's life when it depends entirely on its mother. Just how critical that period is depends on the species, location, patterns of maternal care, and environmental conditions, and can be quite predictable. For example, pinnipeds that disperse for much of the year regroup to give birth and mate again at a certain time and place. While obviously effective for the population, the resulting crowding on the rookery poses a risk of serious injury, super-infection by parasites, and disease transmission—ironically more often in the pups than the adults.

## II. What Can Go Wrong?

Body systems work together, all complementing one another. Impairment of one system can disturb the entire equilibrium, leading to secondary problems, which then threaten the animal's health (Gulland *et al.*, 2001; Geraci and Lounsbury, 2005). For example, blubber is a protecting blanket, hydrodynamic shield, and a source of energy, insulation, water reserves, and buoyancy. In the simple case where food is scarce and blubber depletion the outcome, all its benefits are under attack. To name a few, the animal gradually becomes less able to rest at the surface, maintain body heat, forage, escape predators, or keep up with a group. The ensuing stress may open the door to disease, further weakening the animal. The situation naturally exacerbates if blubber loss was caused by illness in the first place.

Injuries and illnesses are not always apparent and are often detected only after analyzing blood or tissue samples from a living animal or dissecting a dead one. Even careful study might not reveal serious biochemical and physiological conditions. Stress is poorly understood and its effects difficult to quantify. What little is known about the process in marine mammals shows that it can disrupt thyroid and adrenal gland function, water and electrolyte balance, and metabolism and reproduction, and can lower the number of certain blood cells that support an animal's immune system and protect it against parasitic infection (St. Aubin and Dierauf, 2001).

### A. Reproductive Failure and Death of the Newborn

Reproduction in animals is an orderly, coordinated progression of biological and behavioral dynamics. Weakness or disruption at any point can lead to failure, evident as abortion, stillbirth, premature birth, or weakness or death of the newborn. The causes of reproductive failure are often obscure, particularly in species that cannot be studied from shore.

In some species, the risk of abortion or stillbirth appears to be greater for first-time mothers. Young mothers are usually smaller and may give birth to smaller offspring that are more vulnerable to hypothermia and injury. The health and nutritional condition of any mother, regardless of age or size, affects the fetus. In some pinnipeds, decrease in prey abundance associated with El Niño events has been linked with decreased fertility, increased abortions, and reduced pup production (Trillmich *et al.*, 1991). The same outcome befell harbor seals (*Phoca vitulina*) in Europe that were swept by a morbillivirus epidemic. Using history as a guide, when launching any study of a population experiencing serious reproductive failure, the compass swings toward environmental disruption, epidemic disease, reduced prey stocks, or high levels of certain anthropogenic contaminants (Geraci *et al.*, 1999).

### B. Starvation

Marine mammals spend much of their time searching for food of the appropriate type, size, and quality to satisfy needs that vary seasonally and with age. Some animals, e.g., dependent young, the sick, and the very old, can starve even when prey are plentiful. Many factors determine how long an animal can survive without food: its age, fat reserves, metabolic rate, energy demands, and general health. Large animals with low metabolic rates survive longer than those with high energy demands, such as small species, newborn, and growing pups. Baleen whales may feed very little for 6–8 months of the year, but a sea otter without food for even 2 days can die from the complications of starvation. Put into perspective, starvation is a major cause of death in pinniped and sea otter pups.

Throughout the period of dependency, a young animal's survival hinges on the health of its mother. Before giving birth, a phocid seal or baleen whale must develop ample fat reserves to carry it through a period of fasting or reduced feeding during lactation. The pup or calf born of a malnourished mother is at risk from the moment of birth and its longevity is compromised early in its development.

The young of species in which females feed continuously during lactation face a different threat. A bottlenose dolphin (*Tursiops truncatus*) calf depends on the state of its mother's nourishment throughout what may be a year or more of nursing. More subtle is an otariid pup's risk of starvation if a shift in prey abundance forces its mother to spend longer periods away from the rookery.

Weaning frees a young animal from dependence to face the challenge of providing for itself. Manatees (*Trichechus* spp.) and some cetaceans and otariids remain with their mothers long enough to learn foraging skills. Not so for all sea otters. Newly independent juveniles, handicapped by their high need for food and inexperience gathering it, often starve. Females are singularly vulnerable because they tend to remain within a prescribed range, even when prey there becomes depleted.

Depletion of food stocks, whether from overgrazing, overfishing, or climatic or oceanographic fluctuation, can affect entire populations. Food scarcity in one area may cause some animals to move elsewhere. When food abundance changed during the El Niño of 1982–1983, California sea lions (*Zalophus californianus*) moved northward, and many northern fur seals (*Callorhinus ursinus*) may have emigrated from San Miguel Island to rookeries in the Bering Sea. Some animals are unable or unwilling to make such excursions, e.g., females with pups, territorial males, or populations in remote ranges. When fish disappeared from surface waters around the Galapagos Islands during the 1982–1983 El Niño, widespread starvation of the islands' fur seals (*Arctocephalus galapagoensis*) soon followed (Trillmich *et al.*, 1991).

Starving animals eventually die—some quickly, as would a pup deprived of milk or a sea otter overcome with hypothermia and exhaustion. Others die after a period of illness triggered by malnutrition and mediated by factors such as hypothermia, dehydration and electrolyte imbalance, hormonal disturbances, and infection by parasites and opportunistic pathogens (Geraci and Lounsbury, 2002). Some starving seal pups may ingest whatever is nearby—gravel, stones, or grass—and consequently die of an impacted stomach.

While a sudden shortage of prey may cause outright, widespread starvation, the more subtle effects of nutritional stress, including low productivity and decreased juvenile survival, may prove equally damaging to a population.

### C. Direct Environmental Effects

Extreme weather conditions can take a toll on all age classes. Intensely cold winters have killed up to 2% of the Florida manatee (*Trichechus manatus latirostris*) population, mostly juveniles. Storms hitting a crowded pinniped rookery during breeding season can leave pups hypothermic, battered on rocks or drowned, separated from their mothers and starving, or victims of adult aggression. Unusual ice conditions can be hazardous even for cold-water species. Sea otters

trapped out of water by heavy ice die of starvation, stress, and shock. An untimely freeze in polar waters can trap cetaceans in ice where they may ultimately suffocate or starve. Severe storms and unseasonable warm weather can fracture ice floes that crush large numbers of breeding seals and their pups, sometimes affecting the entire year's production. The early retreat of Arctic sea ice observed in recent years may force nutritionally stressed Pacific walrus mothers—deprived of the platform essential for rearing young and reaching vital feeding grounds—to abandon their calves (Cooper *et al.*, 2006).

### D. Trauma

For most marine mammals, the risk of injury is continual, whether from natural sources, such as storms, predators, and aggressive encounters, or human activities, such as fishery operations and recreational boating. For example, injuries are common on pinniped rookeries where pups often are trampled accidentally or attacked by adults, fall into gullies or crevices, or wash off unprotected beaches into pounding surf. Adults can be victims as well, as bulls compete for territories and females, and females compete for space.

Historically, commercial hunting had serious impacts on certain species or stocks of marine mammals. Today, more animals die in accidents. Interaction with fishers is a leading cause of death and injury: pelagic odontocetes die in purse seines and drift nets, coastal cetaceans and pinnipeds in gill net and trawl fisheries, and some river dolphins from fishing methods that use electricity and explosives. Marine mammals thought to compete with commercial operations may be deliberately shot.

Discarded net fragments, ropes, packing bands, monofilament line, and other debris also put animals at risk (Laist *et al.*, 1999). The effects of entanglement on populations vary: some suffer significant impact, others do not. For the individual victim, the problem is always serious. An animal that does not drown immediately may escape with fractures and internal injuries or may carry net fragments, ropes, or bands that increase drag, impede swimming ability, or become snagged. A seal pup growing into its packing-band "collar" will eventually die, either from suffocation or from deep cuts and infection.

Coastal dwellers are especially prone to certain injuries from human activities. For example, dugongs (*Dugong dugon*) in Queensland (Australia) waters have died in shark nets set to protect public beaches. Right whales (*Eubalaena glacialis*) in the Northwest Atlantic and manatees in Florida are injured or killed by collisions with vessels in coastal channels at rates that jeopardize their populations. Cetaceans—especially beaked whales—may be injured and strand after exposure to high-intensity underwater sound, such as that associated with naval sonar operations. Studies of cetaceans involved in such events show evidence consistent with acute decompression sickness, or "the bends" (Jepson *et al.*, 2003)—a condition previously thought to have dropped out of the cetacean gene pool.

### E. Predation

There are times in a marine mammal's life when it draws the attention of predators. Probably the easiest meal is a small, inexperienced animal that can be found in a particular place on schedule—criteria often met by young pinnipeds, whether on land or ice or at sea. As some examples, arctic foxes (*Alopex lagopus*) and polar bears (*Ursus maritimus*) break into ringed seal (*Pusa hispida*) birth lairs to take pups and, sometimes, their mothers. Steller sea lions (*Eumetopias jubatus*) on the Pribilof Islands eat young northern fur seals (*Callorhinus ursinus*) that venture into the water. Southern sea lions (*Otaria flavescens*) raid South American fur seal (*Arctocephalus australis*) rookeries, driving away the adults and killing pups. Leopard seals (*Hydrurga leptonyx*) consume large numbers of crabeater seals (*Lobodon carcinophaga*) from the time the weaned pups leave the safety of the ice until they are several months old and large enough to escape attack. Killer whales (*Orcinus orca*), working in teams, patrol pinniped rookeries, washing seals into the water or even chasing them onto a beach; other pods may attack baleen or sperm (*Physeter macrocephalus*) whales. In the Aleutian Islands, killer whales facing reduced abundance of pinniped prey may have turned to sea otters, perhaps contributing to the dramatic decline in local sea otter populations over the past few decades. Sharks pose a danger to many species or populations, including the Hawaiian monk seal (*Monachus schauinslandi*).

The impact of a predator can extend beyond its effect on the individual prey. Killing a pregnant mother with a dependent young removes not one, but three animals from the population. A female northern elephant seal (*Mirounga angustirostris*) may recover rapidly from a shark attack, as many seem to do, but may be less able to nurse her pup and is unlikely to mate in the compressed breeding season. In this case, a single attack, while only injuring the mother, may have cost the population two pups.

### F. Parasites

Almost all marine mammals are infected by parasites by the time they are weaned or shortly afterward. Most of these parasites have evolved with their hosts and, under normal circumstances, cause little damage to otherwise healthy animals. Among these are the amphipods and copepods that eat bits of whale skin, seal lice that normally occur in small numbers and consume insignificant amounts of blood, and gastrointestinal helminths ("stomach worms"). Others are harmful enough to affect the well-being of individuals and even segments of a population. For pinnipeds, these include heartworms, some lungworms, and the hookworm *Uncinaria lucasi*; and in cetaceans, the nematodes *Crassicauda* spp. (in the mammary glands, cranial sinuses, and kidneys) and the trematodes *Nasitrema* spp. (in the cranial sinuses) and *Campula* spp. (in the liver and pancreas).

Consequences can be unexpectedly serious when individuals are exposed to parasites for which they have evolved no tolerance. Two such parasites have been implicated in about 40% of California sea otter deaths in recent years (Kreuder *et al.*, 2003). Marine mammal infections with the protozoan *Toxoplasma gondi*, a parasite of cats, may be one example of "pathogen pollution," as oocysts shed in cat feces enter coastal waters in runoff and sewage. In sea otters, infection may have little apparent effect or may lead to encephalitis, heart disease, or abnormal behavior that increases the risk of shark attack. Infection with the acanthocephalan *Profilicollis* spp. is acquired through ingestion of infected mole crabs (*Emerita* spp. and *Blepharipoda* spp.), items traditionally not included among the sea otter's preferred prey. This parasite actively penetrates the intestinal wall, causing fatal peritonitis.

Any parasite can become destructive when the mechanisms that maintain the host–parasite balance break down, as they do when an animal is ill or starving. Prolonged stress, by retarding wound healing and destroying protective blood cells, can set the stage for a parasite to do real harm. It is no surprise that debilitated animals that strand ashore often suffer from serious parasitic conditions.

An animal's parasite burden can offer clues to its overall health and to changes in its environment, such as alterations in prey abundance. Certain seal lice that transmit heartworm feed and proliferate on the animal only on land. Any illness that causes the seal to spend more time ashore assures that both the lice and the heartworms will flourish by sapping what is left of the animal's energy. To find the cause of the animal's original illness, one must go deeper than the tempting diagnosis of louse and heartworm infestation. A fast-swimming odontocete offers barnacles little opportunity to attach; the presence of species such as *Lepas* spp. or *Xenobalanus* spp. on a dolphin's flukes or dorsal fin suggests that the animal has been moving unusually slowly, a common sign of illness. Differences in parasite fauna can indicate differences in feeding habits or prey availability: walruses feeding on benthic invertebrates have few if any nematodes in their stomachs, whereas those that eat fish have more. The relationship between diet and parasitism is predictable enough that variations in parasite burden are used to distinguish populations and help identify segregated social groups.

### G. Microorganisms

Microorganisms of all kinds—bacteria, fungi, and viruses among them—abound in the sea. Some are of the types found on land and in land dwellers; others, including certain *Vibrio* bacteria, thrive only in aquatic habitats. Like terrestrial mammals, marine mammals harbor many organisms that are considered normal. Few of these are necessarily pathogenic, meaning they do not always cause infectious disease, but some are more threatening than others (Dunn *et al.*, 2001; Kennedy-Stoskopf, 2001). The fine line between infection and infectious disease depends on both the aggressiveness of the organism and the susceptibility of the host, that is determined by the condition of its immune system. Age is also a factor. A very young animal may benefit from maternal antibodies, that protect it against organisms with which the mother has earlier come into contact. The pup or calf then develops its own active immune capability, which affords increasing protection until its declining years, when immune function once again weakens. For these reasons the very young and the very old are more likely to acquire infections. Of course, natural and human-related stresses may compromise immune function in animals of all ages.

*1. Bacteria*   The nature and severity of bacterial infections can be influenced by the animal's behavior and age, and environmental conditions. Habitat also plays a determining role. A phocid pup born on clean sand is less likely to acquire a serious navel infection as it drags its unhealed umbilicus across the rookery than a pup born in areas fouled by feces, stagnant water, or decaying vegetation. For pups in fouled environments, bite wounds provide another route for infection by bacteria such as *Streptococcus* and *Corynebacterium* (Baker, 1984).

Infections are sometimes predictable. During molt, seals slough skin and hair. In northern elephant seals the process is exaggerated and large sheets of epidermis are lost; many animals, particularly yearlings, come ashore with skin infections during this time. Weddell seals (*Leptonychotes weddellii*), which use their teeth to maintain breathing holes in ice, and sea otters that feed on hard-shelled prey grind down their teeth to such an extent that they develop abscesses and bone infection.

A few bacteria are inherently pathogenic (Geraci *et al.*, 1999). Leptospirosis, caused by the spirochete *Leptospira* spp., occurs in domestic and wild animals worldwide. Infection in California sea lions has caused kidney disease in juvenile and subadult males and abortion in females. A concern for human health arose with the discovery of disease outbreaks associated with *Mycobacteria* of the complex associated with tuberculosis (*M. bovis*, *M. tuberculosis*) in captive New Zealand fur seals (*Arctocephalus forsteri*) and Australian sea lions (*Neophoca cinerea*) in the early 1990s. The disease has since been found in

**H**

free-ranging otariids from Australia, New Zealand, and South America and may be endemic in certain wild populations. Bacteria representing an apparently new strain or species of *Brucella* have been found in many marine mammal populations. The implications for the animals and the persons who work with them are as yet unknown. To date, infection has been linked to brain lesions in striped dolphins (*Stenella coeruleoalba*) and abortion in dolphins and possibly pinnipeds.

The impact of bacterial infection on animal health depends on the organ involved. An isolated abscess in a muscle may have little apparent effect, while a similar infection in the lung can be seriously debilitating. Bacterial pneumonia, often associated with lungworms, can be serious enough to cause death or stranding. The same is true of a condition like gastroenteritis that can quickly alter water and electrolyte balance and lead to a host of secondary problems.

*2. Mycotic Infections*   Fungal organisms rank low on the list of primary pathogens of marine mammals. They tend to infect animals that are weakened, perhaps by other chronic debilitating disease. After a long, terminal illness, an animal can be literally riddled with systemic fungi. The usual source is soil, dust, or water. The wide variety of fungi found in marine mammals includes *Candida, Aspergillus, Coccidioides, Blastomyces, Histoplasma, Fusarium, Nocardia,* and *Lacazia*.

Lobomycosis, a skin infection caused by the yeast *Lacazia loboi*, has an unusual range. The disease occurs in bottlenose dolphins from Florida waters and in tucuxi (*Sotalia fluviatili*) in South America. Curiously, other than in cetaceans, Lobo's disease occurs only in people inhabiting low-lying wetlands of Central and South America.

Coccidioidomycosis, caused by the soil fungus *Coccidioides immitis* and generally associated with terrestrial animals, was historically considered rare in marine mammals. What might be described as outbreaks of infection in California sea lions and sea otters between 1986 and 1994 coincided with a rise in human infections, attributed to unusual environmental conditions. The sea lions and otters presumably inhaled fungal spores transported from land by offshore winds.

*3. Viruses*   First recognized in the late 1960s, viral infections in marine mammals have emerged as the greatest cause of large-scale mortality (Geraci *et al.*, 1999). To spread rapidly, a virus requires a naive host population of a minimum density, which can arise either through population growth or changes in social behavior. Once infected, a migrating or wandering animal may carry the virus to new habitats.

More than 450 harbor seals died in a disease outbreak in New England during the winter of 1979–1980. The cause was found to be an influenza virus of avian origin that had infected the seals, probably as they hauled out on the rookeries of Cape Cod, Massachusetts. Seals of all ages developed pneumonia, which forced many out of the water and onto crowded beaches where the virus could spread easily from seal to seal by aerosol transmission. This was the first marine mammal die-off of demonstrated viral origin.

Morbilliviruses—a group that includes human measles virus and canine distemper virus (CDV)—have since proved to be an ominous threat to marine mammals (Van Bressem *et al.*, 2001). Phocine distemper virus (PDV), closely related to CDV, killed more than 18,000 harbor seals and a few hundred gray seals (*Halichoerus grypus*) in the North Sea in 1988; another outbreak of similar scale and pattern occurred in 2002. CDV killed several thousand Baikal seals (*Pusa sibirica*) in 1987 and thousands of Caspian seals (*P. caspica*) in 1997 and 2000. Between 1990 and 1992, another morbillivirus (a cetacean morbillivirus or CeMV) killed thousands of striped dolphins in the

Mediterranean Sea. Infected animals developed pneumonia, fever, and neurological disorders associated with encephalitis. The immunosuppressive effect of these viruses led to the development of secondary, often overwhelming, infections by bacteria, fungi, and other viruses.

Morbillivirus infection, often without recognized illness, is now known to be common in many cetacean and pinniped populations, and present in some long before the European epidemics. Retrospective studies indicate that morbillivirus infection, which was observed in some bottlenose dolphin carcasses examined during a die-off along the US mid-Atlantic coast in 1987–1988, played an important role in that event, and that outbreaks have occurred sporadically in coastal bottlenose dolphins populations along the southeast United States since the early 1980s. Pilot whales (*Globicephala* spp.), pelagic social species with large populations in which the virus may be endemic, may be a vector for CeMV in the North Atlantic. Similarly, harp seals (*Pagophilus groenlandicus*) are a likely reservoir for PDV and may have introduced the virus into immunologically naive European seal populations. Terrestrial carnivores are the likely source of CDV. Serological studies have tentatively linked a 1955 die-off of crabeater seals along the Antarctic Peninsula to CDV, perhaps transmitted from sled dogs (*Canis lupus familiaris*).

A number of viruses are associated with less serious health conditions (Kennedy-Stoskopf, 2001). Poxviruses, e.g., commonly cause skin lesions in pinnipeds and cetaceans; pox disease can appear and disappear in conjunction with other illnesses or stress. Herpesviruses are also common in cetaceans and pinnipeds and, although not usually serious, they have been associated with fatal pneumonia and hepatitis in harbor seal pups and encephalitis in a stranded harbor porpoise (*Phocoena phocoena*). Calicivirus infection is common among many marine mammals in the North Pacific; clinical disease, which in California sea lions appears as vesicular lesions on the skin of the flippers and mouth, may accompany stress, debilitation, or other infectious conditions, particularly leptospirosis.

Numerous other viruses have been found in marine mammals, many without recognized effect. The number of viruses and other pathogens continues to grow. Some may represent new or emerging diseases in marine mammal populations, while other "discoveries" reflect the growing intensity of research and advances in pathogen isolation and identification.

## H. Metabolic Disorders

Metabolic processes sometimes break down (St. Aubin and Dierauf, 2001). Environmental and biological factors that control hormonal regulation may fail to become synchronized, demands on the system may be overtaxing, and organ function, under the influence of a genetic clock, deteriorates with age and illness. The animal becomes incapacitated, but the underlying reason may be evident only at the molecular level and therefore difficult to detect. Not surprisingly, little is known about metabolic diseases in aquatic species.

In marine mammals, salt and water balance is regulated in part by the adrenal gland. Aldosterone, secreted from the adrenal cortex, normally acts on the kidney tubules to conserve sodium and thereby maintain salt and water balance. In pinnipeds, conditions that lead to prolonged stress, including molt, malnutrition, and disease, can exhaust the gland of aldosterone, resulting in loss of sodium from the body, a condition known as hyponatremia. Affected animals lose their appetite, become weak and disoriented, and eventually die. Aldosterone features in the stress response of cetaceans as well, only it does not become depleted and the animals do not develop hyponatremia. Quite the contrary, in severe stress following a stranding, a cetacean may eventually begin to drink seawater and develop a condition of salt overload, or fatal hypernatremia, that dehydrates tissues, including the brain.

## I. Tumors

Marine mammals develop all kinds of tumors, from benign lipomas that are little more than fatty lumps in the great whales to highly malignant lymphomas in young seals. As studies on marine mammals have increased, so have the numbers and variety of reported tumors (Gulland *et al.*, 2001).

In other mammals, tumors have been associated with a variety of factors, including hormones, viruses, congenital and hereditary defects, and physical and chemical agents. Establishing these links has generally required years of investigation on large populations and a systematic consideration and elimination of other possible contributors. These requirements are difficult to meet in marine mammal studies (O'Shea *et al.*, 2003). Hence it may never be possible to prove the assumption that environmental contaminants are responsible for the unusually high incidence of tumors in beluga whales (*Delphinapterus leucas*) in the St. Lawrence River, however, plausible or tempting the link. One study has been more fruitful: a herpesvirus, in combination with exposure to PCB contaminants, has been linked to the high rate of urogenital cancer in California sea lions.

## J. Biotoxins

Of the thousands of known species of marine phytoplankton, at least 40 can produce toxins that are harmful to humans and other top predators. Only since the late 1980s have we begun to realize the potential impact of such toxins on marine mammal populations (Geraci *et al.*, 1999; Van Dolah *et al.*, 2003). These compounds are difficult to detect and may leave little evidence of their presence. Thus their role in marine mammal mortality is often uncertain and may have gone unrecognized in the past. As one example, ciguatoxin, a dinoflagellate neurotoxin, was implicated in the illness of about 50 Hawaiian monk seals on Laysan Island in 1978; the weak, lethargic seals eventually became emaciated, suffered from severe parasitic infections, and died. Fourteen humpback whales (*Megaptera novaeangliae*) died in Cape Cod Bay (Massachusetts) in the winter of 1987 after eating mackerel containing saxitoxin, a neurotoxin that even in minute quantities causes respiratory paralysis.

In 1988, brevetoxin, a neurotoxin produced by the dinoflagellate *Karenia brevis*, the organism responsible for "red tides," was implicated in a die-off of several hundred Atlantic bottlenose dolphins along the US mid-Atlantic coast. Although the role of brevetoxin in that event remains unclear, this toxin has since been linked to mortality of bottlenose dolphins and Florida manatees in the Gulf of Mexico, where red tides are a recognized threat to the manatee population. Red tide outbreaks in southwest Florida in 1982, 1996, 2003, and 2005 killed about 37, 150, 98, and 81 manatees, respectively; these animals died of acute and chronic poisoning after consuming and inhaling *K. brevis* toxins.

In 1998, California sea lions along central California were poisoned by domoic acid, a neurotoxin produced by the diatom *Pseudonitzschia* sp. It caused convulsions, loss of coordination, and vomiting. While more than half of the stranded sea lions died, others were brought to rehabilitation centers and recovered. This discovery clarified some previously unexplained sea lion strandings and die-offs in this region. Subsequent outbreaks have affected sea lions and other marine mammals from central California to Mexico.

Marine mammals may be particularly susceptible to the neurological action of biotoxins for several reasons: (1) during a dive, blood is channeled to the heart and brain, effectively concentrating toxin there, and away from the liver and kidney where it is normally metabolized and excreted; (2) a short period of disorientation may be enough to impede an animal's ability to reach the surface for a vital breath of air (or to evade an oncoming vessel); and (3) animals that remain in the area of a bloom may be subject to the cumulative effects of toxins ingested or toxic aerosols inhaled over a period of days or weeks.

## K. Strandings

Stranding is defined as having run aground. The term here describes any marine mammal that falters ashore ill, weak, or simply lost. Most animals die at sea and only a fraction reach the shore. Those that do come ashore generally reflect the age, sex, and density of the animals in the area (Geraci and Lounsbury, 2005). Any change in the expected profile may be a signal that something unusual is happening, such as a toxic event, a disease outbreak, intensive local fisheries operations, or a change in prey abundance.

Pinnipeds and, to a lesser extent, sea otters normally spend time ashore, but only those unwilling or unable to return to sea are considered stranded. These would include pups that become separated from their mothers prematurely or fail to make a successful transition to independence. Most strand in the vicinity of the rookery, although some may stray far from their normal range. Other than in spring, when pups may come ashore in large numbers, and in the absence of unusual events such as disease outbreaks, pinnipeds normally strand alone.

Many cetaceans that strand singly are debilitated in some way. Some offshore species strand with characteristic illnesses. Short-beaked common dolphins (*Delphinus delphis*) along California, e.g., develop parasite-related brain damage, and dwarf (*Kogia sima*) and pygmy (*K. breviceps*) sperm whales along the southeastern US coast often come ashore with heart disease of uncertain cause and with impacted stomachs after ingesting plastic bags and other debris.

A mass stranding can be defined as two or more cetaceans, excluding mother–calf pairs, that come ashore alive at the same time and place. Highly social species of odontocetes [e.g., sperm whales, pilot whales, false killer whales (*Pseudorca crassidens*), and Atlantic white-sided dolphins (*Lagenorhynchus acutus*)] are the most probable victims. Many explanations have been proposed, but the only common link seems to be the strong social nature of these species. Once one or more animals strand, for whatever reason, the compulsion to stay together brings others ashore.

A stranded animal's chances of surviving diminish by the hour. Sea otters and pinnipeds risk hyperthermia, injury from terrestrial predators, and starvation. A cetacean has difficulty shedding heat even in cold weather, and a larger one may develop respiratory fatigue and distress as the chest cavity is compressed under its own weight. Within a few hours of stranding, some cetaceans begin to show evidence of shock or vascular collapse, which leads to poor circulation and impaired organ function. The onset of shock further impairs the whale's health and may prevent its recovery, even if it is returned to sea in what appears to be good condition.

## L. Habitat Alteration and Disturbance

Marine mammals have adapted over millions of years to the often harsh conditions of the marine environment. In the past few decades, environmental change has proceeded at a rate far exceeding the slow pace of evolution. How well can marine mammals cope with urban and industrial wastes, coastal dredging, undersea construction, vessel traffic, noise, and intense competition from humans for food resources? As with other influences on health, the effects—if they can be determined with any degree of certainty—will vary depending on species, sex, age, individual tolerance and behavior, and a host of other factors.

*1. Contaminants*    As long-lived predators at the top of the food chain, marine mammals accumulate contaminants in their tissues (O'Shea *et al.*, 2003). The concentrations and distribution within tissues depend on the type of contaminant and the animal's age and sex. Because most compounds accumulate over time, older animals generally have more. Fat-soluble substances, such as the persistent DDT, PCBs, and related organochlorines, reside in fatty tissues like blubber, liver, and brain; heavy metals are found in liver but also distribute in muscle, kidney, and other organs. Pregnant and lactating females produce milk using stored fat and the chlorinated hydrocarbons that came with it. While the suckling offspring loads up with contaminated milk, the female depletes her stores and, over time, has proportionally less and less than a male of equivalent size and age. What concentrations are eventually harmful to the male, or to the female as she loads and unloads the compounds with each reproductive cycle, or to the pup that may be even more sensitive? What happens to an animal of any age that becomes ill, stops eating, and uses stored fat, which releases these potentially toxic compounds into the bloodstream where, in increasingly higher concentrations, they are carried to other tissues?

As yet, no clear picture emerges, and broad differences in effects among species continue to invite speculation. In Baltic seals, organochlorine levels seem to be associated with low pregnancy rates and uterine pathology, as well as a disease complex characterized by metabolic disorders, hormonal imbalance, cranial bone lesions, and reduced immune function. The nature of marine mammals and the environment they live in pose serious challenges to conducting investigations that require tight controls and sophisticated technology. Meanwhile, we rely on empirical observations and preliminary studies that offer clues. Experimental studies are, nevertheless, yielding data supporting the link between exposure to certain chlorinated hydrocarbons and impaired immune function—and to tumors—in at least some species. A better understanding of the influence of contaminants on health will likely emerge from continued laboratory investigations.

*2. Oil Spills*    Oil spills are visible and unsightly, and sea otters show us how quickly fatal one can be. The 1989 *Exxon Valdez* incident in Prince William Sound, Alaska, was dramatic and beyond the proportion of other spills that have affected marine mammals (Loughlin, 1994). Until that event, relatively few marine mammals were known to have been killed by oil.

The impact of spilled oil depends on its composition, environmental conditions, and the species involved (Geraci and St. Aubin, 1990). During the first few hours or days after a spill, low molecular weight fractions are the most acutely toxic. They irritate and harm tissues, especially the sensitive membranes of the eyes and mouth; they can be ingested during feeding or when a fouled animal is grooming; or their vapors can be inhaled and damage the lungs. Light fractions are absorbed into the blood where they can attack the liver, nervous system, and blood-forming tissues. Sea otters caught in the *Exxon Valdez* spill showed signs of lethargy, respiratory distress, and diarrhea, and evidence of liver damage, kidney failure, and endocrine imbalance. Between 3500 and 5500 otters were estimated

to have died. Three hundred harbor seals also died; many had brain lesions, probably resulting from inhalation of vapors from fresh oil.

Evaporation of the low molecular weight fractions leaves heavy residues and thick, foamy emulsions called "mousse." By sticking tenaciously to vital insulating hairs of sea otters, polar bears, and some species of pinnipeds (e.g., fur seals), these substances can destroy the animals' ability to maintain thermal balance. The sea otter is especially vulnerable because its survival depends on a well-groomed hair coat.

Except for the sea otter, there is no real evidence that marine mammals ingest much oil. They may be able to deal with small quantities of fresh oil or that premetabolized by their prey because they, like other mammals, have liver enzymes required to metabolize and excrete such compounds.

*3. Ingesting Debris*    Some marine mammals become entangled in fishing nets and debris. Others are as likely to ingest various types of discarded items and trash that enter the oceans—mostly from land sources—at a rate of over 6 million metric tons each year. Florida manatees, e.g., face increasing risks of ingesting fishing line and hooks, wire, plastic bags, and other rubbish trapped in floating mats of vegetation (O'Shea *et al.*, 2001). Some cetaceans, including pygmy sperm whales and some beaked whales, share a tendency to ingest plastics. Some items are small and inconsequential; others may block or perforate the gastrointestinal tract, leading to slow starvation or sudden death.

*4. Other Disturbing Influences*    Habitat degradation can take many other forms: prey depletion, nutrient enrichment that leads to toxic algal blooms, underwater drilling noise, heavy vessel traffic, and disturbance of pupping or calving areas, to name a few. The potential range of effects is immense. A boat traveling through one of Florida's canals might collide with a manatee and kill it or raise the turbidity and inhibit the growth of water plants that are vital to its diet. Individuals might respond to food shortage or disturbance by moving to marginally suitable environments, e.g., northward to colder waters, where risks of cold stress are increased. A harp seal wandering far from its normal range following a collapse of prey stocks might introduce a pathogenic virus into a susceptible population. A sudden, unusual noise near a crowded pinniped rookery might cause animals to panic and stampede, trampling or abandoning their young. Intense underwater sound offshore might cause deep-diving beaked whales to ascend rapidly to the surface in panic and consequently suffer "the bends."

Other reactions to disturbances may be more subtle. In terrestrial mammals, intense noise alone can cause disorders ranging from long-term hearing loss to physiological stress, hypertension, hormonal imbalance, and lowered resistance to disease. Such effects are nearly impossible to document in marine mammals. It can be assumed that animals are generally unlikely to become habituated physiologically to any disturbances that are associated with threatening situations.

## III. The Future

We have a growing understanding of the range of pathogens in the sea and the mechanisms marine mammals have evolved to counter their effects. However, the expression of illness, whether in an individual or a population, is governed by dynamic environmental conditions, some of which are within our ability to control. For example, reductions in pollution, coastal habitat loss, and harmful fishing practices would have both direct and indirect health benefits for some marine mammal populations. Other environmental conditions are more difficult or impossible to control.

Harmful algal blooms have increased in frequency and distribution in the past few decades. Blooms in some areas are linked to human activities, in others to oceanographic anomalies that may be associated with climate change. There is also growing evidence that changing environmental conditions can influence the prevalence or virulence of pathogens (Harvell *et al.*, 1999). Thus it may be no coincidence that several major outbreaks of viral disease in marine mammals during the past few decades occurred in during periods of unusually warm weather.

If it is true that changes in our environment are occurring at an accelerated pace—and if the few years since the first edition of this encyclopedia are any indication of future trends—the ocean environment and every species dependent upon it, from benthic bacteria to blue whales (*Balaenoptera musculus*), and ourselves, are facing challenging times (Smetacek and Nicol, 2005; Grebmeier *et al.*, 2006). New chapters on health will be written with every technological leap in oceanography, remote sensing, medical diagnostics, and data integration. What pathogens will accompany the increasing traffic across oceans of people, animals, and goods? The effects of global warming on weather patterns, ocean dynamics, nutrient transport, prey distribution and abundance, and the spread of (and emergence of new) toxic organisms and pathogens will dramatically change what we have written here. Adaptation by some marine mammal species or populations may become evident as changes in general distribution, migratory patterns, and feeding areas or prey. Species already pressed to the edge of their "preferred" habitat may not be able to bear much more change. Others on the margin, like bowhead whales (*Balaena mysticetus*), may prosper as retreating sea ice opens the way for greater primary production and prey availability. Barring a catastrophic event, many—if not most—species will, as they have throughout their evolution, muster their pre-adaptive genes to secure a niche that will assure their survival, and their health.

## See Also the Following Articles

Endangered Species and Populations ■ Energetics ■ Fishing Industry, Effects of ■ Habitat Pressure ■ Mass Mortality

## References

Baker, J. R. (1984). Mortality and morbidity in grey seal pups (*Halichoerus grypus*): Studies on its causes, effects of environment, the nature and sources of infectious agents and the immunological status of pups. *J. Zool. Lond.* **203**, 23–48.

Cooper, L. W., Ashjian, C. J., Smith, S. L., Codispoti, L. A., Grebmeier, J. M., Campbell, R. G., and Sherr, E. B. (2006). Rapid seasonal sea-ice retreat in the Arctic could be affecting Pacific walrus (*Odobenus rosmarus divergens*) recruitment. *Aquat. Mamm.* **32**, 98–102.

Dunn, J. L., Buck, J. D., and Robeck, T. R. (2001). Bacterial diseases of cetaceans and pinnipeds. *In* "CRC Handbook of Marine Mammal Medicine" (L. A. Dierauf, and F. M. D. Gulland, eds), 2nd Ed., pp. 309–335. CRC Press, Boca Raton.

Elsner, R. (1999). Living in water: Solutions to physiological problems. *In* "Biology of Marine Mammals" (J. E. Reynolds, and S. A. Rommel, eds), pp. 73–116. Smithsonian Institution Press, Washington, DC.

Geraci, J. R., and St. Aubin, D. J. (1990). Summary and conclusions. *In* "Marine Mammals and Oil: Confronting the Risks" (J. R. Geraci, and D. J. St. Aubin, eds), pp. 253–256. Academic Press, San Diego.

Geraci, J. R., and Lounsbury, V. J. (2002). Marine mammal health: Holding the balance in an ever-changing sea. *In* "Marine Mammals: Biology and Conservation" (P. G. H. Evans, and J. A. Raga, eds), pp. 365–383. Kluwer Academic/Plenum Publishers, London.

Geraci, J. R., and Lounsbury, V. J. (2005). "Marine Mammals Ashore: A Field Guide for Strandings." National Aquarium in Baltimore, Baltimore.

Geraci, J. R., Harwood, J., and Lounsbury, V. J. (1999). Marine mammal die-offs: Causes, investigations, and issues. *In* "Conservation and Management of Marine Mammals" (J. R. Twiss, Jr., and R. R. Reeves, eds), pp. 367–395. Smithsonian Institution Press, Washington, DC.

Grebmeier, J. M., *et al.* (10 authors) (2006). A major ecosystem shift in the northern Bering Sea. *Science* **311**, 1461–1464.

Gulland, F. M. D., Lowenstine, L. J., and Spraker, T. R. (2001). Noninfectious diseases. *In* "CRC Handbook of Marine Mammal Medicine" (L. A. Dierauf, and F. M. D. Gulland, eds), 2nd Ed., pp. 521–547. CRC Press, Boca Raton.

Harvell, C.D., *et al.* (13 authors) (1999). Emerging marine diseases—climate links and anthropogenic factors. *Science* **285**, 1505–1510.

Jepson, P. D., *et al.* (18 authors) (2003). Gas-bubble lesions in stranded cetaceans. *Nature* **425**, 575–576.

Kennedy-Stoskopf, S. (2001). Viral diseases. *In* "CRC Handbook of Marine Mammal Medicine" (L. A. Dierauf, and F. M. D. Gulland, eds), 2nd Ed., pp. 285–307. CRC Press, Boca Raton.

Kreuder, C., *et al.* (9 authors) (2003). Patterns of mortality in southern sea otters (*Enhydra lutris nereis*) from 1998–2001. *J. Wildl. Dis.* **39**, 495–509.

Laist, D. W., Coe, J. M., and O'Hara, K. J. (1999). Marine debris pollution. *In* "Conservation and Management of Marine Mammals" (J. R. Twiss, Jr., and R. R. Reeves, eds), pp. 342–366. Smithsonian Institution Press, Washington, DC.

Loughlin, T. R. (ed.) (1994). "Marine Mammals and the *Exxon Valdez*." Academic Press, San Diego.

O'Shea, T. J., Lefebvre, L. W., and Beck, C. A. (2001). Florida manatees: Perspectives on populations, pain, and protection. *In* "CRC Handbook of Marine Mammal Medicine" (L. A. Dierauf, and F. M. D. Gulland, eds), 2nd Ed., pp. 31–43. CRC Press, Boca Raton.

O'Shea, T. J., Bossart, G. D., Fournier, M., and Vos, J. G. (2003). Conclusions and perspectives for the future. *In* "Toxicology of Marine Mammals" (J. G. Vos, G. D. Bossart, M. Fournier, and T. J. O'Shea, eds), pp. 595–613. Taylor & Francis, New York.

Smetacek, V., and Nicol, S. (2005). Polar ocean ecosystems in a changing world. *Nature* **437**, 362–368.

St. Aubin, D. J., and Dierauf, L. A. (2001). Stress and marine mammals. *In* "CRC Handbook of Marine Mammal Medicine" (L. A. Dierauf, and F. M. D. Gulland, eds), 2nd Ed., pp. 253–269. CRC Press, Boca Raton.

Trillmich, F., *et al.* (11 authors) (1991). The effects of El Niño on pinniped populations in the Eastern Pacific. *In* "Pinnipeds and El Niño: Responses to Environmental Stress" (F. Trillmich, and K. A. Ono, eds), pp. 247–270. Springer-Verlag, Berlin.

Van Bressem, M.-F., *et al.* (17 authors) (2001). An insight into the epidemiology of dolphin morbillivirus worldwide. *Vet. Microbiol.* **81**, 287–304.

Van Dolah, F. M., Doucette, G. J., Gulland, F. M. D., Rowles, T. L., and Bossart, G. D. (2003). Impacts of algal toxins on marine mammals. *In* "Toxicology of Marine Mammals" (J. G. Vos, G. D. Bossart, M. Fournier, and T. J. O'Shea, eds), pp. 247–269. Taylor & Francis, London and New York.

# Hearing

## SIRPA NUMMELA

Animals use sound and hearing for communication, especially signaling different behaviors related to reproduction, breeding, territory marking, as well as for detection and localization of prey and predators and navigation. Cetaceans have succeeded superbly in aquatic hearing, and have also become crucially dependent on their hearing while adapting to the aquatic world. Sirenians have not invested so fully to their auditory sense as whales, but their hearing is sufficiently functional even though it does not have a very prominent role among their senses. Pinnipeds are dependent on their audition in both air and water, and their requirement to hear in both media poses demands beyond those on the whales and manatee ear. Good hearing ability in air does not necessarily imply an equally good hearing ability in water, and it is impossible to optimize the auditory functions in both media; compromises are made regarding hearing sensitivity and frequency ranges.

## I. Acoustics

### A. Sound Velocity

For pure tones sound velocity $c$ is the product of sound frequency $f$ and sound wavelength $\lambda$; $c = f\lambda$, and is nearly 5 times higher in water than in air. This means that for a given sound frequency, the wavelength is nearly 5 times longer in water than in air. Shorter wavelengths have better spatial resolution, and hence high frequencies are better suitable for detecting small objects than are low frequencies, and are thus used in echolocation. However, high frequencies attenuate rapidly and do not carry very far.

### B. Impedance

The characteristic acoustic impedance of a medium is $Z = p/v$, where $p$ is the sound pressure, $v$ is the particle velocity. A given sound pressure gives air molecules a larger particle velocity than water molecules, leading to lower impedance of air than of water. In addition, the specific acoustic impedance of the fluid-filled cochlea is approximately one-tenth of the characteristic acoustic impedance of water, $Z_c = 150\,\text{kPas/m}$. Hence in hearing, airborne sounds travel from a medium of low impedance to one with much higher impedance in the ear. Waterborne sounds travel from a medium with higher impedance to one with slightly lower impedance. The impedance mismatch causes a reflection of sounds at the interface, and to overcome this, an impedance matching device is needed. This device adjusts the sound pressure and particle velocity, either by increasing the pressure, and/or decreasing (airborne sound) or increasing (waterborne sound) the particle velocity between the outer medium and the cochlea.

### C. Sound Intensity and Sound Pressure

Sound loudness is measured with pressure meters and is given in decibels, generally expressed using sound pressure level: $L_p = 20\,\text{dB}$ $\log p/p_o$. The reference pressure value $p_o$ for sounds in air is $20\,\mu\text{Pa}$, and in water $1\,\mu\text{Pa}$. For the cochlear sensitivity, the incident sound energy is a relevant parameter, and sound intensity should be used when comparing the hearing of a terrestrial vs an aquatic animal.

For a plane wave, the sound intensity $I = p^2/Z$, where $Z$ is the characteristic acoustic impedance of the medium, $Z_{\text{air}}$ or $Z_{\text{water}}$. The ratio $Z_{\text{water}}/Z_{\text{air}}$ is approximately 3700. This means that when plane waves with equal intensities in air and water are compared, the sound pressure in water is larger. With the conventional reference pressure $p_o$ values of $20\,\mu\text{Pa}$ (air) and $1\,\mu\text{Pa}$ (water), the waves have equal intensities when the sound pressure level $L_p$ of the aquatic wave is $61.8\,\text{dB}$ larger than sound pressure level $L_p$ of the wave in air. Hence, for audiograms from different habitats, a terrestrial animal and an aquatic animal have equal sensitivities (equally sensitive hearing) if the threshold $L_p$ value of the aquatic animal is $61.8\,\text{dB}$ larger than the threshold $L_p$ of the terrestrial animal.

### D. Directional Hearing

Interaural time difference and interaural intensity difference are two methods that are used by mammals to determine the direction of a sound. The interaural time difference is a usable method when the interaural distance (head size) is relatively large when compared to the wavelength of the sound; mostly at lower frequencies. Measuring experimentally the temporal resolution or the minimum audible angle gives indications of the sound localization abilities of an animal.

The interaural intensity difference is caused by the shadowing effect of the head, sound intensity is generally larger in the ear where it arrives first. This means of binaural hearing is important when the interaural time difference is small, usually at higher frequencies. Additionally, the intensity difference between the ears increases with the sound frequency (Fig. 1). Hence, echolocation using high-frequency hearing not only gives better spatial resolution than using low frequencies, but it also improves directional hearing.

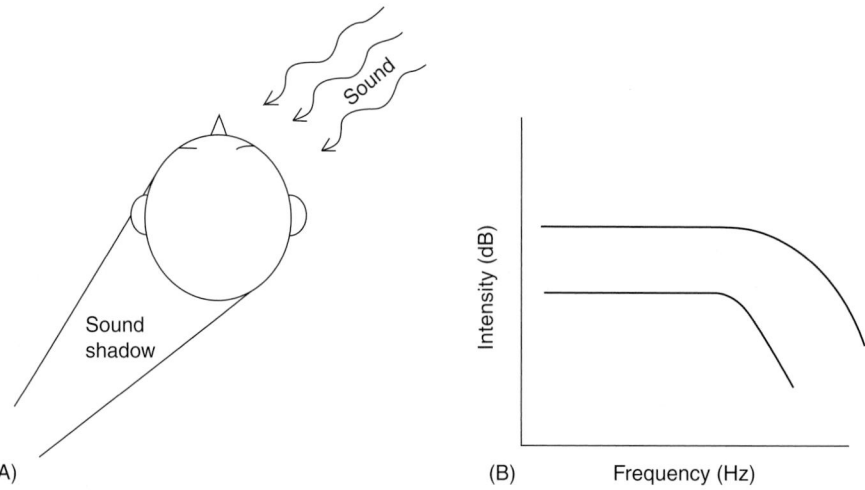

(A)

(B) Frequency (Hz)

**Figure 1** (A) *Diagram of a mammalian head with sound arriving obliquely from the right, showing that the left ear is in a sound shadow. (B) Approximate sound intensities at left and right ear for mammal in (A). From Nummela and Thewissen (2008), with permission.*

## II. Ear Anatomy and Hearing Mechanisms

### A. Land Mammals

In a land mammal ear, sound is collected by the outer ear pinna, and guided through the external auditory meatus (outer ear canal) to the tympanic membrane, which is set into vibration (Fig. 2A). These vibrations are carried further by the ossicular chain, situated in the middle ear cavity and consisting of three small bones, malleus, incus, and stapes. The malleus is attached to the tympanic membrane with its handle, the manubrium. The middle ear ossicles together form a swing which vibrates between the tympanic membrane and the oval window of the cochlea, setting the inner ear fluid into motion. The tympanic bone and the periotic bone have several contacts with each other, and are also in close contact with the squamosum and other skull bones. In land mammals, the ear is not acoustically isolated from the skull, and this makes bone conduction a possible hearing mechanism, in addition to the general land mammal mechanism.

While hearing in air, sound energy is transmitted from areas with lower impedance of air to much higher impedance of the inner ear (discussed earlier). In such a case, the pressure should be increased, and the particle velocity should decrease (as $Z = p/v$). This is in fact what happens. The pressure is increased when sound is transmitted from a larger area (tympanic membrane) to a smaller area (oval window), and the particle velocity is decreased when the malleus lever arm is longer than the incus lever arm. These two arrangements together, the area ratio and the lever ratio, contribute to the sound energy transmission in the middle ear, and the middle ear functions as an impedance matching device between the surrounding air and the inner ear fluid, decreasing the reflection of sound at the tympanic membrane. Additionally, the middle ear functions as an intensity amplifier.

The mammalian cochlea is a coiled structure with channels, filled with inner ear fluid (endolymph and perilymph). The basilar membrane supports the organ of Corti which contains the auditory sensory cells, along with a row of supporting cells. The vibrations of the stapes at the oval window are transferred to the movements of the inner ear fluid, this causes the bending of the auditory hair cells and leads to a nerve impulses in the cochlear nerve. The basilar membrane is tonotopic so that the base of the membrane is most sensitive to high frequencies, and the apex (helicotrema) is most sensitive to low frequencies. The number of turns in the cochlea varies, but is related to the sound frequencies heard by the animal; animals that hear high frequencies have more turns than animals that specialize on hearing low frequencies.

The hearing range is in general determined by the ear structures. In the middle ear, low-frequency transmission is constrained by the elasticity of the system; large tympanic membrane and middle ear volume together with non-stiff ossicular chain improve low-frequency hearing. High-frequency transmission is constrained by the mass of the system, mainly by the middle ear ossicular mass; for high-frequency transmission small middle ear is needed. The inner ear can be a constraining factor too, and this is often seen in the audiogram as a very steep rise at the high-frequency part.

### B. Bone Conduction

Bone conduction occurs when sound energy is transmitted from a surrounding medium to the cochlea through vibrations of the soft tissues and bony parts of the head directly, instead of the outer ear and the rotating ear ossicles. Functionally, there are two types of bone conduction: compressional and the inertial. In the compressional type, a pressure differential develops across the cochlear partition of the inner ear. In the inertial type, relative motion between the ossicular chain and the temporal bone leads to cochlear stimulation much the same way as in air-conducted hearing of land mammals. Both of these types lead to displacement of the basilar membrane and (eventually) bursts of neural impulses.

In general, bone conduction is disadvantageous to land-living mammals, as it interferes with airborne sounds entering the ear.

**H**

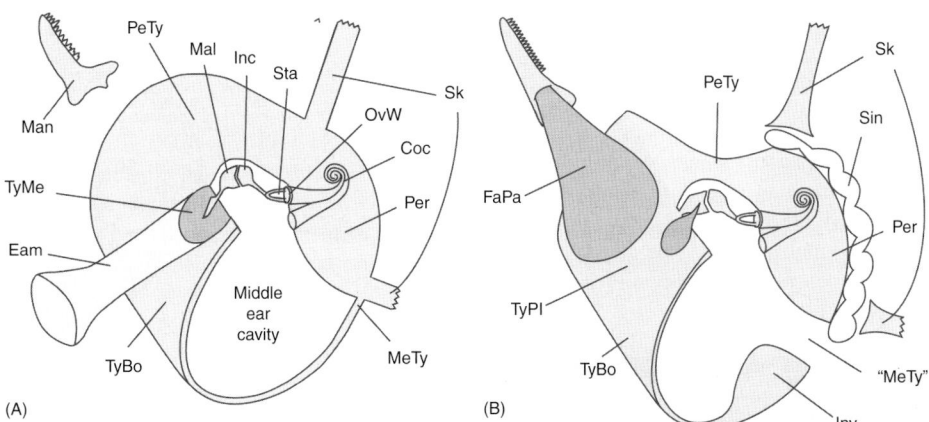

**Figure 2**  *(A) Diagram of the land mammal ear. (B) Diagram of the modern odontocete ear. For technical reasons, the mandibular foramen and the mandibular fat pad are shown on the lateral side of the mandible, although they in reality are situated on the medial side. Coc, cochlea; EAM, external acoustic meatus; FaPa, mandibular fat pad; Inc, incus; Inv, involucrum; Mal, malleus; Man, mandible; MeTy, medial synostosis between periotic and tympanic bone, in cetaceans this synostosis is absent and is homologous to a gap between these bones ("MeTy"); OvW, oval window; Per, periotic bone; PeTy, joint between periotic and tympanic; Sin, air sinuses; Sk, skull; Sta, stapes; TyBo, tympanic bone; TyMe, tympanic membrane; TyPl, tympanic plate. Reprinted by permission from MacMillan Publishers Ltd: Nature (Nummela et al., 2004).*

Bone conduction can be minimized if the mass center point of the ossicular chain coincides with its rotational axis. However, bone conduction in marine mammals can be useful because the density of water is similar to the density of the body tissues allowing for efficient transfer of energy. A disadvantage of bone conduction is that it does not allow for directional hearing. Marine mammals such as cetaceans avoid bone conduction by inserting air cushions between ears and skull in order to acoustically isolate the ears.

### C. Generalized Odontocetes

Odontocetes have excellent hearing; they are able to hear very high frequencies, exceeding even 150 kHz, and their hearing sensitivity in water is very acute. Additionally, as an adaptation to the aquatic environment, odontocetes have developed rapid auditory temporal processing which facilitates sound localization and echolocation.

The ears of modern odontocetes are exclusively adapted for underwater hearing (Fig. 2B). Odontocetes lack an outer ear pinna, and their external auditory meatus is very thin and partly occluded and not functional in hearing. In most species, the tympanic and periotic bones are connected to each other through small, usually non-bony contacts, and together these bones form a tympano-periotic complex that houses the middle ear and the inner ear. The tympanic bone is bowl-shaped, and houses the middle ear cavity. The lateral wall of the tympanic bone is thin, like in mammals in general; in odontocetes this wall is called the tympanic plate. The medial part of the tympanic is thick and bulky, a massive counterpart to the thin lateral part. The tympanic membrane has become a long conical structure, sometimes called the tympanic ligament, it attaches with its medial tip to a small process in the malleus. The odontocete tympanic membrane is not functional in hearing, but rather in the pressure regulation of the middle ear cavity. The three middle ear ossicles are situated between the tympanic plate and the oval window. The malleus has lost its manubrium but has a long thin anterior process, processus gracilis, which connects it to the anterior rim of the tympanic ring. The malleus joints the incus, and these two ossicles together form most of the mass of the chain. The incus in turn joins with the stapes. The footplate of the stapes sits at the cochlear oval window (Fig. 3 and 4).

The tympano-periotic complex is surrounded by air sinuses and thus it is acoustically isolated from the skull. The odontocete lower jaw contains fatty tissue called mandibular fat pad in the mandibular canal, and through the large mandibular foramen this fatty structure reaches up to the tympanic plate (Figs. 3 and 4). The mandibular fat pad is composed of triacylglycerols that are known to conduct vibrations efficiently. Experimental evidence has shown that the odontocete lower jaw is very sensitive to sound, and it has been

suggested that the odontocetes use their lower jaw as an outer ear that collects sound energy that is then guided forward by the mandibular fat pad to the tympanic plate. The tympanic plate vibrations are moved forward to the ossicular chain and to the cochlea. It should be noted that in water, ossicles of high mass can transmit high frequencies which in air can only be transmitted by very light ossicles. Killer whales and mice can hear equally high frequencies, but the ossicles differ hugely in their size.

For an odontocete in water, sound energy travels from higher impedance to somewhat lower impedance. Then, to minimize the reflection of sound energy, pressure should be decreased and particle velocity should be increased. However, when the vibration moves from the larger tympanic plate to the smaller oval window, the pressure increase is inevitable. On the one hand it is advantageous; it increases the intensity, resulting in better hearing sensitivity. But the increase of pressure is disadvantageous in an odontocete ear and

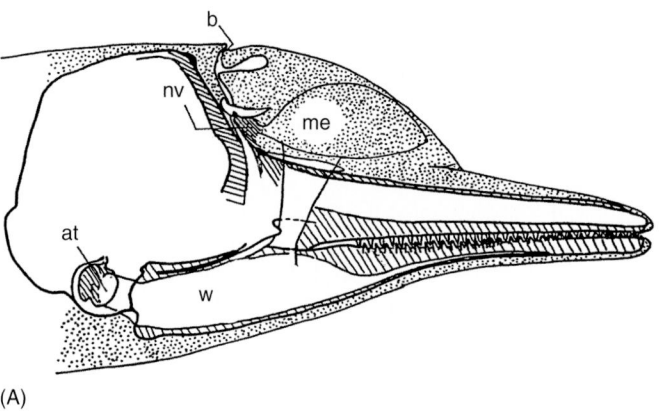

(A)

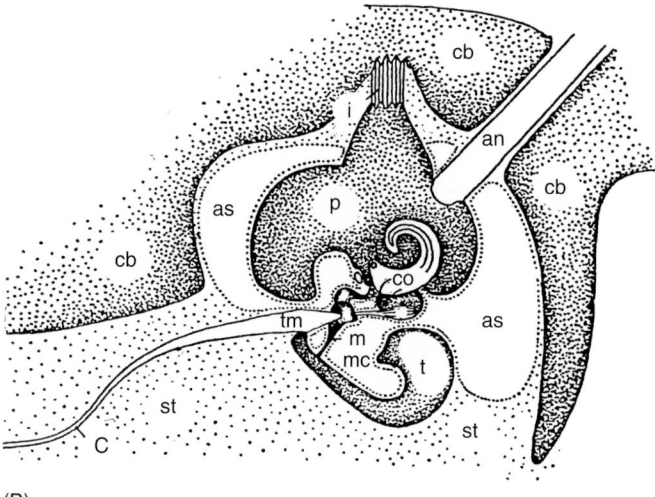

(B)

Figure 4  (A) Structures in the odontocete head (modified from Norris, 1968). (B) Odontocete middle ear region (modified from Reysenbach de Haan, 1957). The nasal valve lips, the blowhole and the melon contribute to sound production and focusing. an, auditory nerve; as, air sinuses; at, attachment of the mandibular wave guide onto the tympanic bulla; b, blow hole; c, auditory canal; cb, cranial bones; co, cochlea; l, ligaments; m, malleus; mc, middle ear cavity; me, melon; nv, nasal valve with lips; p, periotic bone; t, tympanic bone; tm, tympanic membrane; st, soft tissues; w, mandibular wave guide.

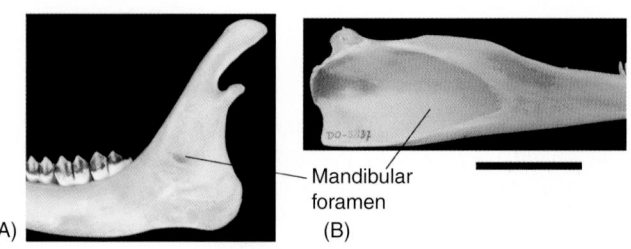

(A)                                              (B)

Figure 3  Medial view of lower jaw showing the mandibular foramen size. (A) Right mandible of deer, Odocoileus. (B) Left mandible of dolphin, Lagenorhynchus. Scale bar = 5 cm. From Nummela et al. (2007).

needs to be compensated by a large increase of particle velocity, hence the lever mechanism (Fig. 5). The lateral wall of the tympanic bone, the tympanic plate, vibrates relatively more than the thick involucrum which stays practically immovable. This structure forms the first of the two middle ear lever mechanisms in the odontocete ear, increasing particle velocity. In the rotation of the ossicular chain the malleus and incus move very little in the epitympanic recess, and the incus long arm moves much more; this forms the second middle ear lever and, again, the particle velocity is increased.

A tympanic plate, made of bone, can be much larger than a tympanic membrane and still vibrate as one single unit. With a large tympanic membrane, when the sound frequency becomes high enough, the vibration of the membrane breaks into several small units, and this may lead to a considerable loss of energy. A bony rigid tympanic plate hence makes it possible for this area to be so large, a condition for the functional success of the odontocete ear. Another factor that helps in rigidity is the increased bone density in the odontocete ear. The tympano-periotic complex together with the ossicles have clearly higher density than the rest of the skull, approximately 2.6–2.7 g/cm$^3$.

The sound velocity is nearly 5 times larger in water than in air means that the wavelength of a given sound is equally much longer in water, reducing considerably the usefulness of the interaural time difference. However, despite this increase in velocity and wavelength, the wavelength of the high frequencies that marine mammals, especially odontocetes, use are sufficiently short to allow binaural hearing. The high frequencies together with a large head compensate for the increase in sound velocity, resulting in good directional hearing in odontocetes.

The auditory input area of the dolphin head was studied by Mohl *et al.* (1999), who found that the best auditory sensitivity is at the middle of the lower jaw (see Figure 3 in Echolocation chapter). Recently Popov *et al.* (2008) presented evidence for double acoustic windows in the dolphin: they were able to identify for the *Tursiops truncatus* at least two different sound-receiving areas (acoustic windows) with different frequency sensitivity. One window was situated 22–26 cm caudal of the melon tip, close to the bulla and auditory meatus, and mainly sensitive to frequencies between 16 and 22.5 kHz. The other window was found 9–13 cm caudal of the melon tip, this place corresponding to a proximal part of the lower jaw, and being most sensitive to frequencies between 32 and 128 kHz. The significance of these multi-receiving areas is not clear yet but it is possible that they provide cues for localization of sound sources and sound pattern recognition, because the best-sensitivity axis direction may be frequency dependent.

Audiograms for odontocetes are presented in the chapter on Echolocation. Comparisons of behavioral and evoked electrophysiological measuring techniques for determining hearing thresholds have been carried out to find out how these two techniques could be replaced by each other. Electrophysiological methods give a relative sensitivity, not the absolute thresholds, but are much less time consuming and in many cases the only possible ones (e.g., with stranded animals). Behavioral methods require time for training, being also much more expensive. In general, the thresholds agree relatively well with each other, between auditory evoked potential and behavioral measurements, the AEP had consistently higher thresholds, with the greatest differences at the lowest frequencies. See Supin *et al.* (2001).

### D. Physeteroidea/Ziphiidae

Beaked whales such as *Ziphius cavirostris* and *Mesoplodon densirostris* produce high-frequency echolocation clicks, in *Z. cavirostris* clicks the energy is centered on 42 kHz, with energy up to about 80 kHz. The temporal resolution of beaked whales is high, similar to that of other cetaceans.

The ear structures of physeteroids and ziphiids are similar to those of other odontocetes, but some differences do exist. The processus gracilis is short, and sometimes hardly discernible, the malleus has a larger and less elastic contact with the tympanic plate rim. Also, the contacts between the tympanic and periotic are less elastic, as there is real synostosis between the bones, although not wide. The malleus lacks a transversal part, being relatively round, and the incus joint facet is flat. There is also a relatively long mastoid process, especially in ziphiids, and this process stays in contact with the skull in a similar way as in mysticetes, although not that deep. The tympano-periotic complex is relatively smaller than in other odontocetes. These morphological differences have functional consequences; the lack of elasticity is the contacts theoretically improves the transmission of higher frequencies. However, no quantitative models have so far been presented for this type of hearing mechanism, and the odontocete model may be applied to these species.

Very little is known of the hearing of these species. Auditory evoked potentials measured on a stranded juvenile beaked whale (*Mesoplodon europaeus*) showed that the animal was most sensitive to high-frequency signals between 40 and 80 kHz, when the lowest tested frequency was 5 kHz, the highest was 80 kHz. The animal was probably able to detect frequencies much higher than 80 kHz (Cook *et al.*, 2006).

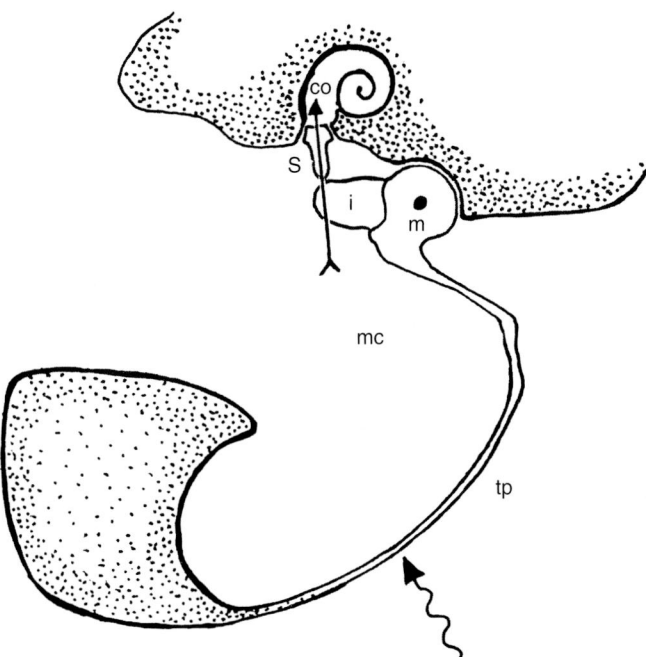

**Figure 5** *Schematic presentation of the odontocete middle ear structure. The incident sound arriving at the tympanic plate is represented by the wave-formed arrow. The black dot indicates the rotational axis of the ossicular chain, and the straight arrow shows the direction of the stapes movement during increasing sound pressure. co, cochlea; i, incus; m, malleus; mc, middle ear cavity; s, stapes; tp, tympanic plate.*

## E. Mysticetes

Low frequencies have long wavelengths, and carry over longer distances as they attenuate more slowly than high frequencies; thus low frequencies are more suitable for long-distance communication. Mysticetes specialize in hearing low frequencies that carry very far with little attenuation. These whales can communicate with each other over hundreds, even thousands of kilometers. The frequency range of sounds that whales utilize ranges from a few hundred hertz (Hz) to well over one hundred kilohertz (kHz).

Mysticetes are specialized to produce frequencies in the lower range, mainly <5 kHz, the lowest frequencies used are 10–15 Hz. Hence the hearing frequency range of mysticetes is clearly different than that of odontocetes. Mostly, the ear anatomy is similar to that of odontocetes, Mysticetes have no outer ear pinna, their external auditory meatus is thin, and hardly functional in hearing. At the medial end of the meatus there is a wax plug attached to the tympanic membrane. The tympanic and periotic together form a tympano-periotic complex, but here the connections between the two bones are true ossifications, although they are relatively small. The bone complex is partly isolated from the skull with sinuses, but a long mastoid process extends posteriorly from the periotic and makes large, although loose contact with the skull. The tympanic membrane is a conical structure, and its medial tip is attached to the malleus. The malleus is fused to the anterior rim of the tympanic ring with its anterior process, the processus gracilis, and this rod-like process is long and robust, not thin and elastic like in odontocetes. The three ossicles form a chain between the lateral wall of the tympanic and the oval window. The term "tympanic plate" is a functional term, and is not used for mysticetes, at least as long as it is unclear whether the lateral wall of the tympanic bone functions as an analog to the land mammal tympanic membrane, as it does in odontocetes.

The size of the ear complex and the ossicles is huge, but still the size relations between different parts coincide with those of odontocetes (Nummela 1999b). Hence, when looking for functionally important relations it would seem that the odontocete and the mysticete ears act similarly. However, the sound route to the ear is unclear in mysticetes. They do not have a mandibular canal, and no fat pad that would conveniently guide sound-induced vibrations further. Mysticetes are highly dependent on their low-frequency hearing in their long-distance communication that can span over tens, even hundreds of kilometers. No behavioral measurements on mysticete hearing exist, but their hearing frequency ranges have been predicted, e.g., on the basis of the detailed inner ear morphology (Ketten, 2000; Parks *et al.*, 2007). This can give relative values for frequency ranges, but does not tell anything about the hearing sensitivity or the sound transmission mechanisms of mysticetes as such.

## F. Pinnipeds

Stimulation of the cochlea by bone conduction can be even enhanced conveniently by morphological changes in the middle ear structures. The mass center point can be moved further away from the rotational axis by placing extra mass in some part of the ossicular chain. This will increase the different phases of vibration between the stapes and cochlear capsule. This kind of morphological evolution has occurred, e.g., in phocids and odobenids, in which the incus is inflated.

Pinnipeds are amphibious and need to hear both in air and in water. In general, they have better hearing sensitivity than terrestrial mammals have in water, but worse sensitivity than odontocetes have in water. All pinnipeds have retained the land mammal type sound transmission mechanism, with some modifications though. All otariids, the eared seals, have a small outer ear pinna, whereas phocids and odobenids do not. The tympanic bone in pinnipeds forms large middle ear cavity, typical of carnivores, and the periotic is in close contact with the skull. The otariid middle ear ossicles are of the same size as normal terrestrial carnivore ossicles, but phocid and odobenid ossicles are enlarged and pachyostotic, the most prominent example being the elephant seal ossicles. The mass of the phocid ossicles is further increased by a somewhat higher ossicular density (2.2–2.3 g/cm³) than the one for land mammals (around 2.0 g/cm³).

Pinnipeds have periotic bones that are fused to the skull, and there may be a fat channel or the enlarged mastoid bone that provides bilateral sound conduction. When diving, the pressure in the middle ear cavity is increased with the help of cavernous tissue in the cavity walls.

For hearing in air, pinnipeds use the normal land mammal hearing mechanism through the tympanic membrane. For hearing in water, no special hearing mechanism is known at least so far, but it is supposed that pinnipeds in water rely on bone conduction. The enlarged ossicular mass of some species limits the high-frequency hearing limit in air, but is apparently advantageous in enhancing the signal in the bone conduction, by creating a larger phase difference between different vibrating structures. As a result, phocids and odobenids have sacrificed part of their high-frequency hearing in air in order to gain better hearing in water.

Both the middle ear and the inner ear can constrain the hearing frequency range, and both may be the main limiting factor in phocids and otariids. As shown in Fig. 6, the high-frequency hearing limits of phocids in air are clearly lower than the limits in water, but for otariids, these limits are similar in air and water (these are experimental results from behavioral audiograms). The mass inertia of the heavy phocid ossicles explains the lower high-frequency hearing limits in air—according to the underwater audiograms the phocid cochlea is sensitive to higher frequencies. The inertia of the normal-sized otariid ossicles should allow the underwater hearing limit to be at higher frequencies than in air. However, their high-frequency hearing limits in air and water are approximately similar, so the limiting factor must be the cochlea alone.

In Fig. 7, the underwater audiogram is shown for walrus, and for two phocids and one otariid. The high-frequency hearing limit is lowest for the walrus, although its middle ear structure is overall similar to that of the elephant seal. Hence, the cochlea is the main limiting factor for the high-frequency hearing of the walrus.

The issue of whether some or all species of pinnipeds possess specialized acoustical abilities for underwater echolocation of the type shown by odontocete cetaceans, has been a controversial issue for decades, and was reviewed by Schusterman *et al.* (2000). The current understanding is that pinnipeds have not developed active biosonar, but rather that their amphibious lifestyle has resulted in relatively non-specialized underwater hearing abilities, which these animals use in combination with alternative senses that are possibly equally prominent, e.g., vision and hydrodynamic reception (see chapters on Vision and Sensory Biology).

In the absence of biosonar, pinnipeds may still have great hearing acuity under water. It has been claimed that due to the amphibious way of life, selection pressures for highly sensitive, acute underwater hearing have not shaped the pinniped auditory system to as great an extent as they have in the dolphins, but instead, pinnipeds would rely largely on other sensory cues (Schusterman *et al.*, 2000). Bodson *et al.* (2006) have shown experimental evidence for underwater auditory localization of *Phoca*, where the seal was able to localize and accurately attain hidden underwater sound sources, and there

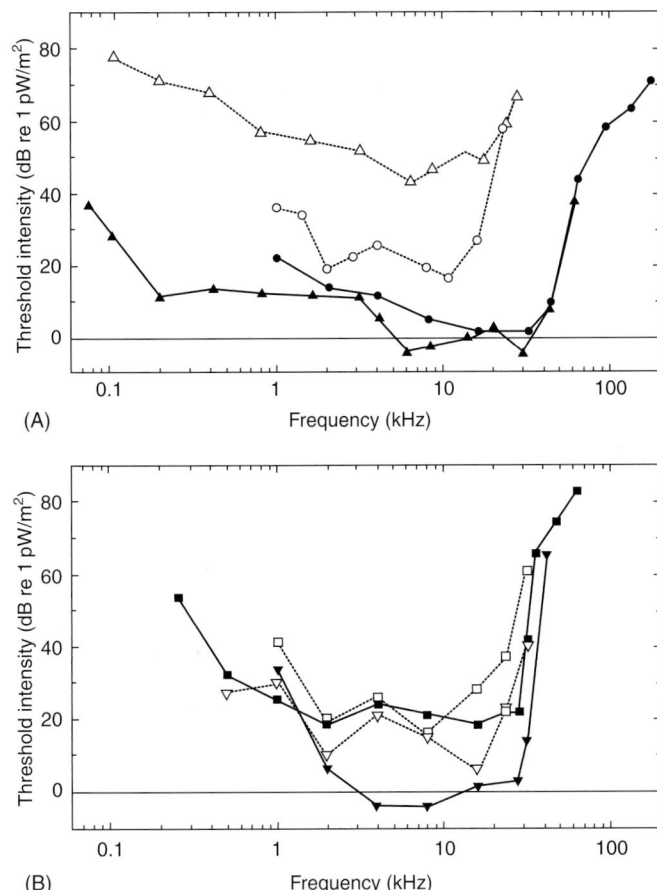

(A)

(B)

**Figure 6** *Audiograms for (A) phocids and (B) otariids. All threshold intensities are given in decibels relative to 1 pW/m², for comparison of thresholds in water and air. Open symbols in-air, filled symbols underwater.* ○, ● *Phoca vitulina;* △,▲ *Mirounga angustirostris;* □,■ *Zalophus californianus;* ▽,▼ *Callorhinus ursinus. From Hemilä et al. (2006), with permission.*

was no indication that the seal was guided by other than acoustical information.

Aerial sound localization abilities of pinnipeds (*Mirounga, Phoca, Zalophus*) are comparable to the domestic cat and rhesus monkey. Factors such as head size and head movements may also help sound localization in pinnipeds (Holt *et al.*, 2004).

### G. Sirenians

Sirenians are obligate herbivores that do not echolocate, but have been reported to have good high-frequency hearing, although their hearing sensitivity is clearly lower than that of odontocetes. Manatees have relatively good localization abilities. The ear anatomy of sirenians is very different from that of any other mammals (Ketten *et al.*, 1992; Chapla *et al.*, 2007), and at the moment it is unclear how the sound reaches the cochlea. No quantitative mechanism for sirenian hearing has yet been presented.

Sirenians have no outer ear pinna, and their external auditory meatus ends in a blind sac. Fatty tissue separates the meatus and the tympanic membrane, this fatty tissue also occurs around the tympanic bone. The tympano-periotic complex is intracranial but is not fused to the skull bones. The tympanic is a ring-shaped bone and does not form a cup-like bulla like in many other mammals, e.g., cetaceans and pinnipeds. The tympanic and periotic are fused to each other at two small locations. The tympanic membrane is large, and thick, and bulges outward, pushed by a cartilaginous keel of the malleus that is an extra structure between the malleus and the tympanic membrane. The middle ear ossicles are massive, and pachyostotic, and their density is the highest known among mammals (*ca.* 2.9 g/cm³; Chapla *et al.*, 2007). The zygomatic process is very fatty. It has been suggested that the sound path to the inner ear would be through this process so that the fat in the process would help in guiding sound to the ear similar to the mandibular fat of odontocetes. However, although experimental evidence for this is lacking, this is the only hearing mechanism suggested for sirenians so far. It is also possible that sirenians use bone conduction hearing to some degree.

**H**

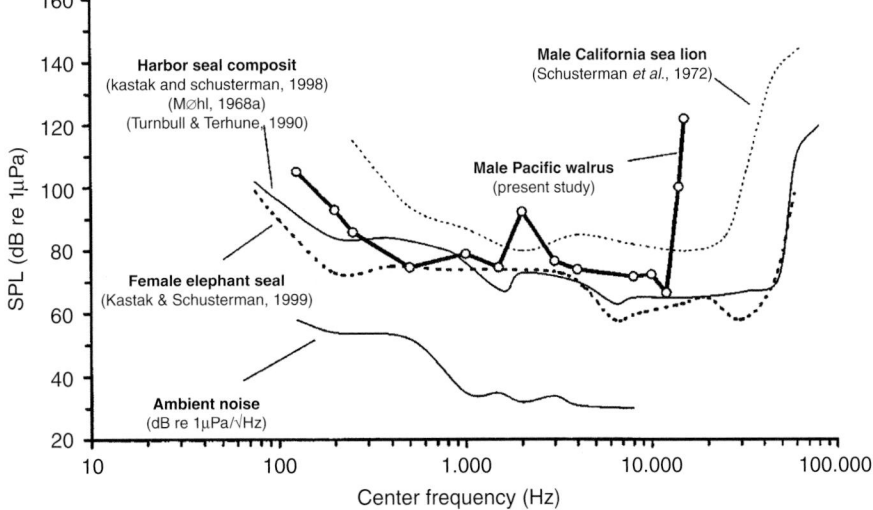

**Figure 7** *Audiograms for the Pacific walrus, harbor seal, northern elephant seal, and California sea lion. The ambient noise is plotted in dB re 1μPa/√Hz (note that this is a different unit than the one along the Y-axis). From Kastelein et al. (2002), with permission.*

It is assumed that sound waves pass directly through the soft tissues of the head to the ears, but little is known about the properties of the manatee's tissues. The fatty tissue is significantly less dense than other soft tissues of the head and the squamosal bone is significantly less dense than the other bones of the head. In contrast to cetaceans where the relationship of the ossicular mass and the tympano-periotic is constant, the sirenian ossicular chain is overly massive where the periosteum separates the bilobed periotic from the squamosal bone.

Evoked potential studies on West Indian manatee (*T. manatus*) in air response up to 35 kHz with the largest peaks in response to stimuli from 1 to 1.5 kHz. AEP measurements on the Amazonian manatee *T. inunguis* in a water-filled bath gave response to stimuli up to 60 kHz. The first underwater behavioral audiogram for the West Indian manatee is of U-shape, like in mammals in general, and the hearing range extends from 0.4 to 46 kHz, with peak sensitivity at 16 and 18 kHz, the range of best hearing is 6–20 kHz (Fig. 8; Gerstein *et al.*, 1999).

Florida manatees have been found to have surprisingly good temporal resolution (600 Hz), roughly 10 times that of humans (50 Hz) and half that of dolphins (1200 Hz). Amazonian manatee was measured to have much lower temporal resolution, but this much reduced response was perhaps be a result of long-term adaptation (Mann *et al.*, 2005).

Manatee vocalizations range from 4 kHz to above 25 kHz. Many of these vocalizations are harmonic, and it is possible that the high temporal resolution of the auditory system provides a useful system for detecting the harmonics. Selective pressures to localize sound under water might also be responsible for the high-frequency sensitivity of the manatee. Higher-frequency sounds will produce larger interaural intensity differences than low-frequency sounds because they are more effectively shadowed by the head.

### H. Polar Bear and Sea Otter

The polar bear (*Ursus maritimus*) and the sea otter (*Enhydra lutris*) are amphibious and apparently have relatively sensitive hearing in their watery habitat. Nachtigall *et al.* (2007) measured auditory evoked potentials for polar bear in air, and made estimations of their hearing sensitivity, based on the background noise. The polar bear hearing was found sensitive over a wide frequency range, with the best sensitivity in the range 11.2–22.5 kHz. No detailed descriptions appear for the ear or hearing mechanisms of these species. Based on their land mammal ear, both these species use the general land mammal hearing mechanism in air, and in water supposedly rely on bone conduction mechanism. Whether these species possess some morphological adaptations that would enhance the bone conduction stimulus, is currently unknown. However, most likely both these amphibious carnivores at least in air hear relatively high frequencies, and have good hearing sensitivity, like carnivores in general.

### III. Evolution of the Aquatic Ear in Cetaceans

### A. General

Evolutionarily, the odontocete ear is best understood. The evolutionary history of cetacean hearing can be divided into three different phases: (1) the modern generalized whale underwater hearing mechanism evolved already among archaeocetes during the Eocene, within less than 10 million years (Nummela *et al.*, 2004, 2007); (2) echolocation evolved around the time of divergence of odontocetes and mysticetes, or alternatively, within the earliest odontocetes; and (3) mysticete low-frequency hearing evolved most likely within the early mysticetes, perhaps as a consequence of the evolution of their large size.

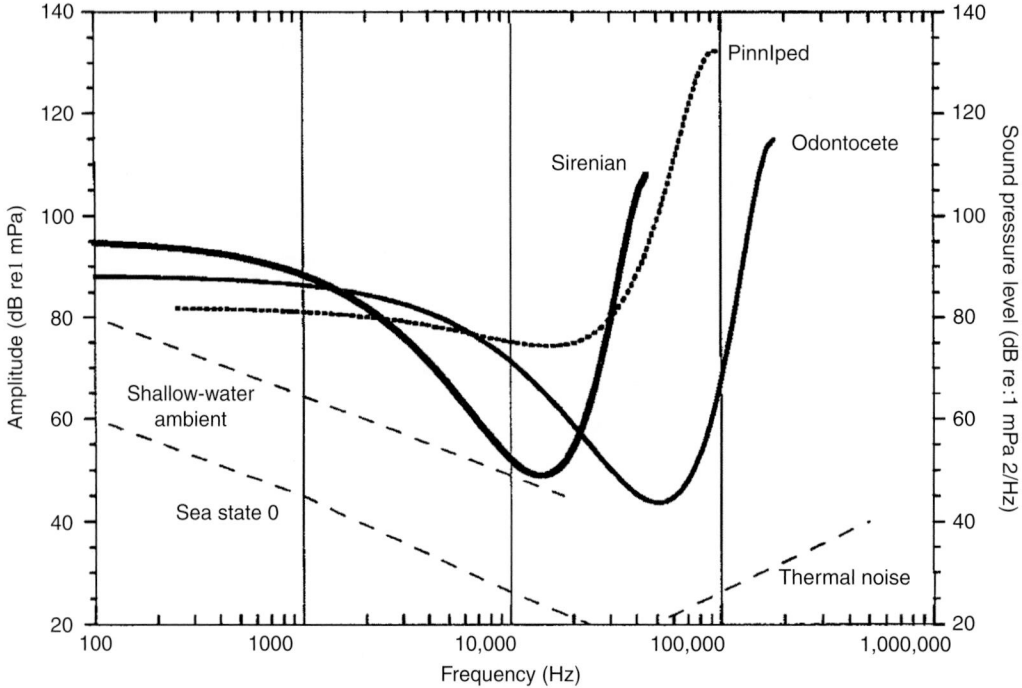

**Figure 8** *Best-fit curves of sirenian, pinniped, and odontocete audiograms. Third-order polynomial curves are accompanied by shallow water and noise curves. From Gerstein* et al. *(1999) with permission.*

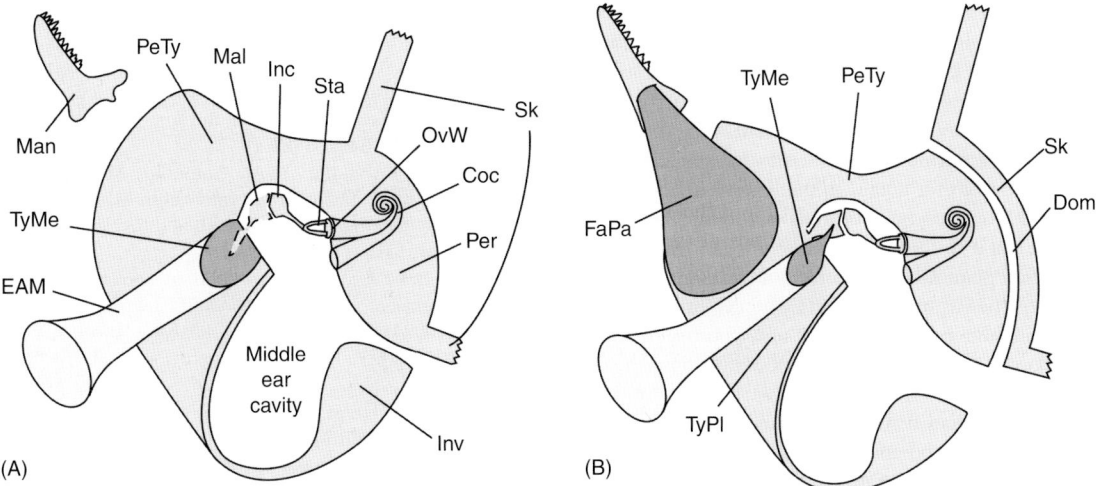

**Figure 9** *(A) Diagram of the pakicetid ear. The malleus is probably of land mammal type, and its position is shown with a dotted line. (B) Diagram of the remingtonocetid/protocetid ear. Coc, cochlea; Dom, dome-shaped depression for periotic; EAM, external acoustic meatus; FaPa, mandibular fat pad; Inc, incus; Inv, involucrum; Mal, malleus; Man, mandible; OvW, oval window; Per, periotic bone; PeTy, joint between periotic and tympanic; Sk, skull; Sta, stapes; TyMe, tympanic membrane; TyPl, tympanic plate. Adapted by permission from Macmillan Publishers Ltd: Nature (Nummela et al., 2004).*

**H**

The ancestral land mammal ear and the modern odontocete ear represent evolutionarily stable configurations. Characteristic of the whale ear evolution from pakicetids to the modern odontocete ear is that the connections between different bone structures, tympanic, periotic, and the rest of the skull, become more delicate and sophisticated. During the Eocene, the size of the tympanic bone and the tympanic plate relatively diminishes. While it is true that a large tympanic plate (just like a large tympanic membrane in land mammals) collects more energy and hence increases the signal/noise ratio, it also has a larger mass which constrains transmission of high frequencies to the ear. With the disconnection between the tympano-periotic and the skull, and the delicate connections between tympanic and periotic themselves, together with a smaller and thinner tympanic plate, the ear became suitable for higher-frequency hearing and in due time, for echolocation.

### B. Archaeocetes

Beginning with the earliest whales, pakicetids, the ear changed gradually and the generalized whale underwater hearing mechanism evolved in less than 10 million years. Pakicetids (Fig. 9A) had an outer ear canal leading to the tympanic membrane, and a small mandibular foramen in their lower jaw, indicating that no mandibular fat pad that would guide vibrations from the lower jaw, as happens in modern odontocetes, was yet present. The tympano-periotic complex stayed in close contact with the skull, giving no acoustic isolation between the two cochleae. A fossil incus of *Pakicetus*, resembling the incus of artiodactyls, indicates that pakicetids still had a land mammal ear with a malleus and tympanic membrane typical of a land mammal. The tympanic bone of pakicetids resembled an artiodactyl tympanic but was unlike that in land mammals, the tympanic was not connected rostro-medially to the periotic. This enabled the medial wall of the tympanic, which was very thick and pachyostotic, to form

a loosely suspended center of mass that could vibrate independently of the periotic. This kind of arrangement could enhance the transmission of bone-conducted sound to the cochlea, and would be a possible mechanism for underwater hearing. A pachyostotic involucrum was earlier thought to be a unique character of cetaceans, but a similarly pachyostotic involucrum has recently described for in *Indohyus*, a fossil artiodactyl and the closest relative of cetaceans (Thewissen *et al.*, 2007).

Pakicetids had two different hearing mechanisms, one in air, and the other in water. In air, they apparently used the land mammal sound transmission mechanism where sound energy reached the inner ear through the tympanic membrane and the middle ear ossicles. Under water, pakicetids used bone conduction mechanism which is not the most sophisticated hearing mechanism but can be combined with the needs of airborne hearing.

Among archaeocetes, the ambulocetid ear structures and their hearing mechanisms are least well known. Ambulocetids had a massive lower jaw with a large mandibular foramen, indicating the presence of a mandibular fat pad for guiding the vibrations received by the lower jaw. The tympanic bone was large, with a thick involucrum, and stayed in close contact with the lower jaw. It is likely that ambulocetids could mainly hear through bone conduction when keeping their lower jaw attached to the ground, like the modern crocodiles do. In water, the same mechanism could be used, but the sensitivity was obviously poor, and the frequency range of hearing was very low.

Remingtonocetids and protocetids had a large mandibular foramen in the lower jaw, indicating the presence of a mandibular fat pad (Fig. 9B). The contact between the tympanic and periotic was further reduced. The shape of the tympanic bone, the ossicles, and the tympanic membrane (as evidenced by the morphology of the malleus and the tympanic ring) was more modern than in earlier archaeocetes. The tympanic membrane had a conical shape, and the malleus had lost its manubrium, being apparently attached to the tip of the conical tympanic membrane from its head. This morphology suggests that the modern generalized whale hearing mechanism made its debut in remingtonocetids and protocetids, sound vibrations in water arriving at the lower jaw were carried further with the mandibular fat pad to the tympanic plate, and the vibrations of the plate were transmitted through the ossicles to the cochlea. With the external auditory meatus still present, it was possible for sound in air to reach the tympanic membrane that way. However, with the malleus lacking the manubrium that would attach it to the tympanic membrane, the efficiency of this mechanism was poor. Due to the lack of air sinuses found in modern odontocetes the tympano-periotic complex was still connected to the skull, and hence the ears were not acoustically isolated from each other, providing these animals with poor directional hearing (Fig. 10).

Basilosaurid cetaceans had a functionally modern ear, with the three parts, outer, middle, and inner ear present. The tympanoperiotic complex was partly isolated from the skull, with air sinuses around, and the ossicles had the morphology resembling that of delphinids. The large mandibular foramen of *Zygorhiza* indicates that the route for sound to reach the ear was through the mandibular fat pad, and although the external auditory meatus is still present, it was hardly patent.

## See Also the Following Articles

Echolocation ■ Sensory Biology, overview ■ Vision

## References

Bodson, A., Miersch, L., Mauck, B., and Dehnhardt, G. (2006). Underwater auditory localization by a swimming harbor seal (*Phoca vitulina*). *J. Acoust. Soc. Am.* **120**, 1550–1557.

Chapla, M. E., Nowacek, D. P., Rommel, S. A., and Sadler, V. M. (2007). CT scans and 3D reconstructions of Florida manatee (*Trichechus manatus latirostris*) head and ear bones. *Hear. Res.* **228**, 123–135.

Cook, M. L. H., Varela, R. A., Goldstein, J. D., McCulloch, S. D., Bossart, G. D., Finneran, J. J., Houser, D., and Mann, D. A. (2006). Beaked whale auditory evoked potential hearing measurements. *J. Comp. Physiol. A.* **192**, 489–495.

Gerstein, E. R., Gerstein, L., Forsythe, S. E., and Blue, J. E. (1999). The underwater audiogram of the West Indian manatee (*Trichechus manatus*). *J. Acoust. Soc. Am.* **105**, 3575–3583.

Hemilä, S., Nummela, S., and Reuter, T. (1999). A model of the odontocete middle ear. *Hear. Res.* **133**, 82–97.

Hemilä, S., Nummela, S., Berta, A., and Reuter, T. (2006). High-frequency hearing in phocid and otariid pinnipeds: An interpretation based on inertial and cochlear constraints. *J. Acoust. Soc. Am.* **120**, 3463–3466.

Holt, M. M., Schusterman, R. J., Southall, B. L., and Kastak, D. (2004). Localization of aerial broadband noise by pinnipeds. *J. Acoust. Soc. Am.* **115**, 2339–2345.

Kastelein, R. A., Mosterd, P., van Santen, P., Hagedoorn, M., and de Haan, D. (2002). Underwater audiogram of a Pacific walrus (*Odobenus rosmarus divergens*) measured with narrow-band frequency-modulated signals. *J. Acoust. Soc. Am.* **112**, 2173–2182.

Ketten, D. R. (2000). Cetacean ears. *In* "Hearing by Whales and Dolphins" (W. W. L. Au, A. N. Popper, and R. R. Fay, eds), pp. 43–108. Springer, New Youk.

Ketten, D. R., Odell, D. K., and Domning, D. P. (1992). Structure, function and adaptation of the manatee ear. *In* "Marine Mammal Sensory Systems" (J. A. Thomas, R. A. Kastelein, and A. Y. Supin, eds), pp. 77–95. Plenum, New York.

Mann, D. A., Colbert, D. E., Gaspard, J. C., Casper, B. M., Cook, M. L. H., Reep, R. L., and Bauer, G. B. (2005). Temporal resolution of the Florida manatee (*Trichechus manatus latirostris*) auditory system. *J. Comp. Physiol. A.* **191**, 903–908.

Møhl, B., Au, W. W. L., Pawloski, J., and Nachtigall, P. E. (1999). Dolphin hearing: Relative sensitivity as a function of point of application of a contact sound source in the jaw and head region. *J. Acoust. Soc. Am.* **105**, 3421–3424.

Nachtigall, P. E., Supin, Y. A., Amundin, M., Röken, B., Möller, T., Mooney, T. A., Taylor, K. A., and Yuen, M. (2007). Polar bear *Ursus maritimus* hearing measured with auditory evoked potentials. *J. Exp. Biol.* **210**, 1116–1122.

Norris, K. S. (1968). The evolution of acoustic mechanism in odontocete cetaceans. *In* "Evolution and Environment" (E. T. Drake, ed.), pp. 297–324. Yale University Press, New Haven, CT.

Nummela, S. (2008). Hearing in aquatic mammals. *In* "Sensory Evolution on the Threshold, Adaptations in Secondarily Aquatic Vertebrates" (J. G. M. Thewissen, and S. Nummela, eds), pp. 211–224. University of California Press, Berkeley, CA.

Nummela, S., and Thewissen, J. G. M. (2008). The physics of sound in air and water. *In* "Sensory Evolution on the Threshold, Adaptations in Secondarily Aquatic Vertebrates" (J. G. M. Thewissen, and S. Nummela, eds), pp. 175–182. University of California Press, Berkeley, CA.

Nummela, S., Thewissen, J. G. M., Bajpai, S., Hussain, S. T., and Kumar, K. (2004). Eocene evolution of whale hearing. *Nature* **430**, 776–778.

Nummela, S., Thewissen, J. G. M., Bajpai, S., Hussain, S. T., and Kumar, K. (2007). Sound transmission in archaic and modern whales: Anatomical adaptations for underwater hearing. *Anat. Rec.* **290A**, 716–733.

Parks, S. E., Ketten, D. R., O'Malley, J. T., and Arruda, J. (2007). Anatomical predictions of hearing of the North Atlantic right whale. *Anat. Rec.* **290**, 734–744.

Popov, V. V., Supin, A. Y., Klishin, V. O., Tarakanov, M. B., and Pletenko, M. G. (2008). Evidence for double acoustic windows in the dolphin, *Tursiops truncatus*. *J. Acoust. Soc. Am.* **123**, 552–560.

Reysenbach de Hann, F. W. (1957). Hearing in Whales. *Acta Oto-Laryngologica, Suppl.* **134**, 1–114.

Schusterman, R. J., Kastak, D., Levenson, D. H., Reichmuth, C. J., and Southall, B. L. (2000). Why pinnipeds don't echolocate. *J. Acoust. Soc. Am.* **107**, 2256–2264.

Supin, A. Y., Popov, V. V., and Mass, A. M. (2001). "The Sensory Physiology of Aquatic Mammals." Kluwer Academic Publishers, Boston, MA.

Thewissen, J. G. M., Cooper, L. N., Clementz, M. T., Bajpai, S., and Tiwari, B. N. (2007). Whales originated from aquatic artiodactyls in the Eocene epoch of India. *Nature* **450**, 1190–1195.

# Hind Limb Anatomy

## PETER J. ADAM

With the development of tail flukes for producing propulsion in whales, manatees (*Trichechus* spp.), and dugongs (*Dugong dugon*), the pelves and hind limbs became vestigial structures that now associate only loosely with the spine. The major role of the pelvic apparatus in these forms, when present, is to serve as attachment points for muscles acting on the genitalia and the abdominal body wall. Marine carnivores, which still maintain close ties with the terrestrial environment, have not had such a dramatic reduction in the pelvis and hind limb structures. Both pinnipeds and the sea otter (*Enhydra lutris*) have united the toes to form flippers. Phocids, which use the hind limbs to generate swimming thrust and which cannot rotate the hind feet under the body while on land, have highly modified hind limbs. Phocid adaptations include increasing the surface areas available for muscles that flex the leg, modifications of limb muscles to aid in undulatory movements of the spine, and a general increase in the muscle mass operating on the hind limb (in particular the muscles acting to flex the limb).

## I. Cetaceans

The known fossil record documenting cetacean evolution shows a progressive reduction and loss of hind limb skeletal elements and dis-association of the pelvic girdle from the vertebral column as whales became less dependent on nearshore environments and developed tail flukes to generate swimming thrust. This trend is most marked with the origin of the basilosaurine whales, in which the tibia and fibula became fused with each other and tarsal elements co-ossified into a single immobile mass. Basilosaurines also mark the point during which the pelves became disassociated from the vertebral column. Among modern forms, only vestiges of the hind limb skeleton can be found, and these are contained within the body wall. Mysticetes may possess fragments of pelvis, femur, and tibia, whereas the occurrence of hind limb and pelvic elements is more variable among both individuals and species of odontocetes. When present, the pelves bear little resemblance to those of terrestrial mammals and, when undeveloped, may exist only as a band of connective tissue connecting spinal muscles to those of the genitalia and abdominal wall. If present as a bony element, each pelvis is typically cigar or sickle shaped, with only the pelvic bone proper contributing to its structure (Fig. 1). Atavistic femora and occasional tibiae have been described from numerous (mysticete and odontocete) taxa, although occurrence of these elements is infrequent. Hind limb buds are present during early embryogenesis of all whale species documented so far, although the mesodermal cells that usually form the internal limb structures die or are reallocated to other functions as limb buds are resorbed later in ontogeny (Sedmera *et al.*, 1997; Thewissen *et al.*, 2006). Retention of a rudimentary pelvis is associated with the attachment of numerous muscles acting on the reproductive organs of both sexes. In males, the pelvis is usually larger relative to that of females. It serves as the site of origin for muscles acting on the genitals and anal region (e.g., the penis retractor and levator ani muscles) and may also serve as a site of attachment for posterior fibers of the rectus abdominis muscle. When present, the pelvis is isolated from the spine (sacral vertebrae are absent) but maintains a soft tissue attachment to the hypaxial spinal musculature. Rearrangements of spinal and pelvic

muscles in association with tail-based locomotion have led to considerable controversy over specific identities of muscles in these regions.

## II. Sirenians

The evolutionary loss of the hind limb in sirenians closely parallels that of cetaceans (Fig. 1), with modern forms possessing only a vestigial pelvis composed of ischium and ilium bones. In dugongs (Domning, 1991), each pelvis is long and stick-like in appearance, and the ilium and ischium are of subequal length, fusing by 5 years of age in both sexes. In manatees, the pelves are more plate-like and cross-shaped in lateral view. The ischium is the largest portion of the manatee pelvis, with the ilium forming a small cap on the anterior surface of the bone complex. As in whales, sirenian pelves lack bony attachment to the vertebral column. In dugongs, the pelves join with the anterior caudal vertebrae by an aponeurosis thought to be homologous with the coccygeus muscle, as well as by the retractor ischii and ischiococcygeus muscles to caudal chevron bones (small bones underneath each tail vertebra). The pelves serve as the origin for muscles acting on the genital organs (e.g., in females, the constrictor vulvae, constrictor vestibuli, and urethralis muscles) as well as some muscles inserting into the skin of the abdominal region (e.g., part of the transversus abdominis). The atavistic appearance of femora has been reported for manatees.

## III. Marine Carnivores
### A. Pinnipeds

The earliest known pinniped, *Enaliarctos* from the late Oligocene of California (Berta and Ray, 1990), possessed a well-developed hind flipper, and intermediate stages in the anatomical progression from a limb used for terrestrial locomotion to one specialized for swimming are undocumented. Anatomical adaptations of the hind limbs of extant pinnipeds largely reflect strategies adopted by each family for swimming and terrestrial locomotion. Phocid seals and walruses (*Odobenus rosmarus*) primarily use the hind limbs to generate thrust while swimming and have relatively more muscle mass in the pelvic region relative to the pectoral region and forelimb. Otariids propel themselves with the forelimbs and have relatively lower pelvic muscle mass. On land, otariids and walruses are able to rotate their hind feet under their body and progress with modified walking motions; phocids lack the ability to rotate their feet under their body and move along the ground with undulatory movements of the body.

Externally, the hind limbs of pinnipeds extend beyond the body contour from the approximate middle or end of the crus. In walruses and phocids the middle digit is the shortest, and digits increase in length both laterally and medially, giving the flipper a crescent shape (more marked in phocids). Thin, extensible interdigital webbing stretches between adjacent digits in these forms. In otariids, the interdigital areas are occupied by thick layers of connective and other tissues, making the hind flipper a much more rigid structure. Emargination of the distal interdigital regions of the flipper confers a scalloped shape to its trailing edge. Claws are reduced in all pinnipeds, although those of the middle three digits tend to be better developed than those of the first and fifth digits. Claws are positioned terminally in phocids and subterminally in walruses, but are located considerably farther proximally in otariids due to the development of distal cartilaginous rods on the ungual phalanges. The presence of these cartilaginous extensions gives the ungual phalanges an hourglass shape and roughened distal ends. The plantar surface of otariid and walrus flippers is hairless, with moderately developed foot pads related to their

Pelvic anatomy

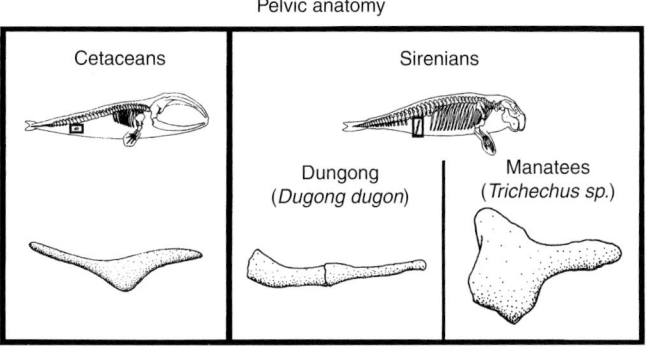

Cetaceans | Sirenians

Dungong
(*Dugong dugon*)

Manatees
(*Trichechus sp.*)

**Figure 1**  *Line drawing of the right pelvis of a cetacean (left) and sirenian (right) in lateral view (anterior toward the right). Position and orientation of the pelvis are indicated by boxes on the skeletal outlines.*

ambulatory terrestrial locomotion. Foot pads are lacking on phocids. With the exception of interdigital regions, which are hairless or have only sparse hair, the dorsal pedal surface typically has a hair density subequal or slightly lower than that of the body.

Departures of the skeletomuscular anatomy from the condition observed in typical terrestrial carnivores are most prevalent in phocid seals due to their highly modified (undulatory) terrestrial locomotion and specialized hind limb swimming. The iliac region of the phocid pelvis is expanded laterally, particularly among phocines. This confers a mechanical advantage to the gluteus muscle complex, which inserts onto the greater trochanter of the femur and functions to flex the leg against the water. The ischiopubic region of phocid pelves, posterior to the acetabulum, is elongate relative to otariids and the walrus. This increases the surface area available for attachment of the strong muscles acting to medially flex the leg during the power stroke of the swimming cycle (e.g., the adductor, gracilis, gemelli, obturatorius, and semimembranosus muscles). The ischial tuberosity, often misidentified as the "ischial spine," is greatly enlarged in phocids, but undeveloped in otariids and the walrus. It serves as the site of origin for the biceps femoris muscle, which inserts broadly onto the tibia. The orientation and widening of the biceps femoris in phocids indicate that it is primarily responsible for lifting the hind limb off the ground during terrestrial locomotion, as well as medially flexing the limb during swimming. The pinniped femur is short and stout, and the distal condyles are inclined relative to the long axis of the shaft. The fovea capitis of the femoral head is lacking. This indicates the loss of the teres ligament, which normally maintains the femoral head within the acetabulum of the pelves in terrestrial mammals that have weight-bearing hip joints. In phocids, the lesser femoral trochanter is either reduced or absent, and the two muscles typically inserting onto it have undergone major changes from their usual orientation and function: (1) the iliacus muscle inserts onto the more distal femoral epicondylar crest or proximal tibia and (2) the psoas major muscle, arising from the posterior thoracic, lumbar, and sacral vertebrae, inserts onto the medial surface of the ilium and thus aids in lateral undulation of the spine during swimming rather than acting on the limb. Proximally, the tibia and fibula of most pinniped species become fused prior to maturity. The posterior tibial fossa is deep in phocids, reflecting enlargement of the tibialis caudalis muscle, which originates from this region and inserts onto the tarsus and first metatarsal, acting to plantar flex the pes during the swimming power stroke. The tendon of the flexor hallucis longus muscle passes over a posterior projection of the astragalus that is unique to phocids, limiting dorsal flexion of the pes. This, in combination with the elongated ischiopubis (which limits anteroventral bending of the spine when phocids are on land), limits the ability of phocids to assume a four-legged stance. Additionally, the tibioastragalar joint of phocids is nearly spherical and would unlikely bear the weight of the animal. This is in contrast to the more rigid, hinge-like joint found in other pinnipeds and terrestrial carnivores. Inserting tendons of the large plantar flexing muscles (i.e., the flexor digitorum longus, flexor hallucis longus, and flexor digitorum superficialis muscles) often combine together in a complex manner at the level of the tarsals. A united tendon of these muscles sends branches out to the digits, although slips extending to the first and fifth digits tend to be larger in phocids. The pedal formula of all pinnipeds is 2-3-3-3-3, the primitive condition for all mammals. In phocids and the walrus, metatarsals and phalanges of the first and fifth digits are more robust than those of the middle three digits. This is associated with hind limb swimming, where both digits may act as leading edges of the flipper during the complex power stroke. Metatarsal–phalangeal and

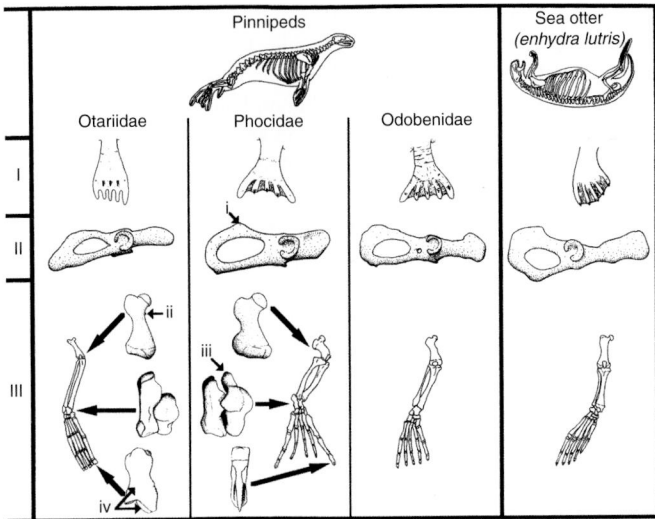

**Figure 2** *Line drawings of the external flipper morphology (row I), right pelvis (row II, lateral view, anterior is toward the right), and right hindlimb skeleton (row III, anterior, or dorsal, view) of representative species of pinnipeds (otariidae, Callorhinus; Phocidae, Monachus; and Odobenidae, Odobenus) and the sea otter (right). Indicated features of the otariid and phocid hind limbs are as follows: i, enlarged ischial tuberosity of phocids; ii, presence of a lesser femoral trochanter in otariids and walrus (absent or reduced in phocids); iii, posterior projection of the phocid astragalus, over which the tendon of the flexor hallucis longus muscle passes; and iv, first ungual phalanx of the otariid pes showing the lack of a claw and distal roughened surface to which cartilage is attached in life.*

interphalangeal joints tend to be of the tongue-and-groove type in phocids and hinge-like in otariids and the walrus. However, all pinnipeds have reduced trochleation of the metatarsals and phalanges (Fig. 2).

Available evidence indicates that blood supplying the crus and pes of phocids passes primarily through the external iliac, femoral, and sapheneous arteries. Maintenance of this primitive condition is related to the efficiency of the system for supplying oxygen and nutrients to the heavily used hind limb musculature. In contrast, otariids have adapted the blood vessels such that most of the blood supplying the distal limb regions passes through the internal iliac and gluteal arteries. The passage of blood via this route is believed to enhance heat dissipation. The walrus is intermediate to these conditions, although detailed description of its anatomy is lacking. The presence of a circulatory countercurrent system (*rete mirabile*) in the hind limbs has been reported for several phocid and otariid species. Anatomy of the spinal nerves serving the hind limb of pinnipeds is poorly known. Available evidence suggests that the lumbosacral plexus has shifted posteriorly by one vertebra, being composed of ventral rami arising from the third through fifth lumbar, first through third sacral, and first caudal vertebrae. Division of this plexus into lumbar and sacral plexi is not possible.

## B. Polar Bears and Sea Otters

Adequate descriptions of polar bear (*Ursus maratimus*) pelvic and hind limb anatomy have not yet been made, but there is little indication that the morphology of this species has diverged appreciably from that of other species of *Ursus*. Departures of sea otter hind limb anatomy from that of other terrestrial mustelids (Tarasoff 1972; Tarasoff *et al.*, 1972), however, are seen more readily. Externally, the leg is enclosed within the loose body skin to the approximate level of the ankle. The digits are bound together by interdigital webbing, although the fourth and fifth digits are bound more closely together than other adjacent digital pairs. The sea otter is unusual in that in overall length the digits decrease in size from the fifth to the first: V > IV > III > II > I. While swimming, sea otters use the hind feet to generate thrust and sweep the leg through the water such that the fifth digit forms the leading edge of the pes. The hair densities for the ankle and interdigital webbing have been estimated at 107,000 and 3300 hairs/cm$^2$, respectively, compared to a density of 125,000 hairs/cm$^2$ for the back. Pads are present on the phalangeal portion of each toe and are variably found ventral to the metatarsals. As with pinnipeds, the fovea capitis is absent from the femur, marking the absence of the teres ligament. The biceps femoris muscle inserts onto the middle of the tibia and maintains the leg in a posterior position. The flexor digit V muscle is very large in the sea otter (relative to other mustelids). This enlargement corresponds to the use of the lateral surface of the pes to lead during the power stroke of the limb. The remaining hind limb anatomy of the sea otter corresponds well with that of terrestrial mustelids (Fig. 2).

## See Also the Following Articles

Forelimb Anatomy ■ Locomotion, Terrestrial ■ Musculature ■ Skeletal Anatomy ■ Swimming

## References

Berta, A., and Ray, C. E. (1990). Skeletal morphology and locomotor capabilities of the archaic piniped *Enaliarctos mealsi. J. Vertebr. Paleontol.* **10**, 141–157.

Bisaillon, A., and Pierard, A. (1981). Osteologie de morse de l'Atlantique (*Odobenus rosmarus*, L. 1758) ceintures et members. zentralbatt Veterinarmedizin. *Reihe C Anat. Histol. Embryol.* **10**, 310–327.

Domning, D. P. (1977). Observations on the myology of *Dugong dugon* (Miller). *Smith. Contrib. Zool.* **226**, 1–57.

Domning, D. P. (1991). Sexual and ontogenetic variation in the pelvic bones of *Dugong dugon. Mar. Mamm. Sci.* **7**, 311–316.

Fay, F. H. (1974). Comparative and functional anatomy of the vascular system in the hind limbs of the Pinnipedia. "Transactions of the First International Theriological Congress." pp. 166–167. Nauka Publishers, Moskow.

Gambarjan, P. P., and Karapetjan, W. S. (1961). Besonderheiten im Baudes Seelowen (Eumetopias californianus), der Baikalrobbe (Phocasibirica) and des Seeotters (*Enhydra lutris*) in Anpassung an die Fortbewegung im Wasser. *Zool. Jahrbucher (Abteilung Anat. On-tog. Tiere)* **79**, 123–148.

Kaiser, H. E. (1974). "Morphology of the Sirenia: A Macroscopic and X-Ray Atlas of the Osteology of Recent Species." Karger, Basel, Switzerland.

Miller, W. C. S. (1888). The myology of the Pinnipedia. *In* "Report on the Scientific Results of the Voyage of H. M. S. Challenger during the Years 1873–76." (C. W. Thomson, and J. Murray), Vol. 26, Order of Her Majesty's Government.

Muizon, C.de (1981). Une interpretation functionelle et phylogene-tique de l'insertion du psoas major chez les Phocidae. *Comp. Ren. Acad. Sci. (Paris)* **292**, 859–862.

Nakanishi, T., Yamamoto, M., and Suenaga, Y. (1978). Comparative anatomical studies on the nerves and muscles of the posterior limb of the northern fur seal and cat. *Okajimas Fol. Anat. Japon.* **54**, 317–340.

Schulte, H. W., and Smith, M.de. F. (1918). The external characters, skeletal muscles, and peripheral nerves of Kogia breviceps (Blainville). A *Bull. Am. Mus. Nat. Hist.* **38**, 7–72.

Sedmera, D., Misek, I., and Klima, M. (1997). On the development of cetacean extremities. I. Hind limb rudimentation in the spotted dolphin. *Eur. J. Morphol.* **35**, 25–30.

Tarasoff, F. J. (1972). Comparative aspects of the hind limbs of the river otter, sea otter and seals. *In* "Functional Anatomy of Marine Mammals" (R. J. Harrison, ed.), Vol. 1, pp. 333–359. Academic Press, New York.

Tarasoff, F. J., Bisaillon, A., Pièrard, J., and Whitt, A. P. (1972). Locomotory patterns and external morphology of the river otter, sea otter, and harp seal (Mammalia). *Can. J. Zool.* **50**, 915–929.

Thewissen, J. G. M., Coh, M. J., Stevens, L. S., Bajpai, S., Heyning, J., and Horton, W. E., Jr. (2006). Developmental basis for hind-limb loss in dolphins and origin of the cetacean body plan. *Proc. Natl. Aca. Sci. USA* **103**, 8414–8418.

Uhen, M. D. (1998). Middle to late Eocene basilosaurines and dorudontines. *In* "The Emergence of Whales: Evolutionary Patterns in the Origin of Cetacea" (J. G. M. Thewissen, ed.), pp. 29–61. Plenum Press, New York.

# History of Marine Mammal Research

### BERND WÜRSIG, WILLIAM F. PERRIN AND J.G.M. THEWISSEN

If research is the gathering of knowledge, then we can think of marine mammal research to have gone on as long as humans have gazed at whales spouting offshore and seals pupping on beaches. But early observations of nature were largely tied up with myths about animals and legends of their capabilities. A common theme appears to have been the changing of humans to dolphins and whales, and the reverse. This theme is recognized in remaining legends of Australian aborigine "dream time," boto (*Inia geoffrensis*) and baiji (*Lipotes vexillifer*) river dolphin folklore (Sangama de Beaver and Beaver, 1989; Zhou and Zhang, 1991, respectively), tales of the god-like killer whales (*Orcinus orca*) of Pacific Northwest indigenous tribes (McIntyre, 1974), and many more.

Some early writings show remarkable insights in marine mammal biology. Well over 2000 years ago, scholars of China's Han Dynasty in the annotated dictionary "Er-Ya," described the baiji as related to marine dolphins, implying that those were known to intellectuals of the time. Even earlier, the Greek philosopher/scientist Aristotle (384–322 BC) differentiated between baleen and toothed whales and described both types in some detail. It is unfortunate but totally understandable in hindsight that he classified cetaceans as fishes, a practice still present in Britain's term "Royal Fishes" under which all whales and dolphins belong by law to the Crown. The Roman writer/lawyer/admiral Pliny the Elder (23–79 AD) published a book on dolphins and whales 400 years after Aristotle's time as part of his 37-volume "Natural History."

Not much scientific inquiry or thought was conducted between Roman times and the western Renaissance, and knowledge, at least written knowledge, of marine mammals languished as well. The modern progression of marine mammal research can perhaps best be described as occurring in four general (and not mutually exclusive) phases: (1) morphological description from beach-cast specimens and fossils; (2) descriptions of behavior, anatomy, and distribution as gathered during hunting and whaling activities; (3) studies of physiology and behavior in captivity; and (4) studies of ecology, habitat use, numbers, life history patterns, behavior, and physiology in nature. A fifth phase may be thought of as an ever-increasing sophistication in integrating knowledge from terrestrial situations as well as from different fields of marine mammal endeavors.

The phases of research mentioned above follow a rough chronology, with morphology and systematics the main topics pre-1900s; hunting-related habitat, morphological, and behavioral research mainly from the 1850s to the 1970s; scientific captive animal descriptions beginning around 1950; and more ecologically oriented descriptions in nature beginning around the 1970s. All phases are ongoing, with electronic devices helping to elevate in-field research on marine mammal lives to a new level of sophistication. A very readable recent account of the history of marine mammal studies is found in Berta and Sumich (1999). Elsewhere, this volume lists some of the major deceased marine mammal researchers of the past and mentions their classic works in the field (see References).

Pierre Bélon was probably the first "modern" marine mammal author since Pliny's time. He published accurate descriptions and woodcuts of some whales, dolphins, and seals (Belloni, 1553), and these (and also, unfortunately, the less accurate ones) were much copied by others in the next two centuries.

The real burst of marine mammal knowledge did not come until later, however. And then it came suddenly, in tune with eighteenth century awakening of scientific thought in the western world. While many authors could be mentioned, three early contemporaries did much to advance cetacean descriptions, taxonomy, and systematics. These were the French zoologist La Cépède (1804) and the Cuvier brothers. Georges Cuvier, who arguably founded modern evolutionary theory, wrote on many topics, including cetaceans; whereas his less-famed brother Frederic published two important works on cetaceans (Cuvier, 1829, 1836). These three were followed by the Belgian zoologist Van Beneden in the latter half of the nineteenth century, with work mainly consisting of compilations of information on fossil whales, and by a host of fine morphologists, taxonomists, systematists, and evolutionary historians in the twentieth century (summaries are provided by Rice, 1998; Thewissen, 1998; Pabst *et al.*, 1999; and Reynolds *et al.*, 1999). While much of the earlier work centered on cetaceans, the British zoologist John Edward Gray described both seals and whales in the British Museum (Gray, 1866), and the American zoologist Joel Allen wrote excellent monographs on whales, pinnipeds, and sirenians (Allen, 1880).

Yamase (1760) began the science of marine mammalogy in Japan at about the same time as serious studies began in the west. He presented accurate figures and descriptions of the external morphology of six toothed and seven baleen whale species and distinguished them from fishes. His work was brought to the west in a marine mammal section of "Fauna Japonica" by Siebold (1842). Otsuki began to describe the internal anatomy of cetaceans of Japan in 1808, but his manuscript remains unpublished.

A second major phase of information gathering, often linked intricately with that just described, involved descriptions of animals as related to HUNTING and WHALING. Morphological information was at the core of these descriptions, but behavior and the basic society structure of whales and pinnipeds—of course much of the time affected by the hunting activities themselves—were recorded as well. One of the earliest accurate accounts consisted of German-born

and Russian-naturalized Georg Steller's descriptions of pinnipeds and the soon-after extinct Steller's sea cow (*Hydrodamalis gigas*), the largest and only cold-water sirenian known (originally published in Latin in 1751, and republished in English as Steller, 1899). Quite a few books related especially to whaling were produced, but perhaps the most enduring one from the nineteenth century was by the North American whaling captain Charles Scammon, who wrote with feeling and accuracy on behavior and life history habits of marine mammals of the North Pacific (Scammon, 1874). In the twentieth century, one of the most famous works largely relying on whaling-accumulated data consists of Everhard Slijper's book "Whales and Dolphins" (published in English in 1976). A very readable account of whaling and the literature derived from whaling can be found in "Men and Whales" by Richard Ellis (1991).

Modern factory whaling itself helped to usher in excellent research on numbers, habitat use, life history patterns, and morphology/physiology. This was so especially during the *Discovery* investigations of 1925–1951, a British research program that was responsible for a wealth of new data, especially on large whales of the southern hemisphere. These investigations consisted in part of extensive long-term tagging ("discovery tags," shot into the blubber and muscle tissues of whales, and later recovered during actual whale kills). In this manner, migrations of great whales were delineated long before modern radio and satellite tags provided such information (e.g., Allen, 1980). Dozens of fine researchers published hundreds of papers that relied on the *Discovery* expeditions, and on other whaling data since then (e.g., Laws, 1959) (see also the section "International Whaling Commission" in this voume). As a counterpoint to early cetacean information, the reader interested in pinniped research from the ancient Greeks to about 1983 can consult an excellent annotated bibliography of over 12,000 publications by Ronald *et al.* (1976, 1983).

While whaling, sealing, and other forms of direct hunting are much abated today as compared to in the 1960s, there are still powerful low-level, oft-indigenous hunts, especially in protein-poor areas of the world (Perrin, 1999). As a result, data are being accumulated and analyzed on morphology, GENETICS, taxonomy, and SYSTEMATICS, life history, prey patterns, and so on. Excellent recent information has become available from results of hunting on, e.g., pilot whales (*Globicephala* spp.), oceanic dolphins (especially of the genus *Stenella*), bowhead whales (*Balaena mysticetus*), sperm whales (*Physeter macrocephalus*), and several seal, fur seal, and sea lion species (summaries in Berta and Sumich, 1999; Reynolds and Rommel, 1999; and Twiss and Reeves, 1999).

A third major research avenue has come about as a result of keeping marine mammals in captivity. Attempts to do so in the early part of the last century usually resulted in the animals' untimely deaths—due to poor water, incorrect or tainted food, disease, and aggression between individuals in confined spaces. Facilities that housed marine mammals simply replaced dead ones by more captures from nature. However, especially since the 1970s, amazing strides in husbandry have been made for all marine mammals (except large whales), and the better aquaria now keep—and breed—animals very well. Unfortunately, there are still many "primitive" facilities, especially in less-developed parts of the world. At present, there are representatives of all major taxonomic groups in captivity, as show animals and for research: toothed whales and dolphins (only two baleen whales, each time young gray whales, *Eschrichtius robustus*, have been successfully kept); pinnipeds of all types, but especially California sea lions (*Zalophus californianus*); sirenians (mainly the West Indian manatee, *Trichechus manatus* and the dugong, *Dugong dugon*); and polar bears (*Ursus maritimus*) and sea otters (*Enhydra lutris*).

Only through holding animals in controlled situations have researchers learned that dolphins echolocate (Au, 1993); that all marine mammals exhibit reduced heart and general metabolic rates during dives (Ridgway, 1972; Pabst *et al.*, 1999); and that both dolphins and sea lions have remarkably advanced cognitive capabilities (Tyack, 1999). Furthermore, it is now fully appreciated that while pinnipeds and cetaceans are finely tuned underwater swimmers and divers with superbly evolved methods of breath holding, avoiding or reducing lactic acid depth during long submergences, and navigating in dark and cold waters, there is no secret "magic" to their energetic capabilities (Costa and Williams, 1999).

One major misstep from studies in captivity took place: the American John Lilly avowed in the 1960s that his research on bottlenose dolphins (*Tursiops truncatus*) proved that these popular show animals have an intelligence superior even to that of the brightest dogs (*Canus lupus familiaris*) and chimpanzees (*Pan troglodytes*), and likely equal to that of humans (Lilly, 1967). Careful studies by others have shown that dolphins are undeniably "smart" (intelligence is very difficult to define and compare, but has something to do with well-developed flexibilities of behavior and of innovative learning), but that there is no reason to believe that dolphins fare better in this "intelligence/cognition" sphere than many other highly social mammals (Herman, 1980, 1986; Tyack, 1999; Wells *et al.*, 1999).

While the study of marine mammals dead from the sea and live in captivity continues and grows, a relatively new approach has become the major research avenue since the 1970s. This consists of our fourth phase, of researchers going out into nature to observe the animals in their own milieu; as the animals associate with conspecifics; eat and are being eaten; and mate, give birth, and raise their young. We are learning more about the lives of these generally social creatures as they face storms, heavy years of sea ice, seasons of poor food resources (e.g., caused by "El Niño" southern oscillation climatic events), parasite infestations, adoring but noisy boatloads of whale-watching tourists, crowded shipping lanes, and habitat degradation near shore and in mighty rivers. This information on ecology of marine mammals is vital if we are to help protect them and their natural ecosystems from the depredations of overfishing, habitat POLLUTION by chemicals, heavy metals, and noise; and the results of global climate change and whole-scale habitat destruction due to the effects of ozone depletion and global warming (Tynan and DeMaster, 1997; Ferguson *et al.*, 2005).

Studies in nature often rely on visual or photographic recognition of individual whales, dolphins, and pinnipeds, often with the help of tags or color marks but also by natural markings (Hammond *et al.*, 1990). Researchers have described movement patterns by tracking animals with surveyor's transits from shore, and from shore and vessels by small radio tags placed on their bodies (Würsig *et al.*, 1991). Since the early 1990s, satellite tags that relay position information to earth-orbiting satellites have become smaller, less expensive, and ever more popular. As a result, we know that northern elephant seals (*Mirounga angustirostris*) swim and dive into deep oceanic waters for months at a time, humpback whales (*Megaptera novaeangliae*) take rapid zigzag courses between their mating and feeding grounds, North Atlantic right whales (*Eubalaena glacialis*) undergo previously unsuspected jaunts between Greenland and New England during the feeding summer, and much more (Wells *et al.*, 1999). Tags are being fitted not only with depth-of-dive measuring and telemetering devices, but also with ways to ascertain geographic position, swimming velocity, angles of dives, water and skin temperature, individual sound production, heart rate, and, in the future, other physiological measures. Recent advances in small and low-light capable video camera/record systems are even giving data on SWIMMING, socializing,

and feeding behavior directly from the animals under water (Davis *et al.*, 1999).

Physiological research, previously entirely within the realm of captivity, is more and more possible with innovative or sophisticated techniques in nature. Samples of stool, urine, blood, and even mother's milk are being collected from pinnipeds resting on land or ice. Trained dolphins have been released at sea, commanded to dive, and then told to exhale into a funnel to ascertain oxygen consumption values and to station themselves so that blood can be drawn. Small darts have been developed that are fired from a crossbow or pneumatic pistol and that obtain skin and blubber samples from free-living cetaceans for analyses of genetics (Dizon *et al.*, 1997), toxin loads, reproductive status, and blubber energy content for relative measurements of health within and between populations. Sloughed skin samples from breaching whales have been successfully collected from the water and genetically sampled for gender, social grouping, and population data. A technique has been developed to harmlessly "skin-swab" bow-riding dolphins, also for genetic analysis (Harlin *et al.*, 1999).

In response to an apparent increase in marine mammal strandings and the emergence of new marine mammal diseases in recent years, studies of wild marine mammal disease and ocean chemical contaminants are on the increase. While studies in nature have yielded data on the presence of deadly viruses and contaminant levels in tissues of beached and dying marine mammals (Aquilar and Borrell, 1997), they have provided little insight into immune defense against disease or the biochemical consequences of contaminants. More recently, e.g., species-specific biomarkers have been developed to assess the dolphin immune system (Romano *et al.*, 1999). Because they are readily available for long-term studies requiring serial sampling of tissues and health and reproductive histories, captive marine mammals afford unique opportunities to provide basic insight into the relationships among contaminants, the immune system, and animal health. Once they are developed and tested on animals in captivity, biomarkers can be used with wild marine mammal populations to assess contaminant exposures and their possible effects on immune systems and neurologic responses (Ridgway and Au, 1999), as well as on reproductive success (Ridgway and Reddy, 1995), growth, and development.

The sensitive hearing of marine mammals has led to concerns that intense sound or noise pollution generated by humans could impede communication, cause physiological stress, or damage hearing. Marine mammal hearing studies currently underway should help to define mitigation criteria for the effects of human-generated sound in the ocean (Schlundt *et al.*, 2000), and ultimately allow us to find a balance between the ecological needs of marine mammals and the role the ocean plays in commerce, exploration, travel, and defense.

Overall, descriptions of marine mammal taxonomy and population biology have shifted from mainly morphological approaches to an increasing reliance on molecular methods. Up through the 1960s, cetologists studied dolphins by harpooning them. For example, the revision of the spotted dolphins (*Stenella attenuata* and *S. frontalis*) by Perrin *et al.* (1987) was based in part on dolphins collected at sea by Francis C. Fraser, Dale W. Rice, William E. Schevill, and Edward D. Mitchell, all eminent scholars and pioneers of modern cetology. Without those specimens, the study would not have been possible; that is the way it was done until protection of marine mammals became the norm in most countries in the 1970s. Another source of specimens has been dolphins that died in oceanaria. The same revision by Perrin *et al.* (1987) included spotted dolphins retrieved from oceanaria by David K. Caldwell through the 1970s and early 1980s. And at that time, there were still a number of accessible directed dolphin fisheries; the spotted dolphin review also included specimens from directed fisheries in the Caribbean, St. Helena, West Africa, Japan, and the Solomon Islands. Today, dolphins, whales, and pinnipeds are stringently protected in the wild in most places. For oceanaria, restrictions have been placed on species and numbers of animals that can be captured for exhibit and the high monetary value of captive marine mammals has resulted in better husbandry and fewer deaths. As a result of these factors, marine mammal biologists practicing morphological approaches became limited to specimens from strandings and bycatch, greatly decreased opportunities for amassing adequate series of specimens for quantitative analysis. But then along came biopsy sampling and ready techniques of amplifying DNA fragments by a technique termed *polymerase chain reaction (PCR)*. Collection of samples by biopsy is legal and doable, so the balance of taxonomic and population studies has shifted from morphology to genetics. And the traditional morphologists have been scrambling to keep up by re-educating themselves in the new techniques or recruiting collaborators who know their way around genetics.

The study of marine mammals has now matured into a fifth phase, characterized by the obliteration of boundaries that separated the previous phases. New studies on marine mammals are often integrative, combining methods and ways of thinking largely gleaned from terrestrial animals. This comparison of ideas and research techniques holds great promise for the understanding of the biology of marine mammals.

As our understanding of their biology increases, marine mammals become appealing subjects for approaches that are at times laboratory heavy and at times nearly biomedical in scope. In turn, these approaches enrich knowledge of marine mammals. For instance, biochemical analyses of body fats, first championed for humans and other terrestrial animals, give new insights into the functions of different fats in cetaceans (Koopman *et al.*, 2003). Immuno-histochemical staining techniques originally used for non-marine mammal studies allow the identification of genes that significantly affected cetacean evolution, such as those genes responsible for the loss of hind limbs (Thewissen *et al.*, 2006). Our understanding of the social systems of terrestrial mammals, with one major aspect being sperm competition at the physiological level (Kenagy and Trombulak, 1986) has begun to inform us about the relatively non-competitive balaenid whales, gray whales, quite a few species of dolphins, and manatees that have polygynous or polygandrous (multi-mate) societies (Reynolds *et al.*, 2004).

Through sophisticated studies with modern techniques, marine mammalogy is beginning to enrich more broad fields of science such as behavioral ecology, physiological ecology, and evolutionary biology. It was recognized long ago that marine mammals represent amazing natural experiments of evolution, and the maturation of the field of marine mammalogy is allowing for these experiments to be explored, and to inform all of biology.

Sophisticated electronic and biochemical techniques have recently been and are being developed to study the lives of marine mammals. However, the "tried and true" methods of looking at fossil bones, dissecting and describing pathologies of a net-entangled animal or one cast on shore after a storm, safely and carefully experimenting with animals in captivity, and the dogged gathering of behavioral information by binoculars and notebook are by no means passé. The greatest change since about the 1960s is the ever wider availability of information. This means that there is now a wealth of background knowledge available to anyone anywhere with a computer and an Internet connection. We are, in this new twenty-first century, in a vibrant phase of marine mammal research, and we see a very bright future for ever-more exciting discoveries in our field.

Although much of the research landscape looks bright, we would be amiss if we did not cite a note of pessimism as well, as it is undeniable that many populations and some entire species are facing reductions and even extinction due to human-caused habitat degradation, including rapid climate change. For example, the Chinese river dolphin, baiji, is very likely extinct (Turvey *et al.*, 2007), and the vaquita (*Phocoena sinus*) and Mediterranean monk sea (*Monachus monachus*) may not be far behind. No amount of modern and multidisciplinary research will be able to wrest information from a species that is gone from the face of the Earth.

## See Also the Following Articles

Hunting of Marine Mammals ■ Marine Protected Areas ■ International Whaling Commission ■ Popular Culture and Literature

## References

Aguilar, A., and Borrell, A. (eds) (1997). "Marine Mammals and Pollutants: An Annotated Bibliography." Foundation for Sustainable Development, Barcelona.

Allen, J. A. (1880). History of North American pinnipeds: A monograph of the walruses, sea-lions, seabears and seals of North America. *US Geol. Surv. Terr. Misc. Publ.* **12**, 1–785.

Allen, K. R. (1980). "Conservation and Management of Whales." University of Washington Press, Seattle, WA.

Au, W. W. L. (1993). "The Sonar of Dolphins." Springer-Verlag, New York.

Belloni, P. (1553). "De Aquatibilis (Book Two)." Stephan Press, Paris.

Berta, A., and Sumich, J. L. (1999). "Marine Mammals: Evolutionary Biology." Academic Press, San Diego.

Costa, D. P., and Williams, T. M. (1999). Marine mammal energetics. *In* "Biology of Marine Mammals" (J. E. Reynolds, III, and S. A. Rommel, eds), pp. 176–217. Smithsonian Institution Press, Washington, DC.

Cuvier, F. (1829). Cétacés. *In* "Histoire Naturelle des Mamifères." Roret Press, Paris.

Cuvier, F. (1836). "De l'Histoire Naturelle des Cétacés." Roret Press, Paris.

Davis, R. W., *et al.* (8 authros) (1999). Hunting behavior of a marine mammal beneath Antarctic fast ice. *Science* **283**, 993–996.

Dizon, A. E., Chivers, S. J., and Perrin, W. E. (eds.) (1997). "Molecular Genetics of Marine Mammals." Special Publication No. 3. The Society for Marine Mammalogy, Allen Press, Lawrence, KS.

Ellis, R. (1991). "Men and Whales." Knopf Press, New York.

Ferguson, S. H., Stirling, I., and McLoughlin, P. (2005). Climate change and ringed seal (*Phoca hispida*) recruitment in western Hudson Bay. *Mar. Mamm. Sci.* **21**, 121–135.

Gray, J. E. (1866). "Catalog of Seals and Whales in the British Museum," 2nd Ed. British Museum Press, London.

Hammond, P. S., Mizroch, S. A., and Donovan, G. P. (eds.) (1990). "Individual Recognition of Cetaceans: Use of Photo Identification and Other Techniques to Estimate Population Parameters." International Whaling Commission, Special Issue No. 12, Cambridge University Press, Cambridge.

Harlin, A. D., Würsig, B., Baker, C. S., and Markowitz, T. M. (1999). Skin swabbing for genetic analysis: Application on dusky dolphins (*Lagenorhynchus obscurus*). *Mar. Mamm. Sci.* **15**, 409–425.

H

Herman, L. M. (1980). Cognitive characteristics of dolphins. *In* "Cetacean Behavior: Mechanisms and Functions" (L. M. Herman, ed.), pp. 363–429. Wiley-Interscience Press, New York.

Herman, L. M. (1986). Cognition and language competencies of bottlenosed dolphins. *In* "Dolphin Cognition and Behavior: A Comparative Approach" (R. J. Schusterman, J. A. Thomas, and F. G. Woods, eds), pp. 221–252. Lawrence Erlbaum Press, Hillsdale, NJ.

Kenagy, G. J., and Trombulak, S. C. (1986). Size and function of mammalian testes in relation to body size. *J. Mammal.* **67**, 1–22.

Koopman, H. N., Iverson, S. J., and Read, A. J. (2003). High concentrations of isovaleric acid in the fats of odontocetes: Stability in the melon vs. variation and patterns of accumulation in blubber. *J. Comp. Physiol.* **173**, 247–261.

La Cépède, Compte de. (1804). "Histoire Naturelle des Cétacé." Paris.

Laws, R. M. (1959). The foetal growth rates of whales with special reference to the fin whale, *Balaenoptera physalus* Linn. *Dis. Rep.* **29**, 281–308.

Lilly, J. C. (1967). "The Mind of the Dolphin." Doubleday Press, New York.

McIntyre, J. (1974). "Mind in the Waters." Charles Scribner's Sons, New York.

Pabst, D. A., Rommel, S. A., and McLellan, W. A. (1999). The functional morphology of marine mammals. *In* "Biology of Marine Mammals" (J. E. Reynolds, III, and S. A. Rommel, eds), pp. 15–72. Smithsonian Institution Press, Washington, DC.

Perrin, W. F. (1999). Selected examples of small cetaceans at risk. *In* "Conservation and Management of Marine Mammals" (J. R. Twiss, Jr., and R. R. Reeves, eds), pp. 296–310. Smithsonian Institution Press, Washington, DC.

Perrin, W. F., Mitchell, E. D., Mead, J. G., Caldwell, D. K., Caldwell, M. C., van Bree, P. J. H., and Dawbin, W. H. (1987). Revision of the spotted dolphins, *Stenella* spp. *Mar. Mamm. Sci.* **3**, 99–170.

Reynolds, J. E., and Rommel, S. A. (1999). "Biology of Marine Mammals." Smithsonian Institution Press, Washington, DC.

Reynolds, J. E., Odell, D. K., and Rommel, S. A. (1999). Marine mammals of the world. *In* "Biology of Marine Mammals" (J. E. Reynolds, III, and S. A. Rommel, eds), pp. 1–14. Smithsonian Institution Press, Washington, DC.

Reynolds, J. E., Rommel, S. A., and Pitchford, M. E. (2004). The likelihood of sperm competition in manatees—explaining an apparent paradox. *Mar. Mamm. Sci.* **20**, 464–476.

Rice, D. W. (1998). "Marine Mammals of the World: Systematics and Distribution." Special Publication No. 4. The Society for Marine Mammalogy, Allen Press, Lawrence, KS.

Ridgway, S., and Reddy, M. (1995). Residue levels of several organochlorines in *Tursiops truncatus* milk collected at varied stages of lactation. *Mar. Pollut. Bull.* **30**, 609–614.

Ridgway, S. E. (1972). "Mammals of the Sea: Biology and Medicine." Charles H. Thomas Press, Springfield, IL.

Ridgway, S. E., and Au, W. W. L. (1999). Hearing and echolocation: Dolphin. *In* "Encyclopedia of Neuroscience" (G. Adelman, and B. Smith, eds), 2nd Ed., pp. 858–862. Springer-Verlag, New York.

Romano, T. A., Ridgway, S. H., Felton, D. L., and Quaranta, V. (1999). Molecular cloning and characterization of CD4 in an aquatic mammal, the white whale, *Delphinapterus leucas. Immunogenetics* **49**, 376–383.

Ronald, K. L., Hanley, L. M., Healey, P. J., and Selley, L. J. (1976). An annotated bibliography of the Pinnipedia. International Council for the Exploration of the Sea. Charlottenlund, Denmark. Also: Supplement 1, 1983.

Sangama de Beaver, M., and Beaver, P. (1989). "Tales of the Peruvian Amazon." AE Publications, Largo, FL.

Scammon, C. M. (1874). "The Marine Mammals of the North-Western Coast of North America Described and Illustrated Together with an Account of the Whale-Fishery." John. H. Carmany, San Francisco, CA.

Schlundt, C. E., Finneran, J. J., Carder, D. A., and Ridgway, S. H. (2000). Temporary shift in masked hearing thresholds (MTTS) of bottlenose dolphins, *Tursiops truncates,* and white whales, *Dephinapterus leucas,* after exposure to intense tones. *J. Acoust. Soc. Am.* **107**, 3496–3508.

Siebold, P. F. von. (1842). "Fauna Japonica: Les Mammiferes Marins." Batavia Press, Jakarta.

Slijper, E. J. (1976). "Whales and Dolphins." University of Michigan Press, Ann Arbor, MI.

Steller, G. W. (1899). The beasts of the sea. *In* "The Fur Seals and Fur Seal Islands of the North Pacific Ocean" (D. S. Jordan, ed.), pp. 179–218. US Government Printing Office, Washington, DC.

Thewissen, J. G. M. (ed.) (1998). "The Emergence of Whales: Evolutionary Patterns in the Origin of Cetacea." Plenum Press, New York.

Thewissen, J. G. M., Cohn, M. J., Stevens, L. S., Bajpai, S., Heyning, J., and Horton, W. E., Jr. (2006). Developmental basis for hind-limb loss in dolphins and the origin of the cetacean bodyplan. *Proc. Natl. Acad. Sci. USA* **103**, 8414–8418.

Turvey, S. T., *et al.* (17 authors) (2007). First human-caused extinction of a cetacean species? *Biol. Lett.* **3**, 537–540.

Twiss, J. R., Jr., and Reeves, R. R. (1999). "Conservation and Management of Marine Mammals." Smithsonian Institution Press, Washington, DC.

Tyack, P. L. (1999). Communication and cognition. *In* "Biology of Marine Mammals" (J. E. Reynolds, III, and S. A. Rommel, eds), pp. 287–323. Smithsonian Institution Press, Washington, DC.

Tynan, C. T., and DeMaster, D. P. (1997). Observations and predictions of Arctic climatic change: Potential effects on marine mammals. *Arctic* **50**, 308–322.

Wells, R. S., Boness, D. J., and Rathbun, G. B. (1999). Behavior. *In* "Biology of Marine Mammals" (J. E. Reynolds, III, and S. A. Rommel, eds), pp. 324–422. Smithsonian Institution Press, Washington, DC.

Würsig, B., Cipriano, F., and Würsig, M. (1991). Dolphin movement patterns: Information from radio and theodolite tracking studies. *In* "Dolphin Societies: Discoveries and Puzzles" (K. Pryor, and K. S. Norris, eds), pp. 79–111. University of California Press, Berkeley, CA.

Yamase, H. (1760). Geishi [Natural History of Whales]. Osaka Shorin, Osaka, Japan.

Zhou, K., and Zhang, X. (1991). "Baiji, the Yangtze River Dolphin, and Other Endangered Animals of China." Yilin Press, Nanjing, China.

**H**

# Hooded Seal
## *Cystophora cristata*

### KIT M. KOVACS

## I. Characteristics and Taxonomy

The hooded seal is a large phocid that is silver-gray in color with irregular black spots covering most of the body; the face is often completely black (Fig. 1). Adult males are about 2.5 m long and weigh an average of 300 kg; large males can be in excess of 400 kg. Adult females are considerably smaller than males, measuring 2.2 m in length and weighing an average of 200 kg. Hooded seal pups are approximately 1 m long when they are born and weigh about 25 kg. They are blue on their backs and silver-gray on their bellies (Fig. 1). This distinctive "blueback" pelage is maintained for about 2 years (Lavigne and Kovacs, 1988).

The most distinctive physical feature of hooded seals is the prominent nasal ornament borne by adult males (Fig. 1). When relaxed the nasal appendage hangs as a loose, wrinkled sac over the front of males' noses. During the breeding season (in March) males inflate this sac to display to females and to other males, forming a tight, bi-lobed "hood" that covers the front of the face and the top of the head. This structure

**H**

**Figure 1** *(A) Hooded seal mother–pup pair with an attending male in the background. (B) Blueback hooded seal newly weaned, 4 days old. (C) Hooded seal male with the nasal septum extruded and hood partially inflated.*

is the source of the species' common name. Males also have the ability to inflate the elastic nasal septum, which when expanded, protrudes through one nostril as a big membranous pink-red balloon; the source of its secondary common name, the bladdernose seal. Both of these secondary sexual characters are used by males to display to females and to other males during the breeding season. The hood is also used as a threat at other times of year.

The taxonomic history of the species is a bit complex. They have been linked with the elephant seals by some authorities (who attributed these animals jointly to the tribe Cystophorini), and the blueback (pups) were actually described originally as a separate species (*Phoca mitrata*, Cuvier) in the early 1800s (Kovacs and Lavigne, 1986). But, for quite some time hooded seals have been classified

within the family Phocidae, subfamily Phocinae, and are the only species in the genus *Cystophora*.

The dentition of hooded seals is I 2/1 C 1/1 PC 5/5. The teeth, other than the canines, are quite small.

## II. Distribution and Abundance

Hooded seals are a migratory species with a range that encompasses a large sector of the North Atlantic (Fig. 2). They follow an annual movement cycle that keeps them in close association with drifting pack ice most of the time. During the spring, the adults concentrate for breeding purposes in three locations: one group forms off the east coast of Canada which is split into two whelping (birthing) patches, one in the Gulf of St. Lawrence and the other north of Newfoundland—an area known as the Front; a second group congregates in the Davis Strait; and a third comes together on the West Ice, east of Greenland. Some weeks after breeding the animals move into traditional molting areas on the southeast coast of Greenland, near the Denmark Strait or in a smaller patch that is found along the northeast coast of Greenland, north of Jan Mayen. After the annual molt, hooded seals disperse broadly for the late summer, autumn, and early winter months, preferring areas along the outer edges of pack ice but ranging quite broadly both toward the north and south of the North Atlantic. Records of hooded seals being found outside their normal range are not uncommon; young animals in particular are great wanderers. Juveniles have been found as far south as Portugal and the Caribbean in the Atlantic Ocean and in California on the Pacific side (e.g., Mignucci-Giannoni and Odell, 2001).

The global population of hooded seals is approximately 660,000 animals. Three stocks are recognized for the purposes of setting harvest quotas: in Canadian waters (including the Front and Gulf breeding areas), the Davis Strait, and the West Ice (west of Jan Mayen Island). The West Ice stock declined markedly from the 1940s to the 1980s but has stabilized during the last two decades at levels (70,000) that are as low as 10–15% of the population size 60 years ago. Annual pup production varies considerably from year to year, in part due to prevailing sea ice conditions.

Genetically speaking, hooded seals are reported to be panmictic.

## III. Ecology

Hooded seals are pack-ice seals that spend much of the year in association with sea ice. However, they can go on pelagic excursions for many consecutive weeks far from ice-filled waters. During such trips they do not haul out. Hooded seals are deep divers; adult animals can dive to depths of over 1000 m and can remain under water for periods of up to almost an hour (Folkow and Blix, 1999). The diel pattern of diving is quite consistent, but they seem to dive deeper and longer during the day compared to at night. Diving in winter also appears to be deeper and longer than during the summer season. Hooded seals feed pelagically on a variety of deep-water fishes, including Greenland halibut, and a variety of redfish species as well as squid. Herring, capelin, sand eels, and various gadoid fishes including Atlantic cod and Arctic cod can be important seasonally. The hooded seal diet appears to be more varied when they are feeding in inshore waters compared to during their pelagic off-shore periods, when they forage on only a few species (Haug *et al.*, 2007).

Polar bears (*Ursus maritimus*) are natural predators of hooded seals, but human exploitation is likely the greatest source of mortality. Killer whales (*Orcinus orca*) are also a likely predator, although this has never been documented conclusively.

**Figure 2**  *Map showing the distribution of hooded seals (pink-shaded area).*

## IV. Behavior and Physiology

Hooded seals are for the most part solitary animals outside the breeding and molting seasons. Even during these two annual phases when they do aggregate into loose herds, they are very aggressive with one another and do not tolerate close contact beyond the mother–offspring bond or a short male–female pairing period. But, even at sea multiple hooded seals can be seen together sometimes, perhaps attracted to a common resource such as abundant food rather than their proximity being a social tie. Their vocal repertoire is quite simple, as would be expected for a species that is not highly social (Ballard and Kovacs, 1995).

The breeding season occurs in late March. It is short, lasting only 2–3 weeks in a given area. Females give birth in loose pack ice areas, preferring quite thick first-year ice floes for whelping. They space themselves out within the herd at intervals of 50 m or more when ice conditions permit, but the form of the herd and inter-female distances are highly variable, depending on the ice conditions. Mothers attend their pups continuously during the 4-day long period of lactation (Bowen *et al.*, 1985). Hooded seals are notable for having the shortest lactation period of any mammal. Pups are born in a very advanced developmental state, having already shed their grayish-white embryonic first coat of hair (which appears as tight little disks

of hair when the placenta is passed) while in the uterus, and having already accumulated a thin layer of subdermal blubber (Lydersen and Kovacs, 1999). During the incredibly short nursing period, pups drink up to 10 l of milk per day that contains an average fat content of 60% (Debier *et al.*, 1999). This energy-rich diet allows pups to more than double their birth mass during the few days that they are cared for by their mothers; they gain approximately 7 kg per day during nursing. Energy assimilation is extremely efficient during lactation; pups store approximately 75% of the energy they ingest (Lydersen *et al.*, 1997). Pups are weaned weighing 50–60 kg. Mothers lose about 10 kg per day during the nursing period; 40 kg are lost over lactation, representing approximately 17% of total maternal body mass (Kovacs and Lavigne, 1992). During the time when mothers are with their pup, a male often attends the pair. Males compete with one another to maintain positions close to a female. The battles are often bloody, and the mating season is energetically costly. Males lose an average of 2.5 kg per day, which represents a seasonal loss of about 44 kg (14% of total body mass; Kovacs *et al.*, 1996). When a mother is ready to leave her offspring, the attending male accompanies her to the water where mating takes place. Males will return to the whelping area after mating with a female, to resume mate searching. Individual males have been recorded with up to eight females in one breeding season; they can be considered polygynous (serially monogamous; Kovacs, 1990). Like all phocid seals, hooded seals have delayed implantation of the embryo, for 3–4 months, and an active gestation period of about 8 months.

The pups remain alone on the ice following weaning for some days or weeks before going to the water and learning to swim, dive, and forage. During their time on the ice they fast, using body reserves stored in their substantial blubber layer to fuel their energy needs. When they do start to eat, pups feed on krill and other invertebrates initially, until they have sufficient aquatic skills to capture fish. Little is known about juvenile hooded seals. They are only infrequently seen among adult breeding or molting aggregations. It is assumed that they spend a lot of their time at sea and in isolated arctic pack ice areas. They dive to depths in excess of 500 m during their first year of life.

Hooded seals molt annually, in mid to late summer at traditional sites (discussed earlier). Molting herds are usually loose aggregations, similar to breeding concentrations. However, ice conditions dramatically affect the geographic locations and the densities at which the animals concentrate.

Most physiological studies of hooded seals have investigated various aspects of their diving capacity. They have elevated hematocrit, large myoglobin stores, high tolerance to $CO_2$, large blood volume, and other features that are typical of deep-diving seals. Adult hooded seals have the largest body oxygen stores reported to date (89.5 ml/kg; Burns *et al.*, 2007). Hooded seals have large, frontally facing eyes that are likely important in light detection at depth (possible detection of prey-emitted phosphorescence). They also have well developed though not elaborate vibrissae.

### V. Life History

Bluebacks go through a gradual transition, changing slowly over a period of several annual molting periods and becoming increasingly spotted until they have the irregularly blotchy pelt of the adult. Male and female pups are similar in size, but by 1 year of age there are distinct differences in length and mass between the sexes that persist throughout life. The marked sexual dimorphism within this species suggests that the sexes are likely to exhibit different distributions, foraging patterns, etc., but few details regarding at-sea portions of the life cycle are available. Only females and young animals have been satellite-tracked to date. Females reach sexual maturity at an age of three years; males are a little bit older when they mature and are probably significantly older before they can compete at a level that permits them to mate. Hooded seals live to be 25–35 years of age.

### VI. Interactions with Humans

Hooded seals have been commercially exploited for centuries; usually in conjunction with hunts whose primary target was the more abundant harp seal. Norway, Russia, Denmark, Greenland, Great Britain, and Canada have taken part in commercial harvesting of hooded seals. Pre-World War II hunting was done for oil and leather, but improved techniques for handling furs meant that the blueback pelt was the most financially lucrative product of the hooded seal harvest following the war. Adults continued to be taken for oil and leather production, but the numbers were reduced because the market demand for these products dropped. Because adult females remain on the ice to defend their pups against hunters, many adult females were killed. Regulations limiting the killing of mothers have become increasingly restrictive and few females are now taken in the whelping patches. Annual catches of hooded seals have always varied dramatically,

depending largely upon ice conditions at the time of breeding. In years of high harvests, up to 150,000 animals have been taken in the North Atlantic. Seal management in international waters has been put under the auspices of the International Commission for the Northwest Atlantic Fisheries (ICNAF), with Canada, Norway, and Denmark being voting members in the early 1960s. Documented population declines of hooded seals lead to the introduction of quota management during the 1970s. A bilateral agreement for East-Atlantic harvesting between Norway and Russia was also formulated. Following Canada's declaration of a 200-mile economic zone in the late 1970s, Norway and Canada also created a bilateral agreement, and ICNAF was transformed into the Northwest Atlantic Fisheries Organization (NAFO). Under this agreement, Canada and Norway cooperate extensively with information exchange regarding hooded seal abundance estimates and quota revisions. The small population of hooded seals breeding in the Gulf of St. Lawrence is protected from harvesting, as is the Denmark Strait molting concentration.

Declines in abundance of West Ice seals have resulted in the termination of hunting in this area since 2005 as a precautionary measure. Subsistence harvesting of hooded seals takes place in Arctic Canada and in Greenland in addition to Canada's commercial harvest. Similar to the case for most pack-ice breeding seals, the most obvious threat to this species is climate change (Kovacs and Lydersen, 2008). Sea-ice predictions suggest that the breeding habitat of hooded seals will decline dramatically in the decades to come, and the precipitous declines in the abundance of the Northeast Atlantic hooded seal stock in recent decades might be linked to changes already taking place in sea-ice conditions and distribution as well as broader ecosystems shifts.

Hooded seals are not a typical aquarium species, but they have been kept in zoological parks and research facilities in Europe and North America.

### *See Also the Following Articles*

Earless Seals (Phocidae) ■ Hunting of Marine Mammals

### *References*

Ballard, K. A., and Kovacs, K. M. (1995). The acoustic repertoire of hooded seals (*Cystophora cristata*). *Can. J. Zool.* **73**, 1362–1374.

Bowen, W. D., Oftedal, O. T., and Boness, D. J. (1985). Birth to weaning in 4 days: Remarkable growth in the hooded seal, *Cystophora cristata*. *Can. J. Zool.* **63**, 2841–2842.

Burns, J. M., Lestyk, K. C., Folkow, L. P., Hammill, M. O., and Blix, A. S. (2007). Size and distribution of oxygen stores in harp and hooded seals from birth to maturity. *J. Comp. Physiol. B* **177**, 687–700.

Debier, C., Kovacs, K. M., Lydersen, C., Mignolet, E., and Larondelle, Y. (1999). Vitamin E and vitamin A contents, fatty acid profiles, and gross composition of harp and hooded seal milk through lactation. *Can. J. Zool.* **77**, 952–958.

Folkow, L. P., and Blix, A. S. (1999). Diving behaviour of hooded seals (*Cystophora cristata*) in the Greenland and Norwegian Seas. *Polar Biol.* **22**, 61–74.

Haug, T., Nilssen, K. T., Lindblom, L., and Linstrom, U. (2007). Diets of hooded seals (*Cystophora cristata*) in coastal waters and drift ice waters along the east coast of Greenland. *Mar. Biol. Res.* **3**, 123–133.

Kovacs, K. M. (1990). Mating strategies in male hooded seals (*Cystophora cristata*)? *Can. J. Zool.* **68**, 2499–2502.

Kovacs, K. M., and Lavigne, D. M. (1986). *Cystophora cristata*. *Mamm. Spec.* **258**, 9.

H

Kovacs, K. M., and Lavigne, D. M. (1992). Mass-transfer efficiency between hooded seal (*Cystophora cristata*) mothers and their pups in the Gulf of St. Lawrence. *Can. J. Zool.* **70**, 1315–1320.

Kovacs, K. M., and Lydersen, C. (2008). Climate change impacts on seals and whales in the North Atlantic Arctic and adjacent shelf seas. *Sci. Prog.*, **91**, 117–150.

Kovacs, K. M., Lydersen, C., Hammill, M., and Lavigne, D. M. (1996). Reproductive effort of male hooded seals (*Cystophora cristata*): Estimates from mass loss. *Can. J. Zool.* **74**, 1521–1530.

Lavigne, D. M., and Kovacs, K. M. (1988). "Harps & Hoods." University of Waterloo Press, UK.

Lydersen, C., and Kovacs, K. M. (1999). Behaviour and energetics of ice-breeding, North Atlantic phocid seals during the lactation period. *Mar. Ecol. Prog. Ser.* **187**, 265–281.

Lydersen, C., Kovacs, K. M., and Hammill, M. O. (1997). Energetics during nursing and early post-weaning fasting in hooded seals (*Cystophora cristata*) pups from the Gulf of St. Lawrence, Canada. *J. Comp. Physiol. B.* **167**, 81–88.

Mignucci-Giannoni, A. A., and Odell, D. K. (2001). Tropical and subtropical records of hooded seals (*Cystophora cristata*) dispel the myth of extant Caribbean monk seals (*Monachus tropicalis*). *Bull. Mar. Sci.* **68**, 47–58.

**Figure 1** *Hourglass dolphins possess two lateral white areas along the flank that are united by a thin white line that resembles an hourglass. Schools of individuals can swell from the small group of 3 pictured, commonly to around 6 or 7 and rarely up to 60 individuals. Photograph by Robert Pitman.*

# Hourglass Dolphin
## *Lagenorhynchus cruciger*

### R. Natalie P. Goodall

### I. Characteristics and Taxonomy

The name *Delphinus cruciger* was based on a drawing from a sighting in the South Pacific in 1820. Synonyms include *D. albigena, D. bivittatus, Electra clancula, E. crucigera, D. superciliosus, Phocoena crucigera, P. d'Orbignyi, Lagenorhynchus wilsoni, L. latifrons,* and *L. Fitzroyi* (with *L. australis* and *L. obscurus*). The complicated synonymy of this species has been reviewed by Goodall *et al.* (1997). The accepted combination, *L. cruciger*, was made by Van Beneden and Gervais in 1880. Common names have included the crucigere, the albigena, grindhval, sea skunks, and springers; the name in Spanish is *delfin cruzado*.

The hourglass dolphin is mainly black or dark with two elongated lateral white areas, in some animals joined with a fine white line, that resembles an hourglass, which gives it its common name (Fig. 1). The forward patch extends onto the face above the eye, which is within the black surface but outlined with a large dark eye spot with a point forward and a thin white line. The dark pigment of the lips is of varying shape; a gape to flipper stripe may be gray to tan, beige, or even rose. One animal had a white half-moon mark outlining the blowhole. On the side below the white flank patch, there is a lobe of white projecting forward, which may form a sharp point, a blunt, curved shape, or a hook. The flank patches on some animals almost meet on the upper tail stock. Part of the underside of the flippers is white. The ventral region is generally white, with some dark areas forward from the tail stock to the genital region. The pigmentation of juveniles has not been described.

The hourglass is a rather stocky dolphin with a large, recurved dorsal fin that is variable in shape from erect to hooked. The tail stock is often keeled, especially in large males. Total lengths (*n* = 13) range from 142 to 187 cm. Females (*n* = 5) measured 142–183 cm, males (*n* = 8) 163–187 cm. This is probably not the total range of length

**Figure 2** *Hourglass dolphin specimen RNP 2366 stranded at Estancia Moat, Tierra del Fuego in January 2005: view of teeth. Photograph by Carolina Navarro.*

for the species. Weights are known for only five specimens. Females of 163.5 and 183 cm weighted 73.5 and 88 kg, respectively. A 174-cm male weighed 94 kg, one of 180 cm 93 kg, and one of 178.5 cm about 100 kg (it had been attacked by birds).

The condylobasal lengths of 12 skulls ranged from 316 to 370 mm. Visible teeth numbered 26–34 upper and 27–35 lower in each jaw (Fig. 2). Vertebral count is Cv7, Th12–13, L18–22, and Ca29–33 for a total of 69–72 (*n* = 9). The first two cervicals are fused (*n* = 6). The vertebrae of *L. cruciger* are smaller than those of *L. australis*, but slightly larger than those of *L. obscurus* and are similar to the latter in shape (illustrated in Goodall *et al.*, 1997). There are 12–13 ribs (*n* = 9); one specimen had seven pairs of sternal ribs, another eight. The phalangeal formula (*n* = 6) is: I = 2–3, II = 8–11, III = 6–8, IV = 2–4, and V = 0–2.

The intestine lengths of three specimens were 18, 18.5, and 19.7 m. One specimen had 670 lobes in the left kidney.

## II. Distribution and Abundance

The hourglass dolphin is pelagic and circumpolar in the Southern Ocean, in both Antarctic and subantarctic waters, from about 45°S to fairly near the ice pack (Fig. 3). Exceptional northern sightings were at 36°14′S in the South Atlantic and 33°40′S in the South Pacific off Valparaiso, Chile. The southernmost sighting was at 67°38′S in the South Pacific. Most specimens were found between 45°S and 60°S, the northernmost from New Zealand and the southernmost from the South Shetland Islands. Hourglass dolphins are often sighted throughout their range. Abundance for south of the Antarctic Convergence in January, based on sighting surveys from 1976/1977 to 1987/1988, was estimated at 144,300 (CV = 0.17) (Kasamatsu and Joyce, 1995).

## III. Ecology

This dolphin is circumpolar in the Southern Hemisphere on both sides of the Antarctic Polar Front and northward in cool currents associated with the West Wind Drift. Recorded water temperatures range from 0.3°C to 13.4°C. Sightings reflect observer effort, with most from the Drake Passage, due to ship traffic between South America and the Antarctic Peninsula, as well as south of New Zealand and Australia.

Although oceanic, this dolphin is often sighted near islands and banks. It is most often seen in areas with turbulent waters. It has been sighted during nearly every cruise in the Southern Ocean, especially in the Drake Passage (A. Walleyn and others, personal communication) and South Georgia. During dedicated bird and mammal surveys, 480 hourglass dolphins were recorded from September to February over the continental slope or in deep waters to the north, east, and south of the Falkland Islands (Isalas Malvinas), but none were seen in July or August (White *et al.*, 1999). Recent sightings have been made in the Beagle Channel (D. Kuntschik, personal communication) and Canals Sarmiento and Ballenero, Chile. On September 24, 2005, one apparently wounded animal (possibly from a boat strike?) swam for several hours among the catamarans tied up at the tourist pier in Ushuaia, Tierra del Fuego. The hourglass pigmentation was clearly visible.

These dolphins were found so often with fin whales (*Balaenoptera physalus*) that whalers used them as cues for finding whales. They have also been seen with sei and minke whales (*B. borealis* and *B. bonaerensis*), large bottlenose whales (*Hyperoodon* and *Berardius*), pilot whales (*Globicephala melas*), and southern right whale dolphins (*Lissodelphis peronii*) (Goodall, 1997; White *et al.*, 1999). Seabirds, mainly great shearwaters and black-browed albatrosses, feed in association with hourglass dolphins (White *et al.*, 1999).

Hourglass dolphins often feed in large aggregations of seabirds and in plankton slicks. The stomachs of a few specimens from different oceans have been examined. Prey items included unidentified small fish (off Chile), the fish *Krefftichtys andersonii* (Myctophidae) of about 2.4 g and a length of 55 mm (east of Cape Horn); small squid, including some from the families Onychoteuthidae and Enoploteuthidae (Indian Ocean); and fish otoliths, squid beaks, and crustaceans (South Pacific) (Goodall *et al.*, 1997). Two males from Chubut, Argentina, contained 10 prey species, including fish, cephalopods, crustaceans, and polychaetes. The most abundant prey in the first dolphin was the lesser shining bobtail squid (*Semirossia tenera*), followed by the small Patagonian squid (*Loligo gahi*) and juvenile Argentine hake (*Merluccius hubbsi*), indicating that the dolphin fed in surface waters. Annelids in the stomachs may have come from prey species. The most abundant prey of the second dolphin was the pelagic fish *Protomyctophum* sp. (Myctophidae); the stomach also contained one Argentine hake and the squid *Illex argentinus* and *L. gahi*.

Recorded parasites include the cestode *Phyllobothrium delphini*, the nematode *Anisakis simplex sensu lato*, and digeneans (Capulidae) of the genera *Hadwenius* and *Oschmariella* (Gasitúa *et al.*, 1999; Fernández *et al.*, 2003).

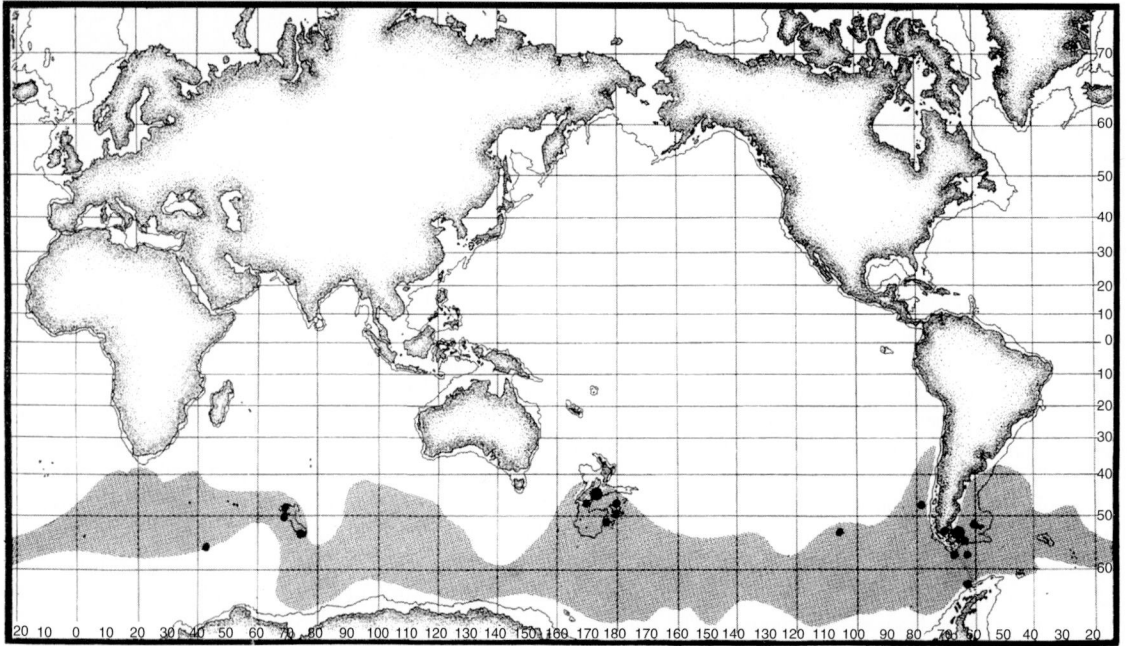

**Figure 3** *Distribution of the hourglass dolphin compiled from incidental and dedicated sighting surveys and the location of specimens from 1824 to 2007. Small circles indicate single specimens, and large circles represent three specimens off New Zealand and six at Tierra del Fuego.*

The largest animal known, a male from the Falkland (Malvinas) Islands, had long-standing gastric trauma, a penetration of some duration, with extreme peritonitis, as adhesions were forming. One of the animals from northern Patagonia had five ulcers in the main stomach, one with an *A. simplex* larva attached (Fernández *et al.*, 2003).

Concentrations of heavy metals in one animal in the blubber were cadmium 0.0022 μg/g damp weight and lead 0.0013 μg/g damp weight (Gazitúa *et al.*, 1999).

No predators are known, although killer whales and leopard seals are possibilities. One animal had wounds in the abdominal area which may have been made by sharks (Fernández *et al.*, 2003).

## IV. Behavior and Physiology

Nothing is known of the migratory movements of this species. Animals of the Antarctic Polar Front zone and the continental shelf break, they may move into subantarctic waters or nearer shore in winter. However, during dedicated surveys, this species was not observed off the Falkland Islands (Las Malvinas) in July and August, where many had been seen in spring and summer (White *et al.*, 1999).

Hourglass dolphins are rapid swimmers with a forward, plunging movement. They commonly bow-ride ships, especially in rough weather. They often leap from the sea, porpoise like penguins, and have been observed to spin. They especially seem to enjoy wave-surfing during very high seas.

During the Southern Hemisphere minke whale assessment cruises conducted by the International Whaling Commission (IWC), school sizes ranged from 1 to 60 animals (mean 7). Other studies reported mean sizes of 4 and 5.7 animals.

During passive acoustic surveys for odontocetes aboard the RV *James Clark Ross* in 1998/1999 and January/February 2000, dolphin-like clicks were confirmed on two occasions when the dolphins were seen near the hydrophone and heard on 22 more in the area north and south of South Georgia and 60°S and southward off the Northwest Antarctic Peninsula, but they could not all be identified with certainty (Leaper *et al.*, 2000).

## V. Life History

Very little is known of growth and reproduction in this species. A 163.5-cm female was sexually immature and one of 183 cm was pubescent. Based on fusion of the vertebral epiphyses, the 163.5-cm female was physically immature. The 183-cm female was subadult.

Males of 174 and 187 cm were sexually mature. A male of 163 cm was physically subadult, one of 187 cm was nearly mature, and one of 174 cm was physically mature. Nothing is known of the young, times of birth, and reproduction rates; only three calves, seen in January and February, have been reported.

Three specimens have been aged through sectioning of teeth. A male of 164 cm had eight growth-layer groups (GLGs), and one of 178 cm had nine; both were sexually mature (Fernández *et al.*, 2003). The Strait of Magellan male of 180 cm had eight GLGs.

The testes of sexually mature males are very large in relation to total length. Those of the 180-cm Strait of Magellan male weighed 420 and 400 g (L-R) and measured 25.5 cm in length and 6.5 cm in width. This animal, collected in November, may not have been sexually active. Those of the Falkland animal (187 cm), stranded in January, weighed 2250 and 2200 g and measured 36 × 11 cm. The testes of the Tierra del Fuego animal of 178.5 cm, also in January, weighed 1900 and 1850 g and measured 42.4 × 8 cm and 42.9 × 10.1 cm. The prostate gland of this animal also seemed large, weighing 429 g (17.7 × 10.8 × 3.4 cm).

## VI. Interactions with Humans

Hourglass dolphins seem to be attracted to ships. They often approach from considerable distances, changing course to do so, and remain with the ship for up to 30 min (Thiele, 1996).

Several hourglass dolphins have been taken for scientific study (Goodall *et al.*, 1997). No other directed catches are known. At least one stranded animal was probably the result of a ship strike or net capture held against the side of a ship, as there were cuts along the anterior dorsal surface and the skin and blubber had been rubbed off near the dorsal fin, flippers, and flukes.

The only INCIDENTAL CATCHES reported were three females from New Zealand and a drift net catch in the southern Pacific Ocean. Hourglass dolphins were seen several times near ships during longline hauling operations for Patagonian toothfish (*Dissostichus eleginoides*) off South Georgia, but "did not appear to interact with fishing operations" (Ashford *et al.*, 1996). Most of the specimens described here were strandings.

No hourglass dolphins have been kept in CAPTIVITY.

## See Also the Following Articles

Antarctic Marine Mammals ■ Delphinids

## References

Ashford, J. R., Rubilar, P. S., and Martin, A. R. (1996). Interactions between cetaceans and longline fishery operations around South Georgia. *Mar. Mamm. Sci.* **12**, 452–457.

Brownell, R. L., Jr. (1999). Hourglass dolphin, *Lagenorhynchus cruciger* (Quoy and Gaimard, 1824). *In* "Handbook of Marine Mammals; the Second Book of Dolphins and Porpoises" (S. H. Ridgway, and R. Harrison, eds), pp. 121–135. Academic Press, San Diego.

Fernández, M., Berón-Vera, B., García, N. A., Raga, J. A., and Crespo, E. A. (2003). Food and parasites from two hourglass dolphins, *Lagenorhynchus cruciger* (Quoy and Gaimard, 1824), from Patagonian waters. *Mar. Mamm. Sci.* **19**, 832–836.

Gazitúa, F., Gibbons, J., and Cárcamo, J. (1999). Descripción de un ejemplar de delfín cruzado, *Lagenorhynchus cruciger* (Delphinidae), encontrado en el Estrecho de Magallanes. *Anales Instit. Patagonia Se. Cs. Nat. (Chile)* **27**, 73–82.

Goodall, R. N. P. (1997). Review of sightings of the hourglass dolphin, *Lagenorhynchus cruciger*, in the South American sector of the Antarctic and sub-Antarctic. *Rep. Int. Whal. Commn* **47**, 1001–1013.

Goodall, R. N. P., Baker, A. N., Best, P. B., Meyer, M., and Miyazaki, N. (1997). On the biology of the hourglass dolphin, *Lagenorhynchus cruciger* (Quoy and Gaimard, 1824). *Rep. Int. Whal. Commn* **47**, 985–999.

International Whaling Commission (1997). Report of the sub-committee on small cetaceans. *Annex H. Rep. Int. Whal. Commn* **47**, 169–191.

Kasamatsu, F., and Joyce, J. (1995). Current status of odontocetes in the Antarctic. *Ant. Sci.* **7**, 365–379.

Kasamatsu, F., Hembree, D., Joyce, G., Tsunoda, L., Rowlett, R., and Nakano, T. (1988). Distribution of cetacean sightings in the Antarctic: Results obtained from the IWC-IDCR minke whale assessment cruises, 1978–79 to 1983–84. *Rep. Int. Whal. Commn* **38**, 449–487.

Leaper, R., Gillespie, D., and Papastavrou, V. (2000). Results of passive acoustic surveys for odontocetes in the Southern Ocean. *J. Cetacean Res. Manage.* **2**, 187–196.

LeDuc, R. G., Perrin, W. F., and Dizon, A. E. (1999). Phylogenetic relationships among the delphinid cetaceans based on full cytochrome *b* sequences. *Mar. Mamm. Sci.* **15**, 619–648.

Thiele, D. (1996). Preliminary results of a Southern Ocean cetacean sightings survey in Areas IV and V. IWC Scientific Committee meeting document SC/48/SH22, 24 pp.

H

Van Waerebeek, K., Goodall, R. N. P., and Best, P. G. (1997). A note on evidence for pelagic warm-water dolphins resembling *Lagenorhynchus*. *Rep. Int. Whal. Commn* **47**, 1015–1017.

White, R. W., Reid, J. B., Black, A. D., and Gillon, K. W. (1999). "Seabird and Marine Mammal Dispersion in the Waters Around the Falkland Islands 1998–1999." Joint Nature Conservation Committee, UK.

# Humpback Dolphins
## *S. chinensis* and *S. teuszii*

### GUIDO J. PARRA AND GRAHAM J.B. ROSS

### I. Characteristics and Taxonomy

Humpback dolphins Fig. 1 are medium-sized delphinids found in coastal waters of the eastern Atlantic (West Africa), Indian, and West Pacific Oceans (Fig. 2). Genetic and morphological information indicate that they are delphinids (family Delphinidae). Initially, humpback dolphins were thought to be related to *Sotalia* spp., small delphinids that inhabit coastal and riverine waters of South America and *Steno bredanensis*, an oceanic dolphin species. However, molecular studies indicate that humpback dolphins are more closely related to tropical oceanic species, including those of the genera *Stenella*, *Delphinus*, *Tursiops*, and *Lagenodelphis* (LeDuc *et al.*, 1999). The taxonomy of the genus *Sousa* is not well established and no study to date has resolved the number of species in the genus. Current views range from recognition of only a single, variable species – *S. chinensis* – to three nominal species: *S. chinensis* (Pacific Ocean), *S. plumbea* (Indian Ocean), and *S. teuszii* (Atlantic Ocean). Other nominal species include *S. lentiginosa* (Owen, 1866) and *S. borneensis* (Lydekker, 1901). Studies on skull morphology support the division of the genus into the *chinensis*, *plumbea*, and *teuszii* forms; however, patterns of cranial variation were thought conservative and no taxonomic revisions were recommended (Jefferson, 2004). Recent phylogenetic studies indicate that Australian humpback dolphins are highly divergent from those in Southeast Asia and may represent a distinct species (Frere *et al.*, 2008). Further morphological and molecular studies are needed to resolve the taxonomy of this highly variable genus. At present, the Scientific Committee of the International Whaling Commission recognizes only two species, *S. teuszii* and *S. chinensis*, the latter comprising all Indo-Pacific populations of *Sousa*.

Humpback dolphins are characterized by a robust and medium-sized body (Jefferson and Karczmarski, 2001; Ross *et al.*, 1994). The melon is moderate in size, slightly depressed and in profile slopes gradually to an indistinct junction with the long, narrow rostrum. Neonates have vibrissae. The gape is straight. The broad flippers are rounded at the tip and the flukes are broad and full, with a deep median caudal notch. Dorsal and ventral ridges on the caudal peduncle are well developed in African and Indian Ocean populations. Overall, humpback dolphins reach a maximum total length of 2.6–2.8 m in different parts of their distribution. A few animals exceeding 3.0 m in length have been recorded in the Arabian and Indian regions. Maximum weights of 250–280 kg have been recorded for humpback dolphins in South Africa and Hong Kong, respectively. Sexual dimorphism in total body length and weight is only apparent in the South African animals where mean lengths and weights

for fully grown males are 2.70 m and 260 kg compared to 2.40 m and 170 kg in females.

Characteristic features of the skull include a long, narrow rostrum strengthened by raised premaxillary bones and increasingly compressed toward the tip, large temporal fossae on which the jaw muscles insert and pterygoid bones that are separated in the midline by up to 11 mm. A broad gap exists between the posterior margin of the maxillary bones and the supraoccipital crest of the skull. The mandibular symphysis is long with each jaw bearing 27–39 teeth, wedge shaped at their base. Skull morphology is similar in all populations, apart from lower tooth counts, a shorter mandibular symphysis and a broader cranium in West African animals (*S. teuszii*). For a thorough review of geographic variation in skull morphology of humpback dolphins, see Jefferson and Van Waerebeek *et al.* (2004). The mean number of teeth per jaw increases eastward from 28 or 29 in West African animals (*teuszii* form) to 35 or 37 teeth in north Indian Ocean populations (*plumbea* form) and 33 or 35 teeth in Southeast Asian and Australian animals (*chinensis* form). The range of vertebral formulae in South African animals was 7 C, 11–12 T, 9–12 L, 20–24 Ca = 49–52. The first and second cervical vertebrae are fused. Vertebral counts in humpback dolphins farther east are similar to those of the South African sample (49–53), while West African humpback dolphins have 52–53 vertebrae.

Regional differences occur in external morphology, especially in coloration and shape and size of dorsal fin and hump (Fig. 1). In Indian humpback dolphins (*plumbea* form), the dorsal fin is smaller, slightly falcate, less triangular in shape and sits atop a prominent and well-developed dorsal hump (Fig. 1A). The dorsal fin of Pacific humpback dolphins (*chinensis* form) is short, triangular in shape, slightly recurved and has a wide base without a basal hump (Fig. 1B). Atlantic humpback dolphins (*S. teuszii*) have a very similar dorsal fin shape and basal hump to Indian humpback dolphins, but the hump tends to be more pronounced and fin more triangular in shape with a rounded tip. Coloration varies greatly according to geographic location and age. Calves throughout the range are mostly dark gray above with a lighter ventral surface. Adults from the western Indian Ocean are usually dark gray; with lighter ventral surface shading to off-white with light spotting sometimes present (Fig. 1A). Atlantic humpback dolphins have a similar appearance to that of Western Indian Ocean animals. Adults from Australia are pale gray in color with flanks shading to off-white and spotted toward the ventral surface (Fig. 1B). Rostrum, melon, and dorsal fin in Australian animals whiten with age. Most adults from Southern China are pure white, often with dark spots on the body and a pinkish tinge resulting from blood flushing during periods of high activity (Fig. 1C).

### II. Distribution and Abundance

Humpback dolphins are tropical to subtropical species found mainly in coastal waters of the eastern Atlantic, Indian and western Pacific Oceans (Fig. 2). Atlantic humpback dolphins are endemic to coastal waters of the eastern Atlantic of West Africa from Morocco to Southern Angola (Van Waerebeek *et al.*, 2004). Indo-Pacific humpback dolphins (including the *plumbea* and *chinensis* forms) occur from South Africa to Central China and northern Australia (Jefferson and Karczmarski, 2001). Recent observations suggest that the *plumbea* form ranges from False Bay in South Africa to at least the Bay of Bengal in India. The *chinensis* form extends from the Gulf of Thailand east to central China and northern Australia. The distributions of the *plumbea* and *chinensis* forms may overlap in the Bay of Bengal. At least one humpback dolphin, most likely *S. plumbea*,

Figure 1    *Regional differences in the external appearance of humpback dolphins. (A) Humpback dolphin from the east coast of South Africa exhibiting dark gray coloration and well-developed dorsal hump typical of animals from the Eastern Atlantic (S. teuszii) and Western Indian Ocean (Sousa plumbea). (B) Humpback dolphin from Cleveland Bay, Australia showing lighter coloration and absence of dorsal hump distinctive of animals from the Eastern Indian Ocean and Pacific Ocean (Sousa chinensis). (C) External appearance of adult humpback dolphin from Hong Kong showing the brilliant white/pink coloration characteristic of animals in southern China which differs from conspecifics elsewhere. Photographs by Brett Atkins (A) and Samuel Hung (C).*

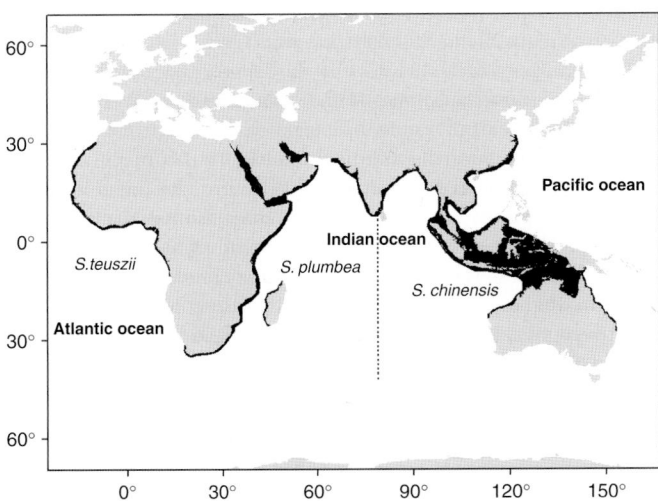

Figure 2    *Map showing the distribution of humpback dolphins in the Eastern Atlantic, Indian and west Pacific Oceans.*

reached the Mediterranean Sea via the Suez Canal (Kerem *et al.*, 2001). Australian humpback dolphins range from approximately the Queensland–New South Wales border (31°27′S, 152°55′E) to western Shark Bay, Western Australia (25° 51′S, E113° 20′E).

Estimates of population size are only available for a few selected areas. At least on a local scale, populations of humpback dolphins are small (usually in the low hundreds) with the exception of the Hong Kong/Pearl River estuary population in China which is estimated to be at about 1500 animals (Jefferson, 2000). In contrast, preliminary surveys of the Xiamen Area to the north of Hong Kong indicate a small population of 80 humpback dolphins (Jefferson and Hung, 2004). Population estimates in Australian waters suggest that there are about 100 individuals inhabiting Moreton Bay, southern Queensland and less than 100 animals in Cleveland Bay, northeast Queensland (Corkeron *et al.*, 1997; Parra *et al.*, 2006a). Populations of South Africa were estimated at 466 (95% CI = 447–485) dolphins in Algoa Bay (Karczmarski *et al.*, 1999b) and 74 (95% CI = 60–88) in Richards Bay (Keith *et al.*, 2002). In Maputo Bay, Mozambique, preliminary estimates indicate that about 105 (95% CI = 30.5–150.9) humpback dolphins inhabit this area (Guissamulo and Cockcroft, 2004). Between 58 and 65 humpback dolphins occur in a small area (26 km²) off the south coast of Zanzibar (Stensland *et al.*, 2006). A very small population (<100 individuals) has been identified off central western Taiwan (Wang *et al.*, 2004).

## III. Ecology

Though key habitats vary with geographical location, humpback dolphins are typically found in shallow waters of less than 20 m, close to the coast and associated with river mouths, mangroves, tidal channels and inshore reefs (Karczmarski *et al.*, 2000a; Atkins *et al.*, 2004; Parra, 2006). In Australia, humpback dolphins occur occasionally offshore but generally in shallow water around islands or reefs as well as in reef lagoons such as Ningaloo Reef, Western Australia (Parra *et al.*, 2004). In Australia, China and India, high-density areas are usually associated with estuarine habitats and deep water channels. In China, dolphins may swim up rivers for tens of kilometers. In the Arabian region, humpback dolphins are mainly found in shallow coastal waters along low-energy sandy shorelines, though in some areas around Oman humpback dolphins are frequently seen along high-energy rocky shorelines in waters over 40 m deep (Baldwin *et al.*, 2004).

Limited quantitative data are available on the feeding ecology of humpback dolphins throughout their range. Based on studies in South Africa and Hong Kong (Barros and Cockcroft, 1991; Barros *et al.*, 2004), it appears that humpback dolphins are opportunistic-generalist feeders, eating a wide variety of coastal, estuarine and reef-associated fishes (and occasionally cephalopods and crustaceans) both on the bottom and within the water column (Fig. 4). Fishes in the families Haemulidae, Sciaenidae, Sparidae, Mugilidae and Clupeidae have been identified as important prey items across the South African and Chinese range of *Sousa*. The most common fish prey species eaten by South African animals were *Mugil cephalus*, *Pomadasys olivaceum*, and *Pachymetopon aeneum* and by Chinese animals, *Johnius* sp. *Collichthys lucida* and *Thryssa* spp. Prey species reported from Senegal include *Pristipoma jubelini*, *Ethmalosa fimbriata*, and *Mugil* spp. Incidence of scars resulting from interactions with sharks has been observed in South African and Australian animals. Adult tiger (*Galeocerdo cuvier*), great white (*Carcharodon carcharias*), and bull (*Carcharinus leucas*) sharks are the most likely predators of humpback dolphins. The effects of predation on humpback dolphins' ecology are uncertain.

## IV. Behavior and Physiology

Humpback dolphins swim slowly at about 5 km/h, surfacing briefly at intervals of up to a minute. Longer dives may last up to 5 min. Typically they avoid boats and rarely bow-ride. Nevertheless, animals in Hong Kong appear to be used to the presence of boats and have been observed bow-riding dolphin-watching boats. When approached, humpback dolphins generally dive, split up into small groups or single animals and often change course underwater, re-appearing unexpectedly some distance away. When a humpback dolphin surfaces, the beak or occasionally the whole head is typically raised clear of the water and the body is arched, showing the upper back and dorsal fin while the rest of the body remains underwater. Flukes are usually exposed at the surface before animals go for a deep dive. Humpback dolphins display a wide variety of aerial displays including vertical leaps, side leaps and forward/backward somersaults (Fig. 3).

The observed daytime behaviors of humpback dolphins include foraging/feeding, traveling, socializing and resting (Parsons, 2004a). Daytime behavior in Algoa and Richards Bays (South Africa), Hong Kong and Cleveland Bay (northeast Queensland, Australia) is dominated by foraging activities followed by traveling and socializing. Foraging

**Figure 3** *Humpback dolphin catching a Mullet (Liza spp.) at the mouth of the Devi River in Orissa, India. Photograph by Dipani Sutaria.*

**Figure 4** *An adult humpback dolphin doing a somersault in Cleveland Bay, Queensland, Australia.*

activities are usually associated with inshore reefs, tidal channels and river mouths. In Algoa Bay, foraging behavior showed tidal, diurnal and seasonal patterns with increased feeding at high tide in the morning and evening and during the winter season. Cooperative feeding appears to be limited. Individuals in foraging schools are usually widely dispersed (50–100-m apart), move in various directions without an obvious pattern, dive frequently and steeply downward [often preceded by fluke (tail fin) up or peduncle (tail stock) arches] and have extended submersion times of more than 2 min. At the surface, individuals often display rapid accelerations and erratic movement while chasing fish. In northeast Queensland, Australia and in the tidal channels of the Bazaruto Archipelago, Mozambique, humpback dolphins have been observed beaching themselves intentionally as they chase fish into shallow waters and sandbanks. Humpback dolphins in Hong Kong are frequently seen feeding in the freshwater/saltwater mixing zone. Schools foraging behind fishing trawlers are common in Australia and Hong Kong and for some individuals this appears to be a major source of food.

Socializing (including mating) in humpback dolphins is characterized by individuals in close proximity showing high levels of physical interaction including body contact (animals touching and biting each other and rubbing their bodies) and frequent aerial behavior such as leaps and somersaults. Fins and flukes often break the surface of the water. Copulation lasting 20–30 sec occurs with one dolphin inverted below its partner. Observations of dolphins rising vertically belly to belly in the Arabian Gulf and the Indus delta have been ascribed to mating behavior. Mating and births occur year round. Allomaternal care of offspring has been suggested for humpback dolphins in Algoa and Plettenberg Bays, South Africa.

Throughout their range, humpback dolphins are most frequently seen in relatively small schools of less than 10 animals. Solitary animals and schools of 2–6 individuals are the most common, although aggregations of 30–100 individuals have been observed along the Arabian Sea coast of Oman. In general, schools consist mainly of adult animals only or combinations of adults, juveniles, and calves. Schools solely composed of juveniles are rare. Little seasonal variation in school size occurs in Australian, Hong Kong, and Mozambique waters. However, significant increases in school size have been documented during summer and winter in Algoa and Plettenberg Bays, South Africa. School size also appears to vary according to behavioral activity. For example, in Cleveland Bay, Australia, socializing schools of humpback dolphins are larger than schools that are foraging or traveling. Additionally, schools foraging behind trawlers are larger than schools foraging independently of trawlers or traveling. Studies in South Africa, Hong Kong, and Australia indicate that humpback dolphins live in a fission–fusion society where individuals associate in schools that change often in size and composition. Long-lasting affiliations among adult animals do occur but are uncommon. Female–calf associations are stable and strong during the first 3–4 years.

Sound production and reception are vital to humpback dolphins in the often murky habitat they occupy. The acoustic repertoire of Pacific humpback dolphins includes a variety of sounds similar to those of other delphinids, ranging in frequency from a minimum of 0.6 kHz to a maximum of at least 200 kHz (Van Parijs and Corkeron, 2001c; Goold and Jefferson, 2004). Sounds produced by Australian humpback dolphins have been classified into five different vocalization categories of variable frequency and length: broad-band clicks (>kHz, 0.1–10 sec); barks (0.6–22 kHz, 0.1–7.4 sec); quacks (0.6–3.7 kHz, 0.08–2.7 sec); and grunts (0.09–1.4 kHz, 0.06–2 sec) (Van Parijs and Corkeron, 2001b, c). High-frequency broadband clicks appear to be used for echolocation and have been recorded mostly during foraging activities and to a lesser extent during socializing and traveling. Barks and quacks are produced predominantly during social and foraging behaviors, while grunts appear to be restricted to social behavior. Additionally, 17 different narrow-band frequency-modulated sounds (whistles) have been described for Australian humpback dolphins. These whistles are mainly heard during social behavior. It has been suggested that humpback dolphins may use their hearing capabilities to locate sound-producing prey by passively listening for the sounds that they make.

Throughout a significant part of their range, humpback dolphins share their coastal habitat with Indo-Pacific bottlenose dolphins (*Tursiops aduncus*), snubfin dolphins (*Orcaella heinsohni*) and finless porpoises (*Neophocaena phocaenoides*). Interactions with bottlenose and snubfin dolphins have been recorded in the wild. Mixed schools of humpback and bottlenose dolphins have been observed in South Africa (Karczmarski *et al.*, 1997), Tanzania (Stensland *et al.*, 2003), Oman (Baldwin *et al.*, 2004) and Australia (Corkeron, 1990). In South Africa, most interactions appear to be non-agonistic, with humpback dolphins remaining in the periphery or at distance of the school of bottlenose dolphins. However, aggressive interactions from bottlenose dolphins toward lone humpback dolphins have been documented in South Africa and Oman. In Tanzania, dolphins in mixed schools are often seen resting, traveling, and socializing, including male bottlenose dolphins herding female humpback dolphins. Mixed schools of bottlenose and humpback dolphins in Moreton Bay, Australia, have only been seen while feeding behind trawlers. During these interactions, bottlenose dolphins were higher in number and appeared to be dominant over humpback dol-

phins. Interactions between humpback and snubfin dolphins have only been observed in northeast Queensland, Australia (Parra, 2006). Here, interspecific interactions are mainly of aggressive-sexual nature with humpback dolphins dominating snubfin dolphins. No interactions have been observed between finless porpoises and humpback dolphins. In Hong Kong these species show spatial and temporal differences in their habitat use. Interspecific interactions between humpback dolphins and other dolphins and porpoises within their range appear to be complex and may be the result of anti-predator and foraging strategies, interspecific mating or competition for resources.

Although humpback dolphins do not undergo large-scale seasonal migrations, seasonal changes in their distribution and abundance have been observed in South Africa (Karczmarski *et al.*, 1999a), Mozambique (Guissamulo and Cockcroft, 2004) and in Hong Kong and adjacent waters (Jefferson, 2000). Long-term observations of individual animals in localized areas in Australia, Hong Kong and South Africa indicate varying degrees of site fidelity, with some animals using local study areas seasonally and some others throughout the year. For example, in Cleveland Bay, northeast Queensland, Australia, most humpback dolphins are not permanent residents, but it was found that individuals did use the same areas within the bay regularly from year to year following a movement model of emigration and re-immigration (Parra *et al.*, 2006a). At Lantau Island, Hong Kong, humpback dolphins are present year round in waters to the north but shift their distribution to the south and east during the summer monsoon season (Parsons, 1998a). These seasonal changes in distribution and abundance are presumably associated with changes in prey availability and increase in the outflow of the Pearl River. Individual linear movements vary from only a few tens of kilometers in Hong Kong up to 120 km along the Natal and Eastern Cape coast in South Africa. Large linear movements seem unlikely, as extensive reviews of photo-identification catalogs from areas wide apart (>500 km) have yielded no matches. The home range size of humpback dolphins is unknown mainly because of the localized nature of the studies conducted and the difficulties in tracking individual animals for long periods of time. Individuals in Hong Kong and the Pearl River estuary region showed range sizes from 24 to 304 km², with an average of 99.5 km² (Hung and Jefferson, 2004). In Cleveland Bay, northeast Queensland, Australia, a representative range of 190 km² and a core area of 17 km² were identified at the population level (Parra, 2006).

## V. Life History

Most of the information available on the life history of humpback dolphins comes from populations in South Africa and Hong Kong (Cockcroft, 1989, Jefferson, 2000, Karczmarski, 1999). Births occur year-round, although there is evidence of seasonality for South Africa and China. In South Africa, calving peaks in spring or summer, the gestation period lasts 10–12 months, lactation may last >2 years, sexual maturity is reached at 10 years of age for females and 12–13 years for males and a 3-year calving interval has been suggested. In Hong Kong, most births occur between January and August, a gestation period of 11 months is presumed, length at birth is assumed to be about 100 cm and females reach sexual maturity at 9–10 years of age.

## VI. Interactions with Humans

The conservation status of almost all populations of humpback dolphins throughout their range is uncertain, primarily because monitoring of population sizes and mortality is lacking in most regions. Humpback dolphins are currently listed as Near-Threatened (*S. chinensis*) and Vulnerable (*S. teuzsii*) by the IUCN and are listed in CITES Appendix I the western Taiwan population of *S. chinensis* is listed as Critically Endangered. Population estimates for a few selected areas indicate that the populations are relatively small (with the exception of Hong Kong) and thus vulnerable to anthropogenic mortality and potentially rapid population declines. Humpback dolphins are vulnerable to human impact because of their dependence on coastal and estuarine habitats which are under increasing pressure from expanding human populations. Anthropogenic threats throughout their range include wildlife tourism, direct takes, incidental captures in gill nets and shark nets and habitat degradation and loss.

In Australia, observations and interactions with free-ranging humpback dolphins occur only in Queensland (Parra *et al.*, 2004). In southern Queensland, up to seven free-ranging humpback dolphins visit Tin Can Bay regularly, where they are fed fish by visitors. Dedicated dolphin-watching trips including humpback dolphins are limited to a handful of boat operators in Moreton Bay and in Hervey Bay. In contrast, dozens of dolphin-watching operations involving humpback dolphins occur in Goa (Parsons, 1998b), Zanzibar (Stensland *et al.*, 2006) and in Hong Kong (Ng and Leung, 2003). If properly managed, marine mammal-watching activities can benefit the animals conservation through promoting increased public awareness of their biology and threats. However, dolphin watching is also recognized as a potential threat to the dolphins. Careful management, official dolphin-watching codes and enforcement are needed in order for the industry to be sustainable. Official dolphin-watching codes have been implemented in Australia and Zanzibar, but enforcement is lacking. A voluntary code has been established in Hong Kong, while no regulations exist in India.

Incidental mortality of humpback dolphins in fishing nets has been reported for almost all areas within their range. Though the data on levels of mortality are lacking for most regions, incidental catch in fishing nets is thought to be one of the most direct sources of human-caused mortality of humpback dolphins. Of 28 humpback dolphins stranded in Hong Kong between May 1993 and March 1998, 21% showed signs of net entanglement and 11% of boat collision (Parsons and Jefferson, 2000). Some of the animals photo-identified in Hong Kong show evidence of scars from fisheries interactions (2.6–6.8%) and boat propellers (1.2–1.9%). Humpback dolphins are also incidentally caught in shark nets set for bather protection in South Africa and Australia (Cockcroft, 1990; Parra *et al.*, 2004)} (Fig. 5). Along the KwaZulu-Natal coast, South Africa, catches are high and shark nets represent a major threat to the small populations of humpback dolphins inhabiting these waters. A total of 129 humpback dolphins were caught in shark nets along the KwaZulu-Natal coast between 1980 and 1998 with the majority being caught at Richards Bay. Humpback dolphins were among the most commonly caught dolphin species in shark nets off Northeast Queensland, Australia. Net attendance rules and gear modifications have been introduced in Queensland's inshore gillnet fishery and most shark nets have been replaced with drumlines to reduce the incidental take of non-target species. However, enforcement is lacking in remote areas and there is no evidence that any of these measures have provided any benefit to the conservation of humpback dolphins.

At present, directed takes of humpback dolphins are rare and are probably restricted to occasional opportunistic hunting. An estimated 22 humpback dolphins were caught intentionally for human consumption between 1986 and 1999 off the east coast of Madagascar (Razafindrakoto *et al.*, 2004). In Zanzibar, dolphins were hunted for shark bait and for local consumption until 1996 (Stensland *et al.*, 2006). This hunt has now been replaced by dolphin-watching tourism which has become an alternative livelihood for the local communities. A total of 36 individuals were taken in Xiamen, China in the early 1960s to determine if leather could be made from the skin (Jefferson and Hung, 2004).

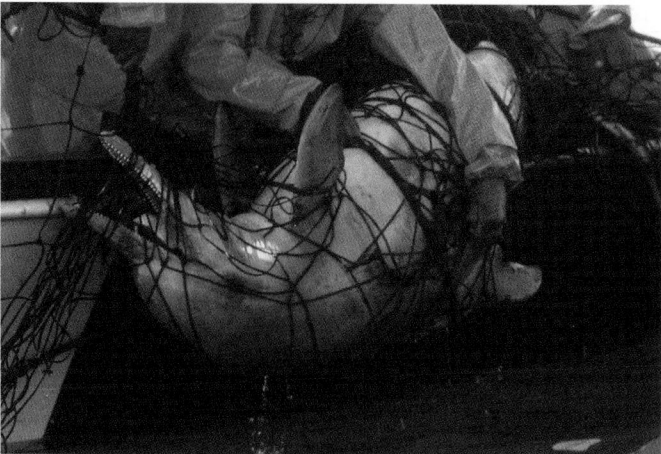

**Figure 5** *Humpback dolphin entangled in a shark net in South Africa. Photograph by Brett Atkins.*

Very few humpback dolphins have been held in captivity. There are reports of live captures of a large number of Indo-Pacific humpback dolphins from the Gulf of Thailand for the oceanarium trade. At least 13 humpback dolphins, most captured in the Tin Can Bay area, have been held in captivity in Australian oceanariums. Humpback dolphins from South Africa, Australia and Thailand have survived in captivity for periods from 3 months to over 30 years.

Because of increasing pressure from expanding human populations (especially in coastal zones throughout the humpback dolphins range), the major threat to all populations is degradation and destruction of coastal habitats. This degradation is mainly being caused by coastal zone development, overfishing of prey, pollution and vessel traffic. For example, high levels of pollutants – particularly mercury and organochlorine contaminants such as DDT – have been found in Hong Kong's population of humpback dolphins (Parsons, 1998c; Parsons, 1999; Parsons, 2004b;Jefferson *et al.*, 2006). The high level of neonatal mortality (53% of strandings) observed in Hong Kong humpback dolphins may be related to organochlorine contamination. Ingestion of contaminated seabed sediments, prey species and transfer of contaminants via lactation are all part of the problem. Studies in Moreton Bay, Australia, indicated that noise from transiting vessels affects group cohesion in humpback dolphins (Van Parijs and Corkeron, 2001a). Moreover, humpback dolphins in Hong Kong tended to dive for longer periods of time in areas of high vessel traffic (Ng and Leung, 2003). In recognition of the numerous risks humpback dolphins face in Hong Kong waters, the Agriculture, Fisheries and Conservation Department (AFCD) of the Hong Kong Government funded several studies to examine the status of the local humpback dolphins. These studies led to the establishment in 1996 of the Sha Chau and Lung Kwu Chau Marine Park, northwest of Lantau Island, as well as the development of a conservation program in 2000 for the protection of humpback dolphins.

The cumulative effect of anthropogenic threats may result in the loss of populations of humpback dolphins already depleted, restricted to certain types of habitats and with small geographic ranges. However, the lack of baseline ecological data for most populations makes determining the effects of habitat loss difficult. Due to their apparent small population sizes, detection of small and progressive population declines is extremely difficult. Thus, detection of population trends should not be the trigger for conservation actions. Precautionary measures to maintain viable populations while minimizing the impacts of management decisions on different stakeholder groups are necessary. The much greater challenge of conserving all forms of humpback dolphins will be the maintenance of high-quality habitat throughout the highly populated developing countries that coincide with their coastal distribution. Improved understanding of humpback dolphins' biology, ecology and taxonomy will be a key element toward their successful conservation and management.

## References

Atkins, S., Pillay, N., and Peddemors, V. M. (2004). Spatial distribution of Indo-Pacific humpback dolphins (*Sousa chinensis*) at Richards Bay, South Africa: Environmental influences and behavioural patterns. *Aquat. Mamm.* **30**, 84–93.

Baldwin, R. M., Collins, M., Van Waerebeek, K., and Minton, G. (2004). The Indo-Pacific humpback dolphin of the Arabian region: A status review. *Aquat. Mamm.* **30**, 111–124.

Barros, N. B. and Cockcroft, V. G. (1991). Prey of humpback dolphins, *Sousa plumbea*, stranded in Eastern Cape Province, South Africa. *Aquat. Mamm.* **17**, 134–136.

Barros, N. B., Jefferson, T. A., and Parsons, E. C. M. (2004). Feeding habits of Indo-Pacific humpback dolphins (*Sousa chinensis*) stranded in Hong Kong. *Aquat. Mamm.* **30**, 179–188.

Cockcroft, V.G. (1989) Biology of Indo-Pacific humback dolphin (*Sousa plumbea*) off Natal, South Africa. *Abstracts of the Biennial Conference on the Biology of Marine Mammals*, 8:13.

Cockcroft, V. G. (1990). Dolphin catches in the Natal shark nets 1980–1988. *S. Afr. J. Wildl. Res.* **20**, 44–51.

Cockcroft, V. G. (1991). Incidence of shark bites on Indian Ocean humpbacked dolphins (*Sousa plumbea*) off Natal, South Africa. *Report of the International Whaling Commission*. 277–282.

Corkeron, P. J. (1990). Aspects of the behavioural ecology of inshore dolphins *Tursiops truncatus* and *Sousa chinensis* in Moreton Bay, Australia. *In* "The Bottlenose Dolphin" (S. Leatherwood, and R. R. Reeves, eds), pp. 285–293. Academic Press, London.

Corkeron, P. J., Morissette, N. M., Porter, L. J., and Marsh, H. (1997). Distribution and status of hump-backed dolphins, *Sousa chinensis*, in Australian waters. *Asian Mar. Biol.* **14**, 49–59.

Frère, C. H., Hale, P. T., Porter, L., Cockcroft, V. G, and Dalebout, M. L. (2008). Phylogenetic analysis of mtDNA sequences suggests revision of humpback dolphin (*Sousa* spp.) taxonomy is needed. *Mar. Freshw. Res.,* **59**, 259-268.

Goold, J. C., and Jefferson, T. A. (2004). A note on clicks recorded from free-ranging Indo-Pacific humpback dolphins, *Sousa chinensis*. *Aquat. Mamm.* **30**, 175–178.

Guissamulo, A., and Cockcroft, V. G. (2004). Ecology and population estimates of Indo-Pacific humpback dolphins (*Sousa chinensis*) in Maputo Bay, Mozambique. *Aquat. Mamm.* **30**, 94–102.

Hung, S. K., and Jefferson, T. A. (2004). Ranging patterns of Indo-Pacific humpback dolphins (*Sousa chinensis*) in the Pearl River Estuary, People's Republic of China. *Aquat. Mamm.* **30**, 159–174.

Hung, C. L. H., *et al.* (2004). A preliminary risk assessment of trace elements accumulated in fish to the Indo-Pacific humpback dolphin (*Sousa chinensis*) in the northwestern waters of Hong Kong. *Chemosphere* **56**, 643–651.

Hung, C. L. H., *et al.* (2006). An assessment of the risks associated with polychlorinated biphenyls found in the stomach contents of stranded Indo-Pacific humpback dolphins (*Sousa chinensis*) and finless porpoises (*Neophocaena phocaenoides*) from Hong Kong waters. *Chemosphere* **63**, 845–852.

Hung, C. L. H., Lau, R. K. F., Lam, J. C. W., Jefferson, T. A., Hung, S. K., Lam, M. H. W., and Lam, P. K. S. (2007). Risk assessment of trace elements in the stomach contents of Indo-Pacific humpback dolphins and finless porpoises in Hong Kong waters. *Chemosphere* **66**, 1175–1182.

Jefferson, T. A. (2000). Population biology of the Indo-Pacific humpbacked dolphin in Hong Kong waters. *Wildl. Monogr.* **144**, 1–65.

Jefferson, T. A. (2004). Geographic variation in skull morphology of humpback dolphins (*Sousa* spp.). *Aquat. Mamm.* **30**, 3–17.

Jefferson, T. A., and Leatherwood, S. (1997). Distribution and abundance of Indo-Pacific hump-backed dolphins (*Sousa chinensis* Osbeck, 1765) in Hong Kong waters. *Asian Mar. Biol.* **14**, 93–110.

Jefferson, T. A., and Karczmarski, L. (2001). Species. *Sousa chinensis*. *Mamm. Species.* **655**, 1–9.

Jefferson, T. A., and Hung, S. K. (2004). A review of the status of the Indo-Pacific humpback dolphin (*Sousa chinensis*) in Chinese waters. *Aquat. Mamm.* **30**, 149–158.

Jefferson, T. A., Hung, S. K., and Lam, P. K. S. (2006). Strandings, mortality and morbidity of Indo-Pacific humpback dolphins in Hong Kong, with emphasis on the role of organochlorine contaminants. *J. Cetacean Res. Manag* **8**, 181–193.

Jefferson, T. A. and Van Waerebeek, K. (2004) Geographic variation in skull morphology of humpback dolphins (*Sousa* spp.). *Aquat. Mamm.,* **30**, 3-17.)

Karczmarski, L. (1999). Group dynamics of humpback dolphins (*Sousa chinensis*) in the Algoa Bay region, South Africa. *J. Zool. (Lond.)* **249**, 283–293.

Karczmarski, L. (2000). Conservation and management of humpback dolphins: The South African perspective. *Oryx* **34**, 207–216.

Karczmarski, L., and Cockcroft, V. G. (1999). Daylight behaviour of humpback dolphins *Sousa chinensis* in Algoa Bay, South Africa. *Mamm. Biol.* **64**, 19–29.

Karczmarski, L., Thornton, M., and Cockcroft, V. G. (1997). Description of selected behaviors of humpback dolphins *Sousa chinensis*. *Aquat. Mamm.* **23**, 127–133.

Karczmarski, L., Cockcroft, V. G., and McLachlan, A. (1999a). Group size and seasonal pattern of occurrence of humpback dolphins *Sousa chinensis* in Algoa Bay, South Africa. *S. Afr. J. Mar. Sci.* **21**, 89–97.

Karczmarski, L., Winter, P. E. D., Cockcroft, V. G., and McLachlan, A. (1999b). Population analyses of Indo-Pacific humpback dolphins *Sousa chinensis* in Algoa Bay, Eastern Cape, South Africa. *Mar. Mamm. Sci.* **15**, 1115–1123.

Karczmarski, L., Cockcroft, V. G., and McLachlan, A. (2000a). Habitat use and preferences of Indo-Pacific humpback dolphins *Sousa chinensis* in Algoa Bay, South Africa. *Mar. Mamm. Sci.* **16**, 65–79.

Karczmarski, L., Thornton, M., and Cockcroft, V. G. (2000b). Daylight occurrence of humpback dolphins *Sousa chinensis* in Algoa Bay, South Africa. *Afr. J. Ecol.* **38**, 86–90.

Keith, M., Peddemors, V. M., Bester, M. N., and Ferguson, J. W. H. (2002). Population characteristics of Indo-Pacific humpback dolphins at Richards Bay, South Africa: Implications for incidental capture in shark nets. *S. Afr. J. Wildl. Res.* **32**, 153–162.

Kerem, D., Goffman, O., and Spanier, E. (2001). Sighting of a single humpback dolphin (*Sousa* sp.) along the Mediterranean coast of Israel. *Mar. Mamm. Sci.* **17**, 170–171.

LeDuc, R. G., Perrin, W. F., and Dizon, A. E. (1999). Phylogenetic relationships among the delphinid cetaceans based on full cytochrome b sequences. *Mar. Mamm. Sci.* **15**, 619–648.

Leung, C. C. M., Jefferson, T. A., Hung, S. K., Zheng, G. J., Yeung, L. W. Y., Richardson, B. J., and Lam, P. K. S. (2005). Petroleum hydrocarbons, polycyclic aromatic hydrocarbons, organochlorine pesticides and polychlorinated biphenyls in tissues of Indo-Pacific humpback dolphins from south China waters. *Mar. Pollut. Bull.* **50**, 1713–1719.

Lydekker, R. (1901). Notice of an apparently new estuarine dolphin from Borneo. *Proc. Zool. Soc. Lond.* **1**, 88–91.

Ng, S. L., and Leung, S. (2003). Behavioral response of Indo-Pacific humpback dolphin (*Sousa chinensis*) to vessel traffic. *Mar. Environ. Res.* **56**, 555–567.

Owen, R. (1866). On some Indian Cetacea collected by Walter Elliot, Esq. *Trans. Zool. Soc. Lond.* **6**, 17–47.

Parra, G. J. (2006). Resource partitioning in sympatric delphinids: Space use and habitat preferences of Australian snubfin and Indo-Pacific humpback dolphins. *J. Anim. Ecol.* **75**, 862–874.

Parra, G. J., Corkeron, P. J., and Marsh, H. (2004). The Indo-Pacific humpback dolphin, *Sousa chinensis* (Osbeck, 1765), in Australian waters: A summary of current knowledge. *Aquat. Mamm.* **30**, 197–206.

H

Parra, G. J., Corkeron, P. J., and Marsh, H. (2006a). Population sizes, site fidelity and residence patterns of Australian snubfin and Indo-Pacific humpback dolphins: Implications for conservation. *Biol. Conserv.* **129**, 167–180.

Parra, G. J., Schick, R. S., and Corkeron, P. J. (2006b). Spatial distribution and environmental correlates of Australian snubfin and Indo-Pacific humpback dolphins. *Ecography* **29**, 396–406.

Parsons, E. C. M. (1998a). The behaviour of Hong Kong's resident cetaceans: The Indo-Pacific hump-back dolphin and the finless porpoise. *Aquat. Mamm.* **24**, 91–106.

Parsons, E. C. M. (1998b). Observations on Indo-Pacific humpbacked dolphins, *Sousa chinensis*, from Goa, Western India. *Mar. Mamm. Sci.* **14**, 166–170.

Parsons, E. C. M. (1998c). Trace metal pollution in Hong Kong: Implications for the health of Hong Kong's Indo-Pacific humpbacked dolphins (*Sousa chinensis*). *Sci. Total Environ.* **214**, 175–184.

Parsons, E. C. M. (1999). Trace element concentrations in the tissues of cetaceans from Hong Kong's territorial waters. *Environ. Conserv.* **26**, 30–40.

Parsons, E. C. M. (2004a). The behavior and ecology of the Indo-Pacific humpback dolphin (*Sousa chinensis*). *Aquat. Mamm.* **30**, 38–55.

Parsons, E. C. M. (2004b). The potential impacts of pollution on humpback dolphins, with a case study on the Hong Kong population. *Aquat. Mamm.* **30**, 18–37.

Parsons, E. C. M., and Jefferson, T. A. (2000). Post-mortem investigations on stranded dolphins and porpoises from Hong Kong waters. *J. Wildl. Dis.* **36**, 342–356.

Razafindrakoto, Y., Andrianarivelo, N., and Rosenbaum, H. C. (2004). Sightings, catches, and other records of Indo-Pacific humpback dolphins in the coastal waters of Madagascar. *Aquat. Mamm.* **30**, 103–110.

Ross, G. J. B., Heinsohn, G. E., and Cockcroft, V. G. (1994). Humpback dolphins – *Sousa chinensis* (Osbeck, 1765). *In* "Handbook of Marine Mammals" (S. H. Ridgway, and R. Harrison, eds), 5, pp. 23–42. Academic Press, London.

Saayman, G. S., and Tayler, C. K. (1972). Observations on inshore and pelagic dolphins on the south-eastern Cape coast of South Africa. *Koedoe.* **15**, 1–24.

Saayman, G. S., and Tayler, C. K. (1973). Social organization of inshore dolphins (*Tursiops aduncus* and *Sousa*) in the Indian Ocean. *J. Mammal.* **54**, 993–996.

Saayman, G. S., and Tayler, C. K. (1979). The socioecology of humpback dolphins (*Sousa* sp.). *In* "Behavior of Marine Animals" (H. E. Winn, and B. L. Olla, eds), 3, pp. 165–226. Plenum Press, New York, USA.

Stensland, E., Angerbjorn, A., and Berggren, P. (2003). Mixed species groups in mammals. *Mammal. Rev.* **33**, 205–223.

Stensland, E., Carlen, I., Särnblad, A., Bignert, A., and Berggren, P. (2006). Population size, distribution, and behavior of indo-pacific bottlenose (*Tursiops aduncus*) and humpback (*Sousa chinensis*) dolphins off the south coast of Zanzibar. *Mar. Mamm. Sci.* **22**, 667–682.

Sutaria, D., and Jefferson, T. A. (2004). Records of Indo-Pacific humpback dolphins (*Sousa chinensis*, Osbeck, 1765) along the coasts of India and Sri Lanka: An overview. *Aquat. Mamm.* **30**, 125–136.

Van Parijs, S. M., and Corkeron, P. (2001a). Boat traffic affects the acoustic behaviour of Pacific humpback dolphins, *Sousa chinensis*. *J. Mar. Biol. Assoc. UK* **81**, 533–538.

Van Parijs, S. M., and Corkeron, P. J. (2001b). Evidence for signature whistle production by a Pacific humpback dolphin, *Sousa chinensis*. *Mar. Mamm. Sci.* **17**, 944–949.

Van Parijs, S. M., and Corkeron, P. J. (2001c). Vocalizations and behaviour of Pacific humpback dolphins *Sousa chinensis*. *Ethology* **107**, 701–716.

Van Waerebeek, K, *et al.* (2004). Distribution, status, and biology of the Atlantic humpback dolphin, *S. teuszii* (Kukenthal, 1892). *Aquat. Mamm.* **30**, 56–83.

Wang, J. Y., Hung, S. K., and Yang, S.-C. (2004). Records of Indo-Pacific humpback dolphins, *Sousa chinensis* (Osbeck, 1765), from the waters of western Taiwan. *Aquat. Mamm.* **30**, 189–196.

# Humpback Whale
*Megaptera novaeangliae*

PHILLIP J. CLAPHAM

The humpback whale (Fig. 1) is one of the best known and easily recognizable of the large whales. It is known for its frequent acrobatic behavior and its occasional tendency to approach vessels. In the last 30 years, thousands of humpback whales have been identified individually from natural markings (notably the pattern on the ventral surface of the tail flukes), and as a result much has been learned about the biology and behavior of this species.

## I. Characteristics and Taxonomy

At close range, humpback whales are easily distinguished from any other large whale by their remarkably long flippers, which are approximately one-third the length of the body. The flippers are ventrally white and can be either white or black dorsally depending on the population and the individual; the flippers of North Atlantic humpbacks tend to be white, while those in the North Pacific are usually black (Fig. 1). The body color is black dorsally, with variable pigmentation on the underside (black, white, or mottled). The head and jaws have numerous knobs called tubercles, which are also diagnostic of the species. The dorsal fin is small but highly variable in shape ranging from low (almost absent) to high and falcate. Like all rorquals, humpbacks have a series of pleats running from the tip of the lower jaw to the umbilicus. The tail is usually raised during a dive; the underside exhibits a pattern that is unique to each individual and which ranges from all white to all black. The presence of white on the ventral surface, and the prominent serration of the trailing edge, distinguishes humpbacks from other whales that "fluke" while diving, such as right, bowhead, blue, gray, and sperm whales.

Adult female humpback whales are typically 1–1.5 m longer than males. Maximum reliably recorded adult lengths are in the 16–17 m range, although 14–15 m is more typical (Clapham and Mead, 1999). Calves are 3.96–4.57 m at birth, and approximately 8–10 m at independence (Clapham *et al.*, 1999), which occurs at the end of the calf's natal year. There are no easily observable differences between male and female humpbacks. Females possess a grapefruit-sized lobe at the rear of the genital slit; this lobe is absent in males (Glockner-Ferrari and Ferrari, 1990). In addition, the spacing between the genital slit and the anus is considerably greater in males.

The skull of the humpback whale is easily distinguished from that of other baleen whales by the narrowness of the rostrum relative to the zygomatic width. The humpback has between 270 and 400 baleen plates on each side of the mouth. The plates are usually black, although those close to the tip of the jaw are sometimes white or partly white.

The genus *Megaptera* is monotypic and is one of two genera in the family Balaenopteridae (the "rorquals"). No subspecies are recognized. The binomial *Megaptera novaeangliae* derives from the Greek for "big wing" (*mega* + *pteron*) and the Latin for "New England" which was the origin of the specimen used by Borowski in his description of the species in 1781.

## II. Distribution and Abundance

Humpback whales are found in all oceans of the world. They are a highly migratory species, spending spring through fall on feeding

Figure 1    *Humpback whale (C. Brett Jarrett).*

grounds in mid- or high-latitude waters, and wintering on calving grounds in the tropics, where they do not eat (Dawbin, 1966). They are typically found in coastal or shelf waters in summer and close to islands or reef systems in winter. Some documented migratory movements of this species represent the longest-known migration of any mammal, being almost 5000 miles one way (Palsbøll *et al.*, 1997). It is possible that some humpbacks do not migrate every year, although the number and sex/age class of these animals remains unclear. Remarkably, the purpose of the migration remains unknown; it may reflect a need to maximize energetic gain by exploiting pulses of productivity in high latitudes in summer, then gaining thermodynamic advantages by over-wintering in warm water. The only non-migratory population is that residing in the Arabian Sea, where monsoon-driven productivity in summer permits the whales to remain in tropical waters year-round (Mikhalev, 1997).

In the North Atlantic, humpbacks return each spring to specific feeding grounds in the Gulf of Maine, Gulf of St. Lawrence, Newfoundland, Labrador, Greenland, Iceland, and Norway. Fidelity to these areas is strong and is determined by where a calf was taken by its mother in the former's natal year. Genetic analysis has indicated that this fidelity is maintained on an evolutionary timescale in at least Iceland and Norway (Palsbøll *et al.*, 1995; Larsen *et al.*, 1996). Despite this fidelity, whales from all feeding grounds migrate to a common breeding area in the West Indies, where they mate and calve (Katona and Beard, 1990). However, genetic data indicate the existence of a second breeding population composed of many of the animals from Iceland and Norway, whose migratory destination is unknown. Historically important breeding areas in the Cape Verde Islands and the southeastern Caribbean appear to be utilized by relatively few whales today.

In the North Pacific, there are at least four separate breeding grounds in Hawaii, coastal Mexico, offshore Mexico (the Revillagigedos Islands), and Japan/Philippines (Calambokidis *et al.*, 1997). Whales from these wintering areas migrate primarily to Alaska, California, possibly the Bering Sea/western Gulf of Alaska, and the western North Pacific, respectively. However, crossover is not unknown and some trans-oceanic movements have been recorded (e.g., British Columbia to Japan and back).

In the Southern Hemisphere, humpbacks feed in circumpolar waters around the Antarctic and migrate to relatively discrete breeding grounds in tropical waters to the north. Seven breeding populations or "management areas" are recognized by the International Whaling Commission (IWC) in the Southern Hemisphere, and these are linked with varying confidence to six feeding areas in the Antarctic. Some movement between these regions is very likely, but the extent of such exchange remains largely unquantified.

## III. Ecology

The humpback whale has a generalist diet, feeding on euphausiids and various species of small schooling fish. The latter include herring (*Clupea* spp.), capelin (*Mallotus villosus*), sand lance (*Ammodytes* spp.), and mackerel (*Scomber scombrus*). Humpbacks appear to be unique among large whales in their use of bubbles to corral or trap schooling fish. Whales blow nets, clouds, or curtains of bubbles around or below schools of fish, then lunge with mouths open into the center of the bubble structure (Jurasz and Jurasz, 1979; Hain *et al.*, 1982). As with other balaenopterids, the ventral pleats expand when a humpback is feeding, allowing the animal to greatly increase the capacity of its mouth.

Rake-mark scars from teeth attest to the fact that humpbacks are commonly attacked by killer whales (*Orcinus orca*). However, it seems likely that fatal attacks are largely confined to young calves (Mehta *et al.*, 2007), and predation does not appear to be a significant factor in the social organization of the humpback (Clapham, 2000).

## IV. Behavior and Physiology

The social organization of the humpback is characterized by small unstable groups, and individuals typically associate with many companions on both feeding and breeding grounds (Clapham, 2000). Longer-term associations (those lasting days or weeks) are occasionally recorded, but their basis is unclear. There appears to be no territoriality in this species.

In winter, male humpback whales sing long, complex songs, the primary function of which is presumably to attract females, although a role in dominance sorting or cooperative behavior among males has also been suggested (Darling *et al.*, 2006). All whales in a given population sing essentially the same song, and although the form and content of all songs change over time, the whales somehow coordinate these changes. Males also compete very aggressively for access to females (Tyack and Whitehead, 1983), and the resulting "competitive groups" can last for hours and involve tail slashing, ramming, or head butting. Males may also form coalitions, but further research is required to assess the significance and composition of such alliances.

In part because of the prominent male display aspect (i.e., singing behavior), the mating system has been compared to a lek (Mobley and Herman, 1985; Clapham, 2000), although it does not possess the rigid territoriality common to such systems. Males almost certainly remain in breeding areas longer than females and attempt to obtain repeated matings, while newly pregnant females return quickly to higher latitudes (Chittleborough, 1965; Dawbin, 1966) where they will feed for many months in order to prepare for the considerable energetic cost of lactation.

H

Humpbacks are well known for their often spectacular aerial behaviors. These include breaching, lobtailing, and flippering. Such behaviors occur at all times of year and in a variety of contexts, and it is clear that they perform a range of functions. These may include play, communication, parasite removal, and expression of excitement or annoyance.

How humpback whales find their food remains largely a mystery. Baleen whales do not appear to possess echolocation, though it is possible that they retain a useful sense of smell which could be used to detect prey patches at the surface. Similarly, we have little understanding of how humpbacks navigate across thousands of miles on their annual migrations, though biomagnetic orientation has been suggested (on little direct evidence) as a component of this ability.

## V. Life History

Breeding in humpback whales is strongly seasonal. Females come into estrus in winter, at which time testosterone production and spermatogenesis also peaks in males (Chittleborough, 1965). The gestation period is about 11.5 months, with the great majority of calves born in mid-winter. Calves probably begin to feed independently at about 6 months of age, but nursing likely continues in many animals until shortly before independence at about a year of age. The age at which sexual maturity is reached appears to vary among populations from 5 to 10 years. Interbirth intervals in females are most commonly 2 years, although annual calving is not unknown (Clapham and Mayo, 1990; Glockner-Ferrari and Ferrari, 1990). Although multiple fetuses have been recorded in dead pregnant females, living twins or multiplets are unknown.

## VI. Interactions with Humans

The humpback whale was heavily exploited by the whaling industry for several centuries. Because of its coastal distribution, it was often the first species to be hunted in a newly discovered area. Last century, more than 213,000 humpbacks were slaughtered in the Southern Hemisphere alone; of these, more than 48,000 were taken illegally by the Soviet Union (Yablokov et al., 1998). Indeed, the USSR killed 25,000 humpbacks in just two Antarctic whaling seasons (1959/1960 and 1960/1961), resulting in the crash of the populations concerned. It is quite likely than more than 90% of the animals in some populations were killed during the most intensive periods of exploitation. As a result, the humpback was long considered an endangered species. Despite this, most studied populations appear to be making a strong recovery, and the species is now rated Least Concern (LC) by the IUCN. However, some populations are still listed as Endangered (Oceauia and Arabian Sea) and others may be in similar trouble. The North Atlantic population was most recently estimated at 11,570 animals (Stevick et al., 2003). An estimate of 6000–8000 for the North Pacific (Calambokidis et al., 1997) is widely regarded as conservative, and new estimates will shortly be available from a major international collaborative study in this ocean. Strong population growth rates have been reported for many areas, ranging from 6.5% in the Gulf of Maine to more than 10% in some Southern Hemisphere populations; however, the IWC recently agreed that annual increase rates above 10.6% are biologically unrealistic for this species (IWC, 2007). Commercial whaling for humpbacks officially ended worldwide in 1966; however, the Soviets continued to hunt this species for some years afterwards, and Japan has recently added a catch quota for humpback whales to its scientific whaling program in the Antarctic. Small aboriginal hunts for humpbacks still occur in a couple of locations, and many more whales die from entanglement in fishing gear or collisions with ships. However, none of these impacts appears to be significant at the population level, and the outlook for this once-overexploited species appears good in most areas.

## References

Calambokidis, J. et al. (1997). Population abundance and structure of humpback whales in the North Pacific basin. Final report to Southwest Fisheries Science Center, La Jolla, CA 92038, 67 pp.

Chittleborough, R. G. (1965). Dynamics of two populations of the humpback whale, Megaptera novaeangliae (Borowski). Aust. J. Mar. Freshw. Res. **16**, 33–128.

Clapham, P. J. (2000). The humpback whale: Seasonal feeding and breeding in a baleen whale. In "Cetacean Societies" (J. Mann, P. L. Tyack, R. Connor, and H. Whitehead, eds), pp. 173–196. University of Chicago Press.

Clapham, P. J., and Mayo, C. A. (1990). Reproduction of humpback whales, Megaptera novaeangliae, observed in the Gulf of Maine. Rep. Int. Whal. Commn,(12), 171–175.

Clapham, P. J., and Mead, J. G. (1999). Megaptera novaeangliae. Mamm. Spec. **604**, 1–9.

Clapham, P. J., Wetmore, S. E., Smith, T. D., and Mead, J. G. (1999). Length at birth and at independence in humpback whales. J. Cetacean Res. Manage. **1**, 141–146.

Darling, J. D., Jones, M. E., and Nicklin, C. P. (2006). Humpback whale songs: Do they organize males during the breeding season? Behaviour **143**, 1051–1101.

Dawbin, W. H. (1966). The seasonal migratory cycle of humpback whales. In "Whales, Dolphins and Porpoises" (K. S. Norris, ed.), pp. 145–170. University of California Press, Berkeley, CA.

Glockner-Ferrari, D. A., and Ferrari, M. J. (1990). Reproduction in the humpback whale (Megaptera novaeangliae) in Hawaiian waters, 1975–1988: The life history, reproductive rates and behaviour of known individuals identified through surface and underwater photography. Rep. Int. Whal. Commn.(12), 161–169.

Hain, J. H. W., Carter, G. R., Kraus, S. D., Mayo, C. A., and Winn, H. E. (1982). Feeding behaviour of the humpback whale, Megaptera novaeangliae, in the western North Atlantic. Fish. Bull. **80**, 259–268.

IWC (2007). Report of the Scientific Committee. J. Cetacean Res. Manage. **9** (supplement) (in press).

Jurasz, C. M., and Jurasz, V. P. (1979). Feeding modes of the humpback whale, Megaptera novaeangliae, in southeast Alaska. Sci. Rep. Whales Res. Inst. Tokyo **31**, 69–83.

Katona, S. K., and Beard, J. A. (1990). Population size, migrations and feeding aggregations of the humpback whale (Megaptera novaeangliae) in the western North Atlantic Ocean. Rep. Int. Whal. Commn.(12), 295–305.

Larsen, A. H., Sigurjónsson, J., Óien, N., Vikingsson, G., and Palsbøll, P. J. (1996). Population genetic analysis of mitochondrial and nuclear genetic loci in skin biopsies collected from central and northeastern North Atlantic humpback whales (Megaptera novaeangliae): Population identity and migratory destinations. Proc. R. Soc. Lond. B. **263**, 1611–1618.

Mehta, A. V., et al. (2007). Baleen whales are not important as prey for killer whales (Orcinus orca) in high latitudes. Mar. Ecol. Prog. Ser. (in press).

Mikhalev, Y. A. (1997). Humpback whales Megaptera novaeangliae in the Arabian Sea. Mar. Eco. Prog. Ser. **149**, 13–21.

Mobley, J. R., and Herman, L. M. (1985). Transience of social affiliations among humpback whales (Megaptera novaeangliae) on the Hawaiian wintering grounds. Can. J. Zool. **63**, 763–772.

Palsbøll, P. J., et al. (1995). Distribution of mtDNA haplotypes in North Atlantic humpback whales: The influence of behavior on population structure. Mar. Ecol. Prog. Ser. **116**, 1–10.

H

Palsbøll, P. J., *et al.* (1997). Genetic tagging of humpback whales. *Nature* **388**, 767–769.

Stevick, P. T., *et al.* (2003). North Atlantic humpback whale abundance and rate of increase four decades after protection from whaling. *Mar. Ecol. Prog. Ser.* **258**, 263–273.

Tyack, P., and Whitehead, H. (1983). Male competition in large groups of wintering humpback whales. *Behaviour* **83**, 1–23.

Yablokov, A. A., Zemsky, V. A., Mikhalev, Y. A., Tormosov, V. V., and Berzin, A. A. (1998). Data on Soviet whaling in the Antarctic in 1947–1972 (population aspects). *Russ. J. Ecol.* **29**, 38–42.

# Hunting of Marine Mammals

## Randall R. Reeves

Ancient middens testify to the importance of marine mammals in the lives of early maritime people around the world. Many of the bones and bone fragments found in such sites probably came from animals that were scavenged from beaches. However, ingenious methods of capturing pinnipeds, sirenians, and cetaceans eventually were developed, and the archaeological refuse came to signify past hunting. The rewards were tempting—large amounts of nutritious meat and fat, hides, ivory, sinews for sewing, and bones for making household implements or weapons. These products eventually came to have high commercial value, fueling the modern global whaling and sealing industries.

No taxonomic group of marine mammals has been spared from hunting pressure. However, some species have been hunted more intensively than others. The great whales (the sperm whale *Physeter macrocephalus* and the baleen whales) have been sought for their oil, meat, and baleen; pinnipeds for their oil and pelts; sea otters (*Enhydra lutris*) for their furs; and sirenians mainly for their flesh and skins. In contrast, some dolphin and porpoise populations have hardly been hunted at all, and they remained secure until the advent and proliferation of unselective fishing methods, which result in the incidental killing of non-target organisms. Marine mammals have also been hunted with the intention of reducing their predation on valued resources such as fish, crustaceans, or mollusks (Bearzi *et al.*, 2004). This culling, often implemented through government-sponsored bounty programs, is similar to that directed at wolves (*Canis lupus*), mountain lions (*Puma concolor*), and other predators in parts of North America, with the outspoken support of ranchers and sport hunters.

## I. Hunting of Whales, Dolphins, and Porpoises

People in the Arctic were hunting bowhead whales (*Balaena mysticetus*) as long ago as the middle of the first millennium, and western Europeans were taking right whales (*Eubalaena glacialis*) by the beginning of the second (Ellis, 1991; McCartney, 1995). The technology and culture of subsistence whaling spread within the Arctic and subarctic from the Bering Strait region, whereas the development and spread of commercial WHALING were driven by the Basques of western Europe (Du Pasquier, 2000). From its beginnings in the Bay of Biscay, this whaling eventually reached all of the world's oceans and involved people of many nationalities. Modern whaling, characterized by engine-driven catcher boats and deck-mounted harpoon cannons firing explosive grenades, began in Norway in the 1860s (Tønnessen and Johnsen, 1982). A key feature of modern whaling was that it

made possible the routine capture of any species, including the blue whale (*Balaenoptera musculus*), fin whale (*B. physalus*), and other fast-swimming balaenopterines. In the first three-quarters of the twentieth century, factory ships from several nations (e.g., Norway, the United Kingdom, Germany, Japan, the United States, and the Soviet Union) operated in the Antarctic, the richest whaling ground on the planet. At its pre-War peak in 1937–1938, the industry's 356 catcher boats, associated with 35 shore stations and as many floating factories, killed nearly 55,000 whales, 84% of them in the Antarctic.

Commercial whaling declined in the 1970s as a result of conservationist pressure and depletion of whale stocks. The last whaling stations in the United States and Canada were closed in 1972, and the last station in Australia ceased operations following the 1978 season. By the end of the 1970s, only Japan, the Soviet Union, Norway, and Iceland were still engaged in commercial whaling. With the decision by the International Whaling Commission (IWC) in 1982 to implement a global moratorium on commercial whaling, Japan and the Soviet Union made their final large-scale factory-ship expeditions to the Antarctic in 1986–1987, and Japan stopped its coastal hunt for sperm whales and Bryde's whales (*Balaenoptera edeni*) in 1988. Iceland closed its whaling station in 1990 and shortly thereafter withdrew its membership in the IWC (only to rejoin in 2002). Contrary to the widespread belief that commercial whaling ended in the 1980s, however, Norway and Japan continued hunting common (*Balaenoptera acutorostrata*) and southern (*B. bonaerensis*) minke whales. By formally objecting to the moratorium decision, Norway reserved its right to carry on commercial whaling. Thus, Norwegian whalers killed hundreds of common minke whales in the North Atlantic annually throughout the 1990s and early 2000s, their self-assigned quota increasing to well over 1000 by 2006. Using a provision in the whaling treaty that allows member states to issue permits to hunt protected species for scientific research, Japan has continued taking hundreds of southern minke whales in the Antarctic and common minke whales in the western North Pacific each year. They now also include in their annual "scientific" sample sei whales (*B. borealis*), Bryde's whales, sperm whales, and fin whales. In 1999, the Icelandic parliament approved the resumption of a shore-based commercial hunt, and whaling for common minke whales resumed in Iceland in 2003. Iceland also has announced plans to follow Japan's lead and expand its efforts to include "scientific" catches of fin whales. Because the main incentive for commercial whaling nowadays is the Japanese demand for whale meat, Norway and Iceland have been eager to reopen the international trade in whale products.

The hunting of smaller cetaceans has generally been confined to coastal waters and conducted on a smaller, or at least localized, scale (Mitchell, 1975). There are, however, some examples of large, well-organized hunts. Fishermen in the Faroe Islands have continued to kill hundreds, and in some years well over a thousand, long-finned pilot whales (*Globicephala melas*) and Atlantic white-sided dolphins (*Lagenorhynchus acutus*) in a centuries-old drive fishery. The method involves a number of small boats that herd the animals into shallow water where they can be killed with lances, long knives, or firearms. There also has been a long-standing drive fishery in Japan, taking a variety of delphinid species, most notably striped dolphins (*Stenella coeruleoalba*) and bottlenose dolphins (*Tursiops* spp.). The Japanese hunt involves other methods as well, including the use of hand-thrown harpoons to take dolphins and Dall's porpoises (*Phocoenoides dalli*) and the use of harpoon guns to take short-finned pilot whales (*Globicephala macrorhynchus*) and other medium-sized cetaceans. A large commercial hunt for short-beaked common dolphins (*Delphinus delphis*), common bottlenose dolphins

H

(*Tursiops truncatus*), and harbor porpoises (*Phocoena phocoena*) was conducted annually in the Black Sea from the nineteenth century through the mid-1960s. The Soviet Union banned dolphin hunting in 1966; hunting with rifles and purse seines continued in the Turkish sector of the Black Sea until at least 1983.

Aboriginal hunters in Russia, the United States (Alaska), Canada, and Greenland kill several tens of bowhead whales, 100–200 gray whales (*Eschrichtius robustus*), and thousands of white whales (*Delphinapterus leucas*), narwhals (*Monodon monoceros*), and harbor porpoises (Greenland only) each year (Fig. 1). This hunting is primarily for food and the products are consumed locally or sold within proscribed markets (see Section V). In recent years, aboriginal whalers in Washington State (USA), British Columbia (Canada), and Tonga (a South Pacific island nation) have expressed interest in reestablishing their own hunts for large cetaceans. In the spring of 1999, the Makah tribe in Washington took their first gray whale in more than 50 years.

## II. Pinnipeds

Sealing began in the Stone Age when people attacked hauled-out animals with clubs (Bonner, 1982). Later methods included the use of harpoons thrown from skin boats and gaff-like instruments for killing pups on ice or beaches. Traps and nets were used as well. The introduction of firearms transformed the hunting of pinnipeds and caused an alarming increase in the proportion of animals killed but not retrieved, especially in those hunts where the animals were shot in deep water before first being harpooned. This problem of "sinking loss" also applies to many cetacean hunts.

In addition to their meat and fat, the pelts of some seals, especially the fur seals and phocids, have had value in the garment industry. Markets for oil and sealskins fueled commercial hunting on a massive scale from the late eighteenth century through the early twentieth century (Busch, 1985). The ivory tusks and tough, flexible hides of walruses (*Odobenus rosmarus*) made these animals exceptionally valuable to both subsistence and commercial hunters. At least 10,000 walruses are still killed every year by the native people of northeastern Russia, Alaska, northeastern Canada, and Greenland. The killing is accomplished mainly by shooting. The meat, BLUBBER, and skin are eaten by people or fed to dogs, while the tusks are either used for carving or sold as curios. Native hunters in the circumpolar north also kill more than a hundred thousand seals each year, mainly ringed seals (*Pusa hispida*) but also bearded (*Erignathus barbatus*), ribbon (*Histriophoca fasciata*), harp (*Pagophilus groenlandicus*), hooded (*Cystophora cristata*), and spotted seals (*Phoca largha*). Seal meat and fat remain important in the diet of many northern communities, and the skins are still used locally in some areas to make clothing, dog traces, and hunting lines. There is also a commercial export market for high-quality sealskins and a strong demand in Oriental communities for pinniped penises and bacula. The sale of these items, along with walrus and narwhal ivory, white whale and narwhal skin (maktak), and polar bear (*Ursus maritimus*) hides and gall bladders, is a significant source of revenue in some local hunting communities.

The scale of commercial sealing, like that of commercial whaling, has declined considerably since the 1960s. It continues, however, in parts of the North and South Atlantic. Norwegian and Russian ships continue to visit the harp and hooded seal grounds in the Greenland Sea ("West Ice") and Barents Sea ("East Ice"). After a period of drastically reduced killing in the 1980s, the Canadian commercial hunt for harp seals has been reinvigorated, at least in part as a result of governmental subsidies (Lavigne, 1999). The total annual commercial catch of harp seals in the North Atlantic is still in the hundreds

**Figure 1** *People living in villages along the rivers of West Africa hunt, trap, and net manatees* (Trichechus senegalensis) *for food and to prevent the animals from raiding their rice crops. In the absence of any kind of management or monitoring, it is impossible to make a meaningful assessment of the status of West African manatee populations. However, with such active, unrestricted persecution, the species has undoubtedly declined, and perhaps even been extirpated, in many parts of its formerly extensive range. Gbandakor, Malen River, Sierra Leone, April 1982. Photograph by Harry Spaling.*

of thousands. Although some molting pups are clubbed to death on the sea ice, most of the killing nowadays is by shooting. Tens of thousands of South African fur seals (*Arctocephalus pusillus pusillus*) are killed annually in Namibia (until 1989 also in South Africa) and large numbers of South American fur seals (*A. australis*) were taken Uruguay until 1991 when the commercial hunt there ended. These hunts are (or were) centuries old, having been driven initially by commercial markets for skins and oil and, more recently, by the Oriental demand for seal penises and bacula. Also, especially in Africa, the hunt has been justified as a response to concerns about competition between seals and fisheries.

## III. Sirenians

Sirenians have been hunted mainly for their delectable meat and blubber and their strong hides. The Steller's sea cow (*Hydrodamalis gigas*) was hunted to extinction within about 25 years of its discovery by European sea otter and fur seal hunters. Much like tortoises on tropical islands, the sea cows were easy to catch and provided local sustenance to ship crews, enabling the men to carry on their pursuit of fur, oil, and other valuable resources. Local people in West Africa and Central and South America used manatee hides to make shields, whips, glue, and plasters for dressing wounds. Large-scale commercial killing of Amazonian manatees (*Trichechus inunguis*) to supply mixira (fried manatee meat preserved in its own fat) took place in Brazil from the 1780s to the late 1950s, and manatee hides were in great demand for making heavy-duty leather products and glue between 1935 and 1954 (Domning, 1982). The total number of manatees (three species, combined) killed by villagers each year in West Africa and South America is probably in the thousands (Fig. 2). Dugongs (*Dugong dugon*), like manatees, have long been a prized food source for seafaring people throughout their (formerly) extensive Indo-Pacific range. Hunting continues in much of that range, including areas of Southeast Asia and Indonesia where few dugongs survive (Marsh *et al.*, 2002).

Sirenians have been captured using many different methods, apart from simply stalking and lancing or harpooning them from boats, or setting nets to enmesh them. People in West Africa and South America developed ingenious fence traps and drop traps to catch manatees. These can be baited to attract the animals or just placed strategically to take advantage of their natural movements through constricted channels. Dugong hunters in some areas used underwater explosives to kill their prey. In Torres Strait between Australia and New Guinea, portable platforms were set up on seagrass beds, and the hunters waited there overnight for opportunities to spear unsuspecting dugongs as they grazed.

## IV. Sea Otters and Polar Bears

Sea otters were cursed by the luxuriance of their pelts, which are among the most desirable of all mammalian furs. They were hunted remorselessly to supply the Oriental market from the 1780s onward—until very few were left and protection came in 1911. As otters were depleted in a region, hunting efforts there would be redirected at fur seals. Although anchored nets were sometimes used to catch sea otters (Kenyon, 1969), most of the hunting was conducted by men in boats, using lances initially and rifles later on. In California, otters were sometimes shot by men standing on shore, and in Washington, shooting towers were erected at the surf line and Indians were employed to swim out and retrieve the carcasses (Busch, 1985). Aboriginal people in Alaska are still allowed to hunt sea otters as long as the furs are used locally to make clothing or authentic handicraft items.

Polar bears have always been prime targets of Eskimo hunters, and non-Eskimo sport hunters have taken large numbers of bears as trophies. At least several hundred polar bears are still killed each year, most of them by Eskimos for meat and the cash value of their hides and gall bladders. Some polar bears are also killed in self-defense each year by Arctic residents or visitors to the North. In Canada, hunting permits allocated to native communities are often sold to sport hunters, on condition that local guides accompany the hunters and that only the heads and hides be exported. These expeditions generally involve dogsled travel, thus reinvigorating a traditional mode of winter transportation. Today, most polar bears are killed by shooting with high-powered rifles, but in the past they were also hunted with baited set-gun traps in Svalbard.

**Figure 2** *Adult white whales (belugas, Delphinapterus leucas) killed by Eskimos in Kasegaluk Lagoon near Point Lay on the Chukchi Sea coast of Alaska in July 1993. Canoes powered by outboard motors are used to drive the whales toward shore before killing them with rifles (top). The flukes, flippers, and skin with adhering blubber (locally called maktak) are saved as a delicacy (bottom). Courtesy of Greg O'Corry-Crowe.*

## V. Market (Commercial) vs Subsistence (Household-Use) Hunting

A distinction is often made between hunting for profit and hunting as a means of survival. This distinction is more than academic. The nature and degree of regulation have often depended on how a given hunter's enterprise was classified. The dichotomy between "commercial" and "subsistence" exploitation has had particular meaning in the context of the worldwide regulation of whaling. The IWC recognizes "aboriginal subsistence" whaling as a special category and has traditionally exempted certain groups of whalers from regulation (Reeves, 2002). Similarly, many national and multilateral restrictions on sealing have applied only or primarily to industrial operations and not to "aboriginal" hunters hunting for "subsistence" (e.g., the US Marine Mammal Protection Act and the North Pacific Fur Seal Convention).

Initially, the reasoning behind such special treatment was that these hunters used less destructive or wasteful gear and methods, and served only local, relatively small markets. However, those criteria are now called into question as aboriginal hunters have adopted modern weaponry and mechanized transport, and increasingly sell their produce for cash. Some products, notably sealskins, polar bear skins and gall bladders, and the ivory obtained from walruses and narwhals, enter a global marketplace.

Anthropologists argue that the term "subsistence" should be broadly defined and not exclude cash-based exchanges when these occur within a context that emphasizes local production and consumption. They point to the fact that modern Eskimos, e.g., are simply adapting to a changing world by hunting marine mammals with rifles, outboard motors, and snowmobiles. Only by selling skins, tusks, and, in the case of polar bear hunting, their own services as guides are these traditional hunters able to obtain the cash needed to live comfortably while continuing to engage in a domestic mode of production, providing highly esteemed and nutritious food for their home communities. Indeed, the IWC still considers Greenland whaling for baleen whales to be "aboriginal subsistence" whaling even though most of the whales are killed with deck-mounted harpoon guns firing explosive grenades and the meat and other products enter a country-wide, cash-based exchange network (Caulfield, 1997). At the same time, the IWC has resisted Japan's efforts to have "small-type coastal" whaling, which also serves a domestic but cash-based market, reclassified as something other than commercial whaling.

The difficulty of distinguishing commercial from subsistence hunting is not unique to situations involving marine mammals. Similar issues have arisen in relation to the trade in "bush meat" in Africa, Asia, and the Neotropics. Unregulated hunting is incompatible with the concept of sustainability. Considering the enormous increases in killing power afforded by firearms and mechanized transport, together with rapid human population growth and the attendant rise in resource consumption, the time is long past when racial or cultural entitlement could be allowed to preclude a vigorously enforced management regime based on conservation principles.

## VI. Future Hunting

For two reasons, the hunting of marine mammals in the foreseeable future is unlikely to approach the scale at which it was pursued throughout the nineteenth and much of the twentieth centuries. First, the populations of many species remain far below the earlier levels. Even if some recovery is achieved, the environmental carrying capacity has almost certainly declined in many instances. Considering the low productivity of these relatively large, long-lived animals, it is unrealistic to expect their numbers to return to "pristine" levels in a world so thoroughly transformed by human endeavor. Second, attitudes toward marine mammals have changed considerably in some parts of the world, and any initiative to expand the scope or scale of hunting is subject to public scrutiny as never before. Many people, particularly in North America, Europe, Australia, and New Zealand, are morally opposed to the killing of cetaceans, if not all marine mammals (Lavigne *et al.*, 1999). Although this does not mean that hunting will stop altogether, it does make it ever more likely that hunters will need to demonstrate that their enterprises are both sustainable (within the productive capacity of the affected animal populations) and humane.

## *See Also the Following Articles*

Conservation Efforts ■ Incidental Catches ■ Inuit and Marine Mammals ■ Polar Bear ■ Steller's Sea Cow ■ Whaling

## *References*

Bearzi, G., Holcer, D., and Notarbartolo di Sciara, G. (2004). The role of historical dolphin takes and habitat degradation in shaping the present status of northern Adriatic cetaceans. *Aquat. Conserv.: Mar. Freshw. Ecosyst.* **14**, 363–379.

Bonner, W. N. (1982). "Seals and Man: A Study of Interactions." University of Washington Press, Seattle.

Busch, B. C. (1985). "The War Against the Seals: A History of the North American Seal Fishery." McGill-Queen's University Press, Kingston.

Caulfield, R. A. (1997). "Greenlanders, Whales, and Whaling: Sustainability and Self-determination in the Arctic." University of New England Press, Hanover.

Domning, D. P. (1982). Commercial exploitation of manatees *Trichechus* in Brazil c. 1785–1973. *Biol. Conserv.* **22**, 101–126.

Du Pasquier, T. (2000). "Les Baleiniers Basques," Editions. S.P.M., Paris, France.

Ellis, R. (1991). "Men and Whales." Alfred A. Knopf, New York.

Kenyon, K. W. (1969). "The Sea Otter in the Eastern Pacific Ocean." United States Department of the Interior, Bureau of Sport Fisheries and Wildlife, North American Fauna No. 68.

Lavigne, D. M. (1999). Estimating total kill of northwest Atlantic harp seals, 1994–1998. *Mar. Mamm. Sci.* **15**, 871–878.

Lavigne, D. M., Scheffer, V. B., and Kellert, S. R. (1999). The evolution of North American attitudes toward marine mammals. *In* "Conservation and Management of Marine Mammals" (J. R. Twiss, Jr., and R. R. Reeves, eds), pp. 99–119. Smithsonian Institution Press, Washington, DC.

Marsh, H., Penrose, H., Eros, C., and Hugues, J. (2002). "Dugong Status Reports and Action Plans for Countries and Territories." UNEP Early Warning and Assessment Report Series.01-1. United Nations Environment Programme, Nairobi, Kenya.

McCartney, A. P. (ed.) (1995). "Hunting the Largest Animals: Native Whaling in the Western Arctic and Subarctic." Canadian Circumpolar Institute, University of Alberta, Edmonton.

Mitchell, E. (1975). "Porpoise, Dolphin and Small Whale Fisheries of the World: Status and Problems." International Union for Conservation of Nature and Natural Resources, Gland. IUCN Monograph No. 3.

Reeves, R. R. (2002). The origins and character of "aboriginal subsistence" whaling: A global review. *Mamm. Rev.* **32**, 71–106.

Tønnessen, J. N., and Johnsen, A. O. (1982). "The History of Modern Whaling." University of California Press, Berkeley, CA.

# Hybridism

## Martine Bérubé

### I. Introduction

Hybridization denotes the successful mating between two individuals, each from different and reproductively isolated gene pools (i.e., species). Hybridization is observed frequently among higher plants but only rarely among vertebrates. Within mammals, hybrids have been recorded in a number of marine as well as terrestrial species (Gray, 1972). The evolutionary consequences of such hybrids vary among incidences depending on the frequency, the degree of genetic differences between the parental species, mating system, and the ecological circumstances (Grant and Grant, 1997).

The examination of hybrids has always attracted much attention, as such incidences and their frequency might provide clues on reproductive behavior, dispersal capabilities, and phylogenetic relationship of species. As might be expected, hybrids are more common within genera where the different species have similar life histories and habitat requirements. When the frequency of hybridization is low, the fitness of the hybrids is generally low as well and hybrids usually are non-viable or sterile, thus not representing a threat to the genetic constitution of the parental species (Mary, 1963). However, as the frequency of hybridization increases, so may the number of viable and reproductive hybrids, which in turn may cause the breakdown of previous reproductive barriers between the two species. One evolutionary consequence of such a scenario is termed introgression; the gradual diffusion of the genes from one species into the gene pool of another. A recent well-documented example of introgression is the high incidence of coyote genes in what morphologically appear to be gray wolves (*Canis lupus*) observed in North America (Lehman *et al.*, 1991). The ultimate evolutionary consequence of introgression is the extinction of the species whose genome is being replaced by that of the other species.

With regard to marine mammals, a total of some 74 cases of putative hybridization events have been described; 53 within Cetacea and 21 within Pinnipedia. Putative hybrids have been observed in captivity and in the wild. Most hybrids among marine mammals reported so far have only been described morphologically. However, recently molecular techniques have been applied to identify hybrids and their parental species.

## II. Evidence of Mating Between Marine Mammal Species

Interspecific sexual interactions between a number of sympatric cetacean species have been described. Attempts at *interspecific mating* have been observed between pinniped species where no hybrids have yet been reported. Such mating appears to be aggressive and usually the *hetero-specific* (of a different species) male is much larger than the female. Often the female does not survive such a mating.

This kind of aggressive *interspecific mating* was first observed between a male gray seal (*Halichoerus grypus*) and a female harbor seal (*Phoca vitulina*) (Wilson, 1975). Later reports of such aggressive behaviors include mating between a male New Zealand sea lion (*Phocarctos hookeri*) and a dead female New Zealand fur seal (*Arctocephalus forsteri*), a South American sea lion (*Otaria flavescens*) and a South American fur seal (*A. australis*), a female California sea lion (*Zalophus californianus*) and a male Steller sea lion (*Eumetopias jubatus*), and between Southern elephant seals (*Mirounga leonina*) and Australian fur seals (*A. pusillus*). The aggressive mating undertaken by sea lions with *hetero-specific* females has been interpreted as "excess of violent sexual selection" (Miller *et al.*, 1996). This aggressive behavior seems to be widespread in the family Otariidae, and the number of hybrids is possibly much higher than reported to date.

## III. Reported Hybridizations in Captivity

Among captive cetaceans, 28 hybrids have been identified, all within the suborder Odontoceti. All hybridizations occurred among seven species of the Delphinoidea superfamily, and the common bottlenose dolphin (*Tursiops truncatus*) was one of the parental species in all cases (Table I). The majority of these hybrids have not survived. However, a viable first-generation hybrid between a bottlenose dolphin and a false killer whale (*Pseudorca crassidens*) has given birth twice after mating with a common bottlenose dolphin (Duffield, 1998). When this was reported in 1998, one of the two calves from the second generation was still alive. A similar event occurred for a hybridization between a common bottlenose dolphin and a long-beaked common dolphin (*Delphinus capensis*), where two first-generation hybrids gave birth to a live-born backcross calf sired by a bottlenose dolphin. In this case, however, the calves died shortly after birth. Within the Pinnipedia hybridization in captivity

has only been observed within the families Phocidae (true seals) and Otariidae (eared seals) (Table I).

## IV. Reported Hybridizations in the Wild

Probably the most impressive cases of hybridization among marine mammals are those identified within the Mysticeti (baleen whales). A total of 11 hybrids between baleen whale species have been reported so far, all were identified in the wild during commercial whaling operations. In all cases, the parents were a blue whale (*Balaenoptera musculus*) and a fin whale (*B. physalus*). The first report was in 1887 by A. H. Cocks, whom recorded six hybrids, or "Bastards," along the Lapland coast. However, this number is likely to be an under-estimate, since, as the author mentioned, hybrids sometimes were entered in the records as a fin whale instead of "Bastard." Later, Doroshenko (1970) reported a hybrid between a blue and a fin whale, taken in 1965 off Kodiak Island (in the Gulf of Alaska); it was identified by its exceptional and intermediate morphological traits.

More recently, three anomalous baleen whales, one female and two males, were caught during Icelandic whaling operations between 1983 and 1989 and described morphologically as fin × blue whale hybrids. The parental species of these specimens were later confirmed by molecular analyses based upon the maternally inherited mitochondrial genome and biparentally transmitted nuclear genes (Arnason *et al.*, 1991; Spilliaert *et al.*, 1991). Interestingly, the female Icelandic fin × blue whale hybrid was in her second pregnancy. Molecular analyses of the fetus revealed the fetus was the result of a mating between the hybrid mother and a male blue whale. Finally, a fin × blue whale hybrid caught off Northwest Spain in 1984 was identified first from its morphology and subsequently by molecular analyses, which identified the maternal and paternal species as blue and fin whale, respectively (Bérubé and Aguilar, 1998).

Within the Odontoceti, the first three hybrids described were from a stranding on the West Coast of Ireland in Blacksod Bay reported by Fraser in 1940. The morphological analysis concluded that the three stranded specimens were hybrids from mating between common bottlenose and Risso's dolphins (*Grampus griseus*). Three hybrid individuals in the same stranding, each a cross of the same parental species, is highly unusual given the overall low rate of hybridization among cetaceans *per se*. For the same reason Fraser himself first thought the hybrids to be a novel species rather than hybrids. Since the stranding in Blacksod Bay, three more cases of hybridization in the wild has been reported within the family Delphinidae identified from intermediate morphological traits. First, a specimen caught by fishermen off the Peruvian coast was determined to be a hybrid between a common (*D. capensis* or *D. delphis*) and a Dusky dolphin (*Lagenorhynchus obscurus*) (Reyes, 1996). Recently, two more interspecific hybrids have been reported, the first one between a spinner dolphin (*Stenella longirostris*) and a pantropical spotted dolphin (*S. attenuata*), the second one between a spinner dolphin (*S. longirostris*) and a Clymene dolphin (*S. clymene*) (Silva *et al.*, 2005).

In 1990, an anomalous whale skull was collected in Disko Bay off West Greenland. The morphological characteristics of this skull were intermediate between those of adult narwhals (*Monodon monoceros*) and belugas (*Delphinapterus leucas*). The authors hypothesized that the specimen was likely a hybrid from a mating between a narwhal and a beluga (Heide-Jørgensen and Reeves, 1993).

Hybridization has also been reported within Phocoenidae between a Dall's (*Phocoenoides dalli*) and a harbor porpoise (*P. phocoena*). The hybrid (identified by its morphology and by molecular

**H**

**TABLE I**
**Reported Hybridization of Marine Mammals in Captivity**

| Family | Species | Parental role | Method of detection | Reported number of hybrids | Reference |
|---|---|---|---|---|---|
| Delphinidae | T. truncatus X G. griseus | Dam Sire | Morphological and molecular | 13 | Sezaki et al. (1984); Sylvestre and Tasaka (1985); Shimura et al. (1986) |
| | T. truncatus X D. delphis | Dam Sire | Morphological | 2 | Duffield (1998) |
| | T. truncatus X D. capensis | Dam Sire | Morphological | 4 | Zornetzer and Duffield (2003) |
| | T. truncatus X P. crassidens | Dam Sire | Morphological | 6 | Nishiwaki and Tobayama (1982); Duffield (1998); Næss et al. (1998) |
| | S. bredanensis X T. truncatus | Dam Sire | Morphological | 1 | Dohl et al. (1974) (See Fig. 1) |
| | G. macrorhynchus X T. truncatus | Dam Sire | Morphological | 2 | Duffield (1998) |
| Phocidae | P. hispida X H. grypus | Dam Sire | Morphological | 1 | Lönnberg (1929) in King (1983) |
| Otariidae | C. ursinus X Z. californianus | Dam Sire | Morphological | 1 | Duffield (1998) |
| | Z. californianus X A. p. pusillus | Dam Sire | Morphological | 1+ | Schlieman (1968) in King (1983) |
| | Z. californianus X O. flavescens | Dam Sire | Morphological | 1 | Kirchschofer (1968) in King (1983) |

**Figure 1** *Hybrid between captive* Tursiops truncatus *and* Steno bredanensis. *Photograph by S. Leatherwood.*

analyses) was a female fetus recovered from a dead Dall's porpoise (Baird *et al.*, 1998; Willis *et al.*, 2004).

In Pinnipedia, most hybrids in the wild have been observed within the families Phocidae and Otariidae. The most common hybridization is between the Subantarctic (*A. tropicalis*) and the Antarctic (*A. gazella*) fur seals (Table II). Based upon the abundance of the two species and the frequency of hybrids, the rate of hybridization between the Subantarctic and the Antarctic fur seal has been estimated at 1% of the population, and the rate of backcrosses to the parental species at 2.4% (Kerley, 1983). A study on Macquarie Island found that although both sexes of each species are present on the island, 91% of the mothers of the Subantarctic and the Antarctic fur

seal hybrids are Antarctic fur seals. The New Zealand fur seal (*A. forsteri*) is also present on the Island, but females are rarely observed. For that reason, it was presumed unlikely that the New Zealand fur seal was involved in hybridization to the degree observed between the Subantarctic and the Antarctic fur seal. However, a recent molecular analysis detected high level of hybridizations (17–30% of the pups were hybrids), not only between *A. tropicalis* and *A. gazella* as expected but also between *A. tropicalis* and *A. forsteri* as well as *A. gazelle* and *A. forsteri* (Lancaster *et al.*, 2006).

## V. Evolutionary Implications of Hybridization

The evolutionary significance of hybridization is unknown, but it provides an opportunity for gene flow between otherwise isolated gene pools (e.g., exchange of adaptive traits). Marine mammals are genetically relatively similar. In comparison, the level of genetic divergence between the fin and the blue whale is similar to that observed between human (*Homo sapiens*), chimpanzee (*Pan troglodytes*, *P. paniscus*), and gorilla (*Gorilla gorilla*) (Arnason and Gullberg, 1993). Most cetacean species have the same number of chromosomes ($2n = 44$, with a few exceptions where $2n = 42$) and similar *karyotypes*, even though mysticetes and odontocetes probably diverged some 40 million years ago. Pinnipeds display a higher degree of variation in chromosome number, with the number of diploid chromosomes varying from 32 to 36 (Arnason, 1990). The relatively similar genetic background and often sympatric existence (in feeding or breeding range) among closely related marine mammals would seem to favor hybridization. However, as mentioned earlier, hybridization is rare, and where several hybrids have been observed (such as between the fin and the blue whale), the genetic integrity of the parental species is intact.

## TABLE II
### Reported Hybridization of Marine Mammals in the Wild

| Family | Species | Parental role | Method of detection | Reported number of hybrids | Reference |
|--------|---------|---------------|---------------------|----------------------------|-----------|
| Balaenopteridae | B. physalus X B. musculus | Sire and dam / Sire and dam | Morphological and molecular | 11+ | Cocks (1887); Doroshenko (1970); Arnason et al. (1991); Spilliaert et al. (1991); Bérubé and Aguilar (1998) |
| Delphinidae | T. truncatus X G. griseus | ? | Morphological and molecular | 3 | Fraser (1940); Shimura et al. (1986) |
| | D. capensis X L. obscurus | ? | Morphological | 1 | Reyes (1996) |
| | S. longirostris X S. attenuata | Dam / Sire | Morphological | 1 | Silva et al. (2005) |
| | S. longirostris X S. clymene | Dam / Sire | Morphological | 1 | Silva et al. (2005) |
| Monodontidae | D. leucas X M. monoceros | ? | Morphological | 1 | Heide-Jørgensen and Reeves (1993) |
| Phocoenidae | P. dalli X P. phocoena | Dam / Sire | Morphological and molecular | 7 | Baird et al. (1998); Willis et al. (2004) |
| Phocidae | C. cristata X P. groenlandica | Dam / Sire | Morphological and molecular | 1 | Kovacs et al. (1997) |
| Otariidae | A. gazella X A. tropicalis | Sire and dam / Sire and dam | Morphological, molecular, and vocal signature | 141 | Condy (1978); Brunner (1998); Page et al. (2001, 2002); Lancaster et al. (2006); Kingston and Gwilliam (2007) |
| | A. gazella X A. forsteri | Dam / Sire | Morphological and molecular | 66 | Lancaster et al. (2006, 2007) |
| | A. tropicalis X A. forsteri | Dam / Sire | Morphological and molecular | 12 | Lancaster et al. (2006, 2007) |
| | O. flavescens X A. australia | ? | Morphological | ? | Miller et al. (1996) |
| | O. byronia X Z. californianus | ? | Morphological | 1 | Brunner (2002) |

Haldane argued in 1922 that hybrids of the heterogametic sex (males in mammals with a single X- and Y-chromosome) was most likely to be sterile or unviable. Since then evolutionary geneticists have attempted to test "Haldane's rule." Among cetaceans, and specifically the family Mysticeti, the only two male blue × fin whale hybrids examined to date were both sexually immature despite their relatively advanced age. Although this observation is consistent with Haldane's rule, which has been supported by a number of studies of hybridization in terrestrial mammals, the small sample size of mysticete hybrids makes it impossible to assess with statistical rigor if the Haldane's rule applies to marine mammals as well.

The cases of anomalous marine mammals reported so far have shown that hybridization does occur in captivity as well as in natural settings. Among cetaceans, hybridization has been shown to occur between a variety of species (Tables I and II). However, the overall rate appears to be quite limited, and no introgression has yet been identified. The apparent scarcity of hybrids may not be a true reflection of the actual rate, i.e., it is possible that hybrids simply are overlooked or not reported (during commercial or subsistence whaling) in order to avoid sanctions for killing protected species (e.g., blue whales). Furthermore, the identification of hybrids so far has relied primarily upon morphological characters, which usually requires that the specimen be killed. However, the recent introduction of non-lethal methods to obtain the necessary tissue for molecular methods as skin biopsies from free-ranging cetaceans makes it a simple task to identify hybrids today.

Within Pinnipedia, multiple cases of hybrids have been carried to term and survived and even reproduced, as observed among the fur seals. It seems that despite the viability of the fur seal hybrids, the rate remains low and some mechanism ensuring species recognition is acting as a barrier to hybridization, keeping the involved species genetically distinct. In the case of captive animals, it is difficult to assess if the seemingly low viability of the offspring is related to their hybrid origin or the general low rate of survival in captive-born cetaceans.

## See Also the Following Articles

Antarctic Fur Seal ■ Southern Fur Seals

## References

Arnason, A., and Gullberg, A. (1993). Comparison between the complete mtDNA sequences of the blue and the fin whale, two species that can hybridize in nature. J. Mol. Evol. **37**, 312–322.

Arnason, U. (1990). Phylogeny of marine mammals—evidence from chromosomes and DNA. *In* "Chromosomes Today." (K. Fredga, B. A. Kihlman, and M. D. Bennet, eds), Vol. 10, pp. 267–278. Routledge, Florence, KY, USA.

Arnason, U., Spillaert, R., Palsdottir, A., and Arnason, A. (1991). Molecular identification of hybrids between the two largest whale species, the blue whale (*Balaenoptera musculus*) and the fin whale (*B. physalus*). *Hereditas* **115**, 183–189.

Baird, R. W., Willis, P. M., Guenther, T. J., Wilson, P. J., and White, B. N. (1998). An intergeneric hybrid in the family Phocoenidae. *Can. J. Zool.* **76**, 198–204.

Bérubé, M., and Aguilar, A. (1998). A new hybrid between a blue whale, *Balaenoptera musculus*, and a fin whale, *B. physalus*: Frequency and implications of hybridization. *Mar. Mamm. Sci.* **14**, 82–98.

Brunner, S. (1998). Cranial morphometrics of the southern fur seals *Arctocephalus forsteri* and *A-pusillus* (Carnivora: Otariidae). *Austral. J. Zool.* **46**, 67–108.

Brunner, S. (2002). A probable hybrid sea lion—*Zalophus californianus* × *Otaria byronia*. *J. Mammal.* **83**, 135–144.

Cocks, A. H. (1887). The fin whale fishery of 1886 on the Lapland coast. *Zoologist* **11**, 207–222.

Condy, P. R. (1978). Distribution, abundance, and annual cycle of fur seals (*Arctocephalus* spp.) on the Prince-Edward-Islands. *S. Afr. J. Wildl. Res.* **8**, 159–168.

Dohl, T. P., Norris, K. S., and Kang, I. (1974). A porpoise hybrid: *Tursiops* × *Steno*. *J. Mammal.* **55**, 217–221.

Doroshenko, N. V. (1970). A whale with features of fin whale and blue whale. *Tinro* **70**, 255–257.

Duffield, D. A. (1998). Examples of captive hybridization and a genetic point of view. *In* "World Marine Mammal Science Conference." (P. G. H. Evans, and E. C. M. Parsons, eds.), Vol. 12, p. 421. UK.

Fraser, F. C. (1940). Three anomalous dolphins from Blacksod Bay, Ireland. *Proc. Roy. Irish Acad.* **45 B**, 413–455, pl. 32–38.

Grant, P. R., and Grant, B. R. (1997). Hybridization, sexual imprinting, and mate choice. *Am. Natural.* **149**, 1–28.

Gray, A. P. (1972). "Mammalian Hybrids. A Check-list with Bibliography." Slough Commonwealth Agricultural Bureaux, Farnham Royal, UK.

Heide-Jørgensen, M. P., and Reeves, R. R. (1993). Description of an anomalous monodontid skull from West Greenland: A possible hybrid? *Mar. Mamm. Sci.* **9**, 258–268.

Kerley, G. I. H. (1983). Relative population sizes and trends, and hybridization of fur seals *Arctocephalus tropicalis* and *A. gazella* at the Prince Edwards Islands, Southern Ocean. *S. Afr. J. Zool.* **18**, 388–392.

Kingston, J. J., and Gwilliam, J. (2007). Hybridization between two sympatrically breeding species of fur seal at Iles Crozet revealed by genetic analysis. *Cons. Gen.* **8**, 1133–1145.

Kovacs, K. M., Lydersen, C., Hammill, M. O., White, B. N., Wilson, P. J., and Malik, S. (1997). A harp seal × hooded seal hybrid. *Mar. Mamm. Sci.* **13**, 460–468.

Lancaster, M. L., Gemmell, N. J., Negro, S., Goldsworthy, S., and Sunnucks, P. (2006). Menage a trois on Macquarie Island: Hybridization among three species of fur seal (*Arctocephalus* spp.) following historical population extinction. *Mol. Ecol.* **15**, 3681–3692.

Lancaster, M. L., Bradshaw, C. J. A., Goldsworthy, S. D., and Sunnucks, P. (2007). Lower reproductive success in hybrid fur seal males indicates fitness costs to hybridization. *Mol. Ecol.* **16**, 3187–3197.

Lehman, N., Eisenhawer, A., Hansen, K., Mech, L. D., Peterson, R. O., Gogan, P. J. P., and Wayne, R. K. (1991). Introgression of coyote mitochondrial DNA into sympatric North American grey wolf populations. *Evolution* **45**, 104–119.

Mayr, E. (1963). "Animal Species and Evolution." Harvard University Press, Cambridge.

Miller, E. H., Ponce de León, A., and DeLong, R. L. (1996). Violent interspecific sexual behavior by male sea lions (*Otariidae*): Evolutionary and phylogenetic implications. *Mar. Mamm. Sci.* **12**, 468–476.

Næss, A., Haug, T., and Nilssen, E. M. (1998). Seasonal variation in body condition and muscular lipid contents in northeast Atlantic minke whale *Balaenoptera acutorostrata*. *Sarsia* **83**, 211–218.

Nishiwaki, M., and Tobayama, T. (1982). Morphological study on the hybrid between *Tursiops* and *Pseudorca*. *Sci. Rep. Whales Res. Inst., Tokyo* **34**, 109–121.

Page, B., Goldsworthy, S. D., and Hindell, M. A. (2001). Vocal traits of hybrid fur seals: Intermediate to their parental species. *An. Behav.* **61**, 959–967.

Page, B., Goldsworthy, S. D., Hindell, M. A., and McKenzie, J. (2002). Interspecific differences in male vocalizations of three sympatric fur seals (*Arctocephalus* spp.). *J. Zool.* **258**, 49–56.

Reyes, J. C. (1996). A possible case of hybridism in wild dolphins. *Mar. Mamm. Sci.* **12**, 301–307.

Sezaki, K., Hirosaki, Y., Watabe, S., and Hashimoto, K. (1984). Electrophoretic characters of the hybrids between 2 dolphins *Tursiops-Truncatus* and *Grampus-Griseus*. *Bull. Japan. Soc. Sci. Fish.* **50**, 1771–1776.

Shimura, E., Numachi, K., Sezaki, K., Hirosaki, Y., Watabe, S., and Hashimoto, K. (1986). Biochemical evidence of hybrid formation between the two species of dolphin *Tursiops truncatus* and *Grampus griseus*. *Bull. Japan. Soc. Sci. Fish.* **52**, 725–730.

Silva, J. M., Jr., Silva, F. J. L., and Sazima, I. (2005). Two presumed interspecific hybrids in the genus *Stenella* (Delphinidae) in the tropical West Atlantic. *Aquat. Mamm.* **31**, 468–472.

Spilliaert, R., Vikingsson, G., Arnason, U., Palsdottir, A., Sigurjonsson, J., and Arnason, A. (1991). Species hybridization between a female blue whale (*Balaenoptera musculus*) and a male fin whale (*B. physalus*): Molecular and morphological documentation. *J. Hered.* **82**, 269–274.

Sylvestre, J.-P., and Tasaka, S. (1985). On the intergeneric hybrids in cetaceans. *Aquat. Mamm.* **11**, 101–108.

Van Gelder, R. G. (1977). Mammalian hybrids and generic limits. *Am. Mus. Nov.* **2635**, 1–25.

Willis, P. M., Crespi, B. J., Dill, L. M., Baird, R. W., and Hanson, M. B. (2004). Natural hybridization between Dall's porpoises (*Phocoenoides dalli*) and harbour porpoises (*Phocoena phocoena*). *Can. J. Zool.* **82**, 828–834.

Wilson, S. C. (1975). Attempted mating between a male grey seal and female harbor seals. *J. Mammal* **56**, 531–534.

Zornetzer, H. R., and Duffield, D. A. (2003). Captive-born bottlenose dolphin × common dolphin (*Tursiops truncatus* × *Delphinus capensis*) intergeneric hybrids. *Can. J. Zool.* **81**, 1755–1762.

# Identification Methods

## Randall S. Wells

Individual identification is an important tool for studies of animal behavior, ecology, and population biology. Much can be learned from recognition of individuals within a population or social unit, or from tracking individuals through time. Repeated observations of a recognizable individual can help to define its ranging patterns or site fidelity, or to quantify habitat use. Behavioral studies benefit greatly from the ability to recognize individuals. Individual identification is essential to understanding group compositions, and this understanding is enhanced when the individual's sex, age, genetic relationships, and reproductive condition are known. Similarly, interpretation of social interactions requires the ability to distinguish between the players. Behavioral descriptions often involve measurements of rates of occurrence of behaviors. These rates are measured most accurately when a selected individual is followed through time, or when the individual's behaviors are recorded at pre-determined intervals, a process referred to as focal animal behavioral observations (Altmann, 1974).

Descriptions of life history patterns and empirical measures of population dynamics can be facilitated by individual identification (Hammond *et al.*, 1990). By following individuals through time it is sometimes possible to determine age at sexual maturity, calving intervals, calf survivorship, and life span, providing measures of reproductive success. Combined, such individual measures can provide population level vital rates, including birth rates, mortality rates, and recruitment (Wells and Scott, 1990). Mark–recapture techniques use individual identification to arrive at abundance estimates. Individual identification provides one of the best tools for documenting exchanges of individuals between populations, allowing estimation of rates of immigration and emigration.

Selection of specific identification techniques depends on the research questions being addressed and the species under study. Frequent monitoring of individuals may require the ability to readily identify animals from a distance at each encounter, whereas other studies may only need to recognize an animal when it is handled subsequently, alive or at the end of its life. Some species exhibit individually specific natural markings that facilitate identification in the field. Other species lack such distinctive markings and require the attachment of artificial marks, or tags, if individual identifications are desired. Some species are visible on land at times, whereas others are entirely aquatic. Morphological, behavioral, and ecological features must be considered in order to determine what kind of tag or attachment is most appropriate, in terms of safety to the animal and effectiveness for the research. It is now also possible to collect small samples of tissues that allow the identification of individuals genetically. Individual identification techniques have been summarized for cetaceans, pinnipeds, and sirenians (Hammond *et al.*, 1990; Scott *et al.*, 1990; Würsig and Jefferson, 1990; Erickson *et al.*, 1993; Wells *et al.*, 1999).

## I. Cetaceans

### A. Natural Markings

Cetaceans exhibit a variety of individually distinctive natural features. In most cases, features appearing above the surface of the water during the respiratory cycle are most useful. In particular, heads, backs, dorsal fins, and flukes are used most frequently for individual identification, with variations occurring in color patterns, skin patches, body scarring, and nicks and notches along fin edges (Hammond *et al.*, 1990). Some individuals of most cetacean species acquire distinctive scars from previous wounds or injuries, which are often used for identifications. Perhaps the most unique features used to identify individual cetaceans are the callosities of the right whales, *Eubalaena* spp. (Payne *et al.*, 1983). These individually distinctive raised patches of roughened skin are present on the rostrum anterior to the blowholes in a pattern referred to as the bonnet, on the chin, lower lips, above the eyes, and near the blowholes. Whale lice (*Cyamus* spp.), cyamid crustaceans that frequently live on the callosities, often give them a white, orange, yellow, or pink appearance. Callosities have allowed for the reliable recognition of individuals over periods of decades.

Color variations, where they exist among cetacean species, have been used with much success for individual identification, especially among the mysticetes (Hammond *et al.*, 1990). Reminiscent of "Moby Dick," a few anomalously white individuals have been noted for several species of large and small cetaceans, offering unusual opportunities for individual identification. Blue whales (*Balaenoptera musculus*) and gray whales (*Eschrichtius robustus*) exhibit individually distinctive mottling on their backs (Fig. 1). The dorsal fin and dorsal ridge, respectively, are used as reference points for locating mottling patterns on these species. Bowhead whales (*Balaena mysticetus*) often have a distinctive pattern of white pigmentation on the chin and/or caudal peduncle. These patterns are readily seen from

**Figure 1** *Distinctive color patterns of a blue whale* (Balaenoptera musculus). *Photo by R. S. Wells.*

aircraft, the most commonly used observation platform for this arctic species. Fin whales (*Balaenoptera physalus*) exhibit strongly asymmetrical body pigmentation, with the lower and upper lips and first third of the baleen on the right side of the head appearing white or pale gray, whereas the left side lips and baleen are dark. A light-colored "blaze" sweeps back on the right side, and a V-shaped light-colored "chevron" occurs on both sides behind the blowhole. Minke whales (*B. acutorostrata* and *B. bonaerensis*) exhibit a pattern of pale lateral pigmentation on the body, often divided into three distinct swaths, with the relative brightness of the three swaths apparently varying consistently between Northern and Southern Hemispheres. The distinctive dark and white patterns of the flippers and ventral surface of the flukes are familiar identification features for humpback whales (*Megaptera novaeangliae*) (Katona *et al.*, 1979).

Some of the smaller cetaceans also exhibit useful color variations, from the perspective of the researcher. Most notable are the light colored saddle patches behind the dorsal fin of the killer whale (*Orcinus orca*), which differ in size and shape. Similar features are used for short-finned pilot whales (*Globicephala macrorhynchus*), though the saddle marks are less distinct. Dorsal fin and/or back pigmentation variation has proven useful in studies of Dall's porpoises (*Phocoenoides dalli*), Pacific white-sided dolphins (*Lagenorhynchus obliquidens*), Risso's dolphins (*Grampus griseus*), and Hector's dolphins (*Cephalorhynchus hectori*); and facial color patterns were used to identify the now-extinct baiji (*Lipotes vexillifer*). Extensive speckling develops with age in spotted dolphins (*Stenella attenuata* and *S. frontalis*). Such speckling has provided much opportunity for individual identification from both above and below the surface of the water in behavioral studies of Atlantic spotted dolphins (*S. frontalis*).

A variation of the color pattern is scarring that results in pigment variations. For example, Risso's dolphins acquire distinctive long-term white scars on their otherwise brown or gray bodies, and belugas (*Delphinapterus leucas*) acquire dark scars on their otherwise white bodies. Bottlenose dolphin (*Tursiops* spp.) scars on the dorsal fin often are white, in contrast to their general gray coloration. Cookie cutter shark (*Isistius* spp.) bite wounds leave permanent small diameter oval-shaped scars which are often depressed and pigmented differently from the rest of many pelagic cetaceans' bodies.

Dorsal fins typically are prominent features that are visible to researchers during most cetacean surfacings. In many cetacean species dorsal fins develop distinctive shapes or acquire nicks and notches, often through intraspecific or interspecific interactions, that allow for individual identification. Among the larger whales, fin, Sei (*Balaenoptera borealis*), Bryde's (*B. edeni*), minke, humpback, and sperm whale (*Physeter macrocephalus*) dorsal fins serve as useful identification features. Building on the pioneering work of Bigg (1982) with killer whales and Würsig and Würsig (1977) with bottlenose dolphins, studies based on dorsal fin identifications of various delphinids and other small cetaceans have blossomed in the last three decades (Hammond *et al.*, 1990; Scott *et al.*, 1990; Würsig and Jefferson, 1990; Wells *et al.*, 1999). Species that have received the most attention include killer whales, bottlenose dolphins, pilot whales, humpbacked dolphins (*Sousa* spp.), Pacific and Atlantic (*Lagenorhynchus acutus*) white-sided dolphins, dusky dolphins (*L. obscurus*), Risso's dolphins, spinner dolphins (*Stenella longirostris*), Atlantic spotted dolphins, Heaviside's dolphins (*Cephalorhynchus heavisidii*), Hector's dolphins, harbor porpoises (*Phocoena phocoena*), Amazon River dolphins (*Inia geoffrensis*), tucuxi (*Sotalia* spp.), long-finned pilot whales (*Globicephala melas*), Irrawaddy dolphins (*Orcaella brevirostris*), and baiji (Fig. 2). The frequency of occurrence of distinctive fin features varies from species to species

**Figure 2** *Killer whale* (Orcinus orca) *dorsal fins and saddle patches provide reliable identification cues. Photo by R. S. Wells.*

**Figure 3** *Distinctive dark and light patterns on the ventral surface of a humpback whale's* (Megaptera novaeangliae) *fluke. Photo by R. S. Wells.*

and in some cases from population to population. Along the central west coast of Florida, approximately 60–80% of bottlenose dolphins are considered to be distinctive based on dorsal fin features. Unlike color patterns that vary from one side of the animal to the other, dorsal fin features are often equally visible from both sides and are distinctive under a broad range of lighting conditions, facilitating data collection in the field.

Some cetacean species regularly lift their flukes from the water prior to a dive, providing predictable opportunities for researchers to note the occurrence of nicks, notches, and other features on the trailing edge of the flukes. Humpback whales offer both distinctive color patterns as well as trailing edge features for identification (Fig. 3). Humpback whale flukes were among the first natural markings on cetaceans to be recognized for their individual specificity, and the technique has achieved extensive application worldwide in studies of population size and structure (Smith *et al.*, 1999). Sperm whales also demonstrate much individual-specific variability in fluke edge features.

Many of the cetacean features used by researchers for individual identification are visible above the surface of the water only briefly during respiratory cycles, or are too subtle to be of use for accurate identification in real time. Most cetacean individual identification research involves the collection of permanent records of the distinctive features for subsequent detailed analysis, through a process generally referred to as photo-identification. As the name indicates, the process frequently involves 35-mm photography of cetaceans. Digital imaging through still cameras and video has greatly expanded the capabilities and possibilities for individual identification, and facilitating image processing, storage, and sharing.

At its most basic level, photo-identification involves trying to obtain high quality, high resolution, full frame images of identifying features (Würsig and Jefferson, 1990). Though photo-identification can sometimes be accomplished from shore, typically scientists in research vessels, whale-watching boats, or aircraft attempt to place themselves in position to be able to obtain an image of the features that is parallel to the photographic plane (lens oriented perpendicular to the feature of interest). Telephoto lenses aid the researcher in enlarging the features to fill the photographic frame. Motor drives or video allow multiple images to be taken in quick succession to optimize capturing fins, backs, or flukes at their greatest perpendicularity and height above the surface of the water, for example. Time and date data printed on the image provide additional assurance that images and data records can be matched correctly during subsequent analyses.

High-resolution digital cameras have now largely replaced film cameras for photo-identification work (Markowitz et al., 2003). In the past, film selection varied with species, lighting conditions, and researcher preference. The film had to have sufficiently fine grain to be able to resolve distinctive features, while allowing a shutter speed setting sufficiently fast to "freeze" the animal but slow enough to optimize depth of field for focus. Some researchers have used black and white film, especially if expense or ease of manual processing were concerns. Color film has often been used when documentation of wounds or the freshness of fin features is desired, for example. High-resolution digital cameras, typically with high-speed auto-focusing, provide high-quality color images in real-time, eliminating delays resulting from the need to wait for film to be developed.

Techniques for image storage, retrieval, and analysis vary greatly across research situations, but the development of digital photography has greatly increased efficiency and decreased costs for image manipulation and storage. Historically, images in the form of slides, prints, or negatives were labeled and stored chronologically in archival plastic sheets in binders, then examined under magnification through a handheld lupe or dissecting microscope. Digital images, obtained either directly from high-resolution digital cameras or from scans of photographs, offer tremendously increased capacity for manipulation, including cropping and enlarging identifying features, electronic storage, or transmission over the Internet.

Digital images also facilitate computer-assisted automated analysis. Previously, photographic matches were made through the laborious process of individual comparison by eye of the image of interest to all possible matches in a catalog of distinctive individuals. Computer software has been developed that can search thousands of images of such animals as sperm whales, humpback whales, bottlenose dolphins, or other odontocetes in a very short time to produce a limited set of potential matches. The researcher can then make the final match using the exceptional resolving capabilities of the human eye. Additional rigor is often incorporated into the process using multiple judges for difficult final identifications. Computer-assisted matching is becoming increasingly important as catalogs are now incorporating many thousands of individuals, and as contributions to centralized catalogs are being made by numerous researchers

in widely dispersed locations (Hillman et al., 2003; Bas et al., 2005; Adams et al., 2006).

Other kinds of "natural markings" that are being used increasingly are genetic markers from skin biopsy samples. Molecular analyses of small samples allow determination of sex and individual identification from genotypes provided by microsatellite loci. This technique was developed for large-scale use during an ocean-basin-wide study of humpback whales, in which photographs were used to identify 2998 individual whales, and microsatellite loci were used to identify 2015 whales (Smith et al., 1999). Based on the results of these initial studies, molecular techniques hold a great deal of promise for studies of a variety of cetaceans.

## B. Temporary Markings

Natural temporary markings include skin lesions on parts of the body visible to researchers (Wilson et al., 1999), and soft-bodied barnacles that attach to dorsal fins, for example. Such markings can be useful for distinguishing between otherwise unmarked animals within a group, but their changeable nature make them less reliable for accurate identifications over long periods. Skin lesions may take weeks to months to fully heal and disappear, but their characteristics change during the healing process. Soft-bodied barnacles favor dorsal fin tips for attachment, leading to low variability in positioning, thus minimizing their value for identification.

Anthropogenic temporary markings have been found to be of limited utility with cetaceans (as reviewed by Scott et al., 1990). Remotely applied paint and tattoos have been tested with small cetaceans, and in all cases the animals were either not re-identified or the markings disappeared within 24 h, due to skin sloughing. In some cases, zinc oxide-based, brightly colored sun protection ointments have been applied to dolphins' dorsal fins prior to release. These have allowed for the short-term identification of animals otherwise lacking in distinctive marks, and transfer of colors between animals can indicate social interactions.

## C. Scarring and Branding

Dorsal fin notching has been attempted in a few cases with killer whales, bottlenose dolphins, pantropical spotted dolphins, and spinner dolphins (Scott et al., 1990). Notching provides the same kinds of features used in photographic identification of natural marks. Such notching requires capturing the animals, which also provides opportunities to learn the sex and age of the marked dolphin, as well as other biological information. One report indicated minor but persistent bleeding as a result of notching, but this has not been reported by others.

Freeze branding, using metal numerals 5–8 cm high applied to the animals' body or dorsal fin for 10–20 sec, has been used safely and successfully with a variety of small cetaceans, including bottlenose dolphins, spinner dolphins, short-beaked common dolphins (*Delphinus delphis*), Pacific white-sided dolphins, short-finned pilot whales, false killer whales (*Pseudorca crassidens*), Amazon River dolphins, and rough-toothed dolphins (*Steno bredanensis*) (Irvine et al., 1982; Scott et al., 1990). Freeze-brand application typically results in little or no reaction by dolphins, but minor skin lesions may occur if brands are applied for too long. Readable white marks usually appear within a few days (Fig. 4). Freeze brands fade over time, but the marks can often still be identified for many years in good quality photographs even if they are not readily visible in the field. Fading appears to be age related, with brands disappearing more rapidly

I

**Figure 4** *Dorsal fin and back of a 41-year-old male bottlenose dolphin* (Tursiops truncatus), *showing a 12-year-old freeze brand ("48") in the center of the dorsal fin, and on the back centered below the caudal insertion of the fin. The healed notch near the base of the fin is from a rototag applied 16 years before. Photo by Sarasota Dolphin Research Program.*

**I**

and more completely on younger animals, but remaining readable on adults for as long as 11 years or more (Irvine *et al.*, 1982; Scott *et al.*, 1990).

### D. Attachment Tags

The use of attachment tags for identification purposes (rather than telemetry, covered elsewhere in this volume) including Discovery tags, spaghetti tags, button tags, and rototags, has been reviewed by Scott *et al.* (1990). Discovery tags are numbered metal cylinders shot into the blubber from whaling ships or research vessels. The tags have been used primarily with baleen and sperm whales, and are recovered when the whales are captured and rendered, providing information on two points within the animals' range. Tagging was initiated in 1932 and continued until the whaling moratorium in 1985. More than 20,000 Discovery tags have been used, but return rates have been low, typically below 15%. Smaller versions of these tags have been used with small whales without notable success, and use with cetaceans less than 4.6-m long has been discouraged because of risk of serious injury.

Streamer or spaghetti tags, originally developed for fish tagging, are colored vinyl-covered strands of wire cable of variable length with steel or metal dart tips that are applied with either a crossbow or a jab stick, with the intent of anchoring the tip between blubber and muscle. Thousands of these tags have been applied to dolphins, porpoises, and belugas, especially in association with the tuna seine net fishery in the eastern tropical Pacific Ocean. Because of poor retention and high risk of injury to the animal, use of spaghetti tags with small cetaceans has been discouraged for many years (Irvine *et al.*, 1982).

Dorsal fins or ridges are commonly used for tag attachment because of their structure, prominence, and regularity of appearance above the surface of the water. Button tags, typically numbered and colored fiberglass or plastic disks or rectangular plates designed after the Peterson disk fish tags have been applied to several species of small cetaceans, including bottlenose dolphins, pantropical spotted dolphins (*Stenella attenuata*), spinner dolphins, common dolphins

(*Delphinus* spp.), Pacific white-sided dolphins, belugas, and harbor porpoises (Evans *et al.*, 1972; Scott *et al.*, 1990). Usually button tags are attached through the dorsal fin by means of one or more plastic (especially delrin) or stainless steel bolts or pins that connect the tag halves on each side of the fin. Although some button tags have lasted for several years on pelagic dolphins, inshore animals often lose the tags within weeks or months, often by breaking them through rubbing on the shallow sea floor. Use of button tags has been largely discontinued due to poor tag retention and the potential for injury to the animals (Irvine *et al.*, 1982).

Small plastic cattle ear tags, or rototags, clipped through the trailing edges of dorsal fins have proved successful for identifying small cetaceans in the field, including bottlenose dolphins, pantropical spotted dolphins, spinner dolphins, common dolphins, rough-toothed dolphins, Pacific white-sided dolphins, short-finned pilot whales, and harbor porpoises (Fig. 4; Norris and Pryor, 1970; Scott *et al.*, 1990). Typically, a small hole is made in the thin tissue of the trailing edge using a sterile technique, and the tag is clipped through the fin with special pliers. Though the written markings are too small to be read at a distance, number of tags, color, and position on the fin provide a useful degree of variation. Rototags have remained in position for a period of years, although often they are lost within months. Rototag halves may separate, leaving a healed hole in the fin, or they migrate through the trailing edge of the fin, leaving a small, healed notch; both pose minimal risks to the animals but offer continuing identification features. Barnacle and algae fouling, and pressure necrosis are infrequent problems. As a modification of this technique, small VHF radio transmitters have been attached to rototags for short-term tracking (up to 30 days), with a modification involving the use of a corrosible nut system to release the tag at that time.

Other attachment techniques, such as the use of tethers or plastic-coated wires or polypropylene or soft rubber tubing have proved to be ineffective and injurious to the animals when attached to the caudal peduncle. Tag loss rates have been high, and abrasions were frequently noted.

### II. Pinnipeds

### A. Natural Markings

Natural body markings have been used in only a few studies of pinnipeds such as gray seals (*Halichoerus grypus*), northern elephant seals (*Mirounga angustirostris*), Steller sea lions (*Eumetopias jubatus*), Hawaiian monk seals (*Monachus schauinslandi*), harbor seals (*Phoca vitulina*), and California sea lions (*Zalophus californianus*). Yochem *et al.* (1990) examined pelage patterns of harbor and largha (*P. largha*) seals to distinguish between populations and individuals. Using black and white photographs they scored the presence or absence of spots, clarity of spots, relative density of spots, complexity of spots, presence of rings, and spacing of rings in selected body areas (especially sides of the head, neck, and chest). Hiby and Lovell (1990) described a computer aided matching system for screening a library of digitized natural mark photographs of gray seals. Their system created a three-dimensional model to locate features on the seal's body, especially using the side of the neck. Harting *et al.* (2004) devised a computer-assisted system for photo-identification of Hawaiian monk seals. For most pinniped species, studies using natural markings are hampered by a lack of distinctive markings, and the large numbers of individuals or pack ice distributions of many species (Erickson *et al.*, 1993). Most pinniped researchers have resorted to the use of artificial markings and tags for individual identification.

## B. Temporary Markings

Techniques for temporary markings of pinnipeds include paints, dyes, bleaches, and pelage clippings (Erickson *et al.*, 1993). These techniques offer the advantages of often being able to be applied without having to restrain the animals and permitting remote identification without disturbance. However, these marks are typically lost upon molting, precluding the continuity of identification beyond a single season. A variety of paints (marine, highway, rubber-based, quick-drying cellulose, aerosol sprays, and house paint) have been used to mark seals and sea lions. Paints have been applied from brushes or rollers on poles, and from plastic bags thrown at the animals. Quick-drying paint has proved relatively effective, with a useful lifespan of about 1 month on average. Northern fur seals (*Callorhinus ursinus*) have been successfully marked for 2–12 months with a fluorescent plastic resin, naptha-based paint. This technique apparently results in matting of guard hairs, which then break off leaving an outline of the mark. High-gloss marine enamel applied from aerosol cans to mark Hooker's sea lions (*Phocarctas hookeri*) has resulted in markings lasting 3 months, even after the animals have been at sea. Carbon dioxide powered paint guns firing small capsules have proved less effective for marking elephant seals due to reliability problems and the small size of the marks.

Dyes have been used with several species of pinnipeds, especially light-colored species (Erickson *et al.*, 1993). Successful dying usually occurs when permanent dyes are used and when the animals are dry and remain out of the water for a period of time following application. Colored dyes and black Nyanzol D have lasted 3–4 months on gray seals, harbor seals, and California sea lions. The addition of alcohol to the Nyanzol D leaves a more distinct marker because it dissolves fur oils, and it also prevents dye solution from freezing. Yellow picric acid in a saturated alcohol solution has been used with gray seals, with results that last through pup molting, appearing on the adults as well. This solution can be applied from a back-pack tree sprayer to wet or dry seals. Fluorescent dye mixed with small quantities of epoxy resin has also been used with success. In some cases, such as southern elephant seals (*Mirounga leonina*) dyes have been less successful.

Bleach offers a very effective and sometimes longer-lasting alternative to paints and dyes (Erickson *et al.*, 1993). Many of the bleach solutions can be applied to sleeping animals via a squeeze bottle, thus minimizing risk, effort, and disturbance. Commercially available products such as Lady Clairol Ultra Blue dye in combination with various chemicals has been used most often, resulting in a white or cream-colored mark that is most visible on dark pelages. Combinations resulting in thicker consistency allow for distinct lines. Bleach marks on elephant seals last until molt, sometimes for 6 months, and have lasted for two seasons on fur seals. Combinations of bleaches and dyes have also been used in some cases such as northern elephant seals (Fig. 5).

Hair clipping is somewhat more difficult than the previous techniques, but effective when the underfur is a different color from the guard hairs (Erickson *et al.*, 1993). This technique involves clipping or singeing the pelage to create a distinctive mark. It has been used with success with northern fur seals, Steller sea lions, and Antarctic fur seals (*Arctocephalus gazella*).

## C. Scarring and Branding

Punch marks and amputations have been used extensively with fur seals, with poor success and concerns about injury to the animals (Erickson *et al.*, 1993). Initial efforts to mark northern fur seals and Antarctic fur seals by punching holes in flippers in unique combinations of numbers and positions found this technique to be unreliable

**Figure 5** *Bleach markings on a northern elephant seal* (Mirounga angustirostris). *"Bilbo" is marked in black dye for identification through the summer molt, and in bleach for the winter breeding season. Photo by C. J. Deutsch.*

due to healing and occlusion. Hair on the flippers of phocids seals precludes utility with these species. Flipper notching was also found to be unreliable due to tissue regrowth. Although ear notching was used successfully for cohort marking in northern fur seals, it is no longer used because of concerns regarding interference with diving abilities.

Both hot branding and freeze branding have been used with great effect with pinnipeds (Erickson *et al.*, 1993). Hot brands have been used since 1912 with thousands of northern fur seals, Cape fur seals (*Arctocephalus pusillus*), southern elephant seals, Weddell seals (*Leptonychotes weddellii*), gray seals, and leopard seals (*Hydrurga leptonyx*). Some marks have remained readable for up to 20 years. The technique seems best suited to colonial seals due to the bulky nature of the branding tools and heat source. Typically, brands are heated to red hot, and applied with firm, even pressure for 2–7 sec, depending on whether the hair has been clipped. Brands are applied to the upper saddle, middle back, or upper shoulder to optimize sightability.

Freeze branding differs from hot branding in that it involves the selective killing of pigment-producing cells through contact with a super-cooled metal numeral or symbol (typically 5-cm high) (Erickson *et al.*, 1993). Brands are cooled with liquid nitrogen or a dry ice and alcohol mixture and applied for about 20 sec to an area where hair has been removed. Correct freeze brand application results in a non-pigmented pelage mark, ranging from dark (elephant seals, California sea lions) to pink (California sea lions, walrus, *Odobenus rosmarus*). Freeze branding has had mixed success. Many freeze brands on pinnipeds have been found to re-pigment within 1–2 years, perhaps as a result of excessive branding. Readable brands have been obtained for elephant seals (up to a year, discernable for 3 years) California sea lions (readable for 1.5 years, discernable for up to 4 years), walrus (readable for many year), and Australian sea lions (*Neophoca cinerea*) (legible on flippers for 7 years, on flanks for 4 years).

## D. Attachment Tags

Plastic or metal attachment tags are used more widely than any other kind of individual identification system with pinnipeds (Erickson *et al.*, 1993). Monel or stainless steel tags such as those

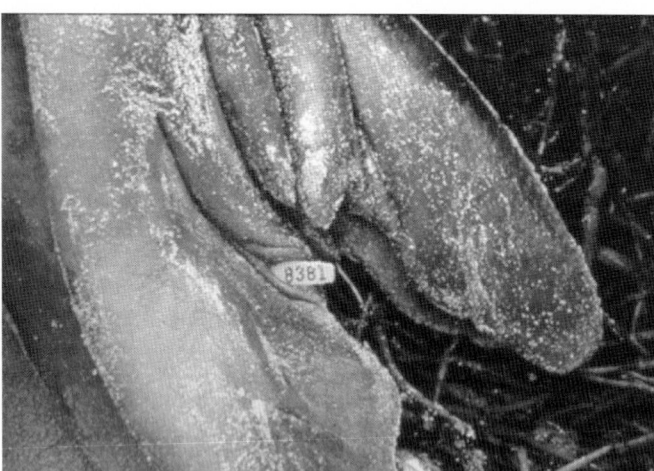

**Figure 6** *Flipper tag on a northern elephant seal* (Mirounga angustirostris). *Photograph by B. J. LeBoeuf.*

used to mark livestock are the most common metal tags. These metal strap tags are self-piercing and are attached by means of special pliers to the trailing edge of the fore flippers of otariids, and to the interdigital web of the hind flippers of phocids. Typically, the tags are stamped with an organization address and serial number. Thousands of metal tags have been attached to phocids. Retention rates on phocid seals are low, with post-attachment tears and cuts sometimes becoming infected. Hundreds of thousands of metal tags have been attached to otariids, with similar poor results.

The use of plastic tags is now much more common than metal tags for identifying pinnipeds (Fig. 6). Two kinds of plastic tags are used commonly, rototags and Allflex tags. Both consist of self-piercing male and female elements that are applied with special pliers, as with the metal tags. Plastic tags are available in a variety of colors, leading to more than 300 unique color combination possibilities. The visibility of both metal and plastic attachment tags can be enhanced using streamer markers such as nylon cloth strips reinforced with vinyl, which may last for a year or more.

Tagging success with both metal and plastic tags is less than desired. Loss rates of the two kinds of plastic tags are variable, but tend to be lower than for metal tags, about 10% annually. However, the long-term durability of metal tags is better than plastic. Wounds from metal tags are more common than from plastic.

## III. Sirenians

### A. Natural Markings

The process of developing new techniques and applying existing technology to studies of sirenians was reviewed by G. Rathbun in Wells *et al.* (1999). Natural marks, including deformities and scars have been used to identify individual Florida manatees (*Trichechus manatus latirostris*) since the 1950s. Among the marks that have proven most useful for individual identification are the scars from collisions with boats, especially propeller scars. Most manatees in Florida waters bear scars from boat collisions, often from more than one event. Boat scars occur over all parts of the manatee's body, but especially the dorsal surface and paddle, where notches may be cut by propellers (Fig. 7). Individual identification progressed from sketches of marks to surface and underwater 35 mm photography. Photography

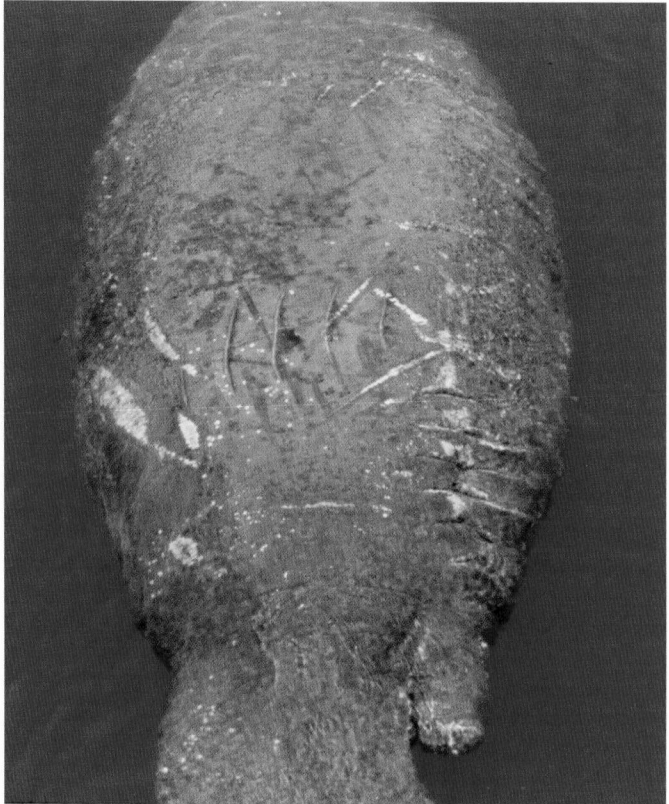

**Figure 7** *Identifying scars from boat collisions on a Florida manatee* (Trichechus manatus). *Photograph by J. K. Koelsch.*

allowed for tracking of changes in identifying characteristics through time, and for distinguishing between manatees with similar markings. Technological advances have resulted in photographic images of scar patterns being saved, cataloged, and searched with the assistance of computers.

### B. Temporary Markings

No widely accepted techniques currently exist for temporarily marking sirenians. Paint, flipper bands, and harnesses have been tested, but have been found to be ineffective (Irvine and Scott, 1984). "Paintstiks," oil-based crayon-like markers, have remained visible for 3–7 days during field tests, though rubbing eventually smears or removes them. Aerosol paint was short-lived, and application startled the animals and polluted the water.

### C. Scarring and Branding

Although not intentional, the most widely used features for identifying individual manatees are propeller scars. In recent years, scientists have begun cutting small notches in the paddles of manatees. The positions of the notches around the paddle are coded to provide information on cohorts. Freeze branding is also used with manatees that have been captured or rehabilitated on occasion, with some success (Irvine and Scott, 1984). Though most brands fade with time, some have remained readable at distances of 15 m for as long as 4 years. Success may vary with whether the manatees are shedding, as well as season, water temperature, and salinity.

## D. Attachment Tags

Lacking dorsal fins, sirenians provide few opportunities for tag attachment. As described for cetaceans, spaghetti tags have been tested with manatees (Irvine and Scott, 1984). These 20-cm-long plastic streamer tags attached to a metal dart have been applied with either a lance or a crossbow, attempting to anchor the tag about 2 cm below the skin. Spaghetti tags demonstrated poor retention, and caused abscesses on some manatees.

The most effective technique for tag attachment involves a breakaway "belt" looped around the animal's peduncle. This belt is designed to minimize chafing, break away if it should become snagged on an obstacle in the environment, and carry a floating very high frequency (VHF) or satellite-linked radio transmitter at the end of a tether. Each transmitter float is color coded to allow for individual identification visually. The tethers can be replaced by swimmers as necessary.

Passive integrated transponders, or PIT tags, have been implanted in nearly every Florida manatee that has been handled in recent years (Wright et al., 1998). These glass-encapsulated microchips are about the size of a rice grain. They are implanted subcutaneously at a depth of about 3.5 cm, dorsal and caudal to the ear, and medial to the scapula. A small incision is made, and the tag is inserted via a 12 ga. needle. Each is programmed with a unique identification code that is activated by a handheld scanner when it passes nearby. PIT tags are relatively easy to implant, last a long time, are reusable, rarely infect the animals, have an unlimited number of potential codes, and allow for easy data recording and transfer, but suffer from the fact that they must be scanned from no more than 15 cm away and the receivers are not waterproof.

## See Also the Following Articles

Behavior, Overview ■ History of Marine Mammal Research ■ Mark and Recapture

## References

Adams, J. D., Speakman, T., Zolman, E., and Schwacke, L. E. (2006). Automating image matching, cataloging, and analysis for photo-identification research. Aq. Mamm. 32, 374–384.

Altmann, J. (1974). Observational study of behavior: Sampling methods. Behaviour 49, 227–267.

Bas, W., Beekmans, P. M., Whitehead, H., Huele, R., Steiner, L., and Steenbeek, A. G. (2005). Comparison of two computer-assisted photo-identification methods applied to sperm whales (Physeter macrocephalus). Aq. Mamm. 31, 243–247.

Bigg, M. (1982). An assessment of killer whale (Orcinus orca) stocks off Vancouver Island, British Columbia. Rep. Int. Whal. Commn 32, 655–666.

Erickson, A. W., Bester, M. N., and Laws, R. M. (1993). Marking techniques. In "Antarctic Seals" (R. M. Laws, ed.), pp. 89–118. Cambridge University Press, Cambridge.

Evans, W. E., Hall, J. D., Irvine, A. B., and Leatherwood, J. S. (1972). Methods for tagging small cetaceans. Fish. Bull. 70, 61.

Hammond, P. S., Mizroch, S. A., and Donovan, G. P. (1990). Report of the workshop on individual recognition and the estimation of cetacean population parameters. In "Individual Recognition of Cetaceans: Use of Photo-Identification and Other Techniques to Estimate Population Parameters" (P. S. Hammond, S. A. Mizroch, and G. P. Donovan, eds), pp. 3–17. Report of the International Whaling Commission, Cambridge, Special Issue 12.

Harting, A., Baker, J., and Becker, B. (2004). Non-metrical digital photo-identification system for the Hawaiian monk seal. Mar. Mamm. Sci. 20, 886–895.

Hillman, G. R., et al. (8 authors) (2003). Computer-assisted photo-identification of individual marine vertebrates: A multispecies system. Aq. Mamm. 29, 117–123.

Hiby, L., and Lovell, P. (1990). Computer aided matching of natural markings: A prototype system for grey seals. In "Individual Recognition of Cetaceans: Use of Photo-Identification and Other Techniques to Estimate Population Parameters" (P. S. Hammond, S. A. Mizroch, and G. P. Donovan, eds), pp. 57–61. Report of the International Whaling Commission, Cambridge, Special Issue 12.

Irvine, A. B., and Scott, M. D. (1984). Development and use of marking techniques to study manatees in Florida. Fla Sci. 47, 12–26.

Irvine, A. B., Wells, R. S., and Scott, M. D. (1982). An evaluation of techniques for tagging small odontocete cetaceans. Fish. Bull. 80, 135–143.

Katona, S., Baxter, B., Brazer, O., Kraus, S., Perkins, J., and Whitehead, H. (1979). Identification of humpback whales from fluke photographs. In "Behavior of Marine Mammals—Current Perspectives in Research" (H. E. Winn, and B. Olla, eds), pp. 33–44. Plenum Press, New York.

Markowitz, T. M., Harlin, A. D., and Würsig, B. (2003). Digital photography improves efficiency of individual dolphin identification. Mar. Mamm. Sci. 19, 217–223.

Norris, K. S., and Pryor, K. W. (1970). A tagging method for small cetaceans. J. Mammal. 51, 609–610.

Payne, R., Brazier, O., Dorsey, E. M., Perkins, J. S., Rowntree, V. J., and Titus, A. (1983). External features in southern right whales (Eubalaena australis) and their use in identifying individuals. In "Communication and Behavior of Whales" (R. Payne, ed.), pp. 371–445. Westview Press, Boulder.

Scott, M. D., Wells, R. S., Irvine, A. B., and Mate, B. R. (1990). Tagging and marking studies on small cetaceans. In "The Bottlenose Dolphin" (S. Leatherwood, and R. R. Reeves, eds), pp. 489–514. Academic Press, San Diego.

Smith, T. D., et al. (12 authors) (1999). An ocean-basin-wide mark–recapture study of the North Atlantic humpback whale (Megaptera novaeangliae). Mar. Mamm. Sci. 15, 10–32.

Wells, R. S., and Scott, M. D. (1990). Estimating bottlenose dolphin population parameters from individual identification and capture-release techniques. In "Individual Recognition of Cetaceans: Use of Photo-Identification and Other Techniques to Estimate Population Parameters" (P. S. Hammond, S. A. Mizroch, and G. P. Donovan, eds), pp. 407–415. Report of the International Whaling Commission, Cambridge, Special Issue 12.

Wells, R. S., Boness, D. J., and Rathbun, G. B. (1999). Behavior. In "Biology of Marine Mammals" (J. E. Reynolds, III, and S. A. Rommel, eds), pp. 324–422. Smithsonian Institution Press, Washington, DC.

Wilson, B., et al. (16 authors) (1999). Epidermal diseases in bottlenose dolphins: Impacts of natural and anthropogenic factors. Proc. R. Soc. Lond., B, Biol. Sci. 266, 1077–1083.

Wright, I. E., Wright, S. D., and Sweat, J. M. (1998). Use of passive integrated transponder (PIT) tags to identify manatees (Trichechus manatus latirostris). Mar. Mamm. Sci. 14, 641–645.

Würsig, B., and Jefferson, T. A. (1990). Methods of photo-identification for small cetaceans. In "Individual Recognition of Cetaceans: Use of Photo-Identification and other Techniques to Estimate Population Parameters" (P. S. Hammond, S. A. Mizroch, and G. P. Donovan, eds), pp. 43–55. Report of the International Whaling Commission, Cambridge, Special Issue 12.

Würsig, B., and Würsig, M. (1977). The photographic determination of group size, composition, and stability of coastal porpoises (Tursiops truncatus). Science 198, 755–756.

Yochem, P. K., Stewart, B. S., Mina, M., Zorin, A., Sadavov, V., and Tablakov, A. (1990). Non-metrical analyses of pelage patterns in demographic studies of harbor seals. In "Individual Recognition of Cetaceans: Use of Photo-Identification and Other Techniques to Estimate Population Parameters" (P. S. Hammond, S. A. Mizroch, and G. P. Donovan, eds), pp. 87–90. Report of the International Whaling Commission, Cambridge, Special Issue 12.

I

# Indo-Pacific Beaked Whale
## *Indopacetus pacificus*

ROBERT PITMAN

## I. Characteristics and Taxonomy

The Indo-Pacific beaked whale, also known as Longman's beaked whale or tropical bottlenose whale, is an uncommon tropical ziphiid that was until very recently one of the least known extant cetaceans. Originally described as *Mesoplodon pacificus* from a beach-worn skull collected in Queensland, Australia in 1882 (Longman, 1926), the validity of the species was initially challenged by researchers who variously suggested that it was a subspecies of True's beaked whale (*Mesoplodon mirus*) or an adult female southern bottlenose whale (*Hyperoodon planifrons*). These allegations were refuted and the validity of the species confirmed by the discovery of a second skull from the coast of Somalia in 1955 (Azzaroli, 1968). After further study, Moore (1968) found it sufficiently distinct to warrant establishing a new genus *Indopacetus*. These two skulls were the only evidence that this species existed until very recently when a series of at-sea sightings of an unidentified beaked whale from the tropical Indian and Pacific oceans were compiled and more closely analyzed. Some of these sightings had previously been tentatively identified as of southern bottlenose whales (*Hyperoodon planifrons*), but color pattern differences ruled out that species and the suggestion was made that it could be the long lost *I. pacificus* (Pitman *et al.*, 1999). This identification was subsequently confirmed by genetically matching stranded animals (that had previously been identified as *H. planifrons*) with the holotype of *I. pacificus* in the Queensland Museum (Dalebout *et al.*, 2003). This has led to the identification of dozens of at-sea sightings, and *I. pacificus* has now become one of the more frequently identified beaked whales.

The Indo-Pacific beaked whale is identified in the field as a relatively large ziphiid with a prominent melon sharply set off from a fairly long beak, a prominent dorsal fin, and a subtle but distinctive color pattern (Fig. 1). The calf is dark gray with a conspicuously pale head. The light color of the melon extends only as far back as the blowhole (this is important because in the otherwise similar-looking *Hyperoodon* spp., the paleness on the melon extends 10 cm or so posterior to the blowhole). Much of the face, lower jaw and throat are also pale. Immediately posterior to the blowhole the dark gray dorsal coloration extends ventrally to form a dark patch around the eye; it also extends ventrally and backward to the insertion of the flipper, forming a broad band. There is a small white "ear spot" embedded in the dark area behind the eye. Immediately posterior to the dark flipper band, and setting it off, is a large white patch formed by the white from the ventral area extending high up on the sides of the animal. This white thoracic coloration apparently darkens with age because it has only been seen on calves.

Adults are similar to calves except that the beak is longer and the color pattern changes somewhat (Fig. 2). Adult females appear brown in good light and grayer with lower light levels. Adult males are similarly colored but often appear lighter than females due, at least in part, to an accumulation of scars from tooth rake marks by other males. Although females have very few of the linear scars found on adult males, both sexes often have numerous white, oval scars that appear to be mainly healed bites from cookie cutter sharks (*Isistius* sp.; Fig. 2). As in calves, the melon of adults is often a pale tan color; this paleness is not evident on all individuals, suggesting

**Figure 1**  Indopacetus pacificus *near the Maldives Islands in the central Indian Ocean showing the pale prominent melon, distinct beak, and conspicuous blow. Photograph by R. C. Anderson.*

**Figure 2**  *An apparently emaciated 6-m female* Indopacetus pacificus *that stranded in Kagaoshima, Japan in 2002; the white spots are bite scars from cookie cutter sharks (*Isistius* sp.). Photograph by Kagoshima City Aquarium, courtesy of T. Yamada.*

that it may become obscured with age. This species is large enough that it produces a clearly visible, low, bushy blow, which is usually angled slightly forward. As in all ziphiids with apical dentition, the gape in both sexes is straight throughout most of its length but turned up slightly at the posterior end. A dip behind the melon (seen in profile) is confluent with the blowhole. The dorsal fin is set approximately two-thirds of the way back on the upper body and it is relatively large, perhaps larger than in any other species of beaked whale. Similar to other ziphiids, the pectoral flippers are small and can fit into depressions that make them flush with the body. *I. pacificus* is the only beaked whale known to exhibit polydactyly.

Adult length measurements are available from only three animals: (1) a pregnant female that stranded in the Maldives in January 2000 was 6 m long (curvilinear measurement) and (2) a physically mature female that stranded in Kagoshima, Japan in July 2002 was 6.5-m long, which may be a maximum for this species, and (3) a 5–65-m female (and her 4.20-m calf) stranded in Taiwan in 2005. A neonate that stranded in South Africa was 2.9 m long.

The skull of the adult (pregnant) female that stranded in the Maldives measured 123 cm. This species has a single pair of teeth set

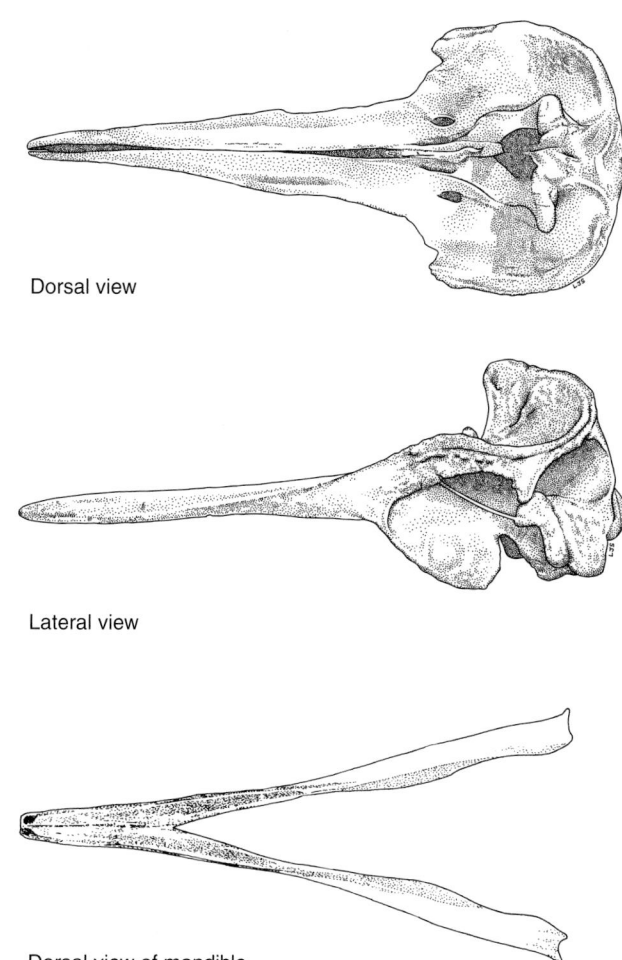

Dorsal view

Lateral view

Dorsal view of mandible

**Figure 3** *Skull of* Indopacetus pacificus. *Courtesy of the Food and Agriculture Organization of the United Nations; from Jefferson* et al. *(1993), Marine Mammals of the World, FAO, Rome.*

near the tip of the lower jaw, which presumably erupt only in adult males (as they do in most species of ziphiids), but no adult male is known to have been examined. A 5.73-m immature male that stranded in the philippines in 2004 had teeth that were just starting to erupt. The teeth are conical in shape and oval in cross section, set in relatively shallow alveoli (sockets), and likely project slightly forward in adult males. In addition to these tooth characteristics, the skull of *I. pacificus* (Fig. 3) is distinguishable from those of *Mesoplodon* and the other living genera of beaked whales by the following features (1) frontal bones occupy an area of the vertex equal to or greater than that of the nasal bones; (2) premaxillary crest with extremely short posterior processes; (3) a deep horizontal groove in the maxillary bone just above the orbit; and (4) about mid-length in the beak, there is a swelling of the lateral margins so that the beak does not grow narrower throughout its entire length (Moore, 1968).

Historically, the validity of the genus *Indopacetus* as distinct from *Mesoplodon* has not been universally accepted. A recent molecular genetics analysis found support for species level differences in *I. pacificus*, but the available samples were too degraded to resolve validity of the genus (Dalebout *et al.*, 2003). The same study, however, was able to verify the morphological characters that Moore (1968) used to diagnose the genus and also identified two other possible diagnostic characters: rib count and number of fused cervical

vertebrae. The more recent study concluded that the genus should be retained pending further evidence to the contrary.

## II. Distribution and Abundance

The Indo-Pacific beaked whale appears to be uncommon or rare throughout much of its range. Before 2003, there were no confirmed live sightings in the wild, and for over 75 years its distribution was inferred from only two skulls collected in east Australia and Somalia, respectively. Since then however, there have been at least 8 new stranding records (including two previously misidentified as *Hyperoodon planifrons*), and at least 65 at-sea sightings. Based on this, *I. pacificus* is now known to inhabit tropical waters throughout the Indo-Pacific, from the west coast of Mexico to the east coast of Africa and the Gulf of Aden (Fig. 4). It is rare in the eastern Pacific but appears to be more common in the western Pacific and is also at least fairly common in the western tropical Indian Ocean, suggesting an affinity for western ocean basins. The only population estimates to date were based on systematic surveys in Hawaiian (EEZ) waters (1007 animals) and the eastern tropical Pacific (291 animals; Barlow *et al.*, 2006).

## III. Ecology

Practically nothing is known about the ecology of *I. pacificus*. Beaked whales are in general mainly squid-eaters and *I. pacificus* appears to be no exception. Stomach contents are known from only two specimens. A stranded animal from Davao, Philippines had only squid beaks in its stomach. Another stranded specimen, from Kagoshima, Japan, contained squid beaks, plastic bits, and nematodes in its stomach; there were no fish remains. Of a total of 69 squid beaks identified, 83% were *Taonius pavo*; other species present included *Moroteuthis loennbergi*, *Onychoteuthis borealijaponica*, *Chiroteuthis imperator*, and *Histioteuthis corona inermis*.

Most sightings have been in deep water (>2000 m), where the sea surface temperatures were ≥26°C. Although a fair number of sightings have been recorded along continental slope areas (200–2000 m), this may reflect a bias for surveys to be more nearshore. It has usually been observed in monospecific groups, but among the 65 sightings recorded to date it has also been associated with short-finned pilot whales (*Globicephala macrorhynchus*; five times), pilot whales and common bottlenose dolphins (*Tursiops truncatus*; two times), and only bottlenose dolphins (one time). Photographs of both stranded and live specimens often show numerous white oval scars, which are probably the healed wounds of cookiecutter shark bites (*Isistius* spp.): the Kagoshima stranding had hundreds of scars giving it a spotted appearance (Fig. 2).

## IV. Behavior and Physiology

Group size tends to be larger than in other beaked whale species except perhaps *Berardius* spp. Evidence to date also suggests there may be regional differences in group size. In the Pacific, mean group size was 18.5 individuals (range 1–100). In the western vs the eastern Pacific, it was 29.2 and 8.6, respectively (Pitman *et al.*, 1999). In the western Indian Ocean, group size averaged 7.2 individuals (range 1–40; Anderson *et al.*, 2006). *I. pacificus* tends to travel in close groups, often with adult males, adult females, and calves present. When traveling fast at the surface, animals bring their head and beak high out of the water, or sometimes porpoise low like large dolphins. Diving and surfacing is largely synchronous within the group. Dive times recorded to date have ranged from 11 to 33 min; one animal was suspected of diving for at least 45 min, and it is probable that maximum dive times may be considerably longer.

I

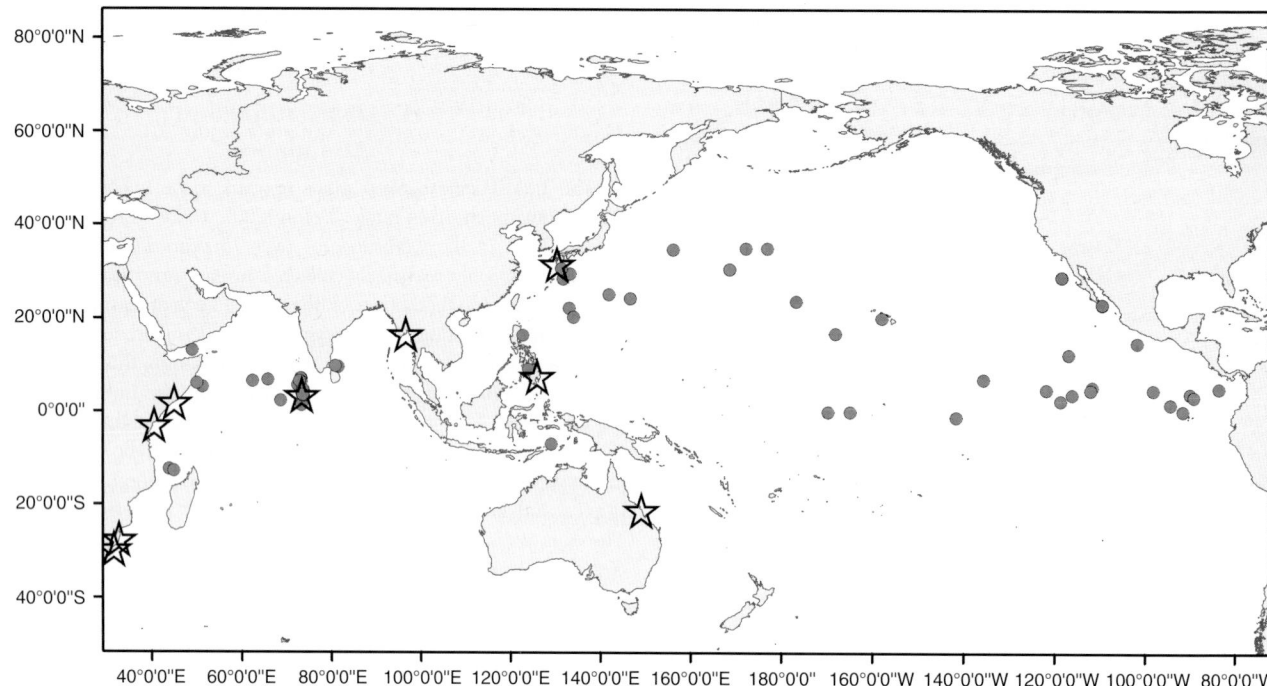

**Figure 4** *Records of* Indopacetus pacificus *from the tropical Indian and Pacific oceans: circles, sightings; stars, strandings. Data from Pitman* et al., *1999; Dalebout* et al., *2003; Anderson* et al., *2006; Watson* et al. *(2008); NOAA Fisheries/SWFSC unpublished data; additional stranding information from J. Acebes (Philippines), T. Yamada (Japan) and T. Tun (Myanmar).*

## V. Life History

Little is known of the life history of this whale: a 6-m female was pregnant; another 5.65-m female with a 4.2-m calf was lactating and had 2 corpora albicantia in the left ovary (Watson *et al.*, 2008).

## VI. Interactions with Humans

Juvenile bottlenose whales that were probably of this species have been caught by gillnet fishermen from Sri Lanka, and the thousands of pelagic gillnet vessels that currently operate across the tropical Indian Ocean are an unknown but potentially significant threat to all whale populations there (Anderson *et al.*, 2006). Unidentified whales that were possibly of this species have also been taken on longline gear in Hawaiian waters, indicating a potential susceptibility to this type of fishing also. Beaked whales in general are known to be particularly vulnerable to loud anthropogenic sounds in the ocean, including active sonar and seismic surveys. An apparently healthy mother and calf *I. pacificus* stranded in Taiwan in 2005 along with numerous other cetaceams perhaps as a result of naval exercises that had just off the coast there. A sub-adult male that stranded in the Philippines reportedly had pieces of plastic bags in its stomach.

## *See Also the Following Articles*

Beaked Whales ■ Mesoplodont Whales

## *References*

Anderson, R. C., Clark, R., Madsen, P. T., Johnson, C., Kiszka, J., and Breysse, O. (2006). Observations of Longman's beaked whale (*Indopacetus pacificus*) in the western Indian Ocean. *Aq. Mamm.* **32**, 223–231.

Azzaroli, M. L. (1968). Second specimen of *Mesoplodon pacificus*, the rarest living beaked whale. *Monitore Zool. Ital. (N. S.)* **2**, 67–79.

Barlow, J., *et al.* (2006). Abundance and densities of beaked and bottlenose whales (family Ziphiidae). *J. Cetacean Res. Manag.* **7**, 263–270.

Dalebout, M. L., *et al.* (2003). Appearance, distribution, and genetic distinctiveness of Longman's beaked whale, *Indopacetus pacificus*. *Mar. Mamm. Sci.* **19**, 421–461.

Jefferson, T.A., Leatherwood, S. and Webber, M.A. (1993). "Marine Mammals of the world." FAO, Rome.

Longman, H. A. (1926). New records of Cetacea, with a list of Queensland species. *Mem. Queensl. Mus.* **8**, 266–278.

Moore, J. C. (1968). Relationships among the living genera of beaked whales with classifications, diagnoses, and keys. *Field. Zool.* **53**, 206–298.

Pitman, R. L., Palacios, D. M., Brennan, P. L. R., Brennan, B. J., Balcomb, K. C., III, and Miyashita, T. (1999). Sightings and possible identity of a bottlenose whale in the tropical Indo-Pacific: *Indopacetus pacificus*? *Mar. Mamm. Sci.* **15**, 531–549.

Watson, A., Kuo, T.-F., Yang, W.-C, Yao, C.-J., and Chou, L.-S. (2008). Distinctive ostcology of distal flipper bones of tropical bottlenose whales. *Indopacetus pacificus* from Taiwan: mother and calf with polydactyly. *Mar. Mamm. Sci.* **24**, 398–410.

# Indo-Pacific Bottlenose Dolphin

## *Tursiops aduncus*

### JOHN Y. WANG AND SHIH CHU YANG

## I. Characteristics and Taxonomy

This dolphin was initially named *Delphinus aduncus. Tursiops* is the combination of the Greek words "Tursio" for dolphin and "ops" for appearance; *aduncus* is from the Latin word

I

meaning hooked (possibly referring to the lower jaw being slightly upturned distally).

## A. Taxonomy

The taxonomic status of *Tursiops aduncus* was uncertain until about 2000 when it gained wide acceptance after studies demonstrated two sympatric forms of *Tursiops* were reproductively isolated, the evidence being the maintenance of congruent and consistent molecular and morphological differences (Wang *et al.*, 1999, 2000a, b). Subsequent studies have also supported the distinct species status of *T. aduncus*. Still, considerable taxonomic uncertainties exist within the genus *Tursiops*, even amongst dolphins that appear to be *T. aduncus*. For example, the dolphins of Shark Bay (Western Australia) are one of the best-known populations of bottlenose dolphins, but their species identity still stymies researchers (note: in this review, the Shark Bay dolphins are considered *T. aduncus*). Although polytypy of *T. aduncus* has been suggested (Natoli *et al.*, 2004), this view is not widely accepted at present. Furthermore, mtDNA analyses have been interpreted as evidence that *T. aduncus* is most closely related to *Stenella frontalis* (LeDuc *et al.*, 1999), which is inconsistent with osteological characters. The osteology of *T. aduncus* (discussed later) is very different from that of *Stenella* spp. and closely resembles *T. truncatus*, to the point that distinguishing between the two species is difficult and was one of the reasons for the long-standing uncertainty about the taxonomic status of *T. aduncus*. Well-designed studies using multiple independent characters are needed to resolve some of these issues.

Other factors contributing to the continuing taxonomic confusion include the wide distribution of *Tursiops* across highly variable environments, thus resulting in great variation exhibited by many locally adapted populations of two (and possibly more) species; the sympatry of the two species in many regions; the lack of *Tursiops* specimens from many regions; and the differences in research methods and designs.

Due to the long-standing taxonomic uncertainties within *Tursiops*, many studies often made no distinction between *T. aduncus* and *T. truncatus* specimens in their analyses. To prevent further confusion with the species' biology, this review only includes information from studies in which *T. aduncus* has been identified and was treated separately (especially in areas where it is sympatric with *T. truncatus*) or in which bottlenose dolphins appear likely to have been *T. aduncus* even though the taxonomy remains to be confirmed. Only when the taxonomy of *Tursiops* becomes clearer worldwide, will our knowledge of *T. aduncus* improve.

## B. External Features

This species appears to be highly plastic and inhabits a variety of coastal habitats through a wide distribution. Nevertheless, *T. aduncus* is generally smaller than *T. truncatus*, reaching a maximum total length of about 2.7 m and about 200 kg (in eastern Asian waters) in weight; the dolphins in some populations do not grow longer than about 2.5 m. In some areas males appear to be slightly larger, sexual dimorphism in the species does not appear to be obvious or consistent. Compared with *T. truncatus*, the appendages (dorsal fin, flipper, and flukes) of *T. aduncus* are generally larger and broader relative to body size, but overall the body appears more slender: the snout is longer and thinner, the melon is less bulbous, and the head has a more pointed profile (Fig. 1). The eye region also appears to bulge out laterally when viewed from above. The length of the rostrum as a proportion of the total body length or relative to the distance from the tip of the rostrum to the middle of the eye is greater than for

**Figure 1** *Indo-Pacific dolphins in the Red Sea. Photograph copyrighted by Beno Steinacher.*

*T. truncatus* (Wang *et al.*, 2000a). In Chinese waters, the rostrum length of *T. aduncus* (excluding young calves) is longer than in *T. truncatus* regardless of body length (Wang *et al.*, 2000a). The overall appearance of dolphins from the northern Philippines, Solomon Islands, Taiwan, and Japan is similar, whereas dolphins from Indonesian, Western Australian, and South African waters look "stubbier" and possess shorter beaks and smaller bodies.

As with *T. truncatus*, the pigmentation of *T. aduncus* is fairly simple, with a dark to medium gray dorsal surface (often appearing as a cape) progressing to light gray on the flanks. The belly is whitish, often with a pinkish hue. A light spinal blaze may be present on some animals. In most regions, *T. aduncus* possess dark spotting on the light ventral half of the body, but some individuals may possess a few spots dorsally as well. The intensity and specific body locations of the ventral spotting appear to be regionally and individually variable, but the general development of spotting is similar. Spotting begins around the onset of sexual maturity and increases in intensity with age (the spots also darken with age). Spotting can be very intense on the oldest animals. The mouth line and tip of the beak also become whiter with age. Calves are generally slightly lighter in overall color and are unspotted. Spotting patterns may be used in individual recognition, to roughly indicate the stage of maturity and also in defining population differences. Dolphins of some populations begin to become spotted as short as 1.6 m long and become intensely spotted at less than 2.2 m, whereas those in other populations may not start to develop spots until at least 2.2 m long. Dolphins of some populations are apparently more or less unspotted (e.g., Jervis Bay and Port Stephens, New South Wales, and southeast Australia).

## C. Osteology

There are only a few, subtle distinguishing features between the skulls of *T. aduncus* and *T. truncatus* (Ross, 1977, 1984; Wang *et al.*, 2000b). The skull of *T. aduncus* is relatively smaller and possesses a narrower rostrum with a prominent premaxillary convexity (when viewed laterally) or "pinch" (when viewed dorsally) that is situated at about one-third of the rostral length anterior of the base of the rostrum (this feature of the rostrum is the attachment point for connective tissues associated with the melon and defines the apex of the melon). The position of the premaxillary convexity is also reflected externally by a longer beak compared with that of *T. truncatus*. The width of the external nares (relative to the parietal width) is also greater in *T. aduncus* than in *T. truncatus*. Teeth of *T. aduncus* are

I

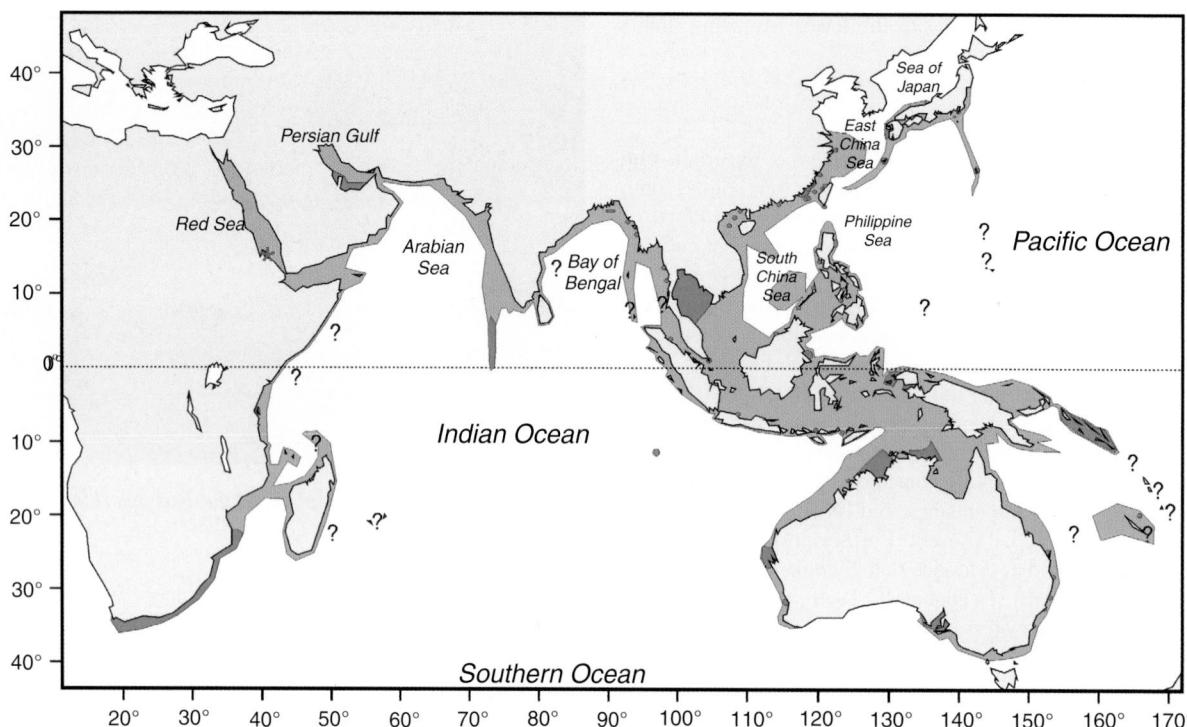

**Figure 2** *Distribution of the Indo-Pacific bottlenose dolphin. Confirmed records are shown in blue and hypothesized distribution in olive. Question marks indicate uncertain records, and red star represents the type locality.*

generally smaller in diameter and more numerous than in *T. truncatus* (in Chinese and South African waters, *T. aduncus* has 23–29 teeth in each row with a total of 97–111 teeth, whereas *T. truncatus* possesses 19–27 teeth per row with a total of 80–106).

The number of vertebrae is fewer than in *T. truncatus*, and although the distributions differ only slightly they do not appear to overlap. The total vertebral count for *T. aduncus* is 59–62, whereas it is 64–67 for *T. truncatus*. There may also be slight differences in the morphology of the cervical vertebrae.

### D. DNA

The mtDNA control regions of *T. aduncus* and *T. truncatus* differ by several fixed nucleotide bases and are highly divergent (at least 4.4%) from each other (Wang *et al.*, 1999). Intraspecific analysis suggested an isolation-by-distance model for the *T. aduncus* of Shark Bay, with females tending to be more philopatric than males (Krützen *et al.*, 2004).

### II. Distribution and Abundance

Because the taxonomic status of *T. aduncus* was only recently accepted and distinguishing this species from *T. truncatus* can be difficult, information about the distribution of *T. aduncus* is patchy, but it clearly indicates that the species is widespread. Confirmed presence of the species in some regions is a reflection of research effort, whereas gaps reflect either absence or lack of cetological studies. It is likely distributed throughout the coastal waters of the Indian and western Pacific oceans from the Solomon Islands and New Caledonia in the east to the southern tip of South Africa in the west and from central Japan to southeastern Australia. However, the level of continuity in the distribution is unknown (Fig. 2). There are

several small apparently isolated, resident populations around a few oceanic islands of Japan and elsewhere, and it is likely that dolphins will be found around more offshore islands throughout the range.

There is no global abundance estimate for the species, and local abundance estimates are relatively few (e.g., Yang *et al.*, 1997; Yang *et al.*, 2000; Shirakihara *et al.*, 2002; Chilvers and Corkeron, 2003; Shirakihara *et al.*, 2003; Kogi *et al.*, 2004; Stensland *et al.*, 2006). It can be the most commonly recorded cetacean in some coastal areas (e.g., southeast coast of South Africa—Ross, 1977, 1984, Arabian Gulf—Preen, 2004) partly because of its near-shore distribution. Because population structure and boundaries are poorly understood in most cases, local estimates must be interpreted cautiously. In Japanese waters, there are an estimated 218 dolphins and at least 160 individuals identified for the Amakusa-Shimoshima and Mikura Island populations, respectively. In Australian waters, there are local estimates for the populations of Shark Bay (>600 dolphins in an area of about 300 km²), Point Lookout (700–1000 dolphins within 150 km²), and Moreton Bay (334 dolphins). In east Africa, a resident population in the waters of Zanzibar (Tanzania) was estimated to contain between 136 and 179 dolphins within a study area of 26 km². There are no abundance estimates for the waters of Taiwan or the Philippines. However, from photo-identification studies, there were at least 24 dolphins within a 110 km² area of southern Taiwan and 44 dolphins in the inshore waters (~200 km²) of the Northern Sierra Madres Natural Park (northeast Philippines). Small populations also appear to exist in the outer part of Malampaya Sound and in and around Tañon Strait in the Philippines. In a small part of the western Taiwan Strait between Xiamen and Dongshan (China), the density was estimated to be less than 5 dolphins per 100 km².

There is little information on population structure. Based on the available information, the species appears to be composed of many relatively small localized populations that are fairly isolated from each other. Dolphins of different regions appear to exhibit strong year-round residency and natal philopatry in both sexes with males being more dispersive than females. There appear to be clear regional differences in the size of the dolphins. The dolphins in the western North Pacific appear to be larger than those of other regions. Spotting intensity also seems to be regionally variable as well (discussed earlier). The great individual and regional variations in size and spotting patterns may be useful for understanding population structure and require more study. Based on cranial and spotting differences, dolphins of Natal and the eastern Cape of South Africa were considered to be of different populations that were year-round residents of those areas (Ross, 1977, 1984). In Japanese waters, dolphin whistles appear to differ amongst populations (Morisaka *et al.*, 2005).

There are indications that this species (like other localized coastal cetaceans) has experienced dramatic declines in the Arabian Gulf and in Vietnamese and Chinese waters. Even in Shark Bay, which is still relatively pristine, a decline in the dolphin population due to activities of tour operators targeting the dolphins has been reported (Bejder *et al.*, 2006). The latter observation further emphasizes the particular vulnerability of this near-shore species to human threats.

## III. Ecology

The species appears to prefer continental shelf waters near shore and in areas with rocky and coral reefs, sandy bottom, or sea grass beds. The dolphins can be found in waters more than 200 m deep but are much more common in water less than 100 m deep. In some areas, estuaries may concentrate dolphins, but they do not seem to frequent muddy estuarine waters. The inshore waters of at least a few small oceanic islands are also home to small populations of dolphins. Although generally considered a coastal species, movement across deep oceanic waters has been reported. However, it is unknown if this behavior is common. The main distribution of the species is in tropical to warm temperate waters of the Indian and western Pacific oceans, but some have been found in cooler waters in Japan, northern China, southern Australia, and South Africa. Sea surface temperatures where *T. aduncus* are found are between 20°C and 30°C, but water temperature can vary greatly by region. The lowest water temperature reported for *T. aduncus* was 12°C in the waters of Amakusa-Shimoshima, Japan.

*Tursiops aduncus* has broadly overlapping distributions with many species but mainly *T. truncatus Sousa chinensis, Neophocaena phocaenoides, Stenella longirostris, S. attenuata,* and *Orcaella* spp. Areas in which sympatry with *T. truncatus* occurs include the waters of central and southern China, western Taiwan, the southern half of Japan, Philippines, Australia, South Africa, and likely other regions. In these areas, *T. aduncus* seems to occupy more coastal waters than *T. truncatus*. In areas where they are sympatric with *S. chinensis*, the distribution of *T. aduncus* is comparatively more offshore.

There are reports of mixed schools with *T. truncatus, Pseudorca crassidens, Delphinus capensis, Sousa chinensis, Delphinus* sp., and *Stenella longirostris*. The interspecific associations may differ regionally (e.g., no schooling with *Sousa chinensis* has been observed in western Taiwan where both species exist).

There is great geographical variability in the species' diet. Throughout most of the distribution of *T. aduncus*, the primary prey species are benthic and reef-dwelling fish and cephalopods of continental shelf waters (Ross, 1977, 1984) but some pelagic and epipelagic species are also consumed. Its prey is usually less than about 300 mm

long and belongs to several families such as Belonidae, Mugilidae, Sciaenidae, Engraulidae, Sepioteuthidae, Sepiidae, Sepiolidae, Loliginidae, and Octopodidae. For dolphins inhabiting waters of oceanic islands, a considerable part of their diet appears to be epi- and mesopelagic fish and cephalopods (Kakuda *et al.*, 2002), but benthic crustaceans can contribute substantially. On the odd occasion, small benthic sharks have also been recorded as prey. At least in some areas, there appears to be little overlap in prey species between sympatric populations of *T. aduncus* and *T. truncatus*. From studies of captive specimens, the daily food required by adults appears to be about 4–5% of their body mass but can increase with lower water temperatures (Ross, 1984).

Little is known about predators in most regions. Sharks are a main cause of mortality for some populations, with the main predators being tiger shark (*Galeocerdo cuvieri*), white shark (*Carcharodon carcharias*), bull shark (*Carcharhinus leucus*), and dusky shark (*Carcharhinus obscurus*). For some populations, shark-bite frequency is very high (e.g., >74% of the non-calf dolphins of Shark Bay bear scars from shark attacks—Heithaus, 2001). Shark-bite scars are much less frequent in Moreton Bay (eastern Australia) and South African waters, at about 37% and 10–20%, respectively (however, it is uncertain if these data include both *T. aduncus* and *T. truncatus*). There are no records of killer whale predation. In addition to predation, dolphins have died as a result of accidental injuries caused by needlefish and the spines of sting rays.

Compared with other species, relatively few parasites and other pathogens have been recorded (Kakuda *et al.*, 2002). However, this is almost certainly due to a lack of research effort. Ectoparasites reported include the cirripede *Xenobalanus globicipitis* and cyamids. There are also some indications that small sharks may behave as ectoparasites by taking small bites out of dolphins living around oceanic islands (e.g., Mikura Island); the dolphins possess scars that have been attributed to "cookie cutter" sharks (*Isistius* sp.). Endoparasites include nematodes in the pterygoid sinuses (possibly *Crassicauda* sp.) and stomachs (likely *Anisakis* sp.), tapeworms in the intestines, *Halocercus lagenorhynchi* in the lungs and respiratory tract, and cestodes of the genus *Phyllobothrium* in the blubber (mainly around the urigenital region) and possibly *Monorygma* internally. Serological tests have shown that dolphins of the western Pacific Ocean are likely exposed to the protozoan *Toxoplasma gondii* (Omata *et al.*, 2005) and dolphin morbillivirus (Van Bressem *et al.*, 2001).

## IV. Behavior and Physiology

Acoustically, *T. aduncus* appears to be similar to *T. truncatus*. However, regional differences in whistle characteristics amongst Japanese populations appear to exist (Morisaka *et al.*, 2005).

Group size tends to be small in most places the species has been studied (note: group size is highly dependent upon how researchers define "group"). In South Africa, up to 2000 individuals were observed on one occasion, but this large aggregation was composed of many smaller groups. The most common group size appears to be between 20 and 50 individuals. In Plettenberg Bay, the mean group size was reported as 140 individuals, but again this was composed of several smaller groups (Ross, 1984). In Japanese waters, more than 100 are commonly seen together (Shirakihara *et al.*, 2002), whereas in Taiwan no group greater than about 35 individuals has been recorded in recent times. Even though past drive-hunting may have caught hundreds in single operations in the Penghu Islands, the last drive in 1993 captured only about 20 individuals; this may be an indication of local depletion after decades of unregulated hunting.

A fission–fusion society best describes the social organization of the species; there are many parallels with that of *T. truncatus* as well as some primates (Connor *et al.*, 2000b). There is great individual variation in social strategies that is dependent upon social context. Males form cooperative alliances (usually as two or three individuals) to challenge other similar alliances for access to females and to help herd them. Females form coalitions, possibly to reduce shark predation, help rear calves, and thwart male coercion. In some regions, associations appear to have correlations with kinship. The mother–calf bond is strong. In the first week after birth, neonates have a strong following response, and their mothers are strongly possessive of their calves (this may be a period of imprinting). Group size tends to be larger when young calves are present, and allomaternal care can be given by other females. Such behavior may reduce shark attacks on small calves as well as their mothers. Mothers have been observed to stay with and protect their deceased calves for extended periods. Social dominance has not been studied directly in wild populations but likely exists (based on captive and some field observations).

In places where the species has been studied extensively, the dolphins appear to exhibit strong year-round residency in fairly limited coastal areas (e.g., around Mikura Island and Shark Bay). Some seasonal movements may occur, but this is uncertain. Based on a few individuals that were marked and followed, minimum home ranges covered more than 200 km distance. At least in some regions, they are philopatric to their natal sites and as with most mammals, males tend to range more widely than females. The normal swimming speed is between 1.5 and 4 km/h, but high-speed bursts can reach at least 16–19 km/h. Maximum dive depth is unknown, but based on their distribution, preferred prey species and typically short submergence times (<5–10 min), dives are mostly, if not always, shallower than 200 m.

A wide variety of foraging techniques is employed depending on prey species. Some techniques are unique and are likely to have been transmitted through learning or may even have been a genetic component. Foraging methods used by dolphins of the Shark Bay population include the use of tail slaps to flush out prey ("kerplunking"), carrying sponges, possibly to protect the tip of the beak from rubbing against the ocean floor, and chasing prey onto beaches and beaching to catch them (Connor *et al.*, 2000a; Krützen *et al.*, 2005; Sargeant *et al.*, 2005).

## V. Life History

Demographic information is limited (Ross, 1984; Connor *et al.*, 2000b; Kogi *et al.*, 2004; Mann and Watson-Capps, 2005). For the Mikura Island population several parameters were estimated: mean annual birth rate = 0.071, mean fecundity rate = 0.239, and mean recruitment rate = 0.068. The sex ratio at birth is about equal, but there are more males at sub-adulthood and then more females at adulthood. This was interpreted to mean that females may have higher survivorship to adulthood than males, but it was uncertain why the sex ratio would be skewed toward males prior to adulthood. Mortality of first-year calves varied from 0.133 (Mikura Island) to 0.29 (Shark Bay). In both the Mikura Island and the Shark Bay populations, 44% of the calves died before weaning and reaching independence (~3-years old). Mortality was especially high for calves of primiparous females.

In a population of at least 160 dolphins, the emigration of five individuals from the Mikura Island population to adjacent waters was documented, but immigration from other regions into this population was not observed. The emigration events seem to have been permanent.

In most regions, the peak calving season is very broad with most births occurring in the months with highest water temperatures after a gestation period of about 12 months. The inter-birth interval is most commonly between 3 and 6 years but can be as low as 2–3 years for captive animals (for the Mikura Island population, the mean is 3.4 years; for the Shark Bay population, 4.1 years). Females that lose calves early can become pregnant soon afterwards and give birth in the following year, and females can be simultaneously pregnant and lactating. Nursing usually lasts 3–5 years (sometimes longer if calves are not as healthy) but can be as short as 18–20 months in captivity, where foraging for solid food requires little learning or practice. Length at birth is about 1 m and the birth weight varies from 9 to 15 kg.

Typically, age of first reproduction for females is about 12–15 years, whereas males reach sexual maturity at between 10 and 15 years (some may reach sexual maturity earlier). The length at sexual maturation varies greatly over the large geographical distribution. Although the smallest mature female reported was less than 1.9 m long, dolphins of Shark Bay more typically reach maturity at about 2.0–2.1 m, whereas South African dolphins are slightly larger with maturity beginning at about 2.1–2.4 m. The largest minimum size at sexual maturation is likely to be found in the waters of eastern Asia (the northern part of the species' distribution).

Ovulation is spontaneous and sporadic. The left ovary is larger (with more corpora) than the right and appears to begin activity earlier. Males have relatively small mature testes compared with other delphinids, but the testes are still very large relative to those of other mammals (combined testes mass about 2 kg).

The maximum age estimated for this species is about 40 years, but some preliminary aging of the teeth of some known old individuals show that they may be 50 or more years of age. Old dolphins can be difficult to age when the pulp cavity of the tooth occludes, which usually coincides with physical maturity (i.e., epiphyseal fusion of all vertebrae). Studies have shown that each dental growth layer group (GLG) is consistent with 1 year.

There is little information on growth curves. Similar to most marine mammals, rapid growth occurs in the earliest years of life. Given the large regional differences in body size, growth curves are sure to vary considerably. For South African animals (Ross, 1984), the length to mass function is $M = 8.0 \times 10^{-9} X^{3.057}$ (where $M$ = mass in kilograms, and $X$ = length in millimeters).

## VI. Interactions with Humans

In many regions, the species is taken directly (usually by some form of spear) for human consumption or bait for shark fishing operations (e.g., Philippines, Taiwan, east Africa). It is also a preferred species of the captive display industry throughout Asia, and with the recent explosion of dolphinarium facilities, particularly in China, over the last decade, there are concerns that local populations may become depleted, if not already so. The infamous drive hunt of the Penghu Islands, which ceased in the early 1990s, and the recent captures of about 100 and several tens of dolphins from the waters of the Solomon Islands in 2003 and 2007, respectively, supplied dolphins for several aquariums and tourist resorts including facilities in Europe, Mexico, and Dubai. These were both from unassessed populations and are likely to be unsustainable. Captures in Southeast Asian waters and other regions with minimal monitoring of the numbers being taken are of great concern.

Mortality due to interactions with fisheries is likely to be the most serious, immediate anthropogenic threat to the species. Bycatch in gillnet (trammel and other bottom set gillnets, driftnets, etc.) fisheries throughout their range is of great concern in many inshore areas but is poorly understood. A large number (possibly >2000 per year) of this species was taken incidentally in northern Australian waters by a Taiwanese drift gillnet fishery (Harwood and Hembree, 1987) that later moved its operation into the neighboring waters of Indonesia after strict regulations were established. The fishery continued largely unmonitored and almost certainly captured dolphins from some of the same populations that straddle political boundaries. In Chinese coastal waters alone, there are more than 3.5 million gillnets in use (not including those of Taiwan) (Zhou and Wang, 1994), and although the level of cetacean mortality is poorly known, the impact of such a large fleet on this coastal species is of concern.

In the South China Sea and parts of eastern Australia, *T. aduncus* has been observed feeding behind bottom trawl nets or on trawl net discards. This behavior increases the risk of incidental capture and in some cases may also expose the animals to higher rates of shark attacks as the dolphins are focused on the trawls, which also attract sharks.

Increasingly, the habitat of coastal cetaceans is being occupied by expansive aquaculture operations (shellfish, finfish, etc.), and dolphins are being displaced physically from parts of their range by the structures (stakes, pens, concrete walls, fences, etc.) associated with this industry (Mann and Watson-Capps, 2005). The impact on local populations being excluded from parts of their habitat (in Australia, western Taiwan, and eastern China) is direct, immediate, and may be very serious.

Heavy industrialization and coastal development in many parts of Asia (e.g., Taiwan and China) with little to no consideration for coastal cetacean species has resulted in severe physical habitat degradation for this and other coastal species. Coastal destruction is still continuing, more or less unabated, with more large-scale development projects being proposed and approved. Along with industrialization in coastal areas comes the associated pollution and degradation of coastal waters. Due to the rapid and massive industrialization of both sides of the Taiwan Strait, pollution is suspected to be especially serious in this region.

Even human activities that are generally considered to have low-level impacts on cetaceans can affect this particularly vulnerable near-shore species. The species is the target of many dolphin-watch tours, and in at least one case a relatively low level of tourism pressure appears to have caused a decline in the dolphin abundance (in Shark Bay—Bejder *et al.*, 2006). Commercial dolphin-swim (e.g., Mikura Island and Zanzibar) and dolphin-feeding tours (Shark Bay) are also likely to have an impact on natural dolphin behavior (Stensland and Berggren, 2007). The feeding of wild dolphins by tourists appears to increase calf mortality and decrease the reproductive success of females in Shark Bay (Mann *et al.*, 2000). Also, noise from dolphin-watch tours, pleasure boating, and commercial vessels can alter the swimming and acoustic behavior of dolphins (Lemon *et al.*, 2006). Vessel collisions have not been identified as an issue yet. In some coastal waters (South Africa and Australia), nets are set around swimming areas to protect the bathers from sharks. Both *T. aduncus* and *Sousa chinensis* are victims of incidental entanglement in these anti-shark nets.

Due to the lack of information needed for assessing the status of the species globally, *T. aduncus* is classified by the IUCN Red List as "Data Deficient." However, this should not be seen as a more favorable status than any of the threatened categories (as is often perceived erroneously and results in less attention being afforded to species classified as such). Once adequate information allows

for an assessment, this species may well meet one of the categories of threat. There are serious concerns about the depletion of local populations because the species appears to be resident (likely with limited exchange with adjacent populations) and exists in relatively small numbers in many areas.

## See Also the Following Articles

Common Bottlenose Dolphin Bycatch

## References

Bejder, L., *et al.* (2006). Decline in relative abundance of bottlenose dolphins exposed to long-term disturbance. *Cons. Biol.* **20**, 1791–1798.

Chilvers, B. L., and Corkeron, P. (2003). Abundance of Indo-Pacific bottlenose dolphins, *Tursiops aduncus*, off Point Lookout, Queensland, Australia. *Mar. Mamm. Sci.* **19**, 85–95.

Connor, R. C., Heithaus, M. R., Berggren, P., and Miksis, J. L. (2000a). "Kerplunking": Surface fluke-splashes during shallow-water bottom foraging by bottlenose dolphins. *Mar. Mamm. Sci.* **16**, 646–652.

Connor, R. C., Wells, R. S., Mann, J., and Read, A. J. (2000b). The bottlenose dolphin. *In* "Cetacean Societies: Field Studies of Dolphins and Whales" (J. Mann, R. C. Connors, P. L. Tyack, and H. Whitehead, eds), pp. 91–126. The University of Chicago Press.

Harwood, M. B., and Hembree, D. (1987). Incidental catch of small cetaceans in the offshore gillnet fishery in northern Australian waters: 1981–1985. *Rep. Int. Whal. Commn* **37**, 363–367.

Heithaus, M. R. (2001). Shark attacks on bottlenose dolphins (*Tursiops aduncus*) in Shark Bay, Western Australia: Attack rate, bite scar frequencies and attack seasonality. *Mar. Mamm. Sci.* **17**, 526–539.

Kakuda, T., Tajima, Y., Arais, K., Kogi, K., Hishii, T., and Yamada, T. K. (2002). On the resident "bottlenose dolphins" from Mikura water. *Mem. Nat. Sci. Mus. Tokyo*, **38**, 255–272.

Kogi, K., Hishii, T., Imamura, A., Iwatani, T., and Dudzinski, K. M. (2004). Demographic parameters of Indo-Pacific bottlenose dolphins (*Tursiops aduncus*) around Mikura Island, Japan. *Mar. Mamm. Sci.* **20**, 510–526.

Krützen, M., Sherwin, W. B., Berggren, P., and Gales, N. (2004). Population structure of an inshore cetacean revealed by microsatellite and mtDNA analysis: Bottlenose dolphins (*Tursiops* sp.) in Shark Bay, western Australia. *Mar. Mamm. Sci.* **20**, 28–47.

Krützen, M., Mann, J., Heithaus, M. R., Connor, R. C., Bejder, L., and Sherwin, W. B. (2005). Cultural transmission of tool use in bottlenose dolphins. *Proc. Nat. Acad. Sci.* **102**, 2943–8939.

Morisaka, T., Shinohara, M., Nakahara, F., and Akamatsu, T. (2005). Geographic variation in the whistles among three Indo-Pacific bottlenose dolphin *Tursiops aduncus* populations in Japan. *Fish. Sci.* **71**, 568–576.

LeDuc, R. G., Perrin, W. F., and Dizon, A. E. (1999). Phylogenetic relationships among the delphinid cetaceans based on full cytochrome *b* sequences. *Mar. Mamm. Sci.* **15**, 619–648.

Lemon, M., Lynch, T. P., Cato, D. H., and Harcourt, R. G. (2006). Response of travelling bottlenose dolphins (*Tursiops aduncus*) to experimental approaches by a powerboat in Jervis Bay, New South Wales, Australia. *Biol. Cons.* **127**, 363–372.

Mann, J., Connor, R. C., Barre, L. M., and Heithaus, M. R. (2000). Female reproductive success in bottlenose dolphins (*Tursiops* sp.): Life history, habitat, provisioning, and group size effects. *Behav. Ecol.* **11**, 210–219.

Mann, J., and Watson-Capps, J. J. (2005). Surviving at sea: Ecological and behavioural predictors of calf mortality in Indian Ocean bottlenose dolphins, *Tursiops* sp. *Anim. Behav.* **69**, 899–909.

I

Natoli, A., Peddemors, V. M., and Hoelzel, A. R. (2004). Population structure and speciation in the genus *Tursiops* based on microsatellite and mitochondrial DNA analyses. *J. Evol. Biol.* **17**, 363–375.

Omata, Y., Hammond, T., Itoh, K., and Murata, K. (2005). Antibodies against *Toxoplasma gondii* in the Pacific bottlenose dolphin (*Tursiops aduncus*) from the Solomon Islands. *J. Parasit.* **91**, 965–967.

Perrin, W. F., Robertson, K. M., van Bree, P. J. H., and Mead, J. G. (2007). Cranial description and genetic identity of the holotype specimen of *Tursiops aduncus* (Ehrenberg, 1832). *Mar. Mamm. Sci.* **23**, 343–357.

Preen, A. (2004). Distribution, abundance and conservation status of dugongs and dolphins in the southern and western Arabian Gulf. *Biol. Cons.* **118**, 205–218.

Ross, G. J. B. (1977). The taxonomy of bottlenosed dolphins *Tursiops* species in South African waters with notes on their biology. *Ann. Cape Prov. Mus. (Nat. Hist.)* **11**, 259–327.

Ross, G. J. B. (1984). The smaller cetaceans of the south east coast of southern Africa. *Ann. Cape Prov. Mus. (Nat. Hist.)* **15**, 173–410.

Ross, G. J. B., and Cockcroft, V. G. (1990). Comments on Australian bottlenose dolphins and taxonomic status of *Tursiops aduncus* (Ehrenberg, 1832). *In* "The bottlenose dolphin" (S. Leatherwood, and R. R. Reeves, eds), pp. 101–128. Academic Press, San Diego.

Sargeant, B. L., Mann, J., Berggren, P., and Krützen, M. (2005). Specialization and development of beach hunting, a rare foraging behavior, by wild bottlenose dolphins (*Tursiops* sp.). *Can. J. Zool.* **83**, 1400–1410.

Shirakihara, M., Shirakihara, K., Tomonaga, J., and Takatsuki, M. (2002). A resident population of Indo-Pacific bottlenose dolphins (*Tursiops aduncus*) in Amakusa, Western Kyushu, Japan. *Mar. Mamm. Sci.* **18**, 30–41.

Shirakihara, M., Yoshida, H., and Shirakihara, K. (2003). Indo-Pacific bottlenose dolphins *Tursiops aduncus* in Amakusa, western Kyushu, Japan. *Fish. Sci.* **69**, 654–656.

Stensland, E., Carlen, I., Sarnblad, A., Bignert, A., and Berggren, P. (2006). Population size, distribution, and behavior of Indo-Pacific bottlenose (*Tursiops aduncus*) and humpback (*Sousa chinensis*) dolphins off the south coast of Zanzibar. *Mar. Mamm. Sci.* **22**, 667–682.

Stensland, E., and Berggren, P. (2007). Behavioural changes in female Indo-Pacific bottlenose dolphins in response to boat-based tourism. *Mar. Ecol. Prog. Ser.* **332**, 225–234.

Van Bressem, M. F., *et al.* (17 authors) (2001). An insight into the epidemiology of dolphin morbillivirus worldwide. *Vet. Microbiol.* **81**, 287–304.

Wang, J. Y., and Yang, S.-C. (2007). "An Identification Guide to the Dolphins and Other Small Cetaceans of Taiwan." Jen Jen Publishing Company and the National Museum of Marine Biology and Aquarium, Taiwan, 208 pp.

Wang, J. Y., Chou, L.-S., and White, B. N. (1999). Mitochondrial DNA analysis of sympatric morphotypes of bottlenose dolphins (genus *Tursiops*) in Chinese waters. *Mol. Ecol.* **8**, 1603–1612.

Wang, J. Y., Chou, L.-S., and White, B. N. (2000a). Differences in the external morphology of two sympatric species of bottlenose dolphins (genus *Tursiops*) in the waters of China. *J. Mammal.* **81**, 1157–1165.

Wang, J. Y., Chou, L.-S., and White, B. N. (2000b). Osteological differences between two sympatric forms of bottlenose dolphins (genus *Tursiops*) in Chinese waters. *J. Zool. (Lond.)* **252**, 147–162.

Yang, G., Zhou, K. Kato, H. and Miyashita, T. (1997). An initial study on the population size and distribution of bottlenose dolphin in the East China Sea. *Act. Ther. Sin.* **17**, 241–247.

Yang, G., Zhou, K., and Xu, X. (2000). Population density, distribution, and incidental catches of bottlenose dolphins in Xiamen-Dongshan waters of the Taiwan Strait. *Act. Ecol. Sin.* **20**, 1002–1008.

Zhou, K., and Wang, X. (1994). Brief review of passive fishing gear and incidental catches of small cetaceans in Chinese waters. *Rep. Int. Whal. Commn* **15**, 347–354, Special Issue.

# Indo-West Pacific Marine Mammals

Peter Rudolph and Chris Smeenk

The Indo-West Pacific is defined here as the tropical and subtropical (warm temperate) waters of the Indian and far western Pacific Oceans, from the Cape of Good Hope in South Africa to the Red Sea and Persian Gulf, and from Australia and Southeast Asia to about 30°N.

The Indo-West Pacific probably offers the greatest diversity of marine mammal species in the world. Within it live representatives of 11 of the 13 families of the Cetacea, with more than 40 of the 85-odd species recognized by most authors, as well as one member of the order Sirenia. We have not included species of the order Carnivora (sea lions, walruses, and seals), nearly all of which normally live in higher latitudes. Only the ranges of the Cape and Tasmanian fur seals (*Arctocephalus pusillus pusillus* and *A. p. doriferus*), the south Australian fur seal (*A. forsteri*), and the Australian sea lion (*Neophoca cinerea*) (family Otariidae) include the southernmost part of the area considered, in South Africa and in southern Australia, respectively.

## I. Endemic Taxa

Many species occurring in the Indo-West Pacific have a cosmopolitan or pantropical distribution, and several Northern and Southern Hemisphere species extend their range to within the confines of the area. However, a relatively large number of species or currently recognized subspecies are endemic to the Indo-West Pacific; the taxonomic position and distribution of some forms are still insufficiently known. Most of these occur mainly in shelf and/or fresh-water ecosystems, although *Tursiops aduncus* is also found in pelagic waters. Two forms of dolphin seem to be largely confined to shelf areas: *Delphinus capensis tropicalis*, and *Stenella longirostris roseiventris*, as well as the small baleen whale *Balaenoptera edeni* (specific status still unclear); five or six species have a decidedly coastal, estuarine, or even partly riverine distribution: *Sousa chinensis/plumbea* (Fig. 1), *Orcaella brevirostris*, *O. heinsohni*, *Neophocaena phocaenoides*, and the sirenian *Dugong dugon*; and two or three are true river dolphins: *Platanista gangetica gangetica*, *P. g. minor*, and *Lipotes vexillifer* (now probably extinct). The distribution and habitat preference of the recently distinguished *B. omurai* are still insufficiently known. In addition, at least two oceanic species: *Mesoplodon ginkgodens* and *Indopacetus pacificus*, appear to be endemic to deep waters of the tropical and subtropical Indo-Pacific at large (including the central and the eastern Pacific), and the same may hold true for the Indo-Pacific form of *Kogia sima*.

## II. Zoogeography

The shelf areas of the Indo-West Pacific show a high primary productivity, the result of monsoon-related currents and strong upwelling. Although the shallow waters of Southeast Asia and Australia would seem to constitute a barrier between the Indian and Pacific Oceans, the deep passages through the eastern Indo-Malayan Archipelago offer suitable dispersion routes for oceanic species, most of which, including sperm whales (*Physeter macrocephalus*), have indeed been recorded from the deeper straits and seas between the islands. Not considering the river dolphins, none of the marine

**Figure 1** *The Indo-West Pacific is home to a wide variety of marine mammals, including many endemic taxa e.g. the Indo-Pacific humpback dolphin. Photograph by Thomas Jefferson.*

mammals in the area is restricted to either the Indian Ocean or the Pacific side of the archipelago, except the western form of the coastal humpback dolphin (*Sousa chinensis/plumbea*).

The center of speciation and distribution in the Indo-West Pacific, containing the greatest diversity of marine mammals, is in the seas of Southeast Asia and northern Australasia, covering the Sunda and Sahul Shelves and neighboring waters. All marine (sub)species endemic to the Indo-West Pacific occur in this area, although they differ in the extent of their range. *S. longirostris roseiventris* (as far as known) and the two species of the genus *Orcaella* appear to be restricted to the Sunda and Sahul Shelves. The deep waters between these shelves form a barrier for some coastal species, which has led to genetic isolation or even speciation within the populations of *Orcaella* and *Sousa* on either shelf. Apart from such shelf forms, the Indo-West Pacific endemics also occur further west and north in the Indian and Pacific Ocean, respectively. In the east, the deep waters of the Pacific form a barrier to further dispersal, although the dugong has penetrated beyond the continental shelves, occurring as far east as Micronesia and, as a vagrant, to Fiji; the eastern confines of *T. aduncus* and of *B. omurai* are not known. In the north, *N. phocaenoides* has extended its range as far as Korea and Japan, and the two baleen whales *B. edeni* and *B. omurai* too have been found off Japan. In the southwest, *S. chinensis/plumbea* and *T. aduncus* occur as far as South Africa. Here, the cold waters off the Cape apparently have prevented their dispersal into the Atlantic Ocean, although the West African *Sousa teuszii* is probably a relatively recent descendant from its Indo-Pacific congener. In the northwest, *S. chinensis/plumbea* occurs in the Suez Canal; this species, as well as the dugong, has even strayed into the Mediterranean Sea.

## III. Annotated Species Accounts

### A. Rorquals, Family Balaenopteridae

Most species of baleen whales undertake extensive seasonal migrations between cold, productive summer feeding grounds in temperate or high latitudes, and winter mating and calving areas in tropical or warm-temperate waters, suggesting that there is only little mixing of Southern and Northern Hemisphere populations. Baleen whales of the Indo-West Pacific, particularly the Indian Ocean, are poorly known.

The modern whaling industry, following the invention of the explosive grenade harpoon, was initially based mainly on rorquals, although sperm and right whales were also taken. In the Indian Ocean, whaling stations existed in South Africa, Mozambique, and Western Australia. Although most pelagic whaling occurred below 40°S, it was practice for whalers to take baleen whales as well as sperm whales in tropical waters on their passage to and from the Antarctic. Newly revealed data on illegal Russian whaling operations between 1947 and 1972 have shown that baleen and sperm whales were taken in the central and northern Indian Ocean, near Madagascar and Australia, as well as in Indonesian waters. These included blue whales (*Balaenoptera musculus*), caught mainly after 1965 when the species had already received protection from the IWC; humpback whales (*Megaptera novae-angliae*) off Oman, Pakistan, and India; as well as southern right whales (*Eubalaena australis*) around the Crozet and Kerguelen Islands and in the central Indian Ocean.

In the period 1963–1967, Russian whalers caught 848 Bryde's whales (probably *Balaenoptera brydei*) in the Gulf of Aden, near the Maldives, and near the Seychelles. During the seasons 1976/77–1978/79, Japanese whalers caught 232 small baleen whales, at the time identified as Bryde's whales (possibly consisting of both *B. brydei* and *B. edeni*) in the western Indian Ocean, the eastern Indian Ocean south of Java and the Lesser Sunda Islands, and in the West Pacific near the Solomon Islands. A small series of specimens caught in the Solomon Sea and near the Cocos Keeling Islands in the eastern Indian Ocean have recently been distinguished as a separate species: *B. omurai*. Relatively small-scale, directed catches of baleen whales are today only known from the Philippines and from Lamakera on Solor Island, Indonesia. In the Philippines, small baleen whales (also collectively called Bryde's whales but including *B. omurai*) were or are still hunted off Pamilacan, Bohol, and Camiguin Islands. The whaling off Pamilacan, now legally prohibited, was seasonal and opportunistic. It started in January and ended in June, with most whales taken in April and May. The animals were caught using a hook of stainless steel, which was driven into the whale by one of the hunters jumping onto the whale's back and using his weight to drive the hook in.

*1. Humpback Whale* Growing evidence shows that discrete populations of humpback whales, *Megaptera novaeangliae*, live year-round in the Arabian Sea, including the Gulf of Oman and Gulf of Aden, the Red Sea, and in the Bay of Bengal, with at least some of these animals feeding and breeding there. A young animal caught in October 2007 in a fishing net at the south coast of Bali, Indonesia, is the first confirmed record for the Indo-Malayan Archipelago. Biological examination of 238 humpback whales caught illegally by Russian whalers in November 1966 off Oman, in the Gulf of Kutch off Pakistan, and west of Bombay, India, showed that they differ significantly from Antarctic humpbacks in size, coloration, body scars, and pathology (Mikhalev, 1997). The song structure of humpback whales recorded off Oman is also different from that in the North Pacific and North Atlantic. Southern Hemisphere humpback whales, which feed in Antarctic waters, winter near Mozambique, around Madagascar, off northwest and northeast Australia, and elsewhere in the tropical Pacific. In Southeast Asian waters, humpback whales have been reported from Vietnam, the Philippines, the South China Sea, and the waters around Taiwan; probably, these are animals that spend the summer in the Northwest Pacific. In the nineteenth and twentieth centuries, humpback whales were hunted off Mozambique, Madagascar, and northwestern Australia.

*2. Minke Whales* Two species of minke whales extend their ranges into the Indo-West Pacific: the common minke whale (*B. acutorostrata*), and the Antarctic minke whale (*B. bonaerensis*). The North Pacific population of the common minke whale has been

I

distinguished as the subspecies *B. a. scammoni*. In summer, it occurs as far south as the East China Sea and the central Pacific at about 30°N; its winter range extends into warm-temperate and tropical waters. Reports of minke whales from the Red Sea, Persian Gulf, and Vietnam are probably all based on confusion with *B. brydei* or *B. edeni* (Andersen and Kinze, 2005). A small, distinctly colored and as yet unnamed southern form, the "dwarf minke whale" (as yet unnamed subspecies), occurs in tropical and warm-temperate waters of the Southern Hemisphere. In the Indo-West Pacific, it is known from South Africa and Australia (Arnold *et al.*, 2005) and has recently been recorded of northwestern New Guinea.

The Antarctic minke whale also ranges into tropical waters; it has been found as far north as about 7°S. Important breeding grounds in the eastern Indian Ocean may lie between 10° and 20°S. The species occurs in part sympatrically with the dwarf minke whale, with the latter seeming to have a more coastal distribution.

*3. Bryde's Whale and Sittang Whale*    These small baleen whales, which until recently were considered conspecific (with the name *B. edeni* taking priority), have a primarily tropical and warm-temperate distribution. The taxonomic status of these whales is still unsettled. The larger form, the "true" Bryde's whale, nominally *B. brydei*, has a cosmopolitan distribution in tropical waters. In the Indo-West Pacific it has been identified from the Indian Ocean, including the Red Sea and Persian Gulf, as well as from the (sub)tropical West Pacific; the limits of its range and the movements of the various populations are still unknown (Kanda *et al.*, 2007).

There are several distinctly smaller, more coastal populations. Of these, the "Sittang whale" has been identified from coastal and shelf areas in the eastern Indian Ocean and (sub)tropical West Pacific: the Andaman Sea, Indo-Malayan Archipelago, Australia, the South China Sea, and southwestern Japan. Molecular evidence, as well as osteological comparisons, have shown that these animals genetically and morphologically do not belong with the larger Bryde's whale and almost certainly form a distinct species. The name *B. edeni* would apply to this species if the holotype, which is from the mouth of the Sittang River in Burma (Myanmar), indeed proves to be of this form.

Despite these differences, several authors still use the name *B. edeni* for both the larger and the smaller forms, pending a molecular-genetic study of the holotype (see chapter Bryde's Whales).

*4. Omura's Whale*    During Japanese scientific whaling operations, two series of small baleen whales were caught near the Solomon Islands (Solomon Sea) in the West Pacific in October 1976 (6 animals) and near the Cocos Keeling Islands in the eastern Indian Ocean in November 1978 (2 animals). They have proved morphologically and genetically distinct from the small *B. edeni* as identified earlier. In 2003, they were described as a new species, *B. omurai*, the holotype being a fresh animal stranded in September 1998 on Tsunoshima Island, in the southern Sea of Japan (Wada *et al.*, 2003; Sasaki *et al.*, 2006). The species has also been identified from a sample taken in the Philippines. All these localities are very far apart and the distribution and ecological requirements of this species are still largely unknown.

*5. Sei Whale*    The sei whale, *Balaenoptera borealis*, appears to be uncommon in tropical waters. None of the published recent records from the northern Indian Ocean is convincing. Sei whales have been confused with Bryde's or fin whales by observers who were not familiar with these species. The sei whale has also been reported from the South China Sea, but this too needs confirmation. Sei whales from the Southern Hemisphere, often distinguished as *B. b. schlegelii*, generally winter as far north as South Africa and Western Australia

(to about 25°S). There is one specimen from the north coast of Java: the holotype of *B. b. schlegelii* (Rudolph *et al.*, 1997). Japanese scouting and research vessels reported seeing sei whales south of Sumatra (5°–10°S) during November in the period 1974/75–1984/85.

*6. Fin Whale*    The fin whale, *Balaenoptera physalus*, also appears to be uncommon in tropical waters. A North Pacific population is thought to winter from the Sea of Japan south to the Philippine Sea, with concentrations in the East China and Yellow Seas. There are a few records from Philippine and Indonesian waters and several reports from the northern Indian Ocean, Persian Gulf, and off the Seychelles. Southern Hemisphere fin whales, which have been distinguished as the subspecies *B. p. quoyi*, winter in the Indian Ocean off South Africa, Madagascar, and Western Australia. Fin whale sightings reported by Japanese vessels in the period 1974/75–1984/85 were concentrated in two longitudinal areas: one west of 50°E, the other at 70–100°E. The northern limits of the species' range in March were found to be at 40–45°S.

*7. Blue Whale*    The blue whale, *Balaenoptera musculus*, is distributed throughout the Indo-West Pacific, in areas with high primary productivity. The species occurs year-round in the Indian Ocean, but some populations undertake seasonal migrations. The taxonomic status and distribution of the various populations of blue whales are still unsettled (Branch *et al.*, 2007). The animals of the northern Indian Ocean, from the Arabian Sea to Southeast Asia, are distinct from Southern Hemisphere blue whales in call pattern and reproductive season. They are often regarded as a separate subspecies *B. m. indica*, although the distinguishing characters of this form are poorly defined. The blue whales of the southern Indian Ocean, off South Africa, Madagascar, the Chagos Archipelago, as well as those from southern Indonesia and off Western Australia and South Australia, are smaller and also have a distinctive call pattern. They have been referred to the subspecies *B. m. intermedia* or, more recently, to the "pygmy blue whale," *B. m. brevicauda*. Finally, Antarctic blue whales probably extend their winter range into the southern confines of the pygmy blue whale. The distribution of blue whales in the northwestern Pacific is poorly known; the species has been recorded from southern China and the Philippines.

Two other species of baleen whale occur in the Indian Ocean. The southern right whale, *Eubalaena australis* (family Balaenidae), extends its range northward to southern Mozambique, Madagascar, and Western Australia. In the nineteenth and early twentieth centuries, right whales were hunted in the Indian Ocean between 30° and 40°S, around Crozet, Kerguelen, and Amsterdam Islands, and on their calving grounds off South Africa and southeastern Australia. The pygmy right whale, *Caperea marginata* (family Neobalaenidae), has been recorded north to 33°S off southwestern Australia. The critically endangered western population of the gray whale, *Eschrichtius robustus* (family Eschrichtiidae), known as the Korea-Okhotsk stock, winters off southern China, with records from the Yellow, East China, and South China Seas, as far south as Hainan. It has been suggested that breeding and calving grounds exist off Guanglong Province in southern China, but this should be investigated.

## B. Sperm Whales, Family Physeteridae

Sperm whales, *Physeter macrocephalus*, occur in all oceans of the world. Concentrations such as found in the traditional whaling grounds appear to be associated with oceanic fronts, steep bottom topography, and high productivity. (Sub)adult males reach temperate or even polar waters and return to lower latitudes to breed, whereas females and immatures are restricted to tropical and warm-temperate seas. The sperm whale is found in deep waters of the Indian Ocean, including the Gulf of Aden, Gulf of Oman and probably the southern Red Sea, and in the West Pacific; its main distribution is well known from records of nineteenth century whalers. The species is sometimes found in shallow waters. The deep passages and seas between the islands of the Indo-Malayan Archipelago have been supposed to form a migration route of sperm whales between the Indian and the Pacific Ocean.

In the western Indian Ocean, exploitation of sperm whales began at about 1800 on the whaling grounds near the Cape of Good Hope, and in later years extended northward to Mozambique, Madagascar, the Comoros, Seychelles, the East African coast, Arabia, and the waters west of Sri Lanka. In the eastern Indian Ocean and western Pacific, sperm whales were taken south of Sumatra and Java, in the Timor Sea, and off Western Australia, in the Moluccan and Sulu Seas, the waters north of New Guinea, and off Japan. Whaling declined for economic reasons toward the close of the nineteenth century. Today, sperm whales are only hunted in subsistence whaling in Indonesia. The whalers of Lamalera on Lembata Island specialize in catching sperm whales and other toothed whales (Rudolph *et al.*, 1997). The animals are caught by using hand harpoons from open rowing boats called "peledang." During the whaling season, mainly from May to October, the boats search an area of up to a few kilometers south of the coast. When a whale is approached, the harpooner leaps from a small platform on the bow and adds his weight to drive the harpoon into the whale's back, similar to the technique used by whalers of the neighboring Solor Island and in the Philippines, who hunt baleen whales.

## C. Pygmy and Dwarf Sperm Whale, Family Kogiidae

Our knowledge of the distribution of the pygmy sperm whale, *Kogia breviceps*, and dwarf sperm whale, *K. sima*, is sketchy and mainly based on animals stranded or caught in fishing gear. A study of *K. sima* has revealed that animals from the Indo-Pacific are genetically distinct from Atlantic ones, perhaps warranting recognition of two species. The waters off the Cape may form the boundary between these two clades, but the taxonomic and ecological situation is still unresolved (Chivers *et al.*, 2005).

Pygmy and dwarf sperm whales occur in tropical and warm-temperate seas, with the pygmy sperm whale extending its range into slightly colder waters. They are often difficult to observe at sea and probably are much more common than sighting records would suggest. Although their diets overlap, prey composition indicates that the pygmy sperm whale has a more oceanic distribution, whereas the dwarf sperm whale prefers shelves and shelf edges.

In the Indo-West Pacific, both species have been recorded from South Africa north to Oman, east to Australia and the Indo-Malayan Archipelago, and north to southern Japan. Dwarf sperm whales appear to be common in the Maldives, particularly over the atoll slopes (Anderson, 2005).

Both species are killed accidentally in Sri Lanka; the dwarf sperm whale is caught in directed fisheries and incidentally in gill net and seine fisheries in the Philippines. In Indonesia, the dwarf sperm whale is taken by subsistence whalers of Lamalera on Lembata Island.

## D. Beaked Whales, Family Ziphiidae

Eleven species of beaked whales have been recorded from the Indo-West Pacific. Only four of these are regularly distributed in tropical seas. The other seven normally live in temperate and cold waters of the Southern Hemisphere, and occasionally migrate or stray into lower latitudes (MacLeod *et al.*, 2006). All are oceanic species, most of them are mainly known from stranded animals.

*1. Cuvier's Beaked Whale* Cuvier's beaked whale, *Ziphius cavirostris*, is the most wide spread of the beaked whales, occurring worldwide in tropical and warm-temperate waters. It has been reported from South Africa, the Comoros and Seychelles, the Arabian Sea, the Maldives, Sri Lanka, Australia, the Indo-Malayan Archipelago, the Philippines, Taiwan, and southern Japan.

*2. Ginkgo-Toothed Beaked Whale* The ginkgo-toothed beaked whale, *Mesoplodon ginkgodens*, is distributed in tropical and warm-temperate waters of the Indian and the Pacific Oceans. In the Indo-West Pacific, there are possible sightings from the Arabian Sea; stranded specimens have been found in the Maldives, Sri Lanka, Malacca, southeast Australia, Guam, northern China, and Japan.

*3. Blainville's or Dense-Beaked Beaked Whale* Blainville's beaked whale, *Mesoplodon densirostris*, occurs in tropical and warm-temperate waters of all oceans. It has been reported from South Africa, Mauritius, the Seychelles, Maldives, Sri Lanka, Australia, the Philippines, China, Taiwan, and Japan.

*4. Longman's or Indo-Pacific Beaked Whale* In recent years, this large beaked whale, *Indopacetus pacificus*, has become clearly defined and its external, morphological, and genetic characters are now well known. Previous sighting records of unidentified "tropical bottlenose whales" and several stranded specimens have now been attributed to this species, and many new records have become known (Dalebout *et al.*, 2003). The Indo-Pacific beaked whale is widely distributed in deep waters of the tropical and warm-temperate Indian and Pacific Oceans, and appears to have a preference for areas over or near deep bottom slopes (Anderson *et al.*, 2006). In the Indo-West Pacific, sightings and (skeletal material from) strandings have been reported from South Africa, the Comoros, Kenya, Somalia, the Gulf of Aden, the Maldives, Sri Lanka, northeastern Australia, the Indo-Malayan Archipelago, the Philippines, and north to southern Japan.

Two species—the strap-toothed whale *Mesoplodon layardii* and the scamperdown whale *M. grayi*—occur perhaps seasonally within the southern confines of the area considered. The remaining five probably are irregular visitors: Arnoux's beaked whale (*Berardius arnuxii*), Andrew's beaked whale (*M. bowdoini*), Hector's beaked whale (*M. hectori*), True's beaked whale (*M. mirus*), and the southern bottlenose whale (*Hyperoodon planifrons*). A skull of the last species (the holotype) was found in the Dampier Archipelago, Western Australia, at about 20°S.

## E. River Dolphins, Families Platanistidae and Lipotidae

Two or three species of river dolphins are represented in the Indo-West Pacific region: the Indian river dolphin or susu, genus *Platanista* (family Platanistidae), and the Yangtze dolphin or baiji *Lipotes vexillifer* (family Lipotidae; See Yan *et al.*, 2005). Two (sub)species of the Indian river dolphin are distinguished: the Ganges dolphin or susu, *P. gangetica gangetica* of the Ganges/Brahmaputra River system of India, Bangladesh, Nepal, and possibly Bhutan, and the Indus dolphin or bhulan, *P. g. minor*, formerly common throughout the Indus River and its tributaries, but now only found in a restricted area in Pakistan. Some authors regard these as distinct species. Both forms are endangered, particularly the Indus dolphin with an estimated population of 1200 in 2001, suffering from habitat degradation (pollution, construction of dams) and mortality in fishing gear.

I

The Yangtze dolphin in China was restricted to the lower Yangtze and Qiantang River systems in China. In the 1990s, its population had decreased to an estimated minimum of 13 animals. A survey in 2006 failed to find the species, which is now considered to be likely extinct (Turvey *et al.*, 2007), although one unconfirmed sighting was reported in 2007. Apart from habitat degradation, bycatch in fishing gear and electric fishing operations are regarded as the proximate causes of extinction.

### F. Dolphins, Family Delphinidae

*1. Rough-Toothed Dolphin* The rough-toothed dolphin, *Steno bredanensis*, inhabits tropical and warm-temperate waters, mainly along the edge of shelf areas. In the Indo-West Pacific, it ranges from South Africa to the Gulf of Aden and Arabian Sea, perhaps the southern Red Sea, and east to the Indo-Malayan Archipelago, Australia, the Solomon Islands, Philippines, and north to southern Japan.

*2. Humpback Dolphins* The taxonomy and geographic history of the genus *Sousa* remains unresolved. In the Indo-West Pacific, several authors have distinguished two species: *S. chinensis* and *S. plumbea*. Recent studies of skull morphology support the division into these two forms, although they may not (yet) be specifically distinct (Jefferson and Van Waerebeek, 2004). Humpback dolphins inhabit tropical and warm-temperate waters. The *plumbea* form occurs from South Africa and Madagascar to the Red Sea including the Suez Canal (it has even strayed into the Mediterranean Sea) and Persian Gulf, west to the Bay of Bengal, perhaps extending as far east as Thailand and northwestern Malaysia. The *chinensis* form (Fig. 1,2B) may occur sympatrically with the *plumbea* form in the Bay of Bengal (Sutaria and Jefferson, 2004) and is further known from the strait of Malacca, off northern Java and Borneo, the Gulf of Thailand, and north along the coast of China as far as the Yangtze, with a possibly isolated population off the west coast of Taiwan. It also occurs off northern Australia, ranging to about 24°S along the west, and 34°S along the east coast (Fig. 2). A recent study has shown that this population may represent a distinct species (Frére *et al.*, 2008).

Humpback dolphins have a coastal distribution and often occur in mangroves, river deltas, and estuaries. They ascend far up the main rivers of Asia as, e.g., the Ganges and the Yangtze. They are reported to be on the decline in the areas where they have been studied in detail. Off South Africa they appear to be killed in anti-shark nets at an unsustainable rate, and a decline is also suspected in Hong Kong waters and off Thailand. Incidental catches in fisheries are known throughout their range. Humpback dolphins have been caught for oceanaria along the coast of Thailand.

*3. Bottlenose Dolphins* The taxonomic situation of bottlenose dolphins, *Tursiops aduncus* and *T. truncatus*, is complicated. The existence of at least two species is now generally accepted. The smaller and longer-beaked form, often with spotted abdomen, is the Indo-Pacific bottlenose dolphin *T. aduncus* (Fig. 3). It is restricted to the tropical and warm-temperate Indo-West Pacific, where it is the more common *Tursiops*. Analysis of mtDNA cytochrome *b* sequences has suggested that this species may be closer related to the genus *Stenella* (LeDuc *et al.*, 1999). It ranges throughout the area, from South Africa to the Red Sea and Persian Gulf, east to Australia, the Solomon Islands and New Caledonia, the Indo-Malayan Archipelago, the Philippines, and north to southern Japan, perhaps including the Ogasawara (Bonin) Islands. It has a predominantly coastal distribution. Genetic analyses of animals from South African and Chinese waters, respectively, have shown that

**Figure 2** *The finless porpoise (A) and the Indo-Pacific humpback dolphin (B) are endemic to the Indo-West Pacific. Photographs by Thomas Jefferson.*

**Figure 3** *The Indo-Pacific bottlenose dolphin is also endemic to the Indo-West Pacific; photographed at Shaab Abu Salama, Hurghada, Egypt, Red Sea. Photograph copyrighted by Bernd Kledt, Eco Diving Society.*

these populations may be specifically distinct, but the situation here and in the intermediate areas is unresolved (Natoli *et al.*, 2004) (Fig. 3).

The larger, cosmopolitan common bottlenose dolphin, *T. truncatus*, also occurs in many parts of the Indo-West Pacific. Genetic analyses have identified it from near Mauritius, the Maldives, China, and Taiwan. It has also been found in South Africa, the Seychelles, the Red Sea, off Oman, the Maldives, eastern and southern Australia, and southern Japan. It occurs in both coastal and oceanic waters, but in general has a more oceanic distribution than *T. aduncus*.

Incidental and directed catches of *Tursiops* sp(p). occur in several kinds of fisheries in the Indo-West Pacific. Direct catches have been reported from Sri Lanka, Thailand, the Philippines, and Taiwan.

*4. Pantropical Spotted Dolphin*   The pantropical spotted dolphin, *Stenella attenuata*, occurs throughout the tropical and warm-temperate Indo-West Pacific, from South Africa to the Red Sea and Persian Gulf, east to Australia and beyond, the Indo-Malayan Archipelago, the Philippines, and north to southern Japan. Mixed groups of spotted and spinner dolphins have been reported from Indonesia, the western Indian Ocean and Red Sea, often associated with concentrations of seabirds and tuna. (See Rudolph *et al.* 1997; Balance and Pitman, 1998).

Spotted dolphins are caught in harpoon fisheries in the Laccadives, Sri Lanka, Indonesia, and the Philippines. A drive fishery at Malaita in the Solomon Islands may take several hundred dolphins a year, spotted dolphins, as well as other species. Spotted dolphins are incidentally caught in gillnet fisheries in Pakistan, Sri Lanka, northern Australia, and the Philippines.

*5. Spinner Dolphin*   The spinner dolphin, *Stenella longirostris*, ranges throughout the tropical and warm-temperate Indo-West Pacific, from South Africa to the Red Sea and Gulf of Oman, east to Australia and beyond, the Indo-Malayan Archipelago, the Philippines, and north to southern Japan. It probably is the most numerous cetacean species in tropical waters of the area. In several places, there are morphological and genetic differences between inshore and offshore populations, the taxonomic implications of which are not clear. Apart from this, two forms of spinner dolphin have been identified in the Indo-West Pacific, based on external, morphological, and behavioral differences, as well as habitat preferences (Perrin *et al.*, 1999). The nominate, pelagic form inhabits the Indian Ocean and West, Central, and South Pacific. The other form, the "dwarf spinner dolphin," has been distinguished as the subspecies *S. l. roseiventris*. It is smaller, with fewer vertebrae and teeth than the pelagic nominate type. It has been identified from shallow inner waters of Southeast Asia: the Gulf of Thailand, off Borneo, the Moluccan Sea, and northern Australia. It appears to feed mainly on benthic and coral reef fishes and invertebrates, whereas the pelagic animals feed primarily on mesopelagic fish and squid.

Off Oman, two types of spinner dolphin have also been recorded, one of them consisting of very small, dark-colored animals (Van Waerebeek *et al.*, 1998); similar dolphins may occur in the Maldives (Anderson, 2005). The taxonomic and ecological position of these animals is still unresolved.

Spinner dolphins are hunted with harpoons in Sri Lanka, Indonesia, and the Philippines. Incidental catches in fishing gear have been reported from Pakistan, India, Sri Lanka, northern Australia, and the Philippines.

*6. Striped Dolphin*   The striped dolphin, *Stenella coeruleoalba*, is primarily a warm-water species, but its range extends into higher latitudes than that of the spotted and spinner dolphins. The species primarily occurs in pelagic waters and mainly approaches the shore where there is deep water close to the coast. It has been reported from South Africa to the Gulf of Aden near the entrance of the Red Sea, east to Australia, and from the Philippines to Japan. Records from the Indo-Malayan Archipelago are scarce. In Southeast Asian waters, the species appears uncommon. It is suspected that animals from Southeast Asia migrate seasonally into Japanese waters, where they are subject to heavy exploitation by drive fishery, which may

seriously have affected their numbers. Striped dolphins are incidentally caught in gillnets, e.g., in Sri Lanka.

*7. Common Dolphins*   Common dolphins, *Delphinus capensis* and *D. delphis*, are distributed in tropical and warm-temperate, mainly coastal, waters of the Indo-West Pacific. The taxonomic situation is complicated. The existence of at least two species is now widely accepted: the long-beaked *D. capensis*, and the short-beaked *D. delphis*, both with a worldwide, but rather disjunct, distribution. However, the affinities of the various populations, particularly within the long-beaked forms, are still unresolved (Natoli *et al.*, 2006).

In the Indian Ocean, only the occurrence of long-beaked animals has been confirmed. *D. capensis* has been identified from coastal waters of South Africa; the situation in East African waters and Madagascar is unknown. In coastal waters of the northern Indian Ocean and West Pacific, a very long-beaked form occurs, with a longer rostrum and a greater number of teeth than in other populations of *D. capensis*. This is now often included in *D. capensis* as a separate subspecies: *D. c. tropicalis* (Jefferson and Van Waerebeek, 2002). It has so far been documented for the Red Sea, Gulf of Aden and the waters off Somalia, the Arabian Peninsula, Gulf of Oman and Persian Gulf, Pakistan, India and Sri Lanka, the southwestern and southeastern Bay of Bengal, Strait of Malacca, off Borneo, Cambodia, the Gulf of Tongking and the South and East China Sea, north to about Shanghai. Long-beaked animals from the East China Sea and southern Japan have been arranged with *D. capensis*, but the taxonomic relation between this population and the animals living off South Africa is still unclear.

The short-beaked common dolphin *D. delphis* mainly lives in (warm-)temperate seas. In the West Pacific it occurs off southern and eastern Australia, even as far as New Caledonia. In the northwestern Pacific, it perhaps extends into the range of *D. capensis*.

*8. Fraser's Dolphin*   Fraser's dolphin, *Lagenodelphis hosei*, was described in 1956 from a skeleton that had been found before 1895 on a beach in Sarawak, Borneo. The species has a worldwide distribution and occurs primarily in deep tropical, and warm-temperate waters. In the Indian Ocean, it has been reported from South Africa, Madagascar, the Seychelles, Maldives, southern India and Sri Lanka, Australia south to about 38°S, the Indo-Malayan Archipelago, Thailand, Vietnam, the Philippines, China, Taiwan, and southern Japan. In Indonesian and Philippine waters, Fraser's dolphin has been observed in mixed groups with other dolphin species.

Directed catches have been reported from Indonesia, the Philippines, and Taiwan and bycatches from the Philippines.

*9. North Pacific White-Sided Dolphin and Dusky Dolphin*   Two species of the genus *Lagenorhynchus* as recognized by most authors, mainly occurring in temperate waters, extend marginally into the Indo-West Pacific. The North Pacific white-sided dolphin *L. obliquidens* has been found as far south as Taiwan. The Southern Hemisphere dusky dolphin *L. obscurus* in some years enters the coastal waters of southern Australia.

*10. Risso's Dolphin*   Risso's dolphin, *Grampus griseus*, is distributed throughout tropical and temperate seas, particularly seaward of steep shelf edges. In the Indian Ocean it is found from South Africa to the Red Sea and Persian Gulf, east to the Bay of Bengal, Australia, the deeper waters of the Indo-Malayan Archipelago, and throughout the West Pacific.

Risso's dolphins are known to be directly caught in the Indonesian subsistence whaling off Lembata Island and in Palawan in the Philippines, Taiwan, and Japan. Between 1983 and 1986, 241 animals

**I**

were reported landed in Sri Lanka gillnet fisheries, but the actual numbers killed here may have been about 1300 per year.

*11. Melon-Headed Whale* The melon-headed whale, *Peponocephala electra*, is mainly found in deep tropical and warm-temperate waters. In the Indo-West Pacific is has been recorded from the Seychelles to the Arabian Sea and Bay of Bengal, Australia, the Indo-Malayan Archipelago, the Philippines, Gulf of Thailand, Taiwan, and southern Japan. There are a few records from South Africa at about 34°S and from southern Australia at about 32°S. Mass strandings are known from the Seychelles, Indonesia, Australia, and Japan.

There is a direct catch of melon-headed whales in the subsistence whaling off Lembata Island in Indonesia, and at Pamilacan Island in the Philippines.

*12. Pygmy Killer Whale* The pygmy killer whale, *Feresa attenuata*, is mainly found in deep tropical and warm-temperate waters. In the Indo-West Pacific, it has been recorded from South Africa to the Gulf of Aden and Gulf of Oman, east to Australia, the Indo-Malayan Archipelago, the Philippines, Taiwan, and southern Japan.

Incidental and directed catches of melon-headed whales have been reported from Sri Lanka with an estimated 300–900 animals taken per year. Although not yet confirmed by a specimen, the species may also be taken in the subsistence whaling off Lembata Island in Indonesia.

*13. False Killer Whale* The false killer whale, *Pseudorca crassidens*, is mainly found in deep tropical and warm-temperate waters. It occurs throughout the Indian Ocean, from South Africa to the Red Sea and Persian Gulf, east to Australia and the Solomon Islands, the Indo-Malayan Archipelago, Thailand, Cambodia, Vietnam, the Philippines, and north to the Yellow Sea and southern Japan. In Australia, mass strandings occur relatively often: since 1970, about once every 2.5 years, on average involving about 100 individuals. Mass strandings have also been reported from (southwestern) South Africa, Tanzania, and Sri Lanka.

In ancient times, false killer whales were hunted for their ivory in the Arabian Sea. Incidental catches have been reported from South Africa, India, Sri Lanka, and northern Australia. The species is sometimes taken in the subsistence whaling off Lembata Island, Indonesia.

*14. Killer Whale* Killer whales, *Orcinus orca*, have generally been considered to form a single, cosmopolitan species, although recently species status has been suggested for different forms/ecotypes that appear reproductively isolated, in the southern oceans and in the North Pacific, based on color pattern, morphological characters, genetic differences, habits, and ecology. The occurrence of different types of killer whales in the Indo-West Pacific is still unknown. Killer whales have been reported throughout the Indian Ocean including the Red Sea, and in the West Pacific there are records for all months and latitudes, although densities in tropical and subtropical waters would seem low.

Usually, killer whales were secondary targets of whalers, but some have been taken by whalers operating from Durban in South Africa, and possibly by Russian pelagic whaling operations in the 1970s, although the localities and size of these catches were not reported to the IWC. Small numbers are caught in the subsistence whaling off Lembata Island, Indonesia. Catches in net-fisheries, though rare, have been reported from Sri Lanka.

*15. Pilot Whales* The two species of pilot whales, *Globicephala macrorhynchus* and *G. melas*, have a largely parapatric distribution. The short-finned pilot whale, *G. macrorhynchus*, mainly occurs in tropical and warm-temperate waters. In the Indo-West Pacific, it occurs from South Africa to the Red Sea and Gulf of Oman, east to Australia and the Solomon Islands, the Indo-Malayan Archipelago, the Philippines, and north to Taiwan and Japan. Mass strandings often occur and have been reported from India, the Indo-Malayan Archipelago, and Australia.

The long-finned pilot whale, *G. melas*, has a circumpolar distribution in temperate waters. Southern Hemisphere animals are distinguished as *G. m. edwardii*. In the Indo-West Pacific, its northern limits extend into the range of the short-finned pilot whale off the Cape Province in South Africa and off southern Australia; occasionally, the species strays further north.

Short-finned pilot whales are caught in coastal fisheries off Pakistan and Sri Lanka and are taken in subsistence whaling off Lembata Island in Indonesia, and in the Philippines.

*16. Snubfin Dolphins* Two species of *Orcaella* are currently recognized, *O. brevirostris* and *O. heinsohni*. The Irrawaddy dolphin, *O. brevirostris*, occurs in the Bay of Bengal, Strait of Malacca, the Indo-Malayan Archipelago, and Gulf of Thailand. It generally lives in muddy, coastal waters and has distinct coastal, estuarine, and riverine populations. Along the Southeast Asian mainland, the species is known from northeastern India, Bangladesh, Burma (Myanmar), Malaysia, and Singapore; Thailand, Cambodia, Laos, and Vietnam. Irrawaddy dolphins have entered the systems of the Ganges/Brahmaputra, Ayeyarwadi (Irrawaddy), Mekong, and several other rivers, with populations originally living as far as about 1300 km upstream in the Ayeyarwadi, and nearly 1000 km up the Mekong/Sekong. In the Indo-Malayan Archipelago, it has been recorded from eastern Sumatra, Belitung, Java, and many places in and around Borneo, including the Barito, Mahakam and Kajan Rivers, and their major tributaries and lakes, and from Malampaya Sound at Palawan in the Philippines. Its occurrence off western Sulawesi needs confirmation. In many coastal areas, numbers have decreased and the distribution appears disjunct. Riverine populations in particular have reached dangerously low levels and are now only found in restricted parts of their original range or may have disappeared altogether. In 1978, the population in the central Mahakam River and lake system was estimated to number 100–150 animals, whereas an estimate made in 1999–2000 over the whole course of the river arrived at 48–55 individuals, with the highest density in the central part, between 180 and 350 km upstream (Kreb, 2004).

The form that is found on the Sahul Shelf is now distinguished as a separate species: the Australian snubfin dolphin, *O. heinsohni*. It was described in 2005, the holotype being an animal caught in 1972 in a shark-net in Horseshoe Bay, Magnetic Island, Queensland. It occurs off northern Australia, on the east coast as far south as Brisbane River (27°S), as well as in southern New Guinea (Beasley et al., 2005). Its presence off northern New Guinea needs confirmation. The species appears restricted to marine environments, up to about 20 km offshore.

The main threats for snubfin dolphins, particularly the riverine populations, are habitat degradation and pollution, mortality in fishing gear, and disturbance by boat traffic.

## G. Porpoises, Family *Phocoenidae*

*Finless Porpoise* The finless porpoise, *Neophocaena phocaenoides* (Fig. 2A), is distributed over a narrow band of shallow water along the coasts of southern and eastern Asia. The species occurs in inshore waters, mangrove zones, and delta areas, including the lower reaches of the major river systems such as the Indus, Ganges/

Brahmaputra, and Mekong. Three subspecies are reasonably well differentiated: *N. p. phocaenoides* occurs from the Persian Gulf to the South China Sea and southern part of the East China Sea, and the western Indo-Malayan Archipelago, east to Java; *N. p. sunameri* is found along the coast of northeastern China, Korea, and southern and eastern Japan. The only population that exclusively inhabits fresh water: *N. p. asiaeorientalis*, occurs in China in the lower and middle course of the Yangtze River and adjacent lakes, originally ranging over an area of almost 1670 km. Some populations have become seriously depleted, mainly due to mortality in fishing gear and habitat degradation.

## H. Dugongs, Family Dugongidae

Dugongs, *Dugong dugon*, live in tropical and subtropical coastal waters of the Indo-West Pacific. Some authors have recognized two subspecies: *D. d. hemprichii* in the Red Sea, and *D. d. dugon* elsewhere in the Indo-West Pacific, discontinuously distributed from Mozambique to the Gulf of Aden, and from the Persian Gulf east to northern Australia. Its range includes many islands in the western and northern Indian Ocean, the Indo-Malayan Archipelago, the Philippines, north to southern China, Taiwan (where probably extinct), and the Ryukyu Islands, east to Guam, Palau, Yap, Pohnpei, the Bismarck Archipelago, Solomon Islands, Vanuatu, New Caledonia, and, as a vagrant, to Fiji. In many areas, the dugong has been reduced to widely separated relict populations, mainly by overhunting. The waters off Papua New Guinea and northern Australia are now the most important stronghold for the species. Dugongs live in areas where there are large quantities ("meadows") of seagrass (family Zosteridae).

Although the dugong is now protected over most of its range, direct hunting for food and other products, as well as indirect catches in fishing gear, are still substantial in many areas: East Africa, India, Sri Lanka, the Indo-Malayan Archipelago, the Philippines, and Australia, but few data on numbers taken are available. Habitat degradation too, constitutes a threat (Marsh *et al.*, 2002).

For extensive reviews of research on marine mammals in Southeast Asia, see the volumes edited by Perrin *et al.* (1996, 2005), Smith and Perrin (1998), and Jefferson and Smith (2002); also see the species accounts.

## *References*

Anderson, C. (2005). Observations of cetaceans in the Maldives, 1990–2002. *J. Cetacean Res. Manage.* **7**, 119–135.

Andersen, M., and Kinze, C. C. (2005). Re-identification of a skeleton of the Bryde's Whale (*Balaenoptera edeni*) from the northern coast of Borneo. *Nat. Hist. Bull. Siam Soc.* **53**, 133–144.

Anderson, R. C., Clark, R., Madsen, P. T., Johnson, C., Kiszka, J., and Breysse, O. (2006). Observations of Longman's Beaked Whale (*Indopacetus pacificus*) in the western Indian Ocean. *Aq. Mamm.* **32**, 223–231.

Arnold, P. W., Birtles, R. A., Dunstan, A., Lukoschek, V., and Matthews, M. (2005). Colour patterns of the dwarf minke whale *Balaenoptera acutorostrata* sensu lato: Description, cladistic analysis and taxonomic implications. *Mem. Queensl. Mus.* **51**, 277–308.

Ballance, L. T., and Pitman, R. L. (1998). Cetaceans of the western tropical Indian Ocean: Distribution, relative abundance, and comparisons with cetacean communities of two other tropical ecosystems. *Mar. Mamm. Sci.* **14**, 429–459.

Beasley, I., Robertson, K. M., and Arnold, P. (2005). Description of a new dolphin, the Australian snubfin dolphin *Orcaella heinsohni* sp. n. (Cetacea, Delphinidae). *Mar. Mamm. Sci.* **21**, 365–400.

Branch, T. A., *et al.* (2007). Past and present distribution, densities and movements of blue whales *Balaenoptera musculus* in the Southern Hemisphere and northern Indian Ocean. *Mamm. Rev.* **37**, 116–175.

Chivers, S. J., LeDuc, R. G., Robertson, K. M., Barros, N. B., and Dizon, A. E. (2005). Genetic variation of *Kogia* spp. with preliminary evidence for two species of *Kogia sima*. *Mar. Mamm. Sci.* **21**, 619–634.

Dalebout, M. L., *et al.* (2003). Appearance, distribution, and genetic distinctiveness of Longman's Beaked Whale, *Indopacetus pacificus*. *Mar. Mamm. Sci.* **19**, 421–461.

Frére, C. H., Hale, P. T., Cockcroft, V. G., and Dalebout, M. L. (2008). Phylogenetic analysis of mtDNA Suggests revision of humpback dolphins (*Sousa* spp.) taxonomy is needed. *Mar. Freshw. Res.* **59**, 259–268.

Jefferson, T. A., and Smith, B. D. (eds) (2002). Facultative freshwater cetaceans of Asia: Their ecology and conservation. *Raffles Bull. Zool.*, **10** (Suppl), 1–187.

Jefferson, T. A., and Van Waerebeek, K. (2002). The taxonomic status of the nominal dolphin species *Delphinus tropicalis* Van Bree, 1971. *Mar. Mamm. Sci.* **18**, 787–818.

Jefferson, T. A., and Van Waerebeek, K. (2004). Geographic variation in skull morphology of humpback dolphins (*Sousa* spp.). *Aq. Mamm.* **30**, 3–17.

Jie, Yan., Kaiya, Zhou., and Guang, Yang. (2005). Molecular phylogenetics of "river dolphins" and the baiji mitochondrial genome. *Mol. Phylogenet. Evol.* **37**, 743–750.

Kanda, N., Goto, M., Kato, H., McPhee, M. V., and Pastene, L. A. (2007). Population genetic structure of Bryde's whales (*Balaenoptera brydei*) at the inter-oceanic and trans-equatorial level. *Conserv. Genet.* **8**, 853–864.

Kreb, D. (2004). Abundance of freshwater Irrawaddy dolphins in the Mahakam River in East Kalimantan, Indonesia, based on mark–recapture analysis of photo-identified individuals. *J. Cetacean Res. Manage.* **6**, 269–277.

LeDuc, R. G., Perrin, W. F., and Dizon, A. E. (1999). Phylogenetic relationships among the delphinid cetaceans based on full cytochrome *b* sequences. *Mar. Mamm. Sci.* **15**, 619–648.

MacLeod, C. D., *et al.* (2006). Known and inferred distributions of beaked whale species (Cetacea: Ziphiidae). *J. Cetacean Res. Manage.* **7**, 271–286.

Marsh, H., Penrose, H., Eros, C., and Hugues, J. (eds) (2002). "Dugong status report and action plans for countries and territories." UNEP, Nairobi, Early Warning and Assessment Report Series 02-1, 1–163.

Mikhalev, Y. A. (1997). Humpback whales *Megaptera novaeangliae* in the Arabian Sea. *Mar. Ecol. Prog. Ser.* **149**, 13–21.

Natoli, A., Peddemors, V. M., and Hoelzel, A. R. (2004). Population structure and speciation in the genus *Tursiops* based on microsatellite and mitochondrial DNA analyses. *J. Evol. Biol.* **17**, 363–375.

Natoli, A., Cañadas, A., Peddemors, V. M., Aguilar, A., Vaquero, C., Fernández-Piqueras, P., and Hoelzel, A. R. (2006). Phylogeography and alpha taxonomy of the common dolphin (*Delphinus* sp.). *J. Evol. Biol.* **19**, 943–954.

Perrin, W. F., Dolar, M. L. L., and Alava, M. N. R., (eds). (1996). "Report of the Workshop on the Biology and Conservation of Small Cetaceans and Dugongs of Southeast Asia". East Asian Seas Action Plan, UNEP, Bangkok.

Perrin, W. F., Dolar, M. L. L., and Robineau, D. (1999). Spinner dolphins (*Stenella longirostris*) of the western Pacific and Southeast Asia: Pelagic and shallow-water forms. *Mar. Mamm. Sci.* **15**, 1029–1053.

Perrin, W. F., Reeves, R. R., Dolar, M. L. L., Jefferson, T. A., Marsh, H., Wang, J. Y., and Estacion, J. (eds). (2005). Report of the Second

I

Workshop on the Biology and Conservation of Small Cetaceans and Dugongs of Southeast Asia. *CMS Technical Series Publication* **9**, 1–161. UNEP/CMS, Bonn.

Rudolph, P., Smeenk, C., and Leatherwood, S. (1997). Preliminary checklist of Cetacea in the Indonesian Archipelago and adjacent waters. *Zool. Verh. Leiden.* **312**, 1–48.

Sasaki, T., Nikaido, M., Wada, S., Yamada, T. K., Ying Cao., Hasegawa, M., and Okada, N. (2006). *Balaenoptera omurai* is a newly discovered baleen whale that represents an ancient evolutionary lineage. *Mol. Phylogenet. Evol.* **41**, 40–52.

Smith, B. D., and Perrin, W. F. (eds) (1998). *Asian Mar. Biol.* **14**, 1–214.

Sutaria, D., and Jefferson, T. A. (2004). Records of Indo-Pacific humpback dolphins (*Sousa chinensis*, Osbeck, 1765) along the coasts of India and Sri Lanka: An overview. *Aq. Mamm.* **30**, 125–136.

Turvey, S. T., *et al.* (2007). First human-caused extinction of a cetacean species? *Biol. Lett.* **3**, 537–540.

Van Waerebeek, K., Gallagher, M., Baldwin, R., Papastavrou, V., and Al-Lawati, S. M. (1999). Morphology and distribution of the spinner dolphin, *Stenella longirostris*, rough-toothed dolphin, *Steno bredanensis*, and melon-headed whale, *Peponocephala electra*, from waters off the Sultanate of Oman. *J. Cetacean Res. Manage.* **1**, 167–177.

Wada, S., Oishi, M., and Yamada, T. K. (2003). A newly discovered species of living baleen whale. *Nature* **426**, 278–281.

Yan, J., Zhou, K., and Yang, G. (2005). Molecular phylogenetics of 'river dolphin' and the baiji mitochondrial genome. *Mol. Phylogen. Evol.* **37**, 743–750.

**I**

# Intelligence and Cognition

## Bernd Würsig

Dolphins and sea lions are wonderful crowd pleasers in oceanaria: they leap, toss balls, swim through hoops or other obstacles, and vocalize on demand. In nature, they race toward boats, surf in the bow wave, and perform amazing acrobatics for—it seems—the pure joy of it. They are highly social, communicate, enjoy contact with humans, and appear to spend much of their time playing. It is therefore easy to understand why one of the most common questions asked by nonmarine mammal researchers is: "They are very intelligent, are they not?" This question is an excellent one, for it forces us to attempt to analyze what we mean by intelligence and how marine mammals might fit our definition of the concept.

Intelligence and cognition go hand in hand. The former refers to the mental capabilities of a human or nonhuman animal and usually is described by assessing problem-solving skills. The latter refers to the information processing within the animal and may be inferred by an analysis of how it appears to plan an action or alter it based on past experience. A "more intelligent" animal responds to an environmental stimulus faster or more accurately than the "less intelligent" one; the "more cognitive" action or animal may indicate more insight and more awareness of the problem than the "less cognitive" one. Unfortunately, past determinations of the concepts tended to be biased by our own human problem-solving skills and sensory systems and, to large degree, still are. However, we now know that indicators of intelligence can even be very different for different human societies or cultural backgrounds, i.e., within species. Can we say that the nature-living Australian aborigine who scores very low on an "intelligence" test designed with problem-solving questions of our modern industrial/electronic society is less intelligent than the student who takes the test in the industrialized world? If we answer "yes," we should be forced to "take the test" on the Aborigine's terms, perhaps by coming up with solutions of survival in the alternately extremely hot and cold, rugged, and food-poor outback. Similarly, it is not reasonable to study intelligence in dolphins and sea lions by asking them to solve problems relative to our linguistic communication or hand manipulation skills (in cognitive psychology, this is called the comparative approach). It is also unreasonable to compare "intelligences" of river dolphins with those of oceanic species by asking them to solve the same problems of space or objects.

An alternative to the comparative approach of describing intelligence and cognition is often called the "absolute method." It involves an attempt to find out how an animal thinks about things. Thinking is defined as mental manipulation of the internal representation of the external world, the stimulus. The cognitive animal is influenced to change its internal manipulations in part by past experience, and the more adept animal does this better than the "less intelligent" one. While it is difficult to judge mental processes, approximate tests and observations to do so have been devised and will be described later on.

One important window into intelligence and cognition for social species (and all marine mammals show a reasonable to very high level of sociality) is certainly communication. The individuals and species that communicate among each other in sophisticated and at times novel and interactive ways are likely the "more intelligent" (by, in this case, the prime criterion of communication) than those whose communication may be structured more rigidly or less complicated. The great US ethologist Donald Griffin has argued persuasively that communication is a major "window into the mind," not only of humans, dolphins, and other mammals, but of ants and honeybees as well (Griffin, 1981). He went on to postulate that it may be more parsimonious to explain the dance language of bees by considering them to be aware of their actions than it is to consider them reacting to complicated chains or sets of stimuli in unthinking ("noncognitive") fashion. This intriguing idea is not yet widely accepted by behavioral researchers and cognitive ethologists. However, most researchers now accept the possibility of "intelligences" and cognition in nonhuman animals, potentially very different in operating modes from our own, and not testable by traditional comparative approaches.

## I. Brain Size and Characteristics

A brain is needed to think and to have the chance of being aware (as a modern book, we need to mention the "brain" of artificially intelligent computers as well). Within a particular taxonomic group, larger and more complex brains tend to show a crude relationship to greater flexibility of behavior, adaptiveness to novel situations, and communication skills, i.e., intelligence. The relationship is imperfect, however, and is notoriously difficult to measure. For example, the entire brain has usually been used for descriptions of size and relative complexity, but there are motor, body function, and sensory parts of the brain that have very little to do with storing, processing, and integrating aspects of memory and thought (the latter occur only in the cerebrum).

Large mammals tend to have larger brains than small ones so brain size to body size ratios have been devised. One of these is the encephalization quotient (EQ), championed by Jerison (1973) and accepted by many researchers, albeit with often slightly different forms of calculation. The EQ is the ratio of brain mass observed to the brain mass predicted from an allometric equation of brain mass/body mass ratio of mammals as a whole. Therefore, an EQ of 1 means that the animal has an "average" brain size. It has been found for terrestrial mammals that EQs tend to be higher for those

## TABLE I
### Brain and Body Weights of Some Marine Mammals as Compared to Humans[a]

| Species | Brain weight (g) | Body weight (ton) | (Brain weight/body weight) × 100 |
|---|---|---|---|
| Pinnipeds | | | |
| Otariids | | | |
| Northern fur seal (*Callorhinus ursinus*) | 355 | 250 (male) | 0.142 |
| California sea lion (*Zalophus californianus*) | 363 | 101 | 0.359 |
| Southern sea lion (*Otaria flavescens*) | 550 | 260 | 0.211 |
| Phocids | | | |
| Bearded seal (*Erignathus barbatus*) | 460 | 281 | 0.163 |
| Gray seal (*Halichoerus grypus*) | 320 | 163 | 0.196 |
| Weddell seal (*Leptonychotes weddellii*) | 550 | 400 | 0.138 |
| Leopard seal (*Hydrurga leptonyx*) | 542 | 222 | 0.244 |
| Walrus (*Odobenus rosmarus*) | 1,020 | 600 | 0.170 |
| Odontocetes | | | |
| Common bottlenose dolphin (*Tursiops truncatus*) | 1,600 | 154 | 1.038 |
| Short-beaked common dolphin (*Delphinus delphis*) | 840 | 100 | 0.840 |
| Pilot whale (*Globicephala* sp.) | 2,670 | 3,178 | 0.074 |
| Killer whale (*Orcinus orca*) | 5,620 | 5,448 | 0.103 |
| Sperm whale (*Physeter macrocephalus*) | 7,820 | 33,596 | 0.023 |
| Mysticetes | | | |
| Fin whale (*Balaenoptera physalus*) | 6,930 | 81,720 | 0.008 |
| Sirenian | | | |
| Florida manatee (*Trichechus manatus latirostris*) | 360 | 756 | 0.047 |
| Human | 1,500 | 64 | 2.344 |

[a]Modified from Berta and Sumich (1999).

species that have few offspring, delayed physical and sexual maturity, long parental care, and generally high behavioral complexity (as estimated by degree of sociality and amount of behavioral flexibility). Examples are primates and social carnivores such as cats and canids. Within the primates, EQs tend to be higher for those in the categories just mentioned than for others, demonstrating that meaningful life history–brain size comparisons can be made at least in that group. Some aspects of general intelligence appear to be correlated with those higher EQs, from tree lemurs at the low end of the scale to the great apes at the pinnacle. Nevertheless, the very concept of EQ represents a general statement for potential comparison within or between taxa but does not represent a fundamental phenomenon *per se*.

Polar bears (*Ursus maritimus*), sea otters (*Enhydra lutris*), and pinnipeds have EQs around 1, as predicted by the overall regression line of brain weight to body weight among mammals. Their brains tend to weigh between 0.1% and 0.3% of their bodies. In other words, there is nothing unusual in brain size of these mammals relative to their terrestrial carnivore cousins. Because brains are energetically expensive, it has been postulated that those of pinnipeds that dive to great depths and hold their breaths for long periods of time might be smaller. At first glance, this appears to be the case for such divers as Weddell (*Leptonychotes weddellii*) and elephant seals (*Mirounga* spp.), but analyses by Worthy and Hickie (1986) for pinnipeds and Marino *et al*. (2006) for cetaceans showed that brain size and dive capability have no clear relationship.

Dolphins and whales have large brains but not all have large brain to body weights or EQs. The sperm whale (*Physeter macrocephalus*) has the largest brain on earth, weighing about 8 kg. This brain is in a body that weighs about 37,000 kg, however. The brain is only 0.02% of the weight of the body, or one-fifth of the size ratio of the smallest-brained pinnipeds. However, at large sizes, a straight-line allometric comparison is probably not fair by any measure, and perhaps the body of the sperm whale simply does not need relatively much brain mass for muscle movement, skin sensation, visceral action, and so on. The common bottlenose dolphin (*Tursiops truncatus*), however, has a brain weighing 1.6 kg in a body that weighs about 160 kg, making it—at 1% of body weight—one of the largest relative brains on earth. This competes only with several other dolphins, great apes, and humans (whose brains are about 1.5 kg in a 65-kg body, or about 2.3%) (Table I).

Baleen whales, like sperm whales, have large absolute brains (about 7 kg in an 80,000-kg fin whale, *Balaenoptera physalus*), but none have brain to body weight ratios as large as even the relatively small ones of the sperm whale. Sirenians have neither absolute nor relatively large brains, with the Caribbean manatee (*Trichechus manatus*) having a 300-g brain in a 750-kg body (0.04% of body weight). It has been postulated that the sirenian, an herbivore, increased body size to house a large gut for processing low-energy food, and a concomitant increase in brain size was not needed to support this size. Similarly, the huge size of baleen whales allows them to have huge mouths and to fast for extended periods. Again, this is a very

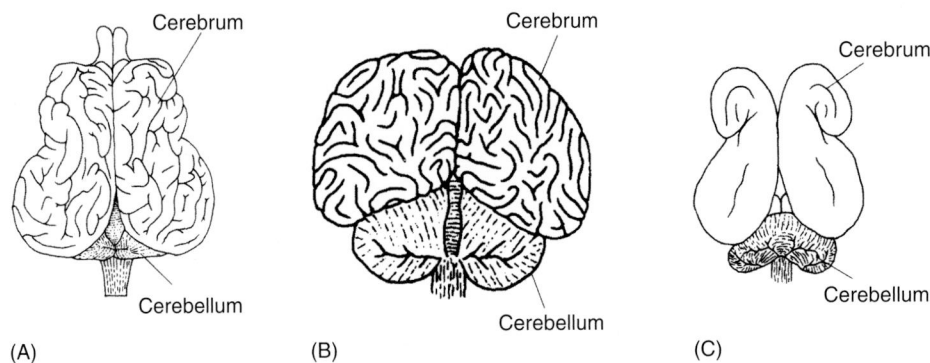

**Figure 1** *Comparison of pinniped,* Otaria flavescens *(A); cetacean,* Tursiops truncatus *(B); and sirenian,* Dugong dugon *(C) brain, dorsal views. Illustrated by P. Adam. From Berta and Sumich (1999).*

different allometric growth than that of a cow, for example, that is "simply" scaled up in size from a sheep (*Ovis* spp.).

Brain weight/body weight relationships are of general interest and have some relationship to relative information-processing capabilities. However, a larger absolute or relatively sized brain than that of another animal does not necessarily serve a "smarter" animal. The concept of intelligence is not a linear one; because there are so many "intelligences" depending on measure or the describer's concept of what is important, intelligence is not definable in absolute terms. All of the marine mammals have well-developed cerebrums. The brains of toothed whales have especially high amounts of neocortical folding and therefore high surface areas (Fig. 1). This quality is believed to be related to thought processes and behavioral flexibility. Polar bears, sea otters, and pinnipeds show a general "terrestrial carnivore" level of folding, whereas baleen whales and sirenians have very smooth cerebrums, with minimal surface areas. Nevertheless, the internal structure of whale and sirenian cerebrums is as well developed as those of other social mammals, and there is no reason to believe that these animals are "dumber" than others based on brain size and gross morphology. Perhaps their ways of finding and securing food, without the need of sophisticated hunting strategies as by toothed whales and carnivores, coupled with some aspects of their communication and society interactions, simply do not require the elaborate neocortical folding seen in many other mammals. Although much more work on brain size and sensory capabilities needs to be done, it is known that toothed whales and dolphins, who echolocate and use sounds intensively for communication, have well-developed auditory processing lobes. Pinnipeds and especially polar bears, however, have well-developed areas for processing smell.

Although brain size and complexity issues used to dominate our thinking about relative intelligence, it is becoming apparent that these can give only vague indicators of complexity of thought. It is likely that brains are structured more along lines of how an animal interacts with others and with its ecology. Higher brain function is a complex mixture of sensory inputs; processing, storing, and reactions to stimuli; innovation; and retrieval and use of previously stored events. Our inability to find clear links of these with measures of brain size and aspects of gross complexity may simply be because of the relatively primitive state of cognitive science, or it could be that clear "all-encompassing" rules of relationships simply do not exist. Promising avenues for future brain studies are noninvasive electrobiological and chemobiological studies from remote sensing of brain tissue while it is undergoing particular tasks. The findings to come

from such work will make our present discussions of brain function seem very primitive indeed.

## II. Learning

We know that dolphins and sea lions do marvelously complex things in captivity, but we also know that most of these behaviors have been reinforced from existing simpler ones and shaped into that dramatic leap to catch a fish. It is positive reinforcement behavior, or operant conditioning, that is at work; the animal gets a food or other reward for having done a good job. Typically, a sea lion or dolphin reward is one to three small fish per performed action. This is not unlike "training" a cat to run into the kitchen when it hears the sound of a can opener or the guppies in a home aquarium all aggregating near the top when a drawer with dried shrimp is opened. Operant conditioning can be performed on just about all animals on earth, and only speed of learning and some aspects of the amount of behavioral shaping can be indicators of a measure of "smartness" or relative intelligence. The animals learn, but there is not necessarily insight to their learning.

### A. Language Studies

It has long been known that dolphins have squeaks and whistles that appear to be used for communication. In captivity, bottlenose dolphins (*Tursiops* spp.) at times appear to imitate or mimic human and other sounds. These observations led an early dolphin communication researcher, John Lilly (Lilly, 1961), to attempt to communicate with dolphins by teaching them human speech. The results were a total failure, with not one clearly definable mimicked human sound; although dolphins are quite good at matching the staccato rhythm, in the form of bursts of sounds emitted in air (or underwater), of human speech. Dolphins do not have the vocal apparatus to produce human speech and may not have the neural wiring for it either. Nevertheless, Lilly's association with dolphins did not stop him from postulating that dolphins have great "extraterrestrial" intelligence. He used their large brains and their purported friendliness as arguments, but could not muster communicative interactions with humans as a part of the argument. Unfortunately, his popular writings have swayed countless laypersons; and a substantial "cult" of believers in extremely high dolphin intelligence and sophisticated human–dolphin communication, even at the nonverbal extrasensory level, has evolved. No other scientists have made similar claims, but the unscientific nature of Lilly's assertions deterred many others from studying dolphin and whale communication, and early on addressing intelligence and cognition in an obviously behaviorally flexible taxonomic order of mammals. By the way, some seals and beluga whales (*Delphinapterus leucas*) do have the ability to mimic human sounds, and one now-deceased harbor seal, *Phoca vitulina*, ("Hoover") at the New England Aquarium used to delight visitors with his rendition of simple sentences mimicked from a human pool cleaner, replete with the pool cleaner's Maine harbor-side accent. This ability does not indicate greater intelligence than in other seals and toothed whales that do not mimic. Instead, the ability (generally found in male pinnipeds) may relate to the way the animals use natural sounds in order to work out dominance relations for mating access to females and for other social interactions.

Two researchers who were not scared off by the unfounded claims of John Lilly and who nevertheless began language communication research, were Lou Herman of the University of Hawaii and Ron Schusterman of the University of California at Santa Cruz. Their studies began in the 1970s and are still ongoing, with a cadre of graduate students and postdoctoral researchers (Herman, 2002; Schusterman *et al.*, 2003).

Lou Herman's work consists of pioneering language studies conducted on dolphins. As in some of the successful studies with chimpanzees (*Pan troglodytes*), who like dolphins also cannot utter sophisticated human sounds, Herman uses a modified form of sign language, with volunteers' arms at poolside "talking to" common bottlenose dolphins. This is thus a gestural, not vocal, language. While Herman and his team have delved into many fascinating aspects of dolphin abilities, the basic study goes somewhat like this. Teach a dolphin a simple sentence, such as "fetch ball hoop," to indicate taking the ball from the hoop and bringing it to poolside. Once this command, reinforced by operant conditioning, is perfected, then the dolphin is presented with new, untrained challenges. Perhaps it is asked to "fetch hoop ball," or either hoop or ball or both objects are replaced with novel items never before put into this context. It is clear that dolphins quickly grasp the basic concept of "object 1," "object 2," and "command" and act correctly a large percentage of the time. These sentence structures have been made more complicated, with similarly positive results. The dolphins are reasonably good at syntactic structure, and they also seem to be able to conceptualize general categories of items. In others words, the ball used in training can be substituted successfully by another ball, and a gestural symbol ("word") can be made to refer to an item very specifically or to be more general, just as in human word use (Herman, 1986, 2006).

Ron Schusterman has repeated many of Herman's studies and invented other experiments of his own, but with California sea lions (*Zalophus californianus*). His results are essentially the same: sea lions are also adept at learning and extrapolating from human-like syntactic structure (Schusterman and Krieger, 1986). Interestingly, the conclusions drawn by these two fine researchers are quite different, indicating the state of knowledge and vibrant nature of the field of animal language and cognition. Herman interprets his findings as the animals using language. "Fetch hoop ball" represents a verb, a direct object, and an indirect object. Schusterman, however, states that there is no reason to believe that the animal perceives this interaction as anything more than an action command and that the linguistic concept "verb" need not enter into the equation. It is true that human children, for example, do not learn language in the structured operant conditioning style as performed here. Instead, we learned (mainly) from people talking around us and from acquiring words and syntactic rules as we went along. It was not until language was already well formed that we were required in school to understand syntactic structure by diagramming or labeling the parts of sentences.

Language acquisition learning in dolphins and sea lions has taught researchers much about imitation, learning, and mental processing abilities. It is undeniable that dolphins learn the basic concepts very rapidly (sea lions a bit less rapidly) and faster than most mammals except for chimpanzees and humans. This by itself indicates a high level of that nebulous and poorly defined "intelligence." However, whether these studies can be called language, or whether that is even an important question, is open to debate. We humans have taken human syntax and foisted it on nonhuman species. Nevertheless, the animals have done remarkably well with what they were given. Perhaps they can do even better as they communicate among each other with signs and symbols and emotive content for which they have evolved.

## B. Inventive Dolphins

Pinnipeds, sea otters, polar bears, and sirenians show elements of learning and play in captivity, but do not show the same kind of quick thinking or innovation as do some dolphins. However, most work has been done with dolphins, so there is some element of bias. Nevertheless, bottlenose dolphins and rough-toothed dolphins (*Steno bredanensis*), both with very large brains, are known as "the best" of performers in oceanaria. It is not clear whether these animals adjust better to captivity than others or whether they are innately more behaviorally flexible than others.

One interesting story of behavioral flexibility comes from a study carried out on two rough-toothed dolphins at Sea Life Park, Hawaii, in the mid-1960s. Karen Pryor, then head trainer at Sea Life Park, introduced a new demonstration into her on-stage performance with one of her dolphins named Malia. The intent was for Pryor to demonstrate to the audience how a previously unconditioned behavior could be reinforced by operant conditioning. In order to do so, she could not use a previously trained repertoire, but each day had to choose a simple behavior (such as a particularly high surfacing or loud blow) that the animal did and then reinforce it. After several days of this, Malia "spontaneously" recognized that "only those actions will be reinforced which had not been reinforced previously" (Pryor *et al.*, 1969). In order to receive rewards rapidly (or for the pure fun of it), Malia "began emitting an unprecedented range of behaviors, including aerial flips, gliding with the tail out of the water, and 'skidding' on the tank floor" (Pryor *et al.*, 1969). None of these behaviors had been shaped, none had even been seen before in the basic repertoire of dolphin behaviors at Sea Life Park! Pryor and her colleagues then repeated the work with an untrained female rough-toothed dolphin named Hou in order to assess experimentally whether creativity could be induced by operant conditioning in another dolphin and how long it would take. The experiment succeeded splendidly, and in a few trials, Hou was also presenting a new "act" after each one that received an operant reward.

Pryor *et al.* (1969) discussed their results very cautiously and reminded the reader that such training for novelty can probably be successful in horses and perhaps even pigeons as well. Many students of animal behavior and intelligence agree and are content to explain the development of novel behavior as simply a trained response. However, others have taken the experimental results further and suggested that much more insight than normal is required for the animal to "learn to learn" (the great philosopher Gregory Bateson called this "deuterolearning") and that the relatively quick manner in which dolphins "caught on" confirms their high intelligence. By the way, similar nonverbal training of reinforcement for novel behaviors has also been conducted for humans; the humans took about as long to realize what was being trained as did the dolphins (Maltzman, 1960). For Hou and for the humans, there was a period of strong frustration (even anger, in the humans) where they had not "caught on." They would be reinforced for a behavior, do it, and then not be rewarded for it ever again. It took some time for the "realization" to come that they then needed to exhibit a new behavior to get a reward. Once realized, the humans expressed great relief at having figured out "the problem," whereas the dolphins raced around the tank excitedly and displayed more and more novel and body-twisting behaviors—to the obvious delight of the researchers.

An interesting observation about dolphins is that they—at least bottlenose dolphins—readily recognize images of humans and of themselves in mirrors (Marino, 2002) and on television screens. Herman *et al.* (1990) were able to elicit correct answers from the televised image of a human giving sign-based directions, even to the point where only white-gloved hands were shown going through the signaling motions. This demonstrates that the animals were able to use representations of the gestural instructions. Several investigators have shown dolphins mirrors and real-time video images of themselves; the dolphins react to the images with curiosity and playfulness, moving their rostrums rapidly and following their own eye movements. Furthermore, the reactions to video images of other dolphins appear to indicate that the viewing animals recognize different individuals on the screen, including themselves. This indicates a "sense of self" and has been described as an important insight into cognition. Interestingly, most chimpanzees and other apes do not have this innate capability to see images on a flat screen as representations of themselves, others, or humans. They can be taught to process the images meaningfully, but only after prolonged exposure.

## III. Behavioral Complexity in Nature
### A. Carnivores and Sirenians

Most marine mammals are highly social, and we would expect that they have sophisticated ways of communicating with each other by showing innovative and variable behaviors in the face of social strategies and interactions. However, the less social species are also behaviorally complex. Examples are polar bears and sea otters. Polar bears have a large repertoire of "sneaking up" on their generally ice-bound prey. They move against the wind, come from the side of the sun glare, and use ice obstructions and stealth in order to surprise their prey. It has been reported that in captivity, they figure out rapidly how to unlatch (and unhinge) doors in order to escape or to move from pen to open enclosure. Sea otters are tool users, prying mussels and abalone from the substrate with rocks or stones they keep cached in an armpit while not in use. At the surface, they retrieve the tool in order to break open their shellfish food; at times using the rock as a hammer and at times laying it on their stomach and using it as an anvil. Individual sea otters have preferred methods of tool use, implying learning and innovation. Polar bears and sea otters are obviously "bright," but few behavioral studies or systematic investigations of learning have been conducted.

Pinnipeds are also behaviorally adept, and—as we have seen—sea lions can learn tricks and some aspects of language in captive training settings. They are all social mammals, especially while hauled out on land in order for males (of most species) to work out dominance relations with each other and for females to mate, give birth, and take care of their altricial (not well developed) young. Vocalizations, body postures, and smell are important aspects of communication. In the sea, most pinnipeds are less social (with the walrus, *Odobenus rosmarus*, being a strong exception), but they likely use more individualized but sophisticated strategies for finding and securing enough prey to survive. We expect that the animals need to periodically adapt to different types of prey, learn which could be physically harmful or poisonous, and learn how to detect and avoid large sharks, killer whales (*Orcinus orca*), and leopard seals (*Hydrurga leptonyx*). Many pinnipeds do not take their young out to sea with them, and therefore all learning to hunt and to survive needs to be without substantial help from more experienced adults. The author suspects, but has no proof for, that the brains of pinnipeds are adapted for relatively quick self-learning to survive and are less adapted or structured for social communication except as that needs to develop for procreation. Polar bears, sea otters, and sirenians would be an exception, although while generally less social than other marine mammals, mothers take prolonged care of their young while the young develop feeding and other skills. We assume, but again have no direct proof for this assertion, that the young learn more easily and completely in the presence of their mother.

### B. Baleen Whales

Baleen whales are social creatures, especially during mating times. Vocal communication is extremely important to them, with drum-like sounds of gray whales (*Eschrichtius robustus*), long low-frequency moans of blue whales (*Balaenoptera musculus*), short low-frequency grunts of fin whales, and the rich repertory of groans, moans, and scream-like sounds of the right (*Eubalaena* spp.) and bowhead (*Balaena mysticetus*) whales. Whereas all whales appear to produce sounds, the most elaborate (and best-studied) sounds are the songs of male humpback whales (*Megaptera novaeangliae*), which likely serve as a male–male (intrasexual) dominance signal, male–female (intersexual) mating advertisement, or both. The songs are copied from listening to each other, are long and complicated, and must require reasonably formidable powers of learning and memory. Baleen whales on the mating grounds also sort out dominance relationships in either aggressive (humpback) or more gentle but highly maneuvering surface-active groups of gray whales (*Eschrichtius robustus*), right whales, and bowhead whales. In the latter gray, right, and bowhead whale groups, it is likely that multiple males allow each other to inseminate a particular female and practice a form of sperm competition instead of physical competition to increase the chances of fathering a young. It is also likely, although behavioral researchers have gathered only incomplete glimpses of the possibility, that female whales make it more difficult for some males than others to mate with them, thereby performing mate choice of preferred partners. If true, it must be important for females to gauge the relative "goodness" of males from the complicated matrix of social sounds and close-up interactions that present themselves. In right and bowhead whales, an adult female has only one young every 2 to 5 years. The calf gestates in her body for 1 year and then is nursed for another. This low-reproductive rate means that she must take very good care of the young to attempt to assure its survival, and researchers would not be surprised at all to find that she also wants to choose the father of her young with care.

Baleen whales tend to be less social on the feeding grounds, although recent behavioral research indicates that at least some long-term bonds of affiliation persist between breeding and feeding grounds. This does not appear to be the norm, however. Generally, blue, humpback, gray, right, and bowhead whales (these five are the best-studied baleen whale species) aggregate at particular areas because of food concentrations. An aggregation due to an outside stimulus is not necessarily a social unit, although it can result in one. Some social interactions do occur, and it is even likely that the whales are paying close attention to each other in order to detect perhaps new or better feeding opportunities somewhere else. As well, blue whales often lunge into their food in tandem, apparently so as to provide a wall next to each other toward which fast-moving krill will not escape. Bowhead and right whales will swim in staggered formations of "echelons," side by side, apparently for the same purpose (Würsig, 1988).

The winner in the baleen whale feeding complexity department must surely be the amazing humpback whale. Humpback whales lunge into their fish food, alone and in coordinated groups up to an observed 22 animals. They are not merely aggregated in such a case, but all lunge (from below and toward the surface) at essentially the same time, coming to the surface within about 6 sec of each other. Apparently, although this is not yet proved, there is a vocal signal at the beginning of these highly coordinated lunges. One whale signals and others follow. Hitting the prey, a huge fish or bait ball, at one time presumably allows for each mouth to be better filled in the resultant prey's confusion than if one or a few mouths attacked. Humpback whales also flick their tails at prey and then circle to engulf it; they flick their long flippers forward as their mouths open, presumably to flash the white undersides of these flippers at the prey and to herd it more efficiently into the mouth. Finally, they release a stream of bubbles from their blowholes while circling around the prey and upward. The rising bubble screen forms an effective net around the prey, and the humpback (alone or with several others) then lunges toward the surface in the center of the "net," filling its capacious mouth with concentrated prey. It is unclear how flexible the several feeding behaviors are, but it is certain that several need social coordination. It is also likely that young humpbacks need to learn and perfect the techniques, and we assume that social learning is the major vehicle to do this.

## C. Toothed Whales

Toothed whales are highly social creatures, except for older adult male sperm whales who tend to be loners, some lone killer whales, and an extremely ("aberrant") low level of singles in many species of dolphins. Some of the deep-ocean beaked whales may be loners as well, but we have no good data on this point. Whereas most species are social, there are very different forms. Hector's dolphins (*Cephalorhynchus hectori*), harbor porpoises (*Phocoena phocoena*), and river dolphins tend to occur in small groups of up to a dozen animals, rarely more. We surmise that in at least some of these dolphins, individuals know each other well. Pantropical spotted (*Stenella attenuata*) and striped (*S. coeruleoalba*) dolphins of the open ocean, however, travel in "herds" of thousands of animals. Although there appear to be subgroups with at least some interindividual fidelity, it is very unlikely that all members of the herd know each other; some may never even meet each other. However, the herd acts as a coordinated unit, traveling at the same speed (which must be near the speed of the slowest animals), turning in essential unison, often diving in synchronized fashion. If a disturbance occurs along a flank or somewhere below, e.g., a shark zooms out of depth, there is rapid information transfer from animal to animal so that the group cascades away from the perceived danger in coordinated fashion. The information transfer is so rapid that we assume that animals are aware not only of their nearest neighbors, but are "looking beyond others," by sight when possible and probably also by echolocation. This is sort of a chorus line effect, where dancers coordinate their movements better by not merely paying attention to their nearest neighbors, but by anticipating the wave of raised legs, for example, as it (the wave) approaches. As well, it is likely that each dolphin

pays attention to the vocalizations and movements of others nearby and thus integrates response information in what the great cetacean researcher Ken Norris called a sensory integration system for dolphins (Norris *et al.*, 1994). Jerison (1986) used the idea of shared echolocation among dolphins to postulate that the animals share sensory inputs in a way that might synergistically enhance an expanded sense of "self." A human analogy would be if several people of a group could know their world and their place in it better by sharing neural data of aggregate visual systems. Jerison postulated that this potential sharing of echolocation data might itself account, at least in part, for large dolphin brains, but we have no direct information on this provocative point.

Coordination of group movements and activities need not be a matter of high intelligence and cognition, of course; and sensory awareness and a collective sensory integration system are well developed in schooling fishes, flocking birds, and so on. Instead, we might do better to look at the complexities of social interactions to gain "a window into the dolphins' minds" (after Griffin, 1981). Unfortunately, we do not yet know very much about the details of communication in delphinid cetaceans, but we do know enough to call it "complex." This complexity, at least for bottlenose dolphins, seems to be on a level of complexity of anthropoid primates such as some monkeys, the great apes, and humans (Marino, 2002).

Dolphins in a group are constantly aware of each other. A flipper touch here, a glance there, a slow echolocation-type click, a whistle. They interact by all sensory modalities available to them. We guess (and it is only a guess) that they are constantly gauging each other, deciding dominance/subservience relationships, seeking the comforting presence of relatives or those that they have found to be helpful in previous encounters, and avoiding those that might be aggressive. We know that there are at least occasions of political intrigue. Indian Ocean bottlenose dolphin (*Tursiops aduncus*) males of Shark Bay, southwestern Australia, have a strong tendency to form alliances to kidnap females. They apparently do so to gain access to reproductive females—access that might otherwise not be available because these males may not be of sufficiently high dominance status or would not be chosen by the females. Interestingly, super alliances of two or more alliances form in order to steal females from another male alliance (Connor *et al.*, 2000). Richard Connor speculates that great cognitive abilities and concomitant large brain evolution in humans, elephants, and some odontocetes may have evolved due to mutual dependence based on threats from predators as well as from members of the same species, i.e. social competition (Connor, 2007).

Toothed whales appear often to be structured along matriarchal (female-based) lines. Sperm whales, killer whales, pilot whales (*Globicephala* spp.), and bottlenose dolphins (of at least several populations) have close ties between mother and female young even after weaning, and in sperm and pilot whales, these ties appear to last for life. This means that potential cultural transmission of knowledge, from generation to generation, is expected to flow especially efficiently along female lines. Mom teaches young, young teaches its children, and so on. In a society of relatively resident killer whales of the US and Canadian Pacific Northwest, female and male offspring stay within the pod for life. This society is thus socially "closed." However, females mate with males outside, and the males mate with females of neighboring pods. Each pod is therefore reproductively matriarchal. These societies of relatively stable long-lived individuals are likely to develop behavioral cultures of their own. We have some evidence: killer whale pods have individually distinctive sound repertoires, or dialects. Individuals of pods can recognize each other easily as of that pod. It is likely, but not proved, that individuals also recognize each other as individuals by sound. In the more open but still matriarchal societies of at least one population of common bottlenose dolphins, studied by Randy Wells and colleagues (Wells 2003), of Sarasota, Florida, male offspring develop signature whistles (individually distinct sounds) more like those of their mothers than do female offspring. The moms and female offspring stay together as daughters mature. The sons, however, leave the natal group, roam elsewhere, and only now and then interact with their natal groups as adults. It is hypothesized that the similar signature sounds of moms and sons may provide an efficient means of recognition and thereby inbreeding avoidance (Sayigh *et al.*, 1995). Signature whistles are also copied by dolphins who are answering the original whistler. This rapid imitation may serve as a societal binding mechanism. It has been postulated that basic greetings and verbal recognition were prerequisite to the development of human language. Dolphins have the signature recognition portion of this capability (Janik, 2000).

Sound has been studied recently relative to kin and others of a society, and much more sound-based learning is likely to come to light as studies progress. This is to be contrasted with the relatively stereotyped sounds of the great apes, for example, that do not change much with age or social association (but then, in all fairness, apes are generally less vocally communicative and more visually based than cetaceans). However, it is also likely that not only vocal evidence for learning and social transmission will come to light. We have some hints, and only hints, here as well. Killer whales of Patagonia, Argentina, beach themselves in order to take South American sea lion (*Otaria flavescens*) and elephant seal pups. The beaching maneuver requires great skill, as the predator needs to gauge exactly where the prey is on the beach, after having seen the prey only from a distance and through murky nearshore waters, beyond the surf zone through which it needs to make its rush to the beach. As well, it needs to beach with such velocity and angle as to be certain that the spilling waves will allow it to reach deep water again. Killer whale adults have been described as making sham rushes at a beach and then waiting along the sides while young killer whales attempt the maneuver, usually clumsily and ineffectively, again and again. Now and then, the adult makes an intervening rush and then retreats to the side again. This behavior was pointed out as probable teaching by Argentine killer whale researchers Juan Carlos and Diana Lopez (1985) and has been studied in greater detail and verified since. It is unclear how well youngsters would learn beaching "on their own," but it is likely that it is transmitted culturally, as killer whale beaching behavior is found in only several populations worldwide. In Galveston Bay, Texas, certain female bottlenose dolphins and their young follow shrimp boats much more so than others, even maneuvering into the shrimp nets to take live fish and then wriggling out again while the shrimper is underway. This activity, video taped underwater, requires skill and dexterity to avoid being entangled in the fishing gear. The dolphins who exhibit this behavior do so "with ease," whereas others do not fish at all in this manner. Again, we wonder whether cultural learning and societal transmission of knowledge is important here. While culture has been explored in birds and nonhuman primates, less has been written on this subject for marine mammals (but see Whitehead, 1998; Rendell and Whitehead, 2001; Laland and Janik, 2006).

Even on an hour-to-hour basis, dolphins of a group are likely to be coordinating their activities superbly well. Although there are many potential examples (and each behavioral observer has his or her favorite ones), the author prefers one that he and his wife have studied for some time. Dusky dolphins (*Lagenorhynchus obscurus*) of the shallow waters of Patagonia, Argentina, coordinate activities to corral fish schools. It appears, and much more work is needed to

properly describe the individual behaviors, that dolphins (circling while vocalizing, tail swiping, and blowing bubbles) surround the prey ball and thereby cause it to tighten. They also herd the prey ball to the surface and then use the surface as a wall through which the prey cannot escape. Interestingly, dolphins do not appear to feed until the prey has been tightened and is at the surface. There may be a form of "temporary restraint," with all animals working toward the common good of getting the prey secured. This coordinated activity stands in stark contrast to taking individual advantage of the prey by grabbing a mouthful here or there and causing the prey to scatter and escape. As an example, sea lions that enter the area work on their own and are highly disruptive to the herding efforts of the dolphins. While we still need to look at the details of this behavior, to see whether kin, for example, help each other more often, we assume that much communication, learning, and individual trust need to go into such coordination. It is likely, but unknown at present, that animals know each other well enough as to have preferred "working" partners and have mechanisms for detecting and effectively ostracizing those cheaters who do not help or are disruptive at critical phases of prey gathering (Würsig *et al.*, 1989).

Such activities require individual recognition, concepts of strategies for dealing with different behaviors of fish schools, coordination, memory of past events, and potential teaching or at least learning from others; in short, considerable behavioral sophistication and flexibility. In New Zealand, far from Argentina and in a different deep-water environment, most dusky dolphins feed not on schooling fishes but on mesopelagic ("midwater, deep ocean") fishes at night. It has been found that a small subsegment of the dolphin population—the same individuals on a regular basis—travels to bays where dolphins take seasonal advantage of fish stocks to herd prey into tight balls as described earlier for Argentina. Because it is apparently the same animals doing so year after year, we believe that there might be cultural transmission of information here as well; only some have learned (or care) to take advantage of this particular foraging style.

A final example of at-times sophisticated-seeming behavior is certainly play. Almost all young mammals play, and this has been interpreted as gaining skills necessary to survive. It certainly seems like much fun as well. In dolphins and a handful of other mammals, adults habitually engage in behavior that is difficult to rationalize as anything but play. Dusky dolphins pull on the legs of floating birds, and individuals of several species perfect the balancing of pieces of kelp or other objects on their rostrums, flippers, dorsal fin, and tail. Play is not the purview of only dolphins, however. Adult baleen whales, sea lions, sea otters, and polar bears play with objects, at times for up to an hour or more. Play seems less common in phocid seals, but the imitative sounds of "Hoover" the harbor seal may have represented a form of vocal play. Play seems more rare (or absent?) in wild adult sirenians, but then long-term studies underwater have not been conducted.

## IV. Conclusions

Marine mammals are not of one taxonomic group and live in many varied ways; we therefore are not surprised to find that they have different brain sizes and ways of adapting to their ecologies, social structures, and behaviors. Because all use marvelously adaptive behaviors to help them survive, they are all "smart." However, such a general definition is not very satisfying. The polar bear, sea otter (and marine otter, *Lontra felina*, of Chile), pinnipeds, sirenians, and baleen whales all have behavioral characteristics and ways of living that might refer to "intelligences" not all that different from terrestrial mammals. Several of the dolphins (not all) stand out as being exceptionally large brained and behaviorally sophisticated; they are quick learners in captivity and have social structures and behaviors that appear to be highly complex and variable. While much of the large brains of these odontocetes may well be taken up by the neural processing required for ECHOLOCATION and other senses, as has often been speculated, it is highly likely that a large part of it also deals with relationships, learning, and long-term memory of events (Schusterman *et al.*, 1986).

Much more needs to be learned about dolphin whistle and click communication. However, it does not seem likely that their combinations of whistles and clicks can be termed "language" in the sense of putting sets of (for example) whistles together as referential communication for different objects or constructs (ideas). Instead, vocalizations seem to carry emotive content, signature information, and may well serve as an important tool for binding social relationships (Janik, 2000). Nevertheless, there are certain to be surprises to be gained from studies on delphinid communication as more information is gleaned. One important avenue for exploration is the extent to which communication and behavior have been transmitted from generation to generation, resulting in distinct cultures in such animals as sperm whales, killer whales, and several species of dolphins (Whitehead, 1998).

While we think of dolphin and other marine mammal "intelligences" and cognitive processes and realize what marvelous animals they are, it is also fair to contemplate their limits. Dolphins are beautifully tuned to the environments in which they have evolved for millions of years, but they do not necessarily have the capability to make behavioral extrapolations that seem to us very simple. A prime example is the fear (or mental incapability) of most wild dolphins to leap over obstructions. This has been a major problem for the tuna purse seining industry—dolphins caught in a net could easily all leap to freedom as the net is pursed. They do not do so because it is not in their repertoire to do so and are caught (and at times entangled and killed) as a result. Only dolphins trained to leap over nets will do so or some animals that seem to have "accidentally" (perhaps the most innovative ones?) discovered the capability in nature. This article ends on this theme of focused mental capabilities because it illustrates two related points: (1) dolphins are not those "super intelligent" beings as claimed by some aspects of the news media and many books and films and (2) dolphins are indeed "intelligent" for those things that they need to solve and interact with in their natural world, but their natural world is very different from ours.

## See Also the Following Articles

Behavior, Overview ■ Brain Size, Evolution ■ Communication ■ Culture ■ Group Behavior ■ Language Learning

## References

Berta, A., and Sumich, J.L. (1999). "Marine Mammals: Evolutionary Biology." Academic Press, San Diego, Calif., USA.

Connor, R. C. (2007). Dolphin social intelligence: Complex alliances in bottlenose dolphins and a consideration of selective environments for extreme brain size evolution in mammals. *Philos. Trans. R. Soc. Lond., B, Biol. Sci.* **362**, 587–602.

Connor, R. C., Read, A. J., and Wrangham, R. (2000). Male reproductive strategies and social bonds. *In* "Cetacean Societies: Field Studies of Dolphins and Whales" (J. Mann, R. C. Connor, P. L. Tyack, and H. Whitehead, eds). University of Chicago Press, Chicago, IL.

Griffin, D. R. (1981). "The Question of Animal Awareness: Evolutionary Continuity of Mental Experience." Rockefeller University Press, NY.

Herman, L. M. (1986). Cognition and language competencies of bottlenosed dolphins. *In* "Dolphin Cognition and Behavior: A Comparative Approach" (R. J. Schusterman, J. Thomas, and F. G. Wood, eds). Lawrence Erlbaum Associates, Hillsdale, NJ.

Herman, L. M. (2002). Vocal, social, and self-imitation by bottlenosed dolphins. *In* "Imitation in Animals and Artifacts" (K. Dautenhahn, and C. L. Nehaniv, eds). MIT Press, Cambridge, MA.

Herman, L. M. (2006). Intelligence and rational behavior in the bottlenosed dolphin. *In* "Rational Animals?" (S. Hurley, and M. Nudds, eds). Oxford University Press, Oxford, UK.

Herman, L. M., Morrel-Samuels, P., and Pack, A. A. (1990). Bottlenosed dolphin and human recognition of veridical and degraded video displays of an artificial gestural language. *J. Exp. Psychol. Gen.* **119**, 215–230.

Janik, V. M. (2000). Whistle matching in wild bottlenose dolphins (*Tursiops truncatus*). *Science* **289**, 1355–1357.

Jerison, H. J. (1973). "Evolution of the Brain and Intelligence." Academic Press, New York.

Jerison, H. J. (1986). The perceptual world of dolphins. *In* "Dolphin Cognition and Behavior: A Comparative Approach" (R. J. Schusterman, J. Thomas, and F. G. Wood, eds). Lawrence Erlbaum Associates, Hillsdale, NJ.

Laland, K. N., and Janik, V. M. (2006). The animal culture debate. *Trends Ecol. Evol.* **21**, 542–547.

Lilly, J. C. (1961). "Man and Dolphin." Doubleday Press, New York.

Lopez, J. C., and Lopez, D. (1985). Killer whales (*Orcinus orca*) of Patagonia, and their behavior of intentional stranding while hunting nearshore. *J. Mammal.* **66**, 181–183.

Maltzman, I. (1960). On the training of originality. *Psychol. Rev.* **67**, 229–242.

Marino, L. (2002). Convergence of complex cognitive abilities in cetaceans and primates. *Brain Behav. Evol.* **59**, 21–32.

Marino, L., Sol, D., Toren, K., and LeFebvre, L. (2006). Does diving limit brain size in ceataceans? *Mar. Mamm. Sci.* **22**, 413–425.

Norris, K. S., Würsig, B., Wells, R. S., and Würsig, M. (1994). "The Hawaiian Spinner Dolphin." University of California Press, Berkeley, CA.

Pryor, K. W., Haag, R., and O'Reilly, J. (1969). The creative porpoise: Training for novel behavior. *J. Exp. Anal. Beh.* **12**, 653–661.

Rendell, L., and Whitehead, H. (2001). Culture in whales and dolphins. *Behav. Brain Sci.* **24**, 309–382.

Sayigh, L. S., Tyack, P. L., Wells, R. S., Scott, M. D., and Irvine, A. B. (1995). Sex differences in signature whistle production of free-ranging bottlenose dolphins, *Tursiops truncatus*. *Behav. Ecol. Sociobiol.* **36**, 171–177.

Schusterman, R. J., Reichmuth-Kastak, C., and Kastak, D. (2003). Equivalence classification as an approach to social knowledge: From sea lions to sirenians. *In* "Animal Social Complexity: Intelligence, Culture, and Individualized Societies" (F. B. M. deWaal, and P. L. Tyack, eds). Harvard University Press, Cambridge, MA.

Schusterman, R. J., and Krieger, K. (1986). Artificial language comprehension and size transposition by a California sea lion (*Zalophus californianus*). *J. Comp. Physiol.* **100**, 348–355.

Schusterman, R. J., Thomas, J. A., and Wood, F. G. (eds) (1986). "Dolphin Cognition and Behavior: A Comparative Approach." Lawrence Erlbaum Associates, Hillsdale, NJ.

Wells, R. S. (2003). Dolphin social complexity: Lessons from long-term study and life history. *In* "Animal Social Complexity: Intelligence, Culture, and Individualized Societies" (F. B. M. deWaal, and P. L. Tyack, eds). Harvard University Press, Cambridge, MA.

Whitehead, H. P. (1998). Cultural selection and genetic diversity in matrilineal whales. *Science* **282**, 1708–1711.

Worthy, G. A. J., and Hickie, J. P. (1986). Relative brain size of marine mammals. *Am. Nat.* **128**, 445–459.

Würsig, B. (1988). The behavior of baleen whales. *Sci. Am.* **256**, 102–107.

Würsig, B., Würsig, M., and Cipriano, F. (1989). Dolphins in different worlds. *Oceanus* **32**, 71–75.

I

# The International Whaling Commission

## G.P. DONOVAN

The International Whaling Commission (IWC) is the intergovernmental body established in 1946 to conserve whale stocks and regulate whaling. Membership is open to any sovereign state. There were 78 member nations (Table I) in December 2007.

## I.  Historical Background

Whaling cannot be put forward as an example of the successful sustainable management of a renewable resource. From the start of

**TABLE I**

**List of Member Nations of the IWC in December 2007, With Dates of Adherence. Some Nations Have Left and Rejoined. The Date of Adherence Refers to Their Most Recent Adherence. Thirty-eight Countries Have Joined Since 2000**

| | |
|---|---|
| Antigua and Barbuda | 21/07/1982 |
| Argentina | 18/05/1960 |
| Australia | 10/11/1948 |
| Austria | 20/05/1994 |
| Belgium | 15/07/2004 |
| Belize | 17/06/2003 |
| Benin | 26/04/2002 |
| Brazil | 04/01/1974 |
| Cambodia | 01/06/2006 |
| Cameroon | 14/06/2005 |
| Chile | 06/07/1979 |
| People's Republic of China | 24/09/1980 |
| Costa Rica | 24/07/1981 |
| Côte d'Ivoire | 08/07/2004 |
| Croatia | 10/01/2007 |
| Cyprus | 26/02/2007 |
| Czech Republic | 26/01/2005 |
| Denmark | 23/05/1950 |
| Dominica | 18/06/1992 |
| Ecuador | 10/05/2007 |
| Finland | 23/02/1983 |
| France | 03/12/1948 |
| Gabon | 08/05/2002 |
| The Gambia | 17/05/2005 |
| Germany | 02/07/1982 |
| Greece | 16/05/2007 |
| Grenada | 07/04/1993 |
| Guatemala | 16/05/2006 |
| Guinea-Bissau | 29/05/2007 |

**TABLE I** (Continued)

| | |
|---|---|
| Republic of Guinea | 21/06/2000 |
| Hungary | 01/05/2004 |
| Iceland | 10/10/2002 |
| India | 09/03/1981 |
| Ireland | 02/01/1985 |
| Israel | 07/06/2006 |
| Italy | 06/02/1998 |
| Japan | 21/04/1951 |
| Kenya | 02/12/1981 |
| Kiribati | 28/12/2004 |
| Laos | 22/05/2007 |
| Luxembourg | 10/06/2005 |
| Republic of Korea | 29/12/1978 |
| Mali | 17/08/2004 |
| Republic of the Marshall Islands | 01/06/2006 |
| Mauritania | 23/12/2003 |
| Mexico | 30/06/1949 |
| Monaco | 15/03/1982 |
| Mongolia | 16/05/2002 |
| Morocco | 12/02/2001 |
| Nauru | 15/06/2005 |
| Netherlands | 14/06/1977 |
| New Zealand | 15/06/1976 |
| Nicaragua | 05/06/2003 |
| Norway | 03/03/1948 |
| Oman | 15/07/1980 |
| Republic of Palau | 08/05/2002 |
| Panama | 12/06/2001 |
| Peru | 18/06/1979 |
| Portugal | 14/05/2002 |
| Russian Federation | 10/11/1948 |
| San Marino | 16/04/2002 |
| St. Kitts and Nevis | 24/06/1992 |
| St. Lucia | 29/06/1981 |
| St. Vincent and The Grenadines | 22/07/1981 |
| Senegal | 15/07/1982 |
| Slovak Republic | 22/03/2005 |
| Slovenia | 20/09/2006 |
| Solomon Islands | 10/05/1993 |
| South Africa | 10/11/1948 |
| Spain | 06/07/1979 |
| Suriname | 15/07/2004 |
| Sweden | 15/06/1979 |
| Switzerland | 29/05/1980 |
| Togo | 15/06/2005 |
| Tuvalu | 30/06/2004 |
| United Kingdom | 10/11/1948 |
| Uruguay | 27/09/2007 |
| United States | 10/11/1948 |

(continues)

the "commercial" exploitation of whales, the story was usually one of eventual overexploitation. Modern commercial whaling began with the invention of the explosive harpoon combined with the development of steam-powered catcher boats in the 1860s (Tønnessen and Johnsen, 1982). This allowed whalers to take the faster-swimming rorquals (e.g., the blue, *Balaenoptera musculus*, and fin, *B. physalus* whales). The promise of large numbers of whales caused whalers to investigate the Antarctic, and the first whaling station was established on South Georgia in 1904 and took 195 whales. By 1913, there were 6 true land stations and 21 floating factories that had to be moored in suitable harbors; the total catch was 10,760 whales. The invention of the stern slipway in 1925 allowed vessels to operate in offshore waters and by 1930/31, 41 factory ships took over 37,000 whales. This overproduction led to a catastrophic decline in the price of whale oil.

It was the fear of low prices rather than the fear of overexploiting whale stocks that was the driving force behind early moves to limit catching. Despite attempts under the auspices of the League of Nations to establish some international control, the production agreements negotiated amongst themselves by the whaling companies produced the first effective limitation of catches in the early 1930s.

World War II caused a world shortage in the supply of fats and several nations had their eyes on profits from pelagic whaling. It was in this light, and the experience gained in developing international agreements just before the war, that discussions were held in London in 1945 and in Washington in 1946 on the international regulation of whaling.

## II. Establishment of the IWC

The International Convention for the Regulation of Whaling was signed at the 1946 Conference. It was a major step forward in the international regulation of natural resources as it was one of the first to place "conservation" at the forefront (Gambell, 1977; Allen, 1980). The Convention was established *to provide for the proper conservation of whale stocks and thus make possible the orderly development of the whaling industry*. This was a laudable aim, but finding the difficult balance between conservation (*achieve the optimum level of whale stocks*) and the interests of the whaling industry' (*without causing widespread economic and nutritional distress*) has dominated the history of the IWC.

An important feature of the Convention was that it established a mechanism whereby regulatory measures included in the *Schedule* to the Convention (catch limits, seasons, size limits, inspections, etc.) could be amended when necessary by a three-quarters majority of members voting (excluding abstentions).

The Convention also formally assigned importance to the need for scientific advice, requiring that amendments to the regulations "shall be based on scientific findings." To this end, the Commission established a Scientific Committee comprising scientists nominated by member governments (and latterly invited experts when appropriate).

Despite this, there are aspects of the Convention that have attracted criticism. For example, any government can "object" to any decision with which it does not agree within a certain time frame. This (along with the right of nations to unilaterally issue permits to catch whales for scientific purposes) has led to accusations that the IWC is "toothless." However, it should be recognized that without these provisions, the Convention would probably have never been signed.

From a management perspective, a more serious flaw was that the IWC could neither restrict operations by numbers or nationality nor allocate quotas per operation. Although it may be questioned whether the IWC could have agreed to national quotas or numbers

of vessels, certainly if such limitations had been reached this would have reduced the management problems associated with increasing numbers of vessels chasing limited quotas.

The Convention formally established the IWC. The IWC comprises one Commissioner from each government who has "one vote and may be accompanied by one or more experts and advisers."

## III. The IWC Before 1972

Perhaps the most serious problem of early management was the use of the *Blue Whale Unit (BWU)*. In terms of oil yield, one blue whale was considered equal to 2 fin, 2.5 humpback (*Megaptera novaeangliae*), or 6 sei (*Balaenoptera borealis*) whales. In 1945, a catch limit of 16,000 BWU was set (suggested by three scientists as being a "reassuring" value in between their estimate of 15–20,000; in fact scientific information on stock status was poor). The flaw in the BWU system is apparent—it allows catching of depleted species below levels at which catching that species alone would be economically unviable. This is apparent from the catch data up to the 1970s, which reveal that as blue whale catches declined, so fin whale catches (the next largest species) increased until they too were overexploited and sei whale catching began.

The lack of national quotas resulted in an "Olympic" system, where it became a race to catch as many whales as possible before the total quota was reached—leading to waste during processing and the use of increasing numbers of catcher boats (129 in 1946/47 and 263 in 1951/52). This neither made economic sense nor encouraged conservation.

Despite the early optimism, as early as 1952 many recognized that the catch quota was too high. The difficulty was in getting all the whaling nations to agree to a reduction—if one nation objected, then all objected. This was the start of a difficult period for the IWC. A combination of short-term economics, greed, and a lack of incontrovertible scientific evidence led to a critical situation for whale stocks; the benefit of the doubt was always given to the industry. Even a single voice on the Scientific Committee (often the Dutch scientist, Slijper) negated its attempts to persuade governments of the desperate need for quota reductions. At one stage, both the Netherlands and Norway withdrew from the Commission and its survival seemed in doubt. The Commission even appointed a group of three, later four scientists—experts in the relatively new science of population dynamics to produce independent advice (Allen, 1980). After considerable argument and controversy, by 1971/72, the catch limit had been reduced to 2300BWU and certain species, including blue and humpback whales had been protected from commercial whaling.

## IV. A Period of Change: 1972 to the "Moratorium"

In 1972, the UN Conference on the Human Environment called for an increase in whale research, a 10-year "moratorium" on commercial whaling and a strengthening of the IWC. Although proposals for a 10-year moratorium were subsequently tabled at the IWC, they failed to reach the required three-quarters majority, largely because the IWC Scientific Committee believed that management on a stock-by-stock basis (Antarctic catches were first set by species in 1972) was the most sensible approach—if required each stock could be independently protected.

The UN resolution was, however, taken seriously by the IWC. By 1976, a permanent Secretariat had been established in Cambridge, an International Decade of Cetacean Research had been declared, and a management procedure (the *New Management Procedure* or *NMP*) had been adopted. The NMP was aimed at bringing all stocks of whales to an optimum level at which the largest number of whales can be taken consistently (the *maximum sustainable yield* or MSY) without depleting the stock. It also gave complete protection to stocks at 54% of their estimated pre-exploitation size, i.e., well before they became endangered.

The NMP was regarded as a major step forward in the management of whaling. It appeared to take the issue of catch limits largely out of the hands of the politicians and into those of the Scientific Committee. In addition, from 1973, the long-awaited international observer scheme was in operation, aimed at ensuring that new catch limits were enforced.

A major feature over this period was the increase in IWC membership. In 1963, there were 18 member nations, of which only 4 were non-whaling countries; in 1978, there were 17 of which 8 were non-whaling and by 1982, membership was 39. Of the 13 whaling nations, 3 had only aboriginal/subsistence operations (Denmark, the United States, and St. Vincent and The Grenadines).

The 1979 meeting was a turning point in the Commission's history. Doubts had been expressed by some over (1) the theoretical and practical application of the NMP and (2) the morality of whaling, irrespective of the status of the stocks. At that meeting, a proposal to end pelagic whaling for all species except minke whales was adopted and a Sanctuary was declared for the Indian Ocean outside the Antarctic. Whereas the onus in the past had been for positive evidence of a decline in stocks before a reduction in catch limits was agreed, positive evidence was now required if a catch limit was to be set. By 1982, a *Schedule* amendment was adopted that implemented a pause in commercial whaling (or to use popular terminology, a "moratorium") from 1986. Originally, four whaling nations, Japan, Norway, Peru, and the USSR, lodged objections to this decision; Peru and Japan subsequently withdrew theirs. In the year 2000, only Norway carried out commercial whaling.

One obvious question to ask as the IWC's moratorium came into effect was whether the Commission been a success. At one level the answer must be no—indeed it could be argued that it had been a disaster. For example, in the Antarctic, the most important area to the IWC initially, (a) blue and fin whales had been reduced to at best 5% and 20% of their original numbers, and possibly much less, respectively—hardly a good example of "conservation of whale resources," and (b) the 1983/84 catch was 6655 minke whales (mainly Antarctic minke whales, *Balaenoptera bonaerensis*), a species not considered worth catching in 1947/48 when the catch in BWU was 25 times greater—hardly "the orderly development of the whaling industry."

So, had the IWC achieved anything? First of all, while it is easy with current levels of knowledge to criticize the IWC's performance, it has to be said that modern whaling had not resulted in the extinction of any species—IWC actions, while insufficient, were better than nothing. Since the 1970s, the trend has been very much toward conservative catch limits based on scientific advice, to a degree probably unparalleled in any fisheries commission. It has been argued by some that this trend reached unreasonable limits with the introduction of the "moratorium." It is indicative of the inherent problems within the Commission that the same decision is hailed by some as its greatest success and others as its most abject failure.

## V. The Commission Today

Since 1976, the IWC has had a full-time Secretariat (of 15–20 people) with headquarters in Cambridge, United Kingdom. Each year,

the Annual Meeting of the Commission is held, either by invitation in any member country, or in the United Kingdom. The Scientific Committee (comprising up to 200 scientists) meets in the 2 weeks immediately before the main Commission meeting, and it may hold special meetings during the year. The information and advice it provides form the basis upon which the Commission develops the regulations for the control of whaling.

### A. Management Issues

The primary function of the IWC is the conservation of whale stocks and the management of whaling. In addition to commercial whaling, the IWC has recognized the discrete nature of aboriginal subsistence whaling and allowed aboriginal catches from stocks that have been reduced to levels at which commercial whaling would be prohibited.

*1. Commercial Whaling* After the moratorium decision, the Scientific Committee recognized the need to develop management procedures that did not repeat past mistakes and the limitations of both the data it had and the data it was likely to obtain. In order to test the possible management strategies, it took the innovative and far-reaching approach of using computer simulations of whale populations over a long (100-year) period.

The most important part of any development process is the determination of *management objectives*. These were set by the Commission and can be summarized as (1) catch limits should be as stable as possible, (2) catches should not be allowed on stocks below 54% of the estimated carrying capacity (as in the NMP), and (3) the highest possible continuing yield should be obtained from the stock. The highest priority was given to objective (2).

After 8 years of intense work, the Committee developed a procedure for determining safe catch limits that required knowledge of only two essential parameters: (1) estimates of current abundance taken at regular intervals and (2) knowledge of past and present catches. Intensive testing of the procedure to numerous assumptions and problems had been undertaken and some of these are summarized in Table II.

The way in which catch limits are calculated from the required information is specified by the *Catch Limit Algorithm (CLA)*. This is a "feedback" procedure—as more information accumulates from sighting surveys (and catches, if taken), then the estimates of necessary parameters are refined. In this way, the procedure constantly monitors itself. Catch limits are set for a period of 5 years. The *CLA*

---

**TABLE II**
**Some Examples of the Trials the Management Procedure Had to Be Able to Cope With**

- Several different population models and associated assumptions
- Different starting population levels, ranging 5–99% of the "initial" population size
- Different MSY levels, ranging 40–80%
- Different MSY rates, ranging 1–7% (including changes over time)
- Various levels of uncertainty and biases in population size
- Changes in carrying capacity (including reduction by half)
- Errors in historic catch records (including underestimation by half)
- Catastrophes (irregular episodic events when the population is halved)
- Various frequencies of surveys

was initially tested on the assumption that it is applied to known biological stocks. To date, testing for specific species and areas has only been carried out for common minke whales (*Balaenoptera acutorostrata*) in the North Atlantic and North Pacific, Antarctic minke whales, and Bryde's whales (*B. edeni*) in the western North Pacific. Unless such testing has occurred and the results indicate otherwise, catch limits under the Revised Management Procedure (RMP) will be zero. It is clear that for very many populations, such as blue whales in the Southern Hemisphere, it will be a very long time before catches would be allowed under the RMP.

The *CLA* plus the rules about, *inter alia*, stock boundaries, allocation of catches to small areas, what to do if many more of one or other sex are caught, form the *Revised Management Procedure* or *RMP*. The RMP sets a standard for the management of all marine and other living resources. It is very conservative and this is a reflection of the relative priorities assigned to the objectives, the level of uncertainty in the information on abundance, productivity and stock identity of whale stocks, and the fact that many years are required before the *CLA* refines its estimates of the required parameters.

Although these scientific aspects were adopted by the IWC in 1994, its actual implementation is a political decision. The IWC will not set catch limits for commercial whaling until it has agreed and adopted a complete Revised Management Scheme (RMS). Any RMS will also include a number of nonscientific issues, including inspection and enforcement, and perhaps humaneness of killing techniques. The importance of an international inspection scheme was highlighted by the recent discovery of widespread falsification of catch data by Soviet whaling operations prior to 1972. There is at present a stalemate in discussions over an RMS. The only commercial whaling being undertaken in 2007/08 is that for common minke whales off Norway, who objected to the moratorium decision.

*2. Aboriginal Subsistence Whaling* Aboriginal subsistence whaling is permitted from Denmark (Greenland: bowhead—*Balaena mysticetus*—, fin and common minke whales), the Russian Federation (Siberia: gray—*Eschrichtius robustus*—and bowhead whales), St. Vincent and The Grenadines (Bequia: humpback whales), and the United States (Alaska: bowhead and gray whales). It is the responsibility of the Committee to provide scientific advice on safe catch limits for such stocks. With the completion of the RMP, the Scientific Committee began the process of developing a new procedure for the management of aboriginal subsistence whaling (AWMP) that takes into account the different objectives for the management of such whaling as compared to commercial whaling. Following the simulation approach used in the RMP development process, the Committee has developed Strike Limit Algorithms for bowhead and gray whales. It is focussing work now on the Greenlandic fisheries. The Commission will be establishing an Aboriginal Whaling Scheme that comprises the scientific and logistical (e.g., inspection/observation) aspects of the management of all aboriginal fisheries. The scientific component will comprise some general aspects common to all fisheries and an overall AWMP within which there will be common components and case-specific components.

*3. Scientific Permit Whaling* A major area of discussion since the moratorium has been the issuance of permits by national authorities for the killing of whales for scientific purposes. The right to issue them is enshrined in Article VIII of the Convention (that furthermore requires that the animals be utilized once the scientific data have been collected) and prior to 1982, over 100 permits had been issued by a number of governments, including Canada, United States, USSR,

South Africa, and Japan. Since the "moratorium," Japan, Norway, and Iceland, have issued scientific permits as part of their research programs. The discussion has centered on accusations that such permits have been issued merely as a way around the moratorium decision contrasted with claims that the catches are essential to obtain information necessary for rational management and other important research needs. All proposed permits have to be submitted for review by the Scientific Committee, but the ultimate responsibility for their issuance lies with the member nation. The Committee has been divided on the value of the programs reviewed to date. Only Japan has issued scientific permits for the year 2007/08 [850 ± 10% Antarctic minke whales and 50 fin whales in the Antarctic; 150 minke, 50 Brydes, 50 sei, and 10 sperm whales (*Physeter macrocephalus*) in the western North Pacific]. As in previous years, a majority of the Commission members urged Japan to refrain from issuing the permits.

*4. Small Cetaceans* It can be argued that no species of large whale is endangered by whaling today and will not be by any resumption of whaling under the RMS or AWMP. Threats to those species, such as the North Atlantic right whale (*Eubalaena glacialis*) and the western gray whale, that remain severely reduced, do not include direct hunting. The most seriously threatened cetaceans (by direct hunting and incidental captures in fisheries) are a number of species and populations of the smaller cetaceans. At present, there is no single international body responsible for their conservation and management. There is considerable disagreement within the IWC as to whether the present Convention is sufficient to allow the IWC to assume such a role. Fortunately, there is general agreement that the IWC Scientific Committee can consider the status of small cetaceans and provide advice to governments even though the IWC cannot set management regulations—it is to be hoped that governments individually and collectively respond. It remains a matter of some urgency that an international agreement or series of regional agreements be reached to ensure the conservation of small cetaceans.

*5. Whalewatching* The IWC is involved (in a monitoring and advisory capacity) with aspects of the management of whalewatching as one type of sustainable use of cetacean resources. It has adopted a series of objectives and principles for managing whalewatching proposed by the Scientific Committee.

## B. Other Scientific issues

The Commission funds and acts as a catalyst for a good deal of cetacean research (in the year 1999/2000 some $400,000 was allocated to scientific research in addition to the IWC-related work undertaken by individual member governments. One major program is a series of Antarctic cruises to estimate abundance that has been carried out since 1978. These are now called SOWER circumpolar cruises (Southern Ocean Whale and Ecosystem Research) and include a component dedicated to blue whales.

With increasing awareness that detrimental environmental changes may threaten whale stocks, the IWC has recently accorded priority to research on the effects of such changes on cetaceans. Whilst the RMP adequately addresses such concerns, the Scientific Committee has agreed that the species most vulnerable to such threats would be those reduced to levels at which the RMP, even if applied, would result in zero catches. It has developed considerable effort into examining the effects of chemical pollutants on cetaceans, the effects of noise, including seismic surveys, and habitat degradation, including the effects of climate change and ozone depletion. It is also increasing collaboration and cooperation with governmental, regional and other international organizations working on related issues.

The work in these areas carried out by the IWC Scientific Committee is recognized worldwide. The Commission has increasingly published scientific reports and papers; this culminated in the launch of the *Journal of Cetacean Research and Management* in 1999.

### C. Politico-ethical Issues

Of prime consideration from both a scientific and an ethical viewpoint is the possibility of extinction of any population due to whaling. No population of whales is currently under threat of extinction from whaling, and it is clear that any acceptable management procedure will ensure that this cannot happen. However, this presumes an acceptance that whales are a natural resource to be harvested. While this is certainly the stated position of many members of the IWC, it is not universally accepted. A wide range of opinions have been expressed, ranging from the belief that whales are such a "special" group of animals that they should not be killed under any circumstances, through the view that they should not be commercially killed as whale products are not essential, to the view that whales are a natural resource to be used like any other.

In this regard, the question of humane killing has once more arisen within the IWC, with some nations stating that even if a safe management procedure is adopted, catch limits should not be set unless a "satisfactorily humane" killing method is available. This subject has been addressed several times during the history of the IWC and the Commission has been active in promoting work on more humane killing techniques for both commercial and aboriginal subsistence whaling. However, obtaining agreement on what comprises a "satisfactorily humane" technique will not be simple. In particular, in the case of aboriginal subsistence whaling, arguments of tradition and culture can clash with the adoption of modern technology.

### VI. Conclusion

Many of the earlier discussed "politico-ethical" issues are linked to questions of culture and freedom; they are complex and almost inevitably will not be resolved unanimously. There is clearly a divergence of opinion within the IWC on such matters to an extent unparalleled in any similar organization. It is, for example, difficult to think of any fisheries organization where some of the members believe it is immoral to catch fish under any circumstances. This is not the place to enter into a philosophical debate over the rights of nations or groups of nations to impose their moral values on others, but merely to point out the necessity of such a debate and the need for a degree of compromise if the IWC is not going to fragment, with potentially serious consequences for the world's whales and other cetaceans (Donovan, 1992).

### References

Allen, K. R. (1980). "Conservation and Management of Whales." University of Washington Press, Seattle and Butterworth & Co, London.
Donovan, G. P. (1992). The International Whaling Commission: Given its past, does it have a future? *In* "Symposium «Whales: Biology—Threats—Conservation" (J. J. Symoens, ed.), pp. 23–44. Royal Academy of Overseas Sciences, Brussels, Belgium.
Gambell, R. (1977). Whale conservation. Role of the International Whaling Commission. *Mar. Policy*, 301–310.
International Whaling Commission. (1950–1998). *Rep. Int. Whal. Commn* 1–48.
International Whaling Commission. (1999a-present). *J. Cetacean Res. Manage*.
International Whaling Commission. (1999b-present). *Ann. Rep. Int. Whal. Commn*.
Tønnessen, J. N., and Johnsen, A. O. (1982). "The History of Modern Whaling." C. Hurst & Co., London.

# Inuit and Marine Mammals

ANNE M. JENSEN, GLENN W. SHEEHAN, AND
STEPHEN A. MACLEAN

Inuit is a northern Alaskan term meaning "people" that has come to include the native "Eskimo" peoples of Chukotka, northern Alaska, Canada, and Greenland (Fig. 1). Inuit represent one extreme of the hunter–gatherer paradigm, relying almost exclusively on hunting to thrive in one of Earth's harshest environments, the Arctic. Most Inuit hunting has focused on marine mammals, with the bowhead whale (*Balaena mysticetus*) making up a central part of the harvest, particularly in the Western and Eastern Arctic coastal areas. Whaling was important to Inuit from Alaska to Greenland and underwrote the formation and survival of permanent sedentary villages on Alaska's arctic coast. When whaling was not feasible, Inuit depended upon caribou (*Rangifer tarandus*) and other marine mammals.

Inuit have hunted marine mammals and caribou for thousands of years. The Birnirk culture (AD 400–900) was the first to successfully incorporate whale hunting into their subsistence regime. Whaling was completely integrated into the succeeding Thule culture starting around AD 900. Around AD 1200, Thule folk and their whaling culture spread out of Alaska and into Canada and Greenland.

The ancestral Inuit tool kit employed raw materials from hunted species plus worked stone and driftwood. Their technology depended heavily on compound (multipart) tools often incorporating several types of raw material. A harpoon might employ a driftwood shaft, a foreshaft made from caribou antler, a socket piece from walrus (*Odobenus rosmarus*) bone, a finger rest made from walrus ivory, lashings made from caribou sinew, a head made from whale bone, a blade made from slate, a line made from walrus hide, and a sealskin float.

The harpoon head toggled, or turned, 90° once it was thrust into the animal, preventing withdrawal. As the head toggled, the shaft fell away, leaving a hide cord running from the head back to the hunter or to a float. The float was a sealskin with all but one of its orifices sewn shut. The remaining orifice was used to inflate the float through an ivory inflation nozzle, which was then plugged with a piece of driftwood. The float marked the prey's location and slowed it down, tiring it as it attempted to swim or dive. The first commercial whalers to enter the northern sea near Greenland in the fourteenth century found Inuit hunting bowhead whales from umiat (skin-covered driftwood framed boats), using compound harpoons with toggling heads. By the early seventeenth century, Greenlandic Inuit were severely impacted by commercial whaling, which decimated the whale stocks, perhaps even eliminating the Svalbard stock upon which the east Greenlanders seem to have depended. In Canada, much commercial

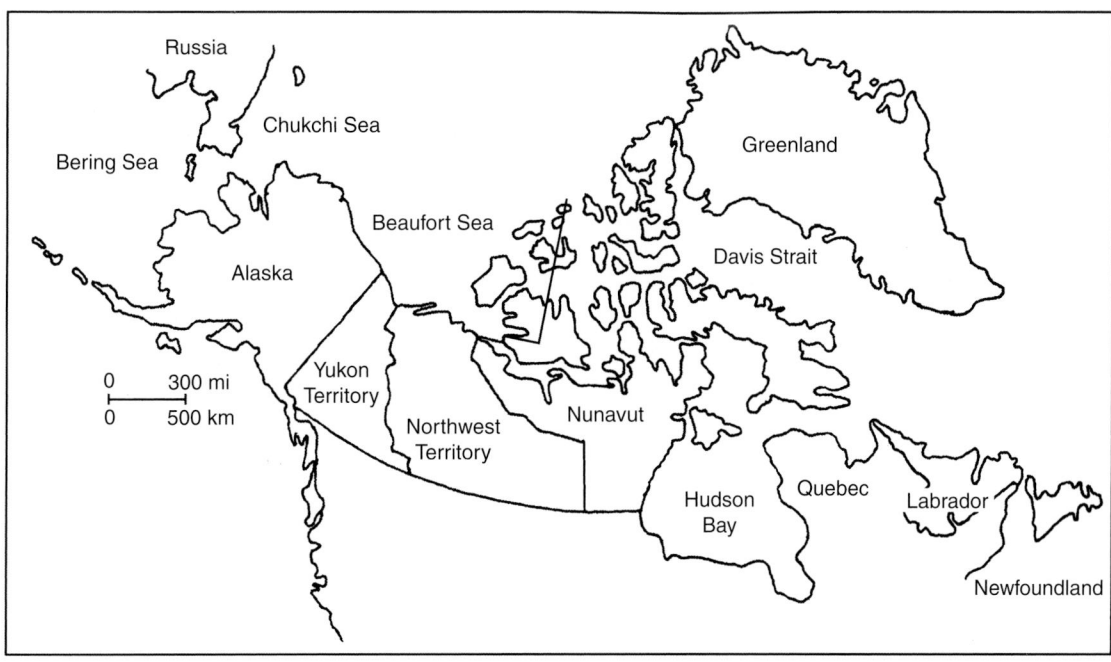

Figure 1    *Coastal Arctic inhabited by the Inuit. Redrawn from Freeman et al. (1998).*

whaling for the European trade came to be shore based and carried out by local Inuit crews, entailing major alterations to Inuit lifestyles compounded by the destruction of the whale stocks.

Westerners first reached northern Alaska in 1826. However, Inuit lifestyles there were relatively unaltered by contact with the West until the second half of the century, when depredation of the bowhead whale stocks by commercial whaling and the spread of European diseases had disastrous consequences.

Inuit clothing was superior to Western cold weather gear and was often sought by Yankee whalers in Alaskan waters. Entire Inuit families were hired to travel aboard commercial whaling ships in the Arctic; women skin sewers made and mended clothing for the crew while the men hunted with the Yankees. By the late nineteenth century, Yankee whalers also adopted toggling harpoons, perhaps based on the Inuit model (Brower, 1942; Bockstoce, 1986).

The Inuit diet relied upon meat and blubber from whales, seals, and polar bears (*Ursus maritimus*). Caribou meat was eaten with seal oil or whale oil. Inland Inuit relied upon traded seal oil for a critical part of their dietary intake (Sheehan, 1997). Skins for boats came from seals and walruses. These, along with caribou and birds, also provided skins for clothing. Whale and seal oil provided fuel for lamps, the only source of heat other than body heat in houses. Alaskan coastal kitchens were separate from the houses and typically had two cooking areas, one fueled by oil, and the other by driftwood.

In Alaska, driftwood semi-subterranean houses incorporated long entrance tunnels made of whale bones, whereas in areas of Canada and Greenland, where driftwood was scarce, even the houses were constructed with whale bones, or with stone and bone. The only prehistoric qargi, or whalers' ceremonial house, that has been excavated in north Alaska was made almost entirely of whale bones.

Pokes (seal skins) filled with seal oil were used to preserve meat. Prehistorically in Alaska, i.e., prior to 1826 and even past the middle of the nineteenth century, seal oil and whale oil pokes were major trade items from coastal areas (Maguire, 1988). Return trade from inland Inuit was primarily caribou skins for clothing and blankets, sinew for sewing, and antler for tools. The economy left nothing to waste, with dog (*Canis lupis familiaris*) teams consuming old clothing as well as any of the harvest not used directly by the Inuit.

Whaling provided a dependable food surplus to the prehistoric coastal Alaskan communities, allowing them to organize their lives around the whale hunt (Sheehan, 1997). This whaling culture was successful for a 1000 years. Whaling remains the organizing focus of Inuit life today in northern Alaska and is still an important part of Inuit ideology in other parts of the Arctic. Marine mammal hunting continues to underpin Inuit subsistence activities and social interactions.

## I. Precontact Whaling

It is commonly believed that indigenous whaling developed in the Bering Sea and Bering Strait region about 2000 years ago with the Okvik and Old Bering Sea cultures. A florescence in the diversity and complexity of tools used for hunting and processing marine mammals took place from approximately 100 BC to AD 600, although such tools and weapons have continued to be a focus for technological innovation. This suggests an increased dependence on large whales and other marine mammals (Stoker and Krupnik, 1993). There appear to be two significant differences between the early groups that hunted whales but did not rely upon them and later groups that were dependent for their survival on the whale hunt. One of these differences was technological, the other social. The introduction of drag float technology may have transformed whale hunting from a "status" activity resulting in lucky "windfalls," into a "normal" activity resulting in a regular and substantial payoff. Transformation of the umialik (whaling captain) from a temporary hunt leader into a permanent political leader responsible for distributing the whaling surplus throughout the community allowed the population to thrive and grow. The combination of technological and social change culminated in the period of the Punuk and the Thule cultures starting at AD 800–900.

It is generally agreed that widespread large whale hunting started when the Thule culture spread across North America to Greenland, but whaling itself may have developed independently in several areas at different times. The earliest of these may be the Maritime Archaic tradition of Labrador and Newfoundland, dating from approximately 3000 BC The Maritime Archaic is believed to be one of the earliest cultures to use the toggling harpoon head. Møbjerg (1999) reported that the Saqqaq culture of Greenland's west coast, part of a broader Arctic small tool tradition, which stretches across the North American Arctic, may have been hunting baleen whales as early as 1600–1400 BC One of the most interesting cases is the so-called Old Whaling culture of Cape Krusenstern, near Kotzebue Sound in Alaska, which appeared suddenly around 1800 BC, but disappeared shortly thereafter. These people used large lance and harpoon points, possibly to hunt for baleen whales. The abundance of whale bones in the area suggested to the original excavators that whaling was practiced, but there is no evidence that the technology was passed to later cultures (Giddings, 1967), and subsequent reanalysis of the materials has called the evidence for whaling into question (Darwent, 2006).

The Thule whaling culture developed in northwestern Alaska around AD 850–900 and a few hundred years later spread very quickly across arctic Alaska and Canada to northern Labrador and Greenland. It then spread into more northerly areas in the central Arctic. The rapid expansion of the Thule whaling culture was perhaps influenced by a period of climatic warming. The warmer weather may have resulted in seasonally open water across the entire coast from northwest Alaska to eastern Canada and Greenland, making Pacific and Atlantic populations of whales contiguous and more numerous. These conditions would encourage the expansion of a shore-based whaling culture.

The climate of the far north did not remain warm and stable for long. Colder weather and a resulting increase in expanse and duration of ice cover reduced the distribution and perhaps the numbers of whales in the Arctic. A concomitant reduction in the geographic range that could sustain a whaling-focused economy made reliance on whales risky in areas that were more marginal. Thule people who could no longer succeed in whaling focused more heavily on smaller marine mammals and other game. Some parts of the central Canadian Arctic were depopulated.

The climatic variations resulted in dramatic changes to the Thule whaling culture throughout its range. The remnant Thule cultures gave rise to the contemporary Inuit cultures of present-day Canada, Greenland, and Alaska. In Alaska, whalers were able to continue their primary reliance on whale hunting by clustering in large permanent villages at points of land, where every spring they could rely on currents and geography to place them within walking distance of nearshore leads in the ice. Whales followed the leads as they went north for the summer. The leads became the foci of the whale harvest, supplemented by fall whaling in open water, as the whales passed the points on their way south.

## II. Mysticetes

### A. Bowhead Whale, agviq

The bowhead whale is the largest animal hunted by any prehistoric or historic hunter–gatherer society. Adults reach at least 20 m and weigh 50,000 kg or more. The slow moving, blubber-rich whale is a particularly suitable target, as it often travels close to shore in predictable migration patterns.

The advent of commercial whaling and the consequential contact with Europeans forever changed the patterns of indigenous bowhead whaling. Commercial whalers reduced bowhead populations to levels too low to support a subsistence hunt in most of the whales' range. The Chukotkan natives continued bowhead whaling until the late 1960s when Soviet authorities replaced the shore-based hunt with a catcher-based hunt, primarily for gray whales (*Eschrichtius robustus*). In 1997, the Alaska Eskimo Whaling Commission (AEWC) began sharing their quota of bowhead whales with Chukotka natives, assigning them five of the total annual strikes allowed under the rules of the International Whaling Commission (IWC). With assistance and training by Alaskan whalers, the Chukotkan natives have begun to hunt bowhead whales again. The Canadian Inuit ceased traditional bowhead hunting around World War I due to low whale numbers and active discouragement by the Canadian government. In 1991, the Canadian Inuit at Aklavik, in the Mackenzie River delta, landed a bowhead for the first time since the early twentieth century. An unsuccessful hunt was carried out in 1994 and a successful hunt in 1996, but they haven't chosen to hunt since then. Greenlandic Inuit hunted bowheads for many centuries before commercial whaling depleted the Atlantic stocks nearly to extinction. Greenlandic Inuit were employed by Danish commercial whalers from the late eighteenth century until 1851, when depleted bowhead numbers brought a halt to commercial hunts.

Currently, the bowhead whale is hunted under the quota system in northern Alaska, in the villages of Savoonga, Gambell, Little Diomede, Wales, Kivalina, Point Hope, Wainwright, Barrow, Nuiqsut, and Kaktovik, along the Bering, Chukchi, and Beaufort Seas.

After commercial whaling ceased in the early twentieth century, Alaskan Inuit returned to a strictly subsistence bowhead hunt. Bockstoce (1986) estimated that an average of 15–20 whales was landed each year from 1914 to 1980. After 1970 there was a significant increase in the number of bowheads landed in Alaska. This was a result of a combination of factors. There was an increase in cultural awareness by Native Americans in general and Alaska Natives in particular, brought about by the passage of the Alaska Native Lands Claim Settlement Act in 1971. The discovery of oil in Prudhoe Bay in 1968 and the construction of the Trans-Alaska pipeline provided significant cash input into the economy of northern Alaska, which prompted a large increase in the number of whaling captains. The position of whaling captain in northern Alaskan Inuit whaling communities has always been one of great respect and authority. Traditionally, only those hunters who demonstrated great hunting success and respect for customs rose to the position of whaling captain. The expense of obtaining whaling gear limited the number of crews and ensured that only experienced whalers rose to the position of captain. The influx of money and employment in the 1970s resulted in a doubling of the whaling crews in northern Alaska from 44 in 1970 to 100 in 1977. The number of whales landed also increased from an average of 15/year to about 30/year from 1970 to 1977. There was also a large increase in the number of whales struck but lost and presumably killed.

The increase in the number of struck but lost whales, combined with a NOAA estimate that only 600–2000 bowheads remained in the Arctic, prompted the IWC to call for a total ban on bowhead whaling. The Inuit reacted strongly to this ban, arguing that the bowhead population was much larger than the IWC scientists were estimating. They formed the AEWC, composed of whaling captains from each whaling village. In 1978 the AEWC, through the US delegation to the IWC, negotiated a quota of 12 bowheads landed or 18 struck for the 9 Alaskan whaling villages. Since then the IWC has established quotas for Alaskan whalers, and the AEWC has distributed strikes to the 10 Alaskan whaling villages (Little Diomede joined AEWC in 1992). Research paid for and conducted through the AEWC and the North Slope Borough (NSB, the regional government in northern

Alaska) Department of Wildlife Management proved the Eskimo whaling captains were correct when they asserted that there were many more whales than estimated by NOAA and presented to the IWC; however, the high mortality rates (>111 strikes in 1977) were probably not sustainable. Careful censuses (George *et al.*, 2004) of the Bering–Chukchi–Beaufort Seas bowhead population estimated that the bowhead population in the western Arctic numbered 10,470 in 2001, (SE = 1351; 95% CI 8100–13,500). The rate of increase (ROI) from 1978 to 2001 is 3.4% per year (95% CI 1.7–5%).

In consequence, the number of strikes allotted to Alaskan whalers was increased to the estimated "need" level based on Inuit population size and patterns of customary and traditional usage of whale products. In 2007, a block quota was set for the years 2008–2012. The quota of 280 whales for that period includes five whales allocated to Chukotka by the AEWC.

Some Alaskan Inuit hunt bowhead during both the spring and fall migration. In spring, bowheads migrate from wintering grounds in the Bering Sea north through the Bering Strait to feeding areas in the eastern Beaufort Sea. Whereas specific dates vary somewhat by year, typically, the whales move along open leads in the ice created when drifting pack ice shears away from the grounded, shore-fast ice. These leads occur in predictable places along the Alaskan coast. Bowheads begin the migration north from the Bering Sea in late March through early April and pass the whaling villages of Gambell and Savoonga soon thereafter. The whales pass by Point Barrow from mid-April to early June and arrive in the eastern Beaufort Sea in May. Bowheads begin the fall migration across the central Beaufort Sea in early September and pass Alaska's north coast from mid-September to early October. Some whales may continue across the northern Chukchi Sea arriving in Chukotka in November, and others may move southward, likely crossing the central Chukchi Sea.

Equipment used in the modern whale hunt is a combination of precontact technology and tools adopted from Yankee whalers. The boat used for the hunt is a skin-covered frame called an umiaq. The frame was traditionally made of driftwood lashed with baleen with some whale bone fittings, but now is made from prepared lumber. The cover is made from the skins of bearded seals (*Erignathus barbatus*) or walrus hunted the previous summer. The skins are left to ferment which softens the skin and allows the hair to be stripped off easily. The skins are sewn together using a special waterproof stitch and stretched over the frame using rawhide thongs or, more recently, jute or nylon line. The average umiaq in Barrow requires six bearded seal skins for the cover, is 6.5–8.5 m long, 1.5–1.8 m across the beam, and weighs approximately 160 kg when dry (Stoker and Krupnik, 1993). The skins are usually replaced every 1 or 2 years, depending on their condition. In some places, aluminum or wooden boats powered with outboard motors have replaced the umiat (plural of umiaq). However, in areas where heavy ice is often encountered, umiat are still used because they are easier to move across and through heavy ice. During fall whaling in Barrow, Nuiqsut, and Kaktovik and during spring whaling in areas where leads are wide and whales travel farther from the lead edge, aluminum or fiberglass boats powered with outboard motors are used.

Weapons used for hunting are essentially the same equipment used by commercial whalers at the end of the nineteenth century. The darting gun and shoulder guns were introduced by Yankee whalers soon after the Civil War and were adopted by Inupiat hunters in the last decades of the nineteenth century (Brower, 1942; Bockstoce, 1986). The harpoon consists of a wooden shaft 1.5–2 m long tipped with a detachable steel harpoon with a brass toggling head attached to a float with 55 m of strong nylon line. The harpoon is tipped with

a plunger trigger-driven gun that fires an 8-gauge, brass bomb simultaneously with the harpoon strike. A second darting gun that resembles the harpoon but without the toggling head harpoon is used to deliver a second bomb. Heavy brass shoulder guns are also used to fire bombs from distances greater than can be attempted with the darting gun. The brass-encased bombs are charged with penthrite, which replaced black powder in 1998. Penthrite bombs deliver a sudden concussion and kill by shock rather than laceration and tissue damage. This reduces the number of whales that are struck but lost. Other equipment includes flensing tools hand-made of steel blades (often from hand saws) attached to long wooden handles, heavy-duty block and tackle to haul the whale onto the sea ice, an aluminum or fiberglass boat used to chase and retrieve a whale after a strike is made from the umiaq, and snowmobiles used to tow equipment to and from camp and to carry meat and maktak back to the village.

Preparations for whaling begin well before the whales arrive. Male members of the crew clean weapons and the ice cellar for storing meat and build sleds and other equipment needed for the camp on the ice. The wives of the captain and crew members sew a new skin cover for the umiaq frame. When the skins are dry, the umiaq is lashed to a sled for the wait until a lead opens.

Sometime before the arrival of the first bowheads the captain will decide where to place his camp. One or several "roads" are built across the ice to the selected sites. The roads are built to smooth the route across the maze of pressure ridges on the ocean ice. Smoothing the route eases the task of hauling sled loads of meat and maktak in the event of a successful hunt and provides a quick escape route if ice conditions become unsafe. Stakes with colors or symbols are often placed along the roads. Camps are located on the ice edge, often in "bays" in anticipation of whales swimming under projecting points and surfacing in those bays, or on points that provide good views of approaching whales.

Inuit believed, and many continue to believe, that whales give themselves willingly to hunters worthy of their sacrifice. Traditionally, many taboos governed activities in whaling camps, and these taboos were strictly followed to ensure a successful hunt. Tents, sleeping gear, and cooking were prohibited in camps. Most taboos have been relaxed or dispensed with, but traditions still govern activity in camps. One tent is set up in camp to allow crew members to sleep in short shifts and for cooking meals. The tent is placed away from the lead and to the right of the boat to prevent approaching whales from seeing the camp. The umiaq is kept ready at the water's edge with a smooth ramp cut into the edge so that it can be launched silently. The harpoon and darting gun are positioned in the bow of the umiaq with the line from the harpoon neatly coiled on the bow. The weapons, lines, and floats are always kept on the right side of the boat, and the strike is always made over the right side of the boat to prevent ENTANGLEMENT in the line. At least one crew member remains on watch at all times, scanning the lead for any sign of an approaching whale.

When a whale comes within range and is determined suitable, the umiaq is launched silently with the harpooner ready in the bow. Two to five paddlers are situated along each side of the umiaq, with a steersman in the stern to steer the umiaq toward the whale. The umiaq is paddled silently, with all crew members stroking in unison. The steersman directs the umiaq to where he or the captain hopes the whale will surface next. The harpooner strikes the whale from as close as possible, often from point-blank range. The preferred target is the postcranial depression just forward of the back. A hit here will often kill the whale instantly. If this target is not available, the spine, heart, or kidney regions are targeted. As soon as the whale is struck, the float is thrown overboard on the starboard (right) side. If possible, a senior crew member other than the harpooner will fire the shoulder gun to plant another bomb into the whale. Other crews, alerted by VHF radio, quickly converge on the site of the strike in aluminum boats powered by outboard motors and may fire another bomb into the whale in an attempt to kill it quickly. Aluminum boats are much faster than umiat and help ensure that a struck whale will not be lost.

Immediately after the whale is killed the captain of the crew that first struck the whale says a prayer (to the Christian God). The prayer is often broadcast over VHF radio and is the first signal of a successful hunt to villagers waiting on shore. The whale's pectoral flippers are then lashed together and the flukes may be removed to reduce drag. A long line is attached to the caudal section forward of the flukes and all available boats attach to the line, with the successful crew at its head, to tow the whale tail-first to the butchering site on the ice. Word of the successful hunt is sent to the village by snowmobile, and the whaling flag of the successful crew is raised over the captain's home. Many members of the community then travel to the butchering site to help with hauling the whale onto the ice and butchering it.

At the butchering site a large block and tackle is attached to the ice and used to haul the whale onto the ice. Every available crew member and community member hauls on the free end of the line running through the block and tackle, pulling on commands from the whaling captains. If the whale is too large to haul onto the ice, some butchering may commence in the water. The tongue or SKULL may be removed to ease the task of hauling the carcass onto the ice. Butchering begins as quickly as possible after the whale is hauled onto the ice because the thick blubber layer retards heat loss and the meat in an unbutchered whale quickly spoils. The whale is butchered according to strict customs governing the distribution of shares (Fig. 2). Parts of the whale are reserved for the captain of the crew that struck the whale. Most of that portion will be shared with the community at feasts and festivals that occur throughout the year. Additional shares are divided among the successful crew and the crews that assisted in killing, towing the whale to the butchering site, hauling the whale onto the ice, and butchering. Individuals not representing a crew are also offered shares of meat and maktak. A group of 20–25 people can butcher an average size bowhead in 6 or 7 h. No shares are distributed until the butchering is complete. Traditionally, following butchering some skulls were rolled into the ocean to allow the spirits of the whales to enter other bodies and again be hunted. The spirit of the whale would remember that the captain treated it well and so sacrifice itself to that captain again. Other skulls were brought ashore and placed at the beginning of the tunnels that led to the entrances of villagers' semi-subterranean homes. These symbolically placed skulls suggested that as you entered the home you also entered the world of the whale. The prehistoric qargi or whalers' ceremonial house was built entirely of whale parts to represent a complete whale (Sheehan, 1990). Today, some skulls are not returned to the ocean but are taken ashore where they are cleaned and displayed in the village. The remainder of the skeleton is left on the ice for gulls, foxes, and polar bears.

Bowhead maktak, served boiled fresh or raw and frozen, is the most prized food in the Arctic. Shares of meat and maktak are widely distributed among family and neighbors, often to family members living in cities who would not receive traditional foods otherwise. Meat is eaten raw and frozen, boiled, or fermented in blood. Many internal organs are also eaten. The kidney, intestines, and heart are boiled. The huge tongue of the bowhead is considered a delicacy when boiled. Baleen was traditionally used to make toboggans, for lashing of umiaq frames, for bird snares, and to make fish nets and seal nets that could easily be freed of the ice that forms on nets immediately as they are removed from the water. A simple snap of the net broke off the ice from this resilient material. Now baleen is crafted into artwork and sold.

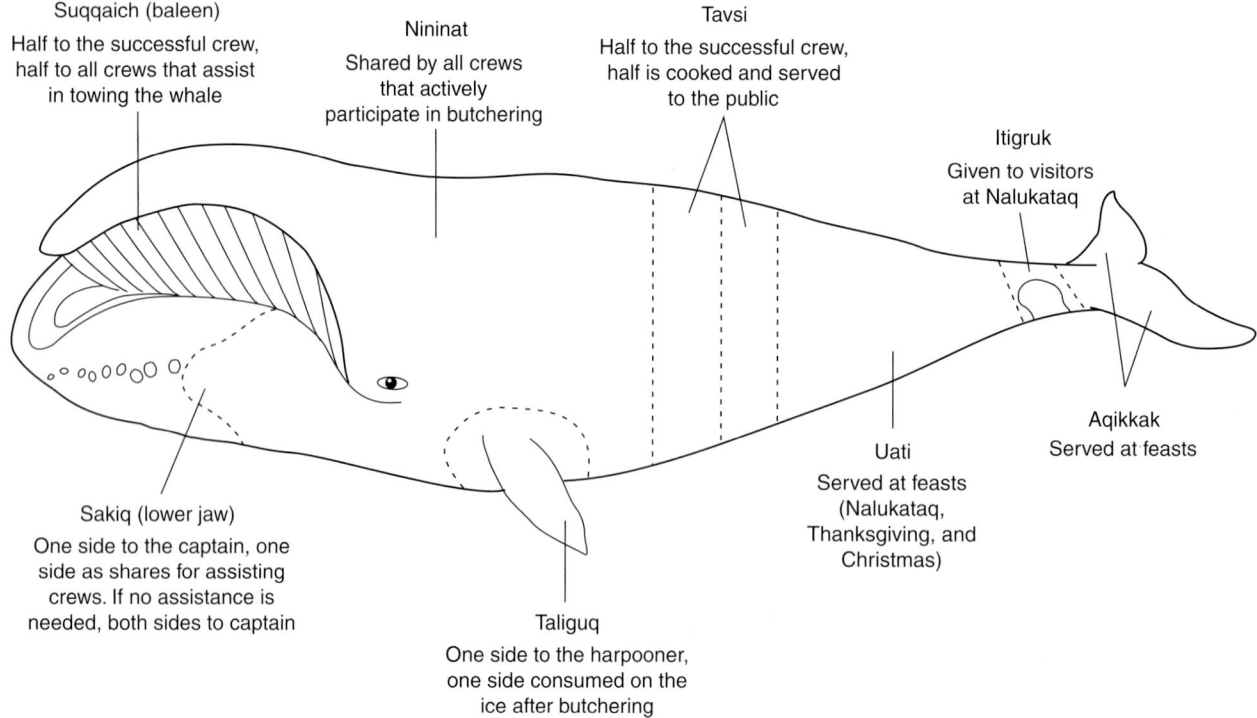

**Figure 2** *Division of bowhead whale* (Balaena mysticetus) *shares in Barrow, Alaska. From Harry Brower, Jr.*

On the day following butchering, the captain of the successful crew opens his home to the community in celebration. All comers are offered food and drink. In early June the umiat of the successful whaling crews are hauled off the ice in ceremonies (apugauti). Once again, the captain supplies food and drink to all who attend. Nalukataq, the formal whaling festival, takes place in June. Each successful crew will have their own nalukataq, or several crews will hold one together. At nalukataq, the members of successful crews distribute the majority of the meat and maktak reserved for the community. The captain and crew also distribute other foods collected during the year, such as caribou meat and soup, duck soup, goose soup, and many other traditional foods. Fruit and candy are also distributed, and coffee, tea, and soft drinks are served throughout the day.

After the food is distributed, the blanket toss begins. Skins from the successful umiaq are removed from the boat and resewn to form a blanket with rope handles along the edge. Community members climb onto the blanket, one at a time, and are thrown into the air by people pulling on the handles in unison. The objective is to jump as high and as many times as possible without falling. Members of successful crews will often climb onto the blanket with bags of candy to fling to the crowd while jumping. After the blanket toss a traditional dance is usually held in the community center. Each successful crew and their families will dance by themselves, but most dances are open to anyone. Nalukataq is one of the most joyful times in the village, and the traditional dance that is the culmination of nalukataq can last late into the night.

### B. Gray Whale, agvigluaq

Only the Chukotkan Inuit of the Russian Far East regularly hunt gray whales. Historically, Chukotkan Inuit hunted both bowhead and gray whales from shore-based stations. The traditional shore-based hunt was banned by the Soviets and replaced by a catcher boat-based hunt in 1954 (Freeman *et al.*, 1998). As a result, the cultural traditions were lost and few people now remember traditional hunting methods. The Soviet catcher boat Zvyozdnyi last hunted in 1992 (Freeman *et al.*, 1998). After the catcher boat stopped whaling, the villagers began to hunt marine mammals again to supplement dwindling food supplies.

The return to traditional, shore-based whaling was a difficult and costly endeavor. The lack of equipment and knowledge had serious consequences in several villages. Hunters from the village of Nunlingran died in several hunting accidents, and one whaling boat from Sireniki was sunk, killing all aboard. Fortunately, material assistance and training by the AEWC and the NSB helped ease the transition. However, the hunters from seven Chukotkan villages landed 51 gray whales in 1994 (Freeman *et al.*, 1998). In Lorino, several experienced marine mammal hunters were able to teach younger hunters the proper use of harpoons, spears, and rifles. Hunters from Lorino landed 38 gray whales in 1994. Several other villages solicited aid from Lorino, and with training from experienced hunters began to successfully hunt gray whales. The hunt is now sanctioned and controlled by the IWC, with a quota of 120 gray whales landed in Chukotkan villages from 1998 to 2002. Gray whale hunting has again become an important part of Chukotkan Inuit cultural and dietary lives.

Gray whale hunting is carried out in the summer when gray whales move into the Bering Sea from their wintering grounds. Whaling is conducted from shore stations using skin boats (baidara) or wooden whaling boats. The harpoon–spear is a special whaling implement traditionally used by the Inuit of Chukotka (Freeman *et al.*, 1998), consisting of a wooden shaft with a detachable metal spear that is attached to a line with a small float. Each boat carries 7–10 of the metal spears and one wooden shaft. The spear is thrown by hand and the metal spear detaches from the wooden shaft. The wooden shaft is retrieved from the water, fitted with another harpoon–spear, and the whale is approached again. The harpooner aims for the back of the whale, trying to hit the main blood vessels or vital organs. Once harpoons have been set, the whales are shot with rifles and darting guns. This form of hunting is often dangerous. Gray whales are known to fight aggressively. Two boats are used to ensure the hunters' safety. The hunters also try to take small- or medium-sized whales.

Gray whales are taken for their meat and blubber. The meat and maktak are eaten frozen, thawed and raw, or boiled. Oil is rendered from the blubber and used as food by itself or added to edible roots, willow leaves, and other vegetables.

In northern Alaska during the early historic period, commercial trade for baleen from bowhead whales created wealth that allowed people to increase the number of dogs in their teams. As a consequence, some gray whales were hunted primarily to feed sled dogs, although some hunters also found the meat to be very tasty. Gray whales are no longer hunted in Alaska.

### C. Humpback Whale (*Megaptera novaeangliae*) and Fin Whales (*Balaenoptera physalus*)

Greenlandic Inuit hunted humpback whales from skin boats in much the same way they hunted bowhead whales. Humpback whales are slow-swimming whales, and the techniques used for bowhead whales were successful for humpbacks as well. Although Greenlandic bowhead hunting ceased in the mid-nineteenth century, humpback whaling continued until the 1980s.

In the 1920s, changing sea ice conditions caused food shortages among the Greenlandic Inuit who could no longer catch seals or humpback whales using traditional means. The Danish government operated a steel catcher boat, the Sonja, with a Danish crew from 1924 to 1949. The Sonja was able to catch larger and faster-swimming whales. In 1927 the Sonja caught 22 fin whales, 9 humpbacks, 7 blue whales (*Balaenoptera musculus*), and 2 sperm whales (*Physeter macrocephalus*). The meat was provided to Inuit of western Greenland and the blubber was shipped to Denmark, where it was rendered to oil and sold. In 1950, the Sonja was replaced with the larger Sonja Kaligtoq. From 1954 onward, the whales were taken to a single flensing station where meat and maktak were frozen for distribution and sale throughout Greenland. In addition to the government catcher boats, in 1948 some local fishermen began installing harpoon cannons on their boats and hunting whales. Fin and humpback whales were taken to the community where meat and maktak were sold. In the late 1980s the IWC eliminated the humpback whale quota, so fin and minke (*Balaenoptera acutorostrata*) whales are currently the only baleen whales that are hunted in Greenland.

### D. Minke Whale

Minke whales have been hunted in Greenland since 1948. The minke whale hunt is now controlled by quotas set by the IWC and administered by the Greenland Home Rule Authority. The variable quotas consider the socioeconomic, cultural, and nutritional needs of the people and the regional abundance of whales. In the 1990s the quota varied from 110 to 175 per year. Minkes are hunted in summer and fall when ice conditions permit.

Hunts from fishing boats and small skiffs are opportunistic. Hunts take place from fishing boats whenever whales are sighted or from skiffs when enough small boat hunters are available. Whalers on fishing boats use deck-mounted harpoon cannons, whereas those aboard skiffs use hand-thrown harpoons and rifles. In each case the whales are towed back to the community for flensing and distribution. Shares are distributed to the vessel owner and crew members, and a large share is reserved for the boat. Little personal share of meat or maktak is sold, but the boat share is sold to contribute to the cost of operating a commercial fishing boat. In the small skiff hunt, shares are divided equally among the participants of the hunt and those helping with the flensing.

## III. Odontocetes

### A. Beluga Whale (*Delphinapterus leucas*), qilalugaq

Beluga whales are hunted across their range in Chukotka, Alaska, Canada, and Greenland. Ancestors to the modern Inuit were involved in beluga hunting as early as 5500 years ago in Alaska (Freeman *et al.*, 1998). The techniques used by the ancestral Inuit are the same as those used in Alaska, Canada, and Greenland before contact with commercial whalers. Entire communities were involved in a collective whale hunt or drive. A shaman typically guided the hunt, which was led by a distinguished hunter from one of the communities involved. Freeman *et al.* (1998) quoted an elder from Escholtz Bay, Alaska, describing a traditional drive from around 1870: "They made a line and moved together. They hollered, splashed their paddles, waved their harpoons to scare them into real shallow water. . . . When a hunter got a beluga, he ties it to his qayaq (kayak) and brought it to shore; if he get two, he'd tie one on each side. . . . If wind came up while men were out hunting, women would take umiaqs (skin boats) off the racks and go to help those hunters who were towing two belugas. People always helped together when they landed and pulled those beluga on the shore." Friesen and Arnold (1995) determined that beluga whales were a focal resource for precontact Inuit of the Mackenzie delta, constituting up to 66% of their meat. Two or more hunters would cooperate in a beluga hunt. The whales were approached by hunters in kayaks who threw harpoons attached to sealskin floats. After the whale tired, it was lanced in the heart with a blade attached to one end of the kayak paddle. In some locations, hunters in kayaks working cooperatively would drive belugas into shallow water where they were killed. In northern Greenland, and possibly elsewhere, belugas were hunted at large cracks in the ice where the whales congregated to breathe.

In the eighteenth, nineteenth, and early twentieth centuries, Canadian Inuit were hired by commercial whaling enterprises to hunt belugas. Skins and blubber from the belugas were shipped to European markets. The Inuit hunters kept the meat and some of the maktak and received trade goods, which often included wooden boats.

Methods changed with the introduction of rifles, fiberglass and aluminum boats, and outboard motors. Today, hunters in Alaska use one of four methods to hunt belugas: harpooning or shooting from the ice edge in spring, shooting from motorized boats in open water, netting, or driving the whales into shallow water. Ice-edge hunting occurs during the northward migration, sometimes concurrently with bowhead whaling. Belugas can also be shot directly from shore if the migrating whales are close enough, as has happened in Barrow in some recent years. Open-water hunting is common in summer and fall when the ocean is free of ice. Netting occurs at headlands where predictable movement patterns make netting practical. Shallow water drives are most common in shallow bays and estuaries, such as Point Lay and Wainwright, Alaska.

Sealskin kayaks were last used to hunt belugas in the 1960s in communities in northern Quebec and the Belcher Islands. Now hunters use skiffs or freighter canoes powered with outboard motors. Harpoons with detachable heads attached to floats are still used, although now floats are made from manmade materials rather than seal skins. Rifles (0.222–30.06 caliber) are used to kill the whales after harpoons have been attached.

Belugas are the most commonly and widely taken whale species in Canada (Freeman *et al.*, 1998). Beluga maktak is highly prized by Canadian Inuit. After a successful hunt the meat and maktak are distributed to family members and neighbors according to traditional customs. In some communities a successful hunt is announced over community radio and all community members are invited to collect a share. Because beluga maktak is so highly prized, very little of it is sold for redistribution through retail outlets in the Canadian Arctic. Beluga maktak is usually eaten raw and fresh, although some now deep-fry it. The meat is usually air dried before being eaten. In some communities, sausages are made by placing meat in sections of intestine that are lightly boiled before being dried or smoked. Beluga oil was used for lamp oil, softening skins, and cleaning and lubricating guns and other equipment.

Beluga hunting in Greenland has followed a history similar to hunting of other larger whale species. For many centuries, local hunters supplied meat and maktak to meet community needs. In colonial times, beluga blubber and oil became an important trade commodity. As a result, the Greenland Trade Department established commercial beluga drives and hired local hunters to carry out the hunt. Commercial drives continued until the 1950s when the European market for whale oil disappeared. Commercial whale drives reappeared in the 1960s when improved coastal communication and refrigeration made it possible to transport beluga meat and maktak from northern hunting communities to southern Greenland. Today, belugas are hunted with rifles (30.06 caliber to 7.62 mm) from small boats. Typically, kayaks and motorized skiffs are used to hunt belugas, often singly or in pairs, but sometimes a larger number of small boats cooperate to hunt belugas swimming together. Meat and maktak are distributed throughout the community, including sale at the local market, and in retail stores throughout Greenland.

Beluga hunting in Russia only occurs in a few villages in Chukotka, and the numbers taken are small. Belugas in Russia are associated with the distribution of fish, especially arctic cod (*Arctogadus glacialis*) and arctic char (*Salvelinus alpinus*). Hunting occurs opportunistically when belugas are encountered during other activities. Hunting occurs either from shore or from the ice edge. Hunters hide behind hummocks of ice and shoot the whales with rifles (7.62 or 9 mm). Meat is dried, frozen, boiled, or fried. Maktak is eaten raw, fresh, boiled, or fried. The skin is used for boot soles, belts, and lines. The oil is used with fish and salad plants. Historically, beluga oil was traded for reindeer (i.e., caribou) meat and skins, although when Soviet state-run fur farms were operating the oil was sold to the farms (Freeman *et al.*, 1998).

### B. Narwhal (*Monodon monoceros*), qilalugaq tuugaalik

Narwhals have been hunted in Greenland and eastern Canada for centuries, and may have brought the Greenlandic Inuit in close contact with the Norse in Greenland beginning in the tenth century. Narwhal ivory was bartered among Inuit long before European contact. Narwhal tusks were highly valued by European traders in the Middle Ages, who sold the tusks in Europe mislabeled as unicorn horn, sometimes for their weight in gold. The royal throne of Denmark, from the fifteenth century, is made almost entirely of narwhal ivory. Narwhal tusks were the basis of trade between Greenlandic Inuit and Europeans from the 10th through the 19th centuries, and were important to Canadian Inuit after the collapse of commercial bowhead whaling in the late nineteenth century. Inuit in Greenland and Canada used the tusks to create durable and functional tools, especially harpoon foreshafts.

Narwhals were hunted from kayaks either along the flow edge, in ice cracks, or in open water. Near ice, the narwhals were harpooned and hauled ashore. In open water, hunters worked together to drive the narwhals into shallow water where they were killed. Another method was to station hunters with rifles on cliffs who would shoot the whales as they swam by. Several hunters in kayaks would wait offshore and harpoon the whales once they were shot. Now, hunting in Canada takes place with small skiffs, rifles, and harpoons attached to floats. Narwhal hunting in northern Greenland is still accomplished with kayaks. Five-meter skiffs or 10–12 m cutters are used in southern Greenland, although occasionally narwhals are shot from shore or netted.

Maktak from narwhals is prized and is eaten fresh raw or aged. Narwhal oil was considered of higher quality than seal oil and was used in lamps for heat and light. The tusk remains the most highly prized product from narwhal. Today tusks are used for artwork or sold. Narwhal ivory sold for an average of $100/foot (30 cm) in 1997 (Freeman *et al.*, 1998). Narwhal meat was used to feed hunters' dog teams.

## IV. Other Small Cetaceans

Small numbers of other cetaceans are taken in eastern Canada and Greenland. The principal species taken in Canada are common bottlenose dolphin (*Tursiops truncatus*) and harbor porpoise (*Phocoena phocoena*). In Greenland, killer whales (*Orcinus orca*), long-finned pilot whales (*Globicephala melas*), northern bottlenose whales (*Hyperoodon ampullatus*), harbor porpoise, white-beaked dolphins (*Lagenorhynchus albirostris*), and Atlantic white-sided dolphins (*Lagenorhynchus acutus*) are taken.

## V. Pinnipeds

### A. Ringed Seal (*Pusa hispida*), natchiq; Bearded Seal, ugruk; and Harp Seal (*Pagophilus groenlandicus*)

Seals are probably the most widely distributed, abundant, and reliable food resource available to coastal Inuit populations. Ringed seals are available nearshore for much of the year. Bearded seals are also important, although less abundant and less widely available than ringed seals. They are important not only for their meat but also as a source of raw materials, particularly their hides (Jensen, 1987). Harp seals are seasonally very abundant in certain areas of Greenland and eastern Canada, and were taken when present. Ribbon seals (*Histriophoca fasciata*), Larga seals (*Phoca largha*), and harbor seals (*P. vitulina*) are only occasionally encountered. All of these pinnipeds are hunted in similar ways and have been combined for the following discussion.

Natchiq (ringed seal) are ice adapted. They are hunted at breathing holes, in subnivean lairs, on drift ice, and in open water. Other seals are not as ice adapted as the natchiq. They can also be hunted on drift ice and in open water. Harbor seals and Larga seals tend to stay away from ice if it is present in significant amounts. Ugruk are common on ice pans and commonly hunted on pans or in open water. Harbor seals tend to be more common than natchiq in more southerly areas (southern Greenland, Labrador), although they have been regarded as shy and also potentially aggressive. Harp seals were generally taken from kayaks in open water or when hauled out on

offshore drift ice, although they could be harpooned from shore or from the ice edge under certain circumstances.

Traditionally, natchiq were hunted at breathing holes on the ice, at pupping dens, while basking in the sun, by netting at the breathing hole, from the ice edge, or from boats in open water. Breathing-hole hunting was most common, as the ocean is ice covered for much of the year. Ringed seals carve out and maintain breathing holes in the ice throughout the winter. In flat ice the breathing holes may be visible from the surface, but often they are covered with snow, and practically invisible. Ringed seals maintain numerous breathing holes, so there was never any guarantee that a seal would visit the hole where the hunter was waiting.

Breathing-hole hunting was a difficult and cold endeavor, and is no longer practiced to any great extent anywhere in the Arctic. Boas (1964) presented an excellent description of pre-rifle seal hunting methods and equipment. A hunter would first locate a breathing hole with the use of one of his sled dogs. Once the hole was found, the hunter set up his equipment around the hole. The hunter sat on an ice block with his feet resting on a piece of fur or stood on the fur with his harpoon in his hand or at his side and waited for the seal to arrive at the breathing hole. There was never any way to determine how long the hunter would have to wait. If the village needed food, it was not uncommon for hunters to wait 24 h or longer for a seal to arrive. Now, more efficient and less strenuous methods are preferred.

When a seal arrives at a breathing hole, the first breath is a short, shallow sniff for any sign of danger. If the seal does not detect danger, the next breath will be deeper. On this second breath, the hunter thrust his harpoon straight down the hole, striking the seal on the head or neck. The toggling head detached, preventing the seal from escaping. The seal was killed and the breathing hole enlarged to pull the seal through. Once rifles became available, seals were shot when they came to the hole, then immediately harpooned to prevent the seal from drifting away or sinking.

After the breeding season, seals enlarge their breathing holes located on large areas of flat ice so they can climb out and bask in the sunshine. Traditionally, Inuit had several methods for hunting seals at this time, described in detail in Nelson (1969) and Boas (1964). A hunter might simply wait near one of the holes for a seal to surface. The water within the hole pulsates when a seal arrives at its hole. When the seal broke the surface of the water, it was speared or shot. Occasionally, hunters placed lines with several hooks along the wall of a breathing hole to catch seals backing into the water after surfacing.

Another traditional seal hunting technique required great stealth and skill. The hunter emulated the behavior of a seal, sliding along the ice on his side, often with a piece of sealskin beneath him to reduce friction and keep his clothing dry. Often hunters would scrape the ice with seal claws attached to a piece of wood to mimic the scratching sound made by resting seals. A skilled hunter could approach very close to a seal basking in the sun. In this way hunters were often able to kill 10–15 seals in 1 day. In a variant of this method, the hunter pushed a small sled with a white shield that hid him from the seals.

Seals could also be netted at their breathing holes. Netting was done at night to prevent the seal from seeing and avoiding the net. This also reduced the hunters' vision and exposed the hunter to many dangers. Four holes were cut around a breathing hole and the net lowered into the water to approximately 10 ft. Seals generally approach breathing holes along the surface, so they did not encounter the net. When the seals dove from the hole, they dove straight down and became entangled in the net. Seal netting was discontinued in the 1960s.

**I**

In spring, pregnant ringed seals hollow a natal den in the snow covering one of their breathing holes. Hunters again use one of their dogs to find the dens. The hunter cut a small hole in the wall of the den through which he could watch for the return of the mother seal. When the seal returned, the hunter jumped through the snow between the seal and its hole, trapping it. Prior to the introduction of rifles the seals were killed with a spear or club; later they were shot through the wall of the den.

Traditionally, ice-edge hunting was accomplished with a small harpoon that was thrown at seals swimming near the edge. A line was attached to the harpoon to retrieve struck seals. Hunters were limited by how far they could accurately throw the harpoons, usually 10–20 ft (Foote, 1992). The introduction of the rifle changed the nature of seal hunting. Hunting seals from the ice edge using rifles is easier and more efficient than breathing-hole hunting, and the range of the hunters has been increased greatly by the rifles. The increased range brought about two new inventions specifically for use in ice-edge rifle hunting: the retrieving hook (manaq or manaqtuun) and a small skin boat (umaiggaluuraq). The manaq consists of a rope up to 200-ft long, attached to a piece of wood with four hooks extruding from the sides. A float is attached to keep the hooks afloat for winter hunting (when seals float after being shot), and a sinker is attached for summer hunting to retrieve seals that sink to the bottom. Once a seal is shot, the hunter grabs his manaq to retrieve the seal from the water. The line is coiled and held in the left hand, while the right hand holds the line 3–5 ft from the hooks. The hook is thrown beyond the seal, the line is slowly drawn in until the hooks are near the seal, a sharp tug sinks the hooks into the hide, and the seal is carefully pulled to the ice edge.

The umaiggaluuraq (literally "small umiaq") is 7–10 ft long and 36–40 in. wide (Nelson, 1969). Two bearded sealskins are used to cover a wooden frame. Once a hunter shoots a seal, he pulls the boat to the ice edge, often with the help of another hunter to prevent damage to the skins by dragging the boat. The boat is rowed to the seal with two short oars lashed to the gunwales. When the hunter reaches the seal, he tows it back to the ice edge with a small hook and line.

Open-water hunting and hunting of seals basking on drift ice became most popular after the introduction of rifles. Before rifles were introduced, hunters occasionally harpooned seals from kayaks, but only in calm water. After rifles and outboard motors became readily available, several men would hunt together from a single umiaq. The hunters were often members of the same whaling crew using the captain's boat. Seals were shot with rifles ranging from 22 to 30.06 caliber and harpooned. Now, aluminum boats have replaced skin boats, but the same methods are used. Open-water hunting from aluminum boats is currently the most popular way to hunt both the ringed and the bearded seal in northern Alaska. Harpoons are still used in the Yukon-Kuskokwim Delta because people feel that shot seals sink too quickly. In Greenland, certain areas still forbid motorized boats in the hunt, although they may be used to travel to the hunting area.

## B. Walrus, aiviq

Walruses are often associated with pack ice and are hunted when the pack ice is close to shore. They do haul out on shore in certain locations, and this may become more common as sea ice diminishes due to climate change. Nelson (1969) reported that hunters in Wainwright, Alaska, only traveled offshore as far as land was still visible on the horizon. However, Spencer (1959) reported that hunters in Barrow often traveled 50–100 miles into the ocean to find walruses. The distances traveled are probably dependent on the proximity of the pack ice to

shore and undoubtedly changed with the introduction of outboard motors.

Hunting walruses was, and remains, a collective hunt. The size of the walrus and the logistics of butchering and transporting the meat back to the village make it necessary for several hunters to work cooperatively. Traditionally, walruses were hunted using large harpoons similar to the harpoons used in bowhead whaling. Long lines, often made of walrus skin, were attached to the harpoons and fastened to a large piece of ice or were held by the hunter who used a smaller spear to drive the end of the line into the ice. Walruses were harpooned while they were lying on the ice. When the harpooned walrus dove, the line prevented it from escaping. When the walrus tired, it was killed with a lance through the heart. Occasionally, walruses were hunted from umiaqs when they were encountered away from the pack ice. In those circumstances, floats were attached to the line or the line was fastened to the umiaq. The walrus was killed with a lance once it tired. Nelson (1969) summarized an elder recounting one traditional method of hunting walruses in which two hunters harpooned two walruses facing opposite directions. The lines from the two harpoons were quickly tied together, and the walruses pulled against each other until they tired enough to be killed with lances through the neck. Now, large rifles are used instead of harpoons, but the methods used to approach the walruses are the same. When a walrus herd is sighted, the ice surrounding the herd is evaluated. There must be enough ice-free water to allow approach and to allow sufficient time for the killed walrus to be butchered before ice closes in.

Walruses are approached slowly with the outboard running. Generally, walruses are approached to within 10 ft before they are shot. All hunters shoot at the same time and continue the volley until enough have been taken or the herd escapes into the water. Walruses must be shot in the brain or the anterior portion of the spinal cord to insure a kill. Walruses will not float once killed, so any dead or seriously wounded walruses that fall into the water are considered struck and lost. Fay et al. (1994) reported that up to 42% of walruses struck in Alaskan hunts from 1952 to 1972 were lost. Wounded walruses are often dangerous, and Nelson (1969) recounted several instances in which wounded walruses damaged boats. In fact, walruses can be so aggressive that they have disrupted mail delivery by kayak and even forced the abandonment of a settlement in Greenland.

Walrus flippers "ripened" in seal oil are considered a delicacy in much of the Arctic. Select portions of meat are eaten, but the bulk of the walrus was used to feed the hunters' dog teams. The skin, bones, and especially the tusks were the most valuable parts of the walrus. Walrus skins often replaced bearded seal skins on umiaqs in places where bearded seals were not abundant. Walrus skins were also used to create strong lines that were attached to harpoons used in seal, walrus, and whale hunting. The bones of walruses were used to make tools and mandibles were used as chocks in house construction. The ivory tusks were often used to make harpoon points and foreshafts. Now, ivory is used in artwork and much is sold to generate a cash income.

## C. Polar Bear, nanuq

Polar bears are found throughout the Arctic and are hunted through much of their range. Polar bears remain on the pack ice for most of the year, and most hunting takes place during the winter on the pack ice. Polar bears are also taken opportunistically when they are encountered on land or in open water.

Polar bear hunting is considered one of the most dangerous hunting activities and successful hunters often enjoy high status in village communities. Traditionally, single hunters using spears, lances, or knives hunted polar bears. Boas (1964) and Nelson (1969) both described polar bear hunts before the introduction of rifles. In the Canadian and Greenlandic Arctic, it was common to release dogs to chase the bears and tire them. Once the bears stopped, they were approached on foot and killed with lances or spears. Dogs were not used commonly in Alaska, but were released if the bears were on young, unsafe ice. Spears and lances were quickly given up once rifles became available.

Hunting for polar bears is now nearly always done on the sea ice, and hunters often travel far offshore to find bears. Walking used to be the preferred method of transportation because it offered the advantages of a silent approach and the ability to hide quickly among the ice hummocks and ridges. Now, snowmobiles are preferred. With snowmobiles, hunters can pull sleds to transport the meat and hide back to the village, eliminating the need to drag the hide and then return with dogs to transport the meat.

Hunters usually find tracks rather than finding the animal itself. From the tracks hunters can tell the size of the animal, its direction and speed, and how long ago the bear passed. Tracks are followed until the bear is sighted. The hunter can then either move quickly to overtake the bear or move ahead to wait in ambush. In either case, it is important to get as close to the bear as possible to ensure a lethal shot. Wounded polar bears are dangerous and sometimes attack the hunter. If the bear is in a position that the hunter cannot reach, the hunter will sometimes try to lure the bear closer by mimicking a sleeping seal. Once the bear stalks close enough, the hunter picks up his rifle and shoots. Sometimes hunters leave seal blood or blubber on the ice and return to the area later to see if any bears have been lured by the smell. When bears venture close to villages or whaling camps they are almost always shot.

Polar bears are hunted for both their meat and their hides, which are divided among the village according to local tradition. In Greenland, the person who sights the bear becomes its "owner" regardless of whether they participate in the hunt. Any other people who shoot the bear or touch it before it is killed also receive shares of the bear. In Alaska and Canada, shares were traditionally distributed widely within the village. A young hunter's first bear was shared among all the people in the hunting party or was distributed to the elders in the village if he was hunting alone. Now, the shares are distributed less formally, but meat is usually shared with family members and others outside the family. The successful hunter usually keeps the hide.

Polar bear meat is prized by many people in the Arctic. Meat is always well cooked to prevent trichinosis, and the liver is never eaten due to high concentrations of vitamin A. In Alaska the sale of polar bear hides is prohibited by the Marine Mammal Protection Act of 1972. Hides are used for clothing such as boots, mittens, or trim for parkas and also for sleeping mats when camping on the ice. In Greenland, polar bear skins were used for warm hunting pants, but now all skins are sold to Greenland's trading department. Since 1994, polar bear hunters in Greenland have been able to sell bear meat to restaurants and hotels.

## VI. Conclusion

Inuit and their ancestors have hunted marine mammals for thousands of years. The technology and techniques of hunting marine mammals evolved in a culture intimately associated with the sea and the creatures that inhabit it. In modern times, the technology and techniques of hunting marine mammals have changed, but the cultural

importance remains, backed up by tradition, beliefs and a web of interlocking obligations. Marine mammal hunting provides access to status within the community and a sense of self-worth for a generation of Inuit struggling to cope with the burdens of cultural assimilation. The product of the hunt provides a sustainable healthy diet. There is every reason to believe that as circumstances continue to change, these subsistence and cultural practices will continue to thrive.

## See Also the Following Articles

Arctic Marine Mammals ■ Whaling, Early and Aboriginal ■ Whaling, Traditional.

## References

Boas, F. (1964). "The Central Eskimo." University of Nebraska Press, Lincoln.

Bockstoce, J. R. (1986). "Whales, Ice, and Men: The History of Whaling in the Western Arctic." University of Washington Press, Seattle.

Brower, C. D. (1942). "Fifty Years Below Zero: A Lifetime of Adventure in the Far North." Dodd Mead and Company, New York.

Darwent, C. M. (2006). Reassessing the old whaling locality at Cape Krusenstern, Alaska. In "Dynamics of Northern Societies" (J. Arneborg, and B. Grønnow, eds), **10**, Proceedings of the SILA/NABO Conference on Arctic and North Atlantic Archaeology, Copenhagen, May 10–14, 2004, pp. 95–102. PNM, Publications from the National Museum, Studies in Archaeology and History, Copenhagen.

Fay, F. H., Burns, J. J., Stoker, S. W., and Grundy, J. S. (1994). The struck-and-lost factor in Alaskan walrus harvests, 1952–1972. Arctic **47**, 368–373.

Foote, B. A. (1992). "The Tigara Eskimos and their environment. North Slope Borough Commission in Inupiat History, Language and Culture." Point Hope, Alaska.

Freeman, M. R. L., Bogoslovskaya, R. A., Caufield, I., Egede, I. I., Krupnik, I., and Stevenson, M. G. (1998). "Inuit, Whaling, and Sustainability." AltaMira Press, Walnut Creek.

Friesen, T. M., and Arnold, C. D. (1995). Zooarchaeology of a focal resource: Dietary importance of beluga whales to the precontact Mackenzie Inuit. Arctic **48**, 22–30.

George, J. C., Zeh, J., Suydam, R., and Clark, C. (2004). Abundance and population trend (1978–2001) of western arctic bowhead whales surveyed near Barrow, Alaska. Mar. Mamm. Sci. **20**, 755–773.

Giddings, J. L. (1967). "Ancient Men of the Arctic." Alfred A. Knopf, New York.

Jensen, A. M. (1987). Patterns of bearded seal exploitation in Greenland. Études/Inuit/Stud. **11**, 91–116.

Maguire, R. (1988). "The Journal of Rochfort Maguire, 1852–1854: Two Years at Point Barrow, Alaska, Aboard H.M.S. Plover in the Search for Sir John Franklin." (J. R. Bockstoce, ed.), 2 Vols, Works issued by the Hakluyt Society, Second Series No. 169. The Hakluyt Society, London.

Møbjerg, T. (1999). New adaptive strategies in the Saqqaq culture of Greenland, c. 1600–1400 BC. World Archaeol. **30**, 452–465.

Nelson, R. K. (1969). "Hunters of the Northern Ice." University of Chicago Press, Chicago.

Sheehan, G. W. (1990). Excavations at Mound 34. In "The Utqiagvik Excavations" (E. S. Hall, Jr., ed.), Vol. 2, pp. 181–325. The North Slope Borough Commission on Iñupiat History, Language and Culture, Barrow, Alaska, 337–353.

Sheehan, G. W. (1997). "In the Belly of the Whale: Trade and War in Eskimo Society." Aurora, Alaska Anthropological Association Monograph Series—VI, Anchorage, Alaska.

Spencer, R. F. (1959). "The North Alaskan Eskimo: A Study in Ecology and Society." Smithsonian Institution Press, Washington, DC.

Stoker, S. W., and Krupnik, I. I. (1993). Subsistence whaling. In "The Bowhead Whale" (J. J. Burns, J. J. Montague, and C. J. Cowles, eds). Academic Press, Lawrence.

I

# Irrawaddy Dolphin
*Orcaella brevirostris*

Brian D. Smith

## I. Characteristics and Taxonomy

### A. Names and Taxonomic History

Vernacular names used for Irrawaddy dolphins include *ikan pesut* or *pesut Mahakam* in Kalimantan, Indonesia; *lumba lumba* in Malaysia; *pa kha* in Laos; *pla loma* (generic word for dolphin); *hooa baht* (monk's bowl, which refers to the resemblance of the dolphin's head) in Thailand; and *Labai* in Myanmar.

*Orcaella* has been placed in the monotypic family Orcellidae (Nishiwaki, 1963), the family Delphinapteridae along with the beluga, *Delphinapterus leucas* (Kasuya, 1973), and in the family Monodontidae that includes both the beluga and narwhal, *Monodon monoceros* (Barnes *et al.*, 1985). Concordant evidence from cladistic analysis of morphology (Arnold and Heinsohn, 1996) and genetics (Arnason and Gullberg, 1996; LeDuc, 1999) place the genus decisively in the family Delphinidae.

The species *Orcaella brevirostris* was recently split into two species based on concordant character differences in external features, osteology, and genetics (Beasley *et al.*, 2002, 2005) between the dolphins occurring in five freshwater systems and nearshore waters of Southeast Asia extending west across the Bay of Bengal and south along the east coast of India to Vishakhapatnam (now considered *O. brevirostris*) and those occurring in coastal waters of northern Australia and southern Papua New Guinea (the newly described Australian snub-fin dolphin *O. heinsohni*). Genetic evidence also supported the existence of two clades within *O. brevirostris*, one occurring in the Mekong River and the other from marine and freshwater sites in Indonesia, Philippines, and Thailand. With systematic examination of a larger sample of specimens there may prove to be additional phylogenetic/taxonomic structure within the species (LeDuc *et al.*, 1999; Beasley *et al.*, 2005).

This account addresses *O. brevirostris*, and caution should be exercised when referring to previous reviews of the "species" (Marsh *et al.*, 1989; Stacey and Arnold, 1999; Arnold, 2002) because much of the information in these pertains to *O. heinsohni*.

The nineteenth century naturalist Dr. John Anderson described Irrawaddy dolphins in the Ayeyarwady (formerly Irrawaddy from where the dolphin takes its common name) River as morphologically distinct from *O. brevirostris* and classified them as a separate species, *Orcella* [*sic.*] *fluminalis* based on an exhaustive lists of differences in anatomical features (Anderson, 1879). However, subsequent authors (Pilleri and Gihr, 1974) rejected Anderson's arguments because the features he described are variable among individuals and his comparisons were limited to two adult males from the Ayeyarwady and two females, one immature and one pregnant, from the Bay of Bengal.

### B. External Characteristics

The Irrawaddy dolphin has a rounded head that overhangs the mouth, which is oriented at a posterior–dorsal angle toward the eye, and a crescent-shaped blowhole positioned to the left of midline (Fig. 1). A posterior neck crease is visible in some individuals, although this feature is apparently less distinct in the Irrawaddy dolphin compared to the Australian snub-fin dolphin. A shallow dorsal

**Figure 1** *Irrawaddy dolphin from Malampaya Sound in the Philippines. Notice the bulbous head, posterior neck crease, and rounded fin. Photograph by M. Matillano.*

groove runs from the neck crease to the dorsal fin, a feature absent in the Australian snub-fin dolphin. In some larger individuals, a conspicuous dorsal hump underlies the dorsal groove; this feature may be sexually dimorphic. The dorsal fin is small, triangular, slightly falcate with a blunt tip, and located about 60% of the body length posterior to the tip of the upper jaw. The flippers have a convex leading edge and are relatively large for delphinids, about one-sixth of the total body length long and about half this length for the width. Span of the flukes is more than one-fourth the total body length, with a concave leading edge and median notch. The species is uniformly dark gray on the dorsal and lateral fields, with variable shading among individuals, and much lighter on the ventral field, which extends from the lower chin to the anus, giving the animals a two-toned appearance. This is in contrast to the tripartite coloration of the Australian snub-fin dolphin (Beasley *et al.*, 2005).

Body lengths of sexually mature Irrawaddy dolphins have been measured at 2.1–2.2 m for females and as large as 2.8 m for males. External morphometrics that differentiate the Irrawaddy dolphin from the Australian snub-fin dolphin include lesser total size and dorsal fin height and longer measurements for the tip of upper jaw to eye, tip of upper jaw to gape, tip of upper jaw to flipper, anterior margin of flipper, and maximum flipper width (Beasley *et al.*, 2005).

### C. Skeletal Characteristics

Irrawaddy dolphins have 62–63 vertebrae (Lloze, 1973; Anderson, 1879), which is slightly less than the number reported for Australian snub-fin dolphins. Only the first two cervical vertebrae are fused and these have greatly reduced transverse processes, giving the head substantial mobility. The acromion process of the scapula is generally larger than the coracoid process, which appears to be the opposite in the Australian snub-finned dolphin (Beasley *et al.*, 2005).

The skull is globe-shaped with an expansive facial region and relatively short rostrum and mandibular symphysis. There are about 68 peg-like teeth, each about 1-cm long: 19 in the upper jaw and 15 in the lower jaw, although all of these may not be erupted. A unique characteristic of the skull is that the tympanoperiotic bones are attached to a triangular ventral pad located on the mastoid portion of the zygomatic arch, rather than within the cavity formed by the squamosal, exoccipital and basioccipital bones, such as the arrangement in other members of the Delphinidae (Stacey and Arnold, 1999).

I

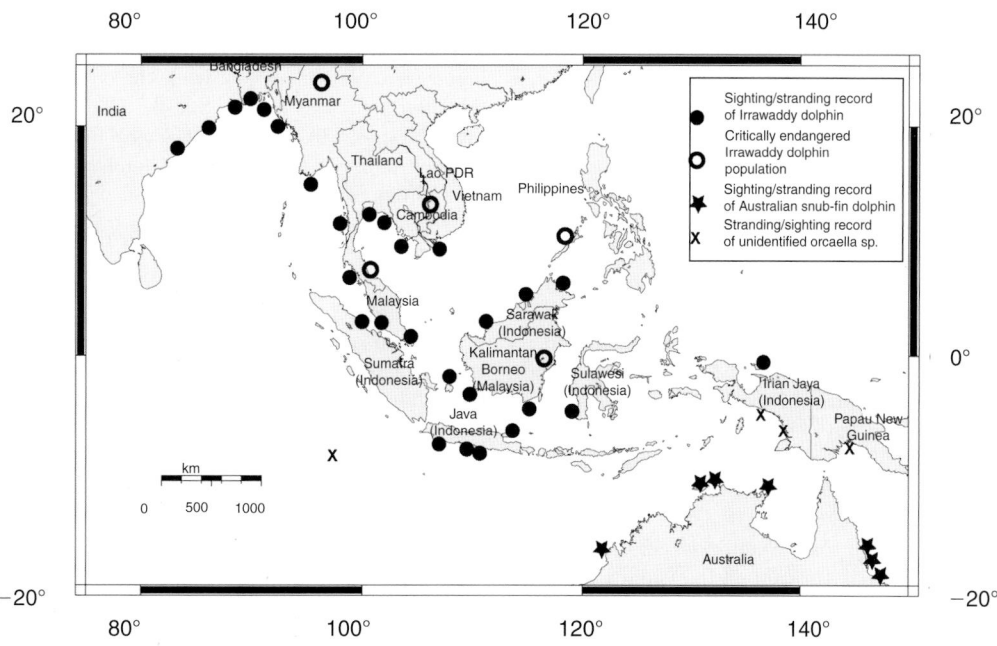

**Figure 2**  *Distribution map for Irrawaddy and Australian snub-fin dolphins.*

Cranial morphometrics differentiating the Irrawaddy dolphin from Australian snub-fin dolphin, in order of decreasing importance, include shorter length of antorbital process, shorter height of temporal fossa, shorter length of rostrum, smaller number of nasal bones/ossicles, less separation between the posterior margin of mesethmoid plate and anteriormost nasal bones/ossicles, shorter length of temporal fossa, smaller condylobasal length, greater minimum distance between pterygoid hamuli, greater depth of pterygoid region, and greater average width of nasal bones/ossicles (Beasley *et al.*, 2002).

## II. Distribution and Abundance

Freshwater populations occur in three river systems: the Ayeyarwady (formerly Irrawaddy) of Myanmar (formerly Burma), Mahakam of Indonesia, and Mekong of Cambodia, Lao PDR and Vietnam. They also inhabit two partially isolated freshwater/brackish lakes or lagoons: Chilika of India and Songkhla of Thailand. All five freshwater populations are believed to be demographically isolated from members of the species occurring in marine waters (Fig. 2).

In the Ayeyarwady River (Fig. 3), the linear extent of Irrawaddy dolphin occurrence has declined by nearly 60% (or 488 km) since the nineteenth century and the animals are presently confined during the dry season to a 370-km river segment located about 1000 km from the sea upstream of Mandalay (Smith and Mya, 2007). Irrawaddy dolphins in the Mahakam River are generally confined to a 195-km river segment of the mainstem starting from about 180 km above the river mouth and inclusive of the lower reaches of the Kedang Rantau, Kedang Kepala, Belayan, Kedang Pahu, and Ratah tributaries and the southern portion of Semayang Lake (Kreb *et al.*, 2007). The range of Irrawaddy dolphins in the Mekong River is a 190-km river segment located about 500 km upstream of the river mouth in Kratie, Cambodia, to Khone Falls slightly upstream of the Lao PDR–Cambodia border (Beasley *et al.*, 2007). Dolphins previously

**Figure 3**  *Irrawaddy dolphin surfacing after cooperating with a cast-net fisherman in the Ayeyarwady River, Myanmar.*

inhabited Tonle Sap (Great Lake) in Cambodia but have apparently been extirpated there. In Chilika Lake, 66% of Irrawaddy dolphins sighted in 2003–2006 were concentrated in 16 km² of the outer channel. (Pattnaik *et al.*, 2007). In Songkhla Lake, 23 of 24 sightings between 2002 and 2004 were within 241 km² while 12 were within 27 km² of the mid-upper portion of Thale Luang, which constitutes the largest part of the lake (Kittiwattanawong *et al.*, 2007).

There are published records of strandings between Vishakhapatnam and Calcutta, India (Owen, 1869; James *et al.*, 1989; Pattnaik *et al.*, 2007) but little is known about the distribution of Irrawaddy dolphins in northeastern India. In Bangladesh, the species occurs in waterways of the Sundarbans Forest (Smith *et al.*, 2006) and in coastal waters offshore of the Sundarbans and Meghna River mouth (Smith *et al.*, in press-a). In Myanmar, Irrawaddy dolphins have been reported from a semi-enclosed bay offshore of the Kyaukpyu and Tennasarim River mouths in the Mergui Archipelago (Smith *et al.* in press-b), in the Ayeyarwady Delta (Smith and Mya, 2007), and in the lower reaches and estuaries of the Myebone, Kalidan, and Kyaukpyu Rivers along the Rakhine (Arakan) coast (Smith *et al.*, 1997). The species occurs in nearshore waters of Thailand, in the Gulf of Thailand at the mouths of the Chao Phraya, Mae Nam Chin, Chanthaburi and Pattani Rivers, and was reported by fishermen to occur in Phang Nga Bay and in certain areas of the Andaman Sea (Chantrapornsyl *et al.*, 1996). Irrawaddy dolphins occur in marine waters of Cambodia along the coast of the Koh Kong Province, Kompong Som Bay, and Raem National Park (Perrin *et al.*, 2005). The species has been recorded in the Belawan Deli River of northeastern Sumatra, and Belitung Island and Cilacap of southern Java, in Surabaya of northeastern Java, in Ujung Pandang or Makassar of Sulawesi, and around Biak Island and various river mouths of the southwestern coast of Irian Jaya (Mörzer Bruyns, 1966). In northern and eastern Borneo of Malaysia and Brunei, Irrawaddy dolphins have been recorded in coastal waters near Muara Island, in Sandakan and Kuching Bays, and in the mouths of the Brunei, Sarawak, Rajang, Kinabatangan, Baram, and Batang Rivers (Mörzer Bruyns, 1966; Beasley and Jefferson, 1997; Dolar *et al.*, 1997). The only records from southern Borneo in Indonesia, outside of the Mahakam River (discussed earlier), are second-hand reports from the Kumay and Kendawangan River mouths (Rudolph *et al.*, 1997). A small isolated population also occurs in Malampaya Sound, Philippines (Dolar *et al.* 2002).

Rigorous abundance estimates are available for only a few portions of the species' range: 77 (CV 27%) in Malampaya Sound, Philippines (Smith *et al.*, 2004a); at least 127 (CV = 7%) in the Mekong River (Beasley *et al.*, 2007); 70 (CV = 10%) in the Mahakam River, Indonesia (Kreb *et al.*, 2007); 58–72 in the Ayeyarwady River, Myanmar (Smith *et al.*, 2007); 62–98 in Chilika Lake, India (Pattnaik *et al.*, 2007); 5383 (CV = 40%) in freshwater affected coastal waters of Bangladesh (Smith *et al.*, in press-a); and 451 (CV = 10%) in waterways of the Sundarbans mangrove forest of Bangladesh (Smith *et al.*, 2006).

## III. Ecology

Irrawaddy dolphins are adapted to relatively rare and patchily distributed ecological conditions—deep pools of large rivers and nearshore marine environments (including appended lakes or lagoons) with freshwater inputs. Sighting data from waterways of the Sundarbans mangrove forest in Bangladesh show clear seasonal movements in response to changes in freshwater input, with the species moving seasonally along a south–west/north–east axis following the salinity gradient (Smith *et al.*, 2006). High salinity does not appear, however, to have direct adverse effects on the species, because in the

Malampaya Sound there was no difference between the mean salinity values recorded for the outer and inner portions when freshwater inputs were particularly low, while the dolphins remained confined to the latter area. This implies that the affinity of the dolphins for low-salinity waters is likely due to ecological preferences (probably related to prey) rather than physiological intolerance to high-salinity conditions (Smith *et al.*, 2004a).

## IV. Behavior and Physiology

Surfacing behavior is relatively inconspicuous, with only the uppermost dorsal surface of the animal generally becoming visible during a slow rolling dive. Leaps are infrequent but occasionally occur when the dolphins are disturbed, socializing or swimming against a strong current. Spyhopping, body rubbing, and tail slaps are also sometimes observed. The animals occasionally engage in water spitting where a narrow, well-directed stream is expelled from the mouth to a distance of 1–2 m. This behavior is believed to be associated with feeding and is sometimes used in the context of social interactions (Smith *et al.*, 1997). Irrawaddy dolphins are not known to bowride.

Reported mean group sizes vary but are normally between 2 and 6 individuals, with up to 15 when two or more groups come together. In the Mahakam River, the average daily home range of 27 groups followed for more than 6 h was 10 linear km, and the ranges for 53 photoidentified dolphins over a 3.5-year period averaged 61 linear km (Kreb *et al.*, 2007).

Vocalizations from two captive Irrawaddy dolphins were fairly basic, with a dominant frequency of about 60 kHz emitted in short pulses of 25–30 μsec. These pulse trains, believed to be used for echolocation, were fairly regular and no audible whistles or pure tones were detected (Kamminga *et al.*, 1983)

In the Ayeyarwady River (Fig. 3), the dolphins engage in a cooperative fishery with cast-net fishermen. During this fishing practice, fishermen search for dolphins and summon them by tapping the sides of their boat with a conical wooden pin called a *labai kway*. One or two lead dolphins then swim in smaller and smaller semi-circles, herding the fish toward the shore. With a wave of their half-submerged flukes, the dolphins deliver a concentrated mass of fish to the fishermen. During cooperative fishing the animals often dive steeply with flukes aloft just after the net is thrown and create turbulence under the surface around the outside of the net. The dolphins appear to benefit from the fishing activity by preying on fish whose movements are confused by the sinking net and those that are momentarily trapped around the edges of the lead line or stuck on the mud bottom just after the net is pulled up (Smith *et al.*, 2007).

## V. Life History

Based on observations in captivity, gestation is believed to last about 14 months and weaning occurs at about 2 years. Births are believed to peak in the pre-monsoon season (April–June) but may take place year-round. A 210-cm female dolphin from the Bay of Bengal near Calcutta was recorded by Anderson (1879) with an 86-cm fetus in June, and a 105-cm newborn was recorded in the Mekong in May (Stacey and Leatherwood, 1997).

## VI. Interactions with Humans
### A. Threats

Irrawaddy dolphins have been documented accidentally caught in fishing nets in almost all areas where they have been studied. In the

Mekong River, 13 of 15 deaths caused by humans between 2001 and 2005 were due to gillnet entanglement (Beasley *et al.*, 2007), and in the Mahakam River 32 of 48 deaths caused by humans between 1995 and 2005 were due to gillnet entanglement (Kreb *et al.*, 2007). Mortality from drifting gillnets has also been documented in nets targeting elasmobranchs in coastal waters of Bangladesh (Smith *et al.*, in press-a) and in bottom-set nylon gillnets used for catching crabs in Malampaya Sound, Philippines (Smith *et al.*, 2004a).

Fishing with electricity is considered the direst threat to Irrawaddy dolphins in the Ayeyarwady River. This destructive technique has been cited as being responsible for the largest number of recent known deaths of the baiji *Lipotes vexillifer* (Zhang *et al.*, 2003), a freshwater dolphin that now may be extinct (Dalton, 2006).

Many dams have been proposed that may adversely affect Irrawaddy dolphins in the Mekong River Basin. Of greatest concern are the large run-of-the-river dams proposed for the Mekong mainstem near Stung Treng and Sambor (Perrin *et al.*, 1996). In waterways of the Sundarbans mangrove forest, the dependence of the Irrawaddy dolphins on relatively deep waters and channel confluences suggests that the animals may be particularly susceptible to habitat loss from increased sedimentation caused by declining freshwater supplies due to upstream withdrawals (Smith *et al.*, in press-c).

Increased sedimentation resulting from deforestation of surrounding watersheds has resulted in decline of water depths in Songkhla, Chilika, and Semayang Lakes. A source of habitat loss and population fragmentation in several areas has also been the proliferation of fixed fishing gears. In the middle and southern portions of Songkhla Lake about 40,000 fixed fishing nets create more than 8000 km of linear barrier in multiple rows. These fishing structures restrict dolphin movements such that their habitat is substantially reduced and the potential for demographic interaction with individuals in the Gulf of Thailand is eliminated (Smith *et al.*, 2004b). Fixed fishing gears also occupy most parts of Semayang Lake and limit dolphin movements to a narrow, dredged channel that is subject to intensive vessel traffic (Kreb *et al.*, 2007).

Removal from the wild for live display is an additional threat. The charismatic appearance of Irrawaddy dolphins and behavioral characteristics they exhibit in the wild (e.g., spitting water, spyhopping, fluke-slapping, etc.) make them especially attractive for live displays. Irrawaddy dolphins are also the subject of nature tourism programs in the Mekong River and Chilika Lake. Although this form of tourism has in some cases been promoted as a substitute for captive displays, in the latter two situations there is concern that collisions with dolphin watching vessels and harassment caused by this activity may threaten the viability of these populations (Smith *et al.*, 2007).

## B. Conservation

The Irrawaddy dolphin is classified in the IUCN Red List as Vulnerable (VU), and five geographically isolated populations (in Malampaya Sound, Philippines, Chilika Lake, India, Ayeyarwady River, Myanmar, Mahakam River, Indonesia, and Mekong River, Lao PDR, Cambodia, and Vietnam) are listed as "Critically Endangered" (IUCN, 2008). Irrawaddy dolphins are included in CITES Appendix I in response to concern about the potential for international trade in live specimens to adversely affect wild populations.

Directed taking of cetaceans is prohibited in Bangladesh, India, Laos, Malaysia, Myanmar, and Thailand. The legal status of Irrawaddy dolphins in Indonesia, and Timor Leste is unclear. In Cambodia, a new fisheries law and royal decree will provide protection

to all cetaceans. In Vietnam, all cetaceans are protected by a decree of the national assembly. Some cetaceans are given legal protection in the Philippines, but as of 2002 Irrawaddy dolphins were not included in the list of species (Perrin *et al.*, 2005).

Although a few areas where the species occurs have been designated as protected, little has been done to conserve dolphin habitat. Malampaya Sound in the Philippines was proclaimed a Protected Seascape in 2000, but this is the lowest possible national prioritization given to a protected area (Smith *et al.*, 2004-a). Portions of Irrawaddy dolphin habitat in the Sundarbans Delta of Bangladesh and India are included within Protected Forests and UNESCO World Heritage Sites but no specific provisions have been implemented for conserving dolphins or their habitat. The Cambodian Department of Fisheries has drafted a Royal Decree for protection of Irrawaddy dolphins in the Mekong River, which includes the designation of eight protected areas (5721 ha) in a 190-km segment of the river above Kratie (Beasley *et al.*, 2007). In December 2005, the Department of Fisheries, Myanmar, announced the establishment of a protected area for Irrawaddy dolphins in a 74-km segment of the Ayeyarwady River between Mingun and Kyaukmyaung. Protective measures in the area include requiring fishermen to immediately release dolphins if found alive and entangled in their nets and prohibiting the catching or killing of dolphins and trade in whole or parts of them and the use of electricity fishing and gill nets that obstruct the water-course, are more than 300-ft long, or spaced less than 600 ft apart (Smith and Mya, 2007).

## See Also the Following Articles

Australian Snubfin Dolphin ■ Indo-west Pacific Marine Mammals

## References

Anderson, J. (1879). "Anatomical and Zoological Researches: Comprising an Account of Zoological Results of the Two Expeditions to Western Yunnan in 1868 and 1875; and a Monograph of the Two Cetacean Genera, *Platanista* and *Orcella* [*sic*.]." Bernard Quaritch, London.

Arnason, U., and Gullberg, A. (1996). Cytochrome *b* nucleotide sequences and the identification of five primary lineages of extant cetaceans. *Mol. Biol. and Evol.* **13**, 407–414.

Arnold, P. W. (2002). Irrawaddy dolphin (*Orcaella brevirostris*). *In* "Encyclopedia of Marine Mammals" (W. F. Perrin, B. Würsig, and J. G. M. Thewissen, eds), pp. 652–654. Academic Press, London, UK.

Arnold, P. W., and Heinsohn, G. E. (1996). Phylogenetic status of the Irrawaddy dolphin *Orcaella brevirostris* (Owen in Gray): A cladistic analysis. *Mem. Queensl. Mus.* **39**, 141–204.

Barnes, L. G., Domning, D. P., and Ray, C. E. (1985). Status of studies on fossil marine mammals. *Mar. Mamm. Sci.* **1**, 15–53.

Beasley, I., and Jefferson, T. A. (1997). Marine mammals of Borneo: A preliminary checklist. *Sarawak Mus. J.* **51**, 193–210.

Beasley, I., Arnold, P., and Heinsohn, G. (2002). Geographical variation in skull morphology of the Irrawaddy dolphin, *Orcaella brevirostris* (Owen in Gray, 1866). *Raff. Bull. Zool. Supplement* **10**, 15–34.

Beasley, I., Robertson, K., and Arnold, P. (2005). Description of a new dolphin, the Australian snubfin dolphin *Orcaella heinsohni* sp. N. (Cetacea, Delphinidae). *Mar. Mamm. Sci.* **21**(3), 365–400.

Beasley, I., Phay, S., Gilbert, M., Phothitay, C., Yim, S., Lor, K. S., and Kim, S. (2007). Status and conservation of Irrawaddy dolphins *Orcaella brevirostris* in the Mekong River of Vietnam, Cambodia and Laos. *In* "Status and Conservation of Freshwater Populations of Irrawaddy Dolphins, WCS Working Paper Series 31" (B. D. Smith,

I

R. G. Shore, and A. Lopez, eds), pp. 67–82. Wildlife Conservation Society, Bronx, NY.

Dolar, M. L. L., Perrin, W. F., Yaptinchay, A. A. S. P., Jaaman, S. A. B. H. J., Santos, M. D., Alava, M. N., and Suliansa, M. S. B. (1997). Preliminary investigation of marine mammal distribution, abundance, and interactions with humans in the southern Sulu Sea. *Asian Mar. Biol.* **14**, 61–81.

Dolar, M. L. L., Perrin, W. F., Gaudiano, J. P., Yaptinchay, A. A. S. P., and Tan, J. M. L. (2002). Preliminary report on a small estuarine population of Irrawaddy dolphins *Orcaella brevirostris* in the Philippines. *Raff. Bull. Zool. Supplement* **10**, 155–160.

Chantrapornsyl, S., Adulyanukosol, K., and Kittiwattanawong, K. (1996). Records of cetaceans in Thailand. *Phuket Mar. Biol. Cent. Res. Bull.* **61**, 39–63.

Dalton, R. (2006). Last hope for river dolphins. *Nature* **440**, 1096–1097.

IUCN. (2008). 2008 IUCN Red List of threatened species. [Available from www.redlist.org].

James, P. S. B. R., Rajagopalan, M., Dan, S. S., Bastian Fernando, A., and Selvaraj, V. (1989). On the mortality and stranding of marine mammals and turtles at Gahirmatha, Orissa from 1983 to 1987. *J. Mar. Biol. Ass. India* **31**, 28–35.

Kasuya, T. (1973). Systematic consideration of recent toothed whales based on the morphology of tympano-periotic bone. *Sci. Rep. Whales Res. Inst. Tokyo* **25**, 1–103.

Kreb, D., Budiono, , and Syachraini, (2007). Status and conservation of Irrawaddy dolphins *Orcaella brevirostris* in the Mahakam River of Indonesia. *In* "Status and Conservation of Freshwater Populations of Irrawaddy Dolphins, WCS Working Paper Series 31" (B. D. Smith, R. G. Shore, and A. Lopez, eds), pp. 53–66. Wildlife Conservation Society, Bronx, NY.

Kamminga, C., Wiersma, H., and Dudok van Heel, W. H. (1983). Sonar sounds in *Orcaella brevirostris* of the Mahakam River, East Kalimantan, Indonesia; the first descriptions of the acoustic behaviour. *Aq. Mamm.* **10**, 125–136.

Kittiwattanawong, K., Chantrapornsyl, S., Ninwat, S., and Chooruk, S. (2007). Status and Conservation of Irrawaddy Dolphins *Orcaella brevirostris* in Songkhla Lake of Thailand. *In* "Status and Conservation of Freshwater Populations of Irrawaddy Dolphins, WCS Working Paper Series 31" (B. D. Smith, R. G. Shore, and A. Lopez, eds), pp. 83–89. Wildlife Conservation Society, Bronx, NY.

Lloze, R. (1973). Contributions à l'étude anatomique, histologique et biologique de l'Orcaelle brevirostris (Gray, 1866) (Cetacea-Delphinidae) du Mekong. Ph.D. Dissertation L'Université Paul Sabatier de Toulouse, France. 598 pp. [In French].

LeDuc, R. G., Perrin, W. F., and Dizon, A. E. (1999). Phylogenetic relationships among the delphinid cetaceans based on full cytochrome b sequences. *Mar. Mamm. Sci.* **15**, 619–648.

Marsh, H., Lloze, R., Heinsohn, G. E., and Kasuya, T. (1989). Irrawaddy Dolphin, *Orcaella brevirostris* (Gray, 1866). *In* "Handbook of Marine Mammals" (S. H. Ridgway, and R. Harrison, eds), pp. 101–118. Academic Press Limited, San Diego, California.

Mörzer Bruyns, W. J. F. (1966). Some notes on the Irrawaddy dolphin, *Orcaella brevirostris* (Owen, 1866). *Zeitsch. Saügetierk.* **31**, 367–370.

Nishiwaki, M. (1963). Taxonomical consideration on genera Delphinidae. *Sci. Rep. Whales Res. Inst. Tokyo* **17**, 93–103.

Owen, R. (1869). On some Indian cetaceans collected by Walter Elliot, Esq. Trans. *Zool. Soc. Lond.* **6**, 17–47.

Pattnaik, A., Sutaria, D., Khan, M., and Behera, B. P. (2007). Status and Conservation of Irrawaddy Dolphins *Orcaella brevirostris* in Chilika Lagoon of India. *In* "Status and Conservation of Freshwater Populations of Irrawaddy Dolphins, WCS Working Paper Series 31" (B. D. Smith, R. G. Shore, and A. Lopez, eds), pp. 41–52. Wildlife Conservation Society, Bronx, NY.

Perrin, W. F., Dolar M. L. L., and Alava, M. N. R. (1996). Report of the Workshop on the Biology and Conservation of Small Cetaceans and Dugongs of Southeast Asia. UNEP(W)/EAS WG 1/2. United Nations Environmental Programme.

Perrin, W. F., Reeves, R. R., Dolar, M. L. L., Jefferson, T. A., Marsh, H., Wang, J. Y., and Estacion, J. (2005). Report of the Second Workshop on the Biology and Conservation of Small Cetaceans and Dugongs of Southeast Asia, Silliman University, Dumaguete City, Philippines 24–26 July, 2002. CMS Technical Series Publication No. 9. Convention on Migratory Species, Bonn. UNEP/CMS.

Pilleri, G., and Gihr, M. (1974). Contribution to the knowledge of the cetaceans of southwest and monsoon Asia (Persian Gulf, Indus Delta, Malabar, Andaman Sea and Gulf of Siam). *Invest. Cetacea* **5**, 95–153.

Rudolph, P., Smeenk, C., and Leatherwood, S. (1997). Preliminary checklist of Cetacea in the Indonesian archipelago and adjacent waters. *Zool. Verh. Leiden.* **312**, 1–48.

Smith, B. D., and Mya, T. T. (2007). Status and conservation of Irrawaddy dolphins *Orcaella brevirostris* in the Ayeyarwady River of Myanmar. *In* "Status and Conservation of Freshwater Populations of Irrawaddy Dolphins, WCS Working Paper Series 31" (B. D. Smith, R. G. Shore, and A. Lopez, eds), pp. 21–40. Wildlife Conservation Society, Bronx, NY.

Smith, B. D., Thant, U. H., Lwin, J. M., and Shaw, C. D. (1997). Investigations of cetaceans in the Ayeyarwady River and Northern coastal waters of Myanmar. *Asian Mar. Biol.* **14**, 173–194.

Smith, B. D., *et al.* (2004a). Status, ecology and conservation of Irrawaddy dolphins *Orcaella brevirostris* in Malampaya Sound, Palawan, Philippines. *J. Cetacean Res. Manag.* **6**, 41–52.

Smith, B. D., Sutaria, D., Piwpong, N., Choorak, S., and Koedpoem, W. (2004b). Can Irrawaddy dolphins survive in Songkhla Lake, Thailand? *Nat. Hist. Bull. Siam Soc.* **52**, 181–194.

Smith, B. D., Ahmed, B. and Mansur, R. (in press-a). Species occurrence and distributional ecology of nearshore cetaceans in the Bay of Bengal, Bangladesh, with abundance estimates for lrrawaddy dolphins *Orcaella brevirostris* and finless porpoises *Neophocaena phocaenoides J. Cetacean Res. Manage.*

Smith, B. D., and Mya, T. T. (in press -b). Species occurrene, distributional ecology and fisheries interactions of cetaceans in the Mergui (Meik) Archipelago, Myanmar. *J. Cetacean Res. Manage.*

Smith, B. D., Braulik, G., Strindberg, S., Mansur, R. Diyan, M. A. A. and Ahmed, B. (in press-c). Habitat selection of freshwater cetaceans and the potential effects of declining freshwater flows and sea-level rise in waterways of the Sundarbans mangrove forest, Bangladesh. *Aquat. Cons. Mar. Freshw. Ecosyst.*

Smith, B. D., Braulik, G., Strindberg, S., Ahmed, B., and Mansur, R. (2006). Abundance of Irrawaddy dolphins (*Orcaella brevirostris*) and Ganges river dolphins (*Platanista gangetica gangetica*) estimated using concurrent counts from independent teams in waterways of the Sundarbans mangrove forest in Bangladesh. *Mar. Mamm. Sci.* **22**, 1–21.

Smith, B. D., Shore, R. G., and Lopez, A. (eds). (2007). Annex 1. Report on the Workshop to Develop a Conservation Action Plan for Freshwater Populations of Irrawaddy Dolphins. *In* "Status and Conservation of Freshwater Populations of Irrawaddy Dolphins," WCS Working Paper Series 31 (B. D. Smith, R. G. Shore, and A. Lopez, eds), pp. 90–109, Wildlife Conservation Society, Bronx, NY.

Stacey, P. J., and Leatherwood, S. (1997). The Irrawaddy dolphin, *Orcaella brevirostris*: A summary of current knowledge and recommendations for conservation action. *Asian Mar. Biol.* **14**, 195–214.

Stacey, P. J., and Arnold, P. W. (1999). *Orcaella brevirostris. Mammal. Spec.* **616**, 1–8.

Zhang, X., *et al.* (2003). The Yangtze River dolphin or baiji (*Lipotes vexilifer*): Population status and conservation issues in the Yangtze River, China. *Aquat. Cons. Mar. Freshw. Ecosyst.* **13**, 51–64.

# Japanese Whaling

TOSHIO KASUYA

Whaling is a fishing activity that targets whales, but the term does not often fit because of the ambiguity of *whale*, sometimes construed to exclude small cetaceans. This is also true in Japanese whaling. This chapter adopts the broadest meaning for the term Japanese whaling to include activities of hunting any cetaceans in Japanese territory, by Japanese companies, or by any companies known to be sponsored by them.

## I. Subsistence Whaling

Numerous bones of gregarious dolphins in a site of the Jomon Era (10,000 BP–3200 BC) on the Noto coasts, Sea of Japan, suggest the presence of a drive fishery. Other sites of similar antiquity on the Pacific coasts of central and northern Japan and on the coasts of northern Kyushu facing the Sea of Japan/East China Sea revealed remains of small cetaceans and detachable harpoon heads. The Okhotsk Sea culture of Hokkaido in the fifth to fourteenth centuries left skeletons, harpoons, and drawings depicting whale harpooning. Ainu people on Uchiura Bay, Pacific coast of southern Hokkaido, opportunistically hunted whales in the late nineteenth century using aconite-poisoned detachable harpoon heads and floats. Skeletal remains from these sites represent at least 13 species of cetaceans: North Pacific right whale (*Eubalaena japonica*), common minke whale (*Balaenoptera acutorostrata*), sei whale (*B. borealis*), humpback whale (*Megaptera novaeangliae*), sperm whale (*Physeter macrocephalus*), false killer whale (*Pseudorca crassidens*), long-finned pilot whale (*Globicephala melas*), Pacific white-sided dolphin (*Lagenorhynchus obliquidens*), common dolphin (*Delphinus* sp.), common bottlenose dolphin (*Tursiops truncatus*), Dall's porpoise (*Phocoenoides dalli*), harbor porpoise (*Phocoena phocoena*), and unidentified beaked whales (Ziphiidae) (Kasuya, 1975). Differentiation of whales hunted from those stranded, however, is often difficult.

## II. Traditional Commercial Whaling

Harpooning whales that are found in a harbor could have taken place widely. Records of such takes in a harbor at Ine, Sea of Japan, included 167 humpback, 148 fin (*Balaenoptera physalus*), and 40 right whales in the period 1656–1913 (Omura, 1984). A village next to Ine took small cetaceans in the same way. A similar fishery was also recorded in the fourteenth century at villages on Tsushima Island, off northern Kyushu. A cooking recipe in 1489 recommended whale meat for noble guests.

Records of "harpoon whaling" started in the 1570s at Morosaki at the entrance of Mikawa Bay, a bay attached to Ise Bay that opens to the Pacific. The whalers first used light harpoons with a detachable head and line. Harpoons with fixed heads and lancing followed. The winter operation continued to the early 1800s and took gray (*Eschrichtius robustus*) and humpback whales for oil and meat. This practice soon spread eastward to Katsuyama at the entrance of Tokyo Bay for Baird's beaked whales (*Berardius bairdii*) and survived until the late nineteenth century (Fig. 1). It also spread westward to the nearby Ise and Kii areas (before 1606), Shikoku (1624), northern Kyushu (1630s), and Nagato (around 1672) (Hashiura, 1969).

Harpoon whaling on the Ise and Kii coasts mostly ceased before 1770. A whaling group at Taiji, Kii was an exception. They modified old harpoon whaling learned from Morosaki in 1606 into new "net whaling" in 1677 (Fig. 2). During the whaling season (winter and spring), harpoon boats waited offshore for a signal from the spotters on cliffs. Receiving signals by flag and smoke indicating species, number, and position of whales, they drove the whales toward the shore where net boats waited to place nets in front of the whales to entangle them. Then the procedures of harpoon whaling followed, i.e., harpooners threw harpoons (fixed head) and lances followed. When a whale became weak, a harpooner swam to the whale to tie ropes through holes made near the blowholes and the back of the body to prevent the carcass from sinking. Boats on each side of the whale towed it, using these ropes and additional ropes that surrounded the body, to the beach for flensing. This method spread to Shikoku (1681) and northern Kyushu (1684). The preference was to harpoon calves first and then their mothers to secure both with ease.

American and European sailing whalers operated off Japan after around 1820. A decline in Japanese coastal whaling became evident in the late nineteenth century, and some whaling groups started modern Norwegian-type whaling, whereas others attempted to improve their traditional method. A few net whalers moved to new grounds in Hokkaido and southern Sakhalin and took gray whales (Hattori, 1887–1888).

Meat and most of the blubber were sold for human consumption fresh or salted. Oil was extracted from chopped bones and some BLUBBER and was used for lighting, for human consumption, and as a pesticide in rice paddies nationwide in Japan.

The traditional whaling was very labor intensive and inefficient. A Tsuro group in Shikoku in the late 1890s used 15 harpoon boats, 2 whale towers, and 14 net boats. The total full-time workers were 356, including 10 whale spotters, 12 flensers, 2 carpenters, 1 cooper, and 2 blacksmiths (Yamada, 1902). Another group in Nagato recorded 587 workers in the early 1800s. Each group took low tens of whales yearly (Table I). Annual expenditures of 12,423–15,864 yen made a profit of 987–25,640 yen (mean 9778) for a group at Kawajiri, Nagato, during 1884–1893 (Tada, 1978).

## III. Modern Coastal Whaling

A Russian, A. Dydymov, started modern Norwegian-type whaling in the western North Pacific in 1889 using a land station east of Vladivostok, Russia. In 1891, the Pacific Whaling Company was established at Vladivostok and operated from the Korean to Sakhalin coasts (Tonnessen and Johnsen, 1982). Large amounts of whale meat sold by Russians at Nagasaki stimulated the Japanese to begin similar operations. After several attempts that caught the first whales in 1898 and survived only for a short period, Nihon Enyo Gyogyo (Japan Far Seas Fishery) founded in 1899 at Senzaki, Nagato established modern whaling in Japan using Norwegian gunners. The company expanded the business and renamed themselves Toyo Gyogyo (Oriental Fishery)

J

TABLE I
Number of Whales Taken by Japanese Net Whaling at Kawajiri (Tada, 1978) and Tsuro (Yamada, 1902)[a]

| Seasons | Humpback | Right | Gray | Blue | Fin | Bryde's | Total |
|---|---|---|---|---|---|---|---|
| *Kawajiri, Nagato* | | | | | | | |
| 1698–1737 | 391(9.8) | 105(2.6) | 60(1.5) | | 22(0.6) | | 518(13.0) |
| 1738–1840[b] | 304(5.4) | 113(2.0) | 72(1.3) | | 3(0.1) | | 492(8.8) |
| 1845–1889 | 198(5.0) | 39(1.0) | 37(0.9) | | 131(3.3) | | 405(10.1) |
| 1894–1901 | 28(3.5) | 7(0.9) | 9(1.1) | | 55(6.9) | | 99(12.4) |
| *Tsuro, Shikoku* | | | | | | | |
| 1849–1865 | 209(12.3) | 19(1.1) | 101(5.9) | 5(0.3) | 0(0) | 35(2.1) | 369(21.7) |
| 1874–1890 | 108(6.4) | 21(1.2) | 82(4.8) | 24(1.4) | 9(0.5) | 41(2.4) | 285(16.8) |
| 1891–1896 | 26(4.3) | 2(0.3) | 18(3.0) | 18(3.0) | 5(0.8) | 31(0.5) | 100(16.7) |

[a]Average annual catches are in parentheses.
[b]Records for 47 years in this period are missing.

in 1904, absorbing other whaling companies. In 1908, a total of 12 modern whaling companies operated using 28 catcher boats. Toyo Gyogyo and five others merged in 1909 to form the new Toyo Hogei (Oriental Whaling), which owned 20 land stations (3 in Korea) and 21 whale catcher boats. Six others remained independent. Data on whales taken by them are not available by species, but comparison with later records throws some light on the species and cetacean fauna during earlier whaling (Table II) (Akashi, 1910).

In November 1909, the Japanese government placed hunting of sperm and baleen whales other than minke whales under its control and limited catcher boats to 30 (Omura *et al.*, 1942). This was further decreased to 25 (1934–1963) and to 5 (1984) in several steps (Kondo, 2001). On June 8, 1938, Japan enacted the protection of certain whales, i.e., cows accompanied by calves and whales below minimum size limits. However, it allowed the taking of right and gray whales, and the size limits, particularly for blue whales (*B. musculus*), were smaller than those existing in international agreements. In November 1945, Japan adopted for all the whaling activities international regulations of the time by the order of the General Headquarters of Allied Forces (GHQ) and joined the International Convention for the Regulation of Whaling of 1946 in April 1951 (before the Peace Treaty).

Postwar coastal whaling started in September 1945 and continued until March 1988 by five major companies using a maximum of 20 land stations (Fig. 1) on the four major islands of Japan. Use of a land station on Hahajima Island, Bonin Islands, started in 1981 to take the then increased Bryde's whale (*B. edeni*) quota and continued to 1987. This whaling was called large-type whaling to distinguish it from small-type whaling established in December 1947 (Ohsumi, 1975). The fishing season and land stations used changed over time. The last season of the fishery (1987/1988) used a land station in the Bonin Islands for 317 Bryde's whales, and four stations at Yamada and Ayukawa (both in Sanriku), Wadaura (Boso), and Taiji (Kii) took 188 sperm whales.

Japan started a national sperm whale quota in 1959. The quota for the North Pacific whaling countries (Canada, Japan, USA, and the USSR) replaced this starting in 1971. The four countries set quotas for fin, sei, and Bryde's whales in 1969; these were replaced by quotas of the International Whaling Commission (IWC) in 1972. The IWC prohibitions by species and dates of enforcement were blue (1965), humpback (1966), and fin and sei whales (1976).

Maximum annual catches by species since 1911 and their dates are 300 blue (1911); 1043 fin, 160 humpback, and 155 gray whales (1914); 14 right whales (1932); 1035 sei whales (1959); 504 Bryde's whales (1962); and 3747 sperm whales (1968). The whaling companies manipulated coastal statistics, particularly for sperm whales and Bryde's whales (Kasuya, 1999).

## IV. Pelagic Whaling in the Antarctic
### A. Before World War II

In 1934, Nihon Hogei ("Japan Whaling", renamed from "Toyo Hogei" in 1934), which merged with Nihon Suisan in 1937, purchased the Norwegian factory ship *Antarctic* (9600 tons) and five catcher boats for £55,000, which was three times the profit from one Japanese Antarctic fleet in 1937/1938. On the way to Japan the *Antarctic* and three catcher boats operated from December 1934 and took 213 whales. This was the first Japanese Antarctic operation. This company built a second fleet in 1936/1937 and a third one in 1937/1938 (Itabashi, 1987).

Hayashikane Shoten, the antecedent of Taiyo Gyogyo (Ocean Fisheries), sent the *Nisshinmaru* fleet to the Antarctic in the 1936/1937 season and had a second fleet in 1937/1938. Kyokuyo Hogei (Polar Sea Whaling) sent the *Kyokuyomaru* fleet in the 1938/1939 season. Thus, the total Japanese Antarctic operation increased to six factory ships in 5 years.

The Japanese government enacted regulations of pelagic whaling on June 8, 1938, including a fishing season from November 1 to March 15; protection of gray whales and right whales (*Eubalaena* spp) (except for the North Pacific north of 20°N) and cows accompanied by calves; minimum size limits; processing within 36h; and full utilization of the catch. This differed from international agreements of the time in a season about 6 weeks longer and the blue whale size limit about 1.4m shorter, allowing taking of blue whales that migrated earlier in the season.

The main product of these operations was whale oil for export. The government strictly limited the importation of Antarctic whale meat until the 1939/1940 season to protect coastal whaling. Some of the whale oil of the last two seasons (1939/1940 and 1940/1941) was landed in northern Korea and was exported to Germany via Siberia (Tokuyama, 1992).

### B. Postwar Operations

In order to feed the starving Japanese population, GHQ issued a permit for Antarctic whaling in August 1946. Taiyo Gyogyo converted an oil tanker into *Nisshinmaru No. 1* and Nihon Suisan another vessel to *Hashidatemaru*. These fleets caught 932 BWU, or 6% of the world catch of the 1946/1947 season, and produced 12,260 tons

## TABLE II
### Expansion of Early Norwegian-type Japanese Whaling from Korean Coasts to Hokkaido via Southwestern Japan and Change in Composition of Species Hunted[a]

| Season | East Korea | NW/Kyushu to Nagato | SE/Kyushu and Shikoku | Kii | Kii plus Boso | Boso | Sanriku | Hokkaido |
|---|---|---|---|---|---|---|---|---|
| 1899/1900 | 15 | — | — | — | | — | — | — |
| 1900/1901 | 42 | — | — | — | | — | — | — |
| 1901/1902 | 58 | — | — | — | | — | — | — |
| 1902/1903 | 89 | — | — | — | | — | — | — |
| 1903/1904 | 182 | — | — | — | | — | — | — |
| 1904/1905 | 336 | — | — | — | | — | — | — |
| 1905/1906 | 294 | 4 | — | 1 | | 74 | 22 | — |
| 1906/1907 | 378 | — | 198 | 199 | | 32 | 88 | — |
| 1907/1908 | 236 | 47 | 289 | 248 | | 160 | 217 | — |
| 1908/1909 | 244 | 59 | 126 | 381 | | 56 | 297 | — |
| 1909/1910 | — | — | — | 58 | | — | 96 | — |
| Total for 11 seasons | 1874 | 110 | 613 | 887 | | 322 | 720 | — |
| 1911 | | | | | | | | |
| Blue | 1 | 4 | 64 | | 177 | | 54 | 0 |
| Fin | 183 | 281 | 7 | | 31 | | 394 | 66 |
| Humpback | 5 | 14 | 4 | | 25 | | 8 | 3 |
| Sei/Bryde's | 0 | 12[b] | 13[b] | | 87[c] | | 260[d] | 1 |
| Gray | 119 | 2 | 0 | | 0 | | 0 | 0 |
| Right | 0 | 0 | 1 | | 1 | | 0 | 0 |
| Sperm | 0 | 0 | 4 | | 9 | | 149 | 0 |
| Total | 308 | 313 | 93 | | 330 | | 865 | 70 |

[a]Statistics for July 1899–April 1910 are from Akashi (1910).
[b]Bryde's whales.
[c]Mostly Bryde's whales.
[d]Mostly sei whales.

of whale oil and 22,167 tons of other edible products. The meat production resulted in total products of 36.9 tons/BWU, almost double the maximum prewar production of 19.0 ton/BWU. Whale meat became an important product of Japanese whaling. Kyokuyo Hogei sent the *Baikarumaru* fleet for sperm whales only in 1951/1952, before it returned to the Antarctic in 1956/1957.

Although world Antarctic fleets recorded an increase from 9 (1946/1947) to 21 (1960/1961–1961/1962) and a subsequent decline, the decline of Japanese fleets was slightly slower, i.e., from a peak of 7 fleets in 1960/1961–1964/1965 to 1 in 1977/1978–1986/1987. In 1956, Japan purchased a foreign fleet to expand its operation. The objective of the purchases changed in 1962, when Japan got a quota allocation of 33%, and a further increase was permitted with the fleet purchase (Tonnessen and Johnsen, 1982). Out of 9 fleets purchased by Japan in the postwar period, 4 were for their quotas.

The total Japanese fleet and number of workers involved varied by quota and species hunted. The *Nisshinmaru No. 1* fleet in the 1950/1951 season, when it processed 631 blue and 1014 fin whales, had 348 persons on the factory ship, 604 on three freezing and salting vessels, and 197 on nine catcher boats. The total was 1149 (Tokuyama, 1992).

Takes of significant numbers of sei whales started in 1949/1950 and reached a maximum of 11,310 in 1965/1966, and that of minke whales (*B. acutorostrata* subsp. and *B. bonaerensis*) started in 1971/1972 and reached 3950 in 1976/1977.

Three Antarctic whaling companies split off their whaling sections to merge them into a new company, Nihon Kyodo Hogei (Japan Union Whaling) in 1976. The new company sent two fleets to the Antarctic in 1976/1977 and one in 1977/1978 to 1986/1987. The last two seasons were operated under objection to the IWC moratorium on commercial whaling.

Southern humpback whales were completely protected as of 1963/1964, "true" blue whales as of 1963/1964, all southern blue whales as of 1964/1965, fin whales as of 1976/1977, sei whales as of 1978/1979, and sperm whales as of 1981/1982.

## V. Pelagic Whaling in the North Pacific
### A. Before World War II

The *Tonanmaru* fleet was sent out in the 1940 and 1941 seasons by Hokuyo Hogei (Northern Sea Whaling), established jointly by three whaling companies, and caught 74 blue, 659 fin, 114 humpback, 9 sei, 333 sperm, 58 gray, and 4 North Pacific right whales in the two seasons off southern Kamchatka and in the Bering and Chukchi Seas (Maeda and Teraoka, 1952).

### B. Postwar, off the Bonin Islands

Whaling had been operated in 1923–1944 using land stations on the Bonin Islands (Ogasawara Islands) for humpback, Bryde's,

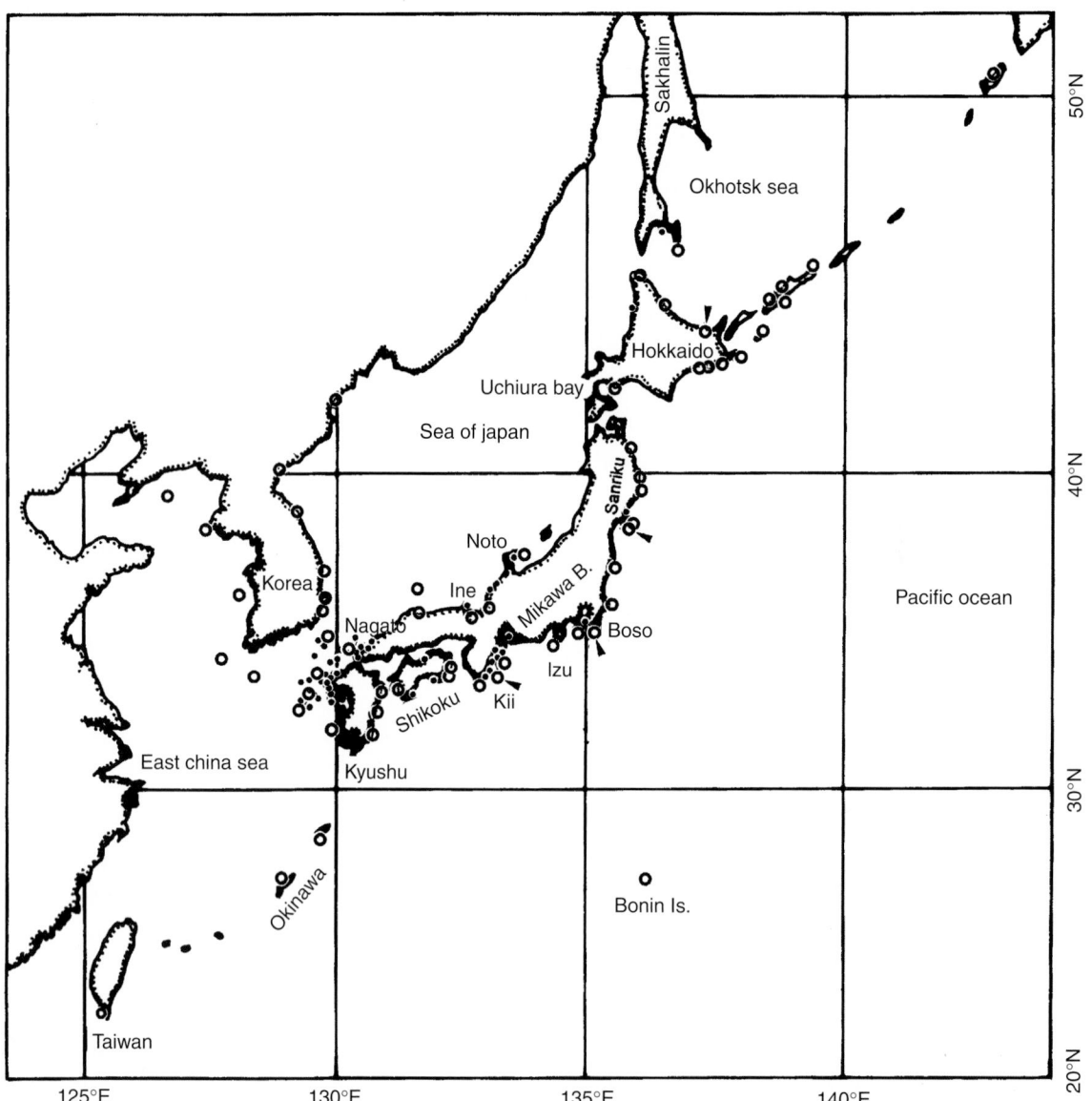

**Figure 1** *Location of major land stations used by Japanese whaling. Closed circles represent harpoon or net whaling, open circles represent large-type whaling during pre- and postwar periods, and arrows indicate five land station currently in use by small-type whaling.*

and sperm whales, but the 1945 permit of the GHQ to whale off the Bonin Islands prohibited the use of land stations. Therefore, Taiyo Gyogyo converted a navy vessel to a factory ship and whaled in March–April 1946. The number of fleets and companies involved subsequently varied by season; the last fleet was sent out in 1951. In 1952, pelagic whaling started in the northern North Pacific, and operations off the Bonin Islands ceased. In the six seasons they took 923 Bryde's, 606 sperm, and 29 other whales. Only 20 humpback whales were taken because of the offshore nature of the operations (Maeda and Teraoka, 1952).

### C. Northern North Pacific

The Peace Treaty came into effect in April 1952, and Japan sent out the *Baikarumaru* fleet to the North Pacific. The fleets increased to two in 1954 and three in 1962, and then in 1976–1979 only the Kyodo Hogei fleet remained. The operation was a joint venture of most of the Japanese whaling companies; Kyokuyo Hogei, Nihon Suisan, and Taiyo Gyogyo were the major ones. Factory ships and quotas changed frequently (Tato, 1985). The IWC ban on pelagic whaling for species other than minke whales came into effect in the 1979/1980 Antarctic and 1980 northern summer season. The first national quota of 350 BWU was for a 1-year test operation by one of the two fleets in 1954. This was followed by a blue whale quota of 70 (1955–1961) or 60 (1962–1965) and quotas of about 800 BWU (1957–1964) and 1000 BWU (1965–1968) for species other than blue whales. Sperm whale quotas were from 1500 to 1800 (1957–1961), 2460 to 2700 (1962–1965), and 3000 (1965–1968).

The North Pacific whaling countries set quotas by species in 1969, which were followed by quotas of the IWC, as of 1971 for baleen whales and 1972 for all large whales. The IWC has protected blue and humpback whales since 1966 and fin and sei whales since 1976.

**Figure 2** *A scene of net whaling (from Oyamada, 1832). A harpooner is climbing on a humpback whale to attach a line to the animal. Boats are ready to kill the animals with lances.*

J

The *Miwamaru*, a whale catcher–factory ship, operated in the 1973–1975 seasons and reported takes of 279 common minke and 6 Baird's beaked whales. The operation was not inspected, and caution should be taken in accepting the statistics.

## VI. Whaling under Foreign Jurisdiction

Since 1957, Japanese whaling has expanded into foreign territories (Tato, 1985), presumably for new whale stocks, efficient vessel allocation, unregulated operation, or for new business opportunities.

### A. Taiwan, 1957–1959

Taiwan was outside the ICRW. Kyokuyo Hogei whaled for two seasons jointly with a local company using a land station in southern Taiwan but took only 29 humpbacks and a sperm whale. The Taiwanese partner operated for a few additional years.

### B. Okinawa, 1958–1965

Under supervision of the US military, the Ryukyu government governed the Ryukyu Islands from the end of World War II to 1972, when the islands were returned to Japan. Hand-harpoon fishermen at Okinawa took humpback whales using harpoon guns beginning around 1950. In 1958, the Ryukyu government introduced IWC regulations. Only a group of Nago fishermen and two other local companies got the new licenses. Two Japanese whaling companies, Taiyo Gyogyo and Nitto Hogei, offered crew and catcher boats to each of the latter. The land stations were at Nago, Sashiki, and Itoman. In addition to catches of 52 humpback whales by the Nago group (1950–1957), the three groups took 788 humpback, 31 sperm, and 1 Bryde's whale in 1958–1965.

### C. Brazil, 1959–1984

Two groups whaled off Brazil, each inviting a Brazilian partner. The Taiyo group whaled in 1960–1963 from a land station at Cabo Frio, and the Nichirei group in 1959–1984 from Costina. Catches were mostly sei whales in 1959–1964 (3214 whales in the six seasons) and then shifted to minke whales (*B.* spp.) with a maximum recorded catch of 1036 in 1975. Sperm, fin, and blue whales were also taken (decreasing order). Some of the supposed sei whales were Bryde's whales.

### D. Canada, 1962–1972

Japanese whalers operated jointly with Canadian partners off Newfoundland and Vancouver Island. The Taiyo group operated in 1962–1967 using a land station at Coal Harbor, Vancouver Island, and caught mostly sei (2153), sperm (1108), and fin (837) whales, but some blue and humpback whales were also taken. Off Newfoundland, the Kyokuyo group operated in 1966–1972 using a land station at Dildo, and the Taiyo group in 1967–1972 using a Williams Port station. Their catch was mostly fin whales (1168) and a few humpback, sei, and sperm whales. The Canadian government closed commercial whaling in 1973.

### E. South Georgia, 1963/1964–1965/1966

Two Japanese expeditions operated using South Georgian land stations leased from the United Kingdom. Their total catches were 1273 fin, 919 sei, and 218 sperm whales. Under international pressure, the United Kingdom agreed at the 1966 IWC conference to voluntarily retain the South Georgian catches at or below the level of the 1964/1965 seasons. This terminated Japanese expeditions.

### F. Chile, 1964–1968

Chile, Peru, and Ecuador jointly regulated whaling in their territorial waters until 1979, when Chile and Peru joined the ICRW. Nitto Whaling and its local partner whaled using one or two land stations and took 516 blue, 582 fin, 1061 sei, and 1221 sperm whales. The catch of blue whales occurred only in 1965 and 1966. Some of the supposed sei whales were probably Bryde's whales.

### G. Peru, 1967–1985

A local company sponsored by Nihon Kinkai Hogei (Japan Coastal Whaling) whaled using a station at Paita. The operation ended in March 1985. The fishing season lasted almost 12 months of the year, with occasional interruptions of 1 or 2 months in winter. The total catch was 291 (1968–1977), 3408 Bryde's (1973–1983), 232 sei (1973–1978), 2304 Bryde's or sei (1968–1972), and 14,331 sperm whales (1968–1981) (Valdivia *et al.*, 1984).

### H. Philippines, 1983–1984

A local company whaled for two seasons using the *Faith No. 1*, the renamed *Miwamaru* catcher–factory ship of Japan. One of the Japanese sponsors had taken part in an earlier *Miwamaru* operation in Japan. A take of 9 Bryde's whales in 1983 and 47 in 1984 was reported with production of 277 tons of meat in 1984. The operation ended due to Japanese rejection of meat import and new regulation for her nationals concerning participation in foreign whaling. The local company operated in 1985 and took 40 Bryde's whales.

### VII. Small-Type Whaling

This is defined as a whaling activity that takes minke whales and toothed whales other than sperm whales using a vessel and a whaling cannon below a certain size limit. This fishery started around the start of the twentieth century, e.g., the Baird's beaked whale fishery off the Boso coast introduced Greener harpoon guns in 1892 and Taiji fishermen 20-mm five-barrel harpoon guns for pilot whales in 1904. The fishery was placed under control of the Minister of Agriculture and Forestry in December 1947 (Ohsumi, 1975). Before this the operation was unregulated except for the Boso coast where the Baird's beaked whale fishery required a license from Chiba Prefecture (since 1920).

About 20 vessels operated the fishery off northern Kyushu, Kii, Boso, and Sanriku before World War II. The number increased to 53 in 1942 and 80 in 1950 and then it declined rapidly to 9 in 1970, 4 in 1988, and 5 since 1992. Conversion from several small vessels to one larger vessel contributed to the earlier decline. During the war the vessel size was 5–20 tons. The size limit was 30 tons in 1947, 40 tons in 1963, and is now 50 tons. The maximum caliber of harpoon gun changed from 40 mm (1947) to 50 mm (1952–present). Other regulations included the prohibition of killing calves and cows accompanied by calves and a fishing season of 6 months. The vessels usually leave port in the morning and return in the evening.

This fishery had no quota until 1977 and took common minke, Baird's beaked, pilot, and killer whales (*Orcinus orca*). Dolphins and porpoises were also taken (Fig. 3). The IWC set a quota for minke whales for the seasons 1978–1987. The government of Japan set a national quota for Baird's beaked whales at 40 (1983–1987), 60 (1988), 54 (1989–1998), 62 (1999–2004), and 66 (2005–2007).

The Japanese government maintains that target species of this fishery other than the minke whale are outside the IWC competence, thus the decision of IWC to cease commercial whaling does not prohibit

**Figure 3** *A hand harpoon fishing vessel operating for Dall's porpoise off Pacific coast of Hokkaido in a summer 1990s (photo by Tomio Miyashita).*

take of these species. Currently five catcher boats operate using five land stations: Abashiri on the Okhotsk Sea coast of Hokkaido, Hakodate on the southern Hokkaido for Sea of Japan operation, Ayukawa on the Sanriku coast, Wadaura on the Boso coast, and Taiji on the Kii coast. Their quota as of 2007 was 66 Bard's beaked whales, 36 short-finned pilot whales (*Globicephala macrorhynchus*, 36 for each of the two populations), and 20 Risso's dolphins (*Grampus griseus*).

### VIII. Dolphin and Porpoise Fisheries

In Japan, dolphins and porpoises are taken by drives, hand harpoon, and small-type whaling. The catch quota for this fishery remained almost unchanged since 1993, but in the 2007/2008 season it had some modification and Pacific white-sided dolphins became a fishing target. Catches are used for human consumption.

At least 52 villages have operated dolphin drive fisheries since the fourteenth century on the Sea of Japan and Pacific coasts, but the number declined throughout the nineteenth and twentieth centuries. When it was placed under the license system of the prefecture governments in 1982, only five groups acquired licenses (Kasuya and Kishiro, 1993). Currently, two groups, at Futo on the Izu coasts and Taiji on the Kii coasts, operate drive fisheries, with quotas of about 3000 dolphins of seven species (170 Pacific white-sided dolphins, 513 striped dolphins, 913 common bottlenose dolphins, 809 pantropical spotted dolphins, 295 Risso's dolphins, 277 southern-stock short-finned pilot whales, and 70 false killer whales). The fishermen drive schools of gregarious dolphins into harbor using several fast boats. Other equipment used is a cone-shaped steel disk welded to one end of a 2-m-long steel pipe. The cone is placed underwater and the other end of the pipe in the air is hammered to scare dolphins acoustically.

Harpoon fisheries started in prehistoric time (see earlier discussion), but large-scale commercial hunts began around 1920 off the Sanriku region for Dall's porpoises accompanied by introduction of motor-driven vessels. Dolphins and porpoises are harpooned when they come to bow ride. An electric shocker is usually connected to the hand harpoon with a detachable head. This fishery came under the control of prefecture governor or regional fishery coordination committees in 1989. In 2007/2008 season, 338 vessels out of Hokkaido, Sanriku, Boso, and Kii operate with a quota of about 18,000 dolphins and porpoises (8707 *dalli*-type Dall's porpoise, 8168 *truei*-type Dall's porpoises, 190 Pacific white-sided dolphins, 172 striped dolphins, 95 common bottlenose dolphins, 70 spotted dolphins (*Stenella attenuata*), and 246 Risso's dolphins).

Okinawa hunters use crossbows to shoot harpoons constructed of steel pipe; their efficiency is superior to hand harpoons for pilot whale hunts. The fishing season is variable among locations. In the 2007/2008 season, 6 crossbow fishermen operated with a quota of 121 dolphins (9 common bottlenose dolphins, 92 southern-stock short-finned pilot whales, and 20 false killer whales).

## IX. Trap Net Fishery

On the coasts of Noto and northern Kyushu, whales were hunted until the end of the nineteenth century by placing small trap nets at whale passages. This fishery has been extinct for some time. However, there are about 20,000 trap nets (also called "set net") of various types now operating in Japan for fish; these occasionally take great whales (Tobayama *et al.*, 1992). In 1990, the Japanese Fisheries Agency prohibited commercial utilization of baleen whales found in the trap net. However, the Ministry of Agriculture and Fisheries Agency changed the rule in July 2001 to make it possible to sell such carcasses. This regulation change resulted in a sudden increase of reported catches to such high level that they cannot be ignored for management purposes. In recent years, the fishery has reported annual take of over 100 minke whales and occasional captures of gray and humpback whales (Kasuya, 2007).

## XI. Scientific Whaling

During 1956–1979, Japan issued several permits to take whales for research purposes based on Article 8 of the ICRW. The scientific collection accompanied the operation of ordinary commercial whaling. It killed a relatively small number of whales or lasted for only a few seasons. This scientific whaling differed from that of the later period.

The ban on commercial whaling by the IWC came in effect in the 1985/1986 Antarctic season and the 1986 coastal season. Japan withdrew its objection to this IWC decision on July 1, 1986, taking effect from May 1, 1987 (pelagic), October 1, 1987 (coastal minke and Bryde's whales), and April 1, 1988 (coastal sperm whales). In November 1987, Nihon Kyodo Hogei dissolved. Half of the staff formed Nihon Kyodo Senpaku (Japan Union Shipping) to operate vessels acquired from Nihon Kyodo Hogei, and the others merged with Geirui Kenkyusho (Whales Research Institute) to establish Nihon Geirui Kenkyusho (Institute of Cetacean Research, ICR).

ICR started to take 300 Antarctic minke whales (*Balaenoptera bonaerensis*) for scientific purposes in the 1987/1988 Antarctic season using a factory ship and catcher boats chartered from Nihon Kyodo Senpaku. This operation moved into the second phase in the 2005/2006 season. In 1994, ICR expanded the project to the North Pacific to take 100 minke whales, which continued in 2000. The numbers of species and individuals to be taken by the projects increased with time. The current projects, both of which are stated to continue for an unlimited period, intend to take annually 850 plus-or-minus 85 Antarctic minke whales, 50 fin whales and 50 humpback whales in the Antarctic and 220 common minke whales, 50 Bryde's whales, 100 sei whales, and 10 sperm whales in the western North Pacific of which 120 North Pacific minke whales are taken by small-type whaling (Kasuya, 2007). Proceeds of products from these operations, about 5 billion yen/year (US\$ = 100–110), and subsistence and contract of about 1 billion yen from the Japanese government financed activities of ICR in the 2003/2004 fiscal year, when it proposed to take 660 plus-or-minus 40 whales.

## See Also the Following Articles

Illegal and Pirate Whaling ■ International Whaling Commission ■ Whaling ■ Traditional

## References

Akashi, K. (1910). "Honpono Noruweshiki Hogeishi [History of Norwegian-Type Whaling in Japan]." Toyohogei, Osaka [In Japanese].

Hashiura, Y. (1969). "Kumano Taijiura Hogeishi (History of Whaling in Taiji, Kumano)." Heibonsha, Tokyo [In Japanese].

Hattori, T. (1887–1888). "Nihon Hogei Iko [Japanese Whaling Miscellanea]." Dainihon Suisankai, Tokyo [In Japanese].

Hawley, F. (1960). "Miscellanea Japonica. II. Whales and Whaling in Japan." Vol. 1, Part 1. Privately published by the author.

Itabashi, M. (1987). "Nanpyoyo Hogeishi [History of Antarctic Whaling]." Chuo Koron, Tokyo [In Japanese].

Kasahara, A. (1950). Nihon Kinkaino Hogei Gyoto Sono Shigen [Whaling and Whale Resources around Japan]. *Rep. Inst. Nihon Suisan.* **4**, 1–103, In Japanese.

Kasuya, T. (1975). Past occurrence of *Globicephala melaena* in the western North Pacific. *Sci. Rep. Whal. Res. Inst., Tokyo* **27**, 95–110.

Kasuya, T. (1999). Examination of reliability of catch statistics in the Japanese coastal sperm whale fishery. *J. Cetacean Res. Manage* **1**, 109–122.

Kasuya, T. (2007). Japanese whaling and other cetacean fisheries. *Env. Sci. Pollut. Res.* **14**, 39–48.

Kasuya, T., and Kishiro, T. (1993). Review of Japanese dolphin drive fisheries and their status. *Rep. Int. Whal. Commn* **43**, 439–452.

Kondo, I. (2001). Nihon Enyo Hogeino Kobo [Rise and Fall of Japanese Coastal Whaling.] Sanyo-sha, Tokyo [In Japanese].

Maeda, K., Teraoka, Y. (1952). "Hogei [Whaling]." Isana Shobo, Tokyo.

Matsubara, S. (1896). "Nihon Hogeishi [History of Japanese Whaling]." Fishery Association of Japan, Tokyo [In Japanese].

Ohsumi, S. (1975). Review of Japanese small-type whaling. *J. Fish. Res. Bd. Can.* **32**, 1111–1121.

Omura, H. (1950). Whales in the adjacent waters of Japan. *Sci. Rep. Whales Res. Inst., Tokyo* **4**, 27–113.

Omura, H. (1984). Nihonkaino Kujira *[Whales in the Sea of Japan] Geiken-Tsushin* **354**, 65–73 [In Japanese].

Omura, H., Matsuura, Y. and Miyazaki, I. (1942). "Kujira [Whales]." Suisansha, Tokyo [In Japanese].

Oyamada, Y. (1832). "Insanatori Ekotoba [Whaling in Words and Pictures]." Tatamiya, Edo. [In Japanese; translated into English in *Invest. Cetacea* **14**, Suppl., 1–119, under erroneous title Yugiotoru Eshi and with wrong spelling of the author's name as Y. Yamada.]

Tada, H. (1978). "Meijiki Yamaguchiken Hogeishino Kenkyu [Study of Whaling History of Yamaguchi in Meiji Era]." Matsuno Shoten, Tokuyama [In Japanese].

Tato, K. (1985). "Hogeino Rekishito Shiryo [Whaling History and Data]." Suisansha, Tokyo [In Japanese].

Tobayama, T., Yanagisawa, F., and Kasuya, T. (1992). Incidental take of minke whales in Japanese trap nets. *Rep. Int. Whal. Commn.* **42**, 433–436.

Tokuyama, N. (1992). "Taiyo Gyogyo Hogei-jigyono Rekishi [History of Whaling Enterprise of Taiyo Gyogyo]." Privately published by the author.

Tonnessen, J. N., and Johnsen, A. O. (1982). "The History of Modern Whaling." Hurst, London.

Valdivia, J., Landa, A., and Ramirez, P. (1984). Peru, progress report on cetacean research 1982–83. *Rep. Int. Whal. Commn.* **34**, 223–228.

Yamada, S. (1902). "Tsuro Hogeishi [History of Whaling at Tsuro]." Tsuro Whaling, Kochi [In Japanese].

**J**

# Killer Whale
## *Orcinus orca*

### JOHN K. B. FORD

## I. Characteristics and Taxonomy

With its striking black and white markings and cosmopolitan range, the killer whale, or orca, is one of the most easily recognized and widely distributed of all cetaceans and is unlikely to be confused with any other species (Fig. 1). It is a large dolphin, attaining maximum body lengths of 9.0 m in males and 7.7 m in females. Maximum measured weights are 6600 kg for a 7.65-m male and 4700 kg for a 6.58-m female (Yamada *et al.*, 2007). In addition to sexual dimorphism in size, mature males develop disproportionately larger appendages than females (Fig. 2). This includes the pectoral flippers, tail flukes (the tips of which curl downward in males), and dorsal fin, which is erect in shape and may attain a height of 1.8 m in males. At birth, neonate killer whales are approximately 2–2.5 m long and weigh approximately 200 kg.

The most distinguishing feature of the killer whale is its striking coloration. Killer whales are generally black dorsally and white ventrally. Above and behind the eye on each lateral side of the whale's head is a conspicuous, elliptically shaped white patch, referred to as the post-ocular patch (or colloquially, the "eye-patch"). On the posterior lateral sides of the whale, the ventral white region continues dorso-posteriorly to form flank patches that extend almost half-way to the dorsal ridge. At the posterior base of the dorsal fin is a gray-pigmented area of variable shape termed the "saddle patch." In neonates, the normally white-pigmented areas on the body have an orange hue, and the saddle patch is indistinct or absent for the first year of life. Considerable variation exists among killer whale populations and individuals in the size and the shape of white and gray patches. In some populations, particularly in the Southern Hemisphere, killer whales have a faint gray pigmentation over much of their body, and a black dorsal "cape" anterior to the dorsal fin.

The skull of the killer whale can be distinguished from those of other odontocetes by its shape, size, dental formula, and large teeth. Typically, 10–12 (up to 14) teeth are found per row, with teeth usually up to 10 cm in length. Upper and lower teeth interlock when the jaws are closed, which may result in considerable wear along their anterior and posterior facets. In some populations, extreme wear of the tooth crowns has been observed, even in young individuals, which may relate to diet.

Taxonomically, the killer whale is the largest species of the family Delphinidae. Only a single species, *Orcinus orca*, is currently

**Figure 2** *Adult male killer whale breaching off Victoria, British Columbia. Note large pectoral flippers and tall dorsal fin typical of mature males. Photo by M. Malleson.*

recognized, though the existence of morphologically, ecologically, and genetically distinct populations indicate that taxonomic revision may be warranted. In the northeastern Pacific, at least two distinct ecotypes—fish-feeding *residents* and mammal-hunting *transients*—co-occur in sympatry but maintain social and reproductive isolation. It has been suggested that residents and transients may constitute subspecies or incipient species. In the Antarctic, three distinct ecotypes—types A, B, and C—have been described from differences in coloration, morphology, and apparent dietary specialization (Fig. 3) (Pitman and Ensor, 2003). Type B and/or C may correspond to one or both of two putative species, *O. nanus* and *O. glacialis*, which were independently proposed in the early 1980s by Soviet researchers for populations of purportedly small individuals in the Antarctic (Mikhalev *et al.*, 1981; Berzin and Vladimirov, 1983). These new proposed species have not received general acceptance due to inadequate

**Figure 3** *Ecotypes of killer whales, known as types A (top), B (middle), and C (bottom), described in waters around Antarctica. Illustration by U. Gorter.*

documentation and the lack of holotype specimens. Recently, aerial photogrammetry studies have determined that type-C killer whales are on average up to 50% smaller than type-A whales, which supports the possibility of species-level variation (Pitman *et al.*, 2007). At present, it remains unclear whether killer whale ecotypes represent a single species, multiple species, or subspecies.

## II. Distribution and Abundance

The killer whale is second only to humans (and possibly our associated pest *Rattus norvegicus*) as the most widely distributed mammal in the world. It has a cosmopolitan distribution, being found in all oceans and most seas, but is most common in coastal, temperate waters, particularly in areas of high marine productivity (Forney and Wade, 2006). Greatest densities occur in waters along the northwestern coast of North America and the Aleutian Islands, along the coast of northern Norway, and in the higher latitudes of the Southern Ocean. In the Antarctic, killer whales are commonly found up to the pack ice edge in many areas and may extend well into ice-covered waters. In the Arctic, killer whales are rarely seen in the vicinity of pack ice but do visit the region during the open-water season in late summer. Information on the species' distribution in most tropical and offshore waters is limited, but numerous scattered records and sightings during cetacean surveys attest to its widespread, if rare, occurrence.

Because of its wide distribution and scarcity in most regions, the killer whale is a difficult species to census. Photo-identification studies in nearshore waters of the northeastern Pacific Ocean from

the Aleutian Islands to California have yielded a total population count of approximately 1600 whales (Ford *et al.*, 2000; Forney and Wade, 2006). Similar studies have identified 450 whales off northern Norway (Similä, 1997), 115 around New Zealand (Visser, 2000), and approximately 900 in waters off the Russian Far East. Line-transect vessel surveys have yielded estimates of 8500 killer whales over an area of 19 million square kilometer in the eastern tropical Pacific, and at least 25,000 in the Southern Ocean. Population counts and estimates from surveys provide a minimum global abundance estimate of 50,000 killer whales, although the total abundance is almost certainly greater than this because estimates are not available for large oceanic areas (Forney and Wade, 2006).

## III. Ecology

The killer whale is the oceans' apex marine predator, capable of preying on a great diversity of vertebrates and invertebrates. It has no natural predators other than humans. Over 140 species have been recorded as killer whale prey. It is the only cetacean that routinely preys upon marine mammals, with attacks or kills documented for 50 different species. Mammalian taxa that are prey of killer whales include other cetaceans–both mysticetes and odontocetes—pinnipeds, sirenians, mustelids and, on rare occasions, ungulates. A variety of fish species are also important food of killer whales, notably salmon (*Oncorhynchus* spp.), herring (*Clupea* spp.), cod (*Gadus* spp.), tuna (*Thunnus* spp.), and various sharks and other elasmobranchs (Ford *et al.*, 1998; Dahlheim and Heyning, 1999; Visser, 1999; Saulitis *et al.*, 2000). Other animals recorded as killer whale prey include squid, octopus, sea turtles, and sea birds.

Although the killer whale is a generalist predator on the global scale, local populations can exhibit remarkable foraging specializations. Best known are two sympatric ecotypes found in coastal waters of the northeastern Pacific, fish-feeding residents and mammal-hunting transients. Residents show strong seasonal movements associated with the coastal migrations of salmon. Observational studies and analyses of stomach contents from beach-cast carcasses have shown that salmon is the principal prey of residents and that they forage selectively for the largest or the fattiest available species [chinook (*Oncorhynchus tshawytscha*] in British Columbia and Washington, and coho (*O. kisutch*) in Prince William Sound, Alaska; [Ford *et al.*, 1998; Saulitis *et al.*, 2000; Ford and Ellis, 2006 (Fig. 4)]. Smaller salmonids such as pink salmon (*O. gorbuscha*) and sockeye salmon (*O. nerka*) are seldom eaten despite their far greater seasonal abundance. Squid and a variety of non-salmonid fish species are also eaten by residents (2006), but there is no evidence that marine mammals are consumed. Foraging groups of residents typically ignore marine mammals in their vicinity, and seldom elicit avoidance responses from those species (Jefferson *et al.*, 1991; Deecke *et al.*, 2002).

Transients show relatively little seasonal change in distribution, most likely because their preferred prey species are present year-round in coastal waters. Harbor seals (*Phoca vitulina*), harbor porpoises (*Phocoena phocoena*), and Dall's porpoises (*Phocoenoides dalli*) are the primary prey of transients, although Steller sea lions (*Eumetopias jubatus*), California sea lions (*Zalophus californianus*), northern elephant seals (*Mirounga angustirostris*), and Pacific white-sided dolphins (*Lagenorhynchus obliquidens*) are also important (Ford *et al.*, 1998). Common minke whales (*Balaenoptera acutorostrata*) are occasionally attacked by transient killer whales in British Columbia and Alaska (Fig. 5), and gray whale (*Eschrichtius robustus*) calves are targeted by transients along the coast of California during their first migration north with their mothers (Ternullo and

**K**

mixing with either the resident or the transient population. They form a genetically distinct group, although they are more closely related to residents than to transients (Barrett-Lennard, 2000). The dietary habits of this population are poorly known, but they have been observed to prey on fishes, including Pacific halibut (*Hippoglossus stenolepis*) and carcharinid sharks (Dahlheim *et al.*, 2008).

Killer whales in other regions may also be highly specialized in feeding habits. Of the three ecotypes described in Antarctic waters, type A is an open-water mammal hunter that may specialize on Antarctic minke whales (*Balaenoptera bonaerensis*), type B feeds on pinnipeds in loose pack ice, and type C is apparently a fish-feeder that inhabits dense pack ice (Pitman and Ensor, 2003). Off the northern coast of Norway, a population of killer whales moves seasonally in relation to the migration pattern of its principal prey, herring (Similä, 1997). It is likely that populations with dietary specializations exist wherever sufficiently abundant and reliable prey resources are available to sustain them year-round. In other regions, more generalist foraging strategies may be expected. For example, in the subantarctic Crozet Islands, killer whales feed seasonally on southern elephant seal (*Mirounga leonina*) pups, but also forage for fish.

As the top predator in the oceans, killer whales have the potential to play important roles in marine ecosystem dynamics. Recently, it has been proposed that prey switching by killer whales drove precipitous declines of marine mammal populations in the North Pacific after commercial whaling depleted their alleged preferred prey, the great whales (Springer *et al.*, 2003). This hypothesis has been rebutted on a variety of grounds, primarily that no compelling evidence has been presented that the large baleen whales targeted by whaling ever represented an important component of the diets of mammal-hunting killer whales (Mizroch and Rice, 2006; Trites *et al.*, in press).

## IV. Behavior and Physiology

### A. Group Structure

Killer whales are social animals that are usually observed traveling in groups containing a few to 20 or more individuals. Reports of larger groups likely involve temporary aggregations of smaller, more stable social units. Long-term photo-identification studies have provided information on the social organization of the species in several regions of the world. The most detailed of these are studies in coastal British Columbia, Washington, and Alaska, particularly for the *resident* ecotype (Bigg *et al.*, 1990; Matkin *et al.*, 1999; Ford *et al.*, 2000).

*Resident* societies can be arranged into a number of groupings based on maternal genealogy, social association, and acoustical relationship. The basic social unit of *residents* is the *matriline*, which is a highly stable group of individuals linked by maternal descent. A typical matriline is comprised of a female, her sons and daughters, and the offspring of her daughters. Because females may live up to 80–90 years of age, and females have their first viable calf at about 15 years of age, a matriline may contain as many as four generations of matrilineally related individuals. Some matrilines contain only one generation, which can result if a matriarch dies and leaves only sons or daughters that have no young of their own. The bonds among members of a matriline are extremely strong, and individuals are seldom seen apart from the group for more than a few hours. No permanent dispersal of individuals has been observed from a resident matriline.

The next level of social organization in resident killer whales is the *pod*, which is a group of related matrilines that likely shared a common maternal ancestor in the recent past. Matrilines within pods are thus more closely related to one another than to matrilines in other pods. Pods are less stable than matrilines, and member matrilines frequently

**Figure 4**  *Resident killer whale with freshly killed salmon, Haro Strait, Washington. Photo by M. Malleson.*

**Figure 5**  *Transients corralling a common minke whale following chase, Ganges Harbour, British Columbia. Photo by D. Ellifrit.*

Black, 2002). Attacks on larger baleen whales by transient killer whales are rare. Transients have not been observed to eat any species of fish, and no fish remains have been found in the stomachs of stranded transients.

Such extreme dietary specialization in sympatric populations is without precedent in mammals. These specializations likely evolved slowly and incrementally by means of increasingly refined and successful foraging strategies that were learned by individuals and passed across generations. Effective foraging for the disparate types of prey of residents and transients may require such divergent skills and tactics that lifestyles dependent on one or the other prey type have become mutually exclusive. Foraging specializations may have played a role in the historical separation of ancestral resident and transient groups, leading to the social and the eventual reproductive isolation of the two populations. Residents and transients are highly distinct in both mitochondrial and nuclear DNA composition (Hoelzel *et al.*, 1998, 2007; Barrett-Lennard, 2000).

More recently, a third sympatric form has been documented in coastal waters off British Columbia south to California (Ford *et al.*, 2000). Provisionally termed "offshores," these whales are seldom encountered in protected inshore waters and have not been observed

K

travel apart for periods of weeks or months. However, these matrilines still tend to travel more often with others from their pod than with matrilines from other pods (Ford et al., 2000). The majority of pods are comprised of 1–3 matrilines. Resident pods in British Columbia, Washington, and Alaska contain a mean of 18 whales (range 2–49).

A further level of social structure is the *clan*, which is defined by the acoustic behavior of pods. All pods within a clan have similar vocal dialects (see *Socializing*) which likely reflect their common matrilineal heritage from an ancestral pod through a process of growth and fragmentation along matrilines. Those pods with many shared features in their dialects are probably more closely related, and have split more recently, than those with more divergent features. Clans are sympatric, and pods from different clans frequently travel together. Clan membership is occasionally—but not usually—reflected in patterns of association. It is not clear how clans are related to each other, as they have no acoustical features in common, nor is the origin of clans known.

The top level of structure in resident killer whale society is the *community*, which is made up of pods that regularly associate with one another. The community is thus defined solely by association patterns rather than maternal genealogy or acoustic similarity. Pods from one community have rarely or never been seen to travel with those from another, although their ranges may partly overlap. Three communities of *residents* have been described in coastal waters of British Columbia, Washington, and Alaska: *southern* (3 pods, 1 clan), *northern* (16 pods, 3 clans), and *southern Alaskan* (11 pods, 2 clans) (Matkin et al., 1999; Ford et al., 2000).

Social organization in mammal-eating transient killer whales is not as well known as in residents. Similar to residents, the basic social unit is the matriline, but unlike residents, offspring often disperse from matrilines for extended periods or permanently, either as juveniles or as adults. As a result, transient matrilines tend to be smaller than those of residents, and lone individuals, particularly males, are often observed. Small group sizes of transients appear to reflect the marine mammal foraging specialization of this population (Baird and Dill, 1996; Ford and Ellis, 1999). Association patterns of transient matrilines are dynamic, and they do not form consistent groupings of matrilines equivalent to resident pods. All transient groups in a community have been observed to interact within this network of associations. Three communities of transients have been described in coastal waters of the northeastern Pacific, the *West Coast Transients* extends from central California north to roughly 56°N latitude in Alaska, the *Gulf of Alaska transients* range from southeastern Alaska to at least Kodiak Island in the west, and the *AT1 transients*, which is a very small population of 11 whales that inhabits Prince William Sound and the Kenai Fjords area in the northern Gulf of Alaska (Ford and Ellis, 1999).

Social organization based on matrilineal descent may be typical of killer whales globally. In other regions where long-term photo-identification studies have been undertaken, close and prolonged associations of mothers and offspring are commonly seen (e.g., Norway, Crozet Islands, Argentina). Temporal persistence of these bonds may be a primary variable determining group sizes and structure.

The activity states of killer whale groups are of four basic types: foraging, traveling, resting, and socializing (Ford, 1989; Saulitis et al., 2000). Minor differences in definitions and classification criteria of activities by different researchers make detailed comparisons difficult, but general patterns are evident. Foraging and traveling are the predominant activity states noted in all populations, although the proportions of the activity budget dedicated to these activities vary. Mammal-eating transients in coastal waters of the northeastern Pacific spend the great majority of their time (~90–95%) foraging

and traveling, whereas fish-eating residents spend only about 60–70% of their time doing so, at least during summer when salmon is abundant. Residents spend considerably more time resting and socializing than do transients. Fish-eating killer whales in northern Norway have activity budgets very similar to those of northeastern Pacific residents (Similä, 1997).

## B. Foraging

Behavior patterns observed during foraging by killer whales vary considerably among populations and prey types. Groups of salmon-hunting residents often disperse over large surface areas while foraging, with members moving at roughly the same speed (mean = 6.0km/h) and direction. Foraging episodes are typically 2–3h in duration, but may last longer. Individual salmon are pursued, captured, and eaten by single animals or shared within small subgroups, usually a mother and a juvenile offspring. Norwegian killer whales feed on herring in a coordinated manner referred to as "carousel feeding" (Similä, 1997). Using percussive actions such as tail lobbing, releasing blasts of bubbles, and flashing the white ventral side of their bodies, the whales herd herring into a tight ball close to the surface. The whales then stun fish by striking the edges of the ball with their tail flukes and eat the debilitated prey. Killer whales in New Zealand have been observed to forage benthically for three species of rays (Visser, 1999). In the Strait of Gibraltar, killer whales catch bluefin tuna (*Thunnus thynnus*) following prolonged directional pursuits that appear to drive these fish to exhaustion (Guinet et al., in press).

Mammal-hunting transient killer whales in the northeastern Pacific typically forage in smaller groups than fish-eating killer whales. Transient groups hunt harbor seals in groups averaging 3–4 individuals (Baird and Dill, 1996; Ford et al., 1998; Saulitis et al., 2000), usually close to shore and near seal haul-out sites. Although foraging, transients remain acoustically quiet, apparently to avoid detection by potential prey and possibly to locate prey by passive listening (Barrett-Lennard et al., 1996). Harbor seals are killed and shared among group members relatively quickly compared to Steller sea lions or California sea lions, which may take over 2h to kill and consume. Sea lions are usually rammed or butted with the whales' heads, and slapped repeatedly with tail flukes, until the animal is debilitated sufficiently to be taken underwater and drowned. When hunting porpoises or dolphins, transients forage in slightly larger groups (averaging five members) that spread out in open water in a rough line abreast. Once an individual porpoise has been singled out, the whales chase it until it tires, then ram it or jump upon it to complete the kill (Ford et al., 1998). Larger schools of Pacific white-sided dolphins are often driven by transients into confined bays where individual dolphins are trapped against the shore and killed.

A variety of specialized tactics have been described for killer whales hunting marine mammals in other regions. In Patagonia, Argentina, killer whales hunt southern sea lion and elephant seal pups in the shallows along sloping pebble beaches and often intentionally strand themselves temporarily in the process (Fig. 6). These whales hunt cooperatively and share their prey after capture. Killer whales in the Crozet Islands hunt elephant seal pups in a manner similar to those in Patagonia, and adults appear to teach this technique to their offspring (Guinet and Bouvier, 1995). In the Antarctic, type-B killer whales have been observed to locate seals hauled out on ice floes by spyhopping, then dislodge them by grouping together and rushing at the ice floe, creating a large wave in the process which tilts and washes over the floe (Fig. 7; Jisser et al., 2008). Attacks on baleen whales or sperm whales often

**K**

**Figure 6**  *Adult female killer whale catching southern sea lion pup, Punta Norte, Argentina. Photo by J. Ford.*

**Figure 7**  *Type B killer whales spyhopping around a weddell seal on an ice floe off the Antarctic Peninsula. A leopard seal (left) looks on. Photo by O. Carlsson.*

**Figure 8**  *Adult male killer whale rubbing on smooth pebbles near Robson Bight, British Columbia. Photo by D. Parer and E. Parer-Cook.*

### E.  Socializing

Socializing activity includes a wide range of physical displays and social interactions. Aerial behaviors are frequent, and may include spyhops, breaches, flipper slaps, tail lobs, and head stands. Juveniles often chase each other, roll and thrash at the surface, and engage in various other forms of play behavior, including playing with objects such as kelp or sea jellies. Sexual interactions involving penile erections are commonly observed, predominantly in all-male play groups. Some individuals may rest quietly at the surface whereas other pod members actively socialize. Rubbing on beaches or kelp is a common behavior observed during socializing in some populations. Killer whales belonging to the northern resident community in British Columbia visit certain beaches repeatedly to rub their bodies on smooth pebbles in shallow water (Fig. 8; Ford, 1989).

### F.  Sound Production

Similar to most delphinids, killer whales are highly vocal. They produce a wide variety of clicks, whistles, and pulsed calls for echolocation and social signaling. Studies of resident killer whales in British Columbia have documented vocal variations associated with activity state and group identity (Ford, 1989, 1991). Vocal exchanges among foraging resident whales are dominated by highly stereotyped, repetitive discrete calls from a repertoire averaging 12 call types (range 7–17 call types) per pod. Resting activity is usually associated with greatly reduced vocal activity, and occasional use of certain calls heard predominantly, but not exclusively, in such contexts. Socializing whales use mainly whistles and non-repetitive, variable pulsed calls, and aberrant versions of discrete calls. Excitement or motivational levels of vocalizing individuals is reflected in minor variations in pitch and in duration of discrete calls.

Call repertoires of resident killer whale pods have features that are distinct, forming systems of group-specific dialects. The entire call repertoire appears to be shared by all pod members. Some portions of a pod's call repertoire may be shared with certain other pods, whereas other portions may be unique. Levels of similarity in these group-specific dialects appear to reflect the degree of relatedness of different pods better than do patterns of travel association. Divergent variations in dialects among related matrilines likely accompany the gradual fission that leads to pod formation. Young whales presumably learn their pod's dialect via mimicry of their mother and siblings, and dialects are retained in the matriline due to the lack of individual

involve groups of 10–20 killer whales working together in a coordinated manner to subdue the prey. Although some group members attempt to grasp the tail flukes or pectoral flippers to immobilize the larger whale, others attack the head and blowhole area, evidently to prevent the whale from breathing. Once a large whale has been killed, killer whales often consume only the tongue, lips, and blubber (Jefferson *et al.*, 1991). Because most baleen whales sink upon death, killer whales may only be able to feed extensively on carcasses of whales killed in shallow waters (Guinet *et al.*, 2000).

### C.  Traveling

Traveling killer whales move in a single direction at a consistent, fast pace, with no evidence of foraging or feeding. Groups often travel in a line abreast, with synchronized dives and surfacings. Resident killer whales have been documented to travel at speeds of over 20 km/h (mean = 10.4 km/h; Ford, 1989).

### D.  Resting

When resting, resident killer whales usually swim tightly together side by side, forming a *resting line*. Group diving and surfacing become closely synchronized and regular, with longer dives of 2–5 min duration separated by 3 or 4 short, shallow dives. Rate of forward progression is slow compared to foraging and traveling, and resting groups may stop altogether and rest motionless at the surface for several minutes (Ford, 1989; Similä, 1997).

dispersal. Thus, dialects likely provide an acoustic means of maintaining group identity and cohesion, and may serve as indicators of relatedness that help in the avoidance of inbreeding between closely related whales (Ford, 1991; Barrett-Lennard, 2000). Dialects have also been documented within a community of pods of killer whales in northern Norway and likely exist elsewhere.

Mammal-eating transient killer whales in the northeastern Pacific have greatly reduced vocalization rates compared to residents. Transients are generally silent when foraging, and even echolocation has been found to be used 27 times less often from foraging transients than foraging residents (using an index adjusted for group size; Barrett-Lennard *et al.*, 1996). Transients are more likely than residents to use individual (or "cryptic") clicks rather than click trains, presumably to avoid alerting potential prey to their approach. Transients often become highly vocal, however, following a successful kill; West Coast transients off the coasts of southeastern Alaska to California produce a number of calls that are shared among all groups in the community. Certain other calls seem exclusive to transient groups in different portions of this range. Group-specific dialects as seen in resident pods are not evident, presumably due to the reduced stability of social structure in transients (Ford and Ellis, 1999).

## V. Life History

Most detailed information on reproduction, mortality, and other life history parameters of killer whales have been derived from long-term photo-identification studies of *resident* killer whales in British Columbia and Washington (Olesiuk *et al.*, 1990, 2005). The reliability and completeness of this information is due to the extremely stable social structure of residents, in which emigration from the natal group does not take place and individual mortalities can be reliably documented (see *Interactions with Humans* for more details). It is not known whether these life history parameters are typical of other populations or regions.

Studies of captive whales indicate that sexually mature females have periods of polyestrous cycling interspersed with noncycling intervals of 3–16 months. The gestation period is 15–18 months (Duffield *et al.*, 1995). In resident killer whales, calving appears to be diffusely seasonal with a peak in the autumn. Neonate mortality may be high, with an estimated 43% dying within the first 6 months (Olesiuk *et al.*, 1990). Calves are nursed for at least a year, but may start taking solid food from the mother while still nursing. Typical age at weaning is not known, but is likely between 1 and 2 years of age.

Females typically give birth to their first viable calf at 12–14 years of age (Olesiuk *et al.*, 2005). Intervals between viable calves average about 5 years (range 2–14 years). Females have an average of about 5 viable calves over a 25-year reproductive life span, which ends at approximately 40 years of age. Females then become reproductively senescent for an average period of 10 years, although this post-reproductive period may extend to more than 30 years. Mean life expectancy for females (calculated at age 0.5 years, following the period of high neonate mortality) is estimated to be approximately 50 years, and maximum longevity is 80–90 years. Males attain sexual maturity at about 15 years of age, as indicated by a rapid growth of the dorsal fin, and continue to grow until they reach physical maturity at about 21 years of age. Mean life expectancy for males (calculated at age 0.5 years) is estimated to be about 30 years, with maximum longevity of about 50–60 years. Mortality curves for both males and females are U shaped, although the male curve is narrower.

## VI. Interactions with Humans

Although the species has long been held in high regard by many aboriginal maritime cultures, other societies feared the killer whale as a reputedly ruthless and a dangerous predator, and the animals were commonly vilified and persecuted. Attitudes toward killer whales have fortunately improved over the past few decades. The killer whale has been an admired display species in aquaria for over 30 years and has been featured in numerous movies, documentaries, and other forms of popular media. It has recently become the focus of commercial whale-watching operations in several regions.

*Orcinus orca* was listed by the IUCN in 2008 as Data Deficient (DD). Should the taxon be revised into two or more species, this will need to be reevaluated. Some regional populations of killer whales are small and highly specialized, and may therefore be vulnerable to over-exploitation and habitat deterioration. The Strait of Gibraltar population of killer whales is critically threatened due to its small population size (<50 individuals) and evidence of recent declines in whale numbers and their primary prey, bluefin tuna. The "southern resident community" or population of killer whales that inhabits the northwest coast of the mainland US and British Columbia is listed as Endangered under the US Endangered Species Act and the Canadian Species-at-Risk Act. Additional small, reproductively isolated populations of killer whales may exist but have not yet been identified, and it is likely that some of these may qualify for a threatened category.

Historically, killer whales in several regions have been the target of directed fisheries, culling, and persecution. An average 43 whales per year were taken by Japanese whalers from their coastal waters during 1946–1981, mostly for human consumption. Norwegian whalers took an average 56 whales per year during 1938–1981 in a government-subsidized hunt aimed at reducing killer whale numbers to reduce competition for other fisheries. The killer whale meat from this fishery was used only for animal consumption. An average of 26 was taken annually by Soviet whalers in the Antarctic from 1939 to 1975, with an exceptionally large take of 916 animals in the 1979/1980 season. Killer whales are still taken in small numbers in coastal fisheries in Japan, Greenland, Indonesia, and the Caribbean islands (Reeves *et al.*, 2003).

Killer whales have long been feared as dangerous predators or vilified as perceived or real threats to fisheries in many regions and were often harassed or shot opportunistically. Although much reduced, some persecution continues today. Killer whales have been shot illegally by fishermen in Alaska to prevent them from taking sablefish (*Anoplopoma fimbria*) from longline fishing operations (Matkin and Saulitis, 1994). Shooting by bluefin tuna fishermen in response to depredation is also considered to be a potential threat to the highly endangered killer whales residing in the Strait of Gibraltar. Killer whale depredation of longline fisheries involving various fish species has also been reported in northern and southern Pacific equatorial waters, the North Atlantic, off Brazil and Tasmania, and in the Southern Ocean.

Live-capture fisheries for killer whales represent another threat to some populations. After the first successful capture and display of the species in Vancouver in 1964, a demand arose for the acquisition of killer whales for public display in aquaria, and a live-capture fishery developed in coastal waters of British Columbia and Washington State in the mid-1960s. During 1964–1977, 63 killer whales were taken in this fishery to supply aquaria in many parts of the world (Olesiuk *et al.*, 1990). There is now evidence that the populations of resident killer whales involved in this fishery were already depressed from shootings related to depredation. During the late 1970s to mid-1980s, live captures shifted to the waters of Iceland, where over 50 whales were taken. Improved success of captive breeding during the

**K**

past decade has reduced the need for capture from wild populations, though periodic live captures continue.

Other conservation concerns include direct effects of oil spills and other forms of toxic pollution on killer whale survival. The *Exxon Valdez* oil spill in Alaska was strongly correlated with the subsequent loss of 14 whales from a pod that was seen swimming through light oil slicks early in the spill, although it was not possible to directly attribute the deaths to this cause. Oil spills may also have indirect effects on killer whales by reducing prey abundance. Their high trophic position in the food web makes killer whales susceptible to bioaccumulation of organochlorine pollutants. Levels of PCBs in resident and, in particular, transient killer whales in British Columbia and Washington state have been shown to be among the highest observed in any cetacean, and are higher than levels found to affect health in European harbor seals. It is not known whether there is a direct impact of PCBs on health in these killer whales, though such effects as immunosuppression and reduced reproductive success are possible (Ross *et al.*, 2000).

Other potential impacts of human activities on killer whale status are reduced prey availability and disturbance caused by vessel traffic. As an example, many stocks of chinook salmon, the principal prey of resident killer whales, have declined significantly in British Columbia and Washington State as a result of overfishing, degradation of spawning grounds, and reduced ocean survival. Vessel disturbance is of particular concern in areas of intensive whale-watching, though many forms of boat traffic have the potential to affect whales. The physical presence of fast moving boats near killer whales can disrupt their activities, particularly during resting, as well as put them at risk of collision. Underwater noise from vessels has the potential to interfere with social or echolocation signals, or to mask passive acoustic cues that may be important in finding prey.

On their own, many of these potential impacts on killer whales are likely insufficient to negatively affect killer whale survival. However, there is a potential for more serious cumulative effects that could displace killer whales from critical habitats or result in reduced survival.

## See Also the Following Articles

Delphinids, Overview ■ Intelligence and Cognition

## References

Baird, R. W., and Dill, M. L. (1996). Ecological and social determinants of group size in *transient* killer whales. *Behav. Ecol.* **7**, 408–416.

Barrett-Lennard, L. G. (2000). "Population Structure and Mating Systems of Northeastern Pacific Killer Whales." Ph.D. dissertation, University of British Columbia, Vancouver.

Barrett-Lennard, L. G., Ford, J. K. B., and Heise, K. A. (1996). The mixed blessing of echolocation: Differences in sonar use by fish-eating and mammal-eating killer whales. *Anim. Behav.* **51**, 553–565.

Berzin, A. A., and Vladimirov, V. L. (1983). A new species of killer whale (Cetacea, Delphinidae) from Antarctic waters. *Zool. Zhurnal* **62**, 287–295.

Bigg, M. A., Olesiuk, P. F., Ellis, G. M., Ford, J. K. B., and Balcomb, K. C., III. (1990). Social organization and genealogy of resident killer whales (*Orcinus orca*) in the coastal waters of British Columbia and Washington State. *Rep. Int. Whal. Commn Spec. Iss.* **12**, 383–405.

Dahlheim, M. E., and Heyning, J. E. (1999). Killer whale *Orcinus orca* (Linnaeus, 1758). *In* "Handbook of Marine Mammals" (S. H. Ridgway, and R. S. Harrison, eds), pp. 281–322. Academic Press, San Diego.

Dahlheim, M. E., Schulman-Janiger, A., Black, N., Ternullo, R., Ellifrit, D., and Balcomb, K. C. III. (2008). Eastern temperate North Pacific Offshore Killer Whales (*Orcinus orca*): occurrence, movements, and insights into seeding ecology. *Mar. Mamm. Sci.* **24**, 719–729.

Deecke, V. B., Slater, P., and Ford, J. K. B. (2002). Selective habituation shapes acoustic predator recognition in harbour seals. *Nature* **420**, 171–173.

Duffield, D. A., Odell, D. K., McBain, J. F., and Andrews, B. (1995). Killer whale (*Orcinus orca*) reproduction at Sea World. *Zoo Biol.* **14**, 417–430.

Ford, J. K. B. (1989). Acoustic behaviour of resident killer whales (*Orcinus orca*) off Vancouver Island, British Columbia. *Can. J. Zool.* **67**, 727–745.

Ford, J. K. B. (1991). Vocal traditions among resident killer whales (*Orcinus orca*) in coastal waters of British Columbia. *Can. J. Zool.* **69**, 1454–1483.

Ford, J. K. B., and Ellis, G. M. (2006). Selective foraging by fish-eating killer whales *Orcinus orca* in British Columbia. *Mar. Ecol. Prog. Ser.* **316**, 185–199.

Ford, J. K. B., Ellis, G. M., Barrett-Lennard, L. G., Morton, A. B., Palm, R. S., and Balcomb, K. C., III. (1998). Dietary specialization in two sympatric populations of killer whales (*Orcinus orca*) in coastal British Columbia and adjacent waters. *Can. J. Zool.* **76**, 1456–1471.

Ford, J. K. B., Ellis, G. M., and Balcomb, K. C. (2000). "Killer Whales: The Natural History and Genealogy of *Orcinus orca* in the Waters of British Columbia and Washington." UBC Press and University of Washington Press, Vancouver, BC and Seattle, WA.

Forney, K. A., and Wade, P. R. (2006). Worldwide distribution and abundance of killer whales. *In* "Whales, Whaling, and Ocean Ecosystems" (J. A. Estes, D. P. Demaster, D. F. Doak, T. M. Williams, and R. L. Brownell, Jr., eds), pp. 145–173. University of California Press, Berkeley.

Guinet, C., and Bouvier, J. (1995). Development of intentional stranding hunting techniques in killer whale (*Orcinus orca*) calves at Crozet Archipelago. *Can. J. Zool.* **73**, 27–33.

Guinet, C., Barrett-Lennard, L. G., and Loyer, B. (2000). Co-coordinated attack behavior and prey sharing by killer whales at Crozet Archipelago: Strategies for feeding on negatively-buoyant prey. *Mar. Mamm. Sci.* **16**, 829–834.

Guinet, C., Domenici, P., de Stephanis, R., Barrett-Lennard, L. G., Ford, J. K. B., and Verborgh, P. (2007). Killer whale predation on bluefin tuna: Exploring the hypothesis of the endurance-exhaustion technique. *Mar. Ecol. Prog. Ser.* **347**, 111–119.

Hoelzel, A. R., Dahlheim, M. E., and Stern, S. J. (1998). Low genetic variation among killer whales (*Orcinus orca*) in the eastern North Pacific, and differentiation between foraging specialists. *J. Hered.* **89**, 121–128.

Jefferson, T. A., Stacey, P. F., and Baird, R. W. (1991). A review of killer whale interactions with other marine mammals: Predation to co-existence. *Mamm. Rev.* **21**, 151–180.

Matkin, C. O., and Saulitis, E. L. (1994). "Killer whale (*Orcinus orca*) biology and management in Alaska." Marine Mammal Commission, Washington, DC, Contract Number T75135023.

Matkin, C. O., Ellis, G., Olesiuk, P., and Saulitis, E. (1999). Association patterns and inferred genealogies of resident killer whales, *Orcinus orca*, in Prince William Sound, Alaska. *Fish. Bull., US* **97**, 900–919.

Mikhalev, Y. A., Ivashin, M. V., Savusin, V. P., and Zelenya, F. E. (1981). The distribution and biology of killer whales in the Southern Hemisphere. *Rep. Int. Whal. Commn* **31**, 551–565.

Mizroch, S. A., and Rice, D. W. (2006). Have North Pacific killer whales switched prey species in response to depletion of the great whale populations? *Mar. Ecol. Prog. Ser.* **310**, 235–246.

Olesiuk, P. F., Bigg, M. A., and Ellis, G. M. (1990). Life history and population dynamics of resident killer whales (*Orcinus orca*) in the coastal waters of British Columbia and Washington State. *Rep. Int. Whal. Comm Spec. Iss.* **12**, 209–242.

Olesiuk, P. F., Ellis, G. M., Ford, J. K. B. (2005). "Life History and Population Dynamics of Northern Resident Killer Whales (*Orcinus orca*) in British Columbia". Canadian Science Advisory Secretariat, Fisheries & Oceans, Canada.

Pitman, R. L., and Ensor, P. (2003). Three forms of killer whales (*Orcinus orca*) in Antarctic waters. *J. Cetacean Res. Manage.* **5**, 131–139.

Pitman, R. L., Perryman, W. L., LeRoi, D., and Eilers, E. (2007). A dwarf form of killer whale in Antarctica. *J. Mammal.* **88**, 43–48.

Reeves, R. R., Smith, B. D., Crespo, E. A., and Notarbartolo di Sciara, G. N. (2003). "Dolphins, Whales and Porpoises: 2002–2010 Conservation Action Plan for the World's Cetaceans." IUCN/SSC Cetacean Specialist Group, IUCN, Gland, Switzerland.

Ross, P. S., Ellis, G. M., Ikonomou, M. G., Barrett-Lennard, L. G., and Addison, R. F. (2000). High PCB concentrations in free-ranging Pacific killer whales, *Orcinus orca*: Effects of age, sex and dietary preference. *Mar. Pollut. Bull.* **40**, 504–515.

Saulitis, E., Matkin, C., Barrett-Lennard, L., Heise, K., and Ellis, G. (2000). Foraging strategies of sympatric killer whale (*Orcinus orca*) populations in Prince William Sound, Alaska. *Mar. Mamm. Sci.* **16**, 94–109.

Similä, T. (1997). "Behavioral Ecology of Killer Whales in Northern Norway." Norwegian College of Fisheries Science. University of Tromso, Tromso.

Springer, A. M., *et al.* (2003). Sequential megafaunal collapse in the North Pacific Ocean: An ongoing legacy of industrial whaling? *Proc. Natl. Acad. Sci. USA* **100**, 12223–12228.

Ternullo, R., and Black, N. (2002). Predation behavior of transient killer whales in Monterey Bay, California. Proceedings of the Fourth International Orca Symposium and Workshop, CEBC-CNRS, France, 156–159.

Trites, A. W., Deecke, V. B., Gregr, E. J., Ford, J. K. B., and Olesiuk, P. F. (2007). Killer whales, whaling, and sequential megafaunal collapse in the North Pacific: A comparative analysis of the dynamics of marine mammals in Alaska and British Columbia following commercial whaling. *Mar. Mamm. Sci.* **23**, 751–765.

Visser, I. (1999). Benthic foraging on stingrays by killer whales (*Orcinus orca*) in New Zealand waters. *Mar. Mamm. Sci.* **15**, 220–227.

Visser, I. N., Smith, T. G., Bullock, I. D., Green, G. D., Carlsson, O. G. I., and Imberti, S. (2008). Antarctic killer whales, *Orcinus orca*, hunt seals and a penguin on floating ice. *Mar. Mamm. Sci.* .**24**, 225–234.

Yamada, T. K., *et al.* (2007). Biological indices obtained from a pod of killer whales entrapped by sea ice off northern Japan. IWC Scientific Committee Meeting Document SC/59/SM12.

# Krill and Other Plankton

ROGER HEWITT AND JESSICA D. LIPSKY

## I. Introduction

Plankton is the collective name given to the assemblage of free-swimming or suspended microscopic organisms considered too small to move independently of ocean currents. Large animals that are able to disperse under their own power are called nekton. The distinction between plankton and nekton, however, is sometimes blurred. For example, larger animals that are capable of limited self-propulsion, such as jellies and salps, are often included in the plankton. Large euphausiids, such as Antarctic krill (*Euphausia superba*), have been referred to as either macroplankton or micronekton. Phytoplankters are plants, and zooplankters are animals.

Phytoplankton consists of microscopic unicellular plants and forms the basis of marine ecosystems; nearly all life in the sea derives from the solar energy fixed in photosynthesis by these plants. Two factors control phytoplankton growth, light irradiance and nutrients.

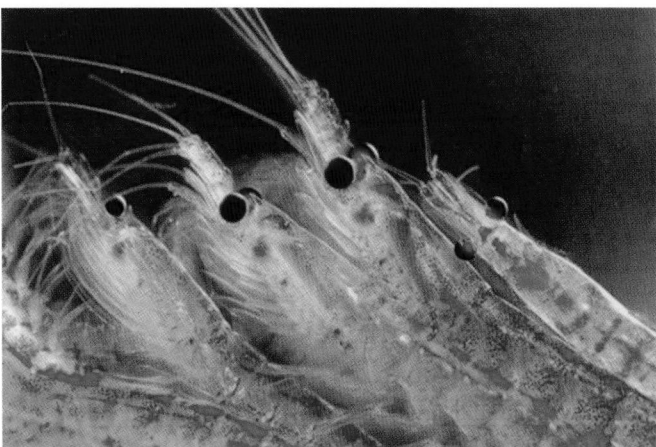

**Figure 1** *Photograph of adult* Euphausia superba. *Photo by Tadashi Mizowaki, Courtesy of Inter-Research Science Center.*

Light is only available in the top layers of the oceans (200 m or less) whereas nutrients are more abundant in the deeper layers. Highest concentrations of phytoplankton occur where light and adequate nutrients are coincident (e.g., areas of coastal upwelling, oceanic fronts, and transition zones). Evolution of small size has enabled phytoplankton to absorb scarce nutrients through maximizing the ratio of surface area to volume. Small size, down to 2 μm, also confers high buoyancy and a low sinking rate, keeping the cells near the surface.

Zooplankton consists of animals from several taxonomic groups from Protozoa to Vertebrata and is a main source of food for many marine mammals. Carnivorous, omnivorous, and herbivorous zooplankters have been found in the stomach of baleen whales. Three groups of crustaceans are the most important: copepods, amphipods, and euphausiids. These planktonic animals have developed a wide variety of specialized mechanisms and techniques for feeding on smaller plankton and suspended particulate matter, including appendicular nets and guiding whorls in copepods, ciliary movements in pteropods, and finely structured appendages used for filtering by euphausiids. Nektonic animals have also developed filters by modifying gillrakers into functional sieves [e.g., basking shark (*Cetorhinus maximus*) and whale shark (*Rhincodon typus*)]. Buccal teeth with well-developed accessory cusps also function as sieves in crabeater seals (*Lobodon carcinophaga*) and leopard seals (*Hydrurga leptonyx*), which feed on krill. The most highly derived filtering system among the vertebrates is that of baleen in whales.

Euphausiids, or krill (Fig. 1), have long been recognized as a critical element of the natural economy of the world's oceans (Sars, 1885; Brinton, 1962; Marr, 1962; Mauchline and Fischer, 1969; Mauchline, 1980). Early fishery biologists repeatedly stressed the importance of various species of euphausiids as food for exploited fish and whale stocks (Lebour, 1924; Hickling, 1927; Hjort and Rund, 1929). Norwegian whalers referred to the euphausiids found in large numbers in the stomachs of whales caught in the North Atlantic as *stor krill* (or large krill, referring to *Meganyctiphanes norvegica*) and *smaa krill* (or small krill, referring to *Thysanoëssa inermis*); the word "krill" is now used in reference to euphausiids in general (Mauchline and Fisher, 1969). Laws (1985) estimated that 190 million tons of Antarctic krill (*Euphausia superba*) were consumed annually by baleen whales in the Southern Ocean prior to their exploitation. It is estimated that current populations of whales, birds, pinnipeds, fish, and squid consume

**K**

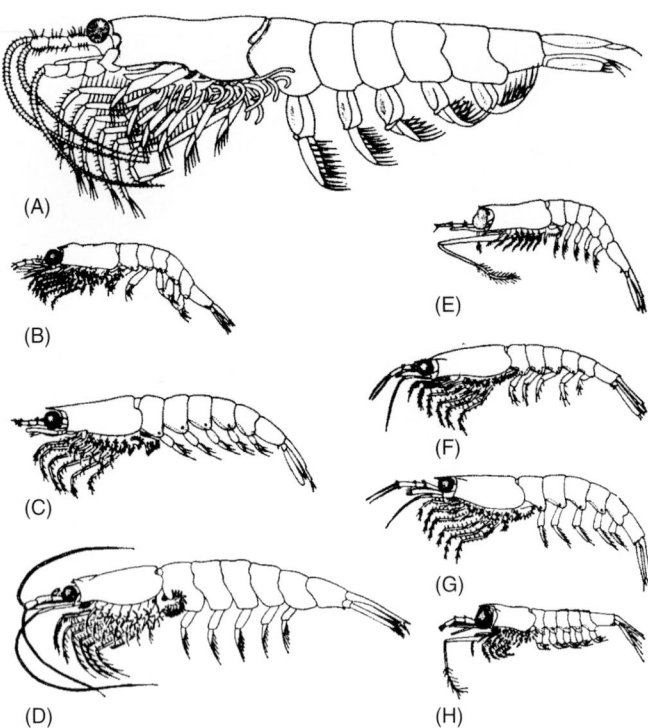

**Figure 2** *Scale drawings of eight most important krill species. (A)* Euphausia superba, *(B)* E. pacifica, *(C)* E. crystallorphias, *(D)* Meganyctiphanes norvegica, *(E)* Thysanoëssa macrura, *(F)* T. inermis, *(G)* T. raschii, *(H)* T. longipes. *From Mauchline and Fischer, 1969.*

250 million tons of Antarctic krill annually (Miller and Hampton, 1989). Of the 85 species of krill, Mauchline and Fisher (1969) list only 8 of primary importance in terms of their distribution range, biomass, and dominance in the diets of vertebrate predators. They note that these species constitute a large fraction of the plankton where they are found and that their biomasses are largest at high latitudes. In addition to their numbers, the habit of euphausiids to form large swarms makes them particularly important as prey to marine vertebrates.

The krill species considered most important to the trophodynamics of marine ecosystems are: *M. norvegica, T. raschii* and *T. inermis* in the North Atlantic Ocean, *E. pacifica, T. inermis, T. raschii, T. longipes,* and *T. inspinata* in the North Pacific Ocean, and *E. superba, E. crystallorophias,* and *T. macrura* in the Southern Ocean (Figs 2 and 3). Mauchline and Fisher (1969) list another seven species of importance in more restricted geographical areas and/or seasons: *Nyctiphanes couchii* in the North Atlantic Ocean, *T. spinifera* and *E. similis* in the North Pacific Ocean, *N. capensis* near the southern part of Africa, *N. australis* and *Pseudoeuphausia latifrons* from western Australia to New Zealand, and *E. vallentini* in the Southern Ocean. In addition, the following seven species are often cited in predator diet samples from restricted locales and time periods: *Nematocelis megalops* in the North Atlantic, *E. recurva, E. lucens, E. hemigibba, T. gregaria, E. spinifera,* and *N. megalops* from western Australia to New Zealand, *E. recurva, E. lucens* and *T. gregaria* near the southern part of Africa, and *N. simplex* along the western coast of North America.

Although many of these species are broadly dispersed, they exhibit their highest densities in areas of enhanced seasonal primary and secondary production. These areas include eastern boundary currents, coastal and oceanic upwelling regions and sea ice edge zones as well as estuaries, fjords, and small-scale eddies where physical mechanisms

may enhance the aggregation of krill. It is not surprising therefore to find krill predators, including baleen whales and crabeater seals, concentrated in these areas as well.

Krill species differ in their geographic distribution, body size (ranging from <1 cm to 14 cm), and longevity (ranging from <1 year to as many as 10 years) but share many other characteristics that contribute to their importance as prey for baleen whales. Furthermore, baleen whales have not shown strong species or size selectivity among krill when foraging in an area where more than one species and/or developmental stage are present. Krill are therefore described here in general terms with species-specific references only where appropriate.

## II. General Morphology and Life History of Krill

The body plan of krill (Fig. 4) is divided into two main regions, the cephalothorax and the abdomen. The cephalothorax, a fused head and thorax, contains the internal organs including the digestive system, the heart, and the gonads. It is about one-third of the body length and is covered by a thin shell or carapace. The muscled abdomen is made up of six segments ending with a telson and two pair of uropods, which together form a fan shape at the tail. At the head there are a pair of eyes and two pair of antennae with tactile and olfactory sensors; excretory organs open near the second set of antennas. The mouth is made up of several parts whose function is to filter, macerate, and manipulate food prior to ingestion. Six to eight pairs of limbs are connected to the thorax and are used to filter particles out of the water and pass them to the mouth. Unlike decapod crustaceans (crabs, lobsters, prawns, shrimps) the gills of krill are exposed, hanging below the carapace. The first five abdominal segments each have a single pair of limbs (pleopods) attached, which are used for swimming; the sixth abdominal segment has no appendages. On a mature adult male the first pair of pleopods is modified to form a petasma which is used during copulation to clasp and transfer spermatophores to the female. The thelycum, or female copulatory organ, is located on the anterior underside of the thorax near the opening of the oviducts.

The exoskeletons of krill are translucent, allowing a view of the internal organs, including the heart, stomach, and hepatopancreas, which is often colored dark green or red. Krill are also luminescent with light-emitting photophores located at the bases of their pleopods, near the thelycum, close to the mouth and in the eye stalks. The photophores are a deep red color but emit electric blue light in the water. Many species are also pigmented with red chromatophores that expand when the animal is stimulated. As a result swarms of krill often appear to be bright red, particularly when under attack by a predator. The guano of krill-eating birds is often pink in color and the feces of krill-eating marine mammals are characteristically dark red.

As krill mature sexually, males elaborate packets of sperm called spermatophores and females develop clusters of eggs or broods. During spawning the male grasps the female with his petasmae and transfers spermatophores to her body where they adhere in the vicinity of her thelycum. Among the various species of krill, brood size ranges from tens of eggs to several thousand and some species have been observed to spawn several broods during a single breeding season. When a female releases a brood of eggs, they are fertilized by spermatozoa now liberated from the spermatophores. For some species the female carries the fertilized eggs in brood pouches until they hatch, thereby protecting them from predation. For most species, however, eggs are released into the open sea. In some cases the eggs are neutrally buoyant, but often they are heavier than water and sink before hatching into nauplius larvae, which in turn develop and molt through a series of larval stages each resembling the adult morphology

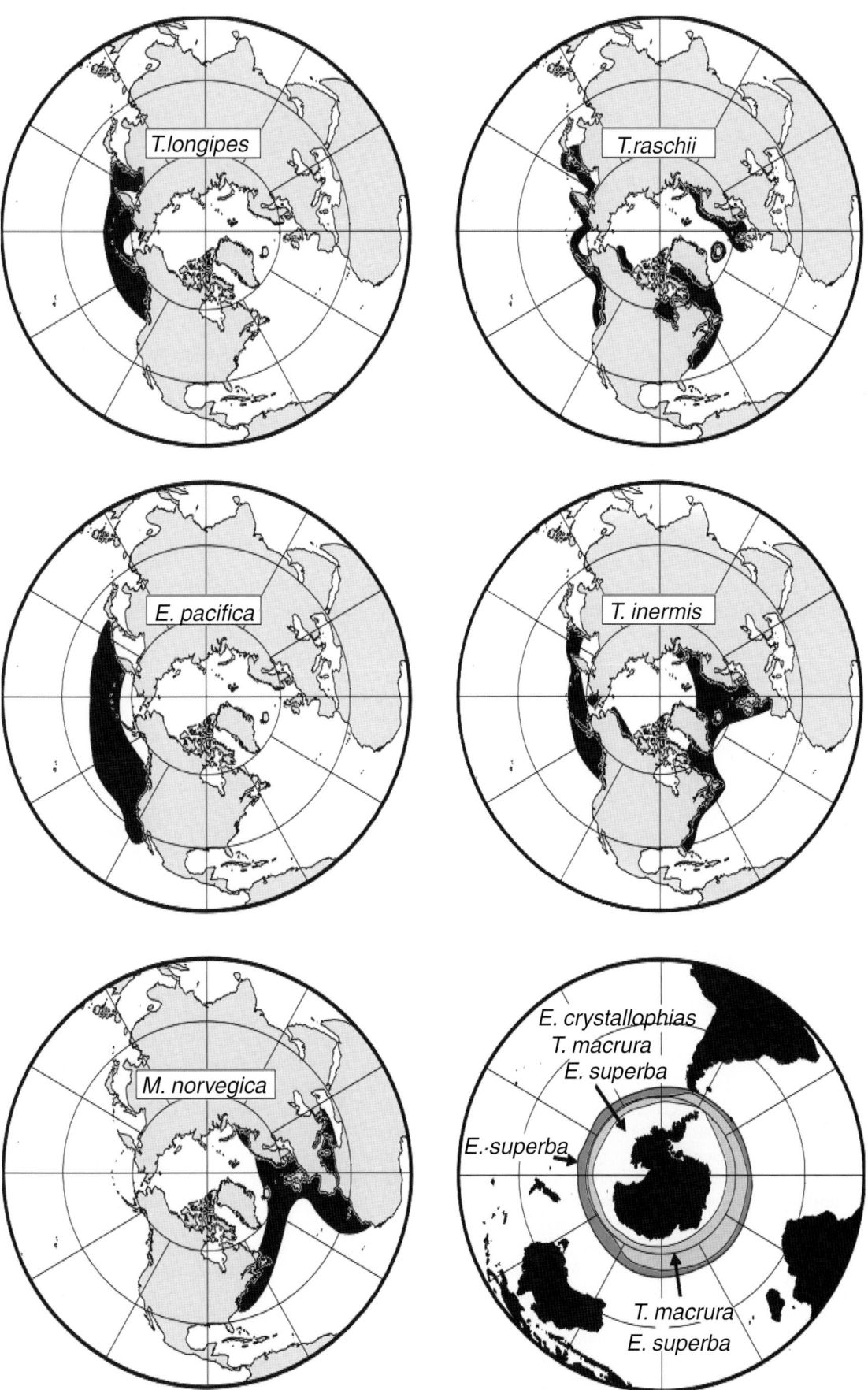

**Figure 3** *Northern and Southern hemisphere maps showing dispersion of important krill species (redrawn from Mauchline and Fischer, 1969).*

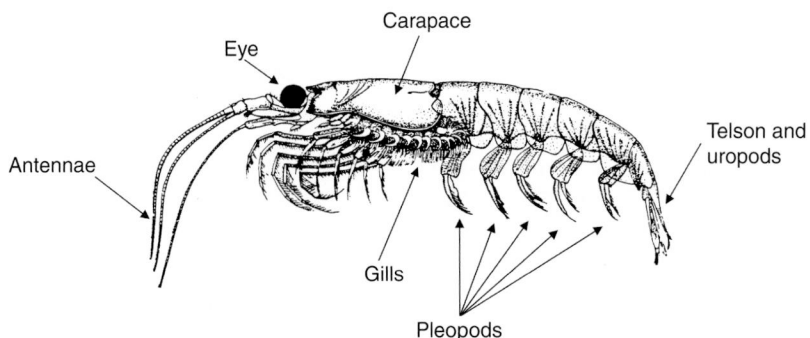

Figure 4    *General krill body plan.*

more than the previous stage. In the case of *E. superba*, a brood of 10,000 fertilized eggs may be released by a single female in a near-surface swarm of spawning adults; the eggs sink to depths of greater than 1000 m, incubate and hatch. The nauplius has no swimming appendages and continues to sink as it grows, molts, and gives rise to more advanced larval forms. Once it can swim, the larva begins its ascent into the surface waters, progressing through several more molts and ultimately emerging as a calyptosis larva. Calyptoses continue to eat, grow, and molt through additional stages in preparation for the winter when food is less available. Sometime in the late winter or the early spring the calyptoses finally metamorphose into juvenile krill, but it may be as long as another year before they are ready to spawn themselves. In the case of *E. pacifica*, this process is compressed to a few months, with spawning occurring during the spring and recruitment into the adult population occurring during the fall.

Except for rich fat stores invested in developing eggs, larval and post-larval krill do not develop high levels of fat reserves. Consequently they must eat constantly in order to offset the energy costs of swimming, growth, and reproduction. In addition, krill periodically shed their exoskeletons throughout their life, adding substantially to their energy requirements. Krill are generally thought to be filter-feeding herbivores, grazing on phytoplankton in the surface layers of the ocean. Many species, however, are reported to be omnivorous, filtering and/or capturing copepods and other small zooplankton. *E. superba* has been observed in the cavities and cracks on the underside of winter sea ice presumably feeding on interstitial ice algae. Krill growth and reproductive activity have been directly linked to available food supplies. Negative growth and regression of sexual characteristics has been observed in several species and related to lowered availability of food.

## III. Swarming

Krill are heavier than water and must continually swim in order to maintain their position. They aggregate into dense swarms, which can take on a variety of shapes from discreet balls to extensive layers. The swarms may range in thickness from one meter to several tens of meters and may extend horizontally tens of meters to several thousand meters. Individual animals appear to be in constant movement, and a sharp gradient in density is often observed at the periphery of a swarm. Within the swarm, volumetric densities may range up to several thousand animals per cubic meter. Near the shelf break surrounding islands in the southwest Atlantic sector of the Southern Ocean it is not uncommon to observe large swarms of *E. superba*, each estimated to contain several thousand tons of krill. Most krill species migrate vertically each day, moving into the upper waters at night and dispersing; just before dawn they move downward and aggregate into denser concentrations. It is generally thought that this behavior is the result of a

trade-off between avoiding predation (dense swarms deep in the water during the day) and maximizing feeding efficiency (dispersed individuals in the more particle-rich surface water at night). Although this is a regular pattern, vertical migration behavior varies among species and within a species depending on location and season. Daytime surface swarms have been observed for several species; they often contain reproductively mature individuals.

Highest densities of krill have been reported near areas of strong vertical mixing and enhanced primary production. These include coastal upwelling zones, ocean frontal boundaries, and topographic features that interrupt or modify currents such as continental shelf breaks, underwater canyons, and escarpments and seamounts. Krill swarms also tend to aggregate in areas of water flow discontinuity such as eddies and sheer zones between opposing currents.

## IV. Recruitment Variability

Recruitment of young animals into adult euphausiid populations is highly variable in space and time. Production of spawn and survival of larvae may vary widely within the distribution range of a species as well as between reproductive events. In his review of euphausiid life histories, Siegel (2000a, b) notes that most species reduce their growth phase and extend their reproductive phase toward the center of their distribution ranges. Closer to their distribution limits, krill put more time into growth and less into reproduction.

There is no apparent relationship between the stock size and the production of new recruits for most species studied. Relatively large adult stocks can produce few new recruits and small adult stocks are capable of producing enough new recruits to increase the stock abundance several fold. The intensity of spawning, survival of eggs and larvae, and the rate of growth have been shown to vary widely between years for several species, resulting in large year-to-year variability in abundance. Interannual variability in abundance has been estimated as 10-fold for *E. pacifica* and *N. simplex* off the west coast of North America, 25-fold for *T. inermis* in the Barents Sea, and 5- to 60-fold for *M. norvegica* at different parts of its range in the North Atlantic.

Recruitment success is affected by exogenous factors, which act to enhance adult reproduction, survival of eggs, and growth of larvae. The best documented of these is the influence of coastal upwelling, which enhances the primary production and the subsequent growth and maturation of young krill. Temperature affects the incubation rate of eggs and growth rate of larvae exposing them to longer or shorter periods of predation. Fluctuations in currents may also transport animals into unfavorable areas. Near the Antarctic Peninsula, *E. superba* spawn earlier in the spring and for a longer period following winters of extensive sea ice development; their larvae enjoy a higher survival rate if sea ice is extensive during the following winter

(Loeb *et al.*, 1997). Four to five year cycles are apparent in the seasonal extent of sea ice and the recruitment of krill in this region of the Southern Ocean. The postulated affect of seasonal sea ice is to provide a refuge, access to a wintertime food source (ice algae), and to inhibit rapid springtime population growth of a potential competitor to krill, *Salpa thompsoni* (Loeb *et al.*, 1997). *Salpa thompsoni* is a pelagic tunicate and obligate filter feeder, which requires open water access to springtime phytoplankton blooms in order to reproduce.

## V. Foraging Tactics of Baleen Whales and Crabeater Seals

The two characteristics of euphausiids described earlier—(1) an immediate response, in terms of individual growth and reproductive output, to favorable conditions and (2) highest densities in predictable locales–allow efficient exploitation of krill by baleen whales. In general, baleen whales migrate between high latitude summer feeding grounds and low latitude winter breeding and calving grounds. Exceptions are bowhead whales, which are restricted to Arctic regions, and Bryde's (*Balaenoptera edeni*) and Omura's whales (*B. omurai*), which usually range from subtropical to temperate waters. Blue (*B. musculus*), fin (*B. physalus*), and sei (*B. borealis*) whales tend to migrate in offshore waters, whereas gray (*Eschrichtius robustus*), right (*Eubalaena* spp.), and humpback (*Megaptera novaeangliae*) whales tend to use a more coastal migration route. Adult whales are thought to feed less during migration than immature or undernourished animals. Off the western coast of North America, blue, fin, Bryde's and humpback whales have been observed feeding on euphausiids aggregated along underwater escarpments and canyons during both winter and summer. The location and timing of whale foraging follows the appearance of high densities of euphausiids and tends to progress from south in the winter to north in the summer. In recent years, aggregations of euphausiids and foraging whales have been a predictable event in the Gulf of California during late winter, near underwater seamounts and canyons off northern California in the summer, and along the shelf break surrounding the Channel Islands in the fall.

Actively feeding whales have been observed to lunge through surface swarms of krill, engulfing large quantities of water and distending their throats, before expelling the water and extruding as much as several hundred kilograms of krill. Similar feeding behavior on subsurface swarms has been inferred from acoustic records of krill layers superimposed with dive tracks simultaneously recorded by instruments attached to foraging whales. There are many reports of humpback and fin whales herding and concentrating their prey before lunging through an aggregation of krill. Bryde's and minke whales (*B. acutorostrata* and *B. bonaerensis*) have also been observed gulping large quantities of aggregated euphausiids. Foraging by right whales has been described as skimming a continuous stream of water rather than gulping; this behavior may be more efficient with dispersed prey (Nemoto, 1970). Sei and gray whales appear to use both methods.

Despite their name, crabeater seals (*Lobodon carcinophaga*) eat very little other than krill. They are found in the sea ice zone in the Southern Ocean and constitute 50% by number (75% by weight) of the world pinniped population. Crabeaters have lobed cusp teeth with spaces between them. It is presumed from the shape of the mouth, tongue, and spacing between the teeth, that crabeater seals engulf a portion of an aggregation of krill and then strain the water similar to a baleen whale. Crabeaters tend to feed at night when krill are in the upper layers and more dispersed than during the day.

## VI. Marine Mammal Diets and Euphausiid Consumption by Ocean Basin

### A. North Pacific

Blue whales in the eastern North Pacific, foraging from the British Columbia to the Californias, feed principally on three species of krill in the California Current; *Euphausia pacifica* and *Thysanoëssa spinifera*, the more inshore species, which is replaced by *Nyctiphanes simplex* moving south. Fin whales have been observed feeding from the Gulf of California to the northern parts of the Bering Sea from April to September respectively. During the late winter and the spring fin whales feed on *N. simplex* in the southern portion of their foraging range; moving north in the summer they feed on *E. pacifica*, *T. raschii*, *T. longipes*, and *T. inermis*. Fin whales have also been observed to feed on copepods, the change in prey type related to changes in local relative densities of prey. During the summer months sei whales consume a variety of euphausiid species including *T. gregaria*, *E. pacifica*, *E. recurva*, *E. diomedeae*, *E. tenera*, *T. inermis*, *T. spinifera*, *N. difficilis*, and *N. gracilis*. South of Japan less than 2% of sei whale diet has been reported to consist of fish. Prey species consumed near the Aleutian Islands include copepods, amphipods, decapods, fishes, and squids. Sei whales, in comparison to blue whales and fin whales, appear to be more opportunistic feeders willing to switch prey type more readily in response to local availability. Bryde's whales have been observed consuming *E. similis*, *N. difficilis*, and *T. gregaria* as well as amphipods, copepods, and fish in the western Pacific and both euphausiids and fish in the Gulf of California. Humpback whales have been observed foraging on euphausiids, including *E. pacifica*, *T. raschii*, *T. longipes*, and *T. spinifera*, from Southeast Alaska to Baja California, although a substantial part of their diet includes clupioid fish as well. Bowhead whales forage primarily on *T. raschii* and *T. inermis* in the Bering and the Beaufort Seas during summer and fall, although copepods, mysids, and amphipods also form a part of their diet. Common minke whales (*B. acutorostrata*) have been observed foraging on euphausiids but appear to prefer fish throughout the North and the northeastern Pacific. Gray whales are thought to consume primarily benthic amphipods in the Bering Sea during the summer months, although there are reports of gray whales consuming *T. raschii* in the Bering Sea and *E. pacifica* off northern California. Prey selectivity among ringed seals appears to be dependent on seasonality and location. Ringed seals (*Pusa hispida*) have been reported to eat *T. raschii*, *T. longipes*, and *T. inermis* in offshore waters in the Northern Hemisphere in spring and summer when krill abundance is greatest; in the winter they consume Arctic cod and other fish species in inshore waters.

### B. South Pacific

Bryde's whales have been observed feeding on *E. diomedeae*, *E. recurva*, and *T. gregaria* and occasionally fish in the Coral Sea (western South Pacific) during the austral spring. In the eastern South Pacific Bryde's whales consume euphausiids during the austral summer between 35° and 40° south latitude. Humpback whales have been observed off the east and west coasts of Australia feeding on euphausiids, including *E. hemigibba*, *P. latifrons*, and *E. spinifera*.

### C. North Atlantic

Fin whales feed primarily on *M. norvegica*, *T. inermis*, and *T. raschii* during the summer months, switching between prey species in response to local availability. Common minke whales consume

*T. inermis* and *M. norvegica* in the North Atlantic, where euphausiids form a much larger portion of the diet than in the Pacific. Northern right whales feed primarily on copepods, although consumption of euphausiids has been observed, particularly when associated with copepods. Harp seals (*Pagophilus groenlandicus*) feed on a variety of prey including decapods, amphipods, euphausiids, and pelagic fishes; however, newly weaned pups and young seals have been reported to feed mainly on *Thysanoëssa* species.

### D. Indian Ocean

Fin and Antarctic minke whales (*B. bonaerensis*) have been observed feeding on euphausiids in the southwest Indian Ocean during their spring and fall migrations to and from the Southern Ocean; prey species include *E. recurva, E. lucens, T. gregaria, E. spinifera, N. capensis,* and *E. diomedeae*. Bryde's whales forage on these euphausiids species as well in the southwest Indian Ocean. Near Durban, South Africa humpback whales have been observed feeding on *E. recurva* and *T. gregaria* and a single pygmy blue whale was reported to be feeding on *E. recurva* and *E. diomedeae*. Sei whales were observed to consume euphausiids as well as copepods, amphipods, pteropods, and fish.

### E. Southern Ocean

Fin and Antarctic minke whales consume several species of krill in the Southern Ocean throughout the austral summer. Species preference appears to be related to local availability, with *T. macrura* and *E. vallentini* more prevalent in the diets of animals foraging in open waters and *E. frigida* and *E. crystallorophias* more prevalent near the continental shelf and ice edge regions. The numerically dominant euphausiid in the Southern Ocean, *E. superba*, is consumed in all areas. Southern right whales (*E. australis*) have been observed foraging on *E. superba* in the Atlantic sector of the Southern Ocean. Humpback whales have been frequently observed foraging on *E. superba* in bays and in fjords along the Antarctic Peninsula. Crabeater seals consume *E. superba* and *E. crystallorophias* in the sea ice zone and in coastal fjords and bays; Antarctic silverfish have been reported as seasonal constituents of their diet but krill has been estimated to provide over 90% of their prey requirements. Much smaller portions of the diets of leopard (*Hydrurga leptonyx*) and Ross (*Ommatophoca rossii*) seals and Antarctic (*Arctocephalus gazella*) and subantarctic (*A. tropicalis*) fur seals have been reported to be composed of krill.

### F. Summary

From these observations some generalizations may be drawn: (1) blue and fin whales appear to have a higher preference for euphausiids than minke, humpback, or bowhead whales; (2) sei and Bryde's whales appear to be more opportunistic feeders; (3) gray whales and northern right whales prefer prey other than euphausiids but will consume them; (4) crabeater seals have a higher preference for euphausiids than other seals in the Southern Ocean; and (5) ringed and harp seals in the Northern Hemisphere include euphausiids in their diets during certain times of the year and life cycle.

Gross estimates of the consumption of euphausiids by marine mammals are summarized in Table I. Estimates of stock abundances were obtained from working papers and reports of the International Whaling Commission, reports from the US National Marine Fisheries Service, and the primary literature. In some cases no reliable estimates are available and broad ranges were used. Daily ingestion rates for baleen

whales during the feeding season were estimated from energetic requirements as a function of body weight following Sigurjonsson and Vikingsson (1997). A daily ingestion rate for seals was estimated as 7% of body weight. Average body weights, the percentages of euphausiids in the diets, and the caloric value of euphausiids (0.93 kcal/g) were taken from the primary literature. The duration of the feeding season was assumed to be 180 days for Northern Hemisphere baleen whales and seals, 120 days for Southern Hemisphere baleen whales and Antarctic and subantarctic fur seals, and 335 days for crabeater, leopard, and Ross seals.

Although Table I is based on several simplifying assumptions, some general conclusions may be drawn. Total consumption of euphausiids by marine mammals is on the order of 10–20 million tons per year in the North Pacific, 15–25 million tons per year in the North Atlantic, and 125–250 million tons per year in the Southern Hemisphere, with the bulk of the latter portion consumed in the Southern Ocean. In the North Atlantic, the largest portion is consumed by fin whales, followed by common minke whales. In the North Pacific, consumption is more evenly distributed, with fin and bowhead whales consuming the most, followed by blue, sei, Bryde's, and common minke whales, all of which consume similar portions. In the Southern Hemisphere, comparable proportions of euphausiids are consumed by crabeater seals and baleen whales. Of the estimated total krill consumption by baleen whales in the Southern Ocean, Antarctic minke whales consume approximately two-thirds. Crabeater seals consume more krill than any other marine mammal population in the world.

These crude calculations suggest that baleen whales consume a substantial amount of euphausiids. Moreover, their food requirements must have been several times higher prior to commercial whaling. Unfortunately, there is little information on which to judge whether krill production was higher prior to the onset of commercial whaling or whether other krill predators (e.g., crabeater seals) benefited as a result of the decline in baleen whale stocks.

What is more apparent is that krill abundance can vary dramatically over relatively short periods of time and that baleen whales have adapted to this variability. Their size and ability to accumulate substantial energy stores allow them to integrate over large distances and periods of time in their search for food. Their longevity allows them to spread reproductive effort over several years. It is reasonable to expect, however, that the supply of euphausiids will not be sufficient in all years to meet total energy requirements, and that reproductive success and population growth among krill-dependent baleen whales may vary from year-to-year in response to the availability of their prey.

### VII. Anthropogenic Affects

The production of euphausiids can be very sensitive to environmental conditions. This raises two concerns with regard to the influence of human activities. The first is that highly productive euphausiid populations may be able to sustain large fisheries (Fig. 5). The second is that climatic change (whether man-induced or not) may affect the frequency of environmental conditions that are favorable for reproductive success. Because these are relatively recent developments, we cite three studies below as entries into a larger body of literature (Nicol and Foster, 2003; Croxall and Nicol, 2004).

Fisheries on euphausiids have the potential of being the largest in the world. In their review of krill fisheries, Nicol and Endo (1999) describe the harvest of *E. pacifica* off the coasts of Japan and

## TABLE I
### Marine Mammal Stock Abundance, Average Body Weight, Daily Ingestion Rate, Duration of Feeding Season, Proportion of Krill in Diet, and Total Krill Consumption by Ocean Basin

| Ocean basin | Whale species | Abundance | Average body weight (t) | Summertime ingestion rate ($10^3$ kcal/day) | Feeding period (days) | Krill in diet (%) | Krill consumed ($10^3$ tons) |
|---|---|---|---|---|---|---|---|
| North Pacific | Blue whale | 3,000–4,000 | 69.2 | 2,136 | 180 | 100 | 1,240–1,654 |
| | Fin whale | 14,600–18,600 | 42.3 | 1,452 | 180 | 80 | 3,282–4,182 |
| | Sei whale | 9,000–13,000 | 19.9 | 805 | 180 | 80 | 1,122–1,621 |
| | Bryde's whale | 34,500–45,500 | 13.2 | 584 | 180 | 40 | 1,559–2,056 |
| | Common minke whale | 30,000–32,000 | 5.3 | 284 | 180 | 70 | 1,153–1,229 |
| | Humpback whale | 5,000–6,000 | 31.8 | 1,161 | 180 | 60 | 674–809 |
| | Bowhead whale | 8,000–10,000 | 80.0 | 2,392 | 180 | 80 | 2,963–3,704 |
| | Gray whale | 25,000–27,000 | 25.0 | 962 | 180 | 5 | 233–251 |
| | North Pacific right whale | 400–600 | 55.0 | 1,784 | 180 | 25 | 35–52 |
| | | | | | | Total | 12,760–15,558 |
| North Atlantic | Blue whale | 750–1,300 | 69.2 | 2,136 | 180 | 100 | 310–538 |
| | Fin whale | 45,000–50,000 | 42.3 | 1,452 | 180 | 80 | 10,117–11,241 |
| | Sei whale | 9,000–13,000 | 19.9 | 805 | 180 | 80 | 1,122–1,621 |
| | Common minke whale | 120,000–182,000 | 5.3 | 284 | 180 | 70 | 4,610–6,992 |
| | Humpback whale | 10,000–11,000 | 31.8 | 1,161 | 180 | 60 | 1,349–1,483 |
| | North Atlantic right whale | 300–350 | 50.0 | 1,656 | 180 | 25 | 24–28 |
| | | | | | | Total | 17,532–21,903 |
| Southern Hemisphere | Blue whale | 600–800 | 83.0 | 3,708 | 120 | 100 | 287–383 |
| | Pygmy blue whale | 2,000–6,000 | 68.9 | 3,205 | 120 | 100 | 827–2,481 |
| | Fin whale | 10,000–20,000 | 48.0 | 2,415 | 120 | 100 | 3,116–6,232 |
| | Sei whale | 35,000–45,000 | 17.5 | 1,096 | 120 | 80 | 3,960–5,091 |
| | Bryde's whale | 78,000–108,000 | 13.2 | 879 | 120 | 40 | 3,538–4,899 |
| | Antarctic minke whale | 650,000–950,000 | 7.0 | 535 | 120 | 100 | 44,858–65,561 |
| | Humpback whale | 15,000–16,000 | 26.5 | 1,517 | 120 | 100 | 2,936–3,131 |
| | Southern right whale | 6,500–7,500 | 55.0 | 2,687 | 120 | 100 | 2,253–2,600 |
| | | | | | | Total | 61,775–90,378 |

| Ocean basin | Seal species | Abundance | Average body weight (kg) | Ingestion rate ($10^3$ kg/day) | Feeding period (days) | Krill in diet (%) | Krill consumed ($10^3$ tons) |
|---|---|---|---|---|---|---|---|
| Northern Hemisphere | Ringed seal | 6,000,000–7,000,000 | 75 | 5.3 | 180 | 25 | 1,418–1,654 |
| | Harp seal | 100,000–400,000 | 130 | 9.1 | 180 | 25 | 41–176 |
| | | | | | | Total | 1,459–1,830 |
| Southern Hemisphere | Crabeater seal | 15,000,000–30,000,000 | 220 | 15.4 | 335 | 94 | 72,742–145,484 |
| | Leopard seal | 300,000–500,000 | 275 | 19.3 | 335 | 37 | 716–1,193 |
| | Antarctic fur seal/ | 1,000,000–1,500,000 | 50 | 3.5 | 120 | 50 | 210–315 |
| | Subantarctic fur seal | 300,000–500,000 | 85 | 6.0 | 120 | 50 | 107–179 |
| | Ross seal | 125,000–225,000 | 175 | 12.3 | 335 | 10 | 51–92 |
| | | | | | | Total | 73,826–14,7263 |

western Canada, *T. inermis* off the coasts off Japan and eastern Canada, *E. nana* off the coast of Japan, *T. raschii* and *M. norvegica* off the coast of eastern Canada, and *E. superba* in the Southern Ocean. In recent years, the harvest of *E. pacifica* off Japan (~60,000 tons per year) and *E. superba* in the Scotia Sea region of the Southern Ocean (~100,000 tons per year) comprised over 90% of the world harvest of euphausiids. Nicol and Endo (1999) note that these yields are well within their theoretical potentials, although expansion of the coastal fisheries is unlikely because of ecological, economic, and political considerations. They further note, however, that as conventional fisheries decline and demand for krill as aquaculture feed increases, fishing pressure is likely to shift to *E. superba* in the Southern Ocean, where current harvests are far below current estimates of sustainable yields.

**Figure 5** *Japanese krill trawlers operating off the north coast of Elephant Island, South Shetland Islands, Antarctica Photograph by (R.P. Hewitt).*

Recent evidence suggests that the production of euphausiids may be affected by long-term climatic change. Warming of the surface waters of the California Current since the mid-1970s has been accompanied by a reduction in the depth of the thermocline, reduced nutrient input via coastal upwelling, reduced primary production and an overall decrease in macrozooplankton biomass by as much as 80% (Roemmich and McGowan 1995; Lavaniegos and Ohman, 2007). Euphausiids are the dominant taxa in the macrozooplankton fauna of the California Current and have shown decreased abundances during warm (El Niño) years and increased abundances during cold (La Niña) years. A 50-year warming trend in the Antarctic Peninsula region has been associated with a decrease in the annual production of sea ice. Loeb *et al.* (1997) and Nicol *et al.* (2000) correlated the reproductive success of *E. superba* with the wintertime extent of sea ice and suggested that the warming trend may cause a decrease in the frequency of strong year classes of Antarctic krill, a decrease in the mean population abundance of krill, and a change in the carrying capacity of vertebrate krill predators in the region.

## See Also the Following Articles

Baleen ▪ Bioluminescence ▪ Cetacean Ecology ▪ Diet ▪ Predator–Prey Relationships

## References

Berta, A., and Sumich, J. L. (1999). "Marine Mammals: Evolutionary Biology." Academic Press, San Diego, CA.

Bowen, W. D., and Siniff, D. B. (1999). Distribution, population biology, and feeding ecology of marine mammals. *In* "Biology of Marine Mammals" (J. E. Reynolds., III, and S. A. Rommel, eds), pp. 422–484. Smithsonian Institution Press, Washington, DC.

Brinton, E. (1962). The distribution of Pacific euphausiids. *Bull. Scripps. Inst. Ocean.* **8**, 51–270.

Clapham, PJ., Young, S. B., and Brownell, R. L., Jr. (1999). Baleen whales: Conservation issues and the status of the most endangered populations. *Mammal Rev.* **29**, 35–60.

Croxall, J. P., and Nicol, S. (2004). Management of Southern Ocean fisheries: Global forces and future sustainability. *Ant. Sci.* **16**, 569–584.

Darling, J. D., Keoghi, K. E., and Steeves, T. E. (1998). Gray whale (*Eschrichtius robustus*) habitat utilization and prey species off Vancouver Island, B. C. *Mar. Mamm. Sci.* **14**, 692–720.

Fraser, J. H. (1969). Plankton resources. *In* "The Encyclopedia of Marine Resources" (E. F. Frank, ed.). Van Nostrand Reinhold.

Gaskin, D. E. (1982). "The Ecology of Whales and Dolphins." Heinemen, London.

Hickling, C. F. (1927). The natural history of the hake. *Fish. Invest. Lond. Ser II* **10**, 1–100.

Hjort, J., and Rund, J. T. (1929). Whaling and fishing in the North Atlantic. *Rapp. P.-v. Reun. Cons. int. Explor. Mer* **41**, 107–119.

Kato, H. (1982). Food habits of largha seal pups in the pack ice area. *Sci. Rep. Whales Res. Inst. Tokyo* **34**, 123–136.

Kawamura, A. (1974). Food and feeding ecology in the southern sei whales. *Sci. Rep. Whales Res. Inst. Tokyo* **26**, 25–144.

Kawamura, A. (1980a). Food habits of the Bryde's whales taken in the South Pacific and Indian Oceans. *Sci. Rep. Whales Res. Inst. Tokyo* **32**, 1–23.

Kawamura, A. (1980b). A review of food of balaenopterid whales. *Sci. Rep. Whales Res. Inst. Tokyo* **32**, 155–197.

Lavaniegos, B. E., and Ohman, M. D. (2007). Coherence of long-term variations of zooplankton in two sectors of the California Current system. *Prog. Oceanogr.* **75**, 42–69.

Laws, R. M. (1984). Seals. *In* "Antarctic Ecology", Vol. 2, pp. 621–715. Academic Press, London.

Laws, R. M. (1985). The ecology of the Southern Ocean. *Am. Sci.* **73**, 26–40.

Lebour, M. V. (1924). The Euphausiidae in the neighborhood of Plymouth and their importance as herring food. *J. Mar. Biol. Assoc. UK* **13**, 810–846.

Loeb, V., Siegel, V., Holm-Hansen, O., Hewitt, R., Fraser, W., Trivelpiece, W., and Trivelpiece, S. (1997). Effects of sea-ice extent and krill or salp dominance on the Antarctic food web. *Nature* **367**, 897–900.

Marr, J. W. S. (1962). The natural history and geography of the Antarctic krill (*Euphausia superba* Dana). *Discov. Rep.* **32**, 33–464.

Mauchline, J. (1980). The biology of mysids and euphausiids. *Adv. Mar. Biol.* **18**, 1–681.

Mauchline, J. (1998). The biology of calanoid copepods. *Adv. Mar. Biol.* **33**, 1–701.

Mauchline, J., and Fischer, L. R. (1969). The biology of euphausiids. *Adv. Mar. Biol.* **7**, 1–454.

Miller, D. G. M., and Hampton, I. (1989). Biology and ecology of the Antarctic krill. *Biomass Sci. Ser.* **9**, 1–166.

Mitchell, E. (1974). Trophic relationships and competition for food in Northwest Atlantic whales. *Proc. Can. Soc. 2001 Ann. Mtg.* **1974**, 123–133.

Nemoto, T. (1970). Feeding pattern of baleen whales in the ocean. *In* "Marine Food Chains" (J. H. Steele, ed.), pp. 241–252. University of California Press, Berkeley and Los Angeles.

Nemoto, T., and Kawamura, A. (1977). Characteristics of food habits and distribution of baleen whales with special reference to the abundance of North Pacific sei and Bryde's whales. *Rep. Int. Whal. Commn.* **1**, 80–87.

Nicol, S., and Endo, Y. (1999). Krill fisheries development, management and ecosystem implications. *Aquat. Liv. Res.* **12**, 105–120.

Nicol, S., and Foster, J. (2003). Recent trends in the fishery for Antarctic krill. *Aquat. Liv. Res.* **16**, 42–45.

Nicol, S., *et al.* (2000). Ocean circulation off East Antarctica affects ecosystem structure and sea-ice extent. *Nature* **406**, 504–507.

Roemmich, D., and McGowan, J. (1995). Climatic warming and the decline of zooplankton in the California Current. *Science* **267**(5202), 1324–1326.

Ryther, J. H. (1969). Photosynthesis and fish production in the sea. *Science* **166**, 72–76.

Sars, G. O. (1885). Report on the Schizopoda collected by H.M.S. "Challenger" during the years 1873–1876. *The Voyage of the H.M.S. "Challenger"* **13**, 1–128.

Siegel, V. (2000a). Krill (Euphausiacea) life history and aspects of population dynamics. *Can. J. Fish. Aquat. Sci.* **57**, 130–150.

Siegel, V. (2000b). Krill (Euphausiacea) demography and variability in abundance and distribution. *Can. J. Fish. Aquat. Sci.* **57**, 151–167.

Sigurjonsson, J., and Vikingsson, G. A. (1997). Seasonal abundance of and estimated food consumption by cetaceans in Icelandic and adjacent waters. *J. Northwest. Atl. Fish. Sci.* **22**, 271–287.

# Language Learning and Cognitive Skills

### Louis M. Herman

No single trait has been linked more closely with the human species than language. However, the definition of language, its relation to animal communication, and its origins continue to be areas of study and debate. One of the most contentious areas is the degree to which there is evolutionary continuity between human language mechanisms, and characteristics and mechanisms extant or tutored in selected animal species. Mechanisms include brain structures and processes as well as articulatory and receptive systems; characteristics include such human language features as semanticity, reference, syntactic structure, and openness as well as the pragmatic characteristics that typify a child's acquisition of language.

The linguist Noam Chomsky (Chomsky, 1972, 1975) viewed human language as a unique development supported by a "language acquisition device" (LAD), whose embodiment in the brain was to be found only in humans. He later elaborated on this idea, postulating the existence of a hard-wired "Universal Grammar" that provides a basic grammatical template that can be modified by exposure to fit the grammar of any language heard by the child. Chomsky asserted that animals do not have a LAD or a universal grammar module, and therefore cannot have language. In these views, human language appeared as a *saltational* event, an abrupt evolutionary development, likely occurring through a genetic mutation and as a by-product of the pattern of growth and development of the human brain. Alternatively, these views maintain, language could have appeared as an *exaptation*, a by-product of a mechanism or function evolved for some purpose other than language. As an example of an exaptation, bird feathers might have evolved originally as an adaptation for thermal insulation but were later co-opted for flight. Pinker (1994), although favoring a discontinuity position—there is no relation between human language and animal communication—believes that the language faculty did not arise *de novo*, but evolved as a unique *adaptation* that followed the laws of Darwinian natural selection, arising relatively gradually. The view that language is a uniquely human "instinct" with no continuity with any form of animal communication is emphatically contested by several other linguists. Lieberman (2000, 2006), a prominent critic of Chomskian notions of a language organ and a linguistic gene, contends that the roots of human language are present in other species. He cites evidence that the human functional language system is distributed widely across many subsystems of the brain, including subcortical areas present in many animal species, and posits that the evolution of a language capacity was a gradual process stretching over 2 million or more years.

Some other recent work appears to attempt to meld properties of human language and animal communication, at least partially by hypothesizing that almost everything essential to human language can be found in other animals, in particular the sensory–motor systems and the cognitive systems ("conceptual–intentional systems") (Hauser *et al.*, 2002). What humans added, these authors contend, is a key computational process, *recursion*, "the capacity to generate an infinite range of expressions from a finite set of elements" (p. 1569). Recursion refers to the hierarchical structure of human language: morphemes combine into words, words into clauses, clauses into sentences, and moreover, for example, sentences can be embedded within sentences. Recursion, coupled with syntactic devices that allow for determining the grammatical relation between words (e.g., as subject vs object), gives language its rich open-ended generative capacity. Jackendoff and Pinker (2005), however, reject recursion as the sole discriminator of human language and animal communication, while at the same time viewing language "as a combination of components, some special to language, others rooted in more general capacities in human or animal cognition" (p. 223). They are therefore supportive of experimental work on animal communicative abilities in efforts to connect theories of grammar to evolutionary principles.

It is within these diverse contexts and disputes that the study of cognitive processes in animals and their competencies for learning forms of communication that have language-like properties take on special meaning and value. As Lieberman (1984, p. 333) stated, "the nature and evolution of the biological basis of language can ultimately be ascertained only by actually studying the cognitive, linguistic, and communicative behavior of human beings and the other animals to whom we are all related." The developing body of work on animal cognition indeed testifies to the depth and the breadth of cognitive skills that may be demonstrated in many large-brained mammals, not only the close relatives of humans, the great apes, but also in the evolutionarily divergent but large-brained common bottlenose dolphin (*Tursiops truncatus*) and several other cetacean species. Within this work, the ability of animals to learn some defining properties of language has been a subject of intense study.

## I. Human Language and Ape Language

The work on teaching language-like systems to apes, by Beatrice and Alan Gardner, David Premack, Duane Rumbaugh, and others, beginning in the mid-1960s and continuing throughout the decade of the 1970s, seemed to provide a genuine link between human and ape in fundamental language competency (see reviews in Ristau and Robbins, 1979; Herman, 1987). This early work reported that common chimpanzees (*Pan troglodytes*) were able to learn to understand and use not only individual words but also words strung together into sentences. Sentences give human language its vast communicative power through the infinite variety of meanings that can be constructed by the combination and recombination of words. To understand a sentence, the human listener must take account not only of the meaning and referents of the words but also their grammatical relationship to one another, as governed by some syntactic device such as word order. This early work on teaching language to apes was thrown into disarray, however, by additional studies and criticisms from other researchers, such as Herbert Terrace (Terrace *et al.*, 1979) and Carolyn Ristau (Ristau and Robbins, 1979). These researchers argued that the putative "sentences" produced by the apes were largely an artifact of context, imitation, or social cueing. Further, although sequences of symbols were indeed produced by the apes, the sequences often had no syntactic structure that enhanced, explained, or modified meaning.

Historically, this work with apes focused primarily on language production and paid scant attention to language comprehension. Investigators attempted to teach the apes to produce requests or statements through learned gestures, or the pressing of keyboard symbols, or the use of other types of artificial symbols, assuming that if the ape produced a gesture or other learned symbol, or a sequence of such productions, that it understood what it was communicating—that it understood what the word or sequence meant or represented. A further assumption was that the ape would understand those same words or sequences when produced by the human partner. These assumptions, when later tested, proved largely false. It was found, instead, that comprehension did not flow automatically from production. The preeminence of language comprehension over language production, only relatively recently appreciated by ape language researchers (Herman and Morrel-Samuels, 1990), has long been appreciated among those studying child language. Language comprehension by young children develops earlier than language production, and even into adulthood, comprehension vocabularies exceed speaking vocabularies (Bloom, 1974).

More recent language work with bonobo chimpanzees (*P. paniscus*), pioneered by Savage-Rumbaugh *et al.* (1993), emphasized language comprehension and has progressed well beyond the findings from the earlier ape language studies. The bonobos have shown an ability to learn to understand instructions given in spoken English sentences, with at least a rudimentary appreciation that sentence structure affects meaning. Further, Savage-Rumbaugh has shown that both common and bonobo chimpanzees can learn to appreciate that symbols (words) of the language can function as references to objects and actions. This understanding that words *refer* is one of the key characteristics of human language. Referential understanding enables us, for example, to discuss objects or events that are not immediately present or that happened at a different place or time.

## II. Dolphins and Language

### A. Natural Language?

Dolphins (including the common bottlenose dolphin) produce various types of sounds including clicks and burst-pulse emissions, and, for some species, whistles. Clicks are used for ECHOLOCATION (Au, 1993). Through echolocation, the dolphin can examine its underwater world by listening to the echoes returning from its emitted clicks as they strike reflective surfaces. Dolphin echolocation apparently can yield an "image" or mental representation of an ensonified object that is functionally similar to the image derived though vision (Herman *et al.*, 1998; Pack *et al.*, 2004). Burst-pulse sounds may indicate the dolphin's emotional state, ranging from pleasure to anger (Herman and Tavolga, 1980; Herzing, 1996). However, burst-pulse sounds have been relatively little studied and much remains to be learned about them. Whistles may be used for COMMUNICATION, but it is still an open question as to whether or how much of whistle production is communicative and intentional (thus, rapidly repeated whistling may be elicited by stress, without any apparent intention to convey that emotional state to others). During the 1960s, researchers attempted to determine the diversity of whistles from dolphins (principally bottlenose dolphins, *Tursiops* spp.), whether these whistles were tied to specific contexts or events, and whether they might support a language (see review in Herman and Tavolga, 1980). Estimates of the number of distinct whistles produced by bottlenose dolphins have varied widely across studies, and the validity and reliability of the different techniques for separating

and classifying samples of whistles still remain an unsettled issue (*cf.* Janik, 1999; McCowan and Reiss, 2001). Some early work pointed to the stereotypy of the whistles from individual dolphins, leading to the hypothesis that the whistle functioned principally as a "signature," with each individual dolphin producing a unique and predominant whistle type (Caldwell and Caldwell, 1965). Presumably, this signature quality enabled that individual to identify itself to others, knowingly or not. McCowan and Reiss (2001), however, contended that there was no individually unique signature. Instead, many different whistles are produced (see also McCowan *et al.*, 1999), but under the conditions of isolation studied by the Caldwells, most dolphins typically produce a single shared whistle that the authors describe as rising in frequency (an upsweep) and which they term "Type 2." Individual recognition was still present, these authors contended, in the distinctive "voice" characteristics of the individual dolphin producing the Type 2 whistle. Janik *et al.* (2006), proponents of the signature whistle hypothesis, have, however, recently presented evidence that individual recognition was not based on voice characteristics but on the contour or shape of the whistle, as it varied in frequency over time. Clearly, additional study is needed to resolve these differences in theory and in evidence.

Richards *et al.* (1984) demonstrated that a bottlenose dolphin could use its whistle mode to imitate a wide variety of sounds of different frequencies and waveforms generated by a computer and broadcast underwater into the dolphin's habitat and could also use learned whistles to "label" objects (see also Reiss and McCowan, 1993). Tyack (1986), studying wild dolphins in Sarasota Bay, Florida, reported that one dolphin could imitate another's typical whistle, thereby possibly referring to or calling that individual. Janik (2000) provided additional evidence for whistle matching in wild dolphins in the Moray Firth in Scotland, and suggested an analogy to the presumptive earliest stage of the development of human language—when a vocalization (or gesture) was invented to refer to some object or event and that symbol was then shared with others. Janik's dolphin analogy was that signature whistles are "invented" individually, and that the imitation of one dolphin's whistle by another dolphin was, in effect, symbol sharing of an invented self-reference. We do not know, however, to what extent the dolphin's imitations are *intentional* references to another or whether whistles, or other dolphin vocalizations, may be used to refer to things other than themselves or another dolphin. This is a fruitful area for additional study.

Although the available evidence strongly suggests that dolphins do not possess a natural language sharing the basic characteristics of semanticity and syntactic structure of human language (Herman, 1980), it is nevertheless important for a deeper understanding of animal communication and its relation to human language to investigate whether they might be able to learn to utilize those fundamental characteristics if tutored in a language-like symbolic system. Any demonstration of language-learning competency by dolphins would bear on questions of the origins of human language, shifting the emphasis from a search for language precursors only in other hominoid species or ancestors to an examination of common or convergent cognitive or social characteristics shared by ape and dolphin that might lead to advanced communicative skills in both species. A large number of behavioral studies of bottlenose dolphins have in fact revealed extensive cognitive skills, both in laboratory settings (reviewed in Herman, 2006) and in the wild (reviewed in Connor, 2007), many of which are shared with the apes. Similarly, the complex social structure of dolphin societies, including higher-order alliances, has also been amply documented (reviewed in Connor *et al.*, 2000, 2007) and has resemblances to some of the complexities of chimpanzee social organization.

## B. Dolphin Cognitive Skills

If language depends in part on rich cognitive structures, then a capacity for learning some elements of a language system might be attainable by large-brained animals exhibiting wide ranging and flexible cognitive skills. The literature demonstrating such skills in the great apes is vast (Tomasello and Call, 1997), and it is no coincidence that these skills combined with the close phylogenetic relationship of humans and apes have motivated the quest for exploration of language-learning capabilities in these species. This quest has met with interesting recent successes, particularly among bonobo chimpanzees, as noted earlier. Bottlenose dolphins share many of the cognitive skills demonstrated in apes (see reviews in Herman, 1980, 1986, 2006).

*1. Memory* Language and indeed all learning skills rest on a strong foundation of memory. Studies of dolphin short-term or "working" memory—the processing of new information and retaining it in conscious memory—show that auditory memory (memory for things heard) and visual memory (memory for things seen, including objects and behaviors) are well developed and robust in bottlenose dolphins (reviewed in Herman, 1980; see also Herman *et al.*, 1989). Dolphin short-term memory is similar in its fidelity and characteristics to demonstrated short-term memory capabilities of nonhuman primates (D'Amato, 1973), or in the case of probe memory—memory for lists of items—comparable to memory characteristics demonstrated in humans (Thompson and Herman, 1977; Herman, 1980).

*2. Rule Learning, Concept Formation, and Representation* Language proficiency is intimately interwoven with rule learning, concept formation, and representation. Bottlenose dolphins can learn a variety of governing rules for solving abstract problems (Herman *et al.*, 1994) including reliably classifying pairs of objects or sounds as "same" or "different" (Mercado *et al.*, 2000). They can learn that arbitrary symbols can represent objects, actions, agents, relationships, and locations. Dolphins can also understand representations of the real world, as illustrated by their spontaneous (untrained) ability to respond to gestural instructions from the small images of televised trainers with the same fidelity that those responses are made to "live" trainers (Herman *et al.*, 1990).

*3. Imitation* Imitation is a key component of the child's early mastery of spoken words. Dolphins (and possibly some other cetacean species) appear to be the only nonhuman animal capable of imitating both arbitrary sounds and arbitrary behaviors including an understanding of the abstract concept of "imitate" (see review in Herman, 2002).

*4. Shared Attention and the Indicative* In the human infant, gazing and pointing as communicative devices often precede more sophisticated language skills. Dolphins tested for "shared attention" reliably choose the object pointed to by a human informant or gazed at with a turn of the head (Tschudin *et al.*, 2001; Pack and Herman, 2006, 2007), and under certain conditions, they will spontaneously use their rostrum and body alignment to point at objects that they desire from a human companion (Xitco *et al.*, 2001).

*5. Creativity* In language, creativity allows for the development of novel sentences and new words. Dolphins have been shown capable of innovating motor behaviors, either individually or in close synchrony with a second dolphin (Pryor *et al.*, 1969; Herman, 2006). Innovations in feedings strategies are also well documented (Smolker *et al.*, 1997).

*6. Cross-modal Transfer* In humans, language can involve transfer from one modality to another, for example the written word to the spoken word. Dolphins have been shown capable of spontaneous transfer of object shape information across the senses of echolocation and vision (Pack and Herman, 1995; Herman *et al.*, 1998; ). An ability to transfer information across sensory modalities has been linked to a variety of intellectual functions in humans and nonhuman primates (Rose and Wallace, 1985; Gunderson *et al.*, 1990).

*7. Cultural Transmission* The child learns the language of its community through such mechanisms as imitation and tutoring within the scope of its predispositions toward language in general. Thus, language is passed from generation to generation. Cultural transmission of group-specific vocalizations or vocal dialects also occurs in some cetacean species, such as the pod-specific vocal dialects of killer whales (*Orcinus orca*) and sperm whales (*Physeter macrocephalus*), and the changing songs of humpback whales (*Megaptera novaeangliae*) that evolve collectively among male humpbacks throughout a winter season in a particular breeding ground. Other non-vocal examples of cultural transmission are the unique sponge-feeding techniques passed on from mother to daughter among subsets of bottlenose dolphins in Shark Bay, Australia, and the cooperative fishing techniques of dolphins and humans in Laguna, Brazil (Rendell and Whitehead, 2001).

There are other demonstrated cognitive abilities that illustrate the breadth of intellectual competence that can be attained, such as the ability to learn a concept of "numerically less" (Kilian *et al.*, 2003; Jaakkola *et al.*, 2005) or to exhibit meta-cognition, "sometimes described as an awareness of one's personal knowledge or lack of knowledge" (Smith *et al.*, 1995).

In summary, this impressive suite of cognitive abilities demonstrates considerable intellectual flexibility that could be applied toward an understanding of some of the elementary features of an imposed language.

## C. Early Attempts at Teaching Language to Dolphins

From the mid-1950s to the mid-1960s, Lilly (1961, 1967) promoted the idea that bottlenose dolphins might possess a natural language. He based this supposition on this species' exceptionally large brain with its richly developed neocortex. He reasoned that the large brain must be a powerful information processor having capabilities for advanced levels of intellectual accomplishment, including the development of a natural language. He set about to uncover the supposed language. Failing in that quest, he then attempted, also without success, to teach human vocal language (English) to dolphins he maintained in his laboratories. Dolphins have a rich vocal repertoire, but not one suited to the production of English phonemes. The procedures used by Lilly and the data he obtained were presented only sketchily, blunting any detailed analysis of his efforts at teaching language. Lilly's language work and his often freewheeling speculations were met with harsh criticisms by several prominent researchers (Wood, 1973; Wilson, 1975).

In the mid-1960s, Duane Batteau developed an automated system that translated spoken Hawaiian-like phonemes into dolphin-like whistle sounds. The sounds were projected underwater into a lagoon housing two bottlenose dolphins (reviewed in Herman, 1980). Batteau attempted to use these sounds as a language for conveying instructions to the dolphins. A major flaw in his approach, however, was that individual sounds were not associated with individual semantic elements, such as objects, actions, or properties, but instead functioned as holophrases (complexes of elements). For example, a particular whistle sound instructed the dolphin to "hit the ball with your pectoral fin." Another sound instructed the dolphins to "swim through a hoop." Unlike a natural language, there was no unique sound to refer to *hit* or *ball*, or *hoop*, or *pectoral fin*, or any other unique semantic element. Hence, there was no way to recombine sounds (semantic

L

elements) to create different instructions, such as "hit the hoop (rather than the ball) with your pectoral fin." As was noted earlier, human language achieves its great communicative power through its ability to combine and recombine words in infinite ways (governed by syntactic structures) to achieve distinct meanings. After several years of effort, Batteau's dolphins were able to reliably follow the holophrastic instructions conveyed by each of 12 or 13 different sounds. However, because of the noted flaw in the approach to construction of a language, the experiment failed as a valid test of dolphin linguistic capabilities.

Bastian (1967) took a different approach, asking whether dolphins might develop arbitrary symbols to refer to some event—the language feature of "openness" (Hockett, 1960; Hauser *et al.*, 2002)—if a situation were structured requiring it. Food reward for each of two dolphins, a male and a female housed in the same pool, was made contingent on an apparent cooperative exchange of information between the pair. Each dolphin had available a pair of paddles and each dolphin was required to press a paddle to its left if an out-of-water light was flashing or a paddle to its right if the light held steady. Both had to press the correct paddle for either to obtain reward, but the male had to press his paddle first. Through a series of steps, Bastian altered the situation so that an opaque screen separated the two and now only the female was able to see the light. Yet, both had to continue to press the appropriate paddle, but with the "blind" dolphin, the male, required, as before, to press his first. Under those conditions, surprisingly, the two dolphins continued to perform almost without error. Bastian noted that the female vocalized at the start of most trials, suggesting that she was intentionally transmitting information to the male about the state of the light or which paddle to press. However, additional tests by Bastian *et al.* (1968) negated this conclusion: the female's vocalizations persisted even when the barrier was removed and the male could again see the light, and also continued even when the male was removed entirely from the pool. Further, subsequent reversal of the contingencies between the state of the light and which paddle to press disrupted joint performance although each dolphin learned the reversal readily. The most likely explanation for the initial high level of joint performance was that the female adopted some stereotyped location in the pool while she vocalized (technically, called a "superstitious" behavior—a behavior correlated with reward but unnecessary for reward) which differed for each state of the light, and the male simply learned to use that unintentional cue (locus of the female's vocalizations) to make his choice. This was confirmed in a replication study carried out in the dolphinarium at Harderwijk by Dudok van Heel (pers. comm.). Evans and Bastian (1969) in a review of the original Bastian study reached a similar conclusion to that of Dudok von Heel. For a fuller description, see Herman and Tavolga (1980).

### D. Kewalo Basin Dolphin Studies of Language Comprehension

The work at the Kewalo Basin Marine Mammal Laboratory in Honolulu examining dolphin language-learning competencies was begun in the mid-1970s and emphasized language comprehension from the start. These researchers tested the capabilities of two female bottlenose dolphins, Akeakamai and Phoenix, to understand instructions given them within artificial languages, including novel requests. Akeakamai was trained in a visually based language and Phoenix in an acoustically based language. Although the vocabularies of the two languages overlapped greatly, the grammars were different, a linear (left-to-right) grammar for Phoenix and an inverse (right-to-left grammar) for Akeakamai. Each dolphin was proficient in its particular language form, as described in detail in Herman *et al.* (1984).

After 1984, the researchers concentrated primarily on analyzing and expanding the language work with Akeakamai. In her language, "words" were produced by gestures of a person's arms and hands. The words referred to floating and fixed objects in the dolphin's habitat, to actions that could be taken to those objects, and to relationships that could be constructed between pairs of objects—taking one object to another or placing one object on top of or inside of another object. There were also location words, *left* and *right*, expressed relative to the dolphin's location, that were used to refer to a particular one of two objects having the same name, e.g., *left hoop* vs *right hoop*. And, there were gestural names for Phoenix, for the dolphin's own body parts, and for abstract concepts, like "mimic," "create," "negate," "tandem," and "question," most of which could be used in combinations with other gestural symbols.

Syntactic rules, primarily word-order rules, governed how words could be combined to extend meaning. For the gestural language, an inverse grammar was used in which the relation between the order of words in a sequence and the order in which response to them were taken were uncorrelated. This required that the dolphin attend to the whole sequence as a unit before responding, as later terms in the sequence could not be reliably predicted from earlier terms. The vocabulary, together with the syntactic rules, allowed for many thousands of unique combinations of words. Sentences up to five words in length could be created, with each sentence a unique instruction to the dolphin. The simplest sentences were instructions to take named actions to named objects. For example, a sequence of two gestures glossed as *surfboard over* directed the dolphin to leap over the surfboard; a sequence of three gestures glossed as *left Frisbee tail-touch* directed the dolphin to touch the Frisbee on her left with her tail. More complex sentences required the dolphin to construct a relationship between two objects, such as taking one named object to another named object or placing one named object in or on another named object. To interpret these "relational" sentences correctly, the dolphin had to account for both word meaning and word order. For example, a sequence of three gestures glossed as *person surfboard fetch* instructed the dolphin to bring the surfboard to the person (who was in the water), but *surfboard person fetch*, the same gestures rearranged, required that the person be carried to the surfboard. By incorporating *left* and *right* into these relational sentences, highly complex instructions could be generated. For example, the sequence of five gestures glossed as *left basket right ball in* asked the dolphin to place the ball on her right into the basket on her left. In contrast, the rearranged sequence *right basket left ball in* meant the opposite, now asking the dolphin to put the ball on her *left* into the basket on her *right*. The results (Herman *et al.*, 1984, Herman, 1986, 1987; Herman *et al.*, 1993b) showed that the dolphin was proficient at interpreting these various types of sentences correctly, as evidenced by her ability to carry out the required instructions, including instructions new to her experience. These were the first published results showing convincingly an animal's ability to process *both* semantic and syntactic information in interpreting language-like instructions.

As a test of Akeakamai's grammatical knowledge of the language she had been taught, Herman *et al.* (1993a) constructed *anomalous* gestural sentences. These were sentences that violated the syntactic rules of the language or the semantic relations among words. The researchers then studied the dolphin's spontaneous responses to these sentences. For example, the researchers compared the dolphin's responses to three similar gestural sequences: *person hoop fetch*, *person speaker fetch*, and *person speaker hoop fetch*. The first sequence is a proper instruction; it violates no semantic or syntactic rule of the learned language. It directs the dolphin to bring the hoop to the person, which the dolphin does easily. The second sequence is a syntactically correct sequence but is a semantic anomaly inasmuch as it directs the dolphin to take the underwater speaker, which is firmly attached to the tank wall, to the person. The dolphin typically rejects sequences like this by not initiating any action. The final sequence is a syntactic anomaly in that there is no sequential structure in the grammar of the language that provides for three object names within a sequence. However, embedded in the four-item anomaly are two semantically and syntactically correct three-item sequences: *person hoop fetch* and *speaker hoop fetch.* The dolphin, in fact, typically extracts one of these subsets and carries out the instruction implicit in that subset by taking the hoop to the person or to the underwater speaker. These different types of responses revealed a rational analysis of the sequences. Thus, the dolphin did not terminate her response when an anomalous initial sequence such as *person speaker* was first detected. Instead, she continued to process the entire sequence, apparently searching mentally backward and forward for proper grammatical structures as well as for proper semantic relationships, until she found something she could act on or not. This analytic type of sequence processing is characteristic of sentence processing by human listeners, as illustrated by so-called "garden-path" sentences that lead the listener into an initially false assumption about meaning, as they wend their way through a sentence, such, for example, "Fat people eat accumulates." For a fuller exposition of dolphin grammatical competencies, see Herman and Uyeyama (1999).

Herman and Forestell (1985) tested Akeakamai's understanding of symbolic references to objects that were not present in her habitat at that time. They constructed a new syntactic frame consisting of an object name followed by a gestural sign glossed as *"question."* For example, the two-item gestural sequence glossed as *basket question* asks whether a basket is present in the habitat. The dolphin could respond *yes* by pressing a paddle to her right or *no* by pressing a paddle to her left. Over a series of such questions, with the particular objects present being changed over blocks of trials, the dolphin responded to approximately 80% of the questions, with correct "absent" responses (83.3%) and correct "present" responses (79.6%) showing no significant statistical difference. These results gave a clear indication that the dolphin understood the gestures assigned to objects referentially, i.e., that the gestures acted as symbolic references to those objects. Further tests revealed the dolphin's creativity with language tasks. When the dolphin was asked to construct a relationship between two named objects, such as "bring the Frisbee to the hoop," but the destination object, the hoop, was absent, she spontaneously created a new response, by taking the indicated transport object (e.g., the Frisbee) to the "no" paddle. This response communicated her knowledge of the presence of the transport object but the absence of the destination object (Herman *et al.,* 1993b).

The television medium can display scenes that are representations of the real world, or sometimes of imagined worlds. As viewers, we understand this and often respond to the displayed content similarly to how we might respond to the real world. We of course understand that it is a representation and not the real world. It appears, however, that an appreciation of television as a representation of the real world does not come easily to animals, even to apes. Sue Savage-Rumbaugh (1986) wrote in her book, "Ape Language," that chimpanzees show at most a fleeting interest in television, and that from their behavior it was not possible to infer that they were seeing anything more than changing patterns or forms. Her own language-trained chimpanzee subjects, Sherman and Austin, learned to attend to and interpret television scenes only after months of exposure in the presence of human companions who reacted to the scenes by exclaiming or vocalizing at appropriate times. Herman *et al.* (1990) tested whether the dolphin Akeakamai might respond appropriately to language instructions delivered by a trainer whose image was presented on a television screen. Akeakamai had never previously been exposed to television of any sort. Then, for the first time, the researchers simply placed a television monitor behind one of the underwater windows in the dolphin's habitat and directed Akeakamai to swim down to the window. On arriving there she saw an image of the trainer on the screen. The trainer than proceeded to give Akeakamai instructions through the familiar gestural language. The dolphin watched and then turned and carried out the first instruction correctly and also responded correctly to 11 of 13 additional gestural instructions given to her at that same testing session. In further tests, Akeakamai was able to respond accurately even to degraded images of the trainer, consisting, for example, of a pair of white hands moving about in black space. The overall results suggested that Akeakamai spontaneously processed the television displays as representations of real-world trainers and of the gestural language she had been exposed to for many years previously.

### E. Studies of Language Production

As noted earlier, studies of language development in humans have shown that comprehension both precedes and exceeds production. A few studies have examined the dolphins' ability to produce symbols that are associated with objects, locations, and outcomes. Richards *et al.* (1984) showed that the dolphin Akeakamai could learn to vocally produce the appropriate acoustic label (in Phoenix's acoustic language system) for each of several objects presented to her. Reiss and McCowan (1993) developed a keyboard with six visual symbols as keys. Each key was associated with a particular acoustic associate and also with a particular outcome (e.g., the presentation of a specific toy like a ball or ring, a reinforcing action like the trainer rubbing the dolphin, etc.). After striking a key or producing the associated sound themselves, a human agent provided the dolphins with the associated outcome. Two of four dolphins showed consistent use of the keyboard in this manner, producing the sounds that gained them the associated item. Finally, Xitco *et al.* (2001) also showed that a dolphin could learn to use an underwater electronic keyboard to make requests of humans, in this case divers present in the dolphin's habitat. Keyboard symbols referred to objects or locations. For example, on seeing fish available at a particular location, the dolphin could press the associated key, alerting the diver to accompany it to that location and to retrieve and offer the fish that was there.

Labeling or naming an object is a necessary but not sufficient act to contend that the name is understood referentially, as a surrogate for or a reference to that object. Further tests for productive naming would be needed, such as establishing that the animal understands the name as a reference to an object even if that object is absent (*cf.* Herman and Forestell, 1985).

### F. Sea Lion Language Competencies

The only other marine mammal that has been trained and tested in a language-learning paradigm is the California sea lion (*Zalophus californianus*). Schusterman and Krieger (1984) tested whether a sea lion named Rocky might be able to learn to understand sentence forms similar to those understood by the dolphin Akeakamai. Rocky was able to carry out gestural instructions effectively for simpler types of sentences requiring an action to an object. The object was specified by its class membership (e.g., "ball") and, in some cases, also by its color (black or white) or size (large or small). In a later study, Schusterman and Gisner (1988) reported that Rocky was able to understand relational sentences requiring that one object be taken to another object. These reports suggested that the sea lion, like the dolphin, was capable of semantic processing of symbols and, to some degree, of syntactic

**L**

processing. As noted by Herman (1989), however, a shortcoming of the sea lion relational work was the absence of contrasting terms for relational sentences such as the distinction between "fetch" (take to) and "in" (place inside of or on top) that was easily understood by the dolphin Akeakamai. Without that contrast the symbol is not necessary and likely not processed semantically, as the sentence form itself tells the animal what to do. Additionally, unlike the dolphin, the sea lion's string of gestures were given discretely, each gesture followed by a pause during which the sea lion looked about to locate specified objects before being given the next gesture in the string. In contrast, gestural strings given to the dolphin Akeakamai were without pause, analogous to the spoken sentence in human language. Further, Rocky did not show significant generalization across objects of the same class (e.g., different balls). Unlike the dolphin, Rocky seemed to regard a gesture as referring to a particular exemplar of the class rather than to the entire class. Thus, although many of the responses of the sea lion resembled those of the dolphin, the processing strategies of the two seemed different, and the concepts developed by the sea lion appeared to be more limited than those developed by the dolphin.

### G. Conclusions

Human language, though unique as a whole, may depend in part on general cognitive processes that are also available to with some nonhuman animals. Language, like other biological traits, is best understood as having its genesis and development within the rules of Darwinian evolution, so that there should be some features that are homologous with features found in other hominoid species, and analogous to features found in some cognitively advanced nonprimate mammals. There is ample evidence for advanced cognitive skills in dolphins, as reviewed earlier. Dolphins can learn concepts and abstract rules and, like apes, are able to learn symbolic systems that have language-like features, particularly semantic and syntactic features. Sea lions can attain the semantic features, but their competency for acquiring syntactic rules has not been well studied. There is evidence that the artificial symbols used in language-like tasks with dolphins are understood referentially, as surrogates for or references to real-world objects, actions, or relationships. There is no evidence for a natural language in dolphins. Although whistle vocalizations have long been targeted as a primary means of vocal communication for bottlenose dolphins, their complexity and function are issues of intense dispute and study. At the same time, vocal learning and vocal mimicry is well developed in dolphins, as illustrated by laboratory behavioral studies and by studies and observations in the wild of vocal imitation by group-specific dialects in some cetacean species.

Finally, the results of the language work with the bonobo chimpanzee and the dolphin show some similarities, especially in the receptivity of the animals to the language formats used and in their proficiency at responding to sequences of symbols. The similarities in these language skills are mirrored by many other convergent cognitive traits in dolphin and in primate (Marino, 2002). An early conclusion by Herman (1980, p. 421) still seems appropriate to accommodate these convergent cognitive and language-learning skills of ape and dolphin: "The major link that cognitively connects the otherwise evolutionarily divergent delphinids and primates may be social pressure—the requirement for integration into a social order having an extensive communication matrix for promoting the well-being and survival of individuals."

### See Also the Following Articles

Brain size, Evolution ■ Communication ■ Intelligence and Cognition

## References

Au, W. W. L. (1993). "The Sonar of Dolphins." Springer-Verlag, New York.

Bastian, J. (1967). The transmission of arbitrary environmental information between bottlenose dolphins. In "Animal Sonar Systems" (R. G. Busnel, ed.), 2, pp. 803–853. Laboratoire de Physiologie Acoustique, Jouy-en-Josas.

Bastian, J., Wall, C., and Anderson, C. L. (1968). "Further investigation of the transmission of arbitrary information between bottlenose dolphins." Naval Undersea Warfare Center, San Diego, pp. 1–40, TP 109.

Bloom, L. (1974). Talking, understanding, and thinking: Developmental relationship between receptive and expressive language. In "Language Perspectives—Acquisition, Retardation, and Intervention" (R. L. Schiefelbusch, and L. L. Lloyd, eds), pp. 285–311. University Park Press, Baltimore.

Caldwell, M. C., and Caldwell, D. K. (1965). Individualized whistle contours in bottlenose dolphins (*Tursiops truncatus*). *Nature Lond.* **207**, 434–435.

Chomsky, N. (1972). "Language and Mind." Harcourt Brace Jovanovich, New York.

Chomsky, N. (1975). "Reflections on Language." Pantheon Books, New York.

Connor, R. C. (2007). Dolphin social intelligence: Complex alliance relationships in bottlenose dolphins and a consideration of selective environments for extreme brain size evolution in mammals. *Phil. Trans. R. Soc. B.* **362**, 587–602.

Connor, R. C., Wells, R., Mann, J., and Read, A. (2000). The bottlenose dolphin: Social relationships in a fission–fusion society. In "Cetacean Societies: Field Studies of Whales and Dolphins" (J. Mann, R. C. Connor, P. Tyack, and H. Whitehead, eds), pp. 91–126. University of Chicago Press, Chicago.

D'Amato, M. R. (1973). Delayed matching and short-term memory in monkeys. In "The Psychology of Learning and Motivation: Advances in Research and Theory" (G. H. Bower, ed.), 7, pp. 227–269. Academic Press, New York.

Evans, W. E., and Bastian, J. (1969). Marine mammal communication: Social and ecological factors. In "The Biology of Marine Mammals" (H. T. Andersen, ed.), pp. 425–475. Academic Press, New York.

Gunderson, V. M., Rose, S. A., and Grant-Webster, K. S. (1990). Cross-modal transfer in high- and low-risk infant pigtailed macaque monkeys. *Dev. Psychol.* **26**, 576–581.

Hauser, M. D., Chomsky, N., and Fitch, W. T. (2002). The faculty of language: What is it, who has it, and how did it evolve? *Science* **298**, 1569–1579.

Herman, L. M. (1980). Cognitive characteristics of dolphins. In "Cetacean Behavior: Mechanisms and Functions" (L. M. Herman, ed.), pp. 363–429. Wiley Interscience, New York.

Herman, L. M. (1986). Cognition and language competencies of bottlenose dolphins. In "Dolphin Cognition and Behavior: A Comparative Approach" (R. J. Schusterman, J. Thomas, and F. G. Wood, eds), pp. 221–251. Lawrence Erlbaum Associates, Hillsdale.

Herman, L. M. (1987). Receptive competencies of language trained animals. In "Advances in the Study of Behavior" (J. S. Rosenblatt, C. Beer, M. C. Busnel, and P. J. B. Slater, eds), 17, pp. 1–60. Academic Press, Petaluma.

Herman, L. M. (1989). In which Procrustean bed does the sea lion sleep tonight? *Psychol. Rec.* **39**, 19–50.

Herman, L. M. (2002). Vocal, social, and self-imitation by bottlenosed dolphins. In "Imitation in Animals and Artifacts" (C. Nehaniv, and K. Dautenhahn, eds), pp. 63–108. MIT Press, Cambridge.

Herman, L. M. (2006). Intelligence and rational behaviour in the bottlenosed dolphin. In "Rational Animals?" (S. Hurley, and M. Nudds, eds), pp. 439–467. Oxford University Press, Oxford.

Herman, L. M., and Forestell, P. H. (1985). Reporting presence or absence of named objects by a language-trained dolphin. *Neurosci. Biobehav. Rev.* **9**, 667–691.

Herman, L. M., and Morrel-Samuels, P. (1990). Knowledge acquisition and asymmetries between language comprehension and production: Dolphins and apes as a general model for animals. In "Interpretation

L

# Male Reproductive Systems

## SHANNON ATKINSON

Studies of the reproductive biology of male marine mammals have not received the attention that has been focused on their female counterparts. In part this is due to the limited numbers of male marine mammals kept in captivity, but also to the difficulty of measuring the anatomical and physiological parameters of free-ranging males. This account focuses on the anatomy of male reproduction, with emphasis on features unique to the various marine mammal groups. Aspects of reproductive life history and behavior, such as endocrine systems, mating systems, territoriality, and sociobiology, are covered elsewhere in this encyclopedia. The effects of environmental and pharmacological factors on reproduction are also briefly reviewed here.

Genitalia are internal in whales, dolphins, and porpoises (Fig. 1; Ommanney, 1932; Harrison Matthew, 1950; Green, 1972; Perrin *et al.*, 1984; and Kenagy and Trombulak, 1986); this contributes to hydrodynamic efficiency. There is no os penis or baculum. The penis is fibroelastic, similar to that in artiodactyls, such as cows, pigs, and antelopes. Cetacea are included in the Artiodactyla, or Cetartiodactyla, by some taxonomists (Slijper 1966; Berta and Sumich, 1999; Rommel *et al.*, 2007). It originates in two crura from the caudal part of the free pelvic bones or from the entire surface of

these bones (Fig. 2; Rommel *et al.*, 2007). The two arms fuse into a long rope-like body, round or oval in cross section. In large rorquals it may be 2.5- to 3-m long and 25–30 cm in diameter. The distal part of the penis tapers smoothly to the tip and is covered with ordinary skin; this may be homologous with the glans penis of some terrestrial mammals. When retracted the penis rests in an S-shaped horizontal loop. Because enlargement through engorgement with blood is limited by the tough tunica albiginosa (modified skin covering) during arousal, this loop allows protrusion of the organ without its lengthening (Ommanney, 1932; Perrin *et al.*, 1984; Rommel *et al.*, 2007). A flat retractor penis muscle runs from its ligamental or rectal wall origin to insert on the ventral surface of the penis; it serves to withdraw the penis back into the penile slit. The prostate is primitive among eutherian mammals, resembling that of marsupials and monotremes in consisting of diffuse urethral glands unlocalized to form a discrete prostate gland.

The testes are intra-abdominal (or cryptic) and mesial/ventral in position, a condition known as secondary testicondy (Rommel *et al.*, 2007). This feature is thought to be synapomorphic in Cetacea but is shared with some marsupials. The position varies among the cetacean taxa, from nearly renal in some odontocetes to ventral in baleen whales (Ommanney, 1932). In *Mesoplodon* spp., the organs are sunk in recesses of the abdominal cavity connected to the main cavity by short vagioperitoneal canals. The organs are long and cylindrical with a smooth shiny white surface. Seminiferous tubules located in the testicle lobe are mostly composed of Sertoli cells. (Honma *et al.*, 2004). In Bryde's and Minke whales, the seminiferous tubules of mature whales only contained a single layer of spermatogonia (Watanabe *et al*, 2004). Relative testis size varies widely among the cetacean groups, being greater in those species thought to engage in sperm competition (Mate *et al.*, 2005), such as right whales (*Eubalaena* spp.) than in polygynous species such as the sperm whale (*Physeter macrocephalus*). Odontocetes in general have testes 7–25 times larger than would be predicted for "average" mammals of their size, ranging to 8% of body weight in the dusky dolphin (*Lagenorhynchus obscurus*), as compared to 0.08% in humans. The largest mammalian testes known are found in right whales, weighing up to 1000 kg (Mate *et al.*, 2005). Testicular activity (and size of some muscles associated with the penis) varies seasonally, and weight increases substantially with breeding

M

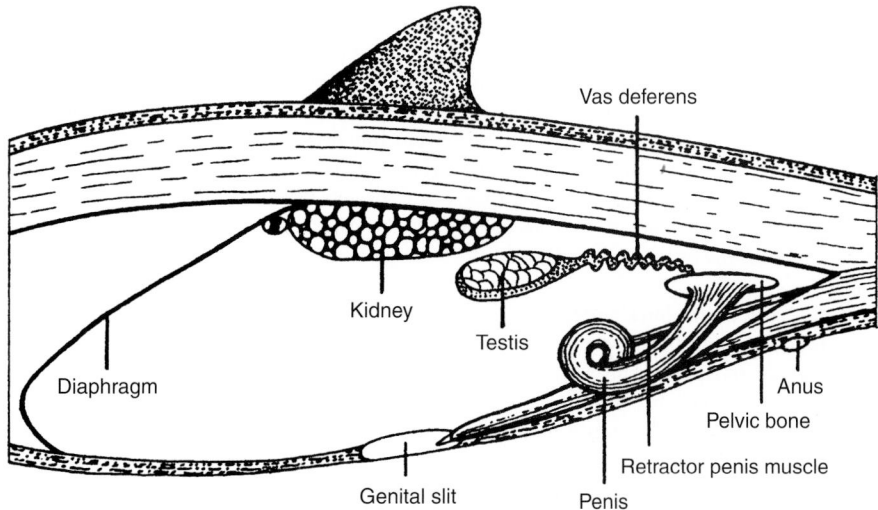

**Figure 1** *Cetacean male reproductive system. From Berta and Sumich (1999), adapted from Slijper (1966).*

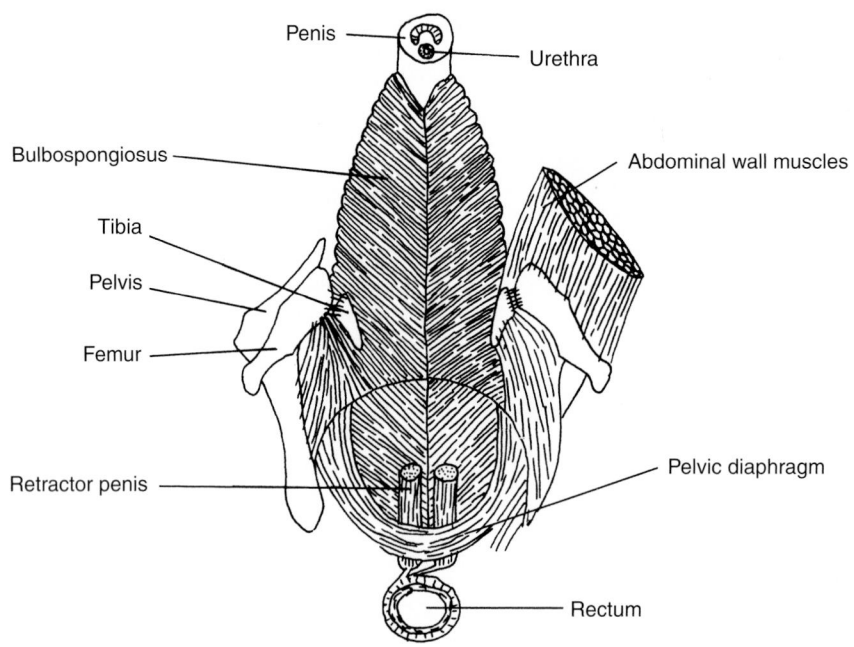

**Figure 2** *Pelvis and male genitals of bowhead whale* (Balaena mysticetus) *in ventral view (top of drawing is anterior). This figure shows the close relation between the internal hind limb bones (pelvis, femur, and tibia), the rectum, and the penis. Retractor penis and bulbospongiosus are penile muscles. Modified after Struthers (1893).*

season in at least some delphinid odontocetes (e.g., the spinner dolphin *Stenella longirostris*). In bowhead whales mean testis mass and mean testis length were correlated as was body length and mean testis mass (O'Hara *et al.*, 2002). The latter correlation had a break point at 12.5–13.0 m suggesting the onset of sexual maturation occurs at that size (O'Hara *et al.*, 2002). In many cetacean species, the spermatic tubules (vasa differentia) are more highly convoluted than in most terrestrial mammals. A distal spiral valve exists in the blue whale (*Balaenoptera musculus*), and mucosal folds have been described in the pygmy sperm whale (*Kogia breviceps*).

The scrotum functions in terrestrial mammals in part to lower ambient temperature to ensure viability of the sperm. Loss of the scrotum in Cetacea in the interest of hydrodynamic efficiency thus introduces a new thermoregulatory problem (Pabst *et al.*, 1998 and 1999). The rete testis is the site where numerous anastomosing tubules are located, whereby cooling through a cardiovascular countercurrent mechanism takes place (Honma *et al.*, 2004). Blood cooled in superficial veins of the dorsal fin and flukes feeds directly into a deep venous plexus closely juxtaposed to a similar arterial plexus that supplies the testis (Rommel *et al.*, 1992). Thus, heat is drawn into the venous blood from the arterial blood before it reaches the testis, cooling the testis to below body core temperature.

All pinnipeds, the polar bear (*Ursus maritimus*), and the sea otter (*Enhydra lutris*) possess a baculum, or penis bone, which is the ossified anterior end of the corpus cavernosum of the penis (Green, 1972; Morejohn, 1975; Atkinson, 1997). The baculum is largest in the walrus (*Odobenus rosmarus*) and smallest in otariid pinnipeds, or fur seals and sea lions. The distal shape varies widely among species and is more complex in phocids (true or earless seals) than in otariids (Morejohn, 1975). As most of the phocids are aquatic copulators, relatively large bacula may function in preventing water damage to sperm after copulation or to facilitate sperm competition in

species where the females mates with more than one male. Bacular size may also be adaptively constrained by a large body size in terrestrial copulators due to the risk of bacular fracture. The penis is vascular, as in terrestrial carnivores (Fig. 3). In phocids, the baculum is surmounted by a fleshy claviform glans. In otariids, it is covered by only a thin layer of epithelium. The prostate gland is bulky, weighing up to 760 g in the southern elephant seal (*Mirounga leonina*) and is similar in anatomy in all seals.

Phocid seals lack a scrotum; the testes are external to the abdominal muscles but covered by the posterior part of a superficial muscle (Laws and Sinha, 1993; Atkinson, 1997). Otariid pinnipeds possess a scrotum, but in some species, e.g., the Antarctic fur seal (*Arctocephalus gazella*), the testes are usually withdrawn into the inguinal position and the scrotum is visible only as two areas of hairless skin; the testes descend into the scrotum only during the need to avoid hyperthermia. Seasonal changes in testis and epididymis size and function occur with breeding season in pinnipeds in which this has been examined (Fig. 4). The testes in the sea otter and polar bear are scrotal (Ramsay and Sterling, 1988).

Sirenians do not possess a baculum. The penis is vascular and retracted when not engorged. The testes are abdominal. The prostate is lacking in the dugong (*Dugong dugon*) and is composed of erectile muscle tissue in the manatees (*Trichechus* spp. Marsh *et al.*, 1984).

Testosterone is the main androgen in male mammals and stimulates spermatogenesis (Atkinson, 1997; Atkinson and Yoshioka, 2007). Testosterone concentrations have been measured in many odontocetes and pinnipeds. In all species for which there are published data, testosterone concentrations increase around the time of sexual maturity, making it a useful diagnostic tool (Desportes *et al.*, 1994; Kita *et al.*, 1999; Kjeld *et al.*, 2003; Robeck and Monfort, 2006; Kjeld *et al.*, 2006; Atkinson and Yoshioka, 2007). A seasonal pattern of circulating testosterone concentrations exists with elevated concentrations during the

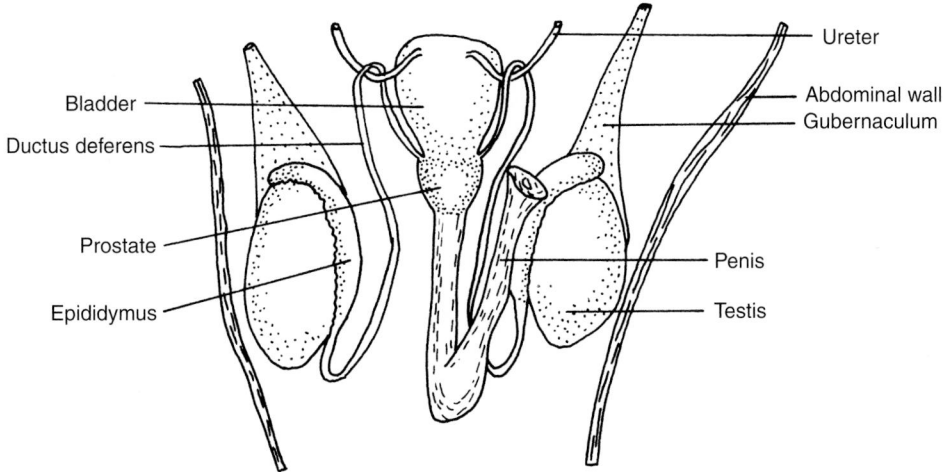

Figure 3    *Male reproductive tract of an otariid in ventral view (top is anterior). Modified after Boyd* et al. *(1999).*

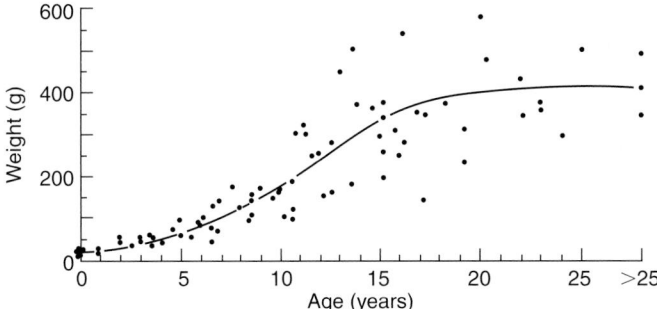

Figure 4    *Weight of nonspermiogenic testes of walrus in relation to age. Modified from Fay (1982).*

breeding season (typically in spring, but a few species are autumnal or multiseasonal breeders). In species with a short, tightly synchronized breeding season, testosterone concentrations are increased for 1–3 months at the start of the season but decline to baseline levels by the time breeding ends. Seasonality is also apparent in most male marine mammals in increased size of the testes and accessory reproductive glands (even muscles in some cetaceans) and increased spermatogenesis. Increased size of the testes is due to an increased diameter of the seminiferous tubules and epididymes, resulting in increases in the volume of sperm. Spermatogenesis usually lags behind testosterone production, as production of testosterone by testicular Leydig cells is necessary for germ cell differentiation in the seminiferous tubules. However, testis weight and circulating testosterone concentrations were correlated in minke whales (Suzuki *et al.*, 2001; Atkinson and Yoshioka, 2007).

A few marine mammals, such as dugongs and sea otters, lack a distinct breeding season. A few older male dugongs that were examined were found to be aspermic, suggesting long periods of sterility (Marsh *et al.*, 1984).

During seasonal quiescence in many cetaceans and pinnipeds, spermatogenesis ceases, although the testes retain relatively large seminiferous tubules with spermatocytes present (Watanabe *et al.*, 2004; Honma *et al.*, 2004). Shrinkage of anterior pituitary cells that produce gonadotrophins is thought to be ultimately responsible for the seasonal testicular regression.

Some environmental factors simply cue physiological events, whereas others have the potential to hasten or disrupt the system. The three most important are photoperiod, nutrition, and social factors.

Seasonal reproduction requires that males have adequate numbers of viable sperm when the females enter estrus. Hence the spermatogenic cycle must be initiated months before breeding (Atkinson, 1997; Atkinson and Yoshioka, 2007). Photoperiod is the most commonly cited environmental cue for synchronizing reproductive processes in both males and females; it appears to function months before the breeding season begins. The pineal gland is responsible for the neuroendocrine communication of photoperiod to the rest of the body (see Endocrine Systems). Melatonin secretion, which is activated during short photoperiods, acts to relay photoperiodic cues to the target organs. In many species, melatonin is inhibitory to the gonadotrophic-releasing hormones (GnRH) that stimulate testosterone production and spermatogenesis. Thus, reproductive processes in most species are stimulated during increasing daylength (i.e., spring). Conversely, increased melatonin concentrations due to a decreased photoperiod lead to inactivation of the testes (see Endocrine Systems).

Both sea otters and sirenians lack a defined breeding season, and the Australian sea lion (*Neophoca cinerea*) has a nonannual, nonseasonal reproductive pattern (Atkinson, 1997). Spermatogenesis in these species may be continuous. The lack of correlation between testicular activity and season in the dugong may relate to the absence of a pineal gland (Marsh *et al.*, 1984). No published studies have accounted for the lack of a defined breeding season in sea otters.

There is little published information of nutritional effects on the reproductive biology of male marine mammals. A high plane of nutrition is known to advance the onset of puberty in females and could be expected to have the same effect in males (Atkinson, 1997). It is also safe to assume that the plane of nutrition of an individual male will affect its position in a dominance hierarchy. For species in which there can be severe natural impacts on food resources, adult males may have lower blubber thickness during years of poor feeding, resulting in reduced stamina during the breeding season. Although the functional or mechanistic nature of the nutrition–reproduction relationship remains unclear, it can safely be concluded that the measurement of body condition and its effects on various reproductive events, especially during natural environmental perturbations, will continue to be important areas of marine mammal research.

**M**

Physiology and the environment influence the development of mating systems by affecting the relative distribution and availability of males and females, thereby altering the reproductive success of an individual male. After sexual maturation, serum testosterone concentrations may vary independent of testis weight, indicating that social factors play a role in reproductive processes (Atkinson and Yoshioka, 2007). It is not uncommon to find captive situations with cetaceans that are of the same age but at very different reproductive states (i.e., one or more males remain sexually immature much longer than the others). Changing the social structure in an enclosure will often stimulate puberty in those lagging behind in sexual development.

The most common reproductive reason for prescribing pharmacological agents is to reduce fertility. The three species for which this has been needed in captivity are the bottlenose dolphin (*Tursiops truncata*) the California sea lion (*Zalophus californianus*), and the harbor seal (*Phoca vitulina*). All of these can be prolific breeders in captivity, and the need regularly arises to control numbers in some facilities. Until recently, physical separation and contraception of females were the only practical methods. GnRH agonists has been demonstrated to be an effective, reliable, and reversible contraceptive in harbor seals (Siebert *et al.*, 2007). Antiandrogens have been tried unsuccessfully with bottlenose dolphins.

A second reason to prescribe pharmacological agents is to suppress aggression among males. The need to control behavior in the captive setting is obvious, especially with adult male bottlenose dolphins during the breeding season. It is less obvious, but equally, if not more important in the management of the Hawaiian monk seal (*Monachus schuainslandi*), a declining species in which males attempt mass matings, usually with a breeding-aged female, sometimes to the point of killing her (Atkinson *et al.*, 1994; Atkinson *et al.*, 1998).

GnRH agonists work by stimulating the anterior pituitary to release GnRH, which stimulates the testes to produce testosterone and initiate spermatogenesis. Paradoxically, the pituitary quickly becomes refractory and ceases its production of GnRH, which inhibits the testes. Injections of GnRH agonists have been used with some success with harbor seals and effectively decrease circulating testosterone concentrations to prepubertal levels in Hawaiian monk seals (Atkinson *et al.*, 1993; Atkinson *et al.*, 1998).

Marine top predators are likely targets for xenobiotic compounds that act either as estrogens or as antiandrogens. The most common of these are the polychlorinated biphenyls (PCBs) and dichlorodiphyenyltrichloroethanes (DDTs). These compounds bioaccumulate up the food chain, making marine mammals highly susceptible to their biological effects. Male marine mammals continue to accumulate organochlorines throughout their lives, whereas females tend to reduce their body burden via transplacental transfer and lactation (Willcox *et al.*, 2004; Meyers *et al.*, 2008). The range of PCB concentrations reported for arctic marine mammals is highest in the walrus (*Odobenus rosmarus*), although the absolute concentrations are highest in polar bears. The effects of organochlorines on male reproductive physiology have not been well studied, as most research has focused on females. The known effects are pathologies related to structural changes and thickening of tubules in organs such as the kidneys, adrenals, and reproductive tract. The most striking possible example has been the occurrence of pseudohermaphroditic polar bears with a normal vaginal opening, a small penis with baculum, and no Y chromosome (Ramsay and Sterling, 1988). The syndrome is hypothesized to be due to either excessive androgen secretion by the mother or endocrine disruption from environmental pollutants.

The impacts of all the detected pathologies are unknown. However, there are widespread reports that xenobiotic compounds are also strongly immunosuppressive, rendering contaminated animals more vulnerable to bacterial and viral infections. Experimental studies using minks (mustelids) indicate that the enzymatic pathways that metabolize steroids are disrupted, but the detailed biosynthetic pathways of the organismal response have not been elucidated.

## References

Atkinson, S., Gilmartin, W. G., and Lasley, B. L. (1993). Testosterone responses to a gonadotrophin-releasing hormone antagonist in Hawaiian monk seals (*Monachus schauinslandi*). *J. Reprod. Fert.* **96**, 35–39.

Atkinson, S., Ragen, T. J., Gilmartin, W. G., Becker, B. L., and Johanos, T. C. (1998). Use of a GnRH agonist to suppress testosterone in wild Hawaiian monk seals (*Monachus schuainslandi*). *Gen. Comp. Endo.* **112**, 178–182.

Atkinson, S. (1997). Reproductive biology of seals. *Rev. Reprod.* **2**, 175–194.

Atkinson, S., and Yoshioka, M. (2007). Endocrinology of reproduction. *In* "Reproductive Biology and Phylogeny of Cetacea" (B. G. M. Jamieson, ed.), Vol. 7, pp. 171–192. Science Publishers, Inc, Enfield, New Hampshire.

Atkinson, S., Becker, B. L., Johanos, T. C., Pietraszek, J. R., and Kuhn, B. C. S. (1994). Reproductive morphology and status of female Hawaiian monk seals fatally injured by adult male seals. *J. Reprod. Fert.* **100**, 225–230.

Berta, A., and Sumich, J. L. (1999). "Marine Mammals: Evolutionary Biology." Academic Press, San Diego.

Desportes, G., Saboureau, M., and Lacroix, A. (1994). Growth-related changes in testicular mass and plasma testosterone concentrations in long-finned pilot whale (*Globicephala melaena*). *J. Reprod. Fert.* **102**, 245–252.

Green, R. F. (1972). Observations on the anatomy of some cetaceans and pinnipeds. *In* "Mammals of the Sea, Biology and Medicine" (S. H. Ridgway, ed.), pp. 247–297. Charles C. Thomas, Springfield, IL.

Harrison Matthew, L. (1950). The male urigenital tract in *Stenella frontalis* (G. Cuvier) [=*S. attenuata*]. *Atlantide Rep.* **1**, 223–247.

Honma, Y., Ushiki, T., Hashizume, H., Takeda, M., Matsushi, T., and Honno, Y. (2004). Histological observations on the reproductive organs of harbor porpoises (*Phocoena phocoena*) incidentally caught in a set net installed off Usujiri, soughern Hokkaido. *Fish. Sci. Tokyo* **70**(1), 94–99.

Kenagy, G. J., and Trombulak, S. C. (1986). Size and function of mammalian testes. *J. Mammal.* **75**, 1–22.

Kita, S., Yoshioka, M., and Kashiwagi, M. (1999). Relationship between sexual maturity and serum and testis testosterone concentrations in short-finned pilot whales (*Globicephala macrorhynchus*). *Fish. Sci.* **65**, 878–883.

Kjeld, M., Olafsson, O., Vikingsson, G. A., and Sigurjonsson, J. (2006). Sex hormones and reproductive status of the North Atlantic fin whales (*Balaenoptera physalus*) during the feeding season. *Aqua. Mammal.* **33**(1), 75–84.

Kjeld, M., Vikingsson, G. A., Alfredsson, A., Olafsson, O., and Arnason, A. (2003). Sex hormone concentrations in the blood of sei whales (*Balaenoptera borealis*) off Iceland. *J. Cetacean Res. Mgmt.* **5**(3), 233–240.

Laws, R. M., and Sinha, A. A. (1993). Reproduction. *In* "Antarctic Seals" (R. M. Laws, ed.), pp. 228–267. Cambridge Univ. Press, Cambridge.

Marsh, H., Heinsohn, G. E., and Glover, T. D. (1984). Changes in the male reproductive organs of the dugong (Sirenia, Dugongidae) with age and reproductive activity. *Aust. J. Zool.* **32**, 721–742.

M

Mate, B., Duley, P., Lagerquist, B., Wenzel, F., Stimpert, A., and
Clapham, P. (2005). Observations of a female North Atlantic right
whale (*Eubalaena glacialis*) in simultaneous copulation with two
males: Supporting evidence for sperm competition. *Aquat. Mammal.*
**31**(2), 157–160.

Morejohn, G. V. (1975). A phylogeny of otariid seals based on morphology of the baculum. *Rapp. Pro. v. Reun. Cons. Int. Expl. Mar.* **169**,
49–56.

Meyers, M. J., Ylitalo, G. M., Krahn, M. M., Boyd, D., Calkins, D.,
Burkanov, V., and Atkinson, S. (2008). Organochlorine contaminants in endangered Steller sea lion pups (*Eumetopias jubatus*) from
western Alaska and the Russian Far East. *Sci. Total Environ.*, 396,
60–69.

O'Hara, T. M., George, J. C., Tarpley, R. J., Burek, K., and Suydam, R.
S. (2002). Sexual maturation in male bowhead whales (*Balaena mys-
ticetus*) of the Bering-Chukchi-Beaufort Seas stock. *J. Cetacean Res.
Mgmt.* **4**(2), 143–148.

Ommanney, F. D. (1932). The urino-genital system of the fin whale
(*Balaenoptera physalus*). *Disc. Rep.* **5**, 363–466, pls. 2–3.

Pabst, D. A., Rommel, S. A., and McLellan, W. A. (1998). Evolution
of thermoregulatory function in cetacean reproductive systems. *In*
"The Emergence of Whales" (J. G. M. Thewissen, ed.), pp. 379–397.
Plenum Press, New York.

Pabst, D. A., Rommel, S. A., and McLellan, W. A. (1999). The functional
morphology of marine mammals. *In* "Biology of Marine Mammals"
(J. E. Reynolds, III, and S. A. Rommel, eds), pp. 15–72. Smithsonian
Institution Press, Washington.

Perrin, W.F., Brownell, R.L., and DeMaster, D.P. (eds). (1984).
Reproduction in whales, dolphins, and porpoises. Int. Whal. Com.
Special Issue 6. Cambridge. 495 pp.

Ramsay, M. A., and Sterling, I. (1986). On the mating system of polar
bears. *Can. J. Zool.* **64**, 2142–2151.

Ramsay, M. A., and Sterling, I. (1988). Reproductive biology and ecology of female polar bears (*Ursus maritimus*). *J. Zool. Lond.* **214**,
601–634.

Robeck, T. R., and Monfort, S. L. (2006). Characterization of male killer
whale (*Orcinus orca*) sexual maturation and reproductive seasonality.
*Theriogenology* **66**(2), 242–250.

Rommel, S.A., Pabst, D.A., McLellan, W.A., Mead, J.G., Potter, C.W.
(1992). Anatomical evidence for a countercurrent heat exchanger
associated with dolphin testes. *Anat. Rec.* 232: 150–156.

Rommel, S. A., Pabst, A., and McLellan, W. A. (2007). Functional anatomy of the cetacean reproductive system, with comparisons to the
domestic dog. *In* "Reproductive Biology and Phylogeny of Cetacean"
(D. Miller, and B. G. M. Jamieson, eds), Vol. 7, pp. 171–192. Science
Publishers, Inc, Enfield, New Hampshire.

Siebert, U., Driver, J., Rosenberger, T., and Atkinson, S. (2007).
Reversible reproductive control in harbour seals (*Phoca vitulina*)
with a gonadotripin releasing hormone agonist. *Theriogenology* **67**,
605–608.

Slijper, E. J. (1966). Functional morphology of the reproductive system
in Cetacea. *In* "Whales, Dolphins, and Porpoises" (K. S. Norris, ed.),
pp. 277–319. Univ. of California Press, Berkely.

Suzuki, T., Mogoe, T., Asada, M., Miyamoto, A., Tetsuka, M., Ishikawa, H.,
Ohsumi, S., and Fukui, Y. (2001). Plasma and pituitary concentrations of gonadotropins (FSH and LH) in minke whales (*Balaenoptera
acutorostrata*) during the feeding season. *Theriogenology* **55**(5),
1127–1141.

Watanabe, H., Mogoe, T., Asada, M., Hayashi, K., Fujise, Y., Isikawa, H.,
Ohsumi, S., Miyamoto, A., and Fukui, Y. (2004). Relationship
between serum sex hormone concentrations and histology of seminiferous tubules of captured baleen whales in the western north Pacific
during the feeding season. *J. Reprod. Devel.* **50**(4), 419–427.

Willcox, M. K., Woodward, L. A., Ylitalo, G. M., Buzitis, J., and Atkinson,
S. (2004). Organochlorines in the free ranging Hawaiian monk seal
(*Monachus schauinslandi*) from French Frigate Shoals, North Pacific
Ocean. *Sci. Total. Environ.* **322**, 81–93.

# Management

Jay Barlow

Management refers to those regulations, laws, treaties, and policies which govern human interactions with marine mammals (Twiss and Reeves, 1999). Marine mammal management may promote a wide variety of human objectives: conservation of marine mammal populations for their intrinsic value, maintenance of marine mammal populations from human exploitation, protection of human health interests, humane treatment of captive animals, reduction of direct or competitive interference with commercial fisheries, and so on. This article concentrates on the general approaches used for marine mammal management.

## I. Management Units

"Management unit" refers to the group of animals that is the target of some management action. It may refer to a colony, subpopulation, population, or species. The term "stock" has traditionally been used instead of "management unit," but this term has evolved to be synonymous with both "population" and "management unit," so, to avoid confusion, "management unit" is preferred. The appropriate definition of a management unit depends on the management objective. Laws to prevent the extinction of a species might have a species or a subspecies as a management unit; however, the likelihood of achieving this management objective may be increased by managing on the basis of populations. Laws may not always explicitly define management units, but the stated goals of that law may give some clues as to how the term should be interpreted. For example, if the goal is to maintain marine mammal populations as functional elements of their ecosystems (one of the goals of the US Marine Mammal Protection Act, MMPA), management units might necessarily be smaller than the entire population to ensure that range contractions would not prevent the attainment of this goal (Taylor, 1997). Knowledge of population structure is critical to defining management units. Population structure has been studied using tagging, radio and satellite tracking, allozymes, DNA fingerprinting, DNA sequencing, photo-identification, morphometrics, and chemical markers. Most of these methods are limited—they can only show that two samples differ and thus that population structure is present, but they cannot be used to demonstrate that population structure is absent. There is almost always some uncertainty in deciding how finely to divide management units, and one of the current challenges in marine mammal management is dealing with this uncertainty.

## II. Methods of Marine Mammal Management

### A. Traditions, Taboos, and Practices

Prior to modern times, management took the form of culturally enforced practices. Ancient Greeks, natives in the Amazon Basin, and many sea-going cultures held dolphins in especially high regard and had proscriptions against killing or eating dolphins. Monk seals (*Monachus monachus*) were considered by the early Greeks to have prophetic powers and to be protected by Poseiden; however, the popular views toward this species included antipathy and hostility. The societies that did harvest whales and seals (including Inuit and Aleut cultures) often had elaborate rules that determined who could hunt these animals and when they could be hunted. It is not known whether traditions and taboos were important in conserving marine mammals, but there

**M**

is no evidence of marine mammal EXTINCTIONS caused by humans prior to that of the Steller's sea cow (*Hydrodamalis gigas*) in the 1700s. Traditions based on superstitions have been increasingly ignored as human populations have increased (Johnson and Lavigne, 1999).

## B. Harvest Bans

The most common method of protecting marine mammals from overexploitation has been a complete ban on harvesting. Most often, this has been practiced after a catastrophic decline that has already occurred. Gray whales (*Eschrichtius robustus*), northern elephant seals (*Mirounga angustirostris*), and Guadalupe fur seals (*Arctocephalus townsendi*) were protected by Mexico after their near extinction. A complete ban on whaling for gray whales and right whales (*Eubalaena* spp.) was instituted early in the history of international whale management. Australia, Mexico, New Zealand, South Africa, and the United States have banned the commercial harvest of all marine mammal species in their waters. The European Union members of ASCOBANS (Agreement on the Conservation of Small Cetaceans of the Baltic and North Seas) have agreed to ban the intentional harvest of all small cetaceans. Exceptions to harvest bans are commonly made for aboriginal or subsistence harvests and for incidental mortality pursuant to other commercial enterprises such as shipping and fishing.

## C. Age/Sex Limitations on Harvests

Age and/or sex limitations on harvest are commonly employed in the management of terrestrial species. The US North Pacific Fur Seal Act of 1910 outlawed the harvest of northern fur seal (*Callorhinus ursinus*) females and pups. This, together with the provisions of the North Pacific Fur Seal Treaty of 1911, effectively reversed the marked declines of the populations that breed on islands in the Bering Sea. Because many marine mammals do not exhibit marked sexual dimorphism (as do fur seals), similar regulations are not practical for all species. The 1931 Convention for Regulation of Whaling and later regulations of the International Whaling Commission (IWC) prohibited the commercial harvest of dependent calves and their mothers. Because most whales have a 2- or 3-year reproductive cycle and are nursing for only 6–12 months, females were not protected for the majority of time. Minimum size limits were also established for various whale species. These and other regulations were not effective in preventing the depletion of most of the world's whale populations.

## D. Seasonal Area Closures

The seasonal closure of certain areas or all areas is another common practice in wildlife management. The 1931 Convention for the Regulation of Whaling established a closed season for factory ships in Antarctic waters from April 7 to December 8. Seasonal area closures have also been used to reduce the number of gill net entanglements of Hector's dolphins (*Cephalorhynchus hectori*) in New Zealand and harbor porpoises (*Phocoena phocoena*) in the US Gulf of Maine.

## E. Restrictions on Methods and Fishing Gear

Regulations may limit the methods by which marine mammals are killed. The Fur Seal Treaty of 1911 eliminated the at-sea pelagic harvest of northern fur seals (*C. ursinus*), which were commonly considered to be wasteful (many carcasses could not be recovered) and which were more difficult to monitor. Methodological restrictions are not limited to direct, intentional harvests. Many gear restrictions have been applied to reduce marine mammal bycatch in commercial fishing operations. Finer mesh panels (Medina panels) were added to tuna purse seine nets in the eastern tropical Pacific to reduce dolphin ENTANGLEMENT. The use of acoustic warning devices (pingers) is required to reduce cetacean bycatch in several US fisheries. Similarly, pinger use in gillnet fisheries was recently mandated by the European Union in the North Sea and English Channel, but compliance and enforcement have been low. Regulations may also address how a particular gear is used; the adoption of a "backdown" procedure greatly reduced the mortality of dolphins in tuna purse seines. In addressing marine mammal bycatch problems, restrictions sometimes take the form of a complete ban on a particular gear type. In 1989 the states and territories of the South Pacific banned the use of large-scale (>2.5 km), drift gill nets in their exclusive economic zones, and in 1992 the United Nations General Assembly extended this ban to all international waters. Drift gill net fishing for tuna is banned in the Mediterranean. In the United States, Florida, Louisiana, Texas, and California have banned gill nets in all or part of their waters in response to marine mammal and other bycatch issues.

## F. Quota-Based Restrictions

The most direct method to manage removals from a wild population is to set a limit on the number of animals that can be taken in a given time period (usually 1 year). Quota-based management was first applied to the directed harvest of marine mammals and was later adapted to regulation of bycatch. This method requires some method for estimating annual mortality, such as from a mandatory program placing observers on whaling or fishing vessels. Whaling on the high seas has been regulated with quotas since 1931, but early quotas were designed only to limit oil production and were based on a "blue whale unit" [the oil equivalent of one blue whale (*Balaenoptera musculus*) being two fin whales (*B. physalus*), six sei whales (*B. borealis*), etc.]. The lack of species- or population-specific whale quotas lasted until 1972 and is widely blamed for the near extinction of most large whale populations. The failure of IWC to effectively manage whaling resulted in an international moratorium on commercial whaling that started in 1986 and continues today. Since 1986, the IWC has devised and adopted a revised management procedure that incorporates a new, well-tested catch limit algorithm (CLA) for setting population-specific quotas. Aboriginal subsistence whaling continues under population-specific quotas that are based on biological considerations and on "cultural and nutritional needs." Quotas were first used in 1976 to limit bycatch in the US tuna purse seine fishery to 78,000 dolphins per year. The US quotas gradually decreased to 20,500 by 1981, but, like the blue whale unit, still had not adequately addressed species- and population-specific conservation concerns. The gradual conversion of the tuna purse seine fishery from a US industry in 1970 to a largely international fleet by 1990 further complicated conservation efforts. The Inter-American Tropical Tuna Commission (IATTC) and several nongovernmental organizations have negotiated with IATTC signatory nations to impose vessel-specific quotas on total dolphin bycatch (1993) and stock-specific quotas (2000). Although the management of dolphin mortality in the tuna fishery has remained a special case, the United States has adopted a more general approach to setting stock-specific quotas on the maximum allowable levels (potential biological removal, PBR) of human-caused mortality for marine mammal populations in its exclusive economic zone. The PBR approach (Wade, 1998) is like the IWC's CLA in that it sets allowable removal rates that are conservative in the face of uncertainty but which can increase as uncertainties are resolved. New Zealand uses a similar approach to setting annual bycatch limits for Hooker's sea lions (*Phocarctos hookeri*) and has closed its squid trawl fishery when this limit was exceeded. Several countries, including the United Kingdom, are investigating similar approaches to setting limits on fishery bycatch.

### G. Market Monitoring and Trade Restrictions

Enforcement of laws on the high seas is often difficult or impossible; therefore, market monitoring and international trade restrictions may be necessary to prevent the illegal harvest and marketing of protected marine mammal. The Convention on International Trade in Endangered Species (CITES) is the primary implement for international trade restrictions and currently bans all trade in whale products, including some species that are not considered "endangered" but whose meat might be confused with that from endangered species. Genetic methods now can distinguish between all species, and CITES is under pressure from pro-whaling countries to lift the "look-alike" ban on non-endangered whales. Surreptitious market surveys by nongovernmental organizations and subsequent genetic analyses have shown that Japanese and Korean markets contain a wide variety of cetacean products (Baker *et al.*, 2002, 2006), many of which are mislabeled and some of which may have been illegally imported (some cetaceans taken within EEZ waters of those countries and whales taken under "scientific whaling" can be legally marketed). Because marine mammal products can be extremely valuable, there will be a strong incentive to cheat. Some IWC member countries are insisting that a system of market monitoring precedes the resumption of commercial whaling, possibly by genetically "fingerprinting" every legally taken whale.

### H. Treatment of Wild and Captive Animals

Marine mammals, especially cetaceans, are regarded by many cultures as deserving special treatment by humans. These attitudes may stem from their similarities to humans (large brain, play behavior, etc.), from their representation in popular media, or from the endangered status of some species. Whatever the reason, the special treatment is often evident in national laws that afford more protection for marine mammals than for similar terrestrial mammals. For example, the US MMPA prohibits "harassment" of marine mammals (defined as any pursuit, torment, or annoyance that has the potential to disrupt the natural behavioral patterns of the animal) unless a specific permit is obtained. Virtually any research on marine mammals (except passive observation) has a potential for harassing the subject and therefore requires an MMPA permit. National laws are also frequently implemented to regulate the public display of marine mammals to ensure that adequate space and care are provided to those animals. Some laws and regulations are expressions of public concern for individual animals (rather than concern about species or populations) and are derived more from the animal rights movement than from a conservation ethic, but this distinction is not clear in many cases. Stranding programs that rehabilitate beached animals may aid individuals and, for endangered species, the survival of the species.

### I. Marine Sanctuaries

There is a long-standing and growing interest in the use of protected areas or sanctuaries as a management tool for marine species. The first marine mammal refuge [for pinnipeds and sea otters (*Enhydra lutris*) on Afognak Island, Alaska] was established in 1892, but most have been established since 1975. Protected areas are a useful management tool because the concept is so simple (easy to understand and to enforce). For marine mammals, established sanctuaries and protected areas are taxonomically limited: the Indian Ocean and Southern Ocean whale sanctuaries (established by the IWC) protect only large whales, the Irish whale and dolphin sanctuary (established by Ireland) protects only cetaceans, and the Banks Peninsula sanctuary (established by New Zealand) was designed to protect only Hector's dolphins. The level of protection varies among sanctuaries; the Irish whale and dolphin sanctuary does not prohibit porpoise and dolphin bycatch in commercial fisheries (although the existence of the sanctuary has focused efforts on reducing bycatch). The utility of protected areas as a management tool depends critically on characteristics of the animals they are designed to protect (residency patterns, home ranges, mating strategies) and on the size of the protected area. The enormous Southern Ocean whale sanctuary (generally, all waters south of 40° but excluding the Indian Ocean sanctuary) is currently recognized as being too small to effectively protect its whales (which migrate out of this area during the southern winter). In contrast, small protected areas are quite effective in sheltering breeding colonies of pinnipeds or essential warm spring habitats of manatees. To conservation biologists, a "marine protected area" refers to an area of complete protection at all ecosystem levels. Existing marine protected areas are too small to afford much protection for marine mammal species, although they may protect some critical habitat.

### J. Pinniped Control Programs

The recovery of many pinniped populations from a legacy of hunting and near extermination is one of the success stories in marine mammal management, but this recovery is hardly viewed as a success by fishermen and aquaculturists who share their waters. Even conservationists are faced with a dilemma in some situations, such as when California sea lions (*Zalophus californianus*) (protected, but now numbering ~250,000) are threatening the survival of a depleted steelhead run in Washington State. Laws protecting marine mammals can and have been modified to deal with such small-scale problems by authorizing the lethal or captive removal of specific problem animals. In some areas, the use of acoustic harassment devices (AHDs) has been authorized to deal with the economic loss to seals by aquaculture facilities or commercial fishermen. Although some of these "fixes" appear to be successful in the short term, their long-term utility is questionable and there is concern about the impact of AHDs on other elements of the ecosystem. Even more controversial are programs designed to reduce entire pinniped populations by culling. Government-sanctioned culling programs to improve fisheries have been practiced in many countries, including Norway and the United States. In Canada, the high annual quota on harp seals (*Pagophilus groenlandicus*) and hooded seals (*Cystophora cristata*) is justified, in part, as a means to reduce seal predation on depleted cod stocks. This approach has been criticized on theoretical grounds because it oversimplifies ecosystem interactions; pinnipeds may feed on a commercially important fish species but may also feed on predators of that species. Management of culling programs would typically fall under national regulations, but the IUCN Marine Mammal Action Plan has established a protocol to evaluate culling proposals.

### K. Ecosystem Management

Ecosystem management refers to approaches ranging from simply considering the impact of a management decision on other elements of the ecosystem to the simultaneous optimization of management strategies to meet management goals of all elements of an ecosystem. There are no examples of the latter approach, although

M

Norway and the signatory nations of the Convention for the Conservation of Antarctic Marine Living Resources are pursuing this goal by promoting multispecies considerations in the management of marine mammal, fish, and seabird resources. Although it is unarguably true that improvements can be made in resource management by considering ecosystem interactions (sometimes called an ecosystem approach to management), it is also true that predicting the implications of even a simple ecosystem perturbation is far beyond our current capabilities. Significant progress in implementing ecosystem management may be left to future generations.

## III.  Trends in Marine Mammal Management

In recent years, there has been a movement toward management procedures that determine quotas for allowable harvests or incidental mortality based on rigid formulae. Both the IWC's CLA and the US PBR approaches are based on formulae that estimate the allowable removals from a management unit based on measurable attributes (such as estimated population size, population growth rates, catch histories, and the precision of the various estimates that are used). The advantage is that all parties can reach *a priori* agreement on the management objectives and on the rules that will be used to reach those objectives without divisive arguments about the effect on anyone's quota. Biological data are inherently imprecise and full of other uncertainties. For both CLA and PBR approaches, computer simulation studies were used to "tune" the quota formulae to achieve their goals even in the presence of imprecision and bias in available data (Wade, 1998). With the increasing emphasis on rigid quota-based management, debates about management practices are changing. Instead of concentrating on which values of biological parameters and which analytical models should be used, managers are now more concerned with how management units should be defined.

Coincident with the movement toward rigid formula-based quota schemes is an increasing reliance on direct approaches to measuring population parameters and a decreasing reliance on industry statistics, such as catch per unit effort. Advances in survey methodology (line-transect and mark–recapture) have greatly improved our ability to estimate the size of cetacean populations. Photo-identification studies, combined with mark–recapture analysis, have refined our understanding of marine mammal life history. Observer programs have increased the reliability of bycatch and harvest estimates. Satellite tagging and the recent revolution in molecular biology have contributed to an explosion of new information on the structure of marine mammal populations. Although all these recent trends promote the potential for effective marine mammal management, the real impediment to effective management is now the lack of collective willpower to implement regulations and to enforce existing regulations.

There has been increasing interests in applying the "precautionary principle" in marine mammal management. In the face of uncertainty, management decisions should be made to minimize the damage caused by being wrong. In most resource protection issues, there are two types of damage: the damage caused to a regulated industry by providing more marine mammal protection than is needed and the damage done to the populations and the industry by providing too little protection. A look at the catastrophic history of marine mammal management illustrates the disastrous economic and ecological results of management approaches that are not precautionary. One way to add precaution is to reverse the legal burden of proof to ensure that any action will *not* adversely affect the population before that action is permitted. Clearly, the future challenge is how to make marine mammal management appropriately precautionary.

## See Also the Following Articles

Conservation Biology ■ Genetics for Management ■ Population Status and Trends ■ Stock Identity ■ Whaling, Modern

## References

Baker, C. S., Dalebout, M. L., and Lento, G. M. (2002). Gray whale products sold in commercial markets along the Pacific coast of Japan. *Mar. Mam. Sci.* **18**, 295–300.

Baker, C. S., Lukoschek, , Lavery, S., Dalebout, M. L., Yong-un, M., Endo, T., and Funahashi, N. (2006). Incomplete reporting of whale, dolphin and porpoise "bycatch" revealed by molecular monitoring of Korean markets. *An. Cons.* **9**, 474–482.

Blix, A. S., Walløe, L., and Ulltang, Ø. (eds) (1994). "Whales, Seals, Fish, and Man." Elsevier, Amsterdam.

Butterworth, D. S., and Punt, A. E. (1999). Experiences in the evaluation and implementation of management procedures. *ICES J. Mar. Sci.* **56**, 985–998.

Johnson, W. M., and Lavigne, D. M. (1999). Monk seals in antiquity. *Netherlands Commission for International Nature Protection.*

Taylor, B. L. (1997). Defining "population" to meet management objectives for marine mammals. *In* "Molecular Genetics of Marine Mammals" (A. E. Dizon, S. J. Chivers, and W. F. Perrin, eds), pp. 49–65. Society for Marine Mammology.

Twiss, J. R., Jr., and Reeves, R. R. (eds) (1999). "Conservation and Management of Marine Mammals." Smithsonian Institution Press, Washington, DC.

Wade, P. R. (1998). Calculating limits to the allowable human-caused mortality of cetaceans and pinnipeds. *Mar. Mam. Sci.* **14**, 1–37.

# Manatees
## *Trichechus manatus, T. senegalensis, and T. inunguis*

### John E. Reynolds, III, James A. Powell and Cynthia R. Taylor

## I.  Characteristics and Taxonomy

The manatees (order Sirenia, family Trichechidae, subfamily Trichechinae) represent one of the most derived groups of extant mammals. Although ancestral forms were terrestrial, descendant forms have occupied aquatic habitats since the Eocene Epoch, providing the group with a long period of time over which to evolve. Apart from their suite of unusual morphological attributes (adaptations) associated with their herbivory and aquatic habitat, manatees have many behavioral and life history traits that are similar to those of other mammals. For most aspects of species biology, the Florida manatee is the best-studied taxon, and without data to the contrary, scientists assume that other manatees may be similar to the Florida subspecies.

The three species and their subspecies are *Trichechus inunguis,* Natterer, 1883 (*Amazonian manatee*); *Trichechus manatus,* Linnaeus, 1758 (West Indian manatee); *T. m. manatus* (Antillean manatee); *T. m. latirostris* (Florida manatee); and *Trichechus senegalensis,* Link, 1795 (West African manatee).

The generic name of the manatees, *Trichechus*, comes from the Greek words *trichos* (hair) and *ekh_* (to have), referring to the sparse body hairs and abundant facial hairs and bristles. *Inunguis* refers to the lack of nails on the pectoral flippers of the Amazonian manatee. At least two possible origins for *manatus* have been suggested: It could refer to the hand (*manus*) since manatees sometimes use their front, or pectoral, limbs to push food into their mouths. More likely, the term comes from the Carib Indian word *manati*, which means woman's breast, perhaps referring to the fact that the manatee's mammary glands are located in the axillary region in approximately the same anatomical location as the breasts of a human female; this particular anatomical feature contributed to the association of the manatee with the mythical mermaid. *Senegalensis* denotes that the West African species is found along the coast of Senegal, although it also occurs in waters of other west-central African countries.

Vernacular names for the manatees vary by region. In English-speaking areas, they are typically referred to as sea cows; similarly, in German, a manatee is referred to as a *Seekuh* or *Manati*, in Dutch as a *zeekoe*, in French as a *lamantin*, in Spanish as a *manati* or *vaca marina*, and in Portuguese as a *peixe-boi*, or ox fish. In some West African countries, a manatee is called a "*mamiwata*," which refers to a water deity. Diverse indigenous names are also in use in Africa and South America.

Unlike some marine mammals (e.g., pinnipeds, polar bears, and sea otters), manatees and the other living member of the order Sirenia, the dugong (*Dugong dugon*), are totally aquatic. They inhabit shallow waterways and feed primarily on plants, a diet that makes the sirenians unique among modern marine mammals.

Although manatees do not dive to great depths or for prolonged periods as many cetaceans and pinnipeds do, they are anatomically well adapted to aquatic habitats (Fig. 1). They lack hind (pelvic) limbs have reduced, paddle-like front (pectoral) limbs, and have fusiform (streamlined, spindle-shaped) bodies with few external protuberances and thick, tough skin, and are very large (an adaptation that facilitates heat conservation). Their heads are somewhat streamlined, and the nostrils are located on the dorsal side of the muzzle. A dorsal fin is lacking. Internally, manatees have extremely thick, heavy (pachyosteosclerotic) bones and an unusual arrangement of the

**Figure 1** *Although manatees may not swim as fast or dive as deep as some cetaceans do, manatees have the fusiform bodies, the reduced or absent limbs, and the powerful locomotory fluke that cetaceans also have. This particular animal has been fitted with a belt attached to a floating canister containing telemetry equipment. Photograph by Patrick Rose.*

diaphragm and lungs that facilitates buoyancy control. Manatees, like other marine mammals, have sensory and other adaptations that enhance diving, osmoregulatory, and thermoregulatory abilities.

Unusual adaptations accommodate the manatee's herbivorous diet. These include (1) enlarged lips (especially the upper lip) equipped with prehensile as well as tactile vibrissae and moved by a muscular hydrostat; (2) the presence of supernumerary (polydont) molariform cheek teeth that are replaced via horizontal migration along the jaws throughout the lifetime of each manatee; and (3) a greatly expanded gastrointestinal tract (specialized for hindgut fermentation, as in horses and elephants) with several unusual gross and microscopic features.

Manatees differ sharply from their close relative, the dugong. Manatees have a rounded fluke, whereas dugongs have split flukes similar to those of cetaceans. Dugongs have tusks, which manatees lack, and the mode of tooth replacement in the two differs. The rostrum of the dugong is much more sharply deflected downward than is the rostrum in any manatee species. In addition, dugong skin is smoother than is the case for West African and West Indian manatees.

The West Indian manatee is the largest living sirenian, with individuals approaching 1500 kg in weight and 4 m in length. Females tend to be somewhat larger than males, but body size cannot be used to determine either the sex or the age of an individual. West Indian manatees are euryhaline (can tolerate both salt and freshwater) but may require periodic access to fresh water to drink. West African manatees are generally very similar to West Indian manatees in terms of their size, general body form, and habitat, but the West African manatee has a blunter snout, somewhat protruding eyes, and a slightly more slender body. The Amazonian manatee is the smallest trichechid, measuring about 3-m long or less and weighing less than 500 kg. Its "rubbery" skin is smoother than that of its congeners, and it lacks nails at the tips of the pectoral flippers, which are proportionately longer than in the other species. In addition, white or pink belly patches are common. The Amazonian manatee is confined to freshwater habitats.

Manatee species also vary in the degree of rostral deflection, corresponding to the predominate location in the water column of food plants in their natural habitats. West African manatees have the least deflected snouts, and Florida manatees the most deflected.

The only species for which subspecies have been identified is the West Indian manatee. The two subspecies differ most obviously in their skeletal (especially skull) morphology.

However, Garcia-Rodriguez *et al.* (1998) examined mitochondrial DNA control regions from 86 individual West Indian manatees from 8 different locations. They found 15 different haplotypes that could be clustered into three, rather than two, distinctive lineages for the species. These authors also noted for three presumed West Indian manatees from Guyana that their mtDNA haplotypes were more consistent with that of the Amazonian manatee. Vianna *et al.* (2006) sequenced the mitochondrial DNA control region from 330 *Trichechus* to compare phylogeographic patterns. In *T. manatus* three haplotype clusters were identified showing distinct spatial distribution. A single expanding population cluster was observed for *T. inunguis*. Analysis revealed a hybrid between the *T. manatus* and *T. inunguis* species at the mouth of the Amazon River in Brazil, extending to the Guyanas and possibly to the mouth of the Orinoco River, Venezuela. Systematics of manatees is a topic that requires additional study.

Sirenians probably arose in the Old World (Eurasia and/or Africa) not later than the early Eocene Epoch, 50–55 million years ago. The oldest fossils are from Jamaica. Within a few million years (i.e., in the

**M**

middle Eocene, 45–50 million years ago), several genera of sirenians existed. Peak diversity of sirenians occurred during the Oligocene and Miocene Epochs (5–35 million years ago).

The first truly manatee-like (i.e., trichechine) sirenian was *Potamosiren*, fossils of which are about 15 million years old (Miocene of Colombia). During the Pliocene Epoch (about 2–5 million years ago), trichechids also inhabited the Amazon Basin and the Caribbean. The Amazonian trichechids gave rise to the Amazonian manatee, and the Caribbean trichechids are thought to have given rise to the West Indian and West African manatees, which are sister taxa.

Due at least in part to their dense bones, sirenians in general are well preserved in the fossil record, but true manatees are rare until the Pleistocene.

Various lines of evidence (e.g., genetic analyses, electrophoresis of serum proteins, and morphological studies) suggest that the order Sirenia (manatees and dugong) is most closely related to a group of mammalian orders called the Paenungulata. The extant paenungulates include the elephants (order Proboscidea) and hyraxes (order Hyracoidea). The sirenians appear to be most closely related to elephants and the extinct, hippopotamus-like desmostylians.

## II. Distribution and Abundance

All extant manatees occupy subtropical and tropical waters (Figs. 2–5).

### A. West Ind ian manatee, *T. manatus*

This species occupies coastal and riverine habitats from the mid-Atlantic region of the United States, throughout the wider Caribbean Sea and Gulf of Mexico, and into coastal parts of northeastern and central-eastern South America. The Florida manatee, *T.m. latirostris*, occurs from eastern Texas to Virginia in the summer with occasional sightings as far north as Massachusetts, but occupies waters of Florida and southeastern Georgia year-round (Fig. 2). Although its distribution is not continuous, the Antillean manatee, *T.m. manatus*, occupies the remainder of the species' range, from southwestern Texas to South America. It occupies the waters of 19 countries (Fig. 3). The range of the Antillean manatee may overlap with that of the Amazonian manatee around the mouth of the Amazon River (Fig. 4).

Scientists estimate that there may be 3000 or more Florida manatees. Some recent analyses of population trends of manatees

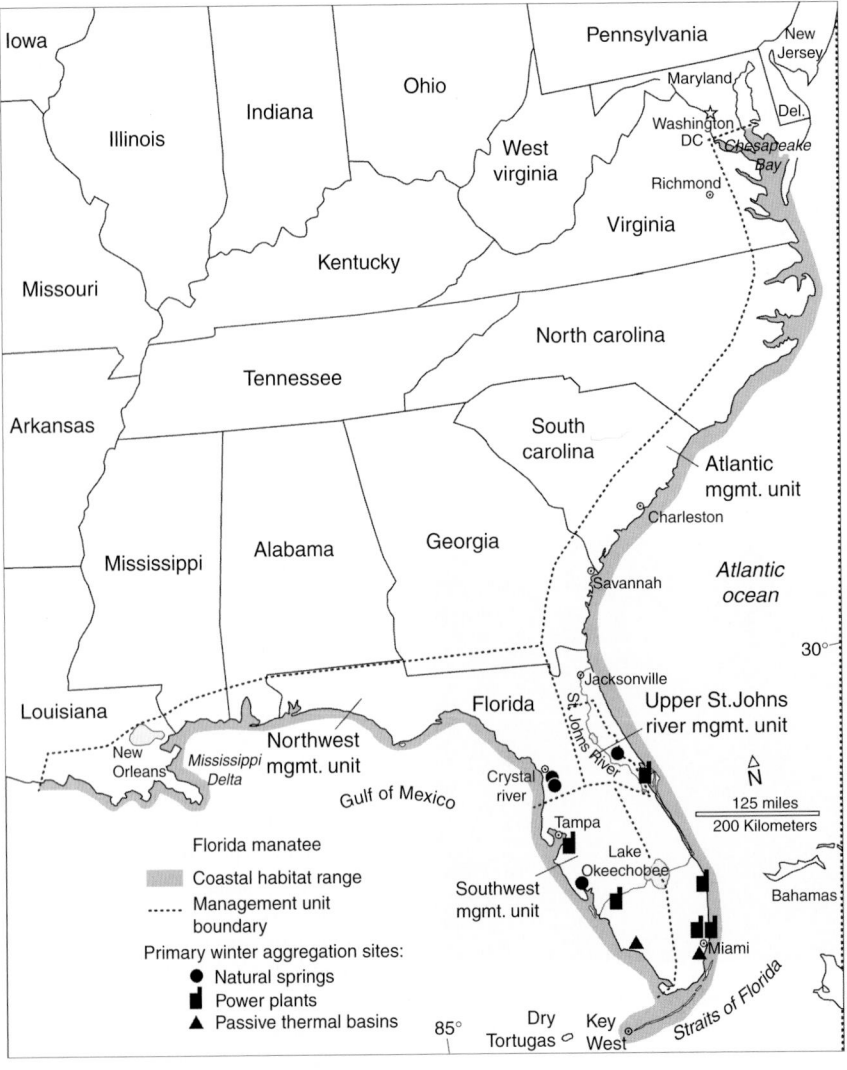

**Figure 2** *Range of the Florida manatee,* T.m. latirostris. *Map produced by Ellen McElhinny.*

United states
Gulf of Mexico
Mexico
Atlantic Ocean
Bahamas
Turks and caicos is.
Cuba
Dominican republic
Cayman Islands
Haiti
Belize
Jamaica
Guatemala
Puerto rico
Honduras
Caribbean Sea
El salvador
Nicaragua
Antilles
Costa rica
Panama
Venezuela
Trinidad & tobago
Pacific Ocean
Colombia
Guyana
Surinam
French guiana
Equator
Equador
Atlantic Ocean
Peru
Brazil
Bolivia
N
500 miles
800 kilometers
60°
Antillean Manatee
Principal Habitat Range
Vagrant Sightings of Individuals
Collaborators:
Mexico - Benjamin Morales, Janneth Padilla
Guatemala - Ester Quintana
Belize - Nicole Auil
Honduras - Daniel Conzalez
Nicaragua - Igancio Jiménez
Costa rica - Igancio Jiménez
Panama - Lenin riquelme, Yolanda Matamoros
Colombia - Dalila Caicedo
Venezuela - Adda Manzanilla, Jaime Bolaños, Esmeranda Mujica
Trinidad - Alêsha Naranjit
Suriname - Benoit de Thoisy
French Guiana - Benoit de Thoisy
Brazil - Carolina Alvite, Regis Pinto
Bahamas - James Reid
Turks and Caicos - Brian Riggs
Cuba - Jose Antonio santos
Dominican Republic - Enrique Pugibet, Monica vega, Haydee Dominguez
Puerto rico - Antonio Mignucci
Virgin Islands - Antonio Mignucci
Curacao - Adolphe Debrot
Cayman Islands - Janice Blumenthal

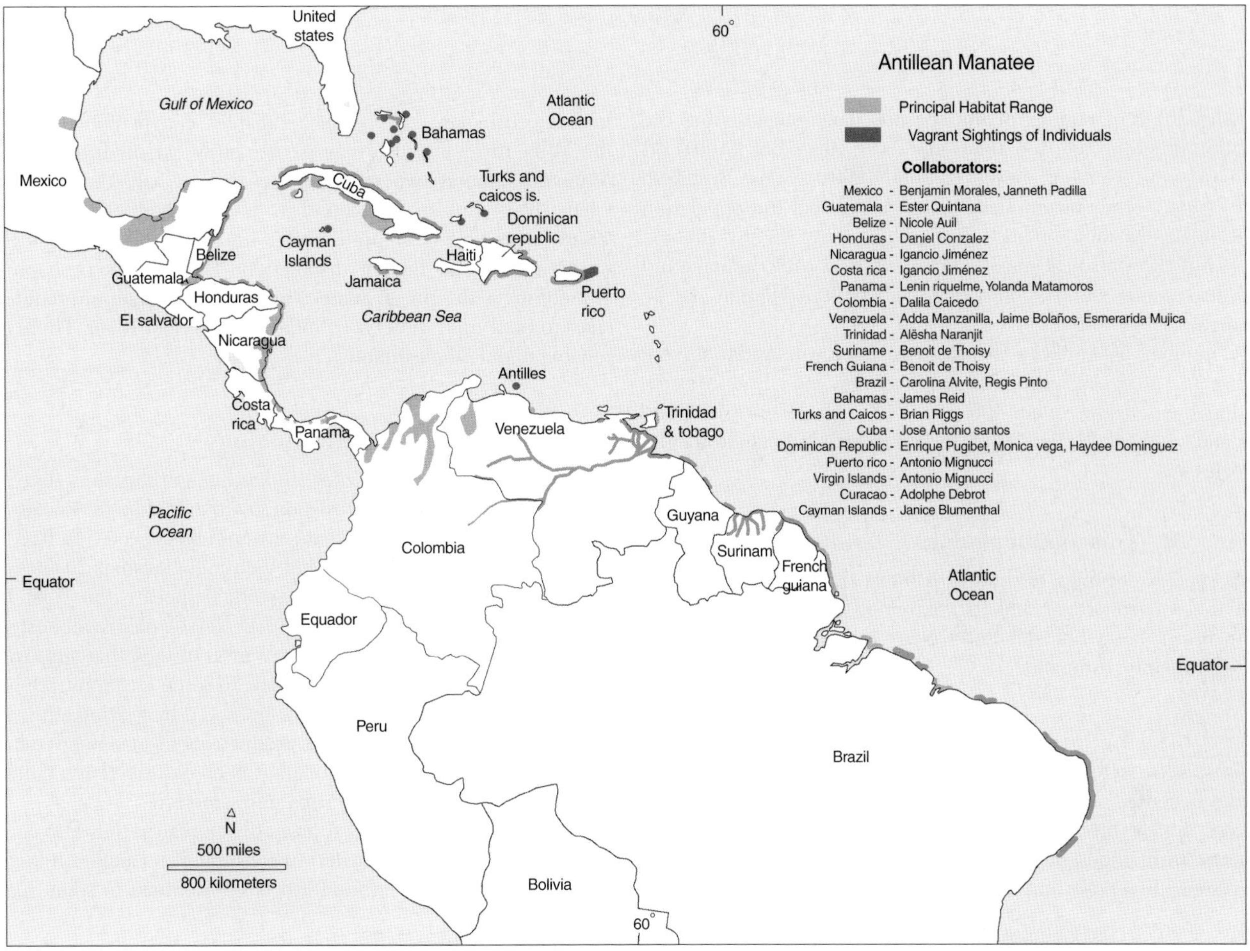

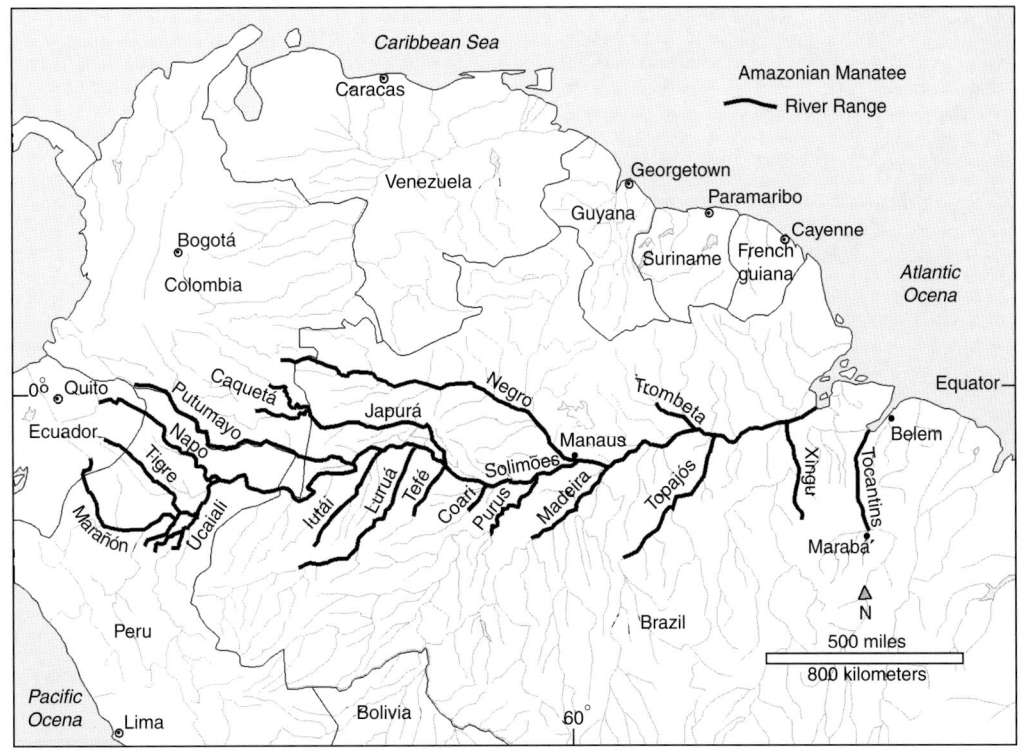

**Figure 3** *Range of the Antillean manatee,* T.m. manatus. *Map produced by Ellen McElhinny.*

Caribbean Sea
Caracas
Amazonian Manatee
River Range
Venezuela
Georgetown
Paramaribo
Guyana
Cayenne
Bogotá
Suriname
French guiana
Colombia
Atlantic Ocena
0°
Quito
Caquetá
Negro
Trombeta
Equator
Putumayo
Japurá
Ecuador
Napo
Manaus
Belem
Tigre
Solimões
Xingu
Tocantins
Iutái
Juruá
Tefé
Coari
Purus
Madeira
Topajós
Marañón
Ucaiali
Marabá
N
Peru
Brazil
500 miles
800 kilometers
Pacific Ocena
Lima
Bolivia
60°

**Figure 4** *Range of the Amazonian manatee,* T. inunguis. *Map produced by Ellen McElhinny.*

occupying different regions of the species' range suggest that the population grew through the 1980s and early 1990s but leveled off in at least some locations during the mid-1990s. As the twentieth century ended, other recent analyses suggested that the population may be relatively stable or may even be increasing slightly in some regions, but the statistical uncertainty associated with the data and models leaves open the possibility that the overall population may be declining.

The number of Antillean manatees is unknown. Aerial surveys of Belizean waters and waters of southern Quintana Roo, Mexico, have documented more than 400 manatees. The corridor between Belize and southern Mexico is considered to be a stronghold for the subspecies.

### B. Amazonian manatee, *T. inunguis*

This species occupies freshwater habitats throughout the drainage of the Amazon River and its tributaries, including rivers and lakes in Brazil, Peru, Ecuador, and Colombia (Fig. 4). There are no reliable population estimates.

### C. West African manatee, *T. senegalensis*

This species is found in the coastal and riverine waters of 21 countries in central and West Africa, from Senegal to Angola (Fig. 5). Manatees inhabit the upper reaches of the Niger River to Guinea and occur throughout the inland delta of Mali. Manatees in the upper Niger River are cut off from the sea by cataracts and a hydroelectric dam. In Ghana, manatees caught above a dam on the Volta River are also now permanently cut off from the ocean. Manatees inhabit two tributaries of Lake Chad, the Logone and Chari rivers, but are not found in the lake itself. The Logone and Chari rivers do not communicate with the sea; during times when water levels were higher, manatees in these rivers probably were able to mix with other manatees by moving through interconnecting lakes to the Benue River, a tributary of the Niger River. There are no reliable population estimates.

### III. Ecology

All manatees are herbivores, and as hindgut digesters (like horses and elephants), they can subsist on low-quality forage. Because they are such large mammals, manatees would be expected to have a low weight-specific metabolic rate, but their metabolism is 20–30% lower than one would expect. The best-studied species, the West Indian manatee, consumes more than 60 species of plants (almost exclusively angiosperms) and may ingest a mass of food that approximates up to 7% of its body weight each day. In some locations, 50–90% of the plant biomass may be eaten or uprooted by grazing animals, but the overall effects on local plant productivity of manatee feeding are not well understood. The dugong has been described as a cultivation grazer, and the manatee may serve the same role.

Although manatees subsist primarily on plants, they also consume flesh. They have been reported to consume fish caught on longlines, and tunicates have been found in large numbers in some manatee

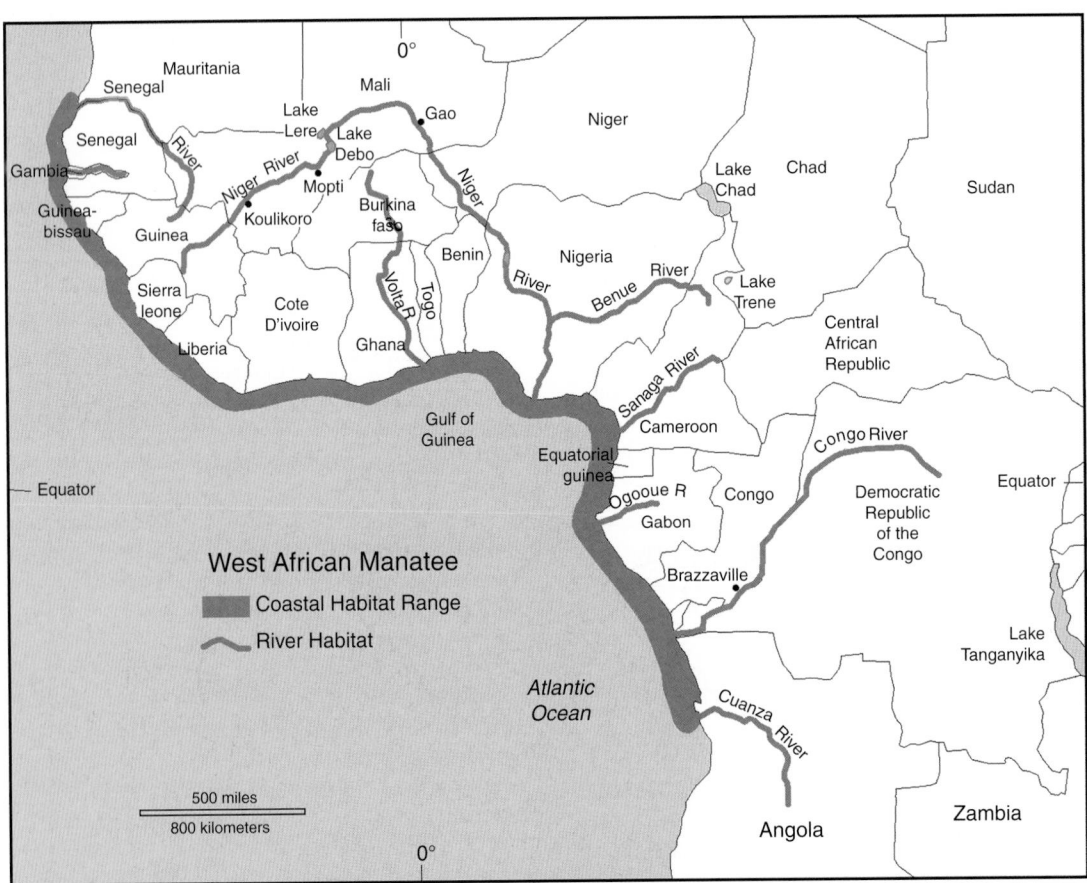

**Figure 5**    *Range of the West African manatee,* T. senegalensis. *Map produced by Ellen McElhinny.*

stomachs. Of course, the plants manatees consume have epiphytic organisms growing on their leaves.

The distribution of the Florida manatee is influenced by temperature and, perhaps, by access to fresh water to drink. In cold weather, manatees tend to migrate south and/or seek refuge at natural and artificial warm-water refugia (Fig. 6). The distribution of other manatee species or subspecies appears to be governed to at least some extent by the availability of water and suitable habitat during the wet and dry seasons. Antillean manatees, for example, may move upstream in coastal rivers during the wet season, when water levels are high, and return to lower reaches of rivers during the dry season. Amazonian manatees occupy lakes during the dry season, when rivers and streams dry up. Because the lakes are murky and lack bottom vegetation, manatees may fast during the dry season for up to 200 days when water levels drop and shoreline vegetation is no longer available for them to eat.

The habitat requirements of West African manatees are similar to those of West Indian manatees. Although manatees along the coast of Africa tend to move up rivers and out of estuaries during the dry season, they can occasionally be found in any aquatic habitat. In the upper reaches of the Niger River and some other large rivers, West African manatees, like Amazonian manatees, may remain in lakes during the dry season, when water levels drop, and stay there until waters rise and they can move back into the rivers.

Predation on manatees has not been well documented, but it appears that they have few natural enemies and that predation levels are very low. There is some evidence to suggest that crocodilians and sharks may account for some manatee mortality in different parts of the world; in Florida, such reports are reinforced by the rare presence on living manatees of wounds caused by alligators or sharks (Falcón-Matos *et al.* 2003). Especially, but not exclusively, when aggregated in lakes during the dry season, Amazonian manatees may also be preyed upon by jaguars.

Natural factors that have been documented to kill large numbers of Florida manatees include cold weather and red tides.

## IV. Behavior and Physiology

Manatees feed on bottom vegetation, plants in the water column, and floating or shoreline vegetation. Their flexible pectoral flippers and prehensile lips, which are equipped with bristles, are used to push vegetation into the mouth. The most striking and well-documented migrations occur in Florida in response to cold weather. Aggregations of more than 300 manatees occur at each of several natural and artificial sources of warm water in winter (Fig. 6). On very cold mornings, the majority of Florida manatees may be found at a few warm-water refugia scattered along the coast. Florida manatees disperse widely in warm weather. Amazonian manatees also migrate seasonally, from rivers and streams in the wet season to deeper waters, such as lakes, in the dry season. In some areas, West African and Antillean manatees show movement patterns similar to those of Amazonian manatees in response to low water and lack of freshwater flow.

Mating herds, composed of a female in estrus and a consort of several (up to 22) males, may remain together for periods of up to a month, typically outside the winter season. The cow is receptive to mating for only a day or two during that time. Although "fighting" does not occur, the males vigorously push and shove one another to gain access to the female. Females mate with several males during the estrous period. The mating system is an example of scramble competition polygamy (specifically polyandry).

Cues that males use to locate estrous females are not known, but it is possible that the males detect olfactory, gustatory, and acoustic signals produced by the females. Males tend to have larger home ranges than females do; thus, wandering males may routinely encounter a number of females. Females in estrus have a larger range of movement than do non-estrous females.

M

**Figure 6**  *Florida manatees aggregate in large numbers around natural and artificial sources of warm water in winter. In this photograph, approximately 230 manatees huddle in the discharge of the Riviera power plant. Photograph by John Reynolds and Florida Power & Light Company.*

Calves tend to stay close to their mothers for some time after birth. Weaning generally occurs when the calf is 1–2 years old, although calves up to 4 years old may still nurse. When traveling, calves swim parallel and close to their mothers, presumably in a position where they are easy to protect, where communication is facilitated, and where hydrodynamic drag is minimized. During the long period of maternal care, calves apparently learn the locations of important resources such as warm-water discharges or fresh water. The learning process causes groups of manatees, including young animals that have recently become independent from their mothers, to use the same areas year after year.

Details of the social structure of manatees are lacking. Florida manatees appear to have a simple fission–fusion society in which individuals come together in a series of temporary groups. G. B. Rathbun has stated that although such societies may be relatively unstructured, the lack of social structure is not the same as a lack of social complexity.

Communication among manatees appears to involve acoustic signals (squeaks and squeals, mostly in the 3–5 kHz range), tactile cues (rubbing and "kissing"), visual cues, and possibly chemical cues (suggested by repeated use of "rubbing posts" and by individuals mouthing one another).

Manatees appear to play. Body surfing and follow-the-leader have been observed.

Manatees have a suite of unusual morphological and physiological adaptations. We note here some features of a variety of organ systems.

The bones, especially the ribs, are dense and heavy (osteosclerotic), and the ribs and some other bones are swollen (pachyostotic). As in other marine mammals, the long bones are shortened and the phalanges in the pectoral appendage are more elongated than are those of "typical" terrestrial mammals. Hyperphalangy, however, does not occur. The first digit is reduced and the fifth digit is enlarged. Pelvic limbs are absent, although vestigial pelvic bones that are sexually dimorphic remain embedded in hypaxial musculature. Erythropoiesis (formation of red blood corpuscles) and granulopoiesis (formation of certain white blood cells) occur primarily in the vertebral bodies. The skull is elongated but not telescoped. Nares and nasal bones have migrated dorsally. The zygomatic arch, which abuts the periotic bone on each side of the skull, is relatively light and porous and is permeated with oil. However, compared to the bones of the zygomatic arches of most marine mammals, the bones of the manatee's arch are huge, reflecting their importance as an attachment for powerful chewing muscles.

The muscle color is of several shades, from almost white to red, apparently due to differences in myoglobin concentration in particular muscles or muscle groups. Axial muscles to the fluke are extremely powerful. The panniculus muscles are very well developed, as are muscles to the pectoral flipper (presumably to facilitate both dexterity and strength).

The skin is extremely heavy and thick and may provide some of the ballast needed for buoyancy control. Body hairs are sparse (~1 every cm$^2$); sweat glands are lacking; and nerve plexuses associated with some hairs suggest that the hairs are important in detecting pressure or in tactile communication. Instead of having one layer of blubber as many other marine mammals do, manatees have alternating layers of panniculus muscle and blubber (somewhat resembling bacon).

The lungs are long and unlobed, and they occupy virtually the entire dorsal region of the trunk. Manatees exchange about 90% of the air in their lungs in a single breath. The branching pattern of the bronchi is monopodial, and the terminal airways are reinforced with cartilage. The diaphragm is large and powerful, is located in a horizontal (coronal) plane, constructed as two independent hemidiaphragms, and instrumental in maintaining buoyancy control.

The large intestine is enormous (>20-m long in large animals), a feature that is not surprising in a hindgut fermenter. The stomach has a large accessory organ of digestion (the cardiac gland), and the capacious duodenum has two prominent diverticulae. The cecum is small and bicornuate. The life-long, horizontally oriented tooth replacement is a very unusual feature and may be an adaptation to facilitate the consumption of the gritty plant material that manatees consume. Histology of the various portions of the gastrointestinal tract shows some unusual cellular arrangements. The accessory organs of digestion (liver, pancreas, salivary glands) are unremarkable. Manatees have taste buds but no vomeronasal organ.

The heart is not unusual except for a persistent interventricular cleft, the presence of notable amounts of cardiac fat, and the large amount of pericardial fluid. Circulatory adaptations (retia, arteriovenous anastomoses, counter-current heat exchangers) facilitate overall heat conservation, while also allowing for the cooling of the reproductive organs and nervous tissues.

The brain is small (the encephalization quotient for *T. manatus* is 0.27), and the cerebral hemispheres lack extensive convolutions. Notably large trigeminal (cranial nerve V) and facial (cranial nerve VII) nerves are associated with the facial vibrissae.

The uterus is bicornuate. The ovaries are rather flattened and diffuse, and in mature individuals, the ovaries have numerous corpora. The penis and testes are located inside the body wall. The testes are relatively small, but the seminal vesicles are very large. The testes abut the kidneys along the caudal part of the diaphragm.

The kidneys are lobular, are located on the ventral surface of the caudal part of the diaphragm, and are often encapsulated in fat. Their microscopic structure suggests an ability to produce concentrated urine and therefore to go for prolonged periods without access to fresh water.

Manatees can remain submerged for more than 20 min but generally dive for much shorter periods of time (2–3 min or less). Because the plants manatees consume grow close to the surface where sunlight is available, dives are usually shallow.

Scientists have historically suggested that temperatures below about 19°C induce sufficient stress to cause at least some manatees to seek warm water as a refuge. Some recent evidence suggests that this temperature may be a little high and that 17°C is perhaps a more realistic point at which stress occurs. Even though scientists may be uncertain of the precise point at which thermal stress occurs, it is clear that both chronic and acute exposure to low temperatures may cause death.

The extent to which manatees physiologically *need* fresh water is unclear. It is clear, however, that Florida manatees *like* fresh water to drink. Functional morphology suggests that the kidney should be able to produce hyperosmotic urine and be able to rid the body of excess salt following seawater ingestion. Manatees, like other marine mammals, are K-strategists when compared to most other animals. In some ways, however, manatees appear to be less K-selected if the comparison group is just the marine mammals. Table I provides life history information on Florida manatees.

## V. Life History

Aspects of life history are known for the Florida manatee based on long-term research (Lefebvre and O'Shea, 1995; Runge *et al.*, 2004). Maximum age was 60 years. The gestation period was 11–13 months. The sex ratio at birth was 1:1. Calf survival to year 1 was 0.81 at Blue Spring and 0.67 at Crystal River. Adult annual survival was

0.937 on the Atlantic coast, 0.96 at Crystal Spring, and Blue Spring and 0.908 on the southwest coast. Earliest age at first reproduction was 3–4 years, mean is 5 year. Thirty-three percent of salvaged female carcasses were pregnant, as were 41% of living adult females at Blue Spring. The mean proportion with nursing calves during the winter season was 0.36. The mean period of calf dependency was 1.2 year. The mean inter-birth interval was 2.5 year. The highest number of births was in May–September, and the highest frequency of mating herds was in February–July.

## VI. Interactions with Humans

Humans have interacted with the various manatee species in a number of ways, most of them harmful to the manatees. The following information includes both well-documented and presumed interactions.

Manatees have historically been hunted throughout their ranges. In Florida, hunting pressure has virtually ceased within the past few decades, although animals are occasionally still taken illegally for meat. The best-documented and most extreme example of manatee hunting occurred in Brazil from 1935 to 1954, when between 80,000 and 140,000 Amazonian manatees were killed for their meat and hides. Primary products included *mixira* (fried manatee meat preserved in its own fat), uncooked meat, lard, and the tough hides, which could be used for a range of products including whips, shields, and machine belts. Although the market for hides diminished after 1954, several thousand manatees were killed each year through the late 1950s, and probably beyond.

In certain countries such as Peru and Ecuador (Amazonian manatee) and possibly in some West African countries, military patrols hunt manatees, or hire local hunters to catch manatees, for food.

Manatees are also hunted for reasons other than the products they provide. In Sierra Leone, the Mende people hunt manatees, in part, to reduce the number of manatees and thereby to keep them from tearing fishing nets, destroying fish that have been netted, and plundering rice fields.

However, local traditions may work in favor of manatees and prevent their harvest in particular areas. In the Korup region of Cameroon, e.g., villagers fear manatees and have no taste for the meat, so they generally do not hunt the animals.

An interesting presumed effect of manatee hunting in tropical America and West Africa is that some manatees have become nocturnal and/or crepuscular.

Manatees are captured accidentally in fishing gear (crab pot lines, trot lines, fishing nets) in the United States and other countries. The extent of serious injury or mortality is unknown.

Collisions with boats and barges account for about 25% of all manatee mortality in the United States (Fig. 7). The number rose at a rate of about 7.5% per year between 1976 and 1996, and currently more than 70 animals die annually in this way. The number of registered boats in Florida alone exceeds 1 million. Based on observations of scarred animals, collisions with boats appear to be occurring with increasing frequency in other parts of the world, but the extent to which those collisions kill manatees outside the United States is unknown. Also unknown is how seriously boat-inflicted injuries debilitate manatees and affect reproduction, without causing immediate death (Fig. 8).

The propeller scars and increased turbidity caused by boats negatively affect the health and distribution of sea grasses and other vegetation eaten by manatees. Boats also make noise, which may affect manatee distribution, habitat use, and energetics. Boats can, therefore, affect manatees both indirectly, by contributing to diminished

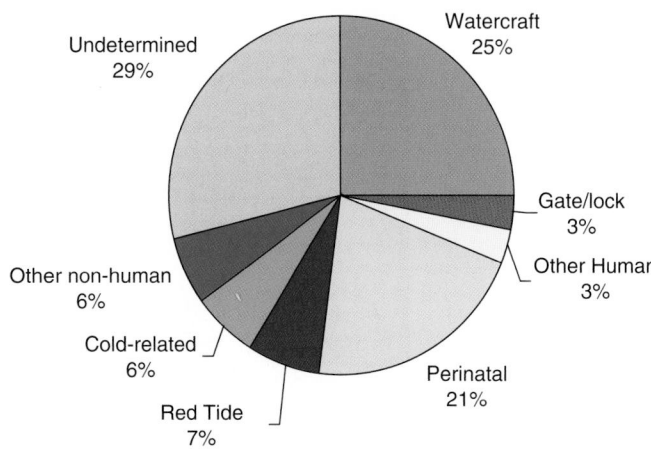

**Figure 7** *Categories of manatee mortality in Florida. This pie chart shows manatee mortality categories based on 6180 carcasses recovered or reported to federal or state agencies from 1974 through 2007. The highest percentage of deaths remains undetermined (n = 1802) and includes unrecovered, badly decomposed, and other carcasses that were not badly decomposed but for which case of death could not be assigned. Total human-related mortality is high (n = 1897, 31% of total) and includes watercraft-related deaths (n = 1551), trauma, or drowning caused by canal locks or flood gates (n = 191), and other human-related factors such as entanglement (n = 155). Non-human-related causes of death (n = 1158, 19% of total) are related to cold exposure (n = 359), red tide outbreaks (n = 460), or other factors (n = 339). Perinatal mortality (n = 1323) refers to the death of a small animal (≤150 cm long) for which cause of death can not be determined; perinatal mortality can be either human or non-human related. Watercraft-related mortality continues to rise annually and is the single highest known category of death. Produced by James Powell/Cynthia Taylor.*

food resources, and directly, by disturbing, injuring, or killing them. In Florida, manatees sometimes become trapped in flood control structures and canal locks and die. About 4% of the manatees known to die between 1974 and 1996 were crushed and drowned in flood gates or canal locks. Increasingly, scientists and environmentalists at the national, state, and regional levels are concerned about the effects of pollution on both the health of individual animals and the status of populations. Levels of certain chemical pollutants have been assessed in some marine mammal tissues, but the effects of these chemicals are unclear. In only one case did scientists experimentally demonstrate a clear cause-and-effect relationship between a toxicant and a reproductive impairment (in harbor seals). Based on toxicological studies of laboratory animals, scientists suspect that chemical pollution is harming the endocrine and immune systems of at least some stocks of marine mammals, but this has yet to be demonstrated.

Scientists have found elevated levels of copper in the tissues of Florida manatees from certain locations, but levels of other metals and of organochlorines have been considered unremarkable, and have not been the focus of many studies. However, Pulster *et al.* (2007) and Wetzel *et al.* (in press) have examined levels of polychlorinated biphenyls (PCBs) and chlorinated pesticides in manatees from southeastern Mexico and southwestern Florida; the scientists found that PCBs in some individuals are surprisingly high, exceeding toxic threshold values for cetaceans (such values are not known for

**Figure 8** *Although several dozen Florida manatees die each year because of collisions with watercraft, many animals survive such encounters, albeit with considerable pain and disfigurement. The extent to which reproduction and longevity of survivors is compromised is unknown. Photograph by Sirenia Project, US Geological Survey.*

sirenians), and that pesticide residues in the blubber are both diverse and higher than expected. No other toxicants have been examined, and the effects of contaminants on manatees are unknown.

In Central America, run off of pesticides and herbicides, and the ingestion of plastic debris, have been suspected of causing the death of Antillean manatees. In the Amazon basin, the water pollution associated with mining activities may be harming manatees.

Noise pollution is a problem of which people around the world have become increasingly aware, but about which few data exist. Many scientists suspect that noise pollution negatively affects manatees, but studies examining this relationship are needed. Underwater noise of anthropogenic origin has been demonstrated or suggested to cause some marine mammals to vary their normal patterns of habitat use, and to expend more energy than usual to avoid disturbance, and these behavioral changes could logically be expected in manatees too. In the coastal waters of Florida, where more than 1 million boats are registered and many additional boats are also found, any cumulative effects of anthropogenic noise are a real cause for concern. Even in Costa Rica, where boat traffic is sparse compared to that in Florida, hunters and scientists have noted that manatees react to the noise of approaching boats and that manatee distribution is inversely related to the amount of boat traffic.

Behavioral and anatomical evidence suggests that manatees hear boat motors, but a recent audiogram suggests otherwise. Nevertheless, the suggestion (unwise in our opinion) has been made to put noise makers or acoustic alarms on boats to alert manatees. Not only may such devices be unnecessary if manatees do, indeed, already hear boats, but because their use would greatly increase existing levels of underwater noise, they would most probably disturb or harm not only manatees but many other organisms as well.

Disturbance can occur in ways besides those associated with boat traffic or boat noise as described above. At Crystal River and nearby springs in Florida, tourists gather in large numbers to swim with manatees when the animals aggregate in winter. Although most people behave responsibly, some stand on, ride, or tie ropes to the manatees. Such behavior by humans could cause manatees to avoid seeking refuge at warm-water springs in winter, which could lead to even higher-than-usual manatee mortality in very cold weather. Disturbance of females accompanied by calves could lead to abandonment of a calf, contributing to escalating levels of "perinatal" mortality.

In Southern Lagoon, Belize, and other locations, ecotourism focusing on manatees has developed. Although the financial benefits to local residents may be significant, such activities should be carefully planned to minimize the negative effects upon the manatees residing in and using the resources of such locations.

As noted above, disturbances from hunting pressure have apparently induced nocturnal or crepuscular behavior in manatees in certain parts of the world.

We have discussed many of the harmful ways that huans have affected manatees—noise, chemical contamination, boat traffic, and ecotourism, e.g.—but in at least one way, we humans may be helping manatees. Most biologists feel that thermal discharges from power plants and other sources have provided winter habitat that has helped the populations of manatees in Florida to survive and even to grow in at least some areas. But these plants have finite lifetimes, and manatee dependence on them creates a long-term dilemma for managers. What will the manatees do if warm water is no longer available at a spot where they have learned to depend on it? Industry (primarily Florida Power & Light Company) and the US Fish and Wildlife Service have initiated discussions to attempt to solve this problem.

Another way in which people have helped manatees is by the introduction of exotic aquatic plants into Florida's waterways. Such plants as *Eichhornia crassipes* (water hyacinths) and *Hydrilla verticillata* have proliferated to such an extent that they provide important and abundant food resources for manatees in certain regions of the state. In fact, the exotics displace native vegetation and may grow so luxuriantly that they create navigation problems in some water-ways, manatees in Florida have been suggested as possible economical weed-clearing agents, a role that they fill well in some canals in Guyana.

Other human-related habitat modifications have not been helpful to manatees. Dams or other structures prevent manatees from pursuing normal migration routes along rivers in South America and West Africa. And finally, the eradication of millions of hectares of rainforest each year in Amazonia cannot help but negatively affect ALL species occupying that area because of factors such as reduced productivity, siltation, and changes in hydrological cycles.

Manatees are maintained in nearly 20 different facilities worldwide. They breed in several of the facilities. In certain facilities in Florida, injured or diseased manatees are routinely rehabilitated and released back into the wild, thereby assisting wild populations.

Without entering the debate about the appropriateness of captivity, we simply note here that facilities that display manatees to the public provide important venues for educating people about manatees and their conservation, as well as for conducting basic and applied research on the different species.

The various manatee species are protected by laws specific to the countries they occupy, but enforcement of these laws is generally minimal. Several countries (e.g., US, Mexico, Belize, Guatemala, Ivory Coast, Cameroon, and Nigeria) have created manatee reserves and sanctuaries, and others (Brazil and Peru) have "protected" areas that include important manatee habitat.

The following list provides examples of broader-scale protection efforts.

| | |
|---|---|
| *Trichechus inunguis*: | CITES: Appendix 1 |
| | World Conservation Union (IUCN): listed as vulnerable |
| | USA, Endangered Species Act (ESA): listed as endangered |
| *Trichechus manatus*: | CITES: Appendix 1 |
| | World Conservation Union (IUCN): listed as vulnerable |
| | USA, Endangered Species Act (ESA): listed as endangered |
| *Trichechus manatus latirostris*: | CITES: Appendix 1 |
| | World Conservation Union (IUCN): listed as endangered |
| | USA, Endangered Species Act (ESA): listed as endangered |
| *Trichechus manatus manatus*: | CITES: Appendix 1 |
| | World Conservation Union (IUCN): listed as endangered |
| | USA, Endangered Species Act (ESA): listed as endangered |
| *Trichechus senegalensis*: | CITES: Appendix 2 |
| | World Conservation Union (IUCN): listed as vulnerable |
| | USA, Endangered Species Act (ESA): listed as threatened |
| | Protected under Class A, African Convention for the Conservation of Nature and Natural Resources |

## See Also the Following Articles

Dugong ■ Sirenian Evolution

## References

Domning, D. P. (1982). Evolution of manatees: A speculative history. *J. Paleo.* **56**, 599–619.

Domning, D. P. (1996). Bibliography and Index of the Sirenia and Desmostylia. *Smithson. Contr. Paleobiol.* **80**, 1–611.

Domning, D. P., and Hayek, L. C. (1986). Interspecific and intraspecific morphological variation in manatees (Sirenia: *Trichechus*). *Mar. Mamm. Sci.* **2**, 87–144.

Falcón-Matos, L., Mignucci-Giannoni, A. A., Toyos-González, G. M., Bossart, G. D., Meisner, R. A., and Varela, R. A. (2003). Evidence of shark attack on a West Indian manatee (*Trichechus manatus*) in Puerto Rico. *J. Neotrop. Mamm.* **10**, 161–166.

Garcia-Rodriguez, A. I., *et al.* (1998). Phylogeography of the West Indian manatee (*Trichechus manatus*): How many populations and how many taxa. *Mol. Evol.* **7**, 1137–1149 (10 authors).

Hartman, D. S. (1979). Ecology and Behavior of the Florida Manatee. *Am. Soc. Mammal. Spec. Pub.* **5**.

Lefebvre, L. W., and O'Shea, T. J. (1995). Florida manatees. *In* "Our Living Resources: A Report to the Nation on the Distribution, Abundance, and Health of US Plants, Animals, and Ecosystems" (E. T. LaRoe, *et al.*, eds), pp. 267–269. US Department of the Interior, National Biological Service, Washington, D.C.

Lefebvre, L. W., O'Shea, T. J., Rathbun, G. B., and Best, R. C. (1989). Distribution, status, and biogeography of the West Indian manatee. *In* "Biogeography of the West Indies" (C. A. Woods, ed.), pp. 567–610. Sandhill Crane Press, Gainesville, Florida, USA.

O'Shea, T. J., Ackerman, B. B., and Percival, H. F. (1995). Population Biology of the Florida Manatee. *US Dep. Int. Nat. Biol. Serv. Inf. Tech. Rep.* **1**.

Powell, J. A. (1996). "The Distribution and Biology of the West African Manatee (*Trichechus senegalensis* Link, 1795)." Report to the United Nations Environment Programme, Nairobi, Kenya.

Pulster, E., Wetzel, D.L., Reynolds, J. E. III, Morales-Vela, B. (2007). "Concentrations of persistent organic pollutants in an endangered species, the West Indian manatee (*Trichechus manatus*) sampled in southeastern Mexico." Abstract, The Society of Environmental Toxicology and Chemistry (SETAC), 11–15 November, Milwaukee, Wisconsin.

Reeves, R. R., Stewart, B. S., and Leatherwood, S. (1992). "The Sierra Club Handbook of Seals and Sirenians." Sierra Club Books, San Francisco.

Reynolds, J. E., III (1999). Efforts to conserve the manatees. *In* "Conservation and Management of Marine Mammals" (J. R. Twiss, Jr, and R. R. Reeves, eds), pp. 267–295. Smithsonian Institution Press, Washington, DC.

Reynolds, J. E., III, and Odell, D. K. (1991). "Manatees and Dugongs." Facts on File, Inc, New York.

Reynolds, J. E., III, and Rommel, S. A. (eds) (1999). "Biology of Marine Mammals." Smithsonian Institution Press, Washington, DC.

Rosas, F. C. (1994). Biology, conservation and status of the Amazonian manatee. *Trichechus inunguis. Mamm. Rev.* **24**, 49–59.

Runge, M. C., Langtimm, C. A., and Kendall, W. L. (2004). A stage-based model of manatee population dynamics. *Mar. Mamm. Sci.* **20**, 361–385.

US Fish and Wildlife Service. (1996). "Florida Manatee Recovery Plan." Second Revision. Prepared by the Florida Manatee Recovery Team for the US Fish and Wildlife Service, Atlanta, Georgia.

US Fish and Wildlife Service. (2001). "Florida Manatee (*Trichechus manatus latirostris*) Recovery Plan." Third Revision. USFWS, Altlanta, Georgia. 144 pp + appendices.

Vianna, J. A., *et al.* (2006). Phylogeography, phylogeny and hybridization in trichechid sirenians: Implications for manatee conservation. *Mol. Ecol.* **15**, 433–447 (14 authors).

Wetzel, D. L., Pulster, E., and Reynolds, J. E. III. in press. Organic contaminants and sirenians. *In* "Sirenian Conservation: Issues and Strategies in Developing Countries" (Hines, E., J.E. Reynolds, III, A. A. Mignucci-Giannoni, L. V. Aragones, and M. Marmontel (eds), University Press of Florida, Gainesville.

M

# Marine Mammal Evolution

### J.G.M. Thewissen and Bobbi Jo Schneider

The front and back inside cover of this volume presents interpretative summaries of the phylogeny of cetaceans, pinnipeds, sirenians, and desmostylians, plotted against the geological timescale. These phylogenies are meant to give the non-specialist an introduction to the confusing array of systematic names of marine mammals by providing a family attribution, an approximate phylogenetic position, and an approximate age range for many genera. They do not represent a conclusive, or even consensus, view of marine mammal evolution.

Most of the topology of these diagrams is based on the entries in this volume (Archaic Archaeocetes, Basilosaurids; Cetacean Evolution; Pinniped Evolution; Sirenian Evolution; Desmostylia; Fossil record), supplemented by data provided by the Paleobiology database (http://paleodb.org) and some technical papers (e.g., Geisler et al., 2005; Fitzgerald, 2006; Steeman, 2007). Modern genera are based on the list of Marine Mammal Species provided in this book, which is mostly consistent with Rice (1987). There are clearly differences of opinion between authorities, for instance in the validity of the right whale genus *Eubalaena* or in whether the dolphin genus *Lagenorhynchus* should be divided into multiple genera (LeDuc et al., 1999). Resolution of these inconsistencies can only be reached with further research.

## See also the Following Articles

Cetacean Evolution ■ Desmostylia ■ Fossil Record ■ Pinniped Evolution ■ Sirenian Evolution

## References

Fitzgerald, E. M. G. (2006). A bizarre new toothed mysticete (Cetacea) from Australia and the early evolution of baleen whales. *Proceedings of the Royal Society B* **273**, 2955–2963.

Geisler, J. H., Sanders, A. E., and Luo, Z. (2005). A new protocetid whale (Cetacea; archaeoceti) from the late middle Eocene of South Carolina. *American Museum Novitates* **3480**, 1–65.

LeDuc, R. G., Perrin, W. F., and Dizon, A. E. (1999). Phylogenetic relations among the delphinid cetaceans based on full cytochrome b sequences. *Marine Mammal Science* **15**, 619–648.

Rice, D. W. (1987). *Marine Mammals of the World: Systematics and Distribution*, pp. 1–231. Society for Marine Mammalogy, Special Publication 4.

Steeman, M. E. (2007). A cladistic analysis and a revised classification of Recent and fossil mysticetes. *Zoological Journal of the Linnean Society* **150**, 875–894.

# Marine Parks and Zoos

### Daniel K. Odell and Loran Wlodarski

## I. The History of Zoological Parks

Humans have exhibited animals from the wild in marine life parks, zoos, and aquaria for hundreds if not thousands of years. The earliest zoos were not meant for the average

## TABLE I
### Worldwide Counts of Zoos and Aquariums with Counts of Those Holding Cetaceans[a]

| Region/Country | Cetacean Facilities[b] | Total Facilities[c] | WAZA[d] |
|---|---|---|---|
| Africa | | 21 | 33 |
|   Cameroon | | | 1 |
|   Ivory Coast | | | 1 |
|   Egypt | 1 | | |
|   Madagascar | | | 1 |
|   South Africa | 2 | | 29 |
|   Uganda | | | 1 |
| Asia/Pacific | | 155 | |
|   Australia | 3 | 10 | 72 |
|   Brunei Darussalam | | | 1 |
|   Cambodia | | | 1 |
|   China | 6 | | 4 |
|   Fiji Islands | | | 1 |
|   French Polynesia | 1 | | |
|   India | | 56 | |
|   Indonesia | 1 | 17 | 18 |
|   Japan | 40 | 140 | 162 |
|   Korea | 1 | | |
|   Malaysia | | | 7 |
|   Myanmar | | | 1 |
|   Nepal | | | 1 |
|   New Caledonia | | | 1 |
|   New Zealand | 1 | | 17 |
|   Papua New Guinea | | | 1 |
|   Phillippines | | | 4 |
|   Singapore | | | 3 |
|   South Korea | | | 2 |
|   Sri Lanka | | | 1 |
|   Taiwan | | 1 | 2 |
|   Thailand | 2 | | 3 |
|   Vietnam | | | 2 |
| North America | | | |
|   Bermuda | 1 | | 1 |
|   Canada | 3 | | 5 |
|   Bahamas | 2 | | |
|   USA | 38 | 230 | 208 |
| Central and South America | | | |
|   Argentina | 2 | 31 | 1 |
|   Belixe | | | 1 |
|   Bolivia | | 1 | |
|   Brazil | | 73 | 1 |
|   Central America | | 8 | |
|   Chile | | 2 | 1 |
|   Colombia | 2 | 1 | 10 |
|   Costa Rica | | | 4 |
|   Cuba | 3 | | 2 |
|   Curaçao | | | 1 |
|   Dominican Republic | | | 2 |
|   El Salvador | | | 3 |
|   Guatemala | | | 7 |
|   Guyana | | 1 | |
|   Honduras | 1 | 1 | 1 |
|   Mexico | 9 | 1 | 24 |
|   Peru | 1 | 1 | |
|   Panama | | | 1 |

(continues)

TABLE I (continued)

| Region/Country | Cetacean Facilities[b] | Total Facilities[c] | WAZA[d] |
|---|---|---|---|
| Puerto Rico | | | 1 |
| Trinidad & Tobago | | | 1 |
| Venezuela | 3 | 4 | 18 |
| Europe and Middle East | | | |
| Austria | 0 | 3 | 8 |
| Bahrain | 1 | | |
| Belgium | 2 | 3 | 8 |
| British Isles | | | 2 |
| Croatia | | | 1 |
| Bulgaria | 1 | | |
| Cyprus | 1 | | |
| Czechoslovakia & Czech Republics | | 15 | |
| Czech Republic | | | 15 |
| Denmark | 1 | 4 | 14 |
| Estonia | | | 1 |
| Finland | 1 | 2 | 2 |
| France | 2 | 7 | 83 |
| Germany | 4 | 42 | 131 |
| Great Britain | 0 | 37 | |
| Greece | | | 2 |
| Hungary | | 9 | 5 |
| Ireland | | | 2 |
| Israel | 1 | 11 | 4 |
| Italy | 5 | 4 | 15 |
| Kazakhstan | | | 1 |
| Kuwait | | | 1 |
| Latvia | | | 1 |
| Lithuania | 1 | | 1 |
| Malta | 1 | | |
| Norway | | | 2 |
| Poland | | 16 | 15 |
| Portugal | 2 | 1 | 9 |
| Russia | | | 25 |
| Russia & Former Soviet Union | 3 | 31 | |
| Serbia | | | 1 |
| Slovakia | | | 4 |
| Slovenia | | | 1 |
| Spain | 8 | 3 | 32 |
| Sweden | 1 | 5 | 17 |
| Switzerland | 1 | 5 | 10 |
| The Netherlands | 1 | 11 | |
| Turkey | | | 2 |
| Ukraine | 4 | | 3 |
| United Arab Emirates | | | 4 |
| United Kingdom | | | 75 |
| Southwest Asia/Middle East | | 29 | |
| South Asia | | 11 | |
| Southeast Asia | | 31 | |
| East Asia | | 11 | |
| Central America | | 9 | |

[a]The counts given are incomplete but do give an idea of the relative distribution and abundance of zoos and aquariums.
[b]From Couquinand-Douaze (1999)
[c]From Kisling (2000)
[d]From World Association of Zoos and Aquariums web page www.waza.org accessed 17 July 2007.

citizen but for the elite, as wealthy rulers collected unusual animals for their enjoyment. Slowly, these private collections turned public, such as when animals gathered at Schloss Schönbrunn, Vienna, Austria, were first exhibited to the public in 1765. This park is considered the first modern zoo. Solski (2006) gives an historical perspective on public aquariums from 1853–1914. Polar bears (*Ursus maritimus*) and various pinnipeds were probably among the first marine mammals to be held by humans. Polar bears may have been held since about 1060, and harbor porpoises (*Phocoena phocoena*) since perhaps as early as the 1400s, but the majority of marine mammals seen more commonly in marine facilities today (2007) were not displayed until the late 1800s and early 1900s. Many species of cetaceans have never been displayed and some have only been seen recently in marine parks. Killer whales (*Orcinus orca*), for example, were first displayed in a sustainable manner in 1961.

## II. Zoos and Marine Parks Worldwide

How many zoos, aquaria, and marine zoological parks exist worldwide? More specifically, how many of these facilities display marine mammals? It is likely that no one has an exact count. New facilities are being built, some facilities close, and some facilities change the animals that they have on display. In 2001, the Conservation Breeding Specialist Group (CBSG) of the World Conservation Union (the IUCN) maintained a global zoo directory on its worldwide web page that listed nearly 1800 institutions but does not give information on marine mammals in the collections (CBSG no longer maintains this directory). A similar list is published in each annual issue of the International Zoo Yearbook (Anonymous, 2006). This list includes postal, phone, fax, and email addresses for the institutions. Kisling (2001) is the most recent review of zoo and aquarium history and includes an admittedly incomplete listing of over 900 zoos and aquaria worldwide. For example, 230 facilities are listed for the United States (The US Department of Agriculture's Animal and Plant Health Inspection Service [APHIS] licenses ~130 facilities to exhibit marine mammals.), 140 for Japan, 56 for India, and 155 for Asia (Table 1). As of July 2007, 216 zoos and aquariums were listed as accredited members of the Association of Zoos & Aquariums (AZA, formerly the American Zoo and Aquarium Association). A full list of these facilities can be found at www.aza.org. A survey by Couquiaud-Douaze (1999) of facilities holding cetaceans lists 166 institutions in 42 countries located on all continents except Antarctica (Table I). If pinnipeds and polar bears were added, the number of institutions would grow considerably. For example, a 1995 survey (Andrews *et al.*, 1997) of the United States and Canada listed 109 facilities that held 1460 marine mammals, including 11 species of cetaceans, 11 species of pinnipeds, the sea otter (*Enhydra lutris*), and the Florida manatee (*Trichechus manatus latirostris*). Polar bears were not included. In 2007, members of the AZA (2007c) listed more than 700,000 animals at their facilities, of which 1260 were marine mammals. Zoological parks can be found through web sites maintained by regional organizations. The following list is representative and not necessarily all inclusive: Alliance of Marine Mammal Parks and Aquariums (Alliance) www.ammpa.org; American Zoo and Aquarium Association www.aza.org; Australian Regional Association of Zoological Parks and Aquariums www.arazpa.org.au; Canadian Association of Zoos and Aquariums www.caza.ca; European Association of Zoos and Aquaria www.eaza.net; Fish Link Central www.fishlinkcentral.com/links/Public_Aquariums/. Japanese Association of Zoological Gardens and Aquariums www.jazga.or.jp; PanAfrican Association of Zoological Gardens, Aquariums and Botanical Gardens www.paazb.com; South East Asian Zoos

**M**

Association www.seaza.org; World Zoo Organization www.waza.org; Zoological Society of London http://www.zsl.org/research/; and Zoos Worldwide www.zoos-worldwide.de.

### III. Challenges

As zoos and aquaria learned more about the marine mammals in their care, there is no question that the quality of the facilities and animal husbandry improved dramatically, especially over the past several decades and especially in developed countries. Enclosure and pool sizes have increased and have gone from caged to cageless exhibits. Governments have enacted (or are considering) legal standards for the care, maintenance, and display of captive animals (e.g., the United States Animal Welfare Act) and these standards are continually evolving. The Animal Welfare Act (including minimal pool size requirements for marine mammals, animal transportation regulation, and so on) is discussed in detail on the United States Department of Agriculture web site www.aphis.usda.gov/. Marginal facilities, including most traveling or temporary exhibits, have been eliminated. However, on a worldwide basis, there remains room for improvement. These improvements and the expert staff necessary for good animal care have a high financial cost and there is often an unavoidable trade-off with funding for other human activities (e.g., health care in the case of publicly funded institutions). If facilities cannot provide proper care for their animals, they should be closed and the animals relocated to responsible facilities. Unquestionably, it is in the best interests of all zoo and aquarium staff to provide the best possible care for the animals in their charge. Whether institutions are formally "for profit" or "not for profit," it still takes large amounts of money to build, operate, maintain, and expand exhibits.

Breeding at zoological parks and aquariums is becoming more important for marine mammal facilities. These programs often require additional separate facilities (e.g., maternity pools) and additional animals. Another challenge is the acquisition of high-quality marine mammal food (primarily fish and squid) on a reliable basis. As fish stocks are depleted around the world, marine mammal facility managers must plan accordingly. Some are considering the development of a mass-produced fish replacement for marine mammal food. Such a product could be produced as needed and would not require the storage of a year's supply.

Facilities exhibiting marine mammals argue that introducing people to living dolphins and other marine mammals is a proven way to promote wildlife conservation, instills an awareness of ecological and conservation issues, and inspires a strong, active commitment to marine mammal conservation. However, the ethics of keeping wild animals (or, for that matter, any animal) in marine life parks, zoos and aquaria for any purpose is an issue for some (Mench and Kreger, 1996). Worldwide, one will encounter any number of animal rights groups dedicated to the elimination of facilities holding cetaceans for any purpose. Interestingly, one seems to see little, if any, opposition to holding pinnipeds. Marine parks and zoos are often targeted by protests when new exhibits are proposed or opened or when new animals are acquired. The effectiveness of these protests and similar activities remains unclear. Ultimately, each person will have to reach her/his own conclusions on the ethics of keeping animals in marine life parks, zoos, and aquaria. One can only ask that people seek factual information even if they choose to ignore it, before making their personal decisions. Zoos and aquariums have an obligation to provide the best available information on the animals in their charge.

The ethics of displaying animals at zoos and aquariums is a topic that more parks are directly discussing. Organizations representing marine life parks, aquariums, and zoos have developed numerous member requirements to assure that the well-being of their animals is the top priority and that they receive state-of-the-art care. For example, members of AZA promote a code of professional ethics that is based on respect and dignity for the care of all animals at these parks (www.aza.org/AboutAZA/CodeEthics/). The Alliance of Marine Mammal Parks and Aquariums (www.ammpa.org) has an accreditation program based on professional standards and guidelines that are updated regularly as the knowledge base expands to integrate advancing science and technologies.

### IV. Research

The mission statements of most zoological parks and aquaria include "recreation, education, conservation, and research" in one form or another. For all institutions, the recreation component is the most visible to the public. The extent to which these institutions are involved in research and conservation programs varies and is, to some extent, dependent on financial resources. However, even the smallest of institutions, in size or in financial resources, can participate in local or multi-partnered national or international research and conservation projects.

Research on wild marine mammals, especially cetaceans, is often expensive and subject to the vagaries of environmental conditions (i.e., weather) among other things. Modern technology (radio and satellite tags, time-depth recorders, GPS tags, "critter cams," hydrophone arrays, etc.) have made huge contributions to our knowledge of free-ranging marine mammals. Nevertheless, it is our opinion that there is still much that cannot be learned from wild animals that can be learned from marine mammals in zoological parks and aquariums. Behavior, including acoustic emissions, can be observed and recorded in 24 h/day if desired. Animals can be trained to hold position for body measurements, collection of body fluids (blood, urine, milk), various medical procedures (i.e., ultrasound examinations), and collection of exhaled breaths for air composition analyses. Animals can be trained for a variety of visual, acoustic (hearing, echolocation), locomotion, and learning studies. The birth, growth, and development (behavioral and physiological) of offspring can be detailed. Therefore, studies on animals in public display facilities are not a replacement for, but a supplement to, studies on free-ranging animals, and the results must be applied with the limits of these "laboratory" studies in mind.

One measure of the involvement of zoos and aquaria in research can be obtained from annual reports of individual institutions and regional organizations. For example, the American Zoo and Aquarium Association's (AZA) annual report on conservation and science for 1996–1997 biennium (Hodskins, 1998) lists over 1100 publications of all types (abstracts, magazine articles, journal articles, etc.) for all animal groups produced by the AZA member institutions during that time period. The Association of Zoos and Aquariums' (2007a) report for CY 2006 lists a total of 866 publications produced by 194 of its 216 member institutions. The Alliance of Marine Mammal Parks and Aquariums publishes a biennial research report that summarizes its members' projects that involve marine mammals (AMMPA, 2005b).

### V. Education

Marine parks, zoos, and aquaria offer a wide variety of education programs, in-park graphics, exhibit narrations, behind-the-scenes tours, curriculum-specific programs for various age and grade levels from preschool through college, camp programs, classroom programs, off-site outreach programs, and, in the electronic age, satellite television and internet (worldwide web) offerings. In fact, US facilities holding marine mammals for public display are required by federal law to have an education program based on professional standards. The US Department of Commerce (1994) published, for reference purposes, the AZA's and Alliance's professionally accepted standards on which their members base their education and conservation programs.

The world we live in might be remarkably different if everyone could travel to the plains of Africa to view cheetahs (*Acinonyx jubatus*) stalking their prey, to view diminutive Humboldt penguins (*Spheniscus humboldti*) basking in the South American shorelines, or perhaps see meandering Florida manatees slowly grazing on vegetation along the coasts of the state of Florida. However, for most people around the planet, such encounters will never occur. In a 1995 Roper poll, 87% of those who participated stated that their only opportunity to see wild animals came from visiting zoological facilities (Roper Starch Worldwide, 1995). Harris Interactive, Inc. [(Rochester, NY) www.harrisinteractive.com] conducted a poll for the Alliance of Marine Mammal Parks and Aquariums (AMMPA) in 2005 (AMMPA, 2005a). Results indicated that 97% of respondents agree that marine life parks, aquariums, and zoos play an important role in educating the public about marine mammals which they might not otherwise have the chance to see.

AZA (2007b) recently completed a 3-year, nationwide study to determine the impact of marine life parks, zoos, and aquaria on the people that visit such facilities. The key results of this study were that visitors felt a stronger connection to the environment after visiting a zoological facility, visitors felt an enhanced public understanding and awareness of conservation issues facing animals in the wild, and overall visitors recognized the value and importance of modern of marine life parks, zoos, and aquaria in the fields of conservation education and animal welfare. Given the fact that most people cannot afford the time and money of a jet-set lifestyle, zoological parks are vital links to connect mankind with the plethora of animals on the planet.

Zoological parks alert people to the increasing threats these animals face. For example, because most people will not travel to view wild cheetahs in Africa, they may not see how these creatures are slaughtered for their hides, how Humboldt penguins are disappearing due to the mining of their guano (feces) deposits where they nest, or how Florida manatees are highly endangered thanks to an ever-increasing presence of humans in their habitat. Zoological facilities may be entertaining, but education, research, and conservation are now cornerstones of major parks. The same Roper poll revealed that 92% of those surveyed agreed that zoological parks are vital educational resources.

Although approximately 71% of the planet is covered by oceans, this realm and its inhabitants remain a mystery to a majority of people. Marine life parks help (1) educate the public about the seas and (2) clear up long-rooted misconceptions about ocean animals. A prime example of this is how killer whales were perceived in the past and how they are viewed today.

Some cultures, like aboriginal tribes of the Pacific Northwest, respected killer whales, although several major whaling countries feared these animals. Indeed, the name *Orcinus* is probably derived from Orcus, an ancient mythological Roman god of the netherworld—a reference to the ferocious reputation of this animal. In 1835, Hamilton wrote that the killer whale "… has the character of being exceedingly voracious and warlike. It devours an immense number of fishes of all sizes …, when pressed by hunger, it is said to throw itself on every thing [*sic*] it meets with …" (Hamilton, 1835). In modern civilization, many still envisioned killer whales as terrifying threats to humans, with a 1973 United States Navy diving manual warning that killer whales "will attack human beings at every opportunity." In the not too distant past, governments such as Japan, Greenland, Canada, and the United States sanctioned the use of lethal force to be used against killer whales. Killer whales were hunted for commercial use and despised by whalers who would "… often carry a rifle expressly for the Killer's benefit," according to Bennett (1932) in his book "Whaling in Antarctica." In 1961, a killer whale was displayed publicly for the first time in recent history, and afterward the perception of these animals began to change. Coupled with a growing environmental awareness in the 1960s, public sentiment rallied to protect cetaceans like killer whales from hunting. Cetaceans are now protected by various national laws and international agreements, and killer whales are generally perceived in a positive way thanks in part to the educational programs of zoological parks.

Conservation programs are linked inextricably with both research and education programs. Zoos, aquaria, and marine parks that hold marine mammals can incorporate conservation messages (i.e., do not feed or swim with wild dolphins or manatees, proper field etiquette, and trash disposal) into static graphics and show and exhibit scripts and narrations, as well as in classroom programs and the electronic media. The American Zoo and Aquarium Association's annual report on conservation and science for the 1996–1997 biennium (Hodskins, 1998) listed over 1200 conservation projects in which the AZA's 185 members were involved during that time. A similar report for CY 2006 lists over 1800 conservation projects reported by 194 member institutions (Association of Zoos and Aquariums, 2007b).

Facility staff experienced in handling marine mammals can provide advice and assistance to field workers, and many of these institutions are actively involved in marine mammal stranding programs. Facilities located near an endangered species' habitat can assist with rescue and rehabilitation of sick or injured animals [e.g., monk seals (*Monachus* spp.) in Hawaii and the Mediterranean/eastern Atlantic, manatees in Florida, sea otters in Alaska and California, and Steller sea lions (*Eumetopias jubatus*) in Canada and Alaska]. No facility is large enough to handle a blue (*Balaenoptera musculus*) or right whale (*Eubalaena* spp.), although SeaWorld California had remarkable success with an orphan gray whale (*Eschrichtius robustus*) calf (Antrim *et al.*, 1998). It may even be possible for facilities to start breeding colonies of severely endangered marine mammals [e.g., the vaquita (*Phocoena sinus*)] if field conservation efforts prove inadequate. In addition, institutions can make direct monetary contributions to conservation programs and encourage their visitors to make their own contributions to *bona fide* programs.

## *See Also the Following Articles*

Captivity ■ Captive Breeding ■ Ethics and Marine Mammals

## *References*

Alliance of Marine Mammal Parks and Aquariums (AMMPA). (2005a). Marine Mammal Poll. Alliance of Marine Mammal Parks and Aquariums, Alexandria, Virginia. Accessed 18 July 2007. http://www.ammpa.org/_docs/HarrisPoll Results.pdf

Alliance of Marine Mammal Parks and Aquariums (AMMPA). (2005b). Research Report 2004–2005. Alliance of Marine Mammal Parks and

M

Aquariums, Alexandria, Virginia. 73p. Accessed 18 July 2007. http://www.ammpa.org/ResearchReport2005.pdf

Andrews, B., Duffield, D. A., and McBain, J. F. (1997). Marine mammal management: Aiming at the year 2000. *IBI Rep.* **7**, 125–130.

Anonymous (2006). Zoos and aquariums of the world. *In* "International Zoo Yearbook" (F. A. Fiskin, and J. Miller, eds), 40, pp. 381–482. Zoological Society of London, London.

Antrim, J., McBain, J., and Parham, D. (1998). Rehabilitation and release of a gray whale calf: J.J's story. *Endang. Spec. Update* **15**, 84–89.

Association of Zoos and Aquariums (AZA). (2007a). 2006 Conservation Impact Report. Association of Zoos and Aquariums. Silver Spring, Maryland. 46p. Accessed 18 July 2007. www.aza.org/ConScience/ARCS/Documents/2006_Conservation_Impact_Report.pdf

Association of Zoos and Aquariums. (AZA) (2007b). Why Zoos Matter. Association of Zoos and Aquariums. Silver Spring, Maryland. 46p. Accessed 18 July 2007. www.aza.org/ConEd/Documents/Why_Zoos_Matter.pdf

Association of Zoos and Aquariums. (AZA) (2007c). Current Statistics. Association of Zoos and Aquariums. Silver Spring, Maryland. 46p. Accessed 18 July 2007. www.aza.org/Newsroom/CurrentStatistics/

Bennett, A. G. (1932). "Whaling in the Antarctic." Henry Holt and Co, New York.

Couquiaud-Douaze, L. (1999). "Dolphins and Whales: Captive Environment Guidebook." National University of Singapore.

de Courcy, C. (1999). The origin and growth of zoos. *Endang. Spec.* **1**, 16–19.

Department of Commerce. (1994). Public Display of Marine Mammals—Notice. Federal Register 59(193). Document ID 092994A. Accessed 20 July 2007. frwebgate2.access.gpo.gov/cgi-bin/waisgate.cgi?WAISdocID = 45861835010 + 0+0 + 0&WAISaction = retrieve

Hamilton, R. (1835). Mammalia: Whales. *In* "The Naturalist's Library" (W. Jardine, ed.), Vol. 26, pp. 228–232. W. H. Lizars, Edinburgh.

Hodskins, L.G. (ed.) (1998). AZA Annual Report on Conservation and Science 1996–97. Volume II. Member Institution Conservation and Research Projects. American Zoo and Aquarium Association, Silver Spring.

Mench, J. A., and Kreger, M. D. (1996). Ethical and welfare issues associated with keeping wild animals in marine life parks, zoos and aquariums. *In* "Wild Mammals in Marine Life Parks, Zoos and Aquariums: Principles and Techniques" (D. G. Kleiman, M. E. Allen, K. V. Thompson, and S. Lumpkin, eds), pp. 5–15. University of Chicago Press, Chicago.

Reeves, R. R., and Mead, J. G. (1999). Marine mammals in marine life parks, zoos and aquariums. *In* "Conservation and Management of Marine Mammals" (J. R. Twiss, Jr., and R. R. Reeves, eds), pp. 412–436. Smithsonian Institution Press, Washington, DC.

Roper Starch Worldwide (1995). "Public Attitudes toward Zoos, Aquariums and Animal Theme Parks." Roper Starch Worldwide, New York.

Solski, L. (2006). Public aquariums 1853–1914: Historical Perspective. *Zoo. Garten N.F.* **75**, 362–397.

# Marine Protected Areas

## ERICH HOYT

A marine protected area is defined by the International Union for the Conservation of Nature (IUCN) as any area of intertidal or subtidal terrain, together with its overlying water and associated flora, fauna, historical, and cultural features, which has been reserved by law or other effective means to protect part or all of the enclosed environment. Marine protected area, or "MPA," is the common generic term, although in various jurisdictions, MPAs are called marine reserves, marine parks, special areas of conservation (SACs), marine wildlife refuges, or national marine sanctuaries. The term "sanctuary," however, in reference to marine mammals, usually refers to the protection of a country's entire EEZ waters in a "national sanctuary" or to an "international sanctuary" on the high seas, e.g., the Indian Ocean Sanctuary. Such national and international sanctuaries typically ban cetacean or marine mammal hunting but rarely have in place detailed conservation measures, or a management plan.

MPAs have been set up to protect vulnerable species and ecosystems, to conserve biodiversity and minimize extinction risk, to re-establish ecosystem integrity, to segregate uses to avoid user conflicts and to enhance the productivity of fish and marine invertebrate populations around a reserve (Pauly *et al.*, 2002; Hooker and Gerber, 2004). MPAs are also useful in terms of providing a public focus for marine conservation (Agardy, 1997). A given MPA may have any one or several of the above goals. A highly protected MPA set aside as a fishery no-take zone, e.g., could be useful for marine mammal conservation by helping predators and prey to recover (Bearzi *et al.*, 2006). Also, setting up an MPA around marine mammals which function as umbrella species can often result in positive effects for many other species (Simberloff, 1998; Hoyt, 2005).

MPAs for marine mammals require targeted management measures to address marine mammal and ecosystem threats either as part of the MPA itself or through existing laws and regulations. Currently, in terms of conservation of most marine mammal populations, MPAs are too small, too few in number, and too weak in terms of protection, and most are "paper reserves"—MPAs in name only (Hoyt, 2005). Yet MPAs hold some promise for marine species and ecosystems when they include substantial highly protected (IUCN Category I) zones, use ecosystem-based management (CBM) principles, and function as part of larger MPA networks.

## I. The Recent Growth and Development of Marine Protected Areas

Even though 71% of the surface of the Earth is ocean, the concept of MPAs is relatively recent, lagging far behind land-based protected areas. The Durban Accord and Action Plan from the V World Parks Congress in 2003 stated that approximately 11.5% of the world's land area has protected status compared to less than 1% of the world ocean and adjacent seas. The first notable MPA of appropriate scale was the Great Barrier Reef Marine Park (GBRMP), established in 1975, although it only achieved a strong level of protection in 2003. Its size, at 340,000 km², makes it one of the largest MPAs in the world managed on a zoned basis. In 2003, nearly a third of it, 111,700 km², became a highly protected, "no take" zone. Although created to protect the world's largest coral reef, GBRMP also contains cetacean populations including mating and calving humpback whales (*Megaptera novaeangliae*) and various dolphins.

The world's first MPA set up specifically for marine mammals was Laguna Ojo de Liebre, or Scammon's Lagoon, established by the Mexican Government in 1971 to protect a prime gray whale (*Eschrichtius robustus*) mating and calving lagoon in Baja California (see Fig. 1). In 1988 the surrounding area of desert and coast was brought together with the San Ignacio and Guerrero Negro lagoons

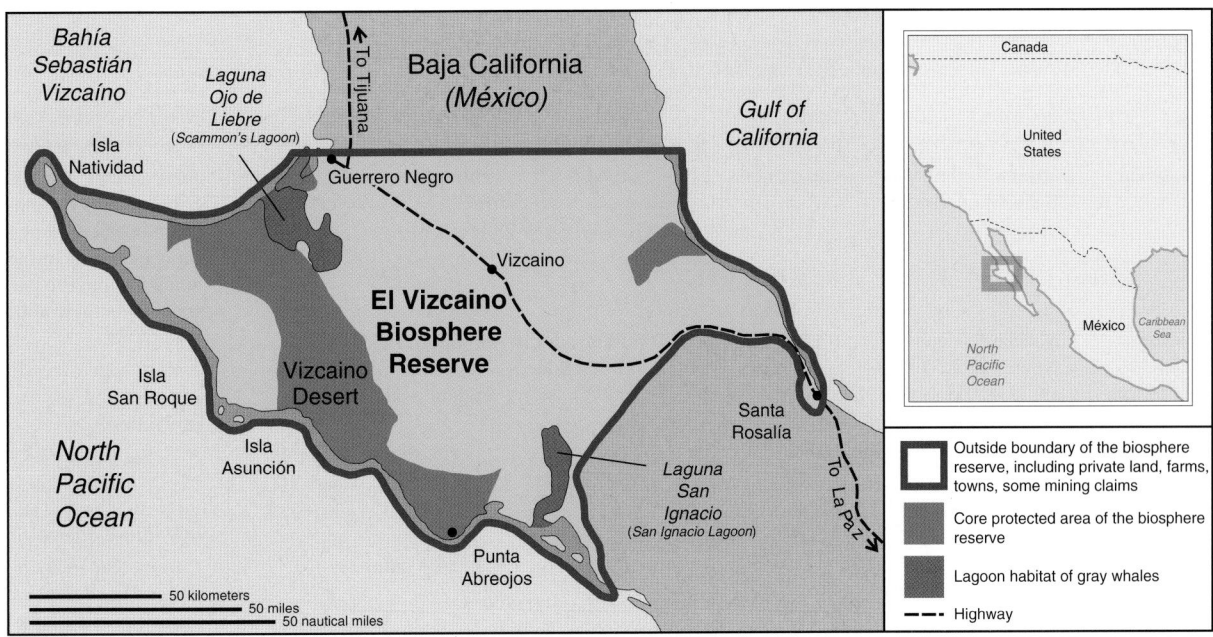

**Figure 1** El Vizcaino Biosphere Reserve. *In 1971 Laguna Ojo de Liebre was made a whale refuge by Presidential decree; protection of Laguna San Ignacio and Laguna Guerrero Negro followed in 1979 and 1980; in 1988, the entire lagoon complex was officially designated a MAB biosphere reserve and UNESCO World Heritage site status followed in 1993. Map by Lesley Frampton.*

to form El Vizcaino Biosphere Reserve. Among cetaceans, gray whales have easy-to-define habitat requirements because they bring their calves every winter to semi-enclosed salt-water lagoons.

Worldwide, as of 2008, there were more than 4500 MPAs, most of them declared in the past two decades. At least 375 MPAs feature or include marine mammals, while a further 200 have been proposed (Hoyt, 2005). Some notable MPAs are shown in Table I. Only an estimated 0.0001 (one ten-thousandth, or 0.01%) of the world ocean is set aside in highly protected IUCN Category I areas.

New Zealand MPA pioneer Bill Ballantine says that we should aim for at least 10% of the world ocean to be in highly protected MPAs (Ballantine, 1995). In a 1998 statement entitled "Troubled Waters: A Call to Action," more than 1600 scientists and conservationists declared that we should aim for 20% of the sea as highly protected MPAs by the year 2020 (Roberts and Hawkins, 2000). Other calls, mainly to address the worldwide collapse of commercial fisheries, have suggested between 20% and 50% of the sea to be protected to enable over-exploited fish stocks to recover. The consensus from MPA practitioners around the world at the V World Parks Congress was that at least 20–30% of each marine and coastal habitat should be in highly protected areas.

## II. Criteria for Selecting MPAs for Marine Mammals

Protecting mobile, wide-ranging marine mammals such as cetaceans and pinnipeds presents unique challenges using the tool of a fixed-boundary protected area. A number of marine mammal species migrate thousands of kilometers twice a year to feeding or breeding areas. Even the feeding or breeding grounds of a population may be spread over a wide area. Although some populations have site fidelity, there may be considerable movement within individual feeding or breeding grounds. Such dispersion is partly due to the peculiarity

of ocean habitat. Besides static topographic features defined largely by slope and depth (the marine equivalent of mountains and valleys), there are persistent yet ever-moving hydrographic features such as currents and frontal systems and ephemeral habitats created by wind- or current-driven upwellings and eddies (Hyrenbach *et al.*, 2000). The option is to conserve large areas and build in flexible boundaries for intra-seasonal protection to accommodate uncertainty, as well as to build networks of MPAs as described in Section VII.

The starting point for establishing marine protected areas should be long-term research of populations of marine species and ecosystems (Simmonds and Hutchinson, 1996; Twiss and Reeves, 1999). "Snapshot" boat or aerial surveys or single season studies are not enough; a several year research period, with dedicated surveys and quantified effort, is ideal. Spatial habitat preference modeling, combining marine mammal sightings, and behavioral observations with oceanographic/environmental and physiographic data, can then be used to characterize cetacean habitats, e.g., as has been done in the western Mediterranean Sea in studies of various dolphins and fin whales (*Balaenoptera physalus*) (Cañadas *et al.*, 2002; Cañadas *et al.*, 2005; Cañadas and Hammond, 2006). The resulting work has lent a strong scientific basis to the choice of marine habitat suitable for protection: the so-called cetacean critical habitat.

However, the dilemma is that time is short for protection and studies can be costly and time-consuming. Partial knowledge must often dictate action to protect populations or ecosystems, with additional research employed as it comes in to refine boundaries and extent of coverage. In Australian waters, governments have taken the approach of seeking to identify critical habitat for marine species, including whales and dolphins, before awarding formal habitat protection (Prideaux, 2003b). This approach, as long as it does not become an excuse for delay, could significantly increase the potential value of future MPAs with cetaceans.

**M**

**TABLE I**
**Notable Marine Protected Areas, National & International Sanctuaries for Marine Mammals**[a]

| Name | Location | Size | Date | Species | Notes |
|---|---|---|---|---|---|
| Abrolhos National Marine Park | 58 km off southern Bahia State, Brazil | 913 km$^2$ | 1984 | Humpback whale, large coral reef | First national marine park protects coral reef & humpback breeding areas; pioneer seismic protection zones around the park have recently been withdrawn |
| Auckland Islands Marine Mammal Sanctuary & Marine Reserve | Auckland Islands, New Zealand subantarctic, 460 km south of New Zealand | 4,840 km$^2$ | 1993; addt'l protection: 2003 | Southern right whale; New Zealand sea lion; various endemic fauna & flora | Important sea lion colony & right whale breeding area; high level of protection |
| Bunaken Marine National Park | off Manado in Minahasa Province, on north Sulawesi, Indonesia | 790.6 km$^2$ | 1991 | Sperm, short-finned pilot & other tropical whales & dolphins; large coral reef | At first a paper park but improving with tourism interest; allows fishing & other development |
| Banks Peninsula Marine Mammal Sanctuary | Banks Peninsula, east coast of South Island, New Zealand | 1140 km$^2$ | 1988 | Hector's dolphin | Commercial gill nets banned but protected area may need to be larger so as not to fragment populations |
| El Vizcaino Biosphere Reserve | Baja California, México | 25,468 km$^2$ | 1971 | Gray whale | Mating & calving grounds |
| Galápagos Marine Resources Reserve & Galápagos Whale Sanctuary | Galápagos Islands, 975 km west of Ecuador, in the equatorial Pacific, Ecuador | 158,000 km$^2$ | 1979; Whale Sanctuary: 1990 | Galápagos sea lion, Galápagos fur seal; sperm, short-finned pilot, humpback whale & tropical dolphins; various endemic fauna & flora | High protection for wildlife but concern about tourism impacts |
| Gerry E. Studds Stellwagen Bank National Marine Sanctuary | Southern Gulf of Maine off Massachusetts, USA | 2,181 km$^2$ | 1993 | Humpback, North Atlantic right, fin, minke, pilot whales | Pop-ups to monitor noise levels (ensonification); no discharge or mining but fishing unrestricted |
| Great Barrier Reef Marine Park & World Heritage Area | Queensland, Australia | 340,000 km$^2$ | 1975 | Humpback whale, dwarf minke whale, dugong; large coral reef | 1/3 highly protected area; intensive management |
| Hawaiian Islands Humpback Whale National Marine Sanctuary | Hawaiian Islands, USA | 3,368 km$^2$ | 1997 | Humpback whale; tropical dolphins | MPA based around humpback whale; no restrictions on fishing or military activities |
| Ilhas Desertas Natural Reserve | Desertas Islands, Madeira | 96.7 km$^2$ | 1990 | Mediterranean monk seal (also: bottlenose dolphin) | High level of protection & small seal population has increased |
| Commander Islands Biosphere Reserve (Zapovednik) | Commander Islands, 50 km east of Kamchatka Peninsula, Russia | 36,490 km$^2$ | 1993 | Sperm & killer whales; fur seal; Steller sea lion; spotted harbor & ringed seals; sea otter | Protected zone extends 50 km around Commander Islands; largest Russian MPA; local fishing, hunting & tourism allowed |
| Marine Mammal Sanctuary of the Dominican Republic | Northeast of the Dominican Republic including Silver Bank, Navidad Bank & part of Samaná Bay, Dominican Republic | 2,500 km$^2$ | 1986, 1996 | Humpback whale; Bryde's, pilot & sperm whales; bottlenose & pantropical spotted dolphins; manatee | Law of the Environment & Natural Resources (2000) strengthens environmental standards & protects coral reefs with breeding areas for humpbacks. From Feb–Apr, Silver Bank has densest concentration of humpbacks in North Atlantic (up to 3,000 present); humpbacks from 5 feeding stocks in the w. North Atlantic aggregate on the bank. |

*(continues)*

M

TABLE I (continued)

| Name | Location | Size | Date | Species | Notes |
|---|---|---|---|---|---|
| Monterey Bay National Marine Sanctuary | North central California coast, USA, adjoining Gulf of the Farallones NMS | 13,802 km² | 1992 | Gray, fin, blue, minke, humpback whales; various dolphins; sea otter; Steller & California sea lions | Largest US NMS (part of growing California state & national MPA network) prohibits oil & gas exploration & waste dumping but no fisheries restrictions |
| Northeast Greenland National Park | Northeast Greenland | Land: 846,100 km²; water: 110,600 km² | 1974, expanded 1988 | Beluga, narwhal, minke whale, white-beaked dolphin, walrus, polar bear | Commercial hunting & mineral development banned but concern remains about high levels of subsistence hunting |
| Papahanaumokuakea Marine National Monument | Northwest Hawaiian Islands, USA | 340,000 km² | 2006 | Humpback whales; tropical dolphins | World's largest highly protected MPA; to allow no commercial fishing & limited tourism |
| PELAGOS Sanctuary for Mediterranean Marine Mammals | Ligurian Corsican & northern Tyrrhenian seas, western Mediterranean Sea, Italy, Monaco, France & High Seas | 87,492 km² | 1999; SPAMI, 2001 | Fin, sperm, minke, Cuvier's beaked whale; bottlenose, striped, common & Risso's dolphin | More than 50% on the high seas of the Mediterranean; first high seas as well as tri-national MPA |
| Saguenay-St. Lawrence Marine Park | St. Lawrence River, Québec, Canada | 1,138 km² | 1998 | Beluga, fin, humpback & minke whales; Atlantic white-sided dolphin & harbor porpoise | Multiple-use zoning with heavy traffic; protects most southerly population of belugas from tour boats |
| Seal Bay Conservation Park | Kangaroo Island, South Australia, Australia | 49.5 km² | 1971 | Australian sea lion (also: New Zealand fur seal) | Intensively managed to protect population & for tourism |
| Shark Bay Marine Park & World Heritage Area | Shark Bay, Western Australia | 23,000 km² | 1990 | Indo-Pacific bottlenose dolpin, humpback whale, dugong, green & loggerhead turtles | High visitor level especially to Monkey Mia; tourism controlled |
| National EEZ Sanctuaries for Marine Mammals | 21 countries & territories (10 in the South Pacific) | 120,000 km² to 16 million km² | Various | Mainly all cetaceans; some include all marine mammals plus turtles | No hunting. These are not MPAs, but with management plans & enhanced protected zones could help conservation. |
| International Whale Sanctuaries | Indian Ocean Sanctuary, Southern Ocean Sanctuary, Eastern Tropical Seascape | IOS: 103.6 million km²; SOS: 50 million km²; ETS: 2.1 million km² | Various | Mainly baleen whales & sperm whale | No commercial hunting |

[a]For a complete directory of more than 600 proposed & existing MPAs & sanctuaries for cetaceans, see www.cetaceanhabitat.org

Defining critical habitat is the crux of the matter. Governments and other agencies have adopted various definitions for critical habitat, but, it is essentially the places, or conditions, where marine mammals feed, socialize, rest, breed, and raise their young as well as where their prey lives. Some times part of migration routes are included, too. The challenge is determining the level of protection needed as well as when areas are essential for day-to-day survival, as well as for maintaining a healthy population growth rate.

The actual selection process for MPAs starts with defining the goals of any proposed MPA in view of marine mammals found and threats to their existence paired with the need to devise the rationale for the proposal (Hoyt, 2005; Notarbartolo di Sciara, 2007). Threats include fishing conflicts (overfishing, bycatch, entanglements), ship collisions, pollution, habitat degradation, and the chronic, high noise levels (ensonification) from shipping traffic as well as acute loud sounds from seismic activities, and low- and mid-frequency Navy sonar. To date, few managent plans for MPAs have addressed fishing conflicts, pollution, or habitat degradation, and noise in a comprehensive way.

Then the question must be asked: Is an MPA the most effective tool—the answer or part of the answer—in terms of addressing threats to marine mammals and ensuring that a favorable conservation status is maintained? At the same time, stakeholders must be brought on board from the start so that the MPA selection process ideally grows out of a community, taking into account socio-economic and other concerns. The most effective MPAs proceed from the bottom up;

M

top-down approaches usually only work if they institute bottom-up procedures early in the process. Even then, it is sometimes impossible to orchestrate public participation and such MPA proposals may ultimately fail.

The next steps are to compile bibliographic information, collect updated scientific data on the animals, human activities, and the threats; and to recommend highly protected zones or core areas as within the MPA. A comprehensive proposal with maps and information on every aspect should then be presented to stakeholders as well as authorities involved in the legal process. This is rarely a one-time process but usually involves a lengthy consultation phase during which stakeholders examine the proposal and help to shape it until conflicts are resolved and acceptable proposals can be formulated.

## III. Designing MPAs for Marine Mammals

MPAs either tend to be managed for multiple- or zoned use. Management for multiple-use is found to a great extent in the flagship US national marine sanctuaries, as well as in the special areas of conservation created under the European Union (EU) Habitats and Species Directive. Multiple-use allows or in some cases tries to regulate a wide variety of uses, from shipping and tour boat traffic to sports and commercial fishing, at "acceptable" levels of use throughout the marine protected area. Of course, some uses may be entirely excluded if deemed too harmful—e.g., oil and gas exploration, waste dumping, and certain kinds of fishing.

In contrast, zoned use, or zoning, attempts to create zones in locations and at sizes appropriate for one or more compatible uses, excluding other uses, but attempting to accommodate all or most uses within a number of zones located within a single MPA. Of course, not every MPA can accommodate every use; many are too small and are most suited to a high level of protection throughout the MPA.

Multiple-use management has had a long history on land, with mixed, often poor results, but land-based protected areas have now employed zoned use successfully for several decades and this is the widely accepted model for many national parks and protected areas. The biosphere reserve concept uses zoning for land-based protected areas and this has also been adapted for MPAs (Batisse, 1990). Biosphere reserves feature a zoned architecture with substantial key core areas reserved for strict protection, surrounding zones for research, tourism, education, and other "light use," and still other zones open for sustainable use of marine resources and as transition areas to the wider community (see Fig. 2).

Highly protected core areas are easy to define for marine mammals spending time on rookeries and haul-outs, and with confined home ranges such as for bottlenose dolphins (*Tursiops* spp.) or humpback and gray whales on winter breeding grounds. But what about marine mammals with less well-defined breeding habitat or on feeding grounds subject more to changing oceanographic conditions? One solution could be to employ adaptable time and area closures such as are used for salmon or other fisheries in various parts of the world. It would be possible to use the biosphere reserve concept to create large overall MPAs with a number of moveable, highly protected "core areas" corresponding to marine mammal critical habitat with boundaries that can be adjusted as needed. Such adjustments would be constantly reviewed and sensitive to seasonal and annual signals from the wider environment. To achieve this fine-grained kind of critical habitat management, however, it is necessary to try to understand ecosystem processes and the impacts that humans have on such processes. An appropriate tool for this is EBM.

## IV. Ecosystem-Based Management

EBM, or ecosystem management, is the management of the uses of ecosystems. An ecosystem *per se* needs no management. It is the escalating human interactions with ecosystems and the damaging human impacts on ecosystems and species that need to be managed. Still, it has become clear that human uses must be accommodated within ecosystem capacities. EBM is a regime that recognizes that ecosystems are dynamic and inherently uncertain yet seeks to manage the human interactions within ecosystems to protect and maintain ecological integrity and to minimize adverse impacts. EBM is widely talked about and is being attempted by some managers but it remains at an embryonic stage, though Australia, e.g., is building its regional marine planning on EBM (Smyth *et al.*, 2003).

To embark on EBM, fundamental shifts in management thinking and research must take place (Hoyt, 2005):

- Management must move from a *reactive* to a *proactive* style. This requires ongoing scientific analysis and the ability to adapt management practice quickly when new information signals the need for a change.
- Research has to re-orient itself to view the ecosystem as a whole, using multiple components such as stability of reef or sea floor, predator presence and water quality as indicators of management success.
- Risk assessments of management choices must be reviewed regularly and adapted to new information.
- Multiple sectoral uses (e.g., commercial and sports fishing) as well as the resulting impacts (e.g., cetacean bycatch), must be viewed as cumulative rather than isolated.
- Managers, policy makers and the public must be alert to the misuse of the term "EBM," particularly by those seeking to justify the culling of predators.
- The ultimate aim is to maintain the ecosystem as it naturally occurs—not to adapt it to human needs but to enable it to accommodate an acceptable level of human use.

Thus, it is important to understand more about the whole ecosystem, rather than focusing on one or other isolated area or species. Without doubt, these are major tasks to undertake in any large marine area, but they are necessary steps to manage human involvement with marine ecosystems.

EBM as a management regime grew out of the widely acknowledged failure of single species management, primarily of fisheries. EBM requires an ongoing research commitment to unravel and model the complex linkages in marine ecosystems. But where knowledge is lacking, it is accepted that a precautionary approach should be invoked to protect ecosystems (Hoyt, 2005). Part of this precautionary approach is creating MPAs as safeguards built into the system from an early stage to secure ecosystem integrity in the absence of scientific certainty.

## V. The Legal Process for Setting up MPAs

To achieve legal status, MPA proposals situated within a country's waters must seek state/provincial/local and/or national approval in law. Such legal status along with appropriate enforcement provisions can be difficult and time-consuming to establish; some governments have only recently approved MPA legislation and others have weak or even no legislation available (Scovazzi, 1999; Hoyt, 2005).

In most parts of the world, regional treaties and international organizations are available to assist with the MPA designation process. These bodies include the IUCN World Commission for

M

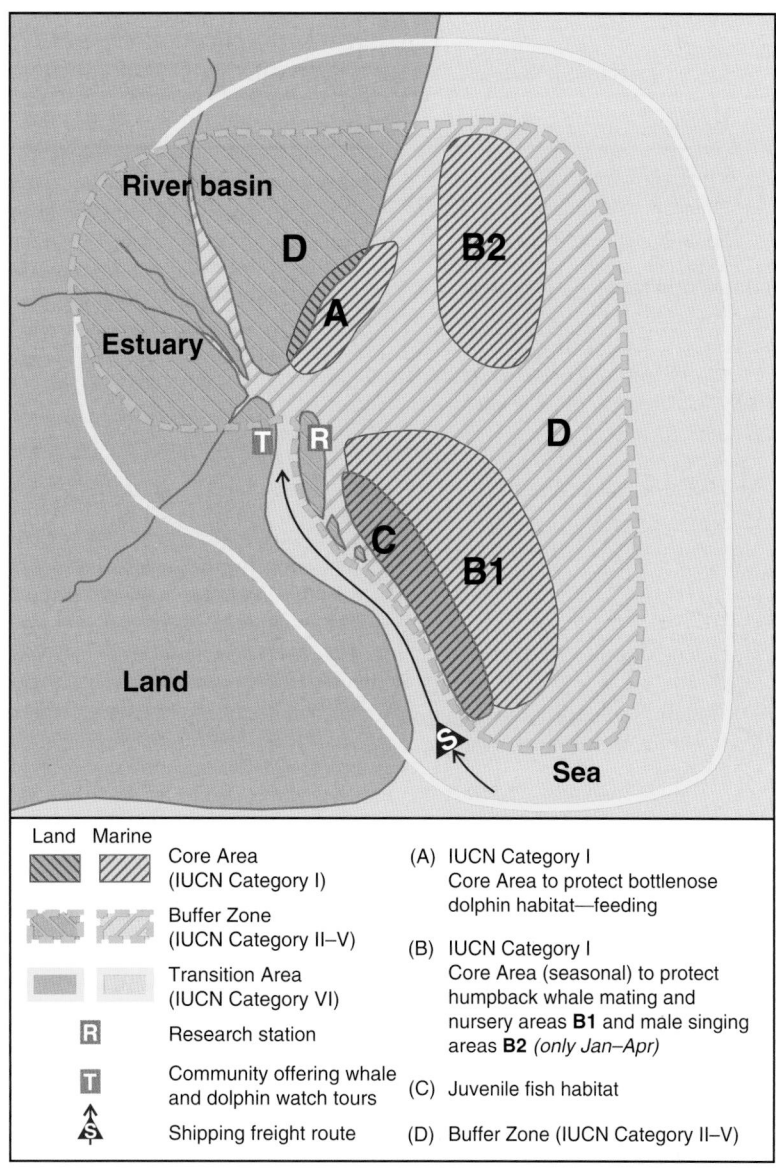

**Figure 2** The Architecture of a Biosphere Reserve. *This map shows the various zones of a hypothetical marine- and land-based biosphere reserve area. Core areas (IUCN Category I) are devoted to strict nature reserve protection; these are surrounded by buffer zones (Category II–V) where activities compatible with the conservation objectives occur, and the buffer zones are in turn surrounded by a more or less defined transition zone (Category VI) which integrates the local people and sustainable resource management into the fabric of the overall reserve. To be effective, the biosphere reserve model must include zoned highly protected areas that are declared and enforced through legislation with management plans formulated by the community, including all stakeholders. Map by Lesley Frampton.*

Protected Areas (WCPA, Marine) with its many regional offices, as well as regional agreements such as ACCOBAMS (the Agreement on the Conservation of Cetaceans of the Black Sea, Mediterranean Sea and contiguous Atlantic Area), the SPAW Protocol in the Caribbean, and the SPREP Convention in the South Pacific.

In addition, international recognition and further protection of an MPA can be valuable. The International Maritime Organization (IMO), e.g., has granted the status of "particularly sensitive sea area" (PSSA) to parts of the GBRMP; such a PSSA requires a compulsory pilotage system through the most sensitive parts of the park. MPAs can also be awarded further protection through World Heritage Site or MAB Biosphere Reserve status (UNESCO), or at the regional level, e.g., designation as a Special Protected Area of Mediterranean Interest (SPAMI). These and other designations are helpful in terms

## TABLE II
### Developing Networks of MPAs for Marine Mammals

| Name | Location | Action | Species |
|------|----------|--------|---------|
| 1. Wadden Sea Conservation Area | SE North Sea off Germany, Denmark, Netherlands | Series of national parks & nature reserves | harbor seal; also harbor porpoise present |
| 2. Sister sanctuaries of Gerry E. Studds Stellwagen Bank National Marine Sanctuary & Marine Mammal Sanctuary of the Dominican Republic | Southern Gulf of Maine off Massachusetts & Caribbean Sea off NE Dominican Republic | Bilateral "sister sanctuary" relationship formally established 2007 with education, research & other planned links | humpback whale on feeding & breeding grounds |
| 3. Natura 2000 network | European Union (EU) waters | Special areas of conservation (SACs) declared for *Tursiops* & *Phocoena* but no protection yet for all other cetaceans | bottlenose dolphin, harbor porpoise |
| 4. MPAs proposed for the ACCOBAMS MPA Work Program (the Agreement on the Conservation of Cetaceans of the Black Sea, Mediterranean Sea & contiguous Atlantic Area) | Mediterranean & Black seas | 18 MPAs proposed 2007 by the Scientific Committee to the Parties to protect cetaceans throughout the ACCOBAMS region; countries have agreed in principle to begin work to establish the MPAs | bottlenose, common, Risso's & striped dolphins; harbor porpoises; fin, sperm, Cuvier's beaked, killer whales |
| 5. 21 manatee sanctuaries | Gulf of Mexico, west coast of Florida | Several protected areas for the critical habitat of manatees | West Indian manatee |
| 6. Various protected areas & recovery plans | Western North Atlantic including Bay of Fundy & Gulf of Maine & approaches | Several marine sanctuaries, conservation zones, species recovery plans & IMO & pilot controls on shipping off NE US & Canada but remains to be seen if this will reduce mortalities from ship strikes. | North Atlantic right whale |
| 7. 9 national (EEZ) marine mammal sanctuaries established among SPREP (South Pacific Region Environment Program) Convention members | South Pacific Ocean: American Samoa, Cook Islands, Fiji, French Polynesia, New Caledonia, Niue, Papua New Guinea, Samoa, Vanuatu | National sanctuaries set up to protect marine mammals & ban whaling in national waters; countries through SPREP plan to devise management &/or zoned conservation plans | Humpback whale, tropical dolphin species including blackfish species; dugongs; turtles in some countries |

**M**

of expanding the remit of an MPA into an important component of an international network.

Whatever designations are obtained, it is useful to consider achieving protection status not as a final goal but as a *first* stage or step toward conservation (Hoyt, 2005).

In some countries, MPAs are situated and must also be considered in the broader context of a general management plan for coastal and marine resources, that is, an umbrella program for conservation of renewable resources as well as implementation of EBM principles (Salm and Clark, 2000; Augustowski and Palazzo, 2003).

## VI. Management Plans, Monitoring, and Evaluation

The management plan is at the heart of the success or failure of an MPA. It is the working plan for what the MPA hopes to do and accomplish along with the time frame for its activities and the schedule for its review.

Many MPAs exist only or mainly on paper. It is fair to say, however, that all MPAs begin as a piece of paper, and it is up to government and stakeholders to devise, put in practice, and enforce their

management plan (Hoyt, 2005). Creating effective protected areas, whether marine- or land-based, is an iterative, participatory process, and is bound to fail if the management plan is seen as set in stone or as a fixed law imposed from the outside. As with MPA design, management must be both top-down and bottom-up. The following are the key steps leading to effective management (Hoyt, 2005; Notarbartolo di Sciara, 2007):

(1) engaging stakeholder involvement from the beginning and throughout the process;
(2) formulating clear management objectives for the proposed MPA;
(3) creating a management body;
(4) developing a management plan, subject to periodic re-examination and revision;
(5) offering management training;
(6) conducting research for baseline numbers, inventory, status and monitoring purposes;
(7) promoting and offering educational programs for the local community and visitors;
(8) developing effective enforcement regimes; and
(9) conducting periodic management review and other evaluations to assess whether objectives are being met.

The last provision is essential to the long-term success of the MPA. Without such evaluations, even MPAs that start out with considerable success may decline in value and fail. An MPA must have clearly defined objectives against which its performance is regularly checked, and a monitoring program to assess management effectiveness and recommend changes (Kelleher, 1999). A number of methods are available for conducting a review (Pomeroy *et al.*, 2004).

Management of an MPA for cetaceans and pinnipeds is similar to managing any other type of MPA (Notarbartolo di Sciara, 2007) but there are several differences that must be kept in mind. MPAs for marine mammals require large sizes to accommodate these highly mobile animals, with all the attendant complications and added problems from size alone. The movement of populations across many national borders and even to opposite ends of an ocean dictate the necessity of creating MPA networks to ensure comprehensive protection. Finally, the use of high seas habitat by many populations of cetaceans and pinnipeds, means that effective legal measures—as well as practical mechanisms for implementation and enforcement—will need to be devised for the high seas (see Section VIII).

## VII. Networks of MPAs

An MPA network can be defined as "an organized collection of individual MPAs operating co-operatively and synergistically, at various spatial scales and with a range of protection levels, to fulfill ecological aims more effectively and comprehensively than individual sites could alone." (WCPA/IUCN, 2006). The idea of creating networks of MPAs is particularly suited to marine mammals. In addition to their long migrations, marine mammals may depend on food webs whose critical habitats are widely separated. Thus, networks are essential to create an effective conservation plan for these wide-ranging species, as well as for the marine ecosystems that help to support them.

A number of MPA networks are beginning to be assembled to confer population-level protection to marine mammals (Reeves, 2000; Hoyt, 2005) (see Table II.) Establishing a network is mainly a "top-down" exercise with governments or regional associations acting as the main initiators and mechanisms. In Table II, individual governments with large territories or undertaking bilateral agreements (United States, Canada, Dominican Republic) have been responsible for nos. 2, 5, and 6. Regional associations, including political and economic unions such as the EU, and conservation agreements and treaties such as ACCOBAMS which draw on the Barcelona Convention are responsible for 1, 3, and 6.

## VIII. High Seas MPAs

Many marine mammal species, including sperm (*Physeter macrocephalus*), beaked, and other toothed whales, large baleen whales and a number of pinnipeds spend part or even most of their life cycles in pelagic waters off the continental shelves and far from the coasts. Large portions of their critical habitats may be in the 50% of the world ocean classed as international waters, or high seas, i.e., outside the 200 nm limits declared by most countries under the United Nations Convention on the Law of the Sea (UNCLOS) (Hoyt, 2005). In such areas—where no single state or authority has the power to designate MPAs, adopt management schemes, or enforce compliance—new strategies must be devised to protect and manage high seas habitats (Thiel and Koslow, 2001).

Various international agreements have the potential to be used to create high seas MPAs. For example, UNCLOS says that States are in a position to take strong conservation measures on the high seas, as long as they cooperate with other States, show that the measures they want to take would enhance the conservation of resources, and that they are based on the best scientific evidence available (de Fontaubert, 2001). Article 194 of UNCLOS establishes a mandate for high seas MPAs by stipulating measures to protect rare and fragile ecosystems as well as the habitat of depleted, threatened, or endangered species and other forms of marine life while Article 197 asks for cooperation on a global basis (Prideaux, 2003a).

Another key treaty, the Convention on Biological Diversity (CBD), has with the work of its scientific advisors, the Subsidiary Body on Scientific, Technical, and Technological Advice (SBSTTA), planned a program of work that includes the creation of high seas MPAs. UNCED: Agenda 21, although it is a "soft-law" instrument, also recognizes the possibility of enacting MPAs on the high seas.

The Convention on the Conservation of Migratory Species of Wild Animals (CMS) may also become an important instrument for high seas critical habitat protection. The harmonization of work plans between CBD and CMS integrates CMS and migratory species into the work program and implementation of CBD with regard to protected areas, as well as the ecosystem approach, and the drive to develop indicators, assessments, and monitoring. In addition, CMS focuses on the establishment of regional agreements (such as ACCOBAMS), which increases its adaptability to regional circumstances. If high seas and multi-jurisdictional cetacean critical habitats are to be protected, CMS and CMS regional agreements may be the most appropriate framework to develop this regime (Prideaux, 2003b).

Besides all of the above approaches (UNCLOS, CBD, CMS, and UNCED: Agenda 21), the IWC whale sanctuaries provide a useful precedent of nations working together to agree on conservation on the high seas. Future IWC agreements could embrace, or even create themselves, highly protected high seas MPAs, though current divisions in the IWC make this unlikely in the near future. In any case, it is important to recognize that those states that are not party to the various conventions and treaties are not bound by them. Yet most states now recognize or are party to at least two of the important conventions for future high seas MPA development: UNCLOS and CBD. Still, it is a huge challenge for the world's nations to come together with the necessary foresight and imagination to create a comprehensive network of MPAs on the new frontier of the high seas.

In 1999, an agreement to create the world's first high seas MPA was signed by France, Monaco, and Italy. The PELAGOS Sanctuary for Mediterranean Marine Mammals, located partly in the national waters of these three countries and partly on the high seas, contains resident populations of sperm, fin, and Cuvier's beaked whales (*Ziphius cavirostris*), as well as striped (*Lagenorhynchus coeruleoalba*), common bottlenose (*Tursiops truncatus*), Risso's (*Grampus griseus*) and short-beaked common dolphins (*Delphinus delphis.*) (see Fig. 3). In 2001, a high seas agreement was forged under the Barcelona Convention, making PELAGOS a SPAMI which confers the official protection of all signatory Mediterranean countries in both national waters and on the high seas (Notarbartolo di Sciara *et al.*, 2008). It could take several years for PELAGOS to come up to speed and to function as a valuable conservation tool. The marine mammals of the Mediterranean are important of course, but no less important is the precedence of both transborder and high seas cooperation by this designation and the implications for other potential areas and cooperation by States. For these reasons, it is hoped that the management plan put in place will employ EBM principles and be effective in terms of identifying and protecting marine mammal critical

M

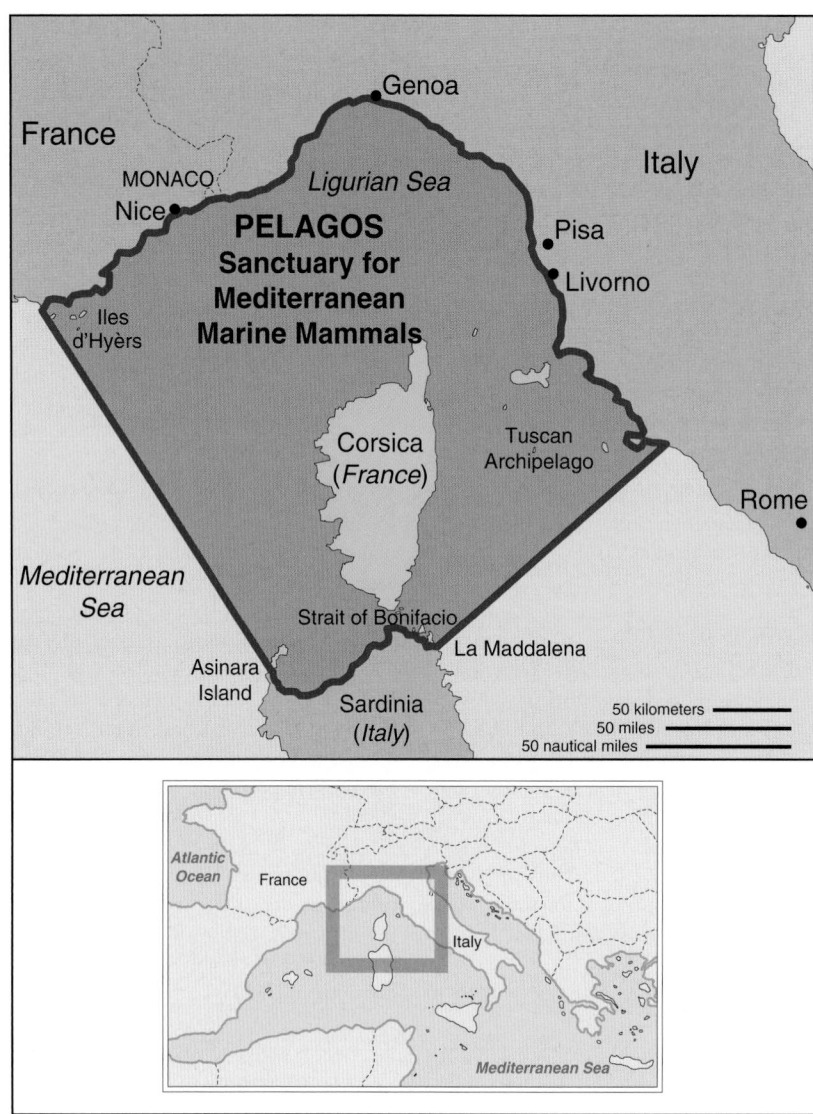

**Figure 3** Map of the PELAGOS Sanctuary for Mediterranean Marine Mammals. *The PELAGOS Sanctuary for Mediterranean Marine Mammals is the world's first high seas MPA. It was designated in 1999 as a transborder MPA in the national waters of France, Italy, and Monaco. In 2001, it was declared a Special Protected Area of Mediterranean Interest (SPAMI) under the Barcelona Convention which gives it protection on the high seas of the Mediterranean. Map by Lesley Frampton.*

habitat core areas as well as responsive to new data and management strategies as they arise in future.

## *See Also the Following Articles*

Conservation Efforts ■ Endangered Species and Populations ■ Habitat Pressure ■ Management

## *References*

Agardy, T. (1997). "Marine Protected Areas and Ocean Conservation." Academic Press, London.

Augustowski, M., and Palazzo, Jr., J. T. (2003). Building a marine protected areas network to protect endangered species: Whale conservation as a tool for integrated management in South America. V World Parks Congress, IUCN, Durban, South Africa, Sept. 2003.

Ballantine, W. J. (1995). Networks of "no-take" marine reserves are practical and necessary. In "Marine Protected Areas and Sustainable Fisheries. Proceedings of a Symposium on Marine Protected Areas and Sustainable Fisheries Conducted at the Second International Conference on Science and the Management of Protected Areas" (N. L. Shackell and J. H. M. Willison, eds), Dalhousie University, Halifax, Nova Scotia, 16–20 May 1994, 13–20.

Batisse, M. (1990). Development and implementation of the biosphere reserve concept and its applicability to coastal regions. *Environ. Conserv.* **17**, 111–116.

Bearzi, G., Politi, E., Agazzi, S., and Azzellino, A. (2006). Prey depletion caused by overfishing and the decline of marine megafauna in eastern Ionian Sea coastal waters (central Mediterranean). *Biol. Conserv.* **127**, 373–382.

Cañadas, A., and Hammond, P. S. (2006). Model-based abundance estimates for bottlenose dolphins off southern Spain: Implications for conservation and management. *J. Cetacean Res. Manag.* **8**, 13–27.

Cañadas, A., Sagarminaga, R., and García-Tiscar, S. (2002). Cetacean distribution related with depth and slope in the Mediterranean waters off southern Spain. *Deep Sea Research I* **49**, 2053–2073.

Cañadas, A., Sagarminaga, R., De Stephanis, R., Urquiola, E., and Hammond, P. S. (2005). Habitat preference modelling as a conservation tool: Proposals for marine protected areas for cetaceans in southern Spanish waters. *Aquatic Conserv: Mar. Freshw. Ecosyst.* **15**, 495–521.

de Fontaubert, A. C. (2001). The status of natural resources on the high-seas—legal and political considerations. WWF/IUCN, Gland, Switzerland, 69–93 [Available online at: www.panda.org/resources/publications/water/highseas.pdf]

Hooker, S. K., and Gerber, L. R. (2004). Marine Reserves as a tool for ecosystem-based management: The potential importance of megafauna. *BioScience* **54**, 27–39.

Hoyt, E. (2005). "Marine Protected Areas for Whales, Dolphins and Porpoises: A World Handbook for Cetacean Habitat Conservation." Earthscan, Sterling. [Available online at www.cetaceanhabitat.org]

Hyrenbach, K. D., Forney, K. A., and Dayton, P. K. (2000). Marine protected areas and ocean basin management. *Aquatic Conserv: Mar. Freshw. Ecosyst.* **10**, 435–458.

Kelleher, G. (1999). "Guidelines for Marine Protected Areas." IUCN, Gland.

Notarbartolo di Sciara, G. (2007). Guidelines for the Establishment and Management of Marine Protected Areas for Cetaceans. Contract RAC/SPA, N° 03/2007:1–29.

Notarbartolo di Sciara, G., Agardy, T., Hyrenbach, D., Scovazzi, T., and Van Klaveren, P. (2008). The Pelagos Sanctuary for Mediterranean Marine Mammals. *Aquatic Conserv: Mar. Freshw. Ecosyst.* **18**, 367–391.

Pauly, D., Christensen, V., Guénette, S., Pitcher, T. J., Sumaila, U. R., and Walters, C. J. (2002). Towards sustainability in world fisheries. *Nature* **418**, 689–695.

Pomeroy, R. S., Parks, J. E., and Watson, L. M. (2004). "How is Your MPA Doing? A Guidebook of Natural and Social Indicators for Evaluating Marine Protected Area Management Effectiveness." IUCN, Gland.

Prideaux, M. (2003a). "Beyond the State: Building Regimes for Species Protection in all Oceans, Hawke Institute." University of South Australia, Adelaide.

Prideaux, M. (2003b). "Small Cetacea and World Politics; Developing Regimes for Species Survival." University of South Australia, Adelaide.

Reeves, R. R. (2000). The value of sanctuaries, parks, and reserves (protected areas) as tools for conserving marine mammals. Report to the Marine Mammal Commission, Contract No. T74465385.

Roberts, C. M., and Hawkins, J. P. (2000). "Fully Protected Marine Reserves: A Guide." WWF-USA, Washington, DC.

Salm, R. V., and Clark, J. R. (2000). "Marine and Coastal Protected Areas: A Guide for Planners and Managers," 3rd Ed. IUCN, Gland.

Scovazzi, T. (1999). "Marine Specially Protected Areas: The General Aspects and the Mediterranean Regional System." Kluwer Law International, Boston.

Simberloff, D. (1998). Flagships, umbrellas, and keystones: Is single-species management passé in the landscape era? *Biol. Conserv.* **83**, 247–257.

Simmonds, M. P., and Hutchinson, J. D. (1996). "The Conservation of Whales and Dolphins: Science and Practice." John Wiley and Sons, Chichester.

Smyth, C., Prideaux, M., Davey, K., and Grady, M. (2003). "Oceans Eleven: The Implementation of Australia's Oceans Policy and Ecosystem-based Regional Marine Planning." Australian Conservation Foundation, Melbourne.

Thiel, H., and Koslow, J. A. (eds) (2001). Managing Risks to Biodiversity and the Environment on the High Sea, including Tools such as Marine Protected Areas—Scientific Requirements and Legal Aspects. Report of the Workshop, 27 Feb–4 Mar, Vilm, Germany.

Twiss, J. R., Jr., and Reeves, R. R. (1999). "Conservation and Management of Marine Mammals." Smithsonian, Washington, DC.

WCPA/IUCN (2006). Establishing networks of marine protected areas: A guide for developing national and regional capacity for building MPA networks. Technical Report.

# Mark–Recapture

PHILIP S. HAMMOND

## I. Introduction

The capture, marking, release, and recapture of individual animals can be used to study movement patterns, the size and structure of populations, survival and recruitment rates (Hammond *et al.*, 1990). It is thus an important method for marine mammal researchers interested in ecology, life history, conservation, and management.

Mark–recapture analysis is widely used in ecology to estimate abundance and survival rates. The basic data required are a set of capture histories of individually identified animals. A capture history is simply a string of 1s and 0s representing whether an animal was (1) or was not (0) captured in a series of sampling occasions. A sampling occasion is a finite period of time during which data are collected, e.g., a survey day, but in studies of marine mammals sampling occasions are often taken to be longer, e.g., a season.

Mark–recapture analyses make a number of assumptions, the violation of which may lead to biased estimates of survival and, especially, abundance. Although it is straightforward to apply mark–recapture analysis to capture history data, great care must be taken to consider the effects of failing to meet these assumptions. This is especially true for studies that were not originally designed for such analyses. Although analytical models exist that allow some assumptions to be relaxed, there is no substitute for a well-designed field study.

## II. Capturing and Identifying Individuals

Mark–recapture methods were initially developed, and have mostly been used, for studies in which individual animals are physically captured in traps of some kind, marked either by the application of a tag or by mutilation, released, and then recaptured or resighted without capture. Pinnipeds can be captured on land during pupping and molting periods and at other times when they are hauled out on land or ice. They can also be captured in nets but are vulnerable to drowning. Typical ways to mark pinnipeds have been the application of flipper tags, and branding methods that are commonly used on terrestrial mammals, and, for short-term studies, gluing plastic numbered "hat-tags" on the heads of animals (Hall *et al.*, 2001).

Cetaceans are difficult to capture and may be vulnerable to being handled, and it has been necessary to develop additional ways of capturing and marking individuals. When whaling was still commonplace, large whales were marked by firing Discovery tags (metal bolts about 30-cm long) into the blubber and recaptured by recovering the tag when the animal was butchered in whaling operations. Some

**M**

more recent studies have physically captured and branded or tagged small cetaceans, but the associated logistical and welfare issues mean that this is not usually a viable method.

## A. Photo-Identification

The most commonly used technique for capturing and marking cetaceans is to take photographs of the natural markings of individual animals (Hammond *et al.*, 1990). Clearly, this is only appropriate if individuals of a species/population are sufficiently well marked. The use of photo-identification has the great advantage of avoiding physical capture, handling, or application of a mark. However, "photo-id" can also make it more challenging to meet the assumptions of analytical methods (see Assumptions). For example, the future recognition of a naturally marked individual may depend on the amount of information in the mark, how long the mark lasts, and the quality of the photograph.

Good photographs are essential for photo-identification; good enough that animals that are considered marked will be recognized with certainty if seen again later. Not all photographs will be of sufficient quality and in most studies photographs are graded according to quality, and only the best are used in analysis.

The list of cetacean species marked through photo-identification is long. It includes humpback (*Megaptera novaeangliae*), right (*Eubalaena* spp.), fin (*Balaenoptera physalus*), blue (*B. musculus*), sperm (*Physeter macrocephalus*), northern bottlenose (*Hyperoodon ampullatus*), killer (*Orcinus orca*) and pilot whales (*Globicephala* spp.), and several species of dolphins, particularly bottlenose dolphins (*Tursiops* spp.). Much of what we know about the biology of these species has come through photo-identification studies. This technique is also used for seals (Hiby and Lovell, 1990; Forcada and Aguilar, 2000; Forcada and Robinson, 2006; Gerondeau *et al.*, 2007).

## B. Genetic Tagging

Individuals can be uniquely identified by genotyping tissue obtained via a biopsy sample. Such genetic tags have been used in a large-scale mark–recapture analysis of North Atlantic humpback whales (Palsbøll *et al.*, 1997; Smith *et al.*, 1999).

## C. Other Marking Methods

Telemetry devices (radio tags) can also be used to mark individuals and provide mark–recapture data (Ries *et al.*, 1998). A recent development involves using mobile telephone technology to send SMS text messages from a tag attached to a seal, which are received when the animal is within range of the coast (McConnell *et al.*, 2004).

## III. Movement Patterns

The simplest information to come from mark–recapture is that an animal marked at a particular place and recaptured in another has clearly moved from one place to the other during the intervening time. This basic information can be used to explore animal movement patterns at a range of spatial and temporal scales from small-scale habitat use to long-distance seasonal migrations (Wilson *et al.*, 1997; Whitehead, 2001; Calambokidis *et al.*, 2001; Stevick *et al.*, 2003a).

## IV. Mark–Recapture Analysis to Estimate Survival Rates and Abundance

Mark–recapture analysis is widely used in ecological studies to estimate survival rates, and arguably it is for this purpose that it is best suited. However, in marine mammal science, although it has been used to estimate survival rates in a variety of species, it is most often considered as a tool for estimating population size (abundance). Time series of data that are sufficiently long to estimate survival rates are relatively rare for marine mammals, but there are many shorter datasets, particularly from relatively small and local photo-identification studies of cetaceans, which are, at least in principle, amenable for estimating abundance.

## A. Survival Rates

To estimate survival rates, cohorts of marked animals are sampled over a period of time, several years if the aim is to estimate annual survival rates. An animal that is not recaptured on a particular sampling occasion may either have died or still be alive but simply not recaptured. The basic task in analysis is to simultaneously estimate two quantities: the probability that an animal survived from one sampling occasion to the next; and the probability of recapturing an animal given that it was alive. Resulting estimates of survival probabilities (rates) are representative of the marked animals in the cohorts. The researcher must then infer population survival rates by assuming that the marked animals are representative of the wider population.

The conventional analytical approach for estimating survival rates is to use the Cormack–Jolly–Seber model. This allows not only time-dependent survival rates to be estimated but can also be collapsed to estimate the average survival rate over a period of time. This model and an expanding range of other more complex models can be implemented in the software MARK http://www.phidot.org/software/mark/, which is accompanied by an excellent guide (Cooch and White, 2007) containing statistical details and references. The other models include the "robust design" for estimating survival rate between primary sampling occasions (e.g., years) and estimating abundance within primary occasions from secondary sampling occasions (e.g., months). This model also allows parameters related to temporary emigration from the study area to be estimated (Bradford *et al.*, 2006). Also useful are models that utilize recaptures additional to those obtained from the designed sampling occasions, such as opportunistic live and dead recoveries, as used by Hall *et al.* (2001) to estimate gray seal (*Halichoerus grypus*) first-year survival rates. If covariate data are available for individuals, e.g., sex, size, age, body condition, analysis can investigate their effects on survival rates.

Survival rates have been estimated using mark–recapture methods for a variety of marine mammal species including humpback, blue, bowhead (*Balaena mysticetus*), right and gray (*Eschrichtius robustus*) whales, elephant (*Mirounga* spp.), monk (*Monachus* spp.), grey and Weddell (*Leptonychotes weddellii*) seals, subantarctic and New Zealand fur seals (*Arctocephalus* spp.), and manatees (*Trichechus manatus*) (Caswell *et al.*, 1999; Pistorius *et al.*, 1999; Schwarz and Stobo, 2000; Hall *et al.*, 2001; Zeh *et al.*, 2002; Bradshaw *et al.*, 2003; Larsen and Hammond, 2004; Langtimm *et al.*, 2004; Mizroch *et al.*, 2004; Cameron and Siniff, 2004; Beauplet *et al.*, 2005; Bradford *et al.*, 2006; Ramp *et al.*, 2006; Baker and Thompson, 2007).

## B. Abundance

Mark–recapture methods estimate the number of animals in a population of individuals that mix together (Hammond, 1986). Obtaining a representative sample of data therefore means sampling individuals representatively (see Assumptions). The estimate is of the population of animals that uses the study area. Note the difference between this and the situation for line-transect sampling, which estimates the density of animals in a defined area at the time of the survey and for which a representative sample of the survey area is required.

The basis of estimating abundance using mark–recapture is as follows. A sample of individuals is captured, marked and released ($n_1$) and on a subsequent occasion, a second sample of individuals is captured ($n_2$) of which a number ($m_2$) are already marked. If a number of assumptions are met (see Assumptions), the proportion of marked individuals in the second sample should equal the proportion of marked animals in the population at large ($N$)

$$\frac{m_2}{n_2} = \frac{n_1}{N}$$

which allows population size to be estimated as:

$$\hat{N} = \frac{n_1 n_2}{m_2}$$

This simple two-sample estimator is known as the Petersen estimator. As well as other basic assumptions, it assumes that the population being estimated is closed to births, deaths, immigration, and emigration. To reduce small sample bias and allow a finite estimate of variance to be estimated, a more commonly used estimator is Chapman's modification of the Petersen estimator,

$$\hat{N} = \frac{(n_1 + 1)(n_2 + 1)}{(m_2 + 1)} - 1$$

with estimated variance

$$\text{var}(\hat{N}) = \frac{(n_1 + 1)(n_2 + 1)(n_1 - m_2)(n_2 - m_2)}{(m_2 + 1)^2(m_2 + 2)}$$

Note that estimates of variance are correlated to the estimates themselves, an undesirable feature of mark–recapture abundance estimates.

In essence, the data on animals that have been captured at least once are used to estimate the number of animals that have never been captured. This inference is what makes mark–recapture abundance estimates so sensitive to the analytical assumptions (see Assumptions). The same idea applies when there are multiple sampling occasions; in this situation each sample (except the first) provides information on recaptures as well as on animals caught for the first time. An important consideration when estimating abundance from multiple samples is whether or not the population being estimated can be considered closed or not. If the sampling occasions are sufficiently close in time that any change in population size is likely to be small, multi-sample closed population models can be used. A variety of models are available in program MARK, including those implemented by the well-known program CAPTURE that allow capture probabilities to vary by time and due to trap dependence and individual heterogeneity (see Assumptions).

If the population is not closed, the most obvious option is to use an open population model to estimate abundance. Program MARK provides a number of options based around the Jolly–Seber model. These models are very flexible and also provide estimates of survival, recruitment, and population growth rates as well as abundance. However, they are not able to take into account individual heterogeneity of capture probabilities, which can be a substantial source of bias (see Assumptions). Another option is to use a series of two-sample estimates (Stevick et al., 2003b).

Not all animals in the population may bear distinguishing markings. For example, bottlenose dolphins acquire nicks as they get older, so younger animals are typically not well marked and will be excluded from a population estimate of well-marked individuals. Data on the proportion of well-marked animals in each school encountered can be used to estimate the proportion of markable animals in the population (Williams et al., 1993; Wilson et al., 1999).

Abundance has been estimated using mark–recapture methods for a number of marine mammal species/populations including humpback, blue, sperm, and northern bottlenose whales; bottlenose, humpback (Sousa spp.), and Hector's dolphins; Weddell, leopard (Hydrurga leptonyx), elephant, monk (Monachus spp.), and gray seals; and Australian and New Zealand fur seals (Arctocephalus forsteri) (Williams et al., 1993; Whitehead et al., 1997a,b; Cerchio, 1998; Smith et al., 1999; Wilson et al., 1999; Forcada and Aguilar, 2000; Shaughnessy et al., 2000; Bejder and Dawson, 2001; Read et al., 2003; Stevick et al., 2003b; Calambokidis and Barlow, 2004; Cameron and Siniff, 2004; Larsen and Hammond, 2004; Garcia-Aguilar and Morales-Bojorquez, 2005; Forcada and Robinson, 2006; Boren et al., 2006; Stensland et al., 2006; Gerondeau et al., 2007).

## C. Assumptions

Analyses to estimate survival rates and abundance make assumptions about the data that, if violated, may lead to bias in estimated population parameters.

There are three fundamental assumptions about the marked individuals:

1. Marks are unique (i.e., there are no twins, triplets, etc.);
2. Marks cannot be lost (to ensure that a marked animal will be recognized on recapture);
3. All marks are correctly recorded and reported.

The uniqueness of a mark can be assured with the application of a numbered tag/brand or the use of genotyping, but it must be inferred for natural markings from the amount of information represented by the marks. For example, the color pattern on the ventral surface and the indentations along the trailing edge of tail flukes of humpback whales, in the Northern Hemisphere at least, are generally believed to contain enough information for the markings to be unique. Similarly, it is usually assumed that the pattern of nicks and notches on the trailing edge of the dorsal fin allows bottlenose dolphins to be uniquely identified. However, not all animals possess such markings so it is important to define what is meant by the markable population.

The issue of mark loss is an important one. Flipper tags may fall off, brands may fade, and natural markings may change. Whether or not mark loss could occur should be determined in any mark–recapture study. If it is, the rate of mark loss should be estimated, e.g., by double tagging (Stobo and Horne, 1994; Bradshaw et al., 2000; Pistorius et al., 2000)

**M**

The correct recording and reporting of marks from captured animals may seem a trivial assumption to meet; however, two aspects of this are important. First, it is surprisingly easy to make mistakes in the field or in transcribing field notes or in laboratory processing of data. Researchers should pay particular attention to avoiding making such errors because they can potentially have a big effect, especially in datasets with few recaptures. Second, the correct identification of recaptures is critical to mark–recapture. The chance of making an error with numbered tags is small, but there are important issues concerning matching individuals using photo-identification, and a strict protocol is essential to minimize false negative errors (missing a match) or false positive errors (calling two different individuals the same). Most researchers implement protocols to avoid making false positive errors. Stevick *et al.* (2001) investigated the rate of false negative errors and found it to increase with decreasing photographic quality. The effects of false negative or false positive errors in matching are again most severe in the analysis of datasets with few recaptures.

The simplest models also make assumptions related to the behavior of the animals or the researcher. These assumptions are often stated as:

1. Marking does not affect future survival or catchability;
2. Animals must have an equal probability of being captured within each sampling occasion.

In many, but not all, circumstances models are available that do not require these assumptions to be made. As a general rule, it is wise to explore datasets with a range of models to determine the level of model complexity needed to minimize bias and maximize precision in population parameter estimates. Model selection and goodness-of-fit testing are important components of mark–recapture analysis; these are well covered in Cooch and White (2007).

Failure of the assumption that marking does not affect future catchability can result in trap dependency where animals become "trap-happy" or "trap-shy" after first capture, which causes bias unless accounted for in estimation models. This is not uncommon in trapping studies of terrestrial animals where capture depends on an animal entering a trap but is less likely to be a problem in studies of marine mammals, which must be actively captured. In particular, animals that have been captured by photo-identification should not become more or less likely to be captured again. However, in analysis, trap dependency may act as a proxy for some other feature of the data (Ramp *et al.*, 2006).

Failure of the assumption that all animals have an equal chance of being captured in each sampling occasion is often referred to as heterogeneity of capture probabilities. This can occur for a variety of reasons: behavioral responses to marking (as described above); failure of animals to mix completely between sampling occasions caused by area preferences of individuals; inability of sampling design to give each animal an equal chance of being captured even within the study area; behavioural aspects making some animals inherently more difficult to catch than others; some animals having more distinguishable marks than others (especially for photo-identification studies).

Heterogeneity of capture probabilities causes bias, which can be severe in abundance estimates. Achieving equal capture probability in each sampling occasion for all individuals in a study population is a practical impossibility; all datasets will suffer from heterogeneity to a greater or lesser extent. The important point is to recognize this and to be prepared to explore the impact in analysis. Notwithstanding this, it is good practice to minimize heterogeneity of capture probabilities in the field as much as possible; this can simplify analysis. The most obvious way to do this is to implement a sampling design that ensures high average capture probability. If a large proportion of the animals in the population can be captured there is less scope for variation in the probability of capture among individuals. How well this can be achieved is a matter of the size and extent of the population, data collection logistics, and the resources available.

## V. Birth Rates

A birth-interval model in which mark–recapture data are used to estimate fecundity and survival rates, and hence population growth rates, was developed by Barlow and Clapham (1997). The model was applied to humpback whales in the Gulf of Maine, but its wider application may be limited because of the requirement for fairly complete data on calving histories of individual females.

## *See Also the Following Articles*

Abundance Estimation ■ Identification Methods

## *References*

Baker, J. D., and Thompson, P. M. (2007). Temporal and spatial variation in age-specific survival rates of a long-lived mammal, the Hawaiian monk seal. *Proc. Roy. Soc. B* **274**, 407–415.

Barlow, J., and Clapham, P. J. (1997). A new birth-interval approach to estimating demographic parameters of humpback whales. *Ecology* **78**, 535–546.

Beauplet, G., Barbraud, C., Chambellant, M., and Guinet, C. (2005). Interannual variation in the post-weaning and juvenile survival of subantarctic fur seals: Influence of pup sex, growth rate and oceanographic conditions. *J. An. Ecol.* **74**, 1160–1172.

Bejder, L., and Dawson, S. (2001). Abundance, residency, and habitat utilisation of Hector's dolphins (*Cephalorhynchus hectori*) in Porpoise Bay, New Zealand. *NZ J. Mar. Freshw. Res.* **35**, 277–287.

Boren, L. J., Muller, C. G., and Gemmell, N. J. (2006). Colony growth and pup condition of the New Zealand fur seal (*Arctocephalus forsteri*) on the Kaikoura coastline compared with other east coast colonies. *Wildl. Res.* **33**, 497–505.

Bradford, A. L., *et al.* (2006). Survival estimates of western gray whales *Eschrichtius robustus* incorporating individual heterogeneity and temporary emigration. *Mar. Ecol. Prog. Ser.* **315**, 293–307.

Bradshaw, C. J. A., Barker, R. J., and Davis, L. S. (2000). Modeling tag loss in New Zealand fur seal pups. *J. Agr. Biol. Env. Stat.* **5**, 475–485.

Bradshaw, C. J. A., Barker, R. J., Harcourt, R. G., and Davis, L. S. (2003). Estimating survival and capture probability of fur seal pups using multistate mark–recapture models. *J. Mamm.* **84**, 65–80.

Calambokidis, J., and Barlow, J. (2004). Abundance of blue and humpback whales in the eastern North Pacific estimated by capture-recapture and line-transect methods. *Mar. Mamm. Sci.* **20**, 63–85.

Calambokidis, J., *et al.* (22 authors) (2001). Movements and population structure of humpback whales in the North Pacific. *Mar. Mamm. Sci.* **17**, 769–794.

Cameron, M. F., and Siniff, D. B. (2004). Age-specific survival, abundance, and immigration rates of a Weddell seal (*Leptonychotes weddellii*) population in McMurdo Sound, Antarctica. *J. Zool.* **82**, 601–615.

Caswell, H., Fujiwara, M., and Brault, S. (1999). Declining survival probability threatens the North Atlantic right whale. *Proc. Nat. Acad. Sci. USA* **96**, 3308–3313.

Cerchio, S. (1998). Estimates of humpback whale abundance off Kauai, 1989–1993: Evaluating biases associated with sampling the Hawaiian Islands breeding assemblage. *Mar. Ecol. Prog. Ser.* **175**, 23–34.

Cooch, E., and White, G. (2007). Program MARK: A gentle introduction. <http://www.phidot.org/software/mark/docs/book/ >

Forcada, J., and Aguilar, A. (2000). Use of photographic identification in capture-recapture studies of Mediterranean monk seals. *Mar. Mamm. Sci.* **16**, 767–793.

Forcada, J., and Robinson, S. L. (2006). Population abundance, structure and turnover estimates for leopard seals during winter dispersal combining tagging and photo-identification data. *Polar Biol.* **29**, 1052–1062.

Garcia-Aguilar, M. D. L., and Morales-Bojorquez, E. (2005). Estimating the haul-out population size of a colony of northern elephant seals *Mirounga angustirostris* in Mexico, based on mark–recapture data. *Mar. Ecol. Prog. Ser.* **297**, 297–302.

Gerondeau, M., Barbraud, C., Ridoux, V., and Vincent, C. (2007). Abundance estimate and seasonal patterns of grey seal (*Halichoerus grypus*) occurrence in Brittany, France, as assessed by photo-identification and capture-mark–recapture. *J. Mar. Biol. Ass. UK* **87**, 365–372.

Hall, A. J., McConnell, B. J., and Barker, R. J. (2001). Factors affecting first-year survival in grey seals and their implications for life history strategy. *J. An. Ecol.* **70**, 138–149.

Hammond, P. S. (1986). Estimating the size of naturally marked whale populations using capture-recapture techniques. *Rep. Int. Whal. Commn (Spec. Iss. 8)*, 253–282.

Hammond, P. S., Mizroch, S. A., and Donovan, G. P. (eds) (1990). Individual recognition of cetaceans: Use of photo-identification and other techniques to estimate population parameters. *Rep. Int. Whal. Commn (Spec. Iss. 12)*.

Hiby, L., and Lovell, P. (1990). Computer-assisted matching of natural markings: A prototype system for grey seals. *Rep. Int. Whal. Commn (Spec. Iss. 12)*, 57–61.

Langtimm, C. A., Beck, C. A., Edwards, H. H., Fick-Child, K. J., Ackerman, B. B., Barton, S. L., and Hartley, W. C. (2004). Survival estimates for Florida manatees from the photo-identification of individuals. *Mar. Mamm. Sci.* **20**, 438–463.

Larsen, F., and Hammond, P. S. (2004). Distribution and abundance of West Greenland humpback whales (*Megaptera novaeangliae*). *J. Zool.* **263**, 343–358.

McConnell, B., Bryant, E., Hunter, C., Lovell, P., and Hall, A. (2004). Phoning home-a new GSM mobile phone telemetry system to collect mark-recapture data. *Mar. Mamm. Sci.* **20**, 274–283.

Mizroch, S. A., *et al.* (2004). Estimating the adult survival rate of central North Pacific humpback whales (*Megaptera novaeangliae*). *J. Mamm.* **85**, 963–972.

Palsbøll, P., *et al.* (19 authors) (1997). Genetic tagging of humpback whales. *Nature (Lond.)* **388**, 767–769.

Pistorius, P. A., Bester, M. N., and Kirkman, S. P. (1999). Survivorship of a declining population of southern elephant seals, *Mirounga leonina*, in relation to age, sex and cohort. *Oecologia* **121**, 201–211.

Pistorius, P. A., Bester, M. N., Kirkman, S. P., and Boveng, P. L. (2000). Evaluation of age- and sex-dependent rates of tag loss in southern elephant seals. *J. Wildl. Manage.* **64**, 373–380.

Ramp, C., Bérubé, M., Hagen, W., and Sears, R. (2006). Survival of adult blue whales *Balaenoptera musculus* in the Gulf of St. Lawrence, Canada. *Mar. Ecol. Prog. Ser.* **319**, 287–295.

Read, A. J., Urian, K. W., Wilson, B., and Waples, D. M. (2003). Abundance of bottlenose dolphins in the bays, sounds, and estuaries of North Carolina. *Mar. Mamm. Sci.* **19**, 59–73.

Ries, E. H., Hiby, L. R., and Reijnders, P. J. H. (1998). Maximum likelihood population size estimation of harbour seals in the Dutch Wadden Sea based on a mark-recapture experiment. *J. Appl. Ecol.* **35**, 332–339.

Schwarz, C. J., and Stobo, W. T. (2000). Estimation of juvenile survival, adult survival, and age-specific pupping probabilities for the female grey seal (*Halichoerus grypus*) on Sable Island from capture-recapture data. *Can. J. Fish. Aquat. Sci.* **57**, 247–253.

Shaughnessy, P. D., Troy, S. K., Kirkwood, R., and Nicholls, A. O. (2000). Australian fur seals at Seal Rocks, Victoria: Pup abundance by mark

recapture estimation shows continued increase. *Wildl. Res.* **27**, 629–633.

Smith, T. D., *et al.* (12 authors) (1999). An ocean-basin-wide mark-recapture study of the North Atlantic humpback whale (*Megaptera novaeangliae*). *Mar. Mamm. Sci.* **15**, 1–32.

Stensland, E., Carlen, I., Sarnblad, A., Bignert, A., and Berggren, P. (2006). Population size, distribution, and behavior of indo-pacific bottlenose (*Tursiops aduncus*) and humpback (*Sousa chinensis*) dolphins off the south coast of Zanzibar. *Mar. Mamm. Sci.* **22**, 667–682.

Stevick, P. T., Palsbøll, P., Smith, T. D., Bravington, M. V., and Hammond, P. S. (2001). Errors in identification using natural markings: Rates, sources and effects on capture-recapture estimates of abundance. *Can. J. Fish. Aquat. Sci.* **58**, 1861–1870.

Stevick, P. T., *et al.* (2003a). Segregation of migration by feeding ground origin in North Atlantic humpback whales (*Megaptera novaeangliae*). *J. Zool.* **259**, 231–237.

Stevick, P. T., *et al.* (2003b). North Atlantic humpback whale abundance and rate of increase four decades after protection from whaling. *Mar. Ecol. Prog. Ser.* **258**, 263–273.

Stobo, W., and Horne, J. K. (1994). Tag loss in grey seals (*Halichoerus grypus*) and potential effects on population estimates. *Can. J. Zool.* **72**, 555–561.

Whitehead, H. (2001). Analysis of animal movement using opportunistic individual identifications: Application to sperm whales. *Ecology* **82**, 1417–1432.

Whitehead, H., Christal, J., and Dufault, S. (1997a). Past and distant whaling and the rapid decline of sperm whales off the Galapagos Islands. *Cons. Biol.* **11**, 1387–1396.

Whitehead, H., Gowans, S., Faucher, A., and McCarrey, S. W. (1997b). Population analysis of northern bottlenose whales in the Gully, Nova Scotia. *Mar. Mamm. Sci.* **13**, 173–185.

Williams, J. A., Dawson, S. M., and Slooten, E. (1993). The abundance and distribution of bottle-nosed dolphins (*Tursiops truncatus*) in Doubtful Sound, New Zealand. *Can. J. Zool.* **71**, 2080–2088.

Wilson, B., Hammond, P. S., and Thompson, P. M. (1999). Estimating size and assessing trends in a coastal bottlenose dolphin population. *Ecol. Appl.* **9**, 288–300.

Wilson, B., Thompson, P. M., and Hammond, P. S. (1997). Habitat use by bottlenose dolphins: Seasonal distribution and stratified movement patterns in the Moray Firth, Scotland. *J. Appl. Ecol.* **34**, 1365–1374.

Zeh, J., Poole, D., Miller, G., Koski, W., Baraff, L., and Rugh, D. (2002). Survival of bowhead whales, *Balaena mysticetus*, estimated from 1981–1998 photoidentification data. *Biometrics* **58**, 832–840.

**M**

# Mass Mortalities

## AILSA HALL AND JOHN HARWOOD

The term "mass mortality" has been used rather imprecisely in the scientific literature. In general, it is an event which involves the death of many hundreds of individuals in a relatively short interval (usually 1–2 months). Mass mortalities of rare species may involve smaller numbers of individuals, but the use of this term can be justified if a large proportion of the population is involved in the die off. Strandings of groups of social cetaceans, such as pilot whales (*Globicephala macrorhynchus* and *G. melas*) and pygmy killer whales (*Feresa attenuata*), should probably not be regarded as mass mortalities (Geraci *et al.*, 1999). However, large numbers of carcasses may wash up along a short section of coast during a mass mortality and this may, initially, resemble a stranding event. Marine mammals spend most of their lives at sea and only a fraction of the number of individuals which die during a mass mortality are likely to be observed. As a result, the true magnitude and effect of these events cannot be assessed from a simple body count. The scale of the mortality is usually best estimated by comparing abundance estimates made before and after the event.

## I. Diagnosis

The fact that marine mammals spend most of their lives at sea not only makes it difficult to determine how many individuals have died during a mass mortality, but it also makes it difficult to diagnose the cause. Many days may elapse between an individual's death and the recovery of its carcass. To make matters worse, many mortalities occur along remote stretches of coastline where access is difficult, further increasing the time between death and examination. Some large cetacean carcasses may float off-shore, necessitating at-sea necropsies. The problem is further complicated because many of the agents which can cause mass mortalities, such as viral infections, also affect immune function and reduce resistance to secondary infections. As a result, the ultimate cause of death may be a pathogen that would otherwise be relatively harmless (Gulland and Hall, 2007).

Despite these problems, it has proved possible to identify the causes of many of the mass mortalities observed since the 1980s. Three causal factors appear to be particularly prevalent: infectious diseases, naturally occurring biotoxins, and environmental events (Harvell *et al.*, 1999; Gulland and Hall, in press). Several of these factors may act together, and the effects of any one of them can be amplified by exposure to anthropogenic factors such as environmental pollutants.

## II. Infectious Disease

Infectious disease of one kind or another is a frequent cause of mortality in marine mammal populations, and highly virulent pathogens (particularly viruses) can cause the death of large numbers of animals in a very short period. One of the first infectious disease outbreaks to be identified involved an influenza virus that probably caused the death of at least 450 harbor seals (*Phoca vitulina*) along the New England coast of the United States in 1979–1980. But the viruses which are most often associated with mass mortalities are those in the Morbillivirus family. *Canine distemper virus* (probably contracted from domestic dogs [*Canis lupis*] or farmed mink [*Mustela* spp.]) caused the death of several thousand Baikal seals (*Phoca sibirica*) in the Russian Federation in 1987–1988. In the spring of 2000, more than 10,000 Caspian seals (*Phoca caspica*) were estimated to have died during an outbreak of canine distemper. Terrestrial carnivores such as wolves or feral dogs appear to be the most likely source of this virus. The closely related *Phocine distemper virus* caused the death of 18,000 harbor seals in the North Sea during 1988, and a second outbreak in the same geographic region in 2002 probably affected almost 30,000 animals. A *Cetacean morbillivirus* caused the death of several thousand striped dolphins (*Stenella coeruleoalba*) in the Mediterranean Sea between 1990 and 1992. The death of more than 700 bottlenose dolphins (*Tursiops truncatus*) along the Atlantic coast of the US in 1987 was initially attributed to poisoning by biotoxins (discussed later), but more recent evidence suggests that this was also caused by *Cetacean morbillivirus*.

Other notable infectious diseases that have caused mass mortalities in pinnipeds include two bacterial infections. Between 2001 and 2003, a marked reduction in the number of New Zealand (Hooker's) sea lion (*Phocarctos hookeri*) pups born on the Auckland island rookeries were followed by very high neonatal mortality. A single highly pathogenic clonal lineage of the opportunistic bacterium *Klebsiella pneumoniae* was identified as the major causative agent of this 2-year mass mortality. Since 1970, California sea lions (*Zalophus californianus*) along the west coast of the USA have been increasingly affected by outbreaks of leptospirosis, caused by *Leptospira interrogans* serovar pomona, that result in death through renal failure.

## III. Biotoxins

Some species of single-celled algae (notably the diatoms and dinoflagellates) produce poisonous compounds (known as phycotoxins) which can accumulate in any fish or invertebrate animals that eat them. When environmental conditions are particularly suitable, these organisms multiply rapidly, creating "blooms" (also sometimes called "red tides" due to the high density of algae in the surface water). The resulting high concentrations of biotoxins can cause mass mortalities of fish and fish predators. The best documented event of this kind involving marine mammals caused the death of over 400 California sea lions along the central California coast during May and June 1998 (Scholin *et al.*, 2000). This coincided with a bloom in the same are of the algal diatom *Pseudo-nitzschia autralis*, which is known to produce domoic acid—a dangerous neurotoxin. Domoic acid was detected in northern anchovies, which are plankton feeders and a well known prey of sea lions, and in the body fluid of sick sea lions. These sea lions also showed many of the neurological symptoms, such as seizures and convulsions, commonly associated with domoic acid poisoning.

Biotoxins produced by dinoflagellate protozoa have been implicated in the deaths of many marine mammals. Ciguatoxin and maitotoxin were the suspected causative agents in the deaths of Hawaiian monk seals (*Monachus schauinslandi*) in 1978, and 14 humpback whales (*Megaptera novaeangliae*) died from saxitoxin exposure in Cape Cod Bay in 1987. Large numbers of Florida manatees (*Trichechus manatus*) have been affected by exposure to brevetoxin produced by the dinoflagellate *Karenia brevis*. The first mass mortality occurred in 1982 and was followed by a second outbreak in 1996. Between 2002 and 2007, this has been an annual occurrence that has also caused the additional death of hundreds of Florida bottlenose dolphins.

However, identifying the primary cause of a mass mortality is not always straightforward. In May and June 1997, the bodies of over 100 Mediterranean monk seals (*Monachus monachus*) were found along a short stretch of the west African coast near the border between Mauritania and the former Western Sahara. Initial investigations revealed that at least some of these individuals had been infected with a morbillivirus, which most closely resembled *Cetacean morbillivirus*, and the mass mortality was originally attributed to this agent. However, most of the seals died quickly with few, if any, overt signs of disease. This was very different from what had been observed in other morbillivirus-induced events. Subsequent analysis provided evidence of the presence of several biotoxins in dead seals (largely saxitoxin and its derivatives), and high concentrations of a dinoflagellate known to produce at least one of these toxins had been observed in local coastal waters. It is possible that both agents were involved in the mortality, but on the basis of the available evidence it is not possible to say with any confidence that either was responsible (Harwood, 1998).

## IV. Environmental Effects

Unusual environmental conditions can cause high levels of mortality, particularly among young animals. For example, severe storms coupled with unusually high tides during the winter of 1982 and 1983 resulted in the death of up to 80% of all northern elephant seal (*Mirounga angustirostris*) pups born at some Californian colonies.

Even more dramatic effects can be caused by changes in oceanographic conditions. For example, El Niño southern oscillation events can dramatically alter the availability of prey species around marine mammal colonies. The severe 1982–1983 El Niño event had wide-ranging effects on fur seal and sea lion colonies throughout the eastern Pacific. Its effects were even evident in populations of seals in the Antarctic.

An intrusion of low oxygen content water into the coastal waters of Namibia in early 1994 resulted in a massive reduction in the availability of fish. Colonies of Cape fur seals (*Arctocephalus pusillus pusillus*) in Namibia suffered the highest levels of pup mortality ever observed: approximately 120,000 pups had died by the end of May 1994. There was also very high mortality among those sub-adult males which remained on the breeding grounds.

In 1992, two floods and a cyclone caused the loss of 1000 km² of sea grass from Hervey Bay in Queensland, Australia, an important habitat for a population of over 2000 dugongs (*Dugong dugon*). By 1993, the dugong population had declined to approximately 600 animals, and many of the recovered carcasses indicated that the animals were severely emaciated and probably died of starvation.

## V. Anthropogenic Effects

Deliberate killing by humans can cause mass mortalities of marine mammals, but the effects of such activities are dealt with elsewhere. However, human activities can also result in mass mortalities through indirect effects. The most obvious of these is the exposure of marine mammals to harmful chemicals which are spilled or discharged into the marine environment. Although marine mammals are probably less vulnerable to the effects of oil spills than seabirds, species that rely on dense fur for insulation, such as the sea otter (*Enhydra lutris*), can be seriously affected. Indeed, it is estimated that 3500–5500 sea otters died in 1989 after the tanker *Exxon Valdez* spilled 42 million liters of oil in Prince William Sound, Alaska. Other chemicals may have a more insidious effect. A wide range of man-made halogenated organic compounds are preferentially soluble in fat and can accumulate at high concentrations in the blubber of predatory marine mammals. These compounds can reduce resistance to disease, and individuals with high tissue levels may be particularly vulnerable during mass mortalities caused by infectious disease agents. High persistent organic and other pollutant levels (DDE, DDT, PCB) may well have been a contributory factor to mortality during the morbillivirus epidemics in the North Sea and Mediterranean.

Between January and May 1992, 220 bottlenose dolphins died in the coastal bays of mid-Texas, USA. Although no definitive cause was established, the low salinity of the coastal waters at that time indicated that there had been unusual levels of agricultural runoff due to record levels of rainfall, and this runoff resulted in elevated concentrations of carbamate pesticides (atrazine and aldicarb). After all other potential causes had been excluded, it was concluded that a combination of pesticide exposure and increased dermal absorption due to low water salinity may have been responsible for the dolphins death.

A combination of environmental factors and fishing activity may also result in mass mortality. For example, large numbers of harp seals (*Phoca groenlandica*) appeared off the north coast of Norway between 1985 and 1988, probably as a result of the collapse of the stocks of capelin, an important prey species, in the Barents and Norwegian Seas. Many of these seals became entangled in fishing nets and subsequently drowned. The Norwegian government compensated fishermen for the damage this caused, and these statistics indicate that at least 79,000 seals died in 1987 and 1988 alone.

## VI. Effects on Populations

Determining the effect of mass mortalities on the marine mammal populations is often difficult, because baseline data on the size and status of the affected populations are lacking. However, their impact can be substantial. For example, the 1988 and 2002 morbillivirus epidemics in the North Sea killed approximately 40% of the harbor seal population, and local mortality rates were as high as 60%. The mass mortality of Mediterranean monk seals in 1997 killed 70% of the local population and about one-third of the world population of this species.

Some marine mammal populations have shown a remarkable ability to recover from the effects of these events. The North Sea harbor seal population returned to its pre-epidemic level within 10 years of the 1988 epidemic, although some local populations (e.g., those along the eastern seaboard of England) are still depleted. However, species whose populations are already small (such as the Mediterranean monk seal, whose world population was less than 1000 individuals at the time of the 1997 mortality) may be reduced to such low levels that they are more susceptible to other problems associated with low population size, such as increased levels of inbreeding and the loss of genetic diversity through genetic drift. As a result, their risk of extinction may be substantially increased.

## VI. Future Trends

There are a number of grounds for predicting that the frequency of mass mortalities will increase during this century (Harvell *et al.*, 1999; Gulland and Hall, 2007). Like many other species, marine mammals will be exposed ever more frequently to novel pathogens as a result of the increased movement of humans and their domestic animals, who can act as vectors for these agents. Climate change is also likely to lead to new movement patterns which will increase the exchange rate of these pathogens.

Exposure to biotoxins, particularly those produced by dinoflagellate algae, is also likely to increase. Levels of nutrients and minerals, which are normally in short supply, are periodically increased in coastal waters by large-scale runoff of rainwater from agriculture land. This occurs more frequently now because of modern drainage techniques and can create conditions that are favorable to algal blooms. In addition, dinoflagellates are particularly well adapted to transportation in the ballast water of large ships, because when conditions are unfavorable they become encased in a protective cyst. As a result, many species that once had a very restricted distribution now have a global one. For example, *Gymnodinium catenatum* (one of the species implicated in the mass mortality of Mediterranean monk seals) was originally confined to the east coast of the US. Since 1970, it has been recorded in Japan and Australia, as well as off the west coast of Africa.

**M**

## See Also the Following Articles

Health ■ Stranding

## References

Geraci, J., Harwood, J., and Lounsbury, A. (1999). Marine mammal die-offs: Causes, investigations and issues. *In* "Conservation and Management of Marine Mammals" (J. Twiss, and R. Reeves, eds), pp. 367–395. Smithsonian Institute Press, Washington, DC.

Gulland, F. M. D., and Hall, A. J. (2007). Is marine mammal health deteriorating? Trends in the global reporting of marine mammal disease. *Eco. Health* **4**, 135–150.

Harvell, C. D., *et al.* (13 authors) (1999). Emerging marine diseases-Climate links and anthropogenic factors. *Science* **285**, 1505–1510.

Harwood, J. (1998). What killed the Mediterranean monk seals? *Nature* **393**, 17–18.

Scholin, C., *et al.* (26 authors) (2000). Mortality of sea lions along the central California coast linked to a toxic diatom bloom. *Nature* **403**, 80–83.

# Mating Systems

SARAH L. MESNICK AND KATHERINE RALLS

## I. Introduction to Mating Systems

Marine mammal mating systems are diverse and variable. The key to understand this variation is the fundamental idea that individuals behave so as to maximize their lifetime reproductive success and that males and females are subject to different selective pressures (Trivers, 1972). In general, female mammals invest heavily in rearing a limited number of young during their lifetimes whereas males tend to maximize reproductive success by mating with as many females as possible to increase the number of offspring they sire. Recent reviews have both questioned this general conceptual framework (Roughgarden *et al.*, 2006), and supported it, although acknowledging that there is more complexity than originally envisioned (Clutton-Brock, 2007). Males and females do not necessarily always cooperate during mating, and sometimes may be in conflict. Together, the mating strategies of males and females define the mating system of a species. This article presents an overview of male and female mating strategies and describes how the different groups of marine mammals solve the problem of finding mates.

Mating systems have traditionally been categorized on the basis of whether the female and male form pair bonds, the duration of those bonds and the number of partners each sex copulates with during a breeding season or lifetime. To facilitate comparisons among marine mammals, in which males do not remain with their mates after mating (with rare exceptions, Danilewicz *et al.*, 2004), we focus on the number of partners for each sex during a breeding season. In *monogamous* systems, each individual has a single partner during the breeding season and in *polygamous* systems, some individuals have multiple partners. There are several types of polygamy: *polygyny*, in which some males have more than one partner; *polyandry* in which some females have more than one partner; and *polygynandry*, in which some females and some males have multiple partners.

Marine mammals, like most mammals, are predisposed to polygyny because of the fundamental disparity in reproductive biology between the sexes (Clutton-Brock, 1989). Females can produce only a limited number of offspring in their lives and they bear the energetic costs of gestation, lactation, and parental care. Males, however, provide little or nothing to the care of offspring and are free to devote their time and energy to competing for access to mates.

The potential for polygyny, and the extent to which that potential is realized, is determined to a great extent by the degree to which receptive females are aggregated in space and time. The distribution of females, in turn, is determined by phylogenetic constraints (such as the retention of terrestrial birthing in pinnipeds) and by ecological and social conditions. Among the most important of these conditions are the distribution of resources necessary for breeding, predation pressure, and the costs and benefits of group living. Several types of polygynous mating systems have been identified in marine mammals. In populations where females are clustered and their movements are limited in space and time, males generally compete to monopolize females. They may defend resources that are vital to females such as parturition sites (territoriality or resource defense polygyny) or they may defend females directly (female defense polygyny). When females are spatially or temporally dispersed, or are highly mobile, males are less able to control access to potential mates. In these situations, males may aggregate on traditional display sites and advertise for females (lekking) or they may search widely and spend little time with females except to mate (roving).

Females of many species also mate with more than one male during a breeding season and they may benefit in many ways by doing so (see below). Polygynandrous (or "promiscuous") mating systems have been described for a number of species. Frequent copulation may have a variety of functions, but not all mating behavior is for procreation; in many species, sexual behavior is an important component of the social fabric and often has little to do with fertilization.

## II. Male Mating Strategies

Male competition for access to mates takes at least three general forms: aggressive interactions to limit the access of other males to females (contest competition), competition to disperse and find sexually receptive females (scramble competition), and competition in courtship to be chosen by the female (mate choice competition). Each of these behavioral strategies has, in turn, generated a number of morphological and physiological adaptations. For example, males are often distinguished by large body size, big canines, or tusks that can be used as weapons in combat with other males (Fig. 1). Among mammals, males generally disperse more widely than females and there is an increasing evidence that the pattern holds in marine mammals [e.g., Dall's porpoise (*Phocoenoides dalli*) and beluga (*Delphinapterus leucas*)]. Males are typically more persistent in courtship and are the more conspicuous and ornamented sex. They may attempt to entice and attract females through visual acoustic, and pheromonal displays (Fig. 2).

Several additional forms of male competition have also been described. Males may attempt to out-compete other males by producing higher quality or greater quantities of sperm or by removing other male's sperm (sperm competition). When a male cannot monopolize access to females by himself, males may cooperate and form alliances. These alliances, comprised of pairs of males or small groups, compete with each other for access to females and have been described in Indian Ocean bottlenose dolphins (*Tursiops aduncus*) from Shark Bay, Australia (Connor *et al.*, 1996), and common

**Figure 1** *Adult male northern elephant seals* (Mirounga angustirostris) *fight for positions in a dominance hierarchy that confers access to receptive females. Photograph by Sarah L. Mesnick.*

**Figure 2** *A singing adult male humpback whale* (Megaptera novaeangliae). *Singing by humpback males presumably acts to attract females, although whether songs contain cues to mate quality remains in dispute. Singing may also function to space males in a breeding area or to aid in the establishment of dominance hierarchies. Maui.* © *Flip Nicklin (Minden Pictures). Photograph obtained under N.M.F.S. Permit No. 987.*

bottlenose dolphins (*Tursiops truncatus*) from Sarasota Bay, Florida (Wells *et al.*, 1987), and are likely to occur in other species as well. Males may also form consortships, maintaining close proximity to a female until she comes into estrus. In common bottlenose dolphins from Sarasota Bay, Florida, such consortships have been found to correlate with the birth of offspring several months later. Species in which males are larger than females, possess dangerous weapons and aggressively pursue copulation, some males may forcibly coerce the female to mate (forced copulation), e.g., northern elephant seals (*Mirounga angustirostris*), Le Boeuf and Mesnick (1990). Males which are not competitive, because of their size, age, or experience, may employ alternative strategies to obtain access to females. Males may sneak copulations when alpha bulls are distracted or abduct females from the territories of dominant males (kleptogyny) as observed in northern fur seals (*Callorhinus ursinus*), Gentry (1998). Although competition between males receives the most attention, the possibility of male choice—the notion that males may prefer to mate with particular females—is an area that deserves further attention.

### III. Female Mating Strategies

Females of most marine mammals produce only a single offspring at a time. The interbirth interval ranges from 1 year in most pinnipeds and small cetaceans to 5, 6, or even 7 years in larger toothed whales such as sperm whales (*Physeter macrocephalus*), killer whales (*Orcinus orca*), and short-finned pilot whales (*Globicephala macrorhynchus*). While females may maximize their reproductive success by being good mothers, they may also enhance their fitness by choosing males as mates that offer resources, protection, or genetic benefits. Multiple mating is also an important female strategy that may function to ensure insemination or to confuse paternity and reduce potential aggression directed toward young. Multiple mating may also enable females to promote sperm competition and exert cryptic female choice.

In other well-studied taxa, such as birds, females are highly discriminating in their choice of sexual partners. Moreover, females often choose in a similar way, so that a few males achieve many copulations and most other males achieve few or none. Females may choose among potential mates directly (based on resources, size, strength, dominance, or display) or indirectly (by mating with the winner of contests for access to females). Some marine mammal females actively seek out particular males and mate. For example, in California sea lions (*Zalophus californianus*), some females were observed to change pupping locations from one year to the next to remain with a territorial male who changed territory location (Heath, 1989). Females may also incite male–male competition. As a result of protesting male sexual advances loudly, female northern elephant seals instigate fights among males, and subsequently mate with the winner of these battles (Cox and Le Boeuf, 1977).

In marine mammals, it is difficult to establish the existence of female choice and even more difficult to determine why females might choose particular males with which to mate or to quantify the benefit to females of exercising choice. Direct benefits to the female, in the form of nutritional resources or parental care, are not known to exist in marine mammals. Females can benefit, however, by choosing males with higher quality territories, which provide better parturition or thermoregulatory sites, or by choosing males that provide protection from harassment by subordinate males (the bodyguard hypothesis; Mesnick, 1997). Females can also benefit by discriminating among potential mates on the basis of genetic benefits (Mays

M

and Hill, 2004). These include choosing males of the correct species, males with immunologically compatible genes, and males with "good genes" who can produce offspring of higher quality. Females can also choose males with better fertilization ability or virility.

Females may make very different decisions regarding which males they associate with, which males they mate with, and which male ultimately sires their offspring. In land-breeding pinnipeds, for example, a female may reside with one dominant or territorial male during lactation but later leave this male's territory to copulate with another male elsewhere (extra-territorial copulation). In some species, such as the common bottlenose dolphin, sexual behavior is a frequent and important component of nonreproductive social life and has little to do with fertilization. As with males, female strategies need not be mutually exclusive and it is likely that different females will utilize different strategies depending on their age, dominance rank, and the number and quality of available mates.

## IV. Taxonomic Descriptions
### A. Pinnipeds

For all pinnipeds studied to date, the data support, or are highly suggestive of, a polygynous mating system (Boness *et al.*, 2002). Pinnipeds are predisposed to polygyny because they give birth on land, which results in the spatial clustering of females, and have an annual birthing cycle, which results in reproductive synchrony among females. The degree of polygyny varies both within and among species with the extent of reproductive synchrony and spatial clustering. Most species have a peak availability of receptive females lasting about 1 month but the availability of receptive females ranges from 10–15 days in harp (*Pagophilus groenlandicus*) and hooded (*Cystophora cristata*) seals to a period of several months for species that breed in tropical habitats such as monk seals (*Monachus* spp.) and Galapagos sea lions (*Zalophus wollebaeki*).

Variation in the degree of spatial clustering within and among species is due to a variety of factors including the spatial distribution of suitable breeding sites, whether mating takes place on land or at sea, the intensity of male harassment, predation pressure, and thermoregulatory pressures. Polygyny and sexual dimorphism are generally much more extreme in species that mate on land than in those that mate in the water and in species that reside in high latitudes than those that reside in lower latitudes (see article on Sexual Dimorphism).

*1. Otariids* Otariid females feed during lactation. Lactation is energetically costly, so females must raise their young on sites near highly productive marine areas. Because these sites are limited, and the season of maximum marine productivity may be short, females typically occur in dense aggregations (numbering from a few individuals to more than one thousand) on beaches or rocky shelves on islands. Mating occurs on land, although evidence of at least some mating at sea exists for a few species [e.g., Juan Fernández fur seal (*Arctocephalus philippii*) and the California sea lion]. The combination of dense female aggregations and terrestrial mating gives some males the opportunity to monopolize mating with many females. Sexual dimorphism among otariids is correspondingly extreme; males are on average 3 times, and up to 10 times, heavier than females and have other traits favored in physical combat over females: large canines, heavy chests, and dense manes.

The northern fur seal is among the most polygynous of the otariids: a single male at the St. George Island rookery mated with 161 females while hundreds of other males had no copulations at all (Gentry, 1998). The lowest levels of polygyny probably occur in species such as the Juan Fernández fur seal, the South American sea lion (*Otaria flavescens*), the Galapagos fur seal (*Arctocephalus galapagoensis*), and Hooker's sea lion (*Phocarctos hookeri*), in which the ratio of sexually active adults ranges from two to six females per male (Boness *et al.*, 1993).

Male otariids typically defend territories containing resources needed by females—parturition and thermoregulatory sites—rather than individual females (Fig. 3). However, female defense has been demonstrated in at least one otariid, the South American sea lion. The two types of polygyny are difficult to distinguish and are not necessarily mutually exclusive. There is some evidence suggestive of lekking in at least three species, the California sea lion, the South American fur seal (*Arctocephalus australis*), and Hooker's sea lion.

A male's ability to acquire and defend a territory depends on his size and age, his ability to compete with other males, and his ability to fast during his tenure (contest competition). Under most circumstances, the boundaries of territories are fixed and are often delineated by natural breaks in the substrate. Males use a species-specific threat display when defending the boundaries of their territory. A male that secures a territory will probably, but not necessarily, mate with many of the females that give birth on his territory. Climate and topography also play an important role in determining a male's mating success. Those males defending territories containing access to the water, tide pools, or shade typically acquire a disproportionately large number of females.

Most otariid bulls fast while maintaining their territories, sometimes for the entire 2–3-month breeding season [e.g., Steller sea lions (*Eumetopias jubatus*) and northern fur seals]. Some males return to the same territory in subsequent years, with remarkably high site fidelity, whereas others move to new territories or are not seen again. Territorial males may try to herd females to prevent them from leaving their territories but in most species, females determined to leave generally can do so. The males of some species, however [e.g., northern fur seal and the South American and Australian (*Neophoca cinerea*) sea lion], are able to prevent females from leaving their territories by threats, herding, and sometimes physical aggression leading to injury.

**Figure 3** *A* Steller sea lion (Eumetopias jubatus) *territory. Adult males defend resource-based territories that encompass female parturition and thermoregulatory sites. Females choose among males in a surprisingly consistent way. As a result, some males holding territories never or rarely mate, while a few males mate with many females. Photograph by Sarah L. Mesnick.*

The importance of male courtship displays in otariids is not well understood. For example, it is not known whether male displays, such as the incessant barking of male California sea lions, function as threat displays directed at other males, as courtship displays to attract females, or both. Alternative male mating strategies are widespread and generally thought to be practiced by subadult or subordinate males. These include gang raids by groups of non-territorial males to abduct or mate with females in the main breeding territories (observed in South American and Australian sea lions), males stealing females from the territories of their neighbors (kleptogyny; northern fur seals) or males trying to sneak copulations or mate at sea (several species). How successful these strategies are in inseminating females is generally not known. In Antarctic fur seals (*Actocephalus gazella*), a large DNA-based paternity study found that females who pupped on the study beach in the year of conception almost certainly conceived to territorial males and males did not appear either to sneak copulations or mate aquatically, although probability of paternity was strongly influenced by maternal state at conception (females who come to the beach but do not pup appear less likely to conceive with males from that beach) and the authors could not entirely preclude the possibility that a few males might employ alternative strategies (Hoffman *et al.*, 2003).

Female mating strategies are less well understood than male mating strategies but several lines of evidence, including genetic studies, suggest that females exercise more choice among males than previously suspected. Female otariids choose which territory to haul out in and usually, but not always, move freely in and out of a male's territory. Estrus occurs within 1–2 weeks postpartum in all but one species (California sea lion, about 21–27 days). When it is time to mate, a female may leave the male's territory in which she gave birth and mate with another male. This behavior has been observed in California sea lions, South American fur seals, and between the territories of sympatric Antarctic fur seals (Goldsworthy *et al.*, 1999).

Climate, rookery topography and the intensity of male herding and harassment influence the ability of females to exercise mate choice. In hot climates, females have more opportunity for mate choice due to their frequent thermoregulatory movements between their birthing site and the water. In contrast, in colder climes, intense male herding restricts female choice and may injure females. Female northern fur seals are thought to successfully reduce the risk of injury from males by forming dense aggregations and competing for central locations within these groups, which minimizes contact with males, and by acting submissive around males. In this species, females do not appear to choose males directly. Rather, by gathering on traditional mating grounds, the result is that males fight for access for these sites, and females subsequently mate with the winners of these contests. Female otariids may also directly solicit and initiate copulation from males. In Steller sea lions, e.g., a female gains the sexual attention of a male by laterally swinging her neck, dragging her hindquarters and sinuous movements of her body against his body. While females tend to direct most solicitation behavior toward the older "proven" territorial males, the extent of female choice remains unclear. Multiple mating is known in 30% of otariids studied (Boness *et al.*, 1993) and suggests an important and variable role for sperm competition and mate choice across species.

*2. Phocids* Most phocid females are thought to fast during a short and concentrated lactation period and are generally thought not to return to sea to feed until after weaning their pup. However, smaller individuals of some species may supplement fat stores by foraging to sustain lactation (Bowen *et al.*, 2001). Many species live in high latitudes, and a number of species give birth on ice. Females of most phocids are spatially dispersed during lactation; they may be solitary or occur in small-to-moderate-sized well-spaced colonies. In 16 of 19 species, the majority of mating takes place in the water near or after the end of lactation (Boness *et al.*, 2002). As a consequence, males have less opportunity to defend and mate with multiple females. Aquatically mating phocids are less polygynous and less sexually dimorphic than terrestrially mating phocids or otariids.

For most species, the breeding season is relatively short and, in species that breed on ice, mating takes place when temperatures are well below freezing. Reverse sexual dimorphism, with females larger than males, occurs in several species. Large female size may help a mother provide greater quantities of fat-rich milk to her pup and protects her from low polar temperatures. Small size in males is also thought to facilitate agility underwater where males may defend aquatic territories, display, and mate with females. Nevertheless, aquatically mating species are considered to be slightly or moderately polygynous. Mating takes place within a few days of the weaning of the pup.

A common feature of the behavior of aquatically mating phocids is the production by males of simple or complex underwater vocalizations and stereotypical dive displays during the breeding season (Van Parijs *et al.*, 2003). These vocalizations are thought to be used predominantly in male–male competition and are also likely to play a role in attracting females. "Eerie but melodious" songs have been described for male bearded seals (*Erignathus barbatus*) and trills, knocks, buzzes, and chirps for male Weddell seals (*Leptonychotes weddellii*). Harbor seal (*Phoca vitulina*) males produce broadband growls. Hooded seals males make numerous sounds underwater and also produce sounds in air as they inflate and deflate their hood and red nasal sac. The number and diversity of underwater vocalizations is correlated with various social and environmental factors, including mating system, female gregariousness and predation intensity, and is geographically variable (Stirling and Thomas, 2003). In leopard seals (*Hydrurga leptonyx*), both males and females produce underwater broadcast calls thought to serve in mate attraction.

In aquatically mating phocids, there is considerable variability in male mating strategies displayed by each species as well as plasticity within and between populations. In some species, such as crabeater seals (*Lobodon carcinophagus*), spotted seals (*Phoca largha*), and hooded seals, males directly defend the lactating female and her immediate vicinity, a strategy akin to roving and mate guarding of a single female or small group of females. A typical group consists of a female and her pup and an adult male who may have to wait before the female comes into estrus and is receptive to mating. Presumably, the male will mate with the female when she enters the water after weaning her pup. This strategy can be described as sequential female defense polygyny since the male may leave after mating to search for another receptive female. Other males may surround these triads and may compete with the attending male for access to the female, typically with threats and sometimes bloody fights.

In other species, males defend aquatic territories (called "maritories") off the beach or ice where females haul out to raise their pups. Males spend considerable time in these territories giving underwater vocal and visual displays and may mate with any receptive female that enters the territory. This characterizes the behavior of Weddell, harp, bearded, and harbor seals, although there is considerable variation within and between breeding colonies. In the geographically wide-spread harbor seal, colonies occur in diverse habitats and males adjust their mating strategies depending on the topography. Some males defend adjacent discrete territories offshore from female

**M**

haulout areas, others defend territories near female foraging areas or along route to foraging areas, and still others do not defend territories but display in the vicinity of female haulouts. Studies from all habitats observe that males perform stereotypic vocal and dive displays during the period the females is in estrus and raise the possibility of a lek-type mating system in which females select males (Boness *et al.*, 2006).

The northern and southern (*Mirounga leonina*) elephant seal and some populations of the gray seal (*Haliochoerus grypus*) are unusual among phocids in that mating takes place on land. These species exhibit forms of female defense polygyny. Harassment by subordinate males is widespread, costly to females, and may be an important factor contributing to spatial and temporal synchrony among breeding females (Le Boeuf and Mesnick, 1990, Boness *et al.*, 1995). In all the three terrestrially breeding species, males that are unable to maintain control over access to females may attempt to "sneak" copulations or capture, sometimes aggressively, and mate with females as they leave the colony.

In elephant seals, males use visual and acoustic threats as well as physical fighting to compete for dominance in a social hierarchy that confers access to females (Le Boeuf, 1974). Female elephant seals may exercise mate choice by competing for central positions in harems where dominant males reside and by inciting male–male competition and subsequently mating with the winner of these battles. Polygyny in elephant seals is among the most extreme recorded in all mammals; lifetime reproductive success of most males is nil or low. Many die before reaching breeding age and higher-ranking males prevent many of those that survive from breeding. DNA paternity analyses confirm that the proportion of pups sired by alpha males is consistent with that expected from observed mating success in southern elephant seals but show that behavioral observations overestimate the success of some northern elephant seal alpha males (Hoelzel *et al.*, 1999). The relatively lower success of northern elephant seal males may be due in part to the behavior of the northern elephant seal females which copulate more frequently than southern females and may mate during departure from the harem, the greater success of non-alpha males, and/or reduced virility of specific alpha males.

Mating behavior among the geographically wide-spread populations of the gray seal is difficult to categorize because of their varied habitat and social organization. Gray seal females typically do not cluster as tightly and are more mobile in the colony than elephant seal females. Males exhibit state-dependent reproductive strategies, a primary mating tactic involving prolonged female defense and three alternative strategies have been identified (Lidgard *et al.*, 2004). Although males exhibiting the primary strategy achieve the greatest mating success, males that exhibit alternative tactics have been shown to achieve some success. At North Rona, Scotland, genetic studies show that over 80% of assigned paternities agree with behavioral observations in which a female in estrus mates with the local dominant male. However, about 10–20% of females do not have their pups sired by local males and may instead seek preferred sires outside the local male's range or mate with unobserved males, during their departure, or at sea (Twiss *et al.*, 2006). Among gray seal colonies, North Rona has a relatively stable social organization; a lower correlation between observational and genetic measures of success would be expected in colonies with a more fluid social structure and increased female mobility (Pemberton *et al.*, 1992).

*3. Walrus (Odobenus rosmarus)* Walruses have the most elaborate courtship displays of all pinnipeds. Walruses show marked sexual size dimorphism and are thought to be strongly polygynous. Atlantic walruses (*O. r. rosmarus*) in the Canadian High Arctic exhibit a mating system that resembles female defense polygyny in which males appear to monopolize access to female herds. Pacific walruses (*O. r. divergens*) in the Bering Sea may have a lek-like mating system. Groups of males cluster around females, which form dense aggregations on pack ice. Males are aggressive toward one another and produce intricate visual and vocal displays, consisting of barks, whistles, growls, and underwater bell-like sounds. The massive tusks of the male walrus appear to play an important role as a symbol of rank used to threaten other males and as a visual signal to females who may choose among males partly by the size of their tusks.

### B. Cetaceans

*1. Odontocetes* In contrast to pinnipeds, which are relatively sedentary and clustered during the breeding season, female odontocetes are mobile and dispersed. This has two important consequences for male mating strategies: males have less opportunity to control access to aggregated females and less assurance of paternity. It is not surprising, therefore, that the basic mating strategy of male odontocetes appears to be one of searching for receptive females and spending little time with them other than to mate. It is likely that mate guarding, or monopolization of females long enough to increase assurance of paternity and to reduce extra pair copulations, also occurs, although the phenomenon has been noted only in bottlenose dolphins and Dall's porpoise.

Female mating strategies in odontocetes are little understood. Given their mobility and three-dimensional habitat, it is generally thought that females are able to exercise choice by out-maneuvering males or by rolling belly-up. Observational and hormonal evidence suggests that females of several species copulate frequently both during and outside the breeding season and may be polyestrous. Frequent copulation may function to ensure fertilization, induce sperm competition, obscure paternity, and help to establish social bonds with potential future partners. In many odontocete species, sexual behavior is an important component of nonreproductive social interactions and often has little to do with fertilization, making it difficult to infer mating strategies from incomplete observations. An example is the intriguing "wuzzling" behavior of Hawaiian spinner dolphins (*Stenella longirostris longirostris*). Wuzzling refers to the interweaving, caressing, and copulating dolphins of both sexes and all ages and is especially common in the summer months, when many females come into estrus. Is the behavior social? Sexual? Or both?

We know little about mating systems in the vast majority of odontocetes. However, there are substantial data on bottlenose dolphins, sperm whales, and killer whales. A mating strategy of female defense or sequential defense polygyny has been suggested for two species of bottlenose dolphin. In Indian Ocean bottlenose dolphins in Shark Bay, Australia, males in pairs or trios cooperate to form temporary consortships with individual females, often through aggressive herding (Connor *et al.*, 1996). "Second-order alliances," teams of pairs or trios, attack other alliances in contests over female consorts. In common bottlenose dolphins in Sarasota Bay, Florida, individual males and members of long-term male associations form temporary consortships with females without obvious aggressive herding (Wells *et al.*, 1999). The extent to which this sequential female defense strategy is successful is uncertain; individual females are polyestrous and associate with several males during the season in which they conceive. These behaviors may facilitate female mate choice, promote sperm competition, and obscure paternity. In Moray Firth, Scotland, common bottlenose dolphin males apparently do not form alliances or aggressively herd females, although single males may accompany groups of females throughout the breeding season. Among these three sites, the level of male bonding may be inversely related to male body size and the degree of sexual dimorphism (Tolley *et al.*, 1995). In Shark Bay, individuals are small in size, not noticeably dimorphic and alliance formation enables males to gain and maintain access to females. In Moray Firth, males are larger than females and can do this on their own.

Most sperm whales in the Galápagos Islands appear to rove between groups of females searching for potential mates. One or more large males may attend a group of females (sometimes simultaneously) for short periods of time ranging from a few minutes to several hours (Fig. 4). Rather than males herding female groups, females were observed to alter course and speed so that they could join a large male hundreds of meters away. Males did not interact aggressively with each other within female groups, despite several accounts in the literature of males fighting outside of groups. Given the apparent roving strategy of males, the role of the tremendously large nose of the male sperm whale and its possible use as a sound generating organ remains unclear. The loud clicks may function in male–male competition or advertisement to attract females.

Pods of resident killer whales in the Pacific Northwest are frequently observed associating with one another in the summer months when prey (and observer) abundance is high. In these multi-pod

**Figure 4** *Adult male sperm whales* (Physeter macrocephalus) *rove among female groups searching for receptive individuals and staying with each group for only a few hours at a time. Dominica. © Flip Nicklin (Minden Pictures).*

groups, there is much sexual activity amongst all pod members, young, and old alike. Since no dispersal of either sex occurs from resident pods, it is thought that mating takes place during these encounters. Considering that the entire pod engages in these ceremonies, it is likely that the function of multi-pod encounters is both sexual and social. Similarly, genetic analyses of long-finned pilot whales captured in a Faroese fishery indicate that males remain in their natal groups but do not mate within them (Amos *et al.*, 1993). Young were sired by males not captured with the group, implying that pilot whales must mate when two or more groups meet or when adult males pay brief visits to other groups.

Very little is known about mating systems in the remaining species of toothed whales. However, we can infer something about the mating strategies of these species from the type and degree of sexual dimorphism and its association with other characteristics such as body scarring and relative testis size (see article on Sexual Dimorphism). For example, relative testis size ranges nearly two orders of magnitude among odontocete species, from less than 0.05% [several Mesoplodon species, the franciscana (*Pontoporia blainillei*), the baiji (*Lipotes vexillifer*), and sperm whale] to nearly 5% [harbor porpoise and dusky dolphin (*Lagenorhynchus obscurus*)] of body weight. These data suggest the importance of sperm competition in several odontocete species, and especially among some of the delphinids and porpoises. The importance of mate choice competition, attempts to entice and attract females through elaborate visual or acoustic displays, is suggested by differences between the sexes in vocalizations and exaggerated visual signals such as the post-anal hump or enlarged dorsal fins. At present, sexually dimorphic acoustic signals are known only in sperm whales. However, because odontocetes produce a wide range of sounds, acoustic displays are likely to occur in several other species as well. The importance of contest competition for access to mates is suggested by sexual dimorphism in size, weaponry (teeth and tusks) and the presence of scarring of conspecific origin (tooth rake marks). Sperm whales, the beaked whales, and bottlenose whale (*Hyperoodon* spp.) exhibit these traits.

*2. Mysticetes* Among the mysticetes, substantial data on breeding behavior exist only for the humpback (*Megaptera novaeangliae*), right (*Eubalaena* spp.), and gray (*Eschrichtius robustus*) whales. Even in these species, little is known about female behavior. The humpback whale has been studied most intensively. Male humpbacks adopt one or more of the three primary strategies: display by singing long, complex songs; direct competition with other males for females in "competitive groups;" and escort of females, including those with newborn calves. It is possible that males escorting females are waiting for mating opportunities or are guarding females after copulation. Two secondary strategies, roving and sneaking, have also been suggested.

Female humpbacks sometimes aggressively reject subadult males and they may incite competition among males. Although molecular analysis of paternity has shown that females are mated by different males between years (Clapham and Palsbøll, 1997), it is unknown whether females mate multiply within a given breeding season. Singing by male humpbacks is an intriguing phenomenon, since songs change over time yet all members of a population sing essentially the same song at any one time. Singing by humpback males may serve to attract females, signal status to other males, or both. Additional suggestions for the function of song include a means to synchronize estrus and a basis for organizing males during breeding season (Darling *et al.*, 2006). Whether the aggregation and displaying

**M**

of humpback whales at specific sites constitutes a lek also remains controversial (see article on Song).

Little else is known about the mating systems of balaenopterid whales. Blue (*Balaenoptera musculus*), fin (*B. physalus*), Bryde's (*B. edeni*), and minke (*B. acutorostrata*) whales have patterned calls, termed songs, which are geographically variable and thought to function in mating, although the mechanism is unclear. Species recognition, male–male competition, and female choice are possibilities. In the Gulf of California, male fin whales may sing while guarding feeding territories (Croll *et al.*, 2002).

Right whales show sexual activity throughout the year, although calving is strongly seasonal. Since the gestation period is 1 year and there is no evidence of diapause, mating leading to conception presumably occurs primarily in the winter. The function of sexual activity during other seasons is unknown. Observations of multiple male right whales mating with single females, together with the huge (1 ton!) testes, strongly suggest that sperm competition is a principal mating strategy in this species, and also probably in bowhead (*Balaena mysticetus*) and gray whales (Brownell and Ralls, 1986). The level of aggression in male–male interactions in these species is low compared to that observed in humpback whales, an observation that is also consistent with the predominance of sperm competition as a mating strategy.

### C. Sirenians, Sea Otters (*Enhydra lutris*) and Polar Bears (*Ursus maritimus*)

Male manatees (genus *Trichechus*) and dugongs (*Dugong dugong*) tend to be solitary and search for potential mates by roaming over large areas that include the home ranges of several females. Groups of males sometimes follow and try to mate with a single female, forming a "mating herd." In both manatee and dugongs, males in these herds threaten and fight with each other but it is still unknown whether this behavior is a form of scramble competition or is more akin to a type of lekking. In Shark Bay, Australia, dugongs associate in a more classical kind of lekking, with several males patrolling exclusive areas and engaging in activities usually indicative of both male competition and mate attraction, including acoustic signaling. In both manatees and dugongs, the mating season extends over several months and sexual dimorphism is slight. Interbirth intervals are at least 2 years and may be as much as 5 years in some cases.

Female sea otters typically give birth annually. Births generally peak in the spring, although females in warmer areas may give birth in any month. Adult males are larger than females. Male sea otters establish territories that contain food resources and sheltered resting places and usually overlap the home ranges of one or more females. Males may defend territories seasonally or all year. Other males congregate in groups outside of the areas occupied by territorial males. Courtship and mating, as are typical for many mustelid species, is physically aggressive and females may be injured or killed by males. Copulation occurs with both the male and female on their backs near the water's surface and the male grasps the female's head or jaws, including the nose, in his own jaws. Recently mated females typically have red, swollen noses. After mating, the pair may stay together for a few days in which they feed, groom, play, and rest in close company.

Polar bears are highly sexually dimorphic and polygynous; adult males may be over twice as heavy as adult females. Females have extensive home ranges and males are forced to travel over large areas when searching for mates. Males apparently fight among themselves for access to females. Specific courtship behaviors are lacking or are as yet undescribed. The largest and strongest males apparently do most of the

mating while other males sometimes wait in the distance. Polar bears are notable among marine mammals in that they are the only species in which females give birth to multiple young (1–2 is the most common litter size and rarely 3–4). The interbirth interval is about 3 years.

## IV. Mating System Studies and the Future

Our knowledge of marine mammal mating systems has increased dramatically in recent years due to technological advances in the use of molecular markers, animal-borne cameras and underwater acoustic recording devices which provide new insights and make it possible to investigate previously inaccessible behaviors and species. Long-term field studies continue to be sources of deepening understanding and new analytical methods such as social network studies and investigations of genetic kinship will likely reveal more complex patterns of social structuring than is currently known. Female mating strategies, including the role of multiple mating and the importance of cryptic female choice are gaining increasing attention. As our understanding of the physiology of female receptivity grows, we will be better able to interpret both female and male mating behavior. At the same time, our increasing ability to hear and see underwater will enable us to tap into the little known realm of underwater acoustic and visual displays.

### *See Also the Following Articles*

Breeding Sites ■ Courtship Behavior ■ Estrus and Estrous Behavior ■ Song ■ Reproductive Behavior ■ Sexual Dimorphism ■ Territorial Behavior

### *References*

Amos, B., Schlötterer, C., and Tautz, D. (1993). Social structure of pilot whales revealed by analytical DNA profiling. *Science* **260**, 670–672.

Boness, D. J., Bowen, W. D., and Francis, J. M. (1993). Implications of DNA fingerprinting for mating systems and reproductive strategies of pinnipeds. *Symp. Zool. Soc., Lond.* **66**, 61–93.

Boness, D. J., Bowen, W. D., and Iverson, S. J. (1995). Does male harassment of females contribute to reproductive synchrony in the grey seal by affecting maternal performance? *Behav. Ecol. Sociobiol.* **36**, 1–10.

Boness, D., Clapham, P. J., and Mesnick, S. L. (2002). Life history and reproductive strategies. *In* "Marine Mammals: An Evolutionary Approach" (R. Hoelzel, ed.), pp. 278–324. Blackwell Science, Ltd.

Boness, D. J., Bowen, W. D., Buhleier, B. M., and Marshall, G. J. (2006). Mating tactics and mating systems of an aquatic-mating pinniped: The harbor seal. *Phoca vitulina. Behav. Ecol. Sociobiol.* **61**, 119–130.

Bowen, W. D., Iverson, S. J., Boness, D. J., and Oftedal, O. T. (2001). Energetics of lactation in harbour seals: Effect of body mass on sources and level of energy allocated to offspring. *Funct. Ecol.* **15**, 325–334.

Brownell, R. L., and Ralls, K. (1986). Potential for sperm competition in baleen whales. *In* "Behaviour of Whales in Relation to Management. Special Issue 8 (G. P. Donovan, ed.), p. 97–112. Reports of the International Whaling Commission, Cambridge.

Clapham, P. J., and Palsbøll, P. J. (1997). Molecular analysis of paternity shows promiscuous mating in female humpback whales (*Megaptera novaeangliae*, Borowski). *Proc. Roy, Soc. Lond.* **264**, 95–98.

Clutton-Brock, T. H. (1989). Mammalian mating systems. *Proc. Roy. Soc. Lond.* **236**, 339–372.

Clutton-Brock, T. (2007). Sexual selection in males and females. *Science* **318**, 1882–1885.

Connor, R. C., Richards, A. F., Smolker, R. A., and Mann, J. (1996). Patterns of female attractiveness in Indian Ocean bottlenose dolphins. *Behavior* **133**, 37–69.

M

Cox, C. R., and Le Boeuf, B. J. (1977). Female incitation of male competition: A mechanism in sexual selection. *Am. Nat.* **111**, 317–335.

Croll, D. A., Clark, C. W., Acevedo, A., Tershy, B., Flores, S., Gedamke, J., and Urbán-Ramírez, J. (2002). Only male fin whales sing loud songs. *Nature* **417**, 809–810.

Danilewicz, D., Claver, J. A., Perez Carrera, A. L., Secchi, E. R., and Fontoura, N. F. (2004). Reproductive biology of male franciscanas (*Pontoporia blainvillei*) (Mammalia: Cetacea) from Rio Grande do Sul, southern Brazil. *Fish. Bull.* **102**, 581–592.

Darling, J. D., Jones, M. E., and Nicklin, C. P. (2006). Humpback whale songs: Do they organize males during the breeding season? *Behavior* **143**, 1051–1101.

Gentry, R. L. (1998). "Behavior and Ecology of the Northern Fur Seal." Princeton University Press, Princeton.

Goldsworthy, S. D., Boness, D. J., and Fleischer, R. C. (1999). Mate choice among sympatric fur seals: Female preference for conphenotypic males. *Behav. Ecol. Sociobiol.* **45**, 253–267.

Heath, C. B. (1989). The behavioral ecology of the California sea lion. Ph.D. Thesis. University of California, Santa Cruz.

Hoelzel, A. R., Le Boeuf, B. J., Reiter, J., and Campagna, C. (1999). Alpha-male paternity in elephant seals. *Behav. Ecol. Sociobiol.* **46**, 298–306.

Hoffman, J. I., Boyd, I. L., and Amos, W. (2003). Male reproductive strategy and the important of maternal status in the Antarctic fur seal. *Arctocephalus gazella. Evolution* **57**, 1917–1930.

Le Boeuf, B. J. (1974). Male-male competition and reproductive success in elephant seals. *Am. Zool.* **14**, 163–176.

Le Boeuf, B. J., and Mesnick, S. L. (1990). Sexual behavior of male northern elephant seals: I. Lethal injuries to adult females. *Behaviour* **116**, 143–162.

Lidgard, D. C., Boness, D. J., Bowen, W. D., McMillan, J. I., and Fleischer, R. C. (2004). The rate of fertilization in male mating tactics of the polygynous grey seal. *Mol. Ecol.* **13**, 3543–3548.

Mays, H. L., and Hill, G. E. (2004). Choosing mates: Good genes versus genes that are a good fit. *Trends Ecol. Evol.* **19**, 554–559.

Mesnick, S. L., and Hill, G.E. (1997). Sexual Alliances: Evidence and Evolutionary Implications. *In* "Feminism and Evolutionary Biology: Boundaries, Intersections and Frontiers" (P. A. Gowaty, ed.), pp. 207–260. Chapman and Hall, New York.

Pemberton, J. M., Albon, S. D., Guinness, F. E., Clutton-Brock, T. H., and Dover, G. A. (2002). Behavioral estimates of male mating success tested by DNA fingerprinting in a polygynous mammal. *Behav. Ecol.* **3**, 66–75.

Roughgarden, J., Oishi, M., and Akçay, E. (2007). Reproductive social behavior: Cooperative games to replace sexual selection. *Science* **311**, 965–969.

Stirling, I., and Thomas, J. A. (2003). Relationships between underwater vocalizations and mating systems in phocid seals. *Aq. Mamm.* **29**, 227–247.

Tolley, K. A., Read, A. J., Wells, R. S., Urian, K. W., Scott, M. D., Irvine, A. B., and Hohn, A. A. (1995). Sexual dimorphism in wild bottlenose dolphins (*Tursiops truncatus*) from Sarasota, Florida. *J. Mammal.* **74**, 11901198.

Trivers, R. L. (1972). Parental investment and sexual selection. *In* "Sexual Selection and the Descent of Man, 1871–1971" (B. Campbell, ed.), pp. 136–179. Aldine, Chicago.

Twiss, S. D., Poland, V. F., Graves, J. A., and Pomeroy, P. P. (2006). Finding fathers: Spatio-temporal analysis of paternity assignment in grey seals (*Halichoerus grypus*). *Mol. Ecol.* **15**, 1939–1953.

Van Parijs, S. M., Lydersen, C., and Kovacs, K. M. (2003). Vocalizations and movements suggest alternative mating tactics in male bearded seals. *Anim. Behav.* **65**, 273–283.

Wells, R. S., Scott, M. D., and Irvine, A. B. (1987). The social structure of free-ranging bottlenose dolphins. *Curr. Mammal.* **1**, 247–305.

Wells, R. S., Boness, D. J., and Rathbun, G. B. (1999). Behavior. *In* "The Biology of Marine Mammals" (J. E. Reynolds, III, and S. Rommell, eds), p. 324–422. Smithsonian Institution Press, Washington, DC.

# Melon-Headed Whale
## *Peponocephala electra*

Wayne L. Perryman

## I. Characteristics and Taxonomy

The melon-headed whale is one of a group of small, dark-colored whales that are often referred to as "blackfish." It is mostly dark gray in color with a faint darker gray dorsal cape that is narrow at the head and dips downward below the tall, falcate dorsal fin (Fig. 1). A faint light band extends from the blowhole to the apex of the melon. A distinct dark eye patch, which broadens as it extends from the eye to the melon, is often present and gives this small whale the appearance of wearing a mask. The lips are often white, and white or light gray areas are common in the throat region and stretching along the ventral surface from the leading edge of the umbilicus to the anus. At sea, this species is difficult to distinguish from the pygmy killer whale (*Feresa attenuata*). It differs externally from the pygmy killer whale by having a more pointed or triangular head and sharply pointed pectoral fins. Both of these characters are difficult to recognize at sea unless these small whales are seen from above. Experienced observers often rely more on behavioral than on physical characters to separate these two blackfish in the field. In stranded specimens, the melon-headed whale can be distinguished from all other blackfish by its high tooth count, 20–26 per row, compared to generally less than 15 teeth per row for pygmy killer whales.

Asymptotic length for males (2.52 m) is greater than for females (2.43 m), and males also have comparatively longer flippers, taller dorsal fins, broader tail flukes, and are more robust (Best and Shaughnessy, 1981; Miyazaki *et al.*, 1998). In addition, some males exhibit a pronounced ventral keel that is found posterior to the anus. The longest specimen reported was a 2.78-m female that stranded in Brazil (Lodi *et al.*, 1990); a 2.64-m male that stranded in Japan weighing 228 kg is the heaviest specimen reported (Miyazaki *et al.*, 1998).

The skull of the melon-headed whale is typically delphinid in shape, with the exception of a very broad rostrum and deep antorbital notches. It is similar to the skull of the common bottlenose dolphin (*Tursiops truncates*), but the teeth of the melon-headed whale are

**M**

**Figure 1** *Melon-headed whales underwater.*

much smaller and more delicate. The high tooth count of this species separates its skull from those of the other small beakless whales.

The melon-headed whale is a member of the subfamily Globicephalinae where it is closely allied with the very similar pygmy killer whale and the larger pilot whales (*Globicephala melas* and *G. macrorhynchus*), false killer whales (*Pseudorca crassidens*), and Risso's dolphin (*Grampus griseus*). Investigations regarding the interrelations of these species have yet to produce definitive results.

## II. Distribution and Abundance

Melon-headed whales are found worldwide in tropical to subtropical waters. They have occasionally been reported from higher latitudes, but these sightings are often associated with incursions of warm water currents (Perryman *et al.*, 1994). They are most often found in offshore, deep waters, and nearshore sightings are generally from areas where deep oceanic waters are found near the coast. Systematic line transect surveys have produced estimates of abundance of 45,000 (CV = 0.47) for the eastern tropical Pacific (Wade and Gerrodette, 1993), 3451 (CV = 0.55) in the northern Gulf of Mexico (Mullin and Fulling, 2004), and in the Philippines, 921 (CV = 0.83) in the eastern Sulu Sea and 1383 (CV = 0.82) in Tañon Strait between the islands of Cebu and Negros (Dolar *et al.*, 2006).

## III. Ecology

Squids appear to be the preferred prey of this species, but small fishes and shrimps have also been found in their stomachs (Jefferson and Barros, 1997). In the eastern tropical Pacific, aggregations of melon-headed whales have frequently been seen in association with Parkinson's petrels (*Procellaria parkinsonii*) which likely feed on scraps of large prey items that the small whales dismember below the surface (Pitman and Balance, 1992).

## IV. Behavior and Physiology

Melon-headed whales are most often found in large aggregations, a behavior that separates them from the very similar pygmy killer whale. They are often seen in large mixed aggregations with Fraser's dolphin (*Lagenodelphis hosei*). They have also been sighted in mixed herds with spinner dolphins (*Stenella longirostris*), common bottlenose dolphins, and rough-toothed dolphins (Dolar *et al.*, 2006). Although they are reported to flee from approaching vessels in the eastern Pacific, it is not uncommon for melon-headed whales to briefly ride the bow wave of passing ships in other areas. They may bow-ride for longer periods if the vessel slows to a speed of a knot or less.

Mass strandings of melon-headed whales have been reported on several occasions and it appears that strandings of this species, at least in the North Pacific, are becoming more common (Brownell *et al.*, 2007). The causes of the strandings are unknown. In two strandings from Japan and one in Brazil, the specimens had high loads of internal parasites, which might have caused some animals to strand. It has been suggested that mass strandings of these highly social animals may be caused by a panic response in the school when a few members accidentally strand (Miyazaki *et al.*, 1998). In a recent stranding event in Hawaii, active sonar transmissions from military vessels were considered as a contributing factor in the stranding (Southall *et al.*, 2006).

## V. Life History

Melon-headed whales are about 1.1 m in length at birth (Perrin, 1976; Bryden *et al.*, 1977) and continue to increase in length until they are 13–14 years old. Females reach sexual maturity at about 11.5 years and a length of about 235 cm. Males appear to mature later, at an age of about 15 years.

## VI. Interactions with Humans

When captured live and transferred to aquariums, melon-headed whales have not thrived and have been difficult to train. They have been aggressive toward keepers and have caused injuries by ramming individuals with their heads or raking them with their teeth. Although melon-headed whales were reported to approached divers in an aggressive manner in Hawaiian waters, several other reports of interactions between these small whales and divers or swimmers found them to be mildly evasive to curious toward people in the water.

Melon-headed whales are taken in small numbers in directed harpoon fisheries and as bycatch in drift net fisheries in the Philippines (Dolar, 1994; Dolar *et al.*, 1994), Indonesia, Malaysia, and in the Caribbean near the island of St. Vincent. Schools of melon-headed whales have been taken in the drive fishery operated from the port of Taiji, Japan. On rare occasions, a member of this species is taken in the purse seine fishery for yellow-fin tuna in the eastern tropical Pacific. Because most of these fisheries are not extensively monitored, the effect of these direct and incidental takes on local populations is unknown.

## See Also the Following Articles

Pilot Whales ■ Pygmy Killer Whales

## References

Best, P. B., and Shaugnessy, P. D. (1981). First record of he melon-headed whale *Peponocephala electra* from South Africa. *Ann. South Afr. Mus.* **83**, 33–47.

Brownell, R. L., Jr., Yamada, T. K., Mead, J. G., and Allen, B. M. (2007). Mass strandings of melon-headed whales, *Peponocephala electra*: A world wide review. IWC Scientific Committee meeting document.

Bryden, M. M., Harrison, R. J., and Lear, R. J. (1977). Some aspects of the biology of *Peponocephala electra* (Cetacea: Delphinidae). I. General and reproductive biology. *Aust. J. Mar. Freshw. Res.* **18**, 703–715.

Dolar, M. L. L. (1994). Incidental takes of small cetaceans in fisheries in Palawan, central Visayas, and northern Mindanao in the Philippines. *Rep. Int. Whal. Commn, Spec. Iss.* **15**, 355–363.

Dolar, M. L. L., Leatherwood, J. S., Wood, C. L., Alava, M. N. R., Hill, C. L., and Aragones, L. V. (1994). *Rep. Int. Whal. Commn* **44**, 439–449.

Dolar, M. L. L., Perrin, W. F., Taylor, B. L., Kooyman, G. L., and Alava, M. N. R. (2006). Abundance and distributional ecology of cetaceans in the central Philippines. *J. Cetacean Res. Manage.* **8**, 93–111.

Jefferson, T. A., and Barros, N. B. (1997). Peponocephala electra. *Mamm. Spec.* **553**, 1–6.

Lodi, L. F., Siciliano, S., and Capistrano, L. (1990). Mass stranding of *Peponocephala electra* (Cetacea Glopicephalinae) on Pirancanga Beach, Bahia, Brazil. *Sci. Rep. Cetacean Res. Inst.* **1**, 79–84.

Miyazaki, N., Yoshihiro, F., and Iwata, K. (1998). Biological analysis of a mass stranding of melon-headed whales (*Peponocephala electra*) at Aoshima, Japan. *Bull. Natl. Sci. Mus. Tokyo Ser.* **A 24**, 31–60.

Mullin, K. D., and Fulling, G. L. (2004). Abundance of cetaceans in the oceanic northern Gulf of Mexico. *Mar. Mam. Sci.* **20**, 787–807.

Perryman, W. L., Au, D. W. K., Leatherwood, S., and Jefferson, T. A. (1994). Melon-headed whale- *Peponocephala electra* (Gray, 1846). *In* "Handbook of Marine Mammals" (S. H. Ridgway, and R. Harrison, eds), Vol. 5, pp. 363–386. Harcourt Brace, London.

Southall, B. L., Braun, R., Gulland, F. M. D., Heard, A. D., Baird, R. W., Wilkin, S. M., and Rowles, T. K. (2006). Hawaiian melon-headed

M

whale (*Peponocephala electra*) mass stranding event of July 3–4, 2004. NOAA Technical Memorandum NMFS-OPR-31. 73 pp.

Wade, P. R., and Gerrodette, T. (1993). Estimates of cetacean abundance and distribution in the eastern tropical Pacific. *Rep. Int. Whal. Commn.* **43**, 477–493.

# Mesoplodont Whales
## (*Mesoplodon* spp.)

### ROBERT PITMAN

Mesoplodont whales are found in nearly all of the deeper, oceanic waters of the world and they are by far the most speciose genus of marine mammal. But despite this, they remain among the least known large animals on the planet. Several species have never been identified alive in the wild, and at least one species is known only from skeletal remains. Although they are fairly common in some parts of the ocean, because of their shyness around vessels and unobtrusive behavior, they are rarely observed, even by those who study cetaceans at sea.

## I. Characters and Taxonomic Relationships

The genus name *Mesoplodon* (Greek: *mesos*, middle: *hopla*, arms; *odon*, tooth; i.e., armed with a tooth in the middle of the jaw) was coined by Gervais in 1850. Ziphiids (beaked whales), including mesoplodonts, appeared suddenly in the fossil record in the lower Miocene (26 Mya), are well represented by the upper Miocene (5 Mya), and their diversity has declined steadily since then. Currently, there are 14 recognized species in *Mesoplodon* (Table I) and new species are still being discovered. In the Pacific Ocean, *M. peruvianus* (Reyes *et al.*, 1991) and *M. perrini* (Dalebout *et al.*, 2002) were recently described; *M. traversii* was resurrected as a distinct species (van Helden *et al.*, 2002), and another unidentified form from the central Pacific is perhaps a new species also (Dalebout *et al.*, 2007). Species limits in the group have been confirmed through a concordance of morphological characters and molecular genetics analyses (Dalebout *et al.*, 2004). Although no subspecies currently recognized within the group, *M. mirus* has disjunct populations in the North Atlantic and the southern Indian Ocean which have markedly different color patterns and they are genetically as distinct as some recognized species of mesoplodonts, suggesting at least a subspecific level of divergence. Longman's beaked whale (*Indopacetus pacificus*) was up until very recently known only from two skulls and has often been included within *Mesoplodon*, but is it now considered to be a separate, monotypic genus.

### TABLE I
#### Living Species of *Mesoplodon*

| Latin name | English name(s) | Length° (m) | Distribution |
|---|---|---|---|
| *M. hectori* | Hector's beaked whale; New Zealand beaked whale | 4.4 | Temperate waters of the Southern Hemisphere |
| *M. mirus* | True's beaked whale | 5.3 | Warm temperate waters of North Atlantic and southern Indian Ocean |
| *M. europaeus* | Gervais' beaked whale, Antillean beaked whale, Gulf Stream beaked whale | 5.2 | Warm temperate and tropical waters of the Atlantic |
| *M. bidens* | Sowerby's beaked whale; North Atlantic beaked whale; North Sea beaked whale | 5.5 | Temperate North Atlantic |
| *M. grayi* | Gray's beaked whale; Haast's beaked whale; scamperdown whale; small-toothed beaked whale | 5.6 | Temperate waters of the Southern Hemisphere |
| *M. peruvianus* | Pygmy beaked whale; Peruvian beaked whale; lesser beaked whale | 3.9 | Mostly tropical waters in the eastern Pacific |
| *M. bowdoini* | Andrew's beaked whale; deep-crest beaked whale | 4.4 | Temperate waters of the Southern Hemisphere |
| *M. traversii* | Spade-toothed beaked whale | ? | Known from 3 strandings: islands off Chile and New Zealand |
| *M. carlhubbsi* | Hubbs' beaked whale; arch-beaked whale | 5.3 | Temperate North Pacific |
| *M. ginkgodens* | Ginkgo-toothed beaked whale; Japanese beaked whale | 5.3 | Tropical and warm temperate waters of Indian and (mainly western) Pacific oceans |
| *M. stejnegeri* | Stejneger's beaked whale; Bering Sea beaked whale; saber-toothed beaked whale | 5.7 | Subarctic and temperate North Pacific |
| *M. layardii* | Layard's beaked whale, strap-toothed beaked whale; long-toothed beaked whale | 6.2 | Temperate and subantarctic Southern Hemisphere |
| *M. densirostris* | Blainville's beaked whale; dense-beaked whale | 4.7 | Circumglobal in warm temperate and tropical waters |
| *M. perrini* | Perrin's beaked whale | 4.4 | Known only from strandings in California, USA |

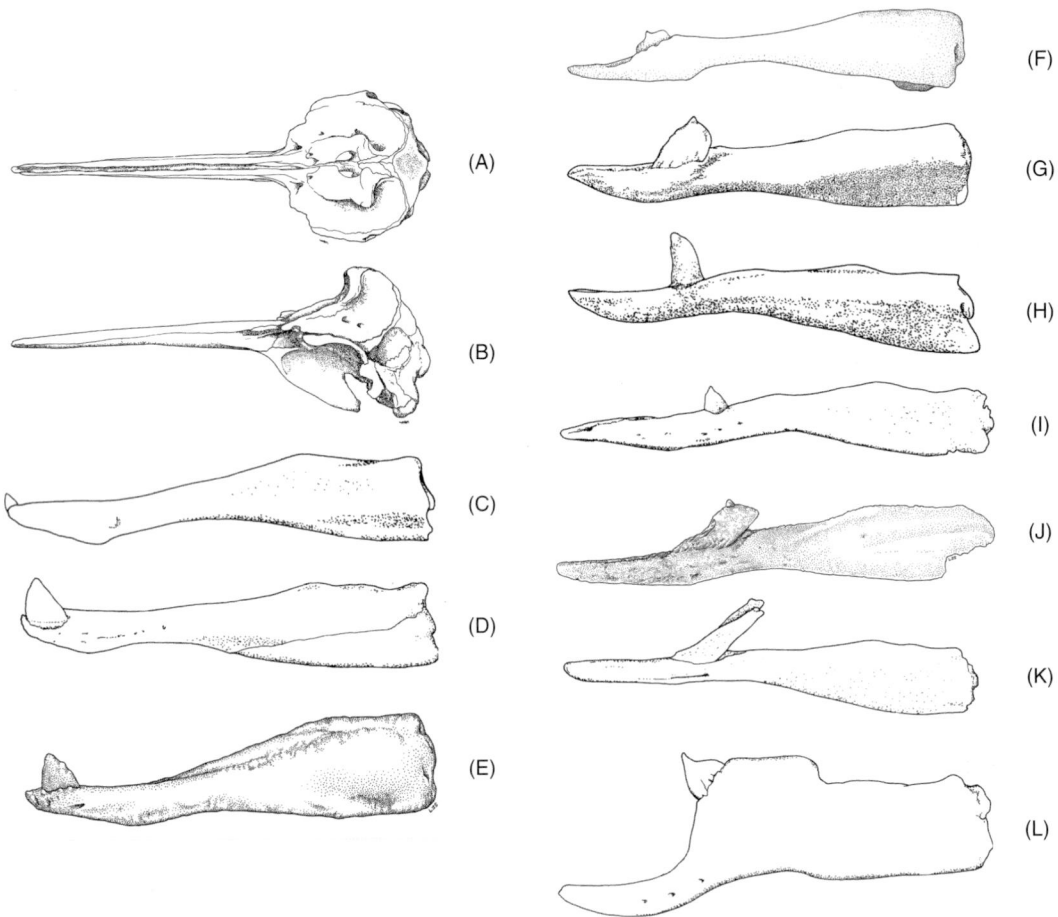

Figure 1 *Selected* Mesoplodon *skulls and mandibles, the latter showing the diversity of tooth shape, size and placement within adult males of this genus (not to scale). (A) Dorsal view of* M. grayi *skull, (B) Lateral view of* M. grayi *skull; lower jaws of (C)* M. mirus, *(D)* M. hectori, *(E)* M. perrini, *(F)* M. grayi, *(G)* M. carlhubbsi, *(H)* M. stejnegeri, *(I)* M. bidens, *(J)* M. traversii, *(K)* M. layardii, *and (L)* M. densirostris. *Modified from Jefferson* et al. *(2008).*

Historically, mesoplodont species have been diagnosed by features of the skull, relying mainly on the length of the rostrum, and the shape, size, and placement of teeth, especially of adult males (Moore, 1968; Fig. 1). However, due to anatomical similarities (especially of females and young), specimens have often been misidentified, even by experts, and molecular genetic analyses have become increasingly important for identifying individual specimens and even uncovering "cryptic" species (Dalebout *et al.*, 2002, 2007; van Helden *et al.*, 2002).

Mesoplodonts are small whales, ranging in size from 3.9 (*M. peruvianus*) to 6.2 m (*M. layardii*) (Table I); there are too few data to determine if there are consistent size differences between males and females, although in at least some cases females appear to be slightly larger, as is the case for most ziphiids. The body is spindle shaped, with a small, usually triangular dorsal fin located approximately two-thirds of the way back on the body (Fig. 2). The flippers are small and narrow and fit into pigmented depressions in the body. The unnotched flukes are usually straight across the trailing edge or even slightly convex. A single pair of external throat grooves is present between the mandibles that apparently assist in suction feeding (discussed later). The head is small and tapered; the melon is small and blends without a crease into

Figure 2 *A rare photograph of a strap-toothed beaked whale* (Mesoplodon layardii), *the most distinctively patterned of all the mesoplodonts, lofting out over the Southern Ocean. This is an especially long-beaked species of* Mesoplodon. *Photography by S. N. G. Howell.*

**Figure 3** *A Hubbs' beaked whale* (Mesoplodon carlhubbsi) *that stranded in Kanagawa, Japan in 2005. The erupted tooth (one on each side) identifies it as an adult male, and the white, linear scars on its head are tooth marks from skirmishes with other males; the red flesh is from beach abrasion. Photo: Courtesy Kanagawa Pref. Museum of Natural History, Japan.*

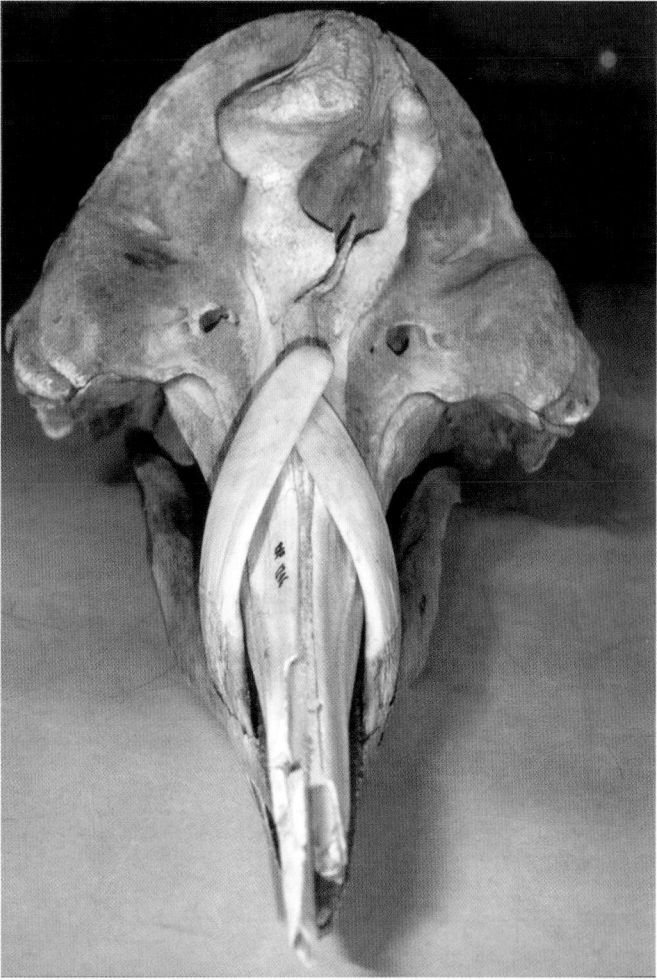

**Figure 4** *The skull of an adult male strap-toothed whale* Mesoplodon layardii *showing assymetrical blowhole and the remarkable way the teeth wrap around the upper jaw.*

**M**

the beak. The blowhole is a half circle with the ends pointed forward and not always symmetric. Beak length is variable depending on the species, ranging from short (e.g., *M. densirostris*) to very long (e.g., *M. grayi*, *M. layardii*, and *M. bidens* Fig. 2).

Most species show three sexually dimorphic traits, all of which relate to male aggressive behavior at the onset of sexual maturity: (1) only adult males have functional teeth (tusks), (2) only adult males have extensive secondary ossification of the mesorostral canal, and (3) only adult males show extensive and conspicuous body scarring. These features are discussed.

Males apparently use their teeth for intraspecific fighting with other males to establish breeding hierarchies (Heyning, 1984). In fully mature males, a single pair of teeth erupts from the mandibles, projecting up, outside of the mouth. Depending on the species, the teeth are located anywhere from the tip of the lower jaw (apical), to about halfway back along the jaw, and they vary markedly in size and shape depending on the species (Fig. 1). The teeth are usually laterally compressed, although in *M. mirus*, which has the most apical teeth, they are oval in cross section. The tooth usually develops a sharp denticle on top for inflicting wounds, but this often worn down smooth in older males. In some species, the tooth is raised up on a high bony arch in the lower jaw (Fig. 3); in other species, the mandible is relatively straight and the tooth is surrounded and supported mainly by gum tissue.

A hypothesized evolutionary trend for mesoplodonts proposes that the occurrence of apical teeth is a primitive condition and more posterior locations are derived. Teeth positioned further back on the mandible apparently allow animals to attack more forcefully with their rostrum with less risk of damage to the mandibles or teeth. However, teeth positioned further back along the lower jaw need to be elevated or elongated so that they are not obstructed by the rostrum, which gets wider and deeper toward the base. Although it has been suggested that the different sizes and shapes of teeth could used for species recognition among the various mesoplodonts,

it is just as likely that tooth position and mandible development within this group correlate with differing amounts of male aggression among the different species. As an extreme example of tooth development, *M. layardii* has some of the most bizarre teeth of any known animal. They are long (at least to 34 cm) and curl back and over the rostrum so that they sometimes overlap each other, clamping the jaws nearly shut (Fig. 4). These teeth have dorsally projecting denticles and, judging by the amount of scarring on adult males, they are still effective for intraspecific fighting. However, in many individuals, tooth wear also results from the rostrum rubbing against the inner sides of the teeth. This clearly indicates that jaw movement is impaired, and the adaptive significance of tooth development in this species has never been adequately explained.

The mesorostral canal is a narrow groove in the midline of the upper rostrum that is filled with cartilage in most cetaceans. This cartilage is continuous with the mesethmoid and homologous to the nasal septum of terrestrial mammals. In adult male mesoplodonts, the cartilage is displaced by expansion of the vomer from below, which is composed of extremely dense bone, in some cases more dense than elephant ivory. When males attack, they make contact with the top of their rostrum and use their teeth with the mouth in the closed position. It has generally been assumed that the heavily reinforced rostrum was selected to allow fighting males to be more forceful with their attacks while reducing the possibility of damage to the rostrum. A recent study, however, suggested that the bone of the rostrum, although very dense, was too brittle to provide mechanical reinforcement and that it might, e.g., have a hydrostatic role to assist in deep diving. Why this would be necessary for males and not females is not clear.

Color patterning among mesoplodonts is gradually becoming known. In general, females and young of most species are indistinguishable either at sea or stranded on the beach because they all tend to be nondescript gray or brown dorsally and somewhat paler ventrally. On the other hand, adult males of most species if seen well are often identifiable in the wild based on tooth shape and location and body coloration. For example, *M. mirus* (Southern Hemisphere form) has an all-white tail stock, dorsal fin, and flukes; *M. layardi* has a very distinctive black and white pattern (Fig. 2), and adult male *M. peruvianus* from the eastern tropical Pacific have a broad white swathe across the body that forms a conspicuous chevron when viewed from above.

The more conspicuous color patterning that most male mesoplodonts exhibit can be due either to normal pigment deposition (e.g., white beak and top of melon of *M. carlhubbsi*; Fig. 3) or to adventitious coloration caused by scarring. Most of the (often numerous) scars on mesoplodonts are attributable to either tooth rake marks from adult males or are the healed scars from the bites of cookiecutter sharks (*Isistius* spp., see below). Both sexes tend to collect the oval cookiecutter shark bite scars as they age but only the adult males acquire the numerous long, linear scars from encounters with other males. All mesoplodonts, except apparently *M. ginkgodens*, form white scar tissue over external wounds so that the body retains a permanent visual record of past injuries. As a result, among groups of mesoplodonts, adult males are usually easily recognizable by their prominent linear scarring (Fig. 3), which may be important for social signaling. In addition to heavy scarring, adult males (and, to a lesser extent, females) of some species have white pigment patches that may serve to highlight important anatomical landmarks, including the beak tip, head, lips, and genital area.

## II. Distribution and Abundance

Mesoplodonts are so difficult to approach and identify under normal conditions that there are several species that have almost never been identified alive in the wild, including, e.g., *M. bowdoini*, *M. perrini*, *M. traversii*, *M. ginkgodens*, and *M. hectori*. Consequently, nearly everything known about their distribution and abundance has been inferred from relatively infrequent stranding events. As a group they are widespread, occurring in all of the world's oceans except for the coldest waters of the Arctic and Antarctica. They normally inhabit deep oceans waters (>2000 m) or continental slopes (200–2000 m), and only rarely stray over the continental shelf. The distribution of most species tends to be somewhat localized (limited to single ocean basins or a hemisphere within an ocean basin; Table I), although *M. densirostris* is found in all tropical and warm temperate oceans and is probably the most widespread and perhaps most abundant mesoplodont. Seasonality of stranding records suggests that at least some high-latitude species (e.g., *M. layardii*) may undergo some limited migration to lower latitudes during the local winter. *M. stejnegeri* in the Bering Sea sometimes have fresh cookiecutter shark bites suggesting that they probably moved up from warmer waters (discussed later).

## III. Ecology

Very little is known about prey preferences of mesoplodonts Based on stomach contents of stranded animals, they appear to feed primarily on relatively small (~500 g), mesopelagic squid (e.g., *Histioteuthis* spp., *Taonis* spp., *Gonatus* spp.). Meso- and benthopelagic fish are also taken (at least some of these fish, however, maybe secondary, i.e., squid prey) and in some species (e.g., *M. bidens*, *M. mirus*, and *M. grayi*), fish may comprise the most important prey items. Most prey are probably caught at depths of well over 200 m or more, using high-frequency echo-location clicks to detect them.

A reduced dentition among various species of odontocetes (toothed cetaceans) is generally interpreted as an adaption for feeding on squid. Most squid-feeding cetaceans, including mesoplodonts, are thought to be suction feeders: prey is sucked in and swallowed whole. Large muscles at the back of the tongue (hyoglossus and styloglossus) are anchored to an enlarged hyoid bone; this allows the tongue to be retracted in a piston-like manner while the paired throat pleats allow distention of the mouth floor, creating the necessary vacuum for sucking in prey. Because ziphiids in general no longer need their teeth for feeding, this has freed them up to evolve for other purposes, i.e., as fighting weapons for males, or to be functionally lost altogether (females). It has been hypothesized that white pigmentation on the anterior floor of the mouth of many beaked whale species may serve as an attractant for bioluminescent squid.

## IV. Behavior and Physiology

Until recently, almost nothing was known about mesoplodont behavior, partly because they are so rarely observed, but also because their behavioral repertoire at the surface is very limited and stereotyped. The most commonly reported behavior has been slow swimming, usually away from a vessel, and often a mile or more away. When undisturbed, they roll quietly several times at the surface and disappear, usually not to be seen again. Most groups surface simultaneously and within a few body lengths of each other, indicating that they are probably communicating as they forage in total darkness of the deep ocean. While at the surface, either traveling or stationary, individuals in groups usually remain within a couple body lengths of each other. Breaching (leaping out of the water; Fig. 2) has been recorded on only a very few occasions. Long-beaked forms (e.g., *M. grayi*, *M. bidens*, and *M. layardii*) often bring their beak up out of the water at a 45° angle when they surface. Normally there is no visible blow, and no species are known to lift their flukes when diving. Tail slapping has been observed on a couple of occasions. Although adult males are assumed to use their erupted teeth as tusks to fight with each other, none of these battles have ever even reported by human observers, even though the extensive scarring found among males of this group suggests that it must be of fairly frequent occurrence (Fig. 3). Photo-identification studies of *M. densirostris* in the Bahamas and around Hawaii indicate long-term site fidelity with individuals re-sighted in the same general area over a period of at least 15 years.

Recent technological advances, including DTAG (acoustic recording tags) and electronic time/depth recorders, have vastly improved our understanding of what these animals are doing during their long absences from the surface. Mesoplodonts are deep divers: suction cup time/depth recorders on *M. densirostris* in Hawaii indicated regular dive times from 48–68 min and reaching a maximum depth of 1408 m. Another study showed that between deep foraging dives, *M. densirostris* spent about twice as much time near the surface undertaking a series of shallow dives, which, it was suggested, allows them to recover from an oxygen debt.

Vocalizations have been reported for only two species: *M. densirostris* and *M. carlhubbsi*. Recently, vocalizations recorded with a suction-cup acoustic recording tag indicate that *M. densirostris* uses two distinct click types associated with different phases of foraging: long duration, frequency-modulated clicks were used for detecting prey and short duration "buzz" clicks used when capturing prey (Johnson *et al.*, 2006). Mesoplodonts are also suspected of producing mid-frequency tonal sounds, similar to those used by other cetacean species for communication. This is apparently rarely used, at least near the surface, which may be a behavioral adaptation for avoiding detection by predatory killer whales ("acoustic crypsis").

## V. Life History

Because of their shy nature, oceanic habitat, and apparent rarity, very little is known about the biology of mesoplodonts, and nearly everything that is known has come from the examination of stranded animals. As in all cetaceans, females give birth to a single calf. Reported mean body length for calves at birth has ranged from 2.1 m for *M. europaeus* to 2.5 m for *M. carlhubbsi*, representing from 40–48% of the adult female body lengths (these are neither the largest nor the smallest mesoplodonts). The smallest mesoplodont calf reported to date was 1.9 m (*M. hectori*), although *M. peruvianus* will probably prove to be smaller. There is no information on gestation or lactation periods.

Longevity data for mesoplodonts are virtually nonexistent, although they may be quite long lived: a count of tooth layers in a specimen of *M. europaeus* suggests that it was at least 27 years old. A female *M. densirostris* estimated to be 9 years old (based on tooth layer counts) had just recently become sexually mature. Very little has been recorded in the way of diseases, parasites, or commensals (Mead, 1989). Osteomyelitis has been reported twice. Endoparasites recorded have included cestodes and occasional heavy infestations of nematodes (*Crassicauda* sp.) in the kidneys. Ectoparasites recorded include *Penella* sp. (a parasitic copepod) and cyamids. The erupted teeth of males often have stalked barnacles (*Conchoderma* sp.) attached to them; bunches of these often appear as "tassels" on the teeth of live animals at sea. A pseudo-stalked barnacle, *Xenobalanus* cf *globicipitis*, sometimes attaches to the trailing edges of the flukes or dorsal fin.

Mesoplodonts occur in small groups typically ranging in size from 1 to 6 animals, although groups of up to 15 have been reported, and a mass stranding of 28 *M. grayi* occurred in New Zealand. Mean school size for 125 *Mesoplodon* sightings (including at least three different species) from the eastern tropical Pacific was 3.0, with 2 being the most common. Although mixed groups of adult males with females and calves have been observed at sea, there is some evidence from both sightings and strandings data that there may be some segregation by sex or age class at times. Predators of mesoplodonts probably include killer whales and large sharks, although direct observations are lacking. Mesoplodonts often have numerous white oval scars (diameter to about 8 cm) caused by the bites of cookiecutter sharks (*Isistius* sp.); one study in Hawaii counted 120 separate bite marks on the visible part of just one side of a *M. densirostris*. These are small (to about 50 cm), mesopelagic sharks that feed by snatching mouthfuls of flesh of larger fish and cetaceans. Although presumably painful, they do not appear to contribute to mesoplodont mortality.

## VI. Interactions with Humans

So few mesoplodonts have been reliably identified at sea that it has been impossible to accurately determine the population status of any species, although based on stranding data, at least some species may not be as rare as the sighting records indicate. *M. grayi* and *M. layardii* appear to be widespread and fairly common in the Southern Ocean, as is *M. densirostris* in tropical oceans; these may be the most abundant mesoplodonts. Most species, however, appear to be neither numerous nor widespread, and some may be quite rare (e.g., *M. bowdoini*, *M. hectori*, *M. perrini*, *M. traversii*). The large number of species in this group suggests a high rate of endemism with naturally small populations and restricted ranges, all of which increases the vulnerability of individual species.

The two main human-caused threats to mesoplodonts are probably bycatch in fishing gear and anthropogenic noise. Although there has never been any directed fishery, a few are occasionally harpooned opportunistically by whalers, and unknown but potentially significant numbers are killed by high seas drift nets and longline fishing gear. Only stranded specimens have ever been kept in captivity and these have usually died within a few days (usually from pre-existing conditions).

Currently, the biggest threat to mesoplodonts, and beaked whales in general, may be anthropogenic noise sources, particularly those associated with airgun arrays (seismic surveys) and military mid-frequency sonar (2–10 kHz). Necropsies of mass-stranded beaked whales exposed to these sound sources have led to a hypothesis that mortality may be caused by gas-bubble disease induced by behavioral responses to acoustic exposure (Cox *et al.*, 2006).

## See Also the Following Articles

Aggressive Behavior, Intraspecific ■ Beaked Whales, Overview ■ Cetacean Life History

**M**

## References

Cox, T. M., *et al.* (35 authors). (2006). Understanding the impacts of anthropogenic sound on beaked whales. *J. Cetacean Res. Manage.* **7**, 177–187.

Dalebout, M. L., Mead, J. G., Baker, C. S., Baker, A. N., and van Helden, A. L. (2002). A new species of beaked whale *Mesoplodon perrini* sp. N. (Cetacea: Ziphiidae) discovered through phylogenetic analyses of mitochondrial DNA sequences. *Mar. Mamm. Sci.* **18**, 577–608.

Dalebout, M. L., Baker, C. S., Cockcroft, V. G., Mead, J. G., and Yamada, T. K. (2004). A comprehensive and validated molecular taxonomy of beaked whales, family Ziphiidae. *J. Hered.* **95**, 459–473.

Dalebout, M. L., *et al.* (8 authors) (2007). Beaked whales: Molecular evidence for a new species in the tropical Pacific? *Mar. Mamm. Sci.* **23**, 954–966.

Heyning, J. E. (1984). Functional morphology involved in intraspecific fighting of the beaked whale, *Mesoplodon carlhubbsi*. *Can. J. Zool.* **62**, 1645–1654.

Jefferson, T. A., Webber, M. A., and Pitman, R. L. (2008). "Marine Mammals of the World." Academic Press, San Diego, CA.

Johnson, M., Madsen, P. T., Ximmer, W. M. X., Aguilar de Soto, N., and Tyack, P. L. (2006). Foraging Blainville's beaked whales (*Mesoplodon densirostris*) produce distinct click types matched to different phases of echolocation. *J. Exp. Biol.* **209**, 5038–5050.

Mead, J. G. (1989). Beaked whales of the genus *Mesoplodon*. In "Handbook of Marine Mammals" (S. H. Ridgway, and R. J. Harrison, eds), Vol. 4, pp. 349–430. Academic Press, London.

Moore, J. C. (1968). Relationships among the living genera of beaked whales with classifications, diagnoses, and keys. *Field. Zool.* **53**, 209–298.

Reyes, J. C., Mead, J. G., and Van Waerebeek, K. (1991). A new species of beaked whale *Mesoplodon peruvianus* sp. n. (Cetacea: Ziphiidae) from Peru. *Mar. Mamm. Sci.* **7**, 1–24.

van Helden, A. L., Baker, A. N., Dalebout, M. L., Reyes, J. C., van Waerebeek, K., and Baker, C. S. (2002). Resurrection of *Mesoplodon traversii* (Gray, 1874), senior synonym of *M. bahamondi* Reyes, Van Waerebeek, Cárdenas and Yañez, 1995 (Cetacea: Ziphiidae). *Mar. Mamm. Sci.* **18**, 609–621.

# Migration and Movement Patterns

## S. Jonathan Stern

Migration and movement are hallmarks of marine mammal behavior. Whether engaged in localized diving or ocean basin-scale displacements, marine mammals use their three-dimensional world in fascinating ways. One of the best ways to view how they use their world is through their movement. The focus of this chapter is to discuss migration and movement patterns in a general sense, using various species as examples, rather than to provide a species by species account.

Movement encompasses a hierarchy of displacements, ranging from a few meters to thousands of kilometers (Fig. 1). A *step* is a relevant distance moved, such as distance between surfacing sites

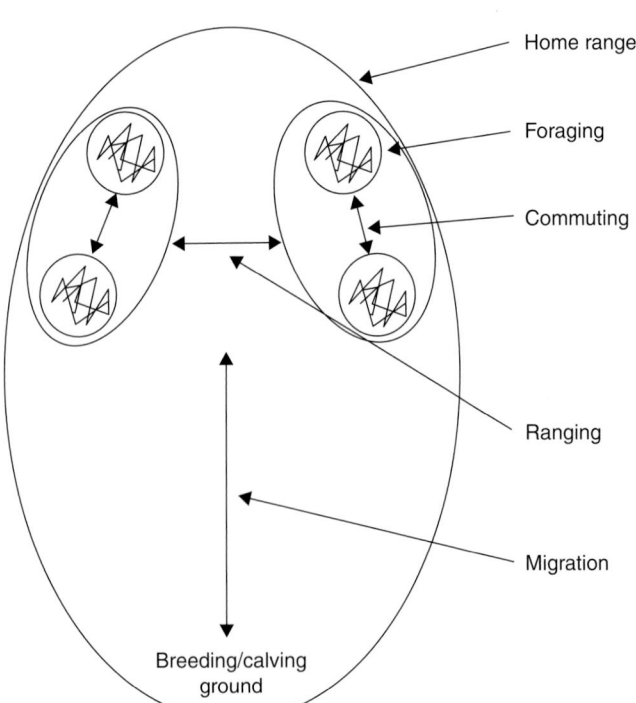

**Figure 1**    *Different types of movements are described in the text.*

or distance traveled between fluke or flipper beats. A step should relate to some relevant distance. *Kinesis* refers to changes in turning or movement rates. *Foraging* is search for resources within a patch. A patch is an area within which resources are randomly distributed. The patch itself may be spatially constant while resource availability and distribution varies temporally. *Commuting* is movement between adjacent patches. *Ranging* is movement between regions. After searching patches in a region, an animal then moves to another region, spatially separated and defined by a different local set of oceanographic conditions. *Dispersal* is permanent movement from one area to another. It is impossible to talk about movement without an environmental context.

As an animal moves greater distances, it experiences greater environmental variability. Variability relates to, among other things, areas of high and low resource concentration, or patchines. Resources include food, conspecifics (including mates), and space. Mates may only be seasonally receptive or available, or sexes may have a different spatial distribution in a particular season. Space is a variable resource because not all habitats are suitable, or available at all times. For example, high tide may cover rocks used for hauling out. In territorial species, once a territory is occupied, other individuals of the same species are excluded. In non-territorial species, crowding often occurs, and while suitable habitat is nearby, individuals crowd.

Foraging animals rely on predictable resources at a specific place at a specific time (scale). Resource variability requires a movement tactics that allows an individual to maximize its fitness. In a constant environment, an individual is likely to find food on most days, while in a high-variance environment, an individual may have difficulty finding food from one day to the next and for considerable stretches of time. This animal is essentially living in an energy sink environment with few areas of energy sources (Stern, 1998). Unless significant energy stores are available, that individual is going to run out of energy.

Marine mammals, by virtue of their large body sizes, require large home ranges (Calder, 1984). And given that resources distribution is variable in space and time, some home ranges are enormous. As home range increases in size, an individual experiences spatial and temporal environmental variability.

Consider a parcel of water with properties of temperature, salinity, dissolved oxygen, and nutrients. Two parcels of water 10 cm apart are more similar, in terms of temperature, salinity, dissolved oxygen, and nutrients than two parcels of water 100 km apart. In terms of depth, one parcel of water at the surface differs from a parcel of water at 100 m. Therefore, a diving animal experiences greater small-scale spatial variability than an individual moving along the surface. In addition, a single parcel of water may have similar oceanographic conditions 2 days apart but may differ significantly 2 weeks apart. And oceanographic conditions in early summer may be very different than those in mid or late summer. The timing of oceanographic conditions favoring primary production must coincide with periods of high hours of sunlight. Thus, resources are variable in space as well as time. Disparity in resource availability results in the necessity to move between places in the home range. Some parcels of water have food, but most do not. The parcels of water with food are rare so an individual has to move to find it. The term *variance* means that food is patchy.

Patchy prey resources require animals adapt to this spatial and temporal variability. The strategy is to conserve energy, since available energy is reduced during parts of the year. One tactic is to stay in a home area and tough out the hard times by somehow reducing the effects of variability. An individual that stays knows the area, reduces the energetic cost of moving, and does not encounter potentially adverse conditions along the way. Survival requires formation of denser fur, thicker BLUBBER layer for thermal protection and energy storage, or adopting a strategy such as hibernation. Female polar bears (*Ursus maritimus*) hibernate over winter, giving birth and feeding cubs until emerging from the den in spring. Other segments of polar bear populations do not hibernate, but make large-scale movements in search of food. For cetaceans, the cost of moving long distances while on migration is probably not very different than moving around in one location. Some pinnipeds conserve energy by hauling out or lying at the surface for extended periods. Energy is also conserved during dives by slow or no tail beats, or gliding. Ranging is a second strategy where an animal moves to another part of their home range where conditions are more favorable, such as a higher resource density, or where environmental conditions are better. Feeding generally occurs along the way. For example, Caribbean manatees (*Trichechus manatus*) move in relation to seasonal changes in water temperature preferring waters warmer than 68°F.

Migration is defined as regular, repeated, and large-scale movement between different sites of the home range (Dingle, 1996). Migration is at the extreme end of a continuum of large-scale displacements. The difference between migration and other large-scale displacements is that migration consists of persistent linear travel not distracted by resources. The goal is to move between sites each serving in a different life history role. Leaving a migration terminus is marked by a decrease in turning angles, resulting in linear travel. Upon arriving at the other migration terminus, turning angles increase in size and frequency resulting in station keeping. In addition, some form of energy re-allocation sustains migration, and time away from feeding grounds (Dingle, 1996). For marine mammals, these destinations are areas for feeding, breeding, birth, lactation, and molting. Pinnipeds generally forage between haul-out sites and feeding grounds. However, some phocids rely on stored energy while hauled out (Table I). Marine mammals have a suite of physiological adaptations for energy allocation and fasting, as well as storage and mobilization of fats to and from blubber stores. Fats are the most important

energy source, as lipids hold more energy per unit weight than other forms. Lipids are also less bulky than protein or carbohydrates because no water is required for storage. During feeding on productive feeding grounds, hyperphagia promotes increased lipid synthesis, fat uptake, and rate of fatty acid synthesis. Fats are stored when supply in blood exceeds metabolic demand and mobilization occurs when the demand for energy in blood exceeds supply.

Feeding in highly productive areas results in an increased fat store to support reproduction. Based on stomach contents analysis of whales killed on winter grounds, mysticetes are thought to feed little on migration routes and breeding grounds, living off stored blubber. However, recent evidence suggests that feeding occurs on the winter grounds, at least opportunistically. Some phocids such as elephant seals (*Mirounga* spp.) move long distances. They feed continuously to store energy to support time spent on breeding and molting haul outs. Stored energy is used for lactation in females and for mating and agonistic displays in males. Elephant seals lose much of their fat when hauled out and need to begin storing energy as rapidly as possible upon reentry into the water. Elephant seal movement can be viewed as migration, though feeding does occur along the migration route. Telemetry studies suggest areas with a higher frequency of deep dives, suggesting intense, localized feeding. Some of these highly productive areas are associated with seamounts. Hauling out for extended periods reduces energy loss to locomotion. Therefore, most of the stored energy can be used for reproductive activities and metabolism.

True migrating marine mammals are the mysticete whales, and their general migration patterns are shown in Fig. 2. However, energetic models suggest that mysticetes might have to feed at a reduced level while away from their feeding grounds (Lockyer and Brown, 1981). Migration was historically assumed to occur based on the occurrence of a particular species at given locations at different times of the year. However, documenting individual movement between locations was the only way to prove migration. Whaling provided the first real evidence for migration in large whales. Numbered darts were fired into dorsal blubber and muscle. If that individual was killed during subsequent whaling operations, tagging and killing dates and locations were compared and some assessment of movements were made. Some pinnipeds are marked with numbered flipper tags. Censuses were conducted on a number of haul-outs that documented tagged individuals moving between haul-out sites. Movement and migration patterns have been described in varying levels of detail using photo-identification and satellite telemetry. For example, locations of individual northern elephant seals (*Mirounga*

**M**

---

TABLE I
**Energy Acquisition**

Fasting

    Mysticetes[a] (except bowhead whales)

    Elephant seals[a]

    Some male otariids[a] (on breeding grounds)

    Harp seals (*Pagophilus groenlandicus*)[a]

    Hooded seals (*Cystophora cristata*)[a]

    Polar bears (females can hibernate)

    Most phocids

    Recently weaned pinnipeds (some exploratory swimming around haul-out site)

Nonfasting

    Odontocetes

    Female otariids

    Some phocids

    Polar bears

    Sirenians

    Otters

[a]A true migratory species.

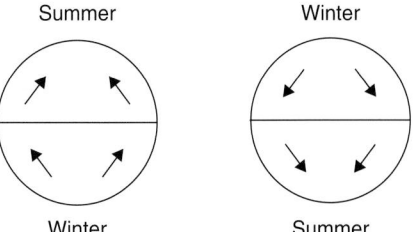

**Figure 2** *Generalized migration patterns of baleen whales in the Northern and Southern Hemispheres. Whales spend summers on productive feeding grounds and then migrate to the winter breeding/calving grounds in warmer waters. Most species of whales are believed to feed little, if at all, while away from their feeding grounds.*

**M**

*angustirostris*) sent by satellites every 2 days revealed fascinating information on movement to and from areas of high food productivity, inferred by persistent signals from a relatively confined location, such as over a seamount. Intricate underwater movement has been described using telemetry devices showing that marine mammals do not swim in straight lines between surfacing locations.

The following scenarios are not migrations but are often labeled as such in literature. Seasonal movement may be a response to changing prey distribution. The occurrence of some groups of killer whales (*Orcinus orca*) in the inland waters of the Pacific Northwest correlates with the seasonal migration of salmon (*Oncorhynchus* spp.). Because these fish ultimately go upstream to die, the whales must find other prey during the winter. It is not clear if they feed on other species during the winter or go offshore to find other salmon schools. Gray seals (*Halichoerus grypus*) move to distinctly different areas on a seasonal basis to feed in productive areas in the Northwest Atlantic. Other large-scale movements are also not truly migration. For example, movement may be in relation to shifting environmental conditions, such as the seasonal advance or retreat of an ice edge. While geographic location of an individual changes, it is essentially maintaining itself in the same general environment.

Dispersal is not migration, as there is no return to the original area. Dispersal is colonization of new or re-colonization of historic breeding habitats. For example, breeding sites of northern elephant seals were historically on islands off the mainland of California, likely due to the presence of terrestrial predators on the mainland. In association with post-exploitation recovery and the decline of terrestrial predators, elephant seals returned to all historic breeding islands and invaded new sites on the mainland. Another example of a dispersal event was observed off California in response to the 1982–1983 El Niño Southern Oscillation event. Common bottlenose

dolphins (*Tursiops truncatus*) moved from southern California to San Francisco with the northward advance of warmer waters. This group of dolphins remained after the warm waters retreated back south.

Baleen whales require large quantities of high quality food. Such resources are found in productive cold waters so it is clear why they migrate to these areas. However, these areas have not been consistent over very long temporal scales. The poleward retreat of productive oceanic waters over the past 20,000 years changed regional productivity, migration routes, destinations, and foraging areas of many marine mammals. Increased ice extent and land emergence of the Pleistocene made current feeding grounds unavailable to gray whales (*Eschrichtius robustus*), bowhead whales (*Balaena mysticetus*), beluga whales (*Delphinapterus leucas*), narwhals (*Monodon monocerus*), walruses (*Odobenus rosmarus*), Northern fur seals (*Callorhinus ursinus*), and other seals. The North Atlantic north of 45° was an ice-bearing polar sea with conditions similar to the Antarctic Convergence, resulting in a larger, more productive sea than at present. The reduced distance between productive cold water and warmer water meant large-scale movement was not necessary (see Fig. 3). In the Southern Ocean, an equatorward shift in isotherms was in response to a northward extension of ice. The Antarctic convergence was 5° north of its current position. As cold, nutrient-rich waters retreated toward the poles, sunlight for photosynthesis became more seasonally variable leading to intense seasonal peaks in production followed by reduced production in winter. Over time, whale distribution followed the poleward retreat of fronts of productive oceans. Along with this the need to store massive amounts of energy for time away from feeding grounds.

A question remains as to why they migrate to warm waters for reproduction. Four reasons have been suggested for migration to

**Figure 3**   *Post-Pleistocene distribution of the polar front in the North Atlantic Ocean. Lines represent the southern extent of sea surface temperature relating to the polar front. Numbers associated with each line are thousands of years before present. Whale movement from warm-water calving/breeding waters to cold productive feeding waters would have been similarly truncated toward the equator in the past. The equator-ward extension of cold water meant that productive seasons were probably longer in the past than at present due to increased hours of daylight.*

warm water breeding and calving areas. The first is to minimize thermal stress on calves. This is likely not a problem for a newborn calf. Smaller mammals with less insulation are able to survive in those conditions. Because of its large body size, a calf is not likely thermally stressed.

The second reason to move to warm waters is resource tracking, i.e., following prey. By definition, this is not migration since foraging occurs along the way. Energetic models suggest that migrating (fasting) baleen whales need to feed, albeit at low levels while away from the feeding grounds. Whales likely feed on whatever they encounter, rather than tracking prey.

Killer whale predation on calves has been a suggested reason for mysticete migration. By migrating to warm, relatively killer whale-free waters to give birth, calf mortality would be reduced. Killer whales have recently been observed feeding on gray whale calves caught along their northward migration. These attacks occur where calves are most vulnerable, far from shore where the mother can lead the calf into the coastal kelp beds. Killer whales exploit this opportunistic but reliable resource.

The final reason to migrate is essentially an evolutionary holdover—individuals migrate because their ancestors did. The evolutionary holdover hypothesis includes feeding and reproduction into a life history strategy. Intense feeding leads to energy storage thus maximizing reproductive success. Natural selection favors individuals that are successful at migrating, feeding, reproducing, and responding to changing environmental conditions.

Resource distribution explains the summer distribution of baleen whales, and for whatever reason, they migrate to warm waters for reproductive activities. How do marine mammals initiate large-scale movement? Many species of marine mammals, and mammals in general, have highly seasonal reproductive strategies to time births relative to optimal environmental conditions. Day length is a cue for seasonal breeding in a number of birds and mammals and is invariant from year to year. The advantage is that an individual can maximize its use of seasonal prey resources as well as seasonally available mates.

Both circadian (daily) and circannual (yearly) cycles use light as a cue; however, the specific cues from light, or *zeitgebers*, vary. Circadian signals are dawn and dusk, whereas circannual signals are perceived as the ratio of number of light to dark hours in a 24-h period. Thus, both cycles are used for seasonal cues. Other cues may act as secondary synchronizers, although these, such as food availability, are more variable.

The pineal gland is responsible for time keeping in birds and mammals via the production of melatonin, as well as other compounds. The number of hours of darkness per 24h is "counted" by the biosynthesis of melatonin, which is produced more in hours of darkness. At seasonal scales, as winter approaches, the hours of daylight decrease and hours of darkness increase. In a given 24-h period, the amount of melatonin produced increases, suppressing gonadal activity. In many mammals, the breeding season of females corresponds to periods of decreasing daylight per 24h. Increasing daylight after the winter solstice is responsible for triggering estrus. Pineal glands are exceptional in size in polar species such as the Weddell seal (*Leptonychotes weddellii*) northern fur seal, and walrus, species that live where day length is most variable.

Testosterone and its metabolites trigger migratory behavior in some animals. Male California sea lions (*Zalophus californianus*) spend the bulk of the year hauled out in large bachelor groups. By June, they have left their haul-outs in central and northern California, Oregon, Washington, and Alaska and migrated to the Channel Islands off southern California. Here, breeding occurs as males set up territories and defend females against other males with whom they spend most of the year in relatively peaceful coexistence. Once migration back to the feeding grounds is triggered in July, an individual is exposed to increasing hours of daylight per 24h as it swims poleward. One benefit from such a signal is that it is invariant from year to year.

Since their entire migration take place close to shore, documenting migration in the Eastern Pacific Gray Whale is relatively easy. Given that the peak migration of gray whales is variable from year to year, environmental cues must be only partly responsible for onset of migration. If endogenous cues play a role, their influence may be at least partly overridden. It is enticing to speculate what will happen to gray whale migration, as well as arctic species as arctic ice retreats.

The mechanisms of orientation, plotting their location at any time, and navigation, directing movement from one location to another, are not known. Individuals are often seen in the same locations from one year to the next. In the interim, they have traveled thousands of kilometers, indicating that marine mammals use some type of cues for orientation and navigation between migratory destinations. Organisms tend to meander if they lack orientation and navigation cues. Therefore, marine mammals must know where they are at a given time (orientation) and where they are going next (navigation).

At the initiation of migration, a direction must be selected. Advancing ice may simply eliminate certain directions as a choice, displacing individuals toward the equator. In higher latitudes, changes in sea conditions influence prey availability, which may also trigger the migratory response. If either or both of these change, individuals will react accordingly.

Cues may vary over time and the course of migration. For example, once migration is initiated, the only cue necessary is which overall direction to travel: north, south, east, or west. Celestial navigation has been suggested as one mechanism of navigation. In the north–south directions, the relative location of the sun in the sky can be monitored. This may be as simple as "keep sunrise on the left side when migrating to the breeding ground and on the right side when migrating to the feeding ground" or as complicated as estimating latitude as a function of position of the sun. Navigation by star location has also been suggested as a mechanism.

Another possible large-scale cue is the direction of a major current if an animal is moving against it. Near the equator in the Northern Hemisphere, western boundary currents move from south to north, while eastern boundary currents move from north to south. Coastal processes and minor currents cause the deformation of major currents, resulting in the formation of gyres and eddies. Migrating whales may use these currents for a free ride and gyres, eddies, and smaller-scale features as landmark.

Magnetite in the brains of some species has been implicated as a mechanism by which individuals could track changes in the earth's magnetic field. Mass STRANDINGS often occur at the same location. These locations may have anomalies in the local earth's magnetic field, which cause whales to become disoriented and strand. A tantalizing example of the possibility of using magnetic cues is seen in humpback whales (*Megaptera novaeangliae*) migrating from Hawaii to Southeast Alaska. Tracks were within 1° of magnetic north.

At smaller scales, other cues could be used. For example, while mysticetes do not have a true sense of smell, they do have a well-developed Jacobsen's organ. This may allow them to "taste" differences

**M**

in water mass composition. For example, freshwater from ice melt or riverine input might provide a "taste trail" to a rich feeding ground, as a lens of fresh water floats on denser salt water.

Routes to and from feeding and breeding grounds may be variable or essentially a retracing of the migratory path. Male humpback whales migrate south from the Gulf of Maine relatively far offshore, while the return trip is much closer to shore. Gray whales along the west coast of North America probably migrate between calving lagoons in Baja California and feeding grounds in the Bering, Chukchi, and Beaufort Seas by following contours of the coastline. Gray whales migrate along the same nearshore corridor in both migratory directions. There is spatial or temporal segregation based on age class or gender. Northern elephant seals disperse from haulout sites into the Gulf of Alaska; however, males go further north than females. Another way to look at this is male and female elephant seals are segregated on their feeding grounds, and migrate to haul-outs for breeding and molting.

In cases where a baleen whale species has a sufficiently long lactation period, offspring can learn migration routes and feeding locations from their mothers. In instances where offspring are weaned prior to reaching the feeding grounds, offspring are left to make exploratory migrations in hopes of finding suitable feeding grounds.

However, little is known about hormonal activity in relation to migratory movements. Prolactin is a hormone responsible for promoting milk production and lactation in mammals. It has the effect of increased fattening in birds. If there were a similar effect on mammals, it would be of importance. For pinnipeds, where lactating females leave pups for one to a few days, increased fat storage for milk is vital. A lactating female mysticete on the feeding ground would store fat relatively faster. This would not only provide for milk for the offspring but also help in restoring the female's blubber layer for subsequent migration back to the breeding-calving grounds.

Within an ocean basin, stocks can be viewed as a metapopulation, which is a number of populations connected by the dispersal of individuals between them. If a species is distributed into populations spread over a sufficiently large area, environmental conditions are more or less independent between areas; therefore, a catastrophe in one area will not affect other populations. Dispersal reduces the risk of population extinction by minimizing the effects of chance environmental changes or changes in population demography. Further, genetic heterozygosity is maintained and the population is less likely to exhibit genetic problems. The probability of dispersal between metapopulations is an important thing to know.

A Steller sea lion (*Eumetopias jubatus*) metapopulation structure has been explored in Alaska. In the Aleutian Islands and Gulf of Alaska, these sea lions have declined by 50% since the 1960s. However, Steller sea lions from Southeast Alaska south to Oregon have remained stable or slightly increased during this time. Evidence from Alaska and the Aleutian Islands suggests that fragmentation will occur, with rookeries being reduced in size and eventually becoming extinct. One reason for hope would be if the population in Southeast Alaska became a source of dispersers into the Aleutian population. The Mediterranean monk seal (*Monachus monachus*) experienced a recent population decline with habitat fragmentation throughout its range. Large expanses of unsuitable habitat separate major pupping sites, with little chance of dispersal between the two remaining large populations, although each separate population may be viable over time.

Movement and migration models can be used to study the effects of changing environmental conditions. Migration in marine mammals evolved within the context of and environment that changes over many spatial and temporal scales. The *rate* of change is as much a concern as the amplitude of the change. If populations are not able to adapt to the rate at which the environment changes, it will not survive. Species had to adapt to deal with novel situations and conditions. A polar front in the Pleistocene as described earlier retreated at a rate that allowed individuals to adapt to its changing distribution. Climatologists predict that global temperatures will increase by as much as 4.5°C in the next century. While it is clear that marine mammals are capable of adapting to changing environments, they might not be able to adapt at a rate commensurate with that of the change in environmental conditions in the near future. The potential implications are profound, and the environmental effects are not entirely clear.

Global warming will vary by latitude and environment, with polar and temperate regions affected to a greater extent than more tropical areas. Because these areas represent feeding grounds for migratory as well as resident species, understanding these effects is of considerable importance. The effect on marine mammals will likely be through changes in the distribution of resources in space and time. Key to survival will be how individuals respond to changes in resource distribution over space and time, and how this affects reproductive success. This hinges on an individual's ability to move in response to changing environmental conditions.

## See Also the Following Articles

Behavior, Overview ■ Breeding Sites ■ Cetacean Ecology ■ Distribution ■ Pinniped Ecology

## References

Baker, R. R. (1978). "The Evolutionary Ecology of Animal Migration." Holmes and Meier Publishers, New York.

Berta, A., and Sumich, J. L. (1999). "Marine Mammals: Evolutionary Biology." Academic Press, San Diego.

Dingle, H. (1996). "Migration: The Biology of Life on the Move." Oxford University Press, New York.

Lockyer, C. L., and Brown, S. G. (1981). The migration of whales. *In* "Animal Migration" (D. J. Aidley, ed.), pp. 105–137. Cambridge University Press, Cambridge.

McMullough, D. R. (1996). "Metapopulations and Wildlife Conservation." Island Press, Washington, DC.

Reidman, M. (1990). "Pinnipeds: Seals, Sea Lions, and Walruses." University of California Press, Berkeley.

Stern, S. J. (1998). Field studies of large mobile organisms: Scale, movement and habitat utilization. *In* "Ecological Scale: Theory and Applications" (D. Peterson, and V. T. Parker, eds), pp. 289–308. Columbia University Press, New York.

# Mimicry

PETER L. TYACK

The words "mimicry" and "imitation" often have a negative connotation in English of being an unoriginal fake. A mimic is often an annoying copycat, while an imitation can be a second-rate copy of a more valuable original. However, mimicry and imitation are based on special cognitive abilities that are rare among animals and that form the basis of culture. Humans learn most cultural traits—from the words in our language to the way we prepare food or hold a tool—through observational learning and imitation. We must learn thousands of these cultural traits through imitation before we can make an original contribution to our culture.

Imitation is a form of social learning—it requires an animal to observe a "demonstrator" performing a behavior and then to be able to perform that behavior itself (Galef, 1988; Whiten and Ham, 1992). Many psychologists distinguish between vocal learning, in which an animal modifies the sounds it produces based on the sounds it hears, and motor imitation, which involves an animal watching a posture or movement of another animal and then copying that movement— "monkey see, monkey do." Some also distinguish between motor imitation of something like clapping, where the animal can watch its hands in the same way that it watches the demonstrator, versus facial gestures or whole body movement, where the actor cannot receive sensory input about its own performance that directly parallels the observation of the demonstrator. However, the discovery of neurons in the premotor cortex of monkeys (Rizzolati and Craighero, 2004) that fire either when the monkey sees a specific action performed by another monkey or when it performs the same action itself suggests that specialized neural mechanisms may facilitate the ability of animals to equate their own actions with observation of the actions of others.

There is strong evidence for vocal learning and imitation among a variety of animals, including marine mammals and birds (Janik and Slater, 1997). Perhaps the simplest evidence involves vocal mimicry, where an animal demonstrates the ability to produce sounds after exposure to model sounds that were not part of its pre-exposure repertoire. Animals such as parrots (Todt, 1975; Pepperberg, 2000), starlings (West *et al.*, 1983), and a harbor seal (*Phoca vitulina*) named Hoover (Ralls *et al.*, 1985) have demonstrated abilities to imitate the sounds of human speech. Other animals such as dolphins have been trained to imitate acoustic features of artificial sounds (Richards *et al.*, 1984).

Humpback whales, *Megaptera novaeangliae*, have not been kept in captivity where one could most easily study imitation of manmade sounds, but their songs have a structure that cannot be explained by any mechanism other than vocal imitation. At any one time, most singing humpback whales within a population sing songs that are similar (Tyack, 1999). These songs change week by week, month by month, and year by year, and individual whales have been shown to track these changes. Humpback songs have been recorded for decades, with no suggestion that the songs repeat. This suggests that humpback whales learn to produce the current song and to track the progressive changes that make up such a distinctive feature of this signal.

Vocal imitation has also been reported for the natural sounds of bottlenose dolphins (*Tursiops truncatus*). Most dolphins studied in either captive or wild settings develop an individually distinctive signature whistle. Dolphins can imitate the whistles of social partners (Tyack, 1986). Three male Indian Ocean bottlenose dolphins (*T. aduncus*) that formed a strong social bond were reported to modify their whistles over 3 years as the bond formed, such that all three converged on a shared whistle (Smolker and Pepper, 1999). Watwood *et al.* (2004) showed that five pairs of male bottlenose dolphins in Sarasota, Florida each had developed whistles more similar to their partner than to the other pairs. The functions of this imitation are currently not known, but similar imitation has been suggested to function as a name for reference (Tyack, 1999), as a threat (Janik and Slater, 1997), or as an affiliative signal (Smolker and Pepper, 1999).

It has been controversial whether animals can perform motor imitation (Galef, 1988). This stems in large part from the difficulty of proving that a display of which the demonstrator was capable was not part of the pre-exposure repertoire of the animal. Ethologists expect that many animal displays represent fixed action patterns that are inherited. If an action pattern is simply triggered by sensing a conspecific performing the same action, that triggering alone does not demonstrate observational learning. There are many anecdotes about motor imitation in marine mammals which are difficult to explain via any mechanism other than observational learning. Tayler and Saayman (1973) provided some of the most interesting examples. They reported captive bottlenose dolphins swimming with postures and motor patterns similar to those of seals, turtles, fishes, and penguins which were housed in the same pool. These postures and swimming patterns are so awkward and different from normal dolphin locomotion that it is scarcely credible that they represent anything other than learned behaviors. The most striking example of imitation involved a calf Indian Ocean bottlenose dolphin that observed through an underwater window a human blowing out a cloud of cigarette smoke. The calf swam over to its mother, suckled, swam back to the window, and expelled a mouthful of milk into a cloud that looked similar to exhaled tobacco smoke! This kind of anecdote clearly suggests that it would be worth conducting careful experimental tests of motor imitation in cetaceans.

A good teacher can shape our behavior in ways that may look like imitation, but only demand associative learning on the part of the student. Animal trainers can shape the behavior of animals in the same way. If wild animals were to train one another in this way, this could create a faulty appearance of imitation, but this is thought not to be a problem for there is little evidence that one animal will train another to perform an action it has mastered (Caro and Hauser, 1992). However, there are some indications of what might be called teaching in cetaceans. Tyack and Sayigh (1997) provided suggestive evidence for possible teaching of signature whistles in wild bottlenose dolphins, and Guinet and Bouvier (1995) suggested that killer whales (*Orcinus orca*) teach the young how to strand to catch pinnipeds on the beach. Rather than being an alternative to passive observational learning, teaching appears at least in our own species to function in tandem with observational learning. This potential synergy between teaching and imitation would be most likely to benefit highly social animal groups, such as carnivores, primates, and cetaceans, in which cultural traditions for foraging provide a strong selective advantage. These kinds of observations in wild cetaceans, coupled with careful experimental tests with captive cetaceans, suggest that cetaceans are promising subjects for the study of mimicry in the development of cultural traits in animals.

## See Also the Following Articles

Culture in Whales and Dolphins ▪ Intelligence and Cognition ▪ Song

## References

Caro, T. M., and Hauser, M. D. (1992). Is there teaching in nonhuman animals? *Q. Rev. Biol.* **67**, 151–174.

Galef, B. G., Jr. (1988). Imitation in animals: History, definitions, and interpretation of data from the psychological laboratory. *In* "Social Learning: Psychological and Biological Perspectives" (T. Zentall, Jr., and B. G. Galef, eds), pp. 3–28. Lawrence Erlbaum Associates, Hillsdale.

Guinet, C., and Bouvier, J. (1995). Development of intentional stranding hunting techniques in killer whale (*Orcinus orca*) calves at Crozet Archipelago. *Can. J. Zool.* **73**, 27–33.

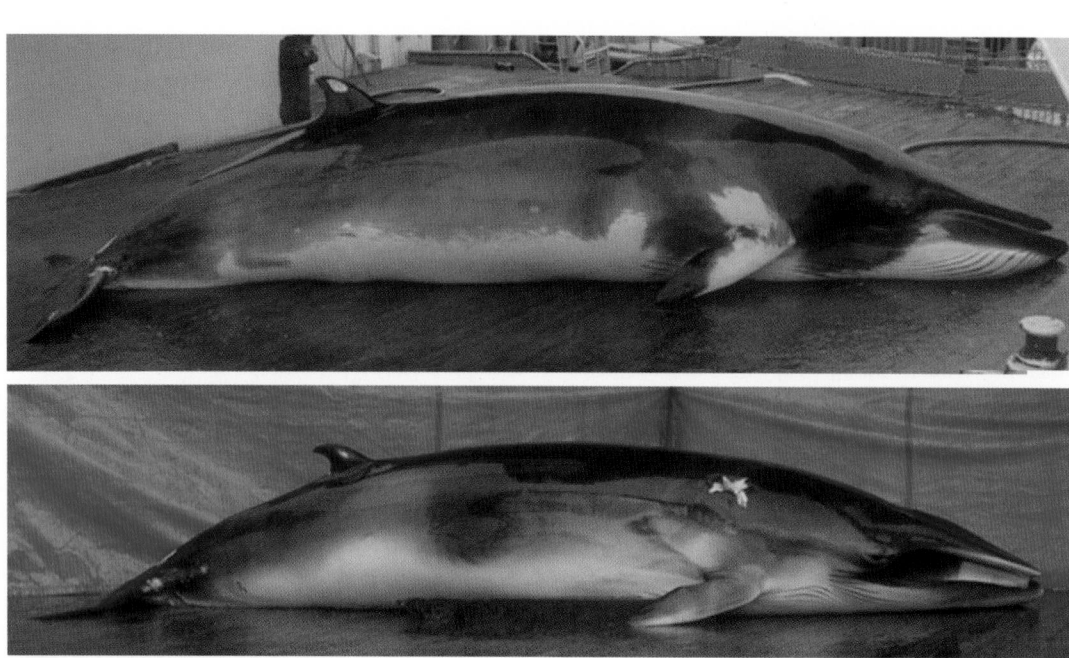

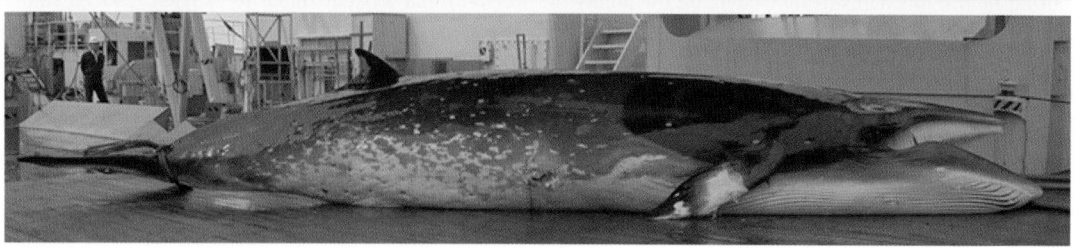

**Figure 1** *(Top) Dwarf minke whale* (Balaenoptera acutorostrata *unnamed subspecies), (middle) Antarctic minke whale* (B. bonaerensis), *and (bottom) North Pacific minke whale* (B. acutorostrata scammoni). *Photographs from Kato and Fujise (2000).*

**M**

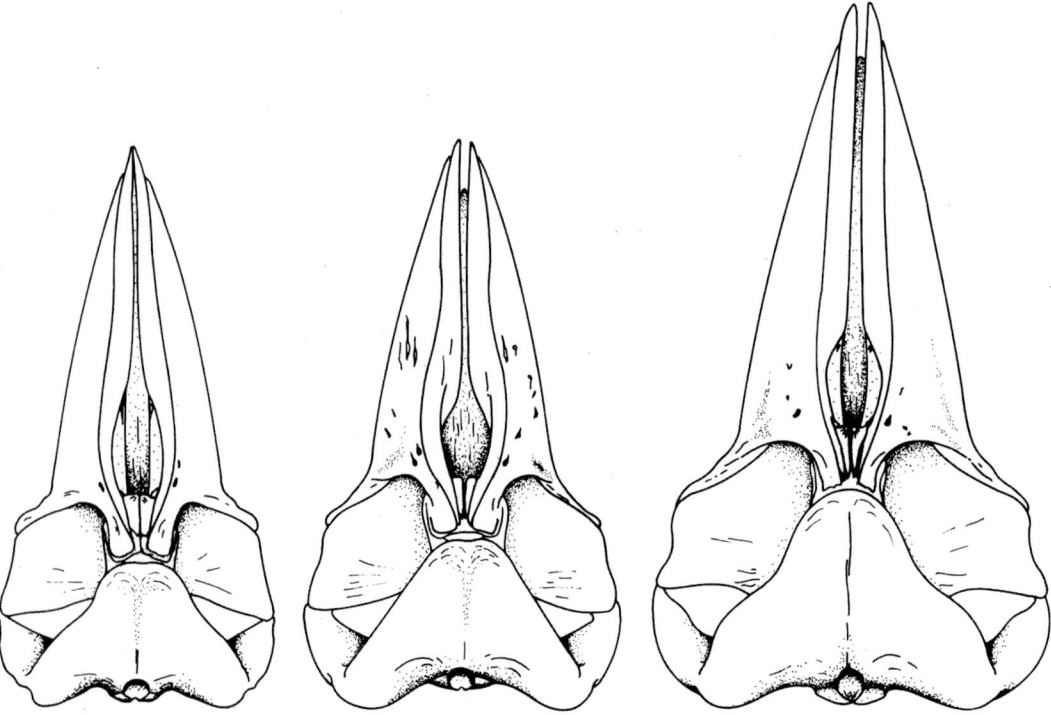

**Figure 2** *Dorsal views of skulls (left to right) North Pacific minke whale* (Balaenoptera acutorostrata scammoni), *dwarf minke whale* (B. acutorostrata *unnamed subspecies) and Antarctic minke whale* (B. bonaerensis). *From Kato and Fujise (2000).*

Janik, V. M., and Slater, P. J. B. (1997). Vocal learning in mammals. *In* "Advances in the Study of Behavior" (P. J. B. Slater, J. S. Rosenblatt, C. T. Snowdon, and M. Milinski, eds), Vol 26, pp. 59–99. Academic Press, New York.

Pepperberg, I. M. (2000). "The Alex Studies: Cognitive and Communicative Abilities of Grey Parrots." Harvard University Press, Cambridge.

Ralls, K., Fiorelli, P., and Gish, S. (1985). Vocalizations and vocal mimicry in captive harbor seals. *Phoca vitulina. Can. J. Zool.* **63**, 1050–1056.

Richards, D. G., Wolz, J. P., and Herman, L. M. (1984). Vocal mimicry of computer-generated sounds and vocal labeling of objects by a bottlenosed dolphin. *Tursiops truncatus. J. Comp. Psychol.* **98**, 10–28.

Rizzolati, G., and Craighero, L. (2004). The mirror-neuron system. *Annu. Rev. Neurosci.* **27**, 169–192.

Smolker, R., and Pepper, J. W. (1999). Whistle convergence among allied male bottlenose dolphins (Delphinidae, *Tursiops* sp.). *Ethology* **105**, 595–617.

Tayler, C. K., and Saayman, G. S. (1973). Imitative behavior by Indian Ocean bottlenose dolphins (*Tursiops aduncus*) in captivity. *Behavior* **44**, 277–298.

Todt, D. (1975). Social learning of vocal patterns and modes of their application in gray parrots (*Psittacus erithacus*). *Zeit. Tierpsychol.* **39**, 179–188.

Tyack, P. (1986). Whistle repertoires of two bottlenosed dolphins, *Tursiops truncatus*: Mimicry of signature whistles? *Behav. Ecol. Sociobiol.* **18**, 251–257.

Tyack, P. L. (1999). Communication and cognition. *In* "Biology of Marine Mammals" (J. E. Reynolds, III, and J. R. Twiss, Jr., eds), Vol 1, pp. 287–323. Smithsonian Institution Press, Washington, DC.

Tyack, P. L., and Sayigh, L. S. (1997). Vocal learning in cetaceans. *In* "Social Influences on Vocal Development" (C. Snowdon, and M. Hausberger, eds), pp. 208–233. Cambridge University Press, Cambridge.

Watwood, S. L., Tyack, P. L., and Wells, R. S. (2004). Whistle sharing in paired male bottlenose dolphins. *Tursiops truncatus. Behav. Ecol. Sociobiol.* **55**, 531–543.

West, M. J., Stroud, A. N., and King, A. P. (1983). Mimicry of the human voice by European starlings: The role of social interaction. *Wilson Bull.* **95**, 635–640.

Whiten, A., and Ham, R. (1992). On the nature and evolution of imitation in the animal kingdom: Reappraisal of a century of research. *In* "Advances in the Study of Behavior" (P. J. B. Slater, J. S. Rosenblatt, C. Beer, and M. Milinski, eds), Vol 20, pp. 239–283. Academic Press, New York.

# Minke Whales
## *Balaenoptera acutorostrata* and *B. bonaerensis*

WILLIAM F. PERRIN AND ROBERT L. BROWNELL, JR.

Until relatively recently, only one species of minke whale was thought to exist (common minke whale); all minke whales were referred to *Balaenoptera acutorostrata* (Lacépède, 1804) (e.g., in Stewart and Leatherwood, 1985). Morphological and genetic evidence of a second species accumulated through the last quarter of the twentieth century, and the Antarctic minke whale, *B. bonaerensis* (Burmeister, 1867) came to be fully recognized in the late 1990s (Rice, 1998; IWC, 2001), but a few workers have withheld judgment about the best taxonomic arrangement pending further studies (Kato and Fujise, 2000). The two species are partially sympatric in the Southern Hemisphere, where a small form (unnamed subspecies) of the common minke whale, the dwarf minke whale, is much smaller than the Antarctic minke whale and possesses the distinct white flipper mark that is characteristic of the species in the Northern Hemisphere and absent in the Antarctic species (Fig. 1). The two species also differ in relative size and shape of several cranial features (Arnold *et al.*, 1987) and in mitochondrial DNA sequences (Pastene *et al.*, 1994). Based on genetic data, the two species are hypothesized to have diverged less than 5 million years ago in the Southern Hemisphere (Pastene *et al.*, 2007).

Phylogenetic relationships of the minke whales to other Cetacea are yet unsettled. In a recent mtDNA control-region tree and in an earlier study including mtDNA and nuclear genes, the two species were sister taxa, in turn sister to the rest of the balaenopterids and the gray whale, *Eschrichtius robustus* (Sasaki *et al.*, 2006; Rychel *et al.*, 2004). In other trees and studies based on cytochrome *b* sequences, nuclear introns and sines, their position was more ambiguous. In a cladogram based on color-pattern characters, they were sister to the fin whale, *B. physalus* (Arnold *et al.*, 2005).

Rice (1998) recognized three subspecies of the common minke whale: the North Atlantic minke whale (*B. a. acutorostrata*), the North Pacific minke whale (*B. a. scammoni*, formerly *B. a. davidsonii*), and the unnamed Southern Hemisphere dwarf minke whale. The dwarf minke whale is genetically closer to the North Atlantic than to the North Pacific form.

The common name comes from Norway. One story has it that a hapless whale spotter named Meincke identified a minke whale as a blue whale, and thereafter small rorquals were called "Minkie's whale." Other common names that have been applied to the minke whales include "lesser rorqual," "little piked whale," "sharpheaded finner," and "lesser finback." Somewhat confusingly, the Antarctic minke is also called "ordinary minke whale" in older IWC literature.

In both species, the rostrum is very narrow and pointed and there is a single ridge on the head. The dorsal fin is relatively tall and falcate and is located relatively far forward on the posterior one-third of the body (in comparison to the larger rorquals). The average length of the common minke whale in the North Atlantic at physical maturity has been estimated variously at about 8.5–8.8 m in females and 7.8–8.2 m in males (Horwood, 1990). One estimate for the North Pacific is 8.5 in females and 7.9 in males. Female Antarctic minke whales are estimated to average 9.0 m at maturity and males 8.5 m. The dwarf minke whale of the Southern Hemisphere is on average about 2 m shorter than the Antarctic minke whale (Kato and Fujise, 2000). The white flipper mark of the common minke whale extends up onto the shoulder in the dwarf subspecies of the Southern Hemisphere (Fig. 1). The BALEEN is white in the northern subspecies of the common minke whale but appears dark-gray or brown posteriorly in the dwarf subspecies due to a narrow dark fringe. In the Antarctic minke whale, the baleen plates are black on the left beyond the first few plates and on the right they are white in the first third and black in the rear two-thirds of the row. The baleen filaments in both species are coarser than in the sei whale but finer than in fin, blue, and Bryde's whales, about 3.0 mm in diameter (Kato and Fujise, 2000). The SKULL (Fig. 2) is larger in *B. bonaerensis* than in both northern and southern forms of *B. acutorostrata*. The modal number of vertebrae in both species is about 49, ranging from 46 to 51.

M

## I. Distribution and Abundance

In the North Atlantic, the common minke whale is found in summer as far north as Baffin Bay in the Canadian Arctic, Denmark Strait, Franz Josef Land, and Novaya Zemlya (IUCN, 2008). The wintering grounds are poorly known but extend at least to the Caribbean in the west and the Straits of Gibraltar in the east. It is considered as a visitor to the Mediterranean. Affinities of minke whales reported from farther south to Senegal are unknown. In the North Pacific, the summer range extends to the Chukchi Sea. In the winter, common minke whales are found south to within 2° of the equator, although those south of central Baja California, Mexico, in the eastern North Pacific are of unknown relationship to the whales farther to the north. It has been seen off Hawaii. It is not known to occur in the northern Indian Ocean. In the Southern Hemisphere, the distribution of the dwarf subspecies is poorly known. It extends as far north as 11°S in the western Pacific off Australia and 2°S off South America in the Atlantic nearly year-round. It has been reported from Chilean waters (Acevedo *et al.*, 2006). In the far south, it is seasonally sympatric with the Antarctic minke whale on the FEEDING grounds during austral summer and occurs off South Africa during the fall and winter. Where sympatric with the Antarctic minke whale, it tends to occur in shallower, more coastal waters over the continental shelf.

Antarctic minke whales are abundant from 60°S to the ice edge during the austral summer, some occurring in the loose ice pack and in polynyas (IUCN, 2008). Some have been recorded to over-winter in the Antarctic. During the austral winter, most retreat to breeding grounds at mid-latitudes: 10–30°S in the Pacific between 170°E and 100°W, off northeastern and eastern Australia, off western South Africa, and off the northeastern coast of Brazil. In these areas their distribution is primarily oceanic, beyond the continental shelf break. One specimen is known from Suriname in the Northern Hemisphere.

Abundance has been estimated for four stocks of common minke whales in the North Atlantic (IUCN, 2008): Northeast Atlantic, about 81,000; Central North Atlantic, about 94,000; West Greenland, about 3500; Canada/US East Coast, about 4000. These total about 182,000. Abundance estimates exist for some areas of the North Pacific (areas may have substructure): Okhotsk Sea (West Pacific), about 25,000; west coast of US, about 1000; central and southeastern Bering Sea, about 2000. Abundance of the Southern Hemisphere dwarf minke whale has not been estimated. Population substructure in the eastern North Pacific is suggested by different calls recorded east and west of 135° (Rankin and Barlow, 2005).

Antarctic minke whale stocks in the Southern Hemisphere have been thought to be still in good condition and stable. However, the older estimates in the neighborhood of 750,000 have been abandoned because a substantial decline may have occurred in recent decades. A new assessment is underway to estimate the current abundance of this species (IWC, 2008).

## II. Ecology

Common minke whales inhabit both coastal and offshore waters. Density of Antarctic minkes whales in the austral summer is greatest near the edge of the pack ice, but they also occur within the pack ice (Shimada and Kato, 2006). Both species of minke whales are catholic feeders but specialize with season and area. In the North Atlantic, reported diet items include sand lance, sand eel, salmon, capelin, mackerel, cod, coal fish, whiting, sprat, wolffish, dogfish, pollack, haddock, herring, euphausiids, and copepods (Stewart and

Leatherwood, 1985); krill (euphausiids) are important off West Greenland, whereas capelin and cod are dominant prey in eastern Newfoundland. In the North Pacific, major food items include euphausiids, Japanese anchovy, Pacific saury, and walleye pollack. In the Antarctic, dwarf minke whales feed mainly on myctophid fishes (Kato and Fujise, 2000), whereas Antarctic minke whales feed mainly on euphausiids.

Killer whales (*Orcinus orca*) prey on minke whales of both species. By one Russian estimate, Antarctic minke whales make up 85% of the diet of killer whales in the Southern Ocean, although this may relate to the diet of only one of the two forms of killer whales found there (Pitman *et al.*, 2007).

## III. Behavior and Physiology

Common minke whales are noted for their curiosity about ships, often coming from afar to cross the bow or run with the vessel for minutes or hours. Their sudden appearance on the bow or alongside has startled many an observer. They are difficult to spot at a distance because of their small inconspicuous blow and brief surfacing behavior. Antarctic minke whales are reported to be easily approachable while feeding. Dwarf minke whales in the Great Barrier Reef region of northeastern Australia readily approach and stay with divers and are the subject of a WHALE-WATCHING tourist operation.

Single animals are often seen and groups are usually small, consisting of two or three individuals, although larger aggregations of up to 400 may form on occasion in high latitudes.

Minke whale sounds recorded in the North Atlantic included grunts, thumps, and frequency down sweeps ranging to 200 Hz. Similar down sweeps have been recorded in the Ross Sea. In the Pacific, the long-mysterious "boing" heard by acousticians has been identified as produced by common minke whales (Rankin and Barlow, 2005).

## IV. Life History

Differential migration by sex and age leads to segregation by sex and breeding condition. Mating behavior has not been directly observed. Breeding is diffusely seasonal in the common minke whale in the Northern Hemisphere, with calves of 2.4–2.7 m appearing approximately 10 months after conception. Lactation lasts 5–6 months. Age at attainment of sexual maturity has been estimated at about 7 years in males and 6 years in females. The pregnancy rate among adult females in some populations approaches 90%; suggesting an annual reproductive cycle. Little is known of the life history of the dwarf minke whale, but limited data available suggest similarity with that of the northern forms. The Antarctic minke whale also exhibits similar life history parameters (Horwood, 1990). Age at attainment of sexual maturity is 8 years in males and 7–8 years in females (although it may have been higher when overall whale densities were much higher earlier in the twentieth century). Pregnancy rates remain at or near 90% for most of the year, again suggesting an annual cycle on average. Generation time is estimated at 23 years (IUCN, 2008). Peak births are in July and August. During the feeding season, mature females are found closer to the ice than immature females, and immature males are more solitary than mature males.

## V. Interactions with Humans

The minke whales in the Southern Ocean were largely ignored in the early days of modern industrialized whaling because of their small size, but as the larger rorquals (blue, fin, and sei) were successively depleted, attention turned to the still abundant minkes in the early 1970s. After 1979, only minke whales were allowed by the IWC to be taken in factory-ship operations. Annual catches in the Antarctic ranged to about 8000 (details on all world catches are given in Horwood, 1990) by Japan and the USSR. Hundreds were also taken from land stations in Brazil and small numbers in South Africa. As of the 1985/1986 Antarctic season, all commercial whaling was banned under an IWC moratorium. Meanwhile, Japan has continued to take Antarctic minke whales annually under a research permit issued under the terms of the whaling convention. The annual research catch limit was increased in the 2005/2006 season to 935, up from 400 (IUCN, 2008).

Common minke whales have also been exploited commercially in the North Pacific and North Atlantic, in both land-based and pelagic whaling operations. Some stocks were depleted and became fully protected under IWC regulations, including the west Greenland, northeastern North Atlantic, and Sea of Japan—Yellow Sea—east China Sea stocks. The main whaling nations involved were Norway and Japan, with catches also by Korea, China, and the USSR. Catches from land stations in Japan continued until the moratorium on commercial whaling in 1987. In the 1990s, Norway recommenced commercial whaling on minke whales in the North Atlantic under an objection to the 1986 moratorium; the take in 2006 was 521 out of a self-imposed quota of 1052 (IUCN, 2008). In the North Pacific, Japan began taking minke whales under a scientific research permit in 1994; the nationally established annual limit for 2006 was 220. Minke whales in the region are also subject to significant bycatch (100s of animals) in set nets in Japan and Korea.

Aboriginal subsistence whaling is exempt from the IWC moratorium on commercial whaling, and localized whaling for North Atlantic minke whales has continued under this provision in West Greenland. In 1999, 165 minke whales were landed, and the current IWC annual catch limit for the years 2003–2007 was 175. The impact of this whale fishery on the West Greenland stock is not known.

In summary, most minke whale stocks are in better condition than most stocks of the other large whales, but questions remain about the status of some populations and the effects of continued whaling. Also, there may be threats to their habitats due to global warming, e.g., a great reduction of extent of sea ice is expected in this century as mean Antarctic temperatures rise faster than the global average (Turner *et al.*, 2006); the implications of this for minke whales are unclear but need to be monitored.

Northern common minke whales are the subject of whalewatching in Europe and the far east, including Japan (Hoyt, 2003, 2005), and dwarf minke whales off northeastern Australia (Arnold *et al.*, 2005).

## See Also the Following Articles

Antarctic Marine Mammals ■ Bow-Riding ■ International Whaling Commission ■ Killer Whale ■ Whaling, Early and Aboriginal

## References

Acevedo, J., Aguayo-Lobo, A., Acuña, P., and Pastene, L. A. (2006). A note on the first record of the dwarf minke whale (*Balaenoptera acutorostrata*) in Chilean waters. *J. Cetacean Res. Manage* **8**, 293–296.

Arnold, P., Marsh, H., and Heinsohn, G. (1987). The occurrence of two forms of minke whales in the east Australian waters with a description of external characters and skeleton of the diminutive or dwarf form. *Sci. Rep. Whales Res. Inst. Tokyo* **38**, 1–46.

Arnold, P., Birtles, R. A., Dunstan, A., Lukoschek, V., and Matthews, M. (2005). Colour patterns of the dwarf minke whale *Balaenoptera acutorostrata* sensu lato: Description, cladistic analysis and taxonomic implications. *Mem. Qld. Mus.* **51**, 277–307.

Horwood, J. (1990). "Biology and Exploitation of the Minke Whale." CRC Press, Boca Raton, FL.

Hoyt, E. (2003). "The Best Whale Watching in Europe." Whale and Dolphin Conservation Society, Unterhaching, Germany, 59p.

Hoyt, E. (2005). "Watching Whales and Dolphins in Japan, Hong Kong, Taiwan and Korea." Whale and Dolphin Conservation Society and International Fund for Animal Welfare, Kochi, Japan, 44p..

IUCN (2008). "The IUCN Red List of Threatened Species." IUCN, Gland, Switzerland.

IWC (2001). Report of the Scientific Committee. *J. Cetacean Res. Manage* **3** (Suppl.).

IWC. (2008). Report of the Scientific Committee. *J. Cetacean Res. Manage.* **8** (Suppl.).

Kato, H., and Fujise, Y. (2000). Dwarf minke whales; Morphology, growth and life history with some analyses on morphometric variation among the different forms and regions. *Int. Whal. Commn* meet. doc. SC/52/OS3, 1–30. Available from IWC, 135 Station Road, Impington, Cambridge CB4 9NP, UK.

Pastene, L. A., Fujise, Y., and Numachi, K. (1994). Differentiation of mitochondrial DNA between ordinary and dwarf forms of southern minke whale. *Rep. Int. Whal. Commn* **44**, 277–281.

Pastene, L. A., et al. (11 authors) (2007). Radiation and speciation of pelagic organisms during periods of global warming: The case of the common minke whale. *Balaenoptera acutorostrata. Mol. Ecol.* **16**, 1481–1495.

Pitman, R. L., Perryman, W. L., LeRoi, D., and Eilers, E. (2007). A dwarf form of killer whale in Antarctica. *J. Mamm.* **88**, 43–48.

Rankin, S., and Barlow, J. (2005). Source of the North Pacific "boing" sound attributed to minke whales. *J. Acoust. Soc. Am.* **118**, 3346–3351.

Reynolds, J. E., III, Odell, D. K., and Rommel, S. A. (1999). Marine mammals of the world. *In* "Biology of Marine Mammals" (J. E. Reynolds III and S. A. Rommel, eds.), pp. 1–14.

Rice, D. W. (1998). Marine mammals of the world. *Soc. Mar. Mamm. Spec. Pub.* **4**, 1–231.

Rychel, A. L., Reeder, T. W., and Berta, A. (2004). Phylogeny of mysticete whales based on mitochondrial and nuclear data. *Mol. Phylo. Evol.* **32**, 892–901.

Sasaki, T., Nikaido, M., Wada, S., Yamada, T. K., Cao, Y., Hasegawa, M., and Okada, N. (2006). *Balaenoptera omurai* is a newly discovered baleen whale that represents an ancient evolutionary lineage. *Mol. Phylo. Evol.* **41**, 40–52.

Shimada, H., and Kato, A. (2006). Tentative population assessment of the Antarctic minke whale within ice field using a sighting data on the ice breaker, *Shirase*, in 2004/2005. IWC Scientific Committee meet. doc. SC/58/IA11.

Stewart, B. S. (1999). Minke whale Balaenoptera acutorostrata. *In* "The Smithsonian Book of North American Mammals" (D. E. Wilson, and S. Ruff, eds), pp. 246–247. Smithsonian Press, Washington, DC.

Stewart, B. S., and Leatherwood, S. (1985). Minke whale *Balaenoptera acutorostrata* Lacépède, 1804. *In* "Handbook of Marine Mammals" (S. H. Ridgway, and R. Harrison, eds), Vol. 3, pp. 91–136. Academic Press, San Diego.

Turner, J., Lachlan-Cope, T. A., Colwell, S., Marshall, G. J., and Connolley, M. (2006). Significant warming of the Antarctic winter troposphere. *Science* **311**(5769), 1914–1917.

Zerbini, A. N., and Simões-Lopes, P. C. (2000). Morphology of the skull and taxonomy of southern hemisphere minke whales. *Int. Whal. Commn* meet. doc. SC/52/OS10, 1–28. Available from IWC, 135 Station Road, Impington, Cambridge, CB4 9NP, UK.

M

# Molecular Ecology

## A. Rus Hoelzel

Molecular ecology explores the interrelationship between ecological and evolutionary processes. From a practical point of view, this often means the application of molecular methods to ecological studies. Applications include individual-based genetics (e.g., assessing kinship to better understand evolutionary strategy), population studies (e.g., to understand the ecological context of population genetic structure), and inter-specific analyses (e.g., to better understand the ecological drivers of speciation). Mammals in the marine environment have adapted to the challenges of aquatic life with some dramatic changes in anatomy, physiology, and behavior. This is especially relevant to molecular ecology when characteristics of the marine habitat, or behaviour enabled by these adaptations, affect patterns of dispersal, reproductive behaviour, and demographics.

For example, while many cetacean species can move great distances and have broad tolerance for different habitats with respect to breeding and foraging, pinnipeds are tied to suitable terrestrial habitat for breeding (e.g., isolated from predators), but must forage at sea. Characteristics of the resource exploited by pinnipeds means that some populations can forage near breeding grounds, while others must travel great distances in search of prey. Resources can also limit the distribution of suitable breeding grounds. These factors affect population genetic diversity and structure through an impact on dispersal range and the "effective size" of populations. The effective population size ($N_e$) is the size of an ideal population (random mating and unaffected by processes such as mutation and selection) that would show the same rate of decay in genetic diversity as the observed population. Population size can be very large at a breeding colony where the resource is local and abundant (and therefore genetic diversity can be high), as seen especially for some otariid species. Phocid seals, on the other hand, often travel great distances on foraging excursions, and breeding colonies are often smaller than for otariids (or restricted to breeding pairs).

The following sections will briefly address the main aspects of molecular ecology in marine mammal species, with a focus on the most abundant taxonomic groups, the cetaceans and the pinnipeds.

## I. Genetic Diversity

Genetic diversity decays by genetic drift more quickly over time in populations with smaller effective population size. Within populations, reproductive strategy can impact the level of diversity, since reproductive skew (such as polygynous mating) reduces $N_e$. The potential for polygyny has been suggested to depend on various aspects of resource exploitation, and in pinnipeds on the consequences this has for the clumping of females (see Boness, 1991). If females are clumped and not too synchronous in estrus, males can monopolize the mating of multiple females to the exclusion of other males (polygyny). Most of the otariid species are polygynous, but only a few of the phocid seals. The phocid elephant seals provide a good illustration of both the potential role of polygyny and the impact of excessive hunting on the loss of genetic diversity. There are two closely related species, the northern (*Mirounga angustirostris*) and the southern (*M. leonina*) elephant seals. Behavioral and genetic studies have shown that these species are among the most polygynous of mammals (LeBoeuf,1972; Hoelzel *et al.*, 1999).

In the ninetieth century elephant seals were exploited heavily for their blubber in both hemispheres. The southern species retained relatively large population numbers, but the more accessible northern species was forced through a severe population bottleneck (Bartholomew & Hubbs, 1960; Hoelzel *et al.*, 1993). Molecular genetic variation in the northern elephant seal is now low at mtDNA, allozyme, immune system, and repetitive DNA loci (see review in Hoelzel, 1999), consistent with predictions based on simulation models given the severity of the bottleneck (Hoelzel *et al.*, 1993; Hoelzel, 1999). The direct loss of diversity, due to the bottleneck, was shown through the comparison of samples collected before and after the bottleneck (Hoelzel *et al.*, 2002). By comparing post-population bottleneck genetic diversity with demographic simulation models and historic data, Hoelzel *et al.* (1993) estimated the severity of the population bottleneck to be fewer than 30 seals over a 20-year period, or a single-year bottleneck of fewer than 20 seals.

Simulation studies illustrate the role of polygyny in further reducing genetic variation during the period of recovery (Hoelzel, 1999). A survey of 54 allozyme loci at an average of 99 individuals per locus revealed no variation (Bonnell and Selander, 1974; Hoelzel *et al.*, 1993). The estimated bottleneck of 20 seals would not have been sufficient to eliminate variation at these loci in a monogamous species, but polygynous mating results in high variance in male reproductive success, and this increases the impact on diversity. The reason is simply that relatively few males are contributing to the gene pool. Figure 1 illustrates how a bottleneck of less than 20 seals can account for the loss of allozyme diversity when you take into account the observed level of polygyny. Only those simulations based on polygynous mating (open bars) predict a level of post-bottleneck diversity that is low enough to be consistent with the measured levels of diversity.

Many other marine mammal species have also been the subject of intensive hunting. Molecular methods can help assess the impact and track recovery. For example, an interspecific comparison of right whale species showed reduced variation in the North Atlantic right whale (*Eubalaena glacialis*) compared to the Southern Hemisphere species (*E. australis*). Waldick *et al.* (2002) analyzed 13 microsatellite DNA loci developed from the *E. glacialis* genome, and found all 13 to be polymorphic in *E. australis*, but only nine were polymorphic in *E. glacialis*. Average heterozygosity was also reduced in the North Atlantic right whale. Long- and short-term estimates of effective population size ($N_e$) suggested an $N_e$ of approximately 100–300 for this species since the eightieth century, and about 4–8 times larger between the elventh and eighteenth centuries. However, Waldick *et al.* (2002) found only weak evidence of a population bottleneck using methods that detect recent events (though these do not reliably detect all bottleneck events). This and a study showing relatively high mtDNA diversity in historical samples (late ninetieth and early twentieth century) suggest that diversity could have been decaying over time prior to the intensive hunting of the ninetieth century (Rosenbaum *et al.*, 2000).

In studies on a related species, also heavily impacted by whaling, Rooney *et al.* (1999, 2001) showed that the modern BCB stock (Bering–Chukchi–Beaufort Seas population) of the bowhead whale (*Balaena mysticetus*) has relatively high diversity at both microsatellite DNA and mtDNA markers. They suggest that the recent intensive whaling on this stock did not significantly reduce genetic variation. This population is understood to have been isolated from the now nearly extinct Spitsbergen stock (Greenland, Norwegian, Barents, and Kara Seas) during periods since the last glaciation by the M'Clintock Channel ice plug. Bjørge *et al.* (2007) investigated mtDNA diversity for the Spitsbergen stock over a Holocene timescale, sequencing 99 historical samples ranging in age from 30–51,000 years

**M**

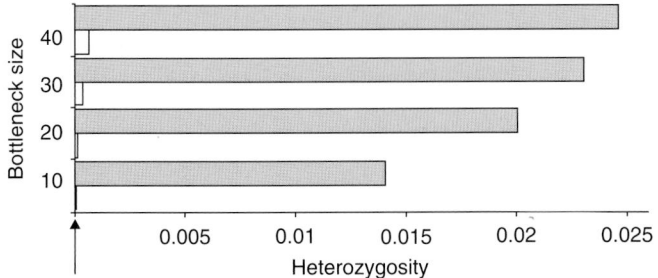

Figure 1 *This graph illustrates for the northern elephant seal the predicted impact of population bottlenecks of various sizes (along the x-axis) on a measure of genetic diversity (heterozygosity) following the bottleneck. The shaded bars are based on simulations that assume monogamous mating, while the open bars are based on simulations assuming observed levels of polygyny. No allozyme heterozygosity was found in the modern population, but the arrow indicates the average heterozygosity that would have been seen (0.00018) if just one individual had been heterozygous at just one locus. After Hoelzel (1999).*

old. (Increasingly studies are incorporating the analysis of ancient DNA, which can greatly enhance our understanding of patterns of change over time.) In this case the samples had been preserved by cold weather and the success rate was comparatively high. They found comparable levels of diversity in the historical Spitsbergen and modern BCB stocks, consistent with the idea that diversity in the BCB stock had not been substantially depleted, and they found a low but significant level of genetic differentiation between these stocks. Comparing early and late Spitsbergen samples suggested little change in this pattern over time.

A direct comparison between samples from before and after periods of hunting and population depletion in the sea otter (*Enhydra lutris*) showed a loss of diversity directly attributable to the anthropogenic bottleneck (Larson *et al.*, 2002), as had a similar study on the northern elephant seal (Hoelzel *et al.*, 2002). Other studies on pinnipeds have either indicated lost diversity due to hunting pressure (e.g., for the Mediterranean monk seal, *Monachus monachus*; Pastor *et al.*, 2004) or suggested that sealing-induced pressure has not had a substantial impact on diversity (e.g., for the Cape fur seal, *Arctocephalus pusillus*; Matthee *et al.*, 2006). The latter study used a coalescent method to detect a much older population expansion in that species and suggested that this was related to abundant food resources in the South Atlantic during the last glacial period. Other studies have used coalescent methods to estimate past population sizes, e.g., pre-whaling populations of baleen whales (Roman & Palumbi, 2003). However, these estimates are complicated by their dependence on parameter estimates (especially mutation and migration rates) which are not well known but can influence $N_e$ estimates substantially (Baker & Clapham, 2004).

## II. Individual Genetics

Molecular methods can also enhance our understanding of the relationships among individuals within populations. At the most basic level, genetic markers can help us identify individual animals. The identification of individuals has been a fundamental method contributing to our better understanding of marine mammal species and has typically been based on acquired or congenital distinguishing markings (e.g., pigmentation patterns and scars) or tags. However, sometimes individuals are hard to tag or are not naturally well marked,

or despite good individual markings numbers of animals are so large as to make visual identification difficult. The latter case is exemplified by a very large-scale survey of humpback whales (*Megaptera novaeangliae*) in the North Atlantic using genetic tags (Palsbøll *et al.*, 1997). This study genetically identified nearly 2400 individual whales (using six polymorphic microsatellite DNA markers) and "recaptured" 692 of them. The 692 repeat-sampling events provided new information on migration patterns and evidence of mixing on feeding grounds, and because it was genetically based, also permitted the identification of the sex of each individual.

Another important application at the individual level is to consider fitness in the context of genetic diversity (measured, e.g., as the proportion of loci heterozygous in that individual). There are many studies showing impact on survival or some indirect measure of fitness (such as fluctuating asymmetry) for a diversity of animal species (see review in Hansson and Westerberg, 2002). Comparatively few such studies exist for marine mammals, but they show consistent results. For example, Acevedo-Whitehouse *et al.* (2003) showed that pathogen susceptibility was correlated to genetic diversity in the California sea lion (*Zalophus californianus*), while less diverse striped dolphins (*Stenella coeruleoalba*) appeared to have been affected more severely by a morbillivirus outbreak (Valsecchi *et al.*, 2004). With respect to indirect measures, fluctuating asymmetry was shown to increase in the northern elephant seal directly after a severe population bottleneck (that also severely reduced genetic diversity; Hoelzel, 1999; Hoelzel *et al.*, 2002).

Perhaps the most common application is toward the assessment of kinship. For pinnipeds the focus has been on testing paternity, e.g., to assess the comparative success of harem holders in the highly polygynous northern and southern elephant seals. Variance in reproductive success among males of the northern species was high and not always consistent with copulatory success (Hoelzel *et al.*, 1999). However, behavioral and genetic assessments of male paternal success were generally consistent for the southern species (Hoelzel *et al.*, 1999; Fabiani *et al.*, 2004). The authors suggest that some northern elephant seal males may have been reproductively compromised following the severe bottleneck event, as has been seen for other species (e.g., lions, *Panthera leo*; Wildt *et al.*, 1987). Although many pinniped species are highly gregarious during the breeding season, there is little evidence that kin typically associate over extended periods. In a study on the southern elephant seal, Fabiani *et al.* (2006) found that females at the colony were more closely related than males, but while there was evidence for some long-term association among female kin, these associations were rare. There were also no indications that females choose harem males to which they are either more or less closely related than expected by chance.

Some of the most interesting studies involving kinship analysis have been on the highly social species in the delphinid radiation. Bottlenose dolphins (*Tursiops* sp.) exist in fission/fusion social groups, but male associations in "alliances" can be long-lasting. Krützen *et al.* (2004) tested the paternity success of alliance members at Monkey Mia in Australia and found that independence from alliances did not preclude paternity, though membership in some alliances could lead to greater success. In a number of species, associated male kin pursue access to females together, perhaps gaining indirect fitness benefits (e.g., in lions; Packer *et al.*, 1991). This may have been the case for some bottlenose dolphin alliances in the Bahamas (Parsons *et al.*, 2003), but apparently not at Monkey Mia (Möller *et al.*, 2001). However, evidence for adult female kin associations has been found for bottlenose dolphins at Monkey Mia (Möller *et al.*, 2006) as well as for striped dolphins in the Mediterranean Sea (Gaspari *et al.*, 2007).

Among the baleen whales, the humpback whale shows highly social behaviors on the breeding grounds, including the association of males into "consorts" pursuing access to females. However, extensive surveys showed no evidence for kin association beyond mother/calf pairs in humpback whales on either the breeding grounds (Pomilla and Rosenbaum , 2006) or during migrations (Valsecchi et al., 2002).

## III. Population Genetics

Dispersal (and hence gene flow) among populations can be restricted by geographic barriers (such as continents) or by habitat specialization and energetic considerations. However, marine mammal species are highly mobile and some are capable of movement over vast distances. For example, the sperm whale (*Physeter macrocephalus*) shows little genetic variation even among different ocean basins (Lyrholm et al., 1999). Other species have been shown to be composed of geographically isolated populations that are genetically differentiated, such as Pacific vs Atlantic populations of humpback whales (Baker et al., 1994) and walrus (*Odobenus rosmarus*; Cronin et al., 1994). This "allopatric" pattern of differentiation is common among both marine and terrestrial species. What is more unusual, but perhaps quite important for the evolution of population structure in marine mammals, is the partitioning of the environment in a way that shows differentiation among sympatric (living in the same geographic region) and parapatric (neighboring regions) populations. This can happen when populations that differentiated in allopatry come back together, or by processes that lead to choosing particular mates in preference to others ("assortative mating") in sympatry.

One mechanism for differentiation in sympatry is called "resource specialization," and this has been described in detail for various terrestrial and aquatic species (see Smith and Skulason, 1996). It means that individuals of a species specialize on habitat or prey choice. Note, however, that these specializations can only lead to genetic differentiation if they also lead to assortative mating. Various studies show genetic differentiation between resource specialists in dolphin species. For example, the killer whale (*Orcinus orca*, the largest dolphin) travels in highly stable social groups ("pods"). Pods tend to specialise on prey resources, with some pods and populations of pods focussing on marine mammal prey and others on fish prey (two different "ecotypes"). Sympatric populations of marine mammal and fish specialists in the eastern North Pacific were found to be genetically differentiated (Hoelzel and Dover, 1991; Hoelzel et al., 1998a). The level of differentiation was great enough to suggest the possibility of two "cryptic" killer whale species, differing especially in foraging behavior. However, the application of non-equilibrium evolutionary models to test for ongoing migration suggested that current dispersal among ecotypes was at a similar level as that seen within ecotypes (Hoelzel et al., 2007), implying multiple populations of a single species. This study further showed that there are clearly two mechanisms differentiating populations of killer whales in the North Pacific: differentiation among ecotypes in sympatry or parapatry, and an "isolation by distance" pattern within ecotypes. The timing of population origins was assessed using a coalescent method, and this suggested that these regional populations had been established after coastal habitat became available when glaciers retreated after the last ice age, emphasizing the likely importance of local habitat in defining genetically isolated populations.

One striking result was the existence of population structure on a geographic scale that is clearly smaller than the dispersion range of the species (see Hoelzel et al., 2007). The southern elephant seal provides another example. In this case male-mediated dispersal can be over a range of thousands of kilometer, and yet significant population genetic structure exists among breeding colonies that are only hundreds of kilometer apart (Fabiani et al., 2003). Satellite tracking studies have further shown that both males and females travel thousands of kilometer on foraging excursions, regularly overlapping the ranges defined by population genetic structure (Biuw et al., 2007). In this case it seems likely that philopatry is driven by the limited availability of suitable breeding habitat. In the case of the killer whales, the key factor may be differences in prey resource among regional habitats. A similar pattern was seen for bottlenose dolphins distributed between the Black Sea and Scotland. Genetic structuring (defined by genotypes without *a priori* assignment to putative populations) matched apparent habitat boundaries between the Black Sea, eastern Mediterranean, western Mediterranean, North Atlantic and Scotland (Natoli et al., 2005; Fig. 2). The implication that local habitat defines population structure in this species was further reinforced by a study that investigated fine-scale population structure around the UK incorporating a now extinct population at the Humber river estuary. The data could rule out an historical range shift or contraction; they instead indicated that this regional population was isolated from neighboring populations and went extinct sometime between 100–1000 years ago.

A number of dolphin species inhabit both coastal and offshore environments, which can differ with respect to the type and distribution of potential prey. Especially in regions where there is upwelling, the habitat in the marine littoral zone can be very different from the offshore habitat. Several studies of nearshore and offshore dolphin populations indicate intraspecific morphological, and in some cases genetic distinctions. For example, nearshore and offshore subspecies of the spotted dolphin (*Stenella attenuata*) can be distinguished by tooth and jaw structure, and the Atlantic spotted dolphin (*S. frontalis*) is also found in nearshore and offshore populations. Two forms of common dolphin (*Delphinus delphis*) have been classified by the proportional length of the beak. In this case, ranges overlap, and both forms are sometimes found in the nearshore habitat. Rosel et al. (1994) found genetic differentiation between these forms; they are now considered by most workers to comprise two species: *D. delphis* and *D. capensis* (but see complications described below).

The best known example is that of the bottlenose dolphin, which occurs in coastal and offshore populations throughout its range. In the eastern North Pacific, nearshore and offshore forms were originally classified as two different species: *T. gilli* (the nearshore form) and *T. nuuanu*, though a reappraisal recognizing extensive overlap in morphotypes later reclassified both as *T. truncatus*. In the western North Atlantic, the nearshore and offshore forms have been described in some detail and show both morphometric (Mead and Potter, 1995) and genetic differentiation (Hoelzel et al., 1998b). There were consistent differences between the two types in feeding behavior with the nearshore form feeding primarily on coastal fish, while the offshore form concentrated on deep water squid (Mead and Potter, 1995). The genetic distinction indicated low levels of gene flow (or no gene flow in the recent past) between the two populations.

Marine mammals are highly mobile, and in many cases they show seasonal differences in distribution, such as the annual migration between breeding and feeding sites seen in baleen whales. This is an important consideration for the identification of populations for protection and management, especially when breeding "stocks"mix on feeding grounds where they may be hunted (see review in Hoelzel, 1998). For example, minke whale (*Balaenoptera acutorostrata*) populations on either side of Japan (off Korean and in the western North

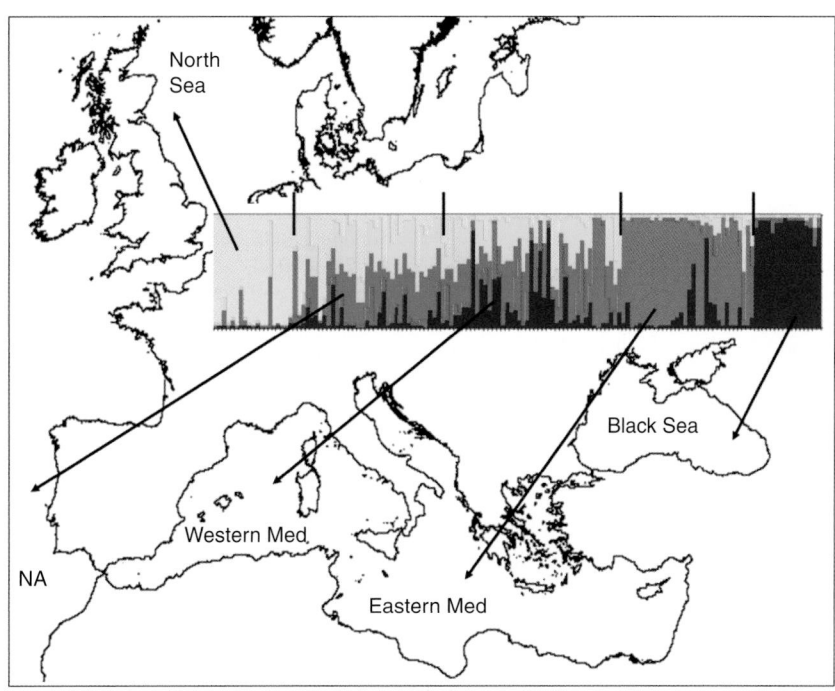

Figure 2    *Assignments to clusters based on equilibrium expectations were generated using the program structure and are shown in the histogram. The colors reflect assignment probability to different putative populations, and the black lines divide sample sets of bottlenose dolphins from different regions. After Natoli* et al. *(2005).*

Pacific) are genetically differentiated at both allozyme (Wada, 1991) and mtDNA loci (Goto and Pastene, 1996). Both studies found evidence of seasonal mixing on feeding grounds to the north in the Okhotsk Sea. In another example involving minke whales, a temporal mixing of two genetically differentiated populations from Antarctic management areas IV and V was described based on mtDNA variation (Pastene *et al.*, 1996).

## IV. Beyond Populations

Sometimes even quite familiar marine mammal "species" become reclassified as multiple cryptic species when molecular genetic methods are employed. One important driver appears the same type of adaptation to different habitats that often defines populations within these species. For example, populations in nearshore and offshore habitat have in some cases been recognized as genetically distinct. For the genus *Tursiops* a nearshore form off China (an "aduncus" type that has a proportionally longer rostrum) was shown to occupy a separate phylogenetic lineage from the offshore form (Wang *et al.*, 1999). However, the coastal *Tursiops aduncus* in South Africa was as genetically differentiated from the coastal aduncus type in Asia as from the *T. truncatus* form, implying at least three species in the genus *Tursiops*, two of which are morphotypically very similar (Natoli *et al.*, 2004; Kurihara and Oda, 2007). As with intraspecific populations of the killer whale, it appears that both ecotype and geographic isolation are important mechanisms promoting differentiation and (in this case) eventual speciation. In the genus *Delphinus* there is also a "long-beaked" form typically found in nearshore habitat (recognized in this volume as *Delphinus capensis*), but only one of several long-beaked populations showed differentiation at a level expected among species (Natoli *et al.*, 2006). In *Sotalia*, another genus where

taxonomic classification has been problematic, a recent phylogenetic study (using mtDNA) proposed the division of *Sotalia fluviatilis* into two species, one adapted to riverine habitat, and the other to marine, coastal waters (Cunha *et al.*, 2005).

## V. Summary

In general, the pattern and degree of genetic differentiation among populations is not easily predicted by geographic patterns for marine mammal species, and instead depends on a complex interaction between life history and habitat. Even for the highly mobile and pelagic species, such as the minke whale, there can be considerable genetic differentiation among regional populations, and for some species (such as the killer whale) among populations in sympatry. What we know of the molecular ecology of these species emphasizes the need for more data and a careful consideration of the mechanisms affecting patterns of diversity within and among populations.

### *See Also the Following Articles*

Genetics, Overview  Poulation Identify

### *References*

Acevedo-Whitehouse, K., Gulland, F., Greig, D., and Amos, W. (2003). Inbreeding-dependent pathogen susceptibility in California sea lions. *Nature* **422**, 35.

Baker, C. S., and. Clapham, P. J. (2004). Modeling the Past and future of whales and whaling. *Trends Ecol. Evol.* **19**, 365–371.

Baker, C. S., *et al.* (1994). Hierarchical structure of mitochondrial DNA gene flow among humpback whales world-wide. *Mol. Ecol.* **3**, 313–327.

M

Biuw, M. *et al* (21 authors). (2007). Variations in behavior and condition of a Southern Ocean top predator in relation to *in situ* oceanographic conditions. PNAS **104**, 13705–13710.

Bartholomew, G. A., and Hubbs, C. L. (1960). Population growth and seasonal movements of the northern elephant seal. *Mirounga angustirostris. Mammalia* **24**, 313–324.

Boness, D. J. (1991). Determinants of mating systems in the Otariidae (Pinnipedia). *In* "Behaviour of Pinnipeds" (D. Renouf, ed.), pp. 1–44. Chapman & Hall, London.

Bonnell, M., and Selander, R. K. (1974). Elephant seals: Genetic variation and near extinction. *Science* **184**, 908–909.

Bjørge, T., Bachmann, L., Bjørnstad, G., and Wiig, O. (2007). Genetic variation in Holocene bowhead whales from Svalbard. *Mol. Ecol.* **16**, 2223–2235.

Cronin, M. A., Hills, S., Born, E. W., and Patton, J. C. (1994). Mitochondrial DNA variation in Atlantic and Pacific walruses. *Can. J. Zool.* **72**, 1035–1043.

Cunha, H. A., *et al.* (2005). Riverine and marine ecotypes of *Sotalia* dolphins are different species. *Mar. Biol.* **148**, 449–457.

Fabiani, A., Hoelzel, A. R., Galimberti, F., and Muelbert, M. M. C. (2003). Long-range paternal gene flow in the southern elephant seal. *Science* **299**, 676.

Fabiani, A., Galimberti, F., Savito, S., and Hoelzel, A. R. (2004). Extreme polygyny among southern elephant seals on Sea Lion Island, Falkland Islands. *Behav. Ecol.* **15**, 961–969.

Fabiani, A., Galimberti, F., Sanvito, S., and Hoelzel, A. R. (2006). Relatedness and site fidelity at the southern elephant seal (*Mirounga leonina*) breeding colony in the Falkland Islands. *Anim. Behav.* **72**, 617–626.

Gaspari, S., Azzellino, A., Airoldi, S., and Hoelzel, A. R. (2007). Social kin associations and genetic structuring of striped dolphin populations (*Stenella coeruleoalba*) in the Mediterranean Sea. *Mol. Ecol.* **16**, 2922–2933.

Goto, M., and Pastene, L.A. (1996). Population genetic structure in the western North Pacific minke whale examined by two independent RFLP analyses of mitochondrial DNA. IWC Scientific Committee meeting document SC/48/NP5.

Hansson, B. and Westerberg, L. (2002). On the correlation between heterozygosity and fitness in natural populations. *Mol. Ecol.* **11**, 2467–2474.

Hoelzel, A. R. (1998). Genetic structure of cetacean populations in sympatry, parapatry, and mixed assemblages: Implications for conservation policy. *J. Hered.* **89**, 451–458.

Hoelzel, A. R. (1999). Impact of population bottlenecks on genetic variation and the importance of life history; a case study of the northern elephant seal. *Biol. J. Linn. Soc.* **68**, 23–39.

Hoelzel, A. R., and Dover, G. A. (1991). Genetic differentiation between sympatric killer whale populations. *Heredity* **66**, 191–195.

Hoelzel, A. R., Dahlheim, M. E., and Stern, S. J. (1998a). Low genetic variation among killer whales in the eastern North Pacific, and genetic differentiation between foraging specialists. *J. Hered.* **89**, 121–128.

Hoelzel, A. R., Potter, C. W., and Best, P. (1998b). Genetic differentiation between parapatric "nearshore" and "offshore" populations of the bottlenose dolphin. *Proc. Royal Soc. B.* **265**, 1–7.

Hoelzel, A. R., LeBoeuf, B. J., Reiter, J., and Campagna, C. (1999). Alpha male paternity in elephant seals. *Behav. Ecol. Sociobiol.* **46**, 298–306.

Hoelzel, A. R., *et al.* (1993). Elephant seal genetic variation and the use of simulation models to investigate historical bottlenecks. *J. Hered.* **84**, 443–449.

Hoelzel, A. R., Fleischer, R. C., Campagna, C., Le Beouf, B. J., and Alvord, G. (2002). Direct evidence for the impact of a population bottleneck on symmetry and genetic diversity in the northern elephant seal. *J. Evol. Biol.* **15**, 567–575.

Hoelzel, A. R., Natoli, A., Dahlheim, M., Olavarria, C., Baird, R. W., and Black, N. (2002). Low world-wide genetic diversity in the killer whale (*Orcinus orca*); Implications for demographic history. *Proc. Roy. Soc. B* **269**, 1467–1475.

Hoelzel, A. R., Hey, J., Dahlheim, M. E., Nicholson, C., Burkanov, V., and Black, N. (2007). Evolution of population structure in a highly social top predator, the killer whale. *Mol. Biol. Evol.* **24**, 1407–1415.

Krützen, M., Barré, L. M., Connor, R. C., Mann, J., and Sherwin, W. B. (2004). O father where art thou?—Paternity assessment in an open fission-fusion society of wild bottlenose dolphins. *Mol. Ecol.* **13**, 1975–1990.

Kurihara, N., and Oda, S. (2007). Cranial variation in bottlenose dolphins *Tursiops* spp. from the Indian and western Pacific Oceans: Additional evidence for two species. *Acta Theriol.* **52**, 403–418.

Larson, S., Jameson, R., Etnier, M., Flming, M., and Bentzen, P. (2002). Loss of genetic diversity in sea otters (*Enhydra lutris*) associated with the fur trade of the 18th and 19th centuries. *Mol. Ecol.* **11**, 1899–1903.

LeBoeuf, B. J. (1972). Sexual behaviour in the northern elephant seal, *Mirounga angustirostris. Behaviour* **41**, 1–25.

Lyrholm, T., Leimar, O., Johanneson, B., and Gyllensten, U. (1999). Sex-biased dispersal in sperm whales: contrasting mitochondrial and nuclear genetic structure of global populations. *Proc. Roy. Soc. B* **266**, 347–354.

Matthee, C. A., Fourie, F., Oosthuizen, W. H., Meyer, M. A., and Tolly, K. A. (2006). Mitochondrial DNA sequence data of the Cape fur seal (*Arctocephalus pusillus pusillus*) suggest that population numbers may be affected by climate shifts. *Mar. Biol.* **148**, 899–905.

Mead, J. G., and Potter, C. W. (1995). Recognizing two populations of the bottlenose dolphin (*Tursiops truncatus*) off the Atlantic coast of North America; morphologic and ecologic considerations. *IBI Rep.* **5**, 31–44.

Möller, L. M., Beheregaray, L. B., Harcourt, R. G., and Krützen, M. (2001). Alliance membership and kinship in wild male bottlenose dolphins (*Tursiops aduncus*) of southeastern Australia. *Proc. Roy. Soc., B* **268**, 1941–1947.

Möller, L. M., Beheregaray, L. B., Allen, S. J., and Harcourt, R. G. (2006). Association patterns and kinship in female Indo-Pacific bottlenose dolphins (*Tursiops aduncus*) of southeastern Australia. *Behav. Ecol. Sociobiol.* **61**, 109–117.

Natoli, A., Peddemors, V., and Hoelzel, A. R. (2004). Population structure and speciation in the genus *Tursiops* based on microsatellite and mitochondrial DNA analyses. *J. Evol. Biol.* **17**, 363–375.

Natoli, A., Birkin, A., Aquilar, A., Lopez, A., and Hoelzel, A. R. (2005). Habitat structure and the dispersal of male and female bottlenose dolphins (*Tursiops truncatus*). *Proc. Roy. Soc. B.* **272**, 1217–1226.

Natoli, A., Cañadas, A., Peddemors, V. M., Aguilar, A., Vaquero, C., Fernández-Piqueras, P., and Hoelzel, A. R. (2006). Phylogeography and alpha taxonomy of the common dolphin (*Delphinus sp*). *J. Evol. Biol.* **19**, 943–954.

Nichols, C., Herman, J., Gaggioti, O.E., Dobney, K. M., Parsons, K. and Hoelzed, A. R. (2007). Genetic isolation of a now extinct population of bottlenose dolphins (Tursiops truncatus). *Proc. Roy. Soc. B* **247**, 1611–1616.

Packer, C., Gilbert, D. A., Pusey, A. E., and O'Brien, S. J. (1991). A molecular genetic analysis of kinship and cooperation in African lions. *Nature* **351**, 562–565.

Palsbøll, J., *et al.* (19 authors). (1997). Genetic tagging of humpback whales. *Nature* **388**, 767–768.

Parsons, K. M., Durban, J. W., Claridge, D. E., Balcomb, K. C., Noble, L. R., and Thompson, P. M. (2003). Kinship as a basis for alliance formation between male bottlenose dolphins, *Tursiops truncatus*, in the Bahamas. *Anim. Behav.* **66**, 185–194.

Pastene, L. A., Goto, M., Itoh, S., and Numachi, K. I. (1996). Spatial and temporal patterns of mitochondrial DNA variation in minke whales from Antarctic areas IV and V. *Rep. Int. Whal. Commn* **46**, 305–314.

**M**

Pastor, T., Garza, J. C., Allen, P., Amos, W., and Aguilar, A. (2004). Low genetic variability in the highly endangered Mediterranean monk seal. *J. Hered.* **95**, 291–300.

Pomilla, C., and Rosenbaum, H. C. (2006). Estimates of relatedness in groups of humpback whales (*Megaptera novaeangliae*) on two wintering grounds of the Southern Hemisphere. *Mol. Ecol.* **15**, 2541–2555.

Rooney, A. P., Honeycutt, R. L., Davis, S. K., and Derr, J. N. (1999). Evaluating a putative bottleneck in a population of bowhead whales from patterns of microsatellite diversity and genetic disequilibria. *J. Mol. Evol.* **49**, 682–690.

Rooney, A. P., Honeycutt, R. L., and Davis, S. K. (2001). Historical population size change of bowhead whales inferred from DNA sequence polymorphism data. *Evolution* **55**, 1678–1685.

Rosel, P. E., Dizon, A. E., and Heyning, J. E. (1994). Genetic analysis of sympatric populations of common dolphins (genus *Delphinus*). *Mar. Biol.* **119**, 159–167.

Rosenbaum, H. C., *et al.* (2000). Utility of North Atlantic right whale museum specimens for assessing changes in genetic diversity. *Cons. Biol.* **14**, 1837–1842.

Smith, T. B., and Skulason, S. (1996). Evolutionary significance of resource polymorphisms in fishes, amphibians, and birds. *Ann. Rev. Ecol. Syst.* **27**, 111–133.

Valsecchi, E., Hale, P., Corkeron, P., and Amos, W. (2002). Social structure in migrating humpback whales (*Megaptera novaeangliae*). *Mol. Ecol.* **11**, 507–518.

Valsecchi, E., Amos, W., Raga, J. A., Podesta, M., and Sherwin, W. (2004). The effects of inbreeding on mortality during a morbillivirus outbreak in the Mediterranean striped dolphin (*Stenella coeruleoalba*). *An. Cons.* **7**, 139–146.

Wada, S., 1991. Genetic distinction between two minke whale stocks in the Okhotsk Sea coast of Japan. IWC Scientific Committee meeting document (SC/43/Mi32).

Waldick, R. C., Kraus, S. S., Brown, M., and White, B. N. (2002). Evaluating the effects of historic bottleneck events: an assessment of microsatellite variability in the endangered, North Atlantic right whale. *Mol. Ecol.* **11**, 2241–2250.

Wang, J. Y., Chou, L. S., and White, B. N. (1999). Mitochondrial DNA analysis of sympatric morphotypes of bottlenose dolphins (genus: *Tursiops*) in Chinese waters. *Mol. Ecol.* **8**, 1603–1612.

Wildt, D. E., Bush, M., Goodrowe, K. L., Packer, C., Pusey, A. E., Brown, J. L. L., Joslin, P., and O'Brien, S. J. (1987). Reproductive and genetic consequences of founding isolated lion populations. *Nature* **329**, 328–331.

# Monk Seals

*Monachus monachus, M. tropicalis,* and *M. schauinslandi*

WILLIAM G. GILMARTIN AND JAUME FORCADA

## I. Characteristics and Taxonomy

Order Carnivora, family Phocidae (Gray 1825). The genus *Monachus* includes two endangered species that live in the world's tropical and subtropical seas of the Northern Hemisphere, the Mediterranean monk seal, *M. monachus*, and the Hawaiian monk seal, *M. schauinslandi*; and the extinct Caribbean monk seal, *M. tropicalis*, once widely distributed around the Caribbean and last seen in 1952.

Monk seals originated in the North Atlantic, with the Hawaiian species of probable Caribbean origin at least 15 million years ago across the Central American seaway (Repenning and Ray, 1977). Unique among pinnipeds in some primitive, unspecialized skeletal, and vascular anatomy, the Hawaiian monk seal is known as the most primitive of living seals.

Mediterranean monk seals are uniformly black at birth with a conspicuous white ventral patch unique to each individual and distinct in shape by sex (Samarach and González, 2000). Newly molted seals of all ages are silvery gray dorsally and lighter ventrally, with the ventral patch pattern persisting through life. Pups are completely molted at 70 days (range 40–108). Juveniles have a medium-to-dark gray pelage, similar to that of adult females but with less coloration disruptions caused by scarring. Pups at birth weigh 20 kg on average and measure between 0.80 and 0.90 m. Near the age of 4, adult males become almost uniformly black, with their white ventral patch; coloration in females is more variable, but not as dark as males dorsally and the ventral fur is also lighter. Adult males may also be slightly longer than females. Adult lengths are 2.3–2.8 m and weights are 240–300 kg, with a maximum recorded weight of 400 kg.

Hawaiian monk seals are also black at birth, with some showing small white patches at various sites. These seals are also silvery gray following molt, with the fur color changing in juveniles to a yellow–brown prior to the next molt and darkening through the year in adults. Hawaiian seals show no differences in fur coloration by sex. Adult females may attain a slightly larger size than males. Adult lengths are 2.1–2.4 m and weights are 170–240 kg. Pups at birth weigh 16–20 kg and measure between 0.80 and 1 m. Both the hair and the epidermis are sloughed and replaced during the annual molt in monk seals. This type of molt is similar to elephant seals, but unlike all other seals. In the Hawaiian monk seal the actual observed molting period is about 10 days, but based on an observed high proportion of time ashore before and after the hair–skin sloughing period, the entire physiological process is probably much longer.

Caribbean monk seals were similar to Hawaiian and Mediterranean monk seals, with uniformly dark brown coloration, lighter in the ventral area, and limited sexual dimorphism. Adult lengths were 2.3 m and weights lower than 200 kg. Pup length at birth was approximately 1 m and weight between 16 and 18 kg. Juveniles reached 1.5 m. The hair was short, about 1 cm in length, and uniformly dark or black in newborn pups. Vibrissae were dark in juveniles, up to 12-cm long, and lighter in adults and up to 10-cm long. The skull had large orbital areas and a smoother part behind these. Adult skull length was approximately 25 cm and similar between sexes. They had 32 teeth, with well-developed canines, like the other *Monachus* species. The other common name of the species is West Indian monk seal.

## II. Distribution and Abundance

Historically, the Mediterranean monk seal inhabited the entire Mediterranean Basin and the southeastern North Atlantic, from the Azores Islands to near the equator (Aguilar, 1999). The current distribution is severely contracted and fragmented, with the largest population approximately 250–300 seals in the Eastern Mediterranean, on the islands in the Aegean and Ionian Seas, and along the coasts of Greece and Turkey. Only two breeding populations are known in the Atlantic, one at the Cap Blanc peninsula in the Western Sahara and Mauritania approximately 120 seals and a smaller group in the Desertas Islands at the Madeira Islands

**M**

approximately 25 seals (Forcada *et al.*, 1999). Sightings are rare now in other areas within the historical range. Genetic variability is low, most likely as a consequence of severe bottlenecks following extensive sealing in the Atlantic and fisheries-related losses that continue today. Mediterranean monk seals have high site-fidelity and tend to occupy only a restricted part of the suitable habitat. The species only preserves colonial aggregations are in the Atlantic, whereas in the Mediterranean it is found in secluded caves, commonly with underwater entrance, and the distribution is greatly fragmented.

Hawaiian monk seals occur only in the Central Pacific in the mostly uninhabited Northwestern Hawaiian Islands (Ragen and Lavigne, 1999). A small breeding population also inhabits the main Hawaiian Islands, and monk seals have been reported on rare occasions south of the Hawaiian Archipelago at Johnston Atoll, Wake Island, and Palmyra Atoll. Hawaiian monk seals have a high fidelity to their island of birth; only 10–30% of seals born at any of the major breeding islands will move to another island during their life. Very low genetic diversity is also evident in this species, which now includes 1200–1300 individuals.

Caribbean monk seals lived in the Caribbean Sea, from the Florida Keys and northeastern Gulf of Mexico to the coast of Guyana. The preferred habitat were small islands, keys, and atolls, surrounded by reefs and shallow, protected waters. Suggested breeding grounds were in the Southern Gulf of Mexico and Bahamas, and Arrecifes Triángulos (Yucatán). Seals occurred at these locations throughout the year (Adam and García, 2003). Historical population numbers are unknown, but following discovery their numbers were reported in the thousands.

### III. Ecology

Both extant monk seal species consume a highly diverse diet of diurnally and nocturnally active fish, octopus, squid, and lobster (Goodman-Lowe, 1999). Proportions of these prey species in the diet vary by location, season, and age of the seals. Although most of the prey species are benthic in the coral reef ecosystem, a few are pelagic. Hawaiian monk seals have a broad prey base of at least 40 species. They forage within their resident atolls and along the fringing reefs; and may, where a population is food stressed, also forage at reefs 60–200 km from their breeding islands and at sea mounts to depths near 500 m. Mediterranean monk seals are mostly benthic feeders, regularly reaching bottom depths in areas of wide continental shelf like the western Sahara. Caribbean monk seals probably preyed upon fish and crustaceans. Killer whales and sharks are probably the only predators of monk seals.

### IV. Behavior, Physiology, and Life History

Both living monk seal species have protracted reproductive seasons, and copulation occurs in the water. Some female Mediterranean monk seals may be reproductively active at 3 years and Hawaiian monk seals at 4 years. However, in the Hawaiian seal, the mean is perhaps 6–7 years and maturity is as late as 10–11 years in females at sites where prey abundance is low, showing nutritional status is a critical factor in maturation. Mediterranean monk seals give birth year-round, but mostly in the summer through early winter months. Caribbean monk seals followed a similar seasonal pattern. Monk seal births in Hawaii usually occur from February to August, peaking in April–June, but births are known in all months. At birth, Mediterranean monk seal pups weigh between 15 and 26 kg and gain weight fast during a suckling time of up to 4 months, with

the females mixing feeding trips with pup attendance. Hawaiian monk seal pups are weaned at 6 weeks, after attaining weights of 50–100 kg, a broad range, reflecting the varying nutritional status of different populations. Hawaiian females do not forage during the lactation period. After weaning, pups of both species survive on their fat reserves while they acquire prey-catching skills. Kenyon and Rice (1959) discussed the life history of monk seals.

Females of both living species have four functional mammary glands, and while they can give birth to single pups in consecutive years, they will also skip some years. Hawaiian females exhibit an average 381-day interval between annual births; thus these females give birth later each year until a year in which they do not give birth, and then they cycle earlier the following year. Some females give birth on a more interrupted schedule. The breeding interval of the Mediterranean monk seal is approximately 1 year with a variability of 15 days and correlates with lactation duration.

When Hawaiian monk seal female–pup pairs are near each other, accidental exchanges of the pups between the females can occur. Serious consequences result when the exchanged pups have suckled for very different times and are of very different size. The mothers will still wean their foster nursing pups after about 6 weeks of lactation, leaving one pup larger than normal size and one smaller, with the latter's survival chances highly compromised. Mediterranean monk seal females may foster pups discarded by other mothers, and also simultaneously nurse two pups, including a pup from a different mother. Milk stealing by pups has also been reported.

A mating tactic of some nondominant male Hawaiian monk seals is an attack by a group of these males, a few to over 20, on an adult female or an immature seal on some occasions. The attention of the dominant male in consort with the female becomes distracted by one or more of the challengers in the group, allowing the others to then breed with the female during bouts that may last over 3 h in the water. The repeated and prolonged biting on the back of the female by males attempting copulation results in extensive trauma and tissue and fluid loss, often leading to her death. This detrimental behavior has occurred primarily at seal colonies where the adult sex ratio was skewed toward males.

### V. Interactions with Humans

Both living species of monk seals have been impacted greatly by human activities, from direct killing to competition for prey with

Figure 1   *Weaned Hawaiian monk seal female pup, showing lighter ventral pelage of younger seals. Photo credit: William G. Gilmartin.*

fisheries and incidental disturbance of seals due to human presence on or near hauling and breeding beaches. Extinction of *M. tropicalis* in the 1950s and the international endangered status of the two remaining *Monachus* species result from this high sensitivity of the genus to direct and indirect human interactions.

The Mediterranean monk seal has been exploited since ancient times, and a significant decline in all of its range occurred in the second half of the twentieth century. The total population was thought to be between 600 and 1000 individuals in the 1970s but at present is estimated at 350–450. Hawaiian monk seals are estimated at 1200–1300 animals. Less than 10% of the population reside in the main Hawaiian Islands where size and survival of young suggest that these seals are doing well. The much larger population in the northwestern islands is currently experiencing very high juvenile losses, primarily due to starvation.

Pup mortality increases sharply for Mediterranean and Hawaiian monk seals when human beach-use of a preferred habitat forces females to give birth at unsuitable pup-rearing sites. Both monk seals also interact with fisheries—commercial bottomfish and longline hooks and hooks from shoreline recreational fisheries in the

**Figure 2** *Interaction between adult Hawaiian monk seals (male, left and female, right). Photo credit: Victoria McCormick.*

**Figure 3** *Mediterranean monk seal adult male (back) showing the typically darker dorsal coloration and juveniles in front. Photo credit: Jaume Forcada.*

main Hawaiian Islands have been observed in Hawaiian seals. Mediterranean seals have been entangled in active fishing gear and shot. Mediterranean monk seals also feed on fish farms which increases their risk of mortality. Entanglement in marine debris is a threat to both monk seals. Usually affecting pup and juveniles, the frequency of observations of entangled Hawaiian monk seals is increasing and, in many cases, where the debris is not removed, the seals are likely to be seriously injured or die.

Although the monk seal colony on the western Sahara coast had been considered the most viable population of the species and characterized by high adult survival rates, it also exhibited high pup mortality and very low recruitment. Then, during May–July 1997, the size of this colony was tragically reduced by two-thirds due to a large-scale mass mortality event. The most probable cause was a phytoplankton paralytic toxin, although a morbillivirus was also detected during the event. The age structure of the surviving population was severely altered because juvenile mortality was insignificant compared to that of adults. This high mortality event severely compromised the recovery potential of the species in the Atlantic.

Hawaiian monk seal numbers have also been reduced due to human activities. Currently, however, all of the major breeding islands of the Hawaiian monk seal occur within federal and state government refuges where access is controlled and the remaining bottomfish fishery is being phased out over 5 years, at which time the foraging habitat in the Northwestern Hawaiian Islands will be completely protected from direct fishery interactions. The Hawaiian monk seal has a well-organized research and recovery effort, with guidance provided by a "recovery team" of scientists. Recovery actions have included disentanglement of seals and two successful programs to enhance female survival. Underweight female pups were collected, rehabilitated, and then released back to the wild during the 1980s and early 1990s. In another effort, adult male seals that were killing females were captured and relocated to areas remote from the main breeding populations. Both projects contributed to population recovery in some colonies during the 1990s.

The remaining populations of both monk seal species are highly vulnerable to random catastrophic events such as die-offs due to introduced disease, biotoxin poisoning, effects of inbreeding depression, human disturbance, and competition with fisheries. The stability of the extant populations relies on high adult female survival rates. Fortunately, most of the Hawaiian monk seal population is now moderately buffered from anthropogenic pressures by the isolation and protected status of its major breeding habitat. The Mediterranean monk seal is not as fortunate. While a few protected areas have been established for the Mediterranean seal by Greece in the Aegean and by Portugal at the Desertas Islands in the Atlantic, only an immediate and significant reduction in anthropogenic pressures on the Mediterranean species and a range-wide coordinated recovery effort will avoid its extinction in the twenty-first century.

## See Also the Following Articles

Earless Seals ■ Extinctions, specific

## References

Aguilar, A. (1999). "Status of Mediterranean Monk Seal Populations." RAC-SPA, United Nations Environment Program, Alos Editions, Tunis.

Adam, P. J., and García, G. G. (2003). New information on the natural history, distribution, and skull size of the extinct (?) West Indian monk seal, *Monachus tropicalis. Mar. Mamm. Sci.* **19**, 297–317.

Forcada, J., Hammond, P. S., and Aguilar, A. (1999). Status of the Mediterranean monk seal *Monachus monachus* in the western Sahara and the implications of a mass mortality event. *Mar. Ecol. Prog. Ser.* **188**, 249–261.

Gerrodette, T., and Gilmartin, W. G. (1990). Demographic consequences of changed pupping and hauling sites of the Hawaiian monk seal. *Conserv. Biol.* **4**, 423–430.

Goodman-Lowe, G. D. (1999). The diet of the Hawaiian monk seal, *Monachus schauinslandi*, from the Northwestern Hawaiian Islands during 1991–1994. *Mar. Biol.* **132**, 535–546.

Kenyon, K. W. (1972). Man versus the monk seal. *J. Mammal.* **53**, 687–696.

Kenyon, K. W., and Rice, D. W. (1959). Life History of the Hawaiian monk seal. *Pacific Sci.* **13**, 215–252.

National Marine Fisheries Service (2007). "Recovery Plan for the Hawaiian Monk Seal (Monachus schauinslandi)," Second Revision. National Marine Fisheries Service, Silver Spring, MD, 165 pp.

Ragen, T. J., and Lavigne, D. M. (1999). The Hawaiian monk seal: Biology of an endangered species. *In* "Conservation and Management of Marine Mammals" (J. Twiss, and R. Reeves, eds), Vol. II, pp. 224–245. Smithsonian Institution Press, Washington, DC.

Repenning, C. A., and Ray, C. E. (1977). The origin of the Hawaiian monk seal. *Proc. Biol. Soc. Wash.* **89**, 667–688.

Samaranch, R., and González, L. M. (2000). Changes in morphology with age in Mediterranean monk seals *Monachus monachus*. *Mar. Mamm. Sci.* **16**, 141–157.

Timm, R. M., Salazar, R. M., and Townsend Peterson, A. (1997). Historical distribution of the extinct tropical seal, *Monachus tropicalis* Carnivora: Phocidae. *Conserv. Biol.* **11**, 549–551.

Wexler, M. (1993). A monk on their backs. *Natl. Wild.* **31**, 44–49.

Winning, B. (1998). The roller coaster ride of the Hawaiian monk seal. *California Wild* **51**, 30–35.

on the forehead (Purves and Pilleri, 1983). As such, they are involved in the production of sound. In many mysticetes, facial muscles also extend between the two halves of the mandible where they assist in squeezing gulps of ingested water through the baleen.

One particular facial muscle, buccinator, is unusual in that it is not near the surface of the skin in land mammals. It forms the wall of the cheek and gives the cheek a rigid wall when suction is produced. As such it is critical for nursing young. In cetaceans, partly as a result of the long snout, this muscle does not give the cheek a rigid wall. Nursing females assist young in suckling by actively squirting milk into their mouths by the contraction of special skin muscles overlying the mammary gland.

The masticatory muscles of pinnipeds are similar to those of terrestrial carnivores, and sirenian masticatory muscles are not unlike those of herbivorous land mammals. In cetaceans, the temporal muscle is greatly reduced and the muscles used to close the jaws are the pterygoids and masseter. Unlike land mammals, in which these muscles direct lateral movements of the lower jaws, simple closing of the jaws is their main function in cetaceans.

The muscles of the throat of most marine mammals do not differ greatly from those of land mammals. The throat of odontocete cetaceans (Fig. 1) is more specialized than that of other marine mammals. The larynx of odontocetes is elongate and its epiglottis projects far anteriorly, reaching the back of the palate and extending into the nasopharyngeal duct. The walls of the nasopharyngeal duct, including the soft palate, consist of a strong annular muscle that encloses the epiglottis and seals the lumen of the larynx functionally from the pharynx.

---

## Musculature

### J.G.M. THEWISSEN

**T**he muscular system of mammals was designed on a single blueprint; there is a remarkable constancy of muscles and associated nerves from the most agile bat to the fastest antelope, and the largest whale. Details differ, and most of these differences reflect adaptations to specific demands of the environment. This article presents an outline of muscular anatomy of marine mammals, emphasizing how pinnipeds, cetaceans, and sirenians differ from terrestrial mammals. General summaries of cetacean muscles can be found in Slijper (1936, 1962), and of manatees in Domning (1978). Howell (1930) discussed locomotor morphology of all marine mammals.

## I. Cranial Muscles

In the cranial muscles, marine mammals differ from terrestrial mammals in the arrangement of their eye muscles, facial muscles, their masticatory muscles, and the muscles of the palate, pharynx, larynx, and tongue.

The facial muscles in land mammals are attached to the skin of the face and moderate facial expressions. In all marine mammals, the facial muscles are involved in closing of the nose opening (or blowhole) to prevent the entry of water during diving. In sirenians, the most important of these muscles insert on the mobile snout and are involved in the manipulation of food. In cetaceans, the facial muscles are greatly rearranged and are positioned around the airsac system

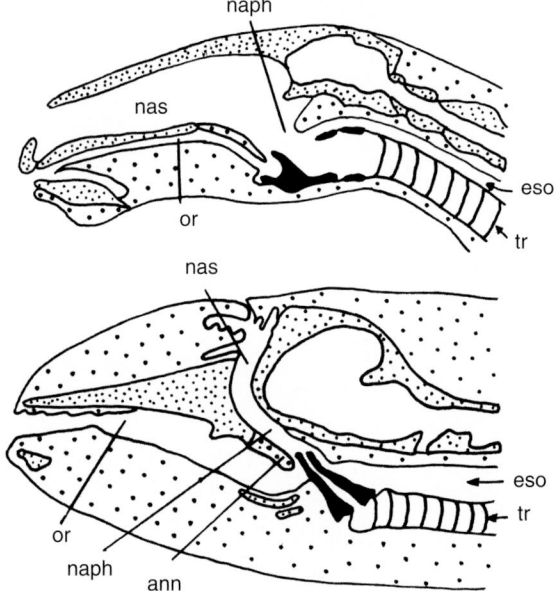

**Figure 1** *Midline sections through the head of a horse (top) and common porpoise* (Phocoena phocoena, *bottom) showing the unique shape of the throat and larynx in odontocete cetaceans. Air in all mammals passes from the nasal cavity (nas) to the nasopharyngeal duct (naph), to the larynx, and to the trachea (tr). Food in all mammals passes from the oral cavity (or), to the pharynx (throat), and to the esophagus (eso). The laryngeal cartilages (colored black); form a spout in odontocetes that fits into the nasopharyngeal duct and can be closed tightly by means of annular muscles (ann). This closure causes a tight separation of the air and food passages. Modified after Slijper (1962).*

The tongue of most mammals is very muscular, and its large size in baleen whales is remarkable. In blue whales (*Balaenoptera musculus*), the tongue is the size of an adult elephant and makes up 2.5% of the whale's entire weight. However, rorqual tongues are not very muscular, consisting mainly of fat and connective tissue.

### A. Axial Muscles

The neck muscles in cetaceans are unremarkable because the neck is short. The neck is long in pinnipeds and may be very muscular. It functions in balancing the body during locomotion and intraspecific fighting.

Muscles extending along the back and tail are the main muscles of propulsion in cetaceans. Epaxial muscles extend along the dorsal side of the transverse processes of the vertebrae. These muscles contract and cause dorsal concavity of the back and tail, pulling the fluke up in the upstroke. These muscles, especially the multifidus and longissimus, are enormous (Fig. 2). The upstroke in cetaceans is powered mainly by the longissimus and extensor caudae lateralis (Pabst, 1993). The latter muscle inserts directly on the dorsal surface of the vertebrae of the fluke, but the longissimus exerts its power by attaching to a subdermal sheath of tendons (Fig. 2) which attaches on spinous and transverse processes along most of the back of the cetacean. It is through the connections of this sheath to the terminal tail vertebrae that the fluke is moved, allowing muscular force to be distributed evenly along the caudal peduncle. The multifidus does not insert on this sheath, instead attaching to the posterior thoracic vertebrae and the lumbar vertebrae. Its main function appears to be to stiffen the back, providing a stable platform of origin for longissimus. The longissimus is also large in sirenians and is probably important in their upstroke. Lumbar epaxial muscles are also important in powering the undulatory movements of *Enhydra lutris* (sea otter).

The downstroke of the fluke in cetaceans is mainly powered by muscles attaching to the ventral side of the thoracic and lumbar vertebrae and inserting, via a tendon sheet, to the ventral side of the caudal vertebrae and chevron bones (Pabst, 1983). These muscles are large and are commonly called hypaxialis lumborum. In sirenians, specialized tail muscles called sacrocaudalis ventralis lateralis and medialis produce depression of the tail.

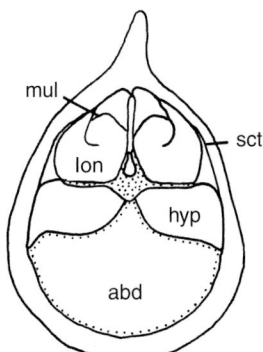

**Figure 2** *Cross section through the lumbar region of a bottlenose dolphin* (Tursiops truncatus). *Note the large development muscles, which, at this level, are cross sectionally larger than the abdominal cavity (abd). Dorsal to the vertebra (stippled area) are epaxial muscles (mul, multifidus; lon, longissimus) and ventral to the vertebra are hypaxial (hyp) muscles. Muscles are closely associated to the subdermal connective tissue sheath (sct). Modified after Pabst (1993).*

Epaxial and hypaxial muscles are large in phocids, where they are used to produce the side-to-side movements that propel the body in swimming. Among the larger of these muscles is the iliocostalis.

### B. Muscles of Thorax, Abdomen, and Limbs

A large superficial skin muscle, the cutaneous trunci (also called panniculus carnosus), covers much of the thorax and abdomen in many mammals. In sirenians, this muscle is especially large and assists in the downstroke of the tail. In cetaceans, part of this muscle is specialized and overlies the mammary gland. It compresses the gland (Fig. 3) and squirts milk into the mouth of nursing young.

Unusual among mammals is the muscle system associated with the penis of cetaceans. Just like in most mammals, erection in cetaceans is not under muscular control. However, unlike most mammals, retraction of the penis into a pouch on the body of the cetacean is caused by contraction of the retractor penis muscles (see Male Reproductive System). These muscles also occur in artiodactyls but are absent in other mammals.

The forelimb of cetaceans (see Forelimb Anatomy) is mainly involved in steering and does not provide propulsive force during rectilinear swimming. Shoulder movements are mainly adduction and abduction; flexion and extension are limited. The shoulder of cetaceans allows less mobility than that of most terrestrial mammals. The clavicle is absent, and tight muscles anchor the scapula to the thorax. These muscles include pectoralis, rhomboids, serratus ventralis, and latissimus dorsi. A large additional muscle, the trapezius, occurs in most mammals but is absent in cetaceans. At the scapulo-humeral joint, the deltoid is a strong abductor, and the latissimus dorsi probably the main adductor, assisted by the subscapularis. The joint between scapula and humerus is a ball and socket joint in cetaceans, as in all mammals, but there are no flexible synovial joints below the cetacean shoulder. Ligaments at the elbow, wrist, and fingers allow for some elastic mobility. A few muscles in cetaceans (e.g., triceps) insert distal to the elbow, but there are no muscle bellies in the forearm and hand.

Just like cetaceans, the forelimb of phocid and odobenid pinnipeds does not provide much of the propulsive force when swimming. It is, however, important in land locomotion. In contrast, otariid pinnipeds use their forelimb as the main propulsor during swimming, and the forelimbs also have an important role in land locomotion.

All pinnipeds lack a functional clavicle, and the shoulder is loosely attached to the chest. Of the shoulder muscles, the pectoralis and latissimus dorsi are the largest and probably provide most propulsive force during swimming in otariids (Fig. 4). The forearm and wrist of otariids contain synovial joints, although mobility at the wrist is reduced. The flippers that form the hands of otariids lack extensive musculature.

The forearm and hand of phocids are relatively mobile, unlike those of otariids. In the northern phocids (phocines), the hands are used in terrestrial locomotion and have well-developed muscles.

Modern Sirenia do not use their forelimbs for propulsion but retain synovial joints at the shoulder, elbow, and wrist. The hands are used in manipulating food and retain many of the muscles that are present in land mammals. In the extinct *Hydrodamalis gigas* (Steller's sea cow), there were no wrist and hand bones and, consequently, no hand muscles.

There are no hind limb muscles in modern cetaceans (see Hind Limb Anatomy), although they were developed in Eocene forms. A rudiment of the pelvis and sometimes the femur and tibia occurs in modern cetaceans, but its main purpose is the attachment of retractor penis. Pelvic bones were also known in early sirenians, but

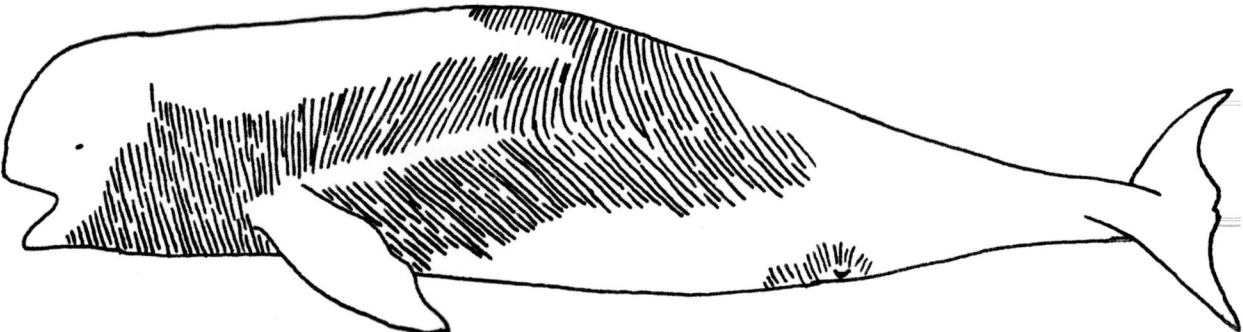

**Figure 3**   *The cutaneous trunci muscle of finless porpoise* (Neophocaena phocaenoides). *Note the muscular tissue overlying the mammany gland on the ventral side near the tail. After Howell (1930).*

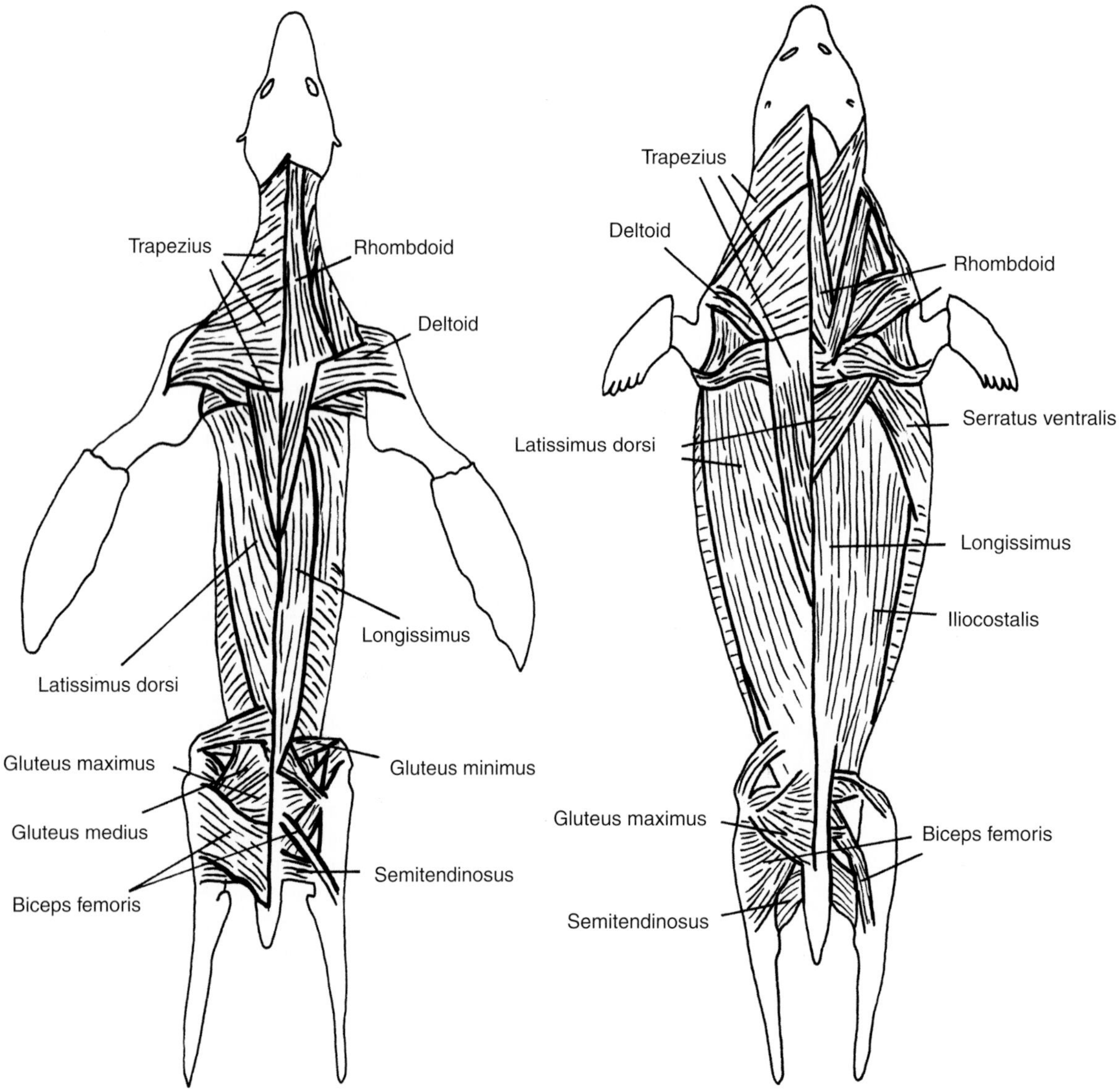

**M**

**Figure 4**   *Dorsal views of a partially dissected sea lion* (Zalophus) *and a harbor seal* (Phoca) *superficial muscles on left side of the animal, deeper muscles on the right. Modified after Howell (1930).*

these bones and all of the hind limb musculature were lost in the extant forms.

Hind limbs are well developed in the pinnipeds. Phocidae and Odobenidae use their hind limbs in aquatic locomotion, making adduction and abduction movements with an inverted foot. Phocidae do not use their hind limbs to support the body while on land. Otariidae trail their hind limbs during swimming, but use them during locomotion on land. Odobenidae also support their body with their hind limbs while on land.

In the phocids, the hind limb flexors (e.g., the hamstrings) are reduced, whereas the extensors are supported by changed insertion of the adductors and obturator externus, which also serve in extension (Fig. 4). Muscles below the knee are present in all pinnipeds, but those crossing the heel are stronger in phocids than in otariids.

## See Also the Following Articles

Cetacean Prenatal Development ■ Skeletal Anatomy ■ Skull Anatomy

## References

Domning, D. P. (1978). The myology of the Amazonian manatee, *Trichechus inunguis* (Natterer)(Mammalia, Sirenia). *Acta Amazon.* **8**(Suppl. 2), 1–80.

Howell, A. B. (1930). "Aquatic Mammals, Their Adaptations to Life in the Water." C.C. Thomas, Baltimore.

Pabst, D. A. (1993). Intramuscular morphology and tendon geometry of the epaxial swimming muscles of dolphins. *J. Zool. Lond.* **230**, 159–176.

Purves, P. E., and Pilleri, G. E. (1983). "Echolocation in Whales and Dolphins." Academic Press, London.

Slijper, E. J. (1936). Die Cetaceen, vergleichend-anatomisch und systematisch. *Cap. Zool.* **6 and 7**, 1–590.

Slijper, E. J. (1962). "Whales." Basic Books, New York.

# Museums and Collections

## JOHN E. HEYNING AND JAMES G. MEAD

The integrative approach to studying biology is similar to constructing a jigsaw puzzle—each discipline and data set contribute in a meaningful way to understand the whole. Individual pieces may contribute more or less to the picture, but nonetheless all pieces are important. In biology, each discipline contributes its own unique set of pieces to the puzzle of life. Research in museums has historically focused on specimen-oriented disciplines and thus has contributed to these suites of puzzle pieces. Specimens are potential sources of data for the disciplines of systematics, paleontology, morphology, histology, genetics, pathology, life history, parasitology, toxicology, and biochemistry. In addition, museums serve as important forums of informal learning for the visitors that peruse the exhibits or engage in an educational program.

## I. Biodiversity and Systematics

Perhaps the most fundamental among the specimen-oriented disciplines is the study of biodiversity, the defining of species and populations within species. Most marine mammalogists working within museums in the nineteenth and early twentieth centuries spent their hours primarily describing new species from the vast array of specimens unloaded from some recent voyage of exploration so characteristic of that time. For instance, from the numerous marine mammal specimens collected by the Southern Hemisphere expeditions of the HMS *Erebus and Terror* during the years 1839–1843, John Gray of the British Museum (Natural History) described numerous new species, including the Ross seal (*Ommatophoca rossii*), the crabeater seal (*Lobodon carcinophaga*), the pygmy right whale (*Caperea marginata*), and the Chilean dolphin (*Cephalorhynchus eutropia*). While the heyday of prolific new species description peaked a century ago, the need for the ongoing study remains very relevant today. Several new species (or resurrected old species) have been defined within recent years, and most populations are just now being understood.

The classical approach of using morphology to define species continues to be relevant. However, analyses of molecular genetic data provide us with additional new tools to help define populations, species, and the relationship among species. Exemplary of this is the recent discovery of a new species of beaked whale. In the mid-1970s, several STRANDINGS occurred of a small species of beaked whale along a restricted section of southern California coastline. Because these specimens morphologically resembled the Southern Hemisphere species *Mesoplodon hectori*, scientists tentatively assigned these California animals to that taxon. A graduate student from New Zealand investigating beaked whale phylogeny sampled the DNA from these specimens along with many others held in museums. To her astonishment, these California specimens clustered nowhere near specimens of *M. hectori* from the Southern Hemisphere (Dalebout *et al.*, 1998), hence providing evidence that they represented a new species.

Determining the evolutionary relationships, or phylogeny, among this diversity of species, both living and extinct, is the study of systematics. Systematics provides an evolutionary framework that becomes the foundation for the comparable biological approach. Phylogenies can be constructed using a variety of data sets, morphological, molecular, and fossils—all of which reside primarily within museums. Hence, researchers today can infer past events from phylogenetic reconstructions of evolutionary relationships. Most modern systematists use a philosophical approach called cladistics. The basic tenets of cladistics are quite simple: organisms are deemed to be related based on shared derived characters called synapomorphies. Derived characters are defined as having arisen in the common ancestor of the taxa and subsequently passed onto their descendant taxa.

Museums have a long-term commitment to house specimens for research. Thus, material collected in the 1700s and 1800s is still available for scientific inquiry today. For many species, it is only through the accumulation of specimens and data over several decades, even over a century that we can obtain the sample sizes needed to begin to understand even the basic biology of these species. For systematic studies, it is crucial to examine a large series of specimens (Fig. 1). To define species or populations, one must first know the limits of variation—individual, ontogenetic, SEXUAL DIMORPHISM—to ascribe that the observed variation is due to limited genetic exchange.

## II. Morphology

How can a blue whale engulf up to 70 tons of water? Why does not a narwhal break its tusk? How can a dolphin cool its testes so that spermatogenesis can occur? All these questions require the detailed examination of anatomical structures. This in turn requires that some

**Figure 1** *Museum workers collec a series of pilot whale specimens* (Globicephala macrorhynchus). *Series such as these allow biologists to define species and to understand populations within species. Defining these biological units is crucial to conservation biology among other disciplines.*

**M**

specimens are readily available. Some studies are limited to hard parts and can be answered by examining osteological material. However, studies of soft anatomy require that these structures be preserved. For most organisms, storage of the whole beast can be accomplished easily by plunking the specimen into a jar of formalin and/or alcohol. Preservation for future study of a good-sized dolphin, let alone a whale, presents far more of a logistical challenge. As the immense specimens typically need to be dissected without preservation, the task can be demanding, as these large, oil-laden mammals produce a rich organic bouquet as they decompose. Fortunately, there is now a renaissance of morphological work requiring innovative ways of preserving and studying cetacean anatomy.

### III. History of Museum Research

The first large collections of marine mammals had their genesis in the grand museums of Europe. Baron von Cuvier amassed and published on a very important collection in the early 1800s, which now resides in the Museum National d'Histoire Naturelle in Paris. By the mid-1800s, the British Museum of Natural History (now the Natural History Museum, London) had built major collections as the British Empire explored the world. Two of the preeminent marine mammalogists of this era, William Henry Flower and John Edward Gray, increased our knowledge considerably by studying the specimens within this venerable museum. Aside from the collections amassed from expeditions, museums in Britain had a distinct advantage for growing their collections. In 1324, stranded whales and dolphins were declared "Royal Fishe" and therefore property of the Crown. The original intent of this decree was to ensure that an economically valuable stranded fresh whale would enhance the coffers of the government. An unforeseen benefit was that the majority of strandings were of the economically non-valuable uneatable variety and therefore available for government supported museums. Hence the first stranding program began (Fraser, 1974). This original decree and subsequent museum-oriented mindset was passed along to the then

British colonies. These former colonies now have museums with major collections including those in Australia, New Zealand, South Africa, and the United States.

Marine mammals as museum specimens are difficult to acquire, store, and maintain. As a result, there are very few large collections for researchers to use. Of the largest collection of land mammals, well over one dozen have more than 100,000 specimens. The majority of specimens in these collections are the taxonomically diverse and numerically abundant rodents and bats. In striking contrast, less than a dozen or so museums have collections of marine mammals numbering over a mere 1000 or so. These include the National Museum of Natural History (Smithsonian), Washington DC, USA; Natural History Museum of Los Angeles County, California, USA; National Science Museum, Tokyo, Japan; The Natural History Museum, London, UK; National Museum of New Zealand (Te Papa), Wellington; American Museum of Natural History, New York, USA; California Academy of Sciences, San Francisco, California, USA; South Australian Museum, Adelaide; Museum National d'Histoire Naturelle, Paris, France; and South African Museum, Cape Town.

For a specimen to be of greatest utility for answering questions, it needs to have as much associated data with it as possible. Such archives provide context for the additional data collected by scientists. Originally, museum curators collected only skulls or skeletons along with occasional sketches of the living beast. Early in this century, following the lead set by the systematic collection of data from whaling stations, museum workers began documenting more data from each specimen. As the number of questions regarding marine mammal biology have increased concurrent with new analytical tools to address these questions, far more is being collected. Now it is not uncommon to collect the complete skeleton, frozen tissues, measurements, fluid-preserved tissues, photographs, and notes.

### IV. Public Display

Over the past century and a half, museums have served an increasing role as important centers for the public to learn about the natural world. Accurately mounted exhibits can convey great biological detail and grand-scale presence that would be difficult for the public to ever experience in the wild.

Many museums have also capitalized on the immensity of whales to create exhibit icons, most notably the model of a living blue whale (*Balaenoptera musculus*). In 1907, a model created from a 74.4-foot blue whale went on display at the American Museum of Natural History. Subsequently, the British Museum of Natural History (now called the Natural History Museum, London) erected its own blue whale model measuring some 88 feet in length. In the early 1960s, the Smithsonian Institution (Washington, DC) unveiled their 92-foot model (Figure 2). Not to be outdone, the American Museum christened their new and anatomically more accurate 94-foot (28.7 m) model in 1969!

Museums hold collections in the public trust so that they are available to scholars in perpetuity. Thus they serve as guardians of the tangible evidence of time past and the archivists of our current natural heritage. In addition, museums serve as important centers at which the public can learn.

### *See Also the Following Articles*

History of Marine Mammal Research ■ Systematics, Overview

**Figure 2** *The full-size model of a blue whale that was mounted at the National Museum of Natural History in Washington, DC.*

### References

Conover, A. (1996). The object at hand. *Smithsonian* **27**(7), 28, 30, 31.

Dalebout, M. L., van Helden, A., van Waerebeek, K., and Baker, C. S. (1998). Molecular genetic identification of southern hemisphere beaked whales (Cetacea: Ziphiidae). *Mol. Ecol.* **7**, 687–694.

Fraser, F. C. (1974). "Report on Cetacea Stranded on the British Coasts from 1948 to 1966." British Museum (Natural History), London.

# Mysticetes, Evolution

Annalisa Berta and Thomas A. Demérè

## I. Introduction

The fossil record of mysticete cetaceans is rapidly improving and the origin and diversification of this highly specialized mammalian group is coming into focus. Crown mysticetes (i.e., extant baleen whales of the families Balaenidae, Neobalaenidae, Balaenopteridae, and Eschrichtiidae) are edentulous as adults, but possess deciduous teeth that are resorbed prior to birth. This ontogenetic pattern reflects an ancestral ontogeny in which fully formed teeth were retained into adulthood. Archaic baleen whales include stem mysticetes, both toothed and toothless, that do not belong to extant families. Toothed mysticetes first evolved in the late Eocene or earliest Oligocene, diversified in the late Oligocene, and appear to have been extinct before the Miocene began. They do not constitute a monophyletic group. Stem edentulous mysticetes are first reported from the late Oligocene coincident with the radiation of toothed forms, but are not diverse until the Miocene. Although contested, it is likely that most, if not all, archaic mysticetes possessed some form of baleen in the upper jaw. This key filter feeding innovation permitted exploitation of a new niche and heralded the evolution of modern baleen whales, the largest animals on Earth.

## II. Toothed Mysticetes

As currently understood, toothed mysticetes are grouped into four families: Llanocetidae, Mammalodontidae, and Janjucetidae from the Southern Ocean and Aetiocetidae, from the North Pacific. To date no toothed mysticetes are known from the Atlantic region. The retention of an adult dentition in toothed mysticetes is the primitive condition seen in basilosaurid "archaeocetes" and stem odontocetes. The degree of telescoping of the skull is also primitive with little interdigitation of rostral and cranial elements. Consequently, there is a long intertemporal exposure of the frontal and parietal on the cranial vertex. In addition, the supraorbital processes of the frontals retain an elevated position on the cranium, and the external narial opening ("blowhole") is only midway between the tip of the rostrum and the orbit. Derived features of toothed and later mysticetes include transverse expansion of the descending process of the maxilla to form an edentulous infraorbital plate, loss of a bony mandibular symphysis, and thin lateral margins of the maxillae.

The geologically oldest purported mysticete is *Llanocetus denticrenatus* from the late Eocene or early Oligocene of the Antarctic Peninsula. Although only a portion of the mandible and an endocranial cast have been described, the holotype also includes a nearly complete skull and partial skeleton under study by Ewan Fordyce. Despite its antiquity, *Llanocetus* was a large whale with a skull length

M

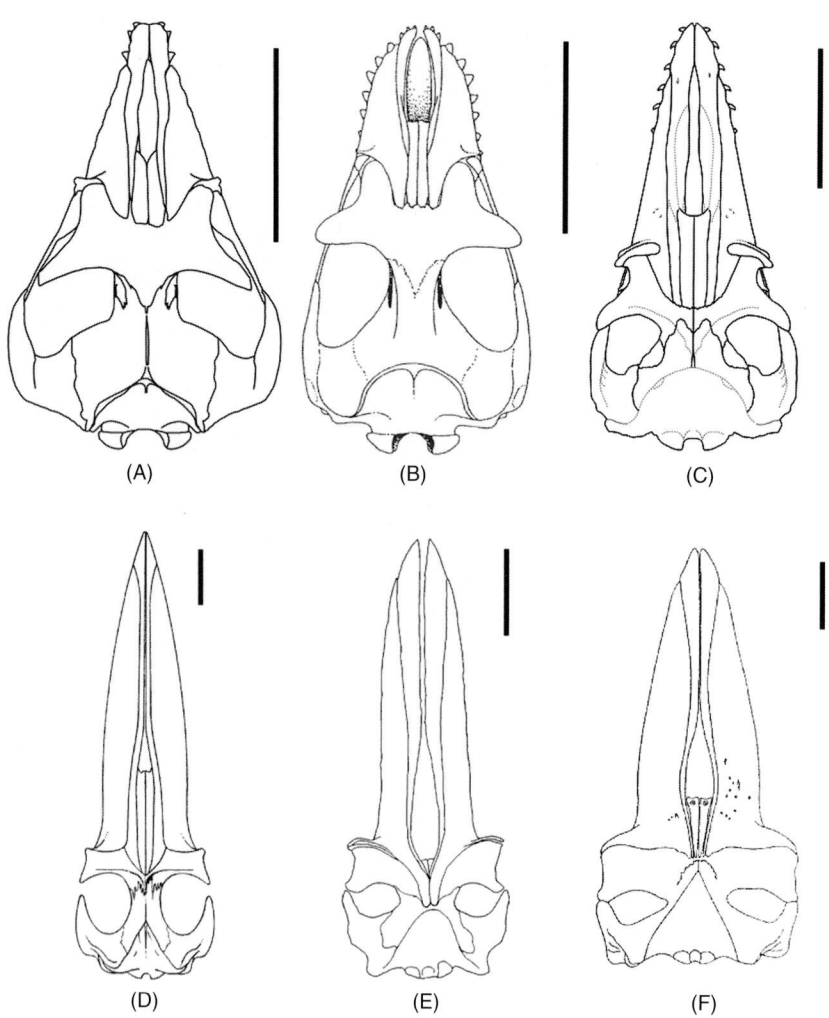

**Figure 1** *Archaic mysticete skulls: (A)* Janjucetus hunderi *(from Fitzgerald, 2006); (B)* Mammalodon collivieri *(from Fordyce and Muizon, 2001); (C).* Aetiocetus weltoni *(from Deméré and Berta, 2008); (D)* Eomysticetus whitmorei *(from Sanders and Barnes, 2002); (E)* Piscobalaena nana *(from Bouetel and Muizon, 2006); (F)* Aglaocetus patulus *(from Kellogg, 1968). Scale bars equal 20 cm.*

**M**

of about 2 m. The distinctly heterodont dentition of *Llanocetus* consisted of widely spaced molariform postcanine teeth with crowns characterized by roughened enamel and large, palmate denticles. Functional comparisons have been made with the palamate teeth of the modern filter-feeding crabeater seal, *Lobodon carcinophagus.*

*Janjucetus hunderi* and *Mammalodon collivieri* from the late Oligocene of Victoria, Australia (Fitzgerald, 2000), were smaller, short-faced toothed mysticetes with closely spaced, heterodont dentitions (Fig. 1). Crown morphology of the postcanine teeth is poorly known for *M. colliveri*, but for *J. hunderi* consists of roughened enamel and moderately sized, closely appressed denticles. The orbits of *J. hunderi* are large relative to skull length (~46 cm), suggesting acute vision. Although both taxa have been assigned to separate monotypic families, character support for this distinction is weak and they eventually may be shown to be sister taxa.

Aetiocetids represent the most diverse clade of toothed mysticetes and include seven nominal species grouped into 3–4 genera. *Aetiocetus* is the most speciose genus, followed by *Chonecetus* and *Morawanocetus.* Overall, aetiocetids were small-bodied cetaceans with skull lengths of about 60–70 cm and an estimated total body length of 2–3 m. Unlike species of *Mammalodon* and *Janjucetus*, aetiocetids had a relatively long rostrum (Fig. 1). Little is known of their postcranial skeleton except that they had elongated necks and relatively long arms with rigid elbow joints. The aetiocetid skull retained numerous primitive features inherited from their archaeocete ancestors (e.g., anteriorly positioned "blowhole," elevated supraorbital processes of the frontals, long intertemporal constriction, large mandibular coronoid process, and large mandibular foramen). However, as mosaic stem mysticetes, aetiocetids also possessed important advanced features (e.g., broad rostrum, vascularized palate, and

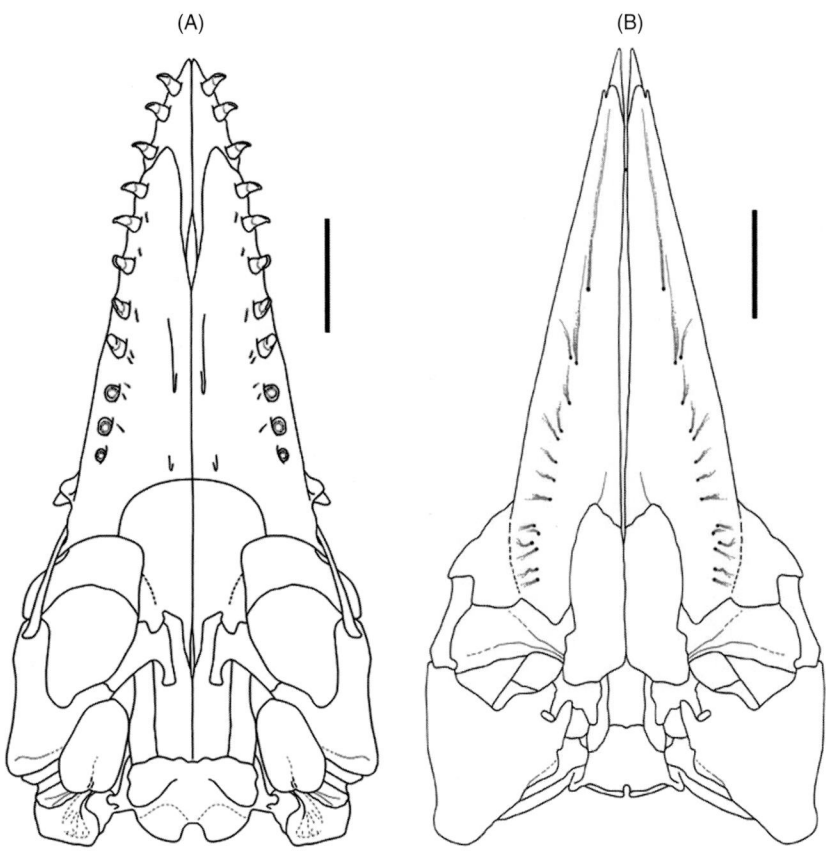

**Figure 2** *Mysticete palates showing lateral foramina and sulci: (A) Toothed mysticete* (Aetiocetus weltoni *~28–25 million years old; scale bar equals 10 cm); (B) Edentulous mysticete* (Balaenoptera acutorostrata; *scale bar equals 20 cm). From Deméré and Berta (2008) and Deméré et al (2008).*

outwardly bowed mandible with prominent groove for the fibrocartilaginous symphysis) that portend the later diversification of fully edentulous baleen whales. Dental morphology varied within the group with *Morawanocetus* possessing a distinctly heterodont dentition and postcanine crowns with roughened enamel and moderately sized denticles. In contrast species of *Aetiocetus* had a more weakly heterodont dentition with postcanine crowns with lightly roughened enamel and diminutive denticles. The teeth were widely spaced and eastern North Pacific species (*A. cotylaveus* and *A. weltoni*) show a tendency toward polydonty, while one western North Pacific species (*A. polydentatus*) was distinctly polydont.

### III. Origin of Baleen

Baleen is a unique mammalian structure consisting of keratinized tubules typically organized into transverse cornified plates suspended from epithelial tissues of the roof of the mouth. The frayed tubules on the medial margin of each plate overlap those of adjacent plates to produce a sieve that entraps prey within the oral cavity. The origin of baleen, although still poorly documented by fossils, appears to have been a stepwise transition from an ancestor with teeth only, to an intermediate state with functional teeth and baleen, to the derived condition with baleen only. Because baleen rarely fossilizes, morphologic evidence for its presence (or absence)

must rely on correlated osteological features. Extant mysticetes have a highly vascularized palate with distinct foramina and associated sulci concentrated along the medial portion of each maxilla (Fig. 2). The blood supply and innervation of the developing baleen apparatus pass through these openings. Thus, the presence of lateral palatal foramina and sulci in fossil mysticetes serves as indirect evidence for the presence of baleen (Figs. 2,3). Importantly, such structures have been reported in some toothed mysticetes.

### IV. Edentulous Mysticetes

Archaic edentulous mysticetes are grouped into two families, Eomysticetidae and "Cetotheriidae." Described eomysteicetids include *Eomysticetus whitmorei* and. *E. carolinensis* from the late Oligocene (30–28 Ma) of South Carolina, USA (Sanders and Barnes, 2002). These earliest edentulous baleen whales were of medium size (skull length ~1.5 m) and possessed a mosaic of primitive (e.g., elongated intertemporal region with long parietal and frontal exposures on the cranial vertex, anteriorly placed "blowholes," elongated nasals, large mandibular coronoid processes, and large mandibular foramina) and derived (e.g., loss of adult dentition, flattened rostrum, and laterally bowed mandibles) features. Other probable eomysticetids include species of *Mauicetus* from the late Oligocene of New Zealand

**Figure 3** *Reconstruction of* Aetiocetus weltoni *by Carl Buell (Deméré et al., 2008).*

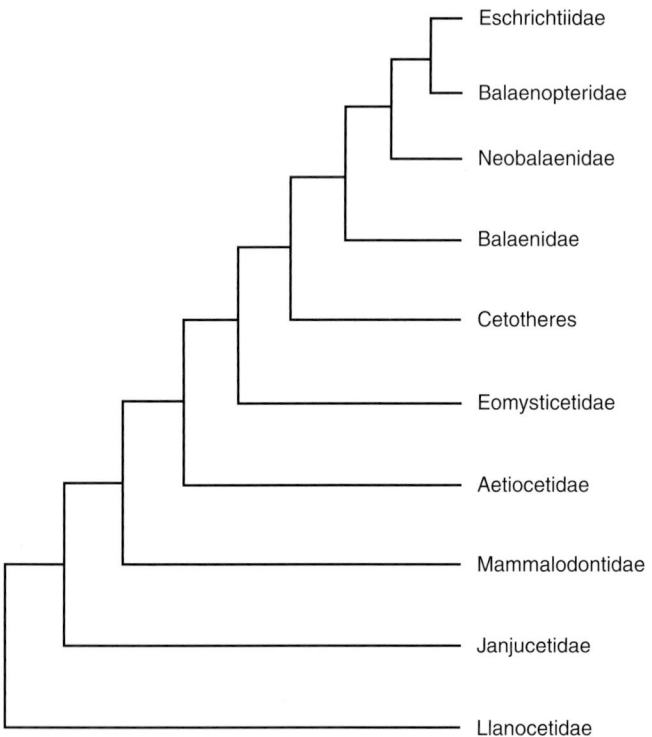

- Eschrichtiidae
- Balaenopteridae
- Neobalaenidae
- Balaenidae
- Cetotheres
- Eomysticetidae
- Aetiocetidae
- Mammalodontidae
- Janjucetidae
- Llanocetidae

**Figure 4** *Phylogeny of mysticetes based on Deméré et al, accepted. (2008).*

have been grouped together primarily based on their lack of synapomorphies of crown mysticetes. "Cetotheres" comprise the greatest taxonomic and morphologic diversity among fossil mysticetes with over 45 described species divided among more than 30 genera. Nominal "cetotheres" range in age from the late Oligocene to the late Pliocene of North and South America, Europe, Japan, Australia, and New Zealand. Monophyly of the "Cetotheriidae" has been questioned since at least the 1920s and the group currently is recognized as paraphyletic or possibly polyphyletic (Bouetel and Muizon, 2006). Several recent cladistic studies have utilized a small number of "cetotheres" to investigate mysticete phylogeny and in the process have recognized two alternative topologies: either "cetotheres" lie outside of crown mysticetes or they are positioned within crown Mysticeti (Fig. 4). With the current increased interest in mysticete paleontology and the description of critical new fossils, it is likely that greater phylogenetic resolution will emerge for "cetotheres." It is also likely that this greater resolution will allow recognition of distinct groups of "cetotheres" that are implicated in the origin of specific crown mysticete clades.

## References

Bouetel, V., and de Muizon, C. (2006). The anatomy and relationships of *Piscobalaena nana* (Cetacea, Mysticeti), a Cetotheriidae *s.s.* from the early Pliocene of Peru. *Geodiversitas* **28**, 319–395.

Deméré, T. A., and A. Berta (2008). Skull anatomy of the Oligocene toothed mysticete *Aetiocetus weltoni* (Mammalia: Cetacea): implications for mysticete evolution and functional anatomy. *Zool. Jour. Linn. Soc.*, 154(2): 308–352.

Deméré, T. A., McGowen, M. R., Berta A., and Gatesy, J. (2008) Morphological and molecular evidence for a stepwise evolutionary transition from teeth to baleen in mysticete cetaceans. *Syst. Biol.* **57**, 15–37.

and undescribed specimens from the late Oligocene of Japan; Baja California, Mexico; and California and Washington, USA.

"Cetotheriidae" is a large, diverse, nonmonophyletic assemblage of extinct small-to-medium-sized toothless mysticetes that traditionally

**M**

Fitzgerald, E. M. (2006). A bizarre new toothed mysticete (Cetacea) from Australia and the early evolution of baleen whales. *Proc. R. Soc. B. London*.

Fordyce, R. E., and Muizon, C.de. (2001). Evolutionary history of cetaceans: A review. *In* "Secondary Adaptation of Tetrapods to Life in Water" (J.-M. Mazin, and V. deBuffrenil, eds), pp. 169–233. Verlag Dr. Frederich Pfeil, Munchen.

Kellogg, A. R. (1968). Fossil marine mammals from the Miocene Calvert Formation of Maryland and Virginia, Part 7: A sharp-nosed cetothere from the Miocene Calvert. Bull. *US Natl. Mus.* **247**, 163–173.

Sanders, A. E., and Barnes, L. G. (2002). Paleontology of the Late Oligocene Ashley and Chandler Bridge Formations of South Carolina, 3: Eomysticetidae, a new family of primitive mysticetes (Mammalia: Cetacea). *Smithsonian Contrib. Paleobiol* **93**, 313–356.

M

# Narwhal
## *Monodon monoceros*

### M.P. HEIDE-JØRGENSEN

### I. Characteristics and Taxonomy

In 1758, Linnaeus used the scientific name *Monodon monoceros* for the whale with one tooth and one horn. Together with the close relative the white whale or beluga, *Delphinapterus leucas*, the narwhal now forms the two-species family of Monodontidae.

Newborn narwhals are evenly gray or dark-brownish gray. While nursing for 1–2 years, the coloration changes gradually to a dark background color with white patches that give a mottled appearance. When adult, the animals are completely mottled on the upper side but with increasing white fields on the ventral side. Old adult males only maintain a narrow dark-spotted pattern on the top of the back, whereas the rest of the body is white. Unlike in other cetaceans, the tail flukes are concave in fully grown narwhals, and a low ridge replaces the dorsal fin.

The most conspicuous feature of the narwhal is the up to 3-m long spiraled tusk (Fig. 1). Six pairs of maxillary and two pairs of mandibulary dental papillae are present in early narwhal embryos, but only two maxillary pairs persist and develop. Of these, the two anterior teeth develop into an elongated tooth that is the start of the tusk. The other two teeth remain vestigial. In males, the left of the two elongated teeth grows and protrudes through the maxillary bones and skin of the rostrum of the whale. During growth the tusk spirals to the left.

In males the right elongated maxillary tooth and in females both maxillary teeth remain inside the skull, sometimes just protruding through an opening in the maxillary bone. Irregularities in the development of tusks are frequently seen; females sometimes attain a tusk, males occasionally have no tusk, and narwhals with two tusks, so-called "double tuskers", are not rare. There are several records of anomalous narwhals with skull characteristics that suggest hybridization with belugas (Heide-Jørgensen and Reeves, 1993). Most of these observations are from Disko Bay in West Greenland where narwhals and belugas occur together during the mating season in late winter (Heide-Jørgensen and Laidre, 2006)

There is a great variability in the shape and the dimensions of the protruding tusk. Some tusks are fairly straight and others are corkscrew-like; some are thin and fragile, whereas others are short and thick. The largest tusk measured was 267-cm long, but a full-grown male usually carries a tusk of about 200 cm. Tusks are sometimes broken, and there are records of a fragment of tusk from another narwhal sitting inside the broken tip, suggesting a head-on collision. The purpose of the tusk has been much disputed, but because both females and males without tusks thrive, they do not seem critical for survival. The tusk is more likely a secondary sexual character that is related to the hierarchy of male narwhals. Displays and crossing of tusks are frequently seen on narwhal summering grounds, and it is likely that this activity determines dominance hierarchies. Narwhals have not been observed using their tusks for fighting or other aggressive behavior.

There are several records of narwhal fragments from Pleistocene deposits in England and Germany. Bones found along the Russian Arctic coasts—both on the mainland and on the Russian Arctic Islands—also suggest a different occurrence or form of narwhals before or during the most recent glaciation. In Canada, bone remains from early postglacial times have also been found both north (Ellesmere Island) and south (Gulf of Saint Lawrence) of present narwhal distribution.

The main reason the narwhal remained a legendary animal for so long may be because of its preference for remote and inaccessible

**Figure 1** *The narwhal,* Monodon monoceros, *occurs in the remote North Atlantic and Arctic Oceans and is conspicuous with a long tusk in males, usually formed from one tooth in the left upper jaw.*

habitats, usually in areas over deep water that is covered with heavy pack ice during dark winter months. Europeans did not visit most of these areas until the nineteenth century, and even though Inuit hunters traded the tusks with whalers, precise descriptions were lacking.

## II. Distribution and Abundance

The narwhal essentially inhabits the Atlantic sector of the Arctic Ocean with few records of stragglers from the Pacific sector (Fig. 2). During the last glaciation, narwhals were restricted to the North Atlantic, but with the retreating ice they inhabited the archipelago of the Canadian high Arctic, northern Hudson Bay, Davis Strait, Baffin Bay, the Greenland Sea, and the Arctic Ocean between Svalbard and Franz Josef Land.

Today, low numbers of narwhals are found offshore in deep-water areas of the Eurasian sector of the Arctic Ocean, where they are seen most frequently around Franz Josef Land and Svalbard. The northernmost recordings of narwhals are from the area between 84°N and 85°N northeast of Franz Josef Land at 70–80°E. In the Greenland Sea, narwhals are widely distributed in the pack ice but probably in low numbers. Along the coast of East Greenland, narwhals are found during the open-water season in fjords from 65°N to 81°N, with particularly large concentrations in Scoresby Sound and Kangerlussuaq. No complete abundance estimates are available from any of the Northeast Atlantic areas, but in 1983 a negatively biased estimate of 300 narwhals was derived from a visual aerial survey in Scoresby Sound (Dietz et al., 1994).

In West Greenland, narwhals visit coastal areas in northwest Greenland (Inglefield Bredning and Melville Bay) during summer and central West Greenland during autumn (Uummannaq) and winter (Uummannaq and Disko Bay). The number of narwhals present at the surface is estimated by visual or photographic aerial surveys (Innes et al., 2002; Heide-Jørgensen, 2004). This number is then corrected for the fraction of the whales that are submerged during passage of the survey plane. Up to 4000 narwhals have been counted in Inglefield Bredning in August and 3000 in Disko Bay in March. Offshore, narwhals are abundant in the heavy consolidated pack ice in northern Davis Strait and Baffin Bay from late November through May, and the number of narwhals wintering in this area has been estimated at 35,000 (NAMMCO, 2006).

Most of the narwhals that spend the winter in Baffin Bay summer in the Canadian high Arctic. During ice break-up, narwhals move from Baffin Bay into the Canadian high Arctic through Lancaster Sound and Pond Inlet. They visit the fjord systems of Eclipse Sound, Admiralty Inlet, Prince Regent Inlet, and Peel Sound during the open-water season from June through September. The abundance in Prince Regent Inlet was estimated at 45,000 in 1996 and in Admiralty Inlet and Eclipse Sound at 15,000 and 3000 narwhals, respectively, in 1984 (NAMMCO, 2006). Prior to the formation of fast ice in October, narwhals move east towards Baffin Bay and Davis Strait.

In northern Hudson Bay and Foxe Basin, a group of 1300 narwhals is found in the summer; hey winter in Davis Strait at the eastern entrance to Hudson Strait.

Narwhals seem to follow relatively strict migration schedules and seem to use the same routes for their annual migrations. It is uncertain how much of the migration is affected by the development of sea ice. The whales arrive at their wintering ground long before pack ice is formed in these areas, and during winter they stay in very heavy consolidated pack ice, usually in leads or holes in areas with less than 3% open water. Before the pack ice breaks up in the spring, the narwhals penetrate north through narrow leads and open-water channels.

Movements from summer through winter have been monitored by tracking of narwhals instrumented with satellite-linked radio transmitters attached to the dorsal ridge or the tusk of the whales. At summering grounds in West Greenland and Canada, narwhals move back and forth between glacier fronts, offshore areas, and neighboring fjords. Before the fast ice forms, the whales move out to deeper water, usually up to 1000 m in depth. In October the whales move southward toward the edge of the continental shelf where the water depth increases over a short distance from 1000 to 2000 m. This slope is also used as a wintering ground, and even though the whales seem stationary in this area, they still conduct shorter movements along this steep slope. Narwhals tracked from Canada and West Greenland were within a few kilometers from each other at these wintering grounds at the deep slope at the edge of the continental shelf in central Baffin Bay. Satellite tracking of a few whales has also confirmed that they return to the summering ground where they were instrumented the year before, and no exchange between summering grounds has so far been found. The mean swimming speed of traveling narwhals is about 60 km per day (Dietz et al., 2001; Heide-Jørgensen et al., 2002).

Studies of mitochondrial DNA have revealed a low level of nucleotide and haplotype diversity in narwhals. This is probably the result of a rapid expansion of the population from a small founding population after the last glaciation. Despite the low variation in narwhal mtDNA, there are still genetic differences between narwhals from different areas. Not so surprisingly, narwhals from East Greenland are different from those inhabiting Baffin Bay, but more surprising is the distinctness of narwhals at two summering grounds (Inglefield Bredning and Melville Bay) and one autumn ground (Uummannaq) in West Greenland (Palsbøll et al., 1997). Apparently, narwhals have annual fidelity to certain summer and autumn feeding localities, but the extent of mixing on the wintering grounds is unknown. The complex stock structure and the apparent overlap in distribution of whales from different areas suggest that narwhals in Baffin Bay and adjacent areas have a metapopulation structure where small populations are connected through movement of at least a few individuals, even if most of the populations remain physically separate.

## III. Ecology

Feeding activity and prey selection by narwhals have been studied by examining stomachs, collected from the Inuit harvest in Canada and Greenland. Stomach samples in the summer were mostly empty with little evidence of recent feeding. Stomachs collected in late fall and winter had considerable amounts of undigested material with evidence of recent feeding. In summer, remains from Arctic cod (Arctogadus glacialis) and polar cod (Boreogadus saida) were occasionally detected. Gonatus squid remains were found in all seasons and in all localities, positively identified as 100% G. fabricii in late fall and winter stomachs. Greenland halibut (Reinhardtius hippoglossoides) was a major part of the diet of narwhals in winter and was often the only prey item identified in a completely full stomach. On a caloric basis, Greenland halibut are lipid-rich and contain higher energy content than Arctic or polar cod. The benefit of making deep dives to the bottom to prey on Greenland halibut may be due to the energy gained from this lipid-rich source (Laidre and Heide-Jørgensen, 2005).

Killer whales and polar bears are the only predators on narwhals. Killer whales feed on narwhals during the summer open-water

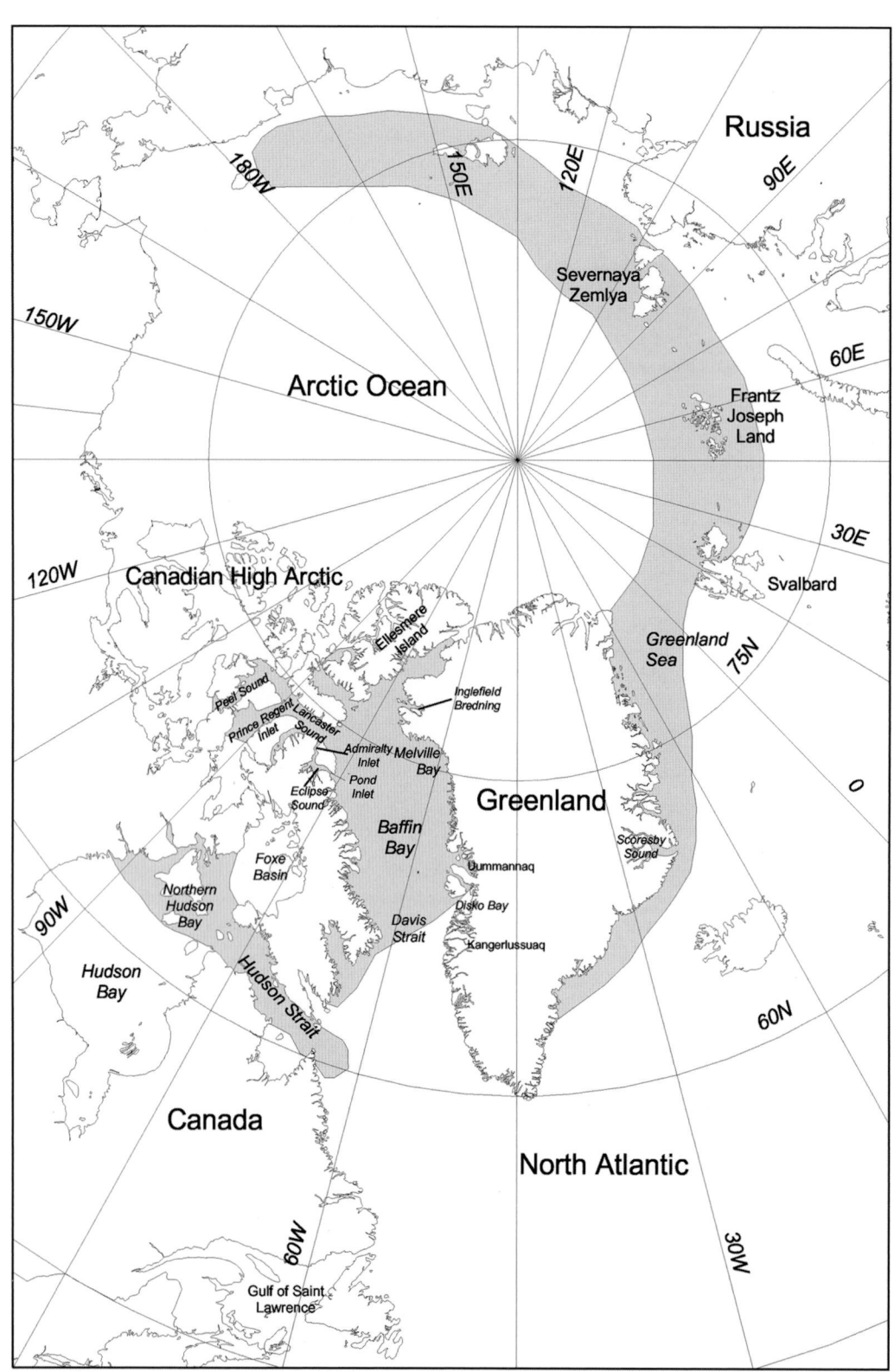

**Figure 2** *Distribution of the narwhal.*

season, when the whales are found in dense aggregations in bays and fjords. Polar bears kill narwhals when they can use the sea ice as a platform to pull the whales out of the water, therefore polar bear predation is limited to winter and spring.

## IV. Behavior and Physiology

Narwhals are usually found in small groups of 5–10 whales, migrating together. Sometimes larger herds are formed that consist of several smaller groups, often all in directional movement along

N

a coastline or toward the head of a fjord. The narwhal groups are usually segregated, with adult males in separate groups and females with calves sometimes together with immature males. Mixed groups occur, especially in large herds, but single animals, particularly males, can also be seen.

Data on narwhal diving have been collected from whales instrumented with dive recorders and satellite transmitters in both Canada and Greenland. Narwhals change their diving habits when moving from coastal and relatively shallow summering grounds to deep offshore wintering grounds. During summer narwhals make few dives below 500 m. This changes gradually during the autumn migration, and by the time they are at the wintering grounds, they make between 5 and 25 dives/day to depths exceeding 800 m, probably in search of deep-water prey. With several dive records exceeding 1500 m, narwhals are among the deepest diving marine mammals, and with a maximum dive record of 1864 m they rival some of the other deep-diving whales (e.g., sperm whales and beaked whales). However, the dives are usually completed within 25 min, and only few dives exceed 30 min, so the whales only have a short time at the bottom as ascent–descent rates for deep dives are 2 and 1 m/sec for shallow dives (Laidre *et al.*, 2003). The accumulated time spent below 800 m can exceed 3 h during a 24-h period on the wintering grounds.

Narwhals make a variety of noises. Clicks that are believed to be used for echolocation have been measured to have their maximum amplitudes at 48 kHz with rates of 3–10 clicks/sec. Faster click rates of 110–150 clicks/sec had maximum amplitudes at 19 kHz. Whistles or pure tones in frequencies from 300 Hz to 18 kHz have also been recorded; they are suspected to serve as social signals among the whales (Ford and Fisher, 1978; Miller *et al.*, 1995).

## V. Life History

Age estimation of narwhals has proven to be difficult. Both the protruding tusk and the embedded teeth contain distinctive growth layers in both dentine and cementum, but with increasing age the growth layers apparently collapse and become unreadable. Also, there has been no empirical way to determine how many growth layers are deposited annually. So far, narwhals have not been kept successfully in captivity.

It has recently been shown that age of narwhals can be estimated by measuring the racemization of l-aspartic acid into d-aspartic acid in the nucleus of the eye lens. Based on this method, it seems likely that female narwhals can attain a maximum age of 115 years and that females have a slightly higher survival rate than males (Garde *et al.*, 2007).

Length at birth is approximately 160 cm. The male tusk erupts at a body length of 260 cm and attains a length of 150 cm at sexual maturity (Fig. 3). Body length at sexual maturity is around 360 and 420 cm for females and males, respectively. Mean length and weight at physical maturity are around 400 cm and 1000 kg for females and 450 cm and 1600 kg for males. Based on age estimates from aspartic acid racemization, age at sexual maturity is likely around 6–7 years for females and 9 years for males (Garde *et al.*, 2007).

The gestation period of the narwhal is subject to some uncertainty, as mating probably occurs in inaccessible areas in April–May. Calving seems to occur in June–August in both Greenland and Canada, and with a mating season early in spring, this implies a gestation period of 13–16 months. Lactation lasts 1–2 years, and females are generally believed to calve every 3 years, but data supporting this seem inadequate.

**Figure 3**  *Narwhals lift their tusks in male-dominance display.*

## VI. Interactions with Humans

The narwhal is the animal behind the legend about the mysterious unicorn: a horse-like creature with a spiraled horn protruding from the forehead. The horn was supposed to have healing abilities, and the wild and shy animal could only be captured with a virgin as bait. Based on narwhal tusks that were brought south from Arctic coasts, this was essentially how narwhals were perceived in western civilization until the seventeenth century when the first descriptions of a fish-like sea monster appeared.

Narwhals have never been a target for commercial whaling, probably because of their skittishness and the difficulties involved in catching them. Inuit hunters in Greenland and Canada hunt narwhals for their valuable tusks and the highly prized skin that is considered a delicacy throughout the Inuit communities. The annual harvest level was on average 535 and 433 during 2000–2004 in West Greenland and Canada, respectively; this is considered small relative to the population size in most areas. However, depending on the population structure, some subpopulations may be overexploited. Concern has been raised about the sustainability of the harvest in West Greenland, and quotas have been installed to reduce the harvest to the recommended level of 135 whales per year. The harvest in East Greenland was on average 100 narwhals per year during 2000–2004; there are no quotas and no assessment of the sustainability of this harvest (NAMMCO, 2006).

Narwhals have high levels of some organochlorines and heavy metals where at least the first are of anthropogenic origin. Possible effects of these contaminants have not been studied in narwhals.

Because of their prevalence for high-density pack ice, narwhals are susceptible to climatic changes that influence the water currents and thereby ice formation in the Arctic. Whether it is naturally occurring or human-induced climate changes, narwhals may become entrapped or lose access to important feeding areas if ice conditions change.

A peculiar feature of the natural history of narwhals is their susceptibility to being entrapped in ice. Because of their preference for heavy pack ice, large schools of narwhals are occasionally caught in ice that freezes rapidly during intense cold, thereby preventing the whales from being able to breathe. This happens particularly often in areas where unpredictable ice conditions persist due to the mixing of warm and cold water masses of variable strength, e.g., Disko Bay in West Greenland. Large numbers of narwhals may succumb during such an ice entrapment, and in January 1915, more than 1000 narwhals died in a well-known ice entrapment in Disko Bay. If the whales are discovered, Inuit hunters may also prey upon them, using the word "sassat" for the event (Siegstad and Heide-Jørgensen, 1994).

N

Narwhals are only hunted in the Canadian Arctic and in East and West Greenland. In Canada the management of the hunt is regulated by a co-management agreement between the Department of Fisheries and Oceans and the Nunavut Wildlife Management Board. In Greenland the hunt is managed by the Greenland Home Rule Government. Scientific and management advice on sustainable harvest levels is generated by the North Atlantic Marine Mammal Commission and the Canada–Greenland Joint Commission for the Conservation and Management of Narwhal and Beluga. Narwhal hunting in both Canada and Greenland is managed by quotas that are based on the latest scientific information on sustainable harvest levels.

International trade in narwhal tusks and other narwhal products is monitored by the Convention on International Trade in Endangered Species of Wild Fauna and Flora (CITES). The narwhals are currently listed under Appendix II of CITES, which requires national permits for import and export of products. Greenland installed a ban on export of narwhal products in 2006 due to concern about the sustainability of the hunt.

Greenland maintains two sanctuaries (Melville Bay and Northeast Greenland) that protect local stocks of narwhals from hunting and disturbances. In Northwest Greenland narwhal hunting is only allowed from kayak, and it is required that the whales are harpooned before being shot.

## *See Also the Following Articles*

Beluga Whale ■ Folklore and Legends ■ Inuit and Marine Mammals

## *References*

Dietz, R., and Heide-Jørgensen, M. P. (1995). Movements and swimming speed of narwhals, *Monodon monoceros*, equipped with satellite transmitters in Melville Bay, Northwest Greenland. *Can. J. Zool*. **73**, 2120–2132.

Dietz, R., Heide-Jørgensen, M. P., Glahder, C., and Born, E. W. (1994). Occurrence of narwhals (*Monodon monoceros*) and white whales (*Delphinapterus leucas*) in East Greenland. *Meddr. Grønland Biosci.* **39**, 69–86.

Dietz, R., Heide-Jørgensen, M. P., Richard, P., and Acquarone, M. (2001). Summer and fall movements of narwhals (*Monodon monoceros*) from Northeastern Baffin Island towards Northern Davis Strait. *Arctic* **54**, 244–261.

Ford, J. K. B., and Fisher, H. D. (1978). Underwater acoustic signals of the narwhal (*Monodon monoceros*). *Can. J. Zool.* **56**, 552–560.

Garde, E., Heide-Jørgensen, M. P., Hansen, S. H., Nachman, G., and Forchhammer, M. C. (2007). Age specific growth and remarkable longevity in narwhals (*Monodon monoceros*) from West Greenland as estimated by aspartic acid racemization. *J. Mammal.* **88**, 49–58.

Hay, K. A., and Mansfield, A. W. (1989). Narwhal *Monodon monoceros* Linnaeus, 1758. *In* "Handbook of Marine Mammals" (S. D. Ridgway, and R. Harrison, eds), Vol. 4, pp. 145–176. Academic Press, San Diego.

Heide-Jørgensen, M. P. (2004). Aerial digital photographic surveys of narwhals, *Monodon monoceros*, in Northwest Greenland. *Mar. Mamm. Sci.* **20**, 58–73.

Heide-Jørgensen, M. P., and Dietz, R. (1995). Some characteristics of narwhal, *Monodon monoceros*, diving behaviour in Baffin Bay. *Can. J. Zool.* **73**, 2106–2119.

Heide-Jørgensen, M. P., and Laidre, K. L. (2006). "Greenland's Winter Whales." Ilinniusiorfik. 124 pp.

Heide-Jørgensen, M. P., and Reeves, R. R. (1993). Description of an anomalous monodontid skull from west Greenland: A possible hybrid? *Mar. Mamm. Sci.* **9**, 258–268.

Heide-Jørgensen, M. P., Dietz, R., Laidre, K., and Richard, P. (2002). Autumn movements, home range and winter density of narwhals (*Monodon monoceros*) from Tremblay Sound, Baffin Island. *Polar Biol.* **25**, 331–341.

Heide-Jørgensen, M. P., Dietz, R., Laidre, K., Schmidt, H. C., Richard, P., and Orr, J. (2003). The migratory behaviour of narwhals (*Monodon monoceros*). *Can. J. Zool.* **81**, 1298–1305.

Innes, S., Heide-Jørgensen, M. P., Laake, J., Laidre, K. L., Cleator, H., Richard, P., and Stewart, R. E. A. (2002). Surveys of belugas and narwhals in the Canadian high Arctic in 1996. *Sci. Pub. No. Atl. Mar. Mamm. Commn.* **4**, 169–190.

Laidre, K. L., and Heide-Jørgensen, M. P. (2005). Winter feeding intensity of narwhals (*Monodon monoceros*). *Mar. Mamm. Sci.* **21**, 45–57.

Laidre, K.. L., Heide-Jørgensen, M. P., and Dietz, R. (2002). Diving behaviour of narwhals (*Monodon monoceros*) at two coastal localities in the Canadian Arctic. *Can. J. Zool.* **80**, 624–635.

Laidre, K. L., Heide-Jørgensen, M. P., Dietz, R., and Hobbs, R. C. (2003). Deep-diving by narwhals, *Monodon monoceros*: Differences in foraging behavior between wintering areas. *Mar. Ecol. Prog. Ser.* **261**, 269–281.

Miller, L., Pristed, J., Møhl, B., and Surlykke, A. (1995). The click-sounds of narwhals (*Monodon monoceros*) in Inglefield Bay, Northwest Greenland. *Mar. Mamm. Sci.* **11**, 491–502.

NAMMCO. (2006). Scientific Committee, Report of the thirteenth meeting. Available from www.nammco.no.

Palsbøll, P., Heide-Jørgensen, M. P., and Dietz, R. (1997). Distribution of mtDNA haplotypes in narwhals, *Monodon monoceros*. *Heredity* **78**, 284–292.

Siegstad, H., and Heide-Jørgensen, M. P. (1994). Ice entrapments of narwhals (*Monodon monoceros*) and white whales (*Delphinapterus leucas*) in Greenland. *Meddr. Grønland Biosci.* **39**, 151–160.

# Neoceti

## R. Ewan Fordyce

## I. Introduction

Neoceti is the taxonomic group comprising the two living clades of Cetacea (Odontoceti and Mysticeti) but excluding the extinct Archaeoceti. The two living groups are quite disparate, each distinguished by a unique combination of anatomical and ecological features. The Odontoceti (toothed whales, dolphins, porpoises) are echolocating macro-predators, whereas Mysticeti (baleen whales) are filter-feeders. Ancient cetaceans from Oligocene times (25 to +30 Ma) show skull structures indicative of echolocation in odontocetes and of filter-feeding in mysticetes, emphasizing the early divergence of feeding habits. Apart from the feeding apparatus, however, basal odontocetes and mysticetes are much more similar to one another than are their modern descendants. Similarities include some evolutionary novelties (synapomorphies) of the skull which are not seen in archaeocetes. Thus, odontocetes and mysticetes are regarded as sister taxa, forming a clade termed crown Cetacea, or Neoceti, or Autoceta. Basal odontocetes and mysticetes also show marked similarities with archaic cetaceans (Archaeoceti), indicating an origin within the archaeocete family Basilosauridae (Fordyce and Muizon 2001; Gingerich 2005).

The qualifiers "crown" and "stem," below, are important in discussing cetacean systematics. The terms ensure that anatomists, molecular biologists, and paleontologists actually discuss the same taxonomic entities.

Figure 1  *W. H. Flower, an influential cetacean systematist from the later 1800s (from Cornish 1904).*

TABLE I
History of First Records for Names of Higher
Divisions of the Cetacea

| Author (date) | Formal name | Status |
|---|---|---|
| Linnaeus (1758) | Cete | A forgotten or little-used name for Cetacea |
| Brisson (1762) | Cetacea | First formal use of name |
| Gray (1864) | Mysticete | Formal precursor to Mysticeti |
| Gray (1864) | Denticete | A forgotten or little-used name for Odontoceti |
| Flower (1865) | Odontocete | Formal precursor to Odontoceti |
| Haeckel (1866) | Autoceta | A forgotten or little-used name for Neoceti |
| Flower (1867) | Odontoceti | First formal use of name |
| Flower (1867) | Mystacoceti | A forgotten or little-used name for Mysticeti |
| Cope (1869) | Mysticeti | First formal use of name |
| Gill (1871) | Zeuglodontia | A forgotten or little-used name for Archaeoceti |
| Flower (1883) | Archaeoceti | First formal use of name |
| Fordyce and Muizon (2001) | Neoceti | Formalization of a name used by J.G. Mead in an oral paper delivered at North American Paleontological Convention, 1996, Washington, DC |

Based partly on Rice (1998).

## II. Changing Concepts of Names

The name Cetacea was first used in a modern sense by Brisson in 1762 for genera and species of living whales, dolphins, and porpoises. Until the mid-1800s, high-level classification was based on superficial features, with no implication that patterns amongst living cetaceans had arisen by evolution. In the 1860s, W. H. Flower (Fig. 1), who was perhaps the first cetologist to use evolutionary principles, was instrumental in establishing formal names for the living cetaceans: Mysticeti and Odontoceti (Table I). Flower implied that these were real groups (in modern terms, clades). The discovery of fossils broadened the concept of Cetacea in the earlier 1800s. Initially, most fossils were recognized as related to living species, and modern generic names (e.g., *Delphinus* and *Balaena*) were often used for such material. The discovery of the archaic Eocene whale *Basilosaurus* in the 1830s eventually led Flower in 1883 to name another formal group of cetaceans, the extinct Archaeoceti. Thus, the concept of Cetacea was expanded to include three suborders, one extinct (Archaeoceti), and two living (Odontoceti and Mysticeti) (Fig. 2).

## III. The Monophyly of Odontoceti

Odontocetes include 74–75 living species in the families Physeteridae, Kogiidae, Ziphiidae, Platanistidae, Delphinidae, Phocoenidae, Monodontidae, Iniidae, Pontoporiidae, and Lipotidae. Phylogenetic analyses based on molecules, anatomy, and fossils indicate that the Odontoceti is monophyletic (Nikaido *et al.*, 2001; Geisler and Sanders, 2003). Strictly defined, the Odontoceti comprises the most-recent common ancestor of all living species, plus all the descendants of that ancestor. The oldest named odontocete is the early Oligocene *Simocetus rayi* (~32 million years), but older unnamed odontocetes are known. Odontocetes arose at the same time as their sister group, Mysticeti (Section IV), by 34.2 million years in late Eocene time. In practice, fossil and recent odontocetes are distinguished by osteological features (Fig. 3), particularly in the skull (Miller, 1923). For example, above the eye, a large supraorbital process in each maxilla rises posteriorly over frontal, forming a voluminous facial fossa in which open dorsal infraorbital foramina for nerves and blood vessels; in living species, this fossa forms the origin for the nasofacial muscles which manipulate diverticula or sacs in the soft nasal passages. In turn, the diverticula help to produce echolocation sounds. Where the rostrum passes into the facial fossa, a vertical antorbital notch transmits the facial nerve to the nasofrontal muscles. Anteriorly, in front of the bony nares, are premaxillary sac fossae, premaxillary foramina, and premaxillary sulci which, in living species, are implicated in sound generation in the nasal passages. Below the face, the infraorbital process is vestigial or absent, and the most posterior tooth lies far forward of the antorbital notch. In the ear region on the skull base, the periotic no longer contributes to the floor of the braincase. Finally, odontocetes have a middle sinus extending laterally from the ear near the jaw joint in the glenoid cavity.

N

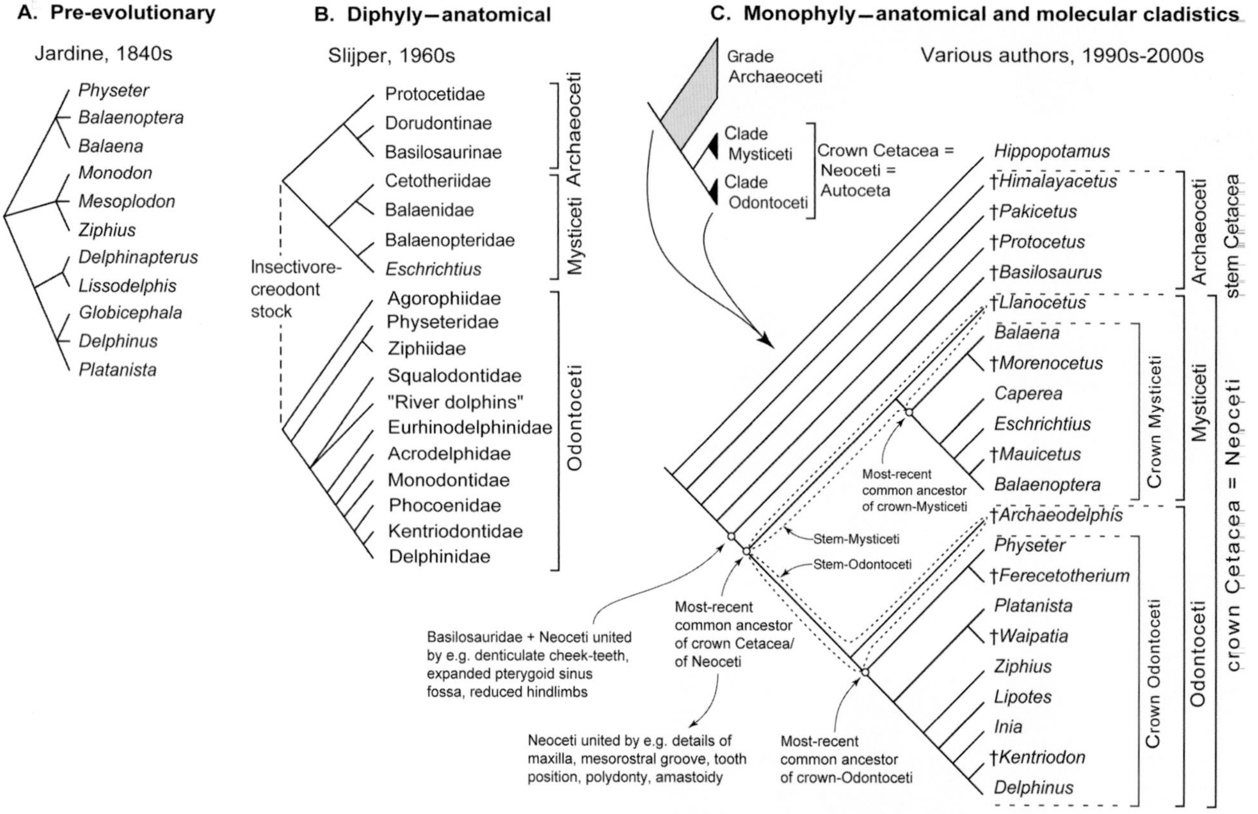

**Figure 2** *Changing concepts of the Cetacea. (A) Pre-evolutionary classification as used by Jardine and others, early to mid-1800s. Species are clustered on the basis of sometimes-superficial features. Genealogical relationships are not particularly implied. (B) Interpretation of cetacean diphyly, as used by Slijper (cladogram based on a phylogeny from Slijper 1979). (C) Composite simple cladogram showing current understanding of Cetacea. Crown-group Cetacea (Neoceti) has two sister taxa, Odontoceti and Mysticeti; included genera represent most of the recognized crown families; only a few fossil genera are included.*

## IV. The Monophyly of Mysticeti

Mysticetes include 13–14 living species in the families Balaenidae, Neobalaenidae, Balaenopteridae, and Eschrichtiidae. Phylogenetic analyses based on molecules, anatomy, and fossils indicate that the Mysticeti is monophyletic. Strictly, the Mysticeti comprises the most-recent common ancestor of all living species, plus all the descendants of that ancestor. The oldest named mysticete is the latest Eocene *Llanocetus denticrenatus* (about 34.2 million years); this fossil is also the oldest named species of Neoceti. Unlike living species, fossil mysticetes are not recognized by the presence of baleen for, though this probably occurred in most extinct species, it preserves rarely. Rather, osteological features (Fig. 3), particularly in the skull (Miller, 1923), distinguish the Mysticeti. The rostrum is relatively large, with thin edges and a smoothly concave and usually broad lower surface. The main bones in the rostrum (vomer, premaxilla, and maxilla) are generally sutured loosely with each other and, posteriorly, with the cranium. The lacrimal is also loosely sutured between the frontal and the preorbital part of the maxilla. Loose sutures between the feeding apparatus and cranium account for the common loss of the rostrum in fossil mysticetes; perhaps, such sutures function in skull kinesis during filter feeding. Ventrally, the toothless maxilla carries baleen; in some archaic forms, the maxilla may have teeth which lie well forward of the orbit. Posteriorly, the maxilla extends towards the orbit, forming a prominent infraorbital plate below the frontal. Finally, the mandibles are joined by ligaments at a short symphysis. Other putative diagnostic features of the skull are seen in most, but not all, mysticetes, as noted below for archaic forms.

## V. The Monophyly of Odontoceti and Mysticeti

In the later 1800s and indeed until the 1960s, the known archaeocetes and fossil mysticetes and odontocetes seemed rather divergent from one another. Slijper (1979) and other influential cetologists doubted a close relationship between the odontocetes and the other two groups, and were uncertain about mysticete origins amongst the archaeocetes. Thus, the two living groups of cetaceans were regarded as diphyletic, of different ancestry. They were sometimes classified as distinct orders.

From the 1970s to 1990s, several major advances overturned notions of diphyly and ultimately changed cetacean nomenclature. The fossil record of Eocene archaeocetes and of Oligocene odontocetes and mysticetes expanded markedly, helping bridge the structural and stratigraphic "gap" between the three groups. It became clear that evolution is not always slow and gradual, and that major structural change can occur in short geological intervals. Developments in deep ocean drilling led to much improved geological correlation, helping to date, and clarify, evolutionary sequences. Molecular and biochemical approaches to phylogeny indicated close relationship between odontocetes and mysticetes. The rise of cladistics (phylogenetic systematics) clarified many concepts of relationship and nomenclature.

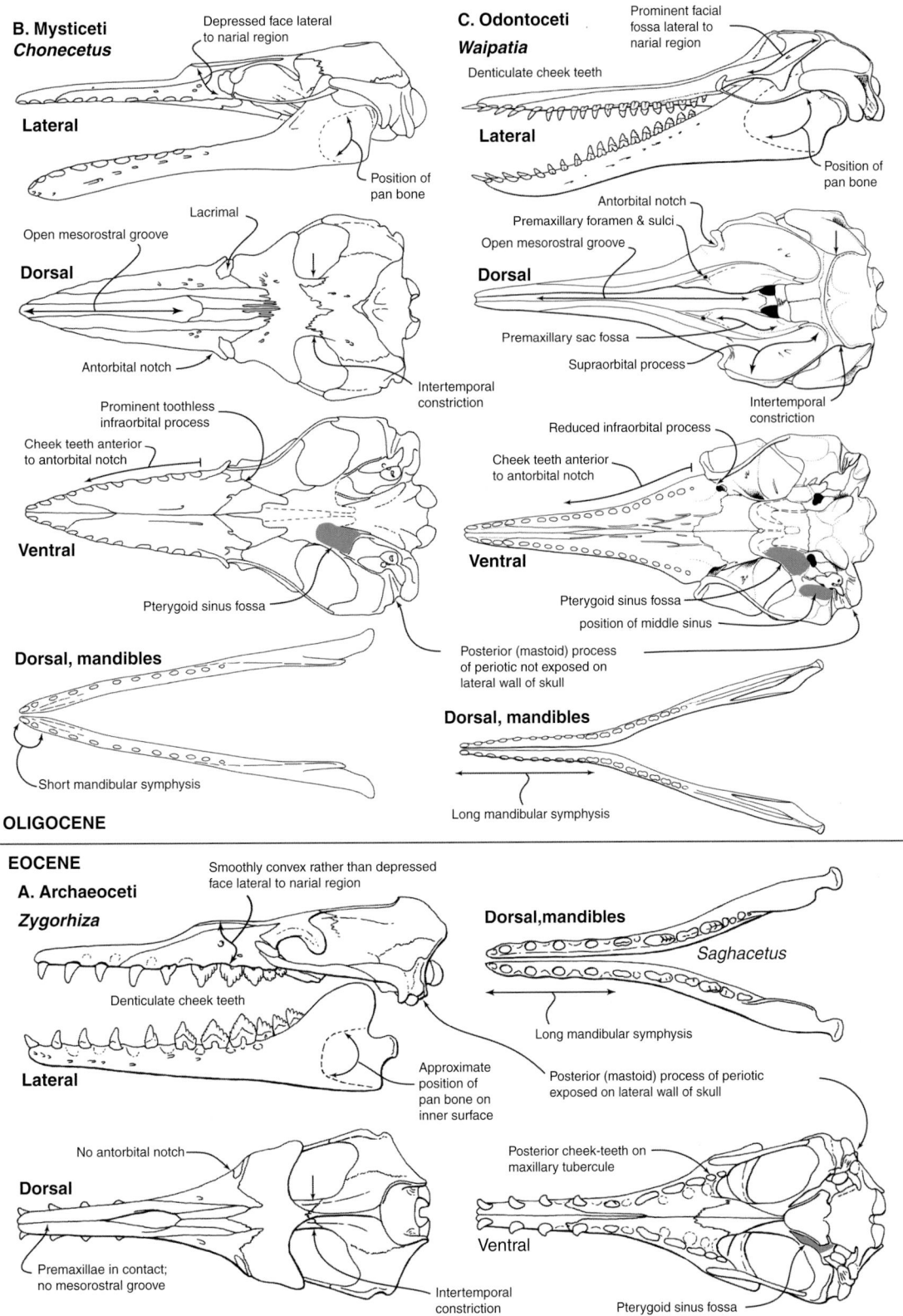

**Figure 3** *Morphological similarities and differences amongst later archaeocetes and basal odontocetes and mysticetes. Originally published as Figure 2, in Neoceti, R. E. Fordyce, Encyclopedia of Marine Mammals, W.F. Perrin et al. (eds), Copyright Elsevier 2002. (A) Archaeocete. Lateral, dorsal and ventral views of the archaeocete skull show the dorudontine Zygorhiza kochii (Basilosauridae: Dorudontinae; Priabonian, latest Eocene), based on Kellogg (1936). Dorsal view of the archaeocete mandibles shows the dorudontine Saghacetus osiris (Basilosauridae: Dorudontinae; Priabonian, latest Eocene), based on Stromer (1908). (B) Archaic mysticete. Lateral, dorsal and ven- tral views of skull, and dorsal view of mandibles show* Chonecetus goedertorum *(Mysticeti: Aetiocetidae; Chattian, Late Oligocene), based on Barnes et al. (1995). In life, the animal had teeth which were lost at burial. (C) Basal platanistoid odontocete. Lateral, dorsal and ventral views of skull, and dorsal view of mandibles show* Waipatia maerewhenua *(Odontoceti, Waipatiidae; Chattian, Late Oligocene), based on Fordyce (1994). Some teeth are in situ on the original fossil, but for simplicity are not shown in other than the lateral view.*

The Odontoceti and Mysticeti are now widely regarded as forming a clade Neoceti, equivalent to the Cetacea in the sense of Brisson and, indeed, of many modern systematists. There is no support for the 1990s molecular diphyly concept of sperm-whales-as-mysticetes. Strictly, the Neoceti is the crown Cetacea; it includes all descendants, living and extinct, of the most-recent common ancestor of Odontoceti and Mysticeti.

Odontocetes and mysticetes do share bony features not seen in archaeocetes, supporting their sister-group relationship. In both, the posterior of the maxilla is at least slightly concave, rather than smoothly convex, and carries multiple dorsal infraorbital foramina that open dorsally rather than anteriorly. On the rostrum, an open mesorostral groove extends far anteriorly, so that the premaxillae have little or no contact in the midline. The posterior-most teeth in odontocetes and toothed mysticetes lie anterior to the antorbital notch. Most (not all) basal species are polydont, with more than the usual mammalian number of cheek-teeth, and a tooth succession is unknown. Compared with basilosaurid archaeocetes, the zygomatic process of squamosal is more robust and anteriorly produced, with a more-delicate jugal. Finally, odontocetes and mysticetes are amastoid, with the posterior (mastoid) process of the periotic not exposed laterally on the skull wall.

Odontocetes and mysticetes are often identified as having a "telescoped" skull, in which bone positions have moved dramatically relative to familiar mammalian landmarks such as the nose and the eye (Miller, 1923). However, "telescoping" is a wide-ranging term applied to at least four different functional shifts involving both the facial region and braincase. It should not be cited to support the monophyly of odontocetes and mysticetes.

## VI. Primitive Features in Basal Odontocetes and Mysticetes

Some early fossil odontocetes and mysticetes have features similar or even identical to those seen in some basilosaurid archaeocetes (Fig. 3) (Fordyce and Muizon, 2001). Examples of such fossils include, amongst odontocetes, *Xenorophus* and *Archaeodelphis*, and amongst mysticetes, the Aetiocetidae, *Mammalodon*, *Janjucetus*, and *Llanocetus*. These examples are stem, rather than crown, members of Odontoceti and Mysticeti respectively. The most-obvious basilosaurid-like features of archaic odontocetes and mysticetes are the prominent intertemporal constriction, formed by elongate parietals dorsally on the braincase, and heterodont teeth. In all, the anterior teeth have single roots and simple crowns clearly distinct from two- or three-rooted cheek-teeth with complex denticulate crowns. Multiple denticles on the crown are an evolutionary novelty linking basilosaurids, odontocetes, and mysticetes. Further, the posterior mandibular cheek-teeth in archaic odontocetes have a distinctive anterior vertical groove, as in basilosaurids. (The loss of the last upper molar, M3, in basilosaurids, has been used to dismiss a basilosaurid origin for odontocetes and mysticetes. However, a widely variable tooth complement in the latter, and the likelihood that polydonty involved increase in the number of mid- to posterior cheek-teeth would allow a basilosaurid origin.)

Other parts of the feeding apparatus are revealing. In basilosaurids and basal odontocetes and mysticetes, the mandible has a large mandibular fossa (reduced in more-crownward mysticetes), and the temporal muscle has a distinct vertical origin on the frontal. (This origin changes dramatically in most odontocetes, becoming overridden by facial bones, and in mysticetes, migrating over the orbit.)

In all groups, the foramina in the orbit are not tightly clustered (as in living species) but are scattered anteroposteriorly. Ventrally,

in the skull base, an enlarged subspherical pterygoid sinus fossa is formed by alisphenoid and pterygoid. Such a fossa is absent in Protocetidae and more basal archaeocetes, and becomes more elaborate in more crownward Odontoceti and Mysticeti. Common features in the ear region include a lack of fusion between the periotic and the tympanic bulla (fusion occurs in later mysticetes), laterally compressed processes on the periotic (becoming more inflated in many later odontocetes), and a rather low squat tympanic bulla (becoming smoothly rounded in later mysticetes, but more elevated and delicate in later odontocetes). Other basilosaurid features in stem Neoceti include persistent postparietal foramina in the braincase, and prominent exoccipital condyles.

## VII. The Paraphyly of Archaeocetes

Because odontocetes and mysticetes (=crown Cetacea) arose from amongst basilosaurid archaeocetes (Uhen, 2004; Gingerich, 2005), the suborder Archaeoceti is paraphyletic. Archaeocetes form an artificial cluster of cetaceans that lack the features of Odontoceti or Mysticeti. Many cladists would not recognize archaeocetes as a formal group and would use the term "stem Cetacea" as an alternative to Archaeoceti. To recognize archaeocetes as cetaceans expands the concept of Cetacea beyond the crown group, and indeed expands it beyond the concept used by Brisson. It is possible that some archaeocete species, in the traditional sense, might be closest to odontocetes, whereas others may be closer to mysticetes.

## See Also the Following Articles

Basilosauridae ■ Cetacean Evolution.

## References

Barnes, L. G., Kimura, M., Furusawa, H., and Sawamura, H. (1995). Classification and distribution of Oligocene Aetiocetidae (Mammalia; Cetacea; Mysticeti) from western North America and Japan. *Isl. Arc* **3**, 392–431.

Cornish, C. J. (1904). "Sir William Henry Flower, K.C.B., F.R.S., LL.D., D.C.L. Late Director of the Natural History Museum, and President of the Royal Zoological Society. A Personal Memoir." MacMillan, London.

Flower, W. H. (1883). On the arrangement of the orders and families of existing Mammalia. *Proc. Zool. Soc. London* **1883**, 178–186.

Fordyce, R. E. (1994). *Waipatia maerewhenua*, new genus and new species (Waipatiidae, new family), an archaic Late Oligocene dolphin (Cetacea: Odontoceti: Platanistoidea) from New Zealand. *Proc. San Diego Mus. Nat. Hist.* **29**, 147–176.

Fordyce, R. E., and Muizon, C. de. (2001). Evolutionary history of whales: A review. In "Secondary Adaptation of Tetrapods to Life in Water. Proceedings of the International Meeting, Poitiers, 1996" (J.-M. Mazin, and V. de Buffrenil, eds), pp. 169–234. Verlag Dr Friedriech Pfeil, München.

Geisler, J. H., and Sanders, A. E. (2003). Morphological evidence for the phylogeny of Cetacea. *J. Mammal. Evol.* **10**, 23–129.

Gingerich, P. D. (2005). Cetacea. *In* "Placental Mammals: Origin, Timing, and Relationships of the Major Extant Clades" (K. D. Rose, and J. D. Archibald, eds), pp. 234–252. Johns Hopkins University Press, Baltimore.

Kellogg, A. R. (1936). A review of the Archaeoceti. *Carnegie Inst. Washington Publ.* **482**, 1–366.

Miller, G. S. (1923). The telescoping of the cetacean skull. *Smithson. Misc. Collect.* **76**, 1–70.

Nikaido, M., Matsuno, F., Hamilton, H., Brownell, R. L., Cao, Y., Ding, W., Zuoyan, Z., Shedlock, A. M., Fordyce, R. E., Hasegawa, M., and Okada, N. (2001). Retroposon analysis of major cetacean lineages:

The monophyly of toothed whales and the paraphyly of river dolphins. *Proc. Natl. Acad. Sci. USA* **98**, 7384–7389.

Rice, D. W. (1998). "Marine mammals of the world. Systematics and distribution." Society for Marine Mammalogy, Lawrence.

Slijper, E. J. (1979). "Whales." Hutchinson, London.

Stromer, E. (1908). Die Archaeoceti des ägyptischen Eocäns. *Beiträge Zur Geologie Und Paläontologie Von Österreich-Ungarn* **21**, 106–178.

Uhen, M. D. (2004). Form, function and anatomy of *Dorudon atrox* (Mammalia: Cetacea): An archaeocete from the middle to late Eocene of Egypt. *Univ. Mich. Papers Paleontol.* **34**, 1–222.

# New Zealand Sea Lion
## *Phocarctos hookeri*

### Nicholas J. Gales

### I. Characteristics and Taxonomy

*Phocarctos hookeri* (Gray, 1844) is named in honor of Sir James Hooker, who was the botanist with the British Antarctic expedition of 1839–1843. It is monotypic and is one of the world's seven extant sea lions.

New Zealand sea lions, like all otariids (eared seals), have marked sexual dimorphism; adult males are 240–350 cm long and weigh 320–450 kg and adult females are 180–200 cm long and weigh 90–165 kg. At birth, pups are 70–100 cm long and weigh 7–8 kg; the natal pelage is a thick coat of dark brown hair that becomes dark gray with cream markings on the top of the head, the nose, tail, and at the base of the flippers. Adult females' coats vary from buff to creamy gray with darker pigmentation around the muzzle and flippers. Adult males are blackish-brown with a well-developed black mane of coarse hair reaching the shoulders (Fig. 1).

### II. Distribution and Abundance

The New Zealand sea lion is endemic to New Zealand and is one of the most regionally localized and rare of the world's pinnipeds. It is classified as a threatened species by the International Union for the Conservation of Nature (IUCN). Prior to human arrival in New Zealand, *P. hookeri* was more widespread and probably more abundant than today. While the pristine breeding range of *P. hookeri* included almost all New Zealand coastal, island, and subantarctic territory, a combination of subsistence hunting by Maori and commercial sealing by Europeans have virtually reduced the current breeding range to two groups of subantarctic islands. Currently about 86% of New Zealand sea lion pups are born at three sites in the remote Auckland Islands; the remaining pups being born at nearby Campbell island (Fig. 2). Occasionally, single pups are born at regular sea lion haul-out sites on the South Island, the most regular of which is at the Otago Peninsula. Total annual pup production has been monitored intensively since 1994. Typically 2800–3000 pups are produced, but since 2004 this has reduced to about 2500, possibly through a combination of disease events and incidental mortality in an adjacent squid fishery (Gales and Fletcher, 1999; Chilvers *et al.*, 2007). Estimates of absolute abundance based on a pup production of 2800 are 12,500 (95% CI 11,100–14,000). Although recent variations in pup production have been recorded, the population appears to have been stable for the past few decades.

### III. Ecology

New Zealand sea lions breed and haul out on a diverse range of terrestrial habitats, including sandy beaches, reef flats, grass and herb fields, dense bush and forests, and solid bedrock. Each site generally has easy access to relatively protected waters (Fig. 3). The sympatric New Zealand fur seal (*Arctocephalus forsteri*) selects rockier, more exposed sites, and the two species rarely interact on land. The marine habitats of New Zealand sea lions have only been described for lactating females from the Auckland Islands. Here they forage on primarily benthic and dermersal habitats in the waters of the adjacent Auckland Island shelf. Other age and sex classes disperse more widely; in particular, adult and subadult males, tagged at the Auckland Islands, have been seen at Macquarie Island and around the southern parts of New Zealand.

Little is known of the normal disease status of New Zealand sea lions. The highly localized distribution of this species makes it particularly vulnerable to the effect of epizootic disease, and indeed in January 1998 an unusual mortality event occurred. At least 53% (*n* = 1600) of the pups of the year died, as well as many juveniles and adults. The principal cause is thought to be the bacteria *Klebsiella pheumoniae*, and several smaller scale outbreaks have been reported since (Wilkinson *et al.*, 2006). The sea lion population was also found to have been exposed previously to phocine distemper virus, a virus that has resulted in many marine mammal deaths worldwide. An undefined suite of environmental factors (including a strong ENSO event, or El Niño) that stressed the sea lion population and decreased its immunity may have influenced these events.

Sharks are likely to be the most significant predator of sea lions; with recent and healed bite wounds being a feature of many animals at the Auckland Islands. Occasional visitors such as leopard seals (*Hydrurga leptonyx*) have been observed to eat small sea lions, and the killer whale (*Orcinus orca*) may also be a predator.

Sea lions at the Auckland Islands forage on a wide variety of prey, with benthic and pelagic organisms being represented. Thirty-three taxa have been identified from analyses of identifiable remains in scats and regurgitations, with fish comprising 59%, cephalopods 22%, and crustaceans 15% of the remains found. The six most abundant prey items [in decreasing order of abundance: opalfish (*Hemerocoetes* sp.), octopus (*Enteroctopus zelandicus*), munida (*Munida gregaria*), hoki

**N**

**Figure 1** *New Zealand sea lions.*

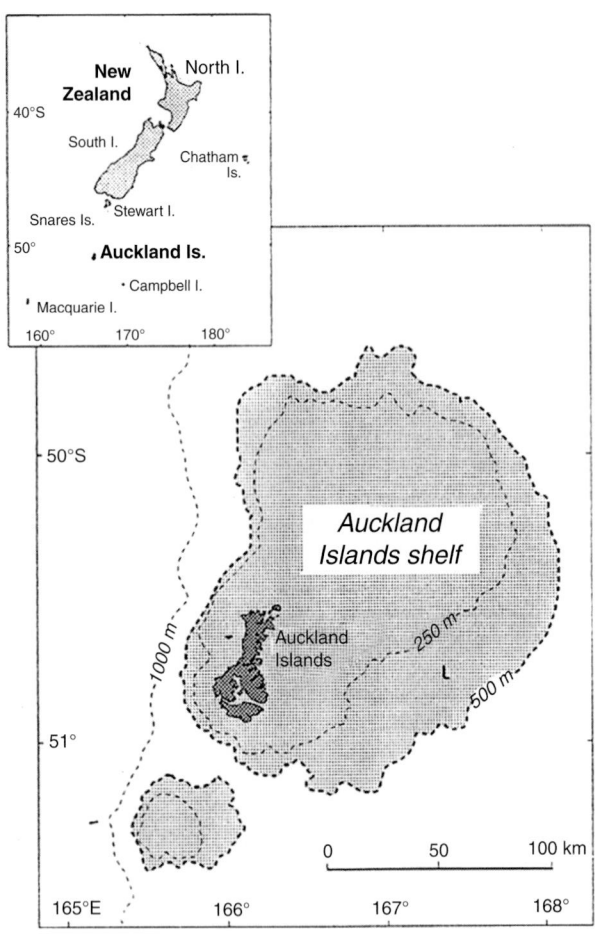
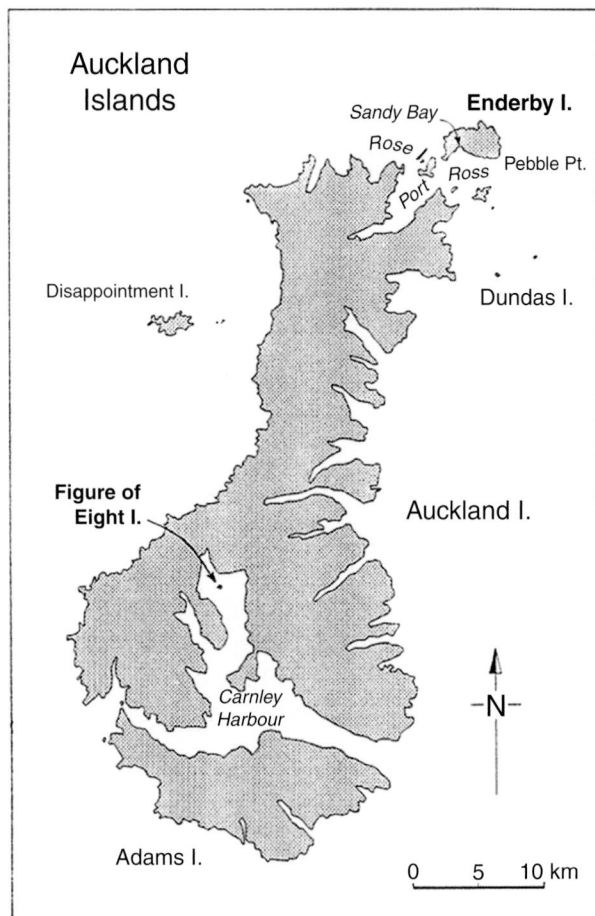

**Figure 2**　*Distribution of the New Zealand sea lion.*

**Figure 3**　*New Zealand sea lion rookery at Enderyb Island.*

(*Macruonus novaezelandiae*), oblique-banded rattail (*Coelorhynchus aspercephalus*), and salps (*Pyrosoma* sp.)] accounted for 90% of the total prey items. The diet of male sea lions on the Otago Peninsula has been found to represent a similar range of prey. Given the problems of bias associated with quantifying diet from scat and regurgitate analysis, further work utilizing newer, more precise tools such as genetic and fatty acid analyses would be most instructive.

## IV. Behavior and Physiology

New Zealand sea lions are among the deepest and longest diving of the otariids. At sea they dive at a mean rate of 7.5 dives/h and spend 45% of the time submerged. They undertake dives to a mean of 123 m (median 124 m, maximum >500 m) and spend an average of 3.9 min (median 4.33 min, maximum 11.3 min) on each dive more than 6 m. Almost half of the dives exceed the calculated *aerobic dive limit*, leading to a hypothesis that the dive behavior of *P. hookeri* reflects either successful physiological adaptation to exploiting benthic prey or a marginal foraging environment in which diving behavior is close to physiological limits (Gales and Mattlin, 1997). It has been shown subsequently that New Zealand sea lions have indeed equipped themselves physiologically for deep diving by having the largest blood volume of any otariid (Costa *et al.*, 1998). Behavioral adaptations, such as "burst and glide" diving, also appear to be used by this species to maximize the time available for foraging on the benthos, while still maintaining and effective energy budget. They have also been shown to be operating at what is likely to be close to their physiological maximum, as the gains in diving performance have been made with $O_2$ storage increase but not through a significant decrease at their at-sea metabolic rate (Costa and Gales, 2000). This degree of physiological and behavioral adaptation may well suggest a marginal habitat for this species (Chilvers *et al.*, 2006).

The breeding behavior of *P. hookeri* is typical of that of a polygynous otariid. Territorial males begin to assemble and defend physical

N

territories at breeding rookeries in late November. Pregnant females begin to arrive in early December and aggregate into harems of up to 25 animals attended by a single dominant bull. Challenges from peripheral males are regular and the tenure of territorial males is short. There are no published studies to elucidate the behavioral mechanisms by which female movement within and between the harems is determined. Females give birth soon after arrival; the pupping season lasting about 35 days. By mid-January most territorial bulls have departed, the harems break up, and females and pups disperse to occupy surrounding areas.

Postparturient females exhibit estrus 7–10 days after the birth of their pup and are mated by the territorial bull. Soon after, they depart on their first foraging trip. During the first few months of lactation trip durations average 1.7 days at sea, interspersed with an average of 1.2 days ashore feeding the pup. Unattended pups gather into groups during maternal absence. At least 10% of pups die during the pupping season. Although the causes of this early mortality have not been studied, it is thought to be caused principally by starvation, parasitism, trauma, and disease. Infanticide and cannibalism by bulls have been observed. A high frequency of cows simultaneously nursing more than one pup (6% of observations) has been reported for *P. hookeri* during the first few weeks of pup rearing. This fostering behavior may be associated with kin selection in small populations. Lactation lasts about 10 months (Gales, 1995).

## V. Life History

Females are thought to become sexually mature at 3 years or older and produce their first pup a year later. The duration of an assumed embryonic diapause has not been measured. The mean reproductive rate of females is 0.67. Males are reported to be sexually mature at 5 years but do not hold territories for a further 3–5 years. Mean adult survival is 0.81 and females have been recorded living to 28 years. The average age of reproductive females is about 11 years. These data show that New Zealand sea lions are among the slowest growing, slowest reproducing, and longest lived sea lion species.

## VI. Interactions with Humans

New Zealand sea lions are subject to incidental drowning in squid and other trawl fisheries that operate around the Auckland Islands. The number of sea lions killed in the squid fishery (estimated to range from 17 to 141 per year for the period 1988–2006) has been the cause of serious concern and has led to a number of management measures. These include the imposition of a 12 nautical-mile marine reserve around the Auckland islands in which commercial fishing is prohibited; the deployment of government observers on trawlers to record the incidence of marine mammal bycatch (7–32% of tows observed each year for 1988–2006; bycatch rate varied from 0.6% to 11.2% of tows during this period); a delay in the opening of the seasonal fishery until February 1 each year; a voluntary code of practice for the industry, which aims to reduce the chance of bycatch; and the imposition by the New Zealand government of a maximum number of sea lions that could be killed in any 1 year. Initially this was a maximum allowable level of fishing related mortality (MALFIRM), the reaching of which leads to the early closure of the fishery for that season. This MALFIRM (set at the beginning of each season on the basis of an approved model at about 60–80 sea lions) was exceeded each year between 1995 and 1998, and the fishery was closed early in the latter 3 years. In 2003 the MALFIRM was superseded by a Bayesian model; the Fisheries-Related Mortality Limit (FRML). This has allowed higher levels of sea lions deaths,

ranging from 97 to 150 each year and has not led to the closure of the fishery. This increase in allowable bycatch has coincided with a substantial decrease in pup production in the past few years. In efforts to mitigate the sea lion–fishery interaction, specially designed escape devices are also being tested and deployed.

Another identified danger to New Zealand sea lions is tourism (principally subantarctic, but also on the Otago Peninsula). These interactions are regulated via a limited-entry permit system on the subantarctic islands and by behavioral protocols for tourists here and elsewhere. Other impacts, such as pollution, entaglement, and direct killing, are not thought to be significant. There are no *P. hookeri* held in captivity. Under the auspices of a population management plan, the New Zealand government aims to monitor and research the sea lion population, mitigate threatening processes and remove the sea lion from its threatened status within 20 years (from 1999).

## See Also the Following Articles

Diving Physiology ■ Eared Seals ■ Management ■ Rookeries

## References

Chilvers, B. L., Wilkinson, I. S., Duignan, P. J., and Gemmell, N. (2006). Diving to extremes: Are New Zealand sea lions (*Phocarctos hookeri*) pushing their limits in a marginal habitat? *J. Zool. Lond.* **269**, 233–241.

Chilvers, B. L., Wilkinson, I. S., and Childerhouse, S. (2007). New Zealand sea lion, *Phocarctos hookeri*, pup production—1995–2005. *N.Z. J. Mar. Freshw. Res.* **41**, 205–213.

Costa, D. P., and Gales, N. J. (2000). Foraging energetics and diving behaviour of lactating New Zealand Sea lions *Phocarctos hookeri*. *J. Exp. Biol.* **203**, 3655–3665.

Costa, D. P., Gales, N. J., and Crcker, D. E. (1998). Blood volume and diving ability of the New Zealand sea lion, *Phocarctos hookeri*. *Phys. Zool.* **71**, 208–213.

Gales, N. J. (1995). Hooker's sea lion recovery plan (*Phocarctos hookeri*). Threatened Species Recovery Plan Ser. No. 17, New Zealand department of Conservation, Wellington.

Gales, N. J., and Mattlin, R. H. (1997). Summer diving behaviour of lactating New Zealand Sea lions, *Phocarctos hookeri*. *Can. J. Zool.* **75**, 1695–1706.

Gales, N. J., and Fletcher, D. J. (1999). Abundance, distribution and status of the New Zealand sea lion, *Phocarctos hookeri*. *Wildl. Res.* **26**, 35–52.

Wilkinson, I. S., Duignan, P. J., Grinberg, A., Chilvers, B. L., and Robertson, B. C. (2006). *Klebsiella pneumoniae* epidemics: Possible impact on New Zealand sea lion recruitment. *In* "Sea Lions of the World" (A. W. Trites, D. P. DeMaster, L. W. Fritz, L. D. Gelatt, L. D. Rea, and K. M. Wynne, eds), pp. 385–404. University of Alaska Fairbanks, Alaska.

N

# Noise, Effects of

BERND WÜRSIG AND W. JOHN RICHARDSON

When we humans submerge our heads, the ocean seems relatively silent. This misconception occurs because our ears are optimized to hear in air and have poor sensitivity in the much denser medium of water. In reality, the oceans are full of sounds. Natural sources of underwater sound include breaking waves and surf, rain striking the sea surface, ice cracking and groaning in the higher latitudes, and the distant rumble of storms and earthquakes. Besides these physical sources, there is also a rich biological repertoire. There are sounds of snapping shrimp, grunting fishes, squeaking and popping sirenians, and the amazingly varied vocalizations of pinnipeds and cetaceans. Walruses (*Odobenus rosmarus*) display by knocks and mews; bearded seals (*Erignathus barbatus*) emit elaborate trills during their breeding season; toothed whales often whistle, send bursts of staccato-like click trains, and echolocate; and large whales moan, groan, and in some cases sing for group cohesion, sexual displays, and communication (Tyack, 2000). Some researchers suspect that strong low-frequency sounds of certain baleen whales may also function as

active sonar, helping them to navigate across wide open ocean spaces or around ice, or to locate silent conspecifics.

Unfortunately, the industrialized world has created other sources of noise underwater (Fig. 1). There is motorized shipping, underwater blasting, and offshore drilling, dredging, and construction. These activities produce underwater sounds incidentally, not purposefully. Several other types of underwater sounds are created purposefully: fathometers and sonars of many types operating at frequencies ranging from very high to low; air gun pulses for oil and gas exploration; pingers used to locate underwater equipment and to alert marine mammals to the presence of fishing nets; electronic acoustic harassment devices to keep marine mammals away

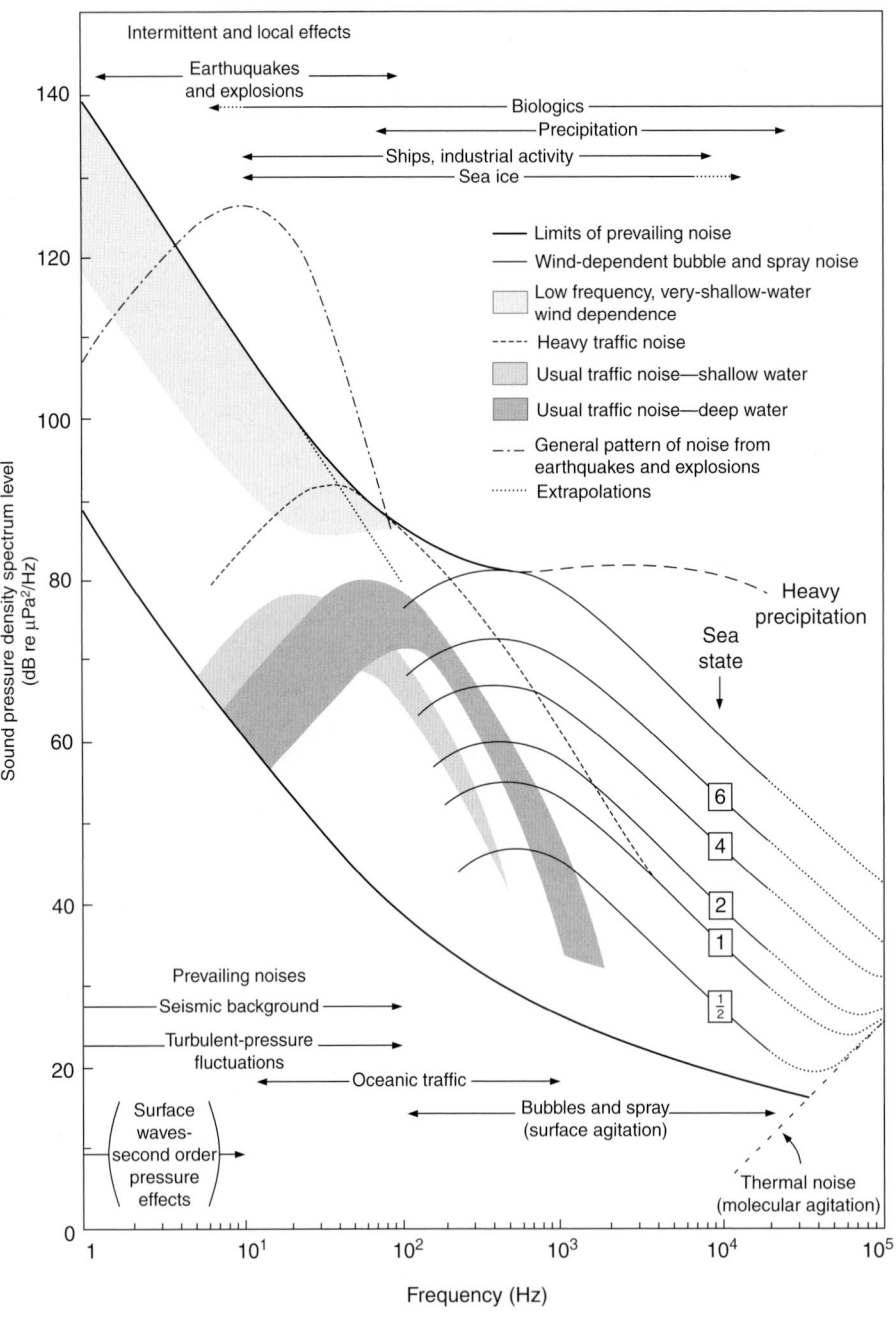

**Figure 1**    *General types of natural and human-made sounds in the world's oceans. Adapted from Wenz (1962).*

from mariculture facilities; and widely varying sounds used for ocean science measurements over short and long distances. Fish and marine mammals have evolved with the rich physical and biological cacophony of nature and are presumably well adapted to those sounds. However, most anthropogenic (human-generated) sounds first appeared in the past about 100 years, and in some parts of the world are increasing in intensity and geographical extent decade by decade (Gisiner *et al.*, 1999; Simmonds *et al.*, 2004; Jasny *et al.*, 2005).

Most marine mammals rely on underwater sound for communicating, finding prey, avoiding predators, and probably navigating. Other senses are available to them (sense organs, overview), but sound is the most important one at distances or in environments where the senses of touch, taste, and sight are not available. However, it is well known that pinnipeds (sea lions, fur seals, seals, and walruses), sirenians (manatees and dugong, *Dugong dugon*), and cetaceans (dolphins, porpoises, and whales) use sound both passively, when listening to the environment, and actively, when communicating. The toothed whales also echolocate to find prey, detect predators, and maneuver in the environment. It is unclear how much sea otters (*Enhydra lutris*) and polar bears (*Ursus maritimus*) rely on sounds underwater.

The acoustic frequencies that are most important vary with the type of marine mammal. See also the chapter on (sense organs, overview). Baleen whales tend to use lower frequencies of sound: predominantly below 1 kHz and reaching down into the infrasonic (below human hearing; <20 Hz) range in fin and blue whales, *Balaenoptera physalus* and *B. musculus*. Toothed whale communication is mainly above 1 kHz, and the echolocation sounds of most species are at very high frequencies, some as high as 150 kHz. The latter sounds are far above the frequency range audible to humans ("ultrasonic"). Pinnipeds and sirenians tend to be intermediate between baleen and toothed whales with respect to the frequencies of their calls (mainly at 0.5–10 kHz) and optimum hearing range. The optimum frequencies for pinnipeds and sirenians are similar to those for humans.

Although sounds of sirenians, some pinnipeds, and some small odontocetes are weak and audible only within a few tens of meters, most cetacean sounds are much stronger (Richardson *et al.*, 1995). Received sound levels and maximum detection distances depend not only on the strength of the sound at the source ("source level"), but also on the frequency, on physical properties of the environment through which the sound propagates, and on background noise levels. Nevertheless, smaller dolphins and porpoises can be heard to several hundred meters. Killer whale (*Orcinus orca*) "screams," social sounds of pilot whales (*Globicephala* spp.) and the staccato click sounds of sperm whales (*Physeter macrocephalus*) reach to several kilometers. Communication sounds of many odontocetes and mysticetes have source levels of 160–180 dB re 1 μPa at 1 m distance (read as "decibels relative to one micropascal at one meter distance"; see noise level descriptions). The clicks of bottlenose dolphins (*Tursiops* spp.) and sperm whales can have much higher peak levels, more than 220 dB in the same units (Au *et al.*, 2000; Møhl *et al.*, 2000). Echolocation clicks of some smaller odontocetes have lower peak levels (Würsig and Au, 2004), and all odontocete echolocation pulses are very brief and do not contain much energy despite the high peak levels sometimes achieved. In a few species of seals, e.g., bearded and Weddell seals (*Leptonychotes weddellii*), the males appear to use complicated tonal sounds as advertising displays to warn off other males (and possibly to attract females); these sounds can also be quite strong. These seals, which defend underwater territories (or "maritories"), are exceptional among the pinnipeds, however.

Anthropogenic sounds can result in a variety of effects whose consequences to marine mammals can range from nil to severe, depending on the type and received level of the sound:

*Tolerance (no overt response).* Mammals exposed to audible levels of human-made sounds often exhibit no obvious responses, continuing their normal activities without moving away. A marine mammal hearing a sound does not always react overtly to it. However, responses can be subtle so a detailed behavioral and physiological study is needed before it is legitimate to conclude that there is no overt response. Marine mammals sometimes tolerate noise in order to remain in a preferred location, such as a feeding area, even if the noise is strong enough to cause reactions when the same species is engaged in other activities.

*Changes in behavior or activity.* Alterations in behavior are common when marine mammals are exposed to human-made sounds. Sometimes the effects are subtle, discernible only by detailed observation and statistical analysis, e.g., changes in surfacing/respiration/dive cycles. More conspicuous effects include changes in activity, e.g., from resting or feeding to alert, facing toward the noise source, and so on.

*Avoidance reactions.* Upon exposure to strong man-made sounds, marine mammals engaged in feeding, social interactions, or other "normal" activities often interrupt those activities and swim away (or occasionally toward). When a noise source operates in the path of migrating whales, they typically deflect a few degrees off their "normal" course and swim to one side or the other side of the noise source.

*Masking.* Masking is the process whereby sounds of interest to a listener are obscured by interfering sounds. If those sounds are important to the listener (e.g., for breeding or predator avoidance), continuous, and prolonged, masking could have a serious effect (Richardson *et al.*, 1995). However, masking is likely to be a problem only if a human-made sound is persistent and at a similar frequency to the sound of interest to the animal, and then mainly if the two sounds are received from similar directions.

*Hearing impairment.* Animals (including humans) exposed to strong sounds can incur a reduction in their hearing sensitivity. This impairment is often temporary, provided the sound levels are not too high or too prolonged. However, repeated exposure to strong sounds, and even a single brief exposure to an extremely strong sound (e.g., nearby explosion), might cause permanent hearing impairment. Temporary threshold shift (TTS) has been demonstrated in captive odontocetes and pinnipeds exposed to strong sounds. Except for nearby explosions, there is as yet no demonstrated case of noise-induced permanent threshold shift (PTS) in a marine mammal.

*Nonauditory physiological effects and stress.* In humans, exposure to very strong underwater sounds can, under special circumstances, cause resonances in lung cavities and other types of nonauditory physiological effects. Chronic exposure to strong noise may also at times cause stress reactions. These phenomena are almost entirely unstudied in marine mammals.

*Mortality of beaked whales.* Several mass strandings and deaths of beaked whales have occurred in various regions upon exposure of these little-known animals to sounds from mid-frequency naval sonars. The involvement of sonar in causing mass strandings of beaked whales in the Mediterranean Sea, Canary Islands, Bahamas, etc., is generally agreed. However, the specific reason(s) for these deaths are controversial. Suggested mechanisms include panic reactions leading to stranding, a form of "the bends" when normal dive patterns are disrupted, resonance in body cavities, and auditory injury (Fernández *et al.*, 2005; Cox *et al.*, 2006).

The term "noise" indicates an unintentional or unwanted sound, possibly disagreeable or noxious. We tend to think of noise as sound we do not like, but this is a subjective impression; a sound that is noise to one person or animal may be an important signal (or music) to another. For example, the songs of humpback whales (*Megaptera novaeangliae*) are undoubtedly important to them. We do not know whether these strong and—in some areas—seasonally persistent sounds are regarded as "music" or as noise by pilot whales and bottlenose, spinner (*Stenella longirostris*), and spotted dolphins (*S. attenuata* and *S. frontalis*) that are attempting to communicate in the presence of this background of whale song. Singing baleen whales and listening/communicating toothed whales and dolphins have co-evolved, and depend primarily on sounds at different frequencies. We expect that neither group is overtly bothered by the sounds of the other. However, industrial sounds are likely to be perceived by marine mammals as noise, especially when the frequencies of those sounds overlap with frequencies used by the mammals. Nevertheless, some anthropogenic noises may themselves be important cues to mammals: a boat is approaching, take evasive action; a seismic vessel is active, do not get too close, etc. Thus, noise can have an important signaling function. In human terms, it is good that we hear an approaching truck or train if we are in its path.

Sound strength can be measured near the source or at some greater distance. By convention, source levels usually are reported for a standard distance of 1 m. The received sound level tends to diminish with increasing distance—often rapidly when close to the source, and then more gradually at longer distances. In interpreting quoted sound levels, it is essential to know whether the value is a source or received level and, for the latter, the distance from the source. The received levels 10, 100, and 1000 m from a source are often about 20, 40, and 50–70 dB, respectively, less than the corresponding source level. However, the initial decrease with distance (close to the source) will be less rapid if the source is large, like a ship or an array of many airguns. In that case, the source level is a somewhat artificial construct, and there is no location in the water—even immediately adjacent to the source—where the sound level is as high as the nominal source level. A further complication is that sound levels are measured in many different and sometimes difficult-to-convert units. Decibel (dB) values are meaningless unless the reference level is also specified. This is typically $1\,\mu\text{Pa}$ for underwater sound and $20\,\mu\text{Pa}$ for airborne sound, but other reference pressures are sometimes used, especially in the older literature. Even when the distance from source and the reference level are stated, other complications can make it difficult or impossible to compare acoustical measurements from different studies. In addition, simple comparisons of levels in water vs air are usually of doubtful validity given a variety of complications.

The levels quoted in this chapter concern underwater sound measured in dB re $1\,\mu\text{Pa}$ and, except for values shown in Fig. 1, are overall broadband levels (i.e., including sound components at a wide range of frequencies). In the case of source levels, the levels are standardized to a distance of 1 m, in which case the units are dB re $1\,\mu\text{Pa} \bullet \text{m}$.

One way of categorizing anthropogenic noises is whether they are transient or continuous. Sounds from explosions and most seismic surveys, sonars, and ocean acoustic studies are inherently transient. The source (and received) levels of some of these transient sources can be high (Richardson *et al.*, 1995). However, because these sounds are brief and intermittent, the average sound level over an extended period is lower than the peak level. Sounds from moving ships and aircraft are continuous, but from the perspective of an individual animal, these sounds will normally be transient. Similarly, an animal traveling past a stationary source of continuous sound will also perceive that sound as a transient. Transient sounds are likely to have less effect on animal behavior and hearing as compared with truly continuous sounds having a similar received level.

The strongest transient sounds in the sea come from explosions, which produce shock waves as well as very strong (but brief) underwater sounds. The shock waves from explosions can injure, stun, and even kill nearby fishes, sea turtles, humans, and marine mammals. The distances within which death or injury will occur depend on the size and depth of the explosion, the type of animal, and other factors. Aside from explosions, some of the strongest sources of underwater sound are pulsed sounds from seismic surveys (typically using air gun arrays to produce noise pulses, mainly at low frequency), military search sonars operating at medium and low frequencies, and some types of pile-driving. Air gun arrays and sonars can produce transient sounds with effective source levels of 230 dB re $1\,\mu\text{Pa} \bullet \text{m}$ or more. Sounds received from these sources diminish rapidly with increasing distance, but often are detectable as much as 100 km away. Under special conditions, they can be detected at considerably greater distances, though whether the weak sounds detected far away from these sources have any biological effects is largely unknown. There are many other types of transient anthropogenic sounds with a wide variety of levels and characteristics.

Continuous sounds are caused by sources such as drillships, dredging, and vessels of all sizes underway, with the aforementioned caveat that sounds from a specific moving ship will be transient for a given mammal. Of these sound producers, large tankers, container ships, and icebreakers while working in ice are among the strongest sources. Their overall source levels (i.e., including all frequencies) are on the order of 200 dB or more. Although that level is not as high as for some transient sources, it is sustained without gaps. Much of this sound energy is at infrasonic frequencies (below 20 Hz), which are important to at least some of the baleen whales, but may be irrelevant to odontocetes that have little or no hearing sensitivity at those low frequencies. Shipping is concentrated along defined shipping lanes, where sound levels will wax and wane as individual ships pass. However, at least in the Northern Hemisphere, the overall din of vessels pervades not only the lanes themselves, but also the major parts of the ocean basins involved (Fig. 2).

The strongest components of sound from many of the major anthropogenic sources are below 1 kHz, in the same frequency band as is used by most of the baleen whales for their communication calls. For seismic surveys and large ships, most sound energy is below 200 Hz. A recently acknowledged source of strong low-frequency sound is the low-frequency active (LFA) sonar used to detect quiet submarines. The US Navy's LFA projectors produce a beam of intense sound (source level up to about 240 dB re $1\,\mu\text{Pa} \bullet \text{m}$) at frequencies in the 100- to 500-Hz band. However, most of the strongest military sonars, some with source levels up to at least 235 dB, operate in the mid-frequency (a few kilohertz) range. Likewise, the sounds from outboard engines operating at high speed, snowmobiles traveling on ice, and acoustic harassment devices at mariculture facilities are in the low kilohertz range. Sounds at medium and especially at low frequencies tend to propagate for longer distances than sounds at high frequencies (e.g., >10 kHz). At progressively higher frequencies, sound is absorbed more rapidly into seawater. Hence, high-frequency pinger and sonar sounds (including dolphin clicks) generally diminish to low or undetectable levels within 1 km or less and thus do not present potential problems except at close range.

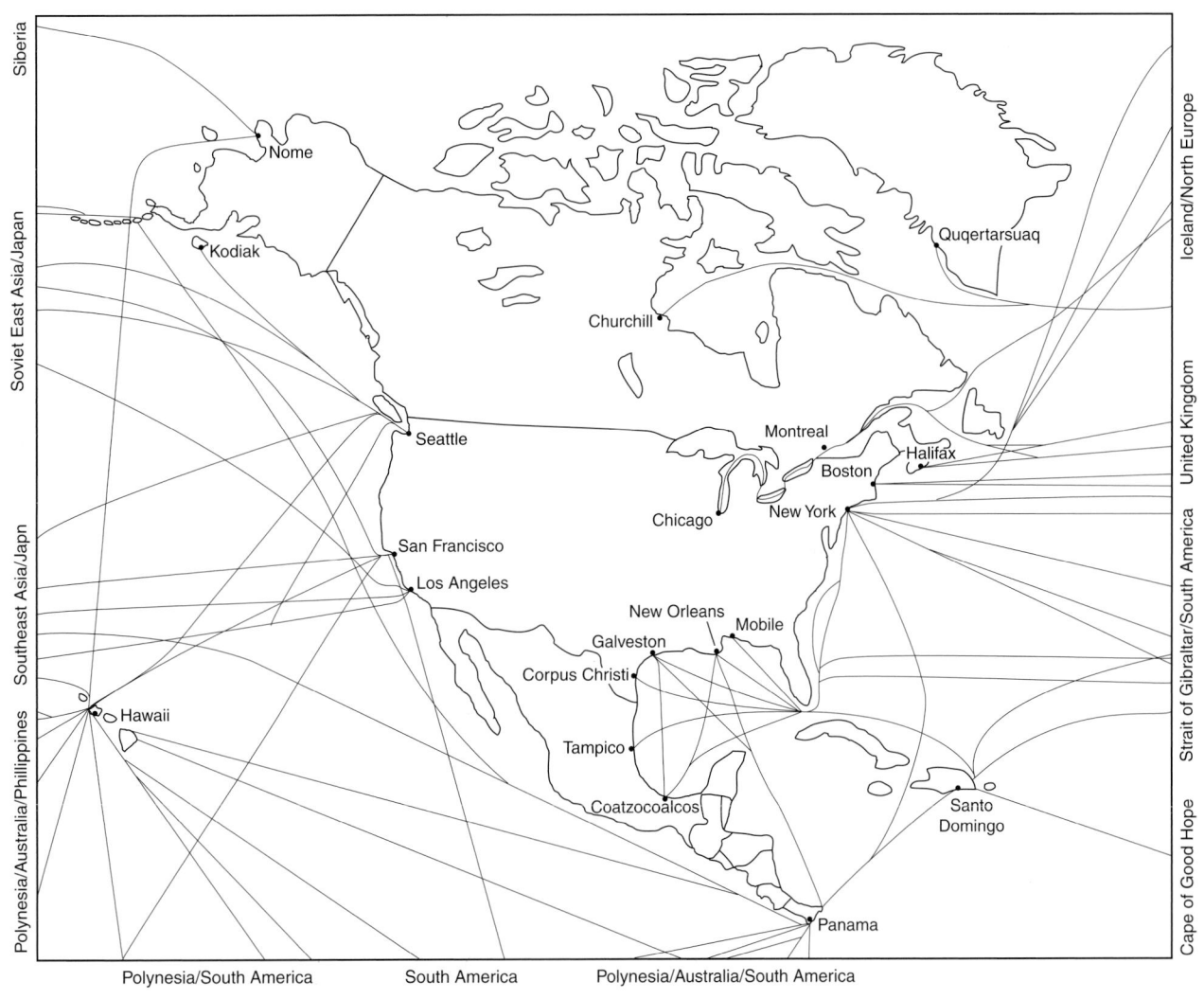

**Figure 2**  *Major international shipping lanes in North American waters. From Jasny* et al. *(1999) and US Department of State.*

N

Because most, if not all, marine mammals rely on underwater sound for various purposes, any strong anthropogenic sounds at relevant frequencies might have an effect. Animals attuned to different sound frequencies will be most affected by different types of sounds. Mid- and high-frequency sounds as produced by small vessels and mid- and high-frequency sonars are likely to affect dolphins, porpoises, pinnipeds, and sirenians. In contrast, large whales are more susceptible to lower frequency sounds such as those from large vessels, drilling operations, and sounds specifically designed to propagate long distances through the water and/or bottom, such as LFA sonar, marine seismic exploration, and large-scale ocean tomography studies.

Although it was once thought that nearshore animals are likely to be most vulnerable to acoustic effects because of the concentration of industrial activities in nearshore waters, this is no longer as clear. Marine mammals on the high seas can be affected as well, given the occurrence in deep water of shipping, military operations, acoustic oceanography projects, and (increasingly) oil industry operations. In general, it is probable that mammals of the southern oceans are still affected less often and less intensively by anthropogenic sounds as compared with those in the Northern Hemisphere, simply because of the preponderance of human occupation, industry, and shipping

in the north. However, anthropogenic underwater noise is now everywhere, at least some of the time, and has the potential to affect all animals that can hear it and that need to communicate through it.

Most, if not all, marine mammals have evolved special adaptations to hear well underwater (Au *et al.*, 2000). However, the adaptations that make their ears sensitive to pressure fluctuations in the water may also increase the risk of damage from exposure to strong waterborne sound and shock waves.

Sounds that are not strong enough to cause profound physical injury or outright death may nonetheless impair hearing, resulting in either temporary or permanent threshold shifts (TTS and PTS, respectively). TTS occurs in humans and animals subjected to intense sounds for short periods of time or to less intense sounds for longer times. The temporary loss of hearing acuity by humans listening to a rock concert is an example. Mild TTS has been demonstrated in common bottlenose dolphins (*Tursiops truncatus*) and beluga whales (*Delphinapterus leucas*) exposed to a single 1 sec pulse of strong sound (192–201 dB) and in dolphins and various pinnipeds exposed to lower sound levels for periods as long as 50 min (Kastak *et al.*, 2005; Finneran *et al.*, 2005). In most experiments, hearing thresholds returned to pre-exposure levels within about 12 h, and none of these exposures caused long-term hearing impairment.

Temporary hearing impairment could be deleterious to animals that rely strongly on their abilities to detect sound. In addition, data on the sound levels and durations at which TTS begins are useful in identifying situations when there is concern about permanent hearing damage (PTS). However, for marine mammals it is not known how much additional exposure (higher levels and/or longer durations) would be necessary before PTS would occur. This is an important data gap. In general, permanent damage typically results from a loss of sensory hair cells within the inner ear. The loss is most pronounced in the part of the inner ear responsible for detecting sound at or near the frequency of the injurious stimulus. In terrestrial mammals, lost hair cells do not grow back, and this is likely the case for marine mammals as well. (Interestingly, at least partial regeneration occurs in birds and fishes.) Older bottlenose dolphins in captivity are known to have reduced hearing sensitivity (especially at the higher frequencies), presumably at least in part because of the cumulative effects of sound (Ridgway and Carder, 1997). This loss is similar to the progressive hearing loss in humans, or similar to presbycusis.

The initial TTS results from small odontocetes and pinnipeds confirm that sound levels necessary to cause TTS are correlated with the duration of exposure. More data are needed to better quantify this relationship and to determine the TTS thresholds for repeated pulsed sounds such as seismic and sonar pulses. Sound levels necessary to cause TTS in baleen whales and sirenians are entirely unknown. In addition, as discussed earlier, it is not known how much additional exposure (above levels causing onset of TTS) any marine mammal can tolerate before there is PTS, but recent work shows that 40 dB of shift is recoverable in dolphins.

Even less is known about potential effects of strong noises on non-hearing physiology than on hearing. Human and animal studies show that strong sounds can affect the vestibular system (and thus sense of balance), air sinuses and adjacent tissues, neural transmission (skin tingling in divers), and reproductive functions. Studies of some terrestrial animals exposed to strong noise have shown reduced sperm production, menstrual irregularities, abortions, and stillbirths. Most of these drastic effects come from intense low-frequency sounds and attendant physical vibrations of body tissues. It is not known whether there are circumstances in which low-frequency underwater sounds from LFA sonars, seismic exploration or ocean tomography studies could elicit similar effects in marine mammals. If this ever occurs, these effects would probably be limited to animals close to the sound source.

It has been suggested that strong and prolonged sound could cause bubbles to form in blood or tissues, and that animals might succumb to a lodging of these bubbles in the brain and elsewhere (Crum and Mao, 1996). This situation is analogous to bubble formation in human divers when they return too rapidly from depth—the condition known as the bends. It is questionable whether this phenomenon would occur under realistic conditions (Cox et al., 2006). However, there has been recent speculation that a different phenomenon, alterations of surfacing-dive cycles of deep-diving species like beaked whales, might lead to accumulation of gas bubbles and cases of "the bends" (Cox et al., 2006).

Stress is one possible outcome of exposure to sounds that are disturbing to animals. Stress could, in theory, inhibit normal social interactions, feeding, reproduction, longevity, and a host of other important functions (Curry, 1999). However, noise-induced stress is not well understood even in humans and terrestrial mammals, and there are few data on noise-induced stress in marine mammals. Research is needed.

Short-term behavioral and avoidance reactions to noise have received more study than damage to hearing and other physiological effects, to the point that there are now at least some data on noise-induced disturbance in every group of marine mammals. For most marine mammals, it is relatively easy to observe some aspects of their distribution and behavior, as they need to come to the surface to breathe. Nevertheless, much of their time is spent below the waves, and our brief glimpses of them at the surface can present a biased view. Further, short-term reactions (or lack thereof) are not necessarily a good indicator of long-term effects, and the latter have rarely been studied directly. This section does not attempt to review the numerous studies on behavioral and distributional responses to underwater noise; for more details, see Richardson et al. (1995) and Southall et al. (in press).

Mammals often appear to become alert to novel sounds when they are received at low levels, but this alertness quickly wanes if there is no danger connected with the sound. Thus, the droning of ships or drilling platforms in the distance often appears to be ignored due to habituation or sensory adaptation. Some mammals may avoid the immediate area around the noise source, but remain within the area where the sounds are at least faintly audible. When sounds are tolerated, it is possible (though unproven) that they may elicit stress or other unseen physiological reactions.

Marine mammals often react more dramatically when received sound levels are high (indicative of a strong or nearby source) or when they are increasing (indicative of an approaching source). Bowhead whales (*Balaena mysticetus*) migrating toward a drillship or a marine seismic operation typically deflect their course so that their closest point of approach as they pass the noise source is at least 20 km away. At that distance, the received level of the strongest air gun pulses was near 130 dB, averaged over pulse duration. Migrating gray whales (*Eschrichtius robustus*) show similar deflections of their migration route as they approach a simulated seismic operation or LFA sonar, but they seem to tolerate higher sound levels than migrating bowheads. These deflections by migrating bowhead and gray whales are not a sudden fright reaction, but an edging away—analogous to a person walking along a sidewalk, seeing a disturbance ahead, and crossing to the sidewalk on the other side of the road. In addition, at least in bowheads, there is a decrease in durations of surfacings and dives, and number of breaths per surfacing. In other words, the whales cycle through their surfacing and dive repertoire more rapidly. There must be at least subtle commensurate changes in activities such as feeding, socializing, or rest.

Cetacean responses to noise are quite variable, depending on the circumstances of exposure and the activities of the animals at the time. For example, bowhead whales exposed to seismic survey sounds are one example; in contrast to the long-range responses of migrating bowheads, summering (feeding) bowheads tolerate much higher levels of airgun pulses from seismic vessels within a few kilometers (Miller et al., 2005). The beluga whale is another example of a species whose responses to anthropogenic noise are highly variable. Belugas sometimes tolerate exposure to large fleets of fishing vessels, but in other circumstances flee when exposed to faint sounds from approaching ships 35–50 km away (Finley et al., 1990). Belugas summering in the Beaufort Sea also show long-range avoidance of an operating seismic vessel (Miller et al., 2005).

A noise that is associated with danger, such as a catcher boat of a whaling fleet or a purse-seine vessel that encircles dolphins to catch the tuna underneath, can trigger strong avoidance reactions at distances of 10 km or more. Reactions may become stronger upon repeated exposure to aversive stimuli; in this case, "sensitization" is said to occur. At least some dolphins seem to distinguish vessels based on their sounds and react differently to boats that habitually harass the animals (such as aggressive tour operators or researchers who tag or collect biopsy samples) vs boats that approach slowly and carefully. In the latter case, the marine mammals may have learned that the vessel is not harmful—a case of behavioral adaptation.

In contrast to the extreme examples just cited, some marine mammals tolerate exposure to high levels of sound. Seals and sea lions attracted to locations where prey fish are concentrated often tolerate exposure to high levels of noise from acoustic harassment devices designed to disperse pinnipeds. Also, during the Heard Island study, hourglass dolphins (*Lagenorhynchus cruciger*) approached to within several hundred meters of a powerful source of low-frequency sound (57 Hz; source level 209–220 dB), whereas long-finned pilot (*Globicephala melas*), southern bottlenose (*Hyperoodon planifrons*), and minke whales (*Balaenoptera* spp.) were seen less often during sound transmissions than during silent periods (Bowles *et al.*, 1994). The hearing systems of the small hourglass dolphins probably were not very sensitive to low-frequency sounds; they may have been curious about but not discomfited by the noise. However, it is also noteworthy that, when visible near the surface, they would not have been exposed to levels much above 160 dB given the nature of the sound field around this particular sound source. It is possible that they exhibited a vertical avoidance response, staying near the surface simply because the noise was less strong there.

Marine mammals sometimes approach and tolerate a sound source for their own benefit. Bottlenose dolphins that feed on prey stirred up by shrimp fishers know the sounds of all aspects of the trawling and net-lifting operations; they move from vessel to vessel according to stage of trawling, even from several kilometers away. Killer and sperm whales and several pinniped species of the northeast Pacific similarly react to net lifting like the gong of a dinner bell. Many species of dolphins and some species of sea lions and fur seals approach vessels to bow ride or wake ride near the vessel, apparently "for fun."

While this discussion centers mainly on underwater sounds, in-air noises can affect pinnipeds on land, occasionally with serious consequences. A low-flying aircraft, the sudden honk of an automobile horn, or the noisy approach of humans on foot can cause animals on land to stampede into the water. If this occurs on a birthing/nursing beach, adults may, on rare occasions, trample pups in their rush to escape the perceived danger; however, it is not always clear whether sound or sight (and, at times, substrate vibration) is more responsible for the stampede and loss of pups.

Because sound is such an important sensory modality for almost all marine mammals, it stands to reason that human-made noise has, at times, the potential to disrupt the efficiency of communication (and echolocation). Indeed, there is some evidence for this. Some animals fall silent when they perceive danger (or a novel sound). During the aforementioned Heard Island study, sperm and pilot whales were heard 24% and 8% (respectively) of 1137 min when there was no sound, but none was heard to vocalize during 2202 min with sound transmissions. Although these data suggest a strong effect of noise on communication, they are not yet sufficient to indicate at what received levels the whales became quiet (Bowles *et al.*, 1994).

Acoustic reactions to intense sounds vary among species and with the activity of the animals when they perceived the sound. Belugas are vocally active when they detect ship sounds, but narwhals (*Monodon monoceros*) fall silent. Dusky dolphins (*Lagenorhynchus obscurus*) resting in small groups fall silent when they perceive danger. However, those same dolphins will continue their social whistling and burst-pulse "chatter" when they are engaged in high-energy social and sexual activities. They seem less easily disturbed at these higher energy times. There is increasing evidence that both toothed

and baleen whales sometimes alter the frequencies or other characteristics of their calls.

Masking of sounds useful to animals by strong anthropogenic noise is another important phenomenon. We humans have the ability to converse with each other even at noisy parties, and can separate useful speech from an amazing background din (the well-known "cocktail party" phenomenon). Marine mammals can probably do likewise, for they have evolved in a world where physical and biological sounds other than their own abound. Nevertheless, studies of captive dolphins and pinnipeds have confirmed that, in marine as well as terrestrial mammals, strong human-made noise can physically "drown out" other sounds at similar frequencies, especially if the two sources are in similar directions. The noise reduces the maximum distance at which one animal can hear calls from another animal or some other environmental sound that may be important (Richardson *et al.*, 1995). There is increasing evidence, e.g., in gray and right (*Eubalaena* spp.) whales, killer whales, belugas, and bottlenose dolphins, that cetaceans sometimes shift the primary frequencies or increase the source levels of their sounds to reduce the effects of masking by background noise.

Marine mammals are generally very social. If noise disrupts social structure, and if the effect is sufficiently strong and long lasting, then it could be detrimental to the well-being and survival of individuals and perhaps ultimately populations. Noise disturbance can cause some degree of social disruption, but the consequences are largely unknown. When a supply ship comes within 2–4 km of a group of bowhead whales, the whales scatter in all directions. They are "socially disrupted" for as much as a few hours. However, we have no idea how important this disruption may be to the well-being and productivity of the whales. We can guess that they will be harmed if this kind of disruption happens day after day to animals that feed more efficiently when together than when apart. Social disruption also includes the accidental separation of a nursing mother from her pup or calf. When this happens, the probability of pup or calf survival may be reduced substantially.

We do not know to what extent the masking of communication sounds by noise can cause social disruption. Blue whales and fin whale calls can, at times, be detected over distances of several hundred kilometers. If there are important social interactions over those long distances (not proven), these interactions may be impeded by the masking effects of background noise, especially if that noise is continuous. For example, these whales may not be able to find each other as efficiently for feeding and mating purposes given the masking effect of the human-made noise. If whale societies are mediated in large part by acoustic contact, as we suspect (Wells *et al.*, 1999; Tyack, 2000), then background noise that diminishes the distances over which whales can communicate also diminishes the spatial scale of their societies.

Although short-term disruption of behavior by noise is known to occur, it is less clear how this may translate to overall use of habitat and to long-term disturbance. It was mentioned earlier that migrating bowhead whales "edge away" from seismic exploration and drillships, in most cases maintaining a minimum distance of about 20 km. It is unlikely that this results in any major harm to the animals. However, if there were a similar and prolonged displacement from a localized area of important feeding habitat, then the potential for harm would be great. At this point, we know that some whales move away from sources of strong noise, but we do not know whether this significantly reduces their yearly food intake or causes other important disruption. There is, however, building evidence that tourism based on motorized vessels going to dolphins can affect overall their

N

occurrence patterns and habitat use, with the best such information coming from bottlenose dolphins in western Australia (Bejder *et al.*, 2006) and southwestern New Zealand (Lusseau, 2003). It is not clear whether it is mainly the noise of the vessels, the physical interaction by the vessel, or both that are responsible. Another area of concern is long-term operation of acoustic harassment devices (AHDs) designed to keep pinnipeds away from aquaculture facilities. Though less than fully successful in deterring pinnipeds from feeding on the fish, AHDs can cause long-term displacement of non-target species such as porpoises and killer whales (Morton and Symonds, 2002).

Similarly, human disturbance of island spinner dolphins has affected their use of nearshore bays that appear to be important to them for daytime rest (after feeding in deep water at night). It is possible to argue that this reduced use of safe havens might affect their survivability as individuals and therefore as populations. However, to date, we have no information on such a population-wide effect.

The concept of habitat available vs habitat disturbed is an important one in conservation biology. The purist would argue that any habitat disrupted is too much, and compromises the natural world. The pragmatist argues that we humans are so overpoweringly disruptive that we must be content with keeping at least a minimum of area available for the animals in question to survive. In the case of marine mammals, that means keeping some critical habitats relatively free of strong noises that create unacceptable disruption in short-term behavior, long-term behavior, aspects of physiology, and hearing. The task is a difficult one and needs further research, monitoring, mitigation, and political enforcement to have a chance of success.

The first step in mitigation is to develop a better knowledge base. Detailed discussions of the research base and needs can be found in NRC (1994, 2000, 2003, 2005), Richardson *et al.* (1995), and Southall *et al.* (in press). We need better information on auditory sensitivity, especially of large whales, and the levels and frequencies of noises that cause TTS, PTS, and (to the extent they occur) nonauditory physiological disruptions in all marine mammals. Researchers have acquired the technical capability to study noise-induced stress and other aspects of physiology, with miniature data loggers and telemetry transmitters attached to animals under field conditions. This is a relatively new avenue of research for marine animals and is likely to grow in the near future. Studies of short-term behavioral disruptions are worthwhile, but it is the long-term behavioral, social, communication, and habitat disruption effects that are in greatest need of study (NRC, 2005). It is our challenge to study and monitor the health of marine mammals as related not only to anthropogenic sound pollution, but also to other aspects of habitat degradation.

Potential mitigation avenues are many. Naval vessels are already engineered to reduce sound emissions, e.g., by physical decoupling of rotating machinery from the hull, propeller design, and emission of screens of air bubbles. This knowledge needs to be taken into the private sector, which is responsible for the vast majority of shipping noise. Bubble screening has been shown to strongly decrease the noise from a stationary pile-driving activity in Hong Kong waters (Würsig *et al.*, 2000), and improved versions of this technique are increasingly being used to dampen some industrial sounds. A controversial technique that is often employed is "ramping up" of sounds in order to alert animals of impending strong noise, essentially to chase them away from the zone of most influence. This approach probably is effective for some species and situations (e.g., baleen whales vs airguns). However, in other situations it may attract curious marine mammals into a zone of danger. Further study is needed to determine the situations when ramping up is a useful mitigation technique.

In projects where sound is emitted purposefully, e.g., marine seismic exploration, ocean science studies, and navy LFA sonar, refinements in equipment and operational procedures may be possible that will reduce the exposure of marine mammals to noise. Some efforts have already been made to reduce source levels to the minimum that will be effective, and to improve beam forming to reduce sound radiation in unnecessary directions. Furthermore, these (and some other) noisy activities are amenable to regulation as to where and when they are used, especially as pertains to the seasonal migration and movement patterns of sensitive baleen whales. It is common for noisy projects to be restricted to certain times of year and/or to certain areas in order to reduce impacts on marine mammals.

It is up to scientists to help provide badly needed information on disturbance reactions and zones of potential influence, and up to politicians and regulators to write and enforce legislation to curb the uncontrolled proliferation of ocean noise. In the absence of adequate information on the levels and durations of sound exposure that cause problems for marine mammals, regulatory actions are often taken based on "best available" information, with or without allowance for the precautionary principle. More specific data are needed to provide adequate protection for marine mammals while avoiding unnecessary restrictions on worthwhile human activities.

## See Also the Following Articles

Behavior ■ Overview ■ Echolocation ■ Habitat Pressures ■ Hearing ■ Pollution and Marine Mammals ■ Sociobiology

## References

Au, W. W. L., Popper, A. N., and Ray, R. R. (2000). "Hearing by Whales and Dolphins." Springer-Verlag, New York.

Bejder, L., *et al.* (10 authors) (2006). Decline in relative abundance of bottlenose dolphins exposed to long-term disturbance. *Conserv. Biol.* **20**(6), 1791–1798.

Bowles, A. E., Smultea, M., Würsig, B., DeMaster, D. P., and Palka, D. (1994). Relative abundance and behavior of marine mammals exposed to transmissions from the Heard Island Feasibility Test. *J. Acoust. Soc. Am.* **96**(4), 2469–2484.

Cox, T. M., *et al.* (36 authors) (2006). Understanding the impacts of anthropogenic sound on beaked whales. *J. Cet. Res. Manage.* **7**(3), 177–187.

Crum, L. A., and Mao, Y. (1996). Acoustically enhanced bubble growth at low frequencies and its implications for human diver and marine mammal safety. *J. Acoust. Soc. Am.* **99**(5), 2898–2907.

Curry, B. E. (1999). "Stress in mammals: The potential influence of fishery-induced stress in dolphins in the Eastern Tropical Pacific Ocean." NOAA Technical Memo, National Marine Fisheries Service, SWFSC, La Jolla, CA.

Fernández, A., *et al.* (9 authors) (2005). "Gas and Fat Embolic Syndrome" involving a mass stranding of beaked whales (Family *Ziphiidae*) exposed to anthropogenic sonar signals. *Vet. Pathobiol.* **42**, 446–457.

Finley, K. J., Miller, G. W., Davis, R. A., and Greene, C. R. (1990). Reactions of belugas, *Delphinapterus leucas*, and narwhals, *Monodon monoceros*, to ice-breaking ships in the Canadian high arctic. *Can. Bull. Fish. Aquat. Sci.* **224**, 97–117.

Finneran, J. J., Carder, D. A., Schlundt, C. E., and Ridgway, S. H. (2005). Temporary threshold shift in bottlenose dolphins (*Tursiops truncatus*) exposed to mid-frequency tones. *J. Acoust. Soc. Am.* **118**(4), 2696–2705.

Gisiner, R., Cudahy, E., Frisk, G., Gentry, R., Hofman, R., Popper, A., and Richardson, W. J. (1999). Proceedings/Workshop on the effects of anthropogenic noise in the marine environment, 10–12 February 1998. Office of Naval Research, Arlington, VA. Available at www.onr.navy.mil/sci_tech/personnel/cnb_sci/Proceed.pdf.

Jasny, M., Reynolds, J., Horowitz, C., and Wetzler, A. (2005). "Sounding the depths II: The rising toll of sonar, shipping and industrial ocean noise on marine life." Natural Resources Defense Council, Los Angeles, CA, Available at http://www.nrdc.org/wildlife/marine/.

Kastak, D., Southall, B. L., Schusterman, R. J., and Reichmuth Kastak, C. (2005). Underwater temporary threshold shift in pinnipeds: Effects of noise level and duration. *J. Acoust. Soc. Am.* **118**(5), 3154–3163.

Lusseau, D. (2003). Effects of tour boats on the behavior of bottlenose dolphins: Using Markov chains to model anthropogenic impacts. *Conserv. Biol.* **17**(6), 1785–1793.

Miller, G. W., Moulton, V. D., Davis, R. A., Holst, M., Millman, P., MacGillivray, A., and Hannay, D. (2005). Monitoring seismic effects on marine mammals—southeastern Beaufort Sea, 2001–2002. *In* "Offshore Oil and Gas Environmental Effects Monitoring/Approaches and Technologies" (S. L. Armsworthy, P. J. Cranford, and K. Lee, eds), pp. 511–542. Battelle Press, Columbus, OH.

MØhl, B., Wahlberg, M., Madsen, P. T., Miller, L. A., and Surlykke, A. (2000). Sperm whale clicks: Directionality and source level revisited. *J. Acoust. Soc. Am.* **107**(1), 638–648.

Morton, A. B., and Symonds, H. K. (2002). Displacement of *Orcinus orca* (L.) by high amplitude sound in British Columbia, Canada. *ICES J. Mar. Sci.* **59**(1), 71–80.

NRC (1994). "Low-frequency sound and marine mammals: Current knowledge and research needs." U.S. Nat. Res. Counc., Ocean Studies Board, Committee on Low-Frequency Sound and Marine Mammals. Natl. Acad. Press, Washington, DC.

NRC (2000). "Marine mammals and low-frequency sound: Progress since 1994." U.S. Nat. Res. Counc., Ocean Studies Board, Committee to Review Results of ATOC's Marine Mammal Research Program. Natl. Acad. Press, Washington, DC.

NRC (2003). "Ocean noise and marine mammals." U.S. Nat. Res. Counc., Ocean Studies Board, Committee on Potential Impacts of Ambient Noise in the Oceans on Marine Mammals. Natl. Acad. Pres., Washington, DC.

NRC (2005). "Marine mammal populations and ocean noise: determining when noise causes biologically significant effects." U.S. Nat. Res. Counc., Ocean Studies Board, Committee on Characterizing Biologically Significant Marine Mammal Behavior. Natl. Acad. Press, Washington, DC.

Richardson, W. J., Greene, C. R., Jr., Malme, C. I., and Thomson, D. H. (1995). "Marine Mammals and Noise." Academic Press, San Diego, CA.

Ridgway, S. H., and Carder, D. A. (1997). Hearing deficits measured in some *Tursiops truncatus*, and discovery of a deaf/mute dolphin. *J. Acoust. Soc. Am.* **101**(1), 590–594.

Simmonds, M., Dolman, S., and Weilgart, L. (2004). "Oceans of Noise 2004." Whale and Dolphin Conservation Society, Chippenham, UK, Available at http://www.wdcs.org/.

Southall, B. L. *et al.* (13 authors) (2007). Marine mammal noise exposure criteria: Initial scientific recommendations. *Aquat. Mamm* **33**, 411–521.

Tyack, P. L. (2000). Functional aspects of cetacean communication. *In* "Cetacean Societies: Field Studies of Dolphins and Whales" (J. Mann, R. C. Connor, P. L. Tyack, and H. Whitehead, eds), pp. 270–307. University of Chicago Press, Chicago, IL.

Wells, R. S., Boness, D. J., and Rathbun, G. B. (1999). Behavior. *In* "Biology of Marine Mammals" (J. E. Reynolds, III, and S. A. Rommel, eds), pp. 324–422. Smithsonian Institution Press, Washington, DC.

Wenz, G. M. (1962). Acoustic ambient noise in the ocean: Spectra and sources. *J. Acoust. Soc. Am.* **34**, 1936–1956.

Würsig, B., and Au, W. W. L. (2004). Echolocation signals of dusky dolphins (*Lagenorhynchus obscurus*) in Kaikoura, New Zealand. *J. Acoust. Soc. Am.* **115**, 2307–2313.

Würsig, B., Greene, C. R., Jr., and Jefferson, T. A. (2000). Development of an air bubble curtain to reduce underwater noise of percussive piling. *Mar. Environ. Res.* **49**(1), 79–93.

# North Atlantic Marine Mammals

Gordon T. Waring, Debra L. Palka and Peter G.H. Evans

Marine mammals are a diverse, widespread, and significant component of North Atlantic marine ecosystems. Four of the five commonly recognized marine mammal taxa reside in the North Atlantic: cetaceans (mysticetes, baleen whales; and odontocetes, toothed whales, dolphins, and porpoises), sirenians [manatees (*Trichechus* spp.)], pinnipeds, [seals and walruses (*Odobenus rosmarus*)], and polar bears (*Ursus maritimus*) (Rice, 1998; Reeves *et al.*, 2002). A fifth taxon [marine (*Lontra felina*) and sea (*Enhydra lutris*) otters] and sea lions and fur seals (family Otariidae) have not inhabited the North Atlantic since at least the late Pleistocene.

The systematics of marine mammals is still being disputed (Rice, 1998). Marine mammals occupy all North Atlantic marine regimes, tropical to polar, although species-specific ranges exist and distribution patterns are not uniform (Tables I and II). The large-scale, nonrandom distribution of marine mammals is influenced by oceanographic features, whereas small-scale distributions are influenced by factors such as the animal's physiology, behavior, and ecology (Bowen and Siniff, 1999). Over geologic time scales, the diversity and ecology of North Atlantic marine mammals reflect adaptation to a dynamic aquatic environment. As elsewhere, North Atlantic marine mammal populations have been impacted significantly by human interactions (Sahrhage and Lundbek, 1992; Kinze, 1995; Gambell, 1999; Reeves *et al.*, 2003). Some species have been, and continue to be, harvested for subsistence and commercial use and for their cultural value. Overexploitation has resulted in extinction [e.g., Caribbean monk seal (*Monachus tropicalis*), Atlantic gray whale (*Eschrichtius robustus*)] and significant population declines [e.g., North Atlantic right whale (*Eubalaena glacialis*)], and has also likely caused significant ecological "changes" (e.g., reduction of top predators and competitive interactions; Rice, 1998; Kraus and Rolland, 2007). Indirect mortality (e.g., fishery bycatch and pollution) has adversely affected numerous species [e.g., harbor porpoise (*Phocoena phocoena*), bottlenose dolphin (*Tursiops truncatus*), beluga whale (*Delphinapterus leucas*), and a ringed seal subspecies (*Phoca hispida botnica*)] (Kinze, 1995; Northridge and Hofman, 1999; O'Shea, 1999; Reijnders *et al.*, 1999; Hall and Donovan, 2002). Climate change is also affecting marine mammal populations, especially species that live in close association with the Arctic ice and/or in the cold temperate to polar seas influenced by Arctic ice [e.g., polar bears and Arctic ringed seal (*Phoca hispida hispida*)] (Learmonth *et al.*, 2006).

## I Physical Environment

The physical characteristics of the North Atlantic ecosystem (Fig. 1) critically influence marine mammal distribution. Although the ocean basin provides marine mammals with an open pathway that extends from the equator northward to the Arctic and includes adjacent bodies of water (e.g., Gulf of Mexico, Caribbean Sea, North Sea, Norwegian Sea, and Bay of Biscay), the North Atlantic has many different ecosystems (Reid *et al.*, 2003; Stenseth *et al.*, 2004). Some adjacent seas, such as the Baltic and Mediterranean, are more isolated from the open ocean and form separate ecosystems. In the

N

## TABLE I
### Occurrence of Marine Mammal Species in the Eastern North Atlantic (including North, Baltic, Barents, and White Seas), by Country

| Cetacean Species | FO | IC | RUS | NO | DK | SE | FI | EBA | PO | DE | NL | BE | UK | IE | FR | ES | PT | MAC | WAFR |
|---|---|---|---|---|---|---|---|---|---|---|---|---|---|---|---|---|---|---|---|
| *(a) Baleen whales and large toothed whales* | | | | | | | | | | | | | | | | | | | |
| Bowhead whale | – | – | RAR | RAR[a] | – | – | – | – | – | – | – | – | – | – | – | – | – | – | – |
| N. Atlantic right whale | VAG | VAG | – | VAG | –° | – | – | – | – | – | VAG | –° | VAG | VAG | –° | VAG | VAG | VAG | VAG |
| Minke whale | COM | COM | COM | COM | COM[b] | RAR | – | – | – | RAR | RAR | VAG | COM[c] | REG[d] | REG | REG | COM | RAR | RAR |
| Sei whale | RAR | REG | RAR | RAR | VAG | RAR | – | – | – | VAG | VAG | VAG | RAR | REG | RAR | RAR | REG | REG | REG |
| Bryde's whale | – | – | – | – | VAG | – | – | – | – | – | –° | – | – | – | – | – | – | REG | REG |
| Blue whale | RAR | REG | RAR | RAR | –° | – | – | – | – | – | –° | –° | RAR | RAR | VAG | RAR | VAG | RAR | RAR |
| Fin whale | REG | REG | REG | REG | VAG | RAR | – | – | – | VAG | VAG | VAG | REG | REG | RAR | REG | REG | REG | REG |
| Humpback whale | REG | REG | RAR | COM | VAG | VAG | –° | – | –° | VAG | VAG | –° | RAR | RAR | VAG | RAR | RAR | RAR | REG |
| Atlantic gray whale[u] | – | – | – | – | – | EXT | – | – | – | – | EXT | EXT | EXT | – | – | – | – | – | – |
| Sperm whale | REG | REG | REG | REG | RAR | RAR | – | – | VAG | VAG | VAG | VAG | RAR | RAR | REG | REG | REG | REG | REG |
| *(b) Small cetaceans* | | | | | | | | | | | | | | | | | | | |
| Pygmy sperm whale | – | – | – | – | – | – | – | – | –° | –° | –° | – | VAG | VAG | RAR | VAG | VAG | RAR | RAR |
| Dwarf sperm whale | – | VAG | – | – | – | – | – | – | – | – | – | – | – | – | VAG | VAG | – | VAG | RAR |
| N. bottlenose whale | REG | REG | REG | REG | VAG | RAR | – | – | –° | VAG | VAG | –° | REG | REG | RAR | REG | – | RAR[l] | VAG |
| Sowerby's beaked whale | RAR | RAR | – | RAR | VAG | RAR | – | – | – | VAG | VAG | VAG | RAR | RAR | RAR | RAR | RAR | RAR[l] | VAG |
| Blainville's beaked whale | – | – | – | – | – | – | – | – | – | – | VAG | – | VAG | – | VAG | VAG | VAG | REG | REG |
| Gervais' beaked whale | – | VAG | – | – | – | – | – | – | – | – | – | – | – | VAG | VAG | – | RAR | VAG[m] | VAG |
| Gray's beaked whale | – | – | – | – | – | – | – | – | – | – | – | – | – | – | – | – | – | – | – |
| True's beaked whale | – | – | – | – | – | – | – | – | – | –° | – | –° | –° | –° | VAG | VAG | – | VAG | VAG |
| Cuvier's beaked whale | – | – | – | – | – | RAR | – | – | – | – | VAG | VAG | RAR | RAR | REG | REG | RAR | REG | REG |
| Beluga | VAG | RAR | COM | RAR | VAG[e] | RAR | – | – | –° | VAG | VAG | VAG | VAG | – | VAG | VAG | – | – | – |
| Narwhal | –° | VAG | RAR[s] | RAR | – | VAG | – | – | – | VAG | –° | – | –° | –° | VAG | VAG | – | – | – |
| Short-bked com. Dolphin | VAG | VAG | – | VAG | REG | RAR | RAR | – | VAG | VAG | RAR | VAG | COM | COM | COM | COM | COM | REG[n] | RAR |
| Long-bked com. Dolphin | – | – | – | – | – | – | – | – | – | – | – | – | – | – | – | – | – | – | COM |
| Pygmy killer whale | – | – | – | – | – | – | – | – | RAR | – | – | – | – | – | VAG | VAG | – | VAG | RAR |
| Short-finned pilot whale | – | – | – | – | – | – | – | – | – | – | – | – | – | – | VAG | VAG | – | COM | COM |
| Long-finned pilot whale | COM | COM | – | COM[f] | RAR | RAR | – | – | – | VAG | VAG | VAG | COM | COM | COM | COM | COM | –[l] | – |
| Risso's dolphin | – | – | – | VAG | –° | RAR | – | – | – | VAG | VAG | RAR | REG | REG | REG | REG | COM | COM | REG |
| Fraser's dolphin | – | – | – | – | – | – | – | – | – | – | – | – | – | – | VAG | – | VAG | VAG | RAR |
| Atlantic White-sided dolphin | COM | COM | – | COM | RAR | RAR | – | – | – | VAG | RAR | VAG | COM | COM | RAR | RAR | RAR | – | – |
| White-beaked dolphin | RAR | COM | – | COM | COM[g] | RAR | – | – | RAR | RAR | REG | RAR | COM | REG | RAR | VAG | – | – | – |
| Killer whale | COM | REG | REG | REG | REG[h] | VAG | – | REG | RAR | VAG | VAG | VAG | REG | REG | RAR | RAR | REG | RAR | RAR |
| Melon-headed whale | –° | – | – | – | –° | –° | – | – | – | –° | –° | –° | –° | – | VAG | – | – | VAGq | REG |
| False killer whale | – | – | – | VAG | –° | – | – | – | – | –° | –° | – | VAG | VAG | VAG | VAG | RAR | RAR | REG |
| Atl. humpback dolphin | – | – | – | – | – | – | – | – | – | – | – | – | – | – | – | – | – | – | REG |

| | | | | | | | | | | | | | | | | | | | | | |
|---|---|---|---|---|---|---|---|---|---|---|---|---|---|---|---|---|---|---|---|---|---|
| **Pantrop. spotted dolphin** | – | – | – | – | – | – | – | – | – | – | – | – | – | – | – | – | – | – | – | – | RAR | REG |
| **Clymene dolphin** | – | VAG | VAG | – | – | – | – | – | – | – | – | – | – | – | – | – | – | – | – | COM | RAR | REG |
| **Striped dolphin** | VAG | VAG | VAG | VAG | VAG | VAG | – | – | VAG | VAG | VAG | VAG | RAR | COM | COM | COM | COM | COM | COM | REG | COM | COM |
| **Atlantic spotted dolphin** | – | – | – | – | – | – | – | – | – | – | – | – | – | – | – | – | – | – | – | – | REG | COM |
| **Spinner dolphin** | – | – | – | – | – | – | – | – | – | – | – | – | – | – | – | – | – | – | – | – | RARr | COM |
| **Rough-toothed dolphin** | – | – | – | – | – | – | – | – | – | – | – | – | – | – | – | – | – | – | – | REG | REG | REG |
| **Bottlenose dolphin** | REG | COM | VAG | VAG | VAG | RAR | VAG | VAG | VAG | VAG | VAG | COM | COM | COM | COM | COM | COM | COM | COM | COM | COM | COM |
| **Harbor porpoise** | COM | RAR | RARt | COM | COM | COM | COM | COM | COM | COM | COM | COM | COM | COM | COM | COM | COM | COM | VAG | RAR | VAG | RAR |
| **(c) Pinnipeds** | | | | | | | | | | | | | | | | | | | | | | |
| **Hooded seal** | REG | REG | REG | RAR | RAR | – | RAR | – | – | – | VAG | VAG | VAG | VAG | VAG | VAG | VAG | – | – | – | – | – |
| **Bearded seal** | RAR | RAR | REG | RAR | RAR | – | RAR | – | – | – | VAG | VAG | VAG | VAG | VAG | VAG | VAG | – | – | – | – | – |
| **Grey seal** | REG | REG | REG | COM | REG | – | REG | – | REG | REG | COM | REG | COM | REG | REG | REGj | – | – | – | – | – | – |
| **Harp seal** | RAR | RAR | RAR | – | COM | – | RAR | – | – | COM | VAG | VAG | – | VAG | VAG | VAG | – | – | – | – | – | – |
| **Ringed seal** | VAG | RAR | COM | REG | REG | REG | REG | RAR | VAG | RAR | VAG | REG | RAR | RAR | – | – | VAG | – | – | – | – | – |
| **Harbor seal** | VAG | COM | REG | COM | COM | COM | REG | COM | COM | REG | COM | COM | COM | COM | RAR | VAG | VAG | – | – | – | VAG | – |
| **Mediterr. monk seal** | – | VAG | REGi | RAR | COM | – | VAG | – | VAG | – | – | VAG | VAG | VAG | VAG | VAG | VAG | RARo | VAG | RARp | – | – |
| **Walrus** | VAG | VAG | REGi | RAR | RAR | COM | – | VAG | – | – | – | – | – | – | – | – | – | – | – | – | – | – |
| **(d) Carnivores** | | | | | | | | | | | | | | | | | | | | | | |
| **Polar bear** | VAG | VAG | REGi | RAR | – | RAR | – | – | – | – | – | – | – | – | – | – | – | – | – | – | – | – |

NOTES: Countries: FO, Faroe Islands; IC, Iceland; NO, Norway; DK, Denmark; SE, Sweden; FI, Finland; RUS, Russia (both Barents Sea and Eastern Baltic); EBA, Eastern Baltic States (Latvia, Lithuania, and Estonia); PL, Poland; DE, Germany; NL, Netherlands; BE, Belgium; UK, United Kingdom; IE, Ireland; FR, (Atlantic) France; ES, (Atlantic) Spain (excl. Canaries); PT, (Atlantic) Portugal (excl. Azores & Madeira/Desertas); MAC, Macaronesia (Azores, Madeira/Desertas, Canaries, Cape Verdes); WAFR, West Africa from Morocco to the Equator. Occurrence (based on records since 1980): VAG, Vagrant; RAR, Rare; REG, Regular (but Uncommon); COM, Common; EXT, Extinct; –, Not Recorded; °, Record(s) before 1980.

RAR in northern Norway only.
REG in Kattegat/Baltic.
But REG in Channel and Southern North Sea.
But COM in Southwest.
But annual, periodically.
But periodic, at other times RAR.
REG in Kattegat/Baltic.
RAR in Kattegat/Baltic.
REG only in arctic islands of Svalbard, Bornøya and Jan Mayen.
COM in N. France, rare further south in France.
But REG in parts of Norway; Includes isolated population in Svalbard.
But VAG in Canaries and Cape Verdes.
But RAR in Canaries.
But RAR in Canaries and Cape Verdes.
RAR in Madeira/Desertas and Canaries, VAG in Cape Verdes, absent from Azores.
RAR only in Morocco and Mauritania, VAG further south.
But RAR in Cape Verdes.
RAR in Canaries, but REG in Cape Verdes and absent from Azores and Madeira/Desertas.
Mainly around Frantz-Josef Land.
But VAG in eastern Baltic.
Records only from Belgium, Netherlands, Sweden, and the United Kingdom.

Table II

Occurrence of Marine Mammal Species in the Western North Atlantic (including Greenland), by Region

| Cetacean SPECIES | CAR | COM | SE-USA | NE-USA | CA-MA | CA-AR | GRE |
|---|---|---|---|---|---|---|---|
| *(a) Baleen whales and large toothed whales* | | | | | | | |
| Bowhead whale | – | – | – | – | – | COM | REG |
| N. Atlantic right whale | RAR | VAG | REG | REG | REG | – | – |
| Minke whale | RAR | RAR | REG | COM | COM | COM | COM |
| Sei whale | COM | VAG | RAR | COM | COM | – | REG |
| Bryde's whale | VAG | REG | RAR | RAR | – | – | – |
| Blue whale | COM[a] | VAG | VAG | RAR | REG | REG | RAR |
| Fin whale | COM | COM | RAR | COM | COM | REG | REG[d] |
| Humpback whale | COM | RAR | COM | COM | COM | COM | COM |
| Sperm whale | COM | COM | COM | COM | COM | COM | REG |
| *(b) Small cetaceans* | | | | | | | |
| Pygmy sperm whale | RAR | COM | REG | VAG | – | – | – |
| Dwarf sperm whale | RAR | COM | REG | VAG | – | – | – |
| N. bottlenose whale | VAG | – | – | RAR | COM | RAR | REG |
| Sowerby's beaked whale | – | VAG | VAG | REG | RAR | RAR | – |
| Blainville's beaked whale | REG | REG | REG | RAR | RAR | – | – |
| Gervais' beaked whale | RAR | REG | REG | VAG | – | – | – |
| True's beaked whale | – | – | REG | REG | RAR | – | – |
| Cuvier's beaked whale | REG | COM | REG | RAR | RAR | – | – |
| Beluga | – | – | – | – | COM | COM | COM |
| Narwhal | – | – | – | – | COM | COM | REG |
| Short-bked com. Dolphin | REG | – | COM | COM | – | – | – |
| Long-bked com. Dolphin | – | COM | – | – | COM | – | – |
| Pygmy killer whale | COM | COM | REG | RAR | – | – | – |
| Short-finned pilot whale | COM | COM | COM | COM | – | – | – |
| Long-finned pilot whale | – | – | REG | COM | COM | – | RAR |
| Risso's dolphin | REG | COM | COM | COM | COM | – | – |
| Fraser's dolphin | REG | REG | REG | – | – | – | – |
| Atl. White-sided dolphin | – | – | – | COM | COM | REG | REG |
| White-beaked dolphin | – | – | – | REG | COM | REG | COM[e] |

| Species | CAR | GOM | SE-USA | NE-USA | CA-MA | CA-AR | CRE |
|---|---|---|---|---|---|---|---|
| Killer whale | REG | REG | RAR | RAR | RAR | REG | REG |
| Melon-headed whale | REG | COM | REG | – | – | – | – |
| False killer whale | REG | COM | REG | – | – | – | – |
| Alt. Hump-backed dolphin | – | – | – | – | – | – | – |
| Pantropical spotted dolphin | COM | COM | COM | RAR | RAR | – | – |
| Clymene dolphin | REG | COM | COM | COM | REG | – | – |
| Striped dolphin | COM[b] | COM | COM | COM | – | – | – |
| Atlantic spotted dolphin | COM | COM | COM | COM | – | – | – |
| Spinner dolphin | COM | COM | REG | – | – | – | – |
| Rough-toothed dolphin | REG | COM | REG | – | – | – | – |
| Bottlenose dolphin | COM | COM | COM | COM | COM | – | COM |
| Harbor porpoise | – | – | VAG | COM | COM | COM | COM |
| *(c) Manatees* | | | | | | | |
| Florida manatee | REG[c] | COM | COM | VAG | – | – | – |
| Antillean manatee | VAG | VAG | – | – | – | – | – |
| *(d) Pinnipeds* | | | | | | | |
| Hooded seal | VAG | – | VAG | REG | COM | COM | COM |
| Atlantic bearded seal | – | – | – | VAG | VAG | COM | COM |
| Gray seal | – | – | VAG | COM | COM | COM | – |
| Harp seal | VAG | – | REG | REG | COM | COM | COM |
| Arctic ringed seal | – | – | – | – | – | COM | COM |
| W. Atlantic harbor seal | – | – | VAG | COM | COM | COM | COM |
| Mediterranean monk seal | EXT | – | – | – | – | – | – |
| Caribbean monk seal | EXT | – | – | – | – | – | – |
| Walrus | – | – | – | – | COM | COM | COM |
| Polar bear | – | – | – | – | VAG | COM | REG[f] |

NOTES "Regions: CAR, Caribbean; GOM, Gulf of Mexico; SE-USA, Southeast USA; NE-USA, Northeast USA; CA-MA, Canadian Maritimes; CA-AR, Canadian Arctic; CRE, Greenland. Occurrence: VAG, Vagrant; RAR, Rare; REG, Regular (but Uncommon); COM, Common; EXT, Extinct; –, Not Recorded.
COM in north but RAR further south,
COM in north, REG in south,
absent from Eastern Caribbean,
COM in southeast,
common near the tip of Greenland and they are more common than Atl. White-sided
common on the east Greenland coast and off shore in West Greenland
rare in the areas with people, common on the east Greenland coast and off shore in West Greenland

N

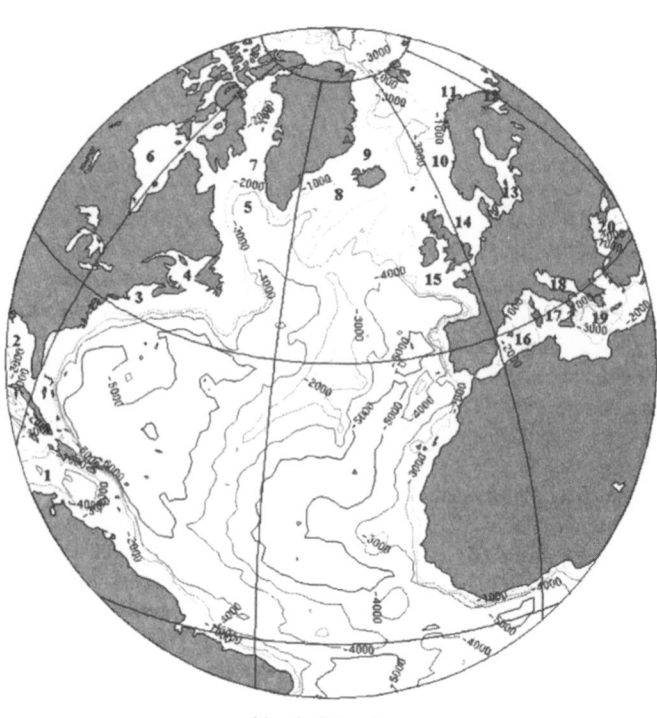

**North Atlantic**

**Figure 1** *Bodies of water in the North Atlantic. Depth contours in meters. 1, Caribbean Sea; 2, Gulf of Mexico; 3, Gulf of Maine; 4, Gulf of St. Lawrence; 5, Labrador Sea; 6, Hudson Bay; 7, Davis Strait; 8, Denmark Strait; 9, Greenland Sea; 10, Norwegian Sea; 11, Barents Sea; 12, White Sea; 13, Baltic Sea; 14, North Sea; 15, Celtic Sea; 16, Mediterranean Sea; 17, Tyrrhenian Sea; 18, Adriatic Sea; 19, Aegean Sea; 20, Black Sea.*

open ocean, water masses define tropical to polar ecosystems that are influenced by circulation patterns of the major ocean currents such as the Gulf Stream, Greenland current, and North Equatorial current. There are broad continental shelf ecosystems defined by basins, banks, channels, ice, submarine canyons, and volcanic islands. Sea mounts and the mid-Atlantic Ridge also define important ecosystems. These types of oceanographic features influence productivity which concentrate prey and create high-use marine mammal habitats (Reid *et al.*, 2003).

## II. Distribution and Habits

Baleen whales are widely distributed in the North Atlantic, with individual species exhibiting preferences for certain ecosystems (Bowen and Siniff, 1999). Some preferences are temperature driven. For example, bowhead whales (*Balaena mysticetus*) occupy only polar waters, whereas Bryde's whales (*Balaenoptera edeni*) are found only in tropical waters. Other preferences are more topography driven. For example, right, humpback (*Megaptera novaeangliae*), and minke (*Balaenoptera acutorostrata*) whales prefer continental shelf ecosystems, whereas blue (*B. musculus*), sei (*B. borealis*), and Bryde's whales are associated with shelf-edge and deeper oceanic water. While fin whale (*Balaenoptera physalus*) habitat preference differs geographically (i.e., shelf ecosystems in the northwest Atlantic and shelf-edge habitats off NW Europe). Large whales, however, are highly mobile and seasonally may occupy different habitats. Baleen whales, except bowhead and Bryde's whales, can undergo the most extensive seasonal migrations of all North Atlantic marine mammals, migrating between

warm low-latitude breeding grounds in winter and cold high-latitude feeding grounds in summer. North Atlantic humpback whales exemplify this migratory behavior (Bowen and Siniff, 1999). In summer, humpback whale stocks feed in Iceland, Greenland, Newfoundland, Gulf of St. Lawrence, and Gulf of Maine/Scotian Shelf and then spend winter on breeding grounds in the Caribbean Sea. A smaller eastern North Atlantic population summers between the Bay of Biscay and Norway, and spends the winter between the British Isles and Cape Verdes. Molecular genetic studies indicate that the feeding stocks are matrilineal groups of related individuals (Baker and Palumbi, 1997). There is little evidence of recent genetic exchange between North Atlantic and South Atlantic populations of baleen whales, due largely to seasonal differences in the migration paths of the two populations (Baker and Palumbi, 1997).

Odontocetes also occupy nearly all marine ecosystems in the North Atlantic, with individual species exhibiting preferences for particular ecosystems (Bowen and Siniff, 1999; Reeves *et al.*, 2002; Reid *et al.*, 2003; Macleod *et al.*, 2006). Continental shelf species found in cool temperate to subpolar waters are harbor porpoises, Atlantic white-sided and white-beaked dolphins (*Lagenorhynchus acutus* and *L. albirostris*), long-finned pilot whales (*Globicephala melas*), and two Arctic species, narwhal (*Monodon monoceros*) and beluga whales. Continental shelf break/pelagic species found in warm-temperate to cooler waters include bottlenose (offshore and coastal forms), short-beaked common (*Delphinus delphis*), Risso's (*Grampus griseus*), striped (*Stenella coeruleoalba*), and Atlantic spotted (*S. frontalis*; coastal form) dolphins, sperm (*Physeter macrocephalus*) and northern bottlenose (*Hyperoodon ampullatus*) whales, and Cuvier's (*Ziphius cavirostris*), Blainville's (*Mesoplodon densirostris*), Sowerby's (*M. bidens*), and True's (*M. mirus*) beaked whales. The range of northern bottlenose and Sowerby's beaked whales extends into subarctic waters. Continental shelf break/pelagic species found in warm-temperate to tropical waters are pantropical spotted (*Stenella attenuata*), Atlantic spotted (offshore form), spinner (*S. longirostris*), Clymene (*S. clymene*), rough-toothed (*Steno bredanensis*), Atlantic humpbacked (*Sousa teuszii*), and Fraser's (*Lagenodelphis hosei*) dolphins and melon-headed (*Peponocephala electra*), false killer (*Pseudorca crassidens*), pygmy killer (*Feresa attenuata*), short-finned pilot (*Globicephala macrorhynchus*), pygmy sperm (*Kogia breviceps*), dwarf sperm (*K. sima*), and Gervais' beaked (*Mesoplodon europaeus*) whales. Within warm temperate to tropical water mass habitats, bottom topography and frontal boundaries are important characteristics that define cetacean distribution. Unlike baleen whales, only a few odontocetes (e.g., sperm and long-finned pilot whales) are known to undergo long-range seasonal migrations (Bowen and Siniff, 1999). Stock structure is largely unknown, except for a few nearshore continental shelf species (e.g., harbor porpoise, beluga). Some oceanic odontocetes likely move between North and South Atlantic waters (e.g., pantropical spotted dolphin and Cuvier's beaked whale).

North Atlantic seals (phocids) include both Northern and Southern Hemisphere species (Tables I and II; Reeves *et al.*, 2002). Northern phocids [harbor (*Phoca vitulina*), and gray seals (*Halichoerus grypus*)] are widely distributed in boreal to polar waters (Bowen and Siniff, 1999). The ice seals [hooded (*Cystophora cristata*), bearded (*Erignathus barbatus*), harp (*Pagophilus groenlandicus*), and ringed (*Pusa hispida*) seals] pup on ice and have seasonal migrations that are strongly associated with seasonal ice fluctuations. Bearded, hooded, and harp seals also utilize pelagic habitats. Ranges change; for example, since the 1990s, the winter/spring distributions of hooded and harp seals extended southward into northeast U.S. coastal waters. Harbor seals are the most widely distributed species, occupying cool

**N**

temperate to Arctic North Atlantic waters. Gray seals have a discontinuous distribution in cold temperate to subarctic coastal waters. Southern phocids include the Mediterranean (*Monachus monachus*) and Caribbean (extinct) monk seals. The Mediterranean monk seal is primarily found in the Mediterranean, adjacent seas, and along northwestern Africa. The Caribbean monk seal previously inhabited the Caribbean Sea and southern portion of the Gulf of Mexico. Stock structure for North Atlantic seals is well defined.

Cetaceans and phocid seals constitute the largest component of North Atlantic marine mammal fauna (Bowen and Siniff, 1999; Reeves *et al.*, 2002). Additional species include walruses, polar bears, and West Indian manatees (*Trichechus manatus*). Walruses and polar bears have a circumpolar distribution. Both species are usually associated with ice habitats but also spend time on coastal land areas. The Florida (*Trichechus manatus latirostris*) and Antillean (*T. m. manatus*) manatees have a tropical to subtropical distribution. The Florida manatee is found in coastal waters of the Gulf of Mexico and southeastern United States. Seasonal extralimital movements northward have been recorded for the Florida manatee. The Antillean manatee is distributed from northern Mexico to central Brazil and throughout the islands of the Caribbean.

## III. Feeding

The taxonomic division of cetaceans into Odontoceti and Mysticeti reflects their different feeding strategies (Rice, 1998; Bowen and Siniff, 1999; Reeves *et al.*, 2002). Baleen whales are strainers who largely feed on planktonic or micronektonic crustaceans and/or relatively small pelagic fish by using visual or passive acoustic techniques. Toothed whales are graspers who capture fish, squid, and other species by hunting using sight, sound, or active echolocation. Pinnipeds and polar bears are carnivores. Pinnipeds consume primarily fish and invertebrates, and some species occasionally eat seabirds, seals, or small whales. Polar bears prey primarily on seals and sometimes feed on fish and other small mammals. In contrast, manatees are herbivores, grazing in shallow waters on vegetation using primarily their sense of touch.

## IV. Human Impact

Centuries of human activities have affected all North Atlantic marine mammal populations. Prehistoric people hunted coastal marine mammals for subsistence use, and in some areas (e.g., Canada, Greenland) aboriginal hunting still exists (Sahrhage and Lundbek, 1992; Kinze, 1995; Gambell, 1999; Heide-Jørgensen and Wiig, 2002; Kraus and Rolland, 2007). Early subsistence hunting, however, was likely insignificant compared to commercial whaling that began in Europe during the tenth century (Slijper, 1979; Sahrhage and Lundbek, 1992). By the beginning of the eighteenth century, European whalers had already depleted bowhead and right (*Eubalaena* spp.) whale stocks in the eastern North Atlantic, so then moved on to hunt these species in the western North Atlantic, from Greenland to the Gulf of St. Lawrence. American whalers also depleted right and humpback whale stocks in coastal waters off the American colonies. Depletion of these stocks initiated pelagic whaling for sperm and humpback whales. Modern whaling, as we know it today, began in the late nineteenth century when Norwegians invented the explosive harpoon and converted from sail to steam vessels. This allowed whaling to expand to the faster swimming blue, fin, and sei whales. By the 1920s, the stocks of North Atlantic large whales had all been over-exploited, and so whaling activities were

redirected into Antarctic waters. Commercial whaling depleted most of these stocks as well. In 1946 the International Convention for the Regulation of Whaling was signed to provide for the CONSERVATION of whale stocks (Gambell, 1999). However, North Atlantic whaling continued until the 1987 INTERNATIONAL WHALING COMMISSION moratorium was enacted. Following the moratorium, fing minke and humpback whales are taken for subsistence in West Greenland, and humpback whales are taken in St. Vincent. Most recently, Norway has been taking minke whales and Iceland minke and fin whales, both commercially under objection to the moratorium. Despite the many years since whaling of most species has stopped, some of the North Atlantic large whales (in particular the North Atlantic right whale) have not yet recovered (Clapham *et al.*, 1999; Kraus and Rolland, 2007). This is probably due to slow growth, low reproductive rates, and other human interactions (Boyd *et al.*, 1999; Evans and Stirling, 2002).

Commercial exploitation of smaller cetaceans began in the fourteenth century when the Danes initiated organized hunts of Baltic Sea harbor porpoises (Kinze, 1995). Although these hunts ended in the mid-twentieth century, there are still very few harbor porpoises in the Baltic Sea. In the 1500s, the Faroese initiated a pilot whale (*Globicephala* spp.) drive fishery that continues to this day. Examples of hunts during the early to mid-1900s include Norwegian hunts of minke, killer (*Orcinus orca*), northern bottlenose, and pilot whales, American bounty hunts on harbor porpoises, and pilot whale drives in Shetland, Orkney, and Newfoundland. The Newfoundland fishery continued through the twentieth century but had to stop in 1971 due to local depletion (Mercer, 1975). In the Atlantic islands of the Azores and Madeira, subsistence hunting of sperm whales continued until as recently as the 1980s. Small-cetacean hunts occurring today are small-scale subsistence fisheries, such as for harbor porpoises in Greenland and belugas in Canada, Greenland, and Russia. It is unknown whether these stocks can sustain these removals. Now, even in the traditional whaling countries, whale and dolphin watching has largely replaced whaling as an economic activity (in the Canary Islands currently estimated to involve more than 1 million tourists a year; Urquiola and de Stephanis, 2000)

North Atlantic walrus populations were similarly exploited (Sahrhage and Lundbek, 1992). In the early 1600s, Britain initiated walrus hunting around Spitzbergen, Jan Mayen, and Norway. Russians, Europeans, and Canadians joined in to expand the hunts further northward. As a result, these walrus populations were severely depleted by the nineteenth century and have not yet recovered.

Seals were first commercially hunted for oil and blubber in Europe and Newfoundland (Sahrhage and Lundbek, 1992). In these areas, large-scale commercial hunts for seal skins started in the early eighteenth century, focusing on harp and hooded seals, although bearded, ring, gray, and harbor seals were also taken. By the late 1800s, hunting expanded to Greenland for harp, hooded, and ringed seals. During the World Wars, hunting slowed down, allowing some populations to recover. However, sealing resumed immediately afterwards. During the 1960s, killing methods raised public opinion against sealing, which then prompted management actions and quotas to reduce hunting. The largest reduction began in 1983, particularly in Canadian waters, when the European community enacted a ban on the importation of seal skins. However, since 1996, the level of Canadian harp sealing has resumed to pre-1970s levels because new markets for skins and meat have opened up (DFO, 2003).

Long-standing conflicts between humans and seals have occurred because seals impact economically valuable fishery resources. Impacts include seals preying on fish species, and seals, particularly gray and harbor seals, infecting many North Atlantic fish species with seal (or cod) worm (*Pseudoterranova decipiens*). These issues have initiated seal bounty programs in Europe and North America, which resulted in regional extirpation (e.g., northeast United States, Baltic Sea), of some gray and harbor seal stocks. Although bounty and other seal removal programs have either ended or been greatly reduced, ecological and fishing gear interactions between seals and fisheries remain a management challenge in the North Atlantic.

Following World War II, technological improvements in fishing gear and vessels led not only to the expansion of coastal and high seas fisheries, but also to the incidental mortality of thousands of marine mammals and rapid depletion of fish resources (Northridge and Hofman, 1999; Hall and Donovan, 2002). By the 1970s, the elevated levels of marine mammal takes, particularly dolphins in the eastern tropical Pacific tuna purse seine fishery, instigated management and conservation measures that were aimed at reducing incidental takes of marine mammals in fisheries (e.g., US Marine Mammal Protection Act of 1972). Over the past two decades, national and international measures have aimed to improve fish stocks and to monitor and reduce fishery-related impacts on marine mammals [e.g., 1991 Agreement on the Conservation of Small Cetaceans of the Baltic and North Sea (ASCOBANS)]. Unfortunately, marine mammal mortality still occurring in many fisheries threatens some marine mammal populations in the North Atlantic, such as right whales, bottlenose and common dolphins (*Delphinus* spp.), harbor porpoises, and Mediterranean monk seals.

Environmental contaminants potentially pose a threat to the health of marine mammals (O'Shea, 1999; Reijnders *et al.*, 1999; Geraci and Lounsbury, 2002). Contaminant levels can become toxic in marine mammals because most feed at high trophic levels and so accumulate low levels of toxins from their contaminated prey. Numerous studies have documented the presence of organochlorine and heavy metals in tissues of marine mammals. The debate is: are these levels dangerous? Potential deleterious biological effects of these contaminants include immunosuppression, endocrine disruption, and reproductive and pathological disorders. Documented cases of deleterious effects include the reproductive failure that has been linked to organochlorine levels in seals from the Baltic and Wadden Seas and to beluga whales from the St. Lawrence Estuary. It has been suggested that some of the large-scale die-off events that have killed thousands of seals and dolphins in northern Europe, the Mediterranean, the US east coast, and Gulf of Mexico are due, at least in part, to high levels of organochlorines (e.g., PCBs; Domingo *et al.*, 2002) or toxic metals (e.g., cadmium, mercury; O'Shea, 1999). Epizootic events and toxic algal blooms have also caused large-scale die-offs (Geraci and Lounsbury, 2002; Härkönen *et al.*, 2006). For example, both the 1988 and the 2002 *Phocine distemper virus* epidemics in Europe killed approximately 56% and 45%, respectively of the European harbor seal populations. In winter 1987/1988, 14 humpback whales died in the vicinity of Cape Cod after consuming Atlantic mackerel (*Scomber scombrus*) containing a dinoflagellate saxitoxin. However, in nearly all cases, it has not been possible to demonstrate a direct link between death and contaminants. Other types of potentially dangerous environmental contaminants include oil spills and acoustic disturbances because these may cause behavioral modifications, prey displacement, or direct mortality. For example, several unusual mass strandings of beaked whales in North Atlantic marine environments (e.g., Bahamas, Canaries, Madeira) have been associated with military sonar activities (Evans and Miller, 2004; Cox *et al.*, 2006).

## V. Status

The current status of North Atlantic marine mammal populations is tightly linked to the population's biological characteristics and their long history of interacting with human activities. Most populations are no longer commercially hunted, but some are still severely depleted (e.g., North Atlantic right whales; Gambell, 1999; Kraus and Rolland, 2007). Human activities, such as hunting, incidental fishing mortality, acoustic activities, vessel strikes, environmental contaminants and climate change continue to directly and indirectly adversely impact marine mammals. Further, human enhanced climate warming is predicted to be detrimental to most marine mammal populations, particularly species associated with Arctic ice (Learmonth *et al.*, 2006). Conservation and research programs, particularly for small cetaceans, are highly variable among countries. Because most marine mammal populations are mobile, the only way to assess the status of and conserve these populations is to ensure that scientific research and conservation programs are effective ocean wide.

## See Also the Following Articles

Cetacea Overview ■ Distribution ■ Fishing Industry ■ Effects of Hunting of Marine Mammals ■ Pinnipedia Overview

## References

Baker, S. C., and Palumbi, S. R. (1997). The genetic structure of whale populations: Implications for management. *In* "Molecular Genetics of Marine Mammals" (A. E. Dizon, S. J. Chivers, and W. F. Perrin, eds), pp. 117–146. The Society for Marine Mammalogy, Lawrence, Special Publication 3.

Bowen, W. D., and Siniff, D. B. (1999). Distribution, population biology, and feeding ecology of marine mammals. *In* "Biology of Marine Mammals" (J. E. Reynolds, III, and S. A. Rommel, eds), pp. 423–484. Smithsonian Institution Press, Washington, DC.

Boyd, I. L., Lockyer, C., and Marsh, H. D. (1999). Reproduction in marine mammals. *In* "Biology of Marine Mammals" (J. E. Reynolds, III, and S. A. Rommel, eds), pp. 218–286. Smithsonian Institution Press, Washington, DC.

Clapham, P. J., Young, S. R., and Brownell, R. L., Jr. (1999). Baleen whales: conservation issues and the status of the most endangered populations. *Mamm. Rev.* **29**, 35–60.

Cox, T. M., *et al.* (2006). Understanding the impacts of anthropogenic sound on beaked whales. *J. Cet. Res. Manage.* **7**, 177–187.

DFO [Dept. of Fisheries and Oceans]. (2003). Atlantic Seal Hunt: 2003-2005 management plan. Available at: http://www.dfo-mpo.gc.ca/seal-phoque/reports-rapports/mgtplan-plangest2003/mgtplan-plangest2003_e.htm.

Domingo, M., Kennedy, S., and Van Bressem, M.-F. (2002). Marine mammal mass mortalities. *In* "Marine Mammals: Biology and Conservation" (P. G. H. Evans, and J. A. Raga, eds), pp. 425–456. Kluwer Academic/Plenum Press, London.

Evans, P. G. H., and Miller, L. A. (2004). *Active Sonar and Cetaceans*. Proceedings of Workshop held at the ECS 17th Annual Conference, Las Palmas, Gran Canaria, March 8, 2003. European Cetacean Society, Kiel, Germany.

Evans, P. G. H., and Stirling, I. (2002). Life history strategies of marine mammals. *In* "Marine Mammals: Biology and Conservation" (P. G. H. Evans, and J. A. Raga, eds), pp. 7–62. Kluwer Academic/Plenum Press, London.

Gambell, R. (1999). The International Whaling Commission and the contemporary whaling debate. *In* "Conservation and Management of Marine Mammals" (J. R. Twiss, Jr., and R. R. Reeves, eds), pp. 179–198. Smithsonian Institution Press, Washington, DC.

Geraci, J., and Lounsbury, V. (2002). Marine mammal health: Holding the balance in an ever changing sea. *In* "Marine Mammals: Biology and Conservation" (P. G. H. Evans, and J. A. Raga, eds), pp. 365–384. Kluwer Academic/Plenum Press, London.

Hall, M., and Donovan, G. (2002). Environmentalists, fishermen, cetaceans and fish: Is there a balance and can science help to find it. *In* "Marine Mammals: Biology and Conservation" (P. G. H. Evans, and J. A. Raga, eds), pp. 491–522. Kluwer Academic/Plenum Press, London.

Härkönen, T., *et al.* (2006). A review of the 1988 and 2002 phocine distemper virus epidemics in European harbour seals. *Dis. Aquat. Org.* **68**, 115–130.

Heide-Jørgensen, M.P., and Wiig, Ø. (eds) (2002). "Belugas in the North Atlantic and the Russian Arctic." NAMMCO Scientific Publications Volume 4. NAMMCO, Tromsø, Norway.

Kinze, C. C. (1995). Exploitation of harbour porpoises (*Phocoena phocoena*) in Danish waters: A historical review. *In* "Biology of the Phocoenids" (A. Bjorge and G. P. Donovan, eds), pp. 141–244. Report of the International Whaling Commission (Special Issue 16), Cambridge.

Kraus, S. D., and Rolland, R. M. (2007). "The Urban Whale: North Atlantic Right Whales at the Crossroads." Harvard University Press, Cambridge.

Learmonth, J. A., Macleod, C. D., Santos, M. B., Pierce, G. J., Crick, H. Q. P., and Robinson, R. A. (2006). Potential effects of climate change on marine mammals. *Oceanogr. Mar. Bio. Annu. Rev.* **44**, 431–464.

Macleod, C. D., *et al.* (2006). Known and inferred distributions of beaked whale species (Cetacea: Ziphiidae). *J. Cet. Res. Manage* **7**, 271–286.

Mercer, M. C. (1975). Modified Leslie-Delury population models of the long-finned pilot whale (*Globicephala melaena*) and annual production of the short-finned squid (*Illex illecebrosus*) based upon their interactions at Newfoundland. *J. Fish. Res. Bd. Can.* **32**, 1145–1154.

Northridge, S. P., and Hofman, R. J. (1999). Marine mammal interactions with fisheries. *In* "Conservation and Management of Marine Mammals" (J. R. Twiss, Jr., and R. R. Reeves, eds), pp. 99–119. Smithsonian Institution Press, Washington, DC.

O'Shea, T. J. (1999). Environmental contaminants and marine mammals. *In* "Biology of Marine Mammals" (J. E. Reynolds, III, and S. A. Rommel, eds), pp. 485–536. Smithsonian Institution Press, Washington, DC.

Reeves, R. R., Smith, B. D., Crespo, E. A., and Notarbartolo di Sciara, G. (compilers) (2003). Dolphins, Whales and Porpoises. 2002–2010 Conservation Action Plan for the World's Cetaceans. IUCN/SSC Cetacean Specialist Group. IUCN, Gland.

Reeves, R. R., Stewart, B. S., Clapham, P. J., and Powell, J. A. (2002). "Sea Mammals of the World." A. C. Black and Sons, New York.

Reid, J. B., Evans, P. G. H., and Northridge, S. P. (2003). "Atlas of Cetacean Distribution in North-west European Waters." Joint Nature Conservation Committee, Peterborough, UK.

Reijnders, P. J. H., Aguilar, A., and Donovan, G.P. (eds) (1999). "Chemical Pollutants and Cetaceans." *J. Cet. Res. Manage.* Special issue 1.

Rice, D. W. (1998). "Marine Mammals of the World: Systematics and Distributions," Special Publication 4. The Society for Marine Mammalogy, Lawrence, KS.

Sahrhage, D., and Lundbek, J. (1992). "A History of Fishing." Springer-Verlag, Berlin.

Slijper, E. J. (1979). "Whales," 2nd Ed. Hutchinson University Press, London (First English Edition, Hutchinson, 1962).

Stenseth, N., Otterson, G., Hurrell, J., and Belgrano, A. (2004). "Marine Ecosystems and Climate Variation. The North Atlantic—A Comparative Perspective." Oxford University Press, Oxford.

Urquiola, E., and de Stephanis, R. (2000). Growth of whale watching in Spain. The success of the platforms in south mainland. New Rules. *Eur. Res. Cetaceans* **14**, 198–204.

# North Pacific Marine Mammals

Sergio Escorza-Treviño

## I. North Pacific Marine and Fresh Water Biomes

The vastness and diversity of the North Pacific Ocean is reflected in the richness of its marine mammal community. Sixteen of the world's 36 species of pinnipeds, 50 of the more than 80 species of cetaceans, and two of the 5 species of sirenians have been reported to occur in the North Pacific, in addition to the polar bear (*Ursus maritimus*) and the sea otter (*Enhydra lutris*). Most of these species are also found in other parts of the world, as is the case of most balenids and delphinids, many ziphiids, and some otariids, phocids, and phocoenids. However, a large proportion of the species found in the North Pacific are endemic to its marine or riverine ecosystems: nine pinnipeds, eleven cetaceans, one sirenian, and the sea otter.

The North Pacific Ocean ranges from about 80°W to 130°E, covering almost 60% of the earth's circumference, and from the Arctic Ocean to the Equator (Fig. 1). The North Pacific encompasses a great number of peripheral basins, as different as the highly evaporative and relatively small Gulf of California (also known as Sea of Cortés) in the east, the large and epicontinental Bering Sea in the north, or the complex region of small, semi-enclosed seas and shallow shelves around the Indo-Pacific Archipelago in the west, where the Pacific and the Indian oceans meet. In addition, there exist a number of large, complex river systems that extend thousands of kilometers upstream, as is the case of the Yangtze River in China.

The geographic distribution of mammal species in the ocean depends on a number of factors, among which temperature, depth, and productivity tend to be the most important. Rice (1998) presents a comprehensive review of the ranges for most species. Some, such as the killer whale (*Orcinus orca*) or the sperm whale (*Physeter macrocephalus*), are considered cosmopolitan. Others, like the vaquita (*Phocoena sinus*), or the now extinct Steller's sea cow (*Hydrodamalis gigas*), have very limited ranges. Many species are circumglobal, but limited to particular climatic zones. For example, some species are pantropical, inhabiting low latitude waters in all the world oceans, whereas others have antitropical (or bipolar) distributions. Species such as the ringed seal (*Pusa hispida*) and polar bear have been sighted as far north as the North Pole. Others can range hundreds of kilometers up the great rivers of both sides of the Pacific Ocean, either permanently or on a seasonal basis.

The North Pacific is dominated by a large subtropical gyre (Fig. 1). This North Pacific central gyre flows clockwise, bounded to the west by the Kuroshio Current, to the north by the North Pacific Current, to the east by the California Current, and along the south by the North Equatorial Current. To the north of the North Pacific central gyre, the cold Oyashio Current flows along the Kamchatka Peninsula and forms the western boundary of a counterclockwise subarctic gyre. The Alaska Current flows counterclockwise along the southeastern coast of Alaska and the Aleutian Peninsula. The convergence zone of these subarctic gyres and the central gyre, known as the Subarctic Boundary, crosses the western and central North Pacific at about 42°N, and marks the steepest change in the abundance of cold-water vs warm-water species. To the south of the central gyre, the equatorial current system

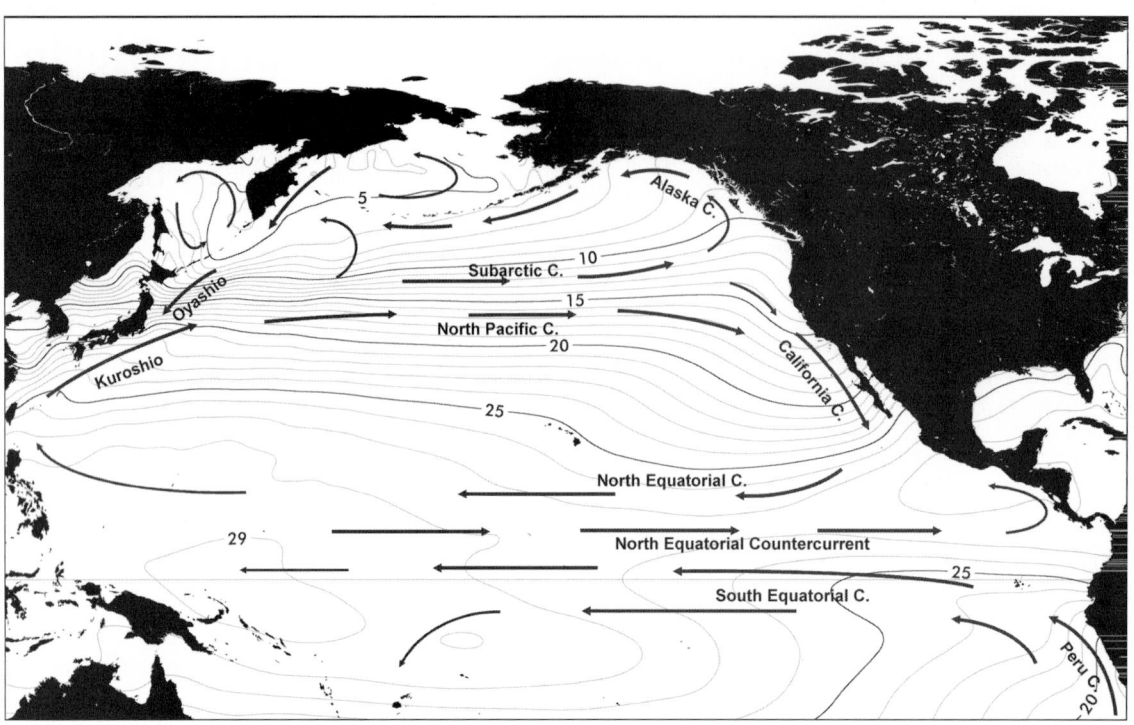

**Figure 1**   *Annual mean surface temperatures and principal currents of the North Pacific Ocean (graphic by Paul Fiedler).*

consists of the North Equatorial Countercurrent between the North and South Equatorial Currents.

An important differentiating characteristic between the Pacific and the other ocean basins is the strong effects of El Niño – Southern Oscillation (ENSO) events. Two distinct modes of Pacific circulation exist, depending on the variability of trade winds stress. When the trade winds weaken and are replaced by westerlies at low latitudes in the western Pacific (an ENSO event), the entire heat balance at low latitudes is perturbed, a situation than can last from a few months to 1 or 2 years. When this occurs, the major oceanographic features usually found in the Eastern Tropical Pacific can weaken or even disappear completely, resulting in an atypical distribution for many marine mammal species.

## II. Marine Mammals of Cold Water Marine Ecosystems

Most of the endemic species of marine mammals in the North Pacific Ocean inhabit its cold temperate, subarctic, or arctic waters. Among the pinnipeds, in the family Otariidae, we can find several of these endemic species. In fact, of the seven species of otariids present in the North Pacific, five are endemic. The other two are also represented in the Southern Pacific, since they occur in the Galapagos archipelago. Arnason *et al.* (2006) hypothesize that pinnipeds originated in North America and that early speciation of otariids took place in the northeast Pacific. The Guadalupe fur seal (*Arctocephalus townsendi*) now breeds only on Guadalupe Island, off Baja California. The northern fur seal (*Callorhinus ursinus*) is a pelagic species that ranges from the Sea of Okhotsk and southern Bering Sea to the Sea of Japan and northern Baja California. The California sea lion (*Zalophus californianus*) includes two geographical divisions, one on the Pacific coast, and one in the Gulf of California (Fig. 2). Some animals have also been sighted on both sides of the Atlantic, but represent escapees

**Figure 2**   *California sea lions (*Zalophus californianus*) occur along much of the coast of southwestern Canada, the western United States, and northern Mexico, as well as on islands nearby. These animals are from the midriff islands of the Mexican Sea of Cortez. Photo by Bernd Würsig.*

from captive facilities. Along the western side of the Pacific, the Japanese sea lion (*Zalophus japonicus*) is, very probably, extinct. The last confirmed sighting took place over 50 years ago. The Steller sea lion (*Eumetopias jubatus*) inhabits the coastal and immediate offshore waters of the cool temperate North Pacific.

Of the eight species of phocids (seals) present in the North Pacific Ocean, four are endemic: the spotted seal (*Phoca largha*) (Fig. 3), the ribbon seal (*Histriophoca fasciata*), the Hawaiian monk seal (*Monachus schauinslandi*), and the northern elephant seal (*Mirounga angustirostris*) (Fig. 4). The former two inhabit the pack ice zone of

**Figure 3** *Spotted and Largha seals* (Phoca largha) *occur throughout the northern area, with this photo taken in the Sea of Okhotsk, far east Russia. Photo by Bernd Würsig.*

**Figure 5** *Walrus* (Odobenus rosmarus) *off Round Island, Alaska. Photo by Bernd Würsig.*

**Figure 4** *The highly sexually dimorphic elephant seals* (Mirounga angustirostris) *can be found on islands, shorelines even quite close to intensive human use, and in the deep sea. These animals are from Año Nuevo, California. Photo by Bernd Würsig.*

**Figure 6** *A gray whale* (Eschrichtius robustus) *migrating south past Point Reyes, California. Photo by Bernd Würsig.*

the arctic North Pacific, although the ribbon seal becomes pelagic during the summer. The Hawaiian monk seal, which probably represents the most primitive of the living phocids, inhabits the Northwestern chain of the Hawaiian Islands, whereas the northern elephant seal, whose population numbers once dwindled down to a few tens, ranges throughout the northeastern Pacific, from the Aleutian Islands to the Gulf of California. Other seals present in the North Pacific are the bearded seal (*Erignathus barbatus*) and the ringed seal (*Pusa hispida*), which are both arctic species, the harbor seal (*Phoca vitulina*), whose range extends from the Bering Sea to Baja California, and the hooded seal (*Cystophora cristata*), present only as a rare vagrant into the Beaufort Sea, Bering Sea, and even southern California. The only living species of the family Odobenidae, the walrus (*Odobenus rosmarus*) also inhabits the arctic North Pacific (Fig. 5).

There are many cetacean species present in the higher latitudes of the North Pacific. Among the mysticeti, three families are found. The balaenids are represented by both the bowhead whale (*Balaena mysticetus*), which inhabits the pack ice zone of the Arctic, and the migrant and endemic North Pacific right whale (*Eubalaena japonica*), whose range extends from the Bering Sea to California in the East

and Taiwan in the West. The single living representative of the family Eschrichtiidae, the gray whale (*Eschrichtius robustus*), occurs only in the North Pacific, as the North Atlantic populations are extinct (Fig. 6). At present, the western, or "Korean," gray whale population is in very low numbers, whereas the eastern, or "California," population is considered to have recovered from past exploitation and ranges from subarctic waters to Baja California. It is not clear whether this is a full recuperation or not, as suggested by Alter *et al.* (2007).

Five species of rorquals are found in the cold waters of the North Pacific. The humpback whale (*Megaptera novaeangliae*) inhabits all the oceans of the world, from high-latitude summer feeding grounds to tropical winter grounds. The sei whale (*Balaenoptera borealis*) also is present in all oceans, but in more temperate waters than other rorquals. The common minke whale (*B. acutorostrata*) can be found from the Chukchi Sea almost to the Equator. The blue whale (*B. musculus*) lives almost everywhere in the world, from the tropics to the arctic pack ice. And the fin whale (*B. physalus*), also almost worldwide in distribution, can be sighted in the North Pacific from the Okhotsk Sea, Bering Sea and Chukchi Sea to Taiwan, Hawaii, and Baja California (including the Gulf of California).

The cosmopolitan sperm whale is present from the Equator to the edges of the pack ice. The ranges of the two species of the family Kogiidae are somewhat smaller, but are present in temperate and tropical waters around the world. The pygmy sperm whale (*Kogia breviceps*) inhabits oceanic waters, whereas the dwarf sperm whale (*K. sima*) lives over the continental shelf and slopes.

The family Ziphiidae is well represented in the North Pacific. Cuvier's beaked whale (*Ziphius cavirostris*) lives in all temperate and tropical waters around the world, whereas the North Pacific bottlenose whale (*Berardius bairdii*) is an endemic species, present only in the temperate and subarctic North Pacific. Also endemic are Hubbs' beaked whale (*Mesoplodon carlhubbsi*), in temperate waters, and the saber-toothed whale (*M. stejnegeri*), which ranges from the subarctic waters to the cold temperate North Pacific. The pygmy beaked whale (*M. peruvianus*) is a recently discovered species, known only from the Gulf of California and Peru. Also newly discovered is the Perrin's beaked whale (*M. perrini*), so far endemic to the North Pacific. In their study, Dalebout *et al.* (2002) indicated that individuals stranded and sighted off southern California, previously considered to be Hector's beaked whales (*M. hectori*), are likely to be *M. perrini*, suggesting that the Hector's beaked whale is confined to the temperate waters of the Southern Hemisphere.

Both members of the family Monodontidae are also present in the North Pacific Ocean. The beluga (*Delphinapterus leucas*) inhabits the Arctic Sea and spreads into its adjacent seas, mainly in shallow shelf waters. The narwhal (*Monodon monoceros*) is present as a vagrant in the Chukchi Sea and Bering Sea.

Several delphinids that can be found in the cold North Pacific are widespread species that mainly inhabit warm waters and will be discussed in the next section. In addition, the cosmopolitan killer whale is present from equatorial regions to the polar pack ice. The long-finned pilot whale (*Globicephala melas*) represents a special case. This species is bipolar in temperate waters and, although common in the North Atlantic, there are no historical records of living animals in the North Pacific. However, skulls were recovered at two archeological sites in Japan, in areas now occupied by the short-finned pilot whale (*Globicephala macrorhynchus*). Two other species present in this area are endemic to the North Pacific: the northern right-whale dolphin (*Lissodelphis borealis*), in subarctic and temperate waters, and the Pacific white-sided dolphin (*Lagenorhynchus obliquidens*), also in cool temperate areas (Fig. 7). Other species of *Lagenorhynchus* occupy this same niche in the Southern Hemisphere. However, genetic studies by LeDuc *et al.* (1999) have indicated that the genus might be polyphyletic, and the Pacific white-sided dolphin might indeed represent a different genus. If this were indeed the case, the correct scientific name for the Pacific white-sided dolphin would be *Sagmatias obliquidens*.

Four species of porpoises are present in the North Pacific. The Dall's porpoise (*Phocoenoides dalli*) is endemic to the cold temperate waters of the North Pacific (Fig. 8). The recently discovered and critically endangered vaquita, with only few hundred individuals left, is endemic to the northern Gulf of California. Its closest relative (Burmeister's porpoise, *Phocoena spinipinnis*), curiously, does not live in the North Pacific, but in southern latitudes, indicating that the vaquita evolved independently in the Gulf of California when it became trapped there after crossing the Equator during the last cooling period. The other two species present in the North Pacific are the harbor porpoise (*Phocoena phocoena*), restricted to shallow coastal temperate waters, and the finless porpoise (*Neophocaena phocaenoides*), which inhabits coastal waters along the mainland of southern and southeastern Asia and Japan, as well as up the Yangtze river system, up to 1600 km inland.

**Figure 7** *Pacific white-sided dolphin* (Lagenorhynchus obliquidens) *in Monterey Bay, California. Photo by Bernd Würsig.*

**Figure 8** *Dall's porpoises* (Phocoenoides dalli) *"rooster tailing" in the southern Bering Sea during rapid surfacings, a very species characteristic behavior. Photo by Bernd Würsig.*

A member of the order Sirenia, the only herbivorous marine mammals, used to be present in the cold waters of the North Pacific. The Steller's sea cow, endemic to shallow subarctic waters of this ocean, has been extinct since 1768. Hopes for the survival of the species went up briefly in 1962, but the alleged sighting has been discredited since.

There are two other marine mammals in the North Pacific that do not belong to the "traditional" marine mammal groups. One is the sea otter, which is endemic to the North Pacific and inhabits the cold temperate and subarctic zones (Fig. 9). This species dipped to near extinction at the end of the nineteenth century and beginnings of the twentieth century. The other is the polar bear, which inhabits the pack ice regions of the Arctic Ocean and contiguous seas and adjacent coastal areas.

## III. Marine Mammals of Warm Water Marine Ecosystems

The marine mammals that inhabit these ecosystems tend to be also present south of the Equator, or even in other oceans. Among the otariids, two Pacific species occur only in tropical waters. The

**Figure 9** *A territorial male sea otter* (Enhydra lutris) *swims within his territory in Simpson Bay, Prince William Sound, Alaska. Sea otters are sexually segregated when not mating, with males and females separated by less than 150 km. During the mating season, male sea otters enter female areas and establish resource-based territories to attract potential mates. Photo by Heidi Pearson.*

Galapagos sea lion (*Zalophus wollebaeki*) was originally confined to this archipelago, but a small rookery exists now on La Plata Island (01°S), which is why the species cannot be considered endemic to the North Pacific. The Galapagos fur seal is also endemic to this equatorial archipelago.

Many of the cetaceans present in warm waters are widespread species, with wide latitudinal ranges, and have already been referred to in the previous section (Table I). This is the case of most rorquals, except for the Bryde's whale (*Balaenoptera brydei*), which prefers tropical and warm temperate waters, and the small type or pygmy Bryde's whale (*B. edeni*), which lives in coastal and shelf waters of the eastern Indian Ocean and the western Pacific. Among the ziphiids, some are found exclusively in warm water. Such is the case of the ginkgo-toothed whale (*Mesoplodon ginkgodens*), which lives in tropical and warm temperate areas of the Indian and Pacific Oceans, and the Blainville's beaked whale (*Mesoplodon densirostris*), also with a tropical and warm temperate distribution, but around the world. Most beaked whales, specially the genus *Mesoplodon*, are rarely spotted at sea and, even when they are, their identification is extremely difficult. It is very probable that some species have not been described yet. Two such undescribed species of *Mesoplodon* have been sighted in the Eastern Tropical Pacific: *Mesoplodon sp. A*, also known as bandolero beaked whale, and *Mesoplodon sp. B*. It has been suggested that the former is in fact the pygmy beaked whale and the latter could possibly be the Bahamonde's beaked whale (*Mesoplodon traversii*). It is also possible that large unidentified "tropical bottlenose whales" seen in the tropical Indian and Pacific Oceans are indeed Longman's beaked whales (*Indopacetus pacificus*), although this will remain a mystery until specimens are collected.

In addition to the widespread delphinids mentioned in the previous section, several species typically found in warm waters can also range into temperate waters. Probably the most familiar is the bottlenose dolphin (*Tursiops truncatus*), which is present in temperate and tropical zones of all oceans. The species usually inhabits coastal areas and shallow offshore banks, although there are pelagic populations that live far offshore. The type specimen of the Indian Ocean bottlenose dolphin (*Tursiops aduncus*) comes from the Red Sea, but some *aduncus*-like *Tursiops* have been sighted throughout the North Pacific. The

striped dolphin (*Stenella coeruleoalba*) is also found worldwide in temperate and tropical waters, as are the Risso's dolphin (*Grampus griseus*) and the false killer whale (*Pseudorca crassidens*). Other species are characteristic of the tropical and warm temperate waters around the world and thus are also found in the North Pacific. Such is the case of the rough-toothed dolphin (*Steno bredanensis*), the pantropical spotted dolphin (*Stenella attenuata*), the spinner dolphin (*Stenella longirostris*), Fraser's dolphin (*Lagenodelphis hosei*), the melon-headed whale (*Peponocephala electra*), the pygmy killer whale (*Feresa attenuata*), and the short-finned pilot whale. Both the short-beaked common dolphin (*Delphinus delphis*) and the long-beaked common dolphin (*D. capensis*) inhabit tropical and warm temperate waters worldwide. The former occupies both offshore and coastal habitats. However, the later shows a disjunct distribution in near-shore waters, which might be due to the fact that these populations had evolved from independent events. The Indo-Pacific humpback dolphin (*Sousa chinensis*) is discontinuously distributed in coastal waters of the western Pacific, as well as in river basins. It is also present in the southern Pacific and Indian Ocean.

The dugong (*Dugong dugon*), the only sirenian found in warm North Pacific waters, has a widespread but discontinuous range along the continental coasts among the islands of the Indian and western Pacific oceans.

## IV. Aquatic Mammals of River Ecosystems

The river basins, systems, and their mouths in southeastern Asia provide a unique habitat for several cetacean species. The most important hydrographic system for marine mammals is the Yangtze River, and several species can be found here. The baiji (*Lipotes vexillifer*) lives exclusively in fresh water, and is endemic to this basin, from its estuary upstream for 1600 km. Unfortunately, it might very well be already extinct. The Indo-Pacific humpback dolphin can ascend as far as 1200 km up this river system as well. The finless porpoise inhabits the coastal waters along the mainland of southern and southeastern Asia and Japan, and also ranges into the middle and lower reaches of the Yangtze River and its tributaries. Finally, the Irrawaddy dolphin (*Orcaella brevirostris*) is discontinuously distributed in shallow and murky waters at the mouth of rivers around the Asian mainland and Australasia, as well as in river systems such as the Mekong.

## V. Conservation Issues

The problems faced by North Pacific marine mammals are common to marine mammals in other areas of the world: pollution, habitat degradation, fishery interactions, directed takes, climatic shifts and lack of sufficient resources for successful protection and management. Although the hunting of marine mammals for oil, meat, or fur has been largely reduced by international agreement, some species are still targeted. The active harpoon fishery for the North Pacific endemic Dall's porpoise, for example, causes the largest fishery-related mortality for any cetacean species in the world. A large number of animals from several species are taken incidentally each year by net entanglements, fishery by-catch, or boat strikes. Two North Pacific marine mammal species have been driven to extinction by humans in the recent past: the Japanese sea lion and Steller's sea cow. A third one, the baiji, is very probably the latest addition to this list, as no authenticated sightings have taken place since 2002, including a dedicated expedition in 2006 by Turvey *et al.* (2007). Several other North Pacific species are dangerously close to following in their wake, as is the case for the vaquita, the western population

## TABLE I
### Marine Mammal Species Present in the North Pacific Ocean, by Taxa, and Waters They Inhabit

| TAXA | Cold | Warm | River |
|---|---|---|---|
| **Order Sirenia** | | | |
| Family Dugongidae | | | |
| *Dugong dugong* (dugong) | | | X |
| **Hydrodamalis gigas (Steller's sea cow)** | | X | |
| **Order Carnivora** | | | |
| Family Ursidae | | | |
| *Ursus maritimus* (polar bear) | X | | |
| Family Mustelidae | | | |
| **Enhydra lutris (sea otter)** | | X | |
| Family Otariidae | | | |
| **Arctocephalus townsendi (Guadalupe fur-seal)** | | X | |
| *Arctocephalus galapagoensis* (Galapagos fur-seal) | | X | |
| **Callorhinus ursinus (northern fur-seal)** | | X | |
| **Zalophus californianus (California sea lion)** | | X | |
| **Zalophus japonicus (Japanese sea lion)** | | X | |
| *Zalophus wollebaeki* (Galapagos sea lion) | | | X |
| **Eumetopias jubatus (Steller sea lion)** | | X | |
| Family Odobenidae | | | |
| *Odobenus rosmarus* (walrus) | X | | |
| Family Phocidae | | | |
| *Erignatus barbatus* (bearded seal) | | X | |
| *Phoca vitulina* (harbor seal) | X | | |
| **Phoca largha (spotted seal)** | X | | |
| *Pusa hispida* (ringed seal) | | X | |
| **Histriophoca fasciata (ribbon seal)** | X | | |
| *Cystophora cristata* (hooded seal) | | X | |
| **Monachus schauinslandi (Hawaiian monk seal)** | | X | |
| **Mirounga angustirostris (northern elephant seal)** | | X | |
| **Order Cetacea** | | | |
| Family Balaenidae | | | |
| **Eubalaena japonica (North Pacific right whale)** | | X | |
| *Balaena mysticetus* (bowhead whale) | X | | |
| Family Eschrichtiidae | | | |
| **Eschrichtius robustus (gray whale)** | X | | |
| Family Balaenopteridae | | | |
| *Megaptera novaeangliae* (humpback whale) | X | X | |
| *Balaenoptera acutorostrata* (minke whale) | X | X | |
| *Balaenoptera borealis* (sei whale) | X | X | |
| *Balaenoptera brydei* (Bryde's whale) | X | X | |
| *Balaenoptera edeni* (small type Bryde's whale) | X | X | |
| *Balaenoptera musculus* (blue whale) | X | X | |
| *Balaenoptera physalus* (fin whale) | X | X | |
| Family Physeteridae | | | |
| *Physeter macrocephalus* (sperm whale) | X | X | |
| Family Kogiidae | | | |
| *Kogia breviceps* (pygmy sperm whale) | X | X | |
| *Kogia sima* (dwarf sperm whale) | X | X | |
| Family Ziphiidae | | | |
| *Ziphius cavirostris* (Cuvier's beaked whale) | X | X | |
| **Berardius bairdii (North Pacific bottlenose whale)** | X | | |
| *Indopacetus pacificus* (Longman's beaked whale) | | X | |
| *Mesoplodon sp. A* | | X | |
| *Mesoplodon sp. B* | | X | |
| **Mesoplodon perrini (Perrin's beaked whale)** | X | | |
| *Mesoplodon peruvianus* (pygmy beaked whale) | | X | |
| **Mesoplodon carlhubbsi (Hubbs' beaked whale)** | X | | |
| *Mesoplodon ginkgodens* (ginkgo-toothed whale) | | X | |
| **Mesoplodon stejnegeri (saber-toothed whale)** | X | | |
| *Mesoplodon densirostris* (Blainville's beaked whale) | | X | |

N

*(continues)*

TABLE I (continued)

| TAXA | Cold | Warm | River |
|---|---|---|---|
| Family Lipotidae | | | |
| **Lipotes vexillifer (baiji)** | | | X |
| Family Monodontidae | | | |
| *Delphinapterus leucas* (beluga) | X | | |
| *Monodon monoceros* (narwhal) | X | | |
| Family Delphinidae | | | |
| *Steno bredanensis* (rough-toothed dolphin) | | X | |
| *Sousa chinensis* (Indo-Pacific humpback dolphin) | X | X | |
| *Tursiops truncatus* (bottlenose dolphin) | X | X | |
| *Tursiops aduncus* | | X | |
| *Stenella attenuata* (pantropical spotted dolphin) | | X | |
| *Stenella longirostris* (spinner dolphin) | | X | |
| *Stenella coeruleoalba* (striped dolphin) | X | X | |
| *Delphinus delphis* (short-beaked common dolphin) | X | X | |
| *Delphinus capensis* (long-beaked common dolphin) | X | X | |
| *Lagenodelphis hosei* (Fraser's dolphin) | | X | |
| **Lagenorhynchus obliquidens (Pacific white-sided dolphin)** | X | | |
| **Lissodelphis borealis (northern right-whale dolphin)** | X | | |
| *Grampus griseus* (Risso's dolphin) | X | X | |
| *Peponocephala electra* (melon-headed whale) | | X | |
| *Feresa attenuata* (pygmy killer whale) | | X | |
| *Pseudorca crassidens* (false killer whale) | X | X | |
| *Orcinus orca* (killer whale) | X | X | |
| *Globicephala melas* (long-finned pilot whale) | X | | |
| *Globicephala macrorhynchus* ((long-finned pilot whale) | X | X | |
| *Orcaella brevirostris* (Irrawaddy dolphin) | | X | X |
| Family Phocoenidae | | | |
| *Neophocaena phocaenoides* (finless porpoise) | X | X | X |
| *Phocoena phocoena* (harbor porpoise) | X | | |
| **Phocoena sinus (vaquita)** | X | | |
| **Phocoenoides dalli (Dall's porpoise)** | X | | |

Endemic species are shown in bold.

of the gray whale and the North Pacific right whale. Sea otters and some pinnipeds, especially some populations of Steller sea lions, have experienced a dramatic decrease in the last few decades whose causing factors are not entirely understood, as demonstrated by the different explanations put forward by Springer *et al.* (2003) or DeMaster *et al.* (2006). However, the recovery of other North Pacific species gives us reason for hope. The northern elephant seal, driven down to a few tens of individuals by hunting in the early 1900s, now numbers in the tens of thousands. The eastern gray whale, whose population size was once dangerously low, has recovered significantly (there exists debate whether to its pre-exploitation stock size) and was taken off the endangered species list in 1995. With continued protection, two species that once faced extinction are now showing positive signs of recovery, the sea otter and the Guadalupe fur seal. Because of their mobility and vast geographic ranges, the distributions of most marine mammal species extend across borders and international waters. For these reasons, effective management and conservation strategies depend greatly on international agreements that coordinate objectives, resources, and efforts.

## See Also the Following Articles

Arctic Marine Mammals ■ Hunting of Marine Mammals ■ Incidental Catches ■ Migration ■ Zoogeography

## References

Alter, S. E., Rynes, E., and Palumbi, S. R. (2007). DNA evidence for historic population size and past ecosystem impacts of gray whales. *Proc. Natl. Acad. Sci. USA* **104**, 15162–15167.

Arnason, U., Gullberg, A., Janke, A., Kullberg, M., Lehman, N., Petrov, E. A., and Väinölä, R. (2006). Pinniped phylogeny and a new hypothesis for their origin and dispersal. *Mol. Phylogenet. Evol.* **41**, 345–354.

Dalebout, M., Mead, J. G., Baker, C. S., Baker, A. N., and van Held, A. L. (2002). A new species of beaked whale *Mesoplodon perrini sp. N.* (Cetacea: Ziphiidae) discovered through phylogenetic analyses of mitochondrial DNA sequences. *Mar. Mamm. Sci.* **18**, 577–608.

DeMaster, D. P., Trites, A. W., Clapham, P., Mizroch, S., Wade, P., Small, R. J., and Ver Hoef, J. (2006). The sequential megafaunal collapse hypothesis: Testing with existing data. *Prog. Oceanogr.* **68**, 329–342.

Favorite, F., Dodimead, A., and Nasu, J. K. (1976). "Oceanography of the subarctic Pacific region." International North pacific Fisheries Commission, Vancouver, Canada.

Green, G. A., Bonnell, M. L., Balcomb, K. C., Bowlby, D. E., Grotefendt, R. A., and Chapman, D. G. (1990). Synthesis of information on marine mammals of the eastern North Pacific, with emphasis on the Oregon and Washington OCS area. *In* "Oregon and Washington Marine Mammal and Seabird Surveys: Information Sythesis and Hypothesis Formulation" (J. J. Bruggeman, ed.), OCS Study MMS 89-0030. U.S. Department of the Interior, Minerals Management Service, Pacific OCS Region, Washington DC

N

Leatherwood, S., Reeves, R. R., Perrin, W. G., and Evans, W. E. (1988). "Whales, Dolphins, and Porpoises of the Eastern North Pacific and Adjacent Arctic Waters. A Guide to their Identification." Dover Publications, Inc., New York.

LeDuc, R. G., Perrin, W. F., and Dizon, A. E. (1999). Phylogenetic relationships among the delphinid cetaceans based on full cytochrome B sequences. *Mar. Mamm. Sci.* **15**, 619–648.

Longhurst, A. (1998). "Ecological Geography of the Sea." Academic press, San Diego.

de Muizon. (1982). Phocid phylogeny and dispersal. *Ann. South African Mus.* **82**, 175–213.

Nishiwaki, M. (1967). Distribution and Migration of Marine Mammals in the North Pacific Area. Bulletin of the Ocean Research Institute University of Tokyo, No 1. Tokyo.

Osborne, R. W., and van Beever, H. (1979). "Sightings of Marine Mammals in Eastern North Pacific Coastal Waters." Ocean Research and Education Society, Boston.

Rice, D. (1998). "Marine Mammals of the World: Systematics and Distribution," Special Publication Number 4. Society for Marine Mammology, Lawrence.

Seed, A. (1971). "Toothed Whales in Eastern North Pacific and Arctic Waters." Pacific Search Books, Seattle.

Seed, A. (1972). "Baleen Whales in Eastern North Pacific and Arctic Waters." Pacific Search Books, Seattle.

Springer, A. M., *et al.* (8 authors) (2003). Sequential megafaunal collapse in the North Pacific Ocean: an ongoing legacy of industrial whaling? *Proc. Natl. Acad. Sci. USA* **100**, 12223–12228.

Turvey, S. T., *et al.* (16 authors) (2007). First human-caused extinction of a cetacean species? *Biol. Lett.* **3**, 537–540.

# Northern Fur Seal
## *Callorhinus ursinus*

### Roger L. Gentry

## I. Characteristics and Taxonomy

*Callorhinus* is the oldest living genus of the family Otariidae, with origins 2–5 million years ago. *Callorhinus* means "beautiful nose" and *ursinus* means "bear-like." Sea Bear was a former common name. This species has a short nose, and lacks the dog-like profile of the other fur seal genus, *Arctocephalus*. Females are small (45 kg) and gray-brown with a light underbelly; males are larger (200–250 kg) and vary from black to reddish (Fig. 1). The young (pups) are generally black with a light belly.

## II. Distribution and Abundance

The northern fur seal inhabits the North Pacific Ocean and Bering Sea (Fig. 2). About 91% of the northern fur seal population breeds at the Pribilof Islands of Alaska (74%) or at the Commander Islands of Russia (17%). The remainder breeds at the Kurile Islands (Kammenye Lovushki and Sredney Rocks in Fig. 2) and Robben Island (Russia), Bogoslof Island (Alaska), and the Farallone and San Miguel Islands (California). The species predates some of the islands on which it now breeds, so some past population redistribution has occurred. For example, bones of pups and adult males in kitchen middens show that breeding colonies once occurred at several mainland sites in the United States. The northern fur seal is primarily a sub-polar species whose adaptations do not preclude breeding at lower latitudes (for general reading about pinniped biology, see Riedman (1990)).

**Figure 1** *Adult male and female northern fur seals and newborn pups. Photo by C.W. Fowler, AFSC, Seattle, WA.*

When the Fur Seal Treaty was resumed after World War II, one of the first decisions was to intentionally reduce the size of the herd. About 331,000 adult females were killed at the Pribilofs and at sea between 1956 and 1968 in the belief that pelt production would increase. From that point onward, the different breeding populations followed different trends. The Commander Islands population increased from 1958 to the mid-1970s, decreased slightly, then remained stable from the mid 1990s to 2005 at about 60,000 pups annually. The Kurile Islands population increased rapidly from 1962 to 1977, decreased slightly until 1988, and increased rapidly again until 2006 to about 30,000 pups annually. Robben Island declined until the mid-1990s, and increased until 2003 at 30,000 pups annually. From 1968 to 1974 the Pribilof Island herd recovered to about 1.25 million animals, declined rapidly (8% per year) until the mid-1980s, stabilized until 1996, then declined rapidly (5.8% per year) to 2004 (Towell *et al.*, 2006). Bogoslof Island began as a new colony in 1980 and increased to 12,600 pups annually by 2005. San Miguel Island, California began as a breeding colony in 1968 and peaked at over 3000 pups 1996. Twice (1982 and 1996) the San Miguel population was driven down by severe El Niño Southern Oscillation events that reduced fur seal food (Angliss and Outlaw, 2007). Overall the species is doing well except the Pribilof Islands seals, which are still declining and producing fewer than half the pups born there in 1968.

## III. Ecology

Ecologically the species is a high-level consumer, taking a variety of fish, cephalopods, crustaceans, and an occasional bird. Seventy-five different prey species have been identified in northern fur seal stomachs. Juvenile walleye pollock, Atka mackerel, capelin, eulachon, herring, and several species of squid predominate. The seals take these prey items at relatively shallow depths (100–200 m for females, <400 m for males) compared to other species of marine mammals that can dive to more than a thousand meters. Northern fur seals feed mostly at night on prey that migrate vertically. A major component of their diet is lantern fish (myctophids) which live in deep water. Northern fur seals are preyed upon by large sharks, killer whales, and (for young only) Steller sea lions (*Eumetopias jubatus*).

This species is not under intense competition for breeding space by other species of eared seals (otariids). California sea lions (*Zalophus californianus*) compete with them for space only at San Miguel. Steller

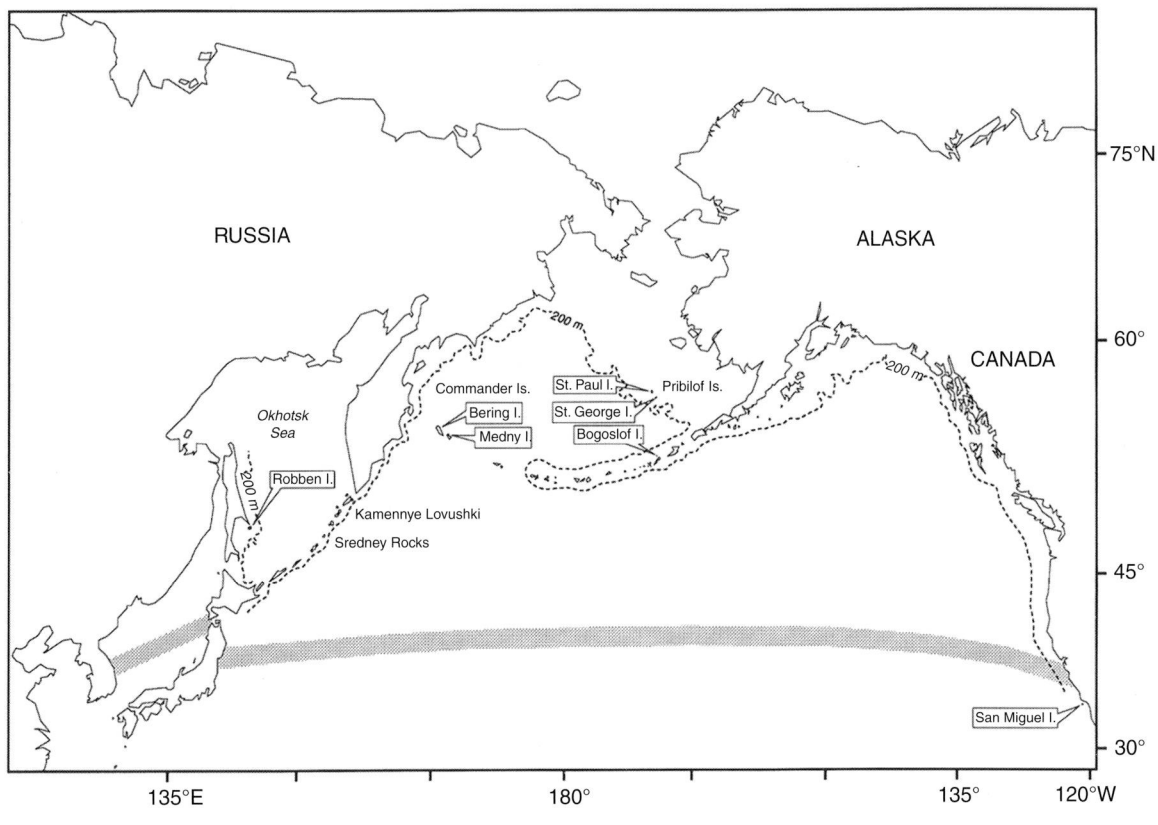

**Figure 2**  *Distribution of the northern fur seal (Shaded line is southern limit of pelagic range).*

sea lions may rest among northern fur seals (and sometimes eat their young at sea) but do not usually mate on the same sites. Guadalupe fur seals (*Arctocephalus townsendi*) are seen on and occasionally pup there, but do not compete with them for space. Furthermore, northern fur seals do not often compete with other otariids for food. They take small fish and squid over deep water, whereas sea lions tend to feed mostly on the continental shelves on larger fish. An exception is that northern fur seals and Steller sea lions in the Bering Sea both take juvenile walleye pollock. Guadalupe fur seals probably prey on small fish and squid like the northern fur seal, but the Oyashio/Kuroshio transition zone effectively separates the two fur seal species at sea.

Like other fur seals and sea lions, northern fur seal females must feed themselves while raising their pups. They are so-called "income" foragers as opposed to "capitol" foragers like true seals, which nurse their young from fat stores laid down well before birth. The need to alternate feeding with nursing means that fur seals can only raise pups where nearby food supplies are abundant. But how they forage and what they eat depend largely on the local environment. The Commander Islands in Russia have a narrow continental shelf that lacks a good fish population. There, females cross the shelf quickly and forage in nearby deep water where small squid rise to the surface at night. The Pribilof Islands have a wide shelf that supports a good fish population. Females there catch fish near the bottom as they cross this shelf going to and returning from more distant deep waters where they feed like the Commander Islands females on deep-water squid at night. The length of time females are away from their pups, the depth of their dives, number of dives per day, and the food they eat all differ among islands, and they change seasonally as the energy needs of the pups increase.

## IV. Behavior and Physiology

Males partition the birthing areas used by females into spaces called territories. They defend the boundaries of these territories (not the females therein) using mainly calls and postural threats, sometimes by fighting. Males arrive on territory about 1 month before the females, and spend their time maintaining boundaries and fasting. Fasts last on average 38–42 days for the population, but uninterrupted fasts of 80 days have been recorded for individuals. Throughout the main breeding season (July) males spend most of their time interacting with females and mating. Each male has exclusive access to all females in his territory. They abandon their territory in late July, and are replaced by large sub-adult males, which perform perhaps 20% of the year's copulations, including with all the virgin females which arrive after pupping has ended. Males are highly specific in the area they will defend, maintaining at least some of the same boundaries from year to year. Very few defend a territory that is more than 10 m distance from the previous year's territory. Males will persist in defending a territory from year to year even if it contains few females.

Females arrive usually one day before giving birth, select a birth site without regard to the male that defends that site, and usually gives birth at night. Females remain with the pup in that location and defend it from other females until mating occurs. Then they leave to feed for a few days before returning to nurse their pup. Returning mothers locate their young by their calls and probably by smell. They will not suckle any pup but their own. Orphans and starvlings sneak suckle but are often bitten and many die of wounds and infection.

Females use the same birthing site from year to year if the population remains the same size. But, if a population declines the females will abandon favored birth sites and move closer to the water to avoid being alone in a territory with an adult male. Males bite and herd females to control their movements. So, by giving birth near other females a mother can reduce being injured by males.

Despite seeking the close company of others, females are quite aggressive toward near neighbors. They bite and frequently threaten all other females, do not allow them to lie in full body contact (like sea lions do), and show no sign of social bonds with other females. That is, females are gregarious but not social. This combination allows each animal to enter and leave the aggregation without loss of social status. The largest females win contests over resting sites, but there is no evidence that individuals form a stable social hierarchy.

Active female mate choice is not apparent. Females mate at nearly the same site for many years whereas the male defending that site as part of his territory changes on average every 1.5 seasons. That is, females do not move around trying to find the highest ranking, most fit male. In captivity, sexually receptive females do not discriminate between high- and low-ranking males, or even between fur seals and humans. Females mate with highly fit males in the population, but by a more indirect means than individual choice. By arriving predictably at a given time and place, receptive females become a resource over which males compete. Males that win this competition remain on shore as potential mates. Females thus indirectly acquire as potential mates the class of males that is highest ranking, but they accomplish this without making a choice among individual males. This appears to explain why females are uniformly receptive to all males when they are in estrus. Individuals show strong preferences for using specific land sites at particular times of year. Despite having many square kilometers of suitable land available for giving birth, individual females bear their young within 8–10 m of the same site in successive years. Males will defend only one territorial location for their reproductive lifetime. This tie to specific sites may be an extreme form of philopatry (mating on the site of their own birth). Furthermore, females and males tend to arrive on a date that is specific to them as individuals despite the fact that births and mating in the population occur over a 6-week period. On the beach, tens of thousands of breeding animals appear to form one large, amorphous mass. However, in reality it is composed of individuals each with fairly specific preferences for breeding at a given location and time of year (Gentry, 1998).

These space and time preferences of individuals create breeding aggregations that are highly predictable in timing and location. Thirty-two breeding sites, or "rookeries," now exist. One in Russia has been occupied yearly since at least 1742. These sites do not change much in size or shape from year to year if the population is stable, mainly because of site fidelity in individual females. The onset and duration of the birth/mating season are quite stable and change little in response to weather or climatic conditions.

The young from all islands except San Miguel leave land at 4 months of age with modest fat reserves and little foraging experience. At sea, they must immediately learn to find and capture solid food alone while migrating in the appropriate direction. This is apparently difficult because some 60% or more of the pups born per year fail to live to age 2 years when their age mates first return to land. Despite this seemingly low survival rate, the species is evolutionarily quite successful; it numbered approximately 2.5 million animals at its peak in the mid-1950s.

The migratory pathways, distribution, diet, and behavior of weaned pups at sea are largely unknown because they are small and difficult to instrument and recapture. It is known that Pribilof pups migrate through the Aleutian passes into the North Pacific Ocean in November of their birth year, and that some of them are seen on the coasts of Washington, British Columbia and Japan by the following

January. Pups return to land during the summer of their second or third year, usually to the island of their birth. Juvenile females usually land on the breeding grounds but juvenile males are excluded from these areas by adult males and instead come ashore on all-male landing areas that old sealers called "hauling grounds." Juvenile males stay around the breeding island for 80 days or so, making periodic feeding trips to sea, visiting different hauling grounds, and playing with other males as they perfect the movements they will later use in fighting. Many juvenile males make brief incursions onto the breeding grounds where they visit sites that some of them will later defend as territories, which suggests philopatry. The hauling grounds are where sealers used to capture and kill animals for pelts. Despite the fact that killing occurred on some of these sites weekly during the breeding seasons for over two centuries, fur seals persist in using them.

Physiology drives some aspects of foraging behavior such as the duration of the deepest dives and the recovery times from these dives neither of which change seasonally. The deepest dives are about 200 m, last about 5 min, and require 20 min of recovery. These are made by large females that are foraging near the bottom. Females that feed in deep water change their dive depths in a way that suggests the prey are deep during the day, rise to the surface at dusk, and remain there all night before migrating to deeper water again at dawn. This pattern fits the behavior of the Deep Scattering Layer, a layer of fish and other animals that forms in deep waters almost globally.

The northern fur seal keeps from losing body heat to cold ocean waters by means of dense underfur that traps small air bubbles. This fur is so dense it stays dry even when submerged. While at sea, and especially after deep dives, fur seals spend long periods grooming the fur and restoring its air content. They lose heat through their naked rear flippers, which are the longest and thinnest of any pinniped. To reduce heat loss while sleeping they often extend one front flipper into the air and hold it between both rear flippers, thereby creating an arch that sealers used to call a "jug handle." To increase heat loss on land in warm weather they fan their rear flippers, keep them damp, or pant. The fur is such an effective insulator that snow falling on sleeping animals does not melt or make the animals shiver.

Oxygen management is as important to fur seals as to any diving mammal. They are apneustic breathers, meaning they take in a breath and hold it as they would during a dive. Even sleeping animals breathe this way. During a dive, they shut off blood circulation to inessential organs and greatly reduce their heart rate to help conserve oxygen stores. While swimming at high speed northern fur seals often leap free of the water to breathe in a "porpoising" motion.

Both vision and hearing are highly effective in both air and under water. In bright light, their pupil closes to a small pinhole showing a brown iris. In dim light, the pupil opens wide to gather more light. An open pupil and a reflective layer in the retina contribute to good vision in dim light where most feeding occurs. The reflective layer is visible as an "eye shine" if a light is trained on animals at night. Their best hearing is between 2 and 29 kHz. The highest frequency they can hear is about 40 kHz (compared to 20 kHz for humans). These modest hearing abilities are related to the fact that fur seals do not find food by echolocation, as dolphins do.

## V. Life History

The species is sexually dimorphic; males are at maximum 4.5 times larger than females. A size difference exists as early in embryonic growth as the sexes can be identified, and it accelerates at sexual maturity, about age 3–5 years of age. Females mate almost immediately after becoming sexually mature, and begin to produce one young per year.

Males do not become socially mature (large enough to gain a territory among females) until they are 8- to 9-years old. Like other otariids, the sex ratio at birth of northern fur seals is nearly 1:1, but due to greater age-specific mortality rates in males than in females, and delayed mating in males, mating is polygynous. The average adult sex ratio is about nine females per male, but it once reached 60:1 when males were being killed for pelts.

Females have the highest pregnancy rates between ages 8 and 13 years, the rates sometimes exceeding 93% of an age class. Females can produce offspring yearly to about age 22 years, although few survive to that age. Older, larger females do not preferentially give birth to male pups, but larger mothers give birth to larger pups. Twinning is rare. For unknown reasons, females in the Commander Islands population have their first pup on average a year younger than females in the Pribilof population (Wickens and York, 1997).

Females usually enter estrus (heat) and mate about 5.3 days after giving birth. About 85% of the females mate only once during estrus; the other 15% mate no more than twice. The reason for this low number of copulations is that the physical act of mating terminates receptivity. Males can cause females to come into estrus by interacting with them (called the Whitten effect). However, if no male is present, females enter estrus spontaneously and become non-receptive again after about 36h if they fail to copulate. No second estrus occurs that year (Gentry, 1998).

Sexually receptive females mate indiscriminately with any male that approaches them. When no adult males are present (as in captivity) females will mate with a male of any size or age. They will arch their backs in a sexually receptive posture when touched on the pelvic area by other females, their own pups, or even a human. Adult males do most of the mating because adults drive juvenile males off the breeding areas, not because females reject juvenile males. Males may mate 115 times or more per year, depending on their size and the location of their territory relative to females. The reproductive life span of males averages 1.5 seasons, but one unusual male held a territory for 10 consecutive seasons.

As in other pinnipeds, the young of northern fur seals are precocial. They can swim on the day of birth but do not voluntarily do so until they are a month old. While their mothers are away foraging, pups withdraw to unoccupied parts of the breeding grounds to avoid being bitten by adult females. There they join other pups in dense "pods" and sleep and play until their mothers return. Males actively avoid trampling pups so these pods are places of relative safety for pups. Intentional infanticide by adult males is unknown in this species. Some of the behavior of pups appears to be instinctual. For example, a 6-day old male pup can mount a female correctly and make coordinated mating motions without ever having observed mating behavior in adult males.

Male pups are on average 0.6 kg heavier than females at birth, and remain larger until weaning. Suckling pups grow slower (1.1–1.3% of maternal mass per day) than in true seals because they fast each time their mothers are away foraging, and most of them face cold wind and rain that requires high metabolic rates to survive. About 10% of pups die on shore before weaning. Starvation, trauma, hookworm, and several infectious diseases account for most of the deaths. Pups are weaned at about 40% of adult female mass. Most pups wean themselves and depart from land before their mothers.

## VI. Human Interactions

Northern fur seals have been studied longer than most other marine mammals. George Wilhelm Steller first described the species in 1749 after he returned from Vitus Bering's 1742 expedition that discovered the Commander Islands, Russia. Heavy exploitation of this colony (for pelts to be made into garments) began almost immediately after its discovery. Killing for pelts expanded considerably in 1786 when the Pribilof population was discovered. To protect the populations from over exploitation the Russians prohibited the killing of females, which was one of the first known wildlife management actions. Despite this protection, the species declined three times after 1742, each time in association with killing for pelts or other management actions. After the biggest decline (1911), the North Pacific Fur Seal Treaty was enacted (Fur Seal Treaty, 1911) which involved Japan, Russia, United States, and Britain (for Canada). Under treaty protection the herd recovered at 8.6% per year from 1912 to 1924. The treaty remained in effect until 1985, except during World War II. Part of the treaty involved a vigorous international research program that made the northern fur seal one of the best known marine mammal species. It produced many of the study methods that are used in modern marine mammal research such as the study of foraging ecology using dive instruments (Gentry and Kooyman, 1986).

The Pribilof Islands herd was declared legally "depleted" in 1988. The stock still contains an estimated 888,120 animals, but the population trend is a major concern because the cause of the ongoing, massive decline is still not known. It may be related to declines in some stocks of harbor seals, sea lions, and sea otters over the past 30 years. The reasons that have been suggested for all these declines include predation by killer whales (Springer *et al.*, 2003), climate change, interaction with fisheries, and combinations of these factors that vary by region (Wade *et al.*, 2007). For more information, consult the chapters on killer whales, ecological effects of marine mammals, and sea otters.

## See Also the Following Articles

Earless Seals ■ Hunting of Marine Mammals

## References

Angliss, R. P., and Outlaw, R. B. (2007). Alaska Marine Mammal Stock Assessments, 2006. *U.S. Dept. Commer., NOAA Tech. Memo.* NMFS-ADFSC-168. (Available from NOAA OPR).

Fur Seal Treaty. (1911). See http://fletcher.tufts.edu/multi/sealtreaty.html#XIII for the on-line text of the treaty.

Gentry, R. L. (1998). "The Northern Fur Seal." Princeton University Press, Princeton, NJ.

Gentry, R. L., and Kooyman, G. L. (eds) (1986). "Fur Seals; Maternal Strategies on Land and at Sea." Princeton University Press, Princeton, NJ.

Ream, R. R., Sterling, J. T., and Loughlin, R. T. (2005). Oceanographic features related to northern fur seal migratory movements. *Deep-Sea Res. II* **52**, 823–843.

Reidman, M. (1990). "The Pinnipeds." University of California Press, Berkeley.

Springer, A. M., *et al.* (2003). Sequential megafaunal collapse in the North Pacific Ocean: An ongoing legacy of industrial whaling? *Proc. Natl. Acad. Sci. USA* **100**, 12223–12228.

Towell, R. G., Ream, R. R., and York, A. E. (2006). Decline in northern fur seal (*Callorhinus ursinus*) pup production on the Pribilof Islands. *Mar. Mamm. Sci.* **22**, 486–491.

Wade, P. R., *et al.* (23 authors) (23 authors) (2007). Killer whales and marine mammal trends in the North Pacific—A re-examination of evidence for sequential megafauna collapse and the prey-switching hypothesis. *Mar. Mamm. Sci.* **23**, 766–802.

Wickens, P., and York, A. E. (1997). Comparative population dynamics of fur seals. *Mar. Mamm. Sci.* **13**, 241–292.

N

# O

# The Ocean Environment

## PAUL C. FIEDLER

Marine mammals live, feed, and reproduce in a vast, three-dimensional fluid environment—the ocean. Air-breathing necessitates frequent attendance to the sea surface for all marine mammals, whereas pinnipeds cannot reproduce in the water and have thus retained a close tie to land or ice. Despite these ties to the boundaries of the ocean environment, oceanography is an important part of the study of marine mammals. Habitat and distribution of marine mammals are affected by the physical and chemical properties of the water through which they swim and communicate, the topography of the ocean bottom where they feed, the physical state of the ocean surface where they breathe and haul out, and numerous factors influencing the distribution of food organisms.

## I. Surface Temperature, Salinity

Temperature at the earth and ocean surface generally decreases from the equator to the poles (Fig. 1), but local processes complicate this simple pattern. Upwelling brings cold water to the surface along the equator and along eastern boundaries between oceans and continents. Ocean currents move cold water equatorward along these eastern boundaries, and warm water poleward along western boundaries.

Salinity of the surface waters of the open ocean varies between 32 practical salinity units (psu) in the subarctic Pacific and 37 psu in subtropical gyres. This variation affects the density of seawater and thus the distribution of mass and the resulting thermohaline circulation. Salinity variations over this range have little or no physiological effect on marine mammals, but can influence availability of food organisms through effects on stratification and circulation. At the coastal and polar limits of the ocean and in marginal seas, processes such as local precipitation and evaporation, river runoff, and ice formation can result in salinities less than 10 and greater than 40 psu. Marine mammals have adapted to tolerate even these extremes.

## II. Surface Currents, Winds

Surface waters of the ocean are constantly in motion due to waves, tides, and currents. Currents move water across ocean basins and thus alter the distribution of temperature, salinity and, indirectly, food organisms. Ocean currents are driven by energy from the sun, both by changes in the distribution of mass due to heating and cooling, and by wind forcing at the surface. The distribution of solar radiation and rotation of the earth set up a basic pattern of winds consisting of (1) easterly trade winds over low latitudes, (2) westerlies over mid-latitudes, and (3) polar easterlies. Heating and cooling of land masses and the seasonal cycle of solar energy input alter this basic pattern.

The surface circulation of most of the area of the world's oceans consists of subtropical gyres which move cyclonically (Fig. 2). Equatorial currents move from east to west, western boundary currents (Gulf Stream, Kuroshio) move warm water from near the equator toward the poles, eastward currents (Antarctic Circumpolar, North Pacific) move water back across the oceans on the poleward side of the gyres, and eastern boundary currents (California, Peru, Canary, Benguela) move cold water towards the equator.

The strength of these currents varies seasonally. Eastern and western boundary currents spin off eddies, on the order of 100 km in diameter, which move both water and prey organisms and alter

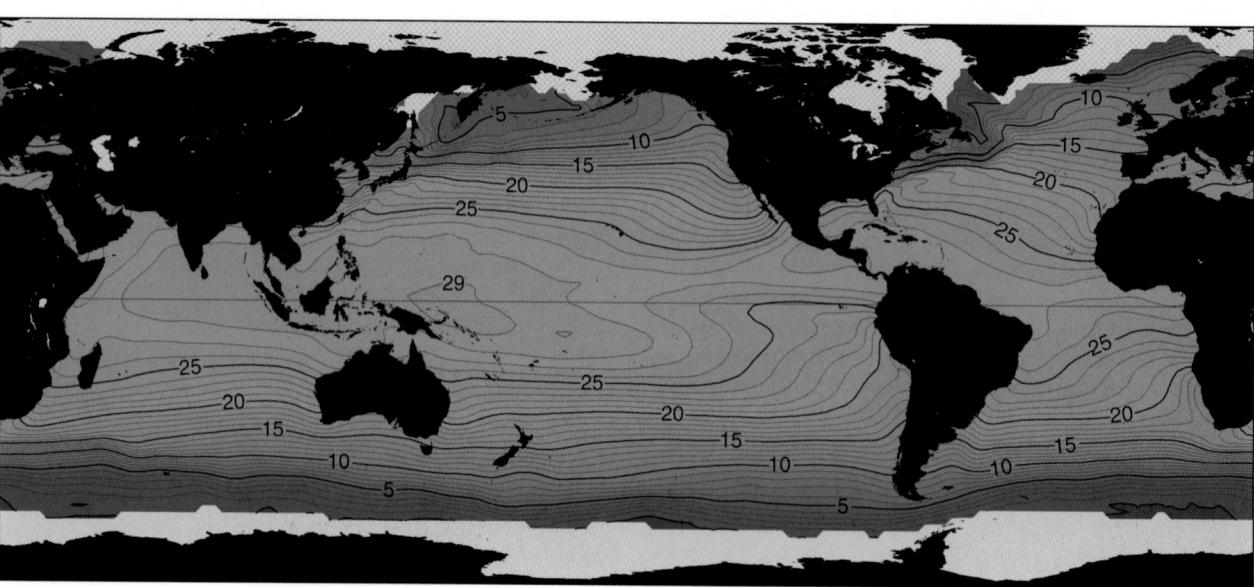

Figure 1   *Mean sea surface temperature (°C, data from Shea et al., 1992).*

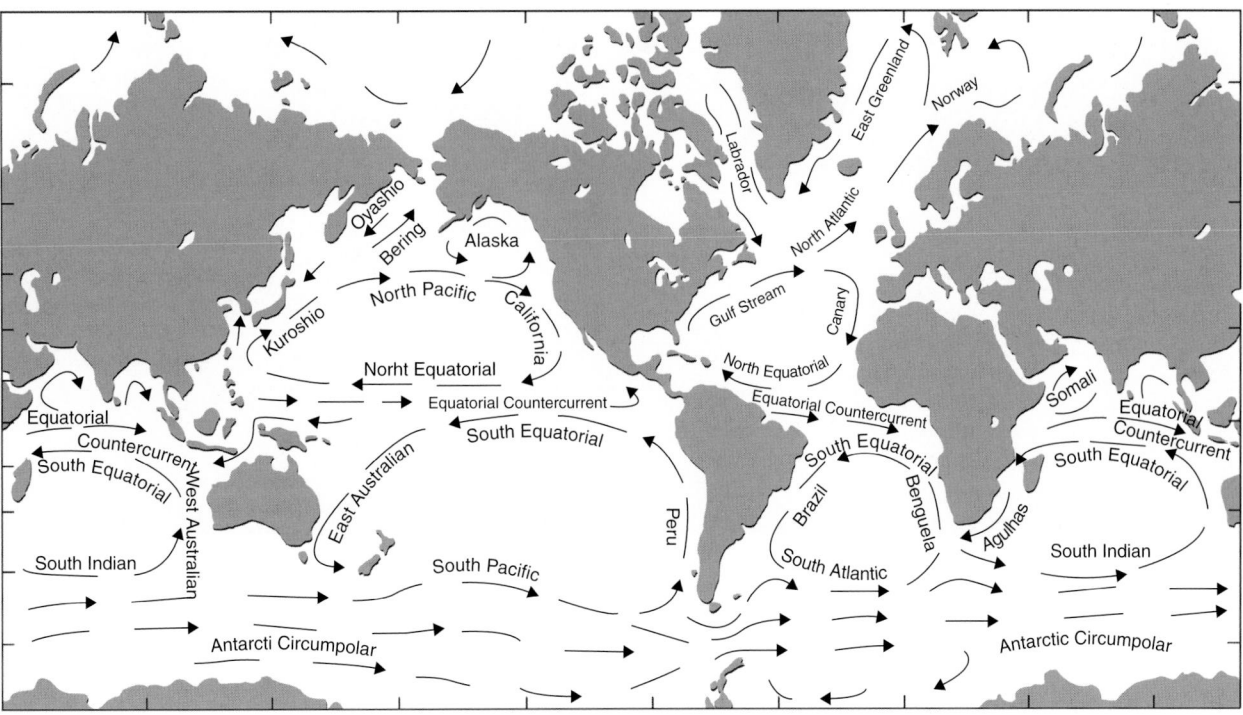

**Figure 2** *Surface currents of the world oceans.*

the distribution of favorable habitat for marine mammals. Eastern boundary currents are sites of seasonal coastal upwelling, when equatorward winds move surface water offshore, and deeper nutrient-rich water is brought to the surface. Westward equatorial currents are also sites of upwelling, due to divergence of surface water caused by the rotation of the earth.

## III. Vertical Structure

Physical (and biological) variability in the ocean environment is generally much greater in the vertical than in the horizontal dimension. Water temperature may be nearly constant for a few meters or tens of meters below the surface in the mixed layer, but then decreases rapidly with depth in the thermocline. Vertical changes in salinity may contribute to this stratification, especially in polar seas. Stratification influences productivity because deeper colder water has higher nutrient concentrations than the warmer water near the surface. Stratification also influences the distribution of food organisms.

## IV. Productivity

Biological productivity is the rate of production of living matter; oceanographers often speak of "production", which is the variable standing stock of living organisms and is closely linked to productivity (Fig. 3). Primary productivity in the ocean is the rate of production of plants (phytoplankton), and is often limited by light and nutrient availability, both of which are controlled by physical oceanographic processes. Light intensity at the sea surface varies with latitude and time of year, as well as time of day. Thus, photosynthesis occurs in polar seas only during summer. Light intensity decreases rapidly with depth, so that photosynthesis is possible within only a few meters or tens of meters of the sea surface.

Plant growth also requires nutrients (nitrate, phosphate and, for some phytoplankton, silicate). The importance of micronutrients, notably iron, in some regions of the open ocean, such as the Southern Ocean around Antarctica where many baleen whales feed, has been recognized in recent years. Nutrient availability is affected by oceanographic processes including stratification, wind and tidal mixing, circulation, and upwelling.

Marine mammals consume zooplankton, fish, squid, and even other marine mammals. Therefore, distribution and foraging are not directly determined by primary production. Food chains involving marine mammals are as short as diatoms–krill–baleen whales. Sperm whales (*Physeter macrocephalus*), however, consume large deep-living squid and are at least four steps removed from primary production at the sea surface.

## V. Ice

Ice is an important component of the habitat of migratory and endemic marine mammals in both the Arctic and the Southern Ocean. Ice cover varies seasonally with temperature and is also subject to the influence of winds and currents. There must be some open water in any ice-covered habitat utilized by marine mammals to allow access to both air and food. Ice on the sea surface is found in a variety of types (pack ice, icebergs, shore ice, fast ice, drift ice, new ice) and forms (small floes, brash ice, pancake ice, etc.).

Sea ice in the Arctic is tightly packed except at the ice edge. Much of the Arctic Ocean is permanently ice covered, although summer ice cover has been decreasing due to global warming (IPCC, 2007). The Antarctic has a broad zone of looser pack ice with many internal leads. Most of this ice melts during summer. Polynyas are areas within the pack ice that are almost always clear of ice. Providing both

**O**

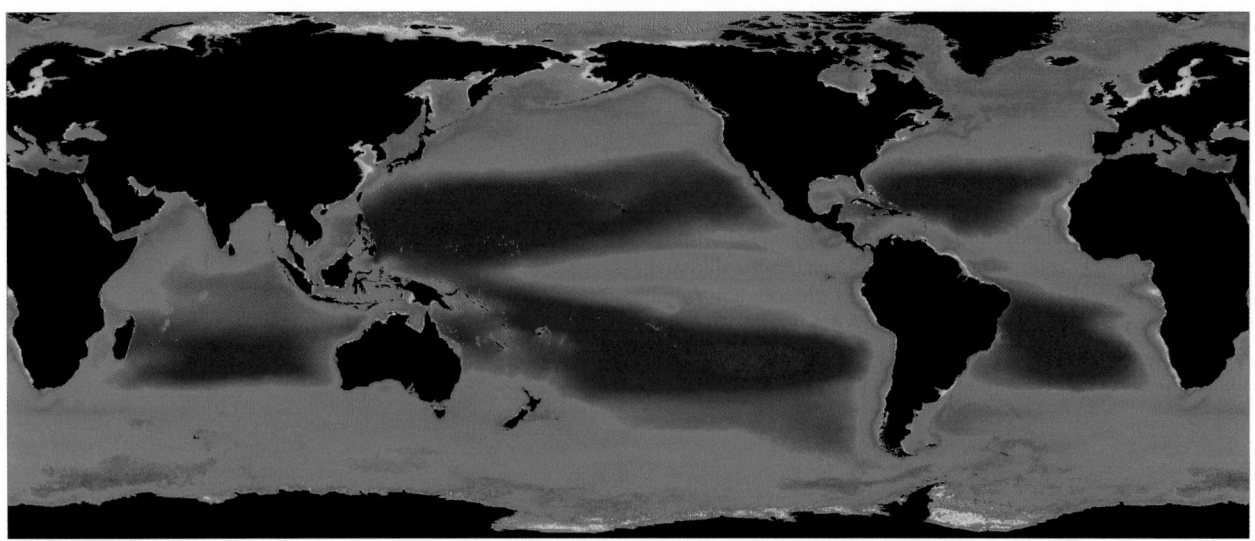

**Figure 3**  *Mean chlorophyll concentration from SeaWiFS satellite data, 1998–2006, increasing from violet and blue to red, black is land and gray is permanent ice (data from Ocean Color Web, SeaWiFS Reprocessing 5, NASA Goddard Space Flight Center, http://oceancolor.gsfc.nasa.gov/).*

access to the water and locally enhanced productivity, they are relatively more important in the Arctic than in the Antarctic.

## VI. Temporal Variability

The ocean environment varies seasonally, especially at higher latitudes where ice closes accessibility to most species during winter. Seasonal variability in wind-driven upwelling causes significant changes during the year, even in the tropics. The ocean environment varies between years as well. The cycle in the ocean–atmosphere system known as the El Niño-Southern Oscillation, or simply "El Niño", causes major changes in the tropics every 3 to 5 years. Both seasonal and interannual variability occur in all marine mammal habitats. However, the relative magnitude of variability at each scale varies (Fig. 4).

Climate variability over periods of decades has been documented in the ocean environment. For example, the Pacific Decadal Oscillation, driven by changes in the strength of the Aleutian Low pressure system results in changes in winds, temperature, and circulation in the North Pacific (Hare and Mantua, 2000). Ecological regime shifts are fundamental changes in the structure and function of ecosystems forced by climate change, fishery mortality or habitat modification, and exhibiting nonlinearities such as irreversibility (Bakun, 2004). Seal and sea lion declines in Alaska have been linked to a major oceanic regime shift that occurred in the late 1970s (Trites *et al.*, 2007).

Marine mammals have experienced climate change—variations in the ocean environment over periods of hundreds to thousands of years—throughout their evolutionary history. Concern is now focused on the possible effects of rapid global warming caused by mankind's input of excess carbon dioxide and other greenhouse gasses into the atmosphere. Scientists generally, although not universally, agree that by 2100 global average temperature will increase by 1.8°C to 4.0°C, sea level will rise by several tens of centimeters, and sea ice coverage will be reduced (IPCC, 2007). Changes in environmental variability on shorter time scales, such as the frequency or magnitude of El Niño events, the amplitude of seasonal cycles, or the frequency or intensity of storms are possible. Regional changes

in precipitation, circulation and upwelling, wind speed, wave conditions, acidity (pH), and UV-B (ultraviolet) radiation are expected.

## VII. Effects on Life History and Function

Temperature can be a critical factor in the energy budget of warm-blooded mammals. However, marine mammals live and breed in the polar ice and even small young harp seals are able to tolerate freezing temperatures. Some baleen whale species tend to be slightly larger in the cold Southern Ocean than in the warmer North Pacific. Adaptations including an insulating blubber layer and an ability to reduce blood circulation to peripheral parts of the body have reduced the direct influence of temperature on distribution.

It has long been believed that large whales migrate from summer polar feeding grounds to warmer tropical breeding grounds because of adverse effects of low temperature on growth and survival of neonates. However, this hypothesis is no longer supported by considerations of energy intake, insulation by blubber, and heat loss. Even though the energy cost of migration may be very low for large whales, adult whales do not optimize their energy budget by migrating to warm tropical or temperate waters.

Cetaceans use underwater sound in several ways. Odontocetes use high-frequency sound to locate prey individuals and patches over short distances. Mysticetes use low-frequency signals for navigation and communication. Although the effects of temperature, salinity, and pressure on the propagation of sound in water are well known, it is not known whether oceanographic variability can affect the use of sound by marine mammals. If blue (*Balaenoptera musculus*) and fin (*B. physalus*) whales use the low-frequency sound channel of the deep ocean to acoustically "visualize" their environment, variations in water properties very likely distort or obscure the image they can obtain.

## VIII. Effects on Feeding

Marine mammals are generally apex predators, or at the top of the food chain, although some are preyed upon by killer whales (*Orcinus orca*) and leopard seals (Hydrurga leptonyx). They are

O

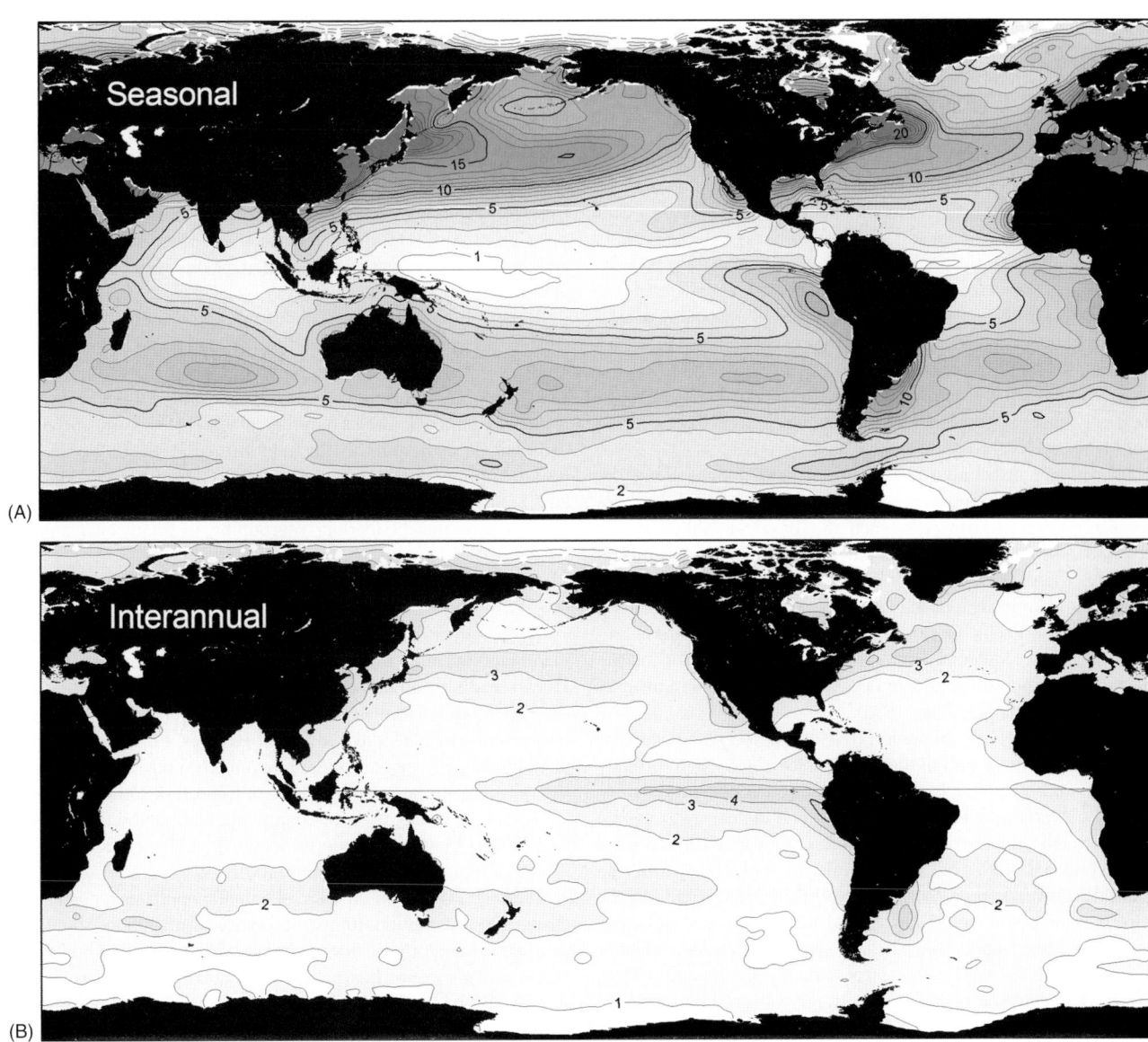

**Figure 4** *Approximate seasonal (A) and interannual (B) ranges of sea surface temperature (UK Meteorological Office HadISST 1.1 global monthly SST, 1950–2004).*

**O**

motile, social, and intelligent, so that they are able to take advantage of prey in locally dense patches. Like all aquatic and terrestrial predators, marine mammals cannot forage efficiently on average concentrations of prey. It has been shown that right whales (*Eubalaena* spp.) and humpback whales (*Megaptera novaeangliae*) require concentrations of prey that are more dense than the maximum densities observed by net sampling. This is an indication of the limitations of oceanographic sampling, as well as the paucity of prey in the ocean environment. Sampling by acoustic backscattering from echo sounders and by photography with tethered cameras has now revealed small patches of prey of the required densities.

Prey consumed by marine mammals include crustaceans (copepods, euphausiids or krill, amphipods, shrimp), cephalopods (squid), and schooling fish [herring (*Clupea* spp.), capelin (*Mallotus villosus*), cod, mackerel, myctophids or lanternfish, and others]. Availability of these prey in dense patches has been linked to oceanographic features including bathymetry, fronts, eddies, and primary productivity.

Except for bathymetry, these features or characteristics of the ocean environment vary over time.

Baleen whales feed on dense patches of zooplankton. Distribution of feeding North Atlantic right whales (*Eubalaena glacialis*) can be predicted by surface temperature and bathymetric variables including depth and slope. These are the simplest oceanographic variables to measure, but they explain the distribution of whales only indirectly through effects on prey. Zooplankton patches are available in productive coastal waters, where upwelling and wind mixing results in high primary production, and in polar seas, where high primary productivity results from summer solar effects and enhanced phytoplankton growth at the melting ice edge and even under the ice. The patches are often located at bathymetric features like the shelf edge, islands, and seamounts, because bathymetry affects circulation and production and serves as a cue for aggregation of krill and fish. Blue whales feed on krill at the ice edge in the Antarctic, but also at the shelf edge off California (Croll *et al.*, 2005).

Along the northeast United States continental shelf, cetaceans tend to frequent distinct regions based on food preferences (Kenney and Winn, 1986). Piscivores—humpback, fin, and minke whales (*Balaenoptera acutorostrata*); and bottlenose (*Tursiops* spp.), Atlantic white-sided (*Lagenorhynchus acutus*), and common dolphins (*Delphinus* spp.)—are most abundant over shallow banks in the western Gulf of Maine and mid-shelf east of Chesapeake Bay. Planktivores—right, blue and sei (*Balaenoptera borealis*) whales—are most abundant in the western Gulf of Maine and over the western and southern portions of Georges Bank. Teuthivores (squid eaters)—sperm and pilot (*Globicephala* spp.) whales and Risso's dolphins (*Grampus griseus*)—are most abundant at the shelf edge. Most of these patterns can be related to availability of specific prey. Right whales feed on dense patches of copepods in the vicinity of the Great South Channel. Humpbacks and fin whales feed on a small schooling fish, the American sand lance (*Ammodytes americanus*), which is very abundant on Stellwagen Bank and Jeffreys Ledge in the Gulf of Maine, and is also abundant off Chesapeake Bay.

Cetaceans are distributed over distinct depth ranges along the continental slope in the northwestern Gulf of Mexico (Davis *et al.*, 2002). Deep-diving teuthivores, such as Risso's dolphins, sperm whales and beaked whales, are found near temperature fronts in deep water, where their squid prey may aggregate. Sperm whales are more abundant in regions of high zooplankton biomass and steep topography in the Pacific Ocean, where deep-living squid are aggregated. Most of the nineteenth-century sperm whaling grounds in the Pacific were in regions of coastal or oceanic upwelling where primary production is high (Jaquet *et al.*, 1996).

In the Antarctic, both Antarctic fur seals (*Arctocephalus gazella*) and minke whales are found in the marginal ice zone, where primary productivity and, presumably, prey availability are enhanced (Ribic *et al.*, 1991). Winter pack ice extends out beyond the narrow Antarctic continental shelf and over the deep ocean. Minke whales, beaked whales and several species of seals live in this ice and feed on krill (the euphausiid *Euphausia superba*), myctophid fishes, and squid. As the pack ice retreats in summer, many of these winter residents and other predators including large baleen whales, male sperm whales, and killer whales, follow the ice edge to feed on abundant plankton and nekton. Right whales, feeding on smaller copepods not concentrated at the ice edge, remain in open waters north of the retreating ice.

Ice is an important habitat component for breeding and shelter of pinnipeds. Antarctic ice floes in the marginal ice zone are a refuge for fur seals from predation by killer whales. Crabeater seals (*Lobodon carcinophaga*), in contrast, are found in the interior of the pack ice where larger, more stable floes provide breeding sites.

In the Arctic, gray (*Eschrichtius robustus*), bowhead (*Balaena mysticetus*), and beluga (*Delphinapterus leucas*) whales migrate into the Chukchi and Beaufort Sea north of Alaska to feed in summer and fall (Moore and DeMaster, 1998). Bottom-feeding gray whales feed in shallow, ice-free waters, piscivorous belugas feed in deep water near the ice edge, whereas planktivorous bowhead whales feed closer to the coast where ice cover is less. These distribution differences are clearly related to prey availability. River plumes formed by summer runoff into the Arctic have been shown to limit the distribution of prey and thus determine the distribution of foraging whales.

## IX. Effects on Distribution and Migration

Distribution of baleen whales on summer feeding grounds and movement on these feeding grounds are associated with oceanographic features, including fronts between water masses, eddies along these fronts, eddies caused by circulation around islands and capes, and oceanic

and coastal upwelling. Prey availability is enhanced by nutrient input and enhanced primary production, or by convergence and aggregation of plankton. Japanese whalers have exploited concentrations of (1) sei whales off Japan where the warm Kuroshio Current meets the cold Oyashio, (2) fin whales in a cold cyclonic eddy and in a mixing zone between cold coastal waters and warmer Bering Sea waters in the western subarctic Pacific, (3) fin whales at shelf edge fronts and upwelling zones off islands in the eastern Bering Sea, (4) blue and fin whales at the Polar Front north of the Antarctic pack ice (Nasu, 1966). Both resident and migrant cetaceans utilize prey aggregations associated with summer upwelling along the California coast (Tynan *et al.*, 2005).

Relatively little is known about migration routes of most baleen whales between summer feeding grounds and winter breeding grounds. Gray whales migrating south along the California coast remain near the coast and apparently use bottom topography for navigation. Humpback whales have been tagged off Hawaii and observed to linger at seamounts in the middle of the North Pacific on the way to summer feeding grounds in the Gulf of Alaska (Mate *et al.*, 2007). Seamounts are known to be sites of aggregation of fishes. These whales may be feeding, regrouping, or simply navigating.

## X. Effects of Climate Change

Direct effects of temperature increase are unlikely for most marine mammals, because of their mobility and thermoregulatory ability. However, direct effects could occur for populations at the limits of their range or with physical limits to migration that preclude distribution shifts to track environmental changes. For example, the Gulf of California harbor porpoise, or vaquita, is a rare species trapped at the warm northern end of the Gulf of California. Other species, such as the beluga whale, with low skin pigmentation and living in a polar region subject to atmospheric ozone depletion, may suffer direct effects of increased UV-B radiation.

Indirect effects on marine mammal populations might include changes in distribution, timing and range of migration, abundance of competitors and/or predators, prey availability, timing of breeding, and reproductive success (Learmonth *et al.*, 2006). Global warming is predicted to be greatest at high latitudes in the Northern Hemisphere. The resulting reduction of ice cover in the Arctic will affect ice-associated species through changes in the distribution of critical habitat such as ice-free areas, polynyas, and the ice edge. Seasonal and geographical shifts in prey availability in both the Arctic and Antarctic may have a variety of indirect effects on nutritional status and reproductive success, geographic range, and timing or patterns of migration. Bowhead whales and narwhals (*Monodon monoceros*), and especially beluga whales, that feed close to the ice edge, will be affected by reduced ice extent in the Arctic. Warming could also reduce the extent of pack ice in the Antarctic and thus affect the distribution and abundance of krill. Declining krill abundance in the region of the Antarctic Peninsula during the early 1990s has been linked to low winter sea ice extent.

Global warming is expected to be less intense at lower latitudes. However, reduction in wind-driven coastal and equatorial upwelling is possible and may reduce prey availability for marine mammals. A decline in zooplankton biomass off California since 1951 has been linked to warming of surface waters, increased stratification, and reduced mixing and upwelling of nutrients. Odontocetes that depend on squid, such as sperm whales, may be more adversely affected because cephalopod populations are known to be highly variable. Rising sea level will change the lagoons and shelf areas where gray whales and humpbacks breed, but past changes in these breeding sites are known to have occurred over longer periods of geologic time.

## *References*

Bakun, A. (2004). Regime shifts. *In* "The Sea" (A. R. Robinson, and K. Brink, eds), pp. 971–1018. Harvard University Press, Cambridge.

Croll, D. A., Marinovic, B., Benson, S., Chavez, F. P., Black, N., Ternullo, R., and Tershy, B. R. (2005). From wind to whales: Trophic links in a coastal upwelling system. *Mar. Ecol. Prog. Ser.* **289**, 117–130.

Davis, R. W., *et al.* (10 authors) (2002). Cetacean habitat in the northern Gulf of Mexico. *Deep-Sea Res. I* **49**, 121–142.

Hare, S. R., and Mantua, N. J. (2000). Empirical evidence for North Pacific regime shifts in 1977 and 1989. *Prog. Oceanogr.* **47**, 103–145.

Intergovernmental Panel on Climate Change (IPCC) (2007). Fourth Assessment Report, Working Group 1: The physical basis of climate change. http:www.//ipcc-wg1.ucar.edu/wg1/wg1-report.html.

Jaquet, N., Whitehead, H., and Lewis, M. (1996). Coherence between 19th century sperm whale distributions and satellite-derived pigments in the tropical Pacific. *Mar. Ecol. Prog. Ser.* **145**, 1–10.

Kenney, R. D., and Winn, H. E. (1986). Cetacean high-use habitats of the northeast United States continental shelf. *Fish. Bull.* **84**, 345–357.

Learmonth, J. A., Macleod, C. D., Santos, M. B., Pierce, G. J., Crick, H. Q. P., and Robinson, R. A. (2006). Potential effects of climate change on marine mammals. *Oceanogr. Mar. Biol. Ann. Rev.* **44**, 431–464.

Mate, B., Mesecar, R., and Lagerquist, B. (2007). The evolution of satellite-monitored radio tags for large whales: One laboratory's experience. *Deep-Sea Res. II* **54**, 224–247.

Moore, S. E., and DeMaster, D. P. (1998). Cetacean habitats in the Alaskan Arctic. *J. Northwest Atl. Fish. Sci.* **22**, 25–69.

Nasu, K. (1966). Fishery oceanographic study on the baleen whaling grounds. *Sci. Rep. Whales Res. Inst. Tokyo* **20**, 157–210.

Ribic, C. A., Ainley, D. G., and Fraser, W. R. (1991). Habitat selection by marine mammals in the marginal ice zone. *Antarct. Sci.* **3**, 181–186.

Shea, D. J., Trenberth, K. E., and Reynolds, R. W. (1992). A global monthly sea surface temperature climatology. *J. Climate* **5**, 987–1001.

Trites, A. W., Deecke, V. B., Gregr, E. J., Ford, J. K. B., and Olesiuk, P. F. (2007). Killer whales, whaling, and sequential megafaunal collapse in the North Pacific: A comparative analysis of marine mammals in Alaska and British Columbia following commercial whaling. *Mar. Mamm. Sci.* **23**, 751–765, doi: 10.1111/j.1748-7692.2006.00076.x.

Tynan, C. T., Ainley, D. G., Barth, J. A., Cowles, T. J., Pierce, S. D., and Spear, L. B. (2005). Cetacean distributions relative to ocean processes in the northern California Current System. *Deep-Sea Res. II* **52**, 145–167.

---

# *Odobenocetops*

## CHRISTIAN DE MUIZON

One of the most unusual fossil cetaceans is a tropical tusked whale related to narwhals—Odobenocetopsidae, *Odobenocetops*—from the early Pliocene beds of the Pisco Formation of Peru (Muizon, 1993; Muizon *et al.*, 2002a, b). As its name indicates, this marine mammal presents several similarities with the living walrus, *Odobenus rosmarus*, in spite of being a cetacean.

*Odobenocetops* is a startling example of convergence with the walrus in its cranial morphology and inferred feeding habits (Muizon, 1993; Muizon *et al.*, 2002a, b). *Odobenocetops* is known from two species, both from the Pliocene Pisco Formation of Peru.

## I. Descriptive Anatomy

*Odobenocetops* (Figs 1 and 2) has lost the elongated rostrum of the other cetaceans. Instead, the rostrum is short, rounded, and blunt and formed by the premaxillae, which are greatly enlarged. It has large, asymmetrical ventral alveolar sheaths holding sexually dimorphic tusks. The right tusk of the male is large and can reach 1 m or more in length. The left tusk is *ca.* 25-cm long, of which a few centimeters only were erupted. Both tusks are straight. In the female both tusks approach the size of the left tusk of the male, although the right tusk is slightly larger than the left. Premaxillary sheath and tusks are oriented posteroventrally. The bony nares are displaced anteriorly (when compared to other odontocetes). The palate is very deep and arched, and its anterior border is U- to V-shaped. It bears no maxillary teeth. The orbits face anterolaterally and dorsally. The portion of the frontal and maxillae which cover the temporal fossae in other odontocetes have been reduced and narrowed in such a way that the temporal fossae are opened dorsally. The periotic and tympanic have the characteristic morphology observed in the other delphinoid cetaceans. The mandible is not known, and there are few postcranial bones. The length of the body could have ranged from 3 to 4 m.

## II. Relationships

In spite of its extremely modified morphology, several characters of the skull indicate that *Odobenocetops* is a delphinoid cetacean. Cetacean characters of *Odobenocetops* are the presence of large air sinuses in the auditory region connected to well-developed pterygoid sinuses, thickened and pachyostotic tympanic bulla, and dorsal opening of the narial fossae. Odontocete affinities of *Odobenocetops* are attested by the maxillae covering the supraorbital processes of the frontals (Muizon, 1994), dorsoventral expansion of the pterygoid sinuses, presence of large premaxillary foramina (in *O. peruvianus* only), and asymmetry of the premaxillae and maxillae in the facial region. The sigmoid morphology of dorsomedial view of the involucrum and the presence of a medial maxilla–premaxilla suture at the anterolateral edge of each narial fossa are delphinoid characters (Muizon, 1988). Several characters of the pterygoid, alisphenoid, and temporal fossa, indicate close affinities of *Odobenocetops* and the Monodontidae (Muizon, 1993; Muizon *et al.*, 2002a). The exceptional specializations of *Odobenocetops* merit that it is referred to a new odontocete family, the Odobenocetopsidae, regarded as the sister group of the Monodontidae.

The occurrence of tusks in *Odobenocetops* is a convergence with *Monodon* as in the latter genus the large tusk of the male is implanted in the left maxilla, whereas in *Odobenocetops* it is implanted in the right premaxilla. Consequently, the tusks of *Monodon* and *Odobenocetops* are not homologous.

## III. Functional Anatomy

*Odobenocetops* and the living walrus are convergent in the large and deep palate; the strong development of a wide, the blunt snout which is highly vascularized and has strong muscular insertions, suggesting

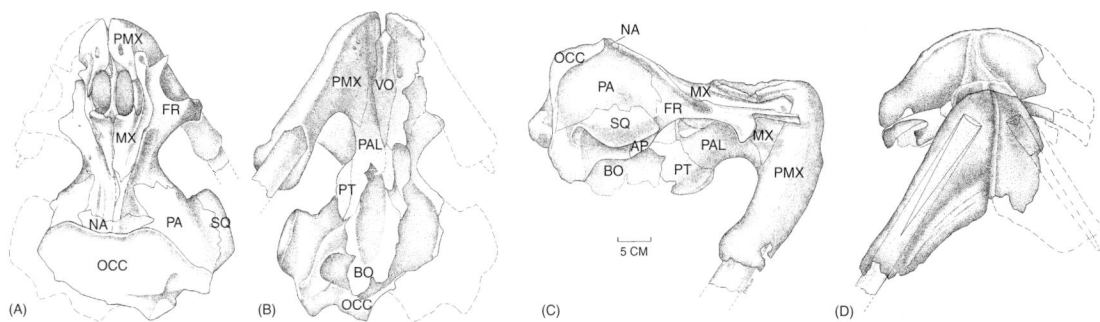

**Figure 1** *Skull of* Odobenocetops peruvianus *in dorsal (A), ventral (B) lateral (C) and rostral (D) view. AP, alisphenoid; BQ, basioccipital; FR, frontal; MX, maxilla; NA, nasal; OCC, occipital; PA, parietal; PAL, palatine; PMX, premaxilla; PT, pterygoid; SQ, squamosal; VO, vomer. Modified from Muizon (1993), reprinted with permission from* Nature (**365**, *745–748, 1993).*

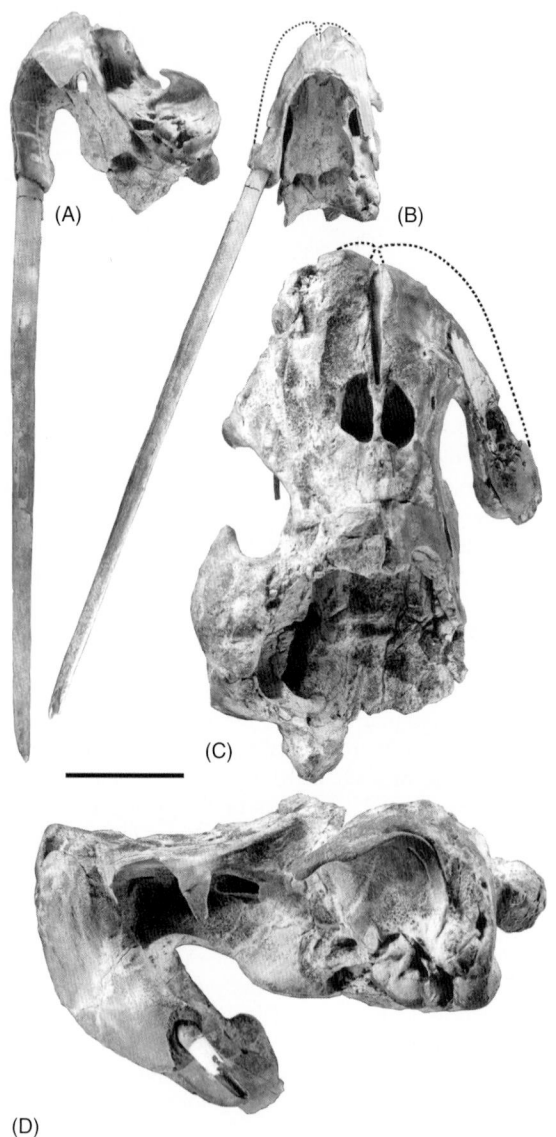

**Figure 2** Odobenocetops leptodon: *skull of a male in left ventro-lateral (A), anteroventral (B), dorsal (C), and left lateral (D) views. Scale bar: 20cm (A, B); 10cm (C, D) from Muizon et al. (1999). Reprinted by permission from the Comptes Rendus de l'Académie des Sciences, Série IIA, **329**, 449–455). Copyright (1999) Elsevier.*

the presence of a powerful tactile upper lip; the presence of tusks, and reduction of the maxillary dentition. The living walrus feeds mainly on benthic invertebrates (Fay, 1982) sucking out the siphon and foot of bivalves or gastropods shells, using the tongue as a piston powered by the very large lingual retractors and depressors. Given the anatomy of its palate, it is likely that *Odobenocetops* used a similar feeding strategy. *Odobenocetops* was, therefore, probably feeding upon benthic invertebrates (bivalves and gastropod mollusks and/or crustaceans), which are abundant in the Pisco Formation. Because of the great vascularization of the inferred upper lip it is possible that, as in the walrus, *Odobenocetops* had vibrissae, which would have had an important tactile role in the search for food. When feeding on the sea floor, the body of *Odobenocetops* was probably held in an oblique position as is observed in the living walrus (Fig. 3).

The visual and the echolocation system of the two species of *Odobenocetops* differed. *O. peruvianus* probably had greatly reduced premaxillary air sacs and melon of most odontocetes, as a result of the extreme shortening of the rostrum and the very short space between the premaxillary foramina and the nares. In contrast, in *O. leptodon* fossae for premaxillary sacs are present and left a clear scar on the premaxillae, and a small melon was probably located anterior to the nares. Thus, *O. peruvianus* probably did not echolocate but *O. leptodon* did.

The anterodorsal edge of the orbit (i.e., the anterior border of the supraorbital process of the frontal) in *O. peruvianus* is deeply notched, allowing overlapping visual fields and binocular vision. In contrast, this region of the frontal in *O. leptodon* is only slightly concave, which probably prevented (or at least reduced) binocular vision.

Therefore, in *O. peruvianus*, the probable lack of echolocation ability was probably compensated for by a good anterodorsal binocular vision. As in the walrus, the latter was especially useful when the animal was searching for food on the sea floor with the body in an oblique position (Kastelein *et al.*, 1993).

The posteroventrally oriented tusks are certainly one of the most unusual characteristics of *Odobenocetops*. Because of its great length and slenderness the right tusk of the male was probably very fragile and it is unlikely that the animal was swimming with the tusk held at a 45° angle with the axis of the body. Comparison of the occipital condyles to the anterior articular facets of the atlas indicates that, when swimming, the dorsal plane of the skull was oriented anterodorsally and the tusk was approximately parallel to the axis of the body. When feeding, the large tusk was dragged on the sea floor, and because of the oblique position of the body, the tusk was at an angle

**Figure 3** *Reconstruction of three male* Odobenocetops leptodon *in swimming and feeding position. From Muizon* et al. *(1999), reprinted with permission. Painting by Mary Parrish.*

of 20° to 40° with the axis of the body (Fig. 3). In other respects, the convexity of the occipital condyles of *Odobenocetops* indicates that its neck was extremely mobile, allowing amplitude of the head movement approaching 90° or more.

In narwhals, the tusk is mainly used in intraspecific behaviors, and the same is probably true for *Odobenocetops*. The tooth was fragile and sexually dimorphic and is unlikely to have been used to exert forces (in digging or fighting), and was probably used in intraspecific displays (Muizon *et al.*, 2002a, b). It is also possible that tusks of walruses and the premaxillary tusks and sheaths of *Odobenocetops* played a role in bottom feeding, as orientation guides to keep the mouth and any vibrissae in stable position relative to the substrate.

## See Also the Following Articles

Walrus ▪ Narwhal ▪ Echolocation

## References

Fay, F. H. (1982). "Ecology and biology of the pacific walrus, *Odobenus rosmarus divergens* Illiger". US Department of Interior; Fish and Wildlife Service. North American fauna, Number **74**, 1–279.

Kastelein, R. C., Zweypfenning, V. J., Spekreijse, H., Dubbeldam, J. L., and Born, E. W. (1993). The anatomy of the walrus head (*Odobenus rosmarus*). Part 3: The eyes and their function in walrus ecology. *Aquat. Mamm.* **19**(2), 61–92.

Muizon, C. de. (1988). Les relations phylogénétiques des Delphinida. *Annales de Paléontologie* **74**, 115–183.

Muizon, C. de. (1993). Walrus-like feeding adaptation in a new cetacean from the Pliocene of Peru. *Nature* **365**, 745–748.

Muizon, C. de. (1994). Are the squalodonts related to the platanistoids? In Contributions in marine mammalogy honoring Frank C. Whitmore, Jr. (A. Berta and T. A. Deméré, eds.). *Proceedings of the San Diego Society of Natural History* **29**, pp. 135–146.

Muizon, C.de., Domning, D. P., and Parrish, M. (1999). Dimorphic tusks and adaptive strategies in the Odobenocetopsidae, walrus-like dolphins from the Pliocene of Peru. *Comptes-Rendus de l'Académie des Sciences Paris, Sciences de la Terre et des planets* **329**, 449–455.

Muizon, C.de., Domning, D. R., and Ketten, D. R. (2002a). *Odobenocetops peruvianus*, the walrus-convergent delphinoid (Cetacea, Mammalia) from the early Pliocene of Peru. *Smithsonian Contrib. Paleobiol.* **93**, 223–261.

Muizon, C.de., and Domning, D. P. (2002b). The anatomy of *Odobenocetops* (Delphinoidea, Mammalia), the walrus-like dolphin from the Pliocene of Peru and its palaeobiologic implications. *Zool. J. Linn. Soc.* **134**, 423–452.

# Omura's Whale
## *Balaenoptera omurai*

### Tadasu K. Yamada

## I. Characteristics and Taxonomy

Omura's whale (Fig. 1) is a medium-sized baleen whale found in lower latitude Indo-Pacific waters of both the hemispheres. Although the species was described quite recently, it has been sighted at sea and a number of specimens have been collected and placed in museums, some only recently identified.

The body length of the species is about 10–12 m. The smallest newborn known was 3.2 m in length. External proportions of the body are similar to those of the Bryde's whales (*B. edeni/brydei* complex) and sei whale (*B. borealis*). Absence of lateral ridges on the dorsal surface of the head is the most diagnostic characteristic for distinguishing it from the Bryde's whales. The lateral contour of the upper jaw is convex as in the blue whale (*B. musculus*). The ventral pleats extend as far as around the umbilicus, separating it from the sei whale. Omura's whale is distinctly counter-shaded, basically black above and off-white below. The dark portion of the color pattern is lighter than in the fin whale (*B. physalus*). The ventral surface of the flukes is off-white with dark irregular margins. Young animals may be lighter in color than adults. The lower jaw area is asymmetrically pigmented, black on the left side and white on the right. The dorsal fin is relatively small, low, and strongly falcate. The number of baleen plates (about 200 on each side) is less than that known for any other species in the genus *Balaenoptera*. The anterior one-third of the baleen plates are yellowish white, more so on the right side. The posterior one-fifth of the plates are all black, whereas the intermediate plates are bi-colored (dark brownish-gray outside and pale inside). The base of a baleen plate is wide relative to the length. The bristles are not coarse but thicker than those in the sei whale.

**O**

Figure 1 *Omura's whale. Illustration by Watanabe Yoshimi.*

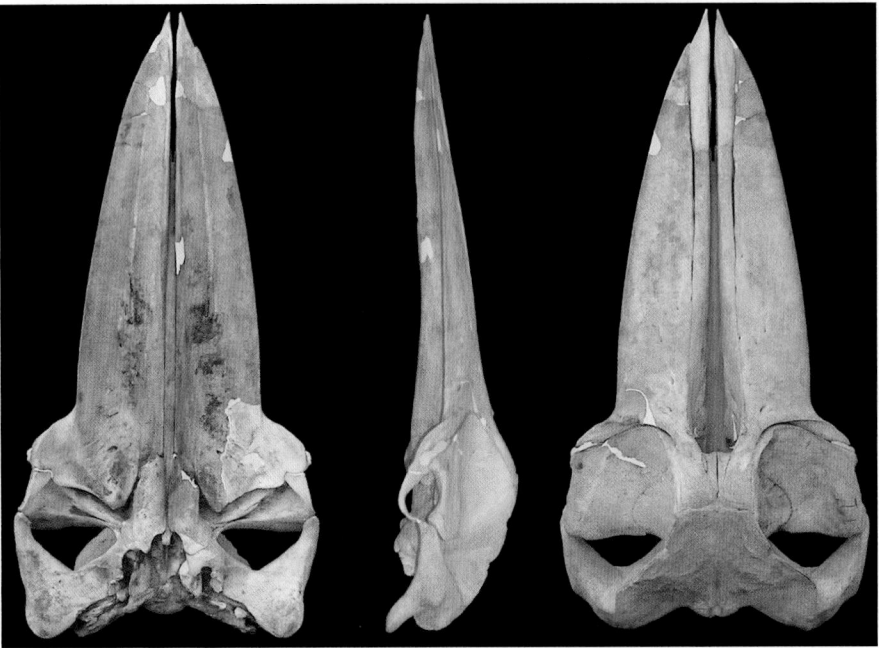

Figure 2 *Skull of Omura's whale.*

There is no record of an anatomical examination of a fresh specimen. Data are available only for the skull and the skeleton (Fig. 2). The most specific feature of the skull is the convex contour of the lateral border in both maxillae. The configuration of the vertex is also unique; the posterior end of the ascending process of the maxilla widens and becomes squarish and the posterior end of the premaxilla becomes thin and sinks between the nasal and the maxilla. A narrow area of the frontal is exposed between the nasals and the occipital shield. There are one or two foramina in the suture line between the parietal and the squamosal on the posterior wall of the temporal fossa, which is unique to this species. In older individuals, however, these foramina are barely discernible. A groove exists in the posterosuperior edge of the orbital process of the frontal. The lateral border of the parietal spreads widely and is visible in dorsal view. The vertebral formula is C7 + Th13 + L12 + Ca21 = 53. The anterior four ribs have both head and tuberculum, whereas the posterior nine are without head. The head of the first rib is not bifurcated (not composed of two fused ribs). The flippers are relatively longer than those of Bryde's whales; the phalangeal formula of the type specimen is I-5, II-7, IV-6, V-3, excluding the metacarpals. The one pair of pelvic bones available for examination have a peculiar spatulate shape, which could be a pathological condition.

Among the species of the genus *Balaenoptera*, *B. omurai* was first considered close to the Bryde's whales. This was only because of the similarity in their size. The morphological characteristics of mainly the skull of *B. omurai* are very different and distinct from those in the Bryde's whales. Sasaki *et al.* (2006) in their molecular work based

on complete mtDNA sequences and short interspersed repetitive element (SINE) insertion patterns suggested that *B. omurai* evolved as an ancient independent lineage that diverged much earlier than the Bryde's whales and sei whale.

The species was first called "small-form Bryde's whale" by Wada and Numachi (1991). "Pygmy Bryde's whale" was also used by several authors. Considering the morphological and molecular differences from the Bryde's whales, common names based on that species or species complex should be avoided. The scientific name, *Balaenoptera omurai*, honors the late Dr. Hideo Omura, who first recognized the significance of the Bryde's whale in global whale fisheries. The Japanese name is related to the type locality Tsunoshima Island in western Japan.

## II. Distribution and Abundance

Current knowledge of the distribution of this whale is very limited and remains tentative (Fig. 3); it is based on sporadic and accidentally collected records. However, the species probably ranges throughout less oceanic waters of tropical/subtropical lower latitudes of the Indo-Pacific region, including the Solomon Sea, Java Sea, Andaman Sea, Gulf of Thailand, Philippines, South China Sea, East China sea, and the western edge of the North Pacific including the seas around Taiwan and southwest Japan. The type locality in the Sea of Japan (34°21′N) and the South Australian record (34°37′S) may be at the northern and southern extremes of the range, respectively. The known longitudinal range is between 90° E and 160°E. There

FAO Fisheries Department

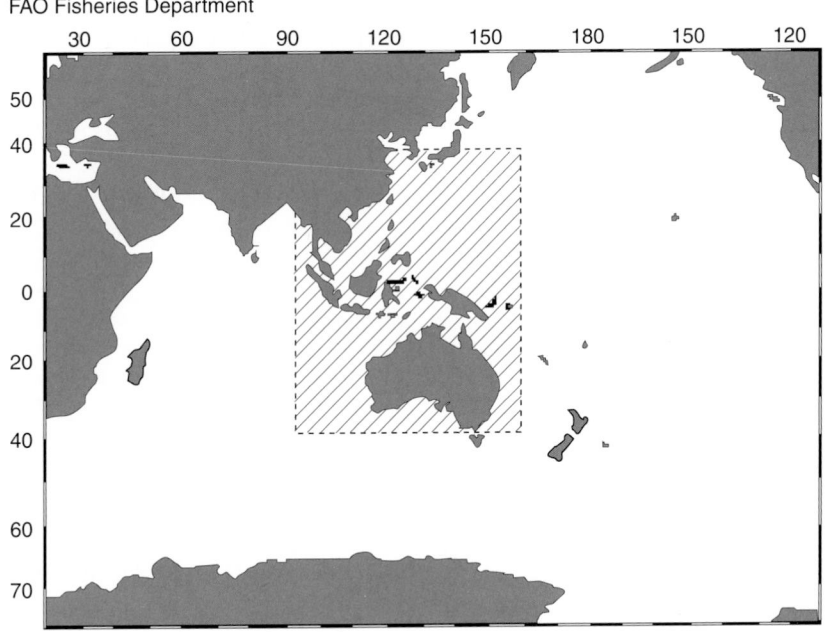

**Figure 3** *Approximate probable range of Omura's whale.*

is no estimate of population size. Strandings have all been of single animals.

## III. Ecology

There is a sighting report of what was possibly a group of Omura's whales chasing tuna (possibly longtail tuna, skipjack tuna, or eastern little tuna) in the Andaman Sea.

## IV. Behavior and Physiology

Nothing is known of the behavior and physiology of this whale.

## V. Life History

Among the six individuals captured in the Solomon Sea, the largest female was 11.5 m and male 9.6 m long. Maximum age among them, based on earplug laminations, was 38 years for a 9.6 m male. The greatest age for a female was 29; she was 11.5 m long. A newborn with fetal folds and folded dorsal fin and flukes was collected off southern Japan on August 30, 2005.

## VI. Interactions with Humans

Apart from the inadvertent take under scientific permit by Japan near the Solomon Islands and Java, there is no record of the species being hunted. Recently, two individuals were incidentally caught in set nets in Japan. In the Philippines sporadic captures by an artisanal land-based fishery in the Bohol Sea existed in the past.

### See Also the Following Articles

Baleen Whales, Overview ■ Bryde's Whales

### References

Sasaki, T., Nikaido, M., Wada, S., Yamada, T. K., Cao, Y., Hasegawa, M., and Okada, N. (2006). *Balaenoptera omurai* is a newly discovered baleen whale that represents an ancient evolutionary lineage. *Mol. Phylogen. Evol.* **41**, 40–52.

Wada, S., and Numachi, K. (1991). Allozyme analyses of genetic differentiation among the populations and species of the *Balaenoptera*. *Rep. Int. Whal. Commn (Spec. Iss.)* **13**, 125–154.

Wada, S., Oishi, M., and Yamada, T. K. (2003). A newly discovered species of living baleen whale. *Nature* **426**, 278–281.

# Osmoregulation

## Daniel P. Costa

### I. Introduction

An organism can be thought of as a large volume of fluid surrounded by the body wall. Mammals must maintain both the concentration and volume of this internal fluid within a very narrow range, and can only tolerate minor deviations. Even though most marine mammals live in an aquatic medium, the animals' internal fluid composition differs from the ambient environment and therefore requires active processes to maintain it. Osmoregulation describes the way in which the internal water and electrolyte concentration of this internal environment is maintained. When animals feed, they take in both water and electrolytes that must be excreted. While they gain water from metabolizing food, they lose water through evaporation when they breathe to obtain the oxygen necessary for metabolism. Maintenance of a constant internal environment requires that whatever comes into the animal must equal what goes out. The easiest way to understand osmoregulation is to account for the ways water and electrolytes enter and leave the organism (Fig. 1). For example, if a dolphin consumes a large volume of water and electrolytes, it must have the capability to excrete an

O

equivalently large volume in the feces and urine, through breathing and in milk during lactation. Conversely, if a seal on the beach does not have access to food or water, it must be able to survive on the water produced from metabolism and have mechanisms in place to reduce water loss. Following the relative rates of water and electrolyte input and output helps us to understand the mechanisms that marine mammals use for osmoregulation.

## II. Water and Electrolyte Ingestion

Water and electrolytes enter the animal through the ingestion of food and water. Water that is consumed in food or actively drunk is called preformed water. Compared to terrestrial mammals, marine mammals consume a water rich diet of fish and marine invertebrates (70–80% water). Prey contains electrolytes and nitrogen that requires water for excretion by the kidney. Ingestion of invertebrate prey (i.e., squid, krill, clams), results in the intake of more electrolytes than vertebrate prey (fish). The internal fluid concentration of invertebrates is essentially the same as seawater, whereas vertebrates contain about one-third the electrolyte content. Thus, a dolphin eating squid will get almost 3 times as much electrolyte than if it consumed fish. Furthermore, an animal like a manatee, *Trichechus* spp., with access to freshwater can drink freshwater to flush electrolytes, whereas an oceanic dolphin can only drink seawater. Water is also produced as a byproduct of metabolism; this is called metabolic

water production (MWP). The amount of MWP varies with the chemical composition of the diet; therefore, different diets produce varying amounts of metabolic water. For example, 1.07 g of water is generated for every gram of fat oxidized, 0.56 g $H_2O$/g of carbohydrate, and only 0.39 g $H_2O$/g of protein.

## III. Water and Electrolyte Output

Both water and electrolytes are excreted in the urine and feces, whereas only water is lost through evaporation. Water is lost via evaporation both across the skin, cutaneous water loss, and through the lungs, respiratory evaporative water loss. Since marine mammals do not sweat there is no loss of salt across the skin (Whittow *et al.*, 1972). Unlike sea birds and marine reptiles, marine mammals lack specialized glands to excrete salts. All salt excretion is through the kidney, and marine mammals have developed a specialized kidney to handle the large volume of electrolytes and water they process (Ortiz, 2001).

## IV. Do Marine Mammals Drink Seawater?

In most cases, marine mammals can derive sufficient water from their diet so that they do not need to ingest seawater. Measurements of the water, electrolyte and nitrogen intake, coupled with measurements of evaporative, urinary and fecal water loss suggest that a feeding seal can get all of the water it needs from its prey (through both preformed and metabolic water) (Pilson, 1970; Depocas *et al.*, 1971; Tarasoff and Toews, 1972; Ortiz, 2001). This is due to the high water content of the prey coupled with the low evaporative water loss of an animal living in a marine environment.

Do animals drink seawater when they become osmotically stressed in environments where the evaporative water loss is high? To determine whether a marine mammal can gain freshwater by drinking seawater we need to know whether the animal can excrete urine that is more concentrated than seawater. The more concentrated the urine, the greater the amount of "freshwater" that can be derived from ingestion of seawater. A simple calculation can show how much water is gained or lost relative to the concentrating ability of the kidney (Table I). For example, if a humpback whale, *Megaptera novaeangliae*, consumed 1000 ml of seawater and its kidney had the ability to excrete urine with a chloride concentration of 820 mmol/l, it could gain 350 ml of freshwater. Whereas humans, who cannot produce urine as concentrated as seawater, would lose 350 ml of freshwater for every liter of seawater they consumed. The maximum urine concentrating ability of marine and terrestrial mammals is presented in Table II.

So, do marine mammals drink seawater? Many species of marine mammals have the capacity to drink seawater, but they do not always

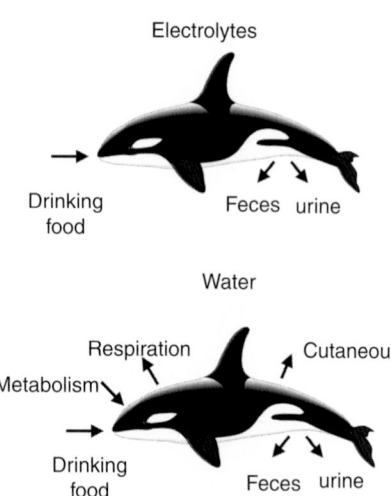

**Figure 1** *Schematic of ways water and electrolytes enter and leave a marine mammal. Excretion of electrolyte and water as milk only occurs when females are lactating.*

TABLE I

Differences in the urine concentrating ability of a humpback whale, and a human given to show a gain or loss of body water after the ingestion of a liter of seawater

| | Seawater consumed volume (ml) | Cl⁻ concentration (mmol/l) | Maximum urine concentration (mmol/l) | Urine volume produced (ml) | Water balance gain or loss (ml) |
|---|---|---|---|---|---|
| Whale | 1000 | 535 | 820 | 650 | +350 |
| Human | 1000 | 535 | 400 | 1350 | −350 |

## TABLE II
The maximum urine chloride concentration and maximum osmolarity measured for marine mammals compared to values of representative terrestrial mammals

|  | Cl⁻ concentration (mEq/l) | Osmolarity (mOsm/l) |
|---|---|---|
| Blue whale, *Balaenoptera musculus* | 340 | 1340 |
| Fin whale, *B. physalus* | 390 |  |
| Sei whale, *B. borealis* | 370 | 1340 |
| Humpback whale, *Megaptera novaeangliae* | 820 |  |
| Bottlenose dolphin, *Tursiops truncates* | 632 | 2458 |
| California sea lion, *Zalophus californianus* | 760 | 2223 |
| Sea otter, *Enhydra lutris* | 555 | 2130 |
| Human, *Homo sapiens* | 400 | 1230 |
| White rat, *Rattus rattus* | 760 | 2900 |
| Camel, *Camelus dromedarius* | 1070 | 2800 |
| Sand rat, *Psammonys obesus* | 1920 | 6340 |
| Sea water | 535 | 1000 |

do so. Isotopically labeled water and/or electrolytes have been used to quantify seawater drinking in a variety of marine mammals (Table III). In these studies, the amount of water and/or electrolytes consumed in the food was added to that produced by metabolism and compared to the total amount of water and/or electrolytes that passed through the animal as measured by isotopic tracers. Using these methods, investigators found that sea otters (*Enhydra lutris*), common bottlenose dolphins (*Tursiops truncatus*), hooded seals (*Cystophora cristata*), and harp seals (*Phoca groenlandica*) that were feeding; and Galapagos fur seals (*Arctocephalus galapagoensis*), short-beaked common dolphins (*Delphinus delphis*), and short-finned pilot whales (*Globicephala macrorhynchus*), that were fasting, consumed seawater (Telfer *et al.*, 1970; Hui, 1981; Costa, 1982; Costa and Trillmich, 1988; Skalstad and Nordoy, 2000; Storeheier and Nordoy, 2001). In contrast, feeding and fasting harbor seals (*Phoca vitulina*), feeding northern fur seals (*Callorhinus ursinus*), and fasting Antarctic fur seals (*Arctocephalus gazella*), all had negligible amounts of seawater ingestion (Depocas *et al.*, 1971; Ortiz *et al.*, 1978; Costa, 1987; Costa and Trillmich, 1988). Weaned northern elephant seal pups (*Mirounga angustirostris*) fast for up to 3 months without any measurable ingestion of seawater (Ortiz *et al.*, 1978). The need to drink seawater varies with climate and habitat. For example, fur seals in warm environments tended to drink seawater whereas those in colder climates did not (Gentry, 1981; Costa and Trillmich, 1988).

## TABLE III
The rate of seawater ingestion measured, using isotopic tracer techniques, in marine mammals

|  | Body mass (kg) | Rate of seawater consumption | | |
|---|---|---|---|---|
|  |  | ml/kg day | ml/day | Proportion of total water influx (%) |
| Pilot whale, *Globicephala macrorhynchus* | 605 | 4.5 | 2720 | n.a. |
| Bottlenose dolphin, *Tursiops truncatus* |  |  |  |  |
| Feeding | 198 | 37.5 | 7420 | 68.8 |
| Common dolphin, *Delphinus delphis* |  |  |  |  |
| Fasting | 57 | 12.5 | 700 | 17 |
| Antarctic fur seal, *Arctocephalus gazelle* |  |  |  |  |
| Fasting | 39.4 | 1.0 | 39 | 15 |
| Galapagos fur seal, *A. galapagoensis* |  |  |  |  |
| Fasting | 37.4 | 18.3 | 684 | 84 |
| Northern fur seal, *Callorhinus ursinus* |  |  |  |  |
| Feeding | 23 | 1.8 | 41 | 2.0 |
| Harbor seal, *Phoca vitulina* |  |  |  |  |
| Feeding | 29.4 | 3.0 | 137 | 9.2 |
| Fasting | 28.6 | 1.3 | 37 | 7.3 |
| Harp seal, *P. groenlandica* |  |  |  |  |
| Feeding | 44.5 | 19 | 900 | 27 |
| Hooded seal, *Cystophora cristata* |  |  |  |  |
| Feeding | 29 | 9 | 300 | 14 |
| Sea otter, *Enhydra lutris* |  |  |  |  |
| Feeding | 24.3 | 62 | 1507 | 23 |

O

## V. Relative Reductions in Water Loss

As described earlier, many marine mammals do not need to drink seawater because they have reduced their evaporative water loss. Amazingly, northern elephant seals can fast for months without access to food or water (Fig. 2). The only water available to fasting seals is MWP from the oxidation of fat and protein in their tissue (Ortiz *et al.*, 1978). Remember that positive water balance requires that water input equals water output. This requires that water lost in the urine, feces, and from evaporation be equal to or less than MWP. How then do elephant seals, and probably other seals and sea lions, reduce their water loss?

### A. Cutaneous Water Loss

Given their aquatic life style, marine mammals have very low evaporative cutaneous (skin) water loss. In water, there would be no evaporative water loss, and on land, as pinnipeds apparently do not sweat, their cutaneous evaporative loss is quite low (Whittow *et al.*, 1972). However, common dolphins (*Delphinus* spp.) and harbor porpoise (*Phocoena phocoena*) appear to lose a substantial amount of water across their skin surface (Hui, 1981; Andersen and

**Figure 2**  *A male northern elephant seal* (Mirounga angustirostris) *fasting on the beach without access to water. Elephant seals undergo fasts of up to 3 months without access to water. Photo by Dan Costa.*

Nielsen, 1983). Common dolphins lose as much as 4 l H$_2$O/day, or 70% of their total water intake. It may be that seawater ingestion is necessary to make up for the water lost across the skin.

### B. Respiratory Evaporative Water Loss

Endotherms lose water through respiration by the simple physics of warming and saturating the air they breathe. Ambient air is inhaled, warmed, and humidified to core body temperature. For example, air fully saturated (100% relative humidity) with water at 10°C contains 10 ml H$_2$O/l of air, whereas fully saturated air in the lungs at 37°C contains 40 ml H$_2$O/l of air. Unless there is a mechanism to recover water, a seal would lose 30 ml of H$_2$O for every liter of 10°C air it inhaled.

Marine mammals employ a few tricks to reduce the water lost through respiration (Lester and Costa, 2006). The first is to breathe periodically; i.e., to inhale, hold their breath and then exhale. This is called apneustic breathing. Apneustic breathing increases the amount of oxygen extracted per liter of air inhaled. While terrestrial animals typically extract 4% oxygen per breath, marine mammals can extract as much as 8% per breath. This allows marine mammals to breathe less frequently and thereby lose less water because they make fewer respirations to obtain an equivalent amount of oxygen. Pinnipeds, sea otters and polar bears (*Ursus maritimus*), further reduce their respiratory evaporative water loss by employing a nasal countercurrent heat exchanger.

*1. Nasal Countercurrent Heat exchanger*  Marine mammals, rodents and desert ungulates have small passageways in their nasal passages that allows them to recover water vapor and heat that was added to the air at inhalation (Huntley *et al.*, 1984; Folkow and Blix, 1987). The nasal turbinates are composed of very small passageways that allow intimate contact between the inhalant air and the nasal membranes (Fig. 3). As the cold air passes across the small nasal passage, it is warmed and water evaporates. Heat and moisture is transferred from the nasal passage to the air so that by the time it leaves the nasal turbinate it is warmed and humidified to body temperature. In the process of warming the inhaled air, the membranes lining the nasal passages have cooled. On the following exhalation the warm moisture laden air is cooled as it passes over the cool membranes. As the air temperature declines, water vapor condenses and is recovered in the nasal passage (Fig. 4).

(A)                          10 cm                                    (B)        1 cm

**Figure 3**  (A) *Sagittal section of a weanling elephant seal skull showing the nasal turbinates.* (B) *Cross-section through one half of the skull at line "X" in A. With permission The Company of Zoologists.*

## C. Fecal Water Loss

Although there are no direct measurements, fecal water loss of feeding cetaceans is probably quite high. Fecal water loss in pinnipeds feeding on fish is comparable to that of terrestrial carnivores. However, it is not clear how marine mammals that ingest seawater avoid the laxative effect of $MgSO_4$. Fasting animals have negligible fecal water loss, as their fecal production is quite low.

## D. Urinary Water Loss

The rate and amount of water lost in the urine is directly related to both the urine concentrating ability of the kidney and the hydration state of the animal. The kidney ultimately regulates the water and electrolyte state of the animal. When there is a surplus

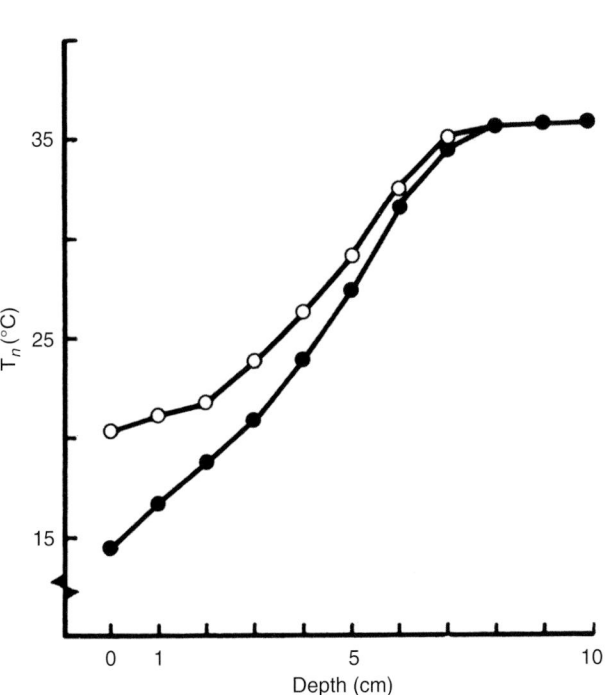

**Figure 4** *Temperature at 1-cm intervals within the nasal passage of a weanling elephant seals where the ambient air temperature was 15°C (open symbols) and 5°C (closed symbols). With permission The Company of Zoologists.*

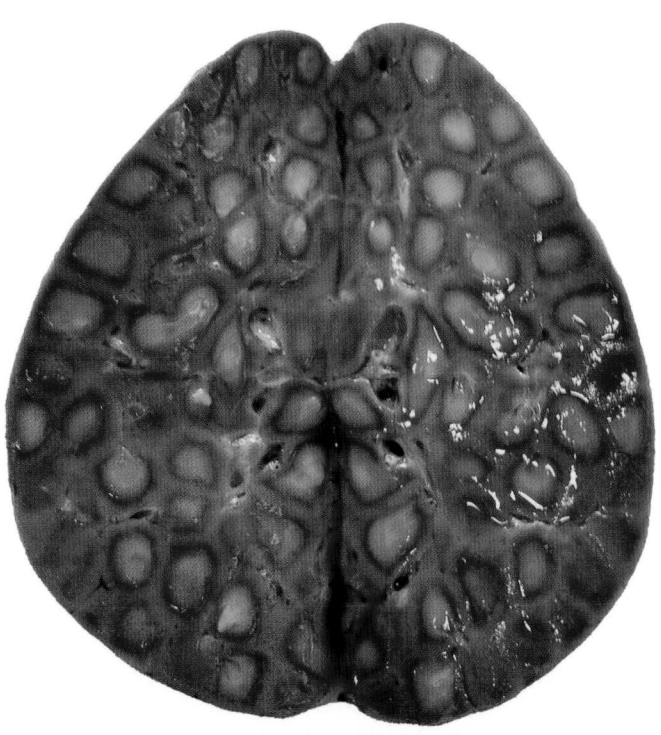

**Figure 5** *A saggital section of a kidney from a California sea lion (Zalophus californianus), showing the two halves. Notice the individual lobules or reniculi that together make up the kidney. Each lobule acts like an individual kidney. Cetaceans, pinnipeds and the sea otter have kidneys constructed this way. Photo by Dave Casper.*

**Figure 6** *A female northern elephant seal and her suckling pup. Over the entire 28 day lactation interval the mother does not eat or drink. All the water and energy contained in the milk provided to the pup must come from the mother. Milk is also the only source of water for the pup. After the pup is weaned it will fast on the beach between 2 and 3 months. During this fasting period the pups does not drink measureable amounts of sea water.*

### TABLE IV
Water, lipid and protein content of marine mammal milk compared to human and cow milk

| | % water | % lipid | % protein |
| --- | --- | --- | --- |
| Blue whale, *Balaenoptera musculus* | 45.4 | 41.5 | 11.9 |
| Minke whale, *B. acutorostrata* | 60.4 | 24.4 | 13.6 |
| Sperm whale, *Physeter macrocephalus* | 64.5 | 24.4 | 9.1 |
| Bottlenose dolphin, *Tursiops truncatus* | 69.6 | 15.3 | 11.5 |
| Galapagos fur seal, *Arctocephalus galapagoensis* | 58.5 | 29.4 | 12.1 |
| Northern fur seal, *Callorhinus ursinus* | 44.3 | 41.5 | 14.2 |
| Australian sea lion, *Neophoca cinerea* | 64.7 | 25.8 | 9.5 |
| Northern elephant seal, *Mirounga angustirostris* | 36.6 | 54.4 | 9.0 |
| Hooded seal, *Cystophora cristata* | 33.7 | 61.4 | 4.9 |
| Grey seal, *Halichoerus grypus* | 36.6 | 52.2 | 11.2 |
| Human, *Homo sapiens* | 87.6 | 3.8 | 1.2 |
| Cow, *Bos Taurus* | 87.3 | 3.7 | 3.3 |

of water, the kidney produces dilute urine, whereas during periods of water stress, the kidney excretes concentrated urine. The kidney must be able to excrete metabolic end products in the form of urea and all excess electrolytes with the water that remains after cutaneous, respiratory, and fecal water loss. While at sea, marine mammals either get all of their water from their prey or they drink seawater. This requires the processing of large urine volumes at moderate to high urine concentrations, and most marine mammals (cetaceans, pinnipeds, sea otters) have a specialized lobulate or reniculate kidney that enables them to do this (Vardy and Bryden, 1981; Costa, 1982; Ortiz, 2001) (Fig. 5; see kidney section).

However, pinnipeds, such as the northern elephant seal, undergo prolonged fasts on land without access to water. These animals are able to stay in water balance by a combination of low rates of evaporative water loss, coupled with low rates of urine production (Adams and Costa, 1993; Lester and Costa, 2006). Elephant seals, *Mirounga* spp., utilize fat almost entirely (96–98%) for their metabolism while fasting. Fat oxidation produces only $CO_2$ and $H_2O$, whereas oxidation of protein results in $CO_2$, $H_2O$, and urea. Urea is the end product de-amination of amino acids and requires water to be excreted by the kidney. Therefore, fat is not only an efficient way to store energy, it is also economical with respect to water balance (see pinniped physiology section).

### VI. Water Balance During Reproduction

Many female pinnipeds do not have access to water while they suckle their young, and thus could become dehydrated during lactation (Fig. 6). However, marine mammal milk is high in lipid and low in water compared to terrestrial mammals (Table IV). This has the advantage of providing the young with the maximum amount of energy with minimal loss of water from the mother. This is likely an advantageous byproduct of the energetics of marine mammal

lactation and not a derived adaptation for water balance (see marine mammal energetics section). Pups also do not have access to water, and therefore must be capable of maintaining water balance entirely from the water provided in the milk.

### See Also the Following Articles

Circulatory System ■ Diving Physiology ■ Thermoregulation

### References

Adams, S. H., and Costa, D. P. (1993). Water conservation and protein metabolism in northern elephant seal pups during the postweaning fast. *J. Comp. Physiol. B* **163**, 367–373.

Andersen, S. H., and Nielsen, E. (1983). Exchange of water between the harbor porpoise, *Phocoena phocoena*, and the environment. *Experientia* **39**, 52–53.

Costa, D. P. (1982). Energy, nitrogen, and electrolyte flux and sea water drinking in the sea otter *Enhydra lutris*. *Physiol. Zool.* **55**, 35–44.

Costa, D. P. (1987). Isotopic methods for quantifying material and energy intake of free-ranging marine mammals. *In* "Approaches to Marine Mammal Energetics," pp. 43–66. Allen Press, Lawrence.

Costa, D. P., and Trillmich, F. (1988). Mass changes and metabolism during the perinatal fast: A comparison between Antarctic (*Arctocephalus gazella*) and Galapagos fur seals (*Arctocephalus galapagoensis*). *Physiol. Zool.* **61**, 160–169.

Depocas, F., Hart, J. S., and Fisher, H. D. (1971). Sea water drinking and water flux in starved and in fed harbor seals. *Phoca vitulina*. *Can. J. Physiol. Pharmacol.* **49**, 53–62.

Folkow, L. P., and Blix, A. S. (1987). Nasal heat and water exchange in gray seals. *Am. J. Physiol.* **253**, R883–889.

Gentry, R. L. (1981). Seawater drinking in eared seals. *Comp. Biochem. Physiol.* **68**, 81–86.

Hui, C. A. (1981). Seawater consumption and water flux in the common dolphin *Delphinus delphis*. *Physiol. Zool.* **54**, 430–440.

Huntley, A. C., Costa, D. P., and Rubin, R. D. (1984). The contribution of nasal countercurrent heat exchange to water balance in the northern elephant seal, *Mirounga angustirostris*. *J. Exp. Bio.* **113**, 447–454.

Lester, C. W., and Costa, D. P. (2006). Water conservation in fasting northern elephant seals (*Mirounga angustirostris*). *J. Exp. Bio.* **209**, 4283–4294.

Ortiz, C. L., Costa, D., and Le Boeuf, B. J. (1978). Water and energy flux in elephant seal pups fasting under natural conditions. *Physiol. Zool.* **51**, 166–178.

Ortiz, R. M. (2001). Osmoregulation in marine mammals. *J. Exp. Bio.* **204**, 1831–1844.

Pilson, M. E. Q. (1970). Water balance in California sea lions. *Physiol. Zool.* **43**, 257–269.

Skalstad, I., and Nordoy, E. S. (2000). Experimental evidence of seawater drinking in juvenile hooded (*Cystophora cristata*) and harp seals (*Phoca groenlandica*). *J. Comp. Physiol. B* **170**, 395–401.

Storeheier, P. V., and Nordoy, E. S. (2001). Physiological effects of seawater intake in adult harp seals during phase I of fasting. *Comp. Biochem. Physiol. A.* **128**, 307–315.

Tarasoff, F. J., and Toews, D. P. (1972). The osmotic and inoic regulatory capacities of the kidney of the harbor seal. *Phoca vitulina*. *J. Comp. Physiol.* **81**, 121–132.

Telfer, N., Cornell, L. P., and Prescott, J. H. (1970). Do dolphins drink water? *J. Vet. Med. Assoc.* **197**, 555–558.

Vardy, P. H., and Bryden, M. M. (1981). The kidney of *Leptonychotes weddelli* Pinnipedia Phocidae with some observations on the kidneys of 2 other southern phocid seals. *J. Morph.* **167**, 13–34.

Whittow, G. C., Matsuura, D. T., and Lin, Y. C. (1972). Temperature regulation in the California sea lion (*Zalophus californianus*). *Physiol. Zool.* **45**, 68–77.

# Otters, Marine

J.A. Estes, J.L. Bodkin, and M. Ben-David

The otters (Mustelidae; Lutrinae) provide an exceptional perspective into the evolution of marine living by mammals. Most extant marine mammals (e.g., the cetaceans, pinnipeds and sirenians) have been so highly modified by long periods of selection for life in the sea that they bear little resemblance to their terrestrial ancestors. Marine otters, in contrast, are more recent expatriates from freshwater habitats and some species still live in both environments. Contrasts among species within the otters, and among the otters, terrestrial mammals, and the more highly adapted pinnipeds and cetaceans provide powerful insights into mammalian adaptations to life in the sea (Estes, 1989). Among the marine mammals, sea otters (*Enhydra lutris*, Fig. 1) provide the clearest understanding of consumer-induced effects on ecosystem function. This is due in part to opportunities provided by history and in part to the relative ease with which shallow coastal systems where sea otters live can be observed and studied. Although more difficult to study than sea otters, other otter species reveal the connectivity among the marine, freshwater, and terrestrial systems. These three qualities of the otters—their comparative biology, their role as predators, and their role as agents of ecosystem connectivity—are what make them most interesting to marine mammalogy.

The following account provides a broad overview of the comparative biology and ecology of the otters, with particular emphasis on those species or populations that live in the sea. Sea otters are featured prominently, in part because they are comparatively well known and in part because they live exclusively in the sea whereas other otters have obligate associations with freshwater and terrestrial environments (Kenyon, 1969; Riedman and Estes, 1990).

## I. Evolution and Phylogeny

Mustelids arose from primitive arctoid carnivores at the Eocene/Oligocene border. Early lutrine phylogenies were based on the morphology of fossil and extant species, from which the otters were viewed as a monophyletic group that diversified into three clades: the fish-eating otters (*Lutra*, *Lontra*, and *Pteronura*), crab-eating otters (*Aonyx*), and the sea otter (*Enhydra*). However, the otters probably have been under strong selection for parallel or convergent evolution, thereby confounding efforts to understand phylogenetic relationships based on morphology. Distinctive features of the three purported clades might thus have resulted from differences among their common ancestors or selective divergence resulting from different prey (fishes vs invertebrates) or habitats (the ocean vs freshwater). Patterns of brain form and function in the otters exemplifies this problem. Sensory and motor function in the mammalian brain maps medio-laterally along the prefrontal gyrus. Architectural and functional differences among otter brains correlate with their principal-foraging modes—invertebrate vs fish feeding. The fish-eating otters, which require precise sensory/motor function of the mouth and facial area, have well-developed proximal regions of the prefrontal gyrus. The invertebrate-feeding otters (*Enhydra* and *Aonyx*), which require precise sensory/motor function of their forelimbs for prey capture, have more highly developed lateral regions of the prefrontal gyrus. But are these features primitive or derived? If primitive, then they might accurately reflect phylogeny; otherwise, they surely do not.

**Figure 1** *Sea otter* (Enhydra lutris). *Photograph courtesy of Randall Davis and students (taken under USFWS permit No. MA078744-2).*

Nucleotide sequence analysis of the mitochondrial cytochrome b gene (Koepfli and Wayne, 1998) has been used to disassociate the confounding effects of adaptation in constructing a lutrine phylogeny for 9 of the 13 extant species (Fig. 2). These data demonstrate that earlier phylogenies based solely on morphology were grossly inaccurate. The molecular analysis indicates three primary clades including the (1) North American otter (*Lontra canadensis*), neotropical otter (*L. longicaudus*), and chungungo (*L. felina*); (2) sea otter (*Enhydra lutris*), Eurasian otter (*Lutra lutra*), spotted-necked otter (*Hydrictis maculicollis*), cape clawless otter (*Aonyx capensis*), and small-clawed otter (*A. cinerea*); and (3) giant otter (*Pteronura brasiliensis*). Fundamental life history differences (e.g., seasonal vs aseasonal reproduction; direct vs delayed implantation) between Eurasian and North American otters, species once thought to have diverged from a common ancestor only after late Pleistocene isolation of Asia and North America, make much more sense based on the molecular phylogeny. The pattern of long terminal branches and short internal branches in the phylogenetic tree further suggests rapid radiation of the otters.

Estimates of divergence time indicate that the clades containing *Pteronura* and *Lontra* had separated by the late Miocene. This phylogeny also suggests that sea otters, the only fully marine otter and the most distinctive of all otter species in terms of morphology, physiology and behavior, diverged recently but have taken on these different characters because of strong selection imposed by life in the sea.

## II. Marine Otters

At least 6 of the 13 extant otter species are fully or partially marine living (Fig. 3). All of the marine-living species or populations occur at high latitudes. Sea otters (North Pacific Ocean, genus *Enhydra*) are the only fully aquatic otter species, with no obligate associations with terrestrial habitats for any life history function. The other otters can be considered semi-aquatic, retaining ties to terrestrial habitats for most behaviors other than hunting. Sea otters and chungungos (*Lontra felina*) are the only lutrines that feed exclusively in the sea. Sea otters range from the northern Japanese archipelago, across the rim of the Pacific to about central Baja California, Mexico. Chungungos range along the west coast of South America from Peru to southern Chile and only spend about 20% of their time feeding in the sea. Chungungos often take their prey ashore to be eaten, and rest, give birth, and rear their young in dens formed in rocky areas just above high water. Sea otters occupy a broad range of habitats, from protected bays to exposed

O

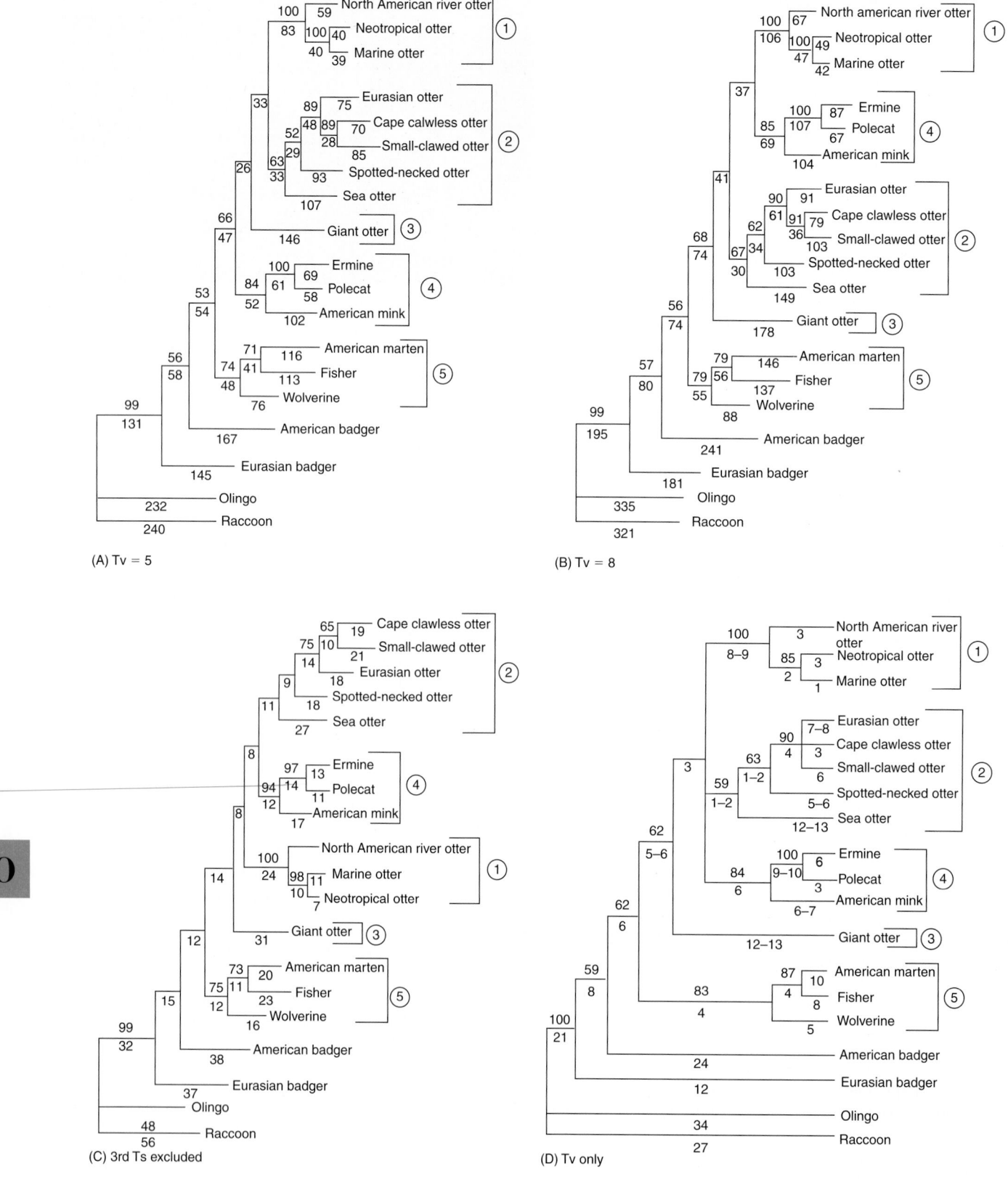

**Figure 2**   *Phylogeny of the otters. From Koepfli and Wayne (1998).*

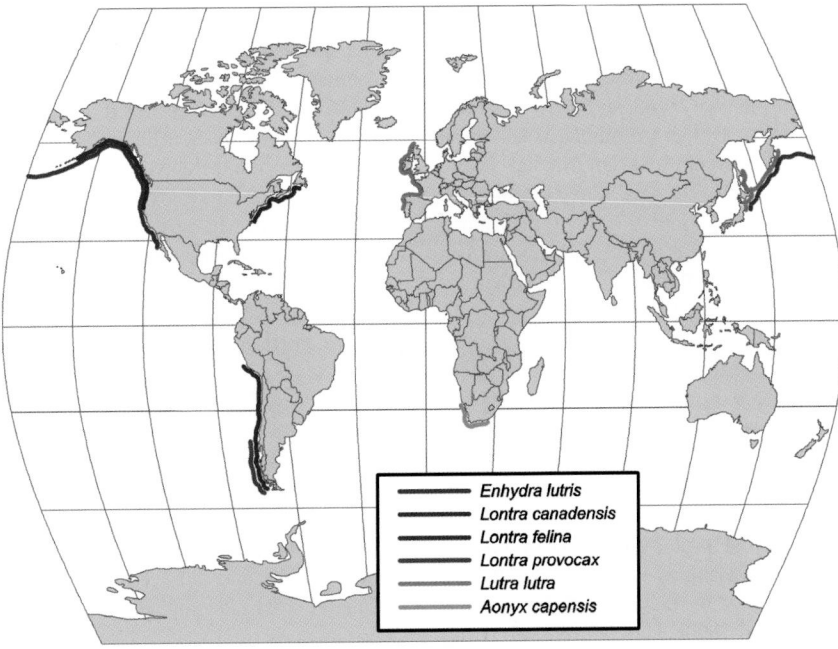

**Figure 3** *The geographical distributions of marine-living otters.*

outer shores while chungungos occur only along exposed shore-lines. Marine-living populations of *Lontra provocax* (southern river otter) occur in the protected inner waters of coastal Chile and southern Argentina. Marine-living populations of North American otters occur northward from about San Francisco in the Pacific and Martha's Vineyard in the Atlantic, although there is anecdotal evidence of marine foraging by river otters in Florida. Marine-living populations of Eurasian otters have a roughly similar latitudinal distribution (from about Portugal northward to northern Scotland and Scandinavia in the North Atlantic; from Japan into Russia in the North Pacific). Marine-living populations of cape clawless otters occur in the southernmost regions of South Africa. Tropical lutrines, in contrast, rarely enter the sea.

While marine-living species or populations of otters occur only at high latitudes, otters inhabiting freshwater habitats occur over a much broader latitudinal range—from the most poleward ice-free environments to the tropics in the northern and southern hemispheres of both the Old and New worlds. These distributional differences between marine- and freshwater-living otters probably relate to latitudinal differences in production between freshwater and coastal marine habitats, and the fact that otters have secondarily entered the sea from freshwater environments. At low latitudes, freshwater production exceeds that of the coastal ocean whereas at high latitudes the pattern is reversed. This production gradient may have drawn the primitively freshwater-living otters into the sea at high latitudes. The most compelling evidence for this proposal is provided by Hans Kruuk's comparative study of freshwater and marine-living populations of river otters in northern Scotland (Kruuk, 2006). Kruuk found that marine-living populations are able to meet their energy requirements by 2–3 h of fishing per day whereas those living in freshwater must double to triple the time investment in foraging to meet their energy requirements. Remarkably different population densities of marine and freshwater-living otters further indicates that production and food availability are superior in coastal marine habitats. Reported densities for Eurasian and North American otters in rivers and streams at high latitudes range between about 1 and 5

individuals per 10 km whereas densities of these same species living in marine habitats are about 5-fold greater. These data indicate that otters at high latitudes maintain higher population densities in the sea than they do in freshwater, and that most low latitude freshwater habitats may not be productive enough to support high otter densities. Some exceptions to this rule are giant otters that live in fairly large family groups in the productive waters of the Amazon and its tributaries, the Asian small-clawed otters (*Aonyx cinerea*) of southeast Asia that forage in productive mangrove forests, and the spotted-necked otters (*Hydrictis maculicollis*) that form large groups and inhabit productive lakes of central Africa.

Different population densities of otters in freshwater and marine environments appear to have influenced the evolution of social behavior. Low-density freshwater environments afford little or no opportunity for males to compete for females; consequently, most freshwater-living otters appear to have monogamous or promiscuous mating systems. The elevated marine production and high female densities may have driven the strongly polygynous mating system of sea otters wherein dominant males defend small territories against other males and attempt to sequester receptive females within their territories. In contrast, female North American otters in coastal marine environments of northwestern North America are usually solitary whereas most males form large groups. Again this discrepancy is likely a result of food availability. While social male otters forage cooperatively on abundant schooling fishes in the nearshore environment, nonsocial females and their cubs forage on the less abundant and spatially dispersed intertidal/demersal fishes.

## III. Status and Trends

Aboriginal maritime hunters developed sophisticated methods to hunt sea otters and may have caused local depletions, as evidenced from faunal remains in Aleut kitchen middens. The species subsequently was hunted to near-extinction in the Pacific maritime fur trade. About a dozen remnant colonies, in total containing no more

O

than a thousand or so individuals, survived into the early twentieth century. Following protection in 1911, these colonies increased with some regaining historical levels by the mid-twentieth century. Reintroductions were undertaken in the 1960s and 1970s to reestablish sea otters in southeast Alaska, British Columbia, Washington and Oregon, and sea otters were relocated to San Nicolas Island (southern California) in 1987. Most of these translocated populations increased at rates of 17–20% per year, about the theoretical maximum ($r_{max}$) for sea otters. All or most of the sea otter's range in Asia is now reoccupied. Several populations in the Kuril and Commander islands had recovered to historical levels by the 1970s or 1980s.

In apparent response to increased killer whale predation, sea otter populations in southwest Alaska declined sharply during the 1990s. This decline ranges from west of the Kodiak archipelago along the Alaska Peninsula and through the Aleutian archipelago. Reintroduced sea otter populations in British Columbia and Washington continue to increase in numbers and range. The introduced population to southeast Alaska appears to have ceased growing in recent years although localized increases are evident in the protected waters of Glacier Bay National Park and Preserve. The California sea otter population, which increased slowly throughout most of the twentieth century, has not changed measurably since the mid-1990s. The reintroduced population at San Nicolas Island, after declining sharply because of emigration, remained roughly stable at 15–25 through the late 1990s but since has increased to more than 40 individuals.

Compared to the sea otter, there is little current information on the status and trends of other marine-living otter populations. The chungungo, listed in IUCN's Red Data Book, is believed by some authorities to be threatened with extinction. Illegal harvesting for fur, habitat destruction in the form of deforestation, mining and pollution, and competition with fisheries are thought to be the species' main threats. Chungungos are rare from Peru southward through central Chile, but there are conflicting reports on their abundance from Chiloe Island southward to Cape Horn (Medina-Vogel *et al.*, 2006).

North American otters (*Lontra canadensis*) are common to abundant along much of the west coast of North America. Their abundance along the Atlantic coast of North America is uncertain and information on population trends is lacking for both areas. Eurasian otters (*Lutra lutra*) are common in marine habitats of Europe and Russia, especially from Norway to Scotland in the eastern North Atlantic. However, as for marine-living river otters in North America, information on the status and trends of these populations is lacking. Coastal populations in Asia are practically unknown. The status and trends of marine-living populations of southern river otters and cape clawless otters are unknown or unreported.

## IV. History

The otter lineage dates back to the early Miocene (approximately 20 million years ago) as indicated by fossils of the genus *Mionictis*. Divergence of this monoplyletic group during the Miocene ("11–14" million years ago) was rapid and resulted in the three main lineages described above. Species resembling modern sea otters (based on body size and dentition) had arisen by the Miocene (Berta and Morgan, 1986). There are two recognized lineages, one of which led to the extinct *Enhydriodon* and the other of which led to *Enhydritherium* and presumably to modern sea otters, *Enhydra* spp. Members of both lineages possessed large flattened molars for crushing the exoskeletons of their invertebrate prey. Fossil remains of these early sea otters are known from North America, Eurasia, and Africa. The distribution of modern sea otters, which apparently arose during the late Pliocene

or early Pleistocene, was restricted to the North Pacific Ocean. One extinct species, *E. macrodonta*, is described from the late Pleistocene of California.

Based on the fossil record it appears that both *Lutra* and *Lontra* groups originated in Southeast Asia from which they expanded into Europe (*Lutra*) and the Americas (*Lontra*). The earliest *Lutra* fossil in Europe is dated to the Pliocene. *Lontra* colonized North America as recently as the Pleistocene, approximately 1.7 million years ago. The relationship between *Lontra* and *Petronura* (giant otters) is unclear but fossil evidence suggest that giant otters evolved from an Asian lineage (*Satherium*) that, as with the ancestors of *Lontra*, emigrated to North America and then spread south.

## V. Morphology and Physiology

Compared with the diversity of size and form in other extant marine mammals, the otters are relatively small and generally similar in overall body plan. Body mass ranges from <5 kg in chungungos to more than 50 kg in sea otters. Their shortened limbs, lengthened and often stout tails, and slim, elongated head, neck and body, and loosely articulated axial skeleton combine to create an almost serpentine appearance. All otter species have dense fur for insulation against heat loss and interdigital webbing of the fore and/or hind limbs to assist in aquatic propulsion. Like pinnipeds, the sea otter's clavicle has been lost or greatly reduced, apparently in response to the reduced need for skeletal support on land and the increased need for flexibility in the water.

The internal morphology of otters is generally unremarkable, except for the sea otter's comparative large lung and blood volumes (which facilitate flotation and oxygen storage), and a large lobular kidney (which facilitates osmoregulation).

Except for sea otters, none of the other marine-living otter species are fully aquatic. Because of their relatively smaller body size and less efficient insulation, these other species or populations are probably incapable of long-term thermal maintenance in high latitude oceans. As the smallest fully aquatic marine mammal, sea otters face an especially severe thermal challenge because they spend all or most of their lives immersed in cold water. This challenge has been met by increasing heat production or reducing heat loss. Like other mustelids, the basal metabolic rate of sea otters is well above that predicted by the Kleiber curve and sea otters gain additional heat from the specific dynamic activity (SDA) of digestion. Sea otters lack the blubber layer that insulates cetaceans and pinnipeds, instead depending upon their dense fur. Although fur is a superior insulator to blubber in air, it has three disadvantages in water: high maintenance costs, the inability to regulate heat flow, and compressibility at depth.

In order to maintain the fur's insulation, sea otters must groom almost continuously, an activity that consumes up to 10% of their time. Because fur (in contrast to blubber) is an inflexible insulator, sea otters require some means of facilitating heat loss during exercise. This apparently is accomplished mainly through the enlarged rear flippers, which are sparsely furred and highly vascularized. The compressibility of fur causes it to loose volume, and hence insulation, with increased depth.

Water conservation also presents a challenge to sea otters because they feed primarily on marine invertebrates, which are isotonic to seawater. They meet this challenge in part with a large and efficient kidney.

Marine-living North American and Eurasian otters also rely on fur for insulation. Unlike sea otters these semi-aquatic otters require rinsing in freshwater to restore the insulative capacity of their coats.

For example, Eurasian otters that were unable to rinse their coats in freshwater spent about 30% longer grooming than animals that washed in freshwater. In general Eurasian otters spend up to 6% of their time grooming. Also, because North American and Eurasian otters have less dense coats than sea otters, they spend much less time in the water. For example, Eurasian otters spend on average only $14.5 \pm 10.7$ min in the water during any given foraging bout.

Like nearly all aquatic mammals, otters must dive to feed. The semi-aquatic otter species usually dive in shallow waters. Eurasian otters may forage as deeply as 15 m but the vast majority of dives occur in depths of 8 m or less. North American otters in Prince William Sound, Alaska, have been observed to dive as deep as 30 m, but again these dives are relatively rare. Although sea otters do not dive to such extreme depths as many other marine mammals, they are considerably better divers than other otter species, attaining depths in excess of 100 m. The respiratory system is appropriately modified for deep diving. For instance, the tracheal length–width ratio is less in sea otters than river otters, thus permitting more rapid and complete air exchange before and after a dive. Like the pinnipeds, sea otters have cartilaginous airways that empty directly into their alveoli, thus insuring patency until alveolar compression collapse during deep dives. Sea otters also have comparatively large lungs that provide both oxygen storage and increased buoyancy. Although the sea otter's blood hemoglobin concentration is similar to that of most terrestrial mammals and less than that of most phocid seals, oxygen–hemoglobin affinity in the sea otter is relatively high, thus increasing their blood-oxygen storage capacity.

## VI. Locomotion

Compared with fully terrestrial or fully aquatic mammals, locomotor efficiency in otters is reduced on both land and in water (Williams, 1999). The otters have retained the general structure of their limbs but all posses webbing between their toes. Among the lutrines, sea otters possess the most extreme modifications for aquatic propulsion. These include enlarged, flipper-like hind limbs; a loosely articulated skeleton that permits increased flexibility, reduced forelimb length, and an increased tendency toward body movement and away from paddling in swimming.

In all otters, two types of terrestrial locomotion—walking and bounding—follow the typical pattern of terrestrial carnivores. Running, described in Eurasian otters as a rapid forward movement in the same pattern as walking, has apparently been lost in the sea otter. In North American otters, stride frequency increases linearly with speed until the animal shifts to bounding. No pattern between speed and bounding strides has been observed. A maximum velocity on land of about 5 m/sec can be achieved in short bursts (Williams *et al.*, 2002).

There are three forms of aquatic locomotion by the sea otter. One involves a sweeping motion of the tail and is used for slow movement while feeding, grooming, and to maintain position while resting. The second is paddling—vertical thrusts and recovery of the hind limbs while in either a supine or pronate position. This type of movement is typically used during long-distance travel and may be interspersed with submerged travel, grooming, or foraging. Paddling velocities range from about 0.5 to 1.0 m/sec. The third means of aquatic locomotion in sea otters is accomplished by craniocaudal thrusts of the pelvic limbs, often including bending of the lumbar, sacral and caudal regions for increased sustained speed up to about 2 m/sec. This method of locomotion is used in foraging dives and rapid surface movement. Other otter species also have flexible spines and use undulation in diving. Many species have laterally flattened tails which assist in controlling the direction and velocity under water. This is especially

pronounced in giant otters. Unlike sea otters, other species of otters do frequently paddle and only Eurasian otters and sea otters float on their back for extended periods of time.

## VII. Diving

In all otter species, diving can occur during traveling and foraging. Because locomotion is more efficient underwater than on the surface, otters frequently make relatively long, shallow dives while traveling. Foraging dives are generally deeper and longer in duration compared to travel dives and include time searching for and capturing prey. Because sea otters forage almost exclusively on benthic prey that are brought to the surface for consumption, classification of dive function is often possible. Visual observations and data from archival time-depth recorders (TDRs) in sea otters indicate substantial individual and sex-related variation in forage dive behavior that is consistent with individual dietary variation. Mean (maximum) forage dive depths ranged from about 5 to 35 m (35–100 m) and dive durations from 57 to 135 sec (162–422 sec) among 31 sea otters that were studied in Alaska (Bodkin *et al.*, 2004). Females generally dive to shallower depths than males, although individuals of both sexes can exhibit strong bimodal forage depth distributions with peaks near 15 and 50 m. Dive times and their interceding surface intervals correlate with water depth although the deepest dives are not necessarily associated with maximum dive times. Surface intervals are highly correlated with prey size and type, with the longest intervals allied with the largest prey, thus reflecting associated increases in handling and consumption times.

Most information on diving in North American and Eurasian otters is derived from captive experiments. These observations were corroborated in the field from telemetry data and direct observations. In general, dive durations of both species are limited to 20–23 sec. The longest voluntary dive in a North American otter was clocked at 3 min. In both North American and Eurasian otters, there is a strong correlation between length of dive and time spent on the surface following the dive.

## VIII. Feeding Ecology

Marine otters feed on a wide array of fish and invertebrate species. With some exceptions, the semi-aquatic otters specialize on fish whereas the fully aquatic sea otter specializes on benthic marine invertebrates. The chungungo, unlike other *Lontra* species feeds extensively on crustaceans, mollusks, and echinoderms and makes relatively low use of fishes (occurrence in about 20% of feces; Medina-Vogel *et al.*, 2004). Cape clawless otters also feed largely on invertebrates. Crabs (*Plagusia chabrus* and *Cyclograpsus punctatus*), and lobster (*Jasus lalandii*) comprise most of the diet of these otters around Cape Town, South Africa. Similarly, the sea otter, typically a predator on benthic invertebrates, occasionally consumes fish in large quantities.

Sea otters feed mainly on echinoderms, mollusks, and arthropods. The enlarged molariform teeth are used to crush the exoskeletons of these invertebrate prey. Tool use has developed in both the sea otter and Cape clawless otter as a further aid to foraging. This is accomplished by using rocks or other hard objects to break open unusually large or well armored invertebrates.

While sea otters are known to consume more than 150 prey species, only a few of these predominate in any particular place, depending on location, habitat type, season and length of occupation. Increasing dietary diversity through time as sea otter populations recolonize new habitats and grow toward resource limitation has been chronicled in the Aleutian Islands, Prince William Sound, and California (Estes

**O**

*et al.*, 1981). These changes are probably the consequence of otters reducing the abundance of their preferred prey.

Studies of marked sea otters in California reveal extreme individual variation in diet and foraging behavior. Most individuals specialize on 1 to 3 prey types. This individual variation does not appear to be directly influenced by access to prey as different individuals often consume different prey at the same time and place. Individual dietary patterns, which appear to be matrilineally inherited, are known to persist for years and may be lifelong characters of individuals. Dietary individuality may develop in response to resource limitation.

Sea otters may also take advantage of episodically abundant prey. Examples include squid (*Loligo* sp.) and pelagic red crabs (*Pleuroncodes planipes*) in California and smooth lumpsuckers (*Aptocyclus ventricosus*) in the Aleutian Islands. Pelagic red crabs appear in coastal waters of southern and central California during strong El Niño events and vast numbers of lumpsuckers appear episodically in coastal waters of the western and central Aleutian Islands to spawn. Sea otters attack and consume various species of sea birds on occasion.

Marine-foraging Eurasian otters usually hunt alone and largely consume benthic fishes such as eelpout (*Zoarces viviparous*), rocklings (*Ciliata mustela*), and sea scorpion (*Taurulus bubalis*). Invertebrates are rarely consumed. Also, there is little difference in the diet of males and females. In contrast, marine-living North American otters consume bivalves and crabs (up to 13% of diet). In addition, diets of males and females differ greatly during the summer. While most females forage alone on benthic fishes, male otters form large groups and cooperatively forage on schooling fishes such as Pacific herring (*Clupea pallasi*), Pacific sandlance (*Ammodytes hexapterus*), capelin (*Mallotus villosus*), and Pacific salmon (*Oncorhynchus* spp).

## IX. Community Ecology

The ecology of sea otters at the community and ecosystem level has been studied extensively whereas the other otter species are poorly known in this regard. Therefore, this section focuses mostly on the interactions between sea otters and coastal marine ecosystems.

### A. Food Web Effects

The sea otter's role in kelp ecosystem preservation, by keeping kelp-eating sea urchin populations in check, was discovered by contrasting otherwise similar areas at which the species was fortuitously present or absent because of exploitation and recovery from the Pacific maritime fur trade. Shallow reef habitats with abundant sea otters supported few sea urchins and well-developed kelp forests whereas abundant sea urchins had destroyed the kelp forests at islands lacking sea otters (Fig. 4). The explanation for this pattern is a straightforward consequence of what has since come to be known as a "trophic cascade". That is, sea urchin populations are regulated by sea otter predation, in turn allowing the kelp forest to flourish in the absence of significant herbivory. When otters were removed, sea urchins increased to such levels that deforestation occurred. These relationships have been documented in Russia, the Aleutian Islands, southeast Alaska, British Columbia, Washington, and California (Estes and Duggins, 1995).

Because sea otters influence the distribution and abundance of kelp forests, they also affect numerous other species via three general ecosystem-level processes: production, habitat modification, and altered flow. Total primary production and the growth rates of filter-feeding invertebrates are significantly greater in otter-dominated compared with otter-free ecosystems (Duggins *et al.*, 1989). Various other indirect effects of sea otter predation are known or suspected.

**Figure 4** *Alternate community states in areas with (kelp forests; Top) and without (sea urchin barrens) sea otters. Photographs were taken of reef habitats at about 10 m depths in the western Aleutian Islands.*

For instance, a reduction in disturbance from herbivory enhances the strength of competitive interactions among kelp species. The sea otter's influence on kelp forest abundance and distribution has strong effects on the abundance and species composition of kelp

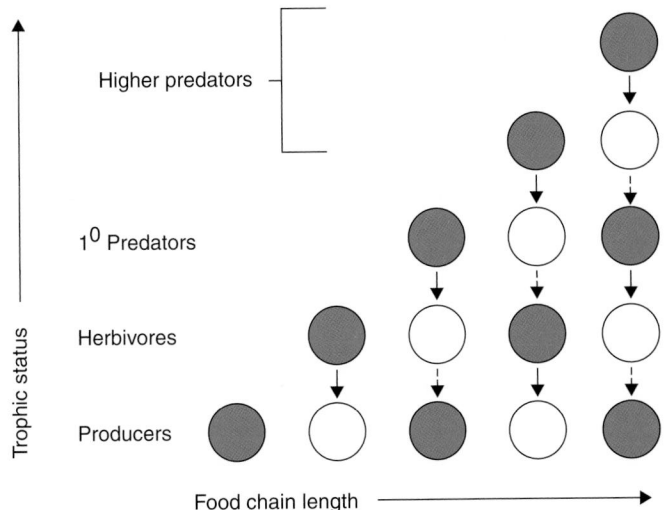

**Figure 5** *Fretwell's (1987) model of alternating plant–herbivore interaction strength with increasing food chain length.*

forest fishes. Rock greenling (*Hexagrammos lagocephalus*) exist at roughly 10-fold higher densities where sea otters are abundant in the Aleutian Islands. Sea otters therefore may facilitate the abundance of solitary river otters as they feed on greenling. If so, trophic cascades in the nearshore environment may have rippling effects onto the adjacent shoreline. Conversely, common eiders (*Somateria mollissima*), which feed mainly on sea urchins, mollusks, and other benthic invertebrates, occur at higher densities where sea otters are rare or absent than where they are abundant. The indirect effects of sea otters also influence the diet and foraging behavior of Glaucous winged gulls (*Larus glaucescens*), which are mainly piscivorous when sea otters are present and invertebrate-feeders when they are absent. Bald eagles (*Haliaetus leucocephalus*) in the Aleutian Islands consume fewer kelp forest fishes and marine mammals and more sea birds when sea otters are lost from the system. Sea otters also influence predator–prey interactions between sea stars and mussels in the Aleutian archipelago.

The top-down influences of apex predators through trophic cascades have been incorporated into a conceptual model relating the strength of plant–herbivore interactions to trophic complexity (Fig. 5). The model predicts that plant–herbivore interactions should be strong in food chains with an even number of trophic levels (e.g., 2, 4, etc.) and relatively weak for odd-numbered food chains (e.g., 1, 3, 5, etc., Fretwell, 1987). The accumulated evidence from three decades of research on sea otters and kelp forests provides empirical evidence for this conceptual model.

### B. Evolutionary Forces

Strong species interactions maintained over sufficiently large scales of space and time should lead to selective responses in the interacting species. This expectation, coupled with predicted differences in the strength of plant–herbivore interactions between odd- vs even-numbered food webs, suggests that kelp forests in the North Pacific Ocean evolved in the absence of intense herbivory whereas Australasian kelp forests evolved in the absence of an apex predator of comparable influence. Striking differences between the two regions in the spatial association between fleshy marine algae and marine invertebrates, the intensity of herbivory, concentrations of

plant secondary metabolites that act as deterrents against herbivory, and the resistance of herbivores to these secondary metabolites have been documented. These patterns are consistent with the hypothesis that intense predation from sea otters and their immediate ancestors decoupled a co-evolutionary arms race between plants and their herbivores in the North Pacific Ocean, perhaps helping to explain why North Pacific kelps are so poorly defended against herbivores and thus why the system changes so markedly in response to the presence or absence of sea otters.

Evolutionary interactions among sea otters, sea urchins, and kelp have also been evoked to explain other seemingly serendipitous patterns in other species of obligate or facultative kelp forest associates. For example, the abundance and high food quality of North Pacific kelps may help explain why Steller's sea cows (*Hydrodamalis gigas*) radiated from the tropics into the North Pacific Ocean, and why body size in North Pacific abalones (genus *Haliotis*) is the largest in the world.

### C. Nutrient Transport

Semi-aquatic otters link aquatic and terrestrial systems by transporting nutrients from sea to land. Most otters use scent marking with feces, urine, and anal gland secretion for social and territorial communication. Otters select specific features of the landscape for establishing communication/latrine sites and these differ among species. For example, North American otters select for high canopy cover of mature trees, large intertidal rocks, and extensive kelp beds of *Laminaria*. There is also high variation among otter species and between the sexes within a species in the use of latrines. For example, in coastal Alaska, social males use latrines for intra-group communication whereas nonsocial otters likely signal mutual avoidance. Females appear to use latrines to defend their territories. Social otters use fewer sites with greater intensity, whereas nonsocial otters use more sites with lower intensity, resulting in high variability in inputs of marine-derived nutrients to different latrine sites (Ben-David *et al.*, 2005). While some sites may receive $2.7 \text{g/m}^2$ year nitrogen (N) and $0.4 \text{g/m}^2$ year phosphorus (P), others may be fertilized with up to $47.6 \text{g m}^2$ year N and $6.7 \text{g/m}^2$ year P. Preliminary studies suggest that these nutrient inputs change the microbial community and increase plant growth and change plant diversity. Although little information on the effects of other otter species exists, it is likely that similar processes occur along all coastlines where otter densities are high and scent marking occurs.

### X. Population Biology

#### A. Genetics

A reasonably well-known history of decline, a recovery period that includes multiple reintroductions, and the relatively sedentary behavior of sea otters combine to provide unique opportunities for the study of population and conservation genetics. Both morphological and genetic study of sea otters indicate some level of population structuring prior to the eighteenth and nineteenth century fur harvests. Isolation and local extinctions caused by Pleistocene ice-sheets extending over large coastal areas probably resulted during glacial extension, whereas levels of gene flow likely increased during periods of glacial retreat. Extensive human harvest reduced and fragmented sea otters into a small number of isolated colonies by 1900, from which all current populations are derived. Additionally, translocations have established viable populations in Washington, British Columbia, and southeast Alaska. The sea otter's dependency on shallow benthic

**O**

habitats and limited diving capacity restricts their distribution to coastal environments where relatively small home ranges (generally <20 km of coast) constrain gene flow over large distances.

Sea otter skull size declines from Russia, across Alaska and into California and variation in skull morphology forms the basis for the three currently designated sub-species–Russia, *E. l. lutris*; Alaska, *E. l. kenyoni*; and California, *E. l. nereis*. Mitochondrial DNA (mtDNA) haplotype frequencies identify at least four distinct groupings of sea otters, including California, Prince William Sound, the Kodiak/Aleutian/Commander Islands, and the Kuril Islands. The magnitude of difference in current mtDNA haplotype frequencies suggests at least some genetic differentiation existed among these groups prior to the population declines and isolation that occurred between 1750 and 1900. The extant California population probably represents a monophyletic mtDNA lineage that contains two unique mtDNA haplotypes.

The extent to which long-term evolutionary processes and recent human exploitation have contributed to genetic differences among and within modern sea otter sub-species is unknown. The degree of difference in mtDNA haplotypes suggests that there has been little gene flow across the range of the species from California to the Kuril Islands. However, low levels of mtDNA sequence divergence in sea otters across their present range also suggests that major phylogenetic breaks or long-term barriers to gene flow have not persisted.

There is concern over the potential loss of genetic diversity stemming from severe population bottlenecks experienced by sea otter populations over the past two centuries. Theoretical analyses suggest that perhaps 23% of the original genetic variation was lost from the California population because of this bottleneck effect. Recent research comparing DNA extracted from pre-fur harvest sea otter bones to contemporary samples suggest significant losses of microsatellite alleles and heterozygosity as a consequence of the fur trade-induced population bottlenecks (Larson *et al.*, 2002). A similar reduction in mtDNA alleles and haplotype diversity was not detected.

Geographic isolation of reintroduced and remnant sea otter populations provide the opportunity to study the effects of the duration and magnitude of bottlenecks on genetic diversity and population growth rates. Genetic diversity was greater in translocated populations that were derived from two source stocks compared with those derived from a single source stock. Haplotype frequencies in populations with estimated founding sizes of 4 and 28 animals (Washington and British Columbia) differed from the source population, probably signaling the effect of genetic drift. This interpretation is supported by the fact that haplotype frequencies from sea otters in southeast Alaska, a translocated population with an estimated founding size of 150 individuals, did not differ from its two source populations. The mtDNA haplotype diversity was inversely correlated with both minimum population size and the number of years a population remained at that minimum size. And although translocated populations demonstrated significantly higher annual growth rates ($\lambda = 1.18$–$1.24$) than remnant populations ($\lambda = 1.06$–$1.09$), no relation was detected between genetic diversity and population growth rates (Bodkin *et al.*, 1999).

As sea otters continue to reoccupy former habitat, and currently isolated populations become contiguous we may be afforded the opportunity to view the process of genetic exchange across the species' range. Genetic differences currently observed among geographically isolated populations may diminish, reducing current levels of genetic population structure.

In contrast with the wealth of genetic information for sea otters, little is known about other marine-living otter species. Several studies in Britain and western Europe indicate that Eurasian otters have relatively high genetic diversity despite large population reductions in most countries. Also, genetic data suggest that the recolonization of extirpated areas in Europe occurred from populations in eastern Europe, and in Britain from remnant populations in Wales and Scotland. In Prince William Sound, North American otters exhibit isolation by distance mainly where dispersal requires the crossing of large bodies of water (Blundell *et al.*, 2002). Recent reintroductions and population supplementations along the Atlantic seaboard resulted in mixing of genetic stocks that previously were distinct.

## B. Demography

Age- and sex-specific rates of reproduction and survival vary markedly among the different otter species. Life history patterns in sea otters are more similar to those of the pinnipeds (with whom they share the ocean as a common environment) than they are to the other lutrines (with whom they share a more recent common ancestor). Perhaps the most remarkable life history feature of sea otters is that they almost invariably conceive and give birth to a single young, a character shared with other marine mammals. Other otter species invariably have multiple young litters, typically varying in size up to five cubs.

Female sea otters become sexually mature at 3 years, occasionally earlier. The reproductive cycle is normally 1 year, with roughly 6 months from conception to birth and another 6 months from birth to weaning. Primiparous females often fail to successfully wean their pups. Adult females apparently enter estrous within several days after mother/pup separation (either from weaning or death). The majority of pre-weaning deaths occur shortly after birth. Thus, in many areas there is a biannual peak in births, the primary peak occurring in spring or early summer and the secondary peak (by females who failed in their previous cycle) in fall or early winter. Females continue to reproduce throughout life, with little evidence for either reproductive senescence or adjustments of fertility rate to environmental variation.

Population regulation in sea otters occurs largely or exclusively through variation in age- and sex-specific mortality rates. The high rates of population growth that have been observed in parts of Alaska, Canada, and Washington (17–20% per year, near $r_{max}$) could only be realized if there were little or no mortality from birth to senescence. As growing populations become limited by resource availability, mortality in young otters increases greatly. Data from the Aleutian Islands and central California indicate that about half of the births fail to reach weaning age, compared to Kodiak where the population was not food limited and pup survival was more than 90%. The probability of mortality during the next 6 months, measured in the Prince William Sound, Alaska population was higher in males (56%) than females (36%) and lower in individuals that were in better condition. Mortality rates from about 1 year of age to physiological senescence (10–15 years) are low, even in food-limited populations. Thus, the principal mechanism of population regulation in sea otters appears to be pup abandonment and starvation-induced mortality early in life.

In Eurasian otters both males and females reach sexual maturity at 2 years of age. Data on age-specific birth rates are few but it is assumed that in coastal areas most females produce young in their third year. Number of cubs is highly correlated with prey abundance. Unlike Eurasian otters, North American otters have delayed implantation and thus while they reach maturity at age 2 they do not have their first litter until age 3. In coastal areas, 55–61% of 3 year olds are pregnant and up to 91% of all adults. Recruitment of Eurasian otters in Shetland, was estimated at 34 2-year-old females per 100 adults. No such information is available for other otter species.

Most semi-aquatic otters die young. In Shetland more than 50% of otters die before they are 3.5 years old. Similarly, the average age of North American otters captured in Prince William Sound, Alaska was 3.5 years. Annual mortality of adults was less than 20% suggesting that animals mostly perish in the first 2 years of life. Causes of mortality vary and in most cases are difficult to document. In Shetland and Alaska, where otters live far from humans, mortality is largely caused by low food availability and less so by predation. Closer to human habitations, other causes are trapping, persecution by owners of fish farms, car collisions, diseases transmitted by pets and livestock, and pollution. Because of their position at the top of the food chain, otters readily accumulate metal and organic pollutants which may lead to reduced survivorship and reproduction (Gaydos *et al.*, 2007).

## XI. Behavior

The behavioral ecology of marine otters strongly reflects life in the sea (Blundell *et al.*, 2002a). Vigorous grooming in sea otters is the species' behavioral hallmark. Sea otters probably spend more time and energy grooming their fur than any other mammal. This is accomplished by rubbing, rolling, blowing, and splashing. Grooming is necessary for cleaning and replenishing air to the under fur.

Sea otters are unusual among both the carnivores and marine mammals in the generally small size of their home ranges, which typically includes no more than 20 km of shoreline. Both males and females occasionally move longer distances for uncertain reasons. Extralimital sightings in central Baja California and near Wrangel Island in the Chuchi Sea demonstrate that, on occasion, individual sea otters can move hundreds of miles.

Adult male sea otters maintain territories that in California average about 0.4 km². Adult females apparently move freely among these territories but the territory holder aggressively excludes other males. Adult males harass females with large pups in an apparent effort to force separation, thus inducing the female to enter estrous and conceive the male's offspring. Copulation occurs repeatedly during brief consorts, after which the pair separates. A male grasps the female's nose in his mouth and rolls vigorously on the surface of the ocean to achieve intromission. Distinctive nose scars in adult females often result from this behavior. In severe cases, trauma to the nose and facial region may result in death to the female. Some males seem especially prone to such brutality. Upon killing their mates, these males usually continue to copulate with the corpses, sometimes for days. Adult male sea otters on occasion will copulate with and kill young harbor seals. These last two behaviors have not been observed outside California.

The sea otter's polygynous mating system (and the resulting high male libido) likely evolved in response to their unusually high population density, thus promoting male competition for females as the limiting resource in sexual reproduction. Polygynous mating systems are typical of all otariids and some phocids in temperate latitude systems but apparently are rare or absent in other species of otters (Blundell *et al.*, 2002a). As is true for other polygynous species, male sea otters provide no parental care.

Mating behavior of other otter species is similar in that the male forcibly holds the female by her nape. Copulation is vigorous, likely because otters are induced ovulators. Male paternal care is absent from most species except for giant otters and possibly the social small-clawed otters. Allo-grooming is common in all social species, likely serving to reinforce social connections.

## XII. Conservation and Management

The recovery and growth of sea otter populations has been heralded as one of conservation's great success stories. However, concerns are now growing over the sea otter's long-term welfare. Several thousand sea otters were killed by the *Exxon Valdez* oil spill in 1989, and while the overall impact of this event on the population is unclear, chronic detrimental effects appear to have persisted for at least 17 years (Ballachey *et al.*, 2003). North American otter populations in Prince William Sound were also negatively affected by the *Exxon Valdez* oil spill. Unlike sea otters, however, North American otters had recovered from the effects of the spill by the end of the 1990s (Ben-David *et al.*, 2002; Bowyer *et al.*, 2003).

Alaska natives are permitted to harvest sea otters for traditional purposes. There are no restrictions on take and large numbers of otters are currently being removed from some populations. Sustainable harvest levels and take quotas need to be developed from life table data and the spatial ecology of individual otters. Quotas also should take into account the status and genetic structure of populations, and the availability of food and space. Population declines in southwest Alaska continue but the ultimate reason for this large-scale collapse remains uncertain and controversial.

As sea otters in British Columbia continue to grow and expand in range, conflicts with various shellfisheries are the primary concern of resource managers. Others believe that the population is still precariously small. Sea otter harvests have been proposed but at present are not permitted under either Canadian national or provincial law. The growing sea otter population in Washington State has spread around Cape Flattery eastward through the Straits of Juan de Fuca. These animals now occupy tribal waters of the Makah Indians and are rapidly depleting red urchin populations, which until recently supported a tribal fishery.

California sea otters have long been the subject of debate and controversy. Fisheries interests have maligned sea otters because they compete with humans for various commercial and recreational resources while conservationists remain concerned over the small size and slow growth rate of the population. The California sea otter's range has recently expanded southeastward around Pt. Conception into the "no-otter zone" established by US PL 625. Shellfishery interests would like these otters removed while conservation groups encourage continued range expansion. Increased trafficking of oil tankers and outer continental shelf oil development along the California coast have heightened concern that a spill might reduce or even extinguish the California sea otter population. Concern has recently developed over land-based effluents and the high frequency of death from various parasites, diseases and biotoxins seen in beach-cast carcasses.

Currently only giant otters (*Pteronura*) are protected internationally through the Convention on International Trade in Endangered Species (CITES). Similarly most populations of Eurasian otters are protected through the European Union, although culling of nuisance otters is allowed in several countries. Of the other otter species, several enjoy local protection although many are still exploited for fur.

**O**

## XIII. Concluding Remarks

After having been hunted to the brink of extinction, sea otters recovered dramatically. Whereas the status of other marine-living otters is less certain, these animals appear better off than their freshwater-living counterparts, many of which are imperiled because of human exploitation and of habitat destruction. The relative well-being of marine otters is thought to result from lower levels of habitat destruction in the sea than on land. Although that may be true, coastal marine environments increasingly are the recipients of terrestrial pollution and are being more extensively exploited by numerous fisheries. Furthermore, scientists are beginning to understand that the coastal zone is linked in important and complex ways with both the open sea on its one side and land on the other. The fact that people live in disproportionately high densities along the land–sea margin is a direct threat to coastal marine ecosystems and many of the species that depend on these systems. All of this does not bode well for the future of marine-living otters. Overall, sea otters are far more abundant today than they were 50 years ago. However, the recent and ongoing declines of otter populations in Alaska and failure of the California population to increase in recent years may be among the first signs from these animals that coastal oceans are in peril.

In the process of recovery, expanding otter populations have come into conflict with fishery interests that developed during their absence. Similar conflicts surround other carnivores, but in few cases is the effect so apparent as it is with sea otters and shellfisheries. Numerous other species and ecosystem-level processes are also influenced by the indirect effects of sea otter predation. These examples contributed significantly to the growing realization that apex predators often play complex but important roles in the maintenance of biodiversity, and thus that effective strategies for biodiversity conservation must include viable populations of apex predators.

## References

Ballachey, B. E., Bodkin, J. L., Howlin, S., Doroff, A. M., and Rebar, A. H. (2003). Correlates to survival of juvenile sea otters in Prince William Sound, Alaska. *Can. J. Zool.* **81**, 1494–1510.

Ben-David, M., Bowyer, R. T., Duffy, L. K., Roby, D. D., and Schell, D. M. (1998). Social behavior and ecosystem processes: river otter latrines and nutrient dynamics of terrestrial vegetation. *Ecology* **79**, 2567–2571.

Ben-David, M., Blundell, G. M., and Blake, J. E. (2002). Post-release survival of river otters: effects of exposure to crude oil and captivity. *J. Wildl. Manage.* **66**, 1208–1223.

Ben-David, M., Blundell, R. M., Kern, J. W., Maier, J. A. K., Brown, E. D., and Jewett, S. C. (2005). Communication in river otters: Creation of variable resource sheds for terrestrial communities. *Ecology* **86**, 1331–1345.

Berta, A., and Morgan, G. S. (1986). A new sea otter (Carnivora: Mustelidae) from the Late Miocene and Early Pliocene (Hemphilian) of North America. *J. Paleontol.* **59**, 809–819.

Blundell, G. M., Ben-David, M., and Bowyer, R. T. (2002a). Sociality in river otters: Cooperative foraging or reproductive strategies? *Behav. Ecol.* **13**, 134–141.

Blundell, G. M., Ben-David, M., Groves, P., Bowyer, R. T., and Geffen, E. (2002b). Characteristics of sex-biased dispersal and gene flow in coastal river otters: implications for natural recolonization of extirpated populations. *Mol. Ecol.* **11**, 289–303.

Bodkin, J. L., Ballachey, B. E., Cronin, M. A., and Scribner, K. T. (1999). Population demographics and genetic diversity in remnant and translocated populations of sea otters. *Conserv. Biol.* **13**(6), 1378–1385.

Bodkin, J. L., Esslinger, G. G., and Monson, D. H. (2004). Foraging depths of sea otters and implications to coastal marine communities. *Mar. Mamm. Sci.* **20**(2), 305–321.

Bowyer, R. T., Blundell, G. M., Ben-David, M., Jewett, S. C., Dean, T. A., and Duffy, L. K. (2003). Effects of the *Exxon Valdez* oil spill on river otters: injury and recovery of a sentinel species. *Wildl. Monogr.* **153**, 1–53.

Duggins, D. O., Simenstad, C. A., and Estes, J. A. (1989). Magnification of secondary production by kelp detritus in coastal marine ecosystems. *Science* **245**, 170–173.

Estes, J. A. (1989). Adaptations for aquatic living in carnivores. In "Carnivore behavior ecology and evolution" (J. L. Gittleman, ed.), pp. 242–282. Cornell University Press Ithaca, New York.

Estes, J. A., and Duggins, D. O. (1995). Sea otters and kelp forests in Alaska: generality and variation in a community ecology paradigm. *Ecol. Monogr.* **65**, 75–100.

Estes, J. A., Jameson, R. J., and Johnson, A. M. (1981). Food selection and some foraging tactics of sea otters, pp. 606–641. In "Worldwide Furbearer Conference Proceedings" (J. A. Chapman and J. D. Pursley, eds.), pp. 3–11, August 1980. Frostburg, Maryland.

Fretwell, S. D. (1987). Food chain dynamics: The central theory of ecology? *Oikos* **50**, 291–301.

Gaydos, J. K., Conrad, P., Gilardi, K. V. K., Blundell, G. M., and Ben-David, M. (2007). Does human proximity affect antibody prevalence in marine-foraging river otters (*Lontra canadensis*)? *J. Wildl. Dis.* **43**, 116–123.

Hairston, N. G., Smith, F. E., and Slobodkin, L. B. (1960). Community structure, population control and competition. *Am. Naturalist* **94**, 421–425.

Larson, S., Jameson, R. J., Etnier, M., Fleming, M. M., and Bentzen, P. (2002). Loss of genetic diversity in sea otters (*Enhydra lutris*) associated with the fur trade of the 18th and 19th centuries. *Mol. Biol.* **11**, 899–1903.

Kenyon, K. W. (1969). "The sea otter in the eastern Pacific Ocean. *North American Fauna* no. 68." Bureau of Sport Fisheries and Wildlife, Dept. of Interior, Washington, DC.

Koepfli, K. P., and Wayne, R. K. (1998). Phylogenetic relationships of otters (Carnivora: Mustelidae) based on mitochondrial cytochrome b sequences. *J. Zool.* **246**(4), 401–416.

Kruuk, H. (2006). "Otters: Ecology, behaviour, and conservation." Oxford University Press Oxford, UK.

Medina-Vogel, G., Delgado Rodriguez, C., Pacheco, R. A., and Bartheld, J. L. (2004). Feeding ecology of the marine otter (*Lutra felina*) in a rocky seashore of the south of chile. *Mar. Mamm. Sci.* **20**(1), 134–144.

Medina-Vogel, G., Bartheld, J. L., Pacheco, R. A., and Delgado Rodriguez, C. (2006). Population assessment and habitat use by marine otter *Lontra felina* in southern Chile. *Wildl. Biol.* **12**, 191–199.

Riedman, M. L., and Estes, J. A. (1990). The sea otter (*Enhydra lutris*): Behavior, ecology and natural history. *US Fish and Wildl. Serv. Biol. Rep.* **90**(14), 12.

Williams, T. M. (1999). The evolution of cost efficient swimming in marine mammals: limits to energetic optimization. *Phil. Trans. R. Soc. Lond. B* **354**, 193–201.

Williams, T. M., Ben-David, M., Noren, S., Rutishauser, M., and McDonald, K. (2002). Running energetics of the North American river otter: do short legs necessarily reduce efficiency on land? *Comparat. Biochem. Physiol.* **A 133**, 203–212.

# Pacific White-Sided Dolphin
## *Lagenorhynchus obliquidens*

### NANCY A. BLACK

## I. Characteristics and Taxonomy

The Pacific white-sided dolphin is a North Pacific endemic and one of six species of the genus *Lagenorhynchus*. It most closely resembles the dusky dolphin (*L. obscurus*) from the southern hemisphere. Theodore Gill originally described this species in 1865 based on three skulls collected by W. P. Trowbridge off San Francisco, California. Adults range from 1.7 to 2.5 m, and weigh 75–198 kg, with males slightly larger than females. They have an average of 30 teeth in each side of the upper and lower jaw (Walker *et al.*, 1986).

These dolphins are boldly marked (Fig. 1) with a dark gray or black dorsal surface, light gray sides with light gray "suspender stripes" originating near the melon and angling toward the blowhole across each side into the light gray flank patch. The beak is dark with a narrow stripe extending to the bicolored flipper. The dorsal fin has a darker leading edge with light gray covering two-thirds of the posterior portion. The flukes are all dark. The white belly is separated from the body by a thin black line. Anomalous or predominately white individuals with small areas of black pigmentation on their sides, heads, and

fins or ones with an unusual white stripe extending up from their flank and widening over each eye are occasionally sighted. Several of these anomalous dolphins have been identified in Monterey Bay, California and Hokkaido, Japan and have been successfully used as "herd markers" (Black, 1994; Tsutsui *et al.*, 2001).

The dorsal fin ranges from falcate to lobate with a more rounded tip. In most cases the shape appears related to age, as both sexes have variable fins. Some dolphins in coastal Japan appear to have a strongly hooked dorsal fin compared to those observed off California. Extremely falcate-finned dolphins also sighted in small subgroups off inland waters of British Columbia may be all males (Morton, 2000).

## II. Distribution and Abundance

The Pacific white-sided dolphin is one of the most abundant pelagic species of dolphin found in the cold-temperate North Pacific. In the western Pacific it occurs coastally from Taiwan to the Sea of Japan and off the Pacific coast of Japan northward to the Kuril and Commander Islands. In the eastern Pacific it occurs as far west as Amchitka in the Aleutian Islands through the Gulf of Alaska and down the west coast to 20°N, just south of Baja, California, and into the southwestern Gulf of California. It occurs continuously across the North Pacific in a band from 38 to 47°N off California; Pacific white-sided dolphins inhabit productive continental shelf and slope waters generally within 185 km of shore.

Although they do not migrate, there are seasonal shifts in distribution found both off Japan and the eastern Pacific that are related to oceanographic conditions, primarily water temperature but also fronts, intensity and location of upwelling regions, currents, and thermocline depths. Changes in distribution are likely prey-related with such changes noted seasonally and interannually depending on warm or cold water years, as evident by increased abundance in Monterey Bay, California, inland waters of British Columbia, and Southeast Alaska in warmer-water years. These dolphins frequent some areas with complex bathymetry such as Monterey Bay, California, an area where a deep submarine canyon approaches shore (Black, 1994; Dahlheim and Towell, 1994; Morton, 2000).

Based on genetic, skull, and body size differences, three populations that are not completely isolated from each other occur in the eastern North Pacific: off Baja California; California to Oregon; and off British Columbia. Additionally these appear separate from dolphins found in offshore waters of the North Pacific. Another two populations occur in coastal Japan, also separate from offshore animals (Walker *et al.*, 1986; Lux *et al.*, 1997; Miyazaki and Shikano, 1997; Hayano *et al.*, 2004).

Accurate population estimates are difficult to ascertain for the entire range, but close to 1,000,000 are estimated for the North Pacific (Buckland *et al.*, 1993; Miyashita, 1993) and 10,000–41,000 for the California Current ecosystem from California to Washington, with the number varying by season and year as animals move into and out of the region (Barlow and Forney, 2007).

## III. Ecology

Killer whales (*Orcinus orca*) appear to be a significant predator, as these dolphins exhibit a strong flight response when killer whales are near, and several direct observations have been made of predation (Fig. 2).

Dominant prey type varies by region. They feed opportunistically on abundant species of fishes (60 species) and cephalopods (20 species) both day and night: schooling epipelagic (0–200 m) fishes and

P

**Figure 2** *A killer whale attacks a Pacific white-sided dolphin in Monterey Bay, California. The killer whale is their main predator Photo by Lori Mazzuca.*

cephalopods in California (northern anchovy, Pacific whiting, and squid), Washington (salmon and squid), and British Columbia (herring, salmon, sardine, cod, shrimp, capelin, and squid) (Stroud *et al.*, 1981; Walker *et al.*, 1986; Black, 1994; Heise, 1997a; Morton, 2000); a large variety of primarily mesopelagic (200–1000 m) species in offshore waters (Walker and Jones, 1993); and predominantly myctophids in the western Pacific (Miyazaki *et al.*, 1991).

## IV. Behavior and Physiology

These highly social dolphins are avid bow riders that commonly occur in groups of less than a hundred but can form herds containing several thousands of individuals. They often associate with northern right whale dolphins (*Lissodelphis borealis*), Risso's dolphins (*Grampus griseus*), and short-beaked common dolphins (*Delphinus delphis*). They occasionally feed in association with humpback whales (*Megaptera novaeangliae*), California sea lions (*Zalophus californianus*), and seabirds in mixed-species aggregations. The cohesiveness of dolphin groups differs according to behavior, with dolphins generally in more dispersed subgroups while milling, socializing, and feeding and usually more tightly grouped while traveling and resting. They often cooperate as a group while herding and feeding on small schooling fish. They are highly acrobatic, often exhibiting percussive repetitive leaps while traveling and single leaps most while feeding and engaged in social behavior. They exhibit a variety of leap types, such as somersaults, dorsal fin slaps, and belly slaps.

Three dolphins radiotracked in Monterey Bay for 2 days exhibited a mean respiration rate of 2.5/min, a mean dive duration of 24 sec, and a maximum dive time of 6.2 min, which is exceptionally long. Dives of around 3 min are more commonly considered long dives (Black, 1994).

## V. Life History

Calving occurs from May to September, corresponding to a mid-to-late summer breeding season. In addition, males exhibit an extreme enlargement in testis size from mid-to-late summer (Ridgway and Green, 1967). Age and length at maturation vary by area, with females becoming sexually mature at 8–11 years and 175–186 cm with a 4 to 5-year calving interval, and males at 9–12 years and 170–180 cm. The mean neonatal length is 94 cm with a gestation of 11–12 months. Males may live to 42 years and females 46 years (Heise, 1997b).

## VI. Interactions with Humans

Until 1992 when large-scale pelagic drift net fishing was banned, a minimum of 4000–8000 of these dolphins were killed each year. Today, dolphins are still taken in Japan by harpoon for human consumption. The drive fishery at Taiji, the last left in Japan, does not generally catch Pacific white-sided dolphins. The long-term effect of the exploitation on the western population has not been evaluated. Dolphins continue to be caught as well to a lesser extent as bycatch in coastal fisheries throughout their range. Pacific white-sided dolphins are found in aquariums in the United States, Canada, and Japan, but generally survival rates are low.

## *See Also the Following Articles*

Delphinids, overview
Dusky Dolphin

## *References*

Barlow, J., and Forney, K. A. (2007). Abundance and population density of cetaceans in the California Current ecosystem. *Fish. Bull. (U.S.)* **105**, 509–526.

Black, N. A. (1994). Behavior and ecology of Pacific white-sided dolphins (*Lagenorhynchus obliquidens*) in Monterey Bay, California. M.S. Thesis, Moss Landing Marine Laboratories (San Francisco State University), Moss Landing, CA.

Buckland, S. T., Cattanach, K. L., and Hobbs, R. C. (1993). Abundance estimates of Pacific white-sided dolphin, northern right whale dolphin, Dall's porpoise and northern fur seal in the North Pacific, 1987/90. *In* "INPFC Symposium: Biology, Distribution and Stock Assessment of Species Caught in the High Seas Driftnet Fisheries in the North Pacific Ocean, 4–6 Nov., 1991." Tokyo, Japan.

Dahlheim, M. E., and Towell, R. G. (1994). Occurrence and distribution of Pacific white-sided dolphins (*Lagenorhynchus obliquidens*) in southeastern Alaska, with notes on an attack by killer whales (*Orcinus orca*). *Mar. Mamm. Sci.* **10**, 458–464.

Hayano, A., Yoshioka, M., Tanaka, M., and Amano, M. (2004). Population differentiation in the Pacific white-sided dolphin (*Lagenorhynchus obliquidens*) inferred from mitochondrial DNA and microsatellite analyses. *Zool. Sci.* **21**, 989–999.

Heise, K. (1997a). Diet and feeding behavior of Pacific white-sided dolphins (*Lagenorhynchus obliquidens*) as revealed through the collection of prey fragments and stomach content analyses. *Rep. Int. Whaling Commn.* **47**, 807–815.

Heise, K. (1997b). Life history and population parameters of Pacific white-sided dolphins (*Lagenorhynchus obliquidens*). *Rep. Int. Whaling Commn.* **47**, 817–825.

Lux, C. A., Costa, A. S., and Dizon, A. E. (1997). Mitochondrial DNA population structure of the Pacific white-sided dolphin (*Lagenorhynchus obliquidens*). *Rep. Int. Whaling Commn.* **47**, 645–652.

Miyashita, T. (1993). Distribution and abundance of some dolphins taken in the North Pacific driftnet fisheries. *Bull. Int. North Pac. Fish. Commn.* **53**, 435–449.

Miyazaki, N., Kukramochi, T., and Amano, M. (1991). Pacific white-sided dolphin (*Lagenorhynchus obliquidens*) off northern Hokkaido. *Mem. Natn. Sci. Mus. (Tokyo)* **24**, 131–139.

Miyazaki, N., and Shikano, C. (1997). Comparison of growth and skull morphology of Pacific white-sided dolphins, *Lagenorhynchus obliquidens*, between the coastal waters of Iki Island and the oceanic waters of the western North Pacific. *Mammalia* **61**, 561–572.

Morton, A. (2000). Occurrence, photo-identification and prey of Pacific white-sided dolphins (*Lagenorhynchus obliquidens*) in the Broughton Archipelago, Canada 1984–1998. *Mar. Mamm. Sci.* **16**, 80–93.

Stroud, R. K., Fiscus, C. H., and Kajimura, H. (1981). Food of the Pacific white-sided dolphin, *Lagenorhynchus obliquidens*, Dall's porpoise, *Phocoenoides dalli*, and Northern fur seal, *Callorhinus ursinus*, off California and Washington. *Fish. Bull. (U.S.)* **78**, 951–959.

Ridgway, S. H., and Green, R. F. (1967). Evidence for a sexual rhythm in male porpoises. *Norw. Whal. Gaz.* **1**, 1–8.

Tsutsui, S., Tanaka, M., Miyazaki, N., and Furuya, T. (2001). Pacific white-sided dolphins (*Lagenorhynchus obliquidens*) with anomalous colour patterns in Volcano Bay, Hokkaido, Japan. *Aquat. Mamm.* **27**, 172–182.

Walker, W. A., and Jones, L. L. (1993). Food habits of northern right whale dolphin, Pacific white-sided dolphin, and northern fur seal caught in the high seas driftnet fisheries of the North Pacific Ocean, 1990. *Int. North Pac. Fish. Commn. Bull.* **53**, 285–295.

Walker, W. A., Leatherwood, S., Goodrich, K. R., Perrin, W. F., and Stroud, R. K. (1986). Geographical variation and biology of the Pacific white-sided dolphin (*Lagenorhynchus obliquidens*) in the north-eastern Pacific. *In* "Research on Dolphins" (M. M. Bryden, and R. Harrison, eds), pp. 441–465. Oxford University Press, Oxford.

**Figure 1** *Pantropical spotted dolphins ride the bow of a research vessel in the eastern tropical Pacific. Juvenile (above) still lacks spots. Photo by Robert L. Pitman.*

# Pantropical Spotted Dolphin
## *Stenella attenuata*

### WILLIAM F. PERRIN

### I. Characteristics and Taxonomy

This dolphin can be identified externally by its long beak sharply demarcated from the melon, slender body, strongly falcate dorsal fin, and (in adults) spots (Fig. 1). The newborn calf is unspotted. Dark spots begin to appear ventrally in large juveniles. Near-adult animals and some young adults have large discrete or overlapping spots both above (light) and below (dark). In adults, the ventral spots fuse and fade to a medium gray, and the dorsal light spots intensify, sometimes to the point of making the animal appear nearly white above. The light spots sweep up behind the dorsal fin. The underlying pattern (observable in calves and juveniles) consists of a dark cape sweeping over the eye to maximum depth below the dorsal fin, a very light to dark gray lateral/ventral field, a narrow well-defined eye stripe to the apex of the melon, a dark band of varying definition extending from the lower corner of the mouth to the flipper, and dorsoventral division of the peduncle into darker upper and lower lighter halves. The last can also be seen in adults. The tip of the beak is white in adults. Details of coloration vary regionally. The large coastal spotted dolphin of the eastern tropical Pacific, *S. g. graffmani*, is extremely heavily spotted, whereas animals around Hawaii tend to be lightly spotted (Perrin and Hohn, 1994; Perrin, 2001).

Sexually mature adults examined range from 166 to 257 cm (*n* > 1650) and weigh up to 119 kg (a 257-cm male from Panama Bay) with a wide GEOGRAPHIC VARIATION. Males are on the average slightly larger than females in body size and most skull characters.

This species may be confused in the tropical Atlantic with the endemic Atlantic spotted dolphin, *S. frontalis*, which is of similar size and may be seen in the same area. A distinguishing characteristic of *S. frontalis* is a light spinal blaze that sweeps up through the cape toward the dorsal fin, but this may be almost absent in some individuals. The dorsoventral division of the peduncle present in *S. attenuata* is absent in *S. frontalis*. The ground pattern is three-part, with a distinct cape, lateral field, and ventral field, as opposed to the two-part pattern in *S. attenuata*. Adults may have dark ventral spots on a very light ground; this is not seen in adults of *S. attenuata* (ventral spots are medium gray and obscured by fusion).

The skull (35.6–46.0 cm long in 183 adults) overlaps with that of *S. frontalis* in tooth counts and all measurements; some specimens can be identified to species only through multivariate analysis (Perrin *et al.*, 1987). The skull in both species has a long narrow rostrum with no palatal grooves, about 35–50 teeth in each row, medium-sized rounded temporal fossae, convergent premaxillae, and arcuate mandibles. However, vertebral counts determined to date have been nonoverlapping: 74–84 (*n* = 75) vs 67–72 (*n* = 52) in *S. frontalis*.

The pantropical dolphin is a member of the delphinid subfamily Delphininae *sensu stricto* (LeDuc *et al.*, 1999). In a cladistic phylogenetic analysis based on cytochrome *b* mtDNA, it shares a strongly supported polytomic clade with *S. clymene* and *S. coeruleoalba* (sister species), *S. frontalis*, *Delphinus* spp., and *Tursiops aduncus* (to the exclusion of *T. truncatus*). In skull characters it is similar to *S. frontalis*, *T. truncatus*, and *T. aduncus* vs another coherent series of species composed of *S. longirostris*, *S. clymene*, and *S. coeruleoalba*. However, these latter groupings are not supported by results of the molecular studies to date. Some of the similarities may be synplesiomorphies, similarities due to the retention of primitive character states.

### II. Distribution and Abundance

Distribution is worldwide in tropical and some subtropical waters, from roughly 30–40°N to 20–40°S (Jefferson *et al.*, 2007). It ranks first in abundance among cetaceans (second only to bottlenose dolphins, *Tursiops truncatus*); in the deeper waters of the Gulf of Mexico (Würsig *et al.*, 2000), second in the eastern tropical Pacific and the Sulu Sea, but only sixth in the tropical Indian Ocean (Ballance and Pitman, 1998; Dolar *et al.*, 2006). The coastal subspecies *S. a. graffmani* occurs only in a narrow coastal band along the Pacific coasts of southern Mexico and south Peru; it may consist of a number of subpopulations (Escorza-Treviño *et al.*, 2005).

An estimated 640,000 northeastern offshore spotted dolphins exist currently in the eastern tropical Pacific, down roughly 80% from original abundance before the tuna purse-seine fishery began killing them in fishing operations (see Section VI below). Other populations of the species in the eastern Pacific have not been as heavily impacted. Estimates of abundance exist for several other regions (IUCN, 2008): Hawaii, about 9000; Japan, about 438,000 in early+ 1990s; northern Gulf of Mexico, about 34,000; east coast of US, about 4400; eastern Sulu Sea, about 15,000; Tañon Strait between Negros and Cebu (Philippines), 640.

## III. Ecology

In the eastern Pacific, the offshore form, *S. a. attenuata*, inhabits the tropical, equatorial, and southern subtropical water masses, being most abundant in waters underlain by a sharp thermocline at depths of 50 m or less, a surface temperature over 25°C, and salinities less than 34 parts per thousand. In these areas, pantropical spotted dolphins commonly occur in large multispecies aggregations, including spinner dolphins (*S. longirostris*) and yellowfin tuna (*Thunnus albacares*) (Perrin and Hohn, 1994; Ballance *et al.*, 2006). A large coastal subspecies, *S. a. graffmani*, occurs in the eastern tropical Pacific from Mexico to northern Peru; this form is replaced ecologically in the Atlantic by a large coastal form (unnamed) of the endemic Atlantic spotted dolphin, *S. frontalis* (Perrin and Hohn, 1994). In the Southwest Atlantic, the species occurs beyond the continental shelf break in waters of >850 m (Moreno *et al.*, 2005). In the Gulf of Mexico it occurs over the lower continental slope and deeper waters (>1000 m) (Baumgartner *et al.*, 2001). In the Philippines it inhabits both shallow and deep waters (Dolar *et al.*, 2006).

Prey of the offshore form include mainly small epipelagic fishes, squids, and crustaceans (Robertson and Chivers, 1997; Wang *et al.*, 2003). Flying fish are a major diet item in some regions. In Hawaii, recorded diving behavior indicates that the dolphins feed primarily at night on organisms associated with the deep-scattering layer as it rises toward the surface (Baird *et al.*, 2001). The diet in the eastern Pacific overlaps greatly with that of the yellowfin tuna, *T. albacares*. Diet of the large coastal form is unknown but may include larger and tougher benthic fishes. Predators include the killer whale, *Orcinus orca* (Pitman *et al.*, 2003) and sharks (Maldini, 2003), probably the pygmy killer whale (*Feresa attenuata*), and possibly the false killer whale (*Pseudorca crassidens*) and the short-finned pilot whale (*Globicephala macrorhynchus*). Parasites may cause direct or indirect mortality.

## IV. Behavior and Physiology

Spotted dolphins exhibit a wide variety of aerial behavior (but not spinning); juveniles (identifiable as such by their lack of spots) make especially high vertical leaps. In areas where they are not harpooned or pursued by purse seiners, they readily come to vessels to ride the bow wave and sometimes can be observed closely for long periods. Burst swimming speed exceeds 12 knt (about 22 km/h), and dives of up to 3.4 min have been observed.

School size may range from a few individuals to several thousand and may be segregated by age or sex or both (Perrin, 2001). Mean school size in the eastern tropical Pacific is about 120, in the western tropical Indian Ocean about 170, in the Philippines about 90, and in the Gulf of Mexico about 70 (Dolar *et al.*, 2006).

Why spotted dolphins associate closely with tuna in the eastern Pacific is unknown, although it is suspected to be related to foraging efficiency at some level (Perrin and Hohn, 1994). Other suggested reasons for the association involve physiological efficiency or protection from predators. Immature and subadult males and females tend to form smaller schools or join larger spinner dolphin schools not associated with tuna.

Physiological and behavioral development of young calves may not be sufficiently advanced to allow them to cope successfully with being chased and captured by tuna purse seiners (Edwards, 2005; Noren and Edwards, 2007). Dolphins have learned to attempt to evade purse seiners (Lennert-Cody and Scott, 2005).

## V. Life History

The breeding system is unknown but may be promiscuous as in the spinner dolphin. Gestation is 11.2–11.5 months (Perrin and Hohn, 1994). Length at birth is 80–85 cm. Females reach sexual maturity at 9–11 years and males at 12–15 years. The calving interval is about 2–3 years but varies with population status. The average age and length at weaning are approximately 9 months and 122 cm, but nursing can continue up to 2 years of age (Archer and Robertson, 2004). Breeding is diffusely seasonal, with multiple peaks in some regions.

## VI. Interactions with Humans

Direct and incidental catches have been substantial. Japanese drive and harpoon fisheries take spotted dolphins for human consumption; the catch in 1 year (1982) was 3799 (Perrin and Hohn, 1994); between 1995 and 2004, the average annual catch was 129 (Kasuya, 2007). Tuna fishermen seek out aggregations of spotted dolphins and tuna in the eastern tropical Pacific and set their nets on them to capture the tuna. Abundance of the northeastern stock of the offshore form has been reduced to a fraction of its original size by incidental kill in tuna purse seines since the early 1960s. While kill in the fishery has been reduced to the low 100s annually for a number of years, the population has not grown toward recovery as rapidly as expected; the continued chase and capture in the fishery may have an indirect effect on fecundity or survival, or there may have been a change in carrying capacity of the ecosystem for this species (Archer *et al.*, 2004; Gerrodette and Forcada, 2005; Wade *et al.*, 2007; Cramer *et al.*, in press). Directed fisheries (legal or illegal) also exist or have existed in the Philippines, Laccadive Islands in the Indian Ocean, Solomon Islands, Indonesia, Taiwan, the Philippines, and St. Helena in the South Atlantic (Perrin and Hohn, 1994), and bycatches also occur in fishing nets of various types in the Philippines, India, Australasia, western North Pacific, Central America, coastal Peru and Ecuador, and Taiwan (Perrin *et al.*, 1994; IUCN, 2008). The impacts of most of these takes on the populations have not been assessed.

Pantropical spotted dolphins have been implicated in depredation on or interference with hook-and-line fisheries for squid and fish in Japan, and 538 were culled during the period 1976–1982 (IUCN, 2008).

The species has not been kept successfully in captivity.

## See Also the Following Articles

Atlantic Spotted Dolphin ■ Incidental Catches ■ Tuna–Dolphin Issue

## References

Archer, F., Gerrodette, T., Chivers, S., and Jackson, A. (2004). Annual estimates of the unobserved incidental kill of pantropical spotted dolphin (*Stenella attenuata attenuata*) calves in the tuna purse-seine fishery of the eastern tropical Pacific. *Fish. Bull. (U.S.)* **102**, 233–244.

Archer, F. I., and Robertson, K. M. (2004). Age and length at weaning and development of diet of pantropical spotted dolphins, *Stenella attenuata*, from the eastern tropical Pacific. *Mar. Mamm. Sci.* **20**, 232–245.

Baird, R. W., Ligon, A. D., Hooker, S. K., and Gorgone, A. M. (2001). Subsurface and nighttime behaviour of pantropical spotted dolphins in Hawaii. *Can. J. Zool.* **79**, 988–996.

Ballance, L. T., and Pitman, R. B. (1998). Cetaceans of the tropical western Indian Ocean: distribution, relative abundance, and comparisons with cetacean communities of two other tropical ecosystems. *Mar. Mamm. Sci.* **14**, 429–459.

Ballance, L. T., Pitman, R. L., and Fiedler, P. C. (2006). Oceanographic influences on seabirds and cetaceans of the eastern tropical Pacific: a review. *Progr. Oceanogr.* **69**, 360–390.

Baumgartner, M. F., Mullin, K. D., May, L. N., and Leming, T. D. (2001). Cetacean habitats in the northern Gulf of Mexico. *Fish. Bull. (U.S.)* **99**, 219–239.

Cramer, K., Perryman, W. L., and Gerrodette, T. (In press). Declines in reproductive indices in two depleted dolphin populations in the eastern tropical Pacific. *Mar. Ecol. Prog. Ser.*

Dolar, M. L. L., Perrin, W. F., Taylor, B. L., Kooyman, G. L., and Alava, M. N. R. (2006). Abundance and distributional ecology of cetaceans in the central Philippines. *J. Cetacean Res. Manage.* **8**, 93–111.

Edwards, E. F. (2005). Duration of unassisted swimming activity for spotted dolphin (*Stenella attenuata*) calves: implications for mother-calf separation during tuna purse-seine sets. *Fish. Bull. (U.S.)* **104**, 125–135.

Escorza-Treviño, S., Archer, F. I., Rosales, M., Lang, A., and Dizon, A. E. (2005). Genetic differentiation and intraspecific structure of eastern tropical Pacific spotted dolphins, *Stenella attenuata*, revealed by DNA analyses. *Conserv. Genet.* **6**, 587–600.

Gerrodette, T., and Forcada, J. (2005). Non-recovery of two spotted and spinner dolphin populations in the eastern tropical Pacific Ocean. *Mar. Ecol. Prog. Ser.* **291**, 1–21.

IUCN (2008). "The IUCN Red List of Threatened Species." IUCN, Gland, Switzerland.

Jefferson, T. A., Webber, M. A., and Pitman, R. L. (2007). "Marine Mammals of the World: A Comprehensive Guide to Their Identification." Academic Press/Elsevier, San Diego, CA.

Kasuya, T. (2007). Japanese whaling and other cetacean fisheries. *Environ. Sci. Poll. Res. (online early)*, 1–10.

LeDuc, R. G., Perrin, W. F., and Dizon, A. E. (1999). Phylogenetic relationships among the delphinid cetaceans based on full cytochrome *b* sequences. *Mar. Mamm. Sci.* **15**, 619–648.

Lennert-Cody, C. E., and Scott, M. D. (2005). Spotted dolphin evasive behavior in relation to fishing effort. *Mar. Mamm. Sci.* **21**, 13–28.

Maldini, D. (2003). Evidence of predation by a tiger shark (*Galeocerdo cuvieri*) on a spotted dolphin (*Stenella attenuata*) off Oahu, Hawaii. *Aquat. Mamm.* **29**, 84–87.

Moreno, I. B., Zerbini, A. N., Danilewicz, D., Santos, M. C. D., Simões-Lopez, P. C., Lailson-Brito, J., and Acevedo, A. F. (2005). Distribution and habitat characteristics of dolphins of the genus *Stenella* (Cetacea: Delphinidae) in the Southwest Atlantic Ocean. *Mar. Ecol. Progr. Ser.* **300**, 229–240.

Noren, S. R., and Edwards, E. F. (2007). Physiological and behavioral development in delphinid calves: implications for calf separation and mortality due to tuna purse-seine sets. *Mar. Mamm. Sci.* **23**, 15–29.

Perrin, W. F. (2001). *Stenella attenuata. Mamm. Species* **633**, 1–8.

Perrin, W. F., Donovan, G. P., and Barlow, J. (eds.). (1994). Cetaceans and gillnets. *Rep. Int. Whal. Commn.* **15**(Special Issue), 629.

Perrin, W. F., and Hohn, A. A. (1994). Spotted dolphin *Stenella attenuata. In* "Handbook of Marine Mammals" (S. H. Ridgway, and R. Harrison, eds), Vol. 5, pp. 71–98. Academic Press, San Diego, CA.

Perrin, W. F., Mitchell, E. D., Mead, J. G., Caldwell, D. K., Caldwell, M. C., Van Bree, P. J. H., and Dawbin, W. H. (1987). Revision of the spotted dolphins, *Stenella* Spp. *Mar. Mamm. Sci.* **3**, 99–170.

Pitman, R. L., O'Sullivan, S., and Mase, B. (2003). Killer whales (*Orcinus orca*) attack a school of pantropical spotted dolphins (*Stenella attenuata*) in the Gulf of Mexico. *Aquat. Mamm.* **29**, 321–324.

Robertson, K. M., and Chivers, S. (1997). Prey occurrence in pantropical spotted dolphins, *Stenella attenuata*, from the eastern tropical Pacific. *Fish. Bull. (U.S.)* **95**, 334–348.

Wade, P. R., Watters, G. M., Gerrodette, T., and Reilly, S. B. (2007). Depletion of northeastern offshore spotted and eastern spinner dolphins in the eastern tropical Pacific and hypotheses for their lack of recovery. *Mar. Ecol. Prog. Ser.* **343**, 1–14.

Wang, M. C., Walker, W. A., Sha, K. T., and Chou, L. S. (2003). Feeding habits of the pantropical spotted dolphin, *Stenella attenuata*, off the eastern coast of Taiwan. *Zool. Stud.* **42**, 368–378.

Würsig, B., Jefferson, T. A., and Schmidly, D. J. (2000). "Marine Mammals of the Gulf of Mexico." Texas A&M University Press, College Station, TX.

# Parasites

J. Antonio Raga, Mercedes Fernández, Juan A. Balbuena and F. Javier Aznar

Beyond their sanitary or economic importance, parasites are an integral part of the biosphere. They are so diverse and pervasive that they virtually infect every free-living organism, potentially influencing, among other things, host health, behavior and population size, food web dynamics, and community structure. These effects are usually undesirable when human health or economy are at stake but confer parasites a paramount importance in nature that should not be neglected.

This entry provides an overview of the diversity of marine mammal parasites. Its aim is to explain concisely what they are and how they have become associated with their hosts. Other aspects, such as the impact of parasites on marine mammal populations, are covered only briefly as they have been dealt with extensively elsewhere (Aznar *et al.*, 2001). Under the term "parasite," we will only consider the protozoons and metazoons (helminths and arthropods) that have adopted this life history strategy. Some representative parasites of marine mammals are shown in Fig. 1.

## I. Parasite Diversity

Protozoan parasites have been reported rarely in marine mammals. Most species have been described only recently thanks to an increasing interest in these organisms and the use of new techniques with fresh samples. So our knowledge of protozoan diversity in marine mammals is likely to increase substantially in the coming years (Raga and Gulland, 2008). As for metazoan parasites, differences in sampling effort are considerable depending on the host group. In particular, beaked whales (Ziphiidae) and fur seals (Otariidae) are speciose taxa for which parasite studies are still very scarce, and therefore diversity is likely to be higher than currently perceived.

Another fundamental factor affecting current diversity estimates is the existence of sibling species. This phenomenon is relatively frequent among marine invertebrates and has been documented extensively in parasitic nematodes of cetaceans and pinnipeds (family Anisakidae). Accordingly, the actual diversity of many other parasite taxa might have been underestimated because of our inability to tell species apart based on morphology. We suspect that many parasites infecting marine mammals that are currently considered as cosmopolitan or widespread may actually represent complexes of sibling or pseudosibling species.

### A. Cetaceans

Four types of coccidians occur in cetaceans: *Cystoisospora delphini* in common bottlenose dolphins (*Tursiops truncatus*) causing enteritis; *Toxoplasma gondii* (or *Toxoplasma* spp.) in four dolphin species, associated with toxoplasmosis; *Sarcocystis* spp. in toothed whales, usually with little or no obvious pathologic effect—fatal hepatic sarcocystosis has been reported only in one striped dolphin (*Stenella coeruleoalba*); and *Neospora caninum* in bottlenose dolphins although its pathological effect is currently unknown (Dailey, 2005). The life cycles of these coccidians have not been elucidated in the aquatic environment, although there are reports of congenital and transplacental transmission of *Toxoplasma* spp. in dolphins. Flagellates have been rarely

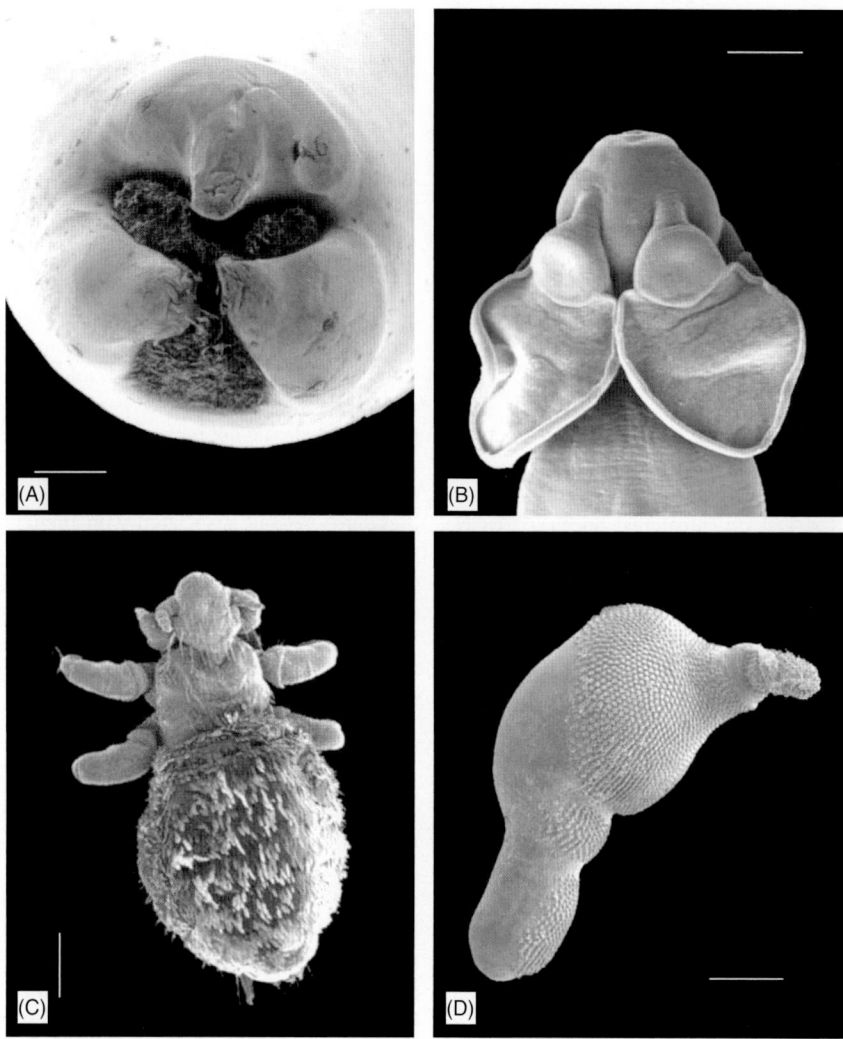

**Figure 1**  *Scanning electron micrographs of some representative parasites of marine mammals. (A) Frontal view of the mouth of* Anisakis physeteris, *a nematode from the stomach of the sperm whale* (Physeter macrocephalus). *(B) Scolex of* Monorygma grimaldii, *a larval cestode from the abdominal mesenteries of the striped dolphin* (Stenella coeruleoalba). *(C) Dorsal view of* Antarctophthirus microchir, *a sucking louse from the fur of the South American sea lion* (Otaria flavescens). *(D) Lateral view of* Corynosoma cetaceum, *an acantocephalan from the stomach of the franciscana* (Pontoporia blainvillei). *Scale bars A: 50 μm; B: 100 μm; C and D: 500 μm.*

P

found in cetaceans. *Jarrellia atramenti* was described in the blowhole mucus of a pygmy sperm whale (*Kogia breviceps*), but whether or not this protozoon is a parasite is still unclear. *Giardia* spp. is found regularly in feces from North Atlantic right whales (*Eubalaena glacialis*) and bowhead whales (*Balaena mysticetus*). In addition, the sarcodinan *Entamoeba* spp. has been recorded in the contents of the colon of the bowhead whale. Parasitic ciliates have been reported in both baleen and toothed whales. *Haematophagus megapterae* feeds on blood cells and attaches to the baleen plates of humpback whales (*Megaptera novaeangliae*), fin whales (*Balaenoptera physalus*), and blue whales (*B. musculus*). *Kyaroikeus cetarius*, *Planilamina ovata*, and *P. magna* (family Kyaroikeidae) occur in the blowhole and lungs of many species of toothed whales. These ciliates are thought to have direct life cycles, requiring perhaps repeated exposure to be successfully transmitted (Dailey, 2005).

Four families of flukes occur typically in cetaceans: species of Brachycladiidae (formerly Campulidae) occur in whales and dolphins; species of Pholeteridae and Brauninidae occur in dolphins, and species of Notocotylidae in whales. In terms of diversity and geographic range, the family Brachycladiidae is the most important. It comprises some 41 species, 35 of which are distributed among the most families of cetaceans. Species of *Campula*, *Oschmarinella*, and *Brachycladium* (formerly *Zalophotrema*) inhabit the hepatic and pancreatic ducts of TOOTHED WHALES; *Hadwenius*, the intestine; *Nasitrema*, the air sinuses; and *Hunterotrema*, the lungs. Baleen whales harbor species of *Brachycladium* (formerly *Lecithodesmus*) in the bile ducts. The life cycles of the brachycladiids are not known, although their widespread occurrence in fish and squid species eaten by cetaceans suggests that these prey act as second intermediate or transport hosts. *Pholeter gastrophilus* (Pholeteridae) parasitizes many toothed whales (mainly delphinids) throughout the world. These flukes bore into the wall of the glandular part of the stomach, and the duodenum, sometimes generating extensive fibrosis. Their distribution among digestive chambers seems to be driven by the diet and digestive physiology of each host species (Aznar *et al.*, 2006). *Braunina cordiformis* (Brauninidae) has a peculiar ovoid morphology and attaches to the stomach and the duodenal ampulla of small toothed whales. Several species of *Ogmogaster* (Notocotylidae) have been found in the large intestine of baleen whales from the Antarctic, Pacific, Atlantic, and Mediterranean. The life cycle is unknown, but information derived from other notocotylids suggests that the cercaria (larvae) might not need a second intermediate host, encysting directly on crustaceans that are preyed upon by whales.

Cetaceans are infected with both adult and larval tapeworms. The adult forms belong to two families, Tetrabothriidae (16 species of 23 in marine mammals) and Diphyllobothriidae (11 of 48 in marine mammals). Their body sizes range from small (a few millimeters) to very large (several meters) and they dwell in the intestine, from the duodenum (e.g., *Tetrabothrius forsteri*) to the terminal colon and rectum (e.g., *Strobilocephalus triangularis*). The Tetrabothriidae have diversified morphologically in cetaceans (represented by the genera *Trigonocotyle*, *Strobilocephalus*, *Priapocephalus*, and *Tetrabothrius*; only the latter two have representatives in baleen whales). The diphyllobothriids belong to the genera *Diphyllobothrium* (infecting toothed whales, more rarely baleen whales), *Diplogonoporus* (in baleen whales and the sperm whale, *Physeter macrocephalus*), *Hexagonoporus* (in the sperm whale), and *Plicobothrium* (typical from pilot whales, *Globicephala* spp., rarely in other dolphins). In general, tapeworms of marine homeotherms use zooplanktonic crustaceans as first intermediate hosts. In some tetrabothriids, euphasiids (the krill) act as intermediate hosts and fish as transport hosts. In contrast, the known cycles of members of the Diphyllobothriidae involve copepods and fish as intermediate hosts. Cetaceans and some pinnipeds also harbor larvae (plerocercoids and merocercoids) of the family Phyllobothriidae worldwide. Recent studies have revealed two types of plerocercoid (termed as "small" and "large") in the digestive tract. Likewise two types of merocercoid, classically designated as *Phyllobothrium delphini* and *Monorygma grimaldii*, occur in the blubber and mesenteries, respectively. A detailed study of these larvae in cetaceans from the Mediterranean, based on morphological, molecular, and ecological data, suggests that "large" plerocercoids, *P. delphini* and *M. grimaldii* represent different species, whereas the "small" plerocercoid is a previous stage of *M. grimaldii*. The adult stage of these tapeworms is not known but all these larvae are closely related to phyllobothriids infecting large sharks that feed on cetaceans. Apparently, cetaceans are intermediate hosts of these tapeworms, raising interesting questions about the evolution of the life cycles of tetraphyllideans (Aznar *et al.*, 2007).

Among the nematodes, the family Anisakidae is probably the most successful in terms of potential for colonizing hosts in many environments. Different studies have revealed the existence of several complexes of sibling species. The anisakids occurring in cetaceans belong to the genera *Pseudoterranova*, *Contracaecum* and *Anisakis*, although only species of the latter are found mainly in cetaceans, hence their vernacular name of "whaleworm." Eight species of *Anisakis* have been reported in at least 35 species of marine cetaceans. Whaleworms occur in the stomach, mainly in the forestomach. The larvae can attach to the stomach walls in aggregates and provoke ulcers. The life cycle is well documented. The eggs, shed in the feces, hatch and release the free-living larvae that are subsequently ingested by planktonic crustaceans. Apparently, the parasite is then ready to infect cetaceans (e.g., mysticetes). However, the larvae are usually transmitted to fish and

squid feeding on infected crustaceans, and most cetaceans become eventually infected by consuming these prey. The worms molt to the adult stage and mate in the stomach of cetaceans, where the cycle is closed when the female nematodes release the fertilized eggs (Dailey, 2005). The large histozoic nematodes of Crassicaudidae (about 11 species exclusive of cetaceans) occur in the kidneys and urogenital organs, placenta, mammary glands, muscles, and pterygoid sinuses, sometimes causing extensive damage. Species of *Crassicauda* infect both whales and dolphins, whereas the up to 9-m-long *Placentonema gigantissima* dwells in the placenta of sperm whales (Geraci and St. Aubin, 1987). Life cycles are largely unknown. It has been speculated that *C. boopis* might infect fin whale calves either by ingestion of larvae shed in their mothers' urine or by transplacental transmission (Dailey, 2005). Pseudaliids (about 17 species exclusive to toothed whales) are distributed among the genera *Pseudostenurus*, *Pharurus*, *Torynurus*, *Stenurus*, *Halocercus*, *Pseudalius*, and *Skrjabinalius*. They occur mainly in the lungs, air sinuses, and heart of phocoenids and monodontids, secondarily delphinids. Prenatal transmission of *Halocercus* in common bottlenose dolphins has been suggested. However, indirect transmission through the food web is probably the main route of infection for most species (Dailey, 2005).

Only species from two genera of acanthocephalans (family Polymorphidae) reproduce in the intestine and occasionally in the stomach of marine mammals. These worms are closely allied to forms infecting aquatic birds. Whales and dolphins are the typical hosts for *Bolbosoma* (some nine species). Pelagic euphasiids and copepods are thought to act as intermediate hosts, and fish as transport hosts (Raga and Gulland, in press). A few members of *Corynosoma* have speciated in cetaceans, but the bulk of diversity in this genus is found in pinnipeds (see later).

Cetaceans harbor a specific and rather diverse fauna of parasitic and nonparasitic crustaceans; interestingly, such associations are rare in sirenians, and do not occur on pinnipeds. Whale lice (Cyamidae, about 26 species) have diversified extensively among dolphins and whales, becoming one of the few groups of parasitic amphipods. Whale lice attach to the skin (especially on natural openings, wounds, and scars), feeding mostly on epidermal tissue. The life cycle takes place completely on the hosts and transmission is by bodily contact (Raga, 1997). Copepods of the genus *Pennella* (family Pennellidae) are primarily parasites of fish. However, about six species have successfully colonized cetaceans. Their life cycle is complex: after two free-swimming stages, they develop, mate, and attain sexual maturity on squid and then females seek a suitable definitive host, e.g., cetaceans, where they burrow on the host's body (mainly on the back and belly) to get their head anchored, and subsequently feed on blood and body fluids.

## B. Sirenians

Four species of Coccidia have been reported in sirenians: *Eimeria manatus*, *E. nodulosa*, and *T. gondii* in the West Indian manatee (*Trichechus manatus*) and *E. trichechi* in the Amazonian manatee (*T. inunguis*). The species of *Eimeria* have been detected as oocysts in feces. Nothing is known about their life cycles, but the oocyst of *E. nodulosa* bears peculiar nodules that are thought to serve for attachment to aquatic vegetation. Thus the cycle might be direct (Raga and Gulland, in press).

The fauna of metazoan parasites in manatees and dugongs is restricted to digeneans and nematodes, but it is fairly diversified relative to the number of extant host species, and very specific. Except *Nudacotyle undicola*, which belongs to a genus typical from land mammals, all species and genera recorded so far (3 species of nematodes belonging to 2 genera; 22 species of digeneans in 16 genera) are exclusive to sirenians (three out of six families also are). Interestingly, the dugong (*Dugong dugon*) exhibits a distinct and richer helminth fauna than the manatees (17 vs 7 species). This difference might simply result from corresponding differences in research effort. However, dugongs inhabit the Tropical Pacific, whose richer marine fauna may offer a larger selection of intermediate hosts and, perhaps more importantly, dugongs have remained longer in the marine habitat of ancestral sirenians. In contrast, the manatees moved and adapted to freshwater habitats, which probably resulted in the loss of some ancestral marine parasites (Beck and Forrester, 1988; Dailey, 2001).

The parasites of sirenians occur in a variety of sites within the host. The roundworms *Paradujardinia halicoris* and *Heterocheilus* spp. inhabit the stomach, more rarely the small intestine. Flukes exhibit a notable morphological diversity, e.g., *Taprobanella bicaudata*, and are found in diverse locations. Species of Paramphistomidae (4), Rhabdiopoeidae (4), and Nudacotylidae (1) and most of the Opisthotrematidae (6) occur in different sites in the digestive system, including the stomach, pyloric ceca, duodenum, ileum, colon, pancreas, and liver. In some cases they form cysts, e.g., *Faredifex clavata* (Rhabdiopoeidae) or *Lankatrema mannarense* (Opisthotrematidae). In contrast, species of *Opisthotrema* (2), *Cochleotrema* (2), and *Pulmonicola* (1) dwell in the ear system or the lungs, whereas *Labicola elongata* (Labicolidae) occurs in the upper lip of dugongs (Lauckner, 1985a; Dailey, 2001). The life cycles have not been determined for any of these parasites. Some authors have suggested that the nematodes of sirenians use crustaceans as intermediate hosts that would be consumed incidentally while feeding on vegetation. Other propose that the eggs might be ingested directly when the hosts feed on contaminated vegetation. Similar conjectures have been advanced for digeneans. *Chiorchis fabaceus* is thought to infect manatees through the incidental ingestion of snails containing metacercariae (larvae), whereas the cercariae of *Lankatrematoides gardneri* use a mollusk as an intermediate host, and perhaps then the larvae escape and encyst on aquatic vegetation, where they wait to infect dugongs (Beck and Forrester, 1988).

## C. Pinnipeds

The most common protozoons found in pinnipeds are coccidians. Species of *Eimeria* have been detected in the intestine of seals, sometimes causing severe disorders. *Eimeria phocae* is typical from harbor seals (*Phoca vitulina*) in the northwestern Atlantic. Although its life cycle is still unclear, experimental evidence shows that oocysts sporulate in feces if incubated in air but not if suspended in seawater, suggesting that transmission occurs on land. Another six species of *Eimeria* occur in Weddell seals (*Leptonychotes weddelli*) and crabeater seals (*Lobodon carcinophagus*). Species of *Sarcocystis* have been reported in harbor seals in California, northern fur seals (*Callorhinus ursinus*) in Alaska, Hawaiian monk seals (*Monachus schauinslandi*), and leopard seals (*Hydrurga leptonyx*) in Antarctica. *Isospora miroungae* and *Cystoisospora israeli* have been described in young Antarctic southern elephant seals (*Mirounga leonina*) and South African fur seals (*Arctocephalus pusillus*), respectively. Toxoplasmosis due to *T. gondii* has been observed in harbor seals, California sea lions (*Zalophus californianus*), ringed seals (*Pusa hispida*), spotted seals (*Phoca largha*), and walruses (*Odobenus rosmarus*) (Raga and Gulland, in press). Experimental infections in gray seals (*Halichoerus grypus*) have demonstrated that oocysts of this parasite can establish viable infection in seals, and that oocysts are probably acquired in surface water runoffs and sewer discharges (see also below). Likewise,

harbor seals, ringed seals, spotted seals, and walruses have been found to be seropositive to *N. caninum* (Dubey *et al.*, 2003). In addition to coccidians, flagellates of *Giardia* spp. have been detected in feces from ringed seals, harp seals (*Pagophilus groenlandicus*), gray seals, and a harbor seal from the arctic, subarctic, and eastern coasts of Canada, and in California sea lions in northern California. It has been suggested that seals and sea lions could serve as reservoirs for *Giardia* spp. but little is known about the transmission of this parasite in the sea (Dailey, 2005).

Flukes are represented by eight families. Within the Brachycladiidae (formerly Campulidae) four species of the genus *Orthosplanchnus* infect arctic and subarctic phocids, and walruses, whereas *O. antarcticus* occurs in Antarctic seals. *Zalophotrema hepaticum* is associated with California sea lions in the northeastern Pacific. Brachyclaidiids of pinnipeds live in the bile ducts, gall bladder, and rarely, the intestine. The family Heterophyidae is widespread in pinnipeds, particularly in the Northern Hemisphere. About 12 intestinal species have been recorded so far, belonging to *Cryptocotyle*, *Phagicola*, *Rossicotrema*, *Galactosomum*, *Mesostphanus*, нeterophyopsis, *Pricetrema*, and *Phocitrema*. Species of the two latter genera are associated more specifically with pinnipeds (in the Pacific region). Nothing is known of the life cycles of these heterophyids. Inferences made from other species strongly suggest that a gastropod would function as first and various species of fish as second intermediate hosts. Five species of opistorchiids from the genera *Opistorchis*, *Metorchis*, and *Pseudamphistomum* inhabit the bile ducts of seals from the Northern Hemisphere. Their life cycles are unknown, but, from other opistorchiids, it might be inferred that fish act as second intermediate hosts. Microphallids (three species) from the genera *Microphallus* and *Maritrema* are intestinal flukes. *Microphallus orientalis* has been reported in immense numbers in walruses and bearded seals (*Erignathus barbatus*) from the Barents Sea. The majority of microphallid larvae encyst in benthic crustaceans, which may explain their occurrence in seals that feed on bottom invertebrates. Finally, *Ogmogaster antarcticus* (Notocotylidae) is known from Weddell and crabeater seals (Lauckner, 1985b). These hosts likely become infected when feeding on benthic invertebrates (see earlier discussion).

The tapeworm fauna of pinnipeds is rich. The majority of species belong to the family Diphyllobothriidae (37 species out of 48 in marine mammals). Of these species, those of *Diphyllobothrium* form the major component, being distributed in pinnipeds worldwide (Lauckner, 1985b). In contrast, species of *Baylisia*, *Baylisiella*, and *Glandicephalus* are exclusive to Antarctic seals, and those of *Diplogonoporus* to boreal pinnipeds. Insights of the life cycles of diphyllobothriids have already been discussed above. One genus of Tetrabothriidae, *Anophryocephalus* (seven species) is associated with arctic and subarctic pinnipeds. Euphasiids and fish appear to be intermediate and transport hosts, respectively, in the life cycle (Lauckner, 1985b).

Pinnipeds harbor a diverse nematode fauna that comprises six families: Ancylostomatidae (2 species), Dipetalonematidae (2), Trichinellidae (1–2), Filaroididae (4), Crenosomatidae (1), and Anisakidae (14). To a great extent, this parasite fauna is related to taxa found in land carnivores. Thus, the cycle of some groups is land-dependent and resembles that of their terrestrial counterparts. For instance, the eggs and the first three larval stages of *Uncinaria* spp. (Ancylostomatidae) develop on the soil of otariid rookeries. Third-stage larvae infect adult hosts by boring the skin (especially in the flippers) and migrate to host tissues, particularly the ventral blubber and mammary glands. Larvae use the milk of adult female seals to infect the pups, where worms become adults in the intestine. Transmission occurs only during the breeding period; infective larvae overwinter on the rookery until the arrival of the hosts (Raga and Gulland, in press). The heartworm, *Acanthocheilonema* (=*Dipetalonema*) *spirocauda* (Dipetalonematidae), likely requires an arthropod vector (a louse) to trigger development and to transmit the microfilaria to North Atlantic seals, although this mode of transmission has yet to be proven. Transplacental or transmammary transmission to pups has been also suggested to occur. Species of *Trichinella* (Trichinellidae) are typical tissue parasites of terrestrial mammals. However, as discussed in the next section, a natural cycle of trichinosis involving pinnipeds seems to exist in the Arctic (Raga and Gulland, in press). With regard to nematodes with aquatic cycles, experiments have demonstrated that the lungworm *Parafilaroides decorus* (Filaroididae) infects fish that consume excrements of Californian sea lions contaminated with the larvae. These fish are, in turn, preyed upon by the sea lions. Recent experimental studies of the lungworm *Otostrongylus circumlitus* (Crenosomatidae), a species infecting boreal seals, have shown that fish seem to act as intermediate hosts (Dailey, 2005). By far the most widespread nematode group in seals is the Anisakidae, dwelling in the stomach and duodenum. Pinnipeds are primary hosts for species of *Pseudoterranova* and *Phocascaris*, whereas species of *Contracaecum* appear primarily in aquatic birds, although some species occur in phocids and otariids. In addition, *Anisakis simplex* may occasionally mature in some phocids, particularly in the gray seal. The life cycle of the sealworm, *Pseudoterranova decipiens*, is the best known. Experiments have shown that the sealworm can infect a wide variety of invertebrates. Under natural conditions, however, it utilizes only benthic and epibenthic organisms as intermediate and transport hosts. The free-living larva emerging from the egg has negative buoyancy and adheres to the substrate by its tail. This behavior favors ingestion by benthic copepods, which, in turn, are consumed by benthic macroinvertebrates. At this point, the larvae have molted twice and are ready to infect seals, but benthophagous fish, or their demersal predators, can be used as transport hosts, enhancing transmission (Aznar *et al.*, 2001).

Acanthocephalans of the genus *Corynosoma* (Polymorphidae, some 20 species in pinnipeds) exhibit a protracted history of association with pinnipeds. They have a cosmopolitan distribution, appearing in the intestine of most species of seals and sea lions, and the walrus. The complete life cycle for *C. strumosum* and *C. pseudohamanni* has been inferred from field collections. Nearshore amphipods acquire the cystacanth larvae, which, following ingestion, encyst in the body cavity of several fish species and await seals to prey on these fish. Species of *Bolbosoma* have also been reported incidentally in Arctic seals (Lauckner, 1985b).

Two main arthropod groups are associated with pinnipeds: sucking lice (Echinophthiriidae) and mites (Halarachnidae). The entire life cycle of sucking lice is spent on the seals and, therefore, they rely on bodily contact for transmission. Echinophthiriids (four genera, nine species) are associated with all major pinniped groups (otariids, odobenids, phocids) worldwide. They are physiologically adapted to the particular hosts' amphibious conditions and are transmitted on land (Lauckner, 1985b). Mites (six species) inhabit the respiratory tract and belong to two genera: *Halarachne* (in phocids) and *Orthohalarachne* (in otariids and odobenids). The life cycle takes place in the same individual host and comprises four stages: a larva, two nymphal stages, and the adult. Transmission occurs on land when active larvae are transferred by nose contact or are sneezed from the nostrils of infested animals. Acari also have some representatives in pinnipeds. *Dermacentor rosmari* (Ixodidae) appears between the fingers of walruses, especially in the hind legs, in the Arctic waters of Russia; nymphal stages are not known. *Demodex zalophi* (Demodicidae) occurs in the hair follicles

of the flippers and genital region of California sea lions. Each follicle usually contains one female and four males. All stages of development take place in the hair follicle, usually in the distal portion of the duct of the sebaceous gland (Lauckner, 1985b).

## D. Sea Otter

Three protozoan species have been reported in the sea otter (*Enhydra lutris*): *T. gondii*, *N. caninum*, and *Sarcocystis neurona*. These species are terrestrial and thus their occurrence in the sea otter and other marine mammals (see above) is intriguing. Recent studies reveal that between 40 and 70% of sea otters analyzed in the North American Pacific coast are seropositive to *T. gondii*. Apparently, oocysts in cat feces are washed into the sea through freshwater runoff. It is assumed that otters are infected either by feeding on filter-feeding invertebrates that retain the oocysts or by direct ingestion in sea water (see above). Interestingly, the most common strain of *T. gondii* infecting sea otters, and some pinnipeds, has not been described in terrestrial hosts. Concerning *S. neurona*, experimental studies have shown that sea otters can support the development of mature sporocysts that are infectious to competent definitive hosts (Miller *et al.*, 2002; Dubey *et al.*, 2003; Kreuder *et al.*, 2003).

Some 20 metazoan parasite species have been reported from sea otters throughout their range. Most are acquired directly from sympatric pinnipeds, for instance, *Orthosplanchnus fraterculus*, *Pricetrema zalophi*, *Phocitrema fusiforme*, *Diplogonoporus* sp., *Diphyllobothrium phocarum*, *Pseudoterranova decipiens*, *P. azarasi*, *Corynosoma strumosum*, *C. villosum*, and *Halarachne miroungae* or from cetaceans (*Anisakis* sp. larvae). Other parasites appear to derive from seabirds, such as the microphallid digeneans *Microphallus pirum*, *M. nicolli*, and *Plenosoma minimum*, and three acanthocephalan species of *Profilicollis* (Polymorphidae) occurring as immature. Apparently, the only parasite specific to the sea otter is *Corynosoma enhydri*, the largest species in this genus (Lauckner, 1985c; Margolis *et al.*, 1997; Mayer *et al.*, 2003).

## E. Polar Bear

Polar bears (*Ursus maritimus*) have been relatively little analyzed for parasites. In wild animals, *Trichinella* spp. (Trichinellidae) represent the most frequent records. Three cestode species, *D. latum*, *Bothriocephalus* sp., and *Taenia ursi-maritimus* have also been reported (Dailey, 2001).

## II. Patterns and Processes in Host–Parasite Associations

To understand how marine mammals and their parasites have become associated, we have to first discuss the principles that regulate the outcomes of host–parasite interactions. The evolutionary fate of every parasite species depends on that of its hosts. If parasites are able to track their hosts' evolution, host and parasite phylogenies will match exactly, resulting in a perfect cospeciation pattern. However, if parasites speciate or go extinct whereas their hosts do not, or if hosts speciate but parasites do not, incongruences between both phylogenies will occur. Another fundamental reason for incongruence is host switching (also called "host capture"), i.e., parasites colonize new hosts through ecological mechanisms. Host switching deserves more detailed comments because of its importance for the development of parasite faunas in marine mammals (Fig. 2) (Aznar *et al.*, 2001).

In order to successfully colonize a new host, the parasite has to encounter the host and be compatible with it. Encounters depend

P

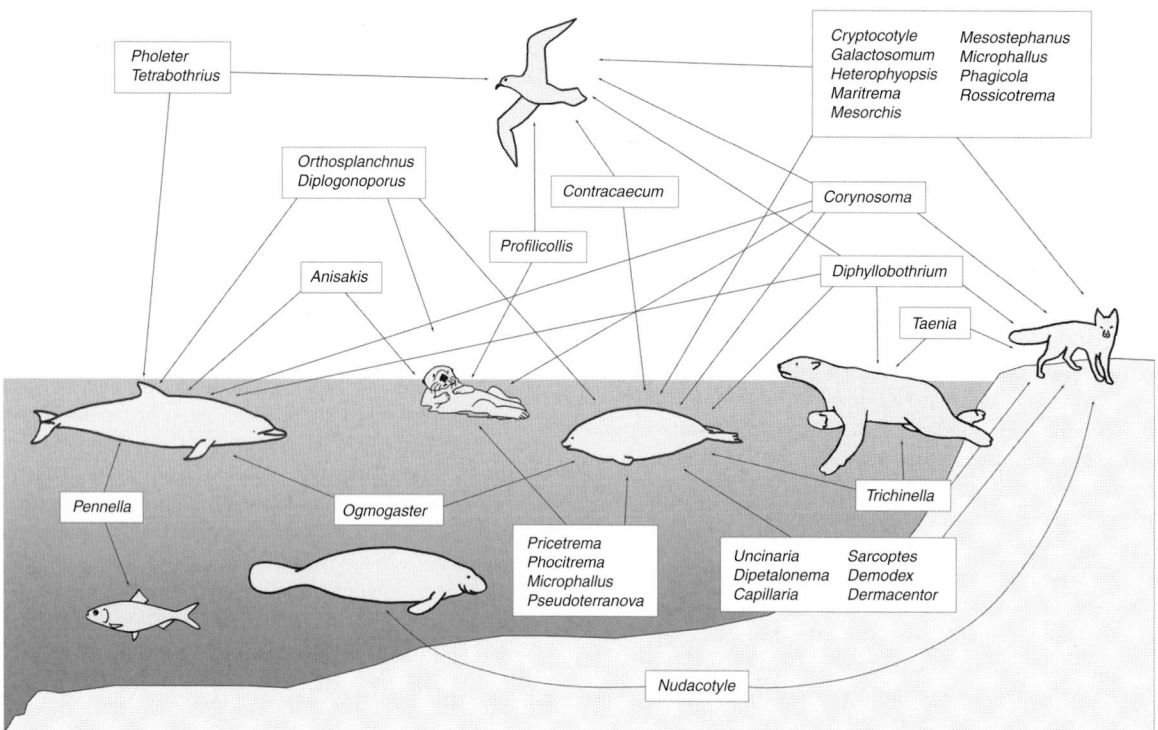

**Figure 2** *Genera of parasitic helminths and arthropods that are shared between marine mammals and between marine mammals and other hosts, i.e., terrestrial mammals (represented by a fox) and seabirds (represented by a gull). Complex patterns of putative colonization involving terrestrial mammals, pinnipeds, fish-eating seabirds, and the sea otter occur in the coastal realm. Pinnipeds also share arthropods and nematodes of terrestrial origin with other mammals. Cetaceans share a smaller number of genera apparently because of the lack of coevolutionary elements of terrestrial origin and their pelagic habits, which would hamper colonization events. Sirenians are isolated from other marine mammals because of their herbivorous diet. Fish are phylogenetically too apart for most parasite exchanges between them and mammals to be successful (see the text for details).*

on behavioral, ecological, or biogeographical factors. For instance, the transmission of digeneans of sirenians appears to be linked to either sea vegetation or the organisms associated to it. Therefore, the chances of transmission to fish-eating marine mammals will be small indeed. If contact occurs, the new host can be "right" or "wrong" depending on compatibility, i.e., those morphological, physiological, or immunological factors that either allow or preclude that a given parasite becomes established, matures, or reproduces in/on the newly contacted host. Since most parasites of marine mammals use food webs for transmission, they cannot exert strong selection on host compatibility, thus ending in both "right" and "wrong" hosts. For instance, species of *Corynosoma* are mainly associated with pinnipeds worldwide, but their larvae frequently end up in sympatric seabirds or cetaceans. However, for unknown physiological and/or immunological reasons, these larvae rarely mature in these nonpinniped hosts.

When encounters with "wrong" hosts are fortuitous, their ecological and evolutionary significance is minimal. For example, sea otters from California seldom acquire larvae of *Polymorphus* spp. from aquatic birds. However, if encounters repeat predictably throughout time, true accommodation or SPECIATION of parasites in initially "wrong" hosts can be promoted, providing that parasites are eventually able to overcome the compatibility filter. This is obviously easier as the target host is phylogenetically closer to the original host. However, the probability of encounter is always of paramount importance. This explains why, for instance, carnivorous marine mammals have more parasitic taxa in common with fish-eating seabirds than with sirenians (Fig. 2).

A typical case of accommodation is that of the ecological facultative hosts, i.e., suitable hosts for a parasite species that originated elsewhere (Hoberg and Adams, 2000). For example, the tapeworm *Anophryocephalus ochotensis* originated in the Steller sea lion (*Eumetopias jubatus*) due to colonization of seals by *Anophryocephalus* spp. However, *A. ochotensis* also infects and reproduces now in sympatric northern fur seals (Hoberg, 1995). Facultative hosts are particularly important when they sustain a significant portion of a parasite population.

## A. General Hypothesis on the Origin of Associations

Based on the above concepts, we can elaborate a general hypothesis on the origin of host–parasite associations. Let us consider the terrestrial ancestors of each marine mammal group (cetaceans, sirenians, pinnipeds, the sea otter, and the polar bear). These ancestors certainly harbored parasites in a variable number, but the subsequent host transitions from land to sea likely posed a barrier for these terrestrial parasites to track their hosts. In other words, parasites were compelled to either precisely adjust their life cycle and/or physiology to the new environment or face extinction. There are beautiful examples of such biological adjustment (e.g., the life cycle of *U. lucasi* explained above), but many parasite extinctions should have occurred, particularly in the two marine mammal groups (cetaceans and sirenians) that eventually severed their ties with land (Aznar *et al.*, 2001; see also Hoberg and Adams, 2000; Hoberg and Klassen, 2002).

When the ancestors of sea mammals made it to the sea, they became literally immersed in an ocean of infective stages of marine parasites. Then the mechanisms favoring host switching began to work. Some host captures involved marine mammals and parasites from nonmammalian hosts. However, due to compatibility limitations, these episodes should have been rare as compared to the parasite exchange between marine mammals themselves. An example can illustrate the intricacies of colonization events. The Brachycladiidae, a family of digeneans occurring in marine mammals worldwide, are the putative sister group of flukes using fish as definitive hosts. Apparently, the brachycladiid ancestors amplified a cycle that formerly ended up in these fish by adding a new step where fish predators (some unknown ancestor of toothed whales) became the new definitive hosts. This initial host switch opened the possibility for brachycladiids of subsequent coevolution (host tracking) of toothed whales and, of course, of new host-switching events. Not surprisingly, the latter involved other marine mammals. Baleen whales and pinnipeds apparently acquired brachycladiids from toothed whales, and sea otters from pinnipeds (Aznar *et al.*, 2001).

It is therefore likely that most parasite taxa infecting cetaceans have a marine origin, whereas coevolved products from land have remained in hosts with a permanent or close contact with this realm. The evidence available indicates that the arthropods and intestinal helminths of cetaceans probably rose from marine ancestors, whereas a sizeable part of parasites of pinnipeds has a clear terrestrial affiliation. To test this prediction thoroughly, we need robust phylogenetic hypotheses dealing with all parasite taxa.

## B. Parasite Exchange in Ecological Scenarios

We have learned from the above discussion that the distribution of parasites in marine mammals can be largely understood from patterns of exchange, either past or present day, within communities of marine vertebrates. This section analyzes with more detail some major features of these patterns within each specific ecosystem.

*1. Terrestrial Ecosystem* According to the previous discussion, we would expect that most monoxenous parasites disappeared when the terrestrial ancestors of cetaceans entered the aquatic environment. Whale lice, coming from free-living amphipods, are among the few aquatic forms that rely exclusively on direct transmission. However, some parasites of pinnipeds, sea otters, and polar bears, derived putatively from terrestrial counterparts, have cycles constrained to develop on land (Aznar *et al.*, 2001). Most of these land parasites have monoxenous life cycles and many are host specific. For instance, echinophthiriid lice and halarachnid mites likely coevolved with the ancestor(s) of pinnipeds and are largely restricted to these hosts today. In this context, the occurrence of *Halarachne miroungae* in sea otters should be regarded as fortuitous. Another interesting example is that of *Trichinella* spp. These extremely generalist parasites occur worldwide in terrestrial CARNIVORES and other mammals. Bearded and ringed seals, walruses, and polar bears are known to serve as hosts of these nematodes in the Arctic. The infections can evidently be traced to a terrestrial origin, but patterns of transmission are intricate and have not been proven definitively. Crustaceans and birds have been suggested to acquire *Trichinella* spp. larvae when feeding on mammal carcasses. Then, seals would be infected by direct ingestion of these crustaceans or while feeding on grounds contaminated with bird feces (Lauckner, 1985b). The nematode might then be transmitted to polar bears and some walruses that feed on seals. Polar bears could also acquire *Trichinella* independently by consuming carrion of other mammals. Perhaps, no other example of terrestrial infections in

marine mammals is as intriguing as that of coccidians. For *T. gondii*, runoff water contaminated with cat excrements has been proposed as the most likely route of transmission, given that felids are the only known hosts that can excrete environmentally resistant oocysts and in high numbers (Miller *et al.*, 2002; Dubey *et al.*, 2003). By contrast, the routes of infection of *N. caninum* and *Sarcocystis* spp. in the marine realm are still largely unknown (Dubey *et al.*, 2003).

*2. Freshwater Ecosystems* River dolphins, Baikal seals (*Pusa sibirica*), and manatees (*Trichechus* spp.) have a comparatively poor and specific helminth fauna. Since these hosts have secondarily colonized freshwater habitats, an interesting question is what kind of parasite fauna they are expected to harbor. Parasites may have followed these hosts from the marine realm or may have colonized them in freshwater habitats. The species of *Hunterotrema* and *Peritrachelius* (=*Anisakis*) *insignis* infecting the Amazonian river dolphin (*Inia geoffrensis*) seem to exemplify ancestral associations of marine origin. In contrast, many other parasites should have been acquired after freshwater colonization. For instance, the congeners of the fluke *Nudacotyle undicola* from West Indian manatees infect freshwater and terrestrial mammals. Likewise, Baikal seals are infected with *Diphyllobothrium dendriticum*, a typical freshwater tapeworm occurring in aquatic birds and land mammals. *Ruffetrema indirae* in the Indian river dolphin (*Platanista gangetica*) provides another example, given that relatives of this fluke infect terrestrial birds and mammals. In this instance, however, infections seem accidental because worms are apparently unable to produce viable eggs, although it might well represent a case of incipient colonization of a new host.

Some other parasites have an uncertain origin. The digenean *Chiorchis fabaceus* and nematodes of the genus *Heterocheilus* are common to species of manatees from both freshwater and coastal habitats, but the origin of these associations is currently unknown. Likewise, the nematode *Contracaecum lobulatum* is exclusive to Indian river dolphins, and species of *Contracaecum* are typical of fish-eating aquatic birds, pinnipeds, and cetaceans. Where might *C. lobulatum* have come from, a marine or a freshwater habitat?

*3. Coastal Ecosystems* Coasts constitute probably the most important realm in terms of historical and current parasite exchange between carnivorous marine mammals, and between these hosts and other vertebrates (Hoberg and Adams, 2000; Hoberg and Klassen, 2002). Perhaps the best example of rich parasite exchange in coastal waters is provided by the parasite fauna of sea otters, which is almost entirely made up of parasites from either sympatric pinnipeds or seabirds (Margolis *et al.*, 1997). The influx of parasites through coastal food webs puts many parasites in contact with incidental hosts, in which, depending on the compatibility filter, parasites are able or not to reproduce at ecological time. Examples of apparent failure are, e.g., *P. decipiens* in sea otters, *A. simplex* in many seals, or most *Corynosoma* spp. in nonpinniped hosts. In contrast, some parasites are notoriously unspecific with respect to the choice of their final hosts: heterophyid, opistorchiid, and microphallid digeneans found typically in terrestrial mammals or aquatic birds infect and reproduce in pinnipeds and the sea otter (rarely coastal cetaceans) in neritic and littoral waters (Lauckner, 1985b). Most of these records are occasional, but in the Caspian seal (*Pusa caspica*) they represent an important portion of the helminth fauna: many species occur also in sympatric aquatic birds or terrestrial carnivores.

There are also many examples of recent accommodation or speciation of parasites in new hosts (although, in most cases, we are not certain of which were the donor and the target hosts). For instance, the brachycladiid *Orthosplanchnus fraterculus* infects and reproduces readily in

**P**

walruses, bearded seals, and sea otters, which share a similar diet on benthic invertebrates (sea otters probably acquired this fluke from pinnipeds) (Margolis *et al.*, 1997). Likewise, diet similarity seems to be responsible for the occurrence of *Ogmogaster antarcticus* in fin whales, blue whales, Weddell seals, and crabeater seals in Antarctic waters. There are also many examples of apparent parasite speciation associated with recent colonization in the coastal environment. Species of *Corynosoma* in cetaceans (e.g., *C. cetaceum* in dolphins from the Southern Hemisphere), otters (*C. enhydri* in sea otters), and seabirds (*C. shackletoni* in the Gentoo penguin, *Pygoscelis papua*) appear to constitute independent host-switching events from *Corynosoma* of pinnipeds.

Finally, some cases represent more archaic colonization events followed by secondary diversification within the new association. It is postulated that the genera *Anophryocephalus* (Tetrabothriidae) and *Orthosplanchnus* (Brachycladiidae), typical from pinnipeds of the arctic and subarctic waters, arose during the Quaternary from ancestors infecting toothed whales (Hoberg and Adams, 2000; Aznar *et al.*, 2001). After the initial host capture, a complex history of parasite diversification apparently occurred associated to pinnipeds themselves. One of the most striking examples is that of *Contracaecum* spp. in pinnipeds: there is phylogenetic evidence that boreal seals and austral otariids acquired species of *Contracaecum* from seabirds independently.

*4. Pelagic Ecosystems* Despite the appearance of continuity between coastal and oceanic domains, the parasitic fauna of pelagic marine mammals is distinctive and comparatively poorer than that of coastal hosts. Several factors appear to contribute to this singularity. First, some parasitic groups are underrepresented. The diversity of digeneans is probably constrained because they almost exclusively require gastropods or bivalves as first intermediate hosts, which are abundant in coastal but not in pelagic waters. Indeed the Brachycladiidae are among the few flukes with representatives in pelagic marine mammals. Second, the probability of parasite exchange is decreased (e.g., there is little chance for terrestrial influence). Third, infective stages are much more "diluted" in the pelagic environment (Hoberg and Adams, 2000; Hoberg and Klassen, 2002). Transmission rates are thus lowered, also reducing the likelihood of host captures.

Some parasitic groups infecting marine mammals are predominantly pelagic and probably originated in this ecosystem. Except for species of the genus *Anophryocephalus*, most tetrabothriid cestodes occur in pelagic birds and cetaceans (Hoberg, 1995). Seabirds have been considered as initial hosts for ancestral tetrabothriids, from which these tapeworms would have switched to cetaceans (seabirds, dolphins, and whales share the genus *Tetrabothrius*). There are two other interesting cases of host switching. First, pelagic cetaceans seem to have acquired species of *Pennella* from oceanic fish. Even, there are records of the same species, *P. crassicornis*, parasitizing both cetaceans and fishes. However, females of *Pennella* can actively select their definitive hosts, in contrast to the majority of examples discussed thus far. How did this host switch to cetaceans happen? Second, some species of *Phyllobothrium* and *Monorygma* (Cestoda) appear to use cetaceans (mostly pelagic) as intermediate hosts (Aznar *et al.*, 2007). Evidently, a shift in the ancestral life cycle of these tapeworms had to occur as other relatives within the class Tetraphyllidea use only invertebrates and fish.

## III. Effects and Applications
### A. Parasitosis

The severity of the parasite-induced damage is related to the type of parasite, its abundance, the host's health status, and the

concurrence of other pathogenic agents. The effects of parasites usually have little relevance on host health, such as local reactions produced by, for instance, the proboscis of *Bolbosoma* or *Corynosoma*. Sometimes, the lesions can be more important, such as ulcers and hemorrhages caused by species of *Anisakis* and *Pseudoterranova*, and on occasion, parasites can compromise the health of its host and even lead to its death. For instance, *Nasitrema* spp. can cause neuropathies that have been related to both single and mass strandings. Likewise, lungworms of cetaceans and pinnipeds cause bronchitis and pneumonia that may result in death, particularly among the youngest individuals. Sometimes, the direct action of the parasite might not be severe, but propitiates more serious viral, bacterial, and other parasite infections, e.g., sucking lice may transmit heartworms while feeding on their hosts (Dailey, 2005; Raga and Gulland, in press).

Sometimes, parasites represent a major cause of mortality in marine mammal populations. For instance, encephalitis by *T. gondii* and intestinal perforation caused by *Profilicollis* spp. accounts for about 38% of the total mortality of southern sea otters off California (Kreuder *et al.*, 2003). Indeed, parasites might contribute significantly to marine mammal POPULATION DYNAMICS by affecting either host reproduction or survival. For instance, many females in a herd of Atlantic white-sided dolphins (*Lagenorhynchus acutus*) suffered from mastitis caused by *Crassicauda* nematodes. The parasites damaged the secretory tissue, affecting the quality and quantity of the milk, which would ultimately compromise the survival of the calves and the reproductive output of the herd (Geraci and St. Aubin, 1987). In other instances, parasites can lead to direct mortality. *Crassicauda* spp. causes cranial damage in pantropical spotted dolphins (*Stenella attenuata*) from the eastern Tropical Pacific. A study of dolphins caught in a tuna purse-seine fishery showed an increase of lesions among individuals up to 5 years old. In contrast, in animals older than 8 years, lesions diminished by 12.3% annually. Assuming that the lesions were irreversible, that there were no reinfections, and that dolphins with lesions had the same probability of being caught as those without, the mortality rate attributable to the parasite was estimated as 1.1%. Since the annual natural mortality rate for the Pacific spotted dolphin population is 10–13%, the study suggests that *Crassicauda* spp. accounts for 11–14% of that rate.

In some cases, parasites might actually regulate marine mammal populations. A density-dependent relationship between parasite-induced mortality and population size has been shown for the hookworm *U. lucasi* in northern fur seals from the North Pacific. These nematodes constitute the most important mortality factor among newborns by causing diarrhea, anemia, and intense intestinal hemorrhages. The population size of pups born in St. Paul Island (Alaska) peaked around 1940 and declined more or less steadily until the present. Data of hookworm mortality from previous surveys showed a decrease in pup numbers, which means that recent mortality is less than in the past. These data suggest a density-dependent relationship and, therefore, a possible regulatory effect.

### B. Economic and Public Health Importance

The parasites of marine mammals can infect valuable animals in aquaria, causing lesions and diseases. For this reason, expensive prophylactic measures must be used routinely. The most important economic impact, however, is due to anisakid nematodes whose larvae occur in commercial fish and squid. These larvae also have public health repercussions (see Box 1) but, from an economic perspective, the problem is cosmetic, rendering the fish unappealing to consumers.

## Box 1 Main parasitic zoonoses

**Trichinosis** occurs mainly among people from the Arctic due to the consumption of raw or undercooked meat, particularly of polar bear, walrus, and some seals. Larvae of *Trichinella* species are found in the striate muscle of many mammals, encapsulated inside individual nurse cells. When the muscles are ingested by another mammal, the larvae are released into the small intestine, where they penetrate the mucosa and develop to adult stage. The females give birth to numerous larvae, which migrate through the circulatory system to the skeletal muscles. The damage to the host is due to both penetration of adult females in the mucosa, migration of juveniles, and penetration in the muscle and nurse cell formation. Infections in humans can be fatal. Traditional arctic dishes based on seal meat, such as igunaq, nikku, raw frozen sausage, and poorly cooked sausage represent a potential source of infection for people (Raga and Gulland, in press).

**Diphyllobotriosis** is caused by tapeworms of the genera *Diphyllobothrium* and *Diplogonoporus*. Human infections occur by ingestion of plerocercoids (larval stages) encysted in fish muscles. These larvae can develop into the adult stage in the human intestine. Usually the infections are not severe, but sometimes can lead to pernicious anemia due to B12 vitamin deficit. Cases of marine transmitted diphyllobothriosis are particularly common in Japan and Peru due to the ingestion of raw fish dishes such as ceviche (Lauckner, 1985b; Oshima and Kliks, 1987).

**Anisakidosis** is produced by anisakid nematodes, particularly those of the genus *Anisakis*. Infections in humans occur when the larvae are eaten with either raw or lightly cooked fish or squid. The larvae cannot develop to the adult stage in the digestive tract of humans, but can make considerable damage to the gastric or intestinal wall. They may produce ulcers and eventually peritonitis and other severe pathologies. Although anisakidosis has been traditionally common in Asian countries, especially in Japan, the popularity of raw fish dishes, as sushi, has spread human infections worldwide (Oshima and Kliks, 1987; Smith, 1999). Allergic reactions due to antigens released by the worms in the fish have been reported both among consumers and workers in fish processing plants. This places the problem of anisakidosis under a whole new light because some of the antigens are thermostable (Moneo *et al.*, 1997). Thus, common prophylactic methods, such as cooking or freezing that kill the larvae and prevent infections are not useful to avoid allergies.

to features of the host and the ecosystem. Despite some limitations, marine mammal parasites have proven themselves as biological indicators of phylogeny, local migration, distribution, disease, stock identity, and social behavior of their hosts.

Regarding host behavior, differences in the occurrence of two whale lice (Cyamidae) species on sperm whales were interpreted as evidence of spatial segregation between the bulls and the rest of the population off South Africa. Assuming that males leave their natal herd at attainment of sexual maturity, parasite information suggested that this should occur at 12 m of length. This prediction was validated later on by analyses of gonadal tissue. Another example is the comparison of intestinal helminth abundance between pods of long-finned pilot whales (*Globicephala melas*) in the North Atlantic. Monte Carlo simulations revealed that adult males were more difficult to allocate into their pods than other individuals. This suggests male exchanges between pods, a conclusion that agrees with genetic studies that adult males do not breed within their natal pods (Balbuena *et al.*, 1995).

Parasites transmitted through the food web have also provided information on past and present host feeding grounds. This has served to reveal geographical differences between areas for Antarctic whales, spotted seal, populations on both sides of the North Pacific, inshore and offshore forms of bottlenose dolphins in both the east coast of the United States and Peru, and geographical segregation of franciscana (*Pontoporia blainvillei*) south and north of the La Plata River Estuary, between Argentina and Uruguay (Aznar *et al.*, 2001).

Finally, marine mammals can be used to evaluate the risk of human infections of protozoan pathogens. In particular, since sea otters are nearshore predators that share the same environment and some food items with humans, the ongoing studies of transmission and spatial dynamics of *T. gondii* in this species may prove useful to tackle toxoplasmosis of marine origin in humans (Miller *et al.*, 2002; Dubey *et al.*, 2003).

### See Also the Following Articles

Health ■ Stock Identity ■ Whale Lice

### References

Aznar, F. J., Fognani, P., Balbuena, J. A., Pietrobelli, M., and Raga, J. A. (2006). Distribution patterns of *Pholeter gastrophilus* (Digenea) in the stomach of four odontocete species: the role of the host digestive physiology. *Parasitology* **133**, 369–380.

Aznar, F. J., Agustí, C., Littlewood, D. T. J., Raga, J. A., and Olson, P. D. (2007). Insight into the role of cetaceans in the life cycle of the tetraphyllideans (Platyhelminthes: Cestoda). *Int. J. Parasitol.* **37**, 243–255.

Aznar, F. J., Balbuena, J. A., Fernández, M., and Raga, J. A. (2001). Living together: the parasites of marine mammals. *In* "Marine Mammals: Biology and Conservation" (P. G. H. Evans, and J. A. Raga, eds). Kluwer Academic/Plenum Publishers, New York.

Balbuena, J. A., Aznar, F. J., Fernández, M., and Raga, J. A. (1995). The use of parasites as indicators of social structure and stock identity of marine mammals. *In* "Whales, Seals, Fish and Man" (A. S. Blix, L. Walløe., and Ø. Ulltang, eds), pp. 133–139. Elsevier Science, Amsterdam.

Beck, C., and Forrester, D. J. (1988). Helminths of the Florida manatee, *Trichechus manatus latirostris*, with a discussion and summary of the parasites of sirenians. *J. Parasitol.* **74**, 628–637.

In 1982, the losses caused by *P. decipiens* in eastern Canada were valued $20 million, only in processing of cod fillets (Aznar *et al.*, 2001).

The parasites with repercussions on public health are those that can infect humans with food (Box 1), when animals consume meat from marine mammals or, more frequently, raw fish or squid containing living infective stages.

### C. Natural Tags

Many parasites are useful natural markers of biological and environmental phenomena because their transmission is linked intimately

**P**

Dailey, M. D. (1985). Diseases of mammalia: cetacea. *In* "Diseases of Marine Animals" (O. Kinne, ed.), vol. 4, pp. 805–847. Biologische Anstalt Helgoland, Hamburg, Part 2.

Dailey, M. D. (2001). Parasitic diseases. *In* "Marine Mammal Medicine" (L. A. Dierauf, and F. M. D. Gulland, eds), 2nd ed., pp. 357–379. CRC Press, Boca Raton, FL.

Dailey, M. D. (2005). Parasites of marine mammals. *In* "Marine Parasitology" (K. Rohde, ed.), pp. 408–414. CSIRO Publishing, Victoria.

Dubey, J. P., *et al.* (12 authors) (2003). *Toxoplasma gondii, Neospora caninum, Sarcocystis neurona,* and *Sarcocystis canis*-like infections in marine mammals. *Vet. Parasitol.* **116**, 275–296.

Geraci, J. R., and St. Aubin, D. J. (1987). Effects of parasites on marine mammals. *Int. J. Parasitol.* **17**, 407–414.

Hoberg, E. P. (1995). Historical biogeography and modes of speciation across high-latitude seas of the Holartic: concepts for host–parasite coevolution among the Phocini (Phocidae) and Tetrabothriidae (Eucestoda). *Can. J. Zool.* **73**, 45–57.

Hoberg, E. P., and Adams, A. (2000). Phylogeny, history and biodiversity: understanding faunal structure and biogeography in the marine realm. *Bull. Scand. Soc. Parasitol.* **10**, 19–37.

Hoberg, E. P., and Klassen, G. J. (2002). Revealing the faunal tapestry: coevolution and historical biogeography of hosts and parasites in marine systems. *Parasitology* **125**, 3–22.

Kreuder, C., *et al.* (9 authors) (2003). Patterns of mortality in southern sea otters (*Enhydra lutris nereis*) from 1998–2001. *J. Wildl. Dis.* **39**, 495–509.

Lauckner, G. (1985a). Diseases of mammalia: sirenia. *In* "Diseases of Marine Animals" (O. Kinne, ed.), vol. 4, pp. 795–803. Biologische Anstalt Helgoland, Hamburg, Part 2.

Lauckner, G. (1985b). Diseases of mammalia: pinnipedia. *In* "Diseases of Marine Animals" (O. Kinne, ed.), vol. 4, pp. 683–793. Biologische Anstalt Helgoland, Hamburg, Part 2.

Lauckner, G. (1985c). Diseases of mammalia: carnivora. *In* "Diseases of Marine Animals" (O. Kinne, ed.), vol. 4, pp. 645–682. Biologische Anstalt Helgoland, Hamburg, Part 2.

Margolis, L., Groff, J. M., Johnson, S. C., McDonald, T. E., Kent, M. L., and Blaylock, R. B. (1997). Helminth parasites of sea otters (*Enhydra lutris*) from Prince William Sound, Alaska: comparisons with other populations of sea otters and comments on the origin of their parasites. *J. Helminthol. Soc. Wash.* **64**, 161–168.

Mayer, K. A., Dailey, M. D., and Miller, M. A. (2003). Helminth parasites of the southern sea otter *Enhydra lutris nereis* in central California: abundance, distribution and pathology. *Dis. Aquat. Org.* **53**, 77–88.

Miller, M. A., *et al.* (10 authors) (2002). Coastal freshwater runoff is a risk factor for *Toxoplasma gondii* infection of southern sea otters (*Enhydra lutris nereis*). *Int. J. Parasitol.* **32**, 997–1006.

Moneo, I., Audicana, M. T., Alday, E., Curiel, G., Del Pozo, M. D., and García, M. (1997). Periodate treatment of *Anisakis simplex* allergens. *Allergy* **52**, 565–569.

Oshima, T., and Kliks, M. (1987). Effects of marine mammal parasites on human health. *Int. J. Parasitol.* **17**, 415–421.

Raga, J. A. (1997). Parasitology of marine mammals. *In* "Marine Mammals, Seabirds and Pollution of Marine Systems" (T. Jauniaux, J. M. Brouquegneau, and F. Coignoul, eds), pp. 67–90. Presses de l'Université de Liège:, Liege.

Raga, J. A., Balbuena, J. A., Aznar, F. J., and Fernández, M. (1997). The impact of parasites on marine mammals: a review. *Parassitologia* **39**, 293–296.

Raga, J. A., and Gulland, F. M. D. (2008). Health, disease and parasites of marine mammals. *In* "Encyclopedia of Life Support Systems." UNESCO-EOLSS Publishers Co. Ltd., Oxford.

Smith, J. W. (1999). Ascaridoid nematodes and pathology of the alimentary tract and its associated organs in vertebrates, including man: a literature review. *Helminthol. Abstr.* **68**, 49–96.

# Parental Behavior

## Janet Mann

Parental behavior in pinnipeds, sirenia, sea otters (*Enhydra lutris*), polar bears (*Ursus maritimus*), and cetaceans shares several features: paternal care is virtually absent, gestation and lactation periods are typically long, females give birth to and nurse one offspring at a time (polar bears excepted), and many marine mammals fast during the early stages of lactation. In sum, marine mammal mothers invest extensively and exclusively in single offspring; this article reviews the diversity and nature of that investment.

Marine mammal maternal strategies vary in important respects. Polar bears, sea otters, and all three families of pinnipeds—Odobenidae (walrus, *Odobenus rosmarus*), Phocidae (earless or "true seals"), and Otariidae (eared seals, sea lions, fur seals)—give birth on land or ice. Twenty-three species of pinnipeds breed on land and 13 breed on ice. Cetacean and sirenian females give birth in the water; this pattern favors precocial swimming and diving. Among many pinnipeds, maternal care is largely restricted to milk transfer, whereas the prolonged association characteristic of many cetaceans, sirenians, otters, polar bears, and some pinnipeds also involves protection and potentially extensive information transfer.

## I. Feeding, Lactation, and Patterns of Association

Lactation strategies in marine mammals generally depend on trade-offs among foraging, predation risk, and reproduction. This trade-off is exemplified by many marine mammal species that fast during the early stages of lactation. That females forgo feeding by breeding on land (i.e., pinnipeds) or in warm coastal waters (i.e., baleen whales) suggests that benefits, such as reduced predation risk and rapid energy transfer from mother to offspring, outweigh the costs of fasting. Larger bodied mammals can withstand fasting for longer periods than smaller, thus able to afford longer fasting periods devoted to offspring care. Fasting and lactation coincide in many marine mammal species and only rarely in terrestrial mammals. Early-weaning marine mammals tend to have milk that is high in fat, investing heavily in offspring for a shorter period. Late-weaning marine mammals tend to have lower fat milk (although still much higher than for terrestrial mammals). This pattern is generally true for comparisons between phocids and otariids, or toothed and baleen whales (Oftedal, 1997; Table I). Other factors, such as the development of pup or calf foraging skills, also contribute to late weaning ages and prolonged association (see Section IV).

Phocids tend to fast and remain near the ROOKERY until their pup is weaned; they rely on fat stores to nurse offspring. Phocid maternal strategies are generally characterized as fasting. A few phocids feed during lactation, notably the harp (*Pagophilus groenlandicus*), harbor (*Phoca vitulina*), ringed (*Pusa hispida*), and Weddell (*Leptonychotes weddellii*) seals. Most remarkable is the hooded seal (*Cystophora cristata*), which breeds on pack ice and nurses her pup for only 4 days, transferring approximately 748,000 J or 178,657 kcal to the pup in that time (Oftedal *et al.*, 1993).

Predator and prey distributions are likely to influence breeding habitat (i.e., pack ice, fast ice, beach, cave, and water) and lactation length. The relationship between breeding habitat and lactation length has been difficult to test using the comparative method because species breeding in similar habitats tend to be close phylogenetically.

TABLE I
Parameters Related to Maternal Care in Marine Mammals

| Taxon | Birth location (land, ice, water) | Sexual dimorphism (high/moderate vs low) | Mother fasts during lactation | Milk fat (%) | Duration of lactation (months) | Association postweaning? |
|---|---|---|---|---|---|---|
| Pinnipedia | | | | | | |
| Otariidae | L | H | Yes | 25–53 | 4–30 | No |
| Phocidae | L, I | H, L | Yes | 47–61 | <3 | No |
| Odobenidae | L, I | H | No | 25–32 | >24 | Yes |
| Sirenia | W | L | No | 13 | 12–36 | No? |
| Ursidae (polar bears) | L | L | Yes | 17–25 | 18–42 | No |
| Mustelidae (sea otters) | L | L | No | 21–26 | 6–12 | Yes |
| Cetacea | | | | | | |
| Platanistoidae | W | L | No | 13 | <12 | Yes |
| Delphinoidae | W | H, L | No | 22–30 | 12–48+ | Yes |
| Ziphoidae | W | H | No | 18–24 | – | – |
| Physeteroidae | W | L | No | 24 | >12 | Yes |
| Balaenidae | W | L | Yes | 22 | <10 | No |
| Neobalaenidae | W | L | Yes | – | <6 | No |
| Eschrichtiidae | W | L | Yes | 53 | <6 | No |
| Balaenopteridae | W | L | Yes | 24–40 | <10 | No |

However a recent analysis using independent contrasts does suggest that breeding substrate selects for lactation length (Schulz and Bowen, 2005). Those breeding on unstable substrates (pack or floe ice) wean early relative to those breeding on land, but other selective pressures (metabolic rate, foraging strategies, predation risk) are likely to be involved (Schulz and Bowen, 2005). Some species with protective lairs or caves [arctic Baikal seal (*Pusa sibirica*), and Mediterranean monk seal (*Monachus monachus*)] wean late for phocids, at 2–3 months. Phocid females tend to be larger than otariid females and can thus fast longer. Since phocid pups are generally more vulnerable to terrestrial predation than otariids, they are adapted to briefer lactation periods.

Otariid mothers leave their young after the first week to hunt for 1–14 days before returning to nurse their pups. When the mother is hunting, the pup is typically fasting. She returns for 1–3 days and fasts while she feeds her pup. The duration of her absence is correlated positively with the milk she provides per pup visit. If food is abundant, then mothers make shorter foraging trips than when food is scarce and feed their young more often. Thus, prey abundance is correlated positively with maternal body condition and pup growth rate. Some otariids, such as sea lions, nurse their pups for 6–11 months. The Galapagos fur seal (*Arctocephalus galapagoensis*) nurses her pup for as long as 3 years.

As pups grow, otariid mothers may spend three-fourths of their time at sea, but each foraging trip tends to vary in duration. Phocids are adapted to store energy and fast for the entire lactation period, whereas otariids must supplement stores with some food. Thus otariid mothers must find breeding breaches that are also near productive feeding grounds. Beach breeding can provide safety to pups, but also constrains how far and how often females can venture to hunt. Otariid pups grow more slowly than phocid pups, but tend to be weaned at

larger weights, 40% of maternal mass compared with 30% in phocids. Recent analyses suggest that weaning and maternal mass are isometric for pinnipeds overall (Schulz and Bowen, 2005). When food abundance is high, otariid mothers can transfer more milk to their offspring than when food is scarce. Phocid females concentrate on feeding offspring rather than expend energy traveling to and from feeding areas. Notably, otariids are the only mammals capable of continuing to lactate despite long periods (occasionally >3 weeks; Georges and Guinet, 2000) of separation from offspring. Molecular data suggest a mutation of the α-LA gene in otariids is involved in the delay of mammary involution (Reich and Arnould, 2007). In sum, most otariids wean their young within the year, but later than most phocids.

Of all pinnipeds, walrus appear to have the longest average period of maternal care, nursing their young for up to 3 years. It is not known whether walrus mothers fast during the initial stage of lactation. They give birth on ice, but the mother nurses her pup in the water. Known for their sociability, females and pups spend most of their time in female herds. Daughters stay in their mothers' herd. Sons leave the herd when 2–3 years old and join all-male herds. Some evidence suggests that adult males return to their natal group after several years (Stewart *et al.*, 2003).

There are few studies of maternal care in Sirenia. Manatees (*Trichechus* spp.) and dugongs (*Dugong dugon*) are thought to have similar gestation lengths (12–13 months), but differ when it comes to other life history traits. Manatees give birth as early as their fourth year, and have interbirth intervals less than 2 years (Koelsch, 2001), whereas dugongs do not breed until age 10, and typically have 3-year intervals (Marsh *et al.*, 1984). Based on tooth wear in dead specimens, Sirenia calves feed on seagrass within months after birth.

Polar bears begin reproduction at 5–6 years of age and nurse their cubs for 2–3 years. Females birth and fast in dens on coastal shores

P

or on multiyear pack-ice up to 300 km offshore (Stirling, 2002). Unlike all marine mammals, polar bear females have the challenge of hunting [mostly ringed and bearded (*Erignathus barbatus*) seals] with two dependent offspring in tow. A long period of dependency is probably essential for cubs to both build body mass, gain protection from aggressive adult male polar bears, and learn hunting and denning skills before they begin breeding themselves.

Similar to the fast vs feed pattern that distinguishes lactating phocids and otariids, the larger baleen whales fast for long periods and toothed whales little or not at all. Baleen whales have higher fat milk and nurse for shorter periods than toothed whales. Mysticete mothers, who typically migrate to warmer waters to birth and rear young, can devote their time and energy almost exclusively to milk transfer, much like phocid mothers. Like pinnipeds, baleen whales are characterized by annual breeding and feeding cycles, including cases of postpartum estrus. Among baleen whales and pinnipeds, females are often attractive to males soon after giving birth. Thus, intense competition between males over access to females could interfere with maternal care, but detailed behavioral studies are needed.

Little is known about the nature of the baleen whale mother–calf relationship. Humpback (*Megaptera novaeangliae*) and right whales (*Eubalaena* spp.) are the best-studied baleen whales, likely due to their tendency to visit protected coastal areas to breed. Interactions between more than one mother–calf pair are infrequent, especially between humpback mother–calf pairs, which appear to actively avoid each other during the initial stage of lactation. The explanation for avoidance is unclear. Association between mother–calf pairs may disrupt the development of maternal or calf recognition. Similar to some of the large ungulate species, mothers may avoid unrelated calves to avoid milk theft. Anecdotal reports of Southern right whale (*E. australis*) behavior indicate that calves occasionally attempt to nurse from other females. With such a tremendous energetic output, milk transfer may be difficult to inhibit if the "wrong" calf attempts to nurse. Additionally, fasting in warm water likely reduces the metabolic demands on the mother. (Warm water would also reduce metabolic demands on the calf.) Fasting terrestrial mammals convert body fat to fatty milk more easily than mammals that eat during lactation, and this pattern is likely to favor fasting in baleen whales that invest in rapid growth of a large calf. Consistent with this, fasting mammals (terrestrial and marine), including baleen whales, have milk low in carbohydrate, protein, and water, but very high in fat. The baleen whale mother can thus fatten in the feeding grounds, transfer that energy efficiently to her calf in the warmer breeding grounds, and assure a safer migration to higher latitudes. Larger calf body size would likely reduce predation risk and other somatic costs (e.g., metabolic). Four balaenopterids [blue (*Balaenoptera musculus*), fin (*B. physalus*), minke (*B. acutorostrata* and *B. bonaerensis*), and sei (*B. borealis*) whales] tend to wean early, before or soon after reaching the feeding grounds (at 6–7 months). Humpback, right, bowhead (*Balaena mysticetus*) and gray (*Eschrichtius robustus*) whales continue to nurse offspring to older ages and are typically feeding during the latter stage of lactation. Although bowhead whales migrate, they do not migrate to warmer waters to calve and apparently do not fast. Given this pattern, bowhead whales are predicted to have milk lower in fat, nurse offspring for longer periods, and have slower growth rates compared to other baleen whales. Existing data on bowhead whales support this, with late age of sexual maturity (20 years) and 4–7 interbirth intervals, but data are still too limited to make explicit comparisons (George *et al.*, 1999).

Little is known about mother–calf behavior during migration, but available data suggest that mothers and calves do remain together during the first migration between the breeding and feeding grounds. Observations of Pacific humpback whales off the coasts of Australia and of gray whales off the coasts of North America indicate that adult males often accompany mothers and calves during the migration. This may reduce killer whale (*Orcinus orca*) predation risk, although the males are likely there to mate with the female, not protect the calf. Several studies have indicated that up to one-third of humpback whale calves bear killer whale tooth marks by the time they reach the feeding grounds with their mothers. Killer whale attacks on gray and humpback whale calves have been observed, but there is no evidence to suggest that males aid calves during such attacks. A recent analysis suggests that killer whale attacks are likely to occur during the baleen whale calf's first migration, but are rare after that (Mehta *et al.*, 2007).

Odontocete life history patterns are slow relative to baleen whales. The baleen whale usually reaches reproductive maturity in 4 or 5 years, but the typical odontocete takes 10 years or more, despite a much smaller body size. A notable exception appears to be the bowhead whale, which reaches sexual maturity much later and may live for 200 years. Only the smallest odontocetes, such as porpoises, tend to reach reproductive maturity by age 5. With a relatively long period of immaturity and calf vulnerability, odontocetes may benefit more by stable patterns of group living compared to mysticetes.

Odontocetes do tend to have stable groups and patterns of association compared to mysticetes. In contrast to the baleen pattern, odontocete mother–calf pairs tend to associate with each other and stable associations with female kin are common. Adult male associations with mothers and young calves are more variable. For example, adult male killer whales consistently associate with female kin and their offspring. Female sperm whales (*Physeter macrocephalus*) and young spend most of their time with female kin and rarely associate with adult males. Shark Bay bottlenose dolphin (*Tursiops* spp.) females and calves appear to avoid juvenile and adult males (Mann and Smuts, 1999; Gibson and Mann, in press).

Close mother–calf association is characteristic of many toothed whales and dolphins, particularly in the delphinid family. Delphinid calves typically swim alongside the mother for the first few weeks in what is known as "echelon position." Within a few months, they begin to swim under the mother regularly in what is known as "infant position." In contrast, humpback and right whale calves tend to swim alongside the mother throughout development and separations are not common during lactation. Bowhead calves occasionally ride on the backs of their mothers during migrations. Odontocete mothers are presented with particular challenges that baleen mothers do not face. Because they feed throughout lactation and hunt single prey, the mother must accelerate or dive deeply to hunt, often leaving her calf alone at the surface. Young calves are not very proficient divers, and they breathe more often, thus they sometimes "wait" at the surface while the mother forages below. This is especially pronounced in deep-diving species such as sperm whales. Mothers may dive for as long as between 30 and 50 min. The calf, often alone at the surface, has a tendency to swim toward any whale that comes up first. Some form of communal care in this deep-diving species may be an important selective force favoring matrilineal sperm whale units. A lone sperm whale mother would be at a distinct disadvantage if she had to abandon her vulnerable calf at the surface for long periods as she hunted for squid at depth. Among pantropical spotted dolphins (*Stenella attenuata*), lactating females feed on flying fish rather than squid (preferred by nonlactating females), which necessitates diving and separation from their calf. In bottlenose dolphins, most mother–calf separations occur during the mother's foraging.

When foraging, bowhead mothers and calves may separate 1–2 km for 30 min or more.

## II. Protection and Predation

Otariids tend to breed on predator-free islands. Phocid pups can be vulnerable on fast ice, where polar bears and killer whales may feed on them. In the North Pacific, killer whales frequently catch harbor seal pups on their first forays into the ocean. South American sea lion (*Otaria flavescens*) pups are picked off beaches by killer whales. Great white sharks (*Carcharon charcharias*) are common predators of elephant seal pups (*Mirounga* spp.) and a variety of other pinniped prey. Mammal-eating killer whales will also prey on most pinniped species they encounter, including the largest species, Steller sea lions (*Eumetopias jubatus*). Maternal protection of pups from predation is not well documented. Adult male and female pinnipeds sometimes pursue predators, such as killer whales and sharks, but nothing about this behavior suggests that they are protecting offspring *per se*.

Mothers and other female whales and dolphins do protect offspring from predatory attacks. Sperm whale females may form a ring "marguerite formation" around a calf and sometimes they place their heads to the center and use their powerful tails to deter a predator. At other times they will face out in a circle, with their tails facing the calf. Indian Ocean bottlenose dolphin mothers and close associates have also defended calves from predation attempts by tiger sharks (*Galeocerdo cuvieri*) (Mann and Barnett, 1999). Gray whale mothers have been observed placing themselves between killer whales and their calves (Black *et al.*, 1999).

Bottlenose dolphin infanticide has been documented from retrieved carcasses at two Atlantic Ocean sites: the Moray Firth in Scotland and Virginia, United States. The perpetrators of calf killing are not known. If the pattern is similar to that found widely in felids, ursids, primates, and rodents, then it is likely that unrelated adult males are the perpetrators and might gain a reproductive advantage by monopolizing and mating with the mother of the victim. Polar bear mothers, like other ursid females, are known for their protectiveness and fierce defense of cubs from infanticidal males.

Among land-breeding SEXUALLY DIMORPHIC pinnipeds, offspring are occasionally killed incidentally by large patrolling males (e.g., elephant seals and sea lions). These killings appear to be by accident, e.g., when the male is preoccupied with defending the breeding beach. However, males also abduct and force matings on pups, sometimes resulting in the pup's death. Among ice-breeding pinnipeds, or when mating occurs in the water, pups are less vulnerable to patrolling males. In these species, sexual dimorphism is also less pronounced and the risk of being squashed is lower than for land-breeding species.

## III. Mother–Offspring Recognition

Mutual vocal and olfactory recognition is common in a number of pinniped species, especially among otariids. This might be expected given that otariid mothers and pups often separate during the mother's foraging trips and the reunion depends on mutual recognition through repeated calling. Otariid pups are also fairly mobile, so the previous location of her pup would not be a reliable cue. In addition, otariids breed colonially, complicating the task of finding the pup or mother among hundreds. Phocid mothers would not necessarily need a recognition system if they remain close to their offspring, but some species that separate during lactation (e.g., harbor seals) exhibit vocal recognition. After feeding trips, it is common for mothers to inspect several pups before finding her own.

Sea otter vocalizations have been compared to pinnipeds in structure and complexity. There is some evidence for short range COMMUNICATION and vocal recognition between mothers and pups. Olfactory cues may also play a role.

Cetaceans have little or no sense of smell (absent in odontocetes; rudimentary in mysticetes), but likely use vocal communication for individual identification. Little is known about mysticete mother–calf recognition and only slightly more is known in odontocetes. Delphinids produce a diverse array of sounds, including echolocation clicks, burst-pulse sounds, and whistles. Dolphin calves are born able to whistle and will whistle often during the first days of life. Bottlenose dolphins and potentially other delphinids exhibit signature whistles, individually distinctive whistles in the first years of life. In bottlenose dolphins, simple whistles are present at birth, but the calf develops a more distinctive contour with age (although the calf may continue to produce simple whistles as well). Mechanisms influencing the development of SIGNATURE WHISTLE contours are not known. Although the calf's signature whistle is not present at birth, the young calf may learn to identify his or her mother's signature whistle soon after birth and mothers may be able to recognize the calf's simple whistle. Field studies suggest that whistles mediate natural separations and reunions between mothers and calves. Calves tend to whistle more than mothers, perhaps indicating the calf's greater motivation to reunite with the mother than vice versa. Experimental work with captive and temporarily captured animals suggests that dolphins can recognize the whistles of others. In the marine environment, depth (water pressure) and other habitat features may alter sound enough to make voice recognition difficult; thus, selection may favor individually distinctive whistle contours in an environment where individuals join and leave each other and frequently lose visual contact. Few studies have sufficiently examined cetacean mother–calf communication in captive or wild settings to elaborate further.

## IV. Maternal Influence on Offspring Behavioral Development

Mothers influence offspring behavioral development by the experiences they provide, including migration and navigation, communication, social interactions, and foraging. To secure the transition from dependence to independence, marine mammal offspring must acquire necessary survival skills. Although data are limited, maternal influence on the development of foraging skills has received the most attention (Mann and Sargeant, 2003).

Some types of marine mammal foraging may not require learning, but other foraging skills clearly require experience. Marine mammal young may develop foraging skills by independent (nonsocial) learning or by social learning, including stimulus or local enhancement (e.g., exposure to foraging areas with the mother), observation of maternal foraging or prey caught, and rarely pedagogy. Teaching requires that the "teacher" modifies his or her behavior in the presence of a naive observer (pupil) at some cost or without benefit to the "teacher," which results in or facilitates acquisition of some skill or knowledge by the "pupil" (Caro and Hauser, 1992). As such, a pup or calf can "learn" from another animal, without teaching occurring. There are good examples of extensive social and nonsocial learning in Cetacea, but only a few possible examples of teaching. There are no examples of teaching in pinnipeds and little evidence for learning from the mother. In sea otters, learning by observing from the mother appears to be important, but teaching probably does not occur (Estes *et al.*, 2003).

P

Pinniped pups theoretically could learn some foraging skills by following and observing the mother. However, there is little evidence to suggest this. Phocid mothers fast for most of lactation and leave the rookery before their pups do. Otariid pups could accompany their mothers on foraging trips, but most studies find little evidence of this. Pup absence from the rookery often coincides with maternal absence (a pup does not leave the rookery while the mother is present), but mother and pup travel and forage independently at sea. Otariid pups have more opportunities to develop swimming, diving, and hunting skills while relying principally on their mothers nutritionally. In the late-weaning Galapagos fur seal, pups catch their own prey for a year or longer before weaning. Some phocid pups gain considerable diving and swimming experience independently before weaning (e.g., bearded and ringed seals). Some harbor seals change haul-out sites during lactation. The pup, then, may follow the mother 30 km or more. Although research with captive pinnipeds indicates that they are adept learners, including vocal learners, there is little to suggest that learning in the context of the mother–pup relationship is particularly important except for mother–pup recognition. Walrus are likely to be an exception among pinnipeds, given their prolonged mother–offspring association and sociability at sea.

Several studies clearly document that otter pups [Eurasian river (*Lutra lutra*) and sea] learn from their mothers. Mothers take their pups on foraging trips and the pup watches the mother as she dives and retrieves prey. Mothers share their catch with pups, and in sea otters, pups even develop the specific tool-using strategy of the mother (use of rocks, soft-drink bottles, or other objects) to open shellfish or other prey (Lyons and Estes, 1989; Estes *et al.*, 2003). It takes time for pups to develop adequate diving and foraging skills, which may help explain the long dependence period in this small marine mammal. Other MUSTELIDS typically wean offspring at much younger ages (i.e., 2–3 months).

To become independent, cetacean calves must be able to forage successfully on their own. They must also catch their first prey on their own. Unlike other hunting species (e.g., felids, canids), cetacean mothers generally do not share prey with young. One exception is the killer whale, where prey sharing between mother and offspring has been well documented. Despite many thousands of hours of observation on bottlenose dolphin calves in Shark Bay by the author, prey sharing has not been observed. Delphinids exhibit a diverse range of foraging strategies and there is some evidence that these are learned. Whether calves learn foraging skills by simple trial and error, social learning, i.e., observing the mother and other conspecifics, or other mechanisms is generally not known. Shark Bay bottlenose dolphin calves begin capturing prey within 4–6 months, even though they continue nursing 3 years or more. Female bottlenose dolphin calves in Shark Bay, Australia, adopt the sponge-foraging tactic of their mother, and those without sponging mothers do not develop the trait. This unusual form of tool use, which is adopted by only a small subset of the population, may be transmitted socially (and culturally) from mother to offspring. Humpback whale juveniles in the North Atlantic have learned new foraging strategies, likely due to shifts in prey density and distribution. These tactics spread rapidly through the population, although data suggest that older animals generally did not adopt the behavior. The humpback example may provide evidence for learning (horizontal transmission), but not from the mother (vertical transmission).

The diversity of foraging strategies that humpback whales exhibit (e.g., bubblenet, cooperative lunge feeding) may help explain their relatively longer nursing period compared to other mysticetes; i.e., the duration of lactation may, in part, be determined by the ability of young animals to forage independently. This might also help explain the longer nursing periods of odontocetes relative to mysticetes. The odontocete's task of capturing single, elusive, and specialized prey is likely to require greater foraging skill than the gulp-feeding techniques characteristic of mysticetes.

Killer whales may provide the only reasonable cetacean example of teaching, but more observations are needed. Experienced beach-feeding whales (who beach to capture pinnipeds) actively help younger animals develop this risky technique by nudging them to shore, partially wounding prey for young to manipulate, and assisting their departure from the beach.

## V. Parity

For many mammals, parity (number of pregnancies/live births experienced) is correlated negatively with offspring mortality. For example, sea otter pups born to older, more experienced mothers tend to have lower mortality than pups born to younger mothers. First-time (primiparous) elephant seal mothers have lower weaning success than experienced multiparous mothers. However, a number of factors may contribute to this difference, including maternal mass (young mothers are smaller than old mothers), experience, dominance, and placement of pups in the rookery. Young, subordinate females relegated to the peripheral area of a breeding beach may have more difficulty relocating their pup and their pup may be more vulnerable to harassment and death.

First-born dolphin calves in captivity have higher mortality than later-born calves, but the cause(s) for the difference is not known. Both maternal experience and body size may be factors. Patterns of first-born mortality in wild populations are not sufficiently documented to draw conclusions, but in Sarasota, Florida, high mortality of first-born offspring might be related to maternal transfer "offloading" of perfluoroalkyl compounds (PFCs) through milk. Later-born offspring receive fewer PFCs from their mothers (Houde *et al.*, 2006).

## VI. Sex-Biased Investment

Several pinniped studies have investigated whether, consistent with the predictions of Trivers and Willard (1973), mothers in good condition preferentially invest in sons over daughters. According to the Trivers–Willard model, in polygynous species, maternal investment is likely to have marginally greater genetic payoff in terms of grand-offspring if the mother can influence offspring quality and thus produce a "super-son" compared to a "super-daughter" because males have greater variance in reproductive success than females. This depends on whether mothers can confer benefits, such as increased body size, to their offspring. Biased investment generally takes two forms: sex ratio and differential investment or maternal care.

Data on biased maternal care or sex ratios are equivocal for both phocids and otariids. The sex ratio does not covary with maternal mass, a good indicator of maternal condition. For example, Northern elephant seal (*M. angustirostris*) males are born heavier than females and young mothers are less likely to successfully wean sons compared to daughters; however, there is little evidence that mothers differentially invest in sons compared to daughters. Males gain most of their size during the 3- to 5-year growth spurt, long after weaning. Maternal expenditure may not contribute significantly to male reproductive success. Gray seal (*Halichoerus grypus*) data are similarly equivocal. Otariid males are born slightly heavier and appear to grow faster than female pups, but none of the pinniped data provides clear evidence that sons exact greater reproductive costs on mothers than daughters.

In cetaceans, much less is known about biases in either sex ratio or maternal care. However, one study found that humpback whales were more likely to give birth to sons than daughters following the close of a long (3-year) birth interval compared to a short (1- to 2-year) one, although they hypothesized that as the mothers were in better condition after a 3-year interval, they were thus more likely to give birth to sons. However, the actual condition of mothers is unknown. Notably, adult female baleen whales are slightly larger than adult males. It thus remains unclear what benefit mothers in good condition might gain by biasing the sex ratio toward sons if larger body size relative to females or other males is unlikely to give them an advantage. Available growth data suggest that weanling males are slightly smaller than weanling females. Other potential cases supporting Trivers and Willard are findings suggesting that male teenage sperm and short-finned pilot whales (*Globicephala macrorhynchus*) are occasionally still nursing, but female offspring nurse no later than age 7. Because both species are markedly sexually dimorphic, these examples may be more likely to fit the Trivers–Willard model.

## VII.  Mother–Offspring Conflict and Weaning

In a number of pinnipeds, weaning can be initiated by either mother or pup. Otariid pups may leave the rookery while their mother continues to return from foraging trips. Phocid pups are typically weaned by the mother. She leaves the rookery permanently. Likely, the proximate cue for both the mother's and the pup's departure from land or ice is hunger. The mother leaves when her fat stores are depleted. The pup leaves when it is no longer receiving enough milk to sustain itself. Among some otariids, such as the Galapagos fur seal, the older sibling (1–2 years) may compete with the newborn for milk, providing one of the few examples of clear sibling competition among marine mammals. The size of the older sibling appears to influence the degree of competition with the younger sibling.

Little is known about the weaning process in whales or dolphins. Weaning may be fairly sudden in some baleen whales that separate feeding and breeding. Some baleen species wean during the migration to higher latitudes or soon after, but other species wean toward the end of the first year. Weaning in toothed whales is likely more gradual than in mysticetes. Evidence for mother–calf conflict comes from studies of Indian Ocean bottlenose dolphins in Shark Bay and southern right whales in Patagonia. At both sites, mothers are less responsible for maintaining proximity to their calves as they age and mothers appear to use the sea floor to prevent calves from nursing. A right whale mother may also roll belly up to prevent her calf from nursing.

## VIII.  Association Postweaning

Like humans and a few other mammals, mother and offspring may continue to associate postweaning and sometimes for life. Evidence for strong mother–offspring association in pinnipeds is weak. Phocids wean abruptly and separate. Among otariids, some postweaning association is possible but difficult to study. Many otariid and walrus females stay in or return to the same breeding areas, and thus may well recognize each other and interact. Walrus males are known to return to their natal area in adulthood and might encounter maternal relatives.

However, among cetaceans, high group stability and patterns of preferential mother–daughter association in fission–fusion societies (fluid patterns of group association, often with some underlying stability) indicate that strong bonds may persist. For example, resident killer whale male and female offspring remain with their mothers for life, a pattern unique to this species. Pilot whale (*Globicephala* spp.) daughters remain with their mothers postweaning, but sons may disperse temporarily or permanently. Sperm whale daughters remain in their matrilineal unit for life, much like elephants. Bottlenose dolphin daughters associate with mothers more postweaning than sons. In contrast, few daughter or son associations postweaning persist in mysticete whales. Among delphinids, the significance of such long-term kin associations is not known, but given the diversity of prey, the complexity of some delphinid social relationships, and other selective pressures on odontocetes, continued association is likely to offer the benefits of group living. Such benefits include protection from predators or conspecifics and sharing of information and/or tasks, such as calf care.

## See Also the Following Articles

Breeding Sites ▪ Aggressive Behavior ▪ Mimicry ▪ Reproductive Behavior

## References

Bachman, K. C., and Irvine, A. B. (1979). Composition of milk from the Florida manatee *Trichechus manatus latirostris*. *Comp. Biochem. Physiol.* **62A**, 873–878.

Bernard, H. J., and Hohn, A. A. (1989). Differences in feeding habits between pregnant and lactating spotted dolphins (*Stenella attenuata*). *J. Mammal.* **70**, 211–215.

Black, N., Ternullo, R., and Schulman-Janiger, A. (1999). Behavior and ecology of transient kills in Monterey Bay, California. 13th Bienniel Conference on the Biology of Marine Mammals (Abstract, pp. 17), Wailea Maui, Hawaii.

Caro, T. M., and Hauser, M. D. (1992). Is there teaching in nonhuman animals? *Q. Rev. Biol.* **67**, 151–174.

Clutton-Brock, T. H. (1991). "The Evolution of Parental Care." Princeton University Press, Princeton, NJ.

Costa, D. P., and Crocker, D. E. (1999). Seals. In "Encyclopedia of Reproduction." Academic Press, San Diego, CA.

Estes, J. A., Riedman, M. L., Staedler, M. M., Tinker, M. T., and Lyon, B. E. (2003). Individual variation in prey selection by sea otters: patterns, causes and implications. *J. Anim. Ecol.* **72**, 144–155.

Gazo, M., Aparicio, F., Cedenilla, M. A., Layna, J. F., and Gonzalez, L. M. (2000). Pup survival in the Mediterranean monk seal (*Monachus monachus*) colony at Cabo Blanco Peninsula (Western Sahara-Mauritania). *Mar. Mamm. Sci.* **16**, 158–168.

Georges, J. Y., and Guinet, C. (2000). Maternal care in the subantarctic fur seals on Amsterdam Island. *Ecology* **81**, 295–308.

George, J. C., Bada, J., Zeh, J., Scott, L., Brown, S. E., O'Hara, T., and Suydam, R. (1999). Age and growth estimates of bowhead whales (*Balaena mysticetus*) via aspartic acid racemization. *Can. J. Zool.* **77**, 571–580.

Gibson, Q. A., and Mann, J. (2008). The size and composition of wild bottlenose dolphin (*Tursiops* sp.) mother-calf groups in Shark Bay, Australia. *Anim. Behav.* **76**, 389–405.

Houde, M., Balmer, B. C., Brandsma, S., Wells, R. S., Rowles, T. K., Solomon, K. R., and Muir, D. C. G. (2006). Perfluoroalkyl compounds in relation to life-history and reproductive parameters in bottlenose dolphins (*Tursiops truncatus*) from Sarasota Bay, Florida, USA. *Environ. Toxicol. Chem.* **25**, 2405–2412.

Janik, V. M., and Slater, P. J. B. (1997). Vocal learning in mammals. *Adv. Study Behav.* **36**, 59–99.

Jenness, R., Williams, T. D., and Mullin, R. J. (1981). Composition of milk of the sea otter *Enhydra lutris*. *Comp. Biochem. Physiol. A.* **70A**, 375–379.

P

Koelsch, J. K. (2001). Reproduction in female manatees observed in Sarasota Bay, Florida. *Mar. Mamm. Sci.* **17**, 331–342.

Le Boeuf, B. J., Condit, R., and Reiter, J. (1989). Parental investment and the secondary sex ratio in Northern elephant seals. *Behav. Ecol. Sociobiol.* **25**, 109–117.

Le Boeuf, B. J., and Laws, R. M. (1994). "Elephant Seals: Population, Ecology, Behavior and Physiology." University of California Press, Berkeley, CA.

Lyons, K. J., and Estes, J. A. (1989). Individual variation in diet and the question of optimal feeding behavior in the female California sea otter. 8th Biennial Conference on the Biology of Marine Mammals (Abstract, pp. 40), Pacific Grove, CA.

Mann, J., and Barnett, H. (1999). Lethal tiger shark (*Galeocerdo cuvier*) attack on bottlenose dolphin (*Tursiops* sp.) calf: defense and reactions by the mother. *Mar. Mamm. Sci.* **15**, 568–575.

Mann, J., Connor, R. C., Barre, L. M., and Heithaus, M. R. (2000). Female reproductive success in wild bottlenose dolphins (*Tursiops* sp.): life history, habitat, provisioning, and group-size effects. *Behav. Ecol.* **11**, 210–219.

Mann, J., and Smuts, B. B. (1998). Natal attraction: allomaternal care and mother–infant separations in wild bottlenose dolphins. *Anim. Behav.* **55**, 1097–1113.

Mann, J., and Smuts, B. B. (1999). Behavioral development in wild bottlenose dolphin newborns (*Tursiops* sp.). *Behaviour* **136**, 529–566.

Mann, J., and Sargeant, B. (2003). Like mother, like calf: the ontogeny of foraging traditions in wild Indian Ocean bottlenose dolphins (*Tursiops* sp.). *In* "The Biology of Traditions: Models and Evidence" (D. Fragaszy, and S. Perry, eds), pp. 236–266. Cambridge University Press, Cambridge.

Mehta, A. V., *et al*. (20 authors) (2007). Baleen whales are not important as prey for killer whales *Orcinus orca* in high-latitude regions. *Mar. Ecol. Prog. Ser.* **348**, 297–307.

Marsh, H., Hensohn, G. E., and Marsh, L. M. (1984). Breeding cycle, life history and population dynamics of the dugong, *Dugong dugon* (Sirenia: Dungongidae). *Aust. J. Zool.* **32**, 767–788.

McShane, L. J., Estes, J. A., Riedman, M. L., and Staedler, M. M. (1995). Repertoire, structure, and individual variation of vocalizations in the sea otter. *J. Mammal.* **76**, 414–427.

Oftedal, O. T. (1993). The adaptation of milk secretion to the constraints of fasting bears, seals, and baleen whales. *J. Dairy Set.* **76**, 3234–3246.

Oftedal, O. T. (1997). Lactation in whales and dolphins: evidence of divergence between baleen- and toothed-species. *J. Mammary Gland Biol. Neoplasia* **2**, 205–230.

Oftedal, O. T., Bowen, W. D., and Boness, D. J. (1993). Energy transfer by lactating hooded seals, nutrient deposition in their pups, during the four days from birth to weaning. *Physiol. Zool.* **66**, 412–435.

Reidman, M. L., Estes, J. A., Staedler, M. M., Giles, A. A., and Carlson, D. R. (1994). Breeding patterns and reproductive success of California sea otters. *J. Wildl. Manage.* **58**, 391–399.

Reich, C. M., and Arnould, J. P. Y. (2007). Evolution of pinnipedia lactation strategies: a potential role for α-lactalbumin? *Biol. Lett.* **3**, 546–549.

Rugh, D. J., Miller, G. W., Withrow, D. E., and Koski, W. R. (1992). Calving intervals of bowhead whales established through photographic identifications. *J. Mammal.* **73**, 487–490.

Sayigh, L. S., Tyack, P. L., Wells, R. S., and Scott, M. D. (1990). Signature whistles of free-ranging bottlenose dolphins *Tursiops truncatus*: stability of mother–offspring comparisons. *Behav. Ecol. Sociobiol.* **26**, 247–260.

Schulz, T. M., and Bowen, W. D. (2005). The evolution of lactation strategies in pinnipeds: a phylogenetic analysis. *Ecol. Monogr.* **75**, 159–177.

Stewart, R. E. A., Outridge, P. M., and Stern, R. A. (2003). Walrus life-history movements reconstructed from lead isotopes in annual layers of teeth. *Mar. Mamm. Sci.* **19**, 806–818.

Smolker, R. A., Mann, J., and Smuts, B. B. (1993). The use of signature whistles during separations and reunions among wild bottlenose dolphin mothers and calves. *Behav. Ecol. Sociobiol.* **33**, 393–402.

Smolker, R. A., Richards, A. F., Connor, R. C., Mann, J., and Berggren, P. (1997). Sponge-carrying by Indian Ocean bottlenose dolphins: possible tool-use by a delphinid. *Ethology* **103**, 454–465.

Stirling, I. (2002). Polar bears and seals in the Eastern Beaufort Sea and Amundsen Gulf: a synthesis of population trends and ecological relationships over three decades. *Arctic* **55**, 59–76.

Trillmich, F. (1996). Parental investment in pinnipeds. *In* "Parental Care: Evolution, Mechanisms and Adaptive Significance" (J. S. Rosenblatt, and C. T. Snowdon, eds), pp. 533–577. Academic Press, San Diego, CA.

Trillmich, F., and Weissing, F. J. (2006). Lactation patterns of pinnipeds are not explained by optimization of maternal energy delivery rates. *Behav. Ecol. Sociobiol.* **60**, 137–149.

Trivers, R. L., and Willard, D. E. (1973). Natural selection of parental ability to vary sex ratio offspring. *Science* **179**, 90–92.

Valsecchi, E., Hale, P., Corkeron, P., and Amos, W. (2002). Social structure in migrating humpback whales (*Megaptera novaeangliae*). *Mol. Ecol.* **11**, 507–518.

Watt, J. (1993). Ontogeny of hunting behaviour of otters (*Lutra lutra* L.) in a marine environment. *Symp. Zool. Soc. Lond.* **65**, 87–104.

Whitehead, H., and Mann, J. (2000). Female reproductive strategies of cetaceans: life histories and calf care. *In* "Cetacean Societies: Field Studies of Dolphins and Whales" (J. Mann, R. C. Connor, P. L. Tyack, and H. Whitehead, eds), pp. 219–246. University of Chicago Press, Chicago, IL.

Würsig, B., and Clark, C. (1993). Behavior. *In* The Bowhead Whale." (J. Burns, J. Montague, and C. Cowles, eds.), Chap. 5, pp. 157–199. Special Pub. No. 2, Society for Marine Mammalogy, Allen Press, Lawrence, Kans, USA.

Würsig, B., Dorsey, E., Fraker, M., Payne, R., and Richardson, J. (1985). Behavior of bowhead whales, *Balaena mysticetus*, summering in the Beaufort sea: a description. *Fish. Bull.* **83**, 357–377.

Würsig, B., Koski, W. R., and Richardson, W. J. (1999). Whale riding behavior: assisted transport for bowhead whale calves during spring migration in the Alaska Beaufort Sea. *Mar. Mamm. Sci.* **15**, 204–210.

# Pathology

## Daniel F. Cowan

## I. Introduction

This article focuses on naturally occurring diseases of marine mammals, and how their behavioral, physiologic, and anatomical adaptations to life spent mainly or entirely in the water influence those disease processes. Pathology is the study of diseases, including their mechanisms, manifestations, and diagnosis. One of the fundamental principles of pathology is that *every disease is a reaction to injury*. This means that an organism responds to an injury within its anatomical and physiological capacity, and that the way the organism responds, that is, the way the disease process shows itself, will be determined or at least influenced by the organism's adaptation to its environment. Thus, a fish can develop an infectious disease of the gills, but cannot get pneumonia, a disease of the lungs. This seems obvious enough, but it may easily be forgotten when making assumptions about marine mammals. For example, the presence or absence of a thick coat of hair makes a difference in the way mercury is handled by an animal. In hairy species, a large portion of the mercury burden accumulated in the diet is eliminated by binding to hair, which is later shed. Cetaceans, however, which may be exposed to at least as much environmental mercury as their hairy fellow marine mammals, the pinnipeds, otters and polar bears, have no hair to bind mercury, and they must find some other way of dealing with it. Their protective mechanism appears to be to combine it with selenium, rendering it insoluble and inert, and tucking it away in storage sites. This seems to serve the cetacean well, but it can be totally misleading to the researcher who finds large tissue burdens, even "lethal levels" of mercury in a healthy dolphin, and immediately casts about for a polluter to blame.

"Injury" is understood broadly to mean any noxious influence, which may include physical trauma; action of physical agents such as heat and cold; infection; intoxication; metabolic disease; nutritional deficiency; genetic disorders and developmental malformations. To this list some would add "stress," the often (but not always) injurious effects of the attempts of an organism to adapt to environmental influences not directly injurious to themselves. This idea allows us to think of concepts of, for example, deleterious effects of excess population density (crowding), even in the presence of an adequate food supply. We can also think of an animal living at the margin of its nutritional support, with enough food available to support day-to-day functions, but unable to respond to an event, such as pregnancy, exposure to unaccustomed cold, and exposure to toxins, among many others, that places an increased energy demand on it. In such circumstances, the factors that comprise immune resistance may be too feeble to prevent heavy parasitic infestation or infection. With this perspective, we can readily understand that the environment in which an organism lives has a profound, if not determining influence on the things that make the organism ill, or cause it to die. By studying the specific diseases and patterns of occurrences of diseases in a group of animals we can gain insight not only into the hazards of the environment, but also the basic physiology of the organism. We also have to consider that animals differ even within species, and that they may differ in their reactions because of experience, so that two animals that seem nearly identical might exhibit wide differences in their response to apparently identical environmental situations or stressors.

Since environments vary, it may be expected that patterns of disease might vary among populations of a single species of animal living in different places, and since physiology and anatomy vary, patterns of disease may also vary among different species living in the same environment. Indeed, this phenomenon is well known in the world of experience. Human populations living in temperate climates and industrialized societies have very different patterns of disease than do people living in tropical agricultural societies, even though both populations would be susceptible to the same noxious agents if only they were exposed to them. Conversely, dogs, cats, and humans all living in the same house will have their own separate infectious diseases, and only rarely do they infect each other.

From the foregoing, it can be seen that study of the diseases of wild marine mammals can offer insights into anatomy, physiology, and environment that might be gained in no other way. Disease represents an interaction between an individual animal and its environment, which demands an understanding of complex, often intricate processes involving several different organisms. For example, it is not enough to simply measure the level of a toxin in water to gauge its effects on animals living in the water, since the toxin itself can be changed not only by the presence of other chemicals in the water, or by bacterial action, but also by sequential processing through metabolic systems of different organisms, and ultimately by the physiology and chemistry of the marine mammal itself. Observations indicate, for example, that adaptation to environmental mercury over millennia makes it possible for dolphins to tolerate tissue burdens of mercury that would be fatal to cattle (Turnbull *et al.*, 1998). Animals with high tissue mercury burdens may show no sign of the tissue lesions associated with mercury toxicity in land animals (Siebert *et al.*, 1999). Diseases may appear as a secondary effect of some primary phenomenon; for example, a population with inadequate food supplies may become debilitated, with lowered resistance, and so be overwhelmed by parasites that ordinarily are held in check. In this instance, the load of parasites is obvious, the lowered resistance to parasitism may be inferred, and the lack of appropriate nutrition recognized by other factors, such as body condition, serum protein levels, and measurements of specific nutrients in body fluids.

It is remarkable that until relatively recent times, almost nothing was known about the diseases of free-ranging marine mammals, and even today not much is known about the diseases of sea otters, polar bears, manatees, and walruses in the wild. Even though many tens of thousands of large whales were taken in the whaling industry, only a few reports of pathology resulted. Simpson and Gardner (1972) compiled a detailed discussion of the histology and histopathology of marine mammals, based mainly on their own experience and an appraisal of the scattered reports then available in the literature. This work remained the standard resource and reference for two decades and is still very valuable. For practical purposes, systematic study of the pathology of marine mammals in the wild is limited to pinnipeds and dolphins, and to a lesser degree manatees, and did not begin until the middle 1960s. The current literature contains many reports of findings in single animals, or small groups of animals, but relatively few investigations resulting from long term, detailed studies. The state of the art and science of pathology of marine mammals is such that new and important observations may still be made using dissection and light microscopy, techniques developed and applied to many species in the nineteenth century, supported by the latest developments in molecular biology and molecular diagnostics.

## II. Sources of information

All marine mammals are protected by law in most advanced countries with notable exceptions being Canada and Japan, which allow commercial hunting of some species. In countries with protective laws, access to marine mammals is carefully limited, and experimentation, with rare exceptions is prohibited. This means that apart from animals maintained in marine aquaria and under the care of specialized veterinarians, all information is derived from animals that strand on beaches, are accidentally killed as by-catch in the fishing industry, or taken by approved subsistence hunters, who may make arrangements with researchers for access. Not much scientific use is made of animals in commercial hunting operations. Apart from the occasional stranding near a population center, or accidental death associated with fishing, it is rare for an animal to be found in an undecomposed condition by a trained observer with appropriate equipment and supplies. Therefore, what we know is based mainly on case reports, often incomplete, and a few studies involving at most a few dozen animals, often of several different species. Stranding and rehabilitation centers are the main source of information. These obviously can describe only what they see, which is mainly coastal species and the occasional pelagic animal that is cast up on the beach.

Despite these limitations, observations made in stranding centers may be the source of powerful insights into the state of the free-ranging, otherwise inaccessible wild populations. Phocine distemper virus (PDV), a morbillivirus responsible for the deaths of at

P

least 18,000 harbor seals in Europe in 1988 was first isolated from stranded seals in that year. Phocine herpesvirus (PhHV1) was identified in 1985. The highly toxic domoic acid, a product of the diatom *Pseudo-nitzschia australis*, was recognized as a cause of death in over 400 California sea lions (*Zalophus californianus*) examined in a rehabilitation center in 1998.

It is not clear just how representative beach-stranded animals are of the population at large, but for the time being this issue is of lesser importance than simply building a reliable database of descriptions of pathological findings, and relating these to age, sex, reproductive status, incidence and type of trauma sustained, identification of infectious diseases, toxin and parasite burdens, etc. Continuing, detailed pathological studies of stranded dolphins are currently taking place in a number of areas, most notably on the North Sea coast of Germany, the St. Lawrence River and estuary in Quebec, coastal New Zealand, the United Kingdom, Italy, Peru, the Canary Islands, and the Texas and Florida coasts of the Gulf of Mexico, among other locations. Pinnipeds are studied in central California, and manatees in Florida. Important but noncontinuing studies of strandings have been done in southern California, in which the focus was solitary strandings, and in the eastern United States, involving mass strandings. Mortality events in which unusually large numbers of deaths of marine mammals, sometimes accompanied by substantial mortality of fish, sea birds and turtles, occur over a period of days to months, are under a special program conducted by the US National Marine Fisheries Service.

### III. Differences between Pathology of Strandings and By-Caught Animals

Spontaneous strandings of single animals are believed to be caused by sickness or some other impairment, such as injury, and are therefore a selected and perhaps nonrepresentative element of the population from which they come, while by-catch animals are snatched from their daily lives, and may be presumed to be healthy, or at least representative of the general condition of the population. Diseases seen in these may be a better indicator of the disease status of the population, or, of disease in an early stage, while strandings might represent a late stage. This is mostly a presumption, based on logic rather than observation, although accumulating experience suggests it is true.

"By-catch" is the term used to refer to animals that are caught by accident during the course of trying to catch something else. In the case of marine mammals, it usually means entanglement of the animal in fishing nets. Commonly this involves one or a few animals, but in the early days of the purse-seine tuna fishery in the eastern tropical Pacific Ocean, it might mean hundreds or even thousands of animals. At least two episodic studies of pathology have been conducted on this group of animals, and a continuing study of by-catch is taking place in northern Germany.

### IV. Mass Strandings vs Mortality Events and Solitary Strandings

Mass strandings, the more or less sudden appearance on the beach of large numbers of whales or dolphins remains a problem to explain. Various theories have been put forth, such as bad luck of animals swimming inshore, in being caught by an ebbing tide. This presupposes several things; complex bottom topography, a gently sloping bottom, a rapidly moving tide, failure of echolocation in shallow water, interference in echolocation by extraneous noise, and

perhaps distraction of attention by feeding. Other theories implicate an impaired leader who misjudges the tide and leads the herd into a strand, or more likely, fails to lead them away from the beach at the right time. One or two observers even see similarities to lemming migrations, and postulate a mass suicide. All of these theories lack supporting experimental evidence. What is known from the few pathological studies of mass stranded animals is that substantial disease is found, mainly parasitism, which would have been present for some time. Some of this may have been severe, but most is of a tolerable level. If the disease causing the strand was present yesterday or the day before, why did they strand today?

Geraci and St. Aubin (1977) examined the naturally occurring diseases in 41 of a herd of approximately 150 stranded Atlantic white-sided dolphins, *Lagenorhynchus obliquidens acutus*. The most prevalent lesions were associated with parasites; mastitis associated with the nematode *Crassicauda grampicola*, and biliary and pancreatic fibrosis associated with biliary flukes *Oschmarinella laevicaecum*. In these conditions, the parasite enters the duct system and provokes inflammation and scarring. Other parasites commonly found were *Phyllobothrium delphini* and *Monorygma grimaldi* cysts in the blubber and abdomen, *Crassicauda* sp in the fascia, all of which were encysted, and long-term infestations. A roundworm, *Stenurus globicephalae* was present in the cranial air sinuses and lungs, where it caused minor inflammation, and the tapeworm *Tetrabothrius forsteri* was found in the duodenum. These findings are all typical of wild dolphins in the North Atlantic, and cannot be invoked to explain the stranding. This is generally the story of mass strandings; many clues, but little definitive evidence.

"Unusual mortality event" is the term applied to excess deaths (over that known to occur in an average period of time) in a relatively limited geographic area. The difference between a mass stranding and a mass mortality is the time scale and circumstances. The mass stranding is an event of a day or so, while the mass mortality evolves slowly over days, weeks, or months. In contrast to mass strandings, the animals involved in a mortality event are very likely to suffer from infectious disease (Schulman *et al.*, 1997) affecting some segment of the local population, or a toxic event such as the brevitoxicosis associated with red tide (Bossart *et al.*, 1998), or domoic acid toxicity associated with bloom of a particular diatom (Scholin *et al.*, 2000). One obvious difference between an infection and an environmental toxicosis is that the infection, typically viral, tends to be limited to one kind of animal, while the toxicosis may affect mammals, birds, fish, and turtles alike. Epizootic infection can occur among terrestrial animals as well, but an environmental bio-toxicosis is a uniquely aquatic event.

Solitary strandings of dolphins, which may be taken as representative of coastal marine mammal strandings, reflect a variety of causes, from trauma (boat strikes, intra-specific and inter-specific aggression, accidents, and a wide variety of disease causes (Haubold *et al.*, 1999). This tends to apply more to juvenile and adult animals, while young calves seem to strand more for social reasons, such as separation from the mother, and although debilitated, are less likely to be sick. This observation does not apply to pinniped pups, however, which tend to be suffering from infection, septicemia, dehydration, starvation, or trauma when found stranded.

### V. Parasitism

Infestation by parasites is nearly universal in wild animals, and marine mammals are no exception. Indeed, the parasites found in pinnipeds and cetaceans are so widespread, and bear such a consistent relationship to their hosts and their environments that they have been used as "tags" to study specific mammal populations (Dailey and Otto, 1982). Parasitism is probably the best recognized disease factor in free-ranging marine mammals, and the variety of lesions caused may be illustrated by a few examples.

Parasitism is by far the leading cause of pathology associated with stranding in wild pinnipeds and cetaceans, affecting the brain and lungs as well as the gastrointestinal tract, liver, and pancreas. Large (baleen) whales may suffer kidney damage from *Crassicauda*, and seals may carry heavy, often fatal burdens of heartworm. While parasites occur in other sites, such as in the blubber and under the peritoneal membranes, they are not usually of any particular significance, and apparently are not an important mortality factor.

The great majority of small cetaceans, approaching 100%, have nematode lung worms. These seem to be relatively innocuous as long as they remain in the airways, but provoke an inflammatory reaction on entering the alveolar spaces, the distal sites of air exchange. These delicate tissues lack the defenses of the airways, which are covered by a cell layer adapted to a passageway, and have a thin coat of protective mucin. The parasite in the alveolus, perhaps associated with bacteria, provokes the formation of small abscesses, which rarely may rupture into the pleural space, but typically subside into fibrous nodules (granulomas), which may mineralize. At this stage, they are inert, and probably of no further consequence, except as a marker of past events. In some delphinids, notably the harbor porpoise, *Phocoena phocoena*, in the North Sea, the airways may be filled to occlusion by large nematodes of several species (Siebert *et al.*, 1995).

The effect of parasites in the lungs illustrates very well the influence of adaptation of the lung to diving on the disease process. The cetacean lung differs from the lung of a typical land mammal in that the diving mammal has structural adaptations designed to keep air in the gas exchange spaces, the alveoli, while also tending to eliminate or reduce "dead space." Dead space is the functional compartment of the lung in which gas exchange does not take place; for practical purposes, the airways. In diving, compression of the animal by surrounding water, which increases with depth, forces incompressible blood into any compressible space. This is familiar to human divers as the "squeeze effect." In the lung, the alveoli can be collapsed by compression of the chest wall, but the airways are held rigidly open by cartilage bars. It appears that reduction of airway dead space, and prevention of forcing of alveolar air back into airways is accomplished by the action of a series of muscular valves or sphincters within the airway. All of this extra tissue (compared with the lung of a terrestrial mammal) makes the cetacean lung dense and heavy. When an agent such as a lungworm, bacteria, or aspirated contaminated water reaches the distal airways and provokes an inflammatory reaction, it appears that the muscular sphincters go into spasm, preventing movement of air, and evacuation of secretions.

The effect of the peculiar anatomy of the cetacean lung is to cause nearly all infections to form abscesses with focal destruction of tissue. The density of the lung tissues in general prevent spread of the inflammation beyond the local focus. In contrast, bacterial infections of the terrestrial mammalian lung may resolve without destruction of tissue, but may spread in an unconfined fashion to involve large regions of the lung.

Delphinids of many species harbor flukes of the genus *Nasitrema* in the pterygoid air sinuses of the skull. These sinuses connect with the external environment by way of the nasal passages, and by way of the eustachian tubes, with the specialized structures of the ear, the ossicles or acoustic bullae. The bullae are connected to the brain by the statico-acoustic nerve, concerned with both balance and hearing, which passes into the skull within a dense fibrous sheath. Characteristics of the anatomy of the air sinus, the acoustic bullae, and the statico-acoustic nerve permit invasion by the worm into the subarachnoid space of the brain, in which the spinal fluid circulates. Taking advantage of this space, they migrate over the surface of the brain, until they reach a point of penetration, at which they burrow through the cortex or gray matter deep into the white matter, laying large numbers of eggs (ova). This migration produces destructive tracts or galleries in the white matter, with hemorrhage and necrosis. It is not clear what induces the migration of the parasite into the brain, as it is a dead-end for the worm as well as damaging to the host. It is a complex problem, seen mainly in beach-stranded dolphins in both the Atlantic and the Pacific Oceans. Dolphins of the same species caught as by-catch in the eastern tropical Pacific tuna fishery also have infestations of the same fluke in the air sinuses, but apparently without invasion of the nervous system (Walker and Cowan, 1981). Why this difference should occur is not known. It is possible that a third factor is involved, such as some agent in the in-shore water that influences the worm; that is, makes it "sick" and disoriented. There is clearly a differential species susceptibility among dolphins, as the air sinus infestation occurs with great frequency in stranded Atlantic bottlenose dolphins, *Tursiops truncatus*, but without nervous system invasion.

Flukes of several species (*Campula rochebruni, C. oblongata*, among others) infest the bile duct and pancreatic duct of dolphins and porpoises. The effect may be a relatively low-grade irritation or inflammatory reaction ranging up to chronic active inflammation with fibrosis and occlusion of the duct. This results in inflammation and fibrosis of the affected organs, a form of hepatitis (cholangitis) and pancreatitis. Serious disease is relatively uncommon, but may be life-threatening when it occurs.

The stomach and intestine of marine mammals are the frequent site of infestation with nematodes, trematodes, and cestodes. A light infestation may be relatively innocuous, but occasionally the stomach or intestine may be perforated by parasites, resulting in peritonitis. In some geographic areas, gastric ulcers associated with the nematode *Anisakis simplex* are common in stranded cetaceans (Abollo *et al.*, 1998). Sea lions may carry heavy burdens of hookworm.

Massive infestation of the heart and great vessels of elephant seals by nematodes (*Otostrongylus circumlitus*) is described from strandings on the central California coast (Gulland *et al.*, 1997). Heavy burdens in the right atrium, right ventricle, and pulmonary artery were associated with pulmonary thromboembolism (blood clots in the lungs) and pulmonary arteritis, an important cause of mortality. Death of infested juvenile elephant seals before the parasite reaches maturity suggests that the host–parasite association is relatively recent. Similar heartworm infestation occurs in seals in the Atlantic Ocean. This disease is very much like the heartworm infestation of dogs seen in parts of the United States.

## VI. Neoplasia

"Neoplasia" is the process of the formation of autonomous new growths in a tissue. The mass of tissue formed is called a tumor. Some of these are benign, meaning limited in capacity to harm, and some are malignant—cancer—having the ability to invade adjacent tissues, and to set up colonies in remote organs and tissues. Some tumors have been shown to be caused by viruses, some by chemicals, and a rare few are associated with parasitic infestation. For most, a definitive "cause" is not known; we can only speak of "associations."

P

Once thought to be rare in marine mammals, neoplasms were found to occur at an incidence of 2.5% in a large surveyed population of marine mammals (Howard *et al.*, 1983b). As of 1987, there were probably only 41 confirmable reports of tumors in cetaceans. Then, in 1994, 21 additional tumors, some benign and some malignant, were reported in 12 of 24 animals from the small, isolated, and highly contaminated population of beluga whales (*Delphinapterus leucas*) in the St. Lawrence River, in Quebec (De Guise *et al.*, 1994). Seven of these animals had more than one tumor; one had as many as three tumors, two malignant. One was as young as 1.5 years, another 3.5 years. The ages ranged up to over 29 years. This particular environment, the estuary of the St. Lawrence River, suffers heavy industrial pollution, with the waters containing polycyclic aromatic hydrocarbons (PAH). High concentrations of organochlorines, heavy metals, and benzo-*a*-pyrene were found in the tissues of the belugas. Thirty-seven percent of all tumors reported from cetaceans to that time were found in this small population of belugas, suggesting the direct carcinogenic effect of the pollutants, or an impairment of resistance to the development of tumors.

Gulland *et al.* (1996) reported an incidence of 66 transitional cell carcinomas in a population of 370 (18%) California sea lions examined over a 15-year period. Transitional cell carcinomas are a particular type of cancer that arise from the lining epithelium or membrane of part of the urogenital tract. The original site and cause of this extraordinary incidence of a particular tumor type was a mystery for a number of years, owing to the advanced state of the disease when recognized. Typically, it had spread to involve most of the lower abdominal lymph nodes and viscera, and the primary site could not be determined. Environmental pollution with a variety of industrial chemicals was speculated to be the cause. Recently, using a variety of modern analytical techniques, strong evidence has been found that the cause of the malignancy is infection with a gamma herpesvirus, and not chemical pollution (Lipscomb *et al.*, 1999). This virus is implicated in the etiology of several animal cancers.

These two studies are very instructive, in that they emphasize that a presumed low incidence of tumors in wild populations may be merely an artifact of not looking. When populations of wild animals are studied carefully over a long time clusters of disease may be revealed. They also illustrate the value of modern technology in evaluating cause. In the case of the transitional cell carcinomas, the cause, a herpesvirus, is demonstrated as well as it can be, while the tumors in Belugas are associated with pollution on epidemiological grounds, but have not been conclusively proven to be caused by pollution.

## VII. Infectious Disease

In their life in the oceans, marine mammals are exposed to a very great array of infectious agents, including dozens, if not hundreds of species of bacteria, viruses, fungi, and protozoa. These are not randomly distributed, and so actual infection will vary with features such as location, water temperature, contamination from terrestrial sources, river effluents, food species, and exposure to other marine mammals. Most organisms are successfully resisted, and a state of more or less peaceful coexistence maintained. Occasionally, however, an individual's defenses are so weakened or breached that one of these microorganisms can gain effective entry and set up a disease state. Sometimes an otherwise perfectly healthy animal comes into contact with a particularly virulent strain of organism, or simply a novel one to which there is no natural or acquired immunity conferred by previous exposure. Under these circumstances, the organism can sweep through a

population, causing epizootic disease with high death rates. Some isolated, small populations may be seriously threatened with extinction by new infections.

Bacterial, viral, and mycotic (fungal) infections are important morbidity and mortality factors in marine mammals, often appearing in outbreaks, such as one that occurred in the endangered New Zealand sea lion (*Phocarctos hookeri*), involving both adults and pups (Gales *et al.*, 1999). At least 53% of the pups born during the 1998 breeding season (total number was not known) had died within the first 2 months of life, with acute necrotizing inflammation of the blood vessels of the skin and lungs in adults, and pneumonia and abscesses in the pups. The causative agent was determined to be a pleomorphic gram negative bacterium, most likely *Campylobacter* sp., an organism not previously associated with this kind of mortality.

Several viral diseases are well known among pinnipeds, including seal pox virus, and the San Miguel calcivirus, which produce contagious crusted lesions on the skin and oral mucosa. These are serious but not fatal diseases, much like human chicken pox. Seals, sea lions, and dolphins are known to be infected with herpesviruses, and dolphins are susceptible to a pox virus, which produces transient "pin hole" lesions of the skin. Rabies virus is uncommon, but has been observed in ringed seals. Herpesvirus is familiar to most people in the context of the infections known as shingles and cold sores. This virus occurs in many types and strains, some relatively innocuous, others capable of producing severe disease of many organs, including the brain, lungs, liver, and heart as well as the skin. Several outbreaks of herpes infection have been recognized among marine mammals, usually producing skin lesions and sometimes pneumonia, as well as the spectacular incidence of cancer in California pinnipeds.

Morbilliviruses, which comprise a large group of viruses that cause measles in humans, and distempers in dogs and many other species also produce respiratory disease, immune deficiency, and neurological injury in seals, dolphins, porpoises, and a wide range of other cetaceans.

Marine mammals are quite susceptible to infection with morbillivirus. Morbillivirus infections, which were not documented in aquatic mammals until 1988, have caused at least five epizootics in these species in recent years (Kennedy, 1998). Disease has been recognized in seals from Europe and Siberia, dolphins in the Mediterranean Sea, the northern Atlantic Ocean, and the Gulf of Mexico, and seropositivity, indicating previous exposure to the virus, has been found in many species of seal, dolphin, porpoise, and whale, from the Antarctic to the Arctic circle, in Florida manatees, and in polar bears in Russia and Canada. The disease in marine mammals is much like distemper in dogs, with destructive and inflammatory lesions in the brain, gastrointestinal tract, lungs, and lymph nodes, with immune suppression, and frequently superinfection of immune impaired animals with fungi. Lung lesions include broncho-interstitial pneumonia, with filling of alveolar spaces by exudates, hyaline membranes formed by protein exudate covering gas exchange surfaces, and hemorrhage. Brain lesions are typically in the form of nonsuppurative encephalitis (nonpus forming inflammation of the brain), with neuronal and glial necrosis, microgliosis (the brain equivalent of scar), and focal demyelination, or loss of the insulating covering of nerve processes. Necrosis of the cerebral cortex is also sometimes found (Kennedy, 1998).

It is not entirely clear whether the observed increase in cases of the morbillivirus distempers is due to actual spread of the infection, or improved case-finding permitted by clear descriptions of the lesions in the literature, accompanied by the development of advanced methods of laboratory diagnosis.

Many bacterial species are found in marine waters and may be recovered from marine mammals, either as primary pathogens, or as part of a complex normal flora. Bacterial infection is thought to be the main cause of disease and death in marine mammals, especially in captivity (Howard *et al.*, 1983a). Certain marine organisms are known to produce severe or fatal infections. These include the halophilic or salt-water *Vibrios* of which there are many species, and *Edwardsiella tarda*. These organisms are common in the marine environment, and are frequently encountered in cetaceans, and less often in pinnipeds. The exact means by which these organisms are acquired by marine mammals is not known, but experience with humans indicates that they can be directly inoculated, infect wounds, and be ingested with food items, such as shellfish. It is reasonable to assume these routes of entry in marine mammals as well.

The *Clostridia* are obligate anaerobic bacilli, ubiquitous in the environment in soil, sewage, marine sediments, decaying animals and plant products, and the intestinal tracts of many animals. More than 80 species are known. Some species are potent toxin producers, causing botulism and gas gangrene. Many species of *Clostridia* have been cultured from blood, lesions and intestinal tract of stranded dolphins in the Gulf of Mexico, but are less common in California pinnipeds. Some *Clostridia* may be merely part of the normal intestinal flora, while some species have been recovered from lungs in cases of pneumonia.

Many species of bacteria have been recovered from cultures of respiratory tract, kidney, and intestinal tract as well as lesions of captive and stranded marine mammals. The majority of these species are known to be pathogenic or potentially pathogenic. There may be a differential distribution of bacteria in different kinds of marine mammals, as in cetaceans *Vibrio*, *Clostridia*, *Pseudomonas*, and *Edwardsiella* tend to predominate, while in pinnipeds from the California coast, for example, the major pathogens encountered in sea lions, elephant seals, and harbor seals are organisms usually associated with the intestine, mainly *Escherichia coli*, *Klebsiella pneumoniae*, *K. oxytoca*, *Proteus spp.*, *Pseudomonas spp.*, *Enterococcus spp.*, and *Salmonella*. Leptospirosis occurs in harbor seals, California sea lions, and northern fur seals.

Brucella infection of the placenta with abortion has been reported in dolphins. The organism, *Brucella delphini*, appears to be readily transmissible among dolphins and has also been cultured from the lung of a bottlenose dolphin at necropsy. *Brucella* meningoencephalitis with hydrocephalus also has been recognized. Brucella infection occurs in other cetaceans and seals. A substantial percentage of marine mammal serum samples (about 30%) react positively on tests used to detect antibody to *Brucella spp.*, and a number of *Brucella* isolates have been obtained from marine mammals. However, only the one first designated *Brucella delphinus* has been associated with reproductive failure. Exactly how many species of marine Brucella exist is under investigation.

The mycobacteria include the organisms that cause tuberculosis and leprosy. They are hardy organisms, and may produce infection across a wide array of warm and cold blooded animals. *Mycobacterium marinum*, originally described from fish, was first recognized as a human pathogen in 1951. *M. marinum* infection has been transmitted to a handler by a dolphin bite. The lesions, when localized resemble abscesses, but are very slow to heal. Infection of deeper tissues and organs, such as heart valves, brain, eye, and joints are very serious. *Mycobacterium bovis*, the agent of bovine tuberculosis, was transmitted from seals to their trainer. It caused similar lesions in both. Six cases of tuberculosis were observed over a 4-year period in stranded sea lions and fur seals from the coast of Argentina

(Bernardelli *et al.*, 1996). Disease was widespread, involving lungs, lymph nodes, liver, spleen, pleura, and peritoneum. The lesions were typical granulomatous inflammation, from which organisms belonging to the *Mycobacterium tuberculosis* complex were isolated. This organism complex was similar to both *M. bovis*, the agent of bovine tuberculosis, and *M. tuberculosis*, the agent of human tuberculosis, but with enough molecular differences to be judged different from both. This suggests that the seals and sea lions did not acquire their infection from contact with humans or cattle. The organisms associated with the pinniped cases all had similar genetic features, suggesting that seal tuberculosis in that geographic area is caused by organisms belonging to a distinct grouping within the *M. tuberculosis* complex. An instance of disseminated *Mycobacterium chelonei* infection in a manatee (*Trichechus inunguis*) has been reported, and cases of cutaneous mycobacteriosis in a manatee and its handler have been attributed to *M. chelonei*. This organism was first identified in a reptile.

Marine mammals are subject to infection with a wide variety of fungi and filamentous bacteria (*actinomyces*, *nocardia*, *dermatophilus*, which produce lesions similar to those caused by fungi) (Migaki and Jones, 1983; Reidarson *et al.*, 1999). Some of these are opportunistic, meaning that they occur in the context of debility or other disease, and some are primary pathogens capable of initiating disease in healthy hosts. Both opportunists and pathogens are species that are familiar in human and veterinary medicine, and are not peculiar to the marine environment.

Animals suffering immune suppression from morbillivirus disease may suffer severe, disseminated fungal infection, frequently from molds such as *Aspergillus* spp., as a terminal event. These organisms may produce disease in one of the two patterns: superficial infections of skin which cause mild disease, or indolent, very chronic processes; and the deep or systemic mycoses, which produce severe pyogranulomatous lesions. A pyogranulomatous lesion is a mixture of granulomatous inflammation, which is produced by persistent infection, with a pus-forming acute inflammation, indicative of a more active injury. Deep infections may involve any organ, especially the lungs and respiratory tract, and the brain. They may produce bulky lesions, which displace normal tissues ("mass lesion") as well as destroy tissues. Some organisms tend to permeate tissues, and invade and block blood vessels.

Dermatophytoses, ringworm-like lesions, occur in manatees and pinnipeds, caused by the same organisms that cause ringworms in land animals and humans. Since infection with fungi requires spores from the environment, rather than the vegetative stages found in marine mammals, direct transmission from animal to animal seems unlikely. Lobo's disease, caused by a fungus *Loboa loboi* (syn. *Lacazia loboi*), is a very unusual disease recognized only in dolphins and humans, and no other species. One instance of transmission from dolphin to human has been reported. Lobo's disease (once called keloidal blastomycosis) is a skin infection producing chronic, treatment-resistant, thick nodular swellings of the superficial dermis and epidermis, occasionally with ulceration. While the lesions in man and dolphin are quite similar, there are subtle morphological differences in the organisms in the lesions, which may represent separate species (Haubold *et al.*, 2000). In man, Lobo's disease is a disease of the Central and South American tropics, while in dolphins it ranges from the Gulf of Mexico, mainly Florida, to South America.

A number of protozoal infections, similar to those seen in land animals are known to occur in marine mammals. These include toxoplasmosis, infection with *Toxoplasma gondii*, found in a West Indian manatee, pinnipeds, stranded Atlantic bottlenose dolphins, *Tursiops truncatus*, a Pacific spinner dolphin, *Stenella longirostris*, Beluga whales, *Delphinapterus leucas*, and a killer whale, *Orcinus orca*. The animals were found both in Atlantic and Pacific Oceans, were captive and free-ranging, and the lesions ranged from incidental to disseminated and fatal. The mode of transmission in these animals is not known. *Cryptosporidium* morphologically, immunologically, and genetically indistinguishable from those obtained from infected calves have been recovered from feces of California sea lions (*Zalophus californianus*) suggesting that the sea lion could serve as a reservoir for environmental transmission of this organism. Giardia have been isolated from fecal material from harp seals (*Phoca groenlandica*), gray seals (*Halichoerus gryphus*), and harbor seals (*Phoca vitulina*) in eastern Canada waters and ringed seals (*Phoca hispida*) from western arctic Canada, and from California sea lions in northern coastal California. The widespread occurrence of these organisms in both terrestrial and marine mammals suggests transmission via rivers and streams, and perhaps carrier species taken as food.

## VIII. Chronic Diseases of Undetermined Cause

Several chronic diseases of unknown cause have been recognized in dolphins and small whales stranding along the Texas coast of the Gulf of Mexico. These are arthritis, pulmonary angiomatosis, and amyloidosis.

A high incidence of arthritis of the synovial joints (neck and flipper joints) was found in several cetacean species, including 10 of 49 (20%) bottlenose dolphins (*T. truncatus*), 1 of 2 (50%) striped dolphins (*Stenella attenuata*), one Fraser's dolphin (*Lagenodelphis hosei*), and one pygmy sperm whale (*Kogia breviceps*) (Turnbull and Cowan, 1999a). Some of these cases were associated with infection of the joint, while others showed features typical of degenerative or osteoarthritis, commonly thought of as a disease of aging. Two such cases however, were in immature animals. The severity of disease ranged from relatively minor loss of articular cartilage, to complete destruction and fusion of the joint.

Angiomatosis, a newly recognized and bizarre disease, initially found only in bottlenose dolphins from the Texas coast (Turnbull and Cowan, 1999b), has now been seen in dolphins of other species from the coast of Florida and southern California. This disease was first recognized in 1991 in a minor form, but over the following 9 years increased in incidence and severity to involve all adult bottlenose dolphins stranding on the Texas coast. It is characterized by the proliferation of small blood vessels in the lungs to form small clusters. It involves all parts of both lungs equally and to the same degree, and is not associated with inflammation. These vascular clusters progressively enlarge and fuse. In advanced stages of the disease, the walls of the new vessels thicken and lumens are reduced. Proliferation of vessels in the bronchial lining mucosa erodes the bronchial cartilages and severely constricts airway diameter. The normally very thin and pliable pleura is similarly involved, and becomes thick, stiff, and opaque, reaching a thickness of 3 mm. The effect on respiratory physiology has not been measured, but the gross and microscopic appearance suggests marked impairment of lung ventilation and restriction of circulation. One animal with advanced disease had massive thickening of the right ventricle of the heart, suggesting marked increase in pulmonary artery blood pressure (pulmonary hypertension). Lymph nodes associated with the lungs, in advanced cases may also show a pronounced vascular proliferation, and hemangiomas (tumors of blood vessels) are a common finding in the lymph nodes and an occasional one in the lungs. The cause of this disease is not known.

Amyloidosis, a disease characterized by a deposition of a dense amorphous waxy proteinaceous material in the interstitial tissues of various organs occurs in bottlenose dolphins stranding along the Texas coast of the Gulf of Mexico at an incidence of about 20%, a remarkable figure considering that the disease had not been reported from any other cetacean species (Cowan, 1995). It has since been found in beaked whales in Japan. Amyloid deposition occurs consistently in the kidneys, but also in small arteries in the spleen, heart, and lungs, and around the acini of the thyroid gland and the palatal salivary gland. In the amounts present, the amyloidosis likely did not result in death, but experience in man and other animals is that it can progress to organ failure. Several causes and associations are known for amyloidosis in man and other animals, including chronic infection and disorders of the lymphoreticular system, but the cause of dolphin amyloidosis remains obscure.

## IX. Stress

Similar recurring patterns of changes in organs and tissues have been observed in many species of beach stranded, net-caught, and captive dolphins at necropsy. They occur across species and appear to be generalizable in cetaceans. The changes are consistent with injury caused by massive release of endogenous catecholamines (adrenalin, noradrenalin) or by spasm of small cardiac arteries, with ischemia and reperfusion. This recurring pattern of pathology includes a particular pattern of necrosis of cardiac and smooth muscle; ischemic injury to the intestinal mucosa, especially the mucosa of the small intestine; and acute necrosis of the proximal tubules of the kidney. The pattern appears to result from a stereotypic stress response, independent of the nature of the provoking stimulus. It may explain the propensity of otherwise hardy animals to die in an otherwise nondamaging stressful situation.

This pattern of injury appears to be rooted in the physiologic adaptations of cetaceans that are associated with a fully aquatic life. In a voluntary dive, these are reflective of exercise, including a reflexive breath holding (apnea), with voluntary override, minimal cardiovascular adjustments, and a general maintenance of aerobic metabolism, the *dive response*. A dolphin in an *involuntary* dive situation undergoes a "dive reflex," better termed as an *alarm reaction*. This includes not only the reflexive apnea, but also decreased heart rate (diving bradycardia), reduction of cardiac output, vasoconstriction with markedly decreased perfusion of gut, liver, kidneys, and skeletal muscle. The clear implication of the distinctive reactions to voluntary and involuntary diving is that the dolphin is responding to the environment *as it is perceived*; the triggering of the alarm reaction is a reaction to a situation interpreted by the dolphin as a dire threat, and is responded to by a marked autonomic reaction. Since the major threats to an aquatic, air breathing mammal are drowning and predation, the alarm reaction is an accentuation of the physiologic dive and escape responses.

Histopathologic findings suggest that the reflexive response of a dolphin to any major perceived threat, the alarm reaction, is to activate all the physiologic adaptations to diving or escape to an extreme or pathological level, resulting, if greatly prolonged, in widespread ischemic injury to tissues. A dolphin in extreme physical or psychological distress will exhibit an extreme, protracted alarm reaction. These observations may explain the mechanism whereby "sensitive" species die abruptly from handling or transportation, why the mortality of highly stressed beach stranded animals is very high.

## X. Conclusions

There is much to be learned from the study of the pathology of marine mammals. Work by many observers over the past 30 years has done much to reveal the causes of sickness and death among these heretofore mysterious animals. We are coming to understand how similar the diseases of marine mammals can be to those of terrestrial mammals, and how different they can be. This is well illustrated by the evolution of inflammatory processes in the lung, which are strongly influenced by the special anatomy of the diving lung, and by the phenomenon of cerebral parasitism, which appears to depend not only on the special anatomy of the nasal passages, air sinuses, the specialized inner ear and their relations to the brain, but also exposure to a very specific parasite.

Recognition of similarities in tissue structure and diseases can be treacherous if we approach the study of the diseases of marine mammals as if they are just like land animals, only wet. What we see, or think we see may mean something very different than it would mean if we saw it in a land animal.

A great deal has been learned about the anatomy and physiology of pinnipeds and cetaceans in particular, and especially about cetaceans, who have taken typical mammalian systems, inherited from what many believe to be terrestrial forbears, back into the water and made them work there. The adaptive changes in the anatomy of the lungs in particular, which influence the development of disease processes in that organ are plain and obvious. These are the adaptations we can see and are relatively easy to puzzle out. We are only now gaining insight into the unseen adaptations, such as the metabolic management of mercury, a protoplasmic poison. The vast majority of mercury (and most other toxic metals) in the environment is from natural, not industrial sources, and comes from rocks and the soil. They are leached out by rain and reach streams, rivers, and ultimately the oceans. These metals have always been present in the environment, perhaps not in the levels now recognized in some geographic regions. However, the adaptive experience of animals who spend their entire lives in the water, and whose food consists of other organisms that spend their entire lives in the water has endowed them with startling abilities. These natural toxins cannot be avoided; they must be dealt with within the metabolic capacities of the animal who is inevitably exposed to them. Who could have predicted that a dolphin can carry a burden of mercury in its tissues sufficient to kill a cow four times over, and still be healthy? The next task is to understand better these unseen adaptations, so that we do not misread or misunderstand the message that Nature is sending us about the state of our oceans, written in the tissues of marine mammals.

## See Also the Following Articles

Pinniped Physiology ■ Diving Physiology

## References

Abollo, E., López, A., Gestal, C., Benavente, P., and Pascual, S. (1998). Long-term recording of gastric ulcers in cetaceans stranded an the Galician (NW Spain) coast. *Dis. Aquat. Org.* **32**, 71–73.

Bernardelli, A., Bastida, R., Louteiro, J. D., Michelis, H., Romano, M. I., Cataldi, I., and Costa, E. F. (1996). Tuberculosis in sea lions and fur seals from the southwestern Atlantic coast. *Revue Scientifique technique Office International Epizooties* **15**, 985–1005.

Bossart, G. D., Baden, D. G., Ewing, R. Y., Roberts, B., and Wright, S. D. (1998). Brevitoxicosis in manatees (*Trichechus manatus latirostris*) from the 1996 epizootic: gross, histologic and immunohistochemical features. *Toxicol. Pathol.* **26**, 276–282.

Cowan, D. F. (1995). Amyloidosis in the bottlenose dolphin *Tursiops truncates. Vet. Pathol.* **34**, 311–314.

Dailey, M. D., and Otto, K. A. (1982). Parasites as biological indicators of the distributions and diets of marine mammals common to the eastern Pacific. NOAA National Marine Fisheries Service, Southwest Fisheries Center, La Jolla, CA, Adman Rept. LJ-82-13C, 44 pp.

De Guise, S., Lagacé, , and Béland, P. (1994). Tumors in St. Lawrence Beluga whales (*Delphinapterus leucas*). *Vet. Pathol.* **31**, 444–449.

Gales, N, Duignan P, Childerhouse, S, and Gibbs, N. (1999). New Zealand sea lion mass mortality event, January/February 1998: (I) Descriptive epidemiology. Society for Marine Mammalogy, 13th Biennial Conference, Wailea, Maui, Hawaii, Nov. 29–Dec. 3, 1999, Abstracts, p. 63.

Geraci, J. R., and St. Aubin, D. J. (1977). Pathologic findings in a stranded herd of Atlantic whitesided dolphins, *Lagenorhynchus obliquidens acutus*. Proceedings of the 8th Annual Conference, International Association for Aquatic Animal Medicine, Boston, MA, May 1977.

Gulland, F. M. D., Trupkiewicz, J. G., Spraker, T. R., and Lowenstine, L. J. (1996). Metastatic carcinoma of probable transitional cell origin in 66 free-living California sea lions (*Zalophus californianus*), 1979 to 1994. *J. Wildl. Dis.* **32**, 250–258.

Gulland, F. M. D., Beckmen, K., Burek, K., Lowenstine, L., Werner, L., Spraker, T., Dailey, M., and Harris, E. (1997). Nematode (*Otostrongylus circumlitus*) infestation of northern elephant seals (*Mirounga angustirostris*) stranded along the central California coast. *Mar. Mamm. Sci.* **13**, 446–459.

Haubold, E. M., Cooper, C. W. Jr., Wen, J., McGinnis, M. R., and Cowan, D. F. (2000). Comparative morphology of *Lacazia loboi* (syn. *Loboa loboi*) in dolphins and humans, Med. Mycol. **38**, 9–14. .

Haubold, E. M., Turnbull, B. S., Clark, L., and Cowan, D. F. (1999). Cycles and trends of pathology of stranded cetaceans in the western gulf of Mexico from 1991–1999. Society for Marine Mammalogy, 13th Biennial Conference, Wailea, Maui, Hawaii, Nov. 29–Dec. 3, 1999, Abstracts, p. 78.

Howard, E. B., Britt, J. O., and Simpson, J. G. (1983a). Neoplasms in marine mammals. *In* "Pathobiology of Marine Mammals" (E. B. Howard, ed.), Vol. II, pp. 95–107. CRC Press, Boca Raton, FL.

Howard, E. B., Britt, J. O., Matsumoto, G. K., Itahara, R., and Nagano, C. N. (1983b). Bacterial diseases. *In* "Pathobiology of Marine Mammals" (E. B. Howard, ed.), Vol. I, pp. 69–118. CRC Press, Boca Raton, FL.

Kennedy, S. (1998). Morbillivirus infections in aquatic mammals. *J. Comp. Pathol.* **119**, 201–225.

Lipscomb, T. P., Scott, D. P., Gulland, F. M. D., Lowenstine, L. J., King, D. P., Hure, M. C., Stott, J. L., and Garber, R. (1999). Metastatic carcinoma of California sea lions: evidence of genital origin and association with gamma herpesvirus infection. Proceedings of the 30th Annual Conference, International Association for Aquatic Animal Medicine, Boston, MA, May 2–5, 1999, Abstracts, p. 102.

Migaki, G., and Jones, S. R. (1983). Mycotic diseases in marine mammals. *In* "Pathobiology of Marine Mammals" (E. B. Howard, ed.), Vol. II, pp. 1–27. CRC Press, Boca Raton, FL.

Reidarson, T. H., McBain, J. F., Dalton, L. M., and Rinaldi, M. G. (1999). Diagnosis and treatment of fungal infections in marine mammals. *In* "Zoo and Wild Animal Medicine: Current Therapy" (M. Fowler, and E. Miller, eds), 4th ed., pp. 478–485. W. B. Saunders Co., Philadelphia, PA.

Schulman, F. Y., Lipscomb, T. P., Moffett, D., Krafft, A. E., Lichy, J. H., Tsai, M. M., Taubenberger, J. K., and Kennedy, S. (1997). Histologic, immunohistochemical, and polymerase chain reaction studies of bottlenose dolphins from the 1987–1988 United States Atlantic coast epizootic. *Vet. Pathol.* **34**, 288–295.

Scholin, C. A., Gulland, F., Doucette, G. J., Benson, S., Busman, M., Chavez, F. P., Cordaro, J., DeLong, R., De Vogelaere, A., Harvey, J., Haulena, M., Lefebvre, K., Lipscomb, T., Loscutoff, S., Lowenstine, L. L., Marin, R., III, Miller, P. E., McLellan, W. A., Moeller, P. D. R., Powell, C. L., Rowles, T., Silvagni, P., Silver, M., Spraker, T., Trainer, V., and Van Dolah, F. M. (2000). Mortality of

P

sea lions along the central California coast linked to a toxic diatom bloom. *Nature* **403**, 80–84.

Siebert, U., Weiss, R., Lick, R., Benke, H., and Frese, K. (1995). Pathology of the respiratory tract of harbour porpoises (*Phocoena phocoena*) from German waters of the North and Baltic Seas. Society for Marine Mammalogy, 11th Biennial Conference, Orlando, FL, Dec. 14–18, Abstracts, p. 105.

Siebert, U., Joiris, C., Holsbeek, L., Benke, H., Failing, K., Frese, K., and Petzinger, E. (1999). Potential relation between mercury concentrations and necropsy findings in cetaceans from German waters of the North and Baltic Seas. *Mar. Pollut. Bull.* **38**, 285–295.

Simpson, J. G., and Gardner, M. B. (1972). Comparative microscopic anatomy of selected marine mammals. *In* "Mammals of the Sea; Biology and Medicine" (S. Ridgway, ed.), pp. 298–418. C. C. Thomas, Springfield, IL.

Turnbull, B. S., Cowan, D. F., Ramanujam, V. M. S., and Alcock, N. W. (1998). Do dolphins have protective mechanisms against mercury toxicity? Proceedings of the 29th Annual Conference, International Association for Aquatic Animal Medicine, San Diego, CA, May 2–6, Abstracts, pp. 163–167.

Turnbull, B. S., and Cowan, D. F. (1999a). Synovial joint disease in wild cetaceans. *J. Wildl. Dis.* **35**, 511–518.

Turnbull, B. S., and Cowan, D. F. (1999b). Angiomatosis, a newly recognized disease in Atlantic bottlenose dolphins (*Tursiops truncatus*) from the Gulf of Mexico. *Vet. Pathol.* **36**, 28–34.

Walker, W. A., and Cowan D. F. (1981). Air sinus parasitism and pathology in free ranging common dolphins (*Delphinus delphis*) in the eastern tropical Pacific. NOAA National Marine Fisheries Service, Southwest Fisheries Center, La Jolla, CA, Adman Rept. LJ-81-23C, 19 pp.

# Peale's Dolphin
## *Lagenorhynchus australis*

R. NATALIE P. GOODALL

### I. Characteristics and Taxonomy

Peale's dolphin (*Lagenorhynchus australis*) is common in inshore waters of southernmost South America, but because it seldom strands, its natural history is not well known. Nearly a century passed before the first descriptions of its pigmentation and a museum skull were discovered to be of the same species. It is the most coastal of the three species of *Lagenorhynchus* inhabiting the Southern Hemisphere.

This species was first described as *Phocaena australis* Peale, 1848 and *Sagmatias amblodon* Cope, 1866. Although lumped with *Delphinus obscurus* and *L. cruciger* at different times, the present combination *Lagenorhynchus australis* was proposed by Kellogg in 1941. *Tursio chiloensis* is a synonym. A review of the early specimens and the history of classification of this species are given by Goodall *et al.* (1997a, b). New work with DNA shows that the genus *Lagenorhynchus* needs revision and that the genus name of one Northern Hemisphere species and the three Southern Hemisphere SPECIES should revert to *Sagmatias*. Although there have been several common names in English, Peale's dolphin is now standard. *Delfin austral* (southern dolphin) is the common name in Argentina and Chile, although *llampa* is sometimes used in the latter.

Peale's dolphins are dark gray or black on the dorsal surface, with two areas of lighter pigmentation on the sides (Fig. 1). A curved white-to-gray flank patch angles forward from the vent, narrowing to a single line ending below or in front of the dorsal fin. The posterior curves of the flank patch almost meet above the tail stock. The larger thoracic patch is light-to-medium gray, outlined with a narrow dark line on its lower surface. Both patches may be flecked with darker gray. A double black eye ring extends forward onto the snout, which is inconspicuous. The black chin or throat patch varies individually in the shape of its posterior border, usually extending backward on the sides to leave a forward-pointing white area in the middle. The white flipper patch is also delineated with gray and may extend onto the ventral part of the flipper. The flippers and dorsal fin are dark with lighter posterior edges. Flippers in older animals may have a series of small knobs on the leading edge. The ventral surface behind the throat patch is white, with a few dark streaks in the genital area. Young animals are lighter gray than adults and have less definition between thoracic and flank patches. Variations in pigmentation have been illustrated by Goodall *et al.* (1997b).

Peale's dolphins can be confused with dusky dolphins, *L. obscurus*, throughout much of their range. The latter are usually a lighter gray and have white on the sides of the face and two light lines running forward from the flank patch, while the Peale's dolphin has a dark chin and only one line running forward along the back.

This is a stocky dolphin with the barest indication of a beak. Total length for 35 specimens ranged from 98 to 218 cm. Females (*n* = 20)

**Figure 1**   *Peale's dolphin*, Lagenorhynchus australis (*C. Brett Jarrett*).

measured 130–210 cm, males ($n = 9$) from 138 to 218 cm, and animals of unknown sex ($n = 5$) from 172 to 213 cm. The collections to date probably do not represent the total size range for the species. The heaviest animal ($n = 5$), a sexually mature female, weighed 115 kg.

The number of teeth is variable, with up to 37 upper and 34 lower teeth in each jaw. The mouth is unusual in having a wide lip outside the tooth line, which may be useful in capturing small squid and octopus (Goodall *et al.*, 1997b).

The condylobasal lengths of 27 skulls ranged from 352 to 359 mm. The vertebral count is CV7 with the first two fused, T14, L13–16, Ca 31–34 for a total of 66–70, normally 67–68. The Peale's dolphin is larger and more robust than the other two southern *Lagenorhynchus*, *L. obscurus*, and *L. cruciger*, with larger, more massive vertebrae. The phalangeal number is I = 2–3, II = 6–9, III = 5–6, IV = 2–4, V = 0–2.

## II. Distribution and Abundance

Peale's dolphin is a southern South American species commonly found from 59°S (the Drake Passage south of Cape Horn) northward to Valdivia, Chile (about 38°S) on the west coast and Golfo San Jorge, Argentina (44°S) on the east, including the Falkland Islands (Islas Malvinas), with exceptional records to 33°S in the southeastern South Pacific and 38°S in the southwestern South Atlantic (Fig. 2). Sightings of animals similar to *L. australis* in 1988 near the Cook Islands, in tropical waters thousands of miles from its normal distribution, have been considered an anomalous occurrence by some authors or perhaps a new species by others (Leatherword *et al.*, 1991).

## III. Ecology

Peale's dolphins occupy two major habitats: (a) open coasts over the shallow continental shelf of the eastern coast of Patagonia and Tierra del Fuego and Chile north of Chiloé and (b) the deep, protected bays, and intricate channels of southern Chile. In the channels, this is an "entrance animal" found in tide rips at the entrance to fjords. They will join a vessel entering a fjord and accompany it to its anchorage; later the dolphins appear when engines are started and accompany the vessel to the mouth of the fjord, where they turn back.

Peale's dolphins are strongly associated with kelp (*Macrocystis pyrifera*) beds. Observed from the coast, they often swim inside, through channels, or on the border of kelp beds. When traveling, they swim outside the kelp. Although their distributions overlap, Peale's dolphins are usually coast hugging, while the similarly pigmented dusky dolphin is mainly found a few miles seaward. There is no information on abundance, stocks, or population size.

Habitat preferences have been studied in the waters of the Chiloé Archipelago (42–43°S), where Chilean dolphins (*Cephalorhynchus eutropia*) select shallow coastal waters (up to 500 m from shore and 20 m in depth) near rivers in southern Chiloé while Peale's dolphins prefer similar areas over shallow shoals in central and southern Chiloé, an area densely occupied by extensive salmon and shellfish (mussel) farms (Heinrich *et al.*, 2008).

Peale's dolphins are noted for feeding in the kelp, where divers collecting sea urchins have observed them picking small octopus from the kelp fronds, probably assisted by their wide, flat lips. They also feed in open waters beyond the kelp on fish, often using the "sunburst" formation for herding. Few stomachs have been examined ($n = 16$) and those only from the southwestern South Atlantic. About 20 prey taxa have been identified, mainly demersal and bottom fish,

octopus, and squid species common over the continental shelf or in kelp beds. The most important prey were bottom fish: hagfish (*Myxine australis*), southern cod (*Salilota australis* and Patagonian grenadier (*Macruronus magellanicus*), followed by red octopus (*Enteroctopus megalocyathus*) and Patagonian squid (*Loligo gahi*). Two very young animals from the eastern Beagle Channel had salps as well as milk in their stomachs (Schiavini *et al.*, 1997).

Peale's dolphins associate closely with other dolphins, especially Commerson's dolphins (*C. commersonii*). We have watched a Peale's dolphin swim with several of these dolphins near shore, moving southward along the Páramo Península for several miles. In Chilean waters, they share shallow bays with Chilean dolphins, occupying areas just slightly seaward of these coastal dolphins.

No predators but humans are known, although killer whales, leopard seals, and sharks are possibilities. In a study of the food of 22 broadnosed seven-gill sharks (*Notorynchus cepedianus*), 30% of the stomach contents were of marine mammals: three species of pinnipeds and two of cetaceans, although no actual attacks on living animals were observed. Parts of three cetaceans were found in three stomachs; one of these may have been the tail of a Peale's dolphin (Crespi-Abril *et al.*, 2003).

## IV. Behavior and Physiology

Peale's dolphins are most often seen swimming slowly in or near the kelp. Dive times range from 3 to 157 sec, with an average of 28 sec ($n = 723$), with three short dives followed by a longer one. They commonly BOW RIDE, with much head movement, rolling, BREACHING, spy hopping, and spinning. They produce a wide splash when surface swimming, earning the name "plough-share" dolphins.

Group size is usually small, from 2 to 5 animals, but aggregations of up to 100 have been seen. Their behavior still needs to be investigated.

Near Isla Chiloé and in the Strait of Magellan, resident groups have been noted throughout the year, although more animals were present during summer. In southern Tierra del Fuego, animals also seem to move inshore during summer, possibly following fish migrations, and are seldom seen in winter. It is not known whether the dolphins in different parts of the range belong to different populations or stocks.

Pulsed sounds were recorded in the Chilean channel region, revealing broadband clicks at 5–12 kHz and narrowband clicks at 1–2 kHz bandwidths. Whistle-like squeals were not recorded, but may have been above the limits of the recording equipment used.

## V. Life History

There is little information on reproduction. A female of 185 cm was sexually immature, one of 193 cm was pubescent, and one of 210 cm was mature. There is no information on sexual maturity in males. Calves have been reported from spring through fall, October to April. Physical maturity, on the basis of epiphyseal fusion, has been recorded for 24 specimens. Neonates measured 98–130 cm, juveniles 138–176 cm, subadults 142–210 cm, and adults over 190 cm. The oldest animal was a physically mature female with 13 growth layer groups (GLGs) in the teeth.

## VI. Interactions with Humans

As far as we know, no Peale's dolphins have been kept in CAPTIVITY.

Peale's (and possibly Commerson's) dolphins are credited with coming to the rescue of Indonesian seamen who jumped overboard

**P**

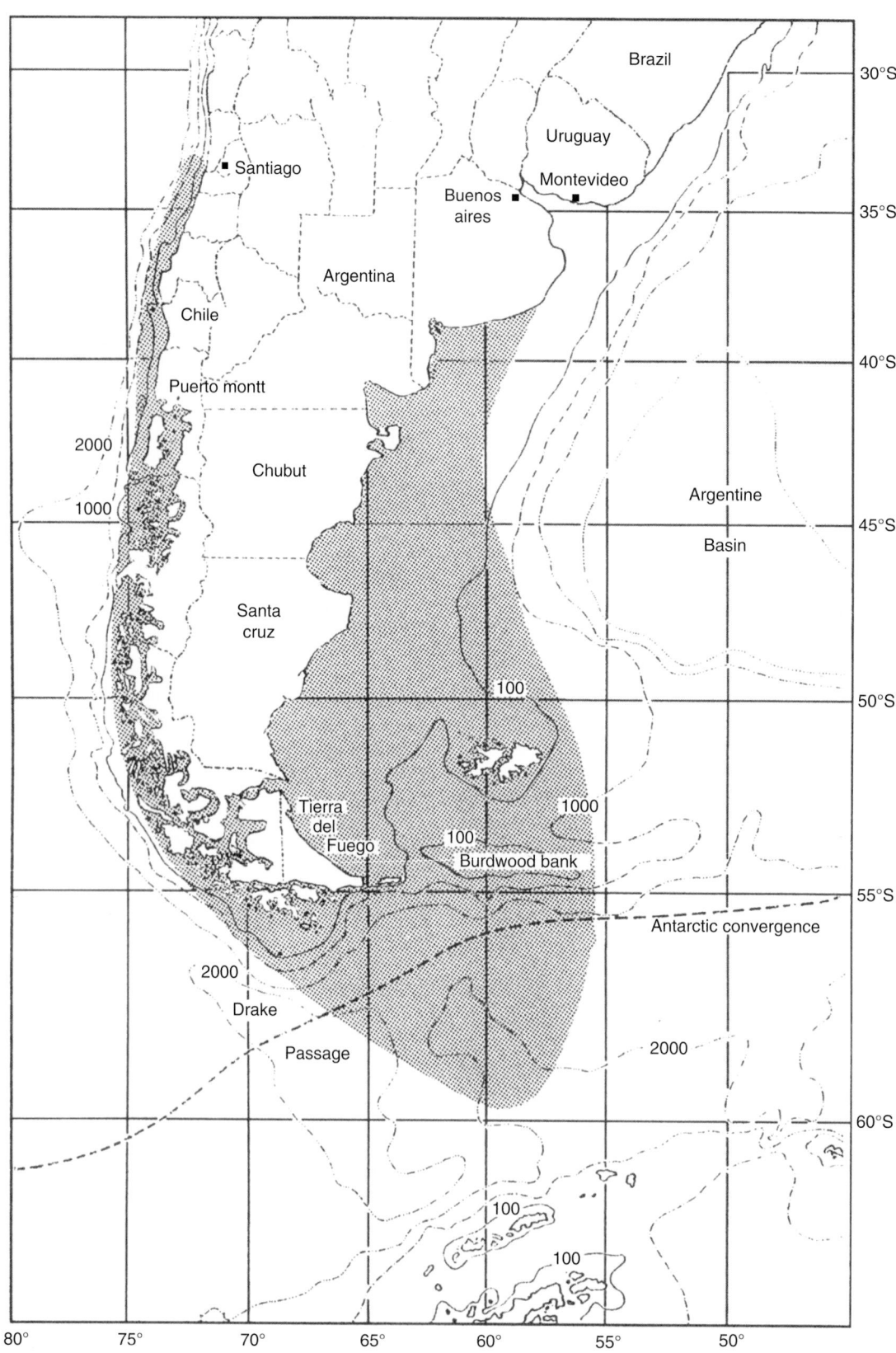

**Figure 2** *The distribution of Peale's dolphin, based on sightings and strandings from 1839 to 2007.*

from their fishing vessel near Port Stanley in the Falkland Islands (Islas Malvinas) in February 2007. The seamen were nudged upward from below by dolphins until they could be picked up by boats. Two seamen drowned, but eight were saved (Anonymous, 2007).

One specimen had a piece of fishing net balled in its first stomach and impacted with food items.

Prehistorically, Peale's dolphins were exploited for food by the canoe people of the channels of southernmost South America; remains have been found in kitchen middens dated at 2500 and 6000 years before present. More recently, a few dolphins were taken for scientific research.

Peale's dolphins have been heavily exploited for crab (centolla, Lithodes santolla) bait in the Magellan region of southern Chile since at least the early 1970s, with highest exploitation in the 1980s (Lescrauwaet and Gibbons, 1994). Overfishing, resulting in greatly reduced populations of crabs, has led fishermen to concentrate on other types of fishing that do not need bait, such as scallops and sea urchins. In addition, crabbing companies now supply bait to their fishermen, so there is less illegal take of dolphins. The full extent of the exploitation of dolphins for crab bait at present in Chile is unknown, but is thought to be much less than formerly. A few Peale's dolphins were taken for crab bait in eastern Beagle Channel in Argentina in the late 1970s and early 1980s, but this has not continued. However, these dolphins have not recolonized the areas of the Beagle Channel with denser human population, but are seen to the east, far west, and southern areas near Cape Horn.

Although common, Peale's dolphins are rarely taken incidentally in nets in the northern part of their range between Valdivia and San Antonio in Chile. A few dolphins have been caught in anti-pinniped nets near the salmon pens in the Isla Chiloé area. A small incidental take occurs in shore-set gill nets off Tierra del Fuego in Argentina, but fishermen claim that Peale's dolphins are usually strong enough to fight and release themselves from the net (Goodall et al., 1994). Likewise, a few Peale's dolphins die in offshore fishing activities south of Golfo San Jorge, but not as many as other species. The extent of this exploitation is unknown; it represents a potential danger that should be monitored.

Kelp forests seem to be a fundamental habitat for Peale's dolphins in coastal ecosystems and their protection may be crucial for the conservation of this species (Viddi and Lescrauwaet, 2005). The coastal habitat preferred by this species is compromised in certain areas of their distribution by extensive salmon and mussel farming activities (Heinrich et al., 2008).

## See Also the Following Articles

Hourglass Dolphin ■ Dusky Dolphin

## References

Anonymous (2007). Dolphins help desperate seamen. *Penguin News Falkland Islands* **18**(43), 1.

Brownell, R. L., Jr. (1999). Peale's dolphin, *Lagenorhynchus australis* (Peale 1848). *In* "Handbook of Marine Mammals; the Second Book of Dolphins and Porpoises" (S. H. Ridgway, and R. Harrison, eds), Vol. 6, pp. 105–120. Academic Press, San Diego, CA.

Crespi-Abril, A. C., García, N. A., Crespo, E. A., and Coscarella, M. A. (2003). Consumption of marine mammals by broadnose sevengill shark *Notorynchus cepedianus* in northern and central Patagonian shelf. *LAJAM* **2**, 101–107.

de Haro, J. C., and Iñíguez, M. A. (1997). Ecology and behaviour of the Peale's dolphin, *Lagenorhynchus australis* (Peale, 1848), at Cabo Vírgenes (52°30′S, 68°28′W), in Patagonia, Argentina. *Rep. Int. Whal. Comm.* **47**, 723–727.

Goodall, R. N. P., de Haro, C., Fraga, F., Iñíguez, M. A., and Norris, K. S. (1997a). Sightings and behavior of Peale's dolphins, *Lagenorhynchus australis*, with notes on dusky dolphins, *L. obscurus*, off southernmost South America. *Rep. Int. Whal. Comm.* **47**, 757–775.

Goodall, R. N. P., Norris, K. S., Schevill, W. E., Fraga, F., Praderi, R., Iñíguez, M. A., and de Haro, C. (1997b). Review and update on the biology of the Peale's dolphin, *Lagenorhynchus australis*. *Rep. Int. Whal. Comm.* **47**, 777–796.

Goodall, R. N. P., Schiavini, A. C. M., and Fermani, C. (1994). Net fisheries and net mortality of small cetaceans off Tierra del Fuego, Argentina. *Rep. Int. Whal. Comm.* **15**(Special Issue), 295–304.

Heinrich, P., Fuentes, R. M., and Hammond, P. S. (2008). Conservation status of small cetaceans in the Chiloé Archipelago, Southern Chile. *IWC Scientific Committee*. Meeting document SC/60/SM23.

International Whaling Commission (1997). Report of the sub-committee on small cetaceans. Annex. H. *Rep. Int. Whal. Comm.* **47**, 169–191.

Leatherwood, S., Grove, J. S., and Zuckerman, A. E. (1991). Dolphins of the genus *Lagenorhynchus* in the tropical South Pacific. *Mar. Mamm. Sci.* **7**, 194–197.

LeDuc, R. G., Perrin, W. F., and Dizon, A. E. (1999). Phylogenetic relationships among the delphinid cetaceans based on full cytochrome *B* sequences. *Mar. Mamm. Sci.* **15**, 619–648.

Lescrauwaet, A.-K. (1997). Notes on the behaviour and ecology of the Peale's dolphin, *Lagenorhynchus australis*, in the Strait of Magellan, Chile. *Rep. Int. Whal. Comm.* **47**, 747–755.

Lescrauwaet, A.-K., and Gibbons, J. (1994). Mortality of small cetaceans and the crab bait fishery in the Magallanes area of Chile since 1980. *Rep. Int. Whal. Comm.* **9**(Special Issue), 103–118.

Schiavini, A. C. M., Goodall, R. N. P., Lescrauwaet, A.-K., and Koen Alonso, M. (1997). Food habits of the Peale's dolphin, *Lagenorhynchus australis*; review and new information. *Rep. Int. Whal. Comm.* **47**, 827–833.

Van Waerebeek, K., Goodall, R. N. P., and Best, P. G. (1997). A note on evidence for pelagic warm-water dolphins resembling *Lagenorhynchus*. *Rep. Int. Whal. Comm.* **47**, 1015–1017.

Viddi, F. A., and Lescrauwaet, A-K. (2005). Insights on habitat selection and behavioural patterns of Peale's dolphins (*Lagenorhynchus australis*) in the Strait of Magellan, southern Chile. *Aquat. Mamm.* **31**, 176–183.

**P**

# Pilot Whales
*Globicephala melas* and
*G. macrorhynchus*

### Paula A. Olson

## I. Characteristics and Taxonomy

Two species are recognized: *Globicephala melas* (long-finned pilot whale) and *G. macrorhynchus* (short-finned pilot whale).

Adult pilot whales reach an average length of approximately 6 m. Males are larger than females. Most pilot whales appear black or dark gray in color. The body is robust with a thick tailstock. The melon is exaggerated and bulbous, and there is either no beak

**Figure 1** *Full-body view of the long-finned pilot whale (top) and the short-finned pilot whale (bottom). Drawings courtesy of P. Folkens.*

**Figure 2** *A group of male and female pilot whales. Males are larger than females and develop exaggeratedly wide dorsal fins. Photo by P. Olson.*

**Figure 3** *This group of long-finned pilot whales with a calf exhibit the external features common to both species including a bulbous melon and a broad-based dorsal fin. Photo by P. Olson.*

or a barely discernible one. A wide, broad-based falcate dorsal fin is set well forward on the body. The flippers are long, slender, and sickle shaped. A faint gray "saddle" patch may be visible behind the dorsal fin as well as a faint postorbital blaze (Fig. 1). On the ventrum, a gray midline extends anteriorly into an anchor-shaped chest patch and widens posteriorly into a genital patch. Calves are paler than adults.

Pilot whales exhibit striking sexual dimorphism in size, similar to that observed in sperm whales (*Physeter macrocephalus*) and killer whales (*Orcinus orca*). Adult males are longer than females, develop a more pronounced melon, and have a much larger dorsal fin (Fig. 2). The function of sexual dimorphism in pilot whales is unknown, although several have been hypothesized. The males' enlarged features may be used for display to other males or females or for increased agility when maneuvering for mate access or for herding females. The males' large size may aid in defense of their school from attacks by killer whales or sharks.

Long-finned and short-finned pilot whales are difficult to distinguish at sea (Fig. 3). The morphological differences between the two species are subtle: length of flippers, differences in skull shape, and number of teeth (Bernard and Reilly, 1999). On average the pectoral flippers of long-finned pilot whales are one-fifth the body length, whereas on short-finned whales they are one-sixth the body length.

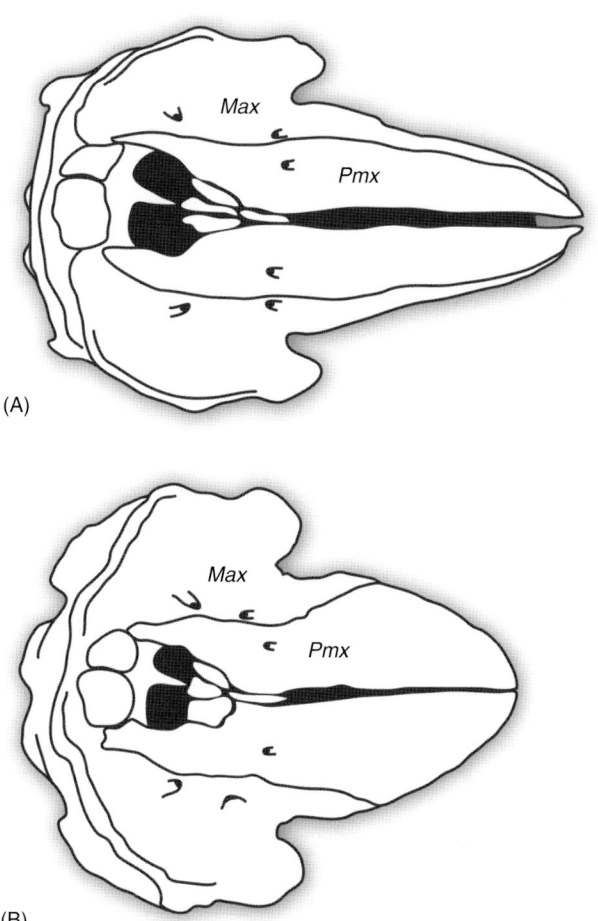

(A)

(B)

**Figure 4** *(A) Dorsal view of the skull of a long-finned pilot whale (G. melas). (B) Dorsal view of the skull of a short-finned pilot whale (G. macrorhynchus). Note the differences in the shape and length of the rostrums and the degree to which the maxillae are covered by the premaxillae. Illustration courtesy of Uko Gorter.*

However, overlap exists between the two species. Long-finned flippers exhibit a noticeable "elbow" whereas short-finned pectorals have a more curved appearance.

Because most size, shape, and color pattern distinctions between the two species are so variable, the shape of the skull is the only definitive characteristic for identification to species. The long-finned pilot whale has a narrower skull, with the premaxillae leaving uncovered 1 cm of the lateral borders of the maxillae (Fig. 4A). There are 9–12 teeth in each row. The skull of the short-finned pilot whale is shorter and broader, and the premaxillae cover the maxillae (Fig. 4B). There are seven to nine teeth in each row. Pilot whales have notably fewer teeth than most other delphiniids. This is similar to the evolutionary reduction of teeth found in Risso's dolphins (*Grampus griseus*) and sperm whales, two other heavy squid consumers.

The extant delphinid groups (to which *Globicephala* belongs) appeared in the Middle to Late Miocene Epoch. Fossils of the genus *Globicephala* dating from the Pleistocene Epoch have been uncovered in Florida (*G. baerreckeii*). Odontocete remains from the Pliocene Epoch uncovered in Tuscany, Italy, have been designated as *Globicephala? eturia.*

The name "pilot whale" originated with an early theory that a school is piloted by a leader. Other common names for these whales

include pothead (after the whales' bulbous melon) and blackfish (a term also used for melon-headed whales, pygmy killer whales, and false killer whales). The genus name, *Globicephala*, is derived from the Latin word *globus*, meaning round ball or globe, and the Greek word *kephale*, meaning head. *Melas* is a Greek word for black. The trivial name *macrorhynchus* is likewise derived from Greek words: *macro*, meaning enlarged and *rhynchus*, meaning snout or beak. For many decades, *malaena* was used as the trivial name for the long-finned pilot whale. However, in 1986 the name was revised to *melas*.

## II. Distribution and Abundance

Pilot whales are wide ranging and abundant (Fig. 5). Generally, short-finned pilot whales have a tropical and subtropical distribution, and long-finned pilot whales are distributed antitropically. There is little overlap in the range of the two species. Areas of marginal overlap include the temperate waters of the North and South Atlantic, in the Pacific off the coast of Peru, and off South Africa. Pilot whales are found in both nearshore and pelagic environments.

Long-finned pilot whales inhabit the cold temperate waters of both the North Atlantic and the Southern Ocean. Whales in the two hemispheres are isolated and are accorded subspecies status: *G. melas melas* in the North Atlantic and *G. melas edwardii* in the Southern Hemisphere. There are slight morphological differences between the subspecies. In the North Atlantic, the range of *G. melas melas* includes the waters of Greenland (Denmark), Iceland, and the Barents Sea south to the Tropic of Cancer. The species is present in the western Mediterranean Sea. *G. melas edwardii* is circumglobal in the Southern Hemisphere, with records as far north as 14°S in the Pacific and farther south than the Antarctic Convergence. Although the long-finned pilot whale does not currently inhabit the North Pacific, skulls dated to the eighth to twelfth centuries have been recovered in Japan. Short-finned pilot whales now inhabit those waters.

Short-finned pilot whales are found worldwide in tropical, subtropical, and warm temperate waters. Their northern range in the Atlantic extends to the mid-coast area of the United States and to France. Short-finned pilot whales are not found in the Mediterranean. Latitude 25°S marks the southernmost record for the Atlantic and the Pacific coasts of South America. Elsewhere in the Pacific, the range of the short-finned pilot whale continues north to Japan and to the west coast of the United States.

Pacific short-finned pilot whales in higher latitudes are generally larger than those in lower latitudes. Two distinct populations of short-finned pilot whales are found off northern and southern Japan. These populations exhibit morphological differences in external and cranial features. The populations are segregated geographically and genetically. However, their exact taxonomic status is undetermined and currently they are both classified as *G. macrorhynchus*.

Estimates of abundance for pilot whales have generally been undertaken in response to management issues. Survey areas are typically determined by management goals rather than natural population boundaries. Most of the quantitatively derived estimates of abundance are for nearshore populations. Estimates using line-transect methods have been made for the eastern US and Canada, Newfoundland/Labrador (Canada), the northeast Atlantic, northern and southern Japan, Hawaiian Islands, the US west coast, the eastern tropical Pacific, the US Gulf of Mexico, and the Antarctic (Table I).

## III. Ecology

Pilot whales are generally nomadic, but resident populations have been documented in a few locations such as coastal California and

**P**

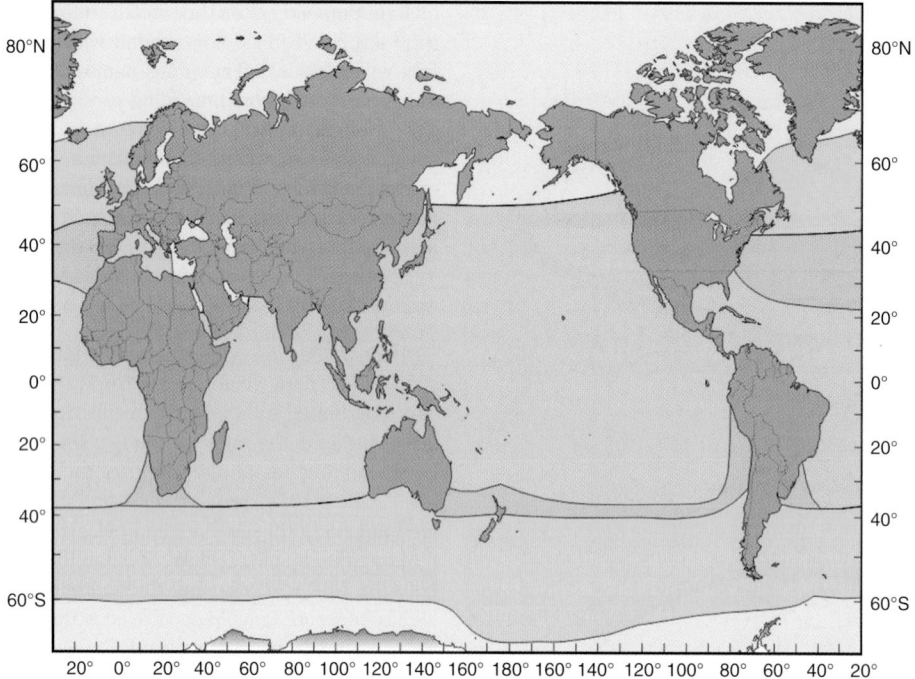

**Figure 5** *Worldwide distribution of pilot whales. The distribution of long-finned pilot whales is indicated by yellow and the distribution of short-finned pilot whales by light green. Dark green indicates areas where the species overlap.*

TABLE I
Estimates of Pilot Whale Abundance using Line-Transect Methods

| Species/subspecies | Area | Population | References |
|---|---|---|---|
| *Globicephala* spp. | East coast US and Canada | 31,139 | Waring *et al.* (2007) |
| *G. melas melas* | Newfoundland/Labrador | 6731–19,603 | Hay (1982) |
| *G. melas melas* | Northeast Atlantic | 778,000 | Buckland *et al.* (1993) |
| *G. melas edwardii* | Antarctic | 200,000 | Kasamatsu and Joyce (1995) |
| *G. macrorhynchus* | Northern Japan | 5300 | Miyashita (1993a) |
| *G. macrorhynchus* | Southern Japan | 53,608 | Miyashita (1993b) |
| *G. macrorhynchus* | Hawaiian Islands | 8806 | Barlow (2006) |
| *G. macrorhynchus* | West coast US | 304[a] | Carretta *et al.* (2007) |
| *G. macrorhynchus* | Eastern tropical Pacific | 160,200 | Wade and Gerrodette (1993) |
| *G. macrorhynchus* | US Gulf of Mexico | 2388 | Waring *et al.* (2007) |

[a] Estimates of abundance for this area vary between years, likely the result of whales moving in and out of the area in response to oceanographic conditions.

Hawaii. Common habitats are the continental shelf break, slope waters, and areas of high topographic relief. Seasonal inshore/offshore movements of pilot whales are related to the distribution of squid, their favorite prey. Studies in Newfoundland and California correlated the seasonal abundance of pilot whales with spawning squid. Individual whales have been recorded moving distances up to 2400 km/month.

The pilot whale diet consists primarily of squid, with lesser amounts of fish. Fish prey in the northwest Atlantic include Atlantic cod (*Gadus morhua*), Greenland turbot (*Rheinhardtius hippoglossoides*), Atlantic mackerel (*Scomber scombris*), Atlantic herring (*Clupea harengus*), hake (*Urophycis* spp.), silver hake (*Merluccius*

*bilinearis*), and spiny dogfish (*Squalus acanthias*). Pilot whales in the northeast Atlantic have been known to take Atlantic Argentine (*Argentina silus*) and blue whiting (*Micromesistius poutassou*).

The ecosystem changes brought about by a strong El Niño event in 1982–1983 affected the distribution of pilot whales off southern California dramatically. With the influx of warm water during the El Niño, squid did not spawn as usual in the area. Pilot whales were virtually absent from the region that year and remained so for another 9 years. It is not known where the whales went during that time or whether the whales sighted there now are returning or are new individuals.

Pilot whales are often observed in mixed species aggregations. They are most commonly sighted in association with common bottlenose dolphins (*Tursiops truncatus*) but have also been seen with short-beaked common dolphins (*Delphinus delphis*) Atlantic white-sided dolphins (*Lagenorhynchus acutus*), Pacific white-sided dolphins (*Lagenorhynchus obliquidens*), striped dolphins (*Stenella coeruleoalba*), Fraser's dolphins (*Lagenodelphis hosei*), melon-headed whales (*Peponocephala electra*), killer whales, fin whales (*Balaenoptera physalus*), sperm whales, and gray whales (*Eschrichtius robustus*). There are accounts of pilot whales behaving aggressively toward humpbacks (*Megaptera novaeangliae*), sperm whales, common dolphins, and dolphins of the genus *Stenella*. Pilot whales have also been reported carrying the carcasses of dead California sea lions (*Zalophus californianus*) and towing a human diver.

Pilot whales in the Atlantic have been affected by the morbilliviruses that have plagued other marine mammals in recent decades (Duignan *et al.*, 1995). Although to date no large-scale outbreak of disease has been reported in pilot whales, high percentages of both species sampled during the 1980s and 1990s carried virus-neutralizing antibodies. It appears that most individuals are immune. Due to their wide-ranging movements and their propensity to mix with other species, it is considered likely that pilot whales act as a vector for morbilliviruses in other cetaceans, such as bottlenose dolphins.

## IV. Behavior and Physiology

Pilot whales are highly social and are usually found in schools, or pods, averaging 20–90 individuals. A variety of group behaviors have been documented. Commonly reported are traveling or foraging in a loose chorus-line formation or collective logging at the surface. The social structure of pilot whale pods is similar to that of killer whales. Pilot whales form stable pods composed of individuals with close matrilineal associations (Amos *et al.*, 1993). All age and sex classes are included, although there is a female bias in adults. The groups are stable; pilot whales grow to maturity in their natal group and most remain there for life. Genetic evidence supports the theory that males breed outside their family group.

Pilot whales are polygynous. Huge aggregations of pilot whales are occasionally reported and it is believed that males move between family groups to mate during these temporary aggregations. This type of social structure where adult males stay with their female kin and mate elsewhere is unusual among mammals.

Pilot whales are one of the most frequently reported cetaceans in mass strandings. Strandings are dramatic events because they usually involve groups of animals and because live whales that are assisted back into the ocean often return to the beach. Pilot whales strand singly as well as in groups; often single animals are diseased.

Curiously, most of the whales in a group event do not show any pathology. It is not understood why apparently healthy animals strand, although there are a variety of hypotheses. The whales may become confused or trapped in shallow areas; geomagnetic anomalies may disorient whales if they are using the earth's magnetic field for navigation; or if an ill animal strands it may be followed by members of its pod. The strong social bonds within a pilot whale pod are likely to play a role in stranding events, whatever the other underlying reasons may be.

Pilot whales are known to make shallower dives (1–16 m) during the day and perform deeper dives (>100 m) at night, presumably foraging. It has been theorized that pilot whales use deeper dives to track the rise of the deep scattering layer (DSL) at night. The deepest dives recorded for pilot whales are over 600 m.

Pilot whales echolocate with a precision similar to that of bottlenose dolphins. Pilot whales also vocalize, the primary purpose probably being to maintain contact among school members. Vocalizations are more complex with active behavior and simpler with less active behavior.

Significant differences have been found between the calls of long-finned and short-finned pilot whales (Rendell *et al.*, 1999). The calls of long-finned pilot whales are of a lower frequency and a narrower frequency range than those of short-finned pilot whales. The mean frequency for long-finned pilot whales is 4480 Hz, for short-finned pilot whales the mean frequency is 7870. Short-finned pilot whales were also found to have distinct group-specific call repertoires, as would be expected for a species with stable matrilineal kinship groups.

## V. Life History

Pilot whales share several features of life history with other large odontocetes: long life span, delayed maturity, different rates of maturation for males and females, seasonal mating, and the production of a single calf in multiyear intervals.

Short-finned pilot whales have a slower growth rate than long-finned pilot whales and reach a shorter body length. The pattern of growth for both species is similar. Rapid neonatal growth is followed by a less rapid but continual growth phase during the juvenile years. Growth slows even further after the attainment of sexual maturity and ceases some years later. Short-finned females become sexually mature at 9 years, males at age 13–16. Long-finned females reach sexual maturity at 8 years, males at about 12. Males become socially mature, e.g., mate successfully, several years after reaching sexual maturity.

In the Northern Hemisphere, mating generally occurs in spring or early summer and parturition in summer or early autumn. An exception to this is the population of pilot whales off northern Japan. Their peak breeding season is autumn with parturition in winter. Gestation is estimated to be 12 months for long-finned pilot whales and 15–16 months for short-finned whales.

The birth interval in pilot whales is one of the longest of all the cetaceans. Lactation lasts for at least 3 years, often longer. Such an extended lactation period probably serves a social rather than nutritional purpose in later years. The capability of a short-finned female to lactate years past final ovulation has been reported. This supports the theory that females may invest more in present offspring as their likelihood to bear more offspring diminishes. Sizable numbers of post-reproductive females have been found in populations of both species.

Estimates of natural mortality for pilot whales are based on data from the directed fisheries in the Faeroe Islands (Denmark) and Japan. These data indicate that male pilot whales have a higher mortality rate at all ages than those of females. Females are known to be longer-lived. Females live past 60 years; males reach 35–45 years.

## VI. Interactions with Humans

Pilot whales are subject to several types of human interactions, including direct exploitation, bycatch in fisheries, and exposure to chemical contaminants (Donovan *et al.*, 1993). The IUCN designates both species as Data Deficient (DD). More research is needed to determine their status.

Because of their cohesive social structure, pilot whales are susceptible to herding by humans. Meat, blubber, and oil are the desired products. Historically, there have been a number of directed fisheries for pilot whales. In the North Atlantic, drive fisheries for long-finned pilot whales were conducted in Newfoundland, Cape Cod, Norway,

**P**

Iceland, Orkney Islands (Scotland), Hebrides Islands (Scotland), Greenland, and the Faeroe Islands. The intensive drive fishery in Newfoundland (1947–1971) is estimated to have taken 54,000 animals and to have reduced the local population substantially. The population may be recovering, but more information is needed. Fisheries for short-finned pilot whales have operated in the Caribbean, Indonesia, and in Japan. The drive fisheries in the Faeroe Islands and in Japan continue today. These fisheries have been in existence for several hundred years. In 2006, the catch of long-finned pilot whales in the Faeroes was 856 animals; in 2004, Japan reported a catch of 63 short-finned pilot whales. The Faeroe fishery is considered sustainable.

The incidental bycatch of cetaceans in fisheries is a worldwide phenomenon. Most bycatch goes unreported because this information is not recorded in many countries. Pilot whales are particularly susceptible to entanglement in driftnets. The effect of such mortality on pilot whale populations is unknown.

Bycatch records are kept in the United States. In northeast US waters, pilot whales have been taken incidentally in a variety of fisheries including pelagic drift gill nets, pelagic long lines, pelagic pair trawls, and trawls for mackerel, herring, and squid. Most takes occurred along the shelf break. Some of these fisheries are now closed. None of the current fisheries exceed the allowable annual take for pilot whales under US law.

Pilot whales off California are taken incidental to driftnet fisheries targeting swordfish and sharks. A take reduction plan was implemented in 1997 and currently the incidental take is lower than the allowable annual limit. Prior to the El Niño of 1982–1983, pilot whales were taken incidentally in the Californian squid purse-seine fishery. Pilot whale redistribution in response to El Niño is the likely reason no mortality was reported for this fishery during the following years. Currently the squid fishery is not monitored, but there have been anecdotal reports of pilot whales seen near squid fishing operations in recent years. Driftnet fisheries similar to those in California operate out of Mexico; any pilot whale takes there are likely from the same population of short-finned pilot whales that occurs off California.

A minimal number of pilot whales have been reported taken incidental to the long-line fisheries based in Hawaii. The mortality and injury sustained by pilot whales interacting with the fishery occurred outside the 200 nmi Exclusive Economic Zone (EEZ) of Hawaii. Less is known about pilot whale populations away from the nearshore areas of Hawaii, so the impact of the takes on these whales is also not known.

Long-finned pilot whales from both sides of the north Atlantic carry high levels of organochlorine contaminants (pesticides such as DDT and PCB) in their tissues. Concentrated organochlorines may impair reproduction or increase susceptibility to disease. Studies are continuing on the effects these compounds have on marine mammals. Accumulations of cadmium and mercury are also present in the tissues of long-finned pilot whales from the Faeroe Islands. As top predators, pilot whales are a repository of these heavy metals accumulating through the marine food chain. Pilot whales seem unusually tolerant to elevated levels of these metals, as studies have not yet revealed a major toxicity problem in these species.

## See Also the Following Articles

See Delphinids, Overview ■ Hunting of Marine Mammals

## References

Amos, B., Schlotterer, C., and Tauz, D. (1993). Social structure of pilot whales revealed by analytical DNA profiling. *Science* **260**, 670–672.

Bernard, H. J., and Reilly, S. B. (1999). Pilot whales. *In* "Handbook of Marine Mammals" (S. H. Ridgway, and R. Harrison, eds), Vol. 6, pp. 245–279. Academic Press, London.

Donovan, G. P., Lockyer, C. H., and Martin, A. R. (eds.). (1993). Biology of Northern Hemisphere pilot whales. *Rep. Int. Whal. Comm.* **14**(Special Issue).

Duignan, P., *et al.* (1995). Morbillivirus infection in two species of pilot whales (*Globicephala* sp.) from the western Atlantic. *Mar. Mamm. Sci.* **11**, 150–162.

Fullard, K. J., Early, G., Heide-Jørgensen, M. P., Bloch, D., Rosing-Asvid, A., and Amos, W. (2000). Population structure of long-finned pilot whales in the North Atlantic: a correlation with sea surface temperature? *Mol. Ecol.* **9**, 949–958.

Nawojchik, R., St. Aubin, D. J., and Johnson, A. (2003). Movements and dive behavior of two stranded, rehabilitated long-finned pilot whales (*Globicephala melas*) in the northwest Atlantic. *Mar. Mamm. Sci.* **19**, 232–239.

Ottensmeyer, C. A., and Whitehead, H. (2003). Behavioural evidence for social units in long-finned pilot whales. *Can. J. Zool.* **81**, 1327–1338.

Rendell, L. E., Matthews, J. N., Gill, A., Gordon, J. C. D., and Macdonald, D. W. (1999). Quantitative analysis of tonal calls from five ondontocete species, examining interspecific and specific variation. *J. Zool. Lond.* **249**, 403–410.

Reynolds, J. E., and Rommel, S. A. (eds) (1999). "Biology of Marine Mammals." Smithsonian Institution Press, Washington, DC.

# Pinniped Ecology

## W.D. Bowen, C.A. Beck, and D.A. Austin

E cology is the study of the interactions between individuals and their environment. In this context, environment is taken broadly to include other organisms and the physical characteristics of habitat. These interactions take place at various spatial and temporal scales, and influence both the abundance and distribution of individuals. However, ecology is also a historical science in that the patterns we see today reflect past events and phylogenetic relationships. Thus, processes acting on both evolutionary and ecological time-scales have undoubtedly influenced many of the characteristics of pinniped ecology we see today. Pinnipeds are large, long-lived, aquatic mammals exhibiting delayed sexual maturity and reduced litter size; a single precocial offspring is the norm. As such, they share many of the demographic features of other large mammals. Population numbers do not change dramatically from year to year, and numbers are most sensitive to changes in adult survival, followed by juvenile survival and fecundity (Eberhardt and Siniff, 1977). We assume that these characteristics are under selection and that variability in foraging success affects survival probability and reproductive performance of individuals. Inevitably, discussions of pinniped ecology and other aspects of pinniped biology will overlap. Here we focus on five aspects of pinniped ecology: abundance, distribution, reproduction, foraging, and the ecological roles of pinnipeds in aquatic ecosystems.

## I. Abundance

Despite the interest in the ecology of pinnipeds, the abundance of many species is poorly known. The abundance of commercially harvested species (e.g., past-northern fur seals [*Callorhinus ursinus*] or present-harp seals [*Pagophilus groenlandicus*]) is generally better known than for those species that have not been exploited. The accuracy and precision attached to the estimates of abundance varies greatly, owing to the difficulty in carrying out surveys or to the lack of effort to obtain good estimates. Good estimates of abundance are important because abundance and trends in abundance are perhaps the most useful indicators of population status.

Commercial exploitation decimated many pinniped species, in some cases to levels nearing extinction (e.g., northern elephant seals [*Mirounga angustirostris*]). Over the past several decades or more, some species have recovered or are continuing to recover. Thus, the present abundance of heavily exploited species may not be a good measure of their preexploitation numbers. Pinniped species range over four orders of magnitude in abundance, from the crabeater seals (*Lobodon carcinophaga*) at about 12 million (probably the most abundant marine mammal in the world) to the Mediterranean monk seal (*Monachus monachus*) at probably fewer than several hundred individuals (Reijnders *et al.*, 1993). Phocid species are generally more abundant than otariids, with 15 of 18 phocid species numbering greater than 100,000 individuals compared with only 8 of 14 otariid species (Bowen and Siniff, 1999). The reasons for this difference are not entirely clear. Over the past 100 years, both families have been commercially exploited and subjected to other human factors that might have influenced abundance. More likely, the greater abundance of phocids is the result of their greater use of high-productivity areas in temperate and polar waters than is the case in most otariid species. The three most abundant otariids, the northern fur seal, Antarctic fur seal (*Arctocephalus gazella*), and South African fur seal (*A. pusillus pusillus*), all forage in seasonally productive, high-latitude ecosystems, a characteristic shared with the most abundant phocid species (i.e., the ringed seal [*Pusa hispida*], the harp seal, and the crabeater seal).

Abundance is determined by the movement of individuals in and out of the population, and by births and deaths. These processes are influenced by ecological factors such as predation, food supply, breeding habitat, disease, competition with other species, and environmental variability, and by both direct and indirect human activities. In the absence of human effects, combinations of these ecological factors determine abundance and some of which, operating in a density-dependent way, will regulate population size about a level known as carrying capacity. With the recovery of populations from earlier periods of exploitation, there is increasing evidence that a number of species have reached or are approaching the current carrying capacity of their environments (e.g., Weddell seal [*Leptonychotes weddellii*], gray seals [*Halichoerus grypus*], harbor seals [*Phoca vitulina*], harp seals). During periods of recovery, species such as Antarctic fur seal, Northern fur seal, gray seal, and harbor seal increased at rates in excess of 12% per year over several decades or more. At Sable Island, Canada, the number of gray seal pups born each year has increased exponentially, with a doubling time of about 6 years, for more than 40 years. Although they can increase rapidly, pinniped populations may decline even more rapidly as a result of epizootics, such as the phocine distemper virus that killed large numbers of harbor seals in the North Sea, and during short-term extreme changes in ocean climate, such as El Niño (see below).

## II. Distribution

Fundamentally, pinniped distributions reflect the need to give birth on solid substrate of land or ice, and to feed at sea. Within these broad constraints, the distribution of pinnipeds is affected by physical (e.g., ice cover, location of remote islands) and biological (e.g., productivity, abundance of predators) characteristics of habitat, demographic factors (e.g., population size, age, sex, and reproductive status), morphological and physiological constraints and human actions (e.g., disturbance). Although each of these factors may influence distribution, combinations of factors are generally responsible for the distribution patterns we observe. Pinniped distribution is also three-dimensional, where the third dimension is water depth and the underlying bathymetry. Although a complete understanding of pinniped distribution must consider this three-dimensional world, this aspect of pinniped behavior is discussed in sections on Energetics, Telemetry, and Diving Behavior.

Pinniped species have a restricted and generally patchy distribution in most aquatic environments: estuaries and continental shelves (e.g., gray seals), tropical seas (e.g., monk seals [*Monachus* spp.], Galapagos fur seals [*Arctocephalus galapagoensis*]), the deep ocean (e.g., elephant seals [*Mirounga* spp.]), Arctic (e.g., ringed seals) and Antarctic polar seas (e.g., crabeater seals, Antarctic fur seals), and freshwater lakes (e.g., Baikal seals [*Pusa sibirica*]) (King, 1983). However, our understanding of the distribution of most species is based primarily on the location of breeding colonies. We know considerably less about where most species forage at sea, and our view of overall distribution is therefore incomplete. For example, based on the location of breeding colonies, northern elephant seals range from Baja California to central California. However, satellite telemetry studies show that this species forages over broad areas of the North Pacific Ocean for much of the year. This new information has dramatically changed our understanding of the ecology of this and a growing number of species (e.g., gray seal, harbor seals, and southern elephant seals [*Mirounga leonina*]).

The distributions of pinniped breeding colonies seem to reflect the evolutionary history of pinnipeds and the distribution of resources. At large scales, both sea lion and fur seal distributions reflect their origins in the Pacific Ocean. Northern fur seals and Steller sea lions (*Eumetopias jubatus*) are widely distributed along both sides of the North Pacific Ocean. The four other species of sea lions occupy colonies along the west coast of South America, southern Australia, and New Zealand. With the exception of the northern fur seal and Guadalupe fur seal, the other six species of fur seals occur in tropical or subtropical southern waters, but also extend into the cool, nutrient-rich waters of the South Atlantic and Indian Oceans. Sea lion and fur seal breeding colonies are usually located on remote islands near areas of high biological productivity (e.g., northern fur seals, Antarctic fur seals), which provide both protection from mainland predators and nearby food sources. These conditions are particularly important for lactating females.

Species of the Family Phocidae are widely distributed in biologically productive temperate and polar seas. Although most abundant in the North Atlantic and Antarctic Oceans, a reflection of their evolutionary origins in the Atlantic basin during the middle Miocene, phocid species have circumpolar distributions in both the Arctic Ocean (e.g., ringed seal, bearded seal [*Erignathus barbatus*]) and Antarctic Ocean (e.g., Weddell seal, crabeater seal), as well as a broad distribution in the North Pacific Ocean (e.g., harbor seal, largha seal [*Phoca largha*], ribbon seal [*Histriophoca fasciata*]). Several endangered species also occur in tropical waters (Hawaiian [*Monachus schauinslandi*] and Mediterranean monk seals).

Pinnipeds must return to a solid substrate (land or ice) to give birth, rear their offspring, and in many species to molt. For most species, these requirements result in seasonal changes in distribution. In the case of species that breed on pack ice, such as harp and hooded (*Cystophora cristata*) seals and the walrus (*Odobenus rosmarus*),

P

seasonal changes in ice cover virtually guarantee changes in distribution. This may partly explain why 7 of 13 (54%) species of pinnipeds that give birth on ice (i.e., most phocid seals and the walrus) are migratory, compared to only 4 of 20 (20%) species that give birth on land (2 of 6 phocids, 2 of 14 otariids; Bowen and Siniff, 1999). However, this difference also may be partly explained by the variable quality of data on the at-sea distribution of pinnipeds.

Migration appears to be a common feature of the ice-breeding phocid species, but this behavior is perhaps best documented in the northern elephant seal. This land-breeding species shows extreme sexual size-dimorphism, with males being about five times heavier than females. Northern elephant seals undertake the longest known migration and some of the deepest dives reported for a mammal (Stewart and DeLong, 1993). Individual elephant seals make two long-distance migrations of 18,000–21,000 km between breeding and molting sites in California and pelagic foraging areas in the North Pacific. Using the California Current as a corridor to areas further north, northern elephant seals leave the breeding beaches in southern California for northern offshore foraging areas. The first migration occurs following the breeding season, in which adult male and female elephant seals travel an average of 11,967 and 6289 km respectively, and remain at sea for an average of 124 and 73 days. After the molt, the seals depart on a second migration; females are at sea for approximately twice as long as males and cover an average distance of 12,264 km compared to an average of 9608 km by males. Males migrate farther north than females, with most males traveling as far as the northern Gulf of Alaska and the eastern Aleutian Islands. These sex differences in foraging distribution, and presumably diet, may have evolved to reduce competition between females and males.

## III. Reproductive Ecology

The reproductive ecology of pinniped species share features that reflect their common ancestry as terrestrial carnivores, and their subsequent adaptation to a predominately aquatic lifestyle. As noted previously, a conserved trait of their terrestrial ancestry is the requirement for all pinniped species to give birth to their offspring on a solid substrate (land or ice). However, pinnipeds must feed at sea, often some distance from the breeding grounds. This spatial and temporal separation of parturition from aquatic foraging is thought to have played a large role in shaping the mating and lactation strategies of pinnipeds. Three general strategies have evolved to deal with the conflict between at-sea foraging and terrestrial parturition (see below); however, the requirement for terrestrial parturition has likely contributed to some common features of pinniped reproduction, such as birth synchrony.

In most pinniped species, reproduction is seasonal and highly synchronous (e.g., harp seals). The evolution of reproductive synchrony is often associated with seasonal resource availability. In ice-breeding species (e.g., harp and hooded seals), the timing of reproduction is linked to the seasonal availability of sea ice. Seasonal changes in prey abundance and environmental conditions can also influence the timing of parturition and mating. The Hawaiian monk seal displays only weak synchrony in reproduction. In this species, births extend over a 6-month period. Given the less variable tropical habitat of this species, reproductive synchrony may not have been under strong selection relative to species in more variable temperate and polar environments. Subtropical populations of California sea lions (*Zalophus californianus californianus*) and Galapagos fur seals also show slightly less temporal synchrony of reproduction relative to more temperate populations (Boness, 1991).

Departures from the annual cycle of reproduction are found in several species. The Australian sea lion (*Neophoca cinerea*) has a cycle lasting 18 months, resulting in a seasonal pattern of births. Similarly, the walrus has a reproductive cycle of 2 years, including a 15-month gestation period, in which the period of births remains seasonal.

Other common features of pinniped reproduction include postpartum mating and delayed implantation. These two characteristics of pinniped reproduction also appear to reflect the terrestrial ancestry of the taxa, with both features occurring in many modern terrestrial carnivores (see Female Reproductive Systems). However, selection for postpartum mating may have continued as pinnipeds adapted to their aquatic environment. Given the wide-ranging and dispersed distribution of pinniped species during the at-sea foraging season, the aggregation of individuals at pupping colonies may have offered one of the few predictable opportunities for males and females to mate.

Another common feature of pinniped reproduction is the production of a single, precocious offspring; litters of two are rare. Offsprings are born with their eyes open and begin to vocalize within minutes of birth. Neonates are also able to move short distances to their mother and begin suckling shortly after birth. Harbor seal females produce extremely precocial offspring that are capable of swimming and diving with their mothers within an hour after birth (Bowen, 1991).

### A. Mating Systems

Within the order Pinnipedia, mating systems range from extreme polygyny (e.g., northern fur seals) to sequential defense by males of individual females. Mating systems are closely associated with the dimensionality and stability of the habitat used, and distribution of females at parturition. Broadly speaking, species can be grouped as land-breeding and aquatic-breeding species.

*1. Land-Breeding Species* Land-breeding pinniped species include all fur seals and sea lions, northern and southern elephant seals, and the gray seal. These species colonize oceanic islands and coastal areas to give birth and mate. The aggregation of individuals during the breeding season has been attributed to the fact that oceanic islands are relatively rare and unevenly dispersed, such that the availability of suitable pupping sites may limit the distribution of females (Boness, 1991). Predation may also select for female clustering, with females being less vulnerable to terrestrial predators and/or harassment by conspecific males when in large groups (dilution effect). Aggregation of females within a stable, two-dimensional habitat has led to the evolution of polygyny in these species, with males defending either resources needed by females (e.g., birth and thermoregulatory sites in otariid species) or the females themselves (e.g., elephant seals and gray seals). By competing with and limiting the access of other males to females, successful males mate with multiple females, thus increasing their reproductive success. The degree of polygyny in land-breeding pinniped species ranges from extreme in the northern fur seal and elephant seals where one male may mate with 16–100 females, to moderate (6–15 females) in gray seals, Hooker sea lions (*Phocarctos hookeri*) and the Galapagos fur seal (Le Boeuf, 1991).

As in other polygynous species, land-breeding pinniped species are sexually size-dimorphic. Males in these species can be much larger than females and often show other secondary sex characteristics. These dimorphic characteristics are the result of sexual selection for traits that increase an individual's ability to monopolize and defend resources needed by females or females themselves. Large body size, and concomitant body energy stores in the form of subcutaneous blubber, permits dominant males to fast and thus remain ashore during the period when females become receptive. The most extreme example of sexual size dimorphism in pinnipeds occurs in elephant seals, where males are 5–6 times heavier than females in the northern species and up to 10 times heavier than females in the southern species.

*2. Aquatic-Breeding Species* Walruses and all other phocid seals (Weddell, Ross [*Ommatophoca rossii*], crabeater, leopard (*Hydrurga leptonyx*), bearded, hooded, ringed, Baikal, Caspian (*Pusa caspica*), spotted, harp and ribbon) give birth on pack ice or fast ice and mate in the water. Although Hawaiian monk seals and harbor seals give birth to their offspring on land, they too mate in the water. In species where pups are born on ice, females tend to be more widely distributed, although access to breathing holes in the ice may promote clumping in some species (e.g., walrus and Weddell seals). This broader distribution of females, on an unstable habitat, limits the number of females a male can monopolize at any given time, and as a result these species typically show reduced levels of polygyny (e.g., harbor seals; Coltman *et al.*, 1999). The fact that mating occurs in the water, a fluid three-dimensional environment, also may limit the ability of males to monopolize females resulting in reduced levels of polygyny.

Wells *et al.* (1999) classified the mating strategies used by ice-breeding species as: scrambling-males search for receptive females and move on to the next, sequential defense-males sequentially defend single females through mating, and lekking-males aggregate and attract females using displays. At present, there is insufficient information on the breeding behavior of most aquatic breeding species to draw firm conclusions about the type of mating system used. Until recently, data on the mating behavior of these species is limited to that which can be observed on ice prior to copulation. For example, observational data suggest that hooded seals use a sequential defense mating system whereby males compete with one another to defend a single female and her pup on the ice. The dominant male remains with the pair until the pup is weaned and then enters the water with the female, presumably to mate. However, the application of newer methods, including genetic paternity assessment, animal-borne video, and positional analysis of vocalizations have clarified the mating systems of harbor seals (Boness *et al.*, 2006) and bearded seals (Van Parijs and Clark, 2006).

In species that mate aquatically, there may be less selective advantage for males to be larger than females because of the limited ability of males to monopolize females in this environment. As a consequence, in most of these species, males and females are of similar size and in some cases females are larger than males. For example, male Weddell seals are slightly smaller than females and it has been suggested that smaller size makes the male more agile during underwater mating activities (Le Boeuf, 1991). Underwater vocalizations also appear to be an important component of the mating behavior in aquatically mating pinniped species. For example, in Pacific walruses, which exhibit a lekking mating system, males perform complex underwater visual and vocal displays in small groups next to female haulout sites to attract females. Male Weddell, harbor, harp, hooded, and bearded seals also produce a range of underwater vocalizations during the breeding season that may be used to attract females or to establish underwater territories or display areas.

## B. Lactation Strategies

Male pinnipeds do not participate in the care of the offspring. Thus, parental care is the exclusive responsibility of the female. Female care involves the transfer of energy-rich milk to the pup, and protection from conspecifics and terrestrial predators (Bowen, 1991).

In some species (e.g., the walrus), females may also teach their young to forage, as young accompany mothers on foraging trips during the lactation period. Female pinnipeds have dealt with the temporal and spatial separation of energy acquisition (aquatic foraging) from high levels of energy expenditure (terrestrial lactation) in different ways, resulting in the three basic lactation strategies: long lactation length and foraging cycle, short lactation length and fasting, and long lactation length and aquatic nursing. Although maternal body size has long been thought to have been an important trait in the evolution of these strategies, on the basis of a comparative analysis of 12 life-history and ecological traits, Schulz and Bowen (2005) concluded that there is little evidence for the influence of body size on lactation length. The patterns we see today appear to reflect an early divergence in body size between otariids and phocids, which influenced their foraging strategies and metabolic rates and subsequently influenced lactation strategies. Abbreviated lactation seems to represent an adaptation for minimizing the relative milk energy expended over lactation, but may also have been selected to reduce terrestrial predation and the uncertainty of breeding on unpredictable pack ice.

*1. Foraging Cycle* All otariids and some of the smaller phocid species (e.g., harbor seals) exhibit this lactation strategy. Females come ashore for parturition with a moderate level of stored body energy. After giving birth, females remain onshore and fast while attending and nursing their young for a perinatal period ranging from a few days to a week. After this initial provisioning period, females leave their pups and return to sea to feed. These trips range from less than 1 day in some species to as long as 23 days in others, depending on the distance to the foraging location and prey abundance. Females then return to land to nurse their pup, after which they repeat the cycle until the pup is weaned. The lactation period in otariid species is quite long, ranging from 4 months to 3 years (Bowen, 1991). Females of these species are considered income breeders, relying on current food intake to support both their own metabolic needs and the energetic cost of milk production. The milk produced by female otariids is relatively energy-dense (24–40% fat) compared to terrestrial mammalian species. Pup growth rates are rather low, ranging from 0.06 kg/day in Galapagos fur seals to 0.38 kg/day in Steller sea lions (Boness and Bowen, 1996).

Harbor seals, a phocid species, also exhibit a form of this lactation strategy—alternating short foraging trips to sea (7–10 h) with terrestrial nursing. The harbor seal is a relatively small phocid species, with females weighing approximately 84 kg at parturition. Given the small quantity of body energy that these females are able to store, female harbor seals are forced to make regular foraging trips to acquire sufficient energy to successfully wean their pups. Compared to otariid species, the length of the lactation period in harbor seals is much shorter (24 days) and the milk produced by females has a relatively higher fat content (50%). Consequently, pup growth rate is higher in harbor seals relative to otariid species (0.6 kg/day). Foraging cycles during lactation also may occur in ringed seals, and other relatively small phocid species. There is evidence that the females of two medium size phocids, the Weddell seal and the harp seal, may also forage during the lactation period. However, as noted above, the intensity of foraging and the degree to which successful weaning of offspring relies on these foraging trips is not clearly understood. Although small body size of some phocid species may limit females to a lactation strategy similar to otariids, difference in mammary gland structure between otariids and phocids may constrain the ability of phocid females from evolving a full-fledged foraging cycle (Schulz and Bowen, 2005).

P

*2. Fasting Strategy* In the larger-bodied phocid species, females fast during lactation. Females arrive at the breeding site with large energy stores in the form of adipose tissue (i.e., blubber). In the western Atlantic, for example, gray seal females arrive at Sable Island weighing an average of 210 kg. Of this body mass, 32% or 67 kg is fat. After parturition, females fast for the entire lactation period (e.g., 16 days in the case of gray seals), using their stored energy to support the energetic cost of milk production and their own maintenance metabolism. For this reason, female phocids are considered capital breeders—having stored energy often months before it is needed. The lactation period in phocids is much shorter than in otariid species ranging from 4 days in hooded seals to 60 days in Weddell seals. Another characteristic feature of the phocid fasting strategy is the production of extremely high fat milk, ranging from 47% fat in southern elephant seals to 61% fat in hooded seals. This energy-dense milk results in a high rate of offspring growth, ranging from 1.4 kg/day in the Hawaiian monk seal to 7.1 kg/day in the hooded seal (Bowen, 1991). Weaning occurs abruptly when mothers return to sea to feed. Pups often fast for weeks following weaning, living off their accumulated fat stores before entering the water and beginning to forage independently. Unstable breeding habitat, reduction in the fraction of energy expenditure devoted to maternal vs offspring requirements, and increased efficiency of milk energy transfer to offspring leading to higher growth rates all appear to have favored the evolution of the fasting strategy (Schulz and Bowen, 2005).

*3. Aquatic Nursing* The walrus is the only pinniped species that exhibits aquatic nursing. Just prior to parturition, pregnant females separate from the herd and give birth to their offspring alone on pack ice. New mothers remain on the ice, fasting for the first few days postpartum, and relying on stored body energy accumulated prior to parturition. Subsequently, females and their young return to the herd to forage. Walrus pups suckle in the water for between 2 and 3 years on relatively low-fat milk (24.1%). As with otariids, weaning is gradual. Young walruses begin to feed on benthic organisms as early as 5 months of age and likely gain valuable foraging experience from their mothers over the remainder of lactation. At weaning, female offspring are assimilated into the mother's herd, whereas male offspring join other male groups.

Lactation strategies are often viewed from the female's perspective. This seems reasonable, but in long-lived species such as pinnipeds, females may trade-off investment in current offspring against investment in future offspring. This may lead to conflicts between females and their offspring over the level of investment received. The transition from nursing pup to nutritionally independent juvenile usually occurs without parental supervision in pinnipeds. This transition is arguably the most important period of a pinniped's young life. As offspring size affects subsequent survival, we should expect that offspring would attempt to obtain as much milk as they can during lactation. Thus, the nutritional requirements and physiological abilities of individual offspring also must play a role in shaping lactation strategies. For example, the fasting ability of offspring constrains the duration of foraging trips by female fur seals and sea lions.

## IV. Foraging

Successful foraging is essential for survival and reproduction, and is therefore a critical determinant of fitness. Pinnipeds are among the largest vertebrate carnivores in marine ecosystems, and yet the foraging behavior of these upper-trophic level predators is generally poorly understood. As noted earlier, pinnipeds inhabit diverse habitats; consequently they forage over highly variable spatial and temporal scales, and in doing so they exploit a wide range of prey.

### A. Methods

As pinnipeds generally feed under water at remote locations, ecologists rely upon indirect methods to gain insight into their foraging behavior and diets. Very high frequency (VHF) radio tags have been used to study the at-sea locations of coastal species such as harbor seals. Acoustic tags have been used to track the underwater movements of gray seals. More recently, microprocessor-based, time-depth recorders (TDRs) have been used to collect information on dive duration, depth, frequency, and temporal distribution and to calculate the at-sea locations of pinnipeds using solar navigation equations. However, the use of TDRs is often limited by the need to recover the instrument to retrieve the stored information and therefore only those species which can be reliably recaptured are used in TDR studies. In contrast, satellite-linked, time-depth recorders (SLTDRs) transmit collected data on diving parameters and surface positions to polar-orbiting satellites operated by Service Argos. This technology has broadened the range of species that have been studied, but the expense of using satellite-linked tags often places limits on the number of individuals studied. A new generation of tags, using fastloc GPS, is providing more accurate locations to permit finer scale studies of foraging behavior and habitat use.

Although we have learned a great deal from the use of location telemetry and dive recorders, these studies have provided little insight into the feeding success rate of pinnipeds. Recent work has demonstrated that estimates of feeding success can be determined using stomach-temperature telemetry and animal-borne video. The body temperature of marine prey is often considerably lower than that of its pinniped predator, thus the stomach temperature of the predator should drop following prey ingestion. This approach has been used successfully on free-living gray seals (Austin *et al.*, 2006a) and several other species. When combined with information on the diving behavior and movement patterns in the same individual, stomach telemetry can provide new insights into the spatial and temporal patterns of foraging success relative to foraging effort (Austin *et al.*, 2006b). Animal-borne video technology has taken our understanding of foraging behavior and diet one step further by providing direct observations of the way in which pinnipeds search for and capture prey, and how foraging behavior changes as a function of prey type (Bowen *et al.*, 2002).

Determining the diet of marine mammals also requires the use of indirect methods. The most common methods rely on the recovery and identification of hard prey structures that are resistant to digestion from the stomach, intestine, or feces of individual animals. Sagittal otoliths, cephalopod beaks, bones, scales, invertebrate exoskeletons, and shells can be used to determine the species consumed, and in some cases, to estimate the size and age of the prey. Fecal samples are increasingly being used for this purpose because they are less expensive to collect; a high proportion of samples contain identifiable prey, and estimates of diet are less affected by differential rates of digestion than are estimates from stomach samples (Bowen and Siniff, 1999). Although the use of hard parts to estimate the diet of pinnipeds is common, this method is subject to a number of biases, which may limit the value of results. Firstly, stomach and fecal contents only provide an estimate of the diet near the point of collection, and as a result, offshore diets cannot easily be sampled. This may seriously bias the diet of wide-ranging species such as elephant seals, harp seals, northern fur seals, and Juan Fernandez fur

P

seals (*Arctocephalus philippii*). Secondly, hard parts are often eroded during digestion or are completely digested, such that prey size may be seriously underestimated and prey identification may not be possible. Finally, dietary analysis based on hard parts is strongly biased against soft-bodied or small prey with fragile structures.

Inevitably, our understanding of the diets of pinnipeds is tied to the development of new methods. One such method is fatty acid signature analysis and its quantitative formulation, called quantitative fatty acid signature analysis (QFASA) (Iverson *et al.*, 2004). Lipids in marine ecosystems are diverse and characterized by long-chain polyunsaturated fatty acids that originate in unicellular phytoplankton. In carnivores such as pinnipeds, ingested fatty acids with a carbon chain length greater than 14 are deposited in body tissues in a predictable way. As a result, the fatty acid composition or signature of the predator reflects (but will never be the same as) the fatty acid composition of the prey species consumed (Iverson *et al.*, 2004, also see Blubber, this volume). By comparing the reference signature of various prey species to the fatty acid signature of the predator, obtained from blubber tissue or milk, diet composition can be estimated. The use of fatty acid signature analysis eliminates the dependence on recovery of hard parts and integrates the diet over a period of weeks to months, such that the location of sampling becomes less important. Nevertheless, QFASA depends on having determined the fatty acid composition of potential prey species; there is always the possibility that some prey species may not be reliably distinguish on the basis of fatty acids, leading to false positive identification of prey consumed.

Stable isotope ratios of carbon and nitrogen found in the muscle, skin, vibrissae, or blood of pinnipeds and other predators are also being used to investigate diet. These ratios reflect a composite of prey species eaten over a broad time scale. By examining the levels of $^{15}N/^{14}N$ found in body tissues, researchers can determine the trophic level at which the pinnipeds fed. The carbon isotope ratio ($^{13}C/^{12}C$) has been found to vary geographically, and thus the level of carbon isotope in the predator's tissues provides insight into foraging location. While this technique is useful in determining trophic level and foraging location, it does not permit the diet composition of individuals to be estimated, except when only a few prey species are consumed.

Finally, recent studies have indicated that the identification of prey species' DNA recovered from pinniped feces can provide qualitative information on prey consumed, and may eventually also yield quantitative diet estimates (Deagle and Tollit, 2007).

## B. Diet

A large number of prey species have been identified in the diet of pinniped species, leading to the view that pinnipeds are generalist predators. This is consistent with the expectation that large wide-ranging predators consume more types of prey as their environment becomes patchier. However, in most cases, a relatively small number of species account for the majority of food eaten (Bowen and Siniff, 1999). For example, gray seals on the Scotian Shelf, Canada, consumed 24 different taxa, but only 2–4 species accounted for over 80% of the energy consumed, depending on the time of the year.

Fish and cephalopod species are the main prey types eaten by pinnipeds (Table I). However, crustaceans also appear to account for a substantial portion of prey consumed by some species. Crustaceans are a major prey of harp seals in the North Atlantic, and of ringed seals and bearded seals in the Bering Sea. In three Antarctic species, Antarctic fur seals, crabeater seals, and leopard seals, krill accounts for up to 50% of the diet. Unlike most pinnipeds, which generally feed on mobile prey (e.g., fishes, cephalopod molluscs, and crustaceans) in pelagic and benthic habitats, the walrus feed almost exclusively on sessile benthic invertebrates in soft-bottom sediments.

Several pinniped species are also known to feed on other pinnipeds (Bowen and Siniff, 1999). There is evidence that adult male South American fur seals (*Arctocephalus australis*) commonly feed on conspecific young. Steller sea lions are known to prey on a variety of pinniped species including harbor seals, ringed seals, bearded seals, young northern fur seals, and spotted seals. Walruses prey on spotted seals, ringed seals, and young bearded seals.

The diet and foraging behavior of pinnipeds are influenced by a number of factors. The ecology and behavior of prey species clearly play a role in shaping the foraging tactics of pinnipeds. Research on the foraging behavior of adult male harbor seals at Sable Island, Canada, using animal-borne video, showed that prey behavior affected both capture technique and profitability of different prey types. Other studies have shown that between-year differences in diet composition of harbor seals were correlated with differences in the distribution and abundance of herring (*Clupea* spp.) and sprat, two important prey species in Scotland.

Intrinsic factors such as age and sex may also play a role in the diet composition of individuals within pinniped species. Given that pinnipeds are long-lived predators, their individual foraging tactics and behavior may change over time to reflect increased physiological capabilities and learning. For example, harbor seal pups feed on pelagic prey such as herring and squid, whereas the diet of adults is dominated by benthic species. Similarly, the contribution of benthic prey (e.g., crabs, clams, and sculpins) to the diet of bearded seals increases with age. Age-specific differences in diet composition have also been found in southern elephant seals and harp seals.

In pinniped species that exhibit sexual body-size dimorphism, the diets of adult males and females may differ due to the relationship between energy requirements and body mass, whereby larger individuals require more total energy per unit time than do smaller individuals. Oxygen storage capacity also increases with body mass due to the larger blood pool in which to store oxygen and the larger muscle (myoglobin) mass. In addition, larger animals have a slower mass-specific metabolic rate, such that they utilize their larger oxygen stores at a slower rate relative to smaller individuals. Thus, larger individuals are capable of longer and deeper foraging dives. These physiological attributes may allow, or require, males (the larger sex) to pursue different prey types (potentially higher quality prey) than females. Males and females may also consume different prey to reduce the effects of intraspecific competition for food, and as a reflection of differences in the timing of reproductive costs. Gray seals are a good example of these effects acting on males and females to produce seasonal differences in both the energy density and diversity of the diet (Beck *et al.*, 2007).

## C. Foraging and Diving Behavior

The foraging ecology of pinnipeds and other air-breathing vertebrates is constrained by the need to surface for oxygen. Dive duration is constrained by the interplay between the amount of oxygen that can be stored and the rate at which the diver expends oxygen. Thus, it is inevitable that the distribution of foraging in time and space will be influenced by the physiological constraints. Other factors, such as prey characteristics, presence of competitors, and predators also play an important role in how pinnipeds forage within these physiological constraints.

Foraging pinnipeds dive repeatedly with relatively short surface intervals between dives; this cluster of dives is called a dive bout. In

**P**

TABLE I
Major Prey of Selected Pinnipeds

| Species | Location | Main prey |
|---|---|---|
| Gray seal (*Halichoerus grypus*) | Eastern Canada | Sandlance, redfish |
| | United Kingdom | Sandlance |
| Harbor seal (*Phoca vitulina*) | Eastern Canada | Herring, Atlantic cod (*Gadus morhua*), pollock (*Pollachius* spp.), squid |
| | Western Canada | Pacific hake (*Merluccius productus*), Pacific herring (*Clupea pallasii*) |
| | Sweden | Atlantic cod, sole, herring, sandlance |
| Harp seal (*Pagophilus groenlandicus*) | Northwest Atlantic | Arctic cod (*Arctogadus glacialis*), herring, capelin (*Mallotus villosus*) |
| | White Sea/East ice | Capelin, sandlance, herring |
| Hooded seal (*Cystophora cristata*) | Greenland | Greenland halibut (*Reinhardtius hippoglossoides*), redfish, Gadidae |
| Ringed seal (*Pusa hispida*) | Bering Sea | Saffron cod (*Eleginus gracilis*), Arctic cod, shrimps |
| Ribbon seal (*Histriophoca fasciata*) | Bering Sea | Pollock, eelpout, Saffron cod |
| Bearded seal (*Erignathus barbatus*) | Bering Sea | Shrimp, crab, clam |
| S. elephant seal (*Mirounga leonina*) | Heard/Macquarie Island | Squids, pelagic and benthic fishes |
| | Heard Island | Squids, pelagic fishes |
| N. elephant seal (*M. angustirostris*) | California | Cephalopods, Pacific whiting (hake) |
| Leopard seal (*Hydrurga leptonyx*) | Southern Ocean | Krill, cephalopods, penguins, seals |
| Northern fur seal (*Callorhinus ursinus*) | North Pacific | Anchovy, herring, capelin, sandlance |
| | Bering Sea | Pollock, capelin, herring, squids |
| South African fur seal (*Arctocephalus pusillus pusillus*) | Benquela Current | Anchovy, hakes, squid |
| Antarctic fur seal (*Arctocephalus gazella*) | South Georgia | Krill, cephalopods, fish |
| Sub-Antarctic fur seal (*A. tropicalis*) | Gough Island | Squids |
| Australian fur seal (*A. pusillus doriferus*) | Tasmania | Squids |
| South American fur seal (*A. australis*) | Peru | Sardine, southern anchovy, jack mackerel (*Trachurus* spp.) |
| Juan Fernández fur seal (*A. philippii*) | Alejando Selkirk Island | Myctophid fishes, squid |
| New Zealand fur seal (*A. forsteri*) | New Zealand | Octopus, squid, barracuda |
| Steller sea lion (*Eumetopias jubatus*) | Gulf of Alaska | Pollock, herring, squids |
| California sea lion (*Zalophus californianus*) | California | Northern anchovy, Pacific whiting, squid |

general, dive bouts are thought to indicate foraging within a prey patch, particularly in otariid species. Theoretically, divers should organize their behavior for optimal use of prey. To organize their behavior in this way, divers should optimize both the time budget of the dive cycle (dive duration and surface interval) and the number of dive cycles to repeat. Both of these factors will influence the amount of prey caught and the energy and time consumed during the dive bout. However, there may be a trade-off between prey depth and profitability, such that prey items that might be exploited when closer to the surface are less likely to be exploited as the depth of that prey increases.

Empirical tests of optimal foraging theory and optimal patch use in diving pinnipeds are uncommon, largely due to the difficulty and expense of studying these wide-ranging predators and their prey. However, it appears that some otariids feed near the surface on vertically migrating prey, such as krill, to maximize energetic efficiency.

Phocids are generally better suited for deep diving and for longer periods of time than are their otariid and odobenid counterparts. This is largely because phocids have a larger blood volume and larger myoglobin content in the muscles and thus store more oxygen

per unit of body mass. Phocids also dive in continuous bouts and are known to spend up to 90% of their time in the water submerged. Thus, unlike otariids and odobenids, phocid seals live at depth, periodically returning to the surface to breathe. Although diving behavior is often considered to be synonymous with foraging in otariids, dive shape analysis in phocids demonstrates that diving may also be used for travel, predator avoidance, and sleep (Wells *et al.*, 1999).

## D. Factors Affecting Foraging Ecology

Pinnipeds are no doubt important consumers of marine species; however, for most species relatively little is still known about the ecological factors affecting diet or foraging success. Knowledge of the at-sea movements of pinnipeds is important because spatial patterns can fundamentally affect the nature and dynamics of species interactions. These interactions largely determine the distribution of foraging. Within the ocean, food is distributed in patches and this distribution can be strongly influenced by the physical properties, such as water temperature and the availability of nutrients. For example, the distribution and migratory patterns of northern elephant seals correspond with the location of three dominant water masses of the North Pacific. The localized biological productivity in these water masses and associated fronts result in a high abundance of cephalopods, an important food of this species.

Seasonal changes in prey distribution and abundance can also influence pinniped foraging patterns. Reduced prey availability leads to changes in foraging behavior that include increased trip duration, trip distance, and increased foraging effort. For example, Antarctic fur seals increase their times at sea, northern fur seals increase diving effort, and California sea lions use both tactics during periods of limited prey resources.

## E. Spatial and Temporal Scales of Foraging

The relative mobility, range, and body size of an animal affects the resolution at which it recognizes environmental heterogeneity. For example, a relatively small-bodied, central place forager, such as a lactating harbor seal, would identify resource patches at a smaller mesoscale than would a highly mobile animal, such as a gray seal. To understand the relationship of an organism to its environment, one must understand the interactions between the intrinsic scales of heterogeneity within the environment and the scales at which the organism can respond to this heterogeneity. Scale issues are critical for effective conservation and management of pinnipeds because of shifts in habitat use and dispersal over ontogeny and a relatively long lifespan.

Large body size and the capacity for storing large amounts of fat in the form of blubber enables some species of pinnipeds to feed irregularly, and thus to exploit distant foraging locations and patchy resources. Austin *et al.* (2006b) found that in adult gray seals, trip duration and total time spent at depth during diving were the best predictors of the number of feeding events per day. However, the predictors differed at different temporal scales. At the 3-h scale, mean bottom time and distance traveled were the best predictors; but at longer scales, only distance traveled predicted the number of feeding events. In contrast, smaller pinnipeds, such as Antarctic fur seals, perceive environmental heterogeneity at a more local scale. For example, fur seals forage at two spatial distributions: (1) fine-scale, represented by short (<5 min) travel durations between patches, and (2) mesoscale, represented by longer periods of travel (>5 min) (Boyd, 1996). Similarly, based on fatty acid signature analysis, harbor seals appear to demonstrate mesoscale partitioning of foraging habitat in Prince William Sound, Alaska. Fatty acid signatures obtained from harbor seal blubber biopsies differed within the Sound at a spatial scale of about 40–50 km, and at a smaller scale of 9–25 km, reflecting fine-scale differences in diet between haul-out sites (Iverson *et al.*, 1997).

Although the patch structure of an environment is expressed in both space and time, temporal variation in predator behavior is likely to provide an insight into the spatial distribution of a highly dynamic prey source that may be difficult to track in other ways. For example, in the Antarctic Ocean, krill is patchily distributed and is the major prey resource of lactating Antarctic fur seals. By using the diving behavior of females obtained from TDR records, it is possible to track the way in which fur seals respond to within-season and inter-annual variation in prey patchiness and abundance. Over a 5-year period, changes in the distribution of travel durations between diving bouts suggested that the spatial distribution of krill swarms varied between years. Although their foraging behavior did not indicate that there was a reduction in the number of krill patches, reduced pup growth rates suggested that patches were of poorer quality, and thus the females had difficulty meeting lactation needs. To compensate,

females spent a greater amount of time at each patch, thereby maximizing their average rate of energy intake (Boyd, 1996).

To maximize fitness during years of reduced prey abundance, pinnipeds must be sufficiently plastic in their foraging strategies to compensate for added foraging costs. To determine the temporal scales at which predators make these behavioral decisions, Boyd and colleagues simulated increased foraging costs in Antarctic fur seals by adding an extra drag to lactating females, thereby increasing energy expenditure. At the scales of individual dives, the treatment group made shorter and shallower dives than did the control (no extra drag added) seals. It appeared that diving behavior was adjusted to maximize the proportion of time spent at the bottom of dives. At the scale of diving bouts, there was no variation between the two groups in bout frequency and duration, or the time spent diving. However, at the scale of complete foraging cycles, the time spent at sea was significantly longer in the treatment group, yet there was no difference in pup growth rate between control and treatment groups.

In contrast to otariids, most phocid seals are able to fast throughout much of the breeding season, owing to their large body size and corresponding energy stores. As a result, behavioral responses of phocids to changes in food availability between years may be more flexible, resulting in less severe effects on their population dynamics. Nevertheless, a change to less profitable prey or increased foraging effort may have energetic consequences that result in impacts at the population level. In the Moray Firth, Scotland, clupeid fishes are the dominant prey of harbor seals. In years when clupeids are absent from inshore waters, seals travel further to feed and use alternative prey. As a consequence, the seals showed evidence of reduced body condition, suggesting that there were energetic consequences to this change in diet. Between-year differences in survival rates suggest that temporal variation in prey abundance and resulting diets also have consequences for the dynamics of phocid populations (Thompson *et al.*, 1996).

## IV. Role of Pinnipeds in Aquatic Ecosystems

Although pinnipeds are one of the more visible components of the marine ecosystems, our understanding of their ecological roles is surprisingly limited. Still, there is some evidence that pinnipeds may have important effects on the structure and functioning of some ecosystems (Bowen, 1997). Given that pinnipeds are large, long-lived animals that are often present in considerable numbers, we might expect some species to exert top-down control on ecosystems. However, conclusive studies are lacking, largely owing to the difficulty of conducting manipulative experiments in the ocean, the fact that interactions occur at quite different spatial and temporal scales, and the inherent indeterminacy in the behavior of complex marine systems.

Ecosystem structure and functioning are influenced by both top-down (i.e., consumer driven) and bottom-up (i.e., producer driven) processes. Pinnipeds likely exert top-down control on some ecosystems through predation and are affected by changes in food available brought about by changes in primary and secondary productivity. An example of a top-down ecosystem perturbation affecting pinniped abundance occurred in the Southern Ocean in the early 1900s. The overexploitation of some species of seals and whales led to an enormous uncontrolled "experiment" in this cold-water ecosystem. A high biomass of Antarctic krill is the cornerstone of the Southern Ocean food web, accounting for half of total zooplankton biomass. Of the six species of pinniped which inhabit the Southern Ocean, crabeater seals, Antarctic fur seals, and leopard seals feed mainly on krill, while southern elephant seals and Ross seals consume mainly

P

cephalopods, and Weddell seals eat primarily fish. Krill is also the main food resource for the resident large baleen whales (blue [*Balaenoptera musculus*], sei [*B. borealis*], minke [*B. acutorostrata*], humpback [*Megaptera novaeangliae*], fin [*B. physalus*], and southern right [*Eubalaena australis*]). As the cetacean biomass declined from exploitation by more than 50% between 1904 and 1973, an estimated 150 million tons of krill were released annually to the remaining predators (Laws, 1985). The abundance of krill-eating species of pinniped, such as the crabeater seal and the Antarctic fur seal, increased substantially following the massive cetacean exploitation.

Bottom-up effects on top predators such as pinnipeds also can occur rapidly over the course of months. Perhaps the most dramatic example of this occurs during El Niño. El Niño events occur approximately every 4 years in the eastern tropical Pacific, resulting in reduced upwelling and a decrease in primary and secondary productivity. During a severe El Niño, the effects of reduced food availability on seabirds and marine mammals can be quite pronounced. Galapagos fur seals have an unusually long lactation period of approximately 2 years that is thought to have evolved to buffer young against minor El Niño events. However, during the severe El Niño event between August 1982 and July 1983, pup production of Galapagos fur seals was only 11% of previous years, and no pups survived past the first 5 months. Adult females responded by increasing foraging trip length, while most adult males did not appear on the breeding site and were unable to hold territories during the breeding season (Trillmich and Ono, 1991).

Top-down and bottom-up processes usually act simultaneously to influence the abundance of pinnipeds, although it may be difficult to determine their relative strengths. One example of this is illustrated by the dramatic declines in western population of Steller sea lion numbers in the Bering Sea, Aleutian Islands, and Gulf of Alaska beginning in the 1970s and continuing through the late 1990s. Both top-down (lethal takes by humans, killer whale [*Orcinus orca*] predation) and bottom-up (regime shift in ocean climate effects on food, commercial fishing effects on food) effects have been advanced to account for the observed declines. Although, specific causes remain uncertain, an important feature of the dynamics of the western population is that both declines and the recent increases in numbers were not spread uniformly throughout the population. This indicates that multiple factors have driven changes in Steller sea lion dynamics, but that both the nature and magnitude of those factors have differed over time and space.

One example of top-down control exerted by a pinniped species comes from a study of lakes in northern Quebec. Lower Seal Lake has a population of landlocked harbor seals, and compared to nine neighboring lakes without seals, supports a different fish community. The relative abundance of lake trout (*Salvelinus fontinalis*) was greater in the nine lakes without seals, whereas brook trout (*S. namaycush*) was the dominant species in Lower Seal Lake. Compared to lake trout in neighboring lakes, those in Lower Seal Lake were on average smaller, younger, grew more rapidly, and matured earlier, all of which represent life history characteristics that are associated with heavy exploitation. Although based on strong inference rather than direct empirical evidence, it appears that seal predation was responsible for both the changes in community structure and life history traits of fish species in Lower Seal Lake (reviewed in Bowen, 1997).

Pinnipeds may also play a role in structuring benthic communities. Walruses disturb bottom sediments during feeding. By selectively feeding on older individuals of a few species of bivalve molluscs, walruses may be responsible for structuring the benthic fauna. Ingestion and defecation by walruses may result in substantial redistribution of bottom sediments, which may favor colonization of some species. In addition, during the process of feeding, walruses produce many pits and furrows in the soft sediments. Thus, walrus feeding appears to affect community structure in three ways: by (1) providing food for scavengers such as sea stars and brittle stars, (2) providing habitat under discarded bivalve shells, and (3) reducing the abundance of macroinvertebrates in feeding pits compared to surrounding sediments. Nonetheless, the effects of walrus feeding behavior on macrobenthic assemblages over periods greater than a few months and at larger spatial scales remain unknown.

## V. Conclusions

Our understanding of the ecology of pinnipeds has increased dramatically over the past several decades, but advances have been rather uneven. For example, the mating systems of terrestrial species and lactation strategies of pinnipeds are reasonably well understood, as are both diving and ranging behavior of many species, but other aspects of foraging ecology, diet, and the ecological role of pinnipeds in aquatic ecosystems remain difficult to study. As in all areas of science, our ability to measure the system under study profoundly influences the rate of progress. New types of data-loggers, telemetry, and methods to estimate the diet of free-ranging pinnipeds have played an important part in recent advances in current understanding. As well, we should not underestimate the importance of collaborative research involving ecologists, oceanographers, and population and ecosystem modelers.

## *See Also the Following Articles*

■ Blubber ■ Diving Behavior ■ Energetics ■ Female Reproductive Systems and ■ Telemetry

## *References*

Austin, D., Bowen, D. W., McMillan, J. I., and Boness, D. J. (2006a). Stomach temperature telemetry reveals temporal patterns of foraging success in a free-ranging marine mammal. *J. Anim. Ecol.* **75**, 408–420.

Austin, D., Bowen, D. W., McMillan, J. I., and Iverson, S. I. (2006b). Consequences of behavior on foraging success: linking movement, diving and habitat to feeding in a large marine predator. *Ecology* **87**, 3095–3108.

Beck, C. A., Iverson, S. I., Bowen, D. W., and Blanchard, W. (2007). Sex differences in grey seal diet reflect seasonal variation in foraging behaviour and reproductive expenditure: evidence from quantitative fatty acid signature analysis. *J. Anim. Ecol.* **76**, 490–502.

Boness, D. J. (1991). Determinants of mating systems in the Otariidae (Pinnipedia). *In* "Behavior of Pinnipeds" (D. Renouf, ed.), pp. 1–44. Chapman and Hall, London.

Boness, D. J., and Bowen, W. D. (1996). The evolution of maternal care in pinnipeds. *BioScience* **46**, 645–654.

Boness, D. J., Bowen, D. W., Bulheier, B. M., and Marshall, G. J. (2006). Mating tactics and mating system of an aquatic-mating pinniped: the harbor seal, *Phoca vitulina*. *Behav. Ecol. Sociobiol.* **61**, 119–130.

Bowen, W. D. (1991). Behavioral ecology of pinniped neonates. *In* "Behavior of Pinnipeds" (D. Renouf, ed.), pp. 66–127. Chapman and Hall and Cambridge University Press, Cambridge.

Bowen, W. D. (1997). Role of marine mammals in aquatic ecosystems. *Mar. Ecol. Prog. Ser.* **158**, 267–274.

Bowen, W. D., Tully, D., Boness, D. J., Bulheier, B., and Marshall, G. (2002). Prey-dependent foraging tactics and prey profitability in a marine mammal. *Mar. Ecol. Prog. Ser.* **244**, 235–245.

Bowen, W. D., and Siniff, D. B. (1999). Distribution, population biology, and feeding ecology of marine mammals. *In* "Biology of Marine Mammals" (J. E. I. Reynolds, and S. A. Rommel, eds), pp. 423–484. Smithsonian Press, Washington, DC.

Boyd, I. L. (1996). Temporal scales of foraging in a marine predator. *Ecology* **77**, 426–434.

Coltman, D. W., Bowen, W. D., and Wright, J. M. (1999). A multivariate analysis of phenotype and paternity in male harbor seals, *Phoca vitulina*, at Sable Island, Nova Scotia. *Behav. Ecol.* **10**, 169–177.

Deagle, B. E., and Tollit, D. J. (2007). Quantitative analysis of prey DNA in pinniped faeces: potential to estimate diet composition. *Conserv. Genet.* **8**, 743–747.

Eberhardt, L. L., and Siniff, D. B. (1977). Population dynamics and marine mammal management policies. *J. Fish. Res. Board Can.* **34**, 183–190.

Iverson, S. J., Frost, K. J., and Lowry, L. F. (1997). Fatty acid signatures reveal fine scale structure of foraging distribution of harbor seals and their prey in Prince William Sound. *Mar. Ecol. Prog. Ser.* **151**, 255–271.

Iverson, S. J., Field, C., Bowen, W. D., and Blanchard, W. (2004). Quantitative fatty acid signature analysis: a new method of estimating predator diets. *Ecol. Monogr.* **74**, 211–235.

King, J. E. (1983). "Seals of the World." Comstock Publishing Associates, Ithaca, NY.

Laws, R. M. (1985). The ecology of the Southern ocean. *Am. Sci.* **73**, 26–40.

Le Boeuf, B. J. (1991). Pinniped mating systems on land, ice and in the water: emphasis on the Phocidae. *In* "Behavior of Pinnipeds" (D. Renouf, ed.), pp. 45–65. Chapman and Hall, Cambridge.

Reijnders, P., Brasseur, S., van der Torn, J., van der Wolf, Boyd, I., Harwood, J., Lavigne, D., *et al.* (1993). Status survey and conservation action plan: seals, fur seals, sea lions, and walrus. IUCN, Gland.

Schulz, T. M., and Bowen, W. D. (2005). The evolution of lactation strategies in pinnipeds: a phylogenetic analysis. *Ecol. Monogr.* **75**, 159–177.

Stewart, B. S., and Delong, R. L. (1993). Seasonal dispersal and habitat use of foraging northern elephant seals. *Symp. Zool. Soc. Lond.* **66**, 179–194.

Thompson, P. M., Tollit, D. J., Greenstreet, S. P. R., MacKay, A., and Corpe, H. M. (1996). Between-year variations in the diet and behavior of harbour seals, *Phoca vitulina* in the Moray Firth; causes and consequences. *In* "Aquatic Predators and their Prey" (S. P. R. Greenstreet, and M. L. Tasker, eds), pp. 44–52. Blackwell Scientific Publishing, Oxford.

Trillmich, F., and Ono, K. A. (1991). "Pinnipeds and El Niño: Responses to Environmental Stress." Springer-Verlag, Berlin.

Van Parijs, S. M., and Clark, C. W. (2006). Long-term mating tactics in an aquatic-mating pinniped, the bearded seal, *Erignathus bartatus*. *Anim. Behav.* **72**, 1269–1277.

Wells, R. S., Boness, D. J., and Rathbun, G. B. (1999). Behavior. *In* "Biology of Marine Mammals" (J. E. I. Reynolds, and S. A. Rommel, eds), pp. 324–422. Smithsonian Press, Washington, DC.

# Pinniped Evolution

## ANNALISA BERTA

The name Pinnipedia was first proposed for fin-footed carnivores more than a century ago. Pinnipeds—fur seals and sea lions, walruses and seals—are one of three major clades of modern marine mammals, having a fossil record going back at least to the late Oligocene (27–25 Ma—millions of years before present).

The earliest pinnipeds were aquatic carnivores with well-developed paddle shaped limbs and feet. A North Pacific origin for pinnipeds has been hypothesized; the group subsequently diversified throughout the world's oceans.

## I. Pinniped Ancestry: Origin and Affinities

There has long been a debate about the relationship of pinnipeds to one another and to other mammals. The traditional view, also referred to as diphyly, proposes that pinnipeds originated from two carnivore lineages, an odobenid (walrus) plus otariids (fur seals and sea lions) grouping affiliated with ursids (bears) and phocids (seals) being related to mustelids (weasels, skunks, otters, and kin) (Fig. 1a). The current view overwhelmingly supported by both morphologic and molecular data confirms pinnipeds as monophyletic (having a single origin). Although the hypothesis presented here positions ursids as the closest relatives of pinnipeds, it is acknowledged that there is difficulty separating the various lineages of arctoid carnivores (mustelids, procyonids, and ursids) at their point of divergence (Fig. 1b). An alternative hypothesis supports pinnipeds as having an ursid-mustelid ancestry (see PINNIPEDIA). With regard to relationships among pinnipeds most current data robustly supports a link between odobenids and otariids.

## II. Divergence of Major Lineages

The broad pattern of evolution within pinnipeds shows divergence of five major lineages. These include the three extant lineages: Otariidae, Phocidae, and Odobenidae and two extinct groups, the Desmatophocidae and a basal lineage *Enaliarctos* (Fig. 2). At times the Odobenidae have been included in the Otariidae, although current studies consistently support these as distinct monophyletic groups that share a sister group relationship.

Within Pinnipedimorpha (living pinnipeds plus their fossil allies) are included archaic pinnipeds *Enaliarctos* and *Pteronarctos* + pinnipeds (Fig. 2). Unequivocal synapomorphies include: large infraorbital foramen, anterior palatine foramina anterior of maxillary-palatine suture, upper molars reduced in size, lower first molar metaconid reduced or absent, humerus short and robust, deltopectoral crest on the humerus strongly developed, digit I on the manus and digit I and V on the pes emphasized. The basal taxon *Enaliarctos* from the late Oligocene and early Miocene (27–18 Ma; Fig. 3) of California is known by five species, one represented by a nearly complete skeleton (Fig. 4). *Enaliarctos* was a small, fully aquatic pinnipedimorph with shearing teeth (as is typical of most terrestrial carnivorans), flexible spine, and fore and hindlimbs modified as flippers. Several features of the hindlimb suggest that *Enaliarctos* was capable of maneuvering on land although probably spent more time near the shore than extant pinnipeds. *Enaliarctos* shows features that are consistent with both fore- and hindlimb swimming, but seems slightly more specialized for forelimb swimming (Berta and Adam, 2001). A later diverging lineage more closely allied with pinnipeds than with *Enaliarctos* is *Pteronarctos* from the late Miocene (19–15 Ma) of coastal Oregon. *Pteronarctos* is recognized as the earliest pinniped to have evolved the unique maxilla diagnostic of modern pinnipeds. The maxilla of pinnipeds makes a significant contribution to the orbital wall. This differs from the condition in terrestrial carnivores in which the maxilla is limited in its posterior extent by contact of several facial bones (jugal, palatine, and/or lacrimal). Ecologically, the earliest pinnipedimorphs were coastal dwellers that evolved a pierce feeding strategy and likely fed on fish and other aquatic prey (Adam and Berta, 2002).

The fur seals and sea lions (eared seals), the Otariidae, are diagnosed by frontals that extend anteriorly between the nasals, large and

**P**

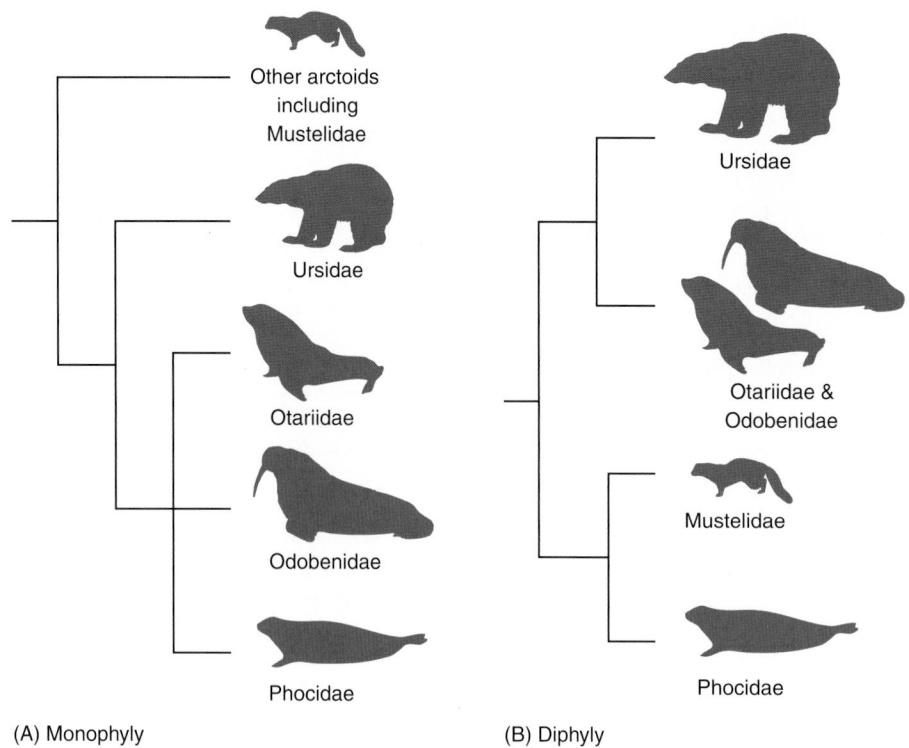

**Figure 1** *Alternative hypotheses for relationships among pinnipeds (A) Monophyly with ursids as the closest pinniped relatives. (B) Diphyly in which phocids and mustelids are united as sister taxa as are otariids, odobenids, and ursids. (From Berta et al., 2006.)*

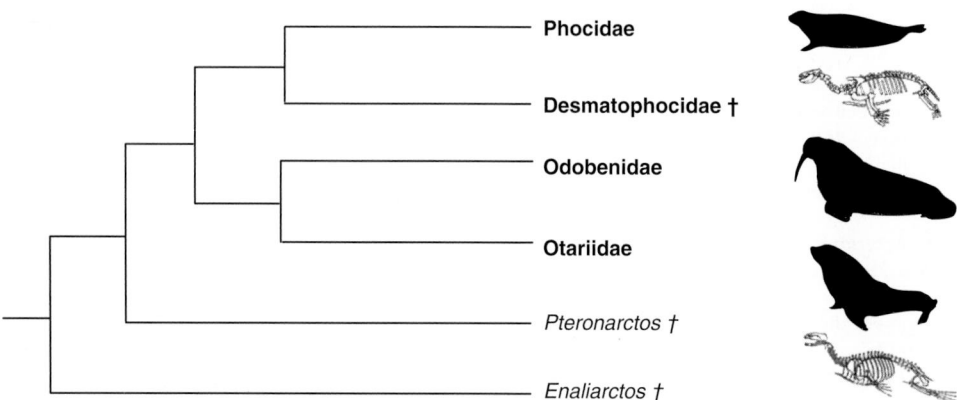

**Figure 2** *A cladogram depicting the relationships of the major clades of pinnipeds. (Modified from Berta et al., 2006.)*

shelf-like supraorbital process of the frontal, secondary spine dividing the supraspinous fossa of the scapula, uniformly spaced pelage units, and by the presence of a trachea with an anterior bifurcation of the bronchi (Fig. 5). Although otariids are often divided into two subfamilies, the Otariinae (sea lions) and the Arctocephaline (fur seals), both groups have been shown to be paraphyletic on the basis of molecular data. The otariids are the earliest diverging pinniped lineage originating approximately 11–12 Ma (Fig. 3) and including the poorly known *Pithanotaria* from the late Miocene of California and three species of *Thalassoleon* described from the late Miocene and early Pliocene of California, Mexico and Japan (Deméré and

Berta, 2005; Fig. 6). An extinct species of the Northern fur seal, *Callorhinus* has been described from the late Pliocene of southern California, Mexico, and Japan. *Hydrarctos* is an extinct fur seal from the Pleistocene of Peru. Several extant species of *Arctocephalus* have a fossil record extending to the Pleistocene in South Africa and North America (California). The fossil record of modern sea lions is poorly known. The following taxa are reported from the Pleistocene: *Neophoca* (New Zealand), *Eumetopias* and *Zalophus* (Japan), and *Otaria* (Brazil).

The basal split between otariids and odobenids based on molecular data is close to the age of the oldest enaliarctid fossils and much

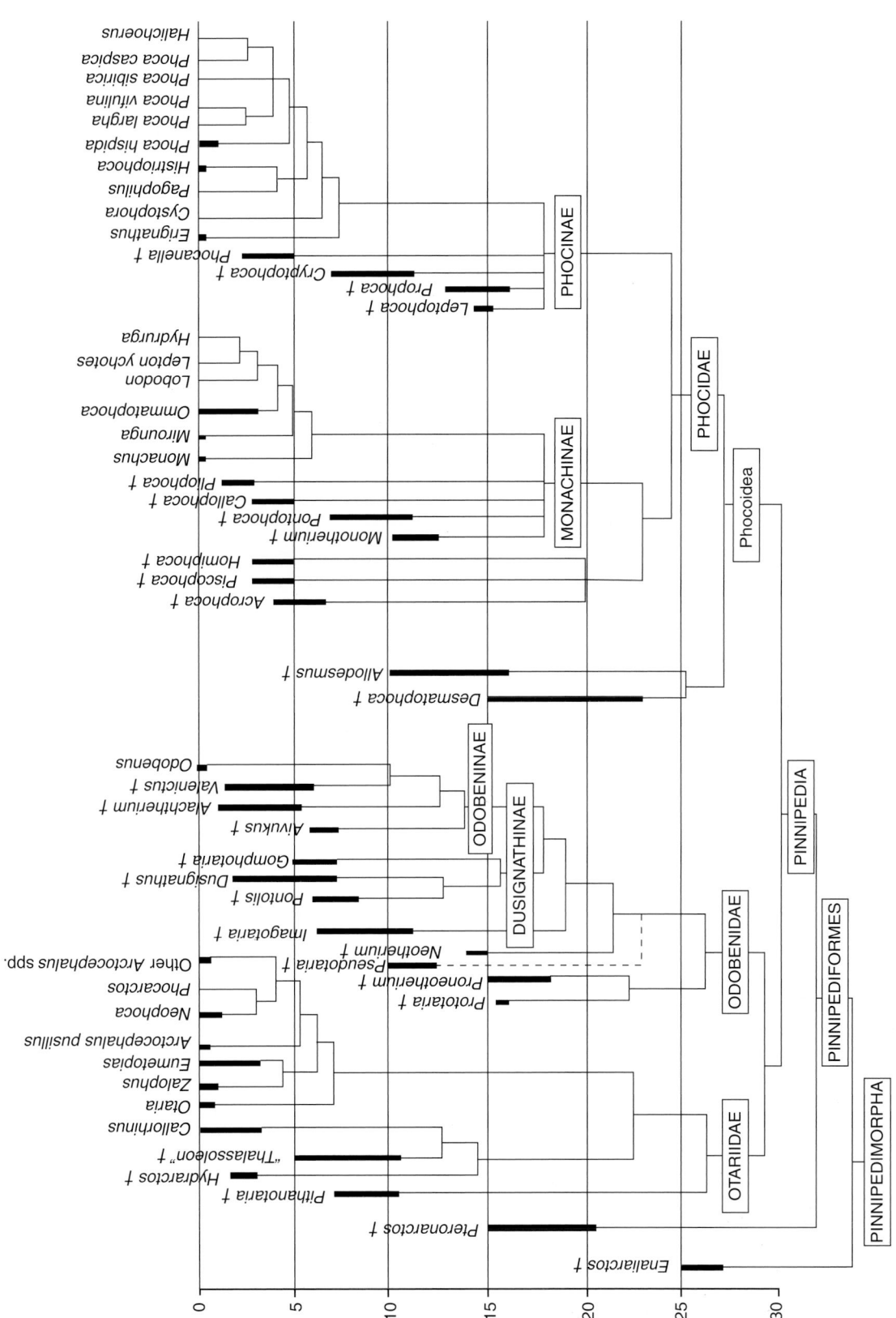

**Figure 3** *Chronologic ranges of major lineages and genera of extinct and living pinnipeds. Time scale in millions of years. Thick lines show stratigraphic ranges of taxa and thin lines indicate their phylogenetic relationships (otariid and phocid phylogeny based on Arnason et al., 2006). Position of fossil walrus Pseudotaria indicated by dashed line is based on Kohno (2006). Branching points are not necessarily correlated with the time scale. (Modified from Deméré et al., 2003.).*

P

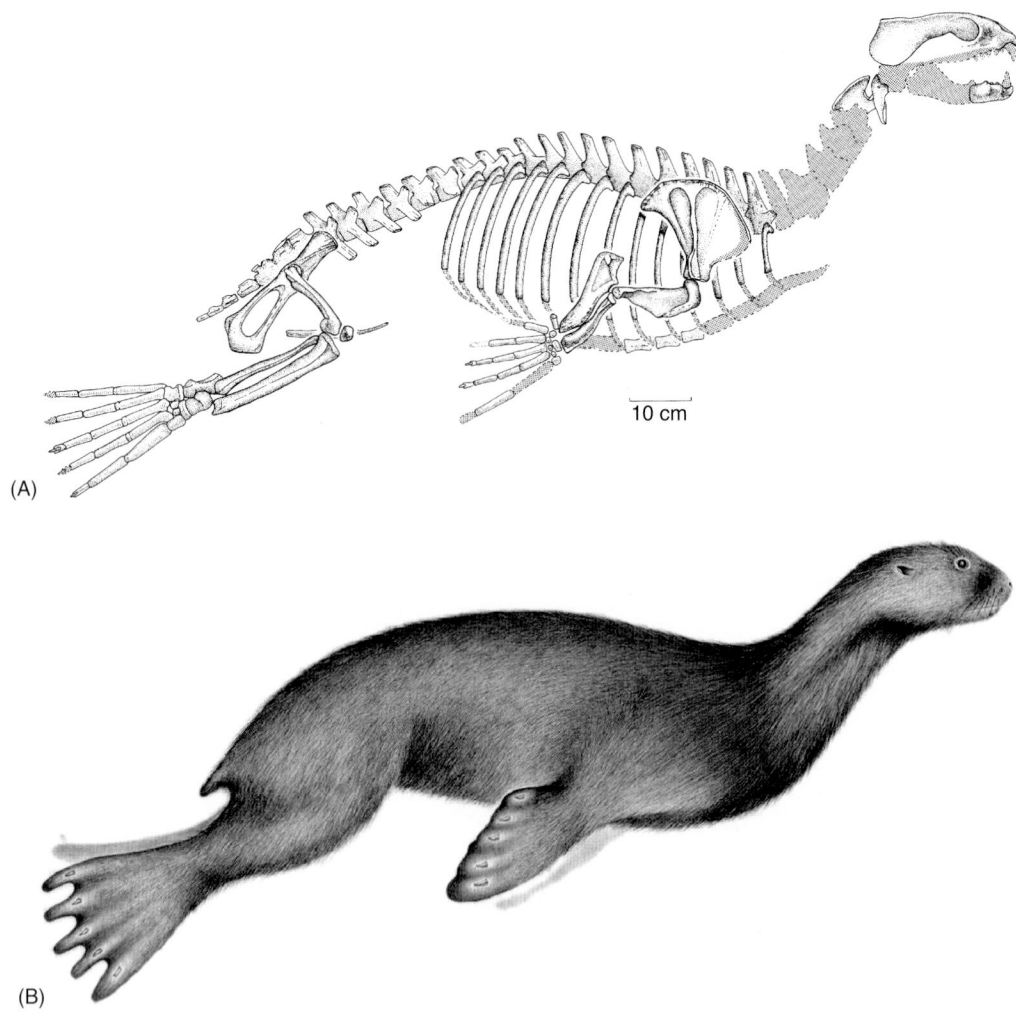

10 cm

(A)

(B)

**Figure 4**    *The pinnipedimorph* Enaliarctos mealsi *showing pinnipedimorph synapomorphies described in text. (From Berta and Ray, 1990.)*

older than the earliest recognized fossils of these lineages (Arnason *et al.*, 2006). Otariids appear to have originated in the North Pacific (Fig. 7). The basal split between *Callorhinus* and other otariids dated to 16 mya is older than the oldest otariid fossil *Pithanotaria* from 11 mya rocks in California. The subsequent divergence of *Otaria*, *Zalophus* + *Eumetopias* from other otariids likely took place in the North Pacific with *Otaria* dispersing along the west coast of South America and later along the Atlantic coast. The diversification of remaining otariids took place in the southern hemisphere with the exception of *Arctocephalus townsendi* which remained in the North Pacific.

The walruses or Odobenidae are diagnosed as a monophyletic group by the presence of a broad, thick pterygoid strut, fourth upper premolar with a strong posterolingually placed protocone shelf, lower first molar with the talonid heel absent, and a calcaneum with a prominent medial tuberosity (Deméré and Berta, 2001 but see also Kohno, 2006; Fig. 6). Morphologic study of the evolutionary relationships among walruses has identified the following taxa: *Prototaria*, *Pseudotaria*, *Proneotherium*, *Neotherium*, *Imagotaria*, dusignathines and odobenines, the latter includes the modern walrus (see PINNIPEDIA).

Fossil walruses first appear in the early Miocene (16–18 Ma; Fig. 3) fossil record with *Prototaria* in Japan and *Proneotherium* in North America (Oregon). These archaic walruses are characterized by unenlarged canines and narrow, multiple rooted premolars with a trend toward molarization (Fig. 8), adaptations suggesting retention of the fish diet hypothesized for archaic pinnipeds rather than evolution of the specialized mollusc diet for the modern walrus. Portions of the axial and hindlimb skeleton of *Proneotherium* preserve evidence of a number of aquatic adaptations including a laterally flexible spine, a broad, shortened femur, and a paddle-shaped foot (Deméré and Berta, 2001). In addition to *Proneotherium*, *Prototaria*, and the recently described *Pseudotaria* (Kohno, 2006) a monophyletic clade of walruses comprised of *Neotherium*, *Imagotaria*, and Dusignathinae and Odobeninae diversified in the middle and late Miocene Fig. 3). Dusignathine walruses which include *Dusignathus*, *Pontolis*, and *Gomphotaria* developed enlarged upper and lower canines. *Gomphotaria pugnax* the most completely known dusignathine is distinct cranially and dentally in its possession of large, procumbent upper lateral incisors and canines; the latter with deeply fluted roots and a small orbit. Odobenines which include *Aivukus*, *Ontocetus* (=*Alachtherium* fide Kohno, 2006), *Valenictus*

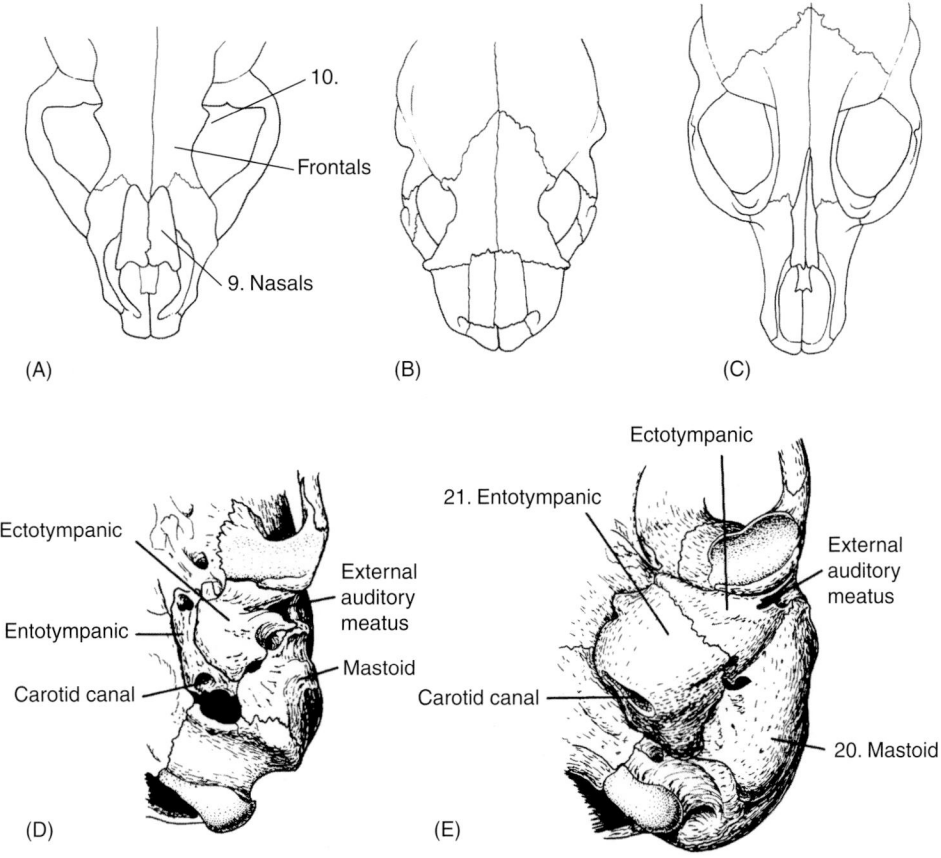

**Figure 5**  *Skulls/ventral view of ear regions of (A, D) otariid, (B) walrus, (C, E) phocid illustrating otariid synapomorphies: frontals extend anteriorly between nasals (contact between these bones in transverse, walrus, or V-shaped, phocids and phocid synapomorphies: pachyostotic mastoid bone (not seen in other pinnipeds) and greatly inflated ectotympanic bone. (From Berta et al., 2006.)*

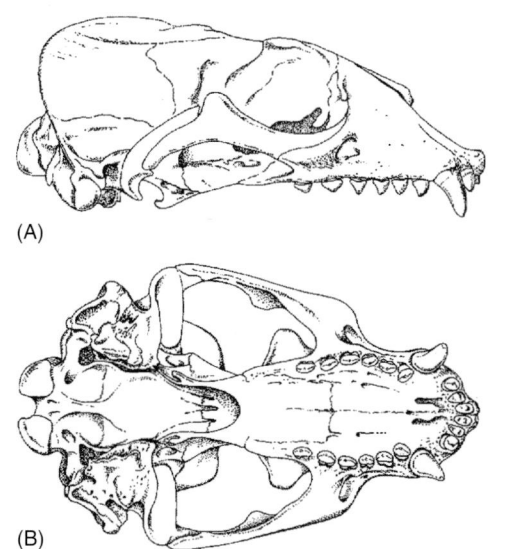

**Figure 6**  *Skull of an early otariid,* Thalassoleon mexicanus *from the late Miocene of western North America in (A) lateral (B) ventral views. Original 25 cm long (From Repenning and Tedford, 1977.)*

and *Odobenus* evolved the enlarged upper tusks seen in the modern walrus. Unique among pinnipeds is the toothlessness of *Valenictus chulavistensis* from the late Pliocene of California, presumably an adaptation for suction feeding.

It has been proposed that the modern walrus lineage (odobenine walruses) entered the Caribbean from the Pacific via the Central American Seaway (between 5 and 8 Ma) and dispersed northward into the North Atlantic (Fig. 7). Less than 1 Ma the living genus *Odobenus* returned to the North Pacific through the Arctic Ocean. Alternatively, on the basis of a new, earlier record of the modern walrus lineage from Japan it has been suggested that this lineage may have evolved in the North Pacific and dispersed instead to the North Atlantic through the Arctic during the Pleistocene.

The earless seals, the Phocidae, are diagnosed as a monophyletic group by pachyostic mastoid region, greatly inflated entotympanic bone, complete absence of the supraorbital process of the frontal, strongly everted ilia and lack of an ability to draw the hindlimbs forward under the body due to a massively developed astragalar process, and greatly reduced calcaneal tuber (Fig. 5). Although phocids have traditionally been divided into two or four major subgroupings, recent molecular studies consistently support monophyly of Monachinae and Phocinae (see also PINNIPEDIA).

**P**

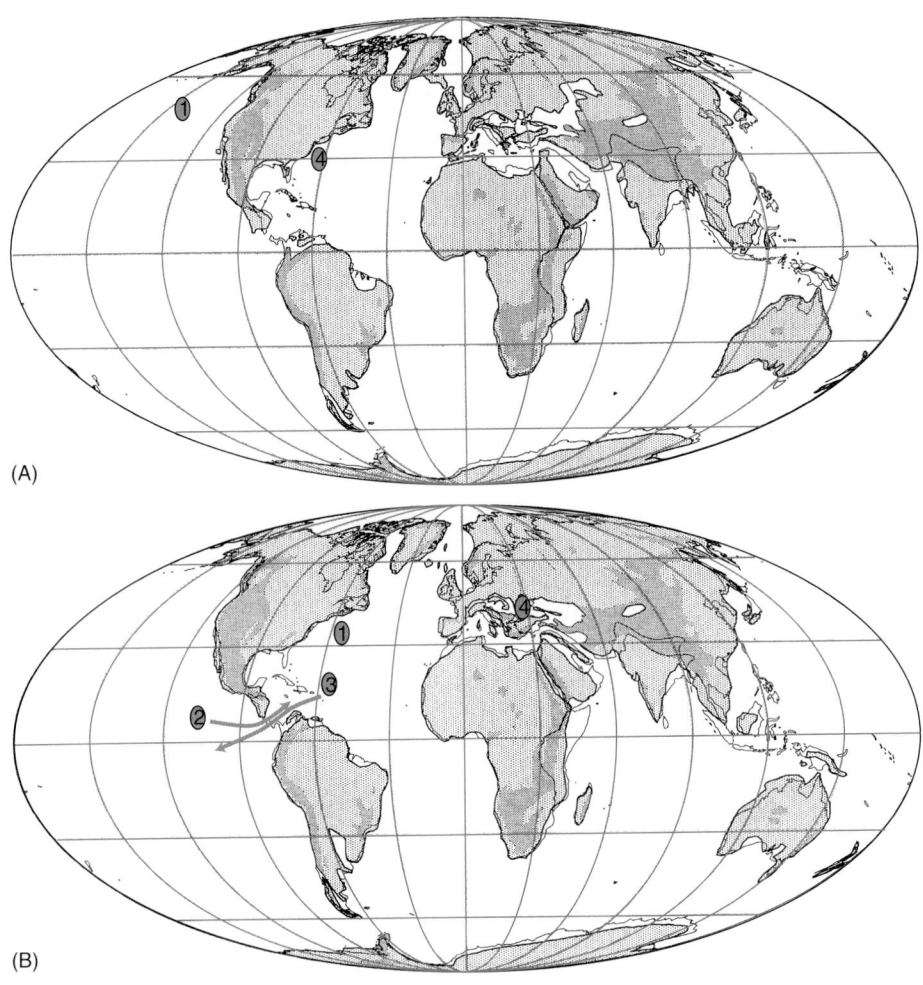

**Figure 7**  *Reconstruction of continents, ocean basins, and paleocoastlines in the (A) early Miocene (20 Ma) (1: 5 early records of archaic pinnipeds, odobenids, and desmatophocids) and (B) middle Miocene (12 Ma) (1: 5 early well documented phocids, 2: 5 dispersal of "monachines" and odobenids to Atlantic, 3: 5 dispersal of phocines to South Pacific, 4: 5 isolation of phocines in remnants of Paratethys Sea and in North Atlantic). (From Berta et al., 2006; base map from Smith et al., 1994.)*

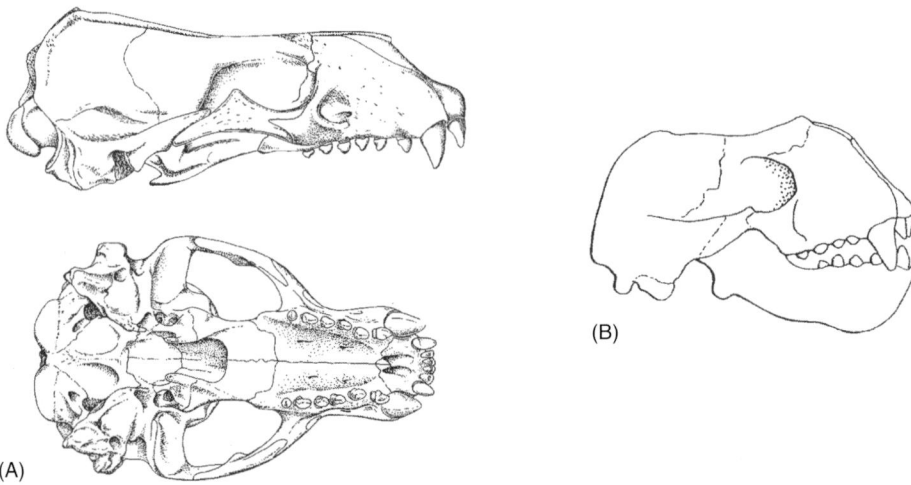

**Figure 8**  *Lateral and ventral views of skulls of fossil odobenids (A)* Imagotaria downsii *from the Miocene of western North America. Original 30 cm long. (From Repenning and Tedford, 1977.) (B) Lateral view of* Protodobenus japonicus *from the early Pliocene of Japan. Original 25 cm. (From Horikawa, 1995.)*

P

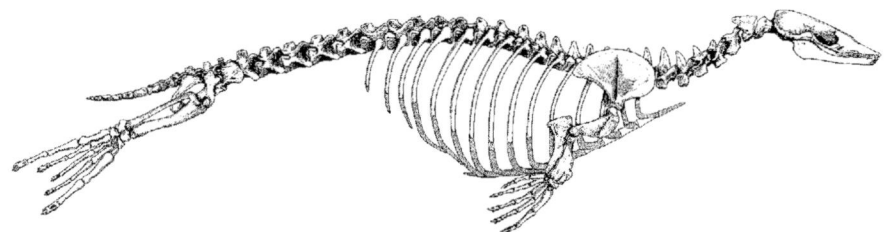

**Figure 9**  *Skeleton of an archaic phocid,* Acrophoca longisrostris *from the Miocene of Peru. (From Muizon, 1981.)*

Although an earlier less well documented record of phocids from the late Oligocene of South Carolina exists, there is undisputed evidence for both monachine and phocine seal lineages from the middle Miocene (approximately 15 Ma) on both sides of the North Atlantic. Molecular estimates place monachine and phocine divergence at approximately 22 mya (Arnason *et al.*, 2006). The earliest phocine *Leptophoca* and the earliest monachine *Monotherium* (Fig. 3) are known from the southeastern North America. Despite the fact that a number of fossil monachines have been described not all are known by comparable elements. In addition to *Monotherium* among the better known taxa from eastern Europe are *Pontophoca* from the middle Miocene and *Callophoca* from the early Pliocene (Koretsky, 2001). Several archaic seals (i.e., *Acrophoca* and *Piscophoca*) represented by complete skeletons are known from the late Miocene and/or early Pliocene of South America (Peru and Chile; Walsh and Naish, 2002) and *Homiphoca* from South Africa (Fig. 3). *Acrophoca* is unique among phocids with its long, slender skull, flexible neck, and elongated body (Fig. 9). Although these fossil monachines have been referred to the lobodontines (see also PINNIPEDIA), new discoveries as well as restudy of material previously referred to these taxa suggests that they may in fact be phocines. The fossil record of extant monachines is poorly known and includes only *Monachus* and *Mirounga* from the late Pleistocene and the lobodontine *Ommatophoca* from the late Pliocene (Fig. 3). With regard to fossil phocines among the best known taxa are *Prophoca* (middle Miocene) in the eastern North Atlantic, *Cryptophoca* (late Miocene) in the Black Sea region, and *Phocanella* (early Pliocene) in both the eastern the western North Atlantic (Fig. 3). Extant phocine genera with a fossil record include *Phoca* from the late Pliocene and *Erignathus* and *Histriophoca* from the late Pleistocene (Fig. 3).

The purported first appearance of phocids in the North Atlantic suggests that the common ancestor of phocids had migrated to the North Atlantic, either northward through the Arctic Basin or southward through the Central American Seaway (Fig. 7). Support for a southern route is based on the hypothesized close relationship of phocids and the extinct desmatophocids, the latter occurring as far south as Mexico, and the fact that the Bering land bridge blocked access to the Arctic through much of the late Oligocene and early Miocene. Fyler *et al.* (2005) confirmed an earlier proposal by Muizon (1982) that monk seals originated in the Mediterranean (*Monachus monachus*) with dispersal first to the Caribbean (*M. tropicalis*) and later to the central North Pacific (Hawaiian monk seal, *M. schauinslandi*). The 4 mya molecular divergence of northern and southern elephant seals (*Mirounga*) may be the result of vicariance following emergence of the Panamanian isthmus (Arnason *et al.*, 2006).

The biogeographic pattern for phocine seals is no less complicated given the different phylogenetic hypotheses proposed. Although it was suggested earlier that phocines were a northern hemisphere radiation, a considerable diversity of phocine seals is known from the southern hemisphere during the late Miocene and/or early Pliocene (assuming that *Acrophoca, Homiphoca,* and *Piscophoca* are phocines rather than lobodontines). In addition other phocine lineages appear to have been isolated in the Paratethys Sea (northern arm of the Tethys Sea stretching across the area now occupied by the Black, Caspian, and Aral Seas of Asia) and the North Atlantic during the late Miocene and Pliocene. Several dispersal routes for phocines seem likely. The deepest phocine split, that between the bearded (*Erignathus*) and hooded seals (*Cystophora*) was dated to 17 mya based on molecular data (Arnason *et al.*, 2006). This suggests an eastward and northward migration of phocines in the North Atlantic. The hooded seal occurs at high latitudes of the Atlantic basin and apparently never successfully dispersed to the Pacific. The bearded seal is presently confined to the Arctic and subarctic around the North Atlantic but Pleistocene records extend as far south as Portugal. A subsequent dispersal route for phocines involved initial migration from Paratethys Sea into the Arctic Basin, followed by an eastward migration to give rise to modern *Phoca/Pusa*. In this scenario the landlocked Baikal seal (*Pusa sibirica*) gained access to Lake Baikal from the Arctic via large lakes at the southern margin of the Siberian ice sheet. A second landlocked species, the Caspian seal (*Pusa caspica*) remained in the Caspian Sea an isolated remnant of Paratethys as judged by the presence of fossils similar to living Caspian seals in this location. Another hypothesis argues for a North Atlantic origin for all phocines with glacial events causing speciation. For example, cyclical fluctuations in glacial maxima (with concomittant variations in sea level) through the Pleistocene mediated range expansions of *Phoca* spp. ultimately leading to isolation of populations in refugial centers (i.e., Arctic, Okhotsk, Aleutian) and divergence of populations (e.g., ribbon seal, *Histriophoca fasciata*, ringed seal, *Pusa hispida*, largha seal, *Phoca largha*, harbor seal, *Phoca vitulina*, harp seal, *Pagophilus groenlandica*).

An extinct family of archaic pinnipeds, the desmatophocids are characterized by elongate skulls, relatively large eyes, mortised contact between two cheekbones, and bulbous cheek teeth (Fig. 10). Cladistic analysis has identified the Desmatophocidae which includes two genera, *Desmatophoca* and *Allodesmus* as the common ancestors of phocid pinnipeds. This clade, phocids + *Allodesmus* and *Desmatophoca*, termed the Phocoidea, is supported by four synapomorphies including: premaxilla–nasal contact reduced, squamosal–jugal contact mortised, marginal process below ascending ramus well-developed (Deméré and Berta, 2002; Fig. 2). This interpretation differs from previous work that recognized desmatophocids as otarioid pinnipeds (a paraphyletic group than includes walruses but excludes phocids). Desmatophocids are known from the early and middle Miocene (23–10 Ma) of western US and Japan (Fig. 3). Newly reported occurrences of *Desmatophoca* confirm the presence of sexual dimorphism and large body size in these pinnipeds (Deméré and Berta, 2002). *Allodesmus* is a diverse taxon (as many

**P**

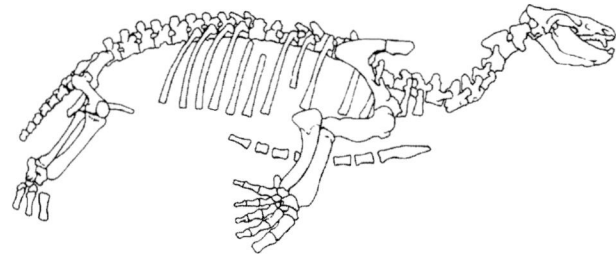

**Figure 10** *Skeleton of the desmatophocid, Allodesmus kernenesis from the Miocene of western North America. Original 2.2 m long. (From Mitchell, 1975.)*

as six species have named) with taxa informally divided into "broad headed" and "long headed" subgroups. The previous suggestion that *Allodesmus* retained a mobile proboscis, much like the modern elephant seal seems unlikely on anatomical grounds. *Allodesmus*, known by several complete skeletons, retains several features consistent with forelimb propulsion but also displays adaptations for hindlimb swimming (Berta and Adam, 2001).

## See Also the Following Articles

Pinnipeds, Overview ■ Fossil Record

## References

Adam, P. J., and Berta, A. (2002). Evolution of prey capture strategies and diet in the Pinnipedimorpha (Mammalia: Carnivora). *Oryctos* **4**, 83–107.

Arnason, U., Gullberg, A., Janke, A., Kullberg, M., Lehman, N., Petrov, E. A., and Vainola, R. (2006). Pinniped phylogeny and a new hypothesis for their origin and dispersal. *Mol. Phylogenet. Evol.* **41**(2), 345–354.

Berta, A., and Adam, P. J. (2001). Evolutionary biology of pinnipeds. *In* "Secondary Adaptation of Tetrapods to Life in Water" (J. M. Mazin, and V. de Buffeenil, eds), pp. 235–260. Verlag Dr. Fredrich Pfeil, Munchen, Germany.

Berta, A., Sumich, J. L., and Kovacs, K. M. (2006). "Marine Mammals: Evolutionary Biology," 2nd ed. Elsevier, San Diego, CA.

Berta, A., and Ray, C. E. (1990). Skeletal morphology and locomotor capabilities of the archaic pinniped *Enaliarctos mealsi*. *J. Vert. Paleontol.* **10**(2), 141–157.

Berta, A., and Wyss, A. R. (1994). Pinniped phylogeny. *Proc. San Diego Soc. Nat. Hist.* **29**, 33–56.

Deméré, T. A., and Berta, A. (2002). The Miocene pinniped *Desmatophoca oregonensis* Condon, 1906 (Mammalia: Carnivora) from the Astoria Formation, Oregon. *Smithsonian Contrib. Paleobiol.* **93**, 113–147.

Deméré, T. A., and Berta, A. (2001). A re-evaluation of *Proneotherium repenningi* from the middle Miocene Astoria Formation of Oregon and its position as a basal odobenid (Pinnipedia: Mammalia). *J. Vertbr. Paleontol.* **21**, 279–310.

Deméré, T. A., and Berta, A. (2005). New skeletal material of *Thalassoleon* (Otariidae: Pinnipedia) from the late Miocene–early Pliocene of California. *Bull. Fla. Mus. Nat. Hist.* **45**(4), 379–411.

Deméré, T. A., Berta, A., and Adam, P. J. (2003). Pinninipedimorph evolutionary biogeography. *Bull. Am. Mus. Nat. Hist.* **279**, 32–76.

Fyler, C. A., Reeder, T. W., Berta, A., Antonelis, G., Aguilar, A., and Androukaki, E. (2005). Historical biogeography and phylogeny of monachine seals (Pinnipedia: Phocidae) based on mitochondrial and nuclear DNA data. *J. Biogeogra.* **32**, 1267–1279.

Horikawa, H. (1995). A Primitive odobenine walrus of early Pliocene age from Japan. *Island Arc.* **3**, 309–329.

Kohno, N. (2006). A new Miocene odobenid (Mammalia: Carnivora) from Hokkaido, Japan, and its implications for odobenid phylogeny. *J. Vert. Paleontol.* **26**(2), 411–421.

Koretsky, I. (2001). Morphology and systematics of Miocene Phocinae (Mammalia: Carnivora) from Paratethys and the north Atlantic region. *Geol. Hung. Ser. Palaeontol. Fasc.* **54**, 1–109.

Muizon, C. de (1982). Phocid Phylogeny and dispersal. *Ann. S. Adr. Mus.* **89**, 175–213.

Mitchell, E. D. (1975). Parallelism and convergence in the evolution of the Otariidae and Phocidae. Conseil international de l'Exploration de la rev. *Rapports et Proces-Verbaux des Reunions* **169**, 12–26.

Muizon, C.de. (1981). Les Vertebres Fossiles de la Formation Pisco (Perou). Part 1. Recherche sur les Grandes Civilisations, Mem. No. 6, Instituts Francais d'Etudes Andines, Paris.

Repenning, C. A., and Tedford, R. H. (1977). Otarioid seals of the Neogene. *Geol. Surv. Prof. Pap. (U.S.)* **992**, 1–93.

Smith, A. G., Smith, D. G., and Funnell, B. M. (1994). "Atlas of Mesozoic and Cenozoic Coastlines." Cambridge University Press, Cambridge.

Walsh, S., and Naish, D. (2002). Fossil seals from late Neogene deposits in South America, a new pinniped (Carnivora: Mammalia) assemblage from Chile. *Palaenotology* **45**(4), 821–842.

Wyss, A. R. (1988). On "retrogression" in the evolution of the Phocinae and the phylogenetic affinities of the monk seals. *Am. Mus. Novit.* **2924**, 1–38.

# Pinniped Life History

## Ian L. Boyd

The life history of an individual is the pattern of allocation of resources to maintenance, growth, and reproduction throughout its lifetime. Life history analysis attempts to explain the scheduling of the allocation process throughout an organism's life. It assumes implicitly that it is appropriate to classify individuals by age because this is a major component of the independent variable representing time that is used to examine variation in resource allocation. However, we know that other properties of an individual, such as its body condition or foraging skill, are also important variables that affect reproduction and ultimately fitness.

Most life history studies involving pinnipeds have assumed that age is the main force in pinniped life histories when, in fact, age *per se* may have relatively little to do with influencing fitness. It is a paradox of life history studies that they are, by definition, time-based approaches to examining variation in the fitness between individuals when time itself probably has less biological importance than other factors. One such factor in pinnipeds is body size, long recognized as a determinant of sexual maturity in pinnipeds. Age at sexual maturity in pinnipeds can be expressed as a decreasing function of growth rate. Expressed at the level of populations, this is interpreted to mean that individuals within pinniped populations that are at a level well below the environmental carrying capacity would experience higher growth rates and would, therefore, become sexually mature at an earlier age (Bengston and Laws, 1985). This was an implicit acknowledgment that age was not the operant factor in pinniped life histories and was at best secondary to the size of the energy reserves of an individual. Nevertheless, despite the considerably greater difficulties that exist with measuring age in pinnipeds than there are with measuring body size (e.g., mass or some other suite of morphometrics), age has continued to be used as the primary independent variable in life history studies.

P

## I. Characteristics of Pinniped Life Histories

Pinniped life histories are characterized by three main features: (1) by mammalian standards, pinnipeds have high annual survival rates, giving potential longevities in the order of two to four decades; (2) the average age at sexual maturity is delayed by 2–6 years depending on the species (Table I); and (3) each adult female normally produces a maximum of one offspring per reproductive cycle. Variations on this theme at the level of individuals and species can provide insight into the evolution of life histories in pinnipeds.

Pinniped life histories are assumed to have evolved to maximize the genetic fitness of individuals. This occurs in pinnipeds within the constraints of a semiaquatic existence and has most probably led to the relatively narrow range of life histories we observe within the taxon. All pinnipeds rely to some degree on ice or land for reproduction, particularly the processes of birth and lactation. Many interacting variables have led to the evolution of pinniped life histories, including the joint and sometimes conflicting needs to avoid predation, to forage with maximum efficiency, and to choose a mate of high quality.

By mammalian standards, pinnipeds are animals with a large body size. However, in terms of their demography and their investment in reproduction, pinnipeds do not appear to differ greatly from other mammals after body size has been taken into consideration. There are also no obvious relationships between body size and life history variables at the species level within the pinnipeds (Table I), although, as we shall see, this is not the case for variation between individuals within species.

Large body size has a cost in that relatively large amounts of resources are invested in tissue growth and maintenance and it takes a relatively long time to reach a body size capable of supporting reproduction. There is also a need to produce precocial young that can defend themselves against predation from an early age or that can forage independently of their mothers within days to weeks of birth. This necessitates greater investment in individual offspring and limits the number of young that can be produced at a single reproductive attempt. It also means that the rate of reproduction (number of young born per unit time) is relatively low The combination of high investment in growth, causing a delay in sexual maturity, and low reproductive rates, even when sexually mature, means that pinnipeds must have relatively high longevities (low rates of mortality). Without this combination of demographic variables individuals could not, on average, replace themselves during their lifetimes.

## II. Methods for Examining Life Histories

Life histories are represented most concisely by demographic models based on empirical measurements of survival and fecundity rates. Demographic variables for pinnipeds are summarized in Table I. Among the 36 species of pinnipeds, some form of demographic information is available for most species, but as seen from Table I, there are very few for which there could be said to be complete information, and, in almost all of these, information is mainly available for females. Very little is known about the life histories of male pinnipeds. It is also perhaps a little misleading to represent these demographic variables in terms of species, as many vary as much between different populations of the same species as they do between the species themselves. Averaging across populations also has the disadvantage that it obscures the variation in life histories between individuals. Therefore, while life histories may, in practice, often be examined at the level of populations using demographic parameters, it is an important tenet of life history analysis that it is based on the demography of individuals. This distinguishes life history analysis from the study of population dynamics, which normally deals with individuals as if they are all identical.

The most complete information about life histories for any population of pinnipeds comes from Weddell seals (*Leptonychotes weddellii*) at McMurdo Sound, Antarctic (Hastings and Testa, 1998), and northern elephant seals (*Mirounga angustirostris*) from Año Neuvo or the Farallon Islands, California (Reiter *et al.*, 1981; Sydeman *et al.*, 1991). These studies were based on the long-term mark-recapture of individuals. Similar studies have been carried out on Antarctic fur seals species (*Arctocephalus gazella*) (Boyd *et al.*, 1995) and gray seals (*Halichoerus grypus*) (Pomeroy *et al.*, 1999). Mark-recapture is probably the only way to examine life histories in pinnipeds to provide the quality of data necessary to understand the complex interactions between factors that influence fitness. However, such studies can only be undertaken in special circumstances where there is particularly easy access to the study population. In most cases, information about population life histories has been derived from cross-sectional samples of populations based on one-off or sequential culls that were often part of a commercial harvest (Bowen *et al.*, 1981; Fowler, 1990). Although some of the disadvantages of this method may be offset by the advantages of a large sample size, it has the potential to lead to misinterpretation of the pattern of life history. Some of these problems are discussed.

## III. Constraints on Life Histories

Pinniped life histories have evolved under a combination of factors that are broadly based around the need for animals to balance their energy budgets. These include the constraints involved with (1) being homeothermic in water that is 25 times more conductive than air and (2) the high temporal and spatial variability in the distribution of resources within the marine environment. Phylogeny may also be seen as a constraint in that the ancestors of pinnipeds may not have possessed an ideal range of characteristics (physiological, anatomical, social, or distributional) for exploiting the marine environment. Therefore, current pinniped life histories may be constrained by difficulties with inherent mechanisms.

An example of such a constraint is the apparent necessity for a terrestrial (or pagophilic) phase during the reproductive cycle. This may be a consequence of the occupancy by ancestral pinnipeds mainly of temperate and polar marine habitats in which small neotates may have difficulty with thermoregulating in cold water, thereby necessitating terrestrial living for young neonates. Pinnipeds may have been locked into this form of reproductive cycle from an early stage in their evolution.

The constraint of the terrestrial phase in reproduction has brought with it other social and life history consequences. The necessity for mothers to find suitable terrestrial habitat (including ice) for parturition has more or less isolated, both spatially and temporally, the reproductive process from the feeding grounds. Species that exploit distant, unpredictable food sources require larger body mass than those that exploit food that is present at relatively close range to the pupping location. This is because there will be a critical duration over which a pup can be left without feeding and with low risk of starvation. If mothers cannot forage profitably during lactation within this critical duration, it is necessary for mothers to carry with them at parturition most of the food reserves required to raise their pup to independence (Boyd, 1998).

The extreme seasonality of food availability in higher latitudes has also led to extreme seasonality of reproduction, resulting in

**P**

TABLE I

Demographic Parameters used to Describe Life Histories of Pinnipeds[a]

| Species | Mean female body mass (kg) | Mean male body mass[b] (kg) | Pup survival rate | Adult female survival rate | Adult male survival rate | Mean age at first parturition (years) | Mean pregnancy rate[c] | References |
|---|---|---|---|---|---|---|---|---|
| Mirounga angustirostris | 393–425 | — | 0.88 | 0.69–0.77[d] | — | 3–4 | 0.80 | Huber et al. (1991); Reiter and Le Boeuf (1991); Le Boeuf et al. (1994) |
| M. leonina | 400–500 | 2100 | 0.98 | 0.67–0.88 | 0.50–0.83 | 4–5 | 0.88 | McCann (1985); Hindell (1991); Galimberti and Boitani (1999) |
| Leptonychotes weddellii | 350–425 | — | 0.80–0.92 | 0.76–0.85[e] | — | 6–8 | 0.46–0.79 | Testa (1987); Testa and Siniff (1987); Testa et al. (1990); Hastings and Testa (1998) |
| Lobodon carcinophaga | 220 | — | 0.21[f] | 0.90–0.97 | — | 2.5 | 0.95–0.98 | Boveng (1993) |
| Pagophilus groenlandicus | 100–140 | — | — | — | — | 4.8 | 0.82–0.97 | Bowen et al. (1981); Kjellqvist et al. (1995) |
| Pusa hispida | 40–50 | — | 0.84[f] | 0.86 | — | 6–8 | 0.88 | Smith (1987) |
| Halichoerus grypus | 160–190 | — | 0.66[f] | 0.93 | — | 5–7 | 0.80–0.98 | Harwood and Prime (1978); Boyd (1985); Pomeroy et al. (1999) |
| Eumetopias jubatus | 250 | — | 0.78[f] | 0.84–0.93 | — | 4–5 | 0.63 | York (1994) |
| Callorhinus ursinus | 29–39 | 97–165 | 0.80–0.96 | 0.86–0.89 | 0.70[g] | 3–4 | 0.69–0.72 | Wickens and York (1997) |
| Arctocephalus townsendi | 49 | — | — | — | — | — | — | Wickens and York (1997) |
| A. galapagoensis | — | 64 | 0.85–0.91 | 0.85 | 0.68 | 5 | — | Wickens and York (1997) |
| A. philippii | — | — | 0.92–0.95 | — | — | — | — | Wickens and York (1997) |
| A. pusillus pusillus | 57 | 247 | 0.65–0.80 | 0.88[h] | 0.70 | 4 | 0.71 | Wickens and York (1997) |
| A. pusillus doriferus | 76 | — | 0.85 | — | — | 4 | 0.73 | Wickens and York (1997) |
| A. forsteri | — | — | 0.40–0.92 | — | — | 5 | 0.67 | Wickens and York (1997) |
| A. australis | 35–58 | — | 0.53–0.90 | — | — | 3 | 0.80–0.82 | Wickens and York (1997) |
| A. tropicalis | 36 | — | 0.85–0.96 | — | — | 5 | 0.79–0.84 | Wickens and York (1997) |
| A. gazella | 45 | 188 | 0.69–0.96 | 0.83–0.92 | 0.50 | 3 | 0.68–0.77 | Wickens and York (1997) |

[a] Rates are expressed per year. Data for fur seals are summaries from tables in Wickens and York (1997); otherwise the original sources are given. Data for male mass were not included if no demographic data were available.

[b] Sexually and socially mature individuals.

[c] Pregnancy and birth rates are assumed to be equivalent.

[d] Juvenile survival rates fall within the same range.

[e] Juvenile survival >1 year old ~0.70.

[f] Survival in first year.

[g] Values for juvenile males aged 4 months–2 years are 0.20–0.50; those for males aged 2–5 years are 0.75–0.90.

[h] Probably negatively biased because of the inclusion of juveniles.

spatially and temporally synchronized reproduction. It is possible that both sexes have used this to affect greater mate choice, which has produced polygynous, highly competitive mating systems. These combined factors have led, in most species to an annual cycle of reproduction.

## IV. Costs vs Benefits of Reproduction

Even though individuals may have the option to reproduce annually, longitudinal studies show that they do not always exercise this option. Even when individuals do reproduce, they may adjust the amount of resources they supply to their offspring. The reasons for this are centered on the decision that individuals must make during their life times in order to maximize their fitness, often measured in terms of number of offspring produced across their whole lifetime and not just one reproductive cycle.

There are obvious fitness gains from reproduction, but there are also costs involved. For example, in Antarctic fur seals (*Arctocephalus gazella*), reproduction in any year carries with it a 40% greater chance of dying in the following year. It also carries a similar cost in terms of reduced probability of breeding in the following year (Boyd *et al.*, 1995). In northern elephant seals (*Mirounga angustirostris*), mothers that reproduce for the first time at age 3 incur greater costs, in terms of reduced survival, than those that breed first at age 4 (Reiter and Le Boeuf, 1991). Female gray seals (*Halichoerus grypus*) that expend more on their offspring in 1 year also have reduced reproductive success in the following year (Pomeroy *et al.*, 1999). Thus, female pinnipeds must find a solution of how best to allocate energy between growth/maintenance and reproduction that optimizes the balance between fitness costs and benefits of reproduction. Male and female pups have different survival rates as a result of maternal investment (Hall *et al.*, 2001). Those individuals that achieve the optimum balance will have greatest lifetime fitness. How pinnipeds make investment decisions in order to optimize this balance is the focus of much recent research (Schulz and Bowen, 2004). In reality, few individuals may actually achieve the optimum, especially in variable environments, but natural selection favors those individuals that make investment decisions that approach the optimum most closely.

## V. Age at First Reproduction

All pinnipeds experience a delay of several years in the time taken to reach sexual maturity (Table I). Several studies have shown that the age at first reproduction is not constant. In harp seals (*Pagophilus groenlandicus*) it is negatively related to population size (Bowen *et al.*, 1981), implying that the age at which individuals mature is density dependent (Trites and York, 1993). Further evidence for a shift in age at sexual maturity with population size exists for crabeater seals (*Lobodon carcinophaga*) (Bengston and Laws, 1985). The speed with which the change occurs shows that this is not an effect mediated by natural selection for individuals with different life history patterns, rather it is almost certainly driven by changes in the growth rates of individuals as population density and, by implication, per capita food availability changes. Consequently, the mean age at sexual maturity in a population may simply be a reflection of the mean growth rate.

Among northern elephant seals, females tend to begin breeding at age 3 or 4. The fitness of individuals that begin to breed at age 4 is greater than those that begin at age 3 because there is a cost, in terms of reduced survivorship, for those that begin breeding at age 3 (Reiter and Le Boeuf, 1991). In Antarctic fur seals there is a similar disadvantage to breeding at an earlier age (Lunn *et al.*, 1994), although, for those individuals that survive, there is no subsequent effect on reproduction through the remainder of life.

These results suggest how age at sexual maturity can be determined by natural selection. In northern elephant seals and Antarctic fur seals there appears to be a trade-off between the fitness costs of breeding early in life and the fitness gains from early reproduction. Although, on average, individuals that begin breeding at age 3 have lower survival, it is possible that those that breed at age 3 and survive have increased fitness mainly because they have, on average, one more reproductive attempt than those that begin breeding at age 4. Animals may opt to take a greater risk by breeding first at age 3 but with the prospect of greater ultimate lifetime fitness. For the trade-off between breeding first at age 3 or 4 to operate and be evolutionarily stable, both strategies must have equal median lifetime fitness.

## VI. Variations in Measures of Fitness

Strictly speaking, fitness should be measured in terms of the number of grandchildren that are produced by an individual. However, no study of pinnipeds has been able to do this, so a variety of fitness indices are used. The simplest and least informative of these is fertility rate, followed by weaning rate, proportion of offspring surviving their first year, and proportion of offspring surviving to reproductive age. There are specific examples of each of these measures from studies of pinnipeds.

Fertility rates in pinnipeds are normally in excess of 0.8 (Table I) and, given other vital rates in pinniped demography, they normally have to be of this order for populations to have the potential to grow. Longitudinal studies of individual pinnipeds show that most females experience fallow reproductive cycles in their lifetimes (Lunn *et al.*, 1994). It remains unclear if the observation of declining fertility with increasing age in crosssectional samples of pinniped populations reflects senescence of individuals. The observation could equally be caused by greater survival rate, and therefore greater representation in older age classes, of individuals with intrinsically low reproductive rates.

Like age at sexual maturity, fertility is probably linked to the attainment of a critical minimum body condition at a specific stage of the reproductive cycle. In fact, physiologically, there may be virtually no difference between the process of puberty and the seasonal recrudescence of the reproductive system, so the two processes could be considered to be controlled by a common mechanism.

Fertility rates are influenced by previous experience of reproduction. In northern elephant seals, it appears that most females that miss a breeding attempt compensate for this by having a higher probability of weaning a pup in the following year, although, early in the reproductive life span, the opposite effect has been observed, i.e., individuals that miss a reproductive cycle have low success in the following year. Therefore, offspring quality may be affected by previous reproductive experience. Antarctic fur seals are significantly less likely to reproduce in a year following a reproductive attempt.

Weaning rates are affected by both age and previous experience of reproduction in northern elephant seals. It appears that although weaning rates increase initially with experience, these begin to decline later in life. This may represent a cumulative cost of reproduction that is manifest as senescence. However, it is still uncertain if this effect is an artifact of sampling caused by greater longevity in individuals that tend to skip reproduction more frequently or invest a smaller proportion of their energy reserves in their offspring.

Weddell seal (*Leptonychotes weddellii*) offspring survival to age 1 and reproductive age both increase with maternal age and experience and, for male offspring, in relation to maternal body length (Hastings and Testa, 1998). Again, this suggests that those individuals that were able to invest more resources in their offspring, by virtue

P

of their larger size and greater experience (perhaps reflecting the occupancy of better habitat), had enhanced fitness.

## VII. Comparing Males and Females

Because females are the limiting sex and because it is much more difficult to study reproductive success in males, more attention has been focused on female than on male pinnipeds. Nevertheless, males may invest large amounts of their energy reserves in reproduction. In general, males have shorter life expectancies than females (shown by lower annual survival rates in Table I), but it is not clear how this is influenced by the investment in reproductive effort. Investment theory would suggest that the shorter life expectancy of males is because of their preparedness to take greater risks with their survival. The potential gains from reproduction, in terms of offspring, in males that are successful competitors because they make a large investment are greater than for females that are restricted to producing a single offspring per season.

There is also confusion in the literature about when males become sexually mature. The age at physiological maturity in males is probably similar to that of females, but many authors make a distinction between physiological and social maturity, which is defined by the age at which individuals are capable of competing for matings. Recent genetic evidence (Amos *et al.*, 1993) is casting doubt on some of the former interpretations of what social maturity actually means because the pattern of mating success in males often does not follow the pattern suggested by the observed social structure. In the near future, we may have to revise our views of the life history patterns of male pinnipeds.

## VIII. Optimal Life Histories: Modeling the Way Forward

Life history analysis in pinnipeds is fraught with difficulties. Longitudinal studies in which individuals are studied throughout their lifetimes can only be carried out on a narrow range of accessible populations and they are expensive and logistically complex to maintain over the time periods (usually decades) required to achieve useful results. Crosssectional studies are extremely limited in what they can tell us about the dynamics of life histories, and commercial harvests, the usual source of these data, are a thing of the past. We have to find a new way forward.

To date, almost all studies of pinniped life histories have been empirically based and, as pointed out in this description, they have highlighted the interactive nature of parameters such as longevity and reproductive rate. A modeling framework is required in order to allow these interactions to be investigated, to make better use of the data sets that already exist, and to identify critical gaps in the empirical data.

If a pinniped is to maximize its lifetime fitness $F$, then it must choose the optimal allocation of resources to reproduction through its lifetime. Thus, $F = f_1 + f_2 + f_3 \ldots f_n$, where $f_a$ is the fitness contribution from year $a$ in the life of the pinniped, which lasts $n$ years. We know that there are certain functional relationships between maternal size or condition and the probability that mothers will reproduce or survive. If we assume that the relationship between offspring condition and its ultimate fitness is asymptotic, then, up to a certain level, the more energy that a female delivers to her offspring the greater will be her fitness. If the energy delivered to an offspring ($ea$) is a proportion $p$ of the energy available to the mother, then from what we know of the growth patterns and the energetic efficiencies of pinnipeds, it is possible to estimate the energy available for reproduction throughout

the life span of an average individual. By setting rules that an individual will only reproduce if it has a sufficient excess of energy above that required for maintenance, we may be able to investigate the life history patterns in different environments as well as the effects of stochastic variability in food availability on life histories.

Many of the dynamic relationships described here should become explicit in the results of such an energy-based life history model. Similarly, such a model could help the interpretation of some of the crosssectional population data in the context of dynamic life history processes. This type of approach seems to be essential if progress is to be made in pinniped life history analysis and for the full implications of life history analysis to be realized. Because the mechanism underlying population trajectories is the sum of individual life histories, understanding the environmental factors that affect life histories is fundamental to understanding population and species viabilities.

## See Also the Following Articles

Cetacean Life History ■ Population Dynamics ■ Sirenian Life History ■ Sociobiology

## References

Amos, W., Twiss, S., Pomeroy, P. P., and Anderson, S. S. (1993). Male mating success and paternity in the grey seal, *Halichoerus grypus*: a study using DNA fingerprinting. *Proc. R. Soc. Lond. B* **252**, 199–207.

Bengston, J. L., and Laws, R. M. (1985). Trends in crabeater seal age at sexual maturity: an insight into Antarctic marine interactions. *In* "Antarctic Nutrient Cycles and Food Webs" (W. R. Siegfried, P. R. Condy, and R. M. Laws, eds), pp. 667–675. Springer, Berlin.

Boveng, P. L. (1993). Variability in a Crabeater Seal Population and the Marine Ecosystem Near the Antarctic Peninsula. Unpublished Ph.D. thesis. Montana State University, Bozeman, MT.

Bowen, W. D., Capstick, C. K., and Sergeant, D. E. (1981). Temporal changes in the reproductive potential of female harp seals (*Pagophilus groenlandicus*). *Can. J. Fish. Aquat. Sci.* **38**, 495–503.

Boyd, I. L. (1985). Pregnancy and ovulation rates in grey seals (*Halichoerus grypus*) on the British coast. *J. Zool. Lond.* **205**, 265–272.

Boyd, I. L. (1998). Time and energy constraints in pinniped lactation. *Am. Nat.* **152**, 717–728.

Boyd, I. L., Croxall, J. P., Lunn, N. J., and Reid, K. (1995). Population demography of Antarctic fur seals: the costs of reproduction and implications for life-histories. *J. Anim. Ecol.* **64**, 505–518.

Fowler, C. W. (1990). Density dependence in northern fur seals (*Callorhinus ursinus*). *Mar. Mamm. Sci.* **6**, 171–195.

Galimberti, F., and Boitani, L. (1999). Demography and breeding biology of a small localized population of southern elephant seals (*Mirounga leonina*). *Mar. Mamm. Sci.* **15**, 159–178.

Hall, A. J., McConnell, B. J., and Barker, R. J. (2001). Factors affecting first-year survival in grey seals and their implications for life history strategy. *J. Anim. Ecol.* **70**, 138–149.

Harwood, J., and Prime, J. H. (1978). Some factors affecting the size of the British grey seal populations. *J. Appl. Ecol.* **15**, 401–411.

Hastings, K. K., and Testa, J. W. (1998). Maternal and birth colony effects on survival of Weddell seal offspring from McMurdo Sound, Antarctic. *J. Anim. Ecol.* **67**, 722–740.

Hindell, M. A. (1991). Some life-history parameters of a declining population of southern elephant seals *Mirounga leonina*. *J. Anim. Ecol.* **60**, 119–134.

Huber, H. R., Rovetta, A. C., Fry, L. A., and Johnston, S. (1991). Age specific natality of northern elephant seals at the South Farrallon Islands, California. *J. Mammal.* **72**, 525–534.

Kjellqwist, S. A., Haug, T., and Øritsland, T. (1995). Trends in age composition, growth, and reproductive parameters of the Barents Sea harp seals, *Phoca groenlandica*. *ICES J. Mar. Sci.* **52**, 197–208.

Le Boeuf, B. J., Morris, P., and Reiter, J. (1994). Juvenile survivorship of northern elephant seals. *In* "Elephant Seals: Population Ecology, Behavior, and Physiology" (B. J. Le Boeuf, and R. M. Laws, eds), pp. 121–136. University of California Press, Berkley, CA.

Lunn, N. J., Boyd, I. L., and Croxall, J. P. (1994). Reproductive performance of female Antarctic fur seals: the influence of age, breeding experience, environmental variation and individual quality. *J. Anim. Ecol.* **63**, 827–840.

McCann, T. S. (1985). Size, status and demography of southern elephant seal (Mirounga leonina) populations. *In* "Sea Mammals in South Latitudes: Proceedings of a Symposium of the 52nd AN-ZAAS Congress in Sydney, May 1982" (J. K. Ling and M. M. Bryden, eds.), pp. 1–17, South Australian Museum, Northfield.

Pomeroy, P. P., Fedak, M. A., Rothery, P., and Anderson, S. S. (1999). Consequences of maternal size for reproductive expenditure and pupping success of grey seals at North Rona, Scotland. *J. Anim. Ecol.* **68**, 235–253.

Reiter, J., and Le Boeuf, B. J. (1991). Life history consequences of variation in age at primipary in northern elephant seals. *Behav. Ecol. Sociobiol.* **28**, 153–160.

Reiter, J., Panken, K. J., and Le Boeuf, B. J. (1981). Female competition and reproductive success in northern elephant seals. *Anim. Behav.* **29**, 670–687.

Schultz, T. M., and Bowen, D. (2004). Pinniped lactation strategies: evaluation of data on maternal and offspring life history traits. *Mar. Mamm. Sci.* **20**, 86–114.

Smith, T. G. (1987). The ringed seal, *Phoca hispida*, of the Canadian Western Arctic. *Can. Bull. Fish. Aquat. Sci.* **216**, 81.

Sydeman, W. J., Huber, H. R., Emslie, S. D., Ribic, C. A., and Nur, N. (1991). Age-specific weaning success of northern elephant seals in relation to previous breeding experience. *Ecology* **72**, 2204–2217.

Testa, J. W. (1987). Long-term reproductive patterns and sighting bias in Weddell seals (Leptonychotes weddelli). *Can. J. Zool.* **65**, 1091–1099.

Testa, J. W., and Siniff, D. B. (1987). Population dynamics of Weddell seals (Leptonychotes weddelli) in McMurdo Sound, Antarctic. *Ecol. Monogr.* **57**, 149–165.

Testa, J. W., Siniff, D. B., Croxall, J. P., and Burton, H. R. (1990). A comparison of reproductive parameters among three populations of Weddell seals (Leptonychotes weddellii). *J. Anim. Ecol.* **59**, 1165–1175.

Trites, A. W., and York, A. E. (1993). Unexpected changes in reproductive rates and mean age at 1st birth during the decline of the Pribilof northern fur seal (Callorhinus ursinus). *Can. J. Fish. Aquat. Sci,* **50**, 858–864.

Wickens, P., and York, A. E. (1997). Comparative population dynamics of fur seals. *Mar. Mamm. Sci.* **13**, 241–292.

York, A. E. (1994). The population dynamics of northern sea lions, 1975–1985. *Mar. Mamm. Sci.* **10**, 38–51.

# Pinniped Physiology

DANIEL E. CROCKER AND DANIEL P. COSTA

## I. Introduction

Pinnipeds are unique among mammals because they feed in the marine environment and reproduce on land or ice, requiring a spatial and temporal separation of feeding from lactation (Costa, 1993). Seals stay at sea for weeks and often months at a time, yet they must spend considerable amounts of time on land. The amphibious nature of pinniped life has necessitated a wide range of physiological adaptations to life in water and on land. Pinnipeds must meet the physiological challenges of marine existence using specialized adaptations that still facilitate existence on land. This life history requires a remarkable plasticity of physiology. Broad categories of physiological adaptation include: (1) aquatic locomotion, (2) apnea and diving physiology, (3) sensory physiology, (4) osmoregulation, (5) thermoregulation, (6) fasting physiology, and (7) lactation physiology.

Pinnipeds have had to overcome numerous problems associated with moving efficiently in the dense aquatic medium and this adaptation has reduced their ability to move about on land. Otariids have hindflippers that can be turned under the body for terrestrial locomotion, whereas phocid seals cannot turn their hindflippers under the body and instead use lunging movements to get around on land.

Perhaps the most complex suite of adaptations required for making a living in the ocean is the physiology associated with breath-hold diving to foraging depths. In addition to adaptations for dealing with great pressures, pinnipeds exhibit physiological adaptation for apnea, increased oxygen storage, bradycardia, hypoperfusion, hypometabolism, and neuronal and hormonal control of cardiac and spleen function.

The sensory systems of pinnipeds enable them to successfully navigate, forage, and communicate in a variety of environments. Seals hear and see relatively well both in the air and underwater. Since the behavior of sound and light in water is markedly different than that in air, this again requires plasticity in their sensory physiology. Ultimately, sensory physiology must provide the appropriate visual and auditory information to facilitate social interactions on land, while allowing detection and capture of prey and detection and avoidance of predators at sea. Adaptations include well-developed underwater directional hearing and visual sensitivity at low light levels.

Living in salt water poses osmoregulatory problems for pinnipeds. In addition pinnipeds must stay in water balance during periods onshore during which they may fast completely from food or water. Since animals also lose water for evaporative cooling, osmoregulatory strategies are linked to thermoregulation.

Pinnipeds are exposed to a remarkably variable range of environmental temperatures. They are able to tolerate frigid ocean temperatures at depth as well as high amounts of thermal radiation encountered when hauled-out on land. Adaptations that help pinnipeds retain heat in the ocean environment, such as thick blubber or dense fur, may also promote overheating on land. Adaptations that may play a role in thermoregulation include large body size, blubber or dense fur, countercurrent heat exchange systems and possibly, high metabolic rates.

## II. Fasting Physiology

### A. Lipid Utilization and Protein Sparing

Many pinnipeds fast for extended periods during their breeding season or during molting (Table I). Mating, giving birth, nursing pups, and for some species, molting all require long periods of time on land. This is particularly true of phocid seals, which undergo voluntary periods of prolonged fasting twice a year. Adult male pinnipeds may abstain from food or water for as long as 3 months while maintaining a territory or competing for dominance rank on the breeding rookery (Costa and Williams, 2000). In many phocid species females fast completely from food and water for over a month, while delivering tremendous amounts of energy to their pups as milk. Offspring are weaned abruptly and in many species the pup then undergoes an extended postweaning fast before departing to sea and initiating foraging. This postweaning fast may be an important developmental time relative to the diving physiology of the offspring.

**P**

## TABLE I
### Duration of Natural Fasts for Pinnipeds Exhibiting Extended Fasts during Breeding

| Females | Duration | Males | Duration |
|---|---|---|---|
| Crabeater seal | ~4 weeks | Crabeater seal | ~4 weeks |
| Gray seal | 2.5–3 weeks | Gray seal | 3–8 weeks |
| Hawaiian monk seal | 5–6 weeks | Hooded seal | ~4 weeks |
| Hooded seal | 1.5–2 weeks | Leopard seal | Unknown |
| Leopard seal | Unknown | Northern elephant seal | 2–3 months |
| Northern elephant seal | 5 weeks | Ross seal | Unknown |
| Ross seal | Unknown | Southern elephant seal | 2–3 months |
| Southern elephant seal | 4 weeks | All fur seals | ~2.5 months |
| Weddell seal[a] | 6–7 weeks[a] | All sea lions[b] | ~2.5 months |

[a]Weddell seal females enter the water frequently and some may feed.
[b]California sea lion males periodically enter the water to feed. Fasting duration in the species is ~2 weeks.

**Figure 1** *An adult female northern elephant seal early and late in the lactation period. Northern elephant seal females lose between 35 and 45% of the body mass during breeding.*

In most cases, these extended fasts are associated with behaviors or processes resulting in considerable energy expenditure (e.g., combat, mating, lactation, molting). Adult animals may lose as much as 35–57% of stored body reserves during these periods (Fig. 1).

The lengths of these voluntary fasts may vary considerably. Fasts can last as long as 3 months in breeding males of both otariids and phocids. Otariid females alternate short on-shore periods with foraging trips. These fasts can last from several days to 1–2 weeks post-partum. In the phocid seals that fast throughout lactation, fasting duration can be as short as 3–5 days in hooded seals, while northern elephant seals nurse a pup for 23–30 days after an additional 1–2 weeks fasting before parturition. Lactating Weddell seals may fast for up to 7 weeks, although there is a growing body of evidence that some females feed during lactation. Unlike most other groups of animals that undergo natural fasts, activity levels remain high. Males expend energy in territorial interaction, dominance interactions, and mating behaviors. Females expend energy in agonistic encounters for breeding space, interaction with males, and for milk synthesis. Pups are also active during their fasts, making daily movements into the

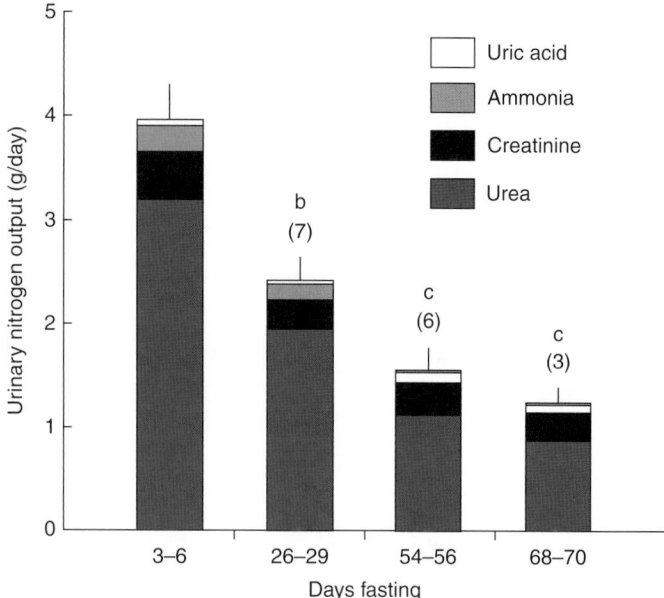

**Figure 2** *Changes in daily urinary nitrogen excretion in fasting elephant seal pups. Letters denote significant differences between periods (P < 0.05). Sample sizes are in parentheses. (From Adams and Costa, 1993.)*

water and often exhibiting high movement rates. In addition pups of many species undergo significant development during fasting including increases on blood volume and synthesis of respiratory pigments.

Despite these high levels of energy expenditure, seals are able to minimize the depletion of lean body mass, with the bulk of energy reserves coming from adipose tissue (Castellini and Rea, 1992). Within the first weeks of the fast, rates of mass loss in nonlactating animals decrease markedly and then remain relatively stable and low for the remainder of the fast. This is accomplished primarily through a reduction in metabolic rate during the fast. This decline is evident in some species on a whole body basis as well as when corrected for changing body size and composition. The key adaptation for extended fasting appears to be the ability to spare protein while fasting and thereby reduce vital organ damage. This stage of fasting, sometimes called Stage II fasting, is characterized by substantial decreases in blood urea nitrogen levels and urinary excretion of nitrogenous wastes. These characteristics are evident throughout the fasts of phocid seals (Fig. 2). This decreased protein degradation is reflected in reduced absolute and proportional declines in the use of protein reserves. Protein contributes as little as 1–6% of total energy utilization by the end of the fast. For example, at the beginning of the postweaning fast, northern elephant seals, *Mirounga angustirostris*, meet around 4% of their energy needs through protein catabolism. By the end of the postweaning fast this value has declined to around 1% (Adams and Costa, 1993).

Nonesterified fatty acids (NEFA) provide the majority of the animal's energy needs during long-term fasts. Increases in both turnover and plasma concentrations of NEFA have been demonstrated in several species. Reported NEFA values are greater than those reported for any other animal (as high as 3.2 mM) and increase over the fast in some species (Houser *et al.*, 2007). Plasma glycerol levels show similar increases and are available as a substrate for gluconeogenesis. There is some evidence that seals can selectively utilize reserves from different parts of the body (such as blubber reserves vs core

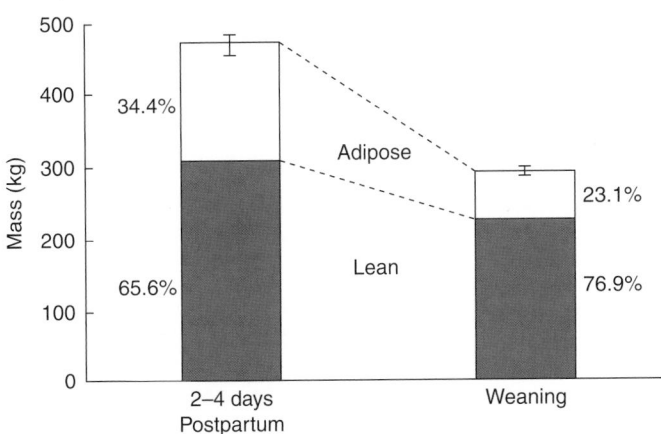

**Figure 3** *Changes in maternal mass and body composition over the lactation period in female northern elephant seals. On average females lose 27% of total body protein stores during lactation.*

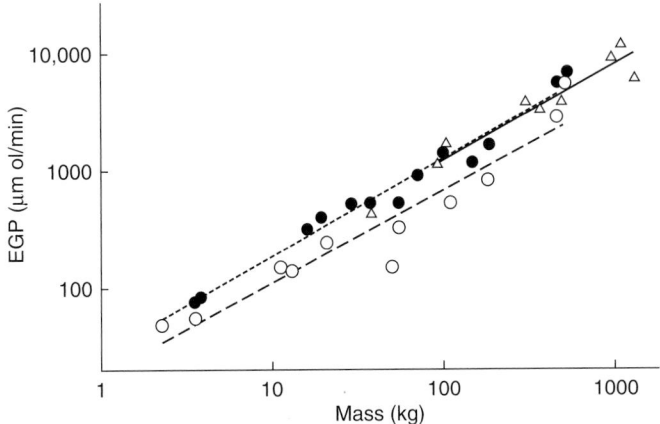

**Figure 4** *Allometric comparison of rates of glucose production in fasting and nonfasting species. Each point represents a mean value for a species or age class from a wide variety of studies on domestic and wild animals. Closed circles represent postabsorptive animals. Open circles represent animals undergoing at least a 48h fast. Note the suppression of glucose production with fasting. Open triangles represent fasting values from several species, sexes, and age classes of phocid seals. Note that fasting phocids exhibit rates of glucose production consistent with nonfasting individuals of other species.*

tissues) during different phases of the fast. Ketone bodies (HBA) accumulate somewhat during the fast in weaned elephant seals and gray seals, *Halichoerus grypus*, and Weddell seals, *Leptonychotes weddelli*, and subsequently decline rapidly as the end of the fast is approached (Castellini and Costa, 1990). This suggests that ketone bodies may contribute to energy metabolism during long-term fasting, although levels are significantly lower than nonfasting adapted species and never reach levels affecting acid–base balance or causing ketosis. Particularly striking in this regard are data from lactating female and breeding adult male northern elephant seals, who, despite the aforementioned high and increasing NEFA levels over the fast, exhibit consistently low HBA values across the breeding fast (<0.80 mM; Champagne *et al.*, 2006). It is also important to note that increasing HBA levels were only demonstrated in juvenile animals of relatively small body size.

Stage III fasting or terminal starvation occurs when 30–50% of total body protein has been used. In nonfasting adapted species this is associated with a decrease in lipid utilization and a decline in circulating ketone bodies. Evidence for entrance into Stage III fasting in seals has been equivocal. The increase and subsequent decline in ketone bodies at the end of the fast in some fasting pups would suggest entry into Stage III (Nordoy *et al.*, 1992; Houser and Costa, 2003). However, only two studies have demonstrated increases in protein catabolism following the period of effective protein sparing. When considering the protein utilization by lactating females it is important to include the loss of body protein for milk synthesis. One study on northern elephant seals demonstrated reductions in protein sparing with the depletion of lipid reserves (Crocker *et al.*, 1998) that together with the nutrient demands of milk synthesis moved females close to the 30% value of body protein loss considered extreme in humans (Fig. 3). It may be that in normal, voluntary fasts, Stage III fasting is never reached, with seals departing to sea before this point. Blubber reserves also play an important role in thermoregulation and blubber depletion for energy needs is limited by the need to thermoregulate (Worthy, 1991).

## B. Glucose Metabolism

The primary constraint on the extended fasting durations exhibited by pinnipeds is the need to provide carbohydrate to glucose-dependent tissues that include the central nervous system,

erythrocytes, renal medulla, and testes. Due to the limited ability to store carbohydrate in the body, fasting pinnipeds are dependent on gluconeogenesis, the production of sugar by the liver and kidney, to provide fuel for these tissues. The general model for fasting species is that protein is spared most effectively by reducing glucose utilization by nondependent tissues and using glycerol mobilized for fats to produce glucose instead of amino acids liberated from muscle. In most species of animals, fasting is associated with a suppression of endogenous glucose production (Fig. 4). In contrast, studies on elephant seals, harbor seals, *Phoca vitulina*, and gray seals, suggest that glucose turnover rates are typical of nonfasting mammals and that there is little or no suppression of glucose production during fasting (Davis, 1983; Nordøy and Blix, 1991; Champagne *et al.*, 2005, 2006). Despite this, the direct contribution of glucose to the total metabolic rate is less than 1% in seals during extended fasts (Keith and Ortiz, 1989). Furthermore, recent studies on the contribution of glycerol to gluconeogenesis in elephant seals suggest that it accounts for less than 3% of the glucose produced while fasting, leaving the primary gluconeogenic substrate as yet unidentified (Houser *et al.*, 2007). These finding suggest high levels of futile cycle activity that may play an important role in regulating fasting metabolism and that recycling of protein and glucose carbon may serve as an important shuttle mechanism for carbon (e.g., synthesis of nonessential amino acids).

As tracer methodologies have been used to examine metabolite flux in fasting pinnipeds, particularly in elephant seals, it has become clear that static values of metabolites in blood samples reflect complex relationships between synthesis, mobilization, and utilization of nutrients and should be interpreted cautiously. For example, dramatic increases in plasma levels of NEFA can occur despite decreases in overall rates of lipolysis, increased protein catabolism can be associated with decreases blood urea levels due to changes in renal filtration, increases in plasma glucose levels can occur despite static levels of glucose production due to impaired glucose clearance, and plasma ketone levels may not reflect levels of lipid mobilization and utilization. Thus the

**P**

traditional methods of examining metabolite flux through blood chemistries may lead to erroneous interpretations.

### C. Hormonal and Fuel Regulation

Studies on hormonal and fuel regulation during fasting have suggested that seals may exhibit the protein conservation and high lipid utilization of Stage II fasting throughout their lives. Fat is an important energy source throughout development and life, including high fat milk, high fat fish, and body fat stores. Studies of the role of insulin in glucose regulation in suckling pups have suggested that seals may be preadapted for fasting (Kirby and Ortiz, 1994). Low insulin concentrations, impaired glucose clearance, and low insulin to glucagon ratios exhibited in both feeding and fasting pups contribute to the mobilization of lipids from body stores. In general, fasting pups are hyperglycemic and hypoinsulemic. Glucose tolerance tests in fasting pups and adults have suggested impaired insulin response and glucose clearance. Most strikingly, lactating females appear to functionally reduce insulin response as lactation progresses. Similarly, hormone challenge studies have suggested deviations from the standard mammalian push–pull model of glucose regulation and important developmental and reproductive impacts on receptor populations for hormones. For example, fasting elephant seal pups exhibit no gluconeogenic response to a pharmacological dose of glucagon. In contrast lactating and molting adult females exhibit a delayed but significant increase in blood glucose levels in response to the same challenge.

In general, studies on fasting pinnipeds have reported significant increases in growth hormone and cortisol associated with fasting (Ortiz *et al.*, 2001; Champagne *et al.*, 2005). However, tracer studies have also suggested an uncoupling between important metabolic regulatory hormones and the processes which they regulate in typical mammals, particularly with respect to carbohydrate metabolism, and a potentially increased importance of cycle activity in regulating metabolite flux. Currently our understanding of the fasting physiology of pinnipeds is biased toward studies on juvenile animals, especially the postweaning fast. Recent investigations have demonstrated dramatic differences in fasting physiology between lactating and molting females as well as fasting pups. The pressures of nutrient delivery for milk synthesis may have significant impacts on the metabolic strategies used by females during extended fasts. Less is known about fasting in adult males, who potentially have some of the highest metabolic rates while fasting. Future studies can benefit from interspecific comparisons of fasting physiology relative to natural fasting durations. Even more instructive may be intraspecific comparisons among sexes and age classes during development, lactation, molting, and breeding. These comparisons will help to demonstrate how the fasting biochemistry of animals responds to the varying energy and nutrient demands of breeding and lactation.

### D. Renal Physiology during Fasting

Water balance during fasting is maintained by input of metabolic derived water from lipid catabolism. Significant reductions in urinary water loss contribute to maintenance of water balance (Adams and Costa, 1993; Fig. 5). The low rate of protein oxidation and an efficient urinary concentrating mechanism in pinnipeds reduces urinary water loss during fasting. Early work on harbor seals demonstrated reductions in glomerular filtration rates (GFR) associated with fasting and hyperfiltration after feeding (Hiatt and Hiatt, 1942). Subsequent investigations have been equivocal, leaving it unclear whether the mechanism underlying reduced urine flow is decreased glomerular filtration or increased tubular resorption (Pernia *et al.*,

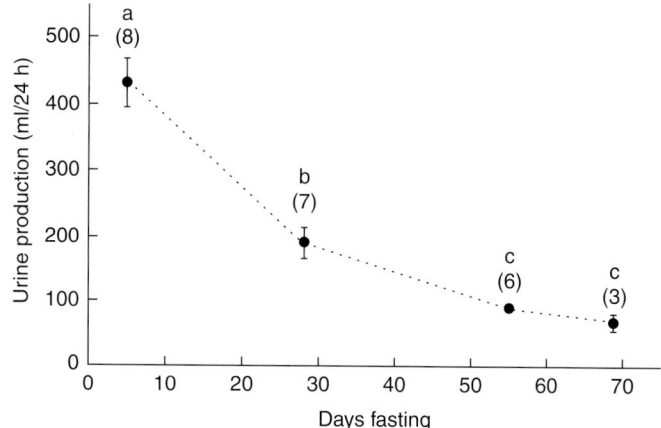

**Figure 5** *The decline in daily urine production of elephant seal pups progressing through the postweaning fast. Letters denote significant differences between the periods (P < 0.05). Sample sizes are in parentheses. (From Adams and Costa, 1993.)*

1989; Houser *et al.*, 2001). Investigations on weaned northern elephant seal pups have revealed no correlation between plasma levels of vasopressin and urinary concentrating ability but suggest that vasopressin may have important effects on solute excretion (Ortiz *et al.*, 2003). Similar investigations have demonstrated increases in aldosterone and angiotensin II during long-term fasts in seals suggesting an important role for aldosterone in regulating urine concentration by its action on sodium resorption in the collecting duct (Ortiz *et al.*, 2006).

Investigations on lactating adult northern elephant seals have demonstrated dramatic increases in GFR across the fast and suggest that these elevated rates could be an adaptation to increasing the efficiency of urea excretion during reduced urine flow (Crocker *et al.*, 1998). This mechanism reduces residency time and passive resorption of urea in the collecting tubules. The efficiency of urea excretion in lactating females declined from 49 to 38% over lactation, suggesting that with a declining urine flow and stable plasma urea concentration, increased GFR is necessary to increase urea excretion as protein catabolism increases.

### E. Lactation Physiology

The physiology of lactation in pinnipeds is significantly impacted by constraints resulting from the temporal separation of foraging and parental investment. Pinnipeds have evolved two general lactation strategies to manage pup provisioning within these constraints (Costa, 1993). For a more detailed discussion of the energetics of lactation, see the section on "Energetics," this volume. After a short perinatal fast, otariids alternate foraging trips with suckling bouts. Initial milk production is synthesized from maternal reserves while subsequent milk nutrients are derived from resources acquired while foraging. Phocids, particularly the larger species, fast during a brief but energy-intensive lactation during which nutrients for milk synthesis are derived exclusively from maternal tissues. Some smaller phocid seals, like harbor seals or ring seals, *Phoca hispida*, forage during lactation. Pinnipeds consistently produce lipid-rich milk, independent of lipid content of the diet (Costa, 1993). Even more amazing is that lactation, an energetically expensive period, occurs while the female is fasting. Long-term fasting is characterized by protein sparing, reductions in metabolic rate, and reductions in water loss for urea nitrogen excretion. Studies on nonlactating fasting phocids have shown that protein stores are spared with the bulk of energy demands being supplied by the oxidation of fatty acids. In contrast, lactation is characterized by dramatic increases in metabolism and significant transfer of nutrients and water to the mammary gland for the synthesis and secretion of milk. The general metabolism of the lactating female is reorganized in a way that ensures the appropriate nutrients are partitioned to the mammary gland. In nonfasting animals, lactation is accompanied by increased levels of food consumption and digestion, with accompanying increases in the absorptive capacity of the gastrointestinal tract. In fasting phocids, regulatory mechanisms override protein and energy sparing mechanisms to make the nutrients necessary for milk synthesis available at the expense of body nutrient reserves. The high demands of lactation coupled with complete abstinence from food and water presents a complex regulatory problem. A recent investigation in northern elephant seals suggested that changes in the energy demands of milk synthesis across lactation may impact fasting physiology and ultimately limit the period of parental investment (Crocker et al., 2001). From this perspective lactating phocid females may be one of the best examples of homeorhesis, "orchestrated changes for the priorities of a physiological state," found in nature.

Studies have suggested changes in milk composition and the nutrient requirements of milk synthesis across lactation in pinnipeds. These patterns are controlled by hormonal and biochemical changes. Of these changes, those that impact mobilization of adipose tissue stores, metabolism of lipids, and utilization of lipids by the mammary gland are the most significant. These changes are also important as pinnipeds transition from periods of nutrient deposition and mobilization for milk synthesis. Decreased insulin levels remove the strong antilipolytic effects of this hormone. Cortisol and other glucocorticoids influence lipid metabolism directly and indirectly. Cortisol stimulates hormone sensitive lipase in adipose tissue and antagonizes the actions of insulin. Hormone sensitive lipase activity increases lipid mobilization from adipose tissue. Lipoprotein lipase (LPL) is the primary enzyme involved in directing triglycerides mobilized from tissue stores to tissues for utilization. LPL is bound to tissues and facilitates the hydrolyzation of triglycerides, allowing uptake of fatty acids by the tissue. Under normal conditions of insulin release, LPL activity in tissues increases and triglyceride is cleared from the blood. During fasting, LPL activity in adipose tissue decreases, while hormone sensitive lipase activity in adipocytes increases. The general pattern found in lactation is a decrease in adipocyte LPL activity before parturition and an increase in mammary gland LPL activity. The hormone prolactin is believed to be primarily responsible for LPL regulation in lactation. Recent investigations on harbor seals and gray seals have suggested a similar pattern. General activity levels of LPL increased 10-fold over lactation in these species and were significantly higher than levels found in humans. The dramatic increase in milk lipid content early in lactation in some phocid seals may in part be explained by a significant increase in mammary gland LPL activity (Iverson et al., 1995).

Very little work has been done on the physiology underlying milk secretion in pinnipeds. The release of milk fat globules occurs by apocrine secretion, in which the apical portion of the cell membrane is sloughed off. The high fat contents of pinniped milk suggest significant increases in the amount of membrane and its turnover. Pinnipeds may partially reduce this requirement by utilizing larger fat globules that require smaller amounts of membrane loss per unit lipid secreted (Tedman, 1983). Data on mammary gland size in phocids has been equivocal, suggesting increased size relative to body mass in some species but not others (Tedman and Bryden, 1981). In any case, phocid seals appear to be particularly efficient at mobilizing and transporting nutrients to the mammary gland, perhaps by reducing the levels of de novo synthesis of milk lipids occurring at the mammary gland (Mellish et al., 1999). This efficiency must in turn be matched by rapid and efficient digestion and assimilation of milk lipids by the offspring.

It is apparent that the regulatory processes involved in the initiation and termination of lactation are quite complex. In most mammals the initiation and eventual termination of lactation is a long complex process. However, all otariids and some phocids are able to intermittently turn milk production on and off. Milk production is downregulated or turned off, while the mother is at sea foraging, but within hours of arriving ashore she can suckle her pup and within hours milk production is in full gear. This ability has received considerable attention by the lactation physiology community and may provide insight into general processes of lactation physiology (Lang et al., 2005; Sharp et al., 2006, 2007).

## See Also the Following Articles

Energetics ■ Endocrine Systems

## References

Adams, S. H., and Costa, D. P. (1993). Water conservation and protein metabolism in northern elephant seal pups during the post-weaning fast. J. Comp. Physiol. B **163**, 367–373.

Castellini, M. J., and Costa, D. P. (1990). Relationships between plasma ketones and fasting duration in neonatal elephant seals. Am. J. Physiol. Regul. Integr. Comp. Physiol. **259**, R1086–R1089.

Castellini, M. A., and Rea, L. D. (1992). The biochemistry of natural fasting at its limits. Experientia **48**, 575–582.

Champagne, C. D., Houser, D. S., and Crocker, D. E. (2005). Glucose production and substrate cycle activity in a fasting adapted animal, the northern elephant seal. J. Exp. Biol. **208**, 859–868.

Champagne, C. D., Houser, D. S., and Crocker, D. E. (2006). Glucose metabolism during lactation in a fasting animal, the northern elephant seal. Am. J. Physiol. Regul. Integr. Comp. Physiol. **291**, R1129–R1137.

Costa, D. P. (1993). The relationship between reproductive and foraging energetics and the evolution of the Pinnipedia. In "Marine Mammals: Advances in Behavioural and Population Biology." (I. L. Boyd, ed.), pp. 293–314. Oxford University Press, Symp. Zool. Soc. Lond.

Costa, D. P., and Williams, T. M. (2000). Marine mammal energetics. In "The Biology of Marine Mammals" (J. Reynolds, and J. Twiss, eds), pp. 176–217. Smithsonian Institution Press, Washington, DC.

Crocker, D. E., Webb, P. M., Costa, D. P., and Le Boeuf, B. J. (1998). Protein catabolism and renal function in lactating northern elephant seals. Physiol. Zool. **71**, 485–491.

Crocker, D. E., Williams, J. D., Costa, D. P., and Le Boeuf, B. J. (2001). Maternal traits and reproductive effort in northern elephant seals. Ecology **82**(12), 3451–3555.

Davis, R. W. (1983). Lactate and glucose-metabolism in the resting and diving harbor seal (Phoca vitulina). J. Comp. Physiol. B **153**, 275–288.

Kirby, V. L., and Ortiz, C. L. (1994). Hormones and fuel regulation in fasting elephant seals. In "Elephant Seals: Population Ecology, Behavior, and Physiology" (B. J. Le Boeuf, and R. M. Laws, eds), pp. 374–386. University of California Press, Los Angeles, CA.

Hiatt, E. P., and Hiatt, R. B. (1942). The effect of food on the glomerular filtration rate and renal blood flow in the harbor seal (Phoca vitulina L.). J. Cell. Comp. Physiol. **19**, 221–227.

Houser, D. S., Crocker, D. E., Webb, P. M., and Costa, D. P. (2001). Renal function in suckling and fasting pups of the northern elephant seal. Comp. Biochem. Physiol. A **129**, 405–415.

Houser, D. S., and Costa, D. P. (2003). Entrance into stage III fasting by starveling northern elephant seal pups. Mar. Mamm. Sci. **19**, 186–197.

Houser, D. S., Champagne, C. D., and Crocker, D. E. (2007). Lipolysis and glycerol gluconeogenesis in simultaneously fasting and lactating

P

northern elephant seals. *Am. J. Physiol. Regul. Integr. Comp. Physiol.* **293**, R2376–R2381.

Iverson, S. J., Hamosh, M., and Bowen, W. D. (1995). Lipoprotein lipase activity and its relationship to high milk fat transfer during lactation in grey seals. *J. Comp. Physiol. B* **165**, 384–395.

Keith, E. O., and Ortiz, C. L. (1989). Glucose kinetics in neonatal elephant seals during postweaning aphagia. *Mar. Mamm. Sci.* **5**, 99–115.

Lang, S. L. C., Iverson, S. J., and Bowen, W. D. (2005). Individual variation in milk composition over lactation in harbour seals (*Phoca vitulina*) and the potential consequences of intermittent attendance. *Can. J. Zool.* **83**, 1525–1531.

Mellish, J. E., Iverson, S. J., Bowen, W. D., and Hammill, M. O. (1999). Fat transfer and energetics during lactation in the hooded seal: the roles of tissue lipoprotein lipase in milk fat secretion and pup blubber deposition. *J. Comp. Physiol. B* **169**, 377–390.

Nordoy, E. S., Stijfhoorn, D. E., Raheim, A., and Blix, A. S. (1992). Water flux and early signs of entrance into phase-iii of fasting in gray seal pups. *Acta Physiol. Scand.* **144**, 477–482.

Nordøy, E. S., and Blix, A. S. (1991). Glucose and ketone body turnover in fasting grey seal pups. *Acta Physiol. Scand.* **141**, 563–571.

Ortiz, R. M., Wade, C. E., and Ortiz, C. L. (2001). Effects of prolonged fasting on plasma cortisol and TH in postweaned northern elephant seal pups. *Am. J. Physiol. Regul. Integr. Comp. Physiol.* **280**, R790–R795.

Ortiz, R. M., Wade, C. E., and Ortiz, C. L. (2003). Body water handling in response to hypertonic-saline induced diuresis in fasting northern elephant seal pups (*Mirounga angustirostris*). *Comp. Biochem. Physiol. A* **134**, 423–428.

Ortiz, R. M., Crocker, D. E., Houser, D. S., and Webb, P. M. (2006). Angiotensin II and aldosterone increase with fasting in breeding adult male northern elephant seals (*Mirounga angustirostris*). *Physiol. Biochem. Zool.* **79**(6), 1106–1112.

Pernia, S. D., Costa, D. P., and Ortiz, C. L. (1989). Glomerular filtration rate in weaned elephant seal pups during natural, long term fasts. *Can. J. Zool. (J. Can. Zool.)* **67**, 1752–1756.

Sharp, J. A., Cane, K. N., Lefevre, C., Arnould, J. P. Y., and Nicholas, K. R. (2006). Fur seal adaptations to lactation: Insights into mammary gland function. *Curr. Top. Dev. Biol.* **72**, 276–308.

Sharp, J. A., Lefevre, C., Brennan, A. J., and Nicholas, K. R. (2007). The fur seal—a model lactation phenotype to explore molecular factors involved in the initiation of apoptosis at involution. *J. Mammary Gland Biol. Neoplasia* **12**, 47–58.

Tedman, R. A. (1983). Ultrastructural morphology of the mammary-gland with observations on the size distribution of fat droplets in milk of the Weddell seal *Leptonychotes weddelli* (Pinnipedia). *J. Zool.* **200**, 131–141.

Tedman, R. A., and Bryden, M. M. (1981). The mammary-gland of the Weddell seal, *Leptonychotes weddelli* (Pinnipedia) 1. Gross and microscopic anatomy. *Anat. Rec.* **199**, 519–529.

Worthy, G. A. J. (1991). Insulation and thermal balance of fasting harp and gray seal pups. *Comp. Biochem. Physiol. A* **100**, 845–851.

# Pinnipedia, Overview

Annalisa Berta

Pinnipeds have always been understood to represent a group distinct from other aquatic mammals. They are recognized as members of the mammalian order Carnivora and include three monophyletic lineages, Otariidae (fur seals and sea lions), Odobenidae (walruses), and Phocidae (true or earless seals).

Pinnipeds comprise slightly more than one-fourth (26%) of the species diversity of marine mammals (approximately 128 species currently recognized). Thirty-three living species of pinnipeds are distributed throughout the world: 18 phocids, 14 otariids, and the walrus. One additional species of phocid (Caribbean monk seal) and one subspecies of otariid (Japanese sea lion) are reported extinct in historical time.

## I. Systematics and Distribution
### A. Otariidae: Fur Seals and Sea Lions

Of the two groups of seals, the otariids are characterized by the presence of external ear flaps or pinnae, and for this reason they are often called "eared" seals (Fig. 1). Otariids (and the walrus) can turn their hindflippers forward and use them to walk (discussed further under Adaptations). The Otariidae traditionally are divided into two subfamilies Otariinae (sea lions) and Arctocephalinae (fur seals) although recent molecular studies have revealed that neither fur seals nor sea lions are monophyletic. Five living genera and species of sea lions are recognized, occurring in both the northern and southern hemispheres: *Eumetopias jubata* (northern or Steller's sea lion), *Neophoca cinerea* (Australian sea lion), *Otaria byronia* (Southern sea lion), *Zalophus californianus* (*Z. c. californianus* California sea lion, *Z. c. japonicus* Japanese sea lion, and *Z. c. wollebacki* Galapagos sea lion), and *Phocarctos hookeri* (New Zealand sea lion; Fig. 2). The fur seals, named for their thick, dense fur, include two genera, the monotypic Northern fur seal (*Callorhinus ursinus*) and the southern fur seals (*Arctocephalus*), consisting of eight species: *A. australis* (South American fur seal), *A. forsteri* (New Zealand fur seal), *A. gazella* (Antarctic fur seal), *A. galapagoensis* (Galapagos fur seal), *A. philippii* (Juan Fernandez fur seal), *A. pusillus* (*A. p. pusillus* South African fur seal and *A. p. doriferus* Australian fur seal), *A. townsendi* (Guadalupe fur seal), and *A. tropicalis* (Subantarctic fur seal). All of the fur seals except the northern and Guadalupe fur seals are found in the southern hemisphere. The northern fur seal is found in subarctic waters of the North Pacific, with the exception of a small population on San Miguel Island off California (Fig. 3).

Otariid monophyly is well supported based on a both morphologic as well as molecular data. Morphologic and molecular studies consistently position *Callorhinus* as the earliest diverging extant otariid (Fig. 4). Extinct otariids *Pithanotaria*, *Thalassoleon*, and *Hydractos* are sequential sister taxa to *Callorhinus*. Molecular results strongly support a branch containing the Guadalupe, South American, and New Zealand fur seals and the Australian and New Zealand sea lion. Good support was also found for South African fur seals as sister to this clade. The relationship between Northern + California sea lions, Southern sea lion and remaining sea lions and *Arctocephalus* species is not conclusively resolved. In support of fur seal and sea lion paraphyly is evidence for hybridization of various sympatric species (e.g., *Arctocephalus gazella/A. tropicalis/A. forsteri, Zalophus californianus/Otaria byronia*). Additionally, the violent sexual behavior by male sea lions toward females of different species my have resulted in more hybridization and introgression than has been typically recognized for the evolutionary history of otariids.

### B. Odobenidae: Walruses

Although tusks are arguably the most characteristic feature of the modern walrus a rapidly improving fossil record indicates that these unique structures evolved in a single lineage of walruses and "tusks do not a walrus make" (Repenning and Tedford, 1977). The living walrus is the sole survivor of what was once a diverse radiation of at least 11

**Figure 1** *Representative otariids (A) southern sea lion,* Otaria byronia *and (B) South African fur seal,* Arctocephalus pusillus, *illustrating pinna. Note also the thick, dense fur characteristic of fur seals. Illustrations by P. Folkens. (From Berta et al., 2006.)*

**Figure 2** *Distribution of sea lions. Based on Riedman (1990). (From Berta et al., 2006.)*

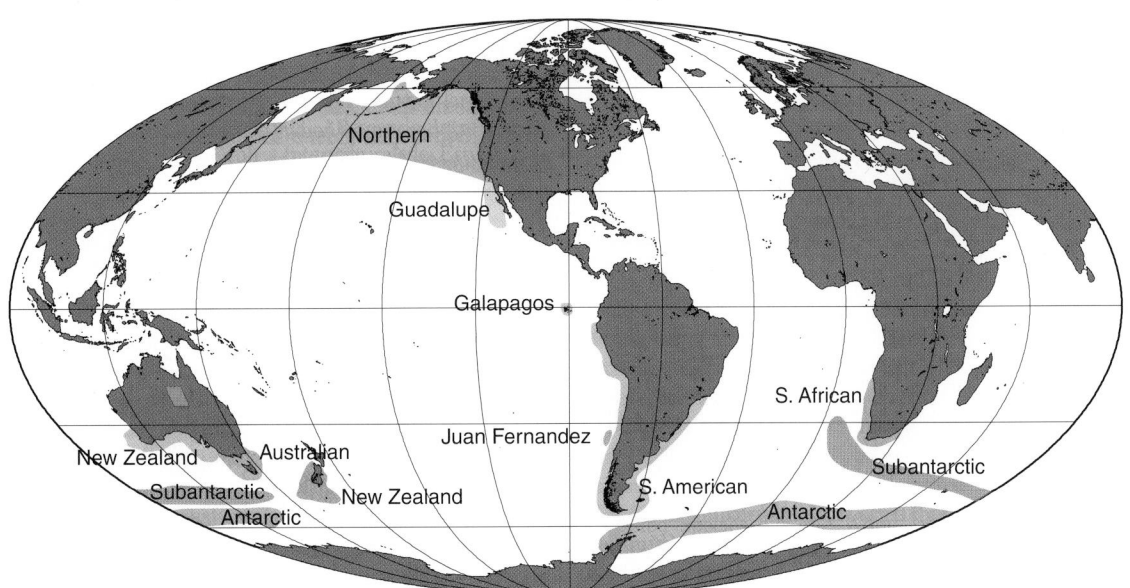

**Figure 3** *Distribution of fur seals. Based on Riedman (1990). (From Berta et al., 2006.)*

genera and 14 species of walruses that lived from the early Miocene to the end of the Pliocene (see also PINNIPED EVOLUTION). Two subspecies of the modern walrus *Odobenus rosmarus* are usually recognized: *O. r. rosmarus* (Atlantic walrus) and *O. rosmarus divergens* (Pacific walrus) although a population from the Laptev Sea has been described as a third subspecies *O. rosmarus. lapteevi* but this has not been substantiated. Pacific walruses are more abundant, are larger, and have longer tusks than Atlantic walruses (Fay, 1981). Walruses inhabit the northern hemisphere in areas with pack ice over shallow water of the continental shelf (Fig. 5). Like phocids, walruses lack external ear flaps. A unique feature of members of the modern walrus lineage are enlarged upper canine tusks that function primarily in breeding and social contexts. Walrus locomotion combines elements of phocid and otariid locomotion (discussed further under Adaptations).

Monophyly of the walrus family is strongly supported although controversy continues regarding whether walruses are more closely related to otariids or to phocids. Although there is morphologic evidence to ally walruses and phocids recent total evidence and molecular studies provide consistent, robust support for an alternative alliance between otariids and odobenids. A survey of genetic variation among Atlantic and Pacific populations of the walrus suggests separation of the subspecies about 500,000–750,000 years ago, supporting the suggestion that *Odobenus* evolved in the Pacific and reached the North Atlantic early in Pleistocene time (Cronin *et al.*, 1994).

### C. Phocidae: Seals

The second major grouping of seals, the phocids, often are referred to as the "true" or "earless" seals for their lack of visible ear pinnae, a characteristic which readily distinguishes them from otariids as well as the walrus (Fig. 6). Among the most distinguishing phocid

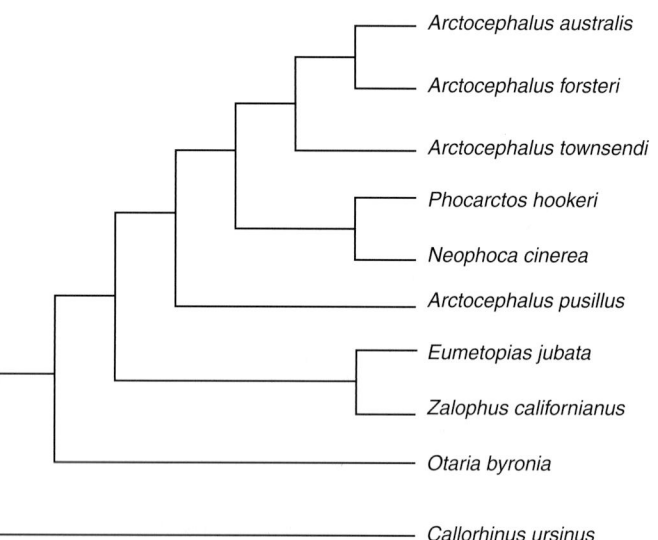

**Figure 4** *Phylogeny of the Otariidae based on molecular data. (From Arnason* et al., *2006.)*

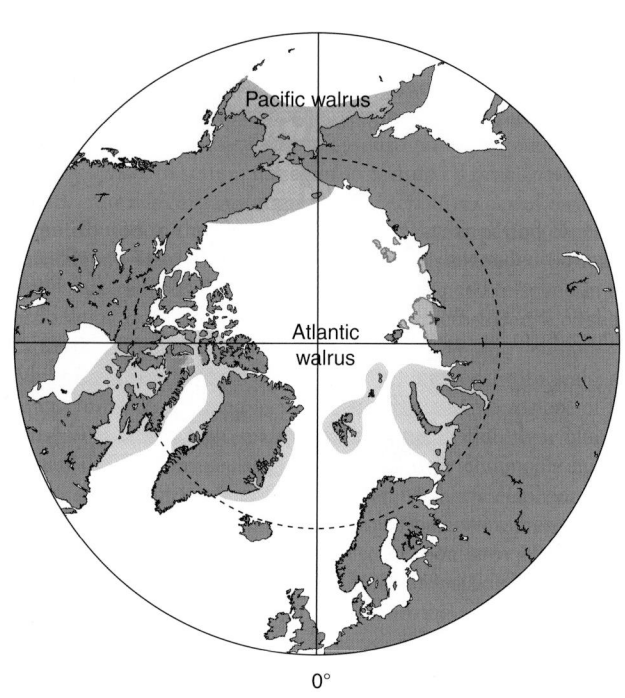

**Figure 5** *Distribution of modern walrus subspecies,* Odobenus rosmarus divergens *(Pacific walrus) and* Odobenus rosmarus rosmarus *(Atlantic walrus).*

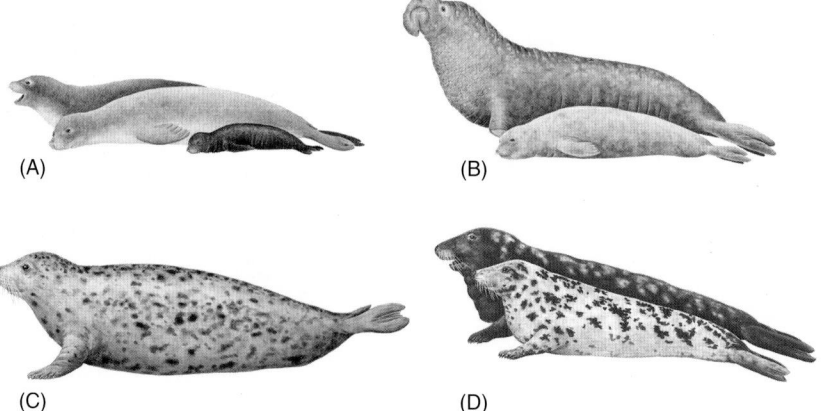

**Figure 6** *Representative monachines (A) Hawaiian monk seal,* Monachus schauinslandi *and (B) northern elephant seal,* Mirounga angustirostris *and phocines (C) Harbor seal,* Phoca vitulina, *and (D) gray seal,* Halichoerus grypus. *Males are shown behind smaller females. Illustrations by P. Folkens. (From Berta* et al., *2006.)*

characteristics is their inability to turn the hindlimbs forward to support the body, resulting in a peculiar crawling locomotion on land (discussed further under Adaptations). Phocids inhabit both northern and southern hemispheres, although they are largely restricted to polar and subpolar regions (Fig. 7). Among pinnipeds, phocids are unique in their ability to survive in estuarine and freshwater habitats (e.g., Caspian and Baikal seals inhabiting landlocked lakes).

Traditionally phocids have been divided into two to four major subgroups (including the Monachinae, Lobodontinae, Cystophorinae, and Phocinae). Recent molecular studies strongly support monophyly of both the Monachinae and Phocinae (Davis *et al.*, 2004; Arnason *et al.*, 2006; Fig. 8).

The Monachinae clade of "southern seals" includes Monachini (monk seals), Miroungini (elephant seals), and the Lobodontini

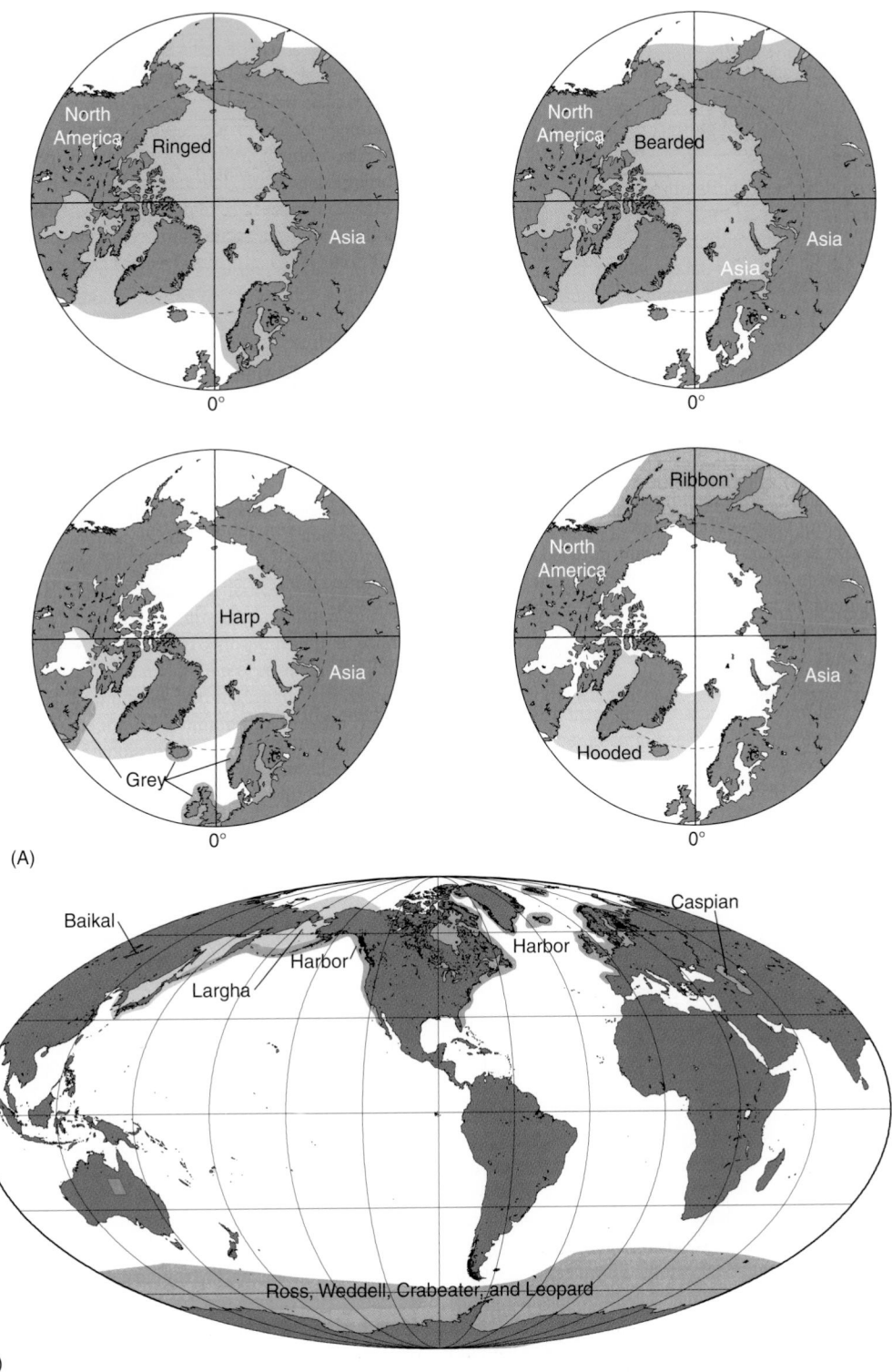

(A)

(B)

**Figure 7** *Distribution of some phocine seals (A), Antarctic phocine seals (B), and monachines (Facing page). Based on Riedman (1990). (From Berta et al., 2006.)*

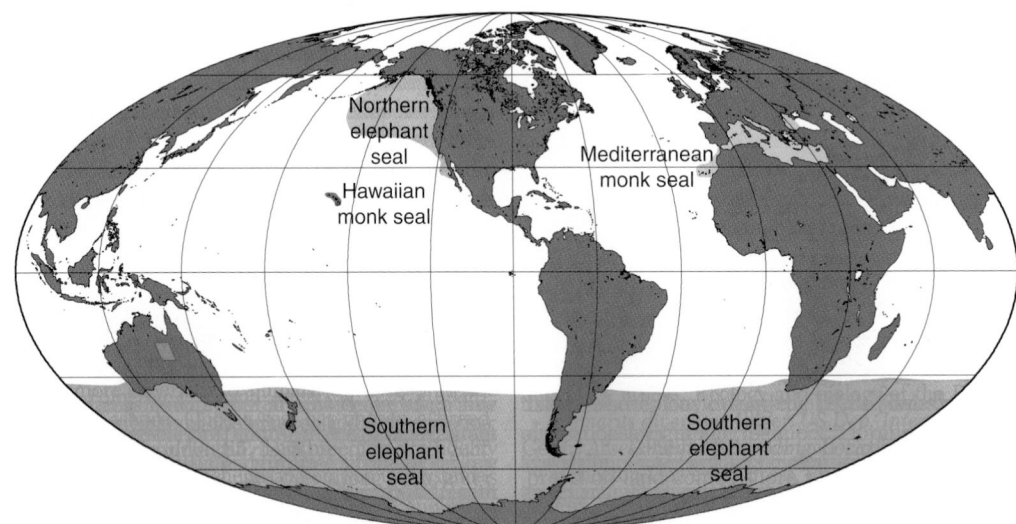

Figure 7 (continued)

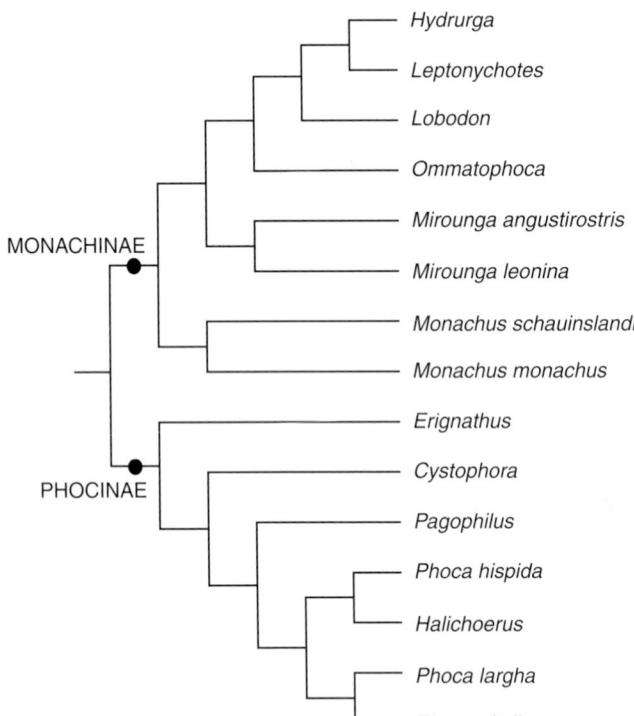

Figure 8 *Phylogeny of the Phocidae based on molecular data (Arnason et al., 2006).*

(Antarctic seals). Within the Monachinae there is strong support for a split between Monachini, the deepest branching lineage and Miroungini and Lobobontini. Three species of *Monachus* have been described, *M. schauinslandi* (Hawaiian monk seal), *M. monachus* (Mediteranean monk seal), and the recently extinct *M. tropicalis* (Caribbean monk seal). Molecular sequence data for extant species of *Monachus* supports a sister group relationship between the Hawaiian and Mediteranean monk seals (Davis *et al.*, 2004; Fyler *et al.*, 2005). Elephant seals, named for their enlarged proboscis, are represented by two species, *Mirounga angustirostris* (northern elephant seal) and *Mirounga leonina* (southern elephant seal).

The Lobodontini which include *Leptonychotes weddelli* (Weddell seal), *Lobodon carcinophagus* (Crabeater seal), *Hydrurga leptonyx* (Leopard seal), and *Ommatophoca rossi* (Ross seal). Molecular studies consistently position the Ross seal as sister to the other three Antarctic seals.

The Phocinae clade of "northern seals" includes *Erignathus barbatus* (bearded seal), *Cystophora cristata* (hooded seal), *Halichoerus grypus* (gray seal), *Phoca* (including among others harbor and spotted seal), *Pusa hispida* (ringed seal), *Histriophoca fasciata* (ribbon seal), and *Pagophilus groenlandica* (harp seal). A basal divergence between the bearded seal and the hooded seal + Phocini (*Pusa, Histriophoca, Pagophilus, Halichoerus,* and *Phoca*) is strongly supported by molecular studies (cited above). Disagreement about relationships among the Phocini, attributed to their rapid radiation, have resulted in taxonomic uncertainty regarding their classification. There is general agreement that the Phocini are divided into an earlier diverging lineage of spotted + harbor seals and remaining seals. It is acknowledged that relationships among the latter group are poorly resolved.

The harbor seal (*Phoca vitulina*) has the most extensive geographic distribution of any seal, with a range spanning over 16,000 km from the east Baltic, west across the Atlantic and Pacific Ocean to southern Japan. The population structure of the harbor seal studied by Stanley *et al.* (1996) revealed that populations in the Pacific and Atlantic Oceans are highly differentiated. The mitochondrial data are consistent with ancient isolation of populations in both oceans coincident with the development of continental glaciers and extensive sea ice. In the Atlantic and Pacific Oceans populations appear to have been established from west to east, with the European populations showing the most recent common ancestry.

## II. Anatomy and Physiology

Pinniped aquatic specializations include their streamlined shape, reduced external ear pinnae, paddle-like limbs and feet, small tail, and genital organs and mammary glands that are retracted beneath the skin. In comparison to most terrestrial carnivorans, pinnipeds are large which helps to conserve warmth. Pinnipeds, particularly phocids show tremendous diversity in size ranging from the smallest pinniped, the Baikal seal reaching a length of just over a meter and a weight of 45 kg to the largest pinniped, the elephant seals nearly 5 m

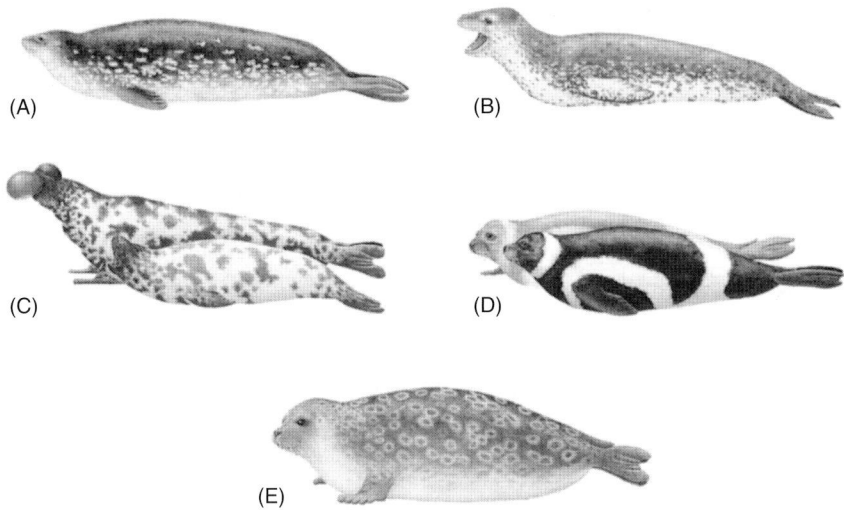

**Figure 9** *Examples of phocid pelage patterns (A) Weddell seal,* Leptonychotes wed-dellii, *(B) leopard seal,* Hydrurga leptonyx, *(C) hooded seal,* Cystophora cristata, *(d) ribbon seal* Histriophoca fasciata, *and (E) ringed seal,* Phoca hispida. *(Illustrations by P. Folkens, in Berta and Sumich, 1999.)*

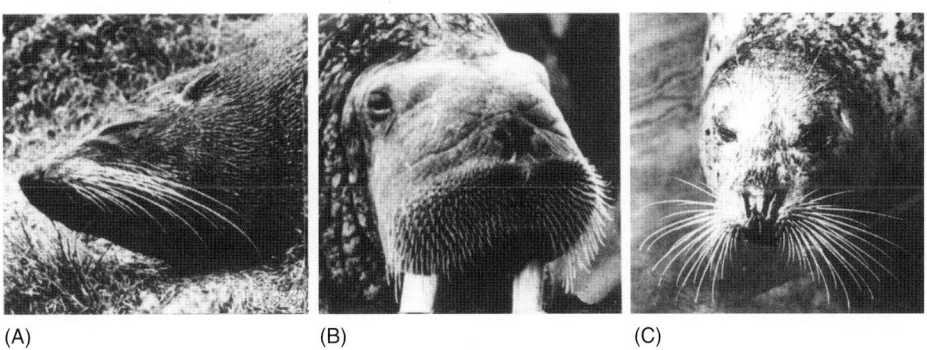

**Figure 10** *Heads of various pinnipeds showing facial vibrissae (A) New Zealand fur seal,* Arctocephalus forsteri, *(B) Walrus,* Odobenus rosmarus, *and (C) Pacific harbor seal,* Phoca vitu-lina richardsi. *(From Ling, 1977 in Berta et al., 2006.)*

in length (adult males) and up to 3200 kg (Bonner, 1990). Pinnipeds are ecologically diverse with habitats ranging from shelf to surface waters in tropical and polar seas, with some species living in fresh water lakes, while others freely move between rivers and the ocean.

Phocids and the walrus have lost much of their hair (fur) and are characterized by thick layers of blubber under the skin. Otariids, especially the fur seals, have retained a thick fur coat. Color patterns in the pelage of pinnipeds occur almost exclusively among phocids. Ice breeding seals (e.g., ribbon seal, harp seal, hooded seal, ringed seal, crabeater seal, Weddell seal, and leopard seal) show contrasting dark and light or disruptive color patterns (Fig. 9). The uniform coloration of some pinnipeds (e.g., white harp seal pups) allows them to readily blend into their arctic environment. Pinnipeds come ashore for birthing and molting. All phocid seals undergo an annual molt. Fur seals and sea lions instead renew their pelt gradually all year.

Vibrissae, or whiskers, are stiff hairs that occur on the face. Most prominent are the mystacial whiskers which range in size from the short stiff bristles of the walrus to the very long, fine bristles of fur seals (Fig. 10). Vibrissae function as sensitive touch receptors. Research on the Baltic ringed seal has shown that they have exceptionally well-developed vibrissae which help them find their way in the dark and often cloudy water beneath the ice. A single vibrissae of the Baltic ringed seal contains more than 10 times the number of nerve fibers typically found in that of a land mammal. Experimental data has also shown that blindfolded harbor seals used vibrations detected by their vibrissae to follow prey (i.e., fish "trails") in the water (Denhardt *et al.*, 2001). Evidence of heat conduction in the vibrissae of harbor seals indicates that they also play a role in thermoregulation (Mauck *et al.*, 2000).

Pinnipeds, like other marine mammals, have evolved ways to accommodate the immense heat loss that occurs in the water. Among these solutions are a spherical body and a resultant decreased surface to volume ratio and increased insulation (thick blubber or fur). In addition, heat exchange systems occurring in the flippers, fins, and reproductive tracts of pinnipeds conserves body heat (see THERMOREGULATION).

Among modern pinnipeds aquatic and terrestrial locomotion are achieved differently (see LOCOMOTION). Three distinct patterns of swimming are recognized: (1) pectoral oscillation (forelimb swimming) seen in otariids where the forelimbs are used in a "flapping" manner to produce thrust, (2) pelvic oscillation (hindlimb swimming)

P

seen in phocids where the hindlimbs are the major propulsors, and (3) a variant of pelvic oscillation exhibited by the walrus where the hindlimbs are the dominant propulsive force and the forelimbs are used as rudders or paddles. There is a major difference in locomotion on land between phocids on the one hand and otariids and walruses on the other. The inability of phocids to turn the hindlimbs forward results in forward progression by vertical undulations of the trunk which do not involve the hindlimbs. In walruses, as in otariids, the hindlimbs can be rotated forward in terrestrial locomotion.

Pinnipeds are carnivores, most are generalists feeding predominately on fish and squid (see also DIET). Several pinnipeds, notably crabeater and leopard seals have highly modified cheek teeth with complex cusps to trap and strain krill. Leopard seals also possess well-developed canines for preying on birds and other pinnipeds. Walruses are specialists feeding almost exclusively on clams using a suction feeding strategy in which the muscular tongue acts as a piston creating low pressure in the mouth cavity. Some pinnipeds, for example elephant seals, rival gray whales in the distances traveled in migration (18,000–21,000 km) to forage offshore between breeding seasons. The advent of microprocessor-based geographic location time and depth recorders (GLTDRs), satellite telemetry, and crittercams has enabled documentation of details of the foraging behavior of these deep diving seals.

Among pinnipeds are found the most extraordinary of marine mammal divers (see DIVING BEHAVIOR and DIVING PHYSIOLOGY). Average dives of small species such as the Ross seal are just under 10 min in duration increasing to over 1 h for the northern elephant seal and the Weddell seal. Maximum depths vary from less than 100 m in the Guadalupe fur seal to more than 1500 m in northern elephant seal males. Some seals (in addition to sperm whales, sea turtles, and some penguins) are "incredible diving machines" with unique ways of budgeting their oxygen supply and responding to pressure.

Sounds produced by pinnipeds include both airborne and underwater vocalizations (see SOUND PRODUCTION). Airborne sounds vary from grunts, snorts, or barks identified as either mother–pup calls or threat calls among seals to the distinctive bell-like sounds produced by male walruses striking throat pouches with their flippers as part of a courting display during the breeding season. Pinnipeds produce a variety of underwater sounds that appear most often related to breeding activities and social interactions. Among these are the whistles, trills, chirps, and buzzes of Weddell seals that are used in territorial defense. These contrast with the soft, lyrical calls of leopard seals that may be related to their solitary social system. In contrast to toothed whales, pinnipeds have not been found to use echolocation in their natural surroundings.

The pinniped eye is adapted for vision both above and under water. The spherical lens, thick retina, and the well-developed tapetum lucidum increase light sensitivity. With the exception of the walrus which has small eyes, seals and sea lions have large eyes in relation to body size. The question of whether pinnipeds have color vision is still debated although behavioral experiments and the presence of both rods and cones in the retina have been documented in some species (e.g., California sea lion, spotted seal, walrus) (see VISION).

### III. Behavior

Unlike other marine mammals, pinnipeds differ in their need to return to land (or to ice) to give birth. Many pinnipeds (e.g., elephant seal) are extremely polygynous, with successful males mating with dozens of females in a single breeding season (see BREEDING SYSTEMS). Species that are polygynous tend to breed in large colonies

on land where males compete for breeding territories (in otariids) or establish dominance hierarchies (in elephant seals). Because these males must compete for access to females associated with extreme polygyny is the strong sexual dimorphism seen in elephant seals including large body size (adult males as much as five times as large as females), elongated proboscis, enlarged canine teeth, and thick skin on the neck. Other phocids (e.g., Weddell seal, harp, ringed, ribbon, bearded, hooded) mate in the water or on ice and show a reduced level of polygyny, that is explained in part by the difficulty in gaining access to females in unstable environments such as pack ice.

Pinnipeds are characterized by sexual bimaturity with females reaching sexual maturity before males. In polygynous species males require several years of physical maturation following sexual maturity before they successfully compete for access to females. Gestation in most species of pinnipeds averages between 10 and 12 months; walruses have the longest gestation period of 16 months. Most species regulate their reproductive cycle (see REPRODUCTIVE BEHAVIORS) by delayed implantation (from 1.5 to 5 months). Delayed implantation prolongs birth until conditions are more favorable for offspring survival. Pinniped females of all species give birth to a single pup. In most species, pupping occurs in spring or summer when food availability is highest.

The maternal behaviors and lactation strategies of pinnipeds are influenced by their breeding habitat whether on ice or land. Most phocids exhibit a fasting strategy where females fast completely and remain out of water for the duration of a relatively short lactation, ranging from less than 1 week in hooded seals to almost 8 weeks in the Weddell seal. It has been suggested that the unstable nature of pack ice has selected for the extremely short lactation periods of some ice breeding phocid seals. To compensate for the brevity of lactation, the milk produced by these species is energy dense, with a fat content up to 60% in some species (i.e., hooded and harp seals). Rapid pup growth is ensured by the richness of the milk. In the foraging cycle strategy of most otariids, mothers fast for only a few days following the birth of pups. Then the mothers begin foraging trips at sea leaving their pups for a few days at a time. The lactation periods of otariids are longer, ranging from several months to more than 1 year, and milk is generally less energy dense than in phocids (e.g., milk fat content averaging between 24 and 40%). Walruses exhibit a variant of the otariid strategy, termed the aquatic nursing strategy, in which walrus pups accompany their mothers on foraging trips into the water. The length of lactation in walruses is the longest among pinnipeds, lasting 2–3 years.

Among generalizations that can be made about pinniped longevity is that females, especially those of polygynous species, tend to live longer than males. In many cases males do not survive even to the delayed age of sexual maturity. Seals have been known to pup successfully at 24–25 years and live as long as 40 years or more. Significant factors implicated in the natural mortality of pinnipeds include disease (especially morbilliviruses), predation (e.g., white sharks, killer whales), starvation, and parasites.

## See Also the Following Articles

Adaptations ■ Diving Physiology ■ Earless Seals ■ Eared Seals

## References

Arnason, U., Gullberg, A., Janke, A., Kullberg, M., Lehman, N., Petrov, E. A., and Vainola, R. (2006). Pinniped phylogeny and a new hypothesis for their origin and dispersal. *Mol. Phylogenet. Evol.* **41**(2), 345–354.

Berta, A., Sumich, J. L., and Kovacs, K. M. (2006). "Marine Mammals: Evolutionary Biology," 2nd ed. Elsevier, San Diego, CA.

Bonner, W. N. (1990). "The Natural History of Seals." Facts on File, New York, NY.

Cronin, M. A., Hillis, S., Born, E. W., and Patton, J. C. (1994). Mitochondrial DNA variation in Atlantic and Pacific walruses. *Can. J. Zool.* **72**, 1035–1043.

Davis, C. S., Delisle, I., Stirling, I., Siniff, D. B., and Strobeck, C. (2004). A phylogeny of the extant Phocidae inferred from complete mitochondrial coding regions. *Mol. Phylogenet. Evol.* **33**, 363–377.

Denhardt, G., Mauck, B., Hanke, W., and Bleckmann, H. (2001). Hydrodynamic trail-following in harbor seals (*Phoca vitulina*). *Science* **293**, 102–104.

Fay, F. H. (1981). Walrus: *Odobenus rosmarus*. *In* "Handbook on Marine Mammals" (S. H. Ridgway, and R. J. Harrison, eds), Vol. 1, pp. 1–23. Academic Press, New York.

Fyler, C. A., Reeder, T. W., Berta, A., Antonelis, G., Aguilar, A., and Androukaki, E. (2005). Historical biogeography and phylogeny of monachine seals (Pinnipedia: Phocidae) based on mitochondrial and nuclear DNA data. *J. Biogeogr.* **32**, 1267–1279.

Ling, J. K. (1977). Vibrissae of marine mammals. *In* "Functional Anatomy of Marine Mammals" (R. J. Harrison, ed.), Vol. 3, pp. 387–415. Academic Press, London.

Mauck, B., Eysel, U., and Dehnhardt, G. (2000). Selective heating of vibrissal follicles in seals (*Phoca vitulina*) and dolphins (*Sotalia fluviatus guianensis*). *J. Exp. Biol.* **203**, 2125–2131.

Repenning, C. A., and Tedford, R. H. (1977). Otarioid seals of the Neogene. *U.S. Geol. Surv. Prof. Pap.*, 992.

Riedman, M. (1990). "The Pinnipeds. Seals, Sea Lions and Walruses." University of California Press, Berkeley, CA.

Stanley, H., Casey, S., Carnahan, J. M., Goodman, S., Harwood, J., and Wayne, R. K. (1996). Worldwide patterns of mitochondrial DNA differentiation in the harbor seal (*Phoca vitulina*). *Mol. Biol. Evol.* **13**(2), 368–382.

# Playful Behavior

## BERND WÜRSIG

Play consists of actions performed for no other apparent purpose than their own enjoyment. However, it is recognized that play occurs in young animals to learn motor and social skills needed to survive. Play as "enjoyment" may have evolved simply because something enjoyable will be sought after, and if needed actions of learning are enjoyable, they will be done. Tussling sibling brown bears, rolling and cuffing each other, are obviously playing. But it has long been a truism that such play in the proximal sense is vital in learning self-defense and in establishing rules of association. Play tends to decrease in frequency as mammals become older and does not often occur in adults. Obvious exceptions are some primates (including humans!) and cetaceans, although as behavioral studies gather details, researchers are learning that play in adults of other species is actually more common than described previously. Play may also be an attempt to relieve boredom, and we would expect play to be especially well developed in the larger brained, behaviorally flexible, mammals (Goodall, 1990; Marino *et al.*, 2007).

## I. Motor Imitations

Many marine mammals seem to be especially good at imitating the actions of their conspecifics or of individuals of other species.

Thus, untrained dolphins in oceanaria have been described as performing a colleagues' trained repertoire essentially flawlessly when called upon to do so. Apparently, the dolphins learned the motor actions simply by observation (Pryor, 1995). While this by itself is not play, the capability of imitation is often expressed as play: dolphins have imitated a diver's movements of cleaning the pool; as well as the grooming and swimming movements of seals and other pool inhabitants. The dolphins would generally approach the object of imitation, slow their own travel to approximate that of the slower coinhabitant, and then move their body in exaggerated imitation of movements of the other individual. The human diver, alternately bending and straightening at the waist as he or she cleaned the aquarium tank with a rubber scraper, was imitated by the dolphin bobbing its head and neck up and down in rapid and jerky fashion. At the same time, the dolphin released clouds of bubbles from its blowhole in synchrony with the bubbles of the diver's air regulator and made a squeaking sound in an apparent attempt to reproduce the squeaks of the rubber on glass. The author has seen similar behavior in nature, with a bottlenose dolphin (*Tursiops truncatus*) adult in the Bahamas imitating a particularly clumsy tourist who had difficulty in descending below the water as she propelled herself with her skin dive flippers and by rotating her arms. The dolphin matched her speed, alternately bobbed at the surface and descended in unison with the woman, jerked its peduncle and tail back and forth in unnatural fashion, while at the same time rotating its short front flippers as if they were flailing arms. The effect looked highly hilarious, and it would be difficult to rationalize the behavior as anything but a bit of malicious fun, or play.

## II. Vocal Imitation

While motor or physical imitation seems to be mainly in the purview of TOOTHED WHALES, other marine mammals also practice vocal imitation. This imitation may be an outgrowth of learning one's own species-specific (and perhaps group or pod-specific, as in killer whales, *Orcinus orca*) vocalizations, but the capability can then become a method of play. In the 1970s, a captive male harbor seal (*Phoca vitulina*), named Hoover by his caretakers, was capable of imitating the voice of a human worker who frequented his area, complete with a New England accent and a bit of a drunk-sounding slur (Ralls *et al.*, 1985). Beluga whales (*Delphinapterus leucas*) can also imitate human sound and will at times use these imitations in apparent mischievous play. While dolphins cannot imitate human sounds very precisely (they seem to lack the vocal capability, not the intelligence, to do so), they will easily imitate clicks, whistles, barks, scrapes, and squeaks (as the forementioned window-washing sounds) that can occur in an aquarium with other animals. A particularly readable account of imitation, innovation, intelligence, and cognition can be found in Tyack (1999).

## III. Examples of Play

When we meet marine mammals underwater, we are apt to be subjects of intense curiosity. Sea lions, fur seals, and harbor seals will dash around us, pirouette in front of us, and gaze at us. Interestingly, these same animals ignore us or become wary if we approach too close to them when they are hauled out onshore. When in their watery milieu, however, fear is gone. Manatees (*Trichechus* spp.) and dugongs (*Dugong dugon*) are similarly attracted to humans underwater, except in those places where they are hunted. This curiosity can turn to play. Just about every researcher who dives with marine

**P**

mammals has tales of sea lions tugging on his or her flippers; manatees and dugongs mouthing an arm or leg in apparent playfulness; and dolphins playing "the leaf game," where they present the human with a bit of flotsam and wait or return until it is given back to them (Johnson and Norris, 1994).

However, although many people prefer to think that dolphins and humans have an especially play-prone bond, dolphins (and pinnipeds) are known to interact in playful fashion with many other animate and inanimate objects in their environment. Thus, they may "tease" a sea turtle by mouthing and pulling on its tail or legs or they may swim beside a like-sized shark, imitating every movement that the shark makes. Some species of dolphins are also known for adroitly balancing pieces of flotsam on their jaws, flippers, dorsal fin, or tail, or carrying a piece of algae, plastic, or other pliable object in a manner to keep it balanced on an appendage simply by force of the forward movement of the body (Silva *et al.*, 2005). This activity tends to take place when the group of animals is being highly social, usually with much sexual activity as well. The dolphin playing with the inanimate object, however, tends to be alone, apparently transferring its sense of socializing to playing with the object (Fig. 1).

While apparent play is often clouded by the possibilities of curiosity or learning, the author has witnessed several clear-cut examples of play in nature, detailed in the next three sections.

## A. The Mischievous Dusky Dolphin

When dusky dolphins (*Lagenorhynchus obscurus*) of Peninsula Valdés, Argentina, have fed and socialized, they are in a very playful mood. They race toward a boat from kilometers away, attempting to ride the bow wave of the vessel. They leap in acrobatic somersaults and flips that appear to show sheer exuberance and orient toward, bite at gently, and otherwise manipulate objects in their environment. There is much social/sexual interaction and apparent play, often in twos and threes, but lone animals carry pieces of kelp on their rostrum or melon and anterior edges of flippers, dorsal fin, or fluke. Most of these behaviors could be interpreted as being a part of learning or honing skills, but the author has repeatedly seen nonequivocal play, as follows.

After dusky dolphins have fed on schooling anchovy near the surface, marine birds who took advantage of the aggregated fish ball sit on the water, digesting. These are of a variety of species, but in Argentina almost always include kelp gulls (*Larus dominicanus*) and brown-hooded gulls (*L. maculipennis*), yellow or red legs, respectively, dangling below a round white rump. The sight seems almost irresistible for some dolphins, who slowly circle one of these birds, approach it at an angle and from the rear, very gently open their jaws around one or both dangling appendages, and then rapidly but not totally close the mouth and surge forward and down. This action drags the bird below water by about 30–60 cm before the mouth is reopened and the bird literally pops back to the surface. The bird frantically flutters and preens before it flies off. The dolphin meanwhile slowly swims off, at times to attempt the same "trick" with another hapless gull. At all times (this has been seen about two dozen times to date), the bird is unharmed, suggesting particular finesse as the dolphin closes its pointy-toothed jaws and surges forward. The intent appears to be to surprise, certainly not to harm, perhaps like a child sneaking up behind a person to pop a balloon (Würsig, 2000). The same attempts have been seen with dolphins orienting toward the dark feet of Magellanic penguins (*Spheniscus magellanicus*) bobbing at the surface, but the penguins, ever aware of their environment below, simply paddle rapidly ahead or dive before the dolphin can complete its action.

## B. Creating a Bow Wave

Dolphins ride or surf on oceanic waves (Fig. 2) and those created by ships. They also ride on the fleeting bow waves created by rapidly surging large whales, and the dolphins appear to "work" particularly hard to get the whales to surge forward. A (generally small) group of dolphins will swim rapidly to the sides and front of the head of a baleen or sperm whale (*Physeter macrocephalus*), close to the eyes. When the whale surges forward abruptly in an apparent attempt to evade the dolphins, the dolphins surf the steep white-water wave so created during that surge. It is believed that the activity is not pleasing to the whale, as the whale exhales forcefully during the surge, with a sonorous "snort" that indicates aggression or anger in other contexts. In right whales (*Eubalaena* spp.), the loud in-air "dolphin blow" of a snorting, surging, whale alerts researchers that *Lagenorhynchus* spp. or bottlenose dolphins are playing with the leviathans. The snort is so loud that on a calm sea it can be heard for a distance of several kilometers. One whale is "good for" anywhere from 5 to 10 surges. It then tires or decides to give up evading the dolphins, and the sport is over for the time or needs to be reinitiated with another whale nearby.

**Figure 1** *A dusky dolphin* (Lagenorhynchus obscurus) *playing with kelp in Patagonia, Argentina. Photo by Bernd Würsig.*

**Figure 2** *A mom and calf common bottlenose dolphin* (Tursiops truncatus) *surf and leap on an oceanic wave off Maui, Hawaii. Photo by Bernd Würsig.*

and Explanation in the Study of Behavior: Vol. 1: Interpretation, Intentionality, and Communication" (M. Bekoff, and D. Jamieson, eds), pp. 283–312. Westview Press.

Herman, L. M., and Tavolga, W. N. (1980). The communication systems of cetaceans. *In* "Cetacean Behavior: Mechanisms and Functions" (L. M. Herman, ed.), pp. 149–209. Wiley Interscience, New York.

Herman, L. M., and Uyeyama, R. K. (1999). The dolphin's grammatical competency: Comments on Kako (1998). *Anim. Learn. Behav.* **27**, 18–23.

Herman, L. M., Richards, D. G., and Wolz, J. P. (1984). Comprehension of sentences by bottlenose dolphins. *Cognition* **16**, 119–129.

Herman, L. M., Hovancik, J. R., Gory, J. D., and Bradshaw, G. L. (1989). Generalization of visual matching by a bottlenosed dolphin (*Tursiops truncatus*): Evidence for invariance of cognitive performance with visual or auditory materials. *J. Exp. Psychol. Anim. Behav. Process.* **15**, 124–136.

Herman, L. M., Morrel-Samuels, P., and Pack, A. A. (1990). Bottlenose dolphin and human recognition of veridical and degraded video displays of an artificial gestural language. *J. Exp. Psychol. Gen.* **119**, 215–230.

Herman, L. M., Kuczaj, S., III, and Holder, M. D. (1993b). Responses to anomalous gestural sequences by a language-trained dolphin: Evidence for processing of semantic relations and syntactic information. *J. Exp. Psychol. Gen.* **122**, 184–194.

Herman, L. M., Pack, A. A., and Morrel-Samuels, P. (1993a). Representational and conceptual skills of dolphins. *In* "Language and Communication: Comparative Perspectives" (H. R. Roitblat, L. M. Herman, and P. Nachtigall, eds), pp. 273–298. Lawrence Erlbaum, Hillside.

Herman, L. M., Pack, A. A., and Wood, A. M. (1994). Bottlenosed dolphins can generalize rules and develop abstract concepts. *Mar. Mamm. Sci.* **10**, 70–80.

Herman, L. M., Pack, A. A., and Hoffmann-Kuhnt, M. (1998). Seeing through sound: Dolphins perceive the spatial structure of objects through echolocation. *J. Comp. Psychol.* **112**, 292–305.

Herzing, D. L. (1996). Vocalizations and associated underwater behavior of free-ranging Atlantic spotted dolphins, *Stenella frontalis* and bottlenose dolphins, *Tursiops truncatus*. *Aquat. Mamm.* **22**, 61–79.

Hockett, C. F. (1960). Logical considerations in the study of animal communication. *In* "Animal Sounds and Communication" (W. E. Lanyon, and W. N. Tavolga, eds), pp. 392–430. American Institute of Biological Science Pub. 7, Washington, DC.

Jaakkola, K., Fellner, W., Erb, L., Rodriguez, M., and Guarino, E. (2005). Understanding of the concept of numerically "less" by bottlenose dolphins (*Tursiops truncatus*). *J. Comp. Psychol.* **119**, 296–303.

Jackendoff, R., and Pinker, S. (2005). The nature of the language faculty and its implications for the evolution of language. (Reply to Fitch, Hauser and Chomsky). *Cognition* **97**, 211–225.

Janik, V. M. (1999). Pitfalls in the classification of behaviour: A comparison of dolphin whistle classification methods. *Anim. Behav.* **57**, 133–143.

Janik, V. M. (2000). Whistle matching in wild bottlenose dolphins (*Tursiops truncatus*). *Science* **289**, 1357–1360.

Janik, V. M., Sayigh, L. S., and Wells, R. S. (2006). Whistle shape conveys identity information to bottlenose dolphins. *Proc. Nat. Acad. Sci.* **103**, 8293–8297.

Kilian, A., Yaman, S., Von Fersen, L., and Gunturkun, O. (2003). A bottlenose dolphin discriminates visual stimuli differing in numerosity. *Learn. Behav.* **31**, 133–142.

Lieberman, P. (1984). "The Biology and Evolution of Language." Harvard University Press, Cambridge.

Lieberman, P. (2000). "Human Language and our Reptilian Brain: The Subcortical Bases of Speech, Syntax, and Thought." Harvard University Press, Cambridge.

Lieberman, P. (2006). "Toward an Evolutionary Biology of Language." Harvard University Press, Cambridge.

Lilly, J. C. (1961). "Man and Dolphin." Doubleday, New York.

Lilly, J. C. (1967). "The Mind of the Dolphin: A Nonhuman Intelligence." Doubleday, New York.

Marino, L. (2002). Convergence in complex cognitive abilities in cetaceans and primates. *Brain Behav. Evol.* **59**, 21–32.

McCowan, B., and Reiss, D. (2001). The fallacy of "signature whistles" in bottlenose dolphins: A comparative perspective of "signature information" in animal vocalizations. *Anim. Behav.* **62**, 1151–1162.

McCowan, B., Hanser, S. F., and Doyle, L. R. (1999). Quantitative tools for comparing animal communication systems: Information theory applied to bottlenose dolphin whistle repertoires. *Anim. Behav.* **57**, 409–419.

Mercado, E. M., III, Killebrew, D. A., Pack, A. A., Macha, I. V. B., and Herman, L. M. (2000). Generalization of same-different classification abilities in bottlenosed dolphins. *Behav. Processes* **50**, 79–94.

Pack, A. A., and Herman, L. M. (1995). Sensory integration in the bottlenosed dolphin: Immediate recognition of complex shapes across the senses of echolocation and vision. *J. Acoust. Soc. Amer.* **98**, 722–733.

Pack, A. A., and Herman, L. M. (2006). Dolphin social cognition and joint attention: Our current understanding. *Aquat. Mamm.* **32**, 443–460.

Pack, A. A., and Herman, L. M. (2007). The dolphin's (*Tursiops truncatus*) understanding of human gaze and pointing: Knowing *what* and *where*. *J. Comp. Psychol.* **121**, 34–45.

Pack, A. A., Herman, L. M., and Hoffmann-Kuhnt, M. (2004). Dolphin echolocation shape perception: From sound to object. *In* "Advances in the Study of Echolocation in Bats and Dolphins" (J. Thomas, C. Moss, and M. Vater, eds), pp. 288–298. University of Chicago Press, Chicago.

Pinker, S. (1994). "The Language Instinct: How the Mind Creates Language." Harper Collins, New York.

Pryor, K., Haag, R., and O'Reilly, J. (1969). The creative porpoise: Training for novel behavior. *J. Exp. Anal. Behav.* **12**, 653–661.

Richards, D. G., Wolz, J. P., and Herman, L. M. (1984). Vocal mimicry of computer generated sounds and vocal labeling of objects by a bottlenose dolphin, *Tursiops truncatus*. *J. Comp. Psychol.* **98**, 10–28.

Reiss, D., and McCowan, B. (1993). Spontaneous vocal mimicry and production by bottlenosed dolphins (*Tursiops truncatus*): Evidence for vocal learning. *J. Comp. Psychol.* **107**, 301–312.

Rendell, L., and Whitehead, H. (2001). Culture in whales and dolphins. *Behav. Brain Sci.* **24**, 309–382.

Ristau, C. A., and Robbins, D. (1979). Language in the great apes: A critical review. *In* "Advances in the Study of Behavior" (J. F. Rosenblatt, R. B. Hinde, C. Beer, and M. C. Busnel, eds), 12, pp. 141–255. Academic Press, New York.

Rose, S. A., and Wallace, I. F. (1985). Cross-modal and intramodal transfer as predictors of mental development in full-term and preterm infants. *Dev. Psychol.* **21**, 949–962.

Savage-Rumbaugh, E. S. (1986). "Ape Language: From Conditioned Response to Symbol." Columbia University Press, New York.

Savage-Rumbaugh, E. S., Murphy, J., Sevcik, R. A., Brakke, K. E., Williams, S. L., and Rumbaugh, D. M. (1993). Language comprehension in ape and child. *Monogr. Soc. Res. Child Dev.* **58**(3–4) 256 pp.

Schusterman, R. J., and Gisiner, R. (1988). Artificial language comprehension in dolphins and sea lions: The essential cognitive skills. *Psychol. Rec.* **34**, 3–23.

Schusterman, R. J., and Krieger, K. (1984). California sea lions are capable of semantic comprehension. *Psychol. Rec.* **38**, 311–348.

Smith, J. D., Schull, J., Strote, J., McGee, K., Egnor, R., and Erb, L. (1995). The uncertain response in the bottlenose dolphin (*Tursiops truncatus*). *J. Exp. Psychol. Gen.* **124**, 391–408.

Smolker, R., Richards, A., Connor, R., Mann, J., and Berggren, P. (1997). Sponge carrying by dolphins (Delphinidae, *Tursiops* sp.): A foraging specialization involving tool use? *Ethology* **103**, 454–465.

Terrace, H. S., Petitto, L. A., Sanders, R. J., and Bever, T. G. (1979). Can an ape create a sentence? *Science* **206**, 891–902.

Thompson, R. K. R., and Herman, L. M. (1977). Memory for lists of sounds by the bottlenosed dolphin: Convergence of memory processes with humans? *Science* **195**, 501–503.

Tomasello, M., and Call, J. (1997). "Primate Cognition." Oxford University Press, New York.

Tschudin, A., Call, J., Dunbar, R. I. M., Harris, G., and van der Elst, C. (2001). Comprehension of signs by dolphins (*Tursiops truncatus*). *J. Comp. Psychol.* **115**, 100–105.

L

Tyack, P. L. (1986). Whistle repertoires of two bottlenose dolphins, *Tursiops truncatus:* Mimicry of signature whistles? *Behav. Ecol. Sociobiol.* **18**, 251–257.

Wood, F. G. (1973). "Marine Mammals and Man: The Navy's Porpoises and Sea Lions." Luce, Washington, DC.

Wilson, E. O. (1975). "Sociobiology." Belknap, Cambridge.

Xitco, M. J., Jr., Gory, J. D., and Kuczaj, II. (2001). Spontaneous pointing by bottlenose dolphins (*Tursiops truncatus*). *Anim. Cogn* **4**, 115–123.

# Locomotion, Terrestrial

### Thomas A. Deméré and Joshua H. Yonas

Among the obligate marine mammals (Cetacea, Sirenia, and Pinnipedia), only pinnipeds are able to locomote on land. They were the last to invade the seas, with the earliest forms known from the latest Oligocene (~24 Ma). Even today, pinnipeds retain a close tie to the land (or ice) for mating, birthing, and molting. In a sense, their bodies represent a compromise between the requirements for aquatic locomotion and the necessity for some form of terrestriality. Some pinnipeds have retained a higher degree of terrestriality than others (i.e., the "walking" otariids and odobenids in contrast to the "wriggling" phocine and monachine phocids).

Extant pinnipeds exhibit a general body plan consisting of a streamlined torso with major portions of the appendicular skeleton concealed within the body contour (i.e., the "armpit" is positioned near the wrist and the "crotch" is positioned near the ankle). The knee joint is held in an obligate flexed posture and proximal limb elements (i.e., humerus and femur) are shortened, whereas distal limb elements (i.e., manus and pes) are elongated. In addition the feet are modified as flippers with retention of five digits and elongation of the "thumb," "big toe," and "little toe." Three patterns of combined aquatic and terrestrial locomotion are recognized: forelimb propulsion and quadrapedal walking (otariids); hindlimb propulsion and quadrapedal walking coupled with sternal lunging (odobenids); and hindlimb propulsion and essentially limb-free undulation (phocids).

Otariids have retained the greatest degree of terrestriality and, surprisingly, represent the latest diverging group of crown pinnipeds. Their fore- and hindlimbs are capable of quadrapedal support and locomotion. The feet are held in a plantigrade stance with the flippers extended laterally. The hind feet are able to rotate under the body. Although some researchers have suggested that Otariids can both walk and gallop, English (1976) has emphasized that in a true gallop there are periods when all feet are off the ground. This condition does not occur in otariids, which never have less than 1 ft on the ground during terrestrial locomotion. Thus otariids can be said to have two gaits; a walk and something between a fast walk and a true gallop. During the otariid walk the forelimbs are alternately in protraction and retraction and rely on flexion and extension of the elbow and dorsoflexion of the wrist. The axial skeleton contributes to the walk with the head and neck swinging from side to side to shift body weight from the side of the raised, flexed, and protracted forelimb to the side of the planted, extended, and retracted forelimb. The hindlimb serves primarily a support role during locomotion, whereas the posterior axial skeleton undergoes flexion on the recovery stroke and extension on the propulsive stroke. The upper ankle joint and transverse tarsal joint remain acutely flexed throughout the walking phase, with the pes turned to the side.

Odobenids also retain a high degree of terrestriality relative to phocids. Like otariids, their fore- and hindlimbs are capable of quadrapedal support and locomotion, their feet are held in a plantigrade stance, and their hind feet are able to rotate under the body. However, rather than being able to completely hold the body off the ground, the limbs of an adult walrus act more like stilts while the venter (primarily represented by the sternum) acts like a "fifth leg" to support the majority of the animal's weight. Odobenids move on land in a slow, blubbery lateral sequence walk punctuated by a forward lunge (Gordon, 1981). First, the forelimb is positioned under the body; with the foreflippers laterally oriented, while the forearm is almost vertical and slightly protracted. The limbs are holding a limited amount of body weight at this stage with the bulk supported by the venter. Next, the forward lunge is facilitated by an initial lift of the body by a "push-up" of the forearm. Here the limb is in full extension and is being used to propel the animal. With the body partially suspended, locomotion occurs by a lateral swing of one forelimb at a time, in the direction of travel. The hindlimb is tucked under the body and pointed cranially, only holding a limited amount of body weight. The hindlimbs then extend and push the body forward. As a forelimb swing goes through completion, and once planted, the hindlimb on the same side is swung laterally and forward in the direction of travel. In this manner the adult walrus swings the left limbs, then the right. The axial skeleton and head also contribute to terrestrial locomotion. With the head aligned with the body axis and the sternum supporting the bulk of the weight, the venter is thrust forward aided by extension of the posterior axial skeleton. Although the forelimbs and hindlimbs swing and are placed in the direction of travel, the head is used as a counterbalance and shifted to the opposite side to allow for less weight on the fore- and hindlimbs, thus acting as a lateral "see-saw."

Phocids have a very divergent mode of terrestrial locomotion. Their fore- and hindlimbs are, for the most part, incapable of supporting the body, the feet are typically held free of the substrate, and the hind feet cannot be rotated forward under the body and instead are held in a relatively hyperextended posture and directed backwards with palmar surfaces opposed. The unique structure of the phocid tarsal bones (e.g., elongate astragalus with hypertrophied calcaneal process) and the position and orientation of the tendon of the flexor hallucis longus prevent the foot from being brought into a plantigrade position (Howell, 1929; Polly, 2008). Phocids have been called "beach maggots" in reference to their inchworm-like undulatory mode of terrestrial locomotion. With the limbs rather useless on land (or ice) most phocids primarily rely on the sternum and pelvis for support and locomotion (Backhouse, 1961). Planting the sternum on the substrate, the posterior part of the axial skeleton is ventroflexed to draw the pelvis forward. When planted the pelvis then serves as an anchor as the axial skeleton is extended forcing the body forward. The forelimbs are generally held against the torso but can sometimes be used to assist in scrambling on land or ice. This is accomplished by flexion of phalanges and extension of metacarpals to form a terminal hook with which the seal can drag its body forward as both forelimbs are retracted (Backhouse, 1961). In elephant seals the forelimbs provide some support of the body and propulsive force as the animal undulates across the land or ice (O'Gorman, 1963). Some Antarctic phocids (e.g., Ross, crabeater, and leopard seals) utilize a sinuous mode of progression to "swim" across soft snow. The forelimbs are alternately retracted whereas the posterior lumbar region, pelvis, and hindlimbs are strongly flexed laterally (O'Gorman, 1963).

Concerning extinct pinnipeds, we are forced to rely on features of fossil bones (e.g., limb proportions, size and position of bony processes, and muscle and ligament attachment sites) to formulate and test hypotheses about ancestral patterns of terrestrial locomotion. Based on fossil limb bones of *Enaliarctos mealsi* from the earliest Miocene of California, United States, it is clear that this stem pinniped had already evolved many of the aquatic limb adaptations of extant otariids (e.g., shortened proximal limb elements and elongate feet), while retaining a greater degree of terrestriality (Berta and Ray, 1990). The same can be said for fossils of the middle Miocene dematopochids, *Desmatophoca* and *Allodesmus*, from the North Pacific region. Extinct odobenids including the stem taxon, *Proneotherium repenningi* from the middle Miocene of Oregon, United States, were also capable of terrestrial quadrapedal walking (Deméré and Berta, 2001). The later evolving dusignathine and odobenine walruses, although diverging in mode of aquatic locomotion (fore- and hindlimb swimming, respectively), retained the ability for terrestrial quadrapedal walking. The phocid fossil record, although in need of rigorous modern analysis, does contain important stem taxa such as the middle Miocene *Leptophoca lenis* from Maryland, United States, that indicate an early evolutionary loss of the ability for terrestrial quadrapedal walking.

## See Also the Following Articles

Forelimb Anatomy ■ Swimming

## References

Backhouse, K. M. (1961). Locomotion of seals with particular reference to the forelimb. *Symp. Zool. Soc. Lond.* **5**, 59–75.

Berta, A., and Ray, C. E. (1990). Skeletal morphology and locomotor capabilities if the archaic pinniped *Enaliarctos mealsi*. *J. Vertebr. Paleont.* **10**, 141–157.

Deméré, T. A., and Berta, A. (2001). A reevaluation of *Proneotherium repenningi* from the Miocene Astoria Formation of Oregon and its position as a basal odobenids (Pinnipedia: Mammalia). *J. Vertebr. Paleont.* **21**, 279–310.

English, A. W. (1976). Limb movements and locomotor function in the California sea lion (*Zalophus californianus*). *J. Zool. (Lond.)* **178**, 341–364.

Gordon, K. R. (1981). Locomotor behaviour of the walrus (*Odobenus*). *J. Zool. (Lond.)* **195**, 349–367.

Howell, A. B. (1929). Contribution to the comparative anatomy of the eared and earless seals (genera *Zalophus* and *Phoca*). *Proc. U.S. Nat. Mus.* 73(15), 1–142.

O'Gorman, F. (1963). Observations on terrestrial locomotion in Antarctic seals. *Proc. Zool. Soc. Lond.* **141**, 837–850.

Polly, P. D. (2008). Adaptive Zones and the Pinniped Ankle: A 3D Quantitative Analysis of Carnivoran Tarsal Evolution. *In* "Mammalian Evolutionary Morphology: A Tribute to Frederick S. Szalay" (E. Sargis and M. Dagosto, eds.), pp. 165–194. Springer: Dordrecht, The Netherlands.

# Leopard Seal
## *Hydrurga leptonyx*

### TRACEY L. ROGERS

## I. Characteristics and Taxonomy

Leopard seals are a large sexually dimorphic species (Kooyman, 1981; Laws, 1984; Bonner, 1994). The females are larger than the males, growing up to 3.8 m in length and weighing up to 500 kg, whereas males grow up to 3.3 m in length and weigh up to 300 kg.

## II. Distribution and Abundance

Although the majority of the leopard seal population remains within the circumpolar Antarctic pack ice, the seals are regular, although not abundant, visitors to the subantarctic islands of the southern oceans and to the southern continents (Erickson *et al.*, 1971; Heller *et al.*, 1977). The most northerly leopard seal sightings are from the Cook Islands. Juveniles appear to be more mobile, moving farther north during the winter. Because it does not need to return to the pack ice to breed, the leopard seal can escape food shortages during winter by dispersing northwards. Every 4–5 years, the number of leopard seals on the subantarctic islands oscillates from a few to several hundred seals. The periodic dispersal could be related to oscillating current patterns or resource shortages in certain years (Testa *et al.*, 1991).

The leopard seal population is estimated to be 222,000–440,000. During summer, leopard seals breed on the outer fringes of the pack ice where they are solitary and sparsely distributed. Their density is inversely related to the amount of pack ice available to the seals as haul-out platforms. Pack ice cover varies with the season, from a maximum between August and October to a minimum between February and March. Population densities are greatest in areas of abundant cake ice (ice floes of 2–20 m in diameter) and brash ice (ice floes greater than 2 m in diameter), whereas they are least in areas with larger floes. Densities range from 0.003–0.151 seals/km², and there is an age-related difference in their spatial behavior. Due to intraspecific aggression there is a greater degree of spatial separation among older seals (Rogers and Bryden, 1997).

## III. Ecology

Leopard seals take a diverse range of prey (Lowry *et al.*, 1988; Hall-Aspland and Rogers, 2004) including fish, cephalopods, sea birds, and seals. Different food sources are used when available or when opportunities to take other more sought- after prey are few. Krill makes up the largest proportion of their diet, particularly during the winter months when other food types are not abundant. At this time the leopard seals must compete directly with krill-feeding specialists, such as the crabeater seal (*Lobodon carcinophagus*) and Adélie penguin (Siniff and Bengtson, 1977). This is believed to be a time of potential food shortage and causes some juvenile leopard seals to move north from the pack ice during the austral winter. The leopard seal is responsible for more predation on warm-blooded prey than any other pinniped. Leopard seals capture and eat juvenile crabeater seals in particular but also prey on Weddell (*Leptonychotes weddellii*) Ross (*Ommatophoca rossii*) and southern elephant (*Mirounga leonina*) seals, subantarctic and Antarctic fur seals (*Arctocephalus tropicalis* and *A. gazella*) and southern sea lions (*Neophoca cinerea* and *Phocarctes hookeri*). Newly weaned crabeater seals are the most vulnerable and are taken from November to February. Crabeater seal survivors bear characteristic parallel paired scars from leopard seal attacks; approximately 78% of adult crabeater seals display such marks. The teeth of the leopard seal have a dual role; the large recurved canines and incisors are designed for gripping and tearing prey, whereas the upper and lower tricuspid (three cusped) molars interlock to provide an efficient krill sieve.

## IV. Behavior and Physiology

Acoustic behavior is important in the mating system of the leopard seal (Rogers *et al.*, 1996; Stirling and Siniff, 1979; Rogers and Cato, 2002). Leopard seals become highly vocal prior to and during

L

**Figure 1** *Leopard seal. Photo by Jaume Forcada, British Antarctic Survey.*

their breeding season (Fig. 1). Lone male leopard seals produce highly stereotyped vocalizations for long periods each day, from early November through January. Female leopard seals also use long-distance acoustic displays during the breeding season. However, female seals vocalize for a brief period only from the beginning of estrus until mating, presumably to advertise sexual receptivity. The calls of the leopard seal are at low-to-medium frequencies and so powerful that they can be heard through the air–water interface and felt through the ice.

## V. Life History

Male leopard seals are sexually mature by 4.5 years and females by 4 years of age (Kooyman, 1981; Bonner, 1994). Females give birth to their pups and wean them on the ice floes of the Antarctic pack ice. Males do not remain with the females; only mother–pup groups are observed on ice floes. Length at birth is approximately 120 cm, with rapid growth through the first 6 months postpartum. Births are believed to occur from October to mid-November and mating from December to early January, after the pups have weaned. Lactation is believed to last for up to 4 weeks. Mating in the wild has been observed rarely, but captive seals mount only when in the water. There is a period of delayed implantation from early January to mid-February. Implanted fetuses are found after mid-February when the corpus luteum (glandular structure in the ovary) has begun to increase in size and the corpus albicans (scar from ovarian glandular structure) from the previous pregnancy has continued to regress.

## VI. Interactions with Humans

There are many reported interactions between humans and leopard seals, mainly involving scientists working in Antarctica (Muir *et al.*, 2006). However, with the number of tourists visiting the Antarctic continent and surrounding islands on the increase, this number is bound to rise. Of all the interactions there has only been one reported death associated with a leopard seal encounter. Generally, the leopard seal behavior during in-water interactions is described as being curious.

Historically leopard seals have never been exploited commercially, however small numbers have been taken for scientific research and for use as pet food.

Leopard seals are currently listed as lower risk, least concern by the International Union for Conservation of Nature (IUCN). The main conservation issue facing leopard seal populations today is the reduction in krill stocks, which is being exacerbated by the rapid increase in climate change.

## *See Also the Following Articles*

Earless Seals (Phocidae) ■ Predation on Marine Mammals

## *References*

Bonner, N. (1994). "Seals and Sea Lions of the World." Blandford, London.

Erickson, A. W., Siniff, D. B., Cline, D. R., and Hofman, R. J. (1971). Distributional ecology of Antarctic seals. *In* "Symposium on Antarctic Ice and Water Masses" (G. Deacon, ed.), pp. 55–76. Scientific Committee on Antarctic Research, Brussels, Belgium.

Hall-Aspland, S. A., and Rogers, T. L. (2004). Summer diet of leopard seals (*Hydrurga leptonyx*) in Prydz Bay, Eastern Antarctica. *Polar Biol.* **27**, 729–734.

Heller and Sons, Cambridge. Gilbert, J. R., and Erickson, A. W. (1977). Distribution and abundance of seals in the pack ice of the Pacific Sector of the Southern Ocean. *In* "Adaptations within Antarctic Ecosystems" (G. A. Llano, ed.), pp. 703–740. Smithsonian Institution, Washington, DC.

Kooyman, G. L. (1981). Leopard seal (*Hydrurga leptonyx* Blainville, 1820). *In* "Handbook of Marine Mammals" (S. Ridgway and R. Harrison, eds), Vol. 2, pp. 261–274. Academic Press, London.

Laws, R. M. (1984). Seals. *In* "Antarctic Ecology" (R. M. Laws, ed.), Vol. 2, pp. 621–715. Academic Press, London.

Lowry, L. F., Testa, J. W., and Calvert, W. (1988). Notes on winter feeding of crabeater and leopard seals near the Antarctic Peninsula. *Polar Biol.* **8**, 475–478.

Muir, S. F., Barnes, D. K. A., and Reid, K. (2006). Interactions between humans and leopard seals. *Ant. Sci.* **18**, 61–74.

Rogers, T., Cato, D. H., and Bryden, M. M. (1996). Behavioural significance of underwater vocalizations of captive leopard seals, *Hydrurga leptonyx*. *Mar. Mamm. Sci.* **12**, 414–427.

Rogers, T. L., and Bryden, M. M. (1997). Density and haul-out behaviour of leopard seals (*Hydrurga leptonyx*) in Prydz Bay, Antarctica. *Mar. Mamm. Sci.* **13**, 293–302.

Rogers, T. L., and Cato, D. H. (2002). Individual variation in the acoustical behaviour of the adult male leopard seal, *Hydrurga leptonyx*. *Behaviour* **139**, 1267–1286.

Rounsevell, D., and Pemberton, D. (1994). The status and seasonal occurrence of leopard seals, *Hydrurga leptonyx*, in Tasmanian waters. *Aust. Mammal.* **17**, 97–102.

Siniff, D. B., and Bengtson, J. L. (1977). Observations and hypothesis concerning the interactions among crabeater seals, leopard seals, and killer whales. *J. Mammal.* **58**, 414–416.

Siniff, D. B., and Stone, S. (1985). The role of the leopard seal in the tropho-dynamics of the Antarctic marine ecosystem. *In* "Antarctic Nutrient Cycles and Food Webs" (W. R. Siegfried, P. R. Condy, and R. M. Laws, eds), pp. 555–559. Springer-Verlag, Berlin.

Stirling, I., and Siniff, D. B. (1979). Underwater vocalizations of leopard seals (*Hydrurga leptonyx*) and crabeater seals (*Lobodon carcinophagus*) near the South Shetland Islands, Antarctica. *Can. J. Zool.* **57**, 1244–1248.

Testa, J. W., Oehlert, G., Ainsley, D. G., Bengtson, J. L., Siniff, D. B., Laws, R. M., and Rounsevell, D. (1991). Temporal variability in Antarctic marine ecosystems: Periodic fluctuations in the phocid seals. *Can. J. Fish. Aquat. Sci.* **48**, 631–639.

## C. Balancing Bowhead Whales (*Balaena mysticetus*)

Baleen whales engage in the surface activities of breaching, tail lobbing, flipper slapping, and holding the tail above the surface of the water. While much of this is certainly play, at least at times, the percussive nature of these activities possibly aids in communication and may also represent outgrowths of anger or frustration. Holding the tail out of the water for many minutes at a time, a habit of some right and bowhead whales, may feel good as a stiff breeze touches the skin and may even be a form of recreational "sailing" with the tail (as suggested by Payne, 1995).

Adult bowhead whales (approximately 18 m long) have been seen interacting with tree trunks, or logs, up to 10 m long on summer–fall feeding grounds in the Beaufort Sea. They nudge and propel the large log, handle it with their flippers and tail, and attempt to push it under water. Several of these activities seem similar to surface social interactions during sexual activity, and it could be argued that log handling is play useful to developing physical social skills. However, the most dramatic part of log handling is balancing the object, quite adroitly, on the back or belly. The author has seen log balancing (by an adult female) wherein the whale rolled gently sideways to compensate for the action of large oceanic swells rolling past the whale's body. The whale was so adept at this balancing that she could briefly keep both ends of the huge log suspended in air, a feat perhaps not unlike a trained sea lion balancing a ball on its snout. It is difficult to imagine how this activity could be anything but play or an "artistic" attempt at perfecting a difficult task (Würsig *et al.*, 1989; Wells *et al.*, 1999) (Fig. 3).

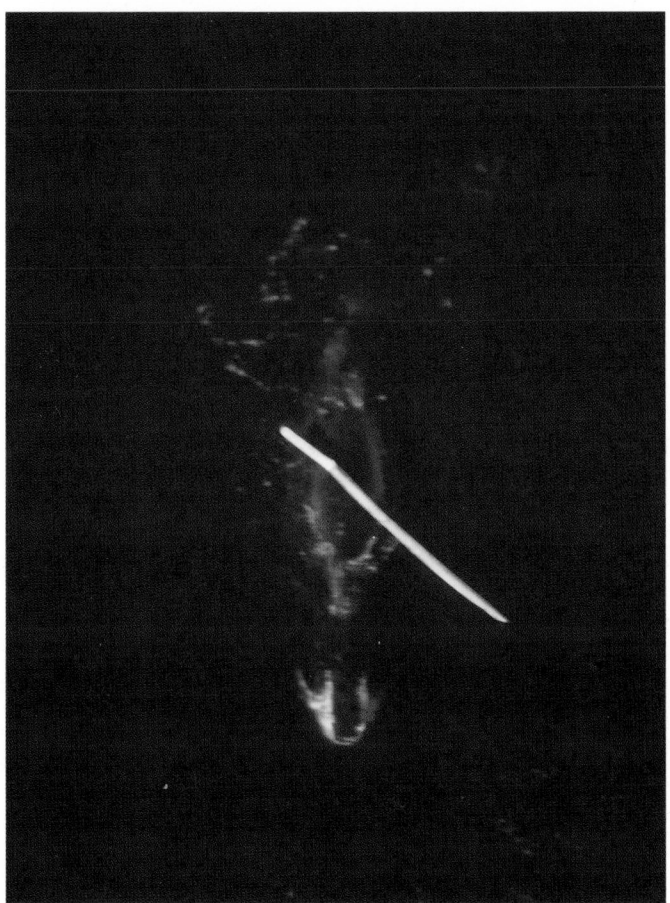

**Figure 3**   *A bowhead whale* (Balaena mysticetus) *playing with a log in the Beaufort Sea in summertime. Photo by Bernd Würsig.*

## D. The Darker Side of Play

While apparent mischievous behavior of pulling on gull or turtle legs or of inciting whales to become aggravated could be potentially AGGRESSIVE or harmful play, it probably is not. Killer whales may use play as a form of learning (and possibly teaching, as well) of youngsters to efficiently hunt (Guinet and Bouvier, 1995; Pitman *et al.*, 2003). However, some interspecies behaviors are definitely harmful and it is unclear whether they represent play or merely aggression. Short-finned pilot whales (*Globicephala macrorhynchus*) have been described aggressively and for almost 2 hr swimming around and into a pod of obviously frightened sperm whales (Weller *et al.*, 1996). Because pilot whales are not known to feed on the much larger sperm whales, it is possible that this action occurred only "for the fun of it," although other explanations, such as perhaps attempting to get whales to regurgitate squid gathered at depth, cannot be ruled out. A large male pilot whale in Hawaii carried a human diver into depth and held onto her with his mouth for tens of seconds. While he could easily have bit into her and killed her, his holding her was relatively gentle (but firm), and play appears to have been the reason (Shane *et al.*, 1993). The woman survived only because she was an expert swimmer and diver and did not panic.

Bottlenose dolphins, the staple of show dolphins in over 100 aquaria worldwide, are known for at times aggressive play. In Scotland, a group of bottlenose dolphins is known to attack and fling about harbor porpoises (*Phocoena phocoena*), not for food and possibly simply for the sport of it. Bottlenose dolphins and harbor porpoises of the area also do not appear to compete with each other for space or food, but it is unknown whether more than a few ("rogue") dolphins are involved in this particularly gruesome sport that killed over one dozen harbor porpoises while being observed by researchers.

## *See Also the Following Articles*

Intelligence and Cognition ■ Mimicry

## *References*

Goodall, J. (1990). "Through a Window: My Thirty Years with the Chimpanzees of Gombe." Houghton Mufflin, Boston, MA.

Guinet, C., and Bouvier, J. (1995). Development of intentional stranding hunting techniques in killer whale (*Orcinus orca*) calves at Crozet Archipelago. *Can. J. Zool.* **73**, 27–33.

Johnson, C. M., and Norris, K. S. (1994). Social behavior. *In* "The Hawaiian Spinner Dolphin" (K. S. Norris, B. Würsig, R. S. Wells, and M. Würsig, eds), pp. 243–286. University of California Press, Berkeley, CA.

Marino, L., *et al.* (16 authors) (2007). Cetaceans have complex brains for complex cognition. *PLoS Biol.* **5**, e139 (online manuscript at www.plosbiology.org).

Payne, R. (1995). "Among Whales." Scribner and Sons, New York.

Pitman, R. L., O'Sullivam, S., and Mase, B. (2003). Killer whales (*Orcinus orca*) attack a school of pantropical spotted dolphins (*Stenella attenuata*) in the Gulf of Mexico. *Aquat. Mamm.* **29**, 321–324.

Pryor, K. (1995). "On Behavior: Essays and Research." Sunshine Books, North Bend, WA.

Ralls, K., Forelli, P., and Gish, S. (1985). Vocalizations and vocal mimicry in captive harbor seals *Phoca vitulina. Can. J. Zool.* **63**, 1050–1056.

Shane, S. H., Tepley, L., and Costello, L. (1993). Life-threatening contact between a woman and a pilot whale captured on film. *Mar. Mamm. Sci.* **9**, 331–335.

Tyack, P. L. (1999). Communication and cognition. *In* "Biology of Marine Mammals" (J. E. Reynolds, III, and S. A. Rommel, eds), pp. 287–323. Smithsonian Institution Press, Washington, DC.

P

Weller, D. W., *et al.* (8 authors) (1996). Observations of an interaction between sperm whales and short-finned pilot whales in the Gulf of Mexico. *Mar. Mamm. Sci.* **12**, 588–594.

Wells, R. S., Boness, D. J., and Rathbun, G. B. (1999). Behavior. *In* "Biology of Marine Mammals" (J. E. Reynolds, III, and S. A. Rommel, eds), pp. 324–422. Smithsonian Institution Press, Washington, DC.

Würsig, B. (2000). In a party mood. *In* "The Smile of a Dolphin" (M. Bekoff, ed.), pp. 188–190. Discovery Book Press, New York, NY.

Würsig, B., Dorsey, E. M., Richardson, W. J., and Wells, R. S. (1989). Feeding, aerial and play behaviour of the bowhead whale, *Balaena mysticetus*, summering in the Beaufort Sea. *Aquat. Mamm.* **15**, 27–37.

# Polar Bear
## *Ursus maritimus*

### Ian Stirling

### I. Characteristics and Taxonomy

The polar bear (*Ursus maritimus* Phipps, 1774) is, on average, the largest of the eight species of bears found worldwide and is completely white. Ecologically, it is the only marine bear. Typically, the body is stocky, lacks a shoulder hump, and has a longer neck in relation to the rest of the body than other ursids. Other English common names for the species are white bear, ice bear, and Nanuk. The taxonomy is order Carnivora, family Ursidae, genus *Ursus*, and species *U. maritimus*. There are no subspecies (Fig. 1).

Polar bears are thought to have originated from brown bears (*Ursus arctos*, also called Grizzly Bear), 1.3 million years ago (Yu *et al.*, 2007). Polar bears and brown bears are capable of interbreeding in zoos and the young are fertile. Although, in the wild, there is little overlap in the habitats during the breeding season, there has been one confirmed

**Figure 1** Ursus maritimus, *the polar bear, is aptly named, as the species is often observed miles from the nearest land on polar pack ice and swimming between ice floes where they hunt ringed seals and sometimes bearded seals. Photo by François Gohier.*

case of a hybrid, in which the mother was a polar bear and the father was a grizzly. The molars and premolars of polar bears are more jagged and sharper than those of other bears, reflecting their rapid evolutionary shift toward carnivory from the flatter grinding teeth of their more omnivorous relatives. The oldest known fossil is from London, England, and is less than 100,000 years old.

Adult males typically measure 200–250 cm in length from the tip of the nose to the tip of the tail and weigh 400–600 kg, although some individuals may reach about 800 kg. The total length of adult females is 180–200 cm and they normally weigh 200–350 kg while pregnant females may occasionally exceed 500 kg. There is some geographic variation in size and growth rates of bears in different populations. Polar bears are plantigrade and have five toes on each foot, with nonretractable claws. The forepaws are large and oar-like, as an adaptation for swimming and walking on thin ice. The hind legs are not used while swimming. The skin is black. Females normally have four functional mammae, although supernumerary nipples have also been reported. The dental formula is I 3/3, C 1/1, P 2–/2–4, M 2/3. The vitamin A content of the liver is about 15,000 units per gram and is toxic to humans.

### II. Distribution and Abundance

Polar bears are distributed in ice-covered waters throughout the circumpolar Arctic. Nineteen individual populations are currently recognized, varying in size from a few hundred to a few thousand bears in each, with a world population estimate in 2005 of 20,000–25,000 (Aars *et al.*, 2006). The southern limit of their distribution in winter varies with the extent of seasonal pack ice in the Bering, Labrador, and Barents seas. In areas where the ice melts completely in summer, such as Hudson Bay or southeastern Baffin Island in Canada, polar bears spend several months onshore fasting on their stored fat reserves until freeze-up in the fall.

### III. Ecology

Although small numbers of animals may be found in the permanent multiyear pack ice of the central arctic basin, their preferred habitat is the annual ice over continental shelves of the continents and islands around the coastline of the Arctic Ocean where overall biological productivity and densities of ringed seals (*Phoca hispida*), their primary prey, are greatest.

Regional concentrations and seasonal movement patterns of polar bears are influenced primarily by the type and distribution of sea ice and by the density and distribution of seals. From freeze-up in the fall until breakup in the spring, polar bears prefer coastal areas and interisland channels lying over the continental shelf, especially active ice areas associated with shore leads or the floe edge. The size of home ranges of bears living within the Canadian Arctic Archipelago may be less than 20,000 km$^2$ or in excess of 250,00 km$^2$ in open ocean areas such as the Chukchi Sea (Amstrup, 2003). In general, polar bears are less abundant in areas of extensive multiyear ice and in the immediate vicinity of polynyas with overwintering walrus (*Odobenus rosmarus*) populations, probably because the density of ringed seals is lower there.

### IV. Behavior and Physiology

Polar bears that have continuous access to sea ice continue to hunt throughout the year. Their hunting methods and rates of success change with the seasons and vary in different areas. The most common method of hunting is to lie beside a breathing hole waiting for a seal

P

to surface and breathe. Contrary to popular myth, they do not cover their noses with a paw when stalking a seal, although when stalking, they do keep their heads low while walking slowly and steadily toward potential prey. The largest proportion of a polar bear's annual caloric intake for the year occurs in spring and early summer, at which time newly weaned ringed seal pups are 50% fat by wet weight (Stirling and Øritsland, 1995). The specific sea ice habitats most hunted by polar bears in spring are stable shore-fast ice with deep snowdrifts along pressure ridges, which are suitable for ringed seal birth lairs and breathing holes; the floe edge where leads are wide, and areas of moving ice with seven-eighths or more of ice cover. After breakup of the annual ice in late spring to early summer, hunting success is reduced and, in areas where open water prevails from late summer through autumn, polar bears seek onshore retreats where they simply fast on their stored fat reserves until freeze-up (Stirling and Parkinson, 2006).

The principal prey species of polar bears are ringed seals and, to a lesser degree, bearded seals (*Erignathus barbatus*). Ringed seals maintain their breathing holes from freeze-up in the fall to breakup in the spring by abrading the ice with the heavy claws on their fore-flippers. Many winter breathing holes are located on the last cracks to close over in the fall and bears are able to locate them by smell, even under a meter or more of compacted wind-blown snow. In areas where wind, water currents, or tidal action causes the ice to continually crack and subsequently refreeze, seals are apparently more accessible to polar bears and the bears are able to hunt them there more successfully. In places where the snow cover in the fiords is deep in spring, large numbers of ringed seals give birth to their pups in sub-nivean liars where they are hunted by polar bears of all ages and sex classes, but especially females with newborn cubs. Bearded seals concentrate where natural cracks and polynyas form through the winter because it is easier to breathe there. Polar bears are also known to prey on harp seals (*Pagophilus groenlandicus*), hooded seals (*Cystophora cristata*) where they are available and occasionally on walruses, belugas (*Dephinapterus leucas*), narwhals (*Monodon monoceros*), harp seals, waterfowl, and seabirds (Stirling, 1988).

Polar bears have a remarkable ability to store large amounts of fat during periods when prey, mainly seals, are available and then fast for protracted periods when food is not available. In Hudson Bay, where the annual ice melts completely by mid-July and does not reform until mid-November, all bears in the population must fast for at least 4 months on their stored fat, whereas pregnant females do not feed for 8 months, during which period they must support themselves as well as nurse their young so the cubs can grow large enough to withstand the rigors of the arctic environment (Ramsay and Stirling, 1988). Only pregnant females go into dens for the winter while bears of all other age and sex classes remain on the ice and hunt throughout the winter, except for brief periods of up to a few weeks during the coldest or most inclement periods when they may occupy temporary dens in the snow. Whenever food is unavailable, the metabolism of a polar bear automatically slows to a hibernation-like physiological state, in which energy is consumed at a lower rate. This change occurs after about 7–10 days of not eating and can occur at any season of the year, rather than only in the autumn prior to winter denning, like other species of bears. Because the digestibility of fat by a polar bear is about 98%, it has the ability to store large amounts of fat quickly when food is available and then switch to a more efficient metabolic state for as long as necessary when food is unavailable. This is probably the polar bear's single most important adaptation (Derocher *et al.*, 1990).

## V. Life History

Breeding occurs from late March to early June. Ovulation is induced so male–female pairs remain together interacting and eventually mating over a period of a week or longer. Dominant adult males try to restrict the movements of adult females and keep them secluded away from areas where competitors are most likely to be until mating is finished. Females have delayed implantation and implant from mid-September to mid-October, probably varying with latitude. The altricial young are born 2 months later at a weight of about 0.6 kg, with fine hair, and closed eyes (Ramsay and Dunbrack, 1986). The family leaves the maternity den to return to the sea ice by late March to mid-April by which time the cubs weigh about 8–10 kg. A litter size of two is most common, followed by singletons and occasionally triplets. One litter of four has been recorded in the wild as it was departing its maternity den.

Most maternity denning takes place in snowdrifts on coastal areas. In western Hudson and James bays, however, pregnant females must enter maternity dens prior to when suitable snowdrifts form in most years. Thus, they dig dens in frozen peat in small banks along the edges of lakes or streams (Ramsay and Stirling, 1988). Individual females show fidelity to denning areas, although not to individual den sites. In the western Beaufort Sea north of Alaska, a large proportion of the adult females den 200 km or more offshore in the multiyear ice of the Beaufort Sea. However, as the climate has warmed and the ice is becoming less stable, the proportion of bears that den on the sea ice in that area has declined (Fischbach *et al.*, 2007).

## VI. Interactions with Humans

Polar bears are important to the culture and economy of aboriginal people in Alaska, Canada, Greenland, and Siberia, who harvest 8–900 bears annually. Offshore development and production of hydrocarbons has the potential to impact negatively on polar bears (Stirling, 1990).

Many anthropogenic contaminants in the Arctic marine ecosystem are lipophilic and bioaccumulate as they move up in the food chain. Because polar bears are at the top of the ecological pyramid, and live predominantly on fat, they are capable of achieving high levels of contaminant loading in some areas. It is not yet known whether these contaminants are high enough to interfere with reproductive functions or the immune system (Norstrom *et al.*, 1998).

In recent years, ecotourism to view polar bears, especially at Churchill, Manitoba, has increased markedly with positive local economic benefits and a significant increase in the worldwide appreciation of the bears themselves. Tourist viewing of polar bears at Churchill in the fall and by ship in various areas of the High Arctic in summer, has not been associated with negative effects on the bears.

Conservation of polar bears and their habitats are mandated by a unique Agreement, signed by the five countries with polar bears [Canada, Denmark (for Greenland), Norway, USA, and USSR], signed in 1973 (Prestrud and Stirling, 1994). Population research continues in many areas to try to ensure that populations are harvested at sustainable levels. Thus, the survival of polar bears is not currently threatened by legal hunting, but this must still be regulated carefully because they have a low reproductive capability and would require 20 or more years to recover from an overharvest (Aars *et al.*, 2006).

In some areas, such as western Hudson Bay and the southern Beaufort Sea, long-term studies have demonstrated that climatic warming is causing the ice to break up earlier, or lie offshore of the continental shelf for extended periods respectively, which correlates with decreasing condition, reproductive rates, and declining population

P

size in both areas. Based on models of the steadily increasing temperatures in the Arctic, as a consequence of increasing concentrations of greenhouse gases, current observations from satellite data and climatic projection models confirm and predict a continuing and significant decline in the total amount of sea ice and an increase in open water in the foreseeable future. This loss of their critical sea ice habitat will likely have a devastating effect on the total population size of polar bears within the next few decades. The IUCN Polar Bear Specialist Group, at their most recent meeting in 2005, voted unanimously to upgrade the classification of polar bears from "species of special concern" to "vulnerable" because of the negative effects of climate warming on their sea ice habitat (Aars *et al.*, 2006).

## See Also the Following Articles

Arctic Marine Mammals ■ Bearded Seals ■ Ringed Seals

## References

Amstrup, S. C. (2003). Polar bear, *Ursus maritimus*. *In* "Wild Mammals of North America: Biology, Management, and Conservation" (G. A. Feldhamer, B. C. Thompson, and J. A. Chapman, eds), 2nd ed., pp. 587–610. Johns Hopkins University Press, Baltimore, MD.

Aars, J., Lunn, N. J., and Derocher, A. E. (eds.). 2006. Polar Bears. Proceedings of the 14th Working Meeting of the IUCN/SSC Polar Bear Specialist Group, 20–24 June 2005, Seattle, Washington, Occasional Paper 32.

Derocher, A. E., Nelson, R. A., Stirling, I., and Ramsay, M. A. (1990). Effects of fasting and feeding on serum urea and serum creatinine levels in polar bears. *Mar. Mamm. Sci.* **6**, 196–203.

Derocher, A. E., Lunn, N. J., and Stirling, I. (2004). Polar bears in a warming climate. *Integr. Comp. Biol.* **44**, 163–176.

Fischbach, A. S., Amstrup, S. C., and Douglas, D. C. (2007). Landward and eastward shift of Alaskan polar bear denning associated with recent sea ice changes. *Polar Biol.* DOI 10.1007/s00300-007-0300-4.

Norstrom, R. J., Belikov, S. E., Born, E. W., Garner, G. W., Malone, B., Olpinski, S., Ramsay, M. A., Schliebe, S., Stirling, I., Stishov, M. S., Taylor, M. K., and Wiig, Ø. (1998). Chlorinated hydrocarbon contaminants in polar bears from eastern Russia, North America, Greenland, and Svalbard: biomonitoring of Arctic pollution. *Arch. Environ. Contam. Toxicol.* **35**, 354–367.

Paetkau, D., Amstrup, S. C., Born, E. W., Calvert, W., Derocher, A. E., Garner, G. W., Messier, F., Stirling, I., Taylor, M., Wiig, Ø., and Strobeck, C. (1999). Genetic structure of the world's polar bear populations. *Mol. Ecol.* **8**, 1571–1585.

Prestrud, P., and Stirling, I. (1994). The International Polar Bear Agreement and the current status of polar bear conservation. *Aquat. Mamm.* **20**, 1–12.

Ramsay, M. A., and Dunbrack, R. L. (1986). Physiological constraints on life history phenomena: the example of small bear cubs at birth. *Am. Nat.* **127**, 735–743.

Ramsay, M. A., and Stirling, I. (1988). Reproductive biology and ecology of female polar bears *Ursus maritimus. J. Zool. Lond.* **214**, 601–634.

Stirling, I. (1988). "Polar Bears." University of Michigan Press, Ann Arbor, MI.

Stirling, I. (1990). Polar bears and oil: ecologic effects. *In* "Synthesis of Effects of Oil on Marine Mammals," pp. 223–234. Academic Press, San Diego, CA.

Stirling, I., and Parkinson, C. L. (2006). Possible effects of climate warming on selected populations of polar bears (*Ursus maritimus*) in the Canadian Arctic. *Arctic* **59**, 261–275.

Stirling, I., and Øritsland, N. A. (1995). Relationships between estimates of ringed seal and polar bear populations in the Canadian Arctic. *Can. J. Fish. Aquat. Sci.* **52**, 2594–2612.

Yu, L., Li, Y.-M., Ryder, O. A., and Zhang, Y.-P. (2007). Analysis of complete mitochondrial genome sequences increases phylogenetic resolution of bears (Ursidae), a mammalian family that experienced rapid speciation. *BMC. Evol Biol.* doi: 10.1186/1471-248-7-198.

# Pollution and Marine Mammals

Peter J.H. Reijnders, Alex Aguilar
and Asuncion Borrell

Awareness of the threat of environmental contaminants to marine mammals is widespread. High concentration of certain compounds in the tissues of these animals has been associated with organ anomalies, impaired reproduction and immune function, and as a consequence of the latter, with the occurrence of large die-offs among seal and cetacean species. This has prompted alertness about the impact of pollution and stimulated research into the relationship between observed effects and pollutants. However, a clear cause and effect relationship between residue levels of contaminants and the observed effects has been demonstrated in only a few studies. This might elicit a serious backlash, because in the absence of evidence, concerns expressed are easily interpreted as fear-mongering. This could lead to inertia in taking appropriate management measures, which is undesirable from a conservation as well as environmental management perspective.

The main reasons for the lack of proof of the impact of pollution on marine mammals are the difficulty or impossibility of experimenting in laboratory conditions with these animals, and the frequent occurrence of confounding factors that hamper the establishment of cause–effect relationships. Examples of these factors are the fact that pollution always occurs as a mixture of a large number of chemical compounds, the lack of data on biological variables influencing tissue levels, quality of samples usually analyzed, the limited information on pathology and occurrence of disease in the specimens studied, the absence of reliable population data, and the lack of information on the influence of other detrimental factors such as the impact of fisheries and of other human-related sources of disturbance.

## I. Substances of Concern

In general, the concept of pollution incorporates many different substances to which marine mammals are exposed and might adversely affect their health. These include chemical compounds, oil-pollution derived substances, marine debris, sewage-related pathogens, excessive amounts of nutrients causing environmental changes, and radionuclides. The influence of oil and petroleum-derived compounds such as the polycyclic aromatic compounds, of marine debris, of sewage-related pathogens and of nutrient-related changes such as the occurrence of biotoxins, has not been the subject of focused research in marine mammalogy. As a consequence, data on these pollutants, either as concentrations in tissue of the affected marine mammals or as effects on them, are extremely limited. This chapter will therefore only address pollution caused by chemical substances.

Traditionally, most laboratories tended to routinely analyze organohalogenated compounds such as DDT, DDE, DDD, polychlorinated biphenyls (PCBs), lindane, dieldrin, endrin, hexachlorobenzene (HCB), heptachloro-epoxide (HEPOX), and mirex, and trace elements such as mercury, lead, selenium, and cadmium. Some laboratories, able to use more sophisticated equipment, have also analyzed polychlorinated dibenzo-dioxins (PCDDs) and polychlorinated dibenzofuranes (PCDFs). Such narrow approach brings the risk of overlooking the impact of other, poorly known compounds. However, the monitoring of all the known synthetic organic chemicals and their metabolites currently in use, would require analysis of about 300,000 compounds. Therefore, criteria have to be developed to identify priority compounds on which to focus monitoring. Criteria for the identification of these compounds should include level of production and release into the environment, bioaccumulation potential, and toxicity. Examples of "novel" compounds that fall into the category of priority compounds are the organotins, polybrominated biphenyls (PBBs), polybrominated diphenyl ethers (PBDEs), and polychlorinated naphthalenes (PCNs).

Because research funds are limited, another issue to be addressed is the choice between monitoring pollutants concentrations or investigating their effects. The latter option is in our view clearly preferable. If an effect is observed, more focused research for the responsible compounds can follow.

Taking into account the two elements discussed above, and without ignoring the potential impact of other compounds, it is currently accepted that a list of compounds of highest priority should include all organohalogens usually referred to as persistent organic pollutants (POPs), particularly including PCBs, DDTs, PCDDs, HCB, HCHs (hexachlorocyclohexanes), dieldrin, endrin, mirex, PCDFs, PBBs, PBDEs, PCNs, polycyclic-aromatic hydrocarbons (PAHs), phenols, and the metals, very particularly including their organic forms such as methyl-mercury and organotins.

## II. Pollution from an Environmental Perspective

Pollution is only one of the many environmental factors that influence the health status of marine mammals. The assessment of the impact of pollution on marine mammals has therefore to be undertaken on a holistic perspective, considering also the potential of pollutants to interfere with their ability to recuperate from stress caused by other environmental forces. As an example, PCBs could cause immune-suppression in a given seal population without directly leading to an increased mortality. However, if such a population is exposed to an introduced virus, the extent of a resulting epidemic is likely to be much aggravated (Aguilar and Borrell, 1994).

Marine ecosystems are complex and environmental forces operating over populations are often multifactorial and produce synergistic or cumulative effects. Therefore, it will be complicated to attribute a given effect to a single factor. To illustrate the complexity of unraveling the impact of pollution, we discuss here some of the environmental factors, natural or anthropogenic, that influence the resilience of marine mammals to pollution.

### A. Prey Depletion

Natural environmental variations such as redistribution of planktonic organisms may bring changes in distribution, abundance, or recruitment of the species that constitute the food of marine mammals. However, depletion of prey may often also be caused by

overfishing by commercial fisheries. Depending on the extent of the depletion, marine mammals may respond to the reduced supply of prey either by switching to other species, or by temporarily moving to another area. But frequently, they undergo an impoverishment of their body condition and their recruitment rates become lower. The resilience of animals in such populations/stocks is negatively affected, potentially increasing the detrimental impact of pollutants.

### B. Habitat Disturbance

Habitat may be disturbed by a wide range of human activities, including recreation, construction works, and many others. For pinnipeds, sea otters (*Enhydra lutris*), and some coastal cetaceans, the physical alteration of the littoral, including the mere presence of humans and their associated infrastructures, may be a significant detrimental factor. Noise pollution is a particular source of concern because many marine mammals rely on sound emission and detection for finding their prey, communication, and navigation. Activities producing noise-related disturbance include shipping, boating, military maneuvers, seismic testing, and oil and gas drilling.

### C. Disease

Natural factors influence incidence of disease. For example, a shift in distribution of prey species may lead to an increased parasite infestation rate likely to affect the resilience of populations to pollutants. Although in general terms the incidence of infectious disease in marine mammals is poorly known, *Morbillivirus* epizootics that have recently affected pinnipeds as well as cetaceans, have elicited extensive research on the effects of viral diseases on marine mammal populations.

Deadly bacterial diseases are generally considered to be secondary to other conditions such as viral disease, parasitic infection, or trauma. However, like some pollutants, bacteria can also interfere with reproduction, as was demonstrated by the finding of *Brucella* organisms in porpoises and dolphins.

In marine mammals, similarly to other better studied groups of vertebrates, disease and the toxic effect of pollutants are often interrelated. This relationship will be discussed in more detail in the next section in this chapter, although we should mention here that diseases can affect metabolic systems and, consequently, alter physiological functions. Chronically diseased females, for example, usually have a poor reproductive performance, as do females affected by some pollutants.

### D. Overall Environmental Changes: Global Warming, Ozone Depletion

Albeit the potential effect of global changes on marine mammals has been very little investigated and its consequences are considered less imminent than those caused by other factors, this matter certainly deserves concern. It is predicted that global rise in temperature will alter marine communities and their productivity, sea level rise, reduce ice cover, and modify rainfall and water current systems. The consequences for marine mammals are unclear but undoubtedly those alterations will affect their behavior and distribution. Increased incidence of epizootics among pinnipeds is also postulated, as higher densities as a result of increased haul-out behavior, shall result in a higher transmission rate of infectious agents.

Despite the long-term character of these threats, changes in distribution and behavior of marine mammals caused by climate variation should be monitored to detect potential relationships at an early stage.

P

The examples mentioned above show clearly that studying the impact of pollution on marine mammals requires a multidisciplinary approach. Therefore we advocate assessing pollution impact not in an isolated way, but always in relation to other environmental factors.

## III. Factors to be Taken into Account in Assessing Pollution Impact

Two sources of information may alert that pollution might affect a given population: high tissue pollutant concentrations in the members of that population, and changes in the population's biological parameters such as physiological condition and changes in reproductive or mortality rates. The latter are often derived from population monitoring and/or pathological investigations. However, a number of biological factors and inconsistencies in the sampling and analytical procedures seriously hamper the establishment of such relationships, sometimes even leading to spurious interpretations of environmental data (Krahn *et al.*, 2003).

### A. Biological Factors Affecting Variability in Pollutant Levels

Some persistent chemicals are bioaccumulative and their concentrations in living organisms undergo a progressive amplification through food chains, a process called biomagnification. However, the increase at each trophic level is usually considerably higher than the 10-fold increase predicted by ecological models. Biomagnification, defined as the ratio of concentrations of a compound in the predator to its prey, can be significantly altered—and often much increased—by a number of variables such as the route of exposure, the physical and chemical properties of the compound, the metabolic capacity of the predator and its physiological constitution (Aguilar *et al.*, 1999).

### B. Diet

Diet composition is a key factor determining resultant tissue concentrations. Because baleen whales feed on planktonic crustaceans, and are thus situated lower in the food web, their tissue organochlorines (OCs) concentrations are almost invariably lower than those in the top-predator toothed whales living in the same ecosystem.

Within a population of the same species, OC levels can also differ because of variations in diet. For example, juvenile pinnipeds often exploit different food resources than adults, and in many species of cetaceans and pinnipeds, adult males prey on different species than adult females. In some marine mammals, differences may even be associated to reproductive status: the diet of lactating females of some dolphin species is different from pregnant or resting females. Also, the geographical region where food is consumed is critical: during most of the year sperm whales (*Physeter macrocephalus*) occupy different geographical regions than females and, as a consequence, their pollutant profiles are quite dissimilar. Differences in diet are also assumed to have an influence on the tissue concentrations of PAHs in marine mammals. Levels of these compounds in marine mammals are generally low, although they tend to be higher in cephalopods-eating marine mammals than in those relying on fish. The explanation appears to be that the ability of fish to metabolize PAHs is better than that of cephalopods.

Tissue levels of metals also appear to be related to the feeding habits and region of exposure. Cadmium, copper, and zinc levels are higher in cetaceans that feed primarily on squid, than in those feeding on fish. This is attributed to the ability of squid to selectively retain these elements. Intraspecific differences in tissue metal concentrations have also been linked to segregation in feeding areas; the levels of lead in kidney and muscle tissue of pilot whales and white beaked dolphins occurring during summer in the same areas, are much different because they segregate geographically—and feed—during the winter.

### C. Age and Sex

The tissue concentration of a pollutant in a marine mammal is a function of the difference between the intake rate and the metabolization and excretion rates. OCs have been found to correlate positively with age; levels are relative low in young animals, increase until a certain age, and then either continue to increase or reach a plateau level or decrease. The leveling-off or decreasing phase is different for males and females, as is addressed further on. Factors that influence the age-related pattern of accumulation of organochlorines are the detoxification ability and the feeding rate. The capacity for detoxification is low in young animals and improves with age; thus, the initial increase during the juvenile stage is slowed down by improved metabolization and excretion rates. The resulting leveling-off of tissue concentrations is further enhanced by reduced feeding rates in adults.

Superimposed on these is the effect of reproduction in females. OCs, as most lipophilic compounds, cross the placenta and reach the fetus, although not all chemicals do it at equal rates. For example, the lower chlorine-substituted (lower weight) congeners of PCBs are more easily transported than the higher chlorinated ones. In addition to placental transport, OCs are also transferred from mother to offspring through milk. Higher chlorinated OCs are less efficiently transferred from the lipid tissue of the mother to her milk and hence to the suckling calve or pup. This process obviously does not start until the females reach sexual maturity and become pregnant for the first time. Therefore, the first pregnancy marks the start of the leveling-off or decrease phase in females. There are differences among species and compounds. But this reproductive discharge in females is not uniform and depends on the characteristics of the reproductive cycle of the species and the physical–chemical properties of the compound. The transfer during lactation is much higher than that occurring through deposition in the tissues of the calf or pup during pregnancy. In cetaceans, discharge of PCBs, expressed as percent transferred in relation to maternal tissue load, ranges from 5 to 96% during lactation and 4–6% during pregnancy. In pinnipeds, the ranges are 23–81% and 1–10%, respectively. Not surprisingly, the length of the lactation period significantly influences the proportion of the OCs' load transferred to the offspring. It has been estimated that this proportion ranged from 3 to 27% in fin whales (*Balaenoptera physalus*), with a lactation period of around 7–8 months, whereas it was around 80% in bottlenose dolphins (*Tursiops* spp.), and 72–91% in striped dolphins (*Stenella coeruleoalba*), two species in which lactation lasts about 14 months. Irrespective of the amount transferred, the reproductive discharge results in lower levels of lipophilic pollutants in reproductively active females as compared to males of the same age. However, there are some exceptions to the general rule. In southern hemisphere minke (*Balaenoptera acutorostrata*) whales, levels of PCBs and DDT were found to be higher in immature males than in mature males, as a result of a shift in diet caused by adult migration to less polluted areas. In the North Atlantic, adult female sperm whales are more polluted than males of comparable age because they feed on more polluted species and distribute year-round in regions where pollutant loads are higher.

Age-related variation in tissue concentrations of trace elements is less homogeneous. Mercury, cadmium, selenium, and lead increase with age, somewhat more steeply in females compared to males. There is no clear leveling-off for any element except for lead, in which a slower increase has been observed at older age. As these elements are not lipophilic, reproductive transfer does not affect their loads in females. It has been suggested that the higher levels of those elements found in females compared to males may be related to differences in metabolic pathways linked to hormone cycles.

Information on other trace elements is scarce. Copper and zinc show no increase with age. In fact, concentrations in newborns are higher than in adults, which are attributed to an age-related decrease in absorption and retention of these essential elements.

### D. Nutritive Condition

Nutritive condition affects the volume of fat in the body and its lipid composition. In some cetaceans and pinnipeds, blubber lipid richness may decrease from 90% in a female near term to 30–35% in females just having weaned their offspring. Though less impressive, males also show changes in blubber layer thickness during the reproductive season. Apart from this reproduction-related change, seasonal variation may also be significant. Variation in blubber layer thickness is lower in toothed whales compared to baleen whales. In some pinnipeds, independently of the reproduction-related changes, blubber layer thickness can vary by as much as 50% (taking the maximum thickness as a reference). This variation has implications for the dynamics of lipophilic contaminants. Because lipids are more readily mobilized from the blubber than lipophilic pollutants, lipid metabolization typically results in an increase in the residue levels. However, it has been found that the increase is less than a kinetic concentrative model would predict. It has been suggested that the more polar fraction of the pollutants is more readily mobilized through the enhanced metabolization and excretory capacity stimulated by a rise in tissue pollutant concentrations subsequent to the lipid metabolization.

It is unclear to what extent changes in nutritive condition affect tissue concentrations of nonlipophilic compounds. Changes in mass and composition of tissues where chemicals (e.g., heavy metals in liver and kidney) are likely to influence the dynamics of these pollutants, but data on these processes in marine mammals is lacking.

Body growth in young animals also influences tissue levels of pollutants. In both pinnipeds and cetaceans it has been found that dilution of contaminants occurs in the early stages of growth due to the rapid deposition of blubber and the amassing of liver and kidney tissue. Calculations of tissue concentrations on a lipid basis instead of a fresh weight basis can partially account for such variation, but it does not account for variation in the qualitative composition of the lipid fraction, which is also likely to affect the retention ability of the tissue.

### E. Body Size

The influence of body size on variation in the accumulation pattern of pollutants is somewhat complex. Generally, elimination rates of xenobiotic compounds per unit of bodyweight are inversely related to bodyweight, a trend that also holds for the activity of detoxifying enzymes. Both would tend to favor accumulation of higher pollutant levels in larger animals. Contrary to that effect, metabolic rate is inversely correlated to body size. Because metabolic rate is correlated with pollutant intake, a higher pollutant accumulation can be expected in smaller species. The influence of metabolic rate has been found to outweigh the counter-effect of elimination and detoxifying activities. The concentration factor in a marine mammal is largely dependent of its daily rate of food consumption—inversely related to body size—and the mean concentration of pollutant in its

prey. Small animals therefore carry generally higher loads of pollutants relative to their bodyweight than do larger animals.

Variation in body size is more dramatic in cetaceans than in pinnipeds. Some dolphin and porpoise species weigh, when adult, about 30–40 kg, while the larger whales can weigh more than 150,000 kg. The range in adult pinnipeds varies from 50 to 4000 kg. An example of variation in pollutant levels between two species of different size is that of two krill eating baleen whale Atlantic species, in which differences in tissue pollutant levels were explained by differences in body mass. It has been proposed that, in species sharing the same waters, the effect of body mass on tissue concentration outweighs that of the small differences in diet or other biological traits.

### F. Body Composition

The distribution pattern of pollutants in the body of an animal depends largely on the physical and chemical properties of the substances involved. For example, much work has been carried out to investigate the influence of the position of H-atoms on the biphenyl ring in all PCBs, which largely determines the possibilities for their metabolization by marine mammals.

Lipophilic pollutants accumulate in fatty tissue, so about 70–95% of lipophilic pollutants end up in the blubber, which in marine mammals is the largest fat compartment. The chemical composition of the blubber also influences pollutant concentrations. In species with thick blubber, pollutants are stratified in the different layers and significant differences may be found between the inner and outer strata. Therefore, the whole blubber layer must be sampled to obtain a representative picture of the individual's load.

Mercury, cadmium, zinc, and other heavy metals accumulate mostly in the liver and kidney, and lead predominantly in bone tissue.

### G. Analysis and Sampling

One of the major handicaps in assessing temporal and spatial trends of contaminants in marine mammals is the poor comparability of data. This holds partly for heavy metals, but it is definitely critical for analyses of OCs. The analytical techniques used, and their accuracy, have changed considerably over time and also vary between different laboratories. This greatly hinders comparison between studies undertaken by different laboratories or time periods. Significant improvement in standardizing procedures has been achieved in the last decade through intercalibration exercises. Quality assurance and quality control are of utmost importance, but this holds also for the sampling procedures. To avoid contamination by the packaging material, clean glass or aluminum foil should be used to preserve samples for OCs analyses, and plastic bags to preserve samples for heavy metal analyses. Each sample should be accompanied by the appropriate biological data, and if possible, also with a detailed pathological examination, to reveal, e.g., incidence of alterations in reproductive biology, early development, and occurrence of diseases.

Detailed field and laboratory protocols taking into account these considerations have to be developed before embarking in any ecotoxicological study.

### IV. Impacts of Pollution on Marine Mammals

Numerous studies have suggested that exposure to pollutants has an impact on marine mammal populations, mainly on reproduction and mortality. However, in most of these studies, the existence of confounding factors prevents reaching conclusive results and only a few have actually succeeded in demonstrating such relationship (O'Shea, 1999; Reijnders *et al.*, 1999).

The effects of pollution, either observed or suggested, can be conveniently grouped under three categories: impaired reproduction, indirect mortality, and direct mortality.

## A. Impaired Reproduction

OCs, particularly PCBs, have been demonstrated to be responsible for impaired reproduction in the harbor seal *Phoca vitulina*. This conclusion was reached by means of a feeding experiment in which 12 female harbor seals were fed diets low in OCs and 12 females received a diet high in OCs, particularly PCBs and DDE. The conclusion was that reproductive success was significantly lower in the more polluted diet group: 4 pups were born instead of 10 born in the control group. The latter figure is similar to what is normally found in free-ranging harbor seals. In addition, the analysis for oestradiol-17β and progesterone in blood samples from these seals revealed that the reproductive failure occurred at the implantation stage, since such failure was accompanied by low levels of oestradiol-17β (Reijnders, 1986). A plausible explanation of this effect is that PCBs enhanced the enzymatic metabolism lowering in this way the circulating levels of estradiol, which in turn led to imperfect endometrial receptivity and prevented successful implementation of the blastocyst.

Elevated OCs concentrations have been associated with reproductive impairment in gray seals, *Halichoerus grypus*, and ringed seals, *Pusa hispida* in the Baltic, and in California sea lions, *Zalophus californianus*. Female Baltic gray and ringed seals exhibited uterine occlusions and stenosis, leading to partial or complete sterility; concentrations of OCs were higher in the affected animals than in normally reproducing females. It has been proposed that pregnancy was interrupted by PCBs (or PCB-metabolites), followed by development of pathological disorders. The epidemiological studies on the involved populations strongly support the hypothesis that PCBs or their metabolites, i.e., methyl-sulfones, are responsible for the observed reproductive impairment. This has been apparently confirmed by the fact that the incidence of pathological conditions in younger but mature age-classes decreased, and OCs levels in seals as well as other Baltic biota, sharply declined between 1970 and 1980 (Olsson *et al.*, 1994). However, unequivocal evidence for a cause–effect relationship has not been provided, although this stage of proof is probably as far as one can get with the constraints of this type of field research.

The case of the California sea lion is even more complex. Initially stillbirths and premature pupping were attributed to high OCs (PCBs and DDE) concentrations. Later studies demonstrated that pathogenic disease agents could also have been responsible of the process. These confounding factors prevented reaching a clear cut conclusion on the causative role of pollution.

The proof for reproductive disorders in cetaceans caused by specific pollutants is weaker than for pinnipeds. There is a risk assessment study indicating a high likelihood of severe reproductive impairment in bottlenose dolphins from the southeast United States coast, caused by chronic exposure to PCBs. Impaired reproductive performance caused by PCBs has been suggested in beluga whales (*Delphinapterus leucas*) in the St. Lawrence River. In 2 out of 120 examined belugas, hermaphroditism was observed. However, the pathological studies were not conclusive, and the lack of sound population data with which to compare the observed findings, made it impossible to conclude on the actual role of pollutants on such abnormalities.

Low levels of testosterone were associated with high levels of PCBs and DDE in Dall's porpoises (*Phocoenoides dalli*). However, the biological significance and underlying mechanism are unclear since both variables are age-related, and further studies are needed to clarify the potential involvement of pollutants.

Abnormal testes, i.e., transformed epididymal and testicular tissue, were observed in North Pacific minke whales. A possible relation with high levels of OCs has been suggested, but not proved.

## B. Disease

A suite of pathological disorders including skull lesions (paradentitis, osteoporosis, exostosis), cortical adenomas, hyperkeratosis, nail malformations, uterine stenosis and occlusions, uterine tumors (leiomyomas), and colonic ulcers, have been observed in Baltic gray and ringed seals and, to a lesser extent, in harbor seals. Pathological and epidemiological investigations revealed that the symptoms observed were part of a disease complex called hyperadrenocorticism, a disease syndrome associated with high levels of PCBs and DDT and their metabolites. Contrary to reproductive impairment, it is not possible to conclusively evaluate which of these substances elicit a response in seals, because of crossed or synergistic effects. Like in the case of the reproductive disorders, the prevalence of uterine lesions, adrenocortical hyperplasia, and skull-bone lesions was found to decrease following a decline of DDT and PCBs in Baltic biota. Conversely, however, the incidence of uterine leiomyoma in Baltic seals has not changed to date. Of even more concern is the increasing incidence during relative recent years of colonic ulcers in young Baltic gray seals, indicating an increasingly compromised immune system in these animals. DDT tissue levels in these animals have decreased strongly between 1969 and 1997, annually by 11–12%, but PCB levels decreased during the same period at a much lower pace, only 2–4% annually. This may suggest a role of PCBs and/or their metabolites in the observed pathologies, although the potential effect of novel, unknown compounds cannot be excluded.

Some studies have shown direct evidence of the immunotoxicity of OCs. Reduced immune responses were correlated with high levels of PCBs and DDT in *in vitro* immune function assays with peripheral blood lymphocytes from free-ranging common bottlenose dolphins, *Tursiops truncatus*. In an experiment with captive harbor seals, *in vitro* and *in vivo* immune function tests showed lower immune function related to higher dietary concentrations of OCs (Swart *et al.*, 1994). While these two studies show that OCs adversely interfere with immune function, the toxicological and biological significance remains unfortunately unclear. An indication for an increased risk of infection from PCBs on harbor porpoises has been reported.

It has been suggested that lowered immuno-competence induced by contaminants aggravated the die-offs of bottlenose dolphins in the Gulf of Mexico (1990, 1991, 1993) and in the east coast of the USA (1987–1988), striped dolphins in the Mediterranean Sea (1990–1992), harbor seals in the North Sea (1988), Baikal seals, *Phoca sibirica*, in Lake Baikal (1987–1988), and Caspian seals, *Phoca caspica*, in the Caspian Sea (2000). In most cases the mortalities were ultimately caused by a *Morbillivirus* infection, but exposure to high levels of OCs was proposed to have played a key role by facilitating viral transmission and increasing susceptibility of individuals to the disease. However, it has been difficult to conclude on the etiology of these mortalities. Different studies have tried to establish links between die-offs and pollution. In the case of the striped dolphin *Morbillivirus* epizootic, animals killed by the disease carried significantly higher PCB concentrations than survivors. This finding could be explained by: (1) immune suppression caused by PCBs leading to higher mortality of the more polluted individuals, (2) mobilization of pollutants stored in depot tissues thinned by the disease, or (3) changes in physiological functions of the affected individuals, leading to increased PCB concentrations.

In two other studies levels of organochlorines were related to mortality. In one study, OCs levels in seals that died during the *Morbillivirus* outbreak were compared with those in surviving seals. In the other study, OCs concentrations in harbor porpoises, *Phocoena phocoena* that died from physical trauma were compared with animals known to have succumbed to an infectious disease. Both studies were inconclusive in establishment of a direct cause–effect relationship between pollutants and susceptibility to disease, because of the existence of confounding factors such as heterogeneous body condition between the groups compared. A recent follow-up study on harbor porpoises from England and Wales has been more conclusive (Jepson *et al.*, 2005). In this study, PCB concentrations in blubber from animals that died due to physical trauma (e.g., bycatch) were compared with those from animals that died because of an infectious disease. A significant association was demonstrated between blubber PCB concentrations and mortality due to infectious disease, suggesting a causal relationship with chronic PCB exposure. Again here, the possibility of additive or synergistic effects of other contaminants must be considered.

Other ecotoxicological studies point toward other effects of pollutants on marine mammals. It has been proposed that OCs produce thyroid hormone and vitamin A deficiency in at least harbor seals. Thyroid hormones are important in the structural and functional development of sex organs and the brain, both intrauterine and postnatal. Vitamin A deficiency may lead to increased susceptibility to microbial infections and retarded growth, as appeared to be indicated by the significant lower birthweights of pups born in the higher contaminated dietary group of the captive harbor seal study discussed earlier.

Another noteworthy example of impaired health status possibly caused by pollution is the case of the St. Lawrence beluga population (Martineau *et al.*, 1994). A range of pathological conditions have been documented in this population, particularly high prevalence of tumors, digestive tract and mammary gland lesions. High tissue levels of OCs, lead, and mercury have been found in these animals. The establishment of a cause–effect relationship between contaminants and the observed effects in this population is hampered by the possible adverse role of other environmental factors such as previous overhunting, high levels of noise pollution, and overall habitat destruction. Any of these factors has a potential for causing most of the observed conditions and the population's small size and slow recovery.

## C. Direct Mortality

There is no clearly documented record of any acute chemical poisoning event affecting marine mammals, apart from one case that affected harbor seals; a small colony had been acutely poisoned by an accidental discharge of mercury-contaminated agricultural disinfectant and several deaths occurred.

## D. Endocrine Disrupting Chemicals

In the last decade, increasing concern has been expressed on xenobiotic-induced endocrine disruption in wildlife and humans. Adverse effects of contaminants on mammalian wildlife through modulation of endocrine systems are predominantly documented in fish-eating (aquatic) mammals. Indeed a large number of xenobiotics with

endocrine disrupting properties, like OCs, have been detected in marine mammal tissue. In previous sections, reproductive and non-reproductive effects, including possible links with the functioning of the immune system, have been discussed in relation to these pollutants. Except for the reproductive toxicity in harbor seals and Baltic seals, the evidence of a causal link between endocrine disruption and observed effects is weak or nonexisting. Most often neither a positive proof nor a dismissal, or simply a negative endocrine-like effect, could be provided. The reasons for lack of proof are the unavailability of reliable population data, the potential interaction between the many pollutants present, the role of disease agents and other environmental factors, the lack of biomarkers to assess endocrine effects, and the little research on early development in marine mammals.

## V. Species Vulnerability

The impact of pollution on marine mammals can occur throughout the entire chain from exposure, uptake, metabolism to excretion. Concentration in prey is a determining factor. Generally, coastal species are exposed to higher environmental levels than more pelagic species, and species occurring in industrialized (including intensive agricultural) areas usually have higher pollutant levels compared to animals in less developed regions. Among marine mammals, coastal seals and dolphins usually carry the highest tissue residue levels. Superimposed to that is the preferred trophic level of feeding. In the same water mass, species feeding at lower trophic levels are exposed to lower levels of pollutants, compared to species feeding higher in the food chain. This is why pollutants levels are almost always lower in baleen whales than in toothed whales. Exceptions to this pattern have been discussed earlier, e.g., for metals in species feeding on squid rather than on fish. However, even species (harbor seals and gray seals from the east coast of Scotland) sharing the same environment, are differentially susceptible to OCs. Immune functions in harbor seals were more affected than in gray seals, the underlying mechanisms are still unclear (Hammond *et al.*, 2005)

As mentioned before, females get rid of pollutants through reproduction. Species that reach sexual maturity at a younger age are in an advantage as compared to those that start reproducing at an older age. Early reproduction is also positive for the offspring. The amount of pollutants descendants receive is lower if mothers initiate reproduction activities early, because they have not yet built up high tissue concentrations. Similarly, an earlier onset of sexual senesce is a disadvantage in this respect, because it halts the discharging process. A protracted lactation period is clearly beneficial for reproductive females, because the amount of lipophilic pollutants that they transfer is high. This latter obviously depends on the time they start to feed again, because then the pollutant uptake will counterbalance the discharge. However, the protracted lactation period may bear adverse effects to the offspring, because often milk is more polluted than the food that descendants will consume once weaned. It is unclear how this resolves at the population level; this is, whether the benefit for the reproductive female is higher or lower than the costs for the offspring.

A factor likely to lead to differential vulnerability between species is body size. Small species have generally higher levels of pollutants relative to their bodyweight, than those of larger body mass.

Metabolization is another operative factor in this context. The P450-enzyme system is the main physiological tool for metabolizing OCs. For example, this system can be induced by PCBs, mediated by the arylhydrocarbon (Ah) receptor, which is found in mammals and birds. But the metabolic ability is not uniform between marine mammals. Overall, cetaceans have a lower metabolization capacity,

P

as measured by Phenobarbital (PB) and methylcholantrene (MC) types of activity. Initially all cetaceans were thought to lack the Phenobarbital (PB)-type of enzyme. However, recent research has shown that several dolphin species posses at least some microsomal PB-type of enzyme. Still, their PB- and MC-type of metabolic activity is usually lower than that of pinnipeds and terrestrial species. At a more specific level, ringed seals and harbor porpoises seem to have metabolic capacities intermediate between those of other seals and cetaceans. In conclusion, apart from the more apparent cetacean–pinniped difference in metabolic capacity, sharp differences exist also among species within any given taxa (Boon *et al.*, 1992).

The critical question in this respect is, however, whether a low activity of, e.g., PB-type and/or MC-type enzymes renders cetaceans more vulnerable to pollution, as has been repeatedly suggested. This may not automatically the case. For example, PCBs can potentially elicit toxicity in at least two ways: as parent compounds (persistent congeners) and as metabolized congeners. The persistent compounds show a PB- and mixed PB- and MC-type of toxicity associated with liver-hyperproliferation, lowered levels of thyroxin and vitamin A, and a dioxin-type of toxicity (MC) resulting in thymic atrophy, dermal disorders, and liver necrosis. Metabolization of parent compounds can result in at least two contrasting effects: a decreased level of dioxin type of toxicity and an increased metabolic specific toxicity such as immunotoxicity. The resultant effect of a lower metabolization capacity therefore depends on the relative contribution of the mitigating influence of a decreased dioxin type of toxicity vs a continued PB and mixed PB/MC induction and the effect of reactive intermediates.

In this respect, attention should be drawn to the often misused concept of toxic equivalency. This concept is based on the structure–activity relationships of contaminants with receptors. Tetrachlorodibenzo-dioxins (TCDD) and PCBs have a structure that fits the Ah-receptor. The degree of induction by TCDD has been correlated with their toxic effects observed in laboratory animals. Given the similarity in structure of PCB congeners, the ability of these latter compounds to induce the Ah-receptor mediated response is expressed as a ratio to the induction by TCDD. This is called the toxic equivalency factor (TEF), which has been extensively used to assess the toxicity of PCB congeners and their mixtures with, e.g., DDT and PCDD. That toxicity is calculated by multiplying the TEF of each compound by its concentration and the sum of the resulting values are considered to be the total toxic equivalent (TEQ) for the mixture of compounds found in the sample.

However, it needs to be stressed that TEFs are based on laboratory animal models. Therefore, the TEQ for a given marine mammal sample only means the effect that the mixture of compounds found in that sample would have on a laboratory animal. Because (1) large differences between species exist in induction of P450-based enzymes, (2) the toxicities of PCB-metabolites are not incorporated into the calculations, and (3) toxicity of modes other than that of a dioxin-type are disregarded, the application of TEQ to assess the toxicological risk to which a particular species is subject to, is not necessarily reliable. The same holds for extrapolating TEQ between species. We would therefore emphasize that the frequently used practice of assessing whether the toxic significance of a certain value of TEQ found in a marine mammal is lower/higher than a TEQ value found in a species where effects were observed, is unfounded and scientifically unsound.

Another issue that remains to be clarified is the occurrence of high levels of some heavy metals, such as mercury, lead, and cadmium, particularly observed in species in the northern-arctic regions (Muir *et al.*, 1999). They apparently possess the ability to tolerate high levels of those compounds. It is known that marine mammals are able to detoxify these metallic compounds by, e.g., de-methylating the highly toxic form of organic mercury into the less dangerous inorganic mercury, by the binding of metals to metallothioneins, or by the binding of selenium to mercury where inactive salts are produced. It is tempting to speculate whether the animals in those areas have evolved responses to mitigate the effects caused by the naturally occurring contaminants.

## VI. Developments in Spatial and Temporal Trends of Pollutants

Data on levels of pollutants in marine mammals are more numerous for western Europe, North America, Canada, and Japan. Limited data are available for many other countries and regions (e.g., Africa, New Zealand, India), and very little information is available for the southern hemisphere. As mentioned in an earlier section, the fish-eating marine mammals from the mid-latitudes (industrialized and intense agricultural use) of Europe, North America, and Japan have the highest loads (Fig. 1). Residue tissue concentrations are lowest in the upper north polar region and the Antarctic. Nearly all of the

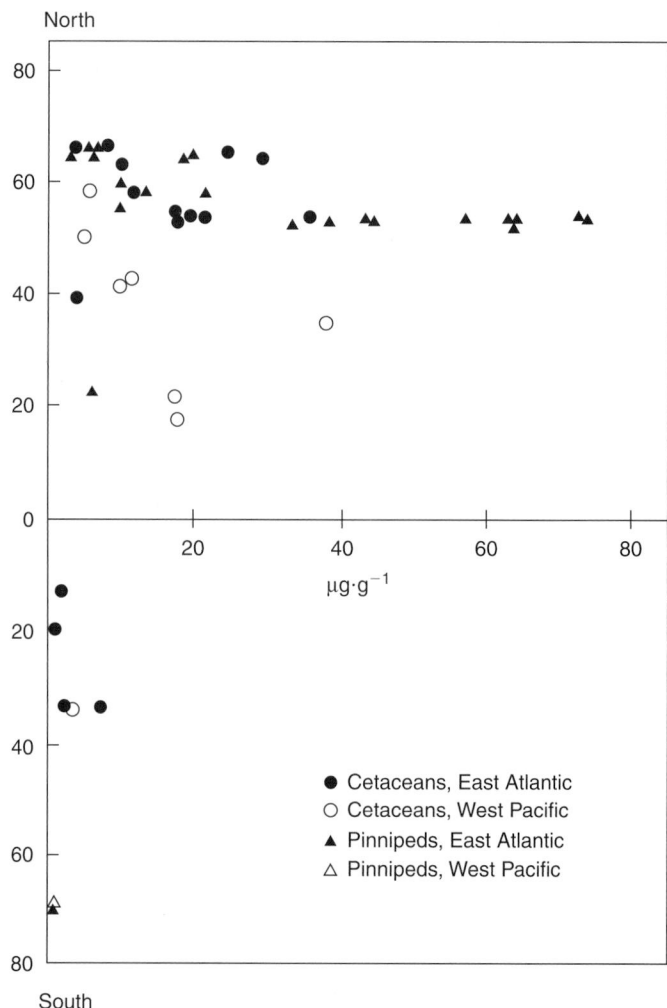

**Figure 1** *Latitudinal ΣPCB concentrations (μg·g⁻¹ wet weight) in blubber tissue of various marine mammals from the western Pacific and the eastern Atlantic.)*

OCs contamination in marine mammals in the Arctic and Antarctic has reached these areas via atmospheric transport. The levels of the more volatile OCs are higher compared to PCBs and DDT and more homogeneously distributed. This pattern of distribution of residue concentrations in marine mammals, however, is gradually changing. Levels of OCs are declining in the mid-latitude areas, whereas they are increasing in regions distant from pollution sources. Particularly the transfer of OCs released in (sub)tropical countries to the atmosphere, causes global redistribution that were predicted to end up in the mid-term future into the Arctic. However, recent studies show that concentrations of most highly persistent OCs (PCBs, DDT, etc.) significantly declined in Canadian Arctic biota from the 1970s to the late 1990s, and today are generally less than half the levels of the 1970s, particularly in ringed seals.

Chlorobenzenes and endosulfan were among the few OCs to show increases during this period while tHCH remained relatively constant in most species. A suite of new-use chemicals previously unreported in Arctic biota (e.g., polybrominated diphenyl ethers (PBDEs), short chain chlorinated paraffins (SCCPs), polychlorinated naphthalenes (PCNs), perfluoro-octane sulfonic acid (PFOS), and perfluorocarboxylic acids (PFCAs) has recently been found. But information to assess the spatial and temporal patterns thereof, is still insufficient (Braune et al., 2005).

For the most widely spread OCs, particularly PCBs and DDT, and the heavy metals mercury, lead, and cadmium, information on temporal trends is increasing. In most heavily industrialized and agricultural regions, the production and use of DDT and PCBs was halted in the early 1970s. From the mid-1970s onwards, levels of DDT and PCBs in marine mammal tissues decreased (Borrell and Reijnders, 1999). The decline in DDT levels was stronger than that of PCBs. In pinnipeds the decline was 80–90% for DDTs and 60% for PCBs. The difference is most likely due to less stringent control measures for PCBs; large quantities of these compounds have remained in use in many applications. The overall time trend for PCBs and DDT levels in marine mammals is that concentrations have decreased since the mid-1970s. The decrease in DDT levels has continued thereafter. However, PCB levels in some areas leveled off at the end of the 1970s/early 1980s, but are now declining.

Figure 2 shows the distribution over different compartments of all the globally produced (1.325.810 tons) PCBs by the industry. In the higher production scenario, 55.4% have been permanently lost through degradation, burial in landfills, or destroyed following various combustion processes; 12% has been emitted into the atmosphere. As a result, only 2.4% of all the PCBs produced has been deposited in soils, 14% is accumulated in dump sites and 15% is still in use (Breivik et al., 2007). It is expected that the observed leveling-off of the decrease in marine mammals will be followed by a strong reduction in the near future.

Trends for heavy metal pollution are less apparent. In general, it is accepted that in the mid-1990s in ringed seals and beluga from the Canadian Arctic and Greenland, the levels of mercury were higher compared to the mid-1970s, whereas for cadmium there was no clear trend (Muir et al., 1999; Breivik et al., 2007). On the contrary, levels of mercury and lead in pinnipeds from the Wadden Sea have considerably decreased.

## VII. A Fundamental Approach to Address Pollution Impact on Marine Mammals

It is clear that a considerable amount of fundamental research is needed before it will be possible to adequately address the impact of pollutants on marine mammals. Realizing this situation, the International Whaling Commission (IWC), through its Scientific Committee, developed recently a comprehensive program to investigate pollutant cause–effect relationships in cetaceans: "Pollution 2000+" (Reijnders et al., 1999). The ultimate objective of pollution studies as related to marine mammal management is to determine a predictive model to link tissue pollutant levels with effects at the population level. This is obviously a longer term goal. It is realized that if any progress is to be made within a reasonable time frame, a multidisciplinary, multinational focused program of research is required, that concentrates on species and areas where there is most chance of success.

Pollution 2000+ focused on PCBs, because these chemicals can be used as model compounds for OCs pollution. Moreover, PCBs are found at extremely high tissue levels in cetaceans, their effects upon mammals are well known and substantial information is available on their patterns of variation, geographical distribution, and tissue kinetics. The focal species in the IWC program are bottlenose dolphin and harbor porpoise, because both species occur in waters extending over a gradient of pollution and are likely to provide reasonable sample sizes.

Because the ultimate aim of the program is to look at potential effects of pollutants at the population level, it was considered necessary to test and develop techniques to feasibly collect data from large numbers of free living animals. Since biopsy techniques allow such type of sampling, an initial step in the project has been to calibrate information obtained from biopsy sampling with that collected from dead animals. It was also considered similarly important to ascertain the influence of postmortem time on levels of contaminants and on indicators of exposure and effect. This calibration is needed to ensure that collected samples are representative of actual pollutant loads.

Phase I of this program is just finalized and a number of significant results were obtained (Reijnders et al., 2007). These results provide a valuable scientific framework for this type of ecotoxicological studies. Worth mentioning publications include a study on pollutant effects at the population level showing that PCB accumulation in Sarasota bottlenose dolphins might be depressing the potential annual growth rate by some 3.6% (Hall et al., 2005), studies in bottlenose dolphins on the effect of OCs and individual biological traits on blubber retinoids, a biomarker of OCs exposure (Tornero et al., 2006b), research on dermal endothelial Cytochrome P450 1A1 expression in biopsies from the same bottlenose dolphin population related to blubber ΣPCB concentration (Wilson et al., 2007), and a study on the postmortem stability of retinol, PCBs, and dioxin-like compounds in by-caught harbor porpoises (Tornero et al., 2005b; Borrell et al., 2007). It is clear that Pollution 2000+ is a core-program to address some fundamental questions. It does not imply that other research on pollutants and marine

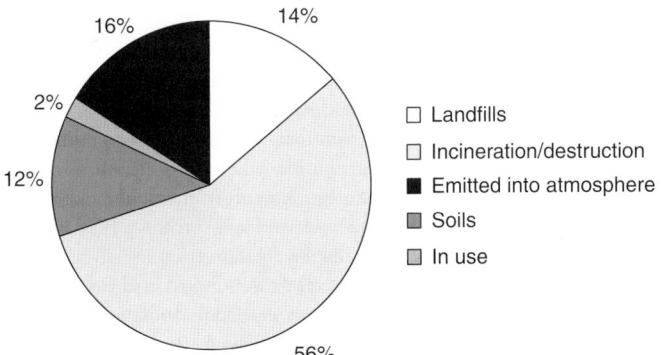

16% 14%
2%
12%
56%

☐ Landfills
☐ Incineration/destruction
■ Emitted into atmosphere
▨ Soils
▨ In use

**Figure 2** *Global budget of produced PCBs.*

mammals is less important. On the contrary, its value is enhanced by cooperation with existing studies and as a context for the development of new programs.

## See Also the Following Articles

Habitat Pressures ■ Health ■ Mass Mortalities

## References

AMBIO (1992). Seals and seal protection: population studies, pathological studies, chemical analysis, experimental studies, comparative studies. *Ambio* **21**, 494–606.

Aguilar, A., and Borrell, A. (1994). Abnormally high polychlorinated biphenyl levels in striped dolphins (*Stenella coeruleoalba*) affected by the 1990–1992 Mediterranean epizootic. *Sci. Total Environ.* **154**, 237–247.

Aguilar, A., Borrell, A., and Pastor, T. (1999). Biological factors affecting variability of persistent pollutant levels in cetaceans. *J. Cetacean Res. Manage.* **1**(Special Issue), 83–116.

Aguilar, A., Borrell, A., and Reijnders, P. J. H. (2002). Geographical and temporal variation in levels of organochlorine contaminants in marine mammals. *Mar. Environ. Res.* **53**, 425–452.

Boon, J. P., *et al.* (9 authors) (1992). The toxicokinetics of PCBs in marine mammals with special reference to possible interactions of individual congeners with the cytochrome P450-dependent mono-oxygenase system—an overview. *In* "Persistent Pollutants in Marine Ecosystems" (C. H. Walker, and D. R. Livingstone, eds), pp. 119–159. Pergamon Press, Oxford.

Borrell, A., and Reijnders, P. (1999). Summary of temporal trends in pollutant levels observed in marine mammals. *J. Cetacean Res. Manage* **1**(Special Issue), 149–157.

Borrell, A., Aguilar, A., Zeljkovic, S., Brouwer, A., Besselink, H. T., and Reijnders, P. J. H. (2007). Post-mortem stability of blubber DLCs, PCB and tDDT in bycaught harbour porpoises (*Phocoena phocoena*). *Mar. Pollut. Bull.* **54**, 1663–1672.

Braune, B. M., *et al.* (14 authors) (2005). Persistent organic pollutants and mercury in marine biota of the Canadian Arctic: an overview of spatial and temporal trends. *Sci. Total Environ.* **351–352**, 4–56.

Breivik, K., Sweetman, A., Pacyna, J. M., and Jones, K. C. (2007). Towards a global historical emission inventory for selected PCB congeners: a mass balance approach-3. An update. *Sci. Total Environ.* **377**, 296–307.

Delong, R. L., Gilmartin, W. G., and Simpson, J. G. (1973). Premature births in California sea lions: association with high organochlorine pollutant residue levels. *Science* **181**, 1168–1170.

Hammond, J. A., Hall, A. J., and Dyrynda, E. A. (2005). Comparison of polychlorinated biphenyl (PCB) induced effects on innate immune functions in harbour and grey seals. *Aquat. Toxicol.* **74**, 126–138.

Hall, A. H., *et al.* (8 authors) (2006). An individual based model framework to assess the population consequences of polychlorinated biphenyl exposure in bottlenose dolphins. *Environ. Health Perspect.* **114**(Suppl. 1), 60–64.

Jepson, P. D., Bennet, P. M., Deaville, R., Allchin, C. R., Baker, J. R., and Law, R. J. (2005). Relationships between polychlorinated biphenyls and health status in harbour porpoises (*Phocoena phocoena*) stranded in the United Kingdom. *Sci. Total Environ.* **243**, 339–348.

Krahn, M. M., Ylitalo, G. M., Stein, J. E., Aguilar, A., and Borrell, A. (2003). Organochlorine contaminants in cetaceans: how to facilitate interpretation and avoid errors when comparing datasets. *J. Cetacean Res. Manage.* **5**, 103–113.

Martineau, D., de Guise, S., Fournier, M., Shugart, L., Girard, C., Lagacé, A., and Béland, P. (1994). Pathology and toxicology of beluga whales from the St. Lawrence Estuary, Québec, Canada. Past, present and future. *Sci. Total Environ.* **154**, 201–215.

Muir, D., *et al.* (11 authors) (1999). Spatial and temporal trends and effects of contaminants in the Canadian Arctic marine ecosystem: a review. *Sci. Total Environ.* **230**, 83–144.

Olsson, M., Karlsson, B., and Ahnland, E. (1994). Diseases and environmental contaminants in seals from the Baltic and the Swedish west coast. *Sci. Total Environ.* **154**, 217–227.

O'Shea, T. J. (1999). Environmental contaminants and marine mammals. *In* "Biology of Marine Mammals" (J. E. Reynolds, III, and S. A. Rommel, eds), pp. 485–564. Smithsonian Institution Press, Washington, DC.

O'Shea, T. J., Reeves, R. R., and Kirk Long, A. (eds.). (1999). Marine mammals and persistent ocean contaminants. Proceedings of the Marine Mammal Commission Workshop, Keystone, Colorado, 12–15 October 1998. Marine Mammal Commission, Bethesda.

Reijnders, P. J. H. (1986). Reproductive failure in harbour seals feeding on fish from polluted coastal waters. *Nature* **324**, 456–457.

Reijnders, P. J. H. (1994). Toxicokinetics of chlorobiphenyls and associated physiological responses in marine mammals, with particular reference to their potential for ecotoxicological risk assessment. *Sci. Total Environ.* **154**, 229–236.

Reijnders, P. J. H., Aguilar, A., and Donovan, G. P. (1999). Chemical pollutants and cetaceans. *J. Cetacean Res. Manage.* **1**(Special Issue), 1–273.

Reijnders, P. J. H., *et al.* (8 authors) (2007). Report from Pollution 2000 + : Phase I. *J. Cetacean Res. Manage.* **9**(Suppl.), 261–274.

Swart, R. D., *et al.* (9 authors) (1994). Impairment of immunological functions in harbour seals (*Phoca vitulina*) feeding on fish from polluted coastal waters. *Ambio* **23**, 155–159.

Tanabe, S., Iwata, H., and Tatsukawa, R. (1994). Global contamination by persistent organochlorines and their ecotoxicological impact on marine mammals. *Sci. Total Environ.* **154**, 263–264.

Tornero, V., Borrell, A., Aguilar, A., Wells, R., Forcada, J., Rowles, T. K., and Reijnders, P. J. H. (2005a). Effect of organochlorine pollutants and individual biological traits on retinoid concentrations in bottlenose dolphins (*Tursiops truncatus*). *J. Environ. Monit.* **7**, 109–114.

Tornero, V., Borrell, A., Pubill, E., Koopman, H., Read, A., Reijnders, P. J. H., and Aguilar, A. (2005b). Post-mortem stability of blubber retinoids in by-caught harbour porpoises (*Phocoena phocoena*): implications for biomarker design studies. *J. Cetacean Res. Manage.* **7**, 147–152.

Wilson, J. Y., *et al.* (8 authors) (2007). Correlates of cytochrome P450 1A1 (CYP1A1) expression in bottlenose dolphin (*Tursiops truncatus*) integument biopsies. *Toxicol. Sci.* **97**, 111–119.

# Popular Culture and Literature

### Paul H. Forestell

## I. Introduction

A review of current human behavior and popular media highlights a contradiction in modern cultural perceptions of marine mammals. On the one hand, it would seem by almost any indicator that human interest in marine mammals is at a fever pitch (Fig. 1). Excursions to view marine mammals in the wild draw countless participants from many cultures and age groups (Hoyt, 2001). Attendance at marine parks and display facilities is at an all time high. The past two decades has seen an explosion in books and videos about marine mammals. The sale of marine mammal "themed" merchandise includes thousands of products worldwide. Campaigns to protect marine mammals and their habitat exist on every continent, and the number of local and national governments involved in legislative

**Figure 1**    *Human interest in marine mammals: Part of the solution or part of the problem? Photo courtesy of Paul Forestell.*

agendas related to marine mammals is growing (Carlson, 2004). In the past 20 years, there has been a strong increase in the number of marine mammal research projects, many involving direct public participation, and the population of marine mammal researchers is growing steadily.

On the other hand, there are at least five species of marine mammal [Chinese river dolphin, or Baiji (*Lipotes vexillifer*), Gulf of California harbor porpoise (*Phocoena sinus*), North Atlantic right whale (*Eubalaena glacialis*), North Pacific right whale (*E. japonica*), and Mediterranean monk seal (*Monachus monachus*)] in danger of becoming extinct because of human activities. Indeed, the Baiji has been deemed to be extinct as of this writing. Attempts to interact with dolphins in the wild may disturb the dolphins and endanger humans (Marsh *et al.*, 2003). Efforts to eliminate the killing of marine mammal fisheries-related by catch are not succeeding. Sealing continues, and efforts to cull seal populations to protect fisheries are on the rise. Commercial whaling (under the guise of "scientific" whaling) has increased dramatically over the past two decades. The oceans are being poisoned: 3 million pounds of trash were cleaned from beaches in the United States in 1999, and that amount must be assumed to represent only a fraction of material floating in the world's oceans.

Lavigne *et al.* (1999) studied North American attitudes toward marine mammals since the early 1960s by reviewing patterns of consumptive exploitation, tourism, media attention, legislation, nongovernmental organization initiatives, and scientific research. They summarized their findings by noting that "With few exceptions… North American attitudes toward marine mammals have shifted from a focus on their killing and material utilization to a more aesthetic interest in observing these creatures in the wild, in captivity, and in various media forms" (Lavigne *et al.*, 1999). The most notable exception to the shift in focus is a general willingness to grant special status to aboriginal groups to hunt marine mammals. The authors further concluded that the change in attitudes has resulted primarily from a rapid urbanization of the human population, increased knowledge of marine mammals through new research and extensive media exposure, and a shift in ethical values.

It seems unarguable that North American attitudes (and those of many nations referred to as "industrialized") have changed from a focus on killing. However, a qualitative analysis of available behavioral evidence (i.e., what people actually do, rather than what they express as a value or attitude in surveys) suggests that humans everywhere still basically regard marine mammals as a commodity. A consideration of how marine mammals are represented in current literature and the popular media; the types of marine mammal issues that win public attention; the types of activities that humans engage in around marine mammals; and the role that government plays in managing such activities, suggest that rather than a shift from material utilization to aesthetic interest, the shift has simply been from one sort of utilization (or exploitation) to another. In brief, while the general trend has been to an increasing effort to protect and conserve, the motivational basis for the change remains highly anthropocentric.

## II. Background: Shaping Cultural Perceptions Prior to the 1950s

The recorded history of human interaction with marine mammals began with small-scale direct exploitation in pursuit of vital resources. From opportunistic scavenging of stranded animals to organized hunting of locally available species, subsistence efforts were focused on obtaining food and by-products needed by the hunters themselves for survival. Subsistence hunting has extended, relatively unchanged, over a considerable period of time in human history. Norwegian petroglyphs (rock drawings) believed to be nearly 4000 years old, show men in boats along with seals, dolphins, and whales. Handheld harpoons, nets, and stranding drives have been employed by indigenous hunters in many parts of the world, including Japan, North America, the South Pacific, and Europe, even into modern times (Fig. 2).

Subsistence hunting of marine mammals until the twelfth century primarily involved hunting for the immediate needs of the community; for food, clothing, and other products of domestic utility. In almost all cases, the traditions of subsistence hunters incorporate

**P**

Figure 2   *Subsistence hunting was conducted for centuries with essentially unchanged techniques. Photo courtesy of the estate of Bill Dawbin.*

The Right Family

Figure 3   *By the middle of the twentieth century whales had become an oddity rather than a direct part of one's survival. Drawing ©Larry Foster, used by permission.*

myths, chants, and rituals honoring the quarry and the skills needed to overcome it. This is an important consideration in understanding current cultural perspectives on marine mammals: a representation of the development of current attitudes as a linear evolution from utilitarian to aesthetic (as presented by Lavigne *et al.*, 1999) is incomplete on two counts. Ancient subsistence attitudes were far more complex than the utilitarian characterization would imply, and current perspectives are far less aesthetic than use of that term would prescribe. In fact, current perspectives on marine mammals probably represent a relatively recent reoccurrence of the same devolution of attitudes from holistic to exploitative that occurred during the transition from subsistence whaling, to commercial whaling, to industrialized whaling. In the present day there has been a parallel transition from "subsistence whale watching," to "commercial whale watching," to "industrialized whale watching" (Forestell and Kaufman, 1996).

The Basques of Spain apparently developed and implemented the first truly "commercial" hunt during the 11th and 12th centuries, initiating a fundamental change in attitude that would sow the seeds of "industrialized" whaling some 600 years later. The Basques probably learned whaling techniques from the Normans, Norwegian settlers who populated the Bay of Biscay during the ninth century. The Basques, tenacious settlers on a critical trade route to the Iberian Peninsula (and onward to Africa), were not subsistence whalers. Originally pastoralists who tended flocks in the mountains, they eventually founded the kingdom of Navarre, which included a number of tiny fishing villages along the Biscayan coast. There they found the North Atlantic right whale, which migrated into the Bay of Biscay each year from autumn through spring. Poised between a vast, untapped resource on the ocean side and a steady flow of traders to the world's centers of commerce on the other, they found a ready market for almost the entire carcass of the whales they brought ashore: oil for lamps and lubricants; baleen for corsets, buggy whips, fishing poles, and brushes; bones for fence posts and portals; skin for footwear; and meat, blubber, and tongue for food. For 300 years Basque whalers pursued the right whale; first across the Bay of Biscay and then to the far side of the Atlantic. So good were they

at their craft that, as the need for whale oil grew in Europe, they became teachers of those who would take commercial whaling to the furthest reaches of the world's oceans.

Preindustrial commercial whaling involved the application of subsistence techniques for the realization of profit rather than direct consumables. As the theater of endeavor moved from the Biscayan coast to North America, Spitsbergen, and the open seas of the Atlantic and Pacific, the method of catching and killing whales changed little, although the processing of whales was carried out variously either onshore or onboard the whaling ship itself. Without benefit of modern machinery, commercial whalers using sailing ships, whale boats, and handheld harpoons successfully pursued right whales, bowhead whales (*Balaena mysticetus*), and gray whales (*Eschrichtius robustus*) to near extinction, and decimated known sperm whale (*Physeter macrocephalus*) populations. A similar fate was visited upon elephant seals (*Mirounga* spp.), walruses (*Odobenus rosmarus*), and sea otters (*Enhydra lutris*), while the Steller sea cow (*Hydrodamalis gigas*) was completely extirpated.

An important change that accompanied commercial hunting was that animals were hunted and killed by humans who were not otherwise connected to the ecosystem or the animals targeted by the enterprise. Strangers came from faraway places for the sole purpose of obtaining as much of the resource as could be most quickly taken away for sale in distant markets. By the middle of the nineteenth century, the cost of whale oil climbed as stocks of whales declined. Alternative products derived from turpentine, cottonseed, and linseed replaced whale oil in a number of uses. And then distillation of petroleum hastened the decline of commercial whaling as practiced by the sailing ships of the American whaling fleets by the end of the nineteenth century.

Industrialized whaling began in 1863 when Svend Føyn, a Norwegian sealer, made his first successful kill of a blue whale (*Balaenoptera musculus*) using a cannon mounted on the bow of a 90-foot steamship. The cannon fired a 100-lb. harpoon fitted with an exploding tip. The combination of cannon, exploding harpoon, and steamship meant that large fast whales could be dispatched with relative speed and efficiency. Readily accessible populations of blue, fin (*B. physalus*), and humpback whales (*Megaptera novaeangliae*), unprofitable targets for sailing ships, gave new life to commercial whaling. For 40 years Norwegian-style whaling spread around the

**Figure 4** *By the 1950s commercial whaling was a highly sophisticated and industrialized operation. Photo courtesy of The New Bedford Whaling Museum.*

world. At the turn of the twentieth century, processes were developed to rid whale oil of its inherently strong taste and odor, and convert it into solid fat. Whale oil once again became a commodity of worldwide importance, particularly for making margarine and soap.

Rapid developments in technology soon made it possible to hunt and kill whales more efficiently and in greater numbers. The hand-held harpoon was replaced with Svend Føyn's cannon, and try-pots gave way to huge pressure cookers. Gigantic hydraulic tail-grabbers dragged entire blue whales out of the ocean, through stern slipways, and onto the flensing decks of pelagic factory ships. Spotter planes, catcher boats, and factory ships coordinated their efforts by radio to increase the extent and efficiency of the catch (Fig. 3). In 1950, a 65-foot whale weighing 80 tons could be hauled up the slipway and turned into oil, frozen meat, and ground-up fertilizer in just over 1 h. Nearly every major nation of the world participated in the decimation of whale stocks—some 30,000–50,000 whales per year were taken in the Antarctic alone between 1930 and 1960.

By the 1950s it could well be said that the human attitude toward marine mammals had become a strictly utilitarian one (Fig. 4). The industrialization of slaughter occurred on a grand scale with regard to whales. At the same time, commercial hunting of elephant seals for oil and sea otters, fur seals, and harp seals (*Pagophilus groenlandicus*) for their coats, had disastrous results on pinniped populations throughout the world and sea otter populations throughout their range. The hunt for sirenians never progressed to anything near an industrialized effort. A combination of subsistence and small-scale commercial hunting has generally been sufficient to reduce known populations to extinction (in the case of Steller's sea cow) or near extinction [in the case of manatees (*Trichechus* spp.) and dugongs

(*Dugong dugon*)]. Hunting of small cetaceans throughout the world has also been marked by the shortsighted exploitation of pilot whales (*Globicephala* spp.), killer whales (*Orcinus orca*), bottlenose whales (*Hyperoodon* spp.), spinner dolphins (*Stenella longirostris*), spotted dolphins [pantropical (*S. attenuata*) and Atlantic (*S. frontalis*)], and common dolphins (*Delphinus* spp.).

What is unique to the history of human interaction with dolphins, compared with our exploitation of other marine mammals, is the massive destruction of pelagic species associated with modern driftnets and purse seining. Industrialized fishing techniques have killed more dolphins in the last 50 years than were killed in the entire prior history of humankind. The late David Gaskin (1982) succinctly described the attitude underlying the industrialization of whaling and fishing as an ethical failure:

> We could have conserved many other resources than just the stocks of large baleen whales, if the dead-end philosophy of the economic industrial growth ethic could be circumvented, and if the aces in the deck of cards with which we play were not invariably short-term economic gain and political expediency. (p. 387)

Little information about marine mammals was available to the public prior to the early nineteenth century. Dewhurst (1834) provides a fairly comprehensive review of scientific information on cetaceans up to that point. He cites 28 writers, beginning with Aristotle (third century BC) and Pliny (first century AD), and gives special praise to the work of Rondèlet (fifteenth century), Sibbald (seventeenth century), La Cépède, and Scoresby (both nineteenth century). A more complete review can be found in Matthews (1978),

P

who incidentally refers to Dewhurst as "…a self-advertising medical quack" (p. 18). Virtually none of the material described by either Dewhurst or Matthews was readily available for public consumption. In 1821 Sir Walter Scott's three volume novel "The Pirate" initiated popular fictional accounts of marine-mammal related literature—followed by James Fenimore Cooper's story "The Pilot" in 1823 and Melville's "Moby Dick" in 1851. The ensuing popular literature of the nineteenth and early twentieth centuries was more about whalers and whaling than it was about whales. Between 1800 and 1950 there were close to 100 published texts focusing on marine mammals. With limited exceptions, they all dealt with popularized accounts of whaling.

## III. The 1950s: Out with the Old

Beginning in the 1950s, there was growing scientific evidence of the complexity of marine mammal behavior. Stimulated by the work of John Lilly in the United States, researchers in other American laboratories, Europe, and Japan conducted neuroanatomical and electrophysiological studies of cetacean brains. This work identified a number of unique characteristics in cortical architecture, structural morphology, degree of lateralization, and brain wave patterns. Philosopher, anthropologist, and ecologist Gregory Bateson explored the intricacies of dolphin social behavior by studying a group of spinner and pantropical spotted dolphins at Sea Life Park in Hawaii. William Schevill and colleagues (including his wife Barbara Lawrence) began collecting acoustic recordings that would eventually document the underwater sounds of nearly three dozen species of marine mammals, including odontocetes, mysticetes, and pinnipeds. David and Melba Caldwell explored the acoustic and social behaviors of whales and dolphins in captivity and the wild.

These researchers heralded an important shift in scientific focus that accompanied the end of the grand age of whaling during the 1950s and the beginnings of "modern" marine mammal studies in the 1960s. Earlier studies of marine mammals had been carried out, for the most part, on the flensing decks of whaling ships, on the slipways of coastal whaling stations, or on the rocky shores of pinniped haul-out and rookery sites. Information gleaned from such studies was applied to quantitative determinations of the size of harvestable stocks and the development of new commercial applications for whale products. Robertson's (1954) account of life aboard an Antarctic factory whaling ship describes such an approach. In one chapter, the author notes some of the questions entertained by the chemist/biologist on board the ship:

> Why has no female sperm whale ever been sighted or killed in the Southern Ocean, though many thousands of old males are killed there every year… At what rate would whale tendon from the flukes be absorbed by the human body if it were put on the market in huge quantities of twenty-foot strands to replace the rare and expensive kangaroo tendon presently used as surgical sutures to repair hernias… What is the nutritive value of a properly cooked fin whale steak… How could one make the whale—as it should be—the world's main supply of the valuable new pituitary and adrenal hormones? (pp. 240–241)

Interest in the management of a commercial harvest prompted a group of American biologists to investigate the life history of gray whales by studying dead ones during the 1960s. From 1959 to 1969, special permits were issued to the Del Monte and Golden Gate Fishing Companies in Richmond, California, by the federally run Marine Mammal Biological Laboratory in Seattle, Washington. The permits allowed the collection of 316 gray whales as they traveled along the west coast of the United States during their annual north–south migration between Mexico and the Bering, Chuckchi, and western Beaufort seas. The description of the whales' life history was based on shipboard and aerial observations of their migration numbers and timing, and documentation of a series of body measurements from the 316 whale carcasses. Data included overall length and weight; width of fluke and length of flippers; thickness of blubber; number and size of baleen plates; degree of vertebral fusion; condition of mammary glands, size of ovaries, and diameter of uterine horns in females; length of penis and weight of testes in males; number of wax ear plug growth layers; number and type of external body scars; number and type of endo- and ectoparasites; and stomach contents. Data were used to determine numbers and migratory patterns of different age/sex classes; prey species; reproductive patterns; and susceptibility to parasites and predators (other than humans). Rice and Wolman (1971) reported the results of the study. The expressed purpose of the work was to provide sound biological knowledge to ensure a wise management program if commercial exploitation, ended by international agreement in 1946, should resume.

Compare the questions of Robertson's chemist/biologist and the Marine Mammal Biological Laboratory's biologists with those posed by the legendary Kenneth Norris (Fig. 5), who began studying wild dolphins in California and Hawaii during the 1960s. In "The Porpoise Watcher" (1974), Norris describes his early musings about the intricacies of social organization in Hawaiian spinner dolphins:

> … most of the subgroups we see in active schools have no clear function. Are they family groups, play groups, are they divided by sex or age… What is it that regulates porpoise school numbers? Why don't their schools grow and grow until every scrap of natural resource is used, as humans are busy doing? What racial wisdom, somehow lost by humans as they adopted civilization, keeps them in balance? (pp. 199–200)

**Figure 5** *Biologist Ken Norris was a pioneer in the study of marine mammals and their behavior in the wild. Photo courtesy of Bernd Würsig.*

Popular Culture and Literature

Norris' approach to the study of live animals represented a significant change in marine mammal science and ultimately helped reshape public perception of marine mammals in many parts of the world.

## IV. The 1960s and 1970s: In with the New

It would be simplistic to point to one central mechanism behind the groundswell of interest in marine mammals (more specifically, cetaceans) that occurred during the second half of the twentieth century. The exotic undersea portraits and exciting escapades of Jacques Cousteau; the new age writings of John Lilly; the increasing exposure to whales and dolphins in zoos, parks, and oceanaria; the growing awareness of conservation issues due to books by (among others) Aldo Leopold (1949), Rachel Carson (1962), and Victor B. Scheffer (1969); and the "protest" mind-set of the 1960s and 1970s all played pivotal roles. The combined effect of these influences was to generate an entirely new look at marine mammals. Four predominant themes emerged quickly and simultaneously throughout North America, the United Kingdom, Europe, and Australia/New Zealand during the 1960s and 1970s: dolphins (and whales) exemplify cognitive capabilities rivaling human capacity; the opportunity to watch dolphins and whales is exciting and entertaining; the worldwide slaughter of whales is unethical and biologically disastrous; and marine mammals demonstrate a number of adaptations to the marine environment that may have important applications to human technology. These have remained the four corners of the box in which humans continue to contain marine mammals: intelligent, entertaining, endangered, and valuable to humans.

The widespread perception of cetaceans as highly intelligent animals was largely due to the writings of Lilly (Fig. 6), beginning with his book "Man and Dolphin" (Doubleday, 1961). Lilly was not the first scientist to note the possibility of complex cognitive capability in cetaceans—Arthur McBride, the first curator of Marine Studios, and Canadian psychologist Donald Hebb published detailed observations of bottlenose dolphin social behavior that served as a landmark reference for more than 20 years (McBride and Hebb, 1948). Lilly used his position as an acclaimed neurophysiologist with the National Institutes of Health to foster an enduring and widespread public perception of whales and dolphins as near-mythical species

with mental capacities rivaling or surpassing those of *Homo sapiens*. His succession of books for the popular press generated a hostile reaction from the scientific community, which disputed the validity of his claims. Overall, however, his characterization of cetaceans as socially and cognitively complex animals with rich emotional lives guided by a code of ethics resonated with the public. Paraphrasing one of his lectures, Lilly (1967) wrote:

I wish to tell you something of what we have learned of a group of uninhibited nudists who have never worn clothes. They have never walked on their own two feet. They have no property. They cannot write their own names. They have no commerce or stores. They have no radio, no TV. They have no fireplaces, nor furnaces, or any fire at all. They have no atomic or nuclear bombs, or power plants. They have no written or printed records. They have no libraries or paintings. In spite of all these handicaps, they are successful. They have big brains and have readily available food supplies. They have the sense to go south in the winter and to go north in summer. They have the ability to out-think, outmaneuver and fight successfully against their enemies. Finally, they think enough of us to save each of us when they find us in trouble. (p. 291)

Sexually liberated, antimaterialistic, antiwar, self-sufficient, intelligent, and altruistic. What suite of characteristics could have more poignantly caught the attention of an up-and-coming generation of "baby boomers" protesting post-World War II materialism, the Vietnam War, restrictive social mores, and cutthroat international capitalism? Lilly's message is clear: cetaceans live a life of sun, surf, and sex—with big brains, and no guilt. Could it get any better?

Lilly's writings and public lectures generated a backdrop against which a new-age philosophy of dolphin as genius, healer, therapist, and spiritual advisor flourished. Although Lilly's interpretations of his findings were frequently based upon gratuitous explanations, he was instrumental in a number of critical discoveries concerning the peculiarities of dolphin acoustic behavior.

One can hardly discuss the work of the new brand of marine mammal scientist that emerged in the 1960s without noting the importance of marine mammal display facilities, beginning with Marine Studios (later called Marineland, Florida). Records of marine mammals in captivity for public amusement date back to ancient times. By far, pinnipeds comprise the majority of display animals. They are relatively easy to take from the wild, adapt well to captivity, and in many cases are trained easily to amuse human spectators. Cetaceans require a degree of knowledge, resources, and commitments that few display facilities are able to meet. Limited, unsuccessful attempts to maintain marine mammals in captivity occurred in the 1800s. However, beginning in 1938, Marine Studios was the first facility to successfully maintain bottlenose dolphins (*Tursiops truncatus*) in captivity for an extended period, in its role as an "oceanarium" rather than aquarium or zoo. Once Marine Studios established that dolphins could be placed successfully on display, the postwar climate fostered the rapid growth of similar facilities, first in the United States and then throughout the world.

From the beginning, those responsible for cetaceans at oceanaria encouraged outside scientists to explore the biology and behavior of the animals at their facilities. Arthur McBride, and later Forrest Wood at Marine Studios; Kenneth Norris, the first curator at Marineland of the Pacific (the west coast "sister" institution of Marine Studios); his graduate students William Evans at Hubbs-Sea World Research Institute and John Prescott at the New England Aquarium; Tap and Karen Pryor at Sea Life Park in Hawaii; and Murray Newman at the Vancouver Public Aquarium all opened their

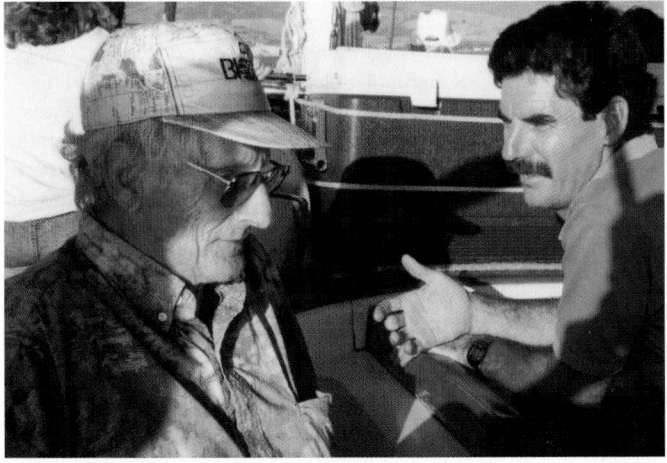

**Figure 6** *Dr. John Lilly, neuro-physiologist (shown here with the author in 1994) had a profound influence on public perception of whales and dolphins. Photo courtesy of Paul Forestell.*

**P**

doors to biologists, neuroscientists, anthropologists, physiologists, psychologists, acoustical engineers, and an ever-widening range of other professionals with an interest in studying live animals. At a time when television and Jacques Cousteau were flooding into American homes, the collaboration between science and the entertainment industry had a tremendous impact on the public perception of marine mammals. Scientists were not used to dealing directly with the public, but those who ran the oceanaria were. As scientists began to discover more and more about the hidden world of whales and dolphins, the oceanaria incorporated much of that knowledge in decisions about what animals to capture, improvements to husbandry programs, and, most importantly, in the subject matter of their shows.

One indication of the "coming of age" of dolphins as a significant cultural phenomenon in the United States was the 1960s production of two movies and a television series based on the adventures of "Flipper," a bottlenose dolphin, and Sandy Ricks, the young son of a fisherman (later changed to a park ranger). The original movie, released in 1963, hinted of the ancient stories of the relationship between a boy and a dolphin, such as described in Iassos, Greece, and the Roman town of Hippo in Africa. In the movie, a dolphin met a young boy and they developed a friendship that led to a bonding that surpassed species distinctions and spawned a sequel. The mythical proportions of the story did not survive long, however, and the subsequent television series seemed to quickly develop into a marine version of "Lassie, Come Home."

While the 1960s initiated a true Renaissance in the portrayal of marine mammals, oceanaria helped emphasize a conceptual "divide" that continues to the present. With limited exceptions, modern public attitudes about marine mammals are strongly predisposed toward whales and dolphins. In reviewing the status of marine mammals in popular literature and culture, it is important to bear the following caveat in mind. The typical representation of "marine mammal" does not reflect the diversity and extent of actual species. Marine mammals include over 120 species, divided into four general categories [whales, dolphins, and porpoises; seals, sea lions, and walruses; manatees and dugongs; sea otters and polar bears (*Ursus maritimus*)]. However, the preponderance of available information on nature excursions, visits to display facilities, and the sales of books, videos, and other merchandise suggest that the prototypical representative of the category "marine mammal" is one of a small set of relatively well-known cetacean species (e.g., bottlenose dolphin, killer whale, humpback whale, sperm whale).

Although occasional interest in other marine mammals has developed with respect to specific issues [e.g., the killing of neonatal harp seals for their white fur coats in Canada; the decimation of Caribbean manatee (*Trichechus manatus latirostris*) populations from boat collisions in Florida; competition between sea otters and abalone (*Haliotis* spp.) fishermen in California], it remains clear that cetaceans have drawn public attention in a way that far outdistances the attention generally paid to pinnipeds, sirenians, sea otters, or polar bears. This difference is despite the fact that about two-thirds of the marine mammals held in captivity since 1973 in North America are either California sea lions (*Zalophus californianus*) or harbor seals (*Phoca vitulina*). The most commonly displayed cetacean is the ubiquitous bottlenose dolphin (Fig. 7), which generally makes up about 20% of the total number of animals on display. Killer whales, although small in number, have generated a huge marketing success for their keepers. Much of the attraction of cetaceans appears to be related to a fascination with the size, beauty, strength, and perceived intelligence of these two most frequently seen examples. While sea lions balance balls on the end of their nose and make hilarious barking sounds on

**Figure 7** *The bottlenose dolphin* (Tursiops truncatus) *is the prototypical marine mammal for most humans. Photo courtesy Pacific Whale Foundation/Paul Forestell.*

cue in oceanaria and marine parks, captive whales and dolphins cooperate in the development of spectacular water shows that train audiences to see them as powerful, beautiful, fun-loving, and intelligent.

An important influence in the perception of marine mammals by the public during the 1970s was a rapid growth in underwater images of whales and dolphins. Although television had introduced the world to the accomplishments of Jacques Cousteau and the crew of the Calypso, their exploits seemed far out of reach for the average individual. Cousteau's books and television programs emphasized the dramatic nature of exploring beneath the surface of the ocean in remote and dangerous settings. Using SCUBA, sophisticated oceanographic equipment, and sheer nerve, Cousteau and his men peeled back the layers of the ocean in a fashion few members of the public thought possible. Until the late 1970s, a sense of the mystique of marine mammals was based on the accounts of explorers, whalers, sealers, and fishers. As late as 1975, an article in Audubon magazine on marine mammals had no photographs, but was illustrated with artwork. National Geographic magazine featured articles on killer whales in 1966, gray whales in 1971, and right whales in 1972 and 1974. The photographs and accompanying stories leave one with the clear impression that such experiences were the prerogative of a highly specialized breed of underwater adventurer/pioneer photographers such as Howard Hall, William Curtsinger, Charles (Chuck) Nicklin, Jen and Des Bartlett, Stan Waterman, and Al Giddings—generally working in the company of field scientists in exotic places.

Then the situation changed, and James Hudnall, a freelance photographer and self-styled whale researcher, helped lead the way. In a 1977 Audubon article, Hudnall wrote of venturing into the winter home of the North Pacific humpback whale offshore of Maui, Hawaii, beginning in 1974. Diving from a small inflatable boat, often by himself, with no more equipment than a mask, snorkel (sometimes a SCUBA tank), fins, and underwater camera, Hudnall revealed the first protracted look at living whales. He described the huge animals as "gentle, clever, passive, and rational beings," and he provided spectacular images to prove it (Hudnall, 1977). By the end of the 1970s the public was sensitized to a perception of marine mammals, especially the great whales, as objects of wonder and mystery; to be held in awe rather than fear; and to be protected rather than hunted. A popular video for National Geographic in 1978 ("The Great Whales"; written, produced, and directed by Nicholas Noxon) represented the changing view rather well. Included in the video was actual footage of modern whaling operations—and it was poignantly contrasted with footage of killer, blue, humpback, and gray whales that mirrored the message in Hudnall's article for Audubon magazine.

Just as Hudnall's photographs softened the public's perception of whales, additional discoveries about the sounds that whales make reinforced the idea that whales and dolphins engage in complex communication behaviors. Sounds made by humpback whales had been documented as early as 1952 by US Navy personnel using deepwater eavesdropping systems developed to detect submarines. In 1971 Roger Payne and Scott McVay reported the first detailed characterization of the sounds in the journal Science. Because of their complex structure and orderly sequence, Payne and McVay (1971) referred to the acoustic displays as songs. Subsequently Payne, his wife Katy, and a number of their students showed that the songs, produced only by males during the breeding season, changed slowly over successive years. In any given season, however, all males within a breeding group appeared to sing the same song. It was shown ultimately that song characteristics could be used to identify separate stocks of whales. The impact of the song on the public was dramatic. Musicians like Paul Winter incorporated the sounds into their recordings. Payne, a creative and charismatic individual, even managed (with Carl Sagan's help) to get a recording of humpback song included in the materials placed aboard the Voyager I spacecraft to provide potential "other intelligences" in the galaxy with artifacts of earth. In January of 1979 a plastic 45-rpm recording of humpback whale song was included as an insert accompanying an article in National Geographic on Payne's work. The hauntingly beautiful sounds of the humpback whale constituted one more piece of evidence that marine mammals demonstrate a high level of cognitive capability.

Lavigne et al. (1999) noted that during the 1970s the rate of marine mammal scientific publications increased from fewer than 10 titles per year to more than 20. As suggested earlier, there was also change in the nature of books published. Mowat's "A Whale for the Killing" (Atlantic Monthly Press, 1972) emphasized the clash between old values and new sensitivities in his fictionalized account of a true incident in Newfoundland, in which locals used a stranded fin whale for target practice. The growing perception that cetaceans are uniquely worthy of our attention because of their intelligence is evident in "Mind in the Waters," edited by McIntyre (Scribners, 1974). The book is an attractive compendium of art, editorial, history, science, and philosophy regarding human association with marine mammals. Scientist Ken Norris (1974) wrote eloquently about the day-to-day lives of dolphins and whales. One of the first "inside" views of a display facility occurred in "Lads Before the Wind" (Harper and Row, 1975) by Karen Pryor, head trainer and cofounder of Sea Life Park in Hawaii. Other books for the popular press included resource texts such as Mathews' "The Natural History of the Whale" (Columbia University Press, 1978) and the fictionalized natural history account of "The Last Blue Whale" (Harper and Row, 1979) by Smith. Fictional writings with marine mammals as central characters were few and largely forgettable.

Television productions featuring marine mammals were few in number, and the most notable movie representation was "The Day of the Dolphin," starring George C. Scott and directed by Mike Nichols (Avco Embassy Pictures, 1973). This box office hit helped solidify public attitudes about the intelligence of dolphins and raised questions about the training of dolphins by the military for a variety of warfare applications.

One of the effects of the exposure to whales and dolphins in oceanaria and in limited television coverage during the 1970s was the realization among the public that at least some species of marine

**Figure 8** *The Save-the-Whales movement was a mediagenic and culturally diverse phenomenon of the 1980s. Photo courtesy of John Perry.*

mammals were very accessible in the wild. Whale watching became a rapidly growing industry. Whale watching first began in California and Mexico in 1955 (gray whales). In the 1970s it developed off the New England coast [humpback, fin, minke (*B. acutorostrata*), and right whales] and in Hawaii (humpback whales). It soon spread to other areas in North America [e.g., killer whale and gray whale watching in the Pacific Northwest; humpbacks in Alaska; blue whales and beluga whales (*Delphinapterus leucas*) in the St. Lawrence River, Canada].

The increased display of marine mammals in oceanaria and the growing opportunity to view them in their own world accelerated the idea that industrialized whaling had become a desecration of nature. The insistence of some nations on continuing to hunt in the face of collapsing populations of whales was viewed by many as an outrage, and efforts to thwart commercial whaling grew increasingly strident. A broad, international coalition of objectors joined forces in launching what later was termed "The Whale War" (Day, 1987). The war was fought on many fronts, and the participants acted more often in independent assaults, but the overall effect was to galvanize the public into a growing chant of "save the whales" (Fig. 8).

The "save the whales" movement was one of the most mediagenic and culturally diverse platforms for environmental activism during the twentieth century. Confrontation between catcher boats and rubber inflatables on the high seas provided a dramatic backdrop to land-based efforts to convince whaling nations to cease the hunt. Saving whales constituted a natural complement to the changing perception of marine mammals generated by oceanaria, whale watching excursions, and media attention.

It was also during the 1970s that the United States incorporated changing public opinion in revised federal legislation. Both the Marine Mammal Protection Act of 1972 and the Endangered Species Act of 1973 became international models of marine mammal protection by

P

Figure 9 *Whale watching grew into a worldwide industry worth hundreds of millions of dollars a year during the 1980s and 1990s. Photo courtesy of Pacific Whale Foundation/Greg Kaufman.*

affording broad powers to federal agencies to prevent not only the killing of marine mammals, but even their disturbance or harassment. These acts were broadly developed to ensure protection of species considered to be of special interest to the public, in part through ecosystem protection, but in general as single species efforts. Both acts provided wide latitude for research and international cooperation in fulfilling conservation agendas. Although revised many times since the original enactment, both these legislative initiatives have received considerable attention around the world as mechanisms for increasing marine mammal protection.

## V. The 1980s and 1990s: Loving Marine Mammals Becomes an Industry

The most significant demonstration of the cultural importance of marine mammals in modern times is the explosion of interest in excursions to view marine mammals in the wild that has taken place over the past three decades (Fig. 9). The growth of the whale and dolphin watching industry has been detailed in a number of places, but the sheer speed and dimension of the changes that took place during the 1980s and 1990s warrant mention here. A detailed report on worldwide trends in whale watching by Erich Hoyt (2001) documents the dramatic growth of whale and dolphin watching from the 1950s. The numbers of people, venues, operators, and targeted species increased exponentially. The number of countries with marine mammal excursions has tripled since 1990 alone, while the number of passengers has more than doubled to some 10 million people each year. Direct revenues from marine mammal watching excursions now exceed 1 billion US dollars a year. If one considers the "multiplier" effect of the other expenditures associated with whale watching (travel, accommodations, food, and souvenirs), the economic impact must be seen as quite substantial.

The media focus on whale and dolphin research, the "taming" of cetaceans at oceanaria, and public access to marine mammals through commercial whale watching trips during the 1980s were accompanied by a unique phenomenon: the use of laypeople as marine mammal research assistants in return for financial support of the research project. One of the first instances of such a program took place aboard the R/V Regina Maris, a three-masted schooner that operated out of Gloucester, Massachusetts, under the direction

of Dr. George Nichols and the nonprofit Ocean Research Expedition Society. This was soon followed by the Center for Field Research in Massachusetts (Earthwatch); the School for Field Studies, also in Massachusetts; the Pacific Whale Foundation in Hawaii; Oceanic Society Expeditions in California; and ultimately a host of organizations throughout the world. Such efforts allowed members of the general public, without experience in conducting whale or dolphin research, the opportunity to participate in field studies of marine mammals. The benefit to the researchers was the financial support of their field work provided by the participants.

Throughout the 1980s, the "discovery" of marine mammals as objects of wonder allowed those with artistic and entrepreneurial abilities the opportunity to develop a wide variety of commercial ventures. Availability of marine mammal books, magazine articles, visual art works, decorated clothing, jewelry, videos, souvenirs and trinkets, marine expeditions, and captive displays increased dramatically. Artistic visual representations of marine mammals benefited enormously from the artists' exposure to live animals (or, perhaps, photographs of live animals). At least two major styles of visual art emerged during the 1980s: one incorporated attention to detail based on biological and morphological data and the second attempted to represent the artist's interpretation of the animal in its world. Prominent examples of attention to detail include the paintings of Larry Foster (Fig. 4), Pieter Folkens, Janet Biondi (all of California), and Richard Ellis (New York). Realism also characterized the work of sculptors John Perry, Randy Puckett, and Steve Aikenhead. More fanciful interpretations of the whale in its world were represented prominently in the works of Robert Lyn Nelson (Fig. 10), Wyland, and Christian Riese Lassen, who reached their pinnacle of commercial success from the island of Maui in Hawaii during the 1990s.

The marketing of marine mammal images grew through the 1980s and 1990s into a multimillion dollar enterprise, driven largely by art galleries catering to wealthy tourists in places like Lahaina Hawaii, Monterey, California, and Provincetown, Massachusetts. New techniques were developed to produce prints, posters, lithographs, and other facsimiles of original works that made art available to everyone. Additional products include T-shirts, coffee mugs, note cards, coffee table books, shopping bags, and stickers. Media attention to public relations extravaganzas such as the air brushing of "Whaling Walls" by Wyland in high-visibility settings in major cities throughout the world helped generate a public perception of marine mammals as beautiful, powerful, friendly, and intelligent. Jewelry, sculptures, and a wide variety of trinkets and other memorabilia were produced in mass quantities for the increasing population of whale watchers and marine mammal lovers with money to spend. A major outlet for such products became the retail centers associated with display facilities. Virtually every aquarium or oceanarium in the world has a retail center strategically located so visitors must move past or through it on the way to the exit. Artifacts related to marine mammals constitute a huge source of related income.

As the market for watching marine mammals in the wild and the purchase of memorabilia grew, the increase in publication of books and videos continued. It is possible to identify three major trends in modern published materials for adult readers: new-age, science fiction, and naturalist accounts. The dolphin as a mystic, healer, ancient communicator shows up in books by Horace Dobbs (British), Patricia St. John (American), Olivia de Bergerac (French), Frank Robinson (New Zealand), and a host of like-minded writers. The popularity of new-age musings on the significance of marine mammals is based on the use of personal transformation stories that foster an emotional response on the part of the reader. In most cases, the new-age perspective on marine mammals emphasizes what they

**Figure 10**  *Marine artist Robert Lyn Nelson is a preeminent figure in the marine art world. Image ©Robert Lyn Nelson, used by permission.*

(most often cetaceans) can do for humans: as teachers, healers, models of a better social order, and indicators of human mistreatment of the oceans.

A second type of modern adult literature provides science-fiction accounts of marine mammals. These works incorporate extensions of scientific studies of communication and other cognitive capabilities of marine mammals into stories based on themes such as military abuse of power, destruction of the environment by corporate greed, mistreatment of marine mammals by egotistical scientists, or cataclysmic events of nature. "Into the Deep" by Grimwood (William Morrow and Co., New York, 1995) weaves a complex plot that has dolphins communicating with humans to warn of impending natural disaster; representatives of each species eventually combine talents to use telepathic communication to prevent massive destruction of both humans and dolphins. "The Secret Oceans" by Ballentine (Bantam Books, New York, 1994) tells of a group of scientists who become captive subjects in a study of humans conducted deep in the ocean by a form of cetacean more intelligent than humans. "Sounding" by Searls (Ballantine Books, New York, 1982) uses two main characters (an aged bull sperm whale and a sonar officer on a Russian submarine) to explore cetacean behavior and the possibilities of interspecies communication.

The third, and largest, category of books includes nonfiction, naturalist accounts of a wide variety of species, including books on all the orders of marine mammals. These books bring together interpretations of scientific findings, stunning art or photography, and personal accounts of fieldwork by accomplished researchers. Two of the earliest are the "Book of Whales" (Alfred A. Knopf, 1980) and "Dolphins and Porpoises" (Alfred A. Knopf, 1982), both by naturalist, painter, and writer Richard Ellis. In "Hawaii's Hump-back Whales" by Kaufman and Forestell (Island Heritage Press, 1986) the authors describe the biology and behavior of humpback whales, based largely on their observations and field studies in Hawaii and Australia. The writing style is meant to model the authors' experiences narrating whale watches and provides a lively, but science-based interpretation of the life of one species. A photographic example of equivalent style is the Japanese photographer Mitsuagi Iwago's "Whales" (Chronicle Books, San Francisco, 1994), which documents his year-long journey to observe humpback whales in Hawaii, Alaska, and the Ogasawara Islands of Japan. Attractive, informative, and entertaining treatments of many species of marine mammals are now available. The National Geographic production of "Whales, Dolphins and Porpoises" (National Geographic Society, Washington, DC, 1995) featuring

**P**

photographs by Flip Nicklin (son of pioneering underwater photographer Chuck Nicklin) and written by several prominent scientists (Darling *et al.*, 1995) is an unsurpassed synthesis of prominent research findings from around the world. Bruemmer's "Seals in the Wild" (Laurel Glen Publishing, San Diego, 1998) offers an elegant collection of facts and photographs about pinnipeds. Sleeper and Foott team up to present an excellent description of the much overlooked manatee in their book "In the Company of Manatees" (Three Rivers Press, New York, 2000). Ken Norris, Roger Payne, Hal Whitehead, Peter Beamish, Jim Darling, and Carol Howard have written intriguing personal accounts of careers spent investigating marine mammals. A number of field guides for whale watchers in many settings throughout the world are also currently available. These books not only provide a great deal of current information on marine mammals, but serve as important vehicles for conservation efforts.

Certainly not all books published during the 1980s and 1990s can be categorized according to the three styles just described. Two books that deserve mention in this regard are Heathcote Williams' "Whale Nation" (Harmony Books, New York, 1988) and "The Delicate Art of Whale Watching" (Sierra Club Books, San Francisco, 1982) by Joana Varawa (formerly Joan McIntyre). "Whale Nation" is a celebration of the history of whale/human interactions through a book-length poem and a collection of whale-related excerpts from other sources. Following its publication, the author did a world tour of highly acclaimed public readings. Varawa's book is a deceptively small but powerful series of meditations on being in the presence of the sea and its creatures.

More than 300 videos and movies on marine mammals have been produced in the last two decades. Most of these are documentaries, generally based on describing the research of one or a few scientists. In some cases, researchers are simply followed in their work while the significance and findings of their efforts are detailed. In other cases, celebrities such as Robin Williams or Christopher Reeves accompany scientists to get a first-hand look at their work. A number of television programs have dealt with controversial issues such as dolphins in captivity ("A Whale of a Business," PBS Frontline, 1997); scientific whaling ("Whale Fever," BBC News and Current Affairs, 1993); captive-dolphin swim programs ("Dying to Please," Biosphere Films, 1990); and bycatch of dolphins by the tuna fishery ("If Dolphins Could Talk," PBS Video, 1990). The invention of a large-format (70-mm) camera by IMAX in the early 1970s has resulted in three big-screen movies that are based on marine mammals. The first was "Nomads of the Deep," featuring humpback whales, released in 1979 by IMAX. In 1996, Destination Cinema produced "Whales," a popular big-screen documentary covering a variety of whale species. In 2000, McGillivray Freeman Films released "Dolphins," a documentary on dolphin communication and behavior that featured the work of Bernd Würsig's students Kathleen Dudzinski and Alejandro Acevedo-Gutiérrez, as well as established researchers Louis Herman and others. It and the artistic and science teams were nominated for an Academy Award.

There have been a handful of feature movies with marine mammals as central characters. A bottlenose dolphin named Snowflake appeared in the comedy "Ace Ventura, Pet Detective" (starring Jim Carrey, 1994). Humpback whales were brought back to the future in the science fiction adventure "Star Trek IV: The Voyage Home" (William Shatner and Leonard Nimoy, 1987). A dog and a dolphin learned to communicate with each other and sparked a romance between the dog's owner and the scientist studying the dolphin in "Zeus and Roxanne" (Steven Guttenberg and Kathleen Quinlan, 1997). "The Secret of Roan Inish" (Jeni Courtney, 1994) was based on a Celtic myth about selkies (seals) and a young girl's search for

her lost brother. None of these movies achieved much success on the basis of their representation of marine mammals, however. In 1993 the movie "Free Willy" became a worldwide sensation and was ultimately followed by two sequels. The series chronicled a friendship between a 12-year-old boy (Jesse) and an adult male killer whale (Willy). When they met in the first movie, the whale was held captive in a small enclosure in a seaside marina. Willy refused to perform for audiences and the owner threatened dire consequences. Jesse arranged for the whale's dramatic release. In the second movie, Jesse and Willy ran into each other in the wild and discovered they must work together to avoid the perils of an offshore oil spill. In the third and final movie, the whale was threatened by an illegal whaling operation, but the whaler's son helped Jesse (who had grown into a 17-year-old whale researcher) save Willy and his family.

The Free Willy movies brought the question of captivity to the forefront of public consciousness, and resulted in a long and expensive campaign to return the killer whale that was used in the first movie to the ocean. That animal, Keiko, had been kept in rather miserable circumstances in a substandard facility in Mexico. With money raised through a publicity campaign sponsored by the movies' producers (Richard Donner and Lauren Shuler-Donner), Keiko went through a reorientation program in an Oregon facility and was then moved to Iceland, where he was reintroduced to the wild. Shortly thereafter the whale died of unknown causes. The public response to Free Willy was indicative of an increase in public questioning throughout the 1990s about the appropriateness of keeping marine mammals in captivity for public display. With increasing access to information about marine mammals in the wild (through excursions, books, and videos), it became clear that life in a captive facility was a poor substitute for nature. In response, the display industry made significant strides in upgrading facilities, improving husbandry, and refocusing shows to emphasize education and the display of animals' natural behaviors. The debate continues on many fronts about the pros and cons of maintaining marine mammals in captivity.

The removal of animals from the wild decreased significantly during the 1990s. Improved health of captive animals and an increase in the number born in captivity provided an ongoing supply of display animals that had no experience in the wild and would be unlikely to survive if released. Some facilities (e.g., the Vancouver Public Aquarium) chose not to replace whales or dolphins that died, and others (Maui Ocean Center; Monterey Bay Aquarium) decided from the outset not to hold cetaceans. In a graphic display of the continuing debate about animals in captivity, the producers of the Free Willy movies threatened a Hollywood-sponsored boycott of film-making on the Island of Maui if a proposal to put dolphins on display at a planned "theme park" on Maui was carried out. County legislators ultimately enacted a ban against keeping cetaceans in captivity anywhere in Maui County.

## VI. Marine Mammals in the Twenty-First Century: The Domestic Dolphin and Urban Whale

The commoditization of marine mammals is now reaching a fever pitch with attempts to resume commercial whaling, continued growth in whale watching, increased competition for ocean resources, space-age technological advances in military and corporate attempts to control the oceans, continued destruction of ocean habitat, and the dire threat of global warming. Each of these concerns has generated an agenda that pits scientists, politicians, environmentalists, and business interests against each other.

Recovery of large whale populations from the depredations of industrial whaling has given whaling interests new-found energy in

P

challenging the two-decade old IWC moratorium. In defending his country's scientific whaling program, a Japanese official referred to minke whales as "cockroaches of the sea." Even while Australian government officials are working to create a sanctuary for whales across the entire South Pacific, one Australian scientist noted that humpback whales are "breeding like rabbits," while others suggest the public's infatuation with whales is the result of gullible susceptibility to "animal protectionists (who) mobilize concerned, yet uninformed citizens to clamour for an end to any animal use" (Marsh *et al.*, 2003). Echoing the claims of prowhaling nations that recovering marine mammal populations are consuming resources needed to feed humans, Corkeron (2004) characterized the "whales as icons" perspective as problematic because it interferes with a rational ability to address such current questions as "… how much of the reduced productivity of the oceans and coasts should remain available to whales" (Corkeron, 2004, p. 848). It may be time, he suggested, to "spread new messages" to whale watchers that "whale populations will fare better under an internationally controlled regime of sustainable hunting rather than under culls instigated by individual nations" (Corkeron, 2004, p. 848). Advocates for an end to the whaling moratorium generally emphasize that "culling" recovered populations may be a necessary antidote to the diminishing availability of the world's fish stocks.

Popular perceptions of marine mammals in the twenty-first century are being shaped by such powerful forces as increased human competition for diminishing resources and the unpredictable impacts of global warming. In the interests of protecting national security and ensuring ready access to world markets, military and commercial interests lobby for exemptions from national and international regulations that protect wildlife and the environment. Despite evidence that military use of powerful underwater acoustic signals harms marine animals, and that ship strikes have been a major factor in the impending demise of the North Atlantic right whale, attempts to curtail the military or redraw shipping lanes have led to several court challenges. As we attempt to engineer the ever-more difficult struggle for survival, human needs trump those of other species. In a recent court case pitting environmental groups against the US Navy over the use of high power SONAR, the court ruled 2-1 in favor of the Navy. Writing for the majority, Appeals Judge Andrew Kleinfeld noted that:

The public does indeed have a very considerable interest in preserving our natural environment and especially relatively scarce whales. But it also has an interest in national defense… The safety of the whales must be weighed, and so must the safety of our warriors. And of our country. (Honolulu Star Bulletin, 2007)

A review of cultural values and literature suggests that our current relationship with marine mammals may best be described as one of "domestication." Oceanaria, whale and dolphin watching operations, and ocean cruise lines provide a wide segment of the world's population the opportunity to interact with groups of a relatively small number of species at close range. Marine mammals are maintained in captivity to entertain, educate and provide "therapy" for the public. Dolphin Swim Programs now offered by most captive display facilities have become a major force in the domestication of bottlenose dolphins, and generate millions of dollars of revenue. Prospective customers are told that captive dolphins are actually better off than those in the wild, as in this statement from the Dolphin Discovery website (http://www.dolphindiscovery.com/teens/info_how-dolphin-learn.asp, retrieved August 22, 2007):

(The dolphins) live in an area protected by fences… where they are safe from predators such as sharks and where they can be with people in a completely safe environment. During the day they play with people several times and they also have free time so they can be alone, play among themselves or do whatever they want. When one of them does not feel well it changes its attitude, it keeps away from the others or does not eat, this is when the veterinarians check them up and give them the right medicine so they can feel better.

These sentiments are echoed on the Dolphin Quest website (http://www.dolphinquest.org/getthefacts/welfare/, retrieved August 22, 2007):

The dolphins and whales in marine life parks, aquariums, oceanaria, and zoos consume consistently high quality nutritional food, receive excellent medical attention, and are kept free of debilitating parasites. This is in stark contrast to the predators, disease, pollution, well-documented commercial fishing and recreational boating dangers, and other stresses they face at sea, resulting in thousands of deaths each year.

In addition to captive swim programs, opportunities to swim with whales and dolphins in the wild may be found in countless locations around the globe—a Google search for "swim with dolphins" will return nearly 2,000,000 hits. The increasing popularity of swim programs may in part be the result of television exposure. The recent advent of High Definition format has resulted in a growing number of truly spectacular and visually pleasing documentaries that include stunning videography of marine mammals. The Discovery Channel and Animal Planet, National Geographic, and BBC have produced a range of programs devoted in whole or in part to marine mammals. BBC's "Blue Planet" is a prime example of the quality and beauty of such efforts.

Books about marine mammals since 2000 reflect a continuing interest in whales and dolphins, and run the gamut of topics. Excellent scientific texts on marine mammal biology and behavior include "Marine Mammal Biology: An Evolutionary Approach" edited by Hoelzel (Blackwell, 2002), and Berta, Sumich, and Kovacs' "Marine Mammals: Evolutionary Biology" (Academic Press, 2006). In-depth accounts of individual species have been written by Whitehead ("Sperm Whales," University of Chicago, 2003), Chadwick ("The Grandest of Whales," Sierra Club, 2006), and Krauss and Rolland ("The Urban Whale," Harvard University, 2007). Impacts of fisheries, tourism and management decisions on marine mammal species is considered in "Marine Mammals: Fisheries, Tourism and Management Issues" edited by Gales, Hindell, and Kirkwood (CSIRO, 2003). "Whales, Whaling, and Ocean Ecosystems" edited by Estes, DeMaster, Doak, Williams, and Brownell (University of California, 2006) considers the ways in which commercial depletion of whales has impacted ocean ecosystems, and how the recovery of depleted species may lead to further unintended consequences. "Leviathan" by Dolin (Norton, 2007) provides a detailed history of American whaling.

Perhaps because of the accessibility of gray whales along the western coastline of the United States, recent popular literature includes a number of books about that species. There is an account by Sullivan of the resumption of whaling by the Makah Indians of Washington state ("A Whale Hunt," Scribner, 2000), Peterson and Hogan's cultural history of human interactions with gray whales ("Sightings," National Geographic, 2002), Russell's documentation of gray whale natural history and migration ("Eye of the Whale," Island Press, 2004), and Thompson's description of gray whale behaviors, drawn from his many years as a whale watch guide ("Whales: Touching the Mystery," New Sage, 2006).

**P**

While personal accounts by those who accompanied whaling ships were popular in the early twentieth century, tales of transformation by personal interaction with whales and dolphins proliferate in the twenty-first century. Some of the work is based on reasonable scientific familiarity with the subject (e.g., Smolker's "To Touch a Wild Dolphin," Random House, 2001; Morton's "Listening to Whales," Ballantine, 2002; Visser's "Swimming with Orca," Penguin, 2005). There are a number of others, however, that derive from the author's conviction they have been chosen to channel cetacean thought processes (e.g., Wyllie's "Adventures Among Spiritual Intelligences," Wisdom Editions, 2001; Getten's "Communicating with Orcas: The Whale's Perspective," Hampton Roads, 2002; Taylor's "Souls in the Sea," Frog, 2003).

The contradiction between public fascination with marine mammals on the one hand, and pursuit of activities we know will destroy them on the other, continues. The twenty-first century may well bring the return of large-scale harvesting of marine mammals, not only for direct consumption, but as a means to reduce competition for commercially important fish species.

## VII.  International Perspectives

Attitudes about consumption of marine mammals differ across cultures that depend on traditional subsistence hunting practices (such as the Inuit, Aborigines, and some areas of the South Pacific); cultures that opportunistically use marine mammals to prop up collapsing fisheries (as is true throughout much of the Indian subcontinent, parts of Asia, Central and South America, and Africa); cultures that pursue relatively modern programs of whaling (including Japan and Norway); and cultures that currently forego whaling (many former whaling nations such as the United States, Canada, Australia, New Zealand, the United Kingdom, and South Africa).

As faulty as simplistic generalizations may be, one can recognize an important dividing line in global perspectives on marine mammals that is based more on economic considerations than fundamental differences in cultural attitudes toward nature. Wealthier countries enjoy the luxury of debating whether marine mammals should be harvested. Japan, Norway, the United States, Britain, Australia, and New Zealand are remarkably alike in their common appreciation of marine mammals as intelligent, amusing, and useful. The difference in those nations' behaviors toward marine mammals is one of time, and perhaps degree, for all have participated in the harvest of marine mammals and each currently reaps the benefits of the worldwide fascination with marine mammals. Less wealthy nations must face immediate considerations of survival before exploring programs devoted to environmental protection or species conservation. Marine mammals are viewed more directly in terms of immediate economic benefit. A major factor in the threat faced by a number of small cetacean populations is the readiness with which incidental takes (accidental entanglement in fishing nets) can be converted into a directed hunt in those areas where there is a market for the meat.

The enthusiasm with which a wide crosssection of nations have begun promoting whale and dolphin watching reinforces the notion that marine mammals are regarded neither as sacred nor profane in most parts of the world; they are primarily a commodity whose particular value can change from time to time. Surveys of attitudes in Australia, New Zealand, and the United States make it clear that, regardless of the qualities that participants may attribute to marine mammals, their primary value is in terms of satisfying human needs. This perspective appears even more entrenched in less affluent areas where marine mammals are harvested directly for food.

One of the most striking demonstrations of the role of marine mammals in modern cultural perspectives in affluent countries is the use of marine mammal images in corporate advertising. A survey by the author of corporate use of marine mammal images in magazines and television commercials (internationally) found over 100 different examples. Almost half the images used were of bottlenose dolphins. The next most frequent use was of killer whale images. Other species included humpback whale, sperm whale, right whale, beluga whale, false killer whale (*Pseudorca crassidens*), sea lion, walrus, spinner dolphin, humpback dolphin, and polar bear. In general, usage appeared to reflect anticipated familiarity on the part of the public for particular species (i.e., bottlenose dolphin, killer whale, humpback whale, and California sea lion), availability of professional quality images, and the extent to which the image symbolized the message of the advertisement (e.g., speed, power, size, intelligence).

The diversity of products marketed with the help of marine mammal images is extremely broad. In some cases, the connection between the product or service and marine mammals is obvious. A number of advertisements use images of marine mammals to promote travel to destinations where marine mammals may be observed (e.g., Olympic Airways in Europe, Air Nippon in Japan). In some cases, there is an identifiable link between the product or service and some perceived feature of the marine mammal. The ruggedness of a Leatherman tool has been highlighted with the help of a humpback whale; the warmth of Eddie Bauer clothing was symbolized by a polar bear; and a "school" of dolphins helped promote educational software for children (Fig. 11). More often the relevance of the image to the product is quite obscure. Marine mammal images have been used to sell aspirin, automobiles, batteries, beer, chewing gum, chocolates, cigarettes, computers, diapers, electronics, film, gold coins, jet skis, life insurance, orange juice, mobile phones, petrochemicals, potato chips, shower stalls, silverware, televisions, tires, video cameras, and a myriad of investment, real estate, interior decorating, and cosmetic services. One can only wonder about a cultural perspective that associates marine mammals with alcohol, tobacco, and petrochemicals.

Not surprisingly, use of marine mammal images in advertising occurred most frequently in more affluent countries, especially in Asia and Europe. North American and Australian use seemed to be more recent and less frequent, but a comprehensive assessment of corporate use of marine mammal images has yet to be conducted. It is clear, however, that over the past quarter century, increasing public attention to marine mammals has created a kind of "rock-star" status that elicits a classically conditioned positive emotional response to marine mammal images that can be exploited by marketing experts.

## VIII.  Conclusion

The one constant element in human perspectives on marine mammals since the beginning of Basque whaling is recognition of their economic significance. We have never stopped making money from them. They are still a resource. We still regard our needs as more critical than theirs. The proof can be seen in the peculiar and contradictory steps that we take to protect them. Consider, for example, the degree of attention that focused on "Humphrey" the "wrong-way whale" that wandered up the Sacramento River in 1985 and would not come out; the three gray whales that became caught in ice in Barrow, Alaska, in 1988 and could not get out; and the mother and calf humpback whale that spent 2 weeks in the Sacramento River in 2007 before getting themselves out. Tremendous resources in time, money, and emotions were devoted to "saving" these animals. Compare those situations with the decision to cull seals hanging out at fish ladders in Seattle a few

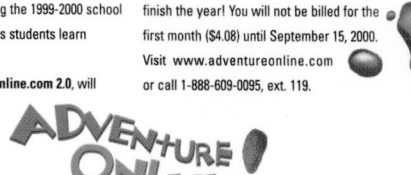
Figure 11 *A wide range of products and services are marketed with the help of marine mammal images. Not all advertisements seem as relevant or appropriate as the one shown here. Image ©Learning Outfitters Inc., used by permission.*

years ago to prevent them from feeding on depleted salmon stocks. No expense was spared to "save" the whales from their own biological predispositions, while the seals were basically held accountable for the precarious status of salmon stocks decimated by human overfishing and habitat degradation. Japanese scientists have been ridiculed for suggesting that Antarctic minke whale (*Balaenoptera bonaerensis*) numbers need to be culled to promote the recovery of blue whales by reducing competition for prey. At the same time, pinniped populations in North America have been culled to promote the recovery of fisheries by reducing prey competition with humans. We pass laws to prevent tourists from approaching marine mammals too closely, but do little to prevent the countless deaths of marine species by a range of human activities. Once we killed marine mammals to get their oil, now we kill them by spilling oil on them. Once we chased and captured them for food, today we chase and capture them for amusement. Instead of using their baleen for corsets, we use their images to sell diapers and real estate.

Nonetheless, there does seem to be a groundswell of respect, hope, and compassion for cetaceans. Thousands of individuals all over the world work tirelessly to assist marine mammals in need. When a humpback whale stranded on a beach near Brisbane, Australia in 1992, more than 5000 people showed up to save it. Our ability to form emotional attachments to individual animals seems boundless (Fig. 12). However, this encourages what right whale biologist Scott Krauss calls "Madison Avenue whale saving"; small-scale, high-profile publicity events that seem to be noble but do little to save marine mammal populations. Our activities will have little real conservation value as long as we ignore the intrinsic value and importance of marine mammals carrying out their natural behaviors in their natural habitat, unimpeded by the presence of humans or the pollution we produce.

We are faced with a fundamental conflict of interest: our need to exploit marine mammals for our own purpose and their need to be protected from our various forms of interference. Our interest in marine mammals is most often based on satisfaction of our own needs. We make them become what we hope they are and ignore what they maybe in reality. We see what we want to see. The sight of a whale or dolphin engaging in high energy activities visible above the surface is exciting for human observers (Fig. 13)—we want to get

**P**

Figure 12  *When a marine mammal strands, the public response is typically immediate and sympathetic. Photo courtesy of Paul Forestell.*

Figure 13  *Our interpretations of animal behavior are frequently clouded by our susceptibility to anthropomorphism. Photo courtesy of Pacific Whale Foundation/Paul Forestell.*

closer and see more. The truth is that in many cases such behavior may signal an angry or otherwise distressed animal. Although the display may not always be related to the presence of a boat or swimmer, there is increasing evidence that such activities frequently indicate the animal does not want us there. Most people simply cannot interpret animal behavior properly, which makes it very likely that they will not know how to avoid disturbing whales and dolphins when they go out on the ocean. It is commendable that many would rather idolize marine mammals in the wild than colonize them in oceanaria, but naive idolization is as questionable as brute colonization.

If we are to achieve the status that Lavigne *et al.* (1999) have described (i.e., an aesthetic interest that reflects increased understanding and a change in ethical values), we must undergo a universal transformation in perspective: the ability to value something independently of its potential for meeting our needs. Until that point, we may continue to find less cruel ways to put marine mammals in our service, but their value will continue to be defined in anthropocentric terms. Philosophically, that may not seem to be a problem of significant proportion, but insofar as our behaviors put our needs ahead of theirs, marine mammals will continue to be at risk.

## Acknowledgments

The ideas expressed reflect nearly 30 years of involvement with the whale and dolphin watching industry in Hawaii, Australia, Costa Rica, Ecuador, and occasional visits to Japan, New Zealand, Alaska, and New England. Many researchers, educators, operators, and members of the public have provided "grist for the mill." I especially thank Greg Kaufman and the Pacific Whale Foundation for support throughout that time. The International Cetacean, Education and Research Center (ICERC Japan) has also provided support and encouragement. Trisha Lamb Furstein's web-based bibliographies have been tremendously helpful to me. Joana Varawa provided critical and encouraging insight and much wisdom through her writings and in conversations we had on the island of Lanai.

## See Also the Following Articles

Captivity ■ Communiation ■ Culture in Whales and Dolphins ■ Ethics and Marine Mammals ■ Marine Parks and Zoos

## References

Carlson, C. (2004). A Review of Whale Watching Guidelines and Regulations around the World. Report for the International Fund for Animal Welfare, Yarmouth Port.

Carson, R. (1962). "Silent Spring." Houghton Mifflin, Boston, MA.

Corkeron, P. J. (2004). Whale watching, iconography, and marine conservation. *Conserv. Biol.* **18**, 847–849.

Darling, J., Nicklin, C., Norris, K. S., Whitehead, H., and Würsig, B. (1995). "Whales, Dolphins, and Porpoises." National Geographic Society Press, Washington, DC.

Day, D. (1987). "The Whale War." Sierra Club, San Francisco, CA.

Forestell, P. H., and Kaufman, G. D. (1996). The development of whale watching in Hawaii and its application as a model for the growth and development of the industry elsewhere. *In* "Encounters with Whales '95" (K. Colgan, ed.), pp. 53–65. Australian Nature Conservation Agency, Canberra.

Gaskin, D. E. (1982). "The Ecology of Whales and Dolphins." Heinemann, Portsmouth.

Dewhurst, H. W. (1834). The Natural History of the Order Cetacea, and the Oceanic Inhabitants of the Arctic Regions. P.P. Thoms, London. Retrieved August 6, 2007 from: http://books.google.com/books?id = gEEAAAAAQAAJ&pg = PA73&dq= Dewhurst + marine + mammals + 1834.

Honolulu Star Bulletin. (2007). Our opinion: appeals court ruling on navy sonar sits on fragile ground. **12**(26). Retrieved September 4 from: (http://starbulletin.com/2007/09/03/editorial/editorial01.html).

Hoyt, E. (2001). "Whale Watching 2001: Worldwide Tourism Numbers, Expenditures, and Expanding Socioeconomic Benefits." International Fund for Animal Welfare, Yarmouth Port, MA.

Hudnall, J. (1977). In the company of great whales. *Audubon* **79**, 62–73.

Lavigne, D. M., Scheffer, V. B., and Kellert, S. R. (1999). The evolution of North American attitudes toward marine mammals. *In* "Conservation of Marine Mammals" (J. R. Twiss, Jr., and R. R. Reeves, eds), pp. 10–47. Smithsonian Institution Press, Washington, DC.

Leopold, A. (1949). "A Sand County Almanac and Sketches Here and There." Oxford University Press, New York.

Lilly, J. C. (1967). "Lilly on Dolphins." Doubleday, New York.

Marsh, H., *et al.* (8 authors) (2003). Strategies for conserving marine mammals. *In* "Marine Mammals: Fisheries, Tourism and Management Issues" (N. Gales, M. Hindell, and R. Kirkwood, eds), pp. 10–18. CSIRO, Collingwood, Victoria.

Matthews, L. H. (1978). "The Natural History of the Whale." Columbia University Press, New York.

McBride, A. F., and Hebb, D. O. (1948). Behavior of the captive bottle-nose dolphin, *Tursiops truncatus*. *J. Comp. Physiol. Psychol.* **41**, 111–123.

P

Norris, K. S. (1974). "The Porpoise Watcher." W. W. Norton, New York.

Payne, R. S., and McVay, S. (1971). Songs of humpback whales. *Science* **173**, 585–597.

Rice, D. W., and Wolman, A. A. (1971). The life history and ecology of the gray whale (*Eschrichtius robustus*). American Society of Mammalogists, Special Publication, **3**, 1–142.

Robertson, R. B. (1954). "Of Men and Whales." Alfred A. Knopf, New York.

Scheffer, V. B. (1969). "The Year of the Whale." Charles Scribner, New York.

# Population Dynamics

## PAUL R. WADE

"*The immediate cause of the increase of population is the excess of the births above deaths*"

When Thomas Malthus wrote his treatise on population growth in 1798, he understood that the study of population dynamics is most simply the study of fundamental life history characteristics such as birth and death rates. However, he also understood that the ultimate explanation for why populations increased or decreased was due to more complicated factors, such as changes in food supply and density-dependent limitations to growth. Much of the research on marine mammal population dynamics has traditionally focused on better definition of life history parameters such as survival and fecundity rates, age of sexual maturity, and longevity. These can, for example, be used to estimate a potential maximum rate of increase for a population. A sign that the study of marine mammal population dynamics has become increasingly sophisticated is the growing number of studies that examine how extrinsic factors can influence populations. These include environmental variability, disease and natural toxins, competition, and predation. There is also much continued interest in the role that density-dependence plays in controlling marine mammal populations. We are currently witnessing the return of many populations to levels limited by their carrying capacity following overexploitation by humans, which means that the study of density-dependence in marine mammals is no longer a theoretical discipline.

## I. Rates of Population Increase

### A. Population Growth of Long-Lived Animals

Most long-lived animals, such as marine mammals and large terrestrial mammals, have relatively slow intrinsic rates of increase compared to most other kinds of animals. These modest population growth rates are the consequence of their life history characteristics. Characteristics such as the age at which females start reproducing, the number of years between births, and how many years a female will live and reproduce, determine how quickly a population can increase. Most marine mammal species take many years to reach sexual maturity and have long gestation periods that result in the production of, at most, only one young a year. In fact, most species give birth only once every several years (Boyd *et al.*, 1999). Even when annual rates of survival are very high, these characteristics cause populations to grow slowly. Low rates of population growth make most species of marine mammal vulnerable to overexploitation, as shown by the rapid depletion of many whale populations by commercial whaling.

## B. Measuring Population Growth

Population growth can be measured in two ways. In general, the most reliable estimates often come from abundance data collected over many years. Population abundance can be estimated from surveys or counts, and when repeated over several years, the trend (percentage change per year) in a population can be estimated (see chapters on ABUNDANCE ESTIMATION and POPULATION STATUS AND TRENDS). Specifically, the slope of a linear regression on the natural logarithm of abundance against time represents the rate of increase ($r$) of a population experiencing exponential growth. Because population growth is slow and population estimates are imprecise, 10 or more years may be required to directly measure population growth rates. Estimates of abundance for some species are so imprecise that it may be difficult to determine trends in abundance (Taylor *et al.*, 2007).

A less direct way of estimating population growth is from life history data. Estimates of age of sexual maturity, birth rate, juvenile and adult survival rate, and maximum age can be compiled in a Leslie matrix or similar model, which can then be used to estimate the rate of increase (usually estimated as $\lambda$, where $\lambda = e^r$). Although such calculations have been useful for exploring potential population growth rates, relatively few estimates of actual population growth have been made in this way. The main hindrance is the lack of direct data on survival rates of marine mammals. Only a few species have been amenable to survival estimation, usually from mark-recapture studies using individuals recognizable from tags, branding, unique scars, or coloration patterns, or other markings (see MARK AND RECAPTURE). Mark-recapture studies of survival or fecundity have been undertaken in California (*Zalophus californianus*) and Steller (*Eumetopias jubatus*) sea lions, bottlenose (*Tursiops truncatus*) and Hector's (*Cephalorhynchus hectori*) dolphins, polar bear (*Ursus maritimus*), manatee (*Trichechus manatus*), and killer (*Orcinus orca*), humpback (*Megaptera novaeangliae*), gray (*Eschrichtius robustus*), bowhead (*Balaena mysticetus*), and right (*Eubalaena* spp.) whales (Fujiwara and Caswell, 2001; Kendall *et al.*, 2004; Taylor *et al.*, 2006).

## C. Taxonomic Differences

Although population growth is comparatively slow for all marine mammal species compared to most other vertebrates, there is a considerable range in life history strategies. For example, just within the cetaceans one can contrast a harbor porpoise (*Phocoena phocoena*) that becomes sexually mature after a few years, can give birth annually, and rarely lives past 12–15 years (Read and Hohn, 1995), with a bowhead whale that becomes sexually mature after perhaps 10–20 years, only gives birth every 3–5 years, and lives for decades, with some individuals apparently living much longer than 100 years. Broad patterns are evident in maximum population growth rates when available information from trend or life history data is examined for different taxonomic groups. Sea otters (*Enhydra lutris*), with maximum observed rates on the order of 17–20% per year (Estes, 1990), are capable of the fastest growth. Next are pinnipeds. Many have the potential to reproduce annually, with observed rates ranging from 8 to 13% for otariids [from northern fur seals (*Callorhinus ursinus*) at the low end to Antarctic fur seals (*Arctocephalus gazella*) at the high end] (Wickens and York, 1997), and 6–13% for phocids [from Hawaiian monk seals (*Monachus schauinslandi*) at the low end to harbor (*Phoca vitulina*) and gray (*Halichoerus grypus*) seals at the high end]. A puzzling species is the Steller sea lion, with the greatest observed increase of only about 3% per year, observed in the eastern North Pacific population; it is uncertain if this population has also

**P**

been affected by factors that have led to the decline and endangered status of the western North Pacific population.

Maximum annual rates of increase for mysticetes likely range from about 4 to 10%, or perhaps even higher (Best, 1993). Observed rates for gray and bowhead whales (3–4%) are at the low end, although the populations that have been studied are likely at high population levels relative to carrying capacity and therefore their growth rate may have already slowed (see Section III below). Model-based estimates suggest bowheads may be capable of 4–5% maximum rates, and gray whales of 5–7%. Southern right whales (*Eubalaena australis*) have shown the ability to increase at 7% per year, and humpback whales have been estimated to increase at 7–10% per year or higher. Estimates at the higher end are surprising given that 10% has been deduced to be the approximate theoretical upper limit for humpback whales from their life history. Estimates for other species, such as blue (*Balaenoptera musculus*) and fin (*Balaenoptera physalus*) whales, are relatively imprecise but mostly fall within a range of 4–8%. Manatees appear to have maximum population growth rates of at least 7% and perhaps higher, whereas dugongs (*Dugong dugon*) are likely only capable of growing at 5–6% per year. Much uncertainty exists about the maximum rates of increase of odontocetes. What little is known suggests this group generally has very low annual population growth rates, 2–3% in some species such as killer whales (Brault and Caswell, 1993), and most species are considered unlikely to have a maximum growth rate of more than 4% (Reilly and Barlow, 1986). There is speculation that a few species, such as the harbor porpoise, may have higher rates of increase because of their life histories, but such higher rates have yet to be documented. Age distributions of by-caught and stranded harbor porpoise support the hypothesis that maximum rates are ~4% for this species.

## II. Extrinsic Factors Affecting Population Size

### A. Environmental Variance

Another aspect of population dynamics is the study of the effects of extrinsic factors on population growth. The difficulty in precisely estimating population size and life history parameters has made the study of variation in population growth rates of marine mammals difficult. However, at least a few conclusions can be made. Long-lived animals with relatively older ages at sexual maturation and relatively slow population growths cannot respond quickly to favorable environmental conditions. Consequently, such species cannot decline too often or too rapidly when conditions are bad, or they would not have persisted on an evolutionary time-scale. Therefore, marine mammals have evolved life-history strategies that keep them relatively buffered from interannual variability in environmental conditions, at least compared to other animals such as small terrestrial mammals. Species with these traits are often referred to as "K-selected species," meaning they have evolved to maintain relatively stable population sizes at or near the carrying capacity (typically represented by the letter "K") of the environment (rather than fluctuating wildly as seen in small mammals such as the lemming, *Dicrostonyx* spp., or snowshoe hare, *Lepus americanus*). Cetaceans, in particular, should be less subject to large fluctuations in survival and fecundity from year to year than would be sea otters or pinnipeds.

Data sufficient to examine such patterns are relatively rare for marine mammals. Studies of pinnipeds provide the best evidence of the effect of changing oceanographic conditions, particularly because of the ability to closely monitor numbers of pups or adults at rookeries from one year to the next. El Niño oceanographic events, through reductions in prey availability, have led to dramatic changes in

survival and reproduction of several species of otariids in places such as California and the Galapagos Islands (Trillmich and Ono, 1991). These changes include lower fecundity, lower pup survival, and even lower adult survival during extreme events. In some cases, such as for the northern fur seal at San Miguel Island, conditions have been bad enough to result in nearly 100% mortality of pups in a given year. Similarly, fluctuations in cohort strength of crabeater seals (*Lobodon carcinophaga*) have been attributed to environmental factors such as sea ice extent and krill production and availability (Boveng and Bengtson, 1997), and survival of juvenile Hawaiian monk seals has been correlated with the location of the chlorophyll front in the North Pacific transition zone (Baker *et al.*, in press). Several populations of pinnipeds have experienced long-term (20–30 year) declines for which the cause has been debated. While some believe these declines are due to oceanographic regime shifts on that time scale, others believe that factors such as direct human-caused mortality or depletion of prey species by commercial fisheries may be at least partially to blame.

Cetaceans probably have similar, though less dramatic, responses to environmental change, but these changes are harder to detect. One of the first examples was that of pregnancy rates of fin whales off Iceland which were correlated with changes in food abundance. Recent studies collecting information about interannual changes in survival or fecundity have led to several other examples of correlations between population dynamics of cetaceans and environmental factors. An at-sea index of chinook salmon (*Oncorhynchus tshawytscha*) abundance is highly correlated with survival rates of fish-eating killer whales in British Columbia (Ford *et al.*, 2005). Production of gray whale calves is correlated with the extent of sea ice in the Bering Sea the previous summer (with greater ice cover presumably restricting access to food resources) (Perryman *et al.*, 2002). Intriguingly, calf production in North Atlantic right whales (*Eubalaena glacialis*) is correlated with the North Atlantic Oscillation index (a physical oceanographic measurement of ocean temperature), apparently through mechanisms that concentrate and favor production of their copepod prey (Greene and Pershing, 2004). Within the sirenians, manatees have been shown to experience higher mortality during years with intense coastal storms (Langtimm and Beck, 2003).

### B. Disease and Natural Toxins

Many species of marine mammal have experienced large mortality events caused by disease or natural toxins. Disease can also affect reproductive rates as well, with a potentially important influence on population dynamics. Disease and toxin related die-offs are likely natural events for marine mammals. It is thought (though not proven) that these events may occur more frequently in populations near or at their carrying capacity due to nutritional stress, enhanced transmission of disease, and other related factors. There is also the suspicion that some of these die-offs have been ultimately triggered by anthropogenic causes, such as degraded habitat or exposure to contaminants, but this has also proven difficult to confirm from field studies. Three natural toxins that cause marine mammal deaths are saxitoxin, brevetoxin, and domoic acid, all of which come from harmful algal blooms. The frequency and severity of harmful algal blooms in some parts of the world have been linked to increased nutrient loading from human activities. The great majority of marine mammal die-offs from disease or toxins have been of coastal or nearshore species; whether this represents a true tendency (perhaps from anthropogenic causes) or is simply due to a greater probability of detection of these events (i.e., coastal species are more likely to strand) is unknown. Such die-offs are particularly dangerous for small populations.

Some mortality events from disease or toxins have been severe enough to be considered a significant influence on the dynamics of the effected population (see MASS MORTALITIES). Seal populations have been most dramatically affected, with die-offs of >18,000 and ~25,000 harbor seals in Europe from phocine distemper virus in 1988 and 2002, respectively, tens of thousands of Baikal seals (*Phoca siberica*) from canine distemper virus (CDV) in 1987–1988, and 10,000 Caspian seals (*Phoca caspica*) from CDV in 2000 (Gulland and Hall, 2007). Even more dramatically, the endangered Mediterranean monk seal (*Monachus monachus*) experienced a loss of 60–70% of its population off Africa in 1997 from a harmful algal bloom and morbillivirus, which has seriously compromised the long-term survival of this species. California sea lions have been effected by at least seven die-offs of hundreds of animals over the last 22 years, caused by leptospirosis or domoic acid. In 1998, 60% of New Zealand sea lion (*Phocarctos hookeri*) pups died, with the suspicion that a bacterial infection was the cause.

In the western North Atlantic Ocean, hundreds of bottlenose dolphins died in 1987–1988 from morbillivirus and possibly brevetoxin, and the event was thought to have caused a significant decline in the population. Other die-offs of hundreds of bottlenose dolphins have occurred along the coasts of Texas and Florida, with brevetoxin the most common cause. Humpback whales in the Atlantic have also experienced mass-mortality events on at least two occasions; in one case at least 14 whales died from saxitoxin poisoning from consuming dinoflagellates concentrated in fish prey. Sixteen humpback whales died in 2003 in Maine, with saxitoxin and domoic acid detected in two and three whales, respectively. It is not known if such mortality is frequent enough to exert a strong influence at the population level. Finally, Florida manatees have experienced at least five die-offs of tens of animals or more from brevetoxin over the last 25 years.

## C. Competition

Competition from other species may influence the population dynamics of marine mammals, although there is little evidence for this. Whether this is due to competition being unimportant or whether it is simply too difficult to demonstrate is an open question. It is important to be clear on what is meant by competition. From an ecological perspective, prey overlap and competition are two different concepts and the latter may or may not occur with the former. The term "competition" is defined to be when one organism has a negative effect upon another by consuming, or controlling access to, a resource that is limited in availability. Therefore, demonstration that two species eat the same prey in the same location does not prove that competition exists between the two species; the prey resource may not be limiting, at least at that time and location, or the species may exploit the prey in different ways (such as at a different depths).

Many claims have been made about competition between whale species and other marine vertebrates in Antarctic waters. There is considerable prey overlap in the Antarctic between cetaceans, pinnipeds, and penguins, but studies are lacking that actually demonstrate competition. An increase in crabeater seals was directly attributed to a release from competition following the severe depletion of several species of baleen whales in the Antarctic, leading to an increased availability of krill. This explanation has recently been reevaluated in light of evidence of environmental influences on the population dynamics of Antarctic pinnipeds. Of many factors considered in a recent review of the decline of southern elephant seal (*Mirounga leonina*) around Antarctica, only competition and oceanographic change were considered plausible hypotheses; the authors concluded the competition hypothesis was not the cause (McMahon *et al.*, 2005). It should be noted that some general textbooks (such as books on oceanography) state that competition for krill with Antarctic minke whales (*Balaenoptera bonaerensis*) has prevented the recovery of depleted blue whales in the Antarctic. However, competition between the two species has not been demonstrated, and recent information indicates the lack of recovery of blue whales in the Antarctic might be explained by previously unknown illegal catches by the former Soviet Union (Clapham *et al.*, 1999).

## D. Predation

Many marine mammals, especially smaller ones, are preyed upon by other animals, but predation in the past had rarely been suggested to be a strong controlling factor in their population dynamics. In recent years, however, increased attention has been paid to the role of top predators in marine ecosystems. It has been suggested that sea otters may have declined in the Aleutian Islands of Alaska because of killer whale predation (Estes *et al.*, 1998). Pinnipeds, particularly pups, are often vulnerable to predation from predators such as leopard seals (*Hydrurga leptonyx*), great white sharks (*Carcharodon carcharias*), and killer whales. While such predation has been shown to affect the growth rate of the subpopulation at local rookeries, it is unclear if it exerts a strong influence on the dynamics of an entire population. For cetaceans, at least some (and perhaps all) pelagic dolphin species experience predation by sharks, and killer whales prey on many cetacean species. Again, even though predation of cetaceans occurs, it is difficult to know whether it is a dominant factor in the dynamics of these populations. From recent studies it appears that killer whale predation on gray whale calves on their first north-bound migration may account for a substantial portion of their natural mortality, though the gray whale population increased for over 20 years while apparently experiencing this source of mortality. In some areas a large proportion of humpback whales show rake-mark scars from killer whale attacks, the great majority of which were apparently acquired as calves. Despite this likely source of natural mortality, humpbacks in both the North Atlantic and Pacific have shown strong recovery from commercial exploitation.

Discussions about the influence of predators on the dynamics of marine species have often been placed in the broader context of an ongoing debate about the relative importance of "top-down" influences (predation or other direct removals such as human-caused kills) vs "bottom-up" influences (from oceanographic productivity) (Ainley *et al.*, 2007; Nicol *et al.*, 2007). For example, it has been hypothesized that killer whale predation caused the decline of southern sea lions (*Otaria flavescens*), southern elephant seals, and Antarctic minke whales in the Southern Ocean, but a review of the southern elephant seal decline concluded that predation was unlikely to be the cause and suggested that oceanographic change or competition were more plausible explanations (McMahon *et al.*, 2005).

An interesting debate has followed the hypothesis that killer whales caused the sequential decline of harbor seals, Steller sea lions, northern fur seals, and sea otters in the Aleutian Islands and Gulf of Alaska; with many articles arguing both for (Springer *et al.*, 2003) and against (Wade *et al.*, 2007) this hypothesis; see Estes *et al.* (2006) for an overview of this issue. In particular, the decline of Steller sea lions has stimulated strong arguments both for top-down causes (from direct takes in fisheries bycatch, intentional shooting, subsistence hunting, or killer whale predation) and for bottom-up processes (from oceanographic regime shift or competition for prey with commercial fisheries). It may never be possible to solve these

P

debates retrospectively, as it is often the case that the key data were not collected during the decline and cannot now be collected.

These extrinsic factors have been discussed independently, but there are likely many interactions between them. As one example, sea otters with moderate to severe encephalitis were estimated to be 3.7 times more likely to be killed by sharks than otters without that condition (Kreuder *et al.*, 2003). Several authors have treated deaths from predation as independent mortality that a population suffers, but it may be that predation mortality is mostly of more vulnerable individuals already known to have lower survival rates, such as sick or very young or old animals and therefore might be correlated to influences from either environmental change or density-dependence (causing nutritional stress). Populations still have the ability to increase while experiencing natural mortality, of which predation is only one component (eastern North Pacific gray whales are a good example of this).

## III. Density-Dependence

### A. Compensation

Another area of great interest is the role of density-dependence in controlling the population dynamics of marine mammals. It is generally accepted that marine mammal populations experience density-dependence. In other words, as populations become relatively larger, they tend to have lower population growth, and eventually stop increasing. This form of density-dependence is termed compensation. The level at which a population stabilizes is called its carrying capacity. Evidence has been found for density-dependence in life history parameters such as the age at sexual maturation. Females from a population at a level well below *K* become sexually mature and start reproducing at an earlier age than do females from a population at a level close to *K*. Presumably this is because of access to greater resources such as prey. For example, the age at sexual maturation apparently became lower for fin and sei whales in the Antarctic as their populations were depleted by commercial whaling.

It has been hypothesized that the mechanisms of the regulation of populations of long-lived mammals would follow a sequence as a population increased, with density-dependence first affecting the rate of immature survival, then the age at sexual maturation and the birth rate, and finally the adult survival rate (Eberhardt, 1977). This hypothesis partially follows from the recognition that a long-lived species which reaches sexual maturity late and has a low intrinsic rate of increase must maximize adult survival in order to persist. Adult females of long-lived species may be able to forgo reproduction to maximize individual survival when conditions are poor, but it is unclear if there is necessarily a specific sequence of effects in how density-dependence operates that is common to all marine mammals.

### B. Linear vs Nonlinear Density-Dependence

It is difficult to assess how these changes in life history translate into changes in population growth, as few direct data are available on changes in population growth at different population sizes. One debate is whether marine mammals experience linear or nonlinear density-dependence. Linear density-dependence is a constant decline in the per capita population growth rate as a population increases, illustrated by the simple logistic population model (Fig. 1). Nonlinear density-dependence is where a population has no decline in the per capita growth rate as it increases until it reaches a level close to *K*, where it then has a rapid decline, illustrated by the θ-logistic population model where the value of θ is greater than 1 (Fig. 1). Both linear and nonlinear density-dependence occurs in single life-history parameters of marine mammals and other large mammals. However,

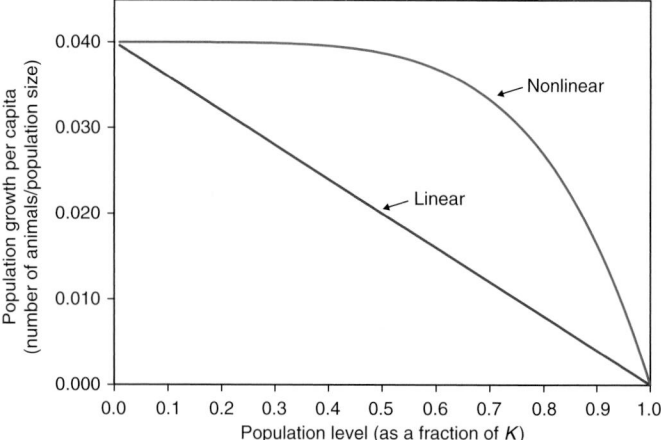

**Figure 1** *An illustration of two forms of density-dependence. "Linear" density dependence represents a constant decline in population growth per capita as the population level approaches carrying capacity (K). Per capita population growth is the number of animals added to the population divided by the total population size, which also represents the population growth rate. "Non-Linear" represents a curved response, where the population growth per capita does not decline until a population level close to K, where it declines rapidly. Both curves represent cases where the maximum population growth rate is 0.04 (4% per year).*

population modeling indicates that nonlinear density-dependence in a single life history trait (such as fecundity) may not translate into nonlinear density-dependence at the population level, particularly if density-dependence in other life history traits is more linear (Taylor and DeMaster, 1993). The form of the density-dependence (linear vs nonlinear) will determine what population size will have the greatest increase in numbers from one year to the next, called the maximum net productivity level (MNPL). For long-lived animals, nonlinear density-dependence will generally lead to populations having their MNPL closer to carrying capacity (Fig. 2). A review concluded that most marine mammal populations likely have their MNPL between 50 and 85% of their carrying capacity. It has been shown, however, that it is very difficult to estimate MNPL for any marine mammal population given the data that are currently available (Ragen, 1995).

### C. Density-Dependence and Management

These concepts of density-dependence have been incorporated into the management and conservation of marine mammals. Both the International Whaling Commission, in its proposed Revised Management Procedure (RMP) for the regulation of whale catches, and the US government, in managing human-caused mortality of marine mammals in US waters, refer to concepts like MNPL, with populations above MNPL considered "healthy."

Many populations of pinnipeds and whales are recovering from unregulated catches that left them at severely depleted population levels. It will be interesting to observe the recovery of these populations over the next few decades, as many are likely to approach previous population levels and are expected to stop growing at some point. Examples of populations that have apparently reached carrying capacity include harbor seals in Washington (Jeffries *et al.*, 2003), harbor seals in Oregon (Brown *et al.*, 2005), and possibly eastern North Pacific gray whales (Moore *et al.*, 2001). It is important to understand when density-dependence is taking place, so that it can

P

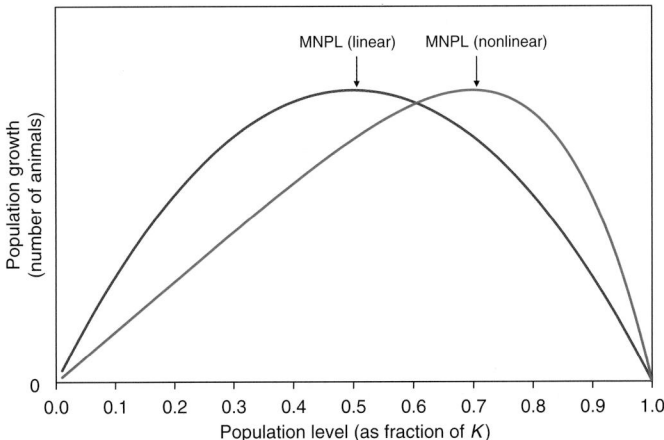

**Figure 2** *Population growth (in numbers of animals added to the population each year) as a function of population level, expressed as a fraction of K. This quantity is also called the "net productivity," and therefore the peak of this curve is referred to as the maximum net productivity level (MNPL). For linear density-dependence the MNPL is at one-half of K. For nonlinear density-dependence the MNPL is closer to K, in this case at 0.7 K.*

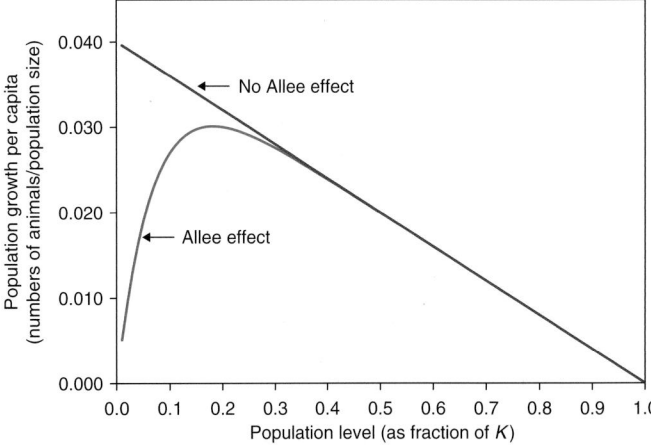

**Figure 3** *An illustration of the Allee effect. One curve represents linear density-dependence without depensation ("No Allee effect"). The second curve represents linear density-dependence with depensation ("Allee effect"). Under depensation, population growth per capita declines at low population levels.*

be distinguished from human-caused effects on population growth that might require conservation action. It is to be expected that populations at or close to K may be subject to more environmental influence, because with populations at the limit of their resources, any drop in resources will potentially have dramatic effects. California sea lions and gray whales may be current examples of this.

### D. Allee Effect (Depensation)

Density-dependence can also work in the opposite direction, where per capita population growth is slowed at very small population sizes (Fig. 3). This is called the Allee effect by ecologists (also called depensation by fisheries biologists). One simple example of the Allee effect would be a reduction in reproduction at very small population sizes due to the inability to find a mate. Allee effects could also occur from inbreeding depression associated with small populations, or from behavioral changes that might accompany a reduction to small numbers of animals, such as decreased foraging success or protection from predators. Unfortunately, the difficulties in studying the population dynamics of marine mammals are compounded by the difficulty of studying small populations, so it may take a long time before much is learned about depensation and its role in the dynamics of marine mammals (Fowler and Baker, 1991). Several severely depleted baleen whale populations have shown little or no recovery despite decades of apparent relief from human exploitation (Clapham *et al.*, in press), raising the possibility that these populations are experiencing Allee effects. However, it has been realized that many of these populations may have continued to experience human-caused mortality from a variety of sources, such as illegal whaling, entanglement in fishery gear, and collisions with ships. This makes it difficult to determine if depensation does play a significant role in the dynamics of small populations of cetaceans.

Similarly, many pinniped species were once hunted to commercial extinction and have since shown recovery, but it is rarely known how small these populations actually were at their low point. One exception is the remarkable story of northern elephant seals (*Mirounga angustirostris*), which were thought to be extinct around 1880 and are now thought to have numbered no more than about

100 individuals around the turn of the century, concentrated in a single colony in Baja California, Mexico (Cooper and Stewart, 1983). The number of northern elephant seals now exceeds 80,000, and they have recolonized many breeding sites in the US and Mexico. However, the apparent lack of depensation in one population does not preclude it from being important to other populations or species. Although difficult to study, a decline in growth rate at small population sizes can substantially increase the risk of extinction for a population, so it will continue to be important to consider depensation despite the lack of solid evidence for its occurrence.

### *See Also the Following Articles*

Abundance Estimation ■ Identification Methods ■ Mass Mortalities ■ Pinniped Life History ■ Predation on Marine Mammals ■ Stock Assessment ■ Sustainability

### *References*

Ainley, D., *et al.* (12 authors) (2007). Paradigm lost, or is top-down forcing no longer significant in the Antarctic marine ecosystem? *Antarct. Sci.* **19**, 283–290.

Baker, J. D., Polovina, J. J., and Howell, E. A. (2007). Effect of variable oceanic productivity on the survival of an upper trophic predator, the Hawaiian monk seal *Monachus schauinslandi*. *Mar. Ecol. Prog. Ser.* **346**, 277–283.

Berta, A., and Sumich, J. L. (2006). "Marine Mammals: Evolutionary Biology." Academic Press, San Diego, CA.

Best, P. B. (1993). Increase rates in severely depleted stocks of baleen whales. *ICES J. Mar. Sci.* **50**, 169–186.

Boveng, P. L., and Bengtson, J. L. (1997). Crabeater seal cohort variation: demographic signal or statistical noise? *In* "Antarctic Communities. Species, Structure and Survival" (B. Battaglia, J. Valencia., and D. W. H. Walton, eds). Cambridge University Press, Cambridge, UK.

Boyd, I. L., Lockyer, C., and Marsh, H. D. (1999). Reproduction in marine mammals. *In* "Biology of Marine Mammals." (J. E. Reynolds, III and S. A. Rommel, eds), pp. 218–286. Smithsonian Institution Press, Washington DC, USA.

Brault, S., and Caswell, H. (1993). Pod-specific demography of killer whales (*Orcinus orca*). *Ecology* **7**, 1444–1454.

Brown, R. F., Wright, B. E., Riemer, S. D., and Laake, J. (2005). Trends in abundance and current status of harbor seals in Oregon: 1977–2003. *Mar. Mam. Sci.* **21**, 657–670.

P

Clapham, P. J., Young, S. B., and Brownell, R. L. (1999). Baleen whales: conservation issues and the status of endangered populations. *Mamm. Rev.* **29**, 35–60.

Clapham, P., Aguilar, A., and Hatch, L. (2008). Determining spatial and temporal scales for management: lessons from whaling. *Mar. Mamm. Sci.* **24**, 183–201.

Cooper, C. F., and Stewart, B. S. (1983). Demography of northern elephant seals, 1911–1982. *Science* **219**, 969–971.

Eberhardt, L. L. (1977). Optimal policies for conservation of large mammals, with special reference to marine ecosystems. *Environ. Conserv.* **4**, 205–212.

Estes, J. A. (1990). Growth and equilibrium in sea otter populations. *J. Anim. Ecol.* **59**, 385–401.

Estes, J. A., DeMaster, D. P., Doak, D. F., Williams, T. M., and Brownell, R. L. (eds) (2006). "Whales, Whaling, and Ocean Ecosystems." University of California Press, Berkeley, CA.

Estes, J. A., Tinker, M. T., Williams, T. M., and Doak, D. F. (1998). Killer whale predation on sea otters linking oceanic and nearshore ecosystems. *Science* **282**, 473–476.

Ford, J. K. B., Ellis, G. M., and Olesiuk, P. F. (2005). Linking prey and population dynamics: did food limitation cause recent declines of 'resident' killer whales (*Orcinus orca*) in British Columbia? Canadian Science Advisory Secretariat Research Document, 2005/042.

Fowler, C. W. (1987). A review of density dependence in populations of large mammals. *In* "Current Mammalogy" (H. H. Genoways, ed.), Vol. 1, pp. 401–441. Plenum Publishing Corporation, New York.

Fowler, C. W., and Baker, J. (1991). A review of animal population dynamics at extremely reduced population levels. *Rep. Int. Whal. Comm.* **41**, 545–554.

Fowler, C. W., and Smith, T. D. (eds) (1981). "Dynamics of Large Mammal Populations." Wiley, New York.

Fujiwara, M., and Caswell, H. (2001). Demography of the endangered North Atlantic right whale. *Nature* **414**, 539–543.

Geraci, J. R., Harwood, J., and Lounsbury, V. J. (1999). Marine mammal die-offs: causes, investigations, and issues. *In* "Conservation and Management of Marine Mammals." (J. R. Twiss and R. R. Reeves, eds.), pp. 367–395. Smithsonian Institution Press, Washington DC, USA.

Greene, C. H., and Pershing, A. J. (2004). Climate and the conservation biology of North Atlantic right whales: the right whale at the wrong time? *Front. Ecol. Environ.* **2**, 29–34.

Gulland, F. M. D., and Hall, A. J. (2007). Is marine mammal health deteriorating? Trends in the global reporting of marine mammal disease. *EcoHealth* **4**, 135–150.

Jeffries, S., Huber, H., Calambokidis, J., and Laake, J. (2003). Trends and status of harbor seals in Washington State: 1978–1999. *J. Wildl. Manage.* **67**, 208–219.

Kendall, W. L., Langtimm, C., Beck, C. A., and Runge, MC. (2004). Capture-recapture analysis for estimating manatee reproductive rates. *Mar. Mamm. Sci.* **20**, 424–437.

Kreuder, C., *et al.* (9 authors) (2003). Patterns of mortality in southern sea otters (*Enhydra lutris nereis*) from 1998–2001. *J. Wildl. Dis.* **39**, 495–509.

Langtimm, C. A., and Beck, C. A. (2003). Lower survival probabilities for adult Florida manatees in years with intense coastal storms. *Ecol. Appl.* **13**, 257–268.

McMahon, C. R., Bester, M. N., Burton, H. R., Hindell, M. A., and Bradshaw, C. J. A. (2005). Population status, trends and a re-examination of the hypotheses explaining the recent declines of the southern elephant seal *Mirounga leonine. Mamm. Rev.* **35**, 82–100.

Moore, S. E., *et al.* (8 authors) (2001). Are gray whales hitting 'K' hard? *Mar. Mamm. Sci.* **17**, 954–958.

Nicol, S., Croxall, J., Trathan, P., Gales, N., and Murphy, E. (2007). Paradigm misplaced: Antarctic marine ecosystems are affected by climate change as well as biological process and harvesting. *Antarct. Sci.* **19**, 291–295.

Perryman, W. L., Donahue, M. A., Perkins, P. C., and Reilly, S. B. (2002). Gray whale calf production 1994–2000: are observed fluctuations related to changes in seasonal ice cover? *Mar. Mamm. Sci.* **18**, 121–144.

Ragen, T. J. (1995). Maximum net productivity level estimation for the northern fur seal (*Callorhinus ursinus*) population of St. Paul Island, Alaska. *Mar. Mamm. Sci.* **11**, 275–300.

Read, A. J., and Hohn, A. A. (1995). Life in the fast lane: the life history of harbor porpoises from the Gulf of Maine. *Mar. Mamm. Sci.* **11**, 423–440.

Reilly, S. B., and Barlow, J. (1986). Rates of increase in dolphin population size. *Fish. Bull. (U.S.)* **84**, 527–533.

Springer, A. M., *et al.* (8 authors) (2003). Sequential megafaunal collapse in the North Pacific Ocean: an ongoing legacy of industrial whaling? *Proc. Natl. Acad. Sci.* **100**, 12223–12228.

Taylor, B. L., and DeMaster, D. P. (1993). Implications of non-linear density dependence. *Mar. Mamm. Sci.* **9**, 360–371.

Taylor, B. L., Martinez, M., Gerrodette, T., Barlow, J., and Hrovat, Y. N. (2007). Lessons from monitoring trends in abundance of marine mammals. *Mar. Mamm. Sci.* **23**, 157–175.

Taylor, M. K., Laake, J., McLoughlin, P. D., Cluff, H. D., and Messier, F. (2006). Demographic parameters and harvest-explicit population viability analysis for polar bears in M'Clintock Channel, Nunavut, Canada. *J. Wildl. Manage.* **70**, 1667–1673.

Trillmich, F., and Ono, K. A. (eds) (1991). "Pinnipeds and El Niño: Responses to Environmental Stress." Springer-Verlag, Berlin.

Wade, P. R., *et al.* (24 authors) (2007). Killer whales and marine mammal trends in the North Pacific—a re-examination of evidence for sequential megafauna collapse and the prey-switching hypothesis. *Mar. Mamm. Sci.* **23**, 766–802.

Wickens, P., and York, A. E. (1997). Comparative population dynamics of fur seals. *Mar. Mamm. Sci.* **13**, 241–292.

# Population Status and Trends

JAY BARLOW AND RANDALL R. REEVES

For marine mammals, status is a measure of the size or general HEALTH of a population relative to some management standard. A trend is a measure of the rate at which a population grows or declines over some (usually long) time period. Taken together, status and trends form the basis for assessing whether management objectives are being met for a given population or management unit.

## I. Status

Inherent in the concept of status is the evaluation of populations relative to some standard or metric. Absolute estimates of population size may be included in an assessment of the population status, but an evaluation of status is incomplete without evaluating the significance of this population size relative to some goal. The standards for evaluating the status of populations are typically related either to a harvest or to a conservation objective.

### A. Harvest Objectives

Traditionally, to evaluate harvest objectives, population size was evaluated relative to the population level (MSYL) that would give the maximum sustainable yield (MSY). Populations are expected to decline as a result of harvests, but as population size decreases, the population growth rate is expected to increase to compensate for this harvest (*density dependence*). The implicit assumption is that populations are resource limited, and as density declines, more per-capita resources are available to support enhanced reproduction, survival, or both. Harvest and growth rates may balance each other over a wide range of possible equilibrium population sizes, but typically harvest will be maximized at one specific equilibrium level (MSYL). Although the MSYL concept has persisted for many years, it was widely recognized by the 1970s that, in practice, it was seriously flawed as a basis for management. Management of populations at their MSYL is a knife-edge balancing act that requires constant conditions and near-perfect data. Usually, when populations slip below that level, the drastic management actions that are required cannot be implemented in time to prevent the collapse of the population as an economic resource. This realization has led to more risk-averse management models that strive to keep populations above their MSYL; this higher target level is sometimes called the optimum sustainable yield level.

Currently, few marine mammal species are managed with explicit harvest objectives. Although a moratorium on commercial whaling has been in place since 1986, the INTERNATIONAL WHALING COMMISSION (IWC) still maintains a harvest-based management framework for large whales [baleen whales, sperm whales (*Physeter macrocephalus*), and bottlenose whales (*Hyperoodon* spp.)]. To determine status, population size is compared to a standard that is based on MSYL. Protected stocks (PS) are less than 0.9 MSYL; sustained management stocks (SMS) are between 0.9 and 1.2 MSYL; and initial management stocks (IMS) are greater than 1.2 MSYL. Acknowledging the risk of managing at the MSYL knife edge, the IWC has, since the mid-1970s, limited harvests to 90% of the estimated MSY. The IWC has been working on a "revised management procedure" (RMP) to replace this MSYL-based management when the moratorium on commercial whaling ends. The performance of the RMP has been tested in simulations where the objective is to maintain populations above 72% of their preexploitation population size.

The Stock Assessment Secretariat of the Canadian Department of Fisheries and Oceans coordinates the production of stock assessment reports for hunted species, notably including harp seals (*Pagophilus groenlandicus*), narwhals (*Monodon monoceros*), and bowhead whales (*Balaena mysticetus*). These reports, prepared by scientists in close consultation with representatives of regional bodies and hunting communities, are intended to provide the basis for managing harvests. However, in the absence of a legal framework equivalent to the US Marine Mammal Protection Act (see later), harvest objectives are *ad hoc*. A stock assessment report recently completed for the Hudson Bay-Foxe Basin stock of bowhead whales, for example, invokes "sustainability" as an implicit management objective, with no reference to recovery or maximizing yield.

## B. Conservation Objectives

The goals of conservation efforts can range from preventing the extinction of species to returning populations to their carrying capacity level. The metrics used to measure the conservation status of populations range over this same spectrum.

When populations become very small or are declining rapidly, their status is often determined by estimating the probability of extinction within a defined time period. For example, the "critically endangered" category of the IUCN Red List (IUCN, 2001, 2006) includes species whose probability of EXTINCTION is estimated to be at least 50% within 10 years or three generations, whichever is longer. Clearly this is nature's intensive care unit for the nearly hopeless cases. In the IUCN categories of "endangered" and "vulnerable," the values change to 20% chance of extinction within 20 years or five generations and 10% chance within 10 years or three generations, respectively. The risk of extinction can be estimated using a technique

developed in conservation biology known as population viability analysis (PVA), which evaluates population size, trends in abundance, life history traits, natural variability, trends in habitat loss, and parameter uncertainty. These onerous data requirements have meant that very few marine mammal species or populations have been evaluated using PVA.

Lacking sufficient data to evaluate the risk of extinction in this manner, several surrogate variables may be measured that are highly correlated with extinction risk. Extinction is obviously correlated with declining abundance, and the IUCN uses total (continuing, irreversible, or not-understood) declines of 80, 50, and 20% over 10 years (or three generations) to classify species as critically endangered, endangered, or vulnerable, respectively. We know that small population size is itself an extinction risk factor, and the IUCN uses population sizes of 50, 250, and 1000 mature individuals to classify species into the same three categories. Other important factors that increase extinction risk include (1) having all individuals in a single location, (2) over-dispersal and the resulting loss of mating opportunities and social facilitation, (3) habitat degradation, and (4) extreme fluctuations in population size. The IUCN recognizes the compounding of risk factors and evaluates some of these factors in tandem with population size or trends. Under domestic legislation, the United States maintains a similar endangered species list with categories of "endangered" and "threatened."

The IUCN's development and adoption of quantitative criteria for Red List classifications prompted various national efforts to take a similar approach. For example, the Mammalogical Society of Japan has applied the IUCN criteria to all marine mammal populations in Japanese waters. The status assigned to many of these populations is worse than that indicated by the IUCN's global listing for the species. There are, for example, critically endangered or endangered populations of finless porpoises (*Neophocaena phocaenoides*), striped dolphins (*Stenella coeruleoalba*), short-finned pilot whales (*Globicephala macrorhynchus*), and dugongs (*Dugong dugon*) in Japan, whereas these species, overall, are listed by IUCN as either "vulnerable (VU)," "data deficient (DD)" (inadequate information to assess risk of extinction), or "least concern (LC)" (out of danger but needing continued conservation measures). In Canada, the task of listing species and populations at risk falls on the Committee on the Status of Endangered Wildlife in Canada (COSEWIC). This group consists of technical experts as well as representatives of government agencies and nongovernmental organizations. Its listing decisions are based on status reports prepared by scientists familiar with the species or populations under review. COSEWIC is in the process of developing its own IUCN-style criteria for status determinations.

For species that are above the size at which extinction is a significant risk, status is usually measured relative to historical abundance or environmental carrying capacity (*K*). Even when the "official" definition of status is based on a percentage of *K*, historical abundance is often substituted because carrying capacity is so difficult to measure. Historical abundance is, itself, poorly known for many species. If there are no direct measures of historical abundance, it can be estimated by a method called "back-calculation" based on current abundance, a time series of annual human-caused mortality, and a model of population growth.

The US Marine Mammal Protection Act of 1972 is probably the most far-reaching and proactive national legislation for the conservation of marine mammals. It has two main conservation goals: to keep populations at their "optimum sustainable population" (OSP) levels and to keep populations as "functioning elements of their ecosystem." OSP has been interpreted to be a range from a population's

**P**

maximum net productivity level (typically thought to be between 60 and 80% of $K$ for marine mammals) up to its carrying capacity. Populations below OSP are considered to be "depleted." The US legislation is significant for having explicit goals to maintain populations near their natural levels instead of protecting them only after they have declined to dangerously low levels.

Similarly, in Europe, the Agreement on the Conservation of Small Cetaceans in the Baltic and North Seas (ASCOBANS) has a conservation goal to restore and/or maintain stocks at a level they would reach when there is the lowest possible anthropogenic influence. ASCOBANS has interpreted this objective to mean restoring and/or maintaining populations at or above 80% of $K$.

## II. Trends

An upward or downward trend in population size is obviously a significant component in evaluating the status of a population; however, it is also one of the more difficult components to determine. Population trends have been directly estimated for only a tiny proportion of all cetacean populations. The primary problems are that population size cannot be estimated very precisely and population growth is typically slow.

Cetacean population size is usually estimated from line-transect SURVEYS. Trends can also be based on an index of relative abundance, such as the number of whales seen per hour on standard transects with consistent survey methods. In either case, the precision of the estimates is measured as a coefficient of variation (CV = standard deviation divided by the mean); CVs of 20% are considered very good and CVs of 30–50% are considered typical. Given their life-history constraints, cetacean populations can grow at a maximum rate of about 10% per year, and, for some slow-reproducing species [such as the killer whale (*Orcinus orca*) and sperm whale], maximum growth rates may be as low as 2–3%. There is no similar constraint on the rate at which populations can decline; however, there have been few instances where long-term rates of decline have been found to exceed 8% per year.

Statistical power is a measure of the probability of detecting a significant change in a population if that population is truly growing or declining. Power increases with the actual rate of change in population size and with the sample size (both the amount of survey effort and, more importantly, the length of the monitoring period). As a rule of thumb for cetaceans, at least 10 annual surveys with good precision (CV < 20%) are required to yield a high probability (>80%) of detecting a 50% change in total population size. Taylor *et al.* (2007) found that in the US, where marine mammal monitoring is probably the best in the world, the majority of populations could decline by 50% without reliable statistical evidence of a decline. The situation is made even more difficult for endangered species [such as the baiji (*Lipotes vexillifer*) and vaquita (*Phocoena sinus*)], which become almost impossible to census as they become increasingly rare. A recent large-scale survey for baiji failed to find even a single individual (Turvey *et al.*, 2007), and the species is now considered "possible extinct" by IUCN. Similar to cetaceans, the detection of population trends for dugongs and manatees is extremely difficult.

The ability of scientists to detect trends in land-hauling pinniped populations is considerably better. Although total population size may be difficult to estimate, a segment of the population predictably returns to land to breed, give birth, or molt, and this segment can be counted accurately by ground-based observers or from aerial photographs. Growth or decline in this segment is taken as an index of trend for the entire population, so this kind of count is sometimes referred to as an index count. Index counts may be based, for example,

on the number of pups born or the number of total seals hauled out at the peak of the molting season. If index counts are used over a wide range of population sizes, ancillary research is typically required to ensure that the fraction of animals counted does not vary in a density-dependent manner (e.g., as per capita resources become scarce, the percentage of time spent foraging away from the beach might increase). Another factor making positive trends in pinniped populations easier to detect is that most are capable of growing at faster rates than cetaceans and sirenians. Abundance and trends are much harder to measure for pinnipeds that haul-out or pup on ice.

## See Also the Following Articles

Abundance Estimation ■ Conservation Biology ■ Endangered Species and Populations ■ Management ■ Population Dynamics ■ Stock Assessment.

## References

Gerrodette, T. (1987). A power analysis for detecting trends. *Ecology* **68**, 1364–1372.

Gerrodette, T. (1993). Trends: software for a power analysis of linear regression. *Wildl. Soc. Bull.* **21**, 515–516, Software available at http://mmdshare.ucsd.edu/trends.html.

IUCN. (2001). IUCN Red List Categories and Criteria. Version 3.1. IUCN, Gland, Switzerland.

IUCN [Standards and Petitions Working Group]. (2006). Guidelines for Using the IUCN Red List Categories and Criteria. Version 6.1. IUCN, Gland, Switzerland. Downloadable from: http://app.iucn.org/webfiles/doc/SSC/RedList/RedListGuidelines.pdf.

IWC (1999). "Schedule of the International Convention for the Regulation of Whaling, as amended by the Commission at the 51st annual meeting 1999, and replacing that dated September 1998." International Whaling Commission, Cambridge, UK.

Klinowska, M. (ed.) (1991). "Dolphins, Porpoises and Whales of the World." IUCN, Gland, Switzerland.

Marsh,, H., and Lefebvre, L. W. (1994). Sirenian status and conservation efforts. *Aquat. Mamm.* **20**, 155–170.

Reijnders, P., *et al.* (8 authors) (1993). "Seals, Fur Seals, Sea Lions, and Walrus Status Survey and Conservation Action Plan. IUCN Species Survival Commission, Seal Specialist Group." IUCN, Gland, Switzerland.

Taylor, B. L., and Gerrodette, T. (1993). The uses of statistical power in conservation biology: the vaquita and northern spotted owl. *Conserv. Biol.* **7**, 489–500.

Taylor, B. L., Martinez, M., Gerrodette, T., Barlow, J., and Hrovat, Y. N. (2007). Lessons from monitoring trends in abundance of marine mammals. *Mar. Mamm. Sci.* **23**, 157–175.

Turvey, S. T., *et al.* (10 authors) (2007). First human-caused extinction of a cetacean species? *Biol. Lett.* **3**, 537–540.

Twiss, J. R., Jr., and Reeves, R. R. (eds) (1999). "Conservation and Management of Marine Mammals." Smithsonian Institute Press, Washington, DC.

# Porpoises, Overview

## ANDREW J. READ

The porpoises are 1 of 10 families that constitute the suborder Odontoceti, or the modern toothed whales. The family Phocoenidae consists of six species, distributed in both hemispheres (Table I). For many years, and in some areas still today, there

## TABLE I
### Living Species of Porpoises

| Common name | Scientific name | Distribution |
|---|---|---|
| Harbor porpoise | *Phocoena phocoena* | Coastal Northern Hemisphere |
| Burmeister's porpoise | *Phocoena spinipinnis* | Coastal South America |
| Vaquita | *Phocoena sinus* | Gulf of California |
| Spectacled porpoise | *Phocoena dioptrica* | Southern Ocean |
| Dall's porpoise | *Phocoenoides dalli* | North Pacific |
| Finless porpoise | *Neophocaena phocaenoides* | Coastal South Asia |

has been confusion regarding the terms "porpoise" and "dolphin." In part, this is because some early taxonomic accounts included porpoises in the family Delphinidae, although all recent accounts separate the two groups. To confuse things further, fishermen have traditionally referred to bottlenose dolphins (*Tursiops* spp.) and other delphinid species as "porpoises" in many areas of the US. Despite their shared vernacular names, porpoises and dolphins are as phylogenetically distinct as are dogs and cats. These differences are manifested in their morphology, ecology, and behavior, as outlined below.

## I. Origins

Our knowledge regarding the evolution of porpoises comes from inferences drawn from the fossil record and from the morphology and genotypes of living species. The earliest porpoise known is *Salumiphocaena stocktoni*, discovered in late Miocene strata of southern California, approximately 11 million years ago (Barnes, 1985). These early porpoises appeared at about the same time as the first dolphins. The dates of the origin of the two families are consistent with estimates of divergence between the cytochrome b genes of phocoenids and delphinids. Most other early fossil porpoises are known from the Pliocene in North and South America. Porpoises, dolphins, and monodontids, all members of the superfamily Delphinoidea, are likely to have descended from the Kentriodontidae, an extinct family of odontocetes. Like modern porpoises, the kentriodontids were rather small animals, approximately 2 m in length. Kentriodontids occurred in both the Atlantic and Pacific Oceans, but disappeared approximately 10 million years ago.

Analyses of the genome of living porpoises suggest that the finless porpoise (*Neophocaena phocaenoides*) is the most primitive, or basal, member of the family (Rosel *et al.*, 1995). This view is consistent with morphological evidence. Gaskin (1982) postulated that the earliest phocoenids radiated into temperate waters of both hemispheres from the tropics, where the finless porpoise is still found today. Other researchers (Barnes, 1985) maintain that phocoenids originated in temperate waters of the North Pacific Ocean, where the oldest fossils are located. With our current knowledge, it is not possible to resolve which of these two scenarios are correct. Molecular evidence indicates that all modern porpoises have evolved within the past few million years.

The nature of phylogenetic relationships among porpoises has been the subject of recent debate. Barnes (1985) suggested that the spectacled porpoise (*Phocoena dioptrica*) of the Southern Ocean is closely related to the Dall's porpoise (*Phocoenoides dalli*) of the North Pacific and that these two species should be classified as a subfamily, distinct from the remaining four species. This view has not been supported by analysis of the molecular genome (Rosel *et al.*, 1995), however, which suggests a close phylogenetic relationship among the vaquita (*Phocoena sinus*), Burmeister's porpoise (*Phocoena spinipinnis*), and spectacled porpoise. Most current practices now separate the family into three genera: *Neophocaena*, *Phocoenoides*, and *Phocoena*. The first two genera are monotypic (the finless porpoise and Dall's porpoise) and the third contains four species (harbor porpoise *Phocoena phocoena*, vaquita, Burmeister's porpoise, and spectacled porpoise).

Periodic cycles of global cooling and warming have had profound effects on the modern biogeography of all cetaceans (Gaskin, 1982), and this holds true for the porpoises. For example, the distribution of the vaquita is currently restricted to the northern Gulf of California. In geographical terms, the closest relative of the vaquita is the harbor porpoise, which is found as far south as the coast of central California. Morphological and genetic evidence, however, indicate that the closest living relative of the vaquita is the Burmeister's porpoise, from the coastal waters of temperate South America. It is likely that, during a cool glacial period of the Pleistocene, ancestors of today's Burmeister's porpoises crossed the equator and were subsequently trapped in the Gulf of California by warming equatorial waters. This isolated population eventually gave rise to the vaquita.

The selective factors shaping the evolution of porpoises are only partially understood, but one important feature is paedomorphosis, the retention of juvenile characters in the adult form. The evolution of small size and early sexual maturation, when compared to most delphinids, is likely driven by paedomorphosis. This phenomenon may also explain convergent evolution with the delphinid genus *Cephalorhynchus*. The morphology, ecology, and behavior of these dolphins are remarkably similar to porpoises in many ways and they appear to have CONVERGED with porpoises in many aspects of their biology.

## II. Morphology

Porpoises share many morphological attributes that distinguish them from other small cetaceans (Fig. 1). All six species are small, with no member of the family exceeding 250 cm in body length. Porpoises are stocky, robust animals, and lack the rostrum common to most delphinids. The appendages of most species are relatively small and, in the case of the finless porpoise, the dorsal fin is missing altogether. Many of these morphological features may be related to the thermal biology of porpoises and, particularly, to the challenges posed by small body size in a cold and conductive medium.

The skulls and postcranial skeletons of porpoises exhibit pronounced paedomorphosis (Barnes, 1985). In particular, the skulls of all porpoises are characterized by short rostra, large and rounded braincases, and delayed fusion of cranial sutures during ontogeny. As a result, the skulls of adult porpoises resemble those of juvenile specimens of other species. The same is true for some features of the postcranial skeleton. For example, the vertebral epiphyses of adult male harbor porpoises rarely display ankylosis, a feature used to diagnose physical maturity in delphinids. Several cranial features are diagnostic of the family, including raised protuberances on the premaxillae and the spatulate shape of the teeth. The latter feature is an easy way to distinguish porpoises from dolphins, which have conical teeth.

An unusual morphological feature of all phocoenid species, except for the Dall's porpoise, is the presence of epidermal tubercules along the leading edge of the dorsal fin. These small, raised protuberances develop shortly after birth and usually occur in several rows. Even the finless porpoise, which lacks a dorsal fin, possesses several rows of tubercules on its dorsal ridge. Tubercules are most prominent in the

**P**

**Figure 1** *This harbor porpoise* (Phocoena phocoena) *illustrates typical external morphology of the family Phocoenidae. Porpoises are small, stocky, robust cetaceans with short rostra and large, rounded braincases. Photograph by Ari Friedlaender.*

Burmeister's porpoise and their presence provides the basis for the specific name *spinipinnis*. The function of these structures is unknown.

The pigmentation patterns of phocoenids vary considerably, although there are several features common to the family. Most porpoises possess patches of dark pigmentation surrounding the eye, although these patches are of varying prominence and contrast. The dark eye patch is particularly well developed in the spectacled porpoise. It is encircled by a narrow white line, which gives the species both its common and specific names. All porpoises (and most dolphins) exhibit a bridle, a system of stripes extending from the eye and blowhole to the apex of the melon. And most porpoises exhibit the general pattern of countershading, common to many odontocetes, of a dark dorsal cape and a lighter abdomen.

## III. Ecology and Behavior

Our general knowledge of the ecology and behavior of porpoises is limited and these areas continue to be fertile subjects for research. Relatively few researchers have had the opportunity to study live vaquita, Burmeister's, or spectacled porpoises. The latter species is known primarily from a small number of STRANDINGS in remote areas of the southern hemisphere. In contrast, our knowledge of the ecology and behavior of harbor and finless porpoises is much more extensive (Read, 1999).

The radiation of phocoenids over the past few million years has allowed porpoises to colonize a variety of habitats in both hemispheres. The harbor porpoise, Burmeister's porpoise, and vaquita are coastal in nature, and the range of the finless porpoise extends well upstream into major river systems. The Dall's porpoise and spectacled porpoise are primarily pelagic animals.

There is little evidence of cooperative feeding in the family and it appears that individual porpoises usually forage alone. Most coastal species, such as the harbor and Burmeister's porpoise, feed on small pelagic fish, such as herring, anchovies, and capelin, and supplement this diet with demersal fishes. In contrast, Dall's porpoises feed primarily on the small mesopelagic fishes and squid that comprise the deep scattering layer. To date, DIVING BEHAVIOR has been studied for only the harbor porpoise, which can dive to depths of over 200 m. Their small size makes it unlikely that any species of porpoise, even the pelagic Dall's porpoise, are exceptional divers. Dall's porpoises likely

take advantage of the daily vertical migration of their prey and feed at night when these mesopelagic fishes and squid are near the surface.

Porpoises are among the smallest and most paedomorphic cetaceans and these aspects of their biology are manifested in their reproductive biology. Compared to many delphinids, porpoises grow rapidly and reach sexual maturity at an early age, features promoted by paedomorphosis. For example, many female harbor porpoises attain sexual maturity in their third year of life, at an age where bottlenose dolphins are still accompanying their mothers. Both Dall's and harbor porpoises are capable of annual reproduction, and females of these species are often simultaneously lactating and pregnant. This demanding reproductive schedule is accompanied by a relatively brief lifespan—very few porpoises live longer than 20 years.

Although our knowledge of the social behavior of phocoenids is limited, we can say that, unlike many pelagic delphinids, porpoises usually occur alone or in small, fluid groups. This is also reflected in patterns of strandings. Porpoises generally strand singly, never in the large groups witnessed in some odontocete species. Occasionally, hundreds of porpoises have been observed together, but these are likely temporary aggregations, rather than stable groups. In general, it appears that porpoises typically exist in fission–fusion societies in which associations among individuals are extremely dynamic. The only long-term association known to occur is between a lactating female and her dependent calf. The duration of lactation varies among species, and perhaps among populations within species, but is unlikely to surpass 2 years in any phocoenid.

With the exception of the Dall's porpoise, porpoises are generally shy, unobtrusive animals that are difficult to sight and follow at sea. Unlike many delphinids, porpoises seldom, if ever, leap clear of the water. A typical surfacing sequence is characterized by a series of gentle rolls at the surface, followed by a longer submergence. Occasionally on calm days, harbor porpoises will lie quietly at the surface for short periods. Individuals seldom approach boats or ride the bow or stern waves of vessels. In contrast, Dall's porpoises are fast swimmers that often make a characteristic splash or "rooster-tail" when they surface. Dall's porpoises are also the only phocoenid that regularly approaches boats to ride the bow wave, a behavior they may have modified from riding the pressure waves produced in front of swimming baleen whales.

## IV. Conservation

Most populations of phocoenids are affected adversely by human activities. Hunting, by-catches in commercial fisheries and habitat degradation have had profound impacts on the health and demography of many affected populations (Bjørge and Donovan, 1995). Only the spectacled porpoise of the Southern Ocean is largely free of the effects of human influences, although even this species has been taken as by-catch in fisheries off Tierra del Fuego. The nature of human activity posing a conservation threat varies from species to species, as does the conservation status of each species.

Porpoises living in coastal areas are affected by the modification, degradation, and destruction of habitat by humans. Particularly affected in this manner is the finless porpoise, which inhabits the temperate and tropical coasts of Asia and the Indian subcontinent, as well as the Yangtze River of China. Human activities, such as dredging, reclamation, pollution, and intense vessel traffic, are felt most keenly in coastal and riverine habitats where the density of humans is high. Thus, particular concern has been expressed for populations of finless porpoises in the South China Sea, adjacent waters, and the Yangtze River. Other coastal species of porpoises are not immune from such effects. Harbor porpoises, for example, may be excluded from portions of their habitat in the North Atlantic and North Pacific by the use of high-intensity acoustic devices designed to keep pinnipeds away from salmon mariculture sites.

Harbor and Dall's porpoises have been hunted for their meat and blubber for many centuries. In the Danish Belt Sea, an annual hunt for harbor porpoises occurred from the fourteenth century until the early twentieth century. More than 1000 porpoises were taken annually between 1834 and 1874 in this hunt. Harbor porpoises have also been harvested from other areas, particularly the Bay of Fundy and western Greenland. A large-scale harpoon fishery for Dall's porpoises still exists off the coast of northern Japan. This hunt supplies meat for the domestic Japanese market, partially offsetting the reduction in the availability in whale meat following the moratorium in commercial whaling that began in 1986. The effects of this exploitation on populations of Dall's porpoises are not fully understood, but the magnitude of catches (more than 40,000 in 1988) is a cause for concern.

The most pressing and pervasive threat to populations of phocoenids is by-catch in commercial fisheries (Jefferson and Curry, 1994). Most porpoises are taken as by-catch in gill nets, often anchored on the bottom to catch demersal species. Porpoises seldom survive entanglement in this type of fishing gear. The causes of by-catches are unclear; it is uncertain, for example why animals with a sophisticated system of ECHOLOCATION do not detect and avoid fishing nets (Nachtigall et al., 1995). Nevertheless, all species of porpoises are taken as by-catch, sometimes in staggering numbers. The estimated annual mortality of harbor porpoises in Danish bottom gill net fisheries in the North Sea was almost 7000 between 1994 and 1998. The consequences of such large by-catches are undocumented, but have likely lead to past depletion. In some areas, such as the coastal waters of Peru, Burmeister's porpoises taken as by-catch gained commercial value for human consumption, so that the distinction between by-catches and directed hunting became blurred. By-catch is the most serious threat to the small endemic population of vaquitas in the upper Gulf of California. Only a few hundred vaquitas remain, and the species is still subject to by-catch in several gill net fisheries. The vaquita is the most endangered marine cetacean; its future depends on current conservation initiatives in Mexico. Unless all entangling nets are removed from the range of this species, it will likely join the Chinese river dolphin, or baiji (*Lipotes vexillifer*), as the second species of cetacean driven to extinction in modern times.

## See Also the Following Articles

Cetacean Ecology ■ Delphinids, Overview

## References

Barnes, L. G. (1985). Evolution, taxonomy and antitropical distribution of the porpoises (Phocoenidae, Mammalia). *Mar. Mamm. Sci.* **1**, 149–165.

Bjørge, A., and Donovan, G. P. (eds) (1995). "Biology of the Phocoenids." Reports of the International Whaling Commission, Cambridge, UK, Special Issue 16.

Gaskin, D. E. (1982). "The Ecology of Whales and Dolphins." Heinemann Educational Books, London.

Jefferson, T. A., and Curry, B. E. (1994). A global review of porpoise (Cetacea: Phocoenidae) mortality in gillnets. *Biol. Conserv.* **67**, 167–183.

Nachtigall, P. E., Lien, J., Au, W. W. L., and Read, A. J. (eds) (1995). "Harbour Porpoises: Laboratory Studies to Reduce Bycatch." De Spil Publishers, Woerden, The Netherlands.

Northridge, S. P., and Hofman, R. J. (1999). Marine mammal interactions with fisheries. *In* "Conservation and Management of Marine Mammals" (J. R. Twiss, Jr., and R. R. Reeves, eds), pp. 99–119. Smithsonian Institution Press, Washington, DC.

Read, A. J. (1999). "Porpoises." Worldlife Library, Colin Baxter Photography, Grantown-on-Spey, Scotland.

Read, A. J., Wiepkema, P. R., and Nachtigall, P. E. (eds) (1997). "The Biology of the Harbour Porpoise." De Spil Publishers, Woerden, The Netherlands.

Ridgway, S. H., and Harrison, R. (eds) (1999). "Handbook of Marine Mammals. Volume 6: The Second Book of Dolphins and the Porpoises." Academic Press, San Diego, CA.

Rosel, P. E., Haygood, M. G., and Perrin, W. F. (1995). Molecular relationships among the true porpoises (Cetacea: Phocoenidae). *Mol. Phylogenet. Evol.* **4**, 463–474.

# Predation on Marine Mammals

## DAVID W. WELLER

## I. Introduction

Although marine mammals are regarded as accomplished and sophisticated hunters, they too are preyed upon by a variety of terrestrial, avian, and aquatic predators. Predation is an ecological factor of significant influence on the behavior and organization of animal societies in general, and the need for protection from predation has likely been an important factor in the evolution of most marine mammal social systems. While the risk of predation is of little or no concern for some species, others exist under high levels of predatory pressure. A large portion of all marine mammals, ranging in size from the enormous blue whale (*Balaenoptera musculus*) to the relatively small sea otter (*Enhydra lutris*), are subjected to varying levels of predation. Responses to predators are complex, and include detection and avoidance, fleeing, seeking habitat features for cover, and active defense by individuals as well as coordinated groups.

While the topic of predation is expansive and multidimensional, the focus of the following review centers on the hunting and consumption of marine mammals by their predators. The definition of predation used here excludes parasitism, filter feeding, scavenging (carrion eating), or browsing, and is limited to situations in which an animal expends time and energy to locate living prey, and exerts additional effort to kill and consume it. Therefore, predation is distinguished from other forms of foraging in that it concludes with the death of an animal that offers some resistance against being discovered and/or being harmed.

## II. Predation on Sirenians

The relatively slow moving and rather lethargic behavior of sirenians (manatees, *Trichechus* spp., and dugongs, *Dugong dugon*) makes them seem particularly vulnerable to predation. However, manatees and dugongs actually have few known natural predators, and appear to experience only occasional mortality due to predation. Although large sharks, crocodiles (*Crocodylus* spp.), and killer whales (*Orcinus orca*) are all considered to be potential predators, few records exist to confirm these suspicions (Anderson and Prince, 1985; Reeves *et al.*, 1992; Wells *et al.*, 1999). Evidence of predation, including tooth-scarring indicative of unsuccessful attacks by predators, has been only rarely observed during long-term field studies on West Indian manatees

P

(*Trichechus manatus*) in Florida and dugongs off Australia. The limited presence of marine predators in the relatively warm and shallow nearshore waters, rivers, and bays where these animals forage on marine vegetation may partially explain the paucity of observed predatory interactions. Further, the particularly thick skin and exceedingly dense bone characteristic of the sirenians may render them rather unpalatable and serve to deter potential predators.

Predation on sirenians does occur, however. For example, in South America, Amazon manatees (*Trichechus inunguis*) have been reported to be preyed on by jaguars (*Panthera onca*) and large sharks, and marine crocodiles may occasionally kill dugongs throughout their distribution (Reeves *et al.*, 1992; Wells *et al.*, 1999). Off western Australia, predation by killer whales on adult dugongs has been reported, including one occasion when 10 killer whales were observed attacking a group of approximately 40 dugongs in shallow water (Anderson and Prince, 1985). During this incident the dugongs were huddled tightly together in an antipredator response, while pieces of flesh and integument floated nearby in blood stained water. Residents of western Australia have also implicated "black porpoises" as predators of dugongs. However, what species these "porpoises" represent is entirely unclear. While some authors suggest that these "porpoise" attacks were likely to be killer whales, such records may also refer to one of several other mammal-killing cetaceans like the false killer whale (*Pseudorca crassidens*). It is conceivable, of course, that predation on sirenians is considerably higher than has been observed and reported. Predatory attacks on young animals, for example, may be particularly successful, and information regarding predator related mortality of species such as the West African (*Trichechus senegalensis*) and Amazon manatees that mainly occur in areas inaccessible to researchers is largely unknown.

### III. Predation on Mustelids

Although sharks and killer whales represent the primary predators of sea otters, several terrestrial and avian predators have also been documented (Reeves *et al.*, 1992; Wells *et al.*, 1999). Coyotes (*Canis latrans*) are known to prey on recently weaned otters in parts of Alaska, and Russian brown bears (*Ursus arctos*) occasionally kill otters that haul-out along the shores of the Kamchatka Peninsula, Russia. Near Amchitka Island, Alaska, bald eagles (*Haliaeetus leucocephalus*) hunt sea otter pups (Sherrod *et al.*, 1975). Pups are particularly vulnerable to avian predation as they float unattended at the surface while their mothers are preoccupied with searching for food. The extraordinary buoyancy of young otter pups prevents them from readily submerging, and greatly reduces their chances of escaping attack by diving. Observations of bald eagles nabbing young otter pups from the surface of nearshore waters confirm that eagles use a hunting strategy similar to that used when capturing large fish. That is, pups are gathered from the water in the talons of an eagle, flown to the nest location, and meticulously devoured. Studies conducted on Amchitka Island between the 1950s and 1970s found that up to 28% of the prey remains in eagle nests were from sea otters. Interestingly, some nests contain high levels of otter remains while other nests have none. This finding suggests that some individual eagles may actually specialize on hunting sea otter pups.

While terrestrial and avian predation account for only a small portion of sea otter mortality, sharks represent a more formidable and common predator (Ames *et al.*, 1996). White shark (*Carcharodon carcharias*) attacks on sea otters along the California coast are thought to account for 9–15% of the total otter mortality recorded in this region. Curiously, there is little evidence from examination of white shark

stomach contents to suggest that sea otters are actually eaten by the individuals that attack them. Instead, otters are stalked and killed by white sharks off California but are apparently abandoned prior to consumption. The absence of sea otter remains in white shark stomachs cannot be considered conclusive at this time however, as only a small number of stomachs have been available for examination. Although other shark species are also suspected to occasionally kill sea otters, few specific details are available.

Killer whales are known predators of sea otters (Estes *et al.*, 1998; Hatfield *et al.*, 1998), but the small number of observed attacks suggests that otters are not preferred prey. Nonetheless, a substantial increase in the number killer whale attacks on sea otters was documented between 1992 and 1996 and corresponded with a notable decline in sea otter population levels over a large part of their western Alaska distribution (Estes *et al.*, 1998). It is unclear if this increase in observed killer whale attacks was due to greater observation effort, or represents a real increase in sea otter predation. If this change is merely related to increased observation effort, than killer whale predation on sea otters may not be as uncommon as previously suggested. However, if this finding represents a true increase in the rate of sea otter attacks it may be related to the relatively recent declines of other killer whale prey, such as Steller sea lions (*Eumetopias jubatus*) and harbor seals (*Phoca vitulina*).

### IV. Predation on Pinnipeds

Of all the marine mammal groups, pinnipeds are probably subjected to the highest level of predation. While some pinniped species experience little or no predation pressure, others are so intensively hunted that important aspects of their natural history, including reproductive strategies, have evolved in response (Trillmich, 1984; Reidman, 1990; Wells *et al.*, 1999). Not even the largest pinnipeds such as the walrus (*Odobenus rosmarus*), bearded seal (*Erignathus barbatus*), and elephant seals (*Mirounga* spp.) are free from predation. Terrestrial predators of pinnipeds are particularly abundant in the subpolar and polar regions of the Northern Hemisphere, usually appearing in the form of polar bears (*Urus maritimus*) and Arctic foxes (*Alopex lagopus*). Southern Hemisphere ice-seals are free from land predators, but instead have fierce aquatic predators such as the leopard seal (*Hydrurga leptonyx*) to contend with. Pinnipeds in temperate and tropical latitudes experience reduced terrestrial predation but are subjected to increased levels of attack by aquatic predators such as sharks and killer whales. When comparing Northern Hemisphere Arctic pinnipeds to Southern Hemisphere Antarctic pinnipeds, clearly divergent predator avoidance tactics are apparent. Arctic pinnipeds escape land predators by fleeing into the water while Antarctic pinnipeds escape aquatic predators by retreating onto ice.

All pinnipeds require a land or ice substrate for pupping, and this facet of their natural history makes them particularly vulnerable to attack in regions where terrestrial predators are present (Reidman, 1990). Golden jackals (*Canis aureus*), for example, are common at a Mediterranean monk seal (*Monachus monachus*) colony on the western coast of Mauritania, and have been reported to consume freshly dead monk seals and are suspected to also prey on living pups. Freshwater pinnipeds in Russia's Lake Baikal and in the Caspian Sea (*Pusa sibirica* and *P. caspica*, respectively) have no aquatic predators, but instead have an unusually high number of terrestrial adversaries. Wolves (*Canis lupus*) and eagles prey on newborn Caspian seals (*Pusa caspica*), and brown bears occasionally hunt Baikal seals (*Pusa sibirica*). Ringed seal pups (*Pusa hispida*) inhabiting Finland's Lake Saimaa and Russia's Lake Ladoga are preyed upon by red foxes

(*Vulpes vulpes*), and suspected to also suffer some level of mortality due to attacks by ravens (*Corvus corax*), wolves, dogs, wolverines (*Gulo gulo*), and other terrestrial predators (Reeves *et al.*, 1992). Similarly, brown bears, wolves, and avian predators including eagles and ravens, sometimes also kill spotted seals (*Phoca largha*) in the Sea of Okhotsk. Glaucous gulls (*Larus hyperboreus*) and ravens may occasionally kill ringed seal pups, and gulls sometimes peck at the eyes of gray seal pups (*Halichoerus grypus*) resulting in some level of mortality (Reeves *et al.*, 1992).

Additional terrestrial predators also hunt pinnipeds at their haul-out sites. Coyotes, for example, prey on harbor seal pups in the Pacific Northwest, and are responsible for at least 16% of the pup mortality within Puget Sound, Washington (Steiger *et al.*, 1985). Similarly, bears and mountain lions (*Felis concolor*) may have historically preyed on northern elephant seals (*Mirounga angustirostris*) along the California coast (Reidman, 1990). In the Southern Hemisphere, pumas (mountain lions) have been reported to prey on southern sea lion pups (*Otaria flavescens*). South African fur seals (*Arctocephalus pusillus*) that breed along the mainland coast of the southern African continent are preyed upon by brown hyenas (*Hyaena brunnea*) and black-backed jackals (*Canis mesomelas*), and South American sea lion pups (*Arctocephalus australis*) are probably attacked by mountain lions (Reeves *et al.*, 1992).

Arctic foxes have been described as hunters of small animals and birds, and as a scavenger of marine mammal remains left by polar bears. However, in parts of the eastern and western expanses of the Beaufort Sea, this fox is considered an active predator of newborn ringed seal pups (Smith, 1976). In early spring, ringed seals give birth and rest in "subnivean birth lairs"—ice caves complete with breathing holes constructed beneath the snow. These lairs provide both shelter from cold temperatures and protection from predators by providing a physical barrier that makes it more difficult for surface predators to detect a newborn pup. Nevertheless, foxes and polar bears enter and kill pups concealed within their subnivean homes with relative frequency. Keen olfaction allows foxes to locate lairs that may be buried under as much as 150 cm of snow. In the Beaufort Sea, Arctic foxes enter about 15% of the birth lairs present within an area. Although the annual average predation rate by Arctic foxes on ringed seal pups is about 26%, rates as high as 58% have been recorded (Smith, 1976). Ringed seals are also preyed upon by polar bears, and may occasionally be attacked by red foxes, wolverines, wolves, dogs, and several avian predators. As such, ringed seals are subjected to perhaps the highest level of predation experienced by any of the marine mammals.

### A. Polar Bears

Throughout their circumpolar range, the major prey of polar bears consists of pinnipeds. Polar bears are versatile predators and are well adapted for catching Arctic pinnipeds (Stirling, 1988). Predation is particularly heavy on pups, as they represent an easily obtained food resource. Foraging strategies employed by polar bears range from sit-and-wait tactics to active stalking and pursuit of seals on ice and in the water. When stalking seals on ice, bears "creep" along with their heads held low, often momentarily hiding behind snowdrifts and irregularities in the ice. Despite their relative stealth and excellent ability to detect prey by olfaction, bears often have little success sneaking up on seals (Stirling, 1988). Observations of bears hunting, and in at least one instance capturing, free-swimming seals in ice-free waters, have also been reported. One of the sit-and-wait strategies employed by polar bears occurs while hunting ringed seals. Ringed seals forage for food under ice-covered waters throughout the winter, and must therefore maintain breathing holes in which to surface. Polar bears seek out such breathing holes, and often patiently await the arrival of an unsuspecting seal. When a seal surfaces in the hole for a breath of air, the bear quickly grasps it and drags it from the water onto the ice.

The ringed seal is a main staple of polar bear diet, although in the Canadian Arctic, bearded seals and harp seals (*Pagophilus groenlandica*) are taken to a lesser extent (Reeves *et al.*, 1992). Harp and hooded seals (*Cystophora cristata*) are particularly vulnerable to predation on the spring pupping grounds, where polar bears may kill more pups than can be consumed. In Alaska, most of the ringed seals attacked by polar bears are over 6 years of age, while in the Canadian Arctic it is mainly 1 to 2-year-old seals that are killed. Polar bears are largely unsuccessful hunting adult ringed seals due to their nearly continuous antipredator vigilance. This vigilance behavior is characterized by constant head-lifting and scanning of the nearby environment for the presence of predators. In late spring, polar bears enter a period of intense feeding that corresponds with the onset of the ringed seal pupping season. During this time, bears prey heavily on pups by digging into birth lairs; adult female seals attempting to protect their pups are also occasionally killed.

Walruses are sometimes preyed upon by polar bears, but this massive obobenid represents a formidable adversary quite capable of killing predatory bears (Stirling, 1984). The extent of polar bear predation on walruses is not well known, and is likely to vary from region to region. Walrus calves, young juveniles, and sick individuals are most vulnerable to polar bear predation. While hunting walruses, bears often cause entire hauled out herds to "stampede" into the water by rushing toward them. Although most individuals in the stampede easily escape approaching bears, calves or young animals may be crushed or injured in the ensuing chaos, making subsequent capture substantially easier.

### B. Pinnipeds

Several pinniped species are recognized as predators of other pinnipeds, and in some locations are responsible for a significant portion of the annual mortality incurred by regional populations (Reidman, 1990). The most ferocious pinniped predators include the leopard seal in the Southern Hemisphere and the walrus in the Northern Hemisphere. In addition, several sea lion species are notorious for feeding on pinnipeds. Two types of pinniped–pinniped predation occur, one at the intraspecific level (cannibalism, within species) and another at the interspecific level (between species). In some cases, particular individuals (usually males) specialize in the predation of pinnipeds. For instance, young male Steller sea lions are known to prey on harbor seals off Alaska, and have been noted to account for approximately 4–8% of the mortality reported for Northern fur seal pups (*Callorhinus ursinus*) at St. George Island, Alaska (Gentry and Johnson, 1981). Adult male Steller sea lions may also prey on other pinnipeds, as was recorded for one individual at Año Nuevo, California, who was observed feeding on a small California sea lion (*Zalophus californianus*). Similarly, Southern sea lions have been observed preying on South American fur seals (*Arctocephalus australis*), and at Punta San Juan, Peru, over 8% of the fur seal pups are killed by marauding sea lions during the breeding season. Off Macquarie Island in the sub-antarctic, one young male Hooker's sea lion (*Phocarctos hookeri*) was thought to be responsible for the mortality of 43% of the fur seal pups (*Arctocephalus gazella* and *A. tropicalis*) from a particular year. At the Snares Islands, New Zealand, Hooker's sea lions have also been observed to prey on New Zealand fur seal pups (*Arctocephalus forsteri*). Finally, gray seals have been reported to consume pups of their own species, but it is unclear if this represents actual predation or merely cannibalistic scavenging of beach cast carcasses.

**P**

*1. Walruses*  Walruses are primarily bottom or benthic feeders whose diet consists largely of bivalve mollusks, a variety of invertebrates, and fish. In addition, they also prey on marine mammals, and are known to feed on bearded seals, ringed seals, largha (spotted) seals, harp seals, and young walruses (Lowry and Fay, 1984). Adult and subadult male walruses are typically responsible for pinniped kills, but females in the Chukchi Sea have also been observed eating seals. Some walruses are habitual predators of other marine mammals. Individuals that regularly attack seals develop massive chest and shoulder muscles, have long and slender tusks, and their upper torsos and normally ivory colored tusks are stained amber from consuming the oil-rich blubber of their prey (Reeves *et al.*, 1992). In general, walruses kill pups and young individuals, but on occasion mature adult pinnipeds are also taken. Observations of attacks on harp seal pups and bearded seals are characterized by walruses impaling the prey with their tusks. Although very little of the skeletal muscle and bone of their mammalian prey are consumed, walruses methodically devour most, if not all, of the highly caloric hide and blubber.

Only since the 1970s have reports of seal-eating walruses become common (Lowry and Fay, 1984). This relatively recent phenomenon has been linked with the almost doubling of the Pacific walrus population between the 1960s and the early 1980s. Although Pacific walrus numbers are currently thought to be in decline, the nearly 20 year increase in population size certainly elevated the probability of contact between walruses and other pinniped species, and may help to explain the greater use of seals as a food source in the past several decades.

*2. Leopard Seals*  The leopard seal is known to prey on penguins, sea birds, fish, squid, krill, and pinnipeds. In certain parts of their range, pinnipeds are an important part of leopard seal diet, while in other areas pinnipeds are rarely taken (Reidman, 1990). Leopard seals commonly hunt a variety of pinnipeds, but young crabeater seal pups (*Lobodon carcinophaga*) are probably the most frequently attacked and form an important part of the leopard seal diet between November and January (Siniff and Bengtson, 1977). After January, crabeater seal pups have physically developed to the point where they are better able to escape leopard seal predation, and the rate at which they are taken declines.

Parallel tooth scars resulting from unsuccessful leopard seal attacks are quite common on crabeater seals. A study of crabeater seals in 1976 reported that 78% of 85 adult seals handled for research purposes had scars likely to have resulted from interactions with leopard seals (Siniff and Bengtson, 1977). Fresh wounds were far more common on subadults than adults, suggesting that immature animals up to the end of their first year were most likely to be attacked, and it is thought that pups younger than 6 months are probably unlikely to survive encounters with leopard seals. The relatively high level of predation on crabeater seals is believed to represent a food source potentially more important to leopard seals than either krill or penguins.

*3. Sharks*  Sharks represent an important predatory threat to a variety of temperate and tropically distributed pinnipeds (Reidman, 1990; Wells *et al.*, 1999). It is probable that all pinniped species, with the exception of inland lake seals, experience some level of shark predation. While the extent of shark attack on pinnipeds is not understood, it is nevertheless thought to play an important role in the population dynamics, life history, and behavior of some pinniped populations. For example, the high incidence of attacks by tiger (*Galeocerdo cuvier*) and white-tip sharks (*Triaenodon obesus*) on Galápagos fur seals (*Arctocephalus galapagoensis*) is thought to

**Figure 1**  *Northern elephant seal* (Mirounga angustirostris) *with shark-inflicted wound near the hindflippers. Photo by Thomas A. Jefferson.*

have contributed to their exceptionally long 3-year period of maternal investment. It has been suggested that this extended period of maternal care reduces the amount of time pups need to spend in the water foraging, which in turn significantly reduces the risk of shark predation (Trillmich, 1984).

White sharks are a common predator of pinnipeds, with seals, sea lions, and fur seals regarded as preferred prey in some regions of the world because of the high lipid stores contained within their blubber (Reidman, 1990). Gray seals and harbor seals are commonly hunted by white sharks off eastern Canada, and northern elephant seals, Steller sea lions, harbor seals, and California sea lions represent common shark prey in the northeastern Pacific. Southern Hemisphere white sharks focus their attacks primarily on fur seals off South Africa, Australia, New Zealand, South America, and the Galápagos Islands, but also occasionally prey on New Zealand and Australian (*Neophoca cinerea*) sea lions (Reidman, 1990).

Off central and northern California, the diet of white sharks consists largely of pinnipeds (Ainley *et al.*, 1981; Le Boeuf *et al.*, 1982). In particular, white sharks are a major predator of northern elephant seals throughout their entire breeding range, with seals of all age and sex classes vulnerable to attack. A large white shark is capable of killing and consuming elephant seals weighing as much as 500 kg and approximately 3.5 m in length (Reidman, 1990).

The hunting behavior of white sharks on northern elephant seals has been well described near the Farallon Islands off northern California (Ainley *et al.*, 1981; McCosker, 1985). White sharks typically attack elephant seals at or near the surface and usually within several kilometers of the islands. In most cases, white sharks approach their prey from below and to the rear, grasp them in their teeth and carry them underwater, release them, and then wait for the prey to die; usually as a result of excessive blood loss. Many shark attacks are unsuccessful, as evidenced by the high incidence of shark-related lacerations and scars on the bodies and appendages of pinnipeds that escape capture (Fig. 1). The nature of these injuries suggests that when attacked from the rear, foreflipper swimming sea lions are more likely to escape than are hindflipper swimming elephant and harbor seals. Most sea lions with evidence of shark-related injuries have lower body and hindflipper injuries, whereas the majority of surviving elephant and harbor seals bear upper body injuries. While sharks sometimes inflict massively disfiguring wounds to pinnipeds, most injured seals that make it to land appear to survive. However, pregnant elephant seals that withstand shark-related injuries usually lose their pups, give birth to a stillborn, abandon the pup shortly after birth, or fail to wean the pup successfully.

In tropical regions, white sharks are less numerous and white-tip reef, gray reef (*Carcharhinus amblyrhynchos*), and especially tiger sharks represent the major pinniped predators. Tiger sharks hunt Hawaiian monk seals (*Monachus schauinslandi*) of all ages off the northwestern Hawaiian Islands (Alcorn and Kam, 1986). Although other predators like hammerhead (*Sphyrna* spp.) and mako sharks (*Isurus* spp.) also occur off northwestern Hawaii, they apparently do not attack monk seals with any regularity. Gray reef sharks are frequently present when tiger sharks kill monk seals, but their presence is thought to represent scavenging rather that direct predation.

A high number of monk seals bear shark-inflicted wounds and scars, indicating that not all predatory attacks are successful. Adult male monk seals seem to have the highest incidence of scarring, suggesting that animals of other age classes are less likely to survive attack (Hiruki *et al.*, 1993). Highly scarred males may also be attributed to the elevated aggressiveness in males during the breeding season, and their propensity to attack or chase approaching tiger sharks. In addition to direct kills of monk seals, the severity and timing of nonfatal injuries to individual females may reduce overall reproductive success. Field observations confirm that female monk seals suffering major, but nonfatal, shark-related injuries have shorter mean lactation periods, and overall lower pup survival. It has been suggested that the combination of lethal and nonlethal tiger shark attacks on Hawaiian monk seals may be hindering the recovery of this endangered population.

Pinnipeds (and cetaceans) are regularly tormented by a diminutive pest called the cookie-cutter shark (*Isistius brasiliensis*). This small squaloid shark, ranges in size from 14 to 50 cm, and inhabits deep tropical and subtropical waters of the Atlantic, Pacific, and Indian Oceans. By use of rasping jaws and teeth well suited to cutting, cookie cutter sharks attach themselves to their marine mammal victims and remove small circular plugs of skin and blubber. While these attacks are nonlethal, they represent a peculiar form of predation that falls outside the definition set forth at the beginning of this review, but is nevertheless of importance to recognize.

*4. Killer Whales* The diet of killer whales varies considerably between and within geographic regions. Some forms of killer whales, termed "transients," are mammal eaters, while others, termed "residents," base their diet on fish (Ford *et al.*, 1994; Ford and Ellis, 1999). Mammal-eating killer whales have been observed to hunt at least 14, and are suspected to take as many as 24, species of pinnipeds (Jefferson *et al.*, 1991). All pinniped species, with the exception of inland lake seals and the monk seals, probably endure some level of killer whale predation.

Pinnipeds comprise a substantial part of the diet for some transient killer whale populations, and observations of predatory events have been witnessed in a variety of locations from around the world. In subpolar and polar areas, where killer whales are most abundant, reports of attacks on pinnipeds are particularly common. Killer whales attack pinnipeds in both offshore and nearshore regions, and often in close proximity to terrestrial haul-out sites (Fig. 2).

Of all the pinniped species hunted by killer whales, southern elephant seals (*Mirounga leonina*), southern sea lions, harbor seals, Steller sea lions, walruses, and California sea lions have been most commonly recorded as prey species. During a 25-year period, approximately 62% of the transient killer whale attacks observed off British Columbia and Washington were on harbor seals (Ford and Ellis, 1999). Harbor

**Figure 2** *Killer whale* (Orcinus orca) *patrolling a harbor seal* (Phoca vitulina) *haul-out site. Photo by Robin W. Baird.*

seals are particularly abundant in this part of the world, and appear to be relatively easy prey for killer whales to capture and kill, perhaps accounting for the apparent dietary preference on this pinniped species. Steller sea lions, which account for about 7% of all observed attacks, are far more difficult to capture, and the large size obtained by adult males combined with their pronounced canines make attack potentially more dangerous. Other less frequently taken pinniped prey includes California sea lions and northern elephant seals.

Pinnipeds are attacked by lone killer whales and by groups ranging in size from 2 to 30 or more, but the majority of reported attacks are by pods of 10 or less. Killer whales have often been referred to as "pack hunters" because of their tendency to forage cooperatively and employ coordinated maneuvers to capture mammalian prey (Baird and Dill, 1995; Baird, 2000). A well-described example of this coordination was witnessed in the Antarctic, where a group of killer whales was observed to work together to generate a wave large enough to sweep a crabeater seal off an ice floe and into the water so that it could be captured (Smith *et al.*, 1981).

Killer whales use a variety of strategies to kill pinnipeds, including ramming them with their rostrums or heads, flinging seals and sea lions high into the air with an abusive slap of their tail flukes, and violently shaking prey while grasped tightly in their mouths. Transient killer whales employ great stealth and remain silent while hunting, so as not to announce their presence to potential prey. Although transients typically hunt for prey near to shore where they are hidden by wave action and turbulence, they also are capable open water foragers. When whales encounter potential prey in open water, one observed hunting strategy is for group members to take turns rushing the prey and striking it with their flukes or ramming it with their heads. Once killed, the pinniped prey is shared among group members, similar to the food sharing behavior observed in social carnivores such as lions (*Panthera leo*) and wolves.

At Peninsula Valdés, Argentina, and on Possession Island, in the Crozet Island Archipelago, killer whales intentionally strand themselves in an effort to hunt sea lions and elephant seals that are on or near the beach. Southern seal lions and southern elephant seals are hunted off Peninsula Valdés (Lopez and Lopez, 1985), while on Possession Island, whales typically take newly weaned southern elephant seal pups (Guinet, 1991). In general, pups and small adult seals are most vulnerable, but adults are also occasionally killed. Once a seal or sea lion has been captured from the beach or nearshore area, they

P

are usually held in the mouth of a killer whale by one of the flippers or taken crossways in the mouth and vigorously shaken. Sometimes, captured pups are exchanged between members of the killer whale pod. Intentional stranding behavior also occurs in the absence of prey, suggesting that adult killer whales may actually teach their youngsters the finer aspects of this foraging strategy.

## V. Predation on Cetaceans

Although killer whales and sharks are responsible for most attacks on whales, dolphins, and porpoises, other cetaceans such as false killer whales, pygmy killer whales (*Feresa attenuata*), and pilot whales (*Globicephala* spp.) also represent potential predatory threat. In addition to these aquatic predators, one terrestrial predator, the polar bear, successfully hunts beluga whales (*Delphinapterus leucas*) and narwhals (*Monodon monoceros*) in Arctic areas (Smith and Sjare, 1990). River dolphins appear to be the only cetacean group free from natural predation, although it has been suggested that freshwater caiman in South America may occasionally take young dolphins (Leatherwood and Reeves, 1983). Finally, killer whales are likely to experience little or no mortality related to predation.

### A. Blackfish

Three members of the delphinid family, including the false killer whale, pygmy killer whale, and short-finned pilot whale (*Globicephala macrorhynchus*), are thought to be hunters of other cetaceans. Each of these species have teeth and jaws suitable for killing and handling large mammalian prey, and all have been observed to at least occasionally prey on other dolphins. Of these three "blackfish," the false killer whale is best known for attacks on small pelagic dolphins, and also has a record of harassing humpback (*Megaptera novaeangliae*) and sperm whales (*Physeter macrocephalus*). A series of observations, mainly by marine mammal observers onboard purse-seine boats fishing for yellowfin tuna (*Thunnus albacares*) in the eastern tropical Pacific (Perryman and Foster, 1980), have detailed false killer whale attacks on pantropical spotted (*Stenella attenuata*) and spinner dolphins (*S. Longirostris*). Although nearly two-dozen attacks were recorded, false killer whale predation on cetaceans outside of the eastern tropical Pacific is rare, suggesting that the high incidence of attack on the yellowfin tuna grounds may be site and circumstance specific. That is to say, false killer whales may be utilizing a prey resource related to tuna fishing operations (i.e., dolphins being released from temporary capture in fishing nets) which is unavailable outside of the eastern tropical Pacific.

Large whales, such as sperm whales and humpback whales, are also subjected to predatory advances by false killer whales. Recently, a group of false killer whales was observed harassing sperm whales off the Galápagos Islands (Palacios and Mate, 1996). In this event, no sperm whale mortality was recorded, but the false killer whales did inflict at least superficial injury to several individuals, and elicited noticeable fear reactions. Similar, albeit uncommon, events have also been suggested for interactions between false killer whales and humpback whales.

Pygmy killer whales have also been observed in predatory attacks on small dolphins during fishery operations in the eastern tropical Pacific, although observations of this nature are less common than those recorded for false killer whales. The predatory habits of pygmy killer whales on other cetaceans are poorly understood. In captivity, this species has been implicated in the death of a young pilot whale

and dusky dolphin (*Lagenorhynchus obscurus*), but it is unclear if these events led to consumption of the prey (Pryor *et al.*, 1965).

Similarly, few records regarding pilot whale predation on other cetaceans are available. Although pilot whales are not generally known to prey on marine mammals, records from the eastern tropical Pacific suggest that this species does chase, attack, and may occasionally eat dolphins during fishery operations (Perryman and Foster, 1980). The incidence at which these predatory events occur, however, is very low. In captivity, pilot whales have been noted to eat stillborn and young dolphins. Short-finned pilot whales have been observed harassing sperm whales in the Gulf of Mexico and off the Galápagos Islands, and although such harassment has been nonlethal, these events nevertheless often elicit a pronounced fear response, called a "marguerite formation", by sperm whale groups (Weller *et al.*, 1996). The marguerite is a defensive formation, in which group members form a heads-in and tails-out circular arrangement resembling the petals of a flower. By placing the powerful flukes, a source of potential danger for predators, toward the outside and containing particularly vulnerable individuals, such as calves, on the inside of the formation, sperm whales can usually defend themselves from harm. This marguerite response has also been noted for sperm whale groups under lethal attack by killer whales (Pitman *et al.*, 2001) and when being hunted by whalers. Therefore, the formation of a marguerite in response to pilot whale harassment suggests that sperm whales do at times appear to be threatened by this species. It remains unclear, however, if such harassment by pilot whales represents actual predatory intent, or if such interactions are merely practice hunting or social play.

As suggested by the accounts presented here, interactions between false killer whales, pygmy killer whales, and pilot whales with other cetaceans are not particularly common. Of the lethal attacks recorded to date for each of these three blackfish species, all have been in relatively unnatural situations. That is, attacks have occurred either in captivity where species that might normally avoid each other are maintained in the same confines, or centered around the eastern tropical Pacific tuna fishing operations where smaller dolphins may become available prey due mainly to capture fatigue associated with the fishing industry. Therefore, it is difficult to assess the regularity of marine mammal predation by these several species, and the scarcity of observed predatory events suggests that marine mammal prey is likely to be secondary to an otherwise fish- and squid-based diet.

### B. Sharks

Sharks represent a significant predatory threat to some populations of dolphins. Crude estimates of predation rates, as determined by the proportion of dolphins within a study population possessing shark-inflicted scars and injuries, vary greatly. Shark-related scars on odontocetes are particularly notable for some populations, while others go seemingly untouched. Results from several long-term photoidentification studies of bottlenose dolphins (*Tursiops truncatus* and *T. aduncus*) have documented shark bite scar rates as low as 1% off southern California, an intermediate rate of 22% in western Florida, and up to about 37% off eastern Australia (Corkeron *et al.*, 1987; Wells *et al.*, 1999; Connor *et al.*, 2000). The frequency of scars may also vary for different dolphin species within the same region. For example, off South Africa, where humpbacked dolphins (*Sousa chinensis*) and bottlenose dolphins overlap in distribution and habitat use, the former species has substantially more scarring related to shark attack than does the later (Cockcroft *et al.*, 1989).

Interestingly, the proportion of individuals bearing crescent shaped shark bite wounds is considerably higher for nearshore species than it is for their offshore counterparts. This apparent discrepancy may be attributable to a variety of factors. To date, most long-term studies on dolphin populations have been conducted nearshore, increasing the opportunity to observe shark scarring. Alternative explanations include the idea that predation on oceanic dolphins is less common overall, or that shark attacks in the open ocean are generally more successful. One theory that may at least partially explain why nearshore dolphins have higher rates of scarring is related to habitat features. The habitat of nearshore cetaceans offers a variety of "cover" features, such as kelp and surf, which may make escape from a predator more successful, while oceanic species have no such cover, and depend solely on fleeing or the protection offered by conspecifics within their social group to escape fatal attack (Wells *et al.*, 1999; Connor *et al.*, 2000).

Tiger sharks, dusky sharks (*Carcharhinus obscurus*), white sharks, and bull sharks (*C. leucas*) are most often implicated in attacks on nearshore dolphins and porpoises. Other sharks, including oceanic white-tip and hammerhead sharks have also been observed to occasionally attack dolphins. Tiger sharks are notorious predators of spinner dolphins (*Stenella longirostris*) off the Hawaiian Islands (Norris *et al.*, 1994), while white sharks prey on a variety of odontocetes ranging in size from the minute harbor porpoise (*Phocoena phocoena*) to more substantial beaked whales, and perhaps even newborn mysticete whales. Evidence of shark predation on baleen whales is relatively uncommon, but a report of a tiger shark attacking a young humpback whale has been recorded. Similarly, large sharks (and killer whales) were observed circling a group of sperm whales in which one adult female was giving birth, but no direct attack was noted. Although the numbers of observations regarding shark attack on large whales are few, it is reasonable to assume that some predatory events probably do occur at least occasionally.

While predation by sharks is of particular concern for cetaceans in the tropics and subtropics, attacks in other regions also occur. The remains of a complete southern right whale dolphin (*Lissodelphis peronii*) fetus as well as the genital region of an adult female were found in the stomach of a sleeping shark (*Somniosus pacificus*) off coastal Chile (Crovetto *et al.*, 1992). In addition, Greenland sharks (*S. squamulosus*) have been reported to prey upon narwhals in the eastern Canadian Arctic, and franciscana dolphins (*Ponotoporia blainvillei*) have been found in the stomachs of seven-gilled (*Heptranchias perlo*) and hammerhead sharks off Brazil (Leatherwood and Reeves, 1983). While each of these accounts is suggestive of predation, they should be considered with caution, as it is unclear if the aforementioned sharks actually attacked living dolphins or if the remains identified from stomach content analyses were attributable to scavenging.

### C. Polar Bears

Although pinnipeds are the principal marine mammal prey of polar bears, they also actively hunt and occasionally consume narwhal and beluga whales (Lowry *et al.*, 1987; Smith and Sjare, 1990). Polar bears off western Alaska, for example, have been observed "fishing" beluga whales and narwhals out of small openings in the ice (Fig. 3), sometimes killing numbers far greater than can possibly be eaten. In one particular event, polar bears killed and dragged onto the ice at least 40 ice-entrapped beluga whales (Lowry *et al.*, 1987), and in a similar episode, a single male polar bear was seen to successfully capture 13 beluga whales from a small opening in the ice over a short period of time.

Beluga whales regularly swim into extremely shallow estuary and river channel areas. On rising tides, whales penetrate far into rivers and creeks, often moving into waters so shallow that they can rest on the bottom while a considerable portion of their body remains above

**Figure 3** *Polar bear* (Ursus maritimus) *hunting beluga whales* (Delphinapterus leucas). *Photo by Sue Flood/BBC Natural History Unit.*

the surface. This behavior can sometimes result in partial stranding, but the animals are typically able to free themselves. On occasion, however, complete stranding occurs accidentally, during which time individual belugas remain beached until the return of the incoming tide. At least some beluga whale mortality results from polar bears feeding on stranded individuals (Smith, 1985). In addition to opportunistic foraging on temporarily beached whales, individual bears have been observed wading into shallow waters and chasing whales passing near to shore (Smith and Sjare, 1990). Predatory polar bears also actively stalk free-swimming belugas from ice edges. In this situation, bears either roam along the ice edge, or remain motionless while awaiting a group of beluga whales to move within striking range. When a whale passes near enough, a polar bear will launch itself from the ice and onto the back of the unsuspecting beluga. In one incident, a single polar bear was observed to use this hunting tactic to capture and kill two beluga whale calves within 24h. This hunting technique requires that bears accurately time their jumps, and more amazingly, handle and debilitate their prey in an aquatic medium. Further, once dead, the beluga must be pulled from the water and dragged onto the ice. In cases where this hunting technique has been directly observed, the captured belugas are generally young, smaller individuals.

Polar bears have also been observed to attempt attacks on belugas while swimming in pursuit of them (Smith, 1985; Smith and Sjare, 1990). Thus far, no successful attacks have been documented for this aquatic hunting tactic, and on at least one occasion, a group of belugas was seen to chase a polar bear out of the water with group coordinated threat behavior including tail lashing and repeated close approaches toward the swimming bear. Aquatic stalks by polar bears are largely unsuccessful due to the greater mobility and speed of whales in the water. In fact, the willingness of belugas to closely approach bears in the water, either out of curiosity or in a possible attempt to harass them, suggests that they have little fear of this predator when it is waterborne.

In contrast to the inshore habits of beluga whales, narwhals prefer deeper water, and are commonly sighted in considerable numbers offshore of beluga groups in the eastern Canadian Arctic. Polar bear predation on narwhals has been rarely observed, with the few attacks reported consisting of narwhals stranded on tidal flats or entrapped by ice. In one incident, three adult female narwhals and one neonate stranded on a tidal flat were being consumed by polar bears (Smith and Sjare, 1990). All three of the adult narwhals bore extensive claw marks and their blubber had been stripped dorsally from the head area back to the tailstock.

**Figure 4** *Killer whale* (Orcinus orca) *attacking a Dall's porpoise* (Phocoenoides dalli). *Photo by Robin W. Baird.*

### D. Killer Whales

In addition to pinnipeds, dugongs, and sea otters, mammal-hunting killer whales (termed transients) also prey upon some species of dolphins and porpoises, and even occasionally attack sperm and baleen whales (Baird, 2000). Transient killer whales are relentless hunters, spending up to 90% of each daylight period searching for food. In addition to marine mammal prey, terrestrial animals such as deer (*Odocoileus hemionus*) and moose (*Alces alces*) are also occasionally taken (Jefferson *et al.*, 1991). In these cases, killer whales opportunistically intercept individual deer and moose as they swim between coastal islands.

More than most other marine mammals, killer whales are social hunters, often working together to capture prey in a coordinated manner resembling that of pack hunting social carnivores like hyenas, wolves, and lions (Baird, 2000). Transients typically form slightly larger groups while hunting dolphins and porpoises, as compared to group sizes observed during pinniped attacks (Baird and Dill, 1995). Most hunts of small cetaceans have some component of chase, making more individuals necessary to prevent prey escape. Sometimes these high-speed chases result in a killer whale leaping free from the water with a dolphin or porpoise in its mouth. When dolphin prey are assembled in relatively large schools, killer whales often attempt to separate one or a few individuals from the group before commencing active pursuit. Once a prey item becomes exhausted, killer whales then attempt to kill the animal by breaching on it, ramming it from below, tossing it into the air, or grasping it in their teeth (Fig. 4).

A variety of dolphins and porpoises are hunted by killer whales. Off New Zealand, common dolphins (*Delphinus delphis*) are most commonly attacked, but bottlenose dolphins and dusky dolphins are also hunted (Visser, 1999). Stomach content analysis of a stranded killer whale off southern Brazil found the remains of three franciscana dolphins. In the Gulf of Mexico, a pod of killer whales chased and killed a pantropical spotted dolphin, and Pacific white-sided dolphins (*Lagenorhynchus obliquidens*), Dall's porpoise (*Phocoenoides dalli*), and harbor porpoises are some of the more commonly hunted small cetaceans off the west coast of North America (Jefferson *et al.*, 1991; Baird, 2000; Pitman *et al.*, 2003). In addition to these relatively small cetaceans, larger prey, including northern bottlenose whales (*Hyperoodon ampullatus*) and long-finned pilot whales (*Globicephala melas*) are also occasionally hunted by killer whales. In Arctic waters, killer whales sometimes herd beluga whales into shallow inlets and creek openings where they then rush into the group to capture young animals. Further, killer whales have been seen feeding on beluga whales and narwhals in open waters and on animals trapped by sea ice (Steltner *et al.*, 1984; Campbell *et al.*, 1988; Smith and Sjare, 1990).

An interesting study of contrasts exists for transient killer whales off British Columbia and in Prince William Sound, Alaska. Although transients in both regions apparently feed exclusively on marine mammals, harbor seals are the most common prey item of whales off British Columbia, while transients in Prince William Sound prey about equally on harbor seals and Dall's porpoises (Ford and Ellis, 1999; Saulitis *et al.*, 2000). Low harbor seal abundance in Prince William Sound may account for the apparent preference for porpoise prey in this region.

Although killer whales tend to focus their predatory attentions on pinnipeds and small odontocetes, numerous reports of attacks on sperm whales have also been recorded (Arnbom *et al.*, 1987; Pitman *et al.*, 2001). In most cases, the sperm whale groups being attacked contained one or more calves. Sperm whales are likely to be difficult for killer whales to kill, as they are excellent deep divers and can escape predation by descending to depth, possess sizable teeth capable of inflicting significant injuries, and actively defend group members when threatened. Regardless of the difficulty in hunting sperm whales, recent field observations from the Pacific noted killer whales successfully killing at least one adult member of a sperm whale group, and fatally injuring at least several others (Pitman *et al.*, 2001).

Killer whales have been noted to hunt all of the mysticete species except for pygmy right whales (*Caperea marginata*), but observations of attacks on baleen whales are not common. As is true for sperm whales, baleen whales are also difficult to kill, requiring extended effort and coordination between pod members. A typical strategy employed by killer whales during large whale hunts consists of first fatiguing the prey by active pursuit, followed then by delivery of debilitating attack. It has been suggested that attacking killer whales may grasp large whales by the flukes and pectoral flippers in an attempt to slow or stop their movement, or perhaps drown their prey by pulling them underwater (Silber *et al.*, 1990).

In the Gulf of California, researchers watched from a small airplane as a group of 15 killer whales attacked and killed a Bryde's whale (*Balaenoptera edeni*). During this event, the killer whales repeatedly swam on to the back and head of the Bryde's whale, a behavior speculated to be useful in hindering life-sustaining respiration of the animal under attack (Silber *et al.*, 1990). A similar incident was recorded off British Columbia (Hancock, 1965) where killer whales were observed to exhaust and kill a fleeing minke whale (*Balaenoptera acutorostrata*). Killer whales also attack humpback whales, but unlike the more passive escape tactics employed by some of the other mysticetes, humpbacks aggressively defend themselves from killer whales by thrashing at them with their tail flukes and flippers (Whitehead and Glass, 1985).

Of all the mysticetes, gray whales (*Eschrichtius robustus*) are probably most frequently attacked by killer whales. On an almost predictable basis each spring, killer whales in Monterey Bay, California, attack gray whales (Fig. 5). Young calves making their first northward migration are particularly vulnerable, even while under the watchful eye of their mothers (Baldridge, 1972; Goley and Straley, 1994). Records from beach cast gray whales along the coast of the Chukchi Sea show a similar pattern to that observed off California; whales with the highest incidence of killer whale-induced injuries (i.e., tooth scarring) were generally under 10 m long, suggesting that killer whales in this region also select young gray whales as their primary predatory target.

P

**Figure 5** *Killer whale* (Orcinus orca) *attack on California gray whales* (Eschrichtius robustus). *Photo by Sue Flood/BBC Natural History Unit.*

**Figure 6** *Young western gray whale* (Eschrichtius robustus) *with evident killer whale* (Orcinus orca) *tooth scarring. Photo by David W. Weller.*

Direct observations of killer whale attacks on large whales are relatively few, but several lines of evidence suggest that predatory interactions may occur more often than suspected. The presence of killer whale tooth rakes on the bodies, flippers, and flukes of many large whales can reach remarkably high proportions (Fig. 6). Photoidentification studies on humpback whales off Newfoundland and Labrador in the north Atlantic found that 33% of the individuals identified had killer whale inflicted tooth rakes on their bodies (Katona *et al.*, 1980). Scars on the flukes of 20–33% of humpback whale calves suggests that predation may be focused on young animals. A similar pattern has also been observed for western gray whales in the Okhotsk Sea, where nearly 34% of all whales photoidentified possess killer whale tooth rakes (Weller, unpublished). In this case, the western gray whale is critically endangered, making any level of killer whale predation significant. Bowhead whales (*Balaena mysticetus*) from the Bering, Chukchi, and Beaufort Sea populations have relatively low rates of killer whale tooth scarring, ranging from about 4 to 8% of the observed individuals (George *et al.*, 1994). In contrast, 31% of bowhead whales in the Davis Strait population show evidence of scars from killer whales.

The relatively high incidence of killer whale tooth scarring on some regional populations of large whales suggests that predatory attempts are probably more regular than indicated by field observations alone, and that many attacks are unsuccessful. Tooth rakes may not be truly indicative of predation attempts by killer whales however, but may instead represent capture practice or instruction of predatory techniques for younger members of the group. Finally, rake marks may also result from killer whales testing large whales to assess the presence of particularly vulnerable individuals that may be easily separated from a group and killed.

Although killer whales exert considerable time and energy in pursuit and capture of large whales, they consume relatively little of their victims (Andrews, 1914). Reports from whaling ship logbooks and more recent field observations, suggest that killer whales often preferentially consume only the tongue, lips, and portions of the ventrum of large whales before abandoning them. This phenomenon is little understood, and stands in stark contrast to the behavior of terrestrial predators that consume all or most of their mammalian prey.

### E. Humans

A review of predation on marine mammals would be incomplete without some mention of humans as predators. No other predator has the ability to harvest marine mammals at the same rate or intensity as man. While killer whales or polar bears may take tens of animals over relatively short periods of time, humans are capable of sometimes killing hundreds of individuals within hours. Although the ecology of the world's oceans is in part maintained by predator–prey interactions, human exploitation of marine mammal populations can have devastating consequences.

## *See Also the Following Articles*

Feeding Strategy and Tactics ■ Predator–Prey relationships

## *References*

Ainley, D. G., Strong, C. S., Huber, H. R., Lewis, T. J., and Morrell, S. H. (1981). Predation by sharks on pinnipeds at the Farallon Islands. *Fish. Bull.* **78**, 941–945.

Alcorn, D. J., and Kam, A. K. H. (1986). Fatal shark attack on a Hawaiian monk seal (*Monachus schauinslandi*). *Mar. Mamm. Sci.* **2**, 313–315.

Ames, J. A., Geibel, J. J., Wendall, F. E., and Pattison, C. A. (1996). White shark-inflicted wounds of sea otters in California, 1968–1992. *In* "Great White Sharks" (A. P. Klimley, and D. G. Ainley, eds), pp. 309–316. Academic Press, New York.

Anderson, P. K., and Prince, R. I. T. (1985). Predation on dugongs: attacks by killer whales. *J. Mammal.* **66**, 554–556.

Andrews, R. C. (1914). Monographs of the Pacific cetacea. I. The California gray whale (*Rhachianectes glaucus* Cope). *Mem. Am. Mus. Nat. Hist.* **1**, 227–287.

Arnbom, T., Papastavrou, V., Weilgart, L. S., and Whitehead, H. (1987). Sperm whales react to an attack by killer whales. *J. Mammal.* **68**, 450–453.

Baird, R. W. (2000). The killer whale: foraging specializations and group hunting. *In* "Cetacean Societies: Field Studies of Dolphins and Whales" (J. Mann, R. C. Connor, P. L. Tyack, and H. Whitehead, eds), pp. 127–153. The University of Chicago Press, Chicago, IL.

Baird, R. W., and Dill, L. M. (1995). Occurrence and behaviour of transient killer whales: seasonal and pod-specific variability, foraging behaviour, and prey handling. *Can. J. Zool.* **73**, 1300–1311.

Baldridge, A. (1972). Killer whales attack and eat a gray whale. *J. Mammal.* **53**, 898–900.

Campbell, R. R., Yurick, D. B., and Snow, N. B. (1988). Predation on narwhals, *Monodon monoceros*, by killer whales, *Orcinus orca*, in the eastern Canadian Arctic. *Can. Field Nat.* **102**, 689–696.

Cockcroft, V. G., Cliff, G., and Ross, G. J. B. (1989). Shark predation on Indian Ocean bottlenose dolphins *Tursiops truncatus* off Natal, South Africa. *S. Afr. J. Zool.* **24**, 305–310.

**P**

Connor, R. C., Wells, R. S., Mann, J., and Read, A. J. (2000). The bottlenose dolphin. *In* "Cetacean Societies: Field Studies of Dolphins and Whales" (J. Mann, R. C. Connor, P. L. Tyack, and H. Whitehead, eds), pp. 91–126. The University of Chicago Press, Chicago, IL.

Corkeron, P. J., Morris, R. J., and Bryden, M. M. (1987). Interactions between bottlenose dolphins and sharks in Moreton Bay, Queensland. *Aquat. Mamm.* **13**, 109–113.

Crovetto, A., Lamilla, J., and Pequeno, G. (1992). *Lissodelphis peronii* Lacépéde 1804 (Delphinidae, Cetacea) within the stomach contents of a sleeping shark, *Somniosus* cf. *Pacificus*, Bigelow and Schroeder 1944, in Chilean waters. *Mar. Mamm. Sci.* **8**, 312–314.

Estes, J. A., Tinker, M. T., Williams, T. M., and Doak, D. F. (1998). Killer whale predation on sea otters linking oceanic and nearshore ecosystems. *Science* **282**, 473–476.

Ford, K. B., Ellis, G. M., and Balcomb, K. C. (1994). "Killer Whales: The Natural History and Genealogy of *Orcinus orca* in British Columbia and Washington State." University of Washington Press, Seattle, WA.

Ford, K. B., and Ellis, G. M. (1999). "Transients: Mammal-Hunting Killer Whales of British Columbia, Washington, and Southeastern Alaska." University of Washington Press, Seattle, WA.

Gentry, R. L., and Johnson, J. H. (1981). Predation by sea lions on northern fur seal neonates. *Mammalia* **45**, 423–430.

George, J. C., Philo, L. M., Hazard, K., Withrow, D., Carroll, G. M., and Suydam, R. (1994). Frequency of killer whale (*Orcinus orca*) attacks and ship collisions based on scarring on bowhead whales (*Balaena mysticetus*) of the Bering-Chukchi-Beaufort Seas stock. *Arctic* **47**, 247–255.

Goley, P. D., and Straley, J. M. (1994). Attack on gray whales (*Eschrichtius robustus*) in Monterey bay, California, by killer whales (*Orcinus orca*) previously identified in Glacier bay, Alaska. *Can. J. Zool.* **72**, 1528–1530.

Guinet, C. (1991). Intentional stranding apprenticeship and social play in killer whales (*Orcinus orca*). *Can. J. Zool.* **69**, 2712–2716.

Hancock, D. (1965). Killer whales kill and eat a minke whale. *J. Mammal.* **46**, 341–342.

Hatfield, B. B., Marks, D., Tinker, M. T., Nolan, K., and Peirce, J. (1998). Attacks on sea otters by killer whales. *Mar. Mamm. Sci.* **14**, 888–894.

Hiruki, L. M., Gilmartin, W. G., Becker, B. L., and Stirling, I. (1993). Wounding in Hawaiian monk seals (*Moachus schauinslandi*). *Can. J. Zool.* **71**, 458–468.

Jefferson, T. A., Stacey, P. J., and Baird, R. W. (1991). A review of killer whale interactions with other marine mammals: predation to co-existence. *Mamm. Rev.* **21**, 151–180.

Katona, S. K., Harcourt, P., Perkins, J. S., and Kraus, S. D. (1980). "Humpback Whales of the Western North Atlantic—A Catalogue of Individuals Identified by Fluke Photographs (2nd edition)." College of the Atlantic, Bar Harbor, ME.

Leatherwood, S., and Reeves, R. R. (1983). "The Sierra Club Handbook of Whales and Dolphins." Sierra Club Books, San Francisco, CA.

Le Boeuf, B. J., Riedman, M. L., and Keyes, R. S. (1982). White shark predation on pinnipeds in California coastal waters. *Fish. Bull.* **80**, 891–895.

Lopez, J. C., and Lopez, D. (1985). Killer whales (*Orcinus orca*) of Patagonia, and their behavior of intentional stranding while hunting nearshore. *J. Mammal.* **66**, 181–183.

Lowry, L. F., and Fay, F. H. (1984). Seal eating by walruses in the Bering and Chukchi Seas. *Polar Biol.* **3**, 11–18.

Lowry, L. F., Burns, J. J., and Nelson, R. R. (1987). Polar bear, *Ursus maritimus*, predation on belugas, *Delphinapterus leucas*, in the Bering and Chukchi seas. *Can. Field Nat.* **101**, 141–146.

McCosker, J. E. (1985). White shark attack behavior: observations of and speculations about predator and prey strategies. *South. Calif. Acad. Sci. Mem.* **9**, 123–135.

Norris, K. S., Würsig, B., Wells, R. S., and Würsig, M. (1994). "The Hawaiian Spinner Dolphin." University of California Press, Berkeley, CA.

Palacios, D. M., and Mate, B. R. (1996). Attack by false killer whales (*Pseudorca crassidens*) on sperm whales (*Physeter macrocephalus*) in the Galapagos Islands. *Mar. Mamm. Sci.* **12**, 582–587.

Perryman, W. L., and Foster, T. C. (1980). Preliminary report on predation by small whales, mainly the false killer whale, *Pseudorca crassidens*, on dolphins (*Stenella spp.* and *Delphinus delphis*) in the eastern tropical Pacific. National Marine Fisheries Service Administrative Report LJ-80-05.

Pitman, R. L., Ballance, L. T., Mesnick, S. I., and Chivers, S. J. (2001). Killer whale predation on sperm whales: observations and implications. *Mar. Mamm. Sci.* **17**, 494–507.

Pitman, R. L., O'Sullivan, S., and Mase, B. (2003). Killer whales (*Orcinus orca*) attack a school of pantropical spotted dolphins (*Stenella attenuata*) in the Gulf of Mexico. *Aquat. Mamm.* **29**, 321–324.

Pryor, T., Pryor, K., and Norris, K. S. (1965). Observations on a pygmy killer whale (*Feresa attenuata* Gray) from Hawaii. *J. Mammal.* **46**, 450–461.

Reeves, R. R., Stewart, B. S., and Leatherwood, S. (1992). "The Sierra Club Handbook of Seals and Sirenians." Sierra Club Books, San Francisco, CA.

Reidman, M. (1990). "The Pinnipeds: Seals, Seal Lions, and Walruses." University of California Press, Berkeley, CA.

Saulitis, E., Matkin, G., Barrett-Lennard, L., Heise, K., and Ellis, G. (2000). Foraging strategies of sympatric killer whale (*Orcinus orca*) populations in Prince William Sound, Alaska. *Mar. Mamm. Sci.* **16**, 94–109.

Sherrod, S. K., Estes, J. A., and White, C. M. (1975). Depredation of sea otters pups by bald eagles at Amchitka Island, Alaska. *J. Mammal.* **56**, 701–703.

Silber, G. K., Newcomer, M. W., and Perez-Cortes, H. (1990). Killer whales (*Orcinus orca*) attack and kill a Bryde's whale (*Balaenoptera edeni*). *Can. J. Zool.* **68**, 1603–1606.

Siniff, D. B., and Bengtson, J. L. (1977). Observations and hypotheses concerning the interactions among crabeater seals, leopard seals, and killer whales. *J. Mammal.* **58**, 414–416.

Smith, T. G. (1976). Predation of ringed seal pups (*Phoca hispida*) by the arctic fox (*Alopex lagopus*). *Can. J. Zool.* **54**, 1610–1616.

Smith, T. G. (1985). Polar bears, *Ursus maritimus*, as predators of belugas, *Delphinapterus leucas*. *Can. Field Nat.* **99**, 71–75.

Smith, T. G., Siniff, D. B., Reichle, R., and Stone, S. (1981). Coordinated behavior of killer whales, *Orcinus orca*, hunting a crabeater seal, *Lobodon carcinophagus*. *Can. J. Zool.* **59**, 1185–1189.

Smith, T. G., and Sjare, B. (1990). Predation on belugas and narwhals by polar bears in nearshore areas of the Canadian high arctic. *Arctic* **43**, 99–102.

Steiger, G. H., Calambokidis, J., Cubbage, J. C., Gribble, D. C., Skilling, D. E., and Smith, A. W. (1985). Comparative mortality, pathology, and microbiology of harbor seal neonates at different sites in Puget Sound, Washington. *In* "Proceedings of the Sixth Biennial Conference on the Biology of Marine Mammals," Nov. 22–26, Vancouver.

Steltner, H., Steltner, S., and Sergeant, D. E. (1984). Killer whales, *Orcinus orca*, prey on narwhals, *Monodon monoceros*: an eyewitness account. *Can. Field Nat.* **98**, 458–462.

Stirling, I. (1984). A group threat display given by walruses to a polar bear. *J. Mammal.* **65**, 352–353.

Stirling, I. (1988). "Polar Bears." University of Michigan Press, Ann Arbor, MI.

Trillmich, F. (1984). The natural history of the Galapagos fur seal (*Arctocephalus galapagoensis*, Heller 1904). *In* "Key Environmental Series: Galapagos" (R. Perry, ed.), pp. 215–223. Pergamon Press, Oxford.

Visser, I. N. (1999). A summary of interactions between orca (*Orcinus orca*) and other cetaceans in New Zealand waters. *N.Z. Nat. Sci.* **24**, 101–112.

Wells, R. S., Boness, D. J., and Rathburn, G. B. (1999). Behavior. *In* "Biology of Marine Mammals" (J. E. Reynolds, and S. A. Rommel, eds), pp. 324–422. Smithsonian Institution Press, Washington, DC.

Weller, D. W., Würsig, B., Whitehead, H., Norris, J. C., Lynn, S. K., Davis, R. W., Clauss, N., and Brown, P. (1996). Observations of an interaction between sperm whales and short-finned pilot whales in the Gulf of Mexico. *Mar. Mamm. Sci.* **12**, 588–593.

Whitehead, H., and Glass, C. (1985). Orcas (killer whales) attack humpback whales. *J. Mammal.* **66**, 183–185.

P

# Predator–Prey Relationships

Andrew W. Trites

Most marine mammals are predators, but some are also preyed upon by other species. Theoretically, the interaction between marine mammals and their prey influences the structure and dynamics of marine ecosystems. Similarly, predators and prey have shaped each other's behaviors, physiologies, morphologies, and life-history strategies. However, there is little empirical evidence of these influences due to the relative scale and complexity of marine ecosystems and the inherent difficulties of observing and documenting marine mammal predator–prey interactions.

## I. Evolutionary Time Scales

Predator–prey relationships have been likened to an evolutionary arms race—the prey become more difficult to capture and eat, while the predators perfect their abilities to catch and kill their prey. Just how strong these selective forces are probably depends on the strength of the interactions between the predators and their prey (Taylor, 1984).

As predators, marine mammals feed primarily upon fish, invertebrates, or zooplankton, which in turn feed primarily upon other species of fish, invertebrates, zooplankton, and phytoplankton (Fig. 1). To capture their prey, marine mammals have evolved special sensory abilities (e.g., vision and hearing), morphologies (e.g., dentition), and physiologies (e.g., diving and breath-holding abilities) (Trites *et al.*, 2006). They have also evolved specialized strategies to capture prey, such as cooperation to corral fish, or the production of curtains of air bubbles used by humpback whales (*Megaptera novaeangliae*) to capture herring. Marine mammals have also evolved specialized

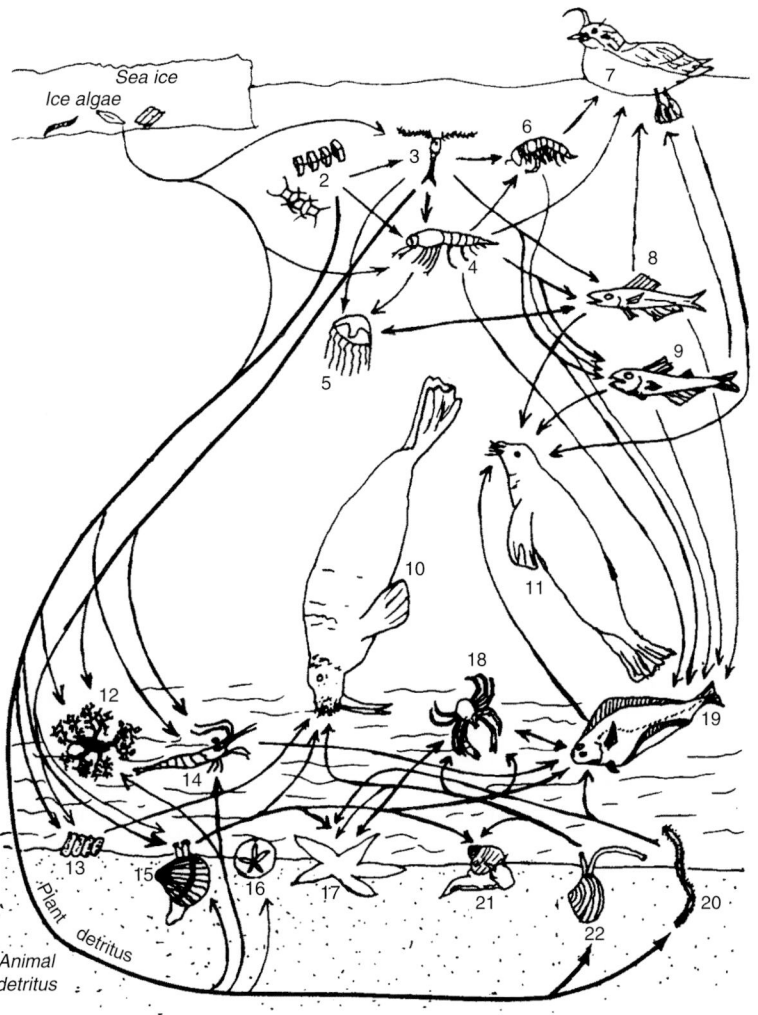

**Figure 1** *A simplified depiction of the Bering Sea food web; (1) ice algae; (2) phytoplankton; (3) copepods; (4) mysids and euphausiids; (5) medusae; (6) hyperid amphipods; (7) seabirds; (8, 9) pelagic fishes; (10) walrus; (11) seals; (12) basket stars; (13) ascideans; (14) shrimps; (15) filter-feeding bivalves; (16) sand dollars; (17) sea stars; (18) crabs; (19) bottom feeding fishes; (20) polychaetes; (21) predatory gastropods; and (22) deposit feeding bivalves. From McConnaughey and McRoy (1976).*

P

feeding behaviors to capture prey that move diurnally up and down the water column or to capture prey that move seasonally across broad geographic ranges. This in turn has likely influenced the life-history strategies of marine mammals and their prey. For example, baleen whales feed for about 6 months when plankton are abundant and concentrated in shallow water, and then fast for the remainder of the year when the plankton are too dispersed to make them worth finding.

As prey, marine mammals have had to escape aquatic and terrestrial predators (Taylor, 1984; Morisaka and Connor, 2007). Porpoise (phocoenidae) for example are preyed upon by killer whales (*Orcinus orca*) and may have evolved an echolocation and communication system through the selective pressures of predation that falls within a range of sounds that killer whales hear poorly or not at all (<2 and >100 kHz). Other species such as pinnipeds can reduce their risk of being eaten by aquatic predators (sharks and killer whales) by hauling out and resting onshore; while species such as Steller sea lions (*Eumetopias jubatus*) and northern fur seals (*Callorhinus ursinus*) reduce their risk of being eaten by terrestrial predators [wolves (*Canis lupus*) and bears] by breeding and hauling out on offshore rocks and islands where terrestrial predators are absent. Other species, such as ringed seals (*Pusa hispida*), give birth in caverns formed between ice and snow to avoid predation by polar bears (*Ursus maritimus*).

Fish and other cold-blooded species of prey have also evolved a number of strategies to increase their chances of survival (Trites *et al.*, 2006). One is cryptic countershading that enables fish to blend in with the bottom when viewed from above, and avoid detection when seen from below against a bright sea surface. Many species of fish, invertebrates, and zooplankton take refuge from predators in the deep, dark waters during the day and move toward the surface to feed under the cover of night. Another strategy evoked by the prey of marine mammals is predator swamping, such as large aggregations of spawning salmon and herring (*Clupea* spp.) that reduce the numerical effect of predators on their prey populations. Schooling is another antipredator behavior that creates confusion through the sheer volume of stimuli from a fleeing school, making it difficult for a marine mammal to actively select and maintain pursuit of single individuals. Scattering and fleeing is yet another option to reduce predation and is used by some prey when attacked by bulk feeders such as baleen whales [e.g., humpback whales and capelin (*Mallotus villosus*)]. The line between feeding and fleeing is undoubtedly fine for species of prey and must be continually evaluated by prey to minimize vulnerability to predation.

Marine mammals may also have indirectly influenced the evolution of nontargeted species in their ecosystems by consuming the predators of these species (Estes, 1996). The best example of this is the apparent influence of sea otters (*Enhydra lutris*) on kelp and other marine algae. Most species of marine algae use secondary metabolites to defend against herbivores. However, marine algae in the North Pacific Ocean have lower levels of chemical defenses where sea otters occur compared to algae species inhabiting the southern oceans where sea otters are not present. Sea otter predation on sea urchins and other herbivores may have removed selective pressure for species of marine algae to defend themselves against herbivores. Because secondary metabolites are expensive to produce, this may have allowed algae, like kelp, to radiate and diversify without the added cost of evolving and producing antigrazer compounds.

## II. Ecological Time Scales

On a shorter time scale than the evolutionary time scale, predators and prey can directly affect the relative abundance of each other, or

they can indirectly affect the abundance of other species. Their interaction may also affect the physical complexity of the marine environment (Katona and Whitehead, 1988; Bowen, 1997; Trites, 1997).

Predation by sea otters on sea urchins is probably the best example of how marine mammals can alter ecosystem structure and dynamics (Estes, 1996). Sea otters were hunted to near extinction in the late 1800s throughout their North Pacific range. Without predation, urchin populations grew unchecked and overgrazed the fleshy algae. Kelp did not replace the underwater barrens until reintroduced sea otters once again began preying upon sea urchins.

Primary production has been estimated to be three times higher in areas where sea otters are present compared to those areas where sea otters are absent, allowing those organisms that feed upon primary production to grow faster and attain larger sizes (e.g., mussels and barnacles). The increase in primary production may even alter settlement patterns of invertebrates. The kelp also provides habitat for fish and suspension feeding invertebrates to spawn, grow, and flourish. It can also change water motion and reduce onshore erosion and may even block the shoreward movement of barnacle larvae. Thus a top predator such as the sea otter can change the structure and dynamics of marine ecosystems.

Gray whales (*Eschrichtius robustus*) and walruses (*Odobenus rosmarus*) are other species of marine mammals whose foraging behavior can also affect community structure. For example, gray whales turn over an estimated 9–27% of the bottom substrate each year in the Bering Sea. The feeding pits created by gray whales draw 2–30 times more scavengers and other invertebrates compared to adjacent sediments. The disturbed sediments may also help maintain the high abundance of gray whale prey and other early colonizing species. Similarly, walruses turn over bottom substrate in their search for clams and other bivalves. There is some evidence that they may feed selectively on certain size classes and certain species and that their defecation may result in the redistribution of sediment. Thus, the interaction of benthic feeding marine mammals with their prey can result in food for scavengers and habitat for other species.

Interactions between predators and prey also influence the shapes of their respective life tables (i.e., age-specific survival and pregnancy rates). In Quebec, Canada, for example, there are a number of freshwater lakes that are home to landlocked harbor seals (*Phoca vitulina*). Studies have found that the trout in these lakes are younger, grow faster, attain smaller sizes, and spawn at younger ages compared to adjacent lakes without seals. As for marine mammals, they typically have elevated mortality rates during their first few years of life. This is likely due to a number of factors, including their relative vulnerability to predators and their inexperience at capturing prey and securing optimum nutrition.

In the Gulf of Alaska and Bering Sea, killer whales have been implicated as a contributing factor, but not the main one, in the decline of Steller sea lions and harbor seals through the 1980s (Williams *et al.*, 2004). Field observations along the Aleutian Islands indicate that these population declines were followed by a decline of sea otters in the 1990s and that this decline was caused by killer whale predation. Some killer whales may have begun supplementing their diet with sea otters because they could not sustain themselves on the low numbers of remaining seals and sea lions. What ultimately caused the decline of Steller sea lions and began this spiraling change of events is a matter of considerable scientific debate. However, it is apparent from mathematical calculations of population sizes and energetic requirements that there are sufficient numbers of killer whales in Alaska to prevent the recovery of pinniped populations. Thus, it is conceivable that populations of pinnipeds and otters may not recover to former levels of abundance until the predation by killer whales is reduced by a reduction in killer whale numbers or by a shift in killer whale diet to other species of mammals such as dolphins and porpoises.

In addition to directly affecting the abundance of their prey, marine mammals can indirectly affect the abundance of other species by outcompeting them or by consuming species the prey upon them (Trites, 1997). A case in point is harbor seals in British Columbia whose diet was about 4% salmon and 43% hake in the 1980s. Contrary to popular opinion, the harbor seals were likely benefiting salmon because they affected the abundance of hake, a species of fish that is one of the largest predators of salmon smolts. Further north in Alaska's Copper River Delta, harbor seals were culled in the 1960s to reduce the predation on salmon. However, the immediate result of the cull was not an increased number of salmon caught, but a decrease and failure of the razor clam (*Siliqua patula*) fishery. It turned out that the seals were primarily eating starry flounder (*Platichthys stellatus*), which fed on the razor clams. Without the seals, the predatory flounder population grew unchecked.

In the Antarctic, commercial whaling systematically removed over 84% of the baleen whales and freed an estimated 150 million tons of krill for other predators to consume each year (Knox, 1994). Species such as crabeater seals (*Lobodon carcinophaga*), Antarctic fur seals (*Arctocephalus gazella*), leopard seals (*Hydrurga leptonyx*), and penguins [chinstrap (*Pygoscelis antarcticus*), Adelie (*P. adeliae*), and macaroni (*Eudyptes chrysolophus*)] increased and moved the Antarctic marine ecosystem to new equilibrium levels. Increases were also observed in Antarctic minke whales (*Balaenoptera bonaerensis*) and squid-eating king penguins (*Aptenodytes patagonicus*) due perhaps to reductions in the respective abundance of blue whales (*Balaenoptera musculus*) and sperm whales (*Physeter macrocephalus*). All of these species appear to have directly benefited from an increase in prey, which was caused by the removal of whales. Penguins and seals may now be hindering the recovery of baleen whale stocks in the Antarctic.

Marine mammals are generally considered to be opportunistic foragers who select from a number of alternative prey according to availability. This is based on the relatively large number of different species that have been reported in the stomachs and feces of marine mammals. Steller sea lions, for example, are known to eat over 50 different species of fish, and even the occasional seabird. However, their diets are typically dominated by five or fewer species, suggesting that they may not be truly opportunistic feeders. Little is yet known about the choices that marine mammals make when foraging. Presumably what marine mammals eat is a function of nutritional value, ease of capture, and digestibility, all of which are invariably linked to the abundance of both predators and prey. These are complex biological interactions about which little is known.

Functional response curves represent rates of predation in relation to the density of prey (Mackinson *et al.*, 2003; Middlemas *et al.*, 2006). In most species, the rate of capture rises with the density of prey to some maximum level. These relationships between prey density and predation rates tend to be sigmoidal (nonlinear and asymptotic), indicating that there are maximum limits to the rate that predators can capture and process prey, which are independent of prey population size. Establishing these functional relationships for different species of prey is fundamental to fully understanding the foraging ecology of marine mammals. Establishing these relationships is beginning to been done for marine mammals and will require further experimentation in captivity or observational studies in the wild.

Ecosystem models are another technique for gaining insight into the effects of predator–prey relationships on ecosystem dynamics and structure (Trites *et al.*, 1999; Morissette *et al.*, 2006). Using a series of mathematical equations to account for the flow of energy from one group of species to another, the models can estimate the extent of competition between species and the effect that changes in abundance of one species will have on other species in the ecosystem. One such ecosystem model describing the Gulf of St. Lawrence revealed that harp seals (*Pagophilus groenlandicus*), gray seals (*Halichoerus grypus*), and hooded seals (*Cystophora cristata*) negatively affect the abundance of the higher trophic level fish they target, which in turn reduces predation pressure on the prey of the species these seals eat. Another ecosystem model constructed for the eastern Bering Sea examined trophic relationships to determine whether the declines of Steller sea lions and forage fishes (such as herring) and the increases in pollock (*Pollachius* spp.) and flatfish between the 1970s and the 1980s were related to the commercial removal of whales.

Removing historic numbers of whales from the simulated Bering Sea ecosystem resulted in an increase in numbers of pollock. However, the increase was only in the order of 10–20%, not the 400% increase believed to have actually occurred. The ecosystem model suggested that the Bering Sea may exist in two alternative states (consisting of two different complexes of species) and that environmental shifts (from periods of cold to warm water years) may ultimately determine when and for how long these shifts occur. The model also suggested that curtailing fishing on pollock (a major prey of Steller sea lions) may affect the Steller sea lion negatively. The explanation for this counterintuitive prediction was that commercial fisheries primarily removed larger pollock than Steller sea lions consumed. Given that pollock are cannibalistic, increasing the size of the adult stock resulted in the increased predation of younger pollock, leaving fewer fish for Steller sea lions to consume. Thus, ecosystem models are useful tools for exploring the influence of predator–prey interactions on one another and on other components of their ecosystems.

## III. Synthesis

Marine mammal predator–prey interactions occur over different spatial and temporal scales, making it difficult to empirically decipher the influences they have on one another and on their ecosystems. However, their coexistence suggests that marine mammal predators and their prey have had profound influences on each other's behaviors, physiologies, morphologies, and life-history strategies. The diversity of niches filled by marine mammals makes it difficult to generalize about the evolutionary consequences of their interactions with prey, beyond stating the obvious: marine mammals have adapted to catch food, while their prey have adapted to avoid being caught.

On the shorter ecological time scale, marine mammals can affect the abundance of other species by consuming or outcompeting them. They can also indirectly affect the abundance of nontargeted species by consuming one of their predators, and can have strong impacts on the overall dynamics and structure of their ecosystems. One of the best tools for understanding marine mammal predator–prey interactions is the ecosystem model. However, more work is required through experimental manipulations and observational studies to evaluate the choices made by marine mammals and the costs of obtaining different species of prey.

## See Also the Following Articles

Feeding Strategies and Tactics ▪ Hearing ▪ Predation on Marine Mammals ▪ Vision

P

## References

Bowen, W. D. (1997). Role of marine mammals in aquatic ecosystems. *Mar. Ecol. Prog. Ser.* **158**, 267–274.

Estes, J. A. (1996). The influence of large, mobile predators in aquatic food webs: examples from sea otters and kelp forests. *In* "Aquatic Predators and their Prey" (S. P. R. Greenstreet, and M. L. Tasker, eds), pp. 65–72. Fishing News Books, Oxford.

Katona, S., and Whitehead, H. (1988). Are cetacea ecologically important? *Oceanogr. Mar. Biol. Annu. Rev.* **26**, 553–568.

Knox, G. A. (1994). "The Biology of the Southern Ocean." Cambridge University Press, Cambridge.

Mackinson, S., Blanchard, J. L., Pinnegar, J. K., and Scott, R. (2003). Consequences of alternative functional response formulations in models exploring whale–fishery interactions. *Mar. Mamm. Sci.* **19**, 661–681.

McConnaughey, T., and McRoy, P. (1976). "Food-web structure and the fraction of carbon isotopes in the Bering Sea. *In* "Science in Alaska 1976." pp. 296–316. Alaska Division of AAAS, Anchorage, AK.

Middlemas, S. J., Barton, T. R., Armstrong, J. D., and Thompson, P. M. (2006). Functional and aggregative responses of harbour seals to changes in salmonid abundance. *Proc. Roy. Soc. B Biol. Sci.* **273**, 193–198.

Morisaka, T., and Connor, R. C. (2007). Predation by killer whales (*Orcinus orca*) and the evolution of whistle loss and narrow-band high frequency clicks in odontocetes. *J. Evol. Biol.* **20**, 1439–1458.

Morissette, L., Hammill, M. O., and Savenkoff, C. (2006). The trophic role of marine mammals in the northern Gulf of St. Lawrence. *Mar. Mamm. Sci.* **22**, 74–103.

Taylor, R. J. (1984). "Predation." Chapman and Hall, New York.

Trites, A. W. (1997). The role of pinnipeds in the ecosystem. *In* "Pinniped Populations, Eastern North Pacific: Status, Trends and Issues" (G. Stone, J. Goebel, and S. Webster, eds), pp. 31–39. New England Aquarium, Conservation Department, Boston, MA.

Trites, A. W, Livingston, P. A., Vasconcellos, M. C., Mackinson, S., Springer, A. M., and Pauly, D. (1999). Ecosystem change and the decline of marine mammals in the eastern Bering Sea: testing the ecosystem shift and commercial whaling hypotheses. Fish. Cent. Res. Rep. **7**(1).

Trites, A. W., Christensen, V., and Pauly, D. (2006). Effects of fisheries on ecosystems: just another top predator? *In* "Top Predators in Marine Ecosystems: Their Role in Monitoring and Management" (I. L. Boyd, K. Camphuysen, and S. Wanless, eds), pp. 11–27. Cambridge University Press, Cambridge.

Williams, T. M., Estes, J. A., Doak, D. F., and Springer, A. M. (2004). Killer appetites: assessing the role of predators in ecological communities. *Ecology* **85**, 3373–3384.

# Pygmy and Dwarf Sperm Whales
## *Kogia breviceps* and *K. sima*

### DONALD F. MCALPINE

## I. Characteristics and Taxonomy

In form, *Kogia* spp. are porpoise-like and robust with a distinctive underslung lower jaw. This latter feature has been described as giving these whales a shark-like appearance (Fig. 1). Although height and position of the dorsal fin have been reported as distinguishing the two currently recognized species, they are probably not separable at sea except under exceptional circumstances. Pygmy sperm whales reach a maximum size of about 3.8 m and a weight of 450 kg. Dwarf sperm whales are smaller at 2.7 m and 272 kg. Adults of both species are dark bluish-gray to blackish-brown dorsally and light below. On the side of the head between the eye and the flipper there is often a crescent-shaped, light colored mark referred to as a "false gill." These whales have the shortest rostrum among living cetaceans, and the skull is markedly asymmetrical. The mandibles are delicate, and the teeth are very sharp, thin, and lack enamel. *K. breviceps* lacks teeth in the upper jaw, but *K. sima* may have up to three pairs of vestigial teeth in this position. Although now recognized as the sole genus within the family Kogiidae, originally these whales were placed within the Physeteridae, with the sperm whale, *Physeter macrocephalus*. Fossil forms of Kogiidae have been described rarely from fragments of teeth, cranium, and lower jaws of late Miocene to early Pliocene age. Most of these may be only distantly related to extant *Kogia* spp. However, *Praekogia cedrosensis*, described from the early Pliocene in the Almejas Formation on Isla Cedros Baja California, Mexico, is reported to clearly be ancestral to living *Kogia*. It is only since 1966 that two species of *Kogia* have been recognized, and no subspecies have been described. On the basis of recent evidence from the mitochondrial cytochrome *b* gene it has been suggested that *K. sima* may consist of two apparently parapatric species occupying the Atlantic and Indo-Pacific Oceans (Chivers *et al.*, 2005). Full recognition of this putative third *Kogia* sp. awaits further supporting evidence.

## II. Distribution and Abundance

Dwarf and pygmy sperm whales occur worldwide in temperate and tropical waters of the Atlantic, Pacific, and Indian Oceans. Although rarely sighted at sea, these whales commonly strand in some regions, and much of the relatively little that is known of their ecology has been gleaned from such stranded animals. In the NE Atlantic most strandings occur in autumn and winter, but more broadly there is little indication for seasonality in the distribution or the migration of these whales. Evidence shows that *K. sima* may prefer warmer seas than *K. breviceps*. The precise at-sea DISTRIBUTION of *Kogia* spp. is unknown, as most records are based on stranded animals, but some evidence suggests *K. sima* may have a more pelagic distribution and feed in deeper water. Analysis of prey in stranded animals suggests that both species of *Kogia* generally inhabit waters along the continental shelf and slope in the epi- and mesopelagic zones.

Although many writers have stated that dwarf and pygmy sperm whales are rare, there is insufficient information to classify the world status of *Kogia* species; neither their population sizes nor trends are known (Baird *et al.*, 1996; Willis and Baird, 1998). The frequency with which *Kogia* strand on certain coasts, especially in southeastern United States and South Africa, suggests that in some regions they may be uncommon rather than rare.

## III. Ecology

*Kogia* spp. feed mostly on mid and deepwater cephalopods but also consume fish and occasionally crustaceans, such as shrimp and crabs (McAlpine *et al.*, 1997; Santos *et al.*, 2006). Stomach contents that have been analyzed have contained cephalopod beaks from at least 55 species representing 15 families, although in NE Atlantic *K. breviceps* squids of the genus *Histioteuthis* predominate. It has been suggested that there may be some competition for prey between adult pygmy sperm whales and juvenile sperm whales. Most feeding seems to takes place on or near the bottom, probably using ECHOLOCATION to find prey. Kogiid hyoid anatomy suggests powerful suction feeding.

P

**Figure 1** *(A) Kogia breviceps. The relatively short and more posteriorly positioned dorsal fin is useful in distinguishing this species from (B) K. sima (C. Brett Jarrett).*

Little is known about disease in *Kogia*. Strandings of unhealthy pygmy and dwarf sperm whales have been attributed to degenerative heart disease, as well as being linked to possible immune system problems associated with the thymus gland. Pneumonia has also been observed in stranded animals. A novel trypanoplasm-like flagellate has been described from the blowhole of several stranded *K. breviceps* (Poynton *et al.*, 2001). *Kogia* specimens are frequently heavily infected with intestinal nematodes (*Anisakis* sp. and *Terranova cetecola*) and blubber-encysted larval cestodes (*Phyllobothrium delphini*). The parasitic crustacean, *Pennella balaenoptera*, has been observed embedded in the epidermal surface of both species of *Kogia*. Scarring indicates that these whales are attacked by lampreys. A white shark (*Carcharadon carcharias*) attack on a pygmy sperm whale has been documented, and pygmy sperm whale remains have been identified in killer whale (*Orcinus orca*) stomachs. Levels of PREDATION on *Kogia* are otherwise unknown. Heavy infestations with larval cestodes, which probably mature in elasmobranchs, suggest that shark attacks may be more common than the single literature report suggests.

## IV. Behavior and Physiology

There have been no comprehensive behavioral studies of dwarf or pygmy sperm whales. Stranded animals that have been maintained in aquaria have usually survived no more than several months and usually only live for a few days. At sea, both species occur individually or in small groups of up to 6 (*K. breviceps*) or 10 (*K. sima*) animals of varying age and sex composition. Strandings usually involve single animals. Dwarf and pygmy sperm whales are reported to spend considerable time lying motionless at the surface with the back of the head exposed and the tail hanging down. *K. breviceps* is easily approached, but is timid and slow moving. Normal swimming speed is thought to be about 3 knt. When surfacing, both species rise slowly, produce an inconspicuous blow, and dive without showing the flukes. Maximum dive durations of nearly 18 min have

been recorded, although most are much shorter (Scott *et al.*, 2001). Neither species is known to be highly vocal (Clarke, 2003).

Like *Physeter macrocephalus*, *Kogia* spp. have a spermaceti organ. However, unlike the sperm whale, current opinion suggests that buoyancy control is not the primary function of the melon. Rather, it has been suggested that its function is to produce, intensify, and actively focus sound used in echolocating prey during deep dives. An elongated balloon-like structure occupies the lower intestine of both species and is usually filled with dark reddish-brown liquid that may be released into the water during foraging or when disturbed.

## V. Life history

*K. sima* reaches sexual maturity at about 2.1 m in length (Caldwell and Caldwell, 1989). In *K. breviceps* males are known to be sexually mature at about 2.7 m and females at a slightly smaller size. Gestation has been cited as 9 or 11 months, with the species about 1 m in length at birth.

## VI. Interactions with Humans

The scarcity of pygmy and dwarf sperm whales and the fact that they are rarely encountered at sea mean that direct effects from humans are probably few. Although by-catch in the pelagic driftnet fishery has been observed, fisheries mortality appears to be very limited. However, there is growing evidence that *Kogia* spp. show a propensity to ingest ocean debris such as plastic bags, latex gloves, and balloons. In several cases such items have been documented to result in intestinal blockage and death in these whales (Stamper *et al.*, 2006). Their habit of lying quietly at the surface seems to have led to occasional ship strikes. Both species are taken infrequently in commercial harpoon fisheries in the Caribbean and Indian Oceans.

## *See Also the Following Articles*

Skull Anatomy ■ Toothed Whales, Overview ■ Sperm Whale

P

## References

Baird, R. W., Nelson, D., Lien, J., and Nagorsen, D. W. (1996). The status of the pygmy sperm whale, *Kogia breviceps*, in Canada. *Can. Field Nat.* **110**, 525–532.

Caldwell, D. K., and Caldwell, M. C. (1989). Pygmy sperm whale, *Kogia breviceps* (de Blainville, 1838): dwarf sperm whale, *Kogia simus*, Owen, 1866. *In* "Handbook of Marine Mammals" (S. H. Ridgway, and R. Harrison, eds), Vol. 4, pp. 235–260. Academic Press, San Diego, CA.

Chivers, S. J., LeDuc, R. D., Robertson, K. M., Barros, N. B., and Dizon, A. E. (2005). Genetic variation of *Kogia* spp. With preliminary evidence for two species of *Kogia sima*. *Mar. Mamm. Sci.* **21**, 619–634.

Clarke, M. R. (2003). Production and control of sound by the small sperm whales, *Kogia breviceps* and *K. sima* and their implications for other Cetacea. *J. Mar. Biol. Ass. U.K.* **83**, 241–263.

McAlpine, D. F., Murison, L. D., and Hob erg, E. P. (1997). New records of the pygmy sperm whale, *Kogia breviceps* (Physeteridae) from Atlantic Canada with notes on diet and parasites. *Mar. Mamm. Sci.* **13**, 701–704.

Poynton, S. L., Whitaker, B., and Heinrich, A. B. (2001). A novel trypanoplasm-like flagellate *Jarrellia altramenti* n. g., n. sp. (Kinetoplastida: Bononidae) and ciliates from the blowhole of a stranded pygmy sperm whale (*Kogia breviceps*) (Physeteridae): morphology, life cycle, and potential pathogenicity. *Dis. Aquat. Org.* **44**, 191–201.

Santos, M. B., Pierce, G. J., López, A., Reid, R. J., and Ridoux, V. (2006). Pygmy sperm whales, *Kogia breviceps* in the Northeast Atlantic: new information on stomach contents and strandings. *Mar. Mamm. Sci.* **22**, 600–616.

Scott, M. D., Hohn, A. A., Westgate, A. J., Nicolas, J. R., Whitaker, B. R., and Campbell, W. B. (2001). A note on the release and tracking of a rehabilitated pygmy sperm whale (*Kogia breviceps*). *J. Cetacean Res. Manage.* **3**, 87–94.

Stamper, M. A., Whitaker, B. R., and Schofield, T. D. (2006). Case study: morbidity in a pygmy sperm whale, *Kogia breviceps*, due to ocean-borne plastic. *Mar. Mamm. Sci.* **22**, 719–722.

Willis, P. M., and Baird, R. W. (1998). Status of the dwarf sperm whale, *Kogia simus*, with special reference to Canada. *Can. Field Nat.* **112**, 114–115.

**Figure 1**   *Pygmy killer whale. Photograph by Robert Pitman.*

# Pygmy Killer Whale
## *Feresa attenuata*

MEGHAN A. DONAHUE AND WAYNE L. PERRYMAN

### I. Characteristics and Taxonomy

The pygmy killer whale has a moderately robust body that narrows posteriorly to the dorsal fin, hence the name *attenuata* from the Latin "to make thin or taper" (Fig. 1). The head is round and blunt and lacks the beak typical of many dolphins. The head does not narrow or appear triangular when viewed from above as with the melon-headed whale (*Peponocephala electra*). The moderately long flippers are rounded at the tips with convex leading and concave trailing edges.

On the back and portions of the flanks and ventral surface, the pygmy killer whale is dark gray to black. A subtle, dark cape (an area of COLORATION extending from the forehead past the dorsal fin) and reaches the greatest distance down the side of the animal below the high, falcate dorsal fin. A paler gray area on each flank is usually present from the tail stock to the eye. Below, the pygmy killer has an irregularly shaped white patch between the flippers, around the genitals, and occasionally on the tail stock. The lips are also edged with white.

The skull is broad and robust. The upper and lower jaws have less than 15 teeth each, a character that distinguishes the pygmy killer whale from the melon-headed whale, which typically has more than 20 teeth per row.

Length measurements from several specimens average 2.31 m (range 2.14–2.59 m). Differences in lengths between males and females have not been observed in measured specimens.

Although called "whale," the pygmy killer whale, like its close relative, the killer whale (*Orcinus orca*), belongs taxonomically to the dolphin family, Delphinidae. Until 1952 this species was only known from two skulls collected in the nineteenth century. Since that time, a number of specimens have been collected from strandings and fishery catches around the world, yet the pygmy killer whale remains one of the least known of the small cetaceans.

### II. Distribution and Abundance

Pygmy killer whales have been recorded in tropical and subtropical waters worldwide. Sightings have been relatively frequent in the eastern tropical Pacific, the Hawaiian Archipelago, and off Japan. The migratory status of this species cannot be determined based on available information. However, incidental catches and observations by fishermen suggest that it is a year-round resident at least in the regions of Sri Lanka and the Lesser Antilles.

Estimates of abundance have been made around the Hawaiian Islands, in the Eastern Tropical Pacific, and the Northern Gulf of Mexico. Between 1986 and 1990, five research vessel surveys were conducted in the eastern tropical Pacific and an abundance of 38,900 (CV = 0.305) pygmy killer whales was estimated for that area (Wade and Gerrodette, 1993). A shipboard line-transect survey of the entire Hawaiian Islands Exclusive Economic Zone (Barlow, 2003) resulted in an abundance estimate of 817 (CV = 1.12). Surveys in the Northern Gulf of Mexico between 1996 and 2001 resulted in an estimate of 408 (CV = 0.60) animals (Mullin and Fulling, 2004).

### III. Ecology

Although the feeding habits of pygmy killer whales are not well known, remnants of cephalopods and small fish have been found in specimens from STRANDINGS and incidental fishery catches. Nothing is known of predators.

### IV. Behavior and Physiology

Pygmy killer whales are found most commonly in small herds, ranging from 12 to 50 animals, although herds of 100 or more have been encountered. This species has been observed BOW RIDING, performing high leaps, and "spyhopping" (raising the head vertically out of the water).

Pygmy killer whales are suspected to be among the small whales that chase, attack, and sometimes eat dolphins (*Stenella* spp. and *Delphinus delphis*) involved in the purse-seine fishery for yellowfin tuna in the eastern tropical Pacific (Perryman and Foster, 1980). AGGRESSIVE BEHAVIOR has also been observed by two pygmy killer whales in captivity in Hawaii and South Africa, but a herd captured off Japan showed no such aggression when placed in an enclosure with other dolphins.

Acoustic recordings from the northern Indian Ocean indicate pygmy killer whale vocalizations have echolocation clicks similar to other comparably sized delphinids and are likely used for detection and classification of prey (Madsen *et al.*, 2004).

### V. Life History

Little is known about this species' growth and reproduction. An estimated length at sexual maturity of 2m based on 85% of the mean length at physical maturity (Laws, 1965) is consistent with data collected from three sexually mature males ranging in length from 2.07 to 2.61m and three pregnant females ranging in length from 2.20 to 2.27m. A lactating female, from a group of animals stranded in the British Virgin Islands, measuring 2.04m constitutes the smallest known sexually mature female to date (Mignucci-Giannoni *et al.*, 1999).

### VI. Interactions with Humans

Pygmy killer whales have been caught directly and incidentally in fisheries. Small-cetacean fisheries in St. Vincent and Indonesia have been known to catch pygmy killer whales, but they comprise a small proportion of the catch and these catches are thought to have little effect on the population in those areas. Monitoring of fisheries in which pygmy killer whales are caught incidentally has not been extensive. Mortality in these fisheries, such as those around Sri Lanka, could be greater than documented and may have a significant impact on stocks in those regions. In Sri Lanka, pygmy killer whales have also been harpooned and used as bait in long-line fisheries for sharks, billfish, and other oceanic fishes (Leatherwood and Reeves, 1989).

### See Also the Following Articles

Delphinids, Overview ■ Melon-headed Whale

### References

Barlow, J. (2003). Cetacean abundance in Hawaiian waters during summer/fall 2002. *NOAA SWFSC Adm. Rep.* **LJ-03-13**.

Laws, R. W. (1965). Growth and sexual maturity in aquatic mammals. *Nature* **178**, 193–194.

Leatherwood, S., and Reeves, R. R. (1989). Marine mammal research and conservation in Sri Lanka 1985–1986. *UNEP Mar. Mamm. Tech. Rep.* **1**.

Madsen, P. T., Kerr, I., and Payne, R. (2004). Source parameter estimates of echolocation clicks from wild pygmy killer whales (*Feresa attenuata*). *J. Acoust. Soc. Am.* **116**, 1909–1912.

Mignucci-Giannoni, A. A., Toyos-González, G. M., Pérez-Padilla, J., Rodríguez-López, M. A., and Overing, J. (1999). Mass stranding of pygmy killer whales (*Feresa attenuata*) in the British Virgin Islands. *J. Mar. Biol. Ass. U.K.* **80**, 759–760.

Mullin, K. D., and Fulling, G. L. (2004). Abundance of cetaceans in the oceanic northern Gulf of Mexico. *Mar. Mamm. Sci.* **20**, 787–807.

Perryman, W. L., and Foster, T. C. (1980). Preliminary report on predation by small whales, mainly the false killer whale, *Pseudorca crassidens*, on dolphins (*Stenella* spp. and *Delphinus delphis*) in the eastern tropical Pacific. *NOAA SWFSC Admin. Rep.* **LJ-80-05**, 9.

Wade, P. R., and Gerrodette, T. (1993). Estimates of cetacean abundance and distribution in the eastern tropical Pacific. *Rep. Int. Whal. Commn.* **43**, 477–493.

# Pygmy Right Whale
## *Caperea marginata*

### CATHERINE M. KEMPER

### I. Characteristics and Taxonomy

The pygmy right whale is the smallest baleen whale and the only member of the family Neobalaenidae (Fig. 1). There is conflicting evidence regarding the evolutionary history of this enigmatic species (Sasaki *et al.*, 2005). Some studies align it more closely to the gray whale (*Eschrichtius robustus*, Eschrichtiidae) and rorquals (Balaenopteridae) than to right (*Eubalaena* spp.) and bowhead (*Balaena mysticetus*) whales (Balaenidae). No fossil neobalaenids have been described. [A reported fossil, *Neobalaena simpsoni*, from Chile is believed to be related to Balaenidae, not Neobalaenidae (Cabrera, personal communications, 2007).] Geographical variation has not been studied and no subspecies are recognized.

Diagnostic features of the pygmy right whale include long, narrow, creamy-white baleen plates with an outer margin of brown or black and very fine bristles (Sekiguchi *et al.*, 1992); a clearly visible band of white gum at the base of the baleen; a moderately arched rostrum that becomes more pronounced as the animal grows; a small, falcate dorsal fin placed about 25–30% of body length from the tail; and shallow throat creases in some animals. The overall body shape of adults is stouter than that of the rorquals but not as broad as in right and bowhead whales. From above, the head is broadest at the eyes and narrows sharply into a long and narrow rostrum on which a medial ridge may be present (Matsuoka *et al.*, 2005). The flukes are very broad and have a deep medial notch. The body is medium to dark gray

**P**

Figure 1 *The smallest baleen whale is the pygmy right whale* (Caperea marginata)*; it achieves a maximum length of about 6.5 m. (C.Brett Jarrett).*

Figure 2 *The pygmy right whale has a distinctively curved jaw line and "throws" its head out of the water while swimming at the surface. These features help to identify it "at sea." Photo credit: Barbara Parker/South Australian Museum.*

above and white to pale gray below. There is some evidence that young animals may not be as dark as older ones. There is a dark eye patch and fairly distinct pale gray chevrons across the back above and anterior to the flipper (Kemper *et al.*, 1997; Matsuoka *et al.*, 2005). The oval scars of cookie-cutter sharks *Isistius* spp. are often present and abundant on large animals. The flippers are small, narrow, and rounded at the tip, and are medium to dark gray above (contrasting with the pale color of the sides of the body) and paler below. Mandibular and rostral hairs persist into adulthood but there are no callosities as in true right whales. The baleen plates number 213–262 on each side (Ivashin *et al.*, 1972; Budylenko *et al.*, 1973).

At sea, pygmy right whales may be confused with minke whales, *Balaenoptera acutorostrata* and *B. bonaerensis*, but close inspection should reveal some of the diagnostic features noted earlier. The blunter rostrum and strongly curved lower jaw of the pygmy right whale and its habit of swimming with its head "thrown" out of the water at an angle should also help identify it (Fig. 2).

The skull and skeleton of the pygmy right whale are unlike those of any other cetacean (Hale, 1964). The supraoccipital bone is very long, extending well forward on the skull. The ear bone has a distinctive wrinkle on its outer surface and is squarish in outline. The mandible is broad and has no coronoid process. The ribs are very broad, flat, and numerous (18 pairs), extending well along the body. All seven cervical vertebrae are fused and the total number of vertebrae is only 44. The flipper has four digits.

Little information on the pygmy right whale's internal anatomy has been published. Ivashin *et al.* (1972) and Budylenko *et al.* (1973) described and weighed body organs, including the reproductive organs, of a few sexually mature whales captured by soviet whaling

ships. The heart and lungs were relatively small, suggesting that the species is not a deep diver (Ross *et al.*, 1975). Reeb and Best (1999) made a detailed study of the larynx and concluded that this organ was considerably different from that of other baleen whales.

## II. Distribution and Abundance

This species is found only in the Southern Hemisphere (Fig. 3). It is circumpolar, between about 30° and 55°S, with records from southern Africa, South America, Australia, and New Zealand (Baker, 1985; Kemper, 2002). It has also been recorded in the vicinity of the Falkland and Crozet islands and in the open ocean of the South Atlantic (Baker, 1985), southwestern Pacific Ocean, and Southern Ocean south of Australia (Matsuoka *et al.*, 2005). There are no estimates of abundance but judging by the number of strandings in Australia and New Zealand, it is likely to be reasonably common in that region.

## III. Ecology

The pygmy right whale lives in temperate and subantarctic regions where water temperatures are between about 5° and 20°C (Baker, 1985). It has been seen in oceanic and neritic environments where some individuals have spent up to 2 months very close to shore, possibly feeding while there (Kemper, 2002). Seasonal movements inshore may be related to the availability of food during spring and summer (Ross *et al.*, 1975; Sekiguchi *et al.*, 1992). Although oceanic feeding has not been observed, animals collected there had full stomachs (Ivashin *et al.*, 1972) or were seen defecating (Matsuoka *et al.*, 2005), both of which indicate recent feeding. The little information available on diet shows that copepods, euphausiids, and possibly other small plankton are eaten. It has been suggested that the subtropical convergence, where sea surface temperatures are 9–13°C and plankton is abundant (Kawamura, 1974), is an important feeding area for pygmy right whales. Strandings and inshore sightings are often in shallow, protected bays. The predators of pygmy right whales are not known.

## IV. Behavior and Physiology

The surface behavior of pygmy right whales is inconspicuous (Ross *et al.*, 1975; Kemper *et al.*, 1997; Matsuoka *et al.*, 2005), but because so few observations have been reported the complete repertoire may not have been recorded. Swimming speeds of 3–8 knt have been noted, and the whale is also capable of very fast acceleration and speed, leaving a conspicuous wake when doing so (Fig. 2). One underwater observation of swimming noted that the body action was very flexed (Ross *et al.*, 1975). When pygmy right whales dive they remain submerged for up to 4 min and surface briefly before diving again (Ivashin *et al.*, 1972; Matsuoka *et al.*, 2005). This behavior

P

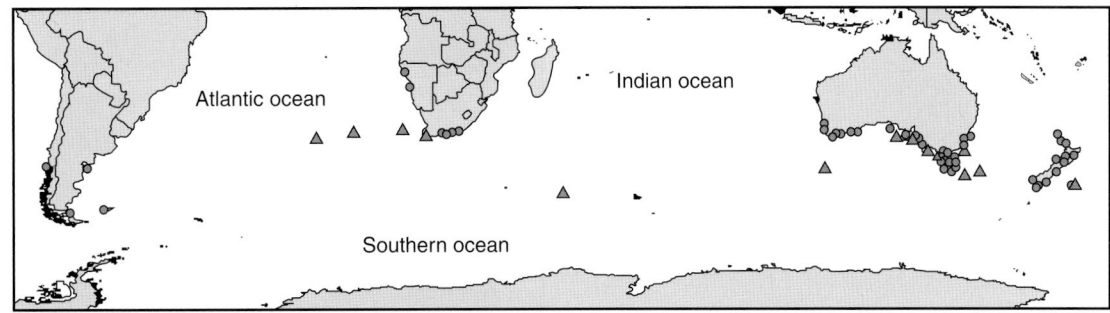

**Figure 3** *Distribution of the pygmy right whale. Symbols show position only, not number, of records. Circle, strandings or entanglements; triangle, sightings (including captures).*

is consistent with shallow dives. The blow is inconspicuous and, when visible, is small and oval. The sounds of one solitary juvenile consisted of pairs or trios of short, thump-like pulses or tone bursts with a down-sweep in frequency and decaying in amplitude. Most energy was between 60 and 120 Hz (Dawbin and Cato, 1992).

Less than 25 sightings of pygmy right whales "at sea" have been recorded. They have been seen with pilot (*Globicephala melas*), sei (*Balaenoptera borealis*), and minke whales and with dolphins. Sightings close to the coast tend to be of one or two animals, and those in the oceanic environment have been of groups up to 80 individuals. Strandings usually involve only one whale but adult female/juvenile pairs have been recorded in some cases (Kemper *et al.*, 1997). Strandings throughout the year in Australia and New Zealand suggest that the species does not migrate north–south as do most other baleen whales.

## V. Life History

Relatively little is known of the life history of pygmy right whales. Length at birth is about 2 m and at weaning, about 3.0–3.5 m (Ross *et al.*, 1975; Kemper, 2002). Most animals are physically mature at around 6 m (Kemper and Leppard, 1999), and maximum length and weight are 6.5 m and 3430 kg (Budylenko *et al.*, 1973). Females are slightly longer than males. Sexual maturity may occur at lengths of greater than 5 m. The calving interval, mating season, and gestation period are not known. The calving season is protracted (Pavey, 1992), possibly year-round. Life expectancy is not known and no age estimates have been made.

## VI. Interactions with Humans

Pygmy right whales were never targeted by whalers, although they were at times taken opportunistically. Intentional killing by inshore fisheries and incidental captures in fishing nets are known (Kemper, 2002). None have been kept in captivity. Toxic contaminants are not believed to be a threat to this species because tissue levels of organochlorines and heavy metals measured in a few animals have been low (Kemper *et al.*, 1994).

## See Also the Following Articles

Baleen Whales (Mysticeti) ■ Right Whales

## References

Arnason, U., Gullberg, A., and Janke, A. (2004). Mitogenomic analyses provide new insights into cetacean origin and evolution. *Gene* **333**, 27–34.

Baker, A. N. (1985). Pygmy right whale *Caperea marginata* (Gray, 1846). *In* "Handbook of Marine Mammals" (S. H. Ridgway, and R. Harrison, eds), Vol. 3, pp. 345–354. Academic Press, London.

Budylenko, G. A., Panfilov, B. G., Pakhomova, A. A., and Sazhinov, E. G. (1973). New data on pygmy right whales *Neobalaena marginata* (Gray, 1848). *Trudy Atlanticheskii Nauchno-Issledovatel'skii Institut Rybnogo Khoz. Okeanogr.* **51**, 122–132.

Dawbin, W. H., and Cato, D. H. (1992). Sounds of a pygmy right whale (*Caperea marginata*). *Mar. Mamm. Sci.* **8**, 213–219.

Hale, H. M. (1964). The pygmy right whale (*Caperea marginata*) in South Australian waters, part 2. *Rec. S. Austral. Mus.* **14**, 679–694.

Ivashin, M. V., Shevchenko, V. I., and Yuchov, V. L. (1972). The pygmy right whale *Caperea marginata* (Cetacea). *Zool. Zh.* **51**, 1715–1723.

Kawamura, A. (1974). Food and feeding ecology in the southern sei whale. *Sci. Rep. Whales Res. Inst. Tokyo* **26**, 25–144.

Kemper, C. M. (2002). Distribution of the pygmy right whale, *Caperea marginata*, in the Australasian region. *Mar. Mamm. Sci.* **18**, 99–111.

Kemper, C. M., and Leppard, P. (1999). Estimating body length of pygmy right whales (*Caperea marginata*) from measurements of the skeleton and baleen. *Mar. Mamm. Sci.* **15**, 683–700.

Kemper, C., Dutton, J., Forster, B., and McGuire, R. (1997). Sightings and strandings of the pygmy right whale *Caperea marginata* near Port Lincoln, South Australia and a review of other Australasian sightings. *Trans. R. Soc. South Aust.* **121**, 79–82.

Kemper, C. M., Gibbs, P., Obendorf, D., Marvanek, S., and Lenghaus, C. (1994). A review of heavy metal and organochlorine levels in marine mammals in Australia. *Sci. Total Environ.* **154**, 129–139.

Matsuoka, K., Yoshihiro, F., and Pastene, L. A. (1996). A sighting of a large school of the pygmy right whale, *Caperea marginata*, in the southeast Indian Ocean. *Mar. Mamm. Sci.* **12**, 594–597.

Matsuoka, K., Pitman, R. L., and Marquez, F. C. (2005). A note on a pygmy right whale (*Caperea marginata*) sighting in the southwestern Pacific Ocean. *J. Cetacean Res. Manage.* **7**, 71–73.

Pavey, C. R. (1992). The occurrence of the pygmy right whale, *Caperea marginata* (Cetacea: Neobalaenidae), along the Australian coast. *Austral. Mamm.* **15**, 1–6.

Reeb, D., and Best, P. B. (1999). Anatomy of the laryngeal apparatus of the pygmy right whale, *Caperea marginata* (Gray 1846). *J. Morphol.* **242**, 67–81.

Ross, G. J. B., Best, P. B., and Donnelly, B. G. (1975). New records of the pygmy right whale (*Caperea marginata*) from South Africa, with comments on distribution, migration, appearance, and behavior. *J. Fish. Res. Board Can.* **32**, 1005–1017.

Sasaki, T., *et al.* (11 authors) (2005). Mitochondrial phylogenetics and evolution of mysticete whales. *Syst. Biol.* **54**, 77–90.

Sekiguchi, K., Best, P. B., and Kaczmaruk, B. Z. (1992). New information on the feeding habits and baleen morphology of the pygmy right whale *Caperea marginata*. *Mar. Mamm. Sci.* **8**, 288–293.

P

# R

## Remoras

DAGMAR FERTL AND ANDRÉ M. LANDRY, JR.

Remora, suckerfish, diskfish, and sucker are some of the names describing eight species of marine fishes in the Family Echeneidae (=Echenedidae) Fischer, 1978; Nelson *et al.*, 2004). Remoras inhabit tropical and subtropical waters worldwide, except for the whitefin sharksucker (=whitefin remora, *Echeneis neucratoides*), which is restricted to the western Atlantic Ocean (Fischer, 1978).

### I. Remora Biology

Remoras use a suction disk to attach to sharks, rays, bony fishes, sea turtles, cetaceans, sirenians, and ships and other floating objects (O'Toole, 2002). When attached to these hosts, remoras appear to swim upside down, but the disk is really on top of their head. The oval-shaped disk is a modified dorsal fin that has split and flattened to form a series of transverse, plate-like fin rays (disk lamellae) that resemble slats of a venetian blind (Fig. 1). When these fin rays are lifted, a strong vacuum is created between the remora's disk and its host (Fulcher and Motta, 2006).

The tenacity with which remoras attach to their hosts is best illustrated by the practice of sea turtle fishing by fishermen in the Caribbean and off China and northern Australia (Gudger, 1919), and in Yemen and Kenya, where it continues to this day. A fisherman ties a line around the tail of a remora and throws the fish into the water. The remora tightly attaches itself to a turtle, and the remora and its "catch" are then hauled ashore.

Suspected benefits of a remora's association with their hosts include transportation, protection from predators, increased courtship/reproduction potential, enhanced respiration, and expanded feeding opportunities (Fertl and Landry, 1999; Silva and Sazima, 2003). Remoras opportunistically feed on parasitic copepods (which constitute the bulk of their diet), zooplankton and smaller nekton, food scraps from meals of their hosts, and sloughing epidermal tissue and feces of the host (Cressey and Lachner, 1970; Williams *et al.*, 2003).

### II. Marine Mammal Hosts

Adult remoras typically attach to the body of a marine mammal (Figs 2, 3). At least three remora species utilize marine mammals as hosts: whalesucker (*Remora australis* = *Remilegia australis*), sharksucker (*Echeneis naucrates*), and whitefin sharksucker (Fertl and Landry, 1999; Williams *et al.*, 2003). Remoras associate with at least 20 cetacean and 2 sirenian species [dugong (*Dugong dugon*) and

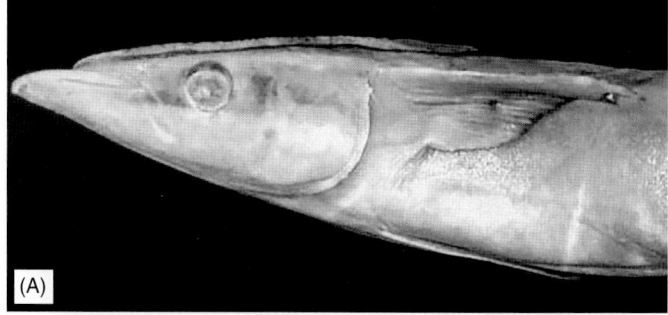

**Figure 1** *(A) Lateral and (B) dorsal view of the head of a remora, with suction disk visible. Photographs by W.H. Dailey.*

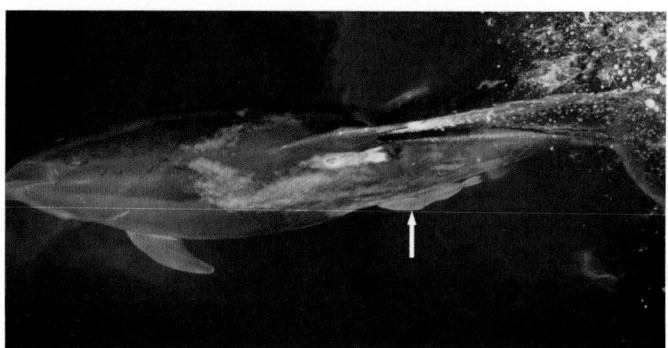

**Figure 2** *Bottlenose dolphin* (Tursiops truncatus) *with remora attached. Photograph by Dagmar C. Fertl.*

West Indian manatee (*Trichechus manatus*)]. The whalesucker has most often been collected and identified from cetaceans, hence, its common name (Rice and Caldwell, 1961; Fertl and Landry, 1999). The sharksucker has also been collected from common bottlenose dolphins (*Tursiops truncatus*) (Fertl and Landry, 1999, 2002; Noke, 2004). Two species of remora have been collected from West Indian manatees; these were positively identified as the whitefin sharksucker and the sharksucker (Williams *et al.*, 2003).

The remora's suction mode of attachment does not hurt the host or leave scars, as has been suggested. However, a temporary mark resembling the disk imprint may be seen. Wounds attributed to remoras are most likely caused by cookiecutter sharks (*Isistius brasiliensis*) or Pacific lampreys (*Lampetra tridentate* = Entosphenus *tridentatus*), which actually bite or rasp their prey or host.

To what degree a remora might irritate its host is uncertain. A remora may slide all over its host's body, possibly tickling the animal.

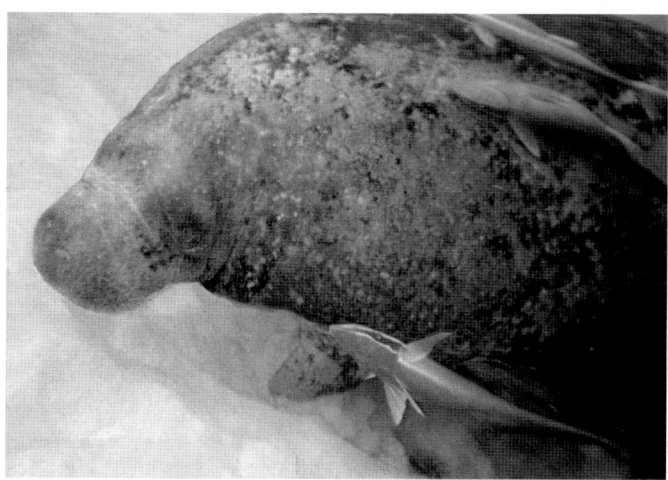

**Figure 3** *Sharksucker* (Echeneis naucrates) *attached to a West Indian manatee* (Trichechus manatus) *off Aguadilla, Puerto Rico. Photo by Edwin Rivera-Colon, La Casa del Buzo.*

Animals observed with remoras sliding over their bodies sometime will jerk and even roll over (Ritter, 2002; Ritter and Brunnschweiler, 2003). Dolphins of various species leap with remoras attached to them, perhaps to dislodge the "hitchhiker" (Fish *et al.*, 2006; Weihs *et al.*, 2006). There are also reports of dolphins dislodging remoras from themselves or their calves and then biting them (Wedekin *et al.*, 2007). Large remoras or multiple remoras on the same host may produce a hydrodynamic drag.

### III. Problems with Remora Identifications

The whalesucker's preference for cetaceans leads many observers to assume that any remora spotted on a cetacean is this species. Most remora–marine mammal associations described in the literature are based on visual or photographic observations of a remote, free-swimming host and its passenger(s) rather than specimens collected from strandings or whaling victims. Host records determined from remote observations should be considered problematic, as the identification of remoras to species is difficult without the actual specimen in hand (Fertl and Landry, 1999; Sazima, 2006).

The whalesucker and other host-specific remoras are typically pelagic forms with a specialized morphology consisting of large disks, short stout bodies, and reduced fin size (when compared to those of inshore counterparts) (Fertl and Landry, 1999). More commonly reported remoras are slender-bodied, inshore forms, such as the sharksucker, that are least particular about their hosts. The possibility that small, slender remoras, as well as more stocky remoras photographed on cetaceans, may represent different life history stages of one species further complicates positive identification from afar.

### See Also the Following Articles

Cetacean Ecology ∎ Parasites

### References

Cressey, R., and Lachner, E. (1970). The parasitic copepod diet and life history of diskfishes (Echeneidae). *Copeia* **2**, 310–318.

Fertl, D., and Landry, A. M., Jr. (1999). Sharksucker (*Echeneis naucrates*) on a bottlenose dolphin (*Tursiops truncatus*) and a review of other cetacean—Remora associations. *Mar. Mamm. Sci.* **15**, 859–863.

Fertl, D., and Landry, A. M., Jr. (2002). Sharksucker (*Echeneis naucrates*) on a bottlenose dolphin (*Tursiops truncatus*) from Sarasota Bay, Florida, with comments on remora-cetacean associations in the Gulf of Mexico. *Gulf Mex. Sci.* **2002**, 151–152.

Fischer, W. (1978). "FAO Species Identification Sheets for Fishery Purposes: Western Central Atlantic (Fishing Area 31), Voulme II." Food and Agricultural Organization of the United Nations, Rome.

Fish, F. E., Nicastro, A. J., and Weihs, D. (2006). Dynamics of the aerial maneuvers of spinner dolphins. *J. Exp. Biol.* **209**, 590–598.

Fulcher, B. A., and Motta, P. J. (2006). Suction disk performance of echeneid fishes. *Can. J. Zool.* **84**, 42–50.

Gudger, E. W. (1919). On the use of the sucking-fish for catching fish and turtles: Studies in Echeneis or Remora, II. *Am. Nat.* **53**, 446–467.

Nelson, J. S., Crossman, E. J., Espinosa-Pérez, H., Findley, L. T., Giltert, C. R., Lea, R. N., and Williams, J. D. (2004). Common and scientific names of fishes from the United States, Canada, and Mexico. *Am. Fish. Soc.*, Spec. Publ. 29, Bethesda.

Noke, W. D. (2004). The association of echeneids with bottlenose dolphins (*Tursiops truncatus*) in the Indian River Lagoon, Florida, USA. *Aquat. Mamm.* **30**, 296–298.

O'Toole, B. (2002). Phylogeny of the species of the superfamily Echeneoidea (Perciformes: Carangoidei: Echeneidae, Rachycentridae, and Coryphaenidae), with an interpretation of echeneid hitchhiking behaviour. *Can. J. Zool.* **80**, 596–623.

Rice, D. W., and Caldwell, D. K. (1961). Observations on the habits of the whalesucker (*Remilegia australis*). *Norsk Hval.–tid.* **5**, 181–189.

Ritter, E. K. (2002). Analysis of sharksucker, *Echeneis naucrates*, induced behavior patterns in the blacktip shark, *Carcharchinus limbatus*. *Environ. Biol. Fishes* **65**, 111–115.

Ritter, E. K., and Brunnschweiler, J. M. (2003). Do sharksuckers, *Echeneis naucrates*, induce jump behaviour in blacktip sharks, *Carcharhinus limbatus*? *Mar. Fresh. Behav. Physiol.* **36**, 111–113.

Sazima, I. (2006). Species records, mistaken identifications, and their further use: The case of the diskfish *Echeneis naucrates* on a spinner dolphin. *Neotrop. Ichthyol.* **4**, 457–460.

Silva, J. M., Jr., and Sazima, I. (2003). Whalesuckers and a spinner dolphin bonded for weeks: Does host fidelity pay off? *Biota Neotropica* **3**(2), published online.

Wedekin, L. L., Freitas, A., Engel, M. H., and Sazima, I. (2004). Rough-toothed dolphins (*Steno bredanensis*) catch diskfishes while interacting with humpback whales (*Megaptera novaeangliae*) off Abrolhos Bank breeding ground, southwest Atlantic. *Aquat. Mamm.* **30**, 327–329.

Weihs, D., Fish, F. E., and Nicastro, A. J. (2007). Mechanics of remora removal by dolphin spinning. *Mar. Mamm. Sci.* **23**, 707–714.

Williams, E. H., Jr., Mignucci-Giannoni, A. A., Bunkley-Williams, L., Bonde, R. K., Self-Sullivan, C., Preen, A., and Cockcroft, V. G. (2003). Echeneid-sirenian associations, with information on sharksucker diet. *J. Fish. Biol.* **63**, 1176–1183.

# Reproductive Behavior

MICHAEL A. FEDAK, BEN WILSON AND
PADDY P. POMEROY

### I. Introduction

Taking a very broad view, the "function" of marine mammals is to convert prey into offspring. Reproductive behavior is an important part of the process by which this is brought about and must serve to create a situation in which the young can safely be born and nurtured, and one which facilitates mating with suitable

R

partners. But in long-lived animals, reproduction has to be linked to the process of gathering the resources for both current and future survival. Because many marine mammals do not feed where they reproduce, they also must locate breeding areas where reproduction and parental care can take place without compromising nutritional requirements before, during, and after the current effort. Here we will consider the basic problems that the animals must solve to reproduce, and give some illustrative examples of their behavior. We will take a broad, strategic view and look at reproductive behavior in a life history context, and consider how animals balance their needs for resources and reproduction.

## A. The Basic Problems to Be Overcome

Although they spend most of their time in the water, seals give birth on ice or on land, and most newborn pups require a period ashore before being able to cope with life at sea. The vulnerability of pinnipeds on land means that suitable breeding sites need to be isolated from potential predators, limiting the choice of suitable ones. Pinnipeds do not feed while ashore. The widely separated and patchy distribution of resources that typifies most marine ecosystems means that animals are often widely separated from one another while foraging, and suitable breeding sites are often few and far between. This necessarily requires the use of stored reserves for periods of days to months. The geographical separation of feeding and breeding sites and the reliance on stored reserves are arguably the most important determinants of seal reproductive strategies and life history patterns.

Whales can give birth, nurse, and mate at sea but conditions suitable for the birth of young are often not suitable for foraging, so these two phases of their annual cycle often take place in widely separated geographical locations. Long migrations between breeding and foraging locations may still be necessary. And while foraging, individuals might be widely separated from potential mates, creating difficulties for locating suitable mates. Little food may be available during the birthing and the mating period, which therefore can require stored energy and materials for its success. Therefore, even though some whales are not constrained to spend time ashore for breeding, in many cases they face some of the same problems as pinnipeds.

Both seals and whales must move to breeding areas and choose a suitable breeding site where they can safely give birth, protect, and feed their young. They must choose a mate, copulate, and produce fertilized eggs. They must protect and feed their young and provide them with the resources and guidance needed to become nutritionally independent and give them a good chance to reach maturity and recruit into the breeding population. Then the adults must reestablish successful foraging patterns to provide resources for their own survival and reproductive success in the following year(s).

The marine habit and the geographic and energetic constraints acting on marine mammals have shaped their life histories and reproductive behaviors to create some of the most dramatic and extreme (some might even say bizarre) reproductive patterns among mammals.

## B. The Importance of Size

Marine mammals as a group contain some of the largest mammals in existence as well as possibly the largest animal to have ever existed. The size adopted by the various species is such an obvious characteristic that we often look past to other features of the animals without considering its fundamental importance to behavior. Yet size stands out as being of fundamental importance as to how they these animals organize their reproductive behaviors. Because of the scaled relationship between body volume or mass (M) and metabolic rate (MR),

where MR $\propto$ M0.75, size has obvious implications for diving and foraging behavior. Larger species and individuals will require more prey each year but they may be able to dive for longer and go longer without food and thus be able to contend with less predictable or more widely distributed food distribution. Large size has equally fundamental implications for variations in reproductive behavior within and between species. Size in large part determines how long animals can fast during reproduction as well as how often they must leave the vicinity of their pups or of potential mates to look for food. In general, bigger animals can maintain their presence on beaches for longer and can breed farther from food sources. Size also sets the relationship between the duration and the efficiency of lactation (energy used in the process divided by energy stored in the pup). It sets the weaning mass of offspring and the relative cost to the mother of achieving offspring of that mass; larger mothers can produce larger pups without putting themselves at risk. The metabolic overheads (i.e., the amount of energy required to support the metabolism of mother and pup during lactation) are relatively lower for large animals in relation to the stored resources available and delivered to offspring. Size can affect the capability of animals (particularly males) to secure mates and, because of its influence on attendance patterns, size can determine the sort of strategies used to gain access to females; larger males can maintain residence for longer on breeding sites. So, size certainly matters in setting the strategies and behaviors used to accomplish reproduction.

Although seals and whales face common problems, the fact that whales do not come ashore to carry out any aspects of reproduction has meant that we have learned about their behavior in very different ways. Behavioral observations of cetaceans are largely confined to activities visible from the surface and "hands-on" techniques are much more difficult to apply. Cetaceans also have grater opportunity for complex social interactions throughout the periods of mating and parental care because parents, offspring, and other members of the social group can remain in contact during the extended times occupied by reproductive activities. The different methodologies used have also lead to separation in the approaches used in the study of the two groups, resulting in emphasis placed on different aspects of behavior. It is therefore expedient to treat the two groups separately for much of the remainder of this discussion, even though we will be considering the same basic strategic goals.

## II. Pinnipeds

We consider the strategies of reproductive behavior within the simple life history model (Fig. 1) in which the animal's mass or condition is viewed as the fundamental state variable that determines the constraints on reproductive success. It considers the life history as an annual cycle of terrestrial and aquatic phases split between foraging, breeding, and molt (see legend for details). Virtually all species can be fit into this conceptual framework and most aspects of reproductive behavior and the links with foraging and molting can be incorporated within it, in terms of how they affect fecundity and offspring quality. As such, the model provides a useful framework within which to describe the requirements of behavior. Some pinnipeds mate exclusively in water, with concomitant underwater displays, often by sound by males; we do not detail these in this general overview.

### A. Transition from Foraging to Breeding: Locating a Suitable Place to Breed

This involves behavior on a wide range of geographical scales, from global to a few meters. Animals must choose a geographical

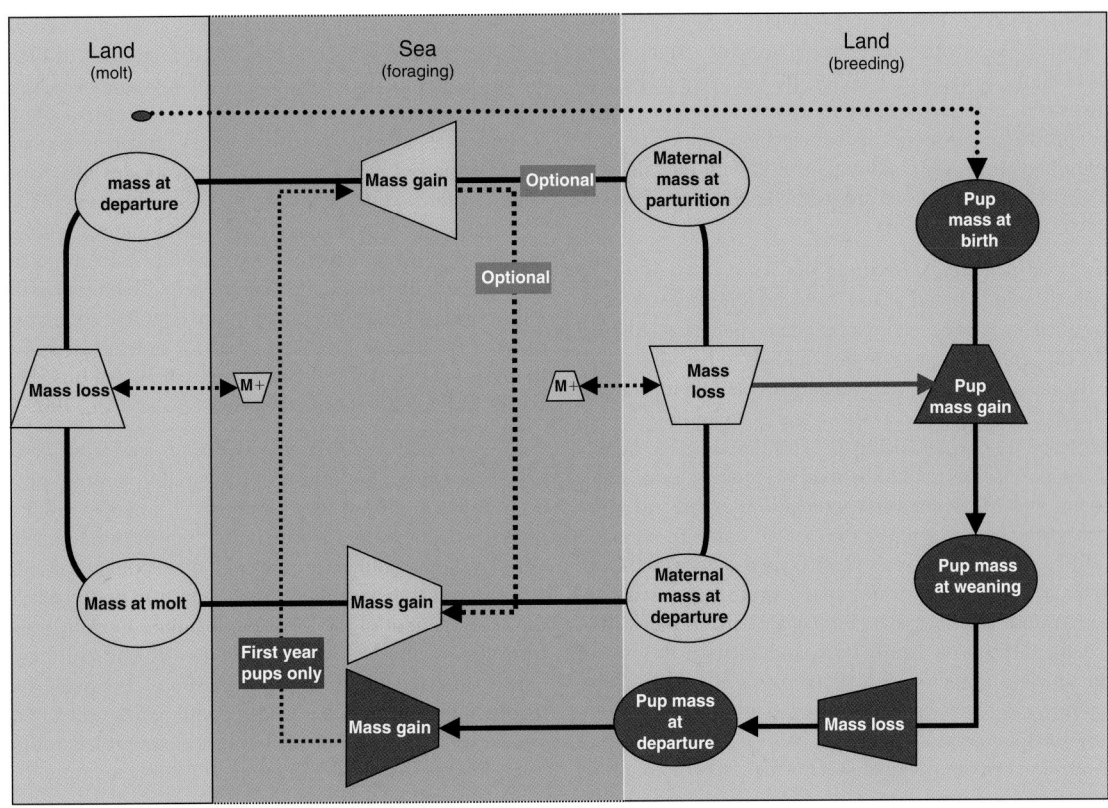

**Figure 1** *A diagram of a model of female pinniped life history with mass (as a surrogate for condition) as a state variable determining reproductive success through fecundity and pup quality. Trapezoids represent mass gain and loss. Ovals denote mass at the start and end of lactation and molt. Dotted lines represent optional paths. Some species and/or individuals (particularly smaller phocids and otariid species and smaller individuals of some phocids) may top up body mass during breeding and molt by returning to the sea to feed. Pups usually bypass molt during their first year and do not breed until they have reached a certain critical mass or condition. Females may opt not to produce a pup in years when they are below a critical mass or condition. Pups that are larger or in better condition at weaning are likely to both be able to remain on the beach longer and depart in better condition. They have a better chance of surviving and to breed earlier. That is, in this model, mass and condition determine the path taken and the resulting reproductive success, rather than age per se (see Arnbom et al., 1997 and Boyd, this volume). In the context of this model, reproductive behavior acts to provide access to suitable mates, influence the transfer of resources to pups and its efficiency, and protect the investment in offspring.*

area, a suitable site within that area (such as a particular beach or ice flow), particular conditions within that site, and a position relative to other animals in the colony. This selection needs to be accomplished in a timely way so that parturition can occur on time, and suitable mates are available.

The transition from foraging to reproduction can be considered to occur as soon as animals switch from a period of net gain of resources to net expenditure of body stores to support the travel to breeding sites. It is not likely to be a sharp boundary as animals may encounter food during their trip but a reduction in resource acquisition is likely because animals must leave prime foraging areas to make their way back to breeding sites. The critical issue for the individual animals is that they arrive at the breeding site in sufficiently good condition to support the breeding expenditure and the activities until successful foraging is reestablished. In the case of many otariids, this means that females must have sufficient reserves to sustain themselves and their pups until the mother's first successful foraging bout, which could be weeks after the animals leave the breeding beaches. Condition in males will in part determine how

long they can remain defending their access to females. In the case of the larger phocids, animals must have sufficient body condition to support the entire breeding effort. Smaller phocids or those with easy access to food may supplement stored reserves with foraging (Bowen *et al.*, in press). It could be argued that the expenditure during this phase should be added to those of reproduction, but we know of no studies that have attempted to do this. Navigational skills and previous experience of suitable sites allow the minimum time and energy to be expended.

*1. Large Scale Movements: Choosing a Geographical Location*
Many species have been shown to have the navigational skills to return to previously used breeding sites from great distances, but the methods they use to accomplish this navigational challenge remain largely unknown. Both southern (*Mirounga leonina*) and northern (*M. angustirostris*) elephant seals have been tracked making directed trips of 1000–3000 km between breeding and foraging locations, arriving on the same beaches they used for breeding the previous season (Le Boeuf *et al.*, 2000; Biuw *et al.*, 2007). Their great

R

size (males may weigh over 3000 kg and females average 500 kg) is important in making such trips energetically feasible. Many other larger species are similarly capable. Smaller species may need a supply of food on-route or feed closer to where they breed as the reproductive season approaches. It seems likely that the evolution of large body size in pinnipeds may have had much to do with enabling the uncoupling of geographical locations of feeding from those of breeding (but see Boyd, 1998 for a contrary view).

*2. The Local Scale: Choosing a Breeding Site Within a Locality*　At a local scale, animals tend to breed where there are other seals present. This aggregative sociality is a key feature of pinniped behavior, although it is modified by the animals' state. The first animals to breed in a season tend to be the older, more experienced animals, and their presence encourages others to come ashore. Once a seal has chosen a breeding site, it tends to be used again and again. This breeding site fidelity means that the same rookeries tend to be used over a long term, and is shown by Weddell seals (*Leptonychotes weddellii*) (Croxall and Hiby, 1983), gray seals (*Halichoerus grypus*) (see later), northern elephant seals (Huber, 1987), Antarctic fur seals (*Arctocephalus gazella*), (Lunn and Boyd, 1991), and is probably widespread. Scottish gray seals are faithful to their previous pupping sites. Females return to pup within an average distance of 55 m on North Rona and 24 m on the Isle of May from their previous breeding sites (Pomeroy *et al.*, 1994, 2000a, b). Males that return also show very similar spatial fidelity at both colonies (Twiss *et al.*, 1994; Pomeroy *et al.*, 2000a, b).

Some seal species display philopatry; i.e., they return to breed at the location where they were born, e.g., fur seals (Gentry, 1999), southern elephant seals (Hindell and Little, 1988). Gray seals also display philopatry, sometimes with surprising accuracy (Pomeroy *et al.*, 2000a, b). It is possible to envisage at least two ways that new colonies may form. First, in a growing population, when space at an established colony has become limiting, pregnant females arriving to breed may be forced to move elsewhere if the available habitat is being used. In this scenario, the new colony should be reasonably close to the original one. Younger, primiparous females that breed later in the season may be forced to use an otherwise unoccupied location, and once there, others join them. The main criteria listed for breeding habitats are isolation (protection from land-based predators) and access to resources nearby. Few seals stray far from the sea. Otariids require a plentiful supply of food within reach of the breeding location, otherwise the breeding attempt may fail, but most phocids are functionally divorced from foraging requirements at breeding time by their ability to store resources, principally as blubber. Thus, phocids may use breeding locations far removed from their foraging grounds (e.g., gray seals, northern and southern elephant seals). Harbor seals (*Phoca vitulina*) represent an intermediate breeding and feeding strategy where mothers supplement their stored reserves with food acquired during foraging trips late in the lactation period (Boness *et al.*, 1994).

*3. Choosing a Position Within a Site: The Individual Scale*　The local topography of the breeding location plays an important role in determining the particular location where pups are born. A featureless surface with unlimited access to the sea such as a sandbank offers the simplest case in which animals choose little except their distance from the sea. Additional resources, such as pools of water, may act as foci for breeding animals. However, seals breeding on many islands or beaches typically arrive at the breeding area through specific access routes. Restricted access produces a radiated pattern of colonization, but also creates thoroughfares where there is continual traffic as animals arrive and leave the colony. Pup mortality in these locations can be substantial (Anderson *et al.*, 1975).

The degree of topographic variation on a breeding colony at a scale relevant to seals is a primary determinant of their distribution within a site. In fact, the degree of topographical variation on the breeding colony probably also defines the scale of site fidelity shown by gray seals at two Scottish colonies, and may explain why fidelity is less apparent at relatively flat open locales like Sable Island (see later, Pomeroy *et al.*, 2000a, b; Boness and James, 1979).

Conservation or management considerations often require information on how animals may use available habitat, particularly where multiple use or potential conflicts occur. Habitat classification within a fine scale Global Information System (GIS) context has been used to identify suitable breeding areas for gray seals, and this was used to make a successful prediction of expansion of breeding areas at the Isle of May colony during the 1990s (Twiss *et al.*, 1999, 2000).

*4. Assessment of Breeding Locations and Site Choice*　Seals that breed on land come ashore for a variable period before parturition occurs. For southern elephant seals this averages 4.5 days; for gray seals the average time ashore before parturition tends to be 2–3 days, although some animals are present at breeding locations for at least a month prior to parturition. This prepartum period appears to involve some assessment of the breeding location. Female gray seals emerge from the water, looking around intently and sniffing continuously, before making tentative moves inshore. Pregnant females collect together in groups near access points where they remain inactive, and any disturbance is likely to make them return to the sea. In some cases final selection of the pupping site occurs immediately prior to parturition when females move inland, and during this movement they can be seen sniffing the ground. Pomeroy *et al.* (1994) found many cases of females returning to the sea before coming ashore again to pup, sometimes in a different location to that chosen originally. Options for changing the pupping site are often limited to the prepartum period, since in most species, once the pup is born, it is not easy for mother and pup to change location together.

## B. Investing in Young After Birth

Parental care in the pinnipeds is the exclusive domain of the mother in all but one species (see later for the single exception). Males take no part in the rearing process and the only part they play in breeding is to contribute sperm during mating. In fact, the process of mating is not without risk to the current offspring. Pups may become separated from their mothers or be the subject of aggressive behavior from males or females at this time, as well as running the risk of being crushed by males. Therefore, maternal care does not consist simply of feeding but includes all behaviors associated with the pup's welfare, such as maintenance of contact, vigilance, and defense against potential aggressors. A mother has finite resources available to service each breeding attempt and, once the pup is born, she has ultimate control over the feeding schedule and its duration. Mothers must gauge their reproductive effort according to the resources they have available, to do enough for the pup to have a good chance of survival without prejudicing the mother's survival or future breeding chances. The costs for mothers that expend too heavily in 1 year are reduced fecundity and lower breeding success in the next year (Pomeroy *et al.*, 1999). Consequently, the fundamental maternal trade-off is one of efficiency: supplying resources to the offspring at a low or acceptable expense. The single most important influence of the efficiency of the process is the "physiological time" it takes to accomplish it (Anderson and Fedak, 1987). For most phocids at least, maternal maintenance during the lactation fast must occur in parallel with the demands of feeding the pup. The shorter this time, the smaller the fraction of maternal resources that are lost as heat (generated by the combined metabolism of mother and pup) and the greater the fraction that can appear as pup growth or remain as maternal condition.

The conflict, between pups' demands for resources and the requirement of mothers to limit expenditure to that which does not incur a threat to themselves, is exemplified by a study of southern elephant seal pups fed as twins. In this case, mothers did not expend resources beyond the level expected for a single pup, so that the cut-off point in this case was fixed by the mother (Arnbom *et al.*, 1997).

*1. Maintaining Contact with Pups*  Seal breeding colonies are typically sensory-rich environments. Many animals are crowded into a restricted area, with the associated sights, smells, noise, and action associated with such a situation. A mother must maintain contact with her pup because, in general, neighbors react aggressively to foreign pups and pups that move away from their mothers may be injured or lose contact with their mothers and starve. At birth, a mother immediately sniffs and interacts with the neonate (Kovacs, 1987). By the time the first feed has been completed, mothers have established a bond with their pup that becomes progressively stronger as lactation proceeds. In most species, pups vocalize almost as soon as they are born, with mothers displaying a varying degree of competence in discriminating their own pup's call (fur seals, reviewed in Gentry, 1998, McCulloch et al., 1999). In gray seals which commute from the breeding beach to the water during lactation, a returning mother looks, sniffs and (presumably) listens when she approaches her pupping site. Often several pups are inspected before one is fed. Reunions involve the approach by the mother, sniffing and flippering of the pup, and finally presenting the nipple to feed. Other pups trying to feed at this time are often dissuaded by aggression, but may also be excluded by the mother turning away. Some gray seal mothers (particularly at the expanding colony on the Isle of May, Scotland) are poor discriminators and feed any pups that approach them. Not surprisingly, these mothers rarely wean large offspring (see earlier).

Otariid mothers leave their pups unattended for several days while they forage for food and must recognize their offspring on their return. The primary mechanisms allowing this to occur successfully are smell and sound recognition.

*2. Providing Protection*  Until the pup is born, many species of seals are tolerant of other conspecifics, so that, for example, large groups of pregnant female gray seals may lie very close to each other, often touching. As soon as the pup is born, this tolerance disappears and the mother becomes fiercely protective of the pup, defending a radius (typically 1.5 body lengths) around it. Any intruders into this space experience an escalating aggressive response, beginning with threats, approaches, then vocalizations, flippering, lunges, and finally at the most extreme, contact involving biting and flippering. Tolerance of conspecifics varies between mothers, but it is not known yet whether this reflects some form of kinship recognition, familiarity based on nonrelated associations, or simply individual variation in response.

There is a single instance of a role for paternal care in pinnipeds, where male Galapagos sea lions (*Zalophus wollebaeki*) mob sharks around colonies, reviewed in Trillmich (1996).

## C.  Lactation and Weaning

The process of lactation is demanding for mothers; most phocid species that fast during lactation lose 30–40% of their postpartum mass, much of it blubber, producing the highest fat milk known in the animal world (up to 60% lipid) in the process. In most phocids the lactation period is short but intense (hooded seals 4 days, pup growth rate 6.0 kg/day; gray seals 18 days, pup growth rate 2.0 kg/day; southern elephant seals 23 days, pup growth rate 4.2 kg/day). Such growth rates can only be achieved by having energy-dense milk, frequent feeds (every 4–5 h in gray seals) and efficient conversion of maternal resources by the offspring. The concentrated milk also conserves water, which may be of short supply. Otariids have less absolute reserves available, although these may be relatively similar to those of phocids (Costa and Trillmich, 1988), and sustain their energy requirements by foraging throughout their extended but less intense lactation periods. This means that otariid pups receive feeds at intervals several days apart.

Weaning is abrupt in most phocids, as females depart from the rookery to return to the sea, leaving the pups on the beach. In most cases, mating has already occurred and indeed observation of a successful mating is a good indication of a female's imminent departure. However, there is considerable individual variation in the time that mothers spend with pups after mating; some may remain for several days before returning to the sea. In many otariid species, a long lactation period allows offspring to develop swimming, diving, and foraging while having the option of maternal milk as a food source. As a result, otariid mothers may have a much more prolonged weaning process, as offspring may still be with their mothers in the second year after birth (Horning, and Trillmich, 1997).

## D.  Locating and Selecting a Mate

Reproduction is the single most important action which individuals of any species carry out in their lifetimes. As such, mate choice is an important consideration. Circumstances dictate the degree of choice likely, in that the distribution of females at breeding time controls the mating patterns seen. For example, solitary hooded seal (*Cystophora cristata*) mothers on fast ice are unlikely to have many options in available mates and may simply mate with the male that has waited persistently beside her until she entered estrus. However, it seems likely that the successful male may have had to defend this position in encounters with others attempting to gain this opportunity. In this situation, males give the appearance of being monogamous. By contrast, large aggregations of female southern elephant seals make it possible for males to attempt to control access for mating, with the result that extreme polygyny occurs. Males in this situation compete vigorously among themselves, as the potential breeding rewards for successful males can be substantial. However, the priorities of each sex are rarely symmetrical. Female elephant seals may be considered to have exercised mate choice just after they arrive at the breeding beach. The 4.5 days spent ashore prior to parturition provide an opportunity to assess the stability and safety of the harems they enter and the qualities of the males guarding it. If pregnant females are disturbed during this pre-parturition period, they often change locations.

## E.  Female Mating Behavior

*1. Resisting Advances*  Female seals are not receptive to males until they enter estrus. In gray seals, this occurs around day 15 of the average 18-day lactation period. Males that attempt to mate before the female is receptive receive a robust and clear message from the

**R**

female indicating her unwillingness. Initially, a female will threaten males that approach and her subsequent vocalizations at a persistent male can alert surrounding females to his presence. Neighboring females may join in this threat display to dissuade the male, although in a very few cases, fights between male and female may develop. Because of the sexual dimorphism common to most polygynous mammals, some males tend to be favored in such encounters. Females dissuade males using the same repertoire of aggressive behaviors as described earlier, but with the additional consideration that males attempting copulation are likely to have tried to mount the female. In this situation, the female's mobility and lack of cooperation, together with the aggressive display, is usually enough to make the male withdraw. Experienced males rarely attempt more than a preliminary investigation into female status and seem particularly adept at gauging a female's receptivity.

The ability of females to resist premature advances is perhaps at its most dramatic in elephant seals, where males can be more than 10 times the mass of the females with which they mate. Even here, in a species in which males are not known for their gentility, females can repel unwanted advances. Males holding harems frequently access the receptivity of females by "heading" them; i.e., they approach and rest their heads on the neck of candidate females. If the females are not receptive, they move their hind flippers rapidly from side to side in a swimming motion, slapping the side of the testing male. Most males take heed and move quickly on to test other females.

*2. Signaling Receptiveness/Estrus*  It is not clear exactly how estrus is signaled in most species. Although the general behavioral indications are simply that a female's initial aggressive response to a mating attempt declines to acceptance and passivity, it is not clear what cues a male uses to judge the situation. Olfaction is probably important as males can be seen sniffing during their approach around females (Gentry, 1998). A successful mating may also signal to other males that a female is receptive. Some females approach males and apparently solicit their attention.

*3. Choosing a Mate*  In reality, mate choice may range from having a single candidate and therefore a passive default, or be an active process involving the assessment of, or competition between a number of candidates. Competition may even be between sperm, where multiple matings occur (e.g., gray seals, elephant seals, some fur seals). The most comprehensive studies so far come from gray seals and elephant seals. Females that occupy prime sites on gray seal breeding colonies tend to have dominant males nearby (Pomeroy *et al.*, 1994; Twiss *et al.*, 1994; Pomeroy *et al.*, 2000a, b), and most mate with the dominant male. However, the number of pups they produce that are fathered by that male does not reflect the male's behavioral dominance or his mating success (Worthington Wilmer *et al.*, 2000). The reasons for this are not yet clear, but may lie in different attendance patterns of individual females at breeding colonies. There is some anecdotal and circumstantial evidence of mate choice in gray seals. At North Rona, where approximately 1200 pups are born each year, the father of a pup born to female J8 in 1986 was seen next to her in 1993, but both were at the other end of the island from where they had been in 1986 (Amos *et al.*, 1995). The pup born to J8 the following year was indeed fathered by the male seen with her in 1993. A more unusual occurrence was observed in 1997, when a known female left her peripheral pupping site to move about 80 m to the center of the colony where she was mated by a dominant male. She then returned to her pup and the attentions of a peripheral male at her pupping site. Females have been seen initiating copulations, but males initiate most of them.

## F. Male Mating Behavior

A male's reproductive success is dependent on the number of offspring he manages to sire and how many of those eventually manage to reproduce as adults. The first part of this requires successful matings and to achieve these, males must be able to take up a place among breeding females, avoid or out-compete other males, and gain a successful copulation. The second part of his reproductive success is less straightforward, since it is possible to achieve many matings without producing any surviving offspring, let alone grand-offspring. Males employ a variety of strategies to achieve success.

*1. Maintaining Access to Females*  The first prerequisite is simply to be around breeding females. Males must coordinate their efforts with the availability of receptive females. One of the most effective ways of gaining success for males is to spend a long time on the breeding colony, but this is costly, both in energetic terms, since males usually fast, and in potential injuries inflicted by competing males (Twiss *et al.*, 1998). For these reasons, large size tends to correlate with male success, so that the largest males tend to have advantages of increased energy reserves and in competitive abilities (Deutsch *et al.*, 1990).

*2. Monogamy—Polygamy*  As already discussed, the potential for polygamy in these animals depends on the distribution of females (2.5 above). Although the terms monogamy and polygamy usually apply to mating patterns of species, they may be applied to the tactics which individuals employ either throughout, or during phases of, their lifetimes. However, without complete knowledge of the reproductive histories of individual animals, it is difficult to make generalizations. Evidence from genetic studies can provide useful insights in these areas. In general, the evidence to date from genetics supports the general observational conclusions on mating patterns, for example in southern elephant seals and in gray seals, although some queries have been raised. One such is the failure of apparently dominant males to account for as many paternities as predicted (Amos *et al.*, 1993a, b). Male reproductive longevities are as important as their within-season success (Worthington Wilmer *et al.*, 1999). Long-lived, reproductively active males may accrue a greater success than live-fast die-young males whose activities are conspicuous (Twiss *et al.*, 2006).

*3. Keeping Other Males Away*  As with many mammals, the risks inherent in engaging in fights over breeding have led to a formalized ritual of aggressive displays in many pinniped species. Dominance hierarchies are common, so that disputes lead to fewer actual fights than otherwise might be expected. Fights do occur, but usually between closely matched opponents where the preliminary assessments could not determine a clear outcome (Arnbom *et al.*, 1997; Deutsch *et al.*, 1990; Twiss *et al.*, 1998). In gray seals, males attempt to control access to groups of females by threatening intruders with open mouth displays, hisses, and vocalizations. Intruders are chased away, but serious challengers may produce fights, which can last up to an hour and leave either or both combatants seriously injured. It is common for losers of such fights to disappear from the breeding colony.

*4. Optional Strategies*  Given the high cost of engaging in the mainstream competition for mating opportunities, it is not surprising that alternative strategies exist. Younger less experienced males are seen around the periphery of breeding colonies and may acquire experience gradually. Some males employ a cryptic tactic, using their similarity to females to gain a position amongst females, making the most of their opportunities when the dominant male is engaged elsewhere.

**R**

It is becoming increasingly evident that aquatic mating may occur to a greater extent than suspected and that the phenotypic qualities that are successful on land may not necessarily be the same for males adopting this mating pattern.

## G. Mating

In most observable species, males usually initiate copulations. Males often act to immobilize the female in some way by biting the back of the neck. On land, the male's weight applied via his body or flippers can help to position the female. In the water, because animals are near neutrally buoyant, the male's weight is less important in restraint. In gray seals breeding on land, a male attempts to mount the female by maneuvering alongside, then throwing his head and shoulders over the female's back. Her response is almost always aggressive, but a female in estrus will accept the male's advances if he persists and manages to grasp the skin of her neck in his jaws. This act is the single best predictor of a female's acquiescence. At the same time, the male tries to achieve intromission by repeated pelvic thrusts, whereas the female either cooperates by lying still, or resists by moving her rear as much as possible. Gray seals also mate underwater. There too, males grasp the females by the back of the neck in their jaws. Because the male cannot restrain the female as easily, she has greater opportunity to avoid the mating. Obviously, both must breathe and move together to the surface when necessary. It is not clear how the need to breathe is communicated to the other partner but cooperation is evident. In both situations, when a successful mating is achieved, the pair remains relatively motionless, for anything as brief as 5 min. or as long as 40 min. (gray seals, Twiss *et al.*, 1998). The function of such long copulations is not known, since males indulging in long copulations are leaving other females unguarded. It is thought that ejaculation occurs toward the end of the copulatory period, and females have been observed to have rhythmic contractions of the lower abdomen in the later stages of copulations. Re-mating of a female may take place soon after a copulation, either by the same or a different male.

## H. Social Structure

Seals are not often regarded as particularly social animals, yet they exhibit philopatry and breeding site fidelity, are long-lived, and often haul out in groups segregated by age and sex. Gray seals breeding on N. Rona Island exhibited spatial structure in average relatedness of neighbors, such that mothers breeding in central areas were more related to the rest of the colony than those breeding at the periphery (Pomeroy *et al.*, 2001). In addition, taking site fidelity into account, some dyads occurred together more often than would be expected, if their typical movement patterns operated at random, leading to the conclusion that these mothers change pupping locations together (Pomeroy *et al.*, 2005). It remains to be seen if such phenomena are widespread.

## I. The Transition to Foraging

*1. Post Weaning Behavior of Mothers* The mothers of most otariids repeatedly leave their pups to feed and gather the resources to support continued lactation. For them, weaning does not therefore represent a dramatic change in behavior. They simply fail to return. Some species such as the Galapagos fur seal (*Arctocephalus galapogoensis*) may give birth to the subsequent pup before the prior pup is weaned, but this pattern is unusual. For most, animals may shift to nonbreeding haul-out locations and engage in longer and more distant foraging trips. But for phocid mothers, weaning occasions can abrupt change in behavior. Typically, soon after mating occurs one or a few times, mothers abandon their pups, leave their position in the colony, and quickly enter the sea. In some species (particularly among otariids) animals may be seen traveling away from breeding sites in groups, but in many phocids, departure appears solitary. Pups are normally left behind. The "decision" to leave is a critical one in relation to the state of body energy reserves. Good foraging areas may be distant from breeding locations, and have changed in position and value while animals were breeding. Mothers must have sufficient stored reserves to enable them to reach these without putting themselves at unacceptable risk.

*2. Post Weaning Behavior of Pups* Weaning occasions dramatic changes in behavior for pups as well. Once pups are weaned, they no longer have the protection of their mothers. They no longer nurse and begin to fast for a time before the transition to nutritional independence. Otariid pups undergo what may be thought of as temporary weaning, when their mothers depart to sea to feed in between bouts of lactation. They have experience of being left unattended prior to true weaning and show some of the behaviors of weaned pups early in development. In either case, pups typically move to areas where they can avoid contact with adults and may often congregate in large groups. In many cases, this movement is stimulated by aggressive encounters with other mothers and adult males within the colony. In elephant seals, mothers leave the weaned pup behind in the harem, possibly at a central location within it. Other mothers will act aggressively to the approach of pups other than their own, and this tends to move unattended pups around with a net movement to the periphery of the harem. Within a day of weaning, pups are usually out of the harems and then move around the beaches in an apparently undirected way. When encountering other pups, they tend to remain with them. The end result of this mobility is that large numbers of pups end up in "crèches" at places on beaches where no harems are present.

Phocid pups often remain in such "crèche" groups, associated with breeding sites, for periods from days to months after weaning. The function of the time spent in these "post weaning fasts" is not understood but is thought to involve a period of physiological, behavioral, and/or social development. Pups may interact with one another, exhibiting some variants of adult behavior. For example, male elephant seal pups often engage in mock fights that involve the rearing up and head strikes seen in the battles between breeding males. Pups of both sexes tend to move into shallow inshore water or fresh water ponds during the night and swim and dive. Dive depths as great as 271 m have been observed in pups from Macquarie Island during this time (Hindell *et al.*, 1999) but in general dives are short and shallow. An increasing fraction of the day is spent in the water until departure on the first foraging trip, typically after about 30–45 days and after about 35% of the mass attained by weaning has been lost. There is some evidence that pups that remain ashore longer have better diving abilities when they enter the sea, which may help to explain the value of the post breeding fast (Noren *et al.*, 2005).

For pups too, the decision as to when to leave is potentially difficult and critical. Phocid pups have no prior experience of foraging locations. They tend to leave as individuals, not forming into groups to avoid predators and in any case, any other pups departing after weaning are similarly naïve. The way they choose to locate foraging sites and the cues they use to help them remain largely unknown.

R

For example, southern elephant seal pups may travel to locations thousands of kilometers away from where they were born before establishing successful foraging. Telemetry devices on elephant seal pups born on Macquarie Island sent back track and dive behavior information, which was used to estimate changes in buoyancy and body condition of pups over their first foraging trip. Pups often traveled 30–50 days before beginning to fatten. These data, combined with survivorship information (McMahon *et al.*, 2000) and estimates of the animals' body condition at weaning and departure suggested that pups that are larger and fatter have the best chance of surviving through their first years of life. Fat content on its own may be less critical than protein stores in determining likelihood of survival to the time successful foraging is established (Bennett *et al.*, 2007).

Interestingly, even in otariids and phocids with unusually precocious young where the opportunity for mothers to lead pups to food seems to exist, it is not exercised. Walrus (*Odobenus rosmarus*) seem to be the only exception to this. Pups travel with mothers prior to weaning and nurse at sea, giving them the opportunity to get geographical information on where to feed. It seems likely that if mothers could direct pups to food, they could obviate the need for a fraction of the material resources given to pups with this information. It seems surprising that this is rarely done.

### III. Odobenids, Sirenians, and Sea Otters (*Enhydra lutris*)

These groups are unique in many ways, showing some particularly interesting variations in reproductive behavior and other life history features. Detailed accounts can be found in Fay (1982) for Walrus, O'Shea *et al.* (1995) for sirenians, and Riedman and Estes (1990) for sea otters. We mention these groups only briefly here, emphasizing a few unusual features of reproductive situation and behavior, and we ignore altogether another marine mammal, the polar bear (*Ursus maritimus*).

Sirenians are unique in being the only group of mammalian marine herbivores. This different lifestyle has led to unusual distribution patterns and some unusual breeding behavior. Individuals of both sexes move about in response to resources such as food and fresh water. In the case of manatees (*Trichechus* spp.) at the extreme northern edge of their range in Florida, warm water sources such as power plants and warm springs cause them to congregate in large groups. They seem dependent on these at times of exceptionally cold weather, but they do not breed in these groups and this opportunistic proximity is not utilized to bring the sexes together for mating. Both males and females can range widely at other times. Locating mates seems to be the result of chance encounters between males and females in estrus. Little is known about how males locate estrus females but increased mobility of estrus females may increase chances of encounters with males. When a female comes into estrus (lasting up to 3–4 weeks), "mating herds" of hopeful males surround her. The normally quite and gentle appearance of the species in general is belied at this time with aggressive behavior between males trying to secure mating opportunities. During estrus period, females may mate with several males. Calves are born underwater and nurse there. They travel with their mothers for 1–4 years. Contact between mothers and young is maintained in part acoustically.

Walrus, too, show some important unique features in their breeding behavior. They are a polygynous group. Males display, interact aggressively, and mate in the water. Although pups are born on ice or land, they nurse in the water. They are the only pinniped group

where maternal foraging and maternal care occur simultaneously. Young move about with mothers when they feed. One of the most common observations of nursing in water is when mothers stay at the surface, oriented vertically with their heads out, the pups nurse upside down with the hind flippers at the surface and their heads down below at the nipples (Miller and Boness, 1983).

We only mention sea otters to the extent of noting that they, too, do not breed colonially and also give birth at sea. We are left with the intriguing question: if the ability to give birth at sea, nurse in the water, and lead pups to food is possible in walrus and sea otters, why do these patterns not occur more often in the "mainstream" pinnipeds. Clearly they are possible options in the pinnipeds as we see later, they are for cetaceans.

### IV. Cetaceans

Unlike pinnipeds, odobenids, and otters, cetaceans have evolved a behavioral and anatomical suite of adaptations allowing them to mate, give birth, suckle, and nurture their young entirely in water. Freed of the spatial and the temporal constraints imposed by reliance on land or ice to breed, the cetaceans have developed a wide diversity of social systems and life history strategies quite unlike those of the pinnipeds. Although some cetaceans, principally the larger mysticetes, compartmentalize breeding to a temporally and spatially discrete component in their lives, the majority breed and acquire resources simultaneously. Further, consecutive breeding attempts themselves may be superimposed upon each other, with females concurrently rearing calves from different breeding attempts and even contributing directly to the survival of their offspring's own offspring.

Having no need of land to reproduce has assuredly led to the success and ubiquitous nature of the cetaceans, but for the same reason has also severely hampered our abilities to understand them. At the most basic level, discrete acts like copulation and birth in most species have never been observed let alone quantified, while comparing the success of different tactics employed by individuals within populations remains somewhat of a Holy Grail. What is known is pieced together from anatomical studies, whaling operations, live captures, individual identification, genetic analyses, and interspecies comparisons (Lockyer, 2007). From these fragments, it is clear that the cetaceans have much to teach us about the ecological determinants of reproductive and social behavior, and even offer potential to broaden our understanding of mammalian reproductive behavior as a whole (Connor *et al.*, 1998).

#### A. Seasonality of Reproduction

For the majority of cetaceans, reproduction has a seasonal component. For the mysticetes [with the possible exception of Bryde's whales (*Balaenoptera edeni*), (Lockyer, 1984)], breeding occurs as a discrete phase of each year with other aspects, principally feeding, being either reduced or halted entirely. The best known species shuttle on an annual basis between productive feeding regions and areas associated with parturition, early nursing, courtship, and mating. The reproductive behavior of gray whales (*Eschrichtius robustus*) is a prime example. Northern Pacific populations migrate from high latitude temperate or polar waters after a summer of feeding, southward along North American and Asian coasts to breed in sheltered coastal waters. Although almost all whales migrate, whether reproducing or not, the pregnant females move south earlier than males and then 80 or so days later return north again following behind the males and newly mated females (Jones and Swartz, 1984).

R

Humpback (*Megaptera novaeangliae*) and right (*Eubalaena* spp.) whales follow similar patterns, but the behavior of rorquals, such as the blue (*Balaenoptera musculus*), fin (*B. physalus*), and minke (*B. acutorostrata*) whale are more poorly known and, though seasonal, it is as yet unclear when and where breeding actually occurs.

The lives of the odontocete cetaceans are less obviously compartmentalized and breeding takes place simultaneously with other activities. Detecting breeding seasons is consequently harder and usually estimated from parameters such as the first appearance of neonates at sea, fetal maturity in stranded or captured animals, and seasonal changes in testes. From such studies, it appears that the majority of odontocetes extend their breeding activities over protracted seasons. Interestingly, those that remain in high latitude areas tend to reproduce at the opposite time of year to neighboring mysticetes. Harbor porpoises (*Phocoena phocoena*) in the north Atlantic, for example, ovulate, mate, and give birth in spring and early summer (Read and Hohn, 1995) whereas seasonally sympatric humpback whales migrate south to breed in winter. Furthermore within species, the specific timing of reproduction may vary by region or population. Bottlenose dolphins (*Tursiops* spp.), for example, show diffuse seasonal peaks in reproduction but these vary in their timing with location (Urian *et al.*, 1996).

The reasons why mysticetes and odontocetes adopt such differing behavioral and physiological strategies toward the seasonality of reproduction remains poorly understood. Body size clearly allows the mysticetes to store sufficient reserves to forego feeding and dedicate a substantial proportion of the year exclusively to breeding. As most odontocetes are smaller, it is tempting to assume that they have less capacity to fast during a discrete breeding season, however they are of similar body size or larger than the highly seasonally breeding pinnipeds. It therefore remains a possibility that odontocetes, and females in particular, have protracted breeding seasons, simply because other aspects of their lives allow it. For all cetaceans, it is likely that food availability, risk of predation, water temperature, and sea or river conditions are important in dictating which season is actually selected to breed (Corkeron and Connor, 1999).

### B. Gathering Resources to Invest in Reproduction

At times outside of specific breeding seasons, mysticetes gravitate toward areas that maximize their potential for prey consumption. Migrations into productive, often high latitude, areas are therefore common. The duration and the rate of energy acquisition appears to be important in determining subsequent reproductive interval, ovulation rate, and fecundity (Lockyer, 1987). For odontocetes, such migrations are less evident, and suggest that their reproductive capabilities allow them to remain in their foraging areas year round. A notable and unusual exception is found in sperm whales (*Physeter macrocephalus*). Females remain in tropical or subtropical waters year round, whereas the sexually dimorphic males migrate from productive high latitude feeding areas toward the equator and their mates to breed. The extreme sexual dimorphism (with males weighing up to 3 times as much as females) may necessitate such migrations to regions especially rich in prey.

### C. Locating a Suitable Place to Breed

Breeding in cetaceans can be broken down into three phases: giving birth, suckling young, and mating. Because, gestation in most cetaceans is close to 12 months these three activities generally occur at a similar time of year and are therefore often considered as if they were one event. However, the factors that influence each differ and thus the choice of breeding habitats may well represent a compromise for the individuals concerned. The processes of giving birth and suckling young may benefit from waters with low predator abundance while these characteristics will be of less importance for mating. Examples of differences in locations for these activities are rare especially in odontocetes where examples of specific sites used for any breeding activities over others are themselves uncommon. One study of harbor porpoises in the North Sea found significantly higher proportions of calves in a specific coastal area relative to neighboring waters, though the reasons why this area was favored is as yet unknown (Sonntag *et al.*, 1999). Mysticetes offer more concrete examples of breeding areas. Those most studied include humpback, gray, and right whales, which typically breed near coasts, with the latter two species favoring sheltered shallow waters. As we might expect, they also offer some evidence of the differing requirements of raising young and mating, with females with newborn calves favoring slightly different areas to the other breeding individuals (Jones and Swartz, 1984; Craig and Herman, 2000). Since the ways in which breeding sites selected by whales actually perform their function are unknown (in contrast to the choice of breeding sites by pinnipeds) we have no information on whether availability of these habitats actually limit the size or behavior of cetacean populations.

Underlying all of the issues associated with mysticete migrations to breeding sites is the controversy over why the mysticetes migrate at all. Sheltered shallow waters are not unique to the tropics, and some mysticetes, such as the bowhead whale (*Balaena mysticetus*), are entirely capable of breeding in the same polar waters in which they feed. Factors that pose direct benefits to adults do not appear to withstand scrutiny while the thermal constraints on calves do not seem likely when neonate mysticetes are larger than most adult odontocetes and are probably already thermoneutral in colder waters. Instead, relatively novel possibilities such as calf predation by killer whales (*Orcinus orca*) might lie as the root cause of such enormous migrations for those species large enough to be capable of making them (Corkeron and Connor, 1999). See also MIGRATION AND MOVEMENT PATTERNS, this volume.

Gray whales appear to navigate to and from their breeding areas by following the coastal margins of their respective continents. In contrast, the north–south migrations of humpback whales, seem to be deflected by coastlines, currents, and underwater topography rather than guided by them (Dawbin, 1966). How these whales find locations such as the Hawaiian Islands each year in waters as large as the Pacific is still unknown. Use of celestial, acoustic, or magnetic markers are distinct possibilities.

### D. Giving Birth

Few cetacean births have been observed in the wild, but in captive odontocetes, most births are accomplished rapidly without direct assistance from conspecifics. However, there are scattered reports of animals seemingly helping in the birth process, either pulling the fetus or placenta clear off the birth canal. The frequency of such activities, if they occur at all in wild populations, is unknown. Because wild births have been so rarely observed, little is known about how females might reduce the risks of predation and separation in the moments after birth. The proficient swimming abilities of newborn calves permit mother and neonate to rapidly vacate an area where the birth occurred and so minimize the attentions of predators whereas the social nature of cetaceans may permit increased

**R**

predator detection and defense. Newborn calves adopt a swimming posture alongside their mothers, permitting tactile communication, camouflage, and slipstreaming.

## E. Investing in Young After Birth

Parental care in cetaceans is predominantly the responsibility of the mother, although male and female kin as well as unrelated females may provide additional care.

*1. Maintaining Contact and Providing Protection*  All cetaceans are born nutritionally dependent on their mothers, but with births occurring without the spatial certainties of land or ice, there is potential for calves to become separated from their mothers and therefore rapidly starve or become prey. For those species that suspend foraging during breeding, females can devote almost continuous attention to their calves. Species that continue foraging, face a problem as the diving abilities of calves may be insufficient to follow their foraging mothers. Babysitting among this latter group appears to be common solution, with an apparent continuum among the odontocetes. This can range from females of similar breeding status schooling with one another and presumably taking turns guarding calves (Wells *et al.*, 1987) or at least acting as a spatial point of reference, through to related females and their adolescent young accompanying mothers and their neonates (Whitehead, 1996). It may even extend to a system demonstrated by killer and pilot (*Globicephala* spp.) whales (apparently unique among mammals) of stable kin groups with neither male nor female dispersal but instead investment in raising their own (females only) and related offspring (females and males) (Bigg *et al.*, 1990; Amos *et al.*, 1993a, b). Despite such behavioral safeguards, however, cetacean mobility makes separations between mother or care-giver and calf inevitable. Individually specific calls (see SIGNATURE WHISTLES, this volume) are thought to be important in reuniting individuals in species like the bottlenose dolphin. However, such mechanisms take time to develop, and neonate mortality, though low compared with other mammals, accounts for a significant fraction of cetacean deaths.

Our understanding of cetacean reproductive behavior is undoubtedly hampered by the potential ambiguity of the behavior that can be observed. This is particularly prevalent in aspects involving spatial proximity of individuals and apparently altruistic or cooperative behavior. The social complexities of cetacean societies and considerable component that appears to be learnt and practiced makes context an essential component of any behavioral observation. Babysitting is an attractive and logical concept that has been frequently described but the appearance of a calf with an adult animal other than its mother may result for other more reproductively selfish motives. Young inexperienced female bottlenose dolphins, for example, may temporarily kidnap calves and thereby improve their own maternal skills (Mann and Smuts, 1998) whereas males or females may benefit from capturing and killing another's young (Patterson *et al.*, 1998).

*2. Lactation and Weaning*  Cetacean calves do not suffer the constraints experienced by phocid seals which need to rapidly transfer resources to the pup in order to resume feeding nor endure the periods of maternal absence experienced by otariid pups. Instead, female cetaceans take their mobile calves with them and are generally only separated for the length of a foraging dive. Thus, calves have the opportunity to suckle frequently and match milk intake with energy expenditure and growth. However, suckling itself poses a behavioral challenge for cetaceans. The two mammary teats in all female cetaceans are located either side of the genital opening on her ventral side, about two-thirds of the way down her body. For pinnipeds, mother and pup can simply lie at right angles to each other on land or ice but for cetaceans, in-water suckling is more complex. To suckle, the calf must either hold its breath, or have the mother roll onto her side and hold hers. Furthermore one member must orient on a different trajectory to the other in order to make firm contact between mother's teat and calf's beak. Thus the pair must slow or halt progress, necessitating suckling in brief bouts lasting only a few seconds, and by default the cooperation of both parties. The presence of bristles on the rostra of neonate cetaceans are thought to help calves orient during suckling whereas a filled margin on their tongues and muscular control of milk ejection by the mother likely aid efficient milk transfer during the abbreviated suckling events. Given the extreme evolutionary sculpting of the anatomy of the cetaceans, it is somewhat perplexing that the mammary teats are universally located in such an awkward position. The sirenians have their mammary teats located posterior to the pectoral flippers, and so a calf can suckle from a cow without either party having to significantly change from a normal orientation. It is intriguing to wonder why an apparently awkward system has been so conserved in the cetaceans.

Mysticete calves are generally weaned within a year of birth, and in migratory species this coincides with the pair reaching high latitude feeding grounds. Weaning may or may not precipitate separation of cow and calf, but in its timing offers opportunities for the mother to train a calf in migration routes, the location of feeding areas, and potentially facilitate membership of feeding assemblages. Lactation in the majority of odontocetes is longer, and in some cases far longer, with weaning appearing to be gradual and occurring over a period of months or years. Lactose, for example has been detected in the stomachs of sperm whales up to 13 years of age (Best *et al.*, 1984). Such long-term maternal investment suggests that many components of odontocete development require a considerable period of learning and training. Foraging tactics in odontocetes (see Foraging Strategies and Tactics, this volume) are often performed in groups and, while it is unknown what proportion are cooperative or exploitative, their complexity is clear, as is the need for a high degree of interindividual coordination and practice (Guinet, 1991). Calves may learn through observation or dedicated tutoring. The prolonged lactation and consequent investment in young allows calves to develop to a high level before facing nutritional independence. Spatial independence may be even longer in coming if it occurs at all. Bottlenose dolphins probably wean around 18 months after birth, but remain closely associated with their mothers for at least 4 years (Wells *et al.*, 1987). Young killer and pilot whales may never separate from their mothers, drawing comparison with elephant matriarchal societies where the eldest animals may function as long-term information stores and guardians for their offspring and offspring's offspring.

## F. Locating and Selecting a Mate

*1. Female Mating Behavior*  The number of offspring that female cetaceans produce in a lifetime varies. Some porpoises and gray and humpback whales may at times give birth on an annual or a biennial cycle, and have the potential to produce between a dozen and twenty offspring in a lifetime. Others produce much fewer. Female killer whales, for example, may only produce five to six young in their entire lives. Whether twenty or five, these numbers are small given the huge investment in time and resources that each calf receives, and therefore the choice of an appropriate mate to father them is a major component of the reproductive fitness for individual females.

*1. Choosing a Mate*  In short, we have little solid information on how female cetaceans choose mates. In any species or population there are many potential junctures where a female may be making behavioral or physiological decisions, both before conception and afterward. In some instances, females may have the opportunity to simply select a particular male with which to copulate from a range of alternatives. Clapham (1996) has proposed such a case for humpback whales, where males engage in communal display behaviors on the breeding grounds without showing any defense of resources. Females would have opportunities to approach males based on the quality of their displays, or choose males who have worked out dominance relationships among themselves. Female bottlenose dolphins are frequently seen being attended by single or alliances of males (Wells *et al.*, 1987; Connor *et al.*, 1996). Although males may have opportunities to herd a female against her will, the females may also have the opportunity to reject or maintain that contact. Females have been observed rebutting the copulation attempts of males by fleeing or rolling upside down at the surface so that males don't have access to their genital opening.

After copulation, females may have a range of behavioral and physiological options to influence the probability of conception. The number of subsequent males with which she mates would influence the probability of a particular male being the father. The large volumes of sperm produced by males of several cetacean species (see later) suggest that females do mate with several males and that competition between the sperm themselves may be a frequent occurrence in such species. Repeatedly mating with a particular male would also significantly bias the odds. Whether or not an egg is available for fertilization is also critical and it appears that ovulation itself may be related to mating and therefore has the potential to be under the female's control.

Even after fertilization has occurred, there are opportunities to select whether or not to continue investment in a particular partner's offspring. These may range from selective abortion, energetic investment in the fetus, and the subsequent level of parental care expended in the calf. At present we have little information to determine whether such behavioral and physiological decisions are actually made, but since such evidence has been found widely in birds and terrestrial mammals it is entirely possible that such options are open to female cetaceans as well.

*2. Male Mating Behavior*  The reproductive behavior displayed by males is a function of the social and the physical environments in which they live and compete. As with the diversity of habitats and lifestyles exhibited by the cetaceans, males of different species and populations show a huge range of tactics to maximize their reproductive potential (see MATING SYSTEMS, this volume). At a most basic level, males should behave to optimize the number of their own sperm competing to fertilize a female's egg and limit the number of those of competitors. Thus males may increase the probability of obtaining copulations by signaling their quality to females and competing males, through physical or acoustic displays (e.g., postural displays in bottlenose dolphins, songs of humpback whales, sperm whale vocalizations, etc.), ornamentation or body scarring (teeth and scars in beaked whales), intermale combat (humpback whales), extreme body size (sperm whales), or simply tracking the long-distance migrations of females (humpback, gray and right whales). Males may attempt to guard receptive or potentially receptive mates to reduce the probability of competitors mating and increase the number of copulations they can obtain themselves. The alliances formed between male bottlenose dolphins may be an example of such behavior (Connor *et al.*, 1996) where pairs or trios of males may trade off their exclusive access to a female in order to ensure that other males cannot gain mating opportunities. The absence of such alliances in some other populations of bottlenose dolphins (Wilson *et al.*, 1993) suggest that such tactics are context specific and influenced by factors such as the relative abundance of receptive females and potential for males to monopolize them.

Even once copulation has occurred, competition between males need not be over. Gray, bowhead and right whales, harbor porpoises, and several species of dolphins all have testes substantially larger than their body size would suggest. Large quantities of sperm and the ability to copulate frequently may allow males to flush away or dilute the sperm and consequent reproductive chances of others males.

Males may also be able to increase the effective pool of receptive females by influencing the fate of other males' calves. Infanticide is common among terrestrial mammals and is often carried out by males in order to force females to switch from investing resources into a previous calf not sired by themselves and become reproductively receptive again. It is unknown if such behavior occurs in cetacean societies but the violent deaths of young bottlenose dolphins in some populations suggest that males may exploit such options (Patterson *et al.*, 1998).

The long lives and intricate social organization of cetaceans, particularly odontocetes, also offers males the opportunity to increase their fitness, not by maximizing their potential to father offspring, but by investing in their kin. The lack of male dispersal in killer and pilot whales (Bigg *et al.*, 1990; Amos *et al.*, 1993a, b) and absence of interbreeding within pod members, suggest that males may remain with their maternally derived relatives in order to provide some degree of care or protection and thus increase their own inclusive fitness.

## G. Mating

Cetaceans live in a three-dimensional environment that facilitates copulation from a variety of orientations. Common positions include ventrum to ventrum with the pair orientated in the same directions or the male may mount the female from a nonparallel position. Intromission may last only a few seconds or far longer and involve vigorous thrusting or a more passive attitude. Mating may be preceded and followed by prolonged periods of courtship and petting and mating episodes (see COURTSHIP, this volume) may be repeated over periods of minutes, hours, or days.

## H. The Transition to the Non-breeding Season

Although copulation and parturition is generally seasonal in cetaceans, investment in reproduction for females is an almost continuous process after reaching sexual maturity. Baleen whales that migrate from breeding grounds with neonate calves, wean them on the feeding grounds, and may either have a few months of intensive feeding before returning to the breeding grounds to mate or be already pregnant following copulation the previous year. Female odontocetes frequently superimpose reproductive events by both being pregnant and lactating or suckling more than one generation of calves at the same time. Perhaps the most intriguing situation is demonstrated by female pilot whales, which appear to cease ovulating after age 40 and yet continue to lactate for well over a decade

R

(Marsh and Kasuya, 1991). In doing so, they have the potential to not only extend long-term care to their own offspring but also have the potential to influence the fate of their offspring's own offspring.

## See Also the Following Articles

Aggressive Behavior ■ Behavior, Overview ■ Breeding Sites ■ Estrus and Estrous Behavior ■ Sociobiology ■ Territorial Behavior

## References

Amos, B., Schlötterer, C., and Tautz, D. (1993a). Social structure of pilot whales revealed by analytical DNA profiling. *Science* **260**, 670–672.

Amos, W. A., Twiss, S. D., Pomeroy, P. P., and Anderson, S. S. (1993b). Male mating success and paternity in the gray seal (*Halichoerus grypus*): A study using DNA fingerprinting. *Proc. R. Soc. Lond., B, Biol. Sci.* **252**, 199–207.

Amos, W. A., Twiss, S. D., Pomeroy, P. P., and Anderson, S. S. (1995). Evidence for mate fidelity in the gray seal. *Science* **268**, 1897–1899.

Anderson, S. S., and Fedak, M. A. (1987). The energetics of sexual success of gray seals and comparison with the costs of reproduction in other pinnipeds. *Symp Zool. Soc. Lond.* **57**, 319–341.

Anderson, S. S., Burton, R. W., and Summers, C. F. (1975). Behavior of gray seals *Halichoerus grypus* during a breeding season at North Rona. *J. Zool. (Lond).* **177**, 179–195.

Arnbom, T., Fedak, M. A., and Boyd, I. L. (1997). Factors affecting maternal expenditure in southern elephant seals during lactation. *Ecology* **78**, 471–483.

Bennett, K. A., Speakman, J. R., Moss, S. E. W., Pomeroy, P., and Fedak, M. A. (2007). Effects of mass and body composition on fasting fuel utilisation in grey seal pups (*Halichoerus grypus fabricius*): An experimental study using supplementary feeding. *J. Exp. Biol.* **210**, 3043–3053.

Best, P. B., Canham, P. A. S., and Macleod, N. (1984). Patterns of reproduction in sperm whales, *Physeter macrocephalus*. *Rept. Int. Whal. Commn.* **8**(Special Issue), 51–79.

Bigg, M. A., Olesiuk, P. F., Ellis, G. M., Ford, J. K. B., and Balcomb, K. C. (1990). Social organization and genealogy of resident killer whales (*Orcinus orca*) in the coastal waters of British Columbia and Washington State. *Rep. Int. Whal. Commn.* **12**(Special Issue), 383–405.

Biuw, M., Boehme, L., Guinet, C., Hindell, M., Costa, D., Charrassin, J. B., Roquet, F., et al. (21 authors) (2007). Variations in behavior and condition of a Southern Ocean top predator in relation to *in situ* oceanographic conditions. *Proc. Natl. Acad. Sci. USA* **104**, 13705–13710, http://www.pnas.org/cgi/doi/10.1073/pnas.0701121104

Boness, D. J., and James, H. (1979). Reproductive behavior of the gray seal (*Halichoerus grypus*) on Sable Island, Nova Scotia. *J. Zool. (Lond).* **188**, 477–500.

Boness, D. J., Bowen, W. D., and Oftedal, O. T. (1994). Evidence of a maternal foraging cycle resembling that of Otariid seals in a small phocid, the harbor seal. *Behav. Ecol. Sociobiol.* **34**, 95–104.

Bowen, W. D., Iverson, S. J., Boness, D. J., and Oftedal, O. T. (2001). Foraging effort, food intake and lactation performance depend on maternal mass in a small phocid seal. *Funct. Ecol.* 15, 325–334.

Boyd, I. L. (1998). Time and energy constraints in pinniped lactation. *Am. Nat.* **152**, 712–728.

Clapham, P. J. (1996). The social and reproductive biology of humpback whales: An ecological perspective. *Mamm. Rev.* **26**, 27–49.

Connor, R. C., Richards, A. F., Smolker, R. A., and Mann, J. (1996). Patterns of female attractiveness in Indian Ocean bottlenose dolphins. *Behavior* **133**, 37–69.

Connor, R. C., Mann, J., Tyack, P. L., and Whitehead, H. (1998). Social evolution in toothed whales. *Trend. Ecol. Evol.* **13**, 228–232.

Corkeron, P. J., and Connor, R. C. (1999). Why do baleen whales migrate? *Mar. Mamm. Sci.* **15**, 1228–1245.

Costa, D. P., and Trillmich, F. (1988). Mass changes and metabolism during the perinatal fast: A comparison between Antarctic

(*Arctocephalus gazella*) and Galapagos fur seals (*Arctocephalus galapogoensis*). *Physiol. Zool.* **61**, 160–169.

Craig, A. S., and Herman, L. M. (2000). Habitat preferences of female humpback whales *Megaptera novaeangliae* in the Hawaiian Islands are associated with reproductive status. *Mar. Ecol. Prog. Ser.* **193**, 209–216.

Croxall, J. P., and Hiby, A. R. (1983). Fecundity, survival and site fidelity in Weddell seals, *Leptonychotes weddelli*. *J. Appl. Ecol.* **20**, 19–32.

Dawbin, W. H. (1966). The seasonal migratory cycle of humpback whales. *In* "Whales, Dolphins and Porpoises" (K. S. Norris, ed.), pp. 145–170. University of California Press, Berkeley.

Deutsch, C. J., Haley, M. P., and Le Boeuf, B. J. (1990). Reproductive effort of male northern elephant seals: Estimates from mass loss. *Can. J. Zool.* **68**, 2580–2593.

Fay, F. H. (1982). Ecology and biology of the Pacific walrus, *Odobenus rosmarus divergens* Illiger. *US Dep. Inter. Fish Wildl. Ser.* **74**.

Gentry, R. (1999). "Behavior and Ecology of the Northern Fur Seal." Princeton University Press, Princeton.

Guinet, C. (1991). Intentional stranding apprenticeship and social play in killer whales (*Orcinus orca*). *Can. J. Zool.* **69**, 2712–2716.

Hindell, M. A., and Little, G. J. (1988). Longevity, fertility and philopatry of two female southern elephant seals (*Mirounga leonina*) at Macquarie Island. *Mar. Mamm. Sci.* **4**, 168–171.

Hindell, M. A., McConnell, B. J., Fedak, M. A., Slip, D. J., Burton, H. R., Reijnders, P. J. H., and McMahon, C. R. (1999). Environmental and physiological determinants of successful foraging by naive southern elephant seal pups during their first trip to sea. *Can. J. Zool.* **77**, 1807–1821.

Horning, M., and Trillmich, F. (1997). Ontogeny of diving behavior in the Galapagos fur seal. *Behavior* **134**, 1211–1257.

Huber, H. R. (1987). Natality and weaning success in relation to age of first reproduction in northern elephant seals. *Can. J. Zool.* **65**, 1311–1316.

Jones, M. L., and Swartz, S. L. (1984). Demography and phenology of gray whales and evaluation of whale-watching activities in Laguna San Ignacio, Baja California Sur, Mexico. *In* "The Gray Whale" (M. L. Jones, S. L. Swartz, and S. Leatherwood, eds), pp. 309–374. Academic Press, San Diego.

Kovacs, K. M. (1987). Maternal behavior and early behavioral ontogeny of gray seals (*Halichoerus grypus*) on the Isle of May, UK. *J. Zool. (Lond).* **213**, 697–715.

Le Boeuf, B. J., Crocker, D. E., Costa, D. P., Blackwell, S. B., Webb, P. M., and Houser, D. S. (2000). Foraging ecology of northern elephant seals. *Ecol. Monogr.* **70**, 353–382.

Lockyer, C. (1984). Review of baleen whale (Mysticeti) reproduction and implications for management. *Rept. Int. Whal. Commn.* **6**(Special Issue), 27–50.

Lockyer, C. (1987). A theoretical approach to the balance between growth and food consumption in fin and sei whales, with special reference to the female reproductive cycle. *Rept. Int. Whal. Commn.* **28**, 243–250.

Lockyer, C. (2007). All creatures great and smaller: A study in cetacean life history energetics. *J. Mar. Biol. Assoc. U.K.* **87**, 1035–1045, doi: 10.1017/S0025315407054720.

Lunn, N. J., and Boyd, I. L. (1991). Pupping-site fidelity of Antarctic fur seals at Bird Island, South Georgia. *J. Mammal.* **72**, 202–206.

Mann, J., and Smuts, B. B. (1998). Natal attraction: Allomaternal care and mother-infant separations in wild bottlenose dolphins. *Anim. Behav.* **55**, 1097–1113.

Marsh, H., and Kasuya, T. (1991). An overview of the changes in the role of a female pilot whale with age. *In* "Dolphin Societies" (K. Pryor, and K. S. Norris, eds), pp. 281–285. University of California Press, Berkeley.

McConnell, B. J., and Fedak, M. A. (1996). Movements of southern elephant seals. *Can. J. Zool.* **74**, 1485–1496.

McConnell, B. J., Fedak, M. A., Burton, H. R., Englehard, G. H., and Reijnders, P. (2002). Movements and foraging areas of naive, recently weaned southern elephant seal pups. *J. Anim. Ecol.* **71**, 65–78.

McCulloch, S., Pomeroy, P. P., and Slater, P. J. B. (1999). Individually distinctive pup vocalizations fail to prevent allo-suckling in gray seals. *Can. J. Zool.* **77**, 716–723.

McMahon, C. R., Burton, H. R., and Bester, M. N. (2000). Weaning mass and the future survival of juvenile southern elephant seals, *Mirounga leonina*, at Macquarie Island. *Antarct. Sci.* **12**, 149–153.

Miller, E. H., and Boness, D. J. (1983). Summer behavior of Atlantic walruses *Odobenus rosmarus rosmarus* (L.) at Coats Island, N. W. T., Canada. *Z. Säugetierkd.* **48**, 298–313.

Noren, S. R., Iverson, S. J., and Boness, D. J. (2005). Development of the blood and muscle oxygen stores in gray seals (*Halichoerus grypus*): Implications for juvenile diving capacity and the necessity of a terrestrial postweaning fast. *Physiol. Biochem. Zool.* **78**, 482–490.

O'Shea, T. J., Ackerman, B. B., and Percival, H. F. (eds.) (1995). Population biology of the Florida manatee. National Biological Service, Information and Technology Report 1, Washington DC.

Patterson, I. A. P., Reid, R. J., Wilson, B., Grellier, K., Ross, H. M., and Thompson, P. M. (1998). Evidence for infanticide in bottlenose dolphins; an explanation for violent interactions with harbor porpoises? *Proc. R. Soc. Lond., B, Biol. Sci.* **265**, 1167–1170.

Pomeroy, P. P., Anderson, S. S., Twiss, S. D., and McConnell, B. J. (1994). Dispersion and site fidelity of breeding female gray seals (*Halichoerus grypus*) on North Rona, Scotland. *J. Zool. (Lond).* **233**, 429–448.

Pomeroy, P. P., Fedak, M. A., Anderson, S. S., and Rothery, P. (1999). Consequences of maternal size for reproductive expenditure and pupping success of gray seals at North Rona, Scotland. *J. Anim. Ecol.* **68**, 235–253.

Pomeroy, P. P., Twiss, S. D., and Duck, C. D. (2000a). Expansion of a gray seal (*Halichoerus grypus*) breeding colony: Changes in pupping site use at the Isle of May, Scotland. *J. Zool. (Lond).* **250**, 1–12.

Pomeroy, P. P., Twiss, S. D., and Redman, P. (2000b). Philopatry, site fidelity and local kin associations within gray seal breeding colonies. *Ethology* **106**, 899–919.

Pomeroy, P. P., Worthington-Wilmer, J., Amos, W., and Twiss, S. D. (2001). Reproductive performance links to fine scale spatial patterns of female grey seal relatedness. *Proc. R. Soc. Lond., B, Biol. Sci.* **268**, 711–717.

Pomeroy, P. P., Redman, P. R., Ruddell, S. J. S., Duck, C. D., and Twiss, S. D. (2005). Breeding site choice fails to explain interannual associations of female grey seals. *Behav. Ecol. Sociobiol.* **57**, 546–556.

Read, A. J., and Hohn, A. A. (1995). Life in the fast lane—The life-history of harbor porpoises from the Gulf of Maine. *Mar. Mamm. Sci.* **11**, 423–440.

Riedman, M. L., and Estes, J. A. (1990). The sea otter (*Enhydra lutris*): Behavior, ecology and natural history. *U.S. Fish and Wildlife Service, Biological Report* **90**(14), Washington, DC.

Sonntag, R. P., Benke, H., Hiby, A. R., Lick, R., and Adelung, D. (1999). Identification of the first harbor porpoise (*Phocoena phocoena*) calving ground in the North Sea. *J. Sea Res.* **41**, 225–232.

Trillmich, F. (1996). Parental Investment in Pinnipeds. *Adv. Study Behav.* **25**, 533–577.

Twiss, S. D., Pomeroy, P. P., and Anderson, S. S. (1994). Dispersion and site fidelity of breeding male gray seals (*Halichoerus grypus*) on North Rona, Scotland. *J. Zool. (Lond).* **233**, 683–693.

Twiss, S. D., Anderson, S. S., and Monaghan, P. (1998). Limited intra-specific variation in male gray seal (*Halichoerus grypus*) dominance relationships in relation to variation in male mating success and female availability. *J. Zool. (Lond).* **246**, 259–267.

Twiss, S. D., Caudron, A., Pomeroy, P. P., Thomas, C. J., and Mills, J. P. (2000). Fine scale topography influences the breeding behavior of female gray seals. *Anim. Behav.* **59**, 327–338.

Twiss, S. D., Pomeroy, P. P., Graves, J. A., and Poland, V. F. (2006). Finding fathers—spatio-temporal analysis of paternity assignment in grey seals (*Halichoerus grypus*). *Mol. Ecol.* **15**, 1939–1953.

Urian, K. W., Duffield, D. A., Read, A. J., Wells, R. S., and Shell, E. D. (1996). Seasonality of reproduction in bottlenose dolphins, *Tursiops truncatus*. *J. Mammal.* **77**, 394–403.

Wells, R. S., Scott, M. D., and Irvine, A. B. (1987). The social structure of free-ranging bottlenose dolphins. *In* "Current Mammalogy" (H. Genoways, ed.), pp. 247–305. Plenum, New York.

Whitehead, H. (1996). Babysitting, dive synchrony, and indications of alloparental care in sperm whales. *Behav. Ecol. Sociobiol.* **38**, 237–244.

Wilson, B., Thompson, P. M., and Hammond, P. S. (1993). An examination of the social structure of a resident group of bottle-nosed dolphins (*Tursiops truncatus*) in the Moray Firth, NE Scotland. *Eur. Res. Cetaceans* **7**, 54–56.

Worthington Wilmer, J., Allen, P. J., Pomeroy, P. P., Twiss, S. D., and Amos, W. (1999). Where have all the fathers gone? An extensive microsatellite analysis of paternity in the gray seal (*Halichoerus grypus*). *Mol. Ecol.* **8**, 1417–1429.

Worthington Wilmer, J., Overall, A. J., Pomeroy, P. P., Twiss, S. D., and Amos, W. (2000). Patterns of paternal relatedness in British gray seal colonies. *Mol. Ecol.* **9**, 283–292.

# Ribbon Seal
## *Histriophoca fasciata*

### LLOYD LOWRY AND PETER BOVENG

## I. Characteristics and Taxonomy

The ribbon seal is one of the least well known of all the world's pinnipeds. The species has been placed by some in the genus *Phoca* based on cranial morphology (Burns and Fay, 1970), but molecular studies (Árnason *et al.*, 1995; Mouchaty *et al.*, 1995) indicate that ribbon seals belong in the separate genus, *Histriophoca* (Carr and Perry, 1997; Rice, 1998; Davis *et al.*, 2004). Its scientific name, *Histriophoca fasciata*, means the banded "actor-seal" in Latin.

Ribbon seals are distinctively marked (Naito and Oshima, 1976). Pups are born with a long, white pelage called lanugo (Fig. 1), and for the first year after the lanugo is shed their coat is silver–gray with a dark blue–black back (Fig. 2). Older seals have a dark background with a set of light bands encircling the head, the posterior trunk, and each front flipper. In males the background color is nearly black and the bands almost white (Fig. 3). Females have a similar pattern with much less contrast (Fig. 1). Adult seals are generally 1.5–1.75 m long (nose to tail) and weigh 70–110 kg. They are considerably more slender than other northern ice-inhabiting seals.

The skull of the ribbon seal is relatively short and broad. Their dentition is similar to that of other phocid seals, usually with 34 small teeth (Burns, 1981).

## II. Distribution and Abundance

The distribution of ribbon seals is restricted to the northern North Pacific Ocean, where they are seen most commonly in the Okhotsk and Bering seas (Fig. 4). In those regions during the late winter and spring, ribbon seals are commonly seen hauled out on sea ice in the ice front region. They stay associated with the ice until it disappears in May–June. During the summer and fall ribbon seals are rarely seen hauled out on ice or land, and the presumption that they spend those months living a pelagic lifestyle (Burns, 1981) has recently been corroborated by tracking with satellite-linked tags. Ribbon seals tagged in the spring of 2005 near the eastern coast of

R

Figure 1 *A ribbon seal mother with her pup in its lanugo pelage. Photo by G. Brady, NOAA.*

Figure 2 *A ribbon seal pup in May, after shedding its lanugo coat. Photo by P. Boveng, NOAA.*

Figure 3 *An adult male ribbon seal in May. Photo by P. Boveng, NOAA.*

Kamchatka spent the summer and fall at sea throughout the Bering Sea and the Aleutian Islands, whereas roughly half of the seals that were tagged in the central Bering Sea in 2007 moved to the Chukchi and Beaufort seas during summer and fall. Reported sightings from

summer–fall have been at sea near the Pribilof Islands, ashore in the eastern Aleutian Islands, and at the ice edge in the northern Chukchi Sea (Kelly, 1988). It is unclear whether or not mixing occurs between seals from the Okhotsk and the Bering Sea regions.

Ribbon seals are difficult to count because of their wide distribution in remote regions, and the fact that they spend little time out of the water. Abundance estimates have been produced by both Russian and US scientists, but they are of unknown accuracy. Burns (1981) presented population estimates for the early 1970s of 140,000 for the Okhotsk Sea and 90–100,000 for the Bering Sea. Fedoseev (2000) gives much higher estimates for the Okhotsk Sea, and indicates an average population size of 370,000 during 1968–1990. In both the Bering and the Okhotsk seas abundance has fluctuated markedly due to increases and decreases in harvests conducted by the former Soviet Union (Burns, 1981; Fedoseev, 2000). There are no recent abundance estimates for either region.

## III. Ecology

The ribbon seal is considered to be a pagophilic (ice-loving) species, and sea ice provides them an essential platform that is used for pupping, nursing their pups, and molting. During the pupping and molting period they primarily use the ice front, where there is a mix of floes and open water (Burns, 1970). The location of the ice front is usually on the continental shelf not far from the continental shelf break.

With their long necks, large eyes, and slender bodies, ribbon seals look quite different from other northern seals. Although the meaning of these characteristics is not fully understood, it seems likely that they are adaptations to a lifestyle that includes spending many consecutive months at sea. Another unusual characteristic is an air sac that connects to the trachea and extends over the ribs (Burns, 1981). When inflated this organ could provide extra buoyancy making it easier for seals to float or rest in the water.

During the spring when they are in the ice front, ribbon seals eat shrimps, squids, and a variety of fishes including arctic cod, saffron cod, pollock, Pacific cod, capelin, and flatfishes (Shustov,1965; Frost and Lowry, 1980). Little is known about feeding habits during the summer, fall, and winter, which are the seasons when they must feed most intensively. A study of winter–spring feeding near Hokkaido, Japan found walleye pollock to be the main prey item (Deguchi *et al.*, 2004).

The ice front region used by ribbon seals is south of the normal range of polar bears (*Ursus maritimus*), so these seals probably rarely encounter bears. During the pelagic phase of their life they undoubtedly encounter killer whales (*Orcinus orca*), and perhaps sharks, but no predation events have been reported (Kelly, 1988).

## IV. Behavior and Physiology

Compared with other seals, the eyes of ribbon seals are quite large. However, they seem to have poor vision in air so this may be an adaptation for improved eyesight under water. When hauled out on the ice they are relatively easy to approach (Burns, 1981), but it isn't clear whether this lack of vigilance is due to their not being subject to polar bear predation or to sensory limitations. Ribbon seals move across the ice with a characteristic snakelike side-to-side motion, unlike other ice seals which hump across the ice. When captured on the ice with a net, many individuals cease to struggle, an apparent "play dead" strategy that is uncharacteristic of pinnipeds and is of unclear adaptive significance.

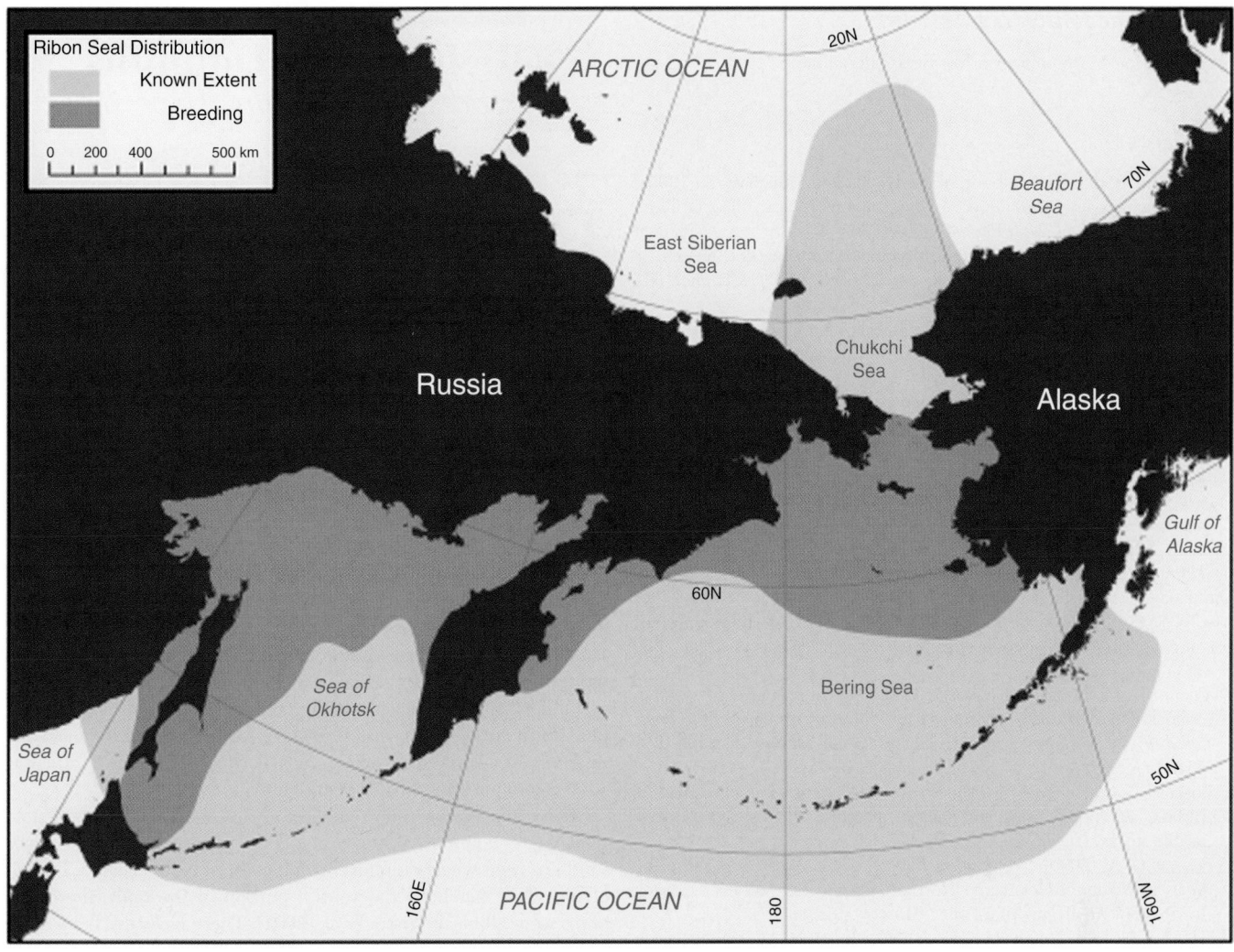

**Figure 4** *The geographic distribution of ribbon seals.*

Two kinds of underwater sounds were recorded from ribbon seals in the ice near St. Lawrence Island in spring (Watkins and Ray, 1977). One was described as a "puffing" sound, and the other a "downward sweeping" sound. Based on analogy with sounds made by other seals, researchers speculated that the ribbon seal sounds probably function in reproductive or territorial behavior.

Ribbon seals have not been the subject of detailed physiological studies that have recently been done on many other pinnipeds. Some studies of respiratory physiology and blood parameters suggest adaptations for a pelagic existence and deep diving (Burns, 1981).

## V. Life History

Ribbon seals give birth in the ice front during March–April. Newborn pups weigh about 10 kg, and their weight doubles during a 3–4 week nursing period. Adult male ribbon seals do not accompany females during the early part of the nursing period, and little is known about their breeding structure. After weaning ribbon seals grow quite rapidly and reach sexual maturity at 3–5 years of age. The peak of breeding occurs in late April and early May, and seals molt shortly thereafter. The maximum lifespan is probably 25–30 years

(Burns, 1981). As mentioned earlier, little is known of the activities of ribbon seals during other months of their life when they are living pelagically.

## VI. Interactions with Humans

Ribbon seals were harvested in substantial numbers in both the Okhotsk and the Bering seas by the former Soviet Union for oil, skins, and animal food. Commercial harvests began in 1961, and in the Bering Sea takes peaked in the mid-1960s at about 22,000 per year (Fedoseev, 1976). With signs of overharvest, quotas were reduced, but 10,000 to 15,000 were taken per year from the sea of Okhotsk in 1990–1993 (Grachev, 2006). More recently, commercial harvests are believed to have virtually ceased.

Their offshore distribution mostly keeps ribbon seals away from coastal hunters, but a few are taken in Alaska by Eskimos at villages in the Bering Strait region (Kelly, 1988). Other than directed harvesting, ribbon seals have few interactions with humans. They are infrequently caught incidental to commercial fishing operations in Alaska and on the high seas (Stewart and Everitt, 1983).

R

## References

Árnason, Ú., Bodin, K., Gullberg, A., Ledje, C., and Mouchaty, S. (1995). A molecular view of pinniped relationships with particular emphasis on the true seals. *J. Mol. Evol.* **40**, 78–85.

Burns, J. J. (1970). Remarks on the distribution and natural history of pagophilic pinnipeds in the Bering and Chukchi seas. *J. Mammal.* **51**, 445–454.

Burns, J. J. (1981). Ribbon Seal. *In* "Handbook of Marine Mammals Vol. 2. Seals" (S. H. Ridgway, and R. J. Harrison, eds), pp. 89–109. Academic Press, London.

Burns, J. J., and Fay, F. H. (1970). Comparative morphology of the skull of the ribbon seal, *Histriophoca fasciata*, with remarks on systematics of the Phocidae. *J. Zool. (Lond).* **161**, 363–394.

Carr, S. M., and Perry, E. A. (1997). Intra- and interfamilial systematic relationships of phocid seals as indicated by mitochondrial DNA sequences. *In* Molecular Genetics of Marine Mammals (A. E. Dizon, S. J. Chivers, and W. F. Perrin, eds)., *The Society for Marine Mammalogy*, Lawrence, KS. *Spec. Pub.* **3**, 277–290.

Davis, C. S., Delisle, I., Stirling, I., Siniff, D. B., and Strobeck, C. (2004). A phylogeny of the extant Phocidae inferred from complete mitochondrial DNA coding regions. *Mol. Phylogenet. Evol.* **33**, 363–377.

Deguchi, T., Goto, Y., and Sakurai, Y. (2004). Importance of walleye pollock (*Theragra chalcogramma*) to wintering ribbon seals (*Phoca fasciata*) in Nemuro Strait, Hokkaido Japan. *Mammal Study* **29**, 55–63.

Fedoseev, G. A. (1976). Morphological-ecological characteristics of ribbon seal populations and factors affecting the conservation of usable stocks. *Izv. TINRO* 86, 158–177.

Fedoseev, G. A. (2000). "Population Biology of Ice-Associated Forms of Seals and Their Roles in the Northern Pacific Ecosystems." Center for Russian Environmental Policy, Moscow, Russia.

Frost, K. J., and Lowry, L. F. (1980). Feeding of ribbon seals (*Phoca fasciata*) in the Bering Sea in spring. *Can. J. Zool.* **58**, 1601–1607.

Grachev, A.I. (2006). Analysis of present day harvest of true seals in the Sea of Okhotsk and its prospects. pp. 502–506 *in* Belkovich, V. M. (ed.) "Marine Mammals of the Holarctic." Magadan, Russia.

Kelly, B. P. (1988). Ribbon seal. *In* "Selected Marine Mammals of Alaska" (J. W. Lentfer, ed.), pp. 97–106. U.S. Marine Mammal Commission, Washington, DC.

Mouchaty, S., Cook, J. A., and Shields, G. F. (1995). Phylogenetic analysis of northern hair seals based on nucleotide sequences of the mitochondrial cytochrome *b* gene. *J. Mammal.* **76**, 1178–1185.

Naito, Y., and Oshima, M. (1976). The variation in the development of pelage of the ribbon seals with reference to the systematics. *Sci. Rep. Whales Res. Inst., Tokyo* **28**, 187–197.

Rice, D. W. (1998). "Marine Mammals of the World: Systematics and Distribution." *The Society for Marine Mammalogy*, Lawrence, KS. *Spec. Pub.* **4**.

Shustov, A. P. (1965). The food of the ribbon seal in the Bering Sea. *Izvest. TINRO* **59**, 183–192.

Shustov, A. P. (1967). The effect of sealing on the state of the population of Bering Sea ribbon seals. *Izvest. TINRO* **59**, 173–178.

Stewart, B. S., and Everitt, W. T. (1983). Incidental catch of a ribbon seal (*Phoca fasciata*) in the central North Pacific. *Arctic* **36**, 369.

Watkins, W. A., and Ray, G. C. (1977). Underwater sounds from a ribbon seal (*Phoca fasciata*). *Fish. Bull., US* **75**, 450–543.

**R**

# Right Whale Dolphins
## *Lissodelphis borealis, L. peronii*

**JESSICA D. LIPSKY**

Right whale dolphins are known for their distinctive black and white color patterns and lack of a dorsal fin. These characteristics make these species easy to identify at sea. Although these species were first described in the beginning of the nineteenth century, their overall biology, life history, taxonomy, and behavior are still poorly known.

## I. Characters and Taxonomic Relationships

The two species of right whale dolphins have very different pigmentation patterns. The northern right whale dolphin (*Lissodelphis borealis* Peale, 1848) is mainly black with a white ventral patch that runs from the fluke notch to the throat region (Fig. 1). This band widens slightly at the urogenital area in males and to a greater extent in females (Leatherwood and Walker, 1979). There is another small white patch on the ventral tip of the rostrum and on the underside of the flippers. The southern right whale dolphin [*L. borealis* (Lacépède, 1804)] has a similar white ventral patch; however, it extends higher on the posterior flanks (Figs 2 and 3). The back of the dolphin is black, and the white area reaches a high point midway along the body, dipping down at the flipper insertion and covering most of the head and rostrum. Newborn calves are usually dark gray or brown, attaining adult coloration after the first year of life.

There have been reported sightings of anomalously pigmented right whale dolphins. Visser *et al.* (2004) observed four melanistic southern right whale dolphins off the coast of New Zealand. Instead of the white patch on the ventral portion of the body, these dolphins were all black (Visser *et al.*, 2004). There have also been other recorded sightings of all white animals (Watson, 1981) and partial whites, darks and gray dolphins, (Newcomer *et al.*, 1996).

Right whale dolphins can grow to lengths of 3 m; males tend to grow larger than females (Leatherwood and Walker, 1979). Weights have been recorded up to 116 kg (Jefferson *et al.*, 1994). The flippers are slender and pointed at the tips. The flukes have a median notch that is moderately deep with concave trailing edges. The teeth are small and sharp, ranging in numbers from 37 to 54 per row in the northern species and 39 to 50 in the southern species, with more teeth in the lower jaw (Jefferson *et al.*, 1994).

Recent classifications have placed the right whale dolphins in a monogeneric delphinid subfamily Lissodelphinae or in the subfamily Delphininae. However, based on an analysis of cytochrome *b* (mtDNA) sequences, LeDuc *et al.* (1999) tentatively placed them in the subfamily Lissodelphinae with *Lagenorhynchus* spp. and *Cephalorhynchus* spp.

## II. Distribution and Ecology

Right whale dolphins are found in cool-temperate and subarctic waters in the North Pacific and circumpolar subantarctic and cool-temperate waters in the Southern Ocean (Fig. 4). Right whale dolphins tend to be offshore oceanic cetaceans, with rare sightings inshore except *L. peronii* which has been observed inshore off the coasts of Chile and Namibia in areas with major upwelling (Aguayo, 1975; Rose and Payne, 1991). In the North Pacific, northern right

Figure 1    *The northern right whale dolphin (C. Brett Jarrett).*

Figure 2    Lissodelphis peronii *incidentally caught on a Japanese squid driftnet vessel in the North Pacific. Photo courtesy of the late Michael Newcomer.*

Figure 3    Lissodelphis peronii *with calf off the coast of South America. Photo courtesy of Laura Morse.*

whale dolphins range from Kuril Islands, Russia, south to Sanriku, Honshu, Japan, extending eastward to the Gulf of Alaska and south to southern California (Rice, 1998). They are distributed approximately from 34°N to 55°N and 145°W to 118°E. Occasional movements south of 30°N are associated with anomalous cold-water temperatures. In the Southern Hemisphere, southern right whale dolphins are found most commonly between 25°S and 55°S in the eastern South Pacific and between 30°S and 65°S in the western South Pacific. They are found most often between the subtropical and the Antarctic convergences, with distributions reflecting the variability in these oceanographic features (Gaskin, 1968). It has also been observed that the range of the southern right whale dolphin often extends northward along eastern cold-water boundary currents. An example of this can be shown with the occurrence of a stranding of the southern species off the coast of Brazil in 1995, where the warm Brazil current meets the cold Malvinas current. A

single male stranded at the Juréia-Itatins Ecological Station in southeastern Brazil (between latitudes 24°17'–24°40'S and longitudes 47°00'–47°30'W) north of its typically observed range. The species had never been seen in this region before (Martuscelli *et al.*, 1996). It is thought that the Malvinas current is a major feature in the occurrence of marine mammals in this area.

Migration of both *Lissodelphis* species is not entirely known or understood. In northern right whale dolphins, there appears to be an inshore shift in winter and spring off California, which coincides with peak abundance of their primary food source (squid) (Leatherwood and Walker, 1979). Forney and Barlow (1998) found that northern right whale dolphin abundance was greatest off the Southern California Bight in winter, whereas in the summer there were no sightings made in this area. In addition, they observed a greater abundance of *L. borealis* offshore in summer and a greater abundance inshore on the Southern California Bight continental shelf in winter (Forney and

R

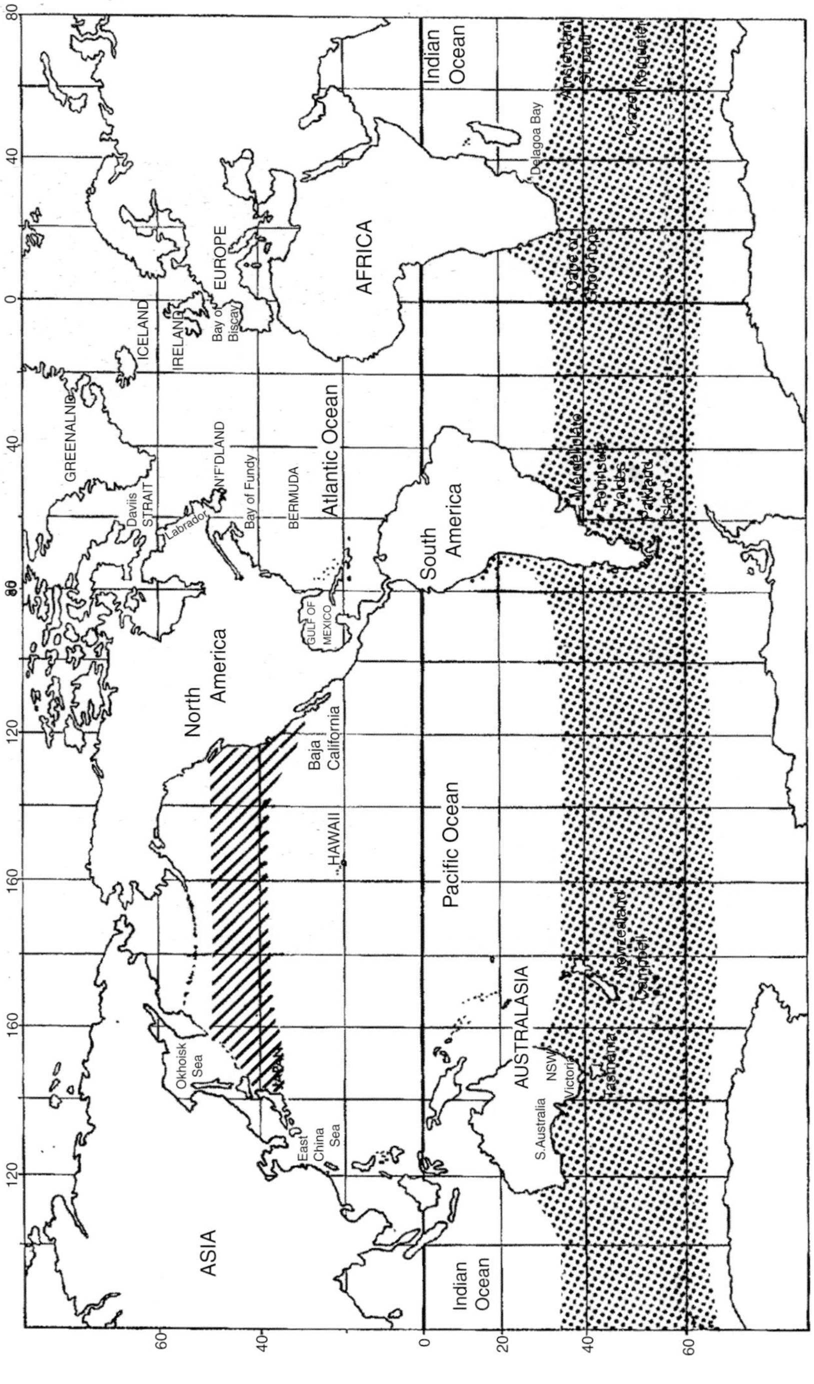

**Figure 4** *Approximate ranges of* Lissodelphis borealis *(hatched) and* L. peronii *(stippled).*

Barlow, 1998). Southern right whale dolphins tend to occur year-round in a localized area off Namibia, Africa, where high-productivity waters prevail (Newcomer *et al.*, 1996). Off the coast of Chile, southern right whale dolphins are present year-round and have been discovered to migrate northward during winter and spring. Food sources, which are affected by changing water temperatures, appear to be a factor in this species' migration. In addition, these two species are commonly found in oceanic, deep waters, on highly productive continental shelves, or sometimes where deep waters approach the coast.

Northern right whale dolphins have been observed to associate with 14 other species of marine mammals in the North Pacific Ocean. They are mainly observed with Pacific white-sided dolphins, *Lagenorhynchus obliquidens*, which share a similar distribution and habitat (Klumov, 1959; Leatherwood and Walker, 1979). They are also commonly found with pilot whales, *Globicephala macrorhynchus* and Risso's dolphins, *Grampus griseus* (Leatherwood and Walker, 1979). Southern right whale dolphins are associated most often with pilot whales (*Globicephala melas*) and dolphins of the genus *Lagenorhynchus* (Jefferson *et al.*, 1994).

Yazdi (2002) described a possible hybrid of southern right whale dolphins with the dusky dolphin (*Lagenorhynchus obscurus*) off the coast of Argentina. Similar phenotypic characteristics were observed from photographic data. These anomalous dolphins have a slender body and the sharp contrasting line between the white ventral patch and the black dorsal part of the body similar to the southern right whale dolphin. Additionally, these possible hybrids have a light gray patch at the peduncle region similar to the dusky dolphin. This anomalous dolphin does have a dorsal fin but one that is much smaller and more triangular than in the dusky dolphin and located approximately two-thirds back on the body. The presence of the modified dorsal fin suggests that this anomalous dolphin is not representative of a color variant of the southern right whale dolphin (Yazdi, 2002).

PREDATION on right whale dolphins is poorly known; however, killer whales and large shark species are occasional predators. There are two records of predation on southern right whale dolphins, a 0.87-m southern right whale dolphin fetus was found in a 3.6-m sleeper shark (*Somniosus* cf. *pacificus*) off the coast of Valdivia, Chile in 1990 (Corvetto *et al.*, 1992) and a 1.7-m Patagonian toothfish taken off central Chile in 1983 had a 0.86-m southern right whale dolphin neonate in its stomach (Van Waerebeek *et al.*, 1991).

The northern right whale dolphin has been observed to feed primarily on squid and lanternfish; however, other prey species include Pacific hake, saury, and mesopelagic fishes (Leatherwood and Walker, 1979). The southern right whale dolphin feeds primarily on various squid and fish species.

Strandings of northern and southern right whale dolphins are uncommon. However, a single mass stranding of three southern right whale dolphins was recorded in New Zealand in 1952 (Fraser, 1955). An apparent increase in strandings of *L. peronii* is possibly the result of discarded animals from a rapidly developing swordfish gillnet fishery off northern Chile (Van Waerebeek *et al.*, 1991).

## III. Behavior and Life History

Right whale dolphins tend to be gregarious animals, often traveling in groups of up to 2000–3000 in the North Pacific (Leatherwood and Walker, 1979) and up to 1000 in the southern species (Gaskin, 1968). Herds are characterized by four main configurations, including V-shaped herds, "chorus line" formation herds, tightly packed herds with no identifiable subgroups, and herds with subgroups within the main group (Leatherwood and Walker, 1979). Both forms have been observed to travel slowly or quickly; this is

associated with surfacing modes, breathing intervals, and travel speeds. Right whale dolphins can travel up to 40 km/h (Leatherwood and Reeves, 1983). In some instances, right whale dolphins will bow ride on vessels, especially in the presence of other species, although sometimes they will actively avoid approaching vessels. Aerial behavior such as breaching, belly flops, and side and fluke slaps are not uncommon, especially in the fast swimming mode.

Information on growth and reproduction for right whale dolphins is limited. Twenty-three specimens have been examined in the eastern North Pacific, and their data suggest that males attain sexual maturity between 212 and 220 cm and females at about 200 cm (Leatherwood and Walker, 1979). In November 1990 and 1991, 229 northern right whale dolphins were obtained from the Japanese squid drift net fishery and examined for total length, age, and sex (Ferrero and Walker, 1993). Ferrero and Walker (1993) found that the average length for sexually mature males is 214.7 cm and 199.8 cm for females in the northern species. In addition, they calculated that the age at the onset of sexual maturity for males is approximately 9.9 years and for females it is approximately 9.7 years (Ferrero and Walker, 1993). In the western North Pacific, other reports indicate that females attain sexual maturity between 206 and 212 cm (Jefferson *et al.*, 1994). Northern right whale dolphin neonates range between 80 and 100 cm at birth (Jefferson *et al.*, 1994). Ferrero and Walker (1993) found that for northern right whale dolphins length at birth ranged between an average of 99.7 and 103.8 cm using three different methods of length estimation. The calving season is unknown; however, small calves are often seen in winter or in early spring. In the Southern Ocean, right whale dolphin reproductive biology is largely unknown. Two females measuring 218 and 229 cm and one male measuring 251 cm have been examined and were all sexually mature (Jefferson *et al.*, 1994).

Sound production in northern right whale dolphins has been recorded (Fish and Turl, 1976). Clicks with high repetition rates were recorded, with few whistles. Sound production in southern right whale dolphins has not been described. There have been few attempts to capture live animals due to the difficulty in maintaining these oceanic species. A northern right whale dolphin was captured live and held for 15 months; however, most live captures have not survived more than 3 weeks (Walker, 1975). There have been no reported attempts to capture live southern right whale dolphins.

## IV. Conservation Status

In the nineteenth century, whalers occasionally took northern right whale dolphins. Although there is currently no direct fishery for right whale dolphins, the northern species is occasionally taken in Japan's harpoon fishery and in the Japanese and Russian purse-seine fisheries (Klumov, 1959). In addition, a few individuals were taken in Japan's salmon gillnet fishery (International Whaling Commission, 1983) and in California's shark and swordfish driftnet fishery. The majority of right whale dolphin bycatches in recent years occurred in the North Pacific squid driftnet fisheries operated by Japan, Korea, and Taiwan (Jefferson *et al.*, 1994). The squid fisheries began in 1978 with small incidental takes of 300–400 dolphins until the mid-1980s when incidental takes were on the order of 15,000–24,000 dolphins per year (Mangel, 1993). It is thought that the stock in this area has been depleted to 24–73% of its pre-exploitation level (Mangel, 1993). In the past southern right whale dolphins have been taken off Chile and Peru for use of their meat and blubber for human consumption or use as crab bait (Newcomer *et al.*, 1996). Since 1989 there has been an increase in southern right whale dolphin bycatches in the developing swordfish gillnet fishery off Chile (Van Waerebeek *et al.*, 1991).

R

The effects of POLLUTION and contaminants on right whale dolphins in general are currently unknown. Given the nature of the pelagic habitat of the northern species, the effects of pollution are probably minimal. However, seasonal shifts in migration and distribution could possibly have a negative impact on these species. Minh *et al.* (2000) estimated the concentrations of polychlorinated biphenyls (PBCs) in cetaceans in the North Pacific. They found that in one northern right whale dolphin individual in their study contained high concentrations of PCBs such that the exceeded levels were associated with immunosuppresion in harbor seals. In another study, Jones *et al.* (1999) found that concentrations of halogenated aromatic hydrocarbons (HAHs) in open ocean cetaceans (southern right whale dolphins) had intermediate levels whereas coastal species had the highest levels of contaminants. So while research shows that *Lissodelphis* species are susceptible to marine pollutants, more research in ecotoxicology is needed.

## See Also the Following Articles

Delphinids, Overview ■ North Pacific Marine Mammals

## References

Aguayo, L. A. (1975). Progress report on small cetacean research in Chile. *J. Fish. Res. Board. Can.* **32**, 1123–1143.

Corvetto, A., Lamilla, J., and Pequeno, G. (1992). *Lissodelphis peronii* Lacépède 1804 (Delphinidae, Cetacea) within the stomach contents of a sleeping shark, *Somniosus* cf. *pacificus*, Bigelow and Schroeder 1944, in Chilean waters. *Mar. Mamm. Sci.* **8**, 312–314.

Ferrero, R. C., and Walker, W. A. (1993). Growth and reproduction of the northern right whale dolphin, *Lissodelphis borealis*, in the offshore waters of the North Pacific Ocean. *Can. J. Zool.* **71**(12), 2335–2344.

Fish, J. F., and Turl, C. W. (1976). Acoustic source levels of four species of small whales. *Naval Undersea Center Tech. Rep., TP* **547**, 1–14.

Forney, K. A., and Barlow, J. (1998). Seasonal patterns in the abundance and distribution of California cetaceans, 1991–1992. *Mar. Mamm. Sci.* **14**(3), 460–489.

Fraser, F. C. (1955). The southern right whale dolphin, *Lissodelphis peronii* (Lacépède) external characteristics and distribution. *Bull. Brit. Mus. (Nat. Hist.) Zool.* **2**(2), 339–346.

Gaskin, D. E. (1968). The New Zealand Cetacea. *Fish. Res. Bull. New Zealand* **1**, 1–92.

International Whaling Commission (1983). Report of the sub-committee on small cetaceans, Annex H. *Rep. Int. Whal. Commn.* **33**, 152–170.

Jefferson, T. A., and Newcomer, M. W. (1993). *Lissodelphis borealis*. *Mamm. Spec.* **425**, 1–6.

Jefferson, T. A., Newcomer, M. W., Leatherwood, S., and Van Waerebeek, K. (1994). Right whale dolphins *Lissodelphis borealis* (Peale, 1848) and *Lissodelphis peronii* (Lacépède, 1804). *In* "Handbook of Marine Mammals" (S. H. Ridgway, and R. Harrison, eds), Vol. 5, pp. 335–362. Academic Press, London.

Jones, P. D., *et al.* (9 authors) (1999). Polychlorinated dibenzo-p-dixions, dibenzofurans and polychlorinated biphenyls in New Zealand cetaceans. *J. Cetacean Res. Manag.*(Special Issue 1), 157–167.

Klumov, S. K. (1959). Commercial dolphins of the far east. *Pacific Sci. Res. Inst. Fish. Econ. Oceanogr. Rep.* **47**, 154–160 [Translated from Russian].

Leatherwood, S., and Reeves, R. R. (1983). "The Sierra Club Handbook of Whales and Dolphins." Sierra Club Books, San Francisco.

Leatherwood, S., and Walker, W. A. (1979). The northern right whale dolphin *Lissodelphis borealis* Peale in the eastern North Pacific. *In* "Behavior of Marine Animals" (H. E. Winn, and B. L. Olla, eds), Vol. 3, pp. 85–141. Plenum Press, New York.

LeDuc, R. G., Perrin, W. F., and Dizon, A. E. (1999). Phylogenetic relationships among the delphinid cetaceans based on full cytochrome b sequences. *Mar. Mamm. Sci.* **15**(3), 619–648.

Mangel, M. (1993). Effects of high-seas driftnet fisheries on the northern right whale dolphin *Lissodelphis borealis*. *Ecol. Appl.* **3**, 221–229.

Martuscelli, P., Olmos, F., Silva e Silva, R., Mazzarella, I. P., Pino, F. V., and Raduan, E. N. (1996). Cetaceans of São Paulo, southeastern Brazil. *Mammalia* **60**(1), 125–139.

Minh, T. B., *et al.* (8 authors) (2000). Isomer-specific accumulation and toxic assessment of polychlorinated biphenyls, including coplanar congeners, in cetaceans from the North Pacific and Asian coastal waters. *Arch. Environ. Contam. Toxicol.* **39**, 398–410.

Newcomer, M. W., Jefferson, T. A., and Brownell, R. L., Jr. (1996). *Lissodelphis peronii*. *Mamm. Spec.* **531**, 1–5.

Rice, D. W. (1998). "Marine Mammals of the World: Systematics and Distribution." *Soc. of Mar. Mamm., Spec. Pub.* **4**. Allen Press. 231 p.

Rose, B., and Payne, A. I. L. (1991). Occurrence and behavior of the southern right whale dolphin *Lissodelphis peronii* off Namibia. *Mar. Mamm. Sci.* **7**(1), 25–34.

Van Waerebeek, K., Canto, J., Gonzalez, J., Oporto, J., and Brito, L. (1991). Southern right whale dolphins, *Lissodelphis peronii*, off the Pacific coast of South America. *Zeitsch. Säugetierk.* **56**, 284–295.

Visser, I. N., Fertl, D., and Pusser, L. T. (2004). Melantistic southern right-whale dolphins (*Lissodelphis peronii*) off Kaikoura, New Zealand, with records of other anomalously all-black cetaceans. *N. Z. J Mar. Fresh. Res.* **38**, 833–836.

Walker, W. A. (1975). Review of the live-capture fishery for smaller cetaceans taken in southern California waters for public display, 1966–73. *J. Fish. Res. Board Can.* **32**, 1197–1211.

Watson, L. (1981). "Sea Guide to Whales of the World." London, Hutchinson & Co. Ltd, 302 p.

Yazdi, P. (2002). A possible hybrid between the dusky dolphin (*Lagenorhynchus obscurus*) and the southern right whale dolphin (*Lissodelphis peronii*). *Aquat. Mamm.* **28**(2), 211–217.

# Right Whales
## *Eubalaena glacialis, E. japonica, and E. australis*

### ROBERT D. KENNEY

## I. Characteristics and Taxonomy

### A. Scientific and Common Names

Three species of right whales are currently recognized: North Atlantic right whale, *Eubalaena glacialis* (Müller, 1776); North Pacific right whale, *E. japonica* (Lacépède, 1818); and southern right whale, *E. australis* (Desmoulins, 1822). The generic name *Eubalaena* means "true whale;" the meanings of specific epithets are: *glacialis* = "of the ice," *japonica* = "Japanese," and *australis* = "southern." Müller based his original description of *E. glacialis* on the "nördcaper" of Norwegian whalers. He did not specify a type locality; Eschricht and Reinhardt (1866) subsequently designated it as North Cape, Norway. Synonyms include *biscayensis* (Eschricht, 1860) and *nordcaper* (Lacépède, 1804). The type specimen of *E. australis* is a skeleton from Algoa Bay, Cape of Good Hope, South Africa in the Museum National d'Histoire Naturelle, Paris. Synonyms include *antarctica* (Lesson, 1828), *antipodarum* (Gray, 1843), and *temminckii* (Gray, 1864).

R

Lacépède's description of *E. japonica* was based upon an illustration by a Japanese artist; the type locality is Japan. A synonym is *sieboldii* (Gray, 1864) (Hershkovitz, 1966; Mead and Brownell, 2005).

Conventional wisdom holds that the common name "right whale" comes from English whalers, who designated this as the "right" (i.e., correct) whale to hunt because it occurred near shore, swam slowly enough to be caught from a small boat propelled by sails or oars, floated when dead, and yielded large amounts of valuable oil and BALEEN. Early writers, however (e.g. Eschricht and Reinhardt, 1866) considered "right" to mean "true" or "proper," meaning showing the characteristics typical of whales generally, as later formally recognized in the Latin generic name *Eubalaena*. Other common names in English include black right whale and black whale.

## B. Systematics and Nomenclature

There has been disagreement on two questions of balaenid systematics and nomenclature—concerning the number of extant species of right whales and whether or not bowhead whales (*Balaena mysticetus*) and right whales are congeneric (see Reeves and Kenney, 2003 and Perrin and Reeves, 2004 for reviews). Northern and Southern Hemisphere right whales were long treated as distinct species (Cummings, 1985). Rice (1998) concluded that there was no evidence for consistent morphological differences and that only one globally distributed species should be recognized. Rosenbaum *et al.* (2000) showed clear, long-established mitochondrial DNA differences among right whale lineages in the North Atlantic, North Pacific, and Southern Hemisphere—concluding that three species should be recognized and that North Pacific and southern right whales are more closely related to one another than either is to North Atlantic right whales. Those results have been confirmed by both nuclear DNA markers (Gaines *et al.*, 2005) and the genetics of whale lice on right whales (Kaliszewska *et al.*, 2005). The conclusions have been accepted by the INTERNATIONAL WHALING COMMISSION's scientific committee (IWC, 2001a) and broadly by marine mammalogists (Mead and Brownell, 2005), but not universally (Baker *et al.*, 2003). The question is more a philosophical one, between phylogenetic and biological species concepts, but under either concept the two-species northern–southern classification is taxonomically invalid.

On the question of the correct generic name for right whales, Eschricht and Reinhardt (1866) published the first widely read, detailed comparison between bowheads and right whales, and maintained both species in the same genus. Rice (1998) reviewed the published comparisons and concluded that there was no scientific evidence from either morphology or genetics for recognizing the separate right whale genus *Eubalaena* (published by J. E. Gray; see Box 1), and that those differences that do exist are less than those between species of *Balaenoptera*. The consensus mitochondrial, nuclear, and combined phylogenetic trees (Gaines *et al.*, 2005) appear to show little genetic divergence between bowheads and the three right whale species. Others (Schevill, 1986; Bannister *et al.*, 1999; Rosenbaum *et al.*, 2000) have argued that maintaining the generic-level separation was justified, in part, for reasons of nomenclatural stability. IWC (2001a), and Mead and Brownell (2005) have accepted that rationale and recognized the right whales as *Eubalaena* spp. That conclusion (although not the underlying justification) has been confirmed by a very recent study of balaenid systematics based on a detailed morphological analysis of all known species, both extant and fossil (Churchill, 2007). The results showed a crown Balaenidae containing two sister clades—one a monophyletic *Eubalaena* and the other containing *Balaena* and *Balaenella*.

## BOX 1: Taxonomic Rules, J. E. Gray, and Right Whale Names

There is a set of very specific rules, the International Code of Zoological Nomenclature, for determining the correct scientific name to apply to any particular animal taxon. One important aspect of the Code is the rule of priority—when determining the valid name for any species (or genus or higher level category), the first published name applied should be used. For example, the original descriptions of all three right whale species included them in the genus *Balaena* (="whale"), a name published by Linnaeus in 1758. If biological evidence supports classifying right whales as a distinct genus from bowheads, the name *Eubalaena*, published by J. E. Gray in 1864, is the oldest name available applicable only to the right whales and therefore the valid name by the rule of priority.

John Edward Gray (1800–1875) was an English zoologist and an important figure in the history of cetacean taxonomy. He began his career at the British Museum at about age 15 as a volunteer insect collector, received a temporary appointment in 1824 to catalog reptiles, and was Keeper of Zoology from 1840 until his retirement only 2 months before his death 35 years later. In his years at the museum, he published well over 1000 papers, including half of the 200 catalogs issued by the museum during his time. *Eubalaena* is not the only cetacean name he coined; 15 or 16 of the currently accepted names of the 85–90 extant species and 40+ genera, 11 of the 14 family names (all except Eschrichtiidae, Kogiidae, and Lipotidae), and many of the subfamily names were authored by Gray. One might conclude that he was particularly knowledgeable about cetacean taxonomy. However, he worked at what was arguably the world's most influential museum of his day and apparently had a penchant for creating new species, genera, and higher taxa. He has been called one of the most notorious taxonomic "splitters" of all time. He also authored at least four other generic names for right whales, and in a single publication in 1871 he counted six genera and 14 species of right whales.

The rule of priority can be set aside under certain circumstances in order to maintain nomenclatural stability. If a taxonomic name is in common usage for a long period and then is discovered to be a junior synonym of an older published name, the International Committee on Zoological Nomenclature can be petitioned to suppress the senior synonym and conserve the junior synonym as the accepted name. For example, the binomial *Tursiops truncatus* (Montagu, 1821) for the common bottlenose dolphin has been conserved by suppression of the senior synonym *T. nesarnack* (Lacépède, 1804) (Rice, 1984, 1998). However, whether right whales and bowheads represent one or two genera is not simply a question of names, but of biological classification. The International Code of Zoological Nomenclature specifically "refrains from infringing upon taxonomic judgment, which must not be made subject to regulation or restraint" (Ride *et al.*, 1985). Invoking the rule of stability in the case of the right whales' generic name is a misapplication of the Code, and rigorous application of the scientific method would suggest that the null hypothesis (i.e., there is no significant generic-level difference between bowhead and right whales) should stand until rejected by evidence. The study by Churchill (2007) now has provided the evidence to

R

conclude that the three living right whale species do comprise a phylogenetic lineage distinct from the bowhead and are rightly classified into a separate genus. One must remember, however, that systematic classifications are scientific hypotheses, subject to revision after further study.

## C. Description

Right whales have an extremely robust body form, bordering on rotund, with a thick blubber layer and the girth at times exceeding 60% of total body length (Omura *et al.*, 1969; Cummings, 1985; Reeves and Kenney, 2003; Fig. 1). The body is mostly black, sometimes with irregular white ventral patches. Some individuals may have a mottled appearance, and calves may sometimes be lighter colored. There is no dorsal fin. The pectoral flippers, which retain all five digits, are large, broad, and blunt. The flukes are very broad (up of 40% of body length), black on both dorsal and ventral surfaces, deeply notched, and smoothly tapered to the tips. Calves are 4.5–6 m long at birth; typical adults are 13–16 m. North Pacific right whales attain larger maximum sizes than the other species, up to 18 m and over 100 metric tons.

The head is relatively large, comprising about one-quarter to one-third of the total body length. The upper jaw is somewhat arched, and the margin of the lower lip forms a very pronounced curve. There are 200–270 baleen plates on each side of the upper jaw, and there is a gap between the two rows of plates at the anterior end of the upper jaw (Nemoto, 1970; Pivorunas, 1979). The baleen plates are relatively narrow and 2–2.8 m long, with very fine fringing hairs. The tongue is massive and solidly muscular.

The most conspicuous external characteristics of right whales are the callosities on the head (Payne and Dorsey, 1983). These are irregular patches of thickened, keratinized tissue, which are inhabited by dense populations of specialized amphipod crustaceans, known as cyamids or "whale lice" (see Box 2). At least three species of whale lice occur on right whales: *Cyamis gracilis*, *C. ovalis*, and *C. erraticus* (Rowntree, 1996; Kaliszewska *et al.*, 2005). In southern right whales the callosities are also inhabited by barnacles, *Tubicinella* sp. The callosities occur at the tip of the snout (called the "bonnet" by whalers), on the lower lips and chin, above the eyes, and in front of and behind the blowholes (Figs 1 and 2). The callosities are congenital and not caused externally, as their beginnings are present in fetuses and neonates, but the pattern is not fully developed and colonized by cyamids until the whale is at least several months old (Fig. 2). The callosity patterns are unique to individuals and are therefore extremely useful as a natural "tag," which allows repeated identification of individuals by photographs (Payne *et al.*, 1983; Kraus *et al.*, 1986).

### BOX 2: Are Whale Lice Parasites?

Two species are said to be symbiotic when they live intimately associated with each other, with the exact nature of the relationship depending on the fitness costs and benefits incurred by each species (Townsend *et al.*, 2003). In the symbiosis involving right whales and cyamids, it can be assumed that the cyamid benefits from the relationship by having a place to live and a ready food supply. If the right whale also benefits, the association is defined as mutualism; if the whale receives neither benefits nor costs, the association is commensalism; and if the whale is harmed, the association is parasitism. Most sources refer to cyamids as parasites or ectoparasites. It is known that they do feed on the whale's skin (Schell *et al.*, 2000), however, there is scant evidence that their presence or feeding causes any harm to the whale. Leung (1976) concluded that "whale-lice cause certain damage to the whale skin when the young begin to maintain their livelihood, and the injury is a result of piercing the tissue in which they shelter for safety and for food," but did not actually show any harm to the whale. Similarly, while it is possible to construct various hypotheses for potential benefits provided to the whale by the cyamids, there is no evidence to test or confirm these theories. It therefore seems that the conservative course would be to consider cyamids as ectocommensals of right whales until convincing evidence demonstrating otherwise has been shown.

The closest relative of the right whale is the bowhead whale. Bowheads may be somewhat longer, are substantially stouter, have relatively larger heads (about 40% of body length) with a more arched appearance, have much longer baleen plates (up to 5.2 m long), and completely lack callosities.

## D. Fossil Record and Evolution

Five fossil species of *Balaena* have been described from deposits of late Miocene, Pliocene, or Pleistocene age (2–10 million years old) from Europe and North America: *B. affinis*, *B. etrusca*, *B. montalionis*,

Figure 1    *The North Atlantic right whale (c. Brett Jarrett).*

**Figure 2** *A mother–calf pair of North Atlantic right whales sighted off the coast of Florida on 9 January 1992. The mother, #1001 ("Fermata"), was the first animal included in the North Atlantic photoidentification catalog, from photographs taken in Cape Cod Bay by W. A. Watkins and W. E. Schevill in March 1978. This was her fifth known calf. A gray area on the calf's rostrum shows where the callosity will develop, and colonizing whale lice can be seen along the curved margin of its lips. "Fermata" was last seen in September of that same year in the Bay of Fundy, and is now presumed dead. Photograph by the author from Airship Shamu (courtesy of Sea World).*

*B. primigenius*, and *B. prisca* (Barnes and McLeod, 1984; McLeod *et al.*, 1993). The last of these is similar enough to modern bowheads (*B. mysticetus*) that it may, in fact, not be a separate species. There is then a long gap in the balaenid FOSSIL RECORD to *Morenocetus parvus*, the oldest known member of the family, found in early Miocene (23 million years old) deposits in South America. Molecular phylogenies generally agree that Balaenidae is the most primitive clade of extant mysticetes (Árnason *et al.*, 1992; Hatch *et al.*, 2006).

## II. Distribution and Abundance

Right whales are found in the middle latitudes of both the Northern and the Southern Hemispheres, between approximately 20° and 60°S and N latitudes (Cummings, 1985; Reeves and Kenney, 2003; Fig. 3). There are three geographically isolated populations currently recognized as separate species: in the North Atlantic, North Pacific, and Southern Ocean. The populations are kept separated by Arctic ice and warm equatorial waters so that there is no interchange between populations, and apparently has not been for millions of years (Rosenbaum *et al.*, 2000; Gaines *et al.*, 2005; Kaliszewska *et al.*, 2005).

### A. North Atlantic Right Whales

The historical range of North Atlantic right whales apparently extended as far south as Florida and northwestern Africa and as far north as Labrador, southern Greenland, Iceland, and Norway (Cummings, 1985; Reeves *et al.*, 2007). The traditional hypothesis has been that there were separate stocks with little interchange on the western and the eastern sides of the basin; however, analysis of some nineteenth century whaling specimens in European museums shows that they do not differ genetically from living western individuals (Rosenbaum *et al.*, 2000), and there have been one or two right whales seen in the eastern North Atlantic in recent years that were known individuals from the western stock (Jacobsen *et al.*, 2004). It is possible that the structure of a right whale population is that a particular ocean basin is inhabited by a single breeding population without long-term genetic isolation of stocks, but where return to traditional habitats learned from the mother (matrilineal habitat fidelity) maintains shorter-term separation between two or more subsets of the population.

The present range of western North Atlantic right whales, from Florida to Nova Scotia with very occasional occurrence beyond those limits, is much reduced from its historical extent (Winn *et al.*, 1986; Kenney *et al.*, 2001; Kraus and Rolland, 2007). The best estimate of present abundance is about 400 animals (NARWC, 2006). In the eastern North Atlantic, there have been only a handful of right whale sightings in the last few decades (reviewed in Reeves and Kenney, 2003). It is not known whether these represent a small remnant eastern stock or whether some or all of them are individuals from the known western population.

### B. North Pacific Right Whales

The historical range in the North Pacific was similarly much more extensive than today. Right whales occurred from Japan and northern Mexico north to the Sea of Okhotsk, Bering Sea, and Gulf of Alaska (Scarff, 1986, 1991; Clapham *et al.*, 2004). Recent sightings are extremely rare, primarily in the Sea of Okhotsk (Brownell *et al.*, 2001) and eastern Bering Sea (Clapham *et al.*, 2004; Shelden *et al.*, 2005; Wade *et al.*, 2006). There are no reliable estimates of abundance, and there may be even fewer whales than in the North Atlantic. Wade *et al.* (2006) reported that 23 individuals had been identified so far in the southeastern Bering Sea. There are also insufficient genetic or resighting data to address whether there is support for the traditional separation into eastern and western stocks (Brownell *et al.*, 2001).

### C. Southern Right Whales

Right whales are known from several areas of the Southern Ocean. Multiple stocks have been hypothesized, including Argentina/Brazil, South Africa, east Africa/Mozambique, western Australia, southeastern Australia, New Zealand, and Chile (Cummings, 1985; Best *et al.*, 2001; IWC, 2001b; Patenaude *et al.*, 2007). Additional stocks have been hypothesized for the central Indian Ocean, the Campbell and Auckland Islands in the southwestern Pacific, and Tristan da Cunha in the central South Atlantic. There have also been suggestions of even finer stock structuring, e.g., between Argentina and Brazil in the western South Atlantic or between Namibia and South Africa in the eastern South Atlantic. There is incomplete genetic isolation separation between stocks, especially between geographically adjacent stocks.

Right whale populations in Argentina, South Africa, and Australia are presently the largest and the best studied. The total abundance of southern right whales was estimated as of 1997 at 7571 animals, with the three well-studied stocks increasing at 7–8% annually (IWC, 2001b). Given that there were no estimates available for some stocks, and a population increasing at 7% doubles every 10 years, total abundance could currently exceed 15,000 animals.

**R**

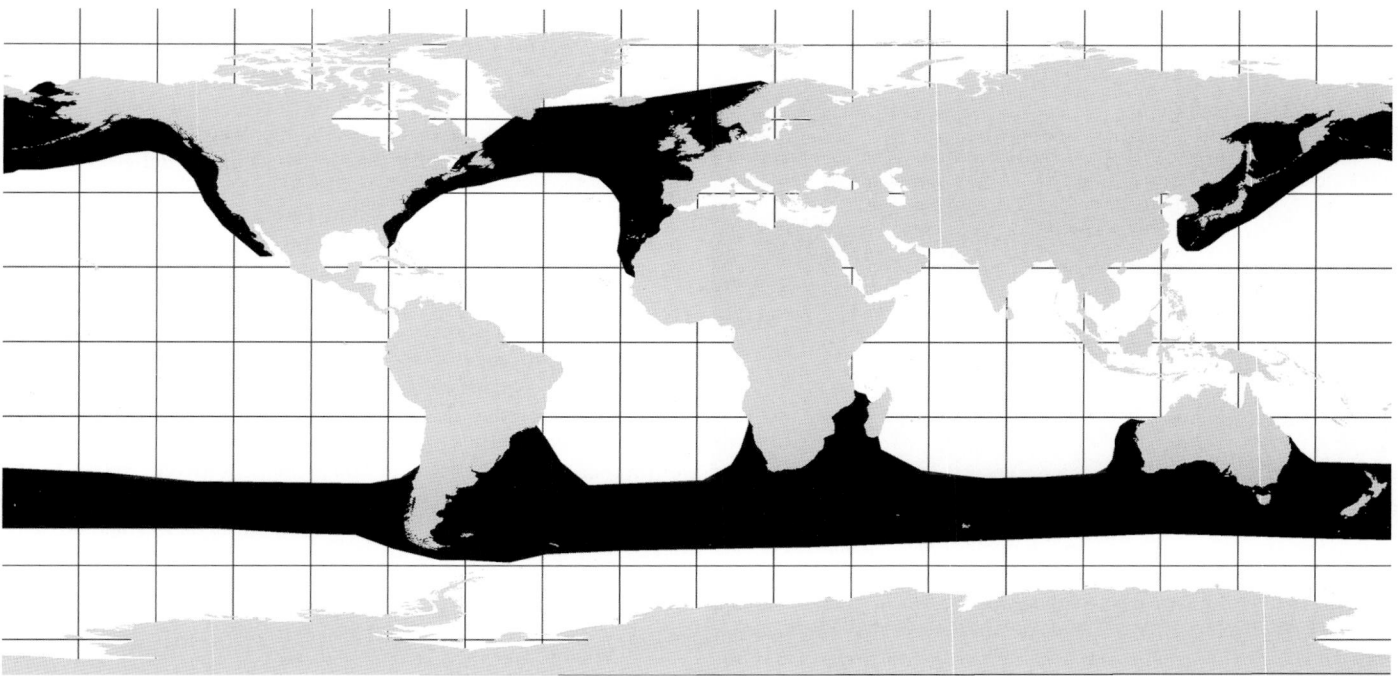

**Figure 3** *Worldwide range of right whales* (Eubalaena glacialis, E. japonica, and E. australis). *Much of what is shown here is relatively speculative based on sparse historical records of whaling catches and the available recent sightings. Most recent data come from areas relatively nearshore, with few or no data for the pelagic areas between the known coastal habitats.*

## III. Ecology

### A. Diet

Right whales feed entirely on zooplankton, especially on large calanoid copepods (crustaceans approximately the size of a grain of rice) (Nemoto, 1970; Cummings, 1985; Baumgartner *et al.*, 2007). At times they also feed on smaller copepods, krill (larger shrimp-like crustaceans), pteropods (tiny planktonic snails), or the planktonic larval stages of BARNACLES and other crustaceans. The details of their diet likely differ between regions, e.g., it is likely that krill comprise a higher proportion of the DIET in southern right whales. It is also likely that right whales can be somewhat opportunistic regarding prey species, FEEDING on any prey of a size that can be filtered efficiently by the baleen, which does not swim strongly enough to escape, and which is concentrated into sufficiently dense patches to trigger feeding behavior. For example, there have been observations of North Atlantic right whales in the Bay of Fundy feeding on aggregations of salps.

### B. Habitat

Right whales migrate annually between high-latitude feeding grounds and low-latitude calving and breeding grounds (Cummings, 1985; Reeves and Kenney, 2003). There are substantial differences in the locations where most research has been conducted between the Northern and the Southern Hemispheres; therefore, there is often a lack of directly comparable information for different populations.

Feeding takes place in spring, summer, and fall in higher-latitude feeding grounds, where ocean temperatures are cooler and overall biological productivity is much higher. The best known right whale feeding grounds are in the western North Atlantic (Winn *et al.*, 1986). These habitats are in nearshore and shelf waters, where some combination of bottom topography, water column structure and stratification, and currents acts to physically aggregate zooplankton into extremely

dense concentrations (Kenney and Wishner, 1995; Baumgartner *et al.*, 2007). The densest zooplankton concentrations measured in the North Atlantic were found by sampling near right whales. There are probably also offshore feeding grounds, in locations not yet known, based on historical whaling records and on the fact that some known whales are often missing from the known habitats for months or years at a time. There must also have been other feeding grounds in the past, when the range of North Atlantic right whales was more extensive.

Feeding grounds for the other species of right whales are much more poorly known. In the North Pacific, based on historical whaling records and the few recent sightings, the principal feeding grounds were most likely in the Sea of Okhotsk, central and eastern Bering Sea, and Gulf of Alaska (Clapham *et al.*, 2004). All of these feeding areas are much more pelagic or offshore than the well-studied North Atlantic habitats. In the Southern Ocean, right whale feeding grounds also appear to be mostly in offshore, pelagic regions. Southern right whale feeding grounds are likely to be found associated with areas of extremely high productivity; limited sighting data available show most whales in the regions between the subtropical and Antarctic convergences (IWC, 2001b).

Calving in right whales occurs during winter. Where the calving grounds are known, they are in shallow coastal regions or bays. The only known current calving ground in the western North Atlantic is in coastal waters near Georgia and northeastern Florida (Winn *et al.*, 1986; Kraus and Rolland, 2007). In that calving ground, right whales show a clear preference for waters in a relatively narrow depth and temperature range—13–19 m and 13–16°C (Keller *et al.*, 2006). It has been speculated that other coastal areas, including Delaware Bay and Cape Cod Bay, may have been calving grounds before the population was depleted by whaling (Reeves *et al.*, 1978). It has been noted, for example, that Cape Cod is similar topographically to Peninsula Valdés in Argentina and is located at about the same latitude (Payne,

1995). In the eastern North Atlantic, Cintra Bay in northwestern Africa is believed to have been a historical right whale calving ground (Reeves, 2001). It is possible that areas near the Azores and Madeira, as well as the Bay of Biscay, were also calving grounds. In the North Pacific, no right whale calving grounds have ever been discovered. In the Southern Hemisphere, shallow coastal waters and bays in many areas are currently known to be southern right whale calving areas or hypothesized to have been calving grounds historically, including Argentina, Brazil, Falkland Islands, Tristan de Cunha, Namibia, South Africa, Mozambique, Kerguelen Island, Australia, New Zealand, Auckland Islands, and Chile (IWC, 2001b).

Breeding or mating also occurs during the winter. Because of the 3-year female reproductive cycle, breeding can take place geographically distant from calving (Knowlton *et al.*, 1994; Kraus *et al.*, 2001, 2007). In the western North Atlantic, the location of the majority of the population during the winter is not known, and adult males are nearly absent from the calving ground (Brown *et al.*, 2001). Breeding must occur wherever the adult population spends the winter, but it is not known whether there are specific, distinct winter habitats or whether the whales are broadly dispersed across wide regions of the North Atlantic.

In southern right whales, at least some mating behavior occurs in or near the calving grounds, although there may be small-scale segregation of breeding adults from females with calves (Payne, 1995). In Argentina, because females are observed infrequently in these breeding groups in the year prior to calving, it is possible either that the mating which actually leads to conception occurs in some other, unknown habitat or that receptive females only visit coastal waters for very brief periods.

Circumstantial evidence suggests that learning is an important component of habitat selection in right whales (Malik *et al.*, 1999; Frasier *et al.*, 2007). Calves apparently learn the locations of feeding grounds by accompanying their mothers during the first year of life and then return to those same habitats for the rest of their lives. This pattern of matrilineal habitat fidelity seems to be common in migratory whale species; resighting and genetic data demonstrate that it is responsible for population structuring in North Atlantic humpback whales (Clapham, 1996).

## C. Predators

Potential predators of right whales include killer whales (*Orcinus orca*) and large sharks, and it is more likely that any predators would attack calves or juveniles. There have been few direct observations of killer whale attacks on right whales, and Mehta *et al.* (2007) concluded that the spacing of tooth rakes on the flukes of North Atlantic right whales was inconsistent with killer whales and more likely from smaller animals such as false killer whales. Reeves *et al.* (2006) reviewed the arguments for and against the hypothesis that migration in large baleen whales to low-latitude calving grounds evolved in response to killer predation on calves in higher latitudes. Predation by sharks similarly may have been one of the selective pressures leading to the evolution of right whales' use of cooler waters in very shallow coastal habitats for calving, since sharks prefer warmer waters and at least white sharks are known to often attack their prey from below (Klimley, 1994).

## IV. Behavior and Physiology

Right whales are observed to frequently perform highly energetic behaviors at or above the surface of the water. These aerial behaviors include BREACHING (jumping partly or almost completely above the surface), LOBTAILING (violently slapping the surface with the flukes), and flippering (slapping the surface with a pectoral flipper). The functions of these behaviors are not known (Whitehead, 1985). They

**Figure 4** *A "skimming" North Atlantic right whale feeding on zooplankton near the surface. The rough black areas of the "bonnet" are the whale's skin, whereas the white areas are comprised of whale lice. Photograph by William Watkins.*

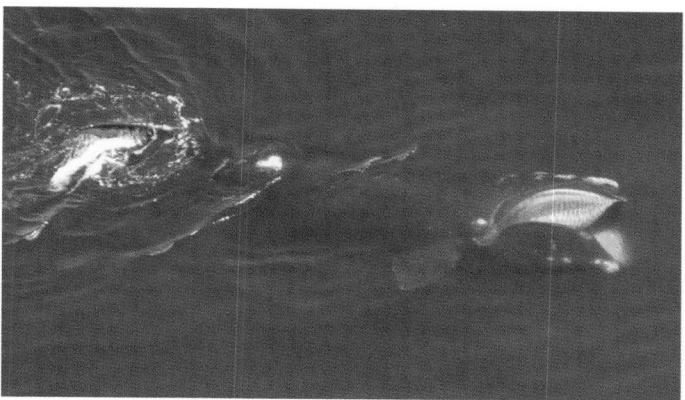

**Figure 5** *Aerial view of a feeding North Atlantic right whale in the Gulf of Maine. Photograph by Peter Duley (courtesy of Northeast Fisheries Science Center).*

all produce very loud sounds, which may sometimes have a communicative and/or aggressive function. Right whales in Argentina and South Africa have been observed to lift their flukes above the surface, where the flukes act like a sail and allow the wind to push the whale horizontally (Payne, 1995). This "tail-sailing" behavior has not been reported in other habitats.

Right whales are "skimmers" (Nemoto, 1970; Pivorunas, 1979; Baumgartner *et al.*, 2007; Figs 4 and 5). They feed by swimming forward with the mouth agape. Water flows into the opening at the front, and out through the baleen, straining their prey from the water. Feeding can occur at or just below the surface, where it can be observed easily, or at depth. At times, right whales apparently feed very close to the bottom, because they are observed to surface at the end of an extended dive with mud on the head. Typical feeding dives last for 10–20 min (Winn *et al.*, 1995; Baumgartner *et al.*, 2007).

Courtship in right whales often involves aggregations of whales termed "surface-active groups" (Donnelly, 1967; Payne and Dorsey, 1983; Kraus and Hatch, 2001; Kraus *et al.*, 2007). These are usually centered around a single female and may involve large numbers of males; groups of more than 20 animals have been observed. Often the female is belly-up at the surface, whereas the males stroke her with their flippers or attempt to push her under. There is evidence that the female

R

initiates the interaction by vocalizing. In the North Atlantic, surface-active groups are observed in all seasons, even though calving is highly synchronous and restricted to winter. Therefore much of the observed activity does not lead to fertilization and may serve a social function. The female may simply use the interactions to assess male quality for later mating. The interactions between males in the group generally involve very little of the violence and aggressiveness seen in humpback whales. One theory is that right whales engage in sperm competition (Brownell and Ralls, 1986), where the volume of semen is important in displacing the sperm of other males mating with the same female. Right whale testes may be the largest of any animal, at 2 m long and 500 kg each.

Right whale vocalizations are primarily low-frequency moans, groans, belches, and pulses (Cummings, 1985; Thomson and Richardson, 1995; Parks and Clark, 2007). Most acoustic energy produced is below 500 Hz, with some sounds up to 1500–2000 Hz. The functions of these sounds are not well understood. Hypothesized functions include maintenance of contact between separated individuals, threats or other aggressive signals, and social signals, including their possible involvement in surface-active group behavior.

## V. Life History

Information on the age at maturity in right whales is not available from whaling data as it is for other whale species taken by twentieth century industrial whaling. The information must be derived from photoidentification studies that track known individuals from birth. The youngest mature female in the western North Atlantic was 4 at maturity and 5 at first calving (Knowlton et al., 1994). From both North Atlantic and Southern Hemisphere data, the average age at first calving is closer to 9 or 10 years (Best et al., 2001; Cooke et al., 2001; Kraus et al., 2001, 2007). Age at maturity is not yet known for males, as there is no external method for identifying paternity. Genetic studies may be better able to identify fathers of calves and begin to provide data for age of maturity in males (Frasier et al., 2007).

Growth is relatively rapid from birth to weaning at about age 1, by which time the calf approximately doubles in body length to 9–11 m (Brown et al., 2001). Available data on growth after age 1 are not entirely consistent. For example, in the North Atlantic, growth also can be relatively rapid in year 2, by which time total length may reach 12–13 m, and thereafter is much slower. However, South African right whales apparently grow little between ages 1 and 4 (Best and Schell, 1996). Growth after age 1 is likely to be dependent on feeding success. The western North Atlantic female that matured at age 4 remained with her mother well into her second year, possibly growing much faster than the typical rate by nursing for a longer period. Right whales are believed to reach sexual maturity at body lengths of 13–16 m.

There are very few data on longevity. Aging baleen whales is an extremely difficult problem. Japanese attempts to use the wax plug found in the auditory canal from the North Pacific right whales taken for research in 1956–1968 were not successful (Omura et al., 1969). The oldest known right whale to date was in the North Atlantic (Kraus and Rolland, 2007). A mother–calf pair near Fort Lauderdale, Florida, was pursued and shot at by fishermen on March 24, 1935. The calf was killed, but the mother escaped. A photograph of her published in the New York Herald Tribune at the time was later matched to photographs taken in Cape Cod Bay (by pioneering right whale researcher William E. Schevill) in April 1959. She was also photographed in 1980, 1985, and 1992. On August 13, 1995 she was photographed offshore, with a large, gaping wound on the head apparently caused by a ship strike. It is unlikely she could have survived that injury. Assuming she was at least 10 years old in 1935, she would have been at least 70 years old in 1995, and may have been

substantially older. Research on bowhead whales suggests that they live substantially longer (George et al., 1999).

The typical reproductive cycle in mature females is 3 years between births (Kraus et al., 2007). The gestation period is approximately 1 year (Best, 1994), and weaning occurs at about 1 year of age (Hamilton et al., 1995). The female then takes a third year to replenish her energy stores, although it is possible for a female who has been especially successful at feeding to skip the resting year and calve after only a 2-year interval (one case observed in the North Atlantic, and at least one in Argentina). An alternative explanation for an observed 2-year interval is calf mortality soon after birth and subsequent avoidance by the mother of the high energetic demands of lactation; documented twice in Australia. This presumes that the mother–calf pair is sighted during the brief interval between birth and death of the calf. Otherwise what would be observed is an apparent 5-year interval, of which 25 were recorded in the North Atlantic between 1980 and 1998 (Kraus et al., 2001). Calving has been observed very rarely; in other instances, known females have been sighted in the calving ground both before and after the calf was born.

## VI. Interactions with Humans

### A. Whaling

North Atlantic right whales were the first whales to be harvested commercially by the Basques along the Atlantic coast of western Europe as early as the eleventh century (Aguilar, 1986; Reeves and Smith, 2006; Reeves et al., 2007). The whales were killed primarily for oil, which was sold across Europe, as the technology of the time did not permit preservation and widespread transportation of meat. As populations nearest shore were depleted, Basque whaling expanded to more distant waters, reaching eastern Canada by 1530. Basque whaling in Canada was centered in the Strait of Belle Isle between Labrador and Newfoundland and took 300–500 whales per year at its peak. Catches were declining by 1610–1620 and ended in 1713, by which time they had taken as many as 40,000 whales. Recent evidence from genetic sampling of whale bones from that era is that nearly all of the whales landed by the Basques in the Strait of Belle Isle were bowheads (Frasier et al., 2007). It is not clear whether or not Basque whaling beyond the Bay of Biscay impacted North Atlantic right whales, and if it did, where that occurred (Reeves et al., 2007).

Local shore-based right WHALING in North America began soon after the establishment of permanent colonies during the early seventeenth century (reviewed in Reeves et al., 2007). Peak catches were in the early eighteenth century (e.g., 86 in Nantucket, Massachusetts in 1726; 111 in Long Island, New York in 1707), and right whales in western North Atlantic waters may have been effectively extinct as a basis for a commercial fishery by the middle of the eighteenth century. The familiar Yankee whaling industry soon developed as a high-seas fishery targeting sperm whales; however, the whalers continued to opportunistically take any right whales encountered. Yankee whaling (including ships from several European nations) spread to the South Atlantic by 1775, into the South Pacific in 1789, and into the North Pacific by 1820. The Japanese had also begun their own shore-based fishery, which took some coastal migrant right whales, in the late sixteenth century.

The traditional high-seas Yankee whale fishery finally ended in the early twentieth century, when it was replaced by modern industrial whaling. Total right whale catches (although records are not complete) were at least 38,000 in the South Atlantic, 39,000 in the South Pacific, 1300 in the Indian Ocean, 15,000 in the North Pacific, and at least a few hundred in the North Atlantic. Some shore-based whaling in the eastern United States persisted into the 1920s, but it was minor, with only 8 taken in Long Island after 1900. In the North Atlantic, the last episode

R

of intensive right whaling was in the late nineteenth and the early twentieth century off Norway, Iceland, and Scotland, and the last right whales taken were at Madeira, 1 in 1952 and 2 in 1967. All right whale populations worldwide were protected from commercial whaling in the 1930s by the first International Convention for the Regulation of Whaling. However, the Japanese took 23 North Pacific right whales in the 1940s and 13 more under special scientific research permits between 1956 and 1968, some illegal takes of right whales along the coast of Brazil were reported in the 1950s, there was a significant amount of illegal Soviet taking of right whales in the North Pacific and Southern Ocean into the 1960s (Brownell *et al.*, 2001; Clapham *et al.*, 2004), and it is possible that there has been illegal right whaling elsewhere in the world.

### B. Ship Strikes

The most significant human-related source of mortality at present in western North Atlantic right whales is collision with large ships (Knowlton and Kraus, 2001; Kraus *et al.*, 2005; Knowlton and Brown, 2007; Waring *et al.*, 2007). Between 1970 and 2005, 24 right whales were known to have been killed by ships, and 3 others were last seen with serious and probably fatal injuries. There are probably additional mortalities that are never discovered because the carcasses are lost at sea. Ship collisions may be less of a mortality factor in other oceans, where right whales spend less time in nearshore habitats or where the level of industrial development is lower, although at least three probable ship-strike mortalities have been recorded in recent years off the Brazilian coast (Greig *et al.*, 2001).

### C. Entanglements

The second most important human-related mortality factor in western North Atlantic right whales is incidental capture in commercial fishing gear (Johnson, 2005; Johnson *et al.*, 2007; Waring *et al.*, 2007). The gear involved is fixed gear (set in one location rather than towed behind a vessel), including sink gill nets, drift nets, and a variety of pot and trap fisheries. In 1970–2005, 6 right whales are known to have been killed by ENTANGLEMENTS and 13 others were seriously injured but disappeared and probably died. It is not always known whether entanglements occur in actively fishing gear or in gear that has been lost, damaged, or moved by storms or vessels (often termed "ghost" gear). There are few data on entanglement mortalities in other populations.

Entanglement seems to be very common in right whales. Many entanglements involve the tail, where the leading edges of the flukes begin, and leave characteristic scars afterward. Over 75% of whales in the western North Atlantic carry such scars, and some individuals have been entangled two or three times (Hamilton *et al.*, 2007). Entanglements are therefore often not lethal. They may be more dangerous in younger animals, who might grow into a relatively benign entanglement until it becomes life-threatening (Moore *et al.*, 2007).

### D. Climate Change

Right whales are feeding specialists, with a relatively narrow range of acceptable prey characteristics and requiring prey to be concentrated in exceptionally high densities. The development of right whale feeding grounds is closely linked to physical phenomena such as water structure, currents, and temperature (Kenney and Wishner, 1995; Baumgartner *et al.*, 2007). This may make right whales more sensitive than other species to impacts from global climate change, with detectable environmentally induced variability in reproduction in recent years (Greene and Pershing, 2004; Leaper *et al.*, 2006; Kenney, 2007).

Any possible impacts may be increased because of matrilineal fidelity to their feeding grounds, and possibly a relatively low ability to locate new feeding grounds when conditions change.

### E. Other Human Impacts

There are a number of other potential human impacts on right whales (reviewed by Katona and Kraus, 1999; Reeves and Kenney, 2003; Kraus and Rolland, 2007).

1. Habitat loss due to high levels of human activity is mentioned frequently as a possible impact. Right whales no longer occur in Delaware Bay, eastern United States; Table Bay, South Africa; Wellington Harbor, New Zealand; or Derwent River, Tasmania. However, a plausible alternative explanation is that they were extirpated by whaling and have never reoccupied the habitat due to matrilineal habitat fidelity.

2. Pollution is another potential impact that is mentioned frequently but where evidence is sparse. Oil spills may be a bigger threat to right whales than to other baleen whales because their very fine baleen might be fouled more easily. Blubber samples show a presence of toxic contaminants, but at lower levels than in cetaceans that feed at higher trophic levels (Woodley *et al.*, 1991; Weisbrod *et al.*, 2000). A recent concern is that some contaminants may act as hormone mimics, affecting reproduction, or as immune system suppressants. They may be exposed via the food chain to naturally occurring biotoxins produced by phytoplankton, such as "red-tide" dinoflagellates (Durbin *et al.*, 2002).

3. Man-made noise may have the potential for interfering with acoustic communication, particularly since the major noise source, shipping, is also concentrated in the lower frequencies (Parks and Clark, 2007).

4. Effects of intensive commercial fisheries may alter ecosystem structure, increasing the abundance of other species that feed on zooplankton, particularly small fishes with lower economic value than the larger species harvested by FISHERIES.

5. The long-term effects of extreme population depletion by whaling might include reduced genetic diversity and associated health and reproductive problems (Frasier *et al.*, 2007).

### See Also the Following Articles

Bowhead Whale ■ Callosities ■ Filter Feeding ■ Species ■ Whale Lice ■ Baleen ■ Mysticetes

### References

Aguilar, A. (1986). A review of old Basque whaling and its effect on the right whales (*Eubalaena glacialis*) of the North Atlantic. *Rep. Int. Whal. Comm., Spec. Iss.* **10**, 191–199.

Árnason, Ú., Grétarsdóttir, S., and Widegren, B. (1992). Mysticete (baleen whale) relationships based upon the sequence of the common cetacean DNA satellite. *Mol. Biol. Evol.* **9**, 1018–1028.

Baker, R. J., *et al.* (2003). Revised checklist of North American mammals north of Mexico, 2003. *Mus. Texas Tech Univ. Occas. Pap.* **229**, 1–23.

Bannister, J. L., Pastene, L. A., and Burnell, S. R. (1999). First record of movement of a southern right whale (*Eubalaena australis*) between warm water breeding grounds and the Antarctic Ocean, south of 60°S. *Mar. Mamm. Sci.* **15**, 1337–1342.

Barnes, L. G., and McLeod, S. A. (1984). The fossil record and phyletic relationships of gray whales. *In* "The Gray Whale, *Eschrichtius robustus*" (M. L. Jones, S. L. Swartz, and S. Leatherwood, eds), pp. 3–32. Academic Press, Orlando, FL.

Baumgartner, M. F., Mayo, C. A., and Kenney, R. D. (2007). Enormous carnivores, microscopic food, and a restaurant that's hard to find. *In*

**R**

"The Urban Whale: North Atlantic Right Whales at the Crossroads" (S. D. Kraus and R. M. Rolland, eds), pp. 138–171. Harvard University Press, Cambridge, MA.

Best, P. B. (1994). Seasonality of reproduction and the length of gestation in southern right whales *Eubalaena australis*. *J. Zool.* **232**, 175–189.

Best, P. B., and Schell, D. M. (1996). Stable isotopes in southern right whales (*Eubalaena australis*) baleen as indicators of seasonal movements, feeding and growth. *Mar. Biol.* **124**, 483–494.

Best, P. B., Brandão, A., and Butterworth, D. S. (2001). Demographic parameters of southern right whales off South Africa. *J. Cetacean Res. Manag., Spec. Iss.* **2**, 161–169.

Brown, M. W., *et al.* (2001). Sighting heterogeneity of right whales in the western North Atlantic: 1980–1992. *J. Cetacean Res. Manag., Spec. Iss.* **2**, 245–250.

Brownell, R. L., Jr., and Ralls, K. (1986). Potential for sperm competition in baleen whales. *Rep. Int. Whal. Comm., Spec. Iss.* **8**, 97–112.

Brownell, R. L., Jr., Clapham, P. J., Miyashita, T., and Kasuya, T. (2001). Conservation status of North Pacific right whales. *J. Cetacean Res. Manag., Spec. Iss.* **2**, 269–286.

Churchill, M. M. (2007). The systematics and biogeography of the Balaenidae (Cetacea: Mysticeti). M.S. Thesis, San Diego State University, San Diego, CA.

Clapham, P. J. (1996). The social and reproductive biology of humpback whales: An ecological perspective. *Mamm. Rev.* **26**, 27–49.

Clapham, P. J., Good, C., Quinn, S. E., Reeves, R. R., Scarff, J. E., and Brownell, R. L., Jr. (2004). Distribution of North Pacific right whales (*Eubalaena japonica*) as shown by 19th and 20th century whaling catch and sighting records. *J. Cetacean Res. Manag.* **6**, 1–6.

Cooke, J., Rowntree, V. J., and Payne, R. (2001). Estimates of demographic parameters for southern right whales (*Eubalaena australis*) observed off Peninsula Valdés, Argentina. *J. Cetacean Res. Manag., Spec. Iss.* **2**, 125–132.

Cummings, W. C. (1985). Right whales *Eubalaena glacialis* (Müller, 1776) and *Eubalaena australis* (Desmoulins, 1822). *In* "Handbook of Marine Mammals" (S. H. Ridgway, and R. Harrison, eds), Vol. 3, pp. 275–304. Academic Press, San Diego, CA.

Donnelly, B. G. (1967). Observations on the mating behaviour of the southern right whale *Eubalaena australis*. *S. Afr. J. Sci.* **63**, 176–181.

Durbin, E., Teegarden, G., Campbell, R., Cembella, A., Baumgartner, M. F., and Mate, B. R. (2002). North Atlantic right whales, *Eubalaena glacialis*, exposed to paralytic shellfish poisoning (PSP) toxins via a zooplankton vector, *Calanus finmarchicus*. *Harmful Algae* **1**, 243–251.

Eschricht, D. F., and Reinhardt, J. (1866). On the Greenland right whale (*Balaena mysticetus* Linn.); with especial reference to its geographical distribution and migrations in times past and present, and to its external and internal characteristics. *In* "Recent Memoirs on the Cetacea by Professors Eschricht, Reinhardt, and Lilljeborg." (W. H. Flower, ed.), pp. 1–150 and plates 1–6. Robert Hardwicke, London. [translated from the 1861 Danish original publication by J. Reinhardt].

Frasier, T. R., McLeod, B. A., Gillett, R. M., Brown, M. W., and White, B. N. (2007). Right whales past and present as revealed by their genes. *In* "The Urban Whale: North Atlantic Right Whales at the Crossroads" (S. D. Kraus and R. M. Rolland, eds), pp. 200–231. Harvard University Press, Cambridge, MA.

Gaines, C. A., Hare, M. P., Beck, S. E., and Rosenbaum, H. C. (2005). Nuclear markers confirm taxonomic status and relationships among highly endangered and closely related right whale species. *Proc R. Soc. B* **272**, 533–542.

George, J. C., Bada, J., Zeh, J., Scott, L., Brown, S. E., O'Hara, T., and Suydam, R. (1999). Age and growth estimates of bowhead whales (*Balaena mysticetus*) via aspartic acid racemization. *Can. J. Zool.* **77**, 571–580.

Greene, C. H., and Pershing, A. J. (2004). Climate and the conservation biology of North Atlantic right whales: The right whale at the wrong time? *Front. Ecol. Environ.* **2**, 29–34.

Greig, A. B., Secchi, E. R., Zerbini, A. N., and Dalla Rosa, L. (2001). Stranding events of southern right whales, *Eubalaena australis*, in southern Brazil. *J. Cetacean Res. Manag., Spec. Iss.* **2**, 157–160.

Hamilton, P. K., Marx, M. K., and Kraus, S. D. (1995). Weaning in North Atlantic right whales. *Mar. Mamm. Sci.* **11**, 386–390.

Hamilton, P. K., Knowlton, A. R., and Marx, M. K. (2007). Right whales tell their own stories: The photoidentification catalog. *In* "The Urban Whale: North Atlantic Right Whales at the Crossroads" (S. D. Kraus and R. M. Rolland, eds), pp. 75–104. Harvard University Press, Cambridge, MA.

Hatch, L. T., Dopman, E. B., and Harrison, R. G. (2006). Phylogenetic relationships among the baleen whales based on maternally and paternally inherited characters. *Mol. Phylogenet. Evol.* **41**, 12–27.

Hershkovitz, P. (1966). "Catalogue of Living Whales." U. S. National Museum, Washington, DC, Bulletin 246.

IWC (International Whaling Commission) (2001a). Annex U. Report of the working group on nomenclature. *J. Cetacean Res. Manag.* **3**(Supplement), 363–367.

IWC (2001b). Report of the workshop on the comprehensive assessment of right whales: A worldwide comparison. *J. Cetacean Res. Manag., Spec. Iss.* **2**, 1–60.

Jacobsen, K.-O., Marx, M., and Øien, N. (2004). Two-way trans-Atlantic migration of a North Atlantic right whale (*Eubalaena glacialis*). *Mar. Mamm. Sci.* **20**, 161–166.

Johnson, A. J., Kraus, S. D., Kenney, J. F., and Mayo, C. A. (2007). The entangled lives of right whales and fishermen: Can they coexist? *In* "The Urban Whale: North Atlantic Right Whales at the Crossroads" (S. D. Kraus and R. M. Rolland, eds), pp. 380–408. Harvard University Press, Cambridge, MA.

Johnson, T. (2005). "Entanglements: The Intertwined Fates of Whales and Fishermen." University Press of Florida, Gainesville, FL.

Kaliszewska, Z. A., *et al.* (2005). Population histories of right whales (Cetacea: *Eubalaena*) inferred from mitochondrial sequence diversities and divergences of their whale lice (Amphipoda: *Cyamus*). *Mol. Ecol.* **14**, 3439–3456.

Katona, S. K., and Kraus, S. D. (1999). Efforts to conserve the North Atlantic right whale. *In* "Conservation and Management of Marine Mammals" (J. R. Twiss and R. R. Reeves, eds), pp. 311–331. Smithsonian Institution Press, Washington, DC.

Keller, C. A., Ward-Geiger, L. I., Brooks, W. B., Slay, C. K., Taylor, C. R., and Zoodsma, B. J. (2006). North Atlantic right whale distribution in relation to sea surface temperature in the southeastern United States calving grounds. *Mar. Mamm. Sci.* **22**, 426–445.

Kenney, R. D. (2007). Right whales and climate change: Facing the prospect of a greenhouse future. *In* "The Urban Whale: North Atlantic Right Whales at the Crossroads" (S. D. Kraus and R. M. Rolland, eds), pp. 436–459. Harvard University Press, Cambridge, MA.

Kenney, R. D., and Wishner, K. F. (eds) (1995). The South Channel Ocean Productivity Experiment: SCOPEX (special journal issue including ten papers). *Cont. Shelf Res.* **15**, 373–611.

Kenney, R. D., Mayo, C. A., and Winn, H. E. (2001). Migration and foraging strategies at varying spatial scales in western North Atlantic right whales: A review of hypotheses. *J. Cetacean Res. Manag., Spec. Iss.* **2**, 251–260.

Klimley, A. P. (1994). The predatory behavior of the white shark. *Am. Sci.* **82**, 122–133.

Knowlton, A. R., and Brown, M. W. (2007). Running the gauntlet: Right whales and vessel strikes. *In* "The Urban Whale: North Atlantic Right Whales at the Crossroads" (S. D. Kraus and R. M. Rolland, eds), pp. 409–435. Harvard University Press, Cambridge, MA.

Knowlton, A. R., Kraus, S. D., and Kenney, R. D. (1994). Reproduction in North Atlantic right whales (*Eubalaena glacialis*). *Can. J. Zool.* **72**, 1297–1305.

Kraus, S. D., and Hatch, J. J. (2001). Mating strategies in the North Atlantic right whale (*Eubalaena glacialis*). *J. Cetacean Res. Manag., Spec. Iss.* **2**, 237–244.

Kraus, S. D., and Rolland, R. M. (2007). Right whales in the urban ocean. *In* "The Urban Whale: North Atlantic Right Whales at the

**R**

Crossroads" (S. D. Kraus and R. M. Rolland, eds), pp. 1–38. Harvard University Press, Cambridge, MA.

Kraus, S. D., Moore, K. E., Price, C. A., Crone, M. J., Watkins, W. A., Winn, H. E., and Prescott, J. H. (1986). The use of photographs to identify individual North Atlantic right whales (*Eubalaena glacialis*). *Rep. Int. Whal. Commn., Spec. Iss.* **10**, 145–151.

Kraus, S. D., Hamilton, P. K., Kenney, R. D., Knowlton, A. R., and Slay, C. K. (2001). Reproductive parameters of the North Atlantic right whale. *J. Cetacean Res. Manag., Spec. Iss.* **2**, 231–236.

Kraus, S. D., *et al.* (2005). North Atlantic right whales in crisis. *Science* **309**, 561–562.

Kraus, S. D., Pace, R. M., III, and Frasier, T. R. (2007). High investment, low return: The strange case of reproduction in *Eubalaena glacialis*. *In* "The Urban Whale: North Atlantic Right Whales at the Crossroads" (S. D. Kraus and R. M. Rolland, eds), pp. 172–199. Harvard University Press, Cambridge, MA.

Leaper, R., Cooke, J., Trathan, P., Reid, K., Rowntree, V., and Payne, R. (2006). Global climate drives southern right whale (*Eubalaena australis*) population dynamics. *Biol. Lett.* **2**, 289–292.

Leung, Y. M. (1976). Life cycle of *Cyamus scammoni* (Amphipoda: Cyamidae), ectoparasite of gray whale, with a remark on the associated species. *Sci. Rep. Whales Res. Inst.* **28**, 153–160.

Malik, S., Brown, M. W., Kraus, S. D., Knowlton, A. R., Hamilton, P. K., and White, B. N. (1999). Assessment of mitochondrial DNA structuring and nursery use in the North Atlantic right whale (*Eubalaena glacialis*). *Can. J. Zool.* **77**, 1–6.

Mead, J. G., and Brownell, R. L., Jr. (2005). Order Cetacea. *In* "Mammal Species of the World, A Taxonomic and Geographic Reference, 3rd edition" (D. E. Wilson and D. M. Reeder, eds), Vol. 1, pp. 723–743. Johns Hopkins University Press, Baltimore, MD.

Mehta, A. V., *et al.* (2007). Baleen whales are not important as prey for killer whales *Orcinus orca* in high latitude regions. *Mar. Ecol. Prog. Ser.* **348**, 297–307.

McLeod, S. A., Whitmore, F. C., Jr., and Barnes, L. G. (1993). Evolutionary relationships and classification. *In* "The Bowhead Whale" (J. J. Burns, J. J. Montague, and C. J. Cowles, eds), pp. 45–70. Society for Marine Mammalogy, Lawrence, KS, Special Publication No. 2.

Moore, M. J., McLellan, W. A., Doust, P.-Y., Bonde, R. K., and Knowlton, A. R. (2007). Right whale mortality: A message from the dead to the living. *In* "The Urban Whale: North Atlantic Right Whales at the Crossroads" (S. D. Kraus and R. M. Rolland, eds), pp. 358–379. Harvard University Press, Cambridge, MA.

NARWC (North Atlantic Right Whale Consortium). (2006). "North Atlantic right whale report card: November 2005–October 2006." North Atlantic Right Whale Consortium, Boston, MA. http://www.rightwhaleweb.org/papers/pdf/NARWC_Report_Card2006.pdf

Nemoto, T. (1970). Feeding patterns of baleen whales in the ocean. *In* "Marine Food Chains" (J. H. Steele, ed.), pp. 241–252. University of California Press, Berkeley, CA.

Omura, H., Ohsumi, S., Nemoto, T., Nasu, K., and Kasuya, T. (1969). Black right whales in the North Pacific. *Sci. Rep. Whales Res. Inst.* **21**, 1–96.

Parks, S. E., and Clark, C. W. (2007). Acoustic communication: Social sounds and the potential impacts of noise. *In* "The Urban Whale: North Atlantic Right Whales at the Crossroads" (S. D. Kraus, and R. M. Rolland, eds), pp. 310–332. Harvard University Press, Cambridge, MA.

Patenaude, N. J., *et al.* (2007). Mitochondrial DNA diversity and population structure among southern right whales (*Eubalaena australis*). *J. Hered.* **98**, 147–157.

Payne, R. (1995). "Among Whales." Dell Publishing, New York.

Payne, R., and Dorsey, E. M. (1983). Sexual dimorphism and aggressive use of callosities in right whales (*Eubalaena australis*). *In* "Communication and Behavior of Whales" (R. Payne, ed.), pp. 295–329. Westview Press, Boulder, CO, AAAS Selected Symposium 76.

Payne, R., Brazier, O., Dorsey, E. M., Perkins, J. S., Rowntree, V. J., and Titus, A. (1983). External features in southern right whales (*Eubalaena australis*) and their use in identifying individuals. *In* "Communication

and Behavior of Whales" (R. Payne, ed.), pp. 371–445. Westview Press, Boulder, CO, AAAS Selected Symposium 76.

Perrin, W. F., and Reeves, R. R. (eds) (2004). Appendix 5. Report of the working group on species- and subspecies-level taxonomy. *In* "Report of the Workshop on Shortcomings of Cetacean Taxonomy in Relation to Needs of Conservation and Management, April 30–May 2, 2004, La Jolla, California" (R. R. Reeves, W. F. Perrin, B. L. Taylor, C. S. Baker, and S. L. Mesnick, eds), NOAA Tech. Memo. NMFS-SWFSC-363, pp. 26–61. National Marine Fisheries Service, La Jolla, CA.

Pivorunas, A. (1979). The feeding mechanisms of baleen whales. *Am. Sci.* **67**, 432–440.

Reeves, R. R. (2001). Overview of catch history, historic abundance and distribution of right whales in the western North Atlantic and in Cintra Bay, West Africa. *J. Cetacean Res. Manag., Spec. Iss.* **2**, 187–192.

Reeves, R. R., and Kenney, R. D. (2003). Baleen whales, *Eubalaena* spp. and allies. *In* "Wild Mammals of North America: Biology, Management, and Economics" (G. A. Feldhamer, B. C. Thompson, and J. A. Chapman, eds), 2nd Ed., pp. 425–453. Johns Hopkins University Press, Baltimore, MD.

Reeves, R. R., and Smith, T. D. (2006). A taxonomy of world whaling eras. *In* "Whales, Whaling, and Ocean Ecosystems" (J. A. Estes, D. P. DeMaster, D. F. Doak, T. M. Williams, and R. L. Brownell, Jr., eds), pp. 82–101. University of California Press, Berkeley, CA.

Reeves, R. R., Mead, J. G., and Katona, S. (1978). The right whale, *Eubalaena glacialis*, in the western North Atlantic. *Rep. Int. Whal. Commn.* **28**, 303–312.

Reeves, R. R., Berger, J., and Clapham, P. J. (2006). Killer whales as predators of large baleen whales and sperm whale. *In* "Whales, Whaling, and Ocean Ecosystems" (J. A. Estes, D. P. DeMaster, D. F. Doak, T. M. Williams, and R. L. Brownell, Jr., eds), pp. 174–187. University of California Press, Berkeley, CA.

Reeves, R. R., Smith, T. D., and Josephson, E. A. (2007). Near-annihilation of a species: Right whaling in the North Atlantic. *In* "The Urban Whale: North Atlantic Right Whales at the Crossroads" (S. D. Kraus and R. M. Rolland, eds), pp. 39–74. Harvard University Press, Cambridge, MA.

Rice, D. W. (1984). *Delphinus truncatus* Montagu, 1821, (Mammalia, Cetacea): Proposed conservation by suppression of *Delphinus nesarnack* Lacépède, 1804. Z. N. (S.) 2082. *Bull. Zool. Nomencl.* **43**, 256–257.

Rice, D. W. (1998). "Marine Mammals of the World: Systematics and Distribution." Society for Marine Mammalogy, Lawrence, KS, Special Publication Number 4.

Ride, W. D. L., *et al.* (1985). "International Code of Zoological Nomenclature," 3rd Ed. University of California Press, Berkeley, CA.

Rolland, R. M., Hunt, K. E., Doucette, G. J., Rickard, L. G., and Wasser, S. K. (2007). The inner whale: Hormones, biotoxins, and parasites. *In* "The Urban Whale: North Atlantic Right Whales at the Crossroads" (S. D. Kraus, and R. M. Rolland, eds), pp. 232–272. Harvard University Press, Cambridge, MA.

Rosenbaum, H. C., *et al.* (2000). World-wide genetic differentiation of *Eubalaena*: Questioning the number of right whale species. *Mol. Ecol.* **9**, 1793–1802.

Rowntree, V. (1996). Feeding, distribution, and reproductive behavior of cyamids (Crustacea: Amphipoda) living on humpback and right whales. *Can. J. Zool.* **74**, 103–109.

Scarff, J. E. (1986). Historic and present distribution of the right whale (*Eubalaena glacialis*) in the eastern North Pacific south of 50°N and east of 180°W. *Rep. Int. Whal. Commn., Spec. Iss.* **10**, 43–63.

Scarff, J. E. (1991). Historic distribution and abundance of the right whale (*Eubalaena glacialis*) in the North Pacific, Bering Sea, Sea of Okhotsk and Sea of Japan from the Maury whale charts. *Rep. Int. Whal. Commn.* **41**, 467–489.

Schell, D. M., Rowntree, V. J., and Pfeiffer, C. J. (2000). Stable-isotope and electron-microscopic evidence that cyamids (Crustacea: Amphipoda) feed on whale skin. *Can. J. Zool.* **78**, 721–727.

Schevill, W. E. (1986). Right whale nomenclature. *Rep. Int. Whal. Commn., Spec. Iss.* **10**, 19.

**R**

Shelden, K. E. W., Moore, S. E., Waite, J. M., Wade, P. R., and Rugh, D. J. (2005). Historic and current habitat use by North Pacific right whales *Eubalaena japonica* in the Bering Sea and Gulf of Alaska. *Mamm. Rev.* **35**, 129–155.

Thomson, D. H., and Richardson, W. J. (1995). Marine mammal sounds. *In* "Marine Mammals and Noise" (W. J. Richardson, C. R. Greene, C. I. Malme, and D. H. Thomson, eds), pp. 159–204. Academic Press, San Diego, CA.

Townsend, C. R., Begon, M., and Harper, J. L. (2003). "Essentials of Ecology," 2nd Ed. Blackwell, Malden, MA.

Wade, P. R., *et al.* (2006). Acoustic detection and satellite-tracking leads to discovery of rare concentration of endangered North Pacific right whales. *Biol. Lett.* **2**, 417–419.

Waring, G. T., Josephson, E., Fairfield, C. P., and Maze-Foley, K. (2007). "U.S. Atlantic and Gulf of Mexico Marine Mammal Stock Assessments—2006." NOAA Technical Memorandum NMFS-NE-201. National Marine Fisheries Service, Woods Hole, MA [Updated annually at http://www.nmfs.noaa.gov/pr/sars/].

Weisbrod, A. V., Shea, D., Moore, M. J., and Stegeman, J. J. (2000). Organochlorine exposure and bioaccumulation in the endangered northwest Atlantic right whale (*Eubalaena glacialis*) population. *Environ. Toxicol. Chem.* **19**, 654–666.

Whitehead, H. (1985). Why whales leap. *Sci. Am.* **252**(3), 84–93.

Winn, H. E., Price, C. A., and Sorensen, P. W. (1986). The distributional ecology of the right whale *Eubalaena glacialis* in the western North Atlantic. *Rep. Int. Whal. Commn., Spec. Iss.* **10**, 129–138.

Winn, H. E., Goodyear, J. D., Kenney, R. D., and Petricig, R. O. (1995). Dive patterns of tagged right whales in the Great South Channel. *Cont. Shelf Res.* **15**, 593–611.

Woodley, T. H., Brown, M. W., Kraus, S. D., and Gaskin, D. E. (1991). Organochlorine levels in North Atlantic right whale (*Eubalaena glacialis*) blubber. *Arch. Environ. Contam. Toxicol.* **21**, 141–145.

# Ringed Seal
## *Pusa hispida*

### M.O. HAMMILL

### I. Characteristics and Taxonomy

The ringed seal (Fig. 1) is among the smallest of pinnipeds, with adults reaching a maximum length of 1.3–1.5 m and weighing up to 100 kg prior to breeding. Males and females are similar in size, with males about 3% larger than females. Pelage descriptions vary slightly depending on the observer's perception. The ventral surface is normally light grey, whereas the dorsal area is variously described as being black with whitish-silvery rings, or silvery grey with black spots producing rings on its back. Its common name is derived from the characteristic ringed pattern on the pelage. The claws on the front flippers are quite rugged and are used to scratch open and maintain holes in the ice, which animals return to repeat to breathe and use to haul out on the ice. During the breeding season, the males emit a strong, pungent odor (Ryg *et al.*, 1992). The odor is produced by modified sebaceous glands that are concentrated in the facial region of the males. On account of this odor, some early descriptions and engravings of ringed seals refer to them as *Pusa* (*Phoca*) *foetida*.

The ringed seal is a member of the subfamily Phocinae of the family Phocidae. Within this subfamily, *Pusa hispida*, along with four other genera, the *Phoca, Halichoerus, Histriophoca,* and *Pagophilus*

**Figure 1** *An adult ringed seal (Pusa hispida) on Scalbard with a VHF transmitter on its back.*

constitute a well marked clade designated as tribe Phocini, which is distinguished from all other phocids by a unique karyotype ($2n = 32$), and a white lanugo (natal fur). Five subspecies are recognized, three are found in marine waters, whereas two subspecies are limited to freshwater areas.

*P.h. hispida* is the most widely distributed subspecies, occurring across northern regions of Canada, Alaska, Greenland, Svalbard, and Russia. This subspecies also includes animals found in freshwater areas in northern Canada such as in Lake Nettling on Baffin Island and in Lake Melville in Labrador, but these animals have received little study.

*P.h. botnica* occurs throughout the northern Baltic Sea, primarily in the Gulf of Bothnia and Gulf of Finland; *P.h. ladogensis* is confined to the freshwater Lake Ladoga in western Russia. Some animals are thought to transit into the Gulf of Finland, and *P.h. saimensis* are found in southeast Finland in a series of landlocked interconnected lakes of Saimaa, Haukivesi, Orivesi, Puruvesi, and Pyhäselkä; *P.h. ochotensis* is found in the western, northern, and northeastern portions of the Sea of Okhotsk ranging south to the northern coast of Hokkaido on the west and to Mys Lopatka, Kamchatka in the east.

### II. Distribution and Abundance

The ringed seal is northern circumpolar in distribution (Fig. 2). Although often considered as an inshore species, they have been observed at the North Pole and large numbers of animals are found in stable offshore pack-ice in polar regions. In North America they occur throughout the Arctic, extending as far south as the Labrador coast of Canada in the east. They are found throughout James and Hudson Bays, the Beaufort Sea, and extend as far south as Norton Sound along the Alaskan coast in the west. Ringed seals are found all along the coast of Greenland, but on the west coast are most abundant north of the Arctic circle. They are also common around Svalbard, but are rare around Iceland. Remnant populations are also found in the Baltic Sea, primarily in the Gulf of Bothnia, in Lake Saimaa in southeastern Finland and Lake Ladoga in western Russia, not far from the Gulf of Finland portion of the eastern Baltic Sea. They also occur along the northern coast of Russia, including the White Sea and are found in the Sea of Okhotsk off eastern Russia.

In general, seal ecology is characterized by marine feeding combined with a need to haul out on a solid substrate for reproduction or molting. Although hauled out under these conditions they are often concentrated in large numbers, which facilitates attempts to estimate abundance using aerial surveys. These surveys often attempt to count total numbers of hauled-out animals such as pups, then use a model incorporating information on haul-out patterns or a combination of

R

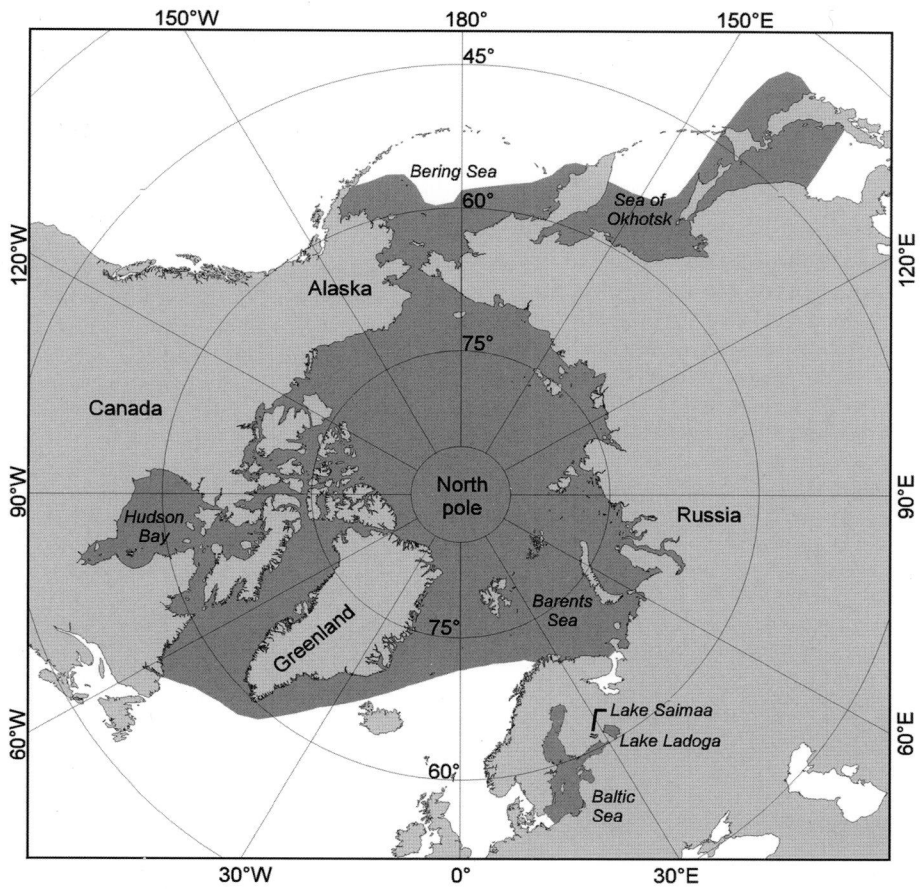

**Figure 2** *Map showing the circumpolar distribution of ringed seals* (Pusa hispida) *in dark gray.*

reproductive rates and survival rates to estimate total population size. Ringed seals present a major challenge to estimating total population size because of their remote location, they are hidden from view in snow caves (lairs) during the breeding season and do not form large aggregations on the ice (compared to other species) during the molt. Unfortunately, the factors that might affect the proportion of animals hauled-out on the ice at any one time, such as ice conditions, age, and sex, stage of molt, and weather conditions are poorly understood. Therefore, any population figure is very likely to be an underestimate and highly uncertain. Very crude estimates of abundance suggest that ringed seal numbers are in the order of 1–1.5 million for Alaska, 1.5 million in Canada, 215,000–2 million around Greenland, 7600 in Svalbard, in eastern Russia, 800,000+ in the Sea of Okhotsk and another 70,000–80,000 in the Bering Sea. Recent surveys estimate ringed seal abundance of around 260 animals in Lake Saimaa, 3000–5000 in Lake Ladoga, and 4000 in the Baltic Sea resulting in a total on the order of 3.6–5.9 million animals.

## III. Ecology

Higher densities of ringed seals are observed in areas that combine stable first-year ice with a certain amount of irregular features such as pressure ridging or frozen in pieces of ice that would encourage snow drift formation, in areas <200-m deep. Within the fast-ice habitat, mature animals occur in the prime areas, with juveniles restricted to less stable ice toward the entrance to bays or toward the offshore edge of the fast-ice.

Ringed seals keep open breathing holes by scratching at the ice using the claws on their foreflippers. As snow builds up over top of these holes, or in areas with snow drifts on the ice, animals may open new holes and dig out caves in the snow using their foreflippers. Also known as subnivean lairs, animals of all age classes and both sexes may haul out in these lairs during winter and early spring to rest. The lairs provide protection from predators and shelter from the wind and cold ambient temperatures.

Over 30 different food species have been identified as ringed seal prey, including both fish and invertebrates prey. Dominant prey among marine animals include Arctic (Polar) cod (*Boreogadus saida*), capelin (*Mallotus villosus*), redfish (*Sebastes* sp.), snailfish (*Liparis* sp.), Greenland halibut (*Reinhardtius hippoglossoides*), and sculpins (Cottidae), and pelagic crustaceans such as amphipods (*Parathemisto libellua*), mysids (*Mysis oculata*), euphausids (*Thysanoessa* sp.), decapods (*Lebbeus* sp.), and cephalopods (*Gonatus* sp.). Among freshwater subspecies, smelt (*Osmarus eperlanus*), vendace (*Coregonus albula*), ruffe (*Gymnocephalus cernuus*), burbot (*Lota lota*), three spined stickleback (*Gasterosteus aculeatus*), perch (*Perca fluviatilis*), roach (*Rutilus rutilus*), whitefish (*Coregonus lavaretus*), trout (*Salmo trutta*), four-horn sculpin (*Triglopsis quadricornis*) are

R

dominant prey for the Ladoga seal. Dominant prey for Lake Saimaa seal include vendace, smelt, whitefish, perch, roach, and burbot as well as the crustacean (*Mysis relicta*). Ringed seals are themselves important prey for polar bears, and to some extent for walrus (*Odobenus rosmarus*), Greenland shark (*Somniosus microcephalus*), and killer whales (*Orinus orca*).

## IV. Behavior and Physiology

Mating occurs in the water, and males are thought to defend underwater territories, around a few females, but this has not been observed directly. There is some evidence of philopatry among mature animals, but juveniles may undertake extensive movements, for example animals captured in the Beaufort Sea in Canada, tagged and released have been recovered from as far away as Siberia, Russia. Although capable of diving to 300 m, most dives go to depths of 100 m or less.

## V. Life History

Ringed seals reach sexual maturity at the age of 4–6 years for females and males respectively. They can be long-lived, with animals up to 45 years old having been recorded. Females give birth to a single, white-coated pup. The pups are born during March–April. At birth they weigh about 5.4 kg (SE = 0.4, N = 6) and are 63 cm (SE = 1.3, N = 11) long. The pups are weaned after 39 days (range = 36–41 days), weighing about 22 kg, for a gain of 0.43 kg/day. The young begin to lose (molt) the white fur or lanugo after 2–3 weeks. The first hair coat has a similar pattern to that of the adult pelage, but the hair is slightly longer in the first year and has a fine texture. These molted young are known as silver jars. As among other phocids, mating occurs at the time of weaning. There is a delay of implantation of 89 days and active gestation of 241 days.

The pups are normally born in a subnivean lair and as the pups grow, they may dig small tunnels off to the sides of the lair. The lower critical temperature for dry ringed seal pups is $-25°C$, but this increases to above $0°C$ for young pups that are wet. Temperatures inside the liars are 0 to $+2°C$, when ambient temperatures may be as low as $-15$ to $-27°C$. Because a dry and exposed pup appears to be capable of coping with ambient conditions, it has been suggested that the lair may play a more important role protecting pups from predators such as polar bears (*Ursus maritimus*), arctic fox (*Alopex lagopus*) and birds such as ravens (*Corvus corax*) or gulls (*Larus hyperboreus*), by hiding the young from view and slowing access by forcing the predator to dig through the lair walls or roof, giving the young some chance to escape. Young ringed seal pups are capable of entering the water soon after birth and diving to depths of as much as 80 m. This precociousness is thought to have evolved in the face of intense predation pressure. During lactation alternative lairs may be used, and if disturbed, both breathing holes and lairs may be abandoned and new lairs formed. The small body size of the ringed seal may be another adaptation to predation pressure, since smaller animals could make use of smaller snow drifts to dig out their subnivean lairs.

## VI. Interactions with Humans

Ringed seals have had a long history of exploitation by humans. In Arctic regions, they are an important component of native subsistence hunts for food, dog food, and also the sale of skins. In the Baltic Sea, they were heavily exploited and are sometimes accused of having a negative impact on commercial fisheries although this is more often attributed to the larger grey seal (*Halichoerus grypus*). Seals are also caught incidentally in fishing gear in Lake Saimaa and Lake Ladoga. In 2002, 480 seals were caught incidentally in Lake Ladoga out of a population of 3000–5000 animals.

Unlike many other seal species, ringed seals tend to be solitary and are widely dispersed, which provides some protection from direct commercial exploitation. The presence of stable ice, and sufficient snow for the snow lairs are important for pup growth and survival. Global warming could reduce snow cover, or may result in lairs melting open earlier and will lead to a decline in stable ice cover. Animals born without the protection of lairs, or where the lairs may open prematurely suffer extremely high mortality from predation by bears, foxes, and avian predators as well. Ringed seal pups need a solid platform during lactation. In areas where the ice is less stable, pups tend to be smaller, and may suffer higher mortality because milk energy must either be diverted to generating heat from spending more time in the water, or early separation (weaning) from the female. Incidental catches in commercial fisheries and coastal development are having a negative impact, particularly on the smaller freshwater subspecies. High catch levels and population declines have already resulted in some areas in Lake Saimaa and Lake Ladoga being abandoned by seals, and pose significant conservation threats to these two subspecies. Hydroelectric development in these areas, resulting in rapidly changing water levels, may lead to the collapse of lairs, particularly along the shoreline, where the largest snow banks are located, resulting in the death of animals. Oil spills would be expected to have only a limited direct impact on ringed seals because they rely on the thick blubber layer, not the fur to keep warm (with the exception of very young, white-coated pups), and the noxious fumes associated with oilspills would result in many animals moving away from the spill area. However, there may be indirect effects if oilspills affect prey abundance. Studies in areas where drilling or production are occurring have not shown a reduction in the number of seal holes or lairs, indicating that animals do not abandon these areas with elevation noise of traffic zones. Seismic exploration activity could cause temporary or permanent hearing damage, and might affect seal behavior as animals may spend more time at the surface to avoid exposure to airgun noise. High contaminants levels, primarily polychlorinated biphenyls (PCBs), were associated with large scale reproductive failure among ringed seals in the Baltic Sea.

## See Also the Following Articles

Gray Seal ■ Earless Seals

## References

Heide-Jørgensen, M.-P., and Lydersen, C. (eds) (1998). Ringed seals in the North Atlantic. NAMMCO, Sci. Publ. 1. 273 p.

McLaren, I. A. (1958). The biology of the ringed seal (*Phoca hispida* Schreber) in the eastern Canadian Arctic. *Bull. Fish. Res. Bd Can.* **118**, 97.

Ryg, M., Solberg, Y., Lydersen, C., and Smith, T. G. (1992). The scent of rutting male ringed seals (*Phoca hispida*). *J. Zool. (Lond).* **226**, 681–689.

Smith, T. G. (1987). The ringed seal, *Phoca hispida*, of the Canadian western Arctic. *Can. Bull. Fish. Aquat. Sci.* **216**, 81.

Sipilä, T., Koskela, J. T., and Kokkonen, T. S. (2005). Spatial differences in the changes of population size of the Saimaa ringed seal. *In* "Kala- ja riistaraportteja No 346 (Symposium of the Biology and Management of seals in Baltic Area)" (E. Helle, O. Stenman, and M. Wikman eds), Helsinki, 48–50.

# Risso's Dolphin
*Grampus griseus*

Robin W. Baird

## I. Characteristics and Taxonomy

Risso's dolphin (*Grampus griseus*) is the fifth largest member of the family Delphinidae, with adults of both sexes reaching up to about 4m in length (Fig. 1). The common name comes from the person (M. Risso) who described the type specimen to G. Cuvier in 1812. Risso's dolphins are unusual looking for a variety of reasons. Their anterior body is extremely robust, tapering to a relatively narrow tail stock, and they have one of the tallest dorsal fins in proportion to body length of any cetacean (Fig. 2). The bulbous head has a distinct vertical crease or cleft along the anterior surface of the melon. Color patterns change dramatically with age. Infants are gray to brown dorsally and creamy-white ventrally, with a white anchor-shaped patch between the pectoral flippers and white around the mouth. Calves then darken to nearly black, while retaining the ventral white patch. As they mature they lighten (except for the dorsal fin, which remains dark in adults in most populations), and the majority of the dorsal and the lateral surfaces of the body become covered with distinctive linear scars, most of which are presumably caused by intraspecific interactions. Older animals can appear almost completely white on the dorsal surface or when swimming just beneath the surface. No evidence of sexual dimorphism has been reported. From a distance Risso's dolphins are most frequently confused with killer whales (*Orcinus orca*) due to the large size of their dorsal fin.

Dentition is unusual, with most individuals having no teeth in the upper jaw and only a small number (two to seven pairs) in the lower jaw. Based on genetic similarity, Risso's dolphins are most closely related to false killer whales (*Pseudorca crassidens*), melon-headed whales (*Peponocephala electra*), pygmy killer whales (*Feresa attenuata*), and pilot whales (*Globicephala* spp.). No subspecies are currently recognized.

## II. Distribution and Abundance

Risso's dolphins are distributed worldwide in temperate and tropical oceans, with an apparent preference for steep shelf-edge habitats between about 400- and 1000-m deep. In the North Pacific they can be found as far north as the Gulf of Alaska and the Kamtchatka Peninsula, in the South Pacific to Tierra del Fuego and New Zealand. In the North Atlantic they have been documented as far north as southern Greenland and southern Norway. They are found throughout the Mediterranean and the Indian Ocean. No worldwide population estimates exist, although a number of regional estimates

**Figure 2** *The Risso's dolphin has one of the tallest dorsal fins in relation to body size of any cetacean. Linear scars on Risso's dolphins can often cover the majority of the body surface. Photo © Annie B. Douglas.*

Adult

Calf

**Figure 1** *The Risso's dolphin is one of the largest dolphins, with a robust body that nevertheless possesses a narrow tail stock (C. Brett Jarrett).*

R

are available. No information on population trends are available. In most areas where their ranges overlap Risso's dolphins are more common than other closely related species, such as pilot whales, false killer whales, melon-headed whales, or pygmy killer whales, although less common than the smaller delphinids. Some evidence of population division exists both between and within ocean basins, based on morphological, genetic, and distribution data, although population boundaries have not been clearly delineated. Seasonal north–south shifts in density have been suggested off the west coast of North America.

## III. Ecology

Risso's dolphins are thought to feed almost entirely on squid (both neritic and oceanic species), and limited behavioral research suggests that they feed primarily at night. Diet may vary by age and sex (Cockroft *et al.*, 1993). No evidence of predation by either killer whales (*O. orca*) or large sharks is available, although both likely prey on Risso's dolphins at least occasionally. Mass strandings of this species are very rare. The range of Risso's dolphins seems to be limited by water temperature, with animals most common in waters between 15°C and 20°C and rarely found in waters below 10°C.

## IV. Behavior and Physiology

Risso's dolphins are relatively gregarious in nature, typically traveling in groups of 10–50 individuals, with the largest group observed estimated to contain over 4000 individuals. Stable groups of adults have been documented within larger aggregations. Based on the age structure of a school killed in a drive fishery in Japan, it has been suggested that mature male Risso's dolphins move among groups. Risso's dolphins frequently travel with other cetaceans. Off southern California they have been documented to "bow ride" on and apparently harass gray whales (*Eschrichtius robustus*), and are often seen "surfing" in swells. Aggressive behavior directed toward short-finned pilot whales (*Globicephala macrorhynchus*) has been observed. No studies on diving behavior have been undertaken.

## V. Life History

Life history information for this species is relatively limited (Amano and Miyazaki, 2004). Gestation has been estimated at 13–14 months and calving interval at 2.4 years. There appears to be a peak in calving seasonality during the winter months in the eastern Pacific and in the summer/fall months in the western Pacific. Age at sexual maturity is thought to be 8–10 years for females and 10–12 years for males. The oldest Risso's dolphin estimated by examining growth layer groups in the teeth was 34.5 years old.

## IV. Interactions with Humans

Interactions with humans are diverse. Although they occasionally bow-ride on vessels, in most cases Risso's seem indifferent to vessels or actively avoid them. Risso's dolphins have been recorded stealing bait from longlines in a number of areas and have been killed as bycatch, as well as being deliberately killed as a result of such interactions. Risso's dolphins are also killed accidentally in gillnet and seine-net fishing around the world, and have been recorded ingesting plastic and with high levels of contaminants in tissues. Small numbers of Risso's dolphins have been killed in small-scale whaling operations around the world, and off Sri Lanka and possibly Japan these takes may seriously jeopardize the local population. Risso's dolphins have been held in aquaria in both Japan and the United States, although they are relatively uncommon in captivity compared to other species of cetaceans. A number of hybrids with bottlenose dolphins (*Tursiops truncatus*) have been documented in Japanese aquaria.

## See Also the Following Article

Delphinids, Overview

## References

Amano, M., and Miyazaki, N. (2004). Composition of a school of Risso's dolphins, *Grampus griseus*. *Mar. Mamm. Sci.* **20**, 152–160.

Baird, R. W., and Stacey, P. J. (1991). Status of the Risso's dolphin, *Grampus griseus*, in Canada. *Can. Field-Nat.* **105**, 233–242.

Baumgartner, M. F. (1997). The distribution of Risso's dolphin (*Grampus griseus*) with respect to the physiography of the northern Gulf of Mexico. *Mar. Mamm. Sci.* **13**, 614–638.

Cockroft, V. G., Haschick, S. L., and Klages, N. T. W. (1993). The diet of Risso's dolphin, *Grampus griseus* (Cuvier, 1812), from the east coast of South Africa. *Z. Säugetierkd.* **58**, 286–293.

Gaspari, S., Airoldi, S., and Hoelzel, A. R. (2007). Risso's dolphin (*Grampus griseus*) in UK waters are differentiated from a population in the Mediterranean Sea and genetically less diverse. *Cons. Gen.* **8**, 727–732.

Kruse, S., Caldwell, D. K., and Caldwell, M. C. (1999). Risso's dolphin *Grampus griseus* (G. Cuvier, 1812). *In* "Handbook of Marine Mammals" (S. Ridgway, ed.), Vol. 6, pp. 183–212. Academic Press, New York.

Leatherwood, S., Perrin, W. F., Kirby, V. L., Hubbs, C. L., and Dahlheim, M. (1980). Distribution and movements of Risso's dolphin, *Grampus griseus*, in the eastern North Pacific. *Fish. Bull. US* **77**, 951–963.

Shane, S. H. (1994). Occurrence and habitat use of marine mammals at Santa Catalina Island, California from 1983–91. *Bull. South. Calif. Acad. Sci.* **93**, 13–29.

Shane, S. H. (1995a). Behavior patterns of pilot whales and Risso's dolphins off Santa Catalina Island, California. *Aquat. Mamm.* **21**, 195–197.

Shane, S. H. (1995b). Relationship between pilot whales and Risso's dolphins at Santa Catalina Island, California, USA. *Mar. Ecol. Prog. Ser.* **123**, 5–11.

Hartman, K. L., Visser, F. and Hendricks, A. J. E. (2008). Social structure of Risso's dolphins (*Grampus griseus*) at the Azores: A stratified community based on highly associated units. *Can. J. Zool.* **86**, 294–306.

# River Dolphins

RANDALL R. REEVES AND ANTHONY R. MARTIN

Most people are surprised to learn that some species of dolphins, and even one porpoise population, live either entirely or partly in freshwater rivers and lakes. Three species have adapted so thoroughly to freshwater habitats, over millions of years, that they now look very different from their marine-dwelling ancestors. A fourth species, the franciscana, looks similar to those three, is most closely related to one of them, and likely once lived in rivers as well, but it is now found only in coastal marine waters. These four dolphin species, collectively (and ambiguously) known as river dolphins, exhibit some extreme characteristics in their morphology and sensory systems. Until quite recently, their similarities persuaded taxonomists that they were closely related, but genetic evidence shows that they have been separated for millions of years,

**Figure 1** *Indus dolphins — mother and calf — near Sehwan Sherif, between Sukkur and Kotri Barrages, Pakistan, April 2006. The mother is shown in lateral profile, swimming toward the left, while the calf is moving directly toward the camera and thus shown face-on. Note the prominent vertical ridge on the top of the calf's head, a typical feature of the species. Photo by Albert Reichert, courtesy of Gill Braulik.*

**Figure 2** *A boto surfaces in its typical riverine habitat, Mamirauá, Brazil, August 2005. Inset: A male boto waves plant material in the air, an action that may function as part of a social or sexual display. Mamiaruá, Brazil, March 2006. Photos by A.R. Martin & P. Gallego..*

and that their morphological similarity is due largely to convergent evolution. In other words, separately and progressively they have adapted their form and function to life in a similar habitat. They are also among the most seriously threatened cetaceans because their habitat and resources must be shared with many millions of people.

## I. Definition and Distribution

In Rice's (1998) evaluation of aquatic mammal SYSTEMATICS, he assigned the term "river dolphin" to the "peculiar long-snouted" dolphins in four single-species genera: *Platanista* (the South Asian river dolphin, known as bhulan in Pakistan, susu in India, and shushuk in Bangladesh), *Lipotes* (the Chinese or Yangtze river dolphin, known as baiji in China), *Inia* (the Amazon and Orinoco river dolphin, or boto), and *Pontoporia* (the franciscana, or La Plata dolphin). Rice placed each genus in a separate family, but Heyning (see River Dolphins, Relationships) recognized two clades: Platanistidae, containing only *Platanista*, and Iniidae, containing the other three genera. Although the genera *Lipotes* and *Pontoporia* are clearly monospecific, it had been customary until recently to recognize two species of *Platanista*—the Indus dolphin (*P. minor*) and the Ganges dolphin (*P. gangetica*). They are now provisionally regarded as subspecies: *P. gangetica minor* in the Indus drainage of Pakistan and *P. g. gangetica* in the Ganges, Brahmaputra, Megna, Karnaphuli, and Sangu drainage systems of India, Bangladesh, and Nepal (Fig. 1). There are three separate populations of the boto (*Inia geoffrensis*): the Bolivian subspecies *I. g. boliviensis* in the Madeira River drainage above the Teotonio Rapids at Porto Velho, the Amazonian subspecies *I. g. geoffrensis* distributed throughout the Amazon drainage basin except the upper Madeira system, and the Orinoco subspecies *I. g. humboldtiana* distributed throughout the Orinoco drainage basin (Fig. 2; see AMAZON RIVER DOLPHIN). The Bolivian form differs from the others in some fundamental morphological characters (e.g., number of teeth), and may deserve recognition as a separate species (Banguera-Hinestroza *et al.*, 2002). The baiji (*Lipotes vexillifer*) is, or was, endemic to China's Yangtze River system. In the past, it also occurred at least seasonally in the two large lakes, Dongting and Poyang, appended to the middle reaches of the Yangtze and in the neighboring Qiantang River. A comprehensive survey of the known current range in 2006 failed to encounter a single baiji, and no

reliable sightings had been reported in several years before then, so the species may already be extinct. If so, its demise was due to conflict with humans sharing its habitat (see later), and it was the first small cetacean to have been wiped out by man.

The outlier in this group, the franciscana (*Pontoporia blainvillei*), lives in coastal marine waters of eastern South America, including the estuary of the Rio de la Plata (River Plate) between Argentina and Uruguay. The modern river dolphins therefore occur in only two continents, South America and Asia. Most questions regarding their origins and how they evolved remain unresolved. In the case of *Inia*, for example, one hypothesis is that their ancestors entered the Amazon basin from the Pacific Ocean approximately 15 million years ago, whereas another is that they entered from the Atlantic Ocean only 1.8–5 million years ago.

## II. Behavior and Ecology

Little is known about river dolphin societies, and they probably vary from species to species. A long-term study of botos in Brazil suggests that they have no long-term affiliations, except in the case of mothers and offspring, and that adult males (which are much larger than females) compete for sexual favors both by fighting and by displaying objects (Fig. 2, inset). Males are often wounded in fights, and are heavily scarred by tooth-rakes, leading to depigmentation of the skin and at least contributing to the bright pink coloration for which the species is well known (Martin and da Silva, 2006). For much of the year, females retreat into the furthest reaches of the flooded forest with their calves, whereas adult males mostly remain on the large rivers. This degree of sexual segregation is unusual in dolphins, and its function is unclear, but it may be related to the needs or protection of the calves (Martin and da Silva, 2004a). River dolphins seem not to be highly social, with observed group sizes rarely exceeding 10 or 15 individuals. Yet the densities at which they exist, expressed in terms of individuals per unit area of water surface, sometimes far exceed those of marine cetaceans. For example, botos and tucuxis in portions of the upper Amazon system typically occur in densities of 1–10 individuals/km$^2$ (Vidal *et al.*, 1997).

Controversy has surrounded the question of whether river dolphins, like their marine counterparts, communicate with high-pitched

R

whistles. Studies of the boto have been complicated by the fact that most of its range is shared with a delphinid (*Sotalia fluviatilis*) that is known to whistle. The question has not yet been answered unequivocally, but very high-frequency whistles of unusual form have been reported from Ecuador in the apparent absence of *Sotalia* (May-Collado and Wartzok, 2007), so it now seems likely that botos do use whistles, as do (did) baijis (Wang and Wang, 2006).

The small-scale distribution of river dolphins is far from random. They tend to congregate at particular points in a river, especially at confluences (where rivers or streams converge), sharp bends, sandbanks, and near the downstream ends of islands. In a study of the distribution of Ganges dolphins in Nepal's Karnali River, Smith (1993) found the animals primarily in eddy countercurrent systems of the main river channel. Such areas of interrupted flow occur when fine sand or silt is deposited as a result of stream convergence. It is not entirely clear why the dolphins are attracted to these sites, but it likely has some relation to prey availability and energy saving. As Smith (1993) points out, positions within eddies "require minimal energy to maintain but are near high-velocity currents where the dolphins can take advantage of passing fish." Large confluences may contain tens of dolphins at a given time, but such concentrations appear to be adventitious rather than formed for social reasons. In other words, noninteracting individuals are found in close proximity due to the clumped nature of resources and refugia in the river systems where they are found. Indus dolphins and botos have similar behavior in terms of habitat preferences and the nature of feeding aggregations.

River dolphins have a number of physical characteristics that set them apart from other cetaceans. All have evolved relatively small eyes, probably because VISION is of limited value in silt-laden water, but the Indus and the Ganges dolphins have moved so far along this path that their eyes now lack a crystalline lens, rendering them functionally blind. At most, they may be able to perceive gross differences between light and dark. These dolphins usually swim on their side, with one flipper (most often the right one) trailing near the river bottom and the body oriented so that the tail end is somewhat higher in the water column than the head. Their head nods constantly as they scan acoustically for prey and obstacles. Indus and Ganges dolphins remain active day and night. All river dolphins are endowed with a sophisticated biosonar system, but those other than the Indus and the Ganges dolphins also have good vision. Other physical adaptations to freshwater environments, where maneuverability may be important in shallow areas and flooded forest, are large flippers and flexible bodies. The river dolphins of the Indus, Ganges, and Amazon/Orinoco all share these characteristics, and at least the boto can even swim backward.

All river dolphins have adapted to living in a highly dynamic environment. Although much of their habitat is silty, they also occur in areas where the water is clear, as in the upper reaches of the Ganges, or "black" (stained by tannic acid), as in many Amazon and Orinoco tributaries. Water levels in the Amazon can vary seasonally by as much as 10–13 m. During the low-water season, the dolphins (and other fauna) are restricted to the deep channels of lakes and rivers, whereas during the flood season they can range widely. Amazon dolphins penetrate into rain forests and venture onto grasslands during the floods. Their DIET seems diverse, with at least 45 fish species from 18 families, in addition to crabs and river turtles, represented in examined stomach contents (Best, 1984). Both schooling and nonschooling fish species are eaten. Botos are the only modern cetaceans with a differentiated dentition. The teeth in the front half of the jaw are conical, whereas those in the latter half have a flange on the inside portion of the crown, more reminiscent of molars (for crushing) than canines or incisors (for biting and holding). Presumably, this feature is related to the hard-bodied or spiny character of some of their prey (e.g., armored catfishes, even turtles); large catfish are often torn into smaller pieces before being eaten.

## III. Threats and Conservation Concerns

Any description of the river dolphins must include a section on their conservation status. They include within their ranks some of the most endangered aquatic mammals (Reeves *et al.*, 2000; Jefferson and Smith, 2002). As mentioned earlier, the baiji is either the most critically endangered cetacean species or it is already extinct (Turvey *et al.*, 2007). Discovered by Western science as recently as 1918, it was apparently still common and widely distributed along the entire Yangtze River, from near the Three Gorges to Shanghai, when China's Great Leap Forward began in the autumn of 1958. From that time, baijis were probably hunted to some extent for meat, oil, and leather. Although protected legally since the early 1980s, they continued to die accidentally in fishing gear, from collisions with powered vessels, and from exposure to underwater blasting during harbor construction, in addition to suffering from the effects of overfishing, pollution, industrial and vessel noise, and the damming of Yangtze tributaries. Efforts to protect the baiji (as well as the Yangtze population of finless porpoises) have been far from adequate. China's commitment to industrial and agricultural development of the Yangtze basin makes its interest in preserving the natural environment pale by comparison.

The Indus and the Ganges dolphins are also classified as endangered, with the former numbering about a thousand and the latter possibly in the low thousands. Indus dolphins occur today only in the main channel of the river, although historically they also inhabited several large tributaries (Sutlej, Ravi, Chenab, and Jhelum). Their population has been fragmented by irrigation dams, and the subpopulations trapped upriver of these dams have progressively gone extinct. Now, only two or three subpopulations of Indus dolphins are large enough to be viable. The Ganges dolphin has also lost large segments of upstream habitat as a result of dam construction, but its generally broader distribution makes it less immediately threatened with extinction. Similar to the baiji, the Indus and the Ganges dolphins have been subjected to incidental capture in fishing gear, especially gill nets. An additional concern for the Ganges dolphin is that fishermen in some parts of India and Bangladesh use dolphin oil as an attractant while fishing for a highly esteemed species of catfish. This means that there is a demand for carcasses and a disincentive for releasing live dolphins found in nets. Also, some tribal people in remote reaches of the Ganges and the Brahmaputra basins still hunt dolphins for food. A proposed seismic profiling survey in the Brahmaputra River in Assam in 2006–2007 brought international attention to the emergent risks to Asian river dolphins from oil and gas development.

Because the Amazon has not yet experienced as much modification or resource extraction as other great rivers, boto populations there appear relatively healthy. The Mamirauá reserve, an area of 11,240 km² of flooded forest in the central Amazon, had an estimated 13,000 animals in the early years of the twenty-first century (Martin and da Silva, 2004b). This density cannot be extrapolated across the species' range because Mamirauá is unusually productive and has an extremely high density of botos. Nonetheless, the total range-wide population is likely in the high tens of thousands at least. This rosy picture is unlikely to last, however. More and more hydroelectric dams are fragmenting the boto's range, riverside human populations are burgeoning, bringing increasing resource extraction, and in recent years large numbers of botos have been killed in Brazil for use as fish bait.

R

Ultimately, all river cetaceans are threatened by the transformation of their habitat to serve human needs. In addition to impeding the natural movements of dolphins and other aquatic organisms, dams in southern Asia divert water to irrigate farm fields and supply homes and businesses in an arid landscape, reducing directly the amount of habitat available to the dolphins. As water becomes an increasingly strategic resource in a warming world with expanding human populations, the prospects for freshwater cetaceans are certain to deteriorate even further.

## See Also the Following Articles

Amazon River Dolphin ■ Baiji ■ Endangered Species and Populations ■ Finless Porpoise ■ Franciscana ■ Irrawaddy Dolphin ■ Susu and Bhulan ■ Tucuxi

## References

Banguera-Hinestroza, E., Cárdenas, H., Ruiz-García, M., Marmontel, M., Gaitán, E., Vázquez, R., and García-Vallejo, F. (2002). Molecular identification of evolutionarily significant units in the Amazon river dolphin *Inia* sp. (Cetacea: Iniidae). *J. Hered.* **93**, 312–322.

Best, R. C. (1984). The aquatic mammals and reptiles of the Amazon. *In* "The Amazon: Limnology and Landscape Ecology of a Mighty Tropical River and its Basin" (H. Sioli, ed.), pp. 371–412. Dr W. Junk, Dordrecht, Dordrecht, The Netherlands.

Best, R. C., and da Silva, V. M. F. (1989). Amazon river dolphin, boto *Inia geoffrensis* (de Blainville, 1817). *In* "Handbook of Marine Mammals" (S. H. Ridgway, and R. Harrison, eds), Vol. 4, pp. 1–23. Academic Press, London.

Brownell, R. L., Jr. (1989). Franciscana *Pontoporia blainvillei* (Gervais and d'Orbigny, 1844). *In* "Handbook of Marine Mammals" (S. H. Ridgway, and R. Harrison, eds), Vol. 4, pp. 45–67. Academic Press, London.

Caldwell, M. C., Caldwell, D. K., and Brill, R. L. (1989). *Inia geoffrensis* in captivity in the United States. *In* "Biology and Conservation of the River Dolphins" Occasional Papers of the IUCN Species Survival Commission No. 3 (W. F. Perrin, R. L. Brownell, Jr., K. Zhou, and J. Liu, eds), pp. 35–41. IUCN, Gland, Switzerland.

Chen, P. (1989). Baiji *Lipotes vexillifer* Miller, 1918. *In* "Handbook of Marine Mammals" (S. H. Ridgway, and R. Harrison, eds), Vol. 4, pp. 25–43. Academic Press, London.

Herald, E. S., Brownell, R. L., Jr., Frye, F. L., Morris, E. J., Evans, W. E., and Scott, A. B. (1969). Blind river dolphins: First side-swimming cetacean. *Science* **166**, 1408–1410.

Jefferson, T. A., and Smith, B.D. (eds.) (2002). Facultative freshwater cetaceans of Asia: Their ecology and conservation. Raffles Bull. Zool. Suppl. 10, 187 pp.

Kasuya, T. (1999). Finless porpoise *Neophocaena phocaenoides* (G. Cuvier, 1829). *In* "Handbook of Marine Mammals" (S. H. Ridgway, and R. Harrison, eds), Vol. 6, pp. 411–442. Academic Press, San Diego.

Martin, A. R., and da Silva, V. M. F. (2004a). River dolphins and flooded forest: Seasonal habitat use and sexual segregation of botos *Inia geoffrensis* in an extreme cetacean environment. *J. Zool. (Lond.)* **263**, 295–305.

Martin, A. R., and da Silva, V. M. F. (2004b). Number, seasonal movements and residency characteristics of river dolphins using an Amazonian floodplain lake system. *Can. J. Zool.* **82**, 1307–1315.

Martin, A. R., and da Silva, V. M. F. (2006). Sexual dimorphism and body scarring in the boto (Amazon river dolphin) *Inia geoffrensis*. *Mar. Mamm. Sci.* **22**(1), 25–33.

May-Collado, L. J., and Wartzok, D. (2007). The freshwater dolphin *Inia geoffrensis geoffrensis* produces high frequency whistles. *J. Acoust. Soc. Am.* **121**(2), 1203–1212.

Perrin, W. F., Brownell, R. L., Jr., Zhou, K., and Liu, J. (eds.) (1989). "Biology and Conservation of the River Dolphins." Occasional Papers

of the IUCN Species Survival Commission No.3. IUCN–The World Conservation Union, Gland, Switzerland.

Reeves, R. R., and Brownell, R. L., Jr. (1989). Susu *Platanista gangetica* (Roxburgh, 1801); and *Platanista minor* Owen, 1853. *In* "Handbook of Marine Mammals" (S. H. Ridgway, and R. Harrison, eds), Vol. 4, pp. 69–99. Academic Press, London.

Reeves, R. R., Smith, B. D., and Kasuya, T. (eds.) (2000). "Biology and Conservation of Freshwater Cetaceans in Asia." Occasional Papers of the IUCN Species Survival Commission No. 23. IUCN–The World Conservation Union, Gland, Switzerland.

Rice, D. W. (1998). "Marine Mammals of the World: Systematics and Distribution." Society for Marine Mammalogy, Lawrence, KS, Special Publication No. 3.

Smith, B. D. (1993). 1990 status and conservation of the Ganges river dolphin *Platanista gangetica* in the Karnali River, Nepal. *Biol. Conserv.* **66**, 159–169.

Smith, B. D., Thant, U. H., Lwin, J. M., and Shaw, C. D. (1997). Investigation of cetaceans in the Ayeyarwady River and northern coastal waters of Myanmar. *Asian Mar. Biol.* **14**, 173–194.

Turvey, S. T., Pitman, R. L., Taylor, B. L., Barlow, J., Akamatsu, T., Barrett, L. A., Zhao, X., Reeves, R. R., Stewart, B. S., Wang, K., Wei, Z., Zhang, X., Pusser, L. T., Richlen, M., Brandon, J. R., and Wang, D. (2007). First human-caused extinction of a cetacean species? *Biol. Letters* **3**, 537–540.

Vidal, O., Barlow, J., Hurtado, L. A., Torre, J., Cendón, P., and Ojeda, Z. (1997). Distribution and abundance of the Amazon river dolphin (*Inia geoffrensis*) and the tucuxi (*Sotalia fluviatilis*) in the upper Amazon River. *Mar. Mamm. Sci.* **12**, 427–445.

Wang, K., and Wang, D. (2006). Estimated detection distance of a baiji's (Chinese river dolphin, *Lipotes vexillifer*) whistles using a passive acoustic survey method. *J. Acoust. Soc. Am.* **120**, 1361–1365.

Zhou, K., and Zhang, X. (1991). "Baiji: The Yangtze River Dolphin and Other Endangered Animals of China." Stone Wall Press, Washington, DC.

# River Dolphins, Evolutionary History and Affinities

## CHRISTIAN DE MUIZON

The term "river dolphins" or Platanistoids has been traditionally used to include the recent odontocetes that live in freshwater and are not members of the other clades of odontocetes: delphinoids, ziphioids, and physeteroids. Their affinities to other groups of odontocetes were unresolved, mainly because they have many plesiomorphic characters (e.g., Slijper, 1936, Simpson, 1945). There are four genera of living "river dolphins" (*Platanista*, *Lipotes*, *Inia*, and *Pontoporia*). Other (partly) freshwater odontocetes include *Orcaella* (Irrawadi River) and *Sotalia* (Amazon River) are not included in the Platanistoidea because they are clearly related to the marine dolphins, Delphinidae. Although it was previously assumed that platanistoids were monophyletic, this is almost certainly not the case, and some of their included taxa have been regarded as closely related to several groups of fossil odontocetes: e.g., the Squalodontidae, the Eurhinodelphinidae, the "Acrodelphinidae." There is now consensus that Platanistoidea is para- or polyphyletic (Muizon, 1984, 1987, 1988, 1991, 1994; Heyning, 1989; Fordyce, 1994; Messenger and McGuire, 1998; Fig. 1).

The genus *Platanista* appears to be an early diverging group of odontocetes, Platanistoidea, and the three other genera (*Lipotes*,

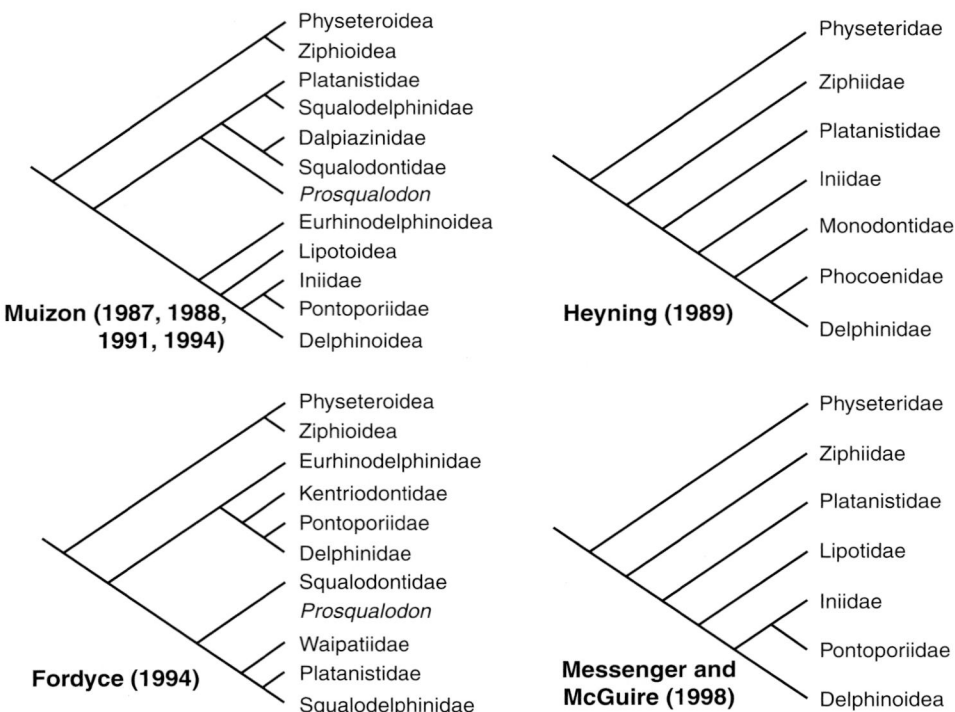

**Figure 1**    *Cladograms of hypotheses on the affinities of "river dolphins."*

*Pontoporia,* and *Inia*) are regarded as closely related to the Delphinoidea. Platanistoidea was a diversified and widely distributed group during the Oligocene and the Miocene. This group includes the modern Platanistidae as well as the fossil families Prosqualodontidae, Squalodontidae, Waipatiidae, Squalodelphinidae, and possibly the Dalpiazinidae. The other "river dolphins" are included with the Delphinoidea within the monophyletic infraorder Delphinida. There is no consensus on their position within the Delphinida, although they are generally regarded as basal taxa.

## I. Platanistoidea

This monophyletic superfamily of odontocetes includes one recent genus (*Platanista*) and approximately 15 fossil taxa (Fordyce and Muizon, 2000). The monophyly of the Platanistoidea is supported by several synapomorphies such as the reduction or loss of the coracoid process of the scapula, the development of articular ridge or peg on the periotic, and the ventral deflection of the anterior process of the periotic (Fordyce, 1994; Muizon, 1994). This superfamily includes five (possibly six) families: the Squalodontidae, the Prosqualodontidae, the Waipatiidae, the Squalodelphinidae, and the Platanistidae. In contrast to their recent representative, all the fossil platanistoids are marine, which indicates that adaptation to freshwater environment is probably a derived condition.

### A. Squalodontidae

Squalodonts (literally shark-toothed) are the most common fossil platanistoids. They have a heterodont dentition, where the posterior teeth are triangular with serrated edges (similar to some sharks). Heterodonty is primitively present in all cetaceans, and not restricted to platanistoids. In the past, this condition was used to include clades in platanistoids, but this view has been abandoned. The squalodontid

genera based on partial or complete skulls are *Squalodon, Kelloggia* (a possible synonym of *Squalodon*), *Eosqualodon,* and *?Phoberodon.* Synapomorphies of Squalodontidae as defined by Fordyce (1994) are essentially based on the morphology of one of the earbones, the periotic, a bone which is unknown in *Phoberodon, Eosqualodon,* and *Kelloggia.* The monophyly and content of the Squalodontidae has still to be evaluated by careful anatomy and further fossil finds. *Patriocetus* was also included in Squalodontidae (Rothausen, 1968), but this needs to be confirmed.

The Squalodontidae are cosmopolitan basal platanistoids. All their remains were found in marine coastal environment. *Squalodon* is present in the Miocene of Europe, Asia, and North America; *Eosqualodon* is present in the Miocene of Europe; *Kelloggia* is present in the late Oligocene of Asia; *Phoberodon* is from the early Miocene of South America. Undescribed squalodontids have also been found in Australia and New Zealand (Fordyce and Muizon, 2000). Squalodontids are relatively large odontocetes approaching the size of the living *Mesoplodon.* They had a long rostrum with strongly procumbent anterior teeth (Fig. 2). In fact, the medial incisors were almost horizontal. The teeth were strongly heterodont. The vertex was low and the skull was symmetrical. As all platanistoids, the Squalodontidae have enlarged and slightly concave premaxillary fossae anterolateral to the nares. These fossae received premaxillary sacs of the nasal tract. Premaxillary sacs are tightly related to the presence of nasal plugs and melon and their presence in the Squalodontidae is an indication of efficient echolocation ability.

### B. Prosqualodontidae

The single genus *Prosqualodon* is included in this family. Initially placed in the Squalodontidae (e.g., Simpson, 1945; Rothausen, 1968), *Prosqualodon* has been removed from this family by Muizon

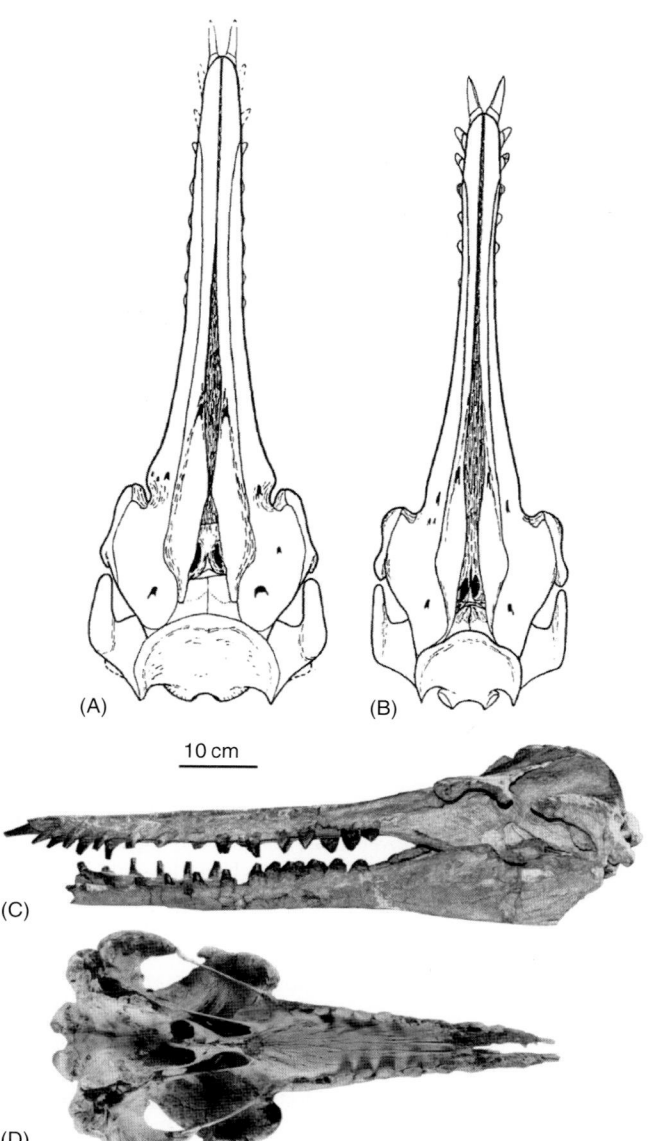

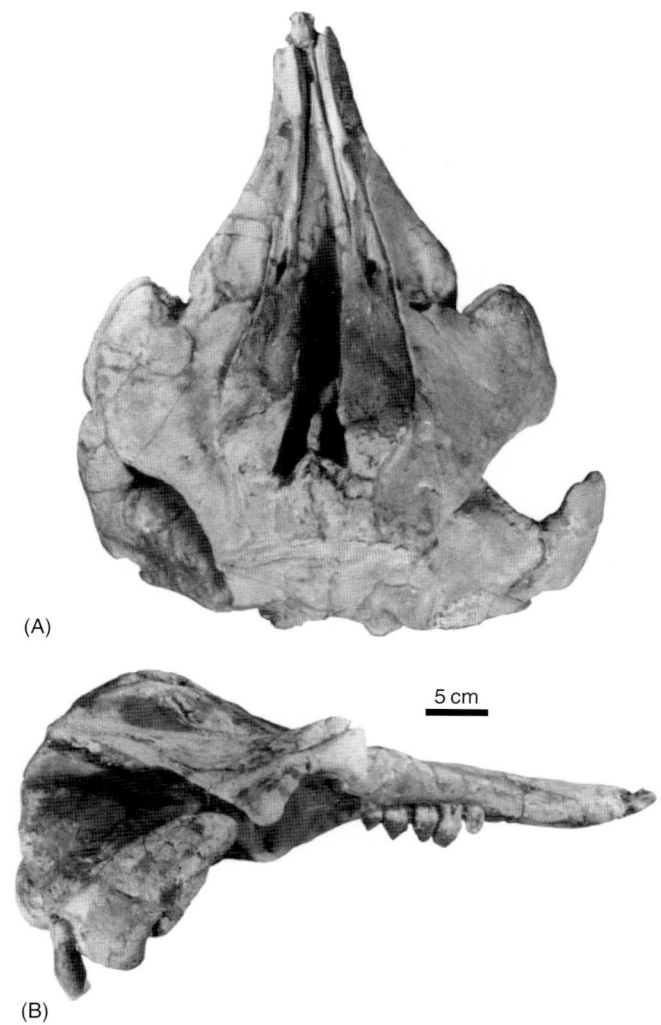

**Figure 2** *Skulls of Squalodontidae. (A) Eosqualodon langewi-eschei (late Oligocene, Germany), reconstruction of the skull in dorsal view (from Rothausen, 1968, modified). (B) Squalodon bellunensis (early Miocene, Italy), reconstruction of the skull in dorsal view (from Rothausen, 1968, modified). (C) S. bellunensis (early Miocene, Italy), skull and mandible (IGUP 26131, 26132, 26133) in lateral view. (D) Squalodon bariensis (early Miocene, France), skull (apex of the rostrum missing) in ventral view (MHNL Dr 15). (A) and (B) are reproduced with permission of Paläontologische Zeitschrift.*

**Figure 3** *Skull (MLP 5–9) of* Prosqualodon australis *(early Miocene, Argentina) in dorsal (A) and lateral (B) views.*

(1991) because it does not possess the synapomorphies of the auditory region observed in the other platanistoids; however, it was maintained in the superfamily because it bears the scapula synapomorphies of the group. This suggests that *Prosqualodon* is the sister group of the other platanistoids. However, *Prosqualodon* has sometimes been included in the infraorder Delphinida on the basis of the presence of a synapomorphy of the palatine (presence of a lateral lamina as defined by Muizon, 1988) but these may not be homologous.

*Prosqualodon* is a southern genus that has been found so far in the early Miocene of Argentina, Australia, and New Zealand. It is a medium-sized odontocete, and its size ranges from a small *Globicephala* to a large *Tursiops*. As squalodontids, it had heterodont teeth. The anterior teeth are elongated conical and project anteroventrally; the posterior teeth are triangular, low, transversely compressed with a rugose enamel and bear several denticles on their anterior and posterior crests. The rostrum is short, the vertex is symmetrical, and the braincase is lower than in the *Squalodon*. Premaxillary fossae are clearly present but they are less developed and shallower than in *Squalodon* (Fig. 3).

## C. Dalpiazinidae

The single known genus of this family, *Dalpiazina*, is probably related to the Platanistoidea given the presence of several similarities with *Squalodon* (Muizon 1991, 1994), although critical synapomorphies are not observable in the preserved fossils. *Dalpiazina* is a medium-sized odontocete (small *Tursiops*). The rostrum is relatively long and bears homodont dentition. It is known from the early

R

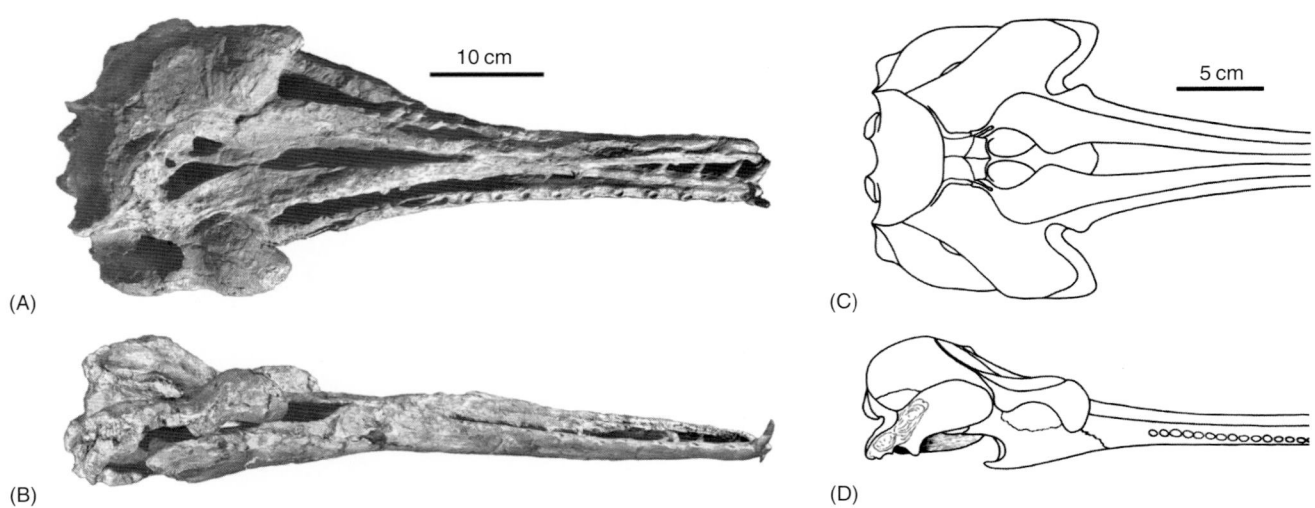

**Figure 4** *Skulls of squalodelphinids. (A)* Squalodelphis fabianii *(early Miocene, Italy), a, skull and mandible (IGUP 26134) in dorsal view. (B) The same in lateral view (note the thickness of the supraorbital region). (C) Reconstruction of the skull of* Medocinia tetragorhina *(early Miocene of France), in dorsal view. (D) The same in lateral view (from Muizon, 1988). Reproduced with permission of the Bulletin du Muséum national d'Histoire naturelle.*

Miocene of Italy and some possible dalpiazinids have been discovered in New Zealand (Fordyce *et al.*, 1994).

### D. Waipatiidae

*Waipatia* is well documented by a relatively complete skull with ear bones and partial skeleton. This genus displays the synapomorphies of the auditory region of the platanistoids and, although its scapula is unknown, is best placed in this superfamily than in any other group of odontocete (Fordyce, 1994; Fordyce and Muizon, 2000). It is a medium-sized platanistoid similar in size to *Tursiops*. The rostrum is long and slender (Fig 4). It bears heterodont teeth but the posterior triangular and double-rooted teeth are smaller than in the Squalodontidae. The incisors are conical and strongly procumbent. The skull roof is very low as in squalodontids. The skull of *Waipatia* shows clear directional asymmetry of the bones. The fossae for the premaxillary sacs are well developed and the premaxillae extend posterior to the nasals and contact the frontals on the vertex as in the other platanistoids. *Waipatia maerewhenua*, the only species unambiguously pertaining to this family, is from the late Oligocene of New Zealand.

*Sulakocetus* is a primitive odontocete from the late Oligocene of the Caucasus (Asia) that has been referred to platanistoids. It bears heterodont dentition but its posterior double-rooted teeth are small as in *Waipatia*. One of its earbones, the tympanic, is squalodont-like, and the scapula bears a small coracoid process, indicating that it is not a platanistoid. However, it is probable that the small (reduced) size of the process represent an incipient development of the platanistoid condition. This genus has been classified by Fordyce and Muizon (2000) as a possible Waipatiidae.

### E. Squalodelphinidae

This family includes the genera *Notocetus*, *Medocinia*, *Phocageneus*, and *Squalodelphis*. The four taxa are based on reasonably well-preserved skulls and/or ear bones. The Squalodelphinidae present the platanistoid synapomorphies of the scapula (loss of the coracoid process, anterior position of the acromion) and of the ear region (e.g., subcircular fossa, articular ridge of the periotic, morphology of the apex of the tympanic; see Muizon 1987, 1994). The Squalodelphinidae are cosmopolitan and marine. *Notocetus* is from the early Miocene of South America and New Zealand; *Squalodelphis* and *Medocinia* are from the early to middle Miocene of Europe; *Phocageneus* is from the early Miocene of North America. The Squalodelphinidae are medium-sized odontocetes similar in size to the living *Tursiops*. The rostrum is of moderate length and slender (Fig. 5). The teeth are more or less homodont: the posterior teeth are single rooted but they are clearly lower and more triangular than the anterior. An interesting characteristic of the Squalodelphinidae is the thickening of the supraorbital region of the skull (maxilla and/or frontal; Fig 5B and D), reminiscent of the specialized supraorbital morphology of Platanistidae (see later, Fig. 6).

### F. Platanistidae

Platanistids are represented in the fossil record by two genera, *Zarhachis* and *Pomatodelphis* (Fig. 6). They both present all the Platanistoid synapomorphies of the ear region, palatine, and scapula. The main characteristic of the Platanistidae is the development of large maxillary (*Platanista*) or maxillofrontal (*Zarhachis*, *Pomatodelphis*) crests, which are already incipiently developed in the Squalodelphinidae (see earlier discussion). A peculiarity of *Platanista* is that the palatine is entirely covered by the maxilla and the pterygoid. In *Zarhachis* and *Pomatodelphis* this condition is incipiently developed since the palatine is partially covered (Muizon 1987, 1994) and the visible portion of the bone is displaced laterally. Both genera have a very long and slender rostrum bearing homodont teeth. *Zarhachis* is slightly larger than *Pomatodelphis* and similar in size to a small *Mesoplodon*. *Allodelphis* has been regarded as a platanistid; however, this genus is still too poorly known to be referred unambiguously to this family. It is regarded here as a possible Platanistoidea, which has

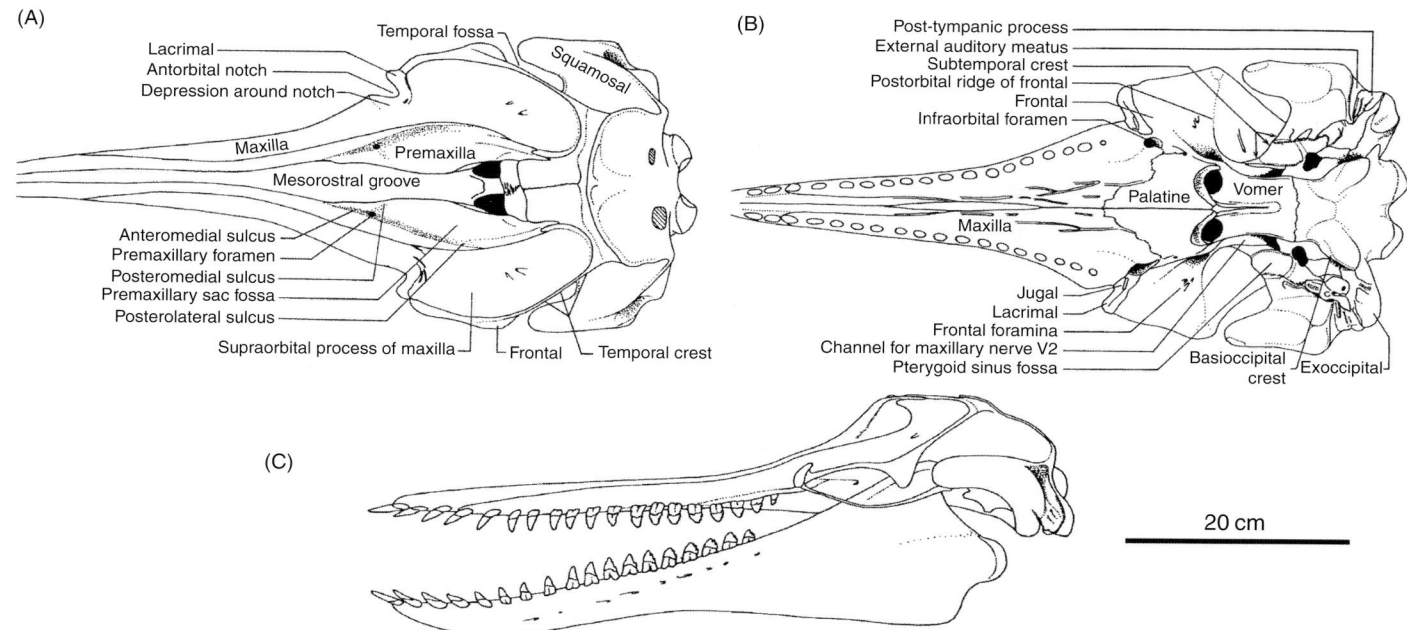

**Figure 5** *Reconstruction of the skull of* Waipatia maerewhenua *(late Oligocene, New Zealand). (A) Dorsal view; (B) ventral view; (C) lateral view (from Fordyce, 1994, modified). Reproduced with permission of the San Diego Society of Natural History.*

to be confirmed by a better knowledge of its anatomy. *Zarhachis* and *Pomatodelphis* were found in marine deposits. They are from the middle Miocene of North America and Europe (*Pomatodelphis* only). No fossil Platanistids have been found, so far, neither in the southern Hemisphere nor in Asia.

## II. Non-platanistoid "River Dolphins"

Non-platanistoid river dolphins are represented by the recent families Lipotidae (*Lipotes*), Iniidae (*Inia*), and Pontoporiidae (*Pontoporia*). Most authors recognize separate families for the three modern genera, but others place all in one family (Iniidae, Heyning, 1989), or two, Pontoporiidae (*Pontoporia* and *Lipotes*) and Iniidae (*Inia*; Fordyce, 1994). Fordyce and Muizon (2000) include these three families in the infraorder Delphinida on the basis of several synapomorphies: the development of a lateral lamina of the palatine, the sigmoid morphology of the involucrum of the tympanic and its posterior excavation, the development of a ventral rim on the ventromedial face of the anterior process of the periotic and the increase in size of the processus muscularis of the malleus. The Lipotidae are the earliest divergent Delphinida. The other Delphinida [Inioidea (Inidae + Pontoporiidae) and Delphinoidea] differ from the Lipotidae in the following synapomorphies: e.g., the reduction of the anterior process of the periotic, which loses the bullar facet; increase in size of the processus muscularis of the malleus, which is distinctly more developed than the manubrium, the presence of a pair of ventral processes on the anterior region of the sternal manubrium (Muizon, 1988). The second diverging clade is the Inioidea. The third clade, the Delphinoidea have an apomorphic thickening of the apex of the anterior process of the periotic and a great reduction of the dorsal portion of the transverse process of the atlas. Therefore, the Delphinida include three superfamilies of odontocetes, the Lipotoidea, the Inoidea, and the Delphinoidea (Muizon, 1988).

### A. Lipotoidea

This superfamily includes a single family with two genera. *Lipotes* (Recent, China) and *Parapontoporia* (Neogene, West coast of North America). *Prolipotes* from the Miocene of China is based on a non-diagnostic mandible fragment and is regarded as an incertae sedis (Fordyce and Muizon, 2000). *Parapontoporia* (Fig. 7) is regarded here as a lipotid, although classified by its author (Barnes 1985) in the Pontoporiidae. The skull of *Parapontoporia* presents a distinct narrowing at the base of the rostrum, which is always present in *Lipotes* and generally absent in pontoporiids (when present in *Pontoporia* it is weak); *Parapontoporia* does not bear the premaxillary eminences which are present in all pontoporiids and iniids; the nasals of *Parapontoporia* tend to be subvertical and not subhorizontal as in pontoporiids.

*Parapontoporia* is the only fossil lipotid for which well-preserved skull material is available. Although its braincase is only slightly larger than that of *Lipotes*, its rostrum is almost twice as long. The asymmetry is less pronounced than in *Lipotes*. The teeth are small, single-cusped, and numerous (~80 on each side). In *Lipotes* the number of teeth varies from approximately 30 to 36. Lipotids are known in the Northern Hemisphere only (China and California) and it is possible that the evolution of the family took place in the Northern Pacific.

### B. Inioidea

This superfamily includes the Iniidae and Pontoporiidae. The two families share synapomorphies of the periotic (great reduction of the anterior and posterior processes), malleus (unique extreme development of the processus muscularis), premaxilla (presence of premaxillary eminences), maxilla (frontomaxillary crests: dorsal inflexion of the postorbital edges of the maxilla and frontal). The superfamily is documented by a few well-established fossil genera mostly from South America.

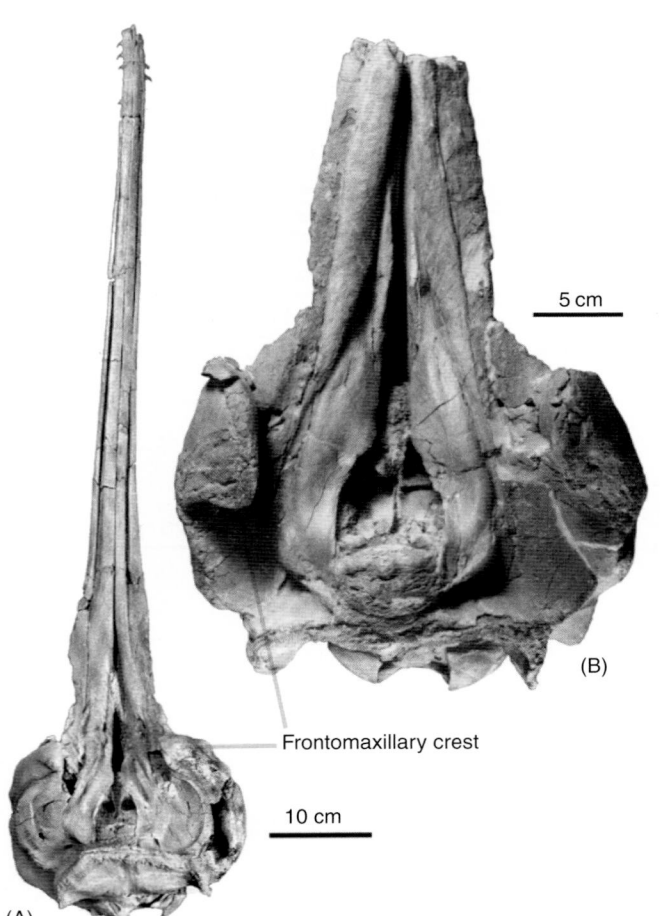

Figure 6 *Skulls of Platanistidae. (A) Pomatodelphis cf. inaequalis (middle Miocene, Maryland, USA), skull (USNM 187414) in dorsal view. (B) Zarhachis flagellator (middle Miocene, Maryland, USA), skull (most of the rostrum missing) in dorsal view (USNM 10911).*

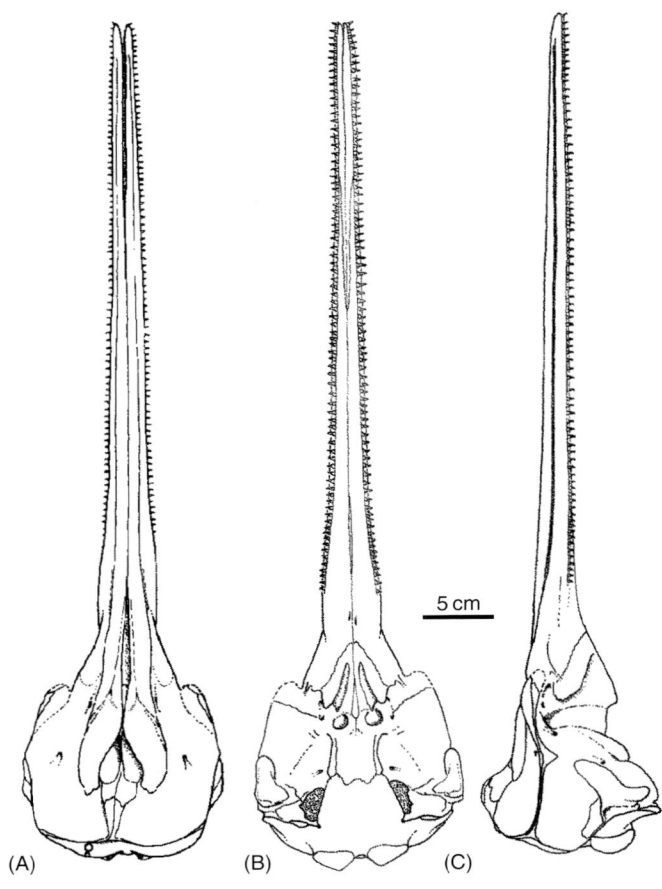

Figure 7 *Reconstruction of the skull of* Parapontoporia sternbergi *(early Pliocene, California, USA). (A) Dorsal, (B) ventral, and (C) lateral views (from Barnes, 1985 modified). Reproduced with permission of the Contribution in Science, Los Angeles County Museum of Natural History.*

The Iniidae are represented in the fossil record by a single genus known by relatively well-preserved cranial remains: *Ischyrorhynchus*. The genera *Saurocetes, Plicodontinia, Hesperocetus, Hesperoinia,* and *Lonchodelphis* that have been related to the Iniidae, are based on non-diagnostic rostra, mandible fragments or isolated teeth, and may or may not pertain to this family. *Goniodelphis,* is based on a partial skull, which probably belongs to an iniid. However, because of its incompleteness, this specimen has been questionably referred to the family by Fordyce and Muizon (2000).

One of the major characteristic of the Iniidae (*Ischyrorhynchus* and *Inia*) is the development of a frontal hump on the vertex, which is expanded at its apex. Iniidae also present an extreme reduction of the posterior process of the periotic.

*Ischyrorhynchus* is approximately 30% larger than *Inia* and its rostrum is relatively longer. Besides these features, *Ischyrorhynchus* is very similar to the recent iniid. It is from the late Miocene of the Paraná Basin (Argentina) and, therefore, was living in a freshwater environment.

The Pontoporiidae are known by two fossil genera based on well-preserved cranial remains (with associated ear bones): *Pliopontos* and *Brachydelphis* (Fig. 8). *Pontistes* is another pontoporiid and is based on a single partial skull with a well-preserved vertex. *Pontivaga*

has been referred to pontoporiids, however this genus, which is based on a partial mandible is regarded as an incertae sedis. The Pontoporiidae share synapomorphies such as a low vertex with flat, more-or-less horizontal nasals and the posterior, blade-like extension of the posterior process of the periotic.

*Pliopontos* is 50% larger than *Pontoporia.* As in the recent taxon, the rostrum is long and slender with sharp small teeth. Except for its size, *Pliopontos* is very similar to *Pontoporia.* It is from the early Pliocene of Peru and was marine. *Brachydelphis* is a much less classical pontoporiid. It has a very short rostrum, which is as long as the braincase. The latter is much larger than in *Pontoporia* and similar in size to that of *Pliopontos.* Because of these unique features for a pontoporiid, *Brachydelphis* has been placed in its own subfamily, the Brachydelphinae. It is from the Middle Miocene of Peru and was living in a marine environment.

### III. Conclusions

There is near consensus that the odontocetes traditionally placed in the "river dolphins" (the "Platanistoidea" of Simpson 1945) belong to two different groups of dolphins and are polyphyletic: the Platanistoidea and the Delphinida (Lipotoidea and Inioidea). The Platanistoidea represent the sister group of a clade, which includes the Delphinida and the

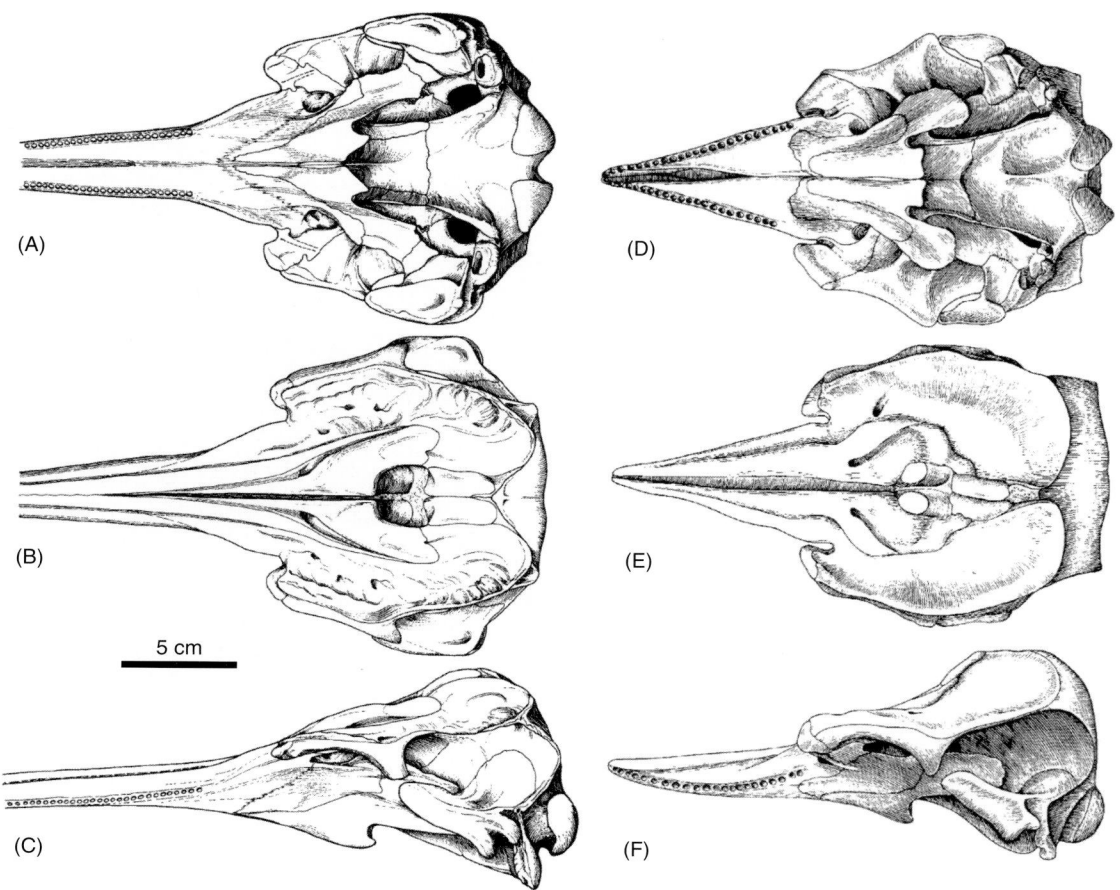

**Figure 8** *Skulls of Pontoporiidae:* Pliopontos littoralis *(early Pliocene, Peru), reconstruction of the skull in (A) dorsal, (B) ventral, (C) and lateral views (from Muizon, 1984);* Brachydelphys mazeasi *(middle Miocene, Peru), reconstruction of the skull in dorsal (D), ventral (E), and lateral (F) views (from Muizon, 1988). Reproduced with permission of the Institut Français d'Études Andines.*

fossil superfamily Eurhinodelphinoidea (Fig. 1). The non-platanistoid "river dolphins" do not represent a monophyletic grouping. In contrast, an exhaustive analysis of cetaceans phylogeny by Geisler and Sanders (2003) did find monophyly of river dolphins but polyphyly of the Delphinoidea. These results are contradicted by all other morphological and molecular analyses (Cassens *et al.*, 2000, Nikaido *et al.*, 2001).

Fossil platanistoids are diverse and distributed into several families. Fossil Lipotids and Inioids are still relatively scarce but can be easily related to one of the three non-Platanistoid families of "river dolphins." Most fossil "river dolphins" are from marine environments and the adaptation to freshwater is a convergence at least in three modern groups: the Platanistidae, the Lipotidae, and the Iniidae. Adaptation to this environment also appeared independently in two delphinoids (*Orcaella* and *Sotalia*).

## See Also the Following Articles

Cetarean Evolution ■ Cetacean Fossil Record

## References

Barnes, L. G. (1985). Fossil pontoporiid dolphins (Mammalia: Cetacea) from the Pacific coast of North America. *Contributions to Science, Natural History Museum of Los Angeles County* **363**, 1–34.

Cassens, I., Vicario, S., Waddell, V. G., Balchowsky, H., Van Belle, D., Ding, W., Fan, C., Mohan, L., Simões-Lopez, P. C., Bastida, R., Meyer, A., Stanhope, M. J., and M.C. Milinkovitch. (2000). *Proc. Natl. Acad. Sci.*, 97: 11343–11347.

Fordyce, R. E. (1994). *Waipatia maerewhenua*, new genus and new species (Waipatiidae, new family), an archaic Late Oligocene dolphin (Cetacea: Odontoceti: Platanistoidea) from New Zealand. *Proceedings of the San Diego Museum of Natural History* **29**, 147–176.

Geisler, J. H., and Sander, A. E. (2003). Morphological evidence for the phylogeny of the Cetacea. *J. Mammal. Evol.* **10**, 23–129.

Heyning, J. E. (1989). Comparative facial anatomy of beaked whales (Ziphiidae) and a systematic revision among the families of extant Odontoceti. *Contributions in science, Natural History Museum of Los Angeles County* **405**, 1–64.

Messenger, S. L., and McGuire, J. A. (1998). Morphology, molecules, and the phylogenetics of Cetaceans. *Syst. Biol.* **47**, 90–124.

Muizon, C. de. (1984). Les Vertébrés fossiles de la Formation Pisco (Pérou). Deuxième partie: Les Odontocètes (Cetacea, Mammalia) du Pliocène inférieur de Sud-Sacaco. *Travaux de l'Institut Français d'Études Andines* **25**, 1–188, *In*: Rech. sur Civ., Mém. 50, A.D.P.F., Paris.

Muizon, C. de. (1987). The affinities of *Notocetus vanbenedeni*, an Early Miocene platanistoid (Cetacea, Mammalia) from Patagonia, southern Argentina. *Am. Mus. Novit.* **2904**, 1–27.

Muizon, C. de. (1988). Les relations phylogénétiques des Delphinida (Cetacea, Mammalia). *Ann. Paléontol.* **74**, 159–227.

R

Muizon, C. de. (1991). A new Ziphiidae (Cetacea) from the Early Miocene of Washington State (USA) and phylogenetic analysis of the major groups of odontocetes. *Bulletin du Muséum national d'Histoire naturelle*, Paris, 4 série 12 C (3–4), 279–326.

Muizon, C. de. (1994). Are the squalodonts related to the platanistoids? *Proceedings Series of the San Diego Society of Natural History* **29**, 135–146.

Nikado, M., Matsuno, F., Hamilton, H., Brownell, R. L., Cao, Y., Ding, W., Zuoyan, Z., Shedlock, A. M., Fordyce, R. E., Hashegawa, M., and Okada, N. (2001). Retroposon analysis of major cetacean lineages: The monophyly of toothed whales and the paraphyly of river dolphins. *Proc. Natl. Acad. Sci.* **98**, 7384–7389.

Rothausen, K. (1968). Die systematische Stellung der europaischen Squalodontidae (Odontoceti: Mamm.). *Paläontologische Zeitschrift* **42**, 83–104.

Simpson, G. G. (1945). The principles of classification and a classification of mammals. *Bull. Am. Mus. Nat. Hist.* **85**, 1–350.

Slijper, E. J. (1936). Die Cetaceen vergleichend-anatomisch undsystematisch. *Capita Zoologica* **7**, 1–590.

# Rookeries

## George Anthony Antonelis

Pinnipeds reproduce in a wide range of marine habitats, including various forms of ice, tidal flats, rock outcroppings, and coastal beaches (Scheffer, 1958). Some species form annual breeding aggregations at traditional locations known as rookeries. These reproductive sites are an integral component of the animals' life history patterns, resulting from a complex suite of adaptive factors involving physiology, morphology, ecology, and distribution. Rookery-breeding pinnipeds exhibit varying forms of polygyny (Boness, 1991); this mode of social organization is believed to have evolved as a consequence of two key traits, parturition on solid substrate and offshore marine foraging (Bartholomew, 1970). The influence of these traits in conjunction with phylogenetic and ecological constraints has likely influenced the development of the polygynous mating systems observed on rookeries today (Emlen and Oring, 1977; Stirling, 1983; Boness, 1991; Boyd, 1991).

Rookery-breeding pinnipeds are subdivided into two families, Otariidae and Phocidae. Rookeries are formed by all otariids (15 species) and three phocids [2 species of elephant seal, *Mirounga angustirostris* and *M. leonina*, and gray seals, *Halichoerus grypus* (Reeves *et al.*, 1992; Rice, 1998)]. This chapter describes the salient social–biological, physical–geographical, and environmental characteristics of pinniped rookeries and provides information on the ecological context in which they occur.

### I. Social–Biological Characteristics

Rookeries are formed at specific times and locations, which optimize the reproductive success and survival of offspring (Bartholomew, 1970; Stirling, 1983; Boyd, 1991). After foraging at sea during the nonbreeding season, adult males return to rookeries and begin establishing territories shortly before or about the same time as the arrival of parturient females. Overt aggression, frequent threat vocalizations, and ritualized boundary displays are common among males when establishing and defending territories (Fig. 1).

**Figure 1**  *Adult male California sea lions* (Zalophus californianus) *compete for territories at San Miguel Island, California* (NMFS, George Antonelis).

Males also attempt to herd females in an effort to keep them within their areas of influence (Fig. 2). Adult females come ashore to find suitable parturition sites and tend to be highly gregarious. Parturient females frequently threaten one another either vocally or visually and are often aggressive toward offspring of other females. Otariid females suckle their pups for about 4–12 months, although longer periods have been documented for some species (Oftedal *et al.*, 1987; Bowen, 1991). Lactation of rookery-breeding phocid females lasts about 0.5 (gray seals) to 1.0 (elephant seals) month. Estrus occurs early in lactation for otariids and late in lactation for phocids (Oftedal *et al.*, 1987). Most copulations occur on land at or near the parturition site, but a few species commonly breed aquatically in the intertidal zone where males maintain aquatic territories. Otariid females intermittently leave the rookery to forage between suckling periods, and phocid females fast on land during the entire lactation period. Thus, some pinniped rookeries may be occupied continuously, but most breeding is completed within a relatively short time period of about 2 months.

Sexual dimorphism is apparent on pinniped rookeries, as adult males usually have distinctly different characteristics and are larger than females. Each sex and species emits stereotypic vocalizations for long- and short-distance communication (Stirling and Warneke, 1971; Miller, 1991). Males emit loud long-distance threat calls toward other males. Lactating otariid females also emit loud pup attraction calls on rookeries to locate their offspring among hundreds of pups. Short-distance threat vocalizations are common on all pinniped rookeries and have less amplitude than long-distance calls. Noise from rookeries initially may be perceived as a cacophony of sounds, but what seems to be chaos is really a well-organized social structure that has evolved over millions of years.

R

Figure 2 *A much larger and darker adult male California sea lion* (Zalophus californianus) *attempts to block an adult female from leaving his territory at San Miguel Island, California (NMFS, George Antonelis).*

Figure 4 *A female California sea lion* (Zalophus californianus) *suckles her pup in a location protected from the surf on San Miguel Island, California (NMFS, George Antonelis).*

Figure 3 *California sea lions* (Zalophus californianus) *are highly polygynous and form dense aggregations on rookeries commonly found along the shoreline at breeding sites on the California Channel Islands (NMFS, George Antonelis).*

## II. Physical–Geographical Characteristics

Pinniped rookeries are typically found on remote offshore islands, although some occur on mainland beaches. Rookeries are formed near shoreline just above the tidal zone in a variety of substrates, including sand, cobble or boulder beaches, rock shelves, and caves. Breeding aggregations usually occur within several hundred meters of the shoreline and also may occur on hillsides or cliffs overlooking the ocean (Fig. 3). Low-growing vegetation such as low grasses or shrubs is common on some rookeries.

The formation of rookeries on substrate near but above the tidal zone provides several functions that reinforce continued use of each site. To insure survival to weaning, neonates are usually born in locations where high tides do not wash them away from their mothers or compromise their ability to thermoregulate (Fig. 4). The gregariousness of females at these nearshore locations facilitates the ability of territorial males to monopolize estrous females, a key component of their complex polygynous mating system (Emlen and Oring,

1977). During anomalous conditions, however, storm surf associated with El Niño events has flooded pinniped rookeries, resulting in significant neonatal mortality and disruption of their polygynous social structure (Trillmich and Ono, 1991). Such events demonstrate the need for rookeries to occur above normal fluctuations in tide height.

Although pinniped rookeries must be located above the tidal zone, they must also be close enough to the water to facilitate access for thermoregulation, foraging trips by lactating females, or escape from terrestrial predators. During high air temperatures in tropical and temperate environments, many rookery-breeding otariids are known to move regularly to the intertidal zone for cooling.

## III. Environmental Characteristics

Environmental characteristics related to the formation, timing, and use of pinniped rookeries vary among species and are likely stimulated by proximate factors such as photoperiod, nutrition, and climate, which ultimately relate to survival and reproductive success (Boyd, 1991). The relative importance of these factors is believed to differ according to species on spatial and temporal scales. Pinniped rookeries occur most commonly during the spring and the summer months when climatic conditions are relatively warm and the frequency of inclement weather diminishes. Such conditions increase the probability of offspring survival, especially in subpolar climates. Rookery-breeding phocids are the exceptions and form aggregations on rookeries in the fall and the winter.

Most rookeries occur in locations where oceanographic conditions result in high productivity. High productivity increases the availability of potential prey resources vital for the foraging success of otariid females, which must feed intermittently during lactation. Rookery-breeding phocid females do not forage during lactation and rely completely on the energy stores obtained before parturition. The availability of prey near rookeries is therefore much more important to otariids than to phocids. However, the availability of prey for pups either before or after weaning may be an essential component for successful transition to foraging self-sufficiency for the young of most pinnipeds born on rookeries.

## See Also the Following Articles

Breeding Sites ■ Estrus and Estrous Behavior ■ Pinniped Life History

## References

Bartholomew, G. A. (1970). A model for the evolution of pinniped polygyny. *Evolution* **24**, 546–559.

Boness, D. J. (1991). Determinants of mating systems in the Otariidae (Pinnipedia). *In* "The Behavior of Pinnipeds" (D. Renouf, ed.), pp. 1–44. Chapman and Hall, London.

Bowen, W. D. (1991). Behavioral ecology of pinniped neonates. *In* "The Behavior of Pinnipeds" (D. Renouf, ed.), pp. 66–117. Chapman and Hall, London.

Boyd, I. L. (1991). Environmental and physiological factors controlling the reproductive cycles of pinnipeds. *Can. J. Zool.* **69**, 1135–1148.

Emlen, S. T., and Oring, L. W. (1977). Ecology, sexual selection, and the evolution of mating systems. *Science* **197**, 215–223.

Miller, E. H. (1991). Communication in pinnipeds, with special reference to non-acoustic signalling. *In* "The Behavior of Pinnipeds" (D. Renouf, ed.), pp. 128–215. Chapman and Hall, London.

Oftedal, O. T., Boness, D. J., and Tedman, R. A. (1987). The behavior, physiology, and anatomy of lactation in Pinnipedia. *Curr. Mamm.* **1**, 175–245.

Reeves, R. R., Stewart, B. S., and Leatherwood, S. (1992). "The Sierra Club Handbook of Seals and Sirenians." Sierra Club Books, San Francisco.

Rice, D. W. (1998). "Marine Mammal of the World, Systematics and Distribution." Allen Press, Lawrence, Special Publication No. 4, Society for Marine Mammalogy.

Scheffer, V. B. (1958). "Seals, Sea Lions, and Walruses." Stanford University Press, Palo Alto, CA.

Stirling, I. (1983). The evolution of mating systems in pinnipeds. *In* "Recent Advances in the Study of Mammalian Behavior" (J. F. Eisenberg, and D. G. Kleiman, eds.), pp. 487–527. Special Publication, American Society of Mammalogists 7.

Stirling, I., and Warneke, R. M. (1971). Implications of a comparison of the airborne vocalizations as some aspects of the behaviour of two Australian fur seals, *Arctocephalus* spp., on the evolution and present taxonomy of the genus. *Aust. J. Zool.* **19**, 227–241.

Trillmich, F., and Ono, K. (1991). "Pinnipeds and El Nino: Responses to Environmental Stress." Springer-Verlag Press, Berlin.

# Ross Seal
## *Ommatophoca rossii*

### JEANNETTE A. THOMAS AND TRACEY ROGERS

## I. Characteristics and Taxonomy

Less is known about the Ross seal than any other pinniped (see Scheffer, 1958; King, 1969; Reeves *et al.*, 1992; Nowak, 1999 for general reviews). It belongs to the family Phocidae and subfamily Monacinae. There is a single species in the genus. Common names include the big-eyed seal and the singing seal. The closest relatives are the other Antarctic seals (crabeater, *Lobodon carcinophagus*; leopard, *Hydrurga leptonyx*; and Weddell, *Leptonychotes weddellii*) and the monk seals (Caribbean, *Monachus tropicalis*; Mediterranean, *M. monachus*; and Hawaiian, *M. schauinslandi*).

As with other phocids, Ross seals crawl on their belly and are not capable of an upright stance or moving the hind limbs forward. The head is proportionally smaller compared to the body than in other Monacinae and the snout is exceptionally short. The neck is thick and short. The Ross seal often assumes a posture with the head raised and mouth open, pointing upward. Because of this posture they have been called the "singing seal" (Fig. 1). However, this is a misnomer, because the seal rarely emits sound in this posture. More likely the open mouth displays teeth, and thrusting-out of the striped chest serves as an aggressive posture.

The skull has an exceptionally large orbit (hence the Greek name *omma* or eye). The zygomatic arch extends well below the palate, supporting part of the skull weight when placed on a table. As with other Antarctic pinnipeds, it is assumed they have a tapetum, which assists in seeing in low-light levels during the austral winter and during deep dives, and a nictitating membrane that protects their eyes from blowing snow and allows opening their eyes in salt water. Condylobasal length has been measured at 244 mm in males and 242 mm in females. The mastoid width of the skull is at 172 and 170 mm in males and in females, respectively. The soft palate is very long, extending posterior to the level of the occipital condyles. The trachea is expanded, and powerful muscles of the tongue and pharynx assist in swallowing large prey. The external ear is absent. The nostrils are normally closed and opened under voluntary control when seals need to respire. They have 15–17 short mystacial whiskers on each lip, only 10–42 mm in length, and superciliary vibrissae. All vibrissae are smooth and not beaded.

The mouth is small. The incisors and canines are small and recurved, an adaptation for holding slippery cephalopods. The front teeth are not procumbent and the seals do not maintain breathing holes by "ice sawing," as the Antarctic Weddell seal does. The cheek teeth are reduced, homodont, and often barely breaking the gum line.

Black claws are reduced and probably useful for gripping the ice or scratching. The phalanges are greatly elongated. The fore and hind flippers are proportionally longer than in other phocids, the latter being 22% as long as the body.

The seals stay warm with a thick fur coat and a layer of subcutaneous fat. They have short, black fur on their back, with grayish silver streaks along their sides that transition into a solid silvery white belly with some spotting along the boundary line. They have a unique color pattern of vivid light silver strips running from the lower jaw to the chest and alongside the neck. A lanugo is present in the newborn, which is long, thick, and soft black on the back, fading into a bright yellow underbelly. This yellowish color may change into silver as the pup grows.

There is no marked sexual dimorphism in body size, but females tend to be slightly larger. Adult males reach 1.7–2.1 m and adult females are 1.9–2.5 m in length. Weight ranges from 129 to 216 kg in adult males and 159 to 201 kg in adult females, with pregnant females obtaining the greatest weights. Post-breeding and newly molted animals average about 1.3 m in girth and weigh about 158 kg and at this time there is no significant difference between the weight of males and females.

## II. Distribution and Abundance

This solitary seal has a circumpolar distribution around Antarctica, occurring in tight pack ice. Ross seals (estimated at 1–2% of Antarctic pinnipeds) are the least abundant of all the Antarctic pinniped species. Logistics for accurate surveys of this species are difficult because it requires icebreaker support in the dense pack-ice habitat and Ross seals generally are located far from permanent research bases, precluding land-based aerial surveys. This species is rarely sighted during other types of Antarctic research, making population estimates difficult.

R

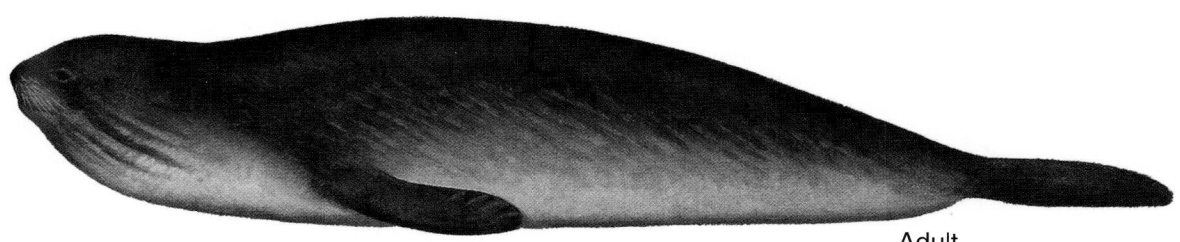

Adult

**Figure 1** *Ross seal (C. Brett Jarrett).*

Estimates range from as low as 20,000–50,000 according to Scheffer (1958), to as high as 220,000 according to Erickson *et al.* (1971). A recent international program, the Antarctic Pack Ice Seal (APIS) study aimed to conduct a systematic, continent-wide census and derive a circumpolar estimate of Ross seal population abundance; however results from different regions within this study are still being compiled. The highest recorded density of Ross seals is 2.9/km².

Distribution of the Ross seal varies with life history stage. During the austral spring and summer when they breed, they are restricted to the heavy, consolidated inner pack ice. In late summer through to early autumn, when they molt, they haul out farther north nearer the ice edge on large stable ice floes. From January to February until the breeding season of the following austral spring, they remain in the open water north or the pack ice. So, although Ross seals gives birth and mate in the pack ice, they spend most of the year living and foraging in the open waters north of the pack ice. Immature and nonbreeding seals spend up to 12 months of the year in open-water pelagic habitats.

Before 1945, there were fewer than 45 sightings of the species. Only with the use of icebreakers into the dense pack ice were more recent sightings obtained, but most often sightings were opportunistic. Ross seals are only seen when hauled out on the ice, which restricts sighting to times when the seals are breeding and molting, because outside this time they are in open water. Sightings usually are of single animals, but small groups occur in local areas; groups have been observed in areas of sparse ice, possibly due to the lack of suitable haul-out platforms. Occasionally, single seals are seen in southern Australia and some subantarctic islands, such as the South Sandwich, Falkland, Scott, and South Orkney Islands and Heard Island.

## III. Ecology

The diet of a Ross seal is primarily cephalopods, even larger species than other Antarctic seals eat. Midwater fish and krill are also eaten. Based on stomach contents analysis, the diet of the Ross seal is 47% squid, 34% fish, and 19% invertebrates. Their specialized diet reduces competition with other Antarctic seals or whales. Their secluded habitat during the breeding season may give them access to food types which are not available to pelagic species of cetaceans or to other Antarctic seals, such as Weddell, leopard, or crabeater seals. Some investigators believe the variable distribution of the species is due to the distribution of their prey, ice type, or both.

Of all the Antarctic phocids, the Ross seal inhabits the densest ice areas, so they have little exposure to predators, such as killer whales and leopard seals, during the breeding season. However, during their open-water swimming they probably are susceptible to predators of other Antarctic seals, like killer whales and leopard seals. Little is known about predation on Ross seals and they do not bear scars that would indicate leopard seal predation as are commonly seen on crabeater seals. Ross seals are sympatric with leopard seals while they breed and molt, so the lack of scarring could indicate total hunting success by leopard seals on this smaller seal or that young Ross seals travel north to open water after weaning to avoid predation. Killer whales (*Orcinus orca*) are known to take other Antarctic seals, so likely would feed on Ross seals if not for habitat segregation.

## IV. Behavior and Physiology

The species is solitary and the seals do no congregate in large breeding colonies. Although mating has not been observed, it is assumed to be in water, the same as with other Antarctic seals.

Ross seals produce up to five call types; one is an explosive noise emitted with the mouth open, but probably through the nares (Rogers, 2003; Stacey, 2006). Underwater calls are made with the mouth closed. They produce a series of pulsed chugs, both on land and in water. In water, they have an unusual two-parted whistle with harmonics that decreases and increases to produce a "W-shaped" sonogram; authors referred to this as a tonal siren call. Ross seals siren calls are heard underwater in the pack ice from mid-October but the greatest number of calls are heard from December through to mid-January. The sounds are low in amplitude compared to other Antarctic phocid calls and generally are not heard in the same areas and times as Weddell, crabeater, or leopard seals, who often call simultaneously.

Ross seals exhibit a distinct diel pattern of haul out in late summer while they are in the pack ice, with most seals hauled out during the midday and in the water during the night. This difference in sightability due to time of day makes accurate population estimates difficult. Stewart (cited in Reeves *et al.*, 1992) documented the diving behavior of one Ross seal with a microprocessor-based dive recorder glued to its back. Most dives were deeper than 100 m and lasted around 6 min. The deepest dive was 212 m and for 10 min.

## V. Life History

No breeding concentrations of Ross seals have been observed, and they do not use traditional hauling out sites. The mating system is promiscuous, with males and females encountering each other for a short time and having no long-term pair bond. Mating occurs after the pup weans in December through to early January, coinciding with a peak in underwater vocalizations and molt that soon follows. Because of the need to synchronize the time of pupping to annual ice conditions, Ross seals exhibit a delayed implantation of 2 months. This delay in pregnancy allows the mother to molt, feed, and recover from the dramatic weight loss associated with lactation before the next fetus starts to develop.

Mother Ross seals haul out in dense pack ice to give birth during the austral spring (November and December), with peak pupping between 3 and 18 November. Typically, a single pup is born. When

**R**

disturbed, a mother vocalizes to her pup and the pup responds with a bawl sound and moves close to the mother. An observation of a newborn pup reported it swimming in icy water from one floe to another. Pups wean after about 1 month of nursing.

Subadult seals are rarely seen as they are presumably in open water north of the pack ice. Longevity is unknown, but is at least 21 years. The exact age of sexual maturity is uncertain, but based on analysis of reproductive tracts is estimated at 3–4 years of age in females and 2–7 years in males. At birth, pups are 105–120 cm in length and weigh 17–27 kg.

## VI. Interactions with Humans

Ross seals have little fear of humans because there are no natural land predators in Antarctica (unlike pinnipeds in the Arctic who adapt to polar bear predators). There is no record of extensive harvest of this species, except for scientific collection. A variety of Antarctic investigators from Britain, France, America, Soviet Union, Australia, and New Zealand reported sightings of Ross seals from ships. However, there have been no ice-camp or land-based studies of this species.

Although no major threats to Ross seal populations have been identified, changes in pack ice due to climate change may influence their pack-ice habitat and prey distribution. The Ross seal is totally protected under the Antarctic Treaty and the Convention for the Conservation of Antarctic Seals. In January 1998, the Environmental Protection Protocol to the Antarctic Treaty was ratified, implementing environmental measures such as the banning of mining and oil drilling in Antarctica for at least 50 years, along with the banning of refuse disposal and the use of pesticides in the region. Because the species inhabits dense pack ice, it is doubtful that ecotourism has an impact on it.

## *See Also the Following Articles*

Antarctic Marine Mammals ▪ Earless Seals

## *References*

Barrett-Hamilton, G. E. H. (1901). Zoologie: Seals. Result. *In* "Voyage S. Y. *Belgica* 1897–1899. Exped. Antarctica Belge." pp. 1–20.

Erickson, A. W., Gilbert, J. R., and Ortis, J. (1971). Distributional ecology of Antarctic seals. *In* "Proceedings of the Symposium on Antarctic Ice and Water Masses" (G. Deacon, ed.), pp. 55–75. SCAR, Cambridge.

Gray, J. E. (1844). Mammalia. 1. The seals of the Southern Hemisphere. *In* "The Zoology of the Voyage of the *HMS Erebus and Terror*, 1839–1843," p. 1. E.W. Jansen, London.

Hofman, R. J., Erickson, A., and Siniff, D. B. (1973). The Ross seal (*Ommatophoca rossi*). *Publ. New Ser.*, Suppl. Paper.

King, J. E. (1969). Some aspects of the anatomy of the Ross seal (*Ommatophoca rossi*). *Brit. Ant. Surv. Sci. Rep.* **63**, 1–54.

Nowak, R. M. (1999). "Walker's Mammals of the World, Vol. II," 6th Ed. The Johns Hopkins University Press, Baltimore.

Øristland, T. (1970). Sealing and seal research in the Southwest Atlantic pack ice, September–October 1964. *In* "Antarctic Ecology" (M. W. Holdgate, ed.), Vol. 1, pp. 367–3765. Academic Press, London.

Rogers, T. L. (2003). Factors influencing the acoustic behaviour of male phocid seals. *Aquat. Mamm.* **29**, 247–260.

Reeves, R. R., Stewart, B. S., and Leatherwood, S. (1992). "The Sierra Club Handbook of Seals and Sirenians." Sierra Club Books, San Francisco.

Scheffer, V. (1958). "Seals, Sea Lions and Walruses: A Review of the Pinnipedia." Stanford University Press, Stanford, CA.

Scheffer, V. (1976). Standard measurements of seals. *J. Mammal.* **48**, 459–462.

Solyanick, G. A. (1964). Experiment in marking seals from small ships. *Sov. Ant. Exped. Bull.* **5**, 212.

Stacey, R. A. (2006). Vocalizations of Antarctic Ross seals (*Ommatophoca rossii*). Masters Thesis, Western Illinois University, Moline, IL.

Stirling, I. (1966). A technique for handling live seals. *J. Mammal.* **47**, 543–544.

Thomas, J. A., DeMaster, D. P., Stone, S., and Andriashek, D. (1980). Observations of a newborn Ross seal pup (*Ommatophoca rossi*). *Can. J. Zool.* **58**, 2156–2158.

Tikhimirov, E. A. (1975). Biology of the ice forms of seals in the Pacific section of the Antarctic. *Rapp. P. V. Reun. Cons. Int. Explor. Mer.* **169**, 409–412.

# Rough-Toothed Dolphin
## *Steno bredanensis*

### Thomas A. Jefferson

## I. Characteristics and Taxonomy

Rough-toothed dolphins are named for the vertical ridges, or wrinkles, on their teeth, which give them a roughened appearance. Other English common names include black porpoise, steno, and slopehead.

The rough-toothed dolphin is very distinctive when seen at close quarters. It is the only long-beaked dolphin with a smoothly sloping melon that does not contain any hint of a crease as it blends into the upper beak (Fig. 1). These dolphins are not particularly slender, and the anterior part of the body may be stocky. The large flippers are set farther back on the body than in most other small cetaceans. They are equal to about 17–19% of the body length. The dorsal fin is tall and generally only slightly falcate. Some large males have a hump of connective tissue posterior to the anus, which gives the appearance of a pronounced keel. Weight is up to 155 kg. Males grow to larger sizes than females (known maximums of 265 and 255 cm, respectively); some evidence suggests they may occasionally reach 280 cm. Females may have proportionately longer beaks (Miyazaki, 1980; Miyazaki and Perrin, 1994).

The color pattern is moderately complex but consists generally of shades of black, white, and gray (Fig. 1). The body is countershaded, with a white belly and black to dark gray back. The sides are a medium shade of gray and are separated from the darker back by the margins of a dorsal cape that is narrow between the blowhole and the dorsal fin and wider behind the fin. The lower sides and mouth area are often dotted with white patches, splotches, and spots. In warm tropical waters, the belly and the lower jaw may be tinged with pink. Some of the white spots are thought to be scars from bites inflicted by cookie-cutter sharks and perhaps squid. Young animals (Fig. 2) have a muted color pattern and generally lack the white spots (Miyazaki and Perrin, 1994).

The skull (adult CBL ~472–555 mm) can be distinguished from those of all other dolphins (except humpback dolphins, *Sousa* spp.) by their size and combination of long beak, concave rostral and maxillary margins, long mandibular symphysis, and large temporal fossae. Tooth counts can be used to distinguish them from humpback dolphins: rough-toothed dolphins have 19–26 teeth in each upper tooth row and 19–28 teeth in each lower row, and humpback dolphins usually have greater than 30. The teeth are ridged. Other differences from the skull of the humpback dolphin are the relatively large orbits

R

and the prominent and long cylindrical ridge on the ventral part of the frontal bones in rough-toothed dolphins (Van Waerebeek *et al.*, 1999). The postcranial skeleton is heavily built, and total vertebral counts generally range from 65 to 67.

Traditionally, morphological characters have been used to infer a close relationship between the rough-toothed dolphin and two other genera of dolphins (*Sotalia* spp., the tucuxi and costero; and *Sousa* spp., the humpback dolphins). Recent genetic analyses (LeDuc *et al.*, 1999) have supported the relationship with *Sotalia* (in the subfamily Stenoninae), but not with *Sousa*, which groups phylogenetically with the Delphininae. In captivity, hybrids between *Steno* and *Tursiops truncatus* (common bottlenose dolphin) have been born (Dohl *et al.*, 1974).

## II. Distribution and Abundance

The rough-toothed dolphin is a tropical to warm temperate species and is found in oceanic waters worldwide (Miyazaki and Perrin, 1994). For instance, it prefers waters greater than 1500-m deep off Hawaii (Baird *et al.*, 2008). However, it can also be found over the continental shelf in some shallow, coastal waters (e.g., Brazil––Flores and Ximinez, 1997). Records from the Atlantic Ocean are mostly from between the southeastern United States and southern

Brazil across to the Iberian Peninsula and tropical West Africa, with some (probably extralimital) records from the English Channel and the North Sea. The normal range includes the Gulf of Mexico, Caribbean Sea, and Mediterranean Sea. In the Pacific, it occurs from central Japan and northern Australia across to southern Baja California, Mexico, and southern Peru. In the eastern tropical Pacific, it is generally associated with warm tropical waters lacking major upwelling. The range includes the southern Gulf of California and the South China Sea. Records from the west coast of the continental United States and New Zealand are considered extralimital. In the poorly studied Indian Ocean, there are only a few scattered records, but the species probably has an extensive distribution there north of about 20°S.

There are no estimates of global abundance, and surveys to estimate abundance have not been conducted in most parts of the species' range. However, about 985 occur in the northern Gulf of Mexico (Mullin and Fulling, 2004) and about 146,000 are found in the eastern tropical Pacific (Wade and Gerrodette, 1993). Almost nothing is known about population or stock structure in this species.

## III. Ecology

The ecology of the species is quite poorly studied (Miyazaki and Perrin, 1994). There have been only a few reports of feeding habits. In the wild, it feeds on a variety of fish and cephalopod species, some coastal and some oceanic. Some large fish may be taken, as suggested by the robust dentition of the species. Algae have been found in the stomachs of stranded specimens, but they may have been ingested incidentally.

Rough-toothed dolphins frequently associate with other species of cetaceans, especially other delphinids in the eastern tropical Pacific, where they also often associate with flotsam. Lone animals have been seen with short-finned pilot whales (*Globicephala macrorhynchus*) and Fraser's dolphins (*Lagenodelphis hosei*) in the Sulu Sea.

Little is known of diseases and pathology, but Japanese animals have been observed with osteopathological conditions. Only a handful of internal parasites have been recorded, although there are undoubtedly others. Externally, cyamid whale lice have been observed, and the cookie-cutter shark is a partial predator.

## IV. Behavior and Physiology

Detailed, long-term behavioral ecology studies on rough-toothed dolphins have only been conducted in the past decade or so, and our

**Figure 1** *A group of rough-toothed dolphins swimming just below the surface in clear waters off Hawaii showing the species' distinctive characteristics. Photo by R. W. Baird.*

**Figure 2** *A rough-toothed dolphin mother and calf surface in Hawaiian waters. The narrow dorsal cape and smoothly sloping forehead show well in this photo. Photo by R. W. Baird.*

R

knowledge is only beginning to accumulate. There now have been such studies in several areas, including the Canary Islands, Caribbean (Honduras), and Hawaii (Ritter, 2002; Gotz et al., 2006; Kuczaj and Yeater, 2007). They are found in moderate-sized groups, most commonly of 10–20, although larger groups have been seen in some areas—up to 50 in the Canary Islands (Ritter, 2002), over 50 in the eastern tropical Pacific, 300 in Hawaii, and 160 in the Mediterranean. Mass STRANDINGS have been recorded in several areas.

These animals are not generally fast swimmers and they often appear rather sluggish in the wild. They do ride bow waves and are known for their habit of skimming along the surface at moderate speed with a distinctive splash. Synchronous swimming in tight formation is common, and recently it has been suggested to facilitate "eavesdropping" on the echolocation clicks of other individuals (Gotz et al., 2006; Kuczaj and Yeater, 2007). Although not highly acrobatic, various leaps and other aerial behaviors have been seen. Photoidentification of individual dolphins has only recently been conducted, and preliminary results suggest that rough-toothed dolphins may have more stability in their associations than do other species of small delphinids (Kuczaj and Yeater, 2007). Few studies of site fidelity have been conducted, but in the Hawaiian Islands, site fidelity appears to be high and inter-island movements relatively rare (Baird et al., 2008).

Although the maximum recorded dive was only to 70 m, rough-toothed dolphins can probably dive much deeper than this. Behavioral and morphological evidence suggests that they are well adapted for long, deep dives. Submergences of up to 15 min have been recorded. A variety of clicks and whistles have been recorded from these dolphins. Highly directional ECHOLOCATION clicks, with some pulses as high as 200 kHz, are known.

The physiology of the rough-toothed dolphin has not been well studied, and not much is known, other than that they have some anatomical adaptations that tend to be associated with deep-diving.

## V. Life History

Detailed studies of life history have only been conducted in Japanese waters. There, males reach sexual maturity at about 14 years and 225 cm, and females at 10 years and 210–220 cm. The maximum age is 32–36 years, although some animals may live significantly longer (Miyazaki, 1980; Miyazaki and Perrin, 1994). Length at birth is thought to be about 1 m.

## VI. Interactions with Humans

Rough-toothed dolphins have been held captive in a number of oceanaria, and some success has been encountered in keeping them alive in the captive environment, especially in Hawaii (Dohl et al., 1974). One lived for over 12 years in captivity. They have been found to be bold and inventive, and one "creative porpoise" at Sea Life Park in Hawaii astounded its trainers by grasping the concept of inventing novel behaviors (Pryor et al., 1969). Although they generally do not survive long, several live-stranded animals have also been kept-captive, and some have been released back to sea.

Although not generally the major target, rough-toothed dolphins have been taken in directed dolphin fisheries in Japan, Sri Lanka, Indonesia, the Solomon Islands, Papua New Guinea, St. Vincent, West Africa, and possibly St. Helena in the South Atlantic (Miyazaki and Perrin, 1994). Probably much more significant is the incidental kill of dolphins in fishing nets. Takes in tuna purse-seine nets are known for the eastern tropical Pacific, and gillnet catches have been documented at least in Sri Lanka, Brazil, and the offshore North

Pacific. This is one of the main species involved in stealing bait from fishermen's hooks off Hawaii, and sometimes animals get caught on the hooks as well (Nitta and Henderson, 1993). Undocumented catches probably occur in most other areas of the range as well.

Habitat degradation impacts and effects of pollutants are probably somewhat less severe for this species than for other, more-coastal small cetaceans. Although little directed work has been done on environmental contaminants, relatively low levels of PCBs and DDTs have been recorded from the few specimens examined so far (O'Shea et al., 1980). However, conservation-oriented studies are almost nonexistent, and therefore the uncertainty that exists about population status for this species should be acknowledged.

## See Also the Following Articles

Captivity ■ Delphinids ■ Skull Anatomy ■ Teeth

## References

Baird, R. W., Webster, D. L., Mahaffy, S. D., McSweeney, D. J., Schorr, G. D., and Ligon, A. D. (2008). Site fidelity and association patterns in a deep water dolphin: Rough-toothed dolphins (*Steno bredanensis*) in the Hawaiian Archipelago. *Mar. Mamm. Sci.* **24**, 535–553.

Dohl, T. P., Norris, K. S., and Kang, I. (1974). A porpoise hybrid: *Tursiops X Steno. J. Mammal.* **55**, 217–221.

Flores, P. A. de C., and Ximinez, A. (1997). Observations on the rough-toothed dolphin *Steno bredanensis* off Santa Catarina Island, southern Brazilian coast. *Biotemas* **10**, 71–79.

Gotz, T., Verfub, U. K., and Schnitzler, H. U. (2006). "Eavesdropping" in wild rough-toothed dolphins (*Steno bredanensis*). *Biol. Lett.* **2**, 5–7.

Kuczaj, S. A., and Yeater, D. B. (2007). Observations of rough-toothed dolphins (*Steno bredanensis*) off the coast of Utila, Honduras. *J. Mar. Biol. Assoc. U.K.* **87**, 141–148.

LeDuc, R. G., Perrin, W. F., and Dizon, A. E. (1999). Phylogenetic relationships among the delphinid cetaceans based on full cytochrome B sequences. *Mar. Mamm. Sci.* **15**, 619–638.

Miyazaki, N. (1980). Preliminary note on age determination and growth of the rough-toothed dolphin *Steno bredanensis*, off the Pacific coast of Japan. *Rep. Int. Whal. Comm. Spec. Issue* **3**, 171–179.

Miyazaki, N., and Perrin, W. F. (1994). Rough-toothed dolphin *Steno bredanensis* (Lesson, 1828). *In* "Handbook of Marine Mammals" (S. H. Ridgway, and R. Harrison, eds), Vol. 5, pp. 1–21. Academic Press, San Diego, California.

Mullin, K. D., and Fulling, G. L. (2004). Abundance of cetaceans in the oceanic northern Gulf of Mexico, 1996–2001. *Mar. Mamm. Sci.* **20**, 787–807.

Nitta, E. T., and Henderson, J. R. (1993). A review of interactions between Hawaii's fisheries and protected species. *Mar. Fish. Rev.* **55**, 83–92.

O'Shea, T. J., Brownell, R. L., Clark, D. R., Walker, W. A., Gay, M. L., and Lamont, T. G. (1980). Organochlorine pollutants in small cetaceans from the Pacific and South Atlantic oceans, November 1968-June 1976. *Pestic. Monit. J.* **14**, 35–46.

Pryor, K. W., Haag, R., and O'Reilly, J. (1969). The creative porpoise: Training for novel behavior. *J. Exp. Anal. Behav.* **12**, 653–661.

Ritter, F. (2002). Behavioural observations of rough-toothed dolphins (*Steno bredanensis*) off La Gomera, Canary Islands (1995–2000), with special reference to their interactions with humans. *Aquat. Mamm.* **28**, 46–59.

Van Waerebeek, K., Gallagher, M., Baldwin, R., Papastavrou, V., and Al-Lawati-Samira, M. (1999). Morphology and distribution of the spinner dolphin, *Stenella longirostris*, rough-toothed dolphin, *Steno bredanensis*, and melon-headed whale, *Peponocephala electra*, from waters off the Sultanate of Oman. *J. Cetacean Res. Manag.* **1**, 167–178.

Wade, P. R., and Gerrodette, T. (1993). Estimates of cetacean abundance and distribution in the eastern tropical Pacific. *Rep. Int. Whal. Commn.* **43**, 477–493.

R

# S

## Scrimshaw

### STUART M. FRANK

Scrimshaw is an occupational handicraft of mariners employing by-products of the whale fishery, often in combination with other found materials. Indigenous to the whaling industry, where it was typically a pursuit of leisure time at sea, it was also adopted in other trades and was occasionally practiced ashore (Flayderman, 1972). It arose among Pacific Ocean whalers circa 1817–1824, persisted throughout the classic "hand-whaling" era of sailing-ship days into the twentieth century, and persisted in degraded form among "modern" whalers on factory ships and shore stations until the industry shut down in the third quarter of the twentieth century (Basseches *et al.*, 1991). Since the early twentieth century, similar materials and techniques have simultaneously been employed by non-mariner artisans for both commercial and hobbyist purposes.

There is no consensus regarding etymology. Plausible and eccentric theories alike have been advanced without any creditable evidentiary basis, whereas academic lexicography (notoriously inconclusive respecting nautical terms) fails to present any convincing hypothesis. The term—also rendered skrimshank, skimshander, skirmshander, and skrimshonting—first appeared in American shipboard usage circa 1826, when the recreational practice of scrimshaw was less than a decade progressed. It originally referred not to whalers' private diversions, but to the fairly common practice whereby crewmen were required to make articles for ship's work (such as tools, tool handles, thole pins, belaying pins, and tackle falls). Sperm whale BONE is ideally suited to such uses: on any "greasy luck" voyage it was in plentiful supply at no cost, its workability is equivalent to the best cabinetmaking hardwoods, its tensile strength is greater than oak, and for many applications its self-lubricating properties were highly desirable. Such was analogously the case regarding the adaptability of cetacean bone and ivory to whales' recreational handicrafts, to which the term "scrimshaw" (and its many variants) came to refer by the 1830s.

## I. Materials and Species

Materials associated most commonly with scrimshaw are the ivory teeth and skeletal bone of the SPERM WHALE (*Physeter macrocephalus*), the ivory tusks of the WALRUS (*Odobenus rosmarus*), and the BALEEN of various mysticete species (the toothless, baleen-bearing whales). In the nineteen century the principal prey species were, roughly in descending order of importance, sperm whale, right whales (*Eubalaena* spp.), Arctic bowhead (*Balaena mysticetus*), gray whale (*Eschrichtius robustus*), and humpback (*Megaptera novaeangliae*). These and the long-finned pilot whale or so-called "blackfish" (*Globicephala melas*), which was hunted primarily from shore, were taken primarily for oil, the mysticetes secondarily for baleen. [The fast-swimming blue whale (*Balaenoptera musculus*) and fin whale (*Balaenoptera physalus*) could not be hunted effectively prior to the introduction of steam propulsion and heavy-caliber harpoon cannons in the late nineteenth century.] From the late sixteenth century, by reason of geographical proximity of Arctic habitats and similar uses of their meat and oil, the hunt for walruses was intimately conjoined with commercial whaling. Later, even when whalers were no longer taking walruses themselves, they characteristically obtained walrus tusks through barter with indigenous Northern peoples.

Commercial uses of walrus ivory were few; there was no significant commercial application for cetacean skeletal bone until the twentieth century (when it was ground and desiccated into industrial-grade meal and fertilizer). The utility and market value of baleen ("whalebone") were subject to mercurial fluctuations of fashion, and sperm whale teeth had little or no commodity value. They thus became available for whalers' recreational use, as did teeth of the Antarctic elephant seal (*Mirounga leonina*), the lower mandibles of various dolphins and porpoises, and tusks of the elusive NARWHAL (*Monodon monoceros*). (Narwhal ivory proved too difficult and brittle for anything much beyond canes and analogous shafts, such as hatracks or bedposts.)

The characteristic pigment for highlighting engraved scrimshaw was lampblack, which is essentially a viscous suspension of carbon particles in oil. (The notion that sailors used tobacco juice for this is a colorful fabrication with no basis in fact.) Lampblack, collected easily from lamps, stoves, and tryworks (shipboard oil-rendering apparatus), was in abundant supply on a whale ship. Colors were introduced almost at the outset: Edward Burdett was using sealing wax and other pigments by 1827 (Fig. 1); full polychrome scrimshaw debuted within the next decade. Sealing wax had the advantages of being universally available, relatively inexpensive, brilliantly colored, and colorfast. Applied properly, it has proven resilient and tenacious, the color as vivid today as when the scrimshaw was new. Improper application—if the cuts are too smooth or insufficiently contoured to grab and hold the wax—results in significant losses from handling and natural desiccation. Sealing wax had the disadvantage of offering only a limited spectrum of colors, all strong. Ambient pigments, however, could be mixed and blended, affording greater subtlety. From the characteristic leeching of pigment into the substrata of some polychrome scrimshaw, a phenomenon that occurs with water- and alcohol-soluble colors but not with waxes or heavy oil-based pigments, it is clear that ambient colors were also favored. Store-bought inks, homemade dyes extracted from berries, and greens from common verdigris seem to predominate; however, their composition has not been investigated comprehensively.

Inlay and other secondary materials—rare on engraved scrimshaw but often encountered on "built" or "architectonic" scrimshaw—were typically obtained at little or no cost, such as other marine byproducts (tortoise shell, mother-of-pearl, sea shells), various woods brought from home or obtained in various ports of call (including exotic tropical species from Africa and Polynesia), and miscellaneous bits of metal (fastenings and finials were often crafted from silver- or copper-alloy coins, typically coins minted in Mexico and South America).

## II. Scrimshaw Precursors

Medieval European artistic productions in walrus ivory and cetacean bone were many, but the whalers themselves had no part

**Figure 1**  *Pan bone plaque by Edward Burdett (1805–1833) of Nantucket, circa 1828. The earliest known scrimshaw artist, Burdett was active from 1824 until he was killed by a whale in 1833. His work is characterized by a bold, confident style, with deep blacks and red sealing-wax highlights. He was serving as a mate in the William Tell when he engraved this section of sperm whale bone, inscribed "William Tell. of New York. homeward bound. in the latitude of. 50 13. S. long[itude] 80. W. got shipwrecked"; "lost her rudder & c."; "by. E. Burdett." 15.7 × 31.8 cm. Kendall Collection, New Bedford Whaling Museum.*

in them beyond gathering the raw materials. Cetacean bone panels and stilettos survive from the Viking era, some incised with rope patterns and animal figures, and cetacean bones served as beams in vernacular buildings in Norway and the Friesian Islands, but even these do not appear to have been made by whalers and are not known to have been part of their occupational culture. Monastic artisans in Denmark and East Anglia carved walrus ivory and cetacean bone into votive art, primarily altar pieces, friezes, and crosses, whereas craftsmen at Cologne and elsewhere produced secular game pieces and chessmen from the same materials. So important was the "Royal Fish" to the Viking economy that a highly sophisticated body of law evolved to regulate whaling itself and the ownership, taxation, distribution, and export of whale products, whether acquired fortuitously (from stranded carcasses) or by hunting. Pliny the Elder (first century C.E.), Olaus Magnus (1555), Conrad von Gessner (1558), and Ambroise Pare (1582) listed the uses of baleen for whips, springs, garment stays, and umbrella ribs; the emergence of pelagic Arctic whaling in the seventeenth century encouraged a search for new applications, especially in Holland. The search, however, proved fruitless and was abandoned by circa 1630, occasioning the appearance of sailor-made baleen objects: there was simply no longer any reason to restrain whalers from using baleen for their private diversions (two centuries later the same principle would make sperm whale teeth available for scrimshaw).

Ditty boxes were the first manifestation of whalers' work. Typically, these have polished baleen sides bent to the oval shape of a wooden bottom 30–35 cm long, to which the baleen is fastened with copper nails and fitted with a wooden top. Two made in 1631 by an anonymous Rotterdam whaling commandeur have baleen sides incised with portraits of whaleships, the wooden tops relief carved with the Dutch lion rampant surrounded by nautical symbols, human figures, and decorative borders. A few North Friesian whalemen—artists of the next generation are known by name. Jacob Floer of Amrum engraved buildings, trees, and geometrical borders on the baleen sides of an oval box, signed and dated 1661. Peter Lorenzen of Sylt signed and dated another in 1687. The form continued for the duration of Arctic whaling and was perpetuated with myriad variations by American and British scrimshaw artists in the nineteenth century.

Another early form was the mangle (paddle for folding cloth). An Amsterdam whaling commandeur decorated one with carved geometric ornaments, signed, dated, and inscribed, "Cornelis Floerensen Bettelem. Niet sonder Godt [Not without God]. Anno 1641." A century later, a North Friesian whaling master, Lødde Rachtsen of Hooge, made one for his daughter's dowry: it has a pierced-work handle and carved geometric and floral decorations, the broadside portrait of a spouting bowhead whale, and a carved inscription dated 1745. Respecting technical aspects of execution and the iconography of their decoration, this kind of piece is the direct ancestor of the sperm whalemen's decorated baleen corset—busks of the nineteenth century.

## III. Origins and Practice

Pictorial engraving on sperm whale teeth—the quintessential manifestation of scrimshaw—resulted from changing circumstances in the fishery in the aftermath of the Napoleonic wars (of which the American theater was the so-called War of 1812). It arose collectively among British and American whalers in the 1820s in the Hawaiian Islands, where (beginning in 1819) the fleets customarily laid over for weeks on end between seasonal cruises, providing ample opportunities for fraternization and foment.

In the late seventeenth century, British colonists on the Atlantic seacoast of New York and Massachusetts hunted right whales along shore—an ancillary day fishery, prosecuted by fishers and farmers in rowboats launched from sandy beaches. In the eighteenth century, expanding markets occasioned offshore cruises in small sailing vessels. The discovery of sperm whales in proximity to New England is ascribed by tradition to Captain Christopher Hussey of Nantucket when he was blown off course while right whaling in 1712. The colonists tuned their technology to accommodate sperm whaling, pioneered the refining of sperm whale oil and the manufacture of spermaceti candles (America's first industry), and developed thriving export networks. Whaling evolved into a full-time occupation, and a distinctive caste of whaler-mariners emerged with its own occupational culture. Scrimshaw would become an integral component of this culture, but it took a whole century for the right circumstances to gel.

Colonial whaling cruises were seasonal, following the Atlantic trade winds on comparatively short passages to and from the grounds. Only a few whales were required to fill the hold before heading home, usually after only a few weeks. The opening of the Pacific grounds in the 1790s changed shipboard dynamics dramatically. Voyages necessarily became longer, as much as 3 or 4 years by the 1840s. The larger catch that was required to make long voyages profitable mandated larger vessels with larger crews so that three–five whaleboats could be launched for the hunt. The effective result was overmanning and an unprecedented abundance of shipboard leisure-long outward and homeward passages, and idle weeks, even months between whales, when there was little to do but maintain the ship and wait. Most sailors worked "watch-on-watch" in 4 h shifts, day and night, whenever the ship was underway, but whaleship crews had most nights off: the hunt could not be prosecuted effectively in darkness, and cutting in (flensing) a carcass with sharp blubber spades was dangerous enough even in daylight. Apart from rendering blubber already on hand, there was little work to do evenings. More than in any other seafaring trade, the nineteenth-century whalers had time to spare. They filled it with reading, journal-keeping, drawing, singing, dancing, gamming (visiting among ships at sea), and a host of other diversions.

At the critical juncture, just when things were ripe for scrimshaw, teeth were in short supply. For in the meanwhile, the China Trade, pioneered in the 1780s, had established a network of Far East destinations and products that involved barter with Pacific islanders to obtain goods for Canton. China traders soon realized that many island cultures placed great value on whale teeth, from which they crafted various totemic and decorative objects. Teeth could be obtained cheaply from whalers (there being no other market), so the China merchants bought them up for barter in the Pacific. Such scrimshaw as there was in the eighteenth century was therefore limited primarily to implements made of skeletal bone-straightedges, hand tools, a few early swifts (yarn-winders), and corset husks; of these, comparatively few were made prior to the florescence of scrimshaw commencing in the 1820s.

Captain David Porter of the US Naval frigate Essex was the inadvertent catalyst for the emergence of scrimshaw. Porter's wartime purpose had been to inflict depredations on British shipping and to disrupt British whaling in the Pacific. His narrative (published in 1815, reissued in an expanded edition in 1821) was valued by mariners for its explicit accounts of conditions in the Marquesas and Galapagos Islands and on the coast of Chile and Peru. It also incidentally revealed the barter value of whales' teeth in Polynesia and disclosed particulars of how they could be gathered cheaply—this at just around the time the vanguard of the whaling fleet reached Hawaii (1819). There was soon a surplus of whales' teeth on the

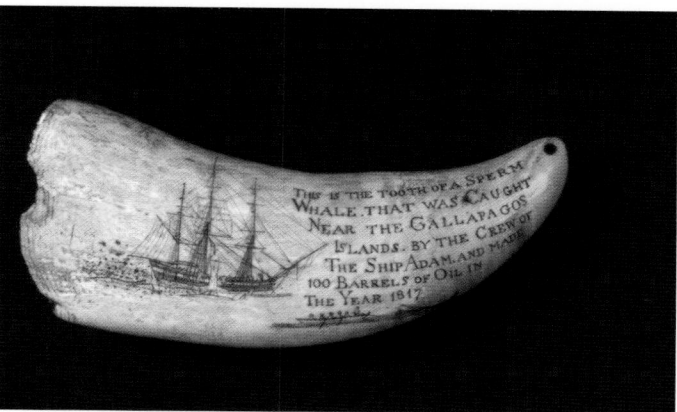

Figure 2  *Genesis of scrimshaw, circa 1817. The oversize tooth is inscribed "This is the tooth of a sperm whale that was caught near the Galapagos Islands by the crew of the ship Adam [of London], and made 100 barrels of oil in the year 1817." Produced in the wake of the Napoleonic Wars, when the British and American whaling fleets were endeavoring to recover their former prowess in the Pacific, it is believed to the earliest full-scale work of engraved pictorial scrimshaw on a sperm whale tooth. Length 23.5 cm. Kendall Collection, New Bedford Whaling Museum.*

Pacific market; as the teeth were no longer valuable as a commodity, they could be relegated to sailors for private use.

Accordingly, the earliest authentic date on any pictorial sperm whale scrimshaw is 1817—a tooth commemorating a whale taken by the ship Adam of London off the Galapagos Islands (Fig. 2); the earliest provisionally identifiable whaleman-engraver of sperm whale ivory is J. S. King, whaling master of London and Liverpool, to whom two teeth are attributed, one perhaps as early as 1821. These suggest a possible British genesis of pictorial scrimshaw; however, the earliest definitively attributable work is by an American, Edward Burdett of Nantucket, who first went whaling from his native port in 1821 and was scrimshandering by 1824. Fellow Nantucketer Frederick Myrick was the first to sign and date his scrimshaw—three dozen teeth produced during 1828–1829 as a seaman aboard the Nantucket ship Susan. Two teeth by Burdett and two "Susan's Teeth" by Myrick were accessioned by the East India Marine Society of Salem, Massachusetts, prior to 1831, while both artists were still living—the first scrimshaw to enter a museum collection. That Myrick's work was listed generically as "Tooth of a Sperm Whale, curiously carved" and "Another, carved by the same hand," with no mention of the exquisitely engraved pictures on them, nor of the artist's name (both are signed), nor of the term "scrimshaw", testifies to the newness of the genre, perhaps also to the low esteem in which sailors' hobby work was held by the great merchants of Salem at the time.

In the 1830s, scrimshaw became widely generalized. On some whaling vessels virtually the entire ship's company participated. In his journal of the New Bedford bark Abigail during 1836–1838, Captain William Hathaway Reynard remarked, "The cooper is going ahead making tools for scrimsham. We had a fracas betwixt the cook and the stewart [sic] … All hands employed in scrimsa." In other ships the best ivory and bone may have been relinquished to some particularly talented member of the crew, such as Joseph Bogart Hersey of Cape Cod on the Provincetown schooner Esquimaux in 1843: "This afternoon we commenced sawing up the large whale's jaws … the bone proved to be pretty good and yielded several canes,

**S**

fids, and busks. I employed a part of my time in engrav[ing] or flow-ering two busks. Being slightly skilled in the art of flowering; that is drawing and painting upon bone; steam boats, flower pots, monu-ments, balloons, landscapes; I have many demands made upon my generosity, and I do not wish to slight any; I of course work for all."

The whaleship labor force was very young on average, with green hands often in their early teens; common seamen and even seasoned harpooners were rarely over 30. Among the greatest scrimshaw art-ists, Frederick Myrick retired from whaling and from scrimshaw at age 21, Edward Burdett was barely 28 when he was killed by a whale, and Welsh ship's surgeon W. L. Roderick left the fishery at 29. Nevertheless, although in the minority, older hands contributed mightily. Seaman Silas Davenport may have been in his forties when he constructed a fine swift of bone and ebony. Former whaleman N. S. Finney was still engraving walrus ivory on commission in San Francisco in his sixties. Ship's carpenters and coopers-trained crafts-men with skills well adapted to scrimshaw, especially architectural pieces were normally older than the average crewman. So, too, whal-ing captains, many of whom were devoted scrimshaw artists: Manuel Enos cut brilliant polychrome portraits into whale ivory right up to the time he was lost at sea at age 55; Frederick Howland Smith was scrimshandering from age 14 until he retired at 61; and the grand old man, Captain Ben Cleveland, was still making napkin rings, man-tle ornaments, gadgets in the 1920s, at age 80.

Scrimshaw was quintessentially a diversion of the whalemen's ample leisure hours, to fill time, and produce mementos, as gifts for loved ones at home. It was occupationally rooted in and wholly indig-enous to the deepwater whaling trade, but was eventually also adopted by merchant sailors, navy tars, and occasionally the seafaring wives and children of whaling masters.

Unfortunately, practitioners in whatever trade rarely signed or dated their work, and family provenance has seldom preserved details of the origins of legacy pieces. Thus, scrimshaw has hitherto been mostly anonymous, the names of only a handful of practition-ers known. However, systematic forensic studies commencing in the 1980s have made stylistic and iconographical attributions increas-ingly possible, and the names and works of a few hundred individual artists are now documented with varying degrees of specificity.

## IV. Taxonomy

Scrimshaw took many forms. Henry Cheever mentions whalers "working up sperm whales' jaws and teeth and right whale [baleen] into boxes, swifts, reels, canes, whips, folders, stamps, and all sorts of things, according to their ingenuity" (The Whale and His Captors, 1850), and Herman Melville alludes to "lively sketches of whales and whaling-scenes, graven by the fishermen themselves on Sperm Whale-teeth, or ladies' busks wrought out of the Right Whale-bone, and other like skrimshander articles, as the whalemen call the numerous little ingenious contrivances they elaborately carve out of the rough mate-rial in the hours of ocean leisure" (Moby-Dick, 1851). Various tools were used for cutting and polishing, but forensic scrutiny corroborates Melville's observation that the ordinary knife predominated: "Some of them [the whaleman-artisans] have little boxes of dentistical-look-ing implements, specially intended for the skrimshandering business. But in general, they toil with their jack-knives alone; and, with that almost omnipotent tool of the sailor, they will turn you out any thing you please, in the way of a mariner's fancy."

Scrimshaw objects intended for practical use included hand tools, kitchen gadgets, sewing implements, toys, and even full-sized furni-ture. Some, such as fids, straightedges, tool handles, seam-rubbers,

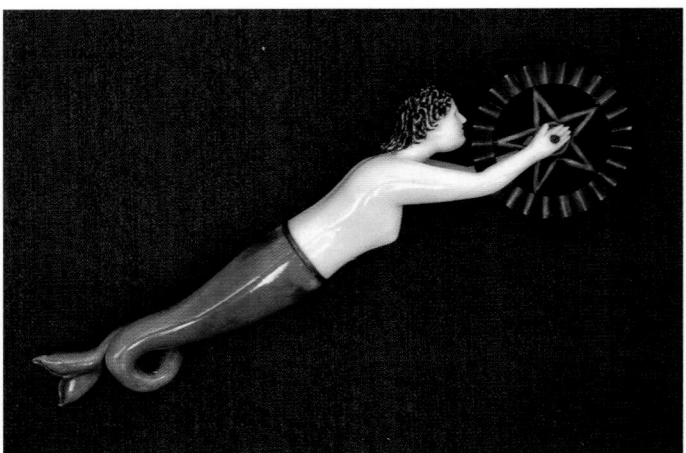

**Figure 3**  *Pie crimper in the form of a mermaid, New Bedford, circa 1875. Practical in origin, these classic kitchen implements inspired some of the scrimshaw's most creative forms and elaborate ornamentation. The jagging wheel was used for crimping pie crusts; they often also had ivory forks to puncture the top of the crust. This one was made aboard the New Bedford ship Europa, Captain James H. McKenzie, during 1871–1876. Length 18 cm. Kendall Collection, New Bedford Whaling Museum.*

napkin rings, and even some canes, could be carved or turned from a single piece of ivory or bone. Although they had a specific func-tion, corset busks (made of bone, walrus ivory, or baleen) were often elaborately engraved; even apart from being products of painstaking labor, as intimate undergarments they were not bestowed casually. Other implements were constructed from two or more pieces joined or hinged together—pie crimpers with rotating jagging wheels and fold-out forks (Fig. 3), canes with shafts of one material, pummels of another, and inlay of a third. The most elaborate forms were truly "built" and may properly be called architectural or architectonic. Swifts (yarn winders) have numerous moving parts, with metal pin-ions and ribbon fastenings (Fig. 4). Bird cages, a labor-intensive technical challenge, could run the gamut of Victorian complexity. Sewing boxes, ditty boxes, chests of drawers, lap secretaries, pocket-watch stands, mantle ornaments, and other composite constructions typically employed combinations of wood, ivory, and bone and may have, hinged lids, internal compartments, legs, finials, handles, draw-ers, drawer pulls, inlay, and all kinds of ornamentation.

The quintessential form of purely decorative scrimshaw is engraved ivory and bone, usually rendered in a single medium—a tooth or pair of teeth; a tusk or pair; or a plaque, strip, or section of sperm whale panbone (jawbone). Finished teeth were sometimes set into wooden, silver, or coin-silver mounts; plaques might be framed by the artist; engraved strips of baleen could become oval boxes. Alternately, teeth and tusks could be carved into stand-alone sculp-tural forms, such as human or animal figures, or could become the components of complex ship models.

There were no rules and few precedents governing the choice of subject matter for pictures on scrimshaw. The earliest work by the anonymous Adam engraver, J. S. King, Edward Burdett, and Frederick Myrick was almost exclusively devoted to ship portraiture and whaling scenes. Figures of Columbia, Liberty, and Britannia appeared by around 1830. The ensuing generation enlarged the vocabulary to include patriotic portraiture (notably of Washington

Figure 4  *Swift of sperm whale ivory and skeletal bone by Captain James M. Clark of Rochester, Massachusetts, circa 1835–1850. Made by a Yankee whaling captain, this exquisite piece typifies the best of the scrimshaw genre. It is inlaid with abalone shell and baleen, fastened with copper, tied with silk ribbons, fitted with two turnscrews in the form of clenched fists, and has a silver presentation plaque inscribed "R W Vose from Jas Clark." Height 40.7 cm. Swifts were a distinctly American form, used for winding the yarn employed in knitting and occasionally other domestic handicrafts and cottage industries. Kendall Collection, New Bedford Whaling Museum.*

Figure 5  *Family album wall hanging, New England, circa 1850. This unusual, elaborate construction features 12 tintype photographic portraits mounted in a triangular framework of walrus ivory and bone. The polychrome engraving on the walrus tusks is particularly interesting, as the woman-and-child portrait pair on the right is no doubt copied from a magazine fashion plate (in typical whalers' fashion), whereas the woman with-telescope on the left appears to be an original image. Height 50 cm. Kendall Collection, New Bedford Whaling Museum.*

and Lafayette), inanimate patriotic devices, female portraiture, landscape, naval engagements, sentimental family scenes, and mortuary motifs (Fig. 5). Gradually, these were canonized as standard genre conventions. Some individual artists developed distinctive styles and themes. George Hilliott's polychrome teeth dialectically juxtapose a Polynesian wahinee in a grass skirt on one side and a New England lady in a fashionable gown on the other. The anonymous "Banknote engraver" did meticulous portraits with banknote-like borders (hence the pseudonym). The "Eagle Artisan" engraved red-and-black American eagles and bold portraits. The "Lambeth Busk Engraver" made busks with London vignettes; a prime example features Lambeth Palace. Much naval scrimshaw is adorned with patriotic devices and naval engagements. Like whalemen's scrimshaw, some examples refer to specific vessels and events. A notable British example is credited to HMS Beagle on the same voyage on which Darwin evolved his theory of natural selection. Edward Yorke McCauley—later an admiral and noted Egyptologist—when he was a young midshipman aboard the U.S. Frigate Powhattan on Perry's

historic Japan expedition in the 1850s, engraved two walrus tusks with portraits of the Powhattan and Susquehanna, exotic Oriental watercraft, and glimpses of Japan, Hong Kong, and Brunei. Even a Confederate infantryman tried his hand: Hampton Wilson, Irish immigrant, North Carolina sharecropper, Confederate draftee, and Union prisoner of war, while recuperating in a military hospital in Kentucky successfully "flowered" a pair of walrus tusks with military and naval vignettes, using materials and methods presumably supplied by a Yankee whaling veteran among his fellow patients.

Most scrimshaw pictures were inspired by or adapted from illustrations in contemporaneous books and periodicals; copying and even direct tracing were standard scrimshaw conventions. Because of their specific functional objectives, scrimshaw implements and architectonic forms were also mostly derivative. However, some of the best pieces—and many of the worst—were truly original creations, drawn from the maker's experience or imagination. A few have authentic stature as significant art, whereas others are little more than mere valentines. In the aggregate, anonymity and quality aside, as an indigenous occupational genre scrimshaw comprises some of the most characteristic and revealing documents of any occupational group, capable of providing profound insights into the life, work, and intentionality of the mariners who made them.

### V. Museum Collections

The famous Kendall Collection (the former Kendall Whaling Museum), which includes what is by far the world's largest and most comprehensive array of scrimshaw, has since 2001 been incorporated into the New Bedford Whaling Museum (Old Dartmouth Historical Society) in New Bedford, Massachusetts. With the two enormous collections now combined, the New Bedford Whaling Museum is the home of the national Scrimshaw Collectors' Guild, houses the world's only Scrimshaw Forensics Laboratory® (to authenticate scrimshaw from institutional and private owners), hosts the annual Scrimshaw Collectors' Weekend, and has taken over the Kendall inventory of scrimshaw publications.

The Nantucket Whaling Museum (Nantucket Historical Association) on Nantucket—the birthplace of sperm whaling and the hometown of scrimshaw pioneers Edward Burdett and Frederick Myrick—holds an eminent collection that was newly installed in a rejuvenated gallery opened in 2006.

Mystic Seaport Museum, Mystic, Connecticut, has a large and comprehensive collection that includes important loan deposits from other private and institutional collection, notable for an informative catalogue and various scrimshaw-related publications.

The Peabody Essex Museum of Salem, Massachusetts, founded in 1799 as the East India Marine Society, was the first institution to include scrimshaw among its nautical relics (circa 1830) and today holds an outstanding collection of American scrimshaw.

The superb collection of the Dietrich American Foundation in Philadelphia is primarily intended for loan exhibitions to qualified institutions.

The Hull Maritime Museum (formerly the Town Docks Museum) in Kingston on Hull, East Yorkshire, a municipal museum in one of England's most historic Arctic whaling ports, holds the most significant scrimshaw collection outside the USA and Australia.

In addition, there are modest but worthwhile collections at the Christensen Whaling Museum (Sandefjord, Norway), the Penobscot Marine Museum (Searsport, Maine), the Providence (Rhode Island) Public Library, the Scott Polar Research Institute (University of Cambridge, England), South Street Seaport Museum (New York City), Whaler's Village (Lahaina, Maui, Hawaii), the Whaling Museum at Cold Spring Harbor (New York), and several state, maritime and municipal museums and libraries in Australia (Sydney, Melbourne, Hobart, and Launceston, Tasmania).

## *See Also the Following Articles*

Folklore and Legends ■ Museums and Collections ■ Popular Culture and Literature ■ Whaling ■ Traditional

## *References*

Basseches, J., and Frank, S. M. (1991). "Edward Burdett (1805–1833): America's First Master Scrimshaw Artist." Kendall Whaling Museum Monograph No. 5, Sharon, MA.

Carpenter, C. H., Jr., and Carpenter, M. G. (1987). "The Decorative Arts and Crafts of Nantucket." Dodd Mead and Co, NY.

Flayderman, E. N. (1972). *In* "Scrimshaw and Scrimshanders, Whales and Whaleman" (R. L. Wilson, ed.), pp. 24–25. N. Flayderman and Co, New Milford, CT.

Frank, S. M. (1991). "Dictionary of Scrimshaw Artists." Mystic Seaport Museum, Mystic, CT.

Frank, S. M. (1992). The origins of engraved pictorial scrimshaw. *Mag. Antiq.* **142**(4), 510–521.

Frank, S. M. (1998). "More Scrimshaw Artists." Mystic Seaport Museum, Mystic, CT.

Frank, S. M. (2000). Scrimshaw: occupational art of the whale-hunters. *Marit. Life Trad.* **7**(March), 42–57.

Frank, S. M. (2006). The whalemen's life. *Ear. Am. Life* **37**, 8–19.

Hellman, N., and Brouwer, N. (1992). "A Mariner's Fancy: The Whaleman's Art of Scrimshaw." South Street Seaport in association with Balsam Press, New York and University of Washington Press, Seattle.

Malley, R. C. (1983). "Graven by the fishermen themselves: Scrimshaw in Mystic Seaport Museum." Mystic Seaport Museum, Mystic, CT.

McManus, M. (1997). "A Treasury of American Scrimshaw: A Collection of the Useful and Decorative." Penguin Books, NY.

Penniman, T. K. [1952] (1984). "Pictures of Ivory and other Animal Teeth, Bone and Antler; With a Brief Commentary on their Use in Identification." Pitt Rivers Museum, Occasional Paper on Technology No. 5, University of Oxford.

Ridley, D. E., *et al.* (2000). "The Scrimshaw of Frederick Myrick (1808–1862): A Catalogue Raisonne and Forensic Survey." The Kendall Whaling Museum, Sharon, MA.

West, J., and Credland, A. G. (1995). "Scrimshaw: The Art of the Whalers." Hull City Museums and Art Galleries and Hutton Press.

# Sea Lions: Overview

## DARYL J. BONESS

Sea lions, like the fur seals, are members of the family Otariidae. There are presently seven sea lion species in five genera, with one genus exclusive to the Northern Hemisphere (Steller sea lion, *Eumetopias jubatus*), one that occurs in both hemispheres [in the north, the California (*Zalophus californianus*) and Japanese (*Z. japonicus*) sea lions, and in the south, the Galapagos sea lion (*Z. wollebaeki*)] and three that are solely in the Southern Hemisphere (southern sea lion, *Otaria flavescens*; Australian sea lion, *Neophoca cinerea*; New Zealand sea lion, *Phocarctos hookeri*).

### I. Origins, Classification, and Size

Sea lions originated in the Northeast Pacific region, sharing a common ancestor with fur seals. Although the fossil record for sea lions is poor, it appears they crossed into the Southern Hemisphere about three million years ago (Berta and Sumich, 1999). For many years, sea lions were thought to be a separate subfamily, the Otariinae, within the family Otariidae. However, genetic analyses now conclusively show a basal split between the Phocidae and the Otaroidea (Otariidae and Odobenidae families) and that the northern fur seal (*Callorhinus ursinus*) is sister to all other sea lions and fur seals (Arnason *et al.*, 2006; Wynen *et al.*, 2001). The fossil record suggests this group splitting off about 6 million years ago. Beyond this level of detail, genetics and fossil record can only show us that there are four sea lion groups (known as clades) and five fur seal ones that evolved rapidly in time. The relationship among these groups, however, is yet to be determined.

The only substantial diagnostic morphological distinction between sea lions and fur seals is the presence of an underhair in fur seals but not in sea lions. Sea lions do tend to be larger than fur seals, with both groups exhibiting substantial differences in body mass, and smaller differences in body length, between males and females, a phenomenon known as sexual dimorphism. Male sea lions are between two and four times heavier than females and up to one and a half times the length. The body mass of males in the different sea lion species ranges from about 250 to 1000 kg and in females from about 75 to 325 kg. In contrast, the heaviest fur seal male is about 300 kg and the heaviest female is about 75 kg. Lengths of male and female sea lions range from 205 to 330 and 180 to 270 cm, respectively.

S

## II. Morphology and Physiology

Sea lions, like fur seals and walruses (*Odobenus rosmarus*), differ anatomically from the true seals (phocids) in several ways. Probably most notable is their ability to walk or run rather than crawl on land (Pabst *et al.*, 1999). Underlying this capability is the ability to rotate the pelvis to a position that allows bringing the hind flippers under the body. As a result, sea lions tend to have more efficient terrestrial locomotion than phocids. Another obvious anatomical feature of sea lions is the extended and flattened fore flippers. This is a feature they have in common with their fur seal cousins, whereby the walrus and phocids have relatively short fore flippers. These differences reflect the different swimming modes of the two groups of seals. The sea lions and fur seals use fore flippers to provide thrusting power while the walrus and phocids use their rear flippers. While one might expect the different swimming styles to yield different swimming speeds, this is not the case. There is marked similarity in the cruising swimming speeds of all seals studied, which is about 2 m/s.

The ability to dive and stay underwater is an important characteristic for all marine mammals, including sea lions. Although large lung volume is important for diving animals, the lung volume for sea lions is equally proportional to their body size as to other marine mammals and even terrestrial mammals. Thus the relatively greater diving capability of sea lions and other marine mammals over terrestrial mammals is controlled by factors more than lung volume. Mechanisms for increased oxygen stores play an important role (Williams and Worthy, 2002). Consistent with being shallow divers, sea lions have lower oxygen stores (40 ml $O_2$/kg) than true seals (60 ml $O_2$/kg), which are generally deep divers, but still much higher stores than humans (20 ml $O_2$/kg), for example. In addition, the relative distribution of oxygen stores is different for sea lions. Sea lions have about 47% of their oxygen in blood, 35% in muscle, and 19% in their lungs. Phocids, however, have 64% of their oxygen in blood, 31% in muscle, and only 5% in their lungs. This larger percentage of oxygen in the lungs of sea lions correlates with the smaller degree to which the lungs collapse from water pressure. In humans, 51% of the oxygen is stored in the lungs.

## III. Life History and Reproduction

Sea lions follow a life style typical to that of all the otariids, with some characteristics common to all seals (Boness *et al.*, 2002). They are long-lived (probably 15–20 years), have delayed sexual maturation, and have physical and social sexual bimaturation, with males maturing more slowly than females. In all species of sea lions, most females give birth for the first time between 4–6 years of age. This may vary to some extent with ecological conditions and population status. Males typically become capable of impregnating a female around the age of 5–8 years of age, but socially cannot achieve a status that gives them access to females for copulation until they are several years older (probably 10–12 years old). For six of the seven sea lion species, there is an annual breeding cycle, but one species, the Australian sea lion, has a unique cycle of just less than 18 months (Gales *et al.*, 1994). The net result of this cycle is that there is a gradual shift in the time of year and season when the breeding period occurs. For example, over a 19-year period, between 1973 and 1991, the median date of pupping occurred in every month of the year. No other species of seal exhibits such a pattern. Why this pattern exists is unclear, but it does link to a lactation length of about 17 months in this species. This may be related to having to forage in marginal ecological conditions that place energetic constraints on females and their pups (Costa and Gales, 2003).

Sea lions generally choose sandy beaches as sites for breeding. This is in contrast to their fur seal cousins, which more often choose beaches with boulders or rocky shelves. The reasons for these differences are not clear, but may be related to methods of behavioral thermoregulation.

The reproductive behavior of female sea lions follows the typical maternal foraging strategy seen in all other otariids (Boness and Bowen, 1996). Almost without exception females give birth to a single pup each breeding season. During what has been termed a perinatal period lasting from about 7–10 days immediately after parturition, the female fasts, remaining with her pup continuously and nursing it frequently. This period of uninterrupted maternal energy transfer allows the pup to rapidly store energy and begin growth. Also during the perinatal period, both female and pup call to each other and sniff one another frequently. This behavior appears to be the basis for establishing a strong bond, which includes mutual vocal and olfactory recognition. This recognition becomes a critical component later in lactation that allows mothers and pups to reunite when the mother comes back from a foraging trip. Without this recognition system in place, females would likely spend critical amounts of milk energy wasted on pups other than their own, potentially compromising the successful weaning of their pup. Because females have been fasting during the perinatal period they will have depleted stores of body fat and need to begin periodic foraging trips to sea. During the foraging trips they will be left behind because they are not adequately developed to make the trips with their mothers. In some species, just before beginning the periodic foraging trips females will come into estrus. However, in other species estrus will occur after foraging trips have begun.

The duration of foraging trips is variable both within and between species (ranging from about 0.5 to 5 days), although they tend to be shorter among the sea lions than among the fur seals (1–12 days). Between foraging trips, females return to their pups on land, nursing them every few hours over a period of 0.5–1.5 days. This cycle is continued throughout lactation, which for most females in each species lasts about a year, except for the Australian sea lion mentioned earlier, which has a 17-month lactation. A small proportion of females in several sea lion species will nurse a pup for longer than a year, and even after they have given birth to a new pup. The number of females that do this may be dependent on the availability of food resources. For at least some sea lion species, the increasing duration of foraging trips over the season appears to be due more to prey movement than the increasing energy demands of growing pups, though both play a role (Boness and Bowen, 1996).

A physiological component of the maternal strategy of sea lions is relatively high-fat milk, which provides the energy needed by the pups as they try to grow and develop during the "feast and famine" situation produced by maternal foraging. We do not have measures of milk fat for all sea lion species, but for those that have been studied the fat content of maternal milk ranges from about 15 to 45%. Most likely the varying levels of fat in milks relate to the typical length of foraging trips. The best example of this is seen in *Zalophus* spp. The California species, which has maternal foraging trips of upward of 3–4 days, produces milk with 43% fat, whereas the Galapagos species, in which females forage for less than a day at a time before returning to pups, has milk containing only 21% fat. Interestingly, the daily growth rates of sea lion pups, after taking the body size of adults into account, are very similar, suggesting that the maternal strategies are finely tuned to ecological conditions.

The reproductive behavior of male sea lions has been investigated unevenly among the various species. We know almost nothing about the New Zealand, and Japanese sea lions but a considerable amount about the California, Galapagos, Steller and southern sea lions. Because females gather on land to give birth and care for their young and estrus is temporally linked to parturition, the conditions

S

are ideal for strong sexual selection through male–male competition. In brief, the tendency for female sea lions to be highly clustered, indeed lying in contact with one another, provides the potential for males to compete for and maintain control over multiple females. The ability to control and mate with multiple females in a given reproductive period is known as polygyny.

As is typical of virtually all otariids, male sea lions return to traditional breeding grounds and vie for positions in areas where females have previously given birth or spent time cooling off during the hottest part of the day. In some species or populations, males may actually defend sites or territories, whereas in others they may be more flexible, defending females directly. Factors that are most important in determining which behavior is typical at a colony are the extent to which females move before they become receptive and the level of competition that exists among males. Female movement is most often associated with the need to cool off because of high ambient temperature. One species for which all studies have shown males to defend territories only is the Steller sea lion. This is likely a result of the high latitude at which this species breeds, eliminating the need for females to engage in thermoregulatory movements.

In contrast, the southern and California sea lions have been shown to behave variably depending on the breeding habitat. At sites where there tend to be large numbers of tidal pools, around which females cluster, males defend territories. However, where there are long narrow sand or pebble beaches, females shift up and down the beach with changes in the state of the tide and air temperatures. Under these conditions, males do not defend territories but shift as females do and defend the females directly.

The level of reproductive success, or number of females mated by the most successful males, may be similar regardless of which pattern of behavior is typical. What seems to constrain the maximum success is the degree to which females are clustered in space and time (Boness, 1991). If receptive females are too dispersed in time, an individual male may not have enough energy stores to remain competitive throughout the entire season. As food sources are usually not close to the breeding grounds, males must fast during the breeding season; rarely are individual males seen leaving their positions on land. This is true even during the hottest part of the day in many breeding colonies, although in some places nearer the equator some sea lion males may either occupy positions in the water or make movements to the water. Minimum estimates of the maximum mating success for the most successful male among the sea lion species are highly variable. The estimate for the most successful Australian sea lion male, a species in which females tend to be quite dispersed, is 7 females mated compared to between 30 and 50 females for California and New Zealand sea lions, species in which females are much more clustered. Recent genetic paternity studies of male success in several species of seals indicate that success levels based on behavioral observations are probably overestimates, however. Some females are fertilized by males other than those with which they are seen mating.

The intense competition among males is what produces the extreme sexual dimorphism we see in sea lions and many other seals (Boness, 1991). At this point it is unclear as to whether the large size of male sea lions is most important in direct competition, i.e., fights and threats with one another, or in the ability to remain ashore for longer periods of time because larger males can store more body fat. In some energetic studies of phocid seals, evidence suggests that it is the amount of energy stores that is more important.

## IV. Feeding Habits

Our understanding of sea lion foraging ecology is much poorer than that of fur seals. What we know comes from both analysis of diving behavior and diet studies based mainly on analysis of food remains in scats (i.e., feces). Most sea lions are relatively shallow divers, diving to much less than 100 m. New Zealand sea lions, however, dive to as deep as 130 m and spend over 50% of their time at sea submerged. All seven species of sea lions eat both fish and cephalopods (e.g., squid and octopus). For most species fish appears to be the predominant food item, but in contrast the Australian sea lion appears to feed more on cephalopods. Special dietary items are found occasionally in the stomachs of some species. These include penguins in the southern, Australian and Galapagos sea lions, all of which live in proximity to penguin populations, and fur seal pups in the two larger sea lions—Steller and southern sea lions. Diets are not necessarily constant. They may change seasonally and during periods of environmental events, such as El Niño and La Niña, which are unusually high warming or cooling of sea surface temperatures, respectively.

## V. Population Status

The status of sea lion populations is variable. The Japanese sea lion has not been sighted since the 1970s and is now considered extinct. According to the International Union for Conservation of Nature (IUCN), three species—the California, Australian and southern sea lions—are currently considered as being at low risk of extinction. The California sea lion population is likely between 250,000 and 300,000, and the Australian sea lion population, while small, appears to be stable. The southern sea lion is a species that may require reassessment since its last one in 1996 because recent surveys of this species at the Falkland Islands and in Argentina suggested population reductions of 93% since its highest levels in the late 1930s. Some of this decline may be due to a severe El Niño in 1997–1998.

The Galapagos and New Zealand sea lions are considered vulnerable to extinction. In the case of the Galapagos sea lion, the major threat continues to be a level of uncontrollable exploitation, whereas in the case of the New Zealand sea lion, although there is currently an increasing trend in the population level, its restricted habitat use and small overall population could lead to rapid depletion under a number of circumstances.

The sea lion species for which there is greatest concern at present and is listed as endangered and at high risk of extinction by the IUCN is the Steller sea lion. Although it is not the smallest population by far (estimated at about 96,000), it declined by about 80% between the 1970s and 1990s. There are two stocks that are managed separately in the United States—the western and eastern stocks. Under the U.S. Endangered Species Act the western stock is considered endangered and the eastern stock threatened. The western stock reached its low point in 2000 and now appears to stable or may even slightly increasing. The eastern stock is increasing overall, but still decreasing slightly at some colonies. Despite substantial research over the past decade, we still are not able to determine conclusively the causes(s) of the decline. One recent hypothesis is that with the historical decline of many large whale species by human hunting killer whales have been driven to shift their primary prey from large whales to other marine mammals, e.g., sea otters (*Enhydra lutris*), harbor seals (*Phoca vitulina*),and sea lions (Williams *et al.*, 2004). Circumstantial evidence based on energetics data, for example, supports the plausibility of this hypothesis, but much more evidence is needed to be confident in the validity of this hypothesis.

## See Also the Following Articles

Eared Seals (Otariidae) ■ Pinnipedia, Overview ■ Pinniped Ecology ■ Pinniped Life History

S

## References

Arnason, U., Gullberg, A., Janke, A., Kullberg, M., Lehman, N., Petrov, E. A., and Vainola, R. (2006). Pinniped phylogeny and a new hypothesis for their origin and dispersal. *Mol. Phylogenet. Evol.* **41**, 345–354.

Berta, A., and Sumich, J. L. (1999). "Marine Mammals." Academic Press, San Diego.

Bininda-Edmonds, O. R. P., Gittleman, J. L., and Purvis, A. (1999). Building large trees by combining phylogenetic information: a complete phylogeny of the extant Carnivora (Mammalia). *Biol. Rev.* **74**, 143–175.

Boness, D. J. (1991). Determinants of otariid mating systems. *In* "Behaviour of Pinnipeds" (D. Renouf, ed.), pp. 1–44. Chapman and Hall, London.

Boness, D. J., and Bowen, W. D. (1996). The evolution of maternal care in pinnipeds. *Bioscience* **46**, 646–654.

Boness, D. J., Clapham, P. J., and Mesnick, S. L. (2002). Life history and reproductive strategies. *In* "Marine Mammal Biology: An Evolutionary Approach" (R. Hoelzel, ed.), pp. 278–324. Blackwell Science, Oxford.

Boyd, I. L. (2002). Energetics consequences for fitness. *In* "Marine Mammal Biology: An Evolutionary Approach" (R. Hoelzel, ed.), pp. 247–378. Blackwell Science, Oxford.

Costa, D. P., and Gales, N. J. (2003). Energetics of a benthic diver: Seasonal foraging ecology of the Australian sea lion, *Neophoca cinerea*. *Ecol. Monogr.* **73**, 27–43.

Ferguson, S. H., and Higdon, J. W. (2006). How seals divide up the world: environment, life history and conservation. *Oecologia* **2**, 318–329.

Gales, N. J., Shaughnessy, P. D., and Dennis, T. E. (1994). Distribution, abundance and breeding cycle of the Australian sea lion, *Neophoca cinerea*; (Mammalia, Pinnipedia). *J. Zool. Lond.* **234**, 353–370.

Pabst, D. A., Rommel, S. A., and McLellan, W. A. (1999). The functional morphology of marine mammals. *In* "Biology of Marine Mammals" (J. E. Reynolds, III, and S. A. Rommel, eds), pp. 15–72. Smithsonian Institution Press, Washington, D.C.

Williams, T. M., Estes, J. A., Doak, D. F., and Springer, A. (2004). Killer appetites: assessing the role of predators on ecological communities. *Ecology* **85**, 3373–3384.

Williams, T. M., and Worthy, G. A. J. (2002). Life history and reproductive strategies. *In* "Marine Mammal Biology: An Evolutionary Approach" (R. Hoelzel, ed.), pp. 73–97. Blackwell Science, Oxford.

Wynen, L. P., *et al.* (11 authors) (2001). Phylogenetic relationships within the eared seals (Otariidae: Carnivora): Implications for the historical biogeography of the family. *Mol. Phylogenet. Evol.* **21**, 270–284.

# Sei Whale
## *Balaenoptera borealis*

### JOSEPH HORWOOD

### I. Characteristics and Taxonomy

The diversity of thought about our great whales is characterized by quotes from the biologist R. C. Haldane on the sei whale. He described the species as the "most graceful of all the whales, as its proportions are so perfect, and wanting the clumsy strength of the two larger *Balaenoptera*, sperms and *Megaptera*." He added, "It is also far the best to eat, the flesh tasting of something between pork and veal and quite tender." The name "sei" whale comes from the Norwegian "*sejhval*," as it would arrive off Norway at the same time as the "*seje*" or saithe (*Pollachius virens*). There are a variety of other common names, but English forms have disappeared from usage in favor of the sei whale.

The sei whale is a typical, sleek rorqual (Fig. 1). It is the third largest whale, reaching a maximum length of almost 20 m. More typically it is 15 m, weighing 20 metric tons (Horwood, 1987). Identification at sea can be difficult. By size alone, adult blue and fin whales are obviously larger and minke whales smaller. The dorsal fin is a useful cue, being relatively taller than that of blue and fin whales. It is also strongly concave on its dorsal edge, similar to that of a minke whale. For a long time it was not distinguished from its close relative, the warm-water Bryde's whales (*B. edeni/brydei*). The Bryde's whales have three distinct ridges, running the length of the head, whereas the sei whale has only one. Nevertheless, it is easy to confuse the sei whale with these other species at sea.

The color helps in identification. It is dark gray dorsally and on the ventral surfaces of the flukes and flippers. There is no whitening of the lower lip as found in fin whales. However, in a few individuals some white baleen plates occur. Often the body can be heavily scarred with healed lamprey bites. Sei whales dive more by sinking than an arched dive, but again other rorquals can also dive in this quiet manner.

A more detailed external inspection, seldom possible at sea, allows a more definite identification. In sei and minke whales the ventral grooves end well before the umbilicus. In other *Balaenoptera* spp., including Bryde's whales, they end at, or posterior to, the umbilicus. The number of ventral grooves varies considerably from whale to whale. In sei whales they vary between 40 and 65 with a mean number of about 50. This is less than in blue (*B. musculus*), fin (*B. physalus*), and minke (*B. acutorostrata* and *B. bonaerensis*) whales, but about the same as in Bryde's whales.

The baleen of sei whales is a dark gray, but often with a yellowing-brown hue, and often with some anterior white plates. The plates number about 350 on each side of the jaw, and the largest is less than 80 cm long. The width of the plate is relatively narrow compared to blue, fin, and Bryde's whales. In the sei whale the length-to-breadth ratio is typically over 2.2, whereas in the Bryde's whale it is always less than 2.2. The bristles of the baleen are particularly fine in sei whales. At their base they are about 0.1 mm in diameter compared with 0.3 mm for the other rorquals.

The sei whales are separated from the other rorquals by the external physical characteristics described above but also by internal characters such as vertebral number, skull and hyoid morphometry, and a bicuspid first rib. Genetically, the rorquals are all very similar having a diploid number of 44 chromosomes. Mitochondrial DNA shows the sei whale is more closely related to the Bryde's whales, and the newly described *B. omurai*, than to the blue, fin and minke whales (Wada *et al.*, 1991, 2003). Genetic differences exist between northern and southern hemisphere forms, and the southern sei whales grow to a larger size.

### II. Distribution and Abundance

The sei whale can be found in all ocean basins. It is an oceanic form and is uncommon in shelf seas. Sei whales undertake extensive, seasonal, latitudinal migrations, spending the summer months feeding in the subpolar higher latitudes and returning to the lower latitudes to calve in winter. Fig. 2 shows the global distribution in summer and winter. In the Southern Hemisphere, they rarely penetrate as far south as blue, fin, and minke whales, with summer concentrations mainly between the sub-tropical and Antarctic convergences.

**S**

**Figure 1**   *The sei whale (C. Brett Jarrett).*

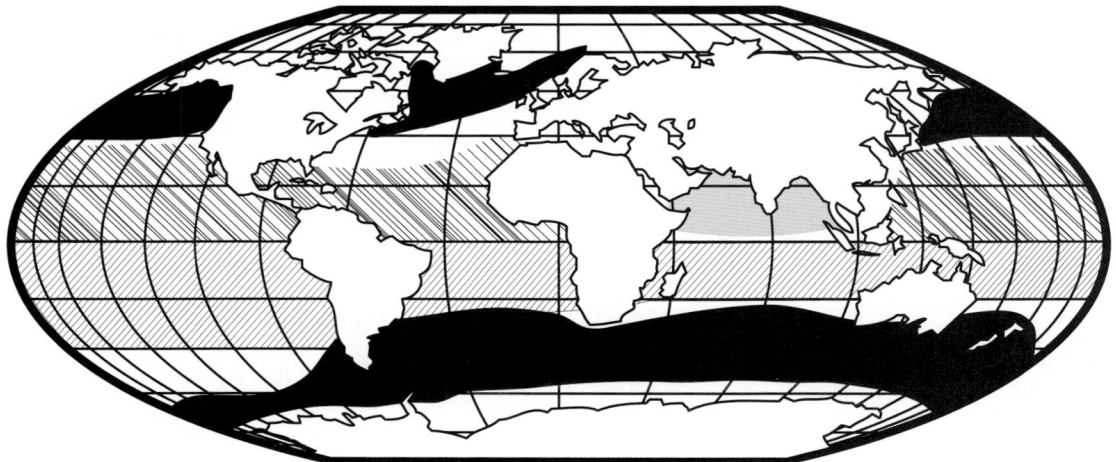

**Figure 2**   *Global distribution of sei whales. Filled areas are the summer feeding distributions, and hatched areas represent breeding areas. The distribution is undetermined in the Indian Ocean.*

Genetic studies have to date not clearly defined different populations of sei whales within the Northern or Southern Hemispheres (Kanda *et al.*, 2006). However, biologists have separated populations, for management purposes, on the basis of different migrations and biological characteristics. In the Southern Hemisphere, six populations are assumed, as for the other rorquals. In the North Pacific, two or three populations have been proposed. In the North Atlantic, as many as eight populations have been suggested, but only three are considered for management purposes.

The size of populations is poorly determined, but whaling significantly depleted populations in all areas. In the Southern Hemisphere the original population was about 100,000, and in 1980 was thought to be 24,000. In the North Pacific, a population of over 60,000 was reduced to about 15,000. By now there may be 70,000 in both areas. The status of the North Atlantic sei whale is more uncertain, but recent sighting surveys indicate about 10,000 sei whales in the central and northeastern North Atlantic. At least in the western North Pacific it does not appear as if the depletion of the population has resulted in any genetic "bottleneck" effects.

## III. Ecology

As with the other rorquals, it is the feeding ecology that determines the great latitudinal migrations between summer, polar feeding areas and winter breeding areas. It is also differences in prey that determine some of the unique features of the sei whale.

The sei whale is grouped with the gray (*Eschrichtius robustus*), and possibly minke, whales as "skimmers and swallowers," in contrast to the right whale (*Eubalaena* spp.) "skimmers" and other rorqual

**Figure 3**   *A feeding sei whale (courtesy of J. Sigurjónsson).*

"swallowers." The fine baleen structure of the sei whales allows them to skim the surface waters for patches of their preferred copepod prey (Fig. 3). The other rorquals would probably find copepod food densities too low. Sei whales also feed on euphausiids, shoals of fish (hence their name) and squid if they are encountered (Kawamura, 1980).

Feeding is predominantly at dawn, lessening in frequency during the day. The behavior appears less bimodal than in other rorquals and is associated with the vertical migrations of the prey species. In the summer about half of the population is feeding on any one day, and stomach contents are about 150 kg. In winter, there is still a high incidence of feeding but consumption is low.

It has been speculated that the wide feeding niche allowed the southern sei whales to expand into the areas occupied by the depleted fin and blue whales; however, it is not possible to substantiate this hypothesis.

## IV. Behavior and Physiology

The key behavior is linked to the annual cycle of migration between the tropical and sub-tropical breeding grounds and the subpolar feeding grounds. The timing of migration differs by sexual status and age. As with the other rorquals, the pregnant females leave for the feeding grounds first, and adults precede immature whales. Migration to the feeding grounds tends to be later than in blue and fin whales (Anon., 1997).

School sizes are poorly determined and vary with the location and migration. In the warmer waters they are most reported in mixed schools of 2–5 in number. In the temperate waters, during migration, they are often solitary. On the feeding grounds they can be solitary or form large aggregations of 20–100.

## V. Life History

Maturity is at about 10 years in both sexes (Masaki, 1976). In most seas, the age of maturity declined by 2–3 years after the populations were depleted by whaling. The peak of conceptions is in June in the Southern Hemisphere and December in the Northern Hemisphere. The young are carried for almost a year and are born at a size of 4.5 m. Most calves are weaned in 7 months, after they have migrated to colder waters with their mothers.

Sei whales reach larger maximum sizes in the Southern than in the Northern Hemisphere. In the south, males mature at about 13–14 m and females at 14 m.

As for most mammals, sei whales can be expected to have increased rates of mortality when very young or old, but actual rates are poorly known. From observations of age compositions, the rate of natural mortality is typically about 5–10% per year. They die naturally from predators, such as killer whales, and weakening from disease and parasites. The population biology of the sei whale is described in detail by Horwood (1987).

## VI. Interactions with Humans

A fast, offshore rorqual, the sei whale was not exploited until the era of modern whaling at the end of the 1800s. Off north Norway, 4000 sei whales were killed between 1885 and 1900. Since then, sei whales were caught in the North Atlantic from land stations in Canada, Faeroes, Iceland, Ireland, Iberia, Norway, and Scotland. In the North Pacific, they were caught from California, Canada, Japan, Kamchatka, and Kuril and by pelagic fleets. In the Southern Hemisphere, they were caught from Brazil, Chile, Peru, South Africa, and South Georgia.

The largest catches were made by the Antarctic pelagic fleets, after the numbers of blue and fin whales had been reduced. Between 1960 and 1970 over 110,000 sei whales were killed.

Whaling is regulated by the IWC. Whaling for sei whales ceased in the North Pacific in 1975 and in the Southern Hemisphere in 1979. Whaling was prohibited in the North Atlantic from 1986. Nevertheless, limited catches continued through subsistence whaling from Greenland and under the IWC scientific special permit. Special permit catches were taken off Iceland, and more recently in the offshore North Pacific.

## See Also the Following Articles

Blue Whale ■ Bryde's Whales ■ Classification ■ Fin Whales ■ Genetics ■ Minke Whales ■ Whaling, Modern

## References

Anonymous (1997). Report of the Special Meeting of the Scientific Committee of the Sei and Bryde's Whales. *Rep. Int. Whal. Commn (Special issue)* **1**, 1–150.
Horwood, J. (1987). "The Sei Whale: Population Biology, Ecology and Management." Croom Helm, London.
Kanda, N., Goto, M., and Pastene, L. A. (2006). Genetic characteristics of western North Pacific sei whales, *Balaenoptera borealis*, as revealed by microsatellites. *Mar. Biotechnol.* **8**, 86–93.
Kawamura, A. (1980). A review of food of balaenopterid whales. *Sci. Rep. Whales Res. Inst. Tokyo* **32**, 155–158.
Masaki, Y. (1976). Biological studies on the North Pacific sei whale. *Bull. Far Sea Fish. Res. Lab.* **14**, 1–104.
Wada, S., and Numachi, K. (1991). Allozyme analyses of genetic differentiation among the populations and species of the Balaenoptera. *Rep. Int. Whaling Commn.* **13**(Special Issue), 125–154.
Wada, S., Oishi, M., and Yamada, T. (2003). A newly discovered species of living baleen whale. *Nature* **426**, 278–281.

# Sensory Biology: Overview

## J.G.M. THEWISSEN

This chapter provides a brief overview of the sense organs of cetaceans, sirenians, and pinnipeds. vision, hearing, and balance are discussed separately in this book and their treatment here is extremely cursory. Instead, this chapter focuses on those sense organs that are not treated elsewhere, namely the chemical senses, haptosense (the sense of touch, also called mechanosense), and electrosense.

## I. Chemical Senses

Chemical cues in water are registered by sense organs (taste buds) that are located on the tongue and in the throat. These stimuli are mediated by Cranial Nerves VII, IX, and X in mammals and this sense is commonly referred to as taste (gustation). The sense of smell (olfaction) is sensitive to airborne chemicals which dissolve in the mucous inside the nasal cavity. These stimuli stimulate Cranial Nerve I. A third chemical sense (absent in humans) is processed by the vomeronasal organ (Organ of Jacobson). The vomeronasal organ is located in the floor of the nasal cavity, but its ducts open just behind the incisors in the oral cavity, and the organ registers stimuli on the palate. The vomeronasal organ is sensitive to large molecules that are caught on the mucous membrane of the palate, and those stimuli are carried to the brain by Cranial Nerve I. Also located in the mouth, but technically not part of gustation are the sensory organs on the tongue that register texture, structure, and temperature of food; these are mediated by Cranial Nerve V, and the process is called chemesthesis.

Pihlström (2008) presented a review of the chemical senses in aquatic mammals. Cetacean tongues are simpler in structure than those of land mammals, with few or no taste buds (Sonntag, 1922).

In spite of this, experimental evidence has shown that, in some cetaceans, thresholds for detecting certain chemicals in water are similar to those of humans (Nachtigall and Hall, 1984). Modern cetaceans lack the vomeronasal organ, and in odontocetes the olfactory bulbs (the parts of the brain that receive olfactory stimuli from the nose) are missing. In mysticetes the olfactory bulbs are present and it is anatomically possible that they have a sense of smell. The presence of a sense of smell in mysticetes could be useful to detect their prey, planktonic invertebrates, which give off a peculiar smell. Experiments show that krill-eating seabirds have olfactory capabilities greater than those of most birds and are attracted by such smells (Verheyden and Jouventin, 1994).

Sirenian tongues have more taste buds than cetacean tongues, but the presence of a vomeronasal organ is disputed. The olfactory bulbs are small but present, suggesting that sirenians have a sense of smell. Pinniped tongue anatomy indicates that they have a sense of taste, but gustation is probably less well developed than in land carnivores. Pinnipeds have a good sense of smell (Kowalewsky *et al.*, 2005) and retain a vomeronasal organ.

## II. Vision

Cetaceans and pinnipeds have well-developed vision, and both groups display adaptations for vision in air and underwater. This is unlike sirenians, which have small eyes used exclusively underwater, and have poor vision (Kröger and Katzir, 2008). Cetaceans living in turbid water, such as river dolphins have reduced vision (Lipotidae) or are nearly blind (Platanistidae). Frequency specializations occur in pinnipeds. In deep-water diving species, such as the elephant seal (*Mirounga spp.*), optimum frequencies of retinal pigments are shifted to the blue-green (high frequency) end of the spectrum, since higher frequency light dominates in deeper waters (Reuter and Peichl, 2008).

## III. Hearing

Hearing in marine mammals was reviewed by Wartzok and Ketten (1999) and Nummela (2008). In odontocete cetaceans, HEARING is the most important sensory modality and is an integral part of ECHOLOCATION. Odontocete sound transmission mechanisms differ greatly from those of land mammals, and are also different from those in mysticetes. Mysticete hearing is poorly understood. Pinnipeds hear well above and underwater, and some pinnipeds (odobenids and phocids) have highly specialized middle ears. Hearing in sirenians is poor.

## IV. Balance

The sense of BALANCE is different from other sense organs in that the stimulus (linear and angular accelerations and decelerations of the head) is basically the same in water and air. However, the organs of balance in cetaceans, and to some extent sirenians and pinnipeds, are very different from those of their land relatives (Spoor and Thewissen, 2008). This may be related to specific means of locomotion or mobility of the neck in these species.

## V. Haptosense

Whereas the other sense organs are concentrated on the head of an animal, haptosense (mechanosense, sense of touch) is diffuse, spread out over much of the body's surface. Unlike the other sensory receptors in the skin (temperature, pain), haptosense in water differs significantly from that in air because of the density of water, which allows animals living in water to detect disturbances in their medium (water currents) at some distance. Haptosense in aquatic mammals was reviewed by Dehnhardt and Mauck (2008).

Very little is known about haptosense in cetaceans, although it is likely that in those species that have bristles on their faces (such as the Amazon River dolphin, *Inia geoffrensis*, and the bowhead whale, *Balaena mysticetus*) deflection of these hairs serves to inform these animals about close-distance disturbances of the water.

In pinnipeds, haptosense is highly developed. Whereas in terrestrial mammals vibrissae (whiskers) are circular in cross section and have a straight shaft, whiskers of pinnipeds are oval in cross section, and, in phocid seals, the shaft of each whisker thickens and thins in an undulating pattern. These hairs are anchored in a vibrissal follicle sinus complex, where bending of the hair shaft causes bowing of its root, and can be registered by an array of sensitive nerve endings. The resulting system can register even weak disturbances in the medium around them. Behavioral experiments have shown that walruses (*O. rosmarus*), California sea lions (*Zalophus californianus*), and harbor seals (*Phoca vitulina*) are capable of identifying objects by touching them with their vibrissae (Dehnhardt and Mauck, 2008). Harbor seals that have been blind-folded can detect the vortices of swimming fishes, and may be able to use this sensory modality to search for and capture prey. This is consistent with anecdotes about blind pinnipeds that appear healthy and well fed. In sirenians, numerous vibrissae are used for haptoreception, and can even bring food items to the mouth.

## VI. Electrosense

Just like other delphinids, the Tucuxi (*Sotalia fluviatilis*) has a row of hair follicles on either side of the snout below the eye. In delphinid embryos, these follicles have hairs, but no hairs occur in adults. The follicles of *Sotalia* are larger than those of other delphinids and are sensitive to weak electric currents (Wilkens and Hofmann, 2008) such as those that would be emitted by the muscles of prey fish buried in the sediment. It is unknown from behavioral work in nature whether they are actually able to use this sensory modality, but it is very likely that they do.

## *See Also the Following Articles*

Balance ▪ Brain ▪ Echolocation ▪ Hearing ▪ Vision

## *References*

Dehnhardt, G., and Mauck, B. (2008). Mechanoreception in secondarily aquatic vertebrates. *In* "Sensory Evolution on the Threshold, Adaptations in Secondarily Aquatic Vertebrates" (J. G. M. Thewissen, and S. Nummela, eds), pp. 295–314. University of California Press, Berkeley.

Kowalewsky, S., Dambach, M., Mauck, B., and Dehnhardt, G. (2005). High olfactory sensitivity for dimethyl sulfide in harbor seals. *Biol. Lett.* **2**, 106–109.

Kröger, R. H. H., and Katzir, G. (2008). Comparative anatomy and physiology of vision in aquatic tetrapods. *In* "Sensory Evolution on the Threshold. Adaptations in Secondarily Aquatic Vertebrates" (J. G. M. Thewissen, and S. Nummela, eds), pp. 121–147. University of California Press, Berkeley.

Nachtigall, P. E., and Hall, R. W. (1984). Taste reception in the bottlenosed dolphin. (*Tursiops truncatus*) Acta Zoologica Fennica **172**, 147–148.

Nummela, S. (2008). Hearing in aquatic mammals. *In* "Sensory Evolution on the Threshold, Adaptations in Secondarily Aquatic

Vertebrates" (J. G. M. Thewissen, and S. Nummela, eds), pp. 211–224. University of California Press, Berkeley.

Pihlström, H. (2008). Comparative anatomy and physiology of chemical senses in aquatic mammals. *In* "Sensory Evolution on the Threshold Adaptations in Secondarily Aquatic Vertebrates" (J. G. M. Thewissen, and S. Nummela, eds), pp. 95–109. University of California Press, Berkeley.

Reuter, T., and Peichl, L. (2008). Structure and function of the retina in aquatic tetrapods. *In* "Sensory Evolution on the Threshold. Adaptations in Secondarily Aquatic Vertebrates" (J. G. M. Thewissen, and S. Nummela, eds), pp. 149–172. University of California Press, Berkeley.

Sonntag, C. F. (1922). The comparative anatomy of the tongues of the Mammalia. VII. Cetacea, Sirenia, and Ungulata. *Proc. Zool. Soc.* **44**, 639–657.

Spoor, F., and Thewissen, J. G. M. (2008). Comparative and functional anatomy of balance in aquatic mammals. *In* "Sensory Evolution on the Threshold, Adaptations in Secondarily Aquatic Vertebrates" (J. G. M. Thewissen, and S. Nummela, eds), pp. 257–284. University of California Press, Berkeley.

Verheyden, C., and Jouventin, P. (1994). Olfactory behavior of foraging procellariiforms. *Auk* **III**, 285–291.

Wartzok, D., and Ketten, D. R. (1999). Marine mammal sensory systems. *In* "Biology of Marine Mammals" (J. E. Reynolds, III, and S. A. Rommel, eds), pp. 117–175. Smithsonian Institution Press, Washington, D.C..

Wilkens, L. A., and Hofmann, M. H. (2008). Electroreception. *In* "Sensory Evolution on the Threshold, Adaptations in Secondarily Aquatic Vertebrates" (J. G. M. Thewissen, and S. Nummela, eds), pp. 325–332. Univ. California Press, Berkeley.

# Sexual Dimorphism

KATHERINE RALLS AND SARAH MESNICK

Dimorphism means two forms. "Sexual dimorphism" means that the two sexes of a species differ in external appearance or other features. Males and females may differ in size, color, shape, the development of appendages (such as horns, teeth, feathers, or fins), and also in scent or sound production. Species in which male and females are identical in appearance or other features are said to be "monomorphic." This article describes the types of dimorphic traits found in marine mammals and explains some of the reasons why these traits might have evolved and what can be inferred about the lives of males and females in a particular species from the pattern of sexual dimorphism. The quality of the information available on sexual dimorphism varies widely across marine mammal species. We know quite a lot about a few species, which are used repeatedly as examples, and virtually nothing about others. Despite the technical difficulties of studying marine mammals, our understanding of the evolution of sexual dimorphism is increasing steadily as observations of rarely encountered species accumulate and new techniques are developed.

## I. Evolution of Sexual Dimorphism

Sexual dimorphism has fascinated biologists since before the time of Darwin. Darwin considered that most sexual dimorphism was due to sexual selection, in which evolutionary forces acted separately on the sexes (Darwin, 1871). For example, females might choose to mate with highly ornamented males (e.g., the peacock's tail) or males might develop characters useful for fighting with other males to win in contests for access to females (e.g., large body size and antlers in deer). Today, these two processes are often referred to as female choice and contest competition, respectively. More recently, scientists have learned that males compete not only by physical fighting and display but also, in species where females mate with more than one male, by sperm competition within the female reproductive tract. Recent reviews have both questioned the very notion of sexual selection (Roughgarden *et al.*, 2006), and reaffirmed it, albeit with acknowledgment of much greater complexity than described by Darwin (Clutton-Brock, 2007).

Although Darwin's ideas about sexual selection have stood the test of time, some cases of sexual dimorphism seem to be best explained by natural selection. For example, males and females in some species of birds [e.g., Galapagos finches (genus *Geospiza*) and the extinct New Zealand huia (*Neomorpha acutirostris*)] have radically different bill morphologies that are best explained by sex differences in foraging habits (Andersson, 1994). In some species, females appear to be larger than males primarily because big mothers are better mothers (more eggs, better at defending their brood; Ralls, 1976). The emerging view is that the degree of sexual dimorphism in a species is the result of the difference between the sum of all the selective pressures (natural selection and sexual selection) affecting the male and the sum of those affecting the female.

## II. Types of Sexual Dimorphism

The adult males and females of a species may differ in size, color, shape, the development of appendages (such as horns, teeth, feathers, or fins), scent, or vocalizations (Figs 1–6). In marine mammals, one of the most striking sexually dimorphic characters is size. In some species, males are dramatically larger than females. For example, in southern elephant seals (*Mirounga leonina*), adult males (maximally at 3700 kg) weigh 4–10 times as much as the adult

**Figure 1** Size. *Adult male South American sea lions* (Otaria flavescens) *are two–three times heavier than females; males grow to 2.8 m and weights of 340 kg; females to 2.2 m and 144 kg. There is extreme sexual dimorphism in body shape and pelage as well as size; males have massive necks, a broad head with a characteristically upturned muzzle, and a thick mane of long guard hairs. Photo by William Conway.*

S

**Figure 2**  Dorsal fins. *A pod of killer whales (Orcinus orca), Alaska. In adult males, the dorsal fin is erect and may grow to 1.8 m in height whereas the dorsal fins of females are less than 0.7 m and distinctly falcate. Sexual dimorphism also occurs in body size, flipper size, and genital pigmentation pattern. Photo by Flip Nicklin (Minden Pictures).*

**Figure 3**  Teeth and Tusks. *Dueling male narwhals (Monodon monoceros), Canada. The unicorn-like tusk of the narwhal is actually a greatly enlarged left upper tooth. The tusk generally erupts only in males and may exceed 3 m in length and 10 kg in weight. Sexual dimorphism also occurs in body size, pigmentation pattern, and the shape of the flukes and pectoral fins. Photo by Flip Nicklin (Minden Pictures).*

**Figure 4**  Teeth and Tusks. *Blainville's beaked whales* (Mesoplodon densirostris) *have a single pair of teeth (which may be better described as tusks) in the lower jaw which appears to primarily function in male–male aggression. Stalked barnacles* (Conchoderma auritum) *often grow in clusters on the erupted teeth, enlarging their effective size and increasing the abrasiveness of the tusks. Adult males also have a distinctive ridge along the dorsum, posterior to the blowhole, which appears to be the area targeted during intra-specific fighting, resulting in heavy scaring from tooth-rakes of other adult males* (Diane Claridge, personal observation.). *The densely ossified rostral bone may function to reinforce the skull when males fight. Photo by Diane Claridge @ Bahamas Marine mammal Research Organization.*

**Figure 5**  Noses. *Threat vocalizations resonate in the greatly enlarged proboscis of adult male northern elephant seals* (Mirounga angustirostris), *Año Nuevo, California. There is extreme dimorphism in body size and shape; males grow maximally to 4 m and 2300 kg and females to 3 m and 360–710 kg. The skin on the neck of adult males is thick, rugose, and scarred by fights and the canine teeth are sexually dimorphic in size and shape. Males are darker brown than females. Photo by Sarah L. Mesnick.*

females (which weigh 350–800 kg). Males in some species also possess greatly enlarged TEETH that are lacking in females and are used in fights between males. The best known example is the unicorn-like tusk of the NARWHAL (*Monodon monoceros*). The tusk, which is actually a greatly enlarged left upper tooth, usually erupts only in males and can grow to an extraordinary size, exceeding 3 m in length and 10 kg in weight. In some odontocete species (e.g., bottlenose whales, genus *Hyperoodon*), males have greatly enlarged and densely ossified heads, which they use to ram other males during fights. In otariids, males have thick necks and massive chests that tend to be covered by a dense mane of hair. The noses of males are sometimes bizarrely modified. For example, the most distinctive feature of the male hooded seal (*Cystophora cristata*) is an inflatable

hood and bright red nasal sac that may function in agonistic and courtship displays. The appendages of males (flippers, flukes, caudal peduncles, and dorsal fins) are sometimes greatly enlarged. The best-known example of dorsal fin enlargement is seen in male killer whales (*Orcinus orca*; Fig. 2).

**Figure 6** Post-anal hump. *The post-anal hump of adult male eastern spinner dolphins (Stenella longirostris orientalis) is tremendously exaggerated. The dorsal fin of adult males is also forward-canted and the tips of the flukes curl up. Photo by Protected Resources Division, Southwest Fisheries Science Center, NOAA Fisheries Service.*

Although sexual dimorphism traditionally referred to differences in morphological traits, the sexes can also produce different vocalizations or odors or be differently colored or patterned. Among marine mammals, differences in color are usually limited to fairly minor differences in pattern or density of pigmentation. For example, in ribbon seals (*Histriophoca fasciata*), the banding pattern is similar in both sexes but paler and less distinct in females. There are numerous examples of sexually dimorphic vocalizations in marine mammals, such as the roars and bellows of male sea lions and fur seals (Otariidae), the songs of male humpback whales (*Megaptera novaeangliae*), and the loud clicks of male sperm whales (*P. macrocephalus*). In terrestrial mammals, males and females often produce different scents via urine, feces, or specialized scent glands. This has not been observed much in marine mammals but may well occur. It is known, for example, that male ringed seals (*Pusa hispida*) produce a strong odor in the breeding season. Male sea otters (*Enhydra lutris*) frequently investigate the anogenital areas of other otters, and male common bottlenose dolphins (*Tursiops truncatus*) investigate the genital region of possibly estrous females with their rostrums.

### III. Taxonomic Distribution
#### A. Baleen Whales

Sexual size dimorphism is "reversed" among the 13 species of baleen whales with females attaining asymptotic lengths that are generally 5% longer than males. Baleen whales typically undertake long-distance migrations between their northern or southern feeding areas and their tropical breeding areas and may not feed while migrating or on the breeding grounds. Females have the added stress of pregnancy and lactation during the non-feeding periods; a larger size may enable them to store more energy resources in the form of BLUBBER to meet their greater reproductive demands.

Sexually dimorphic vocalizations are well known in humpback whales. Male humpbacks sing lengthy, elaborate songs, the function of which has been the subject of much speculation. Songs might function to attract females, signal status to other males, space males on the breeding grounds, synchronize estrus in females, or some combination of these. The humpback song is particularly intriguing because songs change over time, yet all members of a population sing similar songs at any one time. Male fin whales (*Balaenoptera physalus*) have a patterned call, which has been termed a breeding display, but observations of COURTSHIP or competitive interactions are sparse. Sexually dimorphic vocalizations may also exist in blue whales (*B. musculus*). There is dimorphism in the shape of the upper jaw of fin whales and, to a lesser extent, Bryde's whales (*B. edeni*), but the function of this dimorphism is unknown.

Observations of clear AGGRESSION between males are known only in humpback and southern right whales (*Eubalaena australis*). Thus, it is not surprising that there are few accounts of sexually dimorphic structures that might be used in contest competition. Male right whales, however, have larger and more numerous callosities (the raised thickened patches of skin on the head) than females, which may function as weapons in contests between males. Male right whales are also scarred more heavily than females.

#### B. Toothed Whales

The relative size of the sexes varies widely among the more than 70 species of toothed whales. Males are larger than females in many species, with the most pronounced dimorphism in sperm whales, killer whales, bottlenose whales, narwhals, belugas (*Delphinapterus leucas*), and pilot whales (genus *Globicephala*). In sperm whales, for example, females reach about 11 m in length and 15 tons, whereas physically mature males are approximately 16 m long and weigh 45 tons. Females are slightly larger than males in Baird's beaked whales (*Berardius bairdii*), the franciscana (*Pontoporia blainvillei*), the Indian river dolphin (*Platanista gangetica*), harbor porpoise (*Phocoena phocoena*), and dolphins in the genus *Cephalorhynchus*. Some species are monomorphic in size, including the Clymene dolphin (*Stenella clymene*), Atlantic spotted dolphins (*S. frontalis*), dwarf and pygmy sperm whales (genus *Kogia*), tucuxi (*Sotalia fluviatilis*), and some dolphins in the genus *Lagenorhynchus*.

Differences between the sexes may occur in the size and shape of the head, teeth, thoracic girth, flukes, flippers, dorsal fin, caudal peduncle, postanal hump, and length of the beak. In general, males tend to have larger appendages than females, the exception being the few species in which females have longer rostra than males [both species of south Asian river dolphin, the franciscana, and the rough-toothed dolphin (*Steno bredanensis*)]. Dimorphism in the size and shape of the head may be a result of enlargement of the nose (in sperm whales) or the forehead [in bottlenose whales (genus *Hyperoodon*), pilot whales, and to a lesser extent in bottlenose dolphins] of adult males. The massive nasal complex of adult male sperm whales may be one-quarter to one-third the length of the animal and is probably used in the generation of sound (Cranford 1999). In bottlenose whales, the forehead is extremely steep and the surface becomes flattened in mature males. Dimorphism in the density of ossification of the bones in the head occurs in bottlenose whales (the cranium) and beaked whales of the genus *Mesoplodon* (the rostrum, which has one of the highest reported densities of any mammalian tissue). Differences between the sexes in the ossification of the head in these species are not surprising given observations of head butting between adult male bottlenose whales and heavy scarring on adult males of several beaked whale species of the genus *Mesoplodon*. The sexually dimorphic pattern of scarring in *Mesoplodon* species is consistent with the idea that

**S**

adult males use their rostrum, and the exposed teeth on the lower jaw, in fights with other males.

Dimorphism in the size, shape, and/or number of teeth is known in the narwhal, sperm whale, Cuvier's beaked whale (*Ziphius cavirostris*), bottlenose whale, and in beaked whales of the genus *Mesoplodon*. In most of these species (exceptions being sperm whales and narwhals) the teeth erupt only in males and only at sexual maturity.

Differences between the sexes are known in flipper length (killer whales and melonheaded whales, *Peponocephala electra*), serration (Heaviside's dolphins, *C. heavisidii*), and shape of the trailing edge (belugas). In some species, including sperm whales and Dall's porpoises (*Phocoenoides dalli*), males have deeper caudal peduncles than females, which may function to give more power to the flukes. Postanal humps (thought to be composed of muscle and connective tissue) are exhibited in mature males of several species, although they have been properly described and correlated with age and sex in only a few. The postanal hump of the male eastern spinner dolphin (*S. longirostris orientalis*) is exaggerated tremendously (Fig. 6). Although the function of the postanal hump remains unknown, it has been suggested to be an anchor for external genitalia and may serve to enhance sexual performance. It may also serve as a visual signal that makes adult males easily recognizable. Dorsal fins may be larger and more erect in males than females, more hooked, or more forward canted (Figs. 2 and 6). The most exaggerated examples of dorsal fin enlargement are seen in male killer whales and pilot whales. The significance of these differences in dorsal fin size and shape is unknown but they may serve a thermoregulatory function and/or as a visual signal. Differences between the sexes also occur in the flukes, which may be longer and broader in males, or differently shaped. In Dall's porpoises and sperm whales, for example, the trailing edge of the flukes of males are convex, and in male eastern spinner dolphins, the tips of the flukes curl up. As is true for mammals in general, the distance between anal and genital openings of odontocetes tends to be greater in males than in females.

Sexual differences in pigmentation patterns are most common in the genital area but are also known to occur on the face, head, and body. Sexual dimorphism in genital pigment patterns is known in several species [killer whales, dolphins in the genus *Cephalorhynchus* and *Lissodelphis*, shortbeaked common dolphins (*Delphinus delphis*), Burmeister's porpoises (*P. spinipinnis*), and Dall's porpoises]. Pigmentation differences may be related to sexual recognition, advertisement (for either males or females), or may help suckling young locate the teats. In most species of beaked whales, the body gets lighter in color with age. The lightening is especially noticeable in adult males and is primarily due to an accumulation of body scars, but may also be due to changes in pigmentation and, in several species, both. In Risso's dolphins (*Grampus griseus*), ontogenetic lightening and an accumulation of body scars make older animals appear almost pure white, and the pattern may be more prevalent in males. Adult male spotted dolphins (*S. attenuata*) bear conspicuous white rostrum tips, visible at a great distance. In Fraser's dolphins (*Lagenorhynchus hosei*), the intensity and thickness of the eye-to-anus stripe becomes more exaggerated (darker and thicker) in adult males. Another type of pigment dimorphism is the occurrence of visible (white or nonpigmented) linear scarring, suggested to result from a lack of pigmentation of damaged tissue from the tooth rake wounds of conspecifics. In some odontocete species, both sexes exhibit heavy scarring [e.g., Baird's beaked whale (genus *Berardius*) and Risso's dolphins]. However, in others (*Mesoplodon* spp., narwhal, and the sperm whale), males are scarred more heavily than females. In these species, scarring is likely the result of wounds inflicted during male fights.

At present, acoustically dimorphic calls are known only in sperm whales. However, because odontocetes produce a wide range of sounds, dimorphic acoustic signals are likely to occur in several other species as well. Because larger animals make larger sounds, it is also reasonable to expect that other sexually dimorphic species, such as pilot whales, will produce acoustically dimorphic calls.

## C. Pinnipeds

The 36 species of pinnipeds show the greatest range in sexual size dimorphism of any higher vertebrate group (Fig. 7). Adult males are up to 10 times as heavy as adult females in some species, whereas females are slightly larger than males in others. For virtually all pinnipeds studied to date, data support, or are highly suggestive of, a polygynous mating system. However, the potential for polygyny varies greatly among species and there is a strong correlation between the degree of polygyny and the degree of dimorphism. The mating system in turn, depends to a large extent on whether breeding takes place on land or at sea. In terrestrially mating pinnipeds [this includes sea lions and fur seals and three species of phocid, the northern and southern elephant seal (genus *Mirounga*) and the gray seal (*Halichoerus grypus*)], extreme polygyny is possible because females gather in dense groups on islands to give birth and mate. Under these conditions, a successful male can defend many females. In these species, males exhibit not only large size but also other characteristics useful in fights over females, such as large canines, massive necks and chests, and dense pelage. Large size may also help males of these species achieve greater reproductive success by enabling them to remain longer on the breeding rookery because larger males have bigger energy reserves in the form of blubber.

In the remaining pinnipeds, the WALRUS (*O. rosmarus*) and nearly all of the phocids, mating takes place in the water. Females of many species give birth on ice (and therefore are not as spatially clumped as terrestrially breeding species in part because they have larger expanses of suitable habitat available) and the mating season is short. Thus, males have less opportunity to defend and mate with multiple females. These characteristics inhibit the development of the extreme polygyny and sexual dimorphism found in terrestrially mating otariids and phocids. In general, males of aquatically mating species are only slightly larger than females or females may be slightly larger than males [bearded (*Erignathus barbatus*), Weddell (*Leptonychotes weddellii*), Ross (*Ommatophoca rossii*), crabeater (*Lobodon carcinophaga*), and leopard (*Hydrurga leptonyx*) seals]. The hooded seal is a notable exception, with males considerably larger than females. In ice-breeding species, large female size may help a mother provide large quantities of fat-rich milk for her pup, and because mating takes place in the water in these species, agility, rather than size or strength, may be important in male contests for females.

In addition to the sexual size dimorphism mentioned earlier, the sexes may also differ in pelage length and color, shape of the head and chest, canine development, and the pattern of scarring on the neck and chest. Adult male otariids tend to be bulkier than females and are distinguished readily by their thicker and more powerful necks and their massive chests. The head, neck, and chest of males tend to be covered by longer, rougher hairs, which gives the impression of a mane [e.g., the South American sea lion (*Otaria flavescens*); Fig. 1]. In older males, the guard hairs are lighter in color and tinged with white, silver, or tan. Adult male California sea lions (*Zalophus californianus*) also develop a pronounced forehead, or sagittal crest, and adult male southern sea lions have a distinctive upturned muzzle. The skin on the necks of adult male elephant seals and gray seals

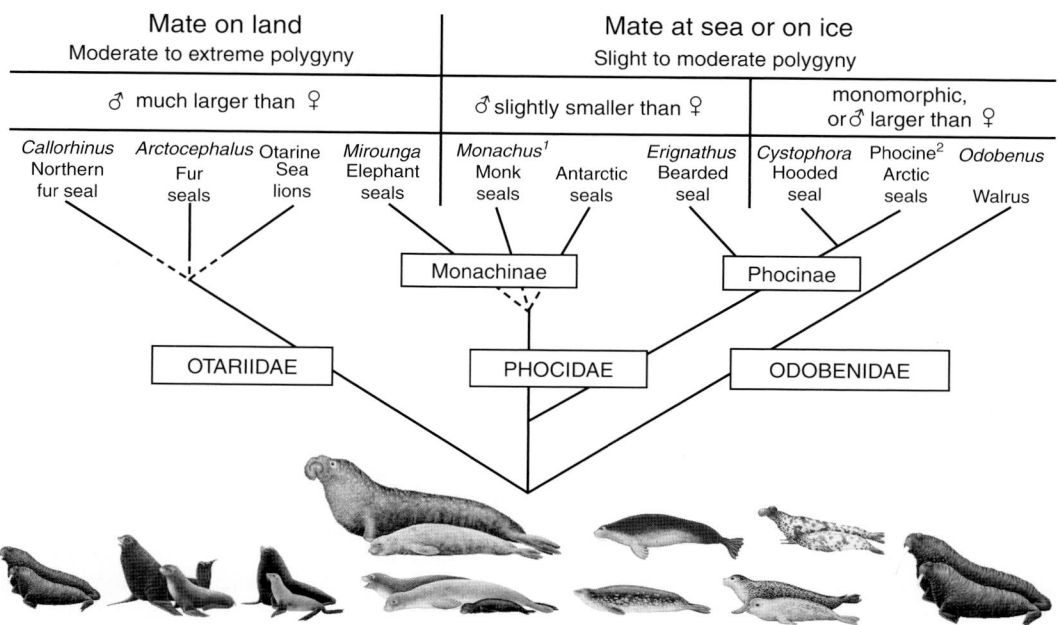

**Figure 7** *Sexual size dimorphism in pinnipeds. A composite phylogenetic tree of the Pinnipedia on which sexual dimorphism, mating location, and degree of polygyny have been overlaid. Sexual size dimorphism varies greatly across pinniped species, and there is a strong correlation between the degree of dimorphism and the mating system. In otariids and three species of phocid (both elephant seals and the gray seal), mating takes place on land and extreme polygyny is possible because a successful male can defend many females. Males are much larger than females (2–10 times larger) and also exhibit other characteristics useful in fights over females, such as large canines. In the remaining pinnipeds, the walrus, and nearly all the phocids, mating takes place in the water or on ice. There is less opportunity for males to mate with multiple females and agility, rather than size or strength, may be important in male contests for females. Males are equal, slightly, or moderately (up to 1.5 times) larger than females or females may be slightly (up to 1.1 times) larger than males in these species. [1]Females are slightly larger than males in the Hawaiian monk seal (Monachus schauinslandi); males are slightly larger than females in the Mediterranean monk seal (Monachus monachus). [2]Among phocines, the gray seal (Halichoerus grypus) represents a notable case because it can exhibit both terrestrial and aquatic mating and males are maximally three times larger than females. Pieter A. Folkens/Higher Porpoise DG.*

is thickened and wrinkled and marked by scars from fights. In general, adult male otariids, as well as adult males in some species of phocids, tend to be more darkly pigmented than females.

Many pinnipeds have sexually dimorphic vocalizations that may function to establish and maintain dominance relationships or to attract females. In some species, the sounds are amplified or resonated in the proboscis (as in hooded and elephant seals; Fig. 5) or an internal air sac (as in ribbon seals, bearded seals, and walruses). Hooded seals produce numerous sounds as they inflate and deflate their hood and bright red nasal sac in response to disturbances and as part of the courtship display. Male elephant seals also have greatly enlarged noses, and snouts of male gray seals are broader and more elongated than those of females. Males of these species establish dominance hierarchies through stereotyped visual and airborne acoustic threats and, less often, physical aggression. Male harbor, bearded, ribbon, Weddell, ringed, and harp (*Pagophilus groenlandicus*) seals are known for their acoustic courtship displays. Male harbor seals engage in complex hierarchical acoustical mating displays in which several subordinate males passively muzzle singing dominant males underwater. Much of the roaring and bellowing of adult male otariids is thought to intimidate rivals but acoustic displays may also be used to advertise to females. The walrus has the most elaborate

courtship display of all pinnipeds involving intricate acoustic and visual components. Vocalizing adult male walruses apparently compete for females in lek-like groups in the water near ice floes on which females gather to pup and rest. Their surface vocal repertoire includes barks, whistles, and growls, and underwater vocalizations sound bell like. It is also thought that the massive tusks may play a role as a visual symbol of rank and as a display to females. Male walruses are larger than females in both body and tusk size. In marine mammals, the only well-documented sexually dimorphic scent of which we are aware occurs in the ringed seal. Male ringed seals give off a strong scent during the breeding season.

### D. Sirenians, Sea Otters, and Polar Bears

Manatees (genus *Trichechus*) are generally monomorphic in size and appearance. Dugongs (*Dugong dugon*) exhibit no obvious sexual dimorphism apart from the short tusks (upper incisors), which usually erupt in adult males, although females may grow to a slightly larger size than males. Male dugongs compete for access to females by patrolling exclusive areas and engaging in threats, fights, and song. Adult male sea otters are larger than adult females. Second only to pinnipeds, polar bears (*Ursus maritimus*) exhibit the greatest

S

sexual size dimorphism among mammals. Male polar bears may be over twice as heavy as females.

## IV. What Can We Infer from Sexual Dimorphism?

The variation in sexual dimorphism among marine mammal taxa is striking. The sexes are visually indistinguishable in some species, whereas in others the differences between the sexes are so extreme that males and females live essentially separate lives except when they meet to mate. This rich variation in sexual dimorphism has prompted scientists to offer a variety of explanations. The various mechanisms of sexual selection—female choice, contest competition, and sperm competition—probably account for a large proportion of sexual dimorphism in marine mammals. However, some dimorphic traits may reflect ecological differences between the sexes (e.g., differences in beak length between the sexes in south Asian river dolphins). Others may be important for females and their young (e.g., larger females make better mothers or urogenital pigment patterns may highlight the mammary glands and help young to find them).

The functional significance of most sexually dimorphic traits in marine mammals remains untested, which is not surprising given the difficulty of observing, let alone experimenting on, most species. In general, the behavior of pinnipeds (which often breed where they can be observed) is better known than the behavior of cetaceans that breed at sea. However, extended observations of behavior have also been possible in a few cetacean species (e.g., bottlenose dolphins, humpbacks, and right whales) that breed close to shore. Nevertheless, we can often infer the functional significance of sexual dimorphism in species whose behavior is poorly known by analogy to well-studied species. The type and degree of sexual dimorphism and its association with other characteristics such as relative testis size and pattern of bodily scarring provide clues to the intensity of sexual selection (the skew in male mating success) in a species and the probable underlying mechanisms of sexual selection.

Based on studies of terrestrial mammals, a positive correlation is generally assumed between the amount of sexual dimorphism in a species and the deviation of the breeding system from monogamy. Thus, in polygynous species, male competition for access to females is severe and males are expected to exhibit traits, such as large size and big canines, favored in fights with other males over access to females. The correlation between sexual size dimorphism and the degree of polygyny has been shown across pinniped taxa (Alexander *et al.*, 1979). For example, among otariids, the northern fur seal (*Callorhinus ursinus*) and Steller sea lion (*Eumetopias jubatus*) show the greatest relative dimorphism in body weight and defend the greatest number of females in their territories as compared to less dimorphic species. Within a species, a large body size has also been shown to be correlated with greater mating success (via dominance rank, endurance, and tenure on rookeries; e.g., elephant seals, gray seals). It is important to note, however, that species that lack sexual size dimorphism do not necessarily lack male–male competition for mates. In these species, sexual selection may be intense, but due to different forms of competition among males for access to mates, and the consequences may not be reflected in size but in other characters, such as song, visual display, or agility.

Sexual dimorphism in size and weaponry (big teeth, enlarged heads, and strong flukes) suggests that contest competition for access to mates plays an important role in the mating strategies of males in many marine mammal species. Contest competition may take the form of ritualized displays (visual or acoustic), by which potential rivals assess relative size or strength, or physical aggression. Among odontocetes, dimorphism in weaponry is correlated to patterns of body scarring and observations of head butting among males. In certain species, such as sperm whales, beaked whales, and narwhals, teeth erupt or are enlarged only in adult males, a pattern that suggests their function has shifted from feeding to use in social interactions. Adult males of many of these same species are heavily scarred, another trait that suggests males use their teeth in physical battles with other males.

Among terrestrial mammals, the relationship between relative testis size and mating system is so strong that the former can be used as a good indicator of the mating system (Gomiendo *et al.*, 1998). In general, where copulatory frequency is high, the testes are large, and where it is low, the testes are small. In right whales, observations of multiple males mating with single females, together with huge (1 ton) testes, strongly suggest that sperm competition is a principal mating strategy in this species, and probably also in bowhead (*Balaena mysticetus*) and gray (*Eschrichtius robustus*) whales (Brownell and Balls, 1986). Odontocete species such as sperm whales and beaked whales that exhibit dimorphic traits associated with intense physical combat (e.g., large size, enlarged teeth) tend to have small testes. The testes of sperm and beaked whales represent less than 0.5% of the body weight, weigh only a few kilograms, and can be held in one hand. At the other extreme, species having the largest testes, suggesting the likelihood of significant sperm competition, do not exhibit the extreme dimorphic traits associated with physical combat. These species tend to be sexually monomorphic or have dimorphic traits that may be associated with agility or visual display. For example, harbor porpoises, finless porpoises (*Neophocaena phocaenoides*), and dusky dolphins (*L. obscurus*) have testes that represent greater than 5% of their body weight. Humans, for comparison, at about the same body mass as these dephinid species, have testes of only 0.08% of body mass (Kenagy and Trombulak, 1986).

In three-dimensional habitats, agility, rather than size or strength, may sometimes determine the outcome of male contests. Agility may be useful in scramble competition for access to mates and it may function as a visual display for female choice. This may be the case in some species, such as the aquatically mating phocids, in which males compete underwater and are smaller than females. Among odontocetes, larger body size typically means that the male's propulsion structures are also proportionally larger than those of the female. The importance of speed and maneuverability is suggested by sexual dimorphism in the flippers, flukes, caudal peduncle, and dorsal fin. Tolley *et al.* (1995) suggested that the larger body size, caudal peduncle, flukes, and dorsal fin of male bottlenose dolphins, and the pattern of dorsal fin scarring, are consistent with males competing for access to dispersed females. Features such as flukes and dorsal fins are used for propulsion, maneuvering, and thermoregulation and in offensive or defensive encounters with other males. More power to the flukes could increase the strength of blows and greater speed could aid in the herding of females.

Traditionally, behavioral ecologists have tended to emphasize the importance of male–male competition in the evolution of exaggerated male traits. More recently, based primarily on bird data, they have found that female choice often plays a critical role. Recordings of male song and the existence of exaggerated morphological traits that make adult males easily recognizable suggest the importance of female choice in marine mammals. The same features that appear to provide advantages in contests between males, such as large size, big canines, or deep roars, may also be used by females to select mates and/or may function to control or intimidate females (Wells *et al.*, 1999). Whether females actually use these traits to assess males or what these traits might signal (e.g., status, fitness, or readiness) is unknown. The enlarged postanal hump of males in some dolphins and porpoises may serve an important biomechanical function for males by facilitating copulation. It may also be important as a visual signal that makes adult males easily recognizable within schools, by both females and other males. Similarly, enlarged dorsal fins, which may have a thermoregulatory function, may also serve as a visual signal. The calls of male pinnipeds may function as male displays to females, in species recognition, and in contests between males. Evidence supporting the idea of lekking in walrus and dugongs suggests an increasing role for female choice in the evolution of vocal mating displays.

Caution is warranted when making inferences about the evolution of sexually dimorphic traits (Roughgarden *et al.*, 2006; Clutton-Brock, 2007). First, our knowledge of sexual dimorphism across marine mammal taxa is incomplete. There are rarely encountered species for which we have very little information about sexual dimorphism. Although our understanding of morphological differences between the sexes is growing, our knowledge of acoustic and pheromonal differences is in its infancy. As we fill in these gaps in our knowledge, our ability to understand the underlying evolutionary patterns and processes will increase. Second, a sexually dimorphic trait may have evolved for different reasons in different species. For example, among odontocetes, males are much larger than females in sperm whales, "resident" killer whales, and long-finned pilot whales (*G. melas*), but it is unlikely that a single explanation fits all three cases. In sperm whales, adult males are solitary and roam great distances searching for females. Males possess large teeth, have massive heads, are scarred, and have been observed ramming each other head-on. It is likely that large size serves male sperm whales well in contest competition over access to females. In contrast, adult male "resident" killer whales and long-finned pilot whales live in stable social groups with their maternal relatives, are not scarred, and we know of no accounts of aggressive interactions between males. In contrast to sperm whales, "resident" male killer whales and long-finned pilot whales may increase their reproductive success, not only by mating with females in other pods, but by assisting kin in their natal pods. At this point, we can only speculate about the advantages that large size confers on males of these species, but assistance in a communal foraging strategy ("resident" killer whales) and protection of the pod (long-finned pilot whales) are possibilities. Females may prefer large males as mates in all these species, but large size may confer different advantages to individuals in each of the three cases.

Despite the technical difficulties of studying marine mammals, our understanding of the evolution of sexual dimorphism is increasing steadily. New techniques, such as scoring molecular genetic markers from tissue samples, are providing insight into social structure and variance in male reproductive success. Video, acoustic recordings, and "critter cams" (small television cameras that can be mounted on individual animals) are providing exciting new data on the behavior and interactions of animals underwater. Clearly, research opportunities abound, and the prospects for increased future understanding of the abundant sexually dimorphic traits in marine mammals are bright.

### *See Also the Following Articles*

Coloration ■ Evolutionary Biology ■ Mating Systems ■ Song ■ Teeth

### *References*

Alexander, R. D., Hoogland, J. L., Howard, R. D., Noonan, K. M., and Sherman, P. W. (1979). Sexual dimorphism and breeding systems in pinnipeds, ungulates, primates, and humans. *In* "Evolutionary Biology and Human Social Behaviour" (N. A. Chagnon, and W. D. Irons, eds), pp. 402–435. Duxbury Press, North Scituate.

Andersson, M. (1994). "Sexual Selection." Princeton University Press, Princeton.

Boness, D. J., Clapham, P. J., and Mesnick, S. L. (2000). Life history and reproductive behaviour. *In* "Marine Mammal Biology: An Evolutionary Approach" (A. R. Hoelzel, ed.). Blackwell Science, Oxford.

Brownell, R. L., and Balls, K. (1986). Potential for sperm competition in baleen whales. *Rep. Int. Whaling Comn.* **8**(Special issue), 97–112.

Clapham, P. J. (1996). The social and reproductive biology of humpback whales: An ecological perspective. *Mamm. Rev.* **26**, 27–49.

Cranford, T. W. (1999). The sperm whale nose: sexual selection on a grand scale? *Mar. Mamm. Sci.* **15**, 1133–1157.

Clutton-Brock, T. (2007). Sexual selection in males and females. *Science* **318**, 1882–1885.

Darwin, C. (1871). "The Descent of Man, and Selection in Relation to Sex." Murray, London.

Gomiendo, M., Harcourt, A. H., and Roldan, E. R. S. (1998). Sperm competition in mammals. *In* "Sperm Competition and Sexual Selection" (T. H. Birkhead, and A. P. Moller, eds), pp. 467–755. Academic Press, San Diego.

Jefferson, T. A. (1990). Sexual dimorphism and development of external features in Dall's porpoise *Phocoenoides dalli*. *Fish. Bull.* **88**, 119–132.

Kenagy, G. J., and Trombulak, S. C. (1986). Size and function of mammalian testes in relation to body size. *J. Mammal.* **67**, 1–22.

Leatherwood, S., and Reeves, R. R. (1983). "The Sierra Club Handbook of Whales and Dolphins." Sierra Club Books, San Francisco.

MacLeod, C. D. (1998). Intraspecific scarring in odontocete cetaceans: an indicator of male "quality" in aggressive social interactions? *J. Zool. (Lond.)* **244**, 71–77.

Morejohn, G. V., Loeb, V., and Baltz, D. M. (1973). Coloration and sexual dimorphism in the Dall porpoise. *J. Mammal.* **54**, 977–982.

Ralls, K. (1976). Mammals in which females are larger than males. *Q. Rev. Biol.* **51**, 245–276.

Reeves, R. R., Stewart, B. S., and Leatherwood, S. (1992). "The Sierra Club Handbook of Seals and Sirenians." Sierra Club Books, San Francisco.

Riedman, M. (1990). "The Pinnipeds: Seals, Sea Lions and Walruses." University of California Press, Berkeley.

Roughgarden, J., Oishi, M., and Akçay, E. (2006). Reproductive social behavior: Cooperative games to replace sexual selection. *Science* **311**, 965–969.

Tolley, K. A., Read, A. J., Wells, R. S., Urian, K. W., Scott, M. D., Irvine, A. B., and Hohn, A. A. (1995). Sexual dimorphism in wild bottlenose dolphins (*Tursiops truncatus*) from Sarasota, Florida. *J. Mammal.* **74**, 1190–1198.

Wells, R. S., Boness, D. J., and Rathbun, G. B. (1999). Behavior. *In* "Biology of Marine Mammals" (J. E. Reynolds, III, and S. A. Rommel, eds). Smithsonian Institution Press, Washington, D.C.

# Shepherd's Beaked Whale
## *Tasmacetus shepherdi*

### JAMES G. MEAD

**S**

## I. Characteristics and Taxonomy

**S**hepherd's beaked whale (*Tasmacetus shepherdi*) (Oliver, 1937) is the only beaked whale to have a full set of functional teeth (Mead, 1989) (Fig. 1). It has between 17 and 21 upper teeth on each side and between 18 and 28 lower teeth. The first tooth in

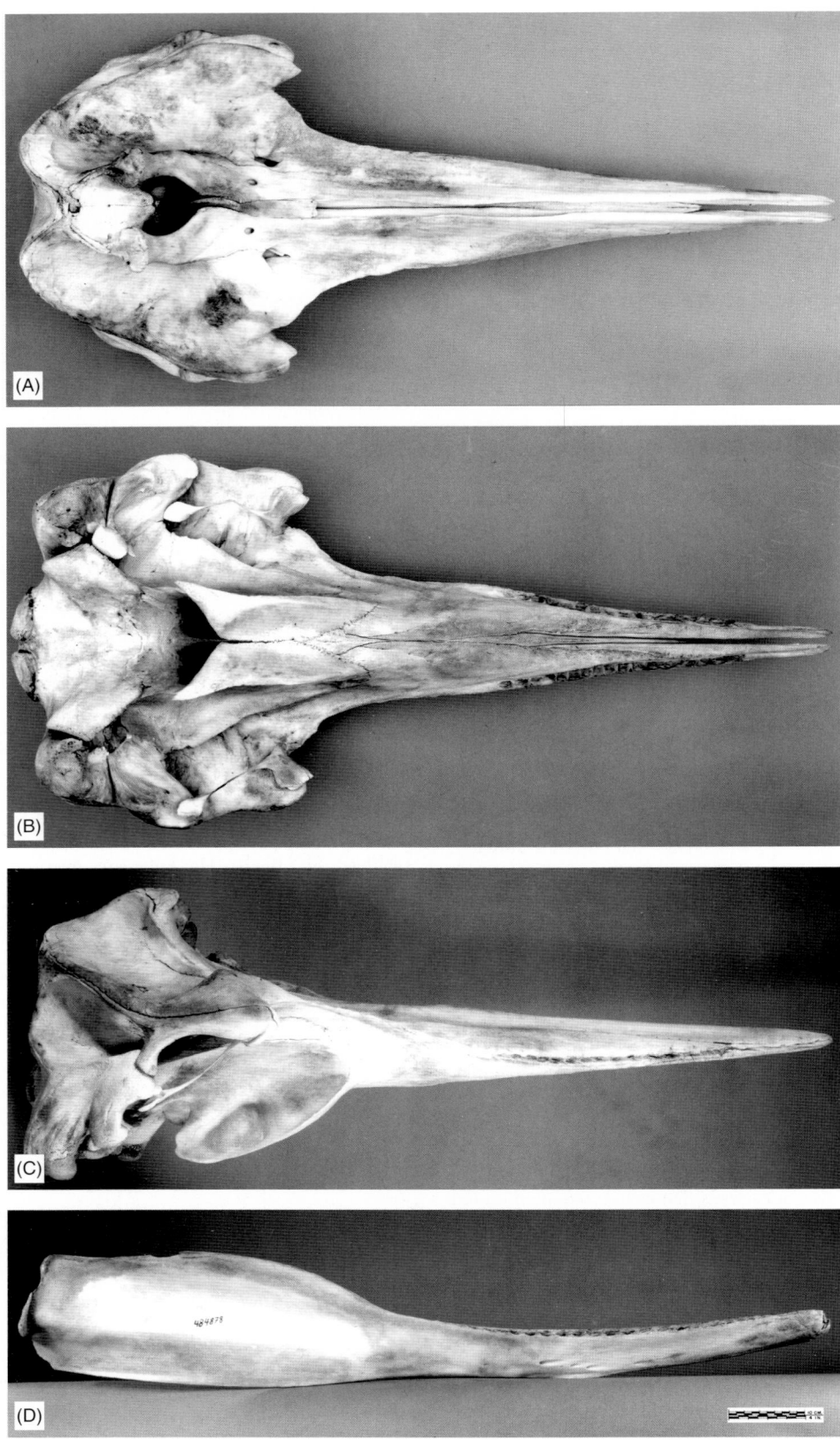

**Figure 1** Tasmacetus shepherdi *USNM 484878 660cm female, 1 February 1973, Peninsula Valdez, Argentina. A. dorsal view of skull; B. ventral view of skull; C. right lateral view of skull; D. right lateral view of mandible.*

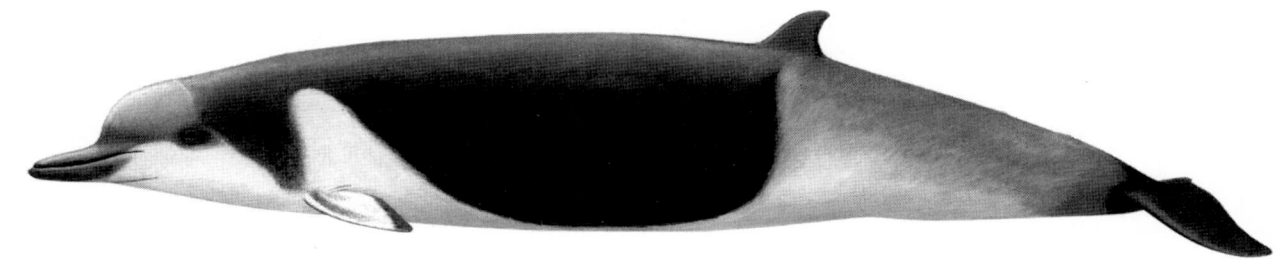

**Figure 2**    *External view of an adult female* Tasmacetus shepherdi *(reconstruction by Pieter Folkens).*

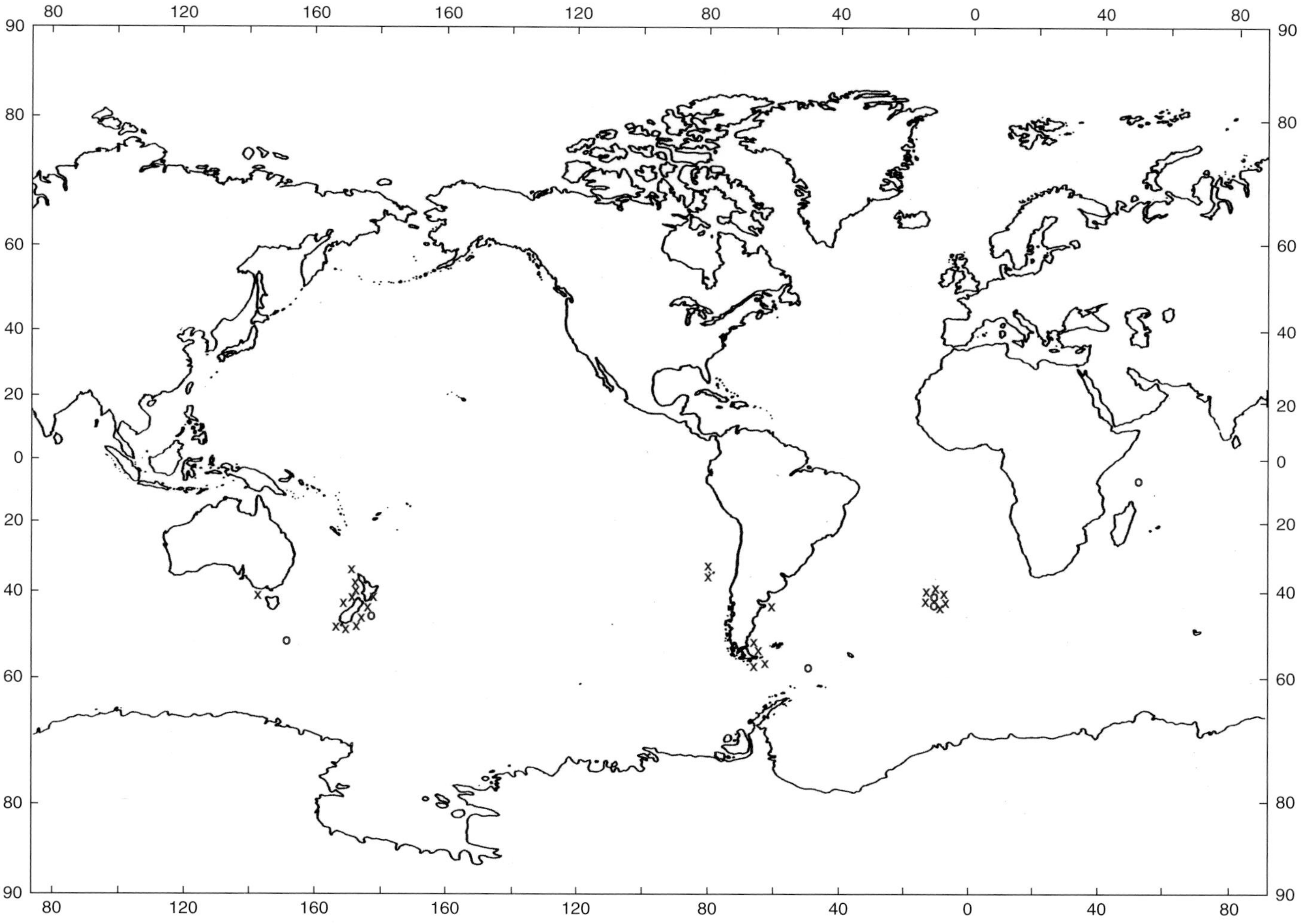

**Figure 3**    *Geographic records of* Tasmacetus shepherdi. *Crosses represent strandings, open circles represent published sightings.*

the mandible is enlarged, and it is thought that size difference makes it homologous to the single mandibular tooth of other beaked whales. It was initially thought to be a primitive member of the beaked whale family (Ziphiidae) on the basis of retention of teeth. There has not been a good study of the evolutionary relationships of the beaked whales, but *Tasmacetus* seems in ways other than the teeth to be just as specialized as the remainder of the family. The pigmentation pattern (Fig. 2) is dark gray dorsally, with a white field ventrally extending dorsally on both the anterior and posterior sides of the flipper (Pitman *et al.*, 2006). The dark dorsal field extends onto the flipper.

Another extension of the white ventral field lies dorsal to the anus. The dorsal surface of the head is pale and the rostrum is darker.

## II.  Distribution and Abundance

The geographic records of *Tasmacetus* are illustrated in Fig. 3. It is known from strandings in New Zealand (13), Australia (1), Argentina (5), Juan Fernandez Islands (2), and Tristan da Cunha (6). There are six published sightings attributed to this species, one from New Zealand, one from the Seychelles Islands, one from the South

**S**

Sandwich Islands, two from Tristan da Cunha and one from south of Tasmania. It is limited to temperate waters and may not ordinarily come as far north as the tip of Africa.

## III. Ecology

This species presumably feeds primarily on squid and fish. One adult female that stranded in Argentina had traces of bottom fish, squid, and one small crab (Mead and Payne, 1975). A calf had plastic debris in its stomach, but it is not known whether it had begun to take normal solid food. Another stranded adult from Tristan da Cunha had only remains of cephalopods in its stomach. All of the beaked whales that are moderately well known feed offshore in deep water.

## IV. Behavior and Physiology

Recent studies have defined the appearance of this species at sea. Groups of up to 10 animals have been sighted, but behavioral observations are rare.

## V. Life History

Adults are between 6 and 7 m in length. The only known calf was 340 cm long. One adult male had 23 growth layer groups in its teeth and was presumably 23 years old.

## VI. Interactions with Humans

Shepherd's beaked whale has not been known to be involved in any fisheries and certainly has not been kept in captivity. It was first known to science by a publication on a stranded adult female which Shepherd found and collected in 1933.

## *See Also the Following Articles*

Beaked whales, Overview ■ Mesoplodont whales

## *References*

Mead, J. G. (1989). Shepherd's beaked whale *Tasmacetus shepherdi* Oliver, 1937. *In* "Handbook of Marine Mammals" (S. H. Ridgway, and R. Harrison, eds), vol. 4, pp. 309–320. Academic Press, London.

Mead, J. G., and Payne, R. S. (1975). A specimen of the Tasman beaked whale, *Tasmacetus shepherdi*, from Argentina. *J. Mammal.* **56**, 213–218.

Pitman, R. L., van Helden, A. L., Best, P. B., and Pym, A. (2006). Shepherd's beaked whale (*Tasmacetus shepherdi*): information on appearance and biology based on strandings and at-sea observations. *Mar. Mamm. Sci.* **22**, 744–755.

# Signature Whistles

LAELA S. SAYIGH AND VINCENT M. JANIK

Signature whistles were first described for several species of captive delphinids by Melba and David Caldwell in the 1960s (Caldwell and Caldwell, 1965) and reviewed in Caldwell *et al.* (1990). They found that individual dolphins, while isolated for medical attention, produced primarily one stereotyped individually distinctive whistle contour (pattern of frequency changes over time). Other researchers had speculated that dolphins would share a specific whistle for each context, such as being in isolation or in distress, or to indicate presence of food. However, it was difficult to support or refute these claims since dolphins do not make any visible movement associated with vocalization, and thus it was not possible to identify which dolphin produced a sound at any given time. By recording isolated dolphins in captivity, the Caldwells were able to get around this problem. The signature whistle is defined as the most common whistle type that an individual uses when in isolation (Fig. 1). Signature whistles have now been documented in more than 300 individual bottlenose dolphins, *Tursiops truncatus*, in a variety of locations, both captive and wild, and there is evidence for them in four other delphinid species (common dolphins, *Delphinus delphis*, Caldwell and Caldwell, 1968; Pacific white-sided dolphins, *Lagenorhynchus obliquidens*, Caldwell and Caldwell, 1971; spotted dolphins, *Stenella plagiodon*, Caldwell *et al.*, 1973; and Pacific humpback dolphins, *Sousa chinensis*, van Parijs and Corkeron, 2001).

Although a study of captive dolphins by McCowan and Reiss (2001) suggested that signature whistles do not exist, more recent studies confirmed that signature whistles are an important component of the bottlenose dolphin vocal repertoire (Janik *et al.*, 2006; Sayigh *et al.*, 2007). In addition, McCowan and Reiss have recently acknowledged that there is evidence for a referential identity-labeling system based on signature whistles, which appears to lay the previous controversy to rest (Marino *et al.*, 2007).

Since the Caldwells' pioneering work, most work on signature whistles has focused on the bottlenose dolphin, and it has focused on four major questions, each of which are addressed below. First, how often do dolphins use signature whistles? Second, how do signature whistles function in the natural communication system of dolphins? Third, what is the role of signature whistle copying in the natural communication system of dolphins? Fourth, how do signature whistles develop in young animals?

The Caldwells had found that isolated animals in captivity tended to produce signature whistles almost exclusively; however, later studies showed that signature whistles are only one component of the whistle repertoire of bottlenose dolphins. For example, Cook *et al.* (2004) used data from brief capture-release studies to identify signature whistles in wild bottlenose dolphins and then followed these animals in other contexts to investigate signature whistle usage in the wild. They found that 52% of whistles produced by undisturbed wild dolphins in Sarasota Bay, FL, were signature or probable signature whistles. Thus, signature whistles are an important component of the whistle repertoire, but dolphins also produce a variety of other whistle types.

Signature whistles function both in individual recognition and in maintaining group cohesion. In playback experiments conducted during brief capture-release events in Sarasota Bay, FL, dolphins responded more strongly to whistles of related than non-related (but familiar) individuals (Sayigh *et al.*, 1999). These experiments were replicated using artificial stimuli resembling the contour shape of whistles of individual dolphins with all voice cues removed, demonstrating that animals actually recognized whistle contours (Janik *et al.*, 2006). The role of signature whistles in maintaining group cohesion was shown by Janik and Slater (1998), who found that captive dolphins were more likely to produce signature whistles when one group member voluntarily isolated itself from the rest of the group. This function of whistles was also supported by Smolker *et al.* (1993), who found that dolphins in Shark Bay, Western Australia, whistled during reunions. Dolphins can also encode information other than identity in their signature whistles (e.g., their own motivational state) by varying whistle parameters while keeping the overall gestalt of the whistle contour constant (Janik *et al.*, 1994).

The first study to address copying, or imitation, of signature whistles was conducted by Tyack (1986). He used a telemetry device called

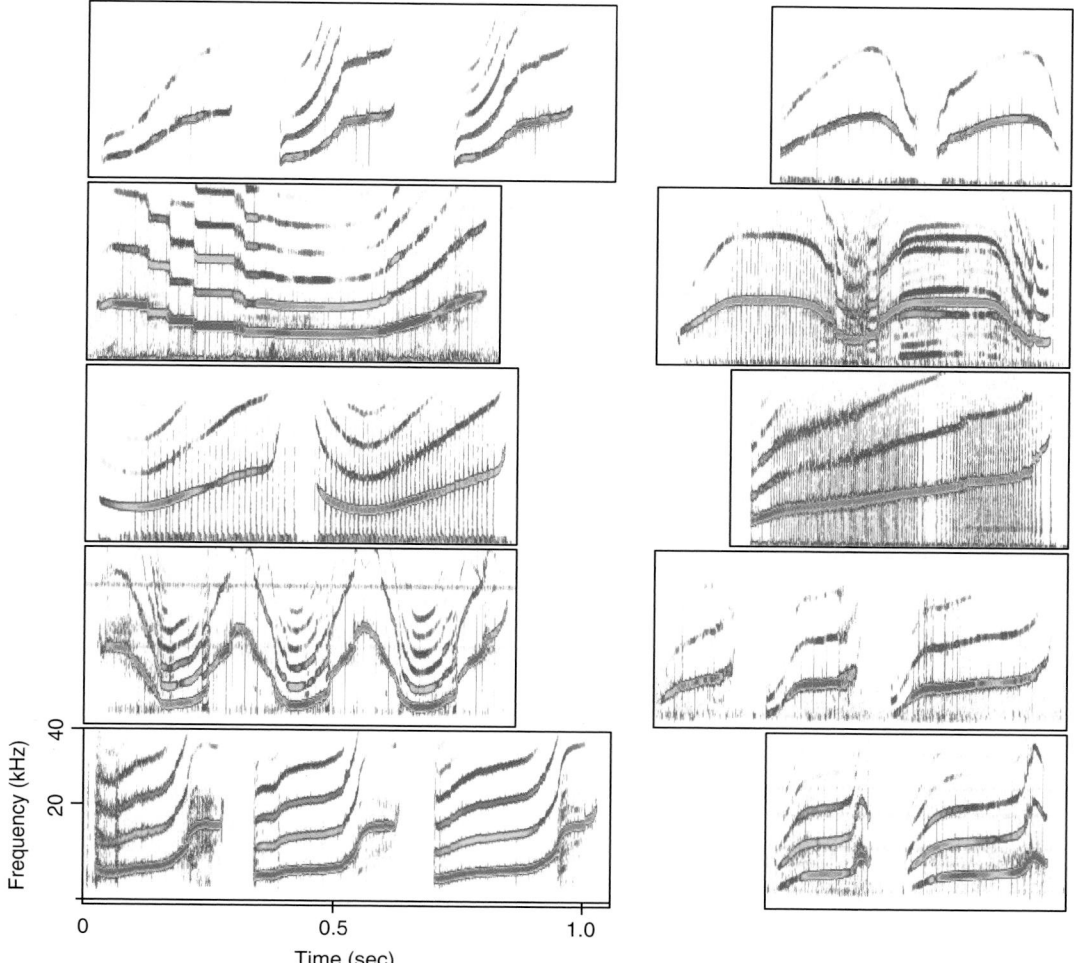

**Figure 1** *Examples of signature whistles of ten different individual bottlenose dolphins* (Tursiops trun-catus) *recorded during temporary capture in Sarasota, Florida. Spectrograms were made with AVISOFT SASLab Pro, at a sample rate of 80,000 Hz and a 256 point FFT.*

a "vocalight," which dolphins wore on their heads and which lit up when a sound was produced, and found that socially interactive dolphins not only produce signature whistles, but that they also produce imitations of one another's signature whistles. Janik (2000) found that free-ranging dolphins in Scotland also engage in such matching whistle exchanges. In addition, in both Shark Bay, Western Australia, and Sarasota, FL, close male associates can converge on a similar signature whistle type that they then use even outside of direct whistle interactions (Smolker and Pepper, 1999; Watwood *et al.*, 1999). Signature whistles of females, on the other hand, usually remain stable for more than a decade (Sayigh *et al.*, 1990, 1995). New technologies that enable localization of vocalizing animals are necessary to study how both signature and non-signature whistles (including whistle copies) are used in the natural communication system of dolphins.

In addition to discovering signature whistles, Caldwell and Caldwell (1979) were the first researchers to investigate vocal development in bottlenose dolphins . They found that young dolphins produce tremulous, quavery whistles, and then gradually converge on a stereotyped whistle contour during their first year of life. Sayigh *et al.* (1990, 1995) found that female dolphin calves in Sarasota, FL, tended

to produce whistles that were highly distinct from those of their mothers, whereas males were more likely than females to produce whistles similar to those of their mothers. This raised the possibility that dolphins may be learning their signature whistles from conspecifics. Much evidence supports the idea that dolphins learn whistles, which sets them apart from many other non-human mammals, in which vocal production is less flexible (Janik and Slater, 1997). Fripp *et al.* (2005) found that free-ranging dolphins in Sarasota appeared to learn their signature whistles from other community members, and Miksis *et al.* (2002) found that captive dolphins can incorporate human-made model sounds into their signature whistles. However, little is known about what factors govern the "choice" of whistle contours by calves. Again, studies that utilize acoustic localization with concurrent behavioral observations promise to shed light on this interesting question.

Signature whistles are an important class of vocalizations produced by bottlenose dolphins. They are learned vocalizations that function both in individual recognition and in maintaining group cohesion. Much work remains to be done in order to learn how these whistles develop and how they are used in the natural communication system of dolphins.

**S**

## See Also the Following Articles

Communication ■ Echolocation ■ Song ■ Sound Production

## References

Caldwell, M. C., and Caldwell, D. K. (1965). Individualized whistle contours in bottlenosed dolphins (*Tursiops truncatus*). *Nature* **207**, 434–435.

Caldwell, M. C., and Caldwell, D. K. (1968). Vocalization of naive captive dolphins in small groups. *Science* **159**, 1121–1123.

Caldwell, M. C., and Caldwell, D. K. (1971). Statistical evidence for individual signature whistles in Pacific whitesided dolphins, *Lagenorhynchus obliquidens*. *Cetology* **3**, 1–9.

Caldwell, M. C., and Caldwell, D. K. (1979). The whistle of the Atlantic bottlenosed dolphin (*Tursiops truncatus*)-Ontogeny. *In* "Behavior of Marine Animals: Current Perspectives in Research. Vol. 3 Cetaceans" (H. E. Winn, and B. L. Olla, eds), pp. 369–401. Plenum Press, New York.

Caldwell, M. C., Caldwell, D. K., and Miller, J. F. (1973). Statistical evidence for individual signature whistles in the spotted dolphin, *Stenella plagiodon*. *Cetology* **16**, 1–21.

Caldwell, M. C., Caldwell, D. K., and Tyack, P. L. (1990). Review of the signature whistle hypothesis for the Atlantic bottlenose dolphin, *Tursiops truncatus*. *In* "The Bottlenose Dolphin" (S. Leatherwood, and R. Reeves, eds), pp. 199–234. Academic Press, New York.

Cook, M. L. H., Sayigh, L. S., Blum, J. E., and Wells, R. S. (2004). Signature-whistle production in undisturbed free-ranging bottlenose dolphins (*Tursiops truncatus*). *Proc. R. Soc. B* **271**, 1043–1049.

Fripp, D., Owen, C., Quintana-Rizzo, E., Shapiro, A., Buckstaff, K., Jankowski, K., Wells, R., *et al.* (2005). Bottlenose dolphin (*Tursiops truncatus*) calves appear to model their signature whistles on the signature whistles of community members. *Anim. Cogn.* **8**, 17–26.

Janik, V. M. (2000). Whistle matching in wild bottlenose dolphins (*Tursiops truncatus*). *Science* **289**, 1355–1357.

Janik, V. M., and Slater, P. J. B. (1997). Vocal learning in mammals. *Adv. Study Behav.* **26**, 59–99.

Janik, V. M., and Slater, P. J. B. (1998). Context-specific use suggests that bottlenose dolphin signature whistles are cohesion calls. *Anim. Behav.* **56**, 829–838.

Janik, V. M., Dehnhardt, G., and Todt, D. (1994). Signature whistle variations in a bottlenosed dolphin, *Tursiops truncatus*. *Behav. Ecol. Sociobiol.* **35**, 243–248.

Janik, V. M., Sayigh, L. S., and Wells, R. S. (2006). Signature whistle shape conveys identity information to bottlenose dolphins. *Proc. Natl. Acad. Sci. U.S.A.* **103**, 8293–8297.

Marino, L., Connor, R. C., Fordyce, R. E., Herman, L. M., Hof, P. R., Lefebvre, L., Lusseau, D., *et al.* (2007). Cetaceans have complex brains for complex cognition. *PLoS Biol* **5**, e139 doi:10.1371/journal.pbio.0050139.

McCowan, B., and Reiss, D. (2001). The fallacy of 'signature whistles' in bottlenose dolphins: a comparative perspective of 'signature information' in animal vocalizations. *Anim. Behav.* **62**, 1151–1162.

Miksis, J. L., Tyack, P. L., and Buck, J. R. (2002). Captive dolphins, *Tursiops truncatus*, develop signature whistles that match acoustic features of human-made model sounds. *J. Acoust. Soc. Am.* **112**, 728–739.

Sayigh, L. S., Tyack, P. L., Wells, R. S., and Scott, M. D. (1990). Signature whistles of free-ranging bottlenose dolphins, *Tursiops truncatus*: stability and mother-offspring comparisons. *Behav. Ecol. Sociobiol.* **26**, 247–260.

Sayigh, L. S., Tyack, P. L., Wells, R. S., Scott, M. D., and Irvine, A. B. (1995). Sex difference in signature whistle production of free-ranging bottlenose dolphins, *Tursiops truncatus*. *Behav. Ecol. Sociobiol.* **36**, 171–177.

Sayigh, L. S., Tyack, P. L., Wells, R. S., Solow, A., Scott, M. D., and Irvine, A. B. (1999). Individual recognition in wild bottlenose dolphins: a field test using playback experiments. *Anim. Behav.* **57**, 41–50.

Sayigh, L. S., Esch, H. C., Wells, R. S., and Janik, V. M. (2007). Facts about signature whistles of bottlenose dolphins (*Tursiops truncatus*). *Anim. Behav.* **74**, 1631–1642.

Smolker, R., Mann, J., and Smuts, B. (1993). Use of signature whistles during separations and reunions between bottlenose dolphin mothers and infants. *Behav. Ecol. Sociobiol.* **33**, 393–402.

Smolker, R., and Pepper, J. (1999). Whistle convergence among allied male bottlenose dolphins (Delphinidae, *Tursiops* sp.). *Ethology* **105**, 595–617.

Tyack, P. L. (1986). Whistle repertoires of two bottlenosed dolphins, *Tursiops truncatus*: mimicry of signature whistles? *Behav. Ecol. Sociobiol.* **18**, 251–257.

van Parijs, S. M., and Corkeron, P. J. (2001). Evidence for signature whistle production by a Pacific humpback dolphin, *Sousa chinensis*. *Mar. Mamm. Sci.* **17**, 944–949.

Watwood, S. L., Tyack, P. L., and Wells, R. S. (2004). Whistle sharing in paired male bottlenose dolphins, *Tursiops truncatus*. *Behav. Ecol. Sociobiol.* **55**, 531–543.

# Sirenian Evolution

## Daryl P. Domning

Sirenia are the order of placental mammals comprising modern seacows (manatees and dugongs) and their extinct relatives. They are the only herbivorous marine mammals now living, and the only herbivorous mammals ever to have become totally aquatic. Sirenians have a known fossil record extending over some 50 million years (early Eocene–Recent). They attained a modest diversity in the Oligocene and Miocene, but since then have declined as a result of climatic cooling, other oceanographic changes, and human depredations. Only two genera and four species survive today: the three species of manatees (*Trichechus*) live along the Atlantic coasts and rivers of the Americas and West Africa; one, the Amazonian manatee, is found only in fresh water. The dugong (*Dugong*) lives in the Indian and southwest Pacific oceans. For comprehensive references to technical as well as popular publications on fossil and living sirenians, see Domning (1996).

## I. Sirenian Origins

The closest living relatives of sirenians are the Proboscidea (elephants). The Sirenia, the Proboscidea, the extinct Desmostylia, and probably the extinct Embrithopoda together make up a larger group called Tethytheria, whose members (as the name indicates) appear to have evolved in the Old World along the shores of the ancient Tethys Sea. Together with the Hyracoidea (hyraces), tethytheres seem to form a more inclusive group long referred to as Paenungulata. The Paenungulata and (especially) Tethytheria are among the least controversial groupings of mammalian orders, and are strongly supported by most morphological and molecular studies. Whereas morphological evidence associates these groups with the ungulates, however, molecular studies have placed them in the Afrotheria, together with insectivores, elephant shrews, and aardvarks. This discrepancy is unresolved (Gheerbrant *et al.*, 2005). In either case, their ancestry is remote from that of cetaceans or pinnipeds; sirenians re-evolved an aquatic lifestyle independently of (though simultaneously with) cetaceans, ultimately displaying strong convergence with them in body form.

## II. Early History, Anatomy, and Mode of Life

Sirenians first appear in the fossil record in the late early Eocene, and the order was already diverse by the middle Eocene (Fig. 1). As inhabitants of rivers, estuaries, and nearshore marine waters, they

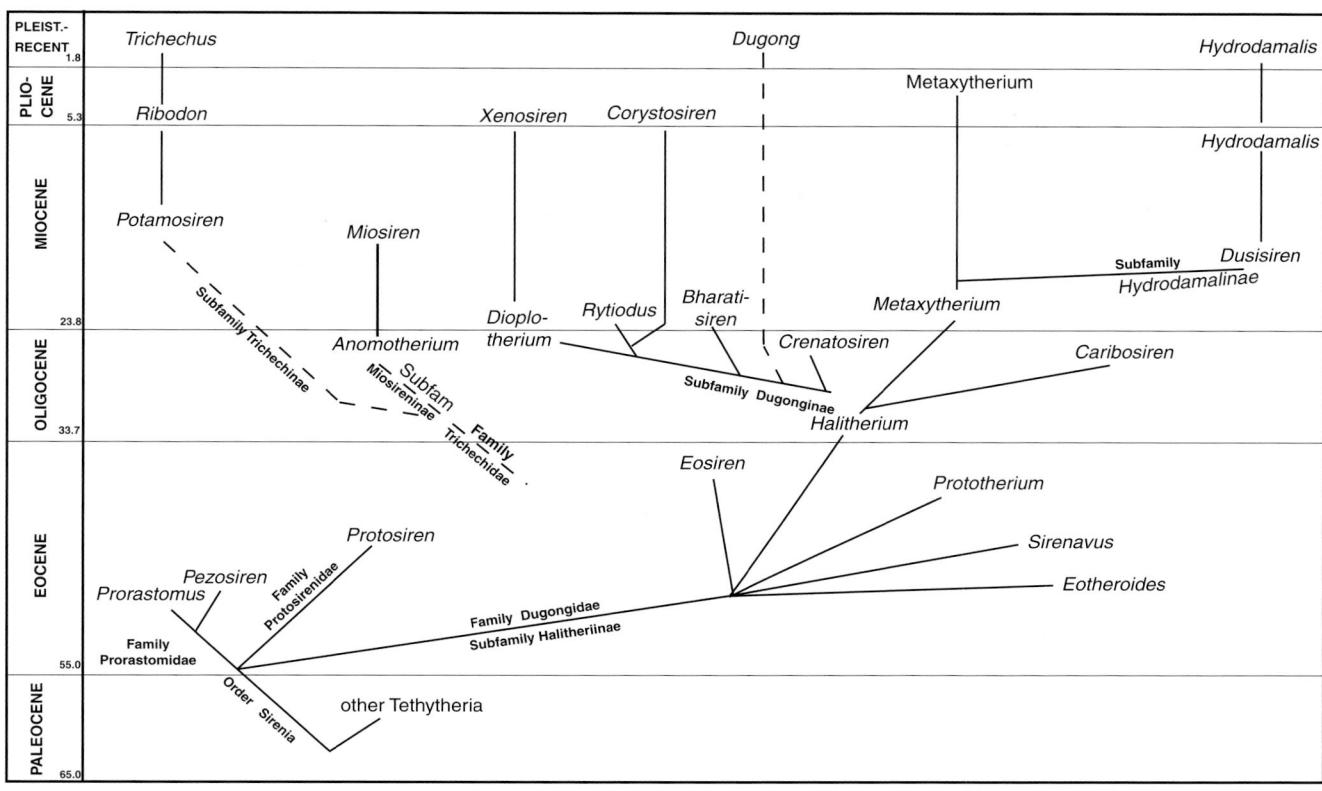

**Figure 1** *Simplified phylogeny of Sirenia, including better-known genera. The time scale (at left) is in millions of years. Undocumented "ghost lineages" that span multiple epochal boundaries are shown as dashed lines.*

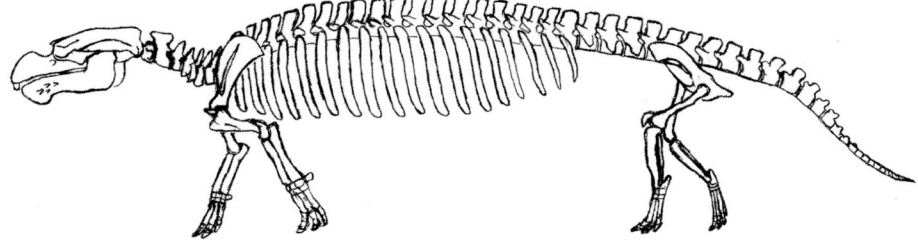

**Figure 2** *Skeleton of* Pezosiren portelli, *an Eocene prorastomid. Total length about 2.1 m.*

were able to spread quickly along the coasts of the world's shallow tropical seas; in fact, the most primitive sirenians known to date (*Prorastomus, Pezosiren*; Fig. 2) were found not in the Old World but in Jamaica (Savage *et al.*, 1994; Domning, 2001a).

The earliest seacows (families Prorastomidae and Protosirenidae, both confined to the Eocene) were pig-sized, four-legged amphibious creatures (Fig. 2). By the end of the Eocene, with the appearance of the Dugongidae, sirenians had taken on their modern, completely aquatic, streamlined body form, featuring flipper-like front legs, no hind legs, and a powerful tail with a horizontal caudal fin, whose up-and-down movements propel them through the water, as in whales and dolphins (Fig. 3). The last-appearing of the four sirenian families (Trichechidae) arose from either protosirenids or primitive dugongids, probably during the Eocene. The sirenian fossil record now documents all the major stages of hindlimb and pelvic reduction from completely "terrestrial" morphology to the extremely reduced condition of the pelvis seen in modern manatees, thereby providing one of the

most dramatic examples of evolutionary change to be seen among fossil vertebrates.

From the outset, sirenians were herbivores, and probably depended on seagrasses and other aquatic angiosperms (flowering plants) for food. To this day, almost all members of the order have remained tropical, marine, and eaters of angiosperms. No longer capable of locomotion on land, sirenians are born in the water and spend their entire lives there. Because they are shallow divers with large lungs, they have heavy skeletons, like a diver's weight belt, to help them stay submerged: their bones are both swollen (pachyostotic) and dense (osteosclerotic), especially the ribs, which are often found as fossils (Domning and de Buffrénil, 1991).

The sirenian skull is characterized by an enlarged and more or less downturned premaxillary rostrum, retracted nasal opening, absence of paranasal air sinuses, laterally salient zygomatic arches, and thick, dense parietals fused into a unit with the supraoccipital. Nasals and lacrimals tend to become reduced or lost, and in most

**S**

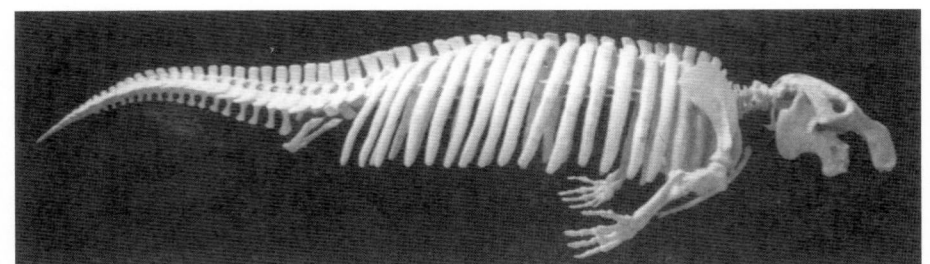

**Figure 3** *Skeleton of* Metaxytherium floridanum, *a Miocene halitheriine dugongid. Total length about 3.2 m. After Domning (1988); reproduced with the permission of the* Journal of Vertebrate Paleontology.

forms the pterygoid processes are large and stout. The periotic is snugly enclosed by a socket in the squamosal and is fused with a ring-shaped tympanic. The mandibular symphysis is long, deep, laterally compressed, and typically fused and downturned; in all but prorastomids the mandibular foramen is enlarged to expose the dental capsule. Incisors, where present, are arranged in parallel, longitudinally aligned rows. In all but the most primitive taxa, the infraorbital and mental foramina are enlarged to accommodate the nerve and blood supply to the large, prehensile, vibrissae-studded lips, which are moved by muscular hydrostats (cf. Marshall *et al.*, 1998).

Eocene sirenians, like Mesozoic mammals but in contrast to other Cenozoic ones, have five instead of four premolars, giving them a 3.1.5.3 dental formula. Whether this condition is truly a primitive retention in the Sirenia is still being debated. The fourth lower deciduous premolar is trilobed, like that of many other ungulates; this raises the further unresolved question of whether the three following teeth (dp5, m1, and m2) are actually the homologs of the so-called m1–3 in other mammals.

Although the cheek teeth are relied on for identifying species in many other mammalian groups, they do not vary much in morphology among the Sirenia, but are almost always low-crowned (brachyodont) with two rows of large, rounded cusps (bunobilophodont). (The most taxonomically informative parts of the sirenian skeleton are the skull and mandible, especially the frontal and other bones of the skull roof; Fig. 4.) Except for a pair of tusk-like first upper incisors seen in most species, front teeth (incisors and canines) are lacking in all but the earliest fossil sirenians, and cheek teeth in adults are commonly reduced in number to four or five on each side of each jaw: one or two deciduous premolars, which are never replaced, plus three molars. As described below, however, all three of the Recent genera have departed in different ways from this "typical" pattern.

### III. Dugongidae

Dugongids comprise the vast majority of the species and specimens that make up the known fossil record of sirenians. The basal members of this very successful family are placed in the long-lived (Eocene-Pliocene) and cosmopolitan subfamily Halitheriinae (Fig. 1). This paraphyletic group included the well-known fossil genera *Halitherium* and *Metaxytherium*, which were relatively unspecialized seagrass eaters.

*Metaxytherium* (Fig. 3) gave rise in the Miocene to the Hydrodamalinae, an endemic North Pacific lineage that ended with Steller's sea cow (*Hydrodamalis*)—the largest sirenian that ever lived (up to 9 m or more in length) and the only one to adapt successfully to temperate and cold waters and a diet of marine algae. It was completely toothless, and its truncated, clawlike flippers, used for gathering plants

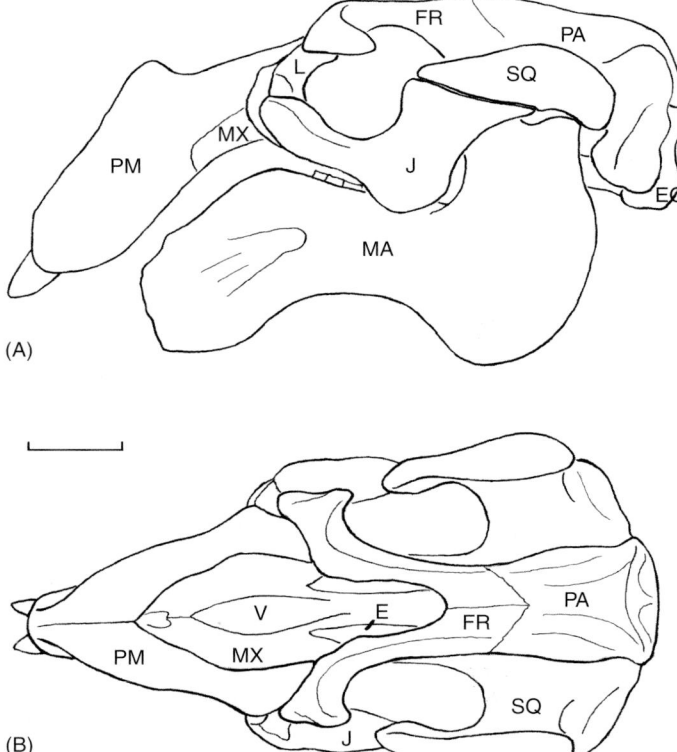

**Figure 4** *Skull of* Crenatosiren olseni, *an Oligocene dugongine dugongid, in* (A) *lateral and* (B) *dorsal views. Note the large incisor tusks in the premaxillae. E, ethmoid; EO, exoccipital; FR, frontal; J, jugal; L, lacrimal; MA, mandible; MX, maxilla; PA, parietal; PM, premaxilla; SQ, squamosal; V, vomer. Scale bar: 5 cm. After Domning (1997); reproduced with the permission of the* Journal of Vertebrate Paleontology.

and fending off from rocks, contained no finger bones (phalanges). It was hunted to extinction for its meat, fat, and hide circa AD 1768.

Another offshoot of the Halitheriinae, the subfamily Dugonginae, appeared in the Oligocene (Fig. 4). Most dugongines were apparently specialists at digging out and eating the tough, buried rhizomes of seagrasses; for this purpose many of them had large, self-sharpening blade-like tusks (Domning, 2001b). The modern *Dugong* is the sole survivor of this group, but it has reduced its dentition (the cheek teeth have only thin enamel crowns, which quickly wear off, leaving simple pegs of dentine) and has (perhaps for that reason) shifted its

S

diet to more delicate seagrasses and ceased to use its tusks for digging (Domning and Beatty, 2007).

## IV. Trichechidae

Trichechidae have a much less complete fossil record than dugongids. Their definition has been broadened by Domning (1994) to include the Miosireninae, a peculiar and little-known pair of genera which inhabited northwestern Europe in the late Oligocene and Miocene (Fig. 1). Miosirenines had massively reinforced palates and dentitions that may have been used to crush shellfish. Such a diet in sirenians living around the North Sea seems less aberrant when we consider that modern dugongs and manatees near the climatic extremes of their ranges are known to consume invertebrates in addition to plants.

Manatees in the strict, traditional sense are now placed in the subfamily Trichechinae. They first appear in the Miocene, represented by *Potamosiren* from freshwater deposits in Colombia. Indeed, much of trichechine history was probably spent in South America, whence they spread to North America and Africa only in the Pliocene or Pleistocene (Domning, 2005).

During the late Miocene, manatees living in the Amazon basin evidently adapted to a diet of abrasive freshwater grasses by means of an innovation still used by their modern descendants: they continue to add on extra teeth to the molar series as long as they live, and as worn teeth fall out at the front, the whole tooth row slowly shifts forward to make room for new ones erupting at the rear. This type of horizontal tooth replacement has often been likened, incorrectly, to that of elephants, but the latter are limited to only three molars. Only one other mammal, an Australian rock wallaby (*Peradorcas concinna*), has truly evolved the kind of tooth replacement seen in manatees (Domning and Hayek, 1984).

## See Also the Following Articles

Desmostylia ▪ Dugong ▪ Manatees ▪ Musculature ▪ Steller's Sea Cow ▪ Tethytheria

## References

Domning, D. (1978). Sirenian evolution in the North Pacific Ocean. *Univ. Calif. Publ. Geol. Sci.* **118**, xi + 176.

Domning, D. (1982). Manatee evolution: a speculative history. *J. Paleontol.* **56**, 599–619.

Domning, D. (1988). Fossil Sirenia of the West Atlantic and Caribbean region. I. *Metaxytherium floridanum* Hay, 1922. *J. Vert. Paleontol.* **8**, 395–426.

Domning, D. (1994). A phylogenetic analysis of the Sirenia. *Proc. San Diego Soc. Nat. Hist.* **29**, 177–189.

Domning, D. (1996). Bibliography and index of the Sirenia and Desmostylia. *Smith. Contrib. Paleobiol.* **80**, 1–611.

Domning, D. (1997). Fossil Sirenia of the West Atlantic and Caribbean region. VI. *Crenatosiren olseni* (Reinhart, 1976). *J. Vert. Paleontol.* **17**, 397–412.

Domning, D. (2001a). The earliest known fully quadrupedal sirenian. *Nature* **413**, 625–627.

Domning, D. (2001b). Sirenians, seagrasses, and Cenozoic ecological change in the Caribbean. *Palaeogeogr. Palaeoclimatol. Palaeoecol.* **166**, 27–50.

Domning, D. (2005). Fossil Sirenia of the West Atlantic and Caribbean region. VII. Pleistocene *Trichechus manatus* Linnaeus, 1758. *J. Vert. Paleontol.* **25**, 685–701.

Domning, D., and Beatty, B. (2007). Use of tusks in feeding by dugongid sirenians: Observations and tests of hypotheses. *Anat. Rec.* **290**, 523–538.

Domning, D., and de Buffrénil, V. (1991). Hydrostasis in the Sirenia: quantitative data and functional interpretations. *Mar. Mamm. Sci.* **7**, 331–368.

Domning, D., and Hayek, L. (1984). Horizontal tooth replacement in the Amazonian manatee (*Trichechus inunguis*). *Mammalia* **48**, 105–127.

Gheerbrant, E., Domning, D., and Tassy, P. (2005). Paenungulata (Sirenia, Proboscidea, Hyracoidea, and relatives). *In* "The Rise of Placental Mammals: Origin and Relationships of the Major Extant Clades" (K. Rose, and J. Archibald, eds), pp. 84–105. Johns Hopkins University Press, Baltimore.

Marshall, C., Clark, L., and Reep, R. (1998). The muscular hydrostat of the Florida manatee (*Trichechus manatus latirostris*): A functional morphological model of perioral bristle use. *Mar. Mamm. Sci.* **14**, 290–303.

Reynolds, J., III, and Odell, D. (1991). "Manatees and Dugongs." Facts on File, New York.

Savage, R., Domning, D., and Thewissen, J. (1994). Fossil Sirenia of the West Atlantic and Caribbean region. V. The most primitive known sirenian, *Prorastomus sirenoides* Owen, 1855. *J. Vert. Paleontol.* **14**, 427–449.

# Sirenian Life History

DANIEL K. ODELL

Sirenians are unique among the marine mammals in that they are herbivores. The mammalian order Sirenia has fossil representatives on the order of 50 million years old (Eocene epoch), making them as old as the cetaceans. While there are similarities between the two groups in terms of adaptation to the aquatic environment (e.g., streamlining, loss of hind limbs, use of the tail for propulsion), the sirenians stayed on the path to herbivory while the cetaceans switched from the herbivorous habits of their terrestrial ancestors to become carnivores. Modern sirenians include three species of manatees (family Trichechidae: Amazonian manatee, *Trichechus inunguis;* West African manatee, *T. senegalensis;* West Indian manatee, *T. manatus;* the latter is divided into two subspecies: Antillean manatee, *T. m. manatus*, and Florida manatee, *T. m. latirostris*). The Sirenia also includes the family Dugongidae: dugong, *Dugong dugon*), and Steller's sea cow, *Hydrodamalis gigas*). The manatees, as their common names suggest, are distributed in the coastal tropical and subtropical regions of the Atlantic Ocean. The dugong is found in the South Pacific and Indian Ocean regions. The manatees and the dugong are considered endangered or threatened under various national laws and international conservation schemes. The Steller's sea cow is the exception, not only is it extinct (eliminated by over-hunting about 25 years after its discovery in 1765) but it inhabited the islands in the western reaches of the Aleutian archipelago in the North Pacific Ocean. Definitely not a tropical clime!

## I. Methodology

Ricklefs (1973) stated that "The life history of an organism can be described in terms of fecundity, growth and development, age at [sexual] maturity, parental care, and longevity." These parameters are closely linked and are acted on by natural selection. While these parameters are easy to state, they are not always easy to measure. It is easy to study a colony of mice in the laboratory but not so with an animal like the Florida manatee that may reach a length of over 4 m

**S**

and a weight in excess of 1500 kg. How, then, does one get this information for Sirenians? Approaches to this question involve both field and "laboratory" (i.e., animals in captivity in marine zoological parks and zoos) studies. In the field, one can learn to identify individual animals using shape, size, and various markings and color patterns or one can mark) using various methods) individual animals. For large, long-lived, aquatic mammals like the sirenians, this becomes an extremely expensive and time-consuming operation. However, in many cases, it is the only way to obtain the desired information. Alternatively, one can collect data from animals that die from both natural and unnatural (i.e., human-related) causes. Animals killed for human consumption or those killed as a result of other human activities provide a sample, albeit potentially biased (nonrandom), of males and females of all age classes. Marine mammal stranding networks collect life history data on animals found dead (i.e., stranded, beach cast, beached). In Florida, the Florida Fish and Wildlife Conservation Commission operates a network dedicated to the collection and examination of all manatees found dead in the state. From these carcasses one can gather information on, among others, reproductive status (immature, mature, pregnant, number of pregnancies) and age. Estimating the age of an animal is critically important in the estimation of life history parameters. Toothed cetaceans and pinnipeds, for example, have permanent teeth with roots that grow continually (but slowly) from birth to death. One can section these teeth and, under the microscope, count growth layer groups (GLGs). If one then knows the frequency with which these GLGs are deposited (e.g., one per year, 2 per year), GLG counts can be directly converted to and estimated age in years. The frequency of GLG deposition can be estimated using known age animals and using chemical markers deposited in the teeth of live animals. There is variability in the reliability of this process from species to species and the only way to know the true age of an animal is to follow it from birth to death. Growth layers are also often deposited in growing bone. The dugong has tusks that erupt through the gums in the male but not in the female. These tusks grow throughout the life of the animal and GLGs accumulate on an annual (assumed) basis. Manatees, however, do not have permanent teeth. While some growth layers are present in the tooth roots in young animals, these are lost through resorption, as manatees have a continual, horizontal (back to front) tooth replacement throughout life. However, because other growing bones accumulate GLGs, scientists discovered that GLGs in manatee ear bones can be used to estimate age. Studies with chemical markers demonstrated that manatee GLGs accumulate on an annual basis.

Studies with dead animals are very useful and relatively inexpensive, but long-term studies on live animals are necessary to fill in the life history blanks. For example, by following a female from year to year, it is possible to document when she has a calf and how much time elapses between successive calves. While radio and satellite tracking allows scientists to follow individual manatees over long periods of time, it is virtually impossible to watch them from minute to minute. However, studies on manatees in captivity (aka "the laboratory") can fill some of these life history gaps, particularly in the areas of reproductive biology and growth and development. In captivity manatees *can* be observed continually if necessary. Blood and urine samples can be collected to document estrous cycling and gestation in females and testicular activity cycling in males. Parental care of the offspring can be documented in great detail.

## II. Longevity

Detailed age estimation studies have only been done for the Florida manatee and for the dugong. Using tetracycline marking, we know that Florida manatees deposit one GLG per year. Based on growth layer groups in the ear bone complex, some Florida manatees may live to be 60 years of age. However, due to bone resorption, 60 years may be a minimum estimate. A manatee born in the Miami Aquarium in 1948 is alive and well today (July 2007) at an aquarium in Bradenton, Florida. This animal provides a good measure of validation for the 60-year estimate. The average longevity for Florida manatees has not been well estimated but is probably on the order of 30 years less. This estimate may be biased downward by manatee deaths resulting from human activities (e.g., watercraft collisions). Average longevity may vary between sexes and among year classes. Longevity data do not exist for Antillean, Amazonian, or West African manatees.

Based on GLGs in their tusks, the oldest dugong examined to date was a female with 73 GLGs, which translates to 73 years if we assume one GLG per year. Tusks in male dugongs erupt through the gums and are worn down, eliminating the early GLGs in older animals. Minimum maximum longevity estimates for male dugongs are about 35 years.

## III. Age at Sexual Maturity

Female Florida manatees reach sexual maturity between 2.5 and 6 years of age. Male Florida manatees reach sexual maturity between 2 and 11 years of age based on the presence of active spermatogenesis in the testes. It is important to note that even if a male manatee is sexually mature at an early age, it may not be physically large enough to compete for mating rights in an estrous (mating herd). However, under circumstances wherein there are not other competing males, small manatees may be able to mate successfully and produce offspring. There are no data on age at sexual maturity for Antillean, Amazonian, or West African manatees.

Even though the estimated maximum life expectancy for the dugong is similar to that of the Florida manatee, dugongs appear to mature at greater ages, and the age at sexual maturity may vary among populations. Female dugongs in Australia and Papua New Guinea mature at 9.5–17.5 years. Male dugongs in the same areas mature at 9–16 years of age. These differences could reflect distinct genetic differences between populations or the effects of density-dependent factors.

## IV. Parental Care

In both manatees and dugongs the male plays no apparent role in the care of the young. After mating, males and females go their separate ways. Florida manatee calves will stay with their mothers for 1 year on average. Most calves (about 70%) stay with their mothers through one winter season and the remainder through two winter seasons. Nutritional weaning is a gradual process but there are few data on the length of lactation in the Florida manatee or in the dugong. Lactation may last 1.5 years in dugongs and is probably similar in the manatee. Some wild and captive manatees appear to lactate for several years under certain circumstances. Most Florida manatee and dugong calves start feeding on vegetation at a few weeks of age and may be nutritionally independent before they reach the age of 1 year. However, at least in the case of the Florida manatee, calves probably learn the locations of feeding areas and warm water refugia by following their mothers through at least one winter season. The latter is extremely important for weaned juvenile survival during cold winters.

S

## V. Growth and Development

Florida manatee calves average about 120 cm in length at birth, but viable calves may have a birth length ranging from about 80 to 160 cm. This, along with variable individual growth rates, results in a highly variable length at age distribution. For example, 2-year-old manatees at Blue Spring (Florida) may range from 210 to 260 cm total length. From a sample of carcasses of Florida manatees in the age class >1 and <2 years, total lengths ranged from about 120 to over 260 cm. Florida manatees grow rapidly during the first few years, and body length becomes asymptotic at about 300 cm and an age of 8-10 years. The birth weight for Florida manatees ranges from 30 to 50 kg and the average adult body weight is about 500 kg. The Antillean and West African manatees are probably similar to the Florida manatee in these respects. The Amazonian manatee is smaller and has a birth length of 85–100 cm and a birth weight of 10–15 kg. Large adults may reach lengths of 280 cm and weights of 480 kg. The dugong has a birth length of 100–130 cm and a birth weight of 25–35 kg. Adult dugongs average 270 cm in length and 250–300 kg body weight.

## VI. Fecundity

As with other aspects of sirenian life history, data on fecundity are limited and based primarily on studies of the Florida manatee and the dugong. A key factor in assessing fecundity is the gestation period, which, despite numerous conceptions and births of Florida manatees, is not known for any sirenian species. However, scientists generally agree that gestation for the Florida manatee and the dugong is in the range of 12–14 months and that the other species of sirenians are probably similar. The known inter-birth interval for wild Florida manatees averages 2.5–2.6 years. The estimated inter-birth interval for Florida manatees ranges from 2.5 to 3.0 years when gestation period estimates of 12, 13, and 14 months are applied. This suggests that the true gestation period may be close to 12 months. Estimated inter-birth intervals for the dugong range from 2.7 to 5.8 years depending on the length of the gestation period assumed (12, 13, or 14 months) and the population of dugongs used. We do know that Florida manatees resume estrous cycling within 1–2 months after the loss of a calf and become pregnant shortly thereafter. Whereas manatees and dugongs (both males and females) display seasonal reproductive activity, there is scant evidence to suggest that they have reproductive senescence as many mammals, including humans, do.

Other factors important in the calculation of fecundity are the number of offspring per birth and the sex ratio of offspring at birth. Wild Florida manatees have been documented producing twin offspring (but no more) in about 1.4–1.8% of births. Estimates of twinning based on Florida manatee carcass studies are as high as 4%. Limited and/or anecdotal data suggest that twinning occurs in the Antillean manatee and in the dugong. The sex ratio of the offspring at birth appears to be 1:1 for both the Florida manatee and the dugong. It is probably reasonable to assume that both Amazonian and West African manatees have similar patterns of twinning and offspring sex ratio.

Applying all of our knowledge of and assumptions about Florida manatee life history, the average fecundity (number of female births per female per year) for age classes 4–29 years was estimated at 0.189, 0.238, and 0.127 using different sets of data obtained from the examination of carcasses. Similarly, estimates of annual pregnancy rate (APR) and gross annual recruitment rate (GARR) ranged from 0.190 to 0.394 and 0.044 to 0.90, respectively, depending on the set of carcass data and gestation period used. Similar calculations for the dugong based on a smaller data set yielded apparent (annual) pregnancy rates of 0.093–0.353 over a series of years. Both the State of Florida (Haubold *et al.*, 2006; Florida Fish and Wildlife Conservation Commission, 2006) and the US Fish and Wildlife Service (2007) have used life history parameters to assess the status of the Florida manatee. These reports include references to several population modeling projects that were in progress and that were not (at the time the reports were released) available for general public consumption in the peer-reviewed or gray literature.

## VII. Summary

Even though the Florida manatee and the dugong have been studied intensively since the 1980s, detailed data on many aspects of their life history are only beginning to be elucidated and there is considerably less information on the other species of sirenians. Given the threatened or endangered status of this unique group of marine mammals, every effort should be made to learn about their biology so that we may ensure their survival.

DKO has been involved with marine mammal studies since 1965 and has worked on pinnipeds, Sirenians and the smaller cetaceans. He has held positions at the University of Miami's Rosenstiel School of Marine and Atmospheric Science, and in Corporate Zoological Operations for the Busch Entertainment Corporation (SeaWorld and Bush Gardens parks). He has served as president of the Society for Marine Mammalogy and president of the Florida Academy of Sciences. He has worked on sirenians, particularly the Florida manatee, since 1974 and was a co-founder of the Florida manatee carcass salvage program.

## See Also the Following Articles

Dugong ■ Manatees

## References

Florida Fish and Wildlife Conservation Commission (2006). "Draft manatee management plan *Trichechus manatus latirostris.*" Florida Fish and Wildlife Conservation Commission, Tallahassee, FL 32399, USA.

Hartman, D. S. (1979). "Ecology and Behavior of the Manatee (*Trichechus manatus*) in Florida." American Society of Mammalogists, Special Publication No. 5. Lawrence, Kansas, USA.

Haubold, E. M., Deutsch, C., and Fonnesbeck, C. (2006). "Final biological status review of the Florida manatee (*Trichechus manatus latirostris*)." Florida Fish and Wildlife Conservation Commission, Fish and Wildlife Research Institute, St. Petersburg, FL.

O'Shea, T. J., Ackerman, B. B., and Percival, H. F. (eds.) (1995). "Population Biology of the Florida Manatee." Information and Technology Report 1. U.S. Department of the Interior, National Biological Service, Washington, DC.

Reynolds, J. E., III, and Odell, D. K. (1990). "Manatees and Dugongs." Facts on File, New York.

Reynolds, J. E., III, and Rommel, S. A. (eds) (1999). "Biology of Marine Mammals." Smithsonian Institution Press, Washington, D.C.

Ricklefs, R. E. (1973). "Ecology." Chiron Press, Portland, OR.

Ridgway, S. H., and Harrison, R. J. (eds) (1985). "Handbook of Marine Mammals," Vol. 3. Academic Press, London.

# Skeleton, Postcranial

SENTIEL ROMMEL AND JOHN E. REYNOLDS, III

The postcranial skeleton includes all the bones and cartilages caudal to the head skeleton; it is subdivided into axial components (the vertebral column, ribs, and sternebrae, which are "on" the midline) and appendicular components (the forelimbs, hindlimbs, and pectoral and pelvic girdles, which are "off" the midline). The skeleton supports and protects soft tissues, controls modes of locomotion, and predominantly determines overall body size and shape. The marrow of some bones may generate the precursors to certain blood cells. Skeletal elements may store lipids (particularly in the Cetacea) and calcium (particularly in the Sirenia) and thus influence buoyancy (Kipps *et al.*, 2002). Because bones are continuously remodeled in response to biochemical and biomechanical demands over the life span of the marine mammal, they offer information that can help interpret life history events after death.

We discuss the skeletons of seven different species of marine mammals: the Florida manatee (*Trichechus manatus latirostris*), the harbor seal (*Phoca vitulina*), the California sea lion (*Zalophus californianus*), the North Atlantic right whale (*Eubalaena glacialis*), the bottlenose dolphin (*Tursiops truncatus*), the polar bear (*Ursus maritimus*), and the sea otter (*Enhydra lutris*). We use the domestic dog skeleton (*Canis familiaris*) to provide a familiar reference. These marine mammal species were chosen, in part, because much is known about them and they provide a wide range of biomechanical morphologies. The skull morphology of these seven species is described in the chapter entitled SKULL ANATOMY of this encyclopedia. The manatee, dolphin, and right whale represent the many species that have lost their hindlimbs and are permanently aquatic; the other species may spend at least some of their life on land.

Visualize the articulated (assembled) skeleton by first considering individuals in an absolute sense (left and lower parts of Fig. 1) and then in a relative sense (upper right of Fig. 1). Consider what decisions morphologists make when comparing features among individuals that differ substantially in size. Contrast the two methods of scaling by using the human swimmer as a reference. Because relational comparisons (such as proportions and percentages) are intuitive, we use the relative scheme for most of this chapter.

We will refer to the manatee when describing individual morphological structures but refer to the seven selected marine mammal species when discussing marine mammals in general (for details on the same structures in the dolphin, see Rommel 1990; for the true seals, see Pierard, 1971; Pierard and Bisaillon, 1978; for the dog, see Evans, 1993; for mammals in general, see Flower, 1885).

## I. Axial Skeleton

### A. Vertebral Structures

A few terms are necessary to better grasp the functional morphology of the vertebral column. Each *vertebra* (plural vertebrae) has several parts (Fig. 2A). The body or *centrum* (plural-centra) of each vertebra forms the primary mechanical support of the vertebral (spinal) column (Fig. 2B). Several projecting parts or *processes* may extend from the centrum and provide connective tissue attachment sites, protection for soft tissues, or both. The largest and most common lateral processes are transverse processes, which tend to be long and robust in the lumbar vertebrae. In the cervical region, there may be two transverse processes extending from each side of a single vertebra.

Dorsal to the centrum is the *neural arch*, an arch of bone that protects the spinal cord. Interestingly, in some marine mammal species, the neural arch is considerably enlarged to accommodate relatively large masses of vascular tissue and/or fat that surround the spinal cord. Because of these enlargements, the neural canal may not reflect the dimensions of the spinal cord as it does in most terrestrial species (Giffen 1992). In addition, each neural arch may extend

dorsally as a *neural spine* (spinous process). Neural spines function as levers to increase the mechanical advantage of the epaxial (epiaxial) muscles to dorsoflex the body (Pabst, 1993).

Relative motion between adjacent vertebrae is controlled, in part, by the size and shape of the *intervertebral disks* (Fig. 2C). Intervertebral disks are flexible yet resist compression; they consist of a fibrous outer ring, the *annulus fibrosus* and a gelatinous inner mass, the *nucleus pulposis*. The jelly-like nucleus forms a dynamic joint that can adjust its shape within the elastic fibers of the annulus. Intervertebral disks are resilient structures that support complex forces of bending. The same design of surrounding a deformable core with elastic fibers is used in baseballs and golf balls.

Flexibility of the vertebral column depends, in part, on the thickness of the disks. Disk thickness varies within and among species, and intervertebral disks represent a substantial proportion (10–30%) of the length of the vertebral column. Relaxed (neutral) curvature of the vertebral column is determined by the shapes of the individual vertebrae, *not* by the shapes of the intervertebral spaces. Intervertebral disks provide flexibility, whereas curvature is provided by (nonparallel) vertebral body faces (Fig. 3).

Adjacent vertebrae may have other surfaces of articulation (facets); *zygapophyses* (singular, zygapophysis) may be located on the neural arch, neural spine, and/or the transverse processes. Zygapophyses are bilaterally paired articular facets, found on the cranial and caudal aspects of each vertebra (Fig. 2D). The cranial pair are termed *prezygapophyses*, and the caudal pair *postzygapophyses*. The region of articulation between prezygapophyses and postzygapophyses is typically a synovial joint if the facets overlap.

The zygapophyses in the neck and cranial thorax tend to be oriented horizontally to allow axial rotation (twisting) and lateral (side-to-side) motions. The zygapophyses in the caudal thorax and tail tend to be oriented vertically to facilitate dorsoventral (up-and-down) motions. In some regions of the vertebral column adjacent vertebrae may have reduced or absent zygapophyses. Test these zygapophyseal constraints on yourself by bending and twisting your body—note how your thorax (rib-bearing region) flexes (most of the axial rotation and some lateral flexion occurs here) when compared with your lumbar (low back) region (where most of the allowable dorsoventral flexion occurs).

### B. Vertebral column

Traditionally, the typical mammalian vertebral column is separated into five regions, each of which is defined by what is or is not attached to the vertebrae. These regions are cervical, thoracic ("dorsal" in the older literature), lumbar, sacral, and caudal (Fig. 3). The *vertebral formula* is an alpha-numerical abbreviation for the numbers of vertebrae in each region. For example C6:T17:L1:S0:Ca25 describes the 6 cervical, 17 thoracic, 1 lumbar, 0 sacral, and 25 caudal vertebrae in the Florida manatee. Individual vertebrae in each region are also referred to by their position using one or two (regiospecific) letters and a number. For example, the third cervical vertebra is designated as C3. The total number of vertebrae, excluding the caudal vertebrae, is surprisingly close to 30 in most mammals (Flower and Lydekker 1891). Using the vertebral formula, contrast the vertebral column of the selected species (Fig. 3).

### C. Cervical Region

Cervical vertebrae (Figs. 3, 4A) are cranial to the rib-bearing vertebrae of the thorax. Most mammals have seven cervical vertebrae; but all the sirenians and the two-toed sloth (*Choloepus*) have six and

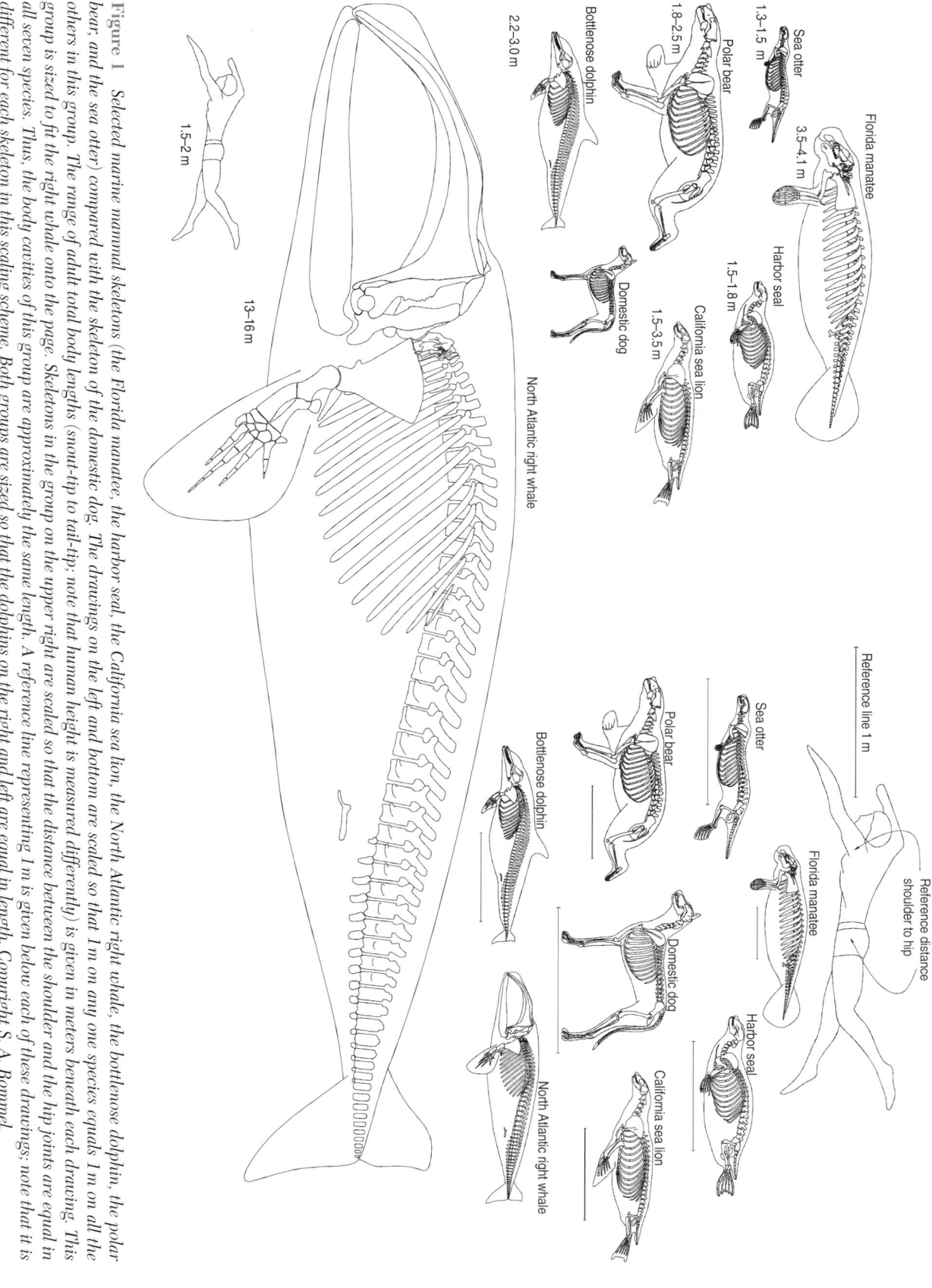

**Figure 1** Selected marine mammal skeletons (the Florida manatee, the harbor seal, the California sea lion, the North Atlantic right whale, the bottlenose dolphin, the polar bear, and the sea otter) compared with the skeleton of the domestic dog. The drawings on the left and bottom are scaled so that 1 m on any one species equals 1 m on all the others in this group. The range of adult total body lengths (snout-tip to tail-tip; note that human height is measured differently) is given in meters beneath each drawing. This group is sized to fit the right whale onto the page. Skeletons in the group on the upper right are scaled so that the distance between the shoulder and the hip joints are equal in all seven species. Thus, the body cavities of this group are approximately the same length. A reference line representing 1 m is given below each of these drawings; note that it is different for each skeleton in this scaling scheme. Both groups are sized so that the dolphins on the right and left are equal in length. Copyright S. A. Rommel.

Sea otter
1.3–1.5 m

Polar bear
1.8–2.5 m

Bottlenose dolphin
2.2–3.0 m

Florida manatee
3.5–4.1 m

Harbor seal
1.5–1.8 m

California sea lion
1.5–3.5 m

Domestic dog

North Atlantic right whale

13–16 m

1.5–2 m

Reference line 1 m

Reference distance shoulder to hip

Sea otter

Polar bear

Bottlenose dolphin

Florida manatee

Domestic dog

Harbor seal

California sea lion

North Atlantic right whale

the three-toed sloth (*Bradypus*) has nine. Serial fusion (ankylosis) of two or more cervical vertebrae is common in cetaceans, but all cetaceans have seven cervical vertebrae. Contrary to what has been reported in some published works, the cervical vertebrae in some

cetaceans (e.g., the narwhal, beluga, and river dolphins) are unfused and provide considerable neck mobility.

The first two cervical vertebrae are named the *atlas* (C1) and the *axis* (C2). Unlike all other vertebrae, the atlas and axis of most

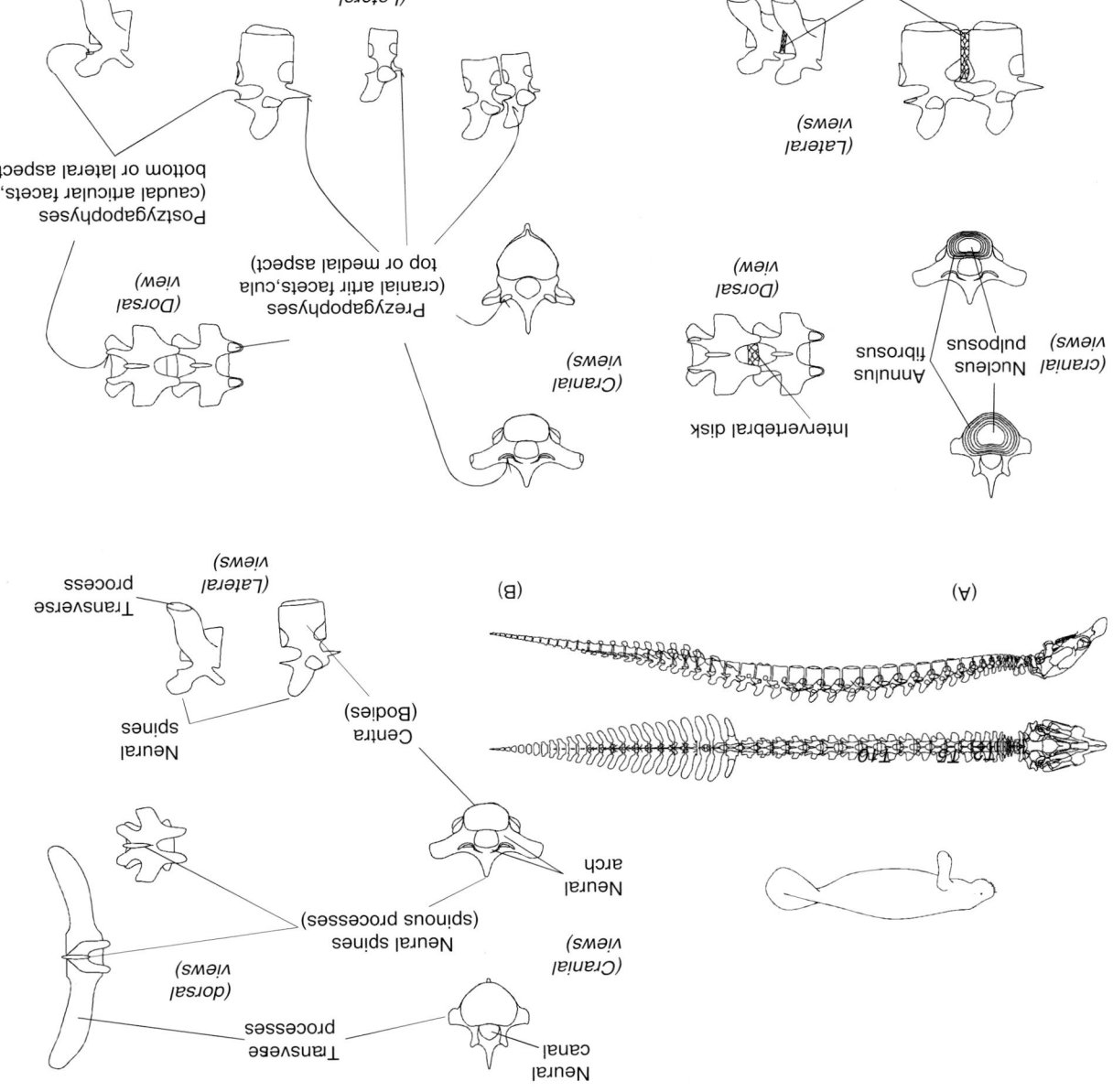

**Figure 2** (A) Dorsal and lateral views of the manatee and its vertebrae. (B) Parts of a vertebra. The centrum (body) of each vertebra is the primary mechanical support of the vertebral column. Vertebral centra help prevent the animal from collapsing when large body muscles contract. Dorsal to the centrum is an arch of bone, the neural arch, within the neural canal. The neural arch protects the spinal cord and it may extend dorsally as a neural spine (spinous process). (C) The flexible region between adjacent centra junctions as a tough joint, the intervertebral disk. These intervertebral fibrocartilage joints have two parts: the inner nucleus pulposus and the outer annulus fibrosus. (D) Zygapophyses are articular facets that constrain motions between adjacent vertebrae. The cranial pair on each vertebra are prezygapophyses and the caudal pair are postzygapophyses. The region of articulation between prezygapophyses and postzygapophyses of adjacent vertebrae is typically a synovial joint if the facets overlap. In some regions of the vertebral column the facets do not overlap and they may be connected by collagen fibers. Copyright S. A. Rommel.

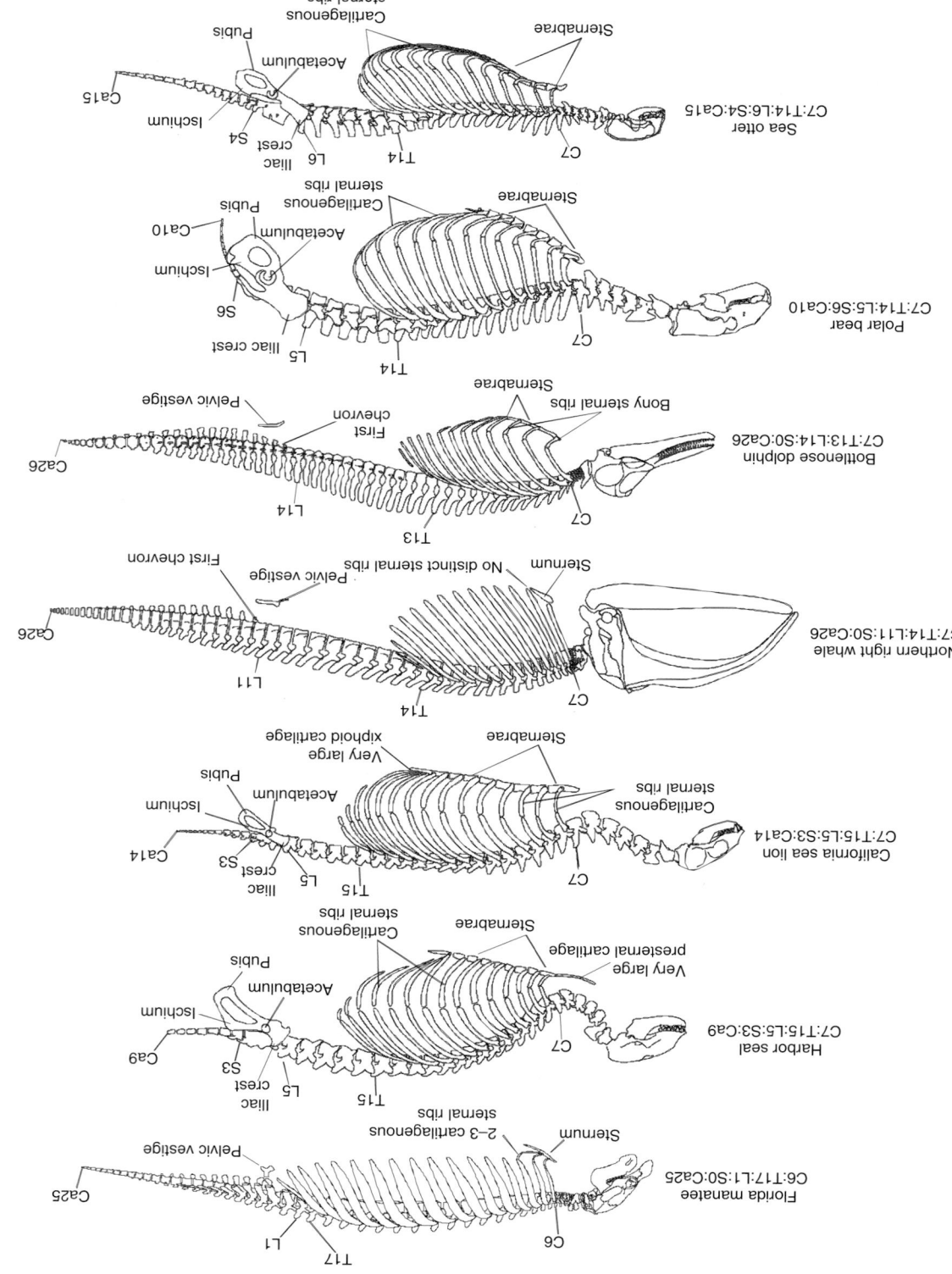

**Figure 3** *The axial skeletons plus pelves of the seven selected marine mammal species. There are typically five separate vertebral regions: cervical, thoracic, lumbar, sacral, and caudal. In this illustration, each skull and vertebral column is scaled to similar shoulder–"hip" distances. Sacral vertebrae are, by definition, associated with an attached pelvis; therefore, the permanently aquatic species (manatee, dolphin, and right whale) have no sacral vertebrae. The permanently aquatic species (manatee, dolphin, and right whale) have pelvic vestiges. The alpha-numerical abbreviation for the numbers of vertebrae in each region is located below each common name; the numbers given are for the specimens illustrated; some of these numbers vary between individuals of each species. Vertebral ribs extend from the vertebral column to join the sternal ribs, which extend from the sternum. The sternum develops from a series of sternebrae. In some species (e.g., humans, right whales, and manatees), the sternebrae fuse into a single bony unit. Copyright S. A. Rommel.*

mammals are not separated by an intervertebral disk but rather by a synovial joint, similar to the skull-atlas joint (Fig. 4A); the shapes of these first two vertebral joints dictate the range of motion between the skull and the vertebral column. Typical skull motions on the atlas are up-and-down and side-to-side (Fig. 4A). The axis has an elongated centrum, the *dens* (odontoid process), which restricts motions between the atlas and axis to rotations. Test these constraints on yourself by bending and twisting your head and neck. Head nodding, as in expressing "yes," is constrained by the skull-atlas joint; head rotations expressing "no," are constrained by the atlas-axis joint.

Typically, the permanently aquatic marine mammals have relatively short necks; in contrast, the pinnipeds have relatively long necks. Comparison of the skeletons of the seal and sea lion reveals that they have similar neck lengths, although external appearance is very different. Seals often hold their heads close to the thorax, which causes the neck to form a deep "S" curve, providing them with a "slingshot potential" for grasping prey (or inattentive handlers).

Cervical vertebrae may have distinct *transverse foramina* that protect blood vessels when the neck is flexed. Transverse foramina are particularly distinct in the cervical vertebrae of the carnivorous marine mammals. A pair of notches (or foramina) in the neural arch of the atlas may be present (Fig. 4A); these openings are for the first spinal nerve pair that exits the neural canal between the back of the skull and the neural arch of C1. Rarely, one finds *cervical ribs*, which are moveable extensions of the ventral transverse processes; cervical ribs do not extend to the sternum.

### D. Thoracic Region

The thoracic region (thorax) is defined by the presence of more or less movable (i.e., not fused to the respective vertebrae) ribs (Figs 3, 4B). The thoracic and lumbar vertebrae support the trunk. By definition, the first thoracic vertebra has a pair of ribs that, unlike cervical ribs, extend to the sternum. Most thoracic vertebrae have ribs that possess two segments. The proximal, dorsal segment is attached to the vertebral column and is termed a *vertebral rib* (Fig. 3). A distal, ventral segment is attached to the sternum and is termed a *sternal rib* (Romer and Parsons, 1986). A joint between the vertebral rib and sternal rib may help accommodate flexibility of the thorax. The distinction between vertebral and sternal ribs is more common in birds and some reptiles, but it is less common in mammals because most mammals have cartilaginous sternal ribs termed costal cartilages. We make this distinction because, unlike most other mammals, some odontocete cetaceans have bony rather than cartilaginous sternal ribs (Rommel 1990; Cotten *et al.*, in press).

As suggested above, the mechanics of motion of the thoracic vertebrae are different from those of either the cervical vertebrae or the post-thoracic vertebrae. Typically, the thorax is relatively more flexible axially and laterally than dorsoventrally. Thoracic vertebrae have dual functions of longitudinal body support and (with their ribs) support of the respiratory muscles. This dual function is reflected in the arrangement, size, and shape of the zygapophyses. The last thoracic vertebrae have ribs associated with them, but these last ribs may not be attached to their respective vertebrae, with little or no indication on their respective vertebrae to indicate their presence. The attachment sites of ribs on the vertebrae are called rib *facets* (costal facets; ribs = costae) if the region of attachment has a distinct articulation (Fig. 4B). Rib facets may be located on the vertebral centrum, on the transverse processes, or on both. Some rib attachments are made via long connective-tissue fibers; these ribs do not actually contact the respective vertebra, and there is no obvious structure on the (cleaned) vertebrae to suggest rib attachment. For this reason, dogmatic adherence to narrow definitions of numbers of thoracic or lumbar vertebrae, or both should be avoided. The cetaceans and pinnipeds have very mobile rib cages. Surprisingly, diving mammals have considerable dorso-cranial to ventro-caudal "tilt" to their ribs. In these species, the double-headed ribs (see below) have joints that are aligned to allow for substantial cranial to caudal swing of the attached vertebral ribs. This extreme mobility of the rib cage can help accommodate the lung volume changes that accompany changes in pressure with depth. It may also have evolved to accommodate rib cage dynamics that occur during lung ventilation (Cotten *et al.*, submitted) caused when abdominal muscle contractions prevent caudal displacement of the diaphragm during inspiration; in this case ribs, could be moved forward to increase lung volume.

Some thoracic vertebrae have ventral vertebral projections called *hypapophyses*—not to be confused with chevron bones (chevrons typically span the intervertebral joints and are not parts of the vertebrae themselves; Figs. 3, 4C). In the manatee, the central tendon of the diaphragm is firmly attached along the midline to hypapophyses. Hypapophyses also occur in some cetaceans (i.e. *Kogia*, the pygmy and dwarf sperm whales) in the caudal thorax and cranial lumbar regions; these hypapophyses are associated with the diaphragm.

Thoracic neural spines are often longer than those in other regions of the body (Figs. 3, 4) to provide mechanical advantage for neck muscles. Interestingly, terrestrial species with large heads tend to have long neural spines; however, the buoyancy of water may negate the need for such long neural spines in the marine mammals—contrast the animals that come onto land with the fully aquatic ones in this regard.

**Figure 4** (A) *The cervical vertebrae.* At the top of this illustration, these vertebrae are filled in on the dorsal and lateral views of the manatee skeleton. The first and second cervical vertebrae are named the atlas and the axis, respectively. Cervical vertebrae may have transverse processes. Typical skull motions on the atlas are up-and-down and side-to-side. The axis has an elongated centrum, the dens, which extends into the large neural canal of the atlas. The shape of the dens restricts motions between the first two vertebrae to rotations parallel to the long axis of the body. (B) *Thoraco-lumbar vertebrae.* Thoracic vertebrae support the ribs. Some ribs are double headed and articulate with their respective vertebrae via capitula and tubercula. Each capitulum articulates with one or more costal facets on the vertebral centrum at or near the intervertebral disk. Each tuberculum articulates with a facet on the transverse process of its respective vertebra. Lumbar vertebrae are trunk vertebrae without ribs—instead they have pronounced transverse processes. In manatees there may be a rib on one side and a transverse process on the other side of the same lumbar vertebra. (C) *Caudal vertebrae are found in the tail.* At the top of this illustration, these vertebrae are filled in on the dorsal and lateral views of the manatee skeleton. In the permanently aquatic marine mammals that have no direct attachment between the pelvic vestiges and the vertebral column, the transition between lumbar and caudal vertebrae is defined by the presence of chevron bones. The chevron bones are central intervertebral ossifications. By definition, each chevron is associated with the vertebra cranial to it. Copyright S. A. Rommel.

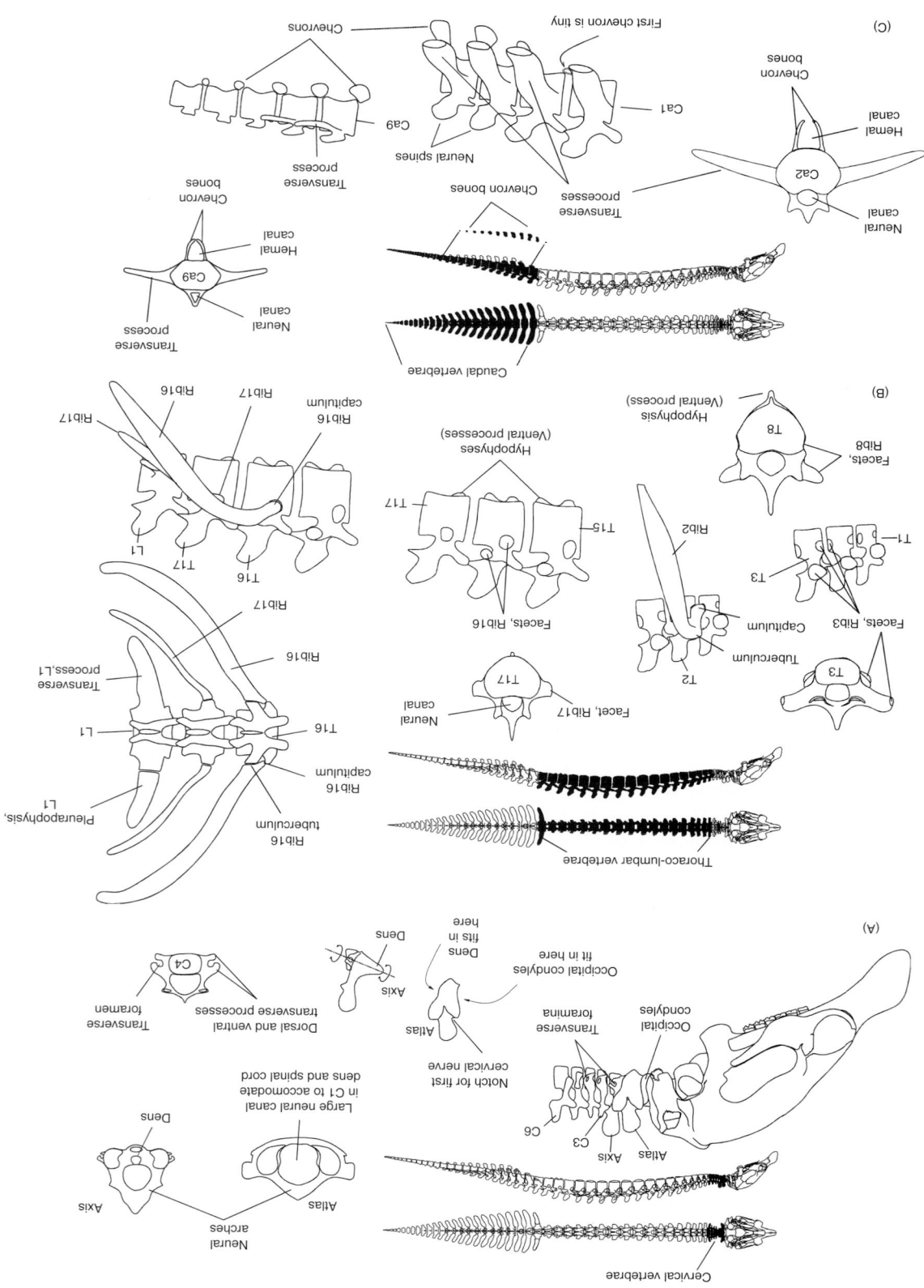

Ribs and transverse processes are homologous structures, but the presence of a movable joint ultimately distinguishes a rib from a transverse process in an adult. An "unfinished" joint may be indicative of developmental age; in manatees, there may be a movable "rib" (pleurapophysis, *pleura* = side or rib) on one side and an attached "transverse process" on the other side of the same (typically the first lumbar) vertebra. Pleurapophyses are more common in some of the lower vertebrates (Romer and Parsons, 1986; Kardong, 1998).

A rib may attach to its respective vertebra (Fig. 4B) at one or more locations. Typically the first few ribs are "double-headed" and have two distinct articulations with their juxtaposed vertebrae. The cranial-most articular part of a double-headed rib is the *capitulum* (plural, capitula) or head, and it articulates (via a synovial joint) with the body (or bodies) of one or two vertebrae. The *tuberculum* (plural, tubercula) is the second attachment site of a double-headed rib; it articulates (either with a synovial joint or by connective tissue fibers) to a transverse process of the respective vertebra. Since the capitula of some double-headed ribs touch two vertebrae, the vertebra with the transverse process attachment is used to identify the respective vertebra.

The caudal-most ribs have single attachments and are referred to as "single-headed". In most mammals, the single-headed ribs have lost their tubercula and are attached to their respective vertebrae by the capitulum at the centrum. In contrast, the single-headed ribs of cetaceans lose their capitula and are attached to their respective vertebrae by their tubercula at the transverse processes of cetacean thoracic vertebrae (Rommel, 1990), as occurs in some reptiles. The transverse processes of cetacean thoracic vertebrae may be longer than those of other mammals. Thus, the single-headed ribs of cetaceans attach to the vertebrae farther from the midline than do the single-headed ribs of other mammals. The last ribs often "float" free at one or both ends, that is, they are attached to neither their associated vertebra nor the sternum.

### E. Sternum

Embryonically, the sternum is formed from serial elements, called sternebrae (the sternal equivalent of vertebrae, Fig. 3). The first and last sternebrae are called the *manubrium* and *xiphisternum*, respectively. In some odontocetes, all mysticetes, and all sirenians, the sternebrae fuse into a single unit. Sternal ribs extend from the sternum to join the vertebral ribs. Seals have an elongated presternal cartilage extending cranially from the manubrium; some taxa, particularly the sea lion, have an enlarged post-sternal cartilaginous extension, the xiphoid cartilage.

### F. Lumbar Region

The lumbar vertebrae are trunk vertebrae that typically do not bear movable ribs (Fig. 3). Recall that ribs and transverse processes develop from homologous embryonic structures. Occasionally, a distinct pleurapophysis may be found in this region. As noted above, pleurapophysis are commonly found in manatees; these "lumbar ribs" are clearly distinct from thoracic ribs (Fig. 4B). Typically, the lumbar vertebrae are more flexible dorsoventrally than they are laterally and they may have no axial flexibility. Some of the mobility of the lumbar vertebrae may be constrained by the ribs in front of them.

As already noted, the number of lumbar vertebrae is usually closely linked to the number of thoracic vertebrae: an increase in number in one section typically means a reduction in the other. For example, there are 19 thoraco-lumbar vertebrae in all species of

Artiodactyla, whereas there are 20 or 21 in the Carnivora; compare these numbers with those of the selected marine mammals (Fig. 3).

### G. Sacral Region

There are at least two commonly accepted definitions for sacral vertebrae: (a) serial fusion of post-lumbar vertebrae, only some of which may attach to the ileum of the pelvis and (b) only those vertebrae that attach to the ileum, whether or not they are serially fused. Both definitions have merit. In species where there is contact between the vertebral column and the pelvis, it is relatively easy to define sacral vertebrae. However, it is not easy in the cetaceans and manatees because there is no direct attachment between the pelvic vestige and the vertebral column and there are no serially fused vertebrae in this region (dugongs have a ligamentous attachment between the vertebral column and the pelvic vestiges). Thus, most of the permanently aquatic species have, by either definition, no sacral vertebrae. In other mammals, the number of sacral vertebrae is commonly 2–5. Within a species, the number of serially ankylosed vertebrae may vary, particularly with age.

### H. Caudal Region

Caudal (cauda = tail) vertebrae are found in the tail (Figs. 3, 4C). The number of caudal vertebrae varies widely. Long tails usually require numerous caudal vertebrae. Florida manatees have between 22 and 27 caudal vertebrae, and bottlenose dolphins 25–28—for cetaceans, the minimum is 13 in *Caperea marginata* (pygmy right whale); the maximum is 49 in *Phocaenoides dalli* (Dall's porpoise). Note that the caudal vertebrae of cetaceans extend to the tip of the tail (fluke notch), whereas manatee caudal vertebrae stop 3–9 % of the total body length from the fluke tip (Fig. 1). Caudal vertebrae range from being robust, important locomotory structures in the permanently aquatic species, to being relatively small vertebrae in the pinnipeds and polar bear.

### I. Chevron Bones

The *chevron bones* or *chevrons* (Fig. 4C) are ventral intervertebral ossifications found only in the caudal region. They are relatively large in the permanently aquatic species but can be found as small ossifications in many other species, including the dog. Chevrons function to increase the mechanical advantage of the hypaxial muscles to flex the tail ventrally. By definition, each chevron is associated with the vertebra cranial to it (note that there is some controversy over which is the "first" caudal Rommel 1990). Pairs of chevron bones form arches, creating a ventral channel called the hemal canal. Within the hemal canal, the blood vessels (caudal arteries and veins) that supply the tail are protected. In sirenians and some cetaceans the chevron bone pairs fuse at the ventral midline.

## II. Appendicular Skeleton

### A. Pectoral Limb Complex

The forelimb (Fig. 5) is referred to as a flipper in the permanently aquatic species (in true seals and sea lions the hindlimb is also referred to as a flipper). The forelimb includes the scapula, humerus, radius and ulna, the multiple bones of the manus, and the clavicle. The scapula is attached to the axial skeleton only by the muscles; there is no functional clavicle in the marine mammals (Klima *et al.* 1980, Strickler, 1978). The scapula consists of an essentially flat (slightly concave medially) blade with an elongate scapular spine extending laterally from it. The distal tip of the spine is the *acromion*.

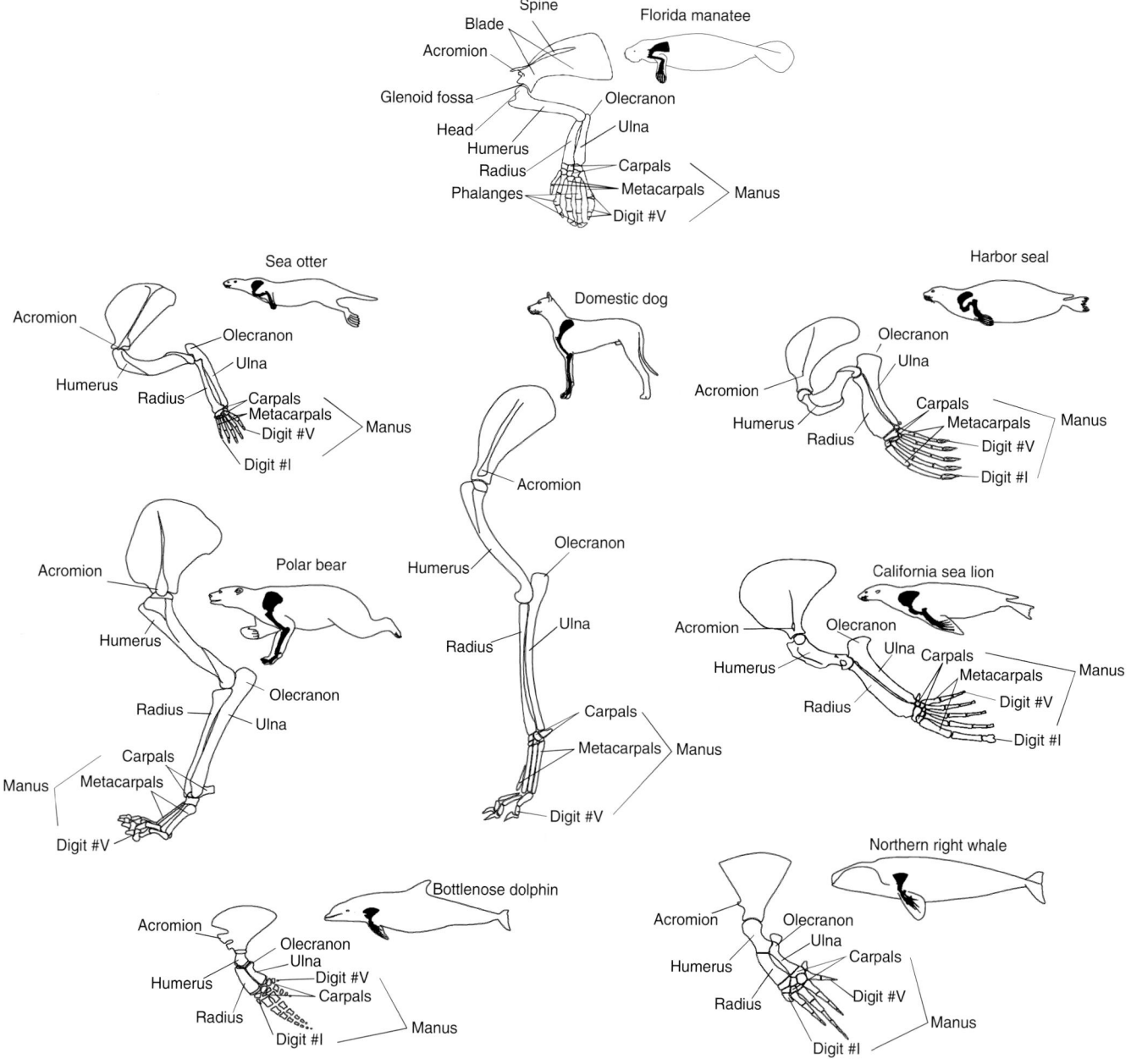

**Figure 5** *The forelimb (pectoral appendage) includes the humerus, radius and ulna, and manus. The forelimb attaches to the pectoral girdle which is made of bilaterally paired scapulae. The scapula is a flat (slightly concave medially) blade with an elongated scapular spine extending laterally from it. The distal tip of the spine is the acromion process. The humerus has a ball-and-socket articulation with the glenoid fossa of the scapula. The humerus articulates with the radius and ulna at the elbow. The radius and ulna articulate with the proximal aspect of the manus at the wrist. The manus includes the carpals, metacarpals, and phalanges. There are five "columns" of phalanges, each "column" is called a digit. The digits are numbered, using Roman numerals, starting from the cranial aspect (associated with the radius, in humans—the thumb). Copyright S. A. Rommel.*

**S**

The scapular spine is roughly in the center of the scapular blade in most mammals. However, in cetaceans and manatees, the scapular spine is close to the cranial margin of the scapular blade, and its acromion process extends beyond the leading edge of the blade.

The proximal humerus has a ball-and-socket articulation in the *glenoid fossa* of the scapula; this is a relatively flexible joint. The humerus articulates distally with the radius and ulna; this is also a flexible joint in most mammals, but it is mechanically constrained in cetaceans. The olecranon is a proximal extension of the ulna that increases the mechanical advantage of the triceps muscles, which extend the forelimb. In species like the polar bear and the sea lion, the olecranon is robust; however, in cetaceans it is relatively small. The radius and ulna

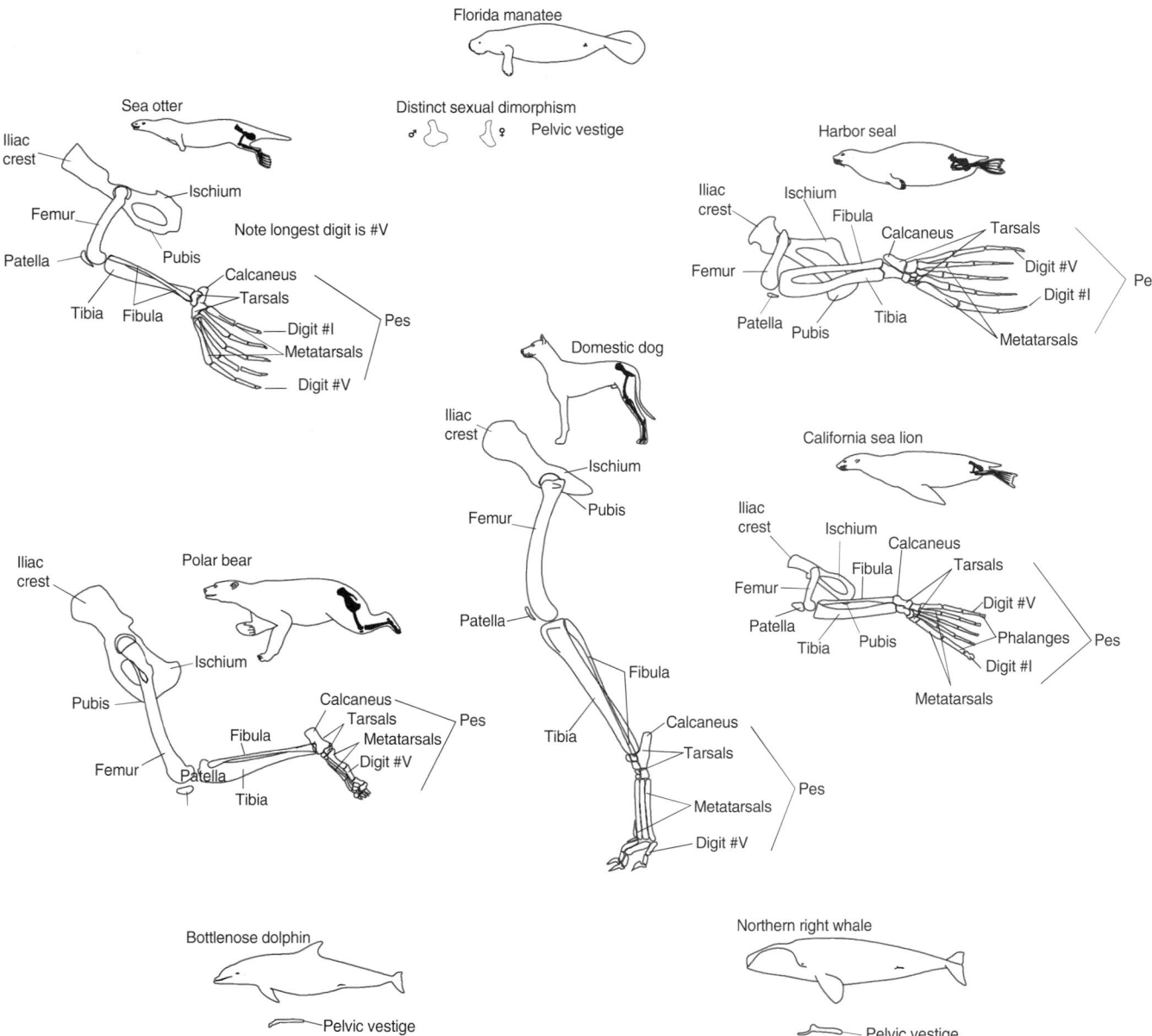

**Figure 6** *The pelvic appendage (hind limb) and pelvic girdle. The typical mammalian pelvis is made of bilaterally paired coxae attached to one or more sacral vertebrae. Each coxa is made of three elements: the ilium, ischium, and pubis. The crest of the ilium is a prominent landmark; it flares forward and outward beyond the region of attachment between the sacrum and the ilium. The two halves of the pelvis join at the ventral midline pubic symphysis. In the permanently aquatic marine mammals there is only a vestige of a pelvis. The hindlimb, if present, is attached to the pelvic girdle via a ball (femoral head) and socket (acetabulum) joint at the hip. The proximal limb bone is the femur. Distally, the femur articulates with the tibia and the fibula at the knee joint. The tibia and fibula distally articulate with the pes at the ankle. Note that the pes is oriented in different "planes" in some of the selected species. The pes consists of the tarsals proximally, the metatarsals medially, and the phalanges distally. Copyright S. A. Rommel.*

**S**

of manatees fuse at both ends as the animal ages. This fusion prevents axial twists that pronate and supinate the manus. The cetacean radius and ulna are also mechanically constrained by the surrounding connective tissues but are not typically fused. The forelimbs of the sea otter are very mobile allowing for the manipulation of food.

The distal radius and ulna articulate with the proximal aspect of the manus. The manus includes the carpals, metacarpals, and phalanges.

There are five "columns" of phalanges, each of which is called a digit. The digits are numbered, using Roman numerals, starting from the cranial aspect (the thumb; associated with the radius). Note that the longest digit may be different in the different species (Fig. 6).

In many of the marine mammals, the "long" bones of the pectoral limb (humerus, radius, and ulna) are relatively short, and the phalanges are elongated. Cetaceans are unique among mammals in

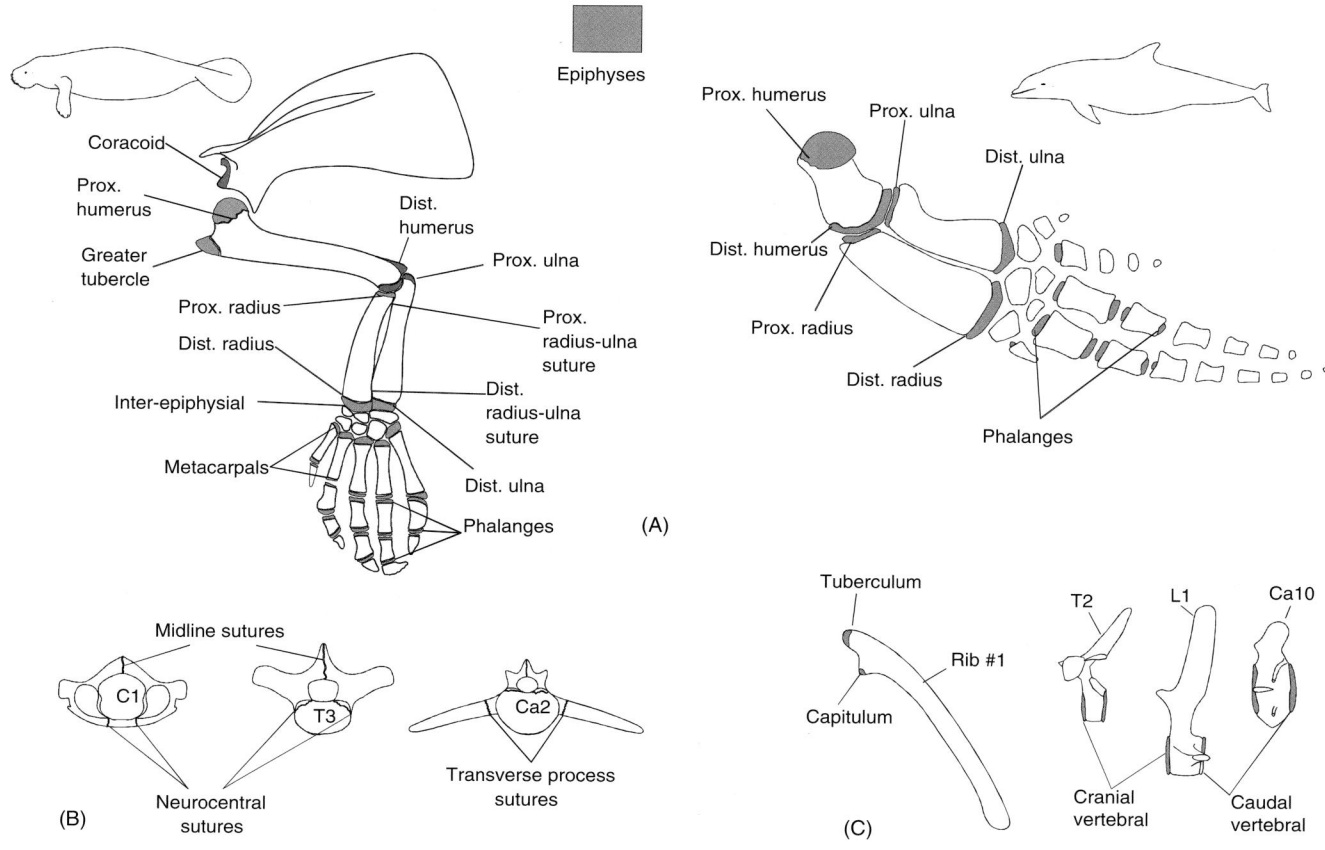

**Figure 7** *(A) Postcranial features that change with age. Some post-cranial elements grow in length by separate ossification centers, the epiphyses, at the margins of the bony element. The humeral head and distal humerus develop as epiphyses (filled parts in illustration). Patterns of fusion of these epiphyses provide information about aging. The epiphyses illustrated here have been used for relative aging of manatees and dolphins. (B) Non-epiphyseal sutures also indicate age in the postcranial skeleton. The transverse processes and the neural arches have sutures that disappear early in adolescence. In addition, the two halves of the neural arch ankylose at the dorsal midline to the base of the neural spine. (C) Epiphyses may occur on the ends of ribs and the tips of transverse processes and neural spines. Most mammals have distinct vertebral epiphyses at each end of their growing vertebrae. Vertebral epiphyses fuse to their respective vertebral centra at physical maturity of the vertebra. Copyright S. A. Rommel.*

that they have more phalanges than do any of the other mammals (Fig. 6); this condition is known as hyperphalangy (Howell 1930, Cooper *et al.*, 2007). The number varies between individuals of each species—the pilot whale has the most (14), the bottlenose dolphin has a maximum of 9, and the right whale has a maximum of 7.

Notice the "plane" in which each manus is oriented (Fig. 6). The sea otter's manus, much like that of humans, is very dexterous. The polar bear manus, much like that of the dog and other terrestrial quadrupeds, is more dexterous. The permanently aquatic species hold the manus roughly parallel to the midsagittal plane of the body or parallel to the nearest body surface.

## B. Pelvic Limb Complex

The typical mammalian pelvis is made of bilaterally paired bones: ilium, ischium, and pubis (Fig. 7). Each half of the pelvis attaches (via the ilium) to one or more sacral vertebrae; the attachment may vary from simple apposition in juveniles to fusion in adults. The crest of the ilium, which is a prominent landmark, flares forward and outward beyond the region of attachment between the sacrum and the

ilium. The two halves of the pelvis join at the pubic symphysis on the ventral midline. In the permanently aquatic marine mammals, there is only a vestige of a pelvis to which portions of the rectus muscles of the abdomen may attach; the male reproductive organs may be supported by these vestiges as well (Pabst *et al.*, 1998). Occasionally, in some of the large whales, a non-functional hindlimb rudiment articulates with the pelvic vestige.

The hindlimb, if present, is attached to the pelvis via a ball and socket joint at the hip. The proximal limb bone is the femur, whose head articulates with the socket of the pelvis, the acetabulum. Distally the femur articulates with the tibia and the fibula at the knee. The tibia and fibula articulate distally with the pes or foot at the ankle (Fig. 7). The pes is composed of the proximal tarsals, the medial metatarsals, and the distal phalanges. Note that the digits of the sea lion terminate a significant distance from the tips of the flipper. Of the chosen group of marine mammals, the limbs of the polar bear are closest in gross appearance to those of the terrestrial mammals. In contrast, many of the other marine mammals have relatively short "long" bones (femur, tibia, and fibula) and relatively long

**S**

phalanges. The femurs of some phocids (e.g., the harbor seal) are so short that the knee cannot contact the ground.

Note that the pes is oriented in different "planes" in some of the selected species (Fig. 7). The pes of the polar bear is oriented in a fashion similar to that of the dog. The sea otter, seal, and sea lion all orient the pes parallel to the sagittal plane when swimming. The sea lion can rotate its pelvis and hind flippers to "walk" on land, and the sea otter can manipulate its hind flippers to hold food. Compare the positions of the first and last digits in the sea otter with those of the seal and sea lion, and note how the sea otter's fibula crosses the tibia to achieve this orientation.

### C. Sexual Dimorphisms

In many mammals the adult males are larger than the adult females, and among marine mammals, this dimorphism is extremely pronounced in certain pinnipeds and the sperm whales. In contrast, in baleen whales and some other species the adult females are larger than the adult males. In the permanently aquatic marine mammals, there may be sexual dimorphisms in the pelvic vestiges (Fagone *et al.*, 2000). The penises of mammals are supported by tough fibrous structures, the crura, which attach to the pubic bones of the pelvis. The muscles that engorge the penis with blood are also attached to the pelvis, and presumably the mechanical forces associated with these muscles influence pelvic vestige size and shape, particularly in manatees. Males in some groups of mammals, particularly the carnivores, have a bone within the penis (the baculum or os penis) that helps support the penis.

### III. Sutures and Epiphyses

Many bony features change in size and shape as the individual develops and matures; these features can be used as milestones in life history studies, and in some cases, they can be fairly accurate estimators of age. Some postcranial elements grow from separate ossification centers at the margins of the bones (Fig. 7A). *Epiphyses* (*epi-*, upon, and *physis-*, grow or generate) are regions at the ends of bones (most often at or near a joint) associated with longitudinal growth in mammals. Epiphyses allow the portion of the bone near a joint to grow at a different rate than the rest of the bone and allow new bone to be laid down away from the articulating surface. Growth occurs at a cartilaginous plate between the epiphysis and the *diaphysis* (*dia-*, between; body or shaft). At the completion of growth, the epiphysis fuses with the rest of the bone, and the cartilage plate disappears. Eventually the suture between the two parts also disappears. Because most marine mammals are large, their epiphyses are more evident than those of the smaller terrestrial mammals.

The best-studied epiphyses are those of the long bones. There are, however, other epiphyses that are less well known. For example, the proximal finger bones may have epiphyses (Fig. 7B). The pattern of epiphysial fusion in the flipper can be used to determine relative age of an individual, when compared to the fusion pattern in other animals of known age. Occasionally epiphyses are found on the proximal ends of ribs, on the scapula located at or near the coracoid (Fig. 7A), or on the tips of neural spines and transverse processes.

Among the epiphyses most important to mammalian osteologists are those located at each end of the vertebral centrum (Fig. 7C). Most mammals have distinct bony plates at each end of their growing vertebrae. These plates are vertebral epiphyses. Manatees (and humans) are among the few species of mammals that have indistinct or no vertebral epiphyses. In manatees, the vertebral epiphyses are

delicate irregular films of bony tissue found within the cartilaginous surfaces at the cranial and caudal ends of the centrum. After these epiphyseal sutures are completely fused, the skeleton cannot grow in length and the individual is considered skeletally mature. Skeletal maturity may or may not coincide with other forms of maturity, such as reproductive maturity.

Non-epiphyseal sutures can also be used to estimate age of parts of the postcranial skeleton (Fig. 7C)—for example, the two halves of the neural arch ankylose at the dorsal midline to form the base of the neural spine. These sutures may persist well into adulthood in cetaceans.

There is much more to learn about the postcranial skeleton. Consult Pabst *et al.* (1999) for additional functional morphology on marine mammals, Young (1975) for information on mammals in general, and Alexander (1994) for vertebrates. For principles on size and scaling try Calder (1966) and Schmidt-Nielsen (1984).

### See Also the Following Articles

Skull Anatomy ■ Forelimb ■ Hind Limb Anatomy

### References

Alexander, R. McN. (1994). "Bones: The Unity of Form and Function." Macmillan, New York.

Calder, W. A., III (1966). "Size, Function, and Life History." Dover Publications, Inc, Mineola, NY.

Cooper, L. N., Berta, A., Dawson, S. D., and Reidenberg, J. S. (2007). Evolution of hyperphalangy and digit reduction in the cetacean manus. *Anat. Rec.* **290**, 654–672.

Cotten, P. B., Piscitelli, M. A., McLellan, W. A., Rommel, S. A., Dearolf, J. L., and Pabst, D. A. Submitted (In press). The gross morphology and histochemistry of respiratory muscles in bottlenose dolphins, *Tursiops truncatus. J. Morphol.*

Evans, H. E. (1993). "Miller's Anatomy of the Dog," 3rd ed. W. B. Saunders Company, Philadelphia, PA.

Fagone, D. M., Rommel, S. A., and Bolen, M. E. (2000). In press. Sexual dimorphism in vestigial pelvic bones of the Florida manatee. *Florida Scientist* **63**(3), 177–181.

Flower, W. H. (1885). "An Introduction to the Osteology of the Mammalia," 3rd ed. Macmillan and Co, London (reprinted by A. Asher & Co., Amsterdam, 1966).

Flower, W. H., and Lydekker, R. (1891). "An Introduction to the Study of Mammals Living and Extinct." Adam and Charles Black, London.

Giffen, E. B. (1992). Functional implications of neural canal anatomy in recent and fossil marine carnivores. *J. Morphol.* **214**, 357–374.

Howell, A. B. (1930). "Aquatic Mammals: Their Adaptions to life in the water." Charles C. Thomas, Springfield, IL.

Kardong, K. V. (1998). "Vertebrates: Comparative Anatomy, Function, Evolution," 2nd ed. McGraw-Hill, Boston.

Kipps, E. K., McLellan, W. A., Rommel, S. A., and Pabst, D. A. (2002). Skin density and its influence on buoyancy in the manatee (*Trichechus manatus latirostris*), harbor porpoise (*Phocoena phocoena*) and bottlenose dolphin (*Tursiops truncatus*). *Mar. Mamm. Sci.* **18**(3), 765–778.

Klima, M., Oeleschlager, H. A., and Wunsch, D. (1980). Morphology of the pectoral girdle in the Amazon dolphin *Inia geoffrensis* with special reference to the shoulder joint and movements of the flippers. *Sonderdeück aus Zeitschrift fur Saugetierkunde* **45**(5), 288–309.

Pabst, D. A. (1993). Intramuscular morphology and tendon geometry of the epaxial swimming muscles of dolphins. *J. Zool. (Lond.)* **230**, 159–176.

Pabst, D. A., Rommel, S. A., and McLellan, W. A. (1999). Functional anatomy of marine mammals. *In* "Marine Mammals" (J. E. Reynolds, and S. A. Rommel, eds), Vol. I, pp. 15–72. Smithsonian Institution Press, Washington, DC.

Pabst, D. A., Rommel, S. A., and McLellan, W. A. (1998). Evolution of thermoregulatory function in cetacean reproductive systems. *In* "The Emergence of Whales" (H. Thewissen, ed.), pp. 379–397. Plenum Press, New York, NY.

Pierard, J. (1971). Osteology and myology of the Weddell seal *Leptonychotes weddelli* (Lesson, 1826). *Antarctic Pinnipedia*. Antarctic Research Series **18**, 53–108.

Pierard, J., and Bisaillon, A. (1978). Osteology of the Ross Seal *Ommatophoca rossi* Gray, 1844. Biology of the Antarctic Seas IX, Antarctic Research Series **31**, 1–24.

Romer, A. S., and Parsons, T. S. (1986). "The Vertebrate Body," 6th ed. Saunders College Publishing, Philadelphia.

Rommel, S. A. (1990). The Osteology of the bottlenose dolphin. *In* "The Bottlenose Dolphin" (R. Reeves, and S. Leatherwood, eds), pp. 29–49. Academic Pr, New York.

Schmidt-Nielsen, K. (1984). "Scaling; Why is Size so Important?." Cambridge University Press, New York.

Strickler, T. L. (1978). Myology of the shoulder of *Pontoporia blainvillei*, including a review of the literature on shoulder morphology in the Cetacea. *The American Journal of Anatomy* **152**(3), 419–431.

Young, J. Z. (1975). "The Life of Mammals, their anatomy and Physiology." Clarendon Pr, Oxford.

# Skull Anatomy

Sentiel A. Rommel, D. Ann Pabst, and William A. McLellan

To appreciate skull anatomy, take a moment and look at your own face in a mirror. The structures above the neck are designed for the acquisition and initial processing of nutrients and respiratory gases, the acquisition of sensory information about light, sound, touch, odor, and taste, and the broadcast of information about your own thoughts and emotions. Sensory and motor information is processed and sent from here to coordinate body functions. Complex signals can be sent to others of our species via vocalizations and/or the contractions of facial muscles. The head is our window for contact, perception, and communication with our world, and the skull provides the framework, the armature, for the head. Thus, the skull is interesting in itself. It is also fundamentally important in our picture of evolutionary biology. This article describes the skull morphology of the evolutionarily diverse group of marine mammals (Reynolds *et al.*, 1999).

We discuss the skulls of seven different species of marine mammals: the manatee (*Trichechus manatus*), the harbor seal (*Phoca vitulina*), the California sea lion (*Zalophus californianus*), the north Atlantic right whale (*Eubalaena glacialis*), the bottlenose dolphin (*Tursiops truncatus*), the polar bear (*Ursus maritimus*), and the sea otter (*Enhydra lutris*). These marine mammal species were chosen, in part, because much is known about them and they illustrate a wide range of morphological adaptations. We use the domestic dog (*Canis familiaris*) to provide a familiar reference.

## I. Defining the Term "Skull"

The term "skull" is inexact. It has been used to describe the entire skeleton of the head. It has also been used to refer to only

the cranium, which is the housing for the brain and sensory organs and the upper jaw. We use the word skull to refer to the entire head skeleton, including the cranium and the derivatives of the first three visceral arches, i.e., the lower jaw (or mandible) and the hyoid apparatus. The mandible and hyoid apparatus of marine mammals have received less attention in the literature, but they are particularly important in adaptations for feeding and in some cases hearing (see below).

The skull acts as a mechanical foundation for the fat, muscle, skin, vascular, and sensory structures that form the head. Thus, the skull alone does not dictate the contours of the head (Fig. 1). For example, odontocete cetaceans have a melon, a fatty facial pad, the shape of which is only partly defined by the underlying bones (Harper *et al.*, 2008; Mead, 1975; Rommel *et al.*, 2006). The relationships between the bones and the soft tissues of the head vary among species, perhaps with major differences in head and skull profiles found in the sperm whales. Contrarily, the dorsal surface of the right whale's head follows closely that of the underlying skull, though the right whale also has huge lower lips that follow the contour of the upper jaw but are not predicted by the outline of the lower jaw. The size and shape of the head may also influence the mechanics of locomotion, balance, and hearing. The completely aquatic species (cetaceans and sirenians) have shorter necks and less need for "anti-gravitational" muscles that support the head than do terrestrial mammals (imagine a right whale moving its head and neck around in the air the way a sea lion does).

## II. Feeding and Swallowing

The specific characteristics of a skull (including dentition) often reflect the animal's methods of feeding (Figs 2, 3). For example, the "typical" heterodont dentition (Kardong, 1998) of terrestrial carnivores, such as the dog, is also found to various degrees in the seal, sea lion, sea otter, and polar bear. Heterodonty is tooth shape differences in different parts of the mouth—incisors and canines rostrally and cheek teeth (premolars and molars) caudally. Although each of these tooth types may vary in shape, their definitions are specific and are related to tooth attachment to the bones of the *upper* jaw (Hildebrand, 1995). Incisors are found only in the premaxillary (incisive) bone; canines are found in the maxilla, in or very near the suture with the premaxilla. Incisors, canines, and premolars are deciduous teeth; molars are not. Developmentally, there are two sets—milk (or "baby") teeth and adult teeth. Premolars erupt from the maxillary bones, they are deciduous cheek teeth that are found rostral to the molars; molars are nondeciduous cheek teeth that erupt from the maxillae. Each tooth shape may perform a distinct function, sort of a "Swiss Army mouth" (Greg Early, personal communication). Incisors, if chisel-like, are for slicing and chipping, and if pointed, for piercing. Long, pointed canines are good for capturing and piercing. Relatively blunt cheek teeth are good for crushing and grinding.

Teeth are also found in the lower jaw (mandible). The mandible is made up of bilaterally paired dentary bones (Fig. 3). The rostral ends of the two dentaries are joined by a mandibular symphysis. The mandibular symphysis ankyloses with age in many mammals making the jaw a single compound bone; this occurs at an early age in manatees whereas it may only fuse in very old dolphins. In contrast, the unfused dentaries of the mysticetes may undergo complex axial rotations, particularly while lunge-feeding in rorquals (Lambertsen, 1983).

In most mammals, each dentary has a horizontal body that presents the teeth. In most mammals the dentary has a vertically

**S**

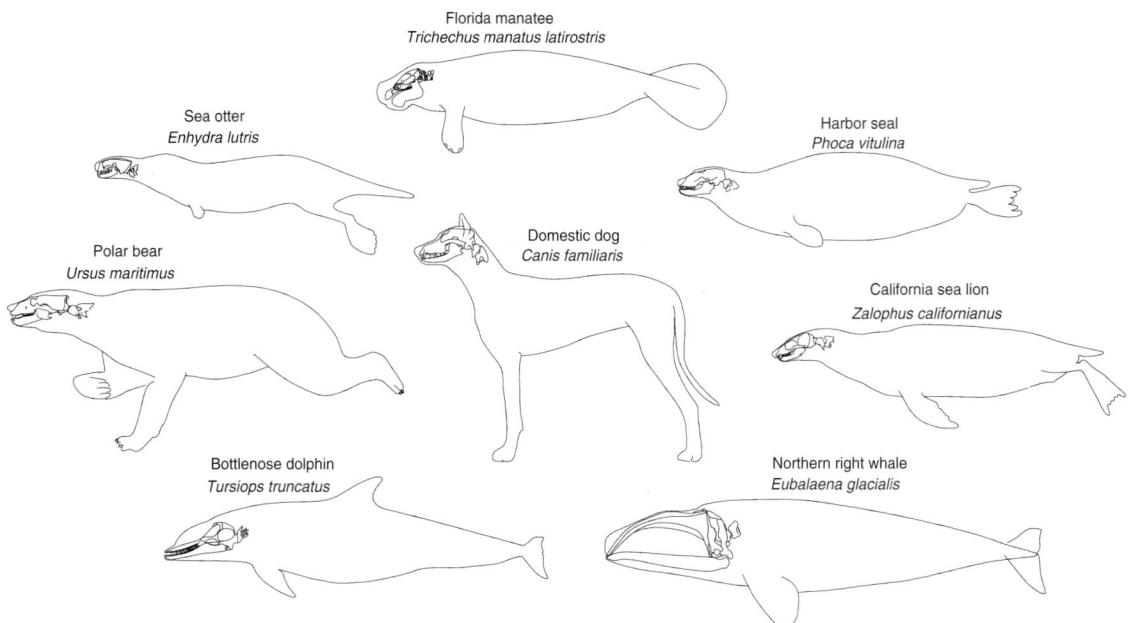

**Figure 1** *Skulls and first two cervical vertebrae (except in the cetaceans illustrated here, in which all, partly fused, vertebrae are fused) of a selection of marine mammals for comparison with those of the dog. Each species is scaled so that the distances between the shoulder and the pelvis are similar; body cavities are therefore roughly similar in length, allowing one to compare head sizes with visceral volumes among species. Note how the outline of the head differs from the midline contour of the skull. Copyright S. A. Rommel.*

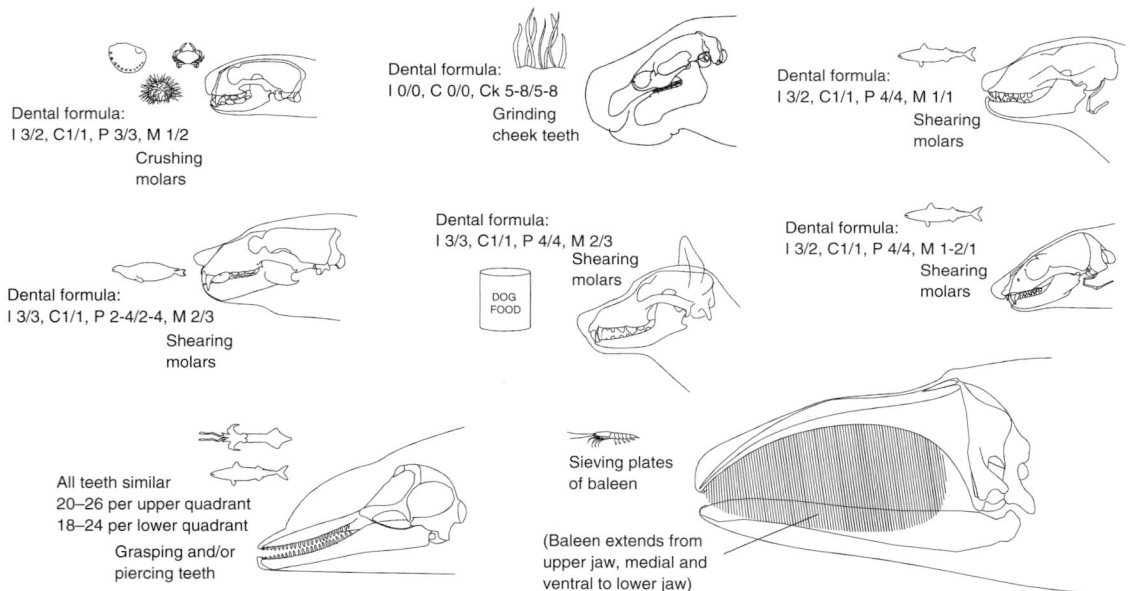

**Figure 2** *Feeding apparatus and typical food. Dominant tooth type is also given. Note that in manatees there are no incisors or canines and that all the cheek teeth (C) are continuously replaced. Also note that the embryonic teeth of the right whale have been replaced with horny plates of baleen in the upper jaw only. Copyright S. A. Rommel.*

directed ramus that projects into the temporal fossa; in dolphins, the ramus is reduced or absent. Typically, the labial surface (lateral aspect) of the dentary has small openings at its rostral end (mental foramina) for the blood vessels and nerves of the chin; manatee mental foramina are relatively large.

Dentition complexity in mammals may be more indicative of food type than is the case for many other vertebrates. The hardness of teeth (which increases their likelihood of preservation in the fossil record) and specificity of dentition contribute significantly to our current understanding of the ecology and evolution of different taxa.

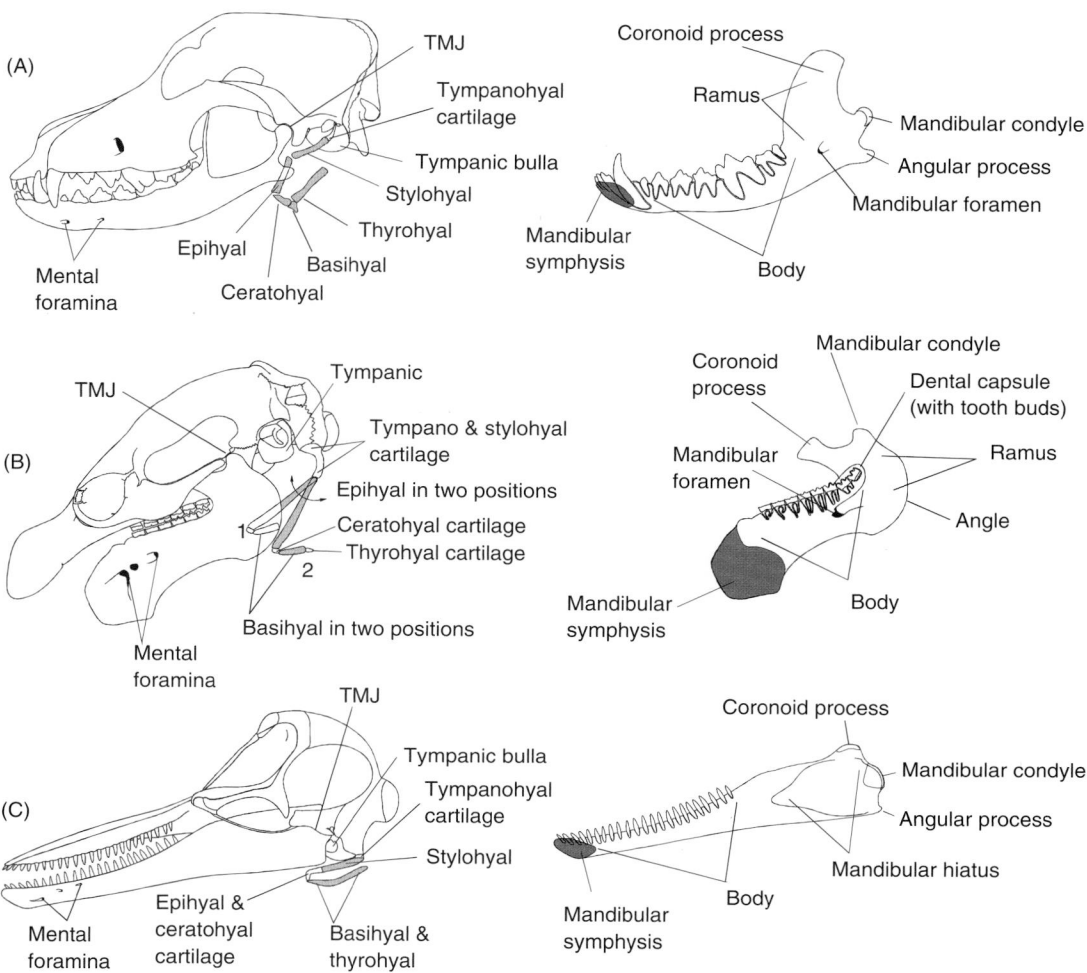

**Figure 3** *Left lateral views of the dog (A), manatee (B), and dolphin (C) skull with attached hyoids and medial views of the isolated right mandible. Bony elements of the hyoids are colored in gray excerpt where they lie deep to the mandible. The manatee hyoid apparatus is presented in two positions to illustrate its motion during swallowing. Muscles between the tongue and basihyal move the hyoid apparatus up and forward (position 1). Muscles between the basihyal and the sternum move the hyoid apparatus down and back (position 2 is exaggerated to illustrate the process). The joint between left and right dentaries (mandibular symphysis) is cross-hatched. The mandibular foramen is enlarged in the manatee and greatly enlarged to form a mandibular hiatus in the dolphin. The mandibular hiatus of the dolphin dentary is a large opening on the medial aspect of the lower jaw. Within this hollow region there is a intramandibular fat body that extends to encompass the ear—this fat functions as an acoustic channel for the reception of sound. TMJ = temporomandibular joint. Teeth and hyoid apparatus of the dog after Evans (1993). Copyright S. A. Rommel*

Typically, deciduous teeth are replaced vertically; the developing permanent teeth are deep to the milk teeth. These teeth are replaced only once. In contrast, manatee (but not dugong!) teeth are continuously replaced. Manatee tooth replacement is horizontal, beginning at the back of the tooth row (Fig. 3). This unusual method of horizontal tooth replacement is found in only a few other mammals such as elephants and kangaroos. In addition, unusual for manatees is the lifelong generation of new tooth buds, which develop in the dental capsule at the caudal end of each tooth row. Each manatee tooth moves forward and the roots grow in length as the crown erupts. The crown of the tooth begins to wear as it occludes with the opposite teeth; simultaneously, the root begins to resorb (Domning and Hayek, 1984); the processes underlying this phenomenon are not yet fully understood. Thus, as each manatee tooth moves forward in the jaw, it becomes smaller at the top and bottom; when it reaches the rostral end of the tooth row the small flattened vestige falls out. Manatees typically do not have incisors, although small rudiments of premaxillary incisors occasionally are observed in fetuses. Baleen whales lose their embryonic teeth and develop baleen (Fig. 2). Postweaning, baleen whales acquire food by sieving plankton, small fish, and (in the case of gray whales), benthic invertebrates, using plates of horny (keratinized) baleen suspended from their upper jaws (Pivorunas, 1979).

In some species, all of the teeth have the same shape—this is the homodont condition. The homodont dentitions of odontocetes and manatees differ in shape and function (they are single-rooted,

**S**

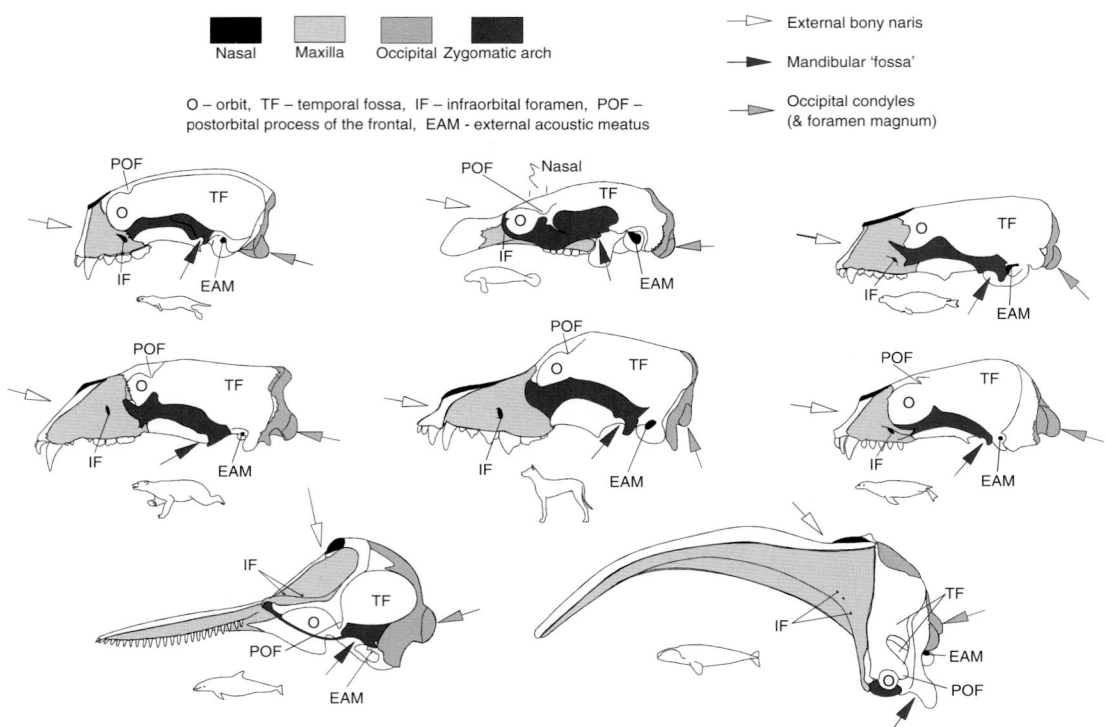

**Figure 4** *Selected bony features of the cranium. The rostrum, composed of the premaxilla, maxilla, and occasionally the nasal, forms the "face" or muzzle of each species. The zygomatic arch, which supports the masseter muscles, may be composed of a single bone (the jugal) or parts of as many as three bones in some species. Arrows indicate directions of air flow at the external bony nares, vertebral column articulation at the occipital condyles, and lower jaw articulation at the mandibular fossa; note that the mandibular fossa of the manatee includes a convex tubercle. Postorbital processes of the frontal may be present; it is absent in the seal, small in the sea otter, and relatively large in the dolphin and right whale. The region spanned by the tympanic membrane is visible in manatees but hidden within the middle ear of the other species. Copyright S. A. Rommel.*

conical, grasping teeth for dolphins and multi-rooted, multi-cusped, grinding teeth for manatees).

The dental formula is an alphanumeric abbreviation for the adult numbers of incisor, canine, premolar, and molar teeth—the number of each tooth type in the upper and lower half of the jaw (Fig. 2). For example, I 3/2, C1/1, P 3/3, M 1/2 describes the dentition of the sea otter. Thus, the adult sea otter has 3 incisors on each side of the upper jaw and 2 incisors on each side of the lower jaw, etc. The lower jaw generally mirrors some of the features of the upper jaw, particularly the dentition (Figs. 2, 3).

Feeding habits may also be reflected in the shapes of the rostrum, the zygomatic arch, and the temporal fossa (Fig. 4). Relatively large temporal muscles and their fossae are typical of carnivorous mammals that tear or shear flesh without finely dividing it in the mouth, and/ or have teeth for killing and temporarily holding prey (Hildebrand, 1995). Carnivorous mammals have upper and lower tooth rows that have little or no horizontal motion but rather occlude with a chopping motion, using a simple hinge joint. Their temporomandibular joint or TMJ is roughly in line with the tooth row. In contrast, the TMJ in herbivores is typically above the tooth line (Figs 2, 3). Relatively large masseter muscles, and robust zygomatic arches that support them, are more typical of herbivorous mammals that use a crushing and rolling action to chew. This feeding style requires a TMJ that can slide horizontally and a large masseter to apply force along the cheek tooth row. Thus, the shape of the TMJ also reflects feeding habits. In carnivores a dorsally-convex mandibular condyle of the dentary bone fits into a distinct, ventrally-concave fossa of the squamosal bone (Figs 3, 4). In herbivores, the TMJ shape is more complex than that of the carnivores. In some ungulates such as the horse and pig, there is a distinct squamosal articular tubercle (tuberculum articulare) that articulates with the mandibular condyle (Nickel *et al.*, 1986; Popesko, 1979).

TMJs of sea otters, seals, sea lions, and polar bears are mechanically constrained, allowing up and down movement but little or no transverse motion of the lower jaw. Of the skulls illustrated, the mandibular fossa of the otter is the most restrictive with a deep concavity that grips the mandibular condyle (in some cleaned skulls the mandible cannot be removed with out damaging the margins of the mandibular fossae). The mandibular fossae of the seal, sea lion, and polar bear are shallower, and more similar to that of the dog. The TMJs of odontocete cetaceans appear to be mechanically less constrained because of their relatively large radii of curvature. Live dolphins, however, exhibit simple up-and-down (opening-and-closing) jaw movements similar to those observed in the dog. As mentioned above, odontocetes have simple fish-and-squid-grasping teeth. The TMJs of rorquals are relatively unconstrained; they can move up and down, forward and back, and rotate along the long axis of the dentary. These relatively unconstrained joints are tough and pliable fibrous

S

structures that can absorb the mechanical shock associated with lunge feeding. Rorqual lower jaws must support the large, pleated gular sac into which flows a large volume of water and prey during lunge feeding. Gular sac contraction forces water out through the relatively short baleen plates trapping the prey (Lambertsen, 1983; Pivorunas, 1979). Right whale TMJs restrict jaw movements to up and down and rotation of the mandible along its long axis (Werth, 2004). The jaws of these skimmers support massive lower lips that guide an almost continuous stream of water past long baleen plates (Pivorunas, 1979; Werth, 2004). Gray whales, which are bottom feeders, have relatively robust lower jaws. The mandibular condyles of manatees are slightly flattened sub-cylinders that articulate with a distinct articular tubercle, which is located rostral to the shallow mandibular fossa (Fig. 3). Manatees must have a relatively mobile TMJ to accommodate grinding their food. The motions of the manatee TMJ include influence of the robust pterygoid process as a pivot, creating a slightly arched transverse travel of the occluding tooth rows. These motions, which include lateral and a small amount of rostral motion, provide the action required for grinding vegetation as well as stimulating the rostral migration of the teeth (Domning, 1978).

The shapes of the dentaries of the three marine mammals that are illustrated in Fig. 3 are dramatically different—that of the dog is roughly similar to those of the sea otter, seal, sea lion, and polar bear (Fig. 1). Note the angular process in these latter species; it is located ventrocaudal to the TMJ—that of the polar bear is much more pronounced than those of the other marine mammals. The dolphin dentary is elongate with a reduced ramus and a very small angular process. The manatee dentary has a forward-directed, robust coronoid process and a relatively flat mandibular condyle; there is no discernable angular process.

Feeding includes swallowing. Chewing involves positioning of the food between the teeth by the tongue; swallowing requires the coordinated action of these muscles and bones as food leaves the oral cavity and moves through the pharynx. How do the bones of the head accommodate swallowing? The hyoid apparatus is an important structure in both feeding and swallowing (Fig. 3); it is a complex of hinged bony and cartilaginous elements that are suspended from the ventral aspect of the cranium and lie between the dentaries. The hyoid bones (labeled with the suffix -hyal to minimize confusion with hyoid muscles, Reidenberg and Laitman, 1994), provide the mechanical support of many of the muscles that act upon the tongue and the larynx. Muscles between the tongue and basihyal move the hyoid apparatus up and forward. Muscles between the basihyal and the sternum move the hyoid apparatus down and back (Fig. 3). The tongue may also help exclude water from food that is swallowed under water.

In most mammals, the hyoid apparatus is attached to the ventral skull at one of the bony elements of the compound temporal bone, at or near the external auditory meatus (Figs 3, 4): in carnivores via the mastoid process of the periotic bone; in man, ruminants, and horses via the styloid process of the tympanic bone; and in the pig via the nuchal process of the squamosal bone (Nickel et al., 1986). The seal, sea lion, and sea otter all have hyoid apparatuses that are similar in configuration and attachment to that of the dog. The dolphin and manatee have relatively robust hyoid apparatuses when compared to the other marine mammals. In suction feeders, such as the squid-eating beaked whales, pilot whales, and kogiids, the hyoid apparatus and its associated muscles are massive (Reidenberg and Laitman, 1994). In contrast to most other mammals, the manatee and the dolphin hyoid apparatus is attached to the ventral skull at the paracondylar (paroccipital) processes of the exoccipital bones (see below) in a position caudolateral to the tympanoperiotic complex. There are

distinct concavities for hyoid attachment in the paracondylar processes (Fraser and Purves, 1960). This attachment helps acoustically isolate the hyoid apparatus from the bones of the tympanoperiotic complex, which are themselves not fused to the rest of the cranium (see below). Interestingly, in the live dolphin there is an air sinus (posterior sinus) at the rostral aspect of the concavity, between the hyoid attachment and the tympanoperiotic (Fraser and Purves, 1960), which would add to the mechanical isolation. In some odontocetes (i.e., *Kogia*, *Ziphius*) there are large, well-developed mastoid process of the tympanoperiotic. These mastoid processes are similar in position to the paracondylar process of the dolphin. There are no deep concavities on these mastoid processes but there are similar but shallow regions, medial to the jugular notches, on the caudolateral margins of the relatively thick crests of the basioccipital bones.

## III. Bony Features and Bones

One approach to studying the skull is to focus on a few specialized bony features (Fig. 4). Bony features are morphological characters or landmarks that make up one or more bones. Size, shape, and positions of bony features reflect evolutionary, developmental, and mechanical pressures in a grossly visible manner. For example, the zygomatic arch, which supports the masseter muscle that helps close the jaws, may be composed of one, two, or three bones depending on the species. The rostrum or muzzle may be elongate and may or may not include the nasal bones. Thus, to characterize individual skulls without having to identify individual bones, biologists use the morphology of bony features such as the postorbital processes; zygomatic arch shape and composition; rostrum length; orbit size, shape, and position; and jaw articulation.

In general, large, forward-facing orbits are characteristic of predators that rely on vision as their primary sensory modality, whereas laterally facing orbits are more typical of non-predatory species (Hildebrand, 1995). Also note that the orbits of most species in Fig. 4 are open caudally (having small postorbital processes), in contrast to those of the fully aquatic mammals. In the species in which the orbit is open, there is a postorbital ligament caudal to the eye that extends between the ventrally projecting postorbital process of the frontal bone and a dorsally projecting postorbital process of the jugal and or the squamosal bones (a bony feature on the dorsal aspect of the zygomatic arch); these postorbital processes are prominent in the polar bear.

In all species, the external bony nares are bordered by the nasal bones. The positions, relative to the rostrum and braincase, of the external bony nares may reflect respiratory adaptations to diving, feeding, and locomotion. The occipital condyles position the head on the neck and influence the flexibility of this joint (Figs. 1, 2). Some marine mammal species have short necks, placing the base of the skull very near the shoulder joint and the thoracic cavity. Species with long necks may have a wide range of neutral head positions and may also have a greater range of movement than the fully aquatic species (King, 1983). In most mammals the joint between occipital condyles and the first cervical vertebra (atlas) is restricted to two degrees of mechanical freedom; an additional degree of freedom is acquired with the rotation between the atlas and axis vertebrae. In cetaceans with fused cervical vertebrae three degrees of freedom are potentially available in the joint between the condyles and the atlas.

## IV. Ground Plan of the Skull Bones

What other factors shape the skull? In all vertebrates, the skull bones develop from ossification centers in a basic pattern that partially or completely encloses the brain and encapsulates the sensory

**S**

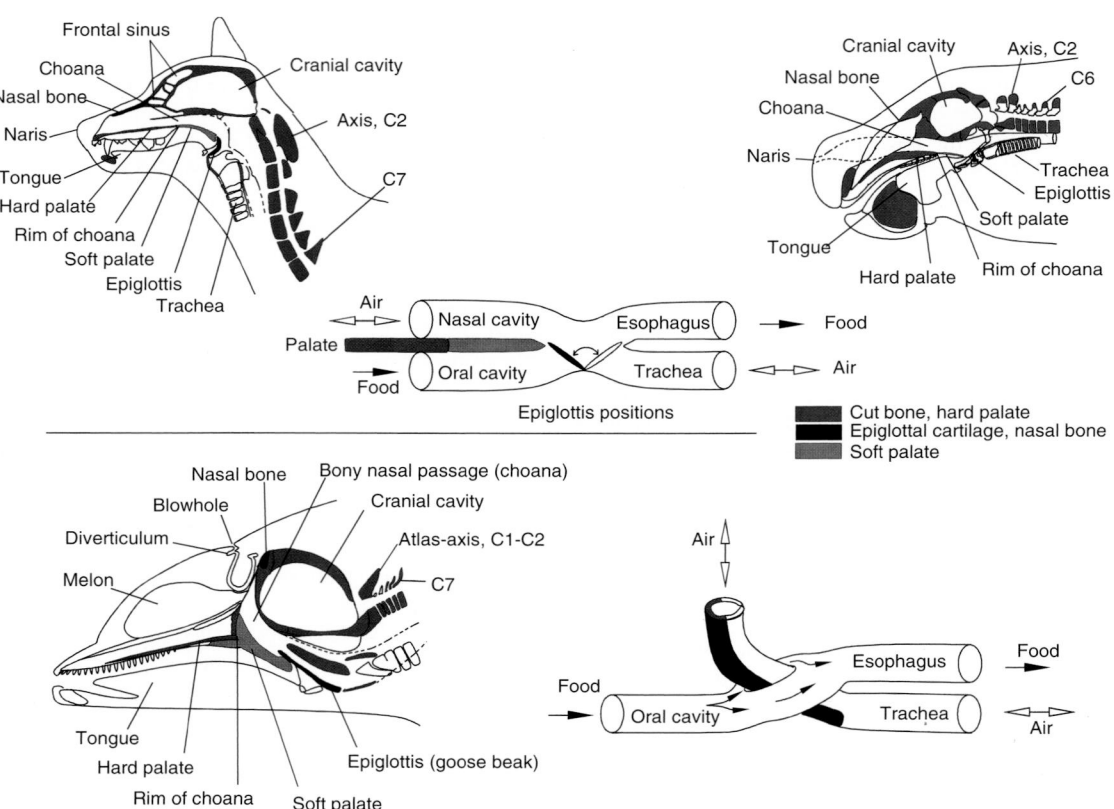

**Figure 5** *Comparisons of the morphological adaptations of the mammalian head that allow respiration while food is in the mouth. Separation of oral and nasal cavities accommodates prolonged chewing and allows teeth to be modified accordingly. The dog and manatee are schematically represented in the upper panel. In the dolphin (lower panel) further modification is shown to accommodate the migration of the respiratory opening to the top of the head. Copyright S. A. Rommel.*

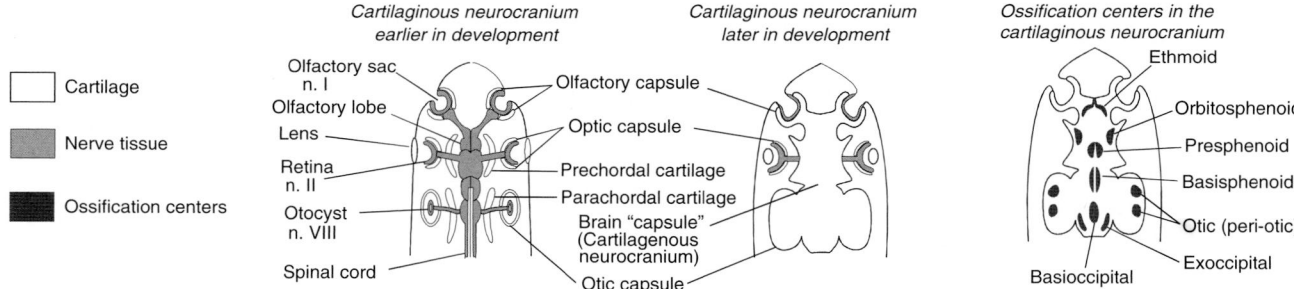

**Figure 6** *Schematic illustration of ventral views of the developing vertebrate skull. Modified after Kent and Miller (1997). The basic plan of encapsulation of the senses and brain is illustrated in the left two drawings. The right drawing illustrates the ossification centers that will eventually become the replacement bones of the cranium.*

organs of olfaction, vision, hearing, and balance (Fig. 6). Tissues that are preformed in cartilage and that eventually develop into bone are referred to as endochondral or replacement bones (these form the chondrocranium); those tissues deposited directly as bone within specialized connective tissue membranes are referred to as dermal or membrane bones (these form the dermatocranium). The distinction between endochondral and dermal bones is valuable in establishing homologies with skull bones of the lower vertebrates; however, once

bony tissues are formed, the two kinds of bone are indistinguishable microscopically (Nickel *et al.*, 1986).

How does one determine which bones of the skull are homologous between species? A systematic but simplified approach that allows one to compare homologous elements is to utilize a generalized schematic skull (Fig. 7). This schematic is a particularly useful way to avoid the "mental indigestion" of having to memorize all the individual bones in several species (Romer and Parsons, 1977).

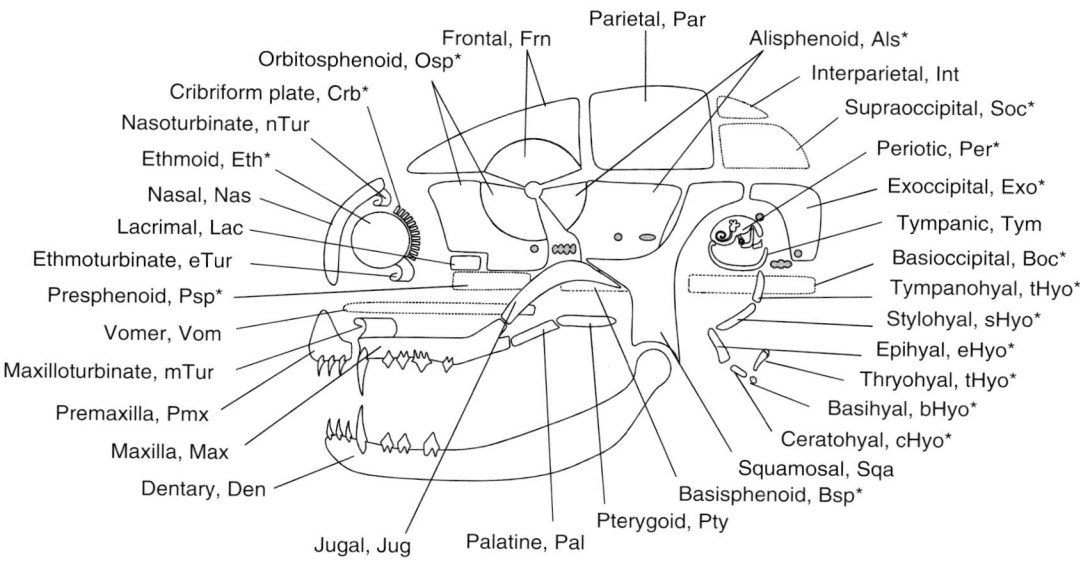

**Figure 7** *Left lateral schematic of the mammalian skull illustrating relative bone positions. Most skull bones are bilaterally paired. This schematic approach has been used for more than 100 years and provides a framework in which to compare a wide variety of mammalian skulls. Modified after Flower (1885), Kent and Miller (1997), and Evans (1993). Recall that the nose, eyes, and ears are encapsulated early in development; these three sensory areas are represented by circular regions in the schematic. Endochondral bones are indicated with an asterisk. Copyright S. A. Rommel.*

In some species, individual bones fuse (ankylose) to form compound bones[1]; these include the occipital, temporal, and sphenoid "bones." Of particular interest is the temporal "bone," which is made up of many separate bony elements and/or ossification centers (Kent and Miller, 1997). These separate bones include the squamosal, the periotic, the tympanic, the middle ear ossicles, and in some species a distinct mastoid. In many mammals, once skeletal maturity is reached, the bulk of each temporal is a single unit with no visible sutures between the bony elements. Thus, it is common with terrestrial mammals to refer to the temporal as a single bone, but this is inappropriate for several of the marine mammals, particularly some of the cetaceans and the sirenians because the earbone complex (periotic and tympanic) do not ankylose to the rest of the skull.

The exploded diagram of the cranium of the Florida manatee illustrates how some bones fit together simply whereas in other places they overlap (Fig. 8). The compound occipital bone, which forms a ring around the foramen magnum, is composed of the basioccipital, exoccipitals, and supraoccipital (Jollie, 1973; Kellogg, 1928; Romer and Parsons, 1977).

Figure 9 is a left lateral view of the individual cranial bones for our representative marine mammals. Contrast this illustration with Fig. 4 to reinforce the distinction between bones and bony features; see Fig. 1 to compare skull size to total body size.

---

[1]We have used three different terms incorporating the word "bone". Bones are discrete ossifications that can be traced phylogenetically. Compound bones are structures that appear to be single units in adults of some species because the joints between the component bones have been resorbed or are not apparent. Bony features are gross features, such as the zygomatic arch, that are made up of discrete and distinct bones or parts of bones (e.g., the zygomatic process of the squamosal bone).

## V. Cranial Joints

Skull bones can meet in several ways and attach to each other by more than one type of material (e.g., cartilage and other connective tissue). Joints between adjacent cranial bones are referred to as sutures or synchondroses. Sutures are fibrous joints between dermal bones; synchondroses are cartilaginous joints between endochondral bones. Sutures and synchondroses are regions of growth between individual bones; in adults they may also function to relieve stresses that are produced in the skull (Gordon, 1988). At parturition, some joints provide the flexibility for a relatively large brain within the cranium to pass through a relatively narrow birth canal. In the illustration of the exploded manatee cranium (Fig. 8), the hatched regions represent (Fig. 9) joints or regions of overlap. In Fig. 10, some of the joints of the dog skull are compared with those of the manatee and dolphin skull. The type of joint generally reflects the mechanical forces acting on the adjacent bones. Different types of joint may be found between the same bones in different species; types may also be different in different parts of the same joint, probably to reflect differences in forces. Interlocking joints can absorb mechanical energy. Movable joints eliminate shearing forces. Butt joints can support little shear but are strong in compression. Squamosal or scarf joints allow more surface contact between adjacent bones and are stronger than simple butt joints. Some bone configurations affect the complexity of joints, and this complexity, in turn, affects the action and strength of the joints. Relative aging of skulls may be determined by evaluating the sequence of ankyloses of sutures and synchondroses, unfortunately few systematic studies have been completed (Moore, 1981).

## VI. Foramina

The development of the skull bones proceeds at a pace different from that of the soft tissues of the head. Bone is constantly being

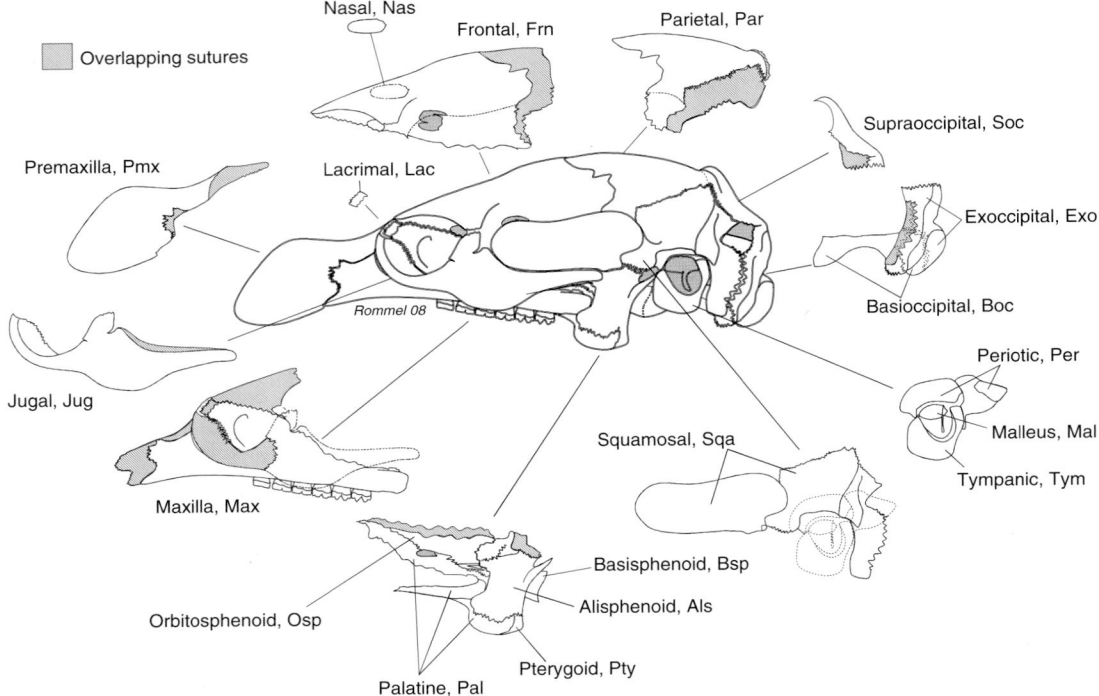

**Figure 8** *Left lateral view of an exploded cranium of the Florida manatee. This figure illustrates the overlapping and/or abutting margins of bones that make up the sutures and synchondroses of the cranium. Copyright S. A. Rommel.*

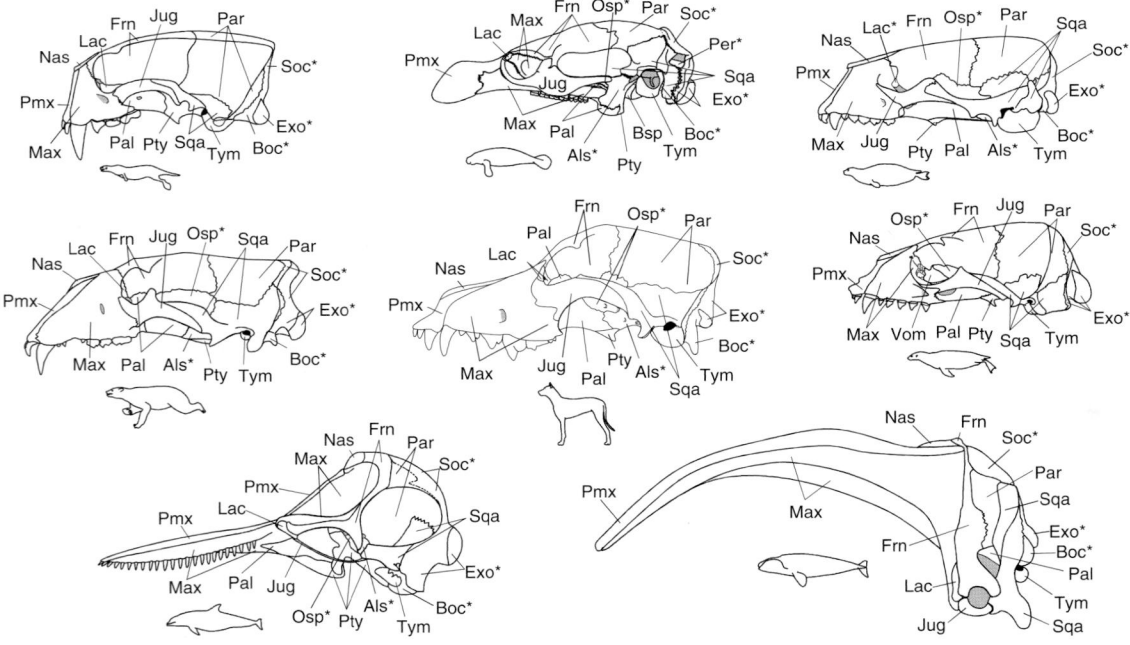

**Figure 9** *Left lateral illustrations of individual cranial bones of selected marine mammals and the dog. Abbreviations are the same as in Figs. 7 and 8. Use Figure 8 to help visualize how the basic plan of mammalian skull morphology is modified in each species. Endochondral bones are indicated with an asterisk. Copyright S. A. Rommel.*

remodeled; this takes place at the level of the individual during its lifetime in response to trauma, nutrition, and localized conditions. Remodeling also takes place at the population level over longer time spans and thus can indicate evolutionary processes. This plasticity is reflected in the way individual skull bones form around vessels and nerves. The resulting openings, or foramina (singular, foramen), are often phylogenetically conserved and so can be used to establish homologies of the same bones across different species. An individual

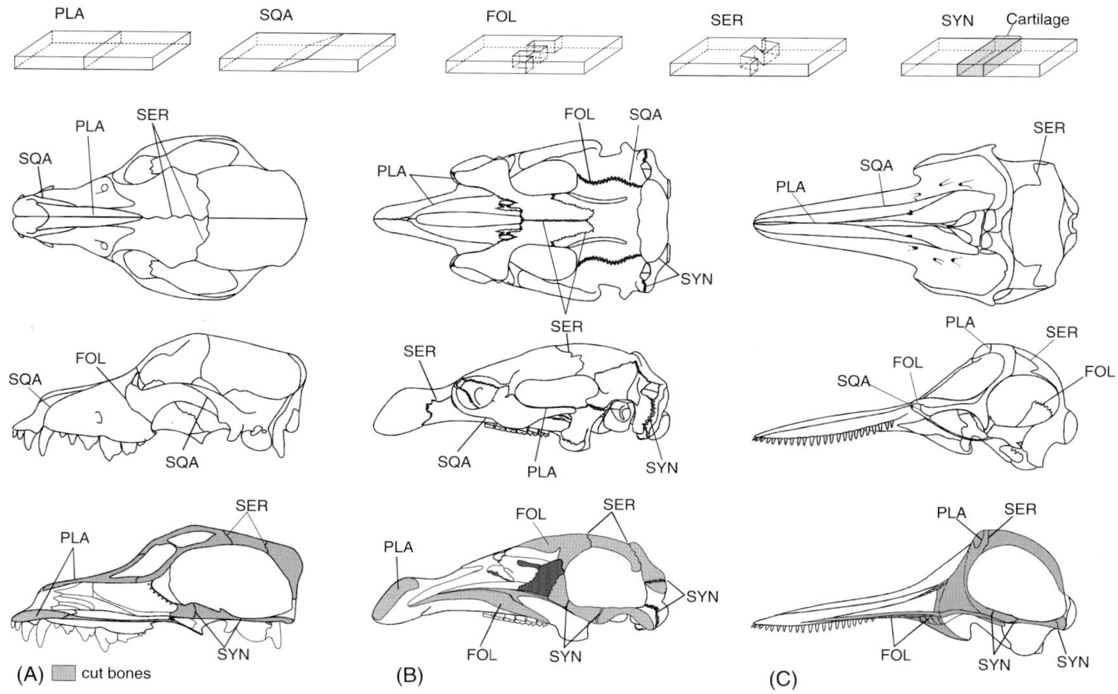

**Figure 10** *Joint types in the cranium of the dog (A), manatee (B), and dolphin (C). Sutures are fibrous joints between dermal bones; synchondroses are cartilaginous joints between endochondral bones. Joints allow growth of adjacent bones, provide limited flexibility and absorb mechanical forces. Suture types are defined by their shape. PLA, plane or butt joint (harmonious suture, sutura plana)—an approximately straight suture with nearly squared-off margins. SQA, squamous or scarf joint (sutura squamosa)—a suture with tapered overlapping margins. FOL, foliate joint (sutura foliata)—a regular suture with small alternating vertical bony plates, in which adjoining bones interleave. SER, serrate joint (sutura serrata)—an irregular suture, in which adjoining bones interlock. A synchondrosis (synchondroses cranii), SYN, has persistent cartilage between bones. Below, simplified, sagittally sectioned crania illustrate those joints that can be observed on the midline. Definitions are from a variety of sources; parenthetical names are from Schaller (1992). Copyright S. A. Rommel.*

nerve or blood vessel may be completely surrounded by a bone or bones of the skull, resulting in a specific foramen. Because this process occurs early in the development of the individual and appears to be similar in all vertebrates, we use cranial nerve foramina to help us identify the skull bones (Figs 11, 12).

In some species, or even in individuals of the same species of different ages, instead of a single foramen for each individual nerve, one or more nerves may exit the braincase through a single opening. Some openings are very large and irregular and are referred to as hiatuses (Fig. 12). The cranial hiatus of the dolphin (Fraser and Purves, 1960) and manatee may include the following nerve openings: optic foramen, orbital fissure (anterior lacerate foramen), and oval foramen—plus the openings for vessels between the last two. The cranial hiatus is not present in the other marine mammals because their skulls have earbone complexes that are firmly attached to the other bones of the skull (see the compound temporal bone above). The tympanic and periotic bones are often referred to as the earbone complex or the tympano-periotic complex. The earbone complexes of the manatee and dolphin have loose connections with the rest of the skull bones, presumably to produce an acoustic isolation from the rest of the skull; in cleaned skulls the earbone complexes may fall out of the skull in these taxa. In life, the odontocete tympanoperiotic is surrounded by peribullar sinuses that add to this acoustic isolation (Fraser and Purves, 1960; Houser *et al.*, 2004). The

cranial hiatus includes the petrooccipital fissure, which is an irregular opening between the tympanic and periotic bones (housing the ear) and the alisphenoid, basisphenoid, basioccipital, and exoccipital bones of skull base (Nickel *et al.*, 1986; Schaller, 1992). In the terrestrial mammals, the margins of the petrooccipital fissure may join to form one or several foramina (foramen ovale, jugular foramen, carotid foramen, hypoglossal foramen, caudal lacerated foramen).

The mandible also has a number of foramina. At its caudal end, the medial aspect (lingual surface) of the dentary has a mandibular foramen, which is the opening of the mandibular canal for the alveolar vessels and nerves. In manatees, the mandibular foramen is relatively large because of the large amount of soft tissues and perioral bristles of the chin supplied and innervated via the mandibular canal. In dolphins, the mandibular foramen is even larger; it is referred to as a hiatus (Fraser and Purves, 1960). The odontocete dentary is almost hollow and is filled with a well-vascularized mandibular fat body, which performs the acoustic function of receiving and guiding sound energy to the earbones (Norris and Harvey, 1974; Koopman *et al.*, 2006).

In some cetaceans (e.g., *Kogia, Ziphius*) the internal auditory meatus is a long narrow canal (Fig. 13) (Rommel *et al.*, 2006). Interestingly, as mentioned above, these cetaceans have distinct mastoid bones. Fraser and Purves (1960) describe secondary growth of the basioccipital, parietal, and squamosal bones around the margins

**S**

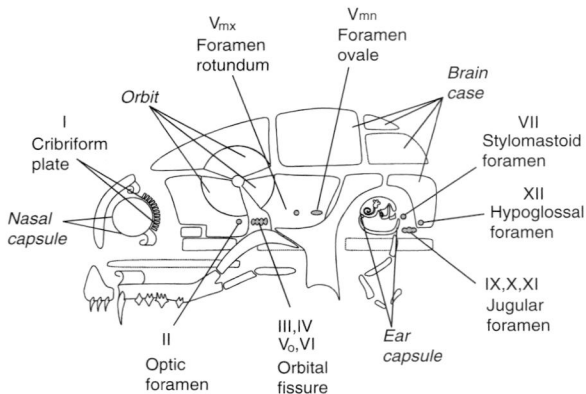

Vmx
Foramen
rotundum

Vmn
Foramen
ovale

*Brain
case*

*Orbit*

I
Cribriform
plate

VII
Stylomastoid
foramen

XII
Hypoglossal
foramen

*Nasal
capsule*

IX,X,XI
Jugular
foramen

II
Optic
foramen

III,IV
V$_o$,VI
Orbital
fissure

*Ear
capsule*

**Figure 11** *Left lateral schematic of openings, or foramina of the skull. Foramina can be used to establish homologies of the same bones in different species; each foramen associated with one or more of the 12 cranial nerves (labeled I through XII) has a name that is used (fairly) consistently by vertebrate morphologists. Thus the cribriform plate, found at the rostral margin of the braincase, is associated with the olfactory nerves (this will be parenthetically referred to as I-olfactory n.) in all of the species that have a sense of smell (even odontocetes, which do not have olfactory nerves as adults, have these perforations (Rommel, 1990). The second cranial nerve passes through the optic foramen and usually perforates the orbitosphenoid bone (II-optic n.). The orbital fissure is usually at the orbitosphenoid bone-alisphenoid bone suture (also anterior lacerate foramen; III-oculomotor n., IV-trochlear n., Vo-ophthalmic branch of trigeminal n., VI-abducens n.). The foramen rotundum (Vmx-maxillary branch of the trigeminal n.) and the foramen ovale (Vmn-mandibular branch of the trigeminal n.) perforate the alisphenoid bone. The stylomastoid foramen is located at the tympanic bone-basioccipital bone suture (VII-facial n.). (Nerve VIII-vestibulocochlear n. is not shown; it perforates the periotic bone through its internal auditory meatus.) The jugular foramen (also caudal lacerate foramen) is at the exoccipital bone-basioccipital bone suture (IX-glossopharyngeal n., X-vagus n., XI-accessory n.). The hypoglossal foramen usually perforates the exoccipital (XII-hypoglossal n.). An additional cranial nerve (O-terminal n.) was discovered after the numbering system was developed. This nerve is found rostral to the olfactory nerve; it has been described only (of the species illustrated) in the odontocetes and is not illustrated here. Copyright S. A. Rommel.*

of the cranial hiatus, which narrows and elongates the channels for the facial and acoustic nerves.

## VII. Skull Cavities

The cranial cavity (Figs 5 and 13) houses the brain, its meninges, and its vasculature (Nickel *et al.*, 1986, Romer and Parsons, 1977). The roof (calvarium) and lateral walls of the braincase are typically made up of the frontal and parietal bones with the caudal wall formed by the supraoccipital and exoccipitals (Figs 7, 8, and 9). Rostrally, there is the ethmoid bone with its perforated cribriform plate, medial extensions of the inner lamina of the frontal bone, and portions of the sphenoids (see below). In odontocetes lateral wings of the vomer also contributes to the rostral wall of the cranial cavity.

The floor of the cranial cavity is formed by the basioccipital, the basisphenoid (including the depression for the pituitary) with its lateral wings (the alisphenoids), and the presphenoid with its lateral wings

(the orbitosphenoids) (Figs 7, and 12). The alisphenoid and orbitosphenoid wings extend dorsolaterally between the more dorsal skull bones in a wide variety of shapes and sizes—these two winged bones (collectively called the sphenoid bone) have most of the foramina for the cranial nerves and may be the most variable and difficult to recognize in different taxa (Figs 7, 9, and 11).

The nasal cavity is separated from the oral cavity by the secondary palate. The secondary palate is formed by the hard and soft palates (Fig. 5). The secondary palate makes a very important contribution to the evolutionary forces that shaped mammal skulls as it allows prolonged chewing while breathing, providing additional time for food processing in the mouth. The nasal cavity extends from the external bony nares to the perpendicular plate of the ethmoid bone. The nasal cavity is typically a roughly tubular structure that occupies the entire length of the rostrum of the skull in most mammals, but not in cetaceans. In some terrestrial species, it may be paralleled on either side by enclosed maxillary sinuses. The bony supports for the nasal cavity are formed by the premaxillae, maxillae, frontals, vomer, palatines, and in some species the lacrimals and jugals (Nickel *et al.*, 1986; Romer and Parsons, 1977). There is often a dorsoventral medial nasal septum in the nasal cavity; it is formed by cartilage rostrally and by a bony extension of the ethmoid caudally. The vomer (an unpaired, relatively thin, ventral midline bone) may also contribute to the ventrolateral aspects of the nasal septum.

The nasal septum divides the nasal cavity into separate left and right air channels called choanae (Fig. 5). Caudally, the choanae may lose their septum near the junction of the soft and hard palates (rim of the choanae; Figs 5 and 12). In many species, a thin layer of the vomer extends caudally along the ventral midline to cover the ventral aspects of the presphenoid and basisphenoid as the animal ages. In some mammals, particularly dolphins, this portion of the thin vomer is often fenestrated, exposing the synchondrosis between the basisphenoid and presphenoid bones (Fig. 12).

Conchae (turbinates, turbinals) are thin lamellae of bone (covered with mucous membrane in life) that project into the nasal cavity. Conchae increase the surface area in the nasal cavity for heat exchange, water balance, and olfaction (Moore, 1981). The more rostral conchae develop as outgrowths of the maxillae and nasals, the more caudal chonchae develop from the ethmoid bones (Nickel *et al.*, 1986). In stark contrast to most other mammals, the nasal cavities of cetacea are not part of the rostrum. Instead, cetacean nasal cavities are almost vertical channels just rostral to the braincase (see telescoping below) and are devoid of conchae; the bones dividing and bordering the nasal cavity are displaced by the vomer and the pterygoid (Mead and Fordyce, 2008; Rommel, 1990). This helps allow for the rapid respiratory cycle of cetaceans (Pabst *et al.*, 1999). The conchae of sea otters, sea lions, and seals are very convoluted, almost filling the nasal cavity with lace-like networks of bone. These structures significantly increase in surface area and have been shown to be important for water and heat conservation in seals and sea otters (Folkow *et al.*, 1988; Huntley *et al.*, 1984).

Skull bones of most mammals are mechanical marvels. Only relatively recently in engineering history has man built composite structures (monocoques) that approach the efficiency of design found in the mammalian skull (Gordon, 1988). In many species, some of the skull bones are made up of a layer of spongy bone (diploe) and/or air-filled sinuses (Fig. 5) sandwiched between two thin "panels" of rigid cortical bone commonly referred to as the internal and external laminae (singular lamina) (Nickel *et al.*, 1986; Schaller, 1992). This excavation of sinuses within bones is called pneumatization (Nickel *et al.*, 1986). The resulting multilayered structure is strong, yet has

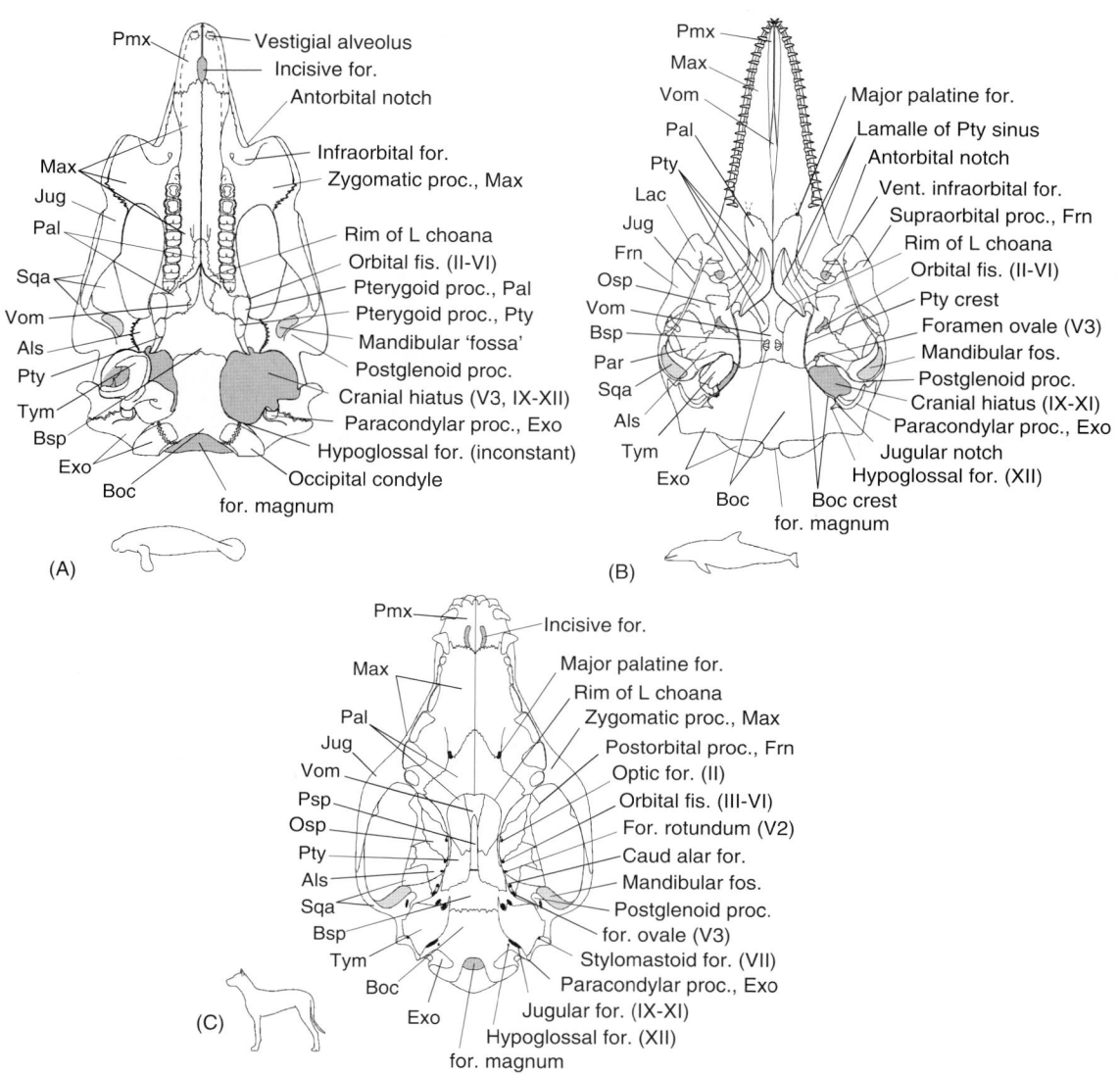

**Figure 12** *Comparison of the basicranial morphology of the manatee, dolphin, and dog. The base of the skull has important morphological features that help with keying out different species. Note the large opening (cranial hiatus), which is illustrated for the manatee and dolphin; it is not present in the other marine mammals because their skulls have earbone complexes that are firmly attached to the other bones of the skull. The earbone complexes of the manatee and dolphin have loose connections with the rest of the skull bones as part of their acoustic isolation from the rest of the skull. The following abbreviations are used: for, foramen; fis, fissure; fos, fossa; proc, process; L, left; R, right. Roman numerals denote cranial nerves. Fenestrations may occur in the vomer ventral to the joint between the basisphenoid and basioccipital bones; these fenestrations make the margins of these bones and this suture visible through the vomer. Copyright S. A. Rommel.*

less weight than other bony structures. In the skull bones of most placental mammals, pneumatized paranasal sinuses develop embryonically as invaginations from the adjacent air spaces. These sinuses vary considerably even within species; they become larger and more numerous as the individual ages (Moore, 1981). In many species, paranasal sinuses help form three-dimensional bracing systems by minimizing the mass of bone that provides the framework needed for mechanical support of different parts of the head[2]. Some paranasal

sinuses increase the available surface area for olfactory epithelium to detect odors (Nickel *et al.*, 1986). Sinuses that are well vascularized may help provide evaporative cooling (Schaller, 1992). In terrestrial mammals, paranasal sinuses may also act as resonators that modify sounds generated by the individual (Moore, 1981). Paranasal sinuses occur in the frontal (Fig. 5) and maxillary bones of the dog and in the frontal, ethmoid, and presphenoid bones of man; in other terrestrial mammals they also occur in the exoccipital, jugal, lacrimal, nasal, palatine, parietal, basisphenoid, and vomer (Moore, 1981; Nickel *et al.*, 1986; Schaller, 1992).

In contrast to most other mammal skulls, manatee skull bones are thick and made of almost solid amedullary bone (Fawcett, 1942).

---

[2]Similar bracing systems are found in the wings and fuselages of aircraft and the hulls of large ships (Gordon 1988).

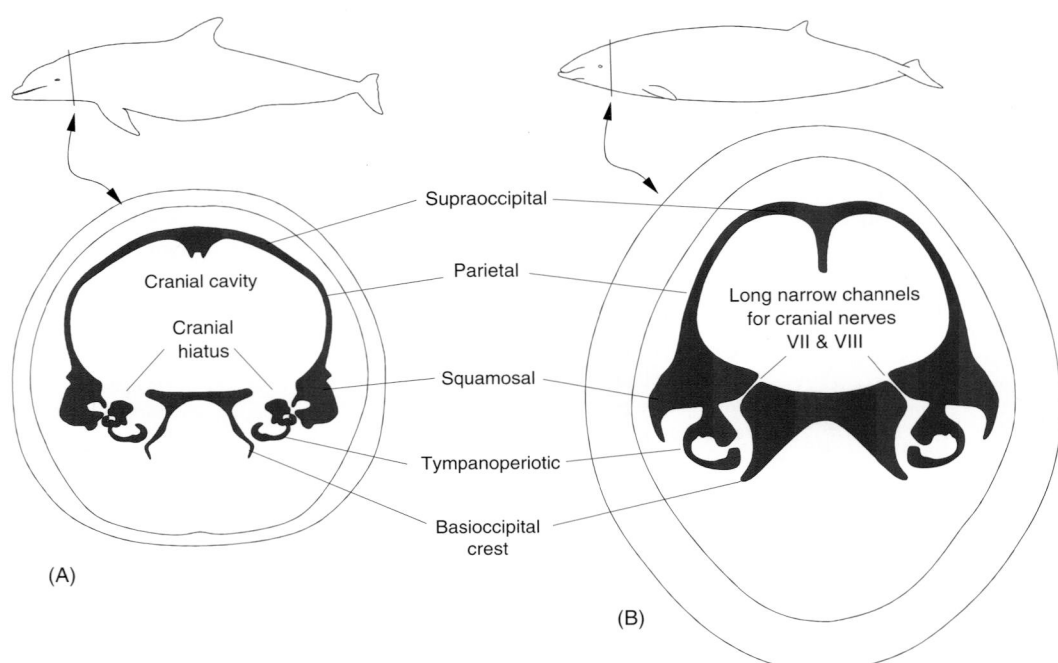

**Figure 13**  *Cross sections of the skulls of* Tursiops *(A) and* Ziphius *(B). The cross sections (at the level of the ear) are scaled to have similar areas of braincase. In* Tursiops, *the pathway out of the braincase for cranial nerves VII and VIII is a short open cranial hiatus bordered by relatively thin bones, whereas in* Ziphius *it is a narrow, relatively long channel. The ziphiid basioccipital bones are relatively massive with thick ventrolateral crests; in contrast, delphinid basioccipital bones are relatively long and tall, but thin and laterally cupped. Note that in contrast to the* Ziphius *calf cross section, the adult head would have a greater amount of bone and the brain size would be relatively smaller. Modified from Rommel et al. (2006).*

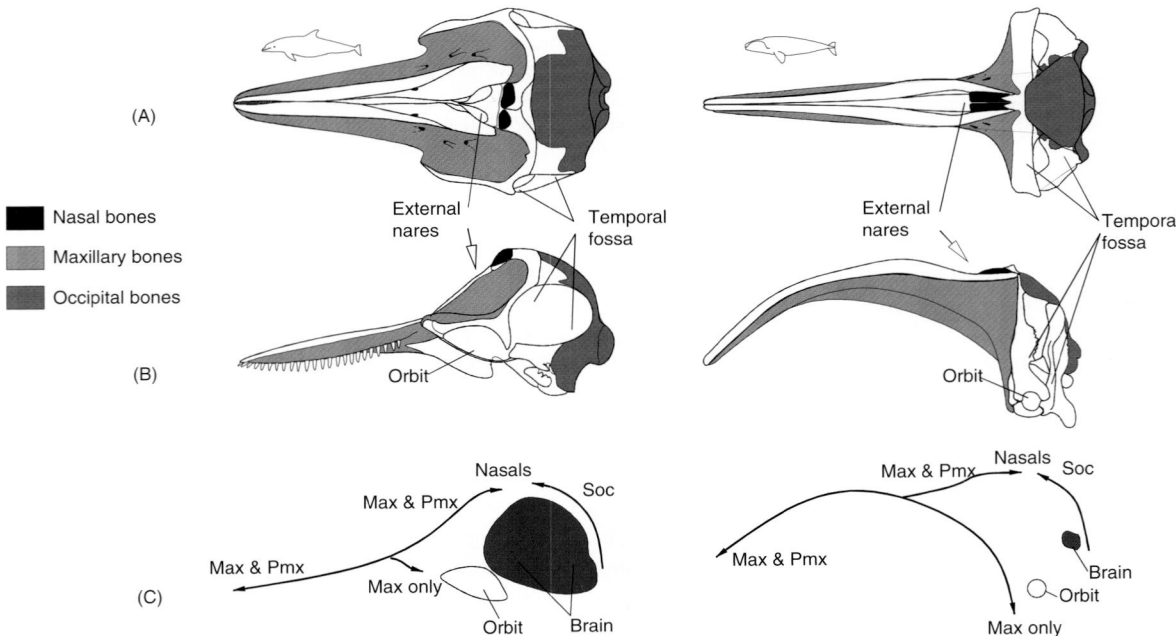

**Figure 14**  *Dorsal (A), lateral (B), and schematic (C) views illustrating telescoping in odontocetes (left, e.g.,* Tursiops*) and mysticetes (right, e.g.,* Eubalaena*). Telescoping refers to the elongation of the rostral elements [both fore and after in the case of the premaxillary and maxillary bones (Pmx and Max), the vomer, and mesorostral cartilage], the dorso-rostral movement of the caudal elements [particularly the supraoccipital bone (Soc)], and the overlapping of the margins of several bones. This overlap or sliding over each other of these elements resembles old-fashioned telescopes.*

In some large terrestrial mammals, such as the elephant the skull is vastly enlarged with pneumatized bones in part to accommodate the large muscles of the head (Nickel *et al.*, 1986). In contrast, the skulls of the larger cetaceans (i.e., *Ziphius* and *Eubalaena*) are enlarged to accommodate the larger muscle masses and larger food processing requirements but are not pneumatized.

In diving mammals, air-filled cavities within rigid enclosures of bone could be damaged during dives when subjected to large rapid pressure changes associated with variations in depth. If air-filled paranasal sinuses are present in divers, they have to be open-ended so that air or other fluids (e.g., blood, lymph, cerebrospinal fluid) in adjacent vascularized structures can move into and out of them to compensate for changes in air volume in response to changes in ambient pressure (Molvaer, 2003); alternatively they must have flexible walls that are capable of collapse. Of the carnivores, terrestrial bears (Moore, 1981) have the most extensive sinuses; to our knowledge, those of polar bears have not been described. Cetaceans particularly odontocetes, have several air-filled regions, those on the ventral aspect of the skull are associated with pneumatized bones. These air-filled sinuses have large openings that are connected to the respiratory system via the Eustacean tube (Fraser and Purves, 1960; Houser *et al.*, 2004). Interestingly, in contrast to the pneumatized bones found in terrestrial mammals, it is the cetacean pterygoid bones that are the most pneumatized. Typically pterygoid bones are small in mammals but in cetaceans they are significantly enlarged (slightly more than 40% of the ventral length of the skull in *Ziphius*; Rommel *et al.*, 2006). In many cetaceans, the pterygoid sinuses have thin medial and lateral bony walls called lamellae. Some species with relatively large pterygoid sinuses (e.g., ziphiids and physeterids) have lost the lateral lamellae and the medial bony wall of the sinus is relatively thick (Fig. 13). In place of bony lateral lamellae, the large pterygoid sinuses of these divers are walled with a tendinous sheath to which the muscles of the palate attach (Fraser and Purves, 1960). In manatees, a large ventromedial pterygoid process of the palatine bone and a large ventrolateral pterygoid process of the alisphenoid both join the relatively small pterygoid bone to produce a robust structure that supports the large pterygoid muscles (Figs 8, 10, and 12).

The walls of the pterygoid sinuses are well vascularized, perhaps to help with adjusting volume as ambient pressures change. The air-filled spaces of live dolphins have been shown to be dynamic structures that function as reflectors to help isolate the earbones from the sound producing apparatus of the head and to help isolate the two ears so that they can have better directional abilities (Houser *et al.*, 2004).

## VIII. Telescoping

Telescoping is a process often discussed when describing the skulls of cetaceans (Figs 14, 15). The term, coined by Miller (1923), refers to the elongation of the rostral elements and the dorsorostral movement of caudal elements (Kellogg, 1928; Miller, 1923; Rommel, 1990). The relative placement of the skull bones in cetaceans results in considerable overlap of some adjacent bones. If the skull is sectioned, one can observe as many as four different bones overlapping each other—this overlap resembles old-fashioned collapsible telescopes. In cetaceans, the external bony nares have been displaced to the dorsal apex of the skull, so the nasal bones are located just caudal to the external bony nares (as in other mammals) but dorsal to the brain case instead of at the apex of the rostrum. The premaxillary and maxillary bones have been extended at their rostral tips; their caudal aspects are pulled up and back over the frontal bones and maintain their relative positions with the nasal bones. The narial passages are essentially vertical in cetaceans, which eliminates the nasal bones as roofing bones of the nasal passages. The nasal bones are, instead, relatively small vestiges that lie in depressions of the frontal bones. Thus, the roof of the cetacean mouth is not the floor of the nasal passages as it is in most other mammals. Caudally, telescoping differs in odontocete and mysticete cetaceans (Fig. 14). The changes in the mysticetes are dominated by a ventrocaudal extension of the maxillary bones, whereas in the odontocetes, the premaxillary and maxillary bones are shifted more dorsocaudally (Kellogg, 1928). Interestingly, whereas the odontocete facial muscles have moved dorsocaudally over the eye, the temporal muscles of the mysticetes have moved dorsorostrally over the eye. Thus, the temporal fossae of the mysticetes are very different from those of the odontocetes.

The remodeling associated with telescoping is reflected in the number and positions of the cranial nerve foramina in the maxilla (Fig. 15). Consider the nerves that are associated with the muscles of the face—the (sensory) trigeminal nerve (V) and the (motor) facial nerve (VII). The right and left facial nerves control such muscle activity as facial expression in the dog, feeding in the manatee, and focusing of sonar pulses in the dolphin. The trigeminal nerves signal the brain to coordinate the muscular activities in the same region. The relatively large sizes of these two nerves in the dolphin and manatee reflect the importance of these neuromuscular actions. The nerve diameters are reflected by the size of the infraorbital foramina through which the trigeminal nerve pierces the maxilla (Fig. 15). In odontocetes, the bones and muscles of the face are reshaped and accommodate the melon and its need for complex mechanical manipulation and the sensory and motor nerves are moved up and over the orbit. The homologous (and thus same-named) opening, which is *infra*-orbital in most species, is now actually dorsal to the orbit (that is, *supra*-orbital) in cetaceans!

In conclusion, we can state that a glance at a skull tells us a great deal about an organism—how it senses its environment, how it feeds, how big its brain is, etc. It is also very important in understanding phylogenetic history and species description. Rather than becoming bogged down in trying to memorize names of bones or bony features, study a skull with an open mind (pun intended) about adaptations and function.

**S**

**Figure 14 (continued)**    *One result of telescoping is the displacement of the external bony nares (and the associated nasal bones) toward the dorsal apex of the skull—up and over the rostral margin of the brain! Telescoping is quite different in odontocete and mysticete cetaceans; in most odontocetes the rostrum is dorsally concave, whereas in mysticetes the rostrum is ventrally concave. The temporal fossae of the mysticetes have moved up and forward over the eye; the temporal fossae in odontocetes are in a more typical mammalian position. Relatively more bone mass is moved up and over the orbit in the odontocetes, whereas relatively more bone mass is moved down and under the orbit in mysticetes. In the lower schematic, arrows indicate the directions of relative movement as each skull is remodeled to accommodate the brain and the respiratory, feeding, and acoustic apparatus of the two types of cetaceans. Note that in C the brains are scaled to fit in the lateral views of the crania in B above them; the odontocete brain makes up a larger percentage of the cranial volume than does the brain of the mysticete. Copyright S. A. Rommel.*

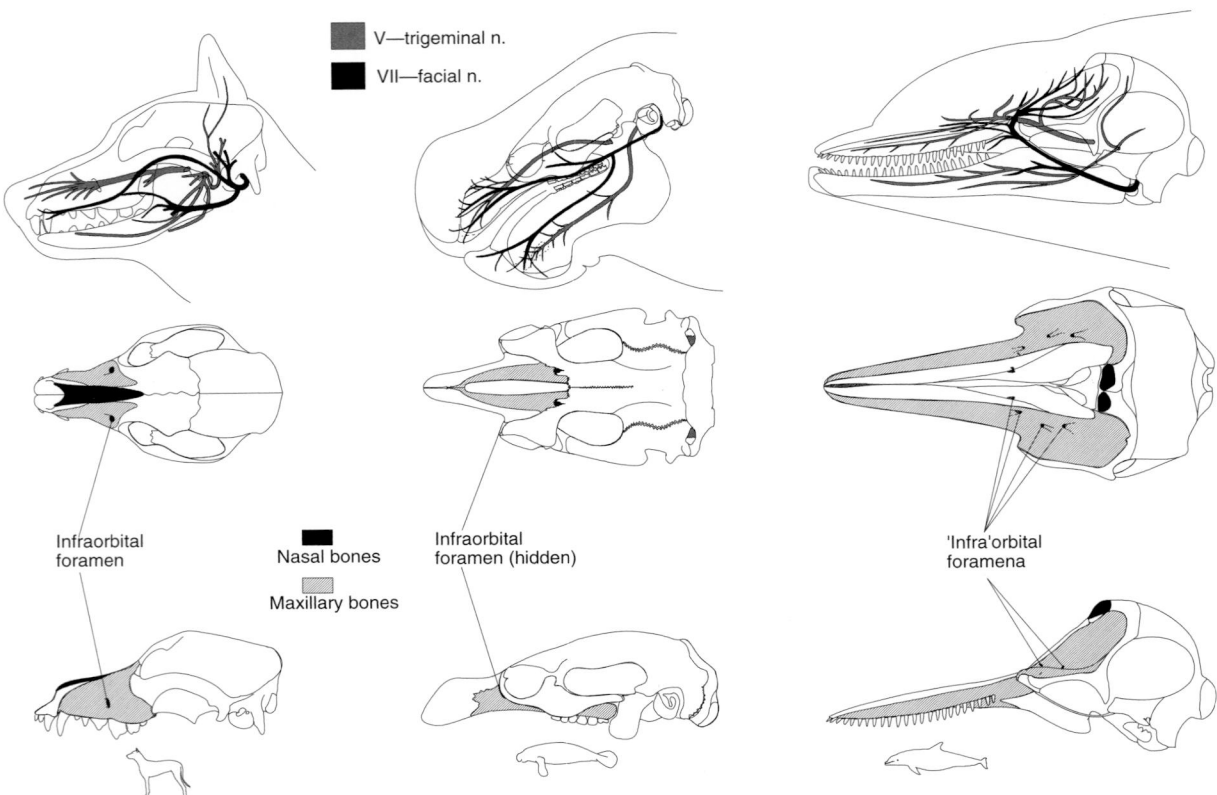

**Figure 15** *The nerves associated with the facial muscles also reflect telescoping. These are the trigeminal and the facial nerves. Here the dog is compared with the manatee and the dolphin. The relatively large sizes of the trigeminal nerves are reflected in the relatively large foramina through the maxillae (infraorbital foramina); the larger the nerve, the more information it can carry. The process of telescoping has remodeled the bones of the rostrum and included in this process are the reshaping of the muscles of the face and the nerves that innervate them. Also note that some odontocete cetaceans have notable bilateral asymmetry in the dorsal elements of the skull. Copyright S. A. Rommel.*

## See Also the Following Articles

Feeding Morphology ■ Sense Organs, Overview

## References

Domning, D. P. (1978). The mycology of the Amazonian manatee, *Trichechus inunguis* (Natterer) (Mammalia: Sirenia). *Acta Amazonica* **VIII**(Supl. 1), 1–81.

Domning, D. P., and Hayek, L.-A. C. (1984). Horizontal tooth replacement in the Amazonian manatee (*Trichechus inunguis*). *Mammalia* **48**(1), 105–127.

Evans, H. E. (1993). "Miller's Anatomy of the Dog," 3rd ed. Saunders, Philadelphia.

Fawcett, D. W. (1942). The Amedullary Bones of the Florida Manatee (Trichechus latirostris). *Am. J. Anat.* **71**, 271–309.

Flower, W. H. (1885). "An Introduction to the Osteology of the Mammalia," 3rd ed. Macmillan, London [Reprinted by A. Asher & Co., Amsterdam, 1966.].

Folkow, L. P., Blix, A. S., and Eide, T. J. (1988). Anatomical and functional aspects of the nasal mucosal and ophthalmic retia of phocid seals. *J. Zool. (Lond.)* **216**, 417–436.

Fraser, F. C., and Purves, P. E. (1960). Hearing in cetaceans—evolution of the accessory air sacs and the structure and function of the outer and middle ear in recent cetaceans. *Bull. Br. Mus.* **7**(1), 1–140, +53 pls.

Gordon, J. E. (1988). "The Science of Structures and Materials." The Scientific American Library, W. H. Freeman, New York.

Harper, C. J., McLellan, W. A., Rommel, S. A., Gay, D. M., Dillaman, R. M. and Pabst, D. A. (in press, 2008) Morphology of the melon and its tendinous connections to the facial muscles in bottlenose dolphins (*Tursiops truncatus*). J. Morphol.

Hildebrand, M. (1995). "Analysis of Vertebrate Structure," 4th ed. Wiley, New York.

Houser, D. S., Finneran, J., Carder, D., Van Bonn, W., Smith, C., Hoh, C., Mattrey, R., and Ridgway, S. (2004). Structural and functional imaging of bottlenose dolphin (*Tursiops truncatus*) cranial anatomy. *J. Exp. Biol.* **207**, 3657–3665.

Huntley, A. C., Costa, D. P., and Rubin, R. D. (1984). The contributions of nasal countercurrent heat exchange to water balance in the northern elephant seal, *Mirounga angustirostris*. *J. Exp. Biol.* **113**, 447–454.

Jollie, M. (1973). "Chordate Morphology." R. E. Krieger Pub. Co, Huntington, NY.

Kardong, K. (1998). "Vertebrates: Comparative Anatomy, Function, Evolution," 2nd ed. McGraw-Hill, New York.

Kellogg, R. (1928). The history of whales: their adaptations to life in the water. *Q. Rev. Biol.* **3**(1), 29–76, and **3**(1), 174–208.

Kent, G. C., and Miller, L. (1997). "Comparative Anatomy of the Vertebrates," 8th ed. Wm. C. Brown, Boston.

King, J. E. (1983). "Seals of the World," 2nd ed. Comstock Publishing Associates, Ithaca, NY.

Koopman, H. N., Budge, S. M., Ketten, D. R., and Iverson, S. J. (2006). Topographical distribution of lipids inside the mandibular fat bodies of odontocetes: remarkable complexity and consistency. *IEEE J. Ocean Eng.* **31**(1), 95–106.

S

Lambertsen, R. H. (1983). Internal mechanism of rorqual feeding. *J. Mammal.* **64**(1), 76–88.

Mead, J. G. (1975). Anatomy of the external nasal passages and facial complex in the Delphinidae (Mammalia: Cetacea). *Smithson. Contrib. Zool.* **207**, 1–72.

Mead, J. G., and Fordyce, R. E. (in press, 2008). The Therian Skull—a lexicon with emphasis on the odontocetes. *Smithson. Contrib. Zool.* **627**.

Miller, G. S., Jr. (1923). The telescoping of the cetacean skull. *Smithson. Miscell. Collect.* **76**(5), 1–71.

Molvaer, O. I. (2003). Otorhinolaryngological aspects of diving. *In* "Bennett and Eliot's Physiology and Medicine of Diving" (A. O. Brubakk, and T. S. Neuman, eds), 5th ed. Saunders, New York.

Moore, W. J. (1981). "The Mammalian Skull." Cambridge University Press, New York.

Nickel, R., Schummer, A., Seiferle, E., Wilkens, H., Wille, K.-H., and Frewein, J. (1986). "The Locomotor System of the Domestic Mammals." Verlag Paul Parey, Berlin.

Norris, K. S., and Harvey, G. W. (1974). Sound transmission in the porpoise head. *J. Acoust. Soc. Am.* **56**(2), 659–664.

Pabst, D. A., Rommel, S. A., and McLellan, W. A. (1999). Functional morphology of marine mammals. *In* "Biology of Marine Mammals" (J. E. Reynolds, and S. A. Rommel, eds), pp. 15–72. Smithsonian Institution Press, Washington, D.C.

Pivorunas, A. (1979). The feeding mechanisms of baleen whales. *Am Sci.* **67**, 432–440.

Popesko, P. (1979). "Atlas of Topographical Anatomy of the Domestic Animals." Saunders, Philadelphia.

Reidenberg, J. S., and Laitman, J. T. (1994). Anatomy of the hyoid apparatus in Odontoceti (toothed whales): specializations of the skeleton and musculature compared with those of terrestrial mammals. *Anat. Rec.* **240**, 598–624.

Reynolds, J. H., Odell, D. K., and Rommel, S. A. (1999). Marine mammals of the world. *In* "Biology of Marine Mammals" (J. E. Reynolds, and S. A. Rommel, eds), pp. 1–14. Smithsonian Institution Press, Washington, D.C.

Romer, A. S., and Parsons, T. S. (1977). "The Vertebrate Body." Saunders College, Philadelphia.

Rommel, S. A. (1990). Osteology of the bottlenose dolphin. *In* "The Bottlenose Dolphin" (S. Leatherwood, and R. R. Reeves, eds), pp. 29–49. Academic Press, San Diego, CA.

Rommel, S. A., Costidis, A. M., Fernandez, A., Jepson, P. D., Pabst, D. A., McLellan, W. A., Houser, D. S., Cranford, T. W., van Helden, A. L., Allen, D. M., and Barros, N. B. (2006). Elements of beaked whale anatomy and diving physiology, and some hypothetical causes of sonar-related stranding. *J Cetacean Res. Manage.* **7**, 189–209.

Schaller, O. (1992). "Illustrated Veterinary Anatomical Nomenclature." Ferdinand Enke Verlag, Stuttgart.

Werth, A. J. (2004). Models of hydrodynamic flow in the bowhead whale filter feeding apparatus. *J. Exp. Biol.* **207**, 3569–3590.

# Sociobiology

## Fritz Trillmich

## I. The Nature of Sociobiological Inquiry

Sociobiology is the study of social behavior and social evolution based on the theory of adaptation through selection. As such it is primarily concerned with the adaptiveness of social behavior and the selective processes producing and maintaining adaptiveness. Understanding the selective processes involved includes studying the ecology, physiology, and behavior, as well as the demography and population genetics, of the species in question. Sociobiological investigation also increasingly attempts to characterize the genetic breeding system, as well as the population dynamics and structure, which importantly influence the effectiveness of evolutionary processes in molding species characteristics.

The sociobiological approach assumes that selection at the individual level is the force producing adaptation. A proper understanding of social phenomena, therefore, needs an understanding of the benefits and costs that the individual derives from its interaction with the social environment. Explicitly, group selection is relegated to a secondary position as in most circumstances selection operates more strongly at the individual than at the group level because fertility, dispersal, and mortality events are more frequent and act much more forceful on individuals than on groups. Explaining social phenomena such as group formation, parental care, and mating systems from the action of selection at the level of the individual forms the core of sociobiological inquiry.

As the majority of sociobiological research in the field of marine mammals has been done on whales and pinnipeds, these two groups form the focus of the following sections. Relevant information on sea otters (*Enhydra lutris*) and manatees (*Trichechus* spp.) is mentioned briefly in Section V.

## II. Group Formation

The most obvious phenomenon of social life is group formation. Suitable feeding or breeding habitat may initially lead to an aggregation of individuals, thus setting the stage for selective processes molding the evolution of elaborate social interactions. In contrast to the term "aggregation," "group" implies that individuals come together to derive benefits from interactions that follow from this proximity. Such a grouping may serve social, foraging, predator avoidance, or defense against predators. Groups may also be established for mating and to share parental care. These kinds of advantages constitute the selective processes that promote group formation in a wide variety of animals. Sociobiology tries to explain groupings from the advantages and disadvantages incurred by individuals (Krause and Ruxton, 2002).

### A. Whales

The open ocean habitat offers few options for hiding from predators. Consequently, predation by large sharks and killer whales (*Orcinus orca*), particularly on newborns, is one important factor selecting for group formation in whales and dolphins. Direct observational evidence for this hypothesis is scarce, but signs of scarring provide evidence of frequent encounters with predators. For example, about one-third of all humpback whale (*Megaptera novaeangliae*) calves carry tooth marks on their flukes when arriving in the foraging areas, presumably from encounters with killer whales or sharks during migration to the feeding grounds (Mann *et al.*, 1999). The most spectacular groupings are found in open ocean species such as spotted dolphins (*Stenella* spp.) which benefit most from the advantages of grouping as protection against predators, but such species may also benefit from group foraging.

Several effects play a role in the protection offered to an individual by a group. The "dilution effect" acts by reducing the probability of an individual to be attacked by a predator who has noticed the group, if the predator takes only one individual out of the group. This effect thus dilutes the chances of an attack on a given individual dramatically (to 1/group size). The "confusion effect," many individuals rushing back and forth through the visual field of an attacking predator, makes it more complicated for the predator to concentrate on one prey to catch. Finally, animals in the middle of a group can in effect use other individuals around them as shields against predatory attack, as there will always be other individuals geometrically closer to a predator that attacks from the edge of a group. This is the so-called "geometry of the selfish herd," which allows an individual to use other group members as protection. These effects may also partly protect individuals in groups against parasites, such as the cookie cutter shark (*Isistius brasiliensis*), which takes little bites out of the side of its victim. The advantages of grouping using dilution, confusion, or cover against predation do not depend on individual recognition or social bonds among individuals, but also function perfectly well in a totally anonymous group.

Eventually, the advantage of gaining protection in the group will be counteracted by increasing food competition among individuals in a large group and will thereby determine the upper size of groups. Because food competition depends on abundance, distribution, and behavior of food organisms, which are little known for whale prey, the importance of food competition is largely unknown. However, signs of territorial behavior between killer whale groups provide at least some evidence for the role of food competition in limiting group size.

For more powerful animals such as killer and sperm whales (*Physeter macrocephalus*), grouping offers the additional option to defend smaller, more vulnerable individuals against predators by taking them into the middle of the group. This kind of group action has a completely different quality than the examples given earlier because it involves bonds among the individuals in the group, which are presumably well cemented by individual recognition. Individuals in these groups actively take some risk to defend others, mostly calves, against predators. This has traditionally been explained by the fact that the individuals involved in such cooperative behaviors are kin to each other (Hamilton, 1964). However, it may also be based on mutualistic cooperation or group augmentation (Kokko *et al.*, 2001). The evolution of mutualism would be eased in kin groups through kin selection, but group augmentation involves a different mechanism. Here individuals attach to a group because survival and reproduction are enhanced or only possible in the protection of a group.

However, grouping can also be of advantage for a predator. Foraging groups have perhaps been best analyzed for killer whales. For the mammal-hunting the so-called *transient* killer whales in British Columbia, groups of three individuals proved most efficient in terms of energy intake per unit time when hunting harbor seals (*Phoca vitulina*) (Baird and Dill, 1996). Optimal efficiency of this group size results because a group of three cooperatively hunting whales seems to be better at detecting and catching prey than single animals or duos. At the same time, competition in such a small group is less than in a larger group, once it comes to sharing a harbor seal carcass. For more evasive prey such as Dall porpoise (*Phocoenoides dalli*), larger groups may be more successful because more animals are better at intercepting fleeing prey. Dusky dolphins (*Lagenorhynchus obscurus*) herd schools of fish cooperatively toward the surface and presumably thereby increase food intake. Such activities have also been observed in other dolphins that herded fish schools into bays or against fishing nets and thereby may increase food intake for all individuals in the group.

Social systems in whales differ widely between mysticetes and odontocetes. Mysticetes often live solitarily and, except for the mother–calf association during the rearing period, show little evidence of long-term social bonds. This is most likely caused by the nature of their food resources and the fact that due to their body size they have largely escaped predation. They may, however, aggregate during

feeding in particularly productive areas and during the mating season (see Section IV).

In contrast, almost all odontocetes are quite social. The social groupings of sperm whales, killer whales, pilot whales (*Globicephala* spp.), and bottlenose dolphins (*Tursiops* spp.) are best documented. In whales, groupings represent matrilines in which male offspring may (in killer and pilot whales) or may not (sperm whales) remain for life. In sperm whales, the grouping of kin serves protection of offspring. During foraging, these animals routinely dive to 400 m depth and stay at depth for 40–60 min. During this period, young would be unprotected, waiting at the surface. This is avoided by adults in a group staggering their diving in such a way that one or more adults almost always attend the calves at the surface. Clearly, such behavior provides indirect fitness gains if the individuals in a group are relatives (such as mother, daughter, granddaughter). Kin selection effects may also explain the lack of dispersal of young in killer whales (in this case of both sexes). In resident, fish-eating groups, male offspring in a matrilineal group may offer protection to relatives as well as help in the defense of foraging areas. Females that stay with their mother may likewise gain fitness from cooperation with close relatives. The size of the group will be limited by food competition, and indeed large groups split up, when they have grown too much, into smaller units along matrilineal kinship lines. Limitation of group size by the potential for food competition is particularly evident, as the so-called *transient*, mammal-eating groups are much smaller (no more than three–five animals) than those of fish-eating *residents*, presumably because otherwise the disadvantage of food competition would offset the advantage of close cooperation with kin. Pilot whales show a similar social structure where male and female offspring stay with mothers (Amos *et al.*, 1993). Advantages and disadvantages of this social structure are less understood for the pilot whale than for the other species mentioned previously.

Bottlenose dolphins and some other inshore odontocetes live in quite open fission–fusion societies in which females associate frequently with many different partners. These associations may be useful in foraging or vigilance and maybe defense against predators, but the last hypothesis seems less likely because females with calves tend to be less gregarious. Whether this is due to food competition is unclear. Females may also sometimes group to avoid harassment by males (see Section IV). Male associations vary between study sites and seem to serve mating purposes (Mann *et al.*, 1999).

## B. Pinnipeds

As a group, pinnipeds are characterized by an amphibious lifestyle. They forage at sea but females need to return to a firm substrate, land or ice, for parturition. A strong selective force during this period of birth and subsequent pup rearing is predation on mother and pup. In land-breeding pinnipeds, mostly otariids but also elephant seals (*Mirounga* spp.), monk seals (*Monachus* spp.), and the largest populations of gray seals (*Halichoerus grypus*), predator avoidance has led to the choice of predator-free oceanic islands for breeding. In comparison to the wide expanse of ocean used for foraging, these islands are limited in area, automatically leading to high densities of females on land. Otariid females come into estrus shortly after parturition, which sets the stage for sexual selection on males trying to station themselves in these female aggregations to breed (see Section IV).

Primarily, the concentration of females on land has all the characteristics of an aggregation, but females of most land-breeding species stay closer together than necessitated by available habitat. Indeed,

S

sea lion colonies were found to show internal social network structure (Wolf *et al.*, 2007). One reason for such active groupings may be to avoid harassment by ardent males trying to mate with females (Bartholomew, 1970; Trillmich and Trillmich, 1984). In addition, subadult males of some species of otariids abduct and herd pups, molesting them to the extent that they may die. In this dangerous situation, grouped pups profit from the dilution effect, which might select for tighter grouping of females (Campagna *et al.*, 1992).

Thermoregulation may also further grouping: Sea lions and the South African (*Arctocephalus pusillus pusillus*) and Australian (*A. p. doriferus*) fur seals tend to rest in body contact after leaving the water, which may serve to keep the body shell warm at minimal metabolic cost to the individual. This is not necessary for a fur seal with its highly insulating, air-trapping fur, and the function of clumping in the South African and Australian fur seal is therefore doubtful.

In contrast, ice-breeding seals tend to be more widely dispersed, be it on pack ice or fast ice. In these species, different factors select for some gregariousness. Most important for many species is predation: in the Arctic by polar bears (*Ursus maritimus*) and in the Antarctic by leopard seals (*Hydrurga leptonyx*) and killer whales. Some kind of a grouping advantage is perhaps operating that might equally benefit animals resting between foraging excursions and animals breeding on the ice. Indeed, harp seals (*Pagophilus groenlandicus*) are well known for breeding—even if dispersed—in well-defined local concentrations. It is not certain whether these concentrations are true social groupings or could also be induced by the characteristics of the ice in combination with the food resources below it, i.e., are aggregations due to resource distribution. It may actually be a combination of both. Weddell seals (*Leptonychotes weddellii*) group around tide cracks in fast ice, which offer holes for entry into the water where the seals forage under the ice. Because suitable holes are limited, this leads to a concentration of animals around entry holes and importantly influences the mating system.

Principally, pinnipeds are solitary foragers. There is little evidence for social foraging except occasional observations of sea lions herding fish into bays and communally preying on the trapped schools of fish. Aggregations of foraging sea lions and fur seals occur quite frequently near large fish schools where pinnipeds, birds, and whales may gather in so-called feeding frenzies. True cooperativity in these aggregations has not been demonstrated. This does not exclude that loose groups of pinnipeds may be more likely to locate food and through signs of foraging activity attract other animals to the site, an advantage frequently exploited by group-foraging birds.

## III. Parental Investment Strategies

There is no convincing evidence of paternal investment in the rearing of young in any whale or pinniped. Lack of paternal care is typical for the majority of mammals and reflects that females alone provide for offspring during pregnancy and lactation, which frees males of parental duties (Clutton-Brock, 1991). Only in a few whale societies, e.g., killer whale matrilineal groups, males—most likely never the fathers of young in the group—act as helpers in defense and perhaps feeding of young.

Maternal strategies in whales and pinnipeds are characterized by massive energetic investment in young through the production of large precocial young and extremely lipid-rich milk. However, parental investment is measured not by energy expenditure, but rather by a reduction in future fitness, a cost incurred by the mother through a loss in subsequent fertility or an increase in mortality due to a expenditure that benefits the offspring by increasing its fitness

(Trivers, 1972). Evidence of parental investment, so defined, is rare in whales and pinnipeds.

The patterning of parental effort differs widely between mysticete whales and odontocetes. Mysticete females gather the material and energy for pregnancy (lasting about 1 year) and lactation during a feeding season of about half a year when food is plentiful and starve for the other half-year. Extraordinary fast fetal growth rates in blue whales (*Balaenoptera musculus*, 27 mm/day) permit even the largest mysticete whale—with the exception of the bowhead (*Balaena mysticetus*), sei (*Balaenoptera borealis*), and gray (*Eschrichtius robustus*) whale—to produce a calf within 1 year. By migrating to warmer oceans in the non-feeding period, mysticete whale females seem to minimize the metabolic overhead for themselves and newborn and sucking calves. Females suckle their young for a short period of 6–8 months on very lipid-rich milk, which is produced from maternal body stores, and wean abruptly. After lactation, mothers need to replenish body reserves, which usually takes a year. Females, therefore, generally breed every other or every third year. Females of the tropical Bryde's whale (*Balaenaptera edeni*) have similar gestation and lactation lengths but show much less of a seasonal breeding pattern. Despite the impressive energy flux involved in pregnancy and lactation in mysticetes, there is no strong evidence that this reproductive effort incurs fitness costs. In other words it does not constitute parental investment, for example, by reducing the probability of a successful pregnancy in the year following lactation.

Large odontocetes have pregnancy periods lasting longer than 1 year and take 8–36 months to wean their young, but reports on 13-year-old sperm whales with milk in their stomachs also exist. All larger odontocetes need more than a year to wean. Females consequently need much longer than 2 years to complete one reproductive cycle. The long period of nursing, even in dolphins, allows young to profit from the milk supply while gradually developing independent hunting skills.

It is presently speculated that the difference in rearing strategy between the two groups depends largely on the difference in hunting strategy. Mysticetes prey on small schooling prey, which are supposedly easy to catch for recently weaned young, particularly since weaning seems to coincide with the annual peak abundance of prey. In contrast, odontocetes hunt large prey individually, which forces them to use more complex foraging tactics. They may need to learn more about prey distribution and behavior before young can feed themselves successfully. This may even involve teaching by mothers as is likely for the technique of beaching used by southern killer whales hunting pinnipeds as observed on Southern sea lion (*Otaria flavescens*) breeding beaches of Argentina.

Another peculiar feature of some odontocetes is the occurrence of menopause. This phenomenon was documented for short-finned pilot whales (*Globicephala macrorhynchus*), and killer whales. One idea about the functional significance of menopause, which finds some support in studies on humans (Mace, 2000), is that menopausal females may help their last born and previous daughters in the group through prolonged maternal and allomaternal care, respectively, and by representing a living memory of how to deal with scarce resources and rare ecological disturbances. In matrilineal societies, menopause could be selected through indirect effects on the fitness of kin if the benefit to kin was larger than the benefit an old female could gain through further reproduction.

Sperm whales, long-finned pilot whales (*Globicephala melas*), false killer whales (*Pseudorca crassidens*), and bottlenose dolphins show evidence of reduced fertility with age, perhaps caused by extended periods of lactation. This change in maternal lactation strategy is expected from life history theory because old females have

**S**

a lower reproductive value and may therefore invest more in current offspring than young females. Alternatively, prolonged lactation in older mothers might be a sign of senescence or it may be caused by a reduction in condition through previous parental care episodes and could then be considered a cost of reproduction.

Despite marked sexual size dimorphism in many species, there is little evidence for sex-biased investment in sons versus daughters. In short-finned pilot whales sons may be suckled for longer than daughters. Males had milk in their stomachs up to an age of 15 years, but females only up to an age of 7 years. This fits with the observation that males grow faster than females and are consistently larger at weaning than female juveniles. Similar observations have been reported for sperm whales. If such a difference in effort expended on the two sexes were consistent in the population, one would predict from models of sex allocation (Frank, 1990) that the sex ratio at birth or at weaning should be biased toward females. There is as yet no evidence for such a sex ratio bias.

## A. Pinnipeds

Pinniped females produce one pup per year, instances of twinning are exceptional. Otariids have a postpartum estrus, whereas phocids copulate around the time of or after weaning of young. Implantation is delayed for 3–4 months and is then followed by an 8–9-month pregnancy. The young are reared on lipid-rich milk and are usually weaned at an age less than 1 year, thus allowing an annual cycle of reproduction. Pinnipeds follow one of two alternative pup-rearing strategies. Females may nurse pups for a short time, a few weeks, from body reserves and then wean abruptly. This is termed the *fasting strategy* and is typical for larger phocids. Smaller phocids and otariids nurse young post-partum from body reserves for a short period, in otariids lasting about 1 week. Thereafter they alternate regularly between foraging close to the colony and nursing ashore; they follow a *foraging cycle strategy* of pup rearing. Phocids wean pups after a shorter duration of lactation, 4–65 days, at a about 25–35% of maternal mass, whereas otariids wean after 120 days to 3 years, at 35–55% of maternal mass (Costa, 1993; Trillmich and Weissing, 2006).

The rearing strategy appears to depend on both female body mass and on phylogenetic grouping (Boyd, 1998; Schulz and Bowen, 2005). The larger a female, the richer a food resource she needs to support pup rearing by foraging during lactation. This is because the energetic costs of traveling to and from the food resource and maintaining the metabolism of mother and pup during the foraging sojourn become too high for large pinnipeds to make foraging during lactation a feasible strategy. Therefore, large species separate foraging completely from lactation, store massive energy reserves during a long foraging period in rich feeding areas often far away from breeding sites, and then fuel lactation out of body reserves. Because phocid females are on average larger (median maternal mass for all species 140 kg) than otariid females (median 55 kg), this might explain why phocids usually follow a fasting strategy and otariids a foraging cycle strategy (Boyd, 1998; Schulz and Bowen, 2005; Trillmich and Weissing, 2006). The largest otariid, the Steller sea lion (*Eumetopias jubatus*) with a female mass around 250 kg, needs to take its unweaned young close to the foraging areas to maintain lactation by alternate foraging and suckling. Similarly, walrus (*Odobenus rosmarus*) females take their young to foraging areas where they are suckled while the mother can forage nearby. Ecological constraints thus play a role in shaping maternal strategy, but phylogenetic constraints also importantly influence presently found lactation strategies.

Evidence is mixed that the energetic effort expended on pup rearing induces fitness costs of reproduction (Boness and Bowen, 1996; Trillmich, 1996). In the northern elephant seal (*Mirounga angustirostris*) primiparous—giving birth for the first time—young females are less likely to bear a pup in the year following birth than older females, thus suffering reduced fertility as a consequence of early onset of reproduction. Also, survival seems to be reduced when females first reproduce at 3 rather than 4 years of age, implying a mortality cost. However, these results were obtained at one but not another site on the Californian islands and the interpretation is not entirely clear. In otariids, Galápagos fur seal (*Arctocephalus galapagoensis*) females incur fitness costs of reproduction in terms of reduced fertility because they frequently lose a newborn pup when still accompanied by their previous young, be it a yearling or a 2-year-old, by the time the next pup is born. This arises because young Galápagos fur seals grow extremely slowly and therefore take unusually long, up to 3 years, to become independent of their mothers. They may suckle for a second or third year if environmental conditions are poor, and thus preclude rearing of another pup by their mother (Trillmich and Wolf, 2007). Clear evidence for a fertility cost of reproduction was also found for Antarctic fur seals (*Arctocephalus gazella*). Parturient females of all ages were significantly less likely to reproduce in the subsequent year than nonparturient females. In addition, females that reproduce in a given year are less likely to survive to the following year than nonreproducing females—a clear mortality cost of reproduction (Boyd *et al.*, 1995). In this species and the northern fur seal (*Callorhinus ursinus*), females older than about 13 years appear to show reproductive senescence. These old females are less fertile and produce smaller newborns than prime-age females. Particularly for otariids, there is thus evidence that the high energetic effort expended by females on pup rearing indeed constitutes maternal investment because it produces fitness costs to the mother.

It has been claimed repeatedly that female pinnipeds of highly polygynous species invest differentially in male and female offspring. Following sociobiological arguments, this would be expected if an increased investment in males, the larger sex showing greater variance in reproductive success would lead to a greater expected reproductive success of such males. In many polygynous pinnipeds, males are born heavier and grow faster than females. This was taken as evidence for greater maternal investment in sons. However, this is no proof of greater investment in male offspring because male pups of some otariids store less fat and produce more lean body mass than female pups. Fat has a higher energy density than lean body mass, and consequently smaller female pups may have taken as much energy from their mothers as the larger, leaner male pups. Also, the most important growth spurt determining later male adult size and probably reproductive success occurs generally after weaning, thus making it less likely that male offspring derive direct benefits for their future reproductive success from increased maternal investment (Trillmich, 1996). Nevertheless, recent data demonstrate a greater cost of male offspring at least for one otariid species (Trillmich and Wolf, 2007).

## IV. Mating Systems

Mammalian females are producers that are limited by the time and resources needed for pregnancy and lactation, as well as by the recovery of condition after a reproductive cycle. This constitutes strong selection to optimize acquisition and efficiency of resource use before and during reproduction. Because the maximal reproductive rate of female mammals is necessarily much lower than that of males, which in the extreme need only the time of one copulation to produce offspring, females become a limiting resource for the reproduction of males. This leads to sexual selection on males for an increased ability to get access to and breed with as many females as possible leading to often extreme sexual size dimorphism. In mammals such as whales and pinnipeds, where males do not contribute to the care of offspring, males are expected to conform to the description of "ardent" males, eagerly searching for females and even harassing them for copulation. Females distribute themselves in relation to the distribution of resources, food, and adequate habitat for reproduction, and males map onto this distribution of females. This difference in the selection on reproductive strategies of males and females leads to the phenomenon that quite often the sexes behave and morphologically look as if they belonged to different, competing species.

Sociobiological reasoning therefore leads to the expectation that observed mating systems represent the compromise arising from the conflict of the sexes. Competition among males for access to females can take the form of aggressive competition, but can also occur via sperm competition when several males copulate with one female, as demonstrated for northern elephant seals. Such sperm competition, if occurring regularly, is expected to lead to the evolution of large testis mass, as larger testes produce more sperm and thus increase a male's chances to fertilize the ova of females in competition with sperm of other males. Such an increase in relative testis mass was documented in other mammalian orders where species in which multiple copulation is frequent have larger testes than species where only one male copulates with a given female, whether the social system is monogamous or polygynous.

### A. Whales

Whale mating systems are still largely unknown partly due to the problem that copulations are hard to observe. Except for a few particularly observable species, this leaves only indirect methods of investigation, such as genetic analyses, to determine the mating pattern in the more elusive species (Connor *et al.*, 1998).

Among mysticete whales, much is known about the humpback whale, so well known for its spectacular songs. During the mating season, males station themselves well spaced out and advertise their position. This is very similar to the lek structures observed in many birds. The song may attract females and keep competing males away, but there is presently no firm evidence for this inference. Alternatively, males follow females, and several males may be doing this simultaneously, leading to competition for proximity to the female. Apparently these males compete over mating access to a female. Because humpback whales have small testes for their size, it is unlikely that females will copulate with several males, thus inducing sperm competition. In contrast, sperm competition is likely to occur regularly in right whales (*Eubalaena* spp.), which—weighing about 50 tons—have testes weighing 1 ton, in strong contrast to blue whales, which weigh about 100 tons but have a testis mass of only about 70 kg. Copulation is observed frequently in right whales but has never been described convincingly in humpbacks, despite much more study of the latter.

Mating patterns in odontocetes are somewhat better known from a few species (Connor *et al.*, 1998). Male strategies vary from singly roving males in the sperm whale to males that cooperate to herd and perhaps force females into copulation in the bottlenose dolphins. Sperm whales show a mating pattern similar to that of elephants in which single fully adult, highly aggressive males rove among female groups in search of receptive females. They stay only briefly with each one of the female groups and then go on. Presumably, the long intervals between estrus in females make it unprofitable for these males to stay with any one group of females waiting for one of the females to enter estrus. Only fully adult males appear able to compete in this system, and

subadult males as well as non-roving males stay at higher latitudes, often in small schools, feeding and maximizing energy intake in this way to grow to a competitive size. It is unclear why these males might stay in small groups as bachelors because they are certainly not endangered by predation and it is unknown which foraging advantages they might derive from grouping.

The killer and pilot whale mating system is the most surprising, least understood of the whale mating systems and has no parallel in terrestrial mammals. Genetic evidence shows that males who remain philopatrically in their maternal group never father offspring in their own group, but apparently in other groups. Some evidence suggests that several related males of one group may mate with several females in another group, presumably during repeated encounters of these pods. This was concluded from the genetic observation that offspring in a group seem to be paternally related.

### B. Pinnipeds

The pattern of mating interactions among individuals depends greatly on the dispersion of females during the breeding season, which in turn reflects the availability of a suitable habitat for pupping and foraging. In pinnipeds, pack ice, fast ice, and land-breeding species differ widely in this respect (Boness, 1991; Le Boeuf, 1991).

Phocids breeding on ice floes seem to have a mating system best described as serial monogamy in which a male stays with a female that has recently pupped until she comes in estrus. He then leaves after copulation to search for another female. The reproductive success for males in such a mating system depends more on their aptness to locate females ready to mate than on fighting abilities. In such species, sexual dimorphism tends to be small, slightly reversed (males smaller than females), or nonexistent.

Some fast-ice breeding species also show reversed sexual dimorphism, which is best analyzed in the mating system of the Weddell seal. Females gather around cracks in the fast ice where they dive for food from holes in the ice. During the reproductive season, they pup near these holes and males claim territories under the ice and defend the holes against other males. Under these conditions, maneuverability is considered more important than sheer size in male–male competition. This may be the reason for the observed reversed sexual size dimorphism. Alternatively, females may be selected for larger size, enabling the production of larger young or the storage of more massive fat reserves for lactation, and males may have remained smaller for lack of such selection. Copulation is underwater and consequently little is known about the reproductive success of males in this mating system.

Phocid seals, such as the harbor seal, which breed in the water close to land areas where females loosely aggregate, seem to engage in fights for the best stations where females have to pass by, and such males are often wounded. Fighting males seem to have the highest reproductive success and, in some places, the mating system of this species may be similar in structure to a lek.

When female otariids or land-breeding phocids come together on predator-free islands, the resulting high female density permits males to station themselves among females and attempt to monopolize access to females. This competitive situation sets up sexual selection for an ever increasing male size, leading to impressive sexual size dimorphism in a few phocid and many otariid species (Lindenfors *et al.*, 2002), such as the northern fur seal in which males weigh six times as much as females. In addition, a larger size also enables males of these species to remain fasting on territory for long periods, thus increasing their chances to mate with females.

S

Otariid mating systems have been described as resource defense, female defense, or leks. The presence of resources important to females with pups, such as shade or access to tide pools, on male territories was demonstrated experimentally for a few species. However, resource distribution is not sufficient to predict the exact location of female aggregations nor does it explain female gregariousness, i.e., an active tendency of females to approach each other. Because high female density correlates with increased pup mortality in breeding colonies, there is a marked cost to female gregariousness, which must be compensated by comparable benefits. Bartholomew (1970) suggested that female choice of genetically superior males was responsible for female gregariousness, but little evidence supporting this view has come forward. New studies suggest a strong selection of female gregariousness through avoidance of interaction with large males, whether territorial or not. In many otariid species and in elephant seals, interactions with a much larger male can be dangerous or even deadly to a female. Females can minimize the probability of interaction with territorial males by aggregating into a "selfish herd." Through this effect and by avoidance of dangerous and sometimes directly fatal harassment of females and pups by marginal males, females are selected to group much more closely than can be explained by resource distribution. Thus, the stationing of large adult males on clustered territories among parturient and estrous females creates a resource "peace from marginal male harassment" (Trillmich and Trillmich, 1984). Except for this socially created resource, the system could also be described as one in which males are lekking on areas where females are forced to stay for a while because they spend the period between parturition and postpartum estrus near-stationary on land. Males may benefit from clustered territories by the reduced chances of losing females when disturbed by marginal subadult or adult males. Within this system, the male defense of access to females varies intra- and interspecifically with habitat and female density, from female defense to territorial site defense with larger or smaller territories reminiscent, respectively, of a resource defense or a lekking system.

Dominating males in these land-based breeding systems gain most copulations and can reach quite extreme reproductive success, sometimes up to 100 copulations in one breeding season. However, genetic studies have shown that observed number of copulations does not always correlate well with actual paternity, suggesting that peripheral, apparently "excluded" males may also gain some reproductive success by keeping close to female groups. There is no evidence of female choice beyond the mechanism that females in dense groups attract the strongest males in the best condition because only these are competitive enough to station themselves in the middle of female groups. Female elephant seals protest when they are attacked and forced into copulation by subadult or peripheral males, thereby attracting the attention of the dominant male who will often chase off the smaller male and copulate with the female. In this way females indirectly choose dominant males as copulation partners.

Gray seal mating systems on land strongly resemble the otariid system. Whether female choice plays a more important role in gray seal colonies remains to be seen. Genetic analysis of subsequent offspring of individual females suggests that females copulate year after year with the same male, even though they may stay in the territories of different males. This would suggest that they actively choose a particular male or, in case of multiple copulations, have mechanisms to choose among sperm of several males. In more dispersed breeding species, such as harbor seals, males have no chance to defend access to females as these are too mobile and not available continually in the same areas, depending on tide level and sea conditions. In situations in which females breed on sandbanks, harbor seal males

were observed to station themselves in areas where females are likely to pass by and make themselves obvious through vocal display. Whether this and similar observations on walrus males that station themselves near females and produce bell-like sounds can be considered a lek display needs further investigation.

## V. Sirenians and Sea Otters

Much less is known about the social life of sirenians and sea otters and, therefore, these two groups are treated only briefly here.

The only clearly recognizable social structure in sirenians is the mother–offspring bond, which may last for up to 3 or 4 years. Other than that, it appears that the dispersion of most sirenians relates directly to the distribution pattern of food, aquatic macrophytes, fresh water for drinking (in the Florida manatee, *Trichechus manatus*) and, particularly in winter, warm water areas. Animals may migrate for large distances between such resources. However, it seems possible that underlying the apparent asocial pattern may be a subtle pattern of individualized relationships. This might be hypothesized from "greeting" displays exchanged between individuals that meet only occasionally at widely distant sites.

Cows in estrus seem to induce male scramble competition. In Florida, manatee herds of up to 20 males may follow an estrous female and compete by pushing to get into a favorable position for mating. In dugong (*Dugong dugon*) males, competition may take a more aggressive form in which males may wound each other with their tusks. For West Australian dugongs, mating competition may lead to a form of lekking (Anderson, 2002). However, the evidence is largely circumstantial.

Sea otter spatial dispersion is related to the need to live close to the coast where they forage relatively shallowly for macroinvertebrates. Females claim year-round foraging territories along the coastline that often overlap. They sometimes aggregate in small groups—so-called "rafts." Young males, and fully adult males outside the reproductive season, also frequently form rafts close to areas of rich feeding resources. Such rafting is presumably related to the reduction of predation risk.

During the reproductive period, fully adult males establish territories that may overlap with more than one female territory. This provides males with a chance for a mild form of polygyny, but hard evidence for paternity of such males is at present missing.

## VI. Concluding Remarks

Much of the sociobiological interpretation of observations on marine mammals is still in an early stage. This situation reflects our lack of detailed knowledge about the marine environment and in particular the macro- and microdistribution of resources and predators of marine mammals. More observation, more comparative studies, and especially more experimental work are urgently needed to understand the sociobiology of these magnificent animals. Obviously, experimental work will be particularly challenging and can only be successful if built on the thorough knowledge of marine mammal natural history. However, a well-founded functional understanding of the social behavior of marine mammals cannot be achieved without experimental tests of our many assumptions. Ingenious instrumentation and molecular genetic tools, developed during the last decade, should prove most helpful in making this summary of marine mammal sociobiology soon outdated.

## *See Also the Following Articles*

Energetics ■ Feeding Strategies and Tactics ■ Group Behavior ■ Mating Systems ■ Parental Behavior ■ Predator–Prey Relationships ■ Thermoregulation

# References

Amos, B., Schlötterer, C., and Tautz, D. (1993). Social structure of pilot whales revealed by analytical DNA profiling. *Science* **260**, 670–672.

Anderson, P. (2002). Habitat, niche, and evolution of Sirenian mating systems. *J. Mamm. Evol.* **9**, 55–98.

Baird, R. W., and Dill, L. M. (1996). Ecological and social determinants of group size in transient killer whales. *Behav. Ecol.* **7**, 408–416.

Bartholomew, G. A. (1970). A model for the evolution of pinniped polygyny. *Evolution* **24**, 546–559.

Boness, D. J. (1991). Determinants of mating systems in the Otariidae, Pinnipedia. *In* "The Behaviour of Pinnipeds" (D. Renouf, ed.), pp. 1–44. Chapman & Hall, London.

Boness, D. J., and Bowen, W. D. (1996). The evolution of maternal care in pinnipeds. *Bioscience* **46**, 645–654.

Boyd, I. L. (1998). Time and energy constraints in pinniped lactation. *Am. Nat.* **152**, 717–728.

Boyd, I. L., Croxall, J. P., Lunn, N. J., and Reid, K. (1995). Population demography of antarctic fur seals: the costs of reproduction and implications for life-histories. *J. Anim. Ecol.* **64**, 505–518.

Campagna, C., Bisioli, C., Quintana, F., Perez, F., and Vila, A. (1992). Group breeding in sea lions: pups survive better in colonies. *Anim. Behav.* **43**, 541–548.

Clutton-Brock, T. H. (1991). "The Evolution of Parental Care." Princeton University Press, Princeton.

Connor, R. C., Mann, J., Tyack, P. L., and Whitehead, H. (1998). Social evolution in toothed whales. *Trends Ecol. Evol.* **13**, 228–232.

Costa, D. P. (1993). The relationship between reproductive and foraging energetics and the evolution of the pinnipedia. *Symp. Zool. Soc. Lond.* **66**, 293–314.

Frank, S. A. (1990). Sex allocation theory for birds and mammals. *Ann. Rev. Ecol. Syst.* **21**, 13–55.

Hamilton, W. D. (1964). The genetical theory of social behaviour. *J. Theor. Biol.* **7**, 1–25.

Kokko, H., Johnstone, R. A., and Clutton-Brock, T. H. (2001). The evolution of cooperative breeding through group augmentation. *Proc. R. Soc. Lond. B* **268**, 187–196.

Krause, J., and Ruxton, G. D. (2002). "Living in Groups." Oxford University Press, New York.

Le Boeuf, B. J. (1991). Pinniped mating systems on land, ice and in the water: emphasis on the Phocidae. *In* "The Behaviour of Pinnipeds" (D. Renouf, ed.), pp. 45–65. Chapman & Hall, London.

Lindenfors, P., Tullberg, B. S., and Biuw, M. (2002). Phylogenetic analyses of sexual selection and sexual size dimorphism in pinnipeds. *Behav. Ecol. Sociobiol.* **52**, 188–193.

Mace, R. (2000). Evolutionary ecology of human life history. *Anim. Behav.* **59**, 1–10.

Mann, J., Connor, R. C., Tyack, P. L., and Whitehead, H. (eds) (1999). "Cetacean Societies: Field Studies of Dolphins and Whales." Chicago University Press, Chicago.

Schulz, T. M., and Bowen, W. D. (2005). The evolution of lactation strategies in pinnipeds: a phylogenetic analysis. *Ecol. Monogr.* **75**, 159–177.

Trillmich, F. (1996). Parental investment in pinnipeds. *Adv. Stud. Behav.* **25**, 533–577.

Trillmich, F., and Trillmich, K. G. K. (1984). The mating systems of pinnipeds and marine iguanas: convergent evolution of polygyny. *Biol. J. Linn. Soc.* **21**, 209–216.

Trillmich, F., and Weissing, F. J. (2006). Lactation patterns of pinnipeds are not explained by optimization of maternal energy delivery rates. *Behav. Ecol. Sociobiol.* **60**, 137–149.

Trillmich, F., and Wolf, J. B. W. (2008). Parent-offspring and sibling conflict in Galápagos fur seals and sea lions. *Behav. Ecol. Sociobiol.* **62**, 363–375.

Trivers, R. (1972). Parental investment and sexual selection. *In* "Sexual Selection and the Descent of Man 1871–1971" (B. Campbell, ed.), pp. 136–179. Aldine Atherton, Chicago.

Wolf, J. B. W., Mawdsley, D., Trillmich, F., and James, R. (2007). Social structure in a colonial mammal: unravelling hidden structural layers and their foundations by network analysis. *Anim. Behav.* **74**, 1293–1302.

# Song

### JIM DARLING

## I. Characteristics

### A. First Descriptions

Although likely heard by sailors for millennia, the first recordings of humpback (*Megaptera novaeangliae*) songs were made via US Navy hydrophones in the 1950s off Hawaii and Bermuda. Scientists first recognized these sounds in the 1960s as coming from humpback whales. The first technical description was published in *Science* in 1971 by Roger Payne and Scott McVay with the revealing subheading, "Humpbacks emit sounds in long, predictable patterns ranging over frequencies audible to humans." Their analysis led to a structural context in which to view the sounds introducing the terms units, phrases, themes, and song. With the observation that humpback whales "produce a series of varied sounds then repeat the same series with considerably precision" these authors called the performance "singing" and the repeated series of sounds the "song."

The biological definition of song is a series of sounds that are repeated over and over. Many animals, therefore, have songs, ranging from the simple "ribet" of frogs, to the huge variety of bird songs, to the loud repetitive signals of the whales. Although whale song was introduced and has been virtually synonymous with the humpback whale, recent studies indicate that other species of baleen whales, including bowheads (*Balaena mysticetus*), blues (*Balaenoptera musculus*), fins (*Balaenoptera physalus*), and minkes (*Balaenoptera acutorostrata*), also repeat patterned sequences of sounds that fit the song definition. Studies of the songs of other mysticetes are young—although this is changing quickly. To date, no known songs are quite as complex and dynamic as those of the humpback whale.

### B. Song Structure

The humpback whale song is composed of a sequence of highly varied sounds ranging from high-pitched squeaks to midrange trumpeting and screeches to lower frequency ratchets and roars, and combinations of all these. This sequence is typically about 10–15 min in duration, although it may range from 5 to 30 min. It is then repeated without a break (Payne and McVay, 1971; Winn and Winn, 1978).

The song has a predictable structure or framework. Discrete sounds are termed *units*. Several different sounds or units in a sequence compose a *phrase*. A phrase is repeated some variable number of times (for example, 10 times), and this series of the same phrase is called a *theme*, say, "theme 1." After several minutes of singing theme 1 the singer changes to a different set of phrases (composed of different units or sounds) and repeats it a number of times. This might be called "theme 2." This pattern repeats until the whale cycles back to its theme 1. A typical song may contain six themes. A singer may sing in order themes 1–2–3–4–5–6 and then start at 1 again. The number of themes in a song varies from population to population and

**S**

from year to year. A song session may continue without a break for hours (Payne and McVay, 1971).

Several characteristics of this song display are notable. First is the precision by which a singer repeats its complex song in any song session. An undisturbed singer may repeat the same themes, phrases, and units faithfully for hours as if on a continuous tape. The second characteristic is that all the singers in a humpback assembly, and there may be hundreds, follow the same structural rules and broadcast essentially the same song (Payne *et al.*, 1983; Noad *et al.*, 2000). At the same time, however, there is some variation in song presentation and some songs are exceptional to the point that they have been termed "aberrant" by investigators (Frumhoff, 1983). The significance of this variability is not yet understood.

## C. Song Progression

One of the more unique characteristics of the humpback song is that it gradually changes over time. Different sounds, and arrangements of sounds, form to create new phrases and themes. These are incorporated into the song gradually, while older patterns are lost. After a period of several years the song my bear little resemblance to the original version (Payne and Payne, 1985; Noad *et al.*, 2000). This is a rare, although not unique, characteristic of biological songs. Several bird species are known to vary their song over time.

The song apparently changes as it is being sung, i.e., the progressive changes occur during the singing or breeding season, not during the off-season. For example, as winter progresses, one unit in a particular phrase may be heard less and less while another becomes more common, or two units in one phrase that had been separate may become joined as one sound. These small changes may eventually lead to new phrases and themes (Payne *et al.*, 1983). During the non-singing or summer season, the song appears to remain relatively stable, to begin to progress again once singing begins the following season. The biological forces or mechanism behind the progression of the song is not known.

The change in the song display occurs in a collective or common way throughout a population of humpback whales, i.e., at any one time all the singers in a population produce fundamentally the same version of the gradually changing song. This is a generalization in that there is certainly some range in when individuals incorporate new elements into their song. However, the majority of singers produce many if not all the same clearly recognizable units, phrases, and themes in the same general order in any one season (Payne *et al.*, 1983; Cerchio *et al.*, 2001). Changes in the song are apparently incorporated throughout the population. The mechanism by which this occurs is not known. All indications are the changes are learned, with some whales making changes and others copying them. This was emphasized by a discovery that, over a 2-year period, the humpback whales off eastern Australia essentially adopted the western Australian humpback song, presumably introduced to that region by the mixing of whales (Noad *et al.*, 2000)

The similarity and collective change of the song may occur not only within one population of humpbacks, but between populations separated by thousands of kilometers. For example, humpback populations that winter in Japan, Hawaii, and Mexico sing similar versions of the song at any one time (Payne and Guinee, 1983; Cerchio *et al.*, 2001). In contrast, the Atlantic humpback song would be substantially different—but at least two populations of humpback whales within the South Atlantic, off Gabon and Brazil, sing similar songs at any one time (Darling and Sousa-Lima, 2005). The means by which this occurs is not known. Some degree of mixing occurs among all populations in the North Pacific in that individual whales may visit

the different winter assembly areas over several years or even in one year. Perhaps this mixing is reflected in the similarity of songs. It is important to emphasize that whereas the generalities described here are striking, much further research is required to conclude on the relationship of songs, singers, and populations.

## II. Singing Behavior

### A. Seasonality

Singing is a seasonal behavior pattern. It occurs primarily during the winter half of the year, throughout the peak of the humpback breeding season. Although a few songs have been recorded during midsummer, they are rare. Songs become increasingly common on the feeding grounds as fall progresses into winter; they are heard regularly during migrations and predominate on the warm water winter breeding grounds. It is not clear when the singing activity declines in the spring, but song has been recorded on early summer feeding grounds.

### B. A Male Communication

The evidence to date indicates that only male humpbacks sing. The sex of a sample of singers has been determined both by photographs of the genital region and by genetic analysis of skin samples, and so far all were males. Also, singing is a behavior pattern common to individual whales that may escort a cow with calf or compete in mating groups—both distinct male behavior patterns (Darling and Berube, 2001). Because there are no confirmed observations of female humpback whales singing, the song may be a male secondary sexual characteristic or display that functions during the breeding season.

A singer on the breeding ground is typically a lone adult male. Also, it is not unusual for an "escort," the adult male accompanying a cow with calf, to be singing. Occasionally a singer has a companion adult or juvenile in the close vicinity (Darling *et al.*, 2006). Often the singer is stationary, remaining in one geographic location for many hours while singing. In these situations the singer generally adopts a motionless, head-down, tail-up posture approximately 50–100 feet beneath the surface. It maintains this posture until it surfaces to breathe and then immediately after diving resumes it. At other times a singer travels steadily as it is singing and may move tens if not hundreds of kilometers across the breeding ground during a song session.

One study in Hawaii indicated that the amount of singing increases significantly at night versus day-time; the reason is unknown (Au *et al.*, 2000). In a study of humpback whale swim speeds during migration off eastern Australia, singers were found to swim significantly slower than the average; the reason for this is also unknown (Noad and Cato, 2007).

### C. Interactions with Singers

Humpback songs are loud and can be heard underwater for at least several kilometers and, in some circumstances in deep ocean basins, possibly hundreds of kilometers. A collection of dozens of singing humpbacks produce a substantial noise, clearly designating the location—on almost an oceanic scale—of an assembly of humpback whales during the breeding season. It is likely that individual humpbacks interact with the collective singing herd over extensive oceanic distances.

On a smaller scale, several studies have focused on the interaction of individual singers with other whales. Generally, singers sing until one of two things happen: First, they are joined by other single adults, the singing stops, and the two interact in some way, ranging from a single pass to rolling, tail lobs, or breaches by one or both animals.

The large majority of these interactions are non-agonistic and brief, with the pair splitting after a few minutes. One or the other may start singing again shortly after the interaction. All evidence to date is that the lone adults that join male singers are also males. Second the singer stops singing without any close approach by another whale and then rushes to approach or join a passing group of whales—often a surface active or competitive group; at times a cow with calf. In these cases, the singer joins a group that includes a potentially breeding female (Darling *et al.*, 2006).

One study carried out by the author, that included extended follows of individual singers, found that the common interaction of a singer joined by another lone male was often just one of a series of similar interactions across the breeding ground. For example, one singer in this study, over a 4 hour period, interacted with 3–4 other males, for 5–6 min each over a distance of 13 km. The singer was joined, stopped singing, then itself joined and split from two separate singers , then began singing again—and was joined again. These observations made it clear that song serves to connect males across a breeding ground, and "singing" and "joining singers" are readily interchangeable male roles. The function of this behavior is not yet known. In several cases in this study, the interacting males (at times joined by a third male) joined either a lone female or a female accompanied by one or more males, as a non-agonistic, and possibly cooperative, unit (Darling *et al.*, 2006). However, such observations around a female are extremely limited, and the behavior pattern is far from understood at this time.

## III. Function

The function of the song has been the subject of much speculation. Much of this has revolved around proposals of the song as a male display that functions to attract females, signal status to other males, or a combination of these. However, other suggestions include that the songs serve as a means to synchronize estrus, a spacing mechanism between males, a migratory beacon, and as sonar by males to find females. A relatively young hypothesis proposes that the song may function as an index of association between males, perhaps as a basis for organizing males during breeding season. Recent research has widened the basis for the speculation but is far from resolving it.

Conservatively, the song is a communication from male humpbacks during the breeding season. It serves to provide the location of the singer, and by association the entire herd, and signals that breeding activity is underway. The song also likely broadcasts information about the individual singer, but what information and who the recipients are remain speculative. The collective, gradual change of the song seems to confound many simple explanations as to its function.

Several early investigations proposed that the song by males serves to attract females, and went further to suggest that changes in the song are driven by female choice (Tyack, 1981). This notion has persisted and predominates in the popular literature; however, there are no data to support it. There is little question that singers join and accompany females, but no evidence that females are attracted by, and voluntarily join, individual singers.

In a related idea, several researchers have proposed that the humpback song functions as part of an aquatic lekking behavior (Clapham, 1996). A lek, as generally defined for land animals, is a seasonal assembly where males display their attributes to attract females for the purpose of mating. This seems to describe the overall breeding assembly of humpbacks, including the male song display; however, if, when, or how females choose a mate is unclear.

Another proposal suggests that the song is a display of males aimed primarily at other males—perhaps signaling social status or position.

In this view the song may enable the male–male interactions that are integral to the establishment and maintenance of the dominance—ordered, polygynous mating system that has been hypothesized for humpback whale (Darling and Berube, 2001). This proposal accounts for the common interactions between male singers and male joiners, and the lack of evidence that females approach, choose, or join singers; however, it does not account for the song characteristics of collective change and similarity.

Regardless of whether the song display is directed at females, males, or both, these potential explanations suggest that song displays should vary with some attribute of the singer. That is, the song should reflect dominance class, fighting ability, or some other measure of the "fitness" of the male. At this stage, research has not revealed any evidence that this is the case. Indeed, the marked similarity of the song between individuals, and maintenance of that similarity, better characterizes the display than individual differences.

An alternative hypothesis, based on the maintenance of song similarity between individuals, and the observations that it mediates non-agonistic male interactions, is that the song may facilitate temporary male associations, cooperative units or coalitions in breeding scenarios. It is clear that non-agonistic and cooperative relations do occur between males on the breeding ground (Clapham *et al.*, 1992), but the extent of this behavior is unknown and its relationship to song is speculative (Darling *et al.*, 2006).

At this time, the function (s) of the song is not clearly understood. With some confidence, we can say that singing is a male behavior, which occurs in the winter season, hence very probably functions in breeding. And, at the least, it is clear that the song functions in mediating brief, non-agonistic male–male interactions. The connection between this behavior and actual mating, or other potential functions of the song remain to be revealed.

## See Also the Following Articles

Behavior, Overview ■ Communication ■ Reproductive Behavior ■ Sociobiology

## References

Au, W. W., Mobley, L. J., Burgess, W. C., Lammers, M. O., and Nachtigall, P. E. (2000). Seasonal and diurnal trends of chorusing humpback whales wintering in waters off western Maui. *Mar. Mamm. Sci.* **16**, 530–544.

Cerchio, S., Jacobsen, J. K., and Norris, T. F. (2001). Temporal and geographical variation in songs of humpback whales, *Megaptera novaeangliae*: synchronous change in Hawaiian and Mexican breeding assemblages. *Anim. Behav.* **62**, 313–329.

Clapham, P. J. (1996). The social and reproductive biology of humpback whales: an ecological perspective. *Mamm. Rev.* **26**, 27–49.

Clapham, P. J., Palsboll, P. J., Mattila, D. K., and Oswaldo, V. (1992). Composition of humpback whale competitive groups in the West Indies. *Behavior* **122**, 182–194.

Darling, J. D., and Berube, M. (2001). Interactions of singing humpback whales with other males. *Mar. Mamm. Sci.* **17**, 570–584.

Darling, J. D., and Sousa-Lima, R. (2005). Songs indicate interaction between humpback whale (*Megaptera novaeangliae*) populations in the western and eastern South Atlantic ocean. *Mar. Mamm. Sci.* **21**, 557–566.

Darling, J. D., Jones, M. E., and Nicklin, C. P. (2006). Humpback whale songs: Do they organize males during the breeding season? *Behavior* **143**, 1051–1101.

Frumhoff, P. (1983). Aberrant songs of humpback whales (*Megaptera novaeangliae*): Clues to the structure of humpback songs. *In* "Communication and Behavior of Whales"

S

(R. Payne, ed.), pp. 81–87. Westview Press, Boulder, AAAS Selected Symposia Series.

Noad, M. J., Cato, D. H., Bryden, M. M., Jenner, M. N., and Jenner, C. S. (2000). Cultural revolution in whale songs. *Nature* **408**, 537.

Noad, M. J., and Cato, D. H. (2007). Swimming speeds of singing and non-singing humpback whales during migration. *Mar. Mamm. Sci.* **23**, 481–495.

Payne, R. S., and Guinee, L. N. (1983). Humpback whale songs as an indicator of "stocks". *In* "Communication and Behavior of Whales" (R. Payne, ed.), pp. 333–358. Westview Press, Boulder.

Payne, R., and McVay, S. (1971). Songs of humpback whales. *Science* **173**, 585–597.

Payne, K., Tyack, P., and Payne, R. (1983). Progressive changes in the songs of humpback whales (*Megaptera novaeangliae*): A detailed analysis of two seasons in Hawaii. *In* "Communication and Behavior in Whales" (R. Payne, ed.), pp. 9–57. Westview Press, Boulder.

Payne, K. P., and Payne, R. S. (1985). Large scale changes over 19 years in songs of humpback whales of Bermuda. *Z. Tierpsychol.* **68**, 89–114.

Tyack, P. L. (1981). Interactions between singing Hawaiian humpback whales and conspecifics nearby. *Behav. Ecol. Sociobiol.* **8**, 105–116.

Winn, H. E., and Winn, L. K. (1978). The song of the humpback whale, *Megaptera novaengliae,* in the West Indies. *Mar. Biol.* **47**, 97–114.

# Sound Production

ADAM S. FRANKEL

Most terrestrial mammals rely heavily on vision and smell. These senses are limited in water by the absorption of light and the slow physical movement of water. As a result, marine mammals have evolved to use sound and hearing as their primary means of communication and sensing their world.

This article briefly reviews the basics of sound and the physical ways sounds are produced by marine mammals. The main focus is on characteristics of sounds made internally by marine mammals. Non-vocal sounds, such as tail slaps, are not treated in detail here. Certain species that stand out as particularly unusual or particularly well studied get special attention. Other species are treated in taxonomic groups, when their acoustic characteristics are similar or less well known. In addition, the odontocetes are organized according to their three main sound types: clicks, pulsed sounds and whistles.

## I. Fundamentals of Sound

Imagine throwing a rock into a calm lake. You can easily picture the ripples, or water waves, that move out in an expanding circle. In these ripples, the water's surface moves up and down, in a smooth progression from crest to trough. Sound waves are similar to these water waves. A sound wave is created by a structure that vibrates, such as a radio loudspeaker or our larynx. The crests of a sound wave are areas of high pressure, and the troughs are areas of low pressure. Like the water wave, there is a smooth progression between these areas of high and low pressure. A sound wave is a propagating (moving forward) alternation of these areas of high and low pressure.

Several terms are used to describe the characteristics of sound. The amount of time it takes for a complete cycle between the highest pressure of a sound wave to the lowest, and back to the highest,

is referred to as the *period* of the sound wave and is measured in seconds. The reciprocal of the period is called the *frequency* of the wave and measured in s$^{-1}$, or more commonly Hertz (Hz). Frequency is a physical characteristic of sound. Our perception of frequency is known as pitch. Humans hear from about 20–20,000 Hz (or 20 kHz). Sounds below 20 Hz are termed infrasound, whereas those above 20 kHz are termed ultrasound.

Marine mammalogists often want to know the loudness of a sound, whether produced by a whale, dolphin, ship, or oil rig. The loudness, or *amplitude*, of a sound is described in decibels (dB). Decibels are defined as a ratio of measured sound pressure level to a reference sound pressure level. The reference sound pressure level in water is one microPascal (μPa). Therefore, in-water decibels are expressed as XX dB re 1μPa. Because sound amplitude decreases with distance, the standard method is to measure the sound level 1 m from the sound source. This measurement is known as the *source level*, and is expressed as XX dB re 1μPa at 1 m. For example, fin whale (*Balaenoptera physalus*) source levels have been measured at 171 dB re 1μPa at 1 m (Charif *et al.*, 2002).

It is also important to remember that in-air and in-water decibels are not the same, due to different reference levels and the physical characteristics of air and water. Generally, adding 61.5 dB to an in-air measurement will convert it to an in-water measurement. Additional information and a technical discussion of sound level measurements can be found in other texts (Hartmann, 2004; Richardson *et al.*, 1995; Urick, 1983).

A sound is said to be frequency modulated (FM) when its frequency changes over time. Dolphin whistles are usually frequency modulated. Amplitude-modulated (AM) signals are those that change loudness, usually rapidly, over time. Many mysticete calls are amplitude modulated and can sound like growls.

A visual record of pressure versus time is known as a *waveform*. Animal sounds are more often represented as a *spectrogram* with frequency on the y-axis and time on the x-axis. Fig. 1 shows both the waveform and spectrogram of two blue whale (*Balaenoptera musculus*) calls. The first is an AM signal, and the second an FM signal. The AM signal shows individual amplitude pulses in the waveform, while the FM signal displays a smooth envelope (or waveform outline). Note also that the AM signal shows the sidebands typical of an AM signal, while the FM signal has a downward sweep from about 20 to 15 Hz. Notice the "extra lines" above the main "line" of the FM signal. This is the lowest frequency or fundamental frequency contour. The additional "lines" above the fundamental frequency are harmonics, which are integer multiples (e.g., 2x, 3x, 4x) of the fundamental frequency. They result from the physical characteristics of the sound-producing structure. Harmonics can occur only with FM calls and are strictly integer multiples of the fundamental frequency. The sidebands of AM signals may resemble harmonics, but are actually the product of the rate of amplitude modulation.

Recently, marine mammal vocalizations have been found to contain spectral features called nonlinear phenomena. These are illustrated in Fig. 2 and include frequency jumps, subharmonics, biphonation, and deterministic chaos. Frequency jumps occur when the frequency of a signal changes almost instantaneously. Subharmonics are additional spectral bands that occur below the fundamental frequency. Animals are also capable of biphonation, or the production of two different harmonically unrelated sounds at the same time. There exists a phenomenon known as deterministic chaos that results from vocal folds oscillating asynchronously. While these signals appear random, they actually are deterministic and repeatable (Fitch *et al.*, 2002).

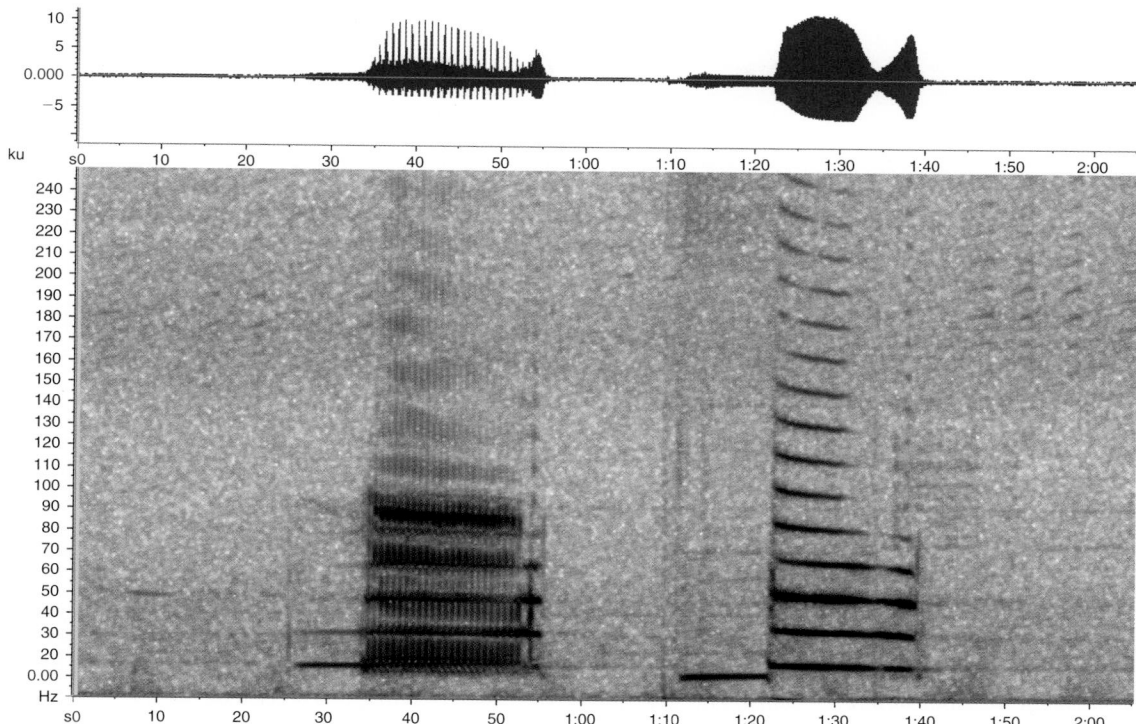

**Figure 1** *Illustration of part of a blue whale* (Balaenoptera musculus) *song. The waveform is shown above and the spectrogram below. One can clearly see the series of loud pulses in the waveform of the first sound, which is amplitude modulated (AM). These produced the sidebands seen in the spectrogram below. The second sound has a smooth envelope of its waveform. The spectrogram of the sound shows change with time, so this is a frequency modulated (FM) signal. The lines of energy above the bottom frequency contour are known as harmonics.*

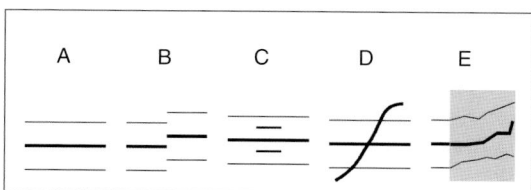

**Figure 2** *Examples of linear and nonlinear signals are shown: (A) typical linear signal (B) frequency jump, (C) subharmonics, (D) biphonation, and (E) deterministic chaos.*

## II. Sound Production Mechanisms

### A. Terrestrial Mammals

When most terrestrial animals vocalize, the lungs push air through the larynx. The vocal folds (more commonly referred to as vocal cords) in the larynx open and close as the air rushes past them, breaking the exhalation up into a series of air puffs or pulses. The resulting undifferentiated sound then passes through the vocal tract of the throat, tongue, mouth, and lips. These structures can move to change the shape of the vocal tract, literally shaping the buzzing sound from the larynx into a vocalization. This mechanism is known as the source-filter model, where the vocal folds are the source of the sound and the vocal tract modifies (or filters) the signal. The source-filter model traditionally has been applied to terrestrial animals, but recently has been employed to investigate sound production in pinnipeds (Sanvito *et al.*, 2007).

### B. Mysticetes

Mysticetes also have a larynx with vocal folds that are believed to be the source of their sound production (Reidenberg and Laitman, 2007). The vocal folds in mysticetes are combined into a single U-shaped fold that is parallel to the airflow (Fig. 3), in contrast to the perpendicular folds in terrestrial mammals. In addition, in mysticetes, a laryngeal sac is found underneath the larynx. The function of the sac is not known. It may serve as an additional acoustic radiating structure, or may help adjust air pressure in the laryngeal structure, or recycle air within the body of the whale. Observations of singing humpbacks (*Megaptera novaeangliae*) show that they do not exhale while singing, indicating a dependence on recycled air.

### C. Odontocetes

Odontocetes can produce clicks, pulsed sounds and whistles. Clicks are often used for echolocation, as described elsewhere in the encyclopedia. Pulsed sounds and whistles are more likely to be used for communication. The production mechanism for these sounds has been the source of considerable debate.

Like mysticetes, delphinid odontocetes have a vocal fold structure in their larynx. These folds may be used in the production of some sounds (Reidenberg and Laitman, 2007). However, it is known that clicks and most (and perhaps all) whistles are produced in the nasal region of the head (Cranford, 2000). All odontocetes, except sperm whales, have a pair of structures known as the "monkey lips"/dorsal bursae (MLDB) complex located in the upper portion of their heads,

S

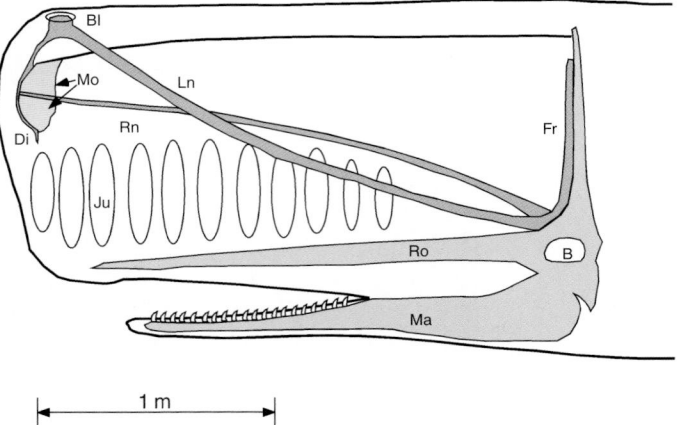

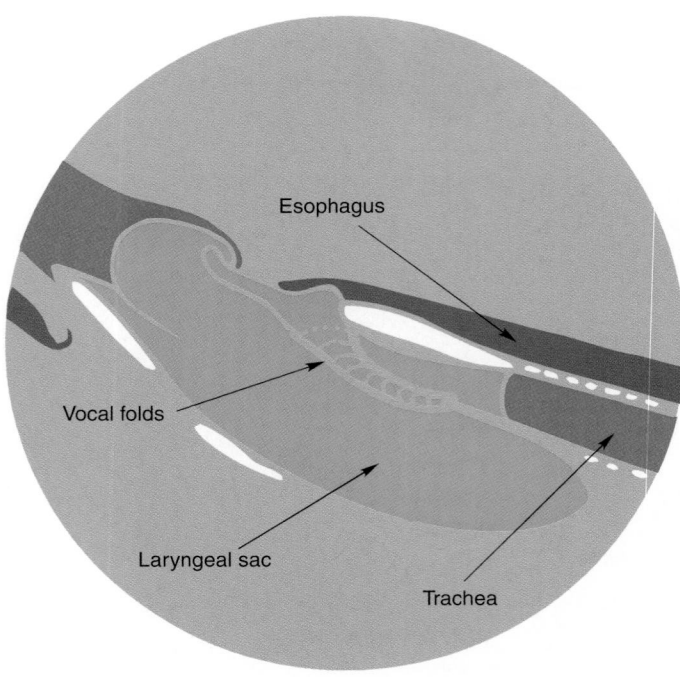

Figure 3   *Details of the mysticete larynx are shown. The vocal folds (in yellow) are parallel to the direction of airflow. The large laryngeal sac below the larynx, unique to the mysticetes, may function to recirculate air, act as resonator, or modulate air pressure in the larynx. Redrawn from Reidenberg and Laitman, 2007 with permission.*

Figure 5   *The anatomy of the sperm whale* (Physeter macrocephalus) *head is shown. B, brain; BL, blowhole; Di, distal air sac; Fr, frontal air sac; Ju, junk; Ln, left naris; Ma, mandible; Mo, monkey muzzle/museau de singe; Ro, rostrum; Rn, right naris; So, spermaceti organ. Clicks are produced on the left at the monkey lips, propagating in all directions. Sound that travels toward the skull is reflected off the frontal air sac and then moves forward out of the head. Some of the sound can be reflected again by the distal sac and frontal air sac again, resulting in a second round trip in the head before being emitted from the forehead. Redrawn from Madsen et al. (2002) with permission.*

The vibration generated by the MLDB is transmitted through the lipid-rich melon, which couples and focuses the sound into the water. In dolphins, the melon sits on top of the skull behind the rostrum (Fig. 4). It also functions as an acoustic lens to focus clicks into a beam. This is similar in function to the focusing of light into a beam by the lens and reflector of a flashlight. The beam pattern (or amount of focusing) is frequency dependent, with a more tightly focused beam at higher frequencies (Au, 1993). Unlike other odontocetes, belugas are able to alter the physical shape of their melon, perhaps to adapt to sound transmission differences due to their movements between the open ocean and less saline estuarine waters (Norris, 1968).

All odontocetes except the sperm whale (*Physeter macrocephalus*) have two MLDB complexes. It appears that they can operate these two structures independently, creating two different sounds simultaneously. Furthermore, in many delphinid species, one of these structures can be larger than the other, thus creating different spectral peaks (Cranford, 2000).

The click production mechanism in sperm whales is related to the dolphins but is somewhat more complex. The huge head of the sperm whale almost certainly evolved as a sound production structure. It appears that the spermaceti organ is the analog of the melon in dolphins, and is used to transmit and focus sound.

Clicks produced by sperm whales make multiple trips through the head (Fig. 5) before being emitted into the water, resulting in a series of pulses that form individual clicks. First, the clicks are produced near the front of the huge head in an organ called the *museau de singe* or monkey's muzzle. A portion of that click goes forward directly into the water while the remainder moves backward, reflecting off the skull into the water as a second sound pulse. Finally, part of this reflected sound can make one more round trip, producing a third pulse of sound by reflecting backward off the distal sac and forward again off the skull (Norris and Harvey, 1972; Møhl, 2001).

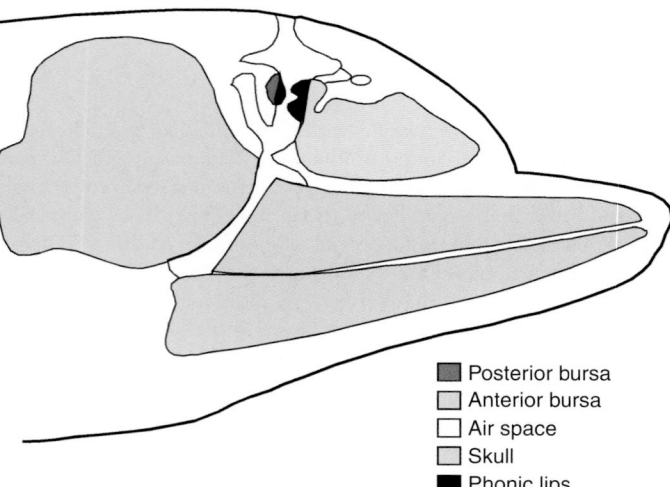

Posterior bursa
Anterior bursa
Air space
Skull
Phonic lips

Figure 4   *A diagram of the MLDB complex and sound production structures in the head of a dolphin. Redrawn from Cranford, 1999 with permission.*

just below the blowhole (Fig. 4). The MLDB complex consists of two lipid-filled sacs or bursae connected to the phonic lips or monkey lips that protrude into the nasal passage. The MLDB creates sound when air pushes past the phonic lips, which then open and slap shut, creating vibration in the dorsal bursae. The air used to create the sound can either be returned to the lower nasal passage or released into the water. Whistling dolphins sometimes, but not always, release air as bubble streams from their blowhole (Fripp, 2005).

S

This pattern of pulses in the sperm whale click has been known for many years, but the relatively low intensity of the early recordings lead some to doubt that clicks were useful for echolocation. However, it has recently been shown that sperm whales can produce directional high intensity clicks that are sufficiently intense for use in echolocation and foraging (Møhl *et al.*, 2000; Miller *et al.*, 2004).

### D. Sirenians and Carnivores [Polar Bear (*Ursus Maritimus*) and Pinnipeds]

Although sound production in sirenians is poorly studied, it is known that manatees (*Trichechus* spp.) and dugongs (*Dugong dugon*) possess a rich repertoire of sounds. Both sirenians and polar bears may produce sound with their larynxes, similar to terrestrial mammals.

Sound production mechanisms in pinnipeds are not well known. When in air, most pinnipeds exhale while vocalizing and probably use their typical mammalian larynx to produce sounds (Tyack and Miller, 2002). However, other sound production mechanisms are found in pinnipeds. For example, male hooded seals (*Cystophora cristata*) have a specialized nasal hood and septum that can be partially extruded out a nostril and then inflated. By alternately deflating and inflating this apparatus, they produce a variety of sounds. Walruses (*Odobenus rosmarus*) also have the ability to create a variety of airborne sounds such as whistles, made by blowing air through the lips, and gong-like sounds, made with their pharyngeal sacs.

Underwater, pinnipeds rarely exhale while vocalizing, indicating that air is shunted between the lungs and mouth/nasal structures. Rather than vocal folds in the larynx, they may use vibrating membranes in the trachea to produce sound underwater. Underwater observations of pulsations in the chest or throat seem to confirm a tracheal mechanism for underwater sound production in many phocids (Tyack and Miller, 2002), as is likely for the long trills of the bearded seal (*Erignathus barbatus*).

## III. Characteristics of Vocalizations by Groups and Selected Species

### A. Pinniped Sounds

Acoustic signal structure varies widely within the 33 species of pinnipeds (Schusterman and Van Parijs, 2003). All pinnipeds are amphibious to some degree, and how they proportion their activities on land and at sea significantly affects their social system, and therefore their acoustic behavior.

Most of the signals of aquatic mating species occur in the water, whereas terrestrial breeding pinnipeds produce a wide variety of vocalizations while hauled out. These signals are often loud, directional, broadband and highly repetitive (Schusterman and Van Parijs, 2003).

Aquatic mating species have evolved a wide variety of signals that are used during the mating season. A review of two phocid species [bearded and Weddell (*Leptonychotes weddellii*) seals], for which there are year-round recordings, found a strong peak in the rate of vocal production associated with breeding (Stirling and Thomas, 2003). Bearded seals became silent after breeding, while Weddell seals continued to vocalize at a low rate, suggesting that they may have been defending breathing holes. This temporal pattern of vocalization indicates, but does not confirm, that these signals are used to attract mates (Van Parijs, 2003).

One notable aspect of life history in many pinniped species is the need for the mother to leave her offspring to forage during lactation. The use of acoustic signals to help reunite mother and pup has been demonstrated in both otariids and phocids. Otariid mothers regularly leave their pups to forage during lactation. This behavior has led, in part, to otariid pups and mothers evolving individually distinct and stereotyped vocal signals that facilitate reunion when the mother returns. These calls tend to be low frequency and about a second in duration. Within a species, there can be significant differences in the fundamental frequency, amplitude modulation and frequency modulation that make the calls distinguishable.

Phocid mothers do not regularly leave their pups during lactation, and their calls are not as individually distinctive. Species with very short lactation periods have no individual recognition mechanisms. Hawaiian monk seals (*Monachus schauinslandi*) are an extreme case. Mothers nurse indiscriminately and do not have individual recognition ability. In general, the amount of individual information in pinniped calls and recognition ability is greater in polygynous societies, those that have a longer lactation period and more crowded breeding colonies (Charrier and Harcourt, 2006).

*1. Phocids* **Ringed seals** (*Pusa hispida*), like many seals, were long thought to be silent while underwater. However, they are now known to produce at least six types of underwater calls. These include clicks, burst pulses, knocks, chirps, yelps and a variety of low-frequency calls similar to those produced in air (Kunnasranta *et al.*, 1996). Clicks are very short, lasting between 6 and 8 msec with energy between 2 and 6 kHz. Burst-pulse sounds are low frequency and narrowband, typically composed of a short sequence of pulses (~5 msec). Knocks are low-frequency sounds, from 150 Hz to 2 kHz that occur in a sequence of increasing frequency pulses. Each knock lasts 40–80 msec, and the entire sequence lasts between 500 and 900 msec. Chirps are FM tonal calls, descending from 1,000 to 500 Hz and lasting between 100 and 300 msec. Yelps show slight frequency modulation around a fundamental frequency of ~1 kHz, and last 100–200 msec (Kunnasranta *et al.*, 1996). The proportion of calls recorded in the Arctic varied seasonally, consistent with a reproductive function (Stirling, 1973).

**Bearded seals** are an aquatic mating species with an extensive repertoire composed of four basic call types: trill, moan, sweep and ascent (Risch *et al.*, 2007). Considerable variation exists within each of these call types. Trills are one of the most distinctive signals produced by marine mammals. They are usually long downsweeps, but some variations include a near-constant frequency contour or alternating increases and decreases of frequency. Typically they start between 3 and 6 kHz, lasting from 30 sec to more than a minute and are characterized by repeated frequency jumps during the signal. In some populations, the long trill is followed by a rapid upsweep to 3 kHz that lasts 1–2 sec (Fig. 6).

Moans are characterized by a short duration, low frequency composition, and lack of frequency modulation. Sweeps resemble a short trill, typically with a rapid decrease in frequency in the second half of the signal. Sweeps are not produced by all populations. Ascents are relatively simple frequency upsweeps lasting from 6 to 25 sec; they are found in those populations that lack sweeps.

The vocalizations of bearded seals show significant variation between individual males of the same populations, as well as significant differences between different populations (Risch *et al.*, 2007). They are thought to function as advertisement displays during the mating season.

**Hooded seal** sounds can be grouped into three major classes. Two of these are produced by normal vocal mechanisms. The third is produced with the hood and septum, a set of specialized anatomical structures. The most common calls are produced both in air and in water. They are pulsed and rarely frequency modulated, with energy

**S**

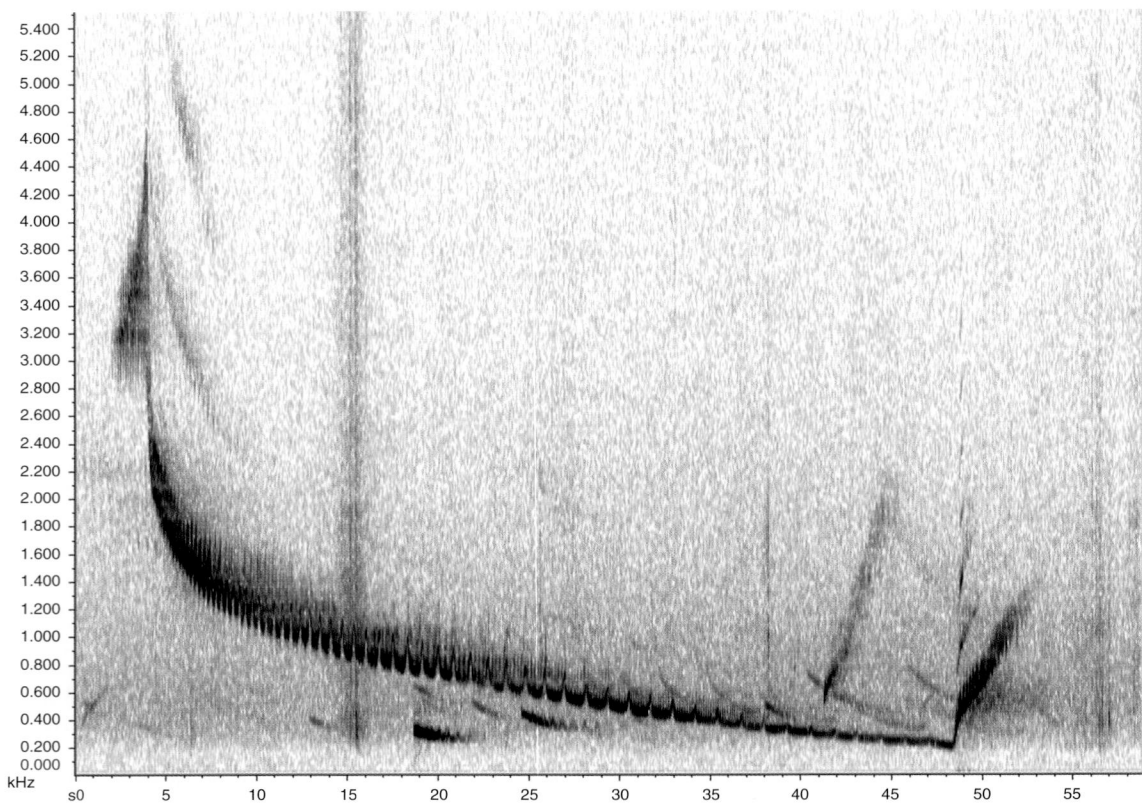

**Figure 6** *The long downward trill and rapid upsweep of a bearded seal from Alaska are shown in this spectrogram.*

ranging from 500 Hz to 6 kHz. These are used in a variety of circumstances, including male displays, female responses, and interactions between females and pups. The second type, growls or roars, appear to be used as low-level threats by both males and females fighting with other males. The third type of signals, produced by the inflation and deflation of the hood and septum, are short-duration, broadband with rapid onsets, and little or no frequency modulation. It is possible that only males make these sounds, described as "bloops", "wooshes", "metallic pings", "clicks", and "knocks" (Ballard and Kovacs, 1995).

**Harp seals** (*Phoca groenlandicus*) aggregate at the ice edge in March in the Northwest Atlantic. At least 19 different call types have been described for this species, ranging from a nearly pure sine wave to pulsed sounds, high-frequency chirps, "broadband warbles," trills, squeaks, and grunts (Terhune, 1994). The maximum source level of these calls is between 135 and 140 dB re 1 μPa at 1 m. Harp seals also produce clicks that are about 25 dB louder than their other calls. Call types have been shown to be stable over a period of tens of years (Serrano and Terhune, 2002). The calls may help individuals locate the herd, and once in the aggregation, may be used to find a mate. Comparisons of different breeding aggregations found many shared calls, but some that were unique to particular breeding areas, suggesting that the populations may be reproductively isolated.

**Ribbon seals** (*Histriophoca fasciata*) are a close relative of harp seals. However, there is only one report on their vocalizations, which describes them as "puff sounds" and frequency downsweeps (Watkins and Ray, 1977). Downsweeps were broken into three categories. Long downsweeps descend from as high as 7100 to 2000 Hz and last up to 4.7 sec. Medium sweeps range from 5300 down to

100 Hz, and last up to 1.8 sec. Finally, short sweeps last less than a second and descend from 2000 to 300 Hz. Puff vocalizations are broadband, below 5 kHz, and last also less than a second.

**Harbor seals** (*Phoca vitulina*) have a mating system with a low degree of polygyny. Females have not been found to vocalize underwater while males produce five types of underwater acoustic displays: roar, bubbly growl, grunt, groan, and creak (Hanggi and Schusterman, 1994). Roars are one of the primary vocalizations of male harbor seals, lasting between 2 and 11 sec, with energy between 20 and 1550 Hz. There is good evidence that the frequency of roars varies between individual males. Roars have been shown to function in territorial behavior (Hayes *et al.*, 2004). Bubbly growls are another long vocalization, lasting between 1 and 8 sec and sounding as if the seal were blowing bubbles underwater. Grunts are short tonal calls from 100 to 4000 Hz lasting less than a second. Groans are a longer version of grunts, lasting between 1 and 5.5 sec. Creaks are tonal calls lasting up to 6 sec with rich harmonic content.

**Spotted seals** (*Phoca largha*) in captivity produced six types of underwater vocalizations: "growls", "drums", "snorts", "chirps", "barks" and "cranky door U" (Beier and Wartzok, 1979). These vocalizations include pulsed and tonal signals. The sounds ranged from 500 Hz to 3.5 kHz and durations from 19–400 msec. Males produce signals at a rate approximately 2.5x times that of females, again suggesting that males are producing signals as part of mating behavior.

**Gray seals** (*Haliochoerus grypus*) recorded in Canada produce calls that have been classified into seven types (Asselin *et al.*, 1993). The most common is a short duration call (~0.2–1.0 sec) that begins with a near-constant low frequency component that then sweeps upward sharply in frequency to about 3 kHz. The second most common call ranges from 0.1 to 2.9 sec in duration, and is composed of a gradual low-frequency downsweep followed by a second component that began at a slightly higher frequency, first sweeping up and then down. Other call types produced much less frequently include growls, knocks, clicks, roars and a pulsed signal called a "trot" (Page *et al.*, 2002).

**Elephant seals** (*Mirounga* spp.), especially males, are known to use acoustic signals in agonistic encounters. They mate on land where males compete for dominance and access to females. Males use threat displays and fights to establish dominance. Male elephant seals make three main types of calls during AGGRESSION: "snores", "snorts", and "clap threats". Snores are used as a low-intensity threat. Dominant males snort more aggressively when approached by a challenging male. Snorts range between 200 and 600 Hz. The clap threat ranges up to 2.5 kHz (Sandegren, 1976; Shipley *et al.*, 1981; Shipley *et al.*, 1986). It is thought that in-air threat calls produced by males are also transmitted through the ground, and can elicit responses from other elephant seals (Shipley *et al.*, 1992). Females produce a low-frequency "belch roar" in aggressive situations and a 500 Hz–1 kHz bark to attract the pup (Bartholomew and Collias, 1962).

Recently the acoustic characteristics of these signals have been compared to the physical attributes of the signaler. While there is only a weak relationship between body size and signal loudness, the signal structure accurately represents the age, size, and resource-holding potential of the signaler (Sanvito *et al.*, 2007). Therefore, in many cases individuals can assess the relative dominance of each other and resolve the interaction without fighting.

Elephant seals are known to make underwater sounds, but these remain poorly described (Poulter, 1968).

**Monk seals** (*Monachus* spp.) are not a very vocal species. They do produce in-air vocalizations composed of soft "liquid bubble sounds", guttural expirations, roars and a "belch-cough". Most of the energy of these calls is below 1 kHz (Miller and Job, 1992).

**Weddell seals** are a very vocal species. Males produce long complex displays that are known as song. Signals appear to vary between different geographic areas. One of the best descriptions of song is based on the Davis Straight population, where seals produced ten different calls consisting of roars, whistles and trills.

Roars can last over 5 sec and vary in frequency from a few hundred to several thousand Hz. Whistles can be both upswept and downswept, ranging to 10 kHz or higher. Trills, produced only by males, may be the most distinctive Weddell seal vocalization (Oetelaar *et al.*, 2003). Trills are long downswept FM calls, beginning above 10 kHz and sweeping down to 1–2 kHz, producing a curved spectrogram. Songs appear to have individually distinct variations in their basic sound units and seem to function in both territorial defense and mate attraction (Terhune and Dell'Apa, 2006).

**Leopard seal** (*Hydrurga leptonyx*) vocalizations show considerable structural variability. Their calls include low-frequency tonal calls, narrowband and wideband high-frequency pulses, and FM calls ranging from 50 Hz to 8 kHz. Males in Prydz Bay, Antarctica, produce a lengthy ordered pattern of five different calls that appears to function as a reproductive display. The structure of these signals varies between individuals (Rogers and Cato, 2002). There is also considerable geographic variation in the signals produced by males of different populations (Thomas and Golladay, 1995). Females in estrus are also known to produce a collection of signals, collectively known as broadcast calls that may serve to indicate their reproductive condition (Rogers *et al.*, 1996). Hauled-out leopard seals that were approached by a researcher made explosive, broadband vocalizations

termed "blasts" and "roars" that may have been used as threats or territorial calls.

**Crabeater seal** (*Lobodon carcinophaga*) vocalizations are poorly known. Only one vocalization has been reported. The groan-like signal had a mean duration of 2.12 sec with most energy below 1.5 kHz (Stirling and Siniff, 1979).

**Ross seal** (*Ommatophoca rossii*) in-air vocalizations include 1–1.5 sec long FM upsweeps and downsweeps between 100 and 1,000 Hz. They also produce sequences of 5–12 short frequency downswept pulses, each lasting 50–100 msec. Sounds produced in water are similar to in-air sounds, except that the frequency range is wider and the longer FM calls possess sidebands not present in air (Watkins and Ray, 1985).

*2. Otariids* The eared seals appear to lack the complex social signals of many phocids. This is probably because phocids tend to mate in the water, and sounds are often, although not always, related to social/sexual interaction. Otariids, however, tend to mate on land, and they are in large part relatively quiet in the sea.

**California sea lions** (*Zalophus californianus*) make bark-like sounds, groans, and grunts in air. The sound of a group of hauled-out California sea lions is loud and far reaching. Barks tend to have most energy below 2 kHz. These sounds and other interactions help produce the structural organization of sea lion society on a beach or headland. Sea lions also bark underwater, in similar fashion as in air (Schusterman and Balliet, 1969).

**Galapagos fur seals** (*Arctocephalus galapagoensis*) have been recorded making few vocalizations while at sea. These include a long growl, lasting 8 sec and composed of a series of discrete low-frequency sounds with energy less than 1 kHz. They also produce knocks, which are single short broadband pulses lasting less than 100 msec with energy above 2 kHz (Merlen, 2000).

**South American sea lion** (*Otaria flavescens*) sound production differs between males and females. During the breeding season, males produce four types of vocalizations: growls, barks, high-pitched calls (HPC), and exhalations, while females produce grunts. Mothers, pups, and yearlings all produce "primary" calls that are used for female–offspring interactions (Fernández-Juricic *et al.*, 1999). Growls are AM calls used during male–female interactions, lasting about one second, with energy between 200 and 2,000 Hz. Barks are a series of short, 0.2 sec, low-frequency pulses produced by males during low-level agonistic interactions. HPCs are short, ~0.4 sec, FM calls with fundamental frequencies between 300 and 500 Hz. Produced during agonistic encounters between males, they are highly directional, suggesting that they can be directed toward specific individuals. Females can produce low intensity, low frequency, 490 Hz, and brief, ~0.4 sec, calls referred to as grunts usually made during female–female agonistic encounters.

**South American fur seals** (*Arctocephalus australis*) make at least 11 types of calls, which appear to fall into 4 functional classes: investigative, threat, submissive, and affiliative (Phillips and Stirling, 2001). Barks are short broadband pulses that occur in a sequence up to several minutes in duration. Barks are used for non-agonistic investigation of other individuals. "Threat calls" are composed of several sounds produced by forceful inhalation or exhalation. Call amplitude can be changed, with louder calls indicating a higher threat level. Threat calls are all amplitude modulated with energy between 400 and 700 Hz. "Submissive calls" are FM tonal calls between 600 and 1600 Hz lasting about a second. Finally, the affiliative calls include pup-attraction calls that last up to 2 sec. These are composed of a pulsed component followed by a tonal component between 700 and 1000 Hz, ending with

**S**

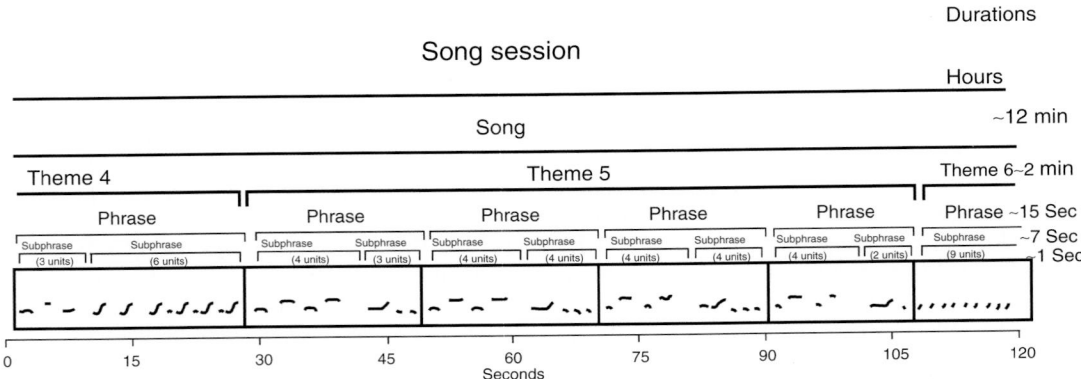

**Figure 7**   *The hierarchical structure of humpback whale* (Megaptera novaeangliae) *song. Redrawn from Payne* et al. *(1983) with permission.*

a downsweep. Female-attraction calls are similar in structure, though they may have more frequency modulation. Both attraction calls have a large degree of inter-individual variability.

**Australian fur seals** (*Arctocephalus pusillus*) produce in-air barks during male–male interactions. Each bark is a short call, ~0.25-sec, with multiple harmonics extending to about 5 kHz and is typically produced in a series. There is enough individual variation to indicate that males may recognize each other with their barks (Tripovich *et al.*, 2005).

*3. Odobenid* **Walrus** adults use roars, grunts, and guttural sounds as threats while hauled out (Miller, 1985). Roars are long, loud mostly low-frequency calls that last a second or more. Grunts vary greatly in amplitude, last between 100 and 400 msec and range from 50 to 250 Hz. The most common acoustic threat is the guttural sound, a series of low frequency, wideband pulses ranging from 13 Hz to 4 kHz. Walruses also bark, usually in a series of short, FM calls ranging from 90 to 260 msec in length and between 300 and 500 Hz. Adult barks indicate submission; louder barking may indicate greater submission. Calves bark in a wide variety of situations. For example, calves separated from their mothers bark loudly and then may continue to bark softly once the two are rejoined. As the calf matures, its bark changes, gradually becoming a longer single call, with both frequency and amplitude modulation. While females do not call to attract the calf, they do produce a short, soft contact call when mother and calf are in close proximity. The call is usually frequency modulated, either as a downsweep or as an alternation of up-and-down sweeps.

Male walruses sing on the mating grounds. Songs are loud, repetitive, and stereotyped sequences of pulsed sounds with bell-like sounds interspersed between them (Sjare *et al.*, 2003). These animals make two types of pulsed sounds: the intense "knock" and the less intense, quickly repeating "tap". Knocks range between 0.2 and 8 kHz and are produced at a rate of 1–3/sec. Taps are produced at a higher rate, 10/sec, and range in frequency from 0.2 to 4 kHz (Stirling *et al.*, 1983). As a courting display, males also make a gong-like sound, both in air and underwater, inflating their throat pouches, sometimes striking their throat with their flippers to augment the sound (Tyack and Miller, 2002). Male walruses also make aggressive nonvocal clacking sounds with their teeth.

### B. Mysticete Sounds

Mysticetes produce a wide variety of sounds, from the blue whale's 7 Hz infrasonic pulses to the humpback whale's songs whose harmonics can range to at least 24 kHz. Individual signals can last less than a second or as long as 30 sec or more. The range of signal types varies from simple growls to loud complex modulated high frequency calls. Four mysticete species are now known to sing: humpback, bowhead (*Balaena mysticetus*), blue and fin whales. Song is typically defined as a repetition of patterned signals. Humpback whale song was first described scientifically in the early 1970s. Bowhead song was recognized in the 1980s. Blue whale song was described in the 1990s and fin whales were recognized as singers in the early 2000s.

**Humpback whale** songs were first described by Roger Payne and Scott McVay in 1971. Only males sing, primarily on the mating grounds in winter, and it is surmised that song is a mating display; intersexual, for males to attract females; intrasexual, as a male dominance display; or both (Frankel *et al.*, 1995). Humpback whale songs have a hierarchical structure, from the shortest utterance to long bouts of singing that can last for days. Individual calls (somewhat analogous to musical notes) are referred to as song units. These units are repeated and combined to form phrases that are then repeated to form longer themes. A song is typically composed of 4–12 themes. Songs can last from 5 to 30 min in length before beginning again. Individual humpbacks are known to sing for as short as a few minutes and as long as 48 h or more. Song units are highly variable, including upsweeps, downsweeps, AM units and complex FM sweeps (Fig. 7). Song units range widely in frequency as well, from fundamental frequencies of about 20 Hz to harmonics reaching 24 kHz or higher. Individual units can last from fractions of a second to several seconds.

Individual whales slowly change the structure of their songs over time (Payne *et al.*, 1983). To illustrate this, consider a theme with two upsweeps and a growl. One type of change would be the addition of a third and then a fourth upsweep. Usually, these changes occur gradually, over a period of about one month. However, the pace of change in song structure is variable as well. In some years the song changes slowly and in other years it evolves rapidly.

Another unique feature is that all humpback whales of a population typically sing essentially the same song, with only minor variations (Payne and Guinee, 1983). It is not known how this uniformity is maintained while the song itself keeps changing. However, it is likely that whales within a discrete area pay attention to each other's song, and some whales copy others. Thus humpback whale song represents an example of cultural evolution.

Once they have evolved, it seems that individual themes are not reused. All themes are created *de novo* (Payne and Payne, 1985). The longest that an individual theme has been known to persist was 5 years (Eriksen *et al.*, 2005).

While most song is produced on the wintering or breeding grounds, it is also heard during migration and on the summer feeding grounds, albeit at a reduced rate (Clark and Clapham, 2004; Gabriele and Frankel, 2002).

Humpback whales on the breeding grounds also produce a rich repertoire of non-song vocalizations, generally termed social sounds. Social sounds can be thought of as a subset of the song units, uttered in a non-patterned fashion. These sounds appear to be used by males as acoustic threat displays in conjunction with visual displays and direct physical contact (Silber, 1986). Humpback calves are also known to produce short, low-frequency calls, although the function of these calls remains unknown (Zoidis *et al.*, 2008).

Finally, humpbacks produce vocalizations during feeding. One call is made during a foraging strategy known as social lunge feeding. This entails a group of 6–12 or more humpbacks all vertically lunging through the surface of the water in a coordinated fashion. The accompanying call is a sequence of "cry" vocalizations. The cry typically begins with a short upsweep, followed by a long near-constant frequency tone, finishing with a short downsweep. There is little variation in the frequency of the call within a sequence, but there is strong variation between different sequences, suggesting the presence of individual variation (Cerchio and Dalheim, 2001). In experiments, these calls were played to herring, which responded by fleeing, suggesting that the most likely function of the call is prey manipulation (Sharpe, 2001).

More recently, humpbacks foraging in the Northwest Atlantic were found to produce click trains and buzzes while feeding at night. Although these signals are reminiscent of odontocete echolocation signals, the function of these "megapclicks" has not yet been established (Stimpert *et al.*, 2007).

**Bowhead whales** were discovered to sing during the acoustic population census of bowhead whales during the spring migration from the Bering Sea to the Chukchi Sea. Songs are usually heard more often at the beginning of the migration and less so at the end. This suggests that most of the singing occurs in the Bering Sea in winter. More recently, songs have also been recorded from the Davis Strait population in Greenland between February and May (Outi *et al.*, 2007).

Song notes are usually longer than non-song moan and grunt-like calls. Bowheads sing between one and three themes, most often two. Unlike humpback whales, bowhead songs regularly show substantial change in structure in successive years. Within a year, all whales within a population sing the same basic version of song, but there is considerable inter- and intra-individual variation. Most of the sound energy of bowhead calls and songs falls below 1000 Hz. The songs are frequently composed of both AM and FM components. Bowhead whales produce fewer units and in a narrower frequency range than humpbacks, yet their signals have great variation in tone, resulting in a wide variety of different sounding song notes.

In addition to songs, bowheads produce a host of calls. There are two main groups: the simple, low-frequency FM calls and complex calls. These FM calls can be categorized by their FM contours, i.e., upsweeps, downsweeps, constant, and inflected contours. FM calls are almost always less than 400 Hz in frequency. The complex calls have been described as pulsive, pulsed tonal, and high. High calls have frequencies above 400 Hz and sound like a whine. The pulsed tonal is a combination of both frequency and amplitude modulation. Pulsive calls are a mixture of pulses, with both frequency and amplitude modulations. Pulsed tonal calls are often below 400 Hz, but pulsive calls can exceed 1000 Hz.

Migrating bowheads will sometimes produce calls for periods ranging from a few minutes to several hours. These are often made in the context of whales counter-calling with each other. Thus it has been suggested that these calls are used to maintain group cohesion. They may also be used to help orient themselves in ice fields. A group of whales was observed approaching a large block of ice. The first whales to encounter the ice only swam around it when they were very close. The following whales deflected much earlier, suggesting that they were listening to the echoes of the early whales and using the acoustic information to avoid the ice (George *et al.*, 1989).

**Fin and blue whales** are the two largest species of extant cetaceans. They both produce the loud and well-known low frequency 20 Hz sounds that are capable of traveling over great distances. In fact, their sounds have been heard at distances covering at least several hundred kilometers. It has been hypothesized that, at least prior to the rise of motorized shipping, whales could hear each other over ocean basin extents (Payne and Webb, 1971). This is possible due to the deep sound channel that propagates low-frequency sound extremely well.

While similar in frequency, blue and fin whale calls differ in length. Most fin whale sounds last about 1 sec, whereas many blue whale signals can be up to 30 sec.

The frequency range of fin whale calls varies in different ocean basins. Most calls from the Atlantic are 1 sec downsweeps from 23 to 18 Hz while those in the Gulf of Mexico sweep down from 42 to 20 Hz (Thompson *et al.*, 1992). These 20 Hz pulses often occur as a long, regularly patterned series of calls. The interval between pulses typically ranges from 6 to 37 sec. In addition to the 20 Hz pulses, fin whales also make tonal pulses at higher frequencies.

In temperate waters, fin whales produce regular sequences of short downswept pulses from about 30 to 15 Hz, predominately during winter. This temporal pattern suggests that these signals may serve a reproductive function (Watkins *et al.*, 1987). Furthermore, only males produce these long patterned sequences, now described as song (Croll *et al.*, 2002). While pulses occur year round, they are less patterned during the non-winter months. It may be that fin whale song is used for reproduction while the less patterned calls may serve another function (Moore *et al.*, 1998).

Male blue whales also produce songs throughout the year (McDonald *et al.*, 2001; Oleson *et al.*, 2007) and it is surmised that they may also serve as a reproductive display. There are at least nine, and probably more, song populations. Songs within a population appear to have a stable structure while song structures differ between populations (McDonald *et al.*, 2006). Some of these song populations have only simple tonal calls, while others exhibit more complex songs that include pulsed as well as tonal units.

As an example, the song of the North Eastern Pacific blue whale population contains both tonal and pulsed components: the pulsed A call, tonal B call, an A–B combination, and downsweeps known as D calls (Thompson *et al.* 1996; McDonald *et al.* 2001). The amplitude-modulated A call typically lasts 17 sec and has a fundamental frequency of 16 Hz. The frequency-modulated B call lasts about 19 sec, sweeping down from ~18 to ~15 Hz and often has strong harmonics. The D call is a short downsweep lasting a few seconds that starts at about 90 Hz descending to about 30 Hz.

Song has been recorded solely from traveling animals. Individual (i.e., non-song) A or B calls were produced from animals in a variety of behavioral states (Oleson *et al.*, 2007), suggesting different functions for non-song calls. Blue whales in the Eastern Tropical Pacific have a higher rate of B call production when their vertically migrating prey is near the surface, suggesting a possible foraging function for the B calls (Stafford *et al.*, 2005). Similarly, the D calls are

**S**

produced by both sexes, frequently during foraging (Oleson *et al.*, 2007).

**Minke whales** (*Balaenoptera acutorostrata*) are known to produce a variety of different calls with a wide frequency range. These include clicks, tonals and FM signals. Minke signals also have significant geographic variation in their call structure. Because minke whales are often difficult to observe in the wild, it has taken many years to associate them with their sounds. Some have only been described in recent years, others almost certainly remain undiscovered.

One notable sound type is the pulse train. These are a repetitive series of pulses, which can either be a narrow-bandwidth (e.g., 80–140 Hz) "growl", or a wider bandwidth "thump" ranging from 100 to 800 Hz (Winn and Perkins, 1976). Thump trains can last over a minute.

Australian dwarf minke whales make a vocalization so unusual that it has been termed the "Star Wars" call (Gedamke *et al.*, 2001). This call consists of a series of three 100 msec pulses ranging between 50 Hz and 9.4 kHz. These pulses are produced simultaneously with harmonically unrelated low frequency, AM pulses between 50 and 750 Hz. Following the pulses, the whale produces a pulsed tone at 1.8 kHz along with a tonal call at 80 Hz. The tonal call shifts up to 140 Hz as the final component of this complex set of vocalizations. The communicative function is unknown.

North Pacific minke whales were recently discovered to be the source of the "boing" sound (Rankin and Barlow, 2005). Boings have been known and recorded since the 1950s (Wenz, 1964). Those recorded in the Eastern Pacific are longer and have a lower pulse repetition rate than those recorded in the Central Pacific (Rankin and Barlow, 2005).

**Sei whales** (*Balaenoptera borealis*) rarely have been recorded. In the Northwest Atlantic they are known to produce two phrases that are each 0.5–0.8 sec long. The phrases are composed of 10–20 FM sweeps between 1.5 and 3.5 kHz, each 30–40 msec in duration (Knowlton *et al.*, 1991; Thompson *et al.*, 1979) with an interval of 0.4 and 1 sec between the two phrases. Recordings off Cape Cod found low frequency downsweeps from 90 to 40 Hz lasting a few seconds in duration (Esch *et al.*, 2007). Sei whales in Antarctica produced low-frequency tonal calls, FM sweeps, and broadband signals. The tonal and FM sweeps ranged from approximately 100 to 1000 Hz, lasting from 0.2 to 1.8 sec and often including a frequency jump. The source levels for these calls ranged from 147 to 156 dB re 1 μPa at 1 m (McDonald *et al.*, 2005).

**Bryde's whales** (*Balaenoptera edeni*) recorded from the Gulf of California contained short low-frequency moans between 70 and 245 Hz and lasting between 0.2 and 1.5 sec. Source levels range between 152 and 174 dB re 1 μPa at 1 m (Cummings *et al.*, 1986). A captive juvenile Bryde's whale produced a pulsed moan, ranging between 100 and 900 Hz and between 0.5 and 51 sec in duration. These moans were similar, though longer, than those recorded in the wild. Finally, wild calves have been recorded making a series of discrete pulses between 700 and 900 Hz (Edds *et al.*, 1993). Bryde's whales in the Eastern Tropical Pacific produce at least six different call types. Most of these are lower in frequency than previous recordings, between 20 and 60 Hz, with one type being a frequency downsweep from 207 to 75 Hz. Durations ranged from 1.1 to 4.9 sec (Oleson *et al.*, 2003). Bryde's whales off New Zealand also produced low-frequency calls, predominately a broadband pulse followed by a 25–22 Hz downsweep and a 5 sec long 22 Hz tonal call (McDonald, 2006).

**Right whales** (*Eubalaena* spp.) from the southern hemisphere produce an "up" call, an upsweep from 50 to 200 Hz that lasts for about a second. This call appears to be used to bring individuals together, because calling stops once the whales rejoin. "Down" calls are downsweeps from 200 to 100 Hz that are also about a second in length. They may serve to maintain acoustic contact, if not physical proximity. "Constant" calls have a nearly constant frequency between 50 and 500 Hz and are 0.5–6 sec in duration. These calls are typically produced by whales that are simply swimming or engaged in a low level of activity (Clark, 1983).

Three call types are associated with surface-active or sexually active right whales: "High" calls are higher in frequency, up to 1 kHz, often with rapid frequency modulations, notably a downsweep at the end. These typically last 0.5–2.5 sec. "Hybrid" calls are similar to the high call, but then end with an AM pulse. "Pulsive signals" are mostly broadband AM signals (Clark, 1983).

Sounds made by North Atlantic right whales (*Eubalaena glacialis*) are similar to their southern counterparts, with one known exception. North Atlantic right whales also make a short broadband vocalization called the "gunshot". Produced only by males, gunshots may be used to attract females. Alternatively, they may be used as an agonistic signal directed toward other males, or may serve both functions (Parks *et al.*, 2005). North Pacific right whales are only known to make calls similar to the southern right whale's up, down, and constant calls.

Right whales feeding with their upper jaw raised above the water's surface can produce a nonvocal sound called "baleen rattle". As water flows through the lower portion of the baleen plates, they apparently rattle together, producing a series of short broadband pulses between 1 and 9 kHz, with most of the energy between 2 and 4 kHz. This sound is audible both in air and underwater, and may be simply a by-product of feeding (Watkins and Schevill, 1976).

**Pygmy right whales** (*Caperea marginata*) have only recently been recorded. A juvenile found in a harbor produced only one sound type. It was a short tonal downsweep that began between 90 and 135 Hz and swept down to 60 Hz. Pulses lasted between 140 and 225 msec in duration and were separated by intervals of 430–510 msec. Source levels were estimated between 153 and 167 dB re 1 μPa at 1 m. These calls were very simple and their function remains unknown (Dawbin and Cato, 1992).

**Gray whales** (*Eschrichtius robustus*) most frequently produce sounds referred to as "knocks" and pulses. These range in frequency from 20 Hz to 3 kHz. Gray whales are fairly vocal while feeding, relatively quiet while migrating, and the most vocal during mating activities. The source levels of gray whale vocalizations range between 167 and 188 dB re 1 μPa at 1 m (Petrochenko *et al.*, 1991).

Six major signal types have been recorded from gray whales on the mating grounds: a series of 2–30 metallic-sounding pulses; a single pulse lower in frequency and longer in duration than the first; a relatively low frequency and long duration "moan"; short "grunts"; a loud "bubble blast" produced by releasing a large amount of air underwater; and finally, a sixth sound type produced by the exhalation of a series of bubbles (Dahlheim *et al.*, 1984). Recordings made on summer feeding grounds contained three signals similar to those heard on the wintering grounds (Moore and Ljungblad, 1984).

## C. Odontocete Sounds

The toothed whales, or odontocetes, are vocal animals *par excellence*. They probably all produce clicks for echolocation, and many produce complicated sets of pulsed sounds and whistles, the latter two for communication. Not all odontocetes make all of these sound types, and there are signals that cross categorical bounds and others, that defy categorization. Nevertheless, these three broad categories can be useful and descriptive. Rather than attempt to consider the signals of each odontocete species individually, this section will focus on the three sound types, with specific examples provided for each.

*1. Click Sounds* Most of the odontocetes are known to produce clicks. In many (if not all) species, these clicks are used for echolocation. An echolocating animal produces a click that travels to and reflects off a target, such as a prey item. The amount of time elapsed between click production and echo reception (i.e., the two-way travel time) provides a measure of the distance to the target. Many species are also able to determine direction to the target. The clicks are usually produced at an interval greater than the two-way travel time. Presumably this is to allow the echo to be processed before the next click is produced. However, some species, such as the beluga (*Delphinapterus leucas*), can produce clicks with intervals shorter than the two-way travel time (Turl and Penner, 1989). When a dolphin approaches an object, it decreases the outgoing level of its clicks to maintain the returning echoes at a near constant sound level (Au and Benoit-Bird, 2003) and increases the rate of click production as the two-way travel time decreases.

Click structure varies between phylogenetic groups. Clicks produced by sperm whales, beaked whales, dolphins, and porpoises can be differentiated by duration, waveform type, and frequency emphasis (Tyack *et al.,* 2006). Sperm whale clicks typically range from 400 Hz to at least 15 kHz (Goold and Jones, 1995). The clicks of Blainville's (*Mesoplodon densirostris*) and Cuvier's (*Ziphius cavirostris*) beaked whales have most of their energy in the 20–50 kHz region (Johnson

*et al.,* 2004). Delphinid clicks generally range from 60–120 kHz. Porpoise clicks are narrowband and usually well above 120 kHz.

**Sperm whales** are the largest toothed mammals on earth and have a disproportionately huge head. It is likely that the evolution of their huge head has, in large part, been driven by the loud and complicated structure of their relatively low-frequency clicks that are used for communication and echolocation.

Sperm whales produce a variety of clicks in a variety of contexts. Clicks can occur singly at different intervals, in a short pattern of distinct clicks called "codas," or in a long sequence of tightly spaced clicks known as "creaks." The frequency content of clicks differs between sexes. It is known that large males have lower frequency content in their clicks than females and young males (Goold and Jones, 1995).

"Usual" clicks are produced in a regular sequence at intervals of 1.17–1.95 sec with a duration between 2 and 24 msec (Goold and Jones, 1995). The click interval varies greatly between individuals (Goold and Jones, 1995), but appears to be stable within the click trains of an individual whale (Fig. 8).

The stereotyped, repetitive patterns of clicks or codas are produced primarily, if not only, by female sperm whales (Marcoux *et al.,* 2006) . While it had been suggested that codas serve the function of individual identification, it is now known that coda repertoires are shared between individuals within a group (Rendell and Whitehead, 2004). This suggests that codas are used in social communication, and perhaps group identity (Weilgart and Whitehead, 1997). In

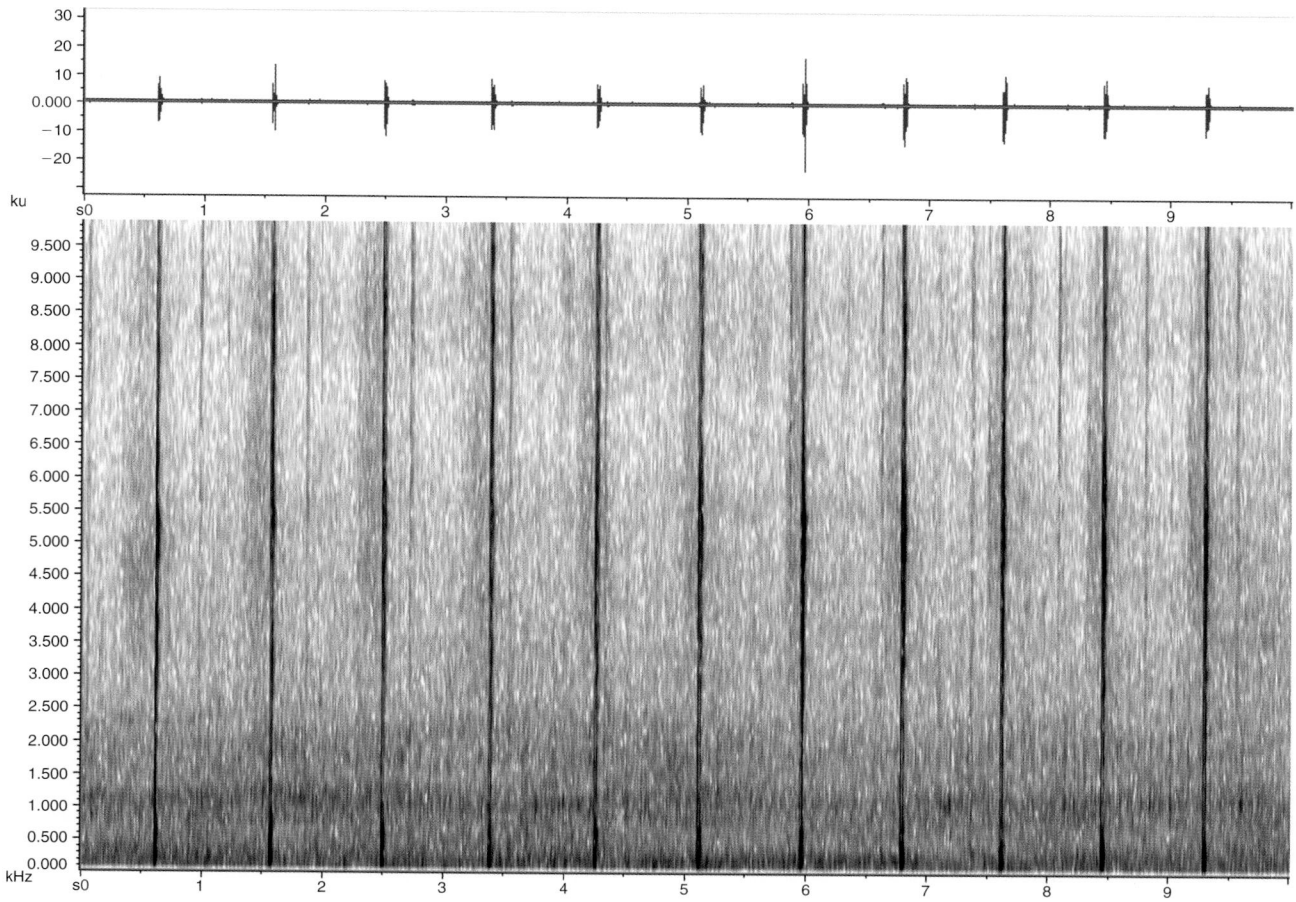

**Figure 8** *The spectrogram and waveform of "usual" sperm whale* (Physeter macrocephalus) *clicks.*

addition, individual social groups of sperm whales appear to belong to clans that share similar coda repertoires. These repertoires may be created and evolve through cultural transmission (Rendell and Whitehead, 2003). This is supported by the finding that the degree of similarity of coda structure is inversely related to the distance between the groups. (Rendell and Whitehead, 2005).

"Creaks" are a rapid sequence of clicks sounding more like a continuous buzzing sound than individual clicks. They occur when a sperm whale is approaching a potential prey item much like a dolphin's terminal buzz (Miller *et al.*, 2004; Watwood *et al.*, 2006).

Sperm whale 'trumpet sounds' are usually produced at the start of a dive in some but not all populations. These sounds are a series of repeated calls, each about 0.2 sec long that occur in sequences lasting between 0.6 and 3.5 sec. Each of these calls is composed of an AM tonal waveform with a complex harmonic structure. The spectrum contains a low frequency component at 500 Hz and a mid-frequency component at 3 kHz (Teloni *et al.*, 2005).

Mature sperm whale males produce another type of click, called a "slow click" for its low repetition rate. These clicks are of longer duration, with a mean of 72 msec compared to 24 msec for usual clicks. Slow clicks have consistent energy concentrations at 1.8 and 2.8 kHz, whereas the energy distribution in the spectra of usual clicks is much more variable. It has been suggested that slow clicks may serve as a long-range social communication signal (Madsen *et al.*, 2002).

**Dolphins**, with the exception of those in the genus *Cephalorhynchus*, produce short clicks lasting only a few cycles. The short duration necessarily produces a broadband signal with most of the energy typically found between 40 and 300 kHz across most species (Au, 1993). The spectrum of white-beaked dolphin (*Lagenorhynchus albirostris*) clicks can span this entire range (Rasmussen and Miller, 2002) while other species may have much narrower bandwidths. The clicks of many species contain two frequency peaks (Au and Würsig, 2004). This may be the result of size differences in the two MLDB sound-generating structures in the dolphin's head (Cranford, 2000). Click source levels can range between 150 and 230 dB re 1 μPa at 1 m (peak to peak).

Dolphins are able to alter the spectral characteristics and source level of their clicks. For example, when a captive beluga was moved from San Diego Bay to Kaneohe, Hawaii, it increased the frequency of its clicks from 40–60 to 100–120 kHz (Au *et al.*, 1985). Furthermore it has been shown that dolphins alter the source level of their clicks while echolocating. They click loudly when far away from the target and more softly as they approach the target. This is apparently done to decrease variation in the level of the returning echo (Au and Benoit-Bird, 2003). Finally, there appears to be a relationship between the source level and the dominant frequency of the click, with louder clicks generally having higher frequency peaks (Au and Würsig, 2004).

Porpoises, the genus *Cephalorhynchus*, the pygmy sperm whale (*Kogia breviceps*) and franciscana (*Pontoporia blainvillei*) produce a very different type of click than the short broadband clicks described above. Called a narrowband high frequency (NBHF) click, they have relatively low power, a narrow bandwidth and a high center frequency (Morisaka and Connor, 2007). NBHF clicks are longer (tens of cycles) than typical delphinid clicks, and have a smooth amplitude envelope (Au, 1993). It may be that the evolution of these higher frequency NBHF clicks was an anti-predation response to killer whales (*Orcinus orca*), who may be unable to hear them (Morisaka and Connor, 2007).

**Beaked whales** have recently been tagged with acoustic recorders. Data from these recordings have provided both a description of beaked whale clicks and evidence that they are used for echolocation (Johnson *et al.*, 2004). The measured source level of the clicks

is as high as 214 dB re 1 μPa at 1 m (peak to peak). The clicks of Blainville's and Cuvier's beaked whales typically show an upward frequency modulation and have very little energy below 20 kHz (Johnson *et al.*, 2004). A similar lack of low-frequency energy has been reported for the clicks of other beaked whale species (Dawson *et al.*, 1998).

2. *Pulsed Sounds*   Pulsed sounds are a series of sound pulses with short intervals. The pulses can occur so quickly, one after the other, that they may be perceived as continuous sound (Watkins, 1967). Short, discrete bursts of broadband sound pulses are referred to as burst-pulses. Typically, most of the energy is in the lower frequencies. However, the burst-pulses of some species can lack the low frequency components and extend up to 60 kHz or higher (Lammers *et al.*, 2003).

Burst-pulses appear to be used for communication between members of the group. Hawaiian spinner dolphins (*Stenella longirostris*) were closer together when they produced burst-pulses than when they used whistles, suggesting a differential function between these two signals (Lammers *et al.*, 2006). Stereotyped patterns of burst-pulse sounds have been described in right whale dolphins (*Lissodelphis borealis*), suggesting that these sounds may have a communicative function similar to stereotyped whistles in other dolphins (Rankin *et al.*, 2007).

In addition to their well-known click sounds, sperm whales produce rapidly pulsed sounds (up to ~1600 pulses/sec) that are described as "squeals" (Weir *et al.*, 2007). These squeals are lower in frequency than their clicks, with most energy below 2 kHz and lasting about 1 sec.

**Killer whales** produce a large number of pulsed signals. The high repetition rate of up to 4000 pulses/sec lends a tonal character to their sounds. Most of the energy in these well-studied signals occurs between 1 and 6 kHz (Ford, 1989). Resident killer whales in the Northeast Pacific share stereotyped pulsed calls or dialects within their stable social pods. Each pod has evolved its own dialect but may share calls with other pods. Pods with similar dialects that share calls are grouped together in acoustic clans (Ford, 1991). Resident pods from different acoustic clans are known to associate, but not to share call types.

Dialects may have evolved through behavioral drift as ancestral groups divided into newer subgroups. The different dialects serve important social functions for group cohesion and inter-group recognition (Ford, 1991). Resident killer whales in Norway may have a similar dialect system (Strager, 1995).

Transient killer whales of the Pacific Northwest (of the United States, and of southern Canada), that feed on marine mammals instead of fish, have dramatically different vocal behavior. They remain silent most of the time, emitting calls for specific functions such as locating other transient killer whales (Saulitis *et al.*, 2005). They also echolocate much less often than resident killer whales (Barrett-Lennard *et al.*, 1996). It is likely that this reduction in acoustic behavior evolved in response to the hearing abilities of their marine mammal prey.

3. *Whistles*   Whistles are narrowband FM signals with a wide variety of contours (*i.e.* frequency-modulation pattern) ranging from short and simple chirps to long complex signals. The frequency range of the fundamental frequency of whistles is typically between 1 and 30 kHz. Whistles can be as short as tens of milliseconds, most often between 500 and 1000 msec, and as long as 3 sec or more. Most dolphin species produce whistles, although some, like those in the genus *Cephalorhynchus*, do not (Morisaka and Connor, 2007). All of the whistles of delphinid species share many characteristics and it

can be difficult to discriminate between them (Oswald *et al.*, 2007). However, additional recordings and analyses are helping to improve the identification process.

There are no universally accepted classifications for whistles, although measures of beginning, ending, minimum and maximum frequency, and kind of frequency modulation are usually described. Whistles in some species can be readily grouped into categories while others form a continuum of structure. In bottlenose dolphins (*Tursiops* spp.), researchers tend to come up with tens of distinct whistle sounds, with many intergrading variations among them (Deecke and Janik, 2006).

*4. Signature Whistles*   Bottlenose dolphins, and probably quite a few other species as well, produce individualized whistle contours called signature whistles. These have been shown to function in maintaining group cohesion (Janik and Slater, 1998) and to convey individual identity information that may be analogous to a "name." This individual identity is encoded in the frequency contour of the whistle and not in the individual characteristics of the vocalization (i.e., the "voice") (Janik *et al.*, 2006). Bottlenose dolphins are the only nonhuman animals yet shown to have this capability. Dolphins are also able to mimic the signature whistles of other individuals in their group, an ability that appears to help maintain social bonds.

The formation of signature whistles is a learned behavior. Dolphin calves learn their signature whistles within their first few months and retain them their entire lives. Interestingly, signature whistles in Florida bottlenose dolphins appear to be more alike in mothers and their male offspring than in mothers and their daughters (Sayigh *et al.*, 1995). The Sarasota, Florida, population of dolphins is generally matriarchal, with daughters being closely affiliated for many years or for life. However, sons leave the natal group as subadults. It has been hypothesized that sons and mothers thereby recognize each other easily after prolonged times apart, perhaps even for years. This recognition may help avoid inbreeding and facilitate other kin-related social behaviors, such as lowered aggression. Further research has shown that the Sarasota calves are more likely to model their signature whistles on those of other Sarasota dolphins, but ones with whom they rarely associate (Fripp *et al.*, 2005).

Debate continues over which dolphin species have signature whistles. It has been suggested that Pacific humpback dolphins (*Sousa* spp.) (Van Parijs and Corkeron, 2001) and Atlantic spotted dolphins (*Stenella frontalis*) (Herzing, 1996) produce them and it is likely that additional species possess them as well.

The complexity of whistle production may well relate to the complexity of behavior or "excitement" level. For example, resting long-finned pilot whales (*Globicephala melas*) make very simple non-wavering whistles. Whistle complexity increases during feeding and bouts of socializing, and variability of whistles and other sounds increases greatly when two pilot whale groups approach each other (Weilgart and Whitehead, 1990). There is a general relationship between the activity level of Hawaiian spinner dolphin groups and their vocalization rate. A resting group produces few sounds, whereas a feeding or socially active group produces many sounds. Higher vocalization rates may result from increased vocalization by all members or from only a subset of individuals within the group. Vocalization behavior may also be related to differences in age, social status, alertness, and gender (Norris *et al.*, 1994).

### D. Sounds of Sirenians and Other Groups

**Manatees** are rather quiet but do produce sounds between 0.15 and 1 sec in duration. Signal structure can be complex, and frequencies

range between 600 Hz and 12 kHz, with most energy found between 2.5 and 5 kHz. The fundamental frequency is at times less intense than the first harmonic. The source level has been estimated at 112 dB re 1 μPa at 1 m (Phillips *et al.*, 2004). Calls consist mainly of "chirp-squeaks," "squeals," and "screams." Their calls contain persistent inter-individual differences, particularly in fundamental frequency, which is inversely correlated with body size (O'Shea and Poche, 2006). Mothers and calves counter-call while rejoining each other, suggesting that they recognize individual call characteristics (O'Shea and Poche, 2006).

**Dugongs** appear to vocalize more often than the manatees, producing three major sound types: chirp-squeaks, barks, and trills (Anderson and Barclay, 1995). They also make intermediate sounds that include components of the three main types. Chirp-squeaks are short FM signals extending upward to 18 kHz. They are about 60 sec in duration, typically trend slightly downward in frequency, and have two to five harmonics. Barks are loud broadband signals that range between 500 Hz and 2.2 kHz, lasting between 0.03 and 0.12 sec. Trills are a series of individual notes, lasting between 100 and 2200 msec. Notes typically begin at about 3.1 kHz and rise to 3.9 kHz. Instead of a linear sweep, the frequency contour has an oscillating character. While the functions of these sounds are uncertain, it is likely that they are used for social communication.

**Sea otters** (*Enhydra lutris*) produce at least 11 different airborne sounds: screams, baby cries, whistles, whines, hisses, snarls, coos, grunts, squeals, squeaks, and whimpers (McShane *et al.*, 1995). These vocalizations are short, lasting less than a second, and extremely variable in frequency and structure. Airborne vocalizations serve to maintain the mother–pup bond. For example, a pup at the surface often vocalizes continuously until the mother resurfaces following a dive. However, if the mother surfaces and does not find the pup, she vocalizes and awaits the pup's response (Sandegren *et al.*, 1973). Pups also vocalize to elicit nursing or grooming. No underwater vocalizations have been recorded.

**Polar bears** may not be quite as vocal as many other carnivores. Males chuff and snort with powerful rapid exhalations, especially in competitive interactions with other males. Females produce low mew-like calls that may be used for mother/pup recognition. Other calls include roars, growls, and bellows (Brown, 1993).

### IV. Conclusions

Marine mammals have a very rich behavioral tapestry of sounds. The basic description here merely hints at this richness in an environment where sight and smell are not transmitted as efficiently as sound. Sound is used for communication and for wresting information from the environment. While only toothed whales are thought to have sophisticated echolocation, it is likely that many sounds give information on depth of water, obstruction ahead, or even silent conspecifics, simply by the alteration of sound reflections in different environments.

Our acceptance that sound is critically important to marine mammals also gives us cause for concern. Since the advent of motorized shipping and more recently industrial seismic, military sonar and other human sources of sound, ambient noise levels in major parts of the oceans are increasing. We do not yet know the details of how this noise may affect the communication and behavior of marine mammals.

*See Also the Following Articles*

Communication ■ Echolocation ■ Intelligence and Cognition ■ Mimicry ■ Noise ■ Effects of ■ Signature Whistles ■ Song

## References

Anderson, P. K., and Barclay, R. M. R. (1995). Acoustic signals of solitary dugongs: physical characteristics and behavioral correlates. *J. Mammal.* **76**, 1226–1237.

Asselin, S., Hammill Mike, O., and Barrette, C. (1993). Underwater vocalizations of ice breeding grey seals. *Can. J. Zool.* **71**, 2211–2219.

Au, W., Carder, D., Penner, R. H., and Scronce, B. L. (1985). Demonstration of adaptation in beluga whale echolocation signals. *J. Acoust. Soc. Am.* **77**, 726–730.

Au, W. W. L. (1993). "The Sonar of Dolphins." Springer-Verlag, New York.

Au, W. W. L., and Benoit-Bird, K. J. (2003). Automatic gain control in the echolocation system of dolphins. *Nature* **423**, 861–863.

Au, W. W. L., and Würsig, B. (2004). Echolocation signals of dusky dolphins (*Lagenorhynchus obscurus*) in Kaikoura, New Zealand. *J. Acoust. Soc. Am.* **115**, 2307–2313.

Ballard, K. A., and Kovacs, K. M. (1995). The acoustic repertoire of hooded seals (*Cystophora cristata*). *Can. J. Zool.* **73**, 1362–1374.

Barrett-Lennard, L. G., Ford, J. K. B., and Heise, K. A. (1996). The mixed blessing of echolocation: differences in sonar use by fish-eating and mammal-eating killer whales. *Anim. Behav.* **51**, 553–565.

Bartholomew, G. A., and Collias, N. E. (1962). The role of vocalization in the social behaviour of the northern elephant seal. *Anim. Behav.* **10**, 7–14.

Beier, J. C., and Wartzok, D. (1979). Mating behavior of captive spotted seals (*Phoca largha*). *Anim. Behav.* **27**, 772–781.

Brown, G. (1993). "The Great Bear Almanac." Lyons and Burford, New York.

Cerchio, S., and Dalheim, M. (2001). Variations in feeding vocalizations of humpback whales (*Megaptera novaeangliae*) from Southeast Alaska. *Bioacoustics* **11**, 277–295.

Charif, R. A., Mellinger, D. K., Dunsmore, K. J., and Clark, C. W. (2002). Estimated source levels of fin whale (*Balaenoptera physalus*) vocalizations: Adjustments for surface interference. *Mar. Mamm. Sci.* **18**, 81–98.

Charrier, I., and Harcourt, R. G. (2006). Individual vocal identity in mother and pup Australian sea lions (*Neophoca cinerea*). *J. Mammal.* **87**, 929–938.

Clark, C. W. (1983). Acoustic communication and behavior of the southern right whale (*Eubalaena* australis). *In* "Communication and Behavior of Whales" (R. Payne, ed.), pp. 163–198. Westview Press, Boulder.

Clark, C. W., and Clapham, P. J. (2004). Acoustic monitoring on a humpback whale (*Megaptera novaeangliae*) feeding ground shows continual singing into late spring. *Proc. R. Soc. Lond., B.* **271**, 1051–1057.

Cranford, T. (2000). In search of impulse sound sources in odontocetes. *In* "Hearing by Whales and Dolphins" (W. W. L. Au, A. N. Popper, and R. R. Fay, eds), pp. 109–155. Springer-Verlag, New York.

Cranford, T. W. (1999). The sperm whale's nose: sexual selection on a grand scale? *Mar. Mamm. Sci.* **15**, 1133–1157.

Croll, D. A., Clark, C. W., Acevedo, A., Tershy, B., Floress, S., Gedamke, J., and Urban, J. (2002). Only Male Fin Whales Sing Loud Songs. *Nature* **417**, 809.

Cummings, W. C., Thompson, P. O., and Ha, S. J. (1986). Sounds from Bryde, *Balaenoptera edeni*, and finback, *B. physalus*, whales in the Gulf of California. *Fish Bull.* **84**, 359–370.

Dahlheim, M. E., Fisher, H. D., and Schempp, J. D. (1984). Sound production by the gray whale and ambient noise levels in Laguna San Ignacio, Baja California Sur, Mexico. *In* "The Gray Whale (*Eschrichtius robustus*)" (M. L. Jones, S. L. Swartz, and S. Leatherwood, eds), pp. 511–541. Academic Press, Inc., London.

Dawbin, W. H., and Cato, D. H. (1992). Sounds of a pygmy right whale *Caperea marginata*. *Mar. Mamm. Sci.* **8**, 213–219.

Dawson, S., Barlow, J., and Ljungblad, D. (1998). Sounds recorded from Baird's beaked whale, *Berardius bairdii*. *Mar. Mamm. Sci.* **14**, 335–344.

Deecke, V. B., and Janik, V. M. (2006). Automated categorization of bioacoustic signals: avoiding perceptual pitfalls. *J. Acoust. Soc. Am.* **119**, 645–653.

Edds, P. L., Odell, D. K., and Tershy, B. R. (1993). Vocalizations of a captive juvenile and free-ranging adult-calf pairs of Bryde's whales, *Balaenoptera edeni*. *Mar. Mamm. Sci.* **9**, 269–284.

Eriksen, N., Miller, L. A., Tougaard, J., and Helweg, D. A. (2005). Cultural change in the songs of humpback whales (*Megaptera novaeangliae*) from Tonga. *Behaviour* **142**, 305–328.

Esch, H. C., Baumgartner, M. F., Wenzel, F., and Van Parijs, S. (2007). Automated detection of stereotyped baleen whale vocalizations: Advantages over traditional methods. 17th Biennial Conference on the Biology of Marine Mammals. Capetown, South Africa.

Fernández-Juricic, E., Campagna, C., Enriquez, V., and Ortiz, C. L. (1999). Vocal communication and individual variation in breeding South American sea lions. *Behaviour* **136**, 495–517.

Fitch, W. T., Neubauer, J., and Herzel, H. (2002). Calls out of chaos: the adaptive significance of nonlinear phenomena in mammalian vocal production. *Anim. Behav.* **63**, 407–418.

Ford, J. K. B. (1989). Acoustic behaviour of resident killer whales (*Orcinus orca*) off Vancouver Island, British Columbia. *Can. J. Zool.* **67**, 727–745.

Ford, J. K. B. (1991). Vocal traditions among resident killer whales (*Orcinus orca*) in coastal waters of British Columbia. *Can. J. Zool.* **69**, 1454–1483.

Frankel, A. S., Clark, C. W., Herman, L. M., and Gabriele, C. M. (1995). Spatial distribution, habitat utilization, and social interactions of humpback whales, *Megaptera novaeangliae*, off Hawaii, determined using Acoustic and Visual Techniques. *Can. J. Zool.* **73**, 1134–1146.

Fripp, D. (2005). Bubblestream whistles are not representative of a bottlenose dolphin's vocal repertoire. *Mar. Mamm. Sci.* **21**, 29–44.

Fripp, D., Owen, C., Quintana-Rizzo, E., Shapiro, A., Buckstaff, K., Jankowski, K., Wells, R., *et al.* (2005). Bottlenose dolphin (*Tursiops truncatus*) calves appear to model their signature whistles on the signature whistles of community members. *Anim. Cogn.* **8**, 17–26.

Gabriele, C. M., and Frankel, A. S. (2002). The occurrence and significance of humpback whale songs in Glacier Bay, Southeastern Alaska. *Arct. Res. U.S.A.* **16**, 42–47.

Gedamke, J., Costa, D. P., and Dunstan, A. (2001). Localization and visual verification of a complex minke whale vocalization. *J. Acoust. Soc. Am.* **109**, 3038–3047.

George, J. C., Clark, C., Carroll, G. M., and Ellison, W. T. (1989). Observations on the ice-breaking and ice navigation behavior of migrating bowhead whales (*Balaena mysticetus*) near Point Barrow, Alaska, Spring 1985. *Arctic.* **42**, 24–30.

Goold, J. C., and Jones, S. E. (1995). Time and frequency domain characteristics of sperm whale clicks. *J. Acoust. Soc. Am.* **98**, 1279–1291.

Hanggi, E. B., and Schusterman, R. J. (1994). Underwater acoustic displays and individual variation in male harbour seals *Phoca vitulina*. *Anim. Behav.* **48**, 1275–1283.

Hartmann, W. H. (2004). "Signals, Sound, and Sensation." American Institute of Physics. Springer-Verlag, New York.

Hayes, S. A., Kumar, A., Costa, D. P., Mellinger, D. K., Harvey, J. T., Southall, B. L., and Le Boeuf, B. J. (2004). Evaluating the function of the male harbour seal, *Phoca vitulina*, roar through playback experiments. *Anim. Behav.* **67**, 1133–1139.

Herzing, D. L. (1996). Vocalizations and associated underwater behavior of free-ranging Atlantic spotted dolphins, *Stenella frontalis* and bottlenose dolphins, *Tursiops truncatus*. *Aq. Mamm.* **22**, 61–79.

Janik, V. M., Sayigh, L. S., and Wells, R. S. (2006). Signature whistle shape conveys identity information to bottlenose dolphins. *Proc. Natl. Acad. Sci. U.S.A.* **103**, 8293–8297.

Janik, V. M., and Slater, P. J. B. (1998). Context-specific use suggests that bottlenose dolphin signature whistles are cohesion calls. *Anim. Behav.* **56**, 829–838.

Johnson, M., Madsen, P. T., Zimmer, W. M. X., Aguilar de Soto, N., and Tyack, P. L. (2004). Beaked whales echolocate on prey. *Proc. R. Soc. Lond., B. Biol. Sci.* **271**, S383–S386.

Knowlton, A., Clark, C. W., and Kraus, S. D. (1991). Sounds recorded in the presence of sei whales, *Balaenoptera borealis*. Ninth biennial conference on the biology of marine mammals, Chicago.

Kunnasranta, M., Heikki, H., and Sorjonen, J. (1996). Underwater vocalizations of Ladoga ringed seals (*Phoca hispida ladogensis* Nordq.) in summertime. *Mar. Mamm. Sci.* **12**, 611–618.

Lammers, M. O., Au, W. W. L., and Herzing, D. L. (2003). The broadband social acoustic signaling behavior of spinner and spotted dolphins. *J. Acoust. Soc. Am.* **114**, 1629–1639.

Lammers, M. O., Schotten, M., and Au, W. W. L. (2006). The spatial context of free-ranging Hawaiian spinner dolphins (*Stenella longirostris*) producing acoustic signals. *J. Acoust. Soc. Am.* **119**, 1244–1250.

Madsen, P. T., Wahlberg, M., and Mohl, B. (2002). Male sperm whale (*Physeter macrocephalus*) acoustics in a high- latitude habitat: implications for echolocation and communication. *Behav. Ecol. Sociobiol.* **53**, 31–41.

Marcoux, M., Whitehead, H., and Rendell, L. (2006). Coda vocalizations recorded in breeding areas are almost entirely produced by mature female sperm whales (*Physeter macrocephalus*). *Can. J. Zool.* **84**, 609–614.

McDonald, M. A. (2006). An acoustic survey of baleen whales off Great Barrier Island, New Zealand. *N. Z. J. Mar. Freshwater Res.* **40**, 519–529.

McDonald, M. A., Calambokidis, J., Teranishi, A. M., and Hildebrand, J. A. (2001). The acoustic calls of blue whales off California with gender data. *J. Acoust. Soc. Am.* **109**, 1728–1735.

McDonald, M. A., Hildebrand, J. A., Wiggins, S. M., Thiele, D., Glasgow, D., and Moore, S. E. (2005). Sei whale sounds recorded in the Antarctic. *J. Acoust. Soc. Am.* **118**, 3941–3945.

McDonald, M. A., Mesnick, S. L., and Hildebrand, J. A. (2006). Biogeographic characterisation of blue whale song worldwide: using song to identify populations. *J. Cetacean Res. Manage.* **8**, 55–65.

McShane, L. J., Estes, J. A., Riedman, M. L., and Staedler, M. M. (1995). Repertoire, structure, and individual variation of vocalizations in the sea otter. *J. Mammal.* **76**, 414–427.

Merlen, G. (2000). Nocturnal acoustic location of the Galapagos Fur Seal *Arctocephalus galapagoensis*. *Mar. Mamm. Sci.* **16**, 248–253.

Miller, E. H. (1985). Airborne acoustic communication in the Walrus (*Odobenus rosmarus*). *Nat. Geogr. Res.* **1**, 124–145.

Miller, E. H., and Job, D. A. (1992). Airborne acoustic communication in the Hawaiian monk seal, *Monachus schauinslandi*. *In* "Marine Mammal Sensory Systems" (J. A. Thomas, R. A. Kastelein, and A. Y. Supin, eds), pp. 485–531. Plenum, New York.

Miller, P. J. O., Johnson, M. P., and Tyack, P. L. (2004). Sperm whale behaviour indicates the use of echolocation click buzzes "creaks" in prey capture. *Proc. R. Soc. Lond., B. Biol. Sci.* **271**, 2239–2247.

Moore, S. E., and Ljungblad, D. K. (1984). Gray whales in the Beaufort, Chukchi and Bering seas: distribution and sound production. *In* "The Gray Whale" (M. L. Jones, S. L. Swartz, and S. Leatherwood, eds), pp. 543–559. Academic Press, New York.

Moore, S. E., Stafford, K. M., Dahlheim, M. E., Fox, C. G., Braham, H. W., Polovin, J. J., and Bain, D. E. (1998). Seasonal variation in reception of fin whale calls at five geographic areas in the North Pacific. *Mar. Mamm. Sci.* **14**, 617–627.

Morisaka, T., and Connor, R. C. (2007). Predation by killer whales (*Orcinus orca*) and the evolution of whistle loss and narrow-band high frequency clicks in odontocetes. *J. Evol. Biol.* **20**, 1439–1458.

Norris, K. S. (1968). The evolution of acoustic mechanisms in odontocete cetaceans. *In* "Evolution and environment" (E. T. Drake, ed.), pp. 297–324. Yale University Press, New Haven.

Norris, K. S., Würsig, B., Wells, R. S., and Würsig, M. (1994). "The Hawaiian spinner dolphin." University of California Press, Berkeley.

O'Shea, T., and Poche, L. B., Jr. (2006). Aspects of underwater sound communication in Florida manatees (*Trichechus manatus latirostris*). *J. Mammal.* **87**, 1061–1071.

Oetelaar, M. L., Terhune, J. M., and Burton, H. R. (2003). Can the sex of a Weddell seal (*Leptonychotes weddellii*) be identified by its surface call? *Aq. Mamm.* **29**, 261–267.

Oleson, E. M., Barlow, J., Gordon, J., Rankin, S., and Hildebrand, J. A. (2003). Low frequency calls of Bryde's whales. *Mar. Mamm. Sci.* **19**, 407–419.

Oleson, E. M., Calambokidis, J., Burgess, W. C., McDonald, M. A., LeDuc, C. A., and Hildebrand, J. A. (2007). Behavioral context of call production by eastern North Pacific blue whales. *Mar. Ecol. Prog. Ser.* **330**, 269–284.

Oswald, J. N., Rankin, S., Barlow, J., and Lammers, M. O. (2007). A tool for real-time acoustic species identification of delphinid whistles. *J. Acoust. Soc. Am.* **122**, 587–595.

Outi, T., Parks, S., and Miller, L. A. (2007). Annual and seasonal changes in the song of the bowhead whales *Balaena mysticetus* in Disko Bay, Western Greenland. 17th Biennial Conference on the Biology of Marine Mammals. Capetown, South Africa.

Page, B., Goldsworthy, S. D., Hindell, M. A., and McKenzie, J. (2002). Interspecific differences in male vocalizations of three sympatric fur seals (*Arctocephalus* spp.). *J. Zool. Lond.* **258**, 49–56.

Parks, S. E., Hamilton, P. K., Kraus, S. D., and Tyack, P. L. (2005). The gunshot sound produced by male North Atlantic right whales (*Eubalaena glacialis*) and its potential function in reproductive advertisement. *Mar. Mamm. Sci.* **21**, 458–475.

Payne, K., and Payne, R. (1985). Large scale changes over 19 years in songs of humpback whales in Bermuda. *Z. Tierpsychol.* **68**, 89–114.

Payne, K., Tyack, P., and Payne, R. (1983). Progressive changes in the songs of humpback whales (*Megaptera novaeangliae*): a detailed analysis of two seasons in Hawaii. *In* "Communication and Behavior of Whales" (R. Payne, ed.), pp. 9–57. Westview Press, Boulder.

Payne, R., and Guinee, L. N. (1983). Humpback whale (*Megaptera novaeangliae*) songs as an indicator of "stocks". *In* "Communication and Behavior of Whales" (R. Payne, ed.), pp. 333–368. Westview Press, Boulder.

Payne, R., and Webb, D. (1971). Orientation by means of long range acoustic signaling in baleen whales. *Ann. N. Y. Acad. Sci.* **188**, 110–142.

Petrochenko, S. P., Potapov, A. S., and Pryadko, V. V. (1991). Sounds, source levels, and behavior of gray whales in the Chukotsoe Sea. *Sov. Phys. Acoust.* **37**, 622–624.

Phillips, A. V., and Stirling, I. (2001). Vocal repertoire of South American fur seals, *Arctocephalus australis*: structure, function, and context. *Can. J. Zool.* **79**, 420–437.

Phillips, R., Niezrecki, C., and Beusse, D. O. (2004). Determination of West Indian manatee vocalization levels and rate. *J. Acoust. Soc. Am.* **115**, 422–428.

Poulter, T. C. (1968). Underwater vocalization and behavior of pinnipeds. *In* "The Behavior and Physiology of Pinnipeds" (R. J. Harrison, R. C. Hubbard, R. S. Peterson, C. E. Rice, and R. J. Schusterman, eds), pp. 69–84. Appleton-Century-Crofts, New York.

Rankin, S., and Barlow, J. (2005). Source of the North Pacific "boing" sound attributed to minke whales. *J. Acoust. Soc. Am.* **118**, 3346–3351.

Rankin, S., Oswald, J., Barlow, J., and Lammers, M. (2007). Patterned burst-pulse vocalizations of the northern right whale dolphin, *Lissodelphis borealis*. *J. Acoust. Soc. Am.* **121**, 1213–1218.

Rasmussen, M. H., and Miller, L. A. (2002). Whistles and clicks from white-beaked dolphins, *Lagenorhynchus albirostris*, recorded in Faxafloi Bay, Iceland. *Aq. Mamm.* **28**, 78–89.

Reidenberg, J. S., and Laitman, J. T. (2007). Discovery of a low frequency sound source in Mysticeti (baleen whales): anatomical establishment of a vocal fold homolog. *Anat. Rec.* **290**, 745–759.

Rendell, L., and Whitehead, H. (2004). Do sperm whales share coda vocalizations?—Insights into coda usage from acoustic size measurement. *Anim. Behav.* **67**, 865–874.

Rendell, L., and Whitehead, H. (2005). Spatial and temporal variation in sperm whale coda vocalizations: stable usage and local dialects. *Anim. Behav.* **70**, 191–198.

Rendell, L. E., and Whitehead, H. (2003). Vocal clans in sperm whales (*Physeter macrocephalus*). *Proc. R. Soc. Lond. B* **270**, 225–231.

**S**

Richardson, W. J., Greene, C. R., Malme, C. I., and Thomson, D. H. (1995). "Marine Mammals and Noise." Academic Press, San Diego.

Risch, D., Clark, C. W., Corkeron, P. J., Elepfandt, A., Kovacs, K. M., Lydersen, C., Stirling, I., and Parijs, S. M. V. (2007). Vocalizations of male bearded seals, *Erignathus barbatus*: classification and geographical variation. *Anim. Behav.* **73**, 747–762.

Rogers, T. L., and Cato, D. H. (2002). Individual variation in the acoustic behaviour of the adult male leopard seal, *Hydrurga leptonyx*. *Behaviour* **139**, 1267–1286.

Rogers, T. L., Cato Douglas, H., and Bryden, M. M. (1996). Behavioral significance of underwater vocalizations of captive leopard seals, *Hydrurga leptonyx*. *Mar. Mamm. Sci.* **12**, 414–427.

Sandegren, F. (1976). Agonistic behavior in the male northern elephant seal. *Behaviour* **57**, 136–158.

Sandegren, F., Chu, E., and Vandevere, J. (1973). Maternal behavior in the California sea otter. *J. Mammal.* **54**, 668–679.

Sanvito, S., Galimberti, F., and Miller, E. H. (2007). Vocal signalling of male southern elephant seals is honest but imprecise. *Anim. Behav.* **73**, 287–299.

Saulitis, E. L., Matkin, C. O., and Fay, F. H. (2005). Vocal repertoire and acoustic behavior of the isolated AT1 killer whale subpopulation in southern Alaska. *Can. J. Zool.* **83**, 1015–1029.

Sayigh, L. S., Tyack, P. L., Wells, R. S., Scott, M. D., and Irvine, A. B. (1995). Sex difference in signature whistle production of free-ranging bottlenose dolphins, *Tursiops truncatus*. *Behav. Ecol. Sociobiol.* **36**, 171–177.

Schusterman, R. J., and Balliet, R. F. (1969). Underwater barking by male sea lions (*Zalophus californianus*). *Nature* **222**, 1179–1181.

Schusterman, R. J., and Van Parijs, S. M. (2003). Pinniped vocal communication: an introduction. *Aq. Mamm.* **29**, 177–180.

Serrano, A., and Terhune, J. M. (2002). Stability of the underwater vocal repertoire of harp seals (*Pagophilus groenlandicus*). *Aq. Mamm.* **28**, 93–101.

Sharpe, F. A. (2001). Social foraging of the Southeast Alaskan humpback whale. Ph.D. Dissertation, Simon Fraser University, Burnaby.

Shipley, C., Hines, M., and Buchwald, J. S. (1981). Individual differences in threat calls of northern elephant seal bulls *Anim. Behav.* **29**, 12–19.

Shipley, C., Hines, M., and Buchwald, J. S. (1986). Vocalizations of northern elephant seal bulls (*Mirounga angustirostris*): development of adult call characteristics during puberty. *J. Mammal.* **67**, 526–536.

Shipley, C., Stewart, B. S., and Bass, J. (1992). Seismic communication in northern elephant seals. *In* "Marine Mammal Sensory Systems" (J. A. Thomas, R. A. Kastelein, and A. Y. Supin, eds), pp. 553–562. Plenum Press, New York.

Silber, G. K. (1986). The relationship of social vocalizations to surface behavior and aggression in the Hawaiian humpback whale *Megaptera novaeangliae*. *Can. J. Zool.* **64**, 2075–2080.

Sjare, B., Stirling, I., and Spencer, C. (2003). Structural variation in the songs of Atlantic walruses breeding in the Canadian High Arctic. *Aq. Mamm.* **29**, 297–318.

Stafford, K. M., Moore, S. E., and Fox, C. G. (2005). Diel variation in blue whale calls recorded in the eastern tropical Pacific. *Anim. Behav.* **69**, 951–958.

Stimpert, A. K., Wiley, D. N., Au, W. W. L., Johnson, M. P., and Arsenault, R. (2007). Megapclicks': acoustic click trains and buzzes produced during night-time foraging of humpback whales (*Megaptera novaeangliae*). *Biol. Lett.* **3**, 467–470.

Stirling, I. (1973). Vocalization in the ringed seal (*Phoca hispida*). *J. Fish. Res. Board Can.* **30**, 1592–1593.

Stirling, I., Calvert, W., and Cleator, H. (1983). Underwater vocalizations as a tool for studying the distribution and relative abundance of wintering pinnipeds in the High Arctic. *Arctic* **36**, 262–274.

Stirling, I., and Siniff, D. B. (1979). Underwater vocalizations of leopard seals (*Hydurga leptonyx*) and crabeater seals (*Lobodon carcinophagus*) near the South Shetland Islands, Antarctica. *Can. J. Zool.* **57**, 1244–1248.

Stirling, I., and Thomas, J. A. (2003). Relationships between underwater vocalizations and mating systems in phocid seals. *Aq. Mamm.* **29**, 227–246.

Strager, H. (1995). Pod-specific call repertoires and compound calls of killer whales, *Orcinus orca* Linnaeus, 1758, in the waters of northern Norway. *Can. J. Zool.* **73**, 1037–1047.

Teloni, V., Zimmer, W. M. X., and Tyack, P. L. (2005). Sperm whale trumpet sounds. *Bioacoustics* **15**, 163–174.

Terhune, J. M. (1994). Geographical variation of harp seal underwater vocalizations. *Can. J. Zool.* **72**, 892–897.

Terhune, J. M., and Dell'Apa, A. (2006). Stereotyped calling patterns of a male Weddell seal (*Leptonychotes weddellii*). *Aq. Mamm.* **32**, 175–181.

Thomas, J. A., and Golladay, C. L. (1995). Geographic variation in leopard seal (*Hydrurga leptonyx*) underwater vocalizations. *In* "Sensory Systems of Aquatic Mammals" (R. A. Kastelein, ed.), pp. 201–222. De Spil, Woerden.

Thompson, P. O., Findley, L. T., and Vidal, O. (1992). 20-Hz pulses and other vocalizations of fin whales, *Balaenoptera physalus*, in the Gulf of California, Mexico. *J. Acoust. Soc. Am.* **92**, 3051–3057.

Thompson, T. J., Winn, H. E., and Perkins, P. J. (1979). Mysticete Sounds. *In* "Behavior of Marine Animals: Current Perspectives in Research" (H. E. Winn and B. L. Olla, Eds.), **3**: Cetaceans, pp. 403–431.

Tripovich, J. S., Rogers, T. L., and Arnould, J. P. Y. (2005). Species-specific characteristics and individual variation of the bark call produced by male Australian fur seals (*Arctocephalus pusillus doriferus*). *Bioacoustics* **15**, 79–96.

Turl, C. W., and Penner, R. H. (1989). Differences in echolocation click patterns of the beluga (*Delphinapterus leucas*) and the bottlenose dolphin (*Tursiops truncatus*). *J. Acoust. Soc. Am.* **86**, 497–502.

Tyack, P., Johnson, M., Zimmer, W. M. X., Aguilar de Soto, N., and Madsen, P. T. (2006). Acoustic behavior of beaked whales, with implications for acoustic monitoring. *Oceans* **September**, 1–6, doi:10.1109/OCEANS.2006.307120.

Tyack, P., and Miller, E. H. (2002). Vocal anatomy, acoustic communication and echolocation. *In* "Marine Mammal Biology: An Evolutionary Approach" (A. R. Hoelzel, ed.), pp. 142–184. Blackwell Publishing, Malden.

Urick, R. J. (1983). "Principles of Underwater Sound." McGraw-Hill Book Company, New York.

Van Parijs, S. M. (2003). Aquatic mating in pinnipeds: a review. *Aq. Mamm.* **29**, 214–226.

Van Parijs, S. M., and Corkeron, P. J. (2001). Evidence for signature whistle production by a Pacific humpback dolphin, *Sousa chinensis*. *Mar. Mamm. Sci.* **17**, 944–949.

Watkins, W. A. (1967). The harmonic interval fact or artifact in spectral analysis of pulse trains. *In* "Marine Bio-Acoustics" (W. N. Tavolga, ed.), **2**, pp. 15–43. Pergamon Press, New York.

Watkins, W. A., and Ray, G. C. (1977). Underwater sounds from ribbon seal, *Phoca fasciata*. *Fish Bull.* **75**, 450–453.

Watkins, W. A., and Ray, G. C. (1985). In-air and underwater sounds of the Ross seal, (*Ommatophoca rossi*). *J. Acoust. Soc. Am.* **77**, 1598–1600.

Watkins, W. A., and Schevill, W. E. (1976). Right whale feeding and baleen rattle. *J. Mammal.* **57**, 58–66.

Watkins, W. A., Tyack, P., and Moore, K. E. (1987). The 20-Hz signals of finback whales (*Balaenoptera physalus*). *J. Acoust. Soc. Am.* **82**, 1901–1912.

Watwood, S. L., Miller, P. J. O., Johnson, M., Madsen, P. T., and Tyack, P. L. (2006). Deep-diving foraging behaviour of sperm whales (*Physeter macrocephalus*). *J. Anim. Ecol.* **75**, 814–825.

Weilgart, L., and Whitehead, H. (1997). Group-specific dialects and geographical variation in coda repertoire in South Pacific sperm whales. *Behav. Ecol. Sociobiol.* **40**, 277–285.

Weilgart, L. S., and Whitehead, H. (1990). Vocalizations of the North Atlantic pilot whale *Globicephala melas* as related to behavioral contexts. *Behav. Ecol. Sociobiol.* **26**, 399–402.

**S**

Weir, C. R., Frantzis, A., Alexiadou, P., and Goold, J. C. (2007). The burst-pulse nature of 'squeal' sounds emitted by sperm whales (*Physeter macrocephalus*). *J. Mar. Biol. Assoc. U.K.* **87**, 39–46.

Wenz, G. M. (1964). Curious noises and the sonic environment in the ocean. *In* "Marine Bio-acoustics" (W. N. Tavolga, ed.), pp. 101–119. Pergamon Press, New York.

Winn, H. E., and Perkins, P. J. (1976). Distribution and sounds of the minke whale, with a review of mysticete sounds. *Cetology* **19**, 1–12.

Zoidis, A. M. *et al.* (8 authors) (2008). Vocalizations produced by humback whale (*Megaptera novaeangliae*) calves recorded in Hawaii. *J. Acoust. Soc. Am.* **123**, 1737–1746.

# South American Aquatic Mammals

### Enrique A. Crespo

## I. South American Marine and Fresh Water Ecosystems

The marine and fresh-water ecosystems of South America are very rich in aquatic mammals. Seventy-one species have been reported to occur within these ecosystems (Table I); most breed locally, and only five species that appear occasionally belong to Antarctic or subantarctic ecosystems (Jefferson *et al.*, 1993). A number of species are found in South America that occur in other parts of the world or the Southern Hemisphere, such as rorquals, ziphiids, and some delphinids. However, 20 species can be considered endemic to the coastal waters or the river systems of South America.

The distribution of marine mammals at sea is related to the distribution pattern of ocean currents that is defined by the characteristics of the major water masses, mainly temperature and salinity. The marine mammal assemblages of South America can be explained in part by the water masses that move around the continent (Fig. 1). However, depth and ocean productivity may also play an important role in the presence, absence, or high concentration of individuals of a given species. Four different water masses each have their own wildlife marine mammal assemblage. These are Humboldt Current, Equatorial Front of the Eastern Tropical Pacific, Malvinas (= Falkland) Current, and Brazil and South Equatorial Atlantic Currents. In addition, a fifth wildlife component is found in continental waters; it is heterogeneous due to the isolation between some of the river basins. Finally, a sixth assemblage that could be defined as "erratic circumpolar" can also be found. However, it is composed of isolated individuals from Antarctic or sub-antarctic populations that breed or live southward of the Polar Front but move erratically in northern water masses.

## II. Marine Mammals of Cold Water Marine Ecosystems

In the extreme south of the continent, the Antarctic Circumpolar Current moves from West to East and splits into two branches: the Malvinas Current in the Atlantic and the Humboldt Current in the Pacific. The cold marine ecosystem in the Pacific almost reaches the equator with waters between 8 and 15°C, but in the Atlantic, the cold-temperate system reaches only to 40°S. Off Perú an upwelling system gives rise to high levels of primary and secondary productivity.

Several cold-water marine mammals are found in both the Humboldt and the Malvinas Currents (Cárdenas *et al.*, 1986). Among those species the most common in coastal waters are two otariids (the South American sea lion *Otaria flavescens* and the South American fur seal *Arctocephalus australis*) and two small cetaceans (the dusky dolphin *Lagenorhynchus obscurus* and Burmeister's porpoise *Phocoena spinipinnis*). Other small cetaceans like the dolphins of the genus *Lagenorhynchus* (*L. australis* and *L. cruciger*, the latter more pelagic and less known) and the southern right whale dolphin (*Lissodelphis peronii*) can be included. Two related species, the Chilean dolphin (*Cephalorhynchus eutropia*) in the Pacific and the Commerson's dolphin (*C. commersonii*) in the Atlantic are endemic to the southern parts of the ecosystems.

One of the most conspicuous species in the southwestern Atlantic is the southern right whale (*Eubalaena australis*). With a geographic distribution between 20 and 55°S, one of the highest breeding concentrations is at Península Valdés (42°S) (Cappozzo *et al.*, 1991). After long-term depletion of its population size; it is now recovering at rates over 7%, like other stocks in South Africa, Australia, and New Zealand. The whales can be observed in several places in the Atlantic: Uruguay, southern Brazil, and Buenos Aires Province in Argentina. At Santa Catarina, Brazil, a new breeding area was established and connected to Península Valdés. On the Pacific side, there are signs of recovery and a possible northward extension of the distribution range. The stocks of Península Valdés and South Africa use the waters around South Georgia as a feeding ground.

The spectacled porpoise (*Phocoena dioptrica*) is known from the eastern coast of South America and several subantarctic islands, and the South American marine otter (*Lontra felina*) is known from Perú to Staten Island in the southern South Atlantic. Two fur seals (*Arctocephalus galapagoensis* and *A. phillippi*) are endemic to the Galápagos and the Juan Fernández Archipelagos, respectively. The latter is also found in a few other places in Perú and Chile. The Galápagos are also home to an endemic sea lion, *Zalophus wollebaecki*.

The long-finned pilot whale (*Globicephala melas*), Risso's dolphin (*Grampus griseus*) and killer whale (*Orcinus orca*) can be included in the cosmopolitan species with locally abundant populations. Eight species of Balaenopteridae and eight ziphiids are common to cold waters of both sides of South America. However, the pygmy beaked whale *Mesoplodon peruvianus* has been recorded only from Peruvian waters of South America (also recorded from México and California) and a recently described species, Bahamonde's beaked whale *M. bahamondi,* from the Juan Fernández Archipelago, Chile.

The dynamics of oceanographic and biological processes that sustain the high productivity of the Peruvian ecosystem can be disturbed by what has been called the El Niño southern oscillation (ENSO event), whose main characteristic is the inflow of tropical waters into the upwelling region around December (Reyes, 1992). The nature of ENSO is irregular and unpredictable, and the impact on the intermediate levels of the food chain (e.g., abundance of anchovies) affects seabirds and marine mammals. Demonstrated effects of ENSO events on fur seals, sea lions, dusky dolphins, and seabirds, have included those on survival, recruitment, and the general condition of the individuals as a consequence of reduced food availability. An ENSO event is part of a more general pattern of oceanographic change affecting not only the Peruvian ecosystem but also the entire Southern Ocean.

**S**

TABLE I
Recorded Presence of Species in South America Marine and Fresh Water Ecosystems

| TAXA | Cold-Temper. Pacific | Tropical Pacific | Tropical Atlantic Caribbean | Cold-Temper. Atlantic | Fresh Waters | Erratic circumpolar |
|---|---|---|---|---|---|---|
| **Sirenidae** | | | | | | |
| 1   Trichechidae | | | | | | |
| 2     *Trichechus manatus* | | | XXX | | | |
| 3[a]   *Trichechus inunguis* | | | | | XXX | |
| **Carnivora** | | | | | | |
|   Pinnipedia | | | | | | |
|     Otariidae | | | | | | |
| 4[a]   *Otaria flavescens* | XXX | | | XXX | | |
| 5[a]   *Zalophus wollebaeki* | XXX | | | | | |
| 6[a]   *Arctocephalus australis* | XXX | | | XXX | | |
| 7[a]   *Arctocephalus philippii* | XXX | | | | | |
| 8     *Arctocephalus gazella* | | | | | | XXX |
| 9[a]   *Arctocephalus galapagoensis* | XXX | | | | | |
| 10   *Arctocephalus tropicalis* | | | | | | XXX |
|     Phocidae | | | | | | |
| 11   *Mirounga leonina* | XXX | | | XXX | | XXX |
| 12   *Lobodon carcinophagus* | | | | | | XXX |
| 13   *Hydrurga leptonyx* | | | | | | XXX |
| 14   *Leptonychotes weddellii* | | | | | | XXX |
| 15   *Monachus tropicalis* | | | **Extinct** | | | |
|   Fissipedia | | | | | | |
|     Mustelidae | | | | | | |
| 16[a]   *Lutra felina* | XXX | | | XXX | | |
| 17[a]   *Lutra provocax* | | | | | XXX | |
| 18[a]   *Lutra longicaudis* | | | | | XXX | |
| 19[a]   *Pteronura brasiliensis* | | | | | XXX | |
| **Cetacea** | | | | | | |
|   Mysticeti | | | | | | |
|     Balaenidae | | | | | | |
| 20   *Eubalaena australis* | XXX | | | XXX | | |
|     Neobalaenidae | | | | | | |
| 21   *Caperea marginata* | XXX | | | XXX | | |
|     Balaenopteridae | | | | | | |
| 22   *Balaenoptera musculus* | XXX | XXX | XXX | XXX | | |
| 23   *Balaenoptera physalus* | XXX | XXX | XXX | XXX | | |
| 24   *Balaenoptera borealis* | XXX | XXX | XXX | XXX | | |
| 25   *Balaenoptera edeni* | XXX | XXX | XXX | XXX | | |
| 26   *Balaenoptera acutorostrata* | XXX | XXX | XXX | XXX | | |
| 27   *Megaptera novaeangliae* | XXX | XXX | XXX | XXX | | |
|   Odontoceti | | | | | | |
|     Physeteridae | | | | | | |
| 28   *Physeter macrocephalus* | XXX | XXX | XXX | XXX | | |
|     Kogiidae | | | | | | |
| 29   *Kogia breviceps* | XXX | XXX | XXX | XXX | | |
| 30   *Kogia sima* | XXX | XXX | XXX | XXX | | |

S

*(continues)*

TABLE I (continued)

| TAXA | Cold-Temper. Pacific | Tropical Pacific | Tropical Atlantic Caribean | Cold-Temper. Atlantic | Fresh Waters | Erratic circun-polar |
|---|---|---|---|---|---|---|
| Ziphiidae | | | | | | |
| 31 Berardius arnuxii | XXX | | | XXX | | |
| 32 Ziphius cavirostris | XXX | XXX | XXX | XXX | | |
| 33 Hyperoodon planifrons | XXX | | | XXX | | |
| 34 Tasmacetus shepherdi | XXX | | | XXX | | |
| 35 Mesoplodon densirostris | XXX | XXX | XXX | XXX | | |
| 36 Mesoplodon grayi | XXX | | | XXX | | |
| 37 Mesoplodon hectori | XXX | | | XXX | | |
| 38 Mesoplodon peruvianus | XXX | | | | | |
| 39 Mesoplodon europaeus | | | XXX | | | |
| 40 Mesoplodon layardii | XXX | | | XXX | | |
| 41[a] Mesoplodon bahamondi | XXX | | | | | |
| 42 Mesoplodon gingkodens | XXX | | | | | |
| Delphinidae | | | | | | |
| 43 Orcinus orca | XXX | XXX | XXX | XXX | | |
| 44 Globicephala melas | XXX | | | XXX | | |
| 45 Globicephala macrorhynchus | XXX | XXX | XXX | XXX | | |
| 46 Pseudorca crassidens | XXX | XXX | XXX | XXX | | |
| 47 Feresa attenuata | XXX | XXX | XXX | XXX | | |
| 48 Peponocephala electra | | XXX | XXX | XXX | | |
| 49[a] Sotalia fluviatilis | | | | | XXX | |
| 50[a] Sotalia guianensis | | | XXX | | | |
| 51 Steno bredanensis | | XXX | XXX | | | |
| 52 Lagenorhynchus obscurus | XXX | | | XXX | | |
| 53 Lagenorhynchus cruciger | XXX | | | XXX | | |
| 54[a] Lagenorhynchus australis | XXX | | | XXX | | |
| 55 Grampus griseus | XXX | XXX | XXX | XXX | | |
| 56 Tursiops truncatus | XXX | XXX | XXX | XXX | | |
| 57 Stenella attenuata | | XXX | XXX | | | |
| 58 Stenella frontalis | | | XXX | | | |
| 59 Stenella longirostris | | XXX | XXX | | | |
| 60 Stenella clymene | | | XXX | | | |
| 61 Stenella coeruleoalba | | XXX | XXX | | | |
| 62 Delphinus delphis | XXX | XXX | XXX | XXX | | |
| 63 Delphinus capensis | XXX | XXX | XXX | XXX | | |
| 64 Lagenodelphis hosei | | XXX | XXX | XXX | | |
| 65 Lissodelphis peronii | XXX | | | XXX | | |
| 66[a] Cephalorhynchus commersonii | | | | XXX | | |
| 67[a] Cephalorhynchus eutropia | XXX | | | | | |
| Phocoenidae | | | | | | |
| 68[a] Phocoena dioptrica | XXX | | | XXX | | |
| 69[a] Phocoena spinipinnis | XXX | | | XXX | | |
| Iniidae | | | | | | |
| 70[a] Inia geoffrensis | | | | | XXX | |
| Pontoporiidae | | | | | | |
| 71[a] Pontoporia blainvillei | | | XXX | XXX | | |

[a]Endemic species.

S

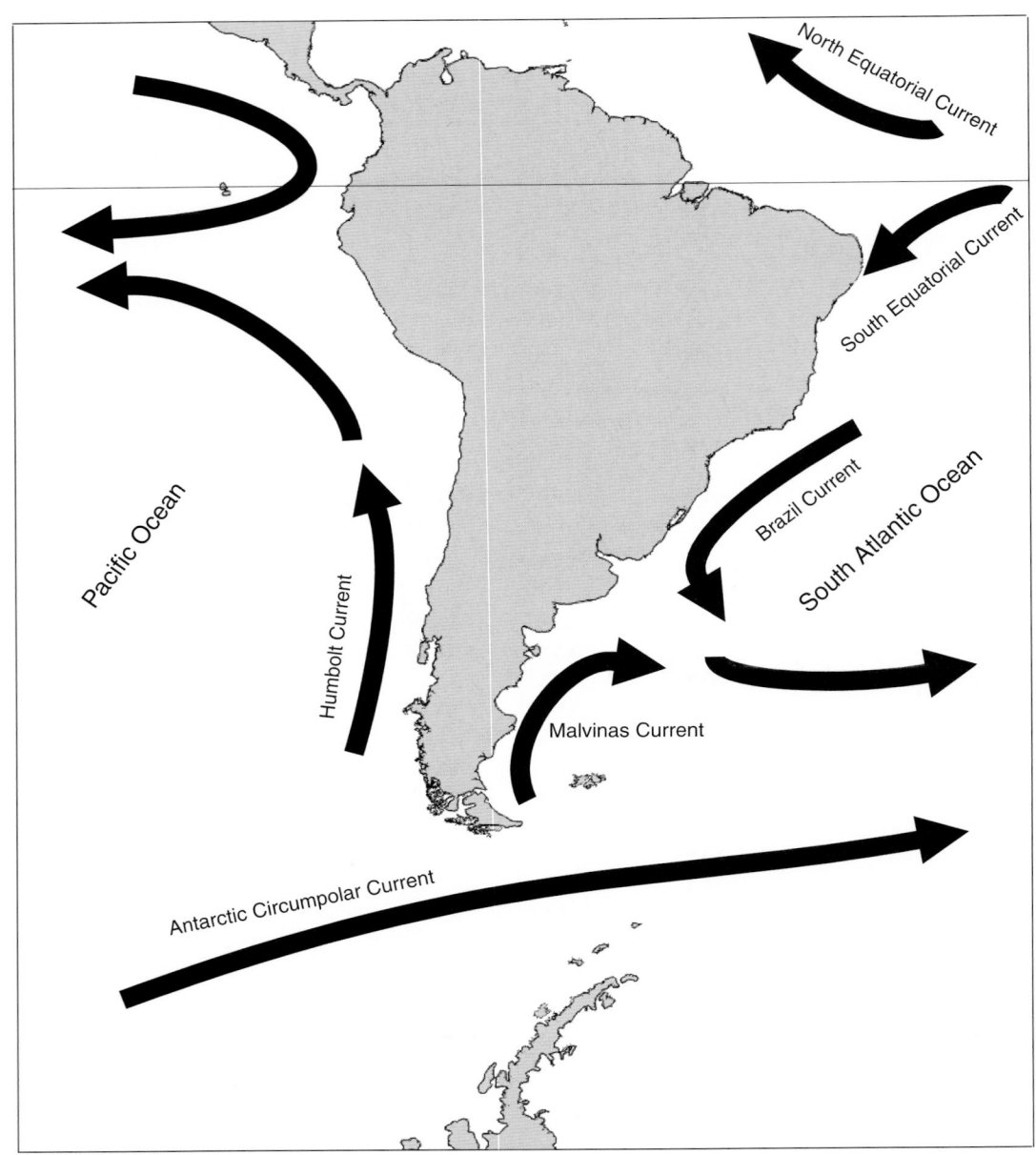

**Figure 1**    *Water currents around the South American continent.*

## III. Marine Mammals in the Tropical Water Ecosystems

On the Atlantic side, the South Equatorial Current moves from east to west and turns southward, becoming the tropical Brazil Current with exclusive influence up to 28°S. The northern Equatorial Current moves northward and turns clockwise in the Caribbean zone. Between 33 and 40°S, the South Atlantic marine ecosystem shows mixed characteristics of the opposing flows of the Malvinas and Brazil Currents. In addition, this region of the Atlantic receives the input of large continental runoffs like La Plata River and Los Patos Lagoon. On the Pacific side of South America, the Humboldt Current turns to the West after forming the Tropical Front around 5°S.

Both the Atlantic and Pacific tropical systems show species assemblages typical of warmer waters globally (Piñedo *et al.,* 1992; Reyes, 1992; Vidal, 1992). They include dolphins of the genus Stenella, (*S. attenuata, S. longirostris* and *S. coeruleoalba),* common dolphins (*Delphinus delphis* and *D. capensis),* melon-headed whale (*Peponocephala electra),* rough-toothed dolphin (*Steno bredanensis),* Fraser's dolphin (*Lagenodelphis hosei),* Bryde's whale (*Balaenoptera borealis),* both species of Kogidae, short-finned pilot whale (*G. macrorhynchus),* pygmy killer whale (*Feresa attenuata),* and false killer whale (*Pseudorca crassidens),* among others. The Clymene dolphin (*Stenella clymene),* Atlantic spotted dolphin (*S. frontalis),* marine tucuxi (*Sotalia guianensis),* and Atlantic manatee (*Trichechus manatus)* are found only in the Atlantic. Some species, such as the common dolphins, false killer whale, and bottle-nosed dolphin (*Tursiops truncatus*), range from the tropics far south in the Atlantic, in the area of mixed waters of the Brazil and Malvinas Currents. Gervais' beaked whale (*Mesoplodon europaeus*) is known from the east coasts of North America and Caribbean islands, and the eastern Atlantic Ocean.

## IV. Aquatic Mammals in the River Basins

South America is very rich in river basins, and all the aquatic mammals in continental waters are endemic (Olson *et al.*, 1998). The most important hydrographic systems are the Amazon, the Orinoco, and the Paraná-Uruguay rivers. However, other small river systems are located in cold and high-latitude parts of the continent. Most of those systems are isolated from each other. The Amazon and Orinoco basins hold important populations of aquatic mammals. The most conspicuous include the boto (*Inia geoffrensis*), a freshwater population of the tucuxi (*Sotalia fluviatilis fluviatilis*) and the Amazon manatee (*Trichechus inunguis*). Two otters are also found in those basins and also in the Paraná–Uruguay River systems: the giant otter (*Pteronura brasiliensis*) and the long-tailed otter (*Lontra longicauda*) (IUCN, 1991). The cold river systems in southern Argentina and Chile sustain populations of the southern river otter (*Lontra provocax*) which is also found on Staten Island at the extreme southern end of the continent.

## V. Occasional Visitors from the Antarctic

Five species of pinnipeds that breed on subantarctic islands or the Antarctic ice (the Antarctic fur seal *Arctocephalus gazella*, subantarctic fur seal *A. tropicalis*, leopard seal *Hydrurga leptonyx*, crabeater seal *Lobodon carcinophagus*, and Weddell seal *Leptonychotes weddellii*) move erratically to more northerly waters of both sides of South America. Probably as a consequence of population increases over the last few decades, it has become more frequent to sight individuals moving as far north as the Equator. The southern elephant seal (*Mirounga leonina*) can be included in this list; while most breeding groups are circumpolar, there is an important and increasing breeding stock at Península Valdés, Patagonia, around 42°S in the Atlantic Ocean.

## VI. Problems Faced by Aquatic Mammals in South America

The problems faced by marine mammals in South America are much like those faced in other parts of the world. They include incidental catch in fisheries, direct exploitation, competition for fishing resources, and habitat loss and degradation (Reeves *et al.*, 2003; Hucke-Gaete *et al.*, 2004). Most of the species in South America are insufficiently known. For most of them, there is no information about abundance and population trends. Exceptions include the pinnipeds, the southern right whale, and several species of small cetaceans such as the dusky and Commerson's dolphins in Patagonia and river dolphins in the Amazon. However, abundance has been estimated for only small parts of their ranges.

While large whales are fully protected, giving them the opportunity to recover, a long list of species of small cetaceans are taken incidentally in coastal and high-seas fisheries both in the tropics and in cold-water ecosystems. The most critical situation is probably that of the franciscana, *Pontoporia blainvillei* (Crespo, this book, CROSS REFERENCE). However, there are other species or local stocks, which interact heavily with different types of fisheries in direct and incidental takes. Examples include the marine tucuxi in some parts of Brazil, local populations of dusky dolphins in Perú and Patagonia, both species of common dolphins and Burmeister's porpoises in Perú, Commerson's dolphins in Patagonia, and Chilean dolphins in southern Chile. The problem of incidental catch of aquatic mammals in fishing gear has not been addressed as part of fishery management in the countries of the region.

Direct takes mostly for food, oil, or bait also include the use of genital organs for aboriginal beliefs in the case of small cetaceans, and illegal small-scale commerce between South American and Asian countries in a black market in the case of otariids. Economic reasons were involved in turning an incidental into a direct catch in Perú during the 1980s. These conditions can be amplified with poverty and hyperinflation. Direct takes are also a matter of concern with other aquatic groups, such as the hunting of river otters for furs in the river systems.

Regarding competition for fishing resources, there is a general perception by fishermen around the world that some marine mammals, mostly pinnipeds, are currently depleting the target species of fisheries. The perception is sustained by the relative increase of marine mammal populations in the past several decades, and the decrease of fishing stocks in the same period. In South America, this is thought to be the case with the South American sea lion. Even though there has been no formal proposal for culling in the countries of the region, with the exception of Perú where the problem has been discussed by the GOVERNMENT, the private sector, and NGOs, sea lions are shot illegally by fishermen in many places throughout their range. There is also a general belief that reducing the populations of competitors will increase the stocks of the target species of the fishery, a supposition that to date has not been supported by data; culling carried out around the world has not increased levels of target species or catches.

Loss of habitat is the most important problem that faces freshwater and coastal species. Indiscriminate clearing of rainforest for the use of land in agriculture and cattle (*Bos taurus*) growing is currently going on, and many species are declining sharply. During the past years, this process was accelerated by large international demand of soya (*Glycine max*), to which the countries of the region responded with increases of the agricultural frontier at the expense of rainforests. Dams and other barriers in large rivers have been or are being planned for hydroelectric power, irrigation, or flood control. These projects have significant impact on regional development and positive benefits for society. Nevertheless, negative consequences in the river ecosystem for river dolphins, manatees (*Trichechus* spp.), river otters, and other wildlife include the isolation of populations of aquatic mammals and their prey, and unnatural water flows and interruption of migratory paths. In addition to dam construction, other threats include pollution, mining, use of dynamite for fishing, and incidental and direct catches in fisheries. In coastal areas, the most important problems include intensive fishing and pollution by hydrocarbons, agrochemical and heavy metal products, and intensive boat traffic. Recently, construction of buildings for tourism has rapidly increased in large portions of coastline, with detrimental consequences on the coastal ecosystem. The importance of each of these varies with the area considered.

S

## VII. The Need for International Agreements in Conservation Policies

With the exception of endemic species, most South American aquatic mammals have a wide distribution and occur or breed passing through political boundaries, e.g., those of Brazil, Perú, Uruguay, Argentina, and Chile. Legal protection or status is not the same in each country and sometimes there is little or no enforcement. For example, the South American sea lion is shot by fishermen in southern Brazil and Uruguay, is one of the tourist targets in Argentine Patagonia, is allowed to be killed if necessary in Chile, and is a potential culling target in Perú. The dusky dolphin is incidentally and directly taken in Perú and northern Chile, and incidentally taken in Patagonia. South American fur seals and Commerson's and Peale's (*Lagnorhynchus australis*) dolphins were used for crab bait for many years in the southern tip of South America. Franciscana cross the boundaries of Brazil, Uruguay, and Argentina, and while they are protected by law in the three countries, there is no enforcement and incidental mortality is a major cause of concern.

At the national level, incidental mortality or bycatch should be considered in fishery management models and decision making. At the international level, agreements should be promoted among the countries of the region in order to give general or particular status of protection for a given species. For example, in 1991, Colombia, Chile, Ecuador, Panamá, and Perú approved the Action Plan for the Conservation of Marine Mammals in the Southeast Pacific, in order to help the governments of the region to agree on appropriate policies for marine mammal conservation and management. This agreement has little or no effective enforcement, and the Atlantic region even still lacks such an agreement.

## *See Also the Following Article*

Distribution

## *References*

Cardenas, J. C., Stutzin, M. E., Oporto, J. A., Cabello, C., and Torres, D. (1986). "Manual de Identificación de los cetáceos chilenos." WWF/CODEFF, Santiago, Chile, Abril 1986.

Capozzo H. L., and Junin, M. (eds.) (1991). Estado de conservación de los mamíferos marinos del Atlántico Sudoccidental. Informes y estudios del Programa de Mares Regionales del PNUMA N° 138, PNUMA.

Hucke-Gaete, R., Crespo, E. A, and Schlatter, R. P. (2004). Aquatic Mammals in Latin America: Proceedings of the Workshop on Identifying High-Priority Conservation Needs and Actions. Valdivia, Chile, 18–19 October, 2002. Convention on the Conservation of Migratory Species of Wild Animals.

IUCN (1991). "Plan de Acción para las Nutrias de Latinoamérica" (P. Foster-Turtley, S. Macdonald, and C. Mason, eds). IUCN/SSC Otter Specialist Group Gland, Switzerland.

Jefferson, T. A., Leatherwood, S., and Weber, M. A. (1993). "FAO Species Identification Guide: Marine Mammals of the World." FAO, Rome.

Olson, D., *et al.* (8 authors) (1998). "Freshwater biodiversity of Latin America and the Caribbean: a conservation assessment." Biodiversity Support Program, Washington, D.C.

Pinedo, M. C., Rosas, F. C. W., and Marmontel, M. (1992). "Cetáceos e pinnípedes do Brasil." UNEP/UEFA, Manaus, Brasil.

Reeves, R. R., Smith, B. D., Crespo, E. A., and Notarbartolo di Sciara, G. (2003). "Dolphins, whales, and porpoises: 2002–2010 Conservation Action Plan for the World's Cetaceans." IUCN/SSC Cetacean Specialist Group, IUCN, Gland.

Reyes, J. C. (1992). Informe nacional sobre la situación de los mamíferos marinos en Perú. Informes y Estudios del Programa de Mares Regionales del PNUMA N° 145, PNUMA.

Vidal, O. (1992). Los mamíferos marinos del Océano Pacífico Sudeste (Panamá, Colombia, Ecuador, Perú y Chile): diagnostico regional. Informes y Estudios del Programa de Mares Regionales del PNUMA N° 142, PNUMA.

# South American Sea Lion
## *Otaria flavescens*

HUMBERTO LUIS CAPPOZZO AND WILLIAM F. PERRIN

## I. Characteristics and Taxonomy

The scientific name of the South American sea lion has been under discussion for many years. Two names were in use until a few years ago: *Otaria flavescens* (Shaw, 1800) and *Otaria byronia* (de Blainville, 1820). More recently, the use of *O. flavescens* has been recommended; this is the name used throughout the distribution area in South America (Rodriguez and Bastida, 1993). The common name changes with location: *lobo marino de un pelo, león marino del sur, lobo marino del sur, lobo común, lobo chusco, leâo marinho,* or *lobo marinho de um pelo.* "South American sea lion" is preferable to "Southern sea lion," as it prevents confusion with Australian and New Zealand sea lions, species also distributed in the Southern Hemisphere.

This species is one of seven that make up the subfamily Otariinae, part of the family Otariidae, usually called otariids, or pinnipeds with ears (Riedman, 1990). It is one of the largest and most dimorphic otariids (Fig. 1). Adult males are much heavier than females. Differences in size between males and females have also been documented among juveniles, and even newborns (Cappozzo *et al.*, 1991). Adult males reach a maximum length of around 3 m and weigh 300–350 kg; adult females are about 2 m long and weigh up to 150 kg. The newborn sea lion usually weighs between 12 and 15 kg and is 0.75–0.86 cm long. Adult males have a characteristic hairy, thick neck. The color is generally brownish, from the very dark brown of adult males to almost yellow in females. Pups are black at birth. After the molt of the embryonic coat (*lanugo*) at 2 months, the first juvenile HAIR is reddish brown, changing in color with age and sex.

## II. Distribution and Abundance

South American sea lions are widely distributed along the Atlantic and Pacific coasts of South America: from Torres in southern Brazil to Cape Horn in the extreme south of the Atlantic coastline, and from Cape Horn to Zorritos in northern Peru, on the Pacific (Reyes *et al.*, 1999; Cappozzo and Rosas, 1991). Total population has been estimated at 110,000 for the southwestern Atlantic coast, concentrated mainly on the Patagonian coast and southern islands. The population in southern Argentina was estimated at 22,157 from surveys in 1992–1997 (Schiavini *et al.*, 2004) and is thought to be increasing (Dans *et al.*, 2004). The population at the Falkland Islands (Las Malvinas) is estimated to have reached less than 1.5% of its pre-sealing size by 1990, with 2034 pups produced in 1995 (Thompson *et al.*, 2005). There is no reliable information concerning the Pacific population, but it is considered smaller than that on the Atlantic side.

In Argentina, the species was widely distributed from the La Plata river to Cape Horn, but at present only two subadult male rookeries

**Figure 1**  *South American sea lion.*

remain in Buenos Aires Province: at Mar del Plata and Quequén harbors, as well as a breeding rookery at Isla Trinidad, 39°S. To the south, they breed along the Patagonian coast, from Punta Bermeja to Tierra del Fuego. There are 18 breeding and non-breeding colonies between Punta Bermeja and Punta Leon, in Northern Patagonia, for which a flow of individuals showing a seasonal pattern has been reported. There is a gap of about 200–300 km between the northern and the central Patagonian stocks. The main concentration of this species occurs in central and southern Patagonia, where there are more than 53 breeding and non-breeding rookeries.

A study analyzed the genetic variability between two rookeries, Isla de Lobos, Uruguay, and Punta Norte, Península Valdés, Argentina; the results suggested that both rookeries belong to the same population (Szapkievich *et al.*, 1999) and that therefore the overall population of the southwestern Atlantic is apparently homogeneous, with movements between rookeries. However, a later analysis of samples from the Atlantic and Pacific suggest that not only do populations in the oceans represent different ESUs but that differences in haplotype frequencies between the two clusters of colonies in Uruguay and Patagonia indicate that they should be considered as different conservation stocks (Tunez *et al.*, 2007).

Variation of the substrate and weather phenomena affect DISTRIBUTION. For example, the occurrence of El Niño on the Peruvian Pacific coast influences the whole marine ecosystem by a drastic superficial thermal inversion that affects the entire trophic web from the plankton up to the top predators. Every time El Niño occurs, there is a drastic population decrease due to higher death rates and migration. Also, on the coasts of southern Brazil and Buenos Aires, Argentina, there has been a great reduction of habitat because of human use of the zone.

## III. Ecology

The South American sea lion eats mainly demersal and benthic species, including fish and squids, but its diet is very variable and it adapts easily to locally abundant prey, including crustaceans and even penguins (Alonso *et al.*, 2000). In central Chile, the prey is mainly pelagic fishes, including jack mackerel (*Trachurus murphyi*) (Hückstadt *et al.*, 2007). Off Peru, a larger diversity of prey species (especially of demersal fishes) was consumed during El Niño, when achovy and lobster were less abundant (Soto *et al.*, 2006). Satellite tracking in Patagonia has revealed that both sexes forage in the temperate waters of the continental shelf, although males traveled farther than females and came closer to the edge of the shelf (Campagna *et al.*, 2001). Trips lasted a mean of 3.4 days and distance traveled averaged 206 km (maximum 864 km). Depths of dives were in the range of 2–30 m.

In some areas of their distribution along the South American coasts, South American sea lions live sympatrically with South American fur seals, but they do not compete with them for breeding space because their respective breeding seasons are out of synchrony. Killer whales prey on sea lions that are at sea and have been seen at Península Valdés preying on groups at the shore by surging out of the water onto the beach and returning with a sea lion in their jaws. They have also been observed preying on sea lions during jack-mackerel fishing operations in central Chile (Huckstadt and Antezana, 2004). Sharks have also been seen eating sea lions in Uruguay.

## IV. Behavior and Physiology

Adult males and females arrive at the breeding rookeries during the first half of December (Campagna, 1985; Campagna and Le Boeuf, 1988a). The males defend a position on the central breeding area, and during the peak of the breeding season they also defend females in ESTRUS. In other rookeries such as Puerto Pirámide (also located at Península Valdés), males defend the territories where females go to mate. In a female-defense polygynous mating system, with males capable of forcing copulation, the female's first priority would be to survive the breeding season and then to mate with high-quality males. Males attempt to mate with as many females as possible. Adult males may maximize their reproductive success through the selective defense of those females that are close to estrus. Adult females develop choice of the male with which they copulate by changing the associated male before giving birth or by mating with more than one male during their maximum sexual receptivity. Thermoregulatory requirements interact with rookery topography to shape mating strategies, variation in mating success, and the mating system type (Campagna and LeBoeuf, 1988b). At Punta Norte, the pebble substrate is homogeneous regarding thermoregulatory advantages. Thus, sites advantageous for THERMOREGULATION

**S**

are not a limited resource that can be used to attract females. Consequently, males acquire mates by selective female defense or abduction. However, at the Puerto Pirámide rookery, a variation in the quality of the substrate with respect to reducing thermal stress favors the development of a territorial system where the best territories contain water or are close to the water. Here, the abduction of females or direct defense of females by males is not required, as females preferentially gather in wet territories. Thus, the topography and substrate of the breeding area, along with thermoregulatory requirements, are driving forces that generate adaptive changes in male mating behavior. Substrate has also been shown to be a factor in degree of agonistic behavior among females (Cassini and Fernando-Juricic, 2007).

Males that are sexually mature but cannot compete gather in "bachelor clubs" close to the main breeding area. These subadult or non-territorial males may develop alternative mating strategies: group raids, solitary breeding (single male with a single female or with a small isolated harem), or female interception (keeping females that leave the main breeding area on the way to and from the water) (Campagna et al., 1988).

A cost–benefit model applied to data from a breeding colony of South American sea lions (Cassini and Fernanez-Juricic, 2003) suggests that gregariousness of females reduces the reproductive costs of interacting with males and that this dilution effect may have played an important role in the evolution of this species' mating system.

There is little information regarding migrations and seasonal movements. To the north of the Uruguayan breeding grounds, in southern Brazil, there are only two non-breeding haul-out rookeries where subadult males predominate, with seasonal movements. Aberrant records for the species have been reported for Rio de Janeiro, Brazil at 23°S, and even to 13°S but these are always solitary individuals.

South American sea lions do not spend long periods away from the coast, as many pinnipeds do, and they gather in groups or colonies. There are both permanent and nonpermanent colonies; the latter are mainly reproductive settlements. At the permanent colonies, during the breeding season, the individuals that do not participate in the reproductive activity stay in the colony segregated in non-reproductive groups.

## V. Life History

The sex ratio at birth is 1:1. Males become sexually mature in their sixth year of age, whereas females produce their first offspring about the fifth year or before. The reproductive behavior of marked individuals was studied for 10 years at Punta Norte rookeries in Peninsula Valdés, Argentina.

The breeding and pupping season begins in mid-December and extends to early February (Campagna, 1985). Most of the pups are born during January, usually 2–3 days after the mother's arrival at the rookery. Copulations occur on land 6 days after parturition, and females spend 2–3 days more with their pups fasting. Mothers begin leaving their pups temporarily and foraging offshore at between 1 and 4 days; each foraging trip is followed by 2 days of nursing bouts on land. Lactation continues for 8–10 months, although it is not unusual to find females with un-weaned yearlings. Lactating females spend around 53% of the time at sea diving, with median and maximum depths ranging from 19 to 62 and from 97 to 175 m, respectively (Werner and Campagna, 1995). The pups wander about and tend to gather in groups; they spend most of the time sleeping or playing. The number of pups in these groups increases as the season goes on. When the mother returns to the colony, she calls her pup and they recognize each other by sound and smell. Each female nurses only her own pup. In 10 years of study at Punta Norte, Península Valdés, only three cases of true adoption were documented, although it is not infrequent to see pups suckling from females other than their mothers, "stealing" milk.

The death rate of pups is usually high, with main reasons being the separation from their mothers, starvation, being crushed by large males where the ground is hard, abuse from juveniles, or predation by other species. Both pup mortality and timing of birth are effected by prey availability influenced by El Niño (Soto et al., 2004).

The degree of dimorphism of South American sea lion pups is similar to that of other otariids. Newborn males are 9–18% heavier than females at Península Valdés. Assuming that there is no sex difference in energy expenditure by fetuses, sexual dimorphism in the size of newborn pups suggests that South American sea lion mothers invest more energy in sons than in daughters during gestation. Contrary to other otariids, in which males increase in mass faster than females, no sex difference was found in growth rates among South American sea lions. The size dimorphism present at birth in this species persists during the nursing period, suggesting that sons continue to be more costly to their mothers than daughters. It is not known if differences in size found at birth in the South American sea lions of Península Valdés persist until weaning, but there is some evidence suggesting that this may be true.

Six-month-old male pups at the Malvinas (Falkland) Islands are longer than females; the difference in length remains in 18-month-old individuals and is even more marked in older juveniles. Data for other otariids show that sexual dimorphism at birth continues through weaning.

## VI. Human Interactions

Human exploitation has been hard on pinnipeds. Some species were hunted down to extinction (Caribbean monk seal, *Monachus monachus*) or barely survive (Mediterranean monk seal, *Monachus schauinslandi*), whereas others were spared thanks to the remoteness of their location, as the Antarctic phocids and the fur seals at Patagonian islands. Pinnipeds in general were killed to obtain oil, fur, and meat either for subsistence or for commercial purposes. The South American sea lion was hunted mainly for oil, as its fur is not as valuable as that of the South American fur seal (*Arctocephalus australis*). In Southern Chile it is still captured by fishermen as bait for the southern king crab. It is also hunted by fishermen who regard it suspiciously, as sea lions are supposed to compete for the same fishes they seek and to damage fishing gear. The South American sea lion endured long periods of exploitation with a severe reduction in ABUNDANCE. Although it is estimated that only 20% of their historical number remains, recent signs of recovery have been reported (Crespo and Pedraza, 1991).

More recently, depredation on fishery catches by sea lions has arisen as a problem in artisanal fisheries in Uruguay (Szteren and Paez, 2002) and Chile (Sepulveda et al., 2007) and commercial purse-seining for jack mackerel off central Chile (Huckstadt and Antezana, 2003; Huckstadt and Krautz, 2004), and attacks on salmon in Chilean fish farms has led to a search for effective mitigation measures (Sepulveda and Oliva, 2005).

## See Also the Following Articles

Mating Systems ■ Rookeries ■ Sexual Dimorphism ■ Southern Fur Seals

## References

Alonso, M. K., Crespo, E. A., Pedraza, , Garcia, N. A., and Coscarella, M. A. (2000). Food habits of the South American sea lion, *Otaria flavescens*, off Patagonia, Argentina. *Fish. Bull. U.S.A.* **98**, 250–263.

Campagna, C. (1985). The breeding cycle of the southern sea lion, *Otaria byronia*. *Mar. Mamm. Sci.* **1**, 210–218.

Campagna, C., and Le Boeuf, B. J. (1988a). Reproductive behaviour of southern sea lions. *Behaviour* **104**, 233–261.

Campagna, C., and Le Boeuf, B. J. (1988b). Thermoregulatory behaviour of southern sea lions and its effects on mating strategies. *Behaviour* **107**, 72–90.

Campagna, C., Le Boeuf, B. J., and Cappozzo, H. L. (1988). Group raids: a mating strategy of male southern sea lions. *Behaviour* **105**, 224–249.

Campagna, C., Werner, R., Karesh, W., Marin, M. R., Koontz, F., Cook, R., and Koontz, C. (2001). Movements and location at sea of South American sea lions (*Otaria flavescens*). *J. Zool.* **255**, 205–220.

Cappozzo, H. L., Campagna, C., and Monserrat, J. (1991). Sexual dimorphism in newborn Southern sea lions. *Mar. Mamm. Sci.* **7**, 385–394.

Cappozzo, H. L., and Rosas, F. C. W. (1991). León Marino Sudamericano, *Otaria flavescens* (Shaw, 1800). *In* "Estado de Conservación de los Mamíferos Marinos del Atlántico Sudoccidental" (H. L. Cappozzo and M. Junin, eds.). *Pub. Cient. PNUMA* **138**, 166–170.

Cassini, M. H., and Fernandez-Juricic, E. (2003). Costs and benefits of joining South American sea lion breeding groups: testing the assumption of a model of female breeding dispersion. *Can. J. Zool.* **81**, 1154–1160.

Cassini, M. H., and Fernandez-Juricic, E. (2007). Intra-sexual female agonistic behaviour of the South American sea lion (*Otaria flavescens*) in two colonies with different breeding substrates. *Acta Ethol.* **10**, 23–28.

Crespo, E. A., and Pedraza, S. N. (1991). Estado actual de la población de lobos marinos de un pelo (*Otaria flavescens*) en el litoral norpatagónico. *Ecol. Austral.* **1**, 87–96.

Dans, S. L., Crespo, E. A., Pedraza, S. N., and Alonso, M. K. (2004). Recovery of the South American sea lion (*Otaria flavescens*) population in northern Patagonia. *Can. J. Fish. Aquat. Res.* **61**, 1681–1690.

Hückstadt, L. A., and Antezana, T. (2003). Behaviour of the southern sea lion (*Otaria flavescens*) and consumption of the catch during purse-seining for jack mackerel (*Trachurus symmetricus*) off central Chile. *ICES J. Mar. Sci.* **60**, 1003–1011.

Hückstadt, L. A., and Antezana, T. (2004). Behaviour of southern sea lions in presence of killer whales during fishing operations in central Chile. *Sci. Mar.* **68**, 295–298.

Hückstadt, L. A., and Krautz, M. C. (2004). Interaction between southern sea lions *Otaria flavescens* and jack mackerel *Trachurus symmetricus* commercial fishery off central Chile: a geostatistical approach. *Mar. Ecol. Prog. Ser.* **282**, 285–294.

Hückstadt, L. A., Rojas, C.p., and Antezana, T. (2007). Stable isotope analysis reveals pelagic foraging by the southern sea lion in central Chile. *J. Exp. Mar. Biol. Ecol.* **347**, 123–133.

Reyes, L. M., Crespo, E. A., and Szapkievich, V. (1999). Distribution and population size of the southern sea lion (*Otaria flavescens*) in central and southern Chubut, Patagonia, Argentina. *Mar. Mamm. Sci.* **15**, 478–493.

Riedman, M. L. (1990). "The Pinnipeds: Seals, Sea Lions and Walruses." University of California Press, Berkeley.

Rodriguez, D. H., and Bastida, R. O. (1993). The southern sea lion, *Otaria byronia* or *Otaria flavescens*? *Mar. Mamm. Sci.* **9**, 372–381.

Schiavini, A. C. M., Crespo, E. A., and Szapkievich, V. (2004). Status of the population of South American sea lion (*Otaria flavescens*) in southern Argentina. *Mamm. Biol.* **69**, 108–118.

Sepulveda, M., and Oliva, D. (2005). Interactions between South American sea lions *Otaria flavescens* (Shaw) and salmon farms in southern Chile. *Aquacult. Res.* **36**, 1062–1068.

Sepulveda, M., Perez, M. J., Sielfeld, W., Oliva, D., Duran, L. R., Rodriguez, L., Araos, V., and Buscaglia, M. (2007). Operation interaction between South American sea lions *Otaria flavescens* and artisanal (small-scale) fishing in Chile: results from interview surveys and on-board observations. *Fish. Res.* **83**, 332–340.

Soto, K. H., Trites, A. W., and Arias-Schreiber, M. (2004). The effects of prey availability on pup mortality and the timing of birth of South American sea lions (*Otaria flavescens*) in Peru. *J. Zool.* **264**, 419–428.

Soto, K. H., Trites, A. W., and Arias-Schreiber, M. (2006). Changes in diet and maternal attendance of South American sea lions indicate changes in the marine environment and prey abundance. *Mar. Ecol. Prog. Ser.* **312**, 277–290.

Szapkievich, V., Cappozzo, H. L., Crespo, E. A., Bernabeu, R. O., Comas, C., and Mudry, M. (1999). Genetic relatedness in two southern sea lion (*Otaria flavescens*) rookeries in the south-western Atlantic. *Z. Saugetierkd.* **64**, 1–5.

Szetern, D., and Paez, E. (2002). Predation by southern sea lions (*Otaria flavescens*) on artisanal fishing catches in Uruguay. *Mar. Freshw. Res.* **53**, 1161–1167.

Thompson, D., Strange, I., Riddy, M., and Duck, C. D. (2005). The size and status of the population of southern sea lions (*Otaria flavescens*) in the Falkland Islands. *Biol. Conserv.* **121**, 357–367.

Tunez, J. I., Centron, D., Cappozzo, H. L., and Cassini, M. H. (2007). Geographical distribution and diversity of mitochondrial DNA haplotypes in South American sea lions (*Otaria flavescens*) and fur seals (*Arctocephalus australis*). *Mamm. Biol.* **72**, 193–203.

Werner, R., and Campagna, C. (1995). Diving behaviour of lactating southern sea lions (*Otaria flavescens*) in Patagonia. *Can. J. Zool.* **73**, 1975–1982.

# Southern Fur Seals
## *Arctocephalus* spp

### JOHN P.Y. ARNOULD

Southern fur seals (genus *Arctocephalus*) are generally recognized as comprising eight species and four subspecies. As the name implies, they occur almost exclusively in the Southern Hemisphere with only one species being found north of the equator. They are circumpolar in distribution, occurring in all the Southern Hemisphere oceans. The generic name, *Arctocephalus*, comes from the Greek words *arktos* and *kephale*, meaning "bear headed," and many of their facial characteristics reflect their terrestrial carnivore ancestry.

The majority of southern fur seals were overexploited during large-scale commercial hunting in the eighteenth and nineteenth centuries, and many species were so depleted in numbers that they were considered extinct. Fortunately, because of the isolated nature of the islands on which many southern fur seal species occur, remnant populations persisted. All known species survived and are now recovering at various rates. Several species have been the focus of extensive research over the last few decades which has greatly improved our understanding of pinniped biology and the influence of environmental variation on the population demography and behavior of top-order marine predators.

## I. Characteristics and Taxonomy

The most obvious diagnostic feature separating fur seals from sea lions is the presence of an underfur layer in their pelage. The

**S**

TABLE I

**Scientific and Common Names of the Southern Fur Seal Species (*Arctocephalus* spp.) with Their Mean Adult Body Mass and Most Recent Estimates of Population Size**

| Scientific name | Common name | Mean adult mass[a] (kg) | | Population size |
| --- | --- | --- | --- | --- |
| | | Female | Male | (× 1000) |
| *A. townsendi* (Merriam, 1897) | Guadalupe fur seal | 49 | 124[b] | >7 |
| *A. galapagoensis* (Heller, 1904) | Galapagos fur seal | 27 | 64 | 12 |
| *A. philipii* (Peters, 1866) | Juan Fernandez fur seal | 48 | 140 | 18 |
| *A. australis* (Zimmerman, 1783) | South American fur seal | 35–58[c] | 75–107[c] | 235–285 |
| *A. forsteri* (Lesson, 1828) | New Zealand fur seal | 39 | 127 | 135 |
| *A. pusillus pusillus* (Schreber, 1775) | Cape or South African fur seal | 57 | 247 | 1700 |
| *A. pusillus doriferus* (Wood Jones, 1925) | Australian fur seal | 76 | 279 | 92 |
| *A. tropicalis* (Gray, 1872) | Subantarctic or Amsterdam Island fur seal | 34–36[c] | 88–131[c] | >310 |
| *A. gazella* (Peters, 1875) | Antarctic or Kerguelen fur seal | 45 | 188 | 1600 |

[a]Means of measurements for each species taken during various seasons and stages of the breeding cycle.
[b]$n = 1$.
[c]Range of means from different populations.

density of hair follicles in the underfur of fur seals is approximately 50 times greater than that in terrestrial mammals and this layer plays an important role in their THERMOREGULATION. Fur seals are generally smaller than their sea lion counterparts. Another difference between the two groups can also be found in the baculum (penis bone), the tip of this bone being narrow in fur seals whereas in sea lions it is broad. In addition, fur seals have six pairs of upper postcanine teeth, compared to five in sea lions, and the third upper incisor is less circular in horizontal cross section.

The southern fur seals can be distinguished from the NORTHERN FUR SEAL, *Callorhinus ursinus*, by the extent to which the fur line extends on the fore flippers. In the northern fur seal the fur stops at the base of the flippers in a sharp line, whereas in southern fur seals it extends across part of the flipper, ending in a line over the metacarpals. A more prominent difference between these species is the shape of the SKULL, with the angle of the slope from the top of the nasal bone to the tip of the premaxilla being much greater in the northern fur seal. This gives the head of the northern fur seal a distinctive shortened-snout appearance in comparison to southern fur seals. Recent phylogenetic analyses (Arnason *et al.*, 2006; Deméré *et al.*, 2003) suggest the southern fur seals are more closely related to sea lions than the northern fur seal.

Within the southern fur seals there is a relative uniformity in appearance. This led to some confusion among early researchers about the classification and nomenclature of the various species in the genus. The history surrounding this has been discussed at length by Bonner (1981, 1994) and Gardner and Robbins (1998). The classification used here is taken from Rice (1998), and the scientific and common names for each species are given in Table I. The discreteness of their distributions aids in the identification of the different species, and in the areas where sympatry occurs morphometric, behavioral, vocalization, and habitat-choice differences can be used to separate the species.

Pelage color among southern fur seals is generally uniform dark brown to dark gray on the dorsal surface with a grizzled appearance caused by the tips of the guard hairs (outer fur layer) being white or pale in color. The fur is a lighter color on the ventral surface, especially

**Figure 1** *Male sub-antarctic fur seal* (A. tropicalis). *Photo courtesy J. Arnould.*

around the abdomen. There can be considerable variation between individuals of a species in the shading of the pelage and the degree to which it appears grizzled depending on age and sex. For example, older Australian fur seal (*A. pusillus doriferus*) and ANTARCTIC FUR SEAL (*A. gazella*) females often have a lighter, more grizzled appearance than younger individuals. The time elapsed since the last molt, which occurs annually in late summer–early autumn, also affects the appearance of the pelage due to wear and soiling.

There are two notable exceptions to the general pelage color within southern fur seals. Subantarctic fur seals (*A. tropicalis*) have a distinctive COLORATION: the chest and face (muzzle and around the eyes to just below the ears) are pale yellow or creamy in color while the top of the head and dorsal surface are dark brown–gray (Fig. 1). The demarcation in coloration is more pronounced in males, which also have a conspicuous tuft of dark hair on the forehead that becomes more erect when the animal is excited. In the Antarctic fur seal population

**Figure 2**   *Female Australian fur seals* (A. pusillus doriferus). *Photo courtesy J. Arnould.*

on South Georgia, 0.1–0.2%; of individuals have a white pelage. These animals are not albinos and have normal skin and eye pigmentation. This white-phase pelage has not been reported in any of the other Antarctic fur seal populations. Incidences where the fur appears uniform golden or tan, however, have also been reported in several other southern fur seal species, albeit rarely and usually only in pups.

Pups of all the southern fur seal species are born with a black natal coat (lanugo) which, with the exception of Cape and Australian fur seals (*A. pusillus* subspp.) where the ventral surface can vary from gray to pale yellow, develops a grizzled appearance soon after birth. In all species, the natal pelage molts 3–4 months after birth to reveal a silky smooth gray or brown fur.

The Galapagos fur seal (*A. galapagoensis*) is the smallest of the southern fur seals (and the Otariidae) with mean female and male adult mass at 27 and 64 kg, respectively (see Table I). Most of the other species are slightly larger, with mean masses ranging from 34 to 58 kg for females and 75 to 188 kg for males. The exceptions to this pattern are the Cape fur seal (*A. pusillus pusillus*) and the Australian fur seal (*A. p. doriferus*), which have mean masses of 58 and 76 kg for females and 247 and 279 kg for males, respectively (Fig. 2). These two subspecies also differ from the remainder of the genus in having some behavioral traits reminiscent of the sea lions. In particular, *A. pusillus* subspp. display very thigmotactic tendencies (tolerance of physical contact between individuals), behavior not seen in other fur seals but common in sea lions. In addition, the mode of terrestrial locomotion and aspects of their vocal repertoire resemble those of sea lions more than those of other fur seals. This is consistent with phylogenetic analyses placing *A. pusillus* as the stem arctocephaline.

Differences in the shape of the flippers can be used as diagnostic characters between several southern fur seal species, especially when their distributions overlap. For example, the fore-flippers of New Zealand seals have a more triangular shape than those of Australian fur seals, which are more paddle shaped and curved. Similarly, Antarctic fur seals have proportionally longer hind flippers than subantarctic fur seals. Snout lengths in southern fur seals also vary, being longest in the Juan Fernandez fur seal (*A. philippii*) and shortest in the Galapagos fur seal. In addition, the rhinarium (soft tissue of the nostrils) is smooth and inconspicuous in Antarctic and Galapagos fur seals, whereas it is inflated and bulbous in Juan Fernandez and New Zealand fur seals.

## II. Distribution and Abundance

The large-scale HUNTING during the commercial sealing era severely depleted the populations of Southern fur seals and it is known to have reduced the breeding distribution of many species. For several species, the pre-sealing distribution and population size are not accurately known, as sealing ships did not always keep detailed records of the number or species taken. This is particularly so for species which have overlapping ranges such as the New Zealand and Australian fur seals in southern Australia and the Antarctic and subantarctic fur seals on several subantarctic island groups.

The current breeding distributions of southern fur seals are shown in Fig. 3 and their population sizes are given in Table I. The Guadalupe fur seal (*A. townsendi*) is the only Southern fur seal species found in the Northern Hemisphere. Its breeding colonies are currently restricted to Isla Guadalupe, situated off the Pacific coast of Mexico, but it once had a wider distribution, including the Channel Islands (of California) and the San Benito Islands (off Baja California). The most recent (1992) estimate of the Guadalupe fur seal population size is >7000 individuals, making this by far the rarest of the southern fur seals.

The distribution of the Galapagos fur seal is limited to the equatorial Galapagos Islands. The population increased following cessation of the extensive sealing in the nineteenth century and it is currently estimated at 12,000 individuals. Periodic El Niño events, however, have been shown to significantly affect the population size by heavily reducing pup production and reproductive success in this species (Trillmich and Ono, 1991).

Further to the south, off the coast of Chile, the Juan Fernandez fur seal is confined to the islands of the same name. This species was once abundant and is estimated to have numbered over four million prior to exploitation. It was thought to have been hunted to extinction until it was rediscovered in 1966. Currently, the Juan Fernandez fur seal occupies four main breeding colonies and has a population size of approximately 18,000, making it the third rarest of the southern fur seals.

S

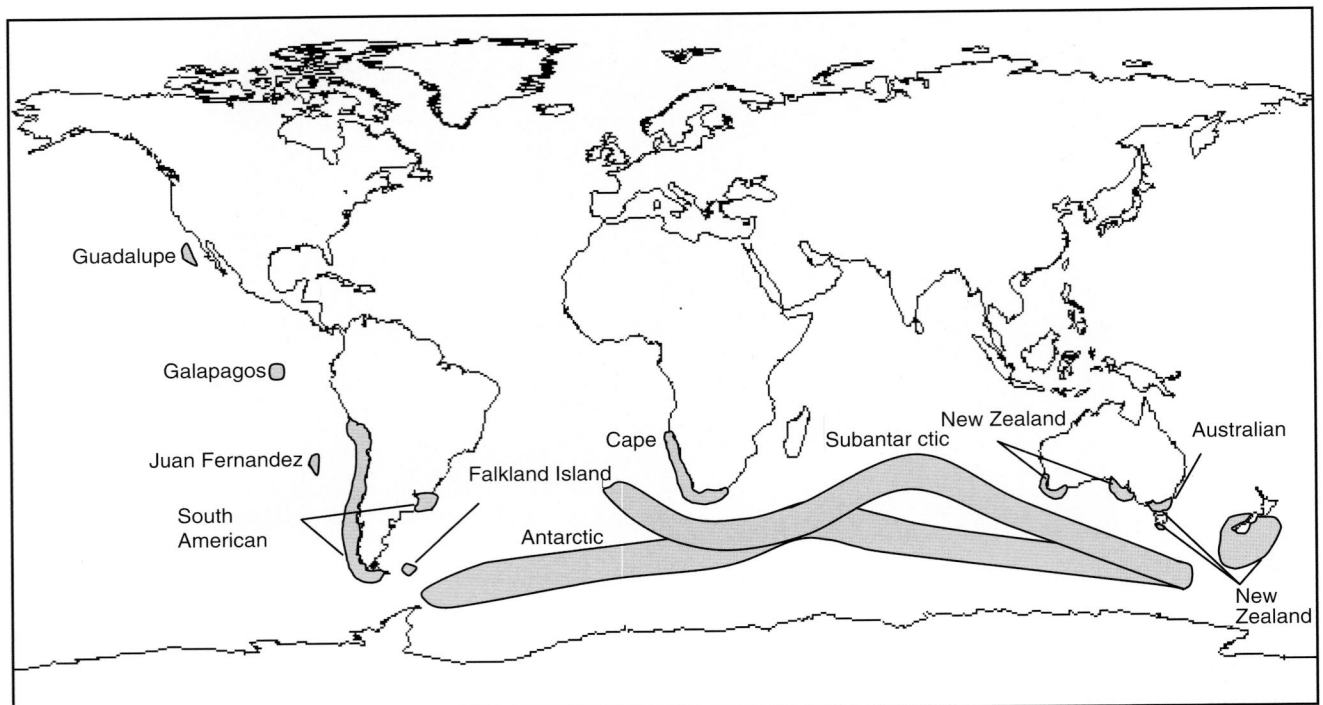

**Figure 3**   *Distribution of southern fur seals.*

The South American fur seal (*A. australis gracilis*) is found at many sites along the Pacific coast of South America, from Peru down to the islands west of Tierra del Fuego and in the Straits of Magellan, and on the Atlantic coast in southern Brazil and Uruguay. The colonies in Uruguay were exploited regularly in a small, controlled harvest until as recently as the early 1990s. The majority of colonies occur on offshore islands except in Peru where the dramatic increase in population since the 1960s (from 40 in the 1950s to over 20,000) has occurred mostly on three mainland colonies at Punta San Juan, Paracas Peninsula, and San Fernando. The Falkland Islands subspecies (*A. a. australis*) is found only on the Falkland Islands where its numbers are still greatly reduced as a result of commercial sealing, which continued into the early part of the last century. The total population size for both subspecies is presently estimated at 235,000–285,000 individuals.

In addition to colonies along the coast of the South Island of New Zealand, the breeding distribution of the New Zealand fur seal (Fig. 4) extends eastward as far as the Bounty, Antipodes, and Chatham Islands and southward to the Snares, Auckland, and Campbell Islands. There are also substantial colonies of New Zealand fur seals along the southern coast of Australia. Colonies in the Recherché Archipelago (Western Australia) and Kangaroo Island (South Australia) have been increasing rapidly in size, and there is the potential for competition to develop between them and the sympatrically breeding AUSTRALIAN SEA LION (*Neophoca cinerea*). Small breeding colonies of New Zealand fur seals (<400 individuals each) have also recently been established at three sites in Bass Strait (southeastern Australia), the most easterly extent of the species in Australia, where it breeds in sympatry with Australian fur seals. Historical evidence suggests both species were once found in larger numbers alongside each other throughout Bass Strait. The total New Zealand fur seal population is estimated at 135,000, with approximately 35,000 being found in Australia. Interestingly, molecular studies

**Figure 4**   *Female New Zealand fur seal* (A. forsteri). *Photo courtesy J. Arnould.*

have shown significant genetic variation, not only between Australian and New Zealand subpopulations of this species, but also within the latter, suggesting deeply divergent lineages "bordering on species-level distinction" (Lento *et al.*, 1997).

Although separated widely in their distributions, the Cape fur seal and the Australian fur seal show morphological and genetic similarities consistent with their status as subspecies. The Cape fur seal breeds at 25 colonies around the coasts of South Africa and Namibia. It is unique among fur seals in that the four largest colonies are located on mainland sites. The fact these colonies border diamond mining areas, where human access is restricted, and the

Namib Desert, which is devoid of large predators, may explain their location. The Cape fur seal has been subject to controlled harvests for over a century, yet the population continued to increase until the early 1990s. The most recent survey (2004) suggests the population has stabilized at around 2 million individuals. In contrast, the Australian fur seal has not been subject to commercial harvesting since 1923, but the population is currently estimated at only 92,000 individuals and breeding colonies are restricted to ten islands in Bass Strait. While the colonies are currently experiencing annual increases of 6–15%, the population size is still well below the estimated 175,000–225,000 individuals present prior to the commercial sealing era. The slower recovery, and smaller pre-exploitation size, of the Australian fur seal population in comparison to the Cape fur seal is thought to be attributed to differences in food availability. Australian fur seal colonies are situated in the nutrient-poor waters of Bass Strait, whereas Cape fur seals forage in the highly productive waters of the Benguela Current.

The subantarctic fur seal has a total population of >310,000 and a widely dispersed breeding distribution with colonies on several subantarctic and subtemperate island groups located north of the Antarctic polar front. The main focus for the species is on Gough Island in the South Atlantic Ocean where colonies total over 200,000 seals. A small population (250) is located to the west at nearby Tristan da Cunha. To the east, in the southern Indian Ocean, substantial colonies exist on the Prince Edward Islands (75,000), the Iles Crozet (1000), and Amsterdam Islands (50,000). A small population also exists to the southeast of Australia on Macquarie Island (<200) where the species breeds sympatrically with the Antarctic fur seal and some hybridization has occurred (Goldsworthy et al., 1999). Hybridization between subantarctic fur seals and Antarctic fur seals has also been recorded at Marion Island (Prince Edward Islands) where the latter is greatly outnumbered (only 1200). These two species also breed sympatrically on Iles Crozet.

The ANTARCTIC FUR SEAL also has a very large breeding range and is the most southerly breeding of the *Arctocephalus* species. With the total population in 1990 estimated at over 1.6 million and growing at an annual rate of 9.8%, the Antarctic fur seal is now likely to be the most numerous of the southern fur seals. Over 95% of the population breeds on South Georgia in the South Atlantic Ocean, but colonies are found on numerous island groups spreading eastward as far as Macquarie Island. All colonies are located south of, or on, the Antarctic polar front with the exception of those at Marion Island, Iles Crozet, and Macquarie Island. The species was considered extinct until a small remnant population of 1000–3000 individuals was discovered in 1950 on Bird Island (northwest tip of South Georgia). From this colony emanated the phenomenal recovery of the species with the recolonization of many sites being attributed to emigration from the expanding South Georgia population. The rapid recovery of the South Georgia population is thought to have been facilitated by its predominant prey, Antarctic krill (*Euphausia superba*), being found in super-abundance in the surrounding waters.

The nonbreeding range of most southern fur seal species is often much greater than the breeding distribution and is determined by the movements of males and juveniles. These segments of the population are not restricted in the same way as adult females, which must return regularly to the natal colony in order to suckle their pup. Males of most species undertake some form of seasonal movement after the breeding season or during the winter months when local food availability is reduced. Whether such movements constitute structured migrations or reflect individual dispersal patterns may vary between species and different age classes.

## III. Ecology

Southern fur seals feed on a variety of prey species, including fish, cephalopods (octopus, squid, and cuttlefish), crustaceans (krill, shrimp, rock lobster), and even penguins and other seabirds. Studies of several species have shown great temporal variation in diet composition, with seals exploiting seasonally abundant prey items such as spawning squid or schools of migrating fish. In addition, substantial differences in diet can be found between populations of the same species. For example, the Antarctic fur seal population at South Georgia feeds almost exclusively on Antarctic krill, whereas at Heard and Kerguelen Islands (southern Indian Ocean), seals of this species feed primarily on myctophid fishes. The majority of detailed studies, however, have concentrated on the diet of adult females because natal colonies have been the easiest to sample for feces (the most commonly used technique for determining diet). Consequently, for most species, there is relatively little information on the diet of adult males or how it varies seasonally. Nonetheless, studies of New Zealand, Juan Fernandez, and Antarctic fur seals have shown significant differences between the sexes in diet which is thought to reflect resource partitioning, differences in nutritional requirements, and physiological diving abilities.

Over the last decade, the foraging zones of several southern fur seal species have been determined through the use of satellite telemetry. Studies of lactating female Cape, Antarctic, subantarctic, and New Zealand fur seals have found that most foraging activity occurs in upwelling zones, oceanic frontal systems, or continental shelf–edge regions, areas generally high in primary productivity. An exception to this trend is the Australian fur seal, which has been found to forage exclusively over the shallow continental shelf region of Bass Strait and surrounding areas in south-eastern Australia. Sex differences in foraging areas have also been observed in New Zealand, Antarctic and Australian fur seals with males generally tending to forage in areas of greater depth. There is little information about the foraging locations for other age and sex classes in southern fur seals.

## IV. Behavior and Physiology

As with diet studies, the majority of research on foraging behavior in southern fur seals has involved lactating females due to their relative ease of handling and recapture in comparison to adult males. Although they are known to be able to dive to much greater depths, it has been shown in most species that foraging activity occurs mainly in the surface mixed layer (<50–60 m) at night. The larger body size of males would be expected to enable them to dive deeper than females, and studies of New Zealand and Antarctic fur seals support this with foraging depths >200 m regularly observed. In most species studied, diving depths change throughout the night, generally reflecting the diel vertical migration of the main prey. The exception to this is the Australian fur seal, the largest of the fur seal species, which has been shown to forage almost exclusively in benthic habitats to depths of up to 200 m during both day and night. This trait, more akin to sea lion foraging behavior, is considered the result of their large body size and the relatively low productivity of their marine habitat.

Like in all otariid seals, female southern fur seals have a lactation pattern characterized by regular nursing periods ashore alternating with foraging trips to sea during which the pup remains on land fasting. However, whereas female sea lions generally only leave their pups for short durations (1–2 days), maternal foraging trips in southern fur seals can be much longer. While in most southern fur seals species maternal foraging trips are 4–6 days long, females of subantarctic, Juan Fernandez, and Australian fur seals have been recorded

S

as regularly undertaking trips of >20 days (Francis *et al.*, 1998; Kirkman *et al.*, 2002). Because of these long separations between mother and pup, lactating southern fur seal females have the longest inter-suckling intervals of any mammal, yet are able to maintain mammary gland function, and their pups endure the longest repeated fasts of any developing mammal. The physiological mechanisms by which these feats are achieved are still poorly understood.

## V. Life History

While definitive data are lacking for some species, southern fur seal females generally reach sexual maturity at 3–5 years of age and thereafter have pregnancy rates of 70–80%. Males reach sexual maturity around the same age but do not attain territorial status and the ability to mate until 7–10 years old. Longevity in the wild has been documented for few species but ranges from 20–23 years for females and 10–18 for males in Australian, subantarctic and Antarctic fur seals.

The breeding biology of the southern fur seals is similar to that of other otariid seals (sea lions). Females give birth to a single pup each year, and the reproductive cycle consists of a perinatal attendance period of 6–10 days, culminating in estrus and copulation; an embryonic diapause lasting 3–4 months; and an active gestation of 8–12 months. Females have two pairs of mammary glands and lactate for 8–12 months, but several species may suckle their pup into part of the second or even third year. The exceptions to this are the Antarctic fur seal, which has a lactation period lasting only 4 months, and the Galapagos fur seal in which weaning may occur at >3 years of age and females can suckle successive offspring at the same time. Differences in food availability and seasonal predictability may explain these divergent strategies.

Timing of the pupping season is fairly consistent, occurring between October and December in the Southern Hemisphere species and in the corresponding late spring–early summer period of June–July in the Northern Hemisphere for the Guadalupe fur seal (Boyd, 1991). Estrus synchrony, measured as the period over which 90% of births occur, varies from 21 days in the Antarctic fur seal to 70 days in the Galapagos fur seal and is negatively related to latitude (Boness, 1991). High estrus synchrony is considered a prerequisite for polygyny, and resource defense polygyny has been demonstrated clearly in New Zealand, Antarctic, and Australian fur seals. As is common in polygynous species, Southern fur seals are sexually dimorphic and have some of the most extreme male:female mass ratios of any mammal. In addition, the operational sex ratio, defined as the average ratio of fertilizable females to sexually active males at any given time, has been shown to be correlated positively to the degree of SEXUAL DIMORPHISM in southern fur seals.

## VI. Interactions with Humans

In addition to experiencing severe population declines during the eighteenth and nineteenth centuries as a result of the commercial sealing era, several southern fur seal species have been subject to culls and harvests, both legal and illegal, at various times since then. The majority of these have been related to commercial fisheries concerns of competition with increasing seal populations. Indeed, as populations continue to recover, and human exploitation of the oceans increases, interaction with commercial fisheries is becoming more prevalent. Currently, only the Cape fur seal in Namibia is legally harvested (>60,000 annually, most which are pups) ostensibly for protecting the Namibian fisheries stocks. However, in addition to competition for resources, direct negative interactions (e.g., entanglement in gear, drowning in nets) can lead to high levels of mortality in some species.

In contrast to the generally negative impacts of interactions with fisheries, populations of several southern fur seal species have enjoyed positive attention in recent decades through eco-tourism activities. The popularity of viewing (and in some locations swimming with) fur seals in their natural habitat has proved to make ecotours focused on these animals important to local economies in numerous places throughout the southern hemisphere. Recent estimates place the annual value of tickets sale alone at >US$8 million.

## References

Arnason, U., Gullberg, A., Janke, A., Kullberg, M., Lehman, N., Petrov, E. A., and Vainola, R. (2006). Pinniped phylogeny and a new hypothesis for their origin and dispersal. *Mol. Phylo .Evol.* **41**, 345–354.

Boness, D. J. (1991). Determinants of mating systems in the Otariidae (Pinnipedia). *In* "Behaviour of Pinnipeds" (D. Renouf, ed.), pp. 1–44. Chapman and Hall, New York.

Bonner, W. N. (1981). Southern fur seals *Arctocephalus* (Geoffroy Saint-Hilaire and Cuvier, 1826). *In* "Handbook of Marine Mammals" (S. H. Ridgway and R. J. Harrison, eds.), Vol. 1: "The Walrus, Sea Lions, Fur Seals, and Sea Otter," pp. 161–208. Academic Press, New York.

Bonner, W. N. (1994). "Seals and Sea Lions of the World." Blandford, London.

Boyd, I. L. (1991). Environmental and physiological factors controlling the reproductive cycles of pinnipeds. *Can. J. Zool.* **69**, 1135–1148.

Deméré, T. A., Berta, A., and Adam, P. J. (2003). Pinnipedimorph evolutionary biogeography. *Bull. Am. Mus. Nat. Hist.*, 32–76.

Francis, J., Boness, D., and Ochoa-Acuña, H. (1998). A protracted foraging and attendance cycle in female Juan Fernandez fur seals. *Mar. Mamm. Sci.* **14**, 552–574.

Gardner, A. L., and Robbins, C. B. (1998). Generic names of northern and southern fur seals (Mammalia: Otariidae). *Mar. Mamm. Sci.* **14**, 544–551.

Goldsworthy, S. D., Boness, D. J., and Fleischer, R. C. (1999). Mate choice among sympatric fur seals: female preference for conphenotypic males. *Behav. Ecol. Sociobiol.* **45**, 253–267.

Kirkman, S. P., Bester, M. N., Hofmeyr, G. J. G., Pistorius, P. A., and Makhado, A. B. (2002). Pup growth and maternal attendance patterns in Subantarctic fur seals. *Afr. Zool.* **37**, 13–19.

Lento, G. M., Haddon, M., Chambers, G. K., and Baker, C. S. (1997). Genetic variation of southern hemisphere fur seals (*Arctocephalus* spp.): investigation of population structure and species identity. *J. Hered.* **88**, 202–208.

Rice, D. W. (1998). Marine mammals of the world: systematics and distribution. *Soc. Mar. Mamm. Spec. Pub.* **4**.

Trillmich, F., and Ono, K. A. (1991). "Pinnipeds and El Niño: Responses to Environmental Stress." Springer-Verlag, Berlin.

# Species

## WILLIAM F. PERRIN

### I. What is a Species?

Although the term "species" is widely used, there is no agreement on exactly what it means beyond the general conception of a basic taxonomic or evolutionary unit. It is important to make progress in our understanding of species and how to define them, not least because much of the legal machinery of international conventions, national legislations, and regulations dealing with wildlife and the conservation of biodiversity uses the term in specifying what they seek to protect. If we cannot agree on what constitutes a wild species, we may not be able to agree on how to conserve and protect it.

There are a number of species concepts in use for vertebrates (Claridge *et al.*, 1997). In the *biological species concept* (BSC) (Mayr, 1963), a species is a group of interbreeding or potentially interbreeding individuals separated from other such groups by intrinsic (genetically fixed) barriers to gene flow. In cases of sympatry or parapatry, direct evidence of non-interbreeding may be available, but more commonly morphological separation has been used as a proxy to indicate the lack of gene flow; in this case the definition works fairly well. In the case of allopatry, however, there is difficulty. While the same operational criterion has been used, in nearly all such cases we do not know whether the groups would interbreed if the chance arose. The barrier to interbreeding may be a broad gap in range or simply a purely physical geographic one such as an isthmus or a climatic zone.

In the more recently developed *phylogenetic species concept* (PSC), a species is "the smallest population or group of populations within which there is a parental pattern of ancestry and descent and which is diagnosable by unique combinations of character states" (Cracraft, 1997), with no inference about potential interbreeding. Indeed, interbreeding is considered a shared primitive character and, as such, is not able to diagnose a lineage, which the PSC requires. In the case of sympatry, this concept does not differ materially from the BSC. Both have difficulties with allopatric populations (biological subspecies under the BSC may equate to phylogenetic species under the PSC) and with paraphyly. Phylogenetic taxonomists insist on reciprocal monophyly (having diagnostic unique characteristics, a single common ancestor, and no reticulation) between most closely related (sister) species, but this may be violated by the real phenomenon of incomplete lineage sorting in the case of some newly evolved but widely recognized species, e.g., because polar bears evolved from brown bears relatively recently, some brown bears are still related more closely genetically to polar bears than they are to other brown bears. In practice, adherents to the biological species concept use morphological characters in exactly the same way as phylogenetic taxonomists, albeit as proxy indicators of reproductive isolation. One difficulty faced by strict phylogenetic species-level taxonomy is that different genes or sequences may have different phylogenies due to differential modes of inheritance or introgression; a gene tree may not equal a species tree.

The two species concepts are actually complementary perspectives on evolution and can reinforce each other. To quote Avise and Wollenberg (1997), "Historical descent and reproductive ties are related aspects of phylogeny and jointly illuminate discontinuity." Furthermore, more important is not the process of specifying units to fit into the simplifying Linnaean binomial system, but rather describing and understanding in detail the coalescent genealogy and demography of discontinuous populations. Where did they come from, how are they different, and where are they likely to go? Whether they are called species (always, under the phylogenetic species concept) or subspecies (sometimes, under the biological species concept) is little more than semantics rooted in perspective. This is generally true also in the conservation of diversity. For example, named subspecies are accorded the same legal protection as species in the U.S. Endangered Species Act and Marine Mammal Protection Act.

There are additional species concepts, e.g., the *cohesion species concept* (Templeton, 1989), under which a species is "the most inclusive group of organisms having the potential for genetic and/or demographic exchangeability." Some pairs or groups of sympatric or parapatric species, such as wolves (*Canis lupus*) and coyotes (*Canis latrans*), may make up *syngameons* (Templeton's term) of species that are fully interfertile and may at times interbreed extensively but nonetheless have persisted separately for geological periods of time. Some of the species of southern fur seals (*Arctocephalus* spp.) may fall into this category (discussed later). Despite sporadic or scattered interbreeding, they would qualify as full species under the cohesion species concept.

It is clear that the current trend is toward pragmatism in addressing species concepts. Agapov and Sluys (2005) noted, "Species are complex things, their identity resting on the methodologies used to diagnose them and on the many different ways that organisms have of belonging to a species. We would do well to embrace plurality, using different species concepts in different circumstances rather than searching for a platonic ideal."

While the various species concepts are necessary (because of the legal need to define units for conservation) and laudable attempts to model the results of actual evolutionary processes, they are human constructs (Hey, 2001), and as for all such constructs, in certain circumstances even the most commonly accepted species concepts fray at the edges. One problem of the biological species concept in its application (mentioned earlier) is that of closely related allopatric species, whether they be defined by morphological or genetic characters. If at some point they became sympatric or parapatric because of some geological or climatic change, would they then freely and effectively interbreed (be one species)? Are they a little different from each other only because of inconsequential genetic drift due to physical separation, or are their differences the result of important differential ecological selection that has led to a permanent parting of their evolutionary paths? Even if their morphological differences are great, the question remains of whether this is due to large genetic differences, or of pleiotrophic effects of a small difference unassociated with genetic differences that would prevent interbreeding. For the most part, of course, answers to these questions will remain unobtainable. However, we can deduce what might happen from looking at the results of some "natural experiments." In one possible example, a zone of hybridization/intergradation between two subspecies of spinner dolphins, *Stenella longirostris*, in the eastern tropical Pacific is thought to perhaps be the result of recontact after separation during an earlier cooler climatic period (Perrin *et al.*, 1991). HYBRIDIZATION in the putative zone of re-contact is evidenced by greater variance in body size and color pattern. The two subspecies, the eastern spinner (*S. l. orientalis*) and Gray's spinner (*S. l. longirostris*), are morphologically so dissimilar that they were once thought to belong to different species (eastern spinners were identified as *S. microps* and Gray's spinners in Hawaii as *S. roseiventris*).

Thus, some of our present allopatric species may be transient taxa that will be absorbed or blended back into a pooled line of descent; i.e., they were never really "good" species at all.

Antitropical species seem especially good candidates for potential recontact and subsequent reticulation; sporadic movement of cool-water forms across the equator during cooler geological periods is thought to have played an important role in speciation, e.g., in the porpoises, including the formation of antitropical pairs of putative species. Species pairs that potentially represent such transient situations include the vaquita (*Phocoena sinus*) and Burmeister's porpoise (*P. spinipinnis*), northern and southern right whale dolphins (*Lissodelphis borealis* and *L. peronii*), the Pacific white-sided dolphin (*Lagenorhynchus obliquidens*) and the dusky dolphin

S

(*L. obscurus*), the North Pacific and southern right whales (*Eubalaena japonica* and *E. australis*), Baird's and Arnoux's beaked whales (*Berardius bairdii* and *B. arnuxii*), North Atlantic and southern bottlenose whales (*Hyperoodon ampullatus* and *H. planifrons*), and the northern and southern elephant seals (*Mirounga angustirostris* and *M. leonina*).

Examples of non-antitropical allopatric species complexes include the four *Cephalorhynchus* species *C. hectori, C. commersonii, C. eutropia,* and *C. heavisidii;* they are sharply distinct morphologically but undoubtedly closely related. The Ganges and Indus river dolphins (formerly *Platanista gangetica* and *P. minor*) are presently separated geographically but are thought to have been in contact in Recent, perhaps historical, time, and Rice (1998) recognized the likely inconsequential nature of any differences by downgrading them to subspecies in *P. gangetica*. Among pinnipeds, the eight currently recognized species of southern fur seals, *Arctocephalus* spp., are almost entirely allopatric, but there are at least four breeding sites where two species (*A. gazella* and *A. tropicalis*) have begun to cooccur as populations have recovered from severe depletion by sealing in the last century (Rice, 1998). At two of these (Marion Island and Heard Island), limited hybridization has occurred, but at the others (Îles Crozet and Bass Strait) it has not been observed. In the North Pacific, there are two allopatric species of sea lions in the genus *Zalophus* (formerly three; the Japanese sea lion *Z. japonicus* is almost certainly extinct). The three seals in the genus *Pusa* are allopatric, albeit strongly differentiated. The allopatric Mediterranean monk seal (*Monachus monachus*) and Caribbean monk seal (*M. tropicalis*) are very similar morphologically (the latter is extinct). The three very similar manatee species *Trichechus manatus, T. senegalensis,* and *T. inunguis* are largely allopatric.

All of these are potential cases of geographical forms within a species being perceived as evolutionary species because of allopatry. In some cases, of course, the question is entirely moot. For example, while the Orinoco and Amazon drainages could merge in the future (bringing *T. manatus* and *T. inunguis* into contact and possible full interbreeding—beyond the occasional marine interbreeding now observed around the mouth of the Amazon—(Vianna *et al.*, 2006), the same could not happen for the Amazon and Congo drainages (inhabited by *T. inunguis* and *T. senegalensis*) under any conceivable geological scenario short of reversal of continental drift; thus the latter two forms are "good" species *perforce* under any species concept.

All of the marine mammal species recognized today, with one exception, were described on morphological grounds, although for the cryptic pairs *Delphinus delphis/D. capensis, Orcaella brevirostris/O. heinsohni* and *Sotalia guianensis/S. fluviatilis*, genetic data provided part of the basis for recognition of more than one species where previously there had been thought to be only one. The exception is Perrin's beaked whale *Mesoplodon perrini*. It was discovered and described primarily based on genetic distinctness, although subsequent research uncovered divergent morphological characters (Dalebout *et al.*, 2002).

The operational criterion of at least one absolutely differentiating character that is commonly used to indicate species-level difference in morphologically based taxonomy has been extended by some to genetic characters. An example of such a potential application concerns Hector's dolphins, *Cephalorhynchus hectori*. Dolphins from east and west sides of the South Island of New Zealand were found to have no mtDNA control–region sequence haplotypes in common (Pichler *et al.*, 1998). They are thus fully diagnosable and reciprocally monophyletic terminal taxa, qualifying as species under the phylogenetic species concept (Cracraft, 1997), if such a decision were

to be based on this single genetic character. Very few would agree with such a decision, although most would probably accept designation of the two populations as *evolutionarily significant units* (ESUs, something worth saving because of unique genetic or morphological diversity) (Ryder, 1986) and agree that they should be managed separately. More recently, the trend has been toward the use of a suite of fixed genetic differences (preferably unlinked) as a criterion for specieshood; this parallels the practice of most morphological taxonomists of not recognizing a species based on a single point of difference but rather on the basis of multiple concordant characters (Avise and Wollenberg, 1997).

In a recent international workshop on cetacean taxonomy (Reeves *et al.*, 2004), it was agreed that a species is a populational entity on an independent evolutionary trajectory and that "the different approaches to species delimitation should be employed in a flexible and pragmatic way, with the basic aim of using proxies to identify irreversible divergence." It was further agreed that multiple lines of evidence are essential. Such kinds of data could include morphological together with genetic data, or data from multiple independent genetic loci.

De Queiroz clarified the "species problem" by pointing out three separate issues subsumed under that rubric: (1) The necessary properties of species, i.e., what is a species? (2) How species come about (the process). (3) How to delimit a species. He suggested that the most reasonable answer to the first question is that a species is a "separately evolving metapopulation lineage," much the same conclusion arrived at by the 2004 cetacean workshop. On the question of the process leading to formation of a species, he proposed that the species category is a cluster concept and that "no single process or set of processes be considered necessary for the existence of species." He concluded that separating out the methodological question of how to delimit a species from the two other more theoretical questions would help to focus research on the important applied problems of estimating the numbers and boundaries of species.

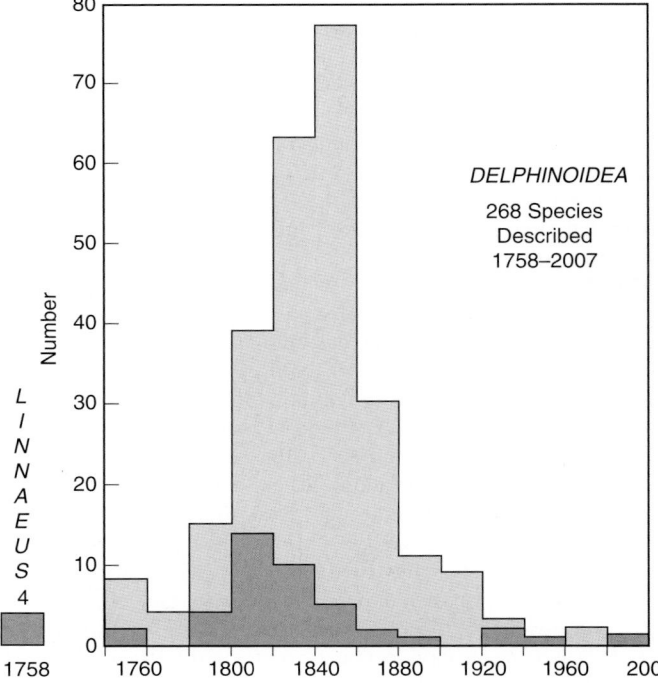

**Figure 1** *The "Great Victorian Radiation" of described (nominal) species of delphinoid cetaceans.*

## II. The Future

The 2004 workshop also emphasized that taxa, including species, are best viewed as hypotheses about evolutionary history rather than rigidly fixed or sacrosanct entities. Thus, while Rice's (1998) classification recognized 127 species of marine mammals, the status of many of these is open to question. For example, Rice "lumped" the northern right whale (*Balaena glacialis*) and the southern right whale (*B. australis*) into a single species *B. glacialis*, but some authors have preferred to continue recognition of two species (under the genus *Eubalena*), one reason being the existence of fixed genetic differences between the forms (but see discussion earlier of allopatric species). The most recent usage is to recognize three species, including the North Pacific *E. japonica* (see Species List). In another example, Rice (1998) provisionally listed three species of humpbacked dolphins, *Sousa teuszii*, *S. plumbea*, and *S. chinensis*, but he discussed the results of taxonomic work in progress that indicate that there likely is only one species, with pronounced geographical variation. In still another example, the number of recognized species of southern fur seals (*Arctocephalus*) may decrease from eight. These and many other examples demonstrate that the *alpha* (species-level) *taxonomy* of marine mammals is still very much in flux: new species are almost certainly yet to be discovered, some species will be lumped (Fig. 1), and others may be split. There is still ample work for the taxonomists.

## See Also the Following Articles

Biogeography ■ Classification ■ Geographic Variation ■ History of Marine Mammal Research ■ Systematics ■ Overview

## References

Agapov, P., and Sluys, R. (2005). The reality of taxonomic change. *TREE* **20**, 278–280.

Avise, J. C., and Wollenberg, K. (1997). Phylogenetics and the origin of species. *Proc. Natl. Acad. Sci. U.S.A.* **94**, 7748–7755.

Claridge, M. F., Dawah, H. A., and Wilson, M. R. (1997). "Species: The Units of Biodiversity." Chapman and Hall, London.

Cracraft, J. (1997). Species concepts in systematics and conservation biology: an ornithological viewpoint. *In* "Species: The Units of Biodiversity" (M. F. Claridge, H. A. Dawah, and M. R. Wilson, eds), pp. 325–339. Chapman and Hall, London.

Dalebout, M. J., Mead, J. G., Baker, C. S., Baker, A. N., and van Helden, A. L. (2002). A new species of beaked whale *Mesoplodon perrini* sp. n. (Cetacea: Ziphiidae) discovered through phylogenetic analysis of mitochondrial DNA sequences. *Mar. Mamm. Sci.* **18**, 577–608.

Hershkovitz, P. (1966). Catalog of living whales. *U.S. Nat. Mus. Bull.* **246**, 1–259.

Hey, J. (2001). "Genes, Categories, and Species. The Evolutionary and Cognitive Causes of the Species Problem." Oxford University Press, Oxford.

Linnaeus, C. (1758). "Systema Naturae", vol. 1. Laurentii Salvii, Stockholm.

Mayr, E. (1963). "Animal Species and Evolution." Harvard University Press, Cambridge, MA.

Perrin, W. F., Akin, P. A., and Kashiwada, J. V. (1991). Geographic variation in external morphology of the spinner dolphin *Stenella longirostris* in the eastern Pacific and implications for conservation. *Fish. Bull. U.S.A.* **89**, 411–428.

Pichler, F. B., Dawson, S. M., Slooten, E., and Baker, C. S. (1998). Geographic isolation of Hector's dolphin populations described by mitochondrial DNA sequences. *Conserv. Biol.* **12**, 676–682.

de Queiroz, K. (2005). Different species problems and their resolution. *Bioessays* **27**, 1263–1269.

Reeves, R. R., Perrin, W. F., Taylor, B. L., Baker, C. S., and Mesnick, S. L. (2004). Report of the Workshop on Shortcomings of Cetacean Taxonomy in Relation to Needs of Conservation and Management, April 30–May 2, 2004 La Jolla, California. NOAA Technical Memorandum NOAA-TM-NMFS-SWFSC-**363**. 94pp.

Rice, D. W. (1998). Marine mammals of the world. *Soc. Mar. Mamm. Spec. Publ.* **4**, 231 pp.

Ryder, O. A. (1986). Species conservation and systematics: the dilemma of subspecies. *Trends Ecol. Evol.* **1**, 9–10.

Templeton, A. R. (1989). The meaning of species and speciation. *In* "Speciation and Its Consequences" (D. Otte, and J. A. Endler, eds), pp. 3–27. Sinauer, Sutherland, MA.

Vianna, J. A., *et al.* (2006). Phylogeography, phylogeny and hybridization in trichechid sirenians: implications for manatee conservation. *Mol. Ecol.* **15**, 433–447.

# Spectacled Porpoise
## *Phocoena dioptrica*

### R. NATALIE P. GOODALL

## I. Characteristics and Taxonomy

The spectacled porpoise is known from a large number of skeletons, mostly from Tierra del Fuego, but from few fresh specimens. As late as 1976, the species was known from only 10 occurrences (nine specimens) all off southeastern South America.

*Phocoena dioptrica* was first described by F. Lahille of Argentina in 1912 from a live, pregnant female stranded near Buenos Aires. A second female and an adult male with an improbably large dorsal fin were examined fresh by Bruch (1916); casts and mounted skeletons of these two animals are in exhibition in the Museo de La Plata. The only synonym is *P. storni*, based on a cranium from Tierra del Fuego. *Phocaena obtusata* Philippi, 1893, sometimes thought to be of this species, has been shown to be a synonym of *Cephalorhynchus eutropia*. In a revision of the taxonomy of the Phocoenidae, *P. dioptrica* was put in a new monotypic genus, *Australophocaena*, but later studies of phocoenid mitochondrial DNA returned it to the genus *Phocoena*. The specific name, *dioptrica*, refers to the double eye patch seen in most specimens. The common name in Spanish is *marsopa de anteojos*.

The spectacled porpoise is a robust animal with a rounded head and no beak (Fig. 1). The gape is short, and the flippers are small and situated well forward on the body. The dorsal fin is broadly triangular and shows strong sexual dimorphism, being much larger and more rounded in adult males. The rather small flukes have rounded tips with a fairly straight posterior border (illustrated in Goodall and Schiavini, 1995). At least some external measurements are available for 29 specimens from Tierra del Fuego. Eleven females ranged from 124.6 (a calf) to 203.5 cm in length. Eleven animals of unknown sex were from 94 cm, the smallest animal found, to 201 cm. Seven males measured from 109 (a neonate) to 224 cm, the largest animal known. This probably does not represent the maximum size for the species. Weights range from a fetus at 1.6 kg, to an 85-kg female, and a 115-kg male. A 135-cm male live-stranded in New Zealand; two juvenile females of 119 cm (17.5 kg) and 127 cm (21.3 kg) stranded in southern Australia (Evans *et al.*, 2001).

The spectacled porpoise is highly distinctive with its unusual pigmentation, small head and facial features, and the large male dorsal

S

Figure 1 *Spectacled porpoise (A) Surfacing female or juvenile, (B) adult male showing large dorsal fin typical of males, and (C) presumed adult female with calf. Photographs by Paula Olson, Southwest Fisheries Science Center, NOAA.*

**S**

fin. Color patterns have been depicted for few animals. Young animals are dark gray dorsally and light gray ventrally with darker gray or brownish streaks, including a well-defined mouth-to-flipper stripe and gray flippers. This pigmentation changes in the adult to shining black on the dorsal surface, sharply separated on the sides from the pure white ventral region, although there is feathering on some specimens. Live animals seen at sea appear lighter (Fig. 1). Some may have a dark patch around the blowhole (Evans *et al.*, 2001) and a paler saddle around the dorsal fin. Females and juveniles appeared lighter in color than adult males; one animal had two lighter stripes

or "bridle" from the blowhole to the apex of the melon (Sekiguchi *et al.*, 2006).

The juvenile flipper stripe seems to disappear with growth. Both young and adults have dark lips surrounded by white; the eye patch is a dark circle, variably outlined by a lighter or white ring. Most animals had a dark upper tail stock and flukes, which are lighter below. The dorsal surface of the flipper varies from white or mottled gray to black, usually ending in a sharp line at the base of the flipper. The pigmentation is illustrated in detail by Goodall and Schiavini (1995) and Sekiguchi *et al.* (2006).

The large dorsal fin of the male is striking. Those from the southern South Pacific may have more pointed fins (Sekiguchi *et al.*, 2006) than those from Tierra del Fuego (Goodall and Schiavini, 1995). The young female from South Australia had 26 small tubercles on the leading edge of the dorsal fin, as do some other phocoenids (Evans *et al.*, 2001). In spite of taking tracings of all dorsal fins, we have not noticed this in Tierra del Fuego animals. The flippers of the two animals from South Australia seemed longer than those from Tierra del Fuego.

The skull has been illustrated and described in several works. Condylobasal lengths of 54 adult skulls ranged from 276 to 424 mm. The rostrum of a specimen from the Auckland Islands was relatively smaller than those of other regions (Perrin *et al.*, 2000). The teeth vary from spade- to peg-shaped and are often hidden in the gums. The posterior teeth are situated in a groove without noticeable alveoli. Tooth counts range from 16 to 26 upper and 17 to 23 lower teeth per side. The vertebral count is C7, T14, L14–16, and Ca 32–33 (66–70). From three to six cervical vertebrae may be fused. There are 13–14 ribs per side, the first 9 bicipital and 8 sternal ribs. The phalangeal count of one specimen was I = 2, II = 7, III = 4, IV = 3, and V = 4. The facial complex system has been studied (A. Purgue, personal communication), but little else is known of the internal anatomy.

## II. Distribution and Abundance

The spectacled porpoise is circumpolar in cool temperate, subantarctic, and Antarctic waters. Specimens are known from southern Brazil (33°S) (Pinedo *et al.*, 2002), Uruguay, and Buenos Aires south to Tierra del Fuego, the Falklands (Malvinas) and South Georgia in the southwestern South Atlantic, Auckland and Macquarie Islands in the southwestern South Pacific; Heard and Kerguelen in the southern Indian Ocean, Burney Island, Tasmania and South Australia (Fig. 2). It is associated with the Falkland (Malvinas) current and the West Wind Drift on both sides of the Antarctic Circumpolar Front.

Sightings are rare but widely distributed: off Uruguay, Patagonia, South Georgia, Kerguelen, the Auckland Islands, south and southeast of New Zealand, Tasmania, and Heard Island. The southernmost sighting was at 64°33.5'S, 176°19'E; there are now nine sightings in Antarctic waters south of 60°S (Sekiguchi *et al.*, 2006). Surface water temperature records to date range from 0.9 to 10.3°C, with most sightings in waters of 4.9–6.2°C. The degree of contact among populations in the different oceans is unknown. Sekiguchi *et al.* (2006) reported 28 live sightings—five of them formerly reported by Kasamatsu and Joyce (1995) and Goodall and Schiavini (1995). Two more recent sightings are from the Drake Passage and east of the Falklands (Malvinas) (A. Walleyn personal communication).

Since it was known mainly from strandings, this species was once thought to be an inshore animal. Offshore sightings show it to be oceanic in cool temperate, subantarctic and Antarctic waters. Although sighted in deep waters far from land, spectacled porpoises

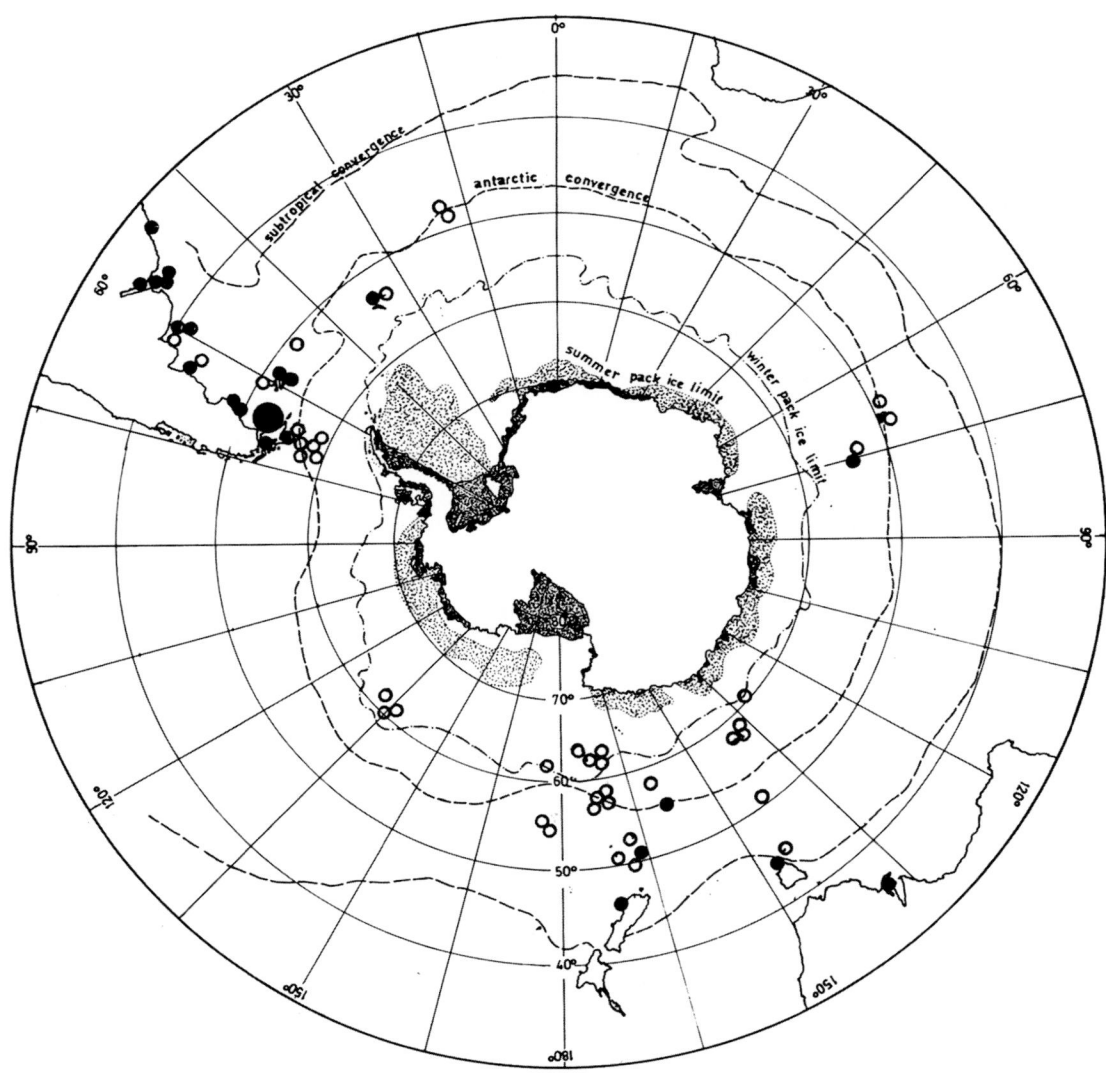

**Figure 2** *Known distribution of the spectacled porpoise. Open circles represent sightings, closed circles are single strandings, and the large closed circle represents over 298 specimens.*

sometimes enter river estuaries and channels. We do not know why so many specimens are found on the beaches of Tierra del Fuego (298 specimens to date) and so few in other areas.

Nothing is known about the abundance of this species. It is the second–most frequent species found stranded in beach SURVEYS in eastern Tierra del Fuego but is uncommon elsewhere.

## III. Ecology

There are no observations of feeding behavior. Only a few stomachs have been examined (stranded animals are soon flensed by birds). One contained a small piece of algae and another meager amounts of anchovy (*Eugraulis* sp.) and stomatopods. The stomach of a 210-cm male from southern Brazil contained remains of Ascidiaceae (Tunicata) (Pinedo *et al.*, 2002). The stomach of the 119-cm juvenile female from Tasmania contained a milk-like liquid (probably still suckling) and two sets of cephalopod beaks (*Sepia* sp.), and a partly digested ornate cowfish (*Aracana ornata*) was found in

the esophagus (Evans *et al.*, 2001). Both young animals from southern Australia were emaciated.

Killer whales (*Orcinus orca*), leopard seals (*Hydrurga leptonyx*), and sharks are possible predators of this species. The stomach of a killer whale stranded in Brazil (33°45'S) contained partial remains of two porpoises, possibly this species, but more probably *P. spinipinnis*. The small animal from Tasmania had fine, incisor-like tooth marks on its side, thought to be caused by a small shark which mouthed the porpoise after death (Evans *et al.*, 2001).

This is mainly an oceanic animal which occasionally comes near shore. Next to nothing is known of its life at sea.

The taphonomy of this species is interesting. Its "yummy factor" is very high. Predators and scavengers, especially birds such as kelp gulls and giant petrels, seem to prefer the flesh of this species; they attack dead animals immediately and nearly skeletonize them within an hour or two. The flippers, being small, are soon lost. Organs and muscle are rapidly consumed and the areas with little flesh soon become dry and hard (head, tail stock, flukes, and dorsal fin) and are

**S**

**Figure 3** *The spectacled porpoise can be sexed by the size of the dorsal fin.*

then ignored. The probable sex of the animal can be determined by the shape and size of the dried dorsal fin (Fig. 3).

## IV. Behavior and Physiology

All sightings to date have been of one–five animals. Mother–calf pairs were recorded stranded twice, but there have been no mass strandings. Five of the seven fresh specimens came ashore alive. Nothing is known of the social behavior, sounds, or associations of this species.

Group size of sightings was small, averaging 2.0 animals. Six cow–calf pairs were accompanied by one–two other animals, including an adult male (Sekiguchi *et al.*, 2006).

Spectacled porpoises are difficult to sight at sea. Swimming behavior has been described as "unobtrusive" and "inconspicuous". Two animals circling in a small bay at South Georgia in March 1995 reacted to boat noise by STRANDING and were pushed back to deeper water. The swimming before stranding was reported as "slow and steady, with dorsal fins breaking the surface together" (A. Martin, personal communication). A re-floated calf had a respiration rate of four to seven breaths a minute. It swam with a slow, rolling motion, arching its back strongly on each dive. Most sightings described "swimming with a slow rolling motion, without any splash." When the vessel approached, the animals swam fast, submerged just under the sea surface, producing almost continuous fluke prints on the surface. There is only one record of porpoising and none of bow-riding (Sekiguchi *et al.*, 2006).

Nothing is known of migrations or seasonal movements of this species. Most sightings have been pelagic, but some animals must move inshore off Tierra del Fuego, where they either strand due to the gently sloping beaches and high tides or die in shore-based nets.

## V. Life History

Twenty-six Tierra del Fuego animals have been aged. The youngest animal that was physically mature (all epiphyses fused) had three growth-layer groups (GLGs), while one animal with five GLGs was not yet mature. Animals of over 189 cm were physically mature. The maximum age to date was 16 GLGs. The smallest neonates examined measured 94 and 109 cm. Teeth were beginning to erupt in two females of 125 (Tierra del Fuego) and 127 cm (South Australia).

One hundred sixty-five Tierra del Fuego animals were classified by physical maturity. Of these, 10 were neonates, 9 were juveniles, 99 subadults, and 47 physically adult.

Little reproductive information is available. Gonads were examined for one male, which at 205 cm and 6 GLGs was sexually mature. Males of 189–224 cm and 4–5 GLGs were physically mature. No ovaries have been examined. Three females of 183–186 cm had calves or fetuses. Female teeth show a change in layering after the second or third GLG, suggesting an early sexual maturation, as found in harbor porpoises.

Calves of 125–163 cm with unerupted teeth stranded from February to May; birth is probably in the late southern spring or summer (November–February). A calf was seen in December off Heard Island. During IWC IDCR and SOWER cruises, six small calves accompanied by a cow were seen from late December to mid-February, confirming the above (Sekiguchi *et al.*, 2006). There is no information on pregnancy rates, gestation, or lactation periods.

No ectoparasites have been reported. Nematodes were found in the blowhole and nasal sacs of the Lahille specimen and in the stomachs of three animals from Tierra del Fuego. The stomach of a 210-cm male from southern Brazil contained fragments of *Anisakis simplex* (Pinedo *et al.*, 2002). No parasites were found in the organs, body cavities, or passages of two young animals from southern Australia. Nothing else is known of parasites or disease.

## VI. Interactions with Humans

One small, emaciated 127-cm female, probably still nursing, which live-stranded on March 24, 1997, at Horseshoe Bay, off Port Elliot, South Australia, was kept in captivity at the RSPCA marine rescue unit, Adelaide. Although it was medicated and force-fed, the animal died early on March 29 (Evans *et al.*, 2001).

The 119-cm female from Tasmania had lacerations near the dorsal fin which may have been from a boat propeller.

Remains of spectacled porpoises were found in 6000- and 1400-year-old kitchen middens of the canoe people in Tierra del Fuego, who harpooned cetaceans for food (Piana *et al.*, 1985). In the twentieth century, whaling captains off South Georgia and fishermen off Uruguay also hunted spectacled porpoises for food. The first four specimens known came ashore alive and were collected by fishermen. There is no known deliberate exploitation of this species at present. At least some (and possibly most) of the specimens found in Tierra del Fuego were taken incidentally in coastal gillnets; the location of the dead animals coincides with that of fishing areas (Goodall *et al.*, 1994; Goodall and Schiavini, 1995). At least one animal died in offshore bottom or mid-water trawls off Patagonia.

None of the biological information needed for conservation and management is available.

## See Also the Following Articles

Burmeister's Porpoise ■ Harbor Porpoise ■ Vaquita

## References

Brownell, R. L., Jr. (1975). Phocoena dioptrica. *Mammal. Spec.* **66**, 1–3.
Brownell, R. L., Jr. (1999). Spectacled porpoise *Australophocaena dioptrica* Lahille, 1912. *In* "Handbook of Marine Mammals: The Second Book of Dolphins and Porpoises" (S. H. Ridgway, and R. Harrison, eds), Vol. 6, pp. 379–391. Academic Press, London.
Bruch, C. (1916). El macho de *Phocaena dioptrica* Lah. *Physis (Buenos Aires)* **2**, 461–462.

Crespo, E. A., Corcuera, J. F., and López Cazorla, A. (1994). Interactions between marine mammals and fisheries in some coastal fishing areas of Argentina. *Rep. Int. Whaling Comm.*(Special Issue 15), 269–281.

Evans, K., Kemper, C., and Hill, M. (2001). First records of the spectacled porpoise *Phocoena dioptrica* in continental Australian waters. *Mar. Mamm. Sci.* **17**, 161–170.

Fordyce, R. E., Mattlin, R. H., and Dixon, J. M. (1984). Second record of spectacled porpoise from subantarctic southwest Pacific. *Sci. Rep. Whales Res. Inst.* **34**, 159–164.

Goodall, R. N. P., and Schiavini, A. C. M. (1995). On the biology of the spectacled porpoise, *Australophocaena dioptrica*. *Rep. Int. Whaling Comm.*(Special Issue 16), 411–453.

Goodall, R. N. P., Schiavini, A. C. M., and Fermani, C. (1994). Net fisheries and net mortality of small cetaceans off Tierra del Fuego, Argentina. *Rep. Int. Whaling Comm.*(Special Issue 15), 195–304.

Kasamatsu, F., and Joyce, J. (1995). Current status of odontocetes in the Antarctic. *Antarct. Sci.* **7**, 365–379.

Lahille, F. (1912). Nota preliminar sobre una nueva especie de marsopa del Río de la Plata (*Phocaena dioptrica*). *An. Mus. Nat. Hist.* **23**, 269–278.

Perrin, W. F., Goodall, R. N. P., and Cozzuol, M. A. (2000). Osteological variation in the spectacled porpoise (*Phocoena dioptrica*). *J. Cetacean Res. Manage.* **2**, 211–215.

Piana, E., Orquera, L., Goodall, R. N. P., Galeazzi, A. R. and Sobral, A. P. (1985). Cetacean remains in Beagle Channel shell middens. Sixth Biennial Conference on the Biology of Marine Mammals, Vancouver, Canada, 22–26 November [Abstract], 64.

Pinedo, M. C., Barreto, A. S., Lammardo, M.. P., Andrade, A. L. V., and Geracitano, L. (2002). Northernmost records of the spectacled porpoise, Layard's beaked whale, Commerson's dolphin and Peale's dolphin in the southwestern Atlantic Ocean. *Aquat. Mamm.* **28**, 32–37.

Rosel, P. E., Haygood, M. G., and Perrin, W. F. (1995). Phylogenetic relationships among the true porpoises (Cetacea: Phocoenidae). *Mol. Phyl. Evol.* **4**, 463–474.

Sekiguchi, K., *et al.* (9 authors) (2006). The spectacled porpoise (*Phocoena dioptrica*) in Antarctic waters. *J. Cetacean Res. Manage.* **8**, 265–271.

# Sperm Whale
## *Physeter macrocephalus*

### HAL WHITEHEAD

Sperm whales (*Physeter macrocephalus*) are animals of extremes. They have unusually large body sizes, sexual dimorphism, brain sizes, home ranges, dive depths, and dive times; they have an ecological role that may be unrivaled in the ocean; and their vocalizations, social structure, and historical relationship with humans are all remarkable. The likelihood of evolutionary or ecological links between these extreme attributes forms one of the scientific attractions of this animal (Whitehead, 2003).

A potential key to the sperm whale's simultaneous possession of a wide range of extreme biological attributes lies in a series of remarkable parallels with the African elephant (*Loxodonta africana*). These include large body sizes, brain sizes, substantial sexual dimorphism, similar life history variables, large ranges, remarkably congruent matrilineally based social systems, and breeding systems in which males roam between groups of females and generally only mate successfully when in their late twenties or older. The highly evolved spermaceti organ may have

paralleled the trunk (another extreme nose) in allowing the animals efficient access to a wide range of resources. Meanwhile, in these animals, large sizes and cooperative societies gave efficient defense against predators, allowing long lives. Long, safe lives, in turn, promote the formation of significant long-term relationships among animals. Thus elephants and sperm whales evolved in highly social populations, near carrying capacity, and become dominant members of their ecological guilds. However, such animals are not well adapted to recovering from sudden depletion. Both the elephant and the sperm whale have been heavily hit, and, because their social structures are so important, exploitation has had consequences beyond animals killed directly.

## I. Characteristics and Taxonomy

In 1758, Linnaeus described four sperm whales in the genus *Physeter*. It soon became clear that all refer to the same species, but there has been a long, and sometimes contentious, debate as to whether *P. catodon* or *P. macrocephalus* has precedence. Currently, most, but not all, authorities prefer *P. macrocephalus*.

The common name, "sperm whale," may have resulted from whalers misinterpreting the function of the spermaceti oil found in the massive forehead of the whale, or the fact that the cooled spermaceti has some physical resemblance to mammalian sperm.

The closest living relatives of the sperm whale are the much smaller dwarf and pygmy sperm whales (*Kogia breviceps* and *K. sima*). Sperm whales seem to have separated from other odontocetes early in modern cetacean evolution, about 20–30 million years ago. See SPERM WHALES: EVOLUTION for more information.

Sperm whales are the largest of the odontocetes (Fig. 1), and the most SEXUALLY DIMORPHIC cetaceans in body length and weight. While adult females reach about 11 m in length and 15 tons, a physically mature male is approximately 16 m and 45 tons (Rice, 1989).

The most distinctive feature of the sperm whale is a massive nasal complex, one quarter to one-third of the length of the animal, situated above the lower jaw and in front of the skull. It principally contains the spermaceti organ, which is enclosed in a muscular "case" (Fig. 2). This is a roughly ellipsoidal-shaped structure made of spongy tissue filled with spermaceti oil and bounded at both ends by air sacs. Between the spermaceti organ and the upper jaws is the "junk," a complex arrangement of spermaceti oil and connective tissue. Spermaceti oil, which has the properties of a wax, differs chemically from the oils found in the "melons" of most other odontocetes.

There is considerable asymmetry in the parts of the skull and air passages that surround the spermaceti organ. This is externally manifested most clearly by a blow which is pointed forward and to the left from the tip of the snout. Compared with the blows of similar-sized baleen whales, the blow of a sperm whale is comparatively weak, low, and hard to see.

Behind the sperm whale's skull lies the largest BRAIN of any animal (mean of 7.8 kg in mature males). However, as a proportion of body size, the sperm whale's brain is not remarkable, and we have no direct information on the sperm whale's cognitive abilities, although its complex social system is consonant with those found in other cognitively advanced mammals.

The sperm whale has 20–26 large conical teeth in each rodlike lower jaw. These teeth do not seem to be necessary for feeding, as they do not erupt until near puberty, and well-nourished sperm whales have been caught that lack teeth, or even lower jaws. The teeth in the upper jaw seem to be vestigial and rarely erupt.

Large corrugations cover most of the body behind the eye, but the surface of the head and the flukes are smooth. The majority

**Figure 1**  *Young male sperm whale* (Physeter macrocephalus) *off Kaikoura, New Zealand. Photo courtesy B. Würsig.*

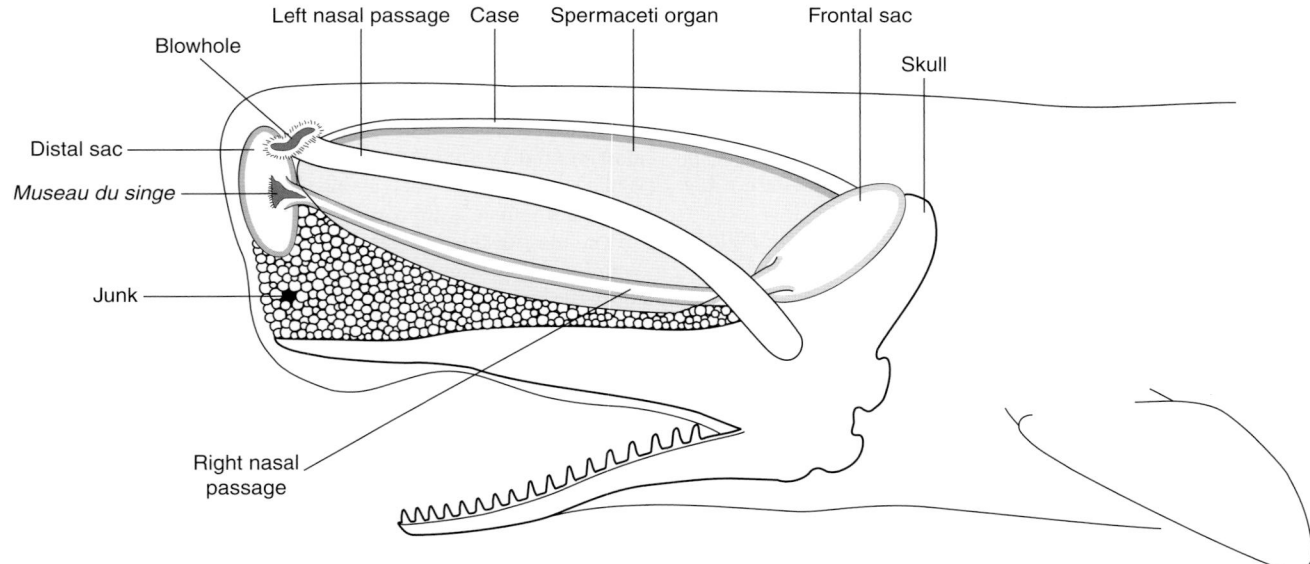

**Figure 2**  *A diagram of the head of a sperm whale* (Physeter macrocephalus). *From Ellis (1980).*

of the body is dark gray in most sperm whales, but there is often a bright white lining to the mouth and sometimes white patches on the belly. The occasional sperm whale has larger white patches, especially in mature males, around the head. The flippers are relatively small and paddle shaped, and the flukes are fairly flat and triangular shaped. The dorsal fin is low, thick, and usually rounded. Especially in mature females, it may be topped by a white or yellowish rough callus. The dorsal ridge, behind the dorsal fin, consists of a series of large crenulations.

## II. Distribution and Abundance

With the exception of humans and killer whales (*Orcinus orca*), few animals on EARTH are as widely distributed as the sperm whale.

They can be seen near the ice-edge in both hemispheres and are also common along the equator, especially in the Pacific. As with so many aspects of sperm whale biology, the sexes have very different distributions.

Although sperm whales have been sighted in most regions of deep water, there are some areas which the whalers called "grounds," where they are more abundant. Many of the grounds coincide with areas of generally higher primary productivity, usually resulting from upwelling, although there are grounds in apparently unproductive waters, such as the Sargasso Sea (Jaquet, 1996).

Concentrations of a few hundred to a few thousand sperm whales can be found in areas a few hundred kilometers across characterized by a relatively high deep water biomass and usually situated within

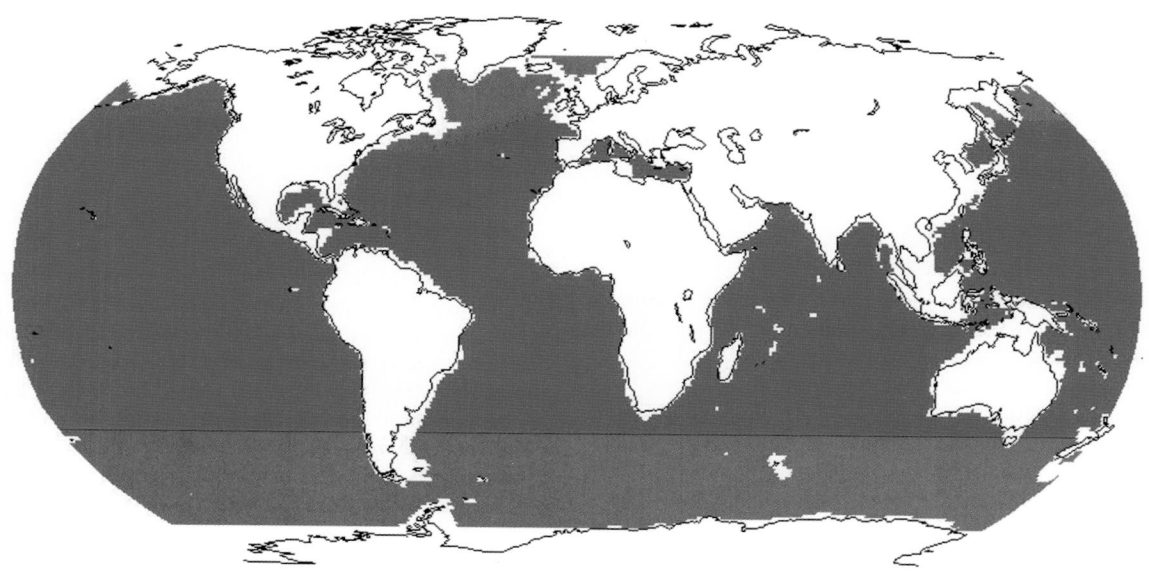

**Figure 3** *Worldwide distribution of sperm whales* (Physeter macrocephalus)*: Females and males in green and just adult males in red.*

grounds. Sometimes aggregations of 50 or more sperm whales can be found within a few kilometers, presumably the result of concentrations of food.

Female sperm whales almost always inhabit water deeper than 1000 m and at latitudes less than 40° (except 50°N in the North Pacific), corresponding roughly to sea surface temperatures greater than 15°C (Fig. 3) (Rice, 1989). Although sometimes seen close to oceanic islands rising from deep ocean floors, most female sperm whales are far from land.

Young male sperm whales accompany the females in tropical and subtropical waters. On leaving their female relatives, sometime between 4 and 21 years of age, the males gradually move to higher latitudes: the larger and older the male, the higher the average latitude. Large males may be found near the edge of pack ice in both hemispheres, although they return to the warm water breeding grounds on an unknown schedule.

The large males of high latitudes can be found over almost any ice-free deep water, but, like the low latitude females, they are more likely to be sighted in productive waters, such as those along the edges of continental shelves. However, in some areas, such as off New York and Nova Scotia, the large males are sighted regularly in waters less than 300 m deep.

Extrapolation from surveys of sperm whale density that covered 24% of their global habitat suggests a current population of about 360,000 animals (CV = 0.36), down from a pre-whaling population of about 1,110,000 whales (Whitehead, 2002).

Information on stock structure in sperm whales is confused, but rather few differences have been found between animals in different ocean areas (Dufault *et al.*, 1999). Genetic studies have been unable to find clearly distinct stocks at geographical scales of less than an ocean basin (Lyrholm *et al.*, 1999). Sperm whales worldwide are remarkably homogeneous in the mitochondrial, maternally transmitted, genome. Explanations for this unexpected result (also found in pilot, *Globicephala* spp., and killer whales) include a historical population bottleneck, a demographic consequence of the result of these whales' matrilineal social system, increased selection on the mitochondrial

genome, or the indirect effects of selection of matrilineally transmitted cultural traits (see CULTURE IN WHALES AND DOLPHINS).

## III. Ecology

The sperm whale has a most catholic DIET consisting of many of the larger organisms that inhabit the deeper regions of the oceans. Females appear to principally eat squids weighing between 0.1 and 10 kg. Favored taxa include Ommastrephidae, Onychoteuthidae, Gonatidae, Pholidoteuthidae, Octopoteuthidae, Histioteuthidae, and Cranchiidae (Kawakami, 1980). Of these, the histioteuthids, mesopelagic gelatinous pelagic cephalopods weighing about 0.1–1 kg, have featured at or near the top of the list of preferred food items in several studies of sperm whales. Females also eat larger prey, such as the giant squid (*Architeuthis* spp.) and the jumbo squid (*Dosidicus* spp.), as well as noncephalopod prey, especially demersal and mesopelagic fish.

Males use the same squid taxa as females, but tend to eat larger individuals. In addition, they eat species that are largely restricted to higher latitudes such as the colossal squid (*Mesonychoteuthis hamiltoni*) of Antarctic waters (Clarke, 1987) and are also more likely to eat demersal fish (including sharks, rays, and gadoids) than females. Off Iceland and in the northern Gulf of Alaska, their primary food is fish.

Sperm whales have competitors for many of these food items—beaked whale (Ziphiidae) and elephant seals (*Mirounga* spp.) also eat mesopelagic squid—and there may be important nonmammal predators of the species that the sperm whales use. For instance, the jumbo squid eats the smaller histioteuthid squids. Although we know virtually nothing of the quantitative ecology of the deep ocean, sperm whales, because of their size and numbers, seem to dominate this trophic level in terms of biomass removed (Clarke, 1987; Whitehead, 2003). Rough estimates of the worldwide consumption of prey by the world's sperm whale population, about 100 million metric tons per year, are comparable with the current annual catch by all human marine fisheries.

Killer whales have been observed attacking sperm whales. They were usually unsuccessful, although at least one sperm whale was

**S**

killed during a well-documented encounter off California (Pitman *et al.*, 2001), and sperm whale remains have, very occasionally, been found in the stomachs of killer whales. Pilot whales have been seen harassing sperm whales on a number of occasions (Fig. 4), but whether they are any real threat to the much larger sperms is uncertain. Large sharks are also potential predators, especially of young animals (Best *et al.*, 1984).

## IV. Life History

The sperm whale is the epitome of the "*K*-selected" mammal, one presumed to have evolved in an environment of competition for resources with members of its own species. It has a very low birth rate, slow growth, slow maturation, and high survival.

Young are born, almost always singly and with an equal sex ratio, following an approximately 14–16-month gestation (Best *et al.*, 1984). Although sperm whales may eat solid food before their first birthday, they continue suckling for several years. The females reach sexual maturity at about age 9 when roughly 9 m long, at which age growth starts to slow (Best *et al.*, 1984). They give birth roughly once every 5 years, although pregnancy rates have been found to vary between areas and in the same area at different stages of exploitation by humans. Female reproductive rates decline with age, and very few give birth after age 40. Females reach physical maturity when growth ceases, at about 30 years old and 10.6 m long.

Males, which are slightly larger than females during the first 10 years of life, continue to grow at a substantial rate until well into their 30s, finally reaching physical maturity at about 16 m long when roughly 50 years old. In males, puberty is prolonged, lasting approximately between ages 10 and 20 years old. However, males seem not to generally take much of an active role in breeding before their late 20s (Best, 1979).

Rates and causes of natural mortality are not well known. Longevity can be at least 50 years. Groups occasionally mass strand on shorelines, usually with fatal results for all members.

## V. Behavior and Physiology

### A. Vocalizations

The principal function of the spermaceti organ and its associated structures (the junk, air sacs and passages, and the *museau de singe*

clapper system at the front of the organ) (Fig. 2) is to form, and focus, the sperm whale's click, making an extremely powerful communication and ECHOLOCATION system (Cranford, 1999; Møhl *et al.*, 2000). The vocalizations of sperm whales consist almost entirely of clicks, although a few relatively quiet "squeals" and "trumpets" are made. However, these clicks are used both for foraging (see below) as well as social signals.

"Codas," stereotyped series of 3–20 clicks lasting 0.2–2 sec, are heard in social situations. Groups of females have distinctive repertoires of codas. In the Pacific, groups are delineated clearly into clans by distinctive coda repertoires (Rendell and Whitehead, 2003). Coda repertoire is probably acquired culturally from within family units.

Whereas codas are heard principally from social females, another form of click, the "slow click" or "clang," in which distinctively ringing clicks are repeated every 6–8 sec, is largely or entirely produced by large males (Weilgart and Whitehead, 1988). The function of slow clicks is not clear, but they may attract females, repel other males, or be used to ECHOLOCATE off other whales.

### B. Movement

Scientists have studied the movements of sperm whales using photoidentification, satellite tags, and artificial marks, as well as by following groups over periods of a few days. Sperm whales usually move through the water at about 4 km/h. When feeding conditions are poor, their tracks are straight, resulting in daily displacements of about 90 km/day. However, when food is plentiful, the animals stay in much smaller areas 10–20 km across. Female home ranges seem to be generally of the order of 2000 km across. Some males roam more widely, but others are found quite consistently over several years in restricted coastal waters.

Migrations of the sperm whale are not as regular or as well understood as those of most baleen whales. In some midlatitudes there appears to be a general seasonal north–south migration, with whales moving poleward in summer, but, in equatorial and some temperate areas, there is no clear seasonal migration (Whitehead, 2003).

### C. Social Structure

The life of a female sperm whale is overwhelmingly social. She is always in the company of other females, some of whom are her

Figure 4 *Sperm whales* (Physeter macrocephalus) *being harassed by a school of several dozen short-finned pilot whales,* Globicephala macrorhynchus, *in the northern Gulf of Mexico. Photo courtesy B. Würsig.*

relatives. The basic element of sperm whale society is a family unit, consisting of about 12 females and their young (Whitehead, 2003). Most females spend their lives in the same unit, with their close female relatives. However, units may contain two or more matrilines and there are recorded instances of females switching units. Two or more units may travel together for a few days as a cohesive group of about 20 animals.

Within the social units there is communal care for the young, with evidence for females suckling calves who are not their own offspring (e.g., Best *et al.*, 1984). Young sperm whale calves do not seem to be able to dive to foraging depths for as long as their mothers. Instead they remain at or near the surface, moving between the members of their group while they are breathing. This babysitting begins soon after birth. Members of groups with young calves seem to intentionally stagger their dives, thus providing better babysitting for the young.

When faced with predators, particularly killer whales, female sperm whales quickly cluster. Two defensive patterns have been described. In the "marguerite" or "wagon wheel" formation, the members of the group place their heads together at the hub, with the bodies radiating out like spokes. In contrast, when adopting the "heads-out" formation, the sperm whales face their attackers, tightly aligned in a rank, and seem to principally use their jaws for defense. Young calves stay toward the center of whichever defensive formation is adopted. Females have been observed risking themselves to assist unit members in peril during a killer whale attack (Pitman *et al.*, 2001).

Young males leave their natal unit when between 4 and 21 years old and then are found in loose aggregations, sometimes called "bachelor schools," with other males of approximately the same size and age (Best, 1979). As the males age and grow they move to generally higher latitudes and the aggregation sizes become smaller, until the largest males are usually alone. Repeat association between males on more than 1 day are rarely observed (Letteval *et al.*, 2002). However, mature and maturing males do strand on beaches together, suggesting significant social relationships on a scale not readily apparent to human boat-based observers.

The large mature males, in their late 20s and older, return to the tropical breeding grounds to mate, although the timing of such visits is largely unknown. When on the breeding grounds, the large males roam between groups of females, usually spending just a few minutes or hours with each (Whitehead, 2003). They are presumably searching for receptive females. There are no clear descriptions of mating itself. The breeding males seem to roam independently and generally avoid one another, although they are sometimes observed within the same group of females, and occasionally fight. These fights are rarely observed, but many large males have deep scars made by the teeth of other males.

## D. Behavioral Modes

Sperm whales possess two quite distinct behavioral modes: foraging and social/resting (Whitehead and Weilgart, 1991). While foraging, the animals make repeated deep dives. Modal dives are about 600 m and for about 45 min (Watwood *et al.*, 2006), but dives can be much deeper (to over 1000 m), shallower (e.g., when in shelf waters 200 m deep), or longer. Between dives the whales come to the surface to breathe for about 9 min. The dive is usually signaled by the raising of flukes out of the water. The descent to depth, as well as the return to the surface, can be nearly vertical.

While foraging, sperm whales generally make regularly spaced clicks at intervals of 0.5–1.0 sec, a searching sonar (Watwood *et al.*, 2006). These are interrupted by creaks, consisting of clicks with accelerating rates, which are assumed to indicate short-range sonar during prey capture events. Small gelatinous squid, such as histioteuthids, are usually both relatively inactive and bioluminescent, so these can be captured using visual or acoustic cues. Larger, more muscular animals may need active chasing.

Groups of females and immatures spread out over 1 km or more of ocean when foraging, often forming a rank aligned perpendicular to the direction of travel. In contrast, males seem generally to forage independently.

Female and young sperm whales spend approximately 75% of their time foraging (Whitehead and Weilgart, 1991). However, during periods of several hours, often in the afternoon, they gather at or near the surface. At these times, their behavior is highly variable. Sometimes the animals may lie still and quiet, closely clustered, for hours at a time, apparently resting (Fig. 5). Particularly at the start or end of non-foraging periods their behavior may be much more active, with breaches, lobtails, animals rolling, maneuvering and touching one another, and codas and creaks being emitted. Large males also lie quietly at the surface for long periods, but they are usually alone unless accompanying a group of females.

**Figure 5** *Sperm whales* (Physeter macrocephalus) *in resting/social mode off the Galapagos Island: one adult male surrounded by immatures and females. Photo courtesy H. Whitehead lab.*

S

Figure 6    *Open-boat sperm whaling. From Beale (1839).*

## VI. Interactions with Humans

In the early eighteenth century, New Englanders began to hunt sperm whales for commercial purposes off their own shores. Over the next two centuries, sperm whaling grew to be a major worldwide industry. By the 1830s, about 5000 sperm whales were being killed each year by whalers from several countries, especially the United States, and the oil produced was a vital element of the burgeoning industrial revolution (Starbuck, 1878). The whalers sailed all oceans of the world in their square-rigged ships. On sighting sperms, open whaleboats were lowered and rowed or sailed to the whales. The whalers threw harpoons into the animals, and then killed them using lances (Fig. 6). Dead animals were towed to the whale ship, where the oil was baled from the spermaceti organ, the BLUBBER stripped from the body and boiled to render the oil, and virtually all the remainder of the carcass discarded. In the nineteenth century, sperm whaling had become such a significant enterprise that it had important effects on fields as diverse as literature (Herman Melville's great novel "Moby Dick") and ocean exploration. The open-boat hunt declined during the latter half of the nineteenth century because of the development of petroleum products as alternatives for sperm oil, an apparent decline in the sperm whale population, and for other reasons.

After about 50 years of relative peace, sperm whale populations were again hit hard following World War II. The whalers of the twentieth century chased sperm whales using mechanized catcher vessels equipped with sonar and killed them using explosive harpoons shot from harpoon guns. This whaling was also widely distributed and took a large number of sperms (up to 30,000 per year). However, unlike their earlier counterparts, the modern whalers used almost all of the whale, and preferentially targeted males. Much, but not all, of this whaling was carried out under the auspices of the International Whaling Commission (IWC). Sperm whale populations, particularly the male portions, were reduced substantially. Commercial sperm whaling declined in the 1970s and 1980s and virtually ceased with the IWC moratorium in 1988. Currently there is a small catch of sperm whales in Lamalera, Indonesia, using primitive methods, and a "scientific" hunt for sperm whales by Japan.

Nowadays, most interactions between sperm whales and humans are more benign. Sperm whales are the principal subjects of WHALE-WATCHING operations in several locations, including Kaikoura (New Zealand), Andenes (Norway), the Azores (Portugal), and Dominica (West Indies). Scientists study their behavior from small vessels in these and a few other locations.

The sperm whale has survived the onslaught of the whalers better than most other large whales. There are still a few hundred thousand sperm whales left in the ocean, sperm whale food is of little interest to human fishers, and their deep water home is further from most sources of pollution than the preferred habitat of most other marine mammals. However, the effects of whaling seem to be lingering. In the southeast Pacific, where modern whaling on males was particularly severe, large breeding males are still scarce and calving seems depressed below the replacement rate (Whitehead *et al.,* 1997). In other parts of the world, the picture appears brighter. For instance, in the northwest Atlantic, where sperm whales were hit hard by whaling in the eighteenth and nineteenth centuries but largely spared by modern whalers, the population appears relatively healthy. However, sperm whales have very low reproductive rates even in the best of times—they do not recover quickly when depleted. They are also killed inadvertently in a range of ways, including entrapment in fishing gear, choking on plastic bags, and collision with ships. The chemical POLLUTION levels in their blubber are generally higher than that of baleen whales, but lower than in inshore odontocetes.

## See Also the Following Articles

Ambergris ■ Culture in Whales and Dolphins ■ Scrimshaw ■ Sperm Whales, Evolution ■ Whaling, Traditional

## References

Beale, T. (1839). "The Natural History of the Sperm Whale." John van Voorst, London.

Best, P. B. (1979). Social organization in sperm whales, *Physeter macrocephalus. In* "Behavior of Marine Animals" (H. E. Winn, and B. L. Olla, eds), 3, pp. 227–289. Plenum, New York.

Best, P. B., Canham, P. A. S., and Macleod, N. (1984). Patterns of reproduction in sperm whales, *Physeter macrocephalus. Rep. Int. Whaling Comm.* **6**(Special Issue), 51–79.

Clarke, M. R. (1987). Cephalopod biomass—estimation from predation. *In* "Cephalopod Life Cycles" (P. R. Boyle, ed.), 2, pp. 221–237. Academic Press, London.

S

Cranford, T. W. (1999). The sperm whale's nose: sexual selection on a grand scale? *Mar. Mamm. Sci.* **15**, 1133–1157.

Dufault, S., Whitehead, H., and Dillon, M. (1999). An examination of the current knowledge on the stock structure of sperm whales (*Physeter macrocephalus*) worldwide. *J. Cetacean Res. Manage.* **1**, 1–10.

Ellis, R. (1980). "The Book of Whales." Knopf, New York.

Jaquet, N. (1996). How spatial and temporal scales influence understanding of sperm whale distribution: a review. *Mammal Rev.* **26**, 51–65.

Kawakami, T. (1980). A review of sperm whale food. *Sci. Rep. Whales Res. Inst.* **32**, 199–218.

Letteval, E., *et al.* (8 authors) (2002). Social structure and residency in aggregations of male sperm whales. *Can. J. Zool.* **80**, 1189–1196.

Lyrholm, T., Leimar, O., Johanneson, B., and Gyllensten, U. (1999). Sex-biased dispersal in sperm whales: contrasting mitochondrial and nuclear genetic structure of global populations. *Proc. R. Soc. Lond. B* **266**, 347–354.

Møhl, B., Wahlberg, M., Madsen, P. T., Miller, L. A., and Surlykke, A. (2000). Sperm whales clicks: directionality and source level revisited. *J. Acoust. Soc. Am.* **107**, 638–648.

Pitman, R. L., Ballance, L. T., Mesnick, S. I., and Chivers, S. J. (2001). Killer whale predation on sperm whales: observations and implications. *Mar. Mamm. Sci.* **17**, 494–507.

Rendell, L., and Whitehead, H. (2003). Vocal clans in sperm whales (*Physeter macrocephalus*). *Proc. R. Soc. Lond. B* **270**, 225–231.

Rice, D. W. (1989). Sperm whale. *Physeter macrocephalus* Linnaeus, 1758. *In* "Handbook of Marine Mammals" (S. H. Ridgway, and R. Harrison, eds), 4, pp. 177–233. Academic Press, London.

Starbuck, A. (1878). History of the American whale fishery from its earliest inception to the year 1876. *In* "United States Commission on Fish and Fisheries, Report of the Commissioner for 1875–1876" Appendix A. Government Printing Office, Washington, D.C.

Watwood, S. L., Miller, P. O., Johnson, M., Madsen, P. T., and Tyack, P. L. (2006). Deep-diving foraging behaviour of sperm whales (*Physeter macrocephalus*). *J. Anim. Ecol.* **75**, 814–825.

Weilgart, L. S., and Whitehead, H. (1988). Distinctive vocalizations from mature male sperm whales (*Physeter macrocephalus*). *Can. J. Zool.* **66**, 1931–1937.

Whitehead, H. (2002). Estimates of the current global population size and historical trajectory for sperm whales. *Mar. Ecol. Prog. Ser.* **242**, 295–304.

Whitehead, H. (2003). "Sperm Whales: Social Evolution in the Ocean." University of Chicago Press, Chicago.

Whitehead, H., and Weilgart, L. (1991). Patterns of visually observable behaviour and vocalizations in groups of female sperm whales. *Behaviour* **118**, 275–296.

Whitehead, H., Christal, J., and Dufault, S. (1997). Past and distant whaling and the rapid decline of sperm whales off the Galápagos Islands. *Conserv. Biol.* **11**, 1387–1396.

# Sperm Whales, Evolution

## GURAM A. MCHEDLIDZE

Sperm whales include the cetacean families Physeteridae (the modern sperm whale) and Kogiidae (the modern pygmy and dwarf sperm whales). Modern sperm whales are represented by two genera only, but sperm whales were much more diverse in past times. Phylogenetically, sperm whales are usually thought to be close to the root of the odontocetes, and they retain many characters that are primitive for odontocetes. They do, however, also have some highly derived features and are not very similar to the primitive Eocene cetaceans (archaeocetes) that they are descended from.

Physeterids and kogiids may be subsequent branches of the basal odontocete phylogenetic tree, or both may be derived from a common sperm whale ancestor, such as *Ferecetotherium* (see Evolution of Marine Mammals).

## I. Physeteridae

Physeterids (including the modern sperm whale *Physeter macrocephalus*) are highly specialized for teutophagy (eating cephalopods such as squid) at great depths, and many parts of sperm whale anatomy show adaptations for this behavior. Some specialized morphologies are already present in the oldest known physeterid, *Ferecetotherium kelloggi* (Mchedlidze, 1984), from the late Oligocene of Azerbaijan, but many fossil sperm whales were probably fisheaters.

The earliest physeterid, *Ferecetotherium*, was small (approximately 5 m long) and had a small head. An increase in size and head size occurred in the Miocene, and modern *Physeter* has a body length of 16 m in females and 21 m in males. The head of *P. macrocephalus* is approximately one-third the size of the body. An increase in body size happened throughout physeterid evolution (Fig. 1).

*Ferecetotherium* had a relatively narrow rostrum. Miocene sperm whales are characterized by the widening of the rostrum. This widening took different paths in different species. In some clades, widening occurred in the maxillae and premaxillae, whereas in other clades only one of these elements enlarged. In the Miocene *Diaphorocetus* and *Orycterocetus*, the premaxilla and maxilla are nearly equal in width near the base of the rostrum, but the tip of the rostrum consists entirely of the premaxillae. In contrast, in *Physeter* the maxillae make up nearly all of the rostrum, and the premaxillae are only exposed near the tip of the rostrum. Widening of the rostrum was probably the result of the enlargement of the spermaceti organ, a large structure housed in a depression (the supracranial basin) on the forehead. The supracranial basin is characteristic of sperm whales, although it is relatively small in older forms. Posteriorly, this basin is bounded by the supraoccipital and the posterior plate of the right premaxilla, and laterally vertical plates of the frontals bound the basin. In modern *Physeter*, the lateral walls of the supraoccipital basin are formed by the maxillae. All physeterids, including *Ferecetotherium*, are also characterized by strong asymmetry of the rostrum, particularly the premaxillae and nasals.

Macroevolutionary changes also occurred in the mandible. In modern sperm whales, the maxilla (upper jaw) is much wider than the mandible as a result of the widening of the rostrum and a narrowing of the mandible. These give the lower jaw a peculiar undersized image. This mismatch evolved gradually in physeterids. *Ferecetotherium* has a primitive mandible, not unlike that of archaeocetes. In late Eocene archaeocetes, the lower margin of the mandible is horizontal, but the mandibular depth increases posteriorly and teeth are present on the ascending ramus. Physeterid specializations in the mandible of *Ferecetotherium* include the parallel edges of the ramus, the superior displacement of the mandibular condyle, and the reduced coronoid process.

As a rule, modern *Physeter* only has teeth in the lower jaw, although occasionally an individual is found with upper teeth. Upper teeth are still present in physeterids from Oligocene through middle Miocene. Teeth in *Physeter* are positioned in the upper edge of the mandible. In *Ferecetotherium*, posterior teeth are rooted in the upper edge of the mandible, but anterior teeth (except the first tooth) are implanted more laterally. The latter condition occurs in all teeth of *Kogia*. All physeterids have homodont and polydont teeth, as do most extant odontocetes. The tooth crowns of physeterids are small, and their roots are more shallow than those of archaeocetes.

**S**

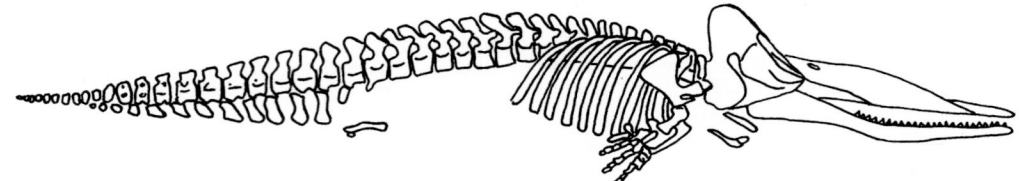

**Figure 1** *The skeleton of the modern sperm whale (Physeter macrocephalus). The concave area on the snout and forehead is the supracranial basin.*

In physeterid evolution, the length of mandibular symphysis increased from one-twelfth of the length of the mandible in the Oligocene to one-fourth in the Miocene, and one-half in *Physeter*. The mandible of *Ferecetotherium* has 27 teeth, whereas that of *Diaphorocetus* has 14. In the middle Miocene *Aulophyseter morricei*, upper teeth, if present at all, were only lodged in the gums, and all upper teeth were lost in the upper Miocene *Placoziphius duboisii*.

The great differences in the shape of the face and jaws between early physeterids and modern ones suggest that the teutophagic specializations of *Physeter* may not have been present in extinct relatives. It is possible that Oligocene and Miocene physeterids were ichthyophagous (fish-eating), as were most cetaceans. Modem sperm whales feed on deep-sea squid, and it is likely early physeterids were not deep-sea animals.

The forelimb of physeterids is very different from that of Eocene cetaceans, but similar to that of other odontocetes. The humerus is shortened and the elbow is immobile. The hand forms a flat, smooth surface with no differentiation of individual fingers, causing the entire limb to be an effective rudder. In early physeterids, the head of the humerus changed considerably. The humeral head of *Ferecetotherium* is deviated externally and faces caudally, but the tubercles are situated anteriorly, as in archaic cetaceans. Whereas *Ferecetotherium* retains a greater and lesser tubercle of the humerus, the lesser tubercle of Miocene physeterids is enlarged greatly and the greater tubercle reduced. *Physeter* has a weak lesser tubercle, set medially on the humerus.

### II. Kogiidae

One genus with two modem species constitutes the Kogiidae (*Kogia brevieeps* and *K. sima*). *Kogia* is similar to *Physeter* but the body *of Kogia* is much smaller (body length approximately 4 m) and the head is smaller (one-sixth to one-eighth of the body). The spermaceti organ is smaller than in *Physeter*, the blowhole more posterior, and the rostrum shorter. Proportions of the head in *Physeter* embryos are similar to those of adult *Kogia*. This suggests that these cetaceans are closely related and that *Physeter* has a more derived facial morphology.

Kogiids are poorly represented in the fossil record, and most specimens are incomplete, although some localities have yielded good material. The genus *Kogiopsis* is known from a single mandible, and fragmentary skulls are known for Miocene *Scaphokogia* and Pliocene *Praekogia*. Another trend in the evolution of kogiids is the reduction of dental enamel. This trend started in the Miocene. In modem *Kogia* some enamel covers the tips of the teeth, whereas Miocene kogiids lack all enamel.

Phylogenetically, *Scaphokogia* is a basal branch of kogiids, retaining primitive morphologies of rostrum, premaxillae, and intermaxillary groove. It is, however, more derived than other kogiids in having a well-developed supracranial basin. *Scaphokogia* may represent an early, specialized branch of kogiids, the subfamily Scaphokogiidae. These went extinct near the end of the Miocene. *Praekogia* is closely related to *Kogia* in that both the nasal passages are anterior due to poorly developed telescoping. The new genus *Aprixokogia* from the late Tertiary of North Carolina is known from a well-preserved skull (Whitmore and Kaltenback, 2008).

### See also the Following Articles

Cetacean Evolution ■ Dental Morphology ■ Evolution of Marine Mammals ■ Fossil Record ■ Pygmy and Dwarf Sperm Whales ■ Sperm Whale

### References

Mchedlidze, G. A. (1984). "General Features of the Paleobiological Evolution of Cetacea." Amerind Publ. Corp. and Smithsonian Institution Libraries.

Whitmore, F. C., Jr., and Kaltenback, J. A. (2008). Neogene Cetacea of the Lee Creek Phosphate Mine, North Carolina. *Virginian Museum of Nattural History Spec. Publ.* **14**, 181–271.

# Spermaceti

## Dale W. Rice

Spermaceti is the term given to the liquid waxes present in the head of the sperm whale (*Physeter macrocephalus*). The word comes from the late Latin *sperma*, "sperm" or "semen," and the classical Latin *ceti*, genitive singular of *cetus*, "sea monster" or "whale"—literally "semen of whale." It was bestowed because of the superficial resemblance of these waxes to semen. The vernacular name "sperm whale" is a contraction of "spermaceti whale."

Spermaceti is present in greatest abundance in the spermaceti organ, or "case," and in the "junk." The case is an elongated barrel-shaped organ, which makes up much of the bulk of the sperm whale's huge head. It consists of soft, white, spongy tissue that is saturated with spermaceti. The junk lies below the case and above the rostrum of the skull; it is also white, but it is more solid than the case and is faintly divided into segments by a series of transverse septa composed of denser tissue. It is also heavily saturated with spermaceti (Clarke, 1978; Rice, 1989). At body temperature, the spermaceti will flow freely as a clear, almost watery, fluid from the case and junk when the latter are slashed with a knife. As soon as it cools to about room temperature, however, it solidifies to a whitish wax similar to paraffin but not as hard. The different fractions may be separated by pressing at various temperatures between 10 and 35°C.

The oils of sperm whales and the other toothed whales (suborder Odontoceti) differ in composition from those of the baleen whales (suborder Mysticeti) (Gilmore, 1951). Those of the baleen whales consist exclusively of triglycerides. Triglycerides are esters compounded from one molecule of the tribasic alcohol glycerol (CH$_2$OH•CHOH•CH$_2$OH) and three molecules of various fatty acids. Lipids of the toothed whales, however, consist of a mixture of triglyceridic oils and wax esters, together with a small proportion (≥2%) of diacyl glyceryl ethers in some species. These waxes are compounds formed from one molecule of a higher monobasic alcohol, mostly cetyl (C$_{16}$H$_{33}$OH) or oleyl (C$_{18}$H$_{35}$OH), and one molecule of a fatty acid.

Although wax esters are present in all odontocetes, only in the giant sperm whale and the Amazon river dolphin (*Inia geoffrensis*) (family Iniidae) do they comprise a majority of the lipids present. In sperm whales the ratio of waxes to triglycerides varies with position in the body. An early study (Hilditch and Lovern, 1928) found that wax esters comprised 74% of the spermaceti in the head, but only 66% of the BLUBBER oil. More detailed studies revealed that even within the head, the proportion of wax esters varies with sex and age as well as position. In the spermaceti case this proportion ranged from 38 to 51% in a calf, 58 to 87% in an adult female, and 71 to 94% in an adult male. Proportions in the junk of the same three whales were 41–62, 65–94, and 91–98%, respectively (Morris, 1975).

Both the triglycerides and the waxes of the sperm whale are characterized by the virtual absence (<2%) of isovaleric acid, a branched, short chain (C$_5$), unsaturated fatty acid. This trait is shared with pygmy sperm whales (*Kogia breviceps*) (family Kogiidae), beaked whales (family Ziphiidae), and some river dolphins (families Platanistidae and Iniidae). This contrasts with the high proportion of isovalerate lipids in delphinoid cetaceans (families Monodontidae, Delphinidae, and Phocoenidae) (Litchfield *et al.*, 1975).

Wax esters of the sperm whale differ from those of all other toothed whales except the Amazon dolphin (family Iniidae) in that they consist predominantly of relatively long chain (C$_{10}$–C$_{22}$) fatty acids (Litchfield *et al.*, 1975). The fatty acid composition of wax esters in the case is mainly shorter chained, saturated acids at the center, with longer chained, more unsaturated acids at the periphery. In the junk, the anterior portion is mainly shorter chained saturated acids, whereas the posterior portion consists of longer chained, more unsaturated acids. The specific proportions of 25 fatty acid moieties in wax esters from the head of an adult female sperm whale were tabulated by Morris (1975).

Because of its heterogeneous lipid composition, and possible internal temperature gradients, the velocity of SOUND through the spermaceti organ varies in a way that could collimate or focus sound waves (Flewellen and Morris, 1978). This accords with the prevailing hypothesis that this organ functions as an acoustic lens to channel acoustic emissions.

In the nineteenth century when the old-style open-boat whale fishery was flourishing, the oil in the head of the sperm whale (the spermaceti proper) was simply ladled from the case and junk, and was kept separate from the oil that was rendered from the blubber (the remainder of each carcass was discarded). Spermaceti was favored for making candles, and the liquid blubber oil was used as a lamp fuel. In the years from 1804 to 1925, a total of 164,073,918 gallons of sperm oil (including spermaceti) was landed at US ports. This quantity was the product of an estimated 262,134 sperm whales killed around the globe. Lesser quantities were taken by vessels out of British, French, and other foreign ports. In the peak year of 1837,

American landings totaled 5,349,138 gallons, representing an estimated kill of 7472 sperm whales, and were worth $4,413,039 at the average going price of 82.5 cents per gallon. Discovery of the first commercially recoverable reserves of petroleum in 1859 triggered the phenomenal growth of the oil industry, so the demand for sperm oil plummeted. The last old-style American sperm-whaling voyage was made in 1925.

In the twentieth century the modern harpoon-cannon whaling industry was carried out from factory ships and shore stations around the world. Oil was rendered from the entire carcass (minus the meat) of each whale, and that from sperm whales ("sperm oil") was kept separate from the oil of baleen whales ("whale oil"). In the early decades of this fishery, whalers took few sperm whales because the edible oil of baleen whales brought higher returns. By the end of World War II, however, baleen whale populations had been severely depleted, and new uses were being discovered for sperm oil. Because of its unique properties, it was in great demand as a high-pressure lubricant, but it also had a diverse array of other specialized industrial uses, such as an ingredient in hydraulic fluids, inks, detergents, and cosmetics, as a plasticizer, and as an agent in the tanning of leathers and the degreasing of wool (Anonymous, 1957). The rising demand led to rapid increases in sperm whale catches, which reached a peak in 1964, when 29,255 animals were killed, yielding 898,257 barrels (=152,703,690 kg) of sperm oil. The International Whaling Commission shut down the entire whale fishery after the 1984 season. Since the cessation of whaling, substitutes for sperm oil have been found—most notably the oil from the seeds of jojoba (*Simmondsia chinensis* of the family Buxaceae), a shrub indigenous to the Sonoran Desert. Jojoba oil has chemical and physical properties similar to those of sperm oil (Harris *et al.*, 1975).

## See Also the Following Articles

Sperm Whale ▪ Toothed Whales ▪ Overview ▪ Whaling ▪ Traditional

## References

Anonymous (1957). Leviathan in the lab. *Laboratory* **26**, 34–38.

Clarke, M. R. (1978). Structure and proportions of the spermaceti organ in the sperm whale. *J. Mar. Biol. Assoc. UK* **58**, 1–17.

Flewellen, C. G., and Morris, R. J. (1978). Sound velocity measurements on samples from the spermaceti organ of the sperm whale (*Physeter catodon*). *Deep-Sea Res.* **A25**, 269–277.

Gilmore, R. M. (1951). The whaling industry: Whales, dolphins, and porpoises. *In* "Marine Products of Commerce" (D. K. Tressler, and J. M. Lemon, eds), pp. 680–715. Reinhod.

Harris, M., Goebel, C. G., Schwartz, A. M., Von Wettberg, E. F., Jr., and Yermanos, D. M. (1975). "Products from Jojoba: A Promising New Crop for Arid Lands." National Academy of Sciences, Washington, D.C, p. 30.

Hilditch, T. P., and Lovern, J. A. (1928). The head and blubber oils of the sperm whale. I. Quantitative determinations of the mixed fatty acids present. *J. Soc. Chem. Ind.* **47**, 105T–t111T.

Litchfield, C., Greenberg, A. J., Caldwell, D. K., Caldwell, M. C., Sipos, J. C., and Ackman, R. G. (1975). Comparative lipid patterns in acoustical and nonacoustical fatty tissues of dolphins, porpoises, and toothed whales. *Comp. Biochem. Physiol. B* **59**, 591–597.

Morris, R. J. (1975). Further studies into the lipid structure of the spermaceti organ of the sperm whale (*Physeter catodon*). *Deep-Sea Res.* **22**, 483–489.

Rice, D. W. (1989). Sperm whale *Physeter macrocephalus* Linnaeus, 1758. *In* "Handbook of Marine Mammals" (S. H. Ridgway, and R. Harrison, eds), 4, pp. 177–233. Academic Press, London.

S

# Spinner Dolphin
*Stenella longirostris*

WILLIAM F. PERRIN

## I. Characteristics and Taxonomy

The spinner dolphin, described by Gray in 1828, is the most common small cetacean in most tropical pelagic waters. It can be seen at a great distance as it spins high in the air and lands in the water with a great splash. It can be identified externally by its relatively long slender beak, color pattern, and dorsal fin (Fig. 1) (Perrin, 1998). The color pattern in most regions is three-part, consisting of a dark-gray cape, light-gray lateral field, and white ventral field. A dark band of even width runs from the eye to the flipper, bordered above by a thin light line. The rostrum is tipped with black or dark gray. Four subspecies are currently recognized: the globally distributed *S. l. longirostris* (Gray's spinner), the eastern tropical Pacific (ETP) endemics *S. l. orientalis* (eastern spinner) and *S. l. centroamericana* (Central American spinner), and *S. l. roseiventris*, the dwarf spinner of central Southeast Asia (Perrin, 1990; Perrin *et al.*, 1999). An additional morphologically different form has been identified off northern Mexico with aerial photogrammetry (Perryman and Westlake, 1998). In the eastern and Central American subspecies (Fig. 2), contrast between

**Figure 1**   *A Gray's spinner dolphin* (Stenella longirostris longirostris) *in the Maldive Islands in the Indian Ocean. Photograph by Robert L. Pitman.*

**Figure 2**   *The eastern spinner dolphin* (Stenella longirostris orientalis) *lacks a bold pattern. This adult male has a forward-canted dorsal fin, unique to this subspecies and the parapatric Central American spinner* (S. l. centroamericana). *Photograph by Robert L. Pitman.*

the cape and the lateral field is very faint to absent, and the ventral field may be restricted to discontinuous axillary and genital-region patches. Various intermediate color patterns are exhibited by "white-belly" spinners in a broad zone of HYBRIDIZATION between the eastern form and the more typically patterned Gray's spinner dolphin, *S. l. longirostris*, in the Central and South Pacific. The dorsal fin in adults in all regions is basically triangular, varying from a slightly falcate right triangle to an erect isosceles triangle. In the adult male of the eastern and Central American subspecies the dorsal fin may lean slightly forward, appearing to "be on backwards." This is correlated with the presence of a large post-anal ventral hump. Both the dorsal fin and the hump appear to be sexual displays, probably important in male–male or male–female communication or both. In calves and juveniles in all regions, the dorsal fin is on average more falcate than in adults.

Sexually mature adults examined ranged from 129 to 235 cm (*n* = 1824) and weighed 23–80 kg (*n* > 33) (Perrin, 1998; Perrin *et al.*, 2005). Males are on average slightly larger than females in body size and most skull characters.

The spinner dolphin may be confused in the tropical Atlantic with the endemic CLYMENE DOLPHIN, *S. clymene,* which is very similar in appearance and has been observed to spin, although not as acrobatically as the spinner dolphin. In the Clymene dolphin, the beak is relatively shorter, the flipper band narrows anteriorly, the lower margin of the cape dips lower toward the ventral field (in the spinner the cape and ventral-field margins are parallel), and there is usually a dark "moustache" mark on the upper side of the beak that has only rarely been observed in spinner dolphins.

The skull (Fig. 3) can be confused with those of *S. coeruleoalba*, *S. clymene*, and *Delphinus* spp.; all have a relatively long and narrow dorsoventrally flattened rostrum, a large number of small slender teeth (about 40–60 in each row), and medially convergent premaxillae and sigmoid ramus (Perrin, 1998). It differs from the skull of *Delphinus* in lacking strongly defined palatal grooves. The rostrum is narrower at the base than in *S. coeruleoalba* (57–84 mm vs. 93–120 mm). It overlaps *S. clymene* in all skull measurements and tooth counts, but the skull (335–464 mm long in 112 adults) is proportionately longer and narrower. The vertebral count is 69–77 (*n* = 90).

The spinner dolphin is a member of the subfamily Delphininae *sensu stricto* (LeDuc *et al.*, 1999). Morphologically and behaviorally it is most similar to the Clymene dolphin, but in a cladistic phylogenetic analysis based on cytochrome *b* mtDNA sequences it was not closely linked to that species, which was the sister species of

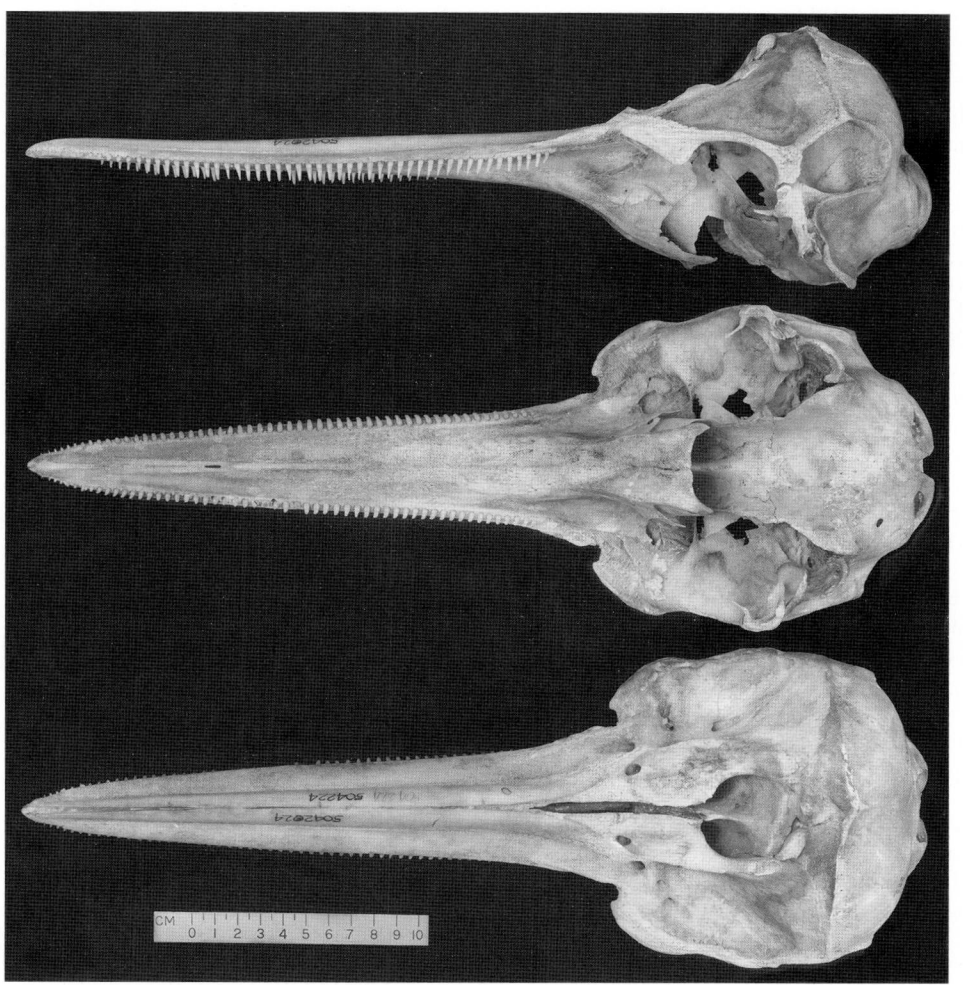

**Figure 3**  *The skull of* Stenella longirostris longirostris; *adult male from Florida. From Perrin and Gilpatrick (1994).*

the striped dolphin, *Stenella coeruleoalba*. The morphological similarly between spinner and Clymene dolphins could be an instance of synplesiomorphy (similarity through retention of primitive characters), but the correlated spinning behavior would seem to be highly derived, a probable synapomorphy (shared derived character). This puzzling discordance between molecular and morphological/behavioral characters shows the need for further molecular and morphological investigation; a hybrid origin of *S. clymene* has been suggested as one possible explanation.

## II. Distribution and Abundance

The spinner dolphin is pantropical, occurring in all tropical and most subtropical waters around the world between roughly 30–40°N and 20–40°S (Jefferson *et al.*, 2007). It is typically thought of as a high-seas species, but coastal populations and races/subspecies exist in the eastern Pacific, Indian Ocean, Southeast Asia, and likely elsewhere (Perrin, 1998; Perrin *et al.*, 1999).

Estimates of abundance exist for several regions (IUCN, 2008): whitebelly spinners in the ETP, about 800,000; eastern spinners in the ETP, about 600,000 (reduced by more than half from original size; see below); northern Gulf of Mexico, about 12,000; Hawaii, about 3,000; southern Sulu Sea, about 4000; and southeastern Sulu Sea, about 31,000. Many other populations that exist in the Pacific, Atlantic, and Indian Oceans have not been surveyed.

## III. Ecology

In the ETP, the habitats of the pelagic eastern subspecies and the eastern/Gray's intergrade/hybrid or "whitebelly spinner" are similar to that of the pantropical spotted dolphin: tropical surface water characterized by a shallow mixed layer, shoal and sharp thermocline, and relatively small annual variation in surface temperature (Ballance *et al.*, 2006). In other tropical waters, spinner dolphins are usually associated with islands and coasts, venturing out to deeper water at night to feed (Benoit-Bird and Au, 2003; Dolar *et al.*, 2003; Karczmarski *et al.*, 2005). Small and semi-isolated island populations scattered across an archipelago may comprise a metapopulation (Karczmarski *et al.*, 2005; Oremus *et al.*, 2007). The Central American spinner inhabits shallower inshore water.

In the eastern and western Pacific, the pelagic form has been shown to prey mainly on small mesopelagic fishes and squids, diving to 600 m or deeper (Perrin and Gilpatrick, 1994; Dolar *et al.*, 2003), but a dwarf subspecies in inner Southeast Asia, *S. l. roseiventris*, inhabits shallow waters in the Gulf of Thailand, Timor Sea, and Arafura Sea and consumes mainly benthic and reef fishes and invertebrates (Perrin *et al.*, 1999). Spinner dolphins are time- and efficiency-limited in their foraging, rather than being limited by available prey; they need to consume 1.25 large prey items per minute during their foraging bouts while the deep scattering layer is close to the surface (Benoit-Bird, 2004). While resting in shallow reef areas during the day, their feces may constitute an important resource for reef fishes (Silva-JR *et al.*, 2005).

Predators include sharks, probably killer whales (*Orcinus orca*) and possibly false killer whales (*Pseudorca crassidens*), pygmy killer whales (*Feresa attenuata*) and short-finned pilot whales (*Globicephala macrorhynchus*). Parasites may cause direct or indirect mortality.

## IV. Behavior and Physiology

Why the spinner spins is unknown. One animal may spin as many as 14 times in quick succession. It has been suggested that the large underwater bubble plume created by the violent spin and reentry may serve as an ECHOLOCATION target for communication across a widely dispersed school (Norris *et al.*, 1994). It is also probable that spinning is an outgrowth of an alert state and as such has a social facilitation function. It could also—at least at times—represent PLAY. Another function proposed is to dislodge remoras; the rotation rates and orientation of the dolphin's body during reentry into the water could produce enough force to accomplish this (Fish *et al.*, 2006).

School size varies greatly, from just a few dolphins to a thousand or more. Spinner dolphins commonly school together with pantropical spotted dolphins, *S. attenuata*, and dolphins and small toothed whales of other species, and sometimes even dugongs (Psarakos *et al.*, 2003; Kiszka, 2007). Social organization in Hawaiian waters is fluid, with schools composed of more or less temporary associations of family units; the associations may vary over days or weeks (Norris *et al.*, 1994). Mating appears to be promiscuous. Adult males form coalitions of up to about a dozen individuals; the function of these is unknown. Maximum recorded movements of individuals are 113 km (over 1220 days) in Hawaii and 275 nmi (over 395 h) in the eastern Pacific (Perrin, 1998).

The whistles and burst pulses produced by spinner dolphins have been intensively studied in several areas (Lammers *et al.*, 2006); they vary in complexity geographically and with activity, serving functions that are only beginning to be understood.

The dwarf spinner dolphin as recently photographed (Perrin *et al.*, 2007) has a pink belly. This accords with the Latin name (*roseiventris*) given to it by its discoverer; however, the COLORATION is not likely a permanent characteristic but may indicate a need to dump excess heat (by dilation of blood vessels near the surface of the skin) generated by its living in very warm shallow reef waters.

## V. Life History

Gestation is about 10 months. Average length at birth is about 75–80 cm. Length of nursing is 1–2 years. Calving interval is about 3 years. Females attain sexual maturity at 4–7 years and males at 7–10 years. Ovulation may be spontaneous. Breeding is seasonal, more sharply so in some regions than in others. The mating system may vary among populations. This is indicated by geographic variation in morphology and testis size. For example, in the ETP eastern spinners are more sexually dimorphic and have smaller testes than whitebelly spinners, likely indicating a greater tendency toward polygyny as opposed to polygynandry (promiscuous mating) (Perrin and Mesnick, 2003).

## VI. Interactions with Humans

Large numbers have been killed incidentally since the early 1960s by tuna purse seiners in the eastern tropical Pacific; the population of *S. l. orientalis* is estimated to have been reduced to less than one-half of its original size, and since reduction of reported purse-seine mortality more than a decade ago it has grown at annual rate of less than 2%, less than the rate expected of about 4% (Gerrodette and Forcada, 2005; Wade *et al.*, 2007). Continued chase and capture in the fishery may adversely affect fecundity or survivorship (Archer *et al.*, 2001; Moore, 2004; Weihs, 2004; Cramer and Gerrodette, 2007; Noren and Edwards, 2007). Bycatches that are also likely unsustainably high occur in drift nets and purse seines in the Philippines (Dolar, 1994, 1999). As is the case for other small cetaceans caught in fishing nets, local human consumption of bycaught animals in several regions has led to the development of markets and large directed catches from unassessed populations (Perrin, 1999; IUCN, 2008), boding ill for CONSERVATION of the species in these regions.

S

Spinner dolphins have been kept in CAPTIVITY only in Hawaii, where some have lived for several years.

Harassment by dolphin-watching boats is emerging as a new threat to spinner dolphins at several localities in Oceania and Southeast Asia (IUCN, 2008).

## See Also the Following Articles

Clymene Dolphin ■ Delphinids, Overview ■ Aerial Behavior ■ Tuna–Dolphin Issue

## References

Archer, F., Gerrodette, T., Dizon, A., Abella, K., and Southern, S. (2001). Unobserved kill of nursing dolphin calves in a tuna purse-seine fishery. *Mar. Mamm. Sci.* **17**, 540–554.

Balance, L. T., Pitman, R. L., and Fiedler, P. C. (2006). Oceanographic influences on seabirds and cetaceans of the eastern tropical Pacific: a review. *Prog. Oceanogr.* **69**, 360–390.

Benoit-Bird, K. J., and Au, W. W. L. (2003). Prey dynamics affect foraging by a pelagic predator (*Stenella longirostris*) over a range of spatial and temporal scales. *Behav. Ecol. Sociobiol.* **53**, 364–373.

Benoit-Bird, K. J. (2004). Prey caloric value and predator energy needs: foraging predictions for wild spinner dolphins. *Mar. Biol.* **145**, 435–444.

Cramer, K. and Gerrodette, T. (2007). Decline in reproductive indices in two depleted dolphin populations in the eastern tropical Pacific. *Mar. Ecol Prog. Ser.*, in press.

Dolar, M. L. L. (1994). Incidental takes of small cetaceans in fisheries in Palawan, Central Visayas and northern Mindanao in the Philippines. *Rep. Int. Whaling Comm.* **15**, 355–363.

Dolar, M. L. L. (1999). "Abundance, Distribution and Feeding Ecology of Small Cetaceans in the Eastern Sulu Sea and Tañon Strait, Philippines," p. 241. Ph.D. dissertation, University of California, San Diego.

Dolar, M. L. L., Walker, W. A., Kooyman, G. L., and Perrin, W. F. (2003). Comparative feeding ecology of spinner dolphins (*Stenella longirostris*) and Fraser's dolphins (*Lagenodelphis hosei*) in the Sulu Sea. *Mar. Mamm. Sci.* **19**, 1–19.

Dolar, M. L. L., Perrin, W. F., Taylor, B. L., Kooyman, G. L., and Alava, M. N. R. (2006). Abundance and distributional ecology of cetaceans in the central Philippines. *J. Cetacean Res. Manage.* **8**, 93–111.

Fish, F. E., Nicastro, A. J., and Weihs, D. (2006). Dynamics of the aerial maneuvers of spinner dolphins. *J. Exp. Biol.* **209**, 590–598.

Jefferson, T. A., Webber, M. A., and Pitman, R. L. (2007). "Marine Mammals of the World: A Comprehensive Guide to Their Identification." Academic Press/Elsevier, San Diego.

Karczmarski, L., Würsig, R., Gailey, G., Larson, K. W., and Vanderlip, C. (2005). Spinner dolphins in a remote Hawaiian atoll: social grouping and population structure. *Behav. Ecol.* **16**, 675–685.

Gerrodette, T., and Forcada, J. (2005). Non-recovery of two spotted and spinner dolphin populations in the eastern tropical Pacific Ocean. *Mar. Ecol. Prog. Ser.* **291**, 1–21.

IUCN (2008). "The IUCN Red List of threatened species." IUCN, Gland Switzerland.

Kiszka, J. J. (2007). Atypical associations between dugongs (*Dugong dugon*) and dolphins in a tropical lagoon. *J. Mar. Biol. Ass. UK* **87**, 101–104.

Lammers, M. O., Schotten, M., and Au, W. W. L. (2006). The spatial context of free-ranging Hawaiian spinner dolphins (*Stenella longirostris*) producing acoustic signals. *J. Acoust. Soc. Am.* **119**, 1244–1250.

LeDuc, R. G., Perrin, W. F., and Dizon, A. E. (1999). Phylogenetic relationships among the delphinid cetaceans based on full cytochrome *b* sequences. *Mar. Mamm. Sci.* **15**, 619–648.

Moore, P. (2004). Examining dolphin hydrodynamics provides clues to calf-loss during tuna fishing. *J. Biol.* **2004**, 6.

Noren, S. R., and Edwards, E. F. (2007). Physiological and behavioral development in delphinid calves: implications for calf separation and mortality due to tuna purse-seine sets. *Mar. Mamm. Sci.* **23**, 15–29.

Norris, K. S., Würsig, B., Wells, R. S., and Würsig, M. (1994). "The Hawaiian Spinner Dolphin." University of California Press, Berkeley.

Perrin, W. F. (1998). Stenella longirostris. *Mammal. Species* **599**, 1–7.

Oremus, M., Poole, M. M., Steel, D., and Baker, C. S. (2007). Isolation and interchange among insular spinner dolphin communities in the South Pacific revealed by individual identification and genetic diversity. *Mar. Ecol. Prog. Ser.* **336**, 275–289.

Perrin, W. F. (1990). Subspecies of *Stenella longirostris* (Mammalia Cetacea Delphinidae). *Proc. Biol. Soc. Wash.* **103**, 453–463.

Perrin, W. F. (1999). Selected examples of small cetaceans at risk. *In* "Conservation and Management of Marine Mammals" (J. R. Twiss Jr., and J. Reynolds, eds). Smithsonian Press, Washington, D.C..

Perrin, W. F., and Mesnick, S. L. (2003). Sexual ecology of the spinner dolphin, *Stenella longirostris*: geographic variation in mating system. *Mar. Mamm. Sci.* **19**, 462–483.

Perrin, W. F., Dolar, M. L. L., and Robineau, D. (1999). Spinner dolphins (*Stenella longirostris*) of the western Pacific and Southeast Asia: Pelagic and shallow–water forms. *Mar. Mamm. Sci.* **15**, 1029–1053.

Perrin, W. F., and Gilpatrick, J. W., Jr. (1994). Spinner dolphin *Stenella longirostris* (Gray, 1828). *In* "Handbook of Marine Mammals" (S. H. Ridgway, and R. Harrison, eds), Vol. 5, pp. 99–128. Academic Press, London.

Perrin, W. F., Dolar, M. L. L., Chan, C. M., and Chivers, S. J. (2005). Length–weight relationships in the spinner dolphin (*Stenella longirostris*). *Mar. Mamm. Sci.* **21**, 765–778.

Perrin, W. F., Aquino, M. T., Dolar, M. L. L., and Alava, M. N. R. (2007). External appearance of the dwarf spinner dolphin *Stenella longirostris roseiventris*. *Mar. Mamm. Sci.* **23**, 464–467.

Perryman, W. L., and Westlake, R. L. (1998). A new geographic form of the spinner dolphin, *Stenella longirostris*, detected with aerial photogrammetry. *Mar. Mamm. Sci.* **14**, 38–50.

Psarakos, S., Herzing, D. L., and Marten, K. (2003). Mixed-species associations between pantropical spotted dolphins (*Stenella attenuata*) and Hawaiian spinner dolphins (*Stenella longirostris*) off Oahu, Hawaii. *Aquat. Mamm.* **29**, 390–395.

Silva-JR, J. M., Silva, F. J. L., and Sazima, I. (2005). Rest, nurture, sex, release, and play: diurnal underwater behaviour of the spinner dolphin at Fernando de Noronha Archipelago, SW Atlantic. *Aquat. J. Ichthyol. Aquat. Biol.* **9**, 161–176.

Van Waerebeek, K., Baldwin, R., Gallagher, M., and Papastavrou, V. (1999). Morphology and distribution of the spinner dolphin, *Stenella longirostris*, rough-toothed dolphin, *Steno bredanensis*, and melon-headed dolphin, *Peponocephala electra*, from waters off the Sultanate of Oman. *J. Cetacean Res. Manage.* **1**, 167–177.

Wade, P. R., Watters, G. M., Gerrodette, T., and Reilly, S. B. (2007). Depletion of northeastern offshore spotted and eastern spinner dolphins in the eastern tropical Pacific and hypotheses for their lack of recovery. *Mar. Ecol. Prog. Ser.*

Weihs, D. (2004). The hydrodynamics of dolphin drafting. *J. Biol.* **2004**, 8.

# Steller's Sea Cow
## *Hydrodamalis gigas*

### PAUL K. ANDERSON AND DARYL P. DOMNING

A very large sirenian once grazed on algae along the shores of the cold North Pacific Ocean. Shipwrecked with Vitus Bering's expedition on Bering Island, the larger of the Kommandorskiye (Commander) Islands, George Wilhelm Steller (1709–1746), the only biologist to observe the species alive before its extinction around

**S**

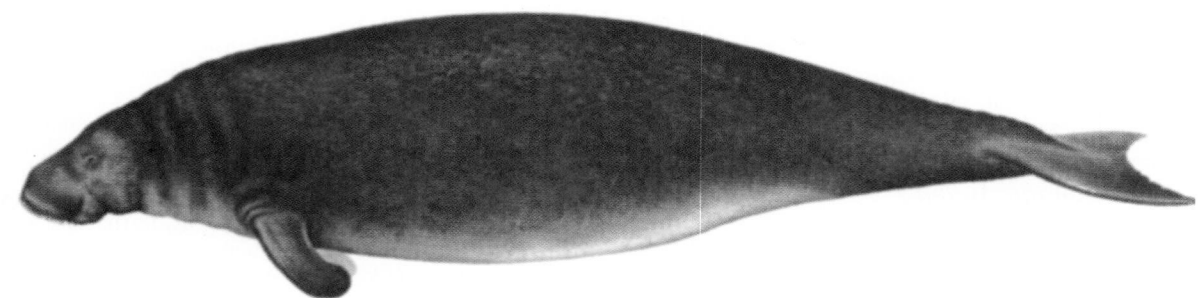

**Figure 1** *A reconstruction of Steller's sea cow. Outstanding features are the relatively small head, bulky body, rough skin, whale-like flukes, and blunt unflipper-like pectoral appendages.*

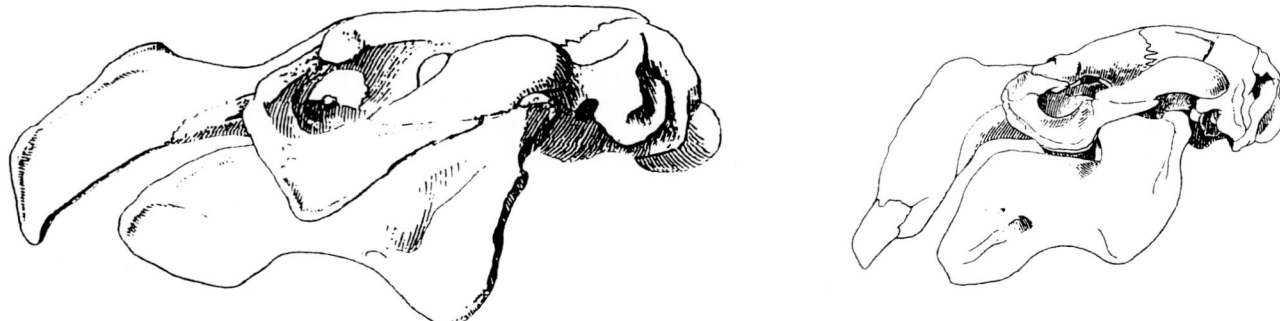

**Figure 2** *Skulls of the two recent dugongids drawn to the same scale. The* Hydrodamalis gigas *skull (left) shows the lack of teeth and the relatively undeflected rostrum compared with the downwardly deflected rostrum of* Dugong dugon *(right). The dugong's tusks and molar teeth are indicated by arrows. The Steller's sea cow's head was relatively small compared with its body and that of the dugong relatively large so that the size of the two skulls does not reflect relative body size.*

1768, observed and described the sea cow and recorded his observations on external and internal anatomy, natural history, and hunting as seen during his forced stay on the island between November 1741 and August 1742.

## I. Characteristics and Taxonomy

Two other common names for Steller's sea cow are the German "Borkentier," referring to its rough bark-like hide, and the Russian "kapustnik" (cabbage eater), which Steller attributed to inhabitants of Kamchatka, where dead sea cows reportedly washed ashore after storms. Because Steller's posthumously published observations (1751) predated Linnean taxonomy, the sea cow was given its formal scientific name by later authors on the basis of Steller's description (no type specimen has ever been designated). Among numerous synonyms, the generic and species names now recognized as correct are *Hydrodamalis* (Retzius, 1794); and *Hydrodamalis gigas* (Zimmerman, 1780). The latter combination was authored by Palmer (1895). The generic synonym used most commonly in the nineteenth century was *Rytina* (Illiger, 1811), improperly emended to *Rhytina*. For full synonymy and comprehensive references, see Domning (1978, 1996).

*Hydrodamalis gigas* (Fig. 1) was a very large toothless sirenian, a member of an order (Sirenia) consisting of completely aquatic, vegetarian, mammals. *H. gigas* was the only sirenian adapted to cool-temperate or cold climates and an algal diet. It differed from all other mammals in feeding predominantly or exclusively on algae and in having a manus composed only of carpals and metacarpals, with the

phalanges vestigial or absent. Its pectoral limb had a densely bristled blunt termination that Steller described as claw-like or hoof-like. It differed from its predecessor, *Hydrodamalis cuestae*, in lacking teeth and from most fossil and living sirenians other than manatees in having a reduced deflection of the rostrum from the occlusal plane (Fig. 2). This characteristic correlated with feeding in the water column above the bottom. Its uniquely large size may have been a factor in its ability to tolerate cold. Surface water temperatures around the Kommandorskiyes range from 0°C in winter to 10°C in summer.

There is no evidence for geographic variation. The northernmost Pacific was likely marginal habitat as members of the Kommandorskiye population may have been stunted. Adults at Bering Island had a body length approximately 750 cm (25 ft.) and a weight of 4500–5900 kg (10,000–13,000 lbs.). Fossil material from farther south suggests maximum body lengths may have reached 9–10 m.

Like other sirenians, Steller's sea cow had paired nostrils, abdominal testes, a pair of axillary mammae, no hindlimbs, horny plates on the occlusal surfaces of the rostrum and mandibular symphysis for macerating vegetation, and ribs and other bones that were swollen (pachyostotic) and completely compact (osteosclerotic). It swam by dorsoventral undulation of the body and its horizontally expanded triangular caudal fluke, resembling those of dugongs and cetaceans. See Domning (1978, 1994) for more detailed morphological diagnoses.

The order Sirenia is agreed to be monophyletic and has been placed in the superorder Tethytheria, which includes the Proboscidea (elephants) as well as an extinct group of marine hippo-like herbivores, the Desmostylia. The cladistic analysis by Domning (1994), based on morphological characters of skull, mandible, and dentition, supports the conclusion that *H. gigas* was the last of two species in the genus *Hydrodamalis*, sharing the subfamily Hydrodamalinae with its ancestral genus *Dusisiren*. The Hydrodamalinae, Dugonginae, and Halitheriinae constitute the family Dugongidae. The Sirenia also include three other families: the extant Trichechidae (manatees) and the extinct Prorastomidae and Protosirenidae.

*Hydrodamalis* evolved from tropical and subtropical, seagrass-eating dugongid ancestors that shifted from bottom feeding to surface feeding and from a seagrass in warm protected waters to an algal diet and a high-energy, low-temperature environment. Seagrasses, which have a large part of their biomass below ground in the form of roots and rhizomatous storage organs, are the preferred forage of the extant dugong (*Dugong dugon*) and apparently of the ancestral dugongid line. Kelps have no belowground biomass. Their softer, more edible, growing portions (the bulk of nutritious biomass of the taller kelps) are at or near the surface, suspended by floats. Fossils record this transition from bottom rooting to surface foraging. *Hydrodamalis* descended from the widely distributed tropical genus *Metaxytherium*, a small-tusked, seagrass-eating Miocene halitherine dugongid. At the time a diversity of herbivorous marine mammals foraging along the east-central Pacific shorelines was presumably supported by a diverse flora of tropical and subtropical seagrasses, as well as algae (perhaps including kelps in cooler, more exposed microhabitats). The sister group and apparent immediate ancestor of hydrodamalines was the species *M. arctodites*, living in southern California and Baja California by about 14–15 Ma. Probably already present were the earliest members of its descendant genus *Dusisiren*, as well as a dugongine, *Dioplotherium allisoni*. These three sympatric sirenians, representing the three dugongid subfamilies, probably occupied distinct feeding niches. The large-tusked *Dioplotherium* with its strongly deflected rostrum was likely a bottom feeder, rooting for rhizomes of the larger seagrasses. The small-tusked *Metaxytherium* probably depended on the smaller rhizomes and leaves of seagrasses. *Dusisiren* and later hydrodamalines, in contrast, progressively gave up bottom and rhizome feeding, specializing on kelps growing higher in the water column, reducing their rostral deflections, and losing their tusks. Also present with these Miocene sirenians were three or more genera of desmostylians, hippopotamus-like bottom feeders, presumably also herbivorous and feeding on intertidal and subtidal seagrasses and kelps.

As the climate cooled after the Middle Miocene, and tectonic uplift rendered the western North American coastline more emergent (replacing protected embayments with more exposed, higher energy habitats), the cold-water plants increased their dominance in the marine flora, and the tropical seagrasses finally disappeared altogether. With them went *Dioplotherium*, *Metaxytherium*, and all the desmostylians. By the Late Miocene, *Dusisiren*, having earlier thrown in its lot with the kelps, had increased its cold tolerance and its morphological specialization for kelp eating (passing through the successive evolutionary stages *D. jordani* and *D. dewana*) and had extended its range to the northwestern Pacific. By the end of the Miocene it had evolved into *Hydrodamalis cuestae*, which gave rise in the Pleistocene to *H. gigas*.

The prehistoric distribution of sea cows and sea otters (*E. lutris*) was coterminous. Anderson (1995) speculated on the evolution of a sea cow–sea otter relationship and sea cow morphology as follows: "As the hydrodamaline range shifted northwards, the adoption of a kelp diet may have contributed (along with the thermoregulatory demands of a cooling environment) to selection for large adult body size, high birth weight, prolonged parental care, and low reproductive rate. At the same time, the more effective chemical defenses of kelps growing below the sea otter foraging zone might have favored the sea cow's specialization for surface feeding and its loss of diving ability."

## II. Distribution

Historically, Steller's sea cow was known only from Bering and Medney (Copper) Islands at the western terminus of the Aleutians, 200 miles to the west of Attu Island and 150 miles east of the Kamchatka Peninsula. Pliocene and Pleistocene fossils demonstrate that the distribution originally extended around the North Pacific from Baja California, Mexico (30° N), along the coast of North America to the Aleutian chain and south in the western Pacific to Honshu, Japan (37° N). Recent evidence implies possible persistence for a few more years in the central Aleutians (Mattioli and Domning, 2006).

## III. Ecology

Bering Island is approximately 50 miles in length and Medney Island 35 miles long. Both are narrow, high, and rocky. Sea cows foraged on softer parts of large marine algae ("kelp") on open coasts in rocky subtidal and intertidal waters. Steller referred to four types of macroalgae (all still undescribed in his day) as preferred foods; these have been variously identified but probably include such forms as *Agarum* spp., *Alaria praelonga, Halosaccion glandiforme, Laminaria saccharina, Nereocystis luetkeana,* and *Thalassiophyllum clathrus.* Tougher stems and holdfasts were not eaten, washing up in heaps on the shore where sea cows fed. Seagrasses have also been suggested as elements of the diet, but this is doubtful. Of the two available forms, *Phyllospadix* spp. are among the toughest seagrasses and *Zostera marina* grows on soft bottoms in protected waters. In any case, *H. gigas* was probably dependent on a small number of forage species. Feeding along the shoreline, with backs above water, the sea cows were not observed to submerge; they moved about using their flukes and maintaining position on the rocks with their bristle-tipped pectoral limbs. Algae growing on rock faces were bitten off or pawed free with the forelimbs, and the softer parts were separated by the bristly lips as if "cut off with a dull knife." If the animals were indeed unable to submerge, they could access food only to depths of about a meter below low tide level. Steller wrote that they fed "in herds," keeping the young in the center of the group. He did not specify herd size but referred to family groups as consisting of a pair with their offspring. In addition to rocky shorelines, Steller stated that sea cows were "fond of shallow sandy places along the seashore, but they like especially to live around the mouths of river and creeks, for they love fresh running water."

As an algivore, *Hydrodamalis* may have been dependent on an ecosystem in which the carnivorous sea otter (*Enhydra lutris*) suppressed invertebrate herbivory in shallow coastal waters. Like Steller's sea cow, the sea otter is confined to the shallows within 1 km of shore where it forages for urchins and other invertebrate herbivores, usually to a maximum depth of about 130 ft. (40 m). Within this zone, otters keep sea urchins and other invertebrate algivores in check. *Hydrodamalis* was dependent on fast-growing kelps on rock faces and on the kelp canopy in deeper waters. Where sea otters are extirpated, grazing by urchins results in "kelp barrens." Thus sea cows depended on sea otter predation that maintained the shallow-water kelp community.

Steller reported both external and internal PARASITES. The crustaceans that he described as infesting the sea cows' skin have been interpreted as cyamids or caprellids, but identification will probably remain uncertain. The only specimens purported to exist

**S**

are ones attached to alleged sea cow skin fragments, but these latter are likely to pertain instead to whales. "White worms half a foot long," observed by Steller in the stomach and intestine of the sea cow, were probably ascarid nematodes; both these and the ectoparasites probably went extinct with the sea cow itself. *Hydrodamalis*, at least the young, may occasionally have been prey for killer whales (*Orcinus orca*) and large sharks.

## IV. Life History

Steller suspected that the animals were monogamous. Mating apparently involved only a pair, with the male following or herding the female during prolonged "amorous preludes." Copulation involved mutual clasping with the pectorals, the female in an inverted position and the male above.

Steller's views as to the annual reproductive cycle present some enigmas. He observed mating behavior and copulation in the early spring, and deduced that gestation was more than 12 months, but stated that young were born "at any time of the year, but most frequently in autumn." Fall births would seem likely to put the young at a disadvantage, especially in view of his observations that "during the winter they are often suffocated by the ice that floats about the shore" and "in the winter the animals become so thin that, besides the bones of the spine, all the ribs show." As Steller did not arrive on the island until late November and left the following August, he could not have had direct knowledge of a concentration of births in the fall. He never saw more than one very small calf with a female, implying that births were single.

## V. Interactions with Humans

These magnificent animals would have been at least as vulnerable to healthy and skilled Pleistocene hunters along mainland shores as they were to the weakened Russians on the Kommandorskiyes. Domning (1978) has suggested that human hunting may have extirpated sea cows from areas within reach of aboriginal hunters, leaving only the population around the remote Kommandorskiyes. Over most of its range the sea cow may have been the only marine mammal to succumb to "Pleistocene overkill."

Steller's sea cows had no fear of humans. Feeding with the head submerged and half of the back above water they were easily approached and could be touched or speared from the rocks. One method of capture was for a swimmer or a boat crew to approach the intended victim with a large hook attached to a line to the shore. When the hook had been driven into the animal's body, a shore crew would attempt to drag it toward the beach. Steller reported that sea cows made no sounds even when wounded by hunters. Of particular interest are the descriptions of both helping behavior and evident pair bonding in this context. Steller reported that while an animal was being dragged ashore other sea cows would attempt to dislodge the hook and/or break the rope. When one female had been dragged ashore, the accompanying male kept station offshore at the site for at least 2 days.

On the basis of his observation of "family groups" and the evidence of strong pair bonds, the vulnerability of the sea cows in their last stronghold made it possible for the weak and scurvy-ridden castaways of Bering's expedition to secure an abundant supply of food and to escape to the mainland in the summer of 1742. Having found ways to capture the sea cows, the survivors were able to divert manpower to salvage materials from the wreck of their ship and build a smaller vessel in which to reach Kamchatka. That voyage, made possible by sea cow vulnerability, carried with it the news of fortunes to be made in hunting sea otters and fur seals. A fur rush followed,

fueled by sea cow meat. The first hunting expedition wintered on Bering Island in 1743–1744. By 1763 several parties had spent up to 9 months on the islands, living almost exclusively on sea cows and salting down barrels of meat to provision the 2–3-year expeditions to the Aleutian chain and the north Asian and North American coasts in search of furs. Sea cow hides were also used to make large skin boats (baidarkas or umiaks). In 1754–1755 a fur-hunting expedition was forced to winter on Bering Island because the sea cows had been extirpated on Medney Island. The last specimen on Bering Island was reported killed in 1766. Steller's biographer, Leonard Stejneger, summarized accounts of the unrestrained killing by the fur hunters and attributed the extinction of the sea cows to ruthless slaughter. Their persistence in the central Aleutians (Domning *et al.*, in press) may have been due to the remoteness, smaller sizes of these islands, and fewer visits by Russian otter hunters.

The drama played out on Bering and Medney Islands suggests a complementary and more complex and instructive story. Anderson (1995), following a suggestion by Delphine Haley, proposed that the final extinction of the sea cow resulted not solely from ruthless harvesting, but from a cascade of events beginning with extirpation of the local sea otter population around the islands in the first rush for furs. Decimation of the otters in all probability triggered a sea urchin population explosion and the disappearance of chemically undefended shallow water kelps that were the sea cows' main food supply. Invasion of the shallow waters by chemically defended deep-water kelps left the hunted remnant of the sea cow population with kelp that was likely toxic. Anderson proposed that this may have been a reenactment of events when North Pacific coastlines were first colonized by humans in the waning Pleistocene.

## *See Also the Following Articles*

Dugong ■ Extinctions, Specific

## *References*

Anderson, P. K. (1995). Competition, predation, and the evolution and extinction of Steller's sea cow, *Hydrodamalis gigas*. *Mar. Mamm. Sci.* **11**, 391–394.

Domning, D. P. (1978). Sirenian evolution in the North Pacific Ocean. University of California Publ. Geol. Sci., 118 xi + 176.

Domning, D. P. (1994). A phylogenetic analysis of the Sirenia. *Proc. San Diego Soc. Nat. Hist.* **29**, 177–189.

Domning, D. P. (1996). Bibliography and index of the Sirenia and Desmostylia. Smithson. Cont. Paleobiol. 80, iii + 611.

Domning, D. P., Thomason, J. and Corbett, D. G. (in press). Steller's sea cow in the Aleutian Islands. Mar. Mamm. Sci.

Golder, F. A. (1922–25). Bering's voyages: An account of the efforts of the Russians to determine the relation of Asia and America. *Am. Geog. Soc. Res. Ser.* **24**, Nos. 1 and No. 2.

Haley, D. (1980). The great northern sea cow. *Oceans* **13**, 7–9.

Mattioli, S., and Domning, D. P. (2006). An annotated list of extant skeletal material of Steller's sea cow *Hydrodamalis gigas* (Sirenia: Dugongidae) from the Commander Islands. *Aquat. Mam.* **32**, 273–288.

Scheffer, V. B. (1972). The last days of the sea cow. *Smithsonian* **3**, 64–67.

Stejneger, L. (1887). How the great northern sea-cow (*Rytina*) became exterminated. *Am. Nat.* **21**, 1047–1054.

Stejneger, L. (1936). "Georg Wilhelm Steller: The Pioneer of Alaskan Natural History." Harvard University Press, Cambridge.

Steller, G. W. (1751). The beasts of the sea. *In* "The Fur Seals and Fur Seal Islands of the North Pacific Ocean" (D. S. Jordan, ed.), pp. 179–218. U.S. Government Printing Office, Washington, D.C., Translated by W. Miller and J. E. Miller.

**S**

# Steller Sea Lion
## *Eumetopias jubatus*

### Thomas R. Loughlin

## I. Characteristics and Taxonomy

The Steller sea lion (or northern sea lion), *Eumetopias jubatus*, is the largest otariid pinniped and one of the more aesthetically appealing sea lions (Loughlin *et al.*, 1987). In Greek, *Eumetopias* means having a well-developed, broad forehead; in Latin *jubatus* means having a mane, as in the male. It is called *qawax* (pronounced ka-wa; *qawan*, plural) by Aleut natives, *sivuch* (*sivuchi*, plural) in Russian, and *todo* in Japanese (*ashika* being the Japanese word for all sea lion species). It exhibits significant sexual dimorphism with males larger (Fig. 1). Average standard length of males is 282 cm and of females 228 cm (maximum of about 325 cm and 290 cm, respectively). Estimated average weight of males is 566 kg and of females 263 kg (maximum of about 1120 kg and 350 kg, respectively). Pup weight at birth is 16–23 kg and may be slightly larger in the western part of their range. Taxonomically the species belongs to the Order Carnivora, Suborder Pinnipedia, Family Otariidae. It can be distinguished from other pinnipeds by a conspicuous diastema between the upper fourth and fifth postcanines. The range overlaps with the California sea lion (*Zalophus californianus*) which is darker color and smaller in size.

The upper postcanine number 5 is double rooted, with the crown directed backward, and does not occlude with lower postcanine 5; all other postcanines are single rooted, slant somewhat forward, and have irregular conical pointed crowns. The diagnostic diastema between upper postcanines 4 and 5 may be caused by rapid growth and extension of the skull rather than by suppression of the 5th and 6th postcanines. Annual growth layers exist in the dentin and cementum with dark layers corresponding to winter and light layers to summer. Dental formula for permanent teeth is i 3/2, c 1/1, pc 5/5, total 34. *E. jubatus* has a double alveolar capillary supply, as in the Cetacea, unique among the pinnipeds.

These are attractive animals. Pups (Fig. 2) are born with a wavy, chocolate brown fur that molts after 3–6 months of age. Adult fur color varies between a light buff to reddish brown with most of the under parts and flippers a dark brown to black; naked parts of the skin are black (Fig. 1). Both sexes become blonder with age. Adult males have long, coarse hair on the chest, neck, and shoulders, which are massive and muscular.

Otariids probably arose in North Pacific temperate waters from the Enaliarctidae; the earliest known otariid, *Pithanotaria starri*, is between 10 and 12 million years old (Mitchell, 1968). The earliest large otariid with most cheek teeth single rooted is from 2 mya Pliocene deposits in Japan and is assigned to the extant genus *Eumetopias*, although probably a different species from *E. jubatus*. A skull, teeth, vertebrae, and other parts of a postcranial skeleton of *E. jubatus* were recovered in California from Pleistocene deposits.

The Steller sea lion has 30 metacentric or submetacentric chromosomes and 4 acrocentric chromosomes; the X chromosome is submetacentric and the Y is acrocentric (2n = 36).

Studies of mitochondrial DNA from throughout the range suggest that at least two stocks exist, an eastern stock (California through southeastern Alaska), and a western stock (Prince William Sound, Alaska, and westward through Russia) (Loughlin, 1997). Designation of a third stock in Russia is equivocal based on examination of nuclear

**Figure 1**  *An* Eumetopias jubatus *rookery with adult females and males. Note the sexual dimorphism between the larger male and the smaller female. (NMFS photograph).*

**Figure 2**  *An Eumetopias jubatus pup at about 1 month of age. (NMFS photograph.)*

DNA. Population declines in western Alaska resulted in the western population being listed as "endangered" in 1997. The eastern population is listed as "threatened", but increases in population numbers in Southeast Alaska and population stability in British Columbia and Oregon may warrant removal from the threatened list for this stock.

## II. Distribution and Abundance

The species was described by the German physician/theologian George Wilhelm Steller based on a specimen that he obtained from the Russian Commander Islands while serving as naturalist on Vitus Bering's fateful voyage to Alaska in 1741–1742. It was also during this voyage that Steller described the northern fur seal (*Callorhinus ursinus*) and THE STELLER'S SEA COW (*Hydrodamalis gigas*, now extinct).

Steller sea lions occur throughout the North Pacific Ocean rim from Japan to southern California (Fig. 3). They abound on numerous breeding sites (rookeries) in the Russian Far East, Alaska, and British Columbia, with lower numbers in Oregon and California. Washington is the only western coastal state that does not contain a Steller sea lion rookery, although pups were observed at one haul-out site in 1997 and 1998. Seal Rocks in Prince William Sound, Alaska, is the northernmost (60°09' N) rookery, and Año Nuevo Island, California, the

**S**

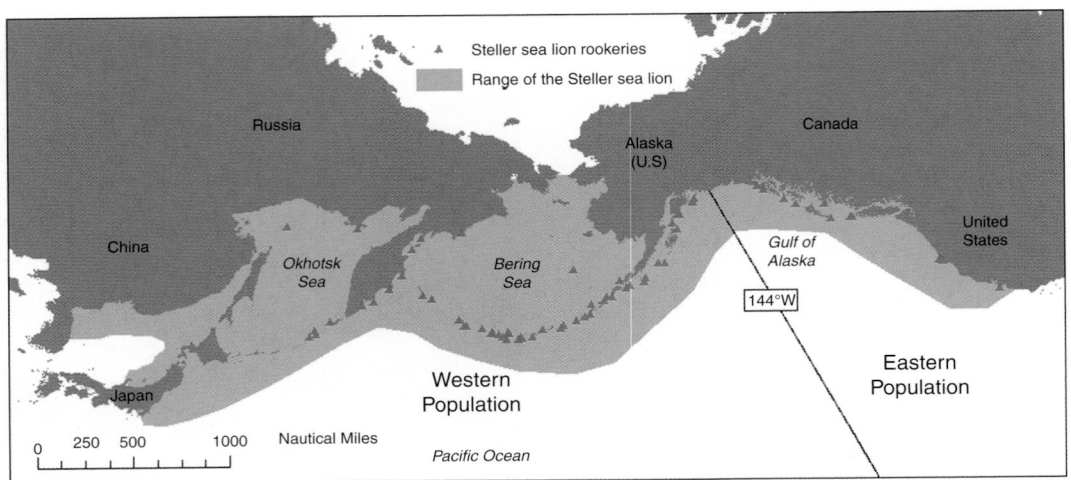

**Figure 3** *A map depicting approximate world distribution (colored area) and rookeries (triangles) of E. jubatus for the eastern and western stocks. Courtesy of NMFS, Seattle, Washington.*

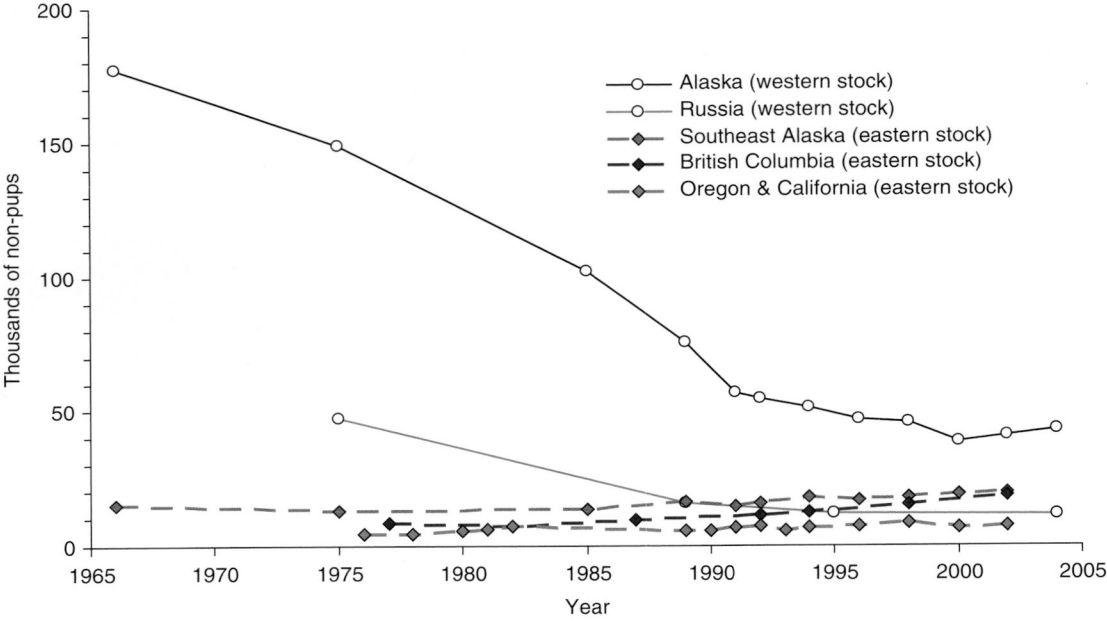

**Figure 4** *The population status of Steller's sea lions throughout their range for major geographic regions with the eastern and western stocks in 1966–2004. Courtesy of NMFS, Seattle, Washington.*

southernmost (37°06′ N). Unlike their more gregarious cousin the California sea lion, Steller sea lions tend to avoid people and prefer isolated offshore rocks and islands to breed and rest. Although rookeries and rest sites occur in many areas, principally on exposed rocky shorelines and wave-cut platforms, the locations used are specific and change little from year to year. Steller sea lions tend to return to their birth island as adults to breed, but they range widely (some yearlings have been seen >1000 km from their birth rookery) during their first few years and during the nonbreeding season.

Population numbers have declined during the last 30 years by over 90% in most of western Alaska and southern California, while populations in Oregon, British Columbia, and southeastern Alaska have remained stable or increased slightly (Fig. 4). The worst declines occurred in the Aleutian Islands and Gulf of Alaska, areas

that historically were the centers of abundance (Loughlin *et al.*, 1992; Trites and Larkin, 1996). Causes for these declines have not been identified but may be related to direct mortality caused by shooting or incidental take in fisheries, disease, and contaminants, or indirectly by reduced food availability through natural changes in the ocean or by fishing pressure and the synergistic effects of commercial fishing and climate change. Studies of blood chemistry and body condition of females and pups during the breeding season are within normal ranges for pinnipeds, but similar studies on juveniles, the age class most likely suffering higher mortality rates, have produced equivocal results. The large population declines resulted in the species being listed as "threatened" by the United States in 1992. A survey in 1989 provided a total estimate of about 116,000 Steller sea lions range-wide, and another in 1994 resulted in about

100,000 of which about 9% of the total was in Russia (Burkanov and Loughlin, 2005), 70% in Alaska, 12% in British Columbia, 8% in Oregon and California, and 1% in Washington. The decline continued throughout the 1990s but seems to have ameliorated in the early 2000s with increases approaching 12% during 2000–2004 in parts of the Aleutian Islands and Gulf of Alaska.

## III. Ecology

Steller sea lions eat a variety of fishes and invertebrates (Calkins, 1998; Merrick *et al.*, 1997; Sinclair and Zeppelin, 2002). In Alaska, the principal prey is walleye pollock (*Theragra chalcogramma*); Pacific cod (*Gadus macrocephalus*), Atka mackerel (*Pleurogrammus monopterygius*), octopus, squid, herring (*Clupea harengus*), flatfishes, sculpins, and a wide variety of other fishes and invertebrates are also consumed. At specific times of the year other prey may be eaten when plentiful (e.g., Pacific salmon, *Onchorynchus* sp.). During the breeding season, females with pups generally feed at night; territorial males eat very little or not at all while on territory. Feeding occurs during all hours of the day once the breeding season ends.

The variety of the sea lion diet has been correlated with population dynamics. In some Alaskan areas where the diet is diverse, sea lion numbers have been stable or increasing slightly. In areas where sea lions primarily depend on one prey item, the sea lion population has declined. Whether population trends are closely associated with diet diversity is still equivocal, but the availability of a variety of prey is optimal for pinnipeds. Steller sea lions are eaten by killer whales (*Orcinus orca*) and sharks, but the possible impact of these predators is unknown.

A cacophony of noise engulfs rookeries and haul-out sites, with animals of both sexes and all ages vocalizing throughout the day and night. Territorial males use low-frequency roars to signal threats to other males and to court females. Females vocalize frequently, calling to their pups and squabbling with other sea lions of all ages. Pups have a bleating, sheep-like cry; their voice deepens with age.

Grooming is performed by bending the head and neck backward and scratching with the claws of the hind flipper. Sea lions also rub themselves on rocks or other animals. While swimming, the fore flippers are used primarily for movement and the rear flippers for braking and turning.

## III. Behavior and Physiology

Observations at sea suggest that large groups usually consist of females of all ages and subadult males; adult males sometimes occur in those groups but are usually found individually. On land, all ages and both sexes occur in large aggregations during the nonbreeding season. Breeding season aggregations are segregated by sexual/territorial status. Steller sea lions are not known to migrate, but they do disperse widely at times of the year other than the breeding season. For example, sea lions marked as pups in the Kuril Islands (Russia) have been sighted near Yokohama, Japan (more than 350 km away) and in China's Yellow Sea (over 750 km away), and pups marked near Kodiak, Alaska, have been sighted in British Columbia, Canada (about 1700 km distant). Generally, animals up to about 4 years of age tend to disperse farther than adults. As they approach breeding age, they have a propensity to stay in the general vicinity of the breeding islands, and, as a general rule, return to their island of birth to breed as adults.

The foraging patterns of adult females vary seasonally. Trip duration for females with young pups in summer is approximately 18–25 h, trip length averages 17 km, and they dive approximately 4.7 h/day.

In winter, females may still have a dependent pup, but a mean trip duration is about 200 h, with a mean trip length of about 130 km, and they dive about 5.3 h/day. Yearling sea lions in winter exhibit foraging patterns intermediate between summer and winter females in trip distance (mean of 30 km) but shorter in duration (mean of 15 h) and with less effort devoted to diving (mean of 1.9 h/day). Estimated home ranges are 320 km$^2$ for adult females in summer, about 47,600 km$^2$ (with large variation) for adult females, and 9200 km$^2$ for winter yearlings in winter. Immature Steller sea lions exhibit long-range trips (>15 km), short-range trips (<15 km), and transits to other sites. Long-range trips start at around 9 months of age and likely occur around the time of weaning, while short-range trips occur almost daily. Transits begin as early as 2.5–3 months of age but occur more often after 9 months of age. Overall, Steller sea lions tend to display two types of distribution when at sea: (1) less than 20 km from rookeries and haul-out sites for adult females with pups, pups, and juveniles, and (2) much larger areas (greater than 20 km) where these and other animals may range to find optimal foraging conditions once they are no longer tied to rookeries and haul-out sites for nursing and reproduction. Large seasonal differences in foraging ranges may be associated with seasonal movements of prey, and seasonal changes in home range are likely related to prey availability.

Compared to some other pinnipeds, Steller sea lions tend to make relatively shallow dives, with few dives recorded to depths greater than 250 m. Maximum depths recorded for individual adult females in summer are in the range from 100 to 250 m; maximum depth in winter is greater than 250 m. The maximum depth measured for yearlings in winter was 72 m and average depths are near 18 m and in shallow nearshore waters.

## IV. Life History

Like most pinnipeds, female Steller sea lions ovulate once a year, and most give birth to a single pup each year; twinning is rare (Pitcher and Calkins, 1981). Males establish territories in May in anticipation of the arrival of females. Viable births begin in late May and continue through early July. The sex ratio at birth is slightly in favor of males. Females breed about two weeks after giving birth. Copulations may occur in the water but most are on land. The mother nurses the pup during the day and after staying with her pup for the first week she goes to sea on nightly feeding trips. Pups generally are weaned before the next breeding season, but it is not unusual for a female to nurse her offspring for a year or more.

Females reach sexual maturity between 3 and 8 years of age and may breed into their early 20s. Females can have a pup every year but may skip years as they get older or when nutritionally stressed. Males reach sexual maturity at about the same age but do not have the physical size or skill to obtain and keep a breeding territory until they are 9 years old or older. Males may return to the same territory from 1 to 7 years, but rarely more than 3 years. While on the territory during the breeding season males may not eat for 1–2 months. The rigors of fighting to obtain and hold a territory and the physiological stress over time during the mating season reduce the life expectancy of these animals. They rarely live beyond their mid-teens, while females may live as long as 30 years.

## V. Interactions with Humans

Steller sea lions are rarely seen in aquaria because of their large size and pugnacious nature. Those in captivity include about 8 in the United States, 12 in Canada, about 4 in Denmark, and about 15 in

**S**

Japan. The 12 in British Columbia at the Vancouver Aquarium and 3 at the Alaska SeaLife Center in Alaska are held for both research and exhibit.

The species is an important subsistence resource for Alaskan natives, who hunt sea lions for food and other uses. Two hundred or more may be taken a year in Alaska.

Steller sea lions may be affected by commercial fishing directly through incidental catch in nets, by entanglement in derelict debris, by shooting, or indirectly through competition for prey, disturbance, or disruption of prey schools (Alverson, 1992). The number of sea lions caught in trawl nets was high during the 1960s and 1970s but has declined since and is presently at very low levels. Incidental entanglement probably contributed to population declines in the Aleutian Islands and western Gulf of Alaska in the 1970s and 1980s, but it is not presently considered an important component in the decline. Entanglement in derelict gear is rare and unlikely to have contributed to the decline. In some areas, Steller sea lions were killed deliberately by fishermen, but it is unclear how such killing affected the world population, especially since declines have occurred in areas uncommonly used by commercial fleets (central and western Aleutian Islands) or where fishermen rarely have guns (Russia). Commercial fisheries target on several of the most important prey eaten by Steller sea lions. In combination, these fisheries remove millions of metric tons of fish. However, the complexity of ecosystem interactions and limitations of data and models make it difficult to determine whether fishery removals, directly or indirectly, have negatively impacted the populations (e.g., see Rosen and Trites, 2000; Fritz and Hinckley, 2005) of steller sea lions.

The U.S. government has implemented numerous measures for the conservation. These include prohibitions on shooting, reductions on allowable incidental take in fisheries, placement of zones around rookeries to restrict commercial fishing, designation of critical habitat, development of a Steller Sea Lion Recovery Plan, and other measures. Research activities have intensified as scientific findings, litigation, and new legislation focused increasing attention on the species' population decline and concerns over possible impacts by commercial fisheries in Alaskan waters. Additional restrictions were placed on these commercial fisheries, resulting in the U.S. congress allocating a seven-fold increase in research funding beginning in 2000 with over 125 individual projects planned or implemented. These studies have provided over 750 primary citations, journal articles, progress and technical reports, contract reports, proceedings of conferences and symposia, books, thesis and other manuscripts. More than 50% of the articles were written since 2004 with the majority pertaining to Steller sea lion life history, foraging ecology and vital rates.

## See Also the Following Articles

Eared Seals (Otariidae) ■ Ecosytem Effects

## References

Alverson, D. L. (1992). A review of commercial fisheries and the Steller sea lion (*Eumetopias jubatus*): The conflict arena. *Rev. Aquat. Sci.* **6**, 203–256.

Burkanov, V. N., and Loughlin, T. R. (2005). Distribution and abundance of Steller sea lions, *Eumetopias jubatus*, on the Asian coast, 1720s–2005. *Mar. Fish. Rev.* **67**, 1–62.

Bickham, J. W., Patton, J. C., and Loughlin, T. R. (1996). High variability for control region sequences in a marine mammal: implications for conservation and biogeography of Steller sea lions (*Eumetopias jubatus*). *J. Mammal.* **77**, 95–108.

Calkins, D. G. (1998). Prey of Steller sea lions in the Bering Sea. *Biosph. Conserv.* **1**, 33–44.

Gentry, R. L. (1970). "Social Behavior of the Steller Sea Lion". *Ph.D. dissertation*, University of California, Santa Cruz, 113 p.

Fritz, L. W., and Hinckley, S. (2005). A critical review of the regime shift "junk food" nutritional stress hypothesis for the decline of the western stock of Steller sea lion. *Mar. Mamm. Sci.* **21**, 476–518.

Holmes, E. E., and York, A. E. (2003). Using age structure to detect impacts on threatened population: a case study with Steller sea lions. *Conserv. Biol.* **17**, 1794–1806.

Loughlin, T. R. (1997). Using the phylogeographic method to identify Steller sea lion stocks. *In* "Molecular Genetics of Marine Mammals" (A. Dizon, S. J. Chivers, and W. F. Perrin, eds.) pp. 159–171. *Soc. Mar. Mamm. Spec. Publ.* **3**, 159–171.

Loughlin, T. R., Perlov, A. S., and Vladimirov, V. A. (1992). Range-wide survey and estimation of total abundance of Steller sea lions in 1989. *Mar. Mamm. Sci.* **8**, 220–239.

Loughlin, T. R., Perez, M. A., and Merrick, R. L. (1987). Eumetopias jubatus. *Mamm. Spec.* **283**, 1–7.

Merrick, R. L. (1995). "The Relationship of the Foraging Ecology of Steller Sea Lions (*Eumetopias jubatus*) to Their Population Decline." *Ph.D. dissertation*, University of Washington, Seattle. 171 p.

Merrick, R. L., Chumbley, M. K., and Byrd, G. V. (1997). Diet diversity of Steller sea lions (*Eumetopias jubatus*) and their population decline in Alaska: a potential relationship. *Can. J. Fish. Aquat. Sci.* **54**, 1342–1348.

Mitchell, E. D. (1968). The Mio-Pliocene pinniped *Imagotaria*. *J. Fish. Res. Board Can.* **25**, 1843–1900.

Pitcher, K. W., and Calkins, D. G.. (1981). Reproductive biology of Steller sea lions in the Gulf of Alaska. *J. Mammal.* **62**, 599–605.

Rea, L. D., Castellini, M. A., Fadely, B. S., and Loughlin, T. R. (1998). Health status of young of the year Steller sea lion pups (*Eumetopias jubatus*) as indicated by blood chemistry and hematology. *Comp. Biochem. Physiol.*, Part A **120**, 617–623.

Rosen, D. A. S., and Trites, A. W. (2000). Pollock and the decline of Steller sea lions: testing the junk-food hypothesis. *Can. J. Zool.* **78**, 1243–1258.

Sinclair, E., and Zeppelin, T. (2002). Seasonal and spatial differences in the diet of western stock of Steller sea lions (*Eumetopias jubatus*). *J. Mammal.* **83**, 973–990.

Springer, A. M., *et al.* (8 authors) (2003). Sequential megafaunal collapse in the North Pacific Ocean: an ongoing legacy of industrial whaling? *Proc. Natl. Acad. Sci.* **100**, 12223–12228.

Trites, A. W., Atkinson, S. K., DeMaster, D. P., Fritz, L. W., Gelatt, T. S., Rea, L. D., and Wynne, K. M. (2006). "Sea Lions of the World." Alaska Sea Grant College Program, University of Alaska, Fairbanks, 644 p.

Trites, A. W., and Larkin, P. A. (1996). Changes in the abundance of Steller sea lions (*Eumetopias jubatus*) in Alaska from 1956 to 1992: How many were there? *Aquat. Mamm.* **22**, 153–166.

York, A. (1994). The population dynamics of northern sea lions, 1975–1985. *Mar. Mamm. Sci.* **10**, 38–51.

# Stock Assessment

JEFFREY M. BREIWICK AND ANNE E. YORK

A marine mammal stock assessment is a process that seeks to estimate the productivity or growth potential of a stock and predict the future growth in conjunction with management objectives and conditions, which often includes removals due to incidental catches, directed harvests or natural causes. It also seeks to measure the capacity of the stock to recover from these removals. The assessment usually encompasses the status of the stock with respect to some reference level, such as the unexploited population size, and the evaluation of the consequences of various management actions. For stocks that are subject to a harvest or experience mortality incidental to fishing operations, a goal is to determine allowable removal levels (e.g., harvests that will allow the population to recover to some desired level during some time frame). A concise definition of a fisheries stock assessment, equally applicable to marine mammals, is that given by Hilborn and Walters (1992): "Stock assessment involves the use of various statistical and mathematical calculations to make quantitative predictions about the reactions of fish populations to alternative management choices."

The components of a stock assessment will vary with the species considered, its stock identification, the quantity and quality of data available, and the methods and mathematical models employed. It is a process whose steps typically include the following: (1) the definition of the geographic and biological extent of the stock, (2) collection of appropriate data, (3) choice of assessment model(s) and parameters, (4) specification of performance criteria and evaluation of alternative actions, (5) estimation of model and other parameters, and (6) presentation of results. While these steps were originally formulated for fisheries stock assessments, they are equally applicable to marine mammal stock assessments.

Marine mammal stock assessments are often carried out to determine what level of mortality a stock can sustain. Several type of information are usually required: current as well as historical abundance, trends in abundance and estimates of biological parameters (such as age at sexual maturity, natural mortality rate, sex ratio, and pregnancy rate), historical harvests, age distribution, maximum sustainable yield level (MSYL), age-specific harvest mortality, sustainable or replacement yield, spatial distribution of the stock in question, and other relevant factors that may vary by species and population (see Population Dynamics). In addition to estimates of these quantities, a measure of parameter uncertainty, such as variance or coefficient of variation, is also necessary.

## I. Productivity

The productivity of a stock is determined by a number of factors, including abundance, rate of increase, population age structure, sex ratio, and the manner in which density-dependence (see Population Dynamics) operates. The productivity is the amount by which the stock increases over a time interval (usually a year) and is the difference between the number of animals added (by reproduction and immigration) and the number that are lost (due to emigration and from natural causes—all causes not due to harvest). Ideally, immigration and emigration are zero or equal, so their effects cancel each other. The amount by which the population increases each year (in the absence of a harvest) is the net production or the replacement yield, and this amount, if harvested, results in the population size at the end of the time period (usually a year) remaining the same size as at the beginning of the year. A related quantity, the sustainable yield, is the productivity when the stock is stable. This occurs when the various population rates (such as natural mortality and reproduction) have remained constant for sufficient time and the environment does not change. Fisheries stock assessments generally determine productivity in terms of biomass, whereas marine mammal stock assessments most often determine productivity in terms of number of animals. This is not only because marine mammals are difficult to weigh but also because they stop growing after reaching physical maturity, whereas most fish grow throughout their life. It thus becomes increasingly difficult to associate age and size for older animals, especially for cetaceans.

## II. Models

Most stock assessments employ a mathematical model of the population to predict the historical trends in abundance as well as future trends under various removal scenarios and choice of model parameters. This approach usually assumes constant environmental conditions and model parameters, and these assumptions are often difficult to evaluate. The model parameters are estimated from the available data or fixed at various plausible values. The more reliable the basic data, the more reliable will be the assessment. A simple population model often used for modeling the dynamics of a population is the discrete, generalized logistic model:

$$N_{t+1} = N_t + R_{max}N_t[1 - (N_t/K)^z] - h_t \tag{1}$$

where $N_t$ is the population size (in numbers) at the start of year $t$, $R_{max}$ is the maximum per capita growth rate, $K$ is the carrying capacity or pre-exploitation abundance of the population, $h_t$ is the harvest in year $t$, and $z$ is a density-dependent exponent which determines at what population level (between 0 and $K$) the productivity is maximum. Equation (1) is a difference equation, with the population size at time $t + 1$ being a function of the population size at time $t$.

The population level at which the productivity curve is maximum, the maximum net productivity level or MNPL, is considered to be greater than 50% of $K$ for marine mammals. When the harvest is random with respect to age and sex, MSYL and MNPL are often used interchangeably. The International Whaling Commission (IWC) has usually adopted a value of 60% for the MSYL, which corresponds to a $z$ of 2.39. A value of $z = 1$ results in a symmetric productivity curve with MSYL at 50% of $K$. This also corresponds to a linear density–dependence relationship between the per capita rate of growth and population density.

This model is simple in that it combines males and females, it ignores age structure and it assumes constant environmental conditions and model parameters. It does, however, capture the basic dynamics of the population, including harvesting or other known removals. It has the desirable feature that the recruitment rate is density dependent; it is greatest at small population sizes and decreases as the population increases toward $K$. The sustainable yield as a function of $N$ for the model shown earlier is given by

$$SY(N) = R_{max}N[1 - (N/K)^z] \tag{2}$$

Productivity, therefore, increases as the population size increases, reaches a maximum when N is equal to the MSYL, a level intermediate between 0 and $K$, and then declines to 0 at $K$. This can be seen in Fig. 1, which shows the sustainable yield curves for $z = 1$ (linear density–dependence with MSYL = 50% of $K$) and $z = 2.39$ (nonlinear density dependence with MSYL = 60% of $K$). By solving for the population size when the productivity is maximum, the MSYL can be shown to be equal to $K(1 + z)^{-1/z}$. Thus, if $z = 1$, MSYL = $K/2$ or 50% of $K$ (see also Population Dynamics).

Equation (1) and similar models, often modified to include sex, age, and spatial structure, have been used to model cetacean, pinniped, and other marine mammal populations. If estimates of $R_{max}$, $z$, and $K$ are available, along with a time series of removals, then the model can be used to project an initial abundance, $N_0$, forward to any particular year. This procedure can be programmed to find an $N_0$ such that the population trajectory "hits" a current abundance estimate. This is a simplification of a technique employed to assess many marine mammal populations. Parameters of the model can be

**S**

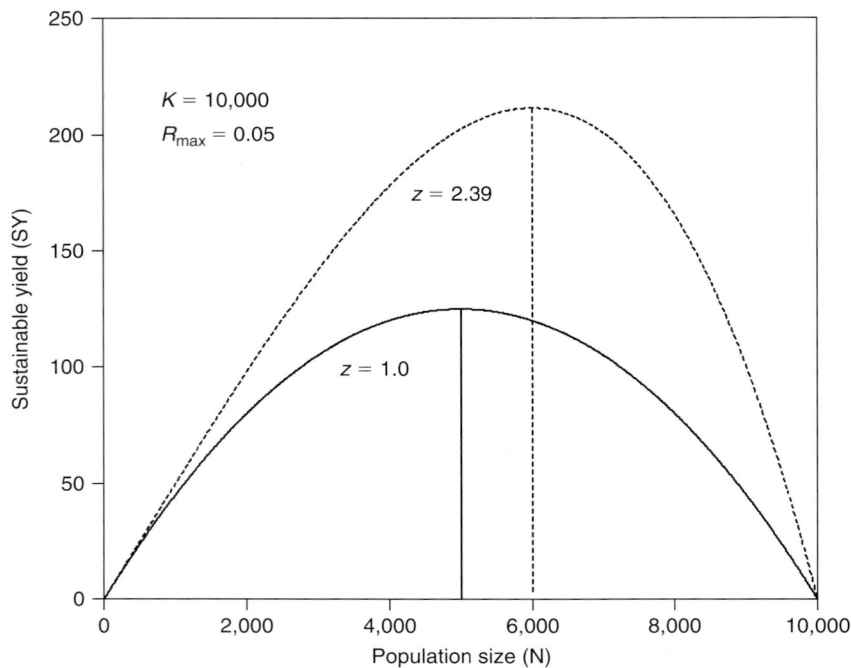

**Figure 1** *A plot showing sustainable yield as a function of population size [see Eq. (2)] for a hypothetical marine mammal population with $R_{max} = 0.05$, $K = 10,000$ and two values for the density-dependent exponent, z.*

estimated by minimizing a measure of discrepancy between observed and model-predicted abundance.

## III. Population Status

The IWC and the US Government (US Marine Mammal Protection Act of 1972: MMPA) have based assessments on classifying stocks by their depletion level with respect to pre-exploitation population size and MNPL. The IWC New Management Procedure (NMP) was based on classifying stocks as Initial Management (IM), Sustained Management (SM), and Protected (P), based on the MSYL and the current depletion level with respect to $K$. However, for stocks classified as IM or SM, the quota required knowledge of the MSY. The MSY, in turn, depends on, among other things, the level of density-dependence (see Fig. 1). The Revised Management Procedure (RMP) addresses shortcomings in the NMP such as the difficulty of determining MSY for a stock and relies primarily on estimates of abundance and their uncertainty and a simple population model such as Eq. (1) requiring few biological parameters.

The US MMPA called for marine mammal populations to be maintained at an "optimum sustainable population" (OSP) level. The US National Marine Fisheries Service defined OSP as a population level between the MNPL and the carrying capacity. The application of this requires the determination of the current population status with respect to the MNPL. In some cases a range of MNPL was used, while in other cases an estimate was made whether the abundance was either less than or greater than the MNPL. The 1994 amendments to the MMPA required that a Potential Biological Removal (PBR) be determined for marine mammal stocks. The PBR is equal to the maximum number of animals that can safely be removed from the population annually. It is calculated as the product of the minimum population estimate of the stock, one-half the maximum theoretical or estimated net productivity rate and a recovery

factor between 0.1 and 1. The assessment of allowable removals therefore hinges on estimation of abundance and the productivity rate. The PBR is a conservative approach whose goal is to allow stocks to reach or maintain their OSP without having to estimate quantities such as MNPL or $K$ (see Management) which can be difficult to estimate.

## IV. Uncertainty and Other Considerations

Uncertainty and how to deal with it is a feature of all stock assessments. The environment and the genetic structure of stocks, as well as the levels of competition and predation may change over time. There is also uncertainty in the underlying population dynamics (termed process error), in the measurement of abundance or indices of abundance (termed measurement error), in the model structure (i.e., is the model correct?), and in the model parameters. Common methods for dealing with some forms of uncertainty include bootstrapping (a method of resampling the data to estimate variability), maximum likelihood (a method for obtaining parameter estimates and their associated variability), model averaging, and Bayesian statistical methods.

Bayesian statistical methods are increasingly being used to deal with uncertainty in stock assessments. Bayesian estimation involves integrating the product of the likelihood of the observed data and the prior probability distribution for parameters of interest to obtain what is termed the posterior distribution for the quantity of interest. Due to the complexity involved in integrating this product it must often be estimated numerically by Monte Carlo methods (based on computer simulations using random numbers). The advantage of the Bayesian methodology is that various sources of information on parameters, including observations from other stocks or species, can be incorporated into the assessment. The end result is not just a simple estimate of the growth rate, for example, but a probability distribution, showing the probability of different values of the parameter. Disadvantages of the Bayesian method are that it is often difficult to construct and obtain agreement on a prior distribution to use in an assessment and the difficulty of detecting inadmissible combinations of parameters that arise in the prior distributions (Borel's Paradox).

Most stock assessments deal with uncertainty in model parameter estimates and management-related quantities, but for the most part they are still based on single-species population models. Some progress has been made in considering marine mammal stocks as part of a larger ecosystem that includes food webs, spatial distribution, and interacting (e.g., competing) species. While it would be preferable to consider a host of interacting factors that influence marine mammal stocks, the paucity of available data often precludes the estimation of the parameters necessary to model the various population interactions. It is likely, however, that future stock assessments will increasingly take into account ecosystem considerations.

Environmental factors are an issue that is becoming more important in fisheries and marine mammal stock assessment. In the North Pacific, for example, decadal oscillations in climate can affect the distribution and abundance of fish and marine mammals (by affecting the distribution and abundance of prey species). Large amounts of long-term data are required to be able to assess the effects of climate on marine mammals. Without these data it is very difficult to determine whether changes in stock abundance are due to climate, harvesting, or a combination of the two.

## V. Examples

### A. Cetaceans

Recent stock assessments carried out for both the eastern North Pacific gray whale (*Eschrichtius robustus*) stock and the Bering-Chukchi-Beaufort Seas stock of bowhead whales (*Balaena mysticetus*) have used Bayesian assessment methods (Wade, 2002; Punt, 2006). Stock definition is not considered an issue for these two stocks, as there is little evidence for sub-stock structure (sub-stock structure has been raised recently with respect to bowhead whales, but the evidence for this is not compelling).

Whales of both of these stocks migrate along the coast where they are counted by shore-based observers. Data collected include counts and sighting distance, which are analyzed to give abundance estimates along with standard errors (measure of uncertainty), and numbers of young and immature animals. Acoustic data are also collected for bowhead whales and integrated with the visual location data to estimate the number of whales passing. Several different assessment models have been used, but all are age- and sex-structured models that incorporate density-dependence in the reproductive rate. A method often used to estimate the posterior probability distribution of management-related parameters (such as MSYL and RY) involves projecting the population model forwards using as inputs parameter values that are randomly chosen from their prior probability distributions (Punt and Hilborn, 1997). A measure of the discrepancy between the observed abundance estimates and the abundances predicted by the population model, the likelihood, is then computed. This is repeated a large number of times, and management-related parameters are calculated for each case. These include historic abundance, current abundance, growth rate, MSYL, and RY. From these repeated model runs, a smaller sample is taken with probability proportional to the total likelihood computed for each trajectory. This second sample is an estimate of the posterior probability distribution of the management-related parameters. From these distributions the median or mean value can

be obtained as well as other statistics of interest, including the answer to questions such as "What is the probability that the RY is less than 100?" or "What is the probability that the current abundance is greater than the MSYL?" In practice, a conservative approach has often been adopted by computing such quantities as the lower 5th or 10th percentile of the RY and other quantities of interest in determining allowable quotas.

Both the US National Marine Fisheries Service and the International Whaling Commission have assessed these two stocks. A Bayesian analysis using an age- and sex-structured model for gray whales resulted in a lower 10th percentile for $R_{max}$ of 0.047. This, coupled with a minimum population estimate of 17,752 [lower 20th percentile of the mean of the 2000/01 and 2001/02 abundance estimates (not significantly different)], and a recovery factor of 1.0, resulted in a potential biological removal (PBR) of 417 animals (Angliss and Outlaw, 2005), well above the average current annual take of less than 180. An assessment based on Bayesian analyses using two different age-structured models (Wade, 2002) resulted in estimates of $K$ of 19,058 and 21,740; $N_{2002}/K$ of 1.04 and 1.00; and $Q_1$, a quantity considered by the IWC to be a more appropriate than RY to use for management advice for populations thought to be above MSYL, of 626 and 669 ($Q_1$ is defined as 0.9MSY for populations above the MSYL, as the minimum of 0.9MSY and the product $N_t°MSYR$ for populations below MSYL, and as zero for populations below $P_{min}$, the population size below which no catches are allowed). These numbers represent medians, but for management advice the IWC used the lower 5th percentile of $Q_1$, 463. This number of whales per year was agreed to be sustainable for at least the medium term (~30 years).

The PBR for the Western Arctic bowhead whale population was based on a minimum population estimate of 9472, a rate of increase of 3.3% (with a harvest, so an $R_{max}$ of 4% was used), and a recovery factor of 0.5 (because the population is increasing in the presence of a known take) (Angliss and Outlaw, 2005). These result in a PBR of 95 animals (9472 × 0.02 × 0.5). The development of a PBR for this stock is required by the MMPA even though the Alaska Eskimo subsistence harvest is managed under the authority of the IWC. Thus, the IWC quota takes precedence over the PBR estimate. An assessment submitted to the IWC Scientific Committee was based on a Bayesian estimation method to fit a density-dependent, age-structured population model to available data on abundance (Punt, 2006). The maximum population growth rate, $R_{max}$, was estimated to be 0.042. The posterior median for K was 11,120 and for $N_{2002}/K$ was 0.86. The posterior distribution for replacement yield (RY) was 172 and for $Q_1$, 243. The difference between these two quantities occurs because the model estimates that the population is above MSLY and close to K.

In the future, gray whale and bowhead whale quotas will be determined by use of a Strike Limit Algorithm (SLA) that has been developed by the IWC Aboriginal Whaling Management Procedure GROUP. The SLA relies mainly on abundance estimates and their variance as well as historic catches. The SLAs for the two species are different, but both of them were developed after a number of years of extensive testing via computer simulations and are robust to uncertainties in abundance, catch, and other factors.

### B. Pinnipeds

The population of Steller sea lions (*Eumetopias jubatus*) in western Alaska has declined sharply since the mid-1970s. No single cause has been implicated in the decline, but human-induced mortality (shootings, incidental takes in fisheries, small directed harvests) and predators (killer whales) are known to have caused sea lion mortality (Loughlin and York, 2000).

A series of modeling papers has attempted to determine the demographic causes of the decline and how they may have changed over time. York (1994) using two age-structure samples with their estimated survival and fecundity rates and counts of Steller sea lions from aerial surveys suggested that the early decline was mostly caused by a decline in juvenile survival. Holmes and York (2003) and Holmes *et al.* (2007) extended this model by additionally using counts of pups and an index of the number of juveniles developed from photographs using pup counts. The later papers estimated time-varying vital rates that were consistent with the non-pup and pup counts and juvenile index data, assuming that the known vital rates were from the sampled animals at time 0. The later models suggested that over the course of the decline, the demographic causes of the decline have changed from low juvenile survival to a combination of low juvenile survival and low adult survival to an increase in survival now in combination with low fecundity.

Commercial harvests of subadult male northern fur seals (*Callorhinus ursinus*) took place on the Pribilof Islands from the time of their discovery in 1786 until 1984. After 1918, the harvest was conducted under the auspices of the Treaty on the Conservation of the Northern Fur Seal (see Northern Fur Seal). The renegotiated Treaty of 1957 provided a vehicle for cooperative research among scientists of the party nations and specified that the population was to be managed to obtain "maximum sustained productivity." Attempts were made to fit spawner-recruit models to fur seal data and to use them to set the harvest, but these methods largely failed, probably due to high variability in year class survival. Using numbers of young of the year (pups) counted earlier in the century and the pattern of harvests, it was estimated that the harvests, on average, took about 30% of the number of male seals born, or about 15% of the total seals born. When managers of the Pribilof herd learned that age at first reproduction of the Russian herd was, on average, 1 year younger than in the Pribilof herd, they justified a large reduction of females in the Pribilof population with the idea that with a lower herd density age at first reproduction would decrease to the level of the Western Pacific population, and a sustainable harvest of the same size would be obtained from a reduced population. That idea was tried and failed, perhaps because the harvesting regime preferentially killed those females that tended to reproduce at a younger age. At present, there is no commercial harvest, but a subsistence take for food is permitted. The maximum size of the subsistence take is set by the PBR approach at about 15,000 animals, well above the mean annual take from 2000 to 2004 of 750 animals.

The Northwest Atlantic harp seal (*Pagophilus groenlandicus*) population in Canada and Greenland is currently subject to a harvest. The current estimate of the population is 5.4 million (Hammill and Stenson, 2007); approximately 18–20% of the population is young of the year. The population model takes into account catches in Canada and Greenland, bycatch in fishing gear and animals struck and lost in the harvests.

Canada uses a management scheme referred to as "Objective Based Fisheries Management." They have identified upper and lower reference points of 30 and 70% of the historical maximum population. The present management objectives are to maximize economic return while maintaining the population over the 70% reference level, with an estimated probability of at least 80%.

The current catches in Canada have averaged 312,000 for the past 5 years. In 2007, the catch was 225,000 (more than 95% are young of the year). An additional 90,000 are taken by Greenland from the same population. The current catch level is about 4–5% of the total population or about 20–25% of the young of year population.

## See Also the Following Articles

Management Population Dynamics

## References

Angliss, R. P., and Outlaw, R. B. (2005). "Alaska Marine Mammal Stock Assessments, 2005." U.S. Dep.Comm, *NOAA-TM-AFSC*-**161**.

Anonymous. (1955). "United States Statement on Estimates of Maximal Sustainable Productivity for the Pribilof Seal Herd." Document 48, presented by the United States during negotiations in Washington, DC, Dec. 19, 1955, preceding ratification of the 1957 Interim Convention on the Conservation of the North Pacific Fur Seal.

Brandon, J. R., and Wade, P. R. (2006). Assessment of the Bering-Chuckchi-Beaufort Seas stock of bowhead whales using Bayesian model averaging. *J. Cetacean Res. Manage.* **8**, 225–239.

Butterworth, D. S., David, J. H. M., McQuaid, L. H., and Xulu, S. S. (1987). Modeling the population dynamics of the South African fur seal, *Arctocephalus pusillus pusillus*. In "Status, Biology, and Ecology of Fur Seals, Proceedings of an International Symposium and Workshop." *NOAA Tech. Rep. NMFS* **51**.

Gerrodette, T., and DeMaster, D. P. (1990). Quantitative determination of optimum sustainable population level. *Mar. Mamm. Sci.* **6**, 1–16.

Goodman, D. (1988). Dynamic response analysis. I. Qualitative estimation of stock status relative to maximum net productivity level from observed dynamics. *Mar. Mamm. Sci.* **4**, 183–195.

Hammill, M. O., and Stenson, G. B. (2007). Application of the precautionary approach and conservation reference points to management of Atlantic seals. *ICES J. Mar. Sci.* **64**, 702–706.

Hilborn, R. (1997). Uncertainty, risk, and the precautionary principle. In "Global Trends: Fisheries Management" (E. K. Pikitch, D. L. Huppert, and M. P. Sissenwine, eds), pp. 100–106. American Fisheries Society, Bethesda, Maryland.

Hilborn, R., and Walters, C. J. (1992). "Quantitative Fisheries Stock Assess-ment: Choice, Dynamics and Uncertainty." Chapman and Hall, New York.

Holmes, E. E., and York, A. E. (2003). Using age structures to detect impacts on threatened populations: a case study using Steller sea lions. *Conserv. Biol.* **17**, 1794–1806.

Holmes, E. E., Fritz, L. W., York, A. E., and Sweeney, K. (2007). Age-structured modeling reveals long-term declines in the natality of western Steller sea lions. *Ecol. Appl.* **17**, 2214–2232.

International Whaling Commission. (1977). Report of the Scientific Committee. *Rep. Int. Whal. Commn.* **27**, 36–51.

Kesteven, G. L. (1999). Stock assessments and the management of fishing activities. *Fish. Res.* **44**, 105–112.

Loughlin, T. R., and York, A. E. (2000). An accounting of the sources of mortality of Steller sea lion, *Eumetopias jubatus*, mortality. *Mar. Fish. Rev.* **62**, 40–45.

National Research Council (1998). "Improving Fish Stock Assessments." National Academy Press, Washington, D.C.

Polacheck, T., Hilborn, R., and Punt, A. E. (1993). Fitting surplus production models: comparing methods and measuring uncertainty. *Can. J. Fish. Aquat. Sci.* **50**, 2597–2607.

Punt, A. E. (2006). Assessing the Bering-Chukchi-Beaufort Seas stock of bowhead whales using abundance data together with data on length or age. *J. Cetacean Res. Manage.* **8**, 127–138.

Punt, A. E., and Hilborn, R. (1997). Fisheries stock assessment and decision analysis: the Bayesian approach. *Rev. Fish Biol. Fish.* **7**, 35–63.

Punt, A. E., and Butterworth, D. S. (2002). An examination of certain of the assumptions made in the Bayesian approach used to assess the eastern North Pacific stock of gray whales (*Eschrichtius robustus*). *J. Cetacean Res. Manage.* **4**, 99–110.

S

Shelton, P. A., Stenson, G. B., Sjare, B., and Warren, W. G. (1996). Model estimates of harp seal numbers-at-age for the Northwest Atlantic. *NAFO Sci. Coun. Stud.* **26**, 1–14.

Smith, T. D. (1983). Changes in size of three dolphin (*Stenella* spp.) populations in the eastern tropical Pacific. *Fish. Bull. (U.S.)* **81**, 1–13.

Wade, P. R. (1998). Calculating limits to the allowable human-caused mortality of cetaceans and pinnipeds. *Mar. Mamm. Sci.* **14**, 1–37.

Wade, P. R. (2002). A Bayesian stock assessment of the eastern pacific gray whale using abundance and harvest data from 1967–1996. *J. Cetacean Res. Manage.* **4**, 85–98.

York, A. E. (1994). Population dynamics of northern sea lions 1975–1985. *Mar. Mamm. Sci.* **10**, 38–51.

York, A. E., and Hartley, J. R. (1981). Pup production following harvest of female northern fur seals. *Can. J. Fish. Aquat. Sci.* **38**, 84–90.

# Stock Identity

JOHN Y. WANG

## I. Importance of Stock Identity

Determining how a species is divided into stocks (the term "stocks" is used here to refer to biological stocks rather than management stocks; see later) is fundamental to the conservation of marine mammals. Because evolutionary processes act at the intraspecific level, genetic differences and locally adaptive characters will accumulate in stocks over time. This reservoir of genetic and phenotypic diversity increases a species' ability to persist through environmental changes. Thus, one of the main goals in conservation is to preserve the evolutionary potential of species by maintaining the diversity found in stocks. Another important goal is to maintain species as functioning elements in their ecosystem by preventing regional overexploitation and depletion. Consequently, knowledge of stock structure of species is integral for developing effective management programs to achieve these goals.

The greatest threats to the survival of marine mammals are human activities. Marine mammals experience various levels and kinds of anthropogenic threats in different regions, and all exhibit life history characteristics (i.e., long-lived, low fecundity, late age of maturity) that make them susceptible to these threats. In order to assess the impact of human activities on marine mammals, it is crucial to identify stocks accurately, establish where the stock boundaries exist, and determine the permeability of the boundaries to genetic exchange with other stocks. This information will influence how the biological data needed for assessments are collected and interpreted and how management plans are designed. Inaccurate stock designations can lead to either unnecessary regulation(s) of fisheries or fallacious management that may result in the depletion of a stock and its accompanying genetic material. For example, if stock structure goes unrecognized and two distinct stocks are incorrectly managed as one, one may inadvertently become depleted.

Understanding stock structure can also help in streamlining the design of other studies, providing insights into evolution and monitoring illegal activities [e.g., DNA analysis of cetacean meat products from Japanese markets found stocks that were prohibited from sale (Baker *et al.*, 2000)]. Therefore, much effort has been directed toward identifying stocks of marine mammals. However, the task remains problematic, with two major difficulties: (1) semantic uncertainty and disagreement in the definition of "stock" and (2) studying stock identity with incomplete biological knowledge.

## II. Definition of Stock

The term "stock" has been used to refer to both biological and management entities (although in many cases, they are combined or inseparable). A management stock is a group of conspecific individuals that are managed separately. The delineation of these stocks is very much dependent on the goals of managers and may not be based on biological discontinuities (e.g., International Whaling Commission management stock designations for baleen whales). With the exception of the definition by Moritz (1994), who described a "management unit" (MU) (which he synonymized with "stock" and appears to be equivalent to management stock) as having significant differences in allele frequencies at nuclear or mitochondrial DNA loci, the criteria for determining management stocks may have little or no biological rationale or consistency and may be influenced greatly by political interests. Nevertheless, management stocks have been used widely due to the paucity of biological information and will likely continue to play an important role in conservation. Developments in management strategies for situations with incomplete biological information should improve the success of conservation programs (Taylor, 1997). Although management stocks offer more flexibility in the sense that they can still be the focus of management programs without evidence of biological distinctiveness, conservation goals (e.g., maintaining genetic diversity) are more likely to be achieved if stocks are based on biological data. Therefore, this article focuses mainly on biological stocks.

Biological stocks are characterized by no or low levels of genetic exchange (which means that members of a biological stock tend to interbreed with each other more often than with other individuals). An entity with this property has also been called a population, subpopulation, evolutionary significant unit (ESU), deme, and subspecies (the only intraspecific taxon recognized by the International Commission on Zoological Nomenclature). When gene flow between two groups is absent, there is usually no disagreement that they represent separate biological stocks. However, it is more typical that some level of genetic exchange exists. Even low levels of genetic exchange can obscure stock boundaries and complicate the task of discriminating biological stocks. Although there is no consensus on the threshold level of gene flow above which stock status is no longer recognized, several approaches have been developed to make the identification of biological stocks more objective and explicit.

## III. Stock Identification Approaches

Defining stocks is linked inextricably with defining species. There are many concepts that propose species definitions, but those advocated most commonly today include biological, evolutionary, and phylogenetic species concepts [for a detailed overview of these and other concepts, see Sites and Crandall (1997) and King (1993)]. However, because these concepts all have major limitations, agreement on the best species definition still eludes biologists. Like the species concepts, each approach to stock identification has limitations and weaknesses. In addition, defining stocks can be influenced, and therefore complicated further, by the goals of conservation and legislation. For example, one of the goals of the U.S. Endangered Species Act (ESA) is to decrease the loss of genetic variation. Thus, for this purpose, defining stocks using genetic criteria [e.g., the ESU of Moritz (1994)] is a reasonable proposal [however, see Pennock and Dimmick (1997) and Dimmick *et al.* (1999)]. Unlike the ESA, the US Marine Mammal Protection Act (MMPA) endeavors to keep biological stocks at or beyond their optimum sustainable levels and functioning in their ecological roles. To accomplish the intent of this

**S**

legislation, defining conservation units also requires demographic information.

There are several operational approaches to stock identification. Whereas some approaches are clear extensions of certain species concepts, the theoretical basis of others may be less explicit or embedded within the methodology. Brief descriptions of the approaches used most commonly are presented here.

Morphological characters have been the main evidence for delineating species. Because differences between stocks are generally less obvious than between species, examination of a large series of specimens is recommended for the identification of stocks using morphology. However, for most marine mammal species, it is difficult, if not impossible, to obtain a large number of specimens for analysis.

The "phylogeographic" approach proposed by Dizon *et al.* (1992) determines the likelihood that a group of organisms is an ESU. The determination is based on distribution, population response (including demography, behavior, vocalizations), and phenotypic and genotypic information, all of which serve as proxies for reproductive isolation (the essence of the biological species concept). Groups most likely to be ESUs have clear geographic and genetic separation and are assigned to "category I." "Category II" units are characterized by clear genetic separation but little to no geographic partitioning. Units with little genetic differentiation but isolated geographically from other conspecifics define "category III." "Category IV" units are least likely to be ESUs because they are separated neither geographically nor genetically from other units. This approach has been described as being unwieldy, but it is explicit, transparent, and has performed well. It also seems to provide the most flexibility in stock delineation, because several kinds of evidence are used and it offers more than a simple dichotomy for the mosaic of variation present. In addition, by considering information on distribution and population responses, it is much better than the other approaches at detecting recently diverged stocks.

Moritz (1994) proposed definitions for ESU and MU that differ from those of the phylogeographic approach. His definition of an ESU requires these entities to exhibit reciprocal monophyly (i.e., to have diagnostic differences) in the mitochondrial DNA (mtDNA) and significant divergence in nuclear DNA allele frequencies, whereas MUs are defined by significant divergence in either mtDNA or nuclear DNA allele frequencies (irrespective of the distinctiveness of the alleles). Because these definitions are based solely on DNA patterns, they cannot be realized with nonmolecular characters and therefore have limited application. Although DNA information may be more direct for determining whether genetic differences exist (some phenotypic characters can be plastic and influenced by environmental factors), it is not always available, and differences in phenotypic characters may be established more rapidly after divergence (e.g., demographic response).

The "phylogenetic" approach for defining ESU was advocated by Vogler and DeSalle (1994). Their procedure for recognizing ESUs is similar to how species are delimited under the phylogenetic species concept. Only heritable genetic, morphological, ecological, or behavioral characters are analyzed. An entity is deemed an ESU if it differs from all other entities in having a unique character or a diagnostic combination of characters. However, it is unclear how the definitions of ESU and species differ with this approach, and the process of determining useful characters may require expert knowledge and can be operationally complex.

The "character concordance" approach (Avise and Ball, 1990; Grady and Quattro, 1999) suggests that a group of individuals sharing a common evolutionary history should share characters that are unique to the group, and the level of concordance among independent, shared characters should increase with increasing divergence time. Thus, high concordance would be strong support for distinctiveness. When concordance is incomplete, the weight of the evidence governs the decision on stock status. Because there are no clear procedural guidelines for interpreting discordant evidence, decisions may be complicated and subjective. Furthermore, many independent characters evolve at different rates, so a lack of concordance may be expected for groups that diverged recently. Therefore this approach may not be effective in identifying recently separated stocks.

A recent workshop on cetacean taxonomy recommended a definition for what should be recognized as a subspecies, the highest category of what could be considered a stock (Reeves *et al.*, 2004): a group of organisms that appear to have been on an independent evolutionary trajectory (with minor continuing flow with other groups) as demonstrated by morphological evidence or at least one line of genetic evidence, with the notes that geographical or behavioral differences can complement morphological and genetic evidence and that subspecies could be geographical forms, incipient species, or even actual species for which data are currently too poor to support elevation to the species level. However, quantitative criteria for determining adequate levels of difference were not specified; it was felt that individual researchers should have the responsibility of convincing their peers of the force of distinctions drawn.

Regardless of the approach one decides to use, it is important that clear hypotheses are stated so that interpretation of the results can be objective and divorced from philosophical or conceptual issues. It is also important that the interpretation of data is within the limitations of the hypotheses being tested. For example, if the results of a study do not support distinct units, then the statistical ability (or power) of the study to detect separate units should be examined. With inadequate power, the appropriate conclusion would be that differences in the characters examined were not detected rather than differences do not exist between the units being studied. Without sufficient power, conclusions regarding stock structure would be premature and should not be made. Finally, it may be tempting to combine units when evidence for separating the units is not found; however, this is neither correct nor does such grouping keep to the precautionary principle.

In situations where essentially no biological stock information exists, the participants of a workshop on the genetics of marine mammals recommended that the smallest area where exploitation occurs be recognized as a stock (management stock). However, they also cautioned that in certain circumstances (e.g., migratory stocks that experience exploitation in several fisheries in different areas or seasons), this strategy may not be precautionary (Dizon *et al.*, 1997). Therefore, the suitability of this approach should be assessed for each case and used only temporarily while immediate attention is directed at studying biological stock structure.

## IV. Analytical Techniques

Several types of information have contributed to our understanding of marine mammal stocks. Which analytical techniques are adopted depends on which stock definition and identification approach are followed. The types of information that have been used in understanding marine mammal stock structure include phenotypic, genetic, distributional, demographic, and ecological information.

Analyses of phenotypic characters have dominated this task. Comparisons of osteology, morphology, and pigmentation have contributed the most to stock identification because these characters provide tangible evidence of distinctness. Also, increased computing capabilities have made multivariate analyses of large data sets simple and quick. However, there are few species (and even fewer stocks) for which data from a large series of specimens can be examined, because specimens are difficult and expensive to obtain, prepare, and store and some characters can be affected greatly by the condition of the specimen (e.g., decomposition and external morphology; postmortem changes in pigmentation).

Increasingly, attention has been shifting toward molecular characters. Protein analyses were important for stock identification but have become obsolete with the development of more efficient DNA technology. Presently, most conclusions regarding stock status are not accepted fully until DNA has been analyzed as well.

Because the properties of mammalian mtDNA are fairly well understood, analyses of mtDNA have dominated molecular studies of marine mammal stocks. For many marine mammal conservation goals, mtDNA evidence is sufficient for designating management stocks [for more details, see Dizon *et al.* (1997)]. For biological stocks, evidence from characters that are heritable from both parents would provide more details about contemporary gene flow and be more convincing. This is because the majority of criteria for interpreting mtDNA data assume that mutation, drift and migration have reached an equilibrium, which can take a long time (particularly in species with long life spans and generation times) so this assumption may not be always valid. Direct analyses of nuclear DNA, especially microsatellites and SNPs (single nucleotide polymorphisms), which is inherited from both parents, are becoming more common.

Most marine mammal species do not have uniform distributions. Areas of high density are usually separated by areas with low or no concentration. Thus, distribution can provide a first approximation of where stock boundaries may exist. Based on heterogeneous distributions, seasonality of occurrence, oceanographic features (e.g., barriers, water currents, temperature), and geographic distance between areas of high abundance, provisional stocks can be proposed for further studies to test. However, distributional data should always be interpreted in conjunction with additional biological knowledge (e.g., daily and seasonal movement patterns, philopatric behavior).

Most distinct stocks are separated geographically or temporally. Therefore, each stock experiences unique ambient conditions (e.g., differential environmental stresses, food quality or availability, exploitation). Adaptations to different conditions may be expressed demographically or ecologically. Different demographic profiles in two groups would be strong evidence of non-interbreeding stocks. Also, demographic differences can reveal recently isolated stocks that have yet to develop genetic or phenotypic distinctiveness. However, to obtain accurate demographic information, a large data set must be analyzed. Because other techniques can address stock identity more directly and efficiently, few studies employ demographic analysis for delineating stocks. If available, demographic information should also be analyzed, especially if stock status, based on molecular and phenotypic evidence, is uncertain.

Many studies have proposed stocks using analyses of ecological differences. Prey preference, parasitology, pollutant loads, stable isotope ratios, and fatty acid signatures are some of the ecological information used most commonly. Although ecological studies provide another line of evidence for understanding stocks, they act only as proxies for genetic and demographic separation.

## V. Study Design and Sampling

Often the people performing analyses of stock structure and identification are distant from the sample collection and may lack knowledge of the local situation. As a result, they are dependent on those collecting the specimens for information, which sometimes may be incomplete. Thus, it is important for studies (and hypotheses) to be designed with a requirement for a set of standard, *a priori* information (e.g., sex, morphology, demography, ecology, oceanography, etc.).

Selection of samples to be analyzed is a critical part of study design, but researchers are commonly handcuffed by the limited availability of samples. As a result, it is common to include samples with limited or no information about their origin, so assumptions are made, often implicitly (e.g., stranded individuals are commonly assumed to be from a local or nearby "population" or designated as being from a certain body of water or area). Inclusion of such specimens and assumptions can unknowingly further complicate or obscure our understanding of stock structure and identity or lead to erroneous conclusions. In addition, political borders can often affect the scope of sampling. Because man-made boundaries are often unconnected with biology, they should not be used for grouping specimens in analyses.

In general, samples of coastal species (e.g., harbor porpoise, franciscana, etc.) are easier to obtain than for offshore species. However, some coastal species have small populations (e.g., *Sousa*, *Orcaella*, etc.) so specimens may be extremely rare. Also, industrial offshore fisheries may result in the capture of large numbers of oceanic species (e.g., the infamous eastern tropical Pacific tuna purse-seine fishery). For coastal species, assuming that stranded specimens are from local or nearby waters is often fairly reasonable, especially if other information exists such as local fauna composition, freshness of carcass, evidence of entanglement in local fisheries, etc.). However, the same assumption is highly questionable for offshore species, which may be less tied to land masses or fixed geographic locations but more related to highly dynamic water masses with boundaries that can vary greatly seasonally or annually. Dead or dying oceanic animals may also be carried by water currents for long distances before stranding. Therefore, great caution must be exercised when attempting to understand stock structure of oceanic species with analyses that include stranded specimens. It is also often assumed implicitly that stranded animals (which are often all that may be available) are representative of a population or stock rather than a specific part of the group.

More and more studies are employing *in situ* sampling (e.g., biopsy darting). Even with such direct sampling methods, it is crucial that extensive information accompany each sample. Standard data such as date, time and geographic location, while vital, are insufficient for species that may be relating to fluid boundaries. Sampling from the same geographical position may result in the sampling of different stocks at different times when water masses shift (Fig. 1). Recording data that can help to characterize water masses from which samples were collected (e.g., water temperature, salinity, turbidity, etc.) will help greatly in understanding stock structure. Photographs of animals sampled may also allow recognition of slight differences in morphology (e.g., pigmentation, etc.) that may help to set *a priori* groupings for tests. Although some analyses do not require *a priori* grouping (e.g., cluster analyses, STRUCTURE for nuclear genetic data), taking a more holistic approach and incorporating ecological, oceanographic and other information into study design and interpretation of results will produce more powerful and convincing conclusions.

## V. Other Complications

Even if there were agreement on a single stock definition and multiple techniques were used, defining stocks would still not be a trivial task. Many situations can obscure and complicate our attempts

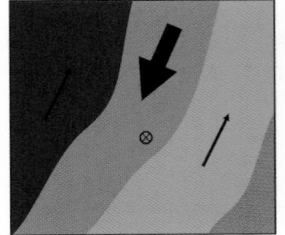 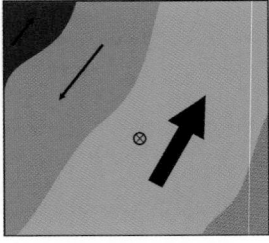 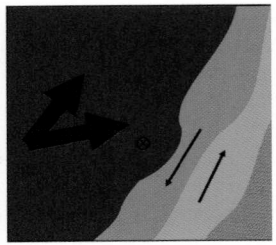

**Figure 1** *Different possible scenarios of shifts in water masses (dark and light blue and green) past a nearby land mass (brown) and through the sampling area (crossed circle) are shown. The sizes of the arrows represent the strength of the currents.*

to delineate stocks, including taxonomic uncertainty; various levels of genetic exchange between stocks; clinal variation; differences between sexes in dispersal and philopatric behavior; diversity in mating strategies; habitat shifts (e.g., occasional environmental fluctuations may bring stocks that are usually separated geographically into contact and allow interbreeding); fragmentation and genetic bottlenecks resulting from exploitation; mixed stocks; social structure; and short-term and seasonal movements, sometimes across international boundaries. Without knowledge of, and consideration for these (and other) attributes, conclusions about stock structure can be compromised. With so many complications, it is not surprising that the biological stock structure of most marine mammal species (even those that were exploited heavily by commercial harvesting) remains uncertain. However, multidisciplinary techniques, technological advancements, and continued attention should allow us to make rapid progress in identifying biological stocks of marine mammals and to design more effective management programs in the absence of essential biological information.

### See Also the Following Articles

Conservation Biology ■ Genetics for Management ■ Molecular Ecology ■ Species

### References

Avise, J. C., and Ball, R. M., Jr. (1990). Principles of genealogical concordance in species concepts and biological taxonomy. *Oxf. Surv. Evol. Biol.* **7**, 45–67.

Baker, C. S., and Palumbi, S. R. (1994). Which whales are hunted? A molecular genetic approach to monitoring whaling. *Science* **265**, 1538–1539.

Dimmick, W. W., Ghedotti, M. J., Grose, M. J., Maglia, A. M., Meinhardt, D. J., and Pennock, D. S. (1999). The importance of systematic biology in defining units of conservation. *Conserv. Biol.* **13**, 653–660.

Dizon, A. E., Lockyer, C., Perrin, W. F., DeMaster, D. P., and Sisson, J. (1992). Rethinking the stock concept: a phylogeographic approach. *Conserv. Biol.* **6**, 24–36.

Dizon, A. E. *et al.* (18 authors). (1997). Report of the Workshop. *In* "Molecular Genetics of Marine Mammals" (A. E. Dizon, S. J. Chivers, and W. F. Perrin, eds.), Special Publication 3, 3–48. The Society for Marine Mammalogy, Lawrence, KS.

Grady, J. M., and Quattro, J. M. (1999). Using character concordance to define taxonomic and conservation units. *Conserv. Biol.* **13**, 1004–1007.

King, M. (1993). "Species Evolution: The Role of Chromosome Change." Cambridge University Press, Cambridge.

Moritz, C. (1994). Defining "evolutionary significant units" for conservation. *Trends Evol. Ecol.* **9**, 373–375.

Pennock, D. S., and Dimmick, W. W. (1997). Critique of the evolutionary significant unit as a definition for "distinct population segments" under the US Endangered Species Act. *Conserv. Biol.* **11**, 611–619.

Reeves, R. R., Perrin, W. F., Taylor, B. L., Baker, C. S., and Mesnick, S. L. (2004). Report of the Workshop on Shortcomings of Cetacean Taxonomy in Relation to Needs of Conservation and Management, April 30–May 2, 2004 La Jolla, California. *NOAA Tech. Mem. NMFS NOAA-TM-NMFS-SWFSC-363*, 94 pp.

Sites, J. W., Jr., and Crandall, K. A. (1997). Testing species boundaries in biodiversity studies. *Conserv. Biol.* **11**, 1289–1297.

Taylor, B. L. (1997). Defining "population" to meet management objectives for marine mammals. *In* "Molecular Genetics of Marine Mammals" (A. E. Dizon, S. J. Chivers, and W. F. Perrin, eds), pp. 49–65. The Society for Marine Mammalogy, Lawrence, KS, Special Publication 3.

Vogler, A. P., and DeSalle, B. (1994). Diagnosing units of conservation management. *Conserv. Biol.* **8**, 354–363.

# Stranding

WILLIAM F. PERRIN AND JOSEPH R. GERACI

Stranded whales have fascinated us through history (Fig. 1). Why do marine mammals strand, what can we learn from their misfortune, and what can we do about it?

### I. Why Do Marine Mammals Strand?

Animals that die or become enfeebled at sea of course may be brought passively to shore by wind and wave action. More intriguing are those cases where marine mammals in distress purposely come ashore. A stranded animal when returned to the water may deliberately strand again. This is very frustrating to those who are trying to "rescue" it. It must be understood that an animal may have stranded because it has decided that it cannot keep itself afloat and survive at sea. Thus, deliberate stranding may represent an effort to keep breathing, whatever the ultimate cost. While this may not be adaptive behavior in evolutionary terms, because nearly all stranded animals die if unassisted, given the alternative of equally certain but earlier death the consideration may be moot. A will to survive is adaptive in general, even if not effective in this circumstance.

**Figure 1** *A sei whale* (Balaenoptera borealis) *stranded in Argentina in 1940. From Leatherwood* et al. *(1982).*

The natural and unnatural causes of death and disablement leading to single strandings are many: environmental conditions such as anomalously low sea temperature or ice entrapment, parasites, disease, biotoxins, ENTANGLEMENT associated with fisheries, starvation due to decreased food supply, collisions with vessels (Laist *et al.*, 2001), contaminants, oil spills, and death or direct injury inflicted by predators, other marine mammals, or at the hands of humans (Geraci and Lounsbury, 2005). The majority of live strandings of pinnipeds are caused by disease or malnutrition (Greig *et al.*, 2005; Colegrove *et al.*, 2005). Determining the cause of a stranding of either a cetacean or a pinniped can be very difficult, even when one is not dealing with a decomposed carcass; symptoms and pathology may be obscure, and two or more factors may be operating simultaneously. Animals that strand in a cluster over a period of a few days may be victims of poisoning, infectious diseases, intensive local fisheries operations (Silva and Sequiera, 2003), or unusual environmental events. By the time the strandings are investigated, the ultimate cause may no longer be evident.

Certain patterns are exhibited in the strandings of particular species related to their distribution, migrations, and reproduction (Geraci and Lounsbury, 2005). For example, newborn gray whales, *Eschrichtius robustus*, are likely to come ashore in the lagoons of Baja California only during the winter calving season. For all cetaceans, the mother–calf bond is very strong and may continue after lactation ceases. If they come ashore together, it may be impossible to determine who led the way. Young juvenile males of highly social pelagic species may strand after being lost or displaced from bachelor schools; this is thought to happen to young Atlantic white-sided dolphins, *Lagenorhynchus acutus*, along the US northeast coast during the fall. Some species follow the migrations of prey. Long-finned pilot whales, *Globicephala melas*, for example, pursue squid into shallow waters of Cape Cod Bay during the autumn and early winter and can be expected to strand at these times.

Mass strandings have always been a puzzle. A large number of whales or dolphins of pelagic species such as sperm whales (*Physeter macrocephalus*), pilot whales (*Globicephala* spp.), false killer whales (*Pseudorca crassidens*), or rough-toothed dolphins (*Steno bredanensis*), may come ashore together for no apparent reason and in seeming good health (Fig. 2). While there undoubtedly is more than one cause

**Figure 2** *A mass stranding of sperm whales in Oregon in 1979. Photo by Robert L. Pitman.*

of mass stranding in these animals, evidence is accumulating that caregiving behavior engendered by tight social bonds may be involved in at least some cases. For example, a herd of 30 false killer whales that semi-stranded in very shallow water in the Dry Tortugas in 1977 included a large male that was moribund due to illness or injury (Connor, 2000). The other whales clustered around this male and did not return to deep water until he died 3 days later. The whales became agitated when would-be rescuers tried to separate them and insisted on remaining in a tight group around the large male. Similar behavior has been reported for other mass strandings, with rescued individuals deliberately rejoining a group on the beach containing one or more severely ill or injured animals (Geraci *et al.*, 1999) or simply joining apparently healthy animals that stranded earlier (Evans *et al.*, 2002). In most cases, in the absence of human intervention, the entire group perishes, victims of a social cohesion that must be highly adaptive in other circumstances. Death may be due to physiological stress or shock, a consequence of lying or struggling on dry land. On a population basis, mass stranding must be a rare event, or it could not persist evolutionarily. Suicidal mass stranding has been suggested

**S**

as a possible mechanism for population regulation (Sergeant, 1982), but this smacks of the group selection hypothesis of Wynne-Edwards (1962), now largely discredited.

In some other instances of mass stranding, careful investigation has uncovered pathological evidence of widespread disease or parasitism that may have been causal or contributory. For example, about half of the mature females in a large stranded group of Atlantic white-sided dolphins were severely infected with a nematode to the extent that reproductive success was likely affected (Geraci *et al.*, 1978). In a stranding of 33 or more short-finned pilot whales, *Globicephala macrorhynchus*, in Florida, all live individuals examined were clinically ill, exhibiting an increased respiratory rate, difficulty in breathing, an elevated heart rate, and a wide range of metabolic and hematological abnormalities (Walsh *et al.*, 1991). As Walsh and colleagues noted: "In such a case, the actual etiology of the stranding event may be unknown because the original inciting factor, such as a virus, may have occurred days or even weeks before."

A number of additional causes for mass stranding have been posited (Geraci and Lounsbury, 2005). Areas with broad tidal flats, strong or unusual currents, or extreme tidal volume may lead to errors of navigation or judgment that result in stranding. It has been suggested that pelagic animals may run aground in shallow water because their echolocation is impaired, or because they encounter acoustical dead zones (Sundaram *et al.*, 2006). Others believe that cetaceans use the earth's magnetic field for navigation and are led astray by magnetic anomalies or disturbances. Links to large-scale climate events (Evans *et al.*, 2005), seasonally changing environmental parameters (Walker *et al.*, 2005), and solar activity (Vanselow and Ricklefs, 2005) have also been suggested. However, the only apparent common factor is strong social cohesion, strong enough that when a single animal comes ashore, for whatever reason, others in the group are likely to follow.

Mass strandings may be increasing in frequency, due possibly to anthropogenic causes. For example, during the period 1981 to 1991, there were 20 mass strandings of long-finned pilot whales on 32 miles of beach in Cape Cod, Massachusetts, whereas only 1 had been reported in the previous 20 years, and epidemiological evidence suggests a possible link with recent morbillivirus outbreaks affecting a number of marine mammal species in the North Atlantic and Mediterranean (Geraci and Lounsbury, 2005; Duignan *et al.*, 1995). The impact of a viral outbreak may be potentiated by organochlorine pollutants that weaken the immune system.

Military sonar can cause whales and dolphins to strand and die singly or *en masse*. Some may die and sink at sea and not wind up on a beach. This has been a controversial issue since a mass stranding of beaked whales in the Bahamas in close conjunction with a nearby naval exercise in 2000 (Balcomb and Claridge, 2001). Use of tactical mid-range-frequency sonar in combination with local oceanographic and topographic conditions was identified as the cause of the stranding event (US Department of Commerce and Secretary of the Navy, 2001). The likely proximate cause of death in sonar-related strandings as determined from necropsies of beaked whales in another mass stranding in the Canary Islands is gas embolic syndrome, similar to decompression sickness ("the bends"), caused by nitrogen coming out of supersaturated blood precipitously, perhaps as the whales are frightened into surfacing faster than usual from a deep dive (Fernández *et al.*, 2005). Other effects may include physical damage to ears and other organs (Wang and Yang, 2006). Another source of loud sounds that may have similar effects is seismic survey operations prospecting for and assessing petroleum deposits (National Research Council, 2003; U.S. Marine Mammal Commission, 2007).

## II. What Can Be Learned From Strandings?

All we know about many species of cetaceans is only what we have learned from strandings. This is true for many of the BEAKED WHALES, the *Mesoplodon* species for example. They are not kept in captivity for exhibit, are hunted only rarely and in remote parts of the world, and are relatively rare, elusive, and notoriously difficult to observe at sea. A rotting carcass on the beach can yield invaluable information on such things as anatomy, life history, genetics, disease, parasites, predators, contaminants, and feeding ecology. A live stranding transported to a holding facility can inform us about physiology, behavior, and cognition. A mass stranding offers a population sample (albeit potentially biased), opening to view parameters such as sex ratio, age structure, pregnancy rate, lactation rate, and relatedness within a group. A released strandling that has been tagged with a transmitter can yield invaluable data on movements, migration, and diving behavior; animals successfully released and tracked have included a pygmy sperm whale (*Kogia breviceps*), common bottlenose dolphins (*Tursiops truncatus*), and long-finned pilot whales (*Globicephala melas*) (Scott *et al.*, 2001; Nawojchik *et al.*, 2003). Every stranding event should be considered a potentially unique opportunity to learn something that cannot be learned any other way.

## III. Stranding Programs and Networks

In order to take full advantage of the scientific opportunities offered by stranded marine mammals, formal stranding response programs have started up in many parts of the world. One of the first was begun by Frederick True, the noted cetologist and one of the first curators of the U.S. National Museum of Natural History (Smithsonian Institution) in Washington, DC, and resulted in the beginnings of the largest collection of marine mammal specimens in the world. The stranding program has continued to the present (Fig. 3) and is a world model for stranding-response procedures and data and specimen collection and curation (Geraci and Lounsbury, 2005). Another early stranding program began in Britain when the board of trade instructed receivers of wrecks to send telegraphic reports of the stranding of whales to the British Museum (Harmer, 1914). As "Royal Fishe," stranded cetaceans are property of the Crown and thus receive special attention and care. This has resulted in a long series of detailed data reports on Cetacea stranded on British coasts and basic knowledge of many North Atlantic species, as well as an immense and irreplaceable collection of specimens (Fraser, 1974; Sheldrick *et al.*, 1994). Perhaps the oldest stranding program, although not scientifically based or motivated, is in Vietnam, where cetaceans are revered as sea-going friends and souls, and those washing ashore have been collected and their skeletons preserved in Buddhist temples for centuries; these accumulations are now yielding information on the cetacean fauna of the region (Smith *et al.*, 1997).

In recent years, formal stranding programs and networks have been established in many countries, including Australia, Japan, New Zealand, Canada, France, Italy, Argentina, Brazil, Thailand, the Philippines, and others. In the United States, a national stranding alert network and officially mandated regional stranding programs arose after passage of the Marine Mammal Protection Act in 1972. These new programs around the world are motivated not only by scientific considerations, but by the desire to achieve humane treatment of live-stranded animals, rescuing them if possible. The goals of these programs are well established, provide for the welfare of live animals, minimize risk to public health and safety, support scientific investigation, and advance public education.

Marine Mammal Salvage Program

# WANTED

INFORMATION CONCERNING
## STRANDED or BEACHED

# WHALES, DOLPHINS & SEALS

Smithsonian Institution
Washington D.C.
Scientific Event Alert Network - SEAN

**Figure 3** *Stranding poster issued by the Smithsonian Institution in the 1970s.*

**Figure 4** *A stranded pygmy killer whale,* Feresa attenuata, *brought into the Miami Seaquarium. From Leatherwood* et al., *1982*

## IV. What Should Be Done With Live-Stranded Marine Mammals?

First, it must be recognized that marine mammals are under legal protection in many countries, and anyone who interferes with them, even though well meaning, may be breaking the law. The first response should be to notify the relevant authorities so that if there is a formal stranding network in the area it can go into action. A telephone call to the nearest natural history MUSEUM, oceanarium, fisheries laboratory, rehabilitation center, or marine wildlife agency will help the information get to the right place. In some areas, hotlines have been established for this purpose and the telephone numbers posted at lifeguard stations, fishery monitoring and landing sites, and other locations.

Not every live animal on the beach needs help. Pinnipeds and sea otters spend time out of the water in the course of their normal affairs. Even some cetaceans may come into very shallow water or ashore for brief periods, e.g., a killer whale to snatch a seal or a bottlenose dolphin to ride a wave. However, certain conditions are unambiguous and do demand attention: a live dolphin obviously in distress on a beach, a sea otter coated with oil, a fur seal too feeble to move, or a porpoise trapped in a fishing weir. Although it is not always possible to judge the health of a cetacean by its outward appearance, coastal animals such as bottlenose dolphins, *Tursiops truncatus*, usually strand singly only when ill and likely will need rehabilitation to survive. Many pelagic cetaceans come ashore in apparent good health, or at least free

of recognizable disease, and have a reasonable chance of withstanding the rigors of being returned to sea, although their long-term survival is undocumented.

Once a decision is made to do something about a live-stranded animal, there are three options: return it to the sea, euthanize it, or transport it to a care facility (oceanarium or marine mammal rescue center) for rehabilitation (Fig. 4). The basic consideration should be to take no action that will only prolong suffering. The basic criteria for making a decision are the following: (1) whether logistical support is available (e.g., a large dolphin or whale cannot be transported without a truck and means to put the animal on it), (2) the number of animals involved (a mass stranding may be a logistical nightmare), (3) the environmental conditions (rough seas, harsh terrain, darkness, or simply a rising tide can increase the risk to the animal and the team, or extremes of heat or cold may affect the animal's ability to thermoregulate), (4) condition of the animal (a healthy animal is resilient, whereas one that is ailing may not survive the ordeal associated with a rescue), (5) risk to wild populations (because of a pathogen), (6) risk to human safety, (7) ease of handling (a very large or struggling animal may be impossible to rescue), and (8) whether care facilities and resources are available.

Immediate return to the sea is an option when the animal is manageable, healthy, and able to function normally; logistical and environmental conditions are favorable; social obligations (e.g., maternal care for the young) can be met; and the area of release is within the normal range of the animal, suitable and navigable. Single-stranded odontocetes, and sea otters or pinnipeds unable to leave the shore, are usually poor candidates for immediate release. Before an animal is released, a plan should be made for monitoring it after release.

Rehabilitation is an option when there is a good chance the animal can be restored to health, facilities are available and equipped for the species and number of animals involved, arrangements can be made for safe and quick transport, the animal is manageable, and, very importantly, there are sufficient funds and staff to provide care for a reasonable period. It should be noted that even where care facilities are increasing in number, more animals come ashore than can be taken into the existing facilities.

Euthanasia is an option when it is necessary to end the suffering of an animal in irreversibly poor condition, no rehabilitation facility is available for orphaned dependent young, rescue is impossible and no care facility is available, animals persistently re-strand, or a distressed cetacean ashore is likely to attract others milling nearby

**S**

to mass strand. The procedure should be carried out humanely by an experienced qualified person and only if essential equipment and materials are available. A clumsy attempt to euthanize an animal without adequate equipment or expertise can cause more suffering than a natural death.

The time between discovery of a stranded cetacean and arrival of a stranding response team can be used by volunteers to relieve stress and improve the animal's chance of recovery. The key is to prevent further injury and keep the animal comfortable while minimizing handling and disturbance. The animal should be protected from blowing sand and kept moist with clean fresh or salt water. Care should be taken to keep water and sand out of the blowhole. In the summer, shade should be provided against the sun. If small, the animal can be positioned on its belly and holes dug to accommodate its flippers. The animal should be out of the surf and protected against lacerations by sharp rocks and seashells.

Moving, release, tagging, transport, rehabilitation, or euthanasia should only be done by qualified, experienced personnel. Detailed guidelines are available in various manuals and government publications.

Mass strandings present a special challenge and can only be coped with effectively by an organized stranding response team backed up with adequate resources. One person, the stranding coordinator or "Incident Commander," must be in charge of all on-site activities. The best procedure is to organize a response team in advance of need to attend to a mass stranding. Volunteers are often indispensable, but their activities must be closely overseen by the coordinator. The most qualified people available should be recruited to the team, including safety officers, veterinarians, and other experienced professionals from aquariums, research stations, veterinary clinics, academic institutions, and wildlife and conservation groups. A list of the kinds of equipment needed to respond effectively to a mass stranding is given in stranding manuals (Geraci and Lounsbury, 2005). The goal should be the swift release of the largest manageable number of animals that have the best chance of surviving. Live animals should be dealt with first. The animals judged to have the greatest prospect of survival should be given priority, not those near death. The integrity of the group may be important to survival of the released animals. A proven approach is to relocate as many animals as possible to a safe place in shallow water where they can rest and become reoriented among their fellows. After tagging and monitoring of condition, the animals can be released together into open water. Because mass-stranded animals returned to sea may re-strand, sometimes immediately but often days or even weeks later, released whales or dolphins should be monitored on a long-term basis, through direct observation (by cooperating fishermen, Coast Guard, sailing clubs, etc.) or telemetry.

## V. What Should Be Collected From a Stranded Carcass?

The condition of a carcass determines much about what can be collected from it and should be specified in field notes. Standard condition codes are (1) alive, (2) freshly dead (i.e., edible), (3) decomposed, but organs basically intact, (4) advanced decomposition (i.e., organs not recognizable, carcass intact), and (5) mummified or skeletal remains only. The quality of information that can be obtained depends on a number of additional factors, including location; size; skills, organization, interests, and morale of the team; adherence to clear, detailed protocols; availability of equipment and supplies; number of animals to be examined; amount of time available; and care maintained in packaging, labeling, shipping, and storing samples. It is well recognized

that it is not possible or practical to collect maximal samples and data in all cases; the effort must be tailored to the conditions. As a rough guide, three levels of collection have been described (modified slightly from Geraci and Lounsbury, 2005):

Level A Data: Basic Minimum Data

1. Name and institutional address of investigator.
2. Reporting source.
3. Species (including preliminary identification and voucher material in the form of photographs in several views, teeth, skulls, and other specimens).
4. Field number.
5. Number of animals, including total and subgroups.
6. Location (preliminary description, plus longitude and latitude and closest named cartographic feature).
7. Date and time of discovery and of specimen recovery.
8. Length (and girth and weight if possible).
9. Condition, using codes above (at time of discovery and at time of recovery).
10. Sex.

Level B Data: Supplementary On-Site Information and Samples

1. Weather and tide conditions.
2. Offshore human/predator activity.
3. Presence of prey species.
4. Behavior before and during stranding, and after return to sea. Also note tag color and number and location of tagging.
5. Samples collected for life history studies (teeth, earplugs, or bone for age determination, reproductive tracts, stomach contents).
6. Samples collected for blood studies.
7. Disposition of carcass.

Level C Data: Necropsy Examination and Parasite Collection

1. Collection of tissues for toxicology, microbiology, and gross histopathology.
2. Collection of parasites.

Detailed protocols for collection of data and specimens are contained in stranding and dissection manuals (Geraci and Lounsbury, 2005). It must be stressed that information has scientific value only when documented carefully. In the case of a mass stranding, it is better, after collecting the minimal basic data for all, to obtain good samples and perform thorough examinations with accurate documentation on a small number of animals than to do a hasty job on many. However, Level-A data should be collected for all animals before moving on successively to Levels B and C.

## See Also the Following Articles

Fishing Industry ■ Mass Die-Offs ■ Pathology ■ Pollution and Marine Mammals ■ Telemetry

## References

Balcomb, K. C., and Claridge, D. E. (2001). A mass stranding of cetaceans caused by naval sonar in the Bahamas. *Bahamas J. Sci.* **5**, 1, 1.

Colegrove, K. M., Greig, D. J., and Gulland, F. M. D. (2005). Causes of live strandings of northern elephant seals (*Mirounga angustirostris*) and Pacific harbor seals (*Phoca vitulina*) along the central California coast, 1992—2001. *Aquat. Mamm.* **31**, 1–10.

Connor, R. C. (2000). Group living in whales and dolphins. *In* "Cetacean Societies: Field Studies of Dolphins and Whales" (J. Mann, R. C. Connor, P. L. Tyack, and H. Whitehead, eds.), pp. 199–218. University of Chicago Press, Chicago, USA.

Duignan, P. J., *et al.* (11 authors) (1995). Morbillivirus infection in two species of pilot whales (*Globicephala* spp.) from the western Atlantic. *Mar. Mamm. Sci.* **11**, 150–162.

Evans, K., Morrice, M., and Hindell, M. (2002). Three mass strandings of sperm whales (*Physeter macrocephalus*) in southern Australian waters. *Mar. Mamm. Sci.* **18**, 622–643.

Evans, K., Thresher, R., Warneke, R. M., Bradshaw, C. J. A., Pook, M., Thiele, D., and Hindell, M. A. (2005). Periodic variability in cetacean strandings: links to large-scale climate events. *Biol. Lett.* **1**, 147–150.

Fernández, A., *et al.* (2005). Gas and fat embolic syndrome" involving a mass stranding of beaked whales (family Ziphiidae) exposed to anthropogenic sonar signals. *Vet. Pathol.* **42**, 446–457.

Fraser, F. C. (1974). "Report on Cetacea Stranded on the British Coasts from 1948 to 1966." Trustees of the British Museum, London.

Geraci, J. R., Dailey, M. D., and St. Aubin, D. J. (1978). Parasitic mastitis in the Atlantic white-sided dolphin, Lagenorhynchus acutus, as a probable factor in herd productivity. *J. Fish. Res. Board Can.* **35**, 1350–1355.

Geraci, J. R., Harwood, J., and Lounsbury, V. J. (1999). Marine mammal die-offs. *In* "Conservation and Management of Marine Mammals" (J. R. Twiss, Jr., and R. R. Reeves, eds), pp. 367–395. Smithsonian Institution Press, Washington, D.C..

Geraci, J. R., and Lounsbury, V. J. (2005). "Marine Mammals Ashore: A Field Guide for Strandings," 2nd ed. National Aquarium in Baltimore.

Greig, D. J., Gulland, F. M. D., and Kreuder, C. (2005). A decade of live California sea lion (*Zalophus californianus*) strandings along the central California coast: causes and trends, 1991–2000. *Aquat. Mamm.* **31**, 11–22.

Harmer, S. F. (1914). "Report on Cetacea Stranded on the British Coasts during 1913." Trustees of the British Museum, London.

Jefferson, T. A., Myrick, A. C. Jr., Chivers, S. J. (1994). Small cetacean dissection and sampling: A field guide. *NOAA Tech. Memo. NMFS NOAA-TM-NMFS-SWFSC*-**198**, 1–54 (Available from NTIS).

Laist, D. W., Knowlton, A. R., Mead, J. G., Collet, A. S., and Podesta, M. (2001). Collisions between ships and whales. *Mar. Mamm. Sci.* **17**, 35–75.

Leatherwood, S., Reeves, R. R., Perrin, W. F., and Evans, W. E. (1982). Whales, dolphins, and porpoises of the eastern North Pacific and adjacent Arctic waters: a guide to their identification. *NOAA Tech. Rep. NMFS Circ.* **444**, 1–245.

National Research Council (2003). "Ocean noise and marine mammals." The National Academies Press, Washington, D.C., pp. 192.

Nawojchik, R., St. Aubin, D. J., and Johnson, A. (2003). Movements and dive behavior of two stranded, rehabilitated long-finned pilot whales (*Globicephala melas*) in the Northwest Atlantic. *Mar. Mamm. Sci.* **19**, 232–239.

Scott, M. D., Hohn, A. A., Westgate, A. J., Nicolas, J. R., Whitaker, B. R., and Campbell, W. B. (2001). A note on the release and tracking of a rehabilitated pygmy sperm whale (*Kogia breviceps*). *J. Cetacean Res. Manage.* **3**, 87–94.

Sergeant, D. E. (1982). Mass strandings of toothed whales (Odontoceti) as a population phenomenon. *Sci. Rep. Whales Res. Inst. Tokyo* **34**, 1–47.

Sheldrick, M. C., *et al.* (1994). Stranded cetacean records for England, Scotland and Wales, 1987–1992. *Investig. Cetacea* **25**, 259–283.

Silva, M. A., and Sequiera, M. (2003). Patterns in the mortality of common dolphins (*Delphinus delphis*) on the Portuguese coast, using stranding records, 1975–1998. *Aquat. Mamm.* **29**, 88–98.

Smith, B. D., *et al.* (1997). Investigations of marine mammals in Vietnam. *Asian Mar. Biol.* **14**, 145–172.

Sundaram, B., Poje, A. C., Veit, R. R., and Nganguia, H. (2006). Acoustical dead zones and the spatial aggregation of whale strandings. *J. Theoret. Biol.* **238**, 764–770.

U.S. Department of Commerce and Secretary of the Navy. (2001). "Joint interim report. Bahamas marine mammal stranding event of 14–16 March 2000."

U.S. Marine Mammal Commission. (2007). "Marine mammals and noise. A sound approach to research and management." A report to Congress from the Marine Mammal Commission, p 187.

Vanselow, K. H., and Ricklefs, K. (2005). Are solar activity and sperm whale *Physeter macrocephalus* strandings around the North Sea related? *J. Sea Res.* **53**, 319–327.

Walker, R. J., Keith, E. O., Yankovsky, A. E., and Odell, D. K. (2005). Environmental correlates of cetacean mass stranding sites in Florida. *Mar. Mamm. Sci.* **21**, 327–335.

Walsh, M. T., Deusse, D. O., Young, W. G., Lynch, J. D., Asper, E. D., and Odell, D. K. (1991). Medical findings in a mass stranding of pilot whales (*Globicephala macrorhynchus*) in Florida. *In* "Marine Mammal Strandings in the United States: Proceedings of the Second Marine Mammal Stranding Workshop, Miami, Florida, December 3–5, 1987" (J. E. Reynolds III and D. K. Odell, eds). *NOAA Tech. Rep. NMFS* **98**, 75–83.

Wang, J. Y., and Yang, S.-C. (2006). Unusual cetacean stranding events of Taiwan in 2004 and 2005. *J. Cetacean Res. Manage.* **8**, 283–292.

Wilkinson, D., and Worthy, G. A. J. (1999). Marine mammal stranding networks. *In* "Conservation and Management of Marine Mammals" (J. R. Twiss, Jr., and R. R. Reeves, eds), pp. 396–411. Smithsonian Institution Press, Washington, D.C..

Wynne-Edwards, V. C. (1962). "Animal Dispersion in Relation to Social Behaviour." Oliver and Boyd, Edinburgh.

# Streamlining

## Frank E. Fish

Streamlining has a major impact on the ecological performance of marine mammals. Because swimming is an integral behavior of marine mammals that forage, mate, escape predation, disperse, and migrate in water, constraints on performance promoted adaptations for effective locomotion by aquatic mammals. To propel itself through the water at a constant swimming speed, a marine mammal needs to generate a forward force (thrust) at the expense of metabolic energy that is equal and opposite to the sum of resistive forces (drag) (Williams, 1987).

## I. Drag

Two major types of drag are experienced by a marine mammal as it swims submerged. These include the pressure or form drag and the viscous or skin friction drag (Williams, 1987). The pressure drag results from the pressure distribution around the body. As water flows about a body, a high pressure is generated at the upstream face and a lower pressure is generated at the downstream face. This difference in pressure produces a force, pressure drag, which opposes forward movement. Viscous drag is a function of the viscosity or stickiness of the water around the body. Water particles adhere to the body surface within a thin layer of water adjacent to the body, called the boundary layer. Friction within the boundary layer and between the boundary layer and the body create a force in the drag direction. The magnitude of the viscous drag will depend on the wetted surface area of the body and the flow conditions within the boundary layer. Boundary flow can be laminar, turbulent, or transitional. A boundary layer with laminar flow produces the lowest viscous drag.

S

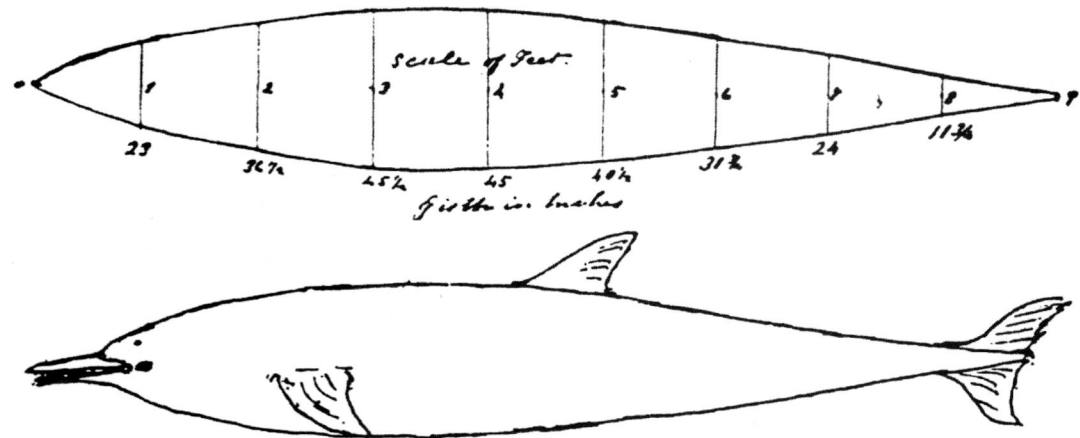

**Figure 1** *A sketch of dolphin and body contours by Cayley.*

Drag is minimized primarily by streamlining the shape of the body and the appendages (Hertel, 1966; Fish, 2006). The streamlined profile of these structures has a fusiform design resembling an elongate teardrop with a rounded leading edge extending to a maximum thickness and a slowly tapering tail. This shape was first investigated in the dolphin by Sir George Cayley (circa 1800) as a solid of least resistance design (Fig. 1). As shown in Fig. 2, marine mammals display a streamlined, fusiform, design. This fusiform shape is sculpted by the distribution of blubber and/or fur covering the body. Modern submarines utilize a fusiform design analogous to the body shape of marine mammals. In addition, the appendages, such as the flukes, flippers, and dorsal fin, have a cross-sectional shape with a fusiform design similar to conventional aircraft wings and hydrofoils.

Streamlining minimizes drag by reducing the magnitude of the pressure difference over the body (Fish, 2006). This reduced pressure difference allows water in the boundary layer to flow without separation from the body surface until near the trailing edge. As separation occurs, a wake is generated downstream. The wake behind the body is narrow, indicating little distortion to the flow and a small pressure drag. Premature separation of the boundary layer occurs because of instabilities in the flow. A laminar boundary flow is inherently less stable and more prone to premature separation than turbulent flow. An animal may pay a higher energetic cost in frictional drag by allowing the development of a turbulent boundary layer, but the pressure and total drags will be substantially lower than with laminar flow where separation transpires.

## II. Body Shape

The bodies of marine mammals are well streamlined (Hertel, 1966; Fish and Hui, 1991). An indicator of the degree of streamlining is the fineness ratio ($FR$ = body length/ maximum diameter). The $FR$ value of 4.5 is considered to provide the least drag and surface area for the maximum volume, although only a 10% increase in drag is realized in the $FR$ range of 3–7 (Fig. 3). However, analysis of streamlined bodies demonstrated minimum drag at $FR$ of 7. In general, cetaceans, pinnipeds, and sirenians have body shapes with $FR$ between 3.3 and 8.0 (Fig. 3). The notable exception for cetaceans is the northern right whale dolphin (*Lissodelphis borealis*) or "snake porpoise" that can have a $FR$ up to 10.9. Despite their bulk and specialization of the head

for filter feeding, the mysticete whales are well streamlined. $FR$ for the Balaenopteridae ranges from 4.8 to 8.1 and Balaenidae ranges from 3.3 to 8.0.

$FR$ is however a crude indicator of the streamlining of the body, because it does not provide information on changes in body contour. Another indicator of body streamlining is the position of the maximum thickness, called the shoulder. Shoulder position is important because this is where transition from laminar to turbulent flow and boundary layer separation are likely to occur. Anterior of the shoulder the pressure distribution favors maintenance of a laminar boundary layer. The position of the shoulder in the most rapidly swimming aquatic mammals is displaced posteriorly which is similar to engineered wings with "laminar" profiles, which reduce drag through maintenance of laminar boundary flow. The shoulder position for dolphins is 34–45% of the body length from the beak.

The shoulder position is located at 40% of body length for otariid seals and 50–60% of body length for phocid seals from the nose. The position can be varied in pinnipeds, because the neck is capable of being retracted and extended. Extension of the neck during rapid swimming could modify the flow over the anterior of the seal and reduce drag by extending the region of laminar flow. Such a drag reduction could aid seals in catching fast swimming, elusive prey.

Experiments on flow visualization using a fluorescent dye applied to a dolphin's melon showed the flow to be laminar over the anterior 32% of the dolphin. Transition began before the dorsal fin with turbulence aft of the fin. Separation of the boundary flow occurred smoothly near the base of the flukes. Flow visualization using bioluminescence within the boundary layer of dolphins and seals similarly indicated a lack of separation from the body surface (Fish, 2006). Flow separation is restricted to the tips of the flukes, flippers, and dorsal fin. The flow separation has been observed as bioluminescent "contrails" (Fig. 4).

The contrails are vortices generated at the tips of the appendages. A tip vortex is generated from pressure differences along the two surfaces of the appendage. The pressure difference produces a lift force similar to the lift produced by airplane wings. Marine mammals utilize lift generated from the appendages to propel the body, increase stability, regulate depth, and maneuver (Fish and Hui, 1991). A consequence of the tip vortices is the loss of energy from the generation of lift, which is referred to as an induced drag.

S

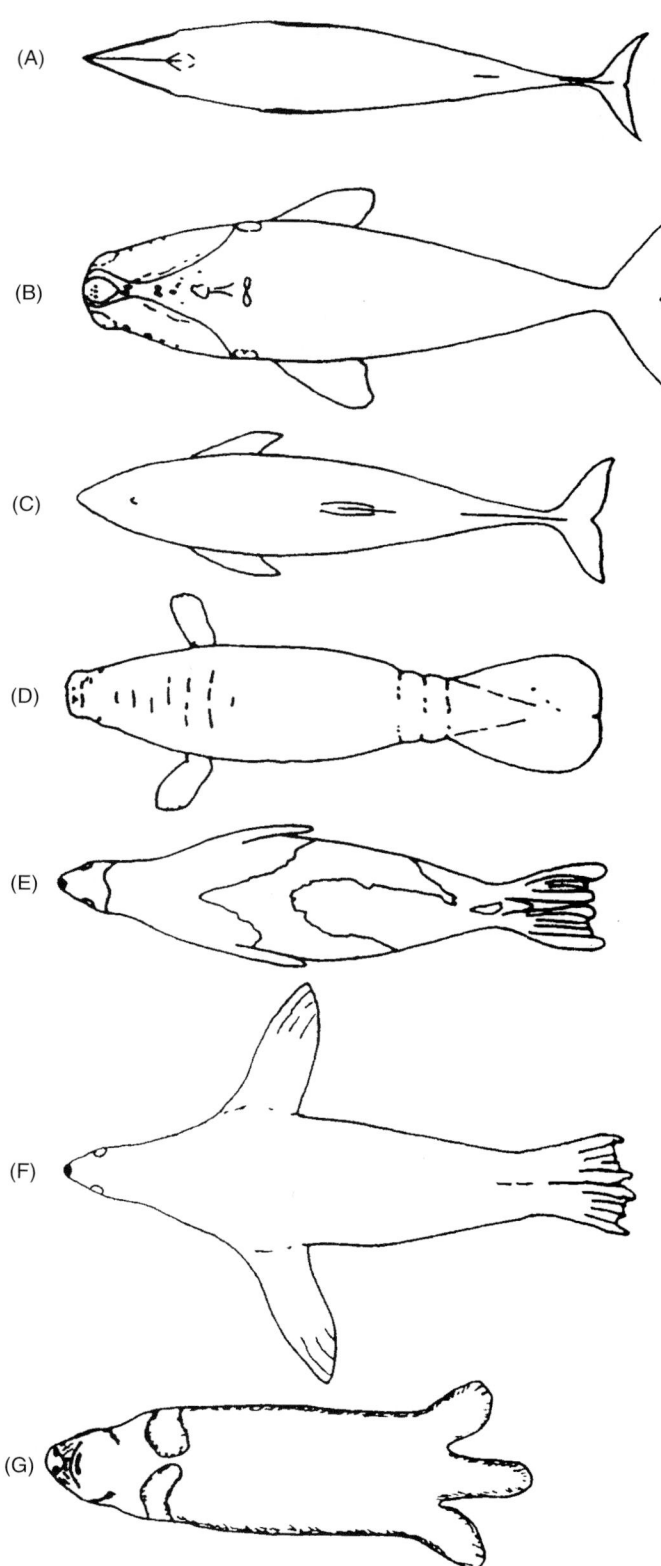

**Figure 2** *Fusiform body shape of marine mammals represented by (A) minke whale,* Balaenoptera acutorostrata, *(B) North Atlantic right whale* Eubalaena glacialis, *(C) harbor porpoise,* Phocoena phocoena, *(D) West Indian manatee,* Trichechus manatus, *(E) harp seal,* Phoca groenlandica, *(F) California sea lion,* Zalophus californianus, *and (G) sea otter* Enhydra lutris.

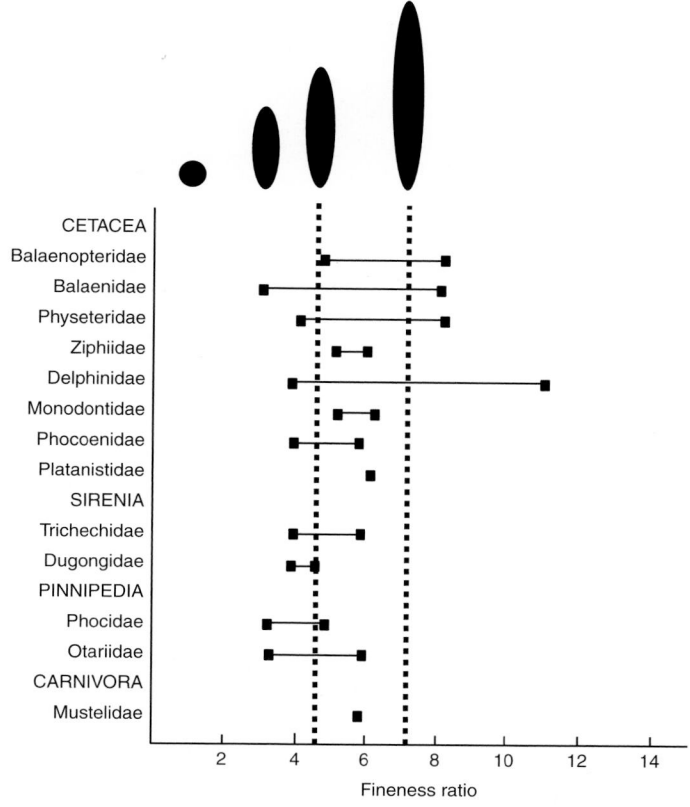

**Figure 3** *A comparison of ranges of Fineness Ratio (FR) for various marine mammal families. The dashed lines indicate FR of 4.5 and 7.0. Minimum drag for streamlined bodies occurs at FR of 7. Silhouettes show the difference in shape in reference to FR from a circular shape (FR = 1) to an elongated form (FR = 7).*

The induced drag and interference to the flow over the body due to the addition of the appendages can affect streamlining. The dorsal fin, pectoral flippers, and flukes comprised only 2.6, 4.2, and 5.6% of the total surface area of the harbor porpoise (*Phocoena phocoena*), respectively; however, these appendages are responsible for 35.7% of the total drag (4.3, 18.0 and 13.4%, respectively). Induced drag is reduced by appendages with a long narrow shape and tips that taper sharply to a point (Fig. 5).

Streamlining also is fostered through buoyancy control. Compared to the center of gravity, the center of buoyancy is closer to the head for terrestrial mammals in water. This relationship produces a torque that causes the body to float at an angle to the horizontal. This orientation would decrease streamlining. Internal and external modifications for buoyancy control provide aquatic mammals with longitudinal trim for better streamlining by presenting a smaller body area to the oncoming flow. For sea otters (*Enhydra lutris*), non-wettable fur provides buoyancy by an entrapped layer of air to maintain a horizontal body orientation when swimming at the surface. The elongate shape of the lungs of marine mammals helps to displace the center of buoyancy posteriorly. This arrangement is possible because the diaphragm of marine mammals is oriented obliquely to almost parallel to the spine.

## III. Drag reduction mechanisms

High swimming speeds attained by many marine mammals have focused attention toward specialized drag reduction mechanisms.

**S**

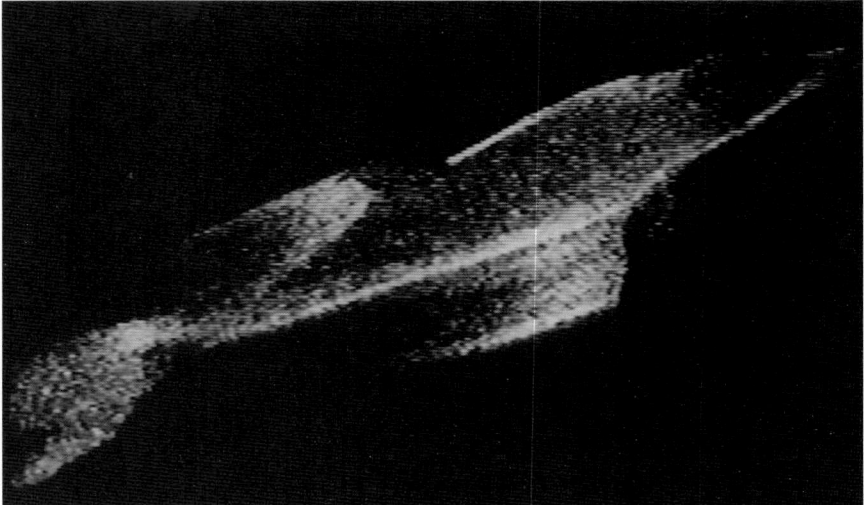

Figure 4 *Bioluminescent image of a gliding bottlenose dolphin* (Tursiops truncatus) *showing lack of separation from the body surface.*

Figure 5 *False killer whale* (Pseudorca crassidens) *displaying highly tapered and elongated appendages.*

In what is known as Gray's Paradox, hydrodynamic estimates of dolphin power output at high speeds were inferred to be greater than the power that could be developed for the mass of muscle available for swimming (Gray, 1936). Resolution of the paradox was believed only possible if the drag was reduced by maintaining laminar flow within the boundary layer, despite a high swimming speed dictating a turbulent boundary flow with increased viscous drag.

While the idea of a special drag reduction mechanism in dolphins has been irresistible, direct evidence of its existence has been elusive. To date, no conclusive evidence has been found of laminar boundary flow over the entire body surface of fast-swimming dolphins (Fish, 2006). The mechanisms examined included compliant skin dampening, secretions, skin cell sloughing, infusion of long-chain polymers into the boundary layer, boundary-layer heating, skin folds, and boundary layer acceleration (Fish and Hui, 1991). Special drag reduction mechanisms are unnecessary to explain Gray's Paradox, which is reconciled when one considers that the calculations of power output were based on burst swimming (10 m/sec for 7 sec) and muscle power output was underestimated, because it was based on sustained performance of dogs and humans.

The naked skin of cetaceans is regarded as a means to maintain a smooth flow with an attached boundary layer over the surface of the body. In addition, the cells of epidermis are produced rapidly, which promotes a high rate of skin sloughing. This increased skin sloughing deters organisms, such as barnacles, from attaching to the skin and thus minimizes drag. The large bumps or tubercles along the leading edge of the flippers of humpback whales (*Megaptera novaeangliae*) produce large vortices, which aid in stabilizing the boundary layer, preventing separation and minimizing drag (Miklosovic *et al.*, 2004). The flippers can then be used to generate large lift forces for maneuvering. The properties of the hair of aquatic mammals are noted to reduce drag by aiding in streamlining of the body. The lack of arrector pili muscles in seals and sea otter permits the pelage to lie flat in water, minimizing resistance to swimming.

Behavioral mechanisms are used also to minimize the drag. Porpoising, which is performed by the fastest swimmers, is a series of rhythmic leaps. By traveling through the air for a given distance, the animal reduces its drag compared to swimming over an equivalent distance underwater (Au and Weihs, 1980). This reduction in drag is due to the lower density and viscosity of air compared to water. In addition, swimming near the water surface to facilitate breathing incurs a large drag increment due to the energy lost for the animal in the formation of surface waves. Porpoising becomes economical only at high swimming speeds when the cost of leaping from the water becomes smaller than the drag on the animal in water. In addition, porpoising is associated with the need to increase ventilation time resulting from greater energetic demands of rapid swimming (Fish and Hui, 1991).

Many dolphins utilize free-riding behaviors to reduce the energy cost of swimming (Williams *et al.*, 1992). In this behavior, the dolphin takes advantage of the pressure field generated by another body and moves along with little or no energetic input. Dolphins have been observed to ride the pressure waves of ships and large whales. By situating itself on the bow wave, the small cetacean can be pushed along or surf down the front slope of the wave (Lang, 1966). This latter mechanism is analogous to human surfing. Sea lions will surf on breakers to get up onto beaches. Even large whales may reduce swimming effort by using the energy of large oceanic waves.

Formation swimming influences the water flow around adjacent individuals. Drafting is the beneficial use of the water flow to reduce drag with a concomitant decrease in the energy cost of locomotion (Weihs, 2004). Large groups of dolphins will organize into side-by-side and echelon formations. Drafting becomes particularly important to young whales to maintain speed with their mothers. A calf swims next to the maximum diameter or near the genital region of the mother. At these locations, the calf is in a hydrodynamically favorable position arising from the interaction of its pressure field with that of the mother's. The calf experiences a thrust component pulling it along with the mother, who experiences increased drag.

## See Also the Following Articles

Energetics ■ Swimming

## References

Au, D., and Weihs, D. (1980). At high speeds dolphins save energy by leaping. *Nature* **284**, 548–550.

Fish, F. E. (2006). The myth and reality of Gray's paradox: implication of dolphin drag reduction for technology. *Bioinsp. Biomim.* **1**, R17–R25.

Fish, F. E., and Hui, C. A. (1991). Dolphin swimming—a review. *Mamm. Rev.* **21**, 181–195.

Gray, J. (1936). Studies in animal locomotion: VI. The propulsive powers of the dolphin. *J. Exp. Biol.* **13**, 192–199.

Hertel, H. (1966). "Structure, Form and Movement." Rheinhold, New York.

Lang, T. G. (1966). Hydrodynamic analysis of cetacean performance. *In* "Whales, Dolphins and Porpoises" (K. S. Norris, ed.), pp. 410–432. University of California Press, Berkeley.

Miklosovic, D. S., Murray, M. M., Howle, L. E., and Fish, F. E. (2004). Leading edge tubercles delay stall on humpback whale (*Megaptera novaeangliae*) flippers. *Phys. Fluids* **16**, L39–L42.

Weihs, D. (2004). The hydrodynamics of dolphin drafting. *J. Biol.* **3**, 8.1–8.16.

Williams, T. M. (1987). Approaches for the study of exercise physiology and hydrodynamics in marine mammals. *In* "Approaches to Marine Mammal Energetics" (A. C. Huntley, D. P. Costa, G. A. J. Worthy, and M. A. Castellini, eds.), pp. 127–145. Spec. Publ. Soc. Mar. Mamm. No. 1.

Williams, T. M., Friedl, W. A., Fong, M. L., Yamada, R. M., Sedivy, P., and Haun, J. E. (1992). Travel at low energetic cost by swimming and wave-riding bottlenose dolphins. *Nature* **355**, 821–823.

# Striped Dolphin
*Stenella coeruleoalba*

FREDERICK I. ARCHER, II

## I. Characteristics and Taxonomy

The striped dolphin (*Stenella coeruleoalba*) is a pelagic small delphinid common in warm-temperate to tropical waters around the world. The specific name, *coeruleoalba*, refers to the diagnostic pattern of blue and white stripes and blazes along the lateral and dorsal sides of their bodies. These dolphins have been well studied in several regions where they occur, such as the western north Pacific, eastern tropical Pacific, and the Mediterranean.

The most striking characteristic of striped dolphins is their bold gray and white color pattern (Fig. 1). The primary features are an eye-to-anus and an eye-to-flipper stripe that are dark gray or bluish black. The dorsal cape is a muted gray or bluish-gray, usually invaded by a white to light-gray spinal blaze that flows from the lateral field. The lateral field is usually darker than the ventral field. A dark stripe branching from the eye-to-anus stripe and continuing into the ventral field is usually present. A faint secondary stripe may also be present in the ventral field.

Striped dolphins, also referred to as "streakers," are similar to most other small oceanic dolphins, having a long beak (well-demarcated from the melon) and falcate dorsal fin. In the field they are most likely to be confused with common dolphins (*Delphinus delphis*), Fraser's dolphin (*Lagenodelphis hosei*), spinner dolphins (*S. longirostris*), or Clymene dolphins (*S. clymene*) but can be easily distinguished by their robust body and coloration. The longest recorded specimen was 2.56 m long and the heaviest specimen weighed 156 kg. Mean maximum body length in the western Pacific is 2.4 m for males and 2.2 m for females (Archer and Perrin, 1999).

Variation in skull and body size has been documented among several geographical regions, and there is genetic evidence that further population subdivision may occur within some of these regions (Bourret *et al.*, 2007). Cranially, *S. coeruleoalba* most closely resembles *S. clymene*, but is significantly larger (Perrin *et al.*, 1981). Sequence analysis of the cytochrome-*b* mitochondrial gene supports a sister species relationship between these two species. The two share an unresolved polytomic clade with *Delphinus* spp., *S. frontalis* and *Tursiops aduncus* (LeDuc *et al.*, 1999). However, as the genus *Stenella* was found to be paraphyletic in this study, the final taxonomic resolution for striped dolphins awaits further work.

## II. Distribution and Abundance

The range is well documented in the western and eastern North and Tropical Pacific, where most records are below about 43°N. Total abundance in this region is estimated to be approximately 570,000 in the western North Pacific and 1.5 million in the eastern North and Tropical Pacific. The species has been recorded from the Atlantic coast of northern South America up to the eastern seaboard of North America, with the northern limit a function of the meanderings of the Gulf Stream. It is found in the eastern North Atlantic south of the United Kingdom and is the most frequently occurring dolphin in the Mediterranean Sea. Abundance in the western Mediterranean Sea has been estimated at approximately 118,000. A recent estimate for the southern Tyrrhenian Sea is 4030 (CV = 0.30) (Fortuna *et al.*,

**S**

Figure 1 *A leaping striped dolphin in the Mediterranean Sea. Photo courtesy of F. Fossa, Delfini Metropolitani, Acquario di Genova.*

2007). It has also been documented from the coast of several countries bordering the Indian Ocean, but its full range in this region is unknown.

## III. Ecology

Striped dolphins can usually be found outside the continental shelf, typically over the continental slope out to oceanic waters and often associated with convergence zones and waters influenced by upwelling (Ballance *et al.*, 2006). The species has been reported in waters from 10–26°C, although most records are from about 18–22°C (Archer and Perrin, 1999).

The species feeds on a variety of pelagic or benthopelagic fish and squid. Off the coast of Japan, South Africa, and in the oceanic waters of the northeast Atlantic, lantern fish (family Myctophidae) are the dominant prey items. In coastal waters in the Northeast Atlantic, their diet shifts more to cod and anchovy, while they primarily feed on squid more in the Mediterranean (Perrin *et al.*, 1994; Ringelstein *et al.*, 2006). To reach potential prey, striped dolphins may be diving from 200 to 700 m. However, analyses of stomach contents suggest that foraging occurs around dusk or early night when prey migrate closer to the surface. Attacks by both sharks and killer whales (*Orcinus orca*) have been reported for this species (Archer and Perrin, 1999).

As a result of a large number of specimens available from the Japanese drive fishery and a morbillivirus epizootic in the western Mediterranean, contaminants and parasites have been studied more intensively in this species than in any other cetacean. High organochlorine loads in the western Mediterranean population have been hypothesized to have caused an immunodepressive state, thus decreasing resistance to infection which in turn led to the deaths of thousands of dolphins from 1990 to 1992. Blubber PCB levels from animals affected by the epizootic may be among the highest recorded values for any mammal (Archer and Perrin, 1999; Aguilar, 2000).

## IV. Behavior and Physiology

Striped dolphins perform a variety of aerial behaviors such as breaching, chin slaps, and a unique behavior termed "roto-tailing," in which they make high arcing jumps while rapidly rotating their tail before reentering the water. They do not commonly ride the bow. School sizes vary between regions, ranging from 10 to 30 to several hundred individuals, but rarely greater than 500 (Archer and Perrin,

1999). In the western Pacific, a complex schooling system has been documented in which individuals move between juvenile, adult, and mixed schools (Miyazaki and Nishiwaki, 1978). The adult and mixed schools are further divided into breeding and nonbreeding schools in what is likely to be a polygynous mating system.

## V. Life History

Mating is seasonal, with gestation lasting from 12 to 13 months. Body length at birth has been estimated to be between 93 and 100 cm. Males enter sexual maturity between 7 and 15 years of age, while females become sexually mature between 5 and 13 years of age. Mean length at sexual maturity is 2.1–2.2 m. Density-dependent changes in several life history parameters have been reported for the western Pacific population, probably a result of large fishery kills. Maximum estimated age for both males and females is 57.5 years (Archer and Perrin, 1999).

## VI. Interactions with Humans

The western north Pacific population(s) of striped dolphins has experienced its heaviest mortality from Japanese drive and hand-harpoon fisheries. Annual catches in one region during the 1940s and 1950s were approximately 8000–9000 animals, reaching as high as 21,000 in some years. However, self-imposed quotas, lower encounter rates, and the dissolution of a fishery cooperative caused annual catches to drop to below ca. 1000 in the early–mid-1980s (Kishiro and Kasuya, 1993). Other directed takes of the species have also occurred in St. Vincent and the Mediterranean for human consumption and protection of fishing gear. Incidental catches have also occurred in the eastern Tropical Pacific tuna purse-seine fishery, pelagic driftnets in the western Mediterranean and northeastern Atlantic, as well as in a variety of gear in fisheries across their range (Archer and Perrin, 1999; Aguilar, 2000). Striped dolphins have not been successfully maintained in captivity, where most have died within 1–2 weeks after not feeding well (Archer and Perrin, 1999).

## See Also the Following Article

Delphinids, Overview

## References

Archer, F. I., and Perrin, W. F. (1999). Stenella coeruleoalba. *Mammal. Species* **603**, 1–9.

Aguilar, A. (2000). Population biology, conservation threats and status of Mediterranean striped dolphins (*Stenella coeruleoalba*). *J. Cetacean Res. Manage.* **2**, 17–26.

Ballance, L. T., Pitman, R. L., and Fiedler, P. C. (2006). Oceanographic influences on seabirds and cetaceans of the eastern tropical Pacific: a review. *Prog. Oceanogr.* **69**, 360–390.

Bourret, V. J. R., Matthias, R. J. M., and Crouau-Roy, M. B. (2007). Genetic variation and population structure of western Mediterranean and northern Atlantic *Stenella coeruleoalba* populations inferred from microsatellite data. *J. Mar. Biol. Assoc. U.K.* **87**, 265–269.

Fortuna, C. M., *et al.* (2007). An insight into the status of the striped dolphin, *Stenella coeruleoalba*, of the southern Tyrrhenian Sea. *J. Mar. Biol. Assoc. U.K.* **87**, 1321–1326.

Kishiro, T., and Kasuya, T. (1993). Review of the Japanese dolphin drive fisheries and their status. *Rep. Int. Whal. Commn.* **43**, 439–452.

Leduc, R. G., Perrin, W. F., and Dizon, A. E. (1999). Phylogenetic relationships among the delphinid cetaceans based on full cytochrome *b* sequences. *Mar. Mamm. Sci.* **15**, 619–648.

Miyazaki, N., and Nishiwaki, M. (1978). School structure of the striped dolphin off the Pacific coast of Japan. *Sci. Rep. Whales Res. Inst. Tokyo* **30**, 65–115.

S

Perrin, W. F., Wilson, C. E., and Archer, F. I. (1994). Striped dolphin *Stenella coeruleoalba* (Meyen, 1833). *In* "Handbook of Marine Mammals, Vol. 5: The First Book of Dolphins" (S. H. Ridgway, and R. Harrison, eds), pp. 129–160. Harcourt Brace and Company, London.

Perrin, W. F., Mitchell, E. D., Mead, J. G., Caldwell, D. K., and van Bree, P. J. H. (1981). *Stenella clymene*, a rediscovered tropical dolphin of the Atlantic. *J. Mamm.* **62**, 583–598.

Ringelstein, J., Pusineri, C., Hassani, S., Meynier, L., Nicolas, R., and Ridoux, V. (2006). Food and feeding ecology of the striped dolphin, *Stenella coeruleoalba*, in the oceanic waters of the north-east Atlantic. *J. Mar. Biol. Assoc. U.K.* **86**, 909–918.

# Surveys

## KARIN A. FORNEY

Surveys are used to address many different marine mammal research questions, including distribution, abundance, trends, and habitat associations. Equipment and methodology vary depending on the species of interest, financial resources, availability of research platforms, and survey objective. Marine mammals at sea are most commonly surveyed aboard ships or large boats (Øien,1991; Wade and Gerrodette, 1993; Barlow, 1995), or from fixed-wing aircraft (Heide-Jørgensen *et al.*, 1993; Forney *et al.* 1995; Garner *et al.*, 1999). Small boats, helicopters, airships, and land-based viewing stations are also used when appropriate (Kraus, *et al.*, 1983; Rathbun, 1988). Line-transect methods (Buckland *et al.*, 2001) are often the most effective for estimating the abundance of marine mammals at sea, although other survey techniques are also used (Hiby and Hammond, 1989; Garner *et al.*, 1999). Marine mammals on or near land, such as pinnipeds, sea otters or walruses, are more commonly counted from land-based viewing points or using aerial photography (Lowry, 1999).

## I. Vessel Surveys

Large oceanographic research vessels (Fig. 1), are the most versatile platform for at-sea surveys of marine mammals. They can carry a dozen or more researchers and remain at sea for weeks at a time, providing the ability to cover extensive marine areas. Search efficiency is greatest aboard these large vessels, because observers can search from a greater height above the water (on the flying bridge, bridge, or in a crow's nest) and use high-power, deck-mounted binoculars ("big eyes") (Fig. 2) when searching for and identifying marine mammals. Large research vessels usually also have equipment for collecting oceanographic data for marine mammal habitat studies (Fiedler and Reilly, 1994), and they may be able to tow hydrophones to detect marine mammals acoustically. Auxiliary studies including photography, biopsy sampling, diving behavior, and prey sampling are also often possible during vessel surveys. A significant disadvantage to large research vessels, however, is their great operating cost: approximately US $10,000 per day. Small- or medium-sized vessels (Fig. 3), including a variety of fishing boats and sail boats, have successfully been used for surveys at a significantly reduced cost (Vidal *et al.*, 1997). This is the most feasible option in many parts of the world, particularly in developing countries (Aragones *et al.*, 1997). On these smaller vessels, searching is usually conducted with hand-held binoculars or by naked eye, from the highest stable deck or platform on the ship. In some shallow bays and rivers, even smaller boats (e.g., rigid-hull inflatable boats, whalers) may be required for safe navigation.

**Figure 1** *NOAA ship McArthur, which has been used for many marine mammal surveys in the eastern Pacific Ocean (Photo: K. Forney).*

**Figure 2** *"Big eyes" used to search for marine mammals on large survey vessels. (Photo courtesy of Protected Resources Division, Southwest Fisheries Science Center, NOAA).*

## II. Aircraft Surveys

The main advantages of aerial surveys are the ability to cover large areas quickly and the lower cost of aircraft compared to large ships. They are particularly useful for rapid assessments and preliminary studies to determine the relative distribution and abundance of species in a particular region. Using strip or line-transect methods, aerial survey can be used to estimate abundance and monitor trends. Aircraft with high wings and bubble windows (Fig. 4) are best suited, because they allow lateral viewing as well as some downward visibility. An additional downward-viewing ("belly") window enhances sighting efficiency considerably, because marine mammals are most easily seen from the air when viewing perpendicular to the water surface. A typical aerial survey observer team consists of two observers searching through the two side windows, one data recorder, and, if possible, a belly observer. Aircraft can also be outfitted with downward viewing instrumentation (e.g., radiometer, hyperspectral imager) to measure ocean surface properties, such as sea surface temperature and ocean color, to provide habitat information during surveys.

S

**Figure 3**  *Small survey vessel used in the Philippines (Photo: W. Perrin).*

**Figure 4**  *NOAA Twin Otter survey aircraft with bubble window (under propeller) and belly window port (above tire) (Photo: K. Forney).*

S

### III. Land-Based Surveys

A few populations of whales reliably migrate close to shore and have successfully been surveyed from land-based stations. These include bowhead whales, *Balaena mysticetus*, at Point Barrow, Alaska, and gray whales, *Eschrichtius robustus*, along the California coast (Buckland *et al.*, 1993). During these surveys, visual and acoustic means may be used to record all whales that travel past the observation point during the migratory period. Adjustments are made for unobservable periods, such as night-time and times of poor weather conditions. Pinnipeds are most commonly counted from land-based viewing stations, such as cliff-tops, although aerial photographs can provide an excellent alternative means of surveying these animals along the coastline (Lowry, 1999).

### IV. Methodological Considerations

Detectability of marine mammals is a key factor in deciding what type of survey platform to use (Hiby and Hammond, 1989; Buckland *et al.*, 2001; Garner *et al.*, 1999). Animals that dive for prolonged periods, such as sperm whales (*Physeter macrocephalus*) and beaked whales (Ziphiidae), will be missed much more frequently during aerial surveys than from ships, because the aircraft travels much faster. For some species, correction factors have been developed to correct for the proportion of animals missed from airplanes or ships. Vessel attraction or avoidance is another concern when designing shipboard marine mammal surveys. For example, harbor porpoise are known to avoid vessels, and if animals are not detected before they react, resulting abundance estimates may be too low. The opposite problem exists for species that are attracted to vessels to "ride the bow"; in these cases abundance estimates may be too high. Both of these problems can be minimized by using a larger vessel that allows viewing from a greater height and with high-power binoculars; animals can then be detected at a greater distance before they react to the vessel.

There is increasing recognition that marine mammal surveys are most effectively interpreted in the context of the habitat conditions at the time of the survey. Marine ecosystems are very dynamic, and the concurrent collection of real-time ecosystem data during surveys provides an ecological context for the observed patterns in marine mammal distribution and abundance. Physical oceanographic measurements and indices of biological productivity can readily be obtained through shipboard sampling, aerial instrumentation, or satellite data. Biological measurements generally require shipboard sampling, such as net tows and hydroacoustic measurements. With the addition of such ecosystem data, survey results can be used to model ecological relationships and evaluate the effect of environmental variability on marine mammal species (Hedley *et al.*, 1999; Forney, 2000; Ferguson *et al.*, 2006).

## See Also the Following Articles

Abundance Estimation Management

## References

Aragones, L. V., Jefferson, T. A., and Marsh, H. (1997). Marine mammal survey techniques applicable in developing countries. *Asian Mar. Biol.* **14**, 15–39.

Barlow, J. (1995). The abundance of cetaceans in California waters. Ship surveys in summer and fall of 1991. *Fish. Bull.* **93**, 1–14.

Buckland, S. T., Breiwick, J. M., Cattanach, K. L., and Laake, J. L. (1993). Estimated population size of the California gray whale. *Mar. Mamm. Sci.* **9**, 235–249.

Buckland, S. T., Anderson, D. R., Burnham, K. P., Laake, J. L., Borchers, D. L., and Thomas, L. (2001). "Introduction to Distance Sampling: Estimating Abundance of Biological Populations." Oxford University Press, New York, 432 p..

Buckland, S. T., Anderson, D. R., Burnham, K. P., Laake, J. L., Borchers, D. L., and Thomas, L. (2004). "Advanced Distance Sampling." Oxford University Press, New York, 434 p..

Ferguson, M. C., Barlow, J., Fiedler, P., Reilly, S. B., and Gerrodette, T. (2006). Spatial models of delphinid (family Delphinidae) encounter rate and group size in the eastern tropical Pacific Ocean. *Ecol. Modell.* **193**, 645–662.

Fiedler, P. C., and Reilly, S. B. (1994). Interannual variability of dolphin habitats in the eastern tropical Pacific. I: Research vessel surveys, 1986–1990. *Fish. Bull.* **92**, 434–450.

Forney, K. A. (2000). Environmental models of cetacean abundance: reducing uncertainty in population trends. *Conserv. Biol.* **14**(5), 1271–1286.

Forney, K. A., Barlow, J., and Carretta, J. V. (1995). The abundance of cetaceans in California waters. Part II. Aerial surveys in winter and spring of 1991 and 1992. *Fish. Bull.* **93**, 15–22.

Garner, G. W., Amstrup, S. C., Laake, J. L., Manly, B. F. J., McDonald, L. L., and Robertson, D. G. (1999). "Marine Mammal Survey and Assessment Methods." A.A. Balkema, Rotterdam.

Hedley, S. L., Buckland, S. T., and Borchers, D. L. (1999). Spatial modeling from line transect data. *J. Cetacean Res. Manage.* **1**(3), 255–264.

Heide-Jørgensen, M. P., Teilmann, J., Benke, H., and Wulf, J. (1993). Abundance and distribution of harbor porpoises *Phocoena phocoena* in selected areas of the western Baltic and the North Sea. *Helg. Meeresunter.* **47**, 335–346.

Hiby, A. R., and Hammond, P. S. (1989). Survey techniques for estimating abundance of cetaceans. *Rep. Int. Whal. Commn.* (Spec. Iss. **11**), 47–80.

Kraus, S. D., Gilbert, J. R., and Prescott, J. H. (1983). A comparison of aerial, shipboard, and land-based survey methodology for the harbor porpoise, *Phocoena phocoena*. *Fish. Bull.* **81**, 910–913.

Lowry, M. S. (1999). Counts of California sea lion (Zalophus californianus) pups from aerial color photographs and from the ground: a comparison of two methods. *Mar. Mamm. Sci.* **15**, 143–158.

Øien, N. (1991). Abundance of the northeastern Atlantic stock of minke whales based on shipboard surveys conducted in July 1989. *Rep. Int. Whal. Commn.* **41**, 433–437.

Rathbun, G. (1988). Fixed-wing airplane versus helicopter surveys of manatees (*Trichechus manatus*). *Mar. Mamm. Sci.* **4**, 71–74.

Vidal, O., Barlow, J., Hurtado, L. A., Torre, J., Cendon, P., and Ojeda, Z. (1997). Distribution and abundance of the Amazon river dolphin (*Inia geoffrensis*) and the tucuxi (*Sotalia fluviatilis*) in the upper Amazon River. *Mar. Mamm. Sci.* **13**, 427–445.

Wade, P. R., and Gerrodette, T. (1993). Estimates of cetacean abundance and distribution in the eastern tropical Pacific. *Rep. Int. Whal. Commn.* **413**, 477–494.

# Sustainability

## Charles W. Fowler and Michael A. Etnier

Sustainability has been elusive in spite of its ubiquitous appearance in the goals for management. Human impacts need to be sustainable, whether they are the harvest of a marine mammal population, the harvest of finfishes in the marine environment, our production of $CO_2$, or the genetic effects we have on other species. Sustainable human interactions with other systems must be established in ways that account for the suite of factors involved in ecosystems and the complexity of the biosphere to include both our direct effects and our indirect effects on such systems. It is unlikely that historical harvests of marine mammal populations are sustainable, partly because of their low productivity levels (Perrin, 1999). Thus, defining sustainability, whether it involves our interactions with marine mammals, fisheries resources, or ecosystems, remains an important objective.

Historically, the concept of Maximum Sustainable Yield (MSY) has played a major role in the management of our utilization of natural resources. This approach has yet to be assessed in its contribution to worldwide problems such as over-harvested fish populations (Rosenberg *et al.*, 1993; Committee on Ecosystem Management for Sustainable Marine Fisheries, 1999). Commercial whaling and sealing have also involved concepts derived from the MSY approach.

The inadequacies of management based on MSY have been recognized [e.g., it is illogical (Fowler and Smith, 2004)]; such approaches are not sustainable. Progress in understanding such problems involve the development of other methodologies, e.g., the Catch Limit Algorithm of the International Whaling Commission (Slooten, 1998) and the Potential Biological Removal approach being used by the National Marine Fisheries Service in the United

**S**

States (Wade, 1998). These alternatives, however, have not escaped the weaknesses of being applicable only to individual species and they do not account for complexity. The ecosystem effects of fishing (Hall, 1999), whaling, or sealing are not adequately considered in current management strategies. The challenges currently facing management are not being met.

It is therefore extremely important to find alternatives that will work. One approach is systemic management (Fowler, 2003) in which empirical examples are used to define and measure sustainability and set sustainable goals. Along with other species, marine mammals are sources of information about sustainability that is broadly applicable and meets the demands being made of management. Using empirical examples of sustainability, the abnormal or pathological can be avoided.

Marine mammals serve as empirical examples of sustainable roles, or niches, within marine ecosystems; they have persisted as parts of such systems to integrate the various factors contributing to their evolution. Resource consumption is an example of an ecosystem relationship for which we need measures of sustainability. Both marine mammals and fisheries consume biomass from resource/prey species, making part of their interaction competitive. The rates of predation by marine mammals exemplify variation in sustainable levels of consumption. The size selectivity of their feeding habits exemplifies sustainability involving genetic impact. Importantly, there are limits to variation in such interspecific interactions as there are with all ecological interactions (Fowler and Hobbs, 2002). The failures of conventional management can be overcome by replacing such management with processes that mimic empirical examples of sustainability (Fowler, 2003). Such management ensures that fisheries catches or selectivity are not abnormal, in comparison to the consumption and selectivity observed among other consumer species. Thus, observed examples of sustainable consumption rates, and size selectivity can be used to regulate the catches taken by fisheries while simultaneously conserving resources and habitat for other species. As such, marine mammals, like other species, are empirical examples of sustainability that provide guiding information. Use of such information prevents the bias of human limitations in converting scientific information to objective management advice (Fowler, 2003).

## I. Management Questions, Empirical Answers

How many tons of whales, seals, fish, cephalopods, or other resources should we harvest each year? What is the appropriate or optimal rate at which to harvest biomass, and when and where should it be harvested? What is an advisable size selectivity for commercial fishing, to deal with one of the many aspects of the genetic effects of harvesting? Such questions can be asked with regard to a single resource species, or with respect to any area of the oceans, an ecosystem, a group of resource species, a season, or the biosphere. How do we answer all such questions so that the answer for one case will not be in conflict with the answer for another? Empirical information is key to answering such questions consistently, and marine mammals are key elements in providing such information for management in marine environments (Fowler, 1999; Fowler, 2003).

Fig. 1 shows frequency distributions (empirical probability distributions) for consumption rates exhibited by various marine mammals. Also shown, for comparison in each case, is the harvest rate for fisheries—the rate at which humans consume biomass. The top panel illustrates consumption from a population of an individual species and the bottom panel total consumption within the biosphere. Intermediate panels depict consumption from a group of resource

species, an ecosystem, and the marine environment. Thus, Fig. 1 represents information for a telescoped series of increasing complexity. In this case, the biosphere contains the marine environment which, in turn, includes the ecosystem (Bering Sea). Within the ecosystem we find populations of resource species (the finfish) among which is the population of walleye pollock (*Theragra chalcogramma*, one species).

Figure 2 is a comparison of the mean size of Atlantic cod (*Gadus morhua*) taken by 19 species of marine mammals with that of commercial fisheries. This represents one measure of the abnormality of commercial fishing in regard to the size composition of catches and its related genetic effects on the resource species compared to the situation normally experienced in the ecosystem (Etnier and Fowler, 2005). The intensity of this selectivity is directly related to the abnormally high harvest rates shown in Fig. 1.

Various requirements are placed on management (e.g., the tenets of management) (Fowler, 2003). It has been made clear that management should avoid abnormality in the components, processes,

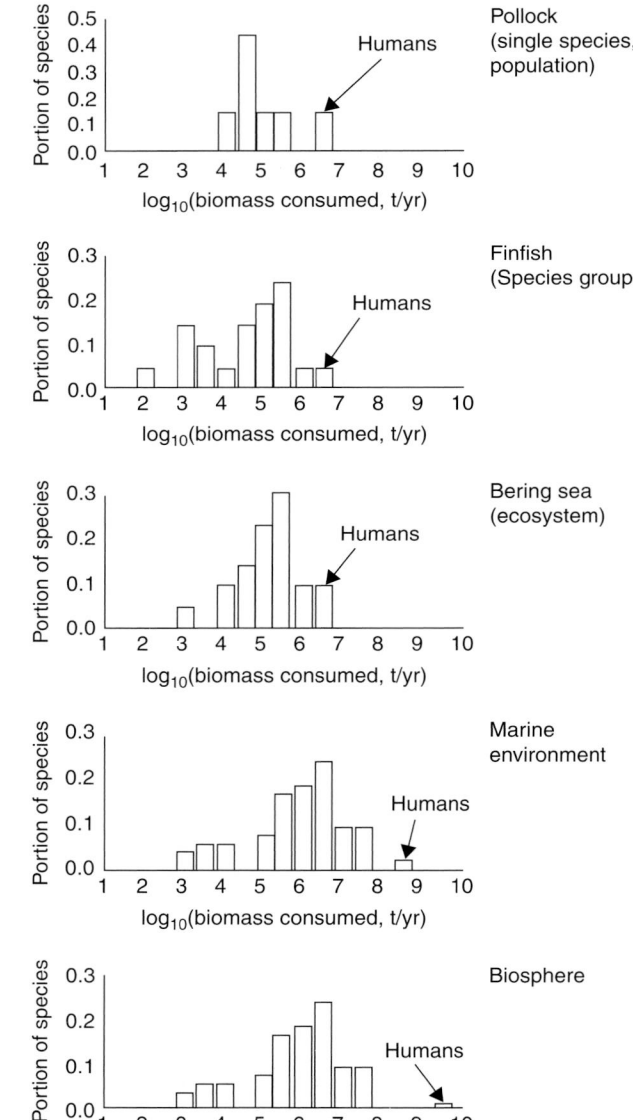

and characteristics of systems (Christensen *et al.*, 1996; Mangel *et al.*, 1996) [see the review of Fowler and Hobbs (2002)]. These include the components and processes that constitute individuals, species, ecosystems, and the biosphere. Humans (through commercial fishing) are, or have been, obvious outliers relative to the natural variation illustrated in Figs. 1 and 2 (Fowler and Hobbs, 2003).

There are things management can not do. The other species within these distributions are largely beyond our control, but not our influence, especially the collective aggregate of species in each distribution. Individual species can undergo change opposite to our intentions if we act to influence them directly. In fact, changes we stimulate in these species may result in unwanted reactions in the rest of the system, whether we purposely manipulate them individually or as a group. These changes include repercussions throughout the food web (e.g., predator/prey dynamics) and domino-effects of genetic consequences in the underlying coevolutionary web. Such reactions cannot be avoided.

Management can move forward, however, by focusing on reducing consumption by humans, or harvesting smaller fish, thus avoiding the abnormal by falling within the range of the variation exhibited by other species. Nonhuman species exhibit predator–prey interactions that occur within the context of complexity—all the things that influence these species to result in what we observe (all the explanatory factors) (Fowler and Crawford, 2004). That is, all the things that have contributed to the observed rates of consumption such as those shown in Fig. 1, or the size selectivity shown in Fig. 2, are taken into account. These factors include all anthropogenic factors to include

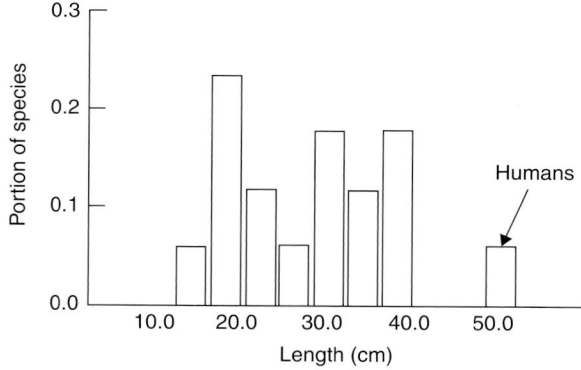

**Figure 2** *The frequency distribution of the mean size of Atlantic cod* (Gadus morhua) *taken by 19 species of marine mammals (1983–1996) compared to that in the take of cod by commercial fisheries prior to their collapse in the Northwest Atlantic. From Etnier and Fowler (2005), with permission.*

the effects of global warming, toxic chemicals, overfishing, introduced species, and oceanic acidification. This complexity is inherent to the patterns we observe.

Figures 1 and 2, then, serve to guide the management of our use of fisheries resources when achieving sustainability is the ultimate goal. The empirical data exemplify what works in the face of the complete suite of factors that set limits on the consumption of resources. Better options are found among the more numerous examples toward the centers of the distributions compared to the lack of examples beyond the tails (Figs. 1 and 2). The risks and constraints that prevent the accumulation of species in such regions are to be avoided. How, then, can we carry out sustainable management?

## II. Management to Achieve Sustainability

It has long been recognized that we need a form of management that applies to ecosystems (Christensen *et al.*, 1996). To develop sustainable management strategies, humans must manage by having sustainable influence on ecosystems. This includes consuming biomass from ecosystems at rates that are sustainable. Sustainable management requires change. For example, reducing consumption by humans to 10% of current harvest rates of fish would place our species squarely within the range of variation shown in relevant ecosystems (third panel of Fig. 1). Such change will require time to accomplish. Maintenance of reduced levels, once achieved, would lead to further challenges for management. These would include responding to changes within ecosystems over seasons, with shifts in climate, or in response to our management.

It has been recognized for some time, in the world of commercial fisheries management, that the historical focus on managing harvests from the individual species point of view has been insufficient, partly because of the need to consider ecosystems. However, management restricted to ecosystems would be similarly insufficient because of the need to account for both the broader marine environment and the biosphere, in addition to individual species.

Management based on empirical examples is an integrated approach that simultaneously helps to define sustainable harvests at the level of individual species and the biosphere. For example, reducing harvests to between 1 and 10% of recent harvest levels would be required for the species represented in the top panel of Fig. 1. We can similarly account for multi-species groups and the entire marine environment (second and fourth panels, respectively, Fig. 1). The biosphere can be involved by reducing human consumption to about 0.1% of current consumption levels (Fig. 1, panel 5). Not only would this account for food-web effects, but it would also deal with the intensity of any selectivity resulting in genetic effects. Selectivity *per se* can also be dealt with directly as a distinct management issue (Fig. 2).

**Figure 1** *The frequency distribution of consumption rates (biomass consumed per year, in* log$_{10}$ *scale) for marine mammals showing optimal consumption rates where most species are concentrated. The rate at which humans harvest biomass is shown for comparison. The top panel shows the natural variation in consumption of walleye Pollock* (Theragra chalcogramma) *as observed for six species of marine mammals in the Bering Sea in comparison to recent takes of pollock by commercial fisheries (Livingston, 1993). The second panel shows consumption of finfish in the Bering Sea by 20 species of marine mammals compared to fisheries takes (predominantly pollock) (Fowler and Perez, 1999). Total biomass consumption is shown for 20 species of marine mammals in the Bering Sea in the third panel, again compared to the commercial take which is predominantly pollock (Fowler and Perez, 1999). Total biomass consumption for the entire marine environment (all oceans combined) is shown in the fourth panel for 55 species of marine mammals, here compared to the take of about 110 million metric tons estimated as the harvest of biomass for human use in the late 1990s (Committee on Ecosystem Management for Sustainable Marine Fisheries, 1999). Worldwide consumption of biomass by humans is compared to that of 55 species of marine mammals in the bottom panel. The last two panels are based on indirect estimates (Fowler and Perez, 1999) using population and body size data from the marine mammal series by Ridgway and Harrison (1981–1999) and equations representing ingestion rates as a function of body size in Peters (1983).*

In all cases shown in Figs. 1 and 2, management would involve change, often measured in orders-of-magnitude, to avoid human abnormality. If the distributions themselves change in response to the reduction of harvested biomass by humans (or other management to deal with factors such as habitat modification, and the production of $CO_2$ and toxic chemicals), then sustainable management would need to respond to the new information. This requires continuous monitoring through concerted scientific effort to observe such changes.

## III. Accounting for Complexity

In management, the list of things to be considered seems endless. For example, we need to account for the effects of all forms of selectivity (evolution), endangered species, and multiple complex processes such as nutrient flow within ecosystems. It is often said that management needs to be interdisciplinary, or an integrated accounting of everything science can study. But we also need to account for the things we do not, or cannot, study or know about. This is accomplished in two ways when we take advantage of empirical examples provided by species such as marine mammals.

First, every species reflects the effects of everything that influences it. These factors are integrated in (inherent to) the distributions shown in Fig. 1 (Fowler and Crawford, 2004). For example, evolution is taken into account through its influence on the position of every species within each distribution. Each species represents a composite of balances among various opposing forces (e.g., those involved in predation, population growth, evolution, or extinction). Each species reflects the constraints of the system—constraints such as competition for the limited availability of energy in its path through the various trophic levels. The relative importance of each factor compared to every other factor is accounted for implicitly. They are integral to what we observe.

Second, we must address other management questions. Distributions similar to those of Fig. 1 can be developed for the allocation of biomass consumption over alternative resource species, seasons, and geographic space. Marine mammals can be used as empirical examples in such an endeavor and through such species we account for the unknowns that influence the position of each species in distributions like those of Fig. 1. For marine mammals themselves, finding the rates at which they can be harvested sustainably would involve information regarding consumption by their predators—including other marine mammals!

## IV. Consistency

An important aspect of empirical examples is their representation of a system that is internally consistent. Advice for management at the ecosystem level (Fig. 1, panel 3) will be consistent with advice at the individual species level (panel 1) when applied simultaneously. Collective application at the individual species level must be constrained to the limits set by application at the ecosystem level. Nutrients, energy, biomass, and species involved in this systemic consistency guarantee freedom of conflict because the conflicting forces of nature result in what we see in distributions derived empirically.

Marine mammals can serve as empirical sources of information about how species fit into marine ecosystems. Through information about nonhuman species, such as marine mammals, we are provided with guidance about sustainable harvest rates, size selectivity, allocation over size, allocation of harvests over time and space, allocation across resource species, and the variety of other management questions left largely unaddressed in current management practices.

## See Also the Following Articles

Abundance Estimation ■ Bycatch ■ Fishing Industry, Effects of ■ Management ■ Population Dynamics

## References

Christensen, N. L., Bartuska, A. M., Brown, J. H., Carpenter, S. R., D'Antonio, C., Francis, R., Franklin, J. F., et al. (1996). The report of the Ecological Society of America Committee on the scientific basis for ecosystem management. Ecol. Appl. **6**, 665–691.

Committee on Ecosystem Management for Sustainable Marine Fisheries (1999). "Sustaining Marine Fisheries." National Academy Press, Washington, D.C..

Etnier, M. A., and Fowler, C. W. (2005). Comparison of size selectivity between marine mammals and commercial fisheries with recommendations for restructuring management policies. U.S. Department of Commerce, NOAA Technical Memorandum NMFS-AFSC-159.

Fowler, C. W. (1999). Natures' Monte Carlo experiments in sustainability. In "Proceedings of the Fifth NMFS Stock Assessment Workshop: Providing Scientific Advice to Implement the Precautionary Approach Under the Magnuson-Stevens Fishery Conservation and Management Act" (V. R. Restrepo, ed.), pp. 25–32. U.S. Department of Commerce, NOAA Technical Memorandum NMFS-F/SPO-40.

Fowler, C. W. (2003). Tenets, principles, and criteria for management: the basis for systemic management. Mar. Fish. Rev. **65**, 1–55.

Fowler, C. W., and Crawford, R. J. M. (2004). Systemic management of fisheries in space and time: tradeoffs, complexity, ecosystems, sustainability. Biosph. Conserv. **6**, 25–42.

Fowler, C. W., and Hobbs, L. (2002). Limits to natural variation: implications for systemic management. Anim. Biodivers. Conserv. **25**, 7–45.

Fowler, C. W., and Hobbs, L. (2003). Is humanity sustainable? Proc. R. Soc. Lond. B. **270**, 2579–2583.

Fowler, C. W., and Perez, M. A. (1999). Constructing species frequency distributions —a step toward systemic management. U.S. Department of Commerce, NOAA Technical Memorandum NMFS-AFSC-109.

Fowler, C. W., and Smith, T. D. (2004). Preface to the 2004 printing. In "Dynamics of Large Mammal Populations" (C. W. Fowler, and T. D. Smith, eds), pp. xiii–xxvi. Blackburn Press, Caldwell.

Hall, S. J. (1999). "The Effects of Fishing on Marine Ecosystems and Communities." Blackwell Science, Oxford.

Livingston, P. A. (1993). Importance of predation by groundfish, marine mammals and birds on walleye pollock Theragra chalcogramma and Pacific herring Clupea pallasi in the eastern Bering Sea. Mar. Ecol. Prog. Ser. **102**, 205–215.

Mangel, M., Talbot, L. M., Meffe, G. K., Agardy, M. T., Alverson, D. L., Barlow, J., Botkin, D., et al. (1996). Principles for the conservation of wild living resources. Ecol. Appl. **6**, 338–362.

Perrin, W. F. (1999). Selected examples of small cetaceans at risk. In "Conservation and Management of Marine Mammals" (J. Twiss, and R. R. Reeves, eds), pp. 296–310. Smithsonian Press, Washington, D.C.

Peters, R. H. (1983). "The Ecological Implications of Body Size." Cambridge University Press, New York.

Ridgway, S. H., and Harrison, R. (eds) (1981–1999). "Handbook of Marine Mammals," pp. 1–6. Academic Press, New York.

Rosenberg, A. A., Fogarty, M. J., Sissenwine, M. P., Beddington, J. R., and Shephard, J. G. (1993). Achieving sustainable use of renewable resources. Science **262**, 828–829.

Slooten, E. (1998). Risk analysis at the International Whaling Commission. In "Statistics in Ecology and Environmental Monitoring 2" (D. J. Fletcher, L. Kavalieris, and B. F. J. Manly, eds), pp. 173–180. University of Otago Press, Dunedin.

Wade, P. R. (1998). Calculating limits to the allowable human-caused mortality of cetaceans and pinnipeds. Mar. Mamm. Sci. **14**, 1–37.

S

# Susu and Bhulan
## *Platanista gangetica gangetica* and *P. g. minor*

BRIAN D. SMITH AND GILLIAN T. BRAULIK

## I. Characteristics and Taxonomy

The susu and bhulan are two river dolphins of the South Asian subcontinent. Some scientists dispute the existence of one species and consider them to be separate, *P. gangetica* and *P. minor* (Pilleri *et al.*, 1982).

A variety of vernacular names are used for the susu and bhulan, mostly connoting the sound made during respiration. For susu, these include *swongsu* and *sus matsya* (*matsya* = fish) in Nepali, *soonse* and *sunsar* in Hindi, *hiho* and *shihu* in Assamese, and *shushuk* and *sishu-*, *foo-*, or *hungmaach* (*maach* = fish) in Bengali. For bhulan these include *bhoolun* in Suraiki and Sindhi and *dolphin muchli* (the name for dolphins generally) in Urdu.

Both subspecies are nearly identical in external appearance (Fig. 1). Their body is supple and robust, attenuating behind the dorsal fin to a narrow tailstock. Coloration is gray or light brown overall, becoming blotchy with age. Bellies of young animals are lighter and often have a pinkish hue. The dolphins have a long snout that becomes thicker at the tip. Mature females generally have a longer rostrum, which sometimes curves upward and to one side at the tip when particularly long. Their numerous narrow pointed teeth (26–39 in each upper jaw and 26–35 in each lower jaw) are curved inward and become longer toward the distal end of the rostrum, sometimes protruding outside the mouth. Teeth become broad, flat, and peg-like in older animals from wear and accumulation of the cement layer. The eyes are extremely small and visible as pin-hole openings slightly above the upturned mouth. The external auditory meatus is larger than the eye opening and located slightly above it, a unique arrangement among odontocetes. The blowhole is a small slit oriented longitudinally, which is a rare but not unique configuration among cetaceans. A distinct median ridge begins anterior of the blowhole and bisects a convex melon, which becomes less rounded as the dolphin approaches adulthood. The dorsal fin is a low triangular hump with a slightly defined knob at the apex located about two-thirds of the body length posterior of the melon. Large, broad flippers sometimes are squared-off at the end with a crenellated margin, and the arm and hand bones are visible beneath the taut dorsal surface. Flukes are broad with a concave margin and distinct median notch. The opening for male genitals is located much closer to the umbilicus in relation to the anus compared with most other cetaceans.

An extraordinary feature of the skull is the projection of a maxillary crest, upward and forward, covering an air sinus that leads to the tympanic cavity. The crest slants to the left, which makes the skull among the most asymmetric of all odontocete skulls. Other skull features unique to *Platanista* include that the palatine is found in the nasal tube and that the pterygoids are external on the palate and enter the temporal fossa (Reeves and Brownell, 1989). The

(A)

(B)

**Figure 1** *Generic appearance of susu and bhulan (top, C. Brett Jarrett) and susu surfacing in the Sundarbans mangrove forest, Bangladesh (bottom; photo by Mowgliz).*

S

mandibular symphysis is laterally compressed, slightly upturned toward the end, and constitutes as much as two-thirds the mandible length of adult females and a little more than half in adult males (Reeves and Brownell, 1989). Cervical vertebrae are unfused, allowing considerable neck movement. Postcranial skeletal features unique to *Platanista* include the location of costal facets of the thoracic vertebrae on the posterior margin of the centrum, a thicker ulna than radius, and the absence of ulnare and pisiform bones in the flippers.

The presence of a cecum between the large and small intestines, a penis with erectile side lobes, nasolaryngeal air sacs that form a diverticulum of the eustachian tube, and a primitive and relatively unlobulated kidney distinguish *Platanista* from all other odontocetes. The length of the intestine is 4–5 times the body length, which is the shortest among the river dolphins (Yamasaki and Kito, 1984). Compared to other dolphins, the brain is small and neocortical development is low; however, sub-cortical components associated with acoustical functions are well developed. Their small eyes have a flat, thick cornea and lack a crystalline lens, giving the dolphins a reputation for being blind. Although the optic nerve is extremely reduced and light can only enter through a pin-hole muscle-controlled sphincter, the retina has densely packed light gathering receptors (Herald *et al.*, 1969).

Fossil and genetic evidence indicates that *Platanista gangetica* is a relict species, the sole surviving member of a primitive and once widespread taxon of archaic cetaceans (Cassens *et al.*, 2000). Although the superfamily Platanistoidea was previously considered to include three other monotypic genera—*Pontoporia*, *Lipotes*, and *Inia* (Kasuya, 1973; Zhou, 1982), the genus *Platanista* is now recognized as the only extant member of the taxon (Rice, 1998). Pilleri *et al.* (1982) recognized two species of *Platanista* based on differences in the prominence of the nasal crests, caudal height of the maxillary crests, length of the lower transverse processes of the sixth and seventh cervical vertebrae, blood protein composition, and free-esterified cholestrin ratio in the lipids. Their arguments were unconvincing, due to the small sample of adult specimens examined and the absence of statistical analyses (Reeves and Brownell, 1989). Based on differences in tail lengths, Kasuya (1972) proposed that the susu and bhulan be considered subspecies; this was followed by Rice (1998).

## II. Distribution and Abundance

A map of historical distribution, charted by the nineteenth-century British naturalist John Anderson (Anderson, 1879), shows the susu occurring throughout the Ganges-Brahmaputra-Megna and Karnaphuli river systems in Nepal, India, and Bangladesh, and the bhulan occurring throughout the Indus River mainstem, and in the Sutlej, Ravi, Chenab, and Jhelum tributaries. Anderson (1879) stated that the range of both species was only limited downstream by increasing salinity in deltas and upstream by rocky barriers or insufficient water. Their distribution has shrunk considerably since then, largely due to water development, which has blocked dolphin movements and degraded their habitat.

With the exception of a few vagrants, bhulans currently occupy less than 700 km of the Indus mainstem (about one-fifth of their historic range), fragmented into three subpopulations by the Chashma, Taunsa, Guddu, and Sukkur barrages (Fig. 2) (Reeves and Chaudhry, 1998). The subspecies has been extirpated from the Sutlej, Ravi, Chenab, and Jhelum Rivers, and from downstream of the Kotri Barrage and upstream of the Jinnah Barrage in the Indus mainstem. The largest subpopulation is located at the downstream end of the dolphins' range, between Guddu and Sukkur barrages, with a reported count of 602 dolphins (3.60/km). Numbers decline progressively in upstream segments, despite their progressively larger geographical size, with counts of 259 (0.74/km) and 84 (0.28/km) individuals between the Guddu and Taunsa, and Taunsa and Chasma barrages, respectively. Correcting for missed animals and extrapolating for un-surveyed channels, the metapopulation was estimated in 2001 to be about 1200 individuals (Brauik, 2006).

Historically, susus were found in the Ganges River as far upstream as Haridwar, about 100 km above their current range, and in the Yamuna River year round as far upstream as Delhi, probably about 400 km above their current low-water range (Fig. 3) (Sinha *et al.*, 2000). In the Ganges mainstem, there are four extant subpopulations isolated between barrages. In the northern tributaries of the Ganges, three of the six subpopulations that have been isolated above or between barrages have been extirpated (Gandak River above the Gandak Barrage and Sarda River above the upper and lower Sarda barrages) (Sinha *et al.*, 2000) and one reduced to insignificant numbers (Kosi River above the Kosi Barrage) (Smith *et al.*, 1994). In the Son River, a southern tributary of the Ganges, a small population of susus has been isolated above the Indrapuri Barrage, and the dolphins no longer occur during the dry season for about 100 km below the barrage until the Ganges confluence (Sinha *et al.*, 2000).

Although no rigorous range-wide surveys of susus have been conducted, the aggregate population is believed to number in the low- to mid-thousands (Smith and Reeves, 2000). In the approximate center of the Ganges mainstem, about 120 susus (1.8/km) occur in the Vikramshila Gangetic Dolphin Sanctuary (Choudhary *et al.*, 2006). In southern Bangladesh, at least 125 susus occur in the Karnaphuli and Sangu Rivers and connecting Sikalbaha-Chandkhali canal. This river system is unconnected to the much larger Ganges-Brahmaputra-Meghna system, and dolphins in the two systems are probably demographically isolated from each other, although it may be possible for there to be movement between them during the high-water season when salinity is low in the coastal area in between. There are occasional reports of dolphins remaining in a reservoir behind Kaptai Dam in the Karnaphuli, but a survey in 1999 produced no sightings (Smith *et al.*, 2001). At the far downstream end of their range in the Sundarbans of Bangladesh, a Huggins conditional likelihood model of double concurrent counts estimated that the population inhabiting waterways of the mangrove forest was 225 individuals (CV = 12.6%) and overlapped the range of Irrawaddy dolphins *Orcaella brevirostris* (Smith *et al.*, 2006).

## III. Ecology

The susu and bhulan are patchily distributed throughout their range, generally occurring in counter-currents and deep pools located downstream of channel confluences and sharp meanders, and upstream and downstream of mid-channel islands. They also occasionally occur in counter-currents induced by engineering structures such as bridge pilings and groynes. The affinity of the species for counter-currents is probably greatest in upstream tributaries where productivity is especially clumped and strong downstream currents restrict occupancy to the hydraulic refuge these areas provide (Smith, 1993; Smith *et al.*, 1998). During the dry season, dolphins are occasionally sighted in secondary braided channels of the Indus River but encounter rates are 62–85% lower compared to the main channel (Brauik, 2006).

## IV. Behavior and Physiology

Captive studies of susus and bhulans have revealed exceptional aspects of their behavior and sensory abilities. The dolphins vocalize

**Figure 2** *Map of the Indus river system of Pakistan showing dams and barrages that have fragmented the population of bhulan and degraded their habitat.*

almost constantly and swim on their sides. The dolphins continuously emit trains of high frequency (15–150 kHz) echolocation clicks, interrupted by short pauses of 1–60 sec (Herald *et al.*, 1969). The click trains are focused in two highly directional fields, the dorsal one emitted directly from the melon and the ventral one reflected downward by the maxillary crest, with an acoustic "scotoma" in front of the rostrum (Pilleri *et al.*, 1976). During a dive, the dolphins spin 90° on their lateral axis and position their head down, sweeping it

back and forth in a scanning motion, while trailing one flipper along or slightly above the bottom. Shortly before surfacing, the dolphins reverse their spin and surface close to their original position and orientation.

In the wild, susus and bhulans are observed alone or in clusters of 2–3, but occasionally as many as 25, individuals. With the exception of mother–young pairs, the attracting force for these clusters may be more related to the patchy distribution of prey and the availability

**S**

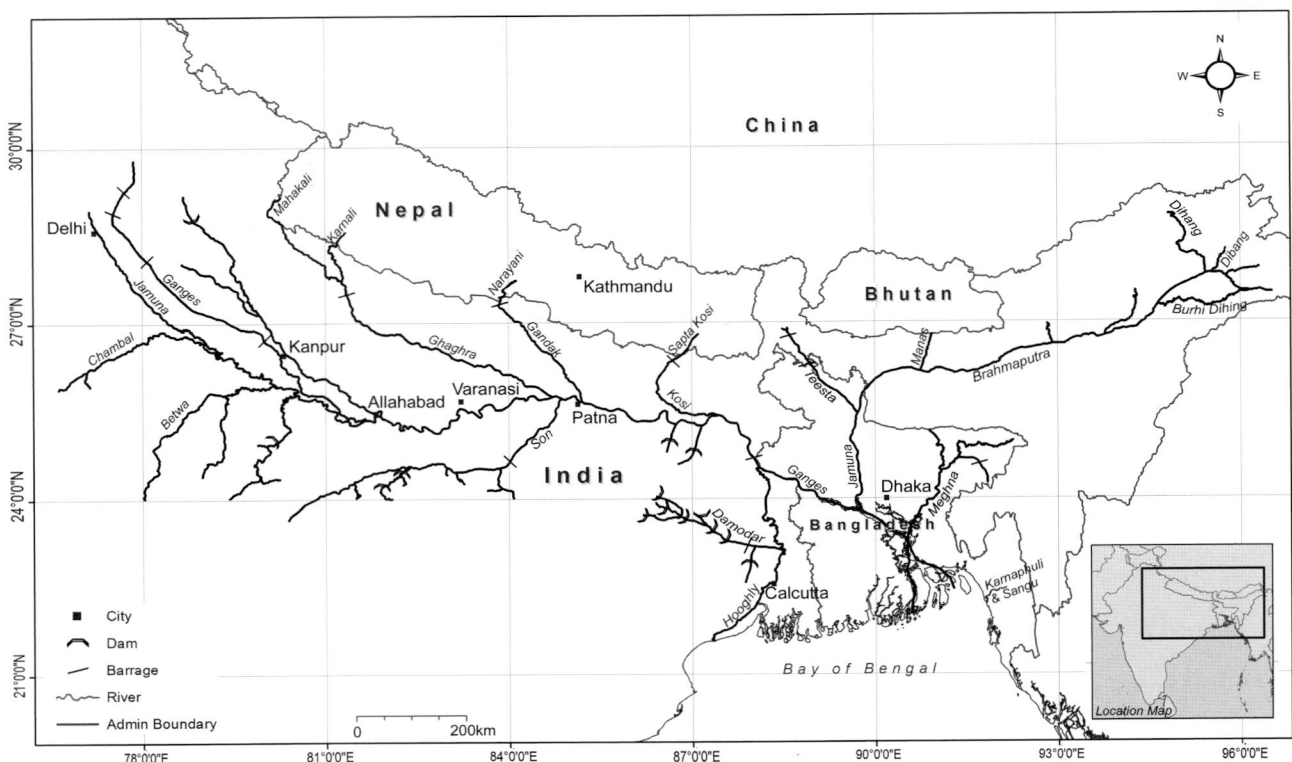

**Figure 3** *Map of the Ganges-Brahmaputra-Megna and Karnaphuli-Sangu river systems of Nepal, India, and Bangladesh showing dams and barrages that have fragmented the population of susu and degraded their habitat.*

of hydraulic refuge and deep pools than to survival or reproductive advantages gained by close social affiliations. Susus have been observed surfacing just inside of the upstream end of counter-current boundaries where the eddies become aligned with main flow. Surfacing at this location may allow the dolphins to minimize energy outputs while monitoring foraging opportunities in the mainstream flow and center pool of the counter-current (Smith, 1993).

## V. Life History

Males attain sexual maturity at a body length of about 170 cm and physical maturity at 200–210 cm. Females attain sexual maturity at similar or slightly larger body lengths but physical maturity at about 250 cm. The generally larger rostrum of females accounts for this sexual dimorphism, which becomes evident at a body length of about 150 cm. Length at birth is estimated to be about 70 cm. Gestation lasts approximately 1 year, with possible peak birthing seasons in early winter and early summer. Young begin feeding on small prey at about 1 or 2 months and are weaned within a year (Kasuya, 1972).

## VI. Interactions with Humans

The susu and bhulan are both classified by the IUCN as endangered due to declines in their range and population size. The most significant threat to their survival is probably the existence of numerous dams and barrages that have severely fragmented populations and reduced the amount and quality of suitable habitat. Dams are absolute barriers to dolphin movements. Subpopulations trapped

above barrages could lose dolphins when they move downstream during high water while the barrage gates are open and probably cannot return due to strong hydraulic forces between the gates (Reeves *et al.*, 1991)—although if flow velocity is sufficiently reduced when discharge approaches its maximum capacity in the downstream channel and there is no head behind the barrage (i.e., water elevations on both sides are equal) it may be possible for there to be upstream movement through the gates. Involuntary attrition due to downstream movement below barrages would exacerbate normal biological problems faced by small isolated populations. Dolphins are also sometimes lost to their metapopulation when they become stranded in irrigation canals where they can apparently survive for long periods before the canals are drained for maintenance. Water diverted by barrages, generally for irrigation and flood control, and abstracted by surface pumps and tube wells also results in dolphins competing with humans for the actual substance of their environment: freshwater. During the low-water season, the Indus and Ganges rivers become virtually dry downstream of the Sukkur and Farakka barrages, respectively. The long-term effects of global climate change will almost certainly hasten the decline of dry season flows as snowmelt from the Himalayan and Karakoram ranges (which feed the Ganges-Brahmaputra-Meghna and Indus River systems, respectively) becomes progressively reduced. An additional threat comes from embankments constructed for flood control. These structures simplify hydraulic complexity and eliminate or reduce the size of counter-current pools where the dolphins most often congregate. The range of the South Asian river dolphins will probably continue to decline as smaller upstream subpopulations are extirpated due to habitat loss related to the combined impacts of escalating water demands, construction of large engineering structures, and long-term climate changes.

Deliberate killing of the bhulan for meat and oil was a traditional practice until at least the early 1970s. Hunting is now banned and no longer common (Pilleri and Zbinden, 1973–1974; Reeves and Chaudhry, 1998). Susus are killed by indigenous people in the upper Brahmaputra for their meat and by fishermen in the middle reaches of the Ganges for their oil, which is used as a fish attractant (Motwani and Srivastava, 1961). Similar to all cetaceans, the susu and bhulan are threatened by entanglement in fishing gear and vessel collisions. Their preferred habitat is often in the same location as primary fishing grounds and ferry crossings, which puts the dolphins at increased risk. The problem of accidental killing in fisheries will undoubtedly worsen as the demand for fish and fishing employment increases. Pollution may also be affecting the survival of both species, especially considering the decline in the flushing benefits of abundant water, the aggregate distribution of river dolphins in areas of intensive human use, and the occurrence of major portions of both subspecies' populations downstream of large cities. As top predators, the dolphins are particularly vulnerable to persistent contaminants (e.g., PCBs and DDTs), some of which are banned or strictly regulated in more developed countries but widely used in industry and agriculture of South Asia.

In 1974, the government of Sindh declared the Indus River between the Sukkur and Guddu barrages a dolphin reserve and the government of Punjab prohibited deliberate killing (Reeves *et al.*, 1991). Enforcement of these measures seems to have stopped the rapid decline of the bhulan between Guddu and Sukkur barrages reported by Pilleri and Zbinden (1973–74), and there is evidence that the population in this segment has been increasing (Braulik, 2006). The susu was perhaps the first cetacean to receive official protection from hunting when it was listed as a protected species in the Moral Edicts of King Asoka in India more than 2000 years ago. Susus currently receive legal protection from deliberate killing in all range states. The Vikramshila Gangetic Dolphin Sanctuary, Bihar, India, located in a ~60 linear kilometer of the Ganges River between Sultanganj and Kahalgaon, was declared a protected area for dolphins in August 1991 and some measures have been taken there to conserve the animals (Choudhury *et al.*, 2006). In a few smaller tributaries, susus receive nominal protection by virtue of small portions of their habitat being included in or adjacent to national parks and sanctuaries.

## References

Anderson, J. (1879). "Anatomical and Zoological Researches: Comprising an Account of Zoological Results of the Two Expeditions to Western Yunnan in 1868 and 1875; and a Monograph of the Two Cetacean Genera, Platanista and Orcella [sic]." Bernard Quaritch, London.

Braulik, G. T. (2006). Status assessment of the Indus River dolphin, *Platanista gangetica minor*, March-April 2001. *Biol. Conserv.* **129**, 579–590.

Cassens, I., *et al.* (12 authors) (2000). Independent adaptation to riverine habitats allowed survival of ancient cetacean lineages. *Proc. Natl. Acad. Sci. U.S.A.* **97**, 11343–11347.

Choudhary, S., Smith, B. D. Dey., Subhasish, Dey. Shushant., and Prakash, S. (2006). Conservation and biomonitoring in the Vikramshila Gangetic Dolphin Sanctuary. *Oryx* **40**, 189–197.

Herald, E. S., Brownell, R. L., Jr., Frye, F. L., Morris, E. J., Evans, W. E., and Scott, A. B. (1969). Blind river dolphins: first side-swimming cetaceans. *Science* **166**, 1408–1410.

Kasuya, T. (1972). Some informations on the growth of the Ganges dolphin with a comment on the Indus dolphin. *Sci. Rep. Whales Res. Inst. Tokyo* **24**, 87–108.

Kasuya, T. (1973). Systematic consideration of recent toothed whales based on the morphology of tympano-periotic bone. *Sci. Rep. Whales Res. Inst. Tokyo* **25**, 1–103.

Motwani, M. P., and Srivastava, C. B. (1961). A special method of fishing for *Clupisoma garua* (Hamilton) in the Ganges River system. *J. Bombay Nat. Hist. Soc.* **58**, 285–286.

Pilleri, G., and Zbinden, K. (1973–74). Size and ecology of the dolphin population (*Platanista indi*) between the Sukkur and Guddu barrages, Indus River. *Investig. Cetacea* **5**, 59–69.

Pilleri, G., Gihr, M., Purves, P. E., Zbinden, K., and Kraus, C. (1976). On the behaviour, bioacoustics and functional morphology of the Indus River dolphin (*Platanista indi* Blyth, 1859). *Investig. Cetacea* **6**, 11–141.

Pilleri, G., Marcuzzi, G., and Pilleri, O. (1982). Speciation in the Platanistoidea: systematic, zoogeographical and ecological observations on recent species. *Investig. Cetacea* **14**, 15–46.

Reeves, R. R., and Brownell, R. L., Jr. (1989). Susu *Platanista gangetica* (Roxburgh, 1801) and *Platanista minor* Owen, 1853. *In* "Handbook of Marine Mammals, River Dolphins and Larger Toothed Whales" (S. H. Ridgeway and R. Harrison, Sir, eds.) Vol. 4, pp. 69–100. Academic Press, London.

Reeves, R. R., Chaudhry, A. A., and Khalid, U. (1991). Competing for water on the Indus Plain: is there a future for Pakistan's river dolphins? *Environ. Conserv.* **18**, 341–350.

Reeves, R. R., and Chaudhry, A. A. (1998). Status of the Indus River dolphin *Platanista minor*. *Oryx* **32**, 35–44.

Rice, D. W. (1998). "Marine Mammals of the World, Systematics and Distribution." Special Publications of the Society of Marine Mammalogy, Lawrence, KS.

Sinha, R. K. *et al.* (8 authors) (2000). Status and distribution of the Ganges Susu, *Platanista gangetica*, in the Ganges river system of India and Nepal. *In* "Biology and Conservation of Freshwater Cetaceans in Asia, IUCN Species Survival Commission Occasional Paper Series 23." (R. R. Reeves, B. D. Smith, and T. Kasuya, eds.), pp. 54–61, Gland, Switzerland.

Smith, B. D. (1993). 1990 status and conservation of the Ganges River dolphin (*Platanista gangetica*) in the Karnali River Nepal. *Biol. Conserv.* **66**, 159–169.

Smith, B. D., Ahmed, B., Edrise, M., and Braulik, G. (2001). Status of the Ganges river dolphin or shushuk *Platanista gangetica* in Kaptai Lake and the southern rivers of Bangladesh. *Oryx* **35**, 61–72.

Smith, B. D., Haque, A. K. M., Hossain, M. S., and Khan, A. (1998). River dolphins in Bangladesh: conservation and the effects of water development. *Environ. Manage.* **22**, 323–335.

Smith, B. D. and Reeves, R. R., eds. (2000). Report of the second meeting of the Asian River Dolphin Committee, 22–24 February 1997, Rajendrapur, Bangladesh. *In* "Biology and Conservation of Freshwater Cetaceans in Asia, IUCN Species Survival Commission Occasional Paper Series 23" (R. R. Reeves, B. D. Smith, and T. Kasuya, eds.), pp. 1–14. Gland, Switzerland.

Smith, B. D., Sinha, R. K., Regmi, U., and Sapkota, K. (1994). Status of Ganges river dolphins (*Platanista gangetica*) in the Karnali, Mahakali, Narayani and Sapta Kosi Rivers of Nepal and India in 1993. *Mar. Mamm. Sci.* **10**, 368–375.

Smith, B. D., Braulik, G., Strindberg, S., Ahmed, B., and Mansur, R. (2006). Abundance of Irrawaddy dolphins (*Orcaella brevirostris*) and Ganges river dolphins (*Platanista gangetica gangetica*) estimated using concurrent counts from independent teams in waterways of the Sundarbans mangrove forest in Bangladesh. *Mar. Mamm. Sci.* **22**, 1–21.

Yamasaki, F., and Kito, K. (1984). A morphological note on the intestine of the boutu, with emphasis on its length and ileo-colic transition compared with other platanistids. *Sci. Rep. Whales Res. Inst. Tokyo* **35**, 165–172.

Zhou, K. (1982). Classification and phylogeny of the superfamily Platanistoidea with notes on evidence for the monophyly of the Cetacea. *Sci. Rep. Whales Res. Inst. Tokyo* **34**, 93–108.

**S**

# Swimming

TERRIE M. WILLIAMS

### I. Introduction

The primary mode of locomotion for marine mammals with the possible exception of polar bears (*Ursus maritimus*) is swimming. For dolphins, porpoises and whales, it is the only form of locomotion. The duration of swimming among these mammals may be as short as several seconds when moving between prey patches or as long as several months during seasonal migrations across entire ocean basins. Although swimming by marine mammals often appears effortless, it is in reality a delicate balance between precise body streamlining, exceptional thrust production by specialized propulsive surfaces, and locomotor efficiency (Fig. 1).

### II. Hydrodynamics and body streamlining

One of the most characteristic features of marine mammals is a streamlined body shape. This is not surprising when one considers the forces that the animal has to overcome in order to move through water. When a swimmer moves through water a force, termed drag, acts backward on it, resisting its forward motion. The equation describing total body drag is given by

$$\text{Total Drag} = 1/2\rho V^2 ACd \qquad (1)$$

where $\rho$ is the density of the fluid, V is the velocity of the fluid relative to the body, A is a characteristic area of the body, and Cd is the drag coefficient (a factor that takes into account the shape of the swimmer). There are four primary types of drag that contribute to total body drag: (1) skin friction drag which is a tangential force resulting from shear stresses in the water sliding by the body, (2) pressure drag which is a perpendicular force on the body associated with the pressure of the surrounding fluid, (3) wave drag that occurs when a swimmer moves on or near the water surface, and (4) induced drag that is associated with water deflection off hydrofoil surfaces such as fins, flukes, or flippers. Of these, pressure drag is the component most influenced by body streamlining in marine mammals. The more streamlined a body, the lower the pressure drag and consequently the lower the total body drag of the swimmer.

Mammals whose lifestyles or foraging habits involve prolonged periods of swimming have streamlined body shapes. In contrast to the lanky appearance and appendages of terrestrial mammals, marine mammals tend to have a reduced appendicular skeleton and characteristic tear drop body profile. External features that may disrupt water flow across the body are also reduced or absent in many species of marine mammal. These features include the pinnae (external ears), limbs, and long fur. In highly specialized swimmers such as dolphins the skin contains microscopic ridges that help to direct the flow of water in a controlled manner down the body. All of these adaptations prevent the onset of turbulence in the water surrounding the swimmer and thereby reduce total body drag.

Hydrodynamic theory describes the streamlined body shape as one in which a rounded leading edge slowly tapers to the tail, and total length is 3–7 times maximum body diameter. The ratio of these

**Figure 1**  *A bottlenose dolphin* (Tursiops truncatus) *swimming at high speed on the water surface. The generation of waves by the dolphin's movements leads to increases in body drag and elevated energetic costs during surface swimming.*

morphological measurements, termed the Fineness Ratio, can be written as

$$\text{Fineness Ratio} = \frac{\text{maximum body length}}{\text{maximum body diameter}} \qquad (2)$$

The optimum fineness ratio that results in minimum drag with maximum accommodation for volume is 4.5. Calculations of the fineness ratio for a wide variety of marine mammals show that many species have body shapes that conform to the ideal hydrodynamic range (Fig. 2). A review paper by Fish (1993) showed that many cetaceans, pinnipeds, and sirenians have body shapes with Fineness Ratios that range from 3.0 to 8.0. The species examined included seals, sea lions, and odontocete whales that are considered by many to typify a streamlined body profile. Even mysticete whales with enlarged heads and jaws specialized for filter feeding maintain a streamlined body profile when swimming (Fig. 2). However, the loss of this hydrodynamic profile when the animals open their mouths quickly results in a marked increase in drag, a reduction in forward speed, and a concomitant elevation in energetic costs (Acevedo-Gutierrez *et al.*, 2002; Goldbogen *et al.*, 2006).

Despite nearly ideal body streamlining, all marine mammals must contend with drag forces when moving through the water. These forces can be a considerable challenge for the swimmer and will influence how quickly the animal will be able to move. It is apparent from Eq. (1) that the velocity of the swimmer will have a large impact on total body drag. As the swimmer moves faster, body drag increases

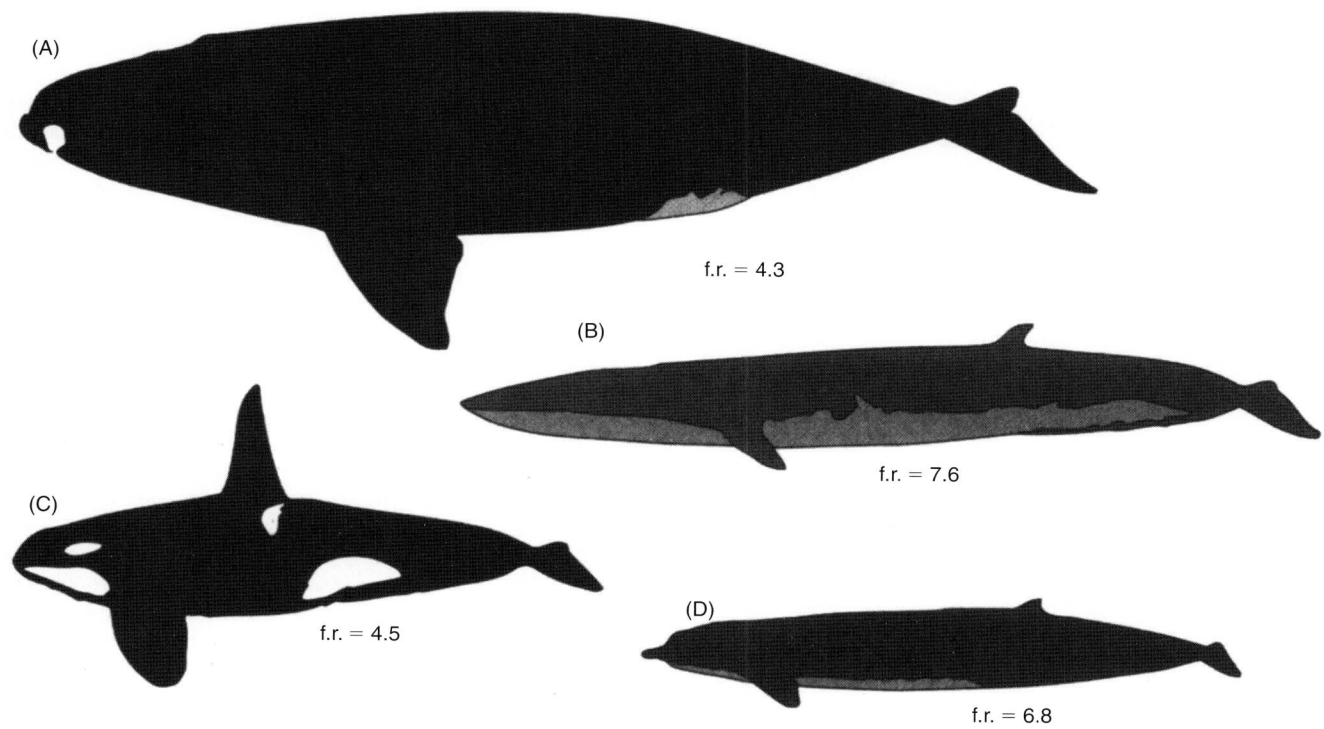

**Figure 2** *Body shapes and fineness ratios for cetaceans. Shapes can range from the robust bowhead whale,* Balaena mysticetus *(A) to the long thin tapered body of the rorqual whales (B) and beaked whales (D). The killer whale* (Orcinus orca) *(C) has the optimum shape in terms of fineness ratio and streamlining. From Berta and Sumich (1999), "Marine Mammals: Evolutionary Biology," Academic Press.*

exponentially. An example of the relationship between total body drag and velocity is presented in Fig. 3 for the sea otter (*Enhydra lutris*) (Williams, 1989). Whether the sea otter swims on the water surface or submerged, body drag increases with velocity. However, body position clearly affects the level of total body drag encountered by the sea otter. At all comparable swimming speeds, body drag is higher for the otter moving on the water surface than when it is swimming submerged. The same results have been found for other swimmers, including humans and harbor seals (*Phoca vitulina*). In general, body drag for a swimmer moving on or near the water surface is 4–5 times higher than the level of drag encountered by the submerged swimmer moving at the same speed (Hertel, 1966). Much of this increase in drag at the water surface is due to energy wasted in the formation waves. This can be avoided if the swimmer is able to submerge to a depth equivalent to three body diameters. For a seal or small whale with a maximum body diameter of 1 m, this would mean changing swimming position to at least three meters in depth to avoid wave drag and the consequent elevation in total body drag. This is one of the reasons that swimming is comparatively difficult for humans—our entire performance takes place on the water surface where wave drag, and hence total body drag, is the highest.

The ability to swim submerged for prolonged periods is one of the most important adaptations for increasing swimming efficiency and performance in marine mammals. The sea otter provides an excellent example of the advantage provided by this adaptation. Sea otters restrict prolonged periods of surface swimming to speeds less than 0.8 m.sec$^{-1}$ and to a maximum body drag of 4.2 N (Fig. 3). For high-speed swimming, sea otters change to a submerged mode of locomotion. In doing so, drag is reduced by 3.5 times and the sea

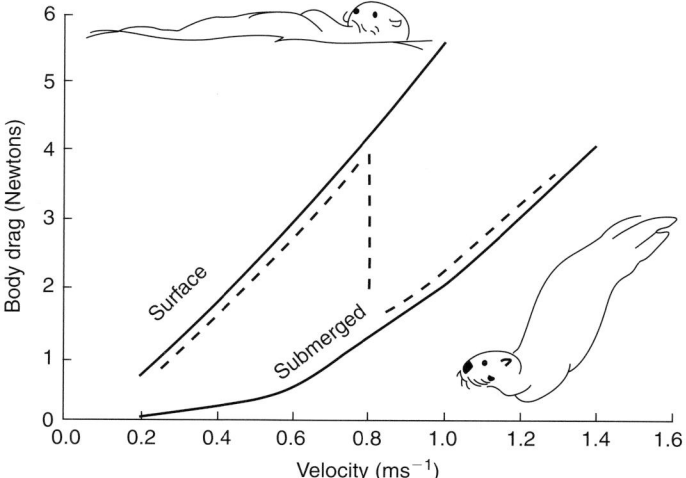

**Figure 3** *A comparison of body drag for surface and submerged sea otters* (Enhydra lutris) *in relation to swimming speed. Note that at all comparable speeds, body drag of the sea otter on the water surface is higher than when the otter is submerged. The dashed line denotes the preferred swimming speeds of surface and submerged sea otters.*

otter is able to reach speeds of 1.4 m.sec$^{-1}$ before body drag once again exceeds 4.0 N. Thus, behavioral changes by the sea otter takes into account the differences in drag associated with body position in the water and allows the animal to extend its range of swimming speeds. Several other behavioral strategies, such as porpoising and

**S**

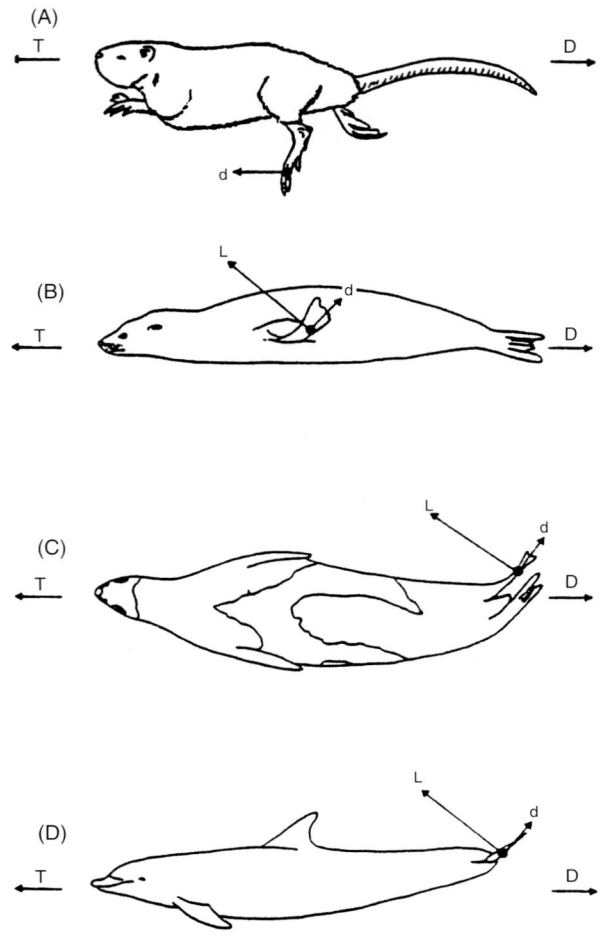

**Figure 4** *Swimming modes for semiaquatic and marine mammals. The muskrat (A) is a semiaquatic mammal that uses drag-based propulsion by paddling its hind feet. Otariids (B), phocid seals (C), and cetaceans (D) use lift-based propulsion that may involve fore flippers (sea lion), lateral body undulation (seal), or dorsoventral undulation (dolphin). Major forces on the animals and propulsive surfaces are shown. T denotes thrust, and D shows the direction of body drag on the animals. L and d illustrate lift and drag forces on the appendages, respectively. From Fish (1993), with permission.*

wave-riding, are also used by marine mammals to avoid elevated body drag while swimming and will be discussed below in Swimming Speed and Behavior.

### III. Kinematics

A hallmark of marine mammal swimming is the use of lift-based propulsion that allows thrust to be generated through the entire stroke cycle. This capability is found in highly adapted marine species such as pinnipeds and cetaceans. It contributes to an increase in locomotor efficiency in marine mammals, especially when compared to the inefficient drag-based swimming styles of humans and terrestrial mammals (Fig. 4).

Marine mammals use a wide variety of swimming styles to move through the water (Table I). The most terrestrial species of this group, the polar bear and sea otter, swim by alternate strokes of the

forelimbs or hind limbs, respectively. Polar bears use a dog style of forelimb paddling with the hind limbs dragged passively behind or used as an aid to steering. Sea otters are unique among marine mammals in their ability to lie on their backs during surface swimming. Propulsion is provided by either the simultaneous or alternate strokes of the hind limbs. When on the surface, sea otters can also swim ventral surface (belly) down using the hind paws for propulsion. The front paws are held against the submerged chest and do not play a role in propulsion during this mode of swimming. Stroke frequency has been measured for swimming sea otters and ranges from approximately 30 to 80 strokes per minute while swimming on the water surface (Williams, 1989).

Polar bears and sea otters are the only marine mammals that rely primarily on drag-based modes of swimming. These modes of swimming have two distinct phases during the stroke cycle, a power phase when thrust is produced, and a recovery phase when the foot or paw is repositioned for the next stroke. During the power phase, the foot is moved backward relative to the body. Drag created by this motion is subsequently translated into thrust and the animal moves forward through the water. The enlarged hind flippers of sea otters and fore paws of polar bears enable the animals to increase propulsive efficiency by moving a large mass of water during this power phase. The recovery phase of the stroke is only used to bring the limb back to its starting position, and occurs without the generation of thrust. Because thrust is produced only during part of the stroke cycle, drag-based modes of swimming are comparatively inefficient.

When sea otters want to move quickly through the water, they switch to an undulatory mode of swimming involving dorsoventral body flexion and simultaneous movements of paired hind flippers. The tail and hind flippers are held straight back and trail the undulatory movements of the trunk. Stroke frequency of sea otters remains relatively constant at 55 strokes per minute during submerged undulatory swimming, which suggests that underwater speed is elevated by increasing stroke amplitude.

As observed for submerged swimming sea otters, dolphins and whales use undulatory modes of propulsion. The primary propulsive movements of all cetaceans occur in the vertical plane with the posterior third of the body undulating in a dorsoventral direction. Termed, thunniform swimming or carangiform swimming with a semi-lunate tail, this mode of locomotion is characterized by an undulatory wave that travels with increasing amplitude down the body, caudal peduncle, and finally the flukes (Fig. 5). "Semi-lunate" refers to the crescent shape of the flukes. This mode of propulsion is shared by other fast swimming vertebrates including tuna, hence the name "thunniform." Undulatory propulsion in cetaceans is considered highly efficient and can generate high levels of thrust on both the upstroke and downstroke. There is no recovery phase and propulsion can be produced throughout the stroke cycle. Stroke frequency using this mode of swimming varies with the speed and size of the cetacean. The range of stroke frequencies for bottlenose dolphins (*Tursiops* spp.) swimming in a pool is 60–180 strokes.min$^{-1}$. Stroke frequency decreases with increasing body size among the cetaceans. Thus, we find that the largest species of swimming mammal, the 100 ton blue whale (*Balaenoptera musculus*), uses stroke frequencies that are only one-tenth of the range observed for bottlenose dolphins. A measurement of the stroke frequency of blue whales ascending during a dive was 6–10 strokes.min$^{-1}$ (Williams *et al.*, 2000).

Swimming by pinnipeds differs markedly between the eared seals, the otariids, and the true seals, the phocids. Otariids use pectoral appendages to generate propulsive forces during swimming, with the hind flippers trailing passively or occasionally used for steering.

**S**

**TABLE I**
**A Comparison of Swimming Characteristics for Four Major Classes of Marine Mammals**

|  | Sea Otter | Otariid | Phocid | Small Cetacean |
|---|---|---|---|---|
| Routine Speed (m/sec) | <0.8 (surface) <1.4 (submerged) | 2.0–3.0 | 1.2–2.0 sprints to 4.0 | 2.0–4.0 sprints to 10.0 |
| Hydrodynamics | Surface/Submerged | Submerged | Submerged | Submerged |
| Kinematics Mode | Paddle, Row (surface) Undulate (submerged) | Pectoral | Lateral Carangiform | Dorso-ventral Thunniform |
| Energetics *(COT measured)* COT predicted | 12.0 (surface) 6.0 (submerged) | 2.3–4.0 | 2.3–4.0 | 2.1–2.9 |

Note: The energetic Cost of Transport (COT) was measured for animals swimming in a flume or freely swimming in open water. The ratio of these values and the predicted values for fish of similar body mass are presented.

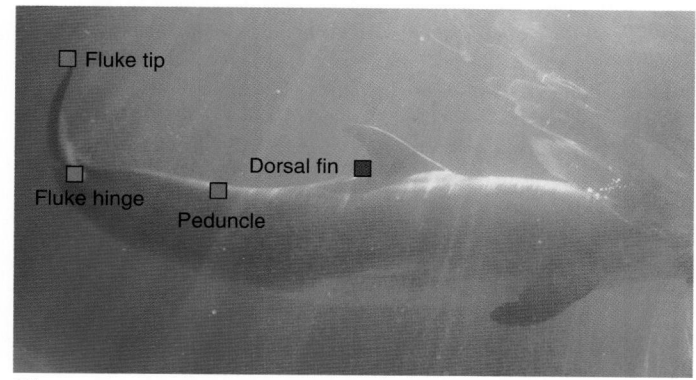

(A)

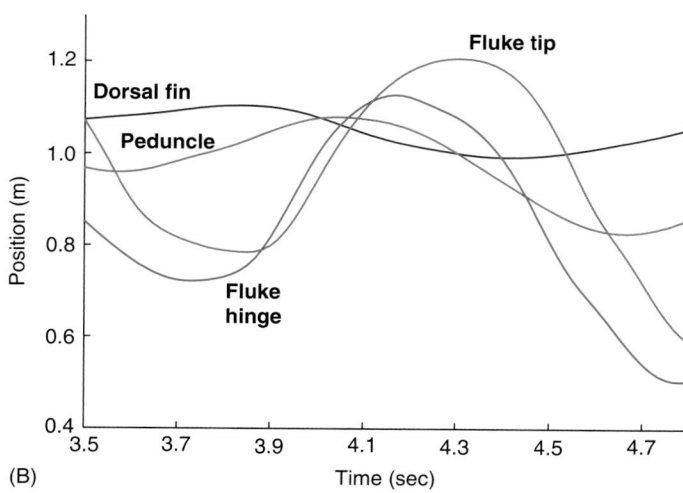

(B)

**Figure 5** *Video image (A) and range of movement (B) of four anatomical sites during a single stroke for a swimming bottlenose dolphin* (Tursiops truncatus). *Colored squares in the picture correspond to the line colors illustrating the movements for each site. Note the traveling wave as it passes from the dorsal fin (dark blue) to the peduncle (red), fluke hinge (green) and finally the fluke tip (pink). From Skrovan et al. (1999), with permission.*

In this way, sea lions and fur seals resemble penguins and sea turtles during swimming. Detailed kinematic analyses have been conducted for California sea lions (*Zalophus californianus*) swimming in a flume (Feldkamp, 1987). These studies revealed three distinct phases to the stroke: (1) the power phase, (2) a paddle phase, and (3) a recovery phase. The majority of thrust is produced during the paddle phase when the fore flippers are quickly and forcibly moved from the water flow to the sides of the animal's body. Stroke frequency for these sea lions increased with swimming speed, and ranged from 15 to 50 strokes.min$^{-1}$ as the animals increased speed from 0.5 to 3.0 m. sec$^{-1}$. In addition to stroke frequency, sea lions increase the amplitude of the fore flipper stroke during high-speed swimming.

When viewed in cross section, the fore flipper of the sea lion resembles a hydrofoil. This specialized shape allows the flipper to produce thrust during both the power and recovery phases of the stroke cycle. As found for cetaceans, the specialized flipper movements of otariids result in thrust production throughout the stroke cycle and contribute to overall locomotor efficiency. Several other advantages are provided by fore flipper propulsion. These include stability at slow speeds and maneuverability at high speeds. Consequently, otariids are champion underwater acrobats and are capable of rapid changes in direction and acceleration.

Phocid seals and walruses (*Odobenus rosmarus*) differ from the otariids in terms of swimming style, and rely on alternate sweeps of the hind flippers for propulsion. In addition to the flippers, the posterior half of the body flexes during each stroke, with the result that body flexion provides nearly 90% of the change in amplitude during the stroke cycle. In phocid seals both hind flippers are swept in the same direction as the posterior portion of the body during each half of the stroke cycle. The leading flipper remains closed and the trailing flipper maximally expands during the sweep to one side. Once the flippers have moved to the maximum lateral position, the flippers switch their open and closed positions in preparation for the reverse lateral sweep. By reversing the role of each flipper during lateral sweeps, one flipper is able to provide thrust while the other flipper recovers. The result, once again, is the ability to produce propulsive thrust during the entire stroke cycle. Stroke frequency in phocids increases linearly with swimming speed. For harbor seals trained to swim at 1.0–1.4 m·sec$^{-1}$ in a water flume stroke frequency ranged from 60 to 78 strokes.min$^{-1}$.

**S**

## IV. Energetics

The energetic cost of swimming has been measured for numerous species of semi-aquatic and marine mammals using a wide variety of techniques. Smaller swimmers such as sea otters, seals, and sea lions have been studied while they swam against a current in water flumes. Similar to placing a human on a treadmill, flume studies have enabled scientists to measure how much energy a swimmer expends while moving at different speeds. Often oxygen consumption is measured during these tests by using a face mask or metabolic hood connected to an oxygen gas analyzer. By training animals to breathe into a metabolic hood, expired respiratory gases can be collected and analyzed for oxygen content. For larger, more powerful swimmers like dolphins and whales, most flumes are not adequate in terms of size or challenging water speeds. Instead, investigators have relied on a variety of novel techniques for determining the energetic cost of swimming in cetaceans. Techniques have included using trained dolphins that match their swimming speed to that of a moving boat in open water (Williams *et al.*, 1992) or having whales swim to metabolic stations where expired gases can be collected for analysis (Worthy *et al.*, 1987).

To compare swimmers of different size, it is useful to convert the metabolic measurements into a cost of transport. Defined as the amount of fuel it takes to transport one unit of body weight over a unit distance, the cost of transport is analogous to the fuel rating of an automobile. In this case, the cost of transport indicates the "gas per mile" used by the swimmer rather than the "miles per gas" achieved by automobiles. The total cost of transport is calculated from the following equation:

$$\text{Total Cost of Transport} = \frac{\text{Oxygen Consumption}}{\text{Swimming Speed}} \quad (3)$$

where oxygen consumption is in $mlO_2 \cdot kg^{-1} \cdot sec^{-1}$ and speed is in $m \cdot sec^{-1}$ which results in a cost of transport in $mlO_2\,kg^{-1} \cdot m^{-1}$. These values are usually converted to an energetic term and expressed as Joules expended per kilogram of body mass per meter traveled ($J \cdot kg^{-1} \cdot m^{-1}$). The conversion calculation assumes a caloric equivalent of 4.8 kcal per liter of oxygen consumed and a conversion factor of $4.187 \times 10^3$ J per kcal.

Comparisons of the cost of transport for a wide variety of mammalian swimmers indicate that swimming is energetically expensive for mammals compared to fish. The total cost of transport for swimming mammals can also be divided into two distinct groups, the semi-aquatic mammals and the marine mammals (Williams, 1999) (Fig. 6). Swimming costs for semiaquatic mammals such as minks (*Mustela* spp.), muskrats (*Ondatra zibethicus*), and humans are 2–5 times higher than observed for marine mammals. These high energetic swimming costs are attributed to a wide variety of factors including elevated body drag associated with a surface swimming position (Fig. 3) and low propulsive efficiency associated with drag-based propulsion.

Mammals specialized for swimming demonstrate comparatively lower energetic costs. Total cost of transport in relation to body mass for swimming marine mammals ranging in size from a 21 kg California sea lion to a 15,000 kg gray whale (*Eschrichtius robustus*) is described by

$$\text{Total Cost of Transport} = 7.79\ \text{mass}^{-0.29} \quad (4)$$

where the cost of transport is in $J \cdot kg^{-1} \cdot m^{-1}$ and body mass is in kilograms. Interestingly, the style of swimming used by marine mammals

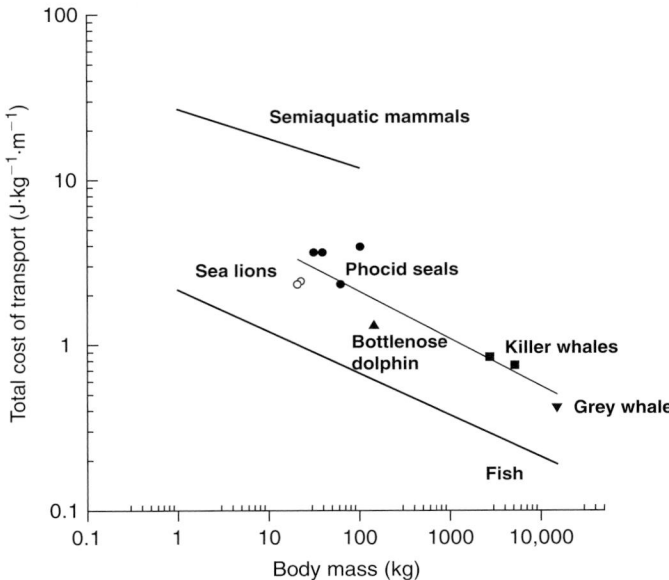

**Figure 6** *Total energetic cost of transport in relation to body mass for different classes of swimmers. Marine mammals include gray seals* (Haliochoerus grypus) *and harbor seals* (Phoca vitulina) *(filled circles), California sea lions* (Zalophus californianus) *(open circles), bottlenose dolphins* (Tursiops truncatus) *(triangle), killer whales* (Orcinus orca) *(squares), and a gray whale* (Eschrichtius robustus) *(downward-pointing triangle). The least squares regression through the data points for marine mammals is presented in the text. This regression is compared to the regressions for swimming semiaquatic mammals (upper solid line), and the predicted regression for salmonid fish (lower solid line). From Williams (1999), with permission.*

did not affect the cost of transport relationship. Species and swimming styles represented in this equation include sea lions using pectoral fins for propulsion, phocid seals using lateral undulation of paired hind flippers, and odontocete and mysticete whales using dorso-ventral undulation of flukes.

As illustrated in Fig. 6, the energetic cost of swimming for marine mammals is greater than predicted for salmonid fish of similar body size. Despite specialization of the body and propulsive surfaces for aquatic locomotion, the cost of transport for swimming by seals and sea lions is 2.3–4.0 times higher than predicted for swimming fish. Values for cetaceans are somewhat lower, and range from 2.1 to 2.9 times values predicted for fish. Differences in the total cost of transport between marine mammals and fish are due in part to the amount of energy expended for maintenance functions, particularly thermoregulation and the support of a high core body temperature. As endotherms, mammals expend more energy to support the production of endogenous heat than ectothermic fish. In addition, many marine mammals show exceptionally high metabolic rates while resting in water in comparison to terrestrial mammals resting in air. A consequence of these high maintenance costs is an overall increase in the total energy expended during swimming, especially when compared to fish.

## V. Swimming Speeds and Behavior

Although body size varies considerably among marine mammals from the 20 kg sea otter to the 122,000 kg blue whale, routine swimming is limited to a surprisingly narrow range of speeds. Many species

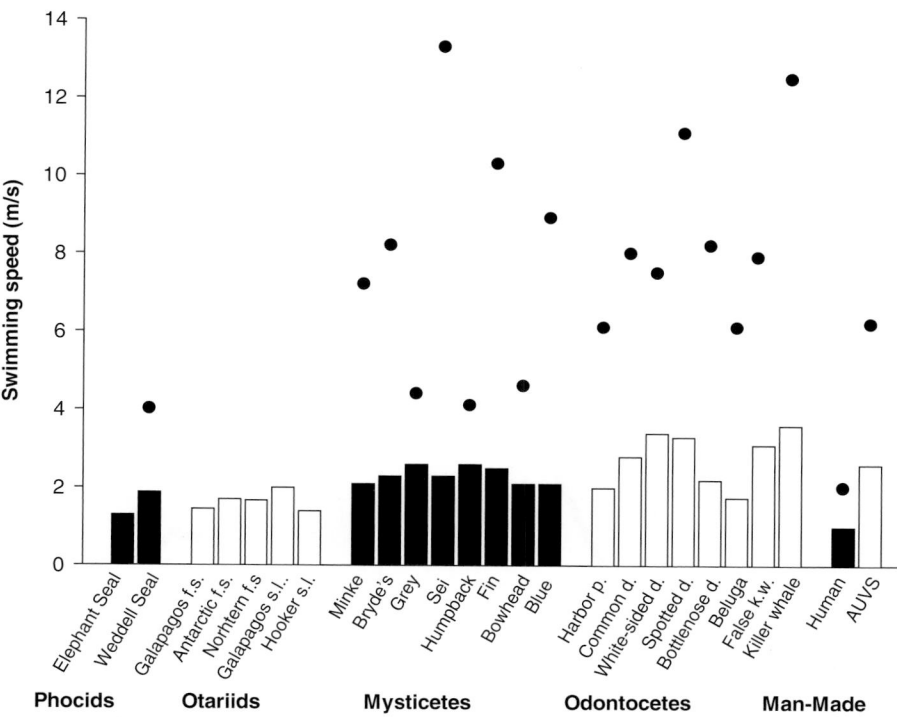

**Figure 7** *Swimming speeds for marine mammals. Routine speeds of phocid seals, otariids, mysticetes, odontocetes, humans and autonomous underwater vehicles (AUVs) are shown by the bars. Filled circles above the bars denote the sprinting speeds recorded for each species. Note the similar range of routine speeds for these marine mammals regardless of body size.*

of marine mammal routinely swim between approximately 1.0 and 3.6 m·sec$^{-1}$ regardless of body size (Fig. 7). Within this range, pinnipeds generally select slower routine traveling speeds than cetaceans, and mysticete whales swim slower than odontocetes. For example, average swimming speeds for a wide variety of otariids and phocids range from 1.3 to 2.0 m·sec$^{-1}$. The massive mysticete whales are only slightly faster; routine speeds for this group of marine mammals ranges from 2.1 to 2.6 m·sec$^{-1}$. Although they are not the largest marine mammals, odontocetes tend to move the fastest during routine travel. The slowest of the odontocetes represented in Fig. 7 was the beluga whale (*Delphinapterus leucas*) with a routine speed of 1.8 m·sec$^{-1}$. In comparison, the killer whale (*Orcinus orca*) demonstrates the fastest routine speed of the marine mammals measured to date and averages 3.6 m·sec$^{-1}$ during casual swimming. These speeds are even more remarkable when compared to the efforts of humans. The routine speed of humans during freestyle swimming is approximately 1.0 m·sec$^{-1}$, about the same speed as a sea otter swimming under water.

As would be expected, the sprinting speeds of marine mammals are considerably faster than routine speeds, and show much variation among the species measured. Most of the information regarding sprint swimming performance in marine mammals is for cetaceans. However, the speed of adult Weddell seals (*Leptonychotes weddellii*) chasing fish beneath the Antarctic sea ice has been measured and was found to exceed 4.0 m·sec$^{-1}$ during the hunt. Among cetaceans, sprint speeds are even higher. The range of sprinting speeds measured for mysticete whales is 4.1–13.3 m·sec$^{-1}$ (Fig. 7); sprint swimming by odontocetes is within the upper end of this range and averages 6.1–12.5 m·sec$^{-1}$. Killer whales remain the fastest of the

odontocetes measured and can sprint at 12.5 m·sec$^{-1}$. This is nearly six times faster than the maximum performance of human swimmers in Olympic sprint competition.

Because marine mammals must periodically surface to breathe, they are subject to high levels of drag associated with the effects of wave formation and splashing especially during high speed swimming. To help minimize body drag and energetic costs during these surface intervals, marine mammals have developed a number of unique behavioral strategies to accommodate breathing while swimming fast. Porpoising is one such highly visible behavioral strategy used by small cetaceans and some pinnipeds moving at high speed near the water surface (Au and Weihs, 1980). Rather than stroke continuously, the animals leap into the air and simply avoid the elevated wave drag that occurs when swimming near the water surface to breathe. Theoretically, this behavior results in an energetic savings to the animal, although the cost of surface swimming versus leaping has yet to be measured. Wave-riding is another strategy that enables the swimmer to avoid the work of continuous stroking while moving near the water surface. In a study involving bottlenose dolphins trained to swim freely or wave-ride next to a moving boat, investigators found that heart rate, respiration rate and energetic cost were reduced for animals riding the bow wave of the boat (Williams *et al.*, 1992). This behavior enabled the dolphins to nearly double their forward traveling speed with only a 13% increase in energetic cost. Consequently, it is not surprising that marine mammals routinely ride waves generated by the wind, surf, the wake of boats and even large whales. What appears to be an amusing activity also provides an energetic benefit to the swimmer.

**S**

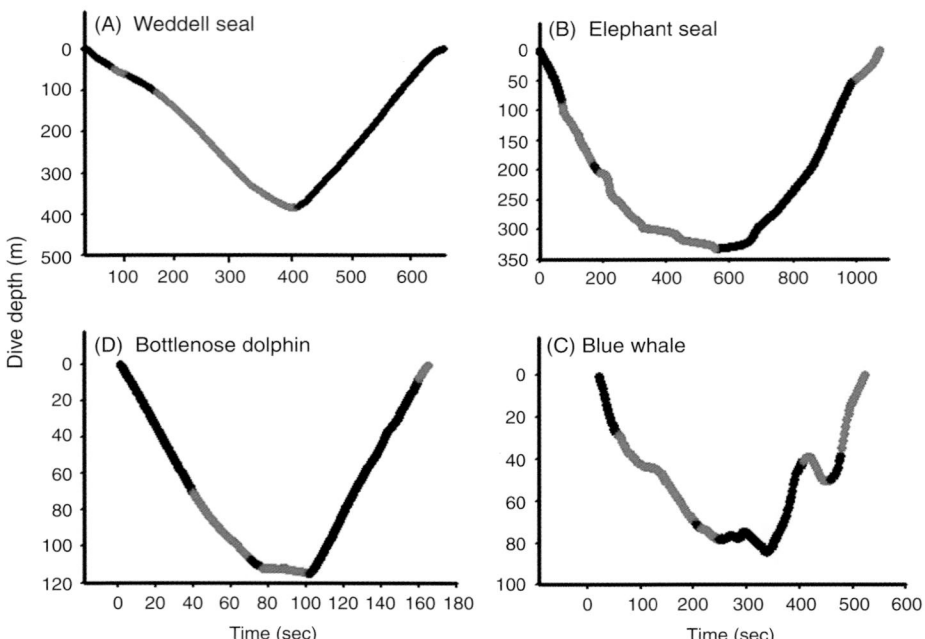

**Figure 8** *Swimming and gliding activity of four species of diving marine mammal. Representative deep dives are presented for the Weddell seal* (Leptonychotes weddellii) *(A), elephant seal* (Mirounga spp.) *(B), bottlenose dolphin* (Tursiops spp.) *(C), and blue whale* (Balaenoptera musculus) *(D). Each curve shows dive depth in relation to time elapsed during the dive. The color of the line corresponds to stroking (black) and gliding (red) periods. For each species the descent was characterized by prolonged periods of gliding. From Williams* et al. *(2000), with permission.*

Although energetically advantageous when swimming near the water surface, both wave-riding and porpoising have been described for only a limited number of marine mammal species moving at high speeds. These locomotor strategies are not possible during slow transit, in large marine mammals such as elephant seals (*Mirounga* spp.) and whales, or in polar regions where ice covers the water surface. Instead, transit swimming is often accomplished by a sawtooth series of sequential dives that allows the animals to remain submerged except for brief surface intervals to breathe (Davis *et al.*, 2001).

## VI. The Special Case of Swimming at Depth

Most of our information about swimming in marine mammals is from animals moving near the water surface. However, the majority of swimming by these animals occurs at depth in conjunction with diving. When descending or ascending during a dive, marine mammals must contend with changes in buoyant forces, hydrostatic pressure, as well as body drag. As discussed above, drag forces resist both forward progression and limb movements of the swimmer. In contrast, buoyant forces act in a vertical direction in the water column and result from the weight, volume, and compressibility of the tissues and air spaces of the animal's body. Hydrostatic pressure results from the weight of the water column above the marine mammal.

The magnitude of buoyant forces and hydrostatic pressure on the swimming marine mammal will depend on where in the water column activity takes place. Hydrostatic pressure progressively increases by 1 ATM for every 10.1 m an animal descends in the water column. This will have a profound effect on compressible spaces or tissues, and hence buoyancy of the animal, especially for marine mammals

that may descend and ascend hundreds of meters during the course of a dive. In addition, seasonal changes in blubber content, pregnancy and lactation will have an effect on the overall buoyancy of the marine mammal.

A consequence of the interrelationships between depth, buoyancy, hydrostatic pressure and body drag is that the physical forces influencing the animal swimming horizontally near the water surface are very different from those encountered by the diving animal moving vertically through the water column. Detailed studies on diving bottlenose dolphins and elephant seals have shown that the animals are positively buoyant near the water surface, and that buoyancy decreases as the animal descends during the dive. For example, the buoyancy of a bottlenose dolphin changes from positive when near the water surface to negative once the animal exceeds 70 m in depth (Skrovan *et al.*, 1999).

These changes in buoyancy are associated with changes in lung compression due to the increase in hydrostatic pressure as marine mammals descend on a dive. Thus, as dolphins, whales and seals dive, hydrostatic pressure increases, the lungs progressively collapse with the result that overall buoyancy is changed. These marked changes in physical forces with depth affect both the locomotor behavior and energetics of the marine mammal as it descends and ascends during a dive (Fig. 8).

Until recently, it was not possible to observe the swimming modes of marine mammals during deep dives. With the development of miniaturized video cameras and instrumentation worn by free-ranging marine mammals, new information about swimming at depth has been obtained (Fig. 9). The videos revealed that bottlenose dolphins, elephant seals, Weddell seals and blue whales switch between different modes of swimming during the dive, much like terrestrial mammals switch between gaits (Williams *et al.*, 2000). Dive descents

**Figure 9** *A bottlenose dolphin* (Tursiops *sp.*) *carries a video camera to record its swimming movements during deep dives. (Courtesy of Kevin McDonnell.)*

usually began with a period of continuous stroking. Once the marine mammals reached 70–80 m in depth they changed to a passive glide for the remainder of the descent. For deep divers such as phocid seals, these gliding periods can be quite long. For example, prolonged gliding periods exceeded 6 min for Northern elephant seals (*Mirounga angustirostris*) traveling to nearly 400 m and Weddell seals descending to 540 m beneath the Antarctic sea ice. Nearly 80% of the descent of diving seals was spent passively gliding rather than actively swimming on dives exceeding 200 m in depth.

The ascent portion of a dive requires more effort by these marine mammals when compared to the descent. The beginning of the ascent represents the period of greatest swimming effort for many mammalian divers. During this period, many species of pinniped and cetacean use sequential, large amplitude strokes to begin moving upward. As the ascent continues, the physical forces impacting the diver are once again altered as they move through the water column. Hydrostatic pressure decreases on ascent. Consequently, the lungs are able to reinflate and the buoyancy of the marine mammal increases. Swimming behavior reflects these changes with the result that the continuous stroking phase is followed by a stroke and glide mode of swimming, and finally a brief glide to the water surface.

An interesting contrast to the seals, dolphins, and blue whales is the right whale (*Eubalaena* spp.). This species, considered the "right" whale to hunt because their carcasses tended to float, are comparatively buoyant. As a result, the right whale displays some of its most powerful fluke strokes at the beginning of descent as it counteracts large positive buoyant forces at the start of a dive (Nowacek *et al.*, 2001). The advantage of this positive buoyancy subsequently occurs during the ascent, when the animals are able to glide to the surface and reduce the number of energetically costly strokes.

By altering the mode of swimming to account for changes in the physical forces that occur during a dive, marine mammals are able to conserve limited oxygen reserves during submergence. Studies investigating the metabolic rates of Weddell seals diving from an ice hole found that the incorporation of prolonged glides enabled seals

to reduce the energetic cost of individual dives by 9–60 % (Williams *et al.*, 2000, 2004). Such an energetic savings could make the difference between completing the dive aerobically or anaerobically, and can increase the time available for hunting or avoiding predators.

In summary, these studies demonstrate that swimming can be energetically expensive for mammals. Marine adapted species including sea otters, pinnipeds, and cetaceans have undergone marked morphological, physiological, and behavioral changes to increase their swimming efficiency. An especially important adaptation that distinguishes marine mammals from semi-aquatic mammals is the ability to remain submerged for prolonged periods when swimming. However, prolonged submergence also requires specialized physiological responses associated with oxygen loading and utilization as described in the chapter "Diving.". A major benefit of these adaptations is a capacity for aquatic performance by marine mammals that far exceeds those of semi-aquatic mammals and the best Olympic efforts of humans.

## See Also the Following Articles

Bow-riding ■ Diving Behavior ■ Energetics ■ Stream lining

## References

Acevedo-Gutierrez, A., Croll, D. A., and Tershy, B. R. (2002). High feeding costs limit dive time in the largest whales. *J. Exp. Biol.* **205**, 1747–1753.

Au, D., and Weihs, D. (1980). At high speeds dolphins save energy by leaping. *Nature* **284**, 548–550.

Davis, R. W., Fuiman, L. A., Williams, T. M., and LeBoeuf, B. J. (2001). Three-dimensional movements and swimming activity of a northern elephant seal. *Comp. Biochem. Physiol. A* **129**, 759–770.

Feldkamp, S. D. (1987). Foreflipper propulsion in the California sea lion, *Zalophus californianus. J. Zool. (Lond.)* **212**(1), 117–136.

Fish, F. E. (1993). Influence of hydrodynamic design and propulsive mode on mammalian swimming energetics. *Aust. J. Zool.* **42**, 79–101.

Goldbogen, J. A., Calambokidis, J., Shadwick, R. E., Oleson, E. M., McDonald, M. A., and Hildebrand, J. A. (2006). Kinematics of foraging dives and lunge-feeding in fin whales. *J. Exp. Biol.* **209**, 1231–1244.

Hertel, H. (1966). "Structure, Form and Movement." Reinhold Publishing Corporation, New York.

Nowacek, D. P., Johnson, M. P., Tyack, P. L., Shorter, K. A., McLellan, W. A., and Pabst, D. A. (2001). Buoyant balaenids: the ups and downs of buoyancy in right whales. *Proc. R. Soc. Lond. B* **268**, 1811–1816.

Skrovan, R. C., Williams, T. M., Berry, P. S., Moore, P. W., and Davis, R. W. (1999). The diving physiology of bottlenose dolphins (*Tursiops truncatus*) II. Biomechanics and changes in buoyancy at depth. *J. Exp. Biol.* **202**, 2749–2761.

Williams, T. M. (1989). Swimming by sea otters: adaptations for low energetic cost locomotion. *J. Comp. Physiol. A* **164**, 815–824.

Williams, T. M. (1999). The evolution of cost efficient swimming in marine mammals: limits to energetic optimization. *Philos. Trans. R. Soc. Lond. B* **354**, 193–201.

Williams, T. M., *et al.* (8 authors) (2000). Sink or swim: strategies for cost-efficient diving by marine mammals. *Science* **288**, 133–136.

Williams, T. M., Friedl, W. A., Fong, M. L., Yamada, R. M., Sedivy, P., and Haun, J. E. (1992). Travel at low energetic cost by swimming and wave-riding bottlenose dolphins. *Nature* **355**, 821–823.

Williams, T. M., Fuiman, L. A., Horning, M., and Davis, R. W. (2004). The cost of foraging by a marine predator, the Weddell seal *Leptonychotes weddellii*: pricing by the stroke. *J. Exp. Biol.* **207**, 973–982.

Worthy, G., Innes, S., Braune, B., and Stewart, R. (1987). Rapid acclimation of cetaceans to an open-system respirometer. *In* "Approaches in Marine Mammal Energetics" (A. Huntly, D. Costa, G. Worthy, and M. Castellini, eds.), pp. 115–126. Soc. Mar. Mammal. Special Publication 1. Society Marine Mammalogy, Lawrence.

**S**

# Systematics

## Annalisa Berta

Systematics is the study of biological diversity that has as its primary goal the reconstruction of phylogeny, the evolutionary or genealogical history of particular group of organisms (e.g., species). Because of its emphasis on phylogeny, this discipline is often referred to as phylogenetic systematics or cladistics. Other related goals of systematics include determination of the times at which species originated and became extinct and the origin and rate of change in their characters. An important component of systematics is taxonomy that involves the identification, description, nomenclature, and classification of organisms. Systematics provides a framework for interpreting patterns and processes in evolution using explicit, testable hypotheses.

The rapid pace of research on marine mammals has resulted in renewed interest in their systematics. Phylogenetic systematic methodology as introduced here has gained near universal acceptance. [For a general introduction see Maddison and Maddison (2000) and for more detailed discussion of methods see Felsenstein (2004).] In addition to their use in elucidating evolutionary relationships, phylogenies are now recognized as powerful tools for unveiling evolutionary patterns of biodiversity in ecological and behavioral settings (Brooks and McLennan, 2002).

## I. Basic Tenets of Phylogenetic Systematics

The recognition of patterns of relationship among species is founded on the concept of evolution. Patterns of relationship among species are based on changes in the characters of an organism. Characters are diverse, heritable attributes of organisms that include DNA base pairs, anatomical and physiological features, and behavioral traits. Two or more forms of a given character are termed the character states. For example, among pinnipeds the character, contact between the maxillary and frontal bones consists of three character states: (1) V-shaped (in bears, extinct desmatophocids, *Enaliarctos*, and phocids), (2) W-shaped (in otariids), and (3) transverse (walruses). In the determination of relationships among groups of organisms phylogenetic systematics emphasizes evolutionary novelties (derived characters) in contrast to ancestral similarities (primitive characters). If derived characters are unique to a particular taxon rather than showing relationships among taxa they are termed autapomorphies. An example of an autapomorphy is the straight contact between the maxillary and frontal bones seen only in walruses among pinnipedimorphs (living pinnipeds and their fossil relatives).

The evolutionary history of a group of organisms can be inferred by sequentially linking species together based on their common possession of derived characters, also known as synapomorphies. These derived characters are considered to be homologous, a similarity that results from common ancestry. For example, the flipper of a seal and that of a walrus are homologous because their common ancestor had flippers. In contrast to homology, a similarity not due to homology is homoplasy. For example the flipper of a pinniped and that of a whale are homoplasious as flippers because their common ancestor lacked flippers. Homoplasy may arise in one of two ways convergence (parallelism) or reversal. Convergence is the independent evolution of a similar feature in two or more lineages. Thus seal flippers and whale flippers evolved independently as swimming appendages,

their similarity is homoplasious by convergent evolution. Reversal is the loss of a derived feature coupled with the reestablishment of an ancestral feature. For example, in phocine seals (e.g., bearded seal, hooded seal, and the Phocini) the development of strong claws, lengthening of the third digit of the foot, and de-emphasis of the first digit of the hand are character reversals because none of them characterize phocids ancestrally but are present in terrestrial arctoid carnivores, common ancestors of pinnipeds.

Relationships among organismal groups are commonly represented in the form of a cladogram, a branching diagram that conceptually represents the best estimate of phylogeny (Fig. 1). Derived characters are used to link monophyletic groups, groups of taxa that consist of a common ancestor plus all descendants of that ancestor (referred to as a clade). For example, currently the best supported hypothesis of relationships among pinnipedimorphs based on both morphologic and molecular characters proposes that phocid seals (Phocidae) and an extinct lineage (Desmatophocidae) are more closely related to each other than either is to other pinnipeds. Fur seals and sea lions (Otariidae) together with their sister taxon walruses (Odobenidae) are positioned as the next closest relatives to this clade with the fossil taxon *Enaliarctos* recognized as the most basal lineage (Fig. 1). According to this hypothesis, relationships among pinnipedimorphs are depicted as sets of nested hierarchies. In this case, four monophyletic groups can be recognized. The most exclusive monophyletic group is that formed by phocid seals and desmatophocids since this clade shares derived synapomorphies not also exhibited by walruses, otariids or *Enaliarctos*. At the other extreme, the most inclusive monophyletic group is that formed by *Enaliarctos* (Otariidae + Odobenidae) (Desmatophocidae + Phocidae).

The task in inferring a phylogeny for a group of organisms is to determine which characters are derived and which are ancestral. If the ancestral condition of a character or character state is established, then the direction of evolution from ancestral to derived, can be inferred, and synapomorphies can be recognized. The methodology for inferring direction of character evolution is critical to cladistic analysis. Outgroup comparison is the most widely used procedure. It relies on the argument that a character state found in close relatives of a group (the outgroup) is likely to be the ancestral or primitive state for the group of organisms in question (the ingroup). Usually more than one outgroup is used in an analysis, the most important being the first or genealogically closest outgroup to the ingroup called the sister group. For example, among pinnipedimorphs the ingroup includes Phocidae (seals), Desmatophocidae (extinct seal relatives), Otariidae (fur seals and sea lions), Odobenidae (walruses), and the fossil taxon *Enaliarctos*. Phocidae is hypothesized as the sister group of Desmatophocidae and these taxa together form the sister taxon to Odobenidae + Otariidae. *Enaliarctos* is positioned as the earliest diverging lineage. Ursids (bears) are hypothesized as the closest pinniped outgroup (Fig. 2) although there is evidence to support an alternative arrangement, an ursid-mustelid ancestry.

## II. Phylogeny Reconstruction

The first step in the reconstruction of phylogeny of a group of organisms is selection and definition of characters and character states for each taxon (e.g., species). Next, the characters and their states are arranged in a data matrix (Table I). Characters can be further distinguished; those with two states are binary, whereas characters with three or more states are multistate. For each character, ancestral and derived states are determined. The determination of character state, whether ancestral or derived (also called polarity assessment) is done using outgroup comparison. For example, if the distribution of character 1, condition of the lacrimal bone, is considered, two character states are recognized, the presence versus the absence (or fusion of the lacrimal such that it does not contact the jugal). The outgroup bears possess a lacrimal bone condition which is the ancestral state (Table I). The ingroup taxa lack the lacrimal bone which is the derived condition. Since this derived state unites pinnipedimorphs to the exclusion of bears it is considered a synapomorphy.

A final step in phylogeny reconstruction is the construction of a cladogram or phylogenetic tree by sequentially grouping taxa based on the common possession of one or more shared derived character states (Fig. 3 A-C) (three possible pinnipedimorph cladograms). An important aspect of phylogeny reconstruction is the principle of

parsimony. The basic tenet of parsimony is that the cladogram that contains the fewest number of evolutionary steps, or changes between character states of a given character summed for all characters, is accepted as being the best estimate of phylogeny. However, multiple equally parsimonious solutions are possible and should be examined as well as suboptimal topologies (i.e., less parsimonious trees). In this example, Fig. 3B is the most parsimonious cladogram. Note that an alternative cladogram for the data set (Fig. 3C) showing a different relationship among the five taxa, requires eight characters state changes, two more than the most parsimonious cladogram (Fig. 3B). It should be noted that this alternative view, an alliance between the walrus and otariids, although less parsimonious on the basis of morphologic characters in the data set employed herein, is consistently and robustly supported by molecular data and total evidence analysis. It also should be pointed out

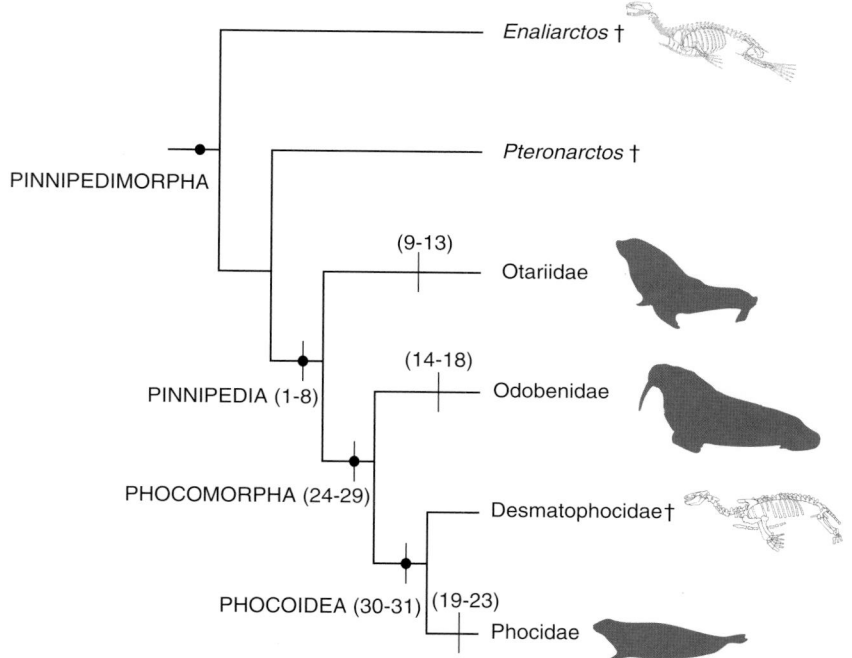

**Figure 1** *Hypothesis of pinnipedimorph relationships modified from Berta et al., 2006. † = extinct taxa.*

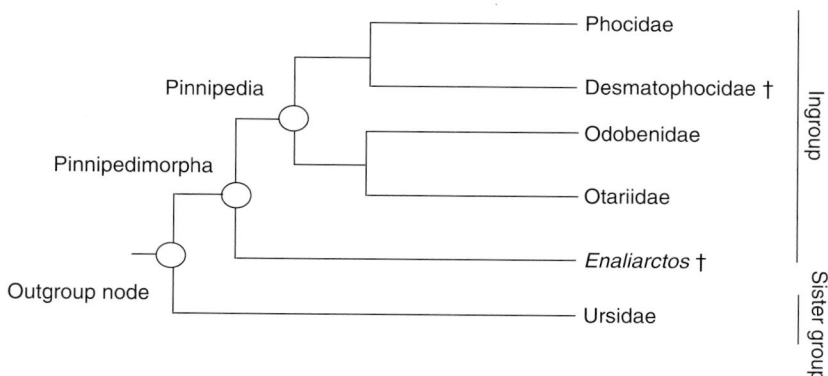

**Figure 2** *Pinniped relationships with ingroup and outgroups identified. † = extinct taxa. Although ursids are hypothesized as the closest outgroup there is evidence to support an ursid-mustelid ancestry for pinnipeds.*

TABLE I
Data Set for Analysis of Pinnipedimorphs Plus an Outgroup Showing Five Characters and Their Character States

| | Character/character states | | | | |
|---|---|---|---|---|---|
| Taxon | Lacrimal bone | Middle ear bones | Orbital/maxilla | Maxilla/frontal | Squamosal/jugal |
| Outgroup | | | | | |
|   Ursids | Present | Small | No | V shape | Overlapping |
| Ingroup | | | | | |
|   *Enaliarctos* | Absent | Small | No | V shape | Overlapping |
|   Otariidae | Absent | Small | Yes | W shape | Overlapping |
|   Desmotophocidae | Absent | Large | Yes | V shape | Interlocking |
|   Phocidae | Absent | Large | Yes | V shape | Interlocking |
|   Odobenidae | Absent | Large | Yes | Straight | Overlapping |

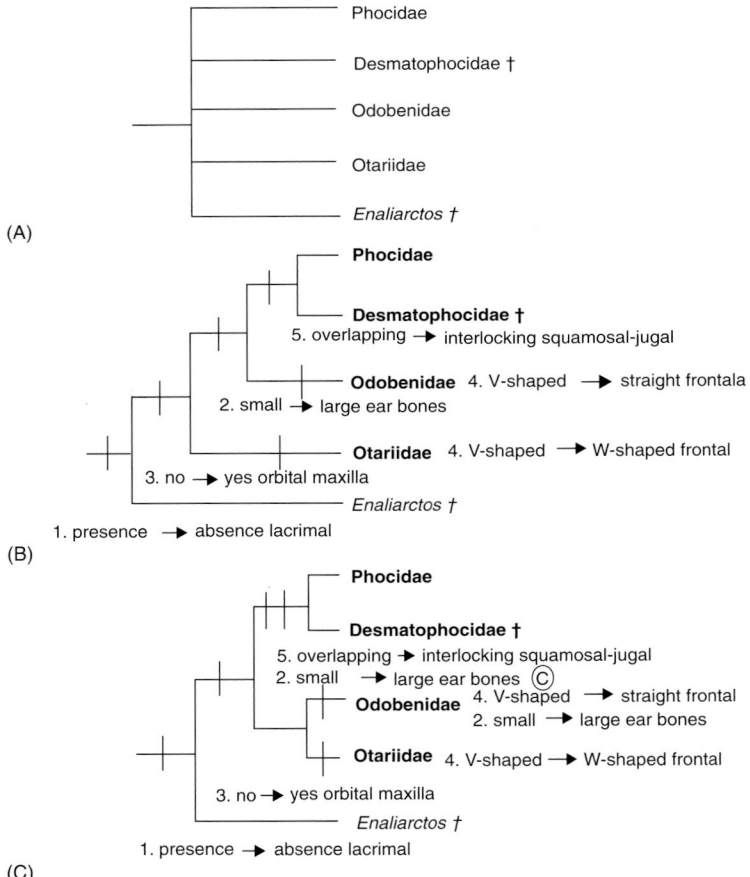

**Figure 3** *Three possible cladograms (A–C) of relationships and character-state distributions for the ingroups listed in Table I. Convergence is denoted by circled C. "C" is the currently best accepted hypothesis (Note: morphologic characters used here do not support this arrangement although it is supported by molecular data and total evidence analysis).*

that parsimony is only one of several methods for reconstructing phylogenetic relationship. Other methods such as maximum likelihood and Bayesian approaches are most often used with molecular or combined data, and perform especially well with large data sets (Felsenstein, 2004).

The methods used to search for the most parsimonious tree depend on the size and complexity of the data matrix. These methods are available in several computer programs, e.g., PAUP (Swofford, 2000) and MacClade (Maddison and Maddison, 2000). The latter is particularly useful in assessing the evolution of characters. Systematists are concerned about the relative accuracy of phylogenetic trees (i.e., how much confidence can be placed in a particular phylogenetic reconstruction). Various measures (e.g., tree length) and indices (e.g., consistency index, rescaled consistency index) have been devised to address this concern (Swofford, 2000). Another approach that is often used to estimate the value of a particular cladogram with respect to the data places confidence limits on the individual branches, techniques known as bootstrapping and Bremer support. Another concern stems from the realization that increase in size of the data sets results in a greater chance of the analysis resulting in more than one equally parsimonious trees. A method of working with multiple trees involves the implementation of consensus trees that are useful in identifying the areas of agreement and conflict among competing trees. A related issue in systematics is how to evaluate different data sets (e.g., morphology, behavior, and DNA sequences) particularly how they should be combined (also referred to as "total evidence" or supermatrix approach) (de Queiroz and Gatesy, 2006). The results of "total evidence" or simultaneous analysis of all character data can then be compared with the results of the separate analyses. Before data sets can be combined, it is necessary to determine if they are congruent, that is the order of branching is not contradictory. Several statistical tests have been developed to test for significant incongruencies among data sets. In summary, many factors should be considered in evaluating phylogenetic hypotheses, among the most important is taxonomic sampling, rigorous analysis including the underlying assumptions of various methods, and computer capabilities.

### III. Phylogenetic Classification

Taxonomy is the language of biology. One aspect of taxonomy is the classification of organisms which allows us to organize and communicate information about life's diversity. Phylogenetic systematists contend that classification should be based on phylogeny and should include only monophyletic groups. In contrast to monophyletic groups, a paraphyletic group (designated by quotation marks) is one that includes a common ancestor and some but not all of the descendants of that ancestor. An example of a paraphyletic taxon is the "Otarioidea," traditionally recognized as group that includes walruses, otariid seals, and their extinct relatives to the exclusion of

---

**TABLE 2**
**Classification of Major Lineages of Pinnipeds**

Pinnipedimorpha (including *Enaliarctos* and all other pinnipeds)

  Pinnipedia (fur seals, sea lions, seals, walruses, and their extinct relatives)

    Otaroidea (walruses, fur seals, sea lions and their extinct relatives)

      Otariidae (fur seals and sea lions)

  Phocomorpha (walruses, seals, and their extinct relatives)

    Phocoidea (including *Allodesmus*, *Desmatophoca*, seals and walruses)

      Phocidae (seals)

      Odobenidae (walruses)

---

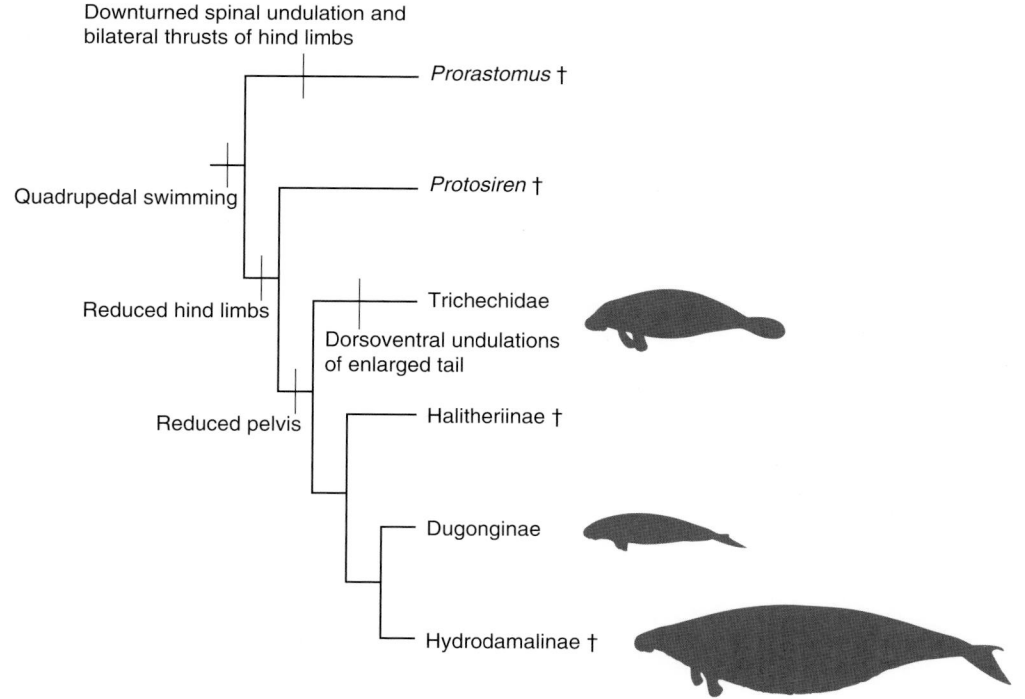

**Figure 4** *Evolution of locomotion among sirenians † = extinct taxa. Based on Domning (2000).*

phocid seals (Fig. 1). The recognition of paraphyletic taxa is to be avoided since by doing so we risk misinterpreting the evolutionary relationships of taxa and their classification. One result of the use of phylogenetic methodology is that traditionally accepted ranks (e.g., phylum, class, order, family, genus, and species) often do not correspond to new information about evolutionary relationships among taxa thus rendering their classification misleading. One method is the elimination of rank altogether and the indication of relative rank by subordination as shown by indentation. An example, the classification of major lineages of pinnipedimorphs, is shown in Table II.

## IV. Uses of a Phylogeny

Once a phylogenetic framework is produced, one of its most interesting uses is to elucidate questions which integrate evolution, behavior, and ecology. One technique used to facilitate such evolutionary studies is optimization or mapping. Once a cladogram is constructed, a feature or condition is selected to be examined in light of the phylogeny of the group. Among several computer programs recently available for reconstruction of ancestral states and character mapping is Mesquite (Maddison and Maddison, 2006). Examples using marine mammals are becoming more numerous. One of these studies, the evolution of locomotion among sirenians, is briefly reviewed here. A cladogram of relationships among sirenians was first established based on morphologic data. Next, using this phylogenetic framework Domning (2000) (Fig. 4) mapped locomotor characters onto the tree. Sirenians passed through three locomotor stages from a terrestrial quadrupedal ancestry. In the first stage exemplified by archaic sirenians, the prorastomids, swimming was accomplished by alternate thrusts of the hind limbs. This was followed by a second stage seen in the extinct taxon *Protosiren* which employed dorsoventral spinal undulation and bilateral thrusts of the hindlimb in swimming. In a final stage, seen in modern sea cows and manatees, sirenians have evolved into completely aquatic animals swimming with the tail only.

## See Also the Following Article

Classification

## References

Berta, A., Sumich, J. L., and Kovacs, K. M. (2006). "Marine Mammals: Evolutionary Biology," 2nd ed. Elsevier, San Diego, CA.

Brooks, D. R., and McLennan, D. A. (2002). "The Nature of Diversity." University of Chicago Press, Chicago.

Davis, C. S., Delisle, I., Stirling, I., Siniff, D. B., and Strobeck, C. (2004). A phylogeny of the extant Phocidae inferred from complete mitochondrial coding regions. *Mol. Phylogenet. Evol.* **33**, 363–377.

Domning, D. P. (2000). The readaptation of Eocene sirenians to life in the water. *Hist. Biol.* **14**, 115–119.

Felsenstein, J. (2004). "Inferring Phylogenies." Sinauer, Sunderland, MA.

Maddison, W. P., and Maddison, D. R. (2000). "MacClade, version 4. Analysis of phylogeny and character evolution." Sinauer Associates, Sunderland, MA.

Maddison, W. P. and Maddison, D. R. (2006). Mesquite: a modular system for evolutionary analysis. Version 1.1 http://mesquiteproject.org

Queiroz, de A., and Gatesy, J. G. (2006). The supermatrix approach to systematics. *Trends Ecol. Evol.* **22**(1), 34–41.

Swofford, D. L. (2000). "PAUP°: "Phylogenetic Analysis Using Parsimony, Version 4." Sinauer Associates, Sunderland, MA.

# Telemetry

## Andrew J. Read

Telemetry is the process of obtaining data remotely, by transmitting information from a marine mammal or by storing it for later retrieval. The field of telemetry includes a number of research approaches, from simple radio tags that allow researchers to relocate a tagged animal, to complex systems that record and transmit data from multiple environmental sensors. Recent advances in computing power, microprocessors, and global telecommunication systems have led to extraordinary insights into the behavior, ecology, and physiology of many marine animals including mammals (Block, 2005). It is now possible to follow the fine-scale behavior of marine mammals for months at a time and from the most remote regions of the world's oceans, from the comfort of one's office. It is also possible to visualize and listen to what a marine mammal sees and hears as it swims through the water column. These and other advances allow researchers to investigate how marine mammals use their three-dimensional world, quantify important physical and biological aspects of their environments, and test potential conservation measures designed to mitigate the effects of adverse human activities. The field continues to develop rapidly, fueled by continuing advances in technology, miniaturization, and data processing.

There are two primary approaches to collect data with telemetry systems. In the first approach, an archival data logger is attached to a marine mammal, records data for a predetermined period, and then it is recovered, allowing researchers to download the information stored in the package. In the second approach, information is transmitted from a marine mammal via radio or acoustic signals.

## I. Archival Tags

The earliest data loggers were simple time-depth recorders that used smoked glass disks rotating past recording needles attached to pressure transducers. These devices were first employed in 1963 by Gerry Kooyman to study the diving behavior of Weddell seals (*Leptonychotes weddellii*; Kooyman, 2006). To record time in these devices, Kooyman incorporated simple kitchen timers. This ingenuity is typical of the field of marine telemetry, which is too small to support much of its own commercial research and development. Instead, biologists have adopted, modified and refined technology developed for other purposes.

Modern data loggers are sophisticated digital devices that are capable of storing large quantities of information. Data are collected from one or more sensors that measure parameters such as depth,

water temperature, light intensity, or swimming velocity. The location of a tagged animal can be determined by several methods, but the most common approach is to record light levels and times of dawn and dusk, and to back-calculate latitude and longitude after the tag is recovered. It is also possible to record physiological data, such as heart rate and body temperature. Researchers routinely record feeding events by transmitting temperature changes in the stomach to external data loggers and jaw movement. A small transmitter, equipped with a temperature sensor, is introduced into the stomach of an animal and transmits data to a data logger mounted on its external surface. Most prey are heterothermic, or cold-blooded, so when the transmitter is swallowed, the temperature of the stomach drops abruptly. Eventually, the transmitter is passed or regurgitated.

One archival tag in particular has yielded considerable new insights into the behavior of marine mammals. This device, known as a digital acoustic tag (or DTAG), was developed by Mark Johnson and Peter Tyack at the Woods Hole Oceanographic Institution. The DTAG records the acoustic environment of the tagged animal, as well as fine-scale data on its three-dimensional orientation (Johnson and Tyack, 2003). The DTAG records sound (at sampling rates up to 200 kHz), depth, temperature, and orientation. In addition to the vocalizations of the animals themselves and the sound of nearby vessels, the loggers have provided unexpectedly rich data on swimming stroke and heart rate.

DTAGs have been used to study the acoustic behavior of marine mammals and to examine their response to anthropogenic sounds. For example, attachment of DTAGs to beaked whales (*Ziphius* and *Mesoplodon*) demonstrated that these deep-diving animals are highly vocal, producing high-frequency echolocation clicks, but only when they are at depths greater than 200 m (Johnson *et al.*, 2004). In addition, the DTAGs were able to detect echoes from prey during foraging events. Such rich detail allows us to understand the behavior of these enigmatic animals, even as they forage near the sea floor in depths of more than 1000 m. Researchers have also used DTAGs to examine the response of North Atlantic right whales (*Eubalaena glacialis*) to alert signals designed to make the animals aware of approaching ships (Nowacek *et al.*, 2004). The whales reacted strongly to the alerting signal by swimming rapidly to the surface, a response likely to increase rather than decrease the risk of collision. This work made it clear that mitigation measures other than alerting sounds will be required to solve the problem of ship strikes with right whales.

Data loggers have several advantages over other types of telemetry systems. First, data storage requires considerably less power than data transmission, so fewer and/or smaller batteries are required. In turn, this means that recoverable loggers can be smaller than transmitting tags. Second, the storage of large quantities of data is possible, particularly with modern digital technology. As noted in Section II, transmitting systems limit the quantity of data that can be relayed from the animal to a receiver.

The primary disadvantage of these systems is the need to recover the data loggers to retrieve stored information. The use of data loggers in studies of pinnipeds is fairly straightforward, because these animals haul out at predictable times and locations. Researchers studying elephant seals (*Mirounga* spp.), for example, are able to recover up to 95% of their loggers because of the strong fidelity of these animals to their rookeries. The use of data loggers with cetaceans, however, is considerably more challenging, as researchers must first attach the package to a dolphin or whale and then recover the tag after it is jettisoned. One solution is to attach the loggers with suction cups using a hand pole or other remote method, and then to recover the buoyant packages after release by homing in on a radio

**T**

signal emitted by a tag in the package. Several research groups have employed this technique with considerable success, although this is possible for only short deployments. The logistics of such field work, in which researchers follow tagged animals at sea, are considerably more complex than the deployment and retrieval of data loggers on pinnipeds.

## II. Transmitting Systems

Transmitting systems have also undergone an extraordinarily rapid development over the past several decades. The earliest transmitters were omnidirectional radio or acoustic transmitters that allowed researchers to relocate a tagged animal but did not provide information on its behavior or physiology. These simple systems have evolved into sophisticated systems in which large quantities of data can be recorded, compressed, and transmitted via orbiting satellite.

Some of the earliest radio transmitters used with marine mammals were developed by Bill Watkins and W.E. Schevill from the Woods Hole Oceanographic Institution. In the 1960s, these researchers developed implantable radio tags fired into the blubber of large whales. Similar tags are still used today to study the movements and behavior of baleen whales. Radio tags have also been attached to the dorsal fins of dolphins and porpoises or glued to the fur of pinnipeds. These tags allow researchers to follow marine mammals at sea and gain insight into their behavior and short-term movements. This labor-intensive field work requires the use of directional receiving antennas to home in the radio signal produced by the transmitter.

The utility of these simple transmitting systems is limited by several factors. First, the high-frequency signals emitted by radio tags attenuate rapidly in salt water, so it is possible to receive signals only when the antenna is above the surface. This complicates the tracking of animals because only a few signals are heard at each surfacing. Acoustic signals propagate for much greater distances underwater, but often overlap with the hearing range of marine mammals, limiting their applicability. Even under ideal circumstances, radio transmitters have effective ranges of only a few tens of kilometers, so researchers are forced to stay in close proximity to tagged animals. Finally, the large size and cumbersome design of many early radio tags created significant hydrodynamic drag and resulted in the premature detachment of the packages and possible harm to the animals wearing them.

Today, there are a wide variety of transmitting systems available to researchers. The most significant advance has been the development of satellite-linked radio transmitters that allow biologists to track the movements and behavior of marine mammals from their offices. The principle underlying these systems is straightforward. Each transmitter emits a stable radio signal to receivers aboard orbiting weather satellites. As a satellite moves across the horizon, the received frequency of the tag changes due to the Doppler shift, allowing estimation of the position of the transmitter. Each transmission also includes the identity of the transmitter and any associated sensor data. Data are processed and relayed to the user by e-mail or made available over the web. This technological advance obviates the need for researchers to follow animals in the field, thus allowing insight into the movements and behavior of marine mammals in the most remote and challenging environments.

Satellite-linked radio transmitters have been coupled with data logging systems, to allow the collection of detailed behavioral or environmental data from marine mammals via satellite. This coupling of data logging and transmitting systems has proven to be very successful, because it precludes the need to recover the logger package to obtain sensor data. Typical data collected by these systems include depth and

swim speed, although in principle any sensor system can be employed. Recent developments in sensor design and data handling have allowed development of a new generation of tags that are deployed as oceanographic sensors on marine mammals (Block, 2005). Thus, marine mammals now collect and transmit large quantities of data on the physical and biological attributes of their environments, greatly facilitating our understanding of their ecology. Marine mammals are adept at exploiting fine-scale oceanographic features that concentrate prey, such as frontal systems, and animals instrumented with appropriate sensors provide considerable information about the location and dynamics of such processes. These marine mammal sentinels are providing a rich picture of the physical structure of their environment to oceanographers in areas where research cruises or the deployment of oceanographic drifters is difficult and prohibitively expensive. It is likely that marine mammals will contribute significantly in this manner to future ocean observing systems (Fedak, 2004).

Satellite-linked data loggers are extremely powerful data acquisition systems, but they do have limitations. Their signals can be received only when the transmitter is above the surface and a satellite receiver is overhead. Energy for signal transmission is a significant limitation with current battery technology, although battery life may be conserved by using a salt-water switch, which suppresses transmissions when the tag is submerged. In addition, the current satellite system limits each transmission to 256 bits, so algorithms are required to compress complex data, such as records of individual dive profiles, prior to transmission. Nevertheless, these tags have revolutionized our understanding of the ecology, behavior, and environments of marine mammals.

## III. Biological Insights

Advances in the field of telemetry have revolutionized our view of marine mammals. As terrestrial observers, we are limited in our ability to study marine mammals and, in the past, have been limited to collecting data from animals at the surface or ashore. While at sea, most marine mammals spend more than 90% of the time submerged, often in remote or harsh environments in which field research is difficult or impossible. Telemetry allows us to peer into the lives of whales, dolphins, sirenians, and pinnipeds as they go about their daily activities of feeding, finding mates, and avoiding predators. We can ask how a beaked whale hunts for food at a depth of 1000 m, or how an elephant seal is capable of such long, repeated submergences, or where blue whales (*Balaenoptera musculus*) go in the winter months. The insights provided by this technology will continue to challenge our thinking about these animals; particularly, as new technological developments improve our ability to collect data at sea.

Elephant seals have proven to be particularly amenable to study with telemetry. These are large animals that haul out to breed and molt at predictable times and locations, but which spend the majority of the year far from shore. Thus it is possible to equip individual elephant seals with fairly large telemetry packages and be confident that most packages will be recovered. Researchers in both hemispheres have equipped a large number of elephant seals with recoverable data loggers and, more recently, satellite-linked data loggers. From this research, we now know that elephant seals in the North Pacific make two long-distance feeding migrations each year, one after breeding and the second after the molt. For example, adult male elephant seals travel from central California to the Gulf of Alaska on each migration, a distance of more than 10,000 km in each round trip. Individuals appear to return to the same feeding area each year and, once on the feeding grounds, forage almost continuously. While

T

feeding, individual elephant seals dive repeatedly, spending more than 90% of their time at sea submerged, and sometimes diving to depths of more than 1500 m. Such behavior is consistent with what we know about diet of these animals; elephant seals feed primarily on mesopelagic squid found at depths of 200–1000 m. These prolonged and continuous dives have raised many physiological questions, particularly with regard to the oxygen storage capacity of these animals (discussed later). The prodigious diving behavior of elephant seal behavior has led some biologists to refer to them as mesopelagic mammals.

Elephant seals are not alone in possessing an impressive diving capacity. Sperm whales (*Physeter macrocephalus*) have been tracked using telemetry to depths of more than 2000 m during dives that may last for more than an hour. Beaked whales are also capable of long, deep dives. Northern bottlenose whales (*Hyperoodon ampullatus*), for example, equipped with time-depth recorders attached with suction cups, have made dives to 1500 m and for over an hour in duration. Studies using satellite-linked data loggers attached to smaller whales, such as belugas (*Delphinapterus leucas*) and narwhals (*Monodon monoceros*), indicate that these species are also capable of prolonged, deep dives under the Arctic ice.

Recent studies with data recorders equipped with video cameras, or crittercams, have provided dramatic findings regarding the behavior of marine mammals. Weddell seals have been videotaped flushing prey from crevices in the ice by blowing bubbles and researchers have watched Hawaiian monk seals (*Monachus schauinslandi*) sleep and forage on the sea floor. This type of research has particular relevance to conservation because it is believed that monk seals may be endangered, in part, due to conflicts with commercial fisheries.

Documenting the availability of prey and the success rate of capture attempts allow us to test such ideas directly for the first time. Backward-mounted crittercams have been used to study the diving behavior of a variety of marine mammals and, in particular, to investigate how whales and seals can make such long dives without exceeding their aerobic capacities. It now appears, for example, that elephant seals and other marine mammals conserve oxygen by gliding extensively during descent. These animals take advantage of the changes in buoyancy brought about by increased pressure at depth and can effectively descend with little extra expenditure of energy or oxygen.

## IV. Challenges and Future Developments

It is difficult to anticipate what surprises the field of telemetry has in hold, but it is clear that these techniques will be an integral component of the toolbox of future marine mammal researchers. In particular, it is likely that more sophisticated sensors will be developed to take advantage of the success of recoverable and satellite-linked data loggers. Such novel sensors will monitor physiological parameters, such as blood oxygen concentration, hormone levels, and blubber thickness (some of these advances have already been tested with other marine vertebrates). Current advances in wireless technology hold great promise for our ability to telemeter data from marine mammals, because many current applications in acoustic and video telemetry are limited by bandwidth—the amount of information that can be transmitted from the animal to a receiver. New telemetry systems based on Global System for Mobile Communications (mobile phone, GSM) technology have been deployed with great success on several species of pinnipeds in Europe and North America. Seals wearing these mobile phone tags send text messages to researchers when they are in areas of GSM coverage, providing information on location and behavior (McConnell *et al.*, 2004). A new generation of low-orbit satellite systems would improve our ability to collect more data from a larger sample of animals. Future readers will, no doubt, find our current suite of data loggers and satellite-based telemetry systems quaint and outdated.

Despite the great promise of telemetry for our understanding of the biology of marine mammals, some significant challenges lie ahead. As the cost of tags has been reduced, and their availability and sophistication increased, the number of individual animals studied has increased exponentially. It is not uncommon for very large samples (hundreds) of tags to be deployed in some of the largest telemetry programs such as the Tagging of Pacific Pelagics (Block, 2005). Our ability to analyze such large data sets has not kept up with the availability of information and new analytical methods are required to gain maximum insight from these large (and costly) field projects. Problems continue to exist with uncertainty in position estimates (although a new generation of GPS tags is helping to resolve some of these issues) and incorporating this uncertainty into an appropriate analytical framework. In particular, it is critical to integrate movement and behavioral data obtained via telemetry with observations of the environment conditions available to the animals (Shaffer and Costa, 2006). Making telemetry information available through publicly available data commons, such as the OBIS-SEAMAP project (Halpin *et al.*, 2006), will facilitate innovative approaches to analysis and integration with environmental data.

Finally, the ready availability of off-the-shelf electronic tags has greatly increased the number of animals instrumented each year. Despite this rapid growth in the field, there have been few attempts to determine what behavioral or physiological effects are caused by carrying a transmitter or archival data logger. Such effects could include mechanical artifacts of the attachment method used, increased energy expenditure caused by elevated rates of hydrodynamic drag, or changes to the thermal biology due to decreased heat flow across the integument. The maturation of the field will, without doubt, bring an evaluation of such potential effects on a variety of marine mammal species.

## See Also the Following Articles

Behavior, Overview ■ Distribution ■ Identification ■ Methods ■ Migration and Movement Patterns

## References

Block, B. A. (2005). Physiological ecology in the 21st century: Advancements in biologging science. *Integr. Comp. Biol.* **45**, 305–320.

Fedak, M. A. (2004). Marine mammals as platforms for oceanographic sampling: A "win/win" situation for biology and operational oceanography. *Mem. Natl Inst. Polar Res. Spec. Iss.* **58**, 133–147.

Halpin, P. N., *et al.* (9 authors) (2006). OBIS-SEAMAP: Developing a biogeographic research data commons for the ecological studies of marine mammals, seabirds, and sea turtles. *Mar. Ecol. Prog. Ser.* **316**, 239–246.

Johnson, M. P., and Tyack, P. L. (2003). A digital acoustic recording tag for measuring the response of wild marine mammals to sound. *IEEE J. Oceanic Eng.* **28**, 3–12.

Johnson, M., Madsen, P. T., Zimmer, W. M. X., Aguilar de Soto, N., and Tyack, P. L. (2004). Beaked whales echolocate on prey. *Proc. R. Soc. Lond. B Suppl.* **271**, S383–S386.

Kooyman, G. L. (2006). Mysteries of adaptation to hypoxia and pressure in marine mammals. *Mar. Mamm. Sci.* **22**, 507–526.

Le Boeuf, B. J., and Laws, R. M. (eds) (1994). "Elephant Seals: Population Ecology, Behavior, and Physiology." University California Press, Berkeley.

McConnell, B., Beaton, R., Bryant, E., Hunter, C., Lovell, P., and Hall, A. (2004). Phoning home—a new GSM mobile phone telemetry system to collect mark-recapture data. *Mar. Mamm. Sci.* **20**, 274–283.

**T**

Nowacek, D. P., Johnson, M. P., and Tyack, P. L. (2004). North Atlantic right whales (*Eubalaena glacialis*) ignore ships but respond to alerting stimuli. *Proc. R. Soc. Lond. B* **271**, 227–231.

Shaffer, S. A., and Costa, D. P. (2006). A database for the study of marine mammal behavior: Gap analysis, data standardization, and future directions. *IEEE J. Oceanic Eng.* **31**, 82–86.

# Territorial Behavior

### Edward H. Miller

## I. Territoriality in Marine Mammals

"Territoriality" refers to the exclusive use of fixed space, which entails obtaining, defending, or advertising occupancy of the space. All agonistic social interactions occur within a spatial framework, but by themselves do not constitute territorial behavior. "Home range" refers to the space used by an individual or group (whether or not it is used exclusively), without attendant behavior in defense or advertisement (Fig. 1). As in other animals, territoriality in marine mammals can evolve if space is defensible, enabling monopolization of resources within that space. Territoriality is absent in most marine mammal species simply because no whales are territorial; the polar bear (*Ursus maritimus*), marine otter (*Lontra felina*), and most Sirenia also are not territorial. The best known examples of territoriality are breeding males in: one population of dugong (*Dugong dugon*; in Shark Bay, Western Australia: Anderson, 2002); sea otter (*Enhydra lutris*); and many pinnipeds. Superficially, territoriality appears to characterize the land-breeding elephant seals (*Mirounga* spp.) and gray seal (*Halichoerus grypus*; some populations also breed on ice), because males are fairly sedentary when hauled out (Fig. 2). All marine mammals that are territorial when breeding are polygynous; polygyny is most extreme in land-breeding pinnipeds, because density of breeding females can be very high (Fig. 3; Boness, 1991; Boness *et al.*, 2002). Territoriality is expressed differently even between closely related species, is rigid or invariant in any species, and varies intraspecifically through short-term opportunistic behavior, throughout development, seasonally, and geographically. Territoriality away from the breeding site and outside the breeding period occurs but has been studied little. Key papers are Bartholomew (1970), Stirling (1983), Boness (1991), Miller (1991), Boness *et al.* (2002), and Tyack and Miller (2002).

## II. Development of Territorial Behavior

Territoriality involves complex behavioral patterns used repeatedly in interaction and communication, which emerge in play early in life. The complex underwater vocal displays of territorial male bearded seals (*Erignathus barbatus*) appeared at 4–5 years of age in several captive males, but not in females. Sexual differences in behavioral development have been documented in most detail for fur seals and sea lions (Otariidae), in which pups segregate by sex, male pups are more aggressive than females, and male pups engage in more play-fighting and territorial displays (Gentry, 1974). Those social interactions occur even in the appropriate topographical context for breeding territoriality; e.g., in the Steller's sea lion (*Eumetopias jubatus*), "Pups located themselves on opposite sides of any available ridge and used boundary display postures, open mouth lunges, and other … behavior characteristic of boundary defense in adult males" (Gentry, 1974, p. 402). Young male otariids and walrus (*Odobenus rosmarus*) of all ages engage extensively in play-fighting (and for walruses, other display forms as well) throughout the year.

Territorial male otariids appear to try to influence females to stay on their territories by "herding" (Section VIII). Such behavior appears early in life, and male pups preferentially direct this behavior toward female pups (e.g., New Zealand fur seal, *Arctocephalus forsteri*). Herding is expressed also by non-territorial or peripheral males during the breeding season; for example when they encounter females outside breeding aggregations. In all otariid species, non-territorial males (including sub-adults) may rush simultaneously into ("raid") breeding aggregations, and herd or interact with females in various ways before they are chased away by territorial males (Peterson, 1968; Campagna *et al.*, 1988a; Section VIII). Outside the breeding season, juvenile and sub-adult male otariids occasionally herd pups or young juveniles, at wintering haulout sites or colony sites (e.g., New Zealand fur seal). An extreme form of this behavior has been noted in several otariid species, in which non-territorial (generally sub-adult) males carry pups away in their mouths following raids on breeding sites (Campagna *et al.*, 1988b; Kiyota and Okamura, 2005). In the southern sea lion (*Otaria flavescens*), males may carry pups to the ocean, and then to non-breeding areas where the males herd and mount them, sometimes over several days; about 6% of pups treated in this manner die as a result (Campagna *et al.*, 1988b).

Breeding territoriality in otariids develops within the context of strong colony and natal-site fidelity (Section V), which become increasingly precise with age [e.g., for male northern fur seals (*Callorhinus ursinus*) aged 2–6 years]. Male northern fur seals start to haul out near the breeding site when they are 2-years old, and hold their first territories at about 7 years of age; first territories tend to be located peripherally, and are occupied late in the breeding season (Gentry, 1998; Kiyota, 2005). Over successive breeding seasons, territories become stabilized in location and size, and are established progressively earlier in the season.

## III. Territorial Functions

Non-breeding territoriality is poorly documented and understood. Male otariid territoriality occurs at some non-pupping sites during the breeding season (Miller, 1991). Many minor disputes over space take place throughout the year at both breeding colonies and non-breeding haulout sites, where individuals of various classes tend to use and interact agonistically repeatedly at the same sites. Non-breeding (winter) territoriality in some individuals of some seal (Phocidae) species may provide exclusive (or priority) access to breath holes in stable ice. The function of another form of territoriality also is not known: mature male otariids may return to occupy their territorial sites for several days in the fall, weeks after abandon them (Section V).

The best known, most dramatic, and most interpretable forms of territoriality in marine mammals are shown by breeding adult males, which establish territories seasonally where females copulate. The ultimate function of male territoriality in all cases is to gain access to estrous females; proximate functions that mediate male reproductive success are more difficult to identify and are both varied and variable. For example, male pinnipeds may change locations in response to female movements, attend and defend isolated lone females, or even defend and copulate with female carcasses.

In many otariids, males hold territories where females give birth and remain on land until they enter estrus; in such cases, male

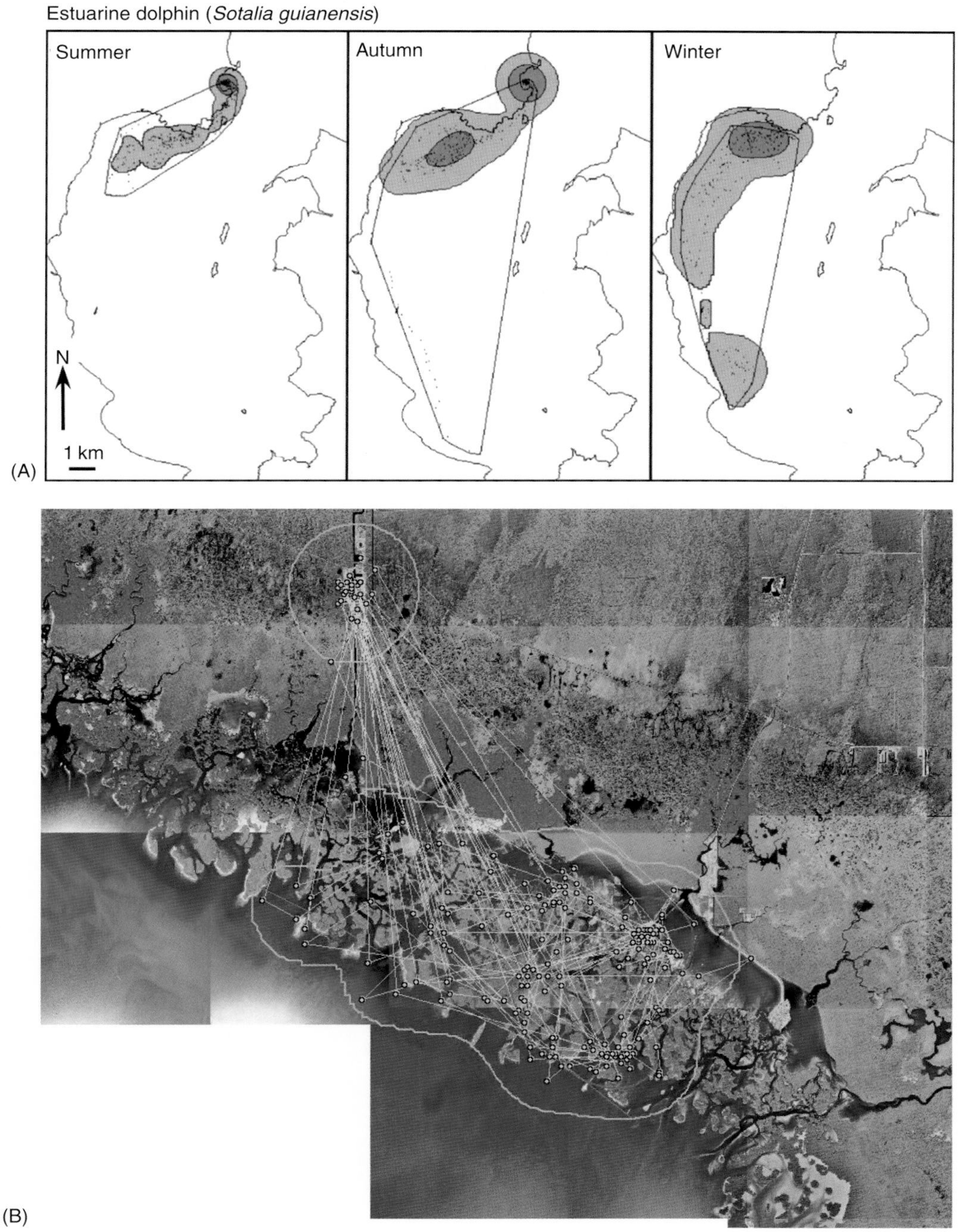

Estuarine dolphin (*Sotalia guianensis*)

(A)

(B)

**Figure 1** *Most marine mammal species are not territorial: they occupy undefended home ranges, which typically incorporate systematic seasonal movements and regular smaller-scale movements between habitats. (A) Group home ranges of estuarine dolphins (Sotalia guianensis) in Norte Bay, southern Brazil in 2002, illustrating size, shape, and seasonal changes in home range. Home ranges were estimated by: (i) the minimum convex polygon method; and (ii) fixed kernel analysis, with dark and light gray areas representing 50% and 95% levels, respectively. (B) Home range of one adult female West Indian manatee (Trichechus manatus) in southern Florida in a dry period (March 2 to May 31, 2001), when it made many trips between marine-coastal feeding areas (seagrass beds), and freshwater rivers and creeks (the species needs regular access to freshwater for drinking and osmoregulation). White lines connect successive high-quality fixes by Argos satellite; colored lines reflect fixed kernel analysis, and depict 95% (orange), 50% (green), and 10% (yellow) levels. (A) After figure 3 of Wedekin et al. (2007); (B) after figure 56 of Stith et al. (2006).*

T

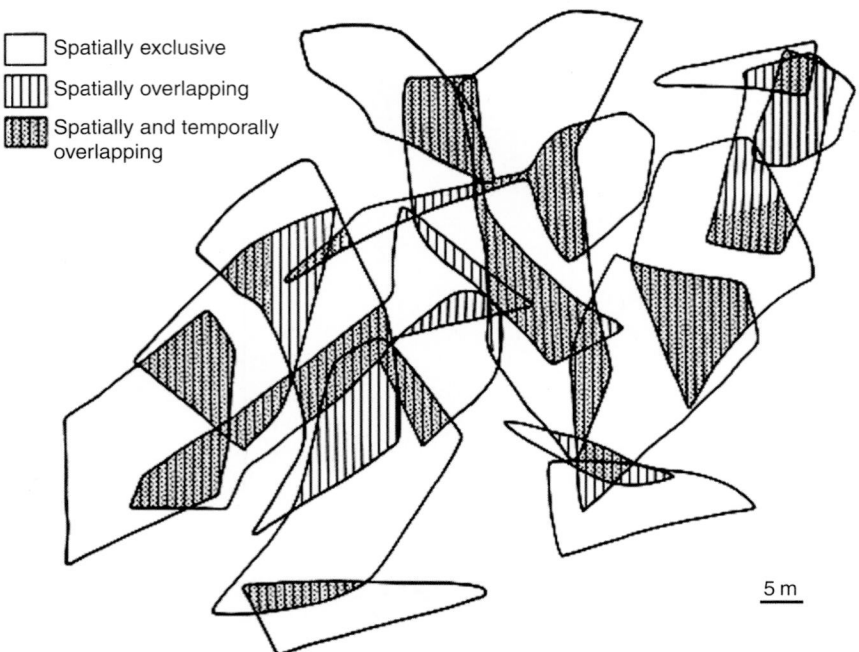

Spatially exclusive

Spatially overlapping

Spatially and temporally overlapping

5 m

**Figure 2**   *The spacing system of breeding male gray seals* (Halichoerus grypus) *on land resembles territoriality but is structured on attendance ("consortship") of females, but not defense of space per se. This figure shows areas used by adult males attending females in a study area on Sable Island, Nova Scotia, in the 1975–1976 breeding season. After figure 9 of Boness and James (1979).*

**Figure 3**   *Female fur seals and sea lions* (Otariidae) *return seasonally to particular terrestrial sites on islands free from predation, to give birth and raise offspring. These factors have enabled the evolution of male territoriality and polygyny; the tendency of females to aggregate when ashore increases polygyny levels further (see text). This photograph illustrates a breeding aggregation of northern fur seals* (Callorhinus ursinus) *at the Little Polovina rookery, Pribilof Islands, Alaska, in July 1948. Photograph by Edward C. Johnston, Fish and Wildlife Service, U.S. Department of the Interior. Photograph courtesy National Marine Mammal Laboratory (NOAA, National Marine Fisheries Service).*

territories hold resources that are needed by females (sites for pupping, reuniting with pups between foraging trips, and nursing). However, territoriality is fluid in many populations, because breeding animals must move regularly for thermoregulatory reasons or

due to tide levels, or because the breeding substrate is homogeneous and hence territorial boundaries are ill defined (Section IV; Boness, 1991; Miller, 1991; Boness *et al.*, 2002). In the Juan Fernández fur seal (*Arctocephalus philippii*), females move to the sea when it is

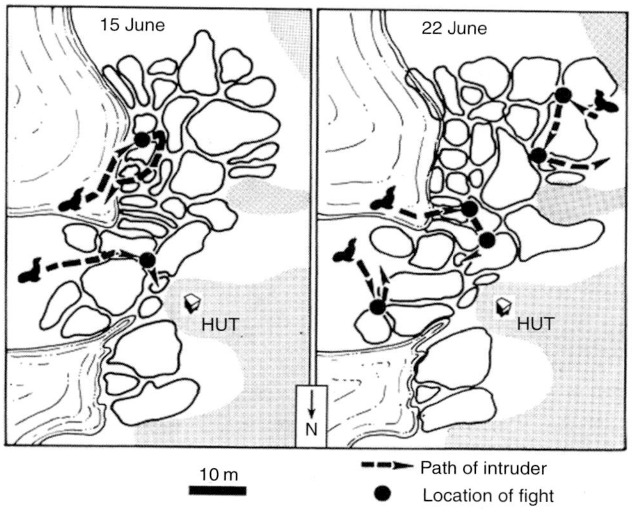

**Figure 4** *Territories are small and fairly fixed in size, shape, and location in many species of fur seals and sea lions (Otariidae). This figure depicts territories (in outlines) and some territorial behavioral events of male northern fur seals* (Callorhinus ursinus) *at Kitovi rookery, Pribilof Islands, Alaska, in 1962. "HUT" is the point of observation. After figure 11 of Peterson (1968).*

hot but remain close to the colony; males may hold inland, shoreline, or completely aquatic territories in this species. The most reproductively successful male southern sea lions hold territories where females can thermoregulate: within tidepools, or along the high water mark where the substrate remains wet. Resource-based territoriality is expressed clearly in sea otters. Adult male sea otters establish territories seasonally in or near areas of high female density. Male territories hold multiple resources used by females, including prey, resting areas, protection from wind and waves, and protection from harassment by other males (Garshelis *et al.*, 1984).

The distinction between territories with and without resources needed by females blurs in many cases. In the Weddell seal (*Leptonychotes weddellii*), mature males set up and defend territories seasonally in traditional areas of stable ice where females congregate, give birth, and later enter estrus; females need access to water for foraging, so these access points can be considered as the key defensible resource on which male territoriality is based (Harcourt *et al.*, 2007). Similarly, male ring seals (*Pusa hispida*) hold underwater territories that are near or encompass the birth lairs of several females (Stirling, 1983; Stirling and Thomas, 2003). Male harbor seals (*Phoca vitulina*) establish territories near female haulouts, or on access routes to and from those haulouts (Boness *et al.*, 2006). At an extreme, male territories may hold no resources needed by females, and male reproductive success depends solely on female visitation for purposes of copulation (e.g., Shark Bay dugongs; walrus; bearded seal). In some situations (e.g., walrus), the presence of males near a small number of females can be construed as defense of females, rather than of space (Sjare and Stirling, 1996).

## IV. Spatial Aspects of Territoriality

Discrete, clearly defined territories are most apparent on small temporal scales, in situations of crowding, in species that have good locomotory abilities so can efficiently patrol or defend their territories, and where environmental features (e.g., topographical irregularities)

occur that can be used by the animals to demarcate territories. Such conditions are lacking in the lives of cetaceans, especially open-ocean species, so territoriality does not occur in that group. River dolphins, with their spatially restricted distributions, or species that feed on concentrated prey that are sedentary or spatially predictable, may prove to be territorial, but this is not known at present. In Scotland's Moray Firth, year-round resident bottlenose dolphins (*Tursiops truncatus*) may be territorial and exclude seasonal (winter) visitants from deep waters, which are most favorable for feeding; group territoriality also has been suggested for this species in Ecuador.

Most species of otariids breed on crowded colony sites and hold small territories; territories of male Hooker's sea lions (*Phocarctos hookeri*) are often no more than 3 m in diameter, for example, and some northern fur seals hold territories that are little larger in diameter than a male's body length (Figs 3 and 4). Larger territories occur in related species (e.g., about 200 m² in male Steller's sea lions). Small aquatic territories are held by Juan Fernández fur seals adjacent to breeding aggregations on land, and by walruses adjacent to mixed herds on ice (Miller, 1991; Sjare and Stirling, 1996). In general however, aquatic territories are large: more than 100 m in length in some male Weddell seals, up to 1 km across in male sea otters (Pearson *et al.*, 2006), and up to 10 km across in some male harbor seals. Many aquatic territories are non-contiguous, but contiguous territories invariably overlap some extent, both in linear and in more complex spatial arrangements (Fig. 5; Boness *et al.*, 2006).

Phocids are specialized for aquatic locomotion, so their locomotion on land is slow and energetically costly. The poor locomotory abilities and large size of the two species of elephant seal usually preclude territoriality, although in small confined areas or rough terrain, defense of space and of females amount to the same thing. In contrast, fur seals and sea lions can move quickly and efficiently on land, so offer many clear examples of terrestrial territoriality.

Precise delimitation of territories occurs at many breeding sites of otariids, because rocks, fracture lines, and other natural features are present; in these situations, territorial boundaries may be stable within and across years. On featureless terrain (e.g., sandy beaches), territories are less clearly defined. Territories of northern fur seals are smaller in exposed terrain, and larger in protected terrain (Peterson, 1968). Aquatic territories also are influenced by physical environmental structures. In species that breed in association with ice, underwater features of ice or fractures or leads in ice may be important in determining territorial density, size, and shape. In the walrus, male territories are established in the water adjacent to mixed herds on ice, which may be stable land-fast ice (e.g., in the Canadian Arctic) or unstable drifting pack ice (e.g., Bering Sea).

## V. Temporal Aspects of Territoriality

Territories of fur seals and sea lions are most clearly defined at the peak of breeding, when territorial density is highest and territorial size is smallest (Figs 3, 4, and 6).

Absence of females from their territories sometimes leads to territorial desertion by male otariids, but more commonly males attempt to acquire a new territory where females are present. In the Hooker's sea lion, males establish territories several times during the breeding season in response to movement of the female aggregation down the beach. Southern sea lion males defend territories early in the breeding season but gradually change to defense of females as the season progresses (Campagna and Le Boeuf, 1988).

Male otariids may haul out at the site of their future territory before territorial behavior begins and, in many species, males

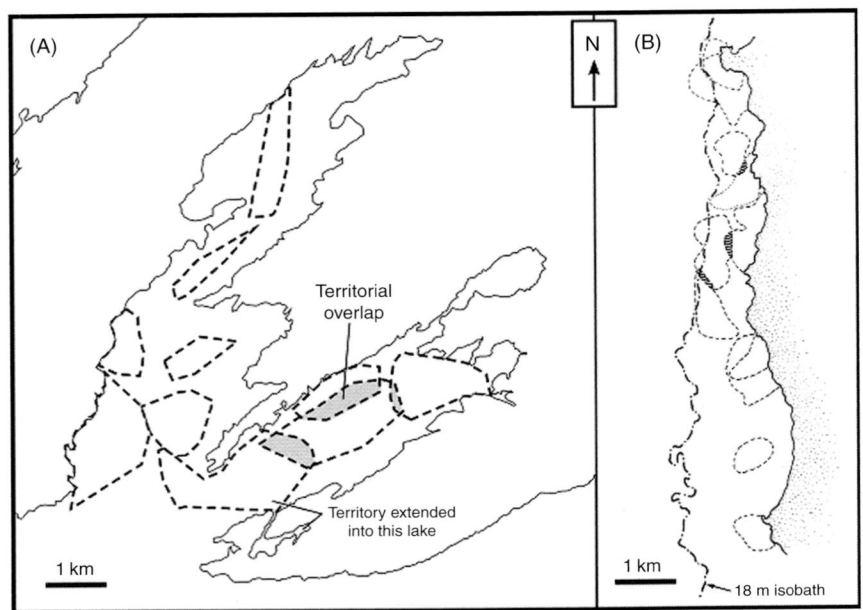

**Figure 5** *Size, shape, and configuration of marine mammal territories are influenced by the physical environment. Aquatic territories held by adult male sea otters (Enhydra lutris) hold multiple resources needed by females. They are large, overlap, and exhibit more complex spatial arrangements in enclosed waters [(A) Prince William Sound, Alaska, 2003; "lake" is a tidal lagoon] than along coasts [(B) central California, 1978–1982; overlapping territories held in different years are shown with areas of overlap not cross-hatched]. (A) After figure 1 of Pearson et al. (2006); (B) after figure 2 of Jameson (1989).*

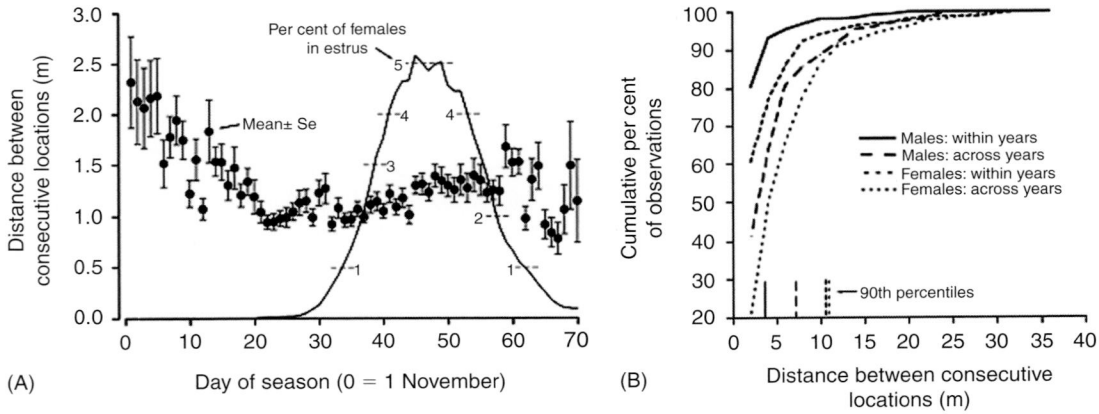

**Figure 6** *Territories are smallest and males move least at the peak of breeding in fur seals and sea lions (otariidae), and males tend to return to the same sites every year, as illustrated by the Antarctic fur seal (Arctocephalus gazella). (A) Movements of territorial males become increasingly localized within a breeding season. (B) Males have high site fidelity both within and across years, and their site fidelity is greater than that of females on both temporal scales. After figures 2 and 1 (respectively) of Hoffman et al. (2006).*

also return to their territories after the breeding season has ended (Fig. 7A). Territorial occupancy may be continuous for weeks or months, or males may leave the territory for periods for thermoregulatory or other reasons, and then return (Fig. 7B; Gentry, 1998). Territorial occupancy is highly variable in the Hooker's sea lion, and males move extensively within and between breeding sites (Fig. 7C);

this is similar to behavior of breeding males in the non-territorial gray seal on Sable Island, Nova Scotia.

Male otariids habituate to neighboring males and engage in fewer and less aggressive interactions with neighbors over time. A similar effect occurs even across years between returning territorial neighbors (e.g., Steller's sea lion; Miller, 1991).

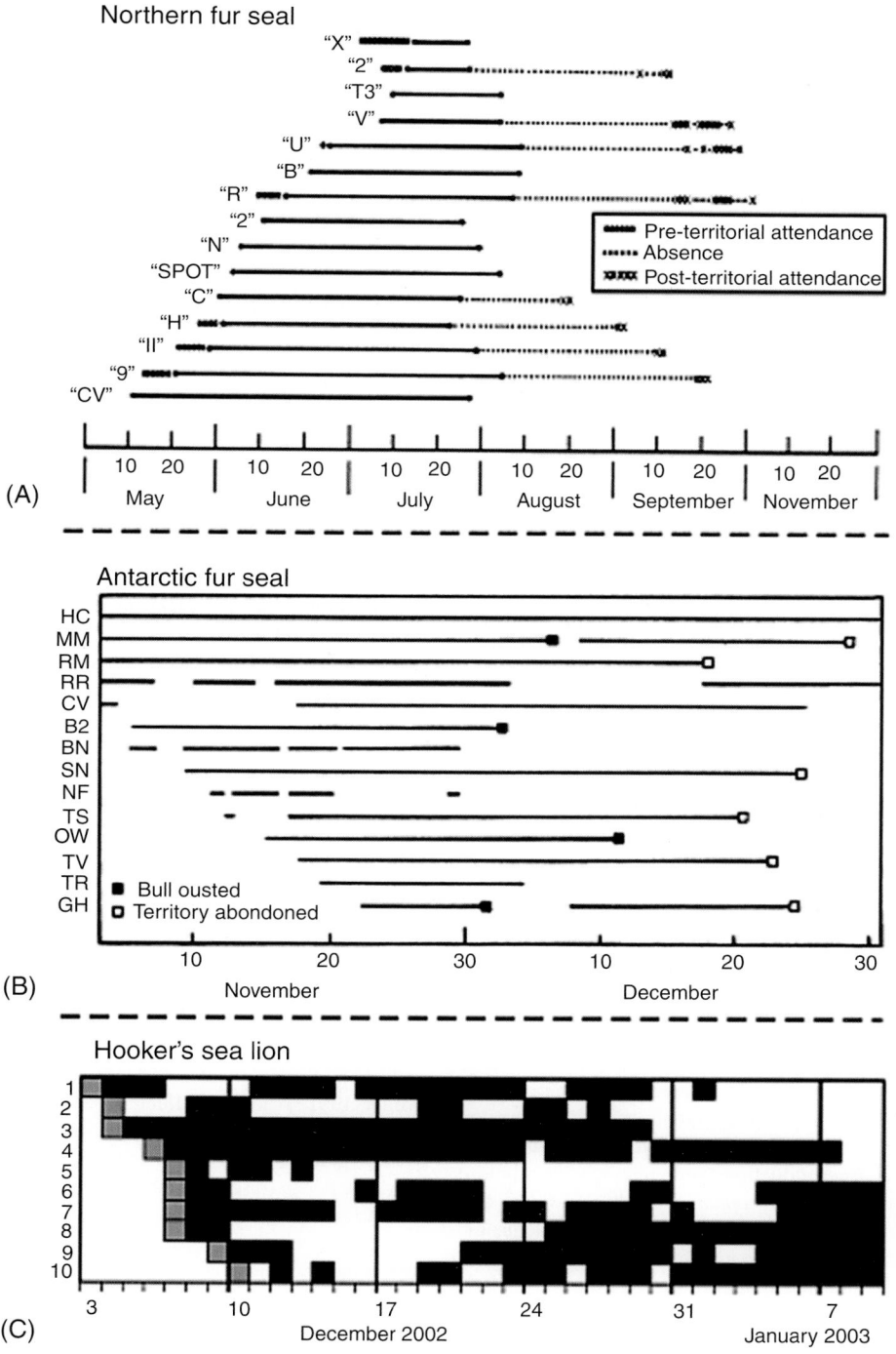

**Figure 7** *Attendance patterns of territorial male fur seals and sea lions (Otariidae) vary within and across species. Territorial attendance patterns are shown for: (A) 15 northern fur seals* (Callorhinus ursinus) *at Kitovi rookery, Pribilof Islands, Alaska, in 1962; (B) 14 Antarctic fur seals* (Arctocephalus gazella) *on Bird Island, South Georgia, in 1975–1976; and (C) 10 Hooker's sea lions* (Phocarctos hookeri) *at Enderby Island, Auckland Islands, in 2002–2003 (shaded squares represent days when males were marked). (A) After figure 12 of Peterson (1968); (B) after figure 2 of McCann (1980); (C) after figure 3 of Robertson et al. (2006), respectively.*

Long-term fidelity (philopatry) to territorial locations has two components. First, there is a tendency for males to return to breed near the site of their birth (natal philopatry). Second, males tend to return in successive years to where they first established a territory. Both forms of site fidelity are well documented in otariids; e.g., about 50% of returning male antarctic fur seals (*Arctocephalus gazella*) occupy a territory that is within half a body length of the territorial site held in the previous year (Fig. 6B). Site fidelity by breeding males occurs also in species that breed in association with land-fast ice (e.g., ring seal, Weddell seal; Harcourt *et al.*, 2007), or near land (e.g., harbor seal, sea otter); male sea otters hold territories in the same location for up to 7 successive years. Site fidelity even occurs in aquatically territorial phocids that display away from land or land-fast ice: some individual male bearded seals returned seasonally to the same territorial display areas every year of the 16-year-long study by Van Parijs and Clark (2006). The extremely strong natal- and territorial-site fidelity shown by otariids, coupled with their high breeding densities, results in kin breeding in proximity to one another.

## VI. Territoriality, Mating Strategies, and Mate Selection

Land-breeding pinnipeds were central to Bartholomew's (1970) model of pinniped polygyny. His model remains as the core paradigm to explain otariid polygyny and territoriality, which assumes lengthy uninterrupted male territoriality, when males fast and expend much energy in territorial display and defense. However, many variations occur in territorial behavior, as noted earlier: thermoregulatory factors influence movements of males and females, and males may hold aquatic or semiaquatic territories (Juan Fernández fur seal, southern sea lion, California sea lion *Zalophus californianus*, etc.); males may vacate their territories for periods ranging from hours to days, even in classically territorial species like the northern and antarctic fur seals (Fig. 7); males may change from territoriality to female defense later in the season (southern sea lion); and so on. Variation in territoriality is paralleled by variation in male mating behavior and success.

Especially in otariids, male reproductive success is closely tied to possession of territory, and territorial males (a) account for most fertilizations and (b) fertilize most females that pup on their territories (Hoffman *et al.*, 2003; Kiyota *et al.*, 2007). Scattered observations over many years have pointed to mating in other situations, however: young male northern fur seals mate with females late in the breeding season, after the main period of reproduction; some mating occurs before pupping in the Cape fur seal (*Arctocephalus pusillus pusillus*) and New Zealand fur seal; and males mature sexually (i.e., physiologically) years before they are large enough to compete for territories. Recent molecular evidence has confirmed that fertilizations occur regularly outside the territorial structure, and involve non-territorial males (Hoffman *et al.*, 2003; Kiyota *et al.*, 2007); furthermore, female antarctic fur seals actively seek and mate with territory holders that are heterozygous and unrelated to them (Hoffman *et al.*, 2007).

Male behavior varies in other territorial species, but its relationship to mating success is not known. Some male bearded seals in Alaska change from being territorial in 1 year to non-territorial in another (Van Parijs and Clark, 2006); in Svalbard, most males are territorial when land-fast ice is extensive. It is not known whether such variability in territorial and mating behavior (a) reflects distinctly different strategies or a continuum, (b) represents evolved adaptations or behavioral plasticity, or (c) is related to male age or other characteristics (e.g., phenotype; in the non-territorial gray seal, males of different body sizes adopt different mating tactics).

## VII. Obtaining, Defending, and Advertising Territories

Dramatic and potentially injurious fights between males occur in all territorial species, particularly when new males attempt to establish themselves. Severe injuries or even death can result, so fighting is uncommon, and most interactions between territorial males instead take the form of display (Peterson, 1968; Miller, 1991; Gentry, 1998).

Territorial male dugongs in Shark Bay repeatedly patrol the margins of their territories along fixed routes (Anderson, 2002). Acoustic displays can travel long distances (especially under water) and are energetically cheap to produce, but long-distance sounds are lacking from the diverse vocal repertoire of this species. Patrolling males perform various distinctive behaviors, e.g., they may swim upside-down or rear high out of the water; fights also occur. Male sea otters likewise patrol their large territories, and announce their presence by exaggerated kicking, splashing, and grooming (Garshelis *et al.*, 1984); males interacting near territorial boundaries may engage in mutual porpoising, and they sometimes fight. Like dugongs, sea otters have a rich vocal repertoire, but also lack sounds for long-distance communication (Miller, 1991). In contrast, walruses and aquatically territorial phocids possess acoustic signals that travel under water for long distances (Fig. 8).

Long-distance underwater sounds of territorial pinnipeds tend to be stereotyped, and to convey little behavioral information; nevertheless, they must be informative to several classes of receiver (e.g., non-territorial males, females), not just other territory holders (Miller, 1991; Tyack and Miller, 2002). Long-distance sounds are not universal in aquatically territorial species; for example, high vocal activity and long-distance sounds probably have been selected against in the ring seal, to minimize detection by polar bears (Stirling, 1983; Stirling and Thomas, 2003). Otariids are highly vocal and have rich vocal repertoires (Phillips and Stirling, 2001). Vocalizations of territorial male otariids are transmitted over medium to short distances; the stereotyped "full-threat call" (Phillips and Stirling, 2001, p. 423) is the loudest, and carries the farthest. Many male calls have been interpreted simply as communication of "threat" but carry much richer behavioral information. Most call types of male otariids are variable, have complicated syntactical arrangements, and are used in multiple social contexts (Miller, 1991; Tyack and Miller, 2002). Many morphological specializations for sound production occur in marine mammals but have evolved to serve general communicative functions, not functions related to territoriality *per se* (discussed later).

Optical signaling, likewise, is complex in territorial marine mammals and involves numerous morphological features. The distinctive appearance of adult male otariids is communicatively important in the context of territoriality. For example, in contrast to females or young males, adult male Hooker's and California sea lions are much darker than females or young males, and adult male southern sea lions have especially distinctive manes (Fig. 10B; Miller, 1991). Many other morphological adaptations for communication occur in marine mammals, including optical patterns of the pelage, sound-producing structures, and weapons (Miller, 1991; Tyack and Miller, 2002). Some structures may have become enlarged, strengthened, or otherwise modified for use in fights (e.g., claws of bearded seal; lower canines of otariids; Miller, 1991), and tusks (upper incisors) have evolved as optical display structures in the walrus. However, no morphological features have evolved specifically to serve territorial functions, as such features (e.g., tusks) have evolved independently in marine mammals with diverse non-territorial systems (Fig. 9A, B).

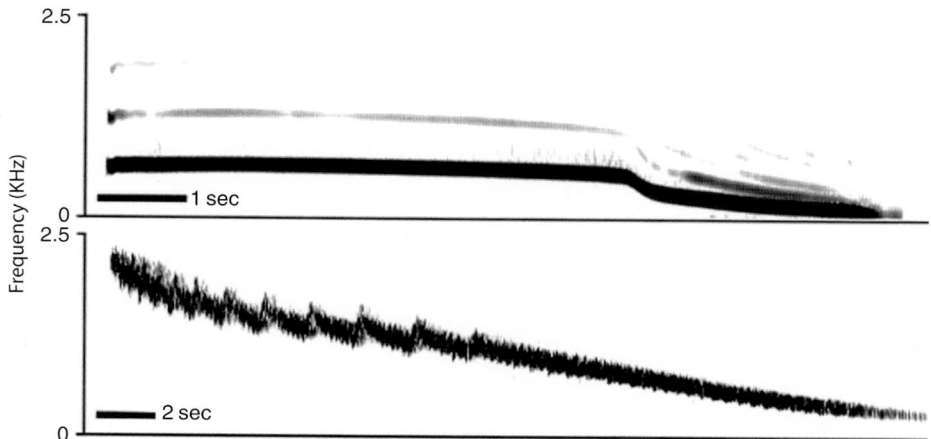

**Figure 8** *Adult males of some species of seals (Phocidae) utter sounds that travel for long distances under water, and function to advertise territorial occupancy. Spectrograms are shown for vocalizations of: (A) wintering territorial male Weddell seal* (Leptonychotes wed-dellii), *at McMurdo Sound, Antarctica; and (B) breeding male bearded seal* (Erignathus bar-batus), *at Svalbard. (A) Courtesy of J.M. Terhune; see Rouget et al. (2007); (B) courtesy of C.M. Lydersen and K.M. Kovacs.*

Some optical displays of otariids are relatively passive and undirected (e.g., the distinctive nose-up upright resting posture of otariids), but most optical displays are active and directed toward specific individuals. In otariids, these involve motor patterns such as distinctive head-and-neck swinging in locomotion, or rapid and complex sequences of feints, oblique stares, sprawls, facial expressions, etc., in displays between neighbors across territorial boundaries (Fig. 10A,B).

Chemical communication is probably important in all land-breeding territorial species of marine mammals but is virtually unstudied. Facial glands are unknown in sea otters, but the species has well developed anatomical (neural) characteristics for olfaction, and individuals often actively smell the air. Facial glands (many associated with vibrissae) occur in walruses, phocids, and otariids, and are known to vary seasonally in size and secretory activity in some species. Male otariids emit distinctive odors during boundary displays between neighbors, apparently from the oral cavity, and have distinctive body odors during the breeding season, but the anatomical source and functions of the smell are unknown. Breeding male ring seals hold underwater territories that are near or overlap with areas used by breeding females. During the breeding season, adult males of this species acquire a strong odor which is the origin of terms like "tiggak" ("stinker") among Inuit, and "gasoline seal" among trappers. Male ring seals may actively deposit secretions from facial glands on entrances to their breathing holes and sub-nivean ("below-snow") resting lairs, as well as within the latter.

The roles of taste or use of the vomeronasal organ (VNO) in marine mammal communication are unknown, although the VNO is absent in Sirenia but present and well developed in pinnipeds (see SENSORY BIOLOGY, OVERVIEW). Otariids commonly exhibit slow, repeated, tongue extrusion following agonistic displays (e.g., boundary displays or fights involving males; females, juveniles, and even pups also express the behavior), suggestive of behavior of other mammals that are known to use the VNO in chemical communication (at the very least, tongue extrusion is a conspicuous optical signal and likely has become ritualized as an optical display).

Tactile communication in marine mammals likewise is important but essentially unstudied (Miller 1991). Breeding males of all species engage physically and contact one another extensively in biting, wrestling, or pushing. Male sea otters and Weddell seals often try to bite the opponent's penis, and fractured bacula (penis bones) of mature male sea otters are relatively common (see "Baculum" article). In phocid species that fight aquatically, the rear flippers (necessary for aquatic locomotion) are frequently bitten and injured in fights; walruses use the tusks in fights at and below the water surface (all sex and age classes of the walrus use tusks in numerous other agonistic contexts, of course).

## VIII. Costs of Territoriality

Territoriality carries benefits and costs. In dispersed aquatically territorial species, most costs fall on territory holders, which must expend energy and time to establish, maintain, and advertise a territory, and must occasionally fight with other males. In fur seals and sea lions, some costs fall also on pups and on breeding females.

Territorial males must balance needs to be vigilant, advertise and defend territories, and signal about their own behavioral and physical attributes. Male sea otters hold large territories, and do not use long-distance advertisement, so must spend much time patrolling: on average, territory holders spent 17% of their time in this activity in Prince William Sound, Alaska (Pearson and Davis, 2005). In one long-term study, territorial male northern fur seals spent 1.5% of their time in overt interactions with neighbors; however, other types of activities also serve territorial functions, such as simply moving around the territory (1.7%), or resting in an upright position (21.1%; Gentry 1998). In the New Zealand fur seal, males spend about 14% of their time upright, compared with 5% for females (Miller, 1991).

The impact on males of time and energy spent in territorial activities is compounded by other factors: effects of fasting or reduced food intake; and time spent as a territory holder. On average, territorial male northern fur seals fast and do not drink for about a month (maximum 87 days), and decline in body mass by about one third

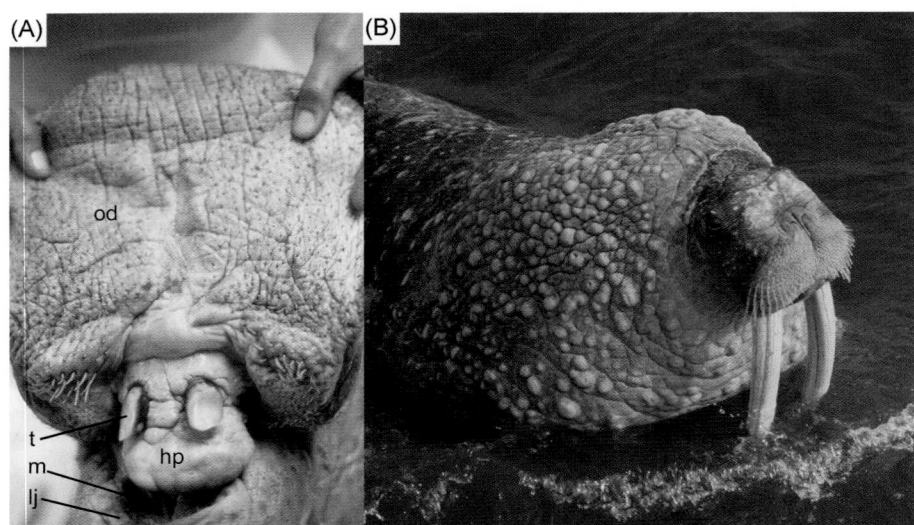

**Figure 9** *Specialized morphology has evolved in marine mammals to serve communication and other social functions, but occurs in diverse social systems and is not restricted to territoriality. (A) Male dugongs* (Dugong dugon) *possess tusks (upper incisors) that are used in fights between territorial males in Shark Bay, Western Australia, and between males elsewhere in the species' range, where the species is non-territorial (hp, horny pad on upper jaw; lj, lower jaw; m, mouth opening; od, oral disc, pulled forward to expose tusks; t, tusk). (B) Adult male walruses* (Odobenus rosmarus; *Alaska), which are territorial in parts of the range, develop conspicuous tubercles on the forequarters with age, which may act as an optical signal of male quality (as suggested for scarring in odontocetes by Macleod (1998). (A) photograph by P.K. Anderson, from figure 4 of Domning and Beatty (2007); (B) photograph by Bill Hickey, courtesy of U.S. Fish and Wildlife Service.*

over that time. This loss in body mass is equivalent to about 0.7% of initial body mass per day, similar to some other species: about 0.8% in the antarctic fur seal and the aquatically territorial Weddell seal. Breeding male harbor seals do not fast completely, and body mass declines more slowly (about 0.4% per day), but overall costs can be high (one male lost 30% of his initial body mass over the breeding season).

Costs and risks of territoriality include the danger of suffering severe physical injury. Most (80–90%) of antarctic fur seals found dead in a study on South Georgia had died as a direct or secondary consequence of fighting injuries, especially to the foreflippers or dorsal surface of the forequarters. Fitness of adult female and pup fur seals and sea lions also is affected by male behavior. For otariids, it is believed that female aggregation on land has been selected for, to reduce harassment by males and to protect pups from males (Chilvers *et al.*, 2005; Kiyota and Okamura, 2005). The significance of harassment is evident in the incidence of injuries and deaths in adult female Hooker's sea lions: greater than 80% of females have permanent scars inflicted by males, and annual mortality of adult females is about 0.5% due to male attacks (Chilvers *et al.*, 2005); infanticide by males also occurs (Wilkinson *et al.*, 2000).

Mortality rates of males are similar to those of females until social maturity is attained and males begin territorial activity. Male mortality increases at that time, and thereafter remains much higher than for adult females: 32% vs 15% in Galápagos fur seals (*Arctocephalus galapagoensis*); 30% vs 10–15% in northern fur seal; 30% vs 12% in Cape fur seals; and 50% vs 8% in Antarctic fur seals. This pattern

is presumed to be due largely to direct and indirect effects of male territoriality.

## References

Anderson, P. K. (2002). Habitat, niche, and evolution of sirenian mating systems. *J. Mammal. Evol.* **9**, 55–98.

Bartholomew, G. A. (1970). A model for the evolution of pinniped polygyny. *Evolution* **24**, 546–559.

Boness, D. J. (1991). Determinants of mating systems in the Otariidae (Pinnipedia). *In* "Behaviour of Pinnipeds" (D. Renouf, ed.), pp. 1–44. Chapman and Hall, London.

Boness, D. J., and James, H. (1979). Reproductive behavior of the grey seal (*Halichoerus grypus*) on Sable Island, Nova Scotia. *J. Zool.* **188**, 477–500.

Boness, D. J., Clapham, P. J., and Mesnick, S. L. (2002). Life history and reproductive strategies. *In* "Evolutionary Biology of Marine Mammals" (A. R. Hoelzel, ed.), pp. 278–324. Blackwell, London.

Boness, D. J., Bowen, W. D., Buhleier, B. M., and Marshall, G. J. (2006). Mating tactics and mating system of an aquatic-mating pinniped: The harbor seal, *Phoca vitulina. Behav. Ecol. Sociobiol.* **61**, 119–130.

Campagna, C., and Le Boeuf, B. J. (1988). Reproductive behaviour of southern sea lions. *Behaviour* **104**, 233–261.

Campagna, C., Le Boeuf, B. J., and Cappozzo, H. L. (1988a). Group raids: A mating strategy of male southern sea lions. *Behaviour* **105**, 224–249.

Campagna, C., Le Boeuf, B. J., and Cappozzo, H. L. (1988b). Pup abduction and infanticide in southern sea lions. *Behaviour* **107**, 44–60.

Chilvers, B. L., Robertson, B. C., Wilkinson, I. S., Duignan, P. J., and Gemmell, N. J. (2005). Male harassment of female New Zealand sea

**Figure 10** *Specialized communicative behavior has evolved in territorial marine mammals, but is based on general mammalian patterns that are not specific to territorial systems. (A and B) Ritualized boundary display; (A) followed by upright "full-neck display"; and (B) between neighboring territorial male Guadalupe fur seals* (Arctocephalus townsendi). *Note the long pelage on the foreguarters.* © *P. Colla.*

lions, *Phocarctos hookeri*: Mortality, injury, and harassment avoidance. *Can. J. Zool.* **83**, 642–648.

Domning, D. P., and Beatty, B. L. (2007). Use of tusks in feeding by dugongid sirenians: Observations and tests of hypotheses. *Anat. Rec.* **290**, 523–538.

Garshelis, D. L., Johnson, A. M., and Garshelis, J. A. (1984). Social organization of sea otters in Prince William Sound, Alaska. *Can. J. Zool.* **62**, 2648–2658.

Gentry, R. L. (1974). The development of social behavior through play in the Steller sea lion. *Am. Zool.* **14**, 391–403.

Gentry, R. L. (1998). "Behavior and Ecology of the Northern Fur Seal." Princeton University Press, Princeton.

Harcourt, R. G., Kingston, J. J., Cameron, M. F., Waas, J. R., and Hindell, M. A. (2007). Paternity analysis shows experience, not age, enhances mating success in an aquatically mating pinniped, the Weddell seal (*Leptonychotes weddellii*). *Behav. Ecol. Sociobiol.* **61**, 643–652.

Hardy, M. H., Roff, E., Smith, T. G., and Ryg, M. (1991). Facial skin glands of ringed and grey seals, and their possible function as odoriferous organs. *Can. J. Zool.* **69**, 189–200.

Hoffman, J. I., Boyd, I. L., and Amos, W. (2003). Male reproductive strategy and the importance of maternal status in the antarctic fur seal *Arctocephalus gazella*. *Evolution* **57**, 1917–1930.

Hoffman, J. I., Trathan, P. N., and Amos, W. (2006). Genetic tagging reveals site fidelity in territorial male Antarctic fur seals *Arctocephalus gazella*. *Mol. Ecol.* **15**, 3841–3847.

Hoffman, J. I., Forcada, J., Trathan, P. N., and Amos, W. (2007). Female fur seals show active choice for males that are heterozygous and unrelated. *Nature*, doi: 10.1038/nature05558.

Jameson, R. J. (1989). Movements, home range, and territories of male sea otters off central California. *Mar. Mamm. Sci.* **5**, 159–172.

Kiyota, M. (2005). Site fidelity, territory acquisition and mating success in male northern fur seals (*Callorhinus ursinus*). *Mammal Study* **30**, 19–27.

Kiyota, M., and Okamura, H. (2005). Harassment, abduction, and mortality of pups by nonterritorial male northern fur seals. *J. Mammal.* **86**, 1227–1236.

Kiyota, M., Insley, S. J., and Lance, S. L. (2007). Effectiveness of territorial polygyny and alternative mating strategies in northern fur seals, *Callorhinus ursinus*. *Behav. Ecol. Sociobiol.*, doi: 10.1007/s00265-007-0499-7.

MacLeod, C. D. (1998). Intraspecific scarring in odontocete cetaceans: An indicator of male 'quality' in aggressive social interactions? *J. Zool.* **244**, 71–77.

McCann, T. S. (1980). Territoriality and breeding behaviour of adult male Antarctic fur seal, *Arctocephalus gazella*. *J. Zool.* **192**, 295–310.

Miller, E. H. (1991). Communication in pinnipeds, with special reference to non-acoustic signalling. *In* "The Behaviour of Pinnipeds" (D. Renouf, ed.), pp. 128–235. Chapman and Hall, London.

de Muizon, C., and Domning, D. P. (2002). The anatomy of Odobenocetops (Delphinoidea, Mammalia), the walrus-like dolphin from the Pliocene of Peru and its palaeobiological implications. *Zool. J. Linn. Soc.* **134**, 423–452.

Pearson, H. C., and Davis, R. W. (2005). Behavior of territorial male sea otters (*Enhydra lutris*) in Prince William Sound, Alaska. *Aq. Mamm.* **31**, 226–233.

Pearson, H. C., Packard, J. M., and Davis, R. W. (2006). Territory quality of male sea otters in Prince William Sound, Alaska: Relation to body and territory maintenance behaviors. *Can. J. Zool.* **84**, 939–946.

Peterson, R. S. (1968). Social behavior in pinnipeds with particular reference to the northern fur seal. *In* "The Behavior and Physiology of

**T**

Pinnipeds" (R. J. Harrison, R. C Hubbard, R. S. Peterson, C. E. Rice, and R. J. Schusterman, eds), pp. 3–53. Appleton-Century-Crofts, New York.

Phillips, A. V., and Stirling, I. (2001). Vocal repertoire of South American fur seals, *Arctocephalus australis*: Structure, function, and context. *Can. J. Zool.* **79**, 420–437.

Robertson, B. C., Chilvers, B. L., Duignan, P. J., Wilkinson, I. S., and Gemmell, N. J. (2006). Dispersal of breeding, adult male *Phocarctos hookeri*: Implications for disease transmission, population management and species recovery. *Biol. Conserv.* **127**, 227–236.

Rouget, P. A., Terhune, J. M., and Burton, H. R. (2007). Weddell seal underwater calling rates during the winter and spring near Mawson Station, Antarctica. *Mar. Mamm. Sci.* **23**, 508–523.

Sjare, B., and Stirling, I. (1996). The breeding behavior of Atlantic walruses, *Odobenus rosmarus rosmarus*, in the Canadian High Arctic. *Can. J. Zool.* **74**, 897–911.

Stirling, I. (1983). The evolution of mating systems in pinnipeds. *In* "Advances in the Study of Mammmalian Behavior. American Society of Mammalogists, Special Publication No. 7" (J. F. Eisenberg, and D. G. Kleiman, eds), pp. 489–527. Allen Press, Lawrence.

Stirling, I., and Thomas, J. A. (2003). Relationships between underwater vocalizations and mating systems in phocid seals. *Aq. Mamm.* **29**, 227–246.

Stith, B. M., Slone, D. H., and Reid, J. P. (2006). "Review and synthesis of manatee data in Everglades National Park." USGS Administrative Report. USGS Florida Integrated Science Center, Gainesville, FL. 126 pp.

Tyack, P., and Miller, E. H. (2002). Vocal communication and anatomy in marine mammals. *In* "Evolutionary Biology of Marine Mammals" (A. R. Hoelzel, ed.), pp. 142–184. Blackwell, London.

Van Parijs, S. M. (2003). Aquatic mating in pinnipeds: A review. *Aq. Mamm.* **29**, 214–226.

Van Parijs, S. M., and Clark, C. W. (2006). Long-term mating tactics in an aquatic-mating pinniped, the bearded seal, *Erignathus barbatus*. *Anim. Behav.* **72**, 1269–1277.

Wedekin, LL., Daura-Jorge, F. G., Piacentini, V. Q., and Simões-Lopes, P. C. (2007). Seasonal variations in spatial usage by the estuarine dolphin, *Sotalia guianensis* (van Bénéden, 1864) (Cetacea; Delphinidae) at its southern limit of distribution. *Braz. J. Biol.* **67**, 1–8.

Wilkinson, I. S., Childerhouse, S. J., Duignan, P. J., and Gulland, F. M. D. (2000). Infanticide and cannibalism in the New Zealand sea lion, *Phocarctos hookeri*. *Mar. Mamm. Sci.* **16**, 494–500.

# Thermoregulation

## Michael Castellini

**M**arine mammals have developed methods to retain heat in cold seas using a suite of physiological, biochemical, anatomical, and behavioral methods. Yet, they must be able to lose excess heat when they are on land or extremely active in the water. If the problem was one of solely evolving methods to stay warm in a cold ocean, it would be easier to wrap themselves in deep blubber and fur or to stay very active and not to have deal with the consequences of heat loading. However, the difficulty of that solution is that the animal would get too warm, which in turn would cause problems with metabolic regulation, reproductive chemistry and neural function. Thus, the thermoregulatory mechanisms that have evolved in marine mammals function not only to conserve heat, but also to dump it when necessary. As poorly insulated humans, we must bring our artificial insulation with us and use exposure suits, wet suits and a variety of human-made insulative materials if we are to spend any significant time in the sea. For marine mammals, the insulation (blubber or fur) is already on board and serves multiple purposes beyond just thermoregulation.

## I. The Physics of Heat and Temperature

The terms "heat" and "temperature" are often incorrectly exchanged for one another, yet they have very different physical aspects. "Heat" is the energy that reflects the molecular motion of atoms and molecules. As energy, heat can flow from an area where the energy is high (an object that is "hot") to an area where the energy is low (something that is "cold"). We quantify how hot something is by using a variety of temperature scales (Kelvin, Celsius, and Fahrenheit). Thus, the temperature of an object is our definition of the level of heat energy contained by that object. The unit of heat energy is the calorie, and a single calorie is defined as the amount of heat necessary to raise 1g of water by 1°C. In common usage in the United States, the calorie associated with food and dieting is actually the kilocalorie (kcal; 1000 calories). In strict scientific terms, a single calorie is defined as 4.184 Joules (J).

As with any energy that flows, there is resistance that impedes the flow of that energy. In the field of thermoregulation, that resistance is insulation (the inverse of insulation is thermal conductance). Thus, poor conductors are excellent insulators. Blubber, for example, makes an excellent insulator and conducts heat poorly. Materials such as silver are poor insulators and thus conduct heat very well. Relatively, water is a better insulator than silver by 1000x. The point that is relevant to marine mammals is that air is a better insulator than water by 25x. In other words, water conducts heat away from a warm body 25x more effectively than air. This becomes an important point for the discussion of how fur works to help keep some marine mammals warm.

One final physics description is the definition of how heat flows from a warm to a cold object. Heat will flow when there is a temperature gradient between two sides of a conducting material in relation to the magnitude of the temperature gradient, the thickness of the material, the inherent thermal conductance of the material and the area that is exposed to the gradient. In biological terms, this means that heat would flow from the interior of a warm blooded mammal through the fat and skin to the cold outside air or water. Since water conducts heat 25x times more effectively than air, this means that heat flows out of a warm object in cold water much more efficiently than it does when that same object is in air. Thus, as humans, we can easily stand around outside in 70°F air, but would find being in 70°F water very cold after a while. This principle states that the thicker the insulator, the less heat flow and the larger the temperature gradient, the greater the driving force. Therefore, to stay warm, an animal would want an effective insulator, a small surface area (reduced appendage size, rolling up into a ball, etc.), a low temperature gradient (seek a warmer area or allow the body temperature to fall, i.e., hibernators) and have a thick insulator. An excellent general discussion of the physics of heat and energy transfer can be found in Schmidt-Nielsen (1997) and in Kooyman (1981) and Pabst *et al.* (1999) for marine mammals.

## II. What Is "Thermoregulation"?

Having discussed the physics of heat, the next step is to define the act of regulating temperature (thermoregulation) in the biological realm. In the broadest sense, animals can be classified as either *endotherms* or *ectotherms*, although some animals cross between those two stages. An endotherm is an animal that generates and controls its internal heat so that its body core temperature can be regulated at a level different than the ambient temperature. Birds and mammals are the most commonly cited examples of endotherms. By contrast, an ectotherm is an animal that allows its body temperature to mimic and follow the ambient temperature. Most fish and invertebrates are ectotherms. All marine mammals are endotherms and regulate their normal body temperature at about 37°C. Animals that hold their body temperature constant are called homeotherms while those that vary body temperature are called *heterotherms*. Most mammals are therefore homeothermic endotherms.

What is the importance of 37°C to a mammal? Perhaps the most critical point is that the actual value of 37°C is not as important as the constancy of that value. There are no biochemical or physical requirements about 37°C that make it the perfect temperature for a mammal, and most isolated biochemical reactions still work above and below that temperature. However, the mammalian body has evolved to balance its myriad of biochemical reactions at 37°C. If an animal gets much warmer or colder, the system comes out of equilibrium and there can be significant failures in metabolic regulation.

The goal of thermoregulation then, is to maintain a constancy of temperature by adjusting all the properties discussed earlier (the temperature gradient, the conductance, the surface area, etc.). Of course, for an endotherm, there is the additional factor that the animal is generating heat through metabolic processes. To maintain a constant body temperature for an endotherm, the produced heat must equal the heat lost to the environment. Put on too much of an insulator and the core temperature goes up if metabolic heat production stays constant. The good example of this concept is a human in Alaska wearing a thick, down jacket outside in the winter to stay warm but getting much too hot on cross country skis wearing that same down jacket. The analogy to marine mammals is straightforward: a whale or a seal may put on large amounts of blubber but as a consequence, overheat when extremely active. A human can simply take off the down jacket while exercising. A whale, however, does not have the option of taking off its blubber layer; it must be able to dump heat and therefore to thermoregulate using other methods.

There are several ways in which heat is transferred to the environment from a warm body. Evaporation is the process of dumping enough heat into a liquid to turn it into a gas. This is the process of cooling down by sweating. Radiation is the movement of heat through the release of electromagnetic energy from the warm body to the cold environment without physical contact (heat energy from the sun traveling through space and warming the earth). Conduction refers to the transfer of heat energy (calories) by physical contact between the warm body and the cold environment (putting your hand into cold water). Finally, convection is a specialized case of conduction where the heat that is transferred from the warm body is moved away from the area by a current of air or water. Thus, the environment provides an infinite sink for the heat (wind chill is the example in air or moving through very cold water as opposed to staying still in the water).

On the whole then, in order to maintain a constant body temperature, the heat that is generated in an endothermic mammal must be balanced by the heat lost or gained through radiation, evaporation, conduction, and convection. This is the fundamental equation of thermoregulatory biology.

There is a long and fascinating history to the study of thermoregulation in mammals. In the modern era, the study of arctic mammals and birds under cold conditions and the definition of the "thermoneutral zone (TNZ)" came about largely from the work of Laurence Irving and Per Scholander (Scholander *et al.*, 1950). The thermoneutral zone is the range of temperature over which an endotherm does not need to regulate its metabolism in order to maintain its body temperature constant. At the lower critical temperature, for example, a mammal would need to increase its heat production by shivering or increasing its activity in order to stay warm. The TNZ can be measured in marine mammals (Rutishauser *et al.*, 2004), but most of this theory was built on detailed studies of the thermal properties of terrestrial mammals (Hammel, 1955).

## III. Thermoregulation in Marine Mammals

So far, most of this discussion could be applied to any biological system, not just marine mammals. Is there anything unique about thermoregulation in this group of marine endotherms? King (1983), Riedman (1990), and Pabst *et al.* (1999) provide excellent summaries of the broad field of thermoregulation in this group of mammals.

A unifying characteristic of most marine mammals is that they spend a great portion of their lives, if not their entire lives, in a liquid environment that is significantly colder than their core temperature of 37EC. Based on the discussion earlier about the fundamental aspects of thermoregulation, it should be clear that this aquatic life represents a significant thermal challenge to these mammals. While radiation and evaporation are probably insignificant sources of heat loss while in the water, conduction and convection are massive. Yet, in the Antarctic, seals will move to relatively warm polar water (at −1.8°C) when the real or wind chill temperature outside falls below about −40°C. Clearly, there is a balance where the extreme cold of the ice covered water, even with its higher thermal conductance, represents less of a thermal challenge than being outside on the surface.

Marine mammals use either fur or blubber for insulation, and like all endotherms, balance their metabolic heat production with various pathways of heat loss. However, the uses of blubber or fur have their own biological costs. Although blubber is used for thermoregulation, it is also a primary source of metabolic fuel for a marine mammal and plays a role in buoyancy regulation. Blubber is a fairly unique tissue for marine mammals and not found outside of that group except for a similar tissue in polar bears (*Ursus maritimus*) and some penguins. Fur, however, is found in both terrestrial and marine mammals, and the highest quality (density) fur is found in the sea otter. Because fur traps air in its hairs, it is a very good insulator as long as it is carefully maintained, groomed, and kept dry on the layer next to the skin. Fur seals and sea otters will spend up to 12% of the daily energy use maintaining their fur coats.

### A. Heat Conservation and Generation

Marine mammals have no unusual heat generating mechanisms or tissues that are not seen in any other mammal. For example, while some large warm-bodied fishes have specialized heat generating tissues behind their eyes, no such organs or tissues exist in marine mammals. Some old data suggest that marine mammals may have an elevated metabolic rate for their mass, but this theory is not generally currently accepted (Lavigne and Kovacs, 1988) with the possible exception of the sea otter. The only heat generating specialized tissue that has ever been found in marine mammals is brown fat in harp seal (*Pagophilus groenlandicus*) pups (Blix *et al.*, 1979). This tissue is thermogenically active via oxidation of lipid compounds, but only for about the first 3 days after birth. This is an important source of heat for these young pups, but not unique since brown fat is found in other terrestrial mammals where it serves the same purpose. As

**T**

noted earlier, marine mammals also have a typical mammalian body core temperature. In fact, upon close examination of the data, there appears to be nothing special about marine mammals that would distinguish them from terrestrial mammals when it comes to heat generating mechanisms or abilities.

Given the particularly nondescript aspects of marine mammal heat generation, there must be something that is different about them since they can live in an extremely cold liquid environment that would be fatal to all terrestrial mammals. Again, given the fundamental balance equation of thermoregulation, this suggests that they must have adapted significant ways to alter the heat loss through reduced conduction and convection. They have done this through the use of blubber, fur, and vascular adaptations.

## B. Blubber

Blubber is often incorrectly assumed to an inert fat layer beneath the skin. However, it actually is a complex, active tissue that consists of a loose, spongy material where the matrix of the sponge is made up of collagen fibers and the volume is made of adipocytes (fat or lipid cells). As the blubber layer increases or decreases, the collagen matrix remains the same and it is the movement of lipid in and out of that matrix which accounts for the change in blubber quality and characteristics. However, all blubber is not the same and it varies from species to species in terms of the ratio of collagen to lipid and it can even vary within the same animal from location to location or with depth. Blubber depth can range from just a millimeter or two in newborn pinniped pups to 50-cm thick in large whales. The key issue here is that blubber, by itself, is a good insulator since it can be up to 93% lipid with very little water content and has roughly the thermal conductance of asbestos. Because lipid has a conductance of only about one-third that of water, it acts as a relatively good insulator. Because blubber is deposited below the skin, it acts as an internal insulator for marine mammals. Therefore, the skin layer itself will be only marginally warmer than the surrounding water. In polar waters for example, the skin of a whale or a seal would be just a degree or two above freezing while the core temperature would remain about 37°C.

In addition to varying between species and with location, blubber quality or thickness can also vary across time in the same animal. This can be seen in the significant seasonal variation in blubber thickness in a seal as it moves between the breeding season (where it is fattest) and the leaner periods associated with molting and mating. For example, northern elephant seals (*Mirounga angustirostris*) can range between 50% and less than 20% body fat depending on the season. Clearly, this temporal change in blubber impacts not only thermoregulation, but also buoyancy and energy reserves during periods of fasting or lactation. Consequently, the role of blubber and its relative thickness as an indicator of nutritional condition has been followed quite closely in studies that seek to address the population health of marine mammals. If a population of marine mammals is nutritionally compromised, one would hypothesize that the blubber layer should be reduced due to consumption of the blubber as a fuel source.

Blubber should be thought of as a very dynamic tissue with multiple stressors and pressures on its biology. Because it is a critical tissue for several different processes in marine mammals, it cannot be modeled in a strictly thermal scenario. For example, during a time of fasting, the animal will be utilizing blubber heavily, which would be inconsistent if it was also being challenged with an increasing thermal demand. Hence, fasting periods associated with breeding occur in warmer months or in warmer water for most marine mammals.

Rosen and Renouf (1997) and Ryg *et al*. (1988) have written about the relationships between blubber seasonal distribution and thermal problems in seals.

## C. Fur

As with terrestrial mammals, fur in marine mammals functions by trapping dry air next to the skin and keeping water (or cold air for a land mammal) away from the skin surface. Thus, the temperature gradient is from the skin outwards with a warm skin surface and cold outer layers of the fur (Boyd, 2000). The most cited example of the use of fur by a marine mammal is that of the sea otter and it provides an excellent example of how this animal lives in a cold environment (Williams *et al*., 1992). The sea otter is faced with a major thermal challenge, as it is a small mammal (large surface area to volume ratio through which to lose heat). It utilizes a dense fur with a series of guard hairs and under furs to keep its skin warm. However, the cost of this luxurious fur coat is a tremendous amount of maintenance with up to 12% of daily energy expenditure being spent on grooming the coat.

Many species of seals utilize blubber for thermal protection as adults but will use a specialized fur, called lanugo, as newborns. Lanugo, or pup fur, is a very effective insulator in the air and is usually both long and very "fluffy." On newborn pups, it functions as protection against the cold air during the time that they are on land or ice for nursing. Lanugo is useless in water and allows the skin to chill to essentially water temperature. A pup must shed its lanugo and develop a significant blubber layer before it can enter the water and be an effective swimmer and diver. Not all species of seal or sea lion pups are born with lanugo, but its purpose is well documented in many cases. Lavigne and Kovacs (1988) provide an excellent description of the first few days of life for harp seals as they adapt from the warm temperature inside the womb to the icy cold of being born on the ice. McCafferty *et al*. (2005) discuss the thermoregulatory problems faced by gray seal (*Halichoerus grypus*) pups while still at their nursing age and Dunkin *et al*. (2005) examined the thermal properties of blubber during development in dolphin calves.

It is the reliance on a high-quality fur in the sea otter and fur seals that makes these mammals particularly vulnerable to oil spills. Oil permeates the fur and destroys the air pockets that provide the thermal insulation for the animal. After the *Exxon Valdez* oil spill (EVOS) in Alaska, there was a massive clean-up operation on the hundreds of sea otters that were brought to rescue and rehabilitation centers. The goal was to clean the fur to restore its thermal insulation properties. However, cleaning the fur of human-made oils also cleans the fur of the natural oils (primarily squalene) that help make the fur water resistant. Therefore, small amounts of lipid had to be added and groomed back into the fur of the otters after they were cleaned of the heavy oil. For a general summary of the impact of the EVOS event on marine mammals, see Loughlin (1994); and for a detailed discussion on otters, see Williams and Davis (1995).

## D. Vascular Adaptations

It is in the area of vascular thermoregulation that marine mammals have evolved several unusual adaptations. The first of these is termed the *rete mirabile*, which is Latin for a "wonderful net." This net, which is a counter-current heat exchanger (Scholander and Schevill, 1955) involves an intertwined network of veins and arteries such that the cold blood returning from the extremities in the veins runs next to the warm blood going out to extremities in the arteries. From the previous discussion on heat flow, it is easy to see how the heat flows from the arteries to the close-by veins thus tending to conserve the heat in the interior and cool the arterial blood going out to the colder regions of the body. Marine mammals have exquisite control of blood flow in their body not only for thermoregulation but also for diving. However, these two demands are themselves inter-related, and the control of one impacts the control of the other. For example, it would do no good for a diving seal to be closely controlling blood flow for oxygen conservation but then to over-ride that control to dump or gain heat. In fact, Elsner and Gooden (1983) discussed some experiments with seals where the diving response inhibited thermoregulatory-driven circulatory adjustments. In another innovative study, divers were able to apply heat flow probes to the skin of dolphins while both divers and dolphins were underwater. The results show that the animals tend to defer heat regulation and favor oxygen conservation vascular adjustments when both must coincide (Noren *et al.*, 1999; Williams *et al.*, 1999).

These circulatory retes are found in several locations in marine mammals (and in some cold adapted birds), with the most cited examples being in the flukes of whales and the flippers of pinnipeds (Tarasoff and Fisher, 1970; Kvadsheim and Folkow, 1997; Meagher *et al.*, 2002). There has been the fascinating description of another rete, but in this case, the rete is used to cool down the reproductive organs of dolphins and seals by bringing in cold blood from the extremities (Rommel *et al.*, 1995).

Another important vascular adjustment seen in marine mammals deals with those mammals that utilize thick blubber as an insulating material. As mentioned several times earlier, having thick blubber is a good technique for staying warm, but can cause serious problems if trying to cool. In fact, the large whales have such a tremendous thermal mass and a low surface area to volume ratio that they may have a much more serious problem dumping heat than conserving it (Hokkanen, 1990). While some fur-bearing marine mammals have been shown to have sweat glands (Rotherham *et al.*, 2005) as a method of dumping heat, these would not be functional underwater. Of greater significance for thermoregulation is that the blubber is not just an inert organic blanket surrounding the animal but is instead vascularized with a series of anastomoses or blood flow shunts. These shunts control the amount of blood moving through the blubber and reaching the skin, thereby controlling the amount of heat lost to the environment. If a seal needs to dump heat, the anastomoses open and warm blood can reach the surface of the skin. When Weddell seals (*Leptonychotes weddellii*) in the Antarctic dump excess heat in this manner, clouds of steam come off the animal as the blood reaches the surface of their skin. In some cases, the seals get so warm that they partially melt their way into the ice and leave perfect "seal shadows." Conversely, when these circulatory shunts are closed, the same seals will be completely covered in snow with no signs of melting at any location except near the eyes and nose.

As mentioned earlier, the balance of blood flow throughout the body of marine mammals can be complex and is controlled by multiple demands: diving, exercise, and heat regulation. Diving requires limited blood circulation, simultaneous underwater exercise requires increased circulation and thermoregulation can require both. How these animals balance those conflicting demands is an area where much more work needs to be done. This can be seen in even simple manipulations of seals and sea lions. When taking blood samples from the flippers of pinnipeds, the flippers must be warm or there is no blood flow out to the periphery. However, if anesthesia or sedation is required to work with the animal, those procedures may also cause a series of vascular adjustments and can dump great amounts of heat quickly. Under these conditions, externally generated heat needs to be added to the animal to keep the core temperature up and blood flow open to the flippers.

The balance between diving and thermoregulation has another interesting aspect if one looks at it from the point of view of diving physiology. One of the central demands for diving is that oxygen must be conserved in order to extend the dive. This can be done by a variety of means and one of those means is to reduce the demand for oxygen by reducing metabolic rate. If a marine mammal were to reduce its body temperature while diving, it would decrease the demand for oxygen, thus extending dive time. There is some evidence from freely diving pinnipeds suggesting that the animals can drop their core temperatures during diving and would thus gain some diving time by reducing metabolic rate. The exact mechanisms by which this is done are not yet known, but temperature drops have been described in freely diving Weddell seals (Kooyman *et al.*, 1980; Hill *et al.*, 1987) and northern elephant seals (Andrews, 1999).

Advances in instrumentation that measure heat flux in free-ranging marine mammals have allowed a suite of experiments on regional measurements of heat flow across the body of the animal under experimental or field conditions. These include studies on body spatial variation in heat flux (Willis *et al.*, 2005), the use of remote infrared thermography to test how instrument attachment to animals may alter heat flow (Mauck *et al.*, 2003; McCafferty *et al.*, 2007) and novel methods for the attachment of heat sensors to marine mammals (Willis and Horning, 2005).

### E. Behavioral Thermoregulation

Most of the mechanisms discussed earlier are biochemical, anatomical, or physiological mechanisms for regulating heat production or loss in a marine mammal. Of course, a marine mammal is not a static system and the animal can alter the demands placed upon it with behavioral modification. For example, sea otters are often seen floating with all four paws out of the water. The paws are highly vascularized but not well insulated with fur. Thus, they would be a tremendous source of heat loss if in contact with the water. The otters keep their paws away from the water if they are trying to stay warm. Similarly, it is not unusual to see rafts of California sea lions (*Zalophus californianus*) floating on the ocean surface with their large fore-flippers extended out of the water. On the beach, both seals and sea lions will move up or down the tidal zone area to either cool off or warm up. When too hot, sea lions will maximize their surface area by spreading out their flippers while if too cold, they will lie on top of their flippers (Beenijes, 2006). As discussed earlier, Weddell seals will head to the water if the actual or convective temperature drops below about −40°C. However, elephant seals will flip cool sand onto their backs to help keep their body temperature down on sunny days and Hawaiian monk seals (*Monachus schauinslandi*) will find shade under bushes or in small ravines out on hot, sandy atolls. These behavioral mechanisms are not unique to marine mammals, except that they have the ability to use the sea to cool down as necessary. A good example of both feeding and thermoregulation are the humpback whales (*Megaptera novaeangliae*) that come into cool Alaskan waters during the summer for feeding, but head south to warm, Hawaiian waters for breeding. Similarly, it has been suggested that thermoregulatory constraints may influence the timing of pupping for seals (Hind and Gurney, 1998). A review of many of these behavior patterns for pinnipeds is found in King (1983).

**T**

## IV. Summary

What are the essential elements of thermoregulation in marine mammals? Like all endotherms, these mammals must obey the physics of heat balance when holding body temperature constant. The methods for producing heat (resting metabolism and exercise) must balance the windows for heat loss (primarily conduction and convection) (Whittow, 1987). Because marine mammals do not appear to have any special adaptations for producing excess heat, most of their ability to thermoregulate comes with their ability to control heat loss. Control of these heat loss mechanisms are via biochemical, anatomical, physiological, and behavioral means. However, as in all levels of adaptation to the environment, systems cannot be considered or modeled in isolation. For example, balancing blood flow for thermoregulation while also controlling blood flow for diving is an excellent example of multiple demands being put on this system.

It is easiest to observe the behavioral means that marine mammals use to stay warm or to cool down: the movement up or down a beach with the tide, the use of shade, flipping of sand, swimming to warmer or colder water, exposing flippers, and so on. Behind all of these behavioral patterns are the physiological or anatomical mechanisms that make the behavioral patterns effective. Counter-current heat exchangers, blood shunts under the blubber, and even the chemistry of the blubber and the microstructure of the fur are all part of the thermoregulatory system. Ultimately, however, we are still left with the paradox of heat balance in marine mammals: they live in a cold, thermally challenging environment that no terrestrial mammal could survive. However, the very means they have utilized to stay warm in cold seas have come at a cost: for many species, they have had to also evolve the means to get rid of the excess heat. The exquisite balance between all these competing demands and systems is what makes the study of thermoregulatory biology in these mammals such a rewarding experience.

## *References*

Andrews, R. A. (1999). Cardio-respiratory metabolism and thermoregulatory physiology of juvenile northern elephant seals (Mirounga angustirostris). Ph.D. Thesis, University of British Columbia, Vancouver, BC.

Beenijes, M. P. (2006). Behavioral thermoregulation of the New Zealand sea lion (Phocarctos hookeri). *Mar. Mamm. Sci.* **22**(2), 311–325.

Blix, A. S., Grav, H. J., and Ronald, K. (1979). Some aspects of temperature regulation in newborn harp seal pups. *Am. J. Physiol.* **236**(3), R188–R197.

Boyd, I. L. (2000). Skin temperatures during free-ranging swimming and diving in Antarctic fur seals. *J. Exp. Biol.* **203**(12), 1907–1914.

Dunkin, R. C., McLellan, W. A., Blum, J. E., and Pabst, D. A. (2005). The ontogenetic changes in the thermal properties of blubber from Atlantic bottlenose dolphin *Tursiops truncatus*. *J. Exp. Biol.* **208**, 1469–1480.

Elsner, R., and Gooden, B. (1983). "Diving and asphyxia. Monographs of the Physiological Society. Number 40." Cambridge University Press, Cambridge.

Hammel, H. T. (1955). Thermal properties of fur. *Am. J. Physiol.* **182**(2), 369–376.

Hill, R. D., Schneider, R. C., Liggins, G. C., Schuette, A. H., Elliot, R. L., Guppy, M., Hochachka, P. W., Qvist, J., Falke, K. J., and Zapol, W. M. (1987). Heart rate and body temperature during free diving of Weddell seals. *Am. J. Physiol.* **253**, R344–R351.

Hind, A. T., and Gurney, W. S. (1997). The metabolic cost of swimming in marine homeotherms. *J. Exp. Biol.* **200**(3), 531–542.

Hind, A. T., and Gurney, W. S. (1998). Are there thermoregulatory constraints on the timing of pupping for harbour seals? *Can. J. Zool.* **76**(12), 2245–2254.

Hokkanen, J. E. (1990). Temperature regulation of marine mammals. *J. Theor. Biol.* **145**(4), 465–485.

King, J. E. (1983). "Seals of the World." Cornell University Press, Ithaca.

Kooyman, G. L. (1981). "Weddell Seal: Consummate Diver." Cambridge University Press, Cambridge.

Kooyman, G. L., Wahrenbrock, E. A., Castellini, M. A., Davis, R. W., and Sinnett, E. E. (1980). Aerobic and anaerobic metabolism during voluntary diving in Weddell seals. Evidence of preferred pathways from blood biochemistry and behavior. *J. Comp. Physiol.* **138**, 335–346.

Kvadsheim, P. H., and Folkow, L. P. (1997). Blubber and flipper heat transfer in harp seals. *Acta Physiol. Scand.* **161**(3), 385–395.

Lavigne, D. M., and Kovacs, K. M. (1988). "Harps and Hoods: Ice Breeding Seals of the Northwest Atlantic." University of Waterloo Press, Waterloo, Ontario, Canada.

Loughlin, T. R. (ed.) (1994). "Marine Mammals and the Exxon Valdez." Academic Press, San Diego.

Mauck, B., Bilgmann, K., Jones, D. D., Eysel, U., and Dehnhardt, G. (2003). Thermal windows on the trunk of hauled-out seals: Hot spots for thermoregulatory evaporation? *J. Exp. Biol.* **206**(10), 1727–1738.

McCafferty, D. J., Moss, S., Bennett, K., and Pomeroy, P. P. (2005). Factors influencing the radiative surface temperature of grey seal (*Halichoerus grypus*) pups during early and late lactation. *J. Comp. Physiol. B, Biochem. Syst. Environ. Physiol.* **175**(6), 423–431.

McCafferty, D. J., Currie, J., and Sparling, C. E. (2007). The effect of instrument attachment on the surface temperature of juvenile grey seals (*Halichoerus grypus*) as measured by infrared thermography. *Deep Sea Res. Part II Top. Stud. Oceanogr.* **54**(3–4), 424–436.

Meagher, E. M., McLellan, W. A., Westgate, A. J., Wells, R. S., Frierson, D., and Pabst, D. A. (2002). The relationship between heat flow and vasculature in the dorsal fin of wild bottlenose dolphins *Tursiops truncatus*. *J. Exp. Biol.* **205**(22), 3475–3486.

Noren, D. P., Williams, T. M., Berry, P., and Butler, E. (1999). Thermoregulation during swimming and diving in bottlenose dolphins, *Tursiops truncatus*. *J. Comp. Physiol. (B)* **169**(2), 93–99.

Pabst, D. A., Rommel, S. A., and McLellan, W. A. (1999). The functional morphology of marine mammals. *In* "Biology of Marine Mammals" (J. E. Reynolds, and S. A. Rommel, eds), pp. 15–72. Smithsonian Institution Press, Washington, DC.

Riedman, M. (1990). "The Pinnipeds. Seals, Sea Lions and Walruses." University of California Press, Berkeley.

Rommel, S. A., Early, G. A., Matassa, K. A., Pabst, D. A., and McLellan, W. A. (1995). Venous structures associated with thermoregulation of phocid seal reproductive organs. *Anat. Rec.* **243**(3), 390–402.

Rosen, D. A. S., and Renouf, D. (1997). Seasonal changes in blubber distribution in Atlantic harbor seals: Indications of thermodynamic considerations. *Mar. Mamm. Sci.* **13**(2), 229–240.

Rotherham, L. S., van der Mewe, M., Bester, M. N., and Oosthuizen, W. H. (2005). Morphology and distribution of sweat glands in the Cape fur seal, *Arctocephalus pusillus pusillus*. *Aust. J. Zool.* **53**(5), 295–300.

Rutishauser, M. R., Costa, D. P., Goebel, M. E., and Williams, T. M. (2004). Ecological implications of body composition and thermal capabilities in young Antarctic fur seals (*Arctocephalus gazella*). *Physiol. Biochem. Zool.* **77**(4), 669–681.

Ryg, M., Smith, T. G., and Øritsland, N. A. (1988). Thermal significance of the topographical distribution of blubber in ringed seals (*Phoca hispida*). *Can. J. Fish. Aquat. Sci.* **45**, 985–992.

Schmidt-Nielsen, K. (1997). "Animal Physiology. Adaptation and Environment." Cambridge University Press, Cambridge.

Scholander, P. E., and Schevill, W. E. (1955). Countercurrent vascular heat exchange in the fins of whales. *J. Appl. Physiol.* **8**, 279–282.

Scholander, P. F., Hock, R., Walters, V., and Irving, L. (1950). Adaptation to cold in arctic and tropical mammals and birds in relation to body temperature, insulation and basal metabolic rate. *Biol. Bull.* **99**(2), 259–271.

T

Tarasoff, F. J., and Fisher, H. D. (1970). Anatomy of the hind flippers of two species of seals with reference to thermoregulation. *Can. J. Zool.* **48**(4), 821–829.

Whittow, G. C. (1987). Thermoregulatory adaptations in marine mammals: Interacting effects of exercise and body mass. A review. *Mar. Mamm. Sci.* **3**(3), 220–241.

Williams, T. D., Allen, D. D., Groff, J. M., and Glass, R. L. (1992). An analysis of California sea otter pelage and integument. *Mar. Mamm. Sci.* **8**(1), 1–18.

Williams, T. M., and Davis, R. W. (eds) (1995). "Emergency care and rehabilitation of oiled sea otters: A guide for oil spills involving fur-bearing marine mammals." University of Alaska Press, Fairbanks.

Williams, T. M., Noren, D., Berry, P., Estes, J. A., Allison, C., and Kirkland, J. (1999). The diving physiology of bottlenose dolphins (*Tursiops truncatus*) – III. Thermoregulation at depth. *J. Exp. Biol.* **202**(20), 2763–2769.

Willis, K., and Horning, M. (2005). A novel approach to measuring heat flux in swimming animals. *J. Exp. Mar. Biol. Ecol.* **315**(2), 147–162.

Willis, K., Horning, M., Rosen, D. A. S., and Trites, A. W. (2005). Spatial variation of heat flux in Steller sea lions: Evidence for consistent avenues of heath exchange along the body trunk. *J. Exp. Mar. Biol. Ecol.* **315**(2), 163–175.

# Tool-Use in Wild Bottlenose Dolphins

## Janet Mann and Brooke Sargeant

Tool-use is generally defined as the exertion of control over a freely manipulateable external object (the tool) with the goal of (1) altering the physical properties of another object, substance, surface, or medium (the target, which may be the tool user or another organism) via a dynamic mechanical interaction, or (2) mediating the flow of information between the tool user and the environment or other organisms in the environment (*sensu* Beck, 1980). Once considered the defining feature of hominids, tool-use is rare in wild animals. Although 10 primate species (van Schaik *et al.*, 1999; Breuer *et al.*, 2005) and 30 bird species (Lefebvre *et al.*, 2002) are known to use tools, only 0.01% of non-primate mammalian species have been documented using tools in the wild (Chevalier-Skolnikoff and Liska, 1993). In Shark Bay, Australia, a subset (11% of adult females) of the Indian Ocean bottlenose dolphin (*Tursiops aduncus*) population uses marine sponges as foraging tools (Mann and Sargeant, 2003). This itself is remarkable because tool-use is typically common to all or none of the individuals within a population. Thus, despite interaction between tool-users and non-tool-users in Shark Bay, only specific individuals adopt tool-use as a foraging method, indicating individual specialization (*sensu* Bolnick *et al.*, 2003) in the use of foraging tactics and probably prey species. Shark Bay is unlikely to be the only place where wild dolphins use tools. There are definitive anecdotal accounts of Indo-Pacific humpback dolphins (*Sousa chinensis*) carrying sponges in other parts of coastal Australia.

## I. History of Research on Dolphin Tool-Use

Rachel Smolker first discovered sponge carrying in 1984, when a dolphin named "Halfluke" was observed with a marine sponge on her rostrum (Smolker *et al.*, 1997). Subsequently several other adult

**Figure 1** *Grand-daughter of "Halfluke" with a sponge. Photograph by J. Mann.*

females were observed carrying sponges on a regular basis (referred to as "spongers"). The cone-shaped sponges, identified as *Echinodictyum mesenterinum*, fit over the dolphin's beak (Smolker *et al.*, 1997). In 1989, Halfluke's 2-year-old daughter, Demi, was observed carrying a small sponge, which she would drop when she moved into "infant position" under her mother (Smolker *et al.*, 1997). Thus, we suspected that the behavior was passed down from mother to offspring, particularly daughters, a pattern later confirmed through extensive behavioral observation (Mann and Sargeant, 2003) and mitochondrial DNA (Krützen *et al.*, 2005). Since 1984, we have been documented over 37 regular spongers in the Eastern Gulf of Shark Bay. Strong vertical transmission of sponging is an evident and has thus been characterized as a tradition (Mann and Sargeant, 2003; Sargeant *et al.*, 2007) or a culture (Krützen *et al.*, 2005). Although we argue that sponging is socially learned, there are strong ecological correlates, including use of deep water channels with high conical sponge density. Such areas are rarely used as much by non-sponging females (Sargeant *et al.*, 2007). Although the transmission mechanism is not known, local enhancement, where mothers provide sponge-habitat, qualifies as some form of social learning, and this is obviously the case for sponging. No individuals are known to develop the tactic unless their mother is also a sponger. Sponge carrying has also been documented in the Western Gulf of Shark Bay, quite far from the Eastern Gulf site. To date, those females are of different maternal lineage than the females in the Eastern Gulf (Krützen and van Schaik, 2007), suggesting at least two independent innovations of the behavior.

## II. Description of the Behavior

In several 100 hours of focal observation of spongers and over 1000 focal hours of non-sponging females, the pattern has proved to be remarkably consistent. Spongers tend to be more solitary than other females in the population. This is largely because they spend more time foraging than other females and over 95% of that foraging time involves sponges (Mann and Sargeant, 2003). Thus, spongers are clearly specialists because it occupies the majority of their foraging budget. Females have only been observed finding sponges in deeper water (>7 m, deep for Shark Bay), and they change sponges approximately every 2–3 h. Typically the base of the sponge, which grows out of rock or shell to anchor it, is left on the sea floor. The sponge is torn from the base, and this usually takes less

T

than 2 min. Searching for sponges (slow swimming with repeated U-turns before diving to the bottom) typically lasts fewer than 20 min. Sponges range in size from about 10–25 cm from base to top and cup the jaw completely when worn (Fig. 1).

When foraging with sponges, dolphins travel in a meandering, undirected manner and nearly all dives are tail-out (fluke-up), reflecting relatively deep dives. Our observations clearly indicate that females are gently disturbing the sandy bottom with the sponge. When dolphins sponge, long dives with multiple breaths at the surface are interspersed with rapid single breaths or leaps, typically without the sponge, when prey chases appear to be underway. Dolphins surfaced with sponges after 76.2% of all dives while foraging with sponges. But, during rapid single-breath surfacings (typically leaps, always lasting less than 3 sec at surface; 10.3% of all surfacing bouts), females wore sponges only 28.2% of the time. During longer surfacings, dolphins carried sponges approximately 80% of the time. Thus, the sponge was used in the search phase and typically not during the chase. Since the sponge cups over the rostrum, it cannot be worn during capture or consumption. In detailed observations of three foraging spongers when water clarity was exceptional, individuals swam slowly along the sandy-bottom habitats with the sponge on, slightly and intermittently disturbing the seafloor. When prey were detected, the dolphin dropped the sponge, accelerated about 5–10 m and then probed the seafloor with her beak. Consistent with all other observations, rapid single breaths or leaps without the sponge were observed before returning to the same spot, indicating that prey may burrow in the sand. Occasionally we have seen spongers with fish at the surface. Fish are always swallowed quickly. Subsequently, the dolphin retrieves the sponge and begins the search process again. Field observations, photographs, and sponging by human divers (with the sponge cupped over one hand) suggest that the prey are small bottom-dwelling fish. Dolphins also transport sponges to foraging areas, and occasionally carry sponges into social groups for use after leaving the group. This indicates "intelligent tool-use" (Beck, 1980; Chevalier-Skolnikoff and Liska, 1993; van Schaik et al., 1999), characterized by planning and social transmission. Chimpanzees (*Pan troglodytes*), for example, carry sticks to termite mounds, transport nuts to areas with appropriate stones to hammer them, and learn socially, particularly by vertical transmission from mother to offspring (Whiten et al., 1999; Lonsdorf et al., 2004). Based on our knowledge of the behavior and substrate where it occurs, sponges are a foraging tool that likely both protects the rostrum from the gritty sand and shell and expands the surface area that can be searched. Fish are probably partially buried in sand and cannot be detected easily any other way, although underwater recordings indicate that echolocation is used intermittently during sponging.

### III. Sex Differences in the Development of Sponging

We documented recurrent sponging in 37 bottlenose dolphins: 26 females (including eight daughters of spongers), five males (four of unknown parentage, one son of a sponger), and six of unknown sex (four born to spongers, two with unknown parentage). Thus, 13 regular spongers (8 females, 1 male, 4 of unknown sex) were definitely born to regular spongers. Critically, no regular sponger is known to have a non-sponging mother. Eighty percent of daughters and 17% of sons born to sponger mothers became regular spongers, but only daughters have been observed using sponges before weaning. From long-term data on all 37 spongers, only two daughters, born to one sponger, failed to follow their mother's tradition. Although most males

did not adopt sponging, the age of onset for those that do is not known. Of offspring born to spongers that were observed in detail during focal follows (216 h, 23 calves: 7 females, 4 males, and 12 unknown sex), all seven daughters carried sponges as calves, typically by their second or third year, but no sons did. Of those not sexed, four carried sponges, one did not, four died too early to assess, and three were not observed enough to classify. Clearly, the explanation for why adult males rarely carry sponges cannot be that males adopt it as calves, but the behavior dwindles with age. With a 3–6 year period of dependency (Mann et al., 2000), social learning likely plays a role in the development of tool-use, but like chimpanzees (Lonsdorf et al., 2004), sex-biased learning is apparent.

It is likely that males do not adopt sponging because of the habitat and time-budget constraints. Male spongers have not been observed in detail, but all female spongers use this foraging technique exclusively and maintain a solitary lifestyle (>80% time alone or with dependent offspring). Adult males form long-term alliances essential for reproductive competition (Connor et al., 1999), with social demands seemingly inconsistent with the solitary demands of sponging. Males would have difficulty finding other sponging males to ally with, and even if they did, they would have little time available to search for females and compete with other male alliances. Further, the habitat constraints of sponging would limit them to consortships in sponging areas (deep water with high sponge density) where there are few females other than spongers. Genetic data suggest that most spongers in the Eastern Gulf are closely related (Krützen et al., 2005), and extreme assortative mating (incest) with female spongers would be problematic. Thus, sponging is unlikely to be advantageous to males. The sex bias in the development of foraging tactics is not limited to sponging. Females exhibit a far wider range of foraging tactics than males and daughters appear to maintain these behaviors long after weaning. How males know not to do what the mother does is not understood.

To date, Shark Bay bottlenose dolphins and sea otters (*Enhydra lutris*) appear to be the only marine mammals that engage in tool-use in the wild. Similarities between dolphins and sea otters include strong matrilineal transmission from mothers to daughters and individual specialization. However, sea otters also develop individual specializations in tool-use itself, with individuals using different tools (e.g., rocks or shells), techniques, and prey (Estes et al., 2003). Although Shark Bay dolphins have individually distinctive foraging profiles and vertical transmission of foraging behaviors appears to be widespread (Mann and Sargeant, 2003), when it comes to tool-use, only sponges have been used and we suspect that they all use sponges in similar ways to capture the same types of prey. In future research, we hope to identify more precisely the mechanisms of transmission, what prey they are eating, and whether this technique confers an adaptive advantage over other foraging tactics.

### See Also the Following Articles

Bottlenose Dolphins ■ Feeding Strategies and Tactics ■ Otters ■ Culture in Whales and Dolphins

### References

Beck, B. (1980). "Animal Tool Behavior." Garland STPM Press, New York.

Bolnick, D. I., Svanbäck, R., Fordyce, J. A., Yang, L. H., Davis, J. M., Hulsey, C. D., and Forister, M. L. (2003). The ecology of individuals: Incidence and implications of individual specialization. *Am. Nat.* **161**, 1–28.

T

Breuer, T., Ndoundou-Hockemba, M., and Fishlock, V. (2005). First observation of tool use in wild gorillas. *PLoS Biol.* **3**, 2041–2043.

Chevalier-Skolnikoff, S., and Liska, J. (1993). Tool use by wild and captive elephants. *Anim. Behav.* **46**, 209–219.

Connor, R. C., Heithaus, M. R., and Barre, L. M. (1999). Superalliance of bottlenose dolphins. *Nature* **397**, 571–572.

Estes, J. A., Riedman, M. L., Staedler, M. M., Tinker, M. T., and Lyon, B. E. (2003). Individual variation in prey selection by sea otters: Patterns, causes and implications. *J. Anim. Ecol.* **72**, 144–155.

Hall, K. R. L., and Schaller, G. B. (1964). Tool-using behavior of the California sea otter. *J. Mammal.* **45**, 287–298.

Krützen, M., and van Schaik, C. P. (2007). "The mother of all cultures: The vertical skill syndrome applied to bottlenose dolphins." Seventeenth Biennial Conference on the Biology of Marine Mammals. Cape Town, South Africa November 29 to December 3. Abstract.

Krützen, M., Mann, J., Heithaus, M. R., Connor, R. C., Bejder, L., and Sherwin, W. B. (2005). Cultural transmission of tool use in bottlenose dolphins. *PNAS* **102**, 8939–8943.

Laidre, K. L., and Jameson, R. J. (2006). Foraging patterns and prey selection in an increasing and expanding sea otter population. *J. Mammal.* **87**, 799–807.

Lefebvre, L., Nektaria, N., and Boire, D. (2002). Tools and brains in birds. *Behaviour* **139**, 939–973.

Lonsdorf, E. V., Eberly, L. E., and Pusey, A. E. (2004). Sex differences in learning in chimpanzees. *Nature* **428**, 715–716.

Mann, J., and Sargeant, B. L. (2003). Like mother, like calf. The ontogeny of foraging traditions in wild Indian Ocean bottlenose dolphins (*Tursiops* sp.). *In* "The Biology of Traditions: Models and Evidence" (D. Fragaszy, and S. Perry, eds), pp. 236–266. Cambridge University Press, Cambridge.

Mann, J., Connor, R. C., Barre, L. M., and Heithaus, M. R. (2000). Female reproductive success in bottlenose dolphins (*Tursiops* sp.): Life history, habitat, provisioning, and group size effects. *Behav. Ecol.* **11**, 210–219.

Mann, J., Sargeant, B. L., Gibson, Q. A., and Singh, L. (2007). "Is sponging a culture?" Seventeenth Biennial Conference on the Biology of Marine Mammals. Cape Town, South Africa November to December 3. Abstract.

Sargeant, B. L., Wirsing, A. J., Heithaus, M. R., and Mann, J. (2007). Can environmental heterogeneity explain foraging variation in wild dolphins (*Tursiops* sp.)? *Behav. Ecol. Sociobiol.* **61**, 679–688.

Smolker, R., Richards, A. F., Connor, R. C., Mann, J., and Berggren, P. (1997). Sponge carrying by dolphins (Delphinidae, *Tursiops* sp.): A foraging specialization involving tool use? *Ethology* **103**, 454–465.

van Schaik, C. P., Deaner, R. O., and Merrill, M. Y. (1999). The conditions for tool use in primates: Implications for the evolution of material culture. *J. Hum. Evol.* **36**, 719–741.

Whiten, A., *et al.* (9 authors) (1999). Cultures in chimpanzees. *Nature* **399**, 682–685.

# Toothed Whales, Overview

SASCHA K. HOOKER

The toothed whales comprise the suborder Odontoceti of the order Cetacea. This suborder includes 10 diverse families, 2 of which contain large numbers of species. There are at least 71 species in all, including the true dolphins, monodontids, river dolphins, porpoises, beaked whales, and sperm whales (Table I). These species occur in three primary clades, the superfamilies Delphinoidea (true dolphins, monodontids, and porpoises), Ziphoidea (beaked whales), and Physeteroidea (sperm whales), whereas the affinities of the river dolphins remain uncertain. With the exception of the sperm whale (males of which reach up to 18 m) and the larger beaked whale species (*Berardius* and *Hyperoodon* spp.), most odontocetes are small to medium-sized cetaceans, ranging in size from the Hector's dolphin (1.5 m) to the killer whale (8.5 m). These species show a range of distributions, with some such as river dolphins, found only in quite specific areas, whereas others such as sperm whales or killer whales show a global distribution.

## I. Diagnostic Characters and Taxonomy

Odontocetes and mysticetes differ fundamentally in three major ways: the way that the bones of the skull have become telescoped, the specialized echolocation system (and associated anatomy) of odontocetes, and the specialized filter-feeding mechanism of the baleen whales (Table II). The name Odontoceti derives from the Greek *odous* or *odontos* for "tooth," and *ketos* for "sea-monster," hence "toothed sea-monster," referring to the presence of teeth (Rice, 1998). In contrast, mysticetes do not possess teeth, but instead have baleen plates, which hang from the upper jaw and are used to filter small prey items from the water. However, although all odontocetes possess teeth, in some species (or sexes) these teeth are much reduced and may not erupt.

Other distinctive features include the possession of a single nares or blowhole, whereas mysticetes have two blowholes. Most odontocetes show some degree of dorsal asymmetry in their skull and facial soft tissue, whereas all mysticetes have a symmetrical skull and facial soft tissue. Odontocetes possess a large ovoid melon in the anterior part of the facial region. This fatty tissue is thought to be an important component of the echolocation system. Although mysticetes possess a fatty structure just anterior to the nasal passages, which may represent a vestigial melon, this is only a fraction of the size of that present in odontocetes.

Odontocetes (with the notable exception of the sperm whale) tend to be smaller in size than mysticetes, although there is some overlap. Odontocetes also show variation in sexual dimorphism. In some species, males are much larger than females (e.g., sperm whale and killer whale), whereas in others there may be reverse sexual dimorphism, in which females are larger than males (e.g., harbor porpoises and Baird's beaked whale). Among mysticetes, adult females are always slightly larger than adult males.

The skull and jawbone of odontocetes also contain a number of diagnostic characteristics. The odontocete mandible is symphyseal (the two jawbones lock together in a bony symphysis anteriorly) and each jawbone spreads into a fat-filled hollow pan at the posterior, non tooth-bearing section, whereas mysticete mandibles are non-symphyseal and are solid. When viewed from above, the odontocete jawbone is relatively straight, whereas the mysticete jawbone curves laterally. The maxilla of both odontocetes and mysticetes has "telescoped," migrating posteriorly to form the long rostrum and dorsal nasal openings. In the odontocetes, the maxillae have extended outwards over the orbits to form an expanded bony supraorbital process of the frontal bone. This process forms an anchoring point for the facial musculature associated with sound production (discussed later). In contrast, the maxillae of mysticetes project under the eye orbit and have developed a bony protuberance on the maxilla anterior to the eye orbit. Odontocetes lack this antorbital process of the maxilla. The earbone of odontocetes is also quite different from that of mysticetes. In odontocetes, the tympanic bulla and periotic bone are fused together, equal size, and not fused to the skull. The

**T**

## TABLE I
### Odontocete Families

| Family | Common names | No. of species |
|---|---|---|
| Kogiidae | Pygmy and dwarf sperm whales | 2 |
| Physeteridae | Sperm whale | 1 |
| Ziphiidae | Beaked whales | 21 |
| Delphinidae | True dolphins | 36 |
| Monodontidae | Narwhal and beluga | 2 |
| Phocoenidae | Porpoises | 6 |
| Iniidae | Boto (Amazon river dolphin) | 1 |
| Pontoporiidae | Fransiscana | 1 |
| Lipotidae | Baiji (Chinese river dolphin) | 1 |
| Platanistidae | South Asian river dolphin | 1 |
| | Total | 72 |

## TABLE II
### Major Differences between Odontocete and Mysticete Suborders of Cetaceans

| Odontocetes | Mysticetes |
|---|---|
| Teeth | Baleen plates |
| Single blowhole | Two blowholes |
| Dorsally asymmetric skull and facial tissue | Symmetrical skull and facial tissue |
| Presence of a melon | No melon |
| Variable size | Always large |
| Variation in sexual dimorphism | Always reverse sexually dimorphic |
| Symphyseal mandible | Nonsymphyseal mandible |
| Hollow pan bone of lower jaw | Lower jawbones solid, no pan bone |
| Maxillae project outward over expanded supraorbital processes | Maxillae project under the eye orbit, and possess bony protuberance anterior to the eye orbit |
| Tympanic and periotic fused and equal sized | Tympanic bulla much larger than periotic from which it separates |

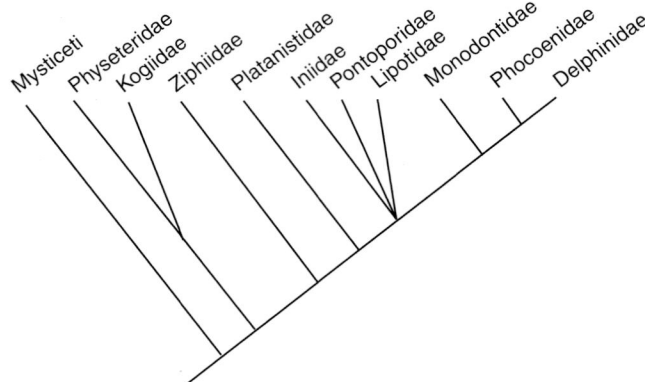

**Figure 1**  *Generally accepted phylogeny extant odontocete families, with mysticetes as sister group.*

sections for further details). Relationships among extant odontocetes are somewhat controversial, but there is a consensus as to the order of branching of the phylogenetic tree, despite controversies over smaller scale relationships (Fig. 1; see RIVER DOLPHINS, EVOLUTIONARY HISTORY).

Odontocetes first appeared in the fossil record in the late Oligocene, approximately 25 Mya (Rice, 1998). Odontocete diversity increased during the warm temperatures of the early Miocene, during which time the earliest ziphiids and platanistids appeared in the eastern North Pacific (see Chapter Cetacean Fossil Record for further details). Middle Miocene diversity was also high and included diversification of these families, together with extensive radiation of the delphinoids, while the platanistoids progressively declined. Modern odontocete families are known from the late Miocene of the eastern North Pacific (monodontids and phocoenids) and western South Atlantic (pontoporiids). Evolutionary trends among odontocetes included the expansion and increase in size of the face, shortening of the intertemporal region, elevation of the cranial vertex posterior to the nasals, increased facial asymmetry and isolation of the earbones from the skull resulting in the diagnostic features described earler. There has also been a trend toward either long, slender jaws with increased number of teeth, or short, robust jaws and reduced number of teeth.

## II. Distribution and Range

Different odontocete species can be found across a wide range of habitats in all oceans of the world. The cyclical changes in sea level over the Quaternary (Pleistocene to Recent) period are thought to be responsible for much of the recent speciation within odontocete families. Sea-level drops associated with cooling are suggested to have isolated populations which then speciated allopatrically. The distributions of modern odontocetes range from species found globally, such as sperm whales and killer whales, to those with more restricted coastal distributions, such as harbor porpoises. Some species are found only in polar regions, including the narwhal and beluga, others only in tropical waters, such as Fraser's dolphin (Jefferson *et al.*, 1993). Segregation by warm tropical waters created antitropical distributions of some species pairs now found only in the Northern and Southern Hemispheres respectively (e.g., the northern and southern right whale dolphin, northern and southern bottlenose whale).

Few odontocetes show the kind of long distance seasonal migration that is found among mysticetes. The only odontocete species

odontocete tympanic is thin-walled and conical, tapering anteriorly. In mysticetes, the tympanic bulla is much larger (nearly twice the volume of the periotic bone) and is thick-walled and spherical. Bony flanges of the periotic wedge it against the skull, such that only the large bulla can be removed.

The clear morphological differences between odontocetes and mysticetes suggest monophyly of the odontocetes. There has been some disagreement about this issue, based on some molecular sequence data that supported a closer relationship between the sperm whales and the baleen whales than between the sperm whales and the other toothed whales. However reanalysis of these data and more recent work on both morphologic and molecular data continue to support odontocete monophyly (see CETACEAN EVOLUTION which is known to show a large (ocean-basin) scale of movement is the sperm whale. At approximately 10 years of age, male sperm whales will leave their natal group in the tropics and subtropics and migrate to cold temperate and sub-polar waters where they feed. Once physically mature they return to the tropics to breed. Other odontocetes may show much smaller migrations as they follow their prey movements (e.g., long-finned pilot whales in the North Atlantic will move from shelf edge up onto shallower banks in response to changes in squid distribution). Other species (e.g., killer whales) show negligible seasonal changes in distribution.

Many odontocete species show variation across their ranges, leading to the definition of several subspecies or races. With increasing scrutiny of both genetic and morphological differences, subspecies and population level differences are continually being identified. In some cases, this is leading to revision of the currently established species (Rice, 1998).

## III. External Appearance

In general, many odontocete species tend to have darker coloration on the dorsal surface (back and cape) and lighter pigmentation on the ventral surface (Fig. 2). Such countershading is relatively common in the marine environment and functions essentially as camouflage, such that when viewed from above the dark dorsal surface is seen against the darker depths, and when viewed from below the lighter ventral surface is seen against the brighter downwelling light from the surface (e.g., harbor porpoise). Exceptions to this include the beluga, which has a uniform color pattern, although this is thought to be related to the bright ice-covered habitat that it lives in. The general tendency toward countershading has become quite elaborate in some species, which also have striped and saddle patterns (e.g., striped dolphins, short-beaked common dolphins), and has become radically modified in others (e.g., killer whale, Dall's porpoise). It has also been suggested that these color patterns may function in signaling between individuals in a group (Norris *et al.*, 1994), in addition to their function in concealment (from both predators and prey).

Odontocetes tend to show greater body flexibility than mysticetes. This is presumably due to differences in prey capture strategies

**Figure 2** *Short-beaked common dolphin* (Delphinus delphis) *pigmentation, shows countershading with dark pigmentation on dorsal surface and light pigmentation on the ventral surface, common to most odontocetes. Photograph by Sascha K. Hooker.*

between odontocetes, which chase fast, mobile, single prey, and mysticetes which engulf less mobile prey schools. This flexibility is quite pronounced in some odontocete species (e.g., belugas).

## IV. Diet and Feeding Strategies

Different odontocete species feed on fish, squid, large crustaceans, birds, and occasionally other marine mammals. They differ from mysticetes in that they generally chase, capture, and swallow single relatively large prey items, rather than filtering and swallowing large quantities of small prey items. This more generalized and more adaptable feeding method is thought to account for much of the diversity of odontocete species and their range of habitats. Many odontocetes also feed on much deeper prey than the surface water plankton diet of many species of mysticetes. Additionally, whereas the diet of baleen whales is highly seasonal due to the seasonal nature of zooplankton biomass and production, that of odontocetes is generally more constant year-round.

Unlike the heterodont condition of most terrestrial mammals, the teeth of odontocetes are uniformly shaped (homodont). There is a wide variation between (and some variation within) species in tooth number, size, and shape (see Dental Morphology). The teeth of most odontocete species tend to be peg-shaped with single, open roots. Exceptions to this include the porpoises, which have spade-shaped teeth, and the beaked whales, which show great variation in size, shape, and location of teeth in the jaw.

Narwhals have only two teeth, both in the upper jaw. In females, these teeth usually remain embedded in the upper jawbones, but in males the left tooth grows out through the front of the head into a tusk up to 3-m long. These tusks are thought to be involved in male–male competition. These are used primarily as a display, although males have been seen sparring with their tusks above the water. The teeth of beaked whales have similarly become adapted for use in male–male competition and generally erupt only in males. In some species, such as the dense-beaked whale (*Mesoplodon densirostris*), these teeth form large structures which protrude above the upper jaw. In the strap-toothed whale (*Mesoplodon layardii*), the teeth, which emerge from the middle of the lower jaw, curl backward and inward, extending over the upper jaw, often preventing it from opening more than a few centimeters.

The diet of a particular species is generally reflected in the morphology of the jaw, and the type and number of teeth. For example, species which feed primarily on fish tend to have more teeth (e.g., spotted dolphin, *Stenella attenuata*), and use these teeth for grasping single prey. Species which feed primarily on squid tend to show reduced dentition (e.g., sperm whales, beaked whales, narwhals, and Risso's dolphin), and are thought to feed by suction. This suction is achieved by using the tongue as a piston in combination with a small gape. The suction feeding mechanism of beaked whales was investigated in detail by Heyning and Mead (1996), and involves distension of the floor of the mouth by expanding the throat grooves together with retraction of the tongue by the styloglossus and hyoglossus muscles. Additionally, it has been suggested that some species may debilitate prey by directing high-intensity sounds at them, prior to capturing them (Norris and Mohl, 1983). Such feeding methods are thought to explain the occasional observation of healthy animals with severe deformity of the jaws (as has been noted in sperm whales).

Notable specializations associated with odontocete feeding are shown by several bottlenose dolphin populations. In Shark Bay, Western Australia, a small number of Indian Ocean bottlenose dolphin females have been observed carrying sponges (Mann *et al.*, 2000).

T

These are thought to function as "tools" to protect the dolphin's rostrum as it roots in bottom coraline sediments in order to flush bottom-dwelling fish out. Another bottlenose dolphin population, in North Carolina, USA, has been observed beaching themselves and the fish they were pursuing up onto the surrounding mudbanks, thus immobilizing their fish prey which they can easily catch, and then slide or wriggle back down into the water (Reynolds and Rommel, 1999). The same behavior has also been observed in killer whales in Patagonia and the Crozet Archipelago in their pursuit of seals up onto steeply shelving beaches.

Foraging specializations of killer whales in British Columbia, Canada, and Washington State, USA, waters, have resulted in two separate forms, "transients" which feed on marine mammals, and "residents" which feed on fish. These two forms are thought to potentially represent a case of incipient speciation. The primary difference between these two forms is in their dietary specialization, which has resulted in their social separation and behavioral, morphological, and genetic differences.

Many odontocetes appear to feed throughout the day and the night (e.g., sperm whales and northern bottlenose whales), but many smaller delphinids show marked diurnal differences in feeding (e.g., spinner dolphin and pantropical spotted dolphin). Since many prey species rise to shallower depths during the darkness of night, it is energetically more efficient for some species to conduct the majority of foraging behavior at night.

As mentioned previously, odontocetes tend to feed at greater depths than mysticetes. In general, however, detailed data on diving and foraging patterns of most species of odontocetes is not yet available. The use of novel technologies such as the attachment of time-depth recorders to monitor dive profiles of these species have lagged behind work done on pinnipeds or mysticetes, primarily due to difficulties in deployment, attachment and recovery (Hooker and Baird, 2001). However, new developments in these technologies and attachment mechanisms are increasingly resulting in the initiation of new studies.

## V. Sound Production and Reception

Toothed whales have developed specialized sound production and reception mechanisms for the use of biosonar. All modern odontocetes are thought to use echolocation, in the same manner as bats, to gain an "image" of their environment. Although only a few species of odontocete are unequivocally known to echolocate, all odontocetes known to produce pulse-like sounds in the wild are assumed to be able to echolocate.

The sound production mechanism of odontocetes consists of a sound generator located in the "monkey lips"/dorsal bursae (MLDB complex) associated with the upper nasal passages. In most odontocetes, there are two bilaterally placed MLDB complexes, but sperm whales have only a single complex. The central components of the MLDB complex are the fatty dorsal bursae, the monkey lips, the bursal cartilages, and the blowhole ligament (Cranford et al., 1996). Sounds are generated as air is forced between the monkey lips, setting the MLDB complex into vibration. Sound is propagated into the water by the melon, a low-density lipid filled structure which acts as an acoustic lens to focus a directional sound beam ahead of the animal (Au, 1993; see ECHOLOGATION). The short-duration clicks produced as a result of this are used primarily for echolocation, although some species appear to use these clicks in a social context (e.g., Hector's dolphin).

Most odontocete species produce broadband echolocation clicks in the ultrasonic sound range, well above the range of human hearing. The pulse duration, frequency, inter-click interval, and source level are adjusted by the animal for optimal performance according to the prevailing conditions of ambient noise, reverberation, target distance, and target characteristics (Au, 1993). With low ambient noise, bottlenose dolphins, belugas, and false killer whales often echolocate using frequencies from 20 to 60 kHz, although at higher noise levels, they emit stronger pulses at 100–130 kHz. Echolocation clicks of porpoises and many small dolphins are greater than 100 kHz, while those of the sperm whale range from less than 100 Hz to 30 kHz, with most energy from 2–4 kHz to 10–16 kHz (Richardson et al., 1995).

In addition to echolocation clicks and loud impulse sounds, many toothed whales also produce other sounds, usually described as whistles, squeals, or less distinct pulsed sounds such as cries, grunts, or barks (Richardson et al., 1995). These tend to be narrow band (sometimes pure tone), frequency modulated sounds, often with a harmonic structure. Most whistles tend to have most of their energy below 20 kHz. These whistles can show a variety of patterns of frequency and amplitude over the duration of the whistle. For many species, the frequency, duration, and level may vary. Observations of captive bottlenose dolphins have shown that individual animals can be identified from the contour of the whistle on a sonogram (a representation of the whistle as sound frequency plotted against time). The distinctive character of a whistle is thought to function in identifying the sound producer to other animals. These whistles are therefore known as signature whistles. Recent evidence also suggests that there may be population or group-level distinctions between whistles of bottlenose dolphins.

Group-specific sounds have also been found among killer whales and sperm whales. Among killer whales, groups appear to have a repertoire of approximately 10 calls. Different groups may share some but not all of the calls within their repertoire, and relationships between groups can be established based on the similarities in their repertoires. In a similar fashion, the codas (rhythmic patterns of clicks) produced by sperm whales also appear to be characteristic of the social group (Whitehead, 2003).

The high-frequency hearing of odontocetes is reflected in the structures of the jaw and ear. The lower jaw of odontocetes is flared out in a thin hollow pan bone at the rear. This is filled with a fat body that directly connects with the bulla of the middle ear. These fat bodies act as low-density sound channels to conduct sounds to the ears. Within the ear, the tympanic bulla is separated from adjacent bones of the skull by peritympanic sinuses filled with an insulating emulsion of mucus, oil, and air. The bulla is suspended in this emulsion

**TABLE III**
**Social Strategies Employed by Odontocete Species Known from Long-Term Studies of Wild Animals**

|  | Social groups | Mating system |
|---|---|---|
| Bottlenose dolphin | Male coalitions<br>Female groups | "Capture" of females by a coalition of males |
| "Resident" killer whales | Matrilineal groups of females and descendents | Inter-group mating thought to occur when groups meet |
| Sperm whale | Matrilineal groups of females and juvenile males; Solitary adult males | Males rove between groups in search of estrus females |

by connective tissue, so that the middle ear functions as a sound receiver isolated from the skull to better localize sound signals. The tympanic bulla membrane of odontocetes is also stiffened with bony ligaments, which appear to be associated with ultrasonic hearing (Reynolds and Rommel, 1999).

## VI. Social Organization and Culture

Toothed whales are particularly well known for their brain size and rich social lives (Mann *et al.*, 2000). The absolute brain size of odontocetes ranges from 840 g in common dolphins to 7820 g in sperm whales (Berta *et al.*, 2006). However, a more useful way to compare brain sizes is to use the ratio of brain size to body size, the encephalization quotient (EQ). The relative brain sizes of odontocetes (ranging from EQ 0.02% for the sperm whale to EQ 1% for the bottlenose dolphin) are much larger than most terrestrial mammals but are similar to those of anthropoid primates (EQ 0.3% in the gorilla). Captive studies of bottlenose dolphins have shown that dolphins may have a "pecking order" similar to that of chimpanzees, in which both males and females have a social hierarchy, but that generally males are dominant to females.

The social systems of only a few odontocete species are known from long-term studies of wild animals, but these suggest some novel adaptations to standard mammalian patterns (Table III; Connor *et al.*, 1998). In Shark Bay, Western Australia, male Indian Ocean bottlenose dolphins form tight associations with one or two other males with whom they co-operate to form aggressively maintained consortships with individual females. These alliances further appear to form moderately strong associations with one or two other alliances and will defend each other in competition over females. A completely different strategy is found among the fish-eating "resident" killer whales living along the coast of British Columbia, Canada, and Washington State, USA. Here, neither males nor females disperse at maturity, but instead they remain with their mothers in stable matrilineal groups. Genetic data suggest that a similar pattern of natal philopatry may exist among long-finned pilot whales. Sperm whales appear to have a different social structure again. Groups of females and immature males are found in the tropical lower latitudes. Juvenile males remain in these natal groups until they reach puberty, at which stage they depart to lead more solitary lives in colder temperate and polar waters. They later return to the tropics when fully mature and search for estrus females with which to mate. Long-term studies of sperm whales have demonstrated strong co-operative bonds between individuals in groups of females and immature males (Fig. 3). These groups were thought to be composed of matrilineally related individuals, however recent evidence suggests that they are not purely matrilineal in structure but are comprised mainly of clusters of related individuals but also of some animals with no close relations. Baird's beaked whales may have one of the most unusual social systems among odontocetes, although this is inferred only from studies of dead animals. It appears that males mature an average of 4 years earlier than females and may live up to 30 years longer (Kasuya *et al.*, 1997). This has led to speculation that males may be providing significant parental care.

An interesting congruence between low levels of mitochondrial diversity and the presence of matrilineal social systems in four species of odontocete may suggest the cultural transmission of matrilineally inherited traits (Whitehead, 1998). Apparent culture in cetaceans includes the learning of particular feeding techniques from the mother or members of the matrilineal group. In fact, many attributes of cetaceans (and odontocetes in particular) favor the evolution of social learning and culture. These include their long lives (~60–90 years),

**Figure 3** *Group of female and immature sperm whales in the Galápagos Islands. Photograph by Sascha K. Hooker.*

advanced cognitive abilities, prolonged parental care, permanent and cohesive groups, and an environment that varies substantially over large spatial and temporal scales (such that individual learning would be costly) (Rendell and Whitehead, 2001). However, claims of animal culture are argued to be subject to weaknesses, in that behavioral differences thought to be cultural may in many cases result from a combination of genetic, ecological, and cultural variation (Laland and Janik, 2006).

A potential offshoot of the advanced sociality observed in some species of odontocetes is the presence of postreproductive care, i.e., "menopause." Pilot whales, killer whales, and sperm whales show similar attributes to human females, with reproduction ceasing at approximately 40 years of age although females live on for several more decades. Short-finned pilot whales (*Globicephala macrorhynchus*) show a decreasing pregnancy rate with increasing maternal age, and a parallel age-related decline in the ovulation rate. Up to 25% of adult females may be postreproductive, ceasing to ovulate after age 40 even though the maximum lifespan exceeds 60 years (Marsh and Kasuya, 1991). However, lactation may continue for up to 15 years after the birth of their last calf. It is unclear whether communal nursing exists.

Another apparent consequence of the strong social bonds found among odontocetes is the phenomenon of mass strandings (Sergeant, 1982; see STRANDING). This tendency for animals to come ashore in groups to die is found only among the most highly social odontocete families. Several explanations for this behavior have been suggested. These include disorientation due to geomagnetic anomalies, panic flight responses from predators or underwater noise, morbillivirus infections, parasitic infections of the respiratory system, brain, or middle ear, or the strong social bonds of a group causing the entire group to follow one intentional strander. It has also been noted that many locations in which mass strandings occur share certain structural characteristics. These sites are often composed of a sandy peninsula or promontory, which extends perpendicular to the coastline and appears to form a "whale-trap," potentially due to the loss of echolocation abilities in the shallow sandy environment.

## VII. Mating System, Reproduction, and Life History

Many odontocete species have promiscuous mating systems, in which several adult males may mate with a female. However, other species such as the bottlenose dolphins in Shark Bay, and Dall's porpoises in the eastern Pacific appear to show a form of mate guarding. Among many beaked whale species males compete, suggesting that there may be a hierarchical nature to their social organization, probably for access to females. However, detailed comparative data on the mating systems of odontocete species will await genetic analyses to determine paternity from offspring within groups.

Gestation periods of odontocetes range between 7 and 17 months and almost all species have interbirth intervals of greater than 1 year (Reynolds and Rommel, 1999; Berta *et al.*, 2006). Length of gestation and fetal growth rate are related to calf size at birth such that larger species tend to require longer periods of gestation. Odontocetes produce a single offspring which is physically well developed (able to swim and surface to breathe), but socially undeveloped. As a result, odontocete species have characteristically long lactation periods, averaging between 32 and 100 weeks (Berta *et al.*, 2006). Females continue to feed throughout this long lactation period.

Odontocetes tend to show extended maternal care, resulting in a strong mother–calf bond. For most species, calves will remain with their mothers for a few years, but for some they will remain in close family groups for their lives (e.g., some killer whale populations, possibly long-finned pilot whales). This high level of investment needed to successfully raise calves may have led to the need to practise mothering behavior. Association between infants and non-mothers, termed allomaternal behavior, has been described for a variety of odontocete cetaceans (Mann *et al.*, 2000). Among bottlenose dolphins, such behavior appears to primarily benefit the non-mother, providing experience in parenting. Similar babysitting has also been documented in sperm whale groups, which stagger their deep-diving behavior such that calves are almost always accompanied by an adult at the surface. However, the function of this behavior in these circumstances appears to relate to increasing calf survival and defense against predation rather than to practise mothering (Mann *et al.*, 2000).

Much of the life history data available for odontocetes has come from studies of dead animals (either from those taken in whaling operations, as bycatch to other fisheries, or from strandings). The age of an odontocetes can be estimated from its teeth using much the same technique as counting the growth rings of a tree trunk. As an individual grows, incremental growth layers are deposited in the teeth and bones. In order to determine the age, the tooth is usually thinly sliced and polished and may often be etched or stained to better resolve the growth layer groups. In most species, each growth layer group is thought to represent an annual increment, but this needs to be independently verified in order to use this method for each species. By establishing the ages of animals, researchers are able to investigate the age-structure of the population, to look at ages at which animals mature, reproduce, the lifespan, etc. Long-term studies of odontocete species in the wild (e.g., those listed in Table III) are gradually allowing life history parameters to be recorded from living animals. In some cases, particularly in establishing lifespan, these are providing records of lifespan to a much greater age than were estimated from catch records.

The life histories recorded in this manner show large differences between different odontocete species. Harbor porpoises have a maximum longevity of 12 years (although some may live up to 17 years); they reproduce at age 3 and become pregnant every year thereafter. In contrast, killer whales have a maximum longevity of about 80 years and first give birth at approximately 15 years of age, with a 5-year interbirth interval. Many species also show sex-specific differences in life history parameters. For example, sperm whale females become sexually mature at approximately age 9–10, whereas sperm whale males do not appear to become sexually mature until approximately age 26–27.

## VIII. Human Interaction and Conservation Status

In the past, the majority of human interaction with odontocetes involved the capture of animals from the wild, either for consumption as part of the whaling industry, or more recently to obtain animals for captive display purposes. Only a few odontocete species were hunted to the same scale as the fisheries for baleen whales in the last two centuries. The main odontocete species taken historically were the sperm whale, some beaked whale species (northern bottlenose whales and Baird's beaked whales), and pilot whales. However, although the moratorium on large whale hunting essentially put an end to the hunting of myticetes, the only odontocete species included in this moratorium was the sperm whale. Thus today, many odontocete species are still taken in large numbers primarily in aboriginal subsistence hunts. Unfortunately, many of these go unmonitored, and so the exact numbers taken are unknown, although they are suspected to be high (Bowen, 2000).

Several odontocete species are or have been maintained in captivity for display, research, or conservation purposes (Twiss and Reeves, 1999). Some species are maintained in captivity for research or public display (e.g., bottlenose dolphin, killer whale, and beluga) whereas others have only rarely been kept in captivity following live stranding (e.g., sperm whales and beaked whales). Some species are successfully bred in captivity, including bottlenose dolphins, Commerson's dolphins, and killer whales, but most other odontocete species do not appear to fare so well in captive situations.

Humans are also increasingly attempting to interact with odontocetes in the wild. Whale-watching operations are increasing in number worldwide. In the San Juan Island area between British Columbia, Canada, and Washington State, USA, the number of whale-watching vessels has increased dramatically over the last two decades. There is currently a good deal of concern about the impact of whale watching on the animals involved. Although there has been little documentation to date of immediate adverse impacts, there has been little investigation of the long-term stresses and consequences for the viability of the populations concerned. Swim-with-dolphin programs are also increasing in frequency. There is also concern about the effect of these, although one of the most pressing concerns is the potential to misinform the public. As a result, of these activities, the public may form the perception that association with wild animals is endorsed. Currently in Florida, many people continue to solicit interactions with wild dolphins and, despite legislation against it, often encourage this by provisioning.

Lone animals of several odontocete species have, at times, been recorded to solicit associations with humans. This behavior has been recorded a number of times all over the world. In all cases, animals have become resident in a small area, where they approach and interact with boats or people in the water. In many cases, the animals involved have been bottlenose dolphins, but this behavior has also been recorded from belugas and rough-toothed dolphins.

Today, conservation problems exist for nearly all odontocete species. In fact, since the International Whaling Commission moratorium on catching large whale species, many of the current conservation threats faced by cetaceans are likely greater for odontocete species than for mysticetes. Their typically smaller size means they are less likely to be able to free themselves when trapped in nets, leading to high incidence of bycatch. The squid diet of many species renders them prone to plastic ingestion and their higher trophic levels magnify their pollutant load. Additionally, the habitat specificity shown by many species means that they have little recourse in the face of habitat destruction (for more details, see Conservation section).

Other less direct threats to odontocetes include noise and disturbance, pollution, and habitat loss, and degradation. In general, noise is thought to be less of a problem for many odontocete species than for mysticetes, since much oceanic anthropogenic noise is low frequency. However, higher frequency noise (such as that created by fishfinder units) is likely to coincide with the hearing range of many odontocetes (Twiss and Reeves, 1999). Additionally, some odontocete species (particularly beaked whales) appear to be highly susceptible to disturbance from mid-frequency sonar, although the mechanism by which this causes mass strandings and death remains unknown (Cox *et al.*, 2006).

## References

Au, W. W. L. (1993). "The sonar of dolphins." Springer-Verlag, New York.

Berta, A., Sumich, J. L., and Kovacs, K. M. (2006). "Marine Mammals: Evolutionary Biology," 2nd Ed. Academic Press, San Diego, San Fransisco, New York, Boston, London, Sydney, Tokyo.

Bowen, B. W. (2000) A field born in conservation's cold war. *Trends Ecol. Evol.* **15**, 1–3.

Carwardine, M., Hoyt, E., Fordyce, R. E., and Gill, P. (1998). "Whales, Dolphins and Porpoises." The Nature Company Guides, Time-Life Books. Weldon Owen Ltd., Sydney, San Fransisco.

Connor, R. C., Mann, J., Tyack, P. L., and Whitehead, H. (1998). Social evolution in toothed whales. *Trends Ecol. Evol.* **13**, 228–232.

Cox, T. M., *et al.* (2006). Understanding the impacts of anthropogenic sound on beaked whales. *J. Cetacean Res. Manag.* **7**, 177–187.

Cranford, T. W., Amundin, M., and Norris, K. S. (1996). Functional morphology and homology in the odontocete nasal complex: Implications for sound generation. *J. Morphol.* **228**, 223–285.

Heyning, J. E., and Mead, J. G. (1996). Suction feeding in beaked whales: Morphological and observational evidence. *Natl Hist. Museum Los Angeles County Contrib. Sci.* **464**, 1–12.

Hooker, S. K., and Baird, R. W. (2001). Diving and ranging behaviour of odontocetes: A methodological review and critique. *Mamm. Rev.* **31**, 81–105.

Jefferson, T. A., Leatherwood, S., and Webber, M. A. (1993). "Marine Mammals of the World." FAO Species Identification Guide. United Nations Environment Programme, Rome.

Kasuya, T., Brownell, R. L., Jr, and Balcomb, K. C. (1997). Life history of Baird's beaked whales off the Pacific coast of Japan. *Rep. Int. Whal. Commn* **47**, 969–979.

Laland, K. N., and Janik, V. M. (2006). The animal cultures debate. *Trends Ecol. Evol.* **21**, 542–547.

Mann, J., Connor, R. C., Tyack, P., and Whitehead, H. (eds) (2000). "Cetacean Societies: Field Studies of Dolphins and Whales." University of Chicago Press, Chicago.

Marsh, H., and Kasuya, T. (1991). An overview of the changes in the role of a female pilot whale with age. *In* "Dolphin Societies: Discoveries and Puzzles" (K. Pryor, and K. S. Norris, eds), pp. 281–285. University of California Press.

McAuliffe, K., and Whitehead, H. (2005). Eusociality, menopause and information in matrilineal whales. *Trends Ecol. Evol.* **20**, 650.

Norris, K. S., and Mohl, B. (1983). Can odontocetes debilitate prey with sound. *Am. Nat.* **122**, 85–104.

Norris, K. S., Wursig, B., Wells, R. S., and Wursig, M. (1994). "The Hawaiian Spinner Dolphin." University of California Press, Berkeley.

Rendell, L., and Whitehead, H. (2001). Culture in whales and dolphins. *Behav. Brain Sci.* **24**, 309–382.

Reynolds, J. E., and Rommel, S. A. (1999). "Biology of Marine Mammals." Smithsonian Institution Press, Washington and London.

Rice, D. W. (1998). "Marine Mammals of the World: Systematics and Distribution." Society for Marine Mammalogy, Special Publication No. 4, Allen Press, Lawrence, KS.

Richardson, W. J., Greene, C. R., Jr, Malme, C. I., and Thomson, D. H. (eds) (1995). "Marine Mammals and Noise." Academic Press, San Diego.

Sergeant, D. E. (1982). Mass strandings of toothed whales (Odontoceti) as a population phenomena. *Sci. Rep. Whale Res. Inst.* **34**, 1–47.

Twiss, J. R., and Reeves, R. R. (eds) (1999). "Conservation and Management of Marine Mammals." Smithsonian Institution Press, Washington and London.

Whitehead, H. (1998). Cultural selection and genetic diversity in matrilineal whales. *Science* **282**, 1708–1711.

Whitehead, H. (2003). "Sperm Whales: Social Evolution in the Ocean." University of Chicago Press, Chicago.

# Training and Behavior Management

TED TURNER, TODD FEUCHT AND TYLER TURNER

## I. Introduction

Marine mammal training is a relatively new field that gained formal recognition in the late 1940s. One of the first institutions to successfully maintain a breeding colony of bottlenose dolphins (*Tursiops* spp.), Marineland of Florida, also became the first institution to begin a formal training program for bottlenose dolphins. As behavioral observations of these animals began to generate questions unanswerable through scientific observation alone, it seemed that further exploration involving dolphin learning abilities could only be facilitated through training. Early training methods were considered industry secrets; yet those techniques were primitive and lacking in their full understanding of current behavior modification principles and procedures. Although the methods used at that time were limited, often consisting of trial and error, the marine mammal training field now employs a range of specific animal learning principles, including operant conditioning, as a means to provide better care, manage breeding groups, and shape specific behaviors to individual animals.

Although the use of operant conditioning is not exclusively the only mechanism used to train marine mammals, it is arguably the most direct. Therefore, the fundamentals of this science, as a minimum, must be assimilated for those who train, or wish to train, marine mammals (Pryor, 1995). Because the field of marine mammal training strongly discourages the use of aversive techniques, it is also highly productive and continues to stand as a model for ethical animal training. This philosophy of focusing on positive proactive behavior management and eliminating aversive and punishing practices has also created much safer environments for both the trainers and the animals by greatly reducing incidents of aggression, between animals and humans as well as within social groups.

The high exposure and excitement that is created when animals perform in shows for the general public have focused attention on

**T**

Figure 1  *Marine mammal training and show performances generate public interest in marine mammals and related issues. Here Pacific white-sided dolphins,* Lagenorhynchus obliquidens, *perform for a crowd.*

the cognitive abilities of these animals (Fig. 1). Though some criticize the use of marine mammals in show performances and other public programs, many who experience these interactions are fascinated by the level of complexity, coordination, and consistency in behavior. Still others are amazed at the degree of cooperation between animal and trainer. The vast majority of guests who participate in controlled swim-with-the-dolphin programs experience a greater sense of appreciation, conservation awareness, and concern for marine mammals (Marine Mammal Poll, 2005). Magical to some and practical to others, it is the training process itself that creates the medium for "communication" between marine mammal behaviorists and the animals with which they closely interact. In most cases, public performances, shows, presentations, and interactions have provided the animals with stimulating enrichment, and the general public with the motivation for further education. As a result, public interest in marine mammals is growing, and quality programs are expanding the use of marine mammal training. In turn, visitors to marine mammal facilities continue to develop a stronger sense of interest, empathy, and awareness, especially when they can interact and observe the results of marine mammal training and behavior management in a variety of contexts. In this and other areas, marine mammal training is making a significant impact on the conservation of these species.

## II. Animal Learning and Behavior

At a fundamental level, learning is a biological process necessary for the survival of an animal. The expression or suppression of certain behaviors enables animals to respond to their environment and adapt to ever-present social and environmental changes. An animal's ability to learn is directly related to its survival.

The study of animal learning involves not only biological processes, but also detailed aspects of psychology such as memory, developmental learning, classical conditioning, operant conditioning, behavior modification, cognition, and other information processing disciplines (Spear and Riccio, 1994). The field of psychology is now experiencing a stronger appreciation for its role in understanding marine mammals,

and although a number of important learning principles have been excluded from this chapter due to their current limited applications for marine mammal training and behavior management, we have attempted to include some of the most widely utilized principles, or principles having significant future impact (Domjan, 1993). When describing the main components of current marine mammal training, three primary learning modes become salient:

### A. Observational Learning

Also known as *modeling*, this vicarious process is defined as learning by observing another (model) engage in a behavior. During observational learning, the animal (observer) need not emit the behavior, experience direct consequences, or receive applied reinforcement for learning to occur. Mimicry seems to be a by-product of observational learning. It is a primary learning element for young animals raised in a complex social environment, and appears to be of significance in developing socialization skills.

The expression of these mimicked behaviors is not relegated to social behaviors only. Young marine mammals attempt to mimic trained behaviors, and often accompany adults in show performances or training sessions (Fig 2.). It is common practice for marine mammal behaviorists to "capture" (reinforce) mimicked behaviors in order to rapidly develop and accelerate the learning process for young animals. However, regression often occurs when the model (adult) is no longer present, a phenomenon that must be considered when zoological exchange, reintroduction programs, or environmental changes occur. Under conditions such as these, expressed behavior (observed while in the presence of the model), may not occur in a changed environment. This may impact behavioral acclimation, foraging skills, socialization patterns, predator avoidance, and other behaviors that may have health or survival ramifications. Training programs that rely on observational learning must often re-train these animals using direct operant conditioning and behavior shaping with individual training sessions. Nonetheless, observational learning plays a critical role in the lives of marine mammals.

### B. Classical Conditioning

Past understanding of Ivan Pavlov's early work investigating anticipatory salivation in dogs led to a restricted view that classical conditioning [the pairing of a conditioned stimulus (CS) in conjunction with an unconditioned stimulus (UCS)], only produced a conditioned response (CR) involving reflex systems. In recent years, groundbreaking research in experimental psychology has replaced this view with a more dynamic understanding of Pavlovian learning. Associations between conditioned and unconditioned stimuli have been linked to learning phenomena such as *sign tracking* (movement towards a stimulus that signals the availability of a positive reinforcer such as food), *conditioned emotional response* (anxiety/fear), and *conditioned taste aversion* (food preference/avoidance) (Cole *et al.*, 1996). These seemingly insignificant components have a measurable and often profound impact on subsequent learning during the training process.

The accidental or deliberate pairing of events, which often occur in any given environment (both wild and zoological); can create learning via classical conditioning that can influence the training of marine mammals. Associating food with a loud noise for example, can eventually diminish the startle effect the noise might otherwise evoke and reduce the likelihood of a negative reaction (such as escape or aggression). However, pairing the same loud noise with a painful medical procedure can exacerbate the negative reaction to that same noise in

**Figure 2** *A killer whale,* Orcinus orca, *calf accompanies its mother during a training session.*

the future, and increase the probability of that negative reaction. A minor avoidance response by the animal could intensify into a severe aggressive or panic response, causing injury to the animal or the animal handlers. In another applied example, pairing food with an aversive medical procedure can cause an animal to avoid eating, leading to weight loss, even though no illness is present. A conditioned food aversion can be an unintended by-product of seemingly unrelated events.

Pavlovian conditioning can be used to help modify behaviors needed for the long-term care of marine mammals, such as introducing new animals into established social groups during animal acclimation procedures. An experienced marine mammal behaviorist understands the principles of Pavlovian conditioning and will manage environmental conditions with behavior to prevent problems, while assisting the animal in the beneficial acquisition of behaviors commensurate to the long-term goals.

### C. Operant Conditioning

Commonly defined as behavior which is modified by consequences, operant conditioning is currently recognized as the most widely employed training program for marine mammals (Ramirez, 1999). The three basic components of instrumental learning (learning that occurs as a result of operant conditioning), however, must occur in precise order for learning to be achieved. These three components are the *antecedent* (a stimulus, cue or hand signal which precedes the behavior), the *behavior* (the resultant observed response emitted by the animal which immediately follows the antecedent), and the *consequence* (a stimulus or applied reinforcement that immediately follows the response and acts to increase or decrease the preceding behavior) (Honig and Staddon, 1977).

Consequences play a key role in operant conditioning as they will cause a behavior to become more or less frequent in the future, and influence the way antecedents set the occasion for repeating or not repeating a behavior (Baldwin, 1998). Reinforcers will strengthen a behavior and increase the likelihood of it occurring in the future, whereas punishers will decrease the frequency of a behavior. The terms "positive" and "negative" are used to indicate if a stimulus is being applied (onset of a stimulus) or removed (termination of a stimulus). Therefore, positive reinforcement would consist of applying a "good" or rewarding stimulus as a consequence of a behavior, whereas negative reinforcement would consist of removing or

terminating a "bad" or aversive stimulus. In both cases, the preceding behavior will have been strengthened. Conversely, positive punishment is the application of a "bad" or aversive stimulus as a consequence of a behavior; while negative punishment is the removal or subtraction of a "good" or rewarding stimulus. In either case, the preceding behavior will decrease in frequency. As previously mentioned, the marine mammal industry concentrates its efforts on the use of positive reinforcement for behavior management plans; however, it is still critical to understand how all consequences affect behavior so one can be fully prepared to manage and understand the many different scenarios that will undoubtedly occur when caring for any animal or group of animals.

During a typical marine mammal training session, a hand signal (antecedent) is usually presented to the animal, followed by the animal's response (behavior). If correct, the behavior is usually reinforced (consequence) by the trainer with either primary reinforcement (food) or secondary reinforcement (touch, toys, play, and activities). Common in marine mammal training, incorrect, or undesired behaviors are usually ignored with minimal or no consequence applied. Again, punishment and negative reinforcement is avoided or eliminated. In the operant process, each behavior is best described as a learning cycle representing three critical elements (Fig. 3). Within a series of behaviors, typical of a formal training session, this cycle is repeated multiple times.

Principles supporting the successful training sessions must be understood by the behaviorist and include *stimulus consistency, stimulus fading, behavioral criterion, behavioral development, delay of reinforcement, schedule of reinforcement, magnitude of reinforcement,* and many others (Kazdin, 1994). Experienced marine mammal trainers understand the implications of precise application. Like other forms of learning, operant conditioning should not be characterized as a "technique" or "system" of training that can be switched on and off conveniently; instead, it is a dynamic and ever-present environmental learning phenomenon that is in continuous action and influences the acquisition, intensity, or extinction of specific behaviors. Behaviors are in the constant process of strengthening or diminishing as a result of the consequences that follow. Not all of the antecedents and consequences experienced by each animal are applied by the marine mammal trainers. Behaviorism views the environment in its' entirety, with many such antecedent stimuli being processed by the animals in context. Even in the most productive marine mammal training programs, it is rare that more than 3h per day, on average, are devoted to actual training sessions, leaving many hours each day to other environmental learning influences.

For large institutions with formal marine mammal training programs, certain principles, and techniques have been refined and "operationalized" to facilitate staff development and expertise, resulting in efficient animal training. Expertise is often gained through practical hands-on application and supplemented with coursework, seminars, and testing. Other mediums such as conference attendance and trade groups provide additional sources of information. The International Marine Animal Trainers Association (IMATA) is an established trade group that offers opportunities to formally exchange specialized marine animal training information. Many marine mammal training programs emphasize the following:

*1. Optimal Learning Conditions* Naïve animals (animals without prior training experience or limited behavior repertoire) often seem cautious and wary of their surroundings. This apprehension can be reduced by minimizing or eliminating those elements that can create anxiety and avoidance such as aversive procedures, unstable environment,

**T**

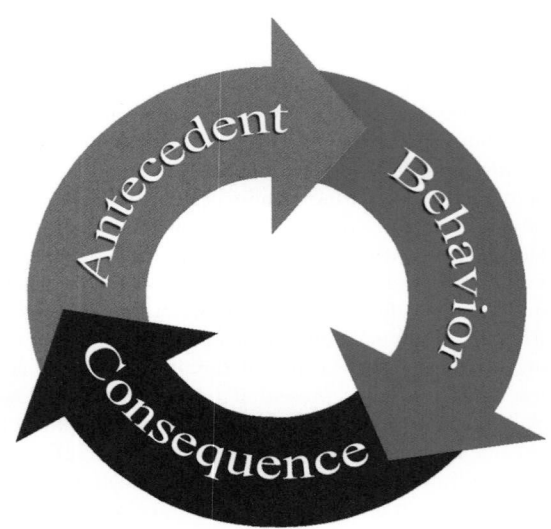

**Figure 3** *A typical operant conditioning cycle where the signal (antecedent) precedes the response (behavior), followed by the reinforcement (consequence). A single training session includes many such cycles.*

incompatible conspecifics, or unfamiliar sounds, to name just a few. The process of extinguishing avoidance behavior is aided by the cumulative positive effect of daily feedings and time spent with the animal, allowing both desensitization and familiarity to influence a more comfortable learning environment. Marine mammals will readily take food from a familiar caregiver and generally allow some calm touching during feeding times. This provides the foundation for positive reinforcement (a stimulus that is applied as a consequence and increases the frequency, intensity, and duration of the preceding behavior). Regression can be minimized if aversive events can be delayed or eliminated. As a reinforcement history develops, unavoidable aversive events (such as routine medical treatment) have less impact and cause minimal regression. The opposite appears to be true as well. Repeated aversive procedures early in the process can have a profound negative effect and will delay the positive response to the trainer.

*2. Conditioned Reinforcers*  We have already discussed briefly and given examples of unconditioned (primary) reinforcers and conditioned (secondary) reinforcers; however, it warrants some additional information. Unconditioned reinforcers satisfy a biological need and therefore are effective without having to be learned or conditioned. Food is a common example of an unconditioned reinforcer.

Conditioned reinforcers are those stimuli that were once neutral but have gained reinforcing properties by repeatedly preceding or predicting other reinforcers (Baldwin, 1998). Some examples of conditioned or secondary reinforcers could include the toys or a rubdown from a trainer. The consistent delivery of an *unconditioned reinforcer* (such as food) as an immediate consequence to a desired behavior establishes the foundation for training a *conditioned reinforcer* (a stimulus that has acquired reinforcing properties through learning) (Chance, 1994). These conditioned reinforcers are an important tool for a marine mammal behaviorist, as they allow a much greater variety of positive consequences to motivate and shape behavior.

Most marine mammal training facilities also use audible whistles called a "bridging" stimulus. This bridging stimulus acts as a cue that signals the imminent application of reinforcement. It is referred to as

a "bridge" because it acts to bridge the time delay between the precise behavior targeted for increase, and delivery of a positive consequence, such as food, toys, play, a favorite activity or another reinforcing behavior. Marine mammal trainers and behaviorists use a variety of tools that act as a bridge. Typically, a high-frequency whistle is used for cetaceans (due in most part to higher hearing ranges), whereas a verbal cue or clicker sound is used for pinnipeds. This bridging stimulus, when consistently followed by the offering of food, quickly becomes recognized by the animal. Within a few training sessions the sound itself evokes a response from the animal and subsequent orientation back to the trainer.

*3. Increasing the Attention Threshold*  In order to maintain the animal's attention sufficient to advance learning and develop a repertoire of beneficial behaviors, an early emphasis on *attention threshold* (the duration that a stimulus will control behavior) is necessary. This is done by systematically increasing the amount of time the animal will continue to watch the trainer prior to bridging and reinforcement delivery. Increments of a few seconds are literally counted by the trainer when shaping an increased attention span. This is particularly important in the training of species such as Commerson's dolphins (*Cephalorhynchus commersonii*), young animals, or animals in high stimulus environments. Eventually, within a few days or weeks, an attention span of several minutes can be conditioned if bridging and reinforcement is precise. Conversely, a lack of precision can interfere with or delay the acquisition of new behaviors. Inexperienced or impatient marine mammal trainers often overlook this critical element.

*4. Following a Target Pole*  If the pairing of food with the bridge has been well timed and effective, the animal will quickly learn to touch a hand prior to the bridge and food delivery. By holding the hand close to the animal, the animal will readily touch the hand. The behavior of touching is then reinforced. Eventually, the hand can be moved back and forth to shape and strengthen the hand-touch response. In time, the animal will move to the hand wherever it is placed. Target poles (a long pole with a small round float, or "target" attached to the end) are often introduced at this stage to simulate an extended reach. Using this target pole, touching and following the target can be reinforced and various high-energy behaviors can now be safely shaped through *successive approximations*.

*5. Achieving Stimulus Control*  As animals begin to learn the relationship between their emitted response (behavior) and the delivery of reinforcement by the trainer (consequence), they also begin to learn which specific behavior will likely lead to reinforcement. Discrimination between learning events occurs when animals respond differently to separate antecedent signals or cues. The trainer offers reinforcement only for specific responses and thereby facilitates discrimination learning. This process is aided with specific hand signals (*discriminative stimuli*) that precede each behavior. When trainers selectively reinforce only desired behaviors, animals learn to recognize the signals and respond appropriately. This is referred to as *stimulus control* where behaviors are reinforced in the presence of a particular signal (or stimulus). *Random sampling* (emitting several behaviors in an attempt to gain access to reinforcement) is a common phenomenon early in the training process. Trainers will ignore all but the requested behavior, thereby aiding the process of stimulus control and discrimination learning.

*6. Successive Approximations*  The most effective trainers break each behavior down into small easily understood and planned steps. Each small step, or approximation, is reinforced until the animal consistently responds to the antecedent cue. Once consistency

T

is established, the trainer will continue in sequence to each new (usually more complex or difficult) step. For example, a high jump begins with the lifting of the head in order to touch the target and receive reinforcement. With each successive approximation, the target is raised slightly until the desired height is achieved. In effect, each behavior can be visualized by the trainer as a systematic process. All behaviors including complex behaviors involved in medical procedures can be broken down in such a manner, creating an easier learning process for the animal.

*7. Schedules of Reinforcement* A schedule of reinforcement refers to the timing and frequency of reinforcement. Any behavioral plan without regard to the precise schedule of reinforcement can lead to training and subsequent learning problems. Specific *schedules of reinforcement* have been empirically derived in learning laboratories and applied in marine mammal training programs. Through practical application, many facilities are beginning to recognize the importance of *variable schedules of reinforcement* including reinforcement variety. This type of reinforcement delivery provides reinforcement and enrichment intermittently and unpredictably. It more closely replicates the type of reinforcement schedule most often associated with the wild environment, helps maintain motivation and interest, and provides enrichment opportunities for each animal while it shapes extended thresholds and increases the number of attempts made.

Only certain components of animal learning disciplines have been described for the purpose of identifying those elements most commonly used to influence the acquisition of trained behaviors, but it is also helpful to recognize that these principles continue to influence animal behavior, whether formally applied or not. The process of "training" through observational, instrumental, or Pavlovian learning is constantly shaping behavior independent of the presence of "trainers" or the existence of a formal training program. Even the normal daily cleaning, feeding, and operational activities in any animal facility can effectively act to shape and modify behaviors. The appearance of the staff veterinarian, for example, might inadvertently act to suppress courtship behavior depending on the timing of appearance. In another example, the appearance of the staff veterinarian could also act as a reinforcer in a different scenario, depending on prior learning experiences, thereby increasing certain behaviors. As many zoological facilities have experienced, the principles of learning and the effects of timing can even affect stereotypic patterns such as anticipatory pacing prior to feeding time, creating an opportunity for *adventitious reinforcement* (coincidental cause and effect reinforcement causing an inadvertent increase in the unwanted behavior). This is why, in addition to animal training objectives, comprehensive behavior management programs are being incorporated into progressive zoological institutions. The Animal Behavior Management Association (ABMA) is a trade group, formed in the United States, to promote the understanding and use of broad-ranged behavior management by zoological professionals. It promotes the systematic use of animal training, environmental enrichment, behavior medicine, exhibit design, and other natural-history behaviors to achieve desired and measurable goals.

As we continue to learn about information processing and marine mammal training, additional learning principles will add to our knowledge of these animals. Investigations into memory retrieval, short and long-term memory capacities, context specific learning (learning in one environment that does not necessarily transfer to another environment), and other principles, may help us understand and predict the success of future reintroduction projects. Awareness of these learning principles enables caretakers, behaviorists, veterinarians, scientists, and trainers to better understand environmental factors influencing the expression of appropriate or inappropriate behaviors. Further investigations into animal emotion, cognition, and stage developmental learning will improve our ability to manage the behavior of marine mammals to their benefit.

## III. Early Development and Learning

For newborn marine mammals, motor skill development occurs rapidly and within a few weeks. Observations of increasing independence (time and distance away from the mother) begin to emerge as motor skill, buoyancy control, visual acuity, echolocation, and auditory learning strengthen. Observational learning becomes evident and mimicry increases until the young animal learns to directly manipulate its environment.

In some cetaceans between 4 and 7 months of age, when teeth begin to erupt and grow, the mother (dam) initiates a weaning process through food sharing (both partially digested and regurgitated or with food that has been ripped into small pieces). This seems to coincide with accelerated learning and may signal the beginning of a *sensitive learning phase* or *critical period* (a developmental stage where rapid acquisition of behaviors occurs). For marine mammals born in zoological facilities, most formal training begins during this stage.

## IV. An Emphasis on Positive Reinforcement

Marine mammals respond readily to operant conditioning and possess a strong aptitude for instrumental learning. The rapid acquisition of behaviors has done little to refute the notion that marine mammals are unusually "intelligent." However, other marine and terrestrial animals trained via operant conditioning sometimes show an equal or superior aptitude and learning curve. Therefore, comparative intelligence remains difficult to define and quantify. Regardless, empirical evidence suggests that positive reinforcement has a strong motivating influence on learning, and this has certainly contributed to the notion that marine mammals can learn quickly.

Positive reinforcement-based training creates an environment that is non-threatening and therefore facilitates learning for marine mammals, although maintaining a strong desire to interact with their trainers. Significantly, the most effective training programs develop a foundation of voluntary cooperation as opposed to compliance. With voluntary cooperation as a prime goal of most marine mammal programs, behaviors once thought impossible to train have been achieved, and positive reinforcement has been the common element for many of the breakthroughs in marine mammal training. These programs emphasize *positive reinforcement* and discount the use of *negative reinforcement* (an increase in a response that causes the removal of an aversive stimulus) and *punishment* (a decrease in a response that is followed by an aversive stimulus). Marine mammal facilities licensed in the United States are regulated by the US Department of Agriculture and abide by the provisions of the Animal Welfare Act. These institutions consider the use of negative techniques as unethical. This philosophy has been a model for other animal management programs and has been drawn acclaim and interest from a variety of sectors. Behavioral consistency and using strictly positive reinforcement may prove to be the marine mammal training industry's most profound contribution to the social sciences.

## V. Enrichment and Variety

Just as we recognize the importance of play in the development of young animals, we also recognize how crucial other physical and

**T**

cognitive challenges are for optimal animal health. An emphasis on behavioral enrichment programs that produce beneficial changes in behavior as well as health and well-being has gained wide acceptance in zoos and aquariums (Shepherdson *et al.*, 1998). In the past, this aspect was considered superfluous to well-designed facilities. It was believed that "natural" behaviors would be an automatic by-product of a naturalistic exhibit. It was soon recognized that these behaviors required more prompting than just high quality food, health care, and naturalistic environments.

A focus on *environmental enrichment* (the application of environmental complexity to stimulate beneficial activity), has become a fixture in the behavior management of mammal species. The goals of environmental enrichment programs clearly fit within the animal training model (i.e., behaviors can be shaped, maintained, and modified using animal learning and training principles). In fact, most successful enrichment programs are now managed by experienced animal trainers and behaviorists. They coordinate environmental change, food placement, toys, games, and other behavioral prompts and reinforcers that target specific behaviors and patterns. When used properly and applied as a consequence of desired behaviors, environmental enrichment devices offer important physical and mental stimulation, while simultaneously maintaining a repertoire of "healthy" behaviors. Daily variety and reinforcement of appropriate behaviors help reduce other inappropriate, destructive, or agonistic behaviors.

## VI. Animal Acclimation

The assumption that animals immediately behave towards a new member with acceptance or at best indifference, is often erroneous. Animals use aggression as a means to acquire and establish territories, food sources, breeding rights, sleeping space, and more. Without employment of simple learning principles, the possibility of any new member being aggressively challenged is highly likely. The use of a reinforcement schedule known as *differential reinforcement* provides selective reinforcement to the group when pro-social behaviors are observed. Through the use of differential reinforcement, inappropriate behaviors are reduced while appropriate behaviors are strengthened.

Differential reinforcement modifies behavior using four main procedures known as; DRO (*Differential Reinforcement of Other behavior*), DRI (*Differential Reinforcement of Incompatible behavior*), DRH (*Differential Reinforcement of High rates of response*), and DRL (*Differential Reinforcement of Low rates of response*) (Kazdin, 1994). Although each differential reinforcement schedule targets specific aspects of behavior, the general procedure is quite simple. Trainers are directed to observe social interactions, and apply reinforcement only when appropriate behaviors are observed. These procedures are commonly used to accelerate acclimation to a new social group, which maintains long-term compatibility quite effectively.

## VII. Behavior Medicine

Achievements in veterinary care, treatments, and procedures have been possible through the training of specific husbandry or medical behaviors. These behaviors provide marine mammal specialists, researchers, and medical personnel access to treatments, procedures, and biological sampling once thought only obtainable through forced restraint, sedation, or anesthesia. Optimal animal health is achieved via routine examinations. For untrained marine mammals however, simple procedures and check-ups can become a risk to both people and animals. Untrained animals usually express fear,

**Figure 4**    *Trainers at the Dolphin Research Center collect a voluntary blood sample from a bottlenose dolphin,* Tursiops truncates.

apprehension, and a "fight or flight" response (*activation syndrome*) that can be dangerous for all involved.

A variety of procedures has been routinely trained through operant conditioning, approximation training, and positive reinforcement. Marine mammal facilities now require the training of behaviors specific to health maintenance for each animal. Commonly referred to as "husbandry behaviors," the effort to begin training medical procedures gained momentum in the early 1980s. Aware of the risks associated with forced restraint or anesthesia, coupled with the need to obtain ongoing biological sampling, marine mammal training took a new direction. Simple procedures, such as allowing a veterinarian to touch a trained animal without restraint, conduct up-close visual examinations, or inspect an open mouth, were the first behaviors trained for health purposes. Within a few years, routine blood sampling (Fig. 4) was trained to dolphins, killer whales (*Orcinus orca*), sea lions, and walrus (*Odobenus rosmarus*), providing valuable data, serum banks, and unprecedented research opportunities. Soon thereafter, trainers began conditioning voluntary urine samples, milk samples, x-rays, sonography, weights and measurements, endoscopy, intubation, and even tooth drilling, all without restraint, sedation, or anesthesia. Marine mammal veterinarians were able to access healthy animals using procedures unimaginable to small animal practitioners, equine, and zoo vets.

Today, these procedures have evolved into comprehensive *behavior medicine* programs. Behavior medicine employs the use of behavior modification to treat chronic and acute medical conditions or to promote and maintain good health through behavior management. A well-known field in psychology, behavior medicine combines traditional medicine and behavior modification techniques to diagnose and successfully treat such complex medical issues as chronic regurgitation, stereotypies, kidney function, eating disorders, conditioned food aversions, osmoregulation, and more. In addition, behavior medicine programs also incorporate the training of behaviors that help maintain optimal health such as exercise, pre/post parturition examinations, dentistry, sonography and EKG, to name a few. Behavior medicine for marine mammals is growing in both understanding and in effective use. Pre-treatment baseline data are normally gathered prior to the onset of behavioral change (treatment) to monitor effectiveness and subsequent health improvement.

Chronic medical conditions can also be treated by engineering a behavioral training program that targets a specific behavioral goal. Acute medical conditions can be treated if prior training was completed. Through training, marine mammal medicine is reaping tremendous benefits from behavioral management. It is no coincidence that dolphins housed in quality aquariums live as long as or longer than their wild counterparts, and that neonate mortality is lower in zoological facilities than in the wild. In fact, the oldest known bottlenose dolphin still lives a productive and healthy life at Marineland in Florida, where she has surpassed her 54th birthday. In addition, recent research by Willis (2007) reports that the life expectancy for dolphins living in accredited facilities of the Alliance of Marine Mammal Parks and Aquariums (AMMPA) exceeds that of the wild. Many zoos are now emulating the success of marine mammal husbandry training and are beginning to experience those same benefits with endangered species. Curators wisely understand that a good animal health program utilizes all available expertise including animal training and behavior management.

## VIII. Research via Training

Research scientists must often be opportunistic when gathering data while investigating marine mammals, since access to these data can be limited. Using a creative approach, scientists can gather information such as food intake, metabolism, respiration, activity levels, and other data during the normal course of operations at marine life facilities. This opportunistic approach has been very useful in producing valuable and factual information. However, the ability to gather data using trained animals has accelerated this process for some investigators, opening up new avenues for researchers and experimental scientists.

New training procedures have given many investigators opportunities for direct experimentation, using non-invasive, reliable, and replicable methodology. Physiological studies using trained animals were pioneered by the US Navy's marine mammal program. This research provided a foundation for understanding the deep diving reflex. Cognitive studies conducted at Kewalo Basin Marine Mammal Laboratory were only possible using trained dolphins. These studies have enhanced our awareness of language use and information processing in animals. More recently, marine mammal energetic studies at West Chester University use trained animals in cooperation with marine mammal programs to measure acceleration and maneuverability in multiple species. Marine mammal training also facilitated signature whistle research for Woods Hole Oceanographic Institution, using a routine pool separation procedure to evoke signature whistle production in bottlenose dolphin calves.

Animals trained to allow voluntary blood samples provided comparative data and titer analysis that helped investigators at the National Marine Fisheries Service identify *morbillivirus* spp., a disease that has devastated some populations of wild marine mammals. In addition, voluntary blood sampling by trained killer whales at SeaWorld enabled scientists to precisely identify gestation in this species. Prior to this, it was speculated that the killer whale gestation period lasted 12 months; but because of this significant research, it is now known to be much longer, at approximately 17 months. Accurate information about reproduction, gestation, and inter-calf interval is vital when calculating the recovery of wild stock. This new information, discovered in the 1980s, had an important impact on stock assessment. These institutions and others, in partnership with marine mammal programs, are providing unprecedented volumes of important information. The utility of marine mammal training continues to have a strong impact on science.

## IX. Interactive Programs

In the early years of keeping marine mammals at oceanaria, scheduled feeding times were billed as a main event that piqued the interest of the visitors. As training advanced, small routines evolved into highly complex behavioral sequences and dramatic shows that featured choreographed behavior, multiple species, and special effects. The significance of marine mammal training became evident as attendance in some marine parks exceeded 5 million guests annually, driven in large part by behavioral programs.

Marine mammal training had become "big business" as the impact of these productions changed marine parks dramatically by generating funds for new and improved exhibits, stadiums, water systems, medical programs, animal rescue, rehabilitation, and research. The most successful marine parks also had the highest quality shows. As new facilities were built, facility design centered on training programs for show production; and attendance at marine parks in the United States soared to over 60 million annual visitors. Millions of park visitors became more interested in marine life than at any other time in history. Vicariously, and from a distance, they were able to experience the excitement of interacting with dolphins, whales, sea lions, and other marine life.

During the early 1980s, Dolphin Research Center in the Florida Keys began a new training program that allowed some guests a hands-on, in-water experience with dolphins (Fig. 5). As with most pioneering efforts, this new program was strongly criticized by anti-captivity groups and even other marine parks, yet they prevailed in developing the first guest interaction program. These early educational programs had a profound impact on the participants, generating a renewed interest in marine life conservation, concern for the oceans and empathy for marine mammals. Just as the early dolphin shows changed public attitudes about these animals in the 1970s by helping to develop a sense of value in protecting wild marine mammals, interactive programs are doing the same. Today, the popularity of these programs once again is dramatically changing marine mammal facility design, animal training programs and public opinion. Since 1999, innovative new facilities have recently been built in Florida, the Bahamas, and the United Arab Emirates, with more large-scale marine mammal training facilities being planned worldwide.

**Figure 5** *Interactive programs are providing a new generation with unique educational experiences.*

T

In addition, new marine mammal legislation has recently been passed in the Bahamas and is being developed in other countries, once again recognizing that marine life facilities and the value of their training programs directly support wildlife conservation. Just as dolphin shows helped in developing and ratifying the US Marine Mammal Protection Act, interactive programs and facilities are the catalyst for new marine mammal legislation that supports professionally supervised interactive programs. At the same time, this particular legislation is beginning to incorporate regulations of wild dolphin interactions. US legislation makes wild dolphin interaction illegal due to the inherent risks involved when attempting to interact with any large, wild mammal. Wild dolphin interactions lack behavioral supervision, pose potential health and safety hazards to animals and people, and adversely affect natural behavior patterns (Spradlin *et al.*, 1999). Supervised programs with trained animals have proven to be a safe and educational alternative.

As well-managed programs in quality institutions continue to evolve, opportunities to interact with species such as beluga whales (*Delphinapterus leucas*), sea lions, false killer whales (*Pseudorca crassidens*), and more species will help shape a future generation of ocean stewardship.

## X. Other Uses for Marine Mammal Training

Marine mammal training and behavior management has always been a fixture in marine mammal programs. With its use in an ever-growing list of projects and applications, this field is rapidly expanding and becoming increasingly sophisticated due to the complexities of each project. In some cases, animal training supplements the project, in others it is of paramount importance.

### A. Military Programs

Marine mammal training has been employed successfully in open-ocean work. The US Navy's marine mammal program has trained mammals a variety of behaviors involved in rescue and recovery, deep diving physiology, echolocation, boat following, harbor security, and object retrievals, to name just a few. Some of these programs actually save human lives. These early and ongoing programs continue to generate enormous benefit to scientists in understanding marine mammal learning capabilities, physiology, and motivation. A number of research institutions now use open-ocean training for interaction programs, research, and reintroduction training.

### B. Rescue and Rehabilitation

Training techniques have also been employed in the rehabilitation process for sick and injured wild animals. Classical conditioning has been successfully used to help condition orienting response, suckling, and bottle feeding for orphaned cetacean calves (Fig. 6) and pinniped pups. The ability to respond, locate, and move toward animal caretakers is not only vital to the survival of the animal, but also becomes a developmental indicator that aids in the diagnosis of neurological disease or injury.

### C. Social Compatibility

The use of body weight maintenance, through trained daily body weights combined with differential reinforcement aids in weight maintenance, avoids radical weight fluctuations and aggression associated with breeding seasons. Through these procedures, social compatibility increases whereas agonistic behaviors are decreased in male California sea lions (*Zalophus californianus*) (Turner and Stafford, 2000). This management strategy has been successfully used in captive bottlenose dolphin populations as well.

### D. Epimeletic Behavior

Epimeletic or care-giving behavior is being investigated as an opportunity to apply training and behavior management to increase the probability of calf survival. During the first few hours of life, the primary challenge for marine mammal newborns is maternal acceptance (vs rejection), followed closely by the need for protection and nutrition. It seems that a dam's natural care-giving behavior should automatically engage, yet healthy newborns are sometimes rejected for reasons not fully understood. Though rare, this phenomenon has been observed in many species (both wild and captive), including humans. It seems evident that many of the skills required for early calf/pup survival and care-giving are acquired through a learning process. In some facilities, training has been applied to this scenario, and although definitive results are not yet available, progress has been made in understanding the process of training to prompt nursing behavior (presentation of the mammary to the soliciting calf), especially for first-time mothers, those without observational models, or those without prior experience raising offspring.

Work has been carried out to apply marine mammal training to accelerate the acquisition of other epimeletic behaviors such as parallel swimming, calf protection, calf retrieval, and other behaviors. Behavioral management is also practiced that encourages calm behaviors and avoids accidental or *adventitious* reinforcement of calf rejection. Fortunately, the majority of newborns calves, *are* accepted (some immediately after birth and some within a few hours) and their dams usually allow them to nurse within 24h. Within a few days, the calf or pup rapidly acquires the motor skills necessary to coordinate successful nursing, while its mother is also acquiring and strengthening her own repertoire of care-giving behaviors, thereby increasing survivability in subsequent offspring. When these young

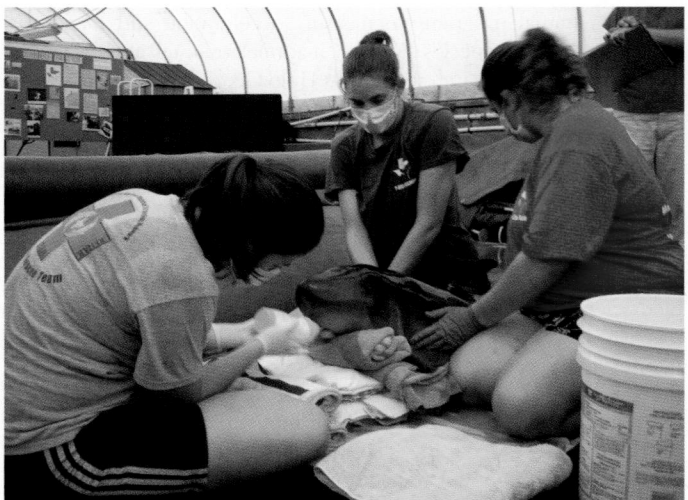

**Figure 6** *Members of the Texas Marine Mammal Stranding Network bottle feed a beached dwarf sperm whale,* Kogia sima. *Photograph supplied by Adrian Dahood.*

survive, opportunities to collect valuable data help scientists better understand the biological and learning developmental processes in wild environments, where observations are difficult or sometimes impossible. These data also act as comparative models for some species, providing like parameters for understanding their development. In some cases, like that of a beached or orphaned killer whale calf, comparing these data can influence the course of treatment, types of supplemental formula, or other behavior management decisions necessary to improve the odds of survival. Having at least some understanding of calf development is particularly critical for other endangered species such as Hector's dolphin (*Cephalorhynchus hectori*) and Vaquita (*Phocoena sinus*).

### E. Feeding and Foraging

An unprecedented opportunity to apply operant conditioning during rehabilitation in a juvenile gray whale calf (*Eschrichtius robustus*) proved successful. In 1997, SeaWorld San Diego assisted in the rescue and managed the recovery and release of this orphaned calf. After initially stabilizing the young animal's health, blood indicators suggested that the animal's level of activity was in need of improvement. This particular animal was placed on a behavioral management plan; and using positive reinforcement the calf was conditioned to move through gates, increase its activity level, and adjust its eating patterns to more closely match the bottom feeding strategies of wild gray whales. Upon reintroduction after a 14 month rehabilitation, the animal had gained over 18,000 pounds (an average of 40 pounds per day), and grew in length from 14 to 30 ft (SeaWorld, 2006). The information gathered during this process has contributed invaluable information on growth rates and may help future beached calves to survive.

### F. Breeding and Reproduction

The training of voluntary blood and urine sampling has enhanced captive breeding programs by providing science with information about progesterone levels and the estrus cycle in cetaceans. Success in conditioning voluntary semen collection in male cetaceans (dolphins and killer whales), and voluntary artificial insemination in females, has resulted in the first successful pregnancies for both species. Artificial insemination using trained procedures is becoming increasingly common. Sonographic evaluations were also trained using positive reinforcement, providing scientific proof of the effectiveness of this training program. The first successful birth of a killer whale calf using artificial insemination training occurred recently at SeaWorld; and a cooperative effort between Miami Seaquarium and Dolphin Research Center resulted in a live birth of a bottlenose dolphin calf in late 2006. This technology provides science with encouraging news for the preservation of endangered marine mammals.

### G. Reintroduction Programs

The translocation of wild animals and the release of marine mammals held for short periods (usually for research or rehabilitation purposes), has provided many opportunities for behavioral management and in some cases, animal training during the holding process. If managed correctly, the success rate of these relatively short-term research projects can be an effective means of gathering data. However, the release of animals born or housed long-term in zoological facilities, should be approached carefully, and include specific behavioral plans as well as health screening and scientific monitoring

of released animals to ensure their well-being and success of the reintroduction.

There have been instances where marine mammal releases have been over-simplified by extremist and anti-captivity groups. Although emotionalizing their concept of "freedom," these groups generally discount the biological need of such experiments, especially for non-endangered species, while often disregarding the very real and complex survival obstacles an animal must confront successfully. Therefore, thorough and unbiased scientific review must be completed before any such permit can be granted in the United States. Attempts to release animals without scientific review and permit authorizations are in violation of the US Marine Mammal Protection Act. Past attempts to do this in the United States has resulted in the needless suffering and death of animals. In one well-publicized case in Florida (1996), this violation led to the prosecution and conviction of those involved (US Department of Commerce, 1999). Unfortunately, not all governments have laws protecting marine mammals from the naive actions of some individuals and claims of success have been reported without proper conditioning, scientific verification, or follow-up, and in most cases it is likely that these animals succumbed to the challenges of life in the wild. However, reintroducing captive-born animals is not without success.

A number of captive-born terrestrial mammal species have been released into historical home ranges with success. Endangered animals such as red wolves (*Canis rufus*), golden lion tamarins (*Leontopithecus rosalia*), and black-footed Ferrets (*Mustela nigripes*) are some examples of well-managed and scientifically sound protocols that are helping in the recovery of wild populations. Due to their endangered status and the biological necessity to make their genetic material available to the wild population, the risks of such programs are deemed acceptable. These animals are carefully managed at reputable facilities, candidate animals are selected based upon specific criteria, population and environmental dynamics carefully reviewed, behavioral repertoires (such as predator avoidance and foraging skills) are strengthened, tracking and follow-up protocols scrutinized, unbiased scientific review completed, and legal permits obtained before such an undertaking begins. Research into the release of captive-born marine mammals, or animals housed long-term, must include an analysis of behavior, a strict behavioral management plan, and the measurable observation of behaviors that enhance survivability in the wild.

In addition, in order for this program to fulfill a biological imperative, the candidate animal must be assimilated into a social group so that breeding and reproduction completes the conservation objective. If a future opportunity to collect endangered marine mammals for captive breeding finds strong support, then using non-endangered marine mammals such as bottlenose dolphins as a concept model for successful captive breeding and reintroduction of endangered species is certainly prudent. Animal training and behavior management will undoubtedly play a role in the future of reintroductions.

### XI. Conservation

Clearly recognizing the valuable contributions made by zoos and aquariums in the conservation of marine mammal species, the US Marine Mammal Protection Act (MMPA) specifically authorizes (and periodically re-authorizes) qualified institutions to collect wild marine mammals for the purpose of public display, education and research (Marine Mammal Permits and Authorizations, 2006). Advances in marine mammal training are contributing to a new wave of public interest among marine life parks. Each year, worldwide attendance continues to grow and in the United States, public support stands firmly behind both the MMPA and marine mammal facilities. More people are now enjoying the benefits of new and exciting training programs, shows, presentations, interaction opportunities, and scientific discoveries, all facilitated through behavior management.

By maintaining a healthy captive population of various marine mammal species, comparative data are generated to assist in understanding wild animals, and these facilities continue to give material support to important research and conservation initiatives. In addition, these facilities act as part of the Marine Mammal Stranding Network, assisting NOAA/NMFS in the rescue, housing, and care of stranded wild animals where expertise in medical care can be applied. These facilities also develop animal management and husbandry skills in staff members who are also able to assist in health assessment studies or during mass strandings, as directed by NMFS. Captive marine mammals and the facilities in which they live provide a template for handling highly endangered animals, or the establishment of a controlled breeding group, when the need arises. Finally, these facilities maintain a living repository of genetic material accessible through behavior management, where candidate animals can be carefully and scientifically selected and trained for introduction into the wild if needed. This successful model has been pioneered in zoological parks, saving numerous species from extinction, including the black-footed ferret and California condor (*Gymnogyps californianus*) (Top 10 Wildlife Conservation Success Stories, 2006). Behavior management plans are an intricate and important part of these programs, helping to generate public interest through fascinating presentations, dynamic behaviors, and invaluable research. These contributions are undoubtedly saving the lives of wild marine mammals.

## XII. Conclusion

As formal behavior management gains wider application, wildlife managers will continue to increase their knowledge and application of animal learning principles, particularly as they relate to the conservation of highly endangered species, shrinking populations, and population recovery. Solutions to man/animal encounters, fisheries depletion, survival skill acquisition, translocation, reintroduction, and relocation efforts will require a specialized understanding of the learning and training processes, pioneered in the learning laboratories, and applied in the field of marine mammal training and behavior management.

## Acknowledgments

We wish to express our sincere gratitude to all the professionals who dedicate their careers to advancements in the care, training, and research of marine mammals. We are especially grateful to Mr. Tom Sanders for his professional input, behavioral review and suggestions for content, as well as his efforts to coordinate file transfers and author communications between Gavutu, Nadi, Los Angeles, Columbus and Cleveland…welcome back.

## See Also the Following Articles

Captivity ■ Behavior, Overview ■ Captive Breeding ■ Ethics and Marine Mammals ■ Marine Parks and Zoos

## References

Baldwin, B., and Baldwin, J. (1998). "Behavior Principles in Everyday Life." Simon & Schuster, New Jersey.

Chance, P. (1994). "Learning and Behavior," 3rd Ed. Brooks/Cole Publishing Company, Belmont.

Cole, K. C., Van Tilburg, D., BurchVernon, A., and Riccio, D. C. (1996). The importance of context in the US preexposure effect in CTA: Novel versus latently inhibited contextual stimuli. *Lear. Motiv.* **27**, 362–374.

Domjan, M. (1993). "The Principles of Learning and Behavior," 3rd Ed. Brooks/Cole Publishing Company, Belmont.

Honig, W. K., and Staddon, J. E. R. (1977). "The Handbook of Operant Behavior." Prentice-Hall, Inc, Englewood Cliffs.

Kazdin, A. E. (1994). "Behavior Modification in Applied Settings," 5th Ed. Brooks/Cole Publishing Company, Belmont.

Marine Mammal Permits and Authorizations. (2006). [Accessed online July 5, 2007]. Available from World Wide Web: http://www.nmfs.noaa.gov/pr/permits/mmpa_permits.htm

Marine Mammal Poll. Harris Interactive. (2005). [Accessed online July 10, 2007]. Available from World Wide Web: http://www.ammpa.org/_docs/HarrisPollResults.pdf

Pryor, K. (1995). "On Behavior: Essays and Research," 1st Ed. Sunshine Books, North Bend.

Ramirez, K. (1999). "Animal Training: Successful Animal Management through Positive Reinforcement." Shedd Aquarium Press, Chicago.

SeaWorld. (2006). The 7th Anniversary of J.J.'s Rescue." Press release, issued spring 2006.

Shepherdson, D. J., Mellen, J. D., and Hutchins, M. (1998). "Second Nature: Environmental Enrichment for Captive Animals." Smithsonian Institution Press, Washington, DC.

Spear, N. E., and Riccio, D. C. (1994). "Memory: Phenomena and Principles." Allyn & Bacon, Needham Heights.

Spradlin, T. R, Drevenak, J. K., Terbush, A. D., and Nitta, E. T. (1999). Interactions between the public and wild dolphins in the United States: Biological concerns and the Marine Mammal Protection Act. *In* "Wild Dolphin Swim Program Workshop," held in conjunction with the 13th Biennial Conference on the Biology of Marine Mammals, November 28, 1999. Silver Spring.

Top 10 Wildlife Conservation Success Stories in 2006. (2006). [Accessed online June 11, 2007]. Available from World Wide Web: http://www.aza.org/Newsroom/PR_TopTenStories2006

Turner, T., and Stafford, G. (2000). Rapid weight fluctuations linked to increased aggression in intact male California sea lions (*Zalophus californianus*). *Mar. Mamm. Pub. Disp. Res.* **4**, 14–20.

U.S. Department of Commerce, National Oceanic and Atmospheric Administration, Office of the Administrative Law Judge In the Matter of: Richard O'Barry United States of America. (1999). [Accessed online June 25, 2007]. Available from World Wide Web: http://www.animallaw.info/cases/caus1999noaalexis1.htm

Willis, K. (2007). Life expectancy of bottlenose dolphins in Alliance of Marine Mammal Parks and Aquariums' North American member facilities: 1990—present. Presented at the 2007 meeting of the Alliance of Marine Mammal Parks and Aquariums.

# Tucuxi and Guiana Dolphin
## *Sotalia fluviatilis and S. guianensis*

Paulo A.C. Flores and Vera M.F. Da Silva

## I. Characteristics and Taxonomy

The genus *Sotalia* of the family Delphinidae was once considered to comprise five species, but in the twentieth century, this was reduced to two, the riverine *Sotalia fluviatilis* and the marine *Sotalia guianensis*. Later these were further lumped into a single species

(*S. fluviatilis*), with marine and riverine ecotypes. Recent morphological and genetic studies, however, concluded that marine and riverine *Sotalia* are different species (Cunha *et al.*, 2005; Caballero *et al.*, 2007). Based on priority criteria, the name *Sotalia guianensis* (Van Bénéden 1864) was assigned to the marine animals, whereas riverine dolphins retained the oldest species name *Sotalia fluviatilis* (Gervais 1853). No fossil record is known.

The common name tucuxi comes from *tucuchi-una* after the Tupi language of the Mayanas Indians from the Amazon region of Brazil, where it is called *boto-tucuxi*, *boto-cinza*, or simply *boto*. In the other Amazon countries it is usually called *delfin* or *bufeo gris del rio*. *S. guianensis* is also known simply as *boto* or *golfinho* and as *boto comum* and *golfinho cinza* along the Brazilian coast; *bufeo gris*, *bufeo blanco*, or *bufeo negro* in Colombia and Peru; *tonina de rio*, *delfin blanco*, or *soplón* in the Venezuela Amazon, *tonina del lago* in Lake Maracaibo, and *bufeo negro*, *bufete*, or *soplón* in the Orinoco River basin in Venezuela; *lam* in Nicaragua; Guyana dolphin or Guiana white dolphin in Guyana; and *profuso* or *dolfijn* in Surinam. There is some controversy about a definitive international common name for *S. guianensis* in English. Various names have been used in the literature, most frequently marine tucuxi, gray dolphin, estuarine dolphin, and recently costero. We avoid the controversy here by using "Guiana dolphin," based on the Scientific name.

The two *Sotalia* species are very similar in coloration, differing mainly in body size and number of teeth, and somewhat resembling a small bottlenose dolphin, *Tursiops* (Fig. 1). They are light gray to bluish gray on the back and pinkish to light gray ventrally, with a distinct line from the mouth gape to the flipper's leading edge. There is a lighter area on the flank between the flippers and the dorsal fin and

another mid-body at the level of the anus. The marine species has another light gray rounded streak on both sides of the caudal peduncle. In both species, the eyes are large, and there is black countershading around the eyes. A case of atypical white coloration was recently reported, although it was not confirmed whether it was albinism or another type of anomalously light pigmentation. The dorsal fin is triangular and sometimes slightly hooked on the tip. The tucuxi has a moderately slender beak, a rounded melon and 26–36 teeth in each mandibular ramus. The Guiana dolphin has more upper teeth and is larger than the tucuxi, with a maximum total length of 220 cm and about 80 kg body mass vs a maximum length of around 152 cm and mass of 55 kg.

## II. Distribution and Abundance

The tucuxi occurs in the main tributaries of the Amazon/Solimões River basin in Brazil as far inland as southeastern Colombia, eastern Ecuador, and northeastern Peru, with records in all three types of water that occur in this region. Several rivers contain impassable falls, rapids, and shallow waters. On the tributaries of the right side of the Amazon basin, the Teotônio and Santo Antônio Falls on the Madeira River, the Santa Isabel Falls on the R. Xingú, and S. Luis Falls on the Tapajós river are impassable barriers, whereas on the left side the falls on the Rio Negro and Raudal La Liberdad on the Caquetá River (Colombia) are also important. The tucuxi does not occur in the Beni/Mamoré River basin in Bolivia and is not known in the upper Rio Negro. The presence of the species in the Orinoco River basin is still controversial, since a stretch of rapids and falls in the Negro River and the 354 km of numerous rapids and outcrops of

Figure 1  *Guiana dolphins*. (Photo by P.A.C. Flores).

the Cassiquiare Channel block the species' movements. Its distribution is influenced by seasonal river level fluctuations, with channels and lakes occupied during rising and high waters but avoided at low water. The tucuxi is abundant in the Solimões and Japurá Rivers as well as in large black water lakes such as Tefé Lake (Brazil) and the El Correo lakes system (Colombia). Tucuxis do not go into flooded forest as does the sympatric boto, *Inia geoffrensis*, but these species share a preference for areas with reduced current and waterway junctions. Mean density along the margins of main rivers in the central Amazon, Brazil within 150 m survey strip of 1,319.7 km was 3.2 individuals per km², with 54% of the individuals occurring within 50 m of the edge (Martin *et al.*, 2004). At the border of Colombia, Brazil, and Peru, Vidal *et al.* (1997) found along about 120 km of the Amazon River a density of 8.6/km² in lakes, 2.8 along main banks and 2.0 around islands (Fig. 2).

The Guiana dolphin is found in the Western Atlantic coastal waters of South and Central America from southern Brazil (27°35'S, 48°35'W) to Nicaragua (14°35'N, 83°14'W), including Colombia, Costa Rica, French Guyana, Guyana, Panama, Suriname, Trinidad, Venezuela, and possibly Honduras (15°58'N, 79°54'W). In the Orinoco River dolphins seen as far up as Ciudad Bolivar may be of this species. The Guiana dolphin is found mostly in estuaries, bays, and other protected shallow coastal waters, although it has also been recorded at the Abrolhos Archipelago, around 70 km off the coast of Bahia State, Brazil. The species' southernmost limit is influenced by the cold waters of the Malvinas current in South Brazil. It is notably recorded throughout the year in many coastal locations such as Baía Norte in Santa Catarina State, Cananéia Estuary and Baía de Guanabara (both in southeastern Brazil), Baía de Todos os Santos and around Fortaleza (northeastern Brazil), Bahia Cispatá and Golfo de Morrosquillo (Colombia), as well as on the Cayos Miskito Coast in Nicaragua. Standard abundance estimates are scarce for the Guiana dolphin, but the species seems to be abundant in various locations along its distribution, mainly in South-Southern Brazil outside of Guanabara Bay. Stocks or significant evolutionary units are evident from residency, site fidelity, genetic and acoustical data.

## III. Ecology

The Guiana dolphin feeds on neritic prey distributed through the water column, mainly on neritic fishes such as clupeids and sciaenids, but cephalopods, shrimps, crabs, and flounders are occasionally taken. Usually young specimens of these teleost fishes, including over 20 species, are the most important diet items. Tucuxis feed mainly on schooling pelagic fish such as characiforms, freshwater clupeids, and sciaenids, no larger than 35 cm. Feeding may occur in pairs, usually mother and calf, and in larger groups or subgroups when different strategies and cooperation among individuals are employed. During feeding activities, Guiana dolphins often associate with birds such as the brown booby (*Sula leucogaster*), terns (*Sterna* spp.), frigate bird (*Fregata magnificens*), and kelp gull (*Larus dominicanus*). Mixed-species flocks of up to a hundred birds can be seen in such associations. In the Amazon, tucuxis may feed occasionally in association with terns (*Phaetusa simplex*). These associations are initiated by the birds and have no impact on the dolphins.

There are no known predators for either species, although bites from sharks of unidentified species have been seen on Guiana dolphins.

Because of the huge regional differences in habitats from temperate waters in the south to the tropical waters, including estuaries such as the Amazon estuary, Guiana dolphins are found in a wide range of water depth, temperature, salinity, and turbidity.

No mass stranding has been reported. Individuals often wash ashore, sometimes due to incidental catch in fisheries in both marine and freshwater environments.

## IV. Behavior and Physiology

*Sotalia* dolphins show a variety of aerial behaviors such as full leaps, somersaults, fluke-ups, spy-hopping, surface rolling, and porpoising. In coastal areas, feeding and traveling are by far the most common behaviors, although resting and milling are rare. Socializing involves various tactile contacts, and herding of females by males has been occasionally seen in southern Brazil. Bow-riding has not been recorded, but Guiana dolphins may surf in waves and wakes produced by passing boats.

Spontaneous swimming interactions with domestic dogs (*Canis familiaris*) and a lone wild Guiana dolphin sociable toward humans were recorded in southeastern Brazil. Epimeletic behavior and hand feeding were also recorded in the same area (Santos *et al.*, 2000). Apparent mating behavior with bottlenose dolphins was recorded off Costa Rica In Baía Norte, South Brazil, at the southernmost distributional limit, Guiana dolphins do not associate with bottlenose dolphins, and rare encounters even result in aggression by bottlenose dolphins or escape behavior by the Guiana dolphins (Flores and Fontoura, 2006). Epimeletic behavior toward an offspring was recorded at that locality.

Dives for Guiana dolphins last about 30–120 sec with shorter dives of 5–10 sec in between. Tucuxis are fast swimmers, spending less than a second at the surface, with an average dive time of about 2 min.

The *Sotalia* species are social dolphins, almost always in cohesive groups engaged in the same activities. Tucuxis are often found in groups of one to six individuals, although larger groups up to 20 individuals are also recorded (Faustino and da Silva, 2006). Groups of up to 50 or 60 Guiana dolphins are common, whereas the average group size is two to six. Large aggregations of up to 200 are reported at Baía de Sepetiba and around 400 individuals in Baía da Ilha Grande on the

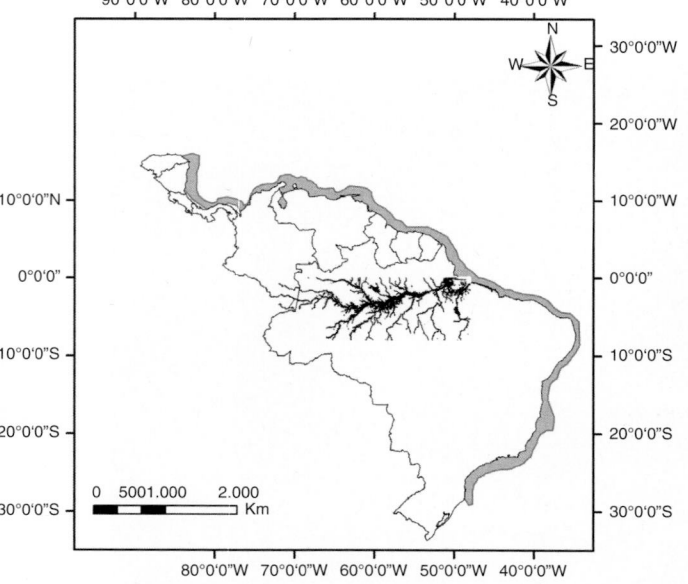

**Figure 2** *Distribution of the tucuxi* Sotalia fluviatilis *and the Guiana dolphin* S. guianensis.

**T**

southeastern Brazilian coast, where these larger aggregations are usually engaged in cooperative feeding. Apparently, larger groups are more common in the south and southeastern Brazilian coast. Mixed groups of adults and calves are common. Individual associations are known only for the Cananéia Estuary population in Brazil; these are weak to moderate, except for a few pairs of individuals with apparently stronger associations, suggesting a relatively fluid society with individuals in fission–fusion (Santos and Rosso, in press).

Photo-identification studies have shown that Guiana dolphins may be resident within and between years for up to 10 consecutive years (e.g., Flores 1999; Flores and Bazzalo, 2004). Home ranges are poorly known and apparently among the smallest for small cetaceans with a mean of about $15 km^2$ in southern Brazil and up to $265 km^2$ in another location. Movement patterns vary among warm and cold seasons in the temperate region, whereas no variation was found in warm waters. In any case, daily movements are small. Freeze-branded tucuxis in Central Amazon were recorded for several years in the same area, suggesting residency and seasonal use of areas.

Comparative analysis of the whistles of Guiana dolphins in different areas along the Brazilian coast revealed significant effects of geographical location. However, is difficult to discriminate between adjacent populations. Guiana dolphins produce mainly upsweep whistles, shorter and less complex in shape than for other species of dolphins. The range of whistle fundamental frequencies recorded was 0.21–24 kHz and durations 38–1064 ms (Azevedo and Van Sluys, 2005).

Research with acoustic pingers in Fortaleza, Brazil, during 345 h of experiment showed that Guiana dolphins avoided areas where pingers were active (Monteiro-Neto et al., 2004).

## V. Life History

Calving is year-round and gestation is estimated to be around 11–12 months for the Guiana dolphin, with calves ranging in size from 90 to 100 cm of total length. Calving interval is believed to be 22–24 months based on photo-identification data. Tucuxi calving occurs between September and November, during low water season, after a gestation time estimated at 11 months, with calves at birth measuring from 71 to 83 cm (da Silva and Best, 1994).

According to tooth growth layer groups (GLGs), life span can reach 30 and 35 years for the Guiana dolphin and tucuxi, respectively. Natural mortality rates are unknown for both species.

## VI. Interactions with Humans

Historically, these species have not been exploited commercially, although incidental mortality in local and commercial fisheries such as those using gillnets and seines are a direct threat to *Sotalia* dolphins. Bottom-set nets for lobsters also occasionally capture Guiana dolphins. On the coasts of Amapá, Maranhão, and Pará States, northern Brazil, Guiana dolphins are killed for shark bait, although they have some protection from myths and legends. This is especially true for the tucuxis in the Amazon (Gravena et al., in press). There, their genital organs and eyes have a local market as love charms, and teeth and bones are used for arts and crafts. Guiana dolphins in some parts of their distribution, mainly on the northern and northeastern coasts of Brazil, may also be used for human consumption. Although these dolphins are fully protected by Federal laws in Brazil, forbidding the harassment, hunting, fishing, or capture of tucuxis and all cetaceans, pinnipeds, and sirenians in national waters, law enforcement may not be effective. In other countries such as Colombia, Ecuador, Peru, and Venezuela, tucuxis are not clearly protected by laws.

Acoustic pingers attached to gill nets may successfully reduce or prevent by catch of Guiana dolphins as suggested by a single study conducted with free ranging dolphins in northeastern Brazil (Monteiro-Neto et al., 2004).

Dams and hydroelectric power facilities in the Amazon region interrupt fish migration, reducing fish abundance, and consequently prey availability for dolphins. Mercury from gold mining, water pollution, seismic activities, oil spills, and boat traffic are other potential threats to tucuxis in the Amazon, while the same factors, except gold mining, plus marine culture farms and destruction of habitats, mainly mangroves and salt marshes, strongly affect the Guiana dolphins. Hand feeding and the behavioral effects caused by boat activities also deserve concern, as these may affect at least populations off the coasts. Bioaccumulation of contaminants and growing pollution outfalls are also concerns.

*Sotalia* dolphins are susceptible to capture stress, quickly become entangled and sometimes suffocate in nets, and are not robust to long periods of transportation or handling after capture. However, Guiana dolphins captured off the coast of Panamá in the late 1970s were kept in captivity in Europe for more than 20 years, and one animal is still alive today. A few Guiana dolphins are still kept in Colombian facilities, although since 2005 it has been illegal to maintain them in captivity.

The separation of the two species is too recent to appear in any of the Species Conservation Status lists, although both *Sotalia* species should be listed as "insufficiently known" by the World Conservation Union (IUCN) as was the status of the unified species. Because of its coastal habits, aggregating in estuaries and bays, and in river channels and lakes, *Sotalia* dolphins are vulnerable to almost all human activities throughout their range. A large proportion of the distributional area of the two species is close to human habitation. Consequently, these habitats are subject to intense fisheries, boat traffic and sewage, industrial waste, and high levels of contaminants. Examples are the Santos and Rio de Janeiro harbors in southeast Brazil, Recife and Rio Grande do Norte in the northeast Brazil, Maracaibo in Venezuela, Golfo de Morrosquillo in Colombia, and Belém, Santarém and Manaus in the Brazilian Amazon.

## See Also the Following Articles

Delphinids, Overview ■ South American Marine Mammals

## References

Azevedo, A. F., and Van Sluys, M. (2005). Whistles of the tucuxi dolphin (*Sotalia fluviatilis*) in Brazil: Comparisons among populations. *J. Acoust. Soc. Am.* **117**, 1456–1464.

Caballero, S., et al. (2007). Taxonomic status of the genus Sotalia: Species level ranking for "tucuxi" (*Sotalia fluviatilis*) and "costero" (*Sotalia guianensis*) dolphins. *Mar. Mam. Sci.* **23**, 358–386.

Cunha, H. A., et al. (2005). Riverine and marine *Sotalia* (Cetacea: Delphinidae) are different species. *Mar. Biol.* **148**, 449–457.

Faustino, C., and da Silva, V. M. F. (2006). Seasonal use of Amazon floodplains by the tucuxi *Sotalia fluviatilis* (Gervais 1853), in the Central Amazon, Brazil. *Lat. Am. J. Aquat. Mamm.* **5**, 95–104.

Flores, P. A. C. (1999). Preliminary results of a photo identification study of the marine tucuxi, *Sotalia fluviatilis*, in southern Brazil. *Mar. Mamm. Sci.* **15**, 840–847.

Flores, P. A. C., and Bazzalo, M. (2004). Home ranges and movements patterns of the marine tucuxi *Sotalia fluviatilis* in Baía Norte, southern Brazil. *Lat. Am. J. Aquat. Mamm.* **3**, 37–52.

Flores, P. A. C., and Fontoura, N. F. (2006). Ecology of marine tucuxi, *Sotalia guianensis*, and bottlenose dolphin, *Tursiops truncatus*, in Baía Norte, Santa Catarina State, southern Brazil. *Lat. Am. J. Aquat. Mamm.* **5**, 105–115.

**T**

Gravena, W., Hrbek, T., da Silva, V. M. F. and Farias, I. P. (2008). Amazonian pink dolphin love fetishes: From folklore to molecular forensics. *Mar. Mamm. Sci.*

Martin, A. R., da Silva, V. M., and Salmon, D. L. (2004). Riverine habitat preferences of botos (*Inia geoffrensis*) and tucuxis (*Sotalia fluviatilis*) in the Central Amazon. *Mar. Mamm. Sci.* **20**, 189–200.

Monteiro-Neto, C., Itavo, R. V., and Moraes, L. E. S. (2003). Concentrations of heavy metals in *Sotalia fluviatilis* (Cetacea: Delphinidae) off the coast of Ceará, northeast Brazil. *Env. Poll.* **123**, 319–324.

Monteiro-Neto, C., *et al.* (2004). Behavioral responses of *Sotalia fluviatilis* (Cetacea, Delphinidae) to acoustic pingers, Fortaleza, Brazil. *Mar. Mamm. Sci.* **20**, 145–151.

Nascimento, L. F. *et al.* (2007). Atypical coloration in a specimen of estuarine dolphin, *Sotalia guianensis*, on the littoral of the state of Rio Grande do Norte, north-east Brazil. *JMBA2 – Biodiv. Rec.* (Published online) 2 pp.

Rossi-Santos, M. R., and Podos, J. (2006). Latitudinal variation in whistle structure of the estuarine dolphin *Sotalia guianensis*. *Behaviour* **143**, 347–364.

Santos, M. C.de. O., and Rosso, S. (2008). Social organization of marine tucuxi dolphins, *Sotalia guianensis*, in the Cananéia estuary, southeastern Brazil. *J. Mammal.* **89**, 347–355.

Santos, M. C. O., Rosso, S., Siciliano, S., Zerbini, A. N., Zampirolli, E., Vicente, A., and Alvarenga, F. (2000). Behavioural observations of the marine tucuxi dolphin (*Sotalia fluviatilis*) in São Paulo estuarine waters, Southeastern Brazil. *Aq. Mamm.* **26**, 260–267.

da Silva, V. M. F., and Best, R. C. (1994). Tucuxi *Sotalia fluviatilis* (Gervais, 1853). *In* "Handbook of Marine Mammals—The First Book of Dolphins" (S. H. Ridgway, and R. J. Harrison, eds), Vol. 5, pp. 43–49. Academic Press, London.

Van Bressem, M.-F., *et al.* (24 authors) (2007). A preliminary overview of skin and skeletal diseases and traumata in small cetaceans from South American waters. *Lat. Am. J. Aquat. Mamm.* **6**, 7–42.

Vidal, O., Barlow, J., Hurtado, L. A., Torre, J., Cendón, P., and Ojeda, Z. (1997). Distribution and abundance of the Amazon River dolphin (*Inia geoffrensis*) and the tucuxi (*Sotalia fluviatilis*) in the upper Amazon River. *Mar. Mamm. Sci.* **13**, 427–445.

Yogui, G. T., Santos, M. C. O., and Montone, R. C. (2003). Chlorinated pesticides and polychlorinated biphenyls in marine tucuxi dolphins (*Sotalia fluviatilis*) from the Cananéia estuary, southeastern Brazil. *Sci. Tot. Env.* **312**, 67–78.

# The Tuna-Dolphin Issue

## Tim Gerrodette

### I. The Problem

In the tropical waters of the Pacific Ocean west of Mexico and Central America, large yellowfin tuna (*Thunnus albacares*) swim together with several species of dolphins: pantropical spotted (*Stenella attenuata*), spinner (*S. longirostris*), and common (*Delphinus delphis* and *D. capensis*) dolphins. This ecological association of tuna and dolphins is not clearly understood, but it has had two important practical consequences: it has formed the basis of a successful tuna fishery, and it has resulted in the deaths of a large number of dolphins. This is the heart of the tuna-dolphin issue.

The bycatch of dolphins in the eastern tropical Pacific (ETP) purse-seine tuna fishery stands apart from marine mammal bycatch in other fisheries, not only in scale but also in the way the dolphins interact with the fishery. Marine mammals interact with most fishing gear only incidentally, but in the ETP tuna fishery the dolphins are an intrinsic part of the fishing operation (Perrin, 1969). The fishermen intentionally capture both tuna and dolphins together, then release the dolphins from the net (National Research Council, 1992). Further, unlike in most other fisheries, the vast majority of dolphins captured by the ETP tuna fishery are released alive; thus, an individual dolphin may be chased, captured, and released many times during its lifetime.

The number of dolphins killed since the fishery began in the late 1950s is estimated to be over 6 million animals, the highest known for any fishery. For comparison, the total number of whales of all species killed during commercial whaling in the twentieth century was about 2 million. The bycatch of dolphins in the ETP tuna fishery has now been successfully reduced by more than 99%, but even at the present level of 1500 dolphins/year, it remains among the largest documented cetacean bycatches in the world.

### II. Purse-Seining for Tuna

Prior to the development of modern purse seines, tropical tuna were caught one at a time using pole-and-line methods. In the late 1950s, the twin technological developments of synthetic netting that would not rot in tropical water and a hydraulically driven power-block to haul the net made it possible to deploy very large purse-seine nets around entire schools of tuna, and thus to catch many tons of fish at a time. Purse-seining for tuna in the ETP can be conducted in one of three ways: the net may be set around schools of tuna associated with dolphins ("dolphin sets," which catch large yellowfin tuna), around schools of tuna associated with logs or other floating objects ("log sets," which catch mainly skipjack but also bigeye and small yellowfin tuna), or around unassociated schools of tuna ("school sets," which catch small yellowfin and skipjack tuna). The proportions of different set types have varied over the history of the fishery, but in recent years, about half have been dolphin sets, one quarter log sets and one-quarter school sets.

Dolphins are killed almost exclusively in dolphin sets. During "porpoise fishing" (the fishermen's term), schools of tuna are located by first spotting the dolphins or the seabird flocks which are also associated with the fish. Speedboats are used to chase down the dolphins and herd them into a tight group; then the net is set around them (Fig. 1). The tuna-dolphin bond is so strong that the tuna stay with the dolphins during this process, and tuna and dolphins are captured together in the net (Fig. 2). Dolphins are released from the net during the backdown procedure (Fig. 3). If all goes well, the dolphins are released alive, but the process requires skill by the captain and crew, proper operation of gear, and conducive wind and sea conditions. As with any complicated procedure at sea, things can go wrong, and when they do, dolphins may be killed.

From an ecosystem perspective, management of the ETP purse-seine tuna fishery poses interesting challenges. The three methods of purse-seining for tuna, log-, school- and dolphin-fishing, catch different mixes of tuna species and sizes, and in addition have different amounts and composition of bycatch. Dolphin sets result in dolphin mortality, but dolphin sets have the least bycatch overall. Log sets have about 30 times the bycatch of school sets by weight per set, which in turn have about 3 times the bycatch of dolphin sets. Most of the bycatch, though, even on dolphin sets, is fish, primarily tuna, marlin, and dorado. These fish have much higher reproductive rates than dolphins, sea turtles, sharks and rays, so the effect of the bycatch is smaller. Although the effects of the fishery on dolphin populations have been strong and are relatively well known, the effects on other marine populations of concern, such as sharks and sea turtles, are mostly unknown.

Figure 1  *Purse-seine being set on tuna and dolphins in the eastern tropical Pacific Ocean. The net is not yet closed, and four speedboats are driving in tight circles near the opening to keep the dolphins from escaping.*

Figure 2  *Spotted dolphins in the purse-seine net. The submerged corkline is being pulled out from beneath them as the boat tows the net in reverse to release the animals.*

Figure 3  *Backdown procedure in progress. As the tuna vessel moves backwards, the net is drawn into a long channel. The corkline at the far end is pulled under water slightly, and the dolphins escape. Speedboats are positioned along the corkline to help keep the net open.*

## III. Actions to Reduce the Dolphin Bycatch

The magnitude of dolphin mortality in the ETP tuna fishery first came to widespread attention in the mid-1960s. The dolphin kill at that time is not known with precision, but without question was very high (Fig. 4). When the US Marine Mammal Protection Act was passed in 1972, it included provisions for reducing the bycatch to "insignificant levels approaching zero" after a 2-year moratorium on regulation during which the tuna industry was expected to solve the problem through development of improved fishing methods. Under this law, scientific studies were initiated, observers were placed on fishing boats, fishing gear was inspected, and boat captains with high dolphin mortality

rates were reviewed. Modifications of fishing gear and procedures were developed to reduce dolphin kill. After much litigation, the first regulations to reduce the dolphin kill on US vessels were promulgated (Gosliner, 1999). By the end of the 1970s, the kill had declined from about 500,000 to about 20,000 dolphins per year (Fig. 4).

As the US tuna fleet decreased in size and the fleets of Mexico, Venezuela, Ecuador, and other Latin American countries increased, the dolphin kill began to grow again, and actions to monitor and reduce the dolphin bycatch became international. The Inter-American Tropical Tuna Commission began a dolphin conservation program in 1979 modeled on the US effort. By 1986, an international

T

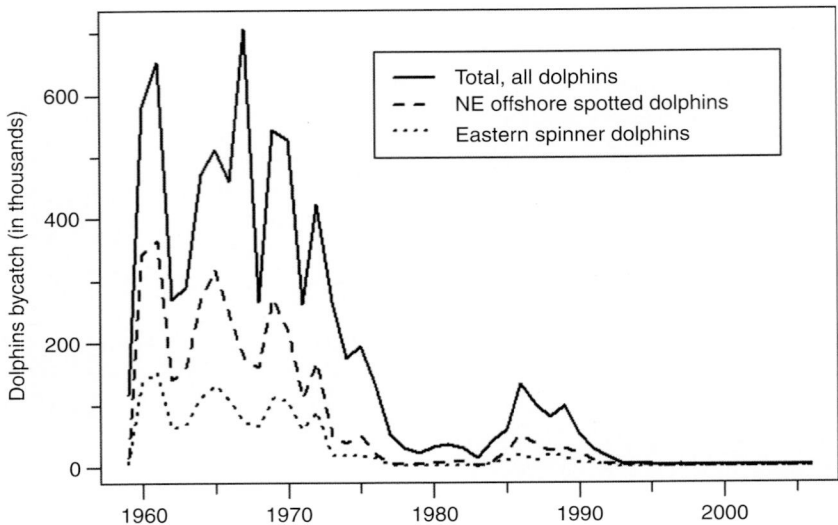

**Figure 4** *Estimated annual number of dolphins killed in the eastern tropical Pacific purse-seine tuna fishery, total for all dolphins and separately for the two dolphin stocks with the highest number killed.*

observer program with all countries participating showed that total dolphin mortality had increased to 133,000/year (Fig. 4). Because US boats operated under restrictions that did not apply to boats of other countries, the United States began requiring that imported tuna be caught at dolphin mortality rates comparable to US boats (Gosliner 1999). The concept of Dolphin-Safe tuna—tuna caught without setting on dolphins (i.e., in log and school sets)—became popular, and by 1994, only Dolphin-Safe tuna could be sold in the United States. These trade actions were important because the United States is the largest market for the canned tuna product of the fishery.

The dolphin kill declined between 1986 and 1993 due to these various political and economic pressures (Fig. 4). Starting in 1993, the ETP fishing countries decided to increase observer coverage, institute skipper review panels, and meet a schedule of decreasing dolphin quotas on an individual boat basis (the La Jolla Agreement). The Declaration of Panama of 1995 carried these ideas further, proposing observers on every boat over 400 tons and strict by-stock dolphin mortality limits. These features became part of the International Dolphin Conservation Program Agreement (Hedley, 2001), a binding document among the major fishing countries that went into force in 1999. By this time total reported dolphin mortality had fallen to fewer than 3000 dolphins/year.

The Declaration of Panama also called for the United States to change its definition of Dolphin-Safe tuna to include tuna caught by setting on dolphins, as long as no dolphins were observed killed or seriously injured *on that set*. Before changing the Dolphin-Safe label, however, the US undertook studies to determine if the process of chasing and encircling dolphins was having a significant adverse impact on depleted dolphin populations. At the conclusion of the studies in 2002, the US National Marine Fisheries Service decided that the fishery was not significantly affecting the dolphin populations, which would have allowed the less restrictive definition of Dolphin-Safe to take effect. The decision was immediately challenged in court by environmental groups, and overturned because the decision was based on "political meddling," not science. Therefore, the original definition of Dolphin-Safe applies to all canned tuna sold in the United States. Tuna caught by setting on dolphins may also be sold but may not be labeled

Dolphin-Safe. The Inter-American Tropical Tuna Commission certifies tuna as Dolphin-Safe under the Declaration of Panama definition, and such tuna is sold in parts of Europe and throughout Latin America.

## IV. Status of the Dolphin Populations

The status of ETP dolphin stocks (management units) is based on two time-series of data: estimates of the number of dolphins killed, based on data from observers on tuna vessels (Fig. 4), and estimates of abundance, based on data from research vessel surveys (Gerrodette and Forcada, 2005; Fig. 5). Combining these data in a population model has indicated that the stocks most affected by the tuna fishery are the northeastern stock of the offshore pantropical spotted dolphin (*S. attenuata attenuata*) and the ETP endemic subspecies, the eastern spinner dolphin (*S. longirostris orientalis*). Both populations declined between 1960 and 1975 during the period of high mortality on US boats but have remained approximately constant since then (Fig. 5). As of 2002, northeastern spotted and eastern spinner dolphins were estimated to be at 19% and 29%, respectively, of population sizes when the fishery began (Wade *et al*., 2007). Other stocks have apparently been less affected, although little is known of the small populations of coastal forms of spotted and spinner dolphins.

Since the early 1990s, reported dolphin mortality has been low enough that the dolphin populations should have started to recover. As of 2002, however, neither dolphin population was recovering at expected rates (Wade *et al*., 2007). Hypotheses to explain the lack of recovery (Gerrodette and Forcada, 2005) have included underreporting of kill by observers, cryptic effects of the fishery not detectable by observers, such as stress, induced abortion or separation of mothers and calves (Archer *et al*., 2004; Noren and Edwards, 2007), long-term ecosystem changes, and a lag in recovery due to interactions with other species. In years with a high number of dolphin sets, there are fewer calves in the spotted dolphin population. Reproduction in both dolphin populations declined between 1993 and 2003, which is at least one reason why recovery has been at a lower-than-expected rate. On the other hand, perhaps pelagic dolphins inherently have low reproductive rates, and our expectation for rate of recovery needs to be

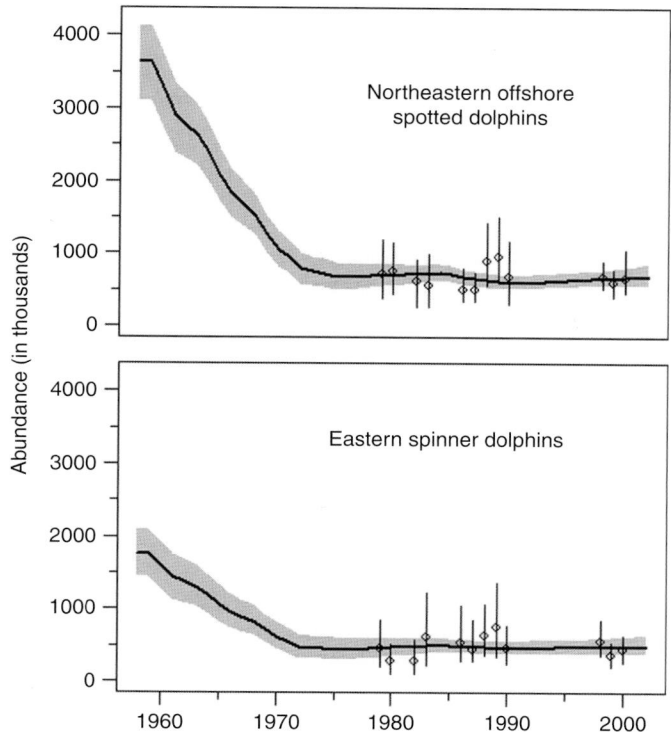

**Figure 5** *Estimated population trajectories of the two dolphin stocks most affected by tuna purse-seine fishing in the eastern tropical Pacific. Estimates of abundance between 1979 and 2000 are shown as points with 95% confidence intervals. The populations declined due to high numbers of dolphins killed in the tuna fishery from 1960 to 1975, as shown in Fig. 4.*

revised. Research is continuing, but until there are clear recoveries of the affected dolphin stocks, the tuna-dolphin issue is likely to remain highly controversial.

## See Also the Following Articles

Fishing Industry, Effects of Management

## References

Archer, F., Gerrodette, T., Chivers, S., and Jackson, A. (2004). Annual estimates of the unobserved incidental kill of pantropical spotted dolphin (*Stenella attenuata attenuata*) calves in the tuna purse-seine fishery of the eastern tropical Pacific. *Fish. Bull. US* **102**, 233–244.

Gerrodette, T., and Forcada, J. (2005). Non-recovery of two spotted and spinner dolphin populations in the eastern tropical Pacific Ocean. *Mar. Ecol. Prog. Ser.* **291**, 1–21.

Gosliner, M. L. (1999). The tuna-dolphin controversy. *In* "Conservation and Management of Marine Mammals" (J. R. Twiss, Jr, and R. R. Reeves, eds), pp. 120–155. Smithsonian Institution Press, Washington, DC.

Hedley, C. (2001). The 1998 Agreement on the International Dolphin Conservation Program: Recent developments in the tuna-dolphin controversy in the eastern Pacific Ocean. *Ocean Dev. Int. Law* **32**, 71–92.

National Research Council (1992). "Dolphins and the Tuna Industry." National Academy Press, Washington, DC.

Noren, S. R., and Edwards, E. F. (2007). Physiological and behavioral development in delphinid calves: Implications for calf separation and mortality due to tuna purse-seine sets. *Mar. Mamm. Sci.* **23**, 15–29.

Perrin, W. F. (1969). Using porpoise to catch tuna. *World Fishing* **18**, 42–45.

Wade, P. R., Watters, G. M., Gerrodette, T., and Reilly, S. B. (2007). Depletion of spotted and spinner dolphins in the eastern tropical Pacific: Modeling hypotheses for their lack of recovery. *Mar. Ecol. Prog. Ser.* **343**, 1–14.

T

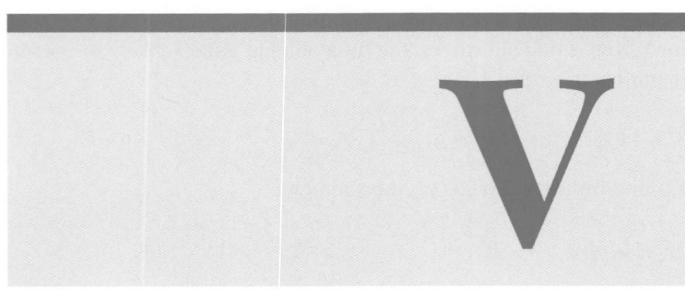

# Vaquita
*Phocoena sinus*

LORENZO ROJAS-BRACHO, ARMANDO M. AND
JARAMILLO-LEGORETTA

The Vaquita (*phocoena Sinus*), or Gulf of California harbor porpoise, is endemic to the upper Gulf of California, Mexico. The Vaquita was first discovered and described in 1958 (Norris and McFarland, 1958), however it was not until the past decades that important advances were made in the knowledge of this porpoise. It is the most critically endangered cetacean species, taken incidentally in fishing nets at an unsustainable rate. Without help it will soon be gone. Currently the Mexican Goernment is putting into action a recovery plan including provision for alternative livlihoods for the fishermen.

## I. Characteristics and Taxonomy

The vaquita is the smallest of all the porpoises (family Phocoenidae). The mean length for females is 140.6 cm; males are slightly smaller (134.9 cm). The vaquita differs from other phocoenids not only in total length, but also in that the flippers of the vaquita are proportionately larger and the dorsal fin is taller and more falcate. Generally, the pigmentation pattern consists of a dark gray cape, pale gray lateral field, and white ventral field. The most conspicuous features of the pigmentation are the relatively large black eye rings and lip patches (Fig. 1). Skulls of adult vaquitas are smaller and have relatively much broader and shorter rostra than those of other members of the genus. The number of teeth in each upper and lower jaw is 16–22 and 17–20, respectively (Brownell *et al.*, 1987; Vidal *et al.*, 1999). Norris and McFarland (1958) suggested that the vaquita had a recent common ancestor with the South American Burmeister's porpoise, *P. spinipinnis*, rather than with the harbor porpoise, *Phocoena phocoena*, its closest phocoenid geographic neighbor. This hypothesis was corroborated almost 37 years later by Rosel *et al.* (1995) using molecular techniques.

## II. Distribution and Abundance

The limited abundance, distribution, and narrow habitat specificity of vaquita makes it one of the rarest marine mammal species (Rojas-Bracho *et al.*, 2006). The most recent estimated population size (Jaramillo-Legorreta *et al.*, 1999) was 567 individuals during the summer of 1997 (CV: 0.51, 95% CI 177–1073). All sightings from systematic ship surveys indicate that its distribution is limited to an area north of 30°45′N (Silber *et al.*, 1994; Gerrodette *et al.*, 1995; Vidal, 1995; Jaramillo-Legorreta *et al.*, 1999) specifically to a small portion of the upper Gulf of California (Fig. 2). Since 1997, acoustic surveys have been carried out as an alternative to the expensive traditional

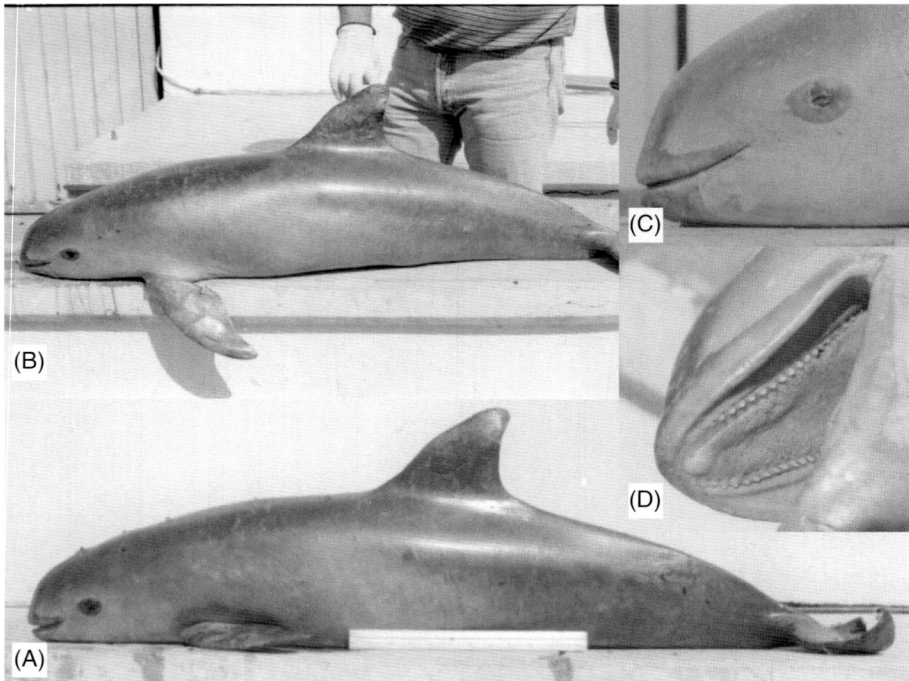

**Figure 1** *Morphology of the vaquita: (A,B) lateral view (note proportionally large dorsal and pectoral fins), (C) lateral view of the head, showing the dark patches surrounding the eye and lips, (D) ventral view of the mouth, showing the palate and the spade-shaped crowns of teeth.*

sighting surveys (Jaramillo-Legorreta *et al.*, 2005). The vaquita emits high frequency narrow clicks (Silber, 1991; Kamminga *et al.*, 1996) that can be automatically detected by specifically designed equipment (Chappell *et al.*, 1996; Gillespie and Chappell, 2002). Using this equipment it has been confirmed that the distribution is very limited, year-round. The core distribution area of acoustic encounters is approximately 1652 km$^2$, which must coincide with its optimal habitat and which overlaps that of the highest concentration of fishing effort with gill and mesh nets defined by Cudney-Bueno and Turk-Boyer (1998, fig. 2). This represents a very high risk for vaquita survival. In a worst-case scenario, having all or the majority of vaquitas in a small area makes the remaining population more vulnerable to a large by-catch event in an area of intense fishing (Rojas-Bracho and Taylor, 1999). The continued increase of fishing effort in the area allowed Jaramillo-Legorreta *et al.* (in press) to estimate that the current population level is, very probably, less than 40% of that estimated in 1997.

## III. Ecology

The vaquita inhabits an area of approximately 4000 km$^2$ (Rojas-Bracho *et al.*, 2006). This area is located toward the west coast of the Upper Gulf of California, where the most turbid waters of the region are encountered (Alvarez and Jones, 2004). The extent of this distribution area is so small that according to measured swimming speeds a vaquita can nearly traverse it in a single tidal cycle (Jaramillo-Legorreta *et al.*, 2005). The acoustic encounters with vaquitas have been constrained to the most turbid zone of the distribution area, which could be part of a *quiet strategy* to avoid predators in clearer waters.

Analysis of stomach contents of 34 vaquitas reveals them to be non-selective feeders (Findley *et al.*, 1995; Pérez-Cortés-Moreno, 1996). Prey consisted of a wide variety of demersal/benthic fishes (21 species), squid (2 species), and crustaceans (1 species, plus 2 that were fish parasites). Several of the fish prey species (such as croakers, family Scianidae) are known to be sound producers so it is possible that vaquitas are in part using passive sound rather than echolocation to find their prey.

Morphological abnormalities in the vertebrate, unusual number of digits (six), and pathological condition of the ovaries (calcification of corpora albicantia) have been reported (Ortega-Ortiz *et al.*, 2000; Torre-Cosio, 1995; Hohn *et al.*, 1996). Taylor and Rojas-Bracho (1999) and Munguia-Vega *et al.* (2007) discounted these and other pathologies as a threat to the survival of the species (endogamic depression) due to a small population size.

## IV. Behavior and Physiology

Little is known of social organization. Mean school size is 2, but groups as large as 8 or 10 individuals have been reported. An important aspect is loose aggregating behavior. The dynamics of these group aggregations are not clear, however it seems that either their duration is short or they shift locations in short periods of time. For example, Jaramillo-Legorreta *et al.* (1999) reported 41 vaquita groups loosely aggregated over several hundred square meters. During the next few days and some weeks later a similar pattern occurred, with several survey days without sightings and few survey days with this kind of aggregations sighted. There are also indications that suggest that shrimp trawlers are related at some extent with this aggregating behavior (Jaramillo-Legorreta *et al.*, 1999), however data on acoustic detection reveals this behavior also in the absence of the trawler fleet (Jaramillo-Legorreta *et al.*, 2005).

As compared with the other phocoenid members, the vaquita has on average relatively larger dorsal fin, flippers, and flukes. It has been confirmed by histological examination that the dorsal fin is highly vascularized. Large arteriole vessels are surrounded with a plexus of thin walled veins (Pérez-Cortés-Moreno, 1996). Some researchers have suggested that these characteristics may be the result of evolutionary adaptations to cope with the extreme water temperatures that prevail in the Upper Gulf of California.

## V. Life History

Hohn *et al.* (1996) noted an unusual age distribution. Most individuals (62%) were found to be between 0 and 2 years of age. The remainder (31%) were between 11 and 16 years of age, with few specimens between 3 and 10 years. There was a complete absence of specimens between the ages of 3 and 6 years. The sample of animals may be biased due to spatial segregation of different age classes or the susceptibility of various ages to be captured in nets. Alternatively, if this were the true age distribution of the population, it would represent a complete recruitment failure in recent years. Demographic models favor the hypothesis of a biased sample.

Like other porpoises, the vaquita is a seasonal reproducer, with most births occurring around March. There are not enough data for a precise estimate, but gestation is probably 10–11 months, as with other porpoises. The maximum observed life span is 21 years, which is higher than the maximum life span known for the harbor porpoise (*Phocoena phocoena*). Age at sexual maturity is difficult to estimate because of the lack of juvenile animals in the sample, but all females less than 3 years were immature and all females older than 6 years were mature. Most female harbor porpoises mature at age 3 and give birth for first time at age 4, which is consistent with observations for the vaquita. Unlike harbor porpoises in the Gulf of Maine, female vaquitas are probably not annual reproducers.

## VI. Interactions with Humans

The vaquita is classified in the most critical conservation categories by the International Union for the Conservation of Nature (IUCN, 2007), the Convention on International Trade in the Endangered Species of Wild Fauna and Flora (UNEP-WCMC, 2007), and the Mexican Government (SEMARNAT, 2002). In 1996, the IUCN concluded that the extinction of the vaquita is likely unless conservation efforts are increased substantially.

To protect the vaquita and other endangered species, the Mexican government created the Upper Gulf of California and Colorado River Delta Biosphere Reserve on June 10, 1993. However, as new knowledge of the vaquita has been gained, it is clear that this measure is insufficient. Results of surveys in 1993 and 1997 indicate that the boundaries of this biosphere reserve do not correspond well to the distribution of vaquitas. A large percentage of the sightings (40%) lie outside the reserve boundary (Fig. 2). Further, no sightings were within the nuclear zone of the reserve, which is the area where all fishing is prohibited (Gerrodette *et al.*, 1995; Jaramillo-Legorreta *et al.*, 1999).

A more specific action has been the creation of the International Committee for the Recovery of the Vaquita (CIRVA) by the Mexican government. Recognized scientists from the United Kingdom, Norway, Canada, the United States, and Mexico make up CIRVA. The mandate of this group is to propose a recovery plan based on the best available scientific information. The plan should also contemplate and consider the socioeconomic impacts of any required regulations on the resource users in the affected areas.

During their first meeting, CIRVA concluded that in the short term, gill nets are the greatest risk to the survival of the vaquita. Estimated incidental mortality in gill nets is 39–84 vaquitas per year (D'Agrosa *et al.*, 2000). This represents from 6% to 14% of the

**V**

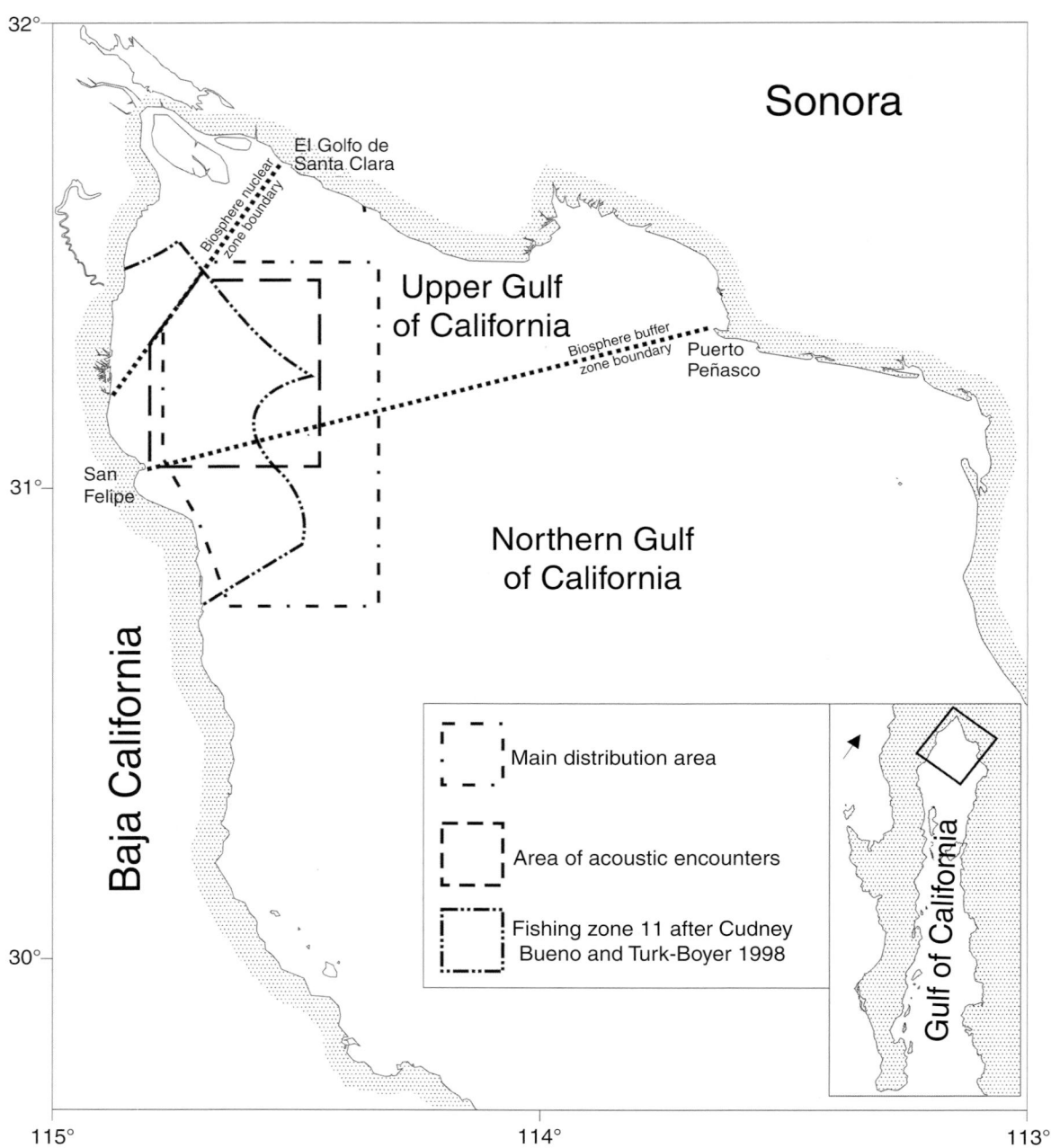

**Figure 2** *Distribution of the vaquita. Main distribution area from sightings data; most of the sightings are from near the coast. Approximately 40% of this area lies outside the Biosphere Reserve. A major portion of the area of acoustic encounters overlaps that of the fishing zone 11 defined by Cudney-Bueno and Turk-Boyer (1998).*

current population size estimate and is from only one fishing port. The committee agreed that inbreeding depression, chlorinated pesticide concentrations in the upper Gulf, and reduced flow from the only freshwater input (the Colorado River) are not risk factors at present (Rojas-Bracho and Taylor, 1999; Taylor and Rojas-Bracho, 1999). It also agreed that, in the long term, changes in vaquita habitat due to reduction of the Colorado River flow are matters of concern and must be investigated. During its second meeting, CIRVA evaluated potential mitigation measures (seasonal closures of specific areas, gear restriction, acoustic deterrents, and marine-protected areas) and strongly recommended that vaquita by-catch should be reduced to zero as soon as possible, the southern boundary of the biosphere reserve be expanded to include the entire range of the vaquita, and gill net and trawlers should be banned in the enlarged biosphere reserve. CIRVA recognized that these protective measures would have significant impacts on the resource users of the upper Gulf of California and therefore it is not possible to implement full protection immediately. Considering this, it was recommended that gill net fishing in the areas inhabited by vaquitas be removed in three stages, starting with large-mesh gill nets. CIRVA also strongly recommended investigating the development of strategies to offset economic hardship imposed by these regulations. Other recommendations were that fishing regulations be effectively enforced; acoustic surveys be started immediately to begin monitoring an index of abundance and to gather data on seasonal movements of vaquitas; development and testing of alternate gear types to replace gill nets be carried out; the design and development of community involvement, education, and public awareness programs go forward; and a description of the critical habitat of vaquita be developed.

Recently, the Scientific Committee of the International whaling Commission indicated that more science is not required to conserve this species. Instead, the Committee *strongly recommended* that resources be found to design and implement a comprehensive programme to eliminate entangling nets from the range of the vaquita (IWC, 2007).

Recently the President of Mexico announced the Conservation Program for Endangered Species (Programa de Conservación de Especies en Riesgo-PROCER) which will instrument specific Species Conservation Action Programs (Programas de Acción para la Conservación de Especies- PACE) for a list of selected species. The vaquita is listed among the top five species. The specific conservation actions will be announced shortly. Mention should be made that an intense effort by the Federal Government is taking place; which includes a coordinated program by the Ministry of Environment, the Ministry of Agriculture and Fisheries (SAGARPA) and all stakeholders, including fishermen (artisanal and industrial), NGOs and NAFTA's Commission for Environmental Cooperation (CEC).

In the absence or ineffectiveness of these kinds of measures the vaquita is very probably doomed to extinction in the near future.

## See Also the Following Articles

Porpoises, Overview ■ Bycatch ■ Fishing Industry, Effect of

## References

Alvarez, L. G., and Jones, S. E. (2004). Short-term observations of suspended particulate matter in a macro tidal inverse estuary: The Upper Gulf of California. *J. Coast. Res.* **20**, 645–654.

Brownell, R. L., Jr, Findley, L. T., Vidal, O., Robles, A., and Manzanilla, S. (1987). External morphology and pigmentation of the vaquita, *Phocoena sinus* (Cetacea: Mammalia). *Mar. Mamm. Sci.* **3**, 22–30.

Chappell, O. P., Leaper, R., and Gordon, J. (1996). Development and performance of an automated harbour porpoise click detector. *Rep. Int. Whal. Commn.* **46**, 587–594.

Cudney-Bueno, R., and Turk-Boyer, P. (1998). "Pescando entre mareas del Alto Golfo de California." Centro Intercultural de Estudios de Desiertos y Océanos, A.C., Sonora, México.

D'Agrosa, C., Lennert, C. E., and Vidal, O. (2000). Preventing the extinction of a small population: Vaquita (*Phocoena sinus*) fishery mortality and mitigation strategies. *Cons. Biol.* **14**, 1110–1119.

Findley, L. T., Nava, J. M., and Torre, J. (1995). Food habits of *Phocoena sinus* (Cetacea: Phocoenidae). *In* "Abstracts Eleventh Biennial Conference on the Biology of Marine Mammals," 14–18 December, Orlando, Florida, USA.

Gerrodette, T., Fleischer, L. A., Pérez-Cortés, H., and Villa-Ramírez, B. (1995). Distribution of the vaquita, *Phocoena sinus*, based on sightings from systematic surveys. *In* "Biology of the Phocoenids" (A. Bjørge and G. P. Donovan, eds), *Rep. Int. Whal. Commn.* Special Issue 16, 273–281.

Gillespie, D., and Chappell, O. (2002). An automatic system for detecting and classifying the vocalisations of harbour porpoises. *Bioacoustics* **13**, 37–61.

Hohn, A. A., Read, A. J., Fernandez, S., Vidal, O., and Findley, L. T. (1996). Life history of the vaquita, *Phocoena sinus* (Phocoenidae, Cetacea). *J. Zool. (Lond.)* **239**, 235–251.

IUCN. (2007). IUCN Red List of Threatened Species. <www.iucnredlist. org>. Downloaded on 26 September 2007.

IWC. (2007). Report of the Scientific Committee IWC/59/Rep 1. 59th Annual Meeting of the International Whaling Commission. Anchorage, Alaska, USA. *Rep. Int. Whal. Commn.*

Jaramillo-Legorreta, A. M., Rojas-Bracho, L., and Gerrodette, T. (1999). A new abundance estimate for vaquitas: First step for recovery. *Mar. Mamm. Sci.* **15**, 957–973.

Jaramillo-Legorreta, A., Rojas-Bracho, L., and Urban, J. (2005). A review of acoustic surveys and conservation actions for the vaquita. IWC Scientific Committee meeting document SC/57/SM10.

Jaramillo-Legorreta, A. M., Rojas-Bracho, L., Brownell, R. L., Jr, Read, A. J., Reeves, R. R., Ralls, K., and Taylor, B. L. (2007). Saving the vaquita: Immediate action, not more data. *Cons. Biol.*.

Kamminga, C., Stuart, A. C., and Silber, G. K. (1996). Investigations on cetacean sonar XI: Intrinsic comparison of the wave shapes of some members of the Phocoenidae family. *Aquat. Mamm.* **22**, 45–55.

Munguia-Vega, A., Esquer-Garrigos, Y., Rojas-Bracho, L., Vazquez-Juarez, R., Castro-Prieto, A., and Flores-Ramirez, S. (2007). Genetic drift *vs.* natural selection in a long-term small isolated population: Major histocompatibility complex class II variation in the Gulf of California endemic porpoise (*Phocoena sinus*). *Mol. Ecol.* **16**, 4051–4065.

Norris, K. S., and McFarland, L. T. (1958). A new harbor porpoise of the genus *Phocoena* from the Gulf of California. *J. Mammal.* **39**, 291–340.

Ortega-Ortiz, J. G., Villa-Ramírez, B., and Gersenowies, J. R. (2000). Polydactyly and other features of the manus of the vaquita, *Phocoena sinus*. *Mar. Mamm. Sci.* **16**, 277–286.

Pérez-Cortés-Moreno, H. (1996). "Contribución al Conocimiento de la Biología de la Vaquita, *Phocoena sinus*." Tesis de Maestría. Instituto de Ciencias del Mar y Limnología. Universidad Nacional Autónoma de México. México, D.F.

Rojas-Bracho, L., and Taylor, B. (1999). Risk factors affecting the vaquita (*Phocoena sinus*). *Mar. Mamm. Sci.* **15**, 974–989.

Rojas-Bracho, L., Reeves, R. R., and Jaramillo-Legorreta, A. (2006). Conservation of the vaquita *Phocoena sinus*. *Mamm. Rev.* **36**, 179–216.

Rosel, P. E., Haygood, M. G., and Perrin, W. F. (1995). Phylogenetics relationship among the true porpoises (Cetacea: Phocoenidae). *Mol. Phylogenet. Evol.* **4**, 463–474.

SEMARNAT. (2002). Norma Oficial Mexicana NOM-059-ECOL-2001, Protección ambiental-Especies nativas de México de flora y fauna silvestres-Categorías de riesgo y especificaciones para su inclusión, exclusión o cambio-Lista de especies en riesgo. *In* "Diario Oficial de la Federación." 6 de marzo de 2002. México.

Silber, G. K. (1991). Acoustic signals of the vaquita (*Phocoena sinus*). *Aquat. Mamm.* **17**, 130–133.

Silber, G. K., Newcomer, M. W., Silber, P. C., Pérez-Cortés, H., and Ellis, G. M. (1994). Cetaceans of the northern Gulf of California: Distribution, occurrence, and relative abundance. *Mar. Mamm. Sci.* **10**, 283–298.

Taylor, B., and Rojas Bracho, L. (1999). Examining the risk of inbreeding depression in a naturally rare cetacean, the vaquita (*Phocoena sinus*). *Mar. Mamm. Sci.* **15**, 1004–1028.

Torre-Cosio, J. (1995). "Descripción del Esqueleto, Dimorfismo Sexual y Crecimiento Alométrico en el Cráneo de la Vaquita, *Phocoena sinus* (Cetacea: Phocoenidae)." Tesis de Maestría. Instituto Tecnológico y de Estudios Superiores de Monterrey, Campus Guaymas. Guaymas, Son., México.

UNEP-WCMC. (2007). UNEP-WCMC Species Database: CITES-Listed Species. On the World Wide Web (26 September, 2007): http://www.unep-wcmc.org/isdb/CITES/Taxonomy/tax-common-result.cfm/isdb/CITES/Taxonomy/tax-common-result.cfm?source = animals&displaylanguage = eng&Common = 19869&tabname = reference.

**V**

Vidal, O. (1995). Population biology and exploitation of the vaquita *Phocoena sinus*. *In* "Biology of the Phocoenids" (A. Bjørge and G. P. Donovan, eds), *Rep. Int. Whal. Commn* Spec. Iss. **16**, 247–272.

Vidal, O., Brownell, R. L., Jr, and Findley, L. T. (1999). Vaquita. *Phocoena sinus*. Norris and McFarland, 1958. *In* "Handbook of Marine Mammals" (S. H. Ridgway, and R. Harrison, eds), Vol. 6, pp. 357–378. Academic Press, San Diego.

# Vision

## ALLA M. MASS AND ALEXANDER YA. SUPIN

The vision of marine mammals has a number of specific features associated with its ability to function in both water and air. Although many marine mammals (cetaceans, sirenians) spend their entire life in water, their aerial breathing confines them to a near-surface layer of water. Other marine mammals (pinnipeds, sea otters) spend a significant part of their life on land. As a result, the organization of their visual system fits requirements of both these different media. Although some aspects of organization of the visual system of marine mammals still remain unstudied, many features of their vision are known already.

## I. Visual Abilities of Marine Mammals

### A. Cetaceans

It was long believed that dolphins—animals with excellent hearing and echolocation—have a poorly developed visual system playing a minor role in their life. However, observations of the visual activity of dolphins have demonstrated the opposite. The ability to catch fish in air, perform precisely aimed jumps to reach targets above the water, and recognize their trainers all show that vision in dolphins is well developed. In conditions of keeping in captivity, dolphins decrease their use of echolocation and, as their interest in events above the water increases, vision takes on a leading role.

Reviews of Madsen and Herman (1980) and Mobley and Helweg (1990) summarize observations of dolphins in captivity and experimental studies which provide a basis for regarding the vision of dolphins as playing an important role in various aspects of their life: in social interactions, discrimination between individuals and species based on their colors and individual marks, the search and discrimination of prey, orientation, reproductive activity, and defense. Only vision provides the ability for rapid and precise assessment of distances to objects in air where echolocation does not operate.

Apart from numerous observations, good visual abilities of cetaceans were demonstrated in behavioral experiments for assessing their visual acuity. Precise behavioral measurements performed by Herman and colleagues (Madsen and Herman, 1980) on the bottlenose dolphin resulted in an estimate of underwater visual acuity of 8.2 arcmin (at the best distance of 1 m) and aerial visual acuity of 12.5 arcmin (at distances of 2.5 m and longer). In general, estimates of visual acuity in dolphins varied from 8 to 27 arcmin in water and from 12 to 18 arcmin in air.

Studies of color vision in cetaceans are very few in number. Only one cone type was found in the bottlenose dolphin, with the best sensitivity at 525 nm; rods are best sensitive to 488 nm. These sensitivity peaks are considerably blue-shifted as compared to those of many terrestrial mammals (Jacobs, 1993). Therefore, the dolphin lacks the common dichromatic vision typical of many terrestrial mammals, which is based on two cone types with different chromatic sensitivity. If color vision is present in dolphins (based on comparison of signals from rods and cones), it is poorly developed and limited to a blue–green region of the spectrum.

In all cetaceans, the eyes are positioned laterally, thus providing a visual field as wide as 120–130° and panoramic vision. Although positioned laterally, the eyes are directed somewhat forward and downward (ventronasally). On viewing visual objects in air, the dolphin eyes can move forward by 10–15 mm, so that the visual fields of the two eyes overlap by 20–30° in the frontal sector, giving a basis for binocular vision. However, uncrossed optic fibers have not yet been demonstrated in dolphins. Therefore, the existence of true binocular (stereoscopis) vision (based on interaction of crossed and uncrossed optic fibers) in dolphins still remains under question.

Dolphins are equally capable of the perception of complex configurations of objects using both vision and echolocation. Besides, there is also a possibility of intermodal transfer between these two modalities: objects known for a dolphin only by visual appearance can be discriminated and recognized by echolocation, and vice versa. The intermodal transfer is equally successful when visual experience is used for echolocation discrimination and when echolocation experience is used for visual discrimination.

Even in riverine cetaceans inhabiting turbid and low-transparent water (the Amazon river dolphin *Inia geoffrensis*, the tucuxi dolphin *Sotalia fluviatilis*), the visual system does not exhibit a significant regression. The only exception is the Indian river dolphin, *Platanista gangetica*, in which the visual system is reduced markedly.

### B. Pinnipeds

Because pinnipeds spend their life partially in water and partially on land, they use both underwater and aerial vision. On land, vision plays an important role during the reproductive period, during birth and feeding of pups, and for maintaining intrapopulation relationships, as well as for orientation. In water, vision is used for prey detection and recognition, avoiding predators, and spatial orientation during migrations.

Because of a great diversity of pinniped species in terms of systematic position and ecology, the role of vision diverges widely as well. Walruses (*Odobenus rosmarus*) rely mainly on their vibrissal sensitivity to identify objects during benthic foraging. Other pinnipeds also have a well-developed vibrissal apparatus; however, in aquatic conditions, most seals use both visual and tactile modalities to search for food. Experiments demonstrated that seals are capable to distinguish rather small objects visually, recognize the shape of figures, and perform a complex analysis of visual images. Data summarized by Fobes and Smock (1981) shows that both otariids and phocids are capable of discriminating objects differing in size from 9% to 24%.

Most pinnipeds (both otariids and phocids) have maximum spectral sensitivity within a range of 496–500 nm. An exception is the southern elephant seal (*Mirounga leonina*), which is sensitive to a shorter wavelength (486 nm).

A possibility of limited color discrimination in a few pinniped species (*Pagophilus groenlandicus*, *Phoca largha*, *Arctocephalus pusillus*, *A. australis*, *Zalophus californianus*) is indicated by their capability to discriminate blue and green objects from gray ones, although they cannot discriminate red and gray objects. The best rod sensitivity in the harbor seal (*Phoca vitulina*) was found at 496 nm, and cone sensitivity at 510 nm; i.e., similarly to dolphins, the spectral sensitivity is blue-shifted as compared to terrestrial mammals. No indication was obtained of more than one cone type in pinnipeds (see Section IV.A).

Measurements of visual acuity based on the use of grids as test stimuli have demonstrated that visual acuity in both water and air is 5–8 arcmin in a few otariide species: *Zalophus californianus, Eumetopias jubatus, Arctocephalus pusillus,* and *A. australis, Phoca vitulina.*

### C. Other Marine Mammals

*1. Sirenians* Little is known of the visual capabilities in sirenians. A few observations summarized by Piggins *et al.* (1983) showed that the Amazon manatee (*Trichechus inunguis*) is capable of visually driven behavior, in particularly, visual tracking of underwater objects. Recently underwater visual acuity was assessed in the Caribbean manatee *Trichechus manatus* as rather poor—from 24 to 56 min of arc, depending on the test gain orientation and media. A capability of this species for dichromatic (blue–green) color vision has been shown. It remains unknown whether the manatee has an ability of good aerial vision.

*2. Sea Otters* Very little is known of the visual abilities of sea otters (*Enhydra lutris*). Inhabiting the coastal zone and feeding under water, sea otters need to have good vision in both air and water. Observations showed that they actively use vision, and experiments have shown their capability to discriminate objects of different sizes. However, quantitative behavioral measurements of their visual abilities are absent.

## II. Eye Anatomy and Optics

### A. Cetaceans

Ocular anatomy in cetaceans is markedly different from that in terrestrial mammals by being adjusted to optical properties of water and to a number of other factors: possibility of eye damage because of high density of water and presence of suspended particles, low temperature and low illumination deep in water, significant light scatter, and so on. Characteristic examples of eye structure in cetaceans are shown in Fig. 1. Remarkable features are a thick sclera (especially so in whales, Fig. 1B), a thickened cornea, a highly developed vascular network forming a typical vascular *rete mirabilia* which fills a significant part of the orbit behind the eyeball, and massive ocular muscles. All these structures take part in protecting the eye from underwater cooling and mechanical damage.

Although in terrestrial mammals the eyeball is almost spherical, in cetaceans its anterior part is flattened, so as the anterior chamber is small and the eyecup is of almost a hemispherical shape. More precisely, the eyecup shape approximates a segment of a sphere of about 150° of arc (Fig. 1A,B), and its naso-temporal diameter slightly exceeds the dorsoventral one.

In terrestrial mammals, the convex outer surface of the cornea is the major refractive element of the eye because it separates media with different refractive indices: air with a refractive index of about 1 and the corneal tissue with a refractive index of more than 1.35. However, the refractive index of water is 1.33–1.34, which is very close to that of the cornea and the intraocular media. As a result, the

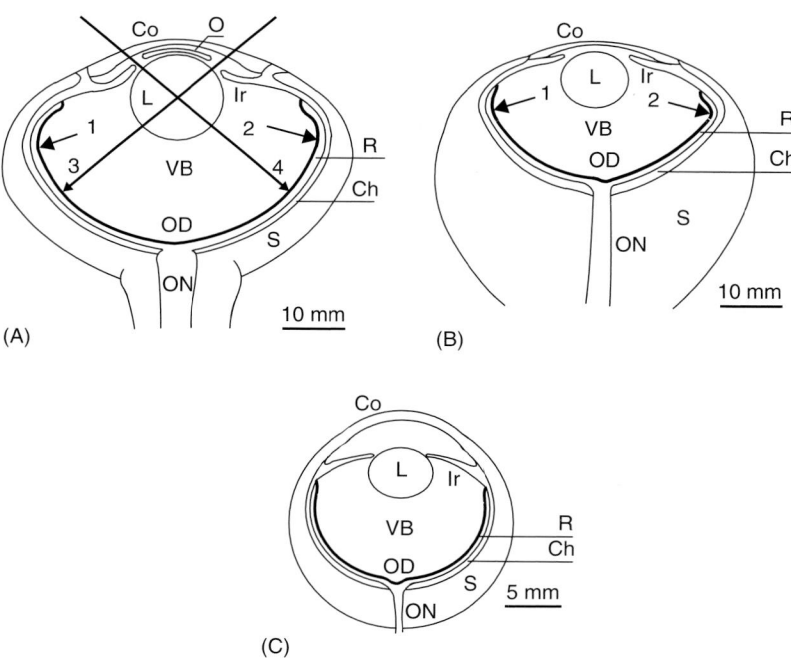

**Figure 1** *Schematic presentation of eye anatomy and optics in some cetaceans: (A) the bottlenose dolphin, (B) the gray whale, and (C) the Amazon river dolphin. Co, cornea; L, lens; Ir, iris; O, operculum; S, sclera; Ch, choroids; R, retina; ON, optic nerve; OD, optic disc; VB, vitreous body. Arrows 1 and 2 delimit a part of the eyecup, which can be approximated by a spherical segment of about 150°. Arrows 3 and 4 show directions of light rays passing through the nasal and temporal holes of the pupil and through the lens center to the high-resolution parts of the retina.*

V

corneal surface plays very little part in underwater light refraction. Therefore, in cetaceans, light refraction and focusing of an image on the retina are almost entirely performed by the lens. This is why the lens in cetaceans is almost spherical or slightly elliptical. The large curvature of the lens surface provides a sufficiently high refractive power of the lens and well-focused images on the retina, despite very weak refractive power of the corneal surface in water. These optics are similar to those in fish, which is not surprising given that in both cases the eye is adjusted to optical properties of the same medium.

A strongly convex (spherical) lens consisting of homogeneous material have a very strong spherical aberration. The cetacean lens is free of this disadvantage due to a heterogeneous structure: outer layers have a lower refractive index than the inner core.

In the cetacean eye, the spherical lens is located in such a way that its center almost coincides with the center of the spherical segment of the eyecup; so light rays coming from any direction are focused almost identically on the retina. This is significantly different from the case in terrestrial mammals, which provides the best focusing on the eye axis.

In terrestrial mammal eyes, accommodation (refraction adjustment to the distance to the object) is performed by change in the shape of the lens due to contraction and relaxation of ciliary muscles. In cetaceans, spherical lens shape and center-symmetric optics of the eye led to loss of this accommodatory mechanism. The ciliary muscles are poorly developed in dolphins and are absent from most whales suggesting that accommodation cannot be achieved by changing the lens shape. It has been suggested that accommodation in cetaceans is performed by another mechanism, namely by axial displacement of the lens due to changes in intraocular pressure. Intraocular pressure can change because of contraction of the massive the massive retractor muscle (*m. retractor bulbi*) which produces axial displacements of the eye in the orbit. When the eye is pulled back into the orbit, intraocular pressure increases, thus shifting the lens forward; when the eye is moved forward, the pressure decreases shifting the lens backward.

The cornea in cetaceans is thicker than in many terrestrial mammals, and this thickness is not uniform: the cornea is thinner in the center and thicker in the periphery. Although major refraction in the cetacean eye is performed by the lens, the refractive role of the cornea is not negligible. Its outer surface is of lower curvature than the inner one; i.e., the cornea has a shape of a divergent lens. Under water, this lens makes minor contribution to the total refraction power, as the media on both sides of the cornea (water outside and the anterior chamber liquid inside) have refractive indices rather close to that of the cornea. However, some difference between the refractive indices of water (1.33) and the cornea (from 1.37 in the central part to 1.53 in the periphery) does exist. Thus, the cornea acts as a weak but nonetheless divergent lens. The total refraction of the cornea and lens makes the cetacean eye well emmetropic within a range of ±1 diopters under water.

Adaptation to underwater vision also affects the cetacean iris and pupil. The cetacean vision functions in conditions of wide and rapid changes of illumination when the animal dives from the well-illuminated water surface into the depth where illumination is very low. This requires the pupil to react in a wide range of illuminations and to have a wide range of sizes. The cetacean pupil is of an unusual shape. The upper part of the iris has a characteristic protuberance, the operculum. At low illumination, the operculum is contracted (raised), so the pupil, similarly to other mammals, is of a round or slightly oval shape; its horizontal diameter in dolphins is of about 10 mm (Fig. 2A). With illumination increase, the operculum advances downward, turning the pupil into a U-shaped slit (Fig. 2B). At high illumination, the

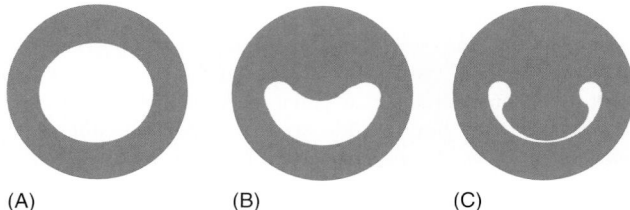

(A)          (B)          (C)

**Figure 2**  *Shape of the pupil in the bottlenose dolphin at various levels of illumination: (A) low illumination, nonconstricted oval pupil; (B) moderate illumination, partially constricted U-shaped pupil; and (C) high illumination, strongly constricted pupil reduced to two pinholes.*

operculum advances so far that the slit becomes closed, leaving only two narrow holes in the temporal and nasal parts of the iris (Fig. 2C). This pupil shape is characteristic for many dolphins, including the bottlenose dolphin *Tursiops truncatus*, harbor porpoise *Phocoena phocoena*, common dolphin *Delphinus* spp., tucuxi dolphin *Sotalia fluviatilis*, and also for a number of whales, although in some whales the operculum is small. A known exception is the Amazon river dolphin which has a round pupil even when it is constricted.

The cetacean eye is well emmetropic in water; however, in air refraction on the outer convex corneal surface adds to the lens refraction. The difference of refractive indices of air and the cornea results in significant refractive power of the central, the most convex part of the corneal surface: about 20 diopters. The addition of this refraction to the emmetropic lens refraction should make the cetacean eye catastrophically myopic (near-seeing) in air. Nonetheless, dolphins have good visual acuity in both water and air.

The solution of the problem is in the presence of flattened (low-curvature) regions of the cornea. A flat corneal surface does not produce additional refraction in air. Even if the surface is not truly flat but a little convex, its refractive power becomes low enough and may be compensated by some additional mechanisms. Keratoscopic studies in common bottlenose dolphins showed a "spoon" shape of the cornea with lower curvature in its nasal and temporal regions.

Aerial myopia can be partially compensated by accommodatory displacements of the lens. For aerial vision, the dolphin eye moves forward thus producing decrease of intraocular pressure; this results in shifting the lens backward and reduced myopia. Additionally, reduction of intraocular pressure decrease the curvature of the cornea. Under water, the eye is retracted into the orbit, which results in increased intraocular pressure and a shift of the lens forward to a position providing underwater emmetropia.

An additional mechanism for the correction of aerial myopia is pupil constriction. Above water, high illumination results in strong pupil constriction; the latter corrects all errors of refraction, including aerial myopia, and provides fairly good depth of field.

Another adaptation of the cetacean eye to low underwater illumination is the well-developed reflective layer, the tapetum (*tapetum lucidum*). It lies behind the retinal pigment epithelium within the choroid. In cetacean, the tapetum is formed with extracellular collagen fibrils (*tapetum fibrosum*). Multiple reflection of light from 50–70 layers of fibrils results in significant light reflection back to the retina, thus increasing visual sensitivity in scotopic conditions.

The tapetum is present in all cetaceans. In most of the investigated cetaceans, particularly in mysticete whales, it covers all of the fundus (although varies in coloration), or at least it covers a large dorsal part of the fundus. Complete coverage of the fundus by the tapetum is

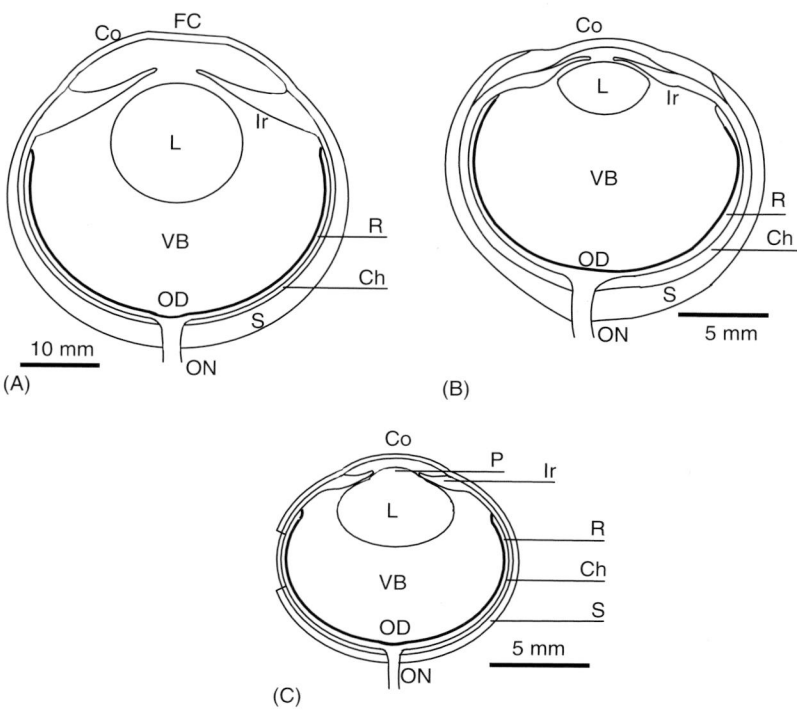

**Figure 3** *Schematic presentation of eye anatomy and optics in some representatives of pinnipeds, sirenians, and otters: (A) the northern fur seal; (B) the manatee; (C) the sea otter. Co, cornea; FC, flattened region of the cornea; L, lens; Ir, iris; S, sclera; Ch, choroids; R, retina; ON, optic nerve; OD, optic disc; P, lens protuberance; VB, vitreous body.*

unique among vertebrates: in terrestrial mammals, the tapetum usually does not extend lower than the horizontal equator of the eyecup.

## B. Pinnipeds

In all pinnipeds (except walruses), both the absolute and the relative sizes of the eyeball are large. Eye structure in pinnipeds (Fig. 3A), despite significant differences from cetaceans, has some common features arising from adaptation to underwater vision (Jamieson and Fisher, 1972). In particular, a characteristic feature is an almost spherical or slightly elliptical lens. Although the eyeball does not appear as shortened in the axial direction, a major part of the eyecup has a shape close to a hemisphere, so a significant part of the retina is almost constantly distant from the lens center. Thus, the eye optics, like in cetaceans, is almost centrally symmetrical. The difference between the eyeball shape in cetaceans and pinnipeds (shorter axial length in cetaceans and longer in pinnipeds) is mainly due to larger size of the anterior chamber in pinnipeds.

Iris in pinnipeds is very muscular and heavily vascularized. The dilator is well developed. Most pinnipeds have a pupil which being constricted becomes pear-shaped. Pupil size can change over a very wide range; at bright illumination, it constricts to a very small hole. In shallow-diving species, the range of pupillary area variation is rather small: 26–70.5 times. In a deep diver, the northern elephant seal, *Mirounga angustirostris*, the pupil area varied within an extremely wide range, from 422 mm$^2$ in dark-adapted conditions to a pinhole of 0.9 mm$^2$ in light-adapted conditions, i.e., almost 470 times.

The ciliary muscle in pinnipeds is well developed, although accommodation is either absent or very weak.

Unlike cetaceans, the central part of the cornea has a clearly delimited region (6–10 mm in diameter) of almost a flat surface. It is located near the center of the cornea, slightly shifted to the nasal direction (FC region in Fig. 3A). Such a flat region of the cornea was found in a number of both otariids and phocids and was demonstrated by precise measurements on the Californian sea lion (*Zalophus californianus*). The flat region of the cornea serves as an emmetropic "window" in which refraction remains almost equal in both water and air. In another pinniped, the hooded seal (*Cystophora cristata*), the flattened part of the cornea does not look like a delimited region but arises because of low curvature of the cornea of the extremely large eyeball.

The existence of a flat region in the central part of the cornea indicates a very specific principle of eye construction in pinnipeds. Indeed, the convex shape of the cornea in most animals is a consequence of excessive intraocular pressure, which is necessary for maintaining the shape and size of the eyeball. Direct data on intraocular pressure in pinnipeds are absent, but their flat cornea suggests that this pressure is very low, perhaps about zero. Anatomical observations on the northern fur seal (*Callorhinus ursinus*) showed that its vitreous body is of a rigid rather than a gelatinous consistency, thus taking a part in maintenance of the eyeball shape and dimensions. This way of maintaining the eye shape is evidently used in a number of pinniped species.

The pinniped tapetum is one of the best developed among both terrestrial and aquatic mammals. Contrary to the tapetum fibrosum

**V**

in cetacean, the tapetum in pinnipeds is formed with intracellular reflective rodlets (tapetum cellulosum). It consists of a large number (20–30 or more) of cell layers and covers all the fundus.

### C. Other Marine Mammals

*1. Sirenians*    Among other marine mammals, the eye anatomy of the manatee (*Trichechus manatus and T. inunguis*) is of interest as an example of the order Sirenia, which, apart of cetaceans, is the only group of completely aquatic mammals.

Both in *Trichechus manatus* and in *T. inunguis* the eye is rather small (13–19 mm diameter) and is set deeply within the ocular fascia. Its general morphology resembles more that of terrestrial mammals than the cetacean eye (Fig. 3B). The eyeball is almost spherical (the axial length differs little from the equatorial diameter), the anterior chamber is shallow, and the lens is set forward and is not true spherical: its axial dimension is markedly shorter than the diameter. The sclera is rather thin. Thus, despite of completely aquatic mode of life of the manatee, its eye anatomy exhibits a number of conservative features (Piggins *et al.*, 1983). Underwater, the eye is almost emmetropic or slightly hyperopic, but in air it is strongly myopic. It remains unknown whether the manatee has some mechanisms to compensate aerial myopia; thus, its capability to aerial vision remains unknown.

A distinctive feature of the manatee's eye is the vascularized cornea which in all other mammals is pathology.

*2. Sea otters*    To a large extent, the eyeball of the sea otter (*Enhydra lutris*) is similar to those of terrestrial mammals (Fig. 3C): it is almost spherical, axial length is only a little shorter than the diameter. Contrary to spherical lenses of cetaceans and pinnipeds, the lens of sea otter is lenticular. However, the front surface of the lens has a protuberance of increased curvature. A characteristic feature of the eye anatomy is that the iris is fastened to the frontal lens surface. Therefore, contraction of iris muscles influence the curvature of the frontal lens surface. This mechanism is capable of providing an accommodation range of up to 60 diopters, thus compensating for the appearance of refraction at the corneal surface in air and its disappearance in water. This accommodation mechanism in the sea otter eye is able to preserve emmetropia in both air and water.

### III. Eye Movements

All dolphins and whales have mobile eyes. However, measurements in the bottlenose dolphin indicated that eye mobility is less than in humans, and eye movements are more slow.

Oculomotor muscles are well developed in dolphins and whales; an exception is the Ganges river dolphin (*Platanista gangetica*), which has reduced eyes and no oculomotor muscles. Other cetaceans have a complete set of muscles known in mammals: four straight and two oblique muscles. These muscles allow eye movements in both the horizontal and the vertical directions. In addition, unlike terrestrial mammals, cetaceans have retractor muscle (*m. retractor bulbi*), which produce axial (in/out) movements of the eye in the orbit. The bottlenose dolphin is capable of moving its eye forward to 10–15 mm and pulling it back. As a rule, forward eye movements (protraction) appear when the dolphin examines an object in air visually. These eye movements may be used for binocular examination of objects. As mentioned earlier, the eye protraction in air can also provide accommodation to avoid the aerial myopia.

Another intriguing feature of oculomotor activity in dolphins is the ability to move the left and right eyes independently. Quantitative measurements in dolphins have shown that correlation of movements of the left and the right eyes are very low; i.e., independent eye movements in dolphins are a rule rather than exception.

In addition to independent eye movements, cetaceans have rather independent pupil reflexes of the two eyes. Moreover, eyelids of the left and the right eyes can also function independently, so one eye can be open while the other is closed. Such observations were made during sleep in dolphins, although similar behavior is also possible in wakefulness: dolphins can swim for long periods with one eye open and the other one closed, with the left and the right eye alternating.

As to pinnipeds and sea otters, there is no significant difference from terrestrial mammals in their oculomotor muscle anatomy and the character of eye movements.

### IV. The Retina and Optic Nerve
### A. Features of the Retina in Cetaceans

The histological structure of the retina has been investigated in a number of cetacean species: the common bottlenose dolphin, short-beaked common dolphin (*Delphinus delphis*), Dall's porpoise (*Phocoenoides dalli*), dwarf sperm whale (*Kogia sima*), Amazon river dolphin (*Inia geoffrensis*), fin whale (*Balaenoptera physalus*), and common minke whale (*B. acutorostrata*). All of these studies have shown that the laminal structure of the cetacean retina is basically similar to that in terrestrial mammals. The retina consists of typical layers as follows (Fig. 4A). The receptor layer (the nearest to the pigment epithelium) is composed of densely packed outer segments of photoreceptors. The outer nuclear layer is composed of receptor pericaria arranged in a multilevel manner. The outer plexiform layer contains cell processes establishing connections between receptors and first-order neurons, bipolar cells. The inner nuclear layer is composed mostly of pericaria of bipolar cells; in addition, this layer contains horizontal and amacrine cells, which establish horizontal connections within the outer and inner plexiform layers. The inner plexiform layer contains processes establishing connections between bipolar and ganglion cells. The ganglion layer contains ganglion cells sending their axons to the optic nerve. Finally, the nerve fiber layer (nearest to the vitreous body) contains optic fibers (axons of ganglion cells), which spread along the inner retinal surface until they reach the optic disk and enter the optic nerve. This laminar structure of the retina is fully developed in all cetaceans. Even in the Ganges river dolphin with strongly reduced eyes, the retina contains all the layers. Being basically similar in cetaceans and terrestrial mammals, the retina has a number of specific features in cetaceans. It is markedly thicker than in terrestrial mammals, ranging from 370 to 425 μm (in terrestrial mammals, the retina is 110–240 μm thick).

The most detailed description of the retina is available for the common bottlenose dolphin (Dral, 1977; Dawson, 1980). Its retinal receptor layer consists predominantly of rods (receptors for achromatic vision). The question of the existence of cones (chromatic-vision receptors) in cetaceans remained debatable for some time. Recent studies of visual pigments have shown that the cetacean retina does contain cone receptors, however rods dominate: cone proportion is in the range of 1–2% (Peichl *et al.*, 2001).

Contrary to the majority of terrestrial mammals which have two types of cones with different pigments providing color vision (short-wave sensitive S-opsin and middle-to-long-wave sensitive L-opsin), only L-opsin containing cones were found in 10 species of odontocetes the cetacean retina. S-opsin have not been reported (Peichl *et al.*, 2001). This corresponds to behavioral data showing poor color vision in dolphins (see Section I.A). As to amacrine, bipolar, and horizontal cells in the cetacean retina, they are generally similar to those in terrestrial mammals.

A marked difference from terrestrial mammals is in the inner plexiform layer and the ganglion layer of the cetacean retina. The ganglion layer looks like a single row of large, sparsely distributed neurons separated by large intercellular spaces. These neurons have large cell bodies with a clearly defined cell membrane, large amount of cytoplasm, a well visible nucleus up to 15 μm in diameter, and clearly defined nucleolus 4–5 μm in diameter. Cell bodies contain clearly visible, well stained Nissl granules.

A remarkable feature of the cetacean retina is large size of ganglion cells, particularly, the presence of giant ganglion cells. Bodies of such cells reach 75–80 μm, sometimes more. Giant ganglion cells were described in a number of odontocete species and in a few mysticete whales. In some dolphins, however, retinal ganglion cells do not reach giant sizes: in the Amazon river dolphin and the Indian river dolphin they do not exceed 40–42 and 20 μm, respectively. However, even these cells are as large as compared to those in many other mammals. The smallest ganglion cells in cetaceans are as large as 10 μm.

Figure 5A,B present ganglion cell size distributions in the retina of the common bottlenose dolphin. The histograms represent samples in different parts of the retina: with high and low concentration of ganglion cells. Despite some difference between the samples (in the area of high cell concentration, cells are a little smaller than in the area of low concentration), both samples demonstrate large cell sizes: the most common size is 20–35 μm, but cells as large as 50–60 μm are also present; there are no cells smaller than 10 μm.

Large cells are not characteristic of all levels of the visual system in cetaceans (the lateral geniculate body, visual cortex); they are typical only in the retina. The largest pyramidal cells in the visual area of the dolphin cerebral cortex are not more than 20–30 μm. In other parts of the dolphin brain, cells do not exceed 20–45 μm either. There is presently no satisfactory explanation why ganglion cells in the cetacean retina are so large. One of possible explanations is that large ganglion cells have thick axons with high velocity of conduction; in a large body, it may be helpful for fast transmission of signals. However, large terrestrial mammals (e.g., the bull or the elephant) have ganglion cells not larger than 25–30 μm.

Apart from large cell sizes, a characteristic feature of the retinal ganglion layer in cetaceans is low cell density. The large neurons are separated by large intercellular spaces.

The question of separation of retinal ganglion cells into different morphological types has not been solved for cetaceans. Large-size ganglion cells in cetaceans resemble large Y-neurons in the visual system of terrestrial mammals, as opposed to smaller X-neurons. However, Y-neurons in terrestrial mammals constitute no more than 1% of ganglion cells, whereas in cetaceans, large ganglion cells predominate.

## B. Optic Nerve Structure in Cetaceans

Retinal ganglion cells send their axons into the optic nerve. Consistent with the large sizes of ganglion cell bodies, the axon diameters in cetaceans are also greater than in terrestrial mammals. In a variety of dolphin species, a significant proportion of optic fibers exceed 15 μm in diameter. For comparison, the maximum fiber diameter in cats and in monkeys, is no more than 8 μm. The only exception is the Chinese river dolphin *Lipotes vexillifer*, which has thin optic fibers, although its retina contains ganglion cells as large as 75 μm.

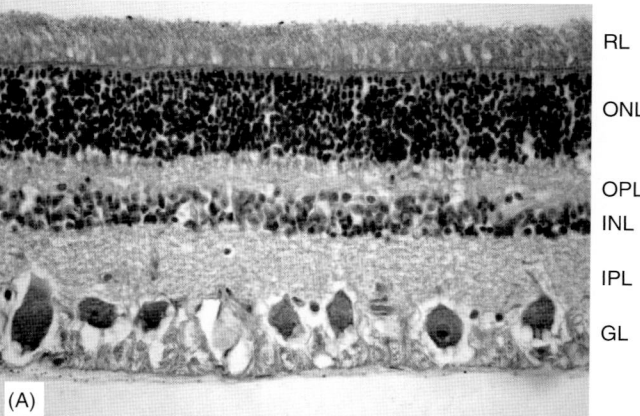

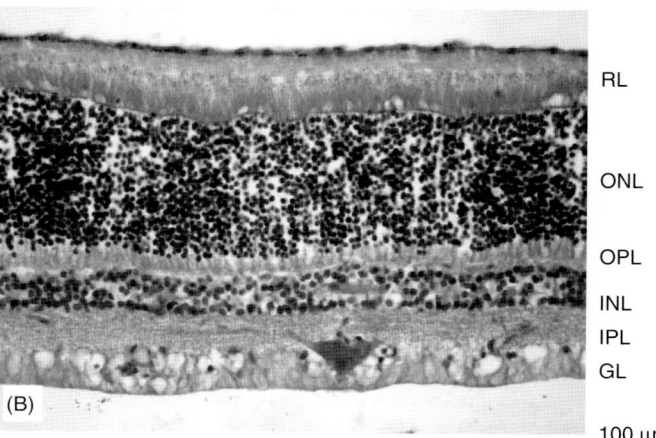

**Figure 4** *Microphotographs of a transverse section of the retina of a bottlenose dolphin (A) and a Steller sea lion (B). RL, receptor layer; ONL, outer nuclear layer; OPL, outer plexiform layer; INL, inner nuclear layer; IPL, inner plexiform layer; GL, ganglion layer.*

The low density of ganglion cells in the retina of cetaceans corresponds to the low density of fibers in the optic nerve. In cross sections of the optic nerve of dolphins, the density of fibers is less than 50,000/mm², whereas in monkeys it exceeds 220,000/mm². Thus, although the optic nerve in cetaceans is of a large diameter, the total number of optic fibers does not exceed than in many terrestrial mammals. More than 50% of the cross-section area of the cetacean optic nerve is occupied by intercellular space (contrary to 12–20% in terrestrial mammals), not by glia.

The total number of optic fibers varies among cetacean species. The smallest number of fibers (14,000–16,000) was found in the Indian river dolphin, *Platanista gangetica*, and the Amazon river dolphin, *Inia geoffrensis*; the number of optic fibers in the Chinese river dolphin, *Lipotes vexillifer*, is a little higher, more than 20,000. In the common bottlenose dolphin, the number of optic fibers is 150,000–180,000. Other odontocetes have an optic fiber number similar to that in the bottlenose dolphin. In mysticetes, the number of optic fibers is within a range of 250,000–420,000.

## C. Features of the Retina in Pinnipeds

In general, the retinal structure in pinnipeds is the same as in terrestrial mammals (Fig. 4B). All layers are present in the pinniped

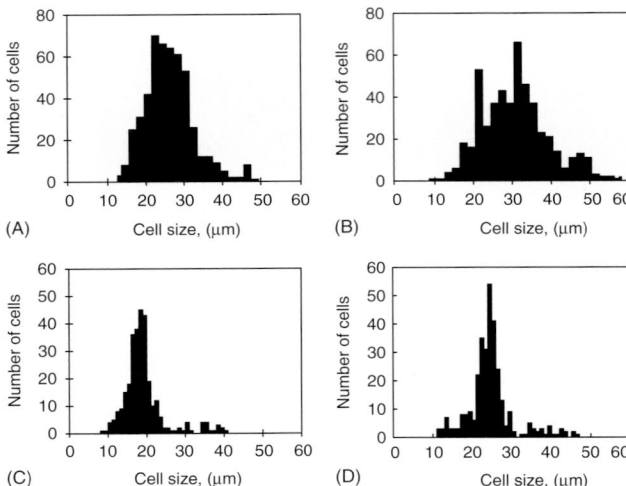

**Figure 5** *Histograms showing size distributions of ganglion cells in the retina of cetaceans and pinnipeds. Abscissa axis, cell size; ordinate axis, number of cells of the present size. (A,B) Data from a common bottlenose dolphin; samples from areas of high (A) and low (B) cell densities. (C,D) The same for a northern fur seal.*

retina, although there are a number of specific features, mainly of the outer nuclear, inner nuclear, and ganglion layers (Jamieson and Fisher, 1972). The very thick outer nuclear layer is characteristic of many pinnipeds. The inner nuclear layer does not have clear margins, in contrast to terrestrial mammals, where this layer is strictly ordered. There are large horizontal cells with very long processes within this layer. The giant horizontal cells are located irregularly among bipolar and amacrine cells, which are also distributed chaotically. Bipolar cells are located mostly in the outer part of the inner nuclear layer while large amacrine cells are located close to the inner plexiform layer.

The ganglion layer in pinnipeds consists of a single row of ganglion cells separated by wide intercellular distances. Ganglion cells have large pericaria, a large amount of Nissl substance in the cytoplasm and long dendrits. Most of these cells are of intermediate size (10–30 μm), although large cells (up to 50 μm) are also encountered (Fig. 5C,D). These sizes are larger than in terrestrial mammals.

All pinnipeds have a predominately rod retina. The question of existence of cones has been a matter of discussion for a long time. However, light and electron microscopy have shown the presence of cones in the harbor seal and harp seal, although photoreceptors of this type are not numerous. Moreover, recently, immunochemical studies in a few pinniped species demonstrated that their retinae contained sparse populations of cones, consisting about 1–2% of photoreceptors (Peichl *et al.*, 2001). However, these studies revealed only one opsin type in the cone receptors, the middle-to-long-wave sensitive L-opsin and did not revealed the short-wave sensitive S-opsin. This feature is common with cetaceans (in spite of the quite different phylogeny of cetaceans and pinnipeds) and distinguishes pinnipeds from majority of terrestrial mammals that have at least two spectrally sensitive cone types (middle- and short-wave sensitive) or three cone types in primates. The existence of some amount of cones corresponds to behavioral data showing a limited capability of color discrimination in pinnipeds (see Section I.B).

## D. The Retina of Other Marine Mammals

*1. Sirenians* The retina of the manatees also features the common laminar organization. Receptors are presented mostly by rods; cones are less numerous. Among specific features, the large size of ganglion cells can be mentioned: up to 60 μm, mostly 15–30 μm, and not less than 10 μm. Thus, the large size of ganglion cells seems to be a common feature of different groups of marine mammals.

*2. Sea Otters* In the sea otter, the retina has many features similar to those in terrestrial rather than in aquatic mammals. The majority of ganglion cells are not of large size: 7–30 μm, mostly 11–15 μm. They can be subdivided into three size groups: large, medium, and small. The retina of the sea otter contains a large number of small amacrine and neuroglial cells.

## V. Retinal Topography and Visual Field Organization

### A. Cetaceans

Ganglion cells are distributed nonuniformly in the mammalian retina: ganglion cell density (number of cells per area unit) is high in some areas and much lower in the remainder of the retina. Regions of ganglion cell concentration (high density) provide the most detailed analysis of visual images. Characteristics of retinal topography in a variety of mammals are presented in a review by Hughes (1977).

In terrestrial mammals, there are two main types of organization of a region with high cell density. In mammals with frontal vision, highest density is in the fovea or area centralis located in the center of the visual field. This retinal area is little vascularized to avoid its shadowing by blood vessels. In mammals with laterally located eyes, the region of high cell density is shaped as a narrow horizontal strip, the visual streak. All terrestrial mammals studied until now have only one, if any, region of the highest ganglion cell density.

The cetacean retina does not have avascular areas that would indicate the presence of fovea or area centralis. Therefore, visual examinations of the eye fundus are not capable of revealing such regions. Data on topography of ganglion cell distribution in the cetacean retina were obtained using retinal whole mounts. Whole mounts are preparation of a total retina flattened on a slide, ganglion layer upward, and stained appropriately. Retinal whole mounts allow to count ganglion cells systematically across all the retina surface, thus constructing a topographic map of ganglion cell distribution. Studies of retinal whole mounts have shown that different regions of the cetacean retina have a very different density of ganglion cells (Fig. 6A,B). Beginning from the pioneering studies of Dral (1977), studies of cetacean retinal whole mounts were performed in a number of dolphin species, particularly, the common dolphin, bottlenose dolphin, harbor porpoise, Dall's porpoise, and Pacific white-sided dolphin (*Lagenorhynchus obliquidens*) (see detail in Supin *et al.*, 2001).

The most characteristic feature of these species is that, unlike terrestrial mammals, all of these marine dolphins do not have a single area of high ganglion cell density but two such areas. They are located near the horizontal diameter of the retina, one in the nasal and the other in the temporal sector (Fig. 7A). In the bottlenose dolphin, both these areas are located at a distance of 15–16 mm from the optic disk, which corresponds to 50–55° of the visual field. Ganglion cell density in each of these areas reaches 700–800 cells/mm², which corresponds to 40–50 cells per squared degree of the visual field (cells/deg²). The two high-density areas are connected by an elongated zone

V

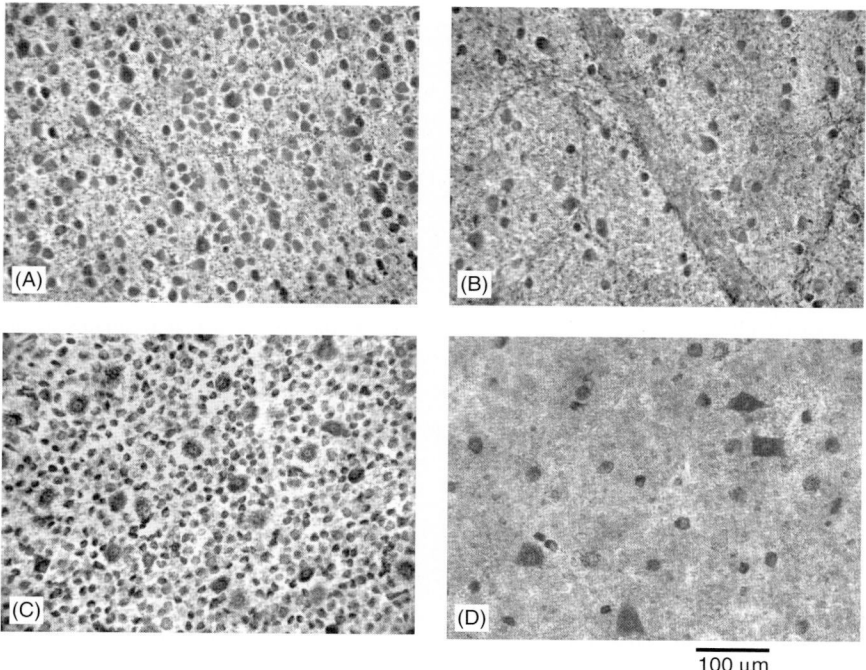

**Figure 6** *Microphotographs of the ganglion layer in retinal whole mounts. (A,B) A bottlenose dolphin, (A) an area of high cell density and (B) an area of low cell density. (C,D) A harp seal, (C) an area of high cell density and (D) an area of low cell density.*

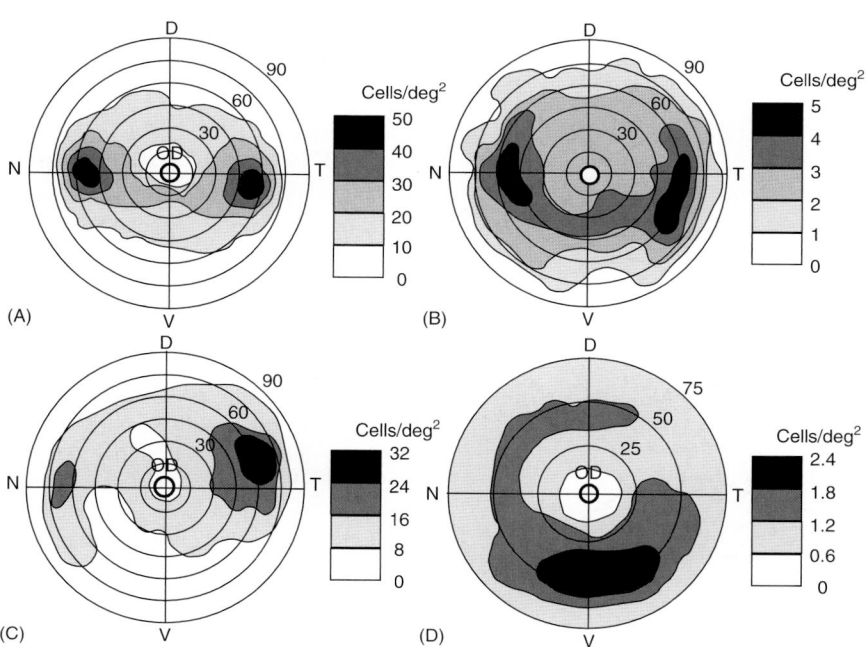

**Figure 7** *Topographic distribution of ganglion cell density in the retina of some cetaceans: (A) the bottlenose dolphin, (B) the tucuxi dolphin, (C) the gray whale, (D) the Amazon river dolphin. Cell density is expressed as a number of cells per squared degree of the visual field and is shown by various shadowing, according to the scales. Concentric circles show angular coordinates on a retinal hemisphere centered on the lens. D, V, N, T, dorsal, ventral, nasal, and temporal poles of the retina, respectively.*

of increased, although somewhat lower, cell density, which runs below the optic disk; this zone looks like a visual streak.

In other dolphin species, the retinal topography is basically similar to that described earlier: there are two areas of high ganglion cell density. Even at low cell density in some cetaceans inhabiting turbid and low-transparent water (e.g., the tucuxi) the retinal topography looks the same (Fig. 7B). However, some quantitative differences do exist. In the bottlenose dolphin, the ganglion cell density is almost equal in the two areas, the nasal and the temporal areas, whereas in the harbor porpoise, the cell density in the temporal area (i.e., the region serving the frontal visual field) is higher than in the nasal region: 28 and 20 cells/deg$^2$, respectively.

The retinal topography of ganglion cells was studied in two mysticete species: the gray whale (*Eschrichtius robustus*) and common minke whale (*Balaenoptera acutorostrata*). Both of them also have ganglion cell distributions with two areas of high cell density, in the nasal and temporal sectors (Fig. 7C). Again, the cell density in the temporal area is higher than in the nasal one: 28 and 21 cells/deg$^2$ in the gray whale.

The significance of the two areas of high ganglion cell density (i.e., of high retinal resolution) is probably associated with the cetacean's capability of good vision both above and under water, in particular, with preventing the aerial myopia. Indeed, the high-resolution areas are located just opposite the two small pupil holes formed when the pupil is constricted in air (see Fig. 1A). Because of the centrally symmetric optics of the cetacean eye, light falls onto each of these areas through the opposite hole of the pupil. The areas of the cornea with minimal curvature are located across from these narrow pupil holes. Both the pinhole pupils and the low cornea curvature are devices

to prevent aerial myopia. Thus, images are projected onto the high-resolution areas of the retina with minimal distortions.

The two high-resolution retinal areas in cetaceans may be used differently for the underwater and aerial vision (Supin *et al.*, 2001). When a dolphin looks at an underwater objects, it takes a position lateral to the object; i.e., the object is placed into the posterolateral part of the visual field, which projects onto the nasal high-resolution area of the retina. On the contrary, when a dolphins looks at an object above water, it places the object into the ventronasal part of the visual field, which projects onto the temporal high-resolution area of the retina (Fig. 8). Of course, the temporal high-resolution area of the retina also participates in underwater vision. This area serves the frontal part of the visual field, which is very important for forward-moving animals. The existence of two high-resolution areas of the retina can also compensate for limited head mobility in many cetaceans. At low head mobility, even at high mobility of the eyes, a single high-resolution area allows the animal to inspect only a limited part of the surrounding space, whereas two such areas can provide almost panoramic vision.

The retina of the Amazon river dolphin is a special case. The visual system of this species is adapted to inhabiting low-transparent turbid water where vision is possible only at short distances. Contrary to all other investigated cetaceans, the retina of the Amazon river dolphin has only one area of higher ganglion cell density. However, this single area is located not in the center or temporal sector, but in the lower part of the retina, i.e., in the region responsible for the upper part of the visual field (Fig. 7D). In turbid low-transparent water, significant illumination exists only near the water surface, i.e., in the upper part of the visual field of a normally oriented animal. Just this part of the visual field is served by the ventral part of the retina where the Amazon river dolphin has higher retinal resolution. The density of ganglion cells in this region reaches 500 cells/mm$^2$; with the small size of the eyeball, this corresponds to a cell density of about 2 cells/deg$^2$.

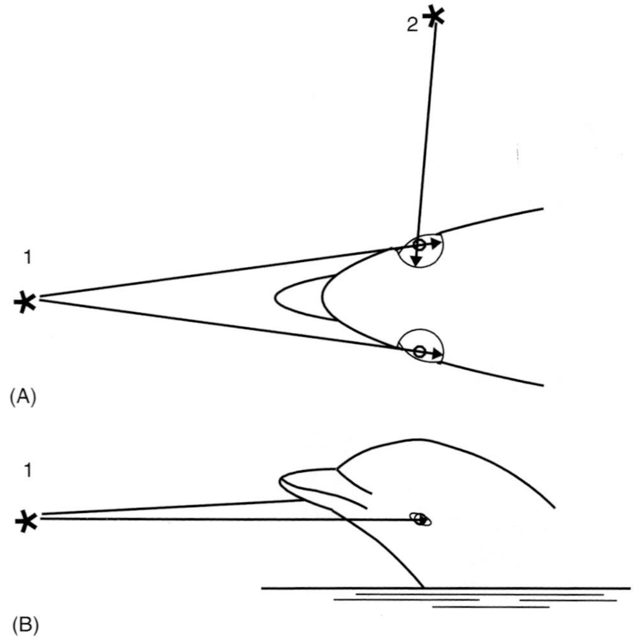

## B. Pinnipeds

Few otariids has been subjects of study of the ganglion cell topography in retinal whole mounts (Supin *et al.*, 2001). All of them have a typical area centralis, well-defined area of high concentration of ganglion cells (Fig. 6C,D). It is located at a distance of 35–40° from the visual field center (Fig. 9A); taking into account the position of the eye in orbit, this place may be at the projection of the vertical meridian of the visual field. Thus, the position of this area is similar to that in terrestrial carnivores. Cell density in this area reaches 1000 cells/mm$^2$, which corresponds to more than 160 cells/deg$^2$.

Quite different is the retinal topography in the walrus. The area of increased ganglion cell density is not defined as clearly as in the northern fur seal. It looks like a horizontally extended oval, resembling the visual streak of terrestrial mammals (Fig. 9B). Within this streak, the highest cell density in its temporal part exceeds 1000 cells/mm$^2$; because of the smaller size of the walrus eye, this cell density corresponds to only about 50 cells/deg$^2$. No phocid seals have been studied successfully studied yet with respect of their retinal topography.

**Figure 8** *Characteristic positions of the dolphin body relative to visually inspected objects: (A) dorsal view (for both underwater and aerial vision) and (B) lateral view (for aerial vision). 1, an above water object; 2, an underwater object. Arrows show directions of light rays from an object to the corresponding high-resolution area of the retina.*

## C. Other Marine Mammals

*1. Sirenians* Ganglion cell distribution in the manatee retina presents an example of low specialization. There is no sharply restricted spot of cell concentration. Ganglion cell distribution is not

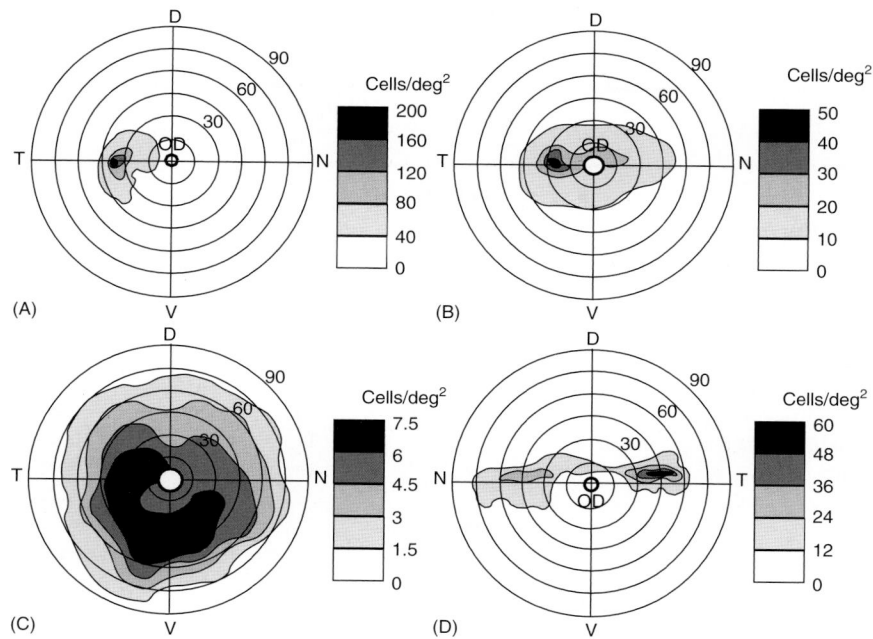

**Figure 9** *Topographic distribution of ganglion cell density in the retina of some pinnipeds, sirenians, and sea otters: (A) the northern fur seal; (B) the walrus; (C) the Caribbean manatee; (D) the sea otter. Cell density is expressed as a number of cells per squared degree of the visual field and is shown by various shadowing, according to the scales. Concentric circles show angular coordinates on a retinal hemisphere centered on the lens. D, V, N, T, dorsal, ventral, nasal, and temporal poles of the retina, respectively.*

uniform but varies smoothly across the retina: cell density is higher in a large center part of the retina (except in the nearest vicinity of the optic disk) and diminishes toward edges (Fig. 9C). The highest cell density is about 250–300 cells/mm²; for a rather small manatee eye, this corresponds to 6–7 cells/deg².

*2. Sea otters* In the sea otter retina ganglion cell topography (Fig. 9D) has a number of features similar to that of terrestrial mammals. The high-density area resembles naso-temporal streak. Within this streak, in its temporal part, there is a narrow and well-defined spot of the highest cells density which is similar to the area centralis in terrestrial mammals. The highest ganglion cell density in the sea otter exceeds 4000 cells/mm²; in the rather small eye of the sea otter, this corresponds to 50–60 cells/deg².

### D. Estimations of Visual Acuity from Ganglion Cell Density

Visual acuity is determined by two factors: quality of the eye optics and the retinal resolution. In normal eyes, these two values are in agreement. Therefore, retinal resolution can be used as a first-order estimate of the visual acuity. The retinal resolution depends on the density of ganglion cells (not of other retinal cells, e.g., photoreceptors, as ganglion cells transmit visual information to the brain). Thus, data on ganglion cell topography can be used to estimate the visual acuity of investigated species. When it is possible to compare estimates of visual acuity obtained by behavioral (psychophysical) methods and those based on ganglion cell topography, these estimates are in good agreement.

Retinal resolution (hence, visual acuity) is defined as the mean angular distance between neighboring ganglion cells; i.e., as $s = 1/\sqrt{D}$, where $s$ is the angular distance between cells and $D$ is the cell density per square degree. The estimation of retinal resolution is different in air and water: if the corneal surface is flat, the retinal image of an object in water is 1.33 times larger than in air (because the ratio of refraction indices of water and air is 1.33). Therefore, retinal resolution in water is 1.33 times better than in air. If the retinal surface is a little convex, this factor is less than 1.33 but still more than 1.

Retinal resolution in the areas of the highest concentration of ganglion cells (i.e., visual acuity in the best-vision areas of the visual field) was estimated in a number of marine mammal species (Table 1). In many cetaceans (except for river dolphins) it varies from 8 to 12 arcmin in water, correspondingly from 11 to 15 arcmin in air. For the bottlenose dolphin, the estimation of visual acuity obtained from the retinal topography almost coincides with that obtained in behavioral experiments. In general, the visual acuity of cetaceans is within a range of visual acuities of many terrestrial mammals, except the foveal vision of primates, where the visual acuity is around 1 arcmin. The Amazon river dolphin has much worse visual acuity: 40–50 arcmin in water; however, this value is adequate for vision in turbid water where objects are visible at best at a few tens of centimeters.

Among pinnipeds, rather acute vision is characteristic for the northern fur seal: better than 5 arcmin in water and better than 7 arcmin in air. This is close to estimates obtained in behavioral experiments in a number of both otariids and phocids: 5–8 arcmin (Schusterman, 1972). In the walrus, the visual acuity is worse than in seals: around 8 arcmin in water and 10 arcmin in air (Table 1).

**V**

#### TABLE I
#### Visual Acuity of Some Aquatic Mammals

| Species | Water | Air | Mode of measurement |
|---|---|---|---|
| **Odontoceti** | | | |
| Bottlenose dolphin (*Tursiops truncatus*) | 8–9 | 11–12 | BR |
| Short-beaked common dolphin (*Delphinus delphis*) | 8 | | R |
| Harbor porpoise (*Phocoena phocoena*) | 11 | 15 | R |
| Tucuxi dolphin (*Sotalia fluviatilis*) | 25 | 33 | R |
| Amazon river dolphin (*Inia geoffrensis*) | 40 | 53 | R |
| Dall's porpoise (*Phocoenoides dalli*) | 11 | | R |
| False killer whale (*Pseudorca crassidens*) | 9 | | R |
| Pacific white-sided dolphin (*Lagenorhynchus obliquidens*) | 11 | | R |
| Beluga whale (*Delphinapterus leucas*) | 12 | | R |
| **Mysticetes** | | | |
| Common minke whale (*Balaenoptera acutorostrata*) | 7 | | R |
| Gray whale (*Eschrichtius robustus*) | 11 | | R |
| **Phocid and Odobenids** | | | |
| Harbor seal (*Phoca vitulina*) | 8 | | B |
| Harp seal (*Pagophilus groenlandicus*) | 4 | 3 | R |
| **Otariids** | | | |
| Northern fur seal (*Callorhinus ursinus*) | 4–5 | 5–7 | R |
| Steller sea lion (*Eumetopias jubatus*) | 6–7 | | B |
| California sea lion (*Zalophus californianus*) | 5–6 | 5–7 | B |
| Cape fur seal (*Arctocephalus pusillus*) | | 6–7 | B |
| Southern fur seal (*A. australis*) | | 7 | B |
| Walrus (*Odobenus rosmarus*) | 8 | 10 | R |
| **Sirenia** | | | |
| Caribbean manatee (*Trichechus manatus*) | 20–24 | | BR |
| **Lutrinae** | | | |
| Sea otter (*Enhydra lutris*) | 7 | | R |

Visual acuity is presented as minimal resolvable distance in minutes of arc, rounded to a whole number of minutes. In some cases, a range of variation is indicated (e.g., 11-12 arc min). Estimates of visual acuity are given for underwater (*Water*) and aerial (*air*) vision. In the column *air*, data are not presented when none of the authors attempted to interpret their results in terms of aerial visual acuity. When several estimates of visual acuity in different conditions are available (e.g., in the nasal and temporal best-vision areas in cetaceans, at various illumination conditions, etc.), the best estimate (i.e., the minimal resolvable distance) is selected. Mode of measurement: B, behavioral data; R, data on retinal resolution.

In the manatee, underwater visual acuity is around 20 arcmin (it remains unknown whether the manatee has good aerial vision). In the sea otter, the visual acuity is around 7 arcmin in water.

## VI. Cerebral Visual Centers

In dolphins, the visual system is well represented in the midbrain (the superior colliculus), thalamus (the lateral geniculate body), and in the cerebral cortex. However, the visual centers (both the superior colliculus and the lateral geniculate body) are several times less in volume than corresponding parts of the auditory system (the inferior colliculus and the medial geniculate body).

In the cerebral cortex of dolphins, visual representation was found by the evoked potential method. This area occupies a part of the cortex named the lateral gyrus (Fig. 10A). The cortical representation of the visual system in dolphins also is not as large as that of the auditory system; nevertheless, it occupies a significant cortical area. There is a differentiation within this area: it contains a zone generating short latency evoked potentials (i.e., the primary projection zone) and another zone generating evoked potentials of longer latency (a nonprimary

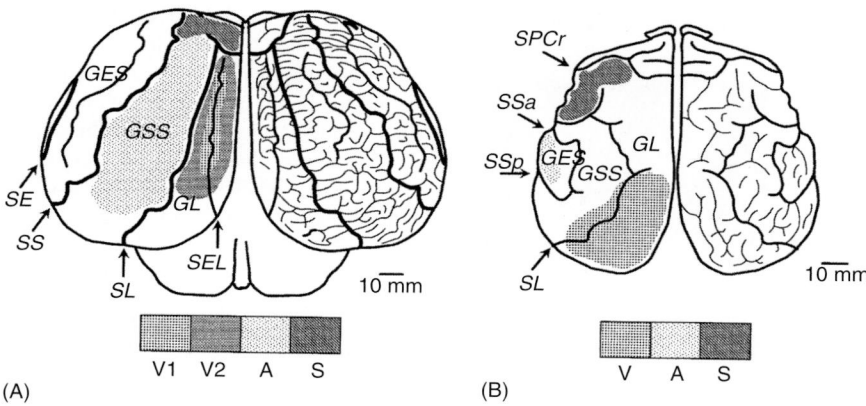

**Figure 10** *Position of projection sensory areas (visual, auditory, and somatosensory) in the cerebral cortex of cetaceans and pinnipeds: (A) the bottlenose dolphin and (B) the northern fur seal. Dorsal view of the cerebral cortex. On the right hemisphere, the pattern of cortical sulci and gyri is shown in more detail. On the left hemisphere, only main cortical sulci are shown and the positions of the visual, auditory, and somatosensory areas are indicated. The main sulci (labeled by arrows at their ends):* SE, *sulcus ectosylvius;* SS, *sulcus suprasylvius;* SL, *sulcus lateralis;* SEL, *sulcus endolateralis;* SSa, *sulcus suprasylvius anterior;* SSp, *sulcus suprasylvius posterior;* SPCr, *sulcus postcruciatus. The main gyri (labeled on their surface):* GES, *gyrus ectosylvius;* GSS, *gyrus suprasylvius;* GL, *gyrus lateralis;* V, *visual area (*V1, *primary projection zone;* V2, *nonprimary zone);* A, *auditory area (only a part of this area is visible in B);* S, *somatosensory area.*

zone). The first of them is located in the depth of the entolateral sulcus (a second-order sulcus within the lateral gyrus); the latter occupies the remainder of the lateral gyrus. These two zones differ in cytoarchitectonic features: the primary projection zone contains an incipient layer IV (the layer where visual thalamocortical afferent fibers end), whereas this layer is absent in the nonprimary zone.

In mysticetes, which do not have echolocation, the sizes of visual and auditory structures in the midbrain and thalamus are comparable. Their cortical sensory areas were not investigated.

Among pinnipeds, the visual representation in the cerebral cortex was found by evoked potential method in one otariid species—the northern fur seal (Fig. 10B), and one phocid species—the Caspian seal, *Pusa caspica*. The location of this area is very similar to that in carnivores: the projection occupies the caudal part of the lateral gyrus.

## VII. Conclusions

In general, the visual system of marine mammals demonstrates a rather high degree of development and performance, in particular, good visual acuity, capabilities to precisely aim visually driven behavior and intermodal transfer, and well-developed visual brain centers. This system also exhibits a number of specific features associated with adaptation to both aquatic and aerial environment, in particular, specific retinal topography (positions of best-vision areas) along with pupil and cornea structure which provide emmetropia in both air and water.

## *See Also the Following Article*

Brain

## *References*

Dawson, W. (1980). The cetacean eye. *In* "Cetacean Behavior: Mechanisms and Functions" (L. Herman, ed.), pp. 53–100. Wiley Interscience, New York.

Dral, A. (1977). On the retinal anatomy of cetacea (mainly *Tursiops truncatus*). *In* "Functional Anatomy of Marine Mammals" (R. Harrison, ed.), pp. 81–134. Academic Press, London.

Fobes, J., and Smock, C. (1981). Sensory capacities of marine mammals. *Psychol. Bull.* **89**, 288–307.

Hughes, A. (1977). The topography of vision in mammals of contrasting life style: Comparative optics and retinal organization. *In* "Handbook of Sensory Physiology: The Visual System in Vertebrates" (F. Crescitelli, ed.), Vol. VII/5, pp. 613–756. Springer, Berlin.

Jacobs, G. H. (1993). The distribution and nature of colour vision among the mammals. *Biol. Rev.* **68**, 413–471.

Jamieson, G. S., and Fisher, H. D. (1972). The pinniped eye: A review. *In* "Functional anatomy of Marine Mammals" (R. J. Harrison, ed.), Vol. 1, pp. 245–261. Academic Press, New York.

Madsen, C., and Herman, L. (1980). Social and ecological correlates of cetacean vision and visual appearance. *In* "Cetacean Behavior: Mechanisms and Functions" (L. Herman, ed.), pp. 101–147. Wiley Interscience, New York.

Mobley, J., and Helweg, D. (1990). Visual ecology and cognition in cetaceans. *In* "Sensory abilities of cetaceans" (J. Thomas, and R. Kastelein, eds), pp. 519–536. Plenum Press, New York.

Peichl, L., Berhmann, G., and Kröger, R. H. H. (2001). For whales and seals the ocean is not blue: A visual pigment loss in marine mammals. *Eur. J. Neurosci.* **13**, 1520–1528.

Piggins, D., Muntz, R., and Best, R. (1983). Physical and morphological aspects of the eye of the manatee *Trichechus inunguis* Natterer 1883 (Sirenia: Mammalia). *Mar. Behav. Physiol.* **9**, 111–130.

Schusterman, R. J. (1972). Visual acuity in pinnipeds. *In* "Behavior of Marine Animals" (H. E. Winn, and B. L. Olla, eds), Vol. 2, pp. 469–492. Plenum Press, New York.

Supin, A. Ya., Popov, V. V., and Mass, A. M. (2001). "The Sensory Physiology of Aquatic Mammals." Kluwer Akademic Publishers, Boston /Dordrecht/London.

**V**

# Walrus
## *Odobenus rosmarus*

### RONALD A. KASTELEIN

The walrus (Fig. 1; Fay, 1982) is the single species of the pinniped family Odobenidae and is distinguished from other pinnipeds by the upper canines in both sexes being prolonged as tusks. Walruses feed mainly on small organisms on the ocean floor, whereas almost all other pinnipeds feed primarily on highly mobile fish and crustaceans. For an animal of such a large size, this predator consumes organisms that are relatively low in the food chain. The diet of the walrus influences its biology: compared to other pinnipeds the walrus has a less streamlined body, swims more slowly, and dives less deep. Its sensory systems are adapted to its benthic foraging technique.

## I. Characteristics and Taxonomy

The latin name *Odobenus rosmarus* means "tooth walking sea horse." The genus *Odobenus* consists of only one species: *O. rosmarus*. Two sub-species are recognized based on morphological characteristics and on mitochondrial DNA divergence: the Pacific walrus (*Odobenus rosmarus divergens* Illiger 1815) and the Atlantic walrus (*Odobenus rosmarus rosmarus* Linnaeus 1758). A potential third sub-species, the Laptev walrus (*Odobenus rosmarus laptevi* Chapski 1940), is dubiously distinct from *Odobenus rosmarus divergens*.

The walrus is the largest pinniped except for the male elephant seal (*Mirounga* spp). The body is rotund; the girth at axilla is almost equal to body length. Males are larger than females of the same age. Adult Pacific walruses are on average slightly larger than adult Atlantic walruses of the same gender. Adult male walruses have an average body length of around 320 cm and weigh 1200–1500 kg; whereas adult females have a body length of around 270 cm and weigh 600–850 kg. Regional differences in body size per gender and age class exist.

Walruses are easily distinguished from other pinnipeds by their flat noses and enlarged upper canines that form huge tusks. Males have longer and thicker tusks than females. The tusks grow throughout life but growth is usually balanced by tooth wear. The large whiskers on the upper lips are translucent and yellowish and are directed forward. The eyes are small relative to body size compared to other pinniped species. They are positioned high on the head and can be protruded and retracted. There are no external pinnae. The color of the skin varies from gray when the walrus has just left cold water to yellowish brown when it is warm and dry. The fur is very short and is absent in some areas of the body. Although much variation exists, walruses generally

molt inconspicuously between May and June and grow new pelage in July and August. The appendages are hairless and the palms and soles are rough. The skin is 2–4 cm thick and very tough. It is thickest around the neck (about 4 cm) and in the area above the whiskers which is used for plowing through the ocean floor. The skin on the neck of adult males is thicker than that of females and is covered with fibrous tubercles. These tubercles are 1 cm thicker than the surrounding skin and protect the underlying tissues against tusk attacks by other males, and are an important visual sexual characteristic (Fig. 2). The blubber layer thickness varies depending on the part of the body and season, and can be up to 10 cm.

The skull (Kastelein and Gerrits, 1990) has some distinct features that relate to the ecology of the walrus. It is very thick and strong, as an adaptation to breaking through ice to make breathing holes. The front of the skull of an adult walrus is much higher and broader than that of other pinnipeds, to accommodate the large tusks. Males, which have thicker tusks than females, also have broader skulls. The tusks are composed of dentine covered with a thin layer of cementum. The ivory of walruses can be distinguished from that of other mammals by its central globular dentine. The heavy weight of the lower jaw probably serves to increase the impact of the tusks. The walrus is able to use its tusks to haul its body upward onto land or ice, because of the strong neck muscles attached to the large mastoid processes and the hinge between these processes: the condyle of the occipital bone. The zygomatic arch below the orbital cavity contains a strip of cartilage, probably to dampen shocks from the tusks to the brain case. In contrast to most pinnipeds, the orbital cavity is not closed on the dorsal side of the head, allowing the walrus to look in the direction it is moving upward during plowing through the substrate at 45°.

The spinal column consists of 7 cervical, 14 thoracic (occasionally 15), 6 lumbar (occasionally 5), 4 sacral, and 8 or 9 caudal vertebrae. There are 14–15 pairs of ribs. The hind flippers of the walrus rotate forward like those of otariids for locomotion on ice and land. The females have a 1- to 2-cm long clitoris bone. The penis bone (baculum) of adult males is up to 62 cm long.

Many of the muscles in the upper lip of the walrus are used to erect the whiskers in unison, although whiskers can be moved individually. The tongue muscles are very strong, and are used to create a low pressure in the mouth to extract the soft parts from clams. The mastication muscles are not very large, as the walrus usually does not chew its prey, but swallows it whole. The cheek muscles are strong, so that the walrus can produce powerful water jets from its mouth to wash sediment away from its prey.

The muscles in the hind limbs are similar to those in otariids. The adult walrus is so heavy and rotund that it does not lift its belly off the substrate when it moves on land or ice. Only calves can walk with just their flippers touching the substrate. In the water, the hind flippers are used for propulsion, and the front flippers mainly for steering.

The trachea is supported by cartilaginous rings throughout its length. It passes between the lungs for a third of their length, before bifurcating into bronchi. The lateral walls of the pharynx of sub-adult and adult males are extremely elastic and, and when inflated, form air sacs (pharyncheal pouches). These sacs are used as resonance chambers for the production of the bell-like sounds underwater, and also for floatation when resting in the water.

The mouth is narrow and bordered by the tusks. The roof of the mouth is very concave, allowing room for the large tongue that acts as a piston when retracted quickly. The force (of up to 119 kPa) generated by this piston is used to extract the soft parts of clams from their shells. The non-tusk teeth are worn down to the gums in wild animals, perhaps by sand moving in the mouth during feeding. The

Pup

Adult Female

Adult Male

**Figure 1**   *Adult male, adult female, and pub walrus (C. Brett Jarrett).*

teeth are pointed in captive walruses that are fed on fish. The dental formula of the Pacific walrus' permanent dentition is as follows:

$$\text{Incisors } \frac{(1)-(2)-3}{(1)-(2)-(3)} \text{ Canines } \frac{1}{1} \text{ Premolar } \frac{1-2-3-(4)}{2-3-4}$$
$$\text{Molar } \frac{(1)-(2)}{(1)-(2)} \times 2 = 18 \text{ to } 38 \text{ teeth}$$

The teeth in parentheses are present in less than 50% of the adult specimens.

The tip of the tongue may be rounded or bifid. The stomach consists of one J-shaped cavity. The intestines are 10–15 times the length of the animal.

A conspicuous aortic bulb is present at the base of the aortic arch. The arteries to the fore limbs are larger than those to the hind limbs. This is related to the division of the muscle masses for the limbs; the muscles of the neck and shoulders are more developed than those of the pelvis area. The testes are situated between the skin and the muscles. The penis is normally retracted into an opening posterior to the umbilicus. The uterus is bicornate and each horn opens separately into the vagina. Walruses usually have four nipples.

## II. Distribution and Abundance

The Pacific walrus is principally found in the Bering Sea south to Bristol Bay and Kamchatka, and in the Chukchi Sea, although in summer it may enter the Beaufort Sea and East Siberian Sea (Fig. 3). Breeding occurs in late winter in the marginal ice zone of the Bering Sea. The location of the main breeding sites varies depending on the state of the ice, but is generally southwest of Nunivak Island and southwest of St. Lawrence Island. Pacific walruses move north in spring with the receding and drifting ice; females give birth on ice floes which drift north. They move south in autumn, following the movement of the pack ice. Their swimming speed is approximately 10 km/h. For most of the year, sexes and age classes live separately. The population presently is estimated at around 200,000 animals.

The Atlantic walrus is found in the western and eastern Atlantic Arctic (Fig. 3; Born *et al.*, 1997). Within their range Atlantic walruses

**W**

**Figure 2** *An adult male Pacific Walrus. Note the large diameter of its tusks, the bulging eye, and the pronounced tubercles of the skin on the neck. The latter feature is only seen in adult males. Photo: Ron Kastelein.*

occur in several (perhaps eight) more or less well defined sub-populations—five to the west and three to the east of Greenland. In the western Atlantic Arctic walruses range from the East Canadian Arctic to West Greenland, including the Davis Strait, Baffin Bay, the archipelago in the Canadian high-Arctic, and Foxe Basin. The western Atlantic population is estimated now at more than 10,000. In the eastern Atlantic Arctic, walruses range from eastern Greenland to Svalbard (Norway), Franz Josef Land (Russia), and the Barents and Kara Seas. The eastern Atlantic population is estimated to be in the low thousands. Movement studies have shown a connection between walruses at Franz Josef Land and Svalbard and between the latter area and eastern Greenland.

The Laptev walrus (if this population is recognized as a sub-species) is found only in the Laptev Sea. The population has been estimated at 4000–5000 animals.

## III. Ecology

The walrus is found in the Arctic, where its distribution is limited by the availability of shallow water foraging grounds and thickness of ice. Walruses prefer relatively shallow water over continental shelves, because they feed on invertebrates which occur on the ocean floor up to a depth of about 80 m. Walruses can break through ice up to about 20 cm thick, but when the ice is thicker than this, they retreat to areas with drift ice. Thus, in winter walruses inhabit those regions of the drifting ice where leads and polynyas (irregular areas of open water surrounded by sea ice) are numerous and where the ice is thick enough to support their weight.

Male Pacific walruses rest in traditional terrestrial haul-out sites (sand, cobble, or boulder beaches), whereas females and calves prefer to haul out on pack ice or ice floes. In summer, males rest and molt while hauled on land out close to their feeding grounds. Females molt when hauled out on ice. Atlantic walruses also molt in the summer and during that time both males and females (sometimes in the same groups) haul out on both land and ice.

The diet of walruses consists mainly of benthic invertebrates. By far the most commonly eaten are bivalve mollusks, which are found buried in the sediment in high-density beds. How walruses find these beds is unknown. When they find a mollusk bed, they plow through the sediment with their snouts, while swimming with their

bodies at a 45° angle to the ocean floor, to find prey items. When foraging by plowing through the sea bed with its snout in search of invertebrates, the walrus uses its tusks and fore flippers as sleds while swimming with its hind flippers. Long furrows in the sediment have been observed in walrus feeding areas.

Once the walrus encounters a potential food item, it is quickly identified by the sensitive whiskers. If it is a bivalve mollusk, the foot or siphon is taken between the mobile lips, and by means of retraction of the large tongue, the soft parts are sucked from the shells and swallowed. The empty shells are discarded and can be found on the ocean floor near the furrows. The walrus can produce strong water jets with its mouth to excavate its prey, but can also remove sediment by producing strong water currents with movements of its fore flippers. The whiskers can be moved individually and are probably used as a tool to manipulate clams so that the foot is directed toward the walrus' mouth. Once a prey item is in front of the mouth, the tongue probably takes over as a touch and manipulation organ.

Compared to most other pinnipeds, walruses consume organisms that are small and low in the food chain. The large walrus has to consume many small organisms and must have a very efficient feeding method. In fact, walruses dive for up to 24 min (average around 5 min), usually spend 80% of that time on the bottom, and generally obtain 40–60 clams per dive. Dive times probably vary depending on water depth, prey type, and prey density. Adult walruses require on average about 25 kg of soft clam parts per day and a walrus has been found with 6000 prey items in its stomach. In captivity, Pacific walruses sometimes eat up to 50 kg of fish per day. Walruses probably store extra fat during summer when they can exploit their inshore foraging grounds, and use this during migration when they swim in waters that are too deep to contain their main prey.

Unusually, individual walruses may take a vertebrate diet. These animals usually kill seals and cetaceans as well as scavenge on their carcasses, from which they suck the blubber and internal organs. Seal-eating walruses thus consume organisms that are higher in the food chain than those consumed by mollusk-eating walruses and therefore carry a relatively high concentration of heavy metals and PCBs. Walruses sometimes also capture and eat sea birds.

## IV. Behavior and Physiology

The walrus has well-developed extrinsic eye muscles. The orbital cavity is not closed on the dorsal side. This allows the walrus binocular vision in the frontal and dorsal direction, as it can protrude its eyes, and does so mostly when excited. The lack of a roof of the orbital cavity suggests that the eyes are vulnerable to mechanical injury. However, the walrus eye has strong retractor muscles. The eyes can be pulled deep into the orbital cavity for protection, and the eye opening can be closed with thick eyelids. Blood vessels and surrounding fat probably serve to keep the eyes warm and functional in cold temperatures. Under high light conditions, the pupil is a vertical slit, during moderate light levels key-hole shaped, and under low light conditions circular. Retinal anatomy suggests that the walrus has color vision, but no psychophysical tests have been carried out, so that is unclear which part of the spectrum it can detect. Visual acuity appears to be less than in other pinnipeds so far investigated, and the eyes of the walrus seem to be specialized for short-range vision.

Because walruses dive up to 130 m and also at night, they often cannot always use vision to detect and process their prey. Instead, they use their sensitive mystacial vibrissae (whiskers). Each walrus has about 450 whiskers, which are highly innervated by sensory and motor nerves. The main sensory nerve of the fare (Trigeminal nerve) is well

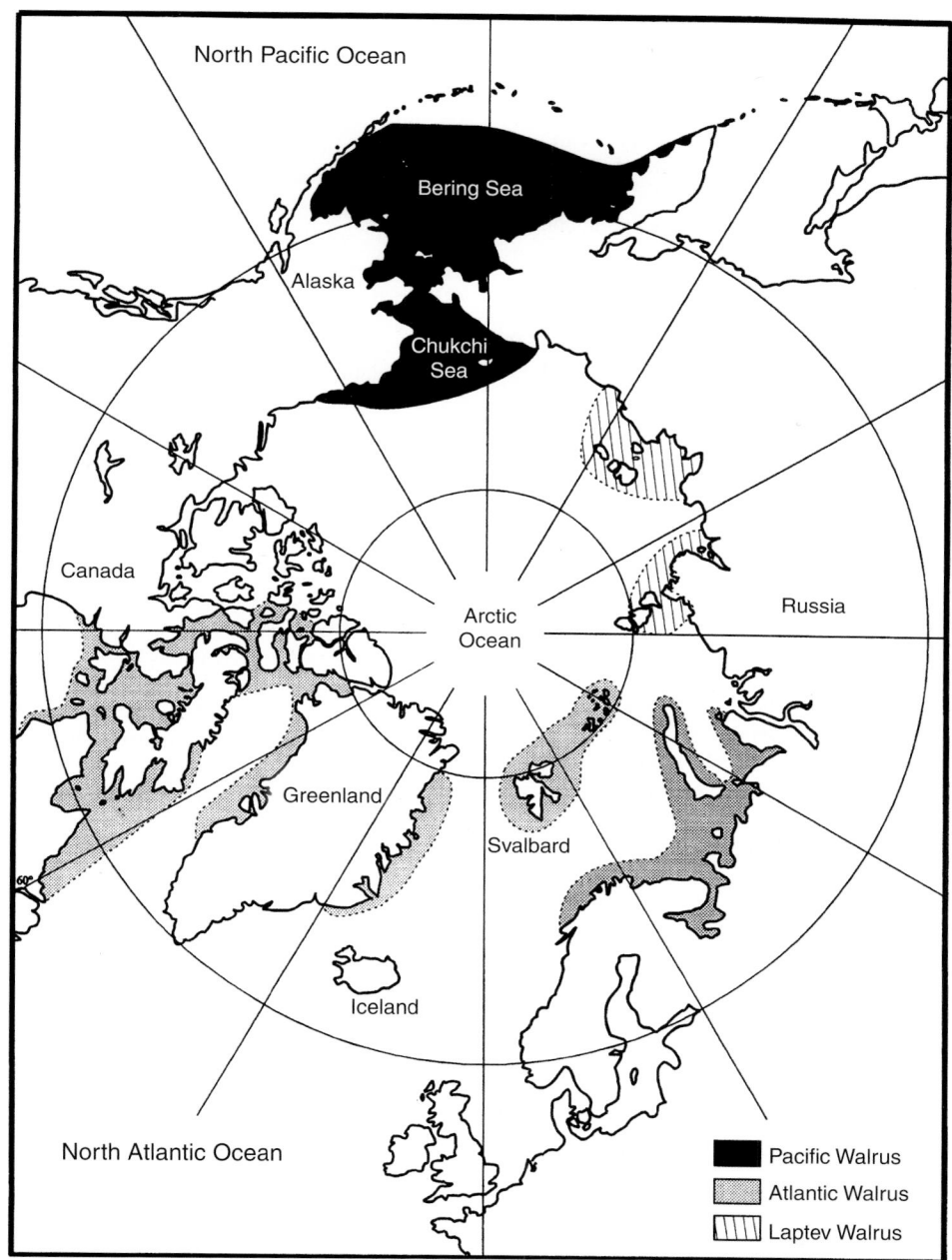

**Figure 3** *The distribution of the three walrus subspecies (the recognition of the Laptev walrus as a subspecies is controversial).*

developed and the infraorbital foramen in the skull is proportionally large to accommodate it. In contrast to most other pinnipeds, which probably use their whiskers to detect vibrations in the water, walruses use their whiskers to examine and manipulate small objects. In a psychophysical test, a captive walrus could distinguish a circle and a triangle which each had a surface area of 0.4 cm². The longer and thicker lateral whiskers are mainly used for the detection of objects, whereas the shorter and thinner ones near the mouth opening are primarily used for identification. The tip of the tongue also contains many mechano-receptors, and can be used to identify or reposition prey.

The behavior of the walrus on land and on ice suggests that it probably relies to a high degree on its sense of smell to obtain information about its surroundings. Anatomical evidence also suggests the

importance of olfaction to walruses. They have large nares that can be closed during dives, and the nasal passage is highly vascularized so that air that passes through it can be heated. However, no conclusive psychophysical tests have been conducted on the olfactory sensitivity of the walrus. Compared to many terrestrial mammals, the walrus has relatively few, but large taste buds. However, not much is known about the taste abilities and discrimination in walruses. Anecdotal information on captive animals suggests that they are not very sensitive to (bitter) flavors that are disgusting to most terrestrial mammals.

The walrus has limited ability to locate the source of sounds, as evidenced by its lack of pinnae. When the walrus is in air, sound reaches the tympanic membrane via a large cartilaginous outer ear tube. The aerial hearing has been tested, and is less acute than that of humans in

**W**

the frequency range tested (125 Hz to 8 kHz). When diving, the walrus closes its auditory meatus, and hears by tissue conduction, probably mainly via the vascular lining of the outer ear tube. The underwater hearing range of the walrus has an upper limit of 16 kHz.

Male and female walruses usually do not have contact with each other for most of the year, but animals of each sex congregate in large numbers both in winter and in summer.

Walruses seem to prefer being in groups. Walruses literally pile on top of each other (probably to conserve heat) in dense aggregations, and spend most of their haul-out time resting. On land, they are usually in deep sleep and are difficult to wake up. They often lie up-side-down, perhaps to reduce the pressure on their lungs. In this position their tusks point upward, they can stretch their necks, and surplus ear wax can drain from their ears. Body size and tusk length play important roles in establishing the hierarchy at haul-out sites. The tusks are used for display and as weapons in fights between walruses, as weapons against polar bears and killer whales (which may prey on walrus calves). They also aid in hauling out on ice floes and shores and serve to enlarge and keep open breathing holes in the ice. In addition, they function to hold heads above water by resting at the edges of ice floes.

Several aerial acoustic signals are used in a social context: barking, coughing and roaring when excited, whistling by males during the reproductive period (source levels of around 120 dB re 1 pW have been recorded), and soft calls from the females toward their calves. Alarm calls and other calls of calves have been described. Underwater, bell-like sounds are produced by the air sacs of adult males. Walruses have well-developed facial muscles and can make many facial expressions, which probably play a role in short-range communication.

## V. Life History

Females generally begin to ovulate at around 7 years of age (but some ovulate at the age of 5 years), and usually give birth for the first time at the age of 9 years. Males become sexually mature between 7 and 10 years of age, but become physically and socially mature, and therefore able to mate, at the age of 15 years. Walruses can reach an age of 30 to 40 years.

Adult male and female walruses congregate during the mating season (January–April). Walruses are polygynous. In the mating season adult males fight intensively in the water, evidently in competition for display sites near females. During courtship, the male walrus emits a stereotyped sequence of underwater sounds consisting of taps, knocks, pulses, and bell-like sounds. This acoustic display probably serves as an advertisement to females and as a warning to other males. Females choose a mate from among the displaying males. Copulation usually occurs in the water, but has been observed to occur on land in captivity, although a pool was available.

After a gestation period of about 15 months (including a period of delayed implantation of 4–5 months), a single calf of around 60 kg and 120 cm in length is born in spring (April–early June). Calves can swim immediately and may sometimes be carried on their mothers' backs for a while. The coat of the calves is slate gray. They are suckled for at least a year (on land, on ice, and in the water), and are usually gradually weaned during their second year, when they begin to forage for invertebrates. Reproductive females can produce about one calf in 3 years. The calves remain near their mothers, in groups of adult females, for several years. Most calves are weaned by the age of 3 years, when juvenile males tend to join male herds. The high degree of maternal care and low predation rate result in low natural mortality in walrus calves. This probably allows the walrus to have fewer offspring than other pinnipeds, most of which produce a calf every year.

## VI. Interactions with Humans

Atlantic walrus stocks were greatly reduced by intense exploitation in the eighteenth and nineteenth centuries by European whalers and sealers (for ivory, oil, and hides), and appear at present not to have recovered fully. The Pacific walrus population recovered from intense hunting, but, for unknown reasons, may have started to decline again.

Atlantic walruses are still hunted for subsistence by native people of Canada and Greenland. For the same purpose, Pacific walruses are taken by indigenous peoples of Russia and the USA (Alaska). Atlantic walruses are fully protected at Svalbard (since 1952) and in the western Russian Arctic (since 1956). Potential threats to walrus populations are: over-hunting, competition with shellfish fisheries, accidental bycatch by trawlers, pollution (PCBs, heavy metals, nuclear radiation), water borne noise at their foraging areas, aerial acoustic disturbances at their resting places (snow mobiles, aircraft, ships, oil and gas exploration), and habitat destruction by bottom trawling.

The first walruses in captivity were in Denmark and Germany. Later, zoological parks in The Netherlands, USA, and Russia began to keep this species. Captive walruses often ingest foreign objects that they cannot digest. A walrus can even die from consuming too many dead leaves that drop in pools in autumn. Another problem is the wearing down of tusks on the pool floor and walls, which, in severe cases, can cause root infections of the tusk.

When properly cared for, walruses are friendly and easily trained. They can perform for the public as well as in psychophysical research projects. Occasionally, males become difficult to handle after they reach maturity. Walruses can be kept in good health on diet of whole fish, which they swallow without chewing. Reproduction in parks and zoos has improved since the mid-90s, but only a few calves have been raised by their mothers, as walrus mothers are very protective of their calves toward both conspecifics and humans. This causes the mother to neglect her calf (she spends more time defending the calf than nursing it) or reduces the production of milk. The main problem in the early years of husbandry was the lack of a good formula to raise calves without milk–producing mothers. This problem has been overcome and calves from the wild or captive born ones can be hand-raised without nutritional deficiencies.

## See Also the Following Article

Pinnipedia, Overview

## References

Andersen, L. W., Born, E. W., Gjertz, I., Wiig, Ø., Holm, L.-E., and Bendixen, C. (1998). Population structure and gene flow of the Atlantic walrus (*Odobenus rosmarus rosmarus*) in the eastern Atlantic Arctic based on mitochondrial DNA and microsatellite variation. *Mol. Ecol.* **7**, 1323–1336.

Born, E. W. (1992). (*Odobenus rosmarus* Linnaeus) 1758—Walross. *In* "Handbuch der Säugetiere Europas. Band 6: Meeressäuger. Teil II: Robben—Pinnipedia" (R. Duguy, and D. Robineau, eds), pp. 269–299. AULA-Verlag, Wiesbaden.

Born, E. W., Gjertz, I., and Reeves, R. R. (1995). "Population assessment of Atlantic walrus (*Odobenus rosmarus rosmarus* L.)." Meddelelser no. 138 of the Norsk Polar Institutt, Oslo, pp. 100.

Born, E. W., Dietz, R., Heide-Jørgensen, M. P., and Knutsen, L.Ø. (1997). Historical and present distribution, abundance and exploitation of Atlantic walruses (*Odobenus rosmarus rosmarus* L.) in eastern Greenland. *Meddelser om Grønland, Bioscience* **46**, 73.

W

Fay, F. H. (1982). "Ecology and biology of the Pacific Walrus, *Odobenus rosmarus divergens*, Illiger. North American Fauna no. 74." United States Dept. of the Anterior, Fish and Wildlife Service, Washington, pp. 279.

Fay, F. H. (1985). Mammalian species, *Odobenus rosmarus*. *Am. Soc. Mammalogists* **238**, 1–7.

Fay, F. H., Eberhardt, L. L., Kelly, B. P., Burns, J. J., and Quakerbush, L. T. (1997). Status of the Pacific walrus population, 1950–1989. *Mar. Mamm. Sci.* **13**(4), 537–565.

Fedoseev, G. A. (1984). Present status of the population of walruses (*Odobenus rosmarus*) in the eastern Arctic and Bering Sea, pp. 73–85. *In* "Marine Mammals of the Far East" (V. E. Rodin, A. S. Perlov, A. A. Berzin, G. M. Gavrilov, A. I. Shevchenko, N. S. Fadeev, and E. B. Kucheriavenko, eds), 149 pp. TINRO, Vladivostok, (Translated by F. H. Fay, 1985).

Murie, J. (1871). Researches upon the anatomy of the Pinnipedia. Part I: On the walrus (*Trichechus rosmarus* Linn.). *Trans. Zool. Soc. (Lond.)* **7**, 411–464.

Gjertz, I. and Wiig, Ø. (1993). Status of walrus research in Svalbard and Franz Josef Land in 1992, A review. Appendix IV. In (R. E. A. Stewart, P. R. Richard, and B. E. Stewart eds), Report of the 2nd Walrus International Technical and Scientific Workshop (WITS), January 11–15, 1993, Winnipeg, Manitoba, Canada. Can. Tech. Rep. Fish. Aquat. Sci. 1940, 68–84.

Kastelein, R. A., and Gerrits, N. M. (1990). The anatomy of the Walrus head (*Odobenus rosmarus*). Part 1: The Skull. *Aquat. Mamm.* **16**(3), 101–119.

Kastelein, R. A., Gerrits, N. M., and Dubbeldam, J. L. (1991). The anatomy of the Walrus head (*Odobenus rosmarus*). Part 2: Description of the muscles and of their role in feeding and haul-out behaviour. *Aquat. Mamm.* **17**(3), 156–180.

Reijnders, P. J. H., Verriopoulos, G., and Brasseur, S. M. J. M. (1997). "Status of Pinnipeds Relevant to the European Union," pp. 160–182. IBN Scientific Contributions 8, The Netherlands.

Richard, P. R., and Campbell, R. R. (1988). Status of the Atlantic Walrus, *Odobenus rosmarus rosmarus* in Canada. *Can. Field Nat.* **102**(2), 337–350.

Wiig, Ø., Gjertz, I., Griffiths, D., and Lydersen, C. (1993). Diving patterns of an Atlantic walrus *Odobenus rosmarus rosmarus* at Svalbard. *Polar Biol.* **13**, 71–72.

# Weddell Seal
## *Leptonychotes weddellii*

### JEANETTE A. THOMAS AND JACK TERHUNE

## I. Characteristics and Taxonomy

The Weddell seal belongs to the family Phocidae, subfamily Monacinae, and tribe Lobodontini (Bertram, 1940; Kooyman, 1981; Nowak, 1991 for general reviews). There is a single species within the genus *Leptonychotes*. The closest relatives are the other Antarctic seals in the tribe Lobodontini, i.e., crabeater, *Lobodon carcinophaga*; leopard, *Hydrurga leptonyx*; and Ross seals, *Ommatophoca rossii*; and the monk seals in the tribe Monachini, i.e., Caribbean, Mediterranean, and Hawaiian monk seals, *Monachus* spp. The subfamily Monachinae evolved in the North Atlantic during the Miocene. A fossil, *Homiphoca*, found off the South African coast could be the intermediate form between monk seals and Antarctic seals.

Weddell seals are black with grayish silver streaks, with individual variations (Fig. 1). The adult pelt does not have underfur. In pups, the light gray lanugo is longer and thicker than adult hair and is shed by 44 days of age. There is no dramatic SEXUAL DIMORPHISM in body size; males reach 2.5–2.9 m and females are 2.6–3.3 m in length. Weights range from 400 to 500 kg, with pregnant females being the heaviest. At birth, pups weigh 22–29 kg and are about 1.5 m in length.

As with other phocids, the Weddell seal has a fusiform body shape and laboriously crawls on its belly or rolls to move on ice. It is not capable of upright stance or moving the hind limbs forward. Under water, the Weddell seal propels itself by moving the hind flippers in a horizontal plane. The manus is a flipper and the pes fully webbed. The first metacarpal is noticeably larger than the others. Black claws on the front flippers are large and useful for gripping the ice or scratching. Claws on the hind flipper are reduced in size. The tail is distinct and free.

The large brown eyes often have wet circles around them because of no lacrimal duct. A tapetum on the back surface of the retina assists in seeing in low-light levels, especially during the austral winter and deep dives. A nictitating membrane protects its eyes from blowing snow and allows its eyes to be open in salt water. The species has excellent VISION under water.

The external ear is absent; the cochlea is about 10 mm high and forms about two and a half turns. The organ of Corti contains one row of inner and three rows of outer hair cells. Microscopic examination of the ear did not reveal obvious specializations for high-frequency hearing. A preliminary attempt was made to measure the in-air hearing of a Weddell seal using auditory evoked potentials and auditory brainstem responses on immobilized Weddell seals, but difficulties in the field precluded completion of an audiogram.

The nostrils are oriented vertically and normally are closed, but open when the seal needs to respire. There are seven rows of mystacial vibrissae and superciliary whiskers that are smooth, not beaded as in some pinnipeds. The tip of the tongue is notched. Testes are abdominal, the penis is retractable, and there is a baculum. The uterus is bipartite. Two mammary glands are present and the milk is exceptionally high in fat and protein content.

The seal has a simple stomach and eats whole fish, heads down first. There are 34 chromosome pairs in this species.

The dental formula is 2/2, 1/1, 4/4, 1/1. The milk teeth disappear before or soon after birth. Cheek teeth have three points and are homodont, but are not the exaggerated tricuspid structure of the crabeater and leopard seals. The outer incisors are larger and procumbent or project forward. The canines also are procumbent and, along with the last incisor, are used in "ice-sawing" behavior. The seal maintains circular breathing holes in the fast ice by turning within the hole and raking ice off the rim of the hole with its teeth. Such behavior allows the seals to maintain holes for breathing and hauling out throughout the year.

## II. Distribution and Abundance

The Weddell seal has a circumpolar distribution around Antarctica, preferring land-fast ice habitats, but the proximity to open water and pack ice also influences the distribution of haul-out sites in fast-ice areas. In the winter, the propensity for the fast ice to crack is the major determinant of Weddell seal distribution. The seals haul out through cracks in the fast ice formed from tidal action. This fast ice provides a stable platform for giving birth to pups, nursing pups, hauling out to avoid predators, and resting. There is no predictable migration in this species, and both males and females exhibit breeding site fidelity

**W**

Figure 1    *Weddell seal.*

(Cameron *et al.*, 2007). Some seals maintain holes in coastal land-fast ice throughout the dark austral winter while others frequent the off-shore pack ice or move back and forth, presumably to forage offshore and return to rest on the stable fast ice. Occasionally, seals are seen at subantarctic islands, including South Shetlands, the South Orkneys, and South Georgia. Single wandering Weddell seals have been found in remote locations such as Heard, Kerguelen, Macquarie, Auckland, Juan Fernandez, and Falkland/Malvinas Islands, as well as Australia, New Zealand, Patagonia, and Uruguay.

No systematic, large-scale population census studies have been conducted, so estimates in the literature are approximate. The Weddell seal is abundant; the estimated range is from 500,000 to 1 million seals. However, population studies following a recent calving of a glacier in McMurdo Sound demonstrate the ability of the species to shift their distribution and adapt to new haul-out sites. Several studies of population dynamics have been carried out (Stirling, 1971; Siniff *et al.*, 1977; Testa, 1987; Testa and Siniff, 1987; Testa *et al.*, 1990; Cameron and Siniff, 2004).

### III. Ecology

The diet includes the large Antarctic cod (*Dissotichus mawsoni*), and smaller fish (*Pleurogramma antarcticum*, *Pagothenia borchgrevinki*, and *Trematomus* spp.). The seals sometimes startle fish by blowing bubbles at them and erect their vibrissae when near a fish (Davis *et al.*, 1999; Plötz *et al.*, 2001).

Weddell seals do not exhibit the scars from leopard seal predation that are seen on crabeater seals. However, killer whales (*Orcinus orca*) are known to take Weddell seals of all ages. When killer whales and leopard seals move into areas of Weddell seals, *Leptonychotes* suddenly and dramatically stops calling underwater, perhaps to avoid detection (Thomas and Kuechle, 1982; Thomas *et al.*, 1983).

### IV. Behavior and Physiology

Weddell seals are extremely good divers, commonly diving to about 600 m for up to 82 min, although shallower dives of <20 min are typical (Elsner *et al.*, 1970; Kooyman, 1981; Wartzok *et al.*, 1992; Harcourt *et al.*, 2000). Some seals range out to 5 km from a breathing hole and returning on a single dive. A great deal about their DIVING BEHAVIOR has been documented using satellite tags with time-depth recorders and video cameras attached to the seal's back. Before a dive, the seal exhales slightly and the nostrils and mouth are closed. Seals usually make a series of short shallow dives before commencing on a longer dive. In one study in the dark of February, the seals foraged almost exclusively in the upper water column at night but often dove to 340–450 m during the day for benthic feeding. Typical swim speed is 8–12 km/h and the descent rate for a dive is 35 m/min, whereas a foraging seal has a swim speed of 0.7–1.8 m/sec. Sometimes the ascending

dive retraces the path of the descending dive. During a dive, the lungs collapse, the trachea compresses, blood is shunted to the extremities, the metabolic rate drops to about 20% of the resting rate, and the heart rate undergoes bradycardia, dropping from about 85 to 16 beats/min. The blood and muscles have a 3–5 times greater oxygen-carrying capacity than those of humans.

Males patrol using loud (up to 193 dB re 1 μPa) trills to advertise and defend their underwater territories, which cover an area 15–50 m wide by 50–400 m long. Sometimes bloody fights occur between males underwater and even continue onto the ice, with the loser evicted from the territory. Territorial males are rarely found more than 1 km from the colony. The underwater repertoire of Weddell seals is elaborate, including over 30 sound types, and many repeated call types have up to seven rhythm patterns. Some researchers suggest the species has song. Their sounds are some of the longest among marine mammals, ranging up to 70 sec. Over 100 calls/min have been recorded at a colony in darkness in the winter and during the height of the breeding season. GEOGRAPHIC VARIATIONS in the repertoire have been documented among McMurdo Sound, Palmer Peninsula, and Davis Station and Mawson Station seals.

The seals stay warm with a thick layer of subcutaneous fat, often remaining hauled out on the ice covered with snow. On warm Antarctic days, steam can be seen rising from their bodies and heat dissipation may be necessary. Weddell seals can dilate blood vessels in the skin to dissipate heat or constrict them to conserve body heat. During winter, however, few seals haul out.

Water is obtained metabolically through the diet; sometimes the seals eat snow. The KIDNEYS are reniculate, adapted for conserving water and removing high salt loads.

### V. Life History

The mating system is likely a variation of harem defense polygamy in which dominant males fight to establish territories at breeding sites, but the territory is underwater (Bartsch *et al.*, 1992). Adult males are not tolerated on the ice in maternal colonies. Occasionally a non-breeding, adult male is seen at the periphery of the colony. Some males establish underwater territories around breathing holes that freeze up in early winter and others arrive and compete for territories after the females give birth. Subadult seals are rarely seen in breeding colonies. They tend to congregate in large groups near the ice edge.

The mother Weddell seals return to tidal cracks each austral spring and give birth to pups from late September to early November, depending on the location (Lugg, 1966; Gelatt *et al.*, 2000; general reviews). Typically, a single pup is born, but twins have been observed and even an albino pup was reported. A few mothers have pups in isolation from other seals, but most give birth in colonies of up to 50 mothers with pups. Mothers and pups maintain

W

individual spacing when hauled out on ice. When disturbed, mothers vocalize to intruders. Pups nurse for a 7- to 8-week period, gaining almost 2 kg per day. At weaning, pups weigh about 125 kg and mothers have lost nearly twice that amount of weight because they do not forage during the nursing period. Pups are first enticed into the water by their mother's calling at 10 to 14 days of age. Pups struggle to swim and stay under the water, popping to the surface. Some pups die in the breathing holes, not being able to crawl out of the slippery, steep hole. Mothers and older pups go into the water progressively more often as the pup grows. They exhibit a distinct diel pattern of haul-out, with most seals hauling out in the colony for several hours around midday.

Mating takes place in the water, with about 80% of females becoming pregnant, at least in the McMurdo Sound area. Once the pups wean, the mother enters the water to feed and mating occurs within a male's territory. Adult seals molt after mating. Implantation of the blastocyst is delayed until mid-January. Because of delayed implantation, pups are born the following spring at approximately the same time of year. Delayed implantation allows the mother to molt, feed, and recover from the dramatic weight loss associated with lactation before another fetus starts to develop.

About the time the last pup in the colony weans in mid-December, the fast ice begins to break up, adults disperse, and newly weaned pups are left to fend for themselves. Some data taken from pups with transmitters indicated pups feed near shore on small fish such as *Pleurogramma* spp. This may be a vulnerable period for Weddell seals because the breakup of the fast ice provides access for predators, like killer whales and leopard seals, to the colonies.

## VI. Interactions with Humans

This species is the most studied of all Antarctic pinnipeds because the seals can be found reliably in breeding colonies at traditional sites. The use of traditional haul-out sites has facilitated research on this species since the early 1960s. Several investigators have maintained research programs near the US bases at McMurdo Sound and Palmer Peninsula, the New Zealand Scott Base, the Australian bases, Davis and Mawson in Eastern Antarctica and the German bases in the Weddell Sea. The seals have little fear of humans because there are no natural land predators like the Arctic polar bears (*Ursus maritimus*).

In 1957, the International Geophysical Year established the importance of conservation and research in the Antarctic. The Antarctic treaty was signed in 1961, establishing the protection of the species and its habitat. During the 1970s and 1980s, inhabitants of the New Zealand station in McMurdo Sound took Weddell seals to feed their sled dogs over the winter. This practice has stopped. Otherwise, there is no record of extensive harvest of this species. Today, Weddell seals encounter humans as various ships resupply research stations and occasionally as passengers on ecotourism vessels disembark on the ice.

## *References*

Bartsch, S. S., Johnston, S. D., and Siniff, D. B. (1992). Territorial behaviour and breeding frequency of male Weddell seals (*Leptonychotes weddellii*) in relation to age, size, and concentrations of serum testosterone and cortisol. *Can. J. Zool.* **70**, 680–692.

Bertram, G. C. L. (1940). The biology of the Weddell and crabeater seals. British Graham Land Expedition 1934–1937. Scientific Reports **1**, 1–39. British Museum of Natural History, London.

Cameron, M. F., and Siniff, D. B. (2004). Age-specific survival, abundance and immigration rates of a Weddell seal (*Leptonychotes weddellii*) population in McMurdo Sound, Antarctica. *Can. J. Zool.* **82**, 601–615.

Cameron, M. F., Siniff, D. B., Proffitt, K. M., and Garrott, R. A. (2007). Site fidelity of Weddell seals: The effects of sex and age. *Antarct. Sci.* **19**, 149–155.

Davis, R. W., Fuiman, L. A., Williams, T. M., Collier, S. O., Hagey, W. P., Kanatous, S. G., and Horning, M. (1999). Hunting behavior of a marine mammal beneath the Antarctic fast ice. *Science* **283**, 993–996.

Elsner, R., Kooyman, G. L., and Drabek, C. M. (1970). Diving duration in pregnant Weddell seals. *In* "Antarctic Ecology" (M. W. Holdgate, ed.), pp. 477–482. Academic Press, New York.

Gelatt, T., Davis, C. S., Cameron, M., Siniff, D. B., and Strobeck, C. (2000). The old and the new: Integrating population ecology and population genetics of Weddell seals. *In* "Antarctic ecosystems: Models for a wider ecological understanding" (W. Davidson, C. Howard-Williams, and P. Broady, eds). Caxton Press, Christchurch.

Harcourt, R. G., Hindell, M. A., Bell, D. G., and Waas, J. R. (2000). Three-dimensional dive profiles of free-ranging Weddell seals. *Polar Biol.* **23**, 479–487.

Kooyman, G. L. (1981). "Weddell Seal, Consummate Diver." Cambridge University Press, Cambridge.

Lake, S., Burton, H., and Wotherspoon, S. (2005). Movements of adult female Weddell seals during the winter months. *Polar Biol.* **29**, 270–279.

Lindsey, A. A. (1937). The Weddell seal in the Bay of Whales, Antarctica. *J. Mammal.* **18**, 127–144.

Lugg, D. J. (1966). Annual cycle of the Weddell seal in the Vestfold Hills, Antarctica. *J. Mammal.* **47**, 317–322.

Moors, H. B., and Terhune, J. M. (2004). Repetition patterns in Weddell seal (*Leptonychotes weddellii*) underwater multiple element calls. *J. Acoust. Soc. Am.* **116**, 1261–1270.

Nowak, R. M. (1991). "Walker's Mammals of the World," 5th Ed. Johns Hopkins University Press, Baltimore.

Plötz, J., Bornemann, H., Knust, R., Schröder, A., and Bester, M. (2001). Foraging behaviour of Weddell seals, and its ecological implications. *Polar Biol.* **24**, 901–909.

Rouget, P. A., Terhune, J. M., and Burton, H. R. (2007). Weddell seal underwater calling rates during the winter and spring near Mawson Station, Antarctica. *Mar. Mamm. Sci.* **32**, 508–523.

Sapin-Jaloustre, J. (1952). Weddell seal, Mammalia. *In* "National Antarctic Expedition 1901–1904." *Nat. His.* **2**, 1–66, British Museum, London.

Schreer, J. F., and Testa, J. W. (1996). Classification of Weddell seal diving behavior. *Mar. Mamm. Sci.* **12**, 227–250.

Siniff, D. B., DeMaster, D. P., Hofman, R. J., and Eberhardt, L. L. (1977). An analysis of the dynamics of a Weddell seal population. *Ecol. Monogr.* **47**, 319–335.

Smith, M. S. R. (1965). Seasonal movements of the Weddell seal in McMurdo Sound, Antarctica. *J. Wildl. Manage.* **29**, 464–470.

Stirling, I. (1971). Population dynamics of the Weddell seal (*Leptonychotes weddellii*) in McMurdo Sound, Antarctica 1966–1968. *Antarct. Res. Ser.* **18**, 141–161.

Terhune, J. M., Burton, H., and Green, K. (1994). Weddell seal in-air call sequences made with closed mouths. *Polar Biol.* **14**, 117–122.

Testa, J. W. (1987). Juvenile survival and recruitment in a population of Weddell seals (*Leptonychotes weddellii*) in McMurdo Sound, Antarctica. *Can. J. Zool.* **65**, 2993–2997.

Testa, J. W., and Siniff, D. B. (1987). Population dynamics of Weddell seals (*Leptonychotes weddellii*) in McMurdo Sound, Antarctica. *Ecol. Monogr.* **57**, 149–166.

Testa, J. W., Siniff, D. B., Croxall, J. P., and Burton, H. R. (1990). A comparison of reproductive parameters among three populations of Weddell seals (*Leptonychotes weddellii*). *J. Anim. Ecol.* **59**, 1165–1175.

**W**

Thomas, J. A., and Kuechle, V. B. (1982). Quantitative analysis of the underwater repertoire of the Weddell seal (*Leptonychotes weddellii*). *J. Acoust. Soc. Am.* **72**, 1730–1738.

Thomas, J. A., Zinnel, K. C., and Ferm, L. M. (1983). Analysis of Weddell seal (*Leptonychotes weddellii*) vocalizations using underwater playbacks. *Can. J. Zool.* **61**, 1448–1456.

Wartzok, D., Elsner, R., Stone, H., and Burns, J. (1992). Under-ice movements and the sensory basis of hold finding in the ringed and Weddell seals. *Can. J. Zool.* **70**, 1712–1722.

Welsh, U., and Ridelsheimer, B. (1997). Histophysiological observations on the external auditory meatus, middle, and inner ear of the Weddell Seal (*Leptonychotes weddellii*). *J. Morph.* **234**, 25–36.

**Figure 1**　　*The life stages of humpback whale cyamids* (Cyamus boopis).

# Whale Lice

### CARL J. PFEIFFER

Whale lice (cyamids) are crustacean ectoparasites living on the skin of some species of cetaceans. Whale lice remain among some of the world's biologically most specialized but least understood crustaceans. The current lack of knowledge about these animals undoubtedly stems from the fact that their natural habitat is limited to the skin surface of primarily slow-moving baleen whales, which themselves are difficult to study because they are generally submerged and constantly moving. Accordingly, it is very difficult for scientists to observe, in their natural setting and over extended periods of time, the behavior and biologic processes of whale lice. Some features have been learned, however, by a small number of scientists who have studied these interesting crustaceans for over a century, and the following brief account reviews these data.

## I. Classification

Whale lice are actually not lice but perhaps they acquired this nickname in the 1800s from whalers who noticed that they crawled, presumably as parasites, on the surface of the whale's skin and that their size, in proportion to that of a great whale, was comparable to the size of louse on a human or dog. Whale lice are arthropods of the subphylum Crustacea, class Malacostraca, order Amphipoda, family Cyamidae, and genus *Cyamus*. As such they are closely related to other more commonly observed amphipods such as sand hoppers and caprellids, i.e., common marine peracarids. Altogether there are more than 5000 crustacean species, but only 23 known species of whale lice or cyamids. At least 15 of these comprise the genus *Cyamus*. Because the practice of WHALING has such a long history, cyamids were historically documented early, including *Cyamus ceti* by Linné in 1758, a genus described further by Latreille in 1796. A number of species were classified in the 1800s, including *C. ovalis* in 1834, *C. delphini* in 1836, *C. boopis* in 1870, and *C. scammoni* in 1872, and these cyamids were found on southern right whales (*Eubalaena australis*), humpback (*Megaptera novaeangliae*), gray (*Eschrichtius robustus*), and other large whales. Species names have occasionally been revised for cyamids and new species were discovered even late in the twentieth century, such as the two new cyamids reported on one of the large beaked whales, the Baird's beaked whale *Berardius bairdii* (Waller, 1989). Genus names for cyamids have also been changed over time and the earlier genus *Paracyamus* is no longer considered valid.

## II. General Ecology and Morphology of Cyamids

Although information does not exist on the early development of cyamids, their evolution has undoubtedly been closely coupled with the long evolution of cetaceans, which first began around 55 million years ago when terrestrial species returned to the marine environment as precursors of the modern whales. Thus, modem cyamids show little resemblance to other crustaceans and have become greatly specialized for their parasitic and lifelong relationship to whales. They have a high degree of host specificity to whales, and one species, *C. boopis*, is found only on humpback whales (Fig. 1) and another species, *C. scammoni*, only occurs on the gray whale. Some other species of whale lice overlap their residence on two to four species of whales. The two isolated Arctic odontocetes, beluga whales (*Delphinapterus leucas*) and narwhals (*Monodon monoceros*), each share the same two cyamids, *C. monodontis* and *C. nodosus*. Table I summarizes the cetacean distribution of cyamids. Cyamids are unable to swim freely in the sea or from whale to whale at any of their developmental stages. Accordingly, they die if they lose their foothold on their host, but can be transferred from cetacean mother to calf or during cetacean mating. Although details on the number of juvenile stages that occur prior to adulthood are not yet established for cyamids, Leung (1976) has estimated that at least seven to eight instar stages exist for *C. scammoni* of the gray whale.

Although the general gross body structure of cyamids has been described elsewhere (Margolis, 1955; Leung, 1967; Berzin and Vlasova, 1982), very few studies have been directed toward the microscopic anatomy of cyamids. Early work on the musculature and very recent work (Levin and Pfeiffer, 1999) on cyamid ocular structure have been reported. Their small, paired eyes appear almost rudimentary. However, ultrastructural analysis of these photoreceptors for *C. ceti* has revealed well-developed sensory organs with each eye containing about 50 visual ommatidial units and an overall organization similar to other amphipod compound eyes. The exoskeleton of cyamids consists of a chitinous cuticle that is similar to that typically observed for other crustaceans. It has an exocuticle with multiple microfibrillar lamellae and an endocuticle traversed by both pore canals and dense fibers, as revealed by electron microscopy (Pfeiffer and Viers, 1998).

Cyamids are SEXUALLY DIMORPHIC. The males are larger and, depending on species, adult cyamids usually range from approximately 6 to 19 mm in length. The most striking features of their appearance are their marked degree of segmentation and prominent gnathopods,

### TABLE I
### Distribution of Cyamid Species on Cetaceans[a]

| Host | Whale lice |
|---|---|
| Mysticeti | |
| Bowhead whale, *Balaena mysticetus* | *Cyamus ceti* |
| Right whales, *Eubalaena* spp. | *C. ceti, C. erraticus, C. gracilis, C. ovalis, C. catadontis* |
| Gray whale, *Eschrichtius, robustus* | *C. ceti, C. kessleri, C. scammoni* |
| Blue whale, *Balaenoptera musculus* | *C. balaenopterae, C. bahamondei* |
| Humpback whale, *Megaptera novaeangliae* | *C. boopis, C. elongatus* |
| Common minke whale, *B. acutorostrata* | *C. balaenopterae* |
| Fin whale, *B. physalus* | *C. balaenopterae, C. bahamondei* |
| Odonticeti | |
| Sperm whale, *Physeter macrocephalus* | *C. ovalis, C. catadontis, C. bahamondei, Isocyamus delphini, Neocyamus physeteris* |
| Baird's beaked whale, *Benardius bairdii* | *Platycyamus flaviscutatus, C. orubraedon* |
| Beluga, *Delphinapteris leucas* | *C. monodontis, C. nodosus* |
| Narwhal, *Monodon monoceros* | *C. monodontis, C. nodosus* |
| Northern and southern bottlenose whales, *Hyperoodon* spp. | *C. thompsoni, I. delphini* |
| Long-finned pilot whale, *Globicephala melas* | *Isocyamus delphini* |
| Short-finned pilot whale, *G. macrorhynchus* | *I. delphini* |
| Short-beaked common dolphin, *Delphinus delphis* | *I. delphini, syncyamus pseudorcae* |
| Risso's dolphin, *Grampus griseus* | *I. delphini* |
| White-beaked dolphin, *Lagenorhynchus albirostris* | *I. delphini, Scutocyamus parvus* |
| Harbor porpoise, *Phocoena phocoena* | *I. delphini* |
| Killer whale, *Orcinus orca* | *I. delphini, C. antarcticensis* |
| False killer whale, *Pseudorca crassidens* | *I. delphini, S. pseudorcae* |
| Rough-toothed dolphin, *Steno bredanensis* | *I. delphini* |
| Gervais' beaked whale, *Mesoplodon densirostris* | *I. delphini* |
| Pantropical spotted dolphin, *Stenella attenuate* | *Syncyamus* sp. |
| Striped dolphin, *S. coeruleoalba* | *Syncyamus* sp. |
| Spinner dolphin, *S. longirostris* | *Syncyamus* sp. |
| Common bottlenose dolphin, *Tursiops truncatus* | *Syncyamus* sp. |

[a]From Margolis (1955), Leung (1967, 1970), Lincoln and Hurley (1974), and Berzin and Vlasova (1982).

**Figure 2**  *Two specimens of* Cyamus ovalis. *The head region faces the top. Note segmented body and antennae.*

or legs, with large dactyli, or hooks, that assure firm attachment to the host. The body is flattened and divided into a small cephalic-cephalon or head with paired, minute eyes, and segmented pereion or body to which are attached two pairs of gills and four pairs of gnathopod-type appendages. Fig. 2 illustrates the general body structure for *C. ovalis*.

Cyamids are mostly found on those areas of the whale surface most protected from the turbulence of water flow, which on baleen whales include regions around barnacles, skin folds or ventral grooves of the head, protected zones around the blowholes, eyes, and flippers, margins of the lips, on callosities, wounds, and genital slit (Fig. 3). In those species of whales that serve as host to several species of cyamids, there may be differences in the spatial distributions of the different species of whale lice, and within one species of cyamid the reproductive status and sex of the cyamid may alter the spatial distribution (Rice and Wolman, 1971; Balbuena and Raga, 1991; Rowntree, 1996). Whale lice do move around on their cetacean hosts; in the case of *C. boopis* of the humpback whale, the larger males may carry their smaller female mates and, in an artificial aquarium setting, were observed to walk at a rate of 4.5 m/h (Rowntree, 1996).

Whale lice breathe by means of two pairs of external gills, which are much reduced in size in early juvenile stages. It has been reported that they can live for up to 3 days out of an aquatic environment, such as on a stranded whale, suggesting that they can also rely on integumentary respiration (Leung, 1976).

### III. Feeding Habits of Cyamids

It was speculated for a long time that whale lice fed on whale skin and hence they were deemed ectoparasites. Some workers suggested that they might be omnivorous and ingest algal filaments or suspended materials or plankton in the water near their attachment

W

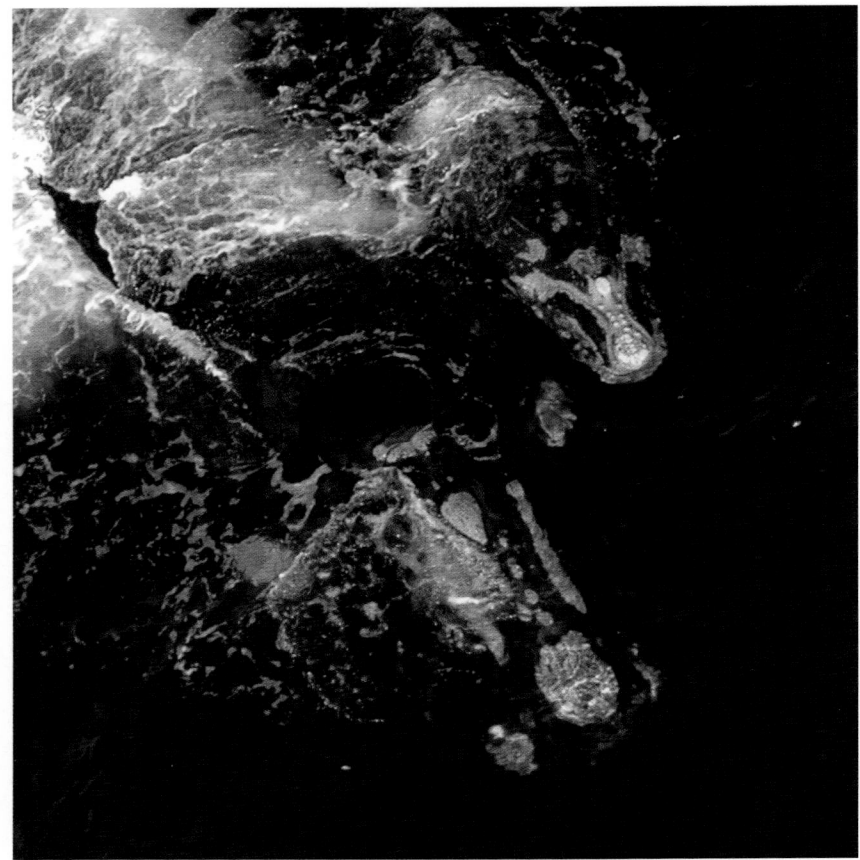

**Figure 3** *Aerial survey photo of mother and calf right whale* (Eubalaena *spp.*) *pair showing orange genital slit cyamids* (Cyamus erraticus) *covering the calf's head and* C. ovalis *(white cyamids) on the mother's callosities. Photo by J. Atkinson.*

site. However, their mouth parts are very small compared to their body size and they do not possess claws such as some other crustaceans or food-gathering cirri such as some predatory, sessile crustacean barnacles (Pfeiffer and Lowe, 1989). Cyamids have poorly developed paired mandibles and incisor processes with strong chitinous teeth that appear well suited for piercing and scraping skin. Rowntree (1983) showed that the color of intestinal contents of cyamids from humpback whales matched the skin color (black or white) from which the cyamids were collected. More recent conclusive evidence has proven that whale skin is a principal dietary material of cyamids. Both electron microscopic proof of whale skin keratinocytes within the upper digestive tract of cyamids and stable isotope evidence have shown that the dietary staple of whale lice is whale skin. Analysis of stable carbon and nitrogen isotope ratios from cyamids and skin from six species of whales have shown that the cyamid stable isotope ratios closely matched those of whale skin, but not those of zooplankton from the sea where the cyamids and whales reside (Schell *et al.*, 2000). Also supporting this conclusion is the evidence of direct damage to the skin by whale lice (Leung, 1976). Thus, cyamids have evolved into the only obligate parasites among the amphipods in distinction to other amphipods, such as caprellids, which are predatory and feed on diatoms, other crustaceans, and so on.

## IV. Reproduction in Cyamids

Reproductive and mating behavior has been less studied in cyamids than in other amphipods and, indeed, has not been investigated in most cyamid species. The males practice mate guarding and consorts are formed, but there appears to be less aggressive territorialism than is evident with some other amphipods. Little is known about cyamid copulation. There is morphological evidence of a secretory product being released on the cuticular surface of amphipods (Pfeiffer and Viers, 1998), but it is not known if this serves as a pheromone-type attractant for mates or serves some other function. Electron microscopic evidence has shown many tactile sensillae on the antennae and head regions of cyamids, some of which are also likely chemoreceptors. One can question if they always sit on their sole food source, why they have evolved so many sensillae. Female cyamids have a brood pouch (four-plated) or marsupium on their ventral surface, and both unhatched eggs and juvenile whale lice are retained in this cavity (Fig. 4). A clutch of 1078 eggs was observed in the marsupium of one female *C. scammoni* (Leung, 1976). The young cyamids measure only about 0.5 mm in length and crawl in and out of the marsupium during development and remain there for at least 2–3 months, when they become about 1.5 mm in length for *C. scammoni*. Several workers have proposed a seasonality for cyamid reproduction, but partly due to the migratory

W

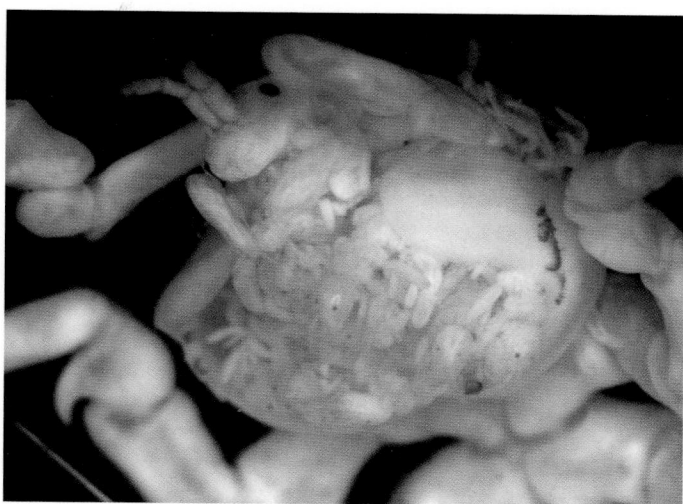

Figure 4  *Sperm whale cyamid with its marsupium full of young. The marsupium is a pouch that is an important adaptation for animals living in a flowing world. Photo by Jon Seger.*

habits of whales, detailed data are not yet available on potential seasonal changes.

## V. Genetic Diversity in Relation to Host Populations

Whale-louse populations are closely coupled to those of their hosts, especially in species that live on only one whale species. The three named species of *Cyamus* found on right whales (*Eubalaena* spp.) occur regularly on no other cetaceans. Their mitochondrial DNA sequence variation shows clearly that the North Atlantic, North Pacific, and southern ocean populations separated roughly 5 mya, near the Miocene–Pliocene boundary (Kaliszewska *et al.*, 2005). This finding supports previous evidence (from the whales' own genes) that right whales themselves speciated at about that time. Because whale lice have been riding on whales for millions of years, have no alternative hosts or free-living life stages, and usually have population sizes that are orders of magnitude larger than those of their hosts, they may be able to teach us about some aspects of whale population history that could not be discerned using other sources of information.

## See Also the Following Articles

Baleen Whales ■ Callosities ■ Parasites

## References

Balbuena, J. A., and Raga, J. A. (1991). Ecology and host relationships of the whale-louse *Isocyamus delphini* (Amphipoda: Cyamidae) parasitizing long-finned pilot whales (*Globecephala melas*) off the Faroe Islands (Northeast Atlantic). *Can. J. Zool.* **69**, 141–145.

Berzin, A. A., and Vlasova, L. P. (1982). Fauna of the Cetacea Cyamidae (Amphipoda) of the world ocean. *Invest. Cetacea* **13**, 149–164.

Kaliszewska, Z. A., *et al.* (18 authors) (2005). Population histories of right whales (Cetacea: *Eubalaena*) inferred from mitochondrial sequence diversities and divergences of their whale lice (Amphipoda: *Cyamus*). *Mol. Ecol.* **14**, 3439–3456.

Leung, Y. M. (1967). An illustrated key to the species of whale-lice (Amphipoda: Cyamidae), ectoparasites of Cetacea, with a guide to the literature. *Crustaceana* **12**, 279–291.

Leung, Y. M. (1970). First record of the whale-louse Genus *Syncyamus* (Cyamidae: Amphipoda) from the western Mediterranean, with notes on the biology of odontocete cyamids. *Invest. Cetacea* **2**, 243–247.

Leung, Y. M. (1976). Life cycle of *Cyamus scammoni* (Amphipoda: Cyamidae), ectoparasite of gray whale, with a remark on the associated species. *Sci. Rep .Whales Res. Inst.* **28**, 153–160.

Levin, M. J., and Pfeiffer, C. J. (1999). Photoreceptor ultrastructure of the amphipod, *Cyamus ceti* (Linn, 1758), an ectoparasite of bowhead, right and gray whales. *J. Submicroscop. Cytol. Pathol.* **31**, 397–405.

Lincoln, R. J., and Hurley, D. E. (1974). *Scutocyamus parvus*, a new genus and species of whale-louse (Amphipoda: Cyamidae) ectoparasitic on the North Atlantic white-beaked dolphin. *Bull. Br. Mus. (Nat. Hist.) Zool.* **27**, 59–64.

Margolis, L. (1955). Notes on the morphology, taxonomy and synonymy of several species of whale-lice (Cyamidae: Amphipoda). *J. Fish. Res. Bd. Canada* **123**, 121–133.

Pfeiffer, C. J., and Viers, V. (1998). Microanatomy of the marsupium, juveniles, eggs and cuticle of cyamid ectoparasites (Crustacea: Amphipoda) of whales. *Aquat. Mamm.* **24**, 83–91.

Pfeiffer, C. J., and Lowe, K. J. (1989). Cirral structure of the pedunculated marine barnacle *Lepas anatifera* L. (Crustacea: Cirripedia). I: Ultrastructure of the neuromuscular apparatus. *Acta Zool.* **70**, 243–252.

Rice, D.L., and Wolman, A.A. (1971). Parasites and epizooites, In "The Life History and Ecology of the Gray Whale (*Eschrichtius robustus*)" (J. N. Lanyne, ed.), Spec. Publ. No. 3, pp. 100–108. Am. Soc. Mammalogists, Provo.

Rowntree, V. (1983). Cyamids: The louse that moored. *Whalewatcher* **17**, 14–17.

Rowntree, V. (1996). Feeding, distribution, and reproductive behavior of cyamids (Crustacea: Amphipoda) living on humpback and right whales. *Can. J. Zool.* **74**, 103–109.

Schell, D. M., Rowntree, V. J., and Pfeiffer, C. J. (2000). Isotopic evidence that cyamids (Crustacea: Amphipoda) feed on whale skin. *Can. J. Zool.* **78**, 721–727.

Waller, G. N. H. (1989). Two new species of whale lice (Cyamidae) from the ziphioid whale *Berardius bairdii*. *Invest. Cetacea* **22**, 292–297.

# Whale Watching

## ERICH HOYT

Whale watching is the human activity of encountering cetaceans in their natural habitat. It can be for scientific, educational, and/or recreational purposes (sometimes all three). Mostly, whale watching refers to a commercial enterprise, although it is sometimes undertaken privately. The wide variety of whale watching activities includes tours lasting from 1 h to 2 weeks, using platforms ranging from kayaks to cruise ships, from land points including cliffs and beaches, from sea planes and helicopters in the air, as well as swimming and diving activities in which the whale watcher enters the water with cetaceans. Whale watching grew out of the traditions of bird watching and, to a lesser extent, other forms of land-based wildlife watching. To this day, the better whale and dolphin trips include sea birds, seals, turtles, and other marine fauna to appeal to more people as well as to give a well-rounded ecological interpretation.

## I. The Birth of Whale Watching

The species originally responsible for the development of whale watching was the gray whale (*Eschrichtius robustus*). Beginning in

**W**

the mid-1940s, students from Scripps Institution of Oceanography, led by Carl L. Hubbs, began participating in annual gray whale counts from university buildings such as Ritter Hall and from coastal headlands and lighthouses. In 1950, the Cabrillo National Monument in San Diego was converted into a public land-based whale watch lookout, attracting 10,000 people the first winter. Year after year, more and more people came to watch whales.

In 1955, the first commercial whale watch operation charged $1 US to see gray whales on their winter migration off San Diego. Although the gray whales passed close to shore, the boat tours sometimes allowed a closer look. By 1959, Raymond M. Gilmore, a US Fish and Wildlife Service biologist who had taken over the gray whale counting chores from Carl Hubbs, began serving as the first naturalist on whale watch trips out of San Diego. Through the 1960s and early 1970s, boat tours and land-based whale watching spread slowly up the coast of California to Oregon and Washington, and the first long-range commercial whale watch trip to the Mexican calving lagoons was organized out of San Diego.

In 1971, the Montreal Zoological Society began offering whale watch tours to go down the St. Lawrence River in Canada to see mainly fin (*Balaenoptera physalus*) and minke whales (*Balaenoptera acutorostrata*), and belugas (*Delphinapterus leucas*). This was the first commercial trip on the east coast of North America. These trips became an annual event.

It was the humpback whale (*Megaptera novaeangliae*), however, that turned commercial whale watching into a big industry. Humpback whales tend to be more active at the surface than gray or other whales, frequently breaching clear of the water—ideal for whale watchers wanting photographs. Added to this is the phenomenon of "friendly" behavior—the tendency of certain individual humpback whales to habituate to the presence of whale watch boats and to approach them regularly. This behavior, first observed commonly in humpback whales, has now also been found in gray whales, particularly in the mating and calving lagoons of Baja California, Mexico; in certain minke whales; and in killer whales, or orcas (*Orcinus orca*), and bottlenose dolphins (*Tursiops truncatus*), among others.

In New England and Hawaii, tours to see humpbacks began in 1975. For more than a decade before, the Wailupe Whale Watchers, a local club on Oahu, sponsored loosely organized, infrequent tours, but when whale watching began in earnest from Lahaina on Maui, where the humpbacks were more numerous and accessible, it immediately became the center of the humpback whale watch industry in the Pacific. Most of the Hawaiian tours were strictly commercial.

In New England, however, operators established their own brand of commercial whale watching with strong scientific and educational components—naturalists on every trip who were often working researchers. Educational programs to introduce school children to wild cetaceans—begun in southern California by such groups as the American Cetacean Society—were expanded in New England. Within a decade, the New England industry would attract even more participants than Californian and Hawaiian whale watching. New England was fortunate to have humpback whales on the feeding grounds centered on Stellwagen Bank, 10 km north of the tip of Cape Cod, as well as North Atlantic right (*Eubalaena glacialis*), fin, minke, and sometimes long-finned pilot whales (*Globicephala melas*), and Atlantic white-sided dolphins (*Lagenorhynchus acutus*). From a commercial point of view, Stellwagen Bank was ideally located close to the large population centers of the US east coast.

## II. Scientific Whale Watching

Whale watching for the purposes of research can be traced back to Aristotle, who spent time on boats and with fishermen in the Aegean Sea. In "Historia Animalium," Aristotle noted that the fishermen would nick the tails of the dolphins and that they could tell them apart. This practice foreshadows the studying of animals by watching them, a key feature of the ethology approach for studying birds and land animals pioneered by Konrad Lorenz, Niko Tinbergen, and others (Hoyt, 1984). It took longer to attempt such research with cetaceans because of the greater difficulties of approaching close and conducting research at sea. The photographic identification (photo-ID) research of cetaceans began in the early 1970s with humpback whales in the North Pacific and North Atlantic, gray whales and killer whales in the eastern North Pacific, and southern right whales (*E. australis*) and bottlenose dolphins off Argentina.

A successful partnership between science and commercial whale watching began in Provincetown, Massachusetts, in 1975, when Al Avellar of the *Dolphin* fleet asked Charles "Stormy" Mayo to be his naturalist. Mayo soon saw the possibilities for using the boat as a platform for studying whales. He set up the Center for Coastal Studies as a research and educational institution, and the close ties with commercial whale watching have been maintained ever since.

The arrangement works as follows: The Center provides naturalist guides for the *Dolphin* fleet. They are paid to help direct the boat to the whales, presenting an informal educational lecture, and answering questions. The Center sells T-shirts and other merchandise on board. Most important, Center researchers can conduct their own photo-ID research, and often collect other data. Sometimes more than one researcher will come aboard to ensure the maximum use of boat time.

This key partnership between science and commerce has determined the course of whale watching, as well as the practice of whale research, throughout southern New England. As of 1995, 18 of the 21 whale watching operators that mainly go to the Stellwagen Bank area had naturalists guiding boats and lecturing whale watchers, while 10 operations were taking and contributing ID photos. Despite the competitive atmosphere of commercial whale watching in New England, the researchers and their representative institutions have cooperated in setting up the North Atlantic Humpback Whale Catalog—a photo catalog and data-base covering more than 10,000 individual whales. As a measure of the scientific value of whale watching, at least 30 published papers in refereed journals have come largely from research aboard whale watching boats on Stellwagen Bank (Hoyt, 1995).

The New England model of successful whale watching and research, like Yankee whaling from an earlier century, has had an impact on the development of whale watching in locales as diverse as the Gulf of St. Lawrence in Quebec, northern Norway, and Dominica in the eastern Caribbean. Of course, a large part of scientific research on cetaceans does not lend itself to being conducted from commercial whale watch trips (such as transect surveys, biopsy darting, and collecting skin and fecal samples). In some cases the research and commercial enterprise operate separately, using different boats and personnel, but the commercial operation supports or contributes to the research. In several areas, whale watch operations have discovered new populations of cetaceans, accessible for study. In all, whale watching worldwide has led to at least 50 cetacean photo-ID programs supported in part or conducted aboard commercial whale watch boats. This has contributed to considerable public support for research through greater familiarity with research programs.

W

## III. The Growth of Commercial Whale Watching

The value of whale watching in 1981 was estimated to be $4.1 million US in direct revenue and $14 million in total revenues (including travel, accommodation, food, and souvenirs), based on approximately 400,000 boat-based whale watchers. By 1988, these numbers had expanded by more than 3 times, with the industry still based largely in New England and California as well as a small amount in Canada, Mexico, and the US Northwest (Table I).

In the late 1980s, whale watching began to spread rapidly to other parts of the world. Between 1987 and 1991, new whale watch industries started up in the Canary Islands, the Azores, Belize, Costa Rica, Dominica, Italy, Madagascar, and New Zealand, whereas existing industries expanded rapidly in Argentina, Australia, South Africa, and in parts of Canada. The diverse opportunities for whale watching included boat tours to view rare species (Heaviside's dolphins, *Cephalorhynchus heavisidii*, in South Africa), observing sperm whales, *Physeter macrocephalus*, from the air (New Zealand), land-based whale watching of southern right whales (South Africa, Australia, Brazil), and glimpsing various beaked whales in the Azores and the Bahamas. However, by the 1990s whale watching meant for the most part going to sea on large, comfortable, purpose-built ships that could take 150 to 400 people to see the whales and return to the dock in 2–4 h.

During this same period, whale watching became important in Norway and Japan, two countries with strong whaling interests. In both countries, the number of whale watchers increased steadily year by year until, in 1998, Norway had more than 21,000 whale watchers spending $6.9 million US, whereas more than 102,000 whale watchers in 20 Japanese communities spent $32.4 million US (Hoyt, 2001). Norway's whale watching industry has about a dozen operators working from four communities and offering sperm and other whales (May–September) or killer whales (October–December); whale watchers (visitors) primarily come from other European countries. Japan's whale watching industry, however, is a 99% domestic industry with diverse attractions including Bryde's whales (*Balaenoptera edeni*) and sperm whales at several warmwater locations from Shikoku Island and adjacent Honshu; humpback whales and tropical dolphins in Okinawa and Ogasawara, both island groups off southern Japan; and minke whales, Dall's porpoises (*Phocoenoides dalli*), and Pacific white-sided dolphins (*Lagenorhynchus obliquidens*) from Hokkaido in the north, as well as multiple locations for watching bottlenose and other dolphins off southern and eastern central Japan.

The compatibility of whaling and whale watching has been debated by whaling and non-whaling countries within the International Whaling Commission (IWC). Whaling can reduce the number of whales available for watching, change whale behavior, diminish the conservation value of whale watching, and potentially affect the larger tourism industry (Hoyt and Hvenegaard, 1999). Despite evidence of such impacts, following a return to whaling in 2003, Iceland has become one of the fastest growing whale watch destinations in the world, with five communities hosting more than 89,000 whale watchers in 2006 and receiving total expenditure of more than $23 million US. Whale watching remains much more popular in non-whaling countries. As of 2006, four countries attracted more than 1 million whale watchers per year: the USA, Canada, Australia, and Spain (including the Canary Islands). According to the most recent worldwide figures (1998), more than 9 million people are going whale watching in 87 countries and overseas territories and spending more than $1 billion US (Hoyt, 2001) (Table I). However, based on the substantial growth (1998–2006) noted in Australia, New Zealand, Iceland, the USA, and certain other countries, it is possible to make a minimum world estimate for 2006 of 12 million whale watchers with total expenditure of $1.5 billion US (IFAW, 2004; IFAW, 2005; Hoyt, 2007).

## IV. Whale Watching Conflicts and Regulations

Such explosive whale watching growth has led to management problems. Typical scenarios include too many boats on the water in a limited area, too many close approaches and sometimes collisions with cetaceans, strain on the infrastructure of local communities from too many visitors, and a lack of guidelines and regulations and/or enforcement of them.

Some operators have formed associations to devise self-imposed guidelines, but most have waited for researchers or NGOs to suggest guidelines or for government to try to impose regulations (Carlson, 2004). Yet even where regulations do exist, enforcement tends to be minimal or absent. In the USA, however, cases have been prosecuted with substantial fines levied against boat operators, as well as researchers and photographers, who approach too close or too aggressively to whales or who operate without a permit. In Hawaii, a film maker was fined for harassing whales when his close-up underwater video of a short-finned pilot whale (*Globicephala macrorhynchus*) mouthing a woman researcher was sold to television.

In 1983, the first whale watch fatality occurred when a mature gray whale overturned a small boat in Laguna Ojo de Liebre (Scammon's Lagoon), Mexico, killing two tourists. Until 1995, this was the only fatal whale watch accident. Then, all in the space of a year, in the Dominican Republic, the upper deck of a crowded boat collapsed after being hit by a wave, killing one tourist and injuring others, and in Kaikoura, New Zealand, a boat overturned, fatally trapping a person underneath. In the same period, on a sightseeing trip near Baffin Island in the Canadian Arctic, a surfacing whale overturned a 5.5 m boat and four tourists died of exposure. Only their guide survived. He was wearing a survival suit.

The number of injuries and fatalities is small considering the millions of people who go whale watching every year. Whale watching is by and large safe for both people and whales. Still, whale watch boats have struck whales, injuring, or even killing them, whereas other boats have been accidentally overturned by whales. Many more accidents have happened due to problems with the boats themselves or

**TABLE I**
**Estimated Growth of Whale Watching Worldwide**

| Year | Number of whale watchers | Direct expenditure (million USD)[a] | Total expenditure (million USD)[b] |
|---|---|---|---|
| 1981 | 400,000 | $4.1 million | $14.0 |
| 1988 | 1,500,000 | $11.0–16.0 | $38.5–56.0 |
| 1991 | 4,046,957 | $77.0 | $317.9 |
| 1994 | 5,425,506 | $122.4 | $504.3 |
| 1998 | 9,020,196 | $299.5 | $1,049.0 million |
| 2006c | 12,000,000+ | $450.0+ | $1,500.0+ |

[a]Direct expenditure = Cost of whale watch tour (ticket price).
[b]Total expenditure = The amount spent by tourists going whale watching from point of decision, including transport, food, accommodation, and souvenirs, as well as ticket price, but not including international air fares.
[c]2006 figures are minimum estimates only.

with overloading—things that are not specific to whale watching but could happen as part of any type of marine tourism. Indeed, most if not all of the accidents to date could have been avoided with due care and precaution.

Perhaps the greatest concern for safety is for the tours involving swimming or diving with cetaceans. Even these have a good safety record with thousands of encounters with dolphins in such places as New Zealand, the Bahamas, and Japan. Despite the generally strict protocol of no touching and approaching cetaceans, some have suggested that swimming with cetaceans should be banned or at least limited to certain dolphin populations or areas and that experienced researchers should always be present as guides to help interpret behavior and ensure safety. Swimming with whales, such as humpback whales on their mating grounds where there is surface active behavior, is potentially more dangerous to humans; despite criticism it continues in several areas of the world, and it remains controversial.

## V. Does Whale Watching Have an Impact on Whale and Dolphin Populations?

Since the late 1990s, researchers have tried to determine whether observed short-term effects on whales and dolphins from whale watching (approaching or avoiding boats, staying down longer, interrupting natural behavior) might be leading to long-term negative impacts (displacement, reduced reproductive success, or reduced survival rate). Concerns have centered around the presence of boats on critical mating, calving, feeding, and resting areas; the amount of time and the number of boats approaching close to the same whales; and the intensive whale watching of certain rare or endangered species. The problems are most evident with small, inshore populations of dolphins living in restricted areas such as Doubtful Sound, New Zealand; eastern Vancouver Island, Canada; and Shark Bay, Australia. In these populations, repeated exposure of individuals to boat-based whale watching is leading to long-term impacts (Report of the Workshop on the Science for Sustainable Whalewatching, 2004; Bejder et al., 2006; Lusseau et al., 2006). These studies highlight the sensitivity of small dolphin populations chronically exposed to whale watching. Yet even large whales sometimes show behavioral changes as a result of whale watching which, for some populations, may represent a threat.

There is much that can be done to manage the development of whale watching to minimize the risk from adverse impacts. In some areas of the world, watching from a large, quiet ship may reduce the pressure exerted by numerous small boats, whereas watching from a land-based lookout can eliminate negative effects on the animals. Well managed whale watching begins with a protective government policy with sensible regulations and an enforcement regime to control the numbers of boats on the water with cetaceans and to limit their approach and the amount of time they spend with cetaceans (IFAW et al., 1995). The Whale and Dolphin Conservation Society has suggested a practical, precautionary plan whereby one-third of every whale and dolphin area and one-third of daylight hours be kept free from any whale watching activity (Hoyt, 2007). Such restrictions on areas and times would also prove useful as controls for researchers doing comparative studies. Equally important for management is the education of whale watch operators, passengers, and other mariners using their boats in whale watch areas. Central to this is the role of the naturalists, or guides, who are the public face of whale watching and marine protected areas, and act as the bridge between the largely urban whale watchers and the sea. Naturalists play an essential conservation role with their ability to shape through their words and actions

### TABLE II
### Brief, Useful Guidelines for Boat Operators[a,b]

1. Do not pursue, overtake, head-off or encircle cetaceans or cause groups to separate.
2. Never approach whales/dolphins head on.
3. Avoid sudden changes in noise level (gear shifts and reverse, unless necessary to back away slowly from a surfacing whale or dolphin group).
4. Reduce speeds in areas where whales may be sighted; approach and leave whales cautiously and slowly.
5. Extreme caution is required when any of the following is present: (a) feeding whales, (b) cow/calf pairs and juveniles, (c) resting, (d) breeding or rowdy groups, or (e) socially active groups. Cetaceans engaged in such behavior are particularly sensitive to disturbance and may be vulnerable to collisions.

[a]This is not a complete list covering every situation but is meant to provide some general suggestions and overall direction for the use of operators offering whale watch tours as well as wildlife managers who are establishing guidelines or regulations on whale watching.
[b]Adapted from IFAW et al. (1997).

the way whales and dolphins and the marine environment are perceived and ultimately respected (IFAW et al., 1997; Hoyt, 1998, 2007).

## VI. Whale Watching and Conservation

In 1983, the IWC co-sponsored the "Whales Alive" conference in Boston, Massachusetts, which examined the "non-lethal" uses of whales. Ten years later, in 1993, the IWC adopted a whale watching resolution prepared by the Whale and Dolphin Conservation Society and successfully proposed by the UK at the IWC annual meeting in Kyoto, Japan. The underlying strength of the argument that the IWC should become involved in whale watching was that, since the IWC moratorium on whaling, the most prevalent "use" of cetaceans among IWC members has been whale watching. However, despite majority agreement on the relevance of whale watching to the IWC, delegates from Japan have repeatedly stated that the IWC has "no competence" concerning whale watching matters.

From 1995 to 2000, a series of six international workshops, organized by the International Fund for Animal Welfare (IFAW), with assistance from the Whale and Dolphin Conservation Society, Tethys Research Institute, World Wide Fund for Nature, and others, brought together more than a 100 cetacean experts from some 25 countries to explore the socioeconomic, educational, and legal aspects of whale watching, as well as the scientific aspects of management. The first of these, the seminal Scientific Aspects of Managing Whale Matching workshop, held in Italy, recommended a precautionary approach to management with a periodic review of regulations based on continuing research and monitoring into possible effects on cetaceans (sample guidelines for boat operators are provided in Table II) (IFAW et al., 1995). The overall impact of this and later workshops has been to focus the debate on the status of whale watching, pointing out that better regulations were needed as well as enforcement, that whale watching had substantial unrealized potential in terms of education and science, and that, economically, whale watching was worth far more than had previously been determined, although some values were difficult to measure in terms of dollars alone (Hoyt, 2005, 2007).

## TABLE III
### Educational Values of Whale Watching[a]

1. Whales are emblems for promoting awareness of endangered species and habitat protection.
2. Whale watching provides the opportunity for people across all ages and cultures to become familiar with environmental issues and to become involved in conservation efforts on a personal, local, regional, national and international level.
3. The development of education programs forges links between the whale watch industry and local communities as well as building bridges between the general public and scientific communities.
4. Natural history knowledge gained through whale watching has intrinsic value.
5. Whale watching provides an opportunity to observe animals in the wild, transmitting factual information and dispelling myths.
6. Whale watching is a model for marine educational programs in adventure travel and ecotourism.
7. Whale watching provides the opportunity for appreciation and understanding of local history, culture and the environment.

[a]Adapted from IFAW *et al.* (1997).

Perhaps the most valuable legacy of whale watching has been the building of a constituency out of the general public that is interested in and sympathetic to marine mammals, the sea, and marine conservation, including marine protected areas. The designation of Stellwagen Bank as a US National Marine Sanctuary in 1993 was largely the result of public interest in whales in New England and in the northeastern US through whale watching. Several million people encountered whales in the wild between 1975 and 1992, saw the research being conducted on whale watching boats, and learned about the whales and problems of the sea, which led to overwhelming popular support for the sanctuary (Hoyt, 2001).

Since the late 1980s as whale watching has expanded, however, it has become less educational in some areas of the world (Table III). A 1998 world survey of whale watch operations found that only 35% of all operators had enlisted naturalists to guide their trips (Hoyt, 1998). In terms of the scientific content, about 9% of operators worldwide had researchers or naturalists on board who conducted regular photo-ID and other research as part of their trips, whereas 57% never conducted scientific research or even offered information to scientists. Most operations were strictly commercial ventures. Clearly, a great deal more could be done to encourage whale watching tours to offer the maximum benefits to local communities and regions in terms of education, science, and conservation, as well as earning tourism dollars, while at the same time protecting the whales and ensuring that they will remain in coastal waters and accessible to whale watchers for generations to come.

## See Also the Following Articles

Conservation Efforts ■ Ethics and Marine Mammals ■ Marine Protected Areas ■ Popular Culture and Literature

## References

Bejder, L., *et al.* (10 authors) (2006). Decline in relative abundance of bottlenose dolphins exposed to long-term disturbance. *Conserv. Biol.* **20**(6), 1791–1798.

Carlson, C. (2004). A review of whale watching guidelines and regulations around the world. International Whaling Commission. [Available from: http://www.iwcoffice.org/conservation/whalewatching. htm#regulations].

Hoyt, E. (1984). "The Whale Watcher's Handbook." Doubleday and Penguin, New York.

Hoyt, E. (1995). Whale watching takes off. *Whalewatcher* 29, 3–7.

Hoyt, E. (1998). Watch a whale; learn from a whale: enhancing the educational value of whale watching. *In* "Proceedings of the 1998 International Forum on Dolphins and Whales," pp. 5–19. Muroran, Japan.

Hoyt, E. (2001). "Whale Watching 2001: Worldwide Tourism Numbers, Expenditures, and Expanding Socioeconomic Benefits." International Fund for Animal Welfare, Yarmouth Port, MA.

Hoyt, E. (2005). Sustainable ecotourism on Atlantic islands, with special reference to whale watching, marine protected areas and sanctuaries for cetaceans. *Biol. Environ. Proc. R. Ir. Acad.* 105B, 141–154.

Hoyt, E. (2007). "A Blueprint for Dolphin and Whale Watching Development." Humane Society International (HSI), Washington, DC [Available from: http://web.mac.com/erich.hoyt/iWeb/www.erichhoyt. com/More%20Info_files/DolphinWhaleManualENG_01_07.pdf].

Hoyt, E., and Hvenegaard, G. (1999). A review of whale watching and whaling with applications for the Caribbean. *Coast. Manage.* **30**, 381–399.

IFAW, Tethys Research Institute, and Europe Conservation (1995). Report of the Workshop on the Scientific Aspects of Managing Whale Watching, Montecastello di Vibio, Italy.

IFAW, WWF, and WDCS (1997). Report of the International Workshop on the Educational Values of Whale Watching, Provincetown.

IFAW (2004). "From Whalers to Whale Watchers: The Growth of Whale Watching Tourism in Australia." Surry Hills, NSW, Australia.

IFAW (2005). "The Growth of the New Zealand Whale Watching Industry." Surry Hills, NSW, Australia.

Lusseau, D., Slooten, E., and Currey, R. J. (2006). Unsustainable dolphin watching activities in Fiordland, New Zealand. *Tourism Mar. Environ.* 3, 173–178.

Report of the Workshop on the Science for Sustainable Whalewatching, Breakwater Lodge, Cape Town, South Africa, 6–9 March 2004. [Available from: http://www.iwcoffice.org/_documents/sci_com/ workshops/WW_Workshop.pdf].

# Whaling, Aboriginal

## RICHARD ELLIS

### I. Man Meets Whale

One of the earliest records of man's interactions with whales can be found in the chronicle of the conquests of Alexander the Great, which took place in the fourth century BC, and was transcribed some 300 years later by the Greek historian Arrian. Because Alexander's empire included the eastern Mediterranean, the northern shore of the Persian Gulf, and the shore of the Indian Ocean from the Strait of Hormuz to the mouth of the Indus River—and also because many of his campaigns were conducted at sea—we can safely assume that he and his army had many opportunities to see whales. The following passage appears in Arrian's description of the officer Nearchus's encounter in the Indian Ocean:

In this foreign sea there lived great whales and other large fish, much bigger than ours in the Mediterranean .... As we set sail we observed that in the sea to the east of us water was blown aloft, as happens with a strong whirlwind. We were terrified and asked our pilots what it was and whence it came. They replied that it was caused by whales, which inhabit this sea. Our sailors were so horrified that the oars fell from their hands .... Then I walked round the fleet and ordered every steersman I met to steer straight at the whales, exactly as if they were going into a naval battle. All the men were to row as hard with as much noise as possible, including yells .... The whales, which could be seen just in front of the ships, dived terrified into the depths. Not long after that, they surfaced behind the fleet, blowing water into the air as before .... Now and again a few of these whales come ashore, having been stranded on the flat beaches at ebb tide. Often, too, they are flung up on dry land by a violent storm. Then they die and rot. When the flesh has mouldered away, the skeletons are left, which the inhabitants of these shores use for building their houses. The large bones at the side form the beams of their houses, the smaller ones the laths. From the jawbones they make doors.

From this description, it is not possible to identify the whales, for there seem to be two types discussed simultaneously. The schooling behavior (including the mass stranding) suggests sperm whales (*Physeter macrocephalus*), but the description of the bones found on the beach would better apply to baleen whales. Whatever the species, it is obvious that stories like these were passed down, modified, embellished, and eventually reconstructed as the stuff of fable and fantasy.

When Jonah fled from Joppa rather than obey the word of the Lord that he go to Nineveh, his ship was caught in a "mighty tempest." To appease the angry God, the mariners cast Jonah into the sea because they knew he had disobeyed the Lord's commands, and immediately "the sea ceased from her raging," a "great fish" swallowed Jonah, and he remained in its belly for 3 days and 3 nights. Inside this uncomfortable sanctuary, Jonah repented, and the creature spewed him up again onto dry land.[1]

---

[1]This is not the first mention of a whale in the Bible: that distinction is found in *Genesis* 21: "And God created great whales." It is, however, the first mention anywhere of a man being swallowed by a whale (or a "great fish"). There are very few animals in the sea large enough to swallow a man, and most of them are whales. Of the sharks, only the whale shark (*Rhincodon typus*), the basking shark (*Cetorhinus maximus*), and the great white shark (*Carcharodon carcharias*) are the requisite size, but the whale shark and the basking shark are plankton-eaters, with the equipment, but not the inclination to swallow a man-sized object. That leaves only the great white, the largest carnivorous shark in the world. This fish has a fully-deserved reputation as a man-eater, but it tends to take great bites of its victims, and while there have been survivors of white shark attacks, there is no record of a human victim having been swallowed whole. The only cetacean with the anatomical equipment required to swallow a human being is the sperm whale, which usually feeds on squid. A giant squid (*Architeuthis* spp.) weighing over 400 pounds was found in the stomach of a sperm whale harpooned off Madeira, so there is no question about the ability of the cachalot to swallow Jonah. Throughout the history of the sperm whale fishery, there have been several tales of whalemen swallowed by the object of their attentions. There are indeed stories of whalemen having been swallowed and recovered alive from the belly of the sperm whale, but under close examination, they begin to resemble the fable of Jonah more than demonstrable fact. An oft-quoted account of a whaler who fell overboard off Newfoundland and was swallowed by a sperm whale is more likely to be true. In this history (published as a letter to the editor of *Natural History* in 1947 by one Edgerton Y. Davis), the man is exhumed from the carcass of the whale, but he is badly crushed, decomposed, and extremely dead. Yes, it is physically possible for a whale to swallow a man, and no, the man would not survive the experience.

In the book of Jonah, there is no mention of a *whale*, but in the book of Job, there are several references to "leviathan," an animal that has been variously interpreted as a crocodile, a shark, and a whale. ("His teeth are terrible round about. His scales are his pride, shut up together as with a close seal. One is so near to another, that no air can come between them .... He maketh the deep to boil like a pot: he maketh the sea like a pot of ointment.") Isaiah 27: "The Lord with his sore and great strong sword shall punish leviathan the piercing serpent, even leviathan that crooked serpent; and he shall slay the dragon that is in the sea." And finally, Psalms 104:25-26: "So is this great and wide sea, wherein are things creeping innumerable, both small and great beasts. There go the ships; there is that leviathan, whom thou hast made to play therein." At best the Bible is ambiguous about the whales; creating them, punishing them, watching them at play, and even feeding Jonah to one of them (maybe). When we really want to know about whales, the Bible ignores them altogether.

When Noah was commanded to build the ark, he "admitted aboard every beast after his kind, and all the cattle after their kind, and every creeping thing that creepeth upon the earth after his kind, and every fowl after his kind, every bird of every sort," but there is no mention of a fish, a dolphin, or a whale. Maybe these creatures were supposed to tag along in the wake of the ark, since they would not be affected by the rains or the "increase in waters." Or perhaps they were not recognized as creatures worth saving, living as they did in an alien environment.

As early as 350 BC, Aristotle recognized that whales were mammals and not fish. He wrote, "The dolphin, the whale and all the rest of the cetacea, all, that is to say, that are provided with a blow-hole instead of gills, are viviparous ... just as in the case of mankind and the viviparous quadrupeds." This knowledge, however, was to prove of little use to mankind for 1000 years, because it was based upon random and infrequent examinations of stranded animals. The observation of cetaceans in the wild only occurred when seafarers spotted dolphins at play in the bow waves of their vessels, or when someone like Nearchus encountered and described living whales. Men would not encounter whales until they began to venture out to explore the oceans, which would not take place for another fifteen centuries. In the meantime, what observations of cetaceans were to be made would be made from land, or in the inland waterways in which men felt more or less secure.

Undaunted by the absence of real whales to describe, ancient authors described them anyway. Pliny the Elder (who often relied upon Aristotle and other authors) included whales in his *Naturalis Historia*, written shortly before he was killed at Pompeii in the AD 79 eruption of Vesuvius. A doctor named Philemon Holland (1552–1637) translated Pliny into Elizabethan English in 1601, and in his discussion of "The Whale," we find the following:

The biggest and most monstrous creature in the Indish Ocean are the whales called Pristis and Balaena. These monstrous Whales named Balaenae, otherwhiles come into our seas also. They say that in the coast of the Spanish Ocean by Gades [Cadiz], they are not seen before midwinter when the daies be shortest: for at their set times they lie close in a certaine calme deepe and large creeke, which they chuse to cast their spawne in, and there delight above all places to breed. The Orcae, other monstrous fishes, know this full well, and deadly enemies they bee unto the foresaid Whales. And verily, if I should portrait them, I can resemble them to nothing else but a mighty masse and lumpe of flesh without all fashion, armed with most terrible, sharp, and cutting teeth.

When the sixth century Irish monk known as St. Brendan set out on his North Atlantic voyages, the result was one of the earliest mentions of a whale–human interaction in the European literature. According to Samuel Eliot Morison (1971), "Brendan was a real person … and his *Navigatio* is based on a real voyage or voyages, enhanced by the Celtic imagination." Brendan's discovery of Iceland, the Canaries or Madeira, while interesting, does not concern us as much as the delightful tale of his encounter with Jasconius the whale. With his crew of seventeen, Brendan came upon "a bare, treeless black island," but when they built a cooking fire, the island sank beneath them. Jasconius told them they could return, but only if they refrained from lighting fires on his back. The story appears in the *Physiologus*, and again in von Mengenberg's translation of Thomas de Cantimpre's *De natura rerum* ("Of things in nature"):

> Some whales are so big that when seen from afar they seem like islands or groves, or resemble great hills. The whale heaps a thick coating of earth upon its back, so that when seamen are driven by the stress of weather upon this earth, they imagine it to be an island and that they have come to land. Rejoiced at this they let down the sails, drop their anchor in the water, build a fire upon the earth and seek to enjoy a little rest. As soon, however, as the whale feels the heat of the fire, it becomes enraged and dives beneath the water, bearing down to the depths both ship and sailors.

It would he another 500 years before men actually went whaling, but whales would continue to inexplicably cast themselves upon the beaches. In his *Historia Animalium* Aristotle wrote, "it is not known for what reason they run themselves aground on dry land, at all events it is said that they do so at times, and for no obvious reason." Whales and dolphins have been running themselves aground for as long as men can remember, and probably long before that. In the twenty-two centuries that have elapsed since Aristotle made his prescient comment, we have come no closer to solving the mystery of whale strandings than the Greeks. Beached whales represented the first important contact between men and whales, one which would set the tone for the interaction of these two mammals for centuries.

Not all knowledge of whales came from those that beached, of course; seafarers encountered all sorts of cetaceans as they plied their trade routes or began their hesitant explorations of distant coasts. Men sailed from the ports of Europe, Asia, and Africa; for conquest, for trade, or to spread the word of their god, but they did not set sail casually. Discovery as an end in itself, exploration in intellectual pursuit of geographical knowledge, or in the romantic pursuit of unusual adventure, is characteristic of a safer, richer, more comfortable society than that of fifteenth century Europe (Parry, 1974). There was no such thing as science for the sake of science, and if men found whales, they took them for what they believed them to be: huge, mysterious, threatening creatures. On their early maps, they figured them as large, scaly animals with a frightening array of unlikely appendages: horns, fringes, crests, armor, lumps, bumps, ridges, horrific dentition, and often twin pipes gushing water into the air.

It was not the intention of these mapmakers to frighten their fellow men; everyone believed that foreign lands harbored all sorts of mysterious animals and equally strange varieties of men. If this was the case, then surely the ocean, home of rite sea serpent and the *kraken* (a giant cephalopod capable of entangling ships in its tentacles and dragging them to the bottom), could be the home of even more terrifying creatures. For medieval man, these superstitions proved to be true, as the sea spewed forth monsters larger and more terrible than any creature imaginable. There are indeed giant squid with arms 50 or 60 ft long, and there certainly are leviathans. Although they did not have scales, horns or twin blowpipes, the leviathans had equally improbable characteristics: giant flattened tails, strange plates where terrestrial mammals had teeth, gigantic reproductive organs (often grotesquely distended in death), and no legs where proper mammals were supposed to have legs. Who could fault the ancients for suspecting that the sea harbored monsters?

We have no way of knowing when or where the first aborigines encountered the first beached whales, but it is obvious that this encounter would eventually lead to whaling. As soon as the inhabitants of what would become Holland, Norway, or Vancouver realized that they did not have to depend on the uncertain generosity of the sea to provide a bounty of meat or oil, they would take to the sea themselves, to hunt the whale.

Many of the earliest descriptions of cetaceans were based on beached animals, and for the scientists of the sixteenth and seventeenth centuries, they were a boon. How else could landlocked Robert Sibbald of Scotland have described a blue whale (*Balaenoptera musculus*) so accurately in his 1692 discussion of the carcass found at Abercorne on the Firth of Forth? At one time or another, every species of whale has come ashore: fin whales (*B. physalus*), minkes (*B. acutorostrata*), right whales (*Eubalaena* spp.), humpbacks (*Megaptera novaeangliae*), gray whales (*Eschrichtius robustus*), and dolphins of every sort. But the most celebrated of all stranders is the sperm whale. With its great square head filled with a mysterious waxy substance, its wrinkled hide and peg-like ivory teeth, a sperm whale appearing on the beach became a *cause célèbre*. In many of the early descriptions of beached whales, the species is open to question, but once you have seen *Physeter*, there is no possibility of confusing it with any other animal on earth, let alone any other whale. (The veneration of the sperm whale reached its apogee in the nineteenth century, when, during the most productive years of the sperm whale fishery, *Moby-Dick* raised the sperm whale to the soaring heights of hyperbole.)

Around 1577, the first engraving of a stranded whale appeared in print. By the turn of the seventeenth century, more whales had stranded on European coasts, and with the heightened interest in popular science, more engravings appeared. Either because the whales preferred the coasts of the Netherlands, or because the Dutch had a particular interest in stranded whales, the majority of the early illustrations of whales were the work of Low Country artists. In these elaborately detailed drawings, the good burghers of Holland are often seen perched upon the carcass, standing around in fashionable attire, or occasionally carrying off a bucket of what may very well have been whale oil.

The North Sea coast of Holland would appear to be one of those places (noteworthy others are in New Zealand and Cape Cod) where whales strand with some degree of regularity. From 1531 to around 1690, some 40 whales of assorted species beached themselves on these shingled coasts. Most of them seem to have been sperm whales, and with their huge heads, a mouthful of ivory teeth, and—in what appear to be a majority of the cases—its male genitalia prominently exposed, the dead whale must have been a wonder of wonders to the Dutchmen who came to view these monsters. It would be another half-century before the whalers of Rotterdam and Delft would head for the icy seas of Spitsbergen, where they would hunt a totally different creature, the Greenland right whale.

One of the best documented of these aliens from the deep was a 54-ft bull sperm whale (*potvisch* to the Dutch) that was discovered floundering helplessly in the shallows of Berckhey in February 1598. When the whale expired, its carcass was sold off for the oil, but its fame lies more in its portrayal than in its products. Drawn by the artist Hendrick Goltzius, the Berckhey whale has appeared in countless versions, often accompanied by a descriptive text that marvels at its leviathanic dimensions. In later years, more whales would strand on the beaches and be immortalized by Dutch engravers, but the Goltzius illustration, repeated and degenerated for 200 years, has probably been employed more often than any other cetacean depiction, before or since. In 1991, German whaling historians Klaus Barthelmess and Joachim Münzig published a three-volume compendium (Barthelmess and Münzig, 1991; *"Horrible Monsters: Whales and Whale Representations in 16th Century Prints, and their Artistic Influence"*) of the illustrations of whales, beached and otherwise, known from European archives.

A dead whale on the beach begins to decompose rather quickly, so often the illustration was erroneous in some of its particulars. In the chapter in *Moby-Dick* on "The Monstrous Pictures of Whales," Melville had the same complaint. He wrote, "Consider! Most of the scientific drawings have been taken from the stranded fish; and these are about as correct as a drawing of a wrecked ship, with broken back, would correctly represent the noble animal itself in all its undashed pride of hull and spars. Though elephants have stood for their full-lengths, the living Leviathan has never yet fairly floated himself for his portrait." Correct or not, these pictures were the best available, and from them, the civilized world began to learn of gigantic animals that lived in the sea, and occasionally appeared on beaches. However, long before scientists would examine, dissect, illustrate, and classify the whales, people with very little concern for their correct nomenclature would be hunting them in the open seas.

Jenkins (1921) writes, "Although the general opinion is that the Basques were the earliest whalers, Noel de la Moriniere says that this is a misapprehension and that the Northmen were really the first in the field." He says that a man called "Ochther" hunted whales and walruses beyond the North Cape, but then the notes that "there is no evidence that it developed into a regular fishery such as that of the Basques."

His "Ochther" was Othere (or Ottar), a Norseman in the service of King Alfred of Wessex around the year AD 890. Alfred (called Alfred the Great) is known for his defense of England against the marauding Danes, and also for the initiation of the *Anglo-Saxon Chronicle*, the first history of England. (Our word "whale" comes from the Anglo-Saxon *hwael* that means "wheel," and probably refers to the shape of the whale's back as it rolls in the water.) Alfred translated many Latin texts, including the one that concerns us here, a description of Europe by one Orosius, who lived four centuries before. To the work of Orosius, Alfred added a description of the northern voyage of Othere, wherein was described the whale- and walrus (*Odobenus rosmarus*)-hunting of a northern people known as the Biarmians. From the location (the White Sea in northern Russia), and description of the whales hunted ("50 ells" in length, which by one calculation works out to 187 ft), it would appear that the larger ones—whose size was greatly exaggerated—were bowheads (*Balaena mysticetus*), whereas Othere's "horsewhales" were walruses. In the history of British *Voyages and Discoveries* compiled in the sixteenth century, Richard Hakluyt, a diplomat and scholar, wrote that the principal purpose of Ochther's expedition was "to increase the knowledge and discovery of these coasts and countries, for the more commodity of fishing for horse-whales, which have in their teeth bones of great price and excellency: whereof he brought some at his return unto the King. Their skins are also very good to make cables for ships, and so used."

In medieval Scandinavia, whales were very much a part of the lives of the people, and were therefore incorporated into their literature. A thirteenth century Icelandic account known as *Konung's skuggsjá* (*Speculum Regale* in Latin; *Konegspiel* in German; "King's Mirror" in English) describes the whales that are found off Iceland, and includes such mysterious creatures as the horse whale, the red whale, and the pig whale, but also discusses recognizable species, such as the killer whale (*Orcinus orca*), the narwhal (*Monodon monoceros*), and the sperm whale. The right whale is described thus:

> People say it does not eat any food except darkness and the rain which falls on the sea. And when it is caught and its intestines opened, nothing unclean is found in its stomach as would be in other fish that eat food, because its stomach is clean and empty. It cannot open its mouth easily, because the baleen that grows there rise up in the mouth when it is opened, and often causes its death because it cannot shut its mouth. It does no harm to ships: it has no teeth, and is a fat fish and edible (Whitaker, 1986.)[2]

There is an almost complete lack of information on Norse whaling, but the waters in which they sailed were then (and are still) among the whale richest in the world. There are right whales, humpbacks, fin whales, sperm whales, belugas (*Delphinapterus leucas*), narwhals, pilot whales (*Globicephala* spp.), and various species of dolphins in the cold, productive waters of the North Atlantic. The Norse sagas are silent on the subject of whales and whaling, but it would be hard to imagine these hardy seafarers ignoring a plentiful source of food and oil as they plied the otherwise inhospitable seas around Iceland, Greenland, and Labrador. There are references, however, to battles royal between various "families" as they dispute the ownership of whale carcasses, which indicates the importance of whales—at least of dead whales—in the lives of the early Norsemen. They left no tryworks, their settlements provide no trace of harpoons or lances, but there are tantalizing hints of Norse whaling in some of the more recent discussions. In his 1928 *History of Whaling*, Sydney Harmer says, "The Icelanders seem to have engaged in whaling ... and the whale known as "Slettibaka" ... is believed to have been the Biscay whale." (The modern Icelandic for the right whale is *sletbag*, which means "smooth back.")

## II. Early Icelandic Whaling

Iceland's early history is to be found in the sagas, tales of the exploits of the island's early heroes. The Vikings of Norway evidently brought to Iceland knowledge of the techniques employed in driving whales (probably pilot whales) into the fjords for slaughter. There are occasional mentions of disputes over stranded whales in the sagas, but as far as we know, there was no active whale fishery. An Icelandic bestiary from about 1200 describes some of the whales (but not accurately enough for modern cetologists to identify them as to species), and the *Konung's skuggsjá* lists no fewer than 21 sea creatures, some of which can be referred to living whales, dolphins and pinnipeds, and some of which—mermaids and mermen, for example—are clearly mythological.

---

[2]The King's Mirror was written in Norwegian, probably as a set of instructions for a king's son. It contains the first description of the ice in the interior of Greenland, a summary of contemporaneous beliefs about the Aurora Borealis, and the most complete inventory of the sea mammals of Greenland and Iceland up to that time.

In a seventeenth century work by an Icelander named Jon Gudmundsson, there is a list of the various whales that might be found in Icelandic waters, including the sperm whale (*Burhvalur*), the narwhal (*Náhvalur*), the right whale (*Slettbakur*), the fin whale (*Geirreydur*), and the blue whale (*Steypireydur*). With the exception of the right whale, which probably refers to the bowhead and was hunted to extinction after this publication appeared, all these whales can still be seen off Iceland. Also included was something that the author referred to as *Sandloegja*, which has been translated as "sandlier" (i.e., one that lies in the sand). Each of the abovementioned whales is accurately illustrated, so there is little doubt as to its identification. The description of the *Sandloegja* is accompanied by a picture of a whale that has not been seen in the Atlantic since commercial whaling began, and if the interpretation is correct, it depicts the only whale to have become extinct in recent history (Lindquist, 2000).

The California gray whale is well known from the North Pacific, where it makes the celebrated round-trip migration from Alaska to Baja California. It was the object of an intense fishery in the nineteenth century which nearly eliminated the species. Fossil remains of a similar—if not identical—species have been found in western Europe (Sweden, England, and the Netherlands), and on the east coast of North America from New Jersey to South Carolina. From the evidence, it appears that there was also an Atlantic gray whale, which probably maintained similar habits to its Pacific cousin; i.e., it fed in cold northern waters (perhaps Iceland and Greenland), and then moved south (Spain, France, England?) to breed and calve. With the exception of the fossil evidence, the only clues to the identity of this whale are found in the work by Gudmundsson and in a debatable reference in a New England work of 1725, where Paul Dudley describes the "scrag whale" with characteristics that are not applicable to any other species except the gray whale. Whether hunted or occasionally appearing on the beach, the gray whale should probably be listed in the Icelandic cetacean fauna, even though no living Icelander has ever seen one in his own country. (In his analysis of the whales of the *Konung's skuggsjá*, Ian Whitaker (1986) wrote that "the gray whale was hunted in the Atlantic between 1100 and 1200, although it has not been found there since the 18th Century." He is unable to correlate this species with any of the thirteenth century Icelandic names, although he indicates that there are two "unallocated" names, which translate as "hog-whale" and "shield-whale.")

Whales were caught by the Norwegians off the Tromsø coast as early as the ninth or tenth century. The oil was used for lighting and the baleen for jewelry, coopering, and boatbuilding. But, as C. B. Hawes wrote in 1924, "with a lamentable lack of foresight, the earliest whaling captains neglected to enlist the services of scholars and historians," so much of the story of early Norse whaling has to be left to conjecture."

One of those who did a lot of conjecturing was Ivan Sanderson. Trained as a zoologist, Sanderson was particularly interested in unusual phenomena, such as the abominable snowman and the Loch Ness monster. He wrote several books on zoological and cryptozoological subjects, but he will probably be best remembered for his *Follow the Whale* (1956). In this book, along with some rather good accounts of the biology of whales and some excellent maps, he re-creates the lives of whalers of the past and the present, "corralling the forgotten and more neglected aspects of whaling history and the new discoveries about the whales themselves, and weaving them into a continuous web of narrative." One of these "neglected aspects" is Norse whaling history, but despite the lack of documentation, Sanderson devotes a whole chapter to the subject, fictionalizing what could not otherwise be ascertained. He has the Norsemen under "Thorvald the Long" trapping

sei whales in the fjords of Norway at an unspecified time, along with "Biarni the Yellow standing in the bow holding a trumpet of cow's (*Bos taurus*) horn in his hand." He also recounts an Icelandic saga of AD 1100, which contains "a delightful passage in which we are told of the stranding of a large rorqual at Rifsker in Iceland and how all the important people who were able went to it." The documentation for this is sparse, but there can be no question that the Norsemen, ranging the North Atlantic from Finnmark to Iceland and from Greenland to North America, had to have encountered whales. Whether they killed the whales in an organized fashion or took them incidentally to their viking and settlement forays may never be known. They did hunt walruses for their skin and ivory tusks, and narwhals for their spiraled ivory tooth, which was passed off as the horn of the fabled unicorn.

## III. Basque Whaling

As far as we can tell, the first people to hunt large whales in an organized and intentional manner were the Basques. As far back as records go—and even further, perhaps as far back as the Stone Age—these men were hunting whales. In his 1820 *Account of the Arctic Regions*, William Scoresby suggests that "the Biscayans were the first who exercised their courage in waging a war of death with the whales," but he attributes their motivation to the protection of their fishing nets, which "would naturally suggest the necessity of driving these intruding monsters from their coasts." Whatever their reasons, the Basques became the paradigms of the whaling industry, establishing the *modus operandi* that would characterize the industry for another 1000 years. "Historians have only recently begun to realize," wrote Farley Mowat (1984) "that it was the Basques who lit the flame that was eventually to consume the mighty hosts of the whale nations." They discovered the "resource," exploited it, and then pursued it so vigorously that it became uneconomical to continue. They probably took their first whales in the shallows, and then, like the bay whalers who were to follow their lead all around the world, realized that it was considerably more expeditious to go after the whales rather than wait for the whales to come to them. The Basques may also have been responsible for the only large cetacean extinction in recorded history.

Somewhere around the year AD 1000, these intrepid hunters of the Bay of Biscay began the slow but systematic eradication of the whales that came into the protected bays in the shadow of the Pyrenees. Obviously, the Basques did not wait for the first millennium to end before beginning their whaling, but most authors cite this as approximately the time they began. Ommanney (1971) wrote: "The industry, founded on the Biscay Right whale, was fully developed by the twelfth century but probably dated from much earlier, possibly from the tenth century when the Basques may have learned the craft from Norse whalers." The Belgian historian W. M. A. De Smet (1981) searched the literature for references to European whaling *before* the Basques, and wrote, "Only a few authors are aware of the fact that whaling existed in still earlier days in other European seas, and that it was practiced in the North Sea and the English Channel during the Middle Ages, certainly from the 9th century onward." Although the species of whale in these early instances was rarely recorded, the likelihood is that it was the right whale that was hunted in the North Sea, and perhaps the gray whale, although the precise date of the disappearance of the Atlantic gray is still being debated.

De Smet cites several instances in which whale meat is mentioned in early texts, and suggests that "it is clear from the regularity with which whale meat occurred in these markets that it cannot have come from stranded animals alone and there must have been regular landings." After providing for themselves, the enterprising

**W**

Basques established markets for the meat and blubber, and even had "consulates" in Holland, Denmark, and England to encourage sales. In French, the blubber was known as *lard de carime*, which means "lenten fat," and Europeans were allowed to eat it on designated meatless calendar days. The oil was used for lighting and the manufacture of soap, wool, leather, and paint; the meat was fed to the poor and to the ships' crews, and the baleen was put to all sorts of uses (including being shredded into plumes for the decoration of knight's helmets), the vertebrae were used for seats, and the ribs were employed as fence-pickets, and beams for cheap housing. The tongue was considered a particular delicacy, and was reserved for the clergy and royalty.

In the unregulated (and largely undocumented) confusion of the Middle Ages, small pockets of Basques lived along the shores of the Bay of Biscay, speaking their own language, about which a contemporaneous cleric wrote, "The Basques speak among themselves in a tongue they say they understand but I frankly do not believe it." In their strongholds in the crook of the elbow of the Iberian peninsula, they were isolated from the turmoil of land wars, fiercely intent upon self-preservation, and coincidentally upon the pursuit of the large black whales (which they called *sarda*) that arrived every autumn in their offshore waters.

It is likely that they also hunted the Atlantic gray whale, although there is no evidence to support this supposition. There is, however, considerable evidence that the Atlantic gray whale (which was called *otta sotta*) was present in the Atlantic during the days of Basque whaling. Remains have been found on both sides of the ocean: in England, Holland, and Sweden in the east; and from New York and New Jersey to North Carolina in the west. An account discovered by Fraser (1970) suggests that a gray whale (called *sandloegja* by the Icelanders) existed as recently as 1640 in the waters off Iceland. With nothing more than the absence of gray whales to substantiate his claim, Mowat (1984) wrote "that by as early as the fourteenth century, the otta sotta had been hunted to virtual extinction in European waters." Mead and Mitchell (1984) recognize only Fraser's *sandloegja*;[3] a 1725 description of the "scrag whale" by Paul Dudley, Esq.,[4] and the 1611 instructions given by the directors of the Muscovy Company to Thomas Edge[5] as "reliable records of gray whales in the North Atlantic." There are no more gray whales in the Atlantic, and while this unfortunate state of affairs might not be directly attributable to the Basques, it is not unreasonable to assign them some part in the disappearance of these whales.

For many years, the most comprehensive study on the subject of Basque whaling was that written by Sir Clements Markham and published in 1881. While working on a study of William Baffin, he learned "that the first English whaling vessels were in the habit of shipping a boat's crew of Basques to harpoon the whales," so he began to investi-

gate, and ended up in Spain. He found that King Sancho the Wise of Navarre had granted petitions to the city of San Sebastian in the year 1150 for the warehousing of certain commodities, among which were *boquinas-barbas de ballenas*, plates of whalebone. Markham traced the fishery through the records of various cities and towns (he found the "Casa de Ballenas" in Asturias), and acknowledged that it was the Basques who taught the British how to kill whales. He sums up the Basque contribution as follows: "Of course the English, in due time, learnt to strike the whales themselves; but the Basques were their instructors; and it is therefore to this noble race that we owe the foundations of our whaling trade" (Markham, 1881).

More recently, the Spanish cetologist Alex Aguilar (1981) has been searching the records for written documentation of Basque whaling and has discovered a reference from Bayona, in the Gulf of Biscay, that dates from the year 1059. From the remains of cetaceans examined at some of the settlements on the shore of the Cantabrian Sea (off the northern coast of Spain), it has been suggested that the Basques occasionally hunted sperm whales, but the predominant object of their fishery was the right whale. Ancient whaling bases have been found along the length of this coastline, which encompasses the provinces of Galicia, Asturias, Santander, and the heart of the Spanish Basque country, Guizpuzcoa. From the western tip of northern Spain the sites have Spanish names (Camariñas, Malpica, Antrellusa, Llanes), but as we move eastward, toward the Basque settlements on the Bay of Biscay, the names take on a decidedly Basque flavor: Lequeitio, Ondarroa, Guetaria, Zarauz. Aguilar quotes several sources (including Markham) for the number of whales killed at Lequeitio from 1517 to 1662, and produces a total of some 62 whales, adults and young, from incomplete records, for a provisional average of something less than 2½ whales per year. Occasional records for Guetaria from 1699 to 1789 provide even lower numbers, suggesting that the Biscayan right whales were on the decline by the eighteenth century.

Along the French and Spanish Biscayan coasts, there are several towns and villages whose seals and coats-of-arms depict whale-fishers, including Bermeo, Ondarroa, Motrico, and Fuenterrabia in Spain, and Biarritz, Hendaye, and Guethary in France. Jenkins (1921) wrote, "in this fishery the Bayonnais took part, and it is one of the most interesting features in the ancient records of the town of Bayonne," For several centuries, the Basques of Biarritz, St. Jean-de-Luz, Bayonne, San Sebastian, and other towns killed the *sarda* in their inshore and offshore waters. This activity did not go unnoticed by the tax collectors. In 1197, King John of England (acting as the Duke of Guienne) collected a tax on the first two whales taken at Biarritz. In 1261, all whales taken at Bayonne were tithed, a continuation of an earlier, voluntary gift of all whales' tongues to the Church. The kings of Castile and Navarre also extracted taxes from the whalers, often in the form of meat or whalebone. Under a 1324 edict known as *De Praerogativa Regis* (The Royal Prerogative), Edward II of England (1307–1327) collected a duty on every whale captured in British waters, and his successors continued to claim the "royal fish" as Crown property.

To this day, we do not know whence the Basques came, or from whom they were descended. (Their blood type distinguishes them from the French and the Spanish, and biologically as well as linguistically they appear to be distinct from any other people now in existence.) As far as we can ascertain from the scanty records and the ruined stone watchtowers (known as *vigías*) that still stand overlooking the bays, they pursued the right whale. Harmer (1928) wrote, "A watchman who tried to use [the towers] for their original purpose would now have an unprofitable occupation, and he would not be likely to see a single whale of this species during his lifetime."

---

[3]Sandloegia. Good eating. It has whiter baleen plates, which project from the upper jaw instead of teeth, as in all other baleen whales, which will be discussed later. It is very tenacious of life and can come to land to lie as a seal like to rest the whole day, But in sand it never breaks up.

[4]The Scrag whale is near a-kin to the Fin-Back, but instead of a Fin on his Back, the Ridge of the After-part of his back is scragged with a half Dozen Knobs or Knuckles; he is nearest the right Whale in Figure and for Quantity of Oil; his Bone is white, but won't split.

[5]The fourth sort of whale is called Otta Sotta, and it is of the same color its the Trumpa having finnes in his mouth all white ball white but not above a yard long, being thicker than the Trumpa but not so long. He yeeldes the best oyle but not above 30 hogs' heads.

Even more significantly, the Basques are said to have invented the on-board tryworks, where whales could be processed at sea, avoiding the time-consuming and arduous process of towing the carcass to shore and then winching it up on the beach for rendering. According to Jenkins, this distinction belongs to "a captain of Cibourre named François Sopite," but surprisingly, in a book heavily footnoted with obscure references, this important fact goes undocumented. In Sanderson's *Follow the Whale*, however, a whole chapter is devoted to a re-creation of Sopite's accomplishments, including a description of him standing "silently on the poop with his hands behind his back peering out from under his curious floppy black hat." Sanderson seems to have consulted many of the same references listed by Jenkins (1921), but he does not tell us where the hat comes from, or how he knows that Sopite was "smiling wryly" at the success of his experiment. Up to that time, whales were flensed and tried-out on shore, which meant that the whalers could never roam too far from their home ports. As we shall see, however, Sopite's "invention" may have been the invention of some creative authors, since real evidence of the Basque whalers has been uncovered, and there is no indication whatsoever of on-board tryworks.

Even though the hunters never took many whales in a given season, the right whale (known as the Biscayan whale to distinguish it from the Greenland right whale) disappeared from Biscayan waters, and the Basques had to look farther afield for oil and bone. Markham (1881) wrote that each of the whaling villages may have taken no more than a couple of whales per year. This would not be enough to decimate the population, but it is possible that the disturbance caused by the whalers drove the whales to other, less perilous breeding grounds.

From Iberia, Basque fishermen crossed the North Atlantic seeking new grounds. Some evidence indicates that they may have fished the Labrador-Newfoundland grounds as early as the fourteenth century, but were driven off by the local Eskimos. The vessels that they used were not known until recently, when a Canadian archaeologist named Selma Barkham (1984) followed up some vague hints in the historical records of Labrador and with the help of divers located the wrecks of several Basque ships in the area known as Red Bay. Found sitting on the bottom of the bay were the remains of a three-masted, 90-ft galleon which is believed to have sunk in a storm in 1565, and the complete hull of one of the frail chalupas.[6] On two of Red Bay's smaller islands, workers found unmistakable evidence of tryworks, where the blubber of the whales was rendered into oil. Since this endeavor took place between the years 1560 and 1570 (ascertained from documents examined in Spanish archives by Barkham), it would appear that Sopite's "invention" of on-board tryworks was either apocryphal or somehow did not extend to the whaling operation at Red Bay.

As the Basques enlarged the scope of their search for whales to the vicinity of Newfoundland and Labrador, they may well have been the first Europeans to fish the Greenland coasts and the Grand Banks, two of the richest cod-fishing grounds in the world. Upon landing, they predated the "official" discoverers of the land known as Terranova, John Cabot, and Jacques Cartier. In their pursuit of the sea's bounty, the adventurous Basques visited Ireland, Iceland, Greenland, and evidently sailed as far north as Spitsbergen. They also crossed the Atlantic to find the right whales that inhabited the inshore waters of

Newfoundland and Labrador, but it is unlikely that they made these voyages without island hopping across the perimeter of the Northern Atlantic, much as the Norse had done before them.

Examination of the bones at Red Bay indicate that bowheads were also processed there by the Baques. This location is considerably south of the known range of the bowhead, which includes eastern Arctic waters and the Bering Sea. (It is likely that the Basques took bowheads farther north, and then brought them back for processing, thereby accounting for bowhead bones in a region where bowheads are not known to have lived.) There are no records of Basques hunting humpbacks, but these whales are found off the Canadian Maritime coasts and Greenland.

The rich days of Newfoundland and Labrador whaling ended for the Basques as the sixteenth century was ending. The destruction of the Spanish Armada in 1588 meant that Spanish ships of war could no longer protect fishing fleets so far from home, and the Basque whalers ventured across the Atlantic unprotected. They had established shore stations at Tadoussac and Sept-Iles on the St. Lawrence, where they hunted humpbacks and probably belugas. By 1738, the last Basques had left Canada. Why bother with the transatlantic crossings and hostile North Americans when there were fat Greenland whales for the taking in Spitsbergen? The Basques participated in the early Dutch and British expeditions in Spitsbergen, bringing with them 500 years of whaling experience.

Six Basque harpooners from St. Jean-de-Luz were part of the crew of the first Muscovy Company expedition to Spitsbergen in 1611. In the early years of the Greenland Fishery (Barendsz had named the Spitsbergen islands "Greenland" when he discovered them in 1596, because he believed they were an extension of the island of that name), the Basques sold their services to whoever was willing to pay, but in addition to their participation in the Dutch/British rivalry, the Spanish Basques also sent their own ships to the northern ice in 1613. No sooner had the Spanish tried to join the fishery on their own than James I of England issued the Muscovy Company an exclusive charter to fish the waters of Spitsbergen, to which the Dutch countered in 1614 by forming their own Noordsche Compagnie with the same objectives.

Although the Spanish Basques had the experience and the expertise, they did not have the naval power to back up their claims, and as the Dutch and the British competed for supremacy in Spitsbergen (the Dutch eventually won the battle because of their more effective management and business practices, but in the end, everybody lost because they ran out of whales), the Basques faded into whaling oblivion. As time and progress passed them by, their domestic whaling capabilities diminished accordingly. According to J.-P. Proulx (1986) when a whale stranded at St. Jean-de-Luz in 1764, the hunters could only find old and rusty implements with which to cut it up.

In many respects, the Basques were the advance guard of what would eventually become an all-out war on the whales, but in those relatively uncomplicated times, they were only aware of the nutritional needs that could be fulfilled by the taking of these large, inoffensive animals. They would, however, establish a pattern with regard to the right whale fishery that would serve as an example for virtually every nation that followed their lead: they took the females and calves because they were the most accessible, and by so doing, guaranteed the catastrophic degeneration of the breeding population. In a review of the available data, Aguilar (1981) estimates that during the period 1530–1610, Basque whalers might have taken as many as 40,000 right whales. Medieval Europeans probably did not have much time to ponder the effects of their actions on future generations, however—certainly not on future generations of whales.

---

[6]A *chalupa* (in French a *chaloupe* and in the British fishery a *shallop*) was a 25-foot-long whaleboat, rowed by six oarsmen, from which the whale was harpooned and towed to shore.

## IV. Aboriginal Whaling

In its early chapters, the story of whaling was a simple one: man against whale. Very infrequently, the whale won. (In *Moby-Dick*, despite the rage of Captain Ahab and the skill of the harpooners, the white whale triumphs.) With the passage of time, the hunters changed the nature of the hunt, and turned it into an industry. The hunted whales remained unevenly matched with their opponents; all they had was the hope of escape in the depth and expanse of the ocean. As the industry grew more economically important, technological innovations were introduced that greatly altered the odds. The introduction of diesel catcher boats, exploding harpoons, spotter planes, sonar and asdic greatly changed the nature of the hunt. No longer remotely equitable, it was not even a hunt any more, but a highly mechanized business. The whale had as much of a fighting chance as a tree had against a chain saw.

There are only a few places in the world where people still *hunt* whales. The Caribbean island of Bequia in the Grenadines, e.g., has a relic humpback whale fishery that the Bequians learned from Yankee whalers in the nineteenth century. The Eskimos of Greenland hunt minke whales and humpbacks (they are given a quota by the International Whaling Commission), as well as belugas and narwhals, which are considered "small cetaceans" and are not under the jurisdiction of the IWC. Alaskan Eskimos "hunt" the bowheads that annually pass their North Slope villages, but because of the complexities of politics and other factors, they have upgraded their weaponry to the point where once again, the whale has hardly any chance of escaping. The only place where whaling takes place in a thoroughly primitive manner (and completely unregulated by the IWC) is in Indonesia.

Lomblen, also known as Lembata, is one of a group of islands that make up the Sunda Archipelago (*Nusa Tenggara Timur* to the Indonesians), which includes the large islands of Timor and Flores, as well as the smaller Solor, Adonara, Pantar, and Alor. Lomblen/Lembata is only one of the 13,000 islands that comprise the 3000 mile long country of Indonesia, but there is something very special about this island. On its southern shore is Lamalarep, one of the few whaling villages in all Indonesia. Lamalarep is the poorest village on the island because it has virtually no industry or agriculture other than whaling, and the success rate of the whalers seems to be rather low. They might capture a whale on 3 trips out of 10; to put it another way, 70% of their trips are unsuccessful. The villagers of Lamalarep do not eat the bulk of the whale meat they take, but dry it in the sun and trade it to other villages for vegetables.

In June of 1979, a research team was sent to Lamalarep by the World Wildlife Fund to investigate the whaling activities there. Unfortunately, on July 17, a giant tsunami inundated Lomblen, causing over 700 casualties and destroying the villages of Wai Teba and Sara Puka. The investigators all survived, however, and remained on Lomblen for 3 weeks. On July 26, on nearby Rote Island, a "giant shark" (species unidentified) was found with the body of what was thought to be a Lomblen fisherman in its stomach. (It is likely that the shark ate one of the victims of the tsunami, rather than taking a swimmer or a fisherman.)

At dawn the Lamalarep fleet sets out for a day's hunting. They may roam as far as 17 miles offshore, but the whales are usually found closer to the islands. The boats (known as *peledang*) are about 30 ft long and brightly painted, often with vigilant eyes on the bows. No nails are used in their construction, only wooden pegs; and the sails are patchwork rattan, a single gaff-rigged square sail for each boat. A crew of 10 to 15 men rows (or sails, if the winds are favorable) the boat out to the whaling grounds, south of the islands in the Savu Sea. They look for the forward-angled spouts of the largest of the toothed whales, the sperm, which they call *ikan pails* in Bahasa Indonesia. (In the language of the islands that was employed before the introduction of this *lingua franca* by President Sukarno, the sperm whale was known as *kotan klema*.) During the 10 weeks that the World Wildlife Fund researchers kept records, the whalers of Lamalarep took sperm whales, killer whales, pilot whales and several species of dolphins. Traditionally, the whalers of Lamalarep do not hunt baleen whales. Although the men of Lamalarep are considered whalers, they will also harpoon any large fish, ray or turtle that they encounter, including sharks, marlin, and ocean sunfish (*Mola mola*).

The whalers of the island are divided into hereditary "corporations," each of which owns a whaling vessel. The vessels—and their names—are passed down from generation to generation, so when a given boat wears out, the next one built by that clan is given the same name.

When a whale is sighted, the *peledang* crews row stealthily upon it, douse the sail, and because they are Christians, they whisper a communal *Pater Noster* for their own protection. The harpooner stands on a narrow platform with his bamboo-shafted harpoon poised. At the critical moment, when he is within striking range of the wrinkled, humped back of the whale, the harpooner launches not only the harpoon, but *himself* through the air, using his strength and his weight to drive the iron deep into the flesh of the whale (Fig. 1).

As the whale is slowed or stopped by the pain of the harpoon in its back, another harpooner throws himself on the whale, and if necessary, another. The iron must be planted in exactly the right place to kill the whale; otherwise the fragile *peledang* will be towed for miles as the whale pulls the whalers on the Indonesian equivalent of the "Nantucket sleigh ride." There are stories of boats being towed all the way to Timor by a maddened whale. (In fact, there are many tales about maddened whales in the Timor Sea. One of the most notorious of all these was a bull sperm whale named "Timor Jack," who savaged whaleboats for years until he was taken by setting out a barrel on a line which he attacked, allowing whalers to lance him.) Or the whale will dive, pulling the line out rapidly and unless it is cut, and pulling the boat down with it. If the right spot is pierced (the heart or lungs), the whale will spout blood from its blow-hole and expire quickly. The dead whale is towed back to the village where it is butchered.

There is a complex system for dividing up the meat of a whale, in which the carcass is portioned out according to rank in the clan and the village. The meat is eaten or bartered to other villages; the oil is used for lamps. The men of the village may carve patterns onto the teeth, like scrimshanders everywhere.

The villagers of Lamalarep kill between 30 and 50 sperm whales every year. They do not take the large bulls, because the big males do not visit these waters. They cannot eat all the meat, so they barter it in neighboring villages. This is in direct contravention of the regulations of the International Whaling Commission, but since Indonesia is not a signatory to the Whaling Convention, the IWC regulations are difficult—if not impossible—to apply.

## See Also the Following Articles

Azorean Whaling ■ Whaling, Illegal and Pirate ■ Whaling, Modern ■ Whaling, Traditional

## References

Aguilar, A. (1981). The black right whale, *Eubalaena glacialis*, in the Cantabrian Sea. *Rep. Intl. Whale Comm.* **31**, 457–459.

**Figure 1** *The whalers of the Indonesian village of Lomblen launch themselves as well as their harpoons.*

Aguilar, A. (1985). A review of old Basque whaling and its effect on right whales (*Eubalaena glacialis*) of the North Atlantic. *Rep. Intl. Whale Comm.* **10**, 191–199 (Special Issue).

Aristotle (nd). "Historia Animalium." Translated by D'Arcy Wentworth Thompson, 1910. Clarendon Press, London.

Barkham, S. (1984). The Basque whaling establishment in Labrador 1536–1632: A summary. *Arctic* **37**, 515–519.

Barthelmess, K., and Münzig, J. (1991). "Monstrum Horrendum: Wale und Waldarstellungen in der Druckgraphik des 16." Jahrhunderts und ihr motivkundlicher Einfluss, Kabel, Hamburg.

De Smet, W. M. A. (1981). Evidence of whaling in the North Sea and English Channel during the Middle Ages. *In* "Mammals in the Seas," FAO Fisheries Series No. 5, Vol. III, pp. 301–309. Food and Agriculture Organization of the United Nations, Rome.

Dudley, P. (1725). An essay upon the natural history of whales, with particular account of the ambergris found in the sperma ceti whale. *Phil. Trans. Royal. Soc. Lond.* **33**, 256–259.

Fraser, F. C. (1970). An early 17th century record of the Californian grey whale in Icelandic waters. *Invest. Cetacea* **2**, 13–20.

Gesner, C. (1551–1558). "Historia Animalium." Zurich.

Harmer, S. F. (1928). The history of whaling. *Proc. Linn. Soc. Lond.* **140**, 51–95.

Hawes, C. B. (1924). "Whaling." Doubleday, Page, Garden City.

Jenkins, J. T. (1921). "A History of the Whale Fisheries." Kennikat Press, Port Washington, Reissued 1971.

Lindquist, O. (2000). The North Atlantic gray whale (*Eschrichtius robustus*): An historical outline based on Icelandic, Danish-Icelandic, English and Swedish sources. Occasional Paper 1. Universities of St. Andrews and Stirling, Scotland.

Markham, C. R. (1881). On the whale fisheries of the Basque provinces of Spain. *Proc. Zool. Soc. Lond.* **62**, 969–976.

Mead, J.G. and Mitchell, E.D. (1984). "Atlantic Gray Whales, in Jones, M.L., Swartz, S.L., and Leatherwood, S. (Eds.). The Gray Whale, *Eschrichtius robustus*. Academic Press, New York, NY.

Melville, H. (1851). "Moby-Dick" (H. Hayford and H. Parker, eds), W. W. Norton, 1967. Norton Critical Edition, New York.

Morison, S. E. (1971). "The European Discovery of America: The Northern Voyages, ad 500–1600." Oxford University Press, New York.

Mowat, F. (1984). "Sea of Slaughter." Atlantic Monthly Press, Boston.

Ommanney, F. D. (1971). "Lost Leviathan." Dodd, Mead, New York.

Parry, J. H. (1974). "The Discovery of the Sea." Dial Press, New York.

Pliny. (n.d.) "Naturalis Historia," Loeb Classical Library, 1933. Harvard University Press, Cambridge.

Proulx, J. P. (1986). "Whaling in the North Atlantic from Earliest Times to the Mid-19th Century." Canadian Printing Service, Ottawa.

Sanderson, I. (1956). "Follow the Whale." Little, Brown, Boston.

Scoresby, W. (1820). An Account of the Arctic Regions with a History and a Description of the Northern Whale-Fishery," David and Charles, 1969 Ed. Archibald Constable, Edinburgh.

Sibbald, R. (1692). "Phalainologia Nova: Sive, Observationes de Rarioribus Quibusdam Balaenis in Scotiae Littus Nuper Ejectis." Edinburgh.

Whitaker, I. (1984). Whaling in classical Iceland. *Polar Rec.* **22**, 249–261.

Whitaker, I. (1985). The king's mirror (*Konung's skuggsjá*). *Polar Rec.* **22**, 615–627.

Whitaker, I. (1986). North Atlantic sea creatures in the king's mirror (*Konung's skuggsjá*). *Polar Rec.* **22**, 3–13.

# Whaling, Illegal and Pirate

ROBERT L. BROWNELL, JR., AND ALEXEY V. YABLOKOV

Illegal whaling occurs in contravention of national laws or internationally agreed quotas, season, area restrictions, and other limitations, whereas "pirate whaling" refers to unregulated whaling conducted outside the aegis of the International Whaling Commission, usually under a flag of convenience. Such activities can lead directly to depletion of whale populations through overexploitation. Furthermore, the lack of catch data, or the reporting of falsified data, can lead to serious error in assessment of the size and status of populations and erroneous management advice ultimately contributing to their collapse.

**W**

## I. Illegal Whaling

Known instances of illegal whaling were conducted by several nations. Because the offenses of the USSR were the most egregious, most of this discussion will focus on what is known of Soviet activities. The Soviet Union conducted massive illegal whaling and falsification of data over a period of decades, with catastrophic consequences for whale CONSERVATION and development of the science of whale MANAGEMENT. The USSR commenced pelagic whaling in the North Pacific with the *Aleut* in June 1933 (Tønnessen and Johnsen, 1982). This floating factory operation continued in the North Pacific through the 1967 season. After this date, four additional Soviet factory ships conducted pelagic whaling operations in the North Pacific. The last season was in 1979 when the Soviets stopped whaling in the region. For more than a decade following the end of World War II, the USSR operated a single whaling factory ship (*Slava*) in the Antarctic. Beginning in 1959, the Soviets began expanding their whaling operations, adding one new factory ship in each of the next three Antarctic seasons (*Sovietskaya Ukraina* in 1959/1960, *Yuri Dologorukiy* in 1960/1961, and *Sovietskaya Rossiya* in 1961/1962). This expansion occurred at a time when there was extensive discussion in the IWC about declining stocks and other countries were decreasing their whaling operations. The USSR voted against the drastic reductions in catch quotas required to meet scientific recommendations and against implementation of an International Observer Scheme (IOS), both of which were eventually put in place. After the breakup of the Soviet Union, a number of Russian and Ukrainian biologists who had served aboard Soviet factory ships and knew of the existence of accurate but unreported catch statistics decided to collect them and make them available to the world scientific community. This section summarizes information on Soviet illegal activities during the two major phases of its whaling in the Southern Hemisphere and North Pacific and briefly notes recent information on illegal whaling activities by other nations.

### TABLE I
**Comparison of Southern Hemisphere USSR Commercial Catch Data (1947–1972)[a]**

| Species | Reported | Actual | Disparity |
|---|---|---|---|
| Blue whale | 3,887 | 3,681 | +206 |
| Pygmy blue whale | 10 | 9,215 | −9,205 |
| Fin whale | 52,860 | 48,889 | +9,971 |
| Sei whale | 29,751 | 53,366 | −23,615 |
| Bryde's whale | 10 | 1,457 | −1,447 |
| Minke whale | 1,246 | 384 | +862 |
| Humpback whale | 2,820 | 42,889 | −43,011 |
| Southern right whale | 4 | 3,368 | −3,364 |
| Sperm whale | 50,715 | 72,372 | −21,657 |
| Killer whale | 482 | 124 | +358 |
| Southern bottlenose whale | 10 | 17 | −7 |
| "Others" | 0 | 29 | −29 |
| Total | 141,795 | 232,733 | −90,938 |

[a]Reported to the IWC with numbers actually taken. Modified from Yablokov *et al.* (1998) and Tormosov *et al.* (1998).

## A. Southern Hemisphere

A summary of the disparity between USSR catch data reported to the IWC and actual takes is given in Table I (modified from Yablokov *et al.*, 1998). These data concern catches made by Soviet fleets working in, or en route to or from Antarctic waters. The period concerned is from the beginning of post-war Soviet whaling in 1947 until the introduction of international observers in the 1972/1973 whaling season, when most illegal activities ceased. During this period, unreported catches totaled 102,335 whales, almost half (44%) of which were humpbacks (*Megaptera novaeangliae*). However, 11,397 animals (primarily fin whales, *Balaenoptera physalus*, which were then an unprotected species) were actually over-reported; this was done to conceal the massive illegal catches of pygmy blue (*B. musculus brevicauda*), sei (*B. borealis*), humpback, and southern right whales (*Eubalaena australis*).

Massive falsification of geographic data began in 1959 and was practiced by all four Soviet Antarctic whaling fleets. The primary areas for illegal catches were numerous sections of the South Atlantic, the Indian Ocean (including as far north as the western Arabian Sea), and the southwestern Pacific (Yablokov, 1994; Zemsky *et al.*, 1995a, b). IWC regulations prohibited the taking of baleen whales (mysticetes) north of latitude 40°S, although the killing of sperm whales (*Physeter macrocephalus*), killer whales (*Orcinus orca*), and bottlenose whales (*Hyperoodon planifrons*) in this region was permissible. Therefore, the Soviets used the pretext of hunting toothed whales to exploit mysticete populations in the prohibited areas. Soviet whaling fleets bound for the Antarctic began search and catcher operations immediately after leaving the Red Sea or after passing either Gibraltar or Portuguese waters.

Virtually all biological data reported to the IWC were "corrected" to disguise the extensive illegal catches. Because of the prohibition on killing mothers and calves, all such cases were either unreported or were concealed with false data. Thus, a fin whale mother and calf might be reported as "two sei whales," whereas a catch of four female sperm whales would be reported as "two males." It has been estimated that at least 80% of all officially reported Soviet data on length, weight, sex ratio, reproduction, and maturational state are false.

The scale of the Soviet catches partly explains the apparent failure to recover that has been evident in some mysticete populations despite their supposedly protected status. The Soviets killed 12,896 blue whales in the Southern Hemisphere, of which more than 9200 were unreported pygmy blues killed after the IWC accorded protected status to both subspecies by the 1965/1966 season. Catches were made over a wide area, including the Indian Ocean north of the equator. Thus, these populations were reduced to much lower levels than the 800–1600 (blue) and 10,000 (pygmy blue) that were estimated at the time. Although recent estimates of 500 and 5000 for current Antarctic populations of blue and pygmy blue whales, respectively, are not statistically robust, it is apparent that the population sizes are smaller than would be expected following three decades of protection.

Humpback whales were even more seriously impacted by illegal catches. In addition to the huge number (45,831) of unreported catches made by the Soviets, it is known that additional illegal takes were made in Antarctic waters by the *Olympic Challenger*, a pirate factory ship owned by Aristotle Onassis (Tønnessen and Johnsen, 1982), discussed later. The Australian biologist Graeme Chittleborough (1965) asserted that large discrepancies in calculated mortality coefficients for humpbacks could be explained only by the occurrence of extensive illegal catches, a view that is now validated by the Soviet catch data reported here. Many of the Soviet catches were made from management division known as Antarctic Area V (Dawbin, 1966), which explains the collapse of shore whaling stations in eastern

Australia and New Zealand in the early 1960s. There is now evidence of strong population growth in some Southern Hemisphere humpback populations (Bannister 1994; Paterson *et al.*, 1994; Findlay and Best, 1996), but the status of other stocks remains unclear.

Additional illegal Soviet pelagic whaling occurred in the Arabian Sea on humpback (Mikhalev, 1997a), blue, Bryde's (*Balaenoptera edeni*), and sperm whales during the 1960s (Mikhalev, 1996, 1997b; personal communication). This whaling occurred while the whaling fleets were en route to the Antarctic whaling grounds. Biological data from humpback catches made in November 1966 off Oman and India have resolved a long-standing issue regarding the identity and boreal breeding cycle of this population, which is unique among humpbacks in that it resides in tropical waters year-around (Mikhalev, 1997a). However, the status of this tropical population of humpbacks remains uncertain.

Southern right whales have always been protected under IWC regulations; this status dates from a League of Nations agreement in 1935. Recovery (increasing populations) is today apparent in only 4 of the 13 putative populations of this species (Best, 1993), although five of the remaining nine stocks are considered impossible to monitor on a regular basis. True data show that the Soviets made large unreported catches (3368) of right whales between 1950 and 1971 (Tormosov *et al.*, 1998). Many of these takes were made around remote islands or in mid-oceanic areas such as Campbell Island, Crozet, Kerguelen, Tristan da Cunha, and the central Indian Ocean.

### B. North Pacific

A single Soviet whaling factory operated in the North Pacific between 1933 and 1967, a small vessel named the *Aleut*. Four additional Soviet factory ships later operated there. A new ship, the *Sovietskaya Rossiya* (built as the sister ship to the *Sovietskaya Ukraina*), operated for four seasons (1962–1965) and then again for three more seasons in 1973, 1978, and 1979. The *Slava*, after working for many years in the Southern Hemisphere, worked in the North Pacific for four seasons (1966–1969). Two sister whaling factory ships were built specifically for the North Pacific (the *Vladivostok* and the *Dalniy Vostok*); both began operations in 1963. The *Vladivostok* operated through 1978 and the *Dalniy Vostok* through 1979. The main whaling grounds for these two fleets were the Bering Sea, Gulf of Alaska, and other more southerly parts of the North Pacific.

The whales in the North Pacific did not fare any better than those in the Southern Hemisphere. However, the available records of the true catches are not as good. Doroshenko (2000a) and others reported numerous illegal catches of North Pacific right whales (*Eubalaena japonica*) in both the western and the eastern North Pacific, catches which devastated the last of the latter population. Brownell and colleagues (2000) provided some data on massive illegal catches of sperm whales, on the order of 180,000 whales, in the North Pacific. Soviet pelagic whaling operations in the North Pacific also illegally took blue, humpback, bowhead, and gray whales (*Eschrichtius robustus*) (Doroshenko, 2000a). A summary of the inconsistencies between data reported by the USSR to the IWC and actual catches is given in Table II.

It is known that illegal whaling on a similarly large scale was also conducted by the USSR throughout the Northern Hemisphere, although the same details of catch data are not available. In light of these revelations about the true catches in both the Southern Hemisphere and the Northern Hemisphere (North Pacific Ocean), it is clear that current views regarding the status and recovery potential of virtually all affected whale populations worldwide need revision. This

is a long-term project that has been underway by the staff of the IWC and others for over a decade and may continue for another decade.

At the 50th IWC meeting in Oman in 1998, the IWC took action on the Scientific Committee's (SC) concern about the falsified Soviet whaling data, mainly for sperm whales, by adopting the SC recommendation to remove the official Soviet Southern Hemisphere whale catches from the IWC database.

The USSR was not alone in the illegal hunting of whales. Recent evidence on the falsification of catch statistics has been reported for various North Pacific land-based operations conducted by the Japanese (Kasuya, 1999; Kasuya and Brownell, 1999, 2001; Kondo, 2001). Suspicions about illegal reporting in Japanese operations are not new and have been presented in the past (Kasuya and Miyazaki, 1997). The scale of these activities, however, was much smaller and the consequences less severe than in the case of the Soviet whaling. Sperm whales catches between 1954 and 1964 were 1.4 to 3 times greater that the numbers Japan reported to the IWC. The total true catches of Bryde's whales taken during the final years of commercial land-based whaling (1981–1987) off the Bonin Islands were 1.6 times the numbers Japan reported to the IWC. Fin whales were reported taken illegally by the Republic of Korea in the 1980s and minke whales were still reported taken illegally in the early 2000s.

During the 1990s and the following decade, numerous reports appeared regarding the sale of "illegal whale products" from protected whales collected in the Japanese market (Baker *et al.*, 2000). It is argued that Japan's scientific whaling program (since 1989) has acted as a cover for undocumented or illegal products from various protected species (fin, sei, humpback, and gray). While this is possible, there are no available data to support the occurrence of any large-scale illegal whaling during the 1990s. The most parsimonious explanation for the whale products from protected species is that they are from (1) whales killed before the 1986 IWC moratorium on commercial whaling, (2) past scientific hunts by Iceland or Norway, (3) by-catches from Japanese fisheries, and (4) STRANDINGS in Japan.

### II. Pirate Whaling

As noted earlier, unregulated whaling conducted under the flags of non-IWC member nations has contributed to the depletion of some whale stocks. The most famous of these operations was that

TABLE II
Comparison of North Pacific Commercial USSR Catch Data
(1961–1972)[a]

| Species | Reported | Actual | Disparity |
|---|---|---|---|
| Blue whale | 517 | 1,205 | −688 |
| Fin whale | 10,613 | 8,621 | +1,992 |
| Sei whale | 9,048 | 4,177 | +4,871 |
| Bryde's whale | 775 | 714 | +61 |
| Humpback whale | 3,043 | 6,793 | −3,750 |
| Gray whale | 0 | 138 | −138 |
| Right whale | 0 | 508 | −508 |
| Bowhead whale | 0 | 133 | −133 |
| Total | 23,996 | 22,289 | +1,707 |

[a]Reported to the IWC with numbers actually taken.

conducted by interests in Norway and Japan from 1968 to 1979 in the North and South Atlantic under the flags of Somalia, Cyprus, Curaçao, and Panama. Meat from the whales was shipped to Japan for human consumption. The *Run* operated mainly in the South Atlantic from January 1968 to February 1972. It was renamed the *Sierra* in 1972 and expanded major operations to the North Atlantic in 1975, where it continued taking whales until it was rammed and sunk by the *Sea Shepherd* (a privately operated vessel dedicated to interference with commercial whaling) in 1979 off Portugal (Watson, 1979). The *Tonna* joined the Sierra in December 1977 and operated until July 1978, when it foundered during processing of a large whale on deck. The *Cape Fisher*, later renamed the *Astra*, operated briefly as a processing vessel for the Sierra in 1979.

The Sierra Fishing Agency submitted its catch statistics to the Bureau of International Whaling Statistics in Norway until 1976, when the practice was discontinued because of a perceived lack of credibility of the data. Data for the remaining years of operation were destroyed, but some information was salvaged through interviews with former crew members (Best, 1992). The catches included blue, fin, sei, Bryde's, humpback, and minke (*Balaenoptera acutorostrata* and/or *B. bonaerensis*). Large catches of fin whales totaling hundreds were made off the coasts of Spain and Portugal (the IWC's "Spain–Portugal–British Isle Management Area") after 1976.

Another notorious episode of pirate whaling occurred in the Southern Hemisphere from 1951 to 1956 by the factory ship *Olympic Challenger* and its fleet of 12 catcher boats (Tønnessen and Johnsen, 1982). The Olympic Whaling Company, an affiliate of the Pacific Tankers Co. of New York, was financed by the Greek-born Argentine citizen Aristotle Onassis. The ownership of the vessel, a converted tanker, was later transferred to the Olympic Whaling Company S.A. in Montevideo, Uruguay. The captain was German and the expedition manager Norwegian. The factory ship and some of the catcher vessels flew the Panamanian flag and the remainder of the catchers the Honduran flag. Neither Panama nor Honduras were members of the IWC at the time, so the whaling operations were completely unregulated. The expedition took thousands of whales in the Antarctic South Pacific sector and off Chile, Peru, and Ecuador, including blue, humpback, sei, right, and sperm. Catch data were reported to the International Bureau of Whaling Statistics, but these have been shown to include extensive falsification of number, species, and sizes of whales caught (Barthelmess *et al.*, 1997). As noted earlier, these unregulated catches in combination with later illegal catches by Soviet fleets contributed in a major way to the catastrophic decline of whales in the Southern Ocean, particularly the humpback.

## See Also the Following Articles

Humpback Whale ■ Hunting of Marine Mammals ■ International Whaling Commission ■ Japanese Whaling ■ Stock Assessment

## References

Baker, C. S., Lento, G. M., Cipriano, F., Dalebout, M. L., and Palumbi, S. R. (2000). Scientific whaling: Source of illegal products for market? *Science* **290**, 1695.

Bannister, J. (1994). Continued increase in humpback whales off Western Australia. *Rep. Int. Whale Comm.* **44**, 309–310.

Barthelmess, K., Kock, K.-H., and Reupke, E. (1997). Validation of catch data of the *Olympic Challenger's* whaling operations from 1950/51 to 1955/56. *Rep. Int. Whale Comm.* **42**, 937–940.

Best, P. B. (1992). Catches of fin whales in the North Atlantic by the M. V. *Sierra* (and associated vessels). *Rep. Int. Whale Comm.* **42**, 697–700.

Best, P. B. (1993). Increase rates in severely depleted stocks of baleen whales. *ICES J. Mar. Sci.* **6**, 93–108.

Brownell, R. L., Jr., Yablokov, A. V., and Zemsky, V. A. (2000). USSR pelagic catches of North Pacific sperm whales, 1949–1979: conservation implications. *In* "Soviet Whaling Data (1949–1979)" (A. V. Yablokov, and V. A. Zemsky, eds), pp. 123–130. Center for Russian Environmental Policy, Moscow.

Chittleborough, R. G. (1965). Dynamics of two populations of the humpbacks whales, *Megaptera novaeangliae* (Borowski). *Aust. J. Mar. Fishwater Res.* **16**, 33–128.

Dawbin, W. H. (1966). The seasonal migratory cycle of humpback whales. *In* "Whales, Dolphins and Porpoises" (K. S. Norris, ed.), pp. 145–170. Univ. California Press, Berkeley and Los Angeles.

Doroshenko, N. V. (2000a). Soviet catches of humpback whales (*Megaptera novaeangliae*) in the North Pacific. *In* "Soviet Whaling Data (1949–1979)" (A. V. Yablokov, and V. A. Zemsky, eds), pp. 48–95. Center for Russian Environmental Policy, Moscow.

Doroshenko, N. V. (2000b). Soviet whaling for blue, gray, bowhead and right whales in the North Pacific, 1961–1979. *In* "Soviet Whaling Data (1949–1979)" (A. V. Yablokov, and V. A. Zemsky, eds), pp. 96–103. Center for Russian Environmental Policy, Moscow.

Findlay, K. P., and Best, P. B. (1996). Estimates of the number of humpback whales observed migrating past Cape Vidal, South Africa, 1988–1991. *Mar. Mam. Sci.* **12**, 354–370.

Kasuya, T. (1999). Examination of the reliability of catch statistics in the Japanese coastal sperm whale fishery. *J. Cetacean Res. Manag.* **1**, 109–122.

Kasuya, T. and Brownell, R. L., Jr. (1999). Additional information on the reliability of Japanese whaling statistics. Unpublished document presented to the 51st Scientific Committee Meeting of the International Whaling Commission, Grenada May 1999. IWC/SC/O7, 15 pp.

Kasuya, T. and Brownell, R. L., Jr. (2001). Illegal Japanese coastal whaling statistics. Unpublished document presented to the 53rd Scientific Committee of the International Whaling Commission, London, UK. IWC/SC/O7, 4 pp.

Kasuya, T., and Miyazaki, N. (1997). Cetacea and Sirenia. *In* "Red List of Japanese Mammals" (T. Kawamichi, ed.), pp. 139–187. Bunichi Sogo Shyuppan, Tokyo [In Japanese with English summary.].

Kondo, I. (2001). "Rise and fall of Japanese coastal whaling." Sanyo-sha, Tokyo, 449. [In Japanese].

Mikhalev Y. A. (1996). Pygmy blue whales of the northern-western Indian Ocean. IWC Scientific Committee. Unpublished Meeting doc. SC/48/SH30, 30 pp. Available from IWC, The Red House, 135 Station Rd, Impington, Cambridge CB4 9NP, UK.

Mikhalev, Y. A. (1997a). Humpback whales *Megaptera novaeangliae* in the Arabian Sea. *Mar. Ecol. Prog. Ser.* **149**, 13–21.

Mikhalev Y. A. (1997b). Bryde's whales of the Arabian Sea and adjacent waters. IWC Scientific Committee. Unpublished. Meeting doc. SC/49/SH30, 10 pp. Available from IWC, The Red House, 135 Station Rd, Impington, Cambridge CB4 9NP, UK.

Paterson, R. A., Paterson, P., and Cato, D. H. (1994). The status of humpback whales *Megaptera novaeangliae* in the east Australia thirty years after whaling. *Biol. Conserv.* **70**, 135–142.

Tønnessen, J. N., and Johnsen, A. O. (1982). "The History of Modern Whaling. [Translated from the Norwegian by R. I. Christophersen]." University of California, Berkeley.

Tormosov, D. D., Mikhalev, Y. A., Best, P. B., Zemsky, V. A., Sekiguchi, K., and Brownell, R. L., Jr. (1998). Soviet catches of southern right whales, *Eubalaena australis*, 1951–1971; Biological data and conservation implications. *Biol. Conserv.* **86**, 185–197.

Watson, P. (1979). Pirate whaler smashed. *Defenders* **54**(6), 338–363.

Yablokov, A. V. (1994). Validity of whaling data. *Nature (Lond.)* **367**(6459), 108.

Yablokov, A. V., Zemsky, V. A., Mikhalev, Y. A., Tormosov, V. V., and Berzin, A. A. (1998). Data on Soviet whaling in the Antarctic in 1947–1972 (population aspects). *Russ. J. Ecol.* **29**, 38–42 [Translated from Ekologiya 1, 43–48 (1998)].

Zemsky, V. A., Berzin, A. A., Mikhalev, Y. A., and Tormosov, D. D. (1995a). Soviet Antarctic pelagic whaling after WWII: Review of actual catch data. *Rep. Int. Whale Comm.* **45**, 131–135.

Zemsky, V. A., Berzin, A. A., Mikhalev, Y. A., and Tormosov, D. D. (1995b). "Soviet Antarctic Whaling Data." Centre for Environmental Policy, Moscow.

# Whaling, Modern

PHILLIP J. CLAPHAM AND C. SCOTT BAKER

*There is no means known to catch the fin whale, or its fast cousins.*

So wrote Herman Melville in the year 1851, echoing the common lament of whalers that the fastest (and some of the largest) species of whales, such as the finback (*Balaenoptera physalus*) and the blue (*B. musculus*), lay beyond contemporary means of capture. At the time that Melville wrote *Moby Dick*, the basic technology of whaling had remained essentially unchanged for centuries. Whaling ships plied their trade under sail, and the small boats that they lowered to pursue whales were also powered by wind, or by the brute strength of their crew's arms at the oars. The killing of whales required men to bring their frail craft alongside the huge quarry, subduing and fastening to it with hand-thrown harpoons. If this dangerous series of actions succeeded, the whale might ultimately be dispatched with a lance thrust deep into some vital organ. Once killed, the whale's carcass would be either towed to shore, or brought alongside the whaling vessel for the time-consuming process of butchering.

## I. The Emergence of Modern Whaling

These methods had been in use in the eleventh century when the Basques began the first sustained commercial whale fishery, on North Atlantic right whales (*Eubalaena glacialis*) in the Bay of Biscay (Ellis, 1991). Although many improvements had been made, the technology available to whalers in the middle of the nineteenth century severely limited the number of animals that could be taken and processed in a working day. Furthermore, as Melville noted, it largely precluded exploitation of the faster species. Blue, fin, sei (*B. borealis*), and Bryde's (*B. edeni/brydei*) whales, all large and desirable targets, were too swift to allow pursuit by oars or sails. Thus, it was the slower species such as the humpback (*Megaptera novaeangliae*), the right, and the sperm whale (*Physeter macrocephalus*) which had borne the brunt of commercial whaling, and they had done so in some cases for almost a 1,000 years.

By 1860, however, all of this was about to change (Tønnessen and Johnsen, 1982). Two men, the Norwegian Svend Føyn and an American named Thomas Welcome Roys, were independently experimenting with explosive harpoons. As patented by Roys in 1861, this device was initially fired from a shoulder gun; Føyn developed a different approach using a bow-mounted cannon, which was to become the industry standard. The *"bomb lance"* was an innovation that was to revolutionize the whaling industry by providing a much more efficient means to dispatch whales, both quickly and from a distance.

About the same time, the use of sail was beginning to give way to steam. This was a key innovation: together with the explosive harpoon, it radically changed the industry, and finally allowed the pursuit and capture of any whale. Suddenly, even the fastest rorquals came under the gun as they were chased down by motorized catcher boats.

A further innovation, the compressor, solved a long-standing problem with regard to the many species which did not float when dead: by pumping air into the carcass immediately after death, whalers could secure it before it sank, thus greatly reducing the loss rate.

For the industry, this transition into the mechanized age could not have come at a more opportune time. By 1900, many populations of the traditionally hunted species were commercially exhausted. In some cases, such as the bowhead (*Balaena mysticetus*) and the North Atlantic right whale, this was the result of exploitation spread over centuries. With some other stocks, decimation had been accomplished in a remarkably short time; the first North Pacific right whale (*Eubalaena japonica*) was not killed until 1835, yet 14 years later the population had already been reduced to the point where many whalers switched their focus to the newly established fishery for bowhead whales in the western Arctic. Another quickly depleted stock was that of the eastern North Pacific gray whale (*Eschrichtius robustus*), made vulnerable by its predictable coastal migration and tendency to concentrate for breeding and calving in the lagoons of Baja California.

The new technology opened up all species to whaling, and did so at a time when the industry, spurred on by steam power, was also expanding geographically. By far the most significant development in this regard was the discovery of the vast stocks of whales in the Southern Ocean. In 1904, the Norwegian whaler C.F. Larsen arrived at the South Atlantic island of South Georgia and reported with astonishment, "I see them in hundreds and thousands." Huge pristine populations of rorquals—notably blues, fin whales, and humpbacks—filled the surrounding waters together with southern right whales and other species. Modern whaling had found its greatest playground, and a slaughter unparalleled in whaling history was about to begin.

Whaling at South Georgia was initially constrained by the need to use land stations for processing of the carcasses. Because of this, spoilage limited the range of the catcher boats; it also left the whaling companies vulnerable to high taxes levied by the British authorities. Despite these difficulties, the industry accomplished the destruction of local stocks of whales with remarkable efficiency. At the height of operations, hundreds of humpback, fin, and blue were taken in a single month. By 1915, the South Georgia population of humpbacks had essentially been extirpated, with a total catch of some 18,557 whales; while occasional catches were made in later years (the largest being one of 238 humpbacks in 1945/1946), the stock had essentially been rendered commercially extinct by the time of the Great War. Blue whales suffered a similar fate; 39,296 were killed at South Georgia between 1904 and 1936, at which point the population had crashed, irretrievably.

The problem of dependence upon land stations was solved, at a stroke, with the introduction of the factory ship. The British vessel *Lansing* was the first such floating factory and began operations in Antarctic waters in 1925. It is difficult to overestimate the importance of this innovation to whaling, or its contribution to the destruction of whale populations in the Antarctic. Factory ships could operate independently far out to sea for months at a time. They maintained round-the-clock processing operations, their huge flensing decks kept constantly supplied by an attendant fleet of catcher boats. Whale carcasses were hauled up the large stern ramp and dismembered with astonishing mechanical efficiency; an adult fin whale of 70 or 80 ft and 100 tons could be rendered

**W**

from whole animal down to bone in half an hour. With the factory ship, all of Antarctic waters became open to whalers, their operations limited only by the constant dangers of weather and ice.

Over the six decades following the opening of the Antarctic grounds in 1904, the whaling industry killed more than 2 million whales in the Southern Hemisphere (Table I). This included 360,000 blue whales, some 200,000 humpbacks, 400,000 sperm whales, and a staggering three-quarters of a million fin whales. By the 1930s, it was apparent even to the whaling nations that some kind of regulation was required. In 1931, the Convention for the Regulation of Whaling was held and adopted worldwide protection for right whales, an action which came into effect in 1935. The Second Convention for the Regulation of Whaling was held in 1937 and provided protection for the much-depleted gray whale. However, neither convention went far enough; among other things, since neither Japan nor the Soviet Union ratified these agreements, both were theoretically free to continue killing the only two species that had been granted protection (Fig. 1).

## II. Advent of the International Whaling Commission

In 1946, following the virtual cessation of whaling that occurred during World War II, the International Convention for the Regulation of Whaling was developed and signed by all major whaling nations (including Japan and the USSR). Among other things, this landmark convention created the International Whaling Commission (IWC), established to regulate whaling and to oversee research on whale stocks. The latter task had as its principal objective learning enough about the abundance, population structure, and life history of the great whales to permit the IWC to set quotas that would allow the highest viable level of exploitation, a concept widely known as Maximum Sustainable Yield (MSY).

Unfortunately, the quota system of the IWC was immediately handicapped by an earlier development. In 1932, the whaling nations had developed the "Blue Whale Unit" (bwu). A single bwu was equivalent to one blue whale, two fin whales, two and a half humpbacks, or six sei whales. As such, quotas set in bwu, by permitting whalers to make their own decisions about which whales to take, made no allowance for the conservation status of a particular species,

### TABLE I
Southern Hemisphere Catch Totals, 1900–2005.

| | |
|---|---|
| Blue | 362,770 |
| Fin | 725,331 |
| Sei | 203,843 |
| Humpback | 213,245 |
| Bryde's | 7,881 |
| Minke | 119,415 |
| Right | 4,424 |
| Sperm | 405,898 |
| Other | 11,835 |
| Total | 2,054,642 |

*Source*: IWC.

**Figure 1** *These blue whales are among those harvested worldwide during the twentieth century. Over 360,000 animals were harvested from the Southern Hemisphere alone. Photo by courtesy of Whales Research Institute, Hideo Omura.*

let alone that of a specific population. It was not until 1949 that a species-specific quota (for humpbacks) was established.

The bwu remained in effect until 1972, despite recommendations from IWC scientists as early as 1963 that it be abolished (also in 1963, the same scientists recommended a halt to all humpback and blue whaling; the IWC responded by setting a quota of 10,000 bwu). In some years, the IWC could not agree on a bwu quota, and whaling nations were left to make their own informal agreements on catch levels. Overall, the bwu arguably represents the most ill-conceived and damaging management strategy in IWC history. However, it was far from the only problem in the Commission's management of whale populations.

## III. Illegal Whaling

*The result of this was that some breeding areas for sperm whales became deserts.*
Soviet biologist Alfred A. Berzin

From its inception, the IWC was hampered by the unwillingness of the whaling nations to pay attention to the mounting evidence of decline in whale populations and by the complete lack of any enforcement or independent inspection measures. That the first humpback whale quota promulgated in 1949 was immediately exceeded in the three subsequent Antarctic seasons pointedly highlighted the latter issue. Additional examples followed, but it was not until the 1990s that the true extent of this problem became apparent, and it was more egregious than anyone could have predicted.

In 1993, following the end of the Cold War, former Soviet biologists revealed that the USSR had conducted a massive campaign of illegal whaling beginning shortly after World War II (Yablokov *et al.*, 1998; Ivashchenko *et al.*, 2006). Soviet factory fleets had killed virtually all whales they encountered, irrespective of size, age, or protected status. The scale of this deception was staggering; in all, the difference between the USSR's reported take and its actual catches was close to a 100,000 animals in the Southern Hemisphere alone (Table II). Of these, the humpback whale was the most heavily

impacted; reporting 2710 catches to the IWC, the Soviets had in fact taken more than 48,000. In the Northern Hemisphere, Soviet activities were on a smaller scale, but were nonetheless extremely damaging in some cases. The virtual disappearance of right whales in the eastern North Pacific in the 1960s was recently explained by revelations of Soviet catches of 372 whales from this already small stock between 1963 and 1967.

In retrospect, it is possible to see clues to this unfolding catastrophe. Beginning in 1959, at a time when there was increasing discussion of declining populations and the need for diminished quotas, the Soviets began adding a new factory ship each year to their Antarctic Fleet. These included the *Sovetskaya Ukraina*, the largest floating factory ever built, with an attendant fleet of 25 catcher vessels. In just two seasons (1959/1960 and 1960/1961), *Sovetskaya Ukraina* and a second factory, the *Slava*, killed almost 25,000 humpbacks, mainly from the high-latitude waters south of Australia and western Oceania. In addition, the intransigence of the USSR in its opposition to a proposed International Observer Scheme (IOS, to permit independent inspections of catches at sea) is now easy to interpret.

In the latter part of the 1960s the IWC finally began to respond to the increasing evidence that whale populations had been exploited well beyond MSY. Blue whales were protected in 1965, and quotas for fin and sei whales were reduced in the late 1960s in response to declines in catches. Nonetheless, any enforcement remained absent, and the Soviets secretly continued to kill whales irrespective of quota or protection until adoption and implementation of the IOS was finally accomplished in 1972.

However, it is now known that extensive falsification of catch data also occurred in Japan's coastal fisheries for Bryde's and sperm whales into the 1980s.

## IV. The Decline of Commercial Whaling

In the following decade, however, a sea change occurred at the IWC. The composition of the Commission slowly shifted as non-whaling nations joined, whereas others ceased whaling and developed instead into advocates for conservation. A whaling moratorium was proposed by the US and Mexico as early as 1974, but this and later proposals were rejected by the IWC until 1982. In that year, a radically changed Commission finally achieved the necessary votes to pass a 10-year moratorium. Predictably, Japan, Norway, and the Soviet Union all objected. The moratorium went into effect in 1986, with a zero catch quota for both pelagic and coastal whaling.

At this point in time Soviet whaling was coming to an end; with aging capital and the imminent dissolution of the USSR, the nation that had wreaked so much havoc on whale populations (a fact still unknown at this time) slowly removed itself from the business of commercial whaling. Japan, Norway, and Iceland, however, remained active, and in 1987 they effectively circumvented the moratorium by beginning "scientific" whaling. This act exploited a provision in the Convention (Article VIII) which allows member nations to issue themselves permits to conduct whaling for scientific research; it was originally included at a time when the only way in which any information could be gathered about whales was to kill them. As opponents of scientific whaling pointed out, the emergence in the 1970s of long-term studies of living whales (frequently based upon the identification of individual animals) provided a much better means to study the biology and behavior of cetaceans.

The stated reason for the moratorium was to permit world whale stocks to recover from the overexploitation to which they had been subject. In the meantime, the IWC's Scientific Committee was

### TABLE II
#### Reported vs. Actual Catches in the Southern Hemisphere by the USSR

| Species | Reported | Actual |
|---|---|---|
| Blue | 3,651 | 3,642 |
| Pygmy blue | 10 | 8,439 |
| Fin | 52,931 | 41,184 |
| Sei | 33,001 | 50,034 |
| Humpback | 2,710 | 48,477 |
| Bryde's | 19 | 1,418 |
| Minke | 17,079 | 14,002 |
| Right | 4 | 3,212 |
| Sperm | 74,834 | 89,493 |
| Other | 1,539 | 1,745 |
| Total | 185,778 | 261,646 |

*Note*: Some catches were actually over-reported; this was to disguise takes of protected species by over-reporting catches of species which could be legally hunted at the time.
*Source*: IWC, Yablokov *et al.* (1998).

charged with developing a new procedure for future management of stocks and setting of quotas. After considerable debate, the so-called Revised Management Procedure (RMP) was accepted by the Scientific Committee in 1994. However, the scheme by which the RMP would actually be implemented had still not been adopted, and there remains considerable resistance to the idea of the IWC endorsing a program which would effectively permit the resumption of commercial whaling. In 1994, Norway preempted such an agreement by resuming commercial whaling under "objection" to the moratorium, and used the RMP to set its own catch quotas for minke whales (*Balaenoptera acutorostrata*) in the northeastern North Atlantic. Norway has recently announced that it intends to unilaterally amend the RMP because it views the procedure as too conservative.

The reluctance of many nations to implement the RMP stems largely from lingering concerns regarding enforcement and transparency (i.e., independent observation and inspection) in whaling operations. These issues are considered by most nations to be essential components of a Revised Management Scheme (RMS), within which the RMP would operate. At present, however, there is little agreement on the general framework or specific inspection mechanisms of an RMS. The whaling nations maintain that adequate measures are now in place to ensure compliance with quotas set under the RMP. Opponents disagree, pointing to the egregious history of deception in modern whaling and noting more recent evidence that such deception continues to exist. In particular, considerable attention has been focused on the use of forensic genetics to test samples of whale meat in Japanese markets; although the only meat that should be found there is that from minke and Antarctic minke whales (*Balaenoptera bonaerensis*) taken in Japanese scientific catches, numerous other species have been detected (Baker and Palumbi, 1994). Although many of these animals certainly represent bycatch (incidental entrapment in fishing gear), their presence reinforces the fact that there is currently no means to adequately track whale products at every stage from catch to market. A DNA Registry of all animals taken in Japanese and Norwegian hunts has been implemented. However, confidence in this system has been compromised by the whaling nations' refusal to permit independent monitoring of the registry and their insistence that any discussion of trade in whale products is outside the competency of the IWC and lies instead within the purview of the Convention on International Trade in Endangered Species (CITES) or the World Trade Organization.

## V. Impact of Whaling on the Stocks of Whales

The impact of modern whaling on the world's stocks of whales has been varied (Clapham *et al.*, 1999). Many species appear to be recovering well in most parts of their range despite exploitation which in many cases may have reduced numbers by 90% or more from pre-exploitation levels (Best, 1993). Humpback whales, which were extensively over-hunted worldwide and which bore the brunt of illegal Soviet catches in the Southern Hemisphere, are showing strong rates of population growth in the North Atlantic, North Pacific and many areas of the Southern Ocean. Eastern gray whales number more than 20,000 animals and in 1994 were removed from the US list of endangered species. Although no reliable estimates of abundance exist, populations of fin and sei whales are assumed to be healthy in the Northern Hemisphere; the status of the extensively exploited Antarctic populations is less clear. Similarly, sperm whales are likely to be generally abundant, although in some areas apparently slow rates of population growth may be attributable to overexploitation of mature males in high latitudes, resulting in insufficient availability of mates

that are "acceptable" to adult females. Furthermore, female sperm whales were extensively exploited in the North Pacific by Soviet illegal whaling, and the status of this population is unknown. Although there is considerable controversy over Japanese and Norwegian minke whaling (and over the associated estimates of abundance), it is generally agreed that the Antarctic minke and some stocks of common minkes are abundant. However, serious concern has been expressed by the IWC Scientific Committee about scientific hunting and unregulated bycatch of minke whales from coastal stocks around Japan and Korea.

In contrast, other populations of whales appear to be struggling to recover from the indiscriminate exploitation to which they were subject. In some extreme cases, local populations appear to have been extirpated, with no recovery evident in the intervening years. Humpback and blue whales at South Georgia were commercially extinct by 1915 and 1936, respectively, and are rarely observed there today. Blue whales were wiped out from the coastal waters of Japan by about 1948, and no members of this species have been recorded there in recent years despite often extensive survey effort. Off Gibraltar, a population of fin whales was extirpated with remarkable rapidity between 1921 and 1927. The population of humpbacks that used the coastal waters of New Zealand as a migratory route crashed in 1960, the result of protracted shore whaling and (in particular) the Soviet catches of some 25,000 humpbacks in the feeding grounds to the south; some sightings have been reported off New Zealand in recent years, perhaps suggesting that a slow recovery is underway. In the North Atlantic, right whales were removed from much of their former range largely by historical whaling prior to 1880, and a remnant population in European waters was extirpated by Norwegians using modern techniques at the beginning of the twentieth century. The demise of at least one stock of whales can be attributed exclusively to pre-modern whaling: the bowhead was commercially extinct from Spitsbergen waters by 1900, and the species is only occasionally observed there today.

In all of these cases, whaling essentially extirpated a stock of whales; the lack of recovery over a timescale ranging from four decades in the case of New Zealand humpbacks to almost four centuries for right whales in some parts of the North Atlantic has important implications for modern management of whale populations. Although it is quite likely that the observed lack of recovery was at least partly due to a simultaneous overexploitation of adjacent populations (i.e., those that might otherwise have provided a source for repopulation), these localized extirpations reinforce the belief that management units should be designed carefully on often smaller spatial scales than has often been the case in the past.

Of those populations which survived, several are critically endangered. Right whales persist in low numbers in the western North Atlantic and the western North Pacific; the present size of the eastern North Pacific population is unknown, but is clearly precariously small following the immense damage done by the Soviets in the 1960s. In sharp contrast to the eastern ("California") gray whale, the outlook for the western North Pacific population of this species is bleak. Whaling on this small stock continued in Korean waters into the 1960s, and only a hundred or so animals may remain today. Furthermore, nothing is known of the location of the breeding grounds for this population; if it is reliant on coastal lagoons for calving (as is a major segment of the eastern stock), the impact of coastal development and other human activities may be severe. Among bowhead populations, that in the Bering/Beaufort/Chukchi Seas is recovering strongly despite continued exploitation by a (well-managed) Inuit hunt. In the eastern Arctic, some hundreds of bowheads remain in Canadian waters (principally Hudson Bay/Foxe Basin and Baffin Bay/Davis Strait), whereas the Spitsbergen stock may be functionally extinct.

Finally, blue whales have fared poorly almost everywhere; the only population which appears to be large and healthy is that which feeds off California in summer. Other blue whale stocks, including all of those in the Southern Ocean, remain small and highly endangered.

## VI. An Uncertain Future and the Rise of Scientific Whaling

It is not clear what the future holds for whaling. With the Moratorium still in place, Japan continues to substantially increase its self-assigned quotas for scientific whaling in both the North Pacific and Antarctic, taking minke, Antarctic minke, sei, sperm, and Bryde's whales. An expanded scientific whaling program in the Antarctic which began in 2005 doubled the catches of minke whales (to about 900 a year) and also added fin and humpback whales to the list of targeted species. The inclusion of the latter two species, considered threatened or endangered by many nations and conventions, has raised further international concern about the lack of control over scientific whaling programs.

All this has fueled arguments that scientific whaling has strayed far from its original purpose of research and is today being used to circumvent the Moratorium (Gales *et al.*, 2005; Clapham *et al.*, 2006). Indeed, since 1987 (the year after the Moratorium came into effect), Japan has killed more than 10,000 whales in its two scientific whaling programs, a figure which is almost 5 times the total number killed for research (2100 whales) by all other nations combined (including Japan) since 1952.

Ultimately, the future of whaling depends upon the outcome of developing geopolitics: put simply, on whether the emerging worldview of commercial whaling as an anachronism prevails or, if it does not, on whether whaling can learn the lessons of its grim past. For now, the outcome remains hung in the balance.

## *See Also the Following Articles*

Forensic Genetics ■ International Whaling Commission ■ Whaling, Illegal and Pirate

## *References*

Baker, C. S., and Palumbi, S. R. (1994). Which whales are hunted? A molecular genetic approach to monitoring whaling. *Science* **265**, 1538–1539.

Best, P. B. (1993). Increase rates in severely depleted stocks of baleen whales. *ICES J. Mar. Sci.* **50**, 169–186.

Clapham, P. J., Young, S. B., and Brownell, R. L., Jr. (1999). Baleen whales: Conservation issues and the status of the most endangered populations. *Mamm. Rev.* **29**, 35–60.

Clapham, P. J., Childerhouse, S., Gales, N., Rojas, L., Tillman, M., and Brownell, B. (2006). The whaling issue: Conservation, confusion and casuistry. *Mar. Policy*. **31**, 314–319.

Ellis, R. (1991). "Men and Whales." Alfred A. Knopf, New York, 542 pp.

Gales, N. J., Kasuya, T., Clapham, P. J., and Brownell, R. L., Jr. (2005). Japan's whaling plan under scrutiny: Useful science or unregulated commercial whaling? *Nature* **435**, 883–884.

Ivashchenko, Y. V., Clapham, P. J., Brownell, R. L. Jr. (eds.) (2006). Scientific Reports of Soviet Whaling Expeditions, 1955–1978. Alaska Fisheries Science Center, Seattle.

Tønnessen, J. N., and Johnsen, A. O. (1982). "The History of Modern Whaling." University of California Press, Berkeley, 798 pp..

Yablokov, A. V., Zemsky, V. A., Mikhalev, Y. A., Tormosov, V. V., and Berzin, A. A. (1998). Data on Soviet whaling in the Antarctic in 1947–1972 (population aspects). *Russ. J. Ecol.* **29**, 38–42.

# Whaling, Traditional

## RICHARD ELLIS

Beginning with the Basques around 1000 AD, there were many peoples that conducted whale hunts, but until the advent of industrial whaling in the twentieth century, there had never been a whale hunt as organized and systematic as the sperm whale fishery out of New England. Founded in the Massachusetts villages of Nantucket and New Bedford, the technology of sperm whale (*Physeter macrocephalus*) hunting—and the profits to be derived therefrom—spread to many locations in and around New England, including Albany, Long Island, New Jersey, Delaware, and Maine. The whale fishery—immortalized in Herman Melville's (1851) *Moby-Dick*—had an enormous effect on the economy of the recently founded USA, and the whaleships spread American culture and customs around the world. The focus here will be on the well-documented American whale fishery, but similar fisheries were operated by many other nations during the same period.

## I. Bowhead (*Balaena mysticetus*) Whaling

The gigantic nose of the sperm whale contains an enormous quantity of valuable oil, and this was the primary object of the sperm whalers. Bowheads are fatter than sperm whales, and their blubber thicker, but it was not so much for the blubber that the bowheads were hunted: their huge, arched mouths contained the longest baleen plates of any whale, and it was this that the whalers were after. It was used for the manufacture of skirt hoops, corset stays, horsewhips, and dozens of other necessities of nineteenth century America.

In July, 1848, Captain Thomas Welcome Roys, a sperm whaler out of Sag Harbor (New York) sailed through Bering Strait, a thousand miles farther north than any whaleship had ever gone in the Pacific, and came upon a thriving population of bowheads that had previously been known only to Eskimos. His crews took 11 bowheads in only 35 days during that eventful summer, and sailed for home with 18,000 barrels of oil, an accomplishment that normally took two or more seasons. According to Bockstoce (1986), the historian of western Arctic whaling, "Roys' cruise was not only the most important whaling discovery of the nineteenth century, it was also one of the most important events in the history of the Pacific …. More than 2700 whaling voyages were made into Arctic waters at a cost of more than 150 whaleships lost and the near extinction of the bowhead whale, as Roys' whales came to be called." Arctic bowhead whalers also changed the way in which whales were hunted; instead of hand-thrown harpoons, they introduced the efficiency of artillery, shooting at the whales first with shoulder-guns, and later killing them with bomb-lances.

Many New England whaleships had been captured or burned by the Confederate raider *Shenandoah* in 1865, and to revive the flagging industry, 39 ships sailed to the Arctic in April, 1971. By August, 32 of them were trapped in the ice off Point Barrow, Alaska. The seven remaining ships took on 1129 men and brought them back to Honolulu, signaling the beginning of the end of Arctic bowhead whaling.

## II. The Beginning of Sperm Whaling

The story of Christopher Hussey's accidental encounter with a school of sperm whales has been told so often that it probably no longer matters whether it really happened. In 1712, Captain Hussey

**W**

**Figure 1** *When a sperm whale* (Physeter macrocephalus) *beached itself at Katwijk in Holland in 1598, Hendrik Goltzius drew it for posterity.*
*Credit: New Bedford Whaling Museum.*

was cruising the Massachusetts coast [there were still right whales (*Eubalaena* spp.) to be caught at that time], when an unexpected storm blew him out to sea. When the clouds cleared, he saw the spouts of whales, but they were forward-angled blows, not the vertical, paired plumes of the right whales. Hussey managed to capture one of these unusual animals and towed it back to Nantucket. Instead of baleen plates, it had ivory teeth in its underslung lower jaw, and in its head was a great reservoir of clear amber oil, which solidified to wax when exposed to the air (Fig. 1).

The first industry practiced by the New England colonists was the export of beaver (*Castor* spp.) pelts and furs to England, but these commodities were quickly exhausted, and given the availability of the easily killed right whales close to their shores, they turned their attention from the forests to the sea. The earliest colonial whaling was practiced in the Indian manner; towers were erected along the shore to enable lookouts to watch for whales, and when one was sighted, the whalers took to the boats. As navigation improved, the whalers began to roam farther offshore, occasionally visiting the rich grounds of Georges Bank, and some vessels even ventured south into the vast oceanic river that would become known as the Gulf Stream. The Yankee whalers also headed north, toward the Gulf of St. Lawrence and the Grand Banks of Newfoundland. By the middle of the eighteenth century, there were some 50 ships bringing oil and bone to England and returning with such things as iron ore, hemp, cloth and other necessities

for the burgeoning new colony. By 1775, Nantucket had a fleet of 150 whalers, which ranged in size from 90 to 180 tons.

The beginning of the sperm whale fishery in 1712 did not automatically spare the remaining right whales. Although sperm oil was enormously desirable for lubrication and candle-making, the need for whalebone had not abated. In the middle of the eighteenth century, European women of fashion still required tight-laced corsets, so the New England whalers captured whatever whales they could find and processed them accordingly. Scammon (1874) tells us that "shore-whaling continued for over fifty years, but eventually it was abandoned, for the same reason that the Spitsbergen and Smeerenburg fisheries were—the scarcity of whales near the coast."

Regardless of the species being hunted, the primary product of the whale fishery was oil (Fig. 2). Earlier, however, the Dutch and English whalers of Spitsbergen and Greenland had concentrated on the whalebone, to the extent that they sometimes cut the slabs of baleen from the mouth of the whale and discarded the carcass. Much of the commerce of whaling was determined by fashion; by the amount of whalebone that would be required to girdle the ladies. In America, by contrast, there was no court, no royalty, and in the mid-eighteenth century Quaker colony of Nantucket, very little fancy dress.

As practiced in the Greenland fishery (and every other whaling operation until that time), the blubber of the whale was cut off in strips (a process known as "making off"), and packed directly into

**W**

**Figure 2** *Casks of oil line the New Bedford wharf. Photo by Old Dartmouth Historical Society—New Bedford Whaling Museum.*

casks for transport to the home port. Scoresby (1820) noted that "in the early ages of the fishery [it was] performed on shore; and even so recently as the middle of the last century, it was customary for ships to proceed into a harbor, and there remain so long as this process was going on." By the middle of the eighteenth century, an innovation that would change the nature of the entire industry had been introduced: iron caldrons set in a brick furnace enabled the whalers to render the oil from the blubber aboard the ship instead of on shore. This method seems to have evolved around the year 1750, but there is no individual whose name is associated with the invention. Although it is not possible to identify the father of the on-board tryworks, there were many mothers, all of whom had "necessity" as part of their names. Among the reasons for its introduction were the unpleasant odors associated with the onshore boiling of the blubber into oil, and the energetic protests of the people who lived downwind of a noisome blubberworks. (Even on board ship, blubber stored in casks tended to spoil quickly, and the stench was overpowering to the whalemen.) On the Spitsbergen and Greenland grounds, the cold climate kept the blubber from turning rancid until the ships could get back to a port, but in New England no such natural refrigeration existed, and the heat often "turned" the oil in the blubber before it could be processed. As long as the whales could be caught within sight—or at most, a couple of days' sail—of shore, the blubber could be casked and stowed, but when the whales became scarcer in the home waters, and longer voyages were required, some other method of processing and stowage was called for. Scoresby (who never employed on-board tryworks, even though

the idea existed during his whaling days) wrote that it was less efficient to carry home the blubber, since "blubber in bulk, notwithstanding every precaution ... generally loses much of its oil."

Now that sperm whales were being processed with some regularity, another change was taking place in the whaling industry. Earlier, all whale oil, casked as blubber or tried-out at sea, was considered usable for lighting and lubrication. (It was often referred to as "train oil," from the Dutch *traan* for "tear" or "drop"). It is a true fat, and impregnates every part of the whale, from the bones to the muscles, but most importantly it is found in the blubber. The fat of right whales and bowheads [and the occasional humpback (*Megaptera novaeangliae*)] provided the whale oil that was extensively used from the tenth century until the middle of the nineteenth for heating, lighting, manufacturing of soap and cosmetics, and lubrication of machinery.

Because spermaceti (from the head of the whale) is quite different from train oil, its processing and utilization were also different. Up to the middle of the nineteenth century, candles were the primary source of indoor light. They were usually made out of wax or tallow, and emitted a smelly black smoke as they burned. The head of the sperm whale contained the mysterious fluid which could be used to make a better kind of candle. This wax, which the whale maintains in a liquid form during its lifetime solidifies when exposed to air, and someone realized that it might be employed in the manufacture of candles. From sometime around 1750, it was used to manufacture smokeless, odorless candles, the best candles known before or since.

**W**

In addition to the liquid oil contained in the case, the sperm whale produced a spongy material also impregnated with oil, known to the whalemen as the "junk." The liquid oil in the case and the oil that was squeezed from the junk were collectively known as "head matter," and were used in the finest lamps and candles. The process of manufacture was a fairly complicated one. Upon delivery, the sludge-like substance was heated in a large copper vat and the impurities drawn off. It was left to congeal in casks and then bagged in woolen sacks, to be pressed later in a large screw press. The oil squeezed from the head matter was the highest quality, and was used in lamps. Further processing produced lower qualities of sperm oil, used for candles. Sperm oil candles were particularly popular in Africa and the Caribbean, but as articles of colonial manufacture they could not be imported into England. Both types of oil were considered superior to the train oil of the right whale and were priced accordingly. As the market developed for the finer qualities of oil, colonial entrepreneurs appeared. In 1751, one Benjamin Crabb of Rehoboth, Massachusetts, applied to the state house of representatives for a monopoly on the manufacture and sale of sperm oil candles, which was granted (Kugler, 1980).

As the right whales became scarcer, the tempo picked up for the sperm whale fishery. Starbuck (1878) called the period from 1750 to 1784 "the most eventful era to the whale fishery that it has ever passed through." New England whaleships were under constant threat of being captured by privateers (the various wars between France and England for control of the North American colonies were going on at the time), and ships that were not commandeered pursued the fishery as far from home as the Grand Banks and the Bahamas. There were also natural disasters attendant upon the nascent whaling industry: ships were lost to storms and occasionally to whales. For reasons of security and increased profitability, the small sloops that had been the mainstay of the fishery were being replaced by larger ships with tryworks aboard; now the whalers could pursue the sperm whale, "the haughty, elusive aristocrat of the high seas." By this time, the method of lowering boats for whales and fastening to them with harpoons attached to the whale-boats had evolved, and would remain the dominant practice for another century. It was this method of whaling—and the great sperm whale—that Melville would immortalize in *Moby-Dick* (Fig. 3). At this time, the fluctuations in the price of whale oil made for a most uneasy market. Good catches would overload the market and depress the price, while in a bad year, the scarcity of the oil would make it dearer.

The failure of the British whaling industry in the early decades of the nineteenth century left the field wide open to the New Englanders, and they were quick to capitalize on it. When the French and Indian War ended in 1763 and France conceded her claims to Canada, the New England whalers moved in. They sailed from Massachusetts and New York to the Gulf of St. Lawrence and the Strait of Belle Isle, and by 1776 they had discovered the whaling grounds off western Africa (Angola and Walvis Bay), the Falkland Islands, and the River Plate

grounds of the South Atlantic. In many of these regions, the whalers occasionally encountered a right whale, but the major object of the fishery by this time had become the sperm whale. These explorers in the name of oil were canvassing the world and perfecting their techniques, but instead of flourishing, they fell deeply into debt. The British government, still trying to support its own collapsing whaling industry, placed a duty on all oil and bone carried to England by colonials. Relations were becoming increasingly strained between the Crown and her rambunctious colony; in the years that followed, the infamous Stamp Act would be enacted and repealed; the Townshend Duties ditto, and finally, the Tea Act was passed in 1773, leading to the Boston Tea Party in the same year. (The East India Company, planning to sell its tea directly to America without having to first sell it to British merchants, shipped 1253 chests of tea from London to Boston on four whaleships: *Beaver, Dartmouth, Eleanor,* and *William and Anne.* It was this tea that the rebels, led by John Hancock and Samuel Adams, dumped into Boston Harbor). The next 2 years saw the first shots fired at Concord Bridge, and for the ensuing decade, most Americans became preoccupied with things other than whaling. (In April 1775, the same time that the "shot heard round the world" was fired, the ship *Amazon,* Captain Uriah Bunker, was discovering the whaling grounds known as the Brazil Banks, some 500 miles off that country.)

At approximately the same time that British whalers were depositing their cargoes of convicts and finding themselves in the middle of the rich whaling grounds of Australia and New Zealand, Yankee whalers were cruising almost everywhere in search of sperm whales. It was dangerous work, and sometimes the voyages seemed to last forever, but there were those who saw it as pleasurable and even romantic, the epitome of the wholesome life.

### III. Life Aboard a Whaler

In retrospect, however, the voyages were often less than romantic and the weather less than benign. There were indeed fresh breezes, tropical sun, and vast herds of sperm whales, or cachalots, but there was also the tedium of years of sailing (the record seems to be the 11 year voyage made by the ship *Nile,* out of New London: 1858–1869), as well as gales, blizzards, typhoons, hurricanes, mountainous seas, and howling winds. The crew's quarters were stinking holes; their food was cheap, coarse, and maddeningly monotonous; the work itself was dirty and dangerous. A voyage aboard a New England whaler was not a luxury cruise.

In the nineteenth century, the hierarchy of officers and men, so important to the successful operation of a whaling vessel, was rigidly observed, and nowhere was the distinction more evident than in their respective living quarters. The captain lived in relative luxury; the ship's officers had smaller cabins; the boatsteerers, the cooper, and the steward occupied the steerage, an irregular compartment fitted with

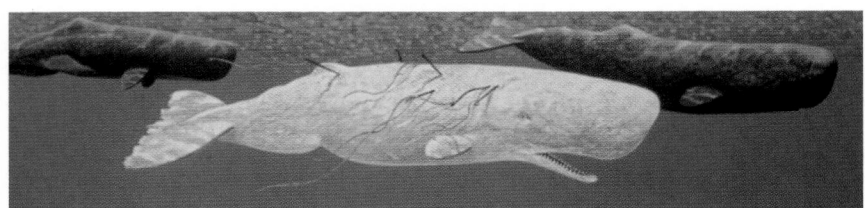

Figure 3  *In this sketch for the 100-foot-long mural in the New Bedford Whaling Museum, Moby Dick is shown with the "harpoons all twisted and wrenched within him."* Credit: Richard Ellis pinxit.

plain bunks. The crew was in the forward section just below the main deck, which followed the shape of the ship: it went from a fairly wide cross-section to a narrow, cramped, triangular warren, where the ship's timbers formed the walls, and the pounding of the waves formed the ambience. The lower portion of the foremast often kept the occupants of the fo'c's'le company, reducing even further their limited space, and the only light that entered this literal and figurative rat hole came from the hatchway cut in the deck for the purpose of giving access to the ladder that allowed the men to climb in and out of their quarters. When the weather turned foul, the hatch was closed, and there was no light but stubby candles, and no ventilation whatever. The number of men that occupied this wretched space often exceeded 20.

No whaleman was ever paid a wage, except in unusual circumstances. If, for instance, a full ship had to take on additional hands on the way home, their share of the profits would be zero (since they had not participated in the whaling), and they were paid a monthly wage. Ordinarily, each man, from the captain to the cabin boy, received a percentage of the profits—called a lay—at the end of the voyage.

The distribution differed from vessel to vessel—larger ships could carry more oil, and therefore the profits to the crew were likely to be proportionately higher—but while a successful voyage could be better for the captain and the officers, it meant precious little indeed to the foremast hands. (On an unsuccessful voyage, where the profits were low or nonexistent, the crew might receive nothing at all.) The captain might earn 1/8 or 1/10 of the net proceeds, whereas a mate could earn 1/15 and a harpooner 1/90. Ordinary seamen could hope at best for 1/150, and there are instances in the records where a green hand signed aboard for 1/350. What did this mean in terms of actual money? On board the *Addison*, First Mate Ebenezer Nickerson, whose lay was 1/18, earned $845. Robert Baxter, the second mate with 1/35, earned $554.83, and a boatsteerer named Narcisco Manuel, with 1/90, got $376.56. Compare these figures to those of the crew: John Martin, at 1/175, earned a total of $31.95, and Francis Finley, got $92.08. During six consecutive voyages totaling 1128 days at sea from 1845 to 1868, the average lay per voyage on the Salem whaler *James Maury* was $321.21, or about 26 cents a day. This compared unfavorably to wages then being paid to unskilled laborers ashore (an average of 90 cents a day), but landlubbers did not get to visit exotic Pacific islands where they might be eaten by cannibals or risk their lives fighting gigantic whales (Hohman, 1928).

Infrequently, the men were paid in the species of whaling; i.e., they received casks of oil that they were then able to sell at the prevailing prices in their port of disembarkation. The cooks often received an added benefit: in addition to their lays, they were permitted to save the grease (known as "slush") from their galleys and sell it to soap-makers ashore.

The whaleman's food and bunk space were generously provided without charge, but throughout the voyage he was docked for various items that he had to buy from the ship's stores. Additional items of clothing, tobacco, knives, needles, and even thread were charged to each man's account, and if he required spending money in a port of call, this too was deducted from the final reckoning. This was a period where the master's voice was law, and if a man needed a new shirt or a pair of boots, he could "either pay up or go naked." Although most of the whalemen signed aboard voluntarily, they usually did not know of the dangers and hardships that lay ahead of them, and the "profit-sharing" which at the outset sounded so attractive often deteriorated into an enforced "risk-sharing," which was invariably uncomfortable, inevitably dirty, and frequently dangerous.

Among the more unusual charges assessed to a whaleman was the cost of desertion. If a man jumped ship, his account included the cost of recapturing him, an expense that was obviously nullified if he remained at large. However, there were captains who rewarded the lookouts with bonuses for the sighting of whales. This exercise was glorified in *Moby-Dick*, where Ahab nails a gold doubloon to the mainmast and exhorts his crew: "Whosoever of ye raises me a whiteheaded whale with a wrinkled brow and a crooked jaw; whosoever of ye raises me that white-headed whale, with three holes punctured in his starboard fluke-look ye, whosoever of ye raises me that same white whale, he shall have this gold ounce, my boys!" The "Spanish ounce" that was offered to the crew was a 16-dollar gold piece.

On a 3- or 4-year voyage, a man might earn $100, but the items billed to him often exceeded this amount, so many hands returned to port not only with no spending money, but in debt. The only thing to do to work off this indebtedness was to sign on for another voyage, thus starting the insidious process all over again. If and when they made it—back to port, the whalemen were set upon by all sorts of "land sharks," eager to assist them in disposing of their wages by enticing them into taverns, brothels, and other iniquitous dens where they could make up for the pleasures they had been denied for the past several years. In an 1860 issue of *Harper's* magazine, an observer describes the arrival of the whalemen in New Bedford:

> A cart rattles by, loaded with recently discharged whalemen—a motley and a savage—looking crew, unkempt and unshaven, capped with the head-gear of various foreign climes and peoples-under the friendly guidance of a land shark, hastening to the sign of the "Mermaid," the "Whale," or the "Grampus," where, in drunkenness and debauchery, they may soonest get rid of their hard-earned wages, and in the shortest space of time arrive at that condition of poverty and disgust of shore life that must induce them to ship for another four years' cruise.

The system of wages aboard a whaler was obviously not conducive to enthusiasm or hard work. In response to the brutal discipline often administered by the captain, there was bound to be apathy, indifference and suspicion on the part of the foremast hands. There was also a profound class distinction between the officers and the men. Despite the abuses, hardships and low earnings which characterized the industry, however, the labor supply was somehow adequate to meet its needs. As Hohman (1928) put it, "The steady stream of men pouring into the forecastles proved sufficient to counteract the continuous labor leakage caused by death, illness, incapacity, discharge and desertion." It was possible (although uncommon) for a dedicated seaman to work his way up through the ranks, and there are instances where a green hand, or even a cabin boy, raised his lay from 1/150 to 1/15, and after perhaps 20 years at sea (in 4- or 5-year increments), a man might even rise to command a whaling vessel.

The *Benjamin Tucker*, a New Bedford whaler, brought back 73,707 gallons of whale oil, 5348 gallons of sperm oil, and 30,012 pounds of whalebone in a voyage that ended in 1851. At the prevailing prices—43 cents a gallon for whale oil, $1.25 a gallon for sperm oil, and 31 cents a pound for bone—the gross value of this cargo was $47,682.73. From this, $2,362.73 was variously deducted, leaving a net of $45,320 to be distributed. But before the profits were divided, the owners took a substantial percentage off the top to compensate for their initial outlay and also because these flinty New Englanders were not in the business for the thrill of the chase. In general, the owners took between 60% and 70% of the profits. On the 1805–1807 cruise of the *Lion*, the various oils yielded a total of $37,661.02. Of this, $24,252.74 went directly to the owners, leaving $13,045.53 to be divided among the captain and the crew for 2 years of work (Hohman, 1928).

**W**

During its heyday, New Bedford was the richest municipality per capita in America, and Melville described it as "perhaps the dearest place to live in all New England … nowhere in America will you find more patrician-like houses, [or] parks and gardens more opulent."

Of course, profits from the whaling industry were not restricted to the owners. They had to repair, refit, and re-provision their ships, which provided work and income for the shipwrights, chandlers, coopers, rope-makers, carpenters, and blacksmiths, and ready markets for the farmers and greengrocers. The entire township of New Bedford benefited from the outfitting and victualing of the armada of ships that annually departed her wharves, loaded with food, clothing and supplies, most of which were bought from local merchants.

The captain had his own cabin, with a proper bunk, a washstand, a table, and perhaps even a sofa and some extra chairs. The captain's quarters of the whaleship *Florida* "opened off the after cabin on the starboard side and extended nearly to the end of the forward cabin. A small room and a toilet room were aft of the stateroom. A large swinging bed was in the captain's cabin instead of the usual fixed berth." The gimballed bed was a special innovation designed by Captain Thomas Williams, because he was planning to bring Mrs. Williams along.

Occasionally a captain took his wife, and even more infrequently, he took his entire family. Captain Williams, of the ship *Florida* out of New Bedford, was accompanied by his wife for a voyage that lasted from September 1858 to October 1861. During the voyage, Eliza Azelia Williams gave birth to two children, who spent the first year of their lives at sea. She also kept a detailed journal of her adventures, which allows us a most unusual perspective of life aboard a whaleship. The voyage commenced on September 7, 1858, in New Bedford, and on January 12 of the next year, Mrs. Williams gave birth to a baby boy, whom they named William. (William's arrival might help to explain her seasickness early in the voyage, when she wrote, "it remains rugged and I remain Sea sick. I call it a gale, but my Husband laughs at me and tells me I have not seen a gale yet. If this is not one I know I do not want to see one.") On August 5, 1859, off the rugged coasts of Sakhalin in the Okhotsk Sea, the *Florida* spoke the *Eliza F. Mason*, and Mrs. Williams visited another "lady ship," where the captain had brought his wife and child, "a Lady Companion, and a little Girl that they brought from the Bay of Islands, New Zealand." On February 27, 1860, Mrs. Williams wrote, "We have had an addition to the *Florida*'s Crew in the form of our little daughter …."

It was US maritime law that a logbook be maintained by the mate or the first officer. (The term "logbook" originated with the practice of casting a log overboard affixed to the ship by a knotted line. The speed at which the line played out—measured in knots—determined the speed of the ship, and the daily records were originally kept in a book reserved for that purpose. Later, the term "logbook" was used to designate the book used for the keeping of all the ship's records). For the most part, logbooks and journals were kept by the masters. Although rarely educated in the classical sense, most of these men could read and write passably well, and their records have given us an enduring picture of life aboard a whaleship. Even though the maintenance of a logbook was mandatory, it obviously served the whalers particularly well, since the appearance of whales at a known latitude and longitude in one season might enable the whalers to predict their reappearance at the same location the following year and thereby avoid aimless wandering.

The more mundane entries consisted of the ship's position, the number of whales caught, and illness and injury aboard ship, but additional dramatic possibilities were vast. In his introduction to the catalog of the logbook collection of Paul Nicholson, Sherman (1965) listed "castaways, mutinies, desertions, floggings, women stowaways, drunkenness, illicit shore leave experiences, scurvy, fever, collisions, fire at sea, stove boats, drownings, hurricanes, earthquakes, tidal waves, shipwrecks, ships struck by lightning, men falling from the masthead, hostile natives, barratry, brutal skippers, escape from Confederate raiders, hard luck voyages and ships crushed by ice." That is not to say that all logbooks read like *Moby-Dick*; dramatic events occurred infrequently, and most of the daily entries—when the ship was not engaged in killing whales—consisted of a remark on the wind direction, the location, and whatever else the keeper of the logbook deemed pertinent.

It is not surprising that few of the foremast hands kept records; their quarters were not conducive to the literary life, and besides, many of them could not write. Francis Olmstead (1841) could. Of the literary aspirations of his fo'c's'le companions, he wrote

> The forecastle of the *North America* is much larger than those of most ships of her tonnage, and is scrubbed out regularly every morning. There is a table and a lamp, so that the men have conveniences for reading and writing if they choose to avail themselves of them; and many of them are practicing writing every day or learning how to write …. When not otherwise occupied, they draw books from the library in the cabin and read; or if they do not know how, get someone to teach them. We have a good library on board, consisting of about two hundred volumes ….

J. Ross Browne, a journalist who shipped aboard the New Bedford whaler *Bruce* in 1842, kept a journal of his experiences that was published, with major revisions, as *Etchings of a Whaling Cruise* in 1846. Browne wanted to do for whaling what Richard Henry Dana had done for merchant sailing in 1840, i.e., exaggerate the problems so that necessary changes would be implemented. Although his account may contain a certain amount of propaganda in the form of negative commentary, he was aboard a whaler for more than a year, and because he is regarded as a reporter and not a writer of fiction, much of the material contained in his book can be taken as fact. Here is Browne's description of the place in which he lived:

> The forecastle was black and slimy with filth, very small and hot as an oven. It was filled with a compound of foul air, smoke, sea-chests, soap-kegs, greasy pans, tainted meat, Portuguese ruffians and sea-sick Americans …. In wet weather, when most of the hands were below, cursing, smoking, singing and spinning yarns, it was a perfect Bedlam. Think of three or four Portuguese, a couple of Irishmen, and five or six tough Americans, in a hole about sixteen feet wide, and as many perhaps, from the bulkheads to the fore-peak; so low that a full-grown person could not stand upright in it, and so wedged with rubbish as to leave scarcely room for a foothold. It contained twelve small berths, and with fourteen chests in the little area around the ladder, seldom admitted of being cleaned. In warm weather it was insufferably close. It would seem like an exaggeration to say, that I have seen Kentucky pig-sties not half so filthy, and in every respect preferable to this miserable hole; such, however, is the fact.

Rats were more numerous on whaleships than on any other vessels, probably because of the profusion of blood and oil that soaked the decks, despite the regular scrubbings. They were more than any ship's cat could cope with, and then as now, there was nothing that could cope with cockroaches. They were endemic aboard the whalers, and for many seamen, the roaches were a more predominant aspect of a whaling voyage than whales. Olmstead wrote that they made "a noise like a flush of quails among the dry leaves of the forest." They are extremely voracious, and destroy almost everything they can find: their teeth are so sharp, the sailors say, that they will eat the edge off a razor."

In *Nimrod of the Sea* William Davis (1874) described roaches as serving a useful purpose: "His chief recommendation is his insane pursuit of the flea … ," but then goes on, "it is a horrible experience to awaken at night, in a climate so warm that a finger-ring is the utmost cover you can endure, with the wretched sensation of an army of cockroaches climbing up both legs in search of some Spanish unfortunate! It reminds me of how many times I have placed my tin plate in the overhead nettings of the forecastle, with a liberal lump of duff reserved from dinner, and on taking it down at supper, have found it scraped clean by the same guerrillas. They leave no food alone, and have a nasty odor, which hot water will scarcely remove. But one becomes philosophical at sea in matters of food."

The crew's rations aboard a whaleship ranged from bad to disgusting, but, Browne says, "a good appetite makes almost any kind of food palatable." He describes the usual fare on board the *Bruce* (which he has, for culinary and other reasons, named the *Styx*): "I had seen the time when my fastidious taste revolted at a piece of good wholesome bread without butter, and many a time I had lost a meal by discovering a fly on my plate. I was now glad enough to get a hard biscuit and a piece of greasy pork; and it did not at all affect my appetite to see the mangled bodies of diverse well-fed cockroaches in my molasses; indeed, I sometimes thought they gave it a rich flavor." Fresh vegetables were taken on at the outset of a voyage, and often picked up when the vessel put in for provisions, but unless they were used quickly, they rotted. (By Browne's time, the causes of scurvy were known, but if the vegetables were used up and the ship was cruising somewhere off the Aleutian Islands, there was not much anyone could do to prevent the dread disease.) Because of their inability to store much water—and to prevent it from spoiling—the whalers hardly ever drank it. (Scammon tells the story of one captain, who, to preserve the dwindling water supply, had the drinking cup hung from the royal-mast head, requiring any man who wanted a drink to climb all the way up after the cup.) They drank "longlick," a mixture of tea, coffee, and molasses, and if the cook was imaginative, he prepared something known as "lobscouse" (or simply "scouse"), which was a hash made of hard biscuits that had been soaked in the greasy water left over after boiling the salted meat. The mainstay of the whaler's diet was salted meat, which was supposed to be pork or beef, but was occasionally horse. In *Omoo* Melville (1847) described the meat on board a whaleship:

When opened, the barrels of pork looked as if preserved in iron rust, and diffused an odor like a stale ragout. The beef was worse yet; a mahogany-colored fibrous substance, so tough and tasteless, that I almost believed the cook's story of a horse's hoof with the shoe on having been fished up out of the pickle of one of the casks.

Nordhoff (1856) described the duff made by a certain cook as "that potent breeder of heartburns, indigestion, and dyspepsia … the very acme of indigestibility," and Ben-Ezra Ely (1849) wrote, "… no swine that gleans the gutters ever subsisted on viler meat and bread than did our crew." The rare opportunity to eat something fresh was a blessing. The cook prepared sea birds, whatever fish they could catch, turtles, dolphins [off the African coast, Nordhoff describes the harpooning and subsequent eating of a hippopotamus(*Hippopotamus amphibius*)], and since they were engaged in the capture of 50- or

60-ton mammals whose carcasses they would otherwise leave for the sharks, they often ate the meat of the whales. On the eating of various parts of the whale, usually during the trying-out, Browne wrote:

About the middle of the watch they get up the bread kid [a kid was a wooden tub] and, after dipping a few biscuits in salt water, heave them into a strainer, and boil them in oil. It is difficult to form any idea of the luxury of this delicious mode of cooking on a long night-watch. Sometimes, when on friendly terms with the steward, they make fritters of the brains of the whale mixed with flour and cook them in the oil. These are considered a most sumptuous delicacy. Certain portions of the whale's flesh are also eaten with relish, though, to my thinking not a very great luxury being coarse and strong ….

It was a different world above decks. On December 28, 1856, the crew of the New Bedford whaler *Addison* caught a porpoise, and Mary Chipman Lawrence (the captain's wife) wrote in her journal, "The meat looks very much like beef. The oil is contained in the skin, which they will boil out tomorrow. Had some of the meat fried for dinner and some made into sausage cakes for supper. They are as nice as pork sausages." If a further demonstration of the disparity between the fare of the men and that of the officers is required, here is Mrs. Lawrence's description of Christmas dinner for that same year: "roast chickens, stuffed potatoes, turnips, onions, stewed cranberries, pickled beets and cucumbers, and a plum duff. For tea I had a tin of preserved grape opened and cut a loaf of fruitcake."

Unlike their British counterparts, American whalers rarely carried any sort of medical man. It commonly fell to the captain to cope with whatever illness or accident befell his crew, and given the master's experience, it was considerably safer to remain healthy. For internal maladies, whaleships were often equipped with medicine chests, which contained various potions and a manual for their dispensation. (Stories were told of masters who, having run out of medicament Number 12, simply administered equal amounts of Numbers 5 and 7.)

Physical injuries were not uncommon, considering the number of sharp-edged tools, whistling whale lines, and hostile natives—not to mention shipboard arguments between men who were almost always armed with knives. Here again, the master served in the role of surgeon, with the same amount of training as he had as apothecary. In *Nimrod of the Sea* (1874), Davis tells the gory tale of a whaleman who was yanked from his boat by a kinked line, and dragged some 125 fathoms from the boat. When he was finally picked up, "it was found that a portion of the hand, including four fingers, had been torn away, and the foot sawed through at the ankle, leaving only the great tendon and the heel suspended to the lacerated stump." Equipped with "his carving knife, carpenter's saw and a fish-hook," the captain "amputated the leg and dressed the hand as best he could."

### IV. Whaleships and Whaleboats

As whaling voyages increased in distance and duration, it became expedient to enlarge the ships. In the early days of the fishery (around 1820), the ships averaged around 280-ton burthen, but within two decades, 400-ton vessels were not uncommon. The move toward bigger whaleships contributed to the decline of Nantucket whaling because there was a prominent sandbar across the harbor, and only the smaller, shallower-draft ships could enter. New Bedford, with its excellent harbor facilities, took up the slack.

All whaling vessels were ships—as opposed to *boats*, which were the smaller vessels that the whalers rowed after their quarry. The literature

W

is replete, however, with references to ships, brigs, brigantines, barks, barkentines, and schooners. These differentiations have to do with the rigging of the masts, and not with the number of masts, although a three-masted, square-rigged vessel was always known as a ship. If the aftermost mast was rigged fore-and-aft, with the sail slung between a gaff and a boom, the vessel was known as a *bark*, the commonest plan, because fewer hands were required to handle the sails, and thus there were more men available for the boats. There were further variations, including the *brig*, where the upper courses of the aftermost mast were rigged with squaresails, but there was also a fore-and-aft sail known as a "spanker." A *barkentine* was square-rigged only on the foremast; the rest fore-and-aft, and a *brigantine* had only two masts, the foremast square-rigged and the mainmast fore-and-aft. A *schooner* had two or more masts, rigged fore-and-aft.

Whaleships differed from merchantmen of the time in that they usually carried less sail. More canvas meant more men aloft, and the whalers needed as many hands as possible for the boats. One further characteristic of the whaler was the presence of masthead hoops, in which the lookouts stood during the daylight hours to watch for whales.

Square-rigged ships, which gave their name to an era of sailing, ran powerfully before the wind, but were not particularly handy in head- or cross-winds. The whalers did not have to perform any smart sailing maneuvers, nor did they have to sail with great speed. All they had to do was get from one location to another and then lower the boats after the whales. Because of the determined, plodding nature of their craft, the masters rarely sailed at night, preferring instead to furl their sails and wait till dawn before continuing.

It was during the heyday of New England whaling, from 1830 to 1860, that the fabulous clipper ships reached the zenith of sailing-ship design, with their graceful lines, sharply raked bows, and opulence of canvas. In marked contrast to these ocean-going greyhounds, the whalers were sturdy, bluff-bowed, flat-bottomed sailers, designed more for durability and storage than for speed. (The *Lagoda* sailed for 50 years, and the all-time record-holder, the *Charles W. Morgan*, sailed for more than 80 years, and earned over a million dollars for her owners. The *Lagoda* was copied at half-scale for the New Bedford Whaling Museum, and the *Morgan*, the last of her kind, is now the proud centerpiece of Mystic Seaport in Connecticut.)

A typical whaler was 100–150 ft long, and especially broad in the beam to accommodate the fixtures of whaling: heavy brick tryworks on deck, iron caldrons, cooling tanks, davits for the boats, and of course, the space required to perform the trying-out of the whale. Ordinary seamen, whose voyages did not take 4 or 5 years, belittled the whale-ships as "built by the mile and cut off in lengths as you want 'em." They were usually painted black, and had mock gun ports painted along the sides, supposedly as a deterrent to pirates or hostile savages.

The naval historian Albert Cook Church (1938) wrote: "Whaleships differed materially from any other type of merchant ship or clipper in model and equipment, and in fact, both sides of a whaleship differed from each other above the waterline." The larger ships were equipped with four boats, one on the starboard quarter, and three on the port (also known as the "larboard") side. This allowed the cutting stages, which were always on the starboard, to be lowered without interference from davits.

When a whale or a group of whales was sighted, the lookout shouted "She blows!" or "Blows!" and when the captain had ascertained "where away," the boats were lowered, and the chase began. All the boats might or might not be lowered, depending upon the number of whales sighted. If only a single whale was seen, the captain might designate one boat to chase it. The starboard boat was reserved for the captain (or the fourth mate, if the captain chose to stay aboard ship during the hunt); the larboard, waist, and bow boats were for the first, second, and third mates, respectively. Each boat contained a regular crew, consisting of five oarsmen, and a boatsteerer/harpooner. Whoever was in command of the whaleboat pulled the steering oar and gave the orders. The boats were double-enders; in case they got turned around in the frenzy of the hunt, they would be able to maneuver, and they were among the most graceful and utilitarian boats ever designed.

All the requisite equipment would be carefully stowed aboard the whaleboats, from the line, which was carefully coiled in a tub so it could be let out rapidly, to the knife that might be required to cut it if a man got his leg entangled. In addition to the six adult men who would be required to man the boat, Scammon lists the contents of a fully equipped whaleboat:

One mast and one yard, one to three sails, five pulling oars, one steering oar, five paddles, three rowlocks, five harpoons, one or two line-tubs, three hand lances, three shortwarps, one boat-spade, three lance-warps, one boat-warp, one boat-hatchet, two boat knives, one boat-waif, one boatcompass, one boat-hook, one drag, one grapnel, one boat-anchor, one sweeping-line, lead, buoy, etc., one boat-keg, one boat-bucket, one piggin, one lantern-keg (containing flint, steel, box of tinder, lantern, candles, bread, tobacco, and pipes), one boat-crotch, one tub-oar crotch, half a dozen chock pins, a roll of canvas, a paper of tacks, two nippers, to which may he added a bomb-gun and four bomb-lances; in all, forty eight articles, and at least eighty-two pieces.

## V. Killing and Processing the Whale

The lowering of the boats took place as the ship was underway; the captain did not come about for the comfort or convenience of his crews. Often in high seas, the graceful whaleboats took off after the whales with the men facing the stern; the boatsteerer was the only man who could see the whales. When they had come within range, the harpooner threw the harpoon. It consisted of a wooden shaft, some 6 ft in length, with a forged iron head. The earliest harpoons had simple fluted arrowhead-shaped heads, but as the fishery developed, more sophisticated designs were introduced. Although the two-flued iron pierced the blubber effectively, its razor edges would occasionally pull out as smoothly as they went in. This led to the introduction of the single-flued iron which held much better. Harpooners and blacksmiths had plenty of time, on board the whalers and in port, to work on harpoon design, and all sorts of elaborate heads with toggles, barbs, and swivels were tried. The most successful of these designs was the double-barbed "Temple" iron, invented in 1848 by a New Bedford blacksmith named Lewis Temple. A graceful, practical device, the Temple iron consisted of a pointed head that was held in the forward position by a wooden shear-pin that broke off when withdrawal forces were applied. This rotated the head 90° in the flesh of the whale, forming a T-shaped device that would not pull out, because the flattened surfaces were pulling against the meat or blubber. The iron was fastened to the shaft of the harpoon by a line which was bent to the heavy manila line. The line, which Melville calls the "magical, sometimes horrible whale line," was originally fashioned of hemp, but was later superseded by manila rope, which was stronger and more elastic. "Hemp is a dusky, dark fellow," Melville wrote, "a sort of Indian, but Manilla is as a golden haired Circassian to behold."

Even though tradition demanded that the harpoon and the lance be thrown separately, some creative whalemen tried to design an iron that would fasten to and kill the whale simultaneously. A

Scottish toxicologist named Robert Christson invented a poison-headed harpoon, equipped with glass cylinders containing prussic acid, one drop of which is lethal enough to kill a man. There is no evidence that prussic-acid harpoons were used in the American fishery, but they were carried on some vessels. The likelihood is that the American harpooners felt that they had enough problems killing the whale without worrying about killing themselves.

If the iron was well placed—the ideal spot was in the flank, forward of the hump—the boat was fast to the whale, and the injured animal took off. Sometimes the whale sounded, taking out the line at such speed that the line smoked as it ran out, and the loggerhead had to be doused with water to keep it from bursting into flame. More often the whale swam at the surface, towing the boat through the waves at a violent clip. Sperm whales are prodigious divers, and no boat could hold enough line for a dive that could be measured in miles. If the whale sounded, another 200-fathom line might be bent to the first, and then another. Eventually, the wounded whale had to surface to breathe.

The lance, also know as the "killing iron," was plunged into the "life" of the whale, a vital artery, the lungs, or the heart. The killing iron consisted of a wooden shaft like that of the harpoon, with a scalpel-sharp head. It was not thrown, but rather stabbed repeatedly into the body of the whale. Melville describes the death-throes of a whale:

> The red tide now poured from all sides, off the monster like brooks down a hill. His tormented body rolled not in brine but in blood, which bubbled and seethed in furlongs behind in their wake. The slanting sun playing upon this crimson pond in the sea sent back its reflection into every face, so that they all glowed to each other like red men … Stubb slowly churned his long sharp lance into the fish and kept it there, carefully churning and churning, as if cautiously seeking to find some gold watch that the whale might have swallowed, and which he was fearful of breaking ere he could hook it out.

The victory did not always go to the whalers. Sperm whales are immensely powerful creatures, and do not take kindly to being stabbed with spears. The most frequent problem occurred when the whale took it into its 20-pound brain to retaliate. A 30-ft whaleboat was no match for an enraged, wounded, 60-ton whale, and the harpooned animal might rise up from the depths and grab the boat in its massive jaws, splintering it into so many matchsticks (Fig. 4). Both ends of a wounded whale are lethal; the triangular flukes, which might measure 20 ft across, could function as a formidable weapon, crashing down upon the whaleboat and dumping the men into the sea. Other perils faced the whalemen, where the whistling line might take a turn around a leg or an arm, surgically severing it, or yanking the man into the water. Even if the boat was not destroyed, it might be upended and its occupants dumped into the ocean. Many of them could not swim, so such a plunge often spelled death.

But more often than not, the world's deadliest predator won the battle, and then faced the problem of bringing whale and ship together. If the conquering whaleboat was downwind of the ship, it was a relatively simple matter to sail the ship to the carcass, but if less propitious conditions prevailed, the tired whalemen might have to tow the whale back to the ship, often for miles. And then, after an exhausting chase and a laborious haul with a 50-ton deadweight in tow, the real work began. What had been a free-swimming, powerful sea mammal was effectively reduced to a disparate assortment of its parts, the reduction accomplished by literally tearing it apart.

**Figure 4** *A painting by Clifford Ashley, the American painter/author/whaleman, showing what happened when the men in a whaleboat got too close to a sperm whale they had harpooned. Credit: New Bedford Whaling Museum.*

As in virtually every other aspect of New England whaling, the cutting-in process was described better by Melville than anybody else. (In the Yankee whale fishery, the process of removing the whale from his outer integuments was known as "cutting-in," and the rendering of the blubber into oil was known as "trying-out." In the English fishery, these operations were known respectively as "flensing" and "making off"). In *Moby-Dick* there is one chapter devoted to the actual process, and several more to the byproducts, including the "blanket," the "funeral," and the "sphynx"—the last referring to the head of the whale after the body and blubber have been separated from it.

The whale was made fast to the ship by lashing heavy chains through its head and around its flukes. The first part of the whale to be brought aboard was the lower jaw, ripped from the head and laid aside to be dealt with later. Then the whale was decapitated, and if it was a small one, the head was brought aboard. But the head of a large whale, often one-third of its 60-ft, 60-ton body, could not be brought on deck (Melville wrote that "even by the immense tackles of the whaler, this were as vain a thing as to attempt weighing a Dutch barn in jeweler's scales"), and had to be processed in the water. The "head matter" was saved for last, however, because the carcass of the whale alongside the ship was threatening to the ship by its weight, and the longer it remained unprocessed, the longer the sharks could wreak havoc on the very outer layer of blubber that was of so much interest to the whalers.

By the use of a complicated series of tackles—described by Melville as "ponderous things comprising a cluster of blocks generally painted green, and which no single man can possibly lift—the cutting stages were lowered, and the process of removing the blubber commenced. Sitting or standing on the lowered cutting stages, men with razor-sharp cutting spades began to slice into the whale's rubbery outer covering. A massive iron hook was inserted in the first piece to come off, and this was hoisted high into the air while the men on the scaffold sliced the blubber. The whale was rotated in the water, and its blubber "stripped off front the body precisely as an orange is sometimes stripped by spiralizing it." The power for this peeling and dismemberment came from the strong backs of the whalemen, who turned the windlass located forward of the foremast.

**W**

As the thick spiral of blubber was peeled from the whale, it was cut into sections approximately 15 ft long and a ton in weight (the "blanket pieces"). These were dropped through a hatch into the blubber room, where they were stored until the carcass of the whale was completely stripped. (With the removal of the blubber and the head, the remainder of the carcass was left for tile sharks.) Workers in the dark, bloody blubber room further reduced the blanket pieces to smaller, more manageable "horse pieces," which were then sliced into "Bible leaves," with cuts almost to the skin making them resemble the splayed pages of a thick-leaved book. (It was believed that the opening of the blubber into "pages" made the oil more accessible). The Bible leaves were then forked back up through the forehatch to the men who would place them in the trypots.

Although the trypot fires were usually started with wood, the unmelted skin of the whale made a wonderful fuel, and the whale was therefore cooked in a fire of its own kindling. As the oil was separated from the blubber, it was carefully ladled into a copper cooling tank, where it rested before being casked. Aside from the obvious danger of a fire spreading, the process was—like almost every aspect of whaling—hard, messy, and dirty. Oil and blood covered the decks and the people, and the smell was often intolerable. J. Ross Browne called the trying-out process "the most stirring part of the whaling business, and certainly the most disagreeable." He described the nighttime scene aboard the "*Styx*":

> Dense clouds of lurid smoke are curling up to the tops, shrouding the rigging from the view. The oil is hissing in the trypots. Half a dozen of the crew are sitting on the windlass, their rough, weather-beaten faces shining in the red glare of the fires, all clothed in greasy duck, and forming about as savage a looking group as ever was sketched by the pencil of Salvator Rosa. The cooper and one of the mates are raking up the fires with long bars of wood or iron. The decks, bulwarks, railing, tryworks, and windlass are covered with oil and slime of black-skin, glistering with the red glare of the try-works. Slowly and doggedly the vessel is pitching her way through the rough seas, looking as if enveloped in flames.

At the end of this description, he wrote, "Of the unpleasant effects of the smoke I scarcely know how any idea can be formed, unless the curious inquirer choose to hold his nose over the smoking wick of a sperm oil lamp, and fancy the disagreeable experiment magnified a 100,000 fold. Such is the romance of life in the whale fishery."

One of the least romantic aspects of the whale fishery was the prospect of fire. Oil-soaked wooden ships upon whose decks fires are being encouraged do not lend themselves to a feeling of security. Care was taken to avoid conflagrations—water was pumped over the decks to keep the planks wet and cool—but occasionally the sails or rigging were ignited by flying sparks, and sometimes the ships burned to the waterline.

When the oil had cooled, it was ladled into the casks that had been assembled by the cooper. Each barrel held 302 gallons, and the figures for the fishery were almost always recorded in barrels. Starbuck's (1878) *History of the Whale Fishery*, which contains the records of every American whaling ship, from every American whaling port, "from its earliest inception to 1876" (insofar as these records were known), lists the result of every whaling voyage in sperm oil (barrels), whale oil (barrels), and whalebone (pounds).

A large female sperm whale might yield 35 barrels of oil, whereas the largest bulls gave up 75–90. As with the sometimes questionable

lengths of large bulls, where there were reports of 90 footers. Ashley (1938) wrote, "If these whalemen's records are accurate, it would appear that the 100-ft Sperm Whale is not an impossibility." Because the reports were invariably made by men whose reputation would be enhanced by overstating the yield of individual whales, many of the whales in the 100–150 barrel range must be questioned.[7]

The amount of oil that could be taken and stored was enormous, but it did not necessarily reflect the success of a voyage. The profits of a voyage could only be calculated when the ship reached port and sold the oil and bone at the prevailing prices. A 31-gallon cask was about 5 ft high and 4 ft in diameter at its bulging middle (Fig. 2). On her maiden whaling voyage, which lasted from October 1841 to September 1843, the *Lagoda* brought home 600 barrels of sperm oil, 2700 barrels of whale oil, and 17,000 pounds of baleen. ("Sperm oil" was the stuff that was ladled out of the whale's "case," and was of a finer quality than "whale oil," which was rendered out of the blubber. Although they were not averse to taking an occasional right whale or humpback, most of the whales hunted by the Yankees were sperm whales). The *Lagoda* was 108 feet long, with a beam of 27 ft. Hunting concluded when there was no more room for the storage of oil, but the whalers sometimes put into port, offloaded some of their greasy cargo, and set out again for the whaling grounds. Some of these sweaty, iron-bound vats were probably stored in the blubber room, but most were stored in the hold.

It was the mysterious "head matter" of the sperm whale that made it the primary object of this globe-girdling enterprise. Other whales were encased in blubber, and some of them had the long "finnes" that could be converted into milady's bodices. But the spermaceti was the *ne plus ultra* of this business, the pot of liquid gold that attracted the whalers to the Azores and the Galapagos, to Zanzibar and the Japan Grounds, to Kamchatka and the Okhotsk Sea. The stuff is as poorly understood today as it was when some early beachcomber presumed that this vast reservoir in the whale's nose was its seminal fluid. Whatever its purpose to the whale (and it certainly is not its seminal fluid), the amber wax that hardened white as it was exposed to air was worth risking life and limb—and sometimes boat and ship—to the whaler. Kept free from contamination by other oils, sperm oil was worth from 3 to 5 times as much as whale oil. In *Nimrod of the Sea*, Davis recorded a whale that yielded 27 barrels of spermaceti from the case, and Ashley's research indicated that the largest bulls gave up something on the order of 30 barrels. At 31.5 gallons per barrel, that works out to 945 gallons of the mysterious liquid wax in the nose of a single whale.

To extract the spermaceti from the head, a much more direct method was employed than the multi-step process of turning blubber into oil. Since the spermaceti already *was* oil, the whalers only had to remove it from the whale and cask it. A hole was cut in the outer fabric of the whale and a man lowered a bucket into it on a long pole, then turned it over to another man on deck who would

---

[7]If only the whaler's stories remained, we would have no way of verifying the size of the largest whales. There is something that they leave behind, however, and Ashley proposes a novel argument for the existence of gigantic bull sperm whales: he examines a particularly large pair of teeth, over 11 in. long, and suggests that "in the days before the Sperm Whale herds were depleted, there must have been exceptional whales, either larger or older than are found today." Mitchell (1983) finds this argument "well taken, but not conclusive," but a look at these teeth, which are on display in the New Bedford Whaling Museum, certainly gives one cause to wonder.

empty the bucket into a waiting tub—or as Melville put it, "Tashtego downward guides the bucket into the Tun, till it entirely disappears; then giving the word to the seamen at the whip, up comes the bucket again, all bubbling like a dairy-maid's pail of new milk."

When the oil had all been casked and the casks stowed, the decks were scrubbed down with lye, which had been leached from the cinders and ashes of the tryworks, and the oily, smoky clothes of the whalemen were also scrubbed down, but the pernicious odor of smoked blubber could never really be removed, and until they could exchange their work clothes for new garments, the whalemen usually smelled like disused tryworks.

## VI. The "Romance" of Whaling

In the nineteenth century, when so much of the world was still unexplored, the whalemen faced even greater hazards than an occasional angry whale. The US Exploring Expedition under Wilkes had visited many of the island groups in the Central Pacific and found that some of the stories of hostile savages, often cannibals, were true. The Fiji Islands were known—more or less accurately—as The Cannibal Isles, and whenever possible the whalers avoided them. In 1835, the whaler *Awashonks*, out of Falmouth, was attacked by the natives of Namarik in the Marshall Islands, and the captain, the first and second mates, and four crew members were killed before an enterprising whaleman dynamited the deck where the would-be conquerors were standing, and the ship was retaken. The *Syren* was recaptured from Palauan natives only after a box of tacks was scattered on deck, driving the barefoot raiders howling overboard.

The need for fresh vegetables and water often outweighed the threat of being attacked, and even though many of the captains knew or had heard stories of cannibalism and "cutoffs" (a whaleship captured by natives and its crew massacred), they could not resist the temptations of cheap provisioning. Whaleships hardly ever carried money; the very same "slop-chest" that provided the foremast hands with their replacement items of clothing also served as a trading bank. "Recruiting ship" was the term used to describe the acquisition of provisions, and the captain would trade cotton cloth, powder, tobacco, knives, and beads for fresh food and water. It was sometimes too much for the parsimonious New Englanders to resist: for a couple of pounds of tobacco or some rusted iron hoops, they could trade for pigs, coconuts, water, wood, and women.

The quality of life aboard a whaler was hardly luxurious, but it was often better than life on the farm. Indeed, many whalemen deserted on the islands, not because they were unduly harassed or flogged, but because life on a lush, green island, with free food and even freer women, was an economic and sociological step upward. By the middle of the nineteenth century, there may have been as many as 3000 deserters from whaleships scattered throughout the coral archipelagoes of Micronesia and Polynesia.

Hawaii, Tahiti, and the Marquesas are picture-book "South Sea Islands," with tall volcanic mountains, tumbling waterfalls, broad white beaches, and swaying palms. When Melville (1846) jumped ship in the Marquesas and subsequently described his experiences in *Typee* he was responsible for many misinterpretations of life on a tropical island. Not all island groups boasted plentiful paw-paws and willing wahines. Many of these tiny specks in the Pacific were mercilessly unforgiving coral atolls; low rings of sand. All along the Equator in the Pacific—the grounds known as "On the Line"— sperm whales occurred in some profusion. Although the vicinity of the Gilbert Islands (then known as the Kingsmill Group) was a good

place to kill sperm whales, it was almost as dangerous for the whalers as it was for the whales. The Gilbertese natives were a particularly aggressive and warlike people, but there were also many beachcombers who had been stranded on these islands, men who were eager to lead the natives in attacks on visiting whalers.

Although an occasional whaler was killed by unfriendly or unreceptive islanders, the effect of the ships' landings on the natives was considerably more severe. Unscrupulous captains would often offer to trade for provisions, and then sail away without giving anything in return, and it is likely that trigger-happy sailors took the lives of many natives without having to worry about punishment so far from any law but the captain's. In *Nimrod of the Sea* (1874), Davis recounts the story of three "kanakas" (probably Maoris from New Zealand) who deserted on a tropical island. After demanding their return from the local natives, "the captain double shorted his 9-pound guns, sent a round into the crowded grass huts of the village, and carried off three natives." From their home ports and pestilent fo'c's'les, the whalemen brought every conceivable communicable disease to the natives, including yaws, influenza, tuberculosis, cholera, syphilis, and the greatest scourge of the unresistant, measles. Before exploration, the population of Tahiti was estimated at some 40,000; by 1830, there were only 9000 Tahitians left. A single measles epidemic in 1875 killed nearly 30,000 Fijians.

There was only so much eating, drinking, holystoning of decks, repairing of sails, and rigging and yarn spinning to occupy the sailors on these seemingly endless voyages. To pass the time, some of them created what Ashley called "the only important indigenous folk art, except for that of the Indians, we have ever had in America; the Art of Scrimshaw." Although there are very few contemporaneous records of scrimshanders at work—probably because the craft was too insignificant to mention—we assume that the whale teeth were carved during periods of sailing or while waiting in port for provisions or repairs. The baleen of the right and bowhead whales was packed into bundles for commerce at home, but occasionally a piece would be shaped into a busk and decorated with contemporary designs. Baleen was colloquially known to the whalers and merchants of the time as *whalebone*, but it is not bone at all; it is made of keratin, the substance of human hair and fingernails. Whales have bones like any other mammals, but with the exception of the lower jaw—known as the "pan bone"—and the teeth, whale bones are too porous for carving. (Other cultures recognized the attractive nature of whale ivory. Certain Polynesian natives made necklaces of dolphin teeth, and the pre-missionary Hawaiians crafted the beautiful *le niho palaoa*, a gracefully carved sperm whale tooth that was worn by royalty on a necklace of braided human hair.)

Despite their discomfort, low wages, and even occasional floggings, the crews of whaleships were remarkably docile. The master's word was law, and when the crew became obstreperous, a "taste of the cat" was not unheard of. Only infrequently did they become so desperate that they rebelled. Since harsh treatment, long hours, uncomfortable quarters, and bad food were expected, the whalemen generally endured these indignities in stoic silence. Also, as with any uprising, a leader is required to galvanize men into action, and on the whaleships, these troublemakers were rare. The story of the *Bounty's* mutiny, which had occurred in 1789—and had nothing whatever to do with whales or whaling—was probably known to every seaman and landlubber on either side of the Atlantic. The fate of Fletcher Christian and the mutineers was not known until 1808, when Captain Mayhew Folger of the Nantucket sealer *Topaz* landed at Pitcairn Island and found the survivors.

**W**

On Nantucket Island there lived a young man named Samuel Comstock, who may or may not have heard the tale told by Captain Folger. At the age of 19, after three previous cruises, he shipped out aboard the Nantucket whaler *Globe*, departing from Edgartown on the neighboring island of Martha's Vineyard on December 15, 1822. The ship rounded the Horn on March 5, and stopped briefly at Hawaii before heading for the newly discovered Japan Grounds. Despite the reports of plentiful whales off Japan, Captain Worth was unable to locate them, and as they sailed in fruitless circles, the crew became increasingly discontented. Rotten meat was an issue, and conditions were so bad that the captain turned back and headed for Hawaii to re-provision. There several members of the crew deserted, and the *Globe's* depleted crew was replenished with beach combers and drunkards. Repeated conflicts between officers and crew increased the tension, and when the captain had one of the men flogged, Comstock decided to initiate a mutiny.

On January 26, 1824, Samuel Comstock led his followers in one of the bloodiest mutinies in American naval history. They murdered Captain Worth with an axe, slaughtered First Mate Beetle with a boarding knife, shot Second Mate Lumbard in the mouth and then bayoneted him, and shot Third Mate Fisher in the back of the head. They heaved the bodies overboard, and with Comstock at the helm, looked for a place where they could land. En route, Comstock decided that one of his crew members was plotting against him, held a "trial" and sentenced him to hang. For 2 weeks they wandered around, uncertain of their location or destination, until they decided to land at tiny Mili Atoll, in what was then known as the Mulgrave Islands and is now known as the Ratak chain of the Marshall Islands. It appears that Comstock's original plan was to arrange things so that he was the only survivor, but the natives and his fellow mutineers conspired against his plan for the perfect mutiny. As Comstock began to give the ship's stores to the natives (to ensure their support), the crew members who had signed on in Hawaii realized that they were in for trouble either from their leader or from the natives, and they shot Comstock dead (Hoyt, 1975).

Those members of the crew of the *Globe* who had not participated in the mutiny managed to gain control of the ship and sailed away, leaving the mutineers stranded on the island. They would not last long. A bloody conflict between the natives and the whalemen resulted in the death of all the latter but two: William Lay, of Saybrook, Connecticut, and Cyrus Hussey, of Nantucket. The *Globe* was sailed to Valparaiso, where the news of the mutiny was made known, and then returned to Nantucket. Her crew was cleared of complicity in the mutiny, and the *Dolphin*, under the command of Lieutenant John ("Mad Jack") Percival, was dispatched to the Pacific to find and bring back the mutineers. Hussey and Lay (1828) had been with the natives for almost a year and a half by the time the *Dolphin* arrived, and they looked more like natives than American whalemen. After considerable tension—the Marshallese chiefs were prepared to kill the newly arrived Americans and take their ship—and confusion about who they were, the last of the *Globe's* crew were transported home. Thus ended the story that Starbuck called "the most horrible mutiny that is recounted in the annals of the whale-fishery from any port or nation."

## See Also the Following Articles

Azorean Whaling ■ Scrimshaw ■ Whaling, Aboriginal ■ Whaling, Modern

## References

Allen, E. S. (1973). "Children of the Light: The Rise and Fall of New Bedford Whaling and the Death of the Arctic Fleet." Little, Brown, New York.

Andrews, R. C. (1916). Shore-whaling: A world industry. *Natl. Geogr.* **22**(5), 411–442.

Ashley, C. W. (1938). "The Yankee Whaler." Riverside Press, New York.

Bockstoce, J. R. (1986). "Whales, Ice and Men: The History of Whaling in the Western Arctic." University of Washington Press, Seattle.

Browne, J. R. (1846). "Etchings of a Whaling Cruise, with Notes of a Sojourn on the Island of Zanzibar. To which is Appended a Brief History of the Whale Fishery, Its Past and Present Condition." Harvard University Press, Cambridge, Reprinted 1968.

Church, A. C. (1938). "Whale Ships and Whaling." Bonanza, New York.

Davis, W. M. (1874). "Nimrod of the Sea, or, the American Whaleman." Christopher, North Quincy.

Ellis, R. (1991). "Men and Whales." Alfred A. Knopf, New York.

Ely, B. E. (1849). "There She Blows: A Narrative of a Whaling Voyage, in the Indian and South Atlantic Oceans." (C. Dahl, ed.), 1971 edition. Wesleyan University Press, Middletown.

Garner, S. (ed.) (1966). "The Captain's Best Mate: The Journal of Mary Shipman Lawrence on the Whaler Addison 1856–1860." Brown University Press, Providence.

Haley, N. C. (1948). "Whale Hunt: The Narrative of a Voyage by Nelson Cole, Haley, Harpooner in the Ship Charles W. Morgan 1849–1853." Ives Washburn, NY.

Henderson, D. A., (Ed.) (1974). Journal Aboard the Bark Ocean Bird on a Whaling Voyage to Scammon's Lagoon, winter of 1858–1859. Dawson's Book Shop, LA.

Hohman, E. P. (1928). "The American Whaleman." Augustus M. Kelley, NJ, Reissued 1972.

Hoyt, E. P. (1975). "Mutiny on the Globe." Random House, New York.

Kugler, R. C. (1980). The whale oil trade, 1750–1775. *Publ. Colonial Soc. Mass.* **52**, 153–173.

Lay, W. and Hussey, C. (1828). A Narrative of the Mutiny on Board the Ship Globe of Nantucket in the Pacific Ocean, January 1824. And the Journal of a Residence of Two Years on the Mulgrave Islands: With Observations on the Manners and Customs of the Inhabitants. New London, CT.

Melville, H. (1846). "Typee." Wiley and Putnam Press, New York, NY.

Melville, H. (1847). "Omoo: A Narrative of the South Seas." Harper and Brothers Press, New York, NY.

Melville, H. (1851). "Moby-Dick." (H. Hayford, and H. Parker, eds.), 1967 Norton Critical Edition. W.W. Norton, NY.

Miller, P. A. (1979). "And the Whale Is Ours: Creative Writing by American Whalemen." David R. Godine, Boston.

Mitchell, E. D. (1983). Potential of logbook data for studying aspects of social structure of the sperm whale *Physeter macrocephalus*, with an example—the ship *Mariner* to the Pacific, 1836–1840. *Rep. Int. Whale Comm.* **5**, 63–80 (Special Issue).

Nordhoff, C. (1856). "Whaling and Fishing." Moore, Wilsatch, Keys and Co, London.

Olmstead, F. A. (1841). "Incidents of a Whaling Voyage." Charles E. Tuttle, Rutland, VT, Reissued 1936.

Scammon, C. M. (1874). "The Marine Mammals of the North-western Coast of North America; Together with an Account of the American Whale Fishery." Carmany and G. P. Putnam's, New York, 1968 Dover edition.

Scoresby, W. (1820). "An Account of the Arctic Regions with a History and Description of the Northern Whale Fishery." David and Charles, Devon, Constable, 1969 edition.

Sherman, S. C. (1965). "The Voice of the Whaleman." Providence Public Library, Providence.

Starbuck, A. (1878). "A History of the American Whale Fishery from Its Earliest Inception to the Year 1876," Part IV, reprinted 1964, Report to the US Commission on Fish and Fisheries, Washington. Argosy-Antiquarian Ltd., New York.

# White-Beaked Dolphin
## *Lagenorhynchus albirostris*

CARL CHRISTIAN KINZE

### I. Characteristics and Taxonomy

#### A. Taxonomy

Traditionally, the genus *Lagenorhynchus* comprises six species: *L. albirostris, L. acutus, L. obliquidens, L. obscurus, L. cruciger,* and *L. australis.* However, recent research (LeDuc *et al.,* 1999) has found this entity to be paraphyletic and suggests a split into three or even four entities instead: the monotypic genus *Lagenorhynchus* containing solely the white-beaked dolphin, another monotypic genus *Leucopleurus* containing *L. acutus,* and a third genus *Sagmatias* for the four Pacific members of the old genus *Lagenorhynchus.* So far, the closest allies of the *L. albirostris* lineage within the delphinids remain obscure. It is not closely related with its present congeners. *L. acutus* shows some affiliation to the generic complex containing *Delphinus, Stenella,* and *Tursiops,* whereas the four Pacific species *L. obscurus, L. cruciger,* and *L. australis* are related with *Lissodelphis* and in particular *Cephalorhynchus. L. cruciger* and *L. australis* may be found to be more closely related to the genus *Cephalorhynchus* than their present congeners.

Recognized only as separate species in 1846, the white-beaked dolphin was among the last of the commonly occurring North Atlantic dolphin species to enter the cetological theater. Earlier finds remain obscure because they have been confused with either bottlenose (*Tursiops truncatus*) or common dolphins (*Delphinus* spp.) and even after its formal description, for many years it was considered a rare species or worse merely a white-beaked type of bottlenose dolphin. It was described twice during 1846: first, in March by the English cetologist John Edward Gray based on an animal that was caught off Great Yarmouth, England in October 1845, and subsequently in November by the Danish cetologist Daniel Frederik Eschricht, who (probably also in 1845) had received another specimen from the northwest coast of Jutland (believed to be Agger Tange) which he named *Delphinus ibsenii* honoring I. P. Ibsen, the donor of the specimen.

As noted, recent DNA studies suggest that the white-beaked dolphin represents its own lineage within the delphinids. However, there is no fossil record to confirm this. Sub-fossil finds of the species exist from the greater North and Baltic Sea area, in particular from Danish and Swedish sites, and document the presence in European waters since the last glacial epoch.

The generic name *Lagenorhynchus* means bottlenose (*lageno* = flask or bottle, *rhynchus* = nose, beak) while *albirostris* refers to the white beak (*albus* = white, *rostrum* = beak).

Vernacular names include Hvidnaese (Danish), Witsnuitdolfijn (Dutch), Kjarthvitur springari (Faroese), dauphin avec bec blanc (French), Weissschnauzendelfin (German), Blettahnydir (Islandic), Kvitnos or Hvitnos (Norwegian), delfin bialosy (Polish), belomordyi del'fin (Russian), delfin de hocicio blanco (Spanish), and ardluasuk (Greenlandic). The names in most languages note the white beak of the species. The Greenlandic name means "killer whale look-alike" obviously due to the marked black and white coloration of the white-beaked dolphin.

#### B. Description

The white-beaked dolphin has a robust appearance. The beak is short, only 5–8 cm long. There is an erect falcate dorsal fin on the middle of the back. Adults grow to between 2.4 and 3.1 m long and may weigh between 180 and 360 kg. Males usually grow larger than females. Newborn animals are about 1.2 m long and weigh approximately 40 kg.

The coloration is typically black on the back with a white saddle behind the dorsal fin and whitish bands on the flanks that vary in intensity from shining white to light gray. The belly and the beak normally are white (Figs 1 and 2). The beak sometimes is ashy gray or even darker, but never all black. The belly may also exhibit a grayish

**Figure 1** *Close-up of the white beak of the species. Photograph courtesy Martin Abramhamsson, Midtrsønderjyllands Museum, Gram, Denmark.*

W

Figure 2　*School of white-beaked dolphins encountered off the Belgian coast in the southern North Sea. Photograph courtesy Christophe Van Driesche.*

coloration leaving only a median narrow band of pure white. Younger animals may exhibit generally a lighter coloration. The whitish bands on the flank will fade shortly after the death of the animal.

There are between 25 and 28 conical teeth in each half of the upper and lower jaws. In very old animals the teeth show considerable wear.

The rostrum of the skull has a broad base and is relatively short. The vertebral column consists of 90–94 vertebrae (C7 Th15–16 L23–24 Ca 45–47). There are either 6 or 7 double-headed ribs anterior in position to 9 or 10 single-headed ribs, the total number of rib pairs usually being 16. The posterior-most pair of ribs may only be present as "floating ribs" without connection to the vertebral column. The phalanges formula is $I_2$, $II_6$, $III_4$, $IV_1$, $V_0$. The pelvic bones in adult specimens are sexually dimorphic and allow both determination of sex and sexual maturity, with male pelvic bones growing longer and heavier.

The blubber thickness varies between 20 and 35 mm. The total blubber weight amounts to between 20% and 30% of the total weight, with the highest figures found in pregnant females. The lungs on their outer edges carry fringes of adipose tissue (missing when animals are in severely deteriorated body condition).

The alimentary tract exhibits the typical delphinid scheme with a stomach consisting of three compartments. The uppermost part of the small intestine may possess a diverticulum sometimes referred to as a "fourth stomach."

Normally, there are two equally well-developed ovaries. Sometimes, however, the right ovary remains sub-mature throughout life, indicating a tendency toward ovarian asymmetry. The mature testes undergo seasonal changes and increase fourfold in weight during the breeding season.

Detailed information on organs and internal anatomy can be found in Reeves *et al.* (1998) and the other references given later.

Based on certain skull measurements significant differences have been found between specimens from either side of the North Atlantic. The use of molecular methods and photo-ID studies may when carried out provide further evidence on the stock identity and possible management stocks.

## II. Distribution and Abundance

The species is endemic to the temperate and subartic North Atlantic (Fig. 3). On the western seaboard it has been documented as far north as the southern tip of Greenland and as far south as Cape Cod on the American east coast. The northernmost record on the eastern side of the North Atlantic originates from the White Sea, the southern most from the Strait of Gibraltar. The species has been reported to stray into the Mediterranean Sea, but these records need confirmation. Being a native species of the northern North Sea and the adjacent Skagerrak, there are frequent intrusions into the Kattegat and western part of the Baltic Sea.

Reflecting the on a gross scale the distribution of shelf waters of the Northern North Atlantic, four principal centers of high density can be identified: (1) the Labrador Shelf including south western Greenland, (2) Icelandic waters, (3) the waters around Scotland including the northern Irish Sea and the North Sea, and (4) the small shelf stretch along the Norwegian coast extending north into the White Sea. The population structure is not known in any detail. So far, comparisons have been carried out only between areas 1 and 3; they revealed significant differences in skull morphology. The size of these designated populations is not known, but abundance estimates are available from the Labrador coast ("at least several thousands" according to Alling and Whitehead, 1987) and the greater North Sea (Hammond *et al.*, 2002): 7856 (95% CI = 4,032–13,301).

## III. Ecology

The diet of the species seems to reflect the local abundance and availability of certain prey species and therefore geographical differences exist. In Danish waters, cod and other gadid fish were found to be the main prey items. White-beaked dolphins prey on larger cods (about 20 cm longer in total length) than do sympatric harbor porpoises. Similar diet analysis from the Netherlands also found codfish as the most common prey, with whiting as the next most common. A German study calculated that by numbers 79% and by mass contribution 94% of the consumed fish were cod (Kinze *et al.*, 1997). Analyses from Scottish waters showed that white-beaked dolphins in addition also consume cephalopods (Santos *et al.*, 1994).

There are no reports of direct attacks on white-beaked dolphins by killer whales (*Orcinus orca*), but schools of white-beaked dolphins have been observed to flee from pods of killer whales. White sharks potentially also could prey on weakened or younger white-beaked dolphins.

The most commonly encountered infestation found in this species is by the nematode *Anisakis simplex* in the first and second stomachs. Severe infestations may cause the development of stomach ulcers. The tape worm, *Plicobotrium globicephalae*, is found in the intestines and another nematode, *Halocercus lagenorhynchi*, in the lungs (Raga, 1994).

Spondylosis deformans, a degenerative disease of the vertebral column, is rather common in physically mature individuals. In general, females seem to be more susceptible, exhibiting a higher number of affected vertebrae and more severe cases of the disease (Galatius *et al.*, 2009).

The species mainly dwells in shelf waters ranging from 150 to 1000 m in depth but also is a facultative coastal species. It is present where prey items concentrate, i.e., where fronts between water masses of different salinity form or where there is up-welling. It prefers water temperatures from 5°C to 15°C.

The white-beaked dolphin is known to occur in mixed groups with white-sided dolphins, e.g., in the North Sea. In the western

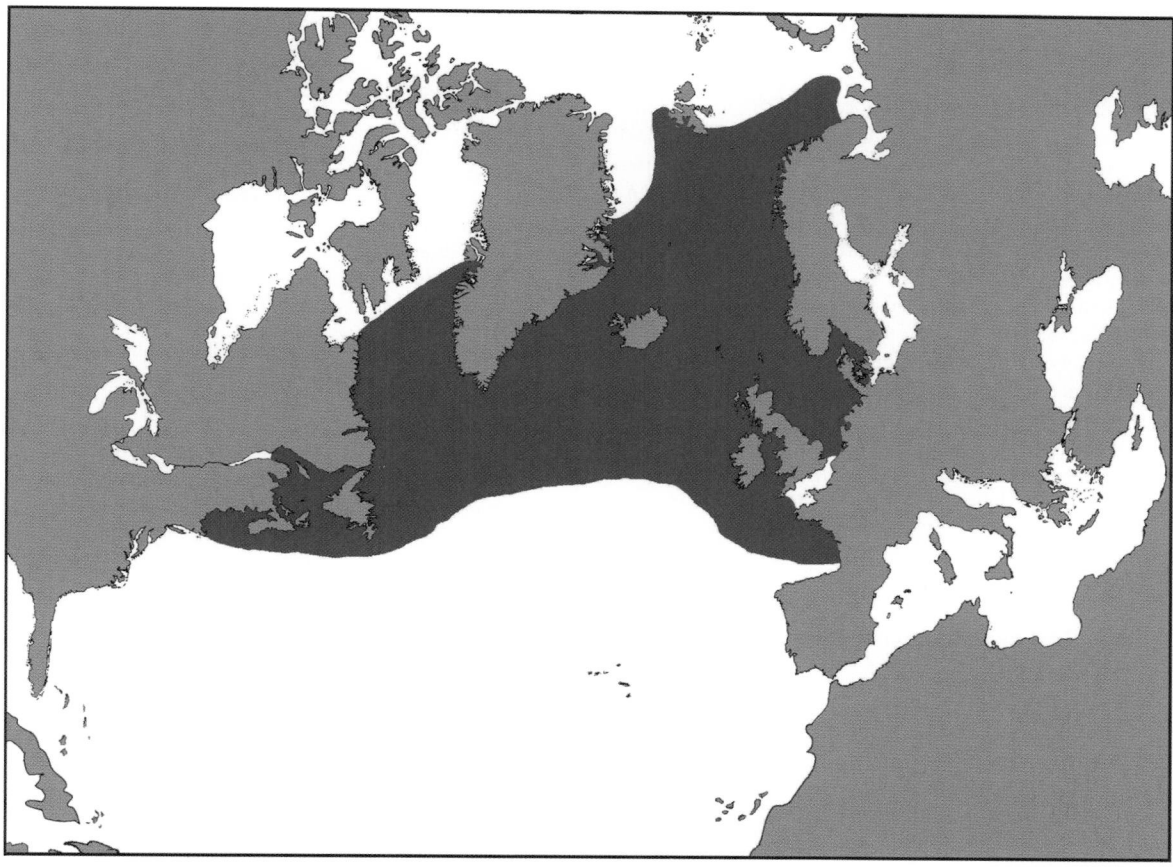

Figure 3 *Distribution of the white-beaked dolphin.*

Baltic, a single white-beaked dolphin paired with a common dolphin for a period of several weeks. In Scottish waters, common bottlenose dolphins (*Tursiops truncatus*) and white-beaked dolphins have been sighted in close vicinity to each other. Hybridization between these two species, however, has not been documented. Other delphinids known to mingle with white beaks are short-beaked common dolphins (*Delphinus delphis*) and Risso's dolphins (*Grampus griseus*). Interactions with larger baleen whales such as fin (*Balaenoptera physalus*), sei (*B. borealis*), and humpback (*Megaptera novaeangliae*) whales have also been documented.

## IV. Behavior and Physiology

White-beaked dolphins readily approach vessels and bow-ride. They frequently leap out of the water, earning them their vernacular name "jumper." They usually travel at speeds of 6–12 km/h, but speeds up to 30 km/h have been recorded (Evans and Smeenk, 2008).

Details on diving behavior only recently have been revealed from a study in Icelandic waters. A 209-cm male tagged in Faxafloi Bay in southwest Iceland performed dives down to 45 m lasting up to 78 sec, with averages of 24 m and 28 sec, respectively (Rasmussen *et al.*, 2007).

Foraging is conducted both in smaller or in larger groups depending on prey availability. The dolphins hunt in a broad front to encircle their prey.

The mean school size is 9 in Icelandic waters and between 4 and 6 in Danish North Sea waters, whereas in offshore waters much larger aggregations have been reported. School composition is known from ice entrapments, strandings, and sightings. All-male groups have been reported. Calves were found in larger groups containing adult animals of both sexes and in cow–calf aggregations.

Migrations over longer distances are not known. Pilot studies conducted in the Skagerrak and northern North Sea applying photo-ID techniques established matches between these areas and the Scottish coast.

Source levels of clicks have been recorded from free-ranging white-beaked dolphins in Icelandic waters. In a single click train they were found to vary from 194 to 211 dB peak-to-peak (p–p) re: 1 μPa. The source levels varied linearly with the log of range. The maximum source levels recorded were 219 dB (p–p) re: 1 μPa. The 3 and 10 dB beam widths were calculated to be 8° and 10°, respectively, indicating a narrower transmission beam for white-beaked dolphins than that reported for common bottlenose dolphins (*Tursiops truncatus*). The beam width was more similar to that found for belugas (*Delphinapterus leucas*). The directivity index was 29 dB. This was the first attempt to estimate the directionality index of dolphins in the field (Rasmussen *et al.*, 2002, 2004).

## V. Life History

The pattern of growth and the age at sexual maturity in the white-beaked dolphin are rather similar to what is known for the bottlenose dolphin. Data on age determination are still in need of further improvement. However, Galatius and Kinze (2007) have aged a sample of 86

**W**

specimens originating from Danish North Sea and Kattegat waters. Females become physically mature on average at lengths of 251 cm, males at lengths of 271 cm, corresponding to mean age of 15.6 years (95% CI = 9.8–23.1) and 11.4 years (95% CI = 7.7–18.1), respectively.

Females attain sexual maturity at a mean age of 8.7 years (95% CI = 5.1–14.6), males at 11.6 years (95% CI = 8.2–16.1). The mean lengths at sexual maturity were found to be 240 cm and 270 cm in females and males, respectively. There is a marked seasonality in the testes' size in mature males. During the mating season (July and August) the combined testes mass will have increased 6 times from 500 g to nearly 3000 g. The gestation period lasts about 11 months. Preliminary studies indicate a rather high annual ovulation rate of 0.7. Maximum recorded age for females is 34 years. Pregnant females are rarely encountered among stranded individuals, possibly indicating long periods of resting.

There are several known cases of ice entrapment of white-beaked dolphins from the northern part of the range. Sub-adult animals may get entangled in gill nets, whereas adults accidentally get drowned when pursuing fish into trawls. There are several reports of live strandings in tidal waters.

## VI. Interactions with Humans

Directed catches or kills are not common today but have occurred off Norway, Iceland, and Newfoundland. There is still opportunistic take in southwestern Greenland. Incidental catches in gillnets mainly affect younger animals.

The first life catch for public display (five animals for the Mystic Aquarium in Connecticut) was conducted in 1983. In Europe live-stranded animals have been brought into rehabilitation facilities.

White-beaked dolphins carry loads of organochlorines in the blubber and heavy metals in the liver and kidneys. Off Newfoundland, the loads were high even in younger animals.

The white-beaked dolphin is probably still one of the most common dolphins of the temperate North Atlantic, but there is a need for further data to ensure proper management.

## References

Alling, A. K., and Whitehead, H. P. (1987). A preliminary study of the status of white-beaked dolphins, *Lagenorhynchus albirostris*, and other small cetaceans off the coast of Labrador. *Can. Field Nat.* **101**, 131–135.

Evans, P. G. H., and Smeenk, C. (2008). White-beaked dolphin. *In* "Handbook of British Mammals" (S. Harris, ed.), 4th Ed. Mammal Society, London.

Galatius, A., and Kinze, C. C. (2007). Aspects of life history of white-beaked dolphins (*Lagenorhynchus albirostris*) from Danish waters. Abstract 21st Annual Conference of the European Cetacean Society.

Galatius, A., Sonne, C., Kinze, C .C., Dietz, R. J, Beck Jensen, J.-E., and Hyldstrup, L. (2009). Occurrence of the degenerative disease *spondylosis deformans* in a museum sample of white-beaked dolphins (*Lagenorhynchus albirostris*) from Danish waters. *J. Wildl. Dis.* **45**, in press.

Hammond, P. S., *et al.* (2002). Abundance of harbour porpoise and other cetaceans in the North Sea and adjacent waters. *J. Appl. Ecol.* **39**, 361–376.

Kinze, C. C., Adddink, M., Smeenk, C., Garcia Hartmann, M., Richards, H. W., Sonntag, R. P., and Benke, H. (1997). The white-beaked dolphin (*Lagenorhynchus albirostris*) and the white-sided dolphin (*Lagenorhynchus acutus*) in the North and Baltic Seas: Review of available information. *Rep. Int. Whale Commn.* **47**, 675–681.

LeDuc, R. G., Perrin, W. F., and Dizon, A. E. (1999). Phylogenetic relationships among the delphinid cetaceans based on full cytochrome *b* sequences. *Mar. Mamm. Sci.* **15**, 619–648.

Raga, J. A. (1994). Parasitismus bei den Cetacea. *In* "Handbuch der Säugetiere Europas 6, Meeressäuger" (R. Duguy, and D. Robineau, eds), pp. 132–179. Teil IA (Wale 1), Wiesbaden.

Rasmussen, M. H., Miller, L. A., and Au, W. W. L. (2002). Source levels of clicks from free-ranging white-beaked dolphins (*Lagenorhynchus albirostris* Gray 1846) recorded in Icelandic waters. *J. Acoust. Soc. Am.* **111**, 1122–1125.

Rasmussen, M. H., Wahlberg, M., and Miller, L. A. (2004). Estimated transmission beam pattern of clicks recorded from free-ranging white-beaked dolphins (*Lagenorhynchus albirostris*). *J Acoust. Soc. Am.* **116**, 1826–1831.

Rasmussen, M. H., Akamatsu, T., Teilmann, J., and Miller, L. A. (2007). First deployment of an acoustic tag on a white-beaked dolphin. Abstract 21st Annual Conference of the European Cetacean Society.

Reeves, R. R., Smeenk, C., Kinze, C.C., Brownell, R.R., Jr. and Lien, J. (1998). Whitebeaked dolphin *Lagenorhynchus albirostris* Gray, 1846. *In* "Handbook of Marine Mammals." (S. H. Ridgway and R. Harrison, eds), Vol. 6, pp. 1–30. Academic Press, San Diego.

Santos, M. B., Pierce, G. J., Ross, H. M., Reid, R. J., and Wilson, B. (1994). Diets of small cetaceans from the Scottish coast. *J. Fish. Res. Bd Can.* **20**, 83–115.

# MARINE MAMMAL SPECIES[1]

## ORDER CARNIVORA

### Family Otariidae

*Arctocephalus pusillus* (Schreber, 1775), Tasmanian and Cape fur seals

*Arctocephalus gazella* (Peters, 1875), Antarctic fur seal

*Arctocephalus tropicalis* (Gray, 1872), Subantarctic fur seal

*Arctocephalus townsendi* Merriam, 1897, Guadalupe fur seal

*Arctocephalus philippii* (Peters, 1866), Juan Fernanádez fur seal

*Arctocephalus forsteri* (Lesson, 1828), South Australian and New Zealand fur seals

*Arctocephalus australis* (Zimmerman, 1783), South American fur seal

*Arctocephalus galapagoensis* Heller, 1904, Galapagos fur seal

*Callorhinus ursinus* (Linnaeus, 1758), Northern fur seal

*Zalophus japonicus* (Peters, 1866), Japanese sea lion

*Zalophus californianus* (Lesson, 1828), California sea lion

*Zalophus wollebaeki* Sivertsen, 1953, Galapagos sea lion

*Eumetopias jubatus* (Schreber, 1776), Steller sea lion, northern sea lion

*Neophoca cinerea* (Péron, 1816), Australian sea lion

*Phocarctos hookeri* (Gray, 1844), New Zealand sea lion, Hooker's sea lion

*Otaria flavescens* (Shaw, 1800), South American sea lion

### Family Odobenidae

*Odobenus rosmarus* (Linnaeus, 1758), Walrus

### Family Phocidae

*Erignathus barbatus* (Erxleben, 1777), Bearded seal

*Phoca vitulina* (Linnaeus, 1758), Harbor seal, common seal

*Phoca largha* (Pallas, 1811), Spotted seal, larga seal

*Pusa hispida* (Schreber, 1775), Ringed seal

*Pusa caspica* (Gmelin, 1788), Caspian seal

*Pusa sibirica* (Gmelin, 1788), Baikal seal

*Halichoerus grypus* (Fabricius, 1791), Gray seal

*Histriophoca fasciata* (Zimmerman, 1783), Ribbon seal

*Pagophilus groenlandicus* (Erxleben, 1777), Harp seal

*Cystophora cristata* (Erxleben, 1777), Hooded seal

*Monachus tropicalis* (Gray, 1850), Caribbean monk seal, West Indian monk seal

*Monachus monachus* (Hermann, 1779), Mediterranean monk seal

*Monachus schauinslandi* Matschie, 1905, Hawaiian monk seal

*Mirounga leonina* (Linnaeus, 1758), Southern elephant seal

*Mirounga angustirostris* (Gill, 1866), Northern elephant seal

*Leptonychotes weddellii* (Lesson, 1826), Weddell seal

*Ommatophoca rossii* Gray, 1844, Ross seal

*Lobodon carcinophaga* (Hombron and Jacquinot, 1842), Crabeater seal

*Hydrurga leptonyx* (Blainville, 1820), Leopard seal

### Family Ursidae

*Ursus maritimus* Phipps, 1774, Polar bear

---

[1]Living and recently extinct. Author(s) and year of description of the species follow the Latin species name; when these are enclosed in parentheses, species was originally described in a different genus. Classification and scientific names follow Rice (1998), with adjustments reflecting more recent literature. The cetaceans genetically and morphologically fall firmly within the artiodactyl clade (Geisler and Uhen, 2005) and therefore we provisionally include them as a suborder in the order Cetartiodactyla, with infraorders Mysticeti and Odontoceti (recognizing that the suborder-level classification of the rest of the cetartiodactyls remains unresolved). The use of *Lontra* rather than *Lutra* for the marine otter follows Larivière (1998) in recognizing the otters of North and South America as a monophyletic taxon distinct from the otters of Eurasia. Recent genetic evidence strongly supports the recognition of three separate phylogenetic species of right whales (Rosenbaum *et al.*, 2000). In addition, the genus *Eubalaena* (rather than *Balaena* as in Rice, 1998) is retained for the right whales as recommended by the Scientific Committee of the International Whaling Commission (IWC, 2002). All Bryde's whales are provisionally considered to comprise a single species, *Balaenoptera edeni*, following the usage of the IWC (IWC, 2001, 2008). We also follow the IWC in listing only two species of *Sousa*; the taxonomy of this group is in flux (see chapter Humpback Dolphins). The Irrawaddy dolphin was recently split into *O. brevirostris* and *O. heinsohni*, the Australian snubfin dolphin (Beasley *et al.*, 2005). The tucuxi has been split into *Sotalia fluviatilis* (tucuxi) and the Guiana dolphin *S. guianensis* (Caballero *et al.*, 2007). *Delphinus tropicalis* is now considered a subspecies of *D. capensis* (Jefferson and Van Waerebeek, 2002). *Mesoplodon traversii* (spade-toothed whale) has been recognized as the senior synonym for *M. bahamondi* (Bahamonde's beaked whale) (van Helden *et al.*, 2002). *Balaenoptera omurai* and *Mesoplodon perrini* are newly described species (Wada *et al.*, 2003 and Dalebout *et al.*, 2002, respectively). The former, previously confounded with Bryde's whales, has been confirmed as having a separate and ancient lineage (Sasaki *et al.*, 2006). For the rest, we do not necessarily endorse Rice's (1998) classification and usage in all their details but have used them as a standard to avoid editorial confusion in the volume.

**Family Mustelidae**

> *Enhydra lutris* (Linnaeus, 1758), Sea otter
> *Lontra felina* (Molina, 1782), Marine otter, chungungo

## ORDER CETARTIODACTYLA

### Suborder Cetacea

### Infraorder Mysticeti

### Family Balaenidae

> *Eubalaena glacialis* (Muller, 1776), North Atlantic right whale
> *Eubalaena japonica* (Lacépède, 1818), North Pacific right whale
> *Eubalaena australis* (Desmoulins, 1822), Southern right whale
> *Balaena mysticetus* Linnaeus, 1758, Bowhead whale, Greenland whale

### Family Neobalaenidae

> *Caperea marginata* (Gray, 1846), Pygmy right whale

### Family Eschrichtiidae

> *Eschrichtius robustus* (Liljeborg, 1861), Gray whale

### Family Balaenopteridae

> *Megaptera novaeangliae* (Borowski, 1781), Humpback whale
> *Balaenoptera acutorostrata* Lacépède, 1804, Common minke whale
> *Balaenoptera bonaerensis* Burmeister, 1867, Antarctic minke whale
> *Balaenoptera edeni* Anderson, 1879, Bryde's whale
> *Balaenoptera omurai* Wada, Oishi and Yamada, 2003, Omura's whale
> *Balaenoptera borealis* Lesson, 1828, Sei whale
> *Balaenoptera physalus* (Linnaeus, 1758), Fin whale
> *Balaenoptera musculus* (Linnaeus, 1758), Blue whale

### Infraorder Odontoceti

### Family Physeteridae

> *Physeter macrocephalus* Linnaeus, 1758, Sperm whale

### Family Kogiidae

> *Kogia breviceps* (Blainville, 1838), Pygmy sperm whale
> *Kogia sima* (Owen, 1866), Dwarf sperm whale

### Family Ziphiidae

> *Ziphius cavirostris* G. Cuvier, 1823, Cuvier's beaked whale, goose-beaked whale
> *Berardius arnuxii* Duvernoy, 1851, Arnoux' beaked whale
> *Berardius bairdii* Stejneger, 1883, Baird's beaked whale
> *Tasmacetus shepherdi* Oliver, 1937, Shepherd's beaked whale, Tasman beaked whale
> *Indopacetus pacificus* (Longman, 1926), Indo-Pacific or Longman's beaked whale
> *Hyperoodon ampullatus* (Forster, 1770), North Atlantic bottlenose whale

> *Hyperoodon planifrons* Flower, 1882, Southern bottlenose whale
> *Mesoplodon hectori* (Gray, 1871), Hector's beaked whale
> *Mesoplodon mirus* True, 1913, True's beaked whale
> *Mesoplodon europaeus* (Gervais, 1855), Gervais' beaked whale
> *Mesoplodon bidens* (Sowerby, 1804), Sowerby's beaked whale
> *Mesoplodon grayi* von Haast, 1876, Gray's beaked whale, scamperdown whale
> *Mesoplodon perrini* Dalebout, Mead, Baker, Baker, and van Helden, 2002, Perrin's beaked whale
> *Mesoplodon peruvianus* Reyes, Mead, and Van Waerebeek, 1991, Pygmy beaked whale, Peruvian beaked whale
> *Mesoplodon bowdoini* Andrews, 1908, Andrew's beaked whale
> *Mesoplodon traversii* (Gray, 1874), Spade-tooted whale
> *Mesoplodon carlhubbsi* Moore, 1963, Hubbs' beaked whale
> *Mesoplodon ginkgodens* Nishiwaki and Kamiya, 1958, Ginkgo-toothed beaked whale
> *Mesoplodon stejnegeri* True, 1885, Stejneger's beaked whale
> *Mesoplodon layardii* (Gray, 1865), Layard's or strap-toothed beaked whale
> *Mesoplodon densirostris* (Blainville, 1817), Blainville's or dense-beaked whale

### Family Platanistidae

> *Platanista gangetica* (Roxburgh, 1801), Susu, bhulan, South Asian river dolphin, Indian river dolphin

### Family Iniidae

> *Inia geoffrensis* (Blainville, 1817), Boto, Amazon river dolphin

### Family Lipotidae

> *Lipotes vexillifer* (Miller, 1918), Baiji, Yangtze river dolphin

### Family Pontoporiidae

> *Pontoporia blainvillei* (Gervais and d'Orbigny, 1821), Franciscana, La Plata river dolphin

### Family Monodontidae

> *Monodon monoceros* Linnaeus, 1758, Narwhal
> *Delphinapterus leucas* (Pallas, 1776), Beluga, belukha, white whale

### Family Delphinidae

> *Cephalorhynchus commersonii* (Lacépède, 1804), Commerson's dolphin
> *Cephalorhynchus eutropia* (Gray, 1846), Chilean dolphin
> *Cephalorhynchus heavisidii* (Gray, 1828), Heaviside's dolphin
> *Cephalorhynchus hectori* (van Bénéden, 1881), Hector's dolphin
> *Steno bredanensis* (G. Cuvier *in* Lesson, 1828), Rough-toothed dolphin
> *Sousa teuszii* (Kükenthal, 1892), Atlantic humpbacked dolphin
> *Sousa chinensis* (Osbeck, 1765), Pacific humpbacked or Chinese white dolphin
> *Sotalia fluviatilis* (Gervais and Deville, 1853), Tucuxi
> *Sotalia guianensis* (Van Bénéden, 1864), Guiana dolphin
> *Tursiops truncatus* (Montagu, 1821), Common bottlenose dolphin
> *Tursiops aduncus* (Ehrenberg, 1833), Indian Ocean or Indo-Pacific bottlenose dolphin

*Stenella attenuata* (Gray, 1846), Pantropical spotted dolphin

*Stenella frontalis* (G. Cuvier, 1829), Atlantic spotted dolphin

*Stenella longirostris* (Gray, 1828), Spinner dolphin

*Stenella clymene* (Gray, 1850), Clymene dolphin

*Stenella coeruleoalba* (Meyen, 1853), Striped dolphin

*Delphinus delphis* Linnaeus, 1758, Short-beaked common dolphin

*Delphinus capensis* Gray, 1828, Long-beaked common dolphin

*Lagenodelphis hosei* Fraser, 1956, Fraser's dolphin

*Lagenorhynchus albirostris* (Gray, 1846), White-beaked dolphin

*Lagenorhynchus acutus* (Gray, 1828), Atlantic white-sided dolphin

*Lagenorhynchus obliquidens* Gill, 1865, Pacific white-sided dolphin

*Lagenorhynchus obscurus* (Gray, 1828), Dusky dolphin

*Lagenorhynchus australis* (Peale, 1848), Peale's dolphin

*Lagenorhynchus cruciger* (Quoy and Gaimard, 1824), Hourglass dolphin

*Lissodelphis borealis* Peale, 1848, Northern right whale dolphin

*Lissodelphis peronii* (Lacépède, 1804), Southern right whale dolphin

*Grampus griseus* (G. Cuvier, 1812), Risso's dolphin, gray grampus

*Peponocephala electra* (Gray, 1846), Melon-headed whale, electra dolphin

*Feresa attenuata* Gray, 1874, Pygmy killer whale

*Pseudorca crassidens* (Owen, 1846), False killer whale

*Orcinus orca* (Linnaeus, 1758), Killer whale, orca

*Globicephala melas* (Traill, 1809), Long-finned pilot whale

*Globicephala macrorhynchus* Gray, 1846, Short-finned pilot whale

*Orcaella brevirostris* (Owen *in* Gray, 1866), Irrawaddy dolphin, pesut

*Orcaella heinsohni* Beasley, Robertson, and Arnold, 2005, Australian snubfin dolphin

## Family Phocoenidae

*Neophocaena phocaenoides* (G. Cuvier, 1829), Finless porpoise

*Phocoena phocoena* (Linnaeus, 1758), Harbor porpoise, common porpoise

*Phocoena sinus* Norris and McFarland, 1958, Vaquita, Gulf of California harbor porpoise

*Phocoena spinipinnis* Burmeister, 1865, Burmeister's porpoise

*Phocoena dioptrica* Lahille, 1912, Spectacled porpoise

*Phocoenoides dalli* (True, 1885), Dall's porpoise, Dall porpoise

## *ORDER SIRENIA*

## Family Trichechidae

*Trichechus manatus* Linnaeus, 1758, Caribbean manatee, West Indian manatee

*Trichechus senegalensis* Link, 1795, African manatee, West African manatee

*Trichechus inunguis* (Natterer, 1883), Amazon manatee

## Family Dugongidae

*Dugong dugon* (Müller, 1776), Dugong, sea cow

*Hydrodamalis gigas* (Zimmerman, 1780), Steller's sea cow, giant sea cow

## *References*

Beasley, I., Robertson, K. M., and Arnold, P. (2005). Description of a new dolphin, the Australian snubfin dolphin *Orcaella heinsohni* sp. n. (Cetacea: Delphinidae). *Mar. Mamm. Sci.* **21**, 365–400.

Caballero, S., Trujillow, F., Vianna, J. A., Barrios-Garrido, H., Montiel, M. G., Beltrán-Pedreros, S., Marmontiel, M., Santos, M. C., Rossi-Santos, M., Santos, F. R., and Baker, C. S. (2007). Taxonomic status of the genus Sotalia: Species-level ranking for "tucuxi" (*Sotalia fluviatilis*) and "costero" (*Sotalia guianensis*) dolphins. *Mar. Mamm. Sci.* **23**, 358–386.

Dalebout, M. L., Mead, J. G., Baker, C. S., Baker, A. N., and van Helden, A. L. (2002). A new species of beaked whale *Mesoplodon perrini* sp. n. (Cetacea; Ziphiidae) discovered through phylogenetic analyses of mitochondrial DNA sequences. *Mar. Mamm. Sci.* **18**, 577–608.

Geisler, J. H., and Uhen, M. D. (2005). Phylogenetic relationships of extinct cetartiodactyls: Results of simultaneous analyses of molecular, morphological, and stratigraphic data. *J. Mammal. Evol.* **12**, 145–160.

van Helden, A. L., Baker, A. N., Dalebout, M. L., Reyes, J. C., Van Waerebeek, K., and Baker, C. S. (2002). Resurrection of *Mesoplodon traversii* (Gray, 1874), senior synonym of *M. bahamondi* Reyes, Van Waerebeek, Cárdenas and Yáñez, 1995 (Cetacea: Ziphiidae). *Mar. Mamm. Sci.* **18**, 609–621.

International Whaling Commission (2001). Report of the Scientific Committee. *J. Cetacean Res. Manag.* **3**(Suppl.), 1–75.

International Whaling Commission (2002). Report of the Scientific Committee. *J. Cetacean Res. Manag.* **4**(Suppl.), 1–78.

International Whaling Commission (2008). Report of the Scientific Committee. *J. Cetacean. Res. Manag.* **10**(Suppl.), 1–74.

Jefferson, T. A., and Van Waerebeek, (2002). The taxonomic status of the nominal species *Delphinus tropicalis* van Bree, 1971. *Mar. Mamm. Sci.* **18**, 787–818.

Larivière, S. (1998). Lontra felina. *Mammalian Species* **575**, 1–5.

Rice, D. W. (1998). Marine mammals of the world: Systematics and distribution. *Soc. Mar Mammalogy Spec. Pub.* **4**, 1–231.

Rosenbaum, H., Brownell, R. L., Jr., Brown, M. W., Schaeff, C., Portway, Y., White, B. N., Malik, S., Pastene, L. A., Patenaude, N. J., Baker, C. S., Goto, M., Best, P. B., Clapham, P. J., Hamilton, P., Payne, R., Rowntree, Y., Tynan, C. T., Bannister, J. L., and DeSalle, R. (2000). World-wide genetic differentiation of Eubalaena questioning the number of right whale species. *Mol. Ecol.* **9**, 1793–1802.

Sasaki, T., Nikaido, M., Wada, S., Yamada, T. K., Cao, Y., Hasegawa, M., and Okada, N. (2006). *Balaenoptera omurai* is a newly discovered baleen whale that represents an ancient evolutionary lineage. *Mol. Phylogenet. Evol.* **41**, 40–52.

Wada, S., Oishi, M., and Yamada, T. K. (2003). A newly discovered species of living baleen whale. *Nature* **426**, 278–281.

# BIOGRAPHIES

*Science is done by individuals, and the student or practicing scientist often wants to know something about those who came before, who developed the field. The brief biographies listed here are meant to meet that need, however sketchily. The list includes those who, in the opinion of the editors, made an important and lasting contribution to the study of marine mammals. Some of their most important works are also listed, and the student is urged to peruse some of these classics. Living persons are not included, as lifetime contributions cannot be fairly weighed until they are completed.*

**Abel, Othenio** (1875–1946) Austrian paleontologist, the main European researcher on fossil whales in the early twentieth century, working on the Belgian Miocene and on *Basilosaurus*. He expanded the perspective of marine mammal paleontologists to include higher level phylogeny.

Abel, O. (1905). Les Odontocétes du Bolderien (Miocene superieure) d 'Anvers. *Mem. Mus. Roy. Hist. Nat. Belg.* **3**, 1–155.

Abel, O. (1907). The genealogical history of the marine mammals. *Ann. Rep. Smithson. Inst.* **1907**, 473–496.

**Allen, Glover M**. (1879–1942) One of the great naturalists of his day; among his enduring works relating to marine mammals are his monograph on the baleen whales of New England and his 1942 book on extinct and endangered mammals, which included the first review of the status of marine mammal species and the conservation problems affecting them.

Allen, G. M. (1916). The whalebone whales of New England. *Mem. Boston Soc. Nat. Hist.* **8**, 109–322.

Allen, G. M. (1942). "Extinct and Vanishing Mammals of the Western Hemisphere with the Marine Species of All the Oceans." American Committee for International Wildlife Protection, Cambridge, Massachusetts. (1972 reprint by Cooper Square, New York).

**Allen, Joel A.** (1838–1921) Completed the first comprehensive bibliography of cetology and sirenology (1495–1840), which is still highly useful today. He also produced a series of important monographs and papers on the marine mammals of North America.

Allen, J. A. (1880). History of North American pinnipeds: A monograph of the walruses, sea-lions, sea-bears and seals of North America. *U.S. Geol. Sum Terr. Misc. Pub*.**12**, 1–785.

Allen, J. A. (1881). Preliminary list of works and papers relating to the mammalian orders Cete and Sirenia. *Bull. U.S. Geol. Geogr. Surv.Terr.* **6**, 399–562.

Allen, J. A. (1908). The North Atlantic right whale and its near allies. *Bull. Am. Mus. Nat. Hist.* **24**, 277–329, pl. 19–24.

**Andrews, Roy Chapman** (1884–1960) Paleontologist, cetologist, educator, and popularizer at the American Museum of Natural History in New York, he traveled the world and authored many important technical papers and monographs as well as best-selling popular books. His cetological work covered porpoises, beaked whales, and baleen whales.

Andrews, R. C. (1916). "Whale Hunting with Gun and Camera." Appleton, New York.

Andrews, R. C. (1916). Monographs of the Pacific Cetacea. II. The sei whale (*Balaenoptera borealis* Lesson). 1. History, habits, external anatomy, osteology, and relationships. *Mem. Am. Mus. Nat. Hist., N.S.* **1**, 293–502.

**Arnold, Peter W.** (1949–2006) Highly respected Australian systematist who applied rigorous cladistic analysis to cranial morphology of delphinoid cetaceans and documented the occurrence of two species of minke whales in Australian waters. Born in Nova Scotia, he spent most of his scientific career in Queensland. He also worked with harbor porpoise parasites, took part in description of the new dolphin species *Orcaella heinsohni*, and pioneered behavioral studies of minke whales subject to whale-watching.

Arnold, P., Marsh, H., and Heinsohn, G. (1987). The occurrence of two forms of minke whales in east Australian waters with a description of external characters and skeleton of the diminutive dwarf form. *Sci. Rep. Whales Res. Inst., Tokyo* **38**, 1–46.

Arnold, P. W., and Heinsohn, G. E. (1996). Phylogenetic status of the Irrawaddy dolphin *Orcaella brevirostris* (Owen in Gray): A cladistic analysis. *Mem. Queensland Mus.* **39**, 141–204.

Beasley, I., Robertson, K. M., and Arnold, P. (2005). Description of a new dolphin, the Australian snubfin dolphin *Orcaella heinsohni* sp. n. (Cetacea, Delphinidae). *Mar. Mamm. Sci.* **21**, 365–400.

**Arsen'ev, Viktor Alexandrovich** (1908–1991) Noted biologist who produced or coauthored several important works on Russian marine mammals; his research focused on the northern fur seal (see also SOKOLOV).

Arsen'ev, V. A., and Zemskiy, V. A. (1954). ["In the Land of Whales and Penguins"]. Moskovkiy Gos. Universitet, Moscow. (In Russian).

Zemskiy, V. A., Arsen'ev V. A., and Studenezskaya, I. S. (1973). ["Marine Mammals"]. Pistchevaya Promyshlennost', Moscow.

**Bélon du Mans, Pierre** (1517–1566) Arguably the first cetologist since Aristotle and Pliny more than a millennium earlier, Bélon was the first to mention cetaceans in the title of a zoological work. The book included an account of the anatomy of a porpoise and listed seven species of whales, dolphins, and porpoises. Figures in his second book were much copied by later workers.

Bélon, P. (1541). "L:Histoire Naturelle des Estranges Poissons Marins, avec la vraie Peincture et Description du Dauphin et de plusieurs autres de son espèce." Regnaud Chaudiere, Paris.

Belloni, P. (1553). "De aquatilibus, Libro duo." Stephan, Paris.

**Bertram, G. Colin L.** (1911–2001) With his wife Kate, founder of modern sirenology. After stints with the British Grahamland Expedition and the Scott Polar Research Institute (as Director), during which he published on Antarctic and Arctic seals, he began investigations of the biology and conservation of manatees and dugongs in the 1960s, becoming the first chairman of the IUCN Sirenia Specialist Group. He drew world attention to the precarious status of sirenian populations globally.

Bertram, C. (1963). "In Search of Mermaids: The Manatees of Guiana." Peter Davies, London.

Bertram, G. C. L., and Bertram, C. K. R. (1973). The modern Sirenia: Their distribution and status. *Biol. J. Linn. Soc.* **5**, 297–338.

**Berzin, Alfred A.** (1930–1996) Latvian whale biologist, born in Rostov-on-Don, Russia; known for his comprehensive research and monograph on the sperm whale. He spent his career at the Pacific Ocean Research Institute for Fisheries and Oceanography (TINRO) in Vladivostock. Most recently, he took part in the recovery and reporting of the true Soviet whaling catch data, which exposed massive unreported kills of protected whales in the 1960s and 1970s that may explain the apparent failure of some stocks to recover.

Berzin, A. A. (1971). "Kashalot" [The sperm whale]. Pischevaya Promyshlennost, Moscow. (Translation by Israel Program for Translation, 1972).

Zemskiy, V. A., Berzin, A. A., Mikhalev, Y. A., and Tormosov, (1995). Soviet Antarctic pelagic whaling after WWII: Review of actual catch data. *Rep. Int. Whal. Commn.* **45**, 131–137.

**Bigg, Michael A.** (1939–1990) Born in London and emigrating to British Columbia as a child, he was the first to use individual identification of killer whales to study long-term social structure, life history, and movements. He discovered the existence of sympatric but nonmixing fish-eating and mammal-eating forms/groups ("residents" and "transients") and established the fact that killer whale pods are matrilineal groups, with all offspring remaining with the maternal pod.

Bigg, M. A., Olesiuk, P. F., Ellis, G. M., Ford, J. K. B., and Balcomb, K. C. (1990). Social organization and genealogy of resident killer whales (*Orcinus orca*) in the coastal waters of British Columbia and Washington State. *Rep. Int. Whal. Commn. (Spec. Iss.)* **12**, 383–405.

**Bonner, W. Nigel** (1928–1994) Biologist at the British Antarctic Survey and a major and primary contributor to the biology, assessment and management of the southern elephant seal, other Antarctic pinnipeds, and British seals. He authored several technical and semipopular books.

Bonner, W. N. (1968). The fur seal of South Georgia. *Brit. Ant. Surv. Sci. Rep.* 56, 1–81.

Bonner, W. N. (1982). "Seals and Man: A Study of Interactions." University of Washington Press, Seattle.

Bonner, W. N. (1990). "The Natural History of Seals." Facts on File, New York.

**Caldwell, David K.** (1928–1990) Cetologist and historian born in Louisville, Kentucky, David Caldwell with his wife and research partner acoustician Melba C. Caldwell discovered that dolphins have individual "signature whistles." He founded and edited the journal *Cetology*.

Caldwell, M. C., and Caldwell, D. K. (1965). Individualized whistle contours in bottlenose dolphins (*Tursiops truncatus*). *Nature* **207**, 434–435.

Caldwell, D. K., and Caldwell, M. C. (1972). "The World of the Bottlenose Dolphin." J. B. Lippincott, Philadelphia.

**Chapman, Douglas G.** (1920–1996) Noted mathematician, wildlife scientist, and teacher, born in Canada and spending his career at the University of Washington in Seattle. He was a member of the famous Committee of Three Scientists established by International Whaling Commission in 1961 as an independent body to advise on status and trends of whale stocks, leading to accurate predictions of collapse of stocks under commercial exploitation. A pioneer in the development and use of quantitative models for marine mammal assessment and management, he worked on pinnipeds as well as cetaceans.

Chapman, D. G. (1961). Population dynamics of the Alaska fur seal herd. *Trans. N. Am. Wildl. Conf.* **26**, 356–369.

Chapman, D. G. (1968). Estimation of fur seal pup populations by randomized sampling. *Trans. Am. Fish. Soc.* **97**, 264–270.

Chapman, D. G. (1971). Status of Antarctic rorqual stocks. *In* "The Whale Problem. A Status Report." (W. E. Schevill, ed.), pp. 218–238. Harvard Univ. Press, Cambridge.

Chapman, D. G. (1981). Evaluation of marine mammal population models. *In* "Dynamics of Large Mammal Populations" (C. W. Fowler and T. D. Smith, eds.), pp. 278–296.

**Chapskiy, Konstantin Konstantinovich** (1906–1984) The preeminent Russian pinniped biologist of the mid-twentieth century, he concentrated on the systematics of the phocid seals but also worked on the ecology and morphology of other groups.

Chapskiy, K. K. (1941). ["Marine Animals of the Soviet Arctic"]. Izdatelstvo Glavsevmorputi,, Moscow.

Chapskiy, K. K. (1955). [An attempt at revision of the systematics and diagnostics of seals of the subfamily Phocinae]. *Trud. Zool. Inst. Akad. Nauk SSSR* **17**, 161–199. (*Fish. Res. Bd Can. Transl. Ser.* **114**, 1–57 (1957)).

Chapskiy, K. K., and Sokolov, V E. (1975). "Morphology and Ecology of Marine Mammals: Seals, Dolphins, Porpoises." Wiley, New York.

**Cuvier, Frederic** (1773–1838) Comparative anatomist and physiologist, brother of Georges Cuvier; Frederic Cuvier synthesized the available information on cetaceans in his landmark volumes, correcting many errors about cetacean natural history accumulated through the centuries.

Cuvier, F. (1829). Cétacés. *In* "Histoire Naturelle des Mamifères" (E. Geoffroy Saint-Hilaire and F. Cuvier, eds). Belin, Paris.

Cuvier, F. (1836). "De l'Histoire Naturelle des Cétacés." Roret, Paris.

**Cuvier, Georges** (1769–1832) Famed paleontologist, zoologist, and evolutionary thinker, widely respected in his time as one of the originators of the idea of evolution and the founder of modern biology and comparative anatomy. He spent his career at the Museum national d'Histoire naturelle, Paris. His enormous treatise on fossils included a volume on cetaceans, and he carried out many basic studies of cetacean anatomy, including the first description of the inner ear, and compiled one of the first stranding reports (1812).

Cuvier, G. (1795). Note sur la découverte de l'oreille interne de Cétacés. *Mag. Encycl.* **6**(2), 130.

Cuvier, G. (1812). Rapport fait à la classe des Sciences mathémathique et physiques, sur divers Cétacés pris sur les côtes de France, principalement sur ceux qui sont échoués près de Paimpol, le 7 janvier 1812. *Ann. Mus. D'Hist. Nat.* **19**, 1–16, pl. l.

Cuvier, G. (1817 and subsequent editions). "Le Règne Animal distribué d'après son organisation, pour servir de base a l'histoire naturelle des animaux et d'introduction à l'Anatomie Comparée." Deterville, Paris.

Cuvier, G. (1821–1825). "Recherches sur les ossemens fossiles, òu l' on rétablit les caractères des plusiers animaux dont les révolutions du globe ont détruit les espèces." Second ed., 10 vol (Vol 5, on Cetacea; 1823). Dufour and d'Ocagne, Paris.

**Dawbin, William H.** (1921–1998) New Zealand-born biologist, historian, and teacher who pioneered much of the early research on distribution and movements of humpback whales and right whales in the Southern Hemisphere. He was launched into marine mammalogy by early years spent in the whaling grounds of the Southern Ocean. His later work focused on catch history of whales in the Southern Hemisphere. He was also a noted herpetologist.

Dawbin, W. H. (1956). The migrations of humpback whales which pass the New Zealand coast. *Trans. Roy. Soc. NZ* **84**, 147–196.

Dawbin, W. H. (1986). Right whales caught in waters around south eastern Australia and New Zealand during the nineteenth and early twentieth centuries. *Rep. Int. Whal. Commn.* (*Spec. Iss.*) **10**, 261–267.

Dawbin, W. H. (1997). Temporal segregation of humpback whales during migration in southern hemisphere waters. *Mem. Queensland Mus.* **42**, 105–138.

**Eschricht, Daniel Frederick** (1798–1863) A pioneer in many aspects of modern cetology: the first to consider cetaceans, including harbor porpoises, narwhals, and belugas, as migrating animals; the first to describe the embryo of a whale based on dissections; the first to distinguish between the bowhead and right whales; the first (with Owen) to see concentric layers of growth in the teeth of cetaceans; and the first to identify different populations of killer whales in the Arctic. Most of his work was in Danish and largely inaccessible to other marine mammal scientists until published in English by Flower (1866). The gray whale, *Eschrichtius robustus*, bears his name.

Eschricht, D. F. (1840–1842). Om undersögelsen of de nordiske Hvaler. *Skand. Naturf. Förhandl.* **2**, 83–108, 203–227.

Flower, W. H. (ed.) (1866). "Recent Memoirs on the Cetacea by Professors Eschricht, Reinhardt and Lilljeborg." For Ray Society by Hardwicke, London.

**Fabricius, Otto** (1744–1822). Danish missionary, cleric, philologist, naturalist, and ethnographer. He published the first account of the fauna of Greenland, including descriptions of 143 new species. This work included precise descriptions of all five pinnipeds inhabiting Greenland. Later works included a monograph on the seals of Greenland and a work on seal hunting implements.

Fabricius, O. (1780). "Fauna Groenlandica." J. G. Rothe, Hafniae and Lipsiae.

Fabricius, O. (1790 and 1791). "Udførlig Besckrivelse over de Grønlandske Saele, Første og Ander Stukke. Skrivter af Naturhistorie-Selskabet, Copenhagen. (In Danish, translated in Kapel, 2005).

Kapel, F. O. (2005). "Otto Fabricius and the Seals of Greenland." Danish Polar Center, Copenhagen.

**Fay, Francis H.** (1927–1994) Born in Massachusetts and spending most of his career at the University of Alaska, "Bud" Fay was the world's leading expert on the walrus. He translated and made accessible to western scientists much of the massive Russian literature on marine mammals of the North Pacific and Arctic.

Fay, F. H. (1982). Ecology and biology of the Pacific walrus, *Odobenus rosmarus divergens Illiger. N. Am. Fauna* **74**, 1–279.

Fay, F. H. (1997). Status of the Pacific walrus population, 1950–1989. *Mar. Mamm. Sci.* **13**, 537–565.

**Flower, William Henry** (1831–1899) Gray's student and his successor as curator of the huge collection of marine mammal specimens at the British Museum in the last quarter of the nineteenth century, Flower was an extremely able zoologist who contributed major works on the beaked whales and the delphinid cetaceans, as well as publications on manatees and pinnipeds.

Flower, W. H. (1871). On the recent ziphioid whales, with a description of the skeleton of *Berardius arnuxi. Trans. Zool. Soc. London* **8**, 203–234, pl. 27–29.

Flower, W. H. (1884). On the characters and divisions of the family Delphinidae. *Proc. Zool. Soc. London* **1883**, 466–513.

**Fraas, Eberhardt** (1862–1915) German paleontologist, one of the first to study the oldest Eocene whales (e.g., *Protocetus*).

Fraas, E. (1904). Neue Zeuglodonten aus dem untern Mitteleozan van Mkattam bei Cairo. *Geol. Palaeot. Abhandl. Jena* **6**(3), 199–220.

**Fraser, Francis C.** CBE, FRS (1903–1978) The main practitioner of British cetology in the mid-twentieth century, which was far ahead of its state in North America. A Scot by birth, he was Keeper of Zoology in charge of the whales, dolphins, and porpoises at the British Museum for several decades. A UK-wide stranding program had started early in the century (stranded animals are "royal fish" and property of the sovereign), and Fraser kept it going and authored elaborate periodic reports up into the 1970s. His classic work with Peter E. Purves on the comparative functional morphology of the cetacean ear resulted in a classification that has endured and proved to be largely consonant with phylogenies generated by the most modern methods. He described Fraser's dolphin, *Lagenodelphis hosei*, from a skeleton that had languished in the BM collection for over 50 years as an "abnormal" or hybrid specimen. Much of today's cetology is built on his work.

Fraser, F. C. (1956). A new Sarawak dolphin. *Sarawak Mus. J.* **7**, 478–503. pl. 22–26.

Fraser, F. C. (1960). Hearing in cetaceans. Evolution of the accessory air sacs and the structure and function of the outer and middle ear in Recent cetaceans. *Bull. Brit. Mus.* (*Nat. Hist.*), *Zool.* **7**, 1–140, 53 pl.

Fraser, F. C. (1974). "Report on Cetacea stranded on the British Coasts from 1948 to 1966." British Museum (Natural History), London.

**Gaskin, David E.** (1939–1998) British biologist; started out as a lepidopterist (and remained one), but a stint as a biologist on a

UK whaling ship in the Antarctic steered him into whale biology and led him to extensive work on the cetaceans of New Zealand and the porpoises of the North Atlantic. For many years, his book on the ecology of whales and dolphins was the only textbook available to teachers of marine mammalogy.

Gaskin, D. E. (1968). The New Zealand Cetacea. N.2. *Mar. Vert Fish. Res. Bull. (New Series)* **1**, 1–92.

Gaskin, D. E. (1982). "The Ecology of Whales and Dolphins." Heinemann, London.

Gaskin, D. E. (1992). Status of the harbour porpoise, *Phocoena phocoena*, in Canada. *Can. Field-Nat.* **106**, 36–54.

**Gervais, Paul Francois Louis** (1816–1879) French zoologist and paleontologist; he described several cetacean species, explored the anatomy and reproduction of dolphins and whales, and collaborated with Van Beneden to produce the classic atlas of cetacean osteology that is still in use today (see also VAN BENEDEN).

Gervais, P. (1853). Description de quelques espèces de phoques et de Cétacés. *Ann. Sci. Nat.* **20**, 281–292.

Gervais, P. (1856). Sur trois espèces de dauphins qui vivent dans la région du Haut Amazone. *Comp. Rend. Acad. Sci.* **42**, 806–808.

**Gilmore, Raymond W.** (1907–1984) The "father" of whale-watching, first naturalist to accompany a whale-watching cruise, on gray whales out of San Diego, California in 1959. Active in early population assessments of gray whales and in observations of gray whales in their breeding lagoons in Baja California. Began his professional career as an entomologist, working on mosquitoes in Central America; a demonstration of how many scientists early on "backed into" marine mammalogy.

Gilmore, R. W. (1960). A census of the California gray whale. *U.S. Fish Wildl. Serv. Spec. Sci. Rep.-Fish.* **342**, 1–30.

Gilmore, R. W. (1976). The friendly whales of Laguna San Ignacio. *Terra* **15**, 24–28.

**Gray, John Edward** (1800–1875) had a plan as a young man to "form the largest and most complete Zoological Collection known" and he succeeded, being personally responsible for the building of the huge collections of the British Museum amassed during the Victorian period of wide zoological exploration. He worked and published prolifically on nearly every animal group. To keep track of the current literature, he started the Zoological Record. Among his 1096 publications (1820–1875) were scores on whales, dolphins, porpoises, seals, otters, and other aquatic mammals. We still recognize 16 of the marine mammal species he described (3 pinnipeds, 1 baleen whale, 2 toothed whales, and 10 dolphins and small toothed whales), as well as many of his higher taxa. *Mesoplodon grayi* is named after him.

Gray, J. E. (1866). "Catalog of Seals and Whales in the British Museum." 2nd Ed. British Museum, London.

Gray, J. E. (1870). The geographical distribution of the Cetacea. *Ann. Mag. Nat. Hist.*, Ser. 4, **6**, 387–394.

**Gulland, John** FRS (1926–1990) Eminent English fisheries mathematician and teacher who was a member of the independent committee of scientists commissioned in the 1960s to assess the status of Antarctic whale stocks; the committee predicted collapse of the stocks due to commercial overexploitation, and that came to pass. He was the mainstay of the fisheries department of the Food and Agriculture Organization (FAO) of the UN in Rome for nearly 20 years, running highly effective training courses in developing countries around the world, and authored several important texts on fishery assessment and management.

Gulland, J. (1965). The plight of the whales. *Oryx* **8**(2), 74–98.

Gulland, J. (1972). The conservation of Antarctic whales. *Biol. Cons.* **4**, 335–344.

Gulland, J. (1983). "Fish Stock Assessment: A Manual of Basic Methods." J. Wiley and Sons, New York.

**Harrison, Sir Richard J.** FRS (1920–1999) English anatomist and functional morphologist who first made his mark in human anatomy and medicine. He became deeply interested and intrigued by marine mammals through discussions with Francis Fraser of the British Museum and went on to produce landmark works on the reproduction and general biology of whales, dolphins, and seals. He received a knighthood for his contributions to marine biology, education, and public service.

Harrison, R., and King, J. E. (1965). "Marine Mammals." Hutchinson University Library, London.

Harrison, R. J. (ed.). (1972–1977). "Functional Anatomy of Marine Mammals," Vol. 1–3. Academic Press, London.

Ridgway, S. H., and Harrison, R. J. (1981–1999). "Handbook of Marine Mammals," Vol. 1–6. Academic Press, San Diego and London.

**Harrison Matthews, Leonard** FRS (1901–1986) An early participant in the British *Discovery* investigations in the Antarctic; authored basic monographs on the elephant seal, humpback whale, sperm whale, and sei whale. Much of his research focused on growth and reproduction. In later life he wrote several semi-popular books on marine mammals and Antarctic research.

Harrison Matthews, L. (1937). The sperm whale, *Physeter catodon. Discovery Rep.* **17**, 93–168, pl. 3–11.

Harrison Matthews, L. (1952). "Sea Elephant – The Life and Death of the Elephant Seal." Macgibbon and Kee, London.

Harrison Matthews, L. (1978). "The Natural History of the Whale." Colombia University Press, New York.

**Hershkovitz, Phillip** (1909–1997) Known mainly for his voluminous works on the systematics and ecology of neotropical terrestrial mammals, Hershkovitz turned to cetaceans at one point and produced his classic *Catalog of Living Whales*, a definitive reference indispensable to any cetacean systematist. While many of his taxonomic "calls" subsequently have been second-guessed, he led the way for cetological systematists in the last half of the twentieth century.

Hershkovitz, P. (1966). Catalog of living whales. *U.S. Nat. Mus. Bull.* **246**, 1–257.

**Heyning, John E.** (1957–2007) Born and reared in southern California, Heyning became Curator and later Deputy Director of the Los Angeles County Museum and built the marine mammal collection there to the second largest in the United States. A dynamic educator and specialist in systematics and functional morphology, he took active roles in the American Cetacean Society and the Natural Science Collections Alliance.

Heyning, J. E., and Perrin, W. F. (1994). Evidence for two species of common dolphins (genus *Delphinus*) from the eastern North Pacific. *Nat. Hist. Mus. Los Angeles Country Contrib. Sci.* **442**, 1–35.

Heyning, J. E. (1995). "Masters of the Ocean Realm". University of Washington Press, Seattle, USA Washington.

Heyning, J. E., and James G. Mead (1997). Thermoregulation in the mouths of feeding gray whales. *Science* **278**, 1138–1139.

**Hubbs, Carl L**. (1894–1979) A polymath naturalist born in Arizona whose career and publications spanned over 65 years and included the study of fishes, birds, marine mammals, archeology, geochronology, geology, climatology, evolution, hybridism, and the practice of systematics. He carried out the first aerial surveys of gray whales, rediscovered the supposedly extinct Guadaloupe and Juan Fernandez fur seals, and was a major force for the conservation of marine animals and their habitats. He persuaded the Mexican Government to establish the first gray whale sanctuary, in Scammon's Lagoon. Hubb's beaked whale, *Mesoplodon carlhubbsi*, was named in his honor; he and other faculty members at the Scripps Institution of Oceanography ate the type specimen during World War II meat rationing (the osteological portions fortunately were saved).

Hubbs, C. L. (1946). First records of two beaked whales, *Mesoplodon bowdoini* and *Ziphius cavirostris*, from the Pacific coast of the United States. *J. Mamm.* **27**, 242–255.

Hubbs, C. L., and Hubbs, L. C. (1967). Gray whale censuses by airplane in Mexico. *Calif. Fish Game* **53**, 23–27.

Hubbs, C. L., and Norris, K. S. (1971). Original teeming abundance, supposed extinction and survival of the Juan Fernandez fur seal. *Ant. Pinnipedia, Ant. Res. Ser.* **18**, 35–52.

**Ichihara, Tadayoshi** (1939–1981) Japanese marine mammalogist known for his description of the pygmy blue whale, *Balaenoptera musculus brevicauda*, and for pioneer research on the telemetry of marine animals, including dolphins and seals. This innovative worker did much in his too-brief career to inject rigor and modern technology into Japanese marine mammalogy.

Ichihara, T. (1957). An application of linear discriminant function to external measurements of fin whale. *Sci. Rep. Whales Res. Inst., Tokyo* **12**, 127–189.

Ichihara, T. (1966). Criteria for determining age of fin whale with reference to ear plug and baleen plate. *Sci. Rep. Whales Res. Inst., Tokyo* **20**, 17–82, 8 pl.

Ichihara, T. (1971). [Ultrasonic, radio tags and various problems in fixing them to marine animal body]. *Rep. Fish. Resource Invest. Scient. Fish. Res. Agency Japanese Gov.* **12**, 29–44.

**Jonsgård, Åge** (1916–1997) Norwegian whale biologist and professor who carried out classical studies on the minke and fin whales and was an influential member of the Scientific Committee of the International Whaling Commission.

Jonsgård, Å. (1951). Studies on the little piked whale or minke whale (*Balaenoptera acutorostrata* Lacépède). Report on Norwegian investigations carried out in the years 1943–1950. *Norsk Hval.-Tid.* **40**, 209–232.

Jonsgård, Å. (1966). Biology of the North Atlantic fin whale *Balaenoptera physalus* (L.). Taxonomy, distribution, migration and food. *Hval. Skr.* **49**, 1–62, fig. 1, map.

**Kamminga, Cees** (1932–2002). Dutch bioacoustician who made many significant contributions to the study of dolphin sonar, applying principles of signal processing to the sound productions of dozens of species.

Kamminga, C. (1975). Remarks on dominant frequencies of cetacean sonar. *Aquat. Mamm.* **7**, 93–101.

Kamminga, C., Cohen, S., and Silber, G. K. (1996). Investigations on cetacean sonar XI: Intrinsic comparison of the wave shapes of some members of the Phocoenidae family. *Aquat. Mamm.* **22**, 45–56.

**Kasamatsu, Fujio** (1950–2001) Highly productive Japanese student of whales and whale ecology in the Antarctic, associated with the Marine Ecology Research Institute in Tokyo. He worked primarily within the context of the International Whaling Commission's Scientific Committee and produced the first rigorous estimates of abundance for many Antarctic small cetaceans.

Kasamatsu, F., and Joyce, G. G. (1995). Current status of odontocetes in the Antarctic. *Ant. Sci.* **7**, 365–370.

Kasamatsu, F. (2000). Species diversity of the whale community in the Antarctic. *Mar. Ecol. Prog. Ser.* **200**, 297–301.

Kasamatsu, F., Masuoka, K., and Hakamada, T. (2000). Interspecific relationships in density among the whale community in the Antarctic. *Polar Biol.* **23**, 466–473.

**Kellogg, Remington** (1892–1969) Paleontologist, mammalogist, and popularizer of cetaceans at the Smithsonian Institution. He produced dozens of works on fossil whales, dolphins, porpoises, and seals, including the classic and massive "Review of the Archeoceti." His doctoral thesis "History of Whales – Their Adaptations to Life in the Water" has served as a primer for cetologists for generations. He took part in the early stages of the organization of international agreements on whaling beginning in the 1930s. The first U.S. Commissioner to the IWC, he battled vigorously but unsuccessfully in the 1960s for limitation of commercial exploitation of whales to sustainable levels. Although he did not win the day, the heat of the controversy led to later independent review of the whale stocks and eventual effective regulation of whaling.

Kellogg, R. (1928). The history of whales – Their adaptation to life in the water. *Quart. Rev. Biol.* **3**, 29–76, 174–208.

Kellogg, R. (1936). A review of the Archeoceti. *Carnegie Inst. Wash. Pub.* **482**, 1–357.

Kellogg, R. (1940). Whales, giants of the seas. *Nat. Geogr.* **77**(1), 35–90.

**Kellogg, Winthrop N.** (1898–1972) A physiologist at Florida State University, discovered ultrasonic hearing in dolphins and was the first to demonstrate cetacean echolocation. The research was accomplished with captive dolphins from Florida Marine Studios. Also worked with dolphin visual learning and sonic size discrimination.

Kellogg, W. N., and Kohler, R. (1952). Reactions of the porpoise to ultrasonic frequencies. *Science* **116**, 250–252.

Kellogg, W. N. (1958). Echo ranging in the porpoise. Perception of objects by reflected sound is demonstrated for the first time in marine animals. *Science* **128**, 982–988.

Kellogg, W. N., and Rice, C. E. (1964). Visual problem-solving in a bottlenose dolphin. *Science* **143**, 1052–1055.

**Kenyon, Karl W.** (1918–2007) A pioneering American naturalist/zoologist who studied marine mammals in the field in the Leeward Islands of Hawaii, Alaska, and the Pacific Northwest. Worked on Hawaiian monk seals, fur seals, sea lions, sea otters, walruses, dolphins, and whales.

Kenyon, K. W., and D. W. Rice. (1959). Life history of the Hawaiian monk seal. *Pac. Sci.* **13**, 215–252.

Kenyon, K. W. (1960). Territoriality and homing in the Alaskan fur seal. *Mammalia* **24**, 431–444.

Kenyon, K. W. (1969). The sea otter in the eastern Pacific Ocean. *North American Fauna* **68**, 1–351.

**Kirkwood, Geoffrey P.** (1947–2006) Australian quantitative modeler who pioneered the application of simulation studies to test the performance of fishery management models. He was pivotal in development of the IWC's Revised Management Procedure for the large whales, serving also as chairman of the IWC Scientific Committee. His last professional berth was at Imperial College, London.

Kirkwood, G. P. (1997). The Revised Management Procedure of the International Whaling Commission. *In* "Global Trends: Fisheries Management". (E. K. Pikitch, D. D. Huppert. and M. P. Sissenwine, eds.) AmericanFisheries Society Symposium 20. Bethesda, Maryland.

**Kleinenberg, Sergei Evgenyevich** (1909–1968) Russian biologist who studied mammals in the field and laboratory; published extensively on the dolphins and porpoises of the Black Sea and, with three colleagues, produced a monograph on the beluga that has been the basis of all subsequent studies of the species. With his student G. A. Klevezal', he contributed greatly to development of the methodology of determining age in mammals through the examinations of layers in hard tissues.

Kleinenberg, S. (1956). "Mlekopitayushchie Cernogo i Azovskogo morei [Marine mammals of the Black and Azov seas]." Izdatelstvo AN SSR. (In Russian).

Kleinenberg, S., Yablokov, A. V., Bel'kovich, B. M., and Tarasevich, M. N. (1964). ["Beluga *(Delphinapterus leucas).* Investigation of the Species"]. Izd. Nauk, Moscow. (Translation by Israel Program for Scientific Translation, Jerusalem, 1969).

Klevezal', G. A., and Kleinenberg, S. E. (1967). ["Age Determination of Mammals from Annual Layers in Teeth and Bones"]. Izd. Nauk, Moscow (Translation by Israel Program for Scientific Trans, Jerusalem, 1969).

**Lacépède** (Bernard Germain Etienne de Laville-sur-Ilon, Compte de Lacépède) (1756–1826) French zoologist who was one of the first modern zoologists to critically review the cetaceans. Colored plates from his book were extensively copied in other authors' works of the nineteenth century. He listed 34 species in the first edition (more in later editions), many more than any previous worker. Working with the collections in the Museum National d'Histoire Naturelle, Paris, beginning in 1784, he was highly honored in his time and worked successfully under the monarchy, the republic, the empire, and the new monarchy.

La Cepède ["Ie Citoyen"]. (1804). "Histoire Naturelle des Cétacés." Paris.

Lacépède ["M. le Comte de"]. (1818). Note sur les Cétacés des mers voisines du Japan. *Acad. Roy. Sci., Paris* **21 septembre 1818**, 119–121.

**Leatherwood, J. Stephen** (1944–1997) Marine mammalogist, naturalist, and conservationist, born in Alabama. He authored or coauthored a series of widely used identification guides and guided the establishment of marine mammal research and conservation programs in many of the less-developed nations in Latin America and Asia.

Leatherwood, S., and Reeves, R. R. (1983). "The Sierra Club Handbook of Whales and Dolphins." Sierra Club Books, San Francisco.

Leatherwood, S., Reeves, R. R., Perrin, W. F., and Evans, W. E. (1988). "Whales, dolphins, and porpoises of the eastern North Pacific and Adjacent Arctic Waters. A Guide to their Identification." Dover Press, New York.

Leatherwood, S., and Donovan, G. P. (eds.). (1991). Cetaceans and cetacean research in the Indian Ocean Sanctuary. *UN Env. Pro. Mar. Mam. Tech. Rep.* **3**, 1–287.

**Linnaeus, Carolus** (1707–1778) Swedish physician and avid naturalist considered the father of modern taxonomy; he worked extensively in the field with birds and plants. Linnaeus named and classified all the then known species of plants and animals in the first edition of his Systema Naturae in 1735, using his binomial system of a species name consisting of genus and trivial name. The official starting point of zoological nomenclature is 1758 (International Commission on Zoological Nomenclature – 1985), the year of publication of the 10th edition of his work. In that edition, he gave us 13 species of marine mammals that are still recognized.

International Commission on Zoological Nomenclature. (1985). "International Code of Zoological Nomenclature." Third Ed. Internat. Trust for Zool. Nomen., Brit. Mus. (Nat. Hist.), London.

Linnaeus, C. (1735). "Systema Naturae." Two vol.

Linnaeus, C. (1758). "Systema Naturae." Laurentii Salvii, Stockholm.

**Mackintosh, Neil A.** (1900–1974) British biological oceanographer and authority on Antarctic whales; his contributions were massive. He was one of the prime organizers of the Discovery investigations that have contributed much to the knowledge of marine mammals and ecology of the Southern Ocean. He authored landmark monographs on whales, plankton, and ice, edited and shepherded the Discovery Reports, launched the international whale-marking scheme, was instrumental in the founding of the British Antarctic Survey, and was a leading player in (and first chairman of the Scientific Committee of) the IWC.

Mackintosh, N. A., and Wheeler, J. F. G. (1929). Southern blue and fin whales. *Discovery Rep.* **1**, 257–540, pl. 25–44.

Mackintosh, N. A. (1942). The southern stocks of whalebone whales. *Discovery Rep.* **22**, 197–300.

Mackintosh, N. A. (1965). "The Stocks of Whales." Fishing News (Books), London.

**Nemoto, Takahisa** (1930–1990) The major authority on the plankton consumed by whales; born in Tokyo and carried out his studies at the Whales Research Institute (1953–1977) and Ocean Research Institute of University of Tokyo (1977–1990). He was instrumental in the organization of the international project BIOMASS to study the ecology of the Southern Ocean.

Nemoto, T (1959). Food of baleen whales with reference to whale movements. *Sci. Rep. Whales Res. Inst., Tokyo* **14**, 149–290.

Nemoto, T (1966). Feeding of baleen whales and krill, and the value of krill as a marine resource in the Antarctic. *In* "Symposium on Antarctic Oceanography" (Scientific Committee on Antarctic Research, Sponsor), 240–253. Polar Research Institute, Cambridge, UK.

Nemoto, T, and Kawamura, A. (1977). Characteristics of food habits and distribution of baleen whales with special reference to the abundance of North Pacific sei and Bryde's whales. *Rep. Int. Whal. Commn. (Spec. Iss.)* **1**, 80–87.

**Nishiwaki, Masaharu** (1915–1984) A "force of nature" for the study and conservation of marine mammals in Asia. After serving as a fighter pilot in World War II, he spent several seasons with the Japanese whaling factory-ship fleet in the Antarctic; this set the direction of his career. He made basic contributions to knowledge of systematics, life history, and reproduction of whales, dolphins, and seals and effectively promoted conservation of cetaceans and sirenians in Japan, China, Bangladesh, India, Pakistan, Nepal, Indonesia, Russia, and several nations in West Africa. He campaigned against overexploitation of whales and dolphins in Japan; his was a lone voice at the time.

Nishiwaki, M. (1965). ["Whales and Dolphins"]. University of Tokyo Press.

Nishiwaki, M. (1972). General biology. *In* "Mammals of the Sea Biology and Medicine." (S. H. Ridgway ed.), pp. 3–204, Thomas, Springfield, Illinois.

Nishiwaki, M., and Marsh, H. (1985). *Dugong – Dugong dugon. In* "Handbook of Marine Mammals, Vol. 3. Sirenians and Baleen Whales." (S. H. Ridgway and R. Harrison, eds.), pp. 1–31. Academic Press, London.

**Norris, Kenneth S**. (1924–1998) The virtual founder of modern cetology in the United States; born in Los Angeles. He organized and chaired the first international conference on cetaceans in 1963 in Washington, D.C. Scientist, teacher, and natural historian, he cofounded the Society for Marine Mammalogy and served as its first president. With William McFarland, he described the vaquita, *Phocoena sinus*, from the Gulf of California in 1958. He demonstrated echolocation in dolphins, developed theories of hearing and sound production in dolphins and the sperm whale, and carried out extensive innovative research on dolphin behavior and ecology. Also an active "biopolitician," he was influential in development of the landmark US Marine Mammal Protection Act of 1972.

Norris, K. S., Prescott, J. H., Asa-Dorian, P. V., and Perkins, P. (1961). An experimental demonstration of echolocation behavior in the porpoise, *Tursiops truncatus* (Montagu). *Biol. Bull.* **120**, 163–176.

Norris, K. S. (ed.). (1966). "Whales, Dolphins and Porpoises." University of California Press, Berkeley.

Norris, K. S. (1974). "The Porpoise Watcher." W. W. Norton and Co., New York.

Norris, K. S., Würsig, B., Wells, R. S., and Würsig, M. (eds). (1994). "The Hawaiian Spinner Dolphin." University of California Press, Berkeley.

**Omura, Hideo** (1906–1993) Considered the dean of Japanese cetology; published detailed monographs on the osteology, systematics, and distribution of baleen whales in Japanese waters and from around the world. He served for decades as Director of the industry-sponsored Whales Research Institute in Tokyo, which fostered and published the bulk of the Japanese research on cetaceans from after World War II until the late 1980s.

Omura, H. (1950). Whales in the adjacent waters of Japan. *Sci. Rep. Whales Res. Inst., Tokyo* **4**, 27–113.

Omura, H., and Ohsumi, S. (1974). Research on whale biology in Japan, with special reference to North Pacific stocks. *In* "The Whale Problem – A Status Report," (W. E. Schevill, G. C. Ray and K. S. Norris, eds.), pp. 196–208. Harvard University Press, Cambridge, MA.

Omura, H. (1984). History of the gray whale in Japan. *In* "The Gray Whale." (M. L. Jones and S. L. Swartz, eds.), pp. 59–77. Academic Press, San Diego.

**Poulter, Thomas C**. (1897–1978) US bioacoustician who worked extensively in the Arctic and Antarctic. He long advocated the idea that pinnipeds, like odontocete cetaceans, use echolocation; this was eventually discounted (see Schusterman, 1967). He is most remembered for organizing a series of conferences on biological sonar and diving mammals in the 1960s and 1970s at Stanford Research Institute; these were the forerunners of the current Biennial Conferences on the Biology of Marine Mammals.

Poulter, T. C. (1963). Sonar signals of the sea lion. *Science* **139**, 753–755.

Schusterman, R. J. (1967). Perception and determination of underwater vocalization in the California sea lion. *In* "Les Systèmes Sonar Animaux. Biologie et Bionique." (R. G. Busnel, ed.), pp. 535–617. Laboratoire de Physiologie Acoustique, Jouy-en-josas, France.

**St. Aubin, David J.** (1952–2002) Canadian pathologist and endocrinologist who explored the role of health in the ecology of cetaceans and seals, what goes wrong and causes epidemics, strandings, and population decline. In his too-short career, he worked in the field from the Arctic to the tropics, probing the effects of contaminants, microbes, and stress.

St. Aubin, D. J. (1979). Strandings: A rare look into the biology of the Atlantic white-sided dolphin, *Lagenorhynchus acutus. In* "The biology of marine mammals: Insights through strandings." (J. R. Geraci and D. J. St. Aubin, eds.). pp. 190–206. *Marine Mammal Commission contract report* **MMC-77/13**, 343pp.

St. Aubin, D. J., Ridgway, S. H., Wells, R. S., and Rhinehart, H. (1996). Dolphin thyroid and adrenal hormones: Circulating levels in wild and semidomesticated *Tursiops truncatus*, and influence of sex, age and season. *Mar. Mamm. Sci.* **12**, 1–13.

**Scammon, Charles M**. (1825–1911) American whaling captain and naturalist who published only one book and one technical paper but greatly influenced cetology and the history of whaling. He discovered the breeding grounds of the gray whale in Laguna Ojo de Liebre (also called Scammon's Lagoon) in Baja California in 1855, which led to the near-extinction of the species in the eastern Pacific. His book (rare and valuable in the original edition) is a detailed chronicle of American whaling and description of the marine mammals of the west coast of North America.

Scammon, C. M. (1869). On the cetaceans of the western coast of North America. *Proc. Acad. Nat. Sci. Philadelphia* **1869**, 13–63.

Scammon, C. M. (1874). "The Marine Mammals of the Northwestern Coast of North America Described and illustrated together with an Account of the American Whale-fishery." John H. Carmany, San Francisco. (Reprinted 1968, Dover, New York).

**Schevill, William E.** (1906–1994) Pioneering bioacoustician who was born in Brooklyn and spent most of his career at Harvard. He made the first recordings of underwater cetacean sounds in the wild and recorded and described sounds of many species. He published the first phonograph record of cetacean sounds. With wife and research partner Barbara Lawrence he described the anatomy of the delphinid nasal passages functional in sound production.

Schevill, W. E., and Lawrence, B. (1950). Underwater listening to the white porpoise (*Delphinapterus leucas*). *Science* **109**, 143–144.

Lawrence, B., and Schevill, W. E. (1956). The functional anatomy of the delphinid nose. *Bull. Mus. Comp. Zool.* **165**, 104–152 + 30 fig.

Schevill, W E., and Watkins, W A. (1962). *Whale and Porpoise Voices. A Phonograph Record.* Woods Hole Oceanographic Institution, Woods Hole, Massachusetts. 24pp. and record.

Schevill, W. E., Ray, G. C., and Norris, K. S. (1974). "The Whale Problem: A Status Report." Harvard University Press, Cambridge.

**Scholander, Per F. ("Pete")** (1905–1980) Norwegian-born physiologist who made many basic discoveries and advances in the respiratory physiology and ecology of diving mammals, including observation of the slowing of heart rate (the "diving response") in submerged seals and description of countercurrent heat-exchange structures in the dorsal fin and flukes of cetaceans. He pioneered direct physiological monitoring and experimentation with captive live marine mammals to explore their respiratory and cardiovascular adaptations.

Scholander, P. F. (1940). Experimental investigations on the respiratory function in diving mammals and birds. *Hval. Skr.* **22**, 1–131.

Scholander, P. F., and Schevill, W. E. (1955). Counter-current vascular heat exchange in the fins of whales. *J. Appl. Physiol.* **8**, 279–282.

Scholander, P. F. (1964). Animals in aquatic environments: Diving mammals and birds. *In* "Handbook of Physiology, Sect. 4: Adaptations to the Environment." (D. B. Dill, ed.), pp. 729–739. Amer. Physiol. Soc., Washington.

**Scoresby, William** (1789–1857) Born in Yorkshire, England; accompanied his father on a whaling expedition when he was 10 years old, an event that shaped his life; he published his first paper on whales at the age of 19. His most important work "An Account of the Arctic Regions and Description of the Northern Whale Fishery" is the best contemporary account of the Arctic and whaling as it existed in the early nineteenth century. Several equally important books followed.

Scoresby, W. (1808). Account of the *Balaena mysticetus* or great northern or Greenland whale. *Edinb. Mem. Wern. Soc.* **1**, 578–586.

Scoresby, W. (1820). "An Account of the Arctic Regions and Description of the Northern Whale Fishery." Constable, Edinburgh.

Scoresby, W. (1860). "The Whaleman's Adventures." Darton, London.

Stamp, T., and Stamp, C. (1975). "William Scoresby Arctic Scientist." Caedmon of Whidby Press, Cambridge, UK.

**Sergeant, David E.** (1927–2008) Born in Hangzhou, China of medical missionary parents and obtained his Ph.D. at Cambridge in 1951. Dedicated natural historian of pinnipeds and cetaceans who produced classical studies of harp seals, belugas, pilot whales and other marine mammals during a long career with the Newfoundland and Canadian Governments. A pioneer in age determination of toothed whales and a creative thinker about population regualtion. Unsung hero for efforts to collect, interpret and report data on harp seals every yera in the face of abuse from all quarters, including the Canadian Government.

Sergeant, D. E. (1957). Age determination in odontocete whales from dentinal growth layers. *Norw. Whale. Gaz.* **6**, 273–288.

Sergeant, D. E. (1962). The biology of the pilot or pothead whale, *Globicephalamelaena* (Traill) in Newfoundland waters. *Fish. Res. Bd Can. Bull.* **132**, 1–84.

Sergeant, D. E. (1973). Biology of white whales (*Delphinapterus leucas*) in western Hudson Bay. *J. Fish. Res Bd Can.* **30**, 1065–1090.

Sergeant, D. E. (1991). "Harp Seals, Man and Ice." *Can. Spec. Pub. Fish. Aquat. Sci.* **114**.

**Slijper, Everhard Johannes** (1907–1968) Like many cetologists of the first half of the twentieth century, E. J. Slijper, the Dutch functional anatomist and student of whales and whaling had his first experience with whale biology on a whaling expedition to the Antarctic. He authored "Walwissen" (1958), which became the famous book "Whales" in English. He analyzed reproduction, locomotion, and other functions in cetaceans and other mammals with a comparative anatomical approach that became standard in such studies.

Slijper, E. J. (1962). "Whales." Hutchinson, London.

Slijper, E. J. (1966). Functional morphology of the reproductive system in Cetacea. *In* "Whales, Dolphins and Porpoises." (K. S. Norris, ed.), pp. 277–319. University of California Press, Berkeley.

**Sokolov, Vladimir Evgenyevich** (1928–1998) An important figure in Russian mammalogy and member of the National Academy of Sciences of the USSR, V. E. Sokolov was a specialist on the microanatomy of the skin. He coauthored a three-volume work on the mammals of the Soviet Union, including a volume on seals and the toothed cetaceans.

Sokolov, V. E. (1962). Adaptations of the mammalian skin to the aquatic mode of life. *Nature* **195**, 464–466.

Sokolov, V. E. (1973). Structure of the skin cover in some cetaceans. *In* "Morphology and Ecology of Marine Mammals: Seals, Dolphins and Porpoises." (K. K. Chapskiy and V E. Sokolov, eds.), pp. 102–118. Wiley, New York.

Geptner, V. G., Chapskiy, K. K., Arseneyev, V. A., and Sokolov, V. E. (1976). ["Mammals of the Soviet Union, Vol. 1. Pinnipeds and Toothed Whales"]. Vyshaya Shkola, Moscow.

**Steller, Georg W.** (1709–1746) German-born botanist who in his relatively short life spent years in the wilds of the Russian Far East with the Great Siberian Expedition and collected specimens and detailed accounts of many new animals, including the Steller's sea lion and the extinct Steller's sea cow of Bering Island.

Steller, G. W. (1751). "De Bestiis Marinus." St. Petersburg Acad. Sci. (posthumous; published in translation as *Von Sonderbaren Meerthieren*, Kummel, Halle, Germany, 1753).

**Tomilin, Avenir Grigoryevich** (1912–2000) Grand old man of Russian cetology who began his 66-year career with a season with the whaling expedition "Aleut" in the Kamchatka, Bering, and Chukchi Seas in 1934. The most famous of his hundreds of technical and semipopular contributions is the massive volume on Cetacea in the series "Mammals of the U.S.S.R. and Adjacent Countries" (1957), which has been translated into English. Tomilin followed a lifelong commitment to popular scientific education, and his books were read by millions.

Tomilin, A. G. (1937). Kity Dal'nego Vostoka [Whales of the Far East]. Ucheniye Zaliski MGU **8**, 1119–1167.

Tomilin, A. G. (1957). Kitoobraznye. Vol. 9, V G. Heptner, ed., Zveri SSSR i prelezhashchikh strano Izdatel'stvo Akademii

Nauk SSSR, Moscow. Translation published as "Cetacea, Vol. 9, Mammals of the U.S.S.R.," V. G. Heptner, ed., by Israel Program for Scientific Translations, Jerusalem (1967).

Tomilin, A. G. (1980). "V mirye kitov i delfinov [In the world of whales and dolphins]." Izdatel'stvo "Znaniye," Moscow.

**True, Frederick W.** (1858–1914) The dominant figure in American cetology at the turn of the century; built the marine mammal collection of the US National Museum into the second largest in the world, after that of the British Museum. Highly prolific, he documented the American dolphin fisheries in New Jersey and North Carolina and produced scores of reports on the systematics of cetaceans, pinnipeds, and other mammals, describing many new species and including the important monographic "Review of the family Delphinidae," which largely delineated the array of dolphin species recognized today, and "Whalebone whales of the North Atlantic."

True, F. W. (1889). Contributions to the natural history of the cetaceans, a review of the family Delphinidae. *Bull. U.S. Nat. Mus.* **36**, 1–191, pl. 1–47.

True, F. W. (1904). The whalebone whales of the North Atlantic. *Smithson. Contrib. Knowledge* **33**, 1–551, pl. 1–50.

True, F. W. (1910). An account of the beaked whales of the family Ziphiidae in the collection of the United States National Museum, with remarks on some specimens in other American museums. *Bull. U.S. Nat. Mus.* **73**, 1–89, pl. 1–42.

**Van Beneden, Pierre Joseph** (1809–1894) Belgian paleontologist and morphologist of the latter half of the nineteenth century; he provided some of the first accurate descriptions of fossil whales. With Paul Gervais, he authored a monumental classic work on the osteology of living and fossil cetaceans that is consulted even today by cetologists needing accurate depictions of dolphin and whale skulls and skeletons.

Van Beneden, P. J. (1864). Recherches sur les ossements provenant du Crag d'Anvers. Les Squalodons. *Acad. Roy. Belg.* **2 April 1864**, 1–85, pl. 1–4.

Van Beneden, P. J., and Gervais, P. (1868–1880). "Ostéographie des Cétacés Vivant et Fossiles Comprenant la Description et !'Iconographie du Squelette et du Système Dentaire de ces Animaux ainsi que des Documents Relatif à leur Histoire Naturelle." Bertrand, Paris.

**Villa-Ramirez, Bernardo** (1912–2006) The father of marine mammal science in Mexico, he studied the endangered vaquita, documented the extirpation of the Caribbean monk seal in Mexico, and launched scores of students into mammalogy. One of the first Honorary Members of the Society for Marine Mammalogy.

Villa-R., B. (1976). Report on the status of *Phocoena sinus*, Norris and McFarland 1958, in the Gulf of California. *An. Inst. Biol. Univ. Nal. Autón. México, Ser. Zoología* **47**, 203–208.

Villa-R., B., Gallo-R., J. P., and Leboeuf, B. (1986). La foca monje *Monachus tropicalis* (Mammalia: Pinnipedia) difinitivament extinguida en México. *Ann. Inst. Biol. Univ. Nat. Autón. México, Ser. Zoología* **56**, 573–588.

**van Utrecht, Willem Lodewijk** (1926–1994) Dutch zoologist who was an important early worker in the study of age and growth structures in both toothed cetaceans and baleen whales, working with teeth in the former and both ear plugs and baleen plates in the latter. He turned to the study of aging of fishes in his later years. Much of his research was in collaboration with his wife Clara N. van Utrecht-Cock.

Utrecht, W. L. van (1965). On the growth of the baleen plate of the fin whale and the blue whale. *Bijdragen tot de Dierkunde* **35**, 1–38.

Utrecht, W. L. van, and Utrecht-Cock, C. N. (1969). Comparison of records of baleen plates and of ear plugs in female in whales, *Balaenoptera physalus* (Linnaeus, 1758). *Bijdragen tot de Dierkunde* **39**, 81–97.

**Watkins, William A.** (1926–2004) Influential and productive researcher at Woods Hole Oceanographic Institution who worked with William Schevill to produce some of the first underwater recordings of marine mammals. A deep background in physics, electronics, and acoustics made him a rigorous mentor to marine mammal bioacousticians and behaviorists. In later years he pioneered telemetry and satellite tracking of large whales.

Schevill, W. E., and Watkins, W. A. (1962). "Whale and Porpoise Voices. A Phonograph Record." Woods Hole Oceanographic Institution, Woods Hole, Massachusetts.

Watkins, W. A. (1963). Portable underwater recording system. *Undersea Technol.* **4**, 23–24.

Watkins, W. A., Sigurjonsson, J., Wartzok, D., Maiefski, R. R., Howey, P. W., and Daher, M. A. (1996). Fin whale tracked by satellite off Iceland. *Mar. Mamm. Sci.* **12**, 564–569.

**Winn, Howard E.** (1926–1995) Oceanographer, whale biologist, and teacher who carried out some of the first acoustic studies of whale songs. He organized and headed the massive Cetacean and Turtle Assessment Program (CETAP), 1978–1982, which provided first estimates of distribution and abundance of whales, dolphins, and porpoises off the eastern United States, and was active in conservation of the right whale in the North Atlantic.

Winn, H. K., and Winn, L. K. (1978). Song of the humpback whale in the West Indies. *Mar. Biol.* **47**, 97–114.

Winn, H. K., and Olla, B. L. (eds.). (1979). "Behavior of Marine Animals. Vol. 3: Cetaceans." Plenum Press, New York.

Winn, H. K. (1982). "A Characterization of Marine Mammals and Turtles in the Mid- and North Atlantic areas of the U.S. outer continental shelf." Cetacean and Turtle Assessment Program of the University of Rhode Island, Kingston.

**Yamada, Munesato** (1921–1994) Noted Japanese anatomist and naturalist, described the cetacean ear in detail. He also produced one of the first analyses of cetacean osteology that took into account individual variation, based on specimens from a mass stranding of false killer whales in Scotland. He rediscovered and described the external appearance of the pygmy killer whale, previously known only from two skulls described by Gray in 1871, and founded the Sea of Japan Cetology Research Group.

Yamada, M. (1953). Contribution to the anatomy of the organ of hearing of whales. *Sci. Rep. Whales Res. Inst., Tokyo* **8**, 1–79.

Yamada, M. (1954). An account of a rare porpoise, *Feresa* Gray from Japan. *Sci. Rep. Whales Res. Inst., Tokyo* **9**, 59–88.

Yamada, M. (1956). An analysis in mass osteology of the false killer whale, *Pseudorca crassidens* (Owen). *Okajimas Folia Anat. Japon.* **28**, 451–463.

# GLOSSARY

*The following section provides definitions for a list of terms that appear in the text of this Encyclopedia. For the most part these are specialized technical terms used in the field of marine mammalogy that are likely to be unfamiliar to a lay reader. Also included are terms that, though likely to be known in their common sense, have a distinctive or highly specific use in marine mammalogy.*

## A

**α-male**  The dominant male which appears to monopolize matings with the females.

**abduction**  The movement of a part of the body (limb) away from the midline of the body (opposite is adduction).

**abuse**  In the context of human interaction with marine mammals, a term for mistreatment involving violence that may result in injuries or death.

**accessory denticles**  Cusps on archaeocete teeth that are not clearly homologous with primitive mammalian tooth cusps.

**accidental reinforcement**  Another term for ADVENTITIOUS REINFORCEMENT.

**acetabulum**  A depression in the pelvis in which the head of the femur is secured.

**acoustic harassment device (AHD)**  A sound-generating device that, because of some combination of intensity, frequency, or other sound characteristic(s), is aversive to marine mammals and keeps or drives them away from an area or structure.

**adduction**  The movement of a part of the body (limb) toward the midline of the body (opposite is abduction).

**adenohypophysis**  A glandular structure, also known as the anterior pituitary, located at the base of the hypothalamus in the brain and producing a variety of hormones primarily responsible for stimulating the function of other endocrine glands.

**adenosine triphosphate**  (ATP) A high-energy compound that, when split by enzymic activity, releases energy in the reaction.

**adipocytes**  The cells of adipose tissue which store fat.

**advanced**  See DERIVED.

**adventitious reinforcement**  Reinforcement that happens to coincide with a particular response even though that particular response was not responsible for delivery of the reinforcement. In effect, the "wrong" behavior is modified. Also, ACCIDENTAL REINFORCEMENT.

**aerial behavior**  Any behavior that takes the animal partly or completely out of the water, for example lobtails, breaches, spy-hops, head, side or back slaps, leaps, flips, and spins.

**aerobic**  Pertaining to activity or metabolism in which oxygen is the final electron acceptor in the breakdown of glucose. This process produces 39 ATPs from 1 mole of glucose.

**aerobic dive limit**  The maximum length of a dive accomplished using mainly aerobic metabolic pathways, and with no net production of anaerobic metabolites.

**AFLPs**  Amplified Fragment Length Polymorphisms. DNA sequences amplified by PCR using generic PCR primers. Mutations in the priming site lead to loss of priming and no PCR amplification, resulting in a bi-allelic dominant genotype (amplification or no amplification).

**aggregation**  A term for a group of individuals that come together because of a common resource, such as a predator-free habitat, rather than by social attraction.

**aggressive behavior** or **aggression**  A set of social interactions ranging from threats to open fights, reflecting in a conflict of interest over limited resources and having the potential to cause injuries and sometimes death to participants. Generally refers to conflict involving members of the same species but may refer to any interaction of this kind.

**agonistic behavior**  See AGGRESSION.

**agviq**  Northern Alaskan Inuit term for a bowhead whale.

**air gun**  In this context, a device that releases a burst of high-pressure air into the water. Groups ("arrays") of air guns are used by the marine seismic industry to create low-frequency, high-level sound pulses that can characterize rock formations below the seafloor.

**aiviq**  Northern Alaskan Inuit term for a walrus.

**aldosterone**  Steroid hormone released by the adrenal gland that increases Na+ retention in the distal convoluted tubule and the collecting duct of the kidney.

**algae**  A general term for various small aquatic organisms, usually single celled, which can synthesize organic matter from carbon dioxide using the energy of sunlight and whose reproductive organs are contained in a single cell. Two groups of algae, diatoms and dinoflagellates, form the basis for all other life in the sea.

**allantois**  Fetal membrane developing from the hindgut or yolk sac that often contributes to the formation of the umbilical cord and placenta in mammals.

**Allee effect**  A form of density dependence in which the population growth rate slows at smaller population sizes, due to factors such as the difficulty of finding suitable mates. (First described by W. C. Allee, 1931).

**allele**  A unique and discernable (haploid) variant of a locus.

**allelic drop-out**  The "loss" of an allele during a PCR amplification resulting in a false homozygote genotype.

**allomatemal care**  Care provided to offspring by individuals other than the mother.

**alloparenting**  Parental behavior by an animal (male or female) shown toward an offspring that is not its own.

**allopatric**  Describing taxa, species, or populations whose ranges are physically separated.

**allopatric speciation**  Speciation when reproductive isolation is caused by geographical isolation.

**allopatry**  The fact of having separate, nonadjacent geographic ranges.

**allozygous**   Describing alleles that are identical by state but not descent.

**allozymes**   Variant forms of an enzyme, differing in amino acid sequence.

**alternate stable states**   The same species assemblage forming different biological communities in a common physical setting.

**alternating carpus and tarsus**   An arrangement of the carpal and tarsal bones in an alternate way, in the manner of bricks in a wall. This is the primitive condition for mammals as seen from various. Mesozoic mammalian groups. *Cf.* SERIAL CARPUS ANDTARSUS.

**altricial**   Being born in a helpless state and depending heavily on adult care. In contrast, precocial young are born in an advanced state of maturation and require relatively little adult care.

**ambergris**   A grayish waxy substance formed in the intestines of sperm whales; once widely used in perfumes.

**amniote**   An air-breathing vertebrate with cleidoic eggs having amniotic membranes, which can be laid outside water; that is, the nonamphibian tetrapods comprising reptiles, birds, and mammals.

**amphibious**   Able to move, feed, and so on, both on land and in water.

**amphipod**   A crustacean of the order Amphipoda, including those on the ocean bottom, fed on by gray whales.

**anadromous**   A fish that swims up a river from the sea to spawn.

**anaerobic**   Relating to activity or metabolism in which the breakdown of glucose occurs in the absence of oxygen.

**anaerobic dive**   A dive during which enough of the required metabolic energy comes from anaerobic pathways to result in net production of anaerobic end products.

**anaerobic metabolism**   Chemical processes that temporarily sustain life in the absence of oxygen.

**anal canal**   The terminal specialized portion of the gastrointestinal system.

**ancestral species**   The species from which a present species has evolved.

**anestrus**   The period of quiescence between two periods of sexual receptivity in cyclically breeding mammals.

**anlage**   (*plural*, **anlagen**)Undifferentiated embryonic cells or tissue from which an organ or part develops; primordium.

**annual pregnancy rate**   The proportion of sexually mature females that are pregnant during any given year, adjusted for the length of the gestation period.

**antecedent**   In operant conditioning, a stimulus (signal or cue) that precedes a behavior emitted by the animal.

**anterior**   Toward the front or head of the body.

**anterior pituitary**   Another term for ADENOHWOPHYSIS.

**anthropogenic**   Created, directly or indirectly, by human activity.

**antidiuretic hormone**   Another term for VASOPRESSIN.

**antilipolytic**   Inhibiting the breakdown of lipids.

**antitropical**   Found in both hemispheres but not in equatorial regions.

**apnea**   A respiratory pause; elapsed time between inspiration and expiration.

**apneustic breathing**   A method of breathing in which the animal rapidly exhales and inhales and then holds its breath for an interval before repeating. This pattern facilitates rapid air exchange while swimming rapidly.

**aponeurosis**   A broad tendon, connecting muscle to muscle or muscle to bone.

**appendicular skeleton**   Bony elements of the limbs.

**Appendix I**   As defined under CITES, this listing includes species that are threatened with extinction.

**Appendix II**   Species listed by CITES as not necessarily threatened with extinction at present, but that may be if trade is not strictly controlled.

**aquatic**   Having to do with water; living in or situated in water, which can be either fresh water or the sea.

**archaeocete**   A cetacean that lacks the synapomorphies of either odontocetes or mysticetes. Archeocetes are restricted to the Eocene.

**Archaeoceti (archaeocetes)**   The suborder of whales that includes the first Cetacea to enter the sea and all their descendants that do not have cranial telescoping.

**Arctoidea**   A phylogenetic grouping that includes the Ursidae (the bear family), Pinnipedia (seals, sea lions, and walruses), Procyonidae (the raccoon family), and Mustelidae (weasels, badgers, otters, and skunks) and that is the sister group to the Cynoidea (the dog family).

**ARGOS**   A joint US–French agency that operates a polar-orbiting satellite system that is used to collect data from and determine locations of transmitters attached to free-ranging wildlife.

**arterio-venous anastomoses (AVAs)**   Vascular shunts between the arterioles and venules in blubber near the body surface.

**artificial insemination**   The introduction of semen into the vagina, cervix, or uterus by artificial means in an attempt to cause conception.

**Artiodactyla**   Even-toed hoofed mammals; the mammalian order that includes hippopotamuses, pigs, peccaries, camels, chevrotains, cattle, antelopes, deer, and giraffes. Among living mammals, the closest relatives of whales.

**asphyxia**   A combination of decreased oxygen (hypoxia), increased carbon dioxide (hypercapnia), and the accumulation of acidic metabolic by-products (acidosis).

**assemblage**   A predictable and particular collection of species within a biogeographic unit.

**assortative mating**   Nonrandom mating in which members of a population pair up with individuals who are either more (positive) or less (negative) like themselves than the average for one or more traits.

**atavistic**   Referring to the reappearance of a characteristic or feature in an individual belonging to a lineage whose members have secondarily lost that characteristic.

**audiogram**   A hearing sensitivity curve drawn as a function of frequency.

**auditory bulla**   The ear bone in odontocetes that houses the middle ear structure.

**auditory meatus**   An opening to the ear.

**autosomal chromosomes**   All chromosomes in the nucleus except the sex chromosomes.

**autozygous**   Alleles identical by state and descent.

**axial skeleton**   The bony elements of the body, excluding the limbs.

# B

**bachelor school**   A term for an apparently loose aggregation of mature or maturing male sperm whales, usually consisting of animals of similar ages.

**backdown procedure**   A procedure used to release dolphins from the purse seine by pulling on the net after it has been pursed.

**baleen**   Plates of dense, hair-like material (keratin) that hang side by side in rows from the roof of the mouth of whales in the order Mysticeti (the baleen whales); forming the "baleen apparatus" for filter feeding on surface plankton. Formerly known as "whale-bone" but bearing no actual resemblance to true bone.

**barnacles** A collective name for various marine crustaceans of the subclass Cirripedia; the adults form a hard outer shell and attach to underwater surfaces such as rocks and ships, as well as to certain whales.

**barrage** In this context, a low gated dam used to divert water for irrigation, flood control, and/or navigation purposes. Normally the gates remain closed during the low-water season and are opened during the high-water season with differing levels of regulation in between.

**basal insectivores** Plesiomorphic representatives of Insectivora (hedgehogs, shrews, moles, etc.). They can be used as a theoretical basis for the calculation of encephalization and size indices and for the derivation of more progressive insectivores and other eutherian mammals.

**basal metabolic rate** Metabolic rate for a adult animal resting within its thermoneutral zone without food in its gut.

**basicranial anatomy** The base of the skull, particularly the size and shape of bones contributing to the auditory bulla; this has proven to be very important for interpreting phylogenetic relationships within Carnivora.

**Bayesian** Referring to a method of drawing inference about unknown parameters in which parameters are assigned a prior distribution independent of new data and a posterior distribution given the new data is calculated. (From Thomas Bayes, 1702–1761, English mathematician.)

**beak** In cetaceans, a term for the elongated anterior portion of the skull that includes both the upper jaw and the lower jaw.

**beaked whales** Members of the family Ziphiidae that includes five current genera: *Berardius*, *Hyperoodon*, *Mesoplodon*, *Tasmacetus*, and *Ziphius*.

**behavior** An observable and measurable event performed by an organism. Can also include nonobservable phenomena (such as emotion, cognition, recall) that can be measurable through other assessment strategies.

**behavioral adaptation** An animal's ability to learn to discern dangerous from nondangerous stimuli, and to react accordingly upon subsequent encounters.

**behavioral ecology** The influence of the environment on behavior.

**behavior medicine** Training programs and behavioral procedures intended to specifically condition behaviors that treat or prevent health disorders.

**bellyflop** A term for a breach in which the animal lands ventrally, on its stomach.

**benthic** Living in or on the ocean floor.

**bicipital groove** A groove on the humerus through which a tendon of the biceps muscle runs.

**binocular vision** The act of seeing an object with two eyes simultaneously, as in humans. Necessary (but not sufficient) for stereoscopic vision.

**binomial name** The scientific Latin name of a species, consisting of two parts: the generic name and the trivial name. Note that the trivial name is not the "species names"; two species can have the same trivial name (e.g., the cetaceans *Stenella attenuata* and *Feresa attenuata*), but each has a unique, two-part species name.

**biodiversity** The diversity of species; a term variously defined but generally construed to mean the relative abundance of different species (and their population size) within a given area, or on earth as a whole.

**bioindicator** A species linked to a particular habitat or biological community.

**bioluminescence** Light emitted by certain marine organisms as the result of a chemical reaction; this may be used to avoid or confuse predators, to attract prey, or to attract mates.

**biomarker** A biological response to a chemical that gives a measure of exposure and sometimes a toxic effect.

**biomass** A measure of the amount of plant or animal matter in a given context or system, expressed in terms of its aggregate weight.

**bipolar** Living or found in both polar regions.

**blow** A term for a cloud of vapor and sea water mixed with air that is exhaled by cetaceans.

**blowhole** The external opening of a whale's ventilatory system. Two blowholes occur in baleen whales and one in toothed whales.

**blue whale unit (BWU)** A measure of baleen whale harvest that reflects oil production. One BWU is equivalent to a blue whale, 2 fin whales, 2.5 humpback whales, or 6 sei whales.

**blubber** A specialized layer of fat that functions as an insulator, found between the skin and underlying muscle of most marine mammals.

**bow-riding** The action or behavior pattern of riding on the pressure wave in front of the bow of a ship.

**bradycardia** A condition of decreased heart rate.

**brainstem-evoked potential** Electrical nervous discharges generated by an external stimulus in a sensory system.

**branchial arches** Embryonic precursors of mammalian jaw, ear, face, and throat structures that are derived from elements homologous to the gill arches of fish.

**breaching** A behavior in which a whale leaps out of the water.

**breeding sites** Traditional land areas (mostly beaches) where adult females give birth and adult males defend territory. In old sealer's jargon, breeding sites were called "rookeries" after breeding aggregations of rooks.

**bridging stimulus** A conditioned stimulus that signals the imminent delivery of reinforcement.

**brine channels** Small pockets of brine in seawater ice that have a lower freezing point than the rest of the ice and often remain liquid.

**bristles** The fine fringe on the inner side of baleen plates that mats to form a sieve for food.

**bristle worm** A marine worm with short legs.

**bunolophodont** Referring to a molar having rounded (bunodont) cusps that are joined to form crests (lophs).

**buoyancy** Upward force exerted on a swimmers body when it is immersed or floating in water.

**burden of proof** The duty of proving a disputed assertion. In ecology, traditionally the burden of proof has been placed on scientists to show that harm to resources will result from some human action. The precautionary principle reverses this standard to require proof that a human action will not harm the resource.

**bycatch** Animals that are caught accidentally in fishing operations; the capture and mortality of all organisms that are not the target species of a fishery.

# C

**calcareous** Composed of calcium carbonate, a chemical compound similar to chalk.

**calf** A young animal dependent on its mother.

**callosity** A patch of thickened, keratinized tissue on the head of a right whale, inhabited by large numbers of whale lice.

**calving interval**   Mean period of time between two successive parturitions for the females of a given population.

**calyptosis larvae**   Developmental series of euphausiid larval stages following the nauplius larvae during which stalked eyes and the abdomen develop.

**cannibalism**   The consumption by an animal of members of its own species.

**carcinoma**   A cancer arising in an epithelial tissue.

**cardiac**   Relating to or describing the portion of the stomach that lies next to the esophageal opening and contains the cardiac glands.

**Carnivora**   An order of living and extinct mammals that includes such species as dogs, cats, bears, and seals.

**carnivoran**   A term applied to all members of the order Carnivora, distinguishing them from other mammals that may be ecological carnivores.

**carnivore**   An organism that feeds on five prey, in the case of marine mammals including fish and cephalopods as well as mammals, birds, and reptiles.

**carnivorous**   Meat-eating; describes the diet of members of the order Carnivora but many other mammals as well.

**carrying capacity**   The maximum population that can be sustained in a given ecosystem without altering the ecosystem in ways that will ultimately reduce the sustainable population.

**catecholamines**   A class of biologically active compounds, including epinephrine (adrenalin) and norepinephrine (nor-adrenalin), produced by cells of the adrenal medulla and other neurological structures.

**caudal**   Having to do with or near the tail or fluke; behind.

**caudal fin**   The fin located on the extremity of the tail of fishes, dolphins, and so on.

**caudal vertebrae**   Vertebrae caudal to the sacral vertebrae in most marine mammals; in cetaceans and sirenia, the vertebrae associated with chevron bones.

**cecum or caecum**   A large blind pouch forming the beginning of the large intestine.

**celomic cavity**   The main body cavity.

**cementum**   A calcified tissue that fastens the roots of teeth to the teeth sockets.

**centrum**   The center part of the vertebra.

**cephalopod**   A member of the group of mollusks including squids, cuttlefish, and octopuses.

**cervical**   Relating to or near the neck region.

**cervical vertebrae**   The vertebrae of the neck.

**cestode**   Any flatworm of the class Cestoda, including tapeworms.

**Cetacea**   An order of living and extinct mammals that includes such species as whales, dolphins, and porpoises.

**cetacean 1.** Any member of the order Cetacea of aquatic, mostly marine mammals that includes whales, dolphins, porpoises, and related forms; among other attributes they have a long tail that ends in two transverse flukes. **2.** Belonging to or describing this order.

**chevron bone**   Any of a number of small bones positioned on the ventral aspect of the caudal intervertebral discs. These bones are common in mammalian tails, but they are particularly large in cetaceans and sirenians. In cetaceans, chevron bones are V-shaped and embrace the arteries that supply the tail.

**chin slap**   A cetacean behavior pattern in which the animal raises its head out of the water and slaps it back down to make a splash.

**chitin**   A tough insoluble polysaccharide, the main constituent of the arthropod exoskeleton.

**Chordata**   One of the phyla (large groupings) of animals. They are characterized by the presence of a chorda (= notochord), a bar that supports the dorsal side of the animal.

**choroid**   The thin, pigmented, vascular layer between the sclera and the retina of the eye. It consists mainly of blood vessels, which provide nourishment to the retina. The choroid also reduces internal reflection of light.

**chromatophore**   Pigment cell or group of cells, which can be altered in shape or color in response to stimuli from the nervous system or hormones.

**cingulum**   A rounded projection at the lower half of the tooth crown.

**circadian**   Describing a biological rhythm with a period of approximately 24 hours.

**circumpolar**   Distributed around the North or South Pole; in this context, referring to the distribution of species living in the cold water masses of the Soudiem Ocean.

**clade**   An evolutionary line; a monophyletic group containing all the descendants of the most recent common ancestor of that group.

**cladistic**   Based on a pattern of descent, with the most closely related (sister) taxa having a common immediate ancestor.

**cladistics**   A method of classifying organisms in which species are delimited in time by branching points on an evolutionary tree (speciation events), with taxa being defined solely on the basis of common ancestry.

**clan**   In this context, killer whale pods that share parts of their vocal repertoire (related dialects), reflecting a probable common matrilineal ancestry.

**classical conditioning**   A type of learning in which a conditioned (learned) stimulus is presented in conjunction with an unconditioned stimulus, creating a conditioned (learned) response. During classical conditioning, learning occurs independent of the ongoing activities of the organism.

**clean-entry leap**   A term for a breach in which the animal returns to the water smoothly, head first.

**click**   In odontoccte cetaceans, a sound of small duration (microseconds) and of narrow band produced at center frequencies between 110 and 150 kHz.

**cline**   Gradual change over a geographic range.

**clupeoid fish**   A member of an order of schooling marine fishes including sardines, anchovies, and herrings that are often observed in the diets of marine mammals.

**coalescence**   Convergence of genetic lineages at some point in the past.

**coda**   A patterned series of 3–20 clicks lasting about 0.5–2.5 seconds, used by sperm whales for communication.

**codon**   The sequence of three nucleotides in DNA or messenger RNA that encodes for a particular amino acid.

**coefficient of variation**   A description of the error associated with a statistical estimation (CV).

**coelom**   The body cavity holding the organs.

**collagen matrix**   A protein web that forms the structure of blubber found in marine mammals.

**collecting duct**   The terminal segment of the mammalian nephron.

**colonization**   Occupation or use of an area for breeding by a group of animals.

**commensalism**   A relationship or association of organisms in which one individual benefits at no expense or damage to the other organism (the host).

**community** An association of interacting groups within a larger population, usually defined by the nature of their interactions or by the place in which they live.

**compensation** A form of density dependence in which the population growth rate slows at larger population sizes.

**competition** Interaction between individuals of the same species (intraspecific competition) or between different species (interspecific competition) such that the simultaneous presence of the two competitors is mutually disadvantageous. In terms of competition between humans and marine mammals, the competitive mechanism may be direct, as when a marine mammal eats a fish that could otherwise have been caught by a fisherman, or indirect, as when a marine mammal consumes a fish that is an important prey species of a commercially desirable fish species.

**competitive group** In this context, a group of three or more humpback whales characterized by a nuclear animal (usually female), a principal escort (male), and one or more other males, who may challenge the principal escort for his position. Competitive groups are assumed to be related to mate choice or mate defense; they are often large and sometimes involve considerable intrasexual aggression.

**competitive release** A phenomenon of competitive interaction that occurs when predation (natural or from increased fishing) causes a reduction in one of two groups that are in competition. The second group then increases in size as a result of this "competitive release."

**computed tomography (CT)** A method of using X-rays to produce visual "slices" through an object.

**conditioned stimulus** A stimulus that attains the ability to "cue" behaviors as a result of learning.

**confidence interval** In statistics, a range of values, expressed as a minimum and a maximum, in which the real value of an estimate can be placed with a desired confidence.

**connecting chamber** A stomach compartment in cetaceans that lies between the main and pyloric chambers; fined with pyloric epithelium.

**conspecific 1.** Of an organism, belonging to the same species as another or others. **2.** An individual of the same species as another or others.

**contest competition** A mechanism of sexual selection in which males compete by fighting with each other to achieve exclusive access to one or more mating females and in which females typically mate with the winners of such battles.

**context-specific learning** Learning that occurs in an environment in which background stimuli and discrete environmental cues are required for the animal to respond correctly.

**continental shelf** The part of a continent that is submerged in relatively shallow waters.

**continental slope** The slope of the sea floor between the continental shelf and the ocean floor, which is steep and 150–4000 m deep.

**convergence zone** An equatorial region where two north–south currents meet, forming an east–west current.

**convergent evolution** The evolution by unrelated organisms of structures that are similar in appearance or function.

**copepod** An organism of the order Copepoda; a small crustacean.

**Copepoda** An order of very small planktonic crustaceans present in a wide variety and great abundance in marine habitats, forming an important basis of ecosystems.

**corpus** Scar on the ovary resulting from ovulation.

**cortex** The outermost layer of tissue of an organ.

**cortisol** A steroid hormone of the adrenal cortex that is involved in the regulation of protein and carbohydrate metabolism.

**cosmopolitan** Found in most parts of the world and under varied ecological conditions.

**costal** Wing-like or lateral.

**cotyledon** A unit of the placenta grossly visible as a lobe circumscribed by a deep cleft.

**cranial** Having to do with the head or nose.

**cranial muscle** A muscle innervated by one of the cranial nerves.

**cranial nerve** A nerve that leaves the central nervous system near its anterior end and enters the periphery of the body by traversing one of the foramina of the skull. There are 12 named pairs of cranial nerves in all mammals.

**cranial vertex** The highest point of the skull immediately caudal to the bony external nares. The nasal, maxillary, pre-maxillary, and frontal bones contribute to this region.

**crepuscular** Having to do with or active at the margins of the day, that is, at dawn and/or dusk.

**Cretaceous** A period of geological time, from about 144 to 65 million years before the present, ending with a mass extinction.

**critically endangered** As defined by the IUCN, a taxon that faces an extremely high risk of extinction in the wild.

**critical period** A period of rapid acquisition that is usually associated with early stage development.

**crossbow whaling** A whaling method using a harpoon that is shot from a rubber-powered crossbow instead of a cannon, originating in Okinawa in 1975 to take advantage of a loophole in whaling regulations.

**cross-sectional studies** Research that captures a snapshot in time.

**crown group** A clade that includes the most recent common ancestor of all living species, plus all its descendants. Examples in Cetacea: Odontoceti and Mysticeti.

**crus** The middle region of the hind limb, consisting of the tibia and fibula and surrounding structures.

**cryopreservation** A process of maintaining the viability of cells, tissues, or organs by storing them at very low subzero temperatures.

**cull 1.** To remove certain individuals from a population to control the overall population. **2.** An individual removed in this manner.

**culling** The process of reducing the population level of a given species. Distinguished from HARVEST, which refers to taking animals for commercial purposes.

**cursorial** Adapted for moving quickly over land; fast-running.

**cusp** A point on the grinding surface of a tooth.

**cutaneous** Referring to the skin surface of an animal.

**cyamid** A term referring to whale lice, all of which belong to the family Cyamidae.

**cyclonic** Describing rotation in the same direction as the earth (clockwise in the Northern Hemisphere and counterclockwise in the Southern Hemisphere).

**cytochrome _b_** A protein involved in respiration, coded for by an 1100-bp region of the mitochondrial genome.

# D

**dactyli** The fingers or claws on appendages of invertebrates.

**deciduous dentition** The first generation of teeth of a mammal that erupt early in life and are later replaced by a second generation of permanent teeth. Unlike other mammals, modern cetaceans only have one tooth generation.

**deciduous tooth**   A tooth that will be replaced, a replacement tooth, as opposed to a permanent tooth.

**deep scattering layer (DSL)**   Organisms associated with the edge of the light, or photic, zone in deep water. The DSL tends to migrate vertically on a daily-nightly (diel) basis, being many hundreds of meters below the surface in daytime and closer to the surface at night.

**definitive host**   The host in which a parasite achieves sexual maturity.

**delayed implantation 1.** A period of suspended development or growth. **2.** Specifically, in pinnipeds, the suspended development of an embryo between shortly after conception and subsequent attachment (implantation) to the uterine wall.

**delphinid**   A dolphin or small whale species belonging to the family Delphinidae.

**demersal**   A fishing term referring to marine resources captured near the bottom.

**demographic connectivity**   Primarily a function of the level of female dispersal between natural groupings.

**demographics**   Population characteristics such as growth rate and age structure.

**demography**   The birth and death rates that determine a population's dynamics.

**density dependence**   The dependence of population growth rate on population density or size.

**density-mediated indirect effects**   Ecological effects that result from the rate of consumption of one species by another.

**dental formula**   A numerical summary of the numbers of teeth of each class (incisor, canine, premolar, molar) in the dentition of a mammal. For example, a dog has three incisors, one canine, four premolars and two molars in the upper jaw, and three incisors, one canine, four premolars, and three molars in its lower jaw; thus its dental formula is 3.1.4.2/3.1.4.3.

**dentinal growth layer group**   A layer of dentine, consisting of one translucent and one opaque zone when examined in the longitudinal, thin section of a tooth (GLG). The number of these growth layer groups is used to indicate the age of toothed whales.

**depleted**   A term used in the US MMPA to indicate that a species or population is below its Optimum Sustainable Population level or that a species or population is listed as endangered or threatened under the US ESA.

**depredation**   An act of facilitated predation, in which an animal raids or plunders something already caught or otherwise claimed by a fishery.

**derived**   Referring to a character or structure of an organism that has been modified relative to its ancestor; an evolutionary "new" feature.

**Desmostylia**   Extinct "seahorses," a group of herbivorous marine mammals characterized by cheek teeth consisting of clusters of enamel/dentine columns.

**diaphragm**   A musculotendinous sheet between the pleural and peritoneal cavities of mammals; generally considered the most powerful muscle associated with breathing.

**diastole**   The period of ventricular relaxation during the contraction–relaxation cycle of the heart.

**diatoms**   Single-celled algae abundant in marine and freshwater environments.

**diel**   Occurring on a 24-hour cycle.

**die-off**   Mortality on a large scale. This may involve more than one species and can occur over days, weeks, or months.

**differential reinforcement**   A procedure in which reinforcement occurs for any behavior other than the target behavior. The effect is to decrease the target behavior while increasing other (more appropriate) behaviors.

**dimethyl sulfoxide (DMSO)**   A chemical compound that increases cell wall permeability.

**dimorphism 1.** A difference in form (body size, shape, or color) between two individuals or two groups of individuals. **2.** see SEXUAL DIMORPHISM.

**dinoflagellates**   Single-celled algae, which, unlike diatoms, are capable of active movement.

**diphyodont**   Developing two successive sets of teeth.

**diphyodonty**   Erupting two sets of teeth, deciduous (milkteeth) and adult.

**diploid**   Two complements of chromosomes (in mammals one paternal and one maternal).

**direct fitness**   The gene contribution of an individual to the ensuing population through its own offspring.

**dispersal**   Outward spreading of organisms from their point of origin.

**display behavior**   Behavior that is evolutionarily specialized for some form of communication, such as bird song. Special structures, colors, or coat patterns commonly evolve as parts of displays.

**distinct population segments or DPS**   A term applied in the US ESA to describe a population of organisms that is discrete from the other populations of its taxon and represents a significant (in terms of numbers or range) or ecologically unique portion of a vertebrate species or subspecies. Also called a "distinct vertebrate population" in the Act.

**diurnal**   Active during the daylight hours, while inactive or sleeping at night.

**diversity 1.** See GENETIC DIVERSITY. **2.** See BIODIVERSITY.

**diving response**   A suite of physiological and biochemical reactions that are activated when a marine mammal dives, in order to conserve oxygen and extend dive time.

**dolphin**   Any of the small cetaceans in the families Delphinidae (under about 3–4 m in length), Iniidae, Pontoporiidae, Lipotidae, and Platanistidae.

**dolphin-assisted therapy**   The use of dolphins to assist in the treatment of such human disorders as depression, autism, cerebral palsy, or mental retardation.

**dominant allele**   At a dominant locus only the dominant allele is detectable, making it impossible to discern between homo- and heterozygote genotypes.

**Doppler shift**   A change in frequency due to the relative motion of a transmitter and observer.

**dorsal**   Toward or relatively near the back and corresponding surface of the head, neck, and tail.

**dorsal cape**   A region of solid color extending along the forward dorsal surface of a dolphin and sweeping up behind the dorsal fin.

**dorsal fin**   The fin on the top midline of the body, as in dolphins.

**dorsoventral**   From top to bottom.

**drag**   The physical force resisting the movement of a body or limb through water or another fluid medium.

**drive fishery**   A style of fishing for dolphins in which speedboats are used to corral a school of dolphins into a bay or shallow water. A net is drawn across the mouth of the bay or around the school, which prevents their escape. The fishermen then wade among the dolphins to kill them. Usually, the entire school is killed.

**Drosophila**   A genus of fruit flies often used as experimental subjects in laboratory studies of chromosome structure and evolution, including *D. mehnogaster*, the familiar fly-in-a-bottle of high school and college genetics laboratories.

**ductus arteriosus** Artery connecting pulmonary arteries and aorta prenatally, shunting oxygen poor blood into the distal part of the arch of the aorta. After birth it becomes a ligament, the ligamentum arteriosum.

**dugong** A strictly marine, herbivorous mammal foraging at the bottom, primarily on seagrasses, and propelling itself with a forked caudal fluke.

**duodenum** The proximal part of the small intestine that is fixed to the dorsal abdominal wall and does not have mesenteries.

**dura mater** The outermost protective layer covering the brain and spinal cord in vertebrate animals.

# E

**easterly** Blowing from the east.

**eared seals** Seals of the suborder Otariidae. They propel themselves with their fore flippers underwater and "waddle" with all four limbs on land. So termed because they have external pinnae, or ear flaps, visible to the casual observer.

**earless seals** Seals of the suborder Phocidae. These "true seals" all use blubber to thermoregulate, do not have external ear pinnae, use their hindlimbs to propel themselves in water, and have a hunching, caterpillar-like mode of locomotion on land.

**echolocation** The production of high-frequency sound waves and reception of echoes to locate objects and investigate the surrounding environment.

**ecology of fear** Ecological effects resulting from the behavioral avoidance of the risk of predation.

**ecophenotypic** Resulting from effects of the environment rather than the genome.

**ecosystem** A biological community and its environment, functioning as a unit in nature.

**ecosystem model** A mathematical representation of an ecological system that attempts to include all the major components of the system.

**ecotype** Ecological form or variation of a species.

**ectoparasite** A type of parasite that resides on the external surface of its host.

**ectotympanic** One of the bones that make up the skull in mammals. In primitive mammals, it is in the form of a horseshoe, and the tympanic membrane (eardrum) is suspended by it; this shape is retained by sirenians. In most mammals, the ectotympanic consists of additional flanges of bone that may form the wall of the middle ear (as in cetaceans and pinnipeds) and that may fuse with other skull bones.

**effective population size** The number of individuals in an ideal population with a level of genetic variation equal similar to that observed, that is, the number of individuals participating in breeding.

**electrolytes** Charged ions such as sodium, chloride, potassium, calcium, and magnesium that are important components of the internal fluid of an organism.

**embryonic diapause** Delayed implantation of the blastocyst (embryo) in the uterine wall and a pause in the process of development. Implantation reactivates the blastocyst and allows placental gestation to proceed.

**emmetropia** The proper eye refraction, as opposed to myopia (near-sight) and hyperopia (far-sight).

**empirical** Describing phenomena that are directly observable or measurable or that require a minimum of models and assumptions to be characterized.

**encephalization** An evolutionary increase in brain size relative to body size.

**encephalization quotient** The numeric comparison of brain size to body size.

**endangered** As defined under the US ESA, "any species which is in danger of extinction throughout all or a significant portion of its range". The IUCN Red List defines the term to mean a taxon that is not critically endangered, but which nonetheless faces a high risk of extinction in the wild in the near future.

**endemic** A species or race which is restricted or peculiar to a locality or region.

**endocranium** The internal space within a cranium that houses the brain and surrounding tissues.

**endocrine** Referring to a ductless gland that secretes hormones.

**endocrine gland** Ductless gland that secretes hormones.

**endotherm** An animal that generates its own body heat for thermoregulation. Mammals and birds are endotherms. Also known as "warm-blooded" in common literature.

**endothermic** Producing its own body heat for temperature regulation; in popular use, warm-blooded.

**endothermy** The physiological condition wherein the body temperature of an animal is controlled by the generating of heat produced in its own body.

**energy reserve** The amount of energy usually stored as fat in the body of the animal that is excess to current requirements.

**entoconid** The cusp (elevated part of a tooth) on the posterolingual side of the posterior heel of a lower cheek tooth.

**environmental carrying capacity** See CARRYING CAPACITY.

**Eocene 1.** The geologic epoch spanning 55–34 million years ago during which most, if not all, of the lineages leading to modern carnivorans evolved. **2.** The sediments deposited in this time period.

**epaxial muscles** Muscles that form, in the embryo on the dorsal side of the vertebrae. In adult mammals these muscles form the back of the neck, the back, and the dorsal surface of the tail.

**epimeletic behavior** Behavior that supports caregiving, nurturing, and survival in young. This also includes the soliciting of caregiving behavior by other animals.

**epipelagic** Occurring at or near the surface in pelagic waters.

**epithelial cells** Cells that form firmly coherent layers of tissue on exposed surfaces.

**epizoic** Referring to nonparasitic animals that attach themselves to the outer surface of another, normally larger, animal.

**epizootic 1.** Referring to a temporarily prevalent and widespread disease in an animal population. **2.** A disease outbreak of this type.

**escort** A term for an adult male humpback whale that accompanies a female with calf on the breeding ground.

**estrous** Relating to the stage of the mammalian sexual cycle when females are receptive to copulation.

**estrus** A phase of the female reproductive cycle when she is receptive to breeding, commonly called "heat" in dogs and other domestic mammals.

**estrus lordosis** See LORDOSIS.

**etiology 1.** Generally, the assignment of a cause or reason for some condition or event. **2.** The field of science that studies the causes of diseases.

**eumustelids** Fossil and extant mustelids that belong to one of the five currently recognized subfamilies; the Mustelinae (weasels), Melinae (badgers), Melivorinae (honey badger), Lutrinae (otters), and Mephitinae (skunks).

**euphausiid** Any of an order (Euphausiacea) of small, usually luminescent shrimp-like crustaceans, forming an important part of the marine zooplankton; commonly known as krill.

**Euphausiacea (euphausiids, krill)** Order of small, usually luminescent shrimp-like crustaceans forming an important part of the marine zooplankton.

**eustachian tube** A slender tube that connects the tympanic cavity with the pharynx and serves to equalize air pressure on either side of the eardrum; the remnant of the embryonic first gill (or pharyngeal) pouch.

**Eutheria** An infraclass of therian mammals including all living forms except the monotremes and marsupials.

**evoked potential** An electric response of the brain to sensory stimulation; reflects an integrated simultaneous activity of many neurons evoked by the stimulus.

**evolution** Descent with modification, with ancestral species giving rise to one or more descendant species.

**evolutionary stable strategy** Patterns of reproduction or behavior that have equal fitness value.

**exon** Part of the DNA sequence at a coding locus that is translated into amino acids.

**exoskeleton** Hard supporting structure made of chitin, secreted by a crustacean and external to the epidermis.

**exponential growth** Population growth for which a constant percentage is added each year, which means that a greater number of animals are added to the population each year.

**ex situ conservation** The process of maintaining and breeding animals or plants outside their natural environment.

**extant** Still existing, not extinct.

**extinct** A taxon for which there is no reasonable doubt that the last individual has died. A taxon is considered to be *extinct in the wild* when the only surviving individuals exist in cultivation, in captivity, or in a naturalized population(s) well outside the past range.

**extinction** The irreversible disappearance of a population, species, or higher taxonomic group.

**extirpation** The extinction of an organism in part of its range; loss or removal of part of a population.

**extradural** Referring to a location superficial to the dura mater, usually within the vertebral canal.

# F

**facultative** "As need dictates;" referring to those animals that can or do live in one particular environment, but are also reasonably well adapted to another environment. For example, facultative river dolphins live in both ocean and river systems. *Cf.* OBLIGATE.

**falcate** Recurved and usually pointed; sickle shaped.

**fast ice** Stable ice that is attached to land.

**fasting** A limited period of not eating or abstaining from food. Differs from starvation in that it is usually a periodic event for which the animal prepares by laying down extra body fat.

**fathometer** A sonar that is directed downward to measure water depth.

**fatty acid** The main constituent of animal storage lipid, composed usually of even-numbered straight chains of 14–24 carbon atoms that are either saturated (i.e., contain no double bonds) or unsaturated (containing from one to six double bonds).

**fecundity 1.** The potential level of female reproductive performance in a population; that is, the average maximum number of offspring that females may bear during their fertile years. **2.** The actual average number of live births (more specifically, of female offspring) per year per female.

**feedback loops** A process through which substances or conditions occurring in the blood stimulate (positive) or inhibit (negative) the release of hormones responsible for affecting changes that modify the physiological state.

**female choice** A mechanism of sexual selection in which females prefer to mate with males exhibiting particular qualities, such as bright colors or vigorous displays, and reject other males.

**female sexual maturity** The stage of female sexual development in which ovulation occurs. Most cetaceans conceive at first ovulation, and past ovulations are identifiable on ovaries.

**fertilization** Union of the sperm and egg.

**fetal folds** Grooves in the skin that appear as stripes on only one lateral surface of the abdominal region and tail of a cetacean fetus, resulting from the curled fetal position.

**fibrosis** A thickening and scarring of connective tissue, as may result from disease or injury.

**filter feeding** Method used by some marine mammals to strain small particles of food from the water, using baleen or serrated teeth.

**fineness ratio** Ratio of body length to maximum diameter.

**fisheries** The enterprise(s) of catching or harvesting fish for human consumption.

**fish ladder** A human-made structure that allows fish to pass upstream with no more effort than they would use to swim against natural river rapids.

**fissipeds** The group of four-legged, mainly carnivorous mammals with toes separate from each other; includes bears, otters, dogs, cats, raccoons, and hyenas.

**fitness** The ability of an organism to survive and to transmit genetic information to future generations.

**flaw zone** An area of labile, fractured or broken ice; caused by various factors that result in impingement or interaction among different types of sea ice, for example, between drifting pack and shore-fast ice.

**flensing** The process of removing the blubber from a marine mammal carcass.

**flipper** The flattened forelimb of a marine mammal.

**flippering** The act of slamming the pectoral fin, or flipper, on the surface of the water.

**flotsam** The traditional term for bits of debris floating at the surface or in the water column, usually but not necessarily of human origin. Contrasted with **jetsam,** which is human debris that has been purposefully discarded.

**flukes** The horizontally spread tail of a whale (in contrast to the vertical tails of fish).

**fluking** The act of raising the tail, or flukes, above the surface of the water during the beginning of a dive.

**food web** The interconnection of species through their diets.

**food web competition** A mutually disadvantageous interaction that occurs due to indirect competition for a shared resource, such as when there is overlap of the trophic flows supporting marine mammals and fisheries.

**foraging** Searching for prey; the process of finding, catching, and eating food.

**foramen ovale** Foramen connecting left and right atrium of the heart prenatally, shunting oxygen-rich blood to the left side of the heart. It usually becoming closed over shortly after birth (then called fossa ovalis).

**foregut**  The anterior portion of the digestive tract, which includes the stomach.

**forestomach**  Cetacean stomach chamber lined with stratified squamous epithelium; lies between the esophagus and the main or fundic stomach.

**fossil**  The prehistoric remains of once-living organisms, either extinct or living species.

**free hormone**  Any hormone in the blood not bound to proteins and therefore more readily available to influence cellular functions.

**freeze brand**  A cryothermic procedure for marking animals with numbers or symbols. Metal brands are cooled in liquid nitrogen or a slurry of dry ice and alcohol, and touched to the skin of an animal, resulting in depigmentation of the area.

**fundic**  The portion of the stomach distal to the cardiac portion, containing the fundic glands.

# G

**gadids**  Fish species of the cod family (Gadidae).

**gape**  The mouth in cetaceans, usually referring to the junction of upper and lower lips.

**gastric**  Relating to or situated in the stomach.

**gene**  The coded information inside cells, which tells the body how to make enzymes, proteins, and other molecules necessary for life.

**gene/allele frequency**  The proportion of gene copies in a population of a particular type; for example, if 40% of the alleles in a population are type W (as opposed to w), the frequency of W in that population is 0.4.

**gene flow**  The exchange of genetic material within and between populations.

**genetic bottleneck**  A severe reduction in genetic diversity.

**genetic diversity**  Variation in the chromosomes of a community of organisms, due to a large number of slightly dissimilar ancestors.

**genetic drift**  Random changes in the gene frequencies of finite populations.

**genetic fingerprinting**  A method of uniquely identifying an individual animal from a sample of tissue.

**genital tubercle**  Swelling and protrusion of tissue in the genital region that appears before differentiation into male or female genitalia.

**genotype**  For a diploid locus, the combined state of both alleles. For a haploid locus the state of the one allele.

**geochronology**  The study of time in relationship to the history of the earth.

**geologic range**  The geologic time span between the first and last appearance of fossil taxa.

**gestation**  The period of time during which the embryo and later the fetus is carried by the mother in the pregnant state.

**gilling**  The process of catching a fish by its gill covers, usually in a mesh of netting, so that the twine, under the gill covers, prevents the fish from moving backwards, while forward movement is prevented by the mesh of the net encircling the fish's head.

**gill net**  A fishing net that is suspended in the water more or less vertically and ensnares fish by the gill covers as they try to swim through its meshes.

**glucagon**  Protein hormone formed in the pancreas that when released to the blood increases the glucose content of the blood by increasing the breakdown of glycogen in the liver.

**glucocorticoid**  Steroids such as cortisol and corticosterone produced by the adrenal cortex and affecting a broad range of metabolic and immunologic processes.

**gluconeogenesis**  The synthesis of glucose from other carbon sources, occuring primarily in the liver.

**glycogen**  The polysaccharide in which carbohydrate is most commonly stored in muscles and liver.

**glycolysis**  The series of reactions that convert glucose to pyruvic acid.

**grade**  A cluster of species that have a common pattern of organization but are not necessarily a clade.

**granuloma**  The product of a type of inflammation, resulting in the formation of distinct nodules of inflammatory tissue.

**gravid**  Pregnant with young (whether live young or eggs).

**gross annual recruitment rate**  The product of the proportion of females in a population, the proportion of sexually mature females, and the annual pregnancy rate.

**ground**  A term for an area of ocean where whalers find abundant sperm whales.

**group selection**  Differences in the attributes of groups are related to differences in group survival, particularly when groups are competing.

**growth layer group (GLG)**  A repeating or serirepeating pattern of adjacent groups of incremental growth layers within the dentine, cement, bone, or other persistent tissue, which is defined as a countable unit. Such a unit must involve at least one change in appearance.

**growth layers**  Incremental layers deposited in structures that persist and reflect growth or physiological processes at the time of deposition.

**groyne**  A transverse structure that deflects water flow to prevent erosion of an embankment.

**gubernaculum**  A fibrous cord connecting the testis to the scrotum that usually plays a significant role to testieular descent.

**gyrification**  The degree of folding of the cerebral cortex of the brain.

# H

**habitnation**  A gradual waning of behavioral (or physiological) responsiveness over time, as an animal learns that a repeated or ongoing stimulus is lacking in significant consequences, *Cf.* SENSITIZATION.

**hair cell**  A special kind of cell that has tiny hairs projecting from its surface into the intercellular space. Movement of the hairs is registered by neurons that contact the hair cell. Hair cells are found in the ear of mammals.

**halogenated organic compounds**  Organic compounds, usually andripogenic or produced by the breakdown of anthropogenic compounds, that contain chlorine or fluorine atoms. Many, including organochlorines such as DDT and the PCBs, break down slowly when released into the environment and are potentially pathogenic.

**hand-harpoon whaling**  A historic Japanese whaling technique that used hand harpoons and lances.

**haploid**  One complement of chromosomes.

**haplotype**  Unique sequence of DNA.

**harem**  A group of females whose breeding is controlled by a single male who seeks to prevent other males from breeding with diem.

**harpoon**   A fishing spear composed of a barbed head, a shaft, and a line that connects the shaft to a speared animal.

**harvest**   The deliberate taking of wild animals, usually by hunting, netting, or trapping (not including live capture or removals for scientific purposes).

**haulout**   The act of a seal leaving the ocean and crawling onto land or ice.

**heat increment of feeding**   Increase in metabolism associated with the processing and digestion of food.

**hemoglobin**   The iron-containing protein that binds to oxygen and is found in red blood cells.

**hemopoietic tissue**   Tissue in which blood cells are formed (e.g., red blood cells are formed in the spleen of developing mammals and in the bone marrow of adults).

**hepatopancreas**   Gland in many invertebrates that secretes digestive enzymes.

**hepatic**   Relating to or occurring in the liver.

**herbivore**   A feeder on plants.

**herbivorous**   Plant-eating.

**hertz**   A measure of frequency corresponding to one cycle per second.

**heterodont**   Having teeth of different types.

**heterodont dentition**   A type of dentition in which there are different tooth shapes in different parts of the mouth: incisors and canines in front and premolars and molars in back. *Cf.* HOMODONT DENTITION.

**heterodonty**   Having teeth that differ in shape.

**heterospecific**   Individual coming from a different species.

**heteroxenous**   Describing a parasite life cycle involving more than one host.

**heterozygosity**   The probability of selecting two different alleles at random from one or more target loci denotes the expected heterozygosity. In contrast the observed heterozygosity describes the proportion of heterozygous individuals.

**heterozygous**   Having one or more pairs of dissimilar alleles (alternative forms of a gene).

**hilus**   The point where the renal artery enters the kidney and the renal vein and ureter exit.

**Hippopotamidae**   The mammalian family that includes extant hippopotamuses and their close extinct relatives.

**histogram**   A graph showing statistical distribution of the probability of events, for example, how frequently cells of one or an other size appear in the overall cell population.

**holotype**   A single specimen laying at the base of the scientific name of a species or subspecies.

**home range**   The total area covered or traversed by an individual animal during the course of normal activities.

**home range**   The area covered by individuals during the course of their normal activities.

**homodont**   Species having teeth of similar type.

**homodont dentition**   A type of dentition in which all of the teeth have similar size and shape. *Cf.* HETERODONT DENTITION.

**homodonty**   See HOMODONT.

**homologous**   Pertaining to a relationship between corresponding parts in different organisms due to evolutionary development from the same structure, such as the wing of a bird and the flipper of a dolphin.

**homoiotherm**   Having a constant body temperature, typically inferred as warm-blooded.

**homoplastic**   Describing a character whose presence in various species is not due to a common origin.

**homoplasy**   Similarity in character state due to other phenomena than inheritance.

**homozygosity**   The probability that two alleles selected at random are identical at the target locus. In contrast the observed homozygosity describes the proportion of homozygous individuals.

**homozygous**   Having one or more pairs of identical alleles (alternative forms of a gene).

**hormone**   An organic compound produced by an endocrine gland that affects other tissues and organs within the body; for example, the hormone progesterone is secreted by the corpus luteum and stimulates the uterus for implantation.

**hormone-sensitive lipase**   An enzyme inside adipose cells that mobilizes free fatty acids stored in triglycerides and makes them available to other cells.

**hydrofoil 1.** An asymmetrical blade whose shape produces lift forces due to differential fluid flow around its upper and lower surfaces. **2.** A wing-like limb flapped underwater to produce lift; for example, penguin "wings."

**hydrophone**   An underwater microphone.

**hyoid apparatus**   A set of bones and/or cartilages attached to the base of the skull that help move the tongue and larynx during feeding and swallowing.

**hyperglycemic**   Describing a condition in which blood glucose concentration is greater than normal.

**hyperoxic**   Describing a gas mixture having a higher concentration of oxygen than in normal air.

**hyperphagia**   Intense feeding.

**hyperventilation**   Increased total ventilation produced by tidal volume and ventilation rate above the resting rate.

**hypoconal crest**   The crest extending to the cheek side from the hypocone on an upper cheek tooth or to the tongue side from the hypoconid on a lower cheek tooth.

**hypocone**   The cusp on the postero-lingual side of an upper cheek tooth of a derived placental mammal having quadritubercular (four-cusped) teeth.

**hypoconid**   The cusp on the postero-labial side of the posterior heel of a lower cheek tooth.

**hypoinsulemic**   Describing a condition in which the level of insulin in the blood is lower than normal.

**hypometabolism**   A metabolic rate that is less than resting or basal metabolism.

**hyponatremia**   An abnormally low blood concentration of sodium.

**hypoxia**   Reduced availability of oxygen.

**hypaxial muscles**   Muscles that form, in the embryo, on the ventral side of the vertebrae. In adult mammals, these muscles are located in front of the neck, the front of the chest and abdomen, the limbs, and the ventral side of the tail.

**Hz**   Short for HERTZ.

# I

**ichthyophagous**   Feeding on fish.

**igneous rock**   Rocks derived from the cooling and solidification of a magma or lava.

**ileum**   The distal part of the small intestine.

**immunodepressive**   Negatively affecting any aspect of the immune response system.

**inbreeding**   Reproduction with a mate more closely related to the individual than the average pair in the population are related to each other.

**inbreeding depression**  Diminution of growth, fertility, or survival as a consequence of inbreeding.

**indirect effects**  Ecological effects of one species on another involving one or more intermediate species.

**indirect fitness**  The gene contribution of an individual to the population through the relatives that it successfully aided in rearing and that would not have existed without its help.

**infanticide**  The killing of young by parents or other members of the same species.

**infrasonic**  Describing sound that is lower in frequency than the minimum audible to humans. Some baleen whales produce infrasonic calls.

**infrasound**  Sound below the human hearing threshold, or 20 Hz (cycles per second). These sounds approximate a very low frequency roar or vibration.

**insectivorous**  Feeding mainly on insects, and often also on other small invertebrates such as scorpions, spiders, worms, and snails. Insectivores are members of the Insectivora, a group of small mammals such as shrews and hedgehogs.

**in situ conservation**  The process of conserving animals or plants within their natural environment, in habitats within which evolutionary processes are maintained.

**instar**  A juvenile stage of invertebrate development, which may be a configuration quite distinct from the adult.

**insulin**  A protein hormone released by the pancreas that stimulates the uptake of glucose by peripheral tissues.

**integument**  The organ system composed of skin, blubber, and pelage. Functions include protection, thermoregulation, buoyancy, and drag reduction or streamlining.

**intermediate host**  A host in which a parasite, develops to some extent but not to sexual maturity. Where more than one intermediate host is involved in a life cycle, they are ordered in chronological sequence as first, second, and so on.

**intermodal transfer**  The transfer of sensory information from one sensory modality to another, for example, from vision to somatic sense and back and from vision to echolocation and back.

**interspecific**  Involving individuals of different species.

**interspecific mating**  Mating occurring between two individuals of different species.

**intraspecific**  Pertaining to a level below or within a species (e.g., subspecies, population, and deme).

**intrinsic rate of increase**  The maximum rate of increase a species is capable of; thought to occur at a relatively small population size where a species has its greatest rates of reproduction and survival.

**introgression**  The incorporation of genes of one species into the genetic constitution of another species, as by means of interspecific hybridization.

**intron**  The part of the nucleotide sequence at a protein-coding locus that is transcribed into messenger RNA but not amino acids.

**Inuit**  The native people of Arctic Chukotka, Alaska, Canada, and Greenland, popularly known as Eskimos.

**Inupiaq** (*plural*, **Inupiat**)  An Inuit person.

**invasive sampling**  Collection of tissue samples that require penetration of the epidermis.

**involucrum**  The internal medial part of the tympanic bulla of cetaceans, which is made of extremely dense and heavy (pachyostotic) bone.

**ischemia**  A blockage or reduction of blood flow that interrupts oxygen supply.

**ischium**  The dorsal posterior part of the hip bone.

**iteroparous**  Describing or referring to animals that reproduce more than once.

# J

**jejunum**  The middle portion of the small intestine. (From a Latin word for dry; this part of the intestine was often empty in cadavers because of postmortem peristalsis.)

**jetsam**  See FLOTSAM.

**junk**  A complex arrangement of whitish tissue beneath the spermaceti organ, saturated with spermaceti oil.

# K

**karyotype**  A display of the chromosomes.

**kelp**  Algae belonging to the division Phaeophyta, which includes the large "macroalgae" commonly known as seaweeds.

**ketone bodies**  Any of three compounds—acetoacetate, acetone, and β-hydroxybutyrate—that can be produced by the liver as a result of lipid oxidization.

**ketosis**  Accumulation of ketone bodies in the blood, usually associated with prolonged fasting or excessive lipid oxidation.

**keystone species**  A species whose presence in (or absence from) a given ecosystem has a significant influence on the structure and function of the system, disproportionate to its numerical abundance.

**killer whale**  The largest species (*Orcinus orca*) of the family Delphinidae, an aggressive cetacean that preys upon other marine mammals.

**kinematics**  The study of motion. Often used in reference to the movement of limbs relative to the body of active animals.

**kin selection**  Individuals engage in activities that benefit their relatives and thus indirectly contribute genes to the population.

**kleptogyny**  Female stealing by males from the territories of other males for the purpose of mating.

**krill**  A general term describing shrimp-like crustaceans in the family Euphausiidae. Krill form an important food resource for filter-feeding marine mammals.

**k-selection**  The reproductive strategy of a species that tends to grow slowly and have relatively few offspring but that has a longer life span and parental involvement in the rearing of young. Certain large mammals are cited as an example of k-selection. Contrasted with an r-selected species, which has more rapid growth and a large number of offspring, but a shorter life span and negligible parental involvement.

**k-selected**  A species is said to be k-selected when it tends to grow and reproduce slowly but has high survival rates, compared to an r-selected species which has higher life history rates and lower survival.

# L

**lactation**  The production of milk by a female mammal for the feeding of young.

**lactose**  A sugar found in some milks.

**Lagrangian study**  The study of individual movement.

**laminar flow**  A smooth and steady flow in which particles travel along a well-defined path.

**lance**  A spear with a blade on both sides, thrust into a harpooned animal for killing.

**land fast ice**  Polar ice that forms large sheets continuous with the shore. It is generally more persistent and stable than pack ice, which floats free of the land.

**landing sites**  Traditional areas used by nonbreeding males for resting, playing, and other activities. In old sealer's jargon, landing sites were called "hauling-out grounds."

**lanugo**  Wooly hair coat of some fetal or neonatal phocids. In some species (e.g., harbor seals) the coat is shed before birth. In other species (e.g., gray seals) the coat is retained for days or weeks after birth and is thought to provide anti-predator and thermoregulatory functions.

**laparoscopy**  The examination of abdominal organs, including the reproductive tract, by means of an illuminated visual instrument (laparascope).

**large-type whaling**  A historic form of Japanese coastal whaling that took sperm and baleen whales other than minke whales using a vessel and harpoon cannon of large size.

**larynx**  The group of cartilages and muscles in the neck that house the vocal cords and are used for sound production in most mammals. In odontocetes, the larynx is not the main sound-producing organ, and it projects far rostrally, reaching the posterior side of the palate.

**lateral**  Away from, or relatively farther from, the middle of the body.

**lead**  A long linear area of open water or thin, newly formed ice; this occurs along shorelines, at the seaward margin of shore-fast ice, or in the drifting pack; often persistent or recurring.

**lead line**  A weighted line onto which netting material has been fixed in order to keep part of the netting relatively low in the water column, typically on or near the seabed.

**least concern**  A category used by the IUCN to describe a taxon which has been evaluated, but which does not satisfy appropriate criteria to be classified as Vulnerable, Endangered, or Critically Endangered (World Conservation Monitoring Centre, 2000).

**lek**  A traditional display site where males gather to defend small individual territories (that lack essential resources) and advertise for mates. (From a Swahili word for a mating ground.)

**lekking**  The social system in which males gather on a communal display ground, and females choose mates by some criteria of "maleness."

**length-specific speed**  Speed measured using the body length of the animal as a scale and calculated as the velocity of the animal divided by the body length.

**Leydig cells**  Testosterone-producing cells between the seminiferous tubules of the testis.

**life history**  The significant features of the life cycle of an organism, particularly the strategies influencing reproduction and survival.

**life span**  Average probability of survival at birth for the individuals of a given population.

**lighting regime**  The ratio of number of hours of light versus number of hours of dark in a 24-hour period (light:dark).

**limiting resource**  A resource whose relative abundance is a determining factor in the size of a given population.

**lineage**  Ancestor-descendant populations of organisms through time.

**lipoprotein lipase**  A tissue-bound enzyme that hydrolyzes circulating triglycerides in the blood, enabling transport of free fatty acids into cells.

**lithification**  The conversion of a newly deposited sediment into solid rock, involving processes of compaction, cementation, and crystallization.

**lithology**  A description of a rock unit based on characteristics such as color, mineralogy, and grain size.

**lithostratigraphy**  The study of rock strata with emphasis on the lithology, succession, and correlation of strata.

**lobster krill**  Larval crustaceans forming dense, often red, shoals in temperate waters.

**lobtailing**  A behavior in which a whale slams its flukes down on the water, usually repeatedly.

**locus**  A specific location in the genome, for example, a gene or DNA sequence.

**longevity**  The average life span of an individual.

**longitudinal studies**  Long-term research that captures temporal variability.

**long line**  A buoyed line onto which are attached numerous branch lines each terminating in a baited hook. Long lines may extend for tens of kilometers and are usually left to drift in the surface waters of the ocean to catch large pelagic fish.

**lordosis**  A reproductive behavior of females in which they arch their back in a manner that raises their rear end upward to make mounting by the male easier.

**low-frequency active sonar (LFA)**  A naval sonar system designed to detect and locate quiet submarines at long range. Operates at frequencies below 1 kHz (typically 100–500 Hz) and at nominal source levels up to at least 230 dB re 1 mPa.

**luciferase**  An enzyme involved in the chemical reaction to create bioluminescence.

**luciferin**  A substrate involved in the chemical reaction to create bioluminescence.

**lumbar**  Describing or located in the region near to the portion of the vertebral column between the thorax and the sacrum.

**lunge**  A term for a thrusting of the forward part of an animal through the water surface, showing less than about 40% of the body (often the result of feeding at the surface).

**lutrine**  Referring to a member of the otter subfamily (the Lutrinae).

# M

**magnetic declination**  The angle between the earth's surface and magnetic lines of force for the earth.

**magnetite**  A magnetic mineral that is composed of iron oxide.

**main stomach**  The glandular portion of the cetacean stomach, lying between the esophagus or forestomach and the connecting chambers.

**maintenance metabolism**  The energetic cost associated with the maintenance of all physiological processes necessary to sustain an animal.

**maktak**  Northern Alaskan Inuit term for the skin and blubber of whales.

**male sexual maturity**  The stage of male sexual development in which all seminiferous tubules at the testicular center produce spermatozoa. Sperm production starts before this and testicular growth continues after.

**management unit**  A group of animals that is the target of a particular management measure.

**manaq** or **manaqtuun**  Northern Alaskan limit term for a retrieval hook used during ice edge seal hunting.

**manatee** A wholly aquatic herbivorous mammal generalize foraging throughout the water column in fresh or salt water, and propelling itself with a broad caudal paddle.

**maritory** Underwater territories held by males.

**mark–recapture** A set of methods for estimating one or more of abundance, survival, and recruitment by recording repeated sightings or captures of animals, some of which are identifiable from marks previously placed on them. Increasingly, natural marks are used, identified from photographs or DNA fingerprinting.

**marsupial** Members of the Marsupialia, mammals found mainly in Australasia and South America, which give birth to very small young that then develop while attached to the mother's teats, often in a pouch, for example, the kangaroo, opossum, wombat.

**mass die-off** Mortality on a large scale that may result in many strandings.

**mass stranding** The simultaneous stranding of two or more animals other than a female and her offspring.

**mastoid process** A component of the periotic bone (which makes the inner ear) that can be seen on the lateral face of the cranium.

**mate choice competition** A mechanism of sexual selection in which males compete among themselves by attempting to entice and attract females with visual, acoustic, or pheromonal displays; females typically mate with the male having the most exaggerated display.

**maternal** Inherited from the mother.

**maternal care** Any form of maternal behavior that appears likely to increase the fitness of her offspring.

**maternally inherited** Inheritance through maternal cell line such as the mitochondrial gene.

**mathematical model** A representation of a system or a synthesis of observations and ideas that is translated into the language of mathematics to facilitate understanding and analysis of the system or process.

**matriline** A descent system in which daughters reproduce in their mothers' social unit, leading to a group of maternally related females.

**matrilineal** Group members are descendants of a single female.

**matrilineal habitat fidelity** The continued return to habitats first learned from the mother during the months before weaning, which may persist for generations.

**maxilla** or **maxillary bone** One of the two major bones found in the upper jaw of mammals, the other being the premaxilla.

**maxillary crest** The anterior portion of the maxilla, which is enlarged in cetaceans due to telescoping of the cranial bones.

**maximum net productivity level** The population size that will result in the maximum growth rate (in number of individuals added to the population per year).

**maximum sustainable yield** The maximum human harvest from a population that can be sustained continuously and, for a non-selective harvest, that is equivalent to the maximum net productivity level.

**medial** Toward, or relatively closer to, the middle of the body.

**mediastinum** In mammals, the region between the lungs.

**melanin** The pigment responsible for dark coloration in the skin and eyes of many animals.

**melon** A lump of fatty tissue that forms the "forehead" of toothed whales and that is thought to function as a means of focusing sound for echolocation.

**Mendelian inheritance** Loci that are inherited from both parents.

**mendelian transmitted nuclear genes** When the parents pass on to their offsprings discrete genes that retain their identity through the generations.

**menopause** The termination of reproductive function in a female, as a result of aging.

**mesentery** The double layer of peritoneum that suspends the abdominal organs from the dorsal wall; it contains blood vessels and nerves that supply the organs.

**mesonephros** Embryonic kidney that functions only until the metanephros is formed and then regresses. Its duct (Wolffian) is retained in the male as the epididymis and ductus deferens.

**mesopelagic** Referring to the middle portion of the water column, generally from 200 to 1000 m.

**mesosalpinx** A special paired mesentery structure, forming part of the broad ligament, and supporting the oviducts.

**Mesozoic** An era of geological time, from about 285 to 65 million years before the present, comprising the Triassic, Jurassic, and Cretaceous periods.

**metabolic overhead** The amount of energy expended by a lactating female for her own maintenance, relative to the total energy expended during lactation.

**metabolic water** Water that is produced as a result of the oxidation of food or stored body tissue.

**metanephros** Precursor of the kidney that will function postnatally.

**metapopulation** An overall population consisting of two or more subpopulations more or less isolated from each other, treated as a single evolutionary or management entity.

**metatarsal** A bone in the middle portion of the hind foot.

**microallopatric** Sharing a geographic range but occurring in different habitats or portions of a habitat, so that there is no mixing on a small spatial scale.

**microsatellite** A subset of genomic locations (loci) that contain variable numbers of very short tandemly repeated DNAs, often highly variable.

**microsatellite loci** Loci that consists of tandem-repeated nucleotide sequences, each of less than six nucleotides.

**minisatellite** Tandem repeated nucleotide sequences with repeats larger than 20+ nucleotides.

**migration** The process of seasonal movement of individuals between different geographic locations.

**migration count** A sample count of migrating whales carried out from a coastal watch point from which population size is estimated.

**mimicry** Behavior resembling the behavior of others occurring in the same locale or within some relationship of time and space; an important component of early social development. Various forms of mimicry are exhibited by many animals for separate survival strategies.

**mineralocorticoids** Steroids such as aldosterone produced primarily in the zona glomerulosa of the adrenal cortex and regulating the metabolism of ions such as sodium and potassium.

**Minoan** Referring to the Bronze Age civilization centered on the Mediterranean island of Crete (c. 3000–1000 B.C.) prior to the Greek empire. (*Mino* was Crete's legendary king.)

**mitochondria** Small cytoplasmic organelles found in almost all living cells.

**mitochondrial DNA** A circular DNA molecule that is found within mitochondria in large numbers of copies. In mammals, mitochondrial DNA is inherited maternally, the gene sequence and contents are much conserved, and the DNA is composed of about 14,000–17,000 nucleotide base pairs.

**molar** One of the rearmost teeth used mainly for crushing and chewing.

**molt** or **molting** The relatively abrupt shedding of an old outer covering (in mammals, the skin) to be replaced by a new layer recently formed underneath.

**monstrous** Having a single estrous cycle once each year.

**monogamy** A mating system in which one male and one female are together for at least one seasonal breeding cycle.

**monomorphic** All of the same type, that is identical (e.g., all sample have the one and same allele).

**monophyletic** Taxa derived from a single common ancestor.

**monophyly** Phylogenetic term that indicates that a group of related species includes an ancestral species and all its descendant.

**monophyodont** Developing only a single set of teeth, rather than deciduous and adult sets.

**monophyodonty** Erupting a single set of teeth, lacking deciduous (milk) teeth.

**monotypic** Being the sole type in its group (genus).

**monoxenous** Describing a parasite life cycle in which only a single host is used.

**monsoon** A constant, seasonal wind within the tropics.

**morphological** Having to do with the appearance or form of an individual or species.

**morphology 1.** The physical appearance or form of an individual or species. **2.** The study of such properties.

**motorboating** A behavior pattern of very rapid tail slapping (two to three slaps a second) in trains up to 15 seconds in length.

**motor pattern** A pattern of coordinated movements, positions, and postures of the body that is recognizable, repeated, and characteristic of particular classes of movements or functions.

**mtDNA** DNA of the closed, circular genome found in mitochondria.

**multiparous** Describing a female that has had more than one pregnancy, as contrasted with nulliparous (no pregnancy) and primiparous (first pregnancy).

**multireniculate** Describing a type of kidney that contains many reniculi.

**multispecies model** A mathematical representation that includes a minimum of two interacting species in the system.

**multivariate analysis** Statistical methods that can be used to simultaneously examine multiple variables and their interactions.

**muscular hydroatat** A three-dimensional array of muscles having constant volume, capable of highly controlled, precise, and varied movements, and not dependent on bony attachments for the support of antagonistic actions of the muscles. Examples include the tentacles of squid, trunks of elephants, and tongues and lips of mammals.

*museau de singe* In sperm whales, a valve-like clapper system at the end of the right nasal passage, thought to be the producer of the whale's clicks.

**mustelids or mustelidae** Members of the family Mustelidae of carnivorous mammals, which includes the otters, stoats, minks, and fishers.

**musteloids** All members of the family Mustelidae and its generally accepted sister taxon Procyonidae, as well as those stem taxa that are thought to have given rise to this clade.

**mutation** An alteration in a gene that causes it to produce a faulty product, or no product at all.

**mutualism** Mutually beneficial interactions between species.

**myoglobin** A protein in mammalian muscle that carries oxygen. Hemoglobin has a similar function in the blood.

**mysid** A shrimp-like crustacean of the order Myscidacea.

**Mysticeti (mysticetes)** The toothless or baleen (whalebone) whales, including the rorquals, gray whale, and right whales; the suborder of whales that includes those that bulk feed and cannot echolocate. Their skulls have an antorbital process of the maxilla, a loose mandibular symphysis, a relatively small pterygoid sinus, and the maxillary bone telescoped beneath the supraorbital process of the frontal.

# N

**nanuq** Northern Alaskan Inuit term for a polar bear.

**nares** Openings of the nasal cavity.

**nasal turbinate** Specialized structures in the nasal passage of mammals that allow for the recovery of water.

**nasolaryngeal air sacs** Air sacs of the larynx that are continuous with the nasal passages.

**natal site** The group of islands or rookery where animals were born.

**natchiq** Northern Alaskan Inuit term for a ringed seal.

**natriuresis** Excretion of sodium ions in the urine.

**natural mark(ing)** Any distinctive feature of an animal that is useful to researchers for individual identification, without any involvement by the researcher in producing the mark.

**natural selection** One of several evolutionary mechanisms; it accounts for an attribute becoming more common (selected) because of some benefit that leads to more offspring being produced.

**Nauplius larvae** Developmental series of free-swimming, nonfeeding planktonic euphausiid larval stages. It is the earliest and the most basic type of euphausiid larva in which only three pairs of appendages are present. At this stage in development the trunk segment has not developed, however an eye, or naupliar, is present on the front of the head.

**necrosis** The death of cells or tissue caused by disease or injury.

**nematodes** Cylindrical worm-shaped invertebrates generally known as roundworms; widespread in diverse habitats. Many are parasitic on vertebrates, including marine mammals.

**neocortical** Referring to the dorsal region of the cerebral cortex, which is the most recently evolved part of the mammalian brain.

**neomorph** A new structure produced by evolution, in a species or lineage in which no comparable structure existed previously.

**neonatal** Referring to or occurring in the period immediately after birth.

**neonate** A newborn.

**neoplasia** The process of formation of a new and abnormal growth of tissue.

**nephron** The functional unit of the kidney. The nephron is a tubule with regions specialized for the reabsorption and secretion of various solutes, processes which ultimately produce urine from fluid filtered from the blood. Each kidney (or reniculus) can contain millions of nephrons.

**neritic** Referring to or occurring in waters above the continental shelf.

**net sounder** An acoustic transmitting device used to try to ensure that fish schools enter a trawl. The device is usually fixed to the trawl head line and provides the ship captain with an indication of the position of the device in relation to tie surface and the seabed, and also the position of fish schools underneath the device.

**net whaling**   A historic whaling method in which whales were first driven into nets and then harpooned.

**neurohypophysis**   Downgrowth of the base of the hypothalamus in the brain, containing the secretory extremities of neurons extending from the brain and producing the hormones oxytocin and vasopressin.

**neutering**   Surgical removal of the testicles; often used for permanent contraception.

**neutral buoyancy**   A state in which the upward force of buoyancy is equal to the downward force of gravity.

**niche**   The ecological space occupied by a species.

**nitrogen narcosis**   A marked decrease in motor coordination and mental capacities, caused by increased concentrations of nitrogen gas in the brain and other nervous tissue.

**nocturnal**   Active at night, while sleeping or at rest during daylight hours.

**nominal species**   A term for any properly named and described (authored) species, whether or not it is recognized today as a valid species.

**nonbasilosaurid archaeocete**   An archaeocete in the families Pakicetidae, Ambulocetidae, Remingtonocetidae, or Protocetidae. Members of this group were formerly included in Protocetidae.

**nonessential amino acids**   Amino acids that can be synthesized.

**nonesterified fatty acids**   A fatty acid that is not attached to a glycerol molecule.

**noninvasive sampling**   The collection of tissue samples in a manner that does not require penetration of the epidermis.

**nonsynonymous mutations**   Single nucleotide substitutions that result in amino acid replacements.

**nontarget species**   A species that is taken in by a fishery even though it is not one that the fishery is primarily concerned with catching.

**Norwegian-type whaling**   A technique of modern whaling, using a harpoon shot from a cannon mounted on a motor-driven vessel.

**nulliparous**   Describing a female that has had no pregnancy, as contrasted with multiparous (more than one pregnancy) and primiparous (first pregnancy).

# O

**obligate**   Those animals consigned to a particular ecological regime by their evolutionary history. Obligate river dolphins can only live in fresh water, *Cf.* FACULTATIVE.

**observational learning**   A process in which animals can acquire changes in their own behavior by observing as another or others engage in some behavior and experience positive or negative consequences.

**oceanic**   Relating to or occurring in deep water off the continental shelves.

**oceanic front**   The point at which two masses of seawater meet that differ in temperature, salinity, or both.

**Odontoceti (odontocetes)**   The toothed whales, including sperm and killer whales, belugas, narwhals, dolphins, and porpoises; the suborder of whales including those able to echolocate. Their skulls have premaxillary foramina, a relatively large pterygoid sinus that extends anteriorly around the nostril passage, 'and the maxillary bone telescoped over the supraorbital process of the frontal.

**offshore killer whales**   A provisional name for a genetically distinct population of killer whales recently identified from British Columbia, California, and southeastern Alaska, which appears to range mostly over continental shelf waters rather than nearshore waters. They are not observed to associate with either resident or transient killer whales, and their feeding habits are unknown.

**ommatidium**   A single functional unit of a compound eye in an arthropod.

**oocyte**   A developing egg cell in one of two stages. The primary oocyte begins differentiation near the time of birth; maturation is arrested until after puberty, when recruitment of the egg causes further differentiation toward potential ovulation.

**operant conditioning**   The process of teaching a specific behavior by rewarding or reinforcing the occurrences of that behavior.

**optimum sustainable population**   Term used in the US MMPA for a population the size of which lies between the presumed carrying capacity of the local environment (K) and the population's maximum net productivity level (MNPL).

**organochlorines**   Synthetic compounds containing chlorine atoms such as PCBs and pesticides.

**organogenesis**   The formation of organs and organ systems.

**osmolality**   A measure of the osmotic pressure of a solution.

**osmolarity**   Changes in the properties of a solution that result from the number of particles that are dissolved in it. The greater the number of dissolved particles, the greater the osmolarity of the solution.

**osmoregulation**   The process by which an organism maintains its internal salt and water composition constant.

**osmotic**   Referring to properties of solutions that depend on concentrations of dissolved salts and other substances.

**osteological**   Relating to the bony structures.

**osteology**   The study of bones and their structures.

**osteoporosis**   Extreme thinning of cortical compact bone, reduction in number of trabeculae, and overall bone loss.

**osteosclerosis**   Increased density of bone due to replacement of spongy or cancellous bone tissue by compact bone tissue. In several groups of marine mammals, this condition is normal and characteristic rather than pathological as it would be in humans.

**Otariidae (otariids)**   The eared seals (sea lions and fur seals), which use their foreflippers for propulsion.

**otolith**   One of three calcareous bodies in the ear of a bony (i.e., teleost) fish, used in orientation.

**overexploitation**   The reduction, as a result of overfishing, of the size of a population to below the level that provides the maximum sustainable yield. Sometimes referred to as "biological overexploitation," because there can be circumstances where it is economically optimal to reduce populations below their maximum sustainable yield levels.

**oxidation**   The "burning" of food items for energy by the body. This biochemical process produces energy by breaking down fats, sugars, and proteins.

**oxygen store**   The total amount of oxygen available for aerobic metabolism. This oxygen can be a gas in the lungs, bound to hemoglobin in the blood, or bound to myoglobm in muscles.

# P

**pachyosteosclerosis**   Combined condition of osteosclerosis and pachyostosis, typified in the thoracic skeleton of sirenians.

**pachyostosis**   Increased density of bone resulting from expansion of bony cross-sectional area through increased deposition of periosteal compact bone.

**pack ice**   Pieces of unstable seawater ice, often called pans, which vary from a few meters to several hundred meters in diameter. These pieces are not attached to land; instead they drift according to local current and wind conditions.

**paedomorphosis**   The retention of juvenile characters in the adult form.

**pagophilic**   Associated with ice.

**PAHs**   Polycyclic aromatic hydrocarbons; large groups of naturally occurring aromatic compounds containing two or more benzene rings fused together. Also, oil compounds derived directly (petrogenic) or indirectly via combustion (pyrogenic). Some of these compounds, such as benzopyrene, are known carcinogens.

**palatine**   One of the two bones that make up the palate.

**pan bone**   Area in the posterior part of the lower jaw, where the bone is very thin.

**panmictic**   Describing the mingling of genetic material throughout a breeding population.

**panmixia**   Random mating, that is, when mate choice is independent of the genotype at the target locus.

**paramesonephric (Mullerian) duct**   Precursor of the uterus and oviduct.

**pantropical**   Occurring or distributed throughout the tropical regions of the world.

**parapatric**   Referring to populations, species, or taxa whose ranges border each other but do not overlap.

**parapatry**   The fact of occurring in adjacent or slightly overlapping geographical areas.

**paraphyletic**   Taxa that are not derived from a single common ancestor.

**paraphyly**   Phylogenetic term that indicates that a group of related species includes an ancestor with some, but not all of its descendants.

**paratenic host**   See TRANSPORT HOST.

**paraxonic**   Describing the hand or foot of a four-footed animal in which the axis of symmetry passes between digits three and four.

**parental investment**   Behavioral and energetic effort expended by a parent on offspring that benefits the offspring at a fitness cost to the parent. Such fitness costs are often measured as a reduction in fertility or as an increase in mortality due to the effort expended on offspring.

**parity**   Number of pregnancies and/or number of times a female has given birth; from the Latin parere "to bear." Nulliparous females have never been pregnant. A primiparous female is pregnant with her first offspring or has had only one offspring. Multiparous females have had more than one pregnancy or offspring.

**partial pressure**   The portion of the total pressure of a gas mixture attributable to one gas component.

**parturition**   Giving birth.

**paternal**   Inherited from the father.

**pathogens**   Organisms or chemicals which can cause disease as a result of infection or ingestion.

**pathogenic**   Causing disease; harmful to an organism.

**PGBs**   polychlqrinated biphenyls; a family of toxic compounds formerly used in industry and manufacturing and frequently discharged into rivers in chemical wastes.

**peduncle**   In cetaceans, the posterior portion of the body bearing the tail, or flukes.

**pelage**   Hair or fur. A hair canal may contain a single, usually coarse, primary or guard hair (cetaceans, sirenians, elephant seals, monk seals). In some species, however, the hair canal may contain up to 100 (sea otter) smaller, softer underhairs or fur hairs in addition to a guard hair.

**pelagic**   Living or occurring in the open sea.

**pelagic trawls**   Bag-like nets that are towed through the upper parts of the water column rather than near the seabed.

**pelagic whaling**   A whaling operation that does not depend on the use of a land station for processing whales, but uses a floating factory ship (mother ship) or catcher boat with factory facilities.

**pelvis**   A bone complex normally formed from the fusion of three bones (ilium, ischium, and pelvic) that forms the pelvic girdle and attaches to the sacral vertebrae.

**perilymphatic duct**   A duct that carries the lymphatic fluids between the inner ear and the otic space of the cranium.

**perinatal**   Relating to or occurring in the period immediately after giving birth.

**permanent ice**   The core area of ice around both poles that does not melt completely in summer.

**periotic**   The bone housing the inner ear of cetaceans. This complex bone is very dense and is loosely attached to the skull in most odontocetes. It is frequently found as an isolated fossil and may be diagnostic as to the level of genus or species.

**Perissodactyla (perissodactyls)**   An order of living and extinct mammals with hooves distinctive for being "odd-toed," a condition in which the weight of the limb is transmitted along digit 3. Examples include horses, tapirs, and rhinos.

**peritoneal**   Refers to the peritoneum, the lining of the abdominal cavity.

**perturbation experiments**   Studies of species interactions based on response to changes in the distribution or abundance of one of the interactors.

**pes**   The foot, consisting of tarsal (ankle) bones, metatarsals, phalanges, and adjacent tissues.

**petasma**   A complex membranous plate on the inner side of a single pair of limbs in the male of several crustaceans. The petasma enlarges during sexual maturation and is used during copulation to aid in the transfer of spermatophores.

**P450 enzyme system**   A group of enzymes induced by and metabolizing a wide range of natural products (e.g., hormones) as well as unnatural chemicals (e.g., dioxins, some PCBs, dibenzofuranes).

**phalanges**   The bones of the toes.

**pharynx**   The anatomical area posterior to the nasal cavity and mouth, where the passage for air crosses that for food. Air passes from the nasal cavity to the nasopharyngeal duct to the pharynx to the larynx to the trachea. Food passes from the oral cavity to the pharynx to the esophagus.

**phenotype** 1. The visible or observable traits of an organism. 2. More generally, all the collective morphological, physiological, or behavioral aspects of the individual; an expression of the genotype as influenced by environmental factors.

**phenotypic**   Relating to the visible body form that is exhibited by an organism.

**pheromone**   A body secretion that affects or influences the behavior of other individuals, such as a sex attractant.

**philopatric**   Describing individuals that tend to return to breed near the site of their birth.

**philopatry**   Mating as an adult at the site of one's own former birth. See site fidelity.

**Phocidae (phocids)**   A family group within the pinnipeds that includes all of the "true" seals (i.e., the "earless" species). Generally used to refer to all recent pinnipeds that are more closely related to *Phoca* than to otariids or the walrus.

**Phocoenidae**  The family of true porpoises.

**Phoresy**  The transportation of one species by another so as to benefit the transported species.

**photic zone**  The depth in the ocean into which natural light penetrates.

**photoidentification**  The use of photographs to identify animals individually.

**photoperiodism**  The fact of an animal's being affected by phase changes in a day–night or light–dark cycle.

**photophore**  Light-emitting organ composed of a cluster of light-producing cells, a reflector, and a lens. The luminescence produced by photophores in euphausiids is presumed to be an adaptation for swarming and reproduction.

**photopic vision**  Vision in conditions of high luminance.

**phycotoxins**  Pathogens produced by algae.

**phylogenetic**  Relating to the evolutionary ancestry or lineage of an organism or group of organisms.

**phylogenetic analysis**  A method of codifying and comparing heritable biological data for a group of organisms, including molecules, soft tissues, bones, and behavioral data, to formulate a hypothesis of how the organisms are related to each other (a **phylogenetic tree**).

**phylogeny**  The evolutionary relationships among different taxa, usually at the species level.

**phylopatry**  See PHILOPATRY.

**phytoplankton**  Collectively, planktonic plant life.

**pigment**  A dark compound that imparts color to hair, eyes, and skin.

**pinna**  A nap of skin supported by cartilage, and sometimes moved by small muscles, that projects from the side of the head in vertebrates and functions as a sound funnel. In popular use, it is called the ear, but scientists reserve the term "ear" for the much larger organ of hearing, which includes many parts not visible externally.

**pinniped**  Seals and sea lions.

**piscivorous**  Feeding primarily on fish.

**pisiform**  A small bone at the junction of the ulna and the carpus (wrist).

**placenta**  The organ of metabolic interchange between fetus and mother.

**placental**  Describing the members of a major group of mammals whose young develop in the womb supplied with nutriment and oxygen from the mother through a placenta; in contrast to the marsupials, in which significant development takes place outside the womb after birth.

**plankton**  Collectively, the passively floating or weakly swimming plant and animal life of a body of water; usually minute.

**plantigrade 1.**  Walking on the sole with the heel touching the ground. **2.** An animal that moves in this manner.

**plantigrady**  A foot posture in which the entire palm and sole of the forefoot and hindfoot are in contact with the ground in the standing position.

**platform transmitter terminal (PTT)**  A radio transmitter that operates on specific radio frequencies to communicate with earth-orbiting satellites to allow collection of diving and environmental data and determination of the geographic location of the transmitter.

**play**  An imprecisely defined category of behavior, especially in juvenile animals; generally including activities that, while having no immediate identifiable value, may benefit the animal in some other situation or at some later stage in life.

**Pleistocene**  The first of two epochs in the Quaternary, extending from 2 million years ago until approximately 10,000 years ago.

**pleomorphic**  Having more than one distinct form within a single species, group, or life cycle.

**pleopod**  Anterior abdominal appendage modified for swimming.

**plesiomorphic**  Describing shared primitive characters inherited from a distant ancestor.

**plesiomorphy**  The existence of shared primitive characters (**plesiomorphies**) inherited from a distant ancestor.

**plicae circulares (circular folds)**  Folds in the epithelium of the small intestine that serve to increase the surface area.

**plicae semilunares (semilunar folds)**  Folds of the epithelium of the large intestine produced by the longitudinal muscles.

**pneumotachograph**  A mechanical device to measure the volume and rate of flow of air moved in and out of a mammal's lungs.

**pod**  In resident killer whales, a group of closely related matrilines that regularly associate with each other in preference to other matrilies within the community.

**poikilotherm**  Animals whose body temperature varies with the surroundings.

**polar front**  The border between the cold waters of the Southern Ocean and the temperate water masses of the south Atlantic, south Pacific, and Indian Oceans (formerly known as the Antarctic convergence).

**polychlorinated biphenyls**  See PCBS.

**polycyclic aromatic hydrocarbons**  See PAHS.

**polydonty**  Having more than the primitive number of teeth (11 per half jaw in placental mammals; 44 total).

**polyestrous**  Having more than one estrous cycle in a single year.

**polygynous**  Describing a mating system in which one male mates with multiple females during a breeding season.

**polygyny**  Mating system in which one male mates with several females within one breeding season. The term usually refers to a mating system in which relatively few successful males breed with most of the females, while others do not mate. The reverse, one female mating with several males, is rare among mammals.

**polynya**  Areas of open water surrounded by heavy pack ice.

**polyphyletic**  Taxonomic group that includes members (as genera or species) from different ancestral lineages.

**polyphyly**  Phylogenetic terms that indicate that a group of related species does not include the last common ancestor of all species.

**polyunsaturated fatty acids (PUFA)**  Fatty acids which contain two to six double bonds. A fatty acid that has 20 carbon atoms with 5 double bonds, where the first double bond is at the third methyl carbon atom, is named 20:5n–3.

**population**  A group of conspecific organisms occupying a specific geographic region at the same time. "Population" tends to be used informally to connote a distinct biological entity, whereas people may tend to use the term "population" as more of a management unit. US MMPA defines this as a group of animals "of the same species or smaller taxa in a common spatial arrangement, that interbreed when mature". This term is used in the MMPA interchangeably with the term "stock."

**population bottleneck**  A rapid reduction in the size of a population.

**population stock**  See stock.

**porpoise**  Any of several small cetaceans in the family Phocoenidae. The dorsal fin may be absent; teeth are spatulate or vestigial.

**porpoising**  The behavior of penguins, dolphins, or whales leaping at least partially clear of the water surface during rapid swimming.

**postcrania**  A collective term for bones that are not part of the skull, for example, the vertebrae, limbs, hands, or feet.

**postpartum**  After birth; especially, shortly after birth.

**postreproductive**  Describing a female of certain toothed whales that ceases ovulating at an advanced age but well before her expected longevity.

**power stroke**  The part of the movement of a locomotor organ (tail, limb) that contributes most to propulsive motion. The opposite is the recovery stroke.

**practical salinity units (psu)**  The measure of salinity of seawater, approximately equal to parts per thousand.

**precautionary principle**  An ecological management philosophy that seeks to maintain natural resources by managing more conservatively when there is greater ignorance about the status of the resource.

**precocial**  A high degree of maturation and independence exhibited by newborns that require relatively little adult care. By contrast, *altricial* young are born in a helpless state and depend heavily on adult care.

**predation**  A situation in which an animal exerts time and energy to locate living prey and expends additional effort to kill and consume it.

**predator**  An animal that obtains its food primarily or exclusively by hunting and killing other animals.

**predatory**  Describing an animal that obtains its food by preying on other animals.

**preening**  In birds, the act of smoothing or cleaning feathers with the beak of the bill.

**preformed water**  Water that exists as water when it is consumed in the food or drunk.

**premaxilla**  One of the two major bones found in the upper jaw of mammals. The other is the maxilla.

**premaxillary sacs**  Premaxillary elements of the air sac system, which is a net of diverticula of the narial passages between the external nares (blowhole) and the bony nares of the skull.

**premolar**  A tooth behind the large canine, used to slice or chew.

**prepartum**  Describing or occurring in the time before birth, especially shortly before birth.

**prey**  Animal that is hunted and killed by another.

**primary production (productivity)**  Organic matter and/or energy produced or captured by organisms (plants) from inorganic compounds and incident solar radiation.

**primiparous**  Describing a female that gives birth for the first time, as contrasted with multiparous (more than one pregnancy) and nulliparous (no pregnancy).

**primordium**  See ANLAGE.

**probability of identity**  The probability that two different individuals have identical genotypes.

**procumbent**  Describing a tooth (generally apical) that tends to acquire a subhorizontal to horizontal orientation.

**protocetid**  A member of the family Protocetidae, a paraphyletic family that is thought to have given rise to the basilosaurids.

**proximate explanation**  An explanation that describes the immediate or short-term causes of a biological incident or occurrence, in contrast to ultimate explanations, which are historical (specifically, evolutionary) in nature. For example, a threat between neighboring territorial male sea otters can be said to have occurred because the two males encountered one another near their common boundary or had high levels of testosterone (proximate explanation). Or, it can be stated that the threats evolved as a low-risk alternative to fighting (ultimate explanation).

**pseudocervix**  Folds of tissue, usually vaginal in origin, that are distal to the true cervix, but play a similar role or barrier as the true cervix.

**psychophysical test**  A test in which an animal is trained to show the experimenter when it detects a physical stimulus, such as light, sound, or a chemical.

**psychosphere**  Global, nutrient-rich, relatively biologically unexploited deep water.

**pteropod**  Small molluscans with two "wings," with which they swim through the water.

**pterygoid**  An anatomical process descending from the sphenoid bone.

**pterygoid sinus**  An airspace adjacent to the flat pterygoid bones of the skull.

**purse seine**  A type of fishing gear that surrounds fish schools with a vertical curtain of netting that is then pursed, or closed off, at the bottom by means of ropes and heavy winches, in order to prevent fish from swimming out of the bottom of the net.

**pyloric stomach**  The most distal portion of the cetacean stomach, adjacent to the pyloric sphincter; contains pyloric glands.

# Q

**qargi**  Northern Alaskan Inuit term for whalers' ceremonial house.

# R

**race**  An interbreeding subgroup of a species that is genetically distinct and usually geographically isolated from other such groups of the same species.

**radius**  A bone extending from the elbow to the wrist on the outside of the forelimb, opposite the ulna.

**range**  The maximum extent of geographic area used by a species.

**rank**  A designated level in a classification scheme.

**raptoria**  Describing a process of feeding by seizing individual food organisms.

**rare**  Not common, widespread, or generalized; not generally found.

**reciprocity**  Individuals engage in helpful actions that will be repaid by the recipient.

**refractive index**  A parameter of a light-transparent medium determining the degree of light refraction at a boundary between media with different indices. The more index difference, the stronger the refraction. The refractive index of a vacuum is 1, that of air is almost the same, and that of water and some eye media (anterior chamber liquid, vitreous body) is 1.33.

**regime shift**  A radical and persistent change in the state of an ocean-atmosphere system or ecosystem.

**regression**  The mathematical relationship between two variables, for example, body mass and brain mass.

**reinforcement**  An event (reward or punishment) that follows a response and that increases or decreases the likelihood that the response will recur.

**reintroduction**  The reestablishment of a species within its natural range.

**release**  The act or technique of moving an animal that has been born in captivity, or that has been in captivity for long enough that it requires retraining in the skills required to survive, back into the wild so that the animal can become free-ranging and no longer under human supervision.

**relict species** Persistent remnants of formerly widespread biotas, typically existing in specific isolated areas or habitats.

**remora** A type of fish that has a modified dorsal fin to suck onto other fish, marine mammals, or turtles; also called "suckerfish."

**renal** Relating to or occurring in the kidney.

**renal pelvis** A funnel-like structure into which urine drains from the nephrons.

**reniculus** In discrete multireniculate kidneys, the anatomically independent functional units into which the kidney is divided. The reniculus resembles a miniature kidney with a cortex, medulla, vascular supply (renicular artery and vein), and ureteral tubule.

**replacement yield** The number of animals, which if removed from a population stock, results in the same number of animals at the end of the year as in the beginning.

**reproductive ecotype** An array of species of plants or animals adapted to the conditions of some breeding site and having inheritable features.

**reproductive success** Number of offspring successfully reared to independence by an individual adopting a particular strategy.

**reproductive value** The number of offspring fiat an individual of a given age can be expected to produce over the rest of its life discounted by the population growth rate.

**resident killer whales** A genetically distinct population of killer whales found in coastal waters of the northeastern Pacific Ocean, which feeds preferentially on fish and squid. *Cf.* OFFSHORE KILLER WHALES, transient killer whales.

**resilience** The recuperative power of a population to recover from or absorb sudden environmental stress and deprivation.

**resource** Any item, factor, or condition that contributes to the ability of an organism to survive and reproduce; for example, food, space, and mates.

**respiratory acidosis** A decrease in blood pH level, caused by an increase in the partial pressure of carbon dioxide that shifts the normal equilibrium of the bicarbonate-carbonic acid buffer system.

**restriction endonuclease** An enzyme that recognizes a specific (typically 4–8 nucleotides) DNA sequence at which point the DNA strands are digested (cut). A mutation resulting in the change of the DNA sequence will then result in failure of digestion by the restriction endonuclease, yielding another detectable allele.

**rete mirabile** (*plural,* **retia mirabilia**) Complex networks of anastomosis, coiled blood vessels.

**retia mirabilia** See RETE MIRABILE

**reticulation** The fact of crossing over or connection between lines of descent.

**retroperitoneal** Outside of but projecting into the body cavity; completely or partially covered by peritoneum.

**rhizome** An elongate, horizontal subterranean stem of a seagrass or other plant, producing shoots above and roots below and serving for the storage of reserve food material.

**ritualization** An evolutionary process through which behavior becomes partly or completely "emancipated" from its original function(s) and acquires new functions for communication. For example, vocalizations presumably represent ritualized breathing.

**riverine** Lving in rivers; found in or relating to a river.

**rookery** A terrestrial breeding area used by a colony of pinnipeds on a seasonal or permanent basis. (From an earlier use of the term to describe the breeding or nesting area of a large number of rooks, a crow-like bird of Europe.)

**rooster tail** A spray of water formed when certain small cetaceans surface at high speed; caused by a cone of water coming off the animals head. (So named because the form of the spray resembles a roosters feathers.)

**rorqual** Any of seven species of baleen whales (the minke, blue, humpback, fin, Bryde's, Omura's or sei whale) belonging to the family Balaenopteridae, characterized by a variable number of pleats that run longitudinally from the chin to near the umbilicus. The pleats expand during feeding to increase the capacity of the mouth.

**rostral** Having to do with the nose or head; toward the front.

**rostral hairs** Small hairs on the outside edge of the rostrum of a cetacean.

**rostrum** The beak-like projection found at the front of the skull or head of a cetacean; the term also refers to the skeletal support of the upper jaw, comprising the anterior parts of the maxillary, premaxiflary, and vomerine bones of the skull.

**r-selection** The reproductive strategy of a species that tends to grow quickly and have a large number of offspring, but that has a relatively short life span and negligible parental involvement in the rearing of young. Certain insects are cited as an example of r-selection. Contrasted with a k-selected species, which has slower growth and a smaller number of offspring, but a longer life span and significant parental involvement.

**ruminant** A hoofed mammal that feeds by grazing or browsing; it has a complex four-chambered stomach (**rumen**) and thus is able to regurgitate and then rechew its food (i.e., it "chews its cud"). Ruminants include domestic cattle, sheep, goats, deer, antelopes, and giraffes.

# S

**sacral vertebrae** Vertebrae associated with the pelves; a feature that is not present in cetaceans and manatees.

**school** An aggregation offish or other animals that regularly swim together as a unit.

**scotoma** An area of diminished vision within the visual field.

**scotopic vision** Vision in conditions of dim luminance.

**scramble competition** A mechanism of sexual selection in which males compete for access to females by searching and locating females more efficiently than their rivals. Females may mate with one or several males.

**sea cow** A strictly marine herbivorous mammal foraging near the surface, primarily on large marine algae, and propelling itself with a forked caudal fluke. The term is also used loosely for any member of the order Sirenia.

**seagrass(es)** Any of various marine flowering plants belonging to the families Potamogetonaceae and Hydrocharitaceae, consisting of 12 genera and some 50 species.

**sea ice** Frozen seawater, which forms ice floes (i.e., pack ice).

**seal bomb** An explosive noise-making device designed to chase marine mammals away from areas where they are regarded as harmful to fisheries, nets, and so on.

**seasonality** The fact of being affected by or dependent on a change of seasons; for example, changes in population distribution that are related to changes in seasons.

**sebaceous glands** Glands located at the base of mammalian hairs that release an oily secretion, sebum.

**secondary sexual characteristic** A feature of a male or female animal that is produced as sexual maturity occurs.

**sedimentology** The scientific study of sediments, solid fragmentary material transported and deposited by wind, water, or ice, or precipitated from solution, or secreted by various organisms.

**selection**   The evolutionary process favoring individuals with genotypes that enable them to raise the greatest number of young which survive to breed.

**selection coefficient**   A mathematical term indicating the decrease in disfavored gene types contributed to the next generation; for example, a selection coefficient of 0.5 for a given genotype indicates a decrease of 50% of that genotype entering the next generation.

**semiaquatic**   Only partly aquatic, for example, an animal that forages for food in water but otherwise lives and breeds on land.

**seminiferous tubules**   Convoluted loops within the testis where spermatozoa are formed.

**sensillum**   A sensory receptor or a receptor complex in the cuticle of an invertebrate.

**sensitization**   An increased behavioral (or physiological) responsiveness occurring over time, as an animal learns that a repeated or ongoing stimulus has significant consequences. *Cf.* HABITUATION.

**sensor**   A device that reacts to a certain physical stimulus in a quantifiable manner.

**sensory adaptation**   The lessening of a response to a stimulus due to physiological fatigue.

**sentience**   The ability to sense; in particular, the capacity to experience pleasure and pain.

**serial carpus and tarsus**   An arrangement of the carpal and tarsal bones in a serial way; that is, the pattern of relations between carpal and tarsal bones is "pillar-like." Each bone of the first row articulates mostly with only one bone just under. This is a derived condition compared to the generalized mammalian condition. *Cf.* ALTERNATING CARPUS AND TARSUS.

**sessile**   Stationary or fixed, for example, plants growing on the sea floor or mussels attached to rocks.

**sex chromosomes**   Nuclear chromosomes that are involved in sex determination, for example, the X and Y chromosomes in mammals.

**sexual bimaturation**   The sexual maturing of one sex significantly earlier than the other.

**sexual dimorphism**   Phenotypic differences between males and females of the same species.

**sexual selection**   Selection due to the advantage some individuals have over others of the same *sex* in exclusive relation to reproduction. This may lead to the evolution of costly secondary sexual characters found only in one sex (such as a male peacocks tail or a stag's antlers), which may decrease the probability of survival and which therefore is opposed by natural selection.

**SINEs**   Short interspersed nuclear elements, that is, an interspersed repetitive element less than 500 nucleotides long of viral origin.

**shelf**   A sea zone located close to land and having the same geological structure. A shelf zone is 100–200 m deep.

**sibling species**   A species that is morphologically very similar to another, sometimes to the point of seeming identity, but demonstrated to be a separate species on the grounds of genetic, behavioral, or cryptic morphological evidence.

**sister group**   The closest relative with which a taxon has a common ancestor.

**site fidelity**   The tendency to return to mate at the same site repeatedly in different breeding seasons. This term is used instead of PHILOPATRY if the animal's original birth site is not known.

**size index**   An analog of the ENCEPHALIZATION INDEX, related to single brain components rather than the total brain.

**small-type whaling**   Japanese coastal whaling that uses a vessel and cannon of small size. Operated since the beginning of the twentieth century for minke and toothed whales (other than sperm whale); came under the control of the Japanese government in 1947.

**SNPs**   Single nucleotide polymorphisms, the substitution of a single nucleotide with another at a specific location in the genome.

**soak time**   A term for the length of time that a piece of fishing gear such as a gill net is left in the water to fish.

**social learning**   The social transfer of information and skill among individuals by processes that include imitation, teaching, exposure, social support, matched dependent learning, stimulus enhancement, observational conditioning, and goal emulation.

**somersault**   A behavior involving an acrobatic stunt in which the body rolls in a complete circle, head over heels, or, in the case of dolphins, head over tail.

**sonar 1.**   An acoustic system or device that emits sound into water or another medium, detects echoes received from objects, and processes those signals to characterize and/or visualize the objects. **2.** In marine mammals, a means of detecting, identifying, and tracking prey by the emission of sound pulses and the analysis of the resulting echo patterns.

**song**   A series of sounds repeated in a specific pattern.

**sonic lens**   An anatomical structure that focuses sound as optical lenses focus light, that is, the melon of odontocete whales.

**sound**   A form of energy manifested by small pressure variations in a medium (such as water or air). Depending on the strength and frequency of the sound, it may be detectable by an organ of hearing.

**source level**   The acoustic pressure that would be measured at a standard distance (usually 1 m) from a point source radiating the same amount of sound as the actual source.

**species interactions**   The effects of one species on another.

**spectral sensitivity**   Selective sensitivity of visual receptors to a certain region of the light wavelength spectrum that corresponds to different light colors.

**spectrogram**   A graphic representation of sound waves per unit of time.

**spermaceti**   A white, translucent substance, obtained from the distinctive form of oil made of wax esters and triglycerides that is found in the spermaceti organ of sperm whales.

**spermaceti organ**   An elongated, barrel-shaped structure making up much of the sperm whale's head, made of soft spongy tissue filled with spermaceti oil.

**spermatogenesis**   The process by which sperm develop and mature from primitive sperm cells (spermatogonia and subsequently spermatids).

**spermatophore**   Collection of spermatozoa enclosed in a sheath of gelatinous material.

**sperm competition**   A mechanism of sexual selection in which a female mates with more than one male, and males compete to fertilize her eggs by producing more sperm per ejaculation, or by mating with her more frequently than their rivals.

**sphincter**   A muscle that forms a ring and contracts upon itself, closing the structure it surrounds (e.g., the external anal sphincter muscles closes the anal canal).

**splanchnic**   Pertaining to the viscera, such as the liver, stomach, and pancreas.

**sporta peiimedullaris**   A sheet-like structure of connective tissue and muscle that lies between the renal cortex and the medulla.

**spout**   A term for a cloud of vapor and sea water mixed with air that is exhaled by cetaceans.

**sprint**   A high-speed, maximum effort or movement of short duration.

**spyhopping** A behavior that involves raising the head vertically out of the water, then sinking below the surface without much splash.

**stable isotope** An alternate form of an element that is more persistent than the common form.

**statistical power** The ability of an analysis to correctly reject the null hypothesis, that is, not to miss a real effect.

**stem group** One extinct species, or several extinct species, that form an extinct sister group, or sequence of sister groups, to the crown group.

**stereoscopic vision** A capability of vision to discriminate the distance to an object based on a small difference of images (disparity) projected on the retinae of two eyes. This ability requires binocular vision.

**stereotyped behavior** A pattern of repetitive movements, such as pacing back and forth or swaying to and fro, that often are performed by captive animals held in impoverished environments, and that are not associated with any behaviors normally performed by these animals in the wild.

**stochasticity** A random occurrence of events in space and/or time; random variability.

**stock** A group of animals of the same species or smaller taxa, existing in the same locale and interbreeding when mature.

**stored reserves** Body components that are supplementary to immediate requirements, most often of fat and blubber.

**stranding** The directed action of beaching.

**strandling** A term for any marine animal that cannot cope with its immediate situation, either ashore or at sea (e.g., a beached whale, an Arctic seal in Florida waters, or a sea otter drenched with oil).

**strategy** Genetically distinctive rules for individual behavior. Selection favors the strategy that yields the highest reproductive success in relation to competing strategies.

**stratification** A vertical density gradient caused by changes in temperature and salinity with depth.

**streamlining** Body shape that causes little distortion to the fluid flow around it and minimizes drag.

**striated muscle** The skeletal or locomotor muscles of the body.

**strip transect sampling** A sample count method in which counts are conducted within long, narrow strips.

**subarachnoid space** A space under the arachnoid membrane, one of the coverings of the brain.

**subarctic** Living in an intermediate zone between arctic and temperate waters.

**subspecies** A formally described and named geographic entity within a species.

**supraoccipital** A bone at the top of the rear of the skull.

**surface-active group** A group of right whales, typically consisting of one female and one to many males, engaged in vigorous physical interactions at the surface, which appear to be courtship or other social behavior.

**surfing** The behavior of riding on a wave or waves. There are two basic patterns: floating on top of the water and falling down the advancing edge of a wave or riding within a wave being created by water being pushed upward by the movement of a denser medium (such as a whale or vessel). Dolphins engage in both types of surfing or "wave riding."

**survival rate** The probability of an animal remaining alive over a given time period; typically expressed as the proportion of individuals in a population surviving from one period (usually defined as a year) to the next.

**sustainability 1.** The combined capacity of a system and its parts to retain their interactions and other natural qualities. **2.** Specifically, the level of human activity in a given ecosystem that can continue without degrading the ecosystem or seriously depleting resources.

**sustained swimming** Long-duration activity generally performed at low speeds.

**sympatric** Occurring together geographically; coexisting in the same place.

**sympatric speciation** Denotes reproductive isolation in the absence of geographical separation.

**sympatry 1.** The fact of occurring or living in the same geographical area. **2.** Specifically, the existence of two or more populations of closely related species in identical or broadly overlapping geographical areas.

**symphysis** A joint between two bones that consists of connective tissue only and lacks a synovial cavity.

**synapomorphic** Sharing a derived character state.

**synapomorphy** Derived character state shared by several taxa, not the ancestral stage.

**synonymous mutations** Single nucleotide substitutions that do not result in amino acid replacements.

**systematics** A method of taxonomy which attempts to ensure that the categories represent evolutionary history and relationships.

**systemic** Broad or universal in application while accounting for everything of importance in proportion to its importance; applying to, and accounting for, complexity.

**systole** The period of ventricular contraction during the contraction–relaxation cycle of the heart.

# T

**tachycardia** A heart rate above the resting rate.

**tag** Any artificial mark attached to animals.

**tail-slapping** Another term for lobtailing, sometimes used when referring to dolphins.

**take** Term used in the MMPA and US ESA. In the MMPA, "take" means "to harass, hunt, capture, or kill, or attempt to harass, hunt, capture, or kill any marine mammal." In the ESA it means approximately the same thing.

**taxon** (*plural*, **taxa**) A natural grouping of related organisms, usually named, and of a given rank (e.g., order, family, genus, species, subspecies).

**taxonomic(al)** Based on or describing the classification of organisms according to species or higher levels.

**taxonomy** The science of classifying organisms into species, genera, and higher categories.

**telemetry** The remote sensing of environmental, behavioral, and physiological data, particularly from animals and via radio frequency signals.

**telescoping** One of two patterns of cranial sutures found in mysticetes and odontocetes. In mysticetes, the occiput is thrust forward and maxilla floors the orbit. In odontocetes the rostral elements are thrust caudally over the cranial elements.

**telson** An impaired terminal abdominal segment of crustaceans.

**temporal fossa** A cavity or depression in the pair of compound bones that form the sides and base of the skull.

**temporary threshold shift** A temporary impairment in hearing capability caused by exposure to strong sounds.

**tension** The partial pressure of a particular gas in solution.

**terrestrial**　On land; occurring or living on land rather than in water or in the air.

**territorialism**　The fact of establishing and defending a territory (see below).

**territory**　Space on the breeding grounds that adult males maintain by calling, threatening, and fighting other males, and in which they mate without harassment from other males.

**Tertiary**　A period of geological time from about 65 to 2 million years before the present day.

**testosterone**　The primary male hormone, produced in the testis.

**teutophagous**　Feeding on squid.

**thelycum**　A female organ to which spermatophores are attached with a sticky fluid.

**thermal insulator**　A material, such as fur or blubber, that serves to retain heat in an animal. The inverse of *insulation* is *conductance* which describes how easily heat flows through a material.

**thermocline**　Steep vertical temperature gradient below a well-mixed surface layer.

**thermohaline**　Concerning the temperature and salinity of seawater.

**thermoneutral**　Falling within the range of temperatures at which thermoregulation can be maintained.

**thermoneutral zone**　Range of environmental temperatures at which an animal does not require an increase in metabolism to maintain its body temperature.

**thermoregulation**　The active process of controlling body temperature in an animal. Sweating to cool down and shivering to warm up are examples of thermoregulation.

**thoracic**　Describing the region within the thorax, or the portion of the vertebral column within the thorax.

**thoracic vertebrae**　Rib-bearing vertebrae.

**thorax**　The part of the vertebrate body that contains the heart and the lungs, in mammals divided from the abdomen by the diaphragm.

**threatened**　As defined under the US ESA, this term describes species likely to become endangered "within the foreseeable future throughout all or a significant portion of its range" (US Fish and Wildlife Service, 1996, p. 4).

**time-activity budget**　A quantitative description of the amount of time spent by animals in different behavioral activities. Energetic costs of different behavioral activities are used to construct such budgets.

**time-depth or time-data recorder (TDR)**　Mechanical or electronic instruments that measure and store information on diving and swimming behavior and environmental variables of aquatic mammals, birds, fish, and turtles.

**toothed whale**　See ODONTOCETI.

**tooth replacement**　The shedding of deciduous (or milk) teeth followed by the eruption of permanent teeth.

**top-down control**　Consumer-mediated regulation of community structure and function.

**trade winds**　Winds blowing almost continually westward and toward the equator from the subtropical high-pressure belts at latitudes near 30° N and 30° S. Trade winds are from the northeast in the Northern Hemisphere and from the southeast in the Southern Hemisphere.

**trait-mediated indirect effects**　Ecological effects that result from the behavior response of one species to another.

**transition**　The substitution of a nucleotide from one purine to another purine (e.g., guanine to adenine) or from a pyrimidine to another pyrimidine (e.g., cytosine to thymidine).

**transient killer whales**　A population of killer whales found in coastal waters of the northeastern Pacific that is sympatric with residents and offshores, but socially and reproductively isolated foam them, and genetically distinct. Transients feed preferentially on marine mammals and seabirds.

**transport host**　A host infected by a larval parasite stage in which the parasite survives, and can be transmitted to another host, but does not develop. Transport hosts are not required to complete the life cycle but enhance the chances of transmission to the definitive host.

**transverse septum**　A membrane that functions as a separator between the heart and the liver.

**transversion**　The substitution of a nucleotide from a purine to a pyrimedine (e.g., guanine with cytosine).

**trawl**　A bag-like net that is dragged through the water in order to catch fish.

**trematode**　Any parasitic flatworm of the class Trematoda, especially a fluke, equipped with hooks or suckers.

**triglyceride**　An ester of glycerol that is combined with three fatty acids. It is the primary chemical form in which lipids are stored in the tissue.

**trophic**　Relating to the process of feeding or feeding patterns.

**trophic cascade**　Interactions among trophic levels through top-down control.

**Trophic flows**　The transfer of food or energy between the different groups of species comprising a food web.

**true seals**　See EARLESS SEALS.

**trying down**　The traditional practice of cooking blubber to extract the valuable oil; rendering.

**tubercule 1.**　One of a series of small, raised epidermal protuberances found on the dorsal fin of most porpoise species. Their distinctness, density, and covering area vary among species. **2.** A knob-like structure of unknown (but probably sensory) function, found on the head and jaws of humpback whales. Each tubercle contains a single hair (vibrissa).

**tumor**　A swelling, especially from an abnormal growth of tissue.

**tuna purse seine**　See PURSE SEINE.

**tunica albuginen**　A white fibrous layer covering an anatomical structure.

**tunicate**　Subphylum of chordates containing the classes Ascidiacea (the sessile sea squirt), the free-swimming tadpole-like Larvacea, and the Thaliacea (free-swimming salps). Chordate features (i.e., notochord and nerve chord) are found only in the larva and are generally lost in the adult. The adult secretes a tough cellulose sac (tunic) in which the animal is embedded.

**turbid**　Exhibiting reduced water clarity because of the presence of suspended matter.

**turbulent flow**　An unsteady and eddying flow of water. Turbulent boundary layers are thicker than laminar boundary layers and likely to separate prematurely.

**tympanic cavity**　A large, irregularly shaped cavity of the middle ear.

**tympanoperiotic earbones**　Very dense bones at the base of the skull consisting of the ventral ectotympanic bone or bulla, which encloses the middle ear cavity, and the more dorsal periotic or petrosal bone, which contains the cochlear apparatus of the inner ear and the balance organ (vestibulum plus semicircular canals). The structure of tympanoperiotic bones is used widely to indicate taxonomic relationships of both extant and fossil cetaceans.

# U

**unequal crossover**   When the breaks in the DNA strands that later recombine during meiosis is not in the exactly homologous position in the pairs of sister chromatids involved. In mini-satellites unequal crossover will result in a mutation, a change in the number of tandem nucleotide repeats.

**ugruk**   Northern Alaskan Inuit term for a bearded seal.

**ultimate explanation**   See PROXIMATE EXPLANTION.

**ultrasonic**   Describing acoustic signals that are above the normal human range of hearing.

**ultrasonography**   The visualization of deep structures of the body by recording and displaying the echoes of pulse of ultrasonic waves directed into the tissues.

**ultrasound 1.**   A sound that is higher in frequency than the maximum audible to humans, usually said to be 20 kHz. Many marine mammals can hear ultrasound. **2.** see ULTRASONOGRAPHY.

**umialik**   Northern Alaskan Inuit term for a whaling captain.

**umiaq** (*plural*, **umiat**)   Northern Alaskan Inuit term for a skin boat used to hunt whales, seals, and walrus.

**underhair**   A layer of shorter, finer hair that provides better insulation than the longer, thicker hair (called guard hairs) covering it.

**ungulate**   Any of a taxonomic group (Ungulata) comprising the hoofed mammals and their derivatives, including the living Artiodactyla, Cetacea, Perissodactyla, Hyracoidea, Sirenia, and Proboscidea, together with the extinct Desmostylia and numerous other fossil groups.

**unguligrade**   Referring to limbs that are elongated and slim with an efficient fore-and-aft drive and nearly no capability of rotating. Unguligrade limbs are typical of ungulate mammals.

**upwelling**   A process by which ocean water rises from a deeper to a shallower level, usually by divergence and offshore currents. This process often occurs along the edges of continental shelves and brings nutrient-rich water to the surface.

**ureter**   The duct that carries urine from the kidney to the urinary bladder.

**urethra**   In anatomy, duct that carries urine from the bladder out of the body.

**uropod**   Fan-shaped paired appendages on the penultimate segment of euphausiids, used for swimming.

**Ursidae**   A family of arctoid carnivorans that includes the subfamilies Ursinae, Hemicyoninae, and Amphicynodontinae, the latter two being represented only by fossil taxa.

# V

**vacuole**   A fluid-filled space in a cell, usually small in animal cells; several vacuoles coalesce in the developing follicle to produce the antrum.

**vaginal process**   Outpocketing of the peritoneal membrane of the abdomen into the scrotum that is associated with testicular descent.

**vascular**   The system of vessels in the body that carry blood. For example, the arteries, veins, capillaries are part of the vascular system.

**vasculitis**   The inflammation of a blood vessel.

**vasopressin**   A peptide hormone (antidiuretic hormone) that is released from the posterior pituitary gland. It increases the permeability of the collecting duct, thereby suppressing urine formation and output.

**venae cavae**   Major veins returning blood from the periphery to the heart.

**ventral**   In anatomy, toward, or relatively near to the belly and the corresponding surface of the head, neck, thorax, and tail.

**ventral grooves**   A series of parallel grooves or pleats running longitudinally on the undersurface of the throat and chest region in balaenopterid whales, allowing great expansion of the mouth during feeding.

**vertebra** (*plural*, **vertebrae**)   A bone of the spine.

**vertebral canal**   The space within the vertebral column that houses the spinal cord.

**vertebral column**   The row of bones forming the longitudinal axis of the body; the backbone.

**vertebrate**   One of the Vertebrata, the subpyhlum including those animals having a backbone; that is, fish, amphibians, reptiles, birds, and mammals.

**vertex**   An elevated conjunction of the nasal, frontal, and supraoccipital bones at the top of the skull.

**vestigial tooth**   A tooth that has no function in food processing. Usually these teeth do not erupt out of the jawbone or do not break through the dental gums.

**vibrissae**   Large, tactile, sensory hairs present on the upper lip of mammals.

**vicariance event**   The development of a natural biogeographical barrier causing groups that previously occupied the same area or adjacent areas to become physically separated.

**vicariant**   Describing a major geological or oceanic split in habitat, such as the geological closure of a strait, thus splitting distributions of many organisms.

**vicariant speciation**   Evolution of species triggered by fragmentation of the ancestral species' distribution area.

**villus** (*plural*, **villi**)   Finger-like projection that greatly increases the surface area and facilitates exchange between two media (e.g. the surface of the small intestine and its contents) or tissues (e.g. the maternal and fetal parts of the placenta).

**visual acuity**   Sharpness of vision; the capability to discriminate visual pattern details.

**vital rates**   Key indicators describing the dynamics of animal populations, including birth rates, recruitment rates, mortality rates, immigration, emigration.

**vomeronasal organ**   An organ for chemical reception situated in the anterior part of the roof of the mouth; anatomically and functionally distinct from the senses of smell and taste; well developed in many species of mammals, in which it is especially important: in reproduction.

**vortex**   The rotating motion of a fluid around a central core.

**vulnerable**   This term, used by the IUCN Red List refers to taxa that are neither endangered nor critically endangered, but which face a high risk of extinction in the medium-term future (World Conservation Monitoring Centre, 2000).

# W

**weaning**   The end of the lactation period; the process of changing from milk to a solid diet in juvenile mammals.

**Westerly**   Blowing from the west.

**whale**   A term commonly used for a subset of the mammalian order Cetacea. In general, it includes the larger members of the order, but the term is inconsistently used for animals of vastly different sizes that are not closely related. In some cases, it refers to the Great Whales, that is, the baleen whales and the (unrelated) sperm whale. Others use whale to include larger members of the dolphin family (such as the killer whale). In that usage, some larger dolphins (such as the bottlenose dolphin) are well above the smallest whales (such as pygmy sperm whales) in size. Paleontologists use the word whale to describe all Eocene forms (some of which were no larger than a dog) and many modern cetaceans. Thus the term whale is scientifically imprecise, and subjects relating to it can be found in tills work under more specific headings.

**whalebone** See BALEEN.

**whale louse** (*plural*, **whale lice**)  An amphipod crustacean of the family Cyamidae; adapted for living in crevices and other secure places on the skin of cetaceans, on which whale lice largely feed.

**whelping**  The process of giving birth; used to refer to seals.

**windkessel**  A capacitance blood vessel, which due to elastic properties in its wall, first expands and accepts blood volume on blood entry and then passes blood onward due to recoil within its wall.

# X

**xenobiotie 1.**  A chemical compound (hat is foreign to biological systems, especially one that is toxic. **2.** Relating to or describing a compound of this type.

# Z

**ziphiid**  Any of at least 19 extant species of beaked whale, of the family Ziphiidae.

**zona glomerulosa**  The outermost layer of the adrenal cortex principally responsible for the production of the mineralocorticoid aldoslerone.

**zoonosis** (*plural*, **zoonoses**)**1.**  A disease of animals that is transmittable to humans. **2.** The process of disease transfer from animals to humans.

**zooplankton**  Collectively, planktonic animal life.

# INDEX

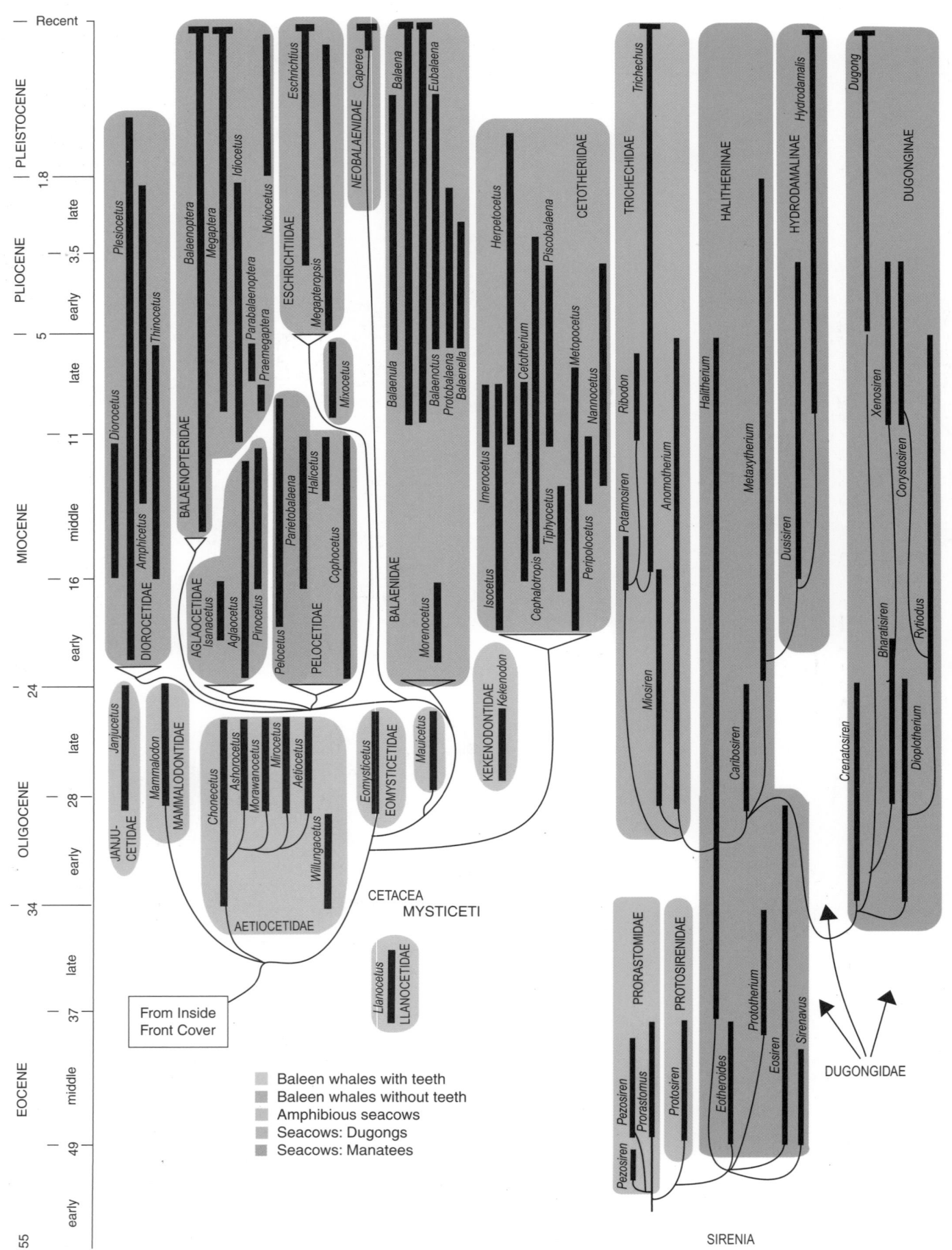